AF696813

Eleventh Edition

The Molecular Probes® Handbook

A GUIDE TO FLUORESCENT PROBES AND LABELING TECHNOLOGIES

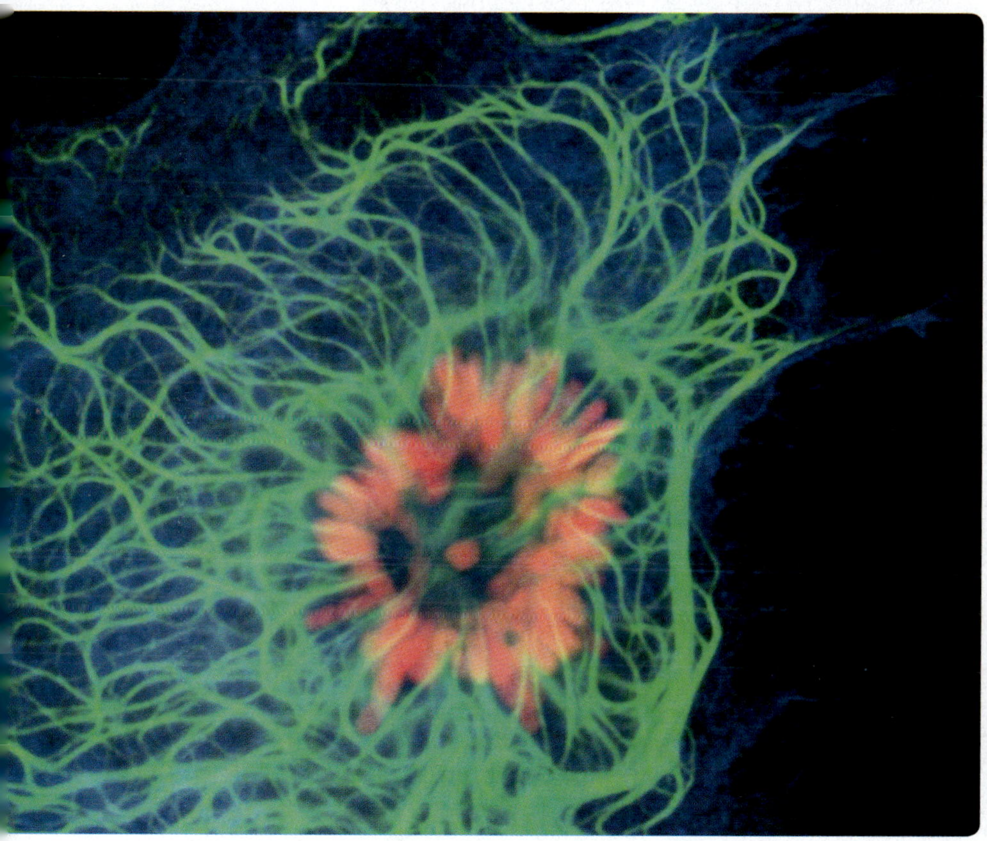

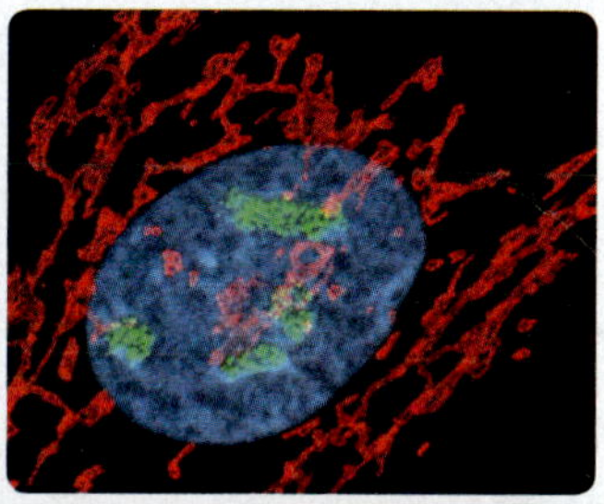

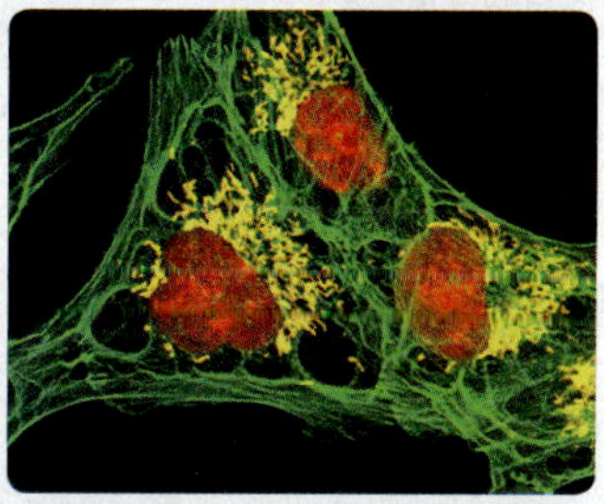

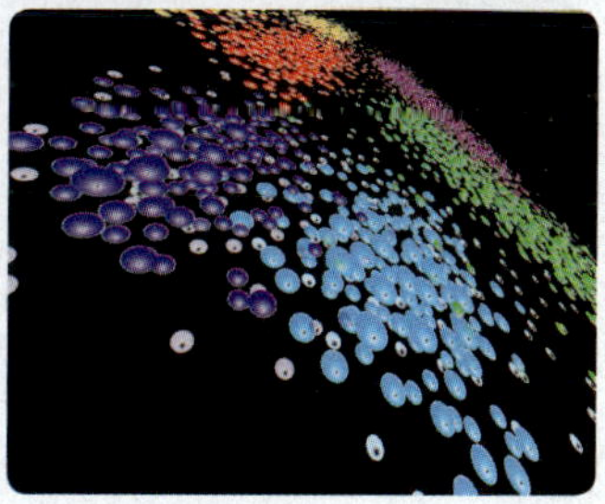

molecular probes® | invitrogen™
by life technologies™

Editors
Iain Johnson, Ph.D.
Michelle T.Z. Spence, Ph.D.

Editorial Assistance
Sarah Adai, M.S., Jay Gregory, Ph.D., Grace Richter, Ph.D., Beth Browne, Ph.D.

Scientific Support
Chip Walker, Ph.D., Gayle Buller, MT(ASCP), Jeff Croissant, Ph.D., Jolene Bradford, MLS(ASCP) CM, Nick Dolman, Ph.D., Kathleen Free, Mike Janes, M.S., Mike Ignatius, Ph.D., Olivia Cholewa, Ph.D., Stephen Full, M.S., Magnus Persmark, Ph.D., Thomas M. Landon, Ph.D.

Content Management
Aaron Basey, Justin Sepulveda, Mary Wisegarver, CPIM, Isabel Melgoza, Nicholas Fargher, Lakshmi Natarajan, Beth Browne, Ph.D.

Publication Design and Production Management
Lynn Soderberg

Graphics Production and Layout
René Nelson, Marilyn Junkins, Heath Thomas, Naomi Valdivia, Kelly Christensen, Wayne Gilbert

Cover Design
Isamu Sato

Legal and Regulatory Affairs
Seri Lee, J.D., Peter Goebel, Ph.D., Mark Stavro

Project Coordination and Marketing
Kerry Lowrie, Ph.D., Chris Scanlon, Dana Bassett, Brian Almond, Ph.D., Jonathan Young

The Molecular Probes® Handbook would not have been possible without the help of our dedicated staff, including those in the following departments: Accounting, Administration Support, Business Development, Commercial, Corporate, Customer Service, Facilities, Human Resources, Cellular Analysis Imaging Group, Information Services, Intellectual Property, Marketing Communications, Operations, Packaging, Product Administration, Program Management, Purchasing, Quality Assurance, Quality Control, Regulatory/Safety, Research and Development, Shipping, Strategic Marketing, Technical Service.

© 2010 by Life Technologies Corporation

First Edition published 1978. Second Edition 1981. Third Edition 1985. Fourth Edition 1989. Fifth Edition 1992. Sixth Edition 1996. Seventh Edition 1999 (on CD only). Eighth Edition 2000 (on CD only). Ninth Edition 2002. Tenth Edition 2005. Eleventh Edition 2010.

All rights reserved. No part of this book may be reproduced or used for commercial purposes in any form or by any means graphic, electronic or mechanical, including photocopying, recording, taping or information storage and retrieval systems without permission of Life Technologies Corporation. All information subject to change without notice.

ISBN 978-0-9829279-0-8

Shaping Discovery, Improving Life

Life Technologies is a global biotechnology tools company that provides premier systems, consumables and services that enable cell biology researchers to accelerate scientific exploration, driving to discoveries and developments that make life even better. Created in 2008 through the combination of Invitrogen Corporation and Applied Biosystems Inc., Life Technologies is committed to changing the future of science and medicine with our innovative product pipeline, and Molecular Probes® products are an important part of that commitment.

The Molecular Probes scientific team has been pioneering fluorescence technologies and cell analysis capabilities, from the Alexa Fluor® dyes to Qdot® nanocrystals, since 1975. A part of the Invitrogen brand since 2003, Molecular Probes® products continue to serve as key tools for investigating every aspect of cell biology. The Molecular Probes site in Eugene, Oregon, serves as the Life Technologies center of excellence for fluorescence labeling and detection, with over 125 chemists (organic, inorganic and analytical), biologists (cell and molecular) and engineers (electrical, mechanical, optical and software) from all over the world. The team in Eugene exemplifies our dedication to the scientific exploration of issues that critically impact our society, encompassing personalized medicine, molecular diagnostics, biofuel development, environmental research and much more.

This significantly revised eleventh edition of *The Molecular Probes® Handbook: A Guide to Fluorescent Probes and Labeling Technologies* embodies Life Technologies' commitment to the continuing legacy of the Molecular Probes® brand, to the superior quality and performance of our products, and to the scientific community we support.

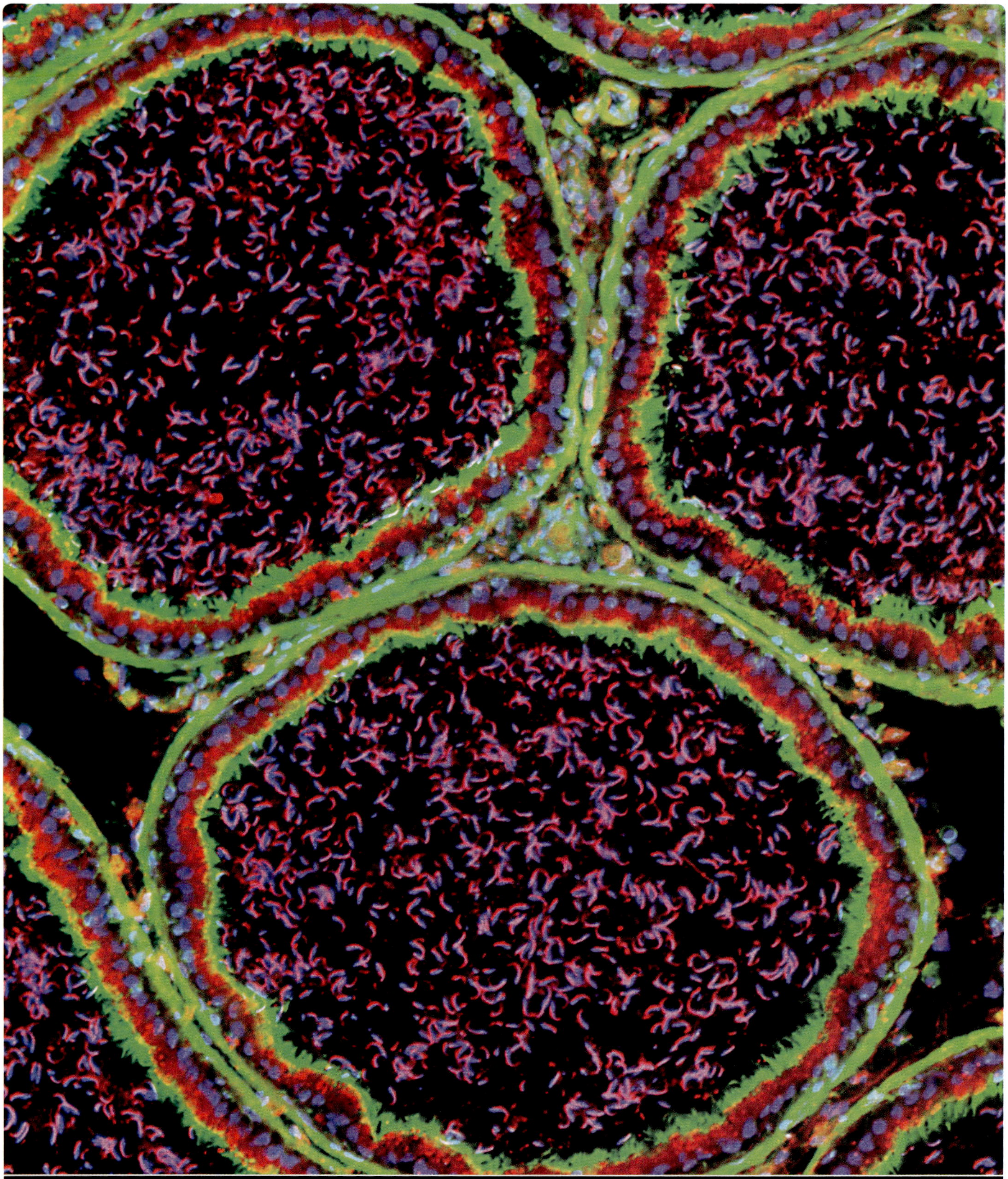

Texas Red® wheat germ agglutinin, Alexa Fluor® 488 phalloidin and Hoechst 33342.

Introduction

PART I — Choosing a Fluorescent Label: Their Properties and Labeling Chemistries

PART II — Detecting Biomolecules: Signal Amplification and Secondary Detection

PART III — Detecting Cell Components: Nucleic Acid and Protein Analysis

PART IV — Probing Cell Structure: From Organelles and Membranes to Whole Cells

PART V — Probing Cell Function: From Cell Viability to Ion Flux

APPENDIX

TECHNICAL NOTES AND PRODUCT HIGHLIGHTS

Ordering and Technical Support

For your convenience, you may place your order by the method that is best for you: online, or by email, mail, phone or fax. Our highly efficient Customer Service representatives will process your order quickly and accurately.

Order Online

Place your order online 24 hours a day, 7 days a week, at our website: **www.invitrogen.com.** Here you will find:

- Real-time pricing and availability
- Secure ordering
- Ability to save "Favorites" for easy future ordering
- Up-to-date access to order status and history
- Simple price checking for any catalog item using your ship-to number
- Rapid checkout with "Quick Order"

Or email your order to **orders@invitrogen.com**.

Order by Mail, Phone or Fax

To place an order by mail, phone or fax, please refer to page 1060 for contact information in your region or visit **www.invitrogen.com/contactus** to obtain contact information for your location. When you order, please include the following information:

1. Institution or company customer number
2. Billing address
3. Shipping address
4. Valid Purchase Order number
5. Name of purchasing agent and department
6. Name and phone number of end user
7. Catalog number and description of product
8. Quantity and size of product

Technical Support

If you have questions about Molecular Probes® products or applications, our Technical Service scientists are ready to help you find the answers. You are our very best source of information about product performance and emerging technologies, and we value your inquiries.

For technical support, email us at **techsupport@invitrogen.com**.

In the US, call us at 800.955.6288.

To better serve your needs, we also offer technical support services around the world. For all locales, visit us at **www.invitrogen.com** and click on the Support link to obtain local contact information and hours of operation, or call your local Invitrogen office to learn more.

Our Promise to You

We seek to exceed your expectations by providing exceptional technical guidance and support on our products, services and technologies that are used in biological research. We will be courteous, helpful, and timely. We strive to build productive, long-lasting partnerships that assist you in reaching your vital research goals.

Commitment to Life Science Research

Passionate about our role as scientific advisors

Service—Second to None

Acknowledged as the #1 life science technical support leader

Training—Cutting Edge, Continuous Learning

Hands-on experience with molecular and cellular applications and technologies

Out Licensing and Commercial Use

The formation of Life Technologies, which includes Molecular Probes® brand products, created the most comprehensive portfolio of products, technologies and capabilities in the life sciences industry. With over 40 years of experience partnering with diagnostics and life sciences companies, we are pleased to provide off-the-shelf solutions, or to develop and manufacture reagents, assays, instruments, software and complete platforms customized to meet your company's needs. We invest heavily in research and development, and are an innovative creator and in-licensor of intellectual property. Our resulting thousands of patents and patent applications, tens of thousands of products and world-class brands are available for licensing and commercial use.

We can assist you with selecting technologies and components that will enhance your product or service offerings, helping to provide you with a competitive edge in your target markets. We can also shorten your development cycle, shortening your time to revenue, and will deliver to your chosen manufacturing or distribution sites around the world, enabling security of supply and smooth, reliable manufacturing processes.

Our multiple manufacturing sites in the US, Europe and Asia let us provide you with products or components made under ISO 9001, ISO 13485 and GMP quality systems, or USDA guidelines. Accurate lot documentation and traceability are standard in our processes. We can synthesize novel molecular entities or particles *de novo*, prepare conjugates, custom-formulate complex buffers and solutions, or provide complete kits. We can custom-package at volumes of single microliters per vial or many liters per container. We can package by the picogram or kilogram, or by activity units using your specified assay. We work with you to define the ideal manufacturing specifications and tolerances to meet your needs. We also can special-label, or provide unlabeled vials, to match your requirements. For your diagnostics kit components, ask about the possibility of safety stock and embedding your forecasts in our manufacturing planning process.

Our portfolio of highly respected technologies that are available for license or commercial use includes:

- Nucleic Acid Amplification Tools
- Cloning and Expression Reagents
- Nucleic Acid Modification Enzymes and Polymerases
- Sequencing Tools
- Sample Preparation and Magnetic Particles
- Microspheres and Nanocrystals
- Fluorescent Labeling and Detection Reagents
- Immunoassays, Antibodies and Protein Detection Reagents
- Cells and Cell Lines
- Recombinant Proteins
- Electrophoresis Reagents and Ladders
- Instruments, Software and Firmware
- Controls and Calibration Standards

We also can provide you with rights to use our industry leading brands, including SYBR®, Alexa Fluor®, Qdot® and others. Please contact us at **outlicensing@lifetech.com** to learn more.

Custom Synthesis and Special Packaging for Research Use

Molecular Probes® services for research use include:

- **Custom synthesis.** From milligram to kilogram quantities, we can synthesize for you a wide selection of customized fluorescent dyes and other small molecules, fluorescent, white and colored microspheres, nanocrystals and other reagents.
- **Custom conjugations.** We offer fast and efficient custom conjugation of antibodies and other proteins, as well as peptides and other molecules to fluorescent dyes, nanocrystals, biotin, haptens, enzymes or other ligands. Our custom conjugation service is confidential, and we guarantee the quality of our work.

We offer discounts on orders of large quantities of our catalog reagents, and can additionally provide special packaging to meet your research needs. Please contact us at custom.services@invitrogen.com or call 1 800 955 6288 x2 to learn more.

For customers outside of the United States, please refer to page 1060 for contact information in your region or visit **www.invitrogen.com/contactus** to obtain contact information for your location.

Molecular Probes® Products and Tools on the Web

For the most up-to-date product information, including pricing and availability, visit the Molecular Probes® brand page at **www.invitrogen.com/handbook/probes**. Here you will find access to a wealth of resources for our broad fluorescence and imaging technology portfolio, spanning everything from basic fluorescence and flow cytometry tutorials to more detailed information on product spectra and structures. From this page, you can easily:

- Search for Molecular Probes® products by technology, application area or brand name
- Explore featured new products and technologies
- Access *The Molecular Probes® Handbook* online
- Stay connected through Twitter, product reviews, the Molecular Probes® Technology Network and much more
- Use the latest tools including the Fluorescence SpectraViewer, Virtual Cell Staining Tool and more

www.invitrogen.com/handbook/probes

Image Gallery

The Molecular Probes® Image Gallery (**www.invitrogen.com/handbook/probesimages**) features an extensive collection of images created with Molecular Probes® fluorescent dyes and probes. You can download hundreds of images, view the products used to create each image and search the gallery for your specific interests.

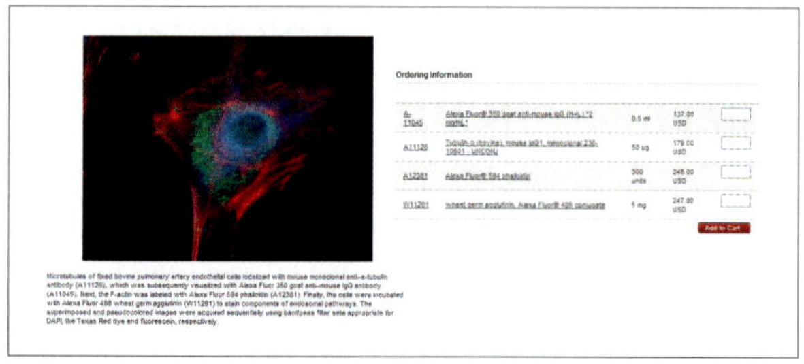

www.invitrogen.com/handbook/probesimages

The Molecular Probes® Handbook—Online

The online version of *The Molecular Probes® Handbook* (**www.invitrogen.com/handbook**) is a comprehensive resource for fluorescence technology that is continually updated. Like the printed version, the online *Handbook* contains detailed information about the thousands of Molecular Probes® products available for fluorescence and imaging applications. The online *Handbook* has product pricing for your region as well as the added advantage of having critical product information at your fingertips. Look for the following icons for easy access to:

- REF Referenced citations
- Product spectra
- Product structures
- Data images

www.invitrogen.com/handbook

Mobile Applications and Widgets

Invitrogen offers an assortment of handy applications and widgets for your everyday mobile needs. Download popular iPhone® apps and widgets from Molecular Probes and other Life Technologies brands, including the DailyCalcs Science Calculator and the Alexa Fluor® Selection Guide widget. Visit **www.invitrogen.com/handbook/apps** to get started. New apps and widgets are introduced on a regular basis, so check back often.

www.invitrogen.com/handbook/apps

Virtual Cell Staining Tool

The Virtual Cell Staining Tool (**www.invitrogen.com/handbook/cellstainingtool**) allows you to select different combinations of cellular structures and fluorophores to create your perfectly labeled fluorescent cell, which you can easily share with your colleagues using our email or print functionality.

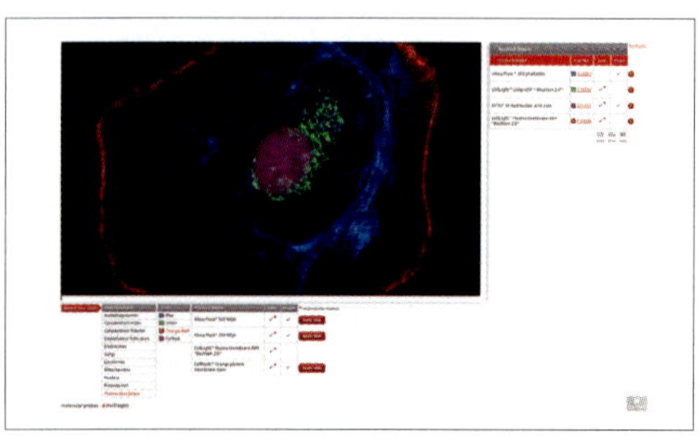

www.invitrogen.com/handbook/cellstainingtool

Introducing the Molecular Probes® Technology Network

The Molecular Probes® Technology Network (**www.invitrogen.com/handbook/mpnetwork**) was designed to connect researchers across the world with a common interest in fluorescence imaging and related applications. You can use this space to share protocols and data with other researchers, and start or join technology groups around your specific niche. You can also search documents and protocols uploaded by other researchers and connect with them individually.

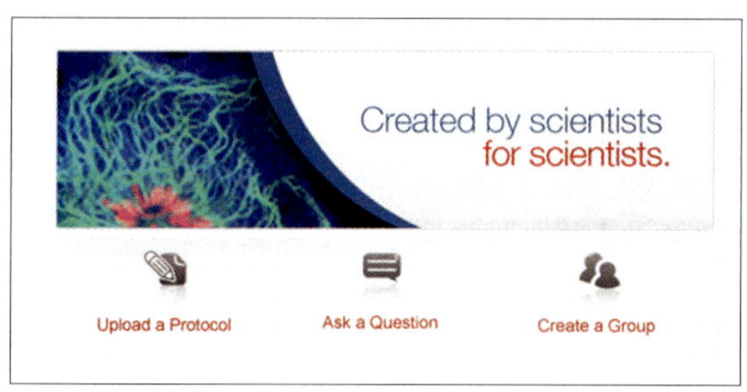

www.invitrogen.com/handbook/mpnetwork

Fluorescence SpectraViewer

The Fluorescence SpectraViewer is an online tool that can be used to plot and compare the excitation and emission spectra of fluorescent dyes, facilitating the design of multicolor fluorescence experiments. As many as five fluorophores can be selected from an extensive list of dyes for simultaneous spectral analysis to determine the suitability of fluorophores for specific instrument configurations, and to evaluate spectral compatibility of different fluorophores (Section 23.1, Using the Flurorescence Spectraviewer—Note 23.1). The SpectraViewer is available at **www.invitrogen.com/handbook/spectraviewer**.

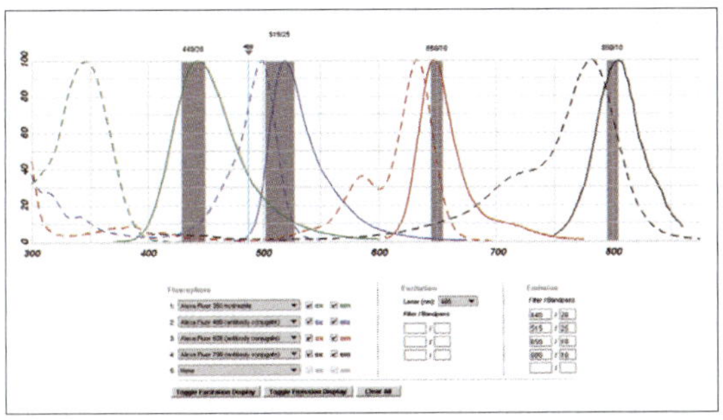

www.invitrogen.com/handbook/spectraviewer

Fluorescence and Flow Cytometry Tutorials

Whether you are new to fluorescence or flow cytometry or just want to brush up on the basics, our flash-based tutorials are a great educational resource. You will find them on our web site at **www.invitrogen.com/handbook/tutorials**.

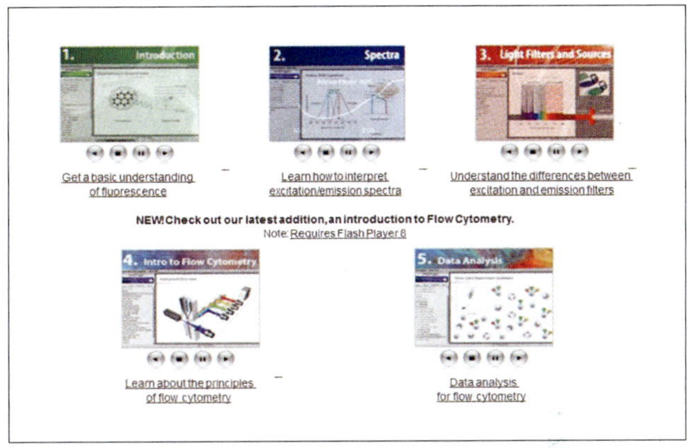

www.invitrogen.com/handbook/tutorials

Molecular Probes® Newsletters

BioProbes® newsletter is published several times each year. This award-winning publication is dedicated to furnishing researchers with the very latest information about cell biology products and their applications, and provides a great way to stay connected with the fluorescence community. You can subscribe and view the latest issues online at **www.invitrogen.com/handbook/bioprobes**. *ProbesOnline* is a monthly email communication that will keep you up to date with the newest Molecular Probes® products and applications to enhance your research. Subscribe and view the latest issues at **www.invitrogen.com/handbook/probesonline**.

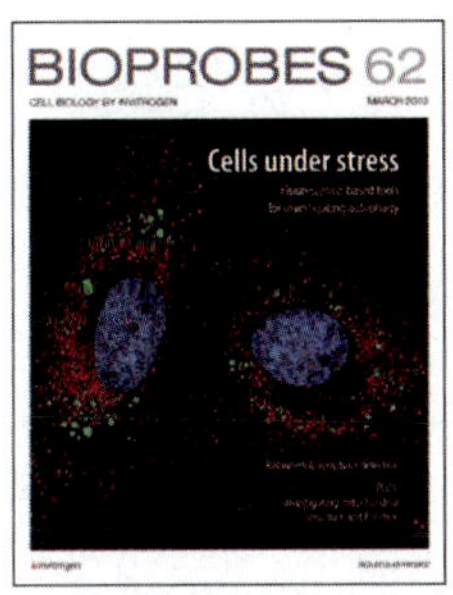

www.invitrogen.com/handbook/bioprobes

www.invitrogen.com/handbook/probesonline

Online Product Selection Guides

Finding primary antibodies, secondary antibodies and reactive dye reagents and kits has never been easier. Our primary antibody (**www.invitrogen.com/handbook/searchprimary**) and secondary antibody (**www.invitrogen.com/handbook/searchsecondary**) selection tools make finding the right antibodies easy. Simply select the type of antibody you're looking for and filter by target, application, reactivity, conjugate or host.

Our Labeling Chemistries Selection Guide (**www.invitrogen.com/handbook/labelingchemistry**) allows you to quickly find the best reactive dye or label for your experiment. Whether you're labeling proteins and antibodies for immunofluorescence, nucleic acids for *in situ* hybridization, or lipids for membrane studies, this tool will help you find the right conjugation product. Access all of our research tools at **www.invitrogen.com/handbook/selectionguides**.

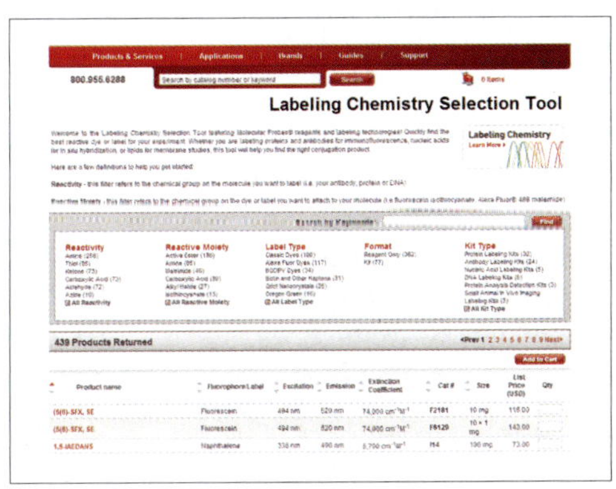

www.invitrogen.com/handbook/selectionguides

Using *The Molecular Probes® Handbook*

Overview

The Molecular Probes® Handbook is is divided into 6 main parts, with a total of 23 chapters, each discussing a product group with common properties or applications. Each chapter is further subdivided into sections, grouping products by more narrowly defined properties or applications. If a product has multiple applications, it may appear in more than one location.

Where possible, we have cited the literature for particular applications. For some products, we have speculated on possible applications. We advise researchers to use appropriate caution when approaching such areas of novel discovery.

Product Lists

There are 3 types of product lists in the *Handbook* to help you find the information you need:

- **Section Product Lists** can be found at the end of each chapter section. They contain the catalog number, product name and unit size of each product discussed in that section. As described above, products can be discussed in multiple sections and will be listed in multiple section product lists.

- **The Master Product List** on page 975 contains the chapter and section locations as well as the Limited Use Label Licenses (LULLs) for all of the products described in the *Handbook*. The products are listed alphabetically by product name.

- **The Master Catalog Number List** on page 1029 contains a list of all of the catalog numbers and their *Handbook* section locations. This list is organized first in alphabetical order using the alpha identifier of the catalog number, and second in numerical order using the numeric identifier of the catalog number.

Please note that pricing is not available in any of these lists but can be found online at **www.invitrogen.com**.

Index

We have also provided an index starting on page 1039. Look for keywords or applications related to the products or data you are interested in to find the page numbers on which the topics are discussed.

Data Tables

Most sections of *The Molecular Probes® Handbook* include tables of technical data, listing the chemical and spectroscopic properties of products described in the section. See page xiii for definitions of data table contents.

More Information Online

For more assistance in finding products for your research, visit the the online version of the *Handbook*. For a review of the additional information available at our Web site, see "Molecular Probes® Products and Tools on the Web" on pages x–xi.

NOTE

Definitions of Data Table Contents

Cat. No.: The alpha-numeric catalog number for a particular product. Data tables are organized first in alphabetical order using the alpha identifier of the catalog number, and second in numerical order using the numeric identifier of the catalog number (e.g., A47, B153, B438, C194, etc.). The Master Product List shows all sections in which products are discussed in the text. In most cases, data tables do not list kits, protein conjugates, dextran conjugates, gel and blot stains, ChromaTide® nucleotides and FluoSpheres® polystyrene microspheres.

MW: Molecular weight (MW) for the anhydrous compound, except in a few cases for which the exact degree of hydration has been determined and is specified in the name of the product. We caution that products may not be completely anhydrous upon receipt, even if they are represented in this form by the chemical structure and MW value. For compounds isolated and sold in salt form, MW is inclusive of counterions unless noted otherwise. In some cases (e.g., nucleotide derivatives, some peptides and proteins), the exact salt form has not been established, and consequently the MW value is approximate, denoted by a preceding "approximately" symbol (~). In general, we recommend measuring the absorbance of a solution and calculating concentration with the extinction coefficient rather than using these MW values.

Storage: Recommended storage conditions for products. Abbreviations are as follows:

A	Material may be air sensitive
AA	Air sensitive, use under a N_2 or Ar atmosphere
D	Desiccation *recommended*
DD	Desiccation *required*
F or FF	Store at ≤–20°C
L	Protect material from long-term exposure to light; may be exposed to light for short periods of time
LL	Protect material from light *at all times*
MIXED	Contains components with incompatible storage requirements; store components separately as indicated
NC	Storage conditions are not critical; store at ≤25°C
RO	Store at ≤6°C; if frozen, AVOID FREEZE-THAW CYCLES
RR	Store at 2–6°C; DO NOT FREEZE
RT	Store at room temperature or below (2–25°C); DO NOT FREEZE
UF	Store at ≤–70°C

Many products may remain at ambient temperature for short periods of time (e.g., during shipment) without any loss of product performance, but should be stored as indicated upon arrival. Storage information is also printed on all product container labels.

Soluble: Recommended solvent for preparing stock solutions of at least 1 mM. Recommendations are based on the best knowledge and experience of our technical staff, but have not necessarily been tested. If a pH is indicated, the product generally ionizes and should be dissolved in aqueous buffer in the specified pH range. Please send any alternate suggestions and corrections to this information based on practical experience to techsupport@invitrogen.com. Abbreviations for solvents are defined below (see Solvent).

Abs: The longest-wavelength (unless noted otherwise) absorption maximum (in nanometers) in the solvent listed in the column headed Solvent.

EC: Molar extinction coefficient (in $cm^{-1}M^{-1}$) determined at the wavelength listed in the column headed Abs. Values above 10,000 $cm^{-1}M^{-1}$ are rounded to the nearest 1000. In most cases, extinction coefficients have not been rigorously determined, and the values may vary somewhat among production lots.

Em: Fluorescence emission maximum (in nanometers) in the solvent listed in the column headed Solvent. Em values are generally not corrected for instrument response characteristics, resulting in small variations when compared to measurements in other laboratories. Considerable environment-dependent variation of Em occurs for some products; when known, this is indicated with a footnote.

Solvent: Solvent used for acquisition of spectroscopic data, including Abs, EC and Em. Solubility of the product in this solvent is not necessarily any greater than is required to obtain an absorption spectrum (i.e., 10 µM or less). In some cases, it is necessary to prepare a stock solution in a different solvent and then to dilute the sample into the indicated solvent to measure spectra. Refer to the column headed Soluble for recommended solvents for preparing stock solutions. Abbreviations are as follows:

$CHCl_3$	Chloroform
DMF	Dimethylformamide
DMSO	Dimethylsulfoxide
EtOAc	Ethyl acetate
EtOH	Ethanol
H_2O	Unbuffered water
pH 7	pH 7 aqueous buffer
H_2O/DNA	Spectra measured for probes bound to DNA in aqueous solution
MeCN	Acetonitrile
MeOH	Methanol
THF	Tetrahydrofuran

Data for ion indicators are usually listed for aqueous solutions both with and without the target ion. The indicator dissociation constant for the target ion, in most cases determined in our laboratories, is listed in a separate column headed $\mathbf{K_d}$ or for pH indicators, $\mathbf{pK_a}$.

Product: Catalog numbers or references to footnotes indicating products that are generated by enzymatic or chemical conversion of the parent compound in the course of standard applications.

Notes: Extensions to the listed data, identified by numbers (1, 2, etc.). For example, in cases where a product exists in two forms (e.g., substrate/enzymatic product, free/bound, unreacted/reacted), data for one form is listed in the table and for the other form in a footnote. Other spectroscopic parameters reported in some footnotes include fluorescence quantum yields (QY) and excited state lifetimes (τ; units = seconds × 10^{-9} = nanoseconds). We do not routinely determine QY and τ for our products, but include this data in the footnotes when it is known.

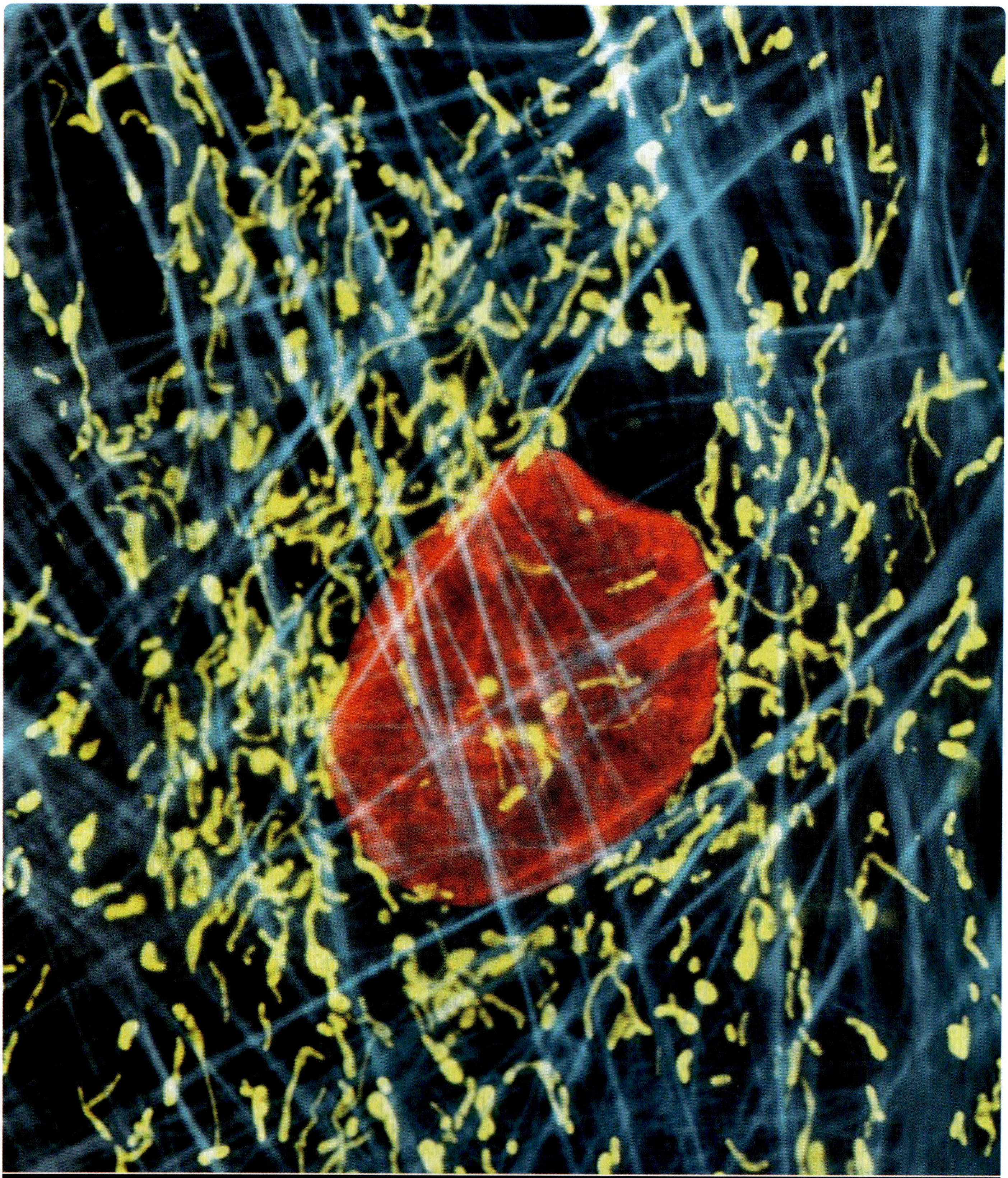

Alexa Fluor® 750 goat anti–mouse IgG, Alexa Fluor® 633 phalloidin and SYTOX® Orange.

PART 1

Choosing a Fluorescent Label: Their Properties and Labeling Chemistries

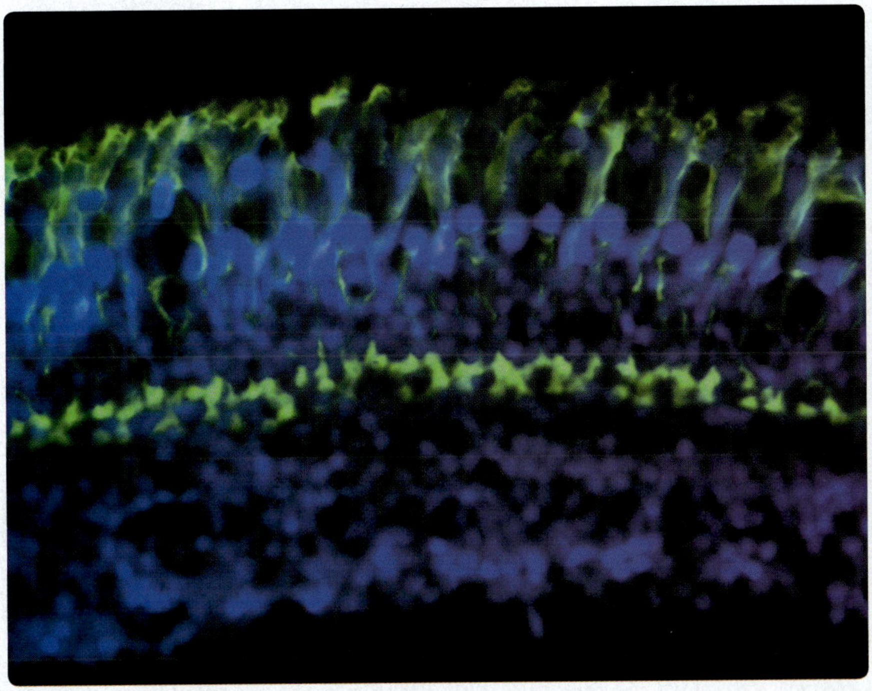

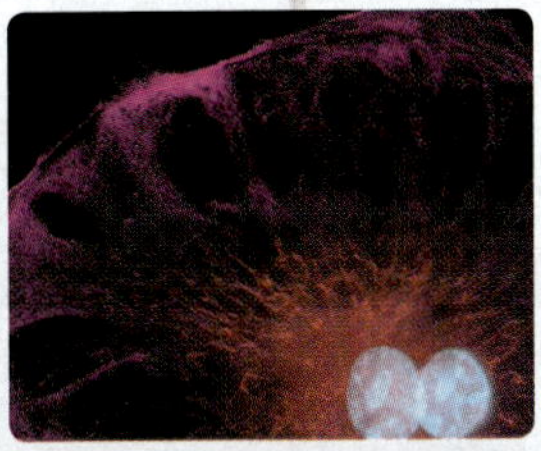

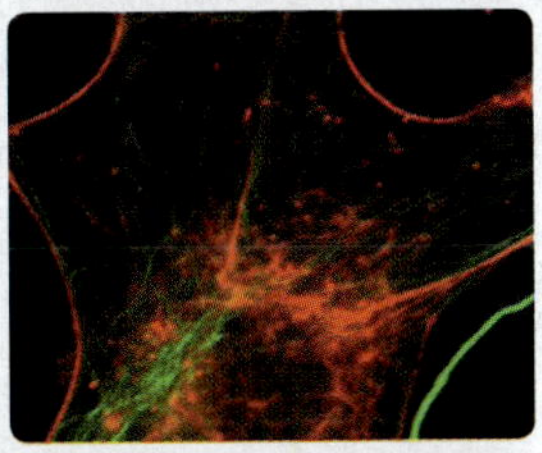

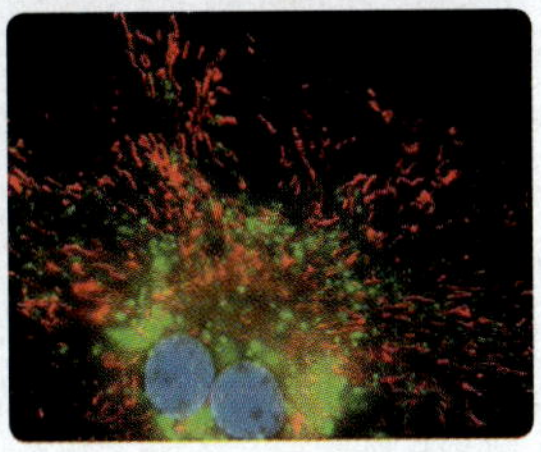

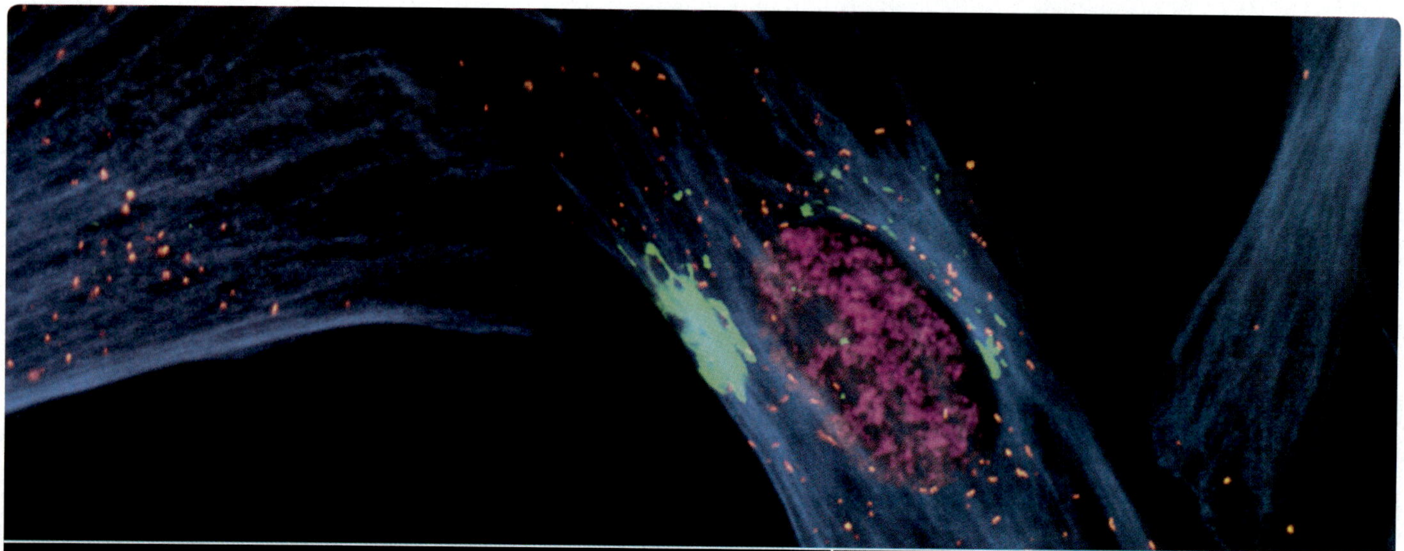

Muntjac cells were treated with 10 µM EdU for 45 minutes. Cells were then fixed and permeabilized, and EdU that had been incorporated into newly synthesized DNA was detected using the far red–fluorescent Click-iT® EdU Alexa Fluor® 647 Imaging Kit (C10085). Tubulin was labeled with a mouse anti-tubulin antibody and visualized with an Alexa Fluor® 350 goat anti–mouse IgG antibody (A11045). The Golgi complex was stained with the green-fluorescent Alexa Fluor® 488 conjugate of lectin HPA from *Helix pomatia* (edible snail) (L11271), and peroxisomes were labeled with a rabbit anti-peroxisome antibody and visualized with an orange-fluorescent Alexa Fluor® 555 donkey anti–rabbit IgG antibody (A31572).

This image appeared on the cover of *BioProbes 55*. *BioProbes®* newsletter is published several times each year. This award-winning publication is dedicated to furnishing researchers with the very latest information about cell biology products and their applications, and provides a great way to stay connected with the fluorescence community. You can subscribe and view the latest issues online at **www.invitrogen.com/handbook/bioprobes**.

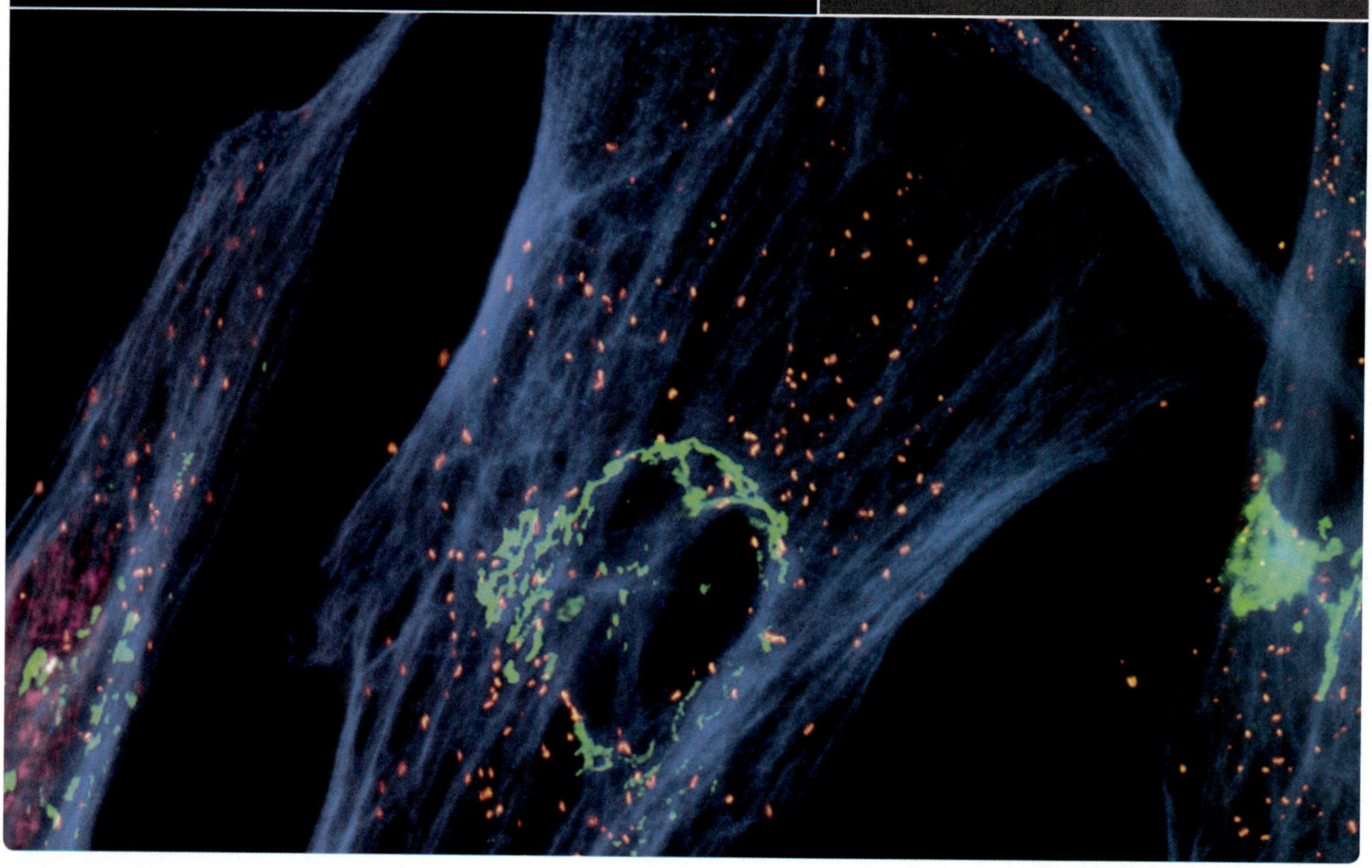

IMPORTANT NOTICE: The products described in this manual are covered by one or more Limited Use Label License(s). Please refer to the Appendix on page 971 and Master Product List on page 975. Products are For Research Use Only. Not intended for any animal or human therapeutic or diagnostic use.

Introduction to Fluorescence Techniques

Fluorescent probes enable researchers to detect particular components of complex biomolecular assemblies, such as live cells, with exquisite sensitivity and selectivity. The purpose of this introduction is to briefly outline fluorescence principles and techniques for newcomers to the field.

The Fluorescence Process

Fluorescence is the result of a three-stage process that occurs in certain molecules (generally polyaromatic hydrocarbons or heterocycles) called fluorophores or fluorescent dyes. A fluorescent probe is a fluorophore designed to respond to a specific stimulus or to localize within a specific region of a biological specimen. The process responsible for the fluorescence of fluorescent probes and other fluorophores is illustrated by the simple electronic-state diagram (Jablonski diagram) shown in Figure 1.

Stage 1: Excitation

A photon of energy $h\nu_{EX}$ is supplied by an external source such as an incandescent lamp or a laser and absorbed by the fluorophore, creating an excited electronic singlet state (S_1'). This process distinguishes fluorescence from chemiluminescence, in which the excited state is populated by a chemical reaction.

Stage 2: Excited-State Lifetime

The excited state exists for a finite time (typically 1–10 nanoseconds). During this time, the fluorophore undergoes conformational changes and is also subject to a multitude of possible interactions with its molecular environment. These processes have two important consequences. First, the energy of S_1' is partially dissipated, yielding a relaxed singlet excited state (S_1) from which fluorescence emission originates. Second, not all the molecules initially excited by absorption (Stage 1) return to the ground state (S_0) by fluorescence emission. Other processes such as collisional quenching, fluorescence resonance energy transfer (FRET) (Fluorescence Resonance Energy Transfer (FRET)—Note 1.2) and intersystem crossing may also depopulate S_1. The fluorescence quantum yield, which is the ratio of the number of fluorescence photons emitted (Stage 3) to the number of photons absorbed (Stage 1), is a measure of the relative extent to which these processes occur.

Stage 3: Fluorescence Emission

A photon of energy $h\nu_{EM}$ is emitted, returning the fluorophore to its ground state S_0. Due to energy dissipation during the excited-state lifetime, the energy of this photon is lower, and therefore of longer wavelength, than the excitation photon $h\nu_{EX}$. The difference in energy or wavelength represented by ($h\nu_{EX} - h\nu_{EM}$) is called the Stokes shift. The Stokes shift is fundamental to the sensitivity of fluorescence techniques because it allows emission photons to be detected against a low background, isolated from excitation photons. In contrast, absorption spectrophotometry requires measurement of transmitted light relative to high incident light levels at the same wavelength.

Fluorescence Spectra

The entire fluorescence process is cyclical. Unless the fluorophore is irreversibly destroyed in the excited state (an important phenomenon known as photobleaching), the same fluorophore can be repeatedly excited and detected. The fact that a single fluorophore can generate many thousands of detectable photons is fundamental to the high sensitivity of fluorescence detection techniques. For polyatomic molecules in solution, the discrete electronic transitions represented by $h\nu_{EX}$ and $h\nu_{EM}$ in Figure 1 are replaced by rather broad energy spectra called the fluorescence excitation spectrum and fluorescence emission spectrum, respectively (Table 1). The bandwidths of these spectra are parameters of particular importance for applications in which two or more different fluorophores are simultaneously detected . The fluorescence excitation spectrum of a single fluorophore species in dilute solution is usually identical to its absorption spectrum. The absorption spectrum can therefore be used as a surrogate excitation spectrum data set. Under the same conditions, the fluorescence emission spectrum is independent of the excitation wavelength, due to the partial dissipation of excitation energy during the excited-state lifetime, as illustrated in Figure 1. The emission intensity is proportional to the amplitude of the fluorescence excitation spectrum at the excitation wavelength (Figure 2).

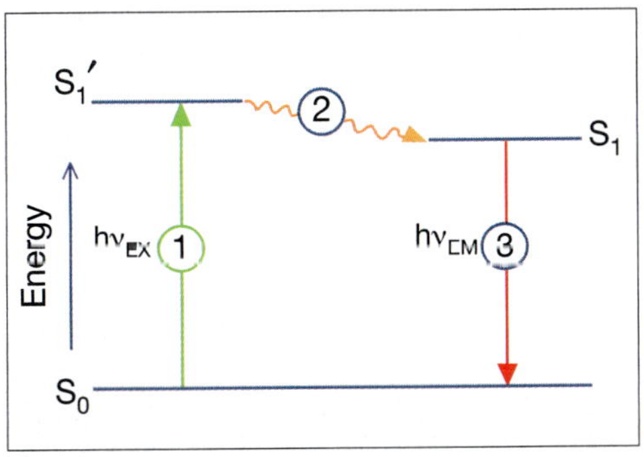

Figure 1 Jablonski diagram illustrating the processes involved in the creation of an excited electronic singlet state by optical absorption and subsequent emission of fluorescence. The labeled stages 1, 2 and 3 are explained in the adjoining text.

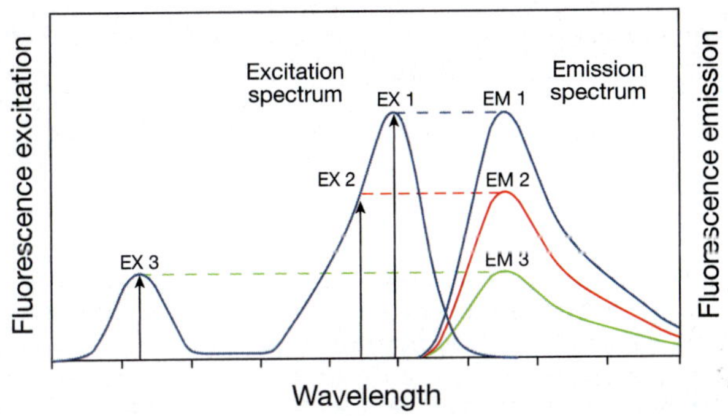

Figure 2 Excitation of a fluorophore at three different wavelengths (EX 1, EX 2, EX 3) does not change the emission profile but does produce variations in fluorescence emission intensity (EM 1, EM 2, EM 3) that correspond to the amplitude of the excitation spectrum.

IMPORTANT NOTICE: The products described in this manual are covered by one or more Limited Use Label License(s). Please refer to the Appendix on page 971 and Master Product List on page 975. Products are For Research Use Only. Not intended for any animal or human therapeutic or diagnostic use.

Fluorescence Detection

Fluorescence Instrumentation

Four essential elements of fluorescence detection systems can be identified from the preceding discussion: 1) an excitation source, 2) a fluorophore, 3) wavelength filters to isolate emission photons from excitation photons, 4) a detector that registers emission photons and produces a recordable output, usually as an electrical signal. Regardless of the application, compatibility of these four elements is essential for optimizing fluorescence detection.

Fluorescence instruments are primarily of four types, each providing distinctly different information:

- Spectrofluorometers and microplate readers measure the *average* properties of bulk (µL to mL) samples.
- Fluorescence microscopes resolve fluorescence as a function of spatial coordinates in two or three dimensions for microscopic objects (less than ~0.1 mm diameter).
- Fluorescence scanners, including microarray readers, resolve fluorescence as a function of spatial coordinates in two dimensions for macroscopic objects such as electrophoresis gels, blots and chromatograms.
- Flow cytometers measure fluorescence per cell in a flowing stream, allowing subpopulations within a large sample to be identified and quantitated.

Other types of instrumentation that use fluorescence detection include capillary electrophoresis apparatus, DNA sequencers [1] and microfluidic devices.[2,3] Each type of instrument produces different measurement artifacts and makes different demands on the fluorescent probe. For example, although photobleaching is often a significant problem in fluorescence microscopy, it is not a major impediment in flow cytometry because the dwell time of individual cells in the excitation beam is short.

Fluorescence Signals

Fluorescence intensity is quantitatively dependent on the same parameters as absorbance—defined by the Beer–Lambert law as the product of the molar extinction coefficient, optical path length and solute concentration—as well as on the fluorescence quantum yield of the dye and the excitation source intensity and fluorescence collection efficiency of the instrument (Table 1). In dilute solutions or suspensions, fluorescence intensity is linearly proportional to these parameters. When sample absorbance exceeds about 0.05 in a 1 cm pathlength, the relationship becomes nonlinear and measurements may be distorted by artifacts such as self-absorption and the inner-filter effect.[4,5]

Because fluorescence quantitation is dependent on the instrument, fluorescent reference standards are essential for calibrating measurements made at different times or using different instrument configurations.[6–8] To meet these requirements, we offer high-precision fluorescent microsphere reference standards for fluorescence microscopy and flow cytometry and a set of ready-made fluorescent standard solutions for spectrofluorometry (Section 23.1, Section 23.2).

A spectrofluorometer is extremely flexible, providing continuous ranges of excitation and emission wavelengths. Laser-scanning microscopes and flow cytometers, however, require probes that are excitable at a single fixed wavelength. In contemporary instruments, the excitation source is usually the 488 nm spectral line of the argon-ion laser. As

Table 1 Spectroscopic proterties of fluorescent dyes.

Property	Definition	Significance
Fluorescence excitation spectrum *	An X,Y plot of excitation wavelength versus number of fluorescence photons generated by a fluorophore.	Optimum instrument setup should deliver excitation light as close to the peak of the excitation spectrum of the fluorophore as possible.
Absorption spectrum	An X,Y plot of wavelength versus absorbance of a chromophore or fluorophore.	To a first approximation, the absorption spectrum of a fluorophore is equivalent to the fluorescence excitation spectrum.† To the extent that this approximation holds, the absorption spectrum can be used as a surrogate for the fluorescence excitation spectrum.
Fluorescence emission spectrum *	An X,Y plot of emission wavelength versus number of fluorescence photons generated by a fluorophore.	Fluorescence emission spectral discrimination is the most straightforward basis for multiplex detection ‡ and for resolving probe fluorescence from background autofluorescence.
Extinction coefficient (EC)	Capacity for light absorption at a specific wavelength.§	Fluorescence output per fluorophore ("brightness") is proportional to the product of the extinction coefficient (at the relevant excitation wavelength) and the fluorescence quantum yield.
Fluorescence quantum yield (QY)	Number of fluorescence photons emitted per excitation photon absorbed.	See "Extinction coefficient."
Quenching	Loss of fluorescence signal due to short-range interactions between the fluorophore and the local molecular environment, including other fluorophores (self-quenching).	Loss of fluorescence is reversible to the extent that the causative molecular interactions can be controlled.**
Photobleaching	Destruction of the excited fluorophore due to photosensitized generation of reactive oxygen species (ROS), particularly singlet oxygen ($^1O_2$).	Loss of fluorescence signal is irreversible if the bleached fluorophore population is not replenished (e.g., via diffusion). Extent of photobleaching is dependent on the duration and intensity of exposure to excitation light.

* Our online Fluorescence SpectraViewer (www.invitrogen.com/handbook/spectraviewer) provides an interactive utility for plotting and comparing fluorescence excitation and emission spectra for over 250 fluorophores (Using the Fluorescence SpectraViewer—Note 23.1).

† Generally true for single fluorophore species in homogeneous solutions but not in more complex heterogeneous samples.

‡ Multiplex detection refers to the process of simultaneously labeling a specimen with two or more fluorescent probes to allow correlation of multiple structural or functional features. As well as specific association with their targets, the probes must have distinctive spectroscopic properties that can be discriminated by the detection instrument.

§ EC (units: $cm^{-1} M^{-1}$) is defined by the Beer-Lambert law A = EC•c•l, where A = absorbance, c = molar concentration, l = optical pathlength. EC values at the absorption maximum wavelength are listed in the Section Data Tables throughout *The Molecular Probes® Handbook.*

** In the case of self-quenching, this can be accomplished by disruption of fluorophore compartmentalization, denaturation or fragmentation of biopolymer conjugates, or functionalization of the fluorophore to produce increased electrostatic repulsion and water solubility.

IMPORTANT NOTICE: The products described in this manual are covered by one or more Limited Use Label License(s). Please refer to the Appendix on page 971 and Master Product List on page 975. Products are For Research Use Only. Not intended for any animal or human therapeutic or diagnostic use.

shown in Figure 3, separation of the fluorescence emission signal (S1) from Rayleigh-scattered excitation light (EX) is facilitated by a large fluorescence Stokes shift (i.e., separation of A1 and E1). Biological samples labeled with fluorescent probes typically contain more than one fluorescent species, making signal-isolation issues more complex. Additional optical signals, represented in Figure 3 as S2, may be due to background fluorescence or to a second fluorescent probe.

Background Fluorescence

Fluorescence detection sensitivity is severely compromised by background signals, which may originate from endogenous sample constituents (referred to as autofluorescence) or from unbound or nonspecifically bound probes (referred to as reagent background). Detection of autofluorescence can be minimized either by selecting filters that reduce the transmission of E2 relative to E1 or by selecting probes that absorb and emit at longer wavelengths. Although narrowing the fluorescence detection bandwidth increases the resolution of E1 and E2, it also compromises the overall fluorescence intensity detected. Signal distortion caused by autofluorescence of cells, tissues and biological fluids is most readily minimized by using probes that can be excited at >500 nm. Furthermore, at longer wavelengths, light scattering by dense media such as tissues is much reduced, resulting in greater penetration of the excitation light.[9]

Multicolor Labeling Experiments

A multicolor labeling experiment entails the deliberate introduction of two or more probes to simultaneously monitor different biochemical functions. This technique has major applications in flow cytometry,[10–12] DNA sequencing,[1] fluorescence *in situ* hybridization[13,14] and fluorescence microscopy.[15–17] Signal isolation and data analysis are facilitated by maximizing the spectral separation of the multiple emissions (E1 and E2 in Figure 3). Consequently, fluorophores with narrow emission spectral bandwidths, such as BODIPY® dyes (Section 1.4) and Qdot® nanocrystals (Section 6.6), are particularly useful in multicolor applications. An ideal combination of dyes for multicolor labeling would exhibit strong absorption at a coincident excitation wavelength and well-separated emission spectra (Figure 3). Unfortunately, it is not easy to find single dyes with the requisite combination of a large extinction coefficient for absorption and a large Stokes shift. Qdot® nanocrystals (Section 6.6) and phycobiliprotein tandem conjugates (Section 6.4) have been developed to meet these requirements and have proven effective in multicolor labeling experiments.[11,18,19]

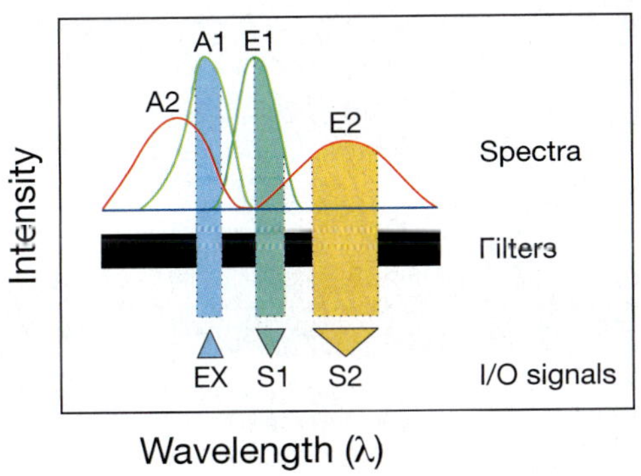

Figure 3 Fluorescence detection of mixed species. Excitation (EX) in overlapping absorption bands A1 and A2 produces two fluorescent species with spectra E1 and E2. Optical filters isolate quantitative emission signals S1 and S2.

Ratiometric Measurements

In some cases—for example the Ca^{2+} indicators fura-2 and indo-1 (Section 19.2) and the pH indicators BCECF and SNARF® (Section 20.2)—the free and ion-bound forms of fluorescent ion indicators have different emission or excitation spectra. With this type of indicator, the ratio of the optical signals (S1 and S2 in Figure 3) can be used to monitor the association equilibrium and to calculate ion concentrations. Ratiometric measurements eliminate distortions of data caused by photobleaching and variations in probe loading and retention, as well as by instrumental factors such as illumination stability.[20,21] (Loading and Calibration of Intracellular Ion Indicators—Note 19.1).

Fluorescence Output of Fluorophores

Comparing Different Dyes

Fluorophores currently used as fluorescent probes offer sufficient permutations of wavelength range, Stokes shift and spectral bandwidth to meet requirements imposed by instrumentation (e.g., 488 nm excitation), while allowing flexibility in the design of multicolor labeling experiments. Our online Fluorescence SpectraViewer (www.invitrogen.com/handbook/spectraviewer) provides an interactive utility for evaluating these factors during the experimental design process (Using the Fluorescence SpectraViewer—Note 23.1). The fluorescence output of a given dye depends on the efficiency with which it absorbs and emits photons, and its ability to undergo repeated excitation/emission cycles. Absorption and emission efficiencies are most usefully quantified in terms of the molar extinction coefficient (EC) for absorption and the quantum yield (QY) for fluorescence. Both are constants under specific environmental conditions. The value of EC is specified at a single wavelength (usually the absorption maximum), whereas QY is a measure of the total photon emission over the entire fluorescence spectral profile. Fluorescence intensity per dye molecule is proportional to the product of EC and QY (Table 1). The range of these parameters among organic dye and autofluorescent protein fluorophores is approximately 5000 to 200,000 $cm^{-1}M^{-1}$ for EC and 0.05 to 1.0 for QY. Phycobiliproteins such as R-phycoerythrin (Section 6.4) have multiple fluorophores on each protein and consequently have much larger extinction coefficients (on the order of 2×10^6 $cm^{-1}M^{-1}$) than low molecular weight fluorophores. Qdot® nanocrystals have even larger extinction coefficients (>2×10^6 $cm^{-1}M^{-1}$), particularly in the blue visible and ultraviolet wavelength regions (Section 6.6).

Photobleaching

Under high-intensity illumination conditions, the irreversible destruction or photobleaching of the excited fluorophore becomes the primary factor limiting fluorescence detectability. The multiple photochemical reaction pathways responsible for photobleaching have been investigated and described in considerable detail.[22–24] Some pathways include reactions between adjacent dye molecules, making the process considerably more complex in labeled biological specimens than in dilute solutions of free dye. In all cases, photobleaching originates from the triplet excited state, which is created from the singlet state (S_1, Figure 1) via an excited-state process called intersystem crossing.[25]

The most effective remedy for photobleaching is to maximize detection sensitivity, which allows the excitation intensity to be

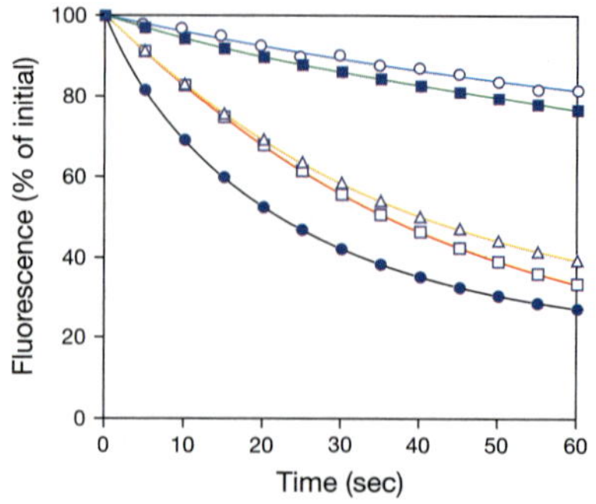

Figure 4 Comparison of photostability of green-fluorescent antibody conjugates. The following fluorescent goat anti–mouse IgG antibody conjugates were used to detect mouse anti–human IgG antibody labeling of human anti-nuclear antibodies in HEp-2 cells on prefixed test slides (INOVA Diagnostics Corp.): Alexa Fluor® 488 (A11001, ○), Oregon Green® 514 (O6383, ■), BODIPY® FL (B2752, △), Oregon Green® 488 (O6380, □) or fluorescein (F2761, ●). Samples were continuously illuminated and viewed on a fluorescence microscope using a fluorescein longpass filter set. Images were acquired every 5 seconds. For each conjugate, three data sets, representing different fields of view, were averaged and then normalized to the same initial fluorescence intensity value to facilitate comparison.

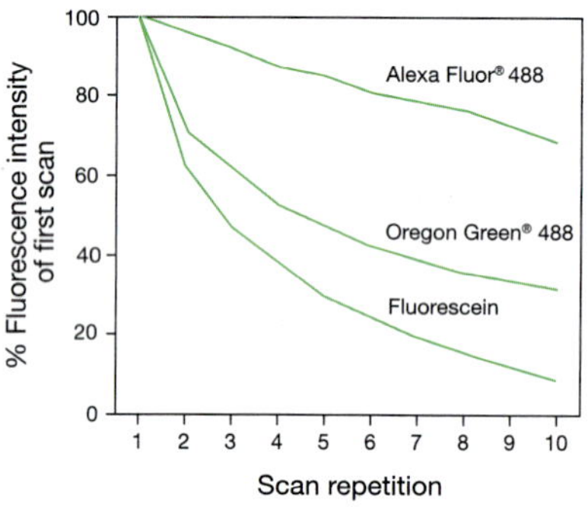

Figure 5 Photobleaching resistance of the green-fluorescent Alexa Fluor® 488, Oregon Green® 488 and fluorescein dyes, as determined by laser-scanning cytometry. EL4 cells were labeled with biotin-conjugated anti-CD44 antibody and detected by Alexa Fluor® 488 (S11223; S32354), Oregon Green® 488 (S6368) or fluorescein (S869) streptavidin. The cells were then fixed in 1% formaldehyde, washed and wet-mounted. After mounting, cells were scanned 10 times on a laser-scanning cytometer; laser power levels were 25 mW for the 488 nm spectral line of the argon-ion laser. Scan durations were approximately 5 minutes, and each repetition was started immediately after completion of the previous scan. Data are expressed as percentages derived from the mean fluorescence intensity (MFI) of each scan divided by the MFI of the first scan. Data contributed by Bill Telford, Experimental Transplantation and Immunology Branch, National Cancer Institute.

reduced. Detection sensitivity is enhanced by low-light detection devices such as CCD cameras, as well as by high-numerical aperture objectives and the widest bandpass emission filters compatible with satisfactory signal isolation. Alternatively, a less photolabile fluorophore may be substituted in the experiment.[26] Alexa Fluor® 488 dye is an important fluorescein substitute that provides significantly greater photostability than fluorescein (Figure 4, Figure 5, Figure 6), yet is compatible with standard fluorescein optical filters. Antifade reagents such as *SlowFade*® and ProLong® reagents (Section 23.1) can also be applied to reduce photobleaching; however, they are usually incompatible with live cells. In general, it is difficult to predict the necessity for and effectiveness of such countermeasures because photobleaching rates are dependent to some extent on the fluorophore's environment.[27–29]

Signal Amplification

The most straightforward way to enhance fluorescence signals is to increase the number of fluorophores available for detection.[30] Fluorescent signals can be amplified using 1) avidin–biotin or antibody–hapten secondary detection techniques, 2) enzyme-labeled secondary detection reagents in conjunction with fluorogenic substrates[31–33] or 3) probes that contain multiple fluorophores such as phycobiliproteins or FluoSpheres® fluorescent microspheres. Our most sensitive reagents and methods for signal amplification are discussed in Chapter 6.

Simply increasing the probe concentration can be counterproductive and often produces marked changes in the probe's chemical and optical characteristics. It is important to note that the effective intracellular concentration of probes loaded by bulk permeabilization methods (Loading and Calibration of Intracellular Ion Indicators—Note 19.1) is usually much higher (>10-fold) than the extracellular incubation concentration. Also, increased labeling of proteins or membranes ultimately leads to precipitation of the protein or gross changes in membrane permeability. Antibodies labeled with more than four to six fluorophores per protein may exhibit reduced specificity and reduced binding affinity. Furthermore, at high degrees of substitution, the extra fluorescence obtained per added fluorophore typically decreases due to self-quenching (Figure 7).

Environmental Sensitivity of Fluorescence

Fluorescence spectra and quantum yields are generally more dependent on the environment than absorption spectra and extinction coefficients. For example, coupling a single fluorescein label to a protein typically reduces fluorescein's QY ~60% but only decreases its EC by ~10%. Interactions either between two adjacent fluorophores or between a fluorophore and other species in the surrounding environment can produce environment-sensitive fluorescence.

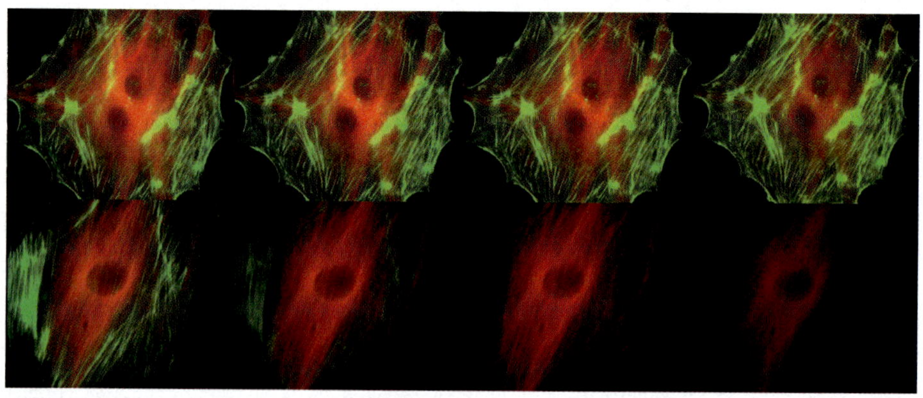

Figure 6 Comparison of the photobleaching rates of the Alexa Fluor® 488 and Alexa Fluor® 546 dyes and the well-known fluorescein and Cy®3 fluorophores. The cytoskeleton of bovine pulmonary artery endothelial cells (BPAEC) was labeled with (top series) Alexa Fluor® 488 phalloidin (A12379) and mouse monoclonal anti–α-tubulin antibody (A11126) in combination with Alexa Fluor® 546 goat anti–mouse IgG antibody (A11003) or (bottom series) fluorescein phalloidin (F432) and the anti–α-tubulin antibody in combination with a commercially available Cy®3 goat anti–mouse IgG antibody. The pseudocolored images were taken at 30-second intervals (0, 30, 90 and 210 seconds of exposure from left to right). The images were acquired with bandpass filter sets appropriate for fluorescein and rhodamine.

IMPORTANT NOTICE: The products described in this manual are covered by one or more Limited Use Label License(s). Please refer to the Appendix on page 971 and Master Product List on page 975. Products are For Research Use Only. Not intended for any animal or human therapeutic or diagnostic use.

www.invitrogen.com/probes

Fluorophore–Fluorophore Interactions

Fluorescence quenching can be defined as a bimolecular process that reduces the fluorescence quantum yield without changing the fluorescence emission spectrum (Table 1); it can result from transient excited-state interactions (collisional quenching) or from formation of nonfluorescent ground-state species. Self-quenching is the quenching of one fluorophore by another;[34] it therefore tends to occur when high loading concentrations or labeling densities are used (Figure 7, Figure 8). DQ™ substrates (Section 10.4) are heavily labeled and therefore highly quenched biopolymers that exhibit dramatic fluorescence enhancement upon enzymatic cleavage[35] (Figure 9).

Fluorescence resonance energy transfer (FRET) (Fluorescence Resonance Energy Transfer (FRET)—Note 1.2) is a strongly distance-dependent excited-state interaction in which emission of one fluorophore is coupled to the excitation of another. Some excited fluorophores interact to form excimers, which are excited-state dimers that exhibit altered emission spectra. Excimer formation by the polyaromatic hydrocarbon pyrene is described in Section 13.2 (Figure 10).

Because they all depend on the interaction of adjacent fluorophores, self-quenching, FRET and excimer formation can be exploited for monitoring a wide array of molecular assembly or fragmentation processes such as membrane fusion (Assays of Volume Change, Membrane Fusion and Membrane Permeability—Note 14.3), nucleic acid hybridization, ligand–receptor binding and polypeptide hydrolysis.

Other Environmental Factors

Many other environmental factors exert influences on fluorescence properties. The three most common are:

- Solvent polarity (solvent in this context includes interior regions of cells, proteins, membranes and other biomolecular structures)
- Proximity and concentrations of quenching species
- pH of the aqueous medium

Fluorescence spectra may be strongly dependent on solvent. This characteristic is most often observed with fluorophores that have large excited-state dipole moments, resulting in fluorescence spectral shifts to longer wavelengths in polar solvents. Representative fluorophores include the aminonaphthalenes such as prodan, badan (Figure 11) and dansyl, which are effective probes of environmental polarity in, for example, a protein's interior.[36]

Binding of a probe to its target can dramatically affect its fluorescence quantum yield (Monitoring Protein-Folding Processes with Environment-Sensitive Dyes—Note 9.1). Probes

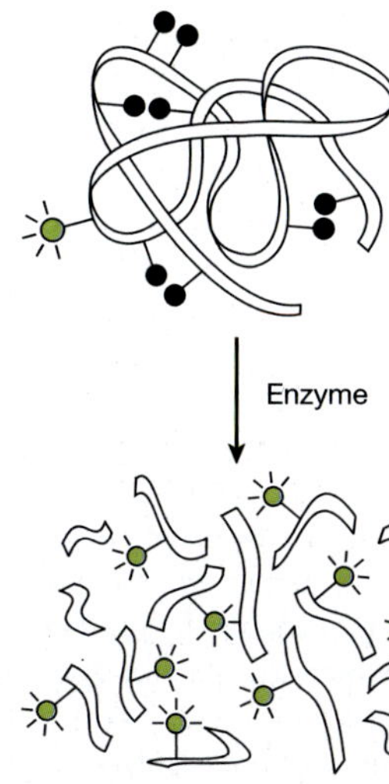

Figure 9 Principle of enzyme detection via the disruption of intramolecular self-quenching. Enzyme-catalyzed hydrolysis of the heavily labeled and almost totally quenched substrates provided in the EnzChek® Assay Kits relieves the intramolecular self-quenching, yielding brightly fluorescent reaction products.

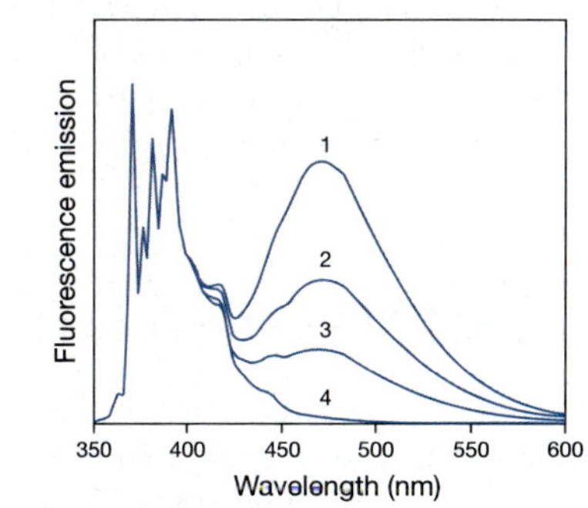

Figure 10 Excimer formation by pyrene in ethanol. Spectra are normalized to the 371.5 nm peak of the monomer. All spectra are essentially identical below 400 nm after normalization. Spectra are as follows: **1)** 2 mM pyrene, purged with argon to remove oxygen; **2)** 2 mM pyrene, air-equilibrated; **3)** 0.5 mM pyrene (argon-purged); and **4)** 2 µM pyrene (argon-purged). The monomer-to-excimer ratio (371.5/470 nm) is dependent on both pyrene concentration and the excited-state lifetime, which is variable because of quenching by oxygen.

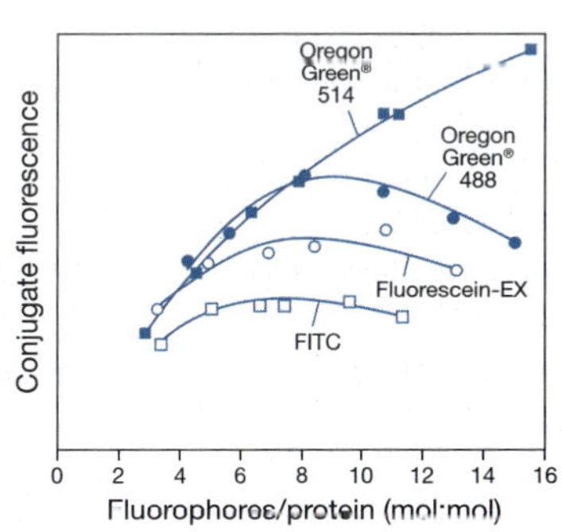

Figure 7 Comparison of relative fluorescence as a function of the number of fluorophores attached per protein for goat anti-mouse IgG antibody conjugates prepared using Oregon Green® 514 carboxylic acid succinimidyl ester (O6139, ■), Oregon Green® 488 carboxylic acid succinimidyl ester (O6147, ●), fluorescein-5-EX succinimidyl ester (F6130, ○) and fluorescein isothiocyanate (FITC; F143, F1906, F1907; □). Conjugate fluorescence is determined by measuring the fluorescence quantum yield of the conjugated dye relative to that of the free dye and multiplying by the number of fluorophores per protein.

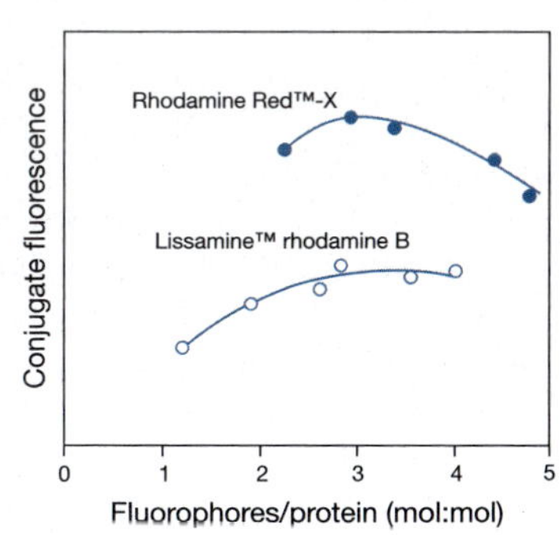

Figure 8 Comparison of the relative fluorescence of goat anti–mouse IgG antibody conjugates of Rhodamine Red™-X succinimidyl ester (R6160, ●) and Lissamine rhodamine B sulfonyl chloride (L20, L1908; ○). Conjugate fluorescence is determined by measuring the fluorescence quantum yield of the conjugated dye relative to that of the free dye and multiplying by the number of fluorophores per protein. Higher numbers of fluorophores attached per protein are attainable with Rhodamine Red™-X dye due to the lesser tendency of this dye to induce protein precipitation.

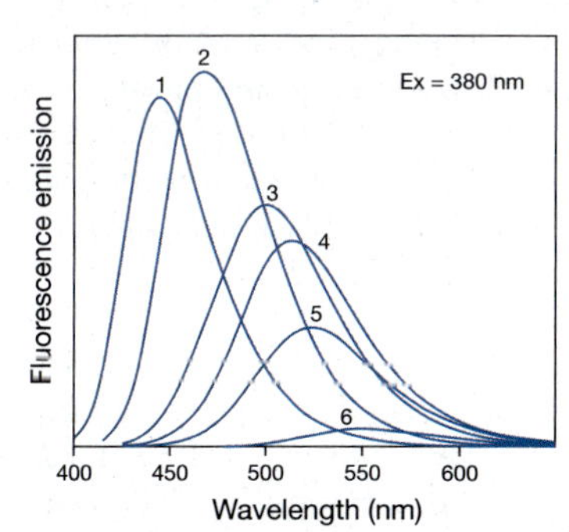

Figure 11 Fluorescence emission spectra of the 2-mercaptoethanol adduct of badan (B6057) in: **1)** toluene, **2)** chloroform, **3)** acetonitrile, **4)** ethanol, **5)** methanol and **6)** water. Each solution contains the same concentration of the adduct. Excitation of all samples is at 380 nm.

IMPORTANT NOTICE: The products described in this manual are covered by one or more Limited Use Label License(s). Please refer to the Appendix on page 971 and Master Product List on page 975. Products are For Research Use Only. Not intended for any animal or human therapeutic or diagnostic use.

that have a high fluorescence quantum yield when bound to a particular target but are otherwise effectively nonfluorescent yield extremely low reagent background signals. The ultrasensitive SYBR®, SYTO®, PicoGreen®, RiboGreen® and OliGreen® nucleic acid stains (Chapter 8) are prime examples of this strategy. Similarly, fluorogenic enzyme substrates, which are nonfluorescent or have only short-wavelength emission until they are converted to fluorescent products by enzymatic cleavage, allow sensitive detection of enzymatic activity (Chapter 10).

Extrinsic quenchers, the most ubiquitous of which are paramagnetic species such as O_2 and heavy atoms such as iodide, reduce fluorescence quantum yields in a concentration-dependent manner. If quenching is caused by collisional interactions, as is usually the case, information on the proximity of the fluorophore and quencher and their mutual diffusion rate can be derived. This quenching effect has been used productively to measure chloride-ion flux in cells (Section 21.2). Many fluorophores are also quenched by proteins. Examples are NBD, fluorescein and BODIPY® dyes, in which the effect is apparently due to charge-transfer interactions with aromatic amino acid residues.[37,38] Consequently, antibodies raised against these fluorophores are effective and highly specific fluorescence quenchers[38] (Section 7.4).

Fluorophores such as BCECF and carboxy SNARF®-1 that have strongly pH-dependent absorption and fluorescence characteristics can be used as physiological pH indicators. Fluorescein and hydroxycoumarins (umbelliferones) are further examples of this type of fluorophore. Structurally, pH sensitivity is due to a reconfiguration of the fluorophore's π-electron system that occurs upon protonation. BODIPY® FL and Alexa Fluor® 488 fluorophores, both of which lack protolytically ionizable substituents, provide spectrally equivalent alternatives to fluorescein for applications requiring a pH-*in*sensitive probe (Section 1.3, Section 1.4).

REFERENCES

1. Annu Rev Genomics Hum Genet (2008) 9:387; **2.** Science (2007) 315:81; **3.** Anal Chem (2008) 80:7063; **4.** Anal Chem (2001) 73:2070; **5.** Analyst (1994) 119:417; **6.** J Microsc (2007) 228:390; **7.** J Res Natl Inst Stand Technol (2001) 106:381; **8.** Methods Cell Biol (1994) 42B:605; **9.** J Microsc (1994) 176:281; **10.** Cytometry B Clin Cytom (2009) 76:295; **11.** Nat Protoc (2009) 4:372; **12.** Nat Rev Immunol (2004) 4:648; **13.** J Immunol Methods (2009) 344:6; **14.** J Neurosci Methods (2007) 162:119; **15.** J Neurosci Methods (2009) 180:116; **16.** BMC Cell Biol (2008) 9:13; **17.** Nat Protoc (2007) 2:1152; **18.** Nat Med (2006) 12:972; **19.** Biophys J (1983) 43:383; **20.** Methods Cell Biol (2007) 81:415; **21.** J Microsc (2009) 233:192; **22.** Nat Methods (2008) 5:197; **23.** J Phys Chem A (2007) 111:429; **24.** Chemphyschem (2008) 9:2019; **25.** Nat Methods (2007) 4:81; **26.** Org Lett (2004) 6:909; **27.** Biophys J (1995) 68:2588; **28.** J Cell Biol (1985) 100:1309; **29.** J Org Chem (1973) 38:1057; **30.** Mol Cell Probes (2008) 22:294; **31.** J Histochem Cytochem (2007) 55:545; **32.** Exp Cell Res (2007) 313:1943; **33.** J Histochem Cytochem (1995) 43:77; **34.** ACS Chem Biol (2009) 4:535; **35.** Anal Biochem (1997) 251:144; **36.** Nature (1986) 319:70; **37.** Bioconjug Chem (2003) 14:1133; **38.** Immunochemistry (1977) 14:533.

Selected Books and Articles

The preceding discussion has introduced some general principles to consider when selecting a fluorescent probe. Application-specific details are addressed in subsequent chapters of *The Molecular Probes® Handbook*. For in-depth treatments of fluorescence techniques and their biological applications, the reader is referred to the many excellent books and review articles listed below.

Principles of Fluorescence Detection

Albani, J.R., *Absorption et Fluorescence: Principes et Applications*, Lavoisier (2001). This book is the first on absorption and fluorescnece to be published in the French language.

Albani, J.R., *Principles and Applications of Fluorescence Spectroscopy*, Wiley-Blackwell (2007).

Brand, L. and Johnson, M.L., Eds., *Fluorescence Spectroscopy (Methods in Enzymology, Volume 450)*, Academic Press (2008).

Gell, C., Brockwell, D. and Smith, A., *Handbook of Single Molecule Fluorescence Spectroscopy*, Oxford University Press (2006).

Goldys, E.M., Ed., *Fluorescence Applications in Biotechnology and Life Sciences*, Wiley-Blackwell (2009).

Guilbault, G.G., Ed., *Practical Fluorescence, Second Edition*, Marcel Dekker (1990).

Joo, C., Balci, H., Ishitsuka, Y., Buranachai, C. and Ha, T., "Advances in single-molecule fluorescence methods for molecular biology," Annu Rev Biochem (2008) 77:51–76.

Lakowicz, J.R., *Principles of Fluorescence Spectroscopy, Third Edition*, Springer (2006).

Mathies, R.A., Peck, K. and Stryer, L., "Optimization of high-sensitivity fluorescence detection," Anal Chem (1990) 62:1786–1791.

Royer, C.A., "Approaches to teaching fluorescence spectroscopy," Biophys J (1995) 68:1191–1195.

Selvin, P.R. and Ha, T., Eds., *Single-Molecule Techniques: A Laboratory Manual*, Cold Spring Harbor Laboratory Press (2007).

Valeur, B., *Molecular Fluorescence: Principles and Applications*, John Wiley and Sons (2002).

Fluorophores and Fluorescent Probes

Berlman, I.B., *Handbook of Fluorescence Spectra of Aromatic Molecules, Second Edition*, Academic Press (1971).

Burry, R.W., *Immunocytochemistry: A Practical Guide for Biomedical Research*, Springer (2009).

Chalfie, M. and Kain, S.R., *Green Fluorescent Protein: Properties, Applications and Protocols, Second Edition*, John Wiley and Sons (2006).

Drexhage, K.H., "Structure and properties of laser dyes" in *Dye Lasers, Third Edition*, F.P. Schäfer, Ed., Springer-Verlag, (1990) p. 155–200.

Green, F.J., *The Sigma-Aldrich Handbook of Stains, Dyes and Indicators*, Aldrich Chemical Company (1990).

Griffiths, J., *Colour and Constitution of Organic Molecules*, Academic Press (1976).

IMPORTANT NOTICE: The products described in this manual are covered by one or more Limited Use Label License(s). Please refer to the Appendix on page 971 and Master Product List on page 975. Products are For Research Use Only. Not intended for any animal or human therapeutic or diagnostic use.

Fluorophores and Fluorescent Probes, *continued*

Hermanson, G.T., *Bioconjugate Techniques, Second Edition*, Academic Press (2008).

Hilderbrand, S.A., "Labels and probes for live cell imaging: Overview and selection guide", Methods Mol Biol (2010) 581:17–45.

Kobayashi, H., Ogawa, M., Alford, R., Choyke, P.L. and Urano, Y., "New strategies for fluorescent probe design in medical diagnostic imaging," Chem Rev (2010) 110:2620–2640.

Kricka, L.J. and Fortina, P., "Analytical ancestry: "Firsts" in fluorescent labeling of nucleosides, nucleotides, and nucleic acids," Clin Chem (2009) 55:670–683.

Liehr, T., Ed., *Fluorescence In Situ Hybridization (FISH): Application Guide*, Springer (2008).

Mason, W.T., Ed., *Fluorescent and Luminescent Probes for Biological Activity, Second Edition*, Academic Press (1999).

Oliver, C. and Jamur, M.C., Eds., *Immunocytochemical Methods and Protocols, Third Edition (Methods in Molecular Biology, Volume 588)*, Humana Press (2010).

Taraska, J.W. and Zagotta, W.N., "Fluorescence applications in molecular neurobiology", Neuron (2010) 66:170–189.

Fluorescence Microscopy

Frigault, M.M., Lacoste, J., Swift, J.L. and Brown, C.M., "Live-cell microscopy: Tips and tools," J Cell Sci (2009) 122:753–767.

Fujimoto, J.G. and Farkas, D., Eds., *Biomedical Optical Imaging*, Oxford University Press (2009).

Goldman, R.D., Swedlow, J.R. and Spector, D.L., Eds., *Live Cell Imaging: A Laboratory Manual, Second Edition*, Cold Spring Harbor Laboratory Press (2009).

Hell, S.W., "Microscopy and its focal switch," Nat Methods (2009) 6:24–32.

Herman, B., *Fluorescence Microscopy, Second Edition*, BIOS Scientific Publishers (1998).

Hibbs, A.R., *Confocal Microscopy for Biologists*, Springer (2004).

Huang, B., Bates, M. and Zhuang, X., "Super-resolution fluorescence microscopy," Annu Rev Biochem (2009) 78:993–1016.

Inoué, S. and Spring, K.R., *Video Microscopy, Second Edition*, Plenum Publishing (1997).

Lichtman, J.W. and Conchello, J.A., "Fluorescence microscopy," Nat Methods (2005) 2:910–919.

Masters, B.R., *Confocal Microscopy and Multiphoton Excitation Microscopy: The Genesis of Live Cell Imaging*, SPIE Press (2006).

Matsumoto, B., Ed., *Cell Biological Applications of Confocal Microscopy, Second Edition (Methods in Cell Biology, Volume 70)*, Academic Press (2003).

Murphy, D.B., *Fundamentals of Light Microscopy and Electronic Imaging*, John Wiley and Sons (2001).

Ntziachristos, V., "Fluorescence molecular imaging," Annu Rev Biomed Eng (2006) 8:1–32.

Patterson, G., Davidson, M., Manley, S. and Lippincott-Schwartz, J., "Superresolution imaging using single-molecule localization," Annu Rev Phys Chem (2010) 61:345–367.

Pawley, J.B., Ed., *Handbook of Biological Confocal Microscopy, Third Edition*, Springer (2006).

Periasamy, A., Ed., *Methods in Cellular Imaging*, Oxford University Press (2001).

Rosenthal, E. and Zinn, K.R., Eds., *Optical Imaging of Cancer: Clinical Applications*, Springer (2009).

Spector, D.L. and Goldman, R.D., *Basic Methods in Microscopy*, Cold Spring Harbor Laboratory Press (2005).

Svoboda, K. and Yasuda, R., "Principles of two-photon excitation microscopy and its applications to neuroscience," Neuron (2006) 50:823–839.

Tsien, R.Y., " Imagining imaging's future," Nat Rev Mol Cell Biol (2003) 4:SS16–SS21.

Yuste, R. and Konnerth, A., Eds., *Imaging in Neuroscience and Development: A Laboratory Manual*, Cold Spring Harbor Laboratory Press (2005).

Flow Cytometry

Darzynkiewicz, Z., Crissman, H.A. and Robinson, J.P., Eds., *Cytometry, Third Edition Parts A and B (Methods in Cell Biology, Volumes 63 and 64)*, Academic Press (2001).

Givan, A.L., *Flow Cytometry: First Principles, Second Edition*, John Wiley and Sons (2001).

Herzenberg, L.A., Parks, D., Sahaf, B., Perez, O., Roederer, M. and Herzenberg, L.A., "The history and future of the fluorescence activated cell sorter and flow cytometry: A view from Stanford," Clin Chem (2002) 48:1819–1827.

Herzenberg, L.A., Tung, J., Moore, W.A., Herzenberg, L.A. and Parks, D.R., "Interpreting flow cytometry data: A guide for the perplexed," Nat Immunol (2006) 7:681–685.

Preffer F. and Dombkowski, D. "Advances in complex multiparameter flow cytometry technology: Applications in stem cell research," Cytometry B (2009) 76:295–314.

Shapiro, H.M., "Optical measurement in cytometry: Light scattering, extinction, absorption and fluorescence," Meth Cell Biol (2001) 63:107–129.

Shapiro, H.M., *Practical Flow Cytometry, Fourth Edition*, Wiley-Liss (2003).

Sklar, L.A., Ed., *Flow Cytometry for Biotechnology*, Oxford University Press (2005)

Other Fluorescence Measurement Techniques

Dorak, M.T., Ed., *Real-Time PCR*, Taylor and Francis (2006).

Gore, M., Ed., *Spectrophotometry and Spectrofluorimetry: A Practical Approach, Second Edition*, Oxford University Press (2000).

Mardis, E.R. "Next-generation DNA sequencing methods," Annu Rev Genomics Hum Genet (2008) 9:387–402.

Patton, W.F., "A thousand points of light: The application of fluorescence detection technologies to two-dimensional gel electrophoresis and proteomics," Electrophoresis (2000) 21:1123–1144.

Shimomura, O., *Bioluminescence: Chemical Principles and Methods*, World Scientific Publishing (2006).

IMPORTANT NOTICE: The products described in this manual are covered by one or more Limited Use Label License(s). Please refer to the Appendix on page 971 and Master Product List on page 975. Products are For Research Use Only. Not intended for any animal or human therapeutic or diagnostic use.

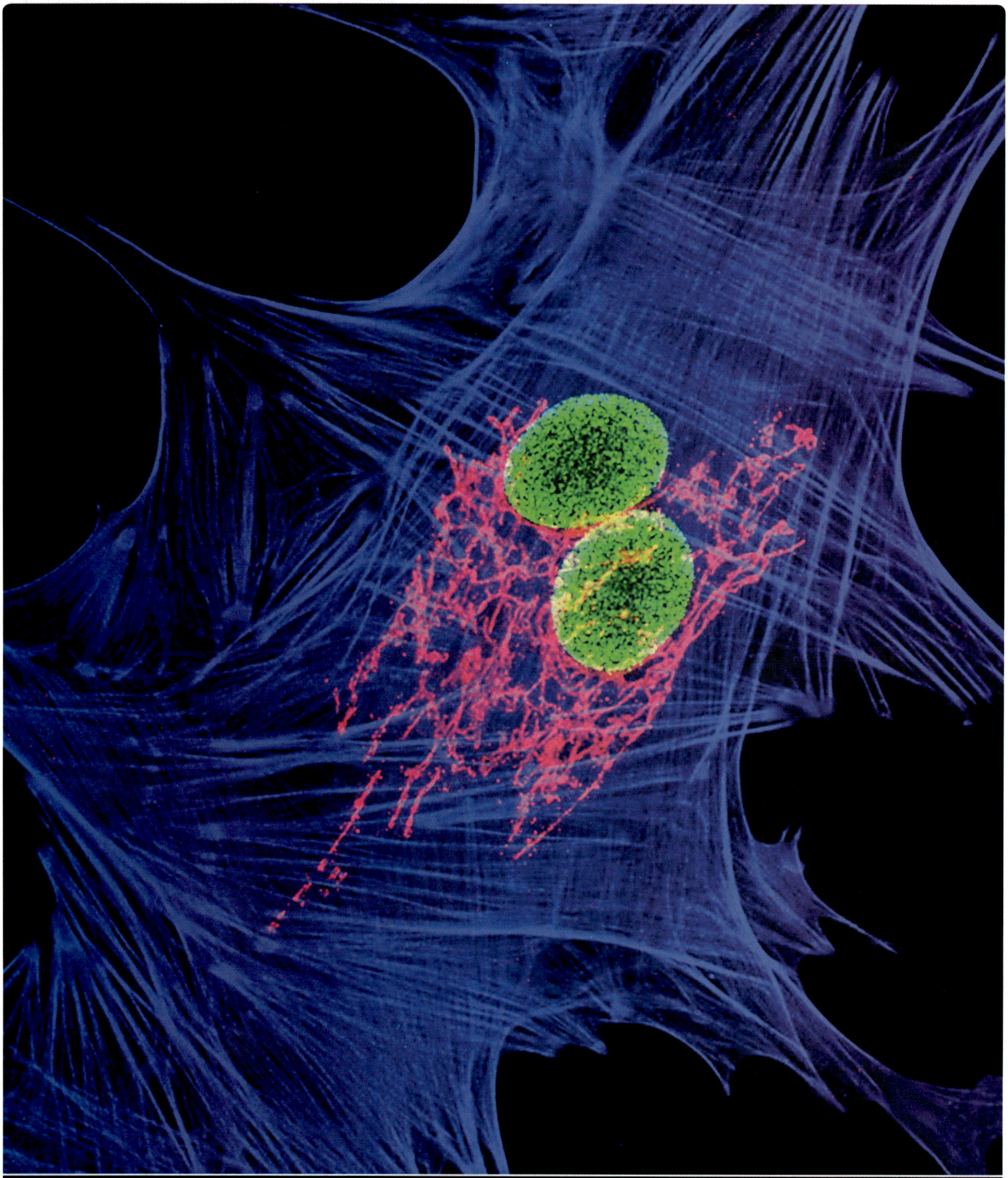

Alexa Fluor® 568 goat anti–rabbit IgG, Alexa Fluor® 488 goat anti–mouse IgG and Alexa Fluor® 350 phalloidin.

IMPORTANT NOTICE: The products described in this manual are covered by one or more Limited Use Label License(s). Please refer to the Appendix on page 971 and Master Product List on page 975. Products are For Research Use Only. Not intended for any animal or human therapeutic or diagnostic use.

ONE

CHAPTER 1

Fluorophores and Their Amine-Reactive Derivatives

IMPORTANT NOTICE: The products described in this manual are covered by one or more Limited Use Label License(s). Please refer to the Appendix on page 971 and Master Product List on page 975. Products are For Research Use Only. Not intended for any animal or human therapeutic or diagnostic use.

IMPORTANT NOTICE: The products described in this manual are covered by one or more Limited Use Label License(s). Please refer to the Appendix on page 971 and Master Product List on page 975. Products are For Research Use Only. Not intended for any animal or human therapeutic or diagnostic use.

IMPORTANT NOTICE: The products described in this manual are covered by one or more Limited Use Label License(s). Please refer to the Appendix on page 971 and Master Product List on page 975. Products are For Research Use Only. Not intended for any animal or human therapeutic or diagnostic use.

IMPORTANT NOTICE: The products described in this manual are covered by one or more Limited Use Label License(s). Please refer to the Appendix on page 971 and Master Product List on page 975. Products are For Research Use Only. Not intended for any animal or human therapeutic or diagnostic use.

1.1 Introduction to Amine Modification

The Molecular Probes® Handbook describes a full spectrum of fluorophores and haptens for covalent derivatization of biopolymers and low molecular weight molecules. Chapters 1–5 describe the chemical and spectroscopic properties of the reactive reagents we offer, whereas the remainder of this book is primarily devoted to our diverse collection of fluorescent probes and their applications in cell biology, neurobiology, immunology, molecular biology and biophysics.

Common Applications for Amine-Reactive Probes

Labeling Biomolecules

Amine-reactive probes are widely used to modify proteins, peptides, ligands, synthetic oligonucleotides and other biomolecules. In contrast to our thiol-reactive reagents (Chapter 2), which frequently serve as probes of protein structure and function, amine-reactive dyes are most often used to prepare bioconjugates for immunochemistry, fluorescence *in situ* hybridization (FISH), cell tracing, receptor labeling and fluorescent analog cytochemistry. In these applications, the stability of the chemical bond between the dye and biomolecule is critical. The bioconjugate will typically be applied to a biochemically complex (and sometimes active) specimen. Furthermore, it will often be subjected to a series of post-processing steps such as washing, permeabilization, fixation and mounting. The integrity of the bioconjugate must be maintained throughout these processes if the fluorescence signal generated by the dye is to have any useful relationship to the abundance or localization of the bioconjugate's molecular target.

Our selection of amine-reactive fluorophores for modifying biomolecules covers the entire visible and near-infrared spectrum (Table 1.1, Table 1.2). Up-to-date bibliographies are available for most of our amine-reactive probes at www.invitrogen.com. Also available are other product-specific bibliographies, as well as keyword searches of the over 60,000 literature references in our product application bibliography database.

Chapter 1 discusses the properties of Molecular Probes® fluorophores, including:

- Alexa Fluor® dyes (Section 1.3)
- BODIPY® dyes (Section 1.4)
- Oregon Green® and Rhodamine Green™ dyes (Section 1.5)
- Rhodamine Red™-X and Texas Red® dyes (Section 1.6)
- UV light–excitable Cascade Blue®, Cascade Yellow™, Marina Blue®, Pacific Blue™ and AMCA-X fluorophores (Section 1.7)

Our essentially nonfluorescent QSY® dyes (Section 1.6, Section 1.8) have strong visible absorption, making them excellent acceptors for fluorescence resonance energy transfer (FRET) applications (Fluorescence Resonance Energy Transfer (FRET)—Note 1.2).

Preparing the Optimal Bioconjugate

The preferred bioconjugate usually has a high fluorescence output (or, in the case of a haptenylated conjugate, a suitable degree of labeling) yet retains the critical functional properties of the unlabeled biomolecule, such as solubility, selective binding to a receptor or nucleic acid, activation or inhibition of a particular enzyme or the ability to incorporate into a biological membrane.[1] Frequently, however, conjugates with the highest degree of labeling precipitate out of solution or bind nonspecifically.[2] Thus there is usually a trade-off between degree of labeling and functional properties that must be resolved through experimental optimization. Lysine residues, the primary targets for amine modification of proteins, are relatively abundant. In mammalian proteins, lysine has the fifth highest occurrence frequency of the 20 naturally occurring amino acids. A typical IgG antibody molecule has about 90 lysine residues, of which about 30 at most can be modified under forcing conditions of high acylating reagent concentration and prolonged incubation. However, maintenance of functional properties (more specifically, antigen binding affinity) typically requires a degree of labeling of <10 dyes per IgG, representing a low fractional modification of available targets. A further consequence of low fractional modification is that, with the exception of small peptides or rare proteins with few lysine residues, bioconjugates prepared by amine modification are polydisperse mixtures containing a range of dye:protein stoichiometries.[3,4]

For the most critical assays, we recommend that researchers consider preparing and optimizing their own conjugates. Our amine-reactive dyes are supplied with a detailed protocol that describes how to use them for labeling biomolecules. This procedure is straightforward and requires no special equipment. Following conjugation, it is very important to remove as much unconjugated labeling reagent as possible, usually by gel filtration, dialysis, bioconjugate precipitation and resolubilization, HPLC or a combination of these techniques. The presence of free dye, particularly if it remains chemically reactive, can greatly complicate subsequent experiments with the bioconjugate. The entire process of labeling reaction and conjugate purification can be completed in little more than two hours, and the main prerequisite is a sufficient amount of purified protein or amine-modified nucleic acid.

With the exception of the phycobiliproteins (Section 6.4, Table 6.2), fluorescent microspheres (Section 6.5, Table 6.7), Qdot® nanocrystals (Section 6.6), Zenon® Antibody Labeling Kits (Section 7.3, Table 7.7) and ULYSIS® Nucleic Acid Labeling Kits (Section 8.2, Table 8.6), virtually all reagents used to prepare Molecular Probes® fluorescent bioconjugates are amine-reactive organic fluorophores, and almost all are described in this chapter.

We have also developed convenient kit formats for labeling proteins and nucleic acids with our most important fluorophores, or alternatively with biotin. Section 1.2 and Table 1.3 include a complete description of these kits. Alternatively, we prepare custom fluorescent conjugates for research use; contact Invitrogen Custom Services for more information. Conjugations with phycobiliproteins, fluorescent polystyrene microspheres and Qdot® nanocrystals require specialized procedures that are described in Section 6.4, Section 6.5 and Section 6.6, respectively.

Derivatizing Low Molecular Weight Molecules

Some amine-reactive probes described in this chapter are also important reagents for various bioanalytical applications, including amine quantitation, protein and nucleic acid sequencing and chromatographic and electrophoretic analysis of low molecular weight molecules. Reagents that are particularly useful for derivatizing low molecular weight amines—including fluorescamine, *o*-phthaldialdehyde, ATTO-TAG™ reagents, NBD chloride and dansyl chloride—are

IMPORTANT NOTICE: The products described in this manual are covered by one or more Limited Use Label License(s). Please refer to the Appendix on page 971 and Master Product List on page 975. Products are For Research Use Only. Not intended for any animal or human therapeutic or diagnostic use.

Table 1.1 Molecular Probes® amine-reactive dyes.

Fluorophore	COOH *	SE *	Other *	Abs (nm)	Em (nm)	Notes
Methoxycoumarin	M1420MP	M1410		340	405	• pH-insensitive alternative to 7-hydroxycoumarins
Dansyl		D6104 (X)	D21 (SC)	340	520 ‡	• Environment-sensitive fluorescence • Large Stokes shift
Pyrene		P130 P6114	P24 (SC)	345	378	• Long excited-state lifetime • Spectral shifts due to excimer emission
Alexa Fluor® 350		A10168		346	442	• Higher fluorescence output than AMCA • Optimally detected with DAPI optical filter sets
AMCA		A6118 (X)		349	448	• Widely used blue-fluorescent labeling dye • Compact structure
Marina Blue® dye		M10165		365	460	• Strongly fluorescent at neutral pH • Optimally detected with DAPI optical filter sets
Dapoxyl® dye		D10161	D10160 (SC)	373	551 ‡	• Environment-sensitive fluorescence • Large Stokes shift
Dialkylaminocoumarin	D126 D1421	D374 D1412		375 430	470 § 475 **	• Longer-wavelength alternatives to AMCA
Bimane	B30250			380	458	• Blue-fluorescent dye • Small size
Hydroxycoumarin	H185 H1428	H1193		385	445 ††	• pH-sensitive fluorescence • Compact structure
Cascade Blue® dye			C2284 (AA)	400	420	• Resistant to quenching upon protein conjugation • Water soluble
Pacific Orange™ dye		P30253		400	551	• Excited with the 405 nm violet diode laser • Compatible with the Pacific Blue™ dye for two-color analysis using the violet diode laser
Alexa Fluor® 405		A30000 A30100		402	421	• Cascade Blue® derivative containing a spacer between the fluorophore and the reactive SE • Near-perfect match to the 405 nm violet diode laser
Cascade Yellow™ dye		C10164		402	545	• Large Stokes shift • High molar absorptivity
Pacific Blue™ dye		P10163		410	455	• Recommended for 405 nm violet diode laser excitation
PyMPO		S6110		415	570	• Large Stokes shift
Alexa Fluor® 430		A10169		434	539	• Large Stokes shift
NBD		S1167 (X)	F486 (AH)	465	535	• Environment-sensitive fluorescence • Compact structure
QSY® 35		Q20133		475	none	• Nonfluorescent quencher • An efficient energy transfer acceptor from blue and green fluorophores
Fluorescein	C1359 C1360 C1904 †	C2210 C6164 C1311 † F6106 (X) F2181 (X) † F6129 (X) † F6130 (EX) C20050 (PA)	D16 (DTA) F143 (ITC) F1906 (ITC) F1907 (ITC)	494	518	• Most widely used green-fluorescent labeling dye • Absorption overlaps the 488 nm spectral line of the argon-ion laser • Prone to photobleaching • pH-sensitive fluorescence between pH 5–8 • Fluorescein-5-EX succinimidyl ester (F6130) is the preferred reactive fluorescein for protein conjugation
Alexa Fluor® 488		A20000 † A20100 †	A30005 (TFP) A30052 (SDP)	495	519	• Bright and photostable fluorescein substitute • Fluorescence output unmatched by any other spectrally similar dye • pH-insensitive fluorescence between pH 4 and 10 • Ideal for excitation by the 488 nm spectral line of the argon-ion laser • As compared with the SE, the TFP ester is less susceptible to spontaneous hydrolysis during conjugation reactions
Oregon Green® 488	O6146	O6147 O6149	O6080 (ITC) †	496	524	• Photostable fluorescein substitute • pH-insensitive fluorescence at pH >6
BODIPY® 493/503		D2191		500	506	• pH-insensitive fluorescence • Narrow spectral bandwidth • Higher 488 nm absorptivity than the BODIPY® FL fluorophore
Rhodamine Green dye		R6107 † R6113 (X) †		502	527	• Photostable fluorescein substitute • pH-insensitive fluorescence

The absorption (Abs) and fluorescence emission (Em) maxima listed in this table are for the goat anti–mouse IgG antibody or dextran conjugates in aqueous buffer. Our online Fluorescence SpectraViewer (www.invitrogen.com/handbook/spectraviewer) provides an interactive utility for plotting and comparing fluorescence excitation and emission spectra for over 250 fluorophores (Using the Fluorescence SpectraViewer—Note 23.1). * COOH = carboxylic acid; SE = succinimidyl ester; (AA) = acetyl azide; (AH) = aryl halide; (C_5) = pentanoic acid; (DTA) = dichlorotriazine; (EX) = seven-atom spacer that is more hydrophilic than X; (ITC) = isothiocyanate; (PA) = photoactivatable; (SC) = sulfonyl chloride; (SDP) = sulfodichlorophenol ester; (SSE) = sulfosuccinimidyl ester; (STP) = 4-sulfotetrafluorophenyl ester; (TFP) = tetrafluorophenyl ester; (X) = an aminohexanoyl spacer between the dye and SE. † Mixed isomers. ‡ Emission spectra of dansyl and Dapoxyl® conjugates may vary considerably depending on the dye attachment site and the degree of labeling. § Spectral maxima for D374. ** Spectral maxima for D1412. †† Spectral maxima for H1193. More information on amine-reactive dyes is available at www.invitrogen.com/handbook/labelingchemistry.

continued on next page

IMPORTANT NOTICE: The products described in this manual are covered by one or more Limited Use Label License(s). Please refer to the Appendix on page 971 and Master Product List on page 975. Products are For Research Use Only. Not intended for any animal or human therapeutic or diagnostic use.

Table 1.1 Molecular Probes® amine-reactive dyes—*continued.*

Fluorophore	COOH *	SE *	Other *	Abs (nm)	Em (nm)	Notes
BODIPY® FL	D2183 ‡ D3834 (C_5)	D2184 D6140 (SSE) D6102 (X) D6184 (C_5)	B10006 (STP)	505	513	• BODIPY® substitute for fluorescein • pH-insensitive fluorescence • Narrow spectral bandwidth
2',7'-Dichloro-fluorescein	C368 †			510	532	• pH-insensitive fluorescence at pH >6
Oregon Green® 514	O6138	O6139		511	530	• Exceptionally photostable • pH-insensitive fluorescence at pH >6
Alexa Fluor® 514		A30002 †		518	540	• Designed to be optically resolved from the Alexa Fluor® 488 dye using spectral imaging instruments with linear-unmixing software • Optimal dye for the 514 nm spectral line of the argon-ion laser
4',5'-Dichloro-2',7'-dimethoxy-fluorescein (JOE)		C6171MP		522	550	• Succinimidyl ester derivative (6-JOE, SE; C6171MP) is widely used for oligonucleotide labeling
Eosin			E18 (ITC)	524	544	• Useful for DAB photoconversion • Phosphorescent
Rhodamine 6G		C6127 C6128 C6157 †		525	555	• Excited by the 514 nm spectral line of the argon-ion laser • Spectra intermediate between those of fluorescein and tetramethylrhodamine
BODIPY® R6G		D6180		528	550	• BODIPY® substitute with spectra similar to rhodamine 6G • pH-insensitive fluorescence • Narrow spectral bandwidth
Alexa Fluor® 532		A20001 A20101MP		531	554	• Bright and photostable dye with spectra intermediate between those of fluorescein and tetramethylrhodamine • Fluorescence output unmatched by any other spectrally similar dye • pH-insensitive fluorescence between pH 4 and 10 • Ideal for excitation by the 532 nm frequency-doubled principal line output of the Nd:YAG laser
BODIPY® 530/550		D2187		534	554	• pH-insensitive fluorescence • Narrow spectral bandwidth
BODIPY® TMR		D6117 (X)		542	574	• BODIPY® substitute for tetramethylrhodamine • pH-insensitive fluorescence • Narrow emission spectral bandwidth
Alexa Fluor® 555		A20009 A20109		555	565	• Red-orange fluorescence • Bright and photostable tetramethylrhodamine or Cy®3 substitute • Spectrally similar to Cy®3
Tetramethyl-rhodamine (TMR)	C6121 C6122 C300 †	C2211 C6123 C1171 † T6105 (X) †	T1480 (ITC) T1481 (ITC) T490 (ITC) †	555	580	• pH-insensitive fluorescence • Good photostability • Prone to aggregation
Alexa Fluor® 546		A20002 † A20102 †		556	575	• Bright and photostable tetramethylrhodamine or Cy®3 substitute • Fluorescence output unmatched by any other spectrally similar dye • pH-insensitive fluorescence between pH 4 and 10 • Less prone to aggregation than tetramethylrhodamine
BODIPY® 558/568		D2219		558	569	• pH-insensitive fluorescence • Narrow spectral bandwidth
QSY® 7		Q10193		560	none	• Nonfluorescent quencher • Broad visible-wavelength absorption • Efficient energy transfer acceptor from green and orange fluorophores
QSY® 9		Q20131		562	none	• Nonfluorescent quencher • Spectrally similar to QSY® 7, but with enhanced water solubility • Efficient energy-transfer acceptor from green and orange fluorophores
BODIPY® 564/570		D2222		565	571	• pH-insensitive fluorescence • Narrow spectral bandwidth
Lissamine rhodamine B			L20 (SC) † L1908 (SC) †	570	590	• Optimal for 568 nm excitation • Photostable
Rhodamine Red dye		R6160 (X) †		570	590	• Conjugates are generally more fluorescent than those of Lissamine rhodamine B sulfonyl chloride, and the succinimidyl ester is more stable in H_2O

The absorption (Abs) and fluorescence emission (Em) maxima listed in this table are for the goat anti-mouse IgG antibody or dextran conjugates in aqueous buffer. Our online Fluorescence SpectraViewer (www.invitrogen.com/handbook/spectraviewer) provides an interactive utility for plotting and comparing fluorescence excitation and emission spectra for over 250 fluorophores (Using the Fluorescence SpectraViewer—Note 23.1). * COOH = carboxylic acid; SE = succinimidyl ester; (AA) = acetyl azide; (AH) = aryl halide; (C_5) = pentanoic acid; (DTA) = dichlorotriazine; (EX) = seven-atom spacer that is more hydrophilic than X; (ITC) = isothiocyanate; (PA) = photoactivatable; (SC) = sulfonyl chloride; (SDP) = sulfodichlorophenol ester; (SSE) = sulfosuccinimidyl ester; (STP) = 4-sulfotetrafluorophenyl ester; (TFP) = tetrafluorophenyl ester; (X) = an aminohexanoyl spacer between the dye and SE. † Mixed isomers. ‡ Emission spectra of dansyl and Dapoxyl® conjugates may vary considerably depending on the dye attachment site and the degree of labeling. § Spectral maxima for D374. ** Spectral maxima for D1412. †† Spectral maxima for H1193. More information on amine-reactive dyes is available at www.invitrogen.com/handbook/labelingchemistry.

continued on next page

IMPORTANT NOTICE: The products described in this manual are covered by one or more Limited Use Label License(s). Please refer to the Appendix on page 971 and Master Product List on page 975. Products are For Research Use Only. Not intended for any animal or human therapeutic or diagnostic use.

Table 1.1 Molecular Probes® amine-reactive dyes—*continued.*

Fluorophore	COOH *	SE *	Other *	Abs (nm)	Em (nm)	Notes
BODIPY® 576/589		D2225		576	590	• pH-insensitive fluorescence • Narrow spectral bandwidth
Alexa Fluor® 568		A20003 † A20103 †		578	603	• Bright and photostable Lissamine rhodamine B substitute • Fluorescence output unmatched by any other spectrally similar dye • pH-insensitive fluorescence between pH 4 and 10
X-rhodamine	C6124 C6156	C6125 C6126 C1309 †	X491 (ITC) †	580	605	• Succinimidyl ester derivative (6-ROX, SE; C6126) widely used for oligonucleotide labeling
BODIPY® 581/591		D2228		584	592	• pH-insensitive fluorescence • Narrow spectral bandwidth
BODIPY® TR		D6116 (X)		589	617	• BODIPY® substitute for the Texas Red® fluorophore • pH-insensitive fluorescence
Alexa Fluor® 594		A20004 † A20104 †		590	617	• Bright and photostable Texas Red® dye substitute • Fluorescence output unmatched by any other spectrally similar dye • pH-insensitive fluorescence between pH 4 and 10 • Ideal for excitation by the 594 nm spectral line of the He-Ne laser
Texas Red® dye		T6134 (X) † T20175 (X)	T353 (SC) † T1905 (SC) †	595	615	• Good spectral separation from green fluorophores • Texas Red®-X succinimidyl ester typically yields higher fluorescence per attached dye than Texas Red® sulfonyl chloride and is more stable in H_2O
Naphthofluorescein	C652 †	C653 †		605	675	• Very long-wavelength excitation and emission • pH-sensitive fluorescence
Alexa Fluor® 610-X		A30050		612	628	• Bright and photostable Texas Red® dye substitute • Fluorescence output unmatched by any other spectrally similar dye • pH-insensitive fluorescence between pH 4 and 10 • Easily differentiated from green fluorophores • Still visible by eye, unlike longer-wavelength fluorophores
BODIPY® 630/650		D10000 (X)		625	640	• pH-insensitive fluorescence • Ideal for excitation by the 633 nm spectral line of the He-Ne laser
Malachite green			M689 (ITC)	630	none	• Nonfluorescent photosensitizer
Alexa Fluor® 633		A20005 † A20105 †		632	647	• Far-red fluorescence • Good spectral separation from green fluorophores • pH-insensitive fluorescence between pH 4 and 10
Alexa Fluor® 635				633	647	• Far-red fluorescence • Good spectral separation from green fluorophores • pH-insensitive fluorescence between pH 4 and 10
BODIPY® 650/665		D10001 (X)		646	660	• pH-insensitive fluorescence • Longest-wavelength BODIPY® dye currently available
Alexa Fluor® 647		A20006 A20106		650	668	• Far-red fluorescence • Produces conjugates that are brighter than those of the Cy®5 dye • pH-insensitive fluorescence between pH 4 and 10
QSY® 21		Q20132		661	none	• Nonfluorescent quencher • Long-wavelength absorption • An efficient energy transfer acceptor from red and near-infrared fluorophores
Alexa Fluor® 660		A20007 A20107		663	690	• Far-red fluorescence • Good spectral separation from green and red-orange fluorophores • pH-insensitive fluorescence between pH 4 and 10
Alexa Fluor® 680		A20008 A20108		679	702	• Far-red fluorescence • Good separation from red fluorophores—useful for three- and four-color applications • pH-insensitive fluorescence between pH 4 and 10
Alexa Fluor® 700		A20010 A20110		702	723	• Far-red fluorescence • Good separation from red fluorophores—useful for three- and four-color applications • pH-insensitive fluorescence between pH 4 and 10
Alexa Fluor® 750		A20011 A20111		749	775	• Far-red fluorescence • Good separation from red fluorophores—useful for three- and four-color applications • pH-insensitive fluorescence between pH 4 and 10 • Spectrally similar to the Cy®7 dye
Alexa Fluor® 790		A30051		782	805	• Spectrally similar to indocyanine green (ICG) dye

The absorption (Abs) and fluorescence emission (Em) maxima listed in this table are for the goat anti–mouse IgG antibody or dextran conjugates in aqueous buffer. Our online Fluorescence SpectraViewer (www.invitrogen.com/handbook/spectraviewer) provides an interactive utility for plotting and comparing fluorescence excitation and emission spectra for over 250 fluorophores (Using the Fluorescence SpectraViewer—Note 23.1). * COOH = carboxylic acid; SE = succinimidyl ester; (AA) = acetyl azide; (AH) = aryl halide; (C_5) = pentanoic acid; (DTA) = dichlorotriazine; (EX) = seven-atom spacer that is more hydrophilic than X; (ITC) = isothiocyanate; (PA) = photoactivatable; (SC) = sulfonyl chloride; (SDP) = sulfodichlorophenol ester; (SSE) = sulfosuccinimidyl ester; (STP) = 4-sulfotetrafluorophenyl ester; (TFP) = tetrafluorophenyl ester; (X) = an aminohexanoyl spacer between the dye and SE. † Mixed isomers. ‡ Emission spectra of dansyl and Dapoxyl® conjugates may vary considerably depending on the dye attachment site and the degree of labeling. § Spectral maxima for D374. ** Spectral maxima for D1412. †† Spectral maxima for H1193. More information on amine-reactive dyes is available at www.invitrogen.com/handbook/labelingchemistry.

IMPORTANT NOTICE: The products described in this manual are covered by one or more Limited Use Label License(s). Please refer to the Appendix on page 971 and Master Product List on page 975. Products are For Research Use Only. Not intended for any animal or human therapeutic or diagnostic use.

discussed in Section 1.8. However, many of the reactive dyes described in Sections 1.2 to 1.7 can also be used as derivatization reagents; likewise, most of the derivatization reagents in Section 1.8 can be utilized for biomolecule conjugation.

Reactivity of Amino Groups

The amine-reactive organic fluorophores described in this chapter are mostly acylating reagents that form carboxamides, sulfonamides or thioureas upon reaction with amines. The kinetics of the reaction depend on the reactivity and concentration of both the acylating reagent and the amine.[5] Of course, buffers that contain free amines such as Tris and glycine must be avoided when using *any* amine-reactive probe. Ammonium sulfate used for protein precipitation must also be removed before performing dye conjugations. In addition, high concentrations of nucleophilic thiols should be avoided because they may react with the amine-reactive reagent to form an unstable intermediate that could consume the dye. Reagents for reductive alkylation of amines are described in Chapter 2 and Chapter 3.

The most significant factors affecting an amine's reactivity are its class (aliphatic or aromatic) and its basicity. Virtually all proteins have lysine residues, and most have a free amine at the N-terminus. Aliphatic amines such as lysine's ε-amino group are moderately basic and reactive with most acylating reagents. However, the concentration of the free base form of aliphatic amines below pH 8 is very low; thus, the kinetics of amine acylation by isothiocyanates, succinimidyl esters or other reagents are strongly pH dependent. A pH of 8.5 to 9.5 is usually optimal for modifying lysine residues. In contrast, the α-amino group at a protein's N-terminus usually has a pK_a of ~7, so it can sometimes be selectively modified by reaction at near neutral pH. Furthermore, although amine acylation should usually be carried out above pH 8.5, the acylation reagents tend to degrade in the presence of water, with the rate increasing as the pH increases.[5] Protein modification by succinimidyl esters can typically be done at pH 8.3, whereas isothiocyanates usually require a pH >9 for optimal conjugations; this high pH may be a factor when working with base-sensitive proteins. DNA and most polysaccharides can be modified at a relatively basic pH if necessary.

Aromatic amines, which are uncommon in biomolecules, are very weak bases and thus unprotonated at pH 7. Modification of aromatic amines requires a highly reactive reagent, such as an isocyanate, isothiocyanate, sulfonyl chloride or acid halide, but can be done at any pH above ~4. A tyrosine residue can be selectively modified to form an *o*-aminotyrosine aromatic amine, which can then be reacted at a relatively low pH with certain amine-reactive probes[6] (Figure 1.1.1).

In aqueous solution, acylating reagents are virtually unreactive with the amide group of peptide bonds and the side-chain amides of glutamine and asparagine residues, the guanidinium group of arginine, the imidazolium group of histidine and the nonbasic amines, such as adenosine or guanosine, found in nucleotides and nucleic acids.

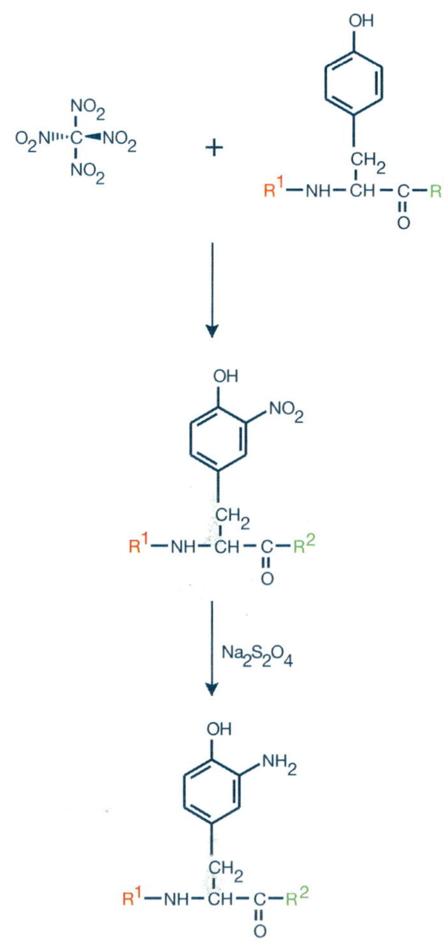

Figure 1.1.1 Nitration of tyrosine by reaction with tetranitromethane, followed by reduction with sodium dithionite, to yield an *o*-aminotyrosine.

Isothiocyanates

Because they are very susceptible to deterioration during storage, we do not sell any isocyanates. Some acyl azides (Section 3.2), however, are readily converted to isocyanates (Figure 1.1.2), which then react with amines to form ureas. As an alternative to the unstable isocyanates, we

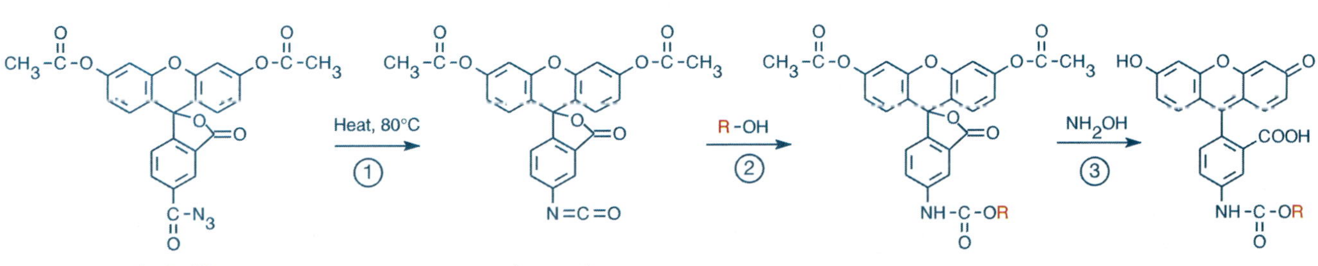

Figure 1.1.2 Derivatization of an alcohol using the diacetate of fluorescein-5-carbonyl azide (F6218). This process consists of three steps: 1) rearrangement of the acyl azide to an isocyanate, 2) reaction of the isocyanate with an alcohol to form a urethane and 3) deprotection of the nonfluorescent urethane derivative using hydroxylamine.

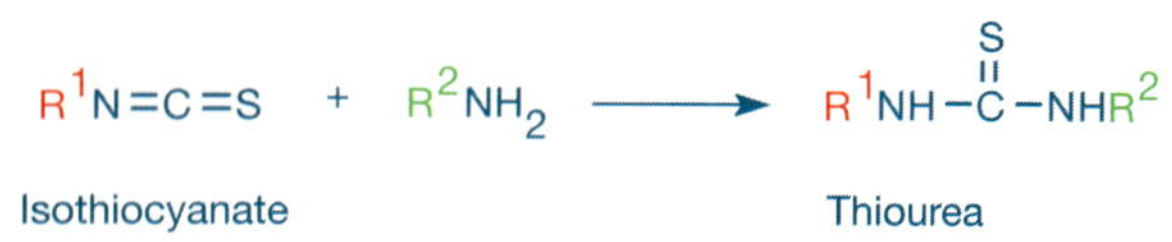

Figure 1.1.3 Reaction of a primary amine with an isothiocyanate.

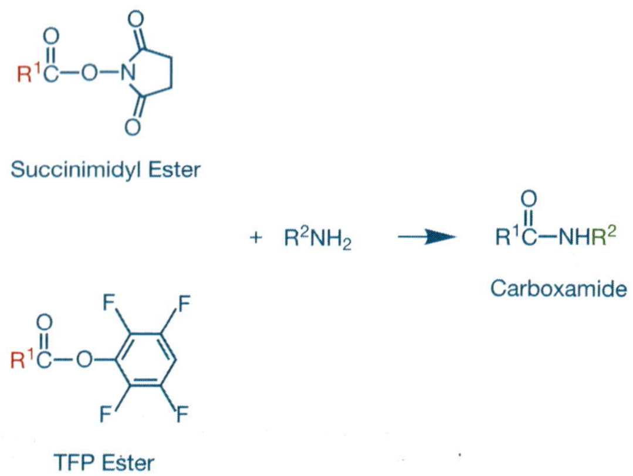

Figure 1.1.4 Reaction of a primary amine with a succinimidyl ester or a tetrafluorophenyl (TFP) ester.

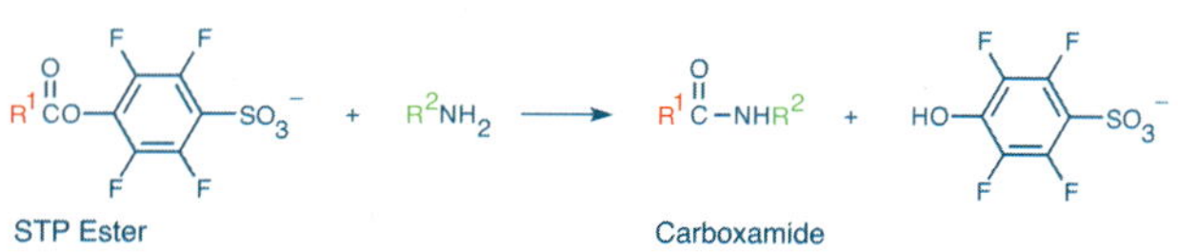

Figure 1.1.6 Reaction of a primary amine with an STP ester.

offer a large selection of isothiocyanates, which are moderately reactive but quite stable in water and most solvents. Isothiocyanates form thioureas upon reaction with amines (Figure 1.1.3). Although the thiourea product is reasonably stable, it has been reported that antibody conjugates prepared from fluorescent isothiocyanates deteriorate over time,[7] prompting us to use fluorescent succinimidyl esters and sulfonyl halides almost exclusively for synthesizing bioconjugates. The thiourea formed by the reaction of fluorescein isothiocyanate with amines is also susceptible to conversion to a guanidine by concentrated ammonia.[8] Despite the growing number of choices in amine-reactive fluorophores, fluorescein isothiocyanate (FITC) and tetramethylrhodamine isothiocyanate (TRITC) are still widely used reactive fluorescent dyes for preparing fluorescent bioconjugates.

Active Esters and Carboxylic Acids

Succinimidyl Esters

Succinimidyl esters are reliable reagents for amine modification because the amide bonds they form (Figure 1.1.4) are as stable as peptide bonds. We provide over 100 succinimidyl esters of fluorescent dyes and nonfluorescent molecules, most of which have been developed within our own laboratories (Table 1.1, Table 1.2). These reagents are generally stable during storage if well desiccated, and they show good reactivity with aliphatic amines and very low reactivity with aromatic amines, alcohols, phenols (including tyrosine) and histidine. Side-reactions of succinimidyl esters with alcohols are generally only observed in applications such as derivatization for mass spectrometry, in which much larger molar excesses of succinimdyl ester reagents and longer reaction times are used than is typically the case in protein labeling for fluorescence detection applications.[9,10] Succinimidyl esters will also react with thiols in organic solvents to form thioesters. If formed in a protein, a thioester may transfer the acyl moiety to a nearby amine. Succinimidyl ester hydrolysis (generating the unreactive carboxylic acid) competes with conjugation, but this side reaction is usually slow below pH 9.

Carboxylic Esters and Their Conversion into Sulfosuccinimidyl Esters and STP Esters

Some succinimidyl esters may not be compatible with a specific application because they can be quite insoluble in aqueous solution. To overcome this limitation, we offer carboxylic acid derivatives of many fluorophores, which can be converted into sulfosuccinimidyl esters or 4-sulfotetrafluorophenyl (STP) esters. These sulfonated esters have higher water solubility than simple succinimidyl esters and sometimes eliminate the need for organic solvents in the conjugation reaction. They are, however, also more polar than succinimidyl esters, which makes them less likely to react with buried amines in proteins or to penetrate cell

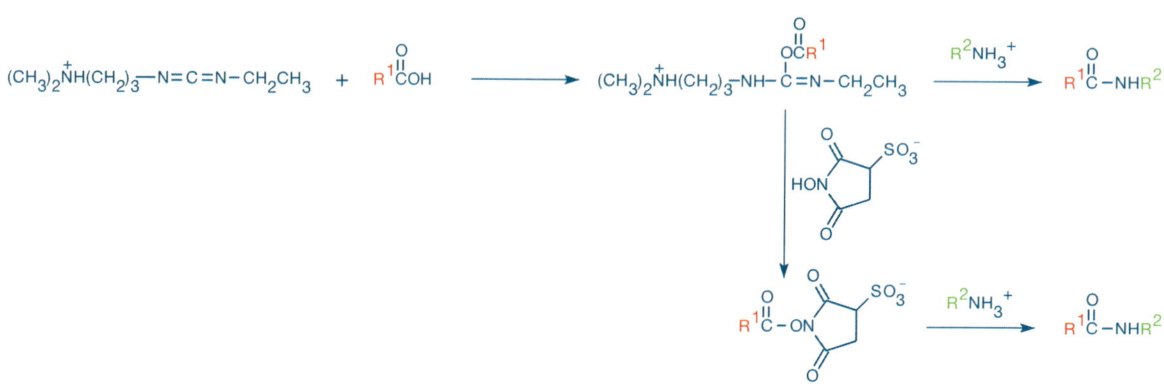

Figure 1.1.5 Stabilization of an unstable *O*-acylisourea intermediate by *N*-hydroxysulfosuccinimide (NHSS, H2249) in a carbodiimide-mediated (EDAC, E2247) modification of a carboxylic acid with a primary amine.

IMPORTANT NOTICE: The products described in this manual are covered by one or more Limited Use Label License(s). Please refer to the Appendix on page 971 and Master Product List on page 975. Products are For Research Use Only. Not intended for any animal or human therapeutic or diagnostic use.

membranes. Because of their combination of reactivity and polarity, sulfosuccinimidyl esters are not easily purified by chromatographic means and thus only a few are currently available. Sulfosuccinimidyl esters can generally be prepared *in situ* simply by dissolving the carboxylic acid dye in an amine-free buffer that contains *N*-hydroxysulfosuccinimide and 1-ethyl-3-(3-dimethylaminopropyl)carbodiimide (NHSS, H2249; EDAC, E2247; Section 3.4). Addition of NHSS to the buffer has been shown to enhance the yield of carbodiimide-mediated conjugations[11] (Figure 1.1.5). STP esters (Figure 1.1.6) are prepared in the same way from 4-sulfo-2,3,5,6-tetrafluorophenol[12] (S10490, Section 3.4), and we find them to be more readily purified by chromatography than their sulfosuccinimidyl ester counterparts. The carboxylic acid derivatives may also be useful for preparing acid chlorides and anhydrides, which, unlike succinimidyl esters, can be used to modify aromatic amines and alcohols.

Tetrafluorophenyl (TFP) Esters

2,4,5,6-Tetrafluorophenyl (TFP) esters (Figure 1.1.4) are more resistant to nonspecific hydrolysis than either succinimidyl esters (Figure 1.1.7) or sulfosuccinimidyl esters, yet they exhibit equal or superior reactivity with amines.[13] At this time, Alexa Fluor® 488 carboxylic acid is the only fluorescent TFP ester we offer (A30005, Section 1.3).

Sulfodichlorophenol (SDP) Esters

The sulfodicholorphenol (SDP) ester is currently the most hydrolytically stable amine-reactive moiety that we offer. As with TFP esters, Alexa Fluor® 488 carboxylic acid is the only fluorescent SDP ester available (A30052, Section 1.3). Conjugates produced with the Alexa Fluor® 488 5-SDP ester produce the same strong amide bond between the dye and the compound of interest as succinimidyl and tetrafluorophenyl (TFP) esters. Because of its improved stability in water and buffers, however, the SDP ester can potentially offer increased control and consistency in reactions as compared with its succinimidyl ester and TFP ester counterparts.

Carbonyl Azides

Section 3.2 describes coumarin, fluorescein and tetramethylrhodamine carbonyl azides (D1446, M1445, F6218, T6219). Like succinimidyl esters, carbonyl azides are active esters that can react with amines to yield amides; however, a more common application of carbonyl azides is thermal rearrangement to a labile isocyanate (which can then react with both aliphatic and aromatic amines to form ureas) for derivatizing alcohols and phenols (Section 3.2, Figure 1.1.2).

Sulfonyl Chlorides

Sulfonyl chlorides, including the dansyl, pyrene, Lissamine rhodamine B and Texas Red® derivatives, are highly reactive but also quite unstable in water, especially at the higher pH required for reaction with aliphatic amines. For example, we have determined that dilute Texas Red® sulfonyl chloride is totally hydrolyzed within 2–3 minutes in pH 8.3 aqueous solution at room temperature.[14] Protein modification by this reagent is therefore best done at low temperature. Once conjugated, however, the sulfonamides that are formed (Figure 1.1.8) are extremely stable; they even survive complete protein hydrolysis (for example, dansyl end-group analysis[15]).

Sulfonyl chlorides can also react with phenols (including tyrosine), aliphatic alcohols (including polysaccharides), thiols (such as cysteine) and imidazoles (such as histidine), but these reactions are not common in proteins or in aqueous solution. Sulfonyl chloride conjugates of thiols and imidazoles are generally unstable, and conjugates of aliphatic alcohols are subject to nucleophilic displacement.[16] Note that sulfonyl chlorides are unstable in dimethylsulfoxide (DMSO) and should never be used in that solvent.[17]

Other Amine-Reactive Reagents

Aldehydes react with amines to form Schiff bases. Notable aldehyde-containing reagents described in Section 1.8 include *o*-phthaldialdehyde (OPA) and naphthalenedicarboxaldehyde (NDA), as well as the 3-acylquinolinecarboxaldehyde (ATTO-TAG™) reagents CBQCA and FQ devised by Novotny and collaborators.[18–20] All of these reagents are useful for the sensitive quantitation of amines in solution, by HPLC and by capillary electrophoresis. In addition, certain arylating reagents such as NBD chloride, NBD fluoride and dichlorotriazines react with both amines and thiols, forming bonds with amines that are particularly stable.

REFERENCES

1. Biotechnol Bioeng (2007) 98:193; 2. J Org Chem (2005) 70:9809; 3. Fresenius J Anal Chem (2000) 366:3; 4. Anal Biochem (1999) 269:312; 5. Bioconjug Chem (2006) 17:501; 6. J Proteome Res (2007) 6:2257; 7. Bioconjug Chem (1995) 6:447; 8. Bioconjug Chem (1998) 9:627; 9. Anal Biochem (2010) 398:123; 10. Anal Bioanal Chem (2008) 392:305; 11. Anal Biochem (1986) 156:220; 12. Tetrahedron Lett (1999) 40:1471; 13. Langmuir (2008) 24:69; 14. Bioconjug Chem (1996) 7:482; 15. Methods Biochem Anal (1970) 18:259; 16. J Phys Chem (1979) 83:3305; 17. J Org Chem (1966) 31:3880; 18. Anal Chem (1991) 63:408; 19. Anal Chem (1991) 63:413; 20. J Chromatogr (1990) 499:579.

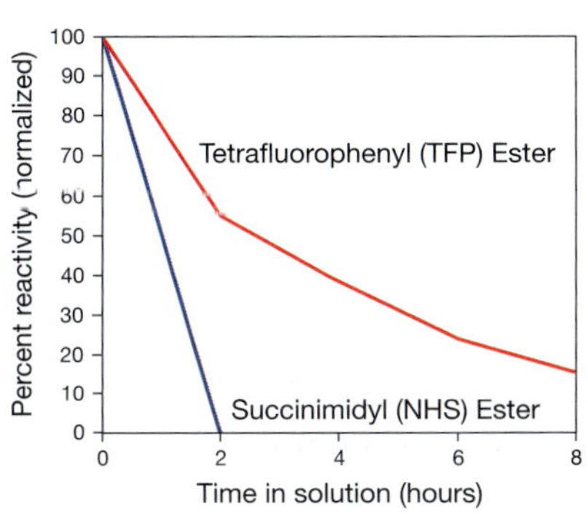

Figure 1.1.7 Stability of the tetrafluorophenyl (TFP) and succinimidyl (NHS) esters at basic pH (8.0–9.0).

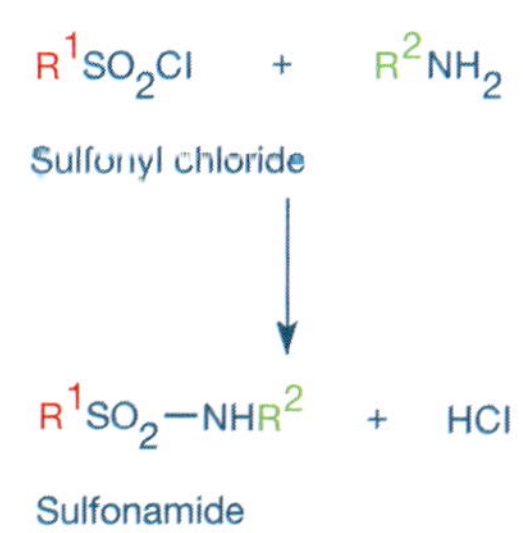

Figure 1.1.8 Reaction of a primary amine with a sulfonyl chloride.

1.2 Kits for Labeling Proteins and Nucleic Acids

We provide a vast number of stand-alone reagents for preparing bioconjugates, most of which are described in detail in other sections of this chapter. This section describes the many specialized kits that we have developed for covalently labeling proteins and nucleic acids with our organic dyes and haptens (Table 1.2, Table 1.3).

As an alternative to direct conjugation of primary antibodies with our reactive dyes and haptens, we recommend using Zenon® technology (Section 7.3) to form labeled antibody complexes. Zenon® labeling can typically be completed in minutes with quantitative yield starting with submicrogram quantities of antibody, and the labeling stoichiometry can be easily adjusted to deliver optimum fluorescence output. Although they do not provide covalent labeling, the Zenon® Antibody Labeling Kits are listed in Table 1.2, as well as in Table 7.7.

Kits for Labeling Proteins with a Fluorescent Dye

APEX® Antibody Labeling Kits

APEX® Antibody Labeling Kits (Table 1.2, Table 1.3) provide a convenient method for covalently labeling small amounts (10–20 µg) of IgG antibody with Alexa Fluor® dyes, Oregon Green® 488 dye or Pacific Blue™ dye. A primary antibody directly labeled with a fluorophore often produces lower background fluorescence and less nonspecific binding than secondary antibodies. Furthermore, multiple primary antibodies of the same isotype or derived from the same species can be used in the same immunostaining experiment if they are directly labeled with compatible fluorophores. Many IgG antibodies, however, are often only available in small quantities and may be mixed with stabilizing proteins, such as BSA, that can react nonproductively with the amine-reactive labeling reagents.

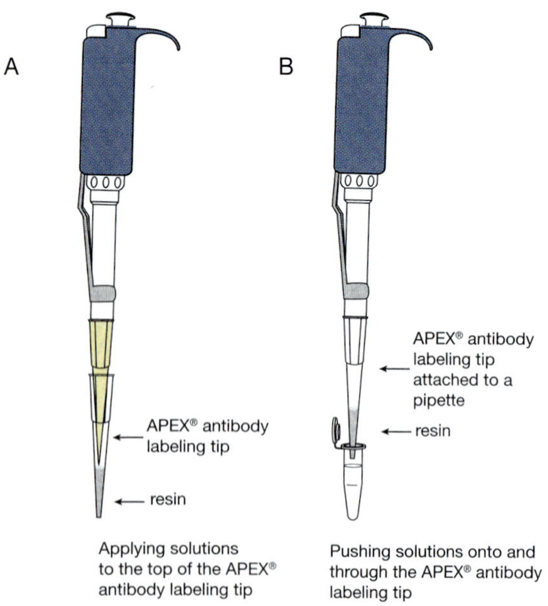

Figure 1.2.1 Illustration of the use of the APEX® antibody labeling tip, provided in APEX® Antibody Labeling Kits. **A)** Applying solutions to the resin in the tip. **B)** Pushing solutions onto the resin in the tip by attaching the APEX® antibody labeling tip to a pipette.

APEX® Antibody Labeling Kits are specifically designed to allow labeling of small amounts of IgG antibody, even in the presence of contaminants. These kits utilize a solid-phase labeling technique that captures the IgG antibody on resin inside an APEX® antibody labeling tip (Figure 1.2.1). Any contaminants, including stabilizing proteins or amine-containing buffers, are eluted through the tip before labeling. After applying the amine-reactive fluorophore to the IgG antibody trapped on the resin, a fluorescent IgG conjugate is formed and subsequently eluted from the resin using elution buffer. The fluorescent IgG conjugate is ready to use in an imaging or flow cytometry assay in as little as 2.5 hours with minimal hands-on time. The typical yield of labeled antibody using this method is between 40 and 80%.

Each APEX® Antibody Labeling Kit provides all reagents required to perform five separate labeling reactions of 10–20 µg IgG antibody, including:

- Five vials of amine-reactive fluorescent dye
- Five APEX® antibody labeling tips
- Wash, labeling, neutralization and elution buffers
- Dimethylsulfoxide (DMSO)
- Labeling protocol

For labeling larger amounts of protein, we recommend the Alexa Fluor® Microscale Protein Labeling Kits, which are optimized for 20–100 µg samples of proteins between 10,000 and 150,000 daltons; the Alexa Fluor® Monoclonal Antibody Labeling Kits, which are optimized for 100 µg samples of mono- or polyclonal antibodies; or the Alexa Fluor® Protein Labeling Kits, which are optimized for 1 mg samples of >30,000-dalton proteins.

Alexa Fluor® Microscale Protein Labeling Kits

Alexa Fluor® Microscale Protein Labeling Kits (Table 1.2, Table 1.3) provide a convenient means for labeling small amounts (20–100 µg) of purified protein with the highly fluorescent Alexa Fluor® dyes. Spin columns are used to purify the labeled protein, with yields between 60 and 90% depending primarily on the molecular weight of the starting material. Labeling and purification can be completed in as little as 30 minutes.

These kits have been optimized for labeling proteins with molecular weights between 12,000 and 150,000 daltons and contain everything needed to perform three labeling reactions and to separate the resulting conjugates from excess dye. Each Alexa Fluor® Microscale Protein Labeling Kit provides:

- Three vials of amine-reactive fluorescent dye (succinimidyl ester or tetrafluorophenyl ester)
- Sodium bicarbonate
- Reaction tubes
- Purification resin and spin filters
- Detailed protocols for conjugation, purification and determination of the degree of labeling

IMPORTANT NOTICE: The products described in this manual are covered by one or more Limited Use Label License(s). Please refer to the Appendix on page 971 and Master Product List on page 975. Products are For Research Use Only. Not intended for any animal or human therapeutic or diagnostic use.
www.invitrogen.com/probes
molecular probes® | invitrogen™ by life technologies™

Table 1.2 Active esters and kits for labeling proteins and nucleic acids.

Label	Fluorescence Color (Abs/Em) *	Succinimidyl Ester, TFP Ester or SDP Ester	Protein Labeling Kits	Zenon® Antibody Labeling Kits	Nucleic Acid Labeling Kits
Alexa Fluor® 350	Blue (346/442)	A10168	A20180 (Mab) A10170 (P)	Z25000 (M IgG_1) Z25100 (M IgG_{2a}) Z25200 (M IgG_{2b}) Z25300 (R IgG) Z25400 (H IgG)	
Marina Blue®	Blue (365/460)	M10165			
Pacific Orange™	Orange (400/551)	P30253	P30014 (Mab) P30016 (P)	Z25256 (M IgG_1) Z25257 (M IgG_{2a})	
Alexa Fluor® 405	Blue (402/421)	A30000 A30100		Z25013 (M IgG_1) Z25113 (M IgG_{2a}) Z25213 (M IgG_{2b}) Z25313 (R IgG)	
Pacific Blue™	Blue (410/455)	P10163	A10478 (APEX®) P30013 (Mab) P30012 (P)	Z25041 (M IgG_1) Z25156 (M IgG_{2a}) Z25341 (R IgG)	
Alexa Fluor® 430	Yellow-green (434/539)	A10169	A10171 (P)	Z25001 (M IgG_1) Z25301 (R IgG)	
Fluorescein-EX	Green (494/518)	F6130	F10240 (P) F6433 (F)	Z25042 (M IgG_1) Z25342 (R IgG)	
FITC	Green (494/518)		F6434 (F)		
Alexa Fluor® 488	Green (495/519)	A20000 † A20100 † A30005 (TFP) A30052 (SDP)	A10468 ‡ (APEX®) A30006 ‡ (Micro) A20181 ‡ (Mab) A10235 ‡ (P)	Z25002 (M IgG_1) Z25090 (TSA™)(M IgG_1) Z25102 (M IgG_{2a}) Z25202 (M IgG_{2b}) Z25302 (R IgG) Z25602 (G IgG) Z25402 (H IgG)	U21650 (ULYSIS®) A21665 (ARES™) F32947 (FT) F32952 (FT) A20191 (Oligo)
Oregon Green® 488	Green (496/524)	O6147 O6149	A10476 (APEX®) O10241 (P) F6153 (F)		U21659 (ULYSIS®)
Oregon Green® 514	Green (511/530)	O6139			
Alexa Fluor® 514	Yellow-green (518/540)	A30002 †			
Alexa Fluor® 532	Yellow (531/554)	A20001 A20101MP	A20182 (Mab) A10236 (P)	Z25003 (M IgG_1) Z25303 (R IgG)	U21651 (ULYSIS®)
Alexa Fluor® 546	Orange (556/573)	A20002 † A20102 †	A20183 (Mab) A10237 (P)	Z25004 (M IgG_1) Z25104 (M IgG_{2a}) Z25204 (M IgG_{2b}) Z25304 (R IgG)	U21652 (ULYSIS®) A21667 (ARES™)
Alexa Fluor® 555	Red-orange (555/565)	A20009 A20109	A10470 (APEX®) A30007 (Micro) A20187 (Mab) A20174 (P)	Z25005 (M IgG_1) Z25105 (M IgG_{2a}) Z25205 (M IgG_{2b}) Z25305 (R IgG) Z25605 (G IgG) Z25405 (H IgG)	A21677 (ARES™) F32948 (FT) F32953 (FT)
Tetramethylrhodamine	Red-orange (555/580)	C2211 C6123 C1171 † T6105 † (X)			
Rhodamine Red™	Red-orange (570/590)	R6160 (X)	F6161 (F)		
Alexa Fluor® 568	Red-orange (578/603)	A20003 † A20103 †	A20184 (Mab) A10238 (P)	Z25006 (M IgG_1) Z25106 (M IgG_{2a}) Z25206 (M IgG_{2b}) Z25306 (R IgG) Z25606 (G IgG)	U21653 (ULYSIS®)
Alexa Fluor® 594	Red (590/617)	A20004 † A20104 †	A10474 (APEX®) A30008 (Micro) A20185 (Mab) A10239 (P)	Z25007 (M IgG_1) Z25107 (M IgG_{2a}) Z25207 (M IgG_{2b}) Z25307 (R IgG) Z25607 (G IgG) Z25407 (H IgG)	U21654 (ULYSIS®) A21669 (ARES™) F32949 (FT) F32954 (FT)

* Approximate absorption (Abs) and fluorescence emission (Em) maxima for conjugates, in nm. † Mixed isomers. ‡ These Alexa Fluor® 488 protein labeling kits contain either the amine-reactive Alexa Fluor® 488 carboxylic acid 5-TFP ester (Micro, Mab, P) or the amine-reactive Alexa Fluor® 488 carboxylic acid 5-SDP ester (APEX®); whereas the Alexa Fluor® 488 nucleic acid labeling kits contain the Alexa Fluor® 488 carboxylic acid succinimidyl ester. § Human vision is insensitive to light beyond ~650 nm, and therefore it is not possible to view the far-red– and near-infrared–fluorescent dyes by looking through the eyepiece of a conventional fluorescence microscope. NA = not applicable. (APEX®) = APEX® Antibody Labeling Kit. (ARES™) = ARES™ DNA Labeling Kit. (D) = DSB-X™ Biotin Protein Labeling Kit. (F) = FluoReporter® Protein Labeling Kit. (FB) = FluoReporter® Biotin-XX Protein Labeling Kit. (FMB) = FluoReporter® Mini-Biotin-XX Protein Labeling Kit. (FT) = FISH Tag™ DNA Kit or FISH Tag™ RNA Kit. (G IgG) = Zenon® Goat IgG Labeling Kit. (H IgG) = Zenon® Human IgG Labeling Kit. (M IgG) = Zenon® Mouse IgG Labeling Kit. (Mab) = Monoclonal Antibody Labeling Kit. (Micro) = Alexa Fluor® Microscale Protein Labeling Kit or Biotin-XX Microscale Protein Labeling Kit. (Oligo) = Alexa Fluor® Oligonucleotide Amine Labeling Kit. (P) = Easy-to-Use Protein Labeling Kit. (R IgG) = Zenon® Rabbit IgG Labeling Kit. (SAIVI™) = SAIVI™ Antibody Labeling Kit. (SDP) = Sulfodichlorophenol ester. (TFP) = Tetrafluorophenyl ester. (TSA™) = Enhanced with TSA™ technology. (ULYSIS®) = ULYSIS® Nucleic Acid Labeling Kit. (X) = An aminohexanoyl spacer between the dye and the SE. More information on amine-reactive dyes is available at www.invitrogen.com/handbook/labelingchemistry.

continued on next page

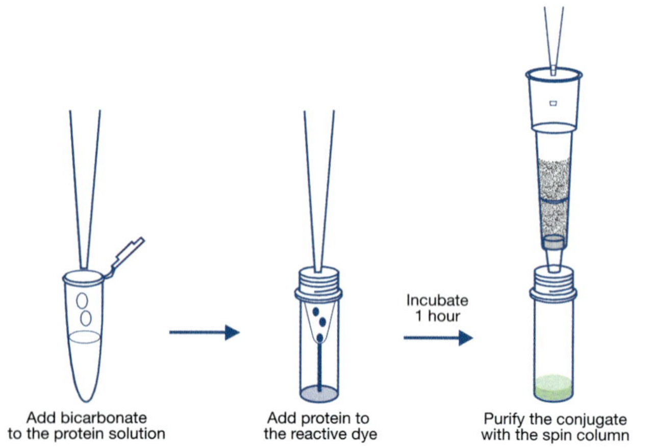

Figure 1.2.2 Illustration of the three simple steps in the protocol for the Monoclonal Antibody Labeling Kits, which provide a convenient method for covalently labeling small amounts of IgG antibodies.

Monoclonal Antibody Labeling Kits

Monoclonal Antibody Labeling Kits (Table 1.2, Table 1.3) provide researchers with a simple yet efficient means of labeling small amounts of IgG antibodies with Alexa Fluor® dyes (Figure 1.2.2). Unlike polyclonal antibodies and most other commercially available proteins, monoclonal antibodies are typically only available in small quantities. These kits contain everything needed to perform five separate labeling reactions containing ~100 µg each of carrier-free monoclonal IgG samples (although other proteins can be labeled).

Each Monoclonal Antibody Labeling Kit provides:

- Five vials of amine-reactive fluorescent dye (succinimidyl ester or tetrafluorophenyl ester)
- Sodium bicarbonate
- Spin columns and collection tubes
- Detailed protocols for conjugation, purification and determination of the degree of labeling

Table 1.2 Active esters and kits for labeling proteins and nucleic acids—*continued.*

Label	Fluorescence Color (Abs/Em) *	Succinimidyl Ester, TFP Ester or SDP Ester	Protein Labeling Kits	Zenon® Antibody Labeling Kits	Nucleic Acid Labeling Kits
Texas Red®	Red (595/615)	T6134 † (X) T20175 (X)	T10244 (P) F6162 (F)	Z25045 (M IgG_1)	
Alexa Fluor® 610	Red (612/628)	A30050 (X)			
Alexa Fluor® 633 §	Far-red (632/647)	A20005 † A20105 †	A20170 (P)		
Alexa Fluor® 635 §	Far-red (633/647)				
Alexa Fluor® 647 §	Far-red (650/668)	A20006 A20106	A10475 (APEX®) A30009 (Micro) A20186 (Mab) S30044 (SAIVI™) A20173 (P)	Z25008 (M IgG_1) Z25108 (M IgG_{2a}) Z25208 (M IgG_{2b}) Z25308 (R IgG) Z25608 (G IgG) Z25408 (H IgG)	U21660 (ULYSIS®) A21676 (ARES™) F32950 (FT) F32955 (FT) A20196 (Oligo)
Alexa Fluor® 660 §	Near-infrared (663/690)	A20007	A20171 (P)	Z25009 (M IgG_1)	
Alexa Fluor® 680 §	Near-infrared (679/702)	A20008 A20108	S30045 (SAIVI™) A20172 (P)	Z25010 (M IgG_1) Z25110 (M IgG_{2a}) Z25210 (M IgG_{2b}) Z25310 (R IgG)	
Alexa Fluor® 700 §	Near-infrared (702/723)	A20010 A20110		Z25011 (M IgG_1)	
Alexa Fluor® 750 §	Near-infrared (749/775)	A20011 A20111	S30046 (SAIVI™)	Z25312 (R IgG)	
Alexa Fluor® 790 §	Near-infrared (782/805)	A30051			
Biotin	NA	B1606 (XX) B6353 (X)	F2610 (FB) F6347 (FMB) B30010 (Micro) B30756 (Micro)	Z25052 (M IgG_1) Z25152 (M IgG_{2a}) Z25252 (M IgG_{2b}) Z25352 (R IgG) Z25452 (H IgG)	
DNP–biotin	NA (364/none)	B2604 (X)	F6348 (F)		
DSB-X™ biotin	NA		D20655 (D)		

* Approximate absorption (Abs) and fluorescence emission (Em) maxima for conjugates, in nm. † Mixed isomers. ‡ These Alexa Fluor® 488 protein labeling kits contain either the amine-reactive Alexa Fluor® 488 carboxylic acid 5-TFP ester (Micro, Mab, P) or the amine-reactive Alexa Fluor® 488 carboxylic acid 5-SDP ester (APEX®); whereas the Alexa Fluor® 488 nucleic acid labeling kits contain the Alexa Fluor® 488 carboxylic acid succinimidyl ester. § Human vision is insensitive to light beyond ~650 nm, and therefore it is not possible to view the far-red– and near-infrared–fluorescent dyes by looking through the eyepiece of a conventional fluorescence microscope. NA = not applicable. (APEX®) = APEX® Antibody Labeling Kit. (ARES™) = ARES™ DNA Labeling Kit. (D) = DSB-X™ Biotin Protein Labeling Kit. (F) = FluoReporter® Protein Labeling Kit. (FB) = FluoReporter® Biotin-XX Protein Labeling Kit. (FMB) = FluoReporter® Mini-Biotin-XX Protein Labeling Kit. (FT) = FISH Tag™ DNA Kit or FISH Tag™ RNA Kit. (G IgG) = Zenon® Goat IgG Labeling Kit. (H IgG) = Zenon® Human IgG Labeling Kit. (M IgG) = Zenon® Mouse IgG Labeling Kit. (Mab) = Monoclonal Antibody Labeling Kit. (Micro) = Alexa Fluor® Microscale Protein Labeling Kit or Biotin-XX Microscale Protein Labeling Kit. (Oligo) = Alexa Fluor® Oligonucleotide Amine Labeling Kit. (P) = Easy-to-Use Protein Labeling Kit. (R IgG) = Zenon® Rabbit IgG Labeling Kit. (SAIVI™) = SAIVI™ Antibody Labeling Kit. (SDP) = Sulfodichlorophenol ester. (TFP) = Tetrafluorophenyl ester. (TSA™) = Enhanced with TSA™ technology. (ULYSIS®) = ULYSIS® Nucleic Acid Labeling Kit. (X) = An aminohexanoyl spacer between the dye and the SE. More information on amine-reactive dyes is available at www.invitrogen.com/handbook/labelingchemistry.

IMPORTANT NOTICE: The products described in this manual are covered by one or more Limited Use Label License(s). Please refer to the Appendix on page 971 and Master Product List on page 975. Products are For Research Use Only. Not intended for any animal or human therapeutic or diagnostic use.
www.invitrogen.com/probes

Table 1.3 Molecular Probes® kits for protein and nucleic acid labeling.

Kit Name	# Labelings	Kit Components	Features
Kits for Labeling Proteins with Fluorescent Dyes			
APEX® Antibody Labeling Kit	5 labelings of 10–20 µg each of IgG antibody	• 5 vials of amine-reactive fluorescent label • Five APEX® antibody labeling tips • Wash, labeling, neutralization and elution buffers • Dimethylsulfoxide (DMSO) • Labeling protocol	APEX® Antibody Labeling Kits utilize a solid-phase labeling technique that captures the IgG antibody on the resin inside the APEX® antibody labeling tip. Any contaminants, including stabilizing proteins, are eluted through the tip prior to labeling with the amine-reactive fluorescent dye. The fluorescent IgG conjugate is ready to use in 2.5 hours, with ~15 minutes hands-on time.
Alexa Fluor® Microscale Protein Labeling Kit	Three 20–100 µg protein samples of a 12,000- to 150,000-dalton protein	• 3 vials of the succinimidyl ester (or tetrafluorophenyl ester) of the fluorescent dye • Sodium bicarbonate • Reaction tubes • Purification resin and spin filters • Detailed protocols for conjugation, purification and determination of the degree of labeling	Alexa Fluor® Microscale Protein Labeling Kits provide a convenient means for labeling small amounts (20–100 µg) of purified protein with our superior Alexa Fluor® dyes and purifying the resulting conjugate. Convenient spin columns are used to purify the labeled protein, with yields between 60 and 90% depending primarily on the molecular weight of the starting material. Labeling and purification can be completed in as little as 30 minutes. Microscale Protein Labeling Kits are also available for biotinylating proteins.
Monoclonal Antibody Labeling Kit	5 labelings of ~100 µg each of carrier-free monoclonal IgG samples	• 5 vials of the succinimidyl ester (or tetrafluorophenyl ester) of the fluorescent dye • Sodium bicarbonate • Spin columns and collection tubes • Detailed protocols for conjugation, purification and determination of the degree of labeling	A buffered solution of the protein is added to one of the five vials of amine-reactive dye. The reactive dye has a succinimidyl ester (or tetrafluorophenyl ester) moiety that reacts efficiently with primary amines of proteins to form stable dye–protein conjugates. The conjugate can be purified on the included size-exclusion spin columns. Labeling and purification can be completed in less than 2 hours.
SAIVI™ Rapid Antibody Labeling Kit	3 labelings of 0.5–3 mg each of carrier-free antibody solution	• 3 vials of amine-reactive Alexa Fluor® dye • Sodium bicarbonate • Regulator solution • Purification resin and purification columns • Phosphate-buffered saline (PBS) • Syringes, syringe filters, column-loading pipettes and catch tubes • Detailed protocols for conjugation, purification and determination of the degree of labeling	SAIVI™ Antibody Labeling Kits provide a convenient means to label antibodies with an optimal degree of labeling for *in vivo* imaging applications over a 6-fold antibody concentration range with no adjustments in reaction volume, dye concentration or antibody concentration necessary. Purification of the dye-labeled antibody is achieved with a simple protocol that can be completed in less than 10 minutes. These optimally labeled antibodies are ready for applications that require azide-free reagents, such as live-cell imaging or direct injection into animals.
Easy-to-Use Protein Labeling Kit	Three ~1 mg protein samples of a 150,000-dalton protein, such as an IgG	• 3 vials of the succinimidyl ester (or tetrafluorophenyl ester) of the fluorescent dye, each containing a magnetic stir bar • Sodium bicarbonate • Gravity-feed columns, a size-exclusion resin and concentrated elution buffer • Column funnels, foam column holders, disposable pipettes and collection tubes • Detailed protocols for conjugation, purification and determination of the degree of labeling	A buffered solution of the protein is added to one of the three vials of the amine-reactive dye. The reactive dye has a succinimidyl ester (or tetrafluorophenyl ester) moiety that reacts efficiently with primary amines of proteins to form stable dye–protein conjugates. Purification of the conjugate can be accomplished on the included gravity-feed size-exclusion columns.
FluoReporter® Protein Labeling Kit	5 to 10 protein samples of 0.2–2 mg each in 200 µL volumes	• 5 vials of the amine-reactive dye • Anhydrous DMSO • Reaction tubes, each containing a stir bar • Spin columns and collection tubes • Detailed protocols for conjugation, purification and determination of the degree of labeling	The amount of dye necessary for the desired protein sample is calculated using the guidelines outlined in the kit protocol. The reactive dye has a succinimidyl ester moiety that reacts efficiently with primary amines of proteins to form stable dye–protein conjugates. Purification of the conjugate can be easily accomplished using the included spin columns.
Qdot® Antibody Conjugation Kit—see Section 6.6			
Zenon® Antibody Labeling Kit—see Section 7.3			
Kits for Labeling Proteins with Biotin or Dinitrophenyl (DNP)			
Biotin-XX Microscale Protein Labeling Kit	Three 20–100 µg protein samples of a 12,000- to 150,000-dalton protein	• 3 vials of biotin-XX sulfosuccinimidyl ester • Sodium bicarbonate • Reaction tubes • Purification resin and spin filters • Detailed protocols for conjugation and purification	The Biotin-XX Microscale Protein Labeling Kit provides a convenient means for biotinylating small amounts (20–100 µg) of purified protein. Convenient spin columns are used to purify the labeled protein with yields between 60 and 90%, depending primarily on the molecular weight of the starting material. Labeling and purification can be completed in as little as 30 minutes. For determining the degree of labeling, the FluoReporter® Biotin Quantitation Assay Kit for proteins is available separately or in combination with the Biotin-XX Microscale Protein Labeling Kit.
FluoReporter® Mini-Biotin-XX Protein Labeling Kit	5 biotinylation reactions of 0.1–3 mg each	• 5 vials of biotin-XX sulfosuccinimidyl ester • Reaction tubes, each containing a stir bar • Purification resin, spin columns and collection tubes • Dialysis tubing • Detailed protocols for conjugation and purification	The biotin-XX sulfosuccinimidyl ester (SSE) is water soluble and reacts with primary amines of proteins or other biomolecules to form stable biotin conjugates. The biotin-XX SSE has a 14-atom spacer that enhances the binding of biotin derivatives to avidin's relatively deep binding sites. Ready-to-use spin columns are included for purification of the biotinylated protein from excess reagents.

continued on next page

Simply dissolve the carrier-free monoclonal antibody at ~1 mg/mL in the provided buffer, then add it to one of the five vials of amine-reactive dye; no organic solvents are required. Purification of the fluorescent conjugate is accomplished on a size-exclusion spin column optimized for proteins with molecular weight greater than 30,000 daltons. Labeling and purification can typically be completed in less than 2 hours.

Mouse monoclonal antibodies in serum, in ascites fluid or diluted with carrier proteins should not be labeled with these kits; however, such antibody preparations can be efficiently labeled with the APEX® Antibody Labeling Kits described above or with the Zenon® Mouse IgG Labeling Kits (Section 7.3, Table 7.7) described below.

SAIVI™ Antibody Labeling Kits

The optimal fluorescent antibody conjugate for *in vitro* detection assays produces an intense fluorescent signal yet retains the binding affinity and specificity of the unlabeled antibody. When preparing a fluorescent antibody conjugate for *in vivo* animal imaging, however, the pharmacokinetics of the labeled probe must also be considered.[1] These additional constraints have led to the development of the SAIVI™ Antibody Labeling Kits for small animal *in vivo* imaging applications (Table 1.2, Table 1.3).

SAIVI™ Antibody Labeling Kits feature reactive far-red and near-infrared Alexa Fluor® dyes, along with a labeling protocol specifically designed to produce a suitable degree of labeling (DOL) for *in vivo*

Table 1.3 Molecular Probes® kits for protein and nucleic acid labeling—*continued.*

Kit Name	# Labelings	Kit Components	Features
FluoReporter® Biotin-XX Protein Labeling Kit	5 biotinylation reactions, each with 5–20 mg of protein	• Biotin-XX succinimidyl ester • DMSO • Gel filtration column • Avidin–HABA complex • Biotinylated goat IgG • Detailed protocols for conjugation, purification and determination of the degree of labeling	The biotin-XX succinimidyl ester (SE) reacts with primary amines of proteins or other biomolecules to form stable biotin conjugates. The biotin-XX SE has a 14-atom spacer that enhances the binding of biotin derivatives to avidin's relatively deep binding sites. A gel filtration column is provided for purifying the labeled proteins from excess biotin reagent. After purification, the degree of biotinylation can be estimated using the included avidin–biotin displacement assay.
FluoReporter® Biotin/DNP Protein Labeling Kit	5 to 10 labeling reactions of 0.2–2 mg of protein each	• 5 vials of DNP-X–biocytin-X succinimidyl ester • DMSO • Reaction tubes • Spin columns and collection tubes • Detailed protocols for conjugation, purification and determination of the degree of labeling	The FluoReporter® Biotin/DNP Protein Labeling Kit is similar to other FluoReporter® Protein Labeling Kits, except that it contains DNP-X–biocytin-X succinimidyl ester as the reactive label. When proteins are labeled with this chromophoric biotin derivative, the degree of biotinylation can be readily assessed from the extinction coefficient of DNP (EC_{360} = 15,000 $cm^{-1}M^{-1}$). An additional feature of the conjugates labeled with DNP-X–biocytin-X succinimidyl ester is that they can be recognized by avidin derivatives (or anti-biotin antibodies) and by anti-DNP antibodies, enabling researchers to choose among several detection techniques suitable for fluorescence and electron microscopy.
DSB-X™ Protein Labeling Kit	5 protein conjugations of 0.5–3 mg each	• 5 vials of DSB-X™ biotin succinimidyl ester • DMSO • Reaction tubes • Purification resin, spin columns and collection tubes • Dialysis tubing for larger-scale separations • Detailed protocols for conjugation and purification	DSB-X™ biotin succinimidyl ester, a derivative of desthiobiotin with an additional seven-atom spacer, reacts with amine groups of biomolecules to form stable amides. The DSB-X™ biotin conjugate can be detected with avidin or streptavidin derivatives. Binding is almost totally reversed by addition of free biotin at neutral pH and normal ionic strength. Materials are included for both small- and large-scale preparations.
Kits for Labeling Nucleic Acids with Fluorescent Dyes			
ULYSIS® Nucleic Acid Labeling Kit	20 labelings of 1 µg DNA	• ULS® labeling reagent and appropriate solvent • Labeling buffer • Deoxyribonuclease I (DNase I), for digesting DNA longer than 1000 base pairs prior to labeling • DNase I storage and reaction buffers • Control DNA from calf thymus • Nuclease-free H_2O • Labeling protocol	The ULS® reagent reacts with the *N*-7 position of guanine residues to provide a stable coordination complex between the nucleic acid and the fluorophore label. Separation of the labeled nucleic acids from the unreacted ULS® complex can be accomplished through a simple procedure using a spin column (not provided).
ARES™ DNA Labeling Kit	10 labelings of 1–5 µg DNA	• 5-(3-Aminoallyl)-dUTP • Amine-reactive fluorescent dye and appropriate solvent • Sodium bicarbonate • Nuclease-free H_2O • Detailed protocol for labeling DNA using reverse transcriptase or nick translation	In the first step, an amine-modified nucleotide, 5-(3-aminoallyl)-dUTP, is incorporated into DNA using conventional enzymatic labeling methods. In the second step, the amine-modified DNA is chemically labeled using an amine-reactive fluorescent dye. The amine-modified DNA can be purified using the PureLink™ PCR Purification Kit (K3100-01).
FISH Tag™ DNA Kit—see Section 8.2			
FISH Tag™ RNA Kit—see Section 8.2			
Oligonucleotide Amine Labeling Kit	3 labelings of 50 µg each of an amine-modified oligonucleotide	• 3 vials of the amine-reactive dye • DMSO • Labeling buffer • Labeling protocol	The reactive dye used in the labeling has an amine-reactive succinimidyl ester moiety that reacts efficiently with an amine-modified oligonucleotide. Following the labeling reaction, the conjugate can be purified from the reaction mixture by preparative gel electrophoresis or reverse-phase HPLC.

IMPORTANT NOTICE: The products described in this manual are covered by one or more Limited Use Label License(s). Please refer to the Appendix on page 971 and Master Product List on page 975. Products are For Research Use Only. Not intended for any animal or human therapeutic or diagnostic use.

molecular probes® | invitrogen™ by life technologies™

imaging applications. When optimally labeled, a fluorescent antibody conjugate produces an intense, targeted fluorescent signal that persists throughout the *in vivo* study, without significant redistribution or clearance of the probe.

The conjugation protocol in the SAIVI™ Rapid Antibody Labeling Kits produces an optimal DOL (~2 fluorophores per antibody) over a 6 fold antibody concentration range with no adjustments in reaction volume, dye concentration or antibody concentration necessary. Purification of the dye-labeled antibody is achieved usng a simple protocol that can typically be completed in less than 10 minutes. With these kits, optimally labeled antibodies are ready for applications that require azide-free reagents, such as live-cell imaging or direct injections into animals. SAIVI™ Rapid Antibody Labeling Kits containing either Alexa Fluor® 680 or Alexa Fluor® 750 dye (S30045, S30046) provide sufficient reagents for three labeling reactions of 0.5–3 mg of protein each, including:

- Three vials of amine-reactive Alexa Fluor® 680 or Alexa Fluor® 750 dye
- Sodium bicarbonate
- Regulator solution
- Purification resin and purification columns
- Phosphate-buffered saline (PBS)
- Syringes, syringe filters, column-loading pipettes and catch tubes
- Detailed protocols for conjugation, purification and determination of the degree of labeling

To control the DOL of the antibody conjugate, the SAIVI™ Alexa Fluor® 647 Antibody/Protein Labeling Kit (S30044) includes a DOL modulating reagent and instructions for decreasing the DOL from its intrinsic highest value by adding specific amounts of this reagent to the labeling reaction. With this method, antibody preparations with varying ratios of dye to protein can be quickly and reproducibly obtained without significant alteration of labeling or purification conditions, allowing more efficient optimization for *in vivo* imaging applications. Each kit provides sufficient reagents for three labeling reactions of 1 mg protein each, including:

- Three vials of amine-reactive Alexa Fluor® 647 dye
- Sodium bicarbonate
- DOL modulating reagent
- Purification resin and purification columns
- Wash buffer
- Column funnels, column holders, disposable pipettes and collection tubes
- Detailed protocols for conjugation, purification and determination of the degree of labeling

Easy-to-Use Protein Labeling Kits

Our Easy-to Use Protein Labeling Kits (Table 1.2, Table 1.3) provide a nearly effortless way to label proteins, especially IgG antibodies, with a fluorescent dye (Figure 1.2.3). Simply add ~1 mg of protein (in a volume of ~500 µL and free of amine-containing buffers such as Tris) to one of the three included vials, which contain a premeasured quantity of amine-reactive dye and a magnetic stir bar. Because the reactive dyes used in these kits are water soluble, no organic solvents are required. Purification is accomplished on a gravity-feed size-exclusion column, which is supplied with the kit. Labeling and purification can typically be completed in about 2 hours, with very little hands-on time.

Each Protein Labeling Kit provides sufficient reagents for labeling three ~1 mg protein samples of a 150,000-dalton protein (such as an IgG), including:

- Three vials of the amine-reactive fluorescent dye (succinimidyl ester or tetrafluorophenyl ester), each containing a magnetic stir bar
- Sodium bicarbonate
- Gravity-feed columns, a size-exclusion resin and concentrated elution buffer
- Column funnels, foam column holders, disposable pipettes and collection tubes
- Detailed protocols for conjugation, purification and determination of the degree of labeling

Researchers have modified α-synuclein using either the Oregon Green® 488 Protein Labeling Kit (O10241) or the Alexa Fluor® 594 Protein Labeling Kit (A10239).[2] They intentionally produced fluorescent conjugates with a low degree of labeling (DOL) so as not to perturb oligomerization. When comparing Oregon Green® 488 synuclein with Alexa Fluor® 594 synuclein, they observed essentially the same results in terms of fibril formation as analyzed by fluorescence polarization. They also detected the Oregon Green® 488 synuclein conjugate with mouse monoclonal anti–fluorescein/Oregon Green® antibody (monoclonal 4-4-20, A6421; Section 7.4) and gold-labeled anti–mouse IgG antibody using electron microscopy.

FluoReporter® Protein Labeling Kits

The FluoReporter® Protein Labeling Kits (Table 1.2, Table 1.3) facilitate research-scale preparation of protein conjugates labeled with some of our brightest fluorescent dyes. Typically, labeling and purifying conjugates with the FluoReporter® Protein Labeling Kits can typically be completed in under 3 hours, with very little hands-on time. First, the amount of dye necessary for the desired protein sample is calculated using the guidelines outlined in the kit protocol. After dissolving the dye in dimethylsulfoxide (DMSO), the calculated amount of dye is added to the protein and the reaction is incubated for 1–1.5 hours. Purification

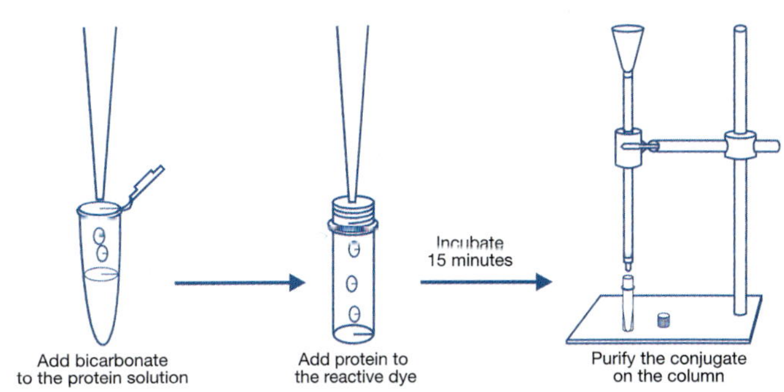

Figure 1.2.3 Illustration of the three simple steps in the protocol for the Easy-to-Use Protein Labeling Kits, which provide a convenient method for covalently labeling most proteins.

is easily accomplished using convenient spin columns designed for use with proteins of molecular weight greater than 30,000 daltons.

Each FluoReporter® Protein Labeling Kits provides sufficient reagents to label 5 to 10 protein samples of 0.2–2 mg each in 200 µL volumes:

- Five vials of the amine-reactive dye
- Dimethylsulfoxide (DMSO)
- Reaction tubes, each containing a stir bar
- Spin columns and collection tubes
- Detailed protocols for conjugation, purification and determination of the degree of labeling

Zenon® Antibody Labeling Kits

The Zenon® Antibody Labeling Kits (Table 1.2, Table 7.7) are useful for the rapid and quantitative labeling of antibodies with dyes (including phycobiliproteins and their tandem conjugates), biotin or enzymes[3] (Figure 1.2.4). Zenon® Antibody Labeling Kits are designed to label intact antibodies in amounts from less than 1 µg to as much as 50 µg, starting with a purified antibody fraction or with a crude antibody preparation such as serum, ascites fluid or a hybridoma supernatant. The Zenon® antibody labeling technology is discussed further in Section 7.3.

Kits for Labeling Proteins with Biotin or Dinitrophenyl (DNP)

Biotin-XX Microscale Protein Labeling Kits

The Biotin-XX Microscale Protein Labeling Kit (B30010; Table 1.2, Table 1.3) provides a convenient means for biotinylating small amounts (20–100 µg) of purified protein. The water-soluble biotin-XX sulfosuccinimidyl ester has a 14-atom spacer (Figure 1.2.5) that enhances the binding of biotin derivatives to avidin's relatively deep binding sites. Spin columns are used to purify the labeled protein, with yields between 60 and 90% depending primarily on the molecular weight of the starting material. Labeling and purification can be completed in as little as 30 minutes.

This kit, which has been optimized for labeling proteins with molecular weights between 12,000 and 150,000 daltons, contains everything needed to perform three labeling reactions and to separate the resulting conjugates from excess reactive biotin. Each Biotin-XX Microscale Protein Labeling Kit provides:

- Three vials of amine-reactive biotin-XX sulfosuccinimidyl ester
- Sodium bicarbonate
- Reaction tubes
- Purification resin and spin filters
- Detailed protocols for conjugation and purification

For determining the degree of labeling, the FluoReporter® Biotin Quantitation Assay Kit for proteins is available separately (F30751) or in combination with the Biotin-XX Microscale Protein Labeling Kit (B30756). When biotinylating larger amounts of protein, we recommend the FluoReporter® Biotin-XX Protein Labeling Kit, which is optimized for 5–20 mg samples, or the FluoReporter® Mini-Biotin-XX Protein Labeling Kit, which is optimized for 0.1–3 mg samples of >30,000-dalton proteins.

FluoReporter® Mini-Biotin-XX Protein Labeling Kit

The FluoReporter® Mini-Biotin-XX Protein Labeling Kit (F6347; Table 1.2, Table 1.3) permits efficient biotinylation of small amounts of antibodies or other proteins. The water-soluble biotin-XX sulfosuccinimidyl ester has a 14-atom spacer (Figure 1.2.5) that enhances the binding of biotin derivatives to avidin's relatively deep binding sites. The ready-to-use spin columns provide a convenient method of purifying the biotinylated protein from excess reagents.

Each FluoReporter® Mini-Biotin-XX Protein Labeling Kit provides sufficient reagents for five labeling reactions of 0.1–3 mg protein each, including:

- Five vials of amine-reactive biotin-XX sulfosuccinimidyl ester
- Reaction tubes, each containing a stir bar
- Purification resin, spin columns and collection tubes
- Dialysis tubing
- Detailed protocols for conjugation and purification

FluoReporter® Biotin-XX Protein Labeling Kit

The FluoReporter® Biotin-XX Protein Labeling Kit (F2610; Table 1.2, Table 1.3) is designed to provide five biotinylation reactions, each containing 5–20 mg of protein. A gel filtration column is used to purify the labeled proteins from excess biotin reagent. Once purified, the degree of biotinylation can be determined using the included avidin–HABA displacement assay; biotinylated goat IgG is provided as a standard.

Each FluoReporter® Biotin-XX Protein Labeling Kit provides sufficient reagents for five labeling reactions of 5–20 mg protein each, including:

- Amine-reactive biotin-XX succinimidyl ester
- Dimethylsulfoxide (DMSO)
- Gel filtration column
- Avidin–HABA complex

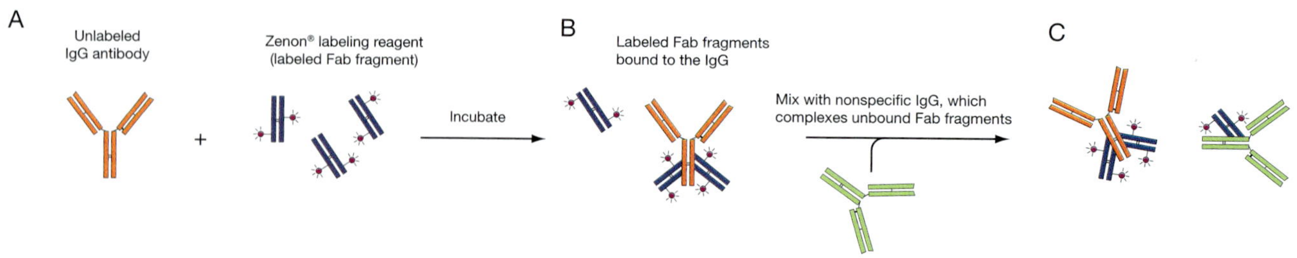

Figure 1.2.4 Labeling scheme utilized in the Zenon® Antibody Labeling Kits. **A)** An unlabeled IgG antibody is incubated with the Zenon® labeling reagent, which contains a fluorophore-labeled, Fc-specific anti-IgG Fab fragment. **B)** This labeled Fab fragment binds to the Fc portion of the IgG antibody. **C)** Excess Fab fragment is then neutralized by the addition of a nonspecific IgG, preventing crosslabeling by the Fab fragment in experiments where primary antibodies of the same type are present. Note that the Fab fragment used for labeling need not be coupled to a fluorophore, but could instead be coupled to an enzyme (such as HRP) or to biotin.

IMPORTANT NOTICE: The products described in this manual are covered by one or more Limited Use Label License(s). Please refer to the Appendix on page 971 and Master Product List on page 975. Products are For Research Use Only. Not intended for any animal or human therapeutic or diagnostic use.

- Biotinylated goat IgG
- Detailed protocols for conjugation, purification and determination of the degree of labeling

FluoReporter® Biotin/DNP Protein Labeling Kit

The FluoReporter® Biotin/DNP Protein Labeling Kit (F6348; Table 1.2, Table 1.3) provides the necessary reagents for labeling proteins with DNP-X–biocytin-X succinimidyl ester (Figure 1.2.6). The degree of biotinylation of proteins labeled with this reagent can be assessed from the optical absorbance of DNP (EC = 15,000 $cm^{-1}M^{-1}$ at ~360 nm). The conjugates are recognized by both avidin derivatives (and anti-biotin antibodies) and by anti-DNP antibodies, permitting a choice of detection techniques.

Each FluoReporter® Biotin/DNP Protein Labeling Kit contains sufficient reagents for 5 to 10 labeling reactions of 0.2–2 mg of protein each, including:

- Five vials of amine-reactive DNP-X–biocytin-X succinimidyl ester
- Dimethylsulfoxide (DMSO) for dissolving the succinimidyl ester
- Reaction tubes
- Spin columns and collection tubes
- Detailed protocols for conjugation, purification and determination of the degree of labeling

DSB-X™ Biotin Protein Labeling Kit

Our unique DSB-X™ biotin technology, which is described in detail in Section 4.1, permits the facile reversal of the biotin–avidin interaction under extremely gentle conditions.[4] DSB-X™ biotin succinimidyl ester, a derivative of desthiobiotin (Figure 1.2.7) with an additional seven-atom spacer, reacts with amine groups of biomolecules to form stable amides. Like biotin conjugates, the DSB-X™ biotin conjugate can be detected with any avidin or streptavidin derivative; with DSB-X™ biotin conjugates, however, this binding is almost totally reversed by addition of free biotin (B1595, B20656; Section 4.2) at neutral pH and normal ionic strength.

The DSB-X™ Biotin Protein Labeling Kit (D20655; Table 1.2, Table 1.3) contains the reagents required for five protein conjugations of 0.5–3 mg each, including:

- Five vials of amine-reactive DSB-X™ biotin succinimidyl ester
- Dimethylsulfoxide (DMSO) for dissolving the succinimidyl ester
- Reaction tubes
- Purification resin, spin columns and collection tubes for small-scale purifications
- Dialysis tubing for larger-scale separations
- Detailed protocols for conjugation and purification

Kits for Labeling Nucleic Acids with a Fluorescent Dye

ULYSIS® Nucleic Acid Labeling Kits

The ULYSIS® Alexa Fluor® Nucleic Acid Labeling Kits (Table 1.2, Table 1.3, Table 8.6) provide a simple, reliable method for producing fluorescent hybridization probes by combining our Alexa Fluor® fluorophores with the versatile, patented Universal Linkage System (ULS®) platinum-based chemistry developed by KREATECH Biotechnology BV. The ULS® technology is based on the use of a platinum dye reagent that forms a stable adduct with the *N*-7 position of guanine and, to a lesser extent, adenine bases in DNA, RNA, peptide–nucleic acid conjugates (PNA) and oligonucleotides (Figure 1.2.8). In protein contexts, ULS® reagents are reactive with cysteine residues and other thiols.[5]

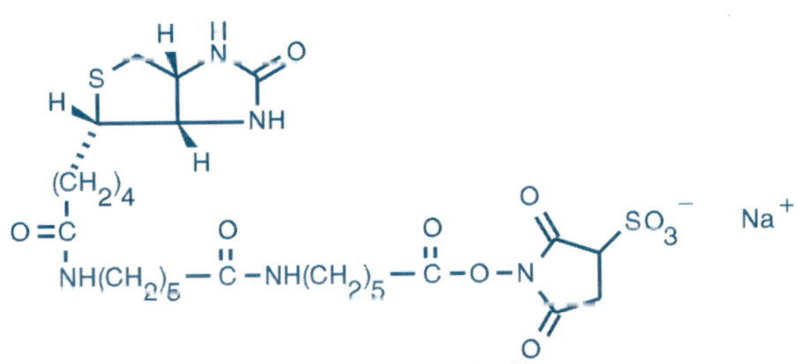

Figure 1.2.5 6-((6-((biotinoyl)amino)hexanoyl)amino)hexanoic acid, sulfosuccinimidyl ester, sodium salt (biotin-XX, SSE; B6352).

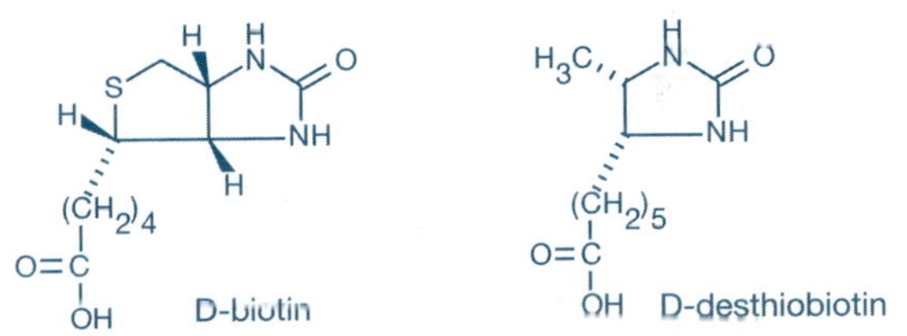

Figure 1.2.7 Comparison of the structures of D-biotin (left) and D-desthiobiotin (right).

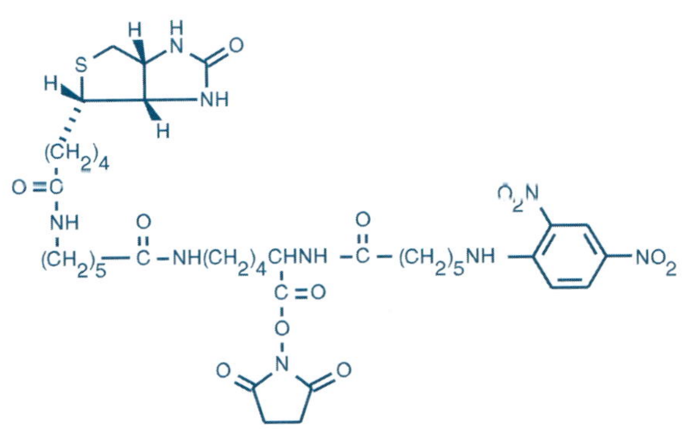

Figure 1.2.6 Biotin-X 2,4-dinitrophenyl-X-L-lysine, succinimidyl ester (DNP-X-biocytin-X, SE; B2604).

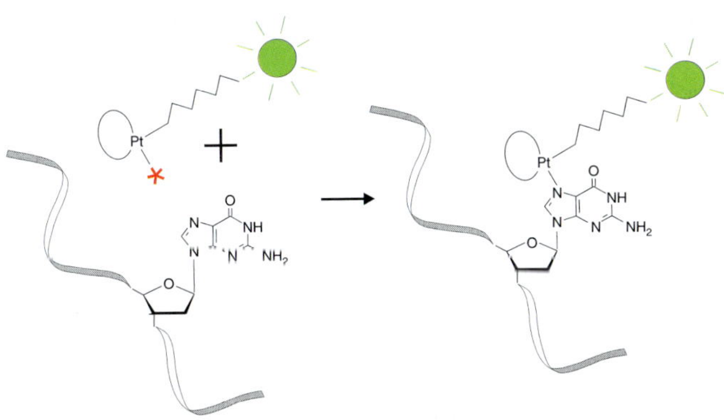

Figure 1.2.8 Schematic diagram of the labeling method provided in our ULYSIS® Nucleic Acid Labeling Kits. The ULS® reagent in the ULYSIS® Nucleic Acid Labeling Kits reacts with the *N*-7 position of guanine residues to provide a stable coordination complex between the nucleic acid and the fluorophore label.

IMPORTANT NOTICE: The products described in this manual are covered by one or more Limited Use Label License(s). Please refer to the Appendix on page 971 and Master Product List on page 975. Products are For Research Use Only. Not intended for any animal or human therapeutic or diagnostic use.

The labeling reaction typically takes only 15 minutes, and separation of the labeled nucleic acids from the unreacted ULS® reagent can be accomplished through use of a simple spin-column procedure (Figure 1.2.9).

In addition to ULYSIS® Alexa Fluor® Nucleic Acid Labeling Kits, we offer the ULYSIS® Oregon Green® 488 Nucleic Acid Labeling Kit (Section 8.2).

Each of these ULYSIS® Kits provides sufficient reagents for 20 labelings of 1 µg DNA each, including:

- ULS® labeling reagent and appropriate solvent
- Labeling buffer
- Deoxyribonuclease I (DNase I), for digesting DNA longer than 1000 base-pairs prior to labeling
- DNase I storage and reaction buffers
- Control DNA from calf thymus
- Nuclease-free H_2O
- Detailed protocols for preparing fluorescent DNA hybridization probes for chromosome *in situ* hybridization and dot-blot hybridization

Probes labeled using the ULYSIS® Kits are stable indefinitely and hybridize effectively to target DNA. The ULS® method has been used to prepare labeled probes for dot, Southern and northern blot analysis, RNA and DNA *in situ* hybridization, multicolor FISH, comparative genome hybridization (CGH) and microarray analysis.

ARES™ DNA Labeling Kits

The ARES™ DNA Labeling Kits (Table 1.2, Table 1.3, Table 8.6) provide a versatile, two-step method for labeling DNA with several of our premier fluorescent dyes[6] (Figure 1.2.10). In the first step, an amine-modified nucleotide, 5-(3-aminoallyl)-dUTP (Figure 1.2.11), is incorporated into DNA using conventional enzymatic labeling methods. This step helps ensure relatively uniform labeling of the probe with primary amine groups. The aminoallyl dUTP substrate used in this reaction is taken up efficiently by reverse transcription or nick translation, for which we provide the protocols; other enzymatic methods are also likely to be compatible. In the second step, the amine-modified DNA is chemically labeled using an amine-reactive fluorescent dye. This chemical reaction varies little in its efficiency from dye to dye, so that it is possible to use any combination of the ARES™ Kits, with their broad selection of the brightest and most photostable dyes, and obtain consistent DNA labeling. The labeling protocols provided generally result in incorporation of about one dye per 12–15 bases, which we have determined to be optimal for fluorescence *in situ* hybridization (FISH) and dot-blot hybridization.

Each ARES™ DNA Labeling Kit provides sufficient reagents for 5 to 10 labelings of 1–5 µg DNA each, including:

- 5-(3-Aminoallyl)-dUTP
- Amine-reactive fluorescent dye and appropriate solvent
- Sodium bicarbonate
- Nuclease-free H_2O
- Detailed protocol for labeling DNA using reverse transcription or nick translation

See Section 8.2 for a complete description of the ARES™ Kits, as well as of the FISH Tag™ DNA and FISH Tag™ RNA Kits, which employ the same aminoallyl nucleotide labeling method but provide a complete workflow solution for fluorescence *in situ* hybridization (FISH) applications. Each FISH Tag™ Kit provides all of the reagents needed for enzymatically incorporating the amine-modified nucleotide (aminoallyl dUTP or aminoallyl UTP) into DNA or RNA, followed by fluorescent labeling with an amine-reactive Alexa Fluor® dye and purification of the labeled probe using PureLink™ nucleic acid purification technology (e.g., PureLink™ PCR Purification Kit, K3100-01). FISH Tag™ DNA and FISH Tag™ RNA Kits are each available as single-color kits containing one of four amine-reactive Alexa Fluor® dyes or in a multicolor kit that contains all four reactive Alexa Fluor® dyes.

Alexa Fluor® Oligonucleotide Amine Labeling Kits

The Alexa Fluor® Oligonucleotide Amine Labeling Kits (Section 8.2; Table 1.2, Table 1.3, Table 8.6) provide the reagents required for labeling synthetic oligonucleotides that have amine groups incorporated at their 5′-terminus. Following purification by standard chromatographic or electrophoretic procedures, these singly labeled oligonucleotides can serve as hybridization or ligation probes for a variety of applications.[7–9]

Each Alexa Fluor® Oligonucleotide Amine Labeling Kit contains sufficient reagents for three labelings of 50 µg each of an amine-modified oligonucleotide, including:

- Three vials of amine-reactive dye
- Dimethylsulfoxide (DMSO)
- Three vials of labeling buffer
- Labeling protocol

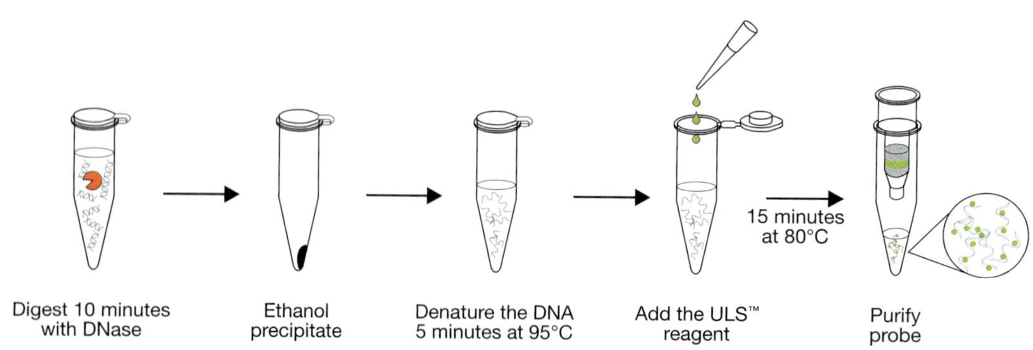

Figure 1.2.9 Nucleic acid labeling method provided in our ULYSIS® Nucleic Acid Labeling Kits.

IMPORTANT NOTICE: The products described in this manual are covered by one or more Limited Use Label License(s). Please refer to the Appendix on page 971 and Master Product List on page 975. Products are For Research Use Only. Not intended for any animal or human therapeutic or diagnostic use.
www.invitrogen.com/probes

Biotin Quantitation Assay Kits

FluoReporter® Biotin Quantitation Assay Kit for Biotinylated Proteins

The FluoReporter® Biotin Quantitation Assay Kit for biotinylated proteins (F30751) provides a sensitive fluorometric assay designed to accurately determine the number of biotin labels on a protein. This assay is based on the displacement of a ligand tagged with a quencher dye from the biotin binding sites of Biotective™ Green reagent.[10] The FluoReporter® biotin quantitation assay can detect from 4 to 80 picomoles of biotin in a sample, providing a 50-fold higher sensitivity than the spectrophotometric HABA biotin binding assay. Furthermore, unlike the HABA biotin binding assay, which requires ~1 mg of protein sample, the FluoReporter® biotin quantitation assay requires a minimum of 600 ng of a singly biotinylated IgG with molecular weight 150,000 daltons. For proteins of lower molecular weight or multiple biotin labels, less protein can be used. To expose any biotin groups in multiply labeled proteins that are sterically restricted and inaccessible to the Biotective™ Green reagent, this kit includes protease and an optional protocol for digesting the protein. With this preliminary digestion, biotin assay values agree well with MALDI-TOF determinations. With excitation/emission maxima of 495/519 nm, this assay is compatible with any fluorescence-based microplate reader capable of detecting fluorescein (FITC) or Alexa Fluor® 488 dye; it can also be scaled up for fluorometer-based experiments.

Each FluoReporter® Biotin Quantitation Assay Kit for biotinylated proteins includes:

- Biotective™ Green reagent
- Biocytin standard
- Protease
- Concentrated phosphate-buffered saline (PBS)
- Biotinylated goat anti–mouse IgG antibody for use as a positive control
- Detailed protocols

Sufficient reagents are provided for assaying 5 samples independently using eight wells in triplicate for the standard curve and three dilutions of the sample in triplicate (totaling 33 wells per assay). However, fewer wells may be used to conserve sample and a single standard curve can be used for multiple samples in the same experimental session.

FluoReporter® Biotin Quantitation Assay Kit for Biotinylated Nucleic Acids

The FluoReporter® Biotin Quantitation Assay Kit for biotinylated nucleic acids (F30755) provides a sensitive fluorometric assay for determining the number of biotin labels on a nucleic acid. This assay is based on the displacement of a quencher dye from the biotin binding sites of Biotective™ Green reagent. The FluoReporter® biotin quantitation assay can detect from 4 to 80 picomoles of biotin in a sample, providing a 50-fold higher sensitivity than the spectrophotometric HABA biotin binding assay. Analysis of multiply biotinylated nucleic acids requires a preliminary nuclease digestion step to avoid underestimation caused by steric restriction of avidin binding. This kit can be applied to as little as 13 ng of biotin-labeled nucleic acid and is ideal for determining the degree of biotinylation of cDNA samples used for microarray expression analysis.

Each FluoReporter® Biotin Quantitation Assay Kit for biotinylated nucleic acids includes:

- Biotective™ Green reagent
- Biotin-dUMP standard
- Nuclease
- Concentrated phosphate-buffered saline (PBS)
- Biotinylated DNA for use as a postive control
- Concentrated nucleic acid digestion buffer
- Detailed protocols

Sufficient reagents are provided for assaying 10 samples independently using eight wells in triplicate for the standard curve and three dilutions of the sample in triplicate (totaling 33 wells per assay). However, fewer wells may be used to conserve sample and a single standard curve can be used for multiple samples in the same experimental session.

REFERENCES

1. Bioconjug Chem (2009) 20:147; 2. Biochemistry (2007) 46:12522; 3. J Mol Diagn (2006) 8:246; 4. Anal Biochem (2002) 308:343; 5. Clin Chem (2002) 48:1352; 6. Biotechniques (2004) 36:114; 7. J Am Chem Soc (2006) 128:11423; 8. Biochemistry (2006) 45:4164; 9. Anal Chem (2003) 75:1664; 10. Biotechniques (2007) 43:503.

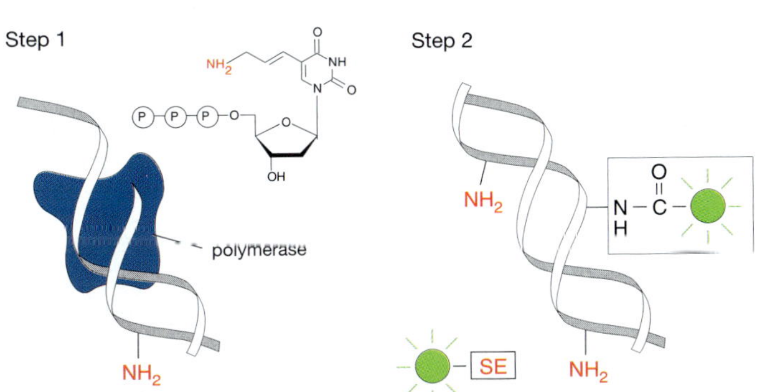

Figure 1.2.10 Schematic diagram of the labeling method provided in our ARES™ DNA Labeling Kits. The ARES™ DNA Labeling Kits use a two-step method to label DNA. Step 1) The aminoallyl dUTP is enzymatically incorporated. Step 2) A reactive fluorophore is used to label the incorporated aminoallyl group.

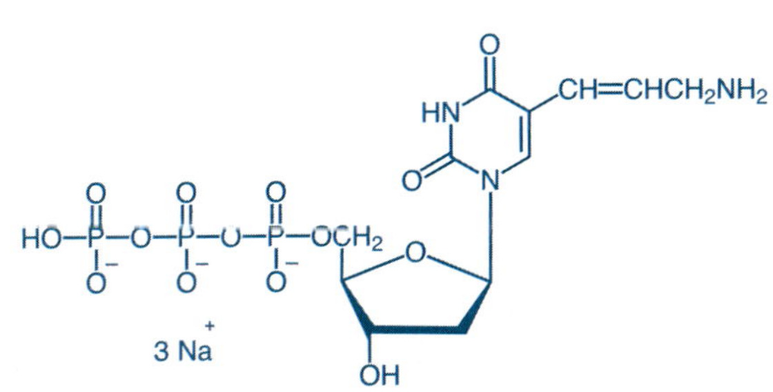

Figure 1.2.11 Aminoallyl-dUTP (5-(3-aminoallyl)-2'-deoxyuridine 5'-triphosphate, trisodium salt, A21664).

PRODUCT LIST 1.2 KITS FOR LABELING PROTEINS AND NUCLEIC ACIDS

Cat. No.	Product	Quantity
A30006	Alexa Fluor® 488 Microscale Protein Labeling Kit *for 20–100 µg protein* *3 labelings*	1 kit
A30007	Alexa Fluor® 555 Microscale Protein Labeling Kit *for 20–100 µg protein* *3 labelings*	1 kit
A30008	Alexa Fluor® 594 Microscale Protein Labeling Kit *for 20–100 µg protein* *3 labelings*	1 kit
A30009	Alexa Fluor® 647 Microscale Protein Labeling Kit *for 20–100 µg protein* *3 labelings*	1 kit
A20180	Alexa Fluor® 350 Monoclonal Antibody Labeling Kit *5 labelings*	1 kit
A20181	Alexa Fluor® 488 Monoclonal Antibody Labeling Kit *5 labelings*	1 kit
A20182	Alexa Fluor® 532 Monoclonal Antibody Labeling Kit *5 labelings*	1 kit
A20183	Alexa Fluor® 546 Monoclonal Antibody Labeling Kit *5 labelings*	1 kit
A20187	Alexa Fluor® 555 Monoclonal Antibody Labeling Kit *5 labelings*	1 kit
A20184	Alexa Fluor® 568 Monoclonal Antibody Labeling Kit *5 labelings*	1 kit
A20185	Alexa Fluor® 594 Monoclonal Antibody Labeling Kit *5 labelings*	1 kit
A20186	Alexa Fluor® 647 Monoclonal Antibody Labeling Kit *5 labelings*	1 kit
A20191	Alexa Fluor® 488 Oligonucleotide Amine Labeling Kit *3 labelings*	1 kit
A20196	Alexa Fluor® 647 Oligonucleotide Amine Labeling Kit *3 labelings*	1 kit
A10170	Alexa Fluor® 350 Protein Labeling Kit *3 labelings*	1 kit
A10171	Alexa Fluor® 430 Protein Labeling Kit *3 labelings*	1 kit
A10235	Alexa Fluor® 488 Protein Labeling Kit *3 labelings*	1 kit
A10236	Alexa Fluor® 532 Protein Labeling Kit *3 labelings*	1 kit
A10237	Alexa Fluor® 546 Protein Labeling Kit *3 labelings*	1 kit
A20174	Alexa Fluor® 555 Protein Labeling Kit *3 labelings*	1 kit
A10238	Alexa Fluor® 568 Protein Labeling Kit *3 labelings*	1 kit
A10239	Alexa Fluor® 594 Protein Labeling Kit *3 labelings*	1 kit
A20170	Alexa Fluor® 633 Protein Labeling Kit *3 labelings*	1 kit
A20173	Alexa Fluor® 647 Protein Labeling Kit *3 labelings*	1 kit
A20171	Alexa Fluor® 660 Protein Labeling Kit *3 labelings*	1 kit
A20172	Alexa Fluor® 680 Protein Labeling Kit *3 labelings*	1 kit
A10468	APEX® Alexa Fluor® 488 Antibody Labeling Kit	1 kit
A10470	APEX® Alexa Fluor® 555 Antibody Labeling Kit	1 kit
A10474	APEX® Alexa Fluor® 594 Antibody Labeling Kit	1 kit
A10475	APEX® Alexa Fluor® 647 Antibody Labeling Kit	1 kit
A10476	APEX® Oregon Green® 488 Antibody Labeling Kit	1 kit
A10478	APEX® Pacific Blue™ Antibody Labeling Kit	1 kit
A21665	ARES™ Alexa Fluor® 488 DNA Labeling Kit *10 labelings*	1 kit
A21667	ARES™ Alexa Fluor® 546 DNA Labeling Kit *10 labelings*	1 kit
A21677	ARES™ Alexa Fluor® 555 DNA Labeling Kit *10 labelings*	1 kit
A21669	ARES™ Alexa Fluor® 594 DNA Labeling Kit *10 labelings*	1 kit
A21676	ARES™ Alexa Fluor® 647 DNA Labeling Kit *10 labelings*	1 kit
B30010	Biotin-XX Microscale Protein Labeling Kit *for 20–100 µg protein* *3 labelings*	1 kit
B30756	Biotin-XX Microscale Protein Labeling Kit with FluoReporter® Biotin Quantitation Assay Kit *includes B30010 and F30751*	1 kit
D20655	DSB-X™ Biotin Protein Labeling Kit *5 labelings*	1 kit
F30755	FluoReporter® Biotin Quantitation Assay Kit *for biotinylated nucleic acids* *10 determinations*	1 kit
F30751	FluoReporter® Biotin Quantitation Assay Kit *for biotinylated proteins* *5 determinations*	1 kit
F6348	FluoReporter® Biotin/DNP Protein Labeling Kit *5–10 labelings*	1 kit
F2610	FluoReporter® Biotin-XX Protein Labeling Kit *5 labelings of 5–20 mg protein each*	1 kit
F6434	FluoReporter® FITC Protein Labeling Kit *5–10 labelings*	1 kit
F6433	FluoReporter® Fluorescein-EX Protein Labeling Kit *5–10 labelings*	1 kit
F6347	FluoReporter® Mini-biotin-XX Protein Labeling Kit *5 labelings of 0.1–3 mg protein each*	1 kit
F6153	FluoReporter® Oregon Green® 488 Protein Labeling Kit *5–10 labelings*	1 kit
F6161	FluoReporter® Rhodamine Red™-X Protein Labeling Kit *5–10 labelings*	1 kit
F6162	FluoReporter® Texas Red®-X Protein Labeling Kit *5–10 labelings*	1 kit
F10240	Fluorescein-EX Protein Labeling Kit *3 labelings*	1 kit
O10241	Oregon Green® 488 Protein Labeling Kit *3 labelings*	1 kit
P30013	Pacific Blue™ Monoclonal Antibody Labeling Kit *5 labelings*	1 kit
P30012	Pacific Blue™ Protein Labeling Kit *3 labelings*	1 kit
P30014	Pacific Orange™ Monoclonal Antibody Labeling Kit *5 labelings*	1 kit
P30016	Pacific Orange™ Protein Labeling Kit *3 labelings*	1 kit
S30044	SAIVI™ Alexa Fluor® 647 Antibody/Protein 1 mg-Labeling Kit *3 labelings*	1 kit
S30045	SAIVI™ Rapid Antibody Labeling Kit, Alexa Fluor® 680 *3 labelings*	1 kit
S30046	SAIVI™ Rapid Antibody Labeling Kit, Alexa Fluor® 750 *3 labelings*	1 kit

IMPORTANT NOTICE: The products described in this manual are covered by one or more Limited Use Label License(s). Please refer to the Appendix on page 971 and Master Product List on page 975. Products are For Research Use Only. Not intended for any animal or human therapeutic or diagnostic use.
www.invitrogen.com/probes
molecular probes® | invitrogen™ by life technologies™

PRODUCT LIST 1.2 KITS FOR LABLING PROTEINS AND NUCLEIC ACIDS—*continued*

Cat. No.	Product	Quantity
T10244	Texas Red®-X Protein Labeling Kit *3 labelings*	1 kit
U21650	ULYSIS® Alexa Fluor® 488 Nucleic Acid Labeling Kit *20 labelings*	1 kit
U21651	ULYSIS® Alexa Fluor® 532 Nucleic Acid Labeling Kit *20 labelings*	1 kit
U21652	ULYSIS® Alexa Fluor® 546 Nucleic Acid Labeling Kit *20 labelings*	1 kit
U21653	ULYSIS® Alexa Fluor® 568 Nucleic Acid Labeling Kit *20 labelings*	1 kit
U21654	ULYSIS® Alexa Fluor® 594 Nucleic Acid Labeling Kit *20 labelings*	1 kit
U21660	ULYSIS® Alexa Fluor® 647 Nucleic Acid Labeling Kit *20 labelings*	1 kit
U21659	ULYSIS® Oregon Green® 488 Nucleic Acid Labeling Kit *20 labelings*	1 kit
Z25400	Zenon® Alexa Fluor® 350 Human IgG Labeling Kit *50 labelings*	1 kit
Z25000	Zenon® Alexa Fluor® 350 Mouse IgG_1 Labeling Kit *50 labelings*	1 kit
Z25100	Zenon® Alexa Fluor® 350 Mouse IgG_{2a} Labeling Kit *50 labelings*	1 kit
Z25200	Zenon® Alexa Fluor® 350 Mouse IgG_{2b} Labeling Kit *50 labelings*	1 kit
Z25300	Zenon® Alexa Fluor® 350 Rabbit IgG Labeling Kit *50 labelings*	1 kit
Z25013	Zenon® Alexa Fluor® 405 Mouse IgG_1 Labeling Kit *50 labelings*	1 kit
Z25113	Zenon® Alexa Fluor® 405 Mouse IgG_{2a} Labeling Kit *50 labelings*	1 kit
Z25213	Zenon® Alexa Fluor® 405 Mouse IgG_{2b} Labeling Kit *50 labelings*	1 kit
Z25313	Zenon® Alexa Fluor® 405 Rabbit IgG Labeling Kit *50 labelings*	1 kit
Z25001	Zenon® Alexa Fluor® 430 Mouse IgG_1 Labeling Kit *50 labelings*	1 kit
Z25301	Zenon® Alexa Fluor® 430 Rabbit IgG Labeling Kit *50 labelings*	1 kit
Z25602	Zenon® Alexa Fluor® 488 Goat IgG Labeling Kit *50 labelings*	1 kit
Z25402	Zenon® Alexa Fluor® 488 Human IgG Labeling Kit *50 labelings*	1 kit
Z25002	Zenon® Alexa Fluor® 488 Mouse IgG_1 Labeling Kit *50 labelings*	1 kit
Z25090	Zenon® Alexa Fluor® 488 Mouse IgG_1 Labeling Kit *enhanced with TSA™ technology* *25 labelings*	1 kit
Z25102	Zenon® Alexa Fluor® 488 Mouse IgG_{2a} Labeling Kit *50 labelings*	1 kit
Z25202	Zenon® Alexa Fluor® 488 Mouse IgG_{2b} Labeling Kit *50 labelings*	1 kit
Z25302	Zenon® Alexa Fluor® 488 Rabbit IgG Labeling Kit *50 labelings*	1 kit
Z25003	Zenon® Alexa Fluor® 532 Mouse IgG_1 Labeling Kit *50 labelings*	1 kit
Z25303	Zenon® Alexa Fluor® 532 Rabbit IgG Labeling Kit *50 labelings*	1 kit
Z25004	Zenon® Alexa Fluor® 546 Mouse IgG_1 Labeling Kit *50 labelings*	1 kit
Z25104	Zenon® Alexa Fluor® 546 Mouse IgG_{2a} Labeling Kit *50 labelings*	1 kit
Z25204	Zenon® Alexa Fluor® 546 Mouse IgG_{2b} Labeling Kit *50 labelings*	1 kit
Z25304	Zenon® Alexa Fluor® 546 Rabbit IgG Labeling Kit *50 labelings*	1 kit
Z25605	Zenon® Alexa Fluor® 555 Goat IgG Labeling Kit *50 labelings*	1 kit
Z25405	Zenon® Alexa Fluor® 555 Human IgG Labeling Kit *50 labelings*	1 kit
Z25005	Zenon® Alexa Fluor® 555 Mouse IgG_1 Labeling Kit *50 labelings*	1 kit
Z25105	Zenon® Alexa Fluor® 555 Mouse IgG_{2a} Labeling Kit *50 labelings*	1 kit
Z25205	Zenon® Alexa Fluor® 555 Mouse IgG_{2b} Labeling Kit *50 labelings*	1 kit
Z25305	Zenon® Alexa Fluor® 555 Rabbit IgG Labeling Kit *50 labelings*	1 kit
Z25606	Zenon® Alexa Fluor® 568 Goat IgG Labeling Kit *50 labelings*	1 kit
Z25006	Zenon® Alexa Fluor® 568 Mouse IgG_1 Labeling Kit *50 labelings*	1 kit
Z25106	Zenon® Alexa Fluor® 568 Mouse IgG_{2a} Labeling Kit *50 labelings*	1 kit
Z25206	Zenon® Alexa Fluor® 568 Mouse IgG_{2b} Labeling Kit *50 labelings*	1 kit
Z25306	Zenon® Alexa Fluor® 568 Rabbit IgG Labeling Kit *50 labelings*	1 kit
Z25607	Zenon® Alexa Fluor® 594 Goat IgG Labeling Kit *50 labelings*	1 kit
Z25407	Zenon® Alexa Fluor® 594 Human IgG Labeling Kit *50 labelings*	1 kit
Z25007	Zenon® Alexa Fluor® 594 Mouse IgG_1 Labeling Kit *50 labelings*	1 kit
Z25107	Zenon® Alexa Fluor® 594 Mouse IgG_{2a} Labeling Kit *50 labelings*	1 kit
Z25207	Zenon® Alexa Fluor® 594 Mouse IgG_{2b} Labeling Kit *50 labelings*	1 kit
Z25307	Zenon® Alexa Fluor® 594 Rabbit IgG Labeling Kit *50 labelings*	1 kit
Z25020	Zenon® Alexa Fluor® 610–R-Phycoerythrin Mouse IgG_1 Labeling Kit *10 labelings*	1 kit
Z25608	Zenon® Alexa Fluor® 647 Goat IgG Labeling Kit *50 labelings*	1 kit
Z25408	Zenon® Alexa Fluor® 647 Human IgG Labeling Kit *50 labelings*	1 kit
Z25008	Zenon® Alexa Fluor® 647 Mouse IgG_1 Labeling Kit *50 labelings*	1 kit
Z25108	Zenon® Alexa Fluor® 647 Mouse IgG_{2a} Labeling Kit *50 labelings*	1 kit
Z25208	Zenon® Alexa Fluor® 647 Mouse IgG_{2b} Labeling Kit *50 labelings*	1 kit
Z25308	Zenon® Alexa Fluor® 647 Rabbit IgG Labeling Kit *50 labelings*	1 kit
Z25021	Zenon® Alexa Fluor® 647–R-Phycoerythrin Mouse IgG_1 Labeling Kit *10 labelings*	1 kit
Z25121	Zenon® Alexa Fluor® 647–R-Phycoerythrin Mouse IgG_{2a} Labeling Kit *10 labelings*	1 kit

continued on next page

IMPORTANT NOTICE: The products described in this manual are covered by one or more Limited Use Label License(s). Please refer to the Appendix on page 971 and Master Product List on page 975. Products are For Research Use Only. Not intended for any animal or human therapeutic or diagnostic use.

PRODUCT LIST 1.2 KITS FOR LABLING PROTEINS AND NUCLEIC ACIDS—*continued*

Cat. No.	Product	Quantity
Z25221	Zenon® Alexa Fluor® 647–R-Phycoerythrin Mouse IgG_{2b} Labeling Kit *10 labelings*	1 kit
Z25009	Zenon® Alexa Fluor® 660 Mouse IgG_1 Labeling Kit *50 labelings*	1 kit
Z25010	Zenon® Alexa Fluor® 680 Mouse IgG_1 Labeling Kit *50 labelings*	1 kit
Z25110	Zenon® Alexa Fluor® 680 Mouse IgG_{2a} Labeling Kit *50 labelings*	1 kit
Z25210	Zenon® Alexa Fluor® 680 Mouse IgG_{2b} Labeling Kit *50 labelings*	1 kit
Z25310	Zenon® Alexa Fluor® 680 Rabbit IgG Labeling Kit *50 labelings*	1 kit
Z25022	Zenon® Alexa Fluor® 680–R-Phycoerythrin Mouse IgG_1 Labeling Kit *10 labelings*	1 kit
Z25011	Zenon® Alexa Fluor® 700 Mouse IgG_1 Labeling Kit *50 labelings*	1 kit
Z25030	Zenon® Alexa Fluor® 700–Allophycocyanin Mouse IgG_1 Labeling Kit *10 labelings*	1 kit
Z25312	Zenon® Alexa Fluor® 750 Rabbit IgG Labeling Kit *50 labelings*	1 kit
Z25031	Zenon® Alexa Fluor® 750–Allophycocyanin Mouse IgG_1 Labeling Kit *10 labelings*	1 kit
Z25350	Zenon® Alkaline Phosphatase Rabbit IgG Labeling Kit *25 labelings*	1 kit
Z25451	Zenon® Allophycocyanin Human IgG Labeling Kit *25 labelings*	1 kit
Z25051	Zenon® Allophycocyanin Mouse IgG_1 Labeling Kit *25 labelings*	1 kit
Z25151	Zenon® Allophycocyanin Mouse IgG_{2a} Labeling Kit *25 labelings*	1 kit
Z25251	Zenon® Allophycocyanin Mouse IgG_{2b} Labeling Kit *25 labelings*	1 kit
Z25351	Zenon® Allophycocyanin Rabbit IgG Labeling Kit *25 labelings*	1 kit
Z25452	Zenon® Biotin-XX Human IgG Labeling Kit *50 labelings*	1 kit
Z25052	Zenon® Biotin-XX Mouse IgG_1 Labeling Kit *50 labelings*	1 kit
Z25152	Zenon® Biotin-XX Mouse IgG_{2a} Labeling Kit *50 labelings*	1 kit
Z25252	Zenon® Biotin-XX Mouse IgG_{2b} Labeling Kit *50 labelings*	1 kit
Z25352	Zenon® Biotin-XX Rabbit IgG Labeling Kit *50 labelings*	1 kit
Z25042	Zenon® Fluorescein Mouse IgG_1 Labeling Kit *50 labelings*	1 kit
Z25342	Zenon® Fluorescein Rabbit IgG Labeling Kit *50 labelings*	1 kit
Z25454	Zenon® Horseradish Peroxidase Human IgG Labeling Kit *25 labelings*	1 kit
Z25054	Zenon® Horseradish Peroxidase Mouse IgG_1 Labeling Kit *25 labelings*	1 kit
Z25154	Zenon® Horseradish Peroxidase Mouse IgG_{2a} Labeling Kit *25 labelings*	1 kit
Z25254	Zenon® Horseradish Peroxidase Mouse IgG_{2b} Labeling Kit *25 labelings*	1 kit
Z25354	Zenon® Horseradish Peroxidase Rabbit IgG Labeling Kit *25 labelings*	1 kit
Z25041	Zenon® Pacific Blue™ Mouse IgG_1 Labeling Kit *50 labelings*	1 kit
Z25156	Zenon® Pacific Blue™ Mouse IgG_{2a} Labeling Kit *50 labelings*	1 kit
Z25341	Zenon® Pacific Blue™ Rabbit IgG Labeling Kit *50 labelings*	1 kit
Z25256	Zenon® Pacific Orange™ Mouse IgG_1 Labeling Kit *50 labelings*	1 kit
Z25257	Zenon® Pacific Orange™ Mouse IgG_{2a} Labeling Kit *50 labelings*	1 kit
Z25455	Zenon® R-Phycoerythrin Human IgG Labeling Kit *25 labelings*	1 kit
Z25055	Zenon® R-Phycoerythrin Mouse IgG_1 Labeling Kit *25 labelings*	1 kit
Z25155	Zenon® R-Phycoerythrin Mouse IgG_{2a} Labeling Kit *25 labelings*	1 kit
Z25255	Zenon® R-Phycoerythrin Mouse IgG_{2b} Labeling Kit *25 labelings*	1 kit
Z25355	Zenon® R-Phycoerythrin Rabbit IgG Labeling Kit *25 labelings*	1 kit
Z25045	Zenon® Texas Red®-X Mouse IgG_1 Labeling Kit *50 labelings*	1 kit
Z25460	Zenon® Tricolor Human IgG Labeling Kit #1 *for green, orange and deep red fluorescence imaging* *3 x 10 labelings*	1 kit
Z25470	Zenon® Tricolor Human IgG Labeling Kit #2 *for blue, green and red fluorescence imaging* *3 x 10 labelings*	1 kit
Z25060	Zenon® Tricolor Mouse IgG_1 Labeling Kit #1 *for green, orange and deep red fluorescence imaging* *3 x 10 labelings*	1 kit
Z25070	Zenon® Tricolor Mouse IgG_1 Labeling Kit #2 *for blue, green and red fluorescence imaging* *3 x 10 labelings*	1 kit
Z25080	Zenon® Tricolor Mouse IgG_1 Labeling Kit #3 *for flow cytometry, 488 nm excitation* *3 x 10 labelings*	1 kit
Z25160	Zenon® Tricolor Mouse IgG_{2a} Labeling Kit #1 *for green, orange and deep red fluorescence imaging* *3 x 10 labelings*	1 kit
Z25170	Zenon® Tricolor Mouse IgG_{2a} Labeling Kit #2 *for blue, green and red fluorescence imaging* *3 x 10 labelings*	1 kit
Z25180	Zenon® Tricolor Mouse IgG_{2a} Labeling Kit #3 *for flow cytometry, 488 nm excitation* *3 x 10 labelings*	1 kit
Z25260	Zenon® Tricolor Mouse IgG_{2b} Labeling Kit #1 *for green, orange and deep red fluorescence imaging* *3 x 10 labelings*	1 kit
Z25270	Zenon® Tricolor Mouse IgG_{2b} Labeling Kit #2 *for blue, green and red fluorescence imaging* *3 x 10 labelings*	1 kit
Z25280	Zenon® Tricolor Mouse IgG_{2b} Labeling Kit #3 *for flow cytometry, 488 nm excitation* *3 x 10 labelings*	1 kit
Z25360	Zenon® Tricolor Rabbit IgG Labeling Kit #1 *for green, orange and deep red fluorescence imaging* *3 x 10 labelings*	1 kit
Z25370	Zenon® Tricolor Rabbit IgG Labeling Kit #2 *for blue, green and red fluorescence imaging* *3 x 10 labelings*	1 kit
Z25380	Zenon® Tricolor Rabbit IgG Labeling Kit #3 *for flow cytometry, 488 nm excitation* *3 x 10 labelings*	1 kit

IMPORTANT NOTICE: The products described in this manual are covered by one or more Limited Use Label License(s). Please refer to the Appendix on page 971 and Master Product List on page 975. Products are For Research Use Only. Not intended for any animal or human therapeutic or diagnostic use.

1.3 Alexa Fluor® Dyes Spanning the Visible and Infrared Spectrum

Overview of the Alexa Fluor® Dyes

The Alexa Fluor® dyes produce exceptionally bright and photostable conjugates (Table 1.4, Note 1.1). The Alexa Fluor® dyes share several significant attributes, including:

- Strong absorption at wavelengths of maximal output of common excitation sources
- Bright and unusually photostable fluorescence of their bioconjugates
- Good water solubility, which makes the reactive dyes easy to conjugate and the conjugates resistant to precipitation and aggregation
- Insensitivity of their absorption and emission spectra to pH over a broad range
- Well-differentiated spectra, providing many options for multicolor detection and fluorescence resonance energy transfer (Fluorescence Resonance Energy Transfer (FRET)—Note 1.2)
- High quantum yields and long fluorescence lifetimes (Table 1.5)
- Extremely high FRET efficiency, with calculated R_0 values of up to 84 Å between pairs of Alexa Fluor® dyes (Table 1.6) and up to 77 Å between Alexa Fluor® dyes and some nonfluorescent quenchers (Table 1.11)

Alexa Fluor® dyes set new standards for fluorophores and the bioconjugates prepared from them. The absorption spectra (Figure 1.3.1, Figure 1.3.2, Figure 1.3.3) of these superior fluorescent dyes—Alexa Fluor® 350, Alexa Fluor® 405, Alexa Fluor® 430, Alexa Fluor® 488, Alexa Fluor® 514, Alexa Fluor® 532, Alexa Fluor® 546, Alexa Fluor® 555, Alexa Fluor® 568, Alexa Fluor® 594, Alexa Fluor® 610, Alexa Fluor® 633, Alexa Fluor® 635, Alexa Fluor® 647, Alexa Fluor® 660, Alexa Fluor® 680, Alexa Fluor® 700, Alexa Fluor® 750 and Alexa Fluor® 790 dyes—span the visible and infrared spectrum (Table 1.4) and match the principal output wavelengths of common excitation sources.[1,2] Because there are so many different Alexa Fluor® dyes, we have had to develop a systematic strategy for naming them. We identify these dyes with the registered trademark Alexa Fluor® followed by the optimal excitation wavelength in nm; for example, Alexa Fluor® 488 dye is optimally excited by the 488 nm spectral line of the argon-ion laser.

With spectra almost identical to those of fluorescein (Figure 1.3.4), but with far greater conjugate fluorescence and significantly better conjugate photostability, Alexa Fluor® 488 dye is indisputably the best green-fluorescent reactive dye available. Spectra of Alexa Fluor® 555 dye are an almost perfect match to those of Cy®3 dye (Figure 1.3.5), but conjugates of Alexa Fluor® 555 dye are more fluorescent (Figure 1.3.6) and more photostable (Figure 1.3.7) than those of Cy®3 dye. Similarly, spectra of Alexa Fluor® 647 conjugates substantially match those of the Cy®5 dye (Figure 1.3.8) and Alexa Fluor® 680 and Alexa Fluor® 750 dyes exhibit spectral properties similar to those of Cy®5.5 and Cy®7 dyes, respectively (Figure 1.3.9, Figure 1.3.10). Tandem conjugates of the long-wavelength Alexa Fluor® dyes with R-phycoerythrin or allophycocyanin (Section 6.4) further expand the utility of this dye series in multiplex flow cytometry applications[3,4] by shifting the phycobiliprotein emission profile relative to the available detection channels (Figure 1.3.11, Figure 1.3.12).

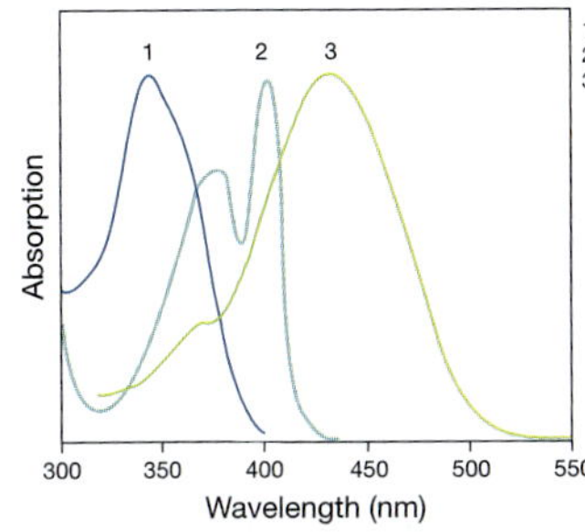

Figure 1.3.1 Absorption spectra of our ultraviolet and blue light-absorbing Alexa Fluor® dyes.

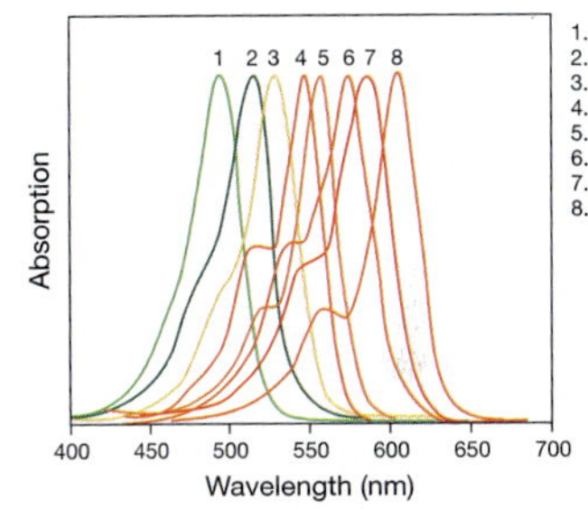

Figure 1.3.2 Absorption spectra of our green, yellow and red light-absorbing Alexa Fluor® dyes.

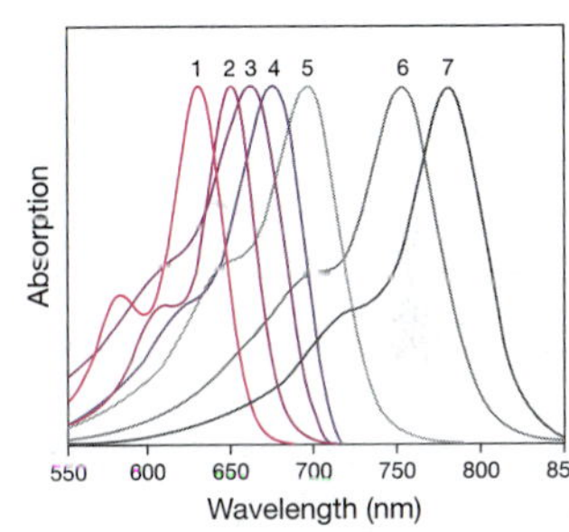

Figure 1.3.3 Absorption spectra of our far-red and infrared light-absorbing Alexa Fluor® dyes. Alexa Fluor® 635 dye, available conjugated to antibodies, streptavidin and phalloidin, is not included here but its absorption spectrum is very similar to that of Alexa Fluor® 633 dye.

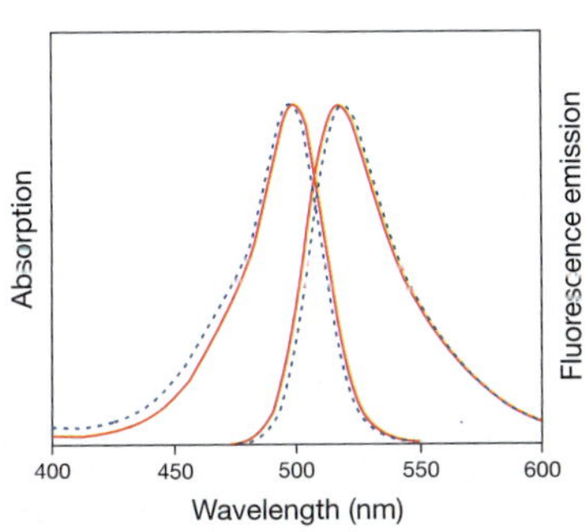

Figure 1.3.4 Absorption and fluorescence emission spectra of fluorescein goat anti-mouse IgG antibody (F2761, (—) and Alexa Fluor® 488 goat anti-mouse IgG antibody (A11001, (- - -)). The fluorescence intensity of the Alexa Fluor® 488 conjugate was significantly higher than that of the fluorescein conjugate. The data are normalized to show the spectral similarity.

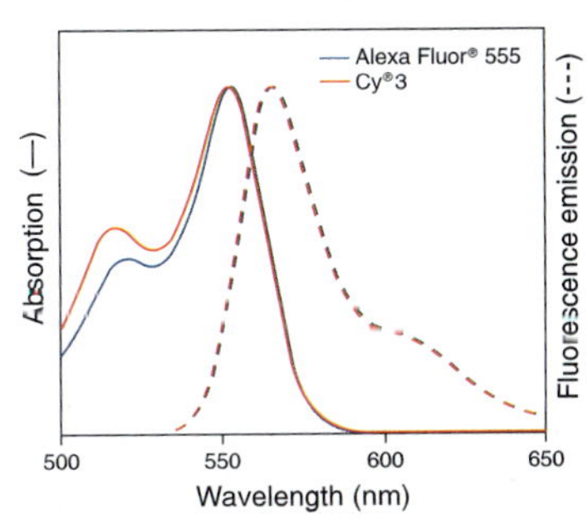

Figure 1.3.5 Comparison of the absorption and fluorescence emission spectra of Alexa Fluor® 555 and Cy®3 dyes. Spectra have been normalized to the same intensity for comparison purposes.

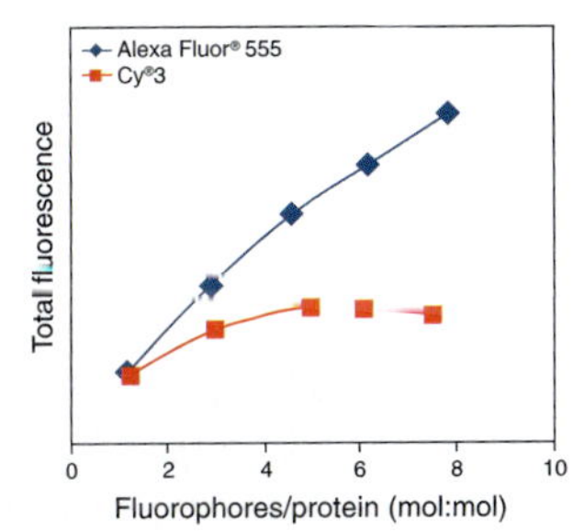

Figure 1.3.6 Comparison of the relative fluorescence of goat anti-rabbit IgG antibody conjugates of Alexa Fluor® 555 and Cy®3 dyes (prepared in our laboratories) at different dye:protein ratios in the conjugate.

IMPORTANT NOTICE: The products described in this manual are covered by one or more Limited Use Label License(s). Please refer to the Appendix on page 971 and Master Product List on page 975. Products are For Research Use Only. Not intended for any animal or human therapeutic or diagnostic use.

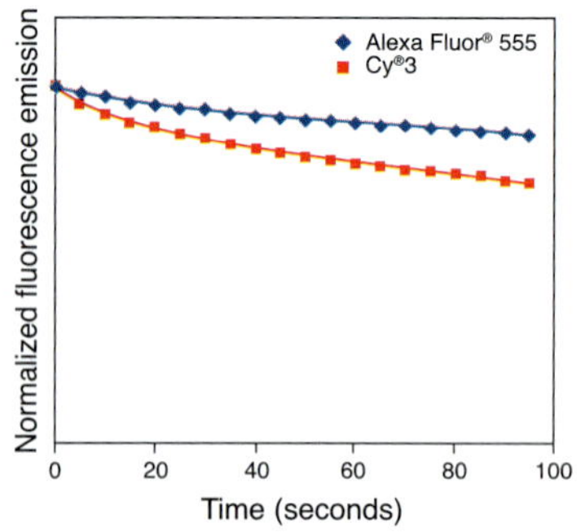

Figure 1.3.7 Photobleaching profiles of Alexa Fluor® 555 and Cy®3 dyes were obtained by placing equal molar concentrations of the free dyes into capillary tubes; the samples were continuously illuminated and data points were collected every 5 seconds. Fluorescence has been normalized to the same initial intensity.

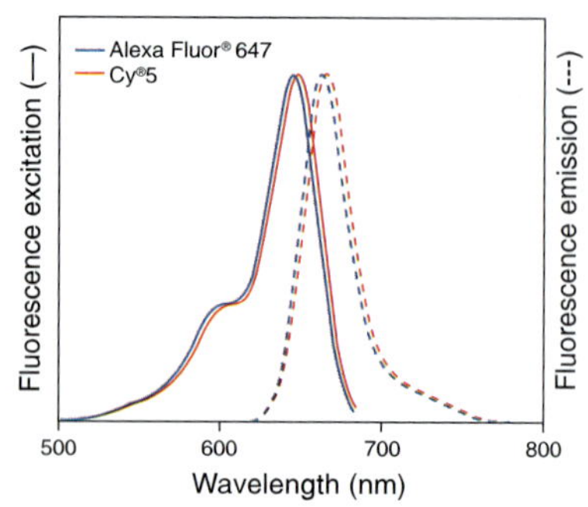

Figure 1.3.8 Comparison of the fluorescence spectra of Alexa Fluor® 647 and Cy®5 dyes. Spectra have been normalized to the same intensity for comparison purposes.

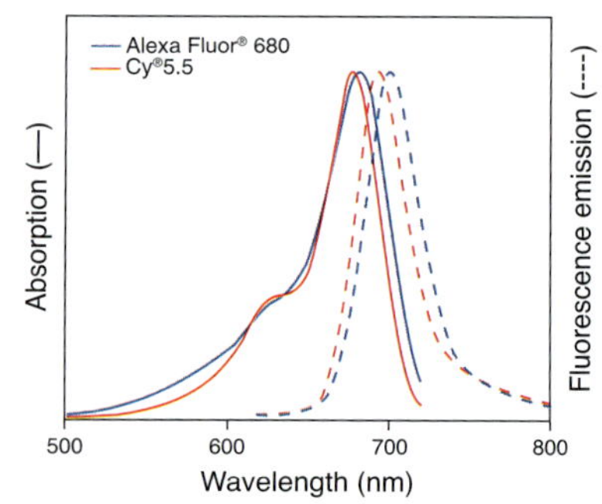

Figure 1.3.9 Comparison of the fluorescence spectra of the unconjugated Alexa Fluor® 680 and Cy®5.5 dyes. Spectra have been normalized to the same intensity for comparison purposes.

Table 1.4 Alexa Fluor® active esters and kits for labeling proteins and nucleic acids.

Alexa Fluor® Dye	Fluorescence Color (Abs/Em) *	Succinimidyl Ester, TFP Ester or SDP Ester	Kits for Labeling Proteins			Kits for Labeling Nucleic Acids and Oligonucleotides
			Protein Labeling Kits	Antibody Labeling Kits	Zenon® Antibody Labeling Kits	
Alexa Fluor® 350	Blue (346/442)	A10168 †	A10170 (P)	A20180 (Mab)	Z25000 (M IgG$_1$) Z25100 (M IgG$_{2a}$) Z25200 (M IgG$_{2b}$) Z25300 (R IgG) Z25400 (H IgG)	
Alexa Fluor® 405	Blue (402/421)	A30000 ‡ A30100 †			Z25013 (M IgG$_1$) Z25113 (M IgG$_{2a}$) Z25213 (M IgG$_{2b}$) Z25313 (R IgG)	
Alexa Fluor® 430	Yellow-green (434/539)	A10169 †	A10171 (P)		Z25001 (M IgG$_1$) Z25301 (R IgG)	
Alexa Fluor® 488	Green (495/519)	A20000 ‡ § A20100 † § A30005 (TFP) A30052 (SDP)	A10235 (P) ** A30006 (Micro) **	A10468 (APEX®) ** A20181 (Mab) **	Z25002 (M IgG$_1$) Z25090 TSA™) (M IgG$_1$) Z25102 (M IgG$_{2a}$) Z25202 (M IgG$_{2b}$) Z25302 (R IgG) Z25602 (G IgG) Z25402 (H IgG)	U21650 (ULYSIS®) A21665 (ARES™) F32947 (FT) F32952 (FT) A20191 (Oligo)
Alexa Fluor® 514	Yellow-green (518/540)	A30002 ‡ §				
Alexa Fluor® 532	Yellow (531/554)	A20001 ‡ A20101MP †	A10236 (P)	A20182 (Mab)	Z25003 (M IgG$_1$) Z25303 (R IgG)	U21651 (ULYSIS®)
Alexa Fluor® 546	Orange (556/573)	A20002 ‡ § A20102 † §	A10237 (P)	A20183 (Mab)	Z25004 (M IgG$_1$) Z25104 (M IgG$_{2a}$) Z25204 (M IgG$_{2b}$) Z25304 (R IgG)	U21652 (ULYSIS®) A21667 (ARES™)
Alexa Fluor® 555	Red-orange (555/565)	A20009 ‡ A20109 †	A20174 (P) A30007 (Micro)	A10470 (APEX®) A20187 (Mab)	Z25005 (M IgG$_1$) Z25105 (M IgG$_{2a}$) Z25205 (M IgG$_{2b}$) Z25305 (R IgG) Z25605 (G IgG) Z25405 (H IgG)	A21677 (ARES™) F32948 (FT) F32953 (FT)
Alexa Fluor® 568	Red-orange (578/603)	A20003 ‡ § A20103 † §	A10238 (P)	A20184 (Mab)	Z25006 (M IgG$_1$) Z25106 (M IgG$_{2a}$) Z25206 (M IgG$_{2b}$) Z25306 (R IgG) Z25606 (G IgG)	U21653 (ULYSIS®)

* Approximate absorption (Abs) and fluorescence emission (Em) maxima for conjugates, in nm. † 5 mg unit size. ‡ 1 mg unit size. § Mixed isomers. ** The APEX® Alexa Fluor® 488 Antibody Labeling Kits contain the Alexa Fluor® 488 carboxylic acid SDP ester, whereas the other Alexa Fluor® 488 protein labeling kits contain the amine-reactive Alexa Fluor® 488 carboxylic acid TFP ester; the Alexa Fluor® 488 kits for labeling nucleic acids and oligonucleotides contain the Alexa Fluor® 488 succinimidyl ester. †† Human vision is insensitive to light beyond ~650 nm, and therefore it is not possible to view the far-red– and near-infrared–fluorescent dyes by looking through the eyepiece of a conventional fluorescence microscope. The Alexa Fluor® Labeling Kits are described in detail in Section 1.2. (APEX®) = APEX® Antibody Labeling Kit. (ARES™) = ARES™ DNA Labeling Kit. (FT) = FISH Tag™ DNA Kit or FISH Tag™ RNA Kit. (G IgG) = Zenon® Goat IgG Labeling Kit. (H IgG) = Zenon® Human IgG Labeling Kit. (M IgG) = Zenon® Mouse IgG Labeling Kit. (Mab) = Monoclonal Antibody Labeling Kit. (Micro) = Alexa Fluor® Microscale Protein Labeling Kit. (Oligo) = Alexa Fluor® Oligonucleotide Amine Labeling Kit. (P) = Easy-to-Use Protein Labeling Kit. (R IgG) = Zenon® Rabbit IgG Labeling Kit. (SAIVI™) = SAIVI™ Antibody Labeling Kit. (SDP) = Sulfodichlorophenol ester. (TFP) = Tetrafluorophenyl ester. (TSA™) = Enhanced with TSA™ technology. (ULYSIS®) = ULYSIS® Nucleic Acid Labeling Kit. (X) = An aminohexanoyl spacer between the dye and the SE. More information on Alexa Fluor® labeling products is available at www.invitrogen.com/handbook/reactivealexa.

continued on next page

IMPORTANT NOTICE: The products described in this manual are covered by one or more Limited Use Label License(s). Please refer to the Appendix on page 971 and Master Product List on page 975. Products are For Research Use Only. Not intended for any animal or human therapeutic or diagnostic use.
www.invitrogen.com/probes
molecular probes® | invitrogen™ by life technologies™

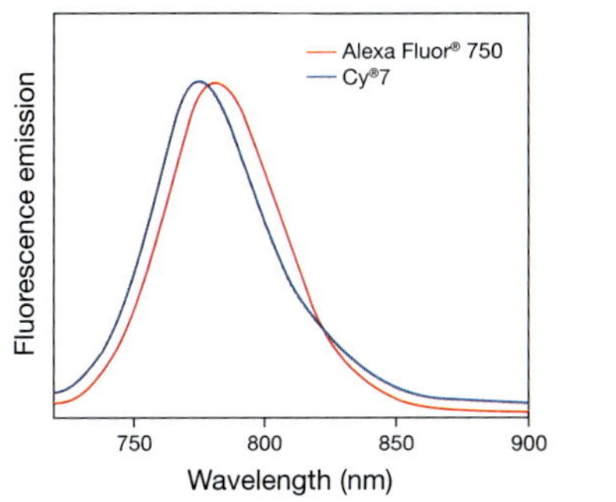

Figure 1.3.10 Comparison of the fluorescence emission spectra of Alexa Fluor® 750 and Cy®7 dyes. Spectra have been normalized to the same intensity for comparison purposes.

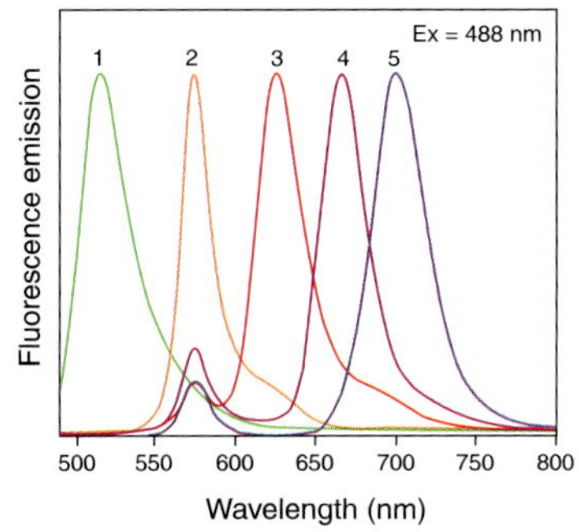

Figure 1.3.11 Normalized fluorescence emission spectra of 1) Alexa Fluor® 488 goat anti-mouse IgG antibody (A11001), 2) R-phycoerythrin goat anti-mouse IgG antibody (P852), 3) Alexa Fluor® 610–R-phycoerythrin goat anti-mouse IgG antibody (A20980), 4) Alexa Fluor® 647–R-phycoerythrin goat anti-mouse IgG antibody (A20990) and 5) Alexa Fluor® 680–R-phycoerythrin goat anti-mouse IgG antibody (A20983). The tandem conjugates permit simultaneous multicolor labeling and detection of up to five targets with excitation by a single excitation source—the 488 nm spectral line of the argon-ion laser.

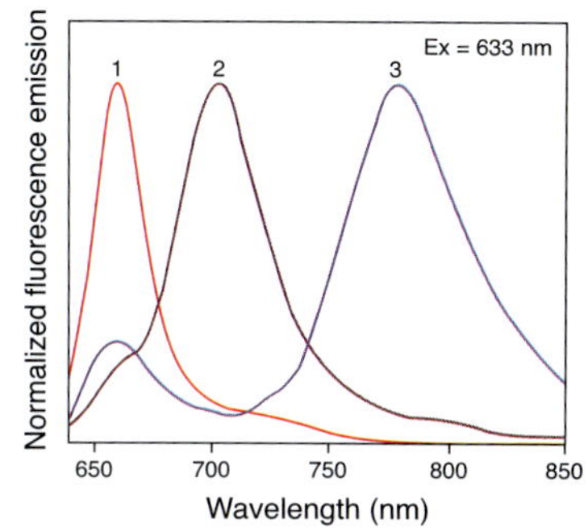

Figure 1.3.12 Normalized fluorescence emission spectra of 1) allophycocyanin goat anti-mouse IgG antibody (A865), 2) Alexa Fluor® 680–allophycocyanin goat anti-mouse IgG antibody (A21000) and 3) Alexa Fluor® 750–allophycocyanin goat anti-mouse IgG antibody (A21006). The tandem conjugates permit simultaneous multicolor labeling and detection of up to three targets with excitation by a single excitation source—the 633 nm spectral line of the He-Ne laser.

Table 1.4 Alexa Fluor® active esters and kits for labeling proteins and nucleic acids—*continued.*

Alexa Fluor® Dye	Fluorescence Color (Abs/Em) *	Succinimidyl Ester, TFP Ester or SDP Ester	Kits for Labeling Proteins			Kits for Labeling Nucleic Acids and Oligonucleotides
			Protein Labeling Kits	Antibody Labeling Kits	Zenon® Antibody Labeling Kits	
Alexa Fluor® 594	Red (590/617)	A20004 ‡ § A20104 † §	A10239 (P) A30008 (Micro)	A10474 (APEX®) A20185 (Mab)	Z25007 (M IgG_1) Z25107 (M IgG_{2a}) Z25207 (M IgG_{2b}) Z25307 (R IgG) Z25607 (G IgG) Z25407 (H IgG)	U21654 (ULYSIS®) A21669 (ARES™) F32949 (FT) F32954 (FT)
Alexa Fluor® 610	Red (612/628)	A30050 (X) ‡				
Alexa Fluor® 633 ††	Far-red (632/647)	A20005 ‡ § A20105 † §	A20170 (P)			
Alexa Fluor® 647 ††	Far-red (650/668)	A20006 ‡ A20106 †	A20173 (P) A30009 (Micro)	A10475 (APEX®) A20186 (Mab) S30044 (SAIVI™)	Z25008 (M IgG_1) Z25108 (M IgG_{2a}) Z25208 (M IgG_{2b}) Z25308 (R IgG) Z25608 (G IgG) Z25408 (H IgG)	U21660 (ULYSIS®) A21676 (ARES™) F32950 (FT) F32955 (FT) A20196 (Oligo)
Alexa Fluor® 660 ††	Near-infrared (663/690)	A20007 ‡	A20171 (P)		Z25009 (M IgG_1)	U21656 (ULYSIS®)
Alexa Fluor® 680 ††	Near-infrared (679/702)	A20008 ‡ A20108 †	A20172 (P)	S30045 (SAIVI™)	Z25010 (M IgG_1) Z25110 (M IgG_{2a}) Z25210 (M IgG_{2b}) Z25310 (R IgG)	
Alexa Fluor® 700 ††	Near-infrared (702/723)	A20010 ‡ A20110 †			Z25011 (M IgG_1)	
Alexa Fluor® 750 ††	Near-infrared (749/775)	A20011 ‡ A20111 †		S30046 (SAIVI™)	Z25312 (R IgG)	
Alexa Fluor® 790 ††	Near-infrared (782/805)	A30051				

* Approximate absorption (Abs) and fluorescence emission (Em) maxima for conjugates, in nm. † 5 mg unit size. ‡ 1 mg unit size. § Mixed isomers. ** The APEX® Alexa Fluor® 488 Antibody Labeling Kits contain the Alexa Fluor® 488 carboxylic acid SDP ester, whereas the other Alexa Fluor® 488 protein labeling kits contain the amine-reactive Alexa Fluor® 488 carboxylic acid TFP ester; the Alexa Fluor® 488 kits for labeling nucleic acids and oligonucleotides contain the Alexa Fluor® 488 succinimidyl ester. †† Human vision is insensitive to light beyond ~650 nm, and therefore it is not possible to view the far-red– and near-infrared–fluorescent dyes by looking through the eyepiece of a conventional fluorescence microscope. The Alexa Fluor® Labeling Kits are described in detail in Section 1.2. (APEX®) = APEX® Antibody Labeling Kit. (ARES™) = ARES™ DNA Labeling Kit. (FT) = FISH Tag™ DNA Kit or FISH Tag™ RNA Kit. (G IgG) = Zenon® Goat IgG Labeling Kit. (H IgG) = Zenon® Human IgG Labeling Kit. (M IgG) = Zenon® Mouse IgG Labeling Kit. (Mab) = Monoclonal Antibody Labeling Kit. (Micro) = Alexa Fluor® Microscale Protein Labeling Kit. (Oligo) = Alexa Fluor® Oligonucleotide Amine Labeling Kit. (P) = Easy-to-Use Protein Labeling Kit. (R IgG) = Zenon® Rabbit IgG Labeling Kit. (SAIVI™) = SAIVI™ Antibody Labeling Kit. (SDP) = Sulfodichlorophenol ester. (TFP) = Tetrafluorophenyl ester. (TSA™) = Enhanced with TSA™ technology. (ULYSIS®) = ULYSIS® Nucleic Acid Labeling Kit. (X) = An aminohexanoyl spacer between the dye and the SE. More information on Alexa Fluor® labeling products is available at www.invitrogen.com/handbook/reactivealexa.

Green-Fluorescent Alexa Fluor® Dyes

Alexa Fluor® 488 Dye: A Superior Fluorescein Substitute

Based on our testing, publications[2,5–7] and results reported by customers, Alexa Fluor® 488 dye is by far the best fluorescein (FITC or FAM) substitute available for most applications (The Alexa Fluor® Dye Series—Note 1.1). It is probably the best dye available for single-molecule detection of bioconjugates, for fluorescence correlation spectroscopy (Fluorescence Correlation Spectroscopy (FCS)—Note 1.3) and for fluorescence polarization measurements[8] (Fluorescence Polarization (FP)—Note 1.4).

This green-fluorescent dye exhibits several unique features:

- Fluorescence spectra almost identical to those of fluorescein, with excitation/emission maxima of 495/519 nm (Figure 1.3.4) and a fluorescence lifetime of ~4.1 nanoseconds (Table 1.5)
- Strong absorption, with an extinction coefficient greater than 65,000 $cm^{-1}M^{-1}$
- Much greater photostability than fluorescein (Figure 1.3.13), allowing more time for observation and image capture (Figure 1.3.14)
- pH-insensitive fluorescence between pH 4 and 10 (Figure 1.3.15)
- Water solubility, with no organic co-solvents required in labeling reactions, suggesting that the succinimidyl ester of Alexa Fluor® 488 carboxylic acid (A20000, A20100) may be the ideal reagent for labeling amines of exposed cell-surface proteins of live cells[9]
- Superior fluorescence output per protein conjugate, surpassing that of any other spectrally similar fluorophore-labeled protein, including fluorescein conjugates (Figure 1.3.16) and Cy®2 conjugates of antibodies (Figure 1.3.17)

NOTE 1.1

The Alexa Fluor® Dye Series

The Alexa Fluor® dyes—a series of superior fluorescent dyes that span the visible spectrum—represent a major breakthrough in the development of fluorescent labeling reagents.[1,2] Benefits of the Alexa Fluor® dyes and their conjugates include:

Brightness—Alexa Fluor® conjugates exhibit more intense fluorescence than other spectrally similar conjugates.

Photostability—Alexa Fluor® conjugates are more photostable than most other fluorescent conjugates, allowing more time for image capture.

Instrument compatibility—Absorption spectra of the Alexa Fluor® conjugates are matched to the principal output wavelengths of common excitation sources.

Color selection—Alexa Fluor® conjugates are available in several distinct fluorescent colors, ranging from blue to red to near-infrared (Figure 1, Table 1).

pH insensitivity—Alexa Fluor® dyes remain highly fluorescent over a broad pH range.

Water solubility—Alexa Fluor® reactive dyes have good water solubility, so protein conjugations can be performed without organic solvents, and the conjugates are relatively resistant to precipitation during storage.

Conventional fluorophores and their conjugates can often be replaced with spectrally similar Alexa Fluor® dyes and conjugates without affecting optical filter choices or other instrumentation considerations (Table 2). Each of the Alexa Fluor® dyes are listed in Table 1 and described in detail in Section 1.3 and in the accompanying Section 1.3 data table.

1. J Histochem Cytochem (2003) 51:1699; **2.** J Histochem Cytochem (1999) 47:1179.

Table 1 Spectral properties of Molecular Probes® Alexa Fluor® dyes.

Alexa Fluor® Dye	Absorption Max (nm) *	Emission Max (nm) *	Emission Color †	Extinction Coefficient ‡
Alexa Fluor® 350	346	442	Blue	19,000
Alexa Fluor® 405	402	421	Blue	35,000
Alexa Fluor® 430	434	539	Yellow-green	15,000
Alexa Fluor® 488	495	519	Green	73,000
Alexa Fluor® 514	518	540	Green	80,000
Alexa Fluor® 532	531	554	Yellow	81,000
Alexa Fluor® 546	556	573	Orange	112,000
Alexa Fluor® 555	555	565	Orange	155,000
Alexa Fluor® 568	578	603	Red-orange	88,000
Alexa Fluor® 594	590	617	Red	92,000
Alexa Fluor® 610	612	628	Red	144,000
Alexa Fluor® 633	632	647	Far-red	159,000
Alexa Fluor® 635	633	647	Far-red	140,000
Alexa Fluor® 647	650	668	Far-red	270,000
Alexa Fluor® 660	663	690	Near-IR §	132,000
Alexa Fluor® 680	679	702	Near-IR §	183,000
Alexa Fluor® 700	702	723	Near-IR §	205,000
Alexa Fluor® 750	749	775	Near-IR §	290,000
Alexa Fluor® 790	782	805	Near-IR §	260,000

* Approximate absorption and emission maxima, in nm, for conjugates. † Typical emission color seen through the eyepiece of a conventional fluorescence microscope with appropriate filters. ‡ Extinction coefficient of the reactive dye at emission maximum in $cm^{-1}M^{-1}$. § Human vision is insensitive to light beyond ~650 nm; it is not possible to directly view far-red– and near-IR–fluorescent dyes.

Table 2 Alexa Fluor® dye alternatives to several common fluorophores.

If you are using...	Try this Alexa Fluor® dye
Coumarin (AMCA)	Alexa Fluor® 350
Cy®2 or Fluorescein (FITC)	Alexa Fluor® 488
Cy®3 or Tetramethylrhodamine (TRITC)	Alexa Fluor® 555
Rhodamine Red™	Alexa Fluor® 568
Texas Red®	Alexa Fluor® 594
Cy®5	Alexa Fluor® 647
Cy®5.5	Alexa Fluor® 680
Cy®7	Alexa Fluor® 750

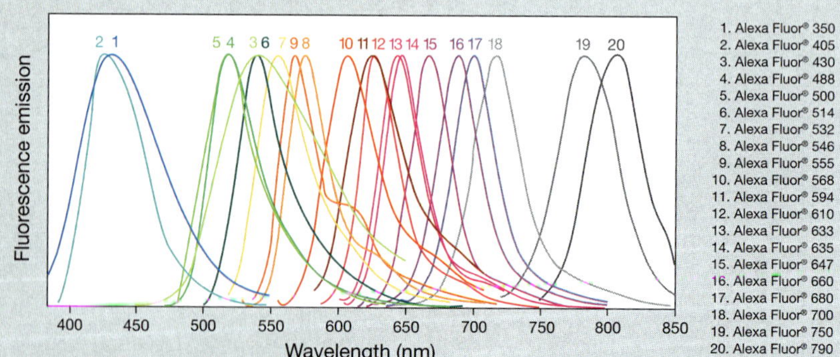

Figure 1 Emission spectra for the Alexa Fluor® dye series.

IMPORTANT NOTICE: The products described in this manual are covered by one or more Limited Use Label License(s). Please refer to the Appendix on page 971 and Master Product List on page 975. Products are For Research Use Only. Not intended for any animal or human therapeutic or diagnostic use.

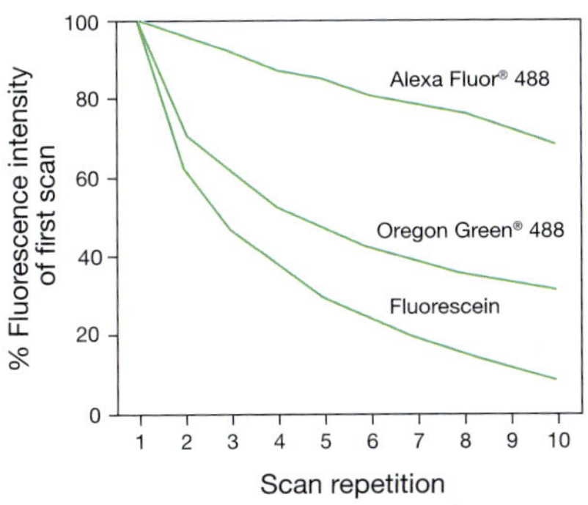

Figure 1.3.13 Photobleaching resistance of the green-fluorescent Alexa Fluor® 488, Oregon Green® 488 and fluorescein dyes, as determined by laser-scanning cytometry. EL4 cells were labeled with biotin-conjugated anti-CD44 antibody and detected by Alexa Fluor® 488 (S11223), Oregon Green® 488 (S6368) or fluorescein (S869) streptavidin. The cells were then fixed in 1% formaldehyde, washed and wet-mounted. After mounting, cells were scanned 10 times on a laser-scanning cytometer; laser power levels were 25 mW for the 488 nm spectral line of the argon-ion laser. Scan durations were approximately 5 minutes, and each repetition was started immediately after completion of the previous scan. Data are expressed as percentages derived from the mean fluorescence intensity (MFI) of each scan divided by the MFI of the first scan. Data contributed by Bill Telford, Experimental Transplantation and Immunology Branch, National Cancer Institute.

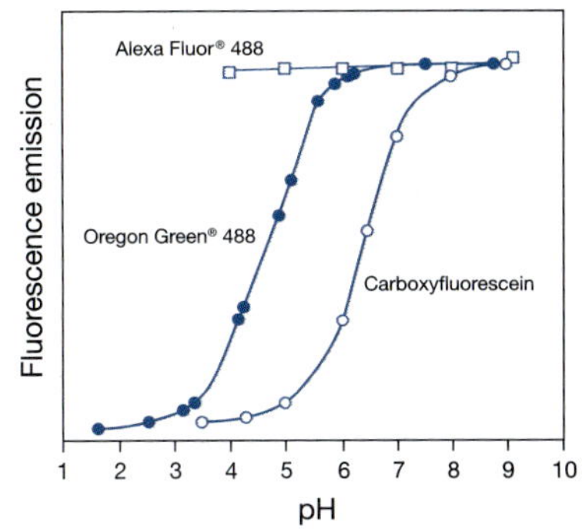

Figure 1.3.15 Comparison of pH-dependent fluorescence of the Oregon Green® 488 (●), carboxyfluorescein (○) and Alexa Fluor® 488 (□) fluorophores. Fluorescence intensities were measured for equal concentrations of the three dyes using excitation/emission at 490/520 nm.

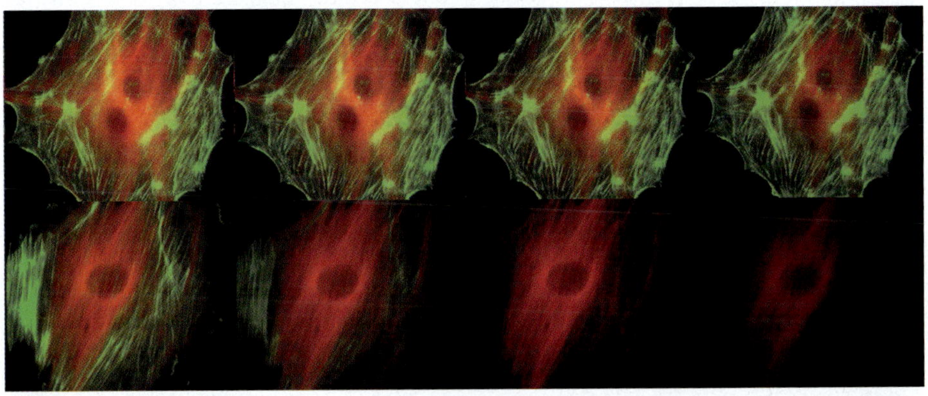

Figure 1.3.14 Comparison of the photobleaching rates of the Alexa Fluor® 488 and Alexa Fluor® 546 dyes and the well-known fluorescein and Cy®3 fluorophores. The cytoskeleton of bovine pulmonary artery endothelial cells (BPAEC) was labeled with (top series) Alexa Fluor® 488 phalloidin (A12379) and mouse monoclonal anti–α-tubulin antibody (A11126) in combination with Alexa Fluor® 546 goat anti–mouse IgG antibody (A11003) or (bottom series) fluorescein phalloidin (F432) and the anti–α-tubulin antibody in combination with a commercially available Cy®3 goat anti–mouse IgG antibody. The pseudocolored images were taken at 30-second intervals (0, 30, 90 and 210 seconds of exposure; left to right). The images were acquired with bandpass filter sets appropriate for fluorescein and rhodamine.

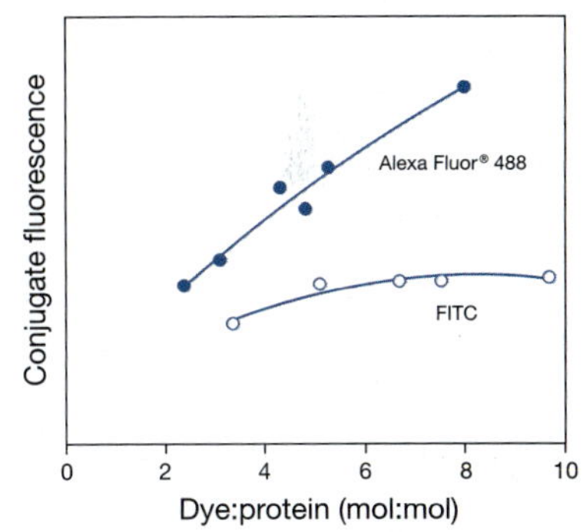

Figure 1.3.16 Comparison of the relative fluorescence of goat anti–mouse IgG antibody conjugates prepared from Alexa Fluor® 488 dye and from fluorescein isothiocyanate (FITC). Conjugate fluorescence is determined by measuring the fluorescence quantum yield of the conjugated dye relative to that of a reference dye and multiplying by the dye:protein labeling ratio.

Table 1.5 Fluorescence quantum yields (QY) and lifetimes (τ) for Alexa Fluor® dyes.

Alexa Fluor® Dye *	QY †	τ (ns) ‡
Alexa Fluor® 488	0.92	4.1 §
Alexa Fluor® 532	0.61	2.5
Alexa Fluor® 546	0.79	4.1
Alexa Fluor® 555	0.10	0.3
Alexa Fluor® 568	0.69	3.6 §
Alexa Fluor® 594	0.66	3.9 §
Alexa Fluor® 647	0.33	1.0
Alexa Fluor® 660	0.37	1.2 **
Alexa Fluor® 680	0.36	1.2
Alexa Fluor® 700	0.25	1.0
Alexa Fluor® 750	0.12	0.7

* Measurements were made on free succinimidyl ester derivatives in aqueous solutiuons. † For Alexa Fluor® 488, Alexa Fluor® 532, Alexa Fluor® 546, Alexa Fluor® 555, Alexa Fluor® 568, Alexa Fluor® 594 and Alexa Fluor® 647 dyes, QY measurements were made in PBS (50 mM potassium phosphate, 150 mM NaCl, pH 7.2) at 22°C relative to fluorescein in 0.01 M NaOH (QY = 0.92). For Alexa Fluor® 660, Alexa Fluor® 680, Alexa Fluor® 700 and Alexa Fluor® 750 dyes, QY measurements were made in PBS (50 mM potassium phosphate, 150 mM NaCl, pH 7.2) at 22°C relative to Alexa Fluor® 647 succinimidyl ester in PBS (QY = 0.33). ‡ Except for the footnoted values, lifetime measurements were made in water at 22°C, data provided by ISS Inc. (Champaign, IL). § Lifetime measurements were provided by the SPEX Fluorescence Group, Horiba Jobin Yvon, Inc. ** Lifetime measurement was made in pH 7.5 buffer at 20°C by Pierre-Alain Muller, Max Planck Institute for Biophysical Chemistry, Göttingen.

Table 1.6 R_0 values for some Alexa Fluor® dyes.

	Acceptor					
Donor	Alexa Fluor® 488	Alexa Fluor® 546	Alexa Fluor® 555	Alexa Fluor® 568	Alexa Fluor® 594	Alexa Fluor® 647
Alexa Fluor® 350	50					
Alexa Fluor® 488	NA	64	70	62	60	56
Alexa Fluor® 546		NA		70	71	74
Alexa Fluor® 555			NA		47	51
Alexa Fluor® 568				NA		82
Alexa Fluor® 594					NA	85
Alexa Fluor® 647						NA

R_0 values in angstroms (Å) represent the distance at which fluorescence resonance energy transfer from the donor dye to the acceptor dye is 50% efficient (Förster radius). Values were calculated from spectroscopic data as outlined (Fluorescence Resonance Energy Transfer (FRET)—Note 1.2). NA = Not applicable.

IMPORTANT NOTICE: The products described in this manual are covered by one or more Limited Use Label License(s). Please refer to the Appendix on page 971 and Master Product List on page 975. Products are For Research Use Only. Not intended for any animal or human therapeutic or diagnostic use.

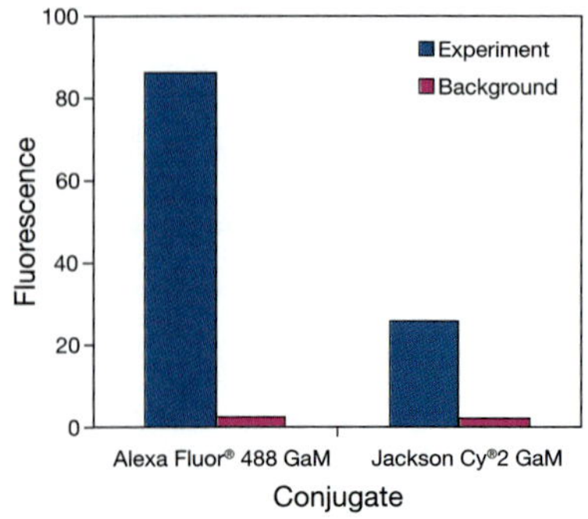

Figure 1.3.17 Brightness comparison of Molecular Probes® Alexa Fluor® 488 goat anti-mouse IgG antibody with Cy®2 goat anti-mouse IgG antibody from Jackson ImmunoResearch. Human blood was blocked with normal goat serum and incubated with an anti-CD3 mouse monoclonal antibody; cells were washed, resuspended and incubated with either Alexa Fluor® 488 or Cy®2 goat anti-mouse IgG antibody at equal concentration. Red blood cells were lysed, and the samples were analyzed with a flow cytometer equipped with a 488 nm argon-ion laser and a 525 ± 10 nm bandpass emission filter.

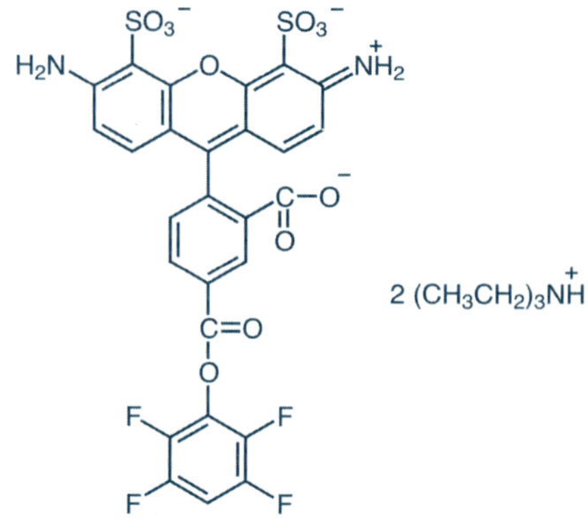

Figure 1.3.18 Alexa Fluor® 488 carboxylic acid, 2,3,5,6-tetrafluorophenyl ester (Alexa Fluor® 488 5-TFP, A30005).

The monosuccinimidyl ester of Alexa Fluor® 488 carboxylic acid is a mixture of two isomers and is available in a 1 mg or 5 mg unit size (A20000, A20100). The isomerically pure 5-isomer of Alexa Fluor® 488 dye is also available as the more hydrolytically stable tetrafluorophenyl (TFP) ester (A30005, Figure 1.3.18) and sulfodichlorophenol (SDP) ester (A30052, Figure 1.3.19).

TFP and SDP esters are an improvement over the succinimidyl ester (NHS ester or SE) chemistry typically used to attach fluorophores or haptens to the primary amines of biomolecules. All three reactive chemistries produce the same strong amide bond between the dye or hapten and the compound of interest (see reaction schemes in Section 1.1), but TFP and SDP esters are less susceptible to spontaneous hydrolysis during conjugation reactions. Both Alexa Fluor® 488 carboxylic

NOTE 1.2

Fluorescence Resonance Energy Transfer (FRET)

Fluorescence resonance energy transfer (FRET) is a distance-dependent interaction between the electronic excited states of two dye molecules in which excitation is transferred from a donor molecule to an acceptor molecule *without emission of a photon*. The efficiency of FRET is dependent on the inverse sixth power of the intermolecular separation,[1] making it useful over distances comparable to the dimensions of biological macromolecules. Thus, FRET is an important technique for investigating a variety of biological phenomena that produce changes in molecular proximity.[2–11] When FRET is used as a contrast mechanism, colocalization of proteins and other molecules can be imaged with spatial resolution beyond the limits of conventional optical microscopy.[12,13]

Primary Conditions for FRET

- Donor and acceptor molecules must be in close proximity (typically 10–100 Å).
- The absorption spectrum of the acceptor must overlap the fluorescence emission spectrum of the donor (Figure 1).
- Donor and acceptor transition dipole orientations must be approximately parallel.

Förster Radius

The distance at which energy transfer is 50% efficient (i.e., 50% of excited donors are deactivated by FRET) is defined by the Förster radius (R_0). The magnitude of R_0 is dependent on the spectral properties of the donor and acceptor dyes (Table 1):

$$R_o = [8.8 \times 10^{23} \cdot \kappa^2 \cdot n^{-4}\, QY_D \cdot J(\lambda)]^{1/6}\ \text{Å}$$

where κ^2 = dipole orientation factor (range 0–4; $\kappa^2 = 2/3$ for randomly oriented donors and acceptors)

QY_D = fluorescence quantum yield of the donor in the absence of the acceptor

n = refractive index

$J(\lambda)$ = spectral overlap integral (see figure) $= \int \varepsilon_A(\lambda) \cdot F_D(\lambda) \cdot \lambda^4 d\lambda\ \text{cm}^3\text{M}^{-1}$

where ε_A = extinction coefficient of acceptor

F_D = fluorescence emission intensity of donor as a fraction of the total integrated intensity

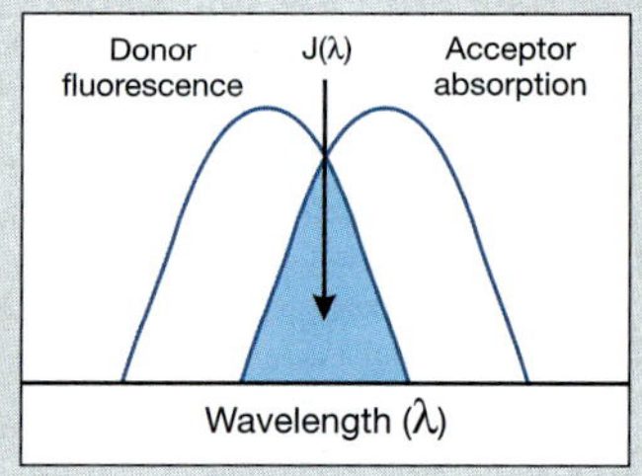

Figure 1 Schematic representation of the FRET spectral overlap integral.

Table 1 Typical Values of R_0.

Donor	Acceptor	R_0 (Å)
Fluorescein	Tetramethylrhodamine	55
IAEDANS	Fluorescein	46
EDANS	Dabcyl	33
Fluorescein	Fluorescein	44
BODIPY® FL	BODIPY® FL	57
Fluorescein	QSY® 7 and QSY® 9 dyes	61

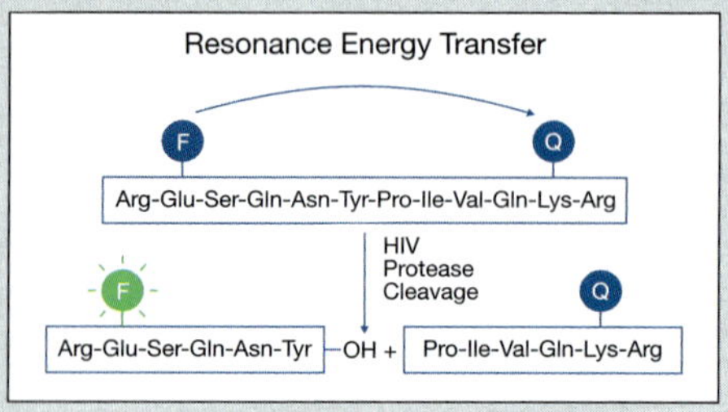

Figure 2 Principle of the fluorogenic response to protease cleavage exhibited by HIV protease substrate 1 (H2930). Quenching of the EDANS fluorophore (F) by distance-dependent resonance energy transfer to the dabcyl quencher (Q) is eliminated upon cleavage of the intervening peptide linker.

IMPORTANT NOTICE: The products described in this manual are covered by one or more Limited Use Label License(s). Please refer to the Appendix on page 971 and Master Product List on page 975. Products are For Research Use Only. Not intended for any animal or human therapeutic or diagnostic use.

acid TFP ester and Alexa Fluor® 488 SDP ester are stable for several hours at the basic pH typically used for reactions—far outlasting succinimidyl esters. Alexa Fluor® 488 carboxylic acid TFP ester is the amine-reactive dye included in most of our labeling kits for proteins, nucleic acids and oligonucleotides (Section 1.2; Table 1.2, Table 1.4), including the Alexa Fluor® 488 Microscale Protein Labeling Kit (A30006), the Alexa Fluor® 488 Monoclonal Antibody Labeling Kit (A20181), the Alexa Fluor® 488 Protein Labeling Kit (A10235), the ARES™ Alexa Fluor® 488 DNA Labeling Kit (A21665) and the Alexa Fluor® 488 Oligonucleotide Amine Labeling Kit (A20191). Alexa Fluor® 488 carboxylic acid SDP ester is the amine-reactive dye included in the APEX® Alexa Fluor® 488 Antibody Labeling Kit (A10468).

Alexa Fluor® 514 Dye: A Perfect Match to the Argon-Ion Laser

Like our Alexa Fluor® 488 dye, Alexa Fluor® 514 dye (excitation/emission maxima ~518/540 nm) is superior to fluorescein in both brightness and photostability and can be detected with standard fluorescein, Oregon Green® dye or Alexa Fluor® 488 dye filter sets. However, Alexa Fluor® 514 dye is spectrally distinguishable from Alexa Fluor® 488 dye and other green fluorophores using spectral imaging instruments with linear-unmixing software.[10,11] Alexa Fluor® 514 dye is one the brightest and most photostable dyes available for excitation by the 514 nm spectral line of the argon-ion laser. Alexa Fluor® 514 dye is available as a succinimidyl ester (A30002) and as antibody (Section 7.2, Table 7.1) and streptavidin (Section 7.6, Table 7.9) conjugates.

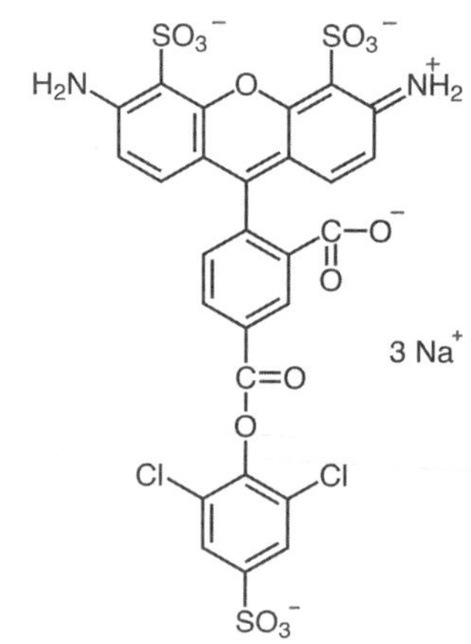

Figure 1.3.19 Alexa Fluor® 488 5-SDP ester (Alexa Fluor® 488 sulfodichlorophenol ester, A30052).

Donor/Acceptor Pairs

In most applications, the donor and acceptor dyes are different, in which case FRET can be detected by the appearance of sensitized fluorescence of the acceptor or by quenching of donor fluorescence. When the donor and acceptor are the same, FRET can be detected by the resulting fluorescence depolarization.[14] Typical values of R_0 for some dye pairs are listed in Table 1 and more extensive compilations are in Table 1.6 and Table 1.11. Note that because the component factors of R_0 are dependent on the environment, the actual value observed in a specific experimental situation is somewhat variable. Extensive compilations of R_0 values can be found in the literature.[3,8,11] Nonfluorescent acceptors such as dabcyl and QSY® dyes (Table 1.10) have the particular advantage of eliminating the potential problem of background fluorescence resulting from direct (i.e., nonsensitized) acceptor excitation. FRET efficiencies from several donor dyes to the QSY® 7 quencher in molecular beacon hybridization probes have been calculated.[15] Probes incorporating fluorescent donor–nonfluorescent acceptor combinations have been developed primarily for detecting proteolysis[16] (Figure 2) and nucleic acid hybridization.

Selected Applications of FRET

- Structure and conformation of proteins [17–22]
- Spatial distribution and assembly of protein complexes [23–27]
- Receptor/ligand interactions [28–32]
- Immunoassays [33,34]
- Probing interactions of single molecules [35]
- Structure and conformation of nucleic acids [36–41]
- Real-time PCR assays and SNP detection [42–47]
- Detection of nucleic acid hybridization [48–53]
- Primer-extension assays for detecting mutations [54]
- Automated DNA sequencing [55–57]
- Distribution and transport of lipids [58–60]
- Membrane fusion assays [61–64] (Lipid-Mixing Assays of Membrane Fusion—Note 13.1)
- Membrane potential sensing [65]
- Fluorogenic protease substrates [16,66–69]
- Indicators for cyclic AMP [70,71] and zinc [72]

1. Proc Natl Acad Sci U S A (1967) 58:719; **2.** Biophys J (2003) 84:3992; **3.** Resonance Energy Transfer: Theory and Data, Van der Meer BW, et al. (1994) p. 133; **4.** J Struct Biol (1995) 115:175; **5.** Photochem Photobiol (1983) 38:487; **6.** Annu Rev Biochem (1978) 47:819; **7.** Methods Enzymol (1995) 246:300; **8.** Anal Biochem (1994) 218:1; **9.** Methods Enzymol (1978) 48:347; **10.** Scanning (1995) 17:72; **11.** J Muscle Res Cell Motil (1987) 8:97; **12.** Methods (2001) 24:289; **13.** Biophys J (1998) 74:2702; **14.** Biophys J (1995) 69:1569; **15.** Nucleic Acids Res (2002) 30:e122; **16.** Science (1990) 247:954; **17.** Biophys J (1998) 74:3111; **18.** Biochemistry (1996) 35:4795; **19.** Biochemistry (1995) 34:8693; **20.** Biochemistry (1995) 34:6475; **21.** J Biol Chem (1998) 273:9119; **22.** J Biol Chem (1993) 268:15588; **23.** Biochemistry (1995) 34:7904; **24.** Biochemistry (1994) 33:13102; **25.** Biochemistry (1994) 33:5539; **26.** J Photochem Photobiol B (1992) 12:323; **27.** J Biol Chem (1989) 264:8699; **28.** J Recept Signal Transduct Res (2002) 22:333; **29.** Biochemistry (1994) 33:11875; **30.** J Cell Physiol (1994) 159:176; **31.** Biophys J (1991) 60:307; **32.** J Biol Chem (1984) 259:5717; **33.** Anal Biochem (1988) 174:101; **34.** Anal Biochem (1980) 108:156; **35.** Proc Natl Acad Sci U S A (1996) 93:6264; **36.** Biochemistry (1998) 37:2979; **37.** Biochemistry (1998) 37:8173; **38.** Anal Biochem (1994) 221:306; **39.** Biophys J (1994) 66:99; **40.** Nucleic Acids Res (1994) 22:920; **41.** Science (1994) 266:785; **42.** Nucleic Acids Res (2000) 28:3752; **43.** Nat Biotechnol (1999) 17:804; **44.** Biotechniques (1999) 27:342; **45.** Genome Res (1996) 6:986; **46.** Nucleic Acids Res (1997) 25:2516; **47.** Genome Res (2001) 11:163; **48.** Nat Biotechnol (1996) 14:303; **49.** Nat Biotechnol (1998) 16:49; **50.** Biochemistry (1995) 34:285; **51.** Nucleic Acids Res (1994) 22:662; **52.** Nonisotopic DNA Probe Techniques, Kricka LR, Ed. (1992) p. 311; **53.** Proc Natl Acad Sci U S A (1988) 85:8790; **54.** Proc Natl Acad Sci U S A (1997) 94:10756; **55.** Anal Biochem (1998) 255:32; **56.** Anal Chem (1995) 67:3676; **57.** Proc Natl Acad Sci U S A (1995) 92:4347; **58.** Biochemistry (1995) 34:4846; **59.** Biochemistry (1992) 31:2865; **60.** J Biol Chem (1983) 258:5368; **61.** Biochemistry (1998) 37:2361; **62.** Biochim Biophys Acta (1994) 1189:175; **63.** Methods Enzymol (1993) 221:239; **64.** Biochemistry (1981) 20:4093; **65.** Biophys J (1995) 69:1272; **66.** FEBS Lett (1997) 413:379; **67.** Techniques in Protein Chemistry V, Crabb JW, Ed. (1994) p. 493; **68.** Biochemistry (1998) 37:11434; **69.** Bioconjug Chem (1993) 4:537; **70.** Nature (1991) 349:694; **71.** Fluorescent and Luminescent Probes for Biological Activity, Mason WT, Ed. (1993) p. 133; **72.** J Am Chem Soc (1996) 118:6514.

IMPORTANT NOTICE: The products described in this manual are covered by one or more Limited Use Label License(s). Please refer to the Appendix on page 971 and Master Product List on page 975. Products are For Research Use Only. Not intended for any animal or human therapeutic or diagnostic use.

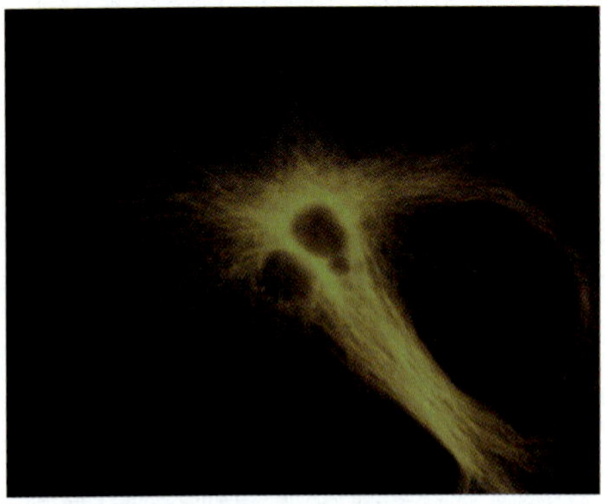

Figure 1.3.20 A bovine pulmonary artery endothelial (BPAE) cell labeled with mouse monoclonal anti–α-tubulin antibody (A11126) in combination with Alexa Fluor® 430 goat anti–mouse IgG antibody (A11063) to stain microtubules. The image was acquired using a longpass filter set allowing excitation at 455 ± 35 nm and emission at wavelengths greater than 515 nm.

Alexa Fluor® 430 Dye: Filling the Spectral Gap Between Green and Yellow

Few reactive dyes that absorb between 400 nm and 450 nm have appreciable fluorescence beyond 500 nm in aqueous solution. Alexa Fluor® 430 dye fills this spectral gap (Figure 1.3.1, Figure 1.3.20). Excitation near its absorption maximum at 431 nm is accompanied by an extremely large Stokes shift and strong yellow-green fluorescence (emission maximum ~541 nm). The coumarin-based amine-reactive succinimidyl ester of Alexa Fluor® 430 carboxylic acid (A10169, Figure 1.3.21) is available, as well as Alexa Fluor® 430 conjugates of secondary antibodies (A11063, A11064; Section 7.2) and streptavidin (S11237, Section 7.6). Alexa Fluor® 430 dye–labeled Fab′ fragments are provided in the Zenon® Alexa Fluor® 430 Antibody Labeling Kits (Section 7.3, Table 7.7). We also offer the Alexa Fluor® 430 Protein Labeling Kit (A10171), which is described in detail in Section 1.2 (Table 1.2, Table 1.4).

Yellow-to-Red–Fluorescent Alexa Fluor® Dyes

As with the green-fluorescent Alexa Fluor® 488 dye, the yellow-, orange- and red-fluorescent Alexa Fluor® dyes exhibit several features that distinguish them from spectrally similar fluorophores:

- Strong absorption, with extinction coefficients greater than 80,000 $cm^{-1}M^{-1}$ for Alexa Fluor® 532, Alexa Fluor® 546, Alexa Fluor® 568 and Alexa Fluor® 594 dyes and greater than 130,000 $cm^{-1}M^{-1}$ for Alexa Fluor® 555 and Alexa Fluor® 610 dyes

NOTE 1.3

Fluorescence Correlation Spectroscopy (FCS)

Fluorescence correlation spectroscopy (FCS) is a technique in which spontaneous fluorescence intensity fluctuations are measured in a microscopic detection volume of about 10^{-15} L (1 femtoliter) defined by a tightly focused laser beam.[1–3] Renewed interest in FCS in recent years has been stimulated by the fact that it is inherently miniaturized and therefore applicable for high-throughput screening applications.[4] Fluorescence intensity fluctuations measured by FCS represent changes in either the number or the fluorescence quantum yield of molecules resident in the detection volume (Figure 1). Small, rapidly diffusing molecules produce rapidly fluctuating intensity patterns, whereas larger molecules produce more sustained bursts of fluorescence.

This situation is in marked contrast to conventional fluorescence photometry carried out in sample volumes of around 0.1–1.0 mL (~10^8 times larger than FCS measurement volumes) that report only the macroscopic average of diffusion-dependent intensity fluctuations. In a typical FCS measurement, fluorescence intensity is recorded for a small number of molecules in the detection volume (e.g., 3 molecules/femtoliter, equivalent to ~5 nM macroscopic concentration) over a time range from about 1 microsecond to 1 second. The time-dependent fluorescence intensity (F(t)) is then analyzed in terms of its temporal autocorrelation function (G (τ)), which compares the fluorescence intensity at time t with the intensity at (t + τ), where τ is a

$$G(\tau) = \frac{\langle \delta F(t) \cdot \delta F(t+\tau) \rangle}{\langle F(t) \rangle^2}$$

variable interval, averaged over all data points in the time series:

The autocorrelation function contains information about equilibrium concentrations, reaction kinetics and diffusion rates of molecules in the sample. The initial amplitude of the autocorrelation function is inversely proportional to the number of molecules in the detection volume. The autocorrelation function decays from its initial value with a time-dependence that is determined by molecular diffusion rates. For example, free fluorescent ligands exhibit faster autocorrelation decay than slower-moving complexed ligands (Figure 2).

Probes and Applications for FCS

FCS is applicable for monitoring a multitude of biomolecular association and dissociation processes (Table 1). Because FCS is intrinsically sensitive to the mass changes occurring in these processes, probe design and selection is generally less critical than it is in assays based on macroscopic fluorescence intensity changes generated by dye–dye interactions (FRET, self-quenching etc.) or environment-dependent fluorescence enhancement. Dyes that perform well in confocal laser-scanning microscopy are usually among the best choices for FCS applications. Laser sources used for excitation in FCS include the 488 nm argon-ion spectral line and the 543 nm and 633 nm He-Ne laser spectral lines. Dyes with appreciable rates of triplet state population via intersystem crossing are generally not well suited for FCS measurement because this process results in an additional submillisecond autocorrelation decay component.[5]

Technical Developments in FCS

Two-photon excitation (TPE) has been applied to FCS for reasons similar to those that have motivated its use in fluorescence microscopy—inherent spatial confinement of excitation, diminished photobleaching and phototoxicity, less scattering and better optical penetration in turbid media.[6,7] Dual-color cross-correlation FCS [8,9] measures the cross-correlation of the time-dependent fluorescence intensities of two spectrally distinct dyes, instead of the conventional autocorrelation for a single dye. This approach has the advantage that cross-correlated fluorescence is only generated by molecules or complexes labeled with both dyes, allowing quantitation of interacting molecules without reference to their diffusion characteristics. In practice, discrimination based on mass in conventional FCS requires that the interacting components should have a molecular weight ratio of at least 1:7. FCS measurements using TPE in combination with dual-color cross-correlation have been reported.[10]

IMPORTANT NOTICE: The products described in this manual are covered by one or more Limited Use Label License(s). Please refer to the Appendix on page 971 and Master Product List on page 975. Products are For Research Use Only. Not intended for any animal or human therapeutic or diagnostic use.

www.invitrogen.com/probes

molecular probes® | invitrogen™ by life technologies™

- Fluorescence that is more photostable than that of other spectrally similar dyes, allowing more time for observation and image capture
- pH-insensitive fluorescence over a broad range
- Good water solubility, permitting labeling reactions to be performed without organic solvents
- Superior fluorescence output per protein or nucleic acid conjugate, surpassing that of any other spectrally similar fluorophore-labeled protein
- Long fluorescence lifetimes (approximately 4.1, 3.6 and 3.9 nanoseconds for Alexa Fluor® 546, Alexa Fluor® 568 and Alexa Fluor® 594 dyes, respectively (Table 1.5))

Alexa Fluor® 532 Dye: Optimal Dye for 532 nm Diode Lasers

The yellow-fluorescent Alexa Fluor® 532 dye (excitation/emission maxima ~532/554 nm) absorbs in the wavelength between the green-fluorescent Alexa Fluor® 488 dye and orange-fluorescent Alexa Fluor® 546 dye (Figure 1.3.2) and provides strong visible fluorescence that contrasts well with these dyes. Five of our Alexa Fluor® dyes—Alexa Fluor® 488, Alexa Fluor® 532, Alexa Fluor® 546, Alexa Fluor® 568 and Alexa Fluor® 594 dyes—have been utilized for simultaneous seven-color fluorescence imaging in tissue samples.[12] The use of Alexa Fluor® 532 dye–labeled actin (instead of the equivalent tetramethylrhodamine conjugate) has been reported to be crucial for studying the polarity of individual filaments during the formation of actin bundles by total internal reflection fluorescence (TIRF) microscopy.[13] The isomer-free, amine-reactive monosuccinimidyl ester of Alexa Fluor® 532 dye is available in either a 1 mg or 5 mg unit size (A20001, A20101MP) and as a

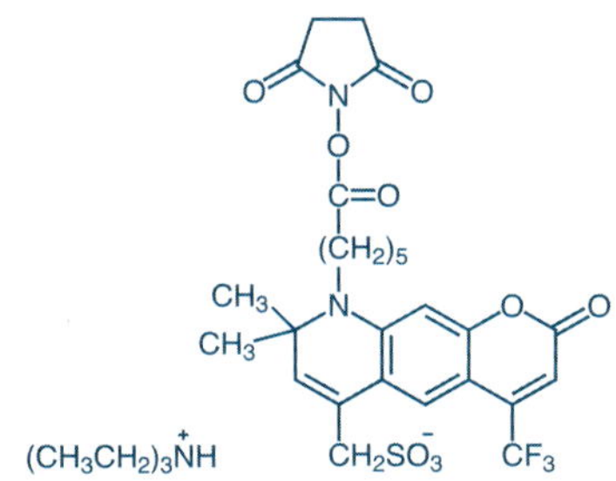

Figure 1.3.21 Alexa Fluor® 430 carboxylic acid, succinimidyl ester (A10169).

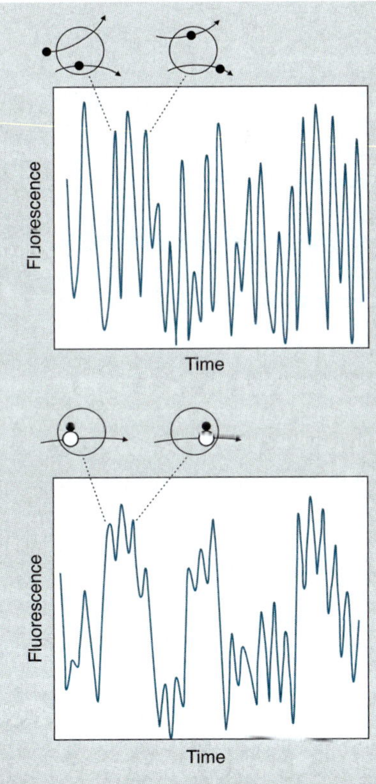

Figure 1 Physical origins of fluorescence correlation spectroscopy data. Free fluorescent ligands move in and out of the detection volume (open circle) and are detected as a series of short, randomized fluorescence bursts (top panel). Macromolecule-bound ligands are less mobile, producing a more slowly fluctuating (i.e., more highly autocorrelated) time-dependent fluorescence pattern (bottom panel).

Table 1 Applications of fluorescence correlation spectroscopy.

Detected Process	References
Nucleic acid fragmentation	Anal Biochem (1998) 260:166; Proc Natl Acad Sci U S A (1998) 95:1416; Proc Natl Acad Sci U S A (1998) 95:1421
Nucleic acid hybridization	Biochemistry (1996) 35:10182; Nucleic Acids Res (1995) 23:1795
PCR product formation	Biochemistry (1998) 37:12971; Biotechniques (1998) 25:706; Proc Natl Acad Sci U S A (1996) 93:12805
Lateral segregation of lipids in bilayer membranes	Cytometry (1999) 36:176; Proc Natl Acad Sci U S A (1999) 96:8461
Molecular diffusion in the nucleus and cytoplasm	Biophys J (1998) 75:2547; Proc Natl Acad Sci U S A (1998) 95:6043
Protein–protein interactions	Biochem Biophys Res Commun (2000) 267:300; Biochemistry (1999) 38:13759; Biochemistry (1999) 38:8402; Chem Biol (1999) 6:53; Cytometry (1999) 36:247; Biophys Chem (1998) 75:151
Binding equilibria for drugs and other low molecular weight ligands	Biochemistry (1999) 38:5082; Biochemistry (1999) 38:8671; J Biomol Screen (1999) 4:355; Biophys J (1997) 73:2195; Biophys Chem (1996) 58:3
Clustering of membrane-bound receptors	Biophys J (1996) 70:2001; Biophys J (1993) 65:1135; Chem Phys Lipids (1989) 50:253

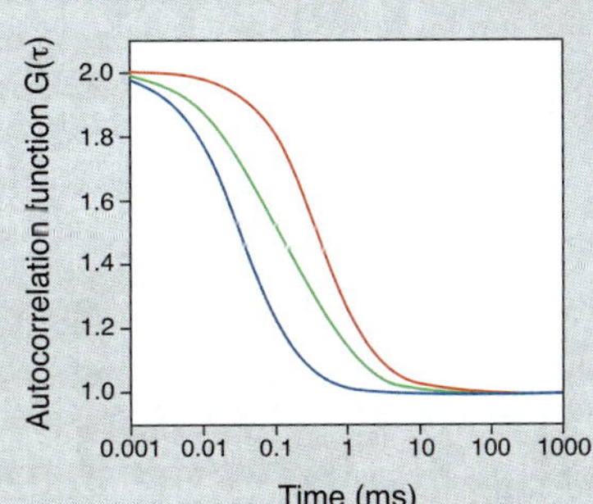

Figure 2 Simulated FCS autocorrelation functions representing a low molecular weight ligand (left curve, blue), macromolecule-bound ligand (right curve, red) and a 1:1 mixture of free and bound ligand (middle curve, green).

1. Biochemistry (2002) 41:697; **2.** Proc Natl Acad Sci U S A (1997) 94:11753; **3.** Ann Rev Biophys Biomol Struct (2007) 36:151; **4.** J Biomol Screen (1999) 4:335; **5.** J Phys Chem (1995) 99:13368; **6.** Biophys J (1999) 77:2251; **7.** Biophys J (1996) 71:410; **8.** Nat Methods (2006) 3:83; **9.** Nat Protoc (2007) 2:2842; **10.** Proc Natl Acad Sci U S A (2000) 97:10377.

IMPORTANT NOTICE: The products described in this manual are covered by one or more Limited Use Label License(s). Please refer to the Appendix on page 971 and Master Product List on page 975. Products are For Research Use Only. Not intended for any animal or human therapeutic or diagnostic use.

NOTE 1.4

Fluorescence Polarization (FP)

Principles

Fluorescence polarization measurements provide information on molecular orientation and mobility and processes that modulate them, including receptor–ligand interactions, protein–DNA interactions, proteolysis, membrane fluidity and muscle contraction (Figure 1).

Because polarization is a general property of fluorescent molecules (with certain exceptions such as lanthanide chelates), polarization-based readouts are somewhat less dye dependent and less susceptible to environmental interferences such as pH changes than assays based on fluorescence intensity measurements. Experimentally, the degree of polarization is determined from measurements of fluorescence intensities parallel and perpendicular with respect to the plane of linearly polarized excitation light, and is expressed in terms of fluorescence polarization (P) or anisotropy (r):

$$P = \frac{(F_\parallel - F_\perp)}{(F_\parallel + F_\perp)} \qquad r = \frac{(F_\parallel - F_\perp)}{(F_\parallel + 2F_\perp)}$$

where $F_\parallel$ = fluorescence intensity parallel to the excitation plane
$F_\perp$ = fluorescence intensity perpendicular to the excitation plane

Note that both P and r are ratio quantities with no nominal dependence on dye concentration. Because of the ratio formulation, fluorescence intensity variations due to the presence of colored sample additives tend to cancel and produce relatively minor inteferences.[1] P has physically possible values ranging from –0.33 to 0.5. In practice, these limiting values are rarely attained. Measured values of P in bioanalytical applications typically range from 0.01 to 0.3 or 10 to 300 mP (mP = P/1000). This measurement range is not as narrow as it might appear to be because very precise measurements (P ± 0.002 or ± 2 mP) are readily obtainable with modern instrumentation.

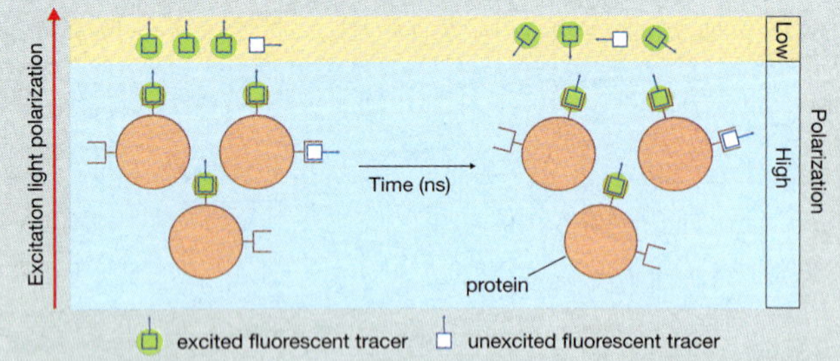

Figure 1 Physical basis of fluorescence polarization assays. Dye molecules with their absorption transition vectors (arrows) aligned parallel to the electric vector of linearly polarized light (along the vertical page axis) are selectively excited. For dyes attached to small, rapidly rotating molecules, the initially photoselected orientational distribution becomes randomized prior to emission, resulting in low fluorescence polarization. Conversely, binding of the low molecular weight tracer to a large, slowly rotating molecule results in high fluorescence polarization. Fluorescence polarization therefore provides a direct readout of the extent of tracer binding to proteins, nucleic acids and other biopolymers.

Dependence of Fluorescence Polarization on Molecular Mobility

Interpretation of the dependence of fluorescence polarization on molecular mobility is usually based on a model derived in 1926 from the physical theory of Brownian motion by Perrin:[2,3]

$$\left(\frac{1}{P} - \frac{1}{3}\right) = \left(\frac{1}{P_0} - \frac{1}{3}\right)\left(1 + \frac{\tau}{\phi}\right)$$

where P_o is the fundamental polarization of the dye (for fluorescein, rhodamine and BODIPY® dyes, P_o is close to the theoretical maximum of 0.5), τ is the excited-state lifetime of the dye and ϕ is the rotational correlation time of the dye or dye conjugate. These relationships can be expressed in terms of fluorescence anisotropy in an equivalent and mathematically simpler manner. For a hydrodynamic sphere, ϕ can be estimated as follows:

$$\phi = \frac{\eta V}{RT}$$

where η = solvent viscosity, T = temperature, R = gas constant and V = molecular volume of the fluorescent dye or dye conjugate. In turn, V can be estimated from the molecular weight of the dye or dye conjugate with appropriate adjustments for hydration. Simulations of these relationships are shown in Figure 2, leading to the following general conclusions:

- Fluorescence polarization increases as molecular weight increases.
- Fluorescence polarization increases as solvent viscosity increases.
- Fluorescence polarization decreases as the excited state lifetime of the dye (τ) increases.

Note that these simulations assume that the dye is rigidly attached to a spherical carrier. When conventional parameter estimates for proteins in aqueous solutions are used, ϕ is found to increase by about 1 ns per 2400 dalton increase of molecular weight.[4]

Dyes for Fluorescence Polarization Assays

Tracers used in fluorescence polarization assays include peptides, drugs and cytokines that are modified by the attachment of a fluorescent dye. Depolarization due to flexibility in the attachment of the dye, sometimes referred to as the "propeller effect," distorts the relationships between P and molecular weight shown in Figure 2. For this reason, it is generally preferable to use reactive dyes without aliphatic linkers between the fluorophore and the reactive group in the preparation of tracers for fluorescence polarization assays.[5,6]

A key factor in the performance of fluorescence polarization assays is the extent to which the biological activity of the tracer is perturbed by the dye modification. BODIPY® dyes generally produce less perturbation of receptor-binding affinity and other activity parameters than conventional dyes such as fluorescein and rhodamine.[7,8] Furthermore, BODIPY® dyes usually have

IMPORTANT NOTICE: The products described in this manual are covered by one or more Limited Use Label License(s). Please refer to the Appendix on page 971 and Master Product List on page 975. Products are For Research Use Only. Not intended for any animal or human therapeutic or diagnostic use.

longer excited-state lifetimes than fluorescein and rhodamine dyes, making their fluorescence polarization sensitive to binding interactions over a larger molecular weight range (Figure 2). The long-wavelength BODIPY® TMR and BODIPY® TR dyes also tend to minimize assay interferences due to intrinsically fluorescent sample contaminants.[7]

Applications

Fluorescence polarization measurements have long been a valuable biophysical research tool for investigating processes such as membrane lipid mobility, myosin reorientation and protein–protein interactions at the molecular level.[9–12] Immunoassays that have been developed and used extensively for clinical diagnostics represent the largest group of bioanalytical applications.[13,14] The more recent advent of microplate readers equipped with polarizing optics has led to the adoption of fluorescence polarization as a readout mode for high-throughput screening.[15–17] Some typical bioanalytical applications of fluorescence polarization–based assays are summarized in Table 1.

1. Anal Biochem (1997) 247:83; **2.** J Phys Radium (1926) 7:390; **3.** Fluorescence and Phosphorescence Analysis, Hercules DM, Ed. (1966) p. 217; **4.** Biophysical Chemistry, Part 2, Cantor CR, Schimmel PR, Eds (1980) p. 454; **5.** Anal Biochem (1997) 249:29; **6.** Anal Biochem (1997) 247:77; **7.** J Biomol Screen (2000) 5:329; **8.** Endocrinology (1997) 138:296; **9.** Methods (1999) 19:222; **10.** Biophys J (1996) 71:3330; **11.** Chem Phys Lipids (1993) 64:99; **12.** Methods Enzymol (1995) 246:283; **13.** Immunochemistry (1970) 7:799; **14.** Immunochemistry (1973) 10:219; **15.** J Biomol Screen (2000) 5:297; **16.** High Throughput Screening: The Discovery of Bioactive Substances, Devlin J, Ed. (1997) p. 389; **17.** J Biomol Screen (2001) 6:275.

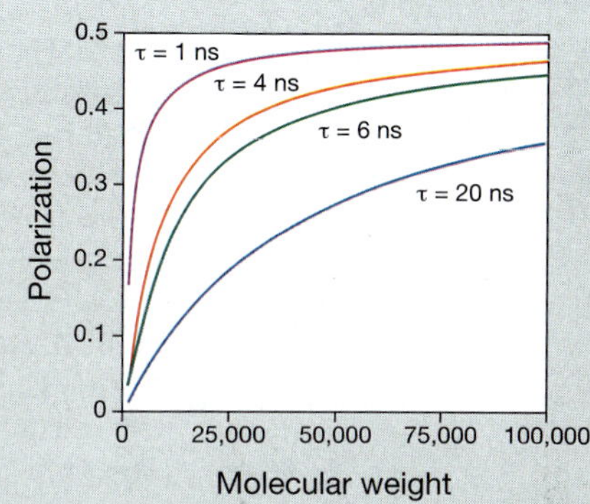

Figure 2 Simulation of the relationship between molecular weight (MW) and fluorescence polarization (P). Simulations are shown for dyes with various fluorescence lifetimes (τ): 1 ns (cyanine dyes) in purple, 4 ns (fluorescein and Alexa Fluor 488 dyes) in red, 6 ns (some BODIPY dyes) in green and 20 ns (dansyl dyes) in blue. At MW = 1000, P = 0.167 for τ = 1 ns, P = 0.056 for τ = 4 ns, P = 0.039 for τ = 6 ns and P = 0.012 for τ = 20 ns. Simulations assume P_0 (the fundamental polarization) = 0.5 and rigid attachment of dyes to spherical carriers.

Table 1 Examples of fluorescence polarization[endash]based assays.

Assay Target	Tracer	References
Ligand binding to neurokinin 1 (NK1) receptor	Fluorescein-labeled substance P	Biochemistry (1994) 33:13079
Ligand binding to melanocortin G-protein–coupled receptors	BODIPY® TMR dye–labeled NDP-αMSH	J Biomol Screen (2000) 5:329
Ligand binding to B2 bradykinin receptor, a G-protein–coupled receptor	BODIPY® TMR dye–labeled HOE140	J Biomol Screen (2002) 7:111
Ligand binding to estrogen receptors	Fluorescein-labeled estradiol	J Biomol Screen (2000) 5:77
Ligand binding to tyrosine kinase Src homology domains	Fluorescein- and BODIPY® TR dye–labeled phosphopeptides	Anal Biochem (1999) 275:62; Anal Biochem (1997) 247:77
Substrate binding to protein farnesyltransferase	Oregon Green® 488 dye–labeled peptide	Biochemistry (1999) 38:13138
β-Lactam antibiotic binding to penicillin-binding proteins	BODIPY® FL dye–labeled penicillin V	Antimicrob Agents Chemother (1999) 43:1124
Protein kinase activity	Fluorescently labeled phosphopeptide	Anal Biochem (2000) 278:206; Methods (2000) 22:61
Nonspecific protease activity	BODIPY® FL dye–labeled casein	Anal Biochem (1996) 243:1
Detection of specific PCR products	Fluorescein-labeled oligonucleotide	Gene (2000) 259:123
Ligation and cleavage of RNA by ribozymes	Fluorescein- or tetramethylrhodamine-labeled oligoribonucleotide	Biotechniques (2000) 29:344
SNP detection by allele-specific primer extension	Fluorescent ddNTP	Genome Res (1999) 9:492
Protein–protein and protein–nucleic acid interactions	Alexa Fluor® 488 dye–labeled human Factor VIIa, Oregon Green® 488 dye–labeled soluble human tissue factor and Oregon Green® 514 dye–labeled oligonucleotide	Anal Biochem (2002) 308:18
Oligomerization and fibril formation of α-synuclein	Oregon Green® 488 dye– and Alexa Fluor® 594 dye–labeled α-synuclein	Biochemistry (2007) 46:12522

IMPORTANT NOTICE: The products described in this manual are covered by one or more Limited Use Label License(s). Please refer to the Appendix on page 971 and Master Product List on page 975. Products are For Research Use Only. Not intended for any animal or human therapeutic or diagnostic use.

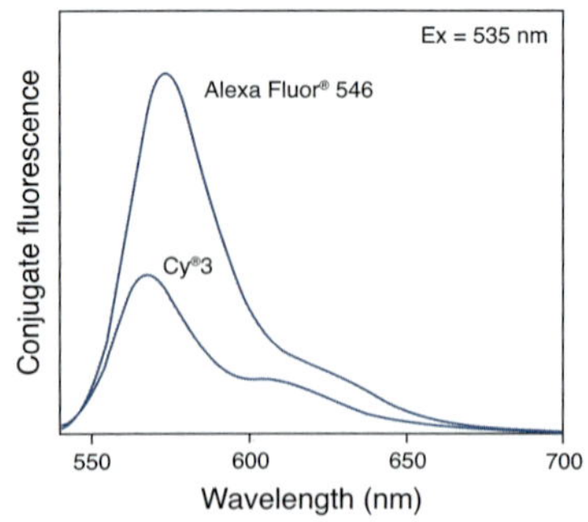

Figure 1.3.22 Fluorescence output from an Alexa Fluor® 546 goat anti–mouse IgG antibody (dye:protein ratio = 5.7) and a commercially available Cy®3 goat anti–mouse IgG antibody (dye:protein ratio = 3.8). Antibody concentrations were adjusted to give equal absorbance at the excitation wavelength (535 nm). The relative fluorescence quantum yield of Alexa Fluor® 546 conjugates is higher than that of Cy®3 conjugates, even at high dye:protein ratios that would typically result in self-quenching effects with most other protein-labeling dyes.

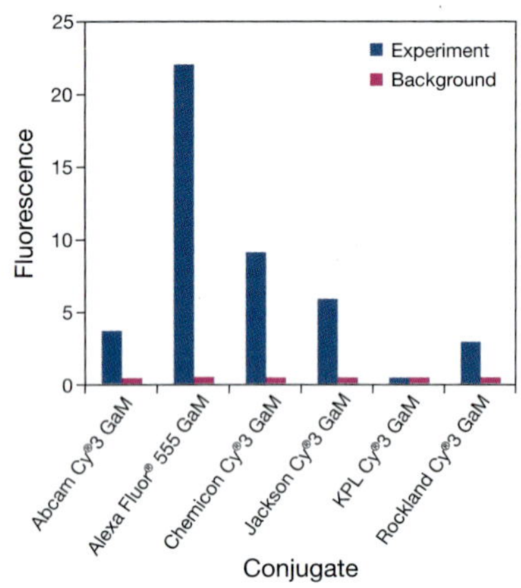

Figure 1.3.23 Brightness comparison of Alexa Fluor® 555 goat anti–mouse IgG antibody with commercially available Cy®3 goat anti–mouse IgG antibody conjugates. Human blood was blocked with normal goat serum and incubated with a mouse monoclonal anti-CD3 antibody; cells were washed, resuspended and incubated with either Alexa Fluor® 555 or Cy®3 goat anti–mouse IgG antibody at equal concentrations. Red blood cells were lysed and the samples were analyzed with a flow cytometer equipped with a 488 nm argon-ion laser and a 585 ± 21 nm bandpass emission filter.

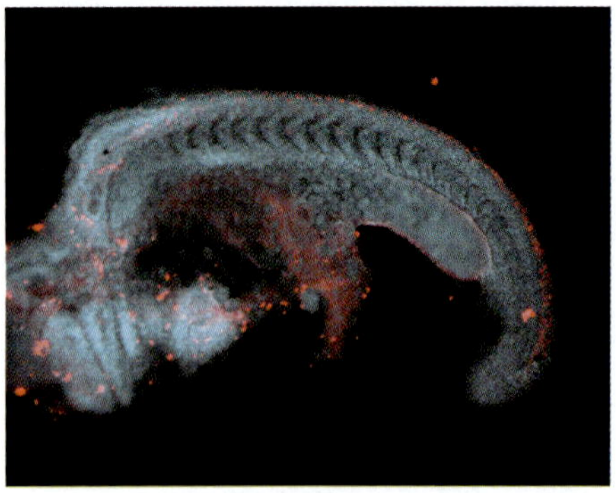

Figure 1.3.24 Neuronal cells in a 22-hour zebrafish embryo were identified with anti-HuC/HuD mouse monoclonal antibody (A21271) and visualized with red-orange–fluorescent Alexa Fluor® 568 goat anti-mouse IgG antibody (A11004). The nuclei were stained with blue-fluorescent DAPI (D1306, D3571, D21490).

component of several labeling kits (Table 1.2, Table 1.4). The contents and utility of these protein and nucleic acid labeling kits are discussed in detail in Section 1.2.

Alexa Fluor® 546 and Alexa Fluor® 555 Dyes: Superior Alternatives to Cy®3 and Tetramethylrhodamine

The orange-fluorescent Alexa Fluor® 546 (excitation/emission maxima ~556/573 nm) and Alexa Fluor® 555 (excitation/emission maxima ~555/565 nm) dyes have spectra that are similar to tetramethylrhodamine and the Cy®3 dye. The spectra of Alexa Fluor® 555 dye are an almost exact match to those of the Cy®3 dye (Figure 1.3.5), and therefore optical filters designed for Cy®3 dye also work with Alexa Fluor® 555 dye. Conjugates of Alexa Fluor® 546 and Alexa Fluor® 555 dyes typically outperfom tetramethylrhodamine (TRITC and TAMRA) and Cy®3 conjugates (Figure 1.3.22, Figure 1.3.23), and Alexa Fluor® 555 conjugates are more fluorescent at a higher degree of substitution (DOS) than are Cy®3 conjugates (Figure 1.3.6). Alexa Fluor® 555 dye is also more photostable than Cy®3 dye (Figure 1.3.7), providing more time for image capture.

We have observed that, unlike most other Alexa Fluor® dyes, antifade reagents provide little protective effect for conjugates of Alexa Fluor® 546 dye; if photobleaching is a limitation, the spectrally similar Alexa Fluor® 555 dye should be used in place of Alexa Fluor® 546 dye. The isomeric mixture of the amine-reactive monosuccinimidyl ester of Alexa Fluor® 546 dye (A20002, A20102) and the isomer-free monosuccinimidyl ester of Alexa Fluor® 555 dye (A20009, A20109) are available in either a 1 mg or 5 mg unit size and as components of several labeling kits (Table 1.2, Table 1.4). The contents and utility of these protein and nucleic acid labeling kits are discussed in detail in Section 1.2.

Alexa Fluor® 568 Dye: A Perfect Match to 561 nm Diode Lasers

The red-orange–fluorescent Alexa Fluor® 568 dye (excitation/emission maxima ~578/603 nm, Figure 1.3.24) is optimally excited by the 561 nm diode lasers used in many confocal laser-scanning microscopes. Although Alexa Fluor® 568 conjugates exhibit absorption and fluorescence emission maxima similar to those of Lissamine rhodamine B conjugates, they are considerably brighter. The isomeric mixture of the amine-reactive monosuccinimidyl ester of Alexa Fluor® 568 dye is available in either a 1 mg or 5 mg unit size (A20003, A20103) and as a component of several labeling kits (Table 1.2, Table 1.4). The contents and utility of these protein and nucleic acid labeling kits are discussed in detail in Section 1.2.

Alexa Fluor® 594 and Alexa Fluor® 610 Dyes: Brighter Red-Fluorescent Dyes

The red-fluorescent Alexa Fluor® 594 dye (excitation/emission maxima ~590/617 nm) has absorption and fluorescence emission maxima similar to those of Texas Red® dye, making it particularly useful for multilabeling experiments in combination with green-fluorescent probes. Alexa Fluor® 594 conjugates are brighter than similarly labeled Texas Red® conjugates, can be labeled to a higher degree of substitution (DOS) (Figure 1.3.25), and are efficiently excited by 561 nm diode lasers and the 594 nm line of the orange He-Ne laser.

The bright and photostable Alexa Fluor® 610 dye (excitation/emission maxima ~612/628 nm) emits an intense red fluorescence that is easily distinguished from green fluorescence and can be visualized with the same optics used for Texas Red® and Alexa Fluor® 594 dyes. Unlike the fluorescence of Alexa Fluor® 633 dye and longer-wavelength fluorophores, Alexa Fluor® 610 fluorescence can still be seen with the human eye.

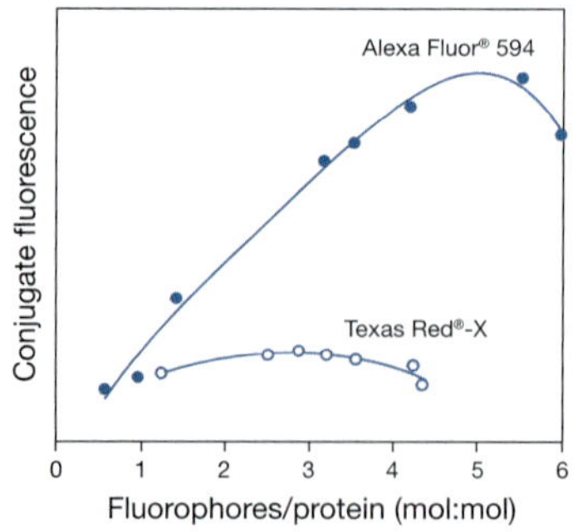

Figure 1.3.25 Comparison of the relative fluorescence of Alexa Fluor® 594 and Texas Red®-X goat anti-mouse IgG antibody F(ab')$_2$ fragment conjugates at different dye:protein ratios.

IMPORTANT NOTICE: The products described in this manual are covered by one or more Limited Use Label License(s). Please refer to the Appendix on page 971 and Master Product List on page 975. Products are For Research Use Only. Not intended for any animal or human therapeutic or diagnostic use.
www.invitrogen.com/probes
molecular probes® invitrogen™ by Life technologies™

The isomeric mixture of the amine-reactive monosuccinimidyl ester of Alexa Fluor® 594 dye is available in either a 1 mg or 5 mg unit size (A20004, A20104) and the 6-isomer of the monosuccinimidyl ester of Alexa Fluor® 610-X dye is available in a 1 mg unit size (A30050). These red-fluorescent Alexa Fluor® dyes are also available as components of several labeling kits (Table 1.2, Table 1.4); the contents and utility of these protein and nucleic acid labeling kits are discussed in detail in Section 1.2.

Far-Red– and Near-Infrared–Fluorescent Alexa Fluor® Dyes

One of our long-term goals has been to develop superior dyes that can be excited by long-wavelength excitation sources, including the red He-Ne laser (at 633 nm), krypton-ion laser (at 647 nm) and diode lasers. Long-wavelength excitation generally provides increased optical penetration of tissues and decreased autofluorescence background, both key enabling prerequisites for *in vivo* imaging applications.[14–16] It has been particularly challenging to prepare long-wavelength reactive dyes whose fluorescence is not significantly quenched upon conjugation. Our far-red– and near-infrared–fluorescent Alexa Fluor® dyes (Figure 1.3.3) meet our goals in several ways:[1,17]

- Very high extinction coefficients—typically >165,000 $cm^{-1}M^{-1}$ but up to 290,000 $cm^{-1}M^{-1}$ for Alexa Fluor® 750 dye
- Excellent spectral match to common long-wavelength excitation sources
- Spectra of Alexa Fluor® 647, Alexa Fluor® 680 and Alexa Fluor® 750 conjugates that virtually match those of the Cy®5 dye (Figure 1.3.8), Cy®5.5 dye (Figure 1.3.9) and Cy®7 dye (Figure 1.3.10), respectively, resulting in an optimal match to optical filters designed for these dyes
- Photostability of Alexa Fluor® 633 and Alexa Fluor® 647 conjugates that exceeds that of Cy®5, allophycocyanin and PBXL-3 conjugates (Figure 1.3.26)
- Unusually low fluorescence quenching upon conjugation to proteins, even at relatively high degrees of substitution (Figure 1.3.27), resulting in protein conjugates that are typically at least three to four times brighter than those of Cy®5, Cy®5.5, Cy®7 and similar dyes[18,19] but that are, in some cases, as much as 40-fold brighter at equal antibody concentrations (Figure 1.3.28, Figure 1.3.29)
- Fluorescence of the nucleotide, oligonucleotide and nucleic acid conjugates of Alexa Fluor® 647 dye that usually exceeds that of the Cy®5 dye conjugates (Section 8.2)

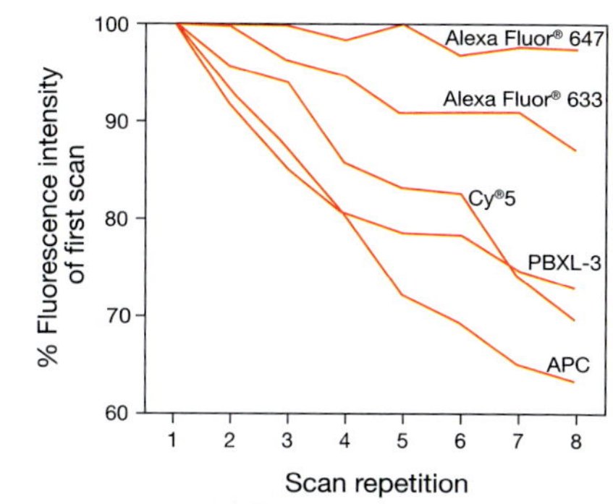

Figure 1.3.26 Photobleaching resistance of the far-red–fluorescent Alexa Fluor® 647, Alexa Fluor® 633, PBXL-3 and Cy®5 dyes and the allophycocyanin fluorescent protein, as determined by laser-scanning cytometry. EL4 cells were labeled with biotin-conjugated anti-CD44 antibody and detected by Alexa Fluor® 647 (S21374), Alexa Fluor® 633 (S21375), PBXL-3, Cy®5 or allophycocyanin (APC, S868) streptavidin. The cells were then fixed in 1% formaldehyde, washed and wet-mounted. After mounting, cells were scanned eight times on a laser-scanning cytometer; laser power levels were 18 mW for the 633 nm spectral line of the He-Ne laser. Scan durations were approximately 5 minutes apiece, and each repetition was started immediately after completion of the previous scan. Data are expressed as percentages derived from the mean fluorescence intensity (MFI) of each scan divided by the MFI of the first scan. Data contributed by Bill Telford, Experimental Transplantation and Immunology Branch, National Cancer Institute.

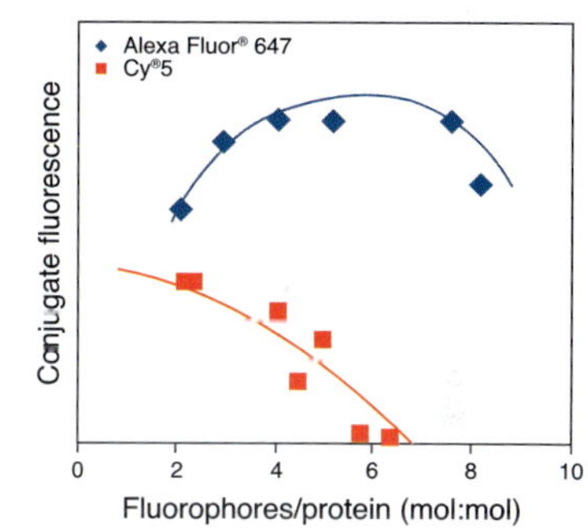

Figure 1.3.27 Comparison of the brightness of Alexa Fluor® 647 and Cy®5 dye antibody conjugates (prepared in our laboratories). More Alexa Fluor® 647 dye molecules can be attached to proteins and nucleic acids without significant quenching, thus yielding conjugates that are much brighter than those possible using the Cy®5 dye.

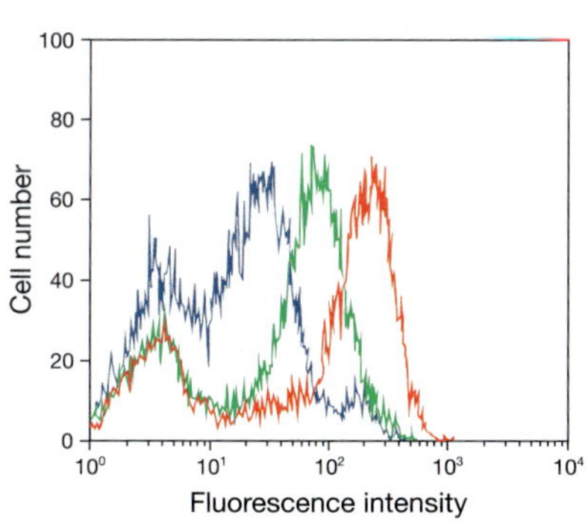

Figure 1.3.28 Flow cytometry was used to compare the brightness of Molecular Probes® Alexa Fluor® 647 goat anti-mouse IgG antibody (red, A21235) with commercially available Cy®5 goat anti-mouse IgG antibody from Jackson ImmunoResearch Laboratories (green) and Amersham® Biosciences Biotech (blue). Human blood was blocked with normal goat serum and incubated with an anti-CD3 mouse monoclonal antibody; cells were washed, resuspended and incubated with either an Alexa Fluor® 647 or Cy®5 goat anti-mouse IgG secondary antibody at equal concentration. Red blood cells were lysed and the samples were analyzed on a flow cytometer equipped with a 633 nm He-Ne laser and a longpass emission filter (>650 nm).

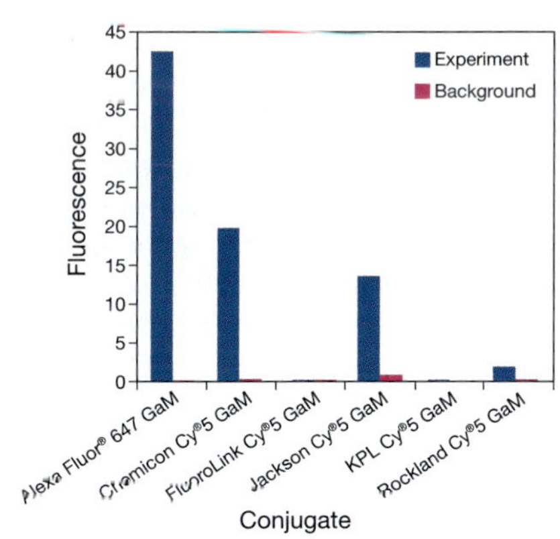

Figure 1.3.29 Brightness comparison of Molecular Probes® Alexa Fluor® 647 goat anti-mouse IgG antibody with Cy®5 goat anti-mouse IgG antibody conjugates commercially available from other companies. Human blood was blocked with normal goat serum and incubated with an anti-CD3 mouse monoclonal antibody; cells were washed, resuspended and incubated with either Alexa Fluor® 647 or Cy®5 goat anti-mouse IgG antibody at an equal concentration. Red blood cells were lysed and the samples were analyzed with a flow cytometer equipped with a 633 nm He-Ne laser and a longpass emission filter (>650 nm).

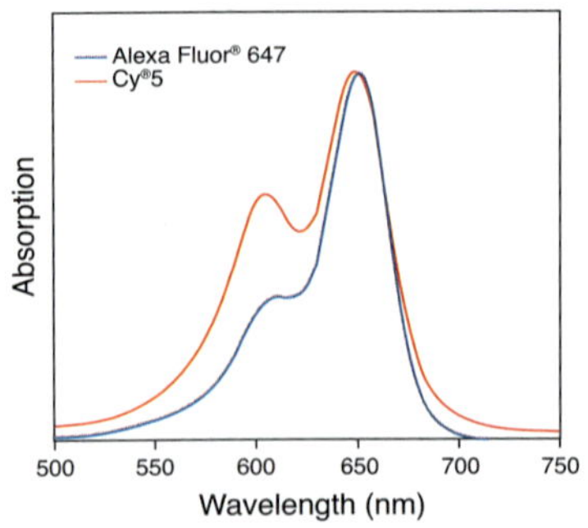

Figure 1.3.30 The absorption spectra of Cy®5 dye conjugates of both proteins and nucleic acids show an additional peak at about 600 nm when compared to the spectrum of the free dye. However, the light absorbed by Cy®5 conjugates at this wavelength does not result in fluroescence. Alexa Fluor® 647 protein conjugates do not exhibit this spectral anomaly. Spectra have been normalized to the same peak intensity for comparison purposes.

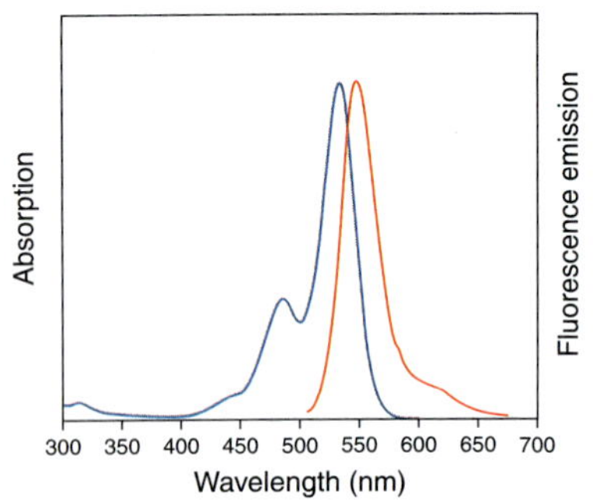

Figure 1.3.31 Absorption and fluorescence emission spectra of Alexa Fluor® 635 goat anti–mouse IgG antibody in pH 7.2 buffer.

- Unlike the Cy®5 dye, very little change in absorption or fluorescence spectra when conjugated to most proteins, oligonucleotides and nucleic acids (Figure 1.3.30), thus yielding significantly greater total fluorescence at the same degrees of substitution (Figure 1.3.27, Figure 1.3.28, Figure 1.3.29)
- Reasonable water solubility of their succinimidyl esters, permitting conjugations to be done without addition of organic solvents if desired
- Chemistry that permits synthesis of pure, monofunctional reactive dyes, thus avoiding crosslinking reactions

The lifetimes and fluorescence quantum yields of far-red and near-infrared Alexa Fluor® dyes are generally shorter and lower, respectively, than the their counterparts in the green-orange-red visible wavelength (488–594 nm) excitation range (Table 1.5). However, the lower quantum yields are compensated by larger extinction coefficients in terms of total fluorescence output. Fluorescence of these long-wavelength Alexa Fluor® dyes is not visible to the human eye but is readily detected by most imaging systems. Pictures of these dyes throughout *The Molecular Probes® Handbook* have been pseudocolored to represent the staining that is observed with sensitive detection equipment.

Alexa Fluor® 633 and Alexa Fluor® 635 Dyes: Optimal Excitation with the He-Ne Laser

These far-red–fluorescent Alexa Fluor® dyes are important labels for fluorescence imaging because their spectra are beyond the range of most sample autofluorescence. With an excitation maximum of 633 nm and 635 nm, respectively, Alexa Fluor® 633 and Alexa Fluor® 635 dyes are a perfect match to the 633 nm spectral line of the He-Ne laser and the 635 nm spectral line of red diode lasers. Although their fluorescence is not visible to the human eye, Alexa Fluor® 633 and Alexa Fluor® 635 conjugates are bright and photostable (Figure 1.3.26), with peak emission at 647 nm.

An isomeric mixture of the amine-reactive succinimidyl ester of Alexa Fluor® 633 dye is available as a stand-alone reagent in either a 1 mg or 5 mg unit size (A20005, A20105) and as a component of several labeling kits (Table 1.2, Table 1.4), which are described in detail in Section 1.2. Alexa Fluor® 635 dye, which is currently only available as antibody (Table 7.1, Figure 1.3.31) and streptavidin (Table 7.9) conjugates, typically produces brighter protein conjugates than does Alexa Fluor® 633 dye because it is less susceptible to self-quenching.

Alexa Fluor® 647 Dye: A Superior Alternative to Cy®5 Dye

Spectra of Alexa Fluor® 647 conjugates are virtually identical to those of the Cy®5 dye (Figure 1.3.8), resulting in an optimal match to optical filters designed for that dye. Total fluorescence of Alexa Fluor® 647 secondary antibody conjugates, however, is significantly higher that that of Cy®5 conjugates commercially available from other suppliers (Figure 1.3.27, Figure 1.3.28, Figure 1.3.29). Also, unlike Cy®5 dye, Alexa Fluor® 647 dye has very little change in absorption or fluorescence spectra upon conjugation to most proteins and nucleic acids (Figure 1.3.30), thus yielding greater total fluorescence at the same degree of substitution.[1] The spectral characteristics of thirteen different red-fluorescent fluorophores, including Alexa Fluor® 647 and BODIPY® 630/660 (Section 1.4) dyes, have been evaluated in different surrounding media to assess the influence of polarity, viscosity and detergent concentration and to facilitate probe choice in fluorescence-based assays.[17] The monosuccinimidyl ester of Alexa Fluor® 647 dye is available as a stand-alone reagent in either a 1 mg or 5 mg unit size (A20006, A20106) and as a component of several labeling kits (Table 1.2, Table 1.4), which are described in detail in Section 1.2.

Alexa Fluor® 660 Dye: A Match for the Krypton-Ion Laser

Alexa Fluor® 660 is optimally excited by the 647 nm spectral line of the krypton-ion laser and well excited by the 633 nm spectral line of the He-Ne laser. Protein conjugates of Alexa Fluor® 660 dye produce bright near-infrared fluorescence, with a peak at 690 nm. This long-wavelength emission is well separated from that of other fluorophores, including Alexa Fluor® 546 and Cy®3 dyes and phycoerythrin conjugates. Alexa Fluor® 660 dye is also the dye of choice as a second label for use with allophycocyanin (APC) conjugates in flow cytometry applications. The monosuccinimidyl ester of Alexa Fluor® 660 dye is available as a 1 mg stand-alone reagent (A20007) and as a component of several labeling kits (Table 1.2, Table 1.4), which are described in detail in Section 1.2.

IMPORTANT NOTICE: The products described in this manual are covered by one or more Limited Use Label License(s). Please refer to the Appendix on page 971 and Master Product List on page 975. Products are For Research Use Only. Not intended for any animal or human therapeutic or diagnostic use.
www.invitrogen.com/probes
molecular probes® | invitrogen™ by Life technologies™

Alexa Fluor® 680 Dye: An Alternative to the Cy®5.5 Dye

With a peak excitation at 679 nm and maximum emission at 702 nm, Alexa Fluor® 680 dye is spectrally similar to the Cy®5.5 dye[14] (Figure 1.3.9). Fluorescence emission of Alexa Fluor® 680 dye is well separated from that of other commonly used red fluorophores, such as the tetramethylrhodamine, Texas Red®, R-phycoerythrin, Alexa Fluor® 594 and Alexa Fluor® 647 dyes, making it ideal for three- and four-color labeling. The monosuccinimidyl ester of Alexa Fluor® 680 dye is available as a stand-alone reagent in either a 1 mg or 5 mg unit size (A20008, A20108) and as a component of several labeling kits (Table 1.2, Table 1.4), which are described in detail in Section 1.2.

Alexa Fluor® 700, Alexa Fluor® 750 and Alexa Fluor® 790 Dyes: Our Longest-Wavelength Dyes

With an absorption maximum at 702 nm, Alexa Fluor® 700 dye can be excited with a xenon-arc lamp, far-red diode lasers or dye-pumped lasers operating in the 675–700 nm range. Alexa Fluor® 700 dye provides near-infrared fluorescence emission, with a peak at 723 nm. Alexa Fluor® 750 dye exhibits fluorescence spectra that are very similar to those of Cy®7 dye (Figure 1.3.10). Its fluorescence emission maximum at 775 nm is well separated from commonly used far-red fluorophores such as Alexa Fluor® 647, Alexa Fluor® 660 or allophycocyanin (APC), facilitating multicolor analysis. With a peak excitation at 749 nm, conjugates of Alexa Fluor® 750 dye are well excited by a xenon-arc lamp or dye-pumped lasers operating in the 720–750 nm range. The monosuccinimidyl esters of Alexa Fluor® 700 (A20010, A20110) and Alexa Fluor® 750 (A20011, A20111) dyes are available as stand-alone reagents in either a 1 mg or 5 mg unit size, and as components of labeling kit (Table 1.2, Table 1.4), which are described in detail in Section 1.2.

Alexa Fluor® 790 dye is the longest-wavelength Alexa Fluor® dye available. With excitation/emission maxima of 784/814 nm, Alexa Fluor® 790 dye has spectral properties similar to those of indocyanine green (ICG) and IRDye® 800 dyes (LI-COR Biosciences). This fluorophore is especially useful for researchers who require an amine-reactive, near-infrared label for small animal *in vivo* imaging (SAIVI) applications, as well as for multicolor analysis with Alexa Fluor® 680 dye and the LI-COR Odyssey® infrared imaging system.[20,21] The monosuccinimidyl ester of Alexa Fluor® 790 dye is supplied in a 100 µg unit size (A30051), enough to label ~1 mg of IgG antibody.

Figure 1.3.32 A10168 Alexa Fluor® 350 carboxylic acid, succinimidyl ester

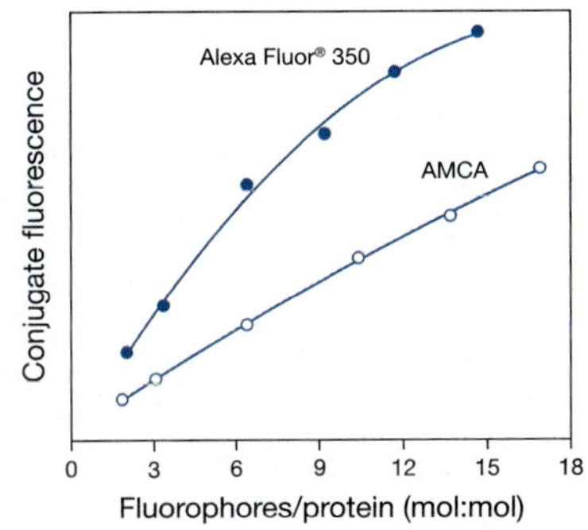

Figure 1.3.33 Comparison of the relative fluorescence of 7-amino-4-methylcoumarin-3-acetic acid (AMCA) streptavidin (○) and Alexa Fluor® 350 streptavidin, a sulfonated AMCA derivative (S11249, ●). Conjugate fluorescence is determined by measuring the fluorescence quantum yield of the conjugated dye relative to that of the free dye and multiplying by the number of fluorophores per protein.

Blue-Fluorescent Alexa Fluor® Dyes

Because their structures are closely related to those of the coumarins and pyrenes, the blue-fluorescent Alexa Fluor® 350 and Alexa Fluor® 405 dyes, as well as the yellow-green–fluorescent Alexa Fluor® 430 dye described above, are also included in Section 1.7. We summarize their properties here to complete our discussion of the Alexa Fluor® dye series.

Alexa Fluor® 350 Dye: Brighter Blue Fluorescence

The blue-fluorescent Alexa Fluor® 350 carboxylic acid succinimidyl ester (A10168, Figure 1.3.32) is a sulfonated coumarin derivative that is more water soluble than either AMCA succinimidyl ester or AMCA-X succinimidyl ester (A6118, Section 1.7) and yields protein conjugates that are typically 50% more fluorescent than those prepared from its nonsulfonated analog (Figure 1.3.33). Alexa Fluor® 350 protein conjugates are optimally excited at 346 nm and exhibit bright blue fluorescence at wavelengths slightly shorter than AMCA or AMCA-X conjugates (442 nm versus 448 nm), which reduces the dye's spectral overlap with the emission of fluorescein. We also prepare Alexa Fluor® 350 conjugates of secondary antibodies (Section 7.2, Table 7.1) and streptavidin (Section 7.6, Table 7.9), as well as several Alexa Fluor® 350 protein labeling kits, which are described in detail in Section 1.2 (Table 1.2, Table 1.4).

Alexa Fluor® 405 Dye: Near-Perfect Match to the Violet Diode Laser

With excitation/emission maxima of 402/421 nm (Figure 1.3.1), the blue-fluorescent Alexa Fluor® 405 dye is well matched to the 405 nm spectral line of violet diode lasers now widely implemented in fluorescence microscopy and flow cytometry. Alexa Fluor® 405 succinimidyl ester is an amine-reactive derivative of our Cascade Blue® dye, which was previously available in amine-reactive form only as its acetyl azide (C2284, Section 1.7). Not only is it offered at higher purity than the alternative Cascade Blue® acetyl azide, but Alexa Fluor® 405 succinimidyl ester also contains a 4-piperidinecarboxylic acid spacer that separates the fluorophore from its reactive moiety

IMPORTANT NOTICE: The products described in this manual are covered by one or more Limited Use Label License(s). Please refer to the Appendix on page 971 and Master Product List on page 975. Products are For Research Use Only. Not intended for any animal or human therapeutic or diagnostic use.

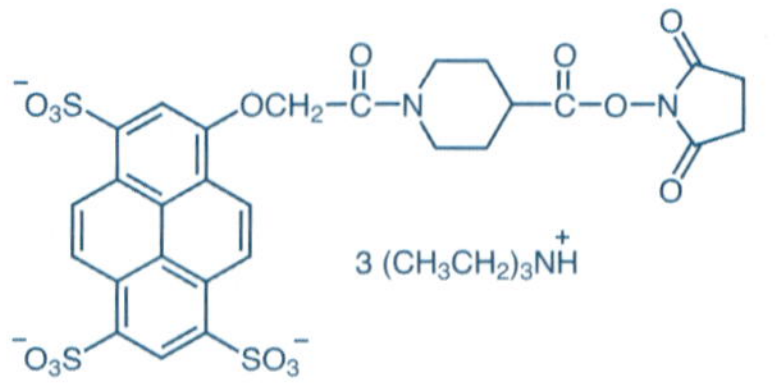

Figure 1.3.34 Alexa Fluor® 405 carboxylic acid, succinimidyl ester (A30000).

(Figure 1.3.34). This spacer enhances the reactivity of the succinimidyl ester and minimizes any interactions between the fluorophore and the biomolecule to which it is conjugated.

As with Cascade Blue® acetyl azide, Alexa Fluor® 405 dye shows minimal spectral overlap with green fluorophores, making it ideal for multicolor applications.[3,22] Alexa Fluor® 405 dye is available as a succinimidyl ester (A30000, A30100) and a cadaverine derivative (A30675, Section 3.4), as well as conjugated to secondary antibodies (Section 7.2, Table 7.1) and streptavidin (Section 7.6, Table 7.9).

Alexa Fluor® 405 dye is also recognized by our anti–Alexa Fluor® 405/Cascade Blue® dye antibody (A5760, Section 7.4). In addition, Alexa Fluor® 405 dye–labeled Fab′ fragments are provided in the Zenon® Alexa Fluor® 405 Antibody Labeling Kits (Section 7.3, Table 7.7).

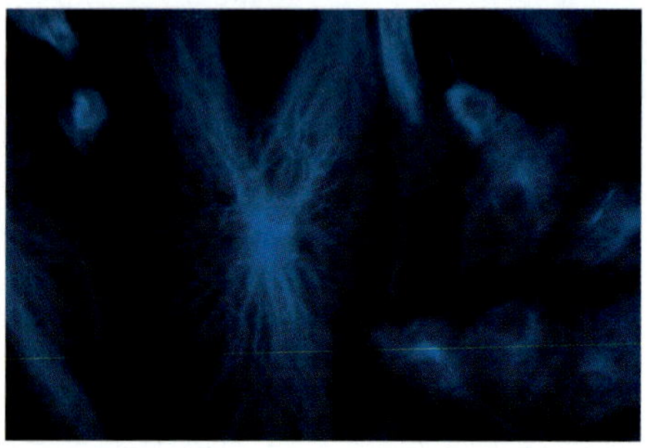

Figure 1.3.36 Fixed and permeabilized bovine pulmonary artery endothelial (BPAE) cells labeled with mouse monoclonal anti-α-tubulin antibody (A11126) and detected using TSA™ Kit #7 with the HRP conjugate of goat anti–mouse IgG antibody and Alexa Fluor® 350 tyramide (T20912).

Alexa Fluor® Protein and Nucleic Acid Labeling Kits

Alexa Fluor® Labeling Kits

As described above, all of our Alexa Fluor® dyes are available as amine-reactive succinimidyl esters (Table 1.2, Table 1.4), and the Alexa Fluor® 488 dye is additionally available as its single-isomer, hydrolysis-resistant tetrafluorophenyl (TFP) ester (A30005). Most of these amine-reactive Alexa Fluor® dyes are also offered as components of several protein and nucleic acid labeling kits (Table 1.2), which are described thoroughly in Section 1.2 and include:

- APEX® Antibody Labeling Kits
- Microscale Protein Labeling Kits
- Monoclonal Antibody Labeling Kits
- SAIVI™ Rapid Antibody Labeling Kits
- Alexa Fluor® Protein Labeling Kits
- Zenon® Antibody Labeling Kits (Section 7.3, Table 7.7)
- ARES™ DNA Labeling Kits (Section 8.2)
- FISH Tag™ DNA Kits and FISH Tag™ RNA Kits (Section 8.2)
- Alexa Fluor® Oligonucleotide Amine Labeling Kits (Section 8.2, Table 8.6)
- ULYSIS® Nucleic Acid Labeling Kits (Section 8.2, Table 8.6)

The purity of the Alexa Fluor® carboxylic acid succinimidyl esters dyes when prepared and packaged in a 5 mg unit size is usually >80–95% by HPLC. However, Alexa Fluor® dyes tenaciously bind water, and packaging of these products in smaller unit sizes—such as the multiple vials used in the Alexa Fluor® labeling kits and the 1 mg stand-alone reagents—may result in some loss of reactivity. Our specifications for Alexa Fluor® carboxylic acid succinimidyl esters provided as kit components or as stand-alone reagents require that the product has ≥50% reactivity after packaging.

Alexa Fluor® Decapacks for Labeling Amine-Modified DNA or RNA

For labeling amine-modified DNA or RNA probes in microarray-based experiments, we offer the Alexa Fluor® 488 reactive dye decapack (A32750), Alexa Fluor® 555 reactive dye decapack

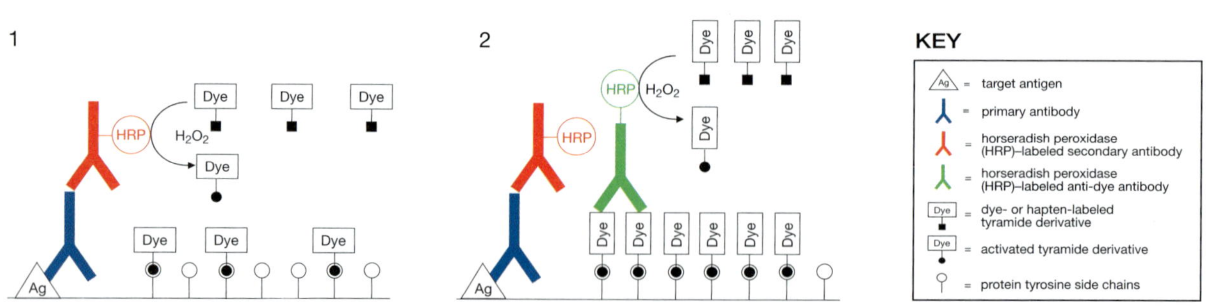

Figure 1.3.35 Schematic representation of TSA™ detection methods applied to immunolabeling of an antigen. The antigen is detected by a primary antibody, followed by a horseradish peroxidase–labeled secondary antibody in conjunction with a dye-labeled (or hapten-labeled) tyramide, resulting in localized deposition of the activated tyramide derivative (Stage 1). Further dye deposition, and therefore higher levels of signal amplification, can be generated by detecting dye deposited in Stage 1 with a horseradish peroxidase–labeled anti-dye antibody in conjunction with a dye-labeled tyramide (Stage 2).

IMPORTANT NOTICE: The products described in this manual are covered by one or more Limited Use Label License(s). Please refer to the Appendix on page 971 and Master Product List on page 975. Products are For Research Use Only. Not intended for any animal or human therapeutic or diagnostic use.

(A32756) and Alexa Fluor® 647 reactive dye decapack (A32757), which provide the corresponding Alexa Fluor® succinimidyl ester conveniently packaged in 10 single-use vials. These specially packaged amine-reactive Alexa Fluor® dyes can be used in conjunction with the aminoallyl dUTP (A21664, Section 8.2) nucleotide[23] or with commercially available aminoallyl nucleotide–based nucleic acid labeling kits.

Each single-use vial contains sufficient Alexa Fluor® succinimidyl ester to optimally label the amount of cDNA produced from reverse transcription of either 20 µg of total RNA or 1–5 µg of poly(A)+ RNA, in the presence of aminoallyl dUTP. For added convenience, we offer a combination set of the Alexa Fluor® 555 and Alexa Fluor® 647 reactive dye decapacks (A32755) that contains 10 vials of each succinimidyl ester and is sufficient for 10 two-color labeling reactions. The Alexa Fluor® 555/Alexa Fluor® 647 dye pair has been shown to display higher signal correlation coefficients than the Cy®3/Cy®5 dye pair in two-color DNA microarray assays.[24–26]

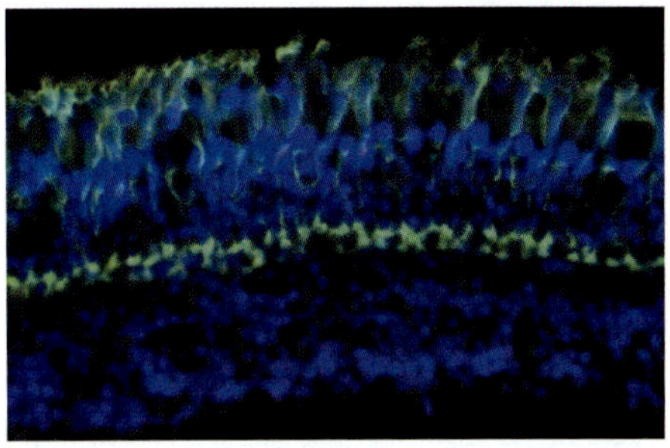

Figure 1.3.37 A zebrafish retina cryosection labeled with the mouse monoclonal antibody FRet 43 and detected using TSA™ Kit #2 with the HRP conjugate of goat anti-mouse IgG antibody and green-fluorescent Alexa Fluor® 488 tyramide (T20912). The nuclei were counterstained with blue-fluorescent Hoechst 33258 (H1398, H3569, H21491).

Other Reactive Alexa Fluor® Derivatives

Several Alexa Fluor® dyes are also available as thiol-reactive maleimides (Section 2.2, Table 2.1), as alkynes and azides for bioorthogonal "click" reactions (Section 3.1) and as aldehyde- and ketone-reactive hydrazides and hydroxylamines (Section 3.3, Table 3.2). The Alexa Fluor® hydrazides and hydroxylamines are also important probes for intracellular tracing (Section 14.3). Although some of the Alexa Fluor® dyes are mixtures of two isomers, all the reactive Alexa Fluor® dyes have only a single reactive substituent linked to each fluorophore.

Alexa Fluor® Tandem Conjugates and Other Bioconjugates

Alexa Fluor® Dye–Phycobiliprotein Tandem Conjugates

We have conjugated R-phycoerythrin with Alexa Fluor® 610, Alexa Fluor® 647 or Alexa Fluor® 680 dye—and in turn conjugated these fluorescent proteins to antibodies or streptavidin, yielding tandem conjugates that permit simultaneous multicolor labeling and detection of multiple targets using a single excitation source (the 488 nm spectral line of the argon-ion laser) and monitoring emission at 628 nm, 668 nm or 702 nm, respectively[3,4] (Section 6.4, Figure 1.3.11). Additionally, we have conjugated allophycocyanin to Alexa Fluor® 680, Alexa Fluor® 700 or Alexa Fluor® 750 dyes to create tandem conjugates for multicolor measurements using excitation sources in the 633 to 650 nm range (Figure 1.3.12). Zenon® Antibody Labeling Kits for the rapid and quantitative labeling of antibodies with the phycobiliprotein tandem dyes are also available (Section 7.3, Table 7.7).

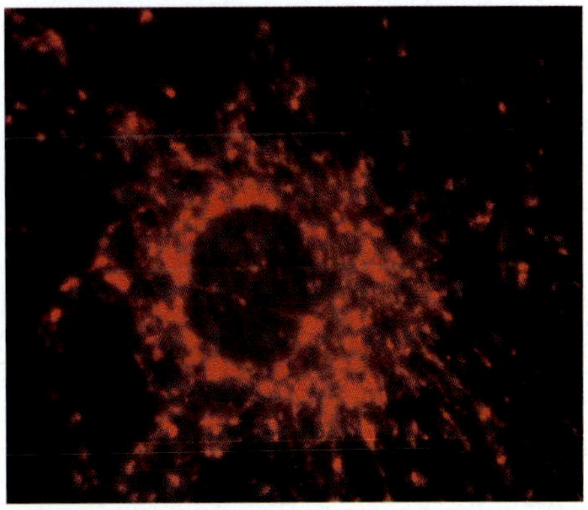

Figure 1.3.38 Fixed and permeabilized bovine pulmonary artery endothelial cell (BPAEC) labeled with anti-OxPhos Complex IV subunit I (human) antibody (anti-cytochrome oxidase subunit I) and detected using TSA™ Kit #4 with the HRP conjugate of goat anti-mouse IgG antibody and Alexa Fluor® 568 tyramide (T20914).

Other Alexa Fluor® Bioconjugates

For immunofluorescence, receptor labeling, nucleic acid synthesis, cell tracing and many other applications, we offer a wide variety of Alexa Fluor® conjugates, including labeled antibodies (Section 7.2, Table 7.1), streptavidin (Section 7.6, Table 7.9) and many other proteins, ligands and nucleotides.

Signal Amplification with Alexa Fluor® Dyes

Tyramide Signal Amplification

Tyramide signal amplification (TSA™) technology, which was developed by NEN (now a part of PerkinElmer Corporation) and licensed for in-cell and in-tissue applications, permits significant amplification of cellular targets by a horseradish peroxidase (HRP)–mediated scheme (Figure 1.3.35). We have introduced several TSA™ Kits (Section 6.2, Table 6.1), including kits that utilize one of the following Alexa Fluor® tyramides:

- Alexa Fluor® 350 tyramide (Figure 1.3.36)
- Alexa Fluor® 488 tyramide (Figure 1.3.37)
- Alexa Fluor® 546 tyramide
- Alexa Fluor® 555 tyramide
- Alexa Fluor® 568 tyramide (Figure 1.3.38)
- Alexa Fluor® 594 tyramide
- Alexa Fluor® 647 tyramide (Figure 1.3.39)

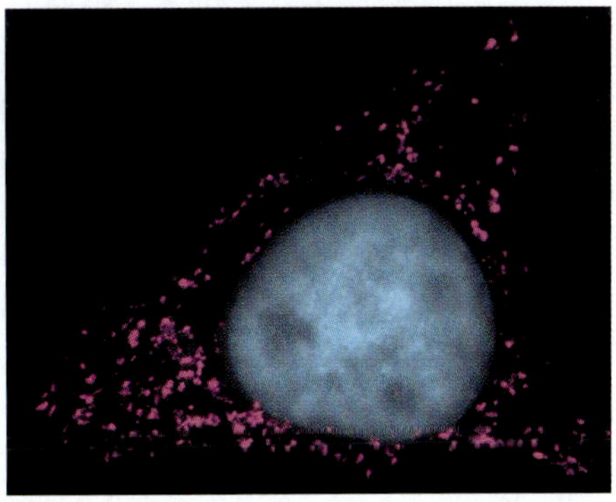

Figure 1.3.39 Fixed and permeabilized bovine pulmonary artery endothelial cell (BPAEC) labeled with anti-OxPhos Complex IV subunit I (human) antibody (anti-cytochrome oxidase subunit I) and detected using TSA™ Kit #6 with the HRP conjugate of goat anti-mouse IgG antibody and Alexa Fluor® 647 tyramide (T20916). The image was deconvolved using Huygens software (Scientific Volume Imaging, http://www.svi.nl/).

IMPORTANT NOTICE: The products described in this manual are covered by one or more Limited Use Label License(s). Please refer to the Appendix on page 971 and Master Product List on page 975. Products are For Research Use Only. Not intended for any animal or human therapeutic or diagnostic use.

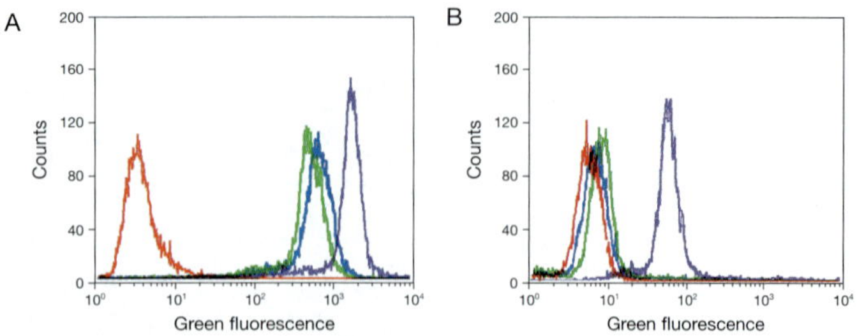

Figure 1.3.40 Detection of epidermal growth factor (EGF) receptors directly or with signal amplification. Cells expressing high (A431 cells, panel **A**) and low (NIH 3T3 cells, panel **B**) levels of EGF receptors were either directly labeled with the preformed Alexa Fluor® 488 complex of biotinylated epidermal growth factor (E13345, blue) or indirectly labeled with biotinylated EGF (E3477) followed by either Alexa Fluor® 488 streptavidin (S11223, green) or HRP-conjugated streptavidin and Alexa Fluor® 488 tyramide (purple), components of our TSA™ Kit #22 (T20932).

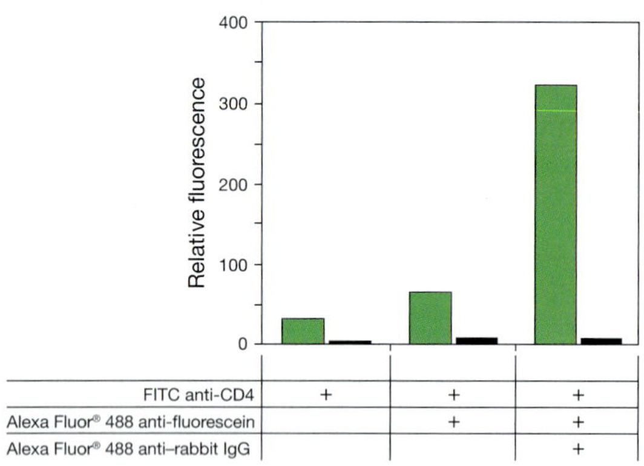

Figure 1.3.41 An example of flow cytometry results obtained using the Alexa Fluor® 488 Signal Amplification Kit for Fluorescein- and Oregon Green® Dye-Conjugated Probes (A11053). Human T cell leukemia cells (Jurkat) were stained with fluorescein (FITC) mouse anti-CD4 antibody and, as indicated, with Alexa Fluor® 488 rabbit anti-fluorescein/Oregon Green® antibody (A11090) and Alexa Fluor® 488 goat anti-rabbit IgG antibody (A11008). The fluorescence values of the negative controls, in which the FITC anti-CD4 antibody was omitted, are shown (black) together with the fluorescence values of the experimental samples (green). The fluorescence values represent the average signals from the population of cells analyzed.

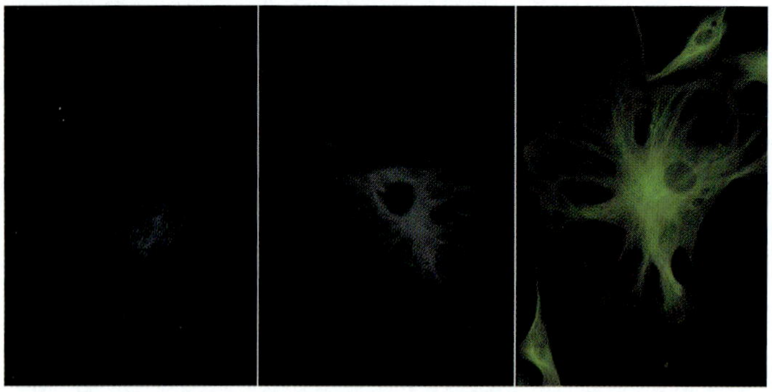

Figure 1.3.42 Demonstration of the amplification obtained with the Alexa Fluor® 488 Signal Amplification Kit for Fluorescein- and Oregon Green® Dye-Conjugated Probes (A11053). Bovine pulmonary artery endothelial cells were labeled with anti-α-tubulin antibody (A11126) in combination with fluorescein goat anti-mouse IgG antibody (F2761) (left panel). The center panel shows the cells after treatment with Alexa Fluor® 488 rabbit anti-fluorescein/Oregon Green® antibody (A11090), and the right panel shows the cells after additional labeling with Alexa Fluor® 488 goat anti-rabbit IgG antibody (A11008). The images were acquired using identical exposure times, and a bandpass filter set appropriate for fluorescein.

The HRP-catalyzed immobilization of a fluorescent tyramide can yield far greater total fluorescence than would ever be possible with direct labeling of the target, enabling detection of very low-abundance receptors (Figure 1.3.40). Furthermore, TSA™ can be used in either live- or fixed-cell preparations, and the increased sensitivity of this signal amplification method often permits use of greatly decreased quantities of antibodies or nucleic acid probes. Our complete selection of TSA™ Kits are listed in Table 6.1 and are extensively discussed in Section 6.2.

Antibody-Based Signal Amplification Kits

Although the direct fluorescence signal of Alexa Fluor® conjugates tends to be significantly greater than that of other dyes with comparable spectra, we have also developed two kits that take further advantage of the superior brightness and photostability of Alexa Fluor® 488 dye– and Alexa Fluor® 594 dye–labeled reagents. These Alexa Fluor® Signal Amplification Kits are designed to substantially increase the signals obtained by immunofluorescence techniques (Figure 1.3.41), thus permitting detection of low-abundance targets.[27] The Alexa Fluor® 488 Signal Amplification Kit for Fluorescein-Conjugated Probes (A11053) provides methods and reagents for two-stage amplification of fluorescein (FITC)–labeled primary antibodies and also overwrites the fluorescein fluorescence signal with that of the brighter and more photostable Alexa Fluor® 488 dye (Figure 1.3.42). The Alexa Fluor® 488 Signal Amplification Kit for Mouse Antibodies (A11054) provides two-stage amplification of imunocytochemical labeling generated by mouse primary antibodies. The similar Alexa Fluor® 568 and Alexa Fluor® 594 Signal Amplification Kits for Mouse Antibodies (A11066, A11067) provide ultrasensitive immunofluorescent detection at longer wavelengths. For additional details about these kits, see Section 7.2.

Alexa Fluor® Conjugates of Anti–Fluorescein/ Oregon Green® Antibody

Alexa Fluor® 488 dye–labeled rabbit anti–fluorescein/Oregon Green® antibody (A11090, Section 7.4) can be used to enhance the green-fluorescent signal of the fluorescein (or Oregon Green®) hapten without changing its fluorescence color. Thus, this conjugate allows researchers to take advantage of the superior photostability of Alexa Fluor® 488 dye, while utilizing existing fluorescein- or Oregon Green® dye–labeled probes and fluorescein-compatible optics. Alexa Fluor® 594 dye–labeled rabbit anti–fluorescein/Oregon Green® antibody (A11091) can be used to convert the green fluorescence of fluorescein or Oregon Green® conjugates into exceptionally photostable red fluorescence (Figure 1.3.43), and to amplify the signal from fluorescein and Oregon Green® conjugates by as much as 100-fold (Figure 1.3.44).

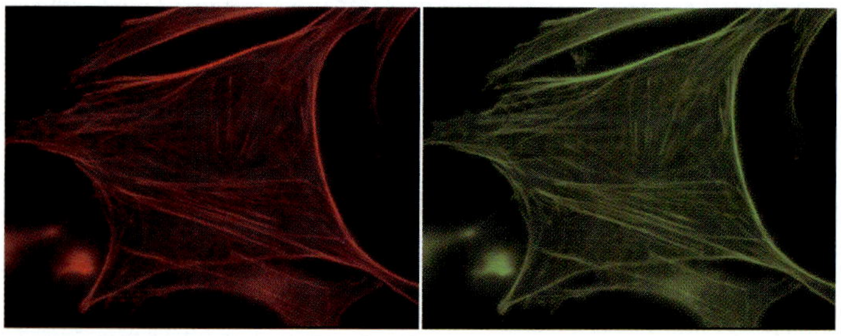

Figure 1.3.43 Fixed and permeabilized bovine pulmonary artery endothelial cells were labeled with the filamentous actin (F-actin) stain, fluorescein phalloidin (F432, right). An Alexa Fluor® 594 anti–fluorescein/Oregon Green® rabbit IgG antibody (A11091) converted the green fluorescence to red (left).

Antibodies to Alexa Fluor® 488 and Alexa Fluor® 405 Dyes

We offer a rabbit polyclonal antibody to Alexa Fluor® 488 dye (A11094, Section 7.4) that quenches the dye's fluorescence and can be used in various signal amplification schemes, including further amplification of the signal from Alexa Fluor® conjugates of proteins or nucleic acids or potentially from Alexa Fluor® 488 tyramide in the corresponding TSA™ Kits (T20912, T20922, T20932; Section 6.2).

As expected, the rabbit polyclonal antibody to the Cascade Blue® dye (A5760, Section 7.4) strongly interacts with Alexa Fluor® 405 dye, making it useful for various fluorescence quenching and amplification schemes. Our Zenon® Rabbit IgG Labeling Kits (Section 7.3, Table 7.7) can also be used to prepare fluorescent dye–, biotin- or enzyme-labeled complexes of these rabbit IgG antibodies for use in various detection and amplification schemes.

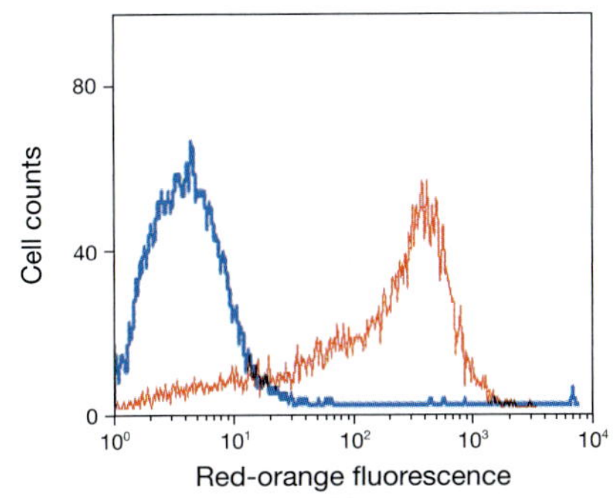

Figure 1.3.44 Color-shifting using a labeled anti–fluorescein/Oregon Green® antibody. Jurkat cells were first stained with a primary mouse anti–human CD3 antibody, followed by fluorescein goat anti–mouse IgG antibody (F2761), with the resultant fluorescence detected in the R-phycoerythrin (red-orange fluorescence) channel of a flow cytometer (blue curve). The weak signal was then shifted to better suit the R-phycoerythrin channel by the addition of an R-phycoerythrin conjugate of anti-fluorescein/Oregon Green® antibody (A21250). The resulting signal intensity is approximately two orders of magnitude greater (red curve) than the direct fluorescence from the first staining step (blue curve).

REFERENCES

1. J Histochem Cytochem (2003) 51:1699; **2.** J Histochem Cytochem (1999) 47:1179; **3.** Nat Protoc (2009) 4:372; **4.** Genes Dev (2008) 22:463; **5.** Cytometry (2000) 41:316; **6.** J Bacteriol (2000) 182:2793; **7.** J Histochem Cytochem (1999) 47:1213; **8.** Anal Biochem (2002) 308:18; **9.** Electrophoresis (2004) 25:779; **10.** Cytometry A (2007) 71:174; **11.** Microsc Res Tech (2005) 68:307; **12.** J Histochem Cytochem (2000) 48:653; **13.** Curr Biol (2006) 16:1924; **14.** Mol Cancer Ther (2009) 8:232; **15.** Clin Cancer Res (2008) 14:4146; **16.** Bioconjug Chem (2008) 19:1186; **17.** Bioconjug Chem (2003) 14:195; **18.** J Immunol Methods (2002) 271:17; **19.** Bioconjug Chem (2000) 11:696; **20.** J Biol Chem (2007) 282:7777; **21.** Proc Natl Acad Sci U S A (2006) 103:15479; **22.** Proc Natl Acad Sci U S A (2005) 102:5346; **23.** Biotechniques (2004) 36:114; **24.** Proteomics (2007) 7:3055; **25.** Nucleic Acids Res (2005) 33:2952; **26.** Anal Biochem (2004) 331:243; **27.** Methods (2003) 30:191.

DATA TABLE 1.3 ALEXA FLUOR® DYES SPANNING THE VISIBLE AND INFRARED SPECTRUM

Cat. No.	MW	Storage	Soluble	Abs	EC	Em	Solvent	Notes
A10168	410.35	F,D,L	H_2O, DMSO	346	19,000	445	pH 7	1
A10169	701.75	F,D,L	H_2O, DMSO	430	15,000	545	pH 7	1
A20000	643.41	F,DD,L	H_2O, DMSO	494	73,000	517	pH 7	1, 2, 3, 4
A20001	723.77	F,DD,L	H_2O, DMSO	530	81,000	555	pH 7	1, 5
A20002	1079.39	F,DD,L	H_2O, DMSO	554	112,000	570	pH 7	1, 6
A20003	791.80	F,DD,L	H_2O, DMSO	578	88,000	602	pH 7	1, 7, 8
A20004	819.85	F,DD,L	H_2O, DMSO	590	92,000	617	pH 7	1, 9, 10
A20005	~1200	F,DD,L	H_2O, DMSO	621	159,000	639	MeOH	1, 11, 12
A20006	~1250	F,DD,L	H_2O, DMSO	651	270,000	672	MeOH	1, 13, 14
A20007	~1100	F,DD,L	H_2O, DMSO	668	132,000	698	MeOH	1, 15, 16
A20008	~1150	F,DD,L	H_2O, DMSO	684	183,000	707	MeOH	1, 17, 18
A20009	~1250	F,DD,L	H_2O, DMSO	555	155,000	572	MeOH	1, 19
A20010	~1400	F,DD,L	H_2O, DMSO	702	205,000	723	MeOH	1, 20, 21
A20011	~1300	F,DD,L	H_2O, DMSO	753	290,000	782	MeOH	1, 22, 23
A30000	1028.26	F,DD,L	H_2O, DMSO	400	35,000	424	pH 7	1, 24
A30002	713.69	F,DD,L	H_2O, DMSO	517	80,000	542	pH 7	1
A30005	884.91	F,DD,L	H_2O, DMSO	494	72,000	520	pH 7	2, 4, 25
A30050	1284.82	F,DD,L	H_2O, DMSO	603	144,000	623	MeOH	1
A30051	~1750	F,DD,L	H_2O, DMSO	784	260,000	814	MeOH	1
A30052	825.46	F,DD,L	H_2O, DMSO	493	73,000	520	pH 7	2, 4, 25

For definitions of the contents of this data table, see "Using *The Molecular Probes® Handbook*" in the introductory pages.

Notes

1. This sulfonated succinimidyl ester derivative is water soluble and may be dissolved in buffer at ~pH 8 for reaction with amines. Long-term storage in water is NOT recommended due to hydrolysis.
2. The fluorescence lifetime (τ) of the Alexa Fluor® 488 dye in pH 7.4 buffer at 20°C is 4.1 nanoseconds. Data provided by the SPEX Fluorescence Group, Horiba Jobin Yvon Inc.
3. The fluorescence quantum yield of Alexa Fluor® 488 carboxylic acid, succinimidyl ester in 50 mM potassium phosphate, 150 mM NaCl pH 7.2 at 22°C is 0.92.
4. Abs and Em of the Alexa Fluor® 488 dye are red-shifted by as much as 16 nm and 25 nm, respectively, on microarrays relative to aqueous solution values. The magnitude of the spectral shift depends on the array substrate material. (Biotechniques (2005) 38:127)
5. The fluorescence quantum yield of Alexa Fluor® 532 carboxylic acid, succinimidyl ester in 50 mM potassium phosphate, 150 mM NaCl pH 7.2 at 22°C is 0.61.
6. The fluorescence quantum yield of Alexa Fluor® 546 carboxylic acid, succinimidyl ester in 50 mM potassium phosphate, 150 mM NaCl pH 7.2 at 22°C is 0.79.
7. The fluorescence lifetime (τ) of the Alexa Fluor® 568 dye in pH 7.4 buffer at 20°C is 3.6 nanoseconds. Data provided by the SPEX Fluorescence Group, Horiba Jobin Yvon Inc.
8. The fluorescence quantum yield of Alexa Fluor® 568 carboxylic acid, succinimidyl ester in 50 mM potassium phosphate, 150 mM NaCl pH 7.2 at 22°C is 0.69.
9. The fluorescence lifetime (τ) of the Alexa Fluor® 594 dye in pH 7.4 buffer at 20°C is 3.9 nanoseconds. Data provided by the SPEX Fluorescence Group, Horiba Jobin Yvon Inc.
10. The fluorescence quantum yield of Alexa Fluor® 594 carboxylic acid, succinimidyl ester in 50 mM potassium phosphate, 150 mM NaCl pH 7.2 at 22°C is 0.66.
11. Alexa Fluor® 633 dye–labeled proteins typically exhibit two absorption peaks at about ~580 and ~630 nm. Fluorescence excitation is more efficient at the 630 nm absorption peak.
12. The fluorescence lifetime (τ) of the Alexa Fluor® 633 dye in H_2O at 20°C is 3.2 nanoseconds. Data provided by LJL BioSystems/Molecular Devices Corporation.
13. The fluorescence lifetime (τ) of the Alexa Fluor® 647 dye in H_2O at 20°C is 1.0 nanoseconds and 1.5 nanoseconds in EtOH. (Bioconjug Chem (2003) 14:195)

continued on next page

IMPORTANT NOTICE: The products described in this manual are covered by one or more Limited Use Label License(s). Please refer to the Appendix on page 971 and Master Product List on page 975. Products are For Research Use Only. Not intended for any animal or human therapeutic or diagnostic use.

DATA TABLE 1.3 ALEXA FLUOR® DYES SPANNING THE VISIBLE AND INFRARED SPECTRUM—*continued*

14. The fluorescence quantum yield of Alexa Fluor® 647 carboxylic acid, succinimidyl ester in 50 mM potassium phosphate, 150 mM NaCl pH 7.2 at 22°C is 0.33.
15. The fluorescence lifetime (τ) of the Alexa Fluor® 660 dye in pH 7.5 buffer at 20°C is 1.2 nanoseconds. Data provided by Pierre-Alain Muller, Max Planck Institute for Biophysical Chemistry, Göttingen.
16. The fluorescence quantum yield of Alexa Fluor® 660 carboxylic acid, succinimidyl ester in 50 mM potassium phosphate, 150 mM NaCl pH 7.2 at 22°C is 0.37.
17. The fluorescence lifetime (τ) of the Alexa Fluor® 680 dye in pH 7.5 buffer at 20°C is 1.2 nanoseconds. Data provided by Pierre-Alain Muller, Max Planck Institute for Biophysical Chemistry, Göttingen.
18. The fluorescence quantum yield of Alexa Fluor® 680 carboxylic acid, succinimidyl ester in 50 mM potassium phosphate, 150 mM NaCl pH 7.2 at 22°C is 0.36.
19. The fluorescence quantum yield of Alexa Fluor® 555 carboxylic acid, succinimidyl ester in 50 mM potassium phosphate, 150 mM NaCl pH 7.2 at 22°C is 0.10.
20. The fluorescence lifetime (τ) of the Alexa Fluor® 700 dye in H_2O at 22°C is 1.0 nanoseconds. Data provided by ISS Inc. (Champaign, IL).
21. The fluorescence quantum yield of Alexa Fluor® 700 carboxylic acid, succinimidyl ester in 50 mM potassium phosphate, 150 mM NaCl pH 7.2 at 22°C is 0.25.
22. The fluorescence lifetime (τ) of the Alexa Fluor® 750 dye in H_2O at 22°C is 0.7 nanoseconds. Data provided by ISS Inc. (Champaign, IL).
23. The fluorescence quantum yield of Alexa Fluor® 750 carboxylic acid, succinimidyl ester in 50 mM potassium phosphate, 150 mM NaCl pH 7.2 at 22°C is 0.12.
24. The Alexa Fluor® 405 and Cascade Blue® dyes have a second absorption peak at about 376 nm with EC ~80% of the 395–400 nm peak.
25. TFP and SDP ester derivatives are water soluble and may be dissolved in buffer at ~pH 8 for reaction with amines. Long-term storage in water is NOT recommended due to hydrolysis.

PRODUCT LIST 1.3 ALEXA FLUOR® DYES SPANNING THE VISIBLE AND INFRARED SPECTRUM

Cat. No.	Product	Quantity
A10168	Alexa Fluor® 350 carboxylic acid, succinimidyl ester	5 mg
A30000	Alexa Fluor® 405 carboxylic acid, succinimidyl ester	1 mg
A30100	Alexa Fluor® 405 carboxylic acid, succinimidyl ester	5 mg
A10169	Alexa Fluor® 430 carboxylic acid, succinimidyl ester	5 mg
A20000	Alexa Fluor® 488 carboxylic acid, succinimidyl ester *mixed isomers*	1 mg
A20100	Alexa Fluor® 488 carboxylic acid, succinimidyl ester *mixed isomers*	5 mg
A30005	Alexa Fluor® 488 carboxylic acid, 2,3,5,6-tetrafluorophenyl ester (Alexa Fluor® 488 5-TFP) *5-isomer*	1 mg
A30002	Alexa Fluor® 514 carboxylic acid, succinimidyl ester *mixed isomers*	1 mg
A20001	Alexa Fluor® 532 carboxylic acid, succinimidyl ester	1 mg
A20101MP	Alexa Fluor® 532 carboxylic acid, succinimidyl ester	5 mg
A20002	Alexa Fluor® 546 carboxylic acid, succinimidyl ester	1 mg
A20102	Alexa Fluor® 546 carboxylic acid, succinimidyl ester	5 mg
A20009	Alexa Fluor® 555 carboxylic acid, succinimidyl ester	1 mg
A20109	Alexa Fluor® 555 carboxylic acid, succinimidyl ester	5 mg
A20003	Alexa Fluor® 568 carboxylic acid, succinimidyl ester *mixed isomers*	1 mg
A20103	Alexa Fluor® 568 carboxylic acid, succinimidyl ester *mixed isomers*	5 mg
A20004	Alexa Fluor® 594 carboxylic acid, succinimidyl ester *mixed isomers*	1 mg
A20104	Alexa Fluor® 594 carboxylic acid, succinimidyl ester *mixed isomers*	5 mg
A30050	Alexa Fluor® 610-X, succinimidyl ester, bis(triethylammonium salt) *6-isomer*	1 mg
A20005	Alexa Fluor® 633 carboxylic acid, succinimidyl ester	1 mg
A20105	Alexa Fluor® 633 carboxylic acid, succinimidyl ester	5 mg
A20006	Alexa Fluor® 647 carboxylic acid, succinimidyl ester	1 mg
A20106	Alexa Fluor® 647 carboxylic acid, succinimidyl ester	5 mg
A20007	Alexa Fluor® 660 carboxylic acid, succinimidyl ester	1 mg
A20008	Alexa Fluor® 680 carboxylic acid, succinimidyl ester	1 mg
A20108	Alexa Fluor® 680 carboxylic acid, succinimidyl ester	5 mg
A20010	Alexa Fluor® 700 carboxylic acid, succinimidyl ester	1 mg
A20110	Alexa Fluor® 700 carboxylic acid, succinimidyl ester	5 mg
A20011	Alexa Fluor® 750 carboxylic acid, succinimidyl ester	1 mg
A20111	Alexa Fluor® 750 carboxylic acid, succinimidyl ester	5 mg
A30051	Alexa Fluor® 790 carboxylic acid, succinimidyl ester, penta(triethylammonium) salt	100 µg
A30006	Alexa Fluor® 488 Microscale Protein Labeling Kit *for 20–100 µg protein* *3 labelings*	1 kit
A30007	Alexa Fluor® 555 Microscale Protein Labeling Kit *for 20–100 µg protein* *3 labelings*	1 kit
A30008	Alexa Fluor® 594 Microscale Protein Labeling Kit *for 20–100 µg protein* *3 labelings*	1 kit
A30009	Alexa Fluor® 647 Microscale Protein Labeling Kit *for 20–100 µg protein* *3 labelings*	1 kit
A20180	Alexa Fluor® 350 Monoclonal Antibody Labeling Kit *5 labelings*	1 kit
A20181	Alexa Fluor® 488 Monoclonal Antibody Labeling Kit *5 labelings*	1 kit
A20182	Alexa Fluor® 532 Monoclonal Antibody Labeling Kit *5 labelings*	1 kit
A20183	Alexa Fluor® 546 Monoclonal Antibody Labeling Kit *5 labelings*	1 kit
A20187	Alexa Fluor® 555 Monoclonal Antibody Labeling Kit *5 labelings*	1 kit
A20184	Alexa Fluor® 568 Monoclonal Antibody Labeling Kit *5 labelings*	1 kit
A20185	Alexa Fluor® 594 Monoclonal Antibody Labeling Kit *5 labelings*	1 kit

IMPORTANT NOTICE: The products described in this manual are covered by one or more Limited Use Label License(s). Please refer to the Appendix on page 971 and Master Product List on page 975. Products are For Research Use Only. Not intended for any animal or human therapeutic or diagnostic use.

PRODUCT LIST 1.3 ALEXA FLUOR® DYES SPANNING THE VISIBLE AND INFRARED SPECTRUM—*continued*

Cat. No.	Product	Quantity
A20186	Alexa Fluor® 647 Monoclonal Antibody Labeling Kit *5 labelings*	1 kit
A20191	Alexa Fluor® 488 Oligonucleotide Amine Labeling Kit *3 labelings*	1 kit
A20196	Alexa Fluor® 647 Oligonucleotide Amine Labeling Kit *3 labelings*	1 kit
A10170	Alexa Fluor® 350 Protein Labeling Kit *3 labelings*	1 kit
A10171	Alexa Fluor® 430 Protein Labeling Kit *3 labelings*	1 kit
A10235	Alexa Fluor® 488 Protein Labeling Kit *3 labelings*	1 kit
A10236	Alexa Fluor® 532 Protein Labeling Kit *3 labelings*	1 kit
A10237	Alexa Fluor® 546 Protein Labeling Kit *3 labelings*	1 kit
A20174	Alexa Fluor® 555 Protein Labeling Kit *3 labelings*	1 kit
A10238	Alexa Fluor® 568 Protein Labeling Kit *3 labelings*	1 kit
A10239	Alexa Fluor® 594 Protein Labeling Kit *3 labelings*	1 kit
A20170	Alexa Fluor® 633 Protein Labeling Kit *3 labelings*	1 kit
A20173	Alexa Fluor® 647 Protein Labeling Kit *3 labelings*	1 kit
A20171	Alexa Fluor® 660 Protein Labeling Kit *3 labelings*	1 kit
A20172	Alexa Fluor® 680 Protein Labeling Kit *3 labelings*	1 kit
A32750	Alexa Fluor® 488 reactive dye decapack *for microarrays* *set of 10 vials*	1 set
A32756	Alexa Fluor® 555 reactive dye decapack *for microarrays* *set of 10 vials*	1 set
A32757	Alexa Fluor® 647 reactive dye decapack *for microarrays* *set of 10 vials*	1 set
A32755	Alexa Fluor® 555 and Alexa Fluor® 647 reactive dye decapacks *for microarrays* *set of 2 x 10 vials* *includes A32756 and A32757 decapacks*	1 set
A30052	Alexa Fluor® 488 5-SDP ester (Alexa Fluor® 488 sulfodichlorophenol ester)	1 mg
A10468	APEX® Alexa Fluor® 488 Antibody Labeling Kit	1 kit
A10470	APEX® Alexa Fluor® 555 Antibody Labeling Kit	1 kit
A10474	APEX® Alexa Fluor® 594 Antibody Labeling Kit	1 kit
A10475	APEX® Alexa Fluor® 647 Antibody Labeling Kit	1 kit
A21665	ARES™ Alexa Fluor® 488 DNA Labeling Kit *10 labelings*	1 kit
A21667	ARES™ Alexa Fluor® 546 DNA Labeling Kit *10 labelings*	1 kit
A21677	ARES™ Alexa Fluor® 555 DNA Labeling Kit *10 labelings*	1 kit
A21669	ARES™ Alexa Fluor® 594 DNA Labeling Kit *10 labelings*	1 kit
A21676	ARES™ Alexa Fluor® 647 DNA Labeling Kit *10 labelings*	1 kit
F32950	FISH Tag™ DNA Far Red Kit *with Alexa Fluor® 647 dye* *10 reactions*	1 kit
F32947	FISH Tag™ DNA Green Kit *with Alexa Fluor® 488 dye* *10 reactions*	1 kit
F32951	FISH Tag™ DNA Multicolor Kit *Alexa Fluor® dye combination* *10 reactions*	1 kit
F32948	FISH Tag™ DNA Orange Kit *with Alexa Fluor® 555 dye* *10 reactions*	1 kit
F32949	FISH Tag™ DNA Red Kit *with Alexa Fluor® 594 dye* *10 reactions*	1 kit
F32955	FISH Tag™ RNA Far Red Kit *with Alexa Fluor® 647 dye* *10 reactions*	1 kit
F32952	FISH Tag™ RNA Green Kit *with Alexa Fluor® 488 dye* *10 reactions*	1 kit
F32956	FISH Tag™ RNA Multicolor Kit *Alexa Fluor® dye combination* *10 reactions*	1 kit
F32953	FISH Tag™ RNA Orange Kit *with Alexa Fluor® 555 dye* *10 reactions*	1 kit
F32954	FISH Tag™ RNA Red Kit *with Alexa Fluor® 594 dye* *10 reactions*	1 kit
S30044	SAIVI™ Alexa Fluor® 647 Antibody/Protein 1 mg-Labeling Kit *3 labelings*	1 kit
S30045	SAIVI™ Rapid Antibody Labeling Kit, Alexa Fluor® 680 *3 labelings*	1 kit
S30046	SAIVI™ Rapid Antibody Labeling Kit, Alexa Fluor® 750 *3 labelings*	1 kit
U21650	ULYSIS® Alexa Fluor® 488 Nucleic Acid Labeling Kit *20 labelings*	1 kit
U21651	ULYSIS® Alexa Fluor® 532 Nucleic Acid Labeling Kit *20 labelings*	1 kit
U21652	ULYSIS® Alexa Fluor® 546 Nucleic Acid Labeling Kit *20 labelings*	1 kit
U21653	ULYSIS® Alexa Fluor® 568 Nucleic Acid Labeling Kit *20 labelings*	1 kit
U21654	ULYSIS® Alexa Fluor® 594 Nucleic Acid Labeling Kit *20 labelings*	1 kit
U21660	ULYSIS® Alexa Fluor® 647 Nucleic Acid Labeling Kit *20 labelings*	1 kit
Z25400	Zenon® Alexa Fluor® 350 Human IgG Labeling Kit *50 labelings*	1 kit
Z25000	Zenon® Alexa Fluor® 350 Mouse IgG_1 Labeling Kit *50 labelings*	1 kit
Z25100	Zenon® Alexa Fluor® 350 Mouse IgG_{2a} Labeling Kit *50 labelings*	1 kit
Z25200	Zenon® Alexa Fluor® 350 Mouse IgG_{2b} Labeling Kit *50 labelings*	1 kit
Z25300	Zenon® Alexa Fluor® 350 Rabbit IgG Labeling Kit *50 labelings*	1 kit
Z25013	Zenon® Alexa Fluor® 405 Mouse IgG_1 Labeling Kit *50 labelings*	1 kit
Z25113	Zenon® Alexa Fluor® 405 Mouse IgG_{2a} Labeling Kit *50 labelings*	1 kit
Z25213	Zenon® Alexa Fluor® 405 Mouse IgG_{2b} Labeling Kit *50 labelings*	1 kit
Z25313	Zenon® Alexa Fluor® 405 Rabbit IgG Labeling Kit *50 labelings*	1 kit
Z25001	Zenon® Alexa Fluor® 430 Mouse IgG_1 Labeling Kit *50 labelings*	1 kit

continued on next page

IMPORTANT NOTICE: The products described in this manual are covered by one or more Limited Use Label License(s). Please refer to the Appendix on page 971 and Master Product List on page 975. Products are For Research Use Only. Not intended for any animal or human therapeutic or diagnostic use.

PRODUCT LIST 1.3 ALEXA FLUOR® DYES SPANNING THE VISIBLE AND INFRARED SPECTRUM—*continued*

Cat. No.	Product	Quantity
Z25301	Zenon® Alexa Fluor® 430 Rabbit IgG Labeling Kit *50 labelings*	1 kit
Z25602	Zenon® Alexa Fluor® 488 Goat IgG Labeling Kit *50 labelings*	1 kit
Z25402	Zenon® Alexa Fluor® 488 Human IgG Labeling Kit *50 labelings*	1 kit
Z25002	Zenon® Alexa Fluor® 488 Mouse IgG_1 Labeling Kit *50 labelings*	1 kit
Z25090	Zenon® Alexa Fluor® 488 Mouse IgG_1 Labeling Kit *enhanced with TSA™ technology* *25 labelings*	1 kit
Z25102	Zenon® Alexa Fluor® 488 Mouse IgG_{2a} Labeling Kit *50 labelings*	1 kit
Z25202	Zenon® Alexa Fluor® 488 Mouse IgG_{2b} Labeling Kit *50 labelings*	1 kit
Z25302	Zenon® Alexa Fluor® 488 Rabbit IgG Labeling Kit *50 labelings*	1 kit
Z25003	Zenon® Alexa Fluor® 532 Mouse IgG_1 Labeling Kit *50 labelings*	1 kit
Z25303	Zenon® Alexa Fluor® 532 Rabbit IgG Labeling Kit *50 labelings*	1 kit
Z25004	Zenon® Alexa Fluor® 546 Mouse IgG_1 Labeling Kit *50 labelings*	1 kit
Z25104	Zenon® Alexa Fluor® 546 Mouse IgG_{2a} Labeling Kit *50 labelings*	1 kit
Z25204	Zenon® Alexa Fluor® 546 Mouse IgG_{2b} Labeling Kit *50 labelings*	1 kit
Z25304	Zenon® Alexa Fluor® 546 Rabbit IgG Labeling Kit *50 labelings*	1 kit
Z25605	Zenon® Alexa Fluor® 555 Goat IgG Labeling Kit *50 labelings*	1 kit
Z25405	Zenon® Alexa Fluor® 555 Human IgG Labeling Kit *50 labelings*	1 kit
Z25005	Zenon® Alexa Fluor® 555 Mouse IgG_1 Labeling Kit *50 labelings*	1 kit
Z25105	Zenon® Alexa Fluor® 555 Mouse IgG_{2a} Labeling Kit *50 labelings*	1 kit
Z25205	Zenon® Alexa Fluor® 555 Mouse IgG_{2b} Labeling Kit *50 labelings*	1 kit
Z25305	Zenon® Alexa Fluor® 555 Rabbit IgG Labeling Kit *50 labelings*	1 kit
Z25606	Zenon® Alexa Fluor® 568 Goat IgG Labeling Kit *50 labelings*	1 kit
Z25006	Zenon® Alexa Fluor® 568 Mouse IgG_1 Labeling Kit *50 labelings*	1 kit
Z25106	Zenon® Alexa Fluor® 568 Mouse IgG_{2a} Labeling Kit *50 labelings*	1 kit
Z25206	Zenon® Alexa Fluor® 568 Mouse IgG_{2b} Labeling Kit *50 labelings*	1 kit
Z25306	Zenon® Alexa Fluor® 568 Rabbit IgG Labeling Kit *50 labelings*	1 kit
Z25607	Zenon® Alexa Fluor® 594 Goat IgG Labeling Kit *50 labelings*	1 kit
Z25407	Zenon® Alexa Fluor® 594 Human IgG Labeling Kit *50 labelings*	1 kit
Z25007	Zenon® Alexa Fluor® 594 Mouse IgG_1 Labeling Kit *50 labelings*	1 kit
Z25107	Zenon® Alexa Fluor® 594 Mouse IgG_{2a} Labeling Kit *50 labelings*	1 kit
Z25207	Zenon® Alexa Fluor® 594 Mouse IgG_{2b} Labeling Kit *50 labelings*	1 kit
Z25307	Zenon® Alexa Fluor® 594 Rabbit IgG Labeling Kit *50 labelings*	1 kit
Z25020	Zenon® Alexa Fluor® 610–R-Phycoerythrin Mouse IgG_1 Labeling Kit *10 labelings*	1 kit
Z25608	Zenon® Alexa Fluor® 647 Goat IgG Labeling Kit *50 labelings*	1 kit
Z25408	Zenon® Alexa Fluor® 647 Human IgG Labeling Kit *50 labelings*	1 kit
Z25008	Zenon® Alexa Fluor® 647 Mouse IgG_1 Labeling Kit *50 labelings*	1 kit
Z25108	Zenon® Alexa Fluor® 647 Mouse IgG_{2a} Labeling Kit *50 labelings*	1 kit
Z25208	Zenon® Alexa Fluor® 647 Mouse IgG_{2b} Labeling Kit *50 labelings*	1 kit
Z25308	Zenon® Alexa Fluor® 647 Rabbit IgG Labeling Kit *50 labelings*	1 kit
Z25021	Zenon® Alexa Fluor® 647–R-Phycoerythrin Mouse IgG_1 Labeling Kit *10 labelings*	1 kit
Z25121	Zenon® Alexa Fluor® 647–R-Phycoerythrin Mouse IgG_{2a} Labeling Kit *10 labelings*	1 kit
Z25221	Zenon® Alexa Fluor® 647–R-Phycoerythrin Mouse IgG_{2b} Labeling Kit *10 labelings*	1 kit
Z25009	Zenon® Alexa Fluor® 660 Mouse IgG_1 Labeling Kit *50 labelings*	1 kit
Z25010	Zenon® Alexa Fluor® 680 Mouse IgG_1 Labeling Kit *50 labelings*	1 kit
Z25110	Zenon® Alexa Fluor® 680 Mouse IgG_{2a} Labeling Kit *50 labelings*	1 kit
Z25210	Zenon® Alexa Fluor® 680 Mouse IgG_{2b} Labeling Kit *50 labelings*	1 kit
Z25310	Zenon® Alexa Fluor® 680 Rabbit IgG Labeling Kit *50 labelings*	1 kit
Z25022	Zenon® Alexa Fluor® 680–R-Phycoerythrin Mouse IgG_1 Labeling Kit *10 labelings*	1 kit
Z25011	Zenon® Alexa Fluor® 700 Mouse IgG_1 Labeling Kit *50 labelings*	1 kit
Z25030	Zenon® Alexa Fluor® 700–Allophycocyanin Mouse IgG_1 Labeling Kit *10 labelings*	1 kit
Z25312	Zenon® Alexa Fluor® 750 Rabbit IgG Labeling Kit *50 labelings*	1 kit
Z25031	Zenon® Alexa Fluor® 750–Allophycocyanin Mouse IgG_1 Labeling Kit *10 labelings*	1 kit

IMPORTANT NOTICE: The products described in this manual are covered by one or more Limited Use Label License(s). Please refer to the Appendix on page 971 and Master Product List on page 975. Products are For Research Use Only. Not intended for any animal or human therapeutic or diagnostic use.

1.4 BODIPY® Dye Series

Overview of the BODIPY® Fluorophores

The BODIPY® fluorophores have spectral characteristics that are often superior to those of fluorescein, tetramethylrhodamine, Texas Red® and longer-wavelength dyes. With derivatives that span the visible spectrum (Figure 1.4.1), BODIPY® dyes are extremely versatile.[1] We use them to generate fluorescent conjugates of proteins, nucleotides, oligonucleotides and dextrans, as well as to prepare fluorescent enzyme substrates, fatty acids, phospholipids, lipopolysaccharides, receptor ligands and polystyrene microspheres.

BODIPY® dyes are unusual in that they are relatively nonpolar and the chromophore is electrically neutral (Figure 1.4.2). These properties tend to minimize dye-induced perturbation of conjugate functional properties. BODIPY® dyes are therefore often the preferred choice for labeling nucleotides, amino acids and other low molecular weight ligands.[2–4] BODIPY® dye conjugates of low molecular weight molecules also tend to be more permeant to live cells than are conjugates of charged fluorophores (Section 14.2). With their high peak intensity, reactive BODIPY® dyes are among the most detectable amine-derivatization reagents available for HPLC and capillary electrophoresis.[5–8] BODIPY® dyes are also more useful than most other long-wavelength dyes, including fluoresceins and carbocyanines, for assays that measure fluorescence polarization[9–12] (Fluorescence Polarization (FP)—Note 1.4), and they have large cross-sections for excitation by multiphoton excitation sources[13,14] (Fluorescent Probes for Two-Photon Microscopy—Note 1.5).

The core structure of the BODIPY® fluorophore is shown in Figure 1.4.2. Solutions of the alkyl-substituted derivatives have a green, fluorescein-like fluorescence. However, when substituents that yield additional conjugation are added to the parent molecule, both the absorption and emission spectra of the resulting derivatives can shift to significantly longer wavelengths, with emission maxima of greater than 750 nm now possible with some BODIPY® derivatives. Our goal has been to develop BODIPY® dyes that are optimal for the major excitation sources and that match the common optical filter sets. Accordingly, our recommended BODIPY® substitutes for the fluorescein, rhodamine 6G, tetramethylrhodamine and Texas Red® fluorophores are named BODIPY® FL, BODIPY® R6G, BODIPY® TMR and BODIPY® TR, respectively (Figure 1.4.3). Because there are so many different BODIPY® dyes, we have had to develop a systematic strategy for naming them. Except for BODIPY® FL, BODIPY® R6G, BODIPY® TMR and BODIPY® TR, we identify these dyes with the registered trademark BODIPY® followed by the approximate absorption/emission maxima in nm (determined in methanol); for example, the BODIPY® 581/591 dye.

Amine-reactive BODIPY® dyes (Table 1.7) are discussed below; thiol-reactive BODIPY® dyes are included in Section 2.2. Other reactive BODIPY® dyes useful for derivatizing aldehydes, ketones and carboxylic acids are described in Section 3.3 and Section 3.4. Applications of some thiol-reactive BODIPY® dyes for cell tracing are discussed in Section 14.2.

BODIPY® FL Dye: A Substitute for Fluorescein

With the most fluorescein-like spectra of the BODIPY® dyes, the green-fluorescent BODIPY® FL fluorophore (Figure 1.4.4) (excitation/emission maxima ~503/512 nm) has several characteristics[15,16] that make it potentially superior to fluorescein in some applications. These include:

- High extinction coefficient (EC >80,000 $cm^{-1}M^{-1}$) and high fluorescence quantum yield (often approaching 1.0, even in water)
- Lack of ionic charge and spectra that are relatively insensitive to solvent polarity and pH[15]
- Narrow emission bandwidth (Figure 1.4.3), resulting in a higher peak intensity than that of fluorescein

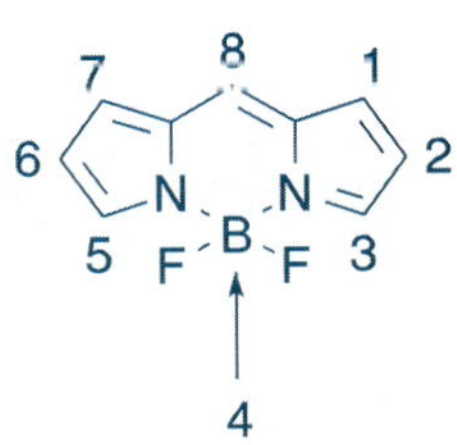

Figure 1.4.2 The structure and numbering of the BODIPY® fluorophore, 4,4-difluoro-4-bora-3a,4a-diaza-s-indacene.

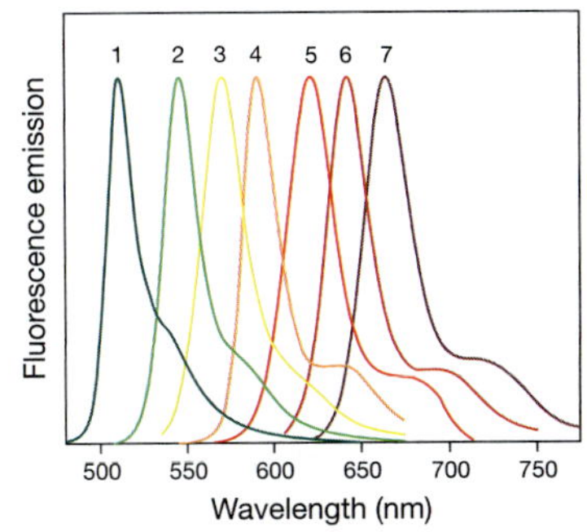

Figure 1.4.1 Normalized fluorescence emission spectra of 1) BODIPY® FL, 2) BODIPY® R6G, 3) BODIPY® TMR, 4) BODIPY® 581/591, 5) BODIPY® TR, 6) BODIPY® 630/650 and 7) BODIPY® 650/665 fluorophores in methanol.

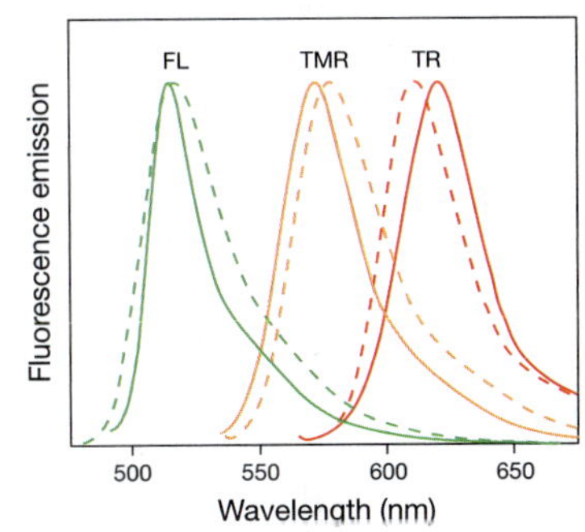

Figure 1.4.3 Normalized fluorescence emission spectra of goat anti-mouse IgG antibody conjugates of fluorescein (FL), tetramethylrhodamine (TMR) and the Texas Red® (TR) dyes, shown by dashed lines (- - -), as compared with goat anti-mouse IgG antibody conjugates of BODIPY® FL, BODIPY® TMR and BODIPY® TR dyes, respectively, shown by solid lines (—).

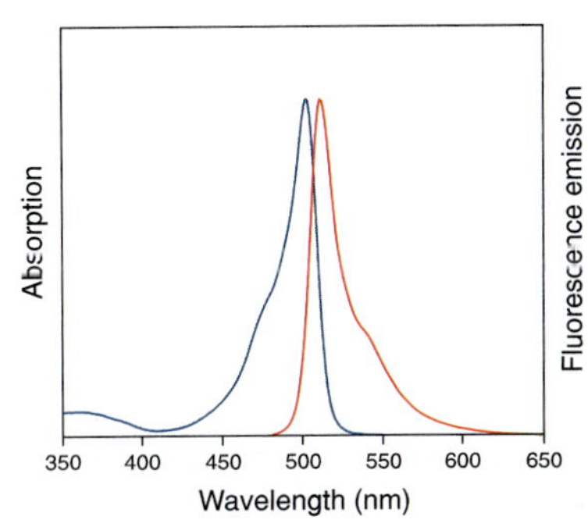

Figure 1.4.4 Absorption and fluorescence emission spectra of BODIPY® FL propionic acid, succinimidyl ester in methanol.

IMPORTANT NOTICE: The products described in this manual are covered by one or more Limited Use Label License(s). Please refer to the Appendix on page 971 and Master Product List on page 975. Products are For Research Use Only. Not intended for any animal or human therapeutic or diagnostic use.

- Red shift in fluorescence emission at high dye concentrations—a property that can be used to detect regions of high probe density[17] (Figure 1.4.5)
- Relatively long excited-state lifetime (typically 5 nanoseconds or longer), which is useful for fluorescence polarization–based assays (Fluorescence Polarization (FP)—Note 1.4)
- Little or no spectral overlap with longer-wavelength dyes such as tetramethylrhodamine and Texas Red® dye (Figure 1.4.3), making BODIPY® FL one of the preferred green-fluorescent dyes for multicolor applications[18]
- Greater photostability than fluorescein in some environments[19]
- Large two-photon cross-section for multiphoton excitation[13,14] (Fluorescent Probes for Two-Photon Microscopy—Note 1.5)

Longer-Wavelength BODIPY® Dyes

It is possible to synthesize BODIPY® fluorophores with altered spectral properties by simply changing the substituents on the parent molecule. This discovery has led to creation of a series of longer-wavelength BODIPY® dyes with fluorescence spectra that span the visible spectrum (Figure 1.4.1). The BODIPY® R6G (excitation/emission maxima ~528/547 nm), BODIPY® TMR (excitation/emission maxima ~543/569 nm) and BODIPY® TR (excitation/emission maxima ~592/618 nm) fluorophores are spectrally similar to the rhodamine 6G, tetramethylrhodamine and Texas Red® fluorophores, respectively, and are thus compatible with standard optical filter sets designed for these important dyes.

The BODIPY® 630/650-X and BODIPY® 650/665-X fluorophores are the longest-wavelength amine-reactive BODIPY® fluorophores currently available. The spectral properties of these longer-wavelength BODIPY® derivatives retain most of the advantages of the BODIPY® FL fluorophore, including narrow bandwidths, high extinction coefficients, good fluorescence quantum yields and relatively long excited-state lifetimes (>3 nanoseconds for the BODIPY® 630/650 dye[20,21]). Like the BODIPY® FL fluorophore, however, most of these dyes have a small Stokes shift, which may require that they be excited or detected at suboptimal wavelengths. The spectral characteristics of 13 different red-fluorescent fluorophores, including the Alexa Fluor® 647 (Section 1.3) and BODIPY® 630/650 dyes, have been evaluated in different surrounding media to assess the influence of polarity, viscosity and detergent concentration and to facilitate probe choice in fluorescence-based assays.[20]

Amine-Reactive BODIPY® Dyes

BODIPY® Succinimidyl Esters

We offer an extensive selection of amine-reactive BODIPY® dyes (Table 1.7). These include succinimidyl esters of several BODIPY® propionic acids and of BODIPY® FL pentanoic acid:

- BODIPY® FL C_3 succinimidyl ester (D2184)
- BODIPY® FL C_5 succinimidyl ester (D6184)
- BODIPY® R6G C_3 succinimidyl ester (D6180)
- BODIPY® 493/503 C_3 succinimidyl ester (D2191)
- BODIPY® 530/550 C_3 succinimidyl ester (D2187)
- BODIPY® 558/568 C_3 succinimidyl ester (D2219)
- BODIPY® 564/570 C_3 succinimidyl ester (D2222)
- BODIPY® 576/589 C_3 succinimidyl ester (D2225)
- BODIPY® 581/591 C_3 succinimidyl ester (D2228)

We have also prepared reactive BODIPY® X succinimidyl esters that contain an additional seven-atom aminohexanoyl spacer ("X") between the fluorophore and the succinimidyl ester group. This spacer helps to separate the fluorophore from its point of attachment, potentially reducing the interaction of the fluorophore with the biomolecule to which it is conjugated and making it more accessible to secondary detection reagents such as anti-dye antibodies.[22–24] These BODIPY® X succinimidyl esters include:

Table 1.7 Amine-reactive BODIPY® dyes.

BODIPY® Dye	Abs *	Em *	COOH	STP	Succinimidyl Ester
BODIPY® 493/503	500	506			D2191
BODIPY® FL	505	513	D2183 (C_3) D3834 (C_5)	B10006	D2184 (C_3) D6140 (SSE) D6102 (X) D6184 (C_5)
BODIPY® R6G	528	550			D6180
BODIPY® 530/550	534	554			D2187
BODIPY® TMR	542	574			D6117 (X)
BODIPY® 558/568	558	569			D2219
BODIPY® 564/570	565	571			D2222
BODIPY® 576/589	576	590			D2225
BODIPY® 581/591	584	592			D2228
BODIPY® TR	589	617			D6116 (X)
BODIPY® 630/650-X †	625	640			D10000
BODIPY® 650/655-X †	646	660			D10001

* Approximate absorption (Abs) and fluorescence (Em) maxima, in nm, for the goat anti-mouse IgG antibody or dextran conjugates in aqueous buffer. † Not recommended for derivatizing proteins. (C_3) = Propionic acid. (C_5) = Pentanoic acid. (COOH) = Carboxylic acid. (SSE) = Sulfosuccinimidyl ester. (STP) = 4-Sulfotetrafluorophenyl ester. (X) = Aminohexanoyl spacer separating the dye and the SE.

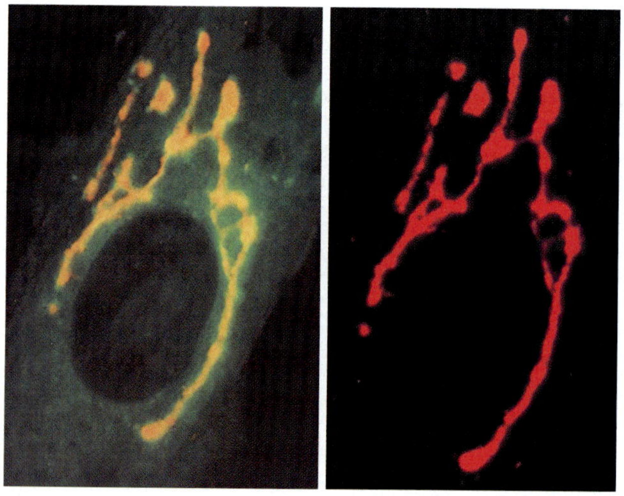

Figure 1.4.5 Selective staining of the Golgi apparatus using the green-fluorescent BODIPY® FL C_5-ceramide (D3521) (left). At high concentrations, the BODIPY® FL fluorophore forms excimers that can be visualized using a red longpass optical filter (right). The BODIPY® FL C_5-ceramide accumulation in the trans-Golgi is sufficient for excimer formation (J Cell Biol (1991) 113:1267). Images contributed by Richard Pagano, Mayo Foundation.

IMPORTANT NOTICE: The products described in this manual are covered by one or more Limited Use Label License(s). Please refer to the Appendix on page 971 and Master Product List on page 975. Products are For Research Use Only. Not intended for any animal or human therapeutic or diagnostic use.
www.invitrogen.com/probes

- BODIPY® FL-X succinimidyl ester (D6102)
- BODIPY® TMR-X succinimidyl ester (D6117)
- BODIPY® TR-X succinimidyl ester (D6116)
- BODIPY® 630/650-X succinimidyl ester (D10000)
- BODIPY® 650/665-X succinimidyl ester (D10001)

BODIPY® succinimidyl esters are particularly useful for preparing conjugates of peptides, nucleotides, oligonucleotides, drugs, toxins, sphingolipids and other low molecular weight ligands that contain aliphatic amines.[2,25] Several BODIPY® succinimidyl esters have been conjugated to aminoacyl tRNAs for metabolic incorporation into proteins through *in vitro* translation.[26–29]

BODIPY® TMR-X SE has been reacted with a series of peptide ligands for use in a high-throughput fluorescence polarization assay of ligand binding to G protein–coupled receptors.[30] The red fluorescence of the BODIPY® 581/591 fluorophore shifts to green fluorescence upon peroxidation, a unique feature that has been exploited for ratiometric measurements of lipid oxidation in live cells (Section 18.2). BODIPY® 630/650-X and BODIPY® 650/665-X succinimidyl esters (D10000, D10001) are quite fluorescent when conjugated to nucleotides[31] and oligonucleotides and can be excited with near-infrared excitation sources.

For amplifying the BODIPY® FL dye signal or converting it into an electron-dense signal, we offer an unlabeled anti–BODIPY® FL rabbit polyclonal antibody (A5770, Section 7.4). This antibody cross-reacts with some other BODIPY® dyes, but not with other fluorophores, and therefore should not be used for simultaneous detection of more than one dye based on the BODIPY® fluorophore.

Water-Soluble BODIPY® Sulfonated Succinimidyl Esters

The moderate lipophilicity of the BODIPY® propionic acid succinimidyl esters discussed above requires their dissolution in an organic solvent before use in conjugations. Although these reactive dyes are very useful for preparing conjugates of amines in organic solvents, they are less suitable for reaction with proteins.

To address the solubility in aqueous solution, we have prepared the sulfosuccinimidyl ester of BODIPY® FL propionic acid (BODIPY® FL, SSE; D6140), as well as the STP ester of BODIPY® FL propionic acid (B10006). STP esters,[32] which are prepared by coupling a carboxylic acid and 4-sulfo-2,3,5,6-tetrafluorophenol (S10490, Section 3.4, Figure 1.4.6), are more readily purified than sulfosuccinimidyl esters but equally amine reactive. BODIPY® FL SSE and BODIPY® FL STP ester are quite soluble in water and more suitable than the corresponding BODIPY® succinimidyl esters for amine conjugation in aqueous solution. These sulfonated succinimidyl esters are useful for preparing conjugates of proteins, amine-modified oligonucleotides and other biomolecules.[33]

BODIPY® Carboxylic Acids

Two green-fluorescent BODIPY® carboxylic acids (D2183, D3834) are available. These carboxylic acid derivatives can be converted to fluorescent esters,[34] acid halides or amides using standard chemical techniques.

BODIPY® Dye Conjugates and Their Applications

The versatility of the BODIPY® fluorophore is demonstrated by its incorporation into literally hundreds of products listed in this *Handbook*, including many of our FluoSpheres® and TransFluoSpheres® microspheres (Section 6.5), enzyme substrates (Chapter 10) and several of our imaging and flow cytometry standards (Section 23.1, Section 23.2). Some examples of our BODIPY® dye conjugates are described below.

BODIPY® Peptide, Protein and Polysaccharide Conjugates

As is common with many fluorescent dyes, conjugation of BODIPY® dyes to proteins is sometimes accompanied by significant fluorescence quenching.[35] Because of this potential problem, we do *not* recommend using the simple BODIPY® propionic acid succinimidyl esters discussed above for preparing most protein conjugates, although peptides labeled with a single BODIPY® dye can be quite fluorescent and are quite useful for fluorescence polarization–based assays[36,37] (Labeling Small Peptides with Amine Reactive Dyes in Organic Solvents—Note 9.2). We prepare conjugates of its BODIPY® dyes with an exceptionally wide variety of peptides, proteins and polysaccharides, including:

- Pepstatin A, a membrane-permeant analog of this important cathepsin D inhibitor (P12271, Section 10.4)
- Antibodies (Section 7.4; Table 7.1)
- Phallacidin and phalloidin for staining F-actin filaments (B607, B3475, B12382; Section 11.1; Table 11.2)
- Bovine serum albumin, for use as a tracer (A2750, Section 14.7)
- Acetylated and non-acetylated low-density lipoproteins (L3485, L3483; Section 16.1)
- *Escherichia coli* and *Staphylococcus aureus* BioParticles® conjugates (E2864, S2854; Section 16.1; Table 16.3)
- Lipopolysaccharide (L23350, Section 16.1; Table 16.1)
- Dextran (D7168, Section 14.5)

In addition, we prepare conjugates of proteins (and of starch) that are so heavily labeled that they are almost nonfluorescent (Figure 1.4.7). Use of the EnzChek® Kits and DQ™ reagents that incorporate these bioconjugates as fluorogenic enzyme substrates is described later in this section and in Section 10.4.

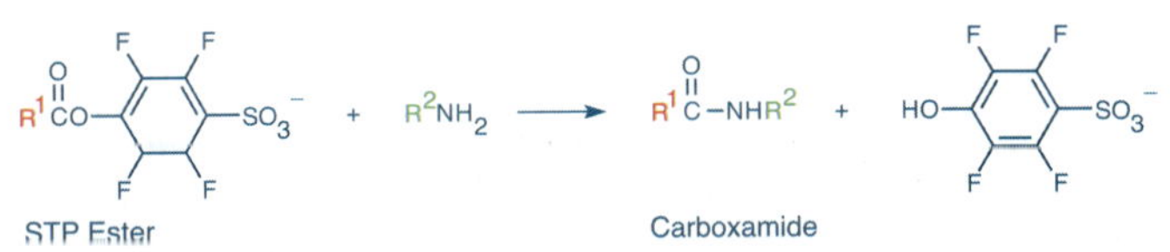

Figure 1.4.6 Reaction of a primary amine with an STP ester.

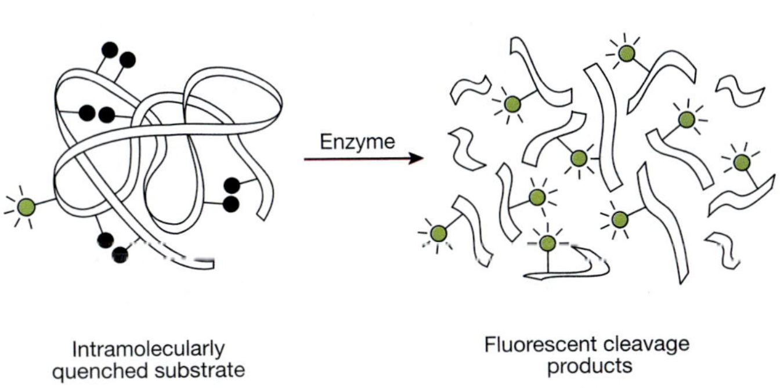

Figure 1.4.7 Principle of enzyme detection via the disruption of intramolecular self-quenching. Enzyme-catalyzed hydrolysis of the heavily labeled and almost totally quenched substrates provided in our EnzChek® Protease Assay Kits relieves the intramolecular self-quenching, yielding brightly fluorescent reaction products.

NOTE 1.5

Fluorescent Probes for Two-Photon Microscopy

Defining Characteristics of Two-Photon Excitation

Two-photon excitation (TPE) is a nonlinear optical process first predicted theoretically by Maria Göppert-Mayer in 1931.[1] Its application to fluorescence microscopy was pioneered much more recently by Denk, Strickler and Webb.[2] In TPE, a fluorophore is excited via near simultaneous absorption of two photons, each having half the energy (twice the wavelength) required for the transition from the ground to the first singlet excited state (Figure 1). The prerequisite for near-simultaneous absorption and the timescale of molecular light absorption (~10–16 seconds) dictates the use of specialized excitation sources; in current instruments, this is typically a mode-locked Ti-Sapphire,Ñ¢ laser delivering infrared light pulses of femtosecond duration at high repetition rates.[3] Two-photon excited fluorescence has a characteristic dependence on the square of the excitation light intensity; doubling the excitation intensity quadruples the fluorescence signal. In contrast, fluorescence derived from conventional one-photon absorption exhibits linear dependence on excitation light intensity.

There are many practical benefits to using TPE, given the transparency of tissues to infrared excitation light:

- Spatial confinement of fluorescence to a very small volume (~0.1 μm^3) defined by the focused excitation light, providing inherent 3D imaging capability (Figure 2)
- Capacity for imaging at increased depths in tissues[4]
- Confinement of photodamage and photobleaching effects to the excitation volume, resulting in increased viability of live specimens.[5,6]

However, in addition to requiring specialized (and therefore fairly expensive) excitation sources, TPE produces photodamage and photobleaching effects within the confined excitation volume that are often more acute than those produced by laser scanning confocal microscopy.[7–10] The advantages of TPE primarily relate to imaging of live specimens. Accordingly, neuroscience—specifically structural and functional imaging of the nervous system—is the largest field of current applications (Table 1). In addition to providing benefits for fluorescence microscopy, TPE offers advantages in other biophotonic techniques such as fluorescence correlation spectroscopy,[11] controlled photoablation,[12] photodynamic therapy[13] and activation of "caged" compounds.[14] To learn more about the technical foundations and applications of TPE microscopy, researchers should consult the growing collection of available review literature.[15–20]

Fluorescence Excitation and Emission Spectra

One-photon and two-photon excitation of a given fluorophore generally result in identical fluorescence emission spectra, as the originating excited state and the photon emission process are the same (Figure 1.1). However, two-photon excitation spectra differ from their one-photon counterparts to an extent that depends on the molecular orbital symmetry of the fluorophore (greater difference for higher symmetry fluorophores).[18] Consequently, most two-photon excitation spectra are blue-shifted and broader compared to the corresponding one-photon spectra plotted on a doubled wavelength axis. Simply stated, a fluorophore with a one-photon excitation peak at 500 nm will probably have a two-photon excitation maximum at <1,000 nm (Figure 3).

Because two-photon excitation spectra are relatively broad, multiplex detection schemes in which two or more fluorophores are excited at a single wavelength and discriminated on the basis of different emission spectra are relatively easy to implement (some examples are included in Table 1). The two-photon absorption cross-section (σ) in units of GM (for Göppert-Mayer; 1 GM = 10^{-50} cm^4 seconds) quantifies the efficiency of TPE for different fluorophores and is plotted on the y-axis of excitation spectra (Figure 3). There are several published collections of two-photon excitation spectra and cross-sections that provide guidance on compatibility of dyes and probes with excitation sources.[21–25] Excitation wavelengths used in selected published TPE microscopy applications are listed in Table 1.

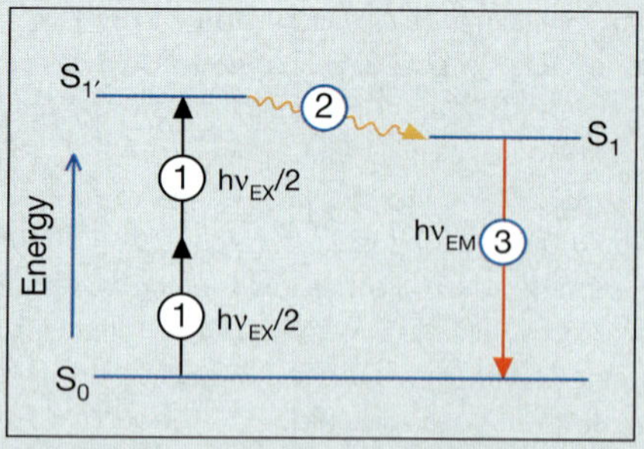

Figure 1 Excited-state energy diagram showing two-photon excitation (1), followed by nonradiative vibrational relaxation (2) and spontaneous fluorescence photon emission (3). In conventional fluorescence detection systems, excitation is achieved by absorption of a single photon of energy ($h\upsilon_{EX}$); processes (2) and (3) are essentially the same.

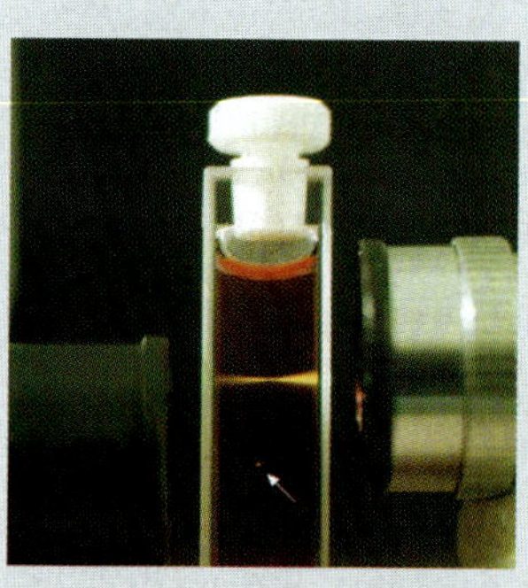

Figure 2 An experiment illustrating ordinary (single-photon) excitation of fluorescence and two-photon excitation. The cuvette contains a solution of the dye safranin O, which normally emits yellow light when excited by green light. The upper lens focuses green (543 nm) light from a CW helium–neon laser into the cuvette, producing the expected conical pattern of excitation (fading to the left). The lower lens focuses pulsed infrared (1046 nm) light from a neodymium–YLF laser. In two-photon absorption, the excitation is proportional to the square of the intensity; thus, the emission is confined to a small point focus (see arrow), which can be positioned anywhere in the cuvette by moving the illuminating beam. Image contributed by Brad Amos, Science Photo Library, London.

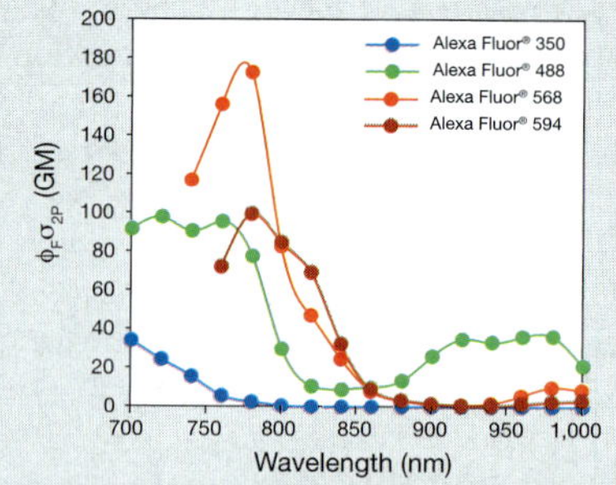

Figure 3 Two-photon excitation spectra of Alexa Fluor® 350, Alexa Fluor® 488, Alexa Fluor® 568 and Alexa Fluor® 594 dyes. The y-axis units are products of fluorescence quantum yields (Φ_F) and two-photon absorption cross-sections (σ). Data courtesy of Warren Zipfel, Cornell University.

IMPORTANT NOTICE: The products described in this manual are covered by one or more Limited Use Label License(s). Please refer to the Appendix on page 971 and Master Product List on page 975. Products are For Research Use Only. Not intended for any animal or human therapeutic or diagnostic use.
www.invitrogen.com/probes
molecular probes® | invitrogen™ by life technologies™

Fluorescent Probes for TPE Microscopy

TPE has added a new spectral dimension to fluorescence microscopy. Probes such as fura-2 (for Ca^{2+}), SBFI (for Na^+), monochlorobimane (for glutathione), and DAPI (for nuclear DNA) were previously of limited utility in confocal microscopy due to their requirements for ultraviolet excitation; now, however, these probes have a new lease on life (Table 1). Furthermore, *in situ* imaging of small endogenous fluorophores, such as serotonin and NADH, that are almost inaccessible to one-photon excitation has now become practicable.[26] Organic fluorophores and fluorescent proteins typically have two-photon absorption cross-sections in the range 1–100 GM. However, fluorophores with smaller cross-sections (e.g., NADH, $\sigma < 0.1$ GM) can still generate sufficient TPE fluorescence for imaging purposes.[26] At the opposite extreme, Qdot® nanocrystals have cross-sections exceeding 10,000 GM, promising even further expansion of TPE imaging, particularly in the area of *in vivo* applications.[27,28]

1. Ann Phys (1931) 9:273; **2.** Science (1990) 248:73; **3.** Micros Res Tech (2005) 67:8; **4.** Nat Methods (2005) 2:932; **5.** Nature Biotech (1999) 17: 763; **6.** Real-Time Imaging (2002) 8:175; **7.** Chemphyschem (2005) 6:791; **8.** Biophys J (2001) 80:2029; **9.** Biophys J (2000) 78:2159; **10.** Appl Phys B (2001) 73:829; **11.** J Cell Biol (2005) 171:527; **12.** Mol Biol Cell (2003) 14:1808; **13.** Photochem Photobiol (2006) 82:443; **14.** Methods Enzymol (1998) 291:356; **15.** Neuron (2006) 50:823; **16.** Q Rev Biophys (2005) 38:97; **17.** Circ Res (2004) 95:1154; **18.** Nature Biotech (2003) 21:1369; **19.** Nat Methods (2005) 2:932; **20.** Handbook of Biological Confocal Microscopy, Third Edition (2006) p. 535; **21.** Proc Natl Acad Sci U S A (1996) 93:10763; **22.** J Microscopy (2002) 208: 108; **23.** J Biomed Opt (2003) 8: 329; **24.** J Neurosci Methods (2005) 148: 94; **25.** Biophys J (2004) 86:1726; **26.** Proc Natl Acad Sci U S A (2003) 100:7075; **27.** Science (2003) 300:1434; **28.** Science (2005) 307:538.

Table 1 Selected applications of fluorescent probes for two-photon excitation (TPE) microscopy.

Probe	TPE Excitation Wavelength	Application	References	Cat. No.
Alexa Fluor® 488 phalloidin	720 nm or 830 nm	Imaging F-actin organization in pancreatic acinar cells	J Biol Chem (2004) 279:37544	A12379
Alexa Fluor® 594 hydrazide	810 nm	Ca^{2+}-insensitive, neuronal tracer *	Neuron (2002) 33:439; www.stke.org/cgi/content/full/sigtrans; 2004/219/pl5	A10438, A10442
Amplex® Red reagent	750 nm or 800 nm	Detection of reactive oxygen species (ROS) associated with amyloid plaques	J Neurosci (2003) 23:2212	A12222, A22177
CFSE, CMTMR	820 nm	Tracking T and B lymphocytes and dendritic cell motility patterns in intact mouse lymph nodes †	Science (2002) 296:1869; Proc Natl Acad Sci U S A (2004) 101:998	C1157, C2927
CM-H_2DCFDA	740 nm	Detection of localized reactive oxygen species release in cardiomyocytes ‡	J Biol Chem (2003) 278: 44735	C6827
DAPI, Hoechst 33342	740 nm	Imaging DNA in nuclei and isolated chromosomes	Micron (2001) 32:679; Histochem Cell Biol (2000) 114:337	D1306, D3571, D21490, H1399, H3570, H21492
DiD	817 nm	Intravital imaging of mouse erythrocytes	Proc Natl Acad Sci U S A (2005) 102:16807	D307, D7757
FM® 1-43	840 nm	Monitoring synaptic vesicle recycling in rat brain slices	Biotechniques (2006) 40:343	T3163, T35356
Fluo-5F §	810 nm	Imaging Ca^{2+} concentration dynamics in dendrites and dendritic spines	Neuron (2002) 33:439; www.stke.org/cgi/content/full/sigtrans; 2004/219/pl5	F14221, F14222
Fura-2	780 nm	Detection of GABA-mediated Ca^{2+} transients in rat cerebellar Purkinje neurons	J Physiol (2001) 536:429	F1200, F1201, F1221, F1225, F6799, F14185
Lucifer yellow CH	850 nm	Identification of gap junctions in rat brain slices	J Neurosci (2003) 23:9254	L453, L682, L1177
Laurdan	800 nm	Detection of ordered membrane lipid domains	Proc Natl Acad Sci U S A (2003) 100:15554; J Cell Biol (2006) 174:725	D250
Monochloro-bimane	780 nm	Imaging glutathione levels in rat brain slices and intact mouse brain	J Biol Chem (2006) 281:17420	M1381MP
MQAE	750 nm	Fluorescence lifetime imaging (FLIM) of intracellular Cl^- concentrations in olfactory sensory neurons	J Neurosci (2004) 24:7931	E3101
Oregon Green® 488 BAPTA-1	880 nm	Imaging spatiotemporal relationships of Ca^{2+} signals among cell populations in rat brain cortex	Proc Natl Acad Sci U S A (2005) 102:14063	O6806, O6807
Qdot® 525, Qdot® 585, Qdot® 655 nanocrystals	750 nm	Multiplexed immunohistochemical analysis of arterial walls **	Am J Physiol (2006) 290:R114	Q11441MP, Q10111MP, Q11621MP, Q11421MP
SBFI	760 nm	Imaging of intracellular Na^+ gradients in rat cardiomyocytes	Biophys J (2004) 87:1360	S1262, S1263, S1264
TMRE	740 nm	Mitochondrial membrane potential sensor ‡	J Biol Chem (2003) 278:44735; Circulation (2006) 114:1497	T669
X-rhod-1	900 nm	Simultaneous imaging of GFP-PHD translocation and Ca^{2+} dynamics in cerebellar purkinje cells	J Neurosci (2004) 24:9513	X14210

* Used in combination with fluo-4, fluo-5F or fluo-4FF to obtain ratio signals that are insensitive to small changes in resting Ca^{2+} and are independent of subcellular compartment volume. † Multiplexed (single excitation/dual channel emission) combination of CFSE and CMTMR. § Techniques also applicable to fluo-4 and fluo-4FF indicators. ‡ Multiplexed (single excitation/dual channel emission) combination of TMRE and CM-H_2DCFDA. ** Multiplexed (single excitation/dual channel emission) combination of Qdot® 585 and Qdot® 655 nanocrystals. PHD = pleckstrin homology domain.

IMPORTANT NOTICE: The products described in this manual are covered by one or more Limited Use Label License(s). Please refer to the Appendix on page 971 and Master Product List on page 975. Products are For Research Use Only. Not intended for any animal or human therapeutic or diagnostic use.

BODIPY® Nucleotide and Oligonucleotide Conjugates

With the exception of guanosine nucleotides, fluorescence quenching is usually not a problem if the BODIPY® derivative is conjugated to nucleotides, oligonucleotides, peptides or low molecular weight amines in which the stoichiometry of modification is 1:1. BODIPY® FL dye–labeled oligonucleotide primers also have lower photodestruction rates than fluorescein-labeled primers, improving the detectability of labeled DNA in sequencing gels.[38] Oligonucleotide conjugates of several of our BODIPY® dyes have been shown to be useful for DNA sequencing[39,40] (Section 8.2, Table 8.7), in part because the dye exhibits minimal effect on the mobility of the fragment during electrophoresis.[41] We also offer a ChromaTide® BODIPY® FL-14-dUTP for enzyme-mediated incorporation into nucleic acids (C7614, Section 8.2, Table 8.5).

In addition, we have prepared BODIPY® FL conjugates of ATP and GTP that are labeled through the ribose moieties and serve as structural probes of nucleotide-binding proteins (A12410, G12411; Section 17.3). The fluorescence of BODIPY® dyes is quenched by photoinduced electron transfer from proximal guanosine bases.[42,43] BODIPY® FL GTP therefore shows significant fluorescence quenching (Figure 1.4.8) that is relieved by binding to GTP-binding proteins (G-proteins). Longer-wavelength BODIPY® TR conjugates of ATP and GTP are also available (A22352, G22351; Section 17.3).

For protein-binding studies that require nonhydrolyzable nucleotides, we offer the BODIPY® FL fluorophore linked through the γ-thiol of ATP-γ-*S* (A22184, Figure 1.4.9) and the BODIPY® FL, BODIPY® 515/530 and BODIPY® TR fluorophores linked through the γ-thiol of GTP-γ-*S*[44] (G22183, G35779, G35780; Section 17.3). Like BODIPY® FL GTP, the fluorescence of the BODIPY® GTP-γ-*S* thioesters is quenched ~90% relative to that of the free dye but is recovered upon protein binding to at least some G-proteins.[44] The green-fluorescent BODIPY® FL GTP-γ-*S* has been used to detect GTP-binding proteins separated by capillary electrophoresis.[45] BODIPY® 515/530 GTP-γ-*S* thioester also exhibits green fluorescence and has a greater fluorescence increase upon protein binding, as compared with the BODIPY® FL GTP-γ-*S* thioester. The BODIPY® TR GTP-γ-*S* thioester is a red-fluorescent analog with spectral properties similar to those of the Texas Red® dye.

We also offer the green-fluorescent BODIPY® FL GTP-γ-*NH* amide (G35778, Section 17.3) as another choice for protein-binding studies. Although this analog exhibits less fluorescence enhancement upon protein binding, it is reportedly the best of the three green-fluorescent GTP-γ analogs for directly monitoring nucleotide exchange.[46] The different linker lengths of the green-fluorescent GTP-γ analogs (six-carbon for BODIPY® FL GTP-γ-*NH* amide, four-carbon for BODIPY® FL GTP-γ-*S* and one-carbon for BODIPY® 515/530 GTP-γ-*S*) may be useful for understanding protein active-site geometries.

In addition to their potential use for binding studies, the BODIPY® FL ATP-γ-*S* and BODIPY® FL GTP-γ-*S* thioesters are important substrates for Fhit (Figure 1.4.10), a member of the histidine triad superfamily of nucleotide-binding proteins that bind and cleave diadenosine polyphosphates.[47–49] Fhit, one of the most frequently inactivated proteins in lung cancer, functions as a tumor suppressor by inducing apoptosis.[48,50,51] These BODIPY® nucleotides should be especially useful for screening potential Fhit inhibitors and activators.

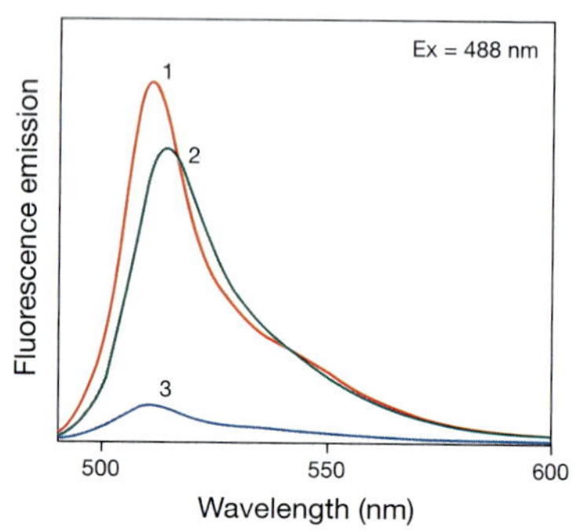

Figure 1.4.8 Fluorescence emission spectra of (1) free BODIPY® FL dye in phosphate-buffered saline, pH 7.2; (2) BODIPY® FL ATP (A12410); and (3) BODIPY® FL GTP (G12411). Samples were prepared with equal absorbance at the excitation wavelength (488 nm). The areas under the curves are therefore proportional to the relative fluorescence quantum yields, clearly showing the quenching effect caused by interaction of the BODIPY® FL fluorophore with the guanine base of GTP.

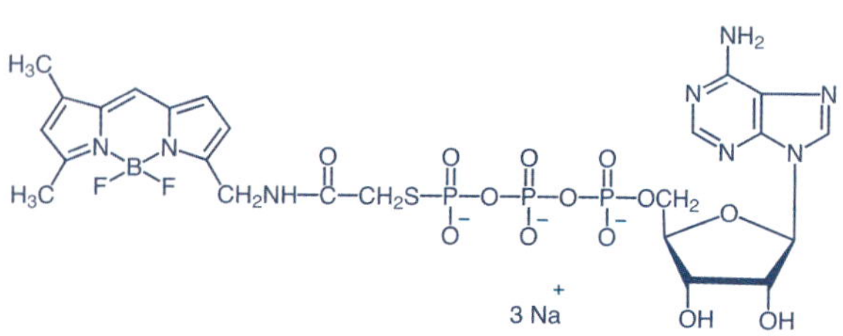

Figure 1.4.9 Adenosine 5'-O-(3-thiotriphosphate), BODIPY® FL thioester, sodium salt (BODIPY® FL ATP-γ-S, thioester, A22184).

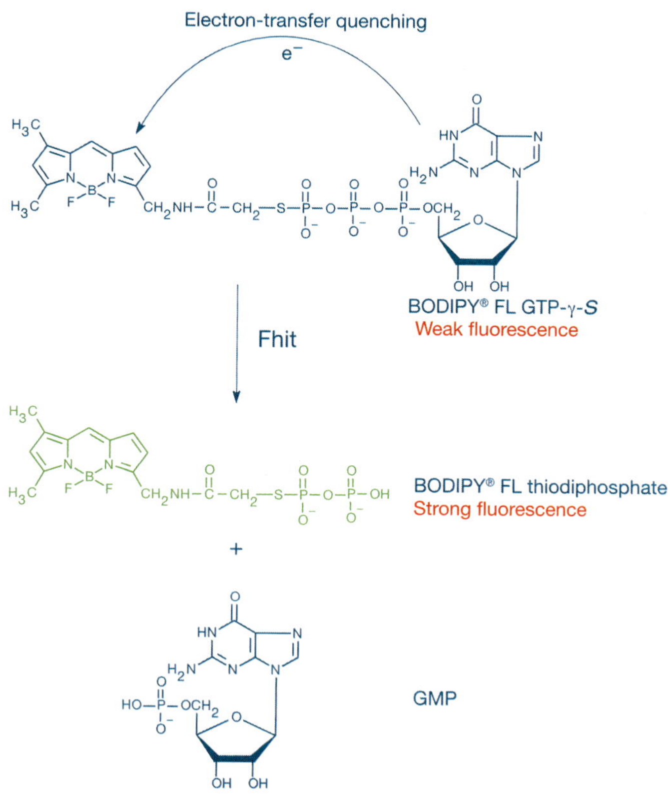

Figure 1.4.10 Principle of fluorescence-based detection of the diadenosine triphosphate hydrolase activity of Fhit using BODIPY® FL GTP-γ-*S* thioester (G22183) as a substrate analog.

IMPORTANT NOTICE: The products described in this manual are covered by one or more Limited Use Label License(s). Please refer to the Appendix on page 971 and Master Product List on page 975. Products are For Research Use Only. Not intended for any animal or human therapeutic or diagnostic use.
www.invitrogen.com/probes
molecular probes® | invitrogen™ by life technologies™

BODIPY® Lipids and BODIPY® Receptor Ligand Conjugates

BODIPY® dye conjugates of lipids, toxins, steroids, drug analogs and receptor probes typically have quantum yields approaching unity, especially in organic solvents. The low polarity of the BODIPY® fluorophore makes probes containing these dyes excellent analogs of biological lipids (Chapter 13). Consequently, these probes are well tolerated by enzymes that metabolize lipids, including phospholipases and sphingomyelinases (Section 17.4, Table 17.3). In most cases, lack of a spectral shift in the metabolic product's fluorescence requires use of an easy extraction and chromatographic separation step to detect product formation, with quantitation possible by photography or with a fluorescence- or absorption-based scanner. BODIPY® derivatives of lipids and low molecular weight ligands include analogs of:

- Phospholipids and fatty acids, numerous versions of which are described in Section 13.2
- Sphingolipids, including several ceramide derivatives (Section 13.3)
- Cholesterol (C3927MP, Section 13.3)
- Cytochalasin D, an actin-polymerization inhibitor (C12377, Section 11.1)
- Paclitaxel (Taxol®), for staining tubulin in isolated preparations (P7500, P7501; Section 11.2)
- Vinblastine, a microtubule-disrupting agent (V12390, Section 11.2)
- Vancomycin, an antibiotic (V34850, Section 15.2)
- Penicillin V (B13233, B13234; Section 15.2)
- Verapamil, for investigating multidrug resistance (B7431, Section 15.6)
- Forskolin, an adenylate cyclase activator (B7469, Section 15.6)
- Prazosin, an α_1-adrenergic receptor probe (B7433, Section 16.2)
- Ouabain, a cardiac glycoside (B23461, Section 16.3)
- Glibenclamide, a probe for potassium channels (E34250, E34251; Section 16.3)
- Dihydropyridine derivative that is selective for Ca^{2+} channels (D7443, S7445; Section 16.3)
- Thapsigargin, which promotes Ca^{2+} release by inhibiting the endoplasmic reticulum Ca^{2+}-ATPase (B7487, B13800; Section 17.2)

In addition to the BODIPY® dye conjugates of receptor ligands in the list above, we have utilized BODIPY® dyes for synthesis of several LysoTracker® and LysoSensor™ dyes, as well as BODIPY® FL histamine (B22461, Figure 1.4.11), which are extremely useful probes for labeling acidic organelles in live cells. These products are discussed in Section 12.3.

DQ™ Reagents: Heavily Labeled BODIPY® Dye Conjugates as Fluorogenic Enzyme Substrates

We have found BODIPY® dye conjugates to be very useful reagents for numerous bioanalytical screening applications. In particular, we have utilized the tendency of BODIPY® dyes to quench their fluorescence on conjugation to certain biopolymers to our advantage (Figure 1.4.7) in the following enzyme-assay kits and reagents:

- EnzChek® Protease Assay Kits, which contain almost nonfluorescent casein derivatives that are heavily labeled with either the green-fluorescent BODIPY® FL dye (E6638; Section 10.4) or red-fluorescent BODIPY® TR-X dye (E6639, R22132; Section 10.4)
- EnzChek® Elastase Assay Kit (E12056, Section 10.4), with DQ™ elastin, a quenched BODIPY® FL conjugate
- EnzChek® *Ultra* Amylase Assay Kit (E33651, Section 10.2), containing a highly quenched BODIPY® FL starch derivative
- DQ™ Green BSA (D12050, Section 10.4) and DQ™ Red BSA (D12051, Section 10.4), heavily labeled and almost nonfluorescent BODIPY® BSA conjugates that yield intense green or red fluorescence upon proteolytic digestion
- DQ™ ovalbumin (D12053, Section 10.4,), a heavily labeled and almost nonfluorescent BODIPY® FL ovalbumin conjugate

Conjugation of either the BODIPY® FL dye (excitation/emission maxima ~500/506 nm) or BODIPY® TR dye (excitation/emission maxima ~589/617 nm) to a biopolymer at high degrees of substitution (DOS) results in almost total quenching of the conjugate's fluorescence; they typically exhibit <3% of the fluorescence of the corresponding free dyes. Enzyme-catalyzed hydrolysis relieves this quenching, yielding brightly fluorescent BODIPY® FL dye– or BODIPY® TR-X dye–labeled peptides (Figure 1.4.7), or, in the case of the BODIPY® FL amylase substrate in the EnzChek® *Ultra* Amylase Assay Kit, BODIPY® FL dye–labeled carbohydrates. The increase in fluorescence, which can be measured with a spectrofluorometer, minifluorometer or fluorescence microplate reader, is proportional to enzymatic activity. The DQ™ BSA and DQ™ ovalbumin substrates are particularly suitable for the study of receptor labeling and antigen processing. DQ™ BSA conjugates can be targeted to Fc receptors after they are complexed with our anti-BSA antibody (A11133, Section 7.5). Ovalbumin is efficiently processed through mannose receptor–mediated endocytosis by antigen-presenting cells and is widely used for studying antigen processing. Upon endocytosis and proteolysis, highly fluorescent peptides are released within intracellular vacuoles. DQ™ ovalbumin appears to be an excellent indicator of macrophage-mediated antigen processing in flow cytometry and microscopy assays.

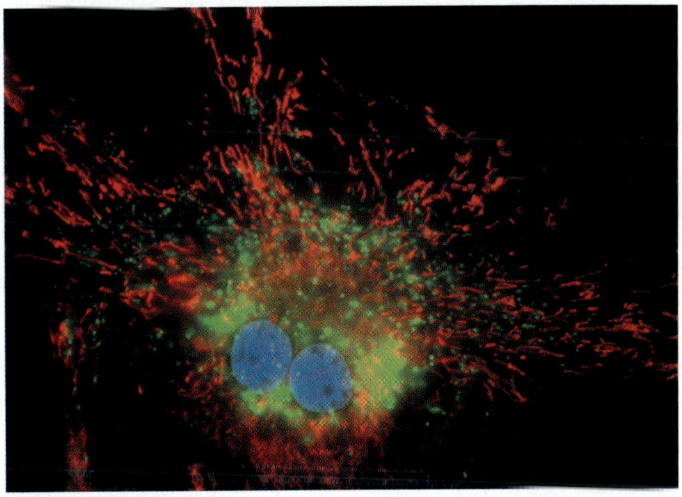

Figure 1.4.11 Viable bovine pulmonary artery endothelial cells simultaneously stained with BODIPY® FL histamine (B22461), MitoTracker® Red CMXRos (M7512) and Hoechst 33342 (H1399, H3570, H21492). Green-fluorescent BODIPY® FL histamine localized to lysosomes, red-fluorescent MitoTracker® Red CMXRos accumulated in the mitochondria and the blue-fluorescent Hoechst 33342 dye stained the nuclei. This multiple-exposure image was acquired with bandpass filters appropriate for fluorescein, the Texas Red® dye and DAPI.

BODIPY® Dye Conjugates for Fluorescence Polarization–Based Assays

When a fluorescent molecule tethered to a protein is excited by polarized fluorescent light, the polarization of fluorescence emission is dependent on the rate of molecular tumbling. Upon proteolytic cleavage of the fluorescently labeled protein, the smaller peptides that result tumble faster and the emitted light is depolarized relative to the light measured from the intact conjugate (Fluorescence Polarization (FP)—Note 1.4). Fluorescence polarization technology is more sensitive than many other nonradioactive assays for proteases and allows measurements to be taken in real time, permitting the collection of kinetic data.

The relatively long fluorescence lifetimes (typically >5 nanoseconds) at visible wavelengths, good anisotropy properties, high molar absorptivity and fluorescence intensity and lack of pH sensitivity in the spectra of the BODIPY® dyes have been shown to make these dyes the preferred fluorophores for high-throughput fluorescence polarization–based assays. The EnzChek® Polarization Assay Kit for Proteases (E6658, Section 10.4) contains green-fluorescent BODIPY® FL casein with an optimal degree of labeling for fluorescence polarization–based protease assays. BODIPY® dye conjugates of nucleotides, peptides and drug analogs are available or are readily prepared from the chemically reactive BODIPY® dyes. Fluorescence polarization–based assays for G-protein–coupled receptors, kinases and phosphatases and for high-affinity receptors are particularly important when screening for new drug candidates.

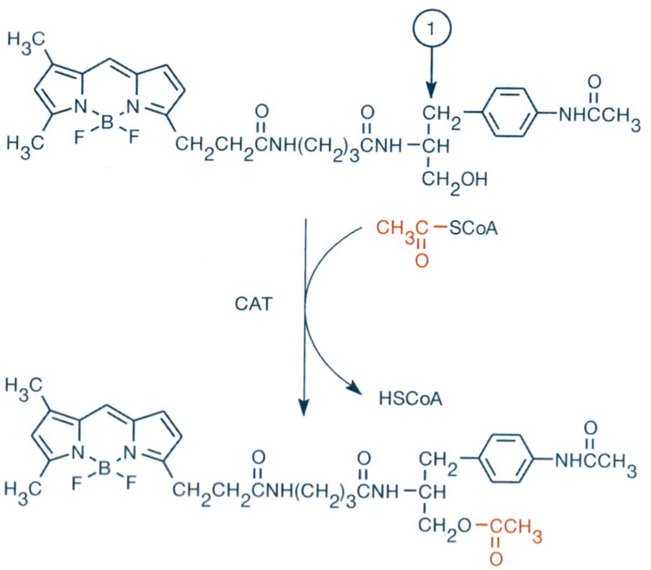

Figure 1.4.12 The green-fluorescent BODIPY® FL 1-deoxychloramphenicol substrate in our *FAST* CAT® Green (deoxy) Chloramphenicol Acetyltransferase Assay Kit (F6616). CAT-mediated acetylation of this substrate and of the BODIPY® TMR 1-deoxychloramphenicol in our *FAST* CAT® Yellow (deoxy) Chloramphenicol Acetyltransferase Assay Kit (F6617) results in single fluorescent products because these substrates contain only one hydroxyl group that can be acetylated. In contrast, the BODIPY® FL chloramphenicol substrate in our original *FAST* CAT® Kit (F2900) contains a second hydroxyl group at the 1-position (indicated by the labeled arrow). This hydroxyl group undergoes a nonenzymatic transacetylation step, restoring the original hydroxyl for a second acetylation. CAT-mediated acetylation of this chloramphenicol substrate produces three fluorescent products, thus complicating the analysis.

BODIPY® Substrates for Chloramphenicol Acetyltransferase

Chloramphenicol acetyltransferase (CAT), an enzyme that is encoded by an important reporter gene, can acetylate chloramphenicol derivatives that incorporate the BODIPY® fluorophore (Figure 1.4.12). The acetylated products are readily separated from the substrate by thin-layer chromatography (Figure 1.4.13) and quantitated by photography, fluorometry or with a plate scanner. Our original *FAST* CAT® Chloramphenicol Acetyltransferase Assay Kit[52,53] (F2900) and improved *FAST* CAT® (deoxy) Chloramphenicol Acetyltransferase Assay Kit (F6616) utilize a green-fluorescent BODIPY® FL substrate, and the *FAST* CAT® Yellow (deoxy) Chloramphenicol Acetyltransferase Assay Kit (F6617) employs a yellow-fluorescent BODIPY® TMR 1-deoxychloramphenicol substrate. These products are described in detail in Section 10.6.

Additional Methods of Analysis Using BODIPY® Dye Conjugates

In addition to their general utility for the intensity-based and fluorescence polarization–based assays described above, the BODIPY® dyes are near optimal for a variety of other bioanalytical techniques:

- The spectral variety and high absorbance of the BODIPY® dyes (Figure 1.4.1) permits their use as efficient donor or acceptor dyes for numerous assays that use fluorescence resonance energy transfer, including internally quenched endopeptidase substrates[54] (Section 10.4), nucleic acid hybridization assays and receptor-binding assays (Fluorescence Resonance Energy Transfer (FRET)—Note 1.2).
- BODIPY® dye conjugates of peptides are readily separated by chromatographic means and can be used to detect the activity of enzymes that catalyze secondary modifications, such as phosphorylation/dephosphorylation, glycosylation/deglycosylation, oxidation/reduction, myristoylation, farnesylation and peptide–peptide crosslinking.
- Hydrolysis of peptides that are singly labeled with BODIPY® dyes to smaller peptides can be detected chromatographically with extremely high sensitivity.
- With their high peak intensity and narrow emission spectra, reactive BODIPY® dyes are among the most detectable amine-derivatization reagents available for HPLC and capillary electrophoresis; thus, amine-containing metabolites can be derivatized with succinimidyl esters of the BODIPY® dyes (Table 1.7) for ultrasensitive analysis.[7,55]

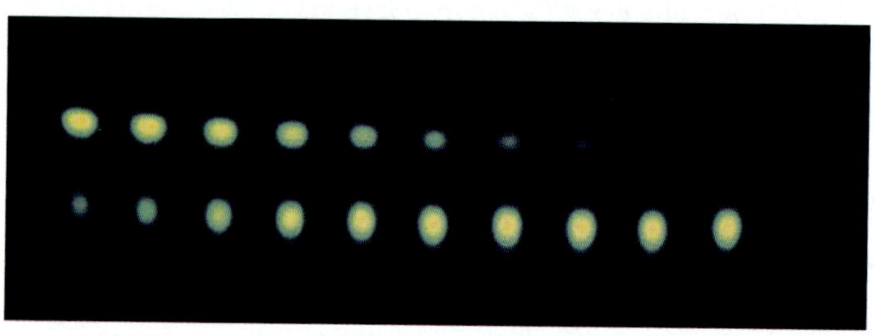

Figure 1.4.13 Chloramphenicol acetyltransferase (CAT) assays using our *FAST* CAT® Yellow (deoxy) Chloramphenicol Acetyltransferase Assay Kit (F6617). Decreasing amounts of purified CAT enzyme (2-fold dilutions) were incubated with the corresponding deoxy substrate in the presence of acetyl CoA; the reaction mixture was then separated with standard thin-layer chromatography (TLC) methods and visualized with 366 nm epi-illumination. The bottom row of fluorescent spots in each TLC represents the substrate; the top, the monoacetylated reaction product.

IMPORTANT NOTICE: The products described in this manual are covered by one or more Limited Use Label License(s). Please refer to the Appendix on page 971 and Master Product List on page 975. Products are For Research Use Only. Not intended for any animal or human therapeutic or diagnostic use.
www.invitrogen.com/probes
molecular probes® | invitrogen™ by life technologies™

REFERENCES

1. Angew Chem Int Ed Engl (2008) 47:1184; **2.** Bioorg Med Chem Lett (2006) 16:3886; **3.** J Med Chem (2007) 50:782; **4.** Pharmacol Ther (2003) 100:101; **5.** Electrophoresis (2007) 28:3901; **6.** Electrophoresis (2005) 26:2599; **7.** Anal Chem (2007) 79:345; **8.** Electrophoresis (2004) 25:839; **9.** J Biochem Biophys Methods (2000) 42:137; **10.** Anal Biochem (1999) 275:62; **11.** Jpn J Physiol (1995) 45:673; **12.** Anal Biochem (1996) 243:1; **13.** Diabetes (2002) 51 Suppl 1:S25; **14.** J Opt Soc Am B (1996) 13:481; **15.** J Am Chem Soc (1994) 116:7801; **16.** Optical Microscopy for Biology, Herman B, Jacobson K, Eds. 1990; p. 143; **17.** J Am Chem Soc (2002) 124:196; **18.** J Microsc (1992) 168:219; **19.** Chemphyschem (2008) 9:2019; **20.** Bioconjug Chem (2003) 14:195; **21.** Biophys J (2002) 83:605; **22.** Biochim Biophys Acta (1992) 1104:9; **23.** Biochim Biophys Acta (1984) 776:217; **24.** Biochemistry (1982) 21:978; **25.** J Am Chem Soc (2005) 127:12766; **26.** Biophys J (2009) 96:226; **27.** ACS Chem Biol (2008) 3:555; **28.** Nat Methods (2006) 3:923; **29.** Nat Biotechnol (2003) 21:1093; **30.** J Recept Signal Transduct Res (2002) 22:333; **31.** Proc Natl Acad Sci U S A (2004) 101:5488; **32.** Tetrahedron Lett (1999) 40:1471; **33.** Bioconjug Chem (2006) 17:1612; **34.** Anal Biochem (1986) 156:220; **35.** Bioconjug Chem (2003) 14:1133; **36.** Endocrinology (1997) 138:296; **37.** Lett Pept Sci (1995) 1:235; **38.** Electrophoresis (1992) 13:542; **39.** Proc Natl Acad Sci U S A (2005) 102:5926; **40.** Proc Natl Acad Sci U S A (2005) 102:5346; **41.** Electrophoresis (1997) 18:2893; **42.** Anal Sci (2001) 17:155; **43.** Nucleic Acids Res (2001) 29:E34; **44.** J Biol Chem (2001) 276:29275; **45.** Anal Chem (2003) 75:4297; **46.** Proc Natl Acad Sci U S A (2004) 101:2800; **47.** Proc Natl Acad Sci U S A (2003) 100:1592; **48.** Curr Biol (2000) 10:907; **49.** J Biol Chem (2000) 275:4555; **50.** Am J Pathol (2000) 156:419; **51.** J Natl Cancer Inst (2000) 92:338; **52.** Biotechniques (1990) 8:170; **53.** Anal Biochem (1991) 197:401; **54.** Nat Biotechnol (2000) 18:1071; **55.** Electrophoresis (2008) 29:4900.

DATA TABLE 1.4 BODIPY® DYE SERIES

Cat. No.	MW	Storage	Soluble	Abs	EC	Em	Solvent	Notes
B10006	542.19	F,D,L	H_2O, DMSO	502	80,000	510	MeOH	1, 2
D2183	292.09	F,L	DMSO, MeCN	505	91,000	511	MeOH	1
D2184	389.16	F,D,L	DMSO, MeCN	502	82,000	510	MeOH	1, 3
D2187	513.31	F,D,L	DMSO, MeCN	534	77,000	551	MeOH	1
D2191	417.22	F,D,L	DMSO, MeCN	500	79,000	509	MeOH	1
D2219	443.23	F,D,L	DMSO, MeCN	559	97,000	568	MeOH	1
D2222	463.25	F,D,L	DMSO, MeCN	563	142,000	569	MeOH	1
D2225	426.19	F,D,L	DMSO, MeCN	575	83,000	588	MeOH	1
D2228	489.28	F,D,L	DMSO, MeCN	581	136,000	591	MeOH	4
D3834	320.15	F,L	DMSO, MeCN	505	96,000	511	MeOH	1
D6102	502.32	F,D,L	DMSO, MeCN	504	85,000	510	MeOH	1
D6116	634.46	F,D,L	DMSO, MeCN	588	68,000	616	MeOH	1, 5
D6117	608.45	F,D,L	DMSO, MeCN	544	60,000	570	MeOH	1
D6140	491.20	F,D,L	H_2O, DMSO	502	75,000	510	MeOH	1, 6
D6180	437.21	F,D,L	DMSO, MeCN	528	70,000	547	MeOH	1
D6184	417.22	F,D,L	DMSO, MeCN	504	87,000	511	MeOH	1
D10000	660.50	F,D,L	DMSO, MeCN	625	101,000	640	MeOH	1, 7
D10001	643.45	F,D,L	DMSO, MeCN	646	102,000	660	MeOH	1

For definitions of the contents of this data table, see "Using *The Molecular Probes® Handbook*" in the introductory pages.

Notes

1. The absorption and fluorescence spectra of BODIPY® derivatives are relatively insensitive to the solvent.
2. This sulfotetrafluorophenyl (STP) ester derivative is water soluble and may be dissolved in buffer at ~pH 8 for reaction with amines. Long-term storage in water is NOT recommended due to hydrolysis.
3. The fluorescence lifetime (τ) of D2184 in MeOH at 20°C is 5.7 nanoseconds. Data provided by the SPEX Fluorescence Group, Horiba Jobin Yvon Inc.
4. Oxidation of the polyunsaturated butadienyl portion of the BODIPY® 581/591 dye results in a shift of the fluorescence emission peak from ~590 nm to ~510 nm. (Methods Enzymol (2000) 319:603, FEBS Lett (1999) 453:278)
5. The fluorescence lifetime (τ) of D6116 in MeOH at 20°C is 5.4 nanoseconds. Data provided by the SPEX Fluorescence Group, Horiba Jobin Yvon Inc.
6. This sulfonated succinimidyl ester derivative is water soluble and may be dissolved in buffer at ~pH 8 for reaction with amines. Long-term storage in water is NOT recommended due to hydrolysis.
7. The fluorescence lifetime (τ) of the BODIPY® 630/650 dye at 20°C is 3.9 nanoseconds in H_2O and 4.4 nanoseconds in EtOH. (Bioconjug Chem (2003) 14:195)

PRODUCT LIST 1.4 BODIPY® DYE SERIES

Cat. No.	Product	Quantity
B10006	BODIPY® FL, STP ester, sodium salt	5 mg
D3834	4,4-difluoro-5,7-dimethyl-4-bora-3a,4a-diaza-s-indacene-3-pentanoic acid (BODIPY® FL C_5)	1 mg
D6184	4,4-difluoro-5,7-dimethyl-4-bora-3a,4a-diaza-s-indacene-3-pentanoic acid, succinimidyl ester (BODIPY® FL C_5, SE)	5 mg
D2183	4,4-difluoro-5,7-dimethyl-4-bora-3a,4a-diaza-s-indacene-3-propionic acid (BODIPY® FL)	5 mg
D2184	4,4-difluoro-5,7-dimethyl-4-bora-3a,4a-diaza-s-indacene-3-propionic acid, succinimidyl ester (BODIPY® FL, SE)	5 mg
D6140	4,4-difluoro-5,7-dimethyl-4-bora-3a,4a-diaza-s-indacene-3-propionic acid, sulfosuccinimidyl ester, sodium salt (BODIPY® FL, SSE)	5 mg
D6102	6-((4,4-difluoro-5,7-dimethyl-4-bora-3a,4a-diaza-s-indacene-3-propionyl)amino)hexanoic acid, succinimidyl ester (BODIPY® FL-X, SE)	5 mg
D6117	6-((4,4-difluoro-1,3-dimethyl-5-(4-methoxyphenyl)-4-bora-3a,4a-diaza-s-indacene-2-propionyl)amino)hexanoic acid, succinimidyl ester (BODIPY® TMR-X, SE)	5 mg
D2187	4,4-difluoro-5,7-diphenyl-4-bora-3a,4a-diaza-s-indacene-3-propionic acid, succinimidyl ester (BODIPY® 530/550, SE)	5 mg
D6180	4,4-difluoro-5-phenyl-4-bora-3a,4a-diaza-s-indacene-3-propionic acid, succinimidyl ester (BODIPY® R6G, SE)	5 mg
D2228	4,4-difluoro-5-(4-phenyl-1,3-butadienyl)-4-bora-3a,4a-diaza-s-indacene-3-propionic acid, succinimidyl ester (BODIPY® 581/591, SE)	5 mg
D2225	4,4-difluoro-5-(2-pyrrolyl)-4-bora-3a,4a-diaza-s-indacene-3-propionic acid, succinimidyl ester (BODIPY® 576/589, SE)	5 mg
D10001	6-(((4,4-difluoro-5-(2-pyrrolyl)-4-bora-3a,4a-diaza-s-indacene-3-yl)styryloxy)acetyl)aminohexanoic acid, succinimidyl ester (BODIPY® 650/665-X, SE)	5 mg
D2222	4,4-difluoro-5-styryl-4-bora-3a,4a-diaza-s-indacene-3-propionic acid, succinimidyl ester (BODIPY® 564/570, SE)	5 mg
D2191	4,4-difluoro-1,3,5,7-tetramethyl-4-bora-3a,4a-diaza-s-indacene-8-propionic acid, succinimidyl ester (BODIPY® 493/503, SE)	5 mg
D2219	4,4-difluoro-5-(2-thienyl)-4-bora-3a,4a-diaza-s-indacene-3-propionic acid, succinimidyl ester (BODIPY® 558/568, SE)	5 mg
D6116	6-(((4-(4,4-difluoro-5-(2-thienyl)-4-bora-3a,4a-diaza-s-indacene-3-yl)phenoxy)acetyl)amino)hexanoic acid, succinimidyl ester (BODIPY® TR-X, SE)	5 mg
D10000	6-(((4,4-difluoro-5-(2-thienyl)-4-bora-3a,4a-diaza-s-indacene-3-yl)styryloxy)acetyl)aminohexanoic acid, succinimidyl ester (BODIPY® 630/650-X, SE)	5 mg

IMPORTANT NOTICE: The products described in this manual are covered by one or more Limited Use Label License(s). Please refer to the Appendix on page 971 and Master Product List on page 975. Products are For Research Use Only. Not intended for any animal or human therapeutic or diagnostic use.

1.5 Fluorescein, Oregon Green® and Rhodamine Green™ Dyes

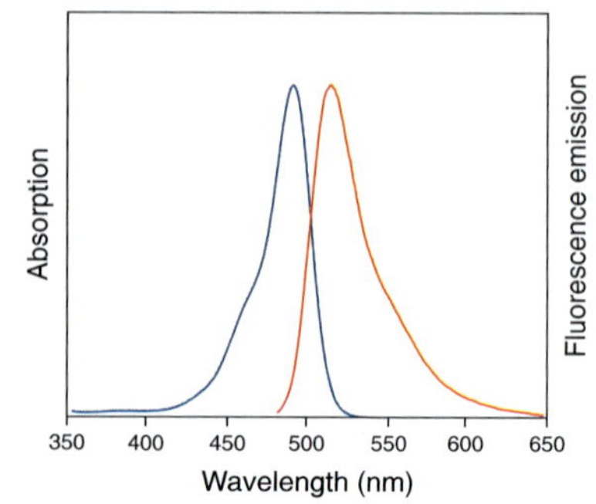

Figure 1.5.1 Absorption and fluorescence emission spectra of fluorescein in pH 9.0 buffer.

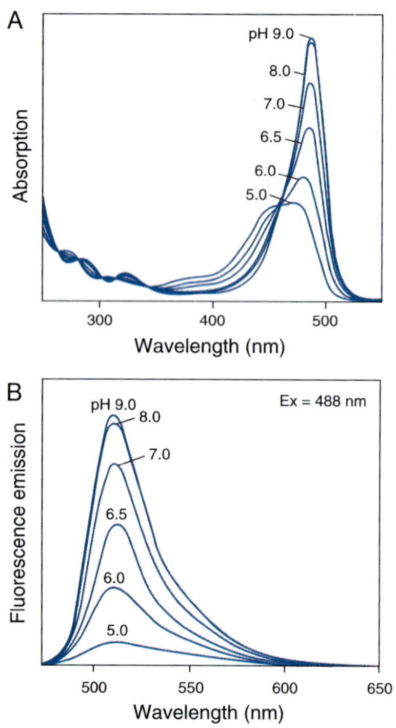

Figure 1.5.2 The pH-dependent spectra of fluorescein (F1300): **A)** absorption spectra, **B)** emission spectra.

Fluorescein

The amine-reactive fluorescein derivatives (Table 1.8) have been the most common fluorescent derivatization reagents for covalently labeling proteins. In addition to its relatively high absorptivity, excellent fluorescence quantum yield and good water solubility, fluorescein (F1300, Figure 1.5.1) has an excitation maximum (494 nm) that closely matches the 488 nm spectral line of the argon-ion laser, making it an important fluorophore for confocal laser-scanning microscopy[1] and flow cytometry applications. In addition, fluorescein's protein conjugates are not inordinately susceptible to precipitation. Because it can be prepared in high purity, fluorescein is one of the five dyes in the Reference Dye Sampler Kit[2] (R14782, Section 23.1). We are also the source of the NIST-traceable fluorescein standard (F36915) described below.

NIST-Traceable Fluorescein Standard

The National Institute of Standards and Technology (NIST) chose a high-grade fluorescein synthesized in our laboratories to create Standard Reference Material 1932 (SRM® 1932), a certified fluorescein solution. We now offer a NIST-traceable fluorescein standard (F36915) that not only meets the stringent criteria established by NIST, but is also directly traceable to SRM® 1932. We supply our NIST-traceable fluorescein standard as a calibrated 50 µM solution of fluorescein in 100 mM sodium borate buffer, pH 9.5; under these conditions, fluorescein is completely ionized[3] and is therefore in its most fluorescent form (Figure 1.5.2), exhibiting an extremely high quantum yield of 0.93 (Section 20.2).

Table 1.8 Amine-reactive xanthene derivatives in this section.

Fluorophore (Abs/Em) *	COOH	Succinimidyl Ester	Other	Protein and Nucleic Acid Labeling Kits	Notes
Eosin (524/544)			E18 (ITC)[5]		• Useful for DAB photoconversion • Phosphorescent
Fluorescein (494/518)	C1359[5] C1360[6] C1904[M]	C2210[5] C6164[6] C1311[M] F6106 (X)[6] F2181 (X)[M] F6129 (X)[M] F6130 (EX)[5] C20050 (C)[5]	F143 (ITC)[5] F1906 (ITC)[5] F1907 (ITC)[5] D16 (DTA)[5]	F6433 (F) F6434 (F) F10240 (P) Z25042 (Z) Z25342 (Z)	• Most widely used green-fluorescent labeling dye • Absorption overlaps the 488 nm spectral line of the argon-ion laser • Prone to photobleaching • pH-sensitive spectra between pH 5 and pH 8 • Common donor in FRET applications that utilize tetramethylrhodamine as the acceptor
JOE (520/548)		C6171MP[6]			• Traditional fluorophore used in automated DNA sequencing • pH-insensitive spectra at pH >6
Oregon Green® 488 (496/524)	O6146[5]	O6147[5] O6149[6] O6185 (X)[6]	O6080[M]	F6153 (F) O10241 (P) A10476 (APEX®) U21659 (U)	• Photostable fluorescein substitute • pH-insensitive spectra at pH >6 • Recognized by anti-fluorescein antibodies
Oregon Green® 514 (511/530)	O6138[M]	O6139[6]			• Photostable fluorescein substitute • pH-insensitive spectra at pH >6
Rhodamine Green™ (502/527)		R6107[M] R6113 (X)[M]			• Photostable fluorescein substitute • pH-insensitive spectra

* The numbers in parentheses reflect the absorption (Abs) and fluorescence emission (Em) maxima, in nm, of the goat anti-mouse IgG antibody or dextran conjugates in aqueous buffer. (APEX®) = APEX® Antibody Labeling Kit (Section 1.2). (C) = Caged; the probe is nonfluorescent until the caging group is removed by UV illumination. (COOH) = Carboxylic acid. (DTA) = Dichlorotriazine. (EX) = A seven-atom spacer that is more hydrophilic than X. (F) = FluoReporter® Protein Labeling Kit (Section 1.2). (ITC) = Isothiocyanate. (5) = 5-Isomer. (6) = 6-Isomer. (M) = Mixture of 5- and 6-isomers. (P) = Easy-to-Use Protein Labeling Kit (Section 1.2). (U) = ULYSIS® Nucleic Acid Labeling Kit (Section 8.2). (X) = Aminohexanoyl spacer separating the dye and SE. (Z) = Zenon® Antibody Labeling Kit (Section 7.3).

IMPORTANT NOTICE: The products described in this manual are covered by one or more Limited Use Label License(s). Please refer to the Appendix on page 971 and Master Product List on page 975. Products are For Research Use Only. Not intended for any animal or human therapeutic or diagnostic use.
www.invitrogen.com/probes
molecular probes® invitrogen™ by life technologies™

Academic researchers and industry scientists alike can use our NIST-traceable fluorescein standard to assess day-to-day or experiment-to-experiment variation in fluorescence-based instrumentation, as well as to determine the Molecules of Equivalent Soluble Fluorophore (MESF) value for an experimental solution. The MESF value is defined not as the actual number of dye molecules present, but rather as the number of fluorophores that would yield a fluorescence intensity equivalent to that of the experimental solution when analyzed on the same instrument under the same conditions.[4–6] Consequently, the MESF value is an important tool for characterizing the fluorescence intensity of a solution containing spectrally similar dye molecules attached to antibodies, nucleic acids, microspheres or other substrates that might enhance or diminish the fluorescence. When its pH is carefully matched with that of the experimental solution, our NIST-traceable fluorescein standard can be used for accurate MESF determinations of a wide range of green-fluorescent dye solutions and on an assortment of fluorescence-based instruments.

Limitations of Fluorescein

Even though fluorescein has been used to derivatize biomolecules for decades, fluorescein-based dyes and their conjugates have several significant drawbacks, including:

- A relatively high rate of photobleaching[7–9] (Figure 1.5.3, Figure 1.5.4)
- pH-sensitive fluorescence[10,11] (pK_a ~6.4) that is significantly reduced below pH 7 (Figure 1.5.5)
- A relatively broad fluorescence emission spectrum, limiting their utility in some multicolor applications
- A tendency toward quenching of their fluorescence on conjugation to biopolymers, particularly at high degrees of labeling (Figure 1.5.6)

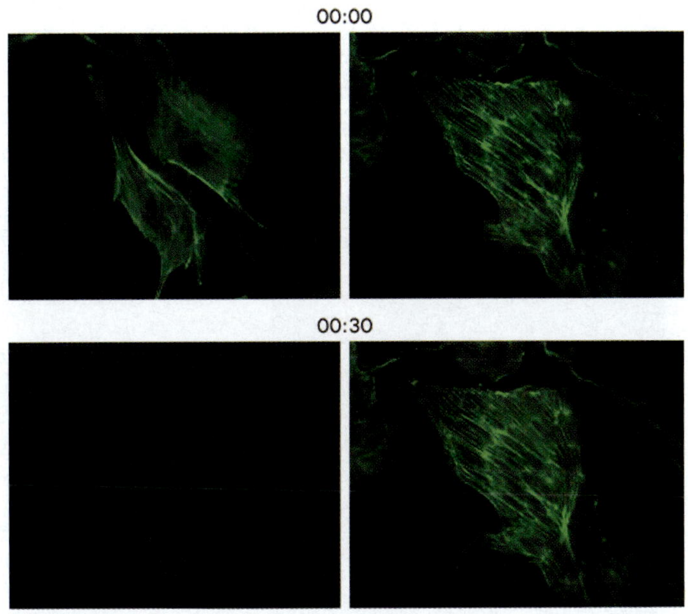

Figure 1.5.4 Bovine pulmonary artery endothelial cells (BPAEC) were labeled with fluorescein phalloidin (left panels, F432), or Alexa Fluor® 488 phalloidin (right panels, A12379), which labels filamentous actin, and mounted in PBS. The cells were placed under constant illumination on the microscope with an FITC filter set using a 60× objective. Images were acquired at 1-second intervals for 30 seconds. Under these illumination conditions, fluorescein photobleached to about 20% of its initial value in 30 seconds; the fluorescence of Alexa Fluor® 488 phalloidin stayed at the initial value under the same illumination conditions.

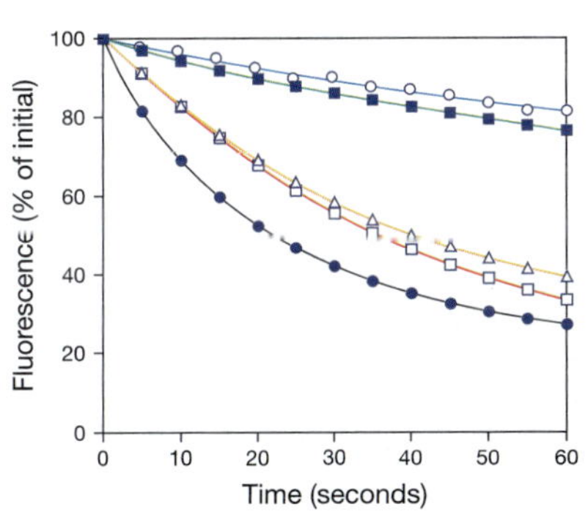

Figure 1.5.3 Comparison of photostability of green-fluorescent antibody conjugates. The following fluorescent goat anti-mouse IgG antibody conjugates were used to detect mouse anti-human IgG antibody labeling of human anti-nuclear antibodies in HEp-2 cells on prefixed test slides (INOVA Diagnostics Corp.): Oregon Green® 514 (O6383, ■), Alexa Fluor® 488 (A11001, ○), BODIPY® FL (B2752, △), Oregon Green® 488 (O6380, □) or fluorescein (F2761, ●). Samples were continuously illuminated and viewed on a fluorescence microscope using a fluorescein longpass filter set. Images were acquired every 5 seconds. For each conjugate, three data sets, representing different fields of view, were averaged and then normalized to the same initial fluorescence intensity value to facilitate comparison.

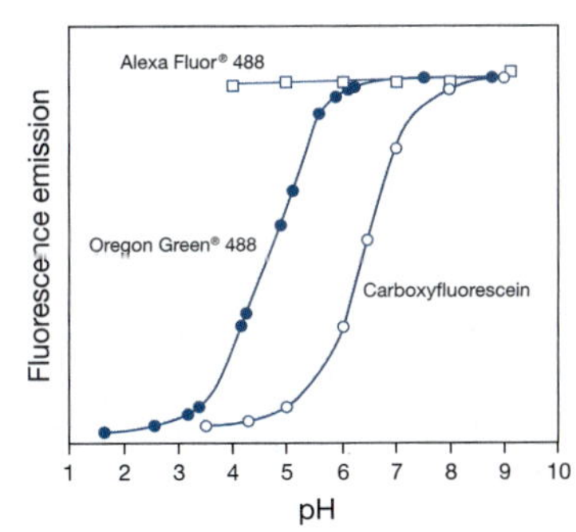

Figure 1.5.5 Comparison of pH-dependent fluorescence of the Oregon Green® 488 (●), carboxyfluorescein (○) and Alexa Fluor® 488 (□) fluorophores. Fluorescence intensities were measured for equal concentrations of the three dyes using excitation/emission at 490/520 nm.

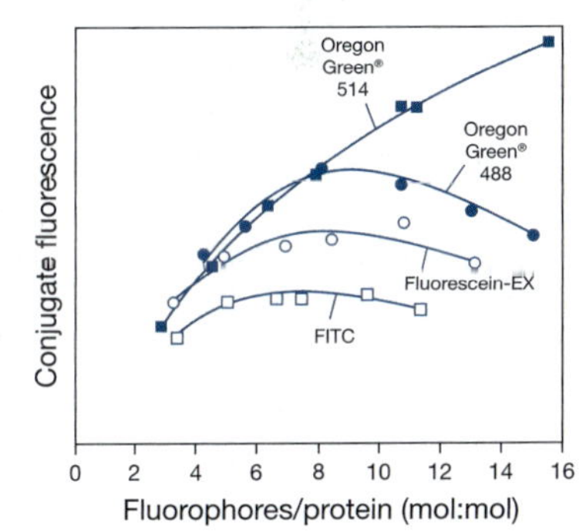

Figure 1.5.6 Comparison of relative fluorescence as a function of the number of fluorophores attached per protein for goat anti-mouse IgG antibody conjugates prepared using Oregon Green® 514 carboxylic acid succinimidyl ester (O6139, ■), Oregon Green® 488 carboxylic acid succinimidyl ester (O6147, ●), fluorescein-5-EX succinimidyl ester (F6130, ○) and fluorescein isothiocyanate (FITC, F143, F1906, F1907, □). Conjugate fluorescence is determined by measuring the fluorescence quantum yield of the conjugated dye relative to that of the free dye and multiplying by the number of fluorophores per protein.

IMPORTANT NOTICE: The products described in this manual are covered by one or more Limited Use Label License(s). Please refer to the Appendix on page 971 and Master Product List on page 975. Products are For Research Use Only. Not intended for any animal or human therapeutic or diagnostic use.

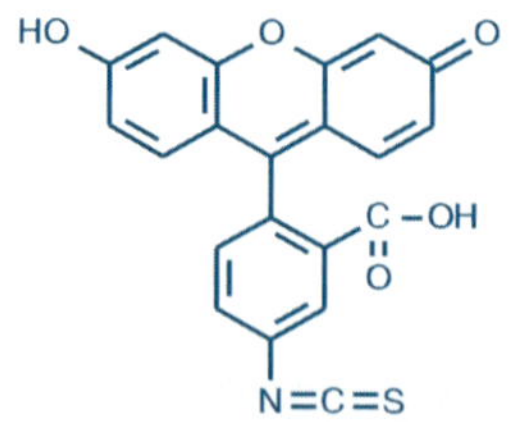

Figure 1.5.7 Fluorescein-5-isothiocyanate (FITC 'Isomer I', F143).

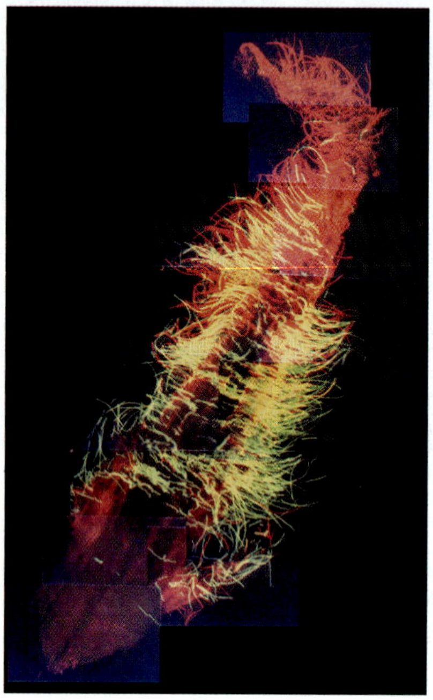

Figure 1.5.8 Two proteobacterial symbionts localized with phylotype-specific 16S rRNA–directed oligonucleotide probes labeled with either fluorescein-5-isothiocyanate (F143, F1906, F1907) or Texas Red® sulfonyl chloride (T353, T1905). The filamentous bacteria are attached to a hair-like structure secreted from a pore on the dorsal surface of the deep-sea hydrothermal vent polychaete *Alvinella pompejana*. Image contributed by M. Cottrell and C. Cary, College of Marine Studies, University of Delaware.

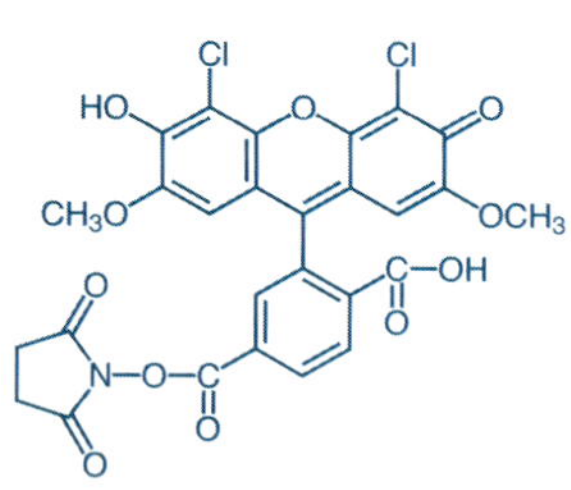

Figure 1.5.9 6-carboxy-4',5'-dichloro-2',7'-dimethoxyfluorescein, succinimidyl ester (6-JOE, SE; C6171MP).

The photobleaching and pH sensitivity of fluorescein make quantitative measurements with this fluorophore problematic. Furthermore, fluorescein's relatively high photobleaching rate limits the sensitivity that can be obtained, a significant disadvantage for applications requiring ultrasensitive detection, such as DNA sequencing, fluorescence *in situ* hybridization and localization of low-abundance receptors. These limitations have encouraged the development of alternative fluorophores.

There are no new dyes available that completely solve fluorescein's photobleaching problems, but we have developed some excellent dyes whose spectra mimic those of fluorescein—the Alexa Fluor® 488 (Section 1.3), BODIPY® FL (Section 1.4), Oregon Green® 488, Oregon Green® 514 and Rhodamine Green™ dyes (this section). These dyes are much more photostable than fluorescein and have less or no pH sensitivity in the physiological pH range. When compared with fluorescein, all of these dyes exhibit the same or slightly longer-wavelength spectra (absorption maxima ~490–515 nm) and comparably high fluorescence quantum yields. Alternatively, where they can be used, our yellow-green fluorescent FluoSpheres® microspheres and our Qdot® nanocrystals (Section 6.5, Section 6.6, respectively) provide a means of preparing bioconjugates that have a combination of fluorescence intensity and photostability far superior to that of any simple dye conjugate.

Single-Isomer Fluorescein Isothiocyanate (FITC) Preparations

Despite the availability of alternative amine-reactive fluorescein derivatives that yield conjugates with superior stability and comparable spectra, fluorescein isothiocyanate (FITC) remains one of the most popular fluorescent labeling reagents.[12] The synthesis of fluorescein isothiocyanate, carboxyfluorescein (FAM) and similar fluorescein-derived reagents yields a mixture of isomers at the 5- and 6-positions of fluorescein's carboxyphenyl ("bottom") ring (Figure 1.5.7). Spectra of the two isomers are almost indistinguishable in both wavelength and intensity. The isomers, however, may differ in the geometry of their binding to proteins, and the conjugates may elute under different chromatographic conditions or migrate differently in an electrophoretic gel. Thus, certain applications may require the single-isomer preparations. Many fluorescein (and rhodamine) probes are available either as a mixture of isomers or as purified single isomers.

The 5-isomer or "isomer I" of FITC (F143, Figure 1.5.7, Figure 1.5.8) is the most widely used FITC isomer, probably because it is easier to isolate in pure form. Because isothiocyanates may deteriorate during storage, we recommend purchasing the 5-isomer of FITC specially packaged in individual vials (F1906, F1907). FITC is readily soluble in aqueous solutions that have a pH above 6. FITC is also available in our FluoReporter® FITC Protein Labeling Kit (F6434, Table 1.2), which is described in Section 1.2.

Mixed-Isomer and Single-Isomer Preparations of FAM and JOE Succinimidyl Esters

Although many other companies still prepare their fluorescein bioconjugates with FITC, we prefer to use amine-reactive succinimidyl esters of carboxyfluorescein (commonly called FAM), which yield carboxamides that are more resistant to hydrolysis. We offer both mixed-isomer and single-isomer preparations of FAM (C1904, C1359, C1360) and FAM succinimidyl esters (C1311, C2210, C6164). A study comparing the relative conjugation rate of several reactive fluorescein derivatives with a protein or L-lysine and the stability of the resulting conjugates concluded that the succinimidyl ester of carboxyfluorescein showed superior performance, followed by fluorescein dichlorotriazine (DTAF). FITC was both the slowest to react and yielded the least stable conjugates;[13] however, the degree of labeling was most easily controlled with FITC.[13] The succinimidyl ester of 5-FAM (C2210) is reported to react much faster than FITC when used to derivatize small biomolecules prior to separation by capillary electrophoresis.[14] We also offer a single-isomer preparation of the succinimidyl ester of 6-carboxy-4′,5′-dichloro-2′,7′-dimethoxyfluorescein (6-JOE, SE; C6171MP; Figure 1.5.9). 6-JOE is one of the four fluorophores (i.e., 5-FAM, 6-JOE, 6-TAMRA and 6-ROX) used in first-generation electrophoretic DNA sequencing instruments.

Succinimidyl Esters of Fluorescein with Spacer Groups

We also prepare succinimidyl esters of fluorescein that contain aliphatic spacers between the fluorophore and the reactive group. These include mixed-isomer (F2181, F6129) and single-isomer

IMPORTANT NOTICE: The products described in this manual are covered by one or more Limited Use Label License(s). Please refer to the Appendix on page 971 and Master Product List on page 975. Products are For Research Use Only. Not intended for any animal or human therapeutic or diagnostic use.
www.invitrogen.com/probes

(F6106) preparations of fluorescein-X succinimidyl ester (SFX), which contains a seven-atom aminohexanoyl spacer ("X") between the FAM fluorophore and the succinimidyl ester (Figure 1.5.10). In addition, we offer fluorescein-5-EX succinimidyl ester (F6130), which contains a seven-atom spacer that is somewhat more hydrophilic than is the spacer in SFX (Figure 1.5.11). These spacers separate the fluorophore from the biomolecule to which it is conjugated, potentially reducing the quenching that typically occurs upon conjugation. We have determined that conjugates of some proteins prepared with fluorescein-5-EX succinimidyl ester are up to twice as fluorescent as the corresponding conjugates labeled with FITC at the same degree of labeling (Figure 1.5.6). Consequently, we now recommend this fluorescein derivative as the preferred dye for preparing most fluoresceinated proteins. Fluorescein-5-EX succinimidyl ester is also available in our convenient FluoReporter® Fluorescein-EX Protein Labeling Kit (F6433) and Fluorescein-EX Protein Labeling Kit (F10240). See Section 1.2 and Table 1.3 for more details about these labeling kits.

The spacers in our SFX and fluorescein-5-EX succinimidyl esters may also make the fluorophore more accessible to secondary detection reagents.[15–17] For example, the spacers should make the fluorescein moiety more available for quenching by our polyclonal and monoclonal anti–fluorescein/Oregon Green® antibodies, a technique used to determine the accessibility of the fluorophore in proteins, membranes and cells.[16,18] Fluorescein is frequently used as a hapten on a primary detection reagent that can be either amplified or converted into a longer-wavelength or electron-dense signal with the appropriate secondary detection reagent. Section 7.4 describes our extensive selection of antibodies to fluorescein and other dyes.

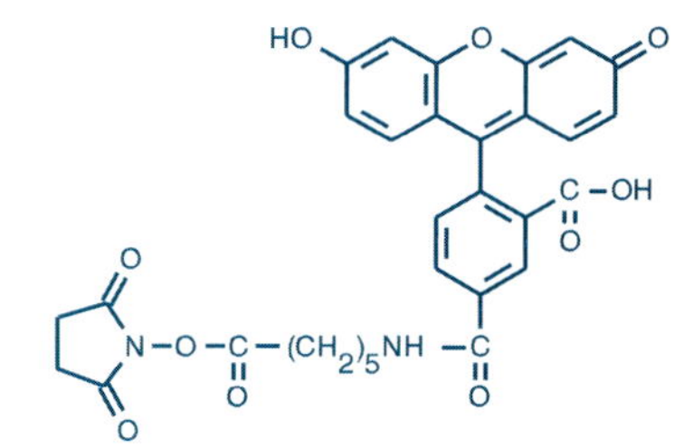

Figure 1.5.10 6-(fluorescein-5-carboxamido)hexanoic acid, succinimidyl ester (5-SFX, F6106).

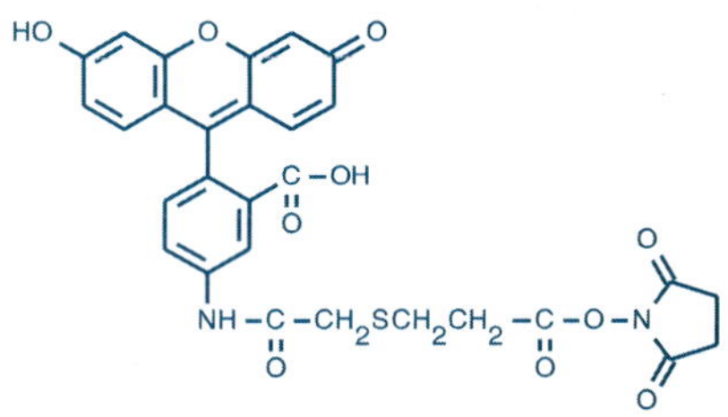

Figure 1.5.11 Fluorescein-5-EX, succinimidyl ester (F6130).

Fluorescein Dichlorotriazine (DTAF)

The 5-isomer of fluorescein dichlorotriazine (5-DTAF, D16) is highly reactive with proteins[19,20] and is commonly used to prepare biologically active fluorescein tubulin.[21] Unlike other reactive fluoresceins, 5-DTAF also reacts directly with polysaccharides and other alcohols in aqueous solution at pH above 9, but cannot be used to modify alcohols in the presence of better nucleophiles such as amines or thiols.[22] Polysaccharides that have been modified by DTAF (or other fluorescein derivatives) are readily radioiodinated.[23]

Caged Fluorescein Succinimidyl Ester

Caged probes are those that can liberate an active species upon illumination with ultraviolet light (Section 5.3). Caged fluorescent dyes can be utilized as polar tracers whose fluorescence can be spatially and temporally activated by illumination. Conjugation of the succinimidyl ester of our water-soluble, caged carboxyfluorescein β-alanine carboxamide (C20050, Figure 1.5.12) to a biomolecule of interest produces an essentially nonfluorescent probe that yields a green-fluorescent fluorescein-labeled product only after ultraviolet photoactivation. Furthermore, caged fluorescein probes are immunochemically cryptic (i.e., the probe is immunoreactive with anti–fluorescein/Oregon Green® antibodies after but not before photoactivation). Caged fluorescein succinimdyl ester is also a key starting material for preparing probes for super-resolution photoactivation microscopy.[24,25]

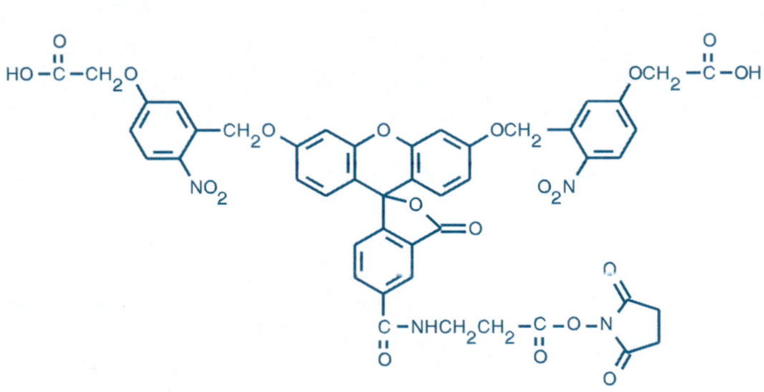

Figure 1.5.12 5-carboxyfluorescein-bis-(5-carboxymethoxy-2-nitrobenzyl) ether, β-alanine-carboxamide, succinimidyl ester (CMNB-caged carboxyfluorescein, SE; C20050).

IMPORTANT NOTICE: The products described in this manual are covered by one or more Limited Use Label License(s). Please refer to the Appendix on page 971 and Master Product List on page 975. Products are For Research Use Only. Not intended for any animal or human therapeutic or diagnostic use.

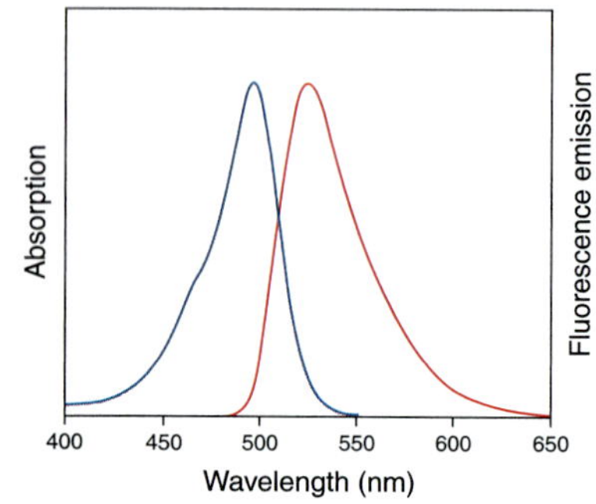

Figure 1.5.13 Absorption and fluorescence emission spectra of Oregon Green® 488 goat anti–mouse IgG antibody in pH 8.0 buffer.

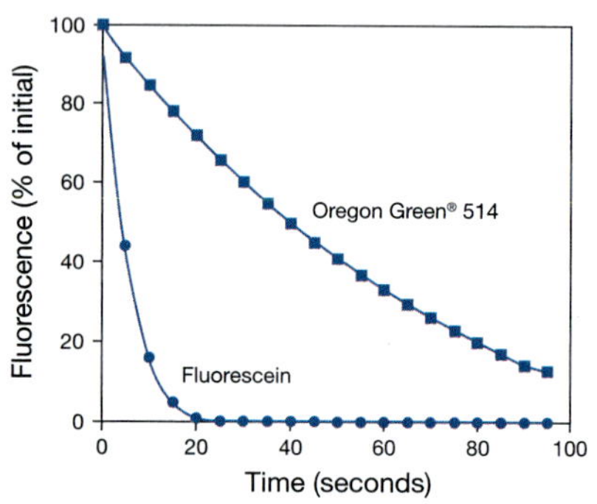

Figure 1.5.14 Photostability comparison for Oregon Green® 514 phalloidin (O7465) and fluorescein phalloidin (F432). CRE BAG 2 fibroblasts were fixed with formaldehyde, permeabilized with acetone and then stained with the fluorescent phallotoxins. Samples were continuously illuminated and images were acquired every 5 seconds using a Star 1 CCD camera (Photometrics®); the average fluorescence intensity in the field of view was calculated with Image-1 software (Universal Imaging Corp.) and expressed as a fraction of the initial intensity. Three data sets, representing different fields of view, were averaged for each labeled phalloidin to obtain the plotted time courses.

Oregon Green® 488 and Oregon Green® 514 Dyes

Oregon Green® 488 and Oregon Green® 514 dyes are fluorinated analogs of fluoresceins.[26] The absorption and emission spectra of Oregon Green® 488 dye (2′,7′-difluorofluorescein; D6145; Figure 1.5.13) is designed to perfectly match those of fluorescein. With additional fluorination of the carboxyphenyl ("bottom") ring of fluorescein, Oregon Green® 514 dye exhibits a moderate shift in its absorption and fluorescence spectra of about 15 nm relative to those of fluorescein or Oregon Green® 488 dye. Because of the near match of their absorption maxima on proteins (~498 nm and ~512 nm) to the strong 488 nm and 514 nm spectral lines of the argon-ion laser, the Oregon Green® 488 and Oregon Green® 514 fluorophores are important dyes for both confocal laser-scanning microscopy and flow cytometry applications. Furthermore, spectral imaging systems with linear-unmixing analysis software[27–29] permit the resolution of Oregon Green® 514 dye from other green-fluorescent dyes.

Bioconjugates prepared from Oregon Green® 488 and Oregon Green® 514 dyes share several advantages over those of other fluorescein dyes. These include:

- Fluorescence of protein conjugates prepared from Oregon Green® 488 and Oregon Green® 514 dyes is not appreciably quenched, even at relatively high degrees of labeling (Figure 1.5.6).
- Conjugates of Oregon Green® 488 and Oregon Green® 514 fluorophores are more photostable than those of fluorescein (Figure 1.5.3, Figure 1.5.14), allowing increased acquisition of photons before photodestruction of the dye and making Oregon Green® dyes particularly useful substitutes for fluoresceins for fluorescence imaging applications (Figure 1.5.15).
- Oregon Green® dyes have a lower pK_a (pK_a = 4.7 versus 6.4 for fluorescein) (Figure 1.5.5), making their fluorescence essentially pH insensitive in the physiological pH range. However, the pH sensitivity of Oregon Green® dyes in the weakly acidic range (pH 4 to 6) also makes these dyes useful as pH indicators for acidic organelles of live cells[30,31] (Section 20.3).
- Oregon Green® dyes are excellent haptens for anti–fluorescein/Oregon Green® antibodies (Section 7.4, Table 4.2), making Oregon Green® bioconjugates useful in a variety of signal amplification schemes.

Both Oregon Green® 488 and Oregon Green® 514 dyes have also proven useful as fluorescence anisotropy probes for measuring protein–protein and protein–nucleic acid interactions.[32]

Reactive Oregon Green® Dyes

We have prepared a variety of amine-reactive derivatives that enable researchers to take advantage of the spectral properties of Oregon Green® 488 and Oregon Green® 514 dyes

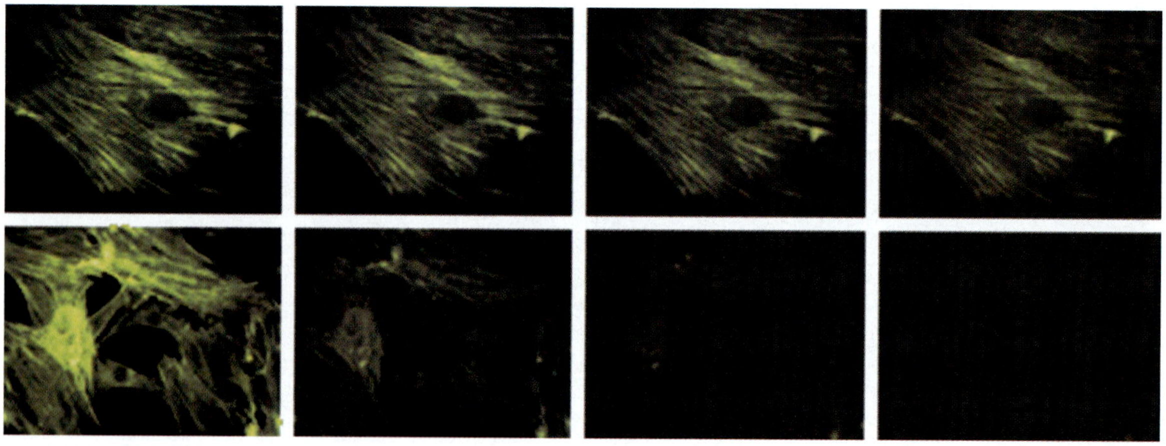

Figure 1.5.15 Photostability comparison of Oregon Green® 514 phalloidin (O7465, upper series) and fluorescein phalloidin (F432, lower series). CRE BAG 2 fibroblasts were fixed with formaldehyde, then permeabilized with acetone and stained with the fluorescent phallotoxin. Samples were illuminated continuously and viewed on a fluorescence microscope equipped with a fluorescein longpass optical filter set. Images acquired at 1, 10, 20 and 30 seconds after the start of illumination (left to right) demonstrate the superior photostability of the Oregon Green® 514 fluorophore.

(Table 1.8). These include the FITC analog, Oregon Green® 488 isothiocyanate (F_2FITC, O6080), and the single-isomer succinimidyl esters of Oregon Green® 488 carboxylic acid (O6147, O6149) and Oregon Green® 514 carboxylic acid (O6139). In addition, we offer the 5-isomer of Oregon Green® 488 carboxylic acid (O6146, Figure 1.5.16) and the mixed-isomer preparation of Oregon Green® 514 carboxylic acid (O6138, Figure 1.5.17). The 6-isomer of Oregon Green® 488-X succinimidyl ester (O6185, Figure 1.5.18) contains a seven-atom aminohexanoyl spacer ("X") between the fluorophore and the succinimidyl ester group. This spacer helps to separate the fluorophore from its point of attachment, reducing the interaction of the fluorophore with the biomolecule to which it is conjugated[33] and making it more accessible to secondary detection reagents.

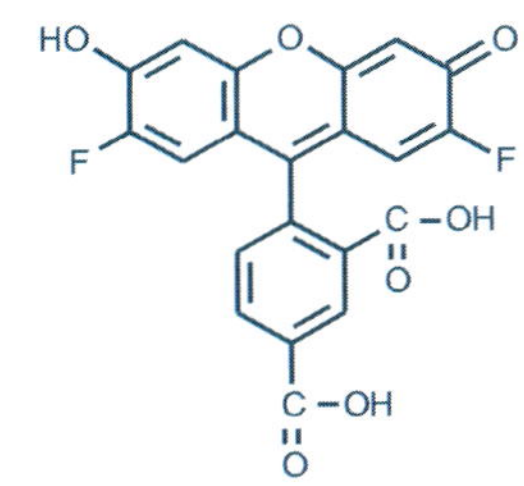

Figure 1.5.16 Oregon Green® 488 carboxylic acid (O6146).

Oregon Green® Protein and Nucleic Acid Labeling Kits

When directly compared with their fluorescein analogs, Oregon Green® 488 and Oregon Green® 514 conjugates typically have higher fluorescence yields and greater resistance to photobleaching. We have used succinimidyl esters of the Oregon Green® 488 and Oregon Green® 514 carboxylic acids to prepare conjugates of antibodies (Section 7.2, Table 7.1), streptavidin (Section 7.6, Table 7.9) and a variety of other proteins and ligands.

To facilitate direct labeling of biomolecules with Oregon Green® dyes, we offer several kits that are easy to use and produce reliable conjugations in minimal time. Our Oregon Green® protein and nucleic acid labeling kits, which are described in detail in the indicated sections, include:

- APEX® Oregon Green® 488 Antibody Labeling Kit (A10476, Section 1.2)
- FluoReporter® Oregon Green® 488 Protein Labeling Kit (F6153, Section 1.2)
- Oregon Green® 488 Protein Labeling Kit (O10241, Section 1.2)
- ULYSIS® Oregon Green® 488 Nucleic Acid Labeling Kit (U21659, Section 8.2)

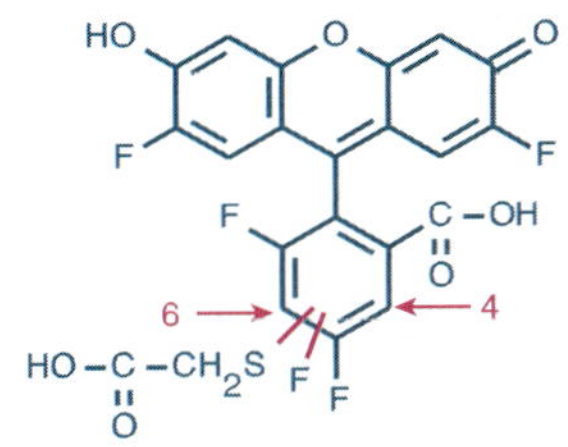

Figure 1.5.17 Oregon Green® 514 carboxylic acid (O6138).

Rhodamine Green™ and Rhodamine Green™-X Dyes

Carboxyrhodamine 110, which we have named Rhodamine Green™ dye, is the nonsulfonated analog of Alexa Fluor® 488 dye. Rhodamine Green™ dye offers a combination of desirable properties, including good photostability, a high extinction coefficient (>75,000 $cm^{-1}M^{-1}$) and a high fluorescence quantum yield, particularly in its nucleotide and nucleic acid conjugates. The Rhodamine Green™ fluorophore is even more photostable than the Oregon Green® 488 dye and about equivalent in photostability to the Oregon Green® 514 dye (Figure 1.5.3). Moreover, the fluorescence of its conjugates is completely insensitive to pH between 4 and 9.

Reactive versions of the Rhodamine Green™ dye (Table 1.8) were originally developed in our laboratories for use in DNA sequencing and other applications. Rhodamine Green™ conjugates can be prepared using the amine-reactive succinimidyl ester of Rhodamine Green™ dye (5(6)-CR 110, SE; R6107) or the succinimidyl ester of the Rhodamine Green™-X dye (R6113), which has an additional seven-atom aminohexanoyl spacer ("X") to reduce interaction of the fluorophore and its reaction site. The absorption and fluorescence emission maxima of Rhodamine Green™ conjugates are red-shifted about 7 nm compared with those of fluorescein; however, they remain compatible with standard fluorescein optical filter sets.

Although the Rhodamine Green™ dye is one of the most photostable of the fluorescein substitutes, its fluorescence when conjugated to proteins is often substantially quenched, and these conjugates also tend to precipitate from solution. Therefore, we do not recommend any of the Rhodamine Green™ succinimidyl esters for preparing protein conjugates. However, when conjugated to dextrans,[34] nucleotides and oligonucleotides,[35] the Rhodamine Green™ fluorophore remains highly fluorescent, and we currently offer two Rhodamine Green™ dextrans (D7153, D7163; Section 14.5, Table 14.4). In addition, Rhodamine Green™ dye–labeled probes have been frequently used for fluorescence correlation spectroscopy[36–38] (Fluorescence Correlation Spectroscopy (FCS)—Note 1.3).

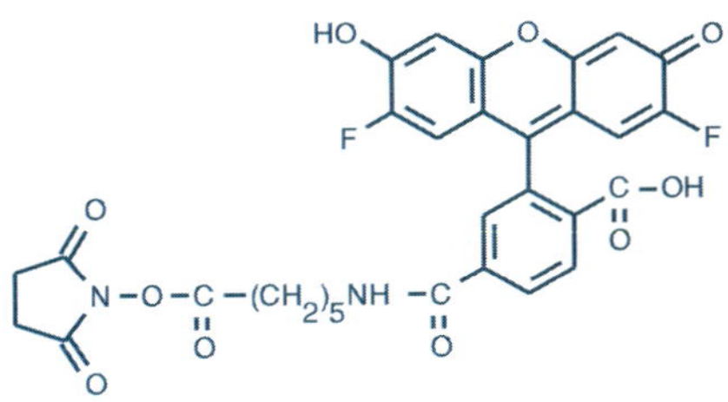

Figure 1.5.18 Oregon Green® 488-X, succinimidyl ester (O6185).

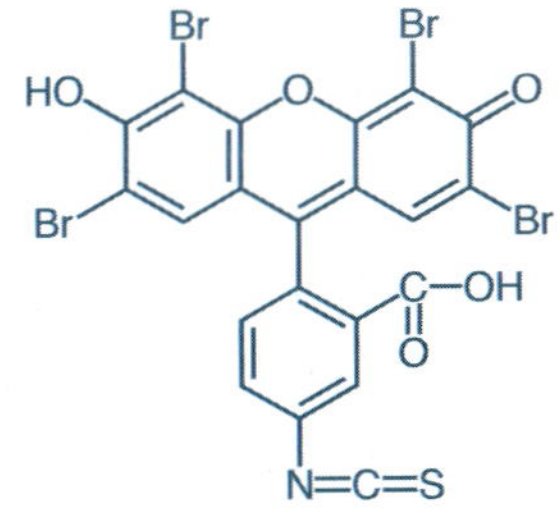

Figure 1.5.19 Eosin-5-isothiocyanate (E18).

Eosin

Eosin (2′,4′,5′,7′-tetrabromofluorescein, Figure 1.5.19) is usually not chosen for its fluorescence properties—the fluorescence quantum yield is typically only about 10–20% that of fluorescein—but rather for its ability to act as phosphorescent probe or as a fluorescence quencher.[39] With its high quantum yield (~0.57) for singlet oxygen generation, eosin and its conjugates can be used as effective photosensitizers of diaminobenzidine (DAB) oxidation in high-resolution electron microscopy studies (Fluorescent Probes for Photoconversion of Diaminobenzidine Reagents—Note 14.2). Like its thiol-reactive maleimide counterpart (E118, Section 2.2), eosin isothiocyanate (E18) is particularly useful as a phosphorescent probe for measuring the rotational properties of proteins, virus particles and other biomolecules in solution and in membranes.[40] In addition, eosin conjugates are employed for fluorescence resonance energy transfer (FRET) studies (Fluorescence Resonance Energy Transfer (FRET)—Note 1.2) and for fluorescence recovery after photobleaching (FRAP) measurements of lateral diffusion.

REFERENCES

1. Three-Dimensional Confocal Microscopy, Stevens JK, Mills LR, Trogadis JE, Eds. (1994); p. 101; **2.** J Fluoresc (2004) 14:465; **3.** J Fluorescence (1996) 6:147; **4.** Cytometry (2009) 75:874; **5.** J Res Natl Inst Stand Technol (2001) 106:381; **6.** J Res Natl Inst Stand Technol (2002) 107:83; **7.** Biophys J (2006) 90:2548; **8.** Cytometry (1997) 27:213; **9.** Biophys J (1995) 68:2588; **10.** Nat Mater (2005) 4:768; **11.** Spectrochim Acta A (1995) 51:7; **12.** Bioconjug Chem (2006) 17:1426; **13.** Bioconjug Chem (1995) 6:447; **14.** J Chromatogr A (1998) 809:203; **15.** Biochim Biophys Acta (1992) 1104:9; **16.** Biochim Biophys Acta (1984) 776:217; **17.** Biochemistry (1982) 21:978; **18.** Biochemistry (1991) 30:1692; **19.** J Immunol Methods (1977) 17:361; **20.** J Immunol Methods (1976) 13:305; **21.** Methods Enzymol (1986) 134:519; **22.** Carbohydr Res (1975) 44:251; **23.** J Biomed Mater Res (1998) 40:275; **24.** Methods Mol Biol (2009) 544:483; **25.** Angew Chem Int Ed Engl (2009) 48:6903; **26.** J Org Chem (1997) 62:6469; **27.** Clin Cancer Res (2007) 13:2936; **28.** Cytometry A (2007) 71:174; **29.** Microsc Res Tech (2005) 68:307; **30.** J Immunol (2006) 176:3070; **31.** J Biol Chem (2009) 284:7681; **32.** Anal Biochem (2002) 308:18; **33.** Biomacromolecules (2006) 7:710; **34.** J Biol Chem (2004) 279:17951; **35.** Biochemistry (2006) 45:10614; **36.** PLoS One (2009) 4:e8074; **37.** J Am Chem Soc (2006) 128:4992; **38.** Proc Natl Acad Sci U S A (2004) 101:105; **39.** Free Radic Biol Med (2007) 43:62; **40.** Proc Natl Acad Sci U S A (1997) 94:4401.

DATA TABLE 1.5 FLUORESCEIN, OREGON GREEN® AND RHODAMINE GREEN™ DYES

Cat. No.	MW	Storage	Soluble	Abs	EC	Em	Solvent	Notes
C1311	473.39	F,D,L	DMF, DMSO	495	74,000	519	pH 9	1
C1359	376.32	L	pH >6, DMF	492	79,000	518	pH 9	1
C1360	376.32	L	pH >6, DMF	492	81,000	515	pH 9	1
C1904	376.32	L	pH >6, DMF	492	78,000	517	pH 9	1, 2
C2210	473.39	F,D,L	DMF, DMSO	494	78,000	520	pH 9	1
C6164	473.39	F,D,L	DMF, DMSO	496	83,000	516	pH 9	1
C6171MP	602.34	F,D,L	DMF, DMSO	520	75,000	548	pH 12	3
C20050	962.79	F,D,LL	DMSO	289	9500	none	MeOH	4, 5
D16	495.28	F,D,L	pH >6, DMF	492	83,000	516	pH 9	1, 6
D6145	368.29	L	pH >6, DMF	490	87,000	514	pH 9	7
E18	704.97	F,DD,L	pH >6, DMF	521	95,000	544	pH 9	8, 9
F143	389.38	F,DD,L	pH >6, DMF	494	77,000	519	pH 9	1, 8, 10
F1300	332.31	L	pH >6, DMF	490	93,000	514	pH 9	1
F1906	389.38	F,DD,L	pH >6, DMF	494	77,000	519	pH 9	1, 8, 10
F1907	389.38	F,DD,L	pH >6, DMF	494	77,000	519	pH 9	1, 8, 10
F2181	586.55	F,D,L	DMF, DMSO	494	74,000	520	pH 9	1
F6106	586.55	F,D,L	DMF, DMSO	494	75,000	521	pH 9	1
F6129	586.55	F,D,L	DMF, DMSO	494	74,000	520	pH 9	1
F6130	590.56	F,D,L	DMF, DMSO	491	86,000	515	pH 9	1
F36915	332.31	RO,L	see Notes	490	93,000	514	pH 9.5	1, 11
O6080	425.36	F,DD,L	DMF, DMSO	493	78,000	520	pH 9	7, 8
O6138	512.36	L	pH >6, DMF	506	86,000	526	pH 9	12, 13
O6139	609.43	F,D,L	DMF, DMSO	506	85,000	526	pH 9	12, 13
O6146	412.30	L	pH >6, DMF	492	85,000	518	pH 9	7, 14
O6147	509.38	F,D,L	DMF, DMSO	495	76,000	521	pH 9	7, 14
O6149	509.38	F,D,L	DMF, DMSO	496	82,000	516	pH 9	7, 14
O6185	622.53	F,D,L	DMF, DMSO	494	84,000	517	pH 9	7
R6107	507.89	F,D,L	DMF, DMSO	504	78,000	532	MeOH	
R6113	621.05	F,D,L	DMF, DMSO	503	74,000	528	MeOH	

For definitions of the contents of this data table, see "Using *The Molecular Probes® Handbook*" in the introductory pages.

Notes

1. Absorption and fluorescence of fluorescein derivatives are pH dependent. Extinction coefficients and fluorescence quantum yields decrease markedly at pH <7.
2. This product is specified to equal or exceed 98% analytical purity by HPLC.
3. Absorption and fluorescence of C6171MP are pH dependent (pK_a ~11.5). Fluorescence is maximal at pH >12.
4. All photoactivatable probes are sensitive to light. They should be protected from illumination except when photolysis is intended.
5. This product is colorless and nonfluorescent until it is activated by ultraviolet photolysis. Photoactivation generates a fluorescein derivative with spectral characteristics similar to C1359.
6. Unstable in water. Use immediately.
7. Absorption and fluorescence of Oregon Green® 488 derivatives are pH dependent only in moderately acidic solutions (pH <5).
8. Isothiocyanates are unstable in water and should not be stored in aqueous solution.
9. Eosin derivatives also exhibit phosphorescence with an emission maximum at ~680 nm. The phosphorescence lifetime is ~1 millisecond for eosin. (Biochem J (1979) 183:561, Spectroscopy (1990) 5:20) Fluorescence lifetime (τ) is 1.4 nanoseconds (QY = 0.2) for eosin. (J Am Chem Soc (1977) 99:4306)
10. The extinction coefficient of fluorescein isothiocyanate decreases about 10% on protein conjugation. (J Immunol Methods (1974) 5:103) The fluorescence lifetime (τ) is 3.8 nanoseconds.
11. F36915 consists of a fluorescein solution in 100 mM sodium borate buffer pH 9.5. The concentration of fluorescein is set spectrophotometrically to be equivalent to that of NIST Standard Reference Material (SRM®) 1932.
12. Absorption and fluorescence of Oregon Green® 514 derivatives are pH dependent only in moderately acidic solutions (pH <5).
13. The fluorescence lifetime (τ) of the Oregon Green® 514 dye in pH 9.0 buffer at 20°C is 4.2 nanoseconds. Data provided by the SPEX Fluorescence Group, Horiba Jobin Yvon Inc.
14. The fluorescence lifetime (τ) of the Oregon Green® 488 dye (O6146) in pH 9.0 buffer at 20°C is 4.1 nanoseconds. Data provided by the SPEX Fluorescence Group, Horiba Jobin Yvon Inc.

IMPORTANT NOTICE: The products described in this manual are covered by one or more Limited Use Label License(s). Please refer to the Appendix on page 971 and Master Product List on page 975. Products are For Research Use Only. Not intended for any animal or human therapeutic or diagnostic use.

www.invitrogen.com/probes

molecular probes® | invitrogen™ by life technologies™

PRODUCT LIST 1.5 FLUORESCEIN, OREGON GREEN® AND RHODAMINE GREEN™ DYES

Cat. No.	Product	Quantity
A10476	APEX® Oregon Green® 488 Antibody Labeling Kit	1 kit
C6171MP	6-carboxy-4',5'-dichloro-2',7'-dimethoxyfluorescein, succinimidyl ester (6-JOE, SE)	5 mg
C1359	5-carboxyfluorescein (5-FAM) *single isomer*	100 mg
C1360	6-carboxyfluorescein (6-FAM) *single isomer*	100 mg
C1904	5-(and-6)-carboxyfluorescein (5(6)-FAM) *FluoroPure™ grade* *mixed isomers*	100 mg
C2210	5-carboxyfluorescein, succinimidyl ester (5-FAM, SE) *single isomer*	10 mg
C6164	6-carboxyfluorescein, succinimidyl ester (6-FAM, SE) *single isomer*	10 mg
C1311	5-(and-6)-carboxyfluorescein, succinimidyl ester (5(6)-FAM, SE) *mixed isomers*	100 mg
C20050	5-carboxyfluorescein-bis-(5-carboxymethoxy-2-nitrobenzyl) ether, β-alanine-carboxamide, succinimidyl ester (CMNB-caged carboxyfluorescein, SE)	1 mg
D16	5-(4,6-dichlorotriazinyl)aminofluorescein (5-DTAF) *single isomer*	100 mg
D6145	2',7'-difluorofluorescein (Oregon Green® 488)	10 mg
E18	eosin-5-isothiocyanate	100 mg
F6434	FluoReporter® FITC Protein Labeling Kit *5–10 labelings*	1 kit
F6433	FluoReporter® Fluorescein-EX Protein Labeling Kit *5–10 labelings*	1 kit
F6153	FluoReporter® Oregon Green® 488 Protein Labeling Kit *5–10 labelings*	1 kit
F36915	fluorescein *NIST-traceable standard* *nominal concentration 50 µM* *special packaging*	5 x 1 mL
F1300	fluorescein *reference standard*	1 g
F6106	6-(fluorescein-5-carboxamido)hexanoic acid, succinimidyl ester (5-SFX) *single isomer*	5 mg
F2181	6-(fluorescein-5-(and-6)-carboxamido)hexanoic acid, succinimidyl ester (5(6)-SFX) *mixed isomers*	10 mg
F6129	6-(fluorescein-5-(and-6)-carboxamido)hexanoic acid, succinimidyl ester (5(6)-SFX) *mixed isomers* *special packaging*	10 x 1 mg
F10240	Fluorescein-EX Protein Labeling Kit *3 labelings*	1 kit
F6130	fluorescein-5-EX, succinimidyl ester	10 mg
F143	fluorescein-5-isothiocyanate (FITC 'Isomer I')	1 g
F1906	fluorescein-5-isothiocyanate (FITC 'Isomer I') *special packaging*	10 x 10 mg
F1907	fluorescein-5-isothiocyanate (FITC 'Isomer I') *special packaging*	10 x 100 mg
O6146	Oregon Green® 488 carboxylic acid *5-isomer*	5 mg
O6147	Oregon Green® 488 carboxylic acid, succinimidyl ester *5-isomer*	5 mg
O6149	Oregon Green® 488 carboxylic acid, succinimidyl ester *6-isomer*	5 mg
O6080	Oregon Green® 488 isothiocyanate (F_2FITC) *mixed isomers*	5 mg
O10241	Oregon Green® 488 Protein Labeling Kit *3 labelings*	1 kit
O6185	Oregon Green® 488-X, succinimidyl ester *6-isomer*	5 mg
O6138	Oregon Green® 514 carboxylic acid	5 mg
O6139	Oregon Green® 514 carboxylic acid, succinimidyl ester	5 mg
R6107	Rhodamine Green™ carboxylic acid, succinimidyl ester, hydrochloride (5(6)-CR 110, SE) *mixed isomers*	5 mg
R6113	Rhodamine Green™-X, succinimidyl ester, hydrochloride *mixed isomers*	5 mg
U21659	ULYSIS® Oregon Green® 488 Nucleic Acid Labeling Kit *20 labelings*	1 kit
Z25042	Zenon® Fluorescein Mouse IgG_1 Labeling Kit *50 labelings*	1 kit
Z25342	Zenon® Fluorescein Rabbit IgG Labeling Kit *50 labelings*	1 kit

IMPORTANT NOTICE: The products described in this manual are covered by one or more Limited Use Label License(s). Please refer to the Appendix on page 971 and Master Product List on page 975. Products are For Research Use Only. Not intended for any animal or human therapeutic or diagnostic use.

1.6 Long-Wavelength Rhodamines, Texas Red® Dyes and QSY® Quenchers

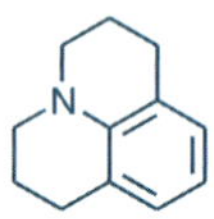

Julolidine

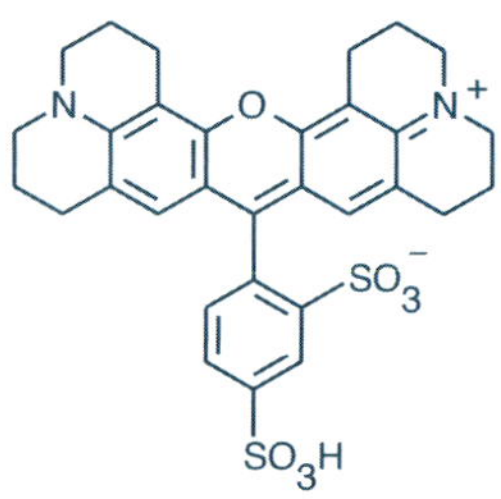

Sulforhodamine 101

Figure 1.6.1 The amine substituents of X-rhodamine, sulforhodamine 101 and Texas Red® dyes are rigidified in a julolidine ring structure.

This section includes dyes that have absorption maxima beyond about 520 nm, extending to nearly 800 nm. Significant exceptions, however, are the long-wavelength Alexa Fluor® dyes, which are all discussed in Section 1.3, the long-wavelength BODIPY® dyes—BODIPY® TMR, BODIPY® TR, BODIPY® 630/650 and BODIPY® 650/665—which are described in Section 1.4 and the 2′,4′,5′,7′-tetrabromofluorescein (eosin) and JOE dyes, which also absorb maximally beyond 520 nm but are discussed with other fluoresceins in Section 1.5. In many applications, the versatile Alexa Fluor® and BODIPY® dyes provide demonstrably superior performance relative to the dyes in this section.

Rhodamine dyes (Table 1.9) are among the most photostable fluorescent labeling reagents available. Moreover, spectra of most of these dyes are not affected by changes in pH between 4 and 10, an important advantage over the fluoresceins for many biological applications.[1] The most common members of this group have been the tetramethylrhodamines—including the reactive isothiocyanate (TRITC) and carboxylic acid (TAMRA) derivatives—as well as the X-rhodamines. The X prefix of the X-rhodamines, which include Texas Red® derivatives, refers to the fluorophore's extra julolidine rings (Figure 1.6.1). These rings prevent rotation about the nitrogen atoms, resulting in a shift in the fluorophore's spectra to longer wavelengths and usually an increase in its fluorescence quantum yield.

QSY® 7, QSY® 9 and QSY® 21 dyes are essentially nonfluorescent diarylrhodamine chromophores with strong absorption in the visible wavelength region, and they have proven to be extremely effective fluorescence quenchers.[2,3] QSY® 7, QSY® 9 and QSY® 21 dyes complement the QSY® 35 dye, a nonfluorescent quencher based on the NBD fluorophore that absorbs maximally near 475 nm, and the dabcyl quencher, both of which are described in Section 1.8.

Table 1.9 Amine-reactive, orange- and red-fluorescent fluorophores in this section.

Fluorophore (Abs/Em) *	Succinimidyl Ester	Other	Protein Labeling Kits	Notes
Lissamine rhodamine B (570/590)		L20 (SC)^{M} L1908 (SC)^{M}		• Optimal for 568 nm excitation • Photostable
Naphthofluorescein (602/672)	C653^{M}	C652 (COOH)^{M}		• Very long-wavelength excitation and emission • pH-sensitive fluorescence, with a high pK_a (~7.6)
Rhodamine 6G (525/555)	C6127^{5} C6128^{6} C6157^{M}			• Excited by the 514 nm spectral line of the argon-ion laser • Spectra intermediate between those of fluorescein and tetramethylrhodamine
Rhodamine Red™-X (580/590)	R6160 (X)^{5}		F6161 (F)	• Conjugates of Rhodamine Red™-X are generally more fluorescent than those of Lissamine rhodamine B, and the succinimidyl ester is more stable in H_2O
Tetramethylrhodamine (555/580)	C2211^{5} C6123^{6} C1171^{M} T6105 (X)^{M}	C6121 (COOH)^{5} C6122 (COOH)^{6} C300 (COOH)^{M} T1480 (ITC)^{5} T1481 (ITC)^{6} T490 (ITC)^{M}		• pH-insensitive fluorescence • Good photostability • Conjugates are prone to aggregation • Succinimidyl ester derivative (6-TAMRA, SE; C6123) is widely used for oligonucleotide labeling
Texas Red® dye (595/615)	T6134 (X)^{M} T20175 (X)^{5}	T353 (SC)^{M} T1905 (SC)^{M} T30200 (DTA)^{M}	F6162 (F) T10244 (P) Z25045 (Z)	• Good spectral separation from green fluorophores • Texas Red®-X succinimidyl ester typically yields greater fluorescence per attached dye than Texas Red® sulfonyl chloride and is more stable in H_2O
X-rhodamine (580/605)	C6125^{5} C6126^{6} C1309^{M}	C6124 (COOH)^{5} C6156 (COOH)^{6} X491 (ITC)^{M}		• Succinimidyl ester derivative (6-ROX, SE; C6126) is widely used for oligonucleotide labeling

* The numbers in parentheses reflect the absorption (Abs) and fluorescence emission (Em) maxima, in nm, of the goat anti–mouse IgG antibody or dextran conjugates in aqueous buffer. (5) = 5-Isomer. (6) = 6-Isomer. (COOH) = Carboxylic acid. (F) = FluoReporter® Protein Labeling Kit (Section 1.2). (ITC) = Isothiocyanate. (M) = Mixed isomers. (P) = Easy-to-Use Protein Labeling Kit (Section 1.2). (S) = Single isomer. (SC) = Sulfonyl chloride. (X) = Aminohexanoyl spacer separating the dye and the SE. (Z) = Zenon® Antibody Labeling Kit (Section 7.3).

IMPORTANT NOTICE: The products described in this manual are covered by one or more Limited Use Label License(s). Please refer to the Appendix on page 971 and Master Product List on page 975. Products are For Research Use Only. Not intended for any animal or human therapeutic or diagnostic use.
www.invitrogen.com/probes
molecular probes® | invitrogen™ by life technologies™

Tetramethylrhodamine

Tetramethylrhodamine (TMR) has been an important fluorophore for preparing protein conjugates, especially the fluorescent antibody and avidin derivatives used in immunochemistry. Under the name TAMRA, the carboxylic acid of TMR has also achieved prominence as a dye for oligonucleotide labeling[3,4] (Section 8.2, Table 8.7) and single-molecule detection applications.[5,6] Because it can be prepared in high purity, the 5-isomer of TAMRA (C6121) is one of the five dyes in our Reference Dye Sampler Kit (R14782, Section 23.1). TMR is efficiently excited by the 543 nm spectral line of the green He-Ne laser, which is increasingly being used for analytical instrumentation; diode lasers with 561 nm output[7] are slightly suboptimal but still effective.

TMR dyes such as TAMRA and TRITC are quite hydrophobic (Figure 1.6.2) when compared with their fluorescein counterparts FAM and FITC. As a result, they have a tendency to aggregate in aqueous solutions under conditions where the labeling density is sufficient to permit dye–dye interactions. A further consequence of these interactions is fluorescence self-quenching, which reduces the fluorescence output of the conjugate. Dye–dye interactions and self-quenching are much less prevalent with the more polar and water-soluble Alexa Fluor® dyes.[8] Another indication of intermolecular interactions of TMR dyes is that the absorption spectrum of TMR-labeled proteins is frequently complex (Figure 1.6.3), usually splitting into two absorption peaks at about 520 and 550 nm,[9] so that the actual degree of labeling is difficult to determine. Excitation at wavelengths in the range of the short-wavelength peak fails to yield the expected amount of fluorescence, indicating that it arises from a nonfluorescent dye aggregate. Furthermore, when the TMR-labeled protein conjugate is denatured by guanidine hydrochloride, the long-wavelength absorption increases, the short-wavelength peak mostly disappears and the fluorescence yield almost doubles[10] (Figure 1.6.3). The absorption spectra of TMR-labeled nucleotides and of other probes such as our rhodamine phalloidin (R415, Section 11.1) do *not* split into two peaks, indicating a labeling ratio of one dye molecule per biomolecule. The emission spectrum of TMR conjugates does not vary much with the degree of labeling.[9] An improved method for estimating the degree of substitution of TRITC conjugates has been described.[10]

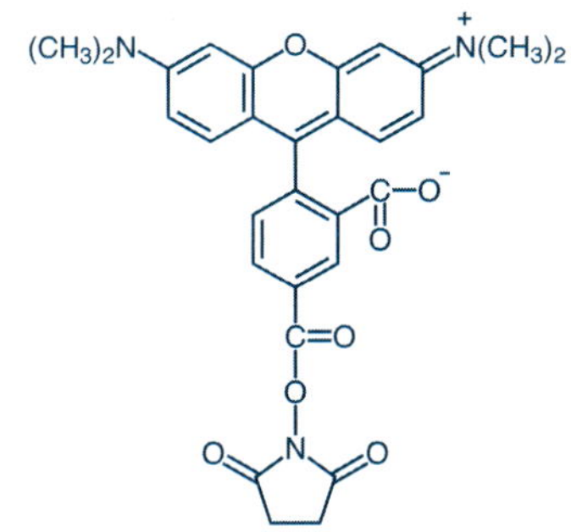

Figure 1.6.2 5-carboxytetramethylrhodamine, succinimidyl ester (5-TAMRA, SE; C2211).

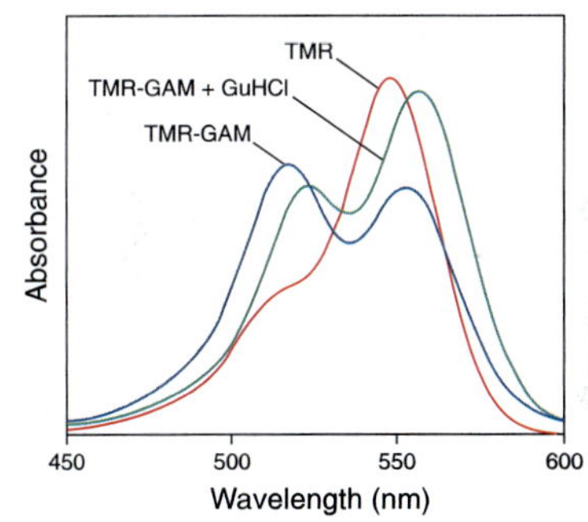

Figure 1.6.3 Effect of protein conjugation on the absorption spectrum of tetramethylrhodamine. The absorption spectrum of tetramethylrhodamine conjugated to goat anti-mouse IgG antibody (TMR-GAM, T2762) shows an additional peak at about 520 nm when compared with the spectrum of the same concentration of the free dye (TMR). Partial unfolding of the protein in the presence of 4.8 M guanidine hydrochloride (TMR-GAM + GuHCl) results in a spectrum more similar to that of the free dye.

Mixed-Isomer and Single-Isomer TRITC Preparations

Our tetramethylrhodamine isothiocyanate (TRITC) is of the highest quality available from any commercial source. Both the mixed-isomer (T490) and single-isomer (T1480, T1481) TRITC preparations typically have extinction coefficients above 80,000 $cm^{-1}M^{-1}$, whereas some competitive sources of TRITC have extinction coefficients reported to be below 50,000 $cm^{-1}M^{-1}$. TRITC is widely used by other companies to prepare most of their so-called "rhodamine" immunoconjugates; however, they also often employ reactive versions of rhodamine B or Lissamine rhodamine B, which have somewhat different spectra, resulting in some confusion in matching the product name to the correct fluorophore.

Succinimidyl Esters of Carboxytetramethylrhodamine

Almost all Molecular Probes® TMR conjugates are prepared using succinimidyl esters of carboxytetramethylrhodamine (TAMRA™ dye), rather than TRITC, because bioconjugates from succinimidyl esters are more stable and often more fluorescent. We offer the mixed-isomer (C300) and single-isomer (C6121, C6122) preparations of carboxymethylrhodamine, as well as the corresponding mixed-isomer (C1171) and single-isomer (C2211, C6123) succinimidyl esters. The single-isomer preparations are most important for high-resolution techniques such as DNA sequencing[11] and separation of labeled carbohydrates by capillary electrophoresis.[12] 6-TAMRA™ dye is one of the traditional fluorophores (5-FAM™, 6-JOE™, 6-TET™, 6-HEX™, 6-TAMRA™ and 6-ROX™ dyes) used in first-generation electrophoresis-based DNA sequencing[11,13–15] (Section 8.2, Table 8.7).

We have also prepared the mixed-isomer TAMRA-X succinimidyl ester (5(6)-TAMRA-X, SE; T6105), which contains a seven-atom aminohexanoyl spacer ("X") between the reactive group and the fluorophore (Figure 1.6.4). This spacer helps to separate the fluorophore from its point of attachment, reducing the interaction of the fluorophore with the biomolecule to which it is conjugated, making it more accessible to secondary detection reagents and facilitating orientational averaging in fluorescence resonance energy transfer (FRET) applications[16] (Fluorescence Resonance Energy Transfer (FRET)—Note 1.2). Polyclonal anti-tetramethylrhodamine and anti–Texas Red® dye antibodies that recognize the tetramethylrhodamine, Rhodamine Red™-X, X-rhodamine and Texas Red® fluorophores are available (Section 7.4).

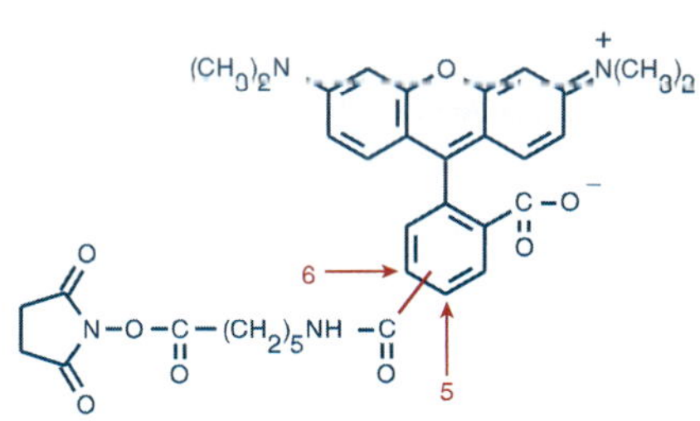

Figure 1.6.4 6-(tetramethylrhodamine-5-(and-6)-carboxamido)hexanoic acid, succinimidyl ester (5(6)-TAMRA-X, SE; T6105).

IMPORTANT NOTICE: The products described in this manual are covered by one or more Limited Use Label License(s). Please refer to the Appendix on page 971 and Master Product List on page 975. Products are For Research Use Only. Not intended for any animal or human therapeutic or diagnostic use.

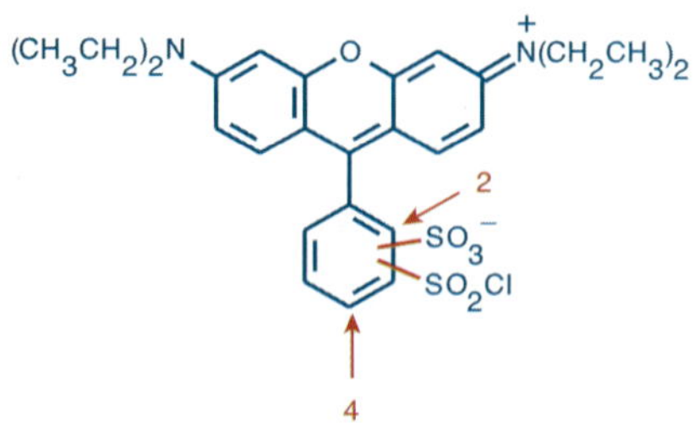

Figure 1.6.5 Lissamine rhodamine B sulfonyl chloride (L20).

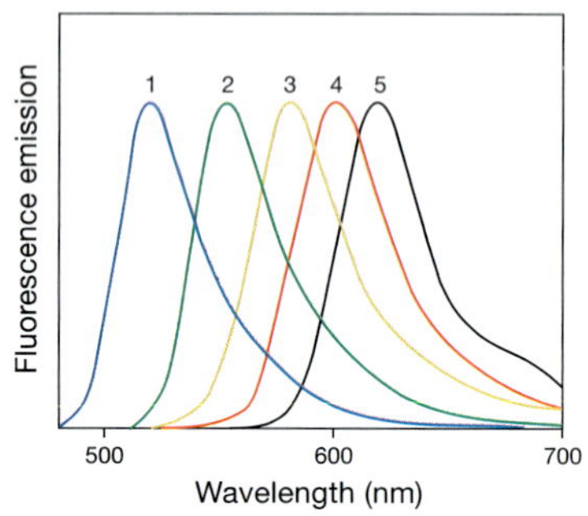

Figure 1.6.6 Normalized fluorescence emission spectra of goat anti–mouse IgG antibody conjugates of 1) fluorescein, 2) rhodamine 6G, 3) tetramethylrhodamine, 4) Lissamine rhodamine B and 5) Texas Red® dyes.

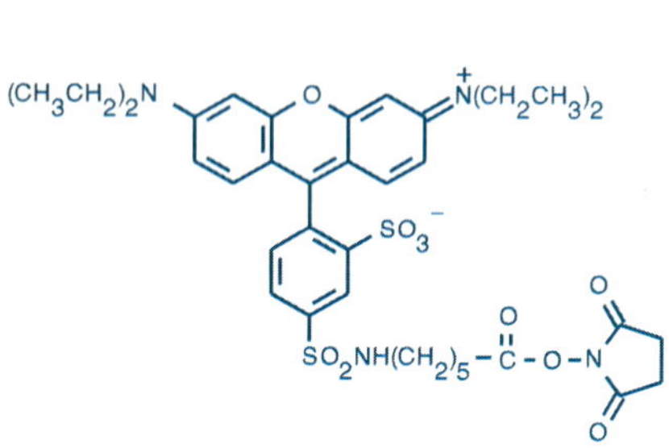

Figure 1.6.7 Rhodamine Red™-X, succinimidyl ester (R6160).

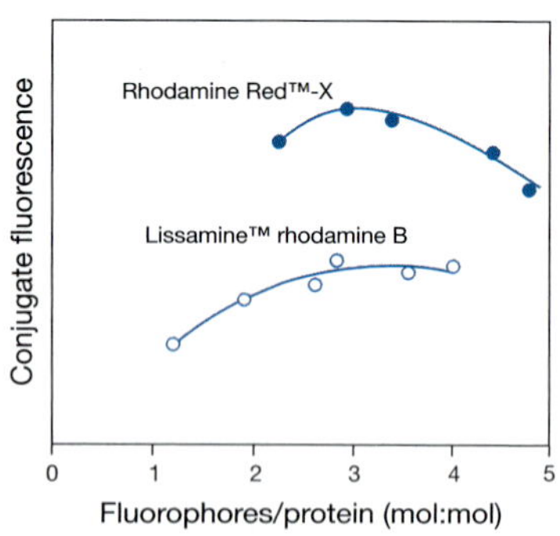

Figure 1.6.8 Comparison of the relative fluorescence of goat anti-mouse IgG antibody conjugates of Rhodamine Red™-X succinimidyl ester (R6160, ●) and Lissamine rhodamine B sulfonyl chloride (L20, L1908, ○). Conjugate fluorescence is determined by measuring the fluorescence quantum yield of the conjugated dye relative to that of the free dye and multiplying by the number of fluorophores per protein. Higher numbers of fluorophores attached per protein are attainable with Rhodamine Red™-X dye due to the lesser tendency of this dye to induce protein precipitation.

Lissamine Rhodamine B and Rhodamine Red™-X Dyes

Lissamine Rhodamine B Sulfonyl Chloride

Lissamine rhodamine B sulfonyl chloride (L20, L1908; Figure 1.6.5) is much less expensive than Texas Red® sulfonyl chloride, and the fluorescence emission spectrum of its protein conjugates lies between those of tetramethylrhodamine and Texas Red® conjugates[17] (Figure 1.6.6). It is more frequently employed as a synthetic precursor for preparing affinity labeling reagents than as a labeling reagent for protein conjugation.[18–20]

Rhodamine Red™-X Succinimidyl Ester

Lissamine rhodamine B sulfonyl chloride is unstable, particularly in aqueous solution, making it somewhat difficult to achieve reproducible conjugations using this dye. Unlike Lissamine rhodamine B sulfonyl chloride, which is a mixture of isomeric sulfonyl chlorides (Figure 1.6.5), Rhodamine Red™-X succinimidyl ester (R6160, Figure 1.6.7) is isomerically pure and is hydrolytically stable for practical purposes at the mild alkaline pH levels typically used for amine-reactive protein conjugation. Rhodamine Red™-X succinimidyl ester incorporates a spacer between the fluorophore and the reactive site, resulting in minimized perturbation of the conjugation partner's functional properties.[21] Moreover, we have found that protein conjugates of Rhodamine Red™-X dye are frequently brighter than those of Lissamine rhodamine B (Figure 1.6.8), and less likely to precipitate during storage.[22] Rhodamine Red™-X succinimidyl ester is used in the FluoReporter® Rhodamine Red™-X Protein Labeling Kit (F6161); see Section 1.2 for further information on preparing red-fluorescent protein conjugates with this kit.

X-Rhodamine

The derivatives of carboxy-X-rhodamine (ROX™ dye)—a dye originally developed in our laboratories in 1986—are widely used for oligonucleotide labeling and DNA sequencing applications (Section 8.2, Table 8.7). Conjugates of this dye and of the similar isothiocyanate (5(6)-XRITC, X491; Figure 1.6.9) have longer-wavelength spectra (Figure 1.6.10) than the spectra of Lissamine rhodamine B, but somewhat shorter-wavelength spectra than those of Texas Red® conjugates. Both the pure 5-isomer (C6124) and 6-isomer (C6156) of carboxy-X-rhodamine are available, as are mixed-isomer (C1309, Figure 1.6.11) and single-isomer (C6125, C6126) preparations of the succinimidyl ester.

Texas Red® and Texas Red®-X Dyes

The Texas Red® fluorophore emits at a longer wavelength than do either tetramethylrhodamine or Lissamine rhodamine B (Figure 1.6.6), making Texas Red® conjugates among the most commonly used long-wavelength "third labels" in fluorescence microscopy (Figure 1.6.12, Figure 1.6.13). Unlike the other rhodamines, the Texas Red® fluorophore exhibits very little spectral overlap with fluorescein (Figure 1.6.6), and its fluorescence can be distinguished from that of

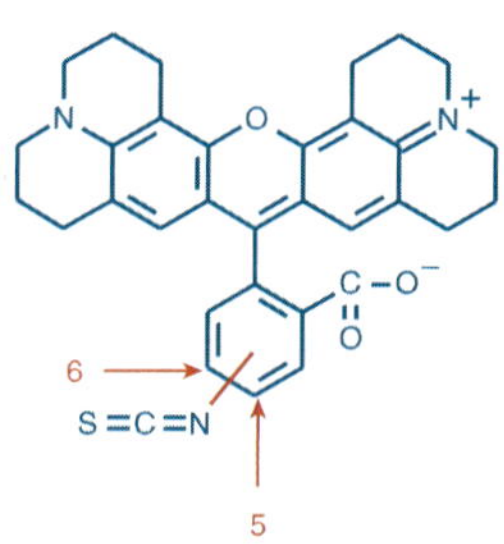

Figure 1.6.9 X-rhodamine-5-(and-6)-isothiocyanate (5(6)-XRITC, X491).

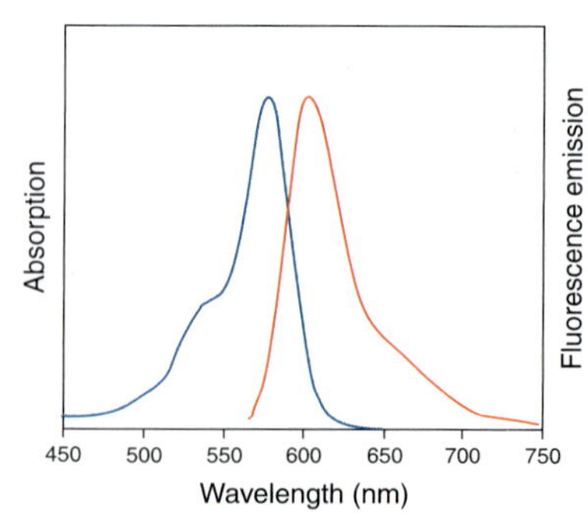

Figure 1.6.10 Absorption and fluorescence emission spectra of 5-carboxy-X-rhodamine (5-ROX) in pH 7.0 buffer.

IMPORTANT NOTICE: The products described in this manual are covered by one or more Limited Use Label License(s). Please refer to the Appendix on page 971 and Master Product List on page 975. Products are For Research Use Only. Not intended for any animal or human therapeutic or diagnostic use.
www.invitrogen.com/probes

phycoerythrins. When the correct optical filter sets are used, Texas Red® conjugates are brighter and have lower background than conjugates of the other commonly used red-fluorescent dyes, with the exception of the Alexa Fluor® 594 dye. Texas Red® conjugates are particularly well suited for excitation by the 594 nm spectral line of the orange He-Ne laser; diode laser excitation at 561 nm is also efficient.

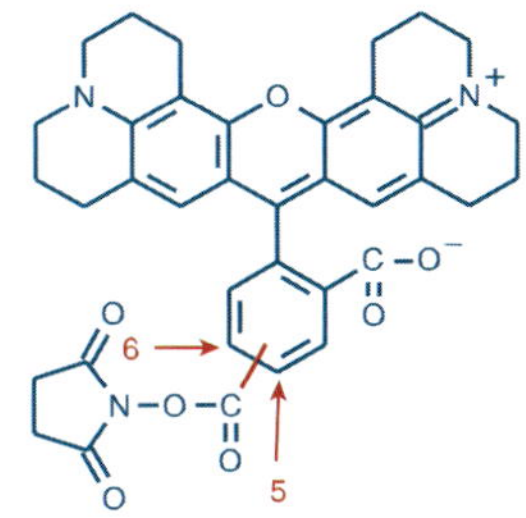

Figure 1.6.11 5-(and-6)-carboxy-X-rhodamine, succinimidyl ester (5(6)-ROX, SE; C1309).

Texas Red® Sulfonyl Chloride

Texas Red® sulfonyl chloride is our trademarked mixture of isomeric sulfonyl chlorides (Figure 1.6.14) of sulforhodamine 101.[23,24] This reagent is quite unstable in water, especially at the higher pH required for reaction with aliphatic amines. For example, dilute solutions of Texas Red® sulfonyl chloride are totally hydrolyzed within 2–3 minutes in pH 8.3 aqueous solution at room temperature.[22] Protein modification by this reagent is best done at low temperature. Once conjugated, however, the sulfonamides that are formed (Figure 1.6.15) are extremely stable; they even survive complete protein hydrolysis.

Because Texas Red® sulfonyl chloride rapidly degrades upon exposure to moisture, we offer this reactive dye specially packaged as a set of 10 vials (T1905), each containing approximately 1 mg of Texas Red® sulfonyl chloride for small-scale conjugations. We also offer the 10 mg unit size packaged in a single vial (T353) for larger-scale conjugations. Each milligram of Texas Red® sulfonyl chloride modifies approximately 8–10 mg of protein. Note that sulfonyl chlorides are unstable in dimethylsulfoxide (DMSO) and should never be used in that solvent.[25] Polyclonal anti-tetramethylrhodamine and anti–Texas Red® antibodies that recognize tetramethylrhodamine, Rhodamine Red™, X-rhodamine and Texas Red® fluorophores are available (Section 7.4, Table 4.2).

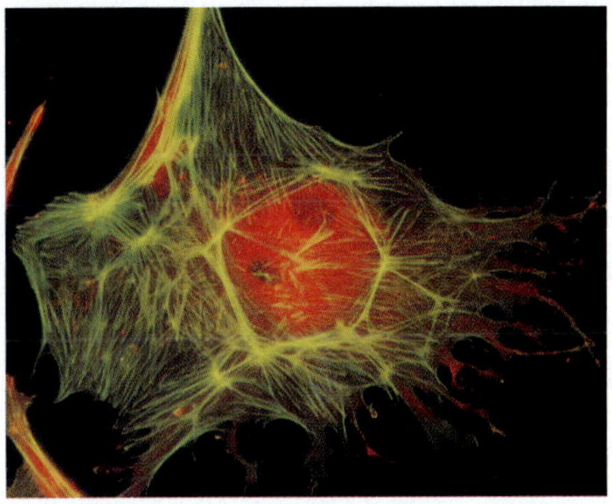

Figure 1.6.12 Simultaneous visualization of F- and G-actin in a bovine pulmonary artery endothelial cell (BPAEC) using F-actin–specific Oregon Green® 488 phalloidin (O7466) and G-actin–specific Texas Red® deoxyribonuclease I. The G-actin appears as diffuse red fluorescence that is more intense in the nuclear region where the cell thickness is greater and stress fibers are less dense. The image was obtained by taking multiple exposures through bandpass optical filter sets appropriate for fluorescein and the Texas Red® dye.

Texas Red®-X Succinimidyl Ester

Texas Red® sulfonyl chloride's susceptibility to hydrolysis and low solubility in water may complicate its conjugation to some biomolecules. To overcome this difficulty, we have developed Texas Red®-X succinimidyl ester, which contains an additional seven-atom aminohexanoyl spacer ("X") between the fluorophore and its reactive group.[22] The single-isomer preparation of Texas Red®-X succinimidyl ester (T20175, Figure 1.6.16) is preferred over the mixed-isomer product (T6134) when the dye is used to prepare conjugates of low molecular weight peptides, oligonucleotides and receptor ligands that are to be purified by high-resolution techniques. Also, because isomers of a reactive dye may differ in their binding geometry, certain applications such as fluorescence resonance energy transfer (FRET) may benefit from the use of single-isomer reactive dyes[26] (Fluorescence Resonance Energy Transfer (FRET)—Note 1.2). Thiol-reactive Texas Red® derivatives that are based on a similar synthetic approach are described in Section 2.2. Texas

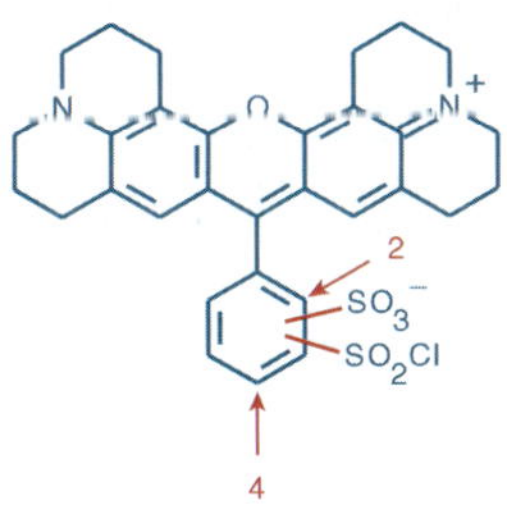

Figure 1.6.14 Texas Red® sulfonyl chloride (T353).

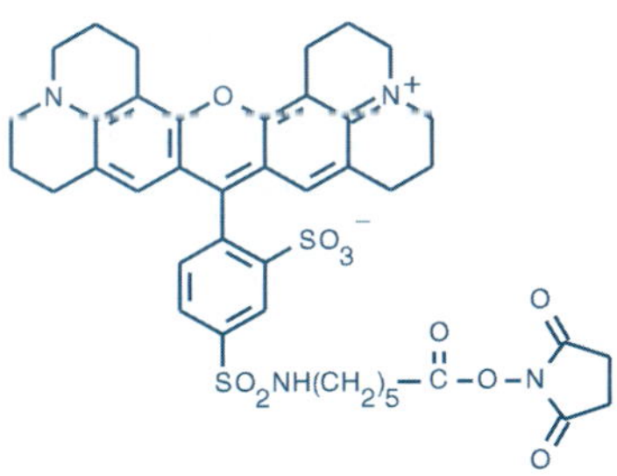

Figure 1.6.16 Texas Red®-X, succinimidyl ester (T20175).

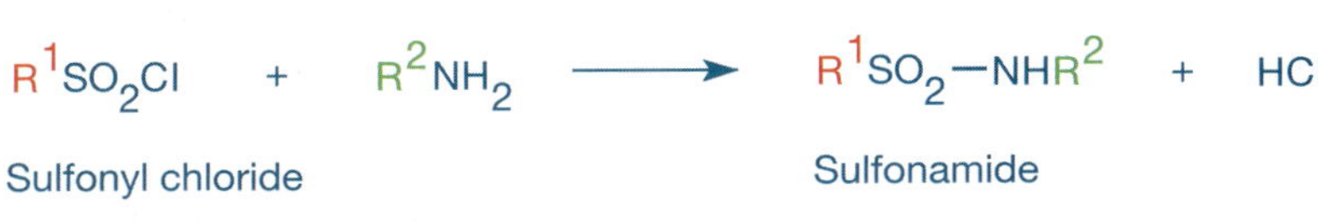

Figure 1.6.15 Reaction of a primary amine with a sulfonyl chloride.

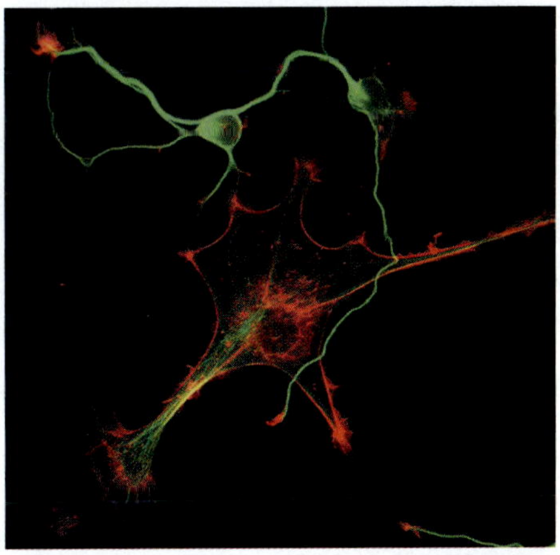

Figure 1.6.13 Confocal micrograph of the cytoskeleton of a mixed population of granule neurons and glial cells. The F-actin was stained with red-fluorescent Texas Red®-X phalloidin (T7471). The microtubules were detected with a mouse monoclonal anti-ß-tubulin primary antibody and subsequently visualized with the green-fluorescent Alexa Fluor® 488 goat anti-mouse IgG antibody (A11001). The image was contributed by Jonathan Zmuda, Immunomatrix, Inc.

IMPORTANT NOTICE: The products described in this manual are covered by one or more Limited Use Label License(s). Please refer to the Appendix on page 971 and Master Product List on page 975. Products are For Research Use Only. Not intended for any animal or human therapeutic or diagnostic use.

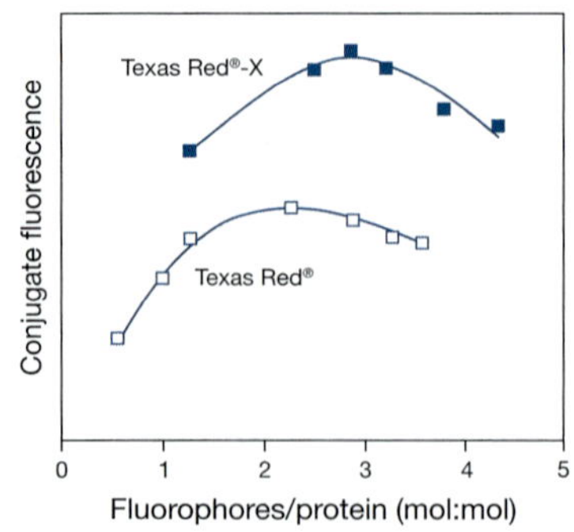

Figure 1.6.17 Comparison of the relative fluorescence of goat anti–mouse IgG antibody conjugates of Texas Red®-X succinimidyl ester (T6134, ■) and Texas Red® sulfonyl chloride (T353, □). Conjugate fluorescence was determined by measuring the fluorescence quantum yield of the conjugated dye relative to that of the free dye and multiplying by the number of fluorophores per protein. Higher numbers of fluorophores attached per protein are attainable with the Texas Red®-X dye due to the lesser tendency of this dye to induce protein precipitation.

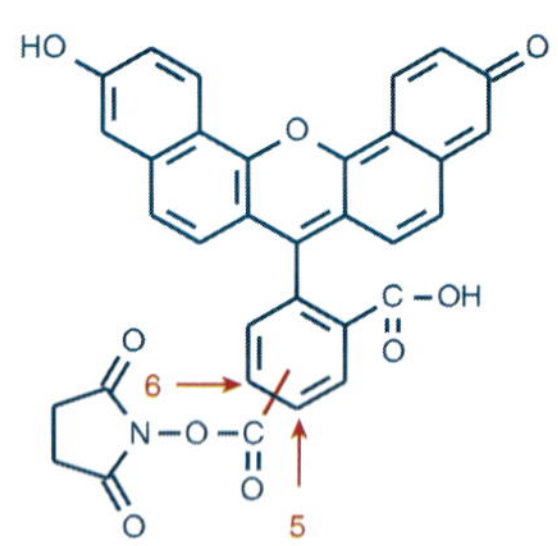

Figure 1.6.18 5-(and-6)-carboxynaphthofluorescein, succinimidyl ester (C653).

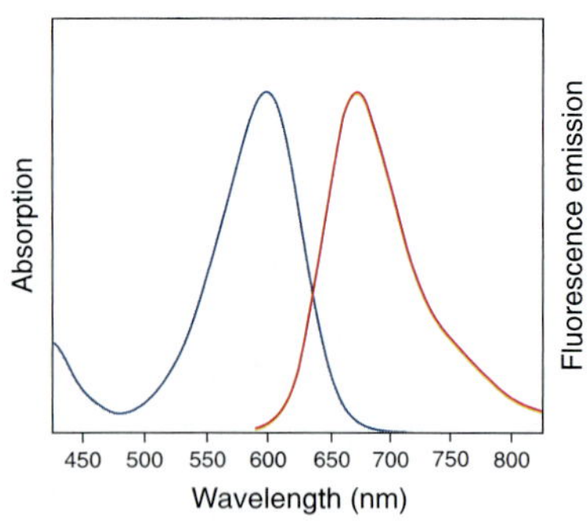

Figure 1.6.19 Absorption and fluorescence emission spectra of 5-(and-6)-carboxynaphthofluorescein in pH 10.0 buffer.

Red®-X succinimidyl ester offers significant advantages over Texas Red® sulfonyl chloride for the preparation of bioconjugates:

- In the absence of amines, greater than 80% of Texas Red®-X succinimidyl ester's reactivity is retained in pH 8.3 solution after one hour at room temperature.[22]
- Much less Texas Red®-X succinimidyl ester (usually half or less of the amount of Texas Red® sulfonyl chloride) is required to yield the same degree of labeling, making the effective costs of these two reagents about the same.
- Conjugations with Texas Red®-X succinimidyl ester are more reproducible.
- Unlike Texas Red® sulfonyl chloride, which can form unstable products with tyrosine, histidine, cysteine and other residues in proteins, the Texas Red®-X succinimidyl ester reacts almost exclusively with amines.
- Protein conjugates prepared with Texas Red®-X succinimidyl ester have a higher fluorescence yield than those with the same labeling ratio prepared with Texas Red® sulfonyl chloride (Figure 1.6.17).
- Texas Red®-X protein conjugates show a decreased tendency to precipitate during the reaction or upon storage.

Texas Red® C_2-Dichlorotriazine

Texas Red® C_2-dichlorotriazine (T30200) is a reactive dye with absorption/emission maxima of ~588/601 nm. Dichlorotriazines readily modify amines in proteins, and they are among the few reactive groups that are reported to react directly with polysaccharides[27] and other alcohols in aqueous solution, provided that the pH is >9 and that other nucleophiles are absent.

Texas Red®-X Conjugates and Texas Red®-X Labeling Kits

Because of the advantages of Texas Red®-X succinimidyl ester, we have converted some of our Texas Red® conjugates to the Texas Red®-X conjugates. We have prepared Texas Red®-X conjugates of:

- Antibodies (Section 7.2, Table 7.1)
- Streptavidin (S6370, Section 7.6, Table 7.9)
- Phalloidin (T7471, Section 11.1, Table 11.2)
- Wheat germ agglutinin (W21405, Section 7.7)
- dUTP (C7631, Section 8.2)

Protein conjugates of the Texas Red®-X dye are readily prepared using our FluoReporter® Texas Red®-X Protein Labeling Kit (F6162) and Texas Red®-X Protein Labeling Kit (T10244); see Section 1.2 for further information on preparing fluorescent protein conjugates with these kits. Zenon® Texas Red®-X Antibody Labeling Kit for mouse IgG_1 antibodies (Z25045, Section 7.3) permits the rapid and quantitative labeling of antibodies from a purified antibody fraction or from a crude antibody preparation such as serum, ascites fluid or a hybridoma supernatant with the Texas Red®-X dye.[28] Polyclonal anti-tetramethylrhodamine and anti–Texas Red® antibodies that recognize tetramethylrhodamine, Rhodamine Red™, X-rhodamine and Texas Red® fluorophores are available (Section 7.4, Table 7.8).

Naphthofluorescein

Naphthofluorescein carboxylic acid and its succinimidyl ester (C652, C653; Figure 1.6.18) have emission maxima of approximately 670 nm in aqueous solution at pH 10 (Figure 1.6.19). However, the fluorescence of naphthofluorescein is pH dependent (pK_a ~7.6), requiring a relatively alkaline pH for maximal fluorescence.

Carboxyrhodamine 6G

The excitation and emission spectra of carboxyrhodamine 6G (CR 6G) fall between those of fluorescein and tetramethylrhodamine (Figure 1.6.6). With a peak absorption at ~520 nm, conjugates prepared from the mixed-isomer (C6157) or single-isomer (C6127, C6128) preparations

IMPORTANT NOTICE: The products described in this manual are covered by one or more Limited Use Label License(s). Please refer to the Appendix on page 971 and Master Product List on page 975. Products are For Research Use Only. Not intended for any animal or human therapeutic or diagnostic use.
www.invitrogen.com/probes

of CR 6G succinimidyl esters are an excellent match to the 514 nm spectral line of the argon-ion laser. They also tend to exhibit a higher fluorescence quantum yield than tetramethylrhodamine conjugates, as well as excellent photostability. As with Rhodamine Green™ dyes, carboxyrhodamine 6G dyes are more suitable for preparing nucleotide and oligonucleotide conjugates than for preparing protein conjugates. Oligonucleotide conjugates of CR 6G have spectroscopic and electrophoretic properties that are superior to the JOE dye (C6171MP, Section 1.5) that is often used for DNA sequencing (Section 8.2, Table 8.7).

QSY® Dyes: Fluorescence Quenchers

Dyes that quench the fluorescence of visible light–excited fluorophores are increasingly important for use in fluorescence resonance energy transfer (FRET) proximity assays (Fluorescence Resonance Energy Transfer (FRET)—Note 1.2), such as those based on DNA hybridization.[29–31] Nonfluorescent acceptors are advantageous in FRET assays because they avoid the complications of proximity-independent signals resulting from direct excitation of fluorescent acceptors. Our QSY® 7, QSY® 9 and QSY® 21 dyes (Table 1.10) are diarylrhodamine derivatives that have several properties that make them superior to the commonly used dabcyl chromophore (Section 1.8) when preparing bioconjugates for use in FRET-based assays:

- Broad absorption in the visible-light spectrum, with an absorption maximum near 560 nm for both the QSY® 7 and QSY® 9 dyes and near 660 nm for the QSY® 21 dye (Figure 1.6.20)
- Extinction coefficients that are typically in excess of 90,000 $cm^{-1}M^{-1}$
- Absorption spectra of the conjugates that are insensitive to pH between 4 and 10
- Fluorescence quantum yields typically <0.001 in aqueous solution (in a few isolated cases, we have observed that some QSY® dyes can exhibit fluorescence when placed in a rigidifying environment such as glycerol.)
- Efficient quenching of the fluorescence emission of donor dyes by the QSY® 7 and QSY® 9 dyes, including blue-fluorescent coumarins, green- or orange-fluorescent dyes, and red-fluorescent Texas Red® and Alexa Fluor® 594 conjugates
- Quenching of red-fluorescent dyes, including Alexa Fluor® 647 dye, by the long-wavelength light–absorbing QSY® 21 dye[32] (Table 1.11)
- Quenching of most green and red fluorophores that is more effective at far greater distances than is possible with dabcyl quenchers (Figure 1.6.21)
- Residual fluorescence of the conjugates, at close spatial separations, that is typically lower than in conjugates that use dabcyl as the quencher
- High chemical stability of the conjugates and very good resistance to photobleaching

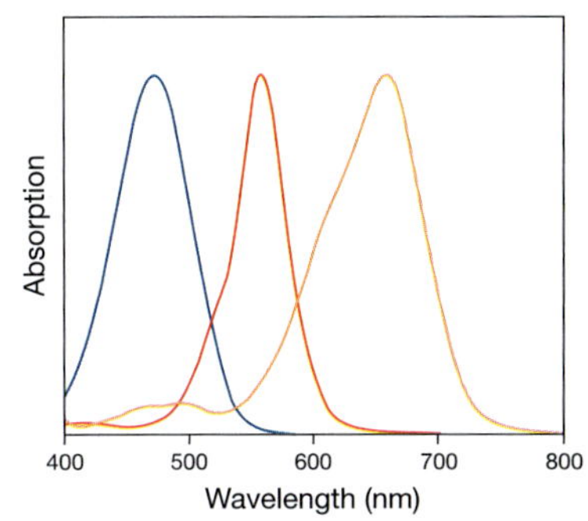

Figure 1.6.20 Normalized absorption spectra of the QSY® 35 (blue), QSY® 7 (red) and QSY® 21 (orange) dyes. The QSY® 7 and QSY® 9 dyes have essentially identical spectra.

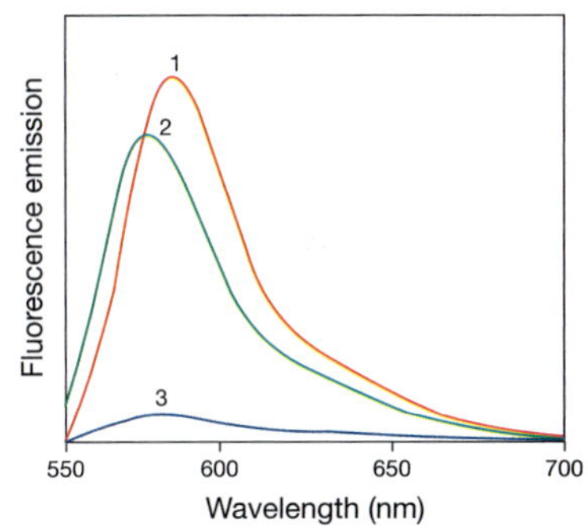

Figure 1.6.21 Fluorescence quenching of 5′-tetramethylrhodamine–labeled M13 primers by nonfluorescent dyes attached at the 3′-end. The comparison represents equal concentrations of oligonucleotides with 1) no 3′-quencher (control), 2) 3′-dabcyl quencer, and 3) 3′-QSY® 7 quencher.

Table 1.10 Molecular Probes® nonfluorescent quenchers and photosensitizers.

Dye	Abs *	Extinction Coefficient †	Amine-reactive Dyes‡	Notes
Dabcyl	453	32,000	D2245 (SE)	• Broad and intense visible-wavelength absorption • Efficient energy transfer acceptor from blue- and green-fluorescent dyes in FRET applications
Malachite green	628	76,000	M689 (ITC)	• Nonfluorescent photosensitizer
QSY® 7	560	90,000	Q10193 (SE)	• Essentially nonfluorescent quencher • Broad visible-wavelength absorption • Efficient energy transfer acceptor from UV light–excited green- and orange-fluorescent dyes in FRET applications
QSY® 9	562	88,000	Q20131 (SE)	• Essentially nonfluorescent quencher • Spectrally similar to QSY® 7, but with enhanced water solubility • Efficient energy transfer acceptor from UV light–excited green- and orange-fluorescent dyes in FRET applications
QSY® 21	661	90,000	Q20132 (SE)	• Essentially nonfluorescent quencher • Long-wavelength absorption • Efficient energy transfer acceptor from red- and near-infrared–fluorescent dyes in FRET applications
QSY® 35	475	23,000	Q20133 (SE) §	• Nonfluorescent quencher • Spectrally similar to dabcyl • Efficient energy transfer acceptor from blue- and green-fluorescent dyes in FRET applications

* Absorption (Abs) maxima, in nm. † Molar extinction coefficient in $cm^{-1}M^{-1}$ determined at the wavelength listed in the column headed Abs. These values may vary with the environment, particularly for the QSY® 35 dye. ‡ (ITC) = Isothiocyanate. (SE) = Succinimidyl ester.

IMPORTANT NOTICE: The products described in this manual are covered by one or more Limited Use Label License(s). Please refer to the Appendix on page 971 and Master Product List on page 975. Products are For Research Use Only. Not intended for any animal or human therapeutic or diagnostic use.

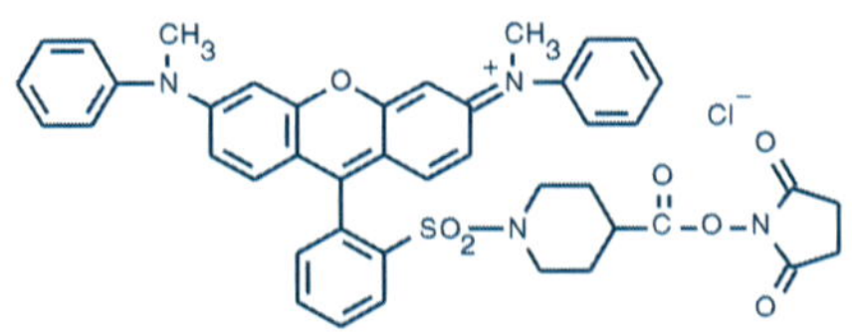

Figure 1.6.22 QSY® 7 carboxylic acid, succinimidyl ester (Q10193).

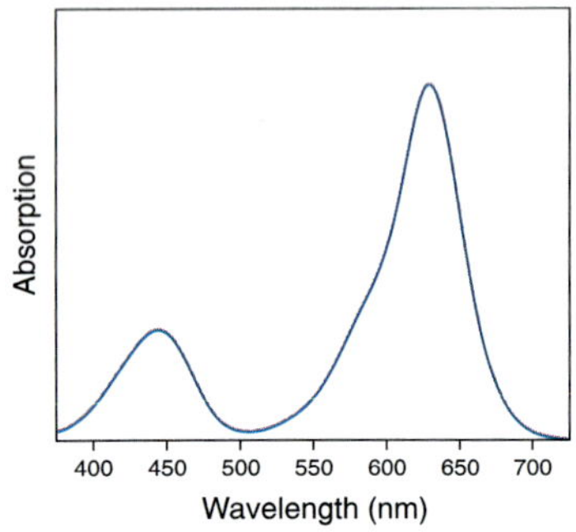

Figure 1.6.23 Absorption spectrum of malachite green isothiocyanate in acetonitrile.

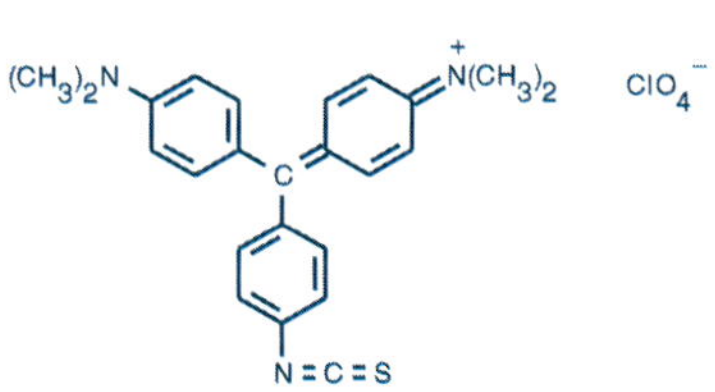

Figure 1.6.24 Malachite green isothiocyanate (M689).

A particularly frequent and effective application of QSY® dyes is as components of fluorogenic protease and peptidase substrates consisting of a fluorescent dye and a QSY® quencher attached to opposite ends of a peptide sequence that is specifically recognized and cleaved by the enzyme.[2,32,33] QSY® 7 dye has been used to create activatable targeted probes comprising a fluorophore–quencher pair and a targeting protein moiety (avidin, which targets the D-galactose receptor, or trastuzumab, a monoclonal antibody that recognizes the human epithelial growth factor receptor type 2 or HER2/neu) for use in *in vivo* tumor imaging. The fluorophore–quencher interaction, in this case between tetramethylrhodamine and QSY® 7 dyes, is disrupted when the probe is internalized in a tumor by receptor-mediated uptake, thereby activating the fluorescence.[34] In this imaging application, QSY® 7 was reported to be superior to azobenzene quenchers (e.g., dabcyl derivatives) because azobenzene dyes are vulnerable to reductive cleavage *in vivo*, producing a false positive signal. Both the amine-reactive and thiol-reactive QSY® 7 derivatives (Q10193, Q10257) have been used to create molecular beacon probes for following the transport of mRNAs in *Drosophila melanogaster* oocytes.[35]

The distance at which energy transfer is 50% efficient (i.e., 50% of excited donors are deactivated by fluorescence resonance energy transfer) is defined by the Förster radius (R_0). The magnitude of R_0 is dependent on the spectral properties of the donor and acceptor dyes. R_0 values calculated for energy transfer from various Alexa Fluor® dyes to QSY® and dabcyl quenchers are listed in Table 1.11. FRET efficiencies from several donor dyes to the QSY® 7 quencher in molecular beacon hybridization probes have also been calculated.[3]

For preparing bioconjugates, we offer several reactive versions of these QSY® dyes:

- Amine-reactive QSY® 7 (Figure 1.6.22), QSY® 9 and QSY® 21 succinimidyl esters (Q10193, Q20131, Q20132)
- Thiol-reactive QSY® 7 C_5-maleimide and QSY® 9 C_5-maleimide (Q10257, Q30457; Section 2.2)
- QSY® 7 aliphatic amine (Q10464, Section 3.4), which can be coupled to carbodiimide-activated carboxylic acids and other functional groups
- α-FMOC-ε-QSY® 7-L-lysine (Q21930, Section 9.5), for automated synthesis of peptides containing the QSY® 7 quencher

In addition to the QSY® 7, QSY® 9 and QSY® 21 dyes, we offer other quenchers that absorb maximally below 500 nm, including the QSY® 35 and dabcyl dyes (Table 1.10). These products are described in Section 1.8.

Nonfluorescent Malachite Green

Malachite green is a nonfluorescent photosensitizer that absorbs at long wavelengths (~630 nm, Figure 1.6.23). Its photosensitizing action can be targeted to particular cellular sites by conjugating malachite green isothiocyanate (M689, Figure 1.6.24) to specific antibodies. Enzymes and other proteins within ~10 Å of the binding site of the malachite green–labeled antibody can then be selectively destroyed upon irradiation with long-wavelength light.[36,37] Studies by Jay and colleagues have demonstrated that this photoinduced destruction of enzymes in the immediate vicinity of the chromophore is apparently the result of localized production of hydroxyl radicals, which have short lifetimes that limit their diffusion from the site of their generation.[38] Earlier studies had supported a thermal mechanism of action.[39–41]

IMPORTANT NOTICE: The products described in this manual are covered by one or more Limited Use Label License(s). Please refer to the Appendix on page 971 and Master Product List on page 975. Products are For Research Use Only. Not intended for any animal or human therapeutic or diagnostic use.

Table 1.11 R_0 values for QSY® and dabcyl quenchers.

	Acceptor			
Donor	**QSY® 35**	**dabcyl**	**QSY® 7 and QSY® 9**	**QSY® 21**
Alexa Fluor® 350	47	50		
Alexa Fluor® 488	44	49	64	
Alexa Fluor® 546	25	29	67	
Alexa Fluor® 555			45	
Alexa Fluor® 568			56	75
Alexa Fluor® 594				77
Alexa Fluor® 647				69

R_0 values in angstroms (Å) represent the distance at which fluorescence resonance energy transfer from the donor dye to the acceptor dye is 50% efficient (Förster radius). Values were calculated from spectroscopic data as outlined (see Fluorescence Resonance Energy Transfer (FRET)—Note 1.2).

REFERENCES

1. Bioconjug Chem (2008) 19:1735; **2.** Nat Chem Biol (2007) 3:668; **3.** Nucleic Acids Res (2002) 30:e122; **4.** J Neurosci Methods (2003) 123:31; **5.** Biophys J (2007) 92:303; **6.** Biophys J (2005) 88:2939; **7.** Cytometry A (2005) 68:36; **8.** ACS Chem Biol (2009) 4:535; **9.** Anal Biochem (1977) 80:585; **10.** J Immunol Methods (1991) 143:263; **11.** Nucleic Acids Res (1992) 20:2471; **12.** J Chromatogr B Biomed Appl (1994) 657:307; **13.** Anal Biochem (1996) 238:165; **14.** Anal Biochem (1994) 223:39; **15.** Proc Natl Acad Sci U S A (1989) 86:9178; **16.** Proc Natl Acad Sci U S A (2010) 107:5453; **17.** J Photochem Photobiol A (2001) 139:151; **18.** Bioconjug Chem (2006) 17:1618; **19.** Bioconjug Chem (2006) 17:735; **20.** Bioconjug Chem (2000) 11:51; **21.** J Med Chem (2005) 48:7513; **22.** Bioconjug Chem (1996) 7:482; **23.** Bioconjug Chem (2001) 12:186; **24.** J Immunol Methods (1982) 50:193; **25.** J Org Chem (1966) 31:3880; **26.** Proc Natl Acad Sci U S A (2000) 97:13021; **27.** J Immunol (2009) 182:3573; **28.** J Biol Chem (2006) 281:18156; **29.** J Virol (2004) 78:4330; **30.** Mol Cell Probes (2004) 18:117; **31.** J Forensic Sci (2003) 48:282; **32.** Bioconjug Chem (2009) 20:702; **33.** Anal Biochem (2004) 332:90; **34.** Mol Pharm (2009) 6:386; **35.** Proc Natl Acad Sci U S A (2003) 100:13308; **36.** J Cell Biol (1996) 134:1197; **37.** Nature (1995) 376:686; **38.** Proc Natl Acad Sci U S A (1994) 91:2659; **39.** Biophys J (1992) 61:956; **40.** Biophys J (1992) 61:631; **41.** Proc Natl Acad Sci U S A (1988) 85:5454.

DATA TABLE 1.6 LONG-WAVELENGTH RHODAMINES, TEXAS RED DYES AND QSY QUENCHERS

Cat. No.	MW	Storage	Soluble	Abs	EC	Em	Solvent	Notes
C300	466.92	L	DMF, DMSO	540	95,000	565	MeOH	1
C652	476.44	L	pH >6, DMF	598	49,000	668	pH 10	2
C653	573.51	F,D,L	DMF, DMSO	602	42,000	672	pH 10	2
C1171	527.53	F,D,L	DMF, DMSO	546	95,000	576	MeOH	1, 3
C1309	631.68	F,D,L	DMF, DMSO	576	80,000	601	MeOH	1
C2211	527.53	F,D,L	DMF, DMSO	546	95,000	579	MeOH	1, 3
C6121	430.46	L	pH >6, DMF	542	91,000	568	MeOH	1
C6122	430.46	L	pH >6, DMF	540	103,000	564	MeOH	1
C6123	527.53	F,D,L	DMF, DMSO	547	91,000	573	MeOH	1, 3
C6124	635.80	F,L	pH >6, DMF	567	92,000	591	MeOH	1
C6125	631.68	F,D,L	DMF, DMSO	574	78,000	602	MeOH	1
C6126	631.68	F,D,L	DMF, DMSO	575	82,000	602	MeOH	1
C6127	555.59	F,D,L	pH >6, DMF	524	108,000	557	MeOH	
C6128	555.59	F,D,L	DMF, DMSO	524	102,000	550	MeOH	
C6156	534.61	F,L	pH >6, DMF	570	113,000	590	MeOH	1
C6157	555.59	F,D,L	DMF, DMSO	524	92,000	552	MeOH	
L20	577.11	F,DD,L	DMF, MeCN	568	88,000	583	MeOH	4
L1908	577.11	F,DD,L	DMF, MeCN	568	88,000	583	MeOH	4
M689	485.98	F,DD,L	DMF, DMSO	629	75,000	none	MeCN	5
Q10193	791.32	F,D,L	DMSO	560	90,000	none	MeOH	
Q20131	951.43	F,D,L	H_2O, DMSO	562	88,000	none	MeOH	6
Q20132	815.34	F,D,L	DMSO	661	90,000	none	MeOH	
R6160	768.90	F,D,L	DMF, DMSO	560	129,000	580	MeOH	
T353	625.15	F,DD,L	DMF, MeCN	588	84,000	601	$CHCl_3$	4
T490	443.52	F,DD,L	DMF, DMSO	544	84,000	572	MeOH	3, 5
T1480	443.52	F,DD,L	DMF, DMSO	543	99,000	571	MeOH	3, 5
T1481	443.52	F,DD,L	DMF, DMSO	544	90,000	572	MeOH	3, 5
T1905	625.15	F,DD,L	DMF, MeCN	587	85,000	602	$CHCl_3$	4
T6105	640.69	F,D,L	DMF, DMSO	543	92,000	571	MeOH	1, 3
T6134	816.94	F,D,L	DMF, DMSO	583	112,000	603	MeOH	
T20175	816.94	F,D,L	DMF, DMSO	587	96,000	602	MeOH	
T30200	796.74	F,D,L	DMF, DMSO	583	87,000	604	MeOH	
X491	547.67	F,DD,L	DMF, DMSO	572	92,000	596	MeOH	5

For definitions of the contents of this data table, see "Using *The Molecular Probes® Handbook*" in the introductory pages.

Notes

1. Abs and Em for TAMRA and ROX dyes in pH 8 buffer are red-shifted approximately 8 nm compared to MeOH, with EC lower by ~10%.
2. Absorption and fluorescence of naphthofluorescein derivatives are pH dependent. Both the absorption and emission spectra shift to much shorter wavelengths at pH <8. Fluorescence quantum yield ~0.14 at pH 9.5.
3. Tetramethylrhodamine protein conjugates often exhibit two absorption peaks at about 520 and 545 nm. The 520 nm peak is due to nonfluorescent dye aggregates. (J Immunol Methods (1991) 143:263, J Phys Chem B (1998) 102:1820)
4. Do NOT dissolve in DMSO.
5. Isothiocyanates are unstable in water and should not be stored in aqueous solution.
6. This sulfonated succinimidyl ester derivative is water soluble and may be dissolved in buffer at ~pH 8 for reaction with amines. Long-term storage in water is NOT recommended due to hydrolysis.

IMPORTANT NOTICE: The products described in this manual are covered by one or more Limited Use Label License(s). Please refer to the Appendix on page 971 and Master Product List on page 975. Products are For Research Use Only. Not intended for any animal or human therapeutic or diagnostic use.

PRODUCT LIST 1.6 LONG-WAVELENGTH RHODAMINES, TEXAS RED DYES AND QSY QUENCHERS

Cat No.	Product	Quantity
C652	5-(and-6)-carboxynaphthofluorescein *mixed isomers*	100 mg
C653	5-(and-6)-carboxynaphthofluorescein, succinimidyl ester *mixed isomers*	25 mg
C6127	5-carboxyrhodamine 6G, succinimidyl ester (5-CR 6G, SE) *single isomer*	5 mg
C6128	6-carboxyrhodamine 6G, succinimidyl ester (6-CR 6G, SE) *single isomer*	5 mg
C6157	5-(and-6)-carboxyrhodamine 6G, succinimidyl ester (5(6)-CR 6G, SE) *mixed isomers*	5 mg
C6121	5-carboxytetramethylrhodamine (5-TAMRA) *single isomer*	10 mg
C6122	6-carboxytetramethylrhodamine (6-TAMRA) *single isomer*	10 mg
C300	5-(and-6)-carboxytetramethylrhodamine (5(6)-TAMRA) *mixed isomers*	100 mg
C2211	5-carboxytetramethylrhodamine, succinimidyl ester (5-TAMRA, SE) *single isomer*	5 mg
C6123	6-carboxytetramethylrhodamine, succinimidyl ester (6-TAMRA, SE) *single isomer*	5 mg
C1171	5-(and-6)-carboxytetramethylrhodamine, succinimidyl ester (5(6)-TAMRA, SE) *mixed isomers*	25 mg
C6124	5-carboxy-X-rhodamine, triethylammonium salt (5-ROX) *single isomer*	10 mg
C6156	6-carboxy-X-rhodamine (6-ROX) *single isomer*	10 mg
C6125	5-carboxy-X-rhodamine, succinimidyl ester (5-ROX, SE) *single isomer*	5 mg
C6126	6-carboxy-X-rhodamine, succinimidyl ester (6-ROX, SE) *single isomer*	5 mg
C1309	5-(and-6)-carboxy-X-rhodamine, succinimidyl ester (5(6)-ROX, SE) *mixed isomers*	25 mg
F6161	FluoReporter® Rhodamine Red™-X Protein Labeling Kit *5–10 labelings*	1 kit
F6162	FluoReporter® Texas Red®-X Protein Labeling Kit *5–10 labelings*	1 kit
L20	Lissamine rhodamine B sulfonyl chloride *mixed isomers*	1 g
L1908	Lissamine rhodamine B sulfonyl chloride *mixed isomers* *special packaging*	10 x 10 mg
M689	malachite green isothiocyanate	10 mg
Q10193	QSY® 7 carboxylic acid, succinimidyl ester	5 mg
Q20131	QSY® 9 carboxylic acid, succinimidyl ester	5 mg
Q20132	QSY® 21 carboxylic acid, succinimidyl ester	5 mg
R6160	Rhodamine Red™-X, succinimidyl ester *5-isomer*	5 mg
T6105	6-(tetramethylrhodamine-5-(and-6)-carboxamido)hexanoic acid, succinimidyl ester (5(6)-TAMRA-X, SE) *mixed isomers*	10 mg
T1480	tetramethylrhodamine-5-isothiocyanate (5-TRITC; G isomer)	5 mg
T1481	tetramethylrhodamine-6-isothiocyanate (6-TRITC; R isomer)	5 mg
T490	tetramethylrhodamine-5-(and-6)-isothiocyanate (5(6)-TRITC) *mixed isomers*	10 mg
T30200	Texas Red® C_2-dichlorotriazine	5 mg
T353	Texas Red® sulfonyl chloride *mixed isomers*	10 mg
T1905	Texas Red® sulfonyl chloride *mixed isomers* *special packaging*	10 x ~1 mg
T10244	Texas Red®-X Protein Labeling Kit *3 labelings*	1 kit
T6134	Texas Red®-X, succinimidyl ester *mixed isomers*	5 mg
T20175	Texas Red®-X, succinimidyl ester *single isomer*	2 mg
X491	X-rhodamine-5-(and-6)-isothiocyanate (5(6)-XRITC) *mixed isomers*	10 mg
Z25045	Zenon® Texas Red®-X Mouse IgG_1 Labeling Kit *50 labelings*	1 kit

1.7 Coumarins, Pyrenes and Other Ultraviolet Light–Excitable Fluorophores

Shorter-wavelength amine-reactive fluorophores are less frequently used for preparing bioconjugates because dyes excited with longer wavelengths, and therefore lower energy, are widely available and less likely to cause photodamage to labeled biomolecules. Moreover, many cells and tissues autofluoresce when excited with ultraviolet (UV) light, producing detection-confounding background signals. However, for certain multicolor fluorescence applications—including immunofluorescence, nucleic acid and protein microarrays, *in situ* hybridization and neuronal tracing—a blue-fluorescent probe provides a contrasting color that is clearly resolved from the green, yellow, orange or red fluorescence of the longer-wavelength probes.

The short-wavelength reactive dyes that we recommend for preparing the brightest blue-fluorescent bioconjugates are the Alexa Fluor® 350, Alexa Fluor® 405, AMCA-X, Marina Blue®, Pacific Blue™ and Cascade Blue® derivatives (Table 1.12). Alexa Fluor® 430, Pacific Orange™ and Cascade Yellow™ dyes fill a spectral void because they exhibit the rare combination of absorption between 400 nm and 450 nm and fluorescence emission beyond 500 nm. The amine-reactive naphthalene, pyrene and Dapoxyl® derivatives are important for the production of environment-sensitive probes in protein structure and function studies (Table 1.13); their thiol-reactive counterparts are discussed in Section 2.3. Many of our UV light–excitable reactive dyes are more commonly employed for such bioanalytical techniques as HPLC derivatization, amino acid sequencing and protein determination and are therefore discussed in Section 1.8.

IMPORTANT NOTICE: The products described in this manual are covered by one or more Limited Use Label License(s). Please refer to the Appendix on page 971 and Master Product List on page 975. Products are For Research Use Only. Not intended for any animal or human therapeutic or diagnostic use.

Table 1.12 Amine-reactive, ultraviolet light–excitable fluorophores for labeling proteins and nucleic acids.

Fluorophore	Abs *	Em *	Amine-Reactive Dyes	Protein and Nucleic Acid Labeling Kits	Notes
Alexa Fluor® 350	346	442	A10168 (SE)	A20180 (Mab) A10170 (P) Z25000 (Z) Z25100 (Z) Z25200 (Z) Z25300 (Z) Z25400 (Z)	• Higher fluorescence per attached dye than AMCA • Protein conjugates emit at slightly shorter wavelengths than AMCA or AMCA-X
Alexa Fluor® 405	402	421	A30000 (SE) A30100 (SE)	Z25013 (Z) Z25113 (Z) Z25213 (Z) Z25313 (Z)	• Cascade Blue® derivative containing a spacer between the fluorophore and the reactive SE • Near-perfect match to the violet diode laser (405 nm)
Alexa Fluor® 430	433	539	A10169 (SE)	A10171 (P) Z25001 (Z) Z25301 (Z)	• Extremely large Stokes shift • Fills spectral gap between green- and yellow-fluorescent dyes
AMCA-X	353	442	A6118 (SE)		• Widely used blue-fluorescent labeling dye • Compact structure
Bimane	380	458	B30250 (COOH)		• Blue-fluorescent dye • Small size
Cascade Blue®	400	420	C2284 (AA)		• Resistant to quenching upon protein conjugation • Trisulfonated pyrene
Dialkylaminocoumarin	375 435	470 [1] 475 [2]	D126 (COOH) D1421 (COOH) D374 (SE) D1412 (SE) D10166 (ITC)		• Longer-wavelength alternatives to AMCA
Hydroxycoumarin	385 360	445 [3] 455 [4]	H185 (COOH) H1428 (COOH) H1193 (SE)		• pH-sensitive fluorescence • Compact structure
Marina Blue®	365	460	M10165 (SE)		• Optimal for 365 nm excitation sources
Methoxycoumarin	358	410	M1420MP (COOH) M1410 (SE)		• pH-insensitive fluorescence • Compact structure
Pacific Blue™	410	455	P10163 (SE)	A10478 (APEX®) P30013 (Mab) P30012 (P) Z25041 (Z) Z25341 (Z)	• Longer-wavelength alternative to the Alexa Fluor® 350 and AMCA-X dyes • Excited with violet diode laser (405 nm)
Pacific Orange™	400	551	P30253 (SE)	P30014 (Mab) P30016 (P) Z25256 (Z) Z25257 (Z)	• Excited with the violet diode laser (405 nm) • Compatible with the Pacific Blue™ dye for two-color analysis using the blue diode laser

* The absorption (Abs) and fluorescence emission (Em) maxima, in nm, listed in this table are for the goat anti-mouse IgG antibody or dextran conjugates in aqueous buffer. (APEX®) = APEX® Antibody Labeling Kit (Section 1.2). (AA) = Acetyl azide. (COOH) = Carboxylic acid. (ITC) = Isothiocyanate. (Mab) = Monoclonal Antibody Labeling Kit (Section 1.2). (P) = Easy-to-Use Protein Labeling Kit (Section 1.2). (SE) = Succinimidyl ester. (Z) = Zenon® Antibody Labeling Kit (Section 7.3). **1.** Spectral maxima for D374; **2.** Spectral maxima for D1412; **3.** Spectral maxima for H1193; **4.** Spectral maxima for 7-hydroxy-4-methylcoumarin-3-acetic acid, succinimidyl ester.

Table 1.13 Amine-reactive, environment-sensitive fluorophores.

Fluorophore	Abs *	Em *	Succinimidyl Ester	Other	Notes
Cascade Yellow™	402	545	C10164		• Fluorescence emission spectrum shifts to shorter wavelengths in nonpolar solvents
Dansyl	335	518	D6104 (X)	D21 (SC)	• Sulfonyl chloride is nonfluorescent until it reacts with amines • Weak fluorescence in aqueous solutions
Dapoxyl®	395	601	D10161	D10160 (SC)	• Very low fluorescence in water • Large Stokes shifts (up to ~200 nm) • Large extinction coefficients of Dapoxyl® derivatives in some solvents [1]
NBD	466	535	S1167 (X)	C20260 (AH) F486 (AH)	• NBD amine derivatives have low fluorescence in water; emission spectra and quantum yields in other solvents are variable [2]
PyMPO	415	570	S6110		• Fluorescence emission spectrum shifts to shorter wavelengths in nonpolar solvents
Pyrene	340	376	P130 P6114 (CASE)	P24 (SC)	• Forms excited-state dimers (excimers) that emit at longer wavelengths (~470 nm) than the lone excited fluorophore • Extremely long fluorescence lifetime (can be >100 ns)

Absorption (Abs) and Emission (Em) maxima, in nm, are for conjugates. (AH) = Aryl halide. (CASE) = Cysteic acid separating the dye and the SE. (SC) = Sulfonyl chloride. (X) = Aminohexanoyl spacer separating the dye and the SE. **1.** Photochem Photobiol (1997) 66:424; **2.** Biochemistry (1977) 16:5150.a

IMPORTANT NOTICE: The products described in this manual are covered by one or more Limited Use Label License(s). Please refer to the Appendix on page 971 and Master Product List on page 975. Products are For Research Use Only. Not intended for any animal or human therapeutic or diagnostic use.

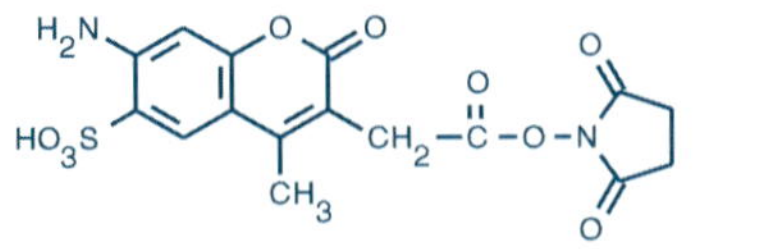

Figure 1.7.1 Alexa Fluor® 350 carboxylic acid, succinimidyl ester (A10168).

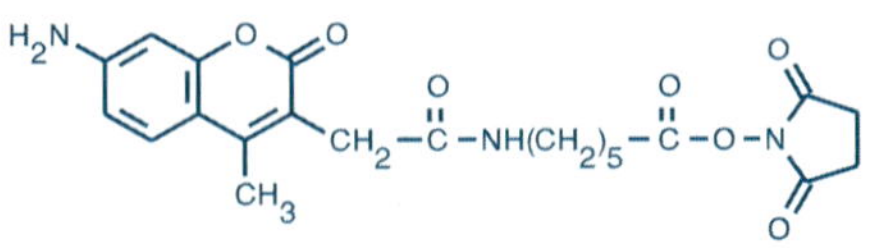

Figure 1.7.2 6-((7-amino-4-methylcoumarin-3-acetyl)amino)hexanoic acid, succinimidyl ester (AMCA-X, SE; A6118).

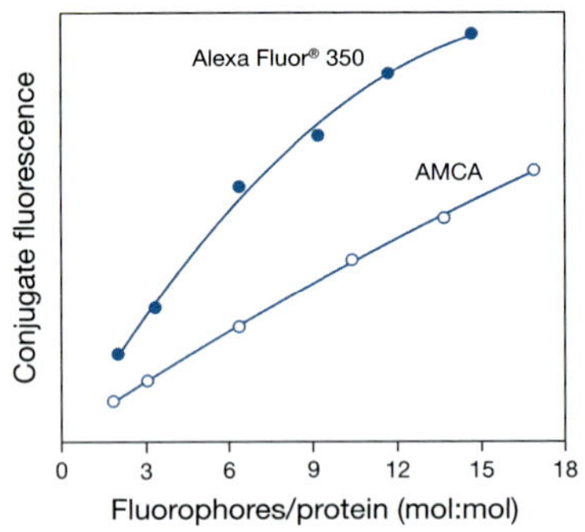

Figure 1.7.3 Comparison of the relative fluorescence of 7-amino-4-methylcoumarin-3-acetic acid (AMCA) streptavidin (O) and Alexa Fluor® 350 streptavidin, a sulfonated AMCA derivative (S11249, ●). Conjugate fluorescence is determined by measuring the fluorescence quantum yield of the conjugated dye relative to that of the free dye and multiplying by the number of fluorophores per protein.

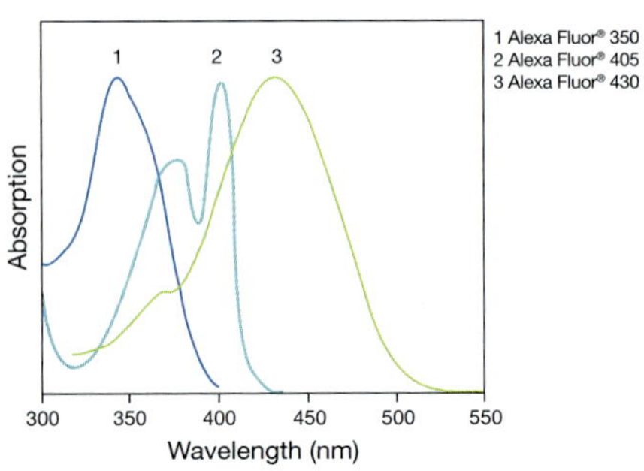

Figure 1.7.4 Absorption spectra of our ultraviolet and blue light-absorbing Alexa Fluor® dyes.

Alexa Fluor® 350 and Other Coumarin Derivatives

Alexa Fluor® 350 and AMCA-X Dyes

Derivatives of 7-aminocoumarin dyes are widely used labeling reagents for preparing protein and nucleic acid conjugates,[1] and we offer two important amine-reactive 7-aminocoumarin derivatives: Alexa Fluor® 350 carboxylic acid succinimidyl ester[2,3] (A10168) and AMCA-X succinimidyl ester (AMCA-X, SE; A6118).

The sulfonated coumarin derivative, Alexa Fluor® 350 carboxylic acid succinimidyl ester (Figure 1.7.1), is more water soluble than either AMCA succinimidyl ester or AMCA-X succinimidyl ester (Figure 1.7.2) and yields protein conjugates that are more fluorescent than those prepared from its nonsulfonated analog (Figure 1.7.3). Alexa Fluor® 350 protein conjugates are optimally excited at 346 nm and have bright blue fluorescence emission (Figure 1.7.4, Figure 1.7.5) at wavelengths slightly shorter than AMCA or AMCA-X conjugates (Figure 1.7.6) (442 nm versus 448 nm), which reduces the dye's spectral overlap with the emission of fluorescein. We offer several reactive versions of Alexa Fluor® 350 dye, including:

- Amine-reactive succinimidyl ester (A10168)
- Thiol-reactive maleimide (A30505, Section 2.3)
- Aldehyde- and ketone-reactive hydrazide and hydroxylamine (A10439, A30627; Section 3.3)
- Aldehyde- and ketone-reactive cadaverine (A30674, Section 3.4)

AMCA-X succinimidyl ester (A6118) contains a seven-atom aminohexanoyl spacer ("X") between the fluorophore and the reactive group. This spacer separates the fluorophore from the biomolecule to which it is conjugated, potentially reducing the quenching that typically occurs upon conjugation and making the dye more available for recognition by secondary detection reagents. Slightly longer-wavelength conjugates can be prepared from the isothiocyanate (DACITC, D10166), succinimidyl esters (D374, D1412) or free acids (D126, D1421) of 7-dialkylaminocoumarins.[4,5]

Alexa Fluor® 430 Dye

Few reactive dyes that absorb between 400 nm and 450 nm have appreciable fluorescence beyond 500 nm in aqueous solution. Alexa Fluor® 430 dye fills this spectral gap.[2] Excitation near its absorption maximum at 431 nm (Figure 1.7.4) is accompanied by strong green fluorescence with an emission maximum at 541 nm. The amine-reactive succinimidyl ester of Alexa Fluor® 430 carboxylic acid (A10169, Figure 1.7.7) is available, as well as Alexa Fluor® 430 conjugates of secondary antibodies (A11063, A11064; Section 7.2, Table 7.1) and streptavidin (S11237, Section 7.6, Table 7.9).

Marina Blue® and Pacific Blue™ Dyes

Marina Blue® and Pacific Blue™ dyes, both of which are based on the 6,8-difluoro-7-hydroxycoumarin fluorophore, exhibit bright blue fluorescence emission near 460 nm[6] (Table 1.12). The Marina Blue® dye is optimally detected using optical filters configured for DAPI, whereas

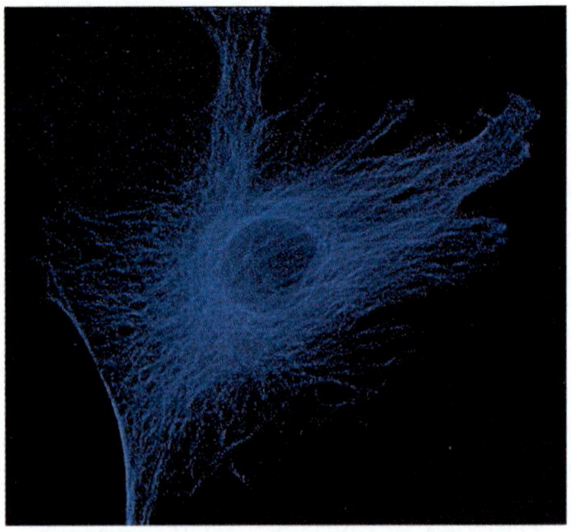

Figure 1.7.5 The microtubules of fixed bovine pulmonary artery endothelial cells (BPAEC) localized with mouse monoclonal anti-α-tubulin antibody (A11126), which was visualized with Alexa Fluor® 350 goat anti–mouse IgG antibody (A11045). The image was acquired using a longpass filter set appropriate for DAPI.

IMPORTANT NOTICE: The products described in this manual are covered by one or more Limited Use Label License(s). Please refer to the Appendix on page 971 and Master Product List on page 975. Products are For Research Use Only. Not intended for any animal or human therapeutic or diagnostic use.
www.invitrogen.com/probes

the Pacific Blue™ dye is ideally suited for 405 nm violet diode laser excitation on the Applied Biosystems® Attune™ Acoustic Focusing cytometer and similarly equipped fluorescence microscopes (Figure 1.7.8). Significantly, the pK_a values of these 6,8-difluoro-7-hydroxycoumarin derivatives are 2–3 log units lower than those of the corresponding 7-hydroxycoumarins (Figure 1.7.9). Thus, the Marina Blue® and Pacific Blue™ dyes yield conjugates that are strongly fluorescent, even at neutral pH. For preparing bioconjugates, we offer amine-reactive succinimidyl esters of the Marina Blue® and Pacific Blue™ dyes (M10165, Figure 1.7.10; P10163, Figure 1.7.11).

Other Hydroxycoumarin and Alkoxycoumarin Derivatives

The hydroxycoumarins (H185, H1193, H1428) exhibit pH-sensitive spectral properties, but the methoxycoumarins (M1410, M1420MP) do not.[7,8] Hydroxycoumarins are often used to prepare reactive intermediates for the synthesis of radioiodinated materials.[9] The spectral properties of the hydroxycoumarins allow their quantitation prior to radioiodination.[9]

Alexa Fluor® 350, Alexa Fluor® 430 and Pacific Blue™ Protein Labeling Kits

For easy and trouble-free labeling of proteins with succinimidyl esters of the Alexa Fluor® 350, Alexa Fluor® 430 and Pacific Blue™ dyes, we offer Alexa Fluor® 350, Alexa Fluor® 430 and Pacific Blue™ Protein Labeling Kits (A10170, A10171, A10171; Table 1.2). These kits, which are described in greater detail in Section 1.2, contain everything that is required to perform three separate labeling reactions and to purify the resulting conjugates (Table 1.3). The Alexa Fluor® 350 and Pacific Blue™ Monoclonal Antibody Labeling Kits (A20180, P30013) can be used to prepare blue-fluorescent conjugates of monoclonal antibodies, as well as of other proteins in limited quantities (five labeling reactions of ~100 µg each). The APEX® Pacific Blue™ Antibody Labeling Kit (A10478) utilizes a solid-phase technique to label 10–20 µg IgG antibody, even in the presence of stabilizing proteins or amine-containing buffers.

The Zenon® Alexa Fluor® 350, Zenon® Alexa Fluor® 430 and Zenon® Pacific Blue™ Antibody Labeling Kits (Table 7.7) permit the rapid and quantitative labeling of antibodies—even submicrogram amounts—using a purified antibody fraction or a crude antibody preparation such as serum, ascites fluid or a hybridoma supernatant. These kits, along with Zenon® technology, are described in detail in Section 7.3.

Pacific Orange™ Dye

The succinimidyl ester of the Pacific Orange™ dye (P30253) yields conjugates with excitation/emission maxima of ~400/551 nm, making it ideal for use with 405 nm violet diode laser–equipped flow cytometers[10] and fluorescence microscopes. Moreover, Pacific Blue™ and Pacific Orange™ conjugates can be simultaneously excited at 405 nm and emit at 455 and 551 nm, respectively, facilitating two-color analysis.

Several of our kits facilitate protein labeling with the Pacific Orange™ succinimidyl ester, including the Pacific Orange™ Protein Labeling Kit (P30016), the Pacific Orange™ Monoclonal Antibody Labeling Kit (P30014) and the Zenon® Antibody Labeling Kits (Table 7.7), all of which are described in greater detail in Section 1.2.

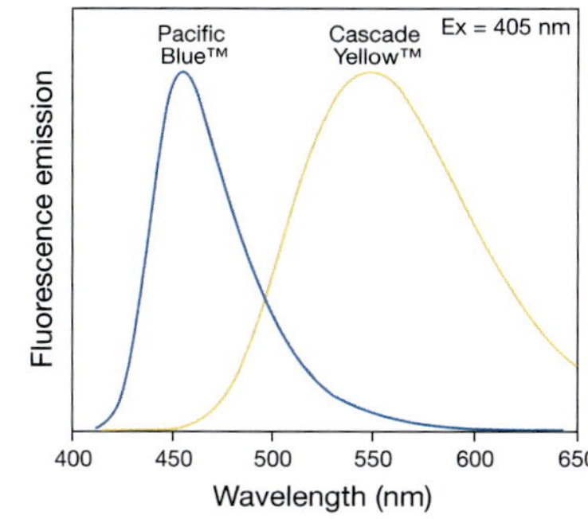

Figure 1.7.8 Normalized fluorescence emission spectra of Pacific Blue™ goat anti–mouse IgG antibody (P10993) and a Cascade Yellow™ goat anti–mouse IgG antibody conjugate prepared with the Cascade Yellow™ succinimidyl ester (C10164). Both fluorescent conjugates are excited at 405 nm. When samples containing equal concentrations of antibody are compared, the peak fluorescence intensity of the Pacific Blue™ conjugate at 456 nm is nine times greater than that of the Cascade Yellow™ conjugate at 548 nm.

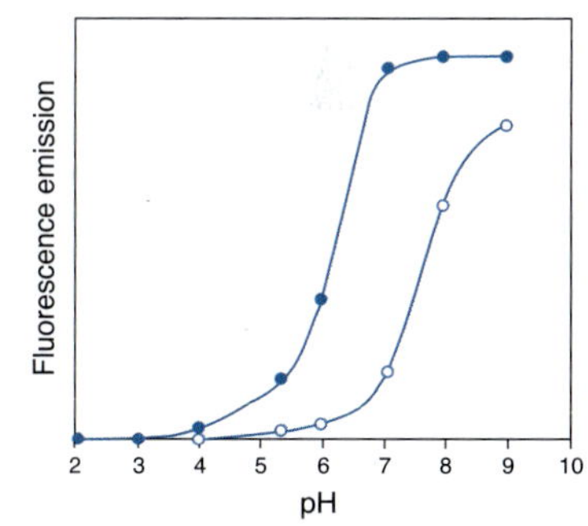

Figure 1.7.9 Comparison of the pH dependent fluorescence changes produced by attachment of electron-withdrawing fluorine atoms to a hydroxycoumarin. 7-Hydroxy-4-methylcoumarin-3-acetic acid (○, H1428) and 6,8-difluoro-7-hydroxy-4-methylcoumarin (●, D6566). Fluorescence intensities were measured for equal concentrations of the two dyes using excitation/emission at 360/450 nm.

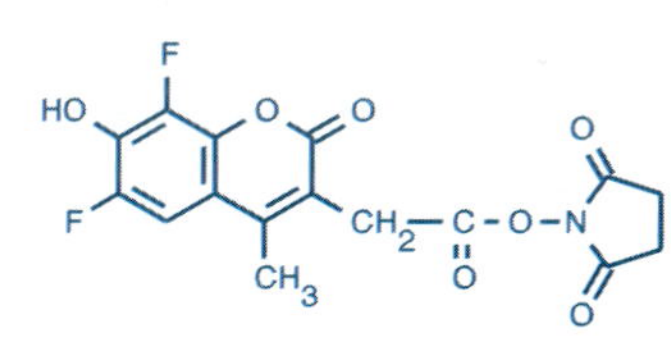

Figure 1.7.10 Marina Blue® succinimidyl ester (M10165).

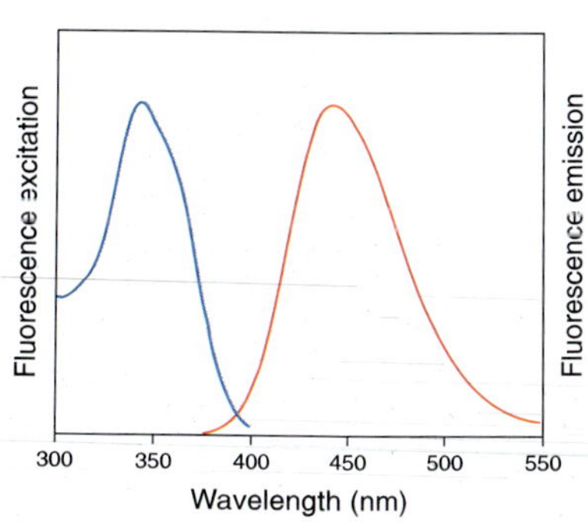

Figure 1.7.6 Fluorescence excitation and emission spectra of Alexa Fluor® 350 goat anti–mouse IgG antibody in pH 8.0 buffer.

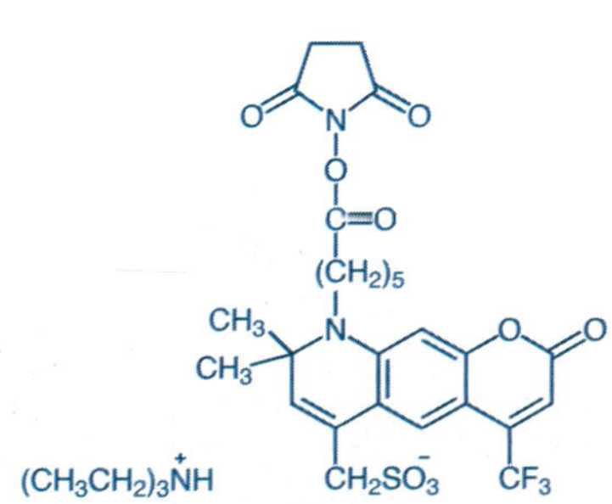

Figure 1.7.7 Alexa Fluor® 430 carboxylic acid, succinimidyl ester (A10169).

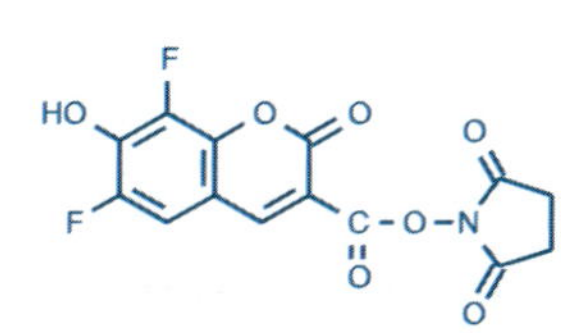

Figure 1.7.11 Pacific Blue™ succinimidyl ester (P10163).

IMPORTANT NOTICE: The products described in this manual are covered by one or more Limited Use Label License(s). Please refer to the Appendix on page 971 and Master Product List on page 975. Products are For Research Use Only. Not intended for any animal or human therapeutic or diagnostic use.

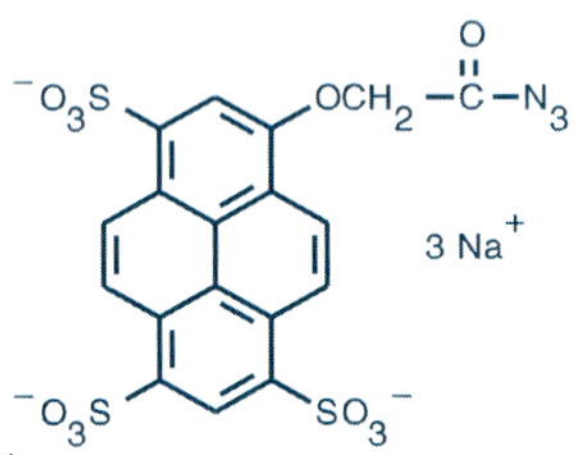

Figure 1.7.12 Cascade Blue® acetyl azide, trisodium salt (C2284).

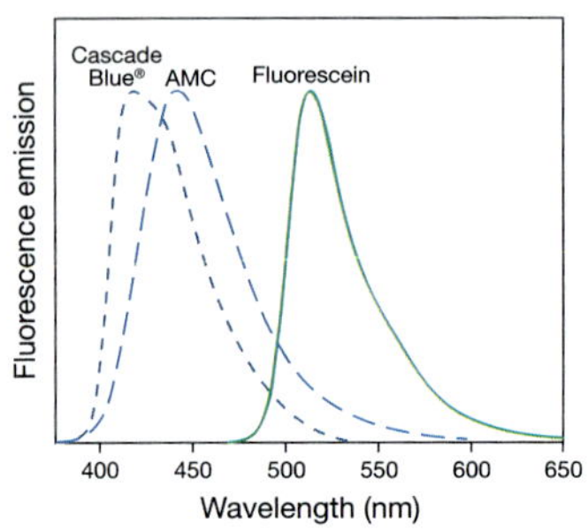

Figure 1.7.13 Normalized fluorescence emission spectra of Cascade Blue® (CB), 7-amino-4-methylcoumarin (AMC) and fluorescein in aqueous solutions.

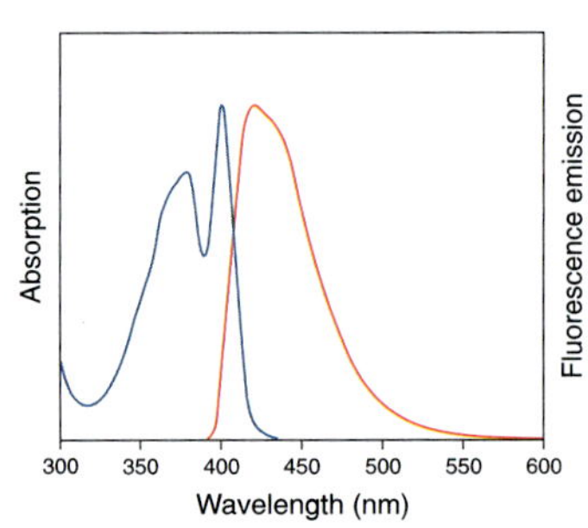

Figure 1.7.14 Absorption and fluorescence emission spectra of Cascade Blue® dye-labeled bovine serum albumin (BSA) in pH 7.0 buffer.

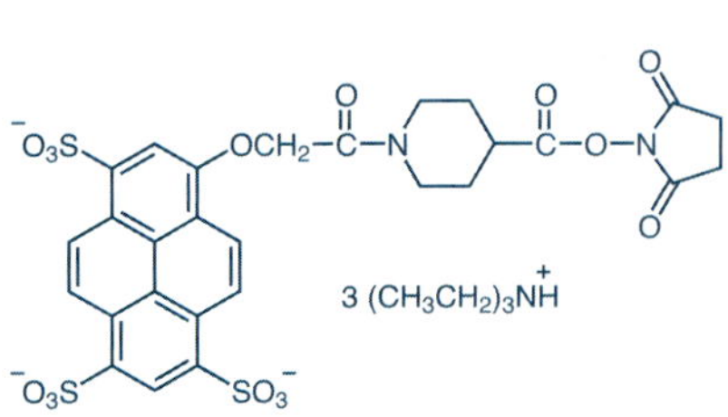

Figure 1.7.16 Alexa Fluor® 405 carboxylic acid, succinimidyl ester (A30000).

Cascade Blue® and Other Pyrene Derivatives

Cascade Blue® Acetyl Azide

Cascade Blue® acetyl azide is the amine-reactive sulfonated pyrene derivative [11] that we use to prepare blue-fluorescent Cascade Blue® dye–labeled proteins and dextrans. The polar nature of this reagent makes it difficult to purify to homogeneity; however, we offer a Cascade Blue® acetyl azide preparation (C2284, Table 1.12, Figure 1.7.12) that is ~60% reactive and packaged according to the net weight of the reactive dye. The remaining constituents are inorganic salts or unreactive forms of the dye that can readily be removed following conjugation.

As compared with the aminocoumarin derivatives, the Cascade Blue® fluorophore shows less spectral overlap with fluorescein (Figure 1.7.13), an important advantage for multicolor applications. In addition, this reactive Cascade Blue® derivative has high absorptivity (Figure 1.7.14), is highly fluorescent and resists quenching upon protein conjugation (Figure 1.7.15). Even at low degrees of labeling, Cascade Blue® conjugates are significantly more fluorescent than are those of 7-amino-4-methylcoumarin-3-acetic acid (AMCA),[11] and they remain preferred reagents for multicolor flow cytometry.[12–14]

Alexa Fluor® 405 Dye

With excitation/emission maxima of 402/421 nm (Figure 1.7.4), Alexa Fluor® 405 dye is well matched to the 405 nm spectral line of violet diode lasers for fluorescence microscopy and flow cytometry.[10,15,16] Alexa Fluor® 405 succinimidyl ester is an amine-reactive derivative of our Cascade Blue® dye. Not only is it offered at higher purity than the alternative Cascade Blue® acetyl azide, but Alexa Fluor® 405 succinimidyl ester also contains a 4-piperidinecarboxylic acid spacer that separates the fluorophore from its reactive moiety (Figure 1.7.16). This spacer enhances the reactivity of the succinimidyl ester and minimizes any interactions between the fluorophore and the biomolecule to which it is conjugated.

As with Cascade Blue® acetyl azide, Alexa Fluor® 405 dye shows minimal spectral overlap with green fluorophores, making it ideal for multicolor applications.[15,17] However, the violet fluorescence of Cascade Blue® and Alexa Fluor® 405 dyes is less visible to the human eye in fluorescence microscopy applications than the blue fluorescence of Alexa Fluor® 350 and AMCA-X dyes. Alexa Fluor® 405 dye is available as:

- Amine-reactive succinimidyl ester (A30000, A30100)
- Aldehyde- and ketone-reactive cadaverine (A30675, Section 3.4)

We also prepare Alexa Fluor® 405 conjugates of secondary antibodies (Section 7.2, Table 7.1) and streptavidin (Section 7.6, Table 7.9). Alexa Fluor® 405 conjugates are recognized by the anti–Alexa Fluor® 405/Cascade Blue® dye antibody (A5760, Section 7.4).

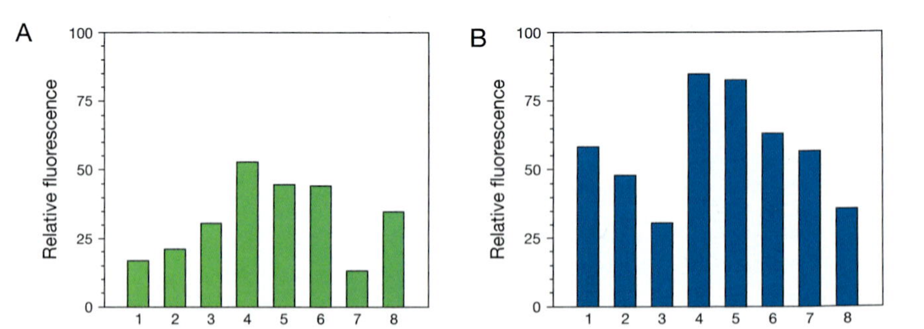

Figure 1.7.15 Histograms showing the fluorescence per fluorophore for **A)** fluorescein and **B)** Cascade Blue® conjugated to various proteins, relative to the fluorescence of the free dye in aqueous solution, represented by 100 on the y-axis. The proteins represented are: **1)** avidin, **2)** bovine serum albumin, **3)** concanavalin A, **4)** goat IgG, **5)** ovalbumin, **6)** protein A, **7)** streptavidin and **8)** wheat germ agglutinin.

IMPORTANT NOTICE: The products described in this manual are covered by one or more Limited Use Label License(s). Please refer to the Appendix on page 971 and Master Product List on page 975. Products are For Research Use Only. Not intended for any animal or human therapeutic or diagnostic use.
www.invitrogen.com/probes

Other Pyrenes

Conjugates of the pyrene succinimidyl esters (P130, P6114) have exceptionally long excited-state lifetimes (sometimes >100 nanoseconds), relatively short-wavelength emission and capacity for proximity-dependent excimer formation (Figure 1.7.17). These amine-reactive pyrene derivatives have primarily been used for labeling and detecting oligonucleotides,[18,19] biogenic amines[20] and polyamines.[21] Pyrene binds strongly to carbon nanotubes via pi-stacking interactions. This property makes 1-pyrenebutanoic acid succinimidyl ester (P130) a valuable reagent for functionalizing these remarkable nanomaterials for coupling to proteins.[22,23]

The long fluorescence lifetime of pyrenebutyric acid (1-pyrenebutanoic acid) permits time-gating of the fluorescence, which is a useful technique for discriminating between the dye signal and sample autofluorescence,[24] and has been exploited for fluorescence immunoassays.[25] For preparing pyrene conjugates with long fluorescence lifetimes, we recommend the more water-soluble succinimidyl ester of *N*-(1-pyrenebutanoyl)cysteic acid (P6114, Figure 1.7.18). The amine-reactive 1-pyrenesulfonyl chloride (P24, Section 1.8) has been used to generate a fluorescent ATP sensor via modification of an ATP-binding ribonucleopeptide.[26]

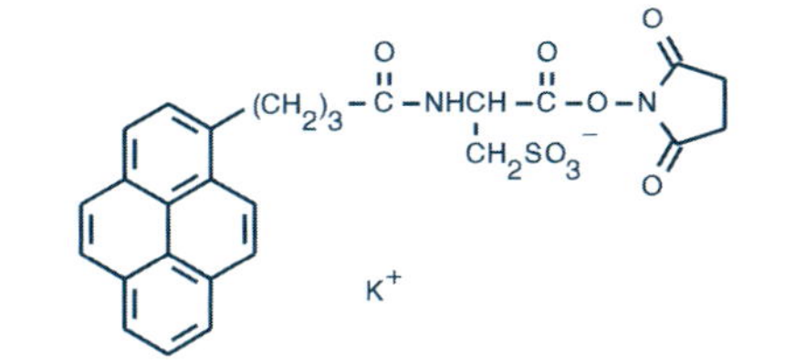

Figure 1.7.18 *N*-(1-pyrenebutanoyl)cysteic acid, succinimidyl ester, potassium salt (P6114).

Cascade Yellow™ and Other Pyridyloxazole Derivatives

Cascade Yellow™ Dye

Like the Alexa Fluor® 430 and Pacific Blue™ dyes described above, the Cascade Yellow™ dye exhibits an excitation maximum that falls between those of the UV light–excited dyes and the fluoresceins. This sulfonated pyridyloxazole (PyMPO) laser dye (Figure 1.7.19) exhibits an absorption maximum near 410 nm and an unusually high Stokes shift, with relatively strong emission at 550–570 nm[27–29] (Figure 1.7.20). The large Stokes shift permits detection at a wavelength well beyond that of most sample autofluorescence, and allows multiple fluorophores to be excited at the same wavelengths and detected at different wavelengths. For example, protein conjugates of Cascade Yellow™ succinimidyl ester (C10164) can be simultaneously excited at 405 nm with Pacific Blue™ conjugates, and then separately detected at longer wavelengths (Figure 1.7.8). Cascade Yellow™ and Cascade Blue® antibody conjugates, along with several phycobiliprotein tandem conjugates, are utilized in an 11-color polychromatic flow cytometry technique.[12,13]

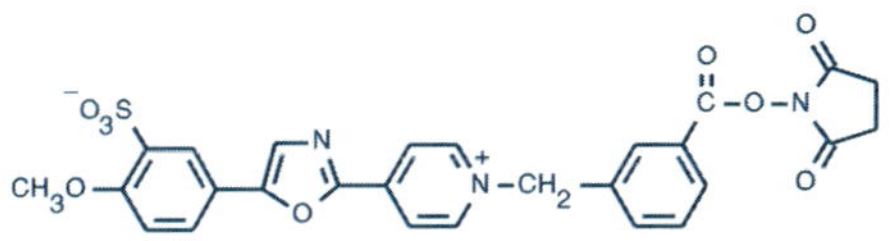

Figure 1.7.19 Cascade Yellow™ succinimidyl ester (C10164).

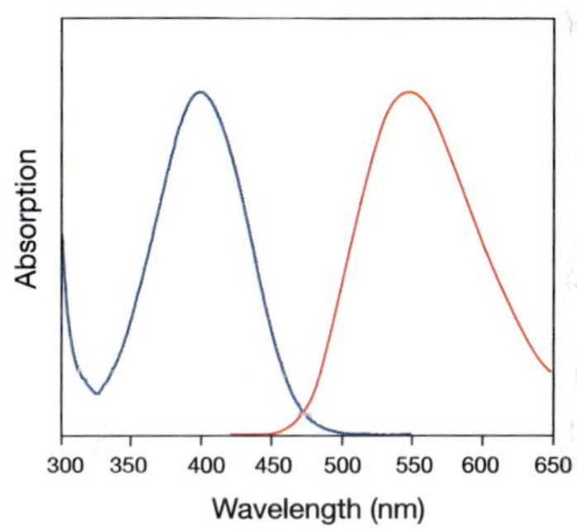

Figure 1.7.20 Absorption and fluorescence emission spectra of Cascade Yellow™ goat anti–mouse IgG antibody in pH 8.0 buffer.

PyMPO Dye

The pyridyloxazole derivatives—including the succinimidyl ester (PyMPO, SE; S6110; Figure 1.7.21) and the thiol-reactive maleimide (M6026, Section 2.2)—fill the spectral gap between UV light–excited dyes and the fluoresceins. These derivatives of the laser dye PyMPO exhibit absorption maxima near 415 nm and unusually high Stokes shifts, with emission at 560–580 nm.[28] Like the naphthalene-based dyes, the pyridyloxazole dyes exhibit environment-sensitive fluorescence spectra. PyMPO SE has been used to synthesize fluorescent gramicidin derivatives for following ion channel–gating processes.[30]

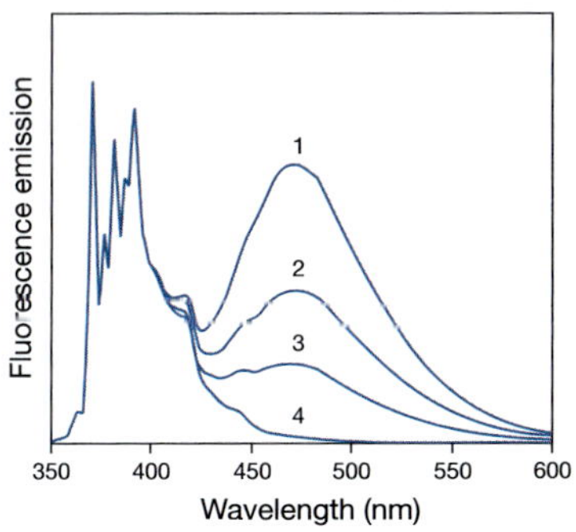

Figure 1.7.17 Excimer formation by pyrene in ethanol. Spectra are normalized to the 371.5 nm peak of the monomer. All spectra are essentially identical below 400 nm after normalization. Spectra are as follows: 1) 2 mM pyrene, purged with argon to remove oxygen; 2) 2 mM pyrene, air-equilibrated; 3) 0.5 mM pyrene (argon-purged); and 4) 2 µM pyrene (argon-purged). The monomer-to excimer ratio (371.5 nm/470 nm) is dependent on both pyrene concentration and the excited-state lifetime, which is variable because of quenching by oxygen.

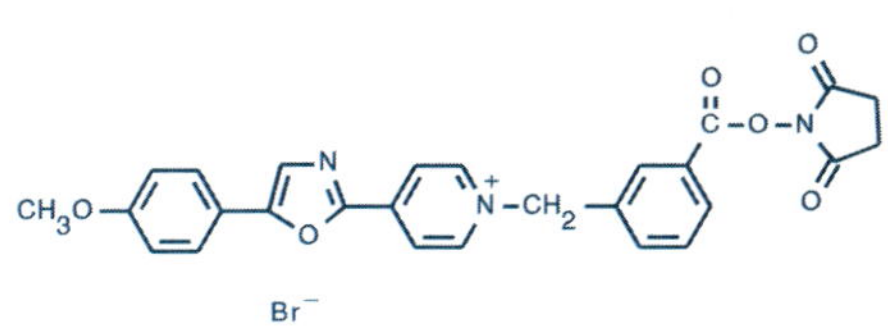

Figure 1.7.21 1-(3-(succinimidyloxycarbonyl)benzyl)-4-(5-(4-methoxyphenyl)oxazol-2-yl)pyridinium bromide (PyMPO, SE; S6110).

IMPORTANT NOTICE: The products described in this manual are covered by one or more Limited Use Label License(s). Please refer to the Appendix on page 971 and Master Product List on page 975. Products are For Research Use Only. Not intended for any animal or human therapeutic or diagnostic use.

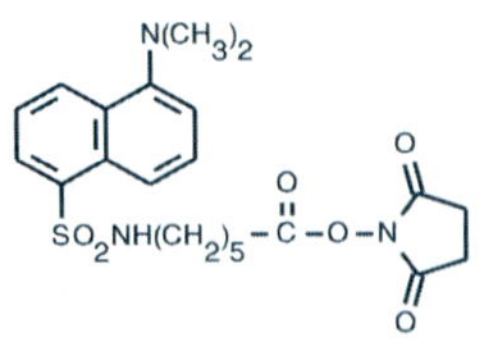

Figure 1.7.22 6-((5-dimethylaminonaphthalene-1-sulfonyl) amino)hexanoic acid, succinimidyl ester (dansyl-X, SE; D6104).

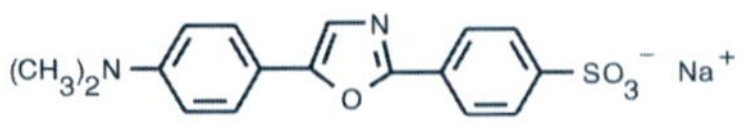

Figure 1.7.23 Dapoxyl® sulfonic acid, sodium salt (D12800).

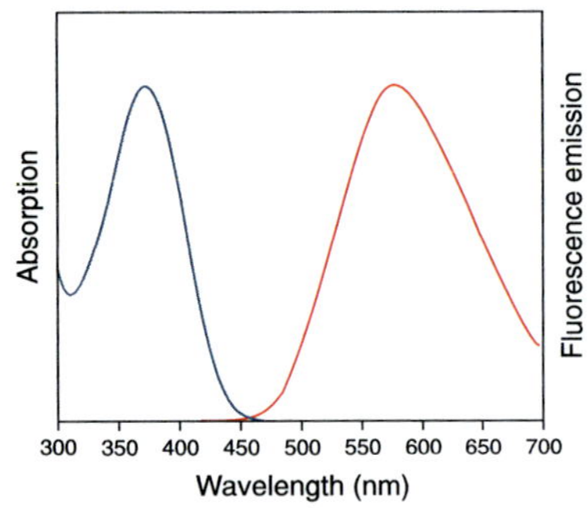

Figure 1.7.24 Absorption and fluorescence emission spectra of Dapoxyl® (2-aminoethyl)sulfonamide in methanol.

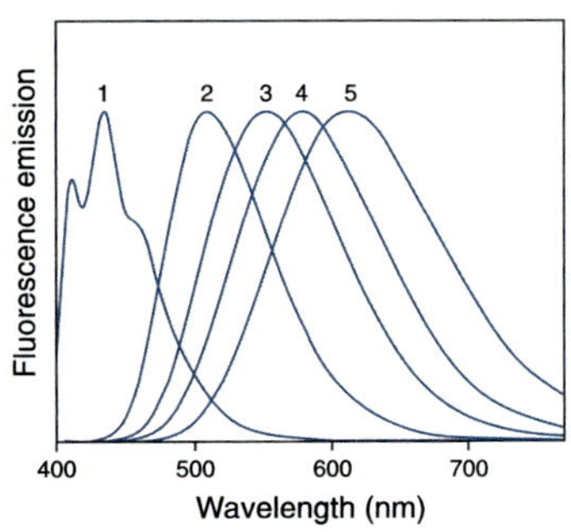

Figure 1.7.25 Normalized fluorescence emission spectra of Dapoxyl® (2-aminoethyl)sulfonamide (D10460) in **1)** hexane, **2)** chloroform, **3)** acetone, **4)** acetonitrile and **5)** 1:1 acetonitrile:water.

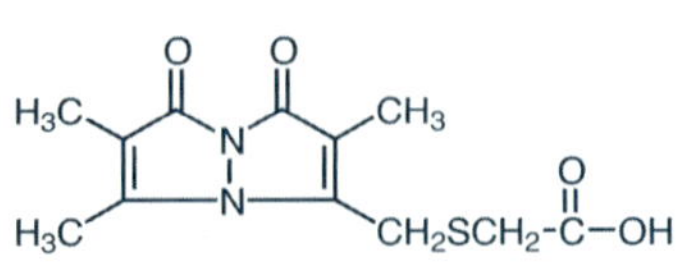

Figure 1.7.26 Bimane mercaptoacetic acid (carboxymethylthiobimane, B30250).

Naphthalenes, Including Dansyl Chloride

Aminonaphthalene-based probes tend to have emission spectra that are sensitive to the environment and to exhibit weak fluorescence in aqueous solution. Spectra of environment-sensitive probes respond to perturbations in the local environment (Table 1.13). For example, changes in solvation that occur because of ligand binding, protein assembly or protein denaturation can often evoke changes in the fluorescence properties of these probes. This property has made dansyl chloride (5-dimethylaminonaphthalene-1-sulfonyl chloride, D21) and other aminonaphthalene-based dyes important tools for protein structural studies.

Dansyl chloride is nonfluorescent until it reacts with amines. The resulting dansyl amides have environment-sensitive fluorescence quantum yields and emission maxima, along with large Stokes shifts. Despite the weak absorptivity (EC ~4000 $cm^{-1}M^{-1}$ at 330–340 nm) and moderate fluorescence quantum yield of dansyl sulfonamides, dansyl chloride is widely used as a derivatization reagent for end-group analysis of proteins, amino acid analysis and HPLC detection (Section 1.8). The succinimidyl ester of dansylaminohexanoic acid (dansyl-X, SE; D6104; Figure 1.7.22) contains a seven-atom spacer ("X") that places the dansyl fluorophore further from its reaction site, potentially reducing the interaction of the fluorophore with the biomolecule to which it is conjugated and enhancing accessibility to antibody binding.[31–33] A rabbit polyclonal antibody to the 1,5-dansyl fluorophore (A6398) that significantly enhances the dye's fluorescence is described in Section 7.4.

Conjugates of two isomers of dansyl chloride (2,5-dansyl chloride, D22; 2,6-dansyl chloride, D23) have smaller Stokes shifts and appreciably longer fluorescence lifetimes (up to ~30 nanoseconds) than conjugates of 1,5-dansyl chloride, making these isomers among the best available probes for fluorescence depolarization studies.[34] These dyes are particularly useful for preparing fluorescent drug or ligand analogs that are expected to bind to hydrophobic sites in proteins or membranes. The lipophilicity of these reagents may also facilitate the labeling of sites within the membrane-spanning portions of cellular proteins.

Dapoxyl® Dye

Dapoxyl® dye (Figure 1.7.23) is a particularly versatile derivatization reagent and precursor to environment-sensitive probes.[35] Like Cascade Yellow™ dye, Dapoxyl® dye exhibits an exceptionally large Stokes shift, with excitation/emission maxima of ~370/580 nm (Figure 1.7.24). Sulfonamides from Dapoxyl® sulfonyl chloride (D10160) have much higher extinction coefficients than those of dansyl chloride (~26,000 $cm^{-1}M^{-1}$ versus about 4000 $cm^{-1}M^{-1}$) and equal or greater quantum yields when dissolved in organic solvents; however, the fluorescence of Dapoxyl® derivatives is very sensitive to the dye environment, and fluorescence in water is very low, making them useful for sensing conformational changes,[36] denaturation and phosphorylation states[37] of proteins.

In addition to Dapoxyl® sulfonyl chloride, we offer the amine-reactive Dapoxyl® succinimidyl ester (D10161) and the carboxylic acid–reactive Dapoxyl® (2-aminoethyl)sulfonamide (D10460, Section 3.4). We have also exploited the environment-sensitive fluorescence of the Dapoxyl® dye (Figure 1.7.25) to develop a highly selective and photostable stain for the endoplasmic reticulum (ER-Tracker™ Blue-White DPX, E12353; Section 12.4).

Bimane Derivative

Bimane mercaptoacetic acid (carboxymethylthiobimane, B30250; Figure 1.7.26) is a blue-fluorescent dye with excitation/emission maxima of ~380/458 nm. It is useful as a reference standard for the fluorogenic monobromobimane and monochlorobimane reagents (Section 2.3) because it is an analog of the thioether product of their reaction with glutathione and other thiols.

IMPORTANT NOTICE: The products described in this manual are covered by one or more Limited Use Label License(s). Please refer to the Appendix on page 971 and Master Product List on page 975. Products are For Research Use Only. Not intended for any animal or human therapeutic or diagnostic use.

REFERENCES

1. Histochem J (1986) 18:497; **2.** J Histochem Cytochem (1999) 47:1179; **3.** Bioorg Med Chem Lett (1999) 9:2229; **4.** J Biol Chem (1996) 271:31160; **5.** Biochemistry (1988) 27:8889; **6.** Bioorg Med Chem Lett (1998) 8:3107; **7.** Anal Chem (1968) 40:803; **8.** Nat Protoc (2007) 2:227; **9.** FEBS Lett (1985) 182:185; **10.** Cytometry A (2006) 69:1153; **11.** Anal Biochem (1991) 198:119; **12.** Nat Med (2001) 7:245; **13.** Nat Biotechnol (2002) 20:155; **14.** Cytometry (1997) 29:328; **15.** Nat Protoc (2009) 4:372; **16.** Proc Natl Acad Sci U S A (2006) 103:12063; **17.** Proc Natl Acad Sci U S A (2005) 102:5346; **18.** Biochemistry (2008) 47:6279; **19.** Nucleic Acids Res (2002) 30:e19; **20.** Anal Sci (2004) 20:557; **21.** J Chromatogr A (2002) 946:265; **22.** Anal Biochem (2008) 381:193; **23.** J Am Chem Soc (2001) 123:3838; **24.** J Biochem Biophys Methods (1994) 29:157; **25.** Anal Biochem (1988) 174:101; **26.** J Am Chem Soc (2006) 128:12932; **27.** Cytometry (1998) 33:435; **28.** IEEE J Quantum Electronics (1980) 16:777; **29.** Cytometry (1999) 36:36; **30.** Bioconjug Chem (2001) 12:594; **31.** Biochim Biophys Acta (1992) 1104:9; **32.** Biochim Biophys Acta (1984) 776:217; **33.** Biochemistry (1982) 21:978; **34.** J Biochem Biophys Methods (1981) 5:1; **35.** Photochem Photobiol (1997) 66:424; **36.** J Am Chem Soc (2006) 128:11736; **37.** Biochim Biophys Acta (2008) 1784:94.

DATA TABLE 1.7 COUMARINS, PYRENES AND OTHER ULTRAVIOLET LIGHT-EXCITABLE FLUOROPHORES

Cat. No.	MW	Storage	Soluble	Abs	EC	Em	Solvent	Notes
A6118	443.46	F,D,L	DMF, DMSO	353	19,000	442	MeOH	
A10168	410.35	F,D,L	H_2O, DMSO	346	19,000	445	pH 7	1
A10169	701.75	F,D,L	H_2O, DMSO	430	15,000	545	pH 7	1
A30000	1028.26	F,DD,L	H_2O, DMSO	400	35,000	424	pH 7	1, 2, 3
B30250	282.31	F,D,L	DMSO	380	5700	458	MeOH	
C2284	607.42	F,D,LL	H_2O, MeOH	396	29,000	410	MeOH	2, 4
C10164	563.54	F,D,L	DMF, DMSO	409	24,000	558	MeOH	5
D21	269.75	F,DD,L	DMF, MeCN	372	3900	none	$CHCl_3$	6, 7
D22	269.75	F,DD,L	DMF, MeCN	403	2900	none	MeOH	7, 8
D23	269.75	F,DD,L	DMF, MeCN	380	16,000	none	$CHCl_3$	7, 8
D126	247.25	L	pH >6, DMF	370	22,000	459	MeOH	
D374	344.32	F,D,L	DMF, MeCN	376	22,000	468	MeOH	
D1412	358.35	F,D,L	DMSO, MeCN	442	64,000	483	pH 9	9
D1421	261.28	L	pH >6, DMF	409	34,000	473	pH 9	
D6104	461.53	F,D,L	DMF, MeCN	335	4200	518	MeOH	
D10160	362.83	F,DD,L	DMF, MeCN	403	22,000	see Notes	MeOH	7, 10
D10161	405.41	F,D,L	DMF, DMSO	395	20,000	601	MeOH	11
D10166	260.31	F,DD,L	DMF, MeCN	400	36,000	476	MeOH	12, 13
H185	206.15	L	pH >6, DMF	386	29,000	448	pH 10	14
H1193	303.23	F,D,L	DMF, MeCN	419	36,000	447	MeOH	
H1428	234.21	L	pH >6, DMF	360	19,000	455	pH 10	
M1410	317.25	F,D,L	DMF, MeCN	358	26,000	410	MeOH	
M1420MP	220.18	L	pH >6, DMF	336	20,000	402	pH 9	
M10165	367.26	F,D,L	DMF, MeCN	362	19,000	459	pH 9	
P24	300.76	F,DD,L	DMF, MeCN	350	28,000	380	MeOH	7, 15
P130	385.42	F,D,L	DMF, DMSO	340	43,000	376	MeOH	16
P6114	574.65	F,D,L	H_2O, DMSO	341	38,000	376	MeOH	1, 16
P10163	339.21	F,D,L	DMF, MeCN	416	46,000	451	pH 9	
P30253	~750	F,D,L	H_2O, DMSO	404	25,000	553	MeOH	1
S6110	564.39	F,D,L	DMF, DMSO	415	26,000	570	MeOH	5

For definitions of the contents of this data table, see "Using *The Molecular Probes® Handbook*" in the introductory pages.

Notes

1. This sulfonated succinimidyl ester derivative is water soluble and may be dissolved in buffer at ~pH 8 for reaction with amines. Long-term storage in water is NOT recommended due to hydrolysis.
2. The Alexa Fluor® 405 and Cascade Blue® dyes have a second absorption peak at about 376 nm with EC ~80% of the 395–400 nm peak.
3. A30100 is an alternative packaging of A30000 but is otherwise identical.
4. Unstable in water. Use immediately.
5. Fluorescence emission spectrum shifts to shorter wavelengths in nonpolar solvents.
6. D21 butylamine derivative has Abs = 337 nm (EC = 5300 $cm^{-1}M^{-1}$), Em = 492 nm in $CHCl_3$. Em and QY are highly solvent dependent: Em = 496 nm (QY = 0.45) in dioxane, 536 nm (QY = 0.28) in MeOH and 557 nm (QY = 0.03) in H_2O. (Biochemistry (1967) 6:3408) EC typically decreases upon conjugation to proteins (EC = 3400 $cm^{-1}M^{-1}$ at 340 nm). (Biochemistry (1986) 25:513) Fluorescence lifetimes (τ) of protein conjugates are typically 12–20 nanoseconds. (Arch Biochem Biophys (1969) 133:263, Arch Biochem Biophys (1968) 128:163)
7. Do NOT dissolve in DMSO.
8. D22 butylamine derivative: Abs = 375 nm (EC = 3100 $cm^{-1}M^{-1}$), Em = 470 nm in MeOH. D23 butylamine derivative: Abs = 375 nm (EC = 13,000 $cm^{-1}M^{-1}$), Em = 419 nm in $CHCl_3$.
9. D1412 reaction product with 1-butylamine has Abs = 427 nm (EC = 48,000 $cm^{-1}M^{-1}$), Em = 478 nm in pH 9 buffer.
10. D10160 fluorescence is very weak. Reaction product with butylamine has Abs = 373 nm (EC = 26,000 $cm^{-1}M^{-1}$), Em = 551 nm.
11. D10161 butylamine derivative: Abs = 367 nm (EC = 25,000 $cm^{-1}M^{-1}$), Em = 574 nm in MeOH. QY of the derivative is approximately 15-fold higher than the unreacted reagent.
12. Isothiocyanates are unstable in water and should not be stored in aqueous solution.
13. D10166 butylamine derivative: Abs = 376 nm (EC = 25,000 $cm^{-1}M^{-1}$), Em = 469 nm in MeOH. QY of the derivative is approximately 6-fold higher than the unreacted reagent.
14. H185 Abs = 339 nm (EC = 19,000 $cm^{-1}M^{-1}$), Em = 448 nm at pH 4.
15. Spectra of the reaction product with butylamine.
16. Pyrene derivatives exhibit structured spectra. The absorption maximum is usually about 340 nm with a subsidiary peak at about 325 nm. There are also strong absorption peaks below 300 nm. The emission maximum is usually about 376 nm with a subsidiary peak at 396 nm. Excimer emission at about 470 nm may be observed at high concentrations.

IMPORTANT NOTICE: The products described in this manual are covered by one or more Limited Use Label License(s). Please refer to the Appendix on page 971 and Master Product List on page 975. Products are For Research Use Only. Not intended for any animal or human therapeutic or diagnostic use.

PRODUCT LIST 1.7 COUMARINS, PYRENES AND OTHER ULTRAVIOLET LIGHT EXCITABLE FLUOROPHORES

Cat. No.	Product	Quantity
A10168	Alexa Fluor® 350 carboxylic acid, succinimidyl ester	5 mg
A30000	Alexa Fluor® 405 carboxylic acid, succinimidyl ester	1 mg
A30100	Alexa Fluor® 405 carboxylic acid, succinimidyl ester	5 mg
A10169	Alexa Fluor® 430 carboxylic acid, succinimidyl ester	5 mg
A20180	Alexa Fluor® 350 Monoclonal Antibody Labeling Kit *5 labelings*	1 kit
A10170	Alexa Fluor® 350 Protein Labeling Kit *3 labelings*	1 kit
A10171	Alexa Fluor® 430 Protein Labeling Kit *3 labelings*	1 kit
A6118	6-((7-amino-4-methylcoumarin-3-acetyl)amino)hexanoic acid, succinimidyl ester (AMCA-X, SE)	10 mg
A10478	APEX® Pacific Blue™ Antibody Labeling Kit	1 kit
B30250	bimane mercaptoacetic acid (carboxymethylthiobimane)	5 mg
C2284	Cascade Blue® acetyl azide, trisodium salt	5 mg
C10164	Cascade Yellow™ succinimidyl ester	5 mg
D10161	Dapoxyl® carboxylic acid, succinimidyl ester	5 mg
D10160	Dapoxyl® sulfonyl chloride	10 mg
D21	5-dimethylaminonaphthalene-1-sulfonyl chloride (dansyl chloride)	1 g
D22	2-dimethylaminonaphthalene-5-sulfonyl chloride	100 mg
D23	2-dimethylaminonaphthalene-6-sulfonyl chloride	100 mg
D6104	6-((5-dimethylaminonaphthalene-1-sulfonyl)amino)hexanoic acid, succinimidyl ester (dansyl-X, SE)	25 mg
D1421	7-diethylaminocoumarin-3-carboxylic acid	100 mg
D1412	7-diethylaminocoumarin-3-carboxylic acid, succinimidyl ester	25 mg
D126	7-dimethylaminocoumarin-4-acetic acid (DMACA)	100 mg
D374	7-dimethylaminocoumarin-4-acetic acid, succinimidyl ester (DMACA, SE)	25 mg
D10166	7-dimethylamino-4-methylcoumarin-3-isothiocyanate (DACITC)	10 mg
H185	7-hydroxycoumarin-3-carboxylic acid *reference standard*	100 mg
H1193	7-hydroxycoumarin-3-carboxylic acid, succinimidyl ester	25 mg
H1428	7-hydroxy-4-methylcoumarin-3-acetic acid	100 mg
M10165	Marina Blue® succinimidyl ester	5 mg
M1420MP	7-methoxycoumarin-3-carboxylic acid	100 mg
M1410	7-methoxycoumarin-3-carboxylic acid, succinimidyl ester	25 mg
P30013	Pacific Blue™ Monoclonal Antibody Labeling Kit *5 labelings*	1 kit
P30012	Pacific Blue™ Protein Labeling Kit *3 labelings*	1 kit
P10163	Pacific Blue™ succinimidyl ester	5 mg
P30014	Pacific Orange™ Monoclonal Antibody Labeling Kit *5 labelings*	1 kit
P30016	Pacific Orange™ Protein Labeling Kit *3 labelings*	1 kit
P30253	Pacific Orange™ succinimidyl ester *triethylammonium salt*	1 mg
P130	1-pyrenebutanoic acid, succinimidyl ester	100 mg
P6114	N-(1-pyrenebutanoyl)cysteic acid, succinimidyl ester, potassium salt	5 mg
P24	1-pyrenesulfonyl chloride	100 mg
S6110	1-(3-(succinimidyloxycarbonyl)benzyl)-4-(5-(4-methoxyphenyl)oxazol-2-yl)pyridinium bromide (PyMPO, SE)	5 mg
Z25400	Zenon® Alexa Fluor® 350 Human IgG Labeling Kit *50 labelings*	1 kit
Z25000	Zenon® Alexa Fluor® 350 Mouse IgG_1 Labeling Kit *50 labelings*	1 kit
Z25100	Zenon® Alexa Fluor® 350 Mouse IgG_{2a} Labeling Kit *50 labelings*	1 kit
Z25200	Zenon® Alexa Fluor® 350 Mouse IgG_{2b} Labeling Kit *50 labelings*	1 kit
Z25300	Zenon® Alexa Fluor® 350 Rabbit IgG Labeling Kit *50 labelings*	1 kit
Z25013	Zenon® Alexa Fluor® 405 Mouse IgG_1 Labeling Kit *50 labelings*	1 kit
Z25113	Zenon® Alexa Fluor® 405 Mouse IgG_{2a} Labeling Kit *50 labelings*	1 kit
Z25213	Zenon® Alexa Fluor® 405 Mouse IgG_{2b} Labeling Kit *50 labelings*	1 kit
Z25313	Zenon® Alexa Fluor® 405 Rabbit IgG Labeling Kit *50 labelings*	1 kit
Z25001	Zenon® Alexa Fluor® 430 Mouse IgG_1 Labeling Kit *50 labelings*	1 kit
Z25301	Zenon® Alexa Fluor® 430 Rabbit IgG Labeling Kit *50 labelings*	1 kit
Z25041	Zenon® Pacific Blue™ Mouse IgG_1 Labeling Kit *50 labelings*	1 kit
Z25341	Zenon® Pacific Blue™ Rabbit IgG Labeling Kit *50 labelings*	1 kit
Z25256	Zenon® Pacific Orange™ Mouse IgG_1 Labeling Kit *50 labelings*	1 kit
Z25257	Zenon® Pacific Orange™ Mouse IgG_{2a} Labeling Kit *50 labelings*	1 kit

IMPORTANT NOTICE: The products described in this manual are covered by one or more Limited Use Label License(s). Please refer to the Appendix on page 971 and Master Product List on page 975. Products are For Research Use Only. Not intended for any animal or human therapeutic or diagnostic use.

1.8 Reagents for Analysis of Low Molecular Weight Amines

Not only are low molecular weight amines abundantly distributed in nature, but numerous drugs, synthetic probes and other molecules of interest also contain amino groups. The sensitive detection, identification and quantitation of amines are important applications of many of the reactive fluorophores in this section. Some of these reagents have also been used to indirectly detect carbohydrates, carboxylic acids, thiols and cyanide.

The preferred reagents for detecting and quantitating amines in solution or on amine-containing polymers are those that are nonfluorescent but form fluorescent conjugates stoichiometrically with amines. It is difficult to compare the sensitivity for amine detection of the different reagents because it depends heavily on the equipment and detection technology used. Many of the assays, however, are rapid, reliable and adaptable to a variety of different sample types and instrumentation.

Fluorescamine

Fluorescamine (F2332, F20261) is intrinsically nonfluorescent but reacts rapidly with primary aliphatic amines, including those in peptides and proteins, to yield a blue-green–fluorescent derivative[1,2] (Figure 1.8.1). Modifications to the reaction protocol permit fluorescamine to be used to detect those amino acids containing secondary amines,[3] such as proline. Excess reagent is rapidly converted to a nonfluorescent product by reaction with water,[4] making fluorescamine useful for determining protein concentrations of solutions.[5,6]

Fluorescamine can also be used to detect proteins in gels and to analyze low molecular weight amines by TLC, HPLC and capillary electrophoresis.[7,8] An optimized procedure that employs fluorescamine for amino acid analysis in microplates has been published.[9] Chiral separation of fluorescamine labeled amino acids has been optimized using capillary electrophoresis in the presence of hydroxypropyl-β-cyclodextrin, a method designed for use in extraterrestrial exploration on Mars.[10] Furthermore, a 200-fold increase in sensitivity and improved resolution in these measurements has been obtained by replacing fluorescamine with Pacific Blue™ succinimidyl ester[11] (P10163, Section 1.7).

Dialdehydes: OPA and NDA

Analyte Detection with OPA and NDA

The homologous aromatic dialdehydes *o*-phthaldialdehyde[12] (OPA, P2331MP) and naphthalene-2,3-dicarboxaldehyde[13] (NDA, N1138) are essentially nonfluorescent until reacted with a primary amine in the presence of a thiol such as 2-mercaptoethanol, 3-mercaptopropionic acid or the less obnoxious sulfite,[14] or in the presence of excess cyanide, to yield a fluorescent isoindole (Figure 1.8.2, Figure 1.8.3). Improved detection sensitivity can be obtained by using SAMSA fluorescein (A685, Section 5.2) as the thiol reagent, thereby incorporating fluorescein as the R^2 substituent of the isoindole product[15] (Figure 1.8.2). Modified protocols that use an excess of an amine and limiting amounts of other nucleophiles permit the determination of carboxylic acids[16] and thiols,[17] as well as of cyanide in blood, urine and other samples.[18–21] Without an additional nucleophile, NDA forms fluorescent adducts with both hydrazine and methylated hydrazines[22] (excitation/emission maxima ~403/500 nm).

Sensitivity of OPA and NDA

Amine adducts of NDA have longer-wavelength spectral characteristics and greater sensitivity than the amine adducts of OPA. The stability and detectability of the amine derivatives of NDA are also superior;[23,24] the detection of glycine with NDA and cyanide is reported to be 50-fold more sensitive than with OPA and 2-mercaptoethanol.[13] The limit for electrochemical detection of the NDA adduct of asparagine has been determined to be as low as 36 attomoles[25,26] (36×10^{-18} moles). An optimized procedure that uses NDA for amino acid analysis in microplates has been published.[9]

Applications for OPA and NDA

OPA and NDA are used extensively for both pre- and post-column derivatization of amines (and thiols) separated by HPLC[27,28] or by capillary electrophoresis.[27,29] The amines in a single cell have been analyzed by capillary electrophoresis using a sequence of on-capillary lysis, derivatization with NDA and cyanide, and laser-excited detection.[30,31]

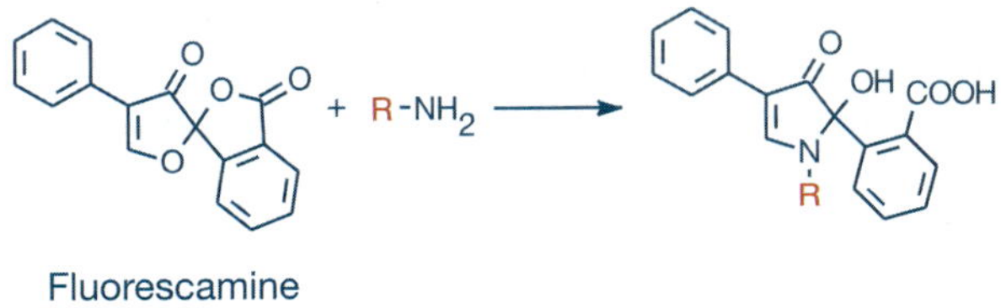

Figure 1.8.1 Fluorogenic amine-derivatization reaction of fluorescamine (F2332, F20261).

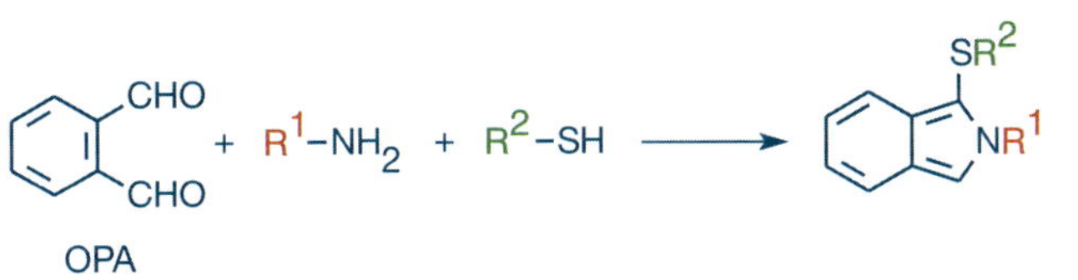

Figure 1.8.2 Fluorogenic amine-derivatization reaction of *o*-phthaldialdehyde (OPA, P2331MP).

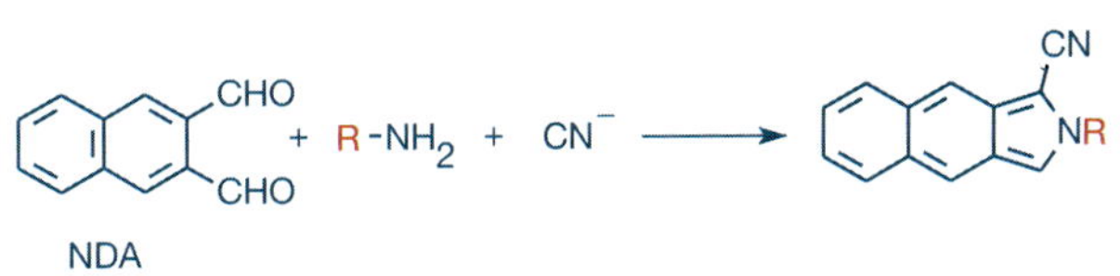

Figure 1.8.3 Fluorogenic amine-derivatization reaction of naphthalene-2,3-dicarboxaldehyde (NDA, N1138).

IMPORTANT NOTICE: The products described in this manual are covered by one or more Limited Use Label License(s). Please refer to the Appendix on page 971 and Master Product List on page 975. Products are For Research Use Only. Not intended for any animal or human therapeutic or diagnostic use.

ATTO-TAG™ Reagents

Sensitivity of ATTO-TAG™ CBQCA and ATTO-TAG™ FQ

ATTO-TAG™ CBQCA (A6222, A2333) and ATTO-TAG™ FQ (A10192, A2334) are designed to provide ultrasensitive detection of primary amines, including those in peptides, aminophospholipids and glycoproteins.[32–35] These reagents combine high sensitivity, visible-wavelength excitation and freedom from background fluorescence, making them useful for research, as well as analytical, forensic and clinical[36] applications. Developed by Novotny and collaborators, the ATTO-TAG™ reagents are similar to OPA and NDA in that they rapidly react with amines in the presence of thiols or cyanide to form highly fluorescent isoindoles[37–47] (Figure 1.8.4).

ATTO-TAG™ CBQCA reagent reacts specifically with amines to form charged conjugates that can be analyzed by electrophoresis techniques. Carbohydrates lacking amines can be detected following reductive amination with ammonia and $NaCNBH_3$.[46,48,49] ATTO-TAG™ CBQCA conjugates are maximally excited at ~456 nm or by the 442 nm spectral line of the He-Cd laser, with peak emission at ~550 nm, whereas ATTO-TAG™ FQ conjugates are maximally excited at ~480 nm or by the 488 nm spectral line of the argon-ion laser, with peak emission at ~590 nm. Ultrasensitive detection of CBQCA-derivatized amino sugars, amino acids and low molecular weight peptides by capillary electrophoresis has been reported.[35,50] In capillary electrophoresis, the sensitivity of amine detection of the laser-induced fluorescence is in the subattomole range ($<10^{-18}$ moles) for ATTO-TAG™ CBQCA and subfemtomole range ($<10^{-15}$ moles) for ATTO-TAG™ FQ.[51,52] Detection sensitivity of reductively aminated glucose using ATTO-TAG™ CBQCA is reported to be 75 zeptomoles[53] (75×10^{-21} moles).

ATTO-TAG™ reagents can, of course, be used in HPLC and other modes of chromatography with either absorption or fluorescence detection. The principal limitation to obtaining ultrasensitive detection using the ATTO-TAG™ reagents and all other chemical derivatization reagents is that relatively high concentrations of the derivatizing reagent are required to obtain adequate kinetics and quantitative modification of the analyte.[54,55] A very sensitive assay that uses ATTO-TAG™ CBQCA for rapid quantitation of protein amines in solution (C6667) is described in Section 9.2. Similarly, ATTO-TAG™ CBQCA has proven useful for *in situ* quantitation of proteins attached to microspheres.[56]

ATTO-TAG™ Reagents and Kits

Because cyclodextrins have been reported to amplify the signal from ATTO-TAG™ CBQCA conjugates up to 10-fold,[34,57] we have included β-cyclodextrin in our ATTO-TAG™ Amine Derivatization Kits (A2333, A2334). These kits contain:

- 5 mg of ATTO-TAG™ CBQCA (in Kit A2333) or ATTO-TAG™ FQ (in Kit A2334)
- Potassium cyanide
- β-Cyclodextrin
- Protocol for amine modification

The ATTO-TAG™ CBQCA and ATTO-TAG™ FQ Amine Derivatization Kits supply sufficient reagents for derivatizing approximately 150 and 100 samples, respectively, depending on the amine concentration and sample volume.

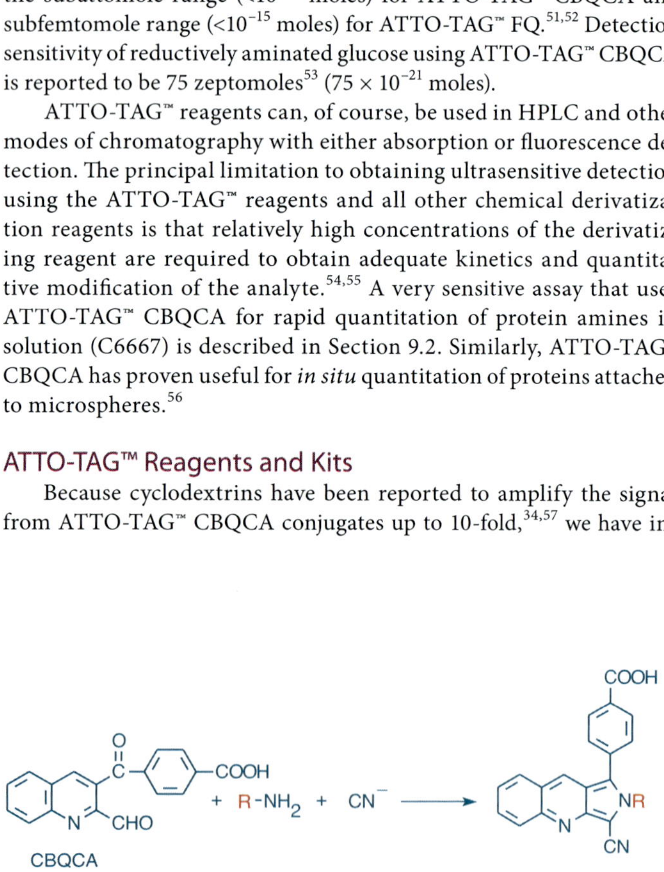

Figure 1.8.4 Fluorogenic amine-derivatization reaction of CBQCA (A6222, A2333).

7-Nitrobenz-2-Oxa-1,3-Diazole (NBD) Derivatives

NBD chloride (C20260, Figure 1.8.5) was first introduced in 1968 as a fluorogenic derivatization reagent for amines.[58] NBD fluoride (F486) usually yields the same products as NBD chloride but is much more reactive;[59] for example, the reaction of NBD fluoride with glycine is reported to be 500 times faster than the reaction of NBD chloride with glycine.[60] Reaction of NBD fluoride with alcohols leads to their utility for derivatizing and detecting lipopolysaccharides[61] (LPS). Unlike OPA and fluorescamine, both NBD chloride and NBD fluoride react with secondary amines and are therefore capable of derivatizing proline and hydroxyproline.[59,62] NBD chloride and NBD fluoride are extensively used as derivatization reagents for chromatographic analysis of amino acids[63] and other low molecular weight amines.[33]

The absorption and fluorescence emission spectra, quantum yields and extinction coefficients of NBD conjugates are all markedly dependent on solvent;[64,65] in particular, the fluorescence quantum yield in water of NBD adducts of amines can be very low (<0.01), particularly of secondary amines. NBD adducts of aromatic amines are essentially nonfluorescent, a property that we have utilized to prepare our QSY® 35 quenchers.

Fluorescence of lysine-modified NBD-labeled actin is sensitive to polymerization.[66] Inactivation of certain ATPases by NBD chloride apparently involves a tyrosine modification followed by intramolecular migration of the label to a lysine residue.[67,68] NBD is also a functional analog of the dinitrophenyl hapten, and its fluorescence is quenched upon binding to anti-dinitrophenyl antibodies[65,69] (Section 7.4).

NBD aminohexanoic acid (NBD-X, N316) and its succinimidyl ester (NBD-X, SE; S1167) are precursors to NBD-labeled phospholipids (Section 13.2), NBD C_6-ceramide (N1154, Section 12.4) and other probes.

Dansyl Chloride and Other Sulfonyl Chlorides

Many of the sulfonyl chlorides described in Section 1.7, including dansyl chloride (D21), 1-pyrenesulfonyl chloride (P24) and Dapoxyl® sulfonyl chloride (D10160), react with amines to yield blue- or blue-green–fluorescent sulfonamides and are particularly useful as chromatographic derivatization reagents. They react with both aliphatic and aromatic amines to yield very stable derivatives. In addition, they are

IMPORTANT NOTICE: The products described in this manual are covered by one or more Limited Use Label License(s). Please refer to the Appendix on page 971 and Master Product List on page 975. Products are For Research Use Only. Not intended for any animal or human therapeutic or diagnostic use.

generally good acceptors for fluorescence resonance energy transfer (FRET) from tryptophan, as well as good donors to longer-wavelength dyes such as QSY® dyes (Section 1.6) (Fluorescence Resonance Energy Transfer (FRET)—Note 1.2). Fluorescence of dansyl conjugates in aqueous solutions can be enhanced by adding cycloheptaamylose.[70]

Although dansyl chloride is the most commonly used of these reagents, the stronger absorption of 1-pyrenesulfonamides and large Stokes shift of Dapoxyl® sulfonamides[71] (Figure 1.8.6) should make these sulfonyl chlorides more sensitive reagents for amine analysis. Note that sulfonyl chlorides are unstable in dimethylsulfoxide (DMSO) and should never be used in that solvent.[72]

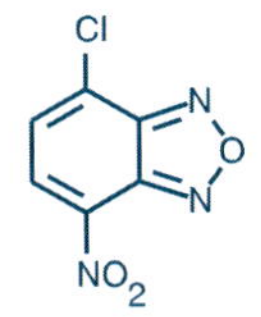

Figure 1.8.5 4-chloro-7-nitrobenz-2-oxa-1,3-diazole (NBD chloride; 4-chloro-7-nitrobenzofurazan, C20260).

Dansyl Chloride

Since its development by Weber in 1951,[73] dansyl chloride (D21, Section 1.7) has been used extensively to determine the N-terminal amino acid residue of proteins and to prepare fluorescent derivatives of drugs, amino acids, oligonucleotides and proteins for detection by numerous chromatographic methods.[74] Nonfluorescent dansyl chloride reacts with amines to form fluorescent dansyl amides that exhibit large Stokes shifts, along with environment-sensitive fluorescence quantum yields and emission maxima.

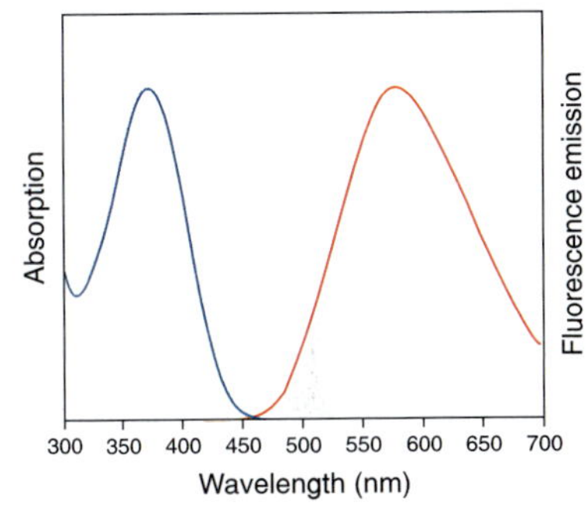

Figure 1.8.6 Absorption and fluorescence emission spectra of Dapoxyl® (2-aminoethyl)sulfonamide in methanol.

Pyrene Sulfonyl Chloride

The absorptivity (and therefore ultimate fluorescence output) of dansyl derivatives is weak compared with that of the more strongly UV light–absorbing fluorophores such as pyrene. Thus, 1-pyrenesulfonyl chloride (P24, Section 1.7) should have greater sensitivity for detection of amines. The fluorescence lifetime of pyrenesulfonamides can also be relatively long (up to ~30 nanoseconds), making them useful for fluorescence anisotropy measurements.[75] Fluorescence polarization measurements of DNA probes labeled with 1-pyrenesulfonyl chloride permit homogeneous detection of hybridization.[76]

Dapoxyl® Sulfonyl Chloride

Sulfonamides derived from Dapoxyl® sulfonyl chloride (D10160, Section 1.7) have much higher extinction coefficients than those of dansyl chloride (~22,000 $cm^{-1}M^{-1}$ versus ~4000 $cm^{-1}M^{-1}$) and equal or greater quantum yields when dissolved in organic solvents; however, Dapoxyl® derivatives have very low fluorescence in water. The huge Stokes shifts (up to ~200 nm) and large extinction coefficients of Dapoxyl® derivatives in some solvents[71] (Figure 1.8.6) make the reactive Dapoxyl® derivatives a good choice for derivatization reagents in chromatographic and electrophoretic analysis.[77]

FITC and Other Isothiocyanates

Isothiocyanates for preparing bioconjugates have been described in several sections of this chapter. However, FITC (F143, F1906, F1907; Section 1.5) can also be used for derivatizing low molecular weight amines[33] and, like phenyl isothiocyanate, for microsequencing of peptides as their thiohydantoins.[78] A method for specific derivatization of the N-terminus of peptides by FITC has been described.[79] FITC-labeled amino acids and peptides have been separated by capillary electrophoresis with a detection limit of fewer than 1000 molecules.[80,81]

Succinimidyl Esters and Carboxylic Acids

Succinimidyl esters have a high selectivity for reaction with aliphatic amines. Most of the succinimidyl ester reagents described elsewhere in this chapter can be used to derivatize low molecular weight amines for subsequent separation by chromatography or capillary electrophoresis. Alexa Fluor®, BODIPY®, Oregon Green® and fluorescein derivatives typically yield the greatest sensitivity, particularly when the conjugate is detected with laser excitation. Use of single isomers of these reactive dyes is essential for all high-resolution analyses. Analysis by capillary electrophoresis shows that carboxyfluorescein succinimidyl ester reacts faster and yields more stable amine conjugates than FITC or DTAF.[82]

IMPORTANT NOTICE: The products described in this manual are covered by one or more Limited Use Label License(s). Please refer to the Appendix on page 971 and Master Product List on page 975. Products are For Research Use Only. Not intended for any animal or human therapeutic or diagnostic use.

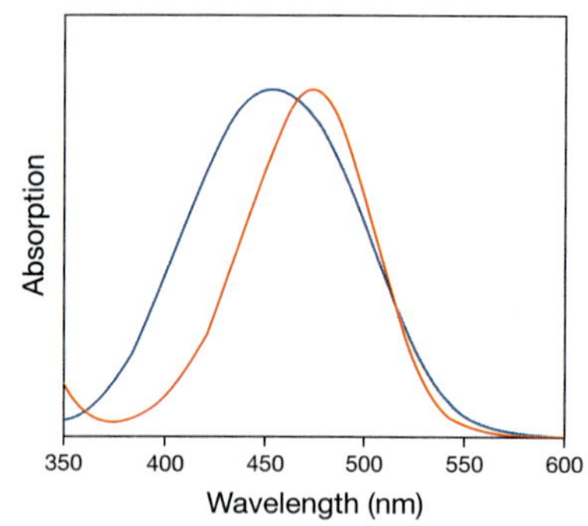

Figure 1.8.7 Normalized absorption spectra of the succinimidyl esters of the dabcyl (D2245, blue) and QSY® 35 (Q20133, red) dyes.

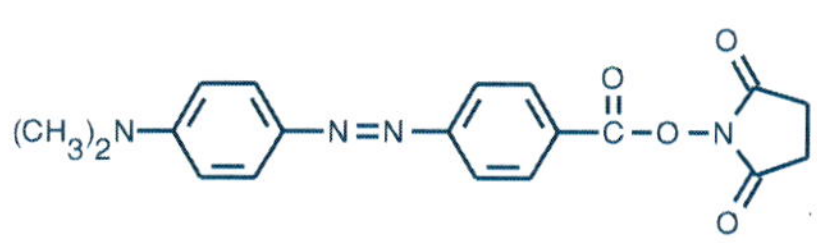

Figure 1.8.8 4-((4-(dimethylamino)phenyl)azo)benzoic acid, succinimidyl ester (dabcyl, SE; D2245).

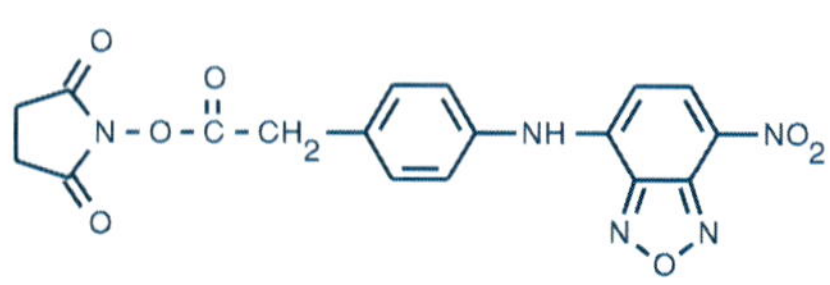

Figure 1.8.10 QSY® 35 acetic acid, succinimidyl ester (Q20133).

The UV light–excitable coumarins described in Section 1.7 have good absorptivity at ~320–420 nm, with purple to bright blue emission at 400–500 nm. Aliphatic polyamines derivatized with 1-pyrenebutanoic acid succinimidyl ester (P130, Section 1.7) have been differentiated from pyrene-labeled monoamines by HPLC using their fluorescent excimer formation.[83]

The Smallest Reactive Fluorophore

N-methylisatoic anhydride (M25) is a useful precursor for preparing esters or amides of the small *N*-methylanthranilic acid fluorophore. The small size of this fluorophore should reduce the likelihood that the label will interfere with the function of the biomolecule, an important advantage when designing site-selective probes. This amine-acylating reagent is often used to prepare fluorescent derivatives of biologically active peptides and toxins[84–87] and, in combination with a quencher, to prepare fluorogenic endoprotease substrates.[88,89]

Chromophoric Succinimidyl Esters: Fluorescence Quenchers

Dabcyl has broad and intense visible absorption (Figure 1.8.7) but no fluorescence, making it useful as an acceptor in FRET applications (Fluorescence Resonance Energy Transfer (FRET)—Note 1.2). Biomolecules double-labeled with dabcyl and the appropriate fluorophore can be used to monitor proteolytic cleavage, conformational changes and other dynamic spatial movements. Dabcyl succinimidyl ester (dabcyl, SE; D2245; Figure 1.8.8) is particularly useful in preparing quenched fluorogenic substrates for proteases, including our HIV protease (Figure 1.8.9) and renin substrates[90–92] (H2930, R2931; Section 10.4), papain,[93,94] Alzheimer disease-associated proteases[95] and others.[96–99] Fluorogenic substrates using this quenching group have also been prepared for interleukin-1β–converting enzyme (ICE),[100] a cysteine protease that is proposed to function in the onset of apoptosis.[101] The dabcyl chromophore has been used as the quencher in donor–acceptor labeled oligonucleotides (molecular beacons);[102–105] unfolding of these probes upon hybridization leads to recovery of the donor dye's fluorescence.

QSY® 35 acetic acid succinimidyl ester (Q20133, Figure 1.8.10) is an essentially nonfluorescent nitrobenzoxadiazole (NBD) derivative. Like the QSY® 7, QSY® 9 and QSY® 21 dyes (Section 1.6), the QSY® 35 dye has absorption at longer wavelengths than does the dabcyl dye (Figure 1.8.7), making it a very good acceptor from most blue-fluorescent dyes. A peptide containing the QSY® 35 quencher paired with the blue-fluorescent 7-hydroxy-4-methyl-3-acetylcoumarin fluorophore has proven useful in a fluorescence resonance energy transfer (FRET) assay for *Bacillus anthracis* lethal factor protease.[106] QSY® 35 iodoacetamide (Q20348, Section 2.2) and an FMOC-protected QSY® 35 amino acid (Q21931, Section 9.5) are available for automated preparation of FRET-based protease substrates.

N-(*t*-BOC)-Aminooxyacetic Acid TFP Ester

The tetrafluorophenyl ester (TFP) of *N*-(*t*-BOC)-aminooxyacetic acid (B30300) is an amine-reactive protected hydroxylamine that is useful for synthesizing new aldehyde- and ketone-reactive probes in an organic solvent. Following coupling to aliphatic amines, the *t*-BOC group can be quantitatively removed with trifluoroacetic acid. The resultant hydroxylamine can then spontaneously react with aldehydes, the reducing ends of saccharides and oligosaccharides and abasic sites in oligonucleotides to form stable adducts (Section 3.3).

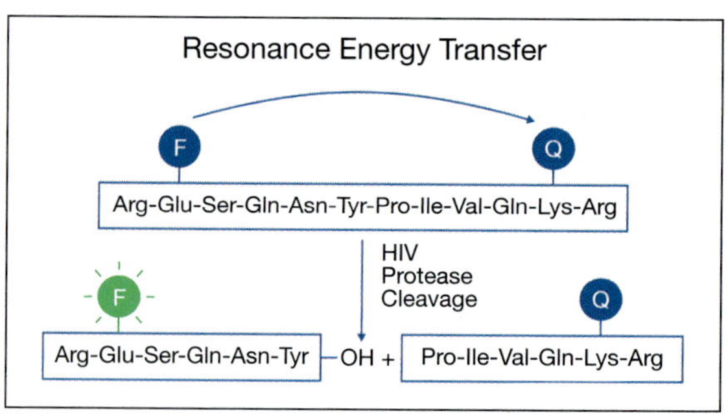

Figure 1.8.9 Principle of the fluorogenic response to protease cleavage exhibited by HIV protease substrate 1 (H2930). Quenching of the EDANS fluorophore (**F**) by distance-dependent resonance energy transfer to the dabcyl quencher (**Q**) is eliminated upon cleavage of the intervening peptide linker.

IMPORTANT NOTICE: The products described in this manual are covered by one or more Limited Use Label License(s). Please refer to the Appendix on page 971 and Master Product List on page 975. Products are For Research Use Only. Not intended for any animal or human therapeutic or diagnostic use.
www.invitrogen.com/probes
molecular probes® | invitrogen™ by life technologies™

REFERENCES

1. Arch Biochem Biophys (1974) 163:390; 2. Science (1972) 178:871; 3. Biochem Biophys Res Commun (1973) 50:352; 4. Arch Biochem Biophys (1974) 163:400; 5. Clin Chim Acta (1986) 157:73; 6. J Lipid Res (1986) 27:792; 7. J Chromatogr (1990) 502:247; 8. J Chromatogr (1991) 548:319; 9. Anal Biochem (2001) 297:128; 10. J Chromatogr A (2003) 1021:191; 11. Anal Chem (2009) 81:2537; 12. Proc Natl Acad Sci U S A (1975) 72:619; 13. Anal Chem (1987) 59:1096; 14. J Chromatogr A (1994) 668:323; 15. Biophys J (2007) 92:2944; 16. Anal Biochem (1990) 189:122; 17. J Chromatogr (1991) 564:258; 18. J Chromatogr (1992) 582:131; 19. Anal Chim Acta (1989) 225:351; 20. Biomed Chromatogr (1989) 3:209; 21. Anal Sci (1986) 2:491; 22. Analyst (1994) 119:1907; 23. Anal Chem (1987) 59:411; 24. J Org Chem (1986) 51:3978; 25. Anal Biochem (1989) 178:202; 26. Anal Chem (1989) 61:432; 27. Electrophoresis (2006) 27:4446; 28. J Neurosci Methods (2007) 160:223; 29. J Chromatogr A (2004) 1022:201; 30. Anal Chem (1995) 67:58; 31. Science (1989) 246:57; 32. Electrophoresis (2007) 29:475; 33. Anal Bioanal Chem (2006) 386:506; 34. Electrophoresis (2002) 23:3071; 35. J Pharm Biomed Anal (2002) 28:475; 36. Clin Chim Acta (2001) 308:147; 37. J Chromatogr B Biomed Sci Appl (1997) 695:67; 38. J Neurosci Methods (1996) 65:33; 39. Anal Chem (1994) 66:3512; 40. Anal Chem (1994) 66:3477; 41. Anal Chem (1993) 65:563; 42. Electrophoresis (1993) 14:373; 43. Anal Chem (1991) 63:413; 44. Anal Chem (1991) 63:408; 45. J Chromatogr (1991) 559:223; 46. Proc Natl Acad Sci U S A (1991) 88:2302; 47. J Chromatogr (1990) 499:579; 48. Anal Chem (1994) 66:3466; 49. Proc Natl Acad Sci U S A (1993) 90:9451; 50. Anal Chim Acta (1995) 299:319; 51. Electrophoresis (1998) 19:2175; 52. Anal Chem (1997) 69:3015; 53. J Chromatogr A (1995) 716:221; 54. J Chromatogr B Analyt Technol Biomed Life Sci (2003) 793:107; 55. Anal Chem (2003) 75:3502; 56. Cytometry A (2005) 65:50; 57. J Chromatogr (1990) 519:189; 58. Biochem J (1968) 108:155; 59. Anal Chim Acta (1981) 130:377; 60. Anal Chim Acta (1985) 170:81; 61. Infect Immun (2005) 73:2321; 62. J Chromatogr (1983) 278:167; 63. Anal Biochem (1981) 116:471; 64. Photochem Photobiol (1991) 54:361; 65. Biochemistry (1977) 16:5150; 66. J Biol Chem (1994) 269:3829; 67. Eur J Biochem (1984) 142:387; 68. J Biol Chem (1984) 259:14378; 69. Biophys J (2007) 92:1988; 70. Chem Pharm Bull (Tokyo) (1974) 22:2413; 71. J Photochem Photobiol A (2000) 131:95; 72. J Org Chem (1966) 31:3880; 73. Biochem J (1952) 51:155; 74. J Liq Chromatogr (1989) 12:2733; 75. J Colloid Interface Sci (1990) 135:435; 76. Anal Biochem (1996) 241:238; 77. Photochem Photobiol (1997) 66:424; 78. Anal Biochem (1984) 141:446; 79. J Chromatogr (1992) 608:239; 80. J Chromatogr (1989) 480:141; 81. Science (1988) 242:562; 82. Bioconjug Chem (1995) 6:447; 83. Anal Chem (2000) 72:4199; 84. Peptides (1992) 13:663; 85. J Neurosci Methods (1985) 13:119; 86. J Biol Chem (1984) 259:6117; 87. J Biol Chem (1983) 258:11948; 88. Anal Biochem (1993) 212:58; 89. Anal Biochem (1987) 162:213; 90. Anal Biochem (1993) 210:351; 91. Science (1990) 247:954; 92. Tetrahedron Lett (1990) 31:6493; 93. Arch Biochem Biophys (1993) 306:304; 94. FEBS Lett (1992) 297:100; 95. Bioorg Med Chem Lett (1992) 2:1665; 96. Drug Des Discov (1997) 15:3; 97. FEBS Lett (1997) 413:379; 98. Anal Biochem (1992) 204:96; 99. J Med Chem (1992) 35:3727; 100. Pept Res (1994) 7:72; 101. Science (1995) 267:1445; 102. Nat Biotechnol (1998) 16:49; 103. Appl Environ Microbiol (1997) 63:1143; 104. Nat Biotechnol (1996) 14:303; 105. Proc Natl Acad Sci U S A (1998) 95:8602; 106. Proc Natl Acad Sci U S A (2002) 99:6603.

DATA TABLE 1.8 REAGENTS FOR ANALYSIS OF LOW MOLECULAR WEIGHT AMINES

Cat. No.	MW	Storage	Soluble	Abs	EC	Em	Solvent	Notes
A2333	305.29	F,D,L	MeOH	465	ND	560	MeOH	1, 2, 3, 4
A2334	251.24	F,D,L	EtOH	486	ND	591	MeOH	4, 5
A6222	305.29	F,D,L	MeOH	465	ND	560	MeOH	1, 2, 3
A10192	251.24	F,L	EtOH	486	ND	591	MeOH	2, 5
B30300	339.24	F,D	DMSO	<300	ND	none		
C20260	199.55	F,D,L	DMF, MeCN	336	9800	none	MeOH	6, 7, 8
D2245	366.38	F,D,L	DMF, DMSO	453	32,000	none	MeOH	9
F486	183.10	F,D,L	MeCN, $CHCl_3$	328	8000	none	MeOH	6, 7
F2332	278.26	F,D,L	MeCN	380	7800	464	MeCN	10
F20261	278.26	F,D,L	MeCN	380	8400	464	MeCN	8, 10
M25	177.16	D	DMF, DMSO	316	3500	386	MeOH	11
N316	294.27	L	DMSO	467	23,000	539	MeOH	7
N1138	184.19	L	DMF, MeCN	419	9400	493	see Notes	12
P2331MP	134.13	L	EtOH	334	5700	455	pH 9	13
Q20133	411.33	F,D,L	DMSO	475	23,000	none	MeOH	
S1167	391.34	F,D,L	DMF, DMSO	466	22,000	535	MeOH	7

For definitions of the contents of this data table, see "Using *The Molecular Probes® Handbook*" in the introductory pages.

Notes

1. Spectral data are for the reaction product with glycine in the presence of cyanide. Unreacted reagent in MeOH: Abs = 254 nm (EC = 46,000 $cm^{-1}M^{-1}$), nonfluorescent.
2. ND = not determined.
3. Solubility in methanol is improved by addition of base (e.g., 1–5% (v/v) 0.2 M KOH).
4. Data represent the reactive dye component of this labeling kit.
5. Spectral data are for the reaction product with glycine in the presence of cyanide. Unreacted reagent in MeOH: Abs = 282 nm (EC = 21,000 $cm^{-1}M^{-1}$), nonfluorescent.
6. Spectra for primary aliphatic amine derivative of NBD chloride in MeOH: Abs = 465 nm (EC = 22,000 $cm^{-1}M^{-1}$), Em = 535 nm (QY = 0.3). Spectra for secondary aliphatic amine derivative in MeOH: Abs = 485 nm (EC = 25,000 $cm^{-1}M^{-1}$), Em = 540 nm (QY <0.1). Aromatic amine derivatives are nonfluorescent. All NBD amine derivatives are almost nonfluorescent in water and have strongly solvent-dependent emission spectra. NBD fluoride yields the same derivatives as NBD chloride but is more reactive.
7. Fluorescence of NBD and its derivatives in water is relatively weak. QY and τ increase and Em decreases in aprotic solvents and other nonpolar environments relative to water. (Biochemistry (1977) 16:5150, Photochem Photobiol (1991) 54:361)
8. This product is specified to equal or exceed 98% analytical purity by HPLC.
9. D2245 is nonfluorescent both before and after reaction with amines. Reaction product with butylamine has Abs = 428 nm (EC = 32,000 $cm^{-1}M^{-1}$) in MeOH.
10. Fluorescamine spectra are for the reaction product with butylamine. The fluorescence quantum yield and lifetime of the butylamine adduct in EtOH are 0.23 and 7.5 nanoseconds, respectively. (Arch Biochem Biophys (1974) 163:390) The unreacted reagent is nonfluorescent (Abs = 234 nm, EC = 28,000 $cm^{-1}M^{-1}$ in MeCN).
11. The amide reaction product of M25 with butylamine has Abs = 353 nm (EC = 5900 $cm^{-1}M^{-1}$), Em = 426 nm in MeOH. Ester reaction products with alcohols have Abs = 350 nm (EC = 5700 $cm^{-1}M^{-1}$), Em = 446 nm in water (pH 8).
12. Spectral data are for the reaction product with glycine in the presence of cyanide, measured in pH 7.0 buffer/MeCN (40:60). (Anal Chem (1987) 59:1102) Unreacted reagent in MeOH: Abs = 279 nm (EC = 5500 $cm^{-1}M^{-1}$), Em = 330 nm.
13. Spectral data are for the reaction product of P2331MP with alanine and 2-mercaptoethanol. The spectra and stability of the adduct depend on the amine and thiol reactants. (Biochim Biophys Acta (1979) 576:440) Unreacted reagent in H_2O: Abs = 257 nm (EC = 1000 $cm^{-1}M^{-1}$).

IMPORTANT NOTICE: The products described in this manual are covered by one or more Limited Use Label License(s). Please refer to the Appendix on page 971 and Master Product List on page 975. Products are For Research Use Only. Not intended for any animal or human therapeutic or diagnostic use.

PRODUCT LIST 1.8 REAGENTS FOR ANALYSIS OF LOW MOLECULAR WEIGHT AMINES

Cat. No.	Product	Quantity
A2333	ATTO-TAG™ CBQCA Amine-Derivatization Kit	1 kit
A6222	ATTO-TAG™ CBQCA derivatization reagent (CBQCA; 3-(4-carboxybenzoyl)quinoline-2-carboxaldehyde)	10 mg
A2334	ATTO-TAG™ FQ Amine-Derivatization Kit	1 kit
A10192	ATTO-TAG™ FQ derivatization reagent (FQ; 3-(2-furoyl)quinoline-2-carboxaldehyde)	10 mg
B30300	N-(t-BOC)-aminooxyacetic acid, tetrafluorophenyl ester	25 mg
C20260	4-chloro-7-nitrobenz-2-oxa-1,3-diazole (NBD chloride; 4-chloro-7-nitrobenzofurazan) *FluoroPure™ grade*	100 mg
D2245	4-((4-(dimethylamino)phenyl)azo)benzoic acid, succinimidyl ester (dabcyl, SE)	100 mg
F2332	fluorescamine	100 mg
F20261	fluorescamine *FluoroPure™ grade*	100 mg
F486	4-fluoro-7-nitrobenz-2-oxa-1,3-diazole (NBD fluoride; 4-fluoro-7-nitrobenzofurazan)	25 mg
M25	N-methylisatoic anhydride *high purity*	1 g
N1138	naphthalene-2,3-dicarboxaldehyde (NDA)	100 mg
N316	NBD-X (6-(N-(7-nitrobenz-2-oxa-1,3-diazol-4-yl)amino)hexanoic acid)	100 mg
P2331MP	*o*-phthaldialdehyde (OPA) *high purity*	1 g
Q20133	QSY® 35 acetic acid, succinimidyl ester	5 mg
S1167	succinimidyl 6-(N-(7-nitrobenz-2-oxa-1,3-diazol-4-yl)amino)hexanoate (NBD-X, SE)	25 mg

IMPORTANT NOTICE: The products described in this manual are covered by one or more Limited Use Label License(s). Please refer to the Appendix on page 971 and Master Product List on page 975. Products are For Research Use Only. Not intended for any animal or human therapeutic or diagnostic use.

TWO

CHAPTER 2

Thiol-Reactive Probes

IMPORTANT NOTICE: The products described in this manual are covered by one or more Limited Use Label License(s). Please refer to the Appendix on page 971 and Master Product List on page 975. Products are For Research Use Only. Not intended for any animal or human therapeutic or diagnostic use.

IMPORTANT NOTICE: The products described in this manual are covered by one or more Limited Use Label License(s). Please refer to the Appendix on page 971 and Master Product List on page 975. Products are For Research Use Only. Not intended for any animal or human therapeutic or diagnostic use.

2.1 Introduction to Thiol Modification and Detection

Common Applications for Thiol-Reactive Probes

Labeling Proteins and Nucleic Acids

Thiol-reactive dyes are principally used to label proteins for the detection of conformational changes, assembly of multisubunit complexes and ligand-binding processes[1] In the case of proteins and peptides, the primary targets of thiol-reactive probes are cysteine residues. In mammalian proteins, the occurrence frequency of cysteine targets (3.3%) is less than half that of lysine targets (7.2%), which are labeled by the amine-reactive reagents described in Chapter 1. Some proteins and many peptides have only a single cysteine residue, enabling site-specific labeling with thiol-reactive probes.[2,3] In proteins with multiple cysteine residues, the multiplicity is often small enough that it is practicable to obtain single-cysteine variants by site-directed mutagenesis without significant disruption of the structure or function of the native protein. Site-specific modification is particularly important for labeling small proteins in applications where the activity or binding affinity of the conjugate is paramount;[3,4] thiol-reactive labeling is the preferred approach over amine-reactive labeling in such cases.

The relatively low abundance of cysteine residues also makes it possible to obtain saturating modification with less risk of incurring the penalties of protein precipitation and fluorescence self-quenching interactions that make high-percentage amine-reactive modification largely impracticable.[5,6] In proteins with multiple cysteine residues, however, the reactivity of an individual cysteine can be very dependent on both its local environment and the hydrophobicity of the reactive dye.[7–10] Site-specific modification strategies involving site-directed cysteine mutagenesis, site-dependent variations in thiol reactivity and functional group protection/deprotection have been developed to double-label proteins with donor and acceptor dyes for fluorescence resonance energy transfer (FRET) applications[2,11–16] (Fluorescence Resonance Energy Transfer (FRET)—Note 1.2). Thiol-reactive dyes can also be reacted with thiolated oligonucleotides for hybridization- or ligation-based nucleic acid detection applications[17] and with thiouridine-modified tRNA for studying its association with protein synthesis machinery.[18,19]

Derivatizing Low Molecular Weight Molecules

Several of the thiol-reactive probes described in this chapter are also useful for derivatizing low molecular weight thiols for various analytical assays that employ chromatographic and electrophoretic separation. An extensive review by Shimada and Mitamura describes the use of several of our thiol-reactive reagents for derivatizing thiol-containing compounds.[20]

Quantitating Thiols

Thiols play a principal role in maintaining the appropriate oxidation–reduction state of proteins, cells and organisms. The susceptibility of thiols to oxidation, however, can lead to the formation of disulfides and higher oxidation products, often with loss of biological activity. Measuring the oxidation state of thiols within live cells is complicated by the high concentration of reduced glutathione in cells, which makes them difficult to assay with reagents that stoichiometrically react with the thiol. Nonetheless, many useful reagents and methods have been developed for the quantitative assay of thiols and disulfides.[21]

Reactivity of Thiol Groups

Reducing Disulfides with DTT or TCEP

In proteins, thiol groups (also called mercaptans or sulfhydryls) are present in cysteine residues. Thiols can also be generated by selectively reducing cystine disulfides with reagents such as dithiothreitol[22] (DTT, D1532) or 2-mercaptoethanol (β-mercaptoethanol), each of which must then be removed by dialysis or gel filtration before reaction with the thiol-reactive probe.[21]

Unfortunately, removal of DTT or 2-mercaptoethanol is sometimes accompanied by air oxidation of the thiols back to the disulfides. Reformation of the disulfide bond can often be avoided by using the reducing agent tris-(2-carboxyethyl)phosphine[23,24] (TCEP, T2556), which usually does not need to be removed prior to thiol modification because it does not contain thiols (Figure 2.1.1). However, there have been several reports that TCEP can react with haloacetamides or maleimides under certain conditions and that labeling in the presence of TCEP is inhibited.[6,25] Carrying out thiol-reactive labeling on ammonium sulfate–precipitated proteins facilitates efficient and rapid removal of DTT after the preparatory reduction step and inhibits thiol reoxidation during the subsequent labeling reaction.[3]

TCEP is more stable at a higher pH and at higher temperatures[26] than is DTT and for a longer period of time in buffers without metal chelators such as EGTA; DTT is more stable than TCEP in solutions that contain metal chelators.[23] TCEP is also more stable in the presence of Ni^{2+} levels that commonly contaminate proteins eluted from Ni^{2+} affinity columns and that rapidly oxidize DTT.[23] Spin labels in TCEP are two to four times more stable than those in DTT, an advantage for electron paramagnetic resonance (EPR) spectroscopy.[23] In addition, TCEP is used to stabilize solutions of ascorbic acid.[27] TCEP is generally impermeable to cell membranes and to the hydrophobic protein core, permitting its use for the selective reduction of disulfides that have aqueous exposure. It has also been reported that TCEP can be used to deplete high-abundance

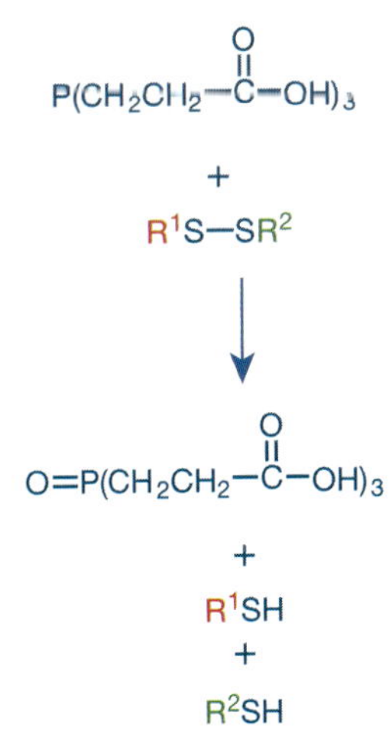

Figure 2.1.1 Reduction of a disulfide using TCEP (tris-(2-carboxyethyl)phosphine, hydrochloride; T2556). Unlike DTT (dithiothreitol, D1532), TCEP does not itself contain thiols, and therefore downstream thiol labeling reactions do not require preliminary removal of the reducing reagent.

plasma proteins (albumins, transferrin, etc.) prior to proteomic analysis because these proteins have a large number of disulfide bridges and are therefore particularly susceptible to reductive denaturation.[28]

Thiol-Reactive Reagents

The primary thiol-reactive reagents, including iodoacetamides, maleimides, benzylic halides and bromomethylketones, react by *S*-alkylation of thiols to generate stable thioether products. Arylating reagents such as NBD halides react with thiols or amines by a similar substitution of the aromatic halide by the nucleophile. Because the thiolate anion is a better nucleophile than the neutral thiol, cysteine is more reactive above its pK_a[21] (~8.3, depending on protein structural context). However, as in the case of amine modification by succinimidyl esters (Chapter 1), reagent stability also decreases with increasing pH (e.g., maleimide hydrolysis to unreactive maleamic acid), and therefore a compromise pH of 7.0–7.5 is typically used for protein modifcation reactions. It has been reported that iodoacetamide and maleimide adducts with intracellular proteins have different degrees of stability and toxicity. Analysis of the intracellular reactivity and toxicity of haloacetyl and maleimido thiol-reactive probes in HEK 293 cells indicates that maleimides are less stable and iodoacetamides are more toxic[29] (putatively because maleimide adducts degrade before they are able to trigger damage-signaling pathways).

Also available are the TS-Link™ series of reagents for reversible thiol modification (Section 2.2). The TS-Link™ reagents are water-soluble thiosulfates that react stoichiometrically with thiols to form mixed disulfides. Thiols also react with many of the amine-reactive reagents described in Chapter 1, including isothiocyanates and succinimidyl esters. However, the reaction products appear to be insufficiently stable to be useful for routine modification of thiols in proteins. Although the thiol–isothiocyanate product (a dithiocarbamate) can react with an adjacent amine to yield a thiourea, the dithiocarbamate is more likely to react with water, consuming the reactive reagent without forming a covalent adduct. In addition to insertion or deletion of cysteine residues by site-directed mutagenesis, several reagents have been developed for introducing thiols into proteins, nucleic acids and lipids. Because the selective introduction of thiols is particularly important for crosslinking two biomolecules, these reagents are discussed in Chapter 5.

Site-Specific Labeling with Thiol-Reactive Reagents

A method for the site-specific double-labeling of a protein containing at least one vicinal diol and another distal thiol has been reported.[11] In this labeling protocol, the vicinal diol is first protected with phenylarsine oxide (PAO) to allow labeling of the unprotected distal thiol with Oregon Green® 488 maleimide (O6034, Section 2.2). The blocked vicinal diol is then deprotected with dithiothreitol (DTT) and labeled with Alexa Fluor® 350 maleimide (A30505, Section 2.3). Target proteins may need to be engineered to contain a vicinal diol and distal thiol in order to employ this labeling strategy.

In a similar double thiol-labeling method, instead of PAO protection/deprotection, the protein's tetracysteine tag was labeled using FlAsH-EDT_2 reagent (T34561, Section 2.2), and the Alexa Fluor® 568 maleimide (A20341, Section 2.2) was used to label a distal cysteine.[30] Fluorescence resonance energy transfer (Fluorescence Resonance Energy Transfer (FRET)—Note 1.2) between the FlAsH label and Alexa Fluor® 568 dye was then used to detect ligand-induced conformational changes in the C-terminal domain of the β_2-adrenoreceptor in SF9 cells.

Iodoacetamides

Iodoacetamides readily react with all thiols, including those found in peptides, proteins and thiolated polynucleotides, to form thioethers (Figure 2.1.2); they are somewhat more reactive than bromoacetamides. When a protein's cysteine residues are blocked or absent, however, iodoacetamides can sometimes react with methionine residues.[13] They may also react with histidine[31] or tyrosine, but generally only if free thiols are absent. Although iodoacetamides can react with the free base form of amines, most aliphatic amines, except the α-amino group at a protein's N-terminus, are protonated and thus relatively unreactive below pH 8.[32] In addition, iodoacetamides react with thiolated oligonucleotide primers, as well as with thiophosphates.[33]

Iodoacetamides are intrinsically unstable in light, especially in solution; reactions should therefore be carried out under subdued light. Adding cysteine, glutathione or mercaptosuccinic acid to the reaction mixture will quench the reaction of thiol-reactive probes, forming highly water-soluble adducts that are easily removed by dialysis or gel filtration. Although the thioether bond formed when an iodoacetamide reacts with a protein thiol is very stable, the bioconjugate loses its fluorophore during amino acid hydrolysis, yielding *S*-carboxymethylcysteine.

Maleimides

Maleimides are excellent reagents for thiol-selective modification, quantitation and analysis. In this reaction, the thiol is added across the double bond of the maleimide to yield a thioether (Figure 2.1.3). Applications of these fluorescent and chromophoric analogs of *N*-ethylmaleimide (NEM) strongly overlap those of iodoacetamides, although maleimides apparently do not react with methionine, histidine or tyrosine. Reaction of maleimides with amines usually requires a higher pH than reaction of maleimides with thiols.

Hydrolysis of the maleimide to an unreactive product can compete significantly with thiol modification, particularly above

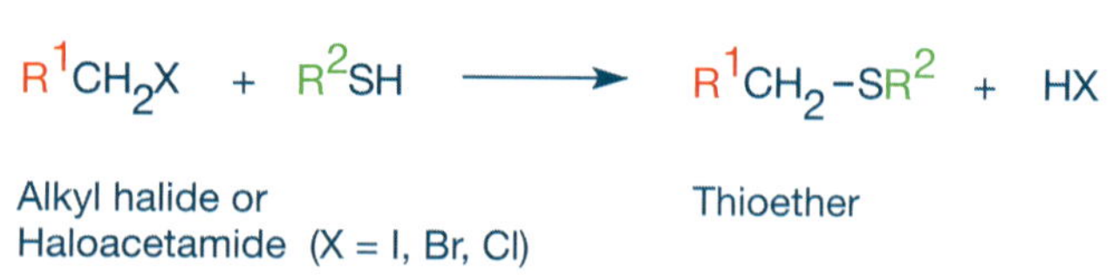

Figure 2.1.2 Reaction of a thiol with an alkyl halide.

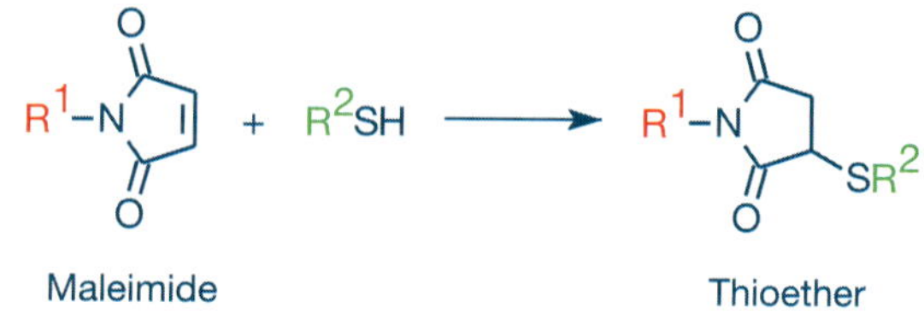

Figure 2.1.3 Reaction of a thiol with a maleimide.

IMPORTANT NOTICE: The products described in this manual are covered by one or more Limited Use Label License(s). Please refer to the Appendix on page 971 and Master Product List on page 975. Products are For Research Use Only. Not intended for any animal or human therapeutic or diagnostic use.
www.invitrogen.com/probes
molecular probes® | invitrogen™ by life technologies™

pH 8. Furthermore, once formed, maleimide-derived thioethers can hydrolyze to an isomeric mixture of succinamic acid adducts, or they can undergo cyclization with adjacent amines to yield crosslinked products. This latter reaction is much less frequently encountered than the former.[34] Deliberate acceleration of the hydrolytic succinimide to succinamide acid ring-opening reaction by molybdate or chromate catalysis provides a strategy for decreasing the heterogeneity of bioconjugates derived from maleimide derivatization of thiols.[35]

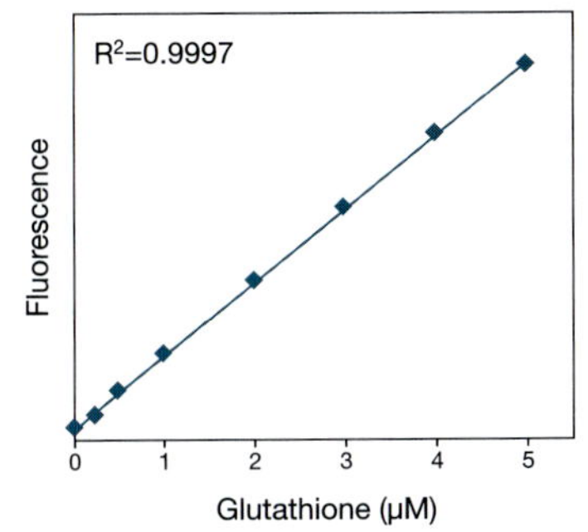

Figure 2.1.6 Linearity and sensitivity of the Measure-iT™ thiol assay. Triplicate 10 µL samples of glutathione were assayed using the Measure-iT™ Thiol Assay Kit (M30550). Fluorescence was measured using excitation/emission of 490/520 nm and plotted versus glutathione concentration. The variation (CV) of replicate samples was <2%.

Reversible Thiol-Reactive Reagents

Several of our thiol-reactive probes can be used to form reversible bonds, including BODIPY® FL L-cystine (B20340), as well as the TS-Link™ BODIPY® thiosulfate and TS-Link™ DSB-X™ biotin C_5-thiosulfate reagents (Section 2.2, Section 4.2).

Symmetric disulfides such as BODIPY® FL L-cystine undergo a thiol–disulfide interchange reaction to yield a new asymmetric disulfide (Figure 2.1.4), a reaction that is freely reversible and thiol-*specific*. This disulfide linkage can be cleaved with reagents such as DTT or TCEP.

Thiosulfates (R–S–SO_3^-), including our water-soluble TS-Link™ reagents, are similar to disulfides in that they stoichiometrically react with thiols to form disulfides (Figure 2.1.5). However, unlike the reaction of the BODIPY® FL cystine probe with a free thiol, no excess of the TS-Link™ reagent is required to drive the equilibrium.

Reagents for Quantitating Thiols

Measure-iT™ Thiol Assay Kit

The Measure-iT™ Thiol Assay Kit (M30550) provides an easy and accurate method for quantitating thiols. This thiol assay has a linear range of 0.05–5 µM thiol (Figure 2.1.6), making it up to 400 times more sensitive than colorimetric methods based on Ellman's reagent.

Each Measure-iT™ Thiol Assay Kit contains:

- Measure-iT™ thiol quantitation reagent (100X concentrate in 1,2-propanediol)
- Measure-iT™ thiol quantitation buffer (50 mM potassium phosphate buffer)
- Measure-iT™ thiol quantitation standard (reduced glutathione)
- Detailed protocols

Simply dilute the reagent 1:100, load 100 µL into the wells of a microplate, add 1–10 µL sample volumes, mix, then read the fluorescence. Maximum fluorescence signal is attained within 5 minutes and is stable for at least 1 hour. The assay is performed at room temperature, and common contaminants are well tolerated in the assay. The Measure-iT™ Thiol Assay Kit provides sufficient materials for 500 assays, based on a 100 µL assay volume in a 96-well microplate format; this thiol assay can also be adapted for use in cuvettes or 384-well microplates.

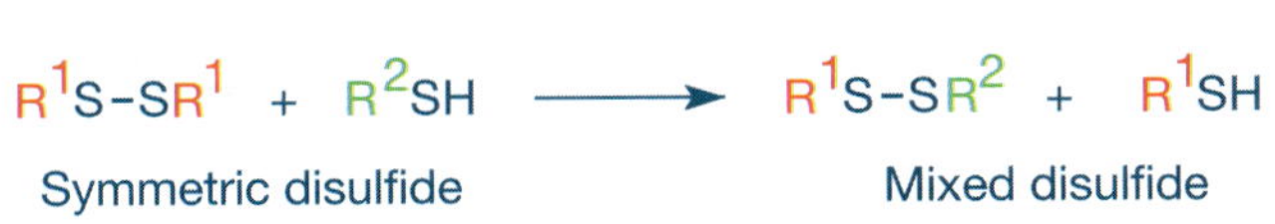

Figure 2.1.4 Reaction of a thiol with a symmetric disulfide.

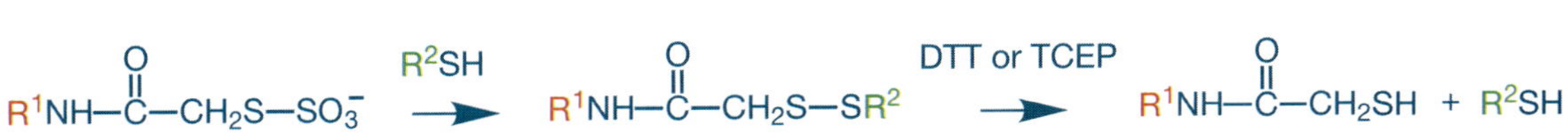

Figure 2.1.5 Reaction of a TS-Link™ reagent with a thiol, followed by removal of the label with a reducing agent.

IMPORTANT NOTICE: The products described in this manual are covered by one or more Limited Use Label License(s). Please refer to the Appendix on page 971 and Master Product List on page 975. Products are For Research Use Only. Not intended for any animal or human therapeutic or diagnostic use.

Thiol and Sulfide Quantitation Kit

Ultrasensitive colorimetric quantitation of both protein and nonprotein thiols can be achieved using the Thiol and Sulfide Quantitation Kit (T6060). In this assay, which is based on a method reported by Singh,[36,37] thiols or sulfides reduce a disulfide-inhibited derivative of papain, stoichiometrically releasing the active enzyme (Figure 2.1.7). Activity of the enzyme is then measured using the chromogenic papain substrate L-BAPNA via spectrophotometric detection of *p*-nitroaniline release at 412 nm.

Although thiols and inorganic sulfides can also be quantitated using 5,5′-dithiobis-(2-nitrobenzoic acid) (DTNB or Ellman's reagent, D8451), the enzymatic amplification step in the Thiol and Sulfide Quantitation Kit enables researchers to detect as little as 0.2 nanomoles of thiols or sulfides—a sensitivity that is about 100-fold better than that achieved with DTNB.[38,39] Thiols in proteins can be detected indirectly by incorporating the disulfide cystamine into the reaction mixture. Cystamine undergoes an exchange reaction with protein thiols, yielding 2-mercaptoethylamine (cysteamine), which then releases active papain. Thiols that are alkylated by maleimides, iodoacetamides or other reagents are excluded from detection and can therefore be assayed subtractively.[40]

The Thiol and Sulfide Quantitation Kit contains:

- Papain–$SSCH_3$, the disulfide-inhibited papain derivative
- L-BAPNA, a chromogenic papain substrate
- DTNB (Ellman's reagent), for calibrating the assay
- Cystamine
- L-Cysteine, a thiol standard
- Buffer
- Detailed protocols for measuring thiols, inorganic sulfides and maleimides

Sufficient reagents are provided for approximately 50 assays using standard 1 mL cuvettes or 250 assays using a microplate format.

Ellman's Reagent (DTNB) for Quantitating Thiols

Ellman's reagent[41] (5,5′-dithiobis-(2-nitrobenzoic acid) or DTNB; D8451) remains an important reagent for spectrophotometric quantitation of protein thiols and, by extension, the analysis of thiol–disulfide exchange reactions[42] and oxidative thiol modifications.[43] It readily forms a mixed disulfide with thiols, liberating the chromophore 5-mercapto-2-nitrobenzoic acid[44] (absorption maximum 410 nm, EC ~13,600 $cm^{-1}M^{-1}$). Only protein thiols that are accessible to this water-soluble reagent are modified.[45,46] Inaccessible thiols can usually be quantitated by carrying out the titration in the presence of 6 M guanidinium chloride. DTNB conjugates of glutathione and other thiols can be separated by HPLC and quantitated based on their absorption.[47]

Other Fluorometric Reagents for Quantitating Thiols

Several maleimides—including 7-diethylamino-3-(4′-maleimidylphenyl)-4-methylcoumarin (CPM, D346; Section 2.3) and *N*-(7-dimethylamino-4-methylcoumarin-3-yl)maleimide (DACM, D10251; Section 2.3)—are not appreciably fluorescent until after conjugation with thiols, and are therefore useful for thiol quantitation. Similarly, fluorescein-5-maleimide (F150, Section 2.2) exhibits an analytically useful 10-fold fluorescence enhancement upon reaction with thiols.[48] Monobromobimane (M1378, M20381; Section 2.3) is also essentially nonfluorescent until it reacts with thiols and can be used to determine thiol levels in cells.

In addition, most of the fluorescent thiol-reactive reagents in this chapter can be used as derivatization reagents for analyzing thiols by techniques such as HPLC that utilize a separation step. 5-(Bromomethyl)fluorescein (B1355, Section 2.2) is the reagent with the greatest intrinsic sensitivity for this application. See Section 15.6 for a further discussion of methods to quantitate reduced glutathione in cells.

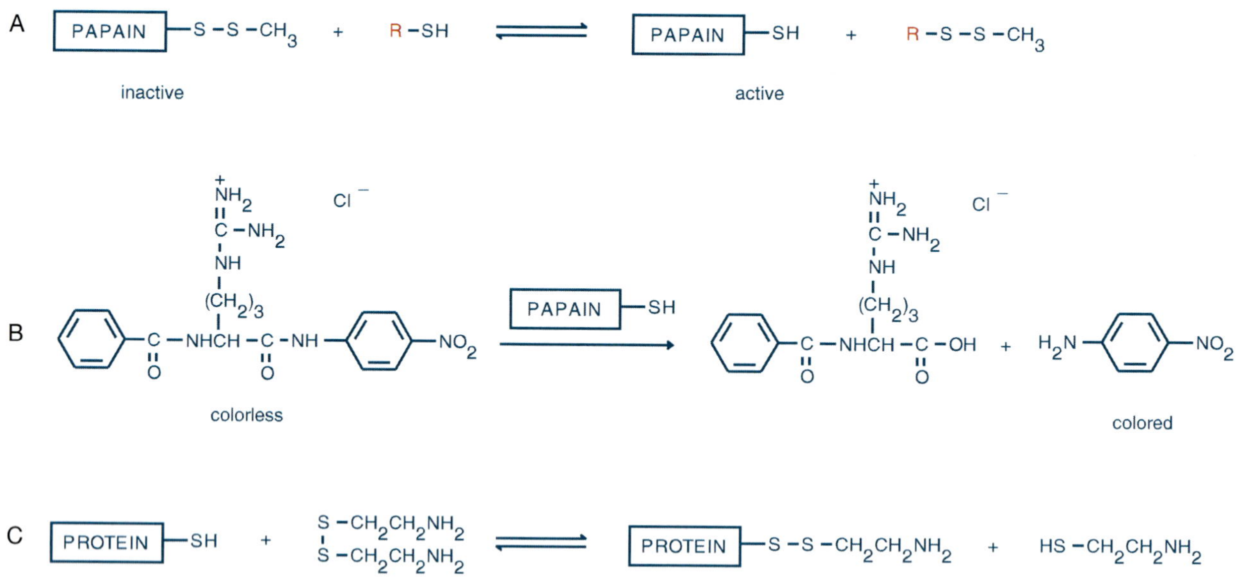

Figure 2.1.7 Chemical basis for thiol detection using the Thiol and Sulfide Quantitation Kit (T6060): **A)** The inactive disulfide derivative of papain, papain–$SSCH_3$, is activated in the presence of thiols; **B)** active papain cleaves the substrate L-BAPNA, releasing the *p*-nitroaniline chromophore; **C)** protein thiols, often poorly accessible, exchange with cystamine to generate 2-mercaptoethylamine (cysteamine), which is functionally equivalent to the thiol R–SH in step **A**.

IMPORTANT NOTICE: The products described in this manual are covered by one or more Limited Use Label License(s). Please refer to the Appendix on page 971 and Master Product List on page 975. Products are For Research Use Only. Not intended for any animal or human therapeutic or diagnostic use.

REFERENCES

1. Neuron (2010) 66:170; **2.** J Am Chem Soc (2008) 130:17664; **3.** Bioconjug Chem (2008) 19:786; **4.** Bioconjug Chem (2008) 19:2527; **5.** Proteomics (2005) 5:1746; **6.** Electrophoresis (2003) 24:2348; **7.** Biochemistry (2008) 47:3615; **8.** J Biol Chem (2007) 282:8667; **9.** J Biol Chem (2002) 277:29018; **10.** J Biol Chem (2004) 279:34913; **11.** Protein Sci (2009) 18:1033; **12.** Bioconjug Chem (2008) 19:1124; **13.** Anal Biochem (2005) 342:271; **14.** Protein Sci (2006) 15:640; **15.** Proc Natl Acad Sci U S A (2001) 98:14464; **16.** Bioconjug Chem (2002) 13:1163; **17.** Org Biomol Chem (2008) 6:908; **18.** Electrophoresis (2005) 26:2384; **19.** J Biol Chem (2000) 275:10727; **20.** J Chromatogr B Biomed Appl (1994) 659:227; **21.** Anal Biochem (2009) 394:147; **22.** Eur J Biochem (1987) 168:169; **23.** Anal Biochem (1999) 273:73; **24.** J Org Chem (1991) 56:2648; **25.** Anal Biochem (2000) 282:161; **26.** Anal Biochem (2004) 325:137; **27.** Anal Biochem (2000) 282:89; **28.** Anal Biochem (2009) 387:184; **29.** Chem Res Toxicol (2008) 21:2361; **30.** J Biol Chem (2007) 282:13895; **31.** Biochemistry (1981) 20:7021; **32.** J Mass Spectrom (2007) 42:233; **33.** Mol Cell Proteomics (2003) 2:242; **34.** Protein Sci (2003) 12:1567; **35.** Bioorg Med Chem Lett (2007) 17:6286; **36.** Anal Biochem (1998) 265:8; **37.** Anal Biochem (1993) 213:49; **38.** Environ Sci Technol (2008) 42:8127; **39.** Anal Chem (2007) 79:1411; **40.** Bioconjug Chem (1994) 5:348; **41.** Biochem Pharmacol (1961) 7:88; **42.** J Virol (2007) 81:2328; **43.** J Biol Chem (2003) 278:9203; **44.** Methods Enzymol (1994) 233:380; **45.** Methods Enzymol (1987) 143:44; **46.** Methods Enzymol (1983) 91:49; **47.** J Chromatogr B Analyt Technol Biomed Life Sci (2002) 781:181; **48.** Anal Biochem (2001) 295:101.

DATA TABLE 2.1 INTRODUCTION TO THIOL MODIFICATION AND DETECTION

Cat. No.	MW	Storage	Soluble	Abs	Em
D1532	154.24	D	H_2O	<300	none
D8451	396.35	D	pH >6	324	none
T2556	286.65	D	pH >5	<300	none

For definitions of the contents of this data table, see "Using *The Molecular Probes® Handbook*" in the introductory pages.

PRODUCT LIST 2.1 INTRODUCTION TO THIOL MODIFICATION AND DETECTION

Cat. No.	Product	Quantity
D8451	5,5′-dithiobis-(2-nitrobenzoic acid) (DTNB; Ellman's reagent)	10 g
D1532	dithiothreitol (DTT)	1 g
M30550	Measure-iT™ Thiol Assay Kit *500 assays*	1 kit
T6060	Thiol and Sulfide Quantitation Kit *50–250 assays*	1 kit
T2556	tris-(2-carboxyethyl)phosphine, hydrochloride (TCEP)	1 g

IMPORTANT NOTICE: The products described in this manual are covered by one or more Limited Use Label License(s). Please refer to the Appendix on page 971 and Master Product List on page 975. Products are For Research Use Only. Not intended for any animal or human therapeutic or diagnostic use.

2.2 Thiol-Reactive Probes Excited with Visible Light

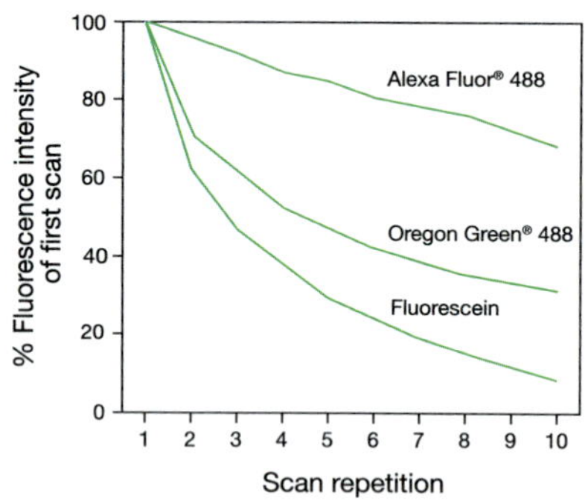

Figure 2.2.1 Photobleaching resistance of the green-fluorescent Alexa Fluor® 488, Oregon Green® 488 and fluorescein dyes, as determined by laser-scanning cytometry. EL4 cells were labeled with biotin-conjugated anti-CD44 antibody and detected by Alexa Fluor® 488 (S11223, S32354), Oregon Green® 488 (S6368) or fluorescein (S869) streptavidin. The cells were then fixed in 1% formaldehyde, washed and wet-mounted. After mounting, cells were scanned 10 times on a laser-scanning cytometer; laser power levels were 25 mW for the 488 nm spectral line of the argon-ion laser. Scan durations were approximately 5 minutes apiece, and each repetition was started immediately after completion of the previous scan. Data are expressed as percentages derived from the mean fluorescence intensity (MFI) of each scan divided by the MFI of the first scan. Data contributed by Bill Telford, Experimental Transplantation and Immunology Branch, National Cancer Institute.

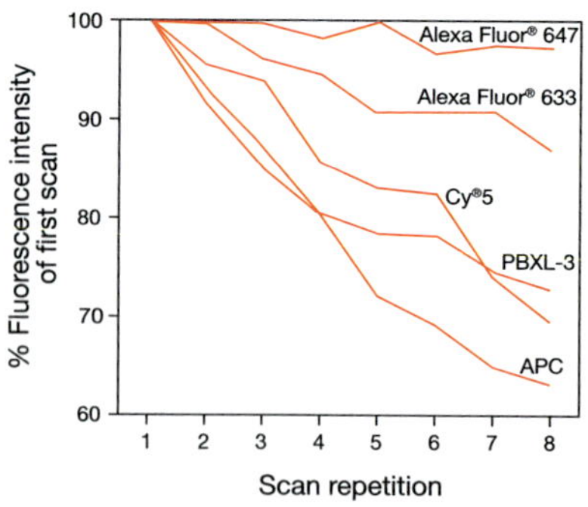

Figure 2.2.2 Photobleaching resistance of the red-fluorescent Alexa Fluor® 647, Alexa Fluor® 633, PBXL-3 and Cy®5 dyes and the allophycocyanin fluorescent protein, as determined by laser-scanning cytometry. EL4 cells were labeled with biotin-conjugated anti-CD44 antibody and detected by Alexa Fluor® 647 (S21374, S32357), Alexa Fluor® 633 (S21375), PBXL-3, Cy®5 or allophycocyanin (APC, S868) streptavidin. The cells were then fixed in 1% formaldehyde, washed and wet-mounted. After mounting, cells were scanned eight times on a laser-scanning cytometer; laser power levels were 18 mW for the 633 nm spectral line of the He-Ne laser. Scan durations were approximately 5 minutes apiece, and each repetition was started immediately after completion of the previous scan. Data are expressed as percentages derived from the mean fluorescence intensity (MFI) of each scan divided by the MFI of the first scan. Data contributed by Bill Telford, Experimental Transplantation and Immunology Branch, National Cancer Institute.

The thiol-reactive Alexa Fluor®, BODIPY®, fluorescein, Oregon Green®, tetramethylrhodamine and Texas Red® derivatives have strong absorptivity and high fluorescence quantum yields. This combination of attributes makes these compounds the preferred reagents for preparing protein and low molecular weight ligand conjugates to study the diffusion, structural properties and interactions of proteins and ligands using techniques such as:

- Fluorescence recovery after photobleaching (FRAP)
- Fluorescence polarization (FP) (Fluorescence Polarization (FP)—Note 1.4)
- Fluorescence correlation spectroscopy (FCS) (Fluorescence Correlation Spectroscopy (FCS)—Note 1.3) and other single-molecule detection techniques
- Fluorescence resonance energy transfer (FRET) (Fluorescence Resonance Energy Transfer (FRET)—Note 1.2)

In this section and in Section 2.3, thiol-reactive reagents with similar spectra, rather than the same reactive group, are generally discussed together. The probes described in this section have visible absorption maxima beyond 410 nm; thiol-reactive probes with peak absorption below 410 nm are described in Section 2.3. Table 2.1 summarizes this section's thiol-reactive probes excited with visible light.

Alexa Fluor® Maleimides

Alexa Fluor® dyes set new standards for fluorescent dyes and the bioconjugates prepared from them (The Alexa Fluor® Dye Series—Note 1.1). Alexa Fluor® dyes exhibit several unique features:

- Strong absorption, with extinction coefficients greater than 65,000 $cm^{-1}M^{-1}$
- Excellent photostability (Figure 2.2.1, Figure 2.2.2), providing more time for observation and image capture than spectrally similar dyes allow (Figure 2.2.3)
- pH-insensitive fluorescence between pH 4 and pH 10
- Superior fluorescence output per protein conjugate, surpassing that of other spectrally similar fluorophore-labeled protein, including fluorescein, tetramethylrhodamine and Texas Red® conjugates, as well as Cy®3 and Cy®5 conjugates [1]

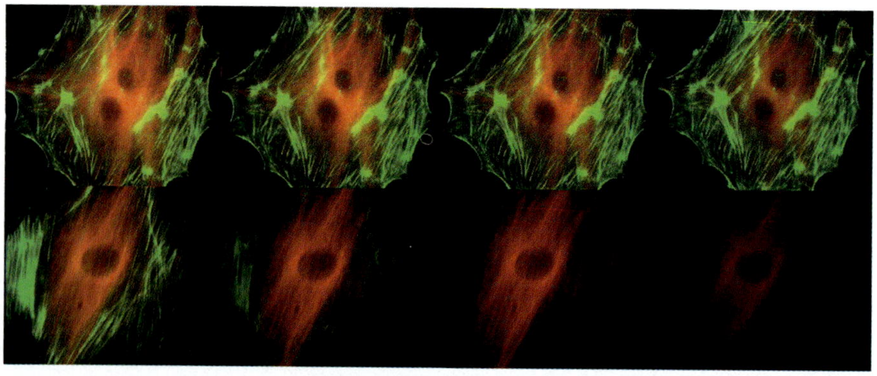

Figure 2.2.3 Comparison of the photobleaching rates of the Alexa Fluor® 488 and Alexa Fluor® 546 dyes and the well-known fluorescein and Cy®3 fluorophores. The cytoskeleton of bovine pulmonary artery endothelial cells (BPAEC) was labeled with (top series) Alexa Fluor® 488 phalloidin (A12379) and mouse monoclonal anti-α-tubulin antibody (A11126) in combination with Alexa Fluor® 546 goat anti-mouse IgG antibody (A11003) or (bottom series) fluorescein phalloidin (F432) and the anti-α-tubulin antibody in combination with a commercially available Cy®3 goat anti-mouse IgG antibody. The pseudocolored images were taken at 30-second intervals (0, 30, 90 and 210 seconds of exposure from left to right). The images were acquired with bandpass filter sets appropriate for fluorescein and rhodamine.

For labeling thiol groups, we offer thiol-reactive Alexa Fluor® dyes that span the visible spectrum:

- Alexa Fluor® 350 C_5-maleimide (A30505, Section 2.3)
- Alexa Fluor® 488 C_5-maleimide (A10254, Figure 2.2.4)
- Alexa Fluor® 532 C_5-maleimide (A10255)
- Alexa Fluor® 546 C_5-maleimide (A10258)
- Alexa Fluor® 555 C_2 maleimide (A20346)
- Alexa Fluor® 568 C_5-maleimide (A20341)
- Alexa Fluor® 594 C_5-maleimide (A10256)
- Alexa Fluor® 633 C_5-maleimide (A20342)
- Alexa Fluor® 647 C_2-maleimide (A20347)
- Alexa Fluor® 660 C_2-maleimide (A20343)
- Alexa Fluor® 680 C_2-maleimide (A20344)
- Alexa Fluor® 750 C_5-maleimide (A30459)

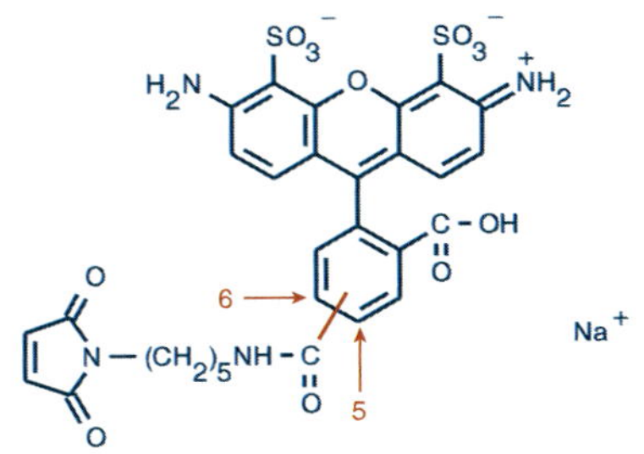

Figure 2.2.4 Alexa Fluor® 488 C_5-maleimide (A10254).

Table 2.1 Molecular Probes® thiol-reactive dyes excited with visible light.

Derivative	Abs *	Em *	Maleimide	Haloacetamide	Bromomethyl	Halide	Cystine^{C} or Thiosulfate^{T}
Alexa Fluor® 488	495	519	A10254^{M}				
Alexa Fluor® 532	532	553	A10255				
Alexa Fluor® 546	556	575	A10258^{M}				
Alexa Fluor® 555	555	565	A20346				
Alexa Fluor® 568	578	603	A20341^{M}				
Alexa Fluor® 594	590	617	A10256^{M}				
Alexa Fluor® 633	632	647	A20342^{M}				
Alexa Fluor® 647	650	665	A20347				
Alexa Fluor® 660	663	690	A20343				
Alexa Fluor® 680	679	702	A20344				
Alexa Fluor® 750	749	775	A30459				
BODIPY® FL	505	513	B10250	D6003			B20340^{C} T30453^{T}
BODIPY® TMR	542	574	B30466				T30454^{T}
BODIPY® TR	589	617					T30455^{T}
BODIPY® 493/503	493	503			B2103		
BODIPY® 499/508	499	508	D20350				
BODIPY® 507/545	508	543		D6004			
BODIPY® 577/618	577	618	D20351				
BODIPY® 630/650	625	640			B22802		T30456^{T}
4-Dimethylamino phenylazophenyl	419	NA	D1521				
Eosin	524	544	E118^{5}				
Fluorescein	494	518	F150^{5}	I30451^{5} I30452^{6}	B1355^{5}		
Lucifer yellow	426	531		L1338			
NBD	478	541		I9 † D2004		F486 F6053 ‡ C20260	
Oregon Green® 488	496	524	O6034^{5}	O6010^{M}			
PyMPO	415	570	M6026				
QSY® 7	560	NA	Q10257				
QSY® 9	562	NA	Q30457				
QSY® 35	475	NA		Q20348			
Rhodamine Red™	570	590	R6029^{M}				
Sulfonerhodamine	555	580		B10621 §			
Tetramethylrhodamine	555	580	T6027^{5} T6028^{6}	T6006^{5}			
Texas Red®	595	615	T6008^{M}		T6009^{M}		

* Absorption (Abs) and emission (Em) maxima, in nm. † Iodoacetate ester. ‡ Like the NBD probes, ABD-F (F6053) is a benz-2-oxa-1,3-diazole, except that it is sulfonated (i.e., an SBD probe) instead of nitrated (i.e., an NBD probe); its reaction product with dimethylaminoethanethiol has Abs/Em maxima of 376/510 nm. § Bifunctional crosslinker. 5 = 5-Isomer. 6 = 6-Isomer. M = Mixed isomers. C = BODIPY® FL L-cystine. T = TS-Link™ fluorescent thiosulfate. NA = Not applicable. More information on thiol-reactive dyes is available at www.invitrogen.com/handbook/labelingchemistry.

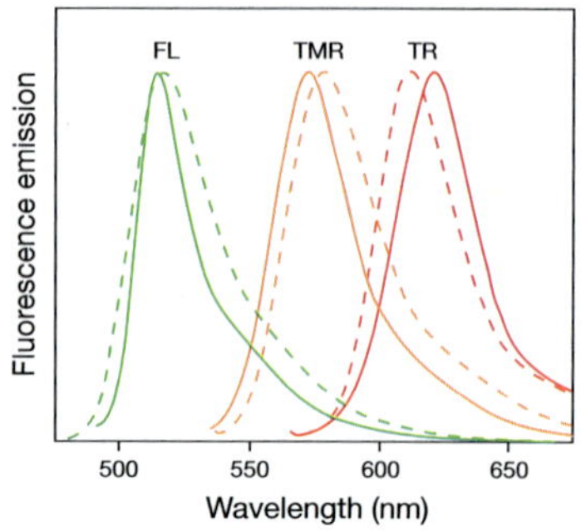

Figure 2.2.5 Normalized fluorescence emission spectra of goat anti–mouse IgG antibody conjugates of fluorescein (FL), tetramethylrhodamine (TMR) and the Texas Red® (TR) dyes, shown by dashed lines (- - -), as compared with goat anti–mouse IgG antibody conjugates of BODIPY® FL, BODIPY® TMR and BODIPY® TR dyes, respectively, shown by solid lines (—).

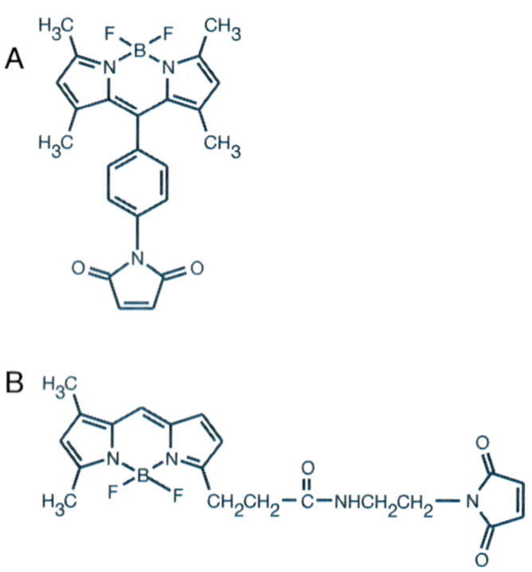

Figure 2.2.6 Comparison of the fluorophore orientation relative to the reactive moiety of two spectrally similar thiol-reactive BODIPY® dyes: **A)** BODIPY® 499/508 maleimide (D20350) and **B)** BODIPY® FL *N*-(2-aminoethyl)maleimide (B10250).

The Alexa Fluor® maleimides are particularly useful for labeling thiol-containing proteins on the surface of live cells, where their polarity permits the sensitive detection of exposed thiols.[2–4] In proteomics applications, Alexa Fluor® protein conjugates can be electrophoretically separated and then detected without additional staining.[5–7] As with their amine-reactive succinimidyl ester counterparts (Section 1.3), Alexa Fluor® 647 maleimide, Alexa Fluor® 750 maleimide and other long-wavelength reactive dyes are frequently used to make conjugates for *in vivo* imaging applications.[8,9] In experiments using Alexa Fluor® 488 maleimide, immunodetection of labeled proteins can be accomplished using our anti–Alexa Fluor® 488 antibody (A11094, Section 7.4).

BODIPY® Derivatives

BODIPY® Iodoacetamides, Maleimides and Methyl Bromides

Like their amine-reactive BODIPY® counterparts (Section 1.4), BODIPY® iodoacetamides, BODIPY® maleimides and BODIPY® methyl bromides yield thiol adducts with several important properties:

- High extinction coefficients (EC >60,000 $cm^{-1}M^{-1}$)
- High fluorescence quantum yields, often approaching 1.0, even in water
- Narrow emission bandwidths (Figure 2.2.5)
- Good photostability[10]
- Spectra that are relatively insensitive to solvent polarity and pH of the medium[11]
- Lack of ionic charge, which is especially useful when preparing membrane probes and cell-permeant reagents

BODIPY® dyes are chemically stable between about pH 3 and pH 10, although they are less stable to extremes of pH than are fluorescein and Alexa Fluor® derivatives. All of the thiol-reactive BODIPY® dyes are suitable for labeling cysteine residues in proteins and thiolated oligonucleotides and for detecting thiol conjugates separated by HPLC and capillary electrophoresis using ultrasensitive laser-scanning techniques.[11] BODIPY® FL iodoacetamide has been shown to be highly selective for cysteine labeling, producing little or no nonspecific labeling even at high dye:thiol ratios; in contrast, tetramethylrhodamine iodoacetamide exhibited nonspecific labeling as dye concentrations increased.[12] Furthermore, actin labeling with BODIPY® FL iodoacetamide (D6003) reportedly does not perturb actin polymerization.[13] BODIPY® FL maleimide is a useful reagent for flow cytometric quantitation and confocal imaging of microparticles released upon agonist-elicited activation of human platelets.[14,15] Labeling can be carried out after activation, avoiding concerns that pre-labeling might interfere with cellular functions involved in the activation process.

Our selection of thiol-reactive BODIPY® reagents includes:

- BODIPY® FL maleimide and BODIPY® FL iodoacetamide (B10250, Figure 2.2.6; D6003), which exhibit spectral characteristics very similar to fluorescein
- BODIPY® 507/545 iodoacetamide (D6004)
- BODIPY® TMR maleimide (B30466)
- BODIPY® 493/503 methyl bromide (B2103)
- BODIPY® 630/650 methyl bromide (B22802), with very long-wavelength spectra

Two additional symmetric maleimidylphenyl BODIPY® derivatives are available with excitation/emission maxima of ~499/508 nm (D20350, Figure 2.2.6) and ~577/618 nm (D20351).

BODIPY® FL L-Cystine

We have attached the BODIPY® FL fluorophore to the amino groups of the disulfide-linked amino acid cystine to create a reagent for reversible, *thiol-specific* labeling of proteins, thiolated oligonucleotides and cells.[16] BODIPY® FL L-cystine (B20340) is virtually nonfluorescent due to interactions between the two fluorophores; however, thiol-specific exchange to form a mixed disulfide results in significant enhancement of the green fluorescence (Figure 2.2.7).

IMPORTANT NOTICE: The products described in this manual are covered by one or more Limited Use Label License(s). Please refer to the Appendix on page 971 and Master Product List on page 975. Products are For Research Use Only. Not intended for any animal or human therapeutic or diagnostic use.
www.invitrogen.com/probes

TS-Link™ BODIPY® Thiosulfate Reagents

The TS-Link™ BODIPY® reagents are water-soluble, fluorescent thiosulfates that react readily and selectively with free thiols to form disulfide bonds (Figure 2.2.8). In contrast to the thioether bonds formed by maleimides and iodoacetamides, the disulfide bond formed by the TS-Link™ reagents is reversible; the TS-Link™ BODIPY® fluorophore can easily be removed using a reducing agent such as dithiothreitol or tris-(2-carboxyethyl)phosphine (DTT, D1532; TCEP, T2556; Section 2.1), leaving the molecule of interest unchanged for downstream processing. These TS-Link™ reagents yield the same disulfide products as methanethiosulfonates (MTS reagents), but they are much more polar and water soluble and may therefore selectively react with residues on the surface of a protein or live cell.[17]

We currently offer:

- TS-Link™ BODIPY® FL C_2-thiosulfate (T30453)
- TS-Link™ BODIPY® TMR C_5-thiosulfate (T30454)
- TS-Link™ BODIPY® TR C_5-thiosulfate (T30455)
- TS-Link™ BODIPY® 630/650 C_5-thiosulfate (T30456)

We also offer TS-Link™ DSB-X™ biotin C_5-thiosulfate (TS-Link™ desthiobiotin-X C_5-thiosulfate, T30754), which is described in Section 4.2.

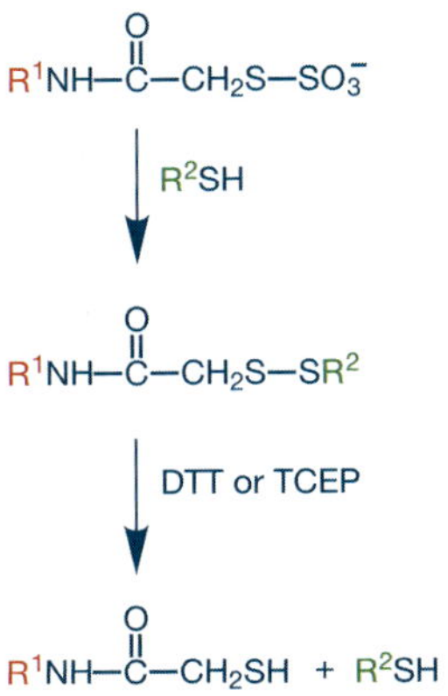

Figure 2.2.8 Reaction of a TS-Link™ reagent (R[1]) with a thiol (R[2]), followed by removal of the label with a reducing agent.

Fluorescein Derivatives, Including Thiol-Reactive Oregon Green® Dyes

Fluorescein Iodoacetamide, Maleimide and Methyl Bromide

The excellent water solubility of the fluorescein iodoacetamide single isomers (I30451, I30452) and fluorescein-5-maleimide (F150, Figure 2.2.9) at pH 7 makes it easy to prepare green-fluorescent thiol conjugates of biomolecules. Fluorescein maleimide and 5-iodoacetamidofluorescein have been the most extensively used visible wavelength–excitable, thiol-reactive dyes for modifying proteins, nucleic acids and other biomolecules. Following conjugation to thiols, fluorescein-5-maleimide (and other fluoresceins) can be radioiodinated.[18]

When compared with these iodoacetamide and maleimide derivatives, 5-(bromomethyl)fluorescein (B1355, Figure 2.2.10) reacts more slowly with thiols of peptides, proteins and thiolated nucleic acids but forms stronger thioether bonds that are expected to remain stable under the conditions required for complete amino acid analysis. With the possible exception of our Alexa Fluor® maleimides and the thiol-reactive BODIPY® dyes described above, 5-(bromomethyl)fluorescein has the highest intrinsic detectability of all thiol-reactive probes, particularly for capillary electrophoresis instrumentation that uses the 488 nm spectral line of the argon-ion laser.[19]

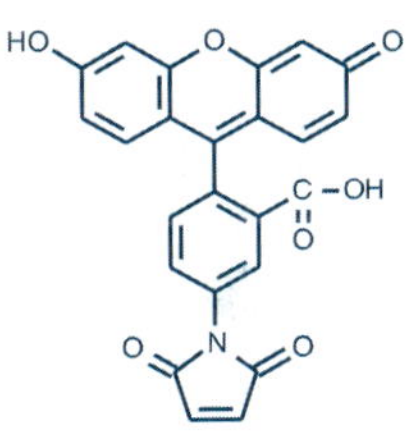

Figure 2.2.9 Fluorescein-5-maleimide (F150).

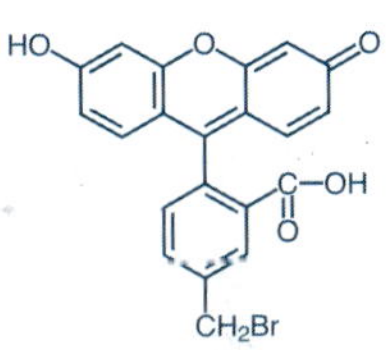

Figure 2.2.10 5-(bromomethyl)fluorescein (B1355).

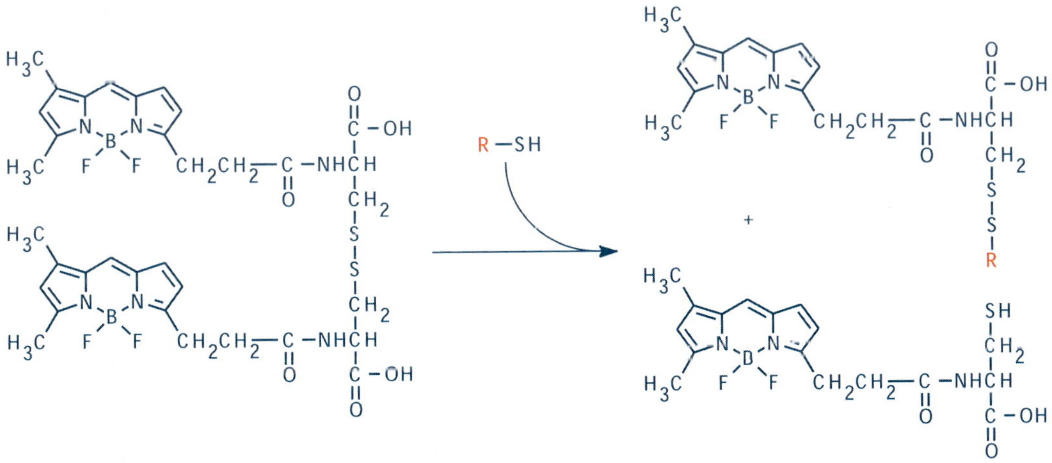

Figure 2.2.7 Reaction of intramolecularly quenched BODIPY® FL L-cystine (B20340) with a thiol, yielding two fluorescent products—a mixed disulfide labeled with the BODIPY® FL dye and a BODIPY® FL cysteine derivative.

IMPORTANT NOTICE: The products described in this manual are covered by one or more Limited Use Label License(s). Please refer to the Appendix on page 971 and Master Product List on page 975. Products are For Research Use Only. Not intended for any animal or human therapeutic or diagnostic use.

Oregon Green® 488 Iodoacetamide and Maleimide

The Oregon Green® 488 dye (2′,7′-difluorofluorescein, D6145; Section 1.5) has absorption and emission spectra that are a perfect match to those of fluorescein. In addition to Oregon Green® 488 isothiocyanate, carboxylic acid and succinimidyl ester derivatives (Section 1.5), we have synthesized the isomeric mixture of Oregon Green® 488 iodoacetamide (O6010) and the single-isomer Oregon Green® 488 maleimide (O6034, Figure 2.2.11). These thiol-reactive probes yield conjugates that have several important advantages when directly compared with fluorescein conjugates, including:

- Greater photostability (Figure 2.2.12)
- A lower pK_a (pK_a of 4.8 for 2′,7′-difluorofluorescein versus 6.4 for fluorescein) (Figure 2.2.13)
- Higher fluorescence and less quenching at comparable degrees of substitution (Figure 2.2.14)
- Utility as fluorescence anisotropy probes for measuring protein–protein and protein–nucleic acid interactions[20] (Fluorescence Polarization (FP)—Note 1.4)

Eosin Maleimide

As compared with the corresponding fluorescein derivative, eosin maleimide (E118, Figure 2.2.15) is less fluorescent but much more phosphorescent and a better photosensitizer.[21] With eosin's high quantum yield of 0.57 for singlet oxygen generation,[22–24] eosin conjugates can be used as effective photooxidizers of diaminobenzidine (DAB) in high-resolution electron microscopy studies[25] (Fluorescent Probes for Photoconversion of Diaminobenzidine Reagents—Note 14.2).

Eosin (excitation/emission maxima ~519/540 nm) derivatives efficiently absorb the fluorescence from fluorescein and other fluorophores such as the BODIPY® FL, Alexa Fluor® 488, Oregon Green® 488, dansyl and coumarin dyes, making them good acceptors in FRET techniques[26] (Fluorescence Resonance Energy Transfer (FRET)—Note 1.2).

Although usually selectively reactive with thiols, eosin maleimide reportedly also reacts with a specific lysine residue of the band-3 protein in human erythrocytes, inhibiting anion exchange in these cells.[27,28] A flow cytometry assay for hereditary spherocytosis (HS), characterized by band-3 protein–deficient erthrocytes, has been developed using this selective binding by eosin maleimide.[29–31] In this assay, HS erythrocytes are identified as the population exhibiting low eosin fluorescence.

Rhodamine Derivatives, Including Thiol-Reactive Texas Red® Dyes

Tetramethylrhodamine Iodoacetamide and Maleimide

Tetramethylrhodamine iodoacetamide (TMRIA) and tetramethylrhodamine maleimide yield photostable, pH-insensitive, red-orange–fluorescent thiol conjugates.[32,33] These iodoacetamide and maleimide derivatives, however, are difficult to prepare in pure form and different batches of our mixed-isomer products have contained variable mixtures of the 5- and 6-isomers. Moreover, certain cytoskeletal proteins

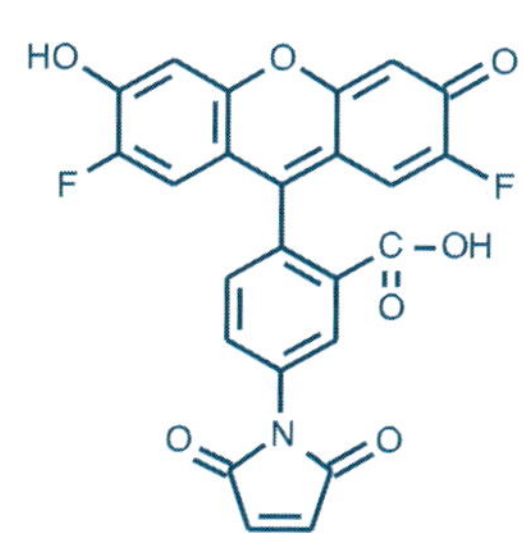

Figure 2.2.11 Oregon Green® 488 maleimide (O6034).

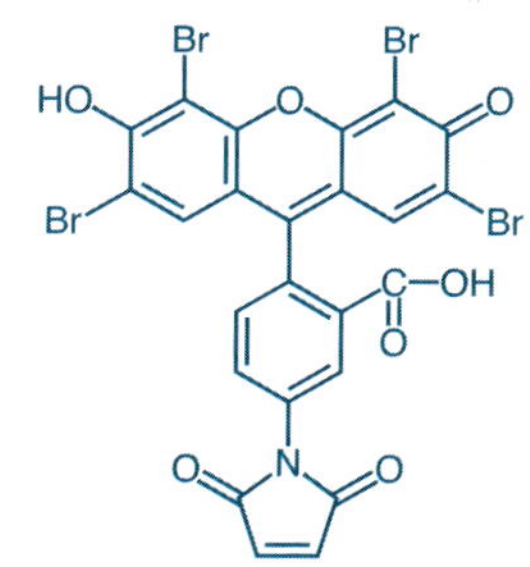

Figure 2.2.15 Eosin-5-maleimide (E118).

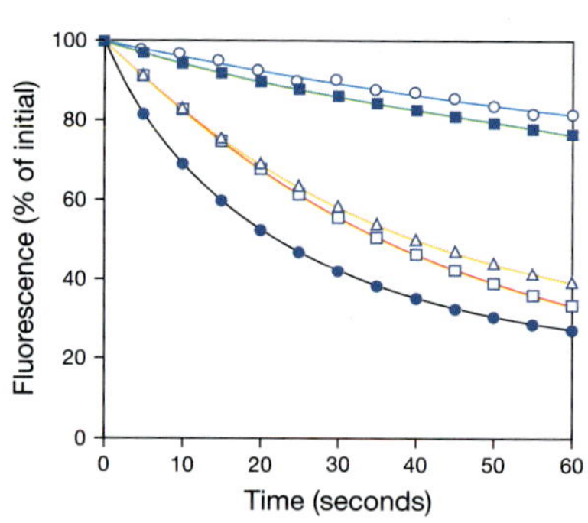

Figure 2.2.12 Comparison of photostability of green-fluorescent antibody conjugates. The following fluorescent goat anti–mouse IgG antibody conjugates were used to detect mouse anti–human IgG antibody labeling of human anti-nuclear antibodies in HEp-2 cells on prefixed test slides (INOVA Diagnostics Corp.): Oregon Green® 514 (O6383, ■), Alexa Fluor® 488 (A11001, ○), BODIPY® FL (B2752, △), Oregon Green® 488 (O6380, □) or fluorescein (F2761, ●). Samples were continuously illuminated and viewed on a fluorescence microscope using a fluorescein longpass filter set; images were acquired every 5 seconds. For each conjugate, three data sets, representing different fields of view, were averaged and then normalized to the same initial fluorescence intensity value to facilitate comparison.

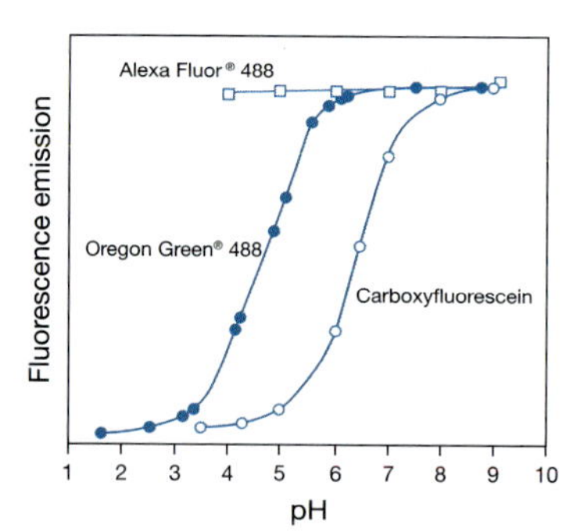

Figure 2.2.13 Comparison of pH-dependent fluorescence of the Oregon Green® 488 (●), carboxyfluorescein (○) and Alexa Fluor® 488 (□) fluorophores. Fluorescence intensities were measured for equal concentrations of the three dyes using excitation/emission at 490/520 nm.

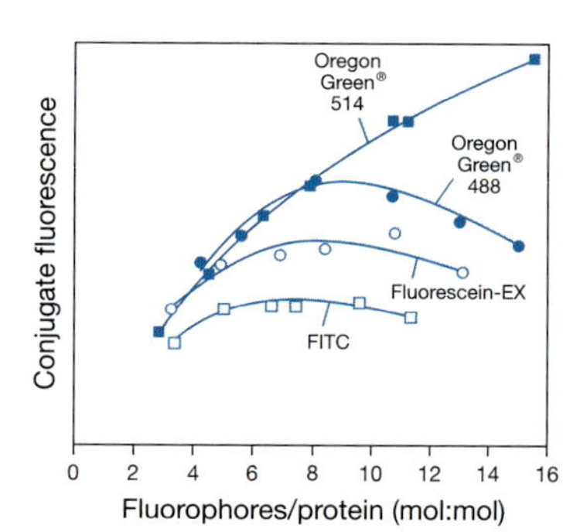

Figure 2.2.14 Comparison of relative fluorescence as a function of the number of fluorophores attached per protein for goat anti–mouse IgG antibody conjugates prepared using Oregon Green® 514 carboxylic acid succinimidyl ester (O6139, ■), Oregon Green® 488 carboxylic acid succinimidyl ester (O6147, ●), fluorescein-5-EX succinimidyl ester (F6130, ○) and fluorescein isothiocyanate (FITC, F143, F1906, F1907, □). Conjugate fluorescence is determined by measuring the fluorescence quantum yield of the conjugated dye relative to that of the free dye and multiplying by the number of fluorophores per protein.

IMPORTANT NOTICE: The products described in this manual are covered by one or more Limited Use Label License(s). Please refer to the Appendix on page 971 and Master Product List on page 975. Products are For Research Use Only. Not intended for any animal or human therapeutic or diagnostic use.

preferentially react with individual isomers, leading to complications in the interpretation of labeling results.[34–37] Consequently, we now prepare the 5-isomer of TMRIA (T6006, Figure 2.2.16) and the 5-isomer (T6027, Figure 2.2.17) and 6-isomer (T6028, Figure 2.2.18) of tetramethylrhodamine maleimide. A fluorogenic ADP biosensor has been described that exploits nucleotide-modulated self-quenching of two TMRIA labels that have been site-specifically attached to *Escherichia coli* ParM nucleotide-binding protein.[38] Tetramethylrhodamine-5-maleimide is often used for voltage-clamp fluorometry,[39] wherein it is attached to cysteine residues in the voltage-sensor domains of ion channels, generating fluorescence signals that are responsive to structural rearrangements associated with channel gating.[40,41] In this context, the dye is sometimes referred to as TMRM, but it should not be confused with tetramethylrhodamine methyl ester (T668, Section 12.2), a structurally similar but functionally quite different dye that is identified by the same acronym.

Rhodamine-Based Crosslinking Reagent

The thiol-reactive, homobifunctional crosslinker bis-((*N*-iodoacetyl)piperazinyl)sulfonerhodamine (B10621, Figure 2.2.19) is derived from a relatively rigid rhodamine dye. It is similar to a thiol-reactive rhodamine-based crosslinking reagent used to label regulatory light-chains of chicken gizzard myosin for fluorescence polarization experiments.[42] Researchers have attached bis-((*N*-iodoacetyl)piperazinyl) sulfonerhodamine to the kinesin motor domain and determined the orientation of kinesin bound to microtubules in the presence of a nonhydrolyzable ATP analog by fluorescence polarization microscopy.[43,44] Images of single molecules of chicken calmodulin crosslinked between two engineered cysteines by bis-((*N*-iodoacetyl)piperazinyl) sulfonerhodamine have been used to generate comparisons of experimental and theoretical super-resolution point-spread functions[45] (PSF). Dibromobimane (D1379, Section 2.3) is a shorter-wavelength alternative for applications requiring a fluorescent homobifunctional thiol crosslinker.

Rhodamine Red™ Maleimide

We offer a maleimide derivative of our Rhodamine Red™ fluorophore (R6029), which is spectrally similar to Lissamine rhodamine B (Figure 2.2.20). The spectral properties of Rhodamine Red™ maleimide have been exploited to improve the light-harvesting efficiency of chlorophyll by site-specific labeling of cysteine residues in the recombinantly expressed apoprotein in order to fill in the "green gap" in the absorption spectrum.[46] Rhodamine Red™ C_2-maleimide is a mixture of two isomeric sulfonamides (Figure 2.2.21).

Texas Red® Bromoacetamide and Maleimide

Conjugates of the bromoacetamide and maleimide derivatives of our Texas Red® fluorophore (T6009, T6008) have very little spectral overlap with fluorescein or Alexa Fluor® 488 conjugates (Figure 2.2.20) and are therefore useful as second labels in multicolor applications or as energy transfer acceptors from green-fluorescent dyes.[47,48] Bromoacetamides are only slightly less reactive with thiols than are iodoacetamides. The Texas Red® bromoacetamide (Figure 2.2.22) and maleimide (Figure 2.2.23) derivatives are mixtures of the corresponding two isomeric sulfonamides.

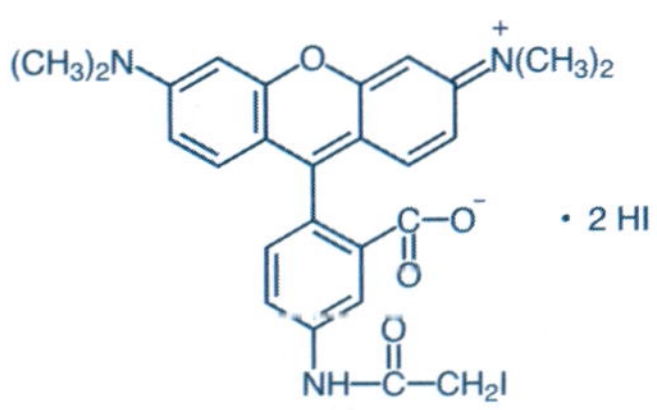

Figure 2.2.16 Tetramethylrhodamine-5-iodoacetamide dihydroiodide (5-TMRIA, T6006)

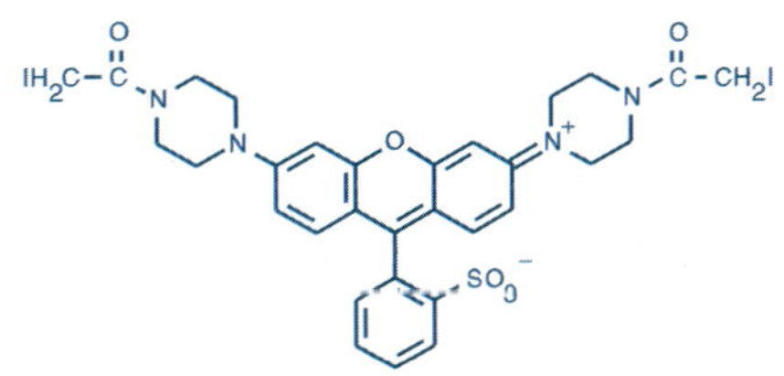

Figure 2.2.19 bis-((*N*-iodoacetyl)piperazinyl)sulfonerhodamine (B10621).

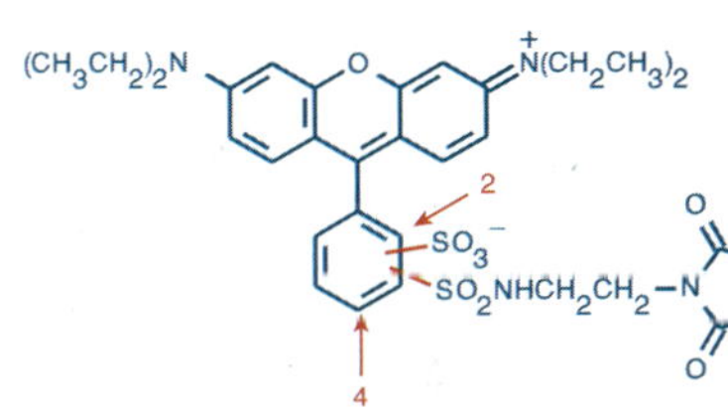

Figure 2.2.21 Rhodamine Red™ C_2-maleimide (R6029).

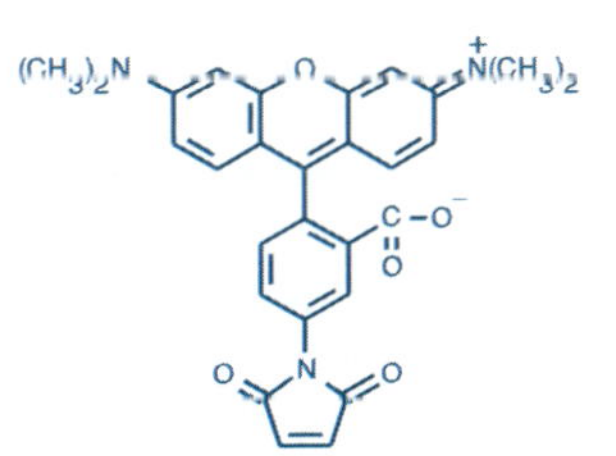

Figure 2.2.17 Tetramethylrhodamine-5-maleimide (T6027).

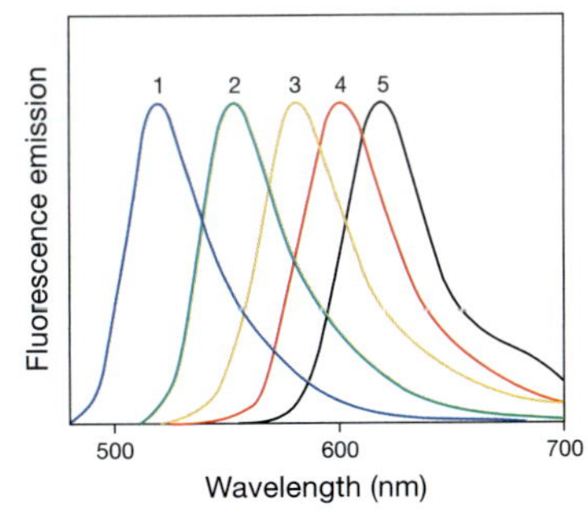

Figure 2.2.20 Normalized fluorescence emission spectra of goat anti-mouse IgG antibody conjugates of 1) fluorescein, 2) rhodamine 6G, 3) tetramethylrhodamine, 4) Lissamine rhodamine B and 5) Texas Red® dyes.

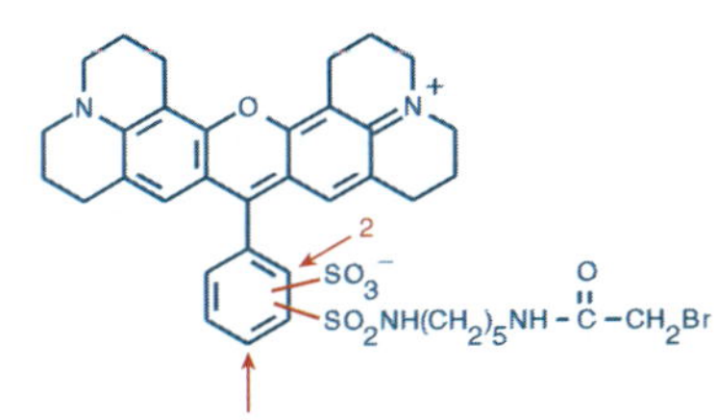

Figure 2.2.22 Texas Red® C_5-bromoacetamide (T6009).

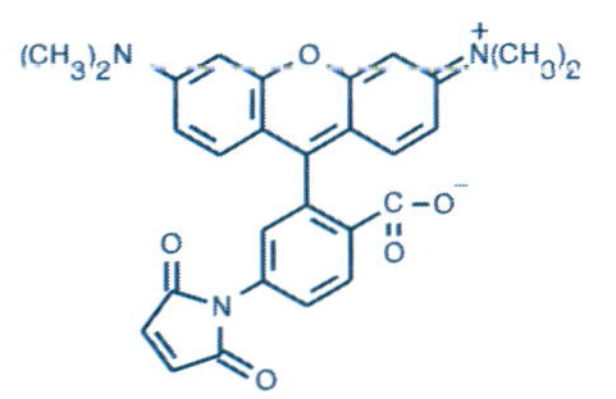

Figure 2.2.18 Tetramethylrhodamine-6-maleimide (T6028).

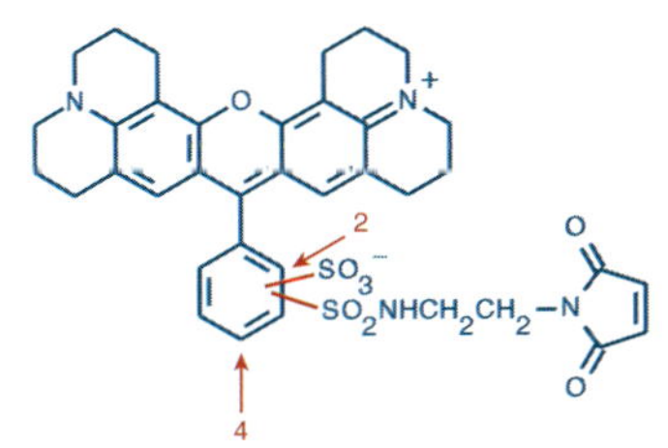

Figure 2.2.23 Texas Red® C_2-maleimide (T6008).

IMPORTANT NOTICE: The products described in this manual are covered by one or more Limited Use Label License(s). Please refer to the Appendix on page 971 and Master Product List on page 975. Products are For Research Use Only. Not intended for any animal or human therapeutic or diagnostic use.

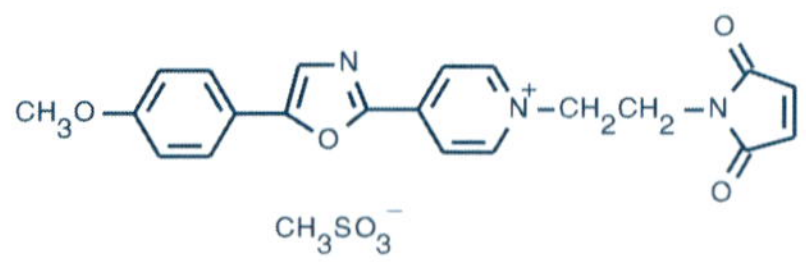

Figure 2.2.24 1-(2-maleimidylethyl)-4-(5-(4-methoxyphenyl)oxazol-2-yl)pyridinium methanesulfonate (PyMPO maleimide, M6026).

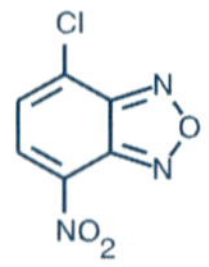

Figure 2.2.25 4-chloro-7-nitrobenz-2-oxa-1,3-diazole (NBD chloride; 4-chloro-7-nitrobenzofurazan, C20260).

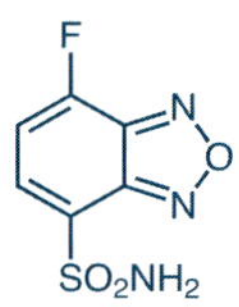

Figure 2.2.26 7-fluorobenz-2-oxa-1,3-diazole-4-sulfonamide (ABD-F, F6053).

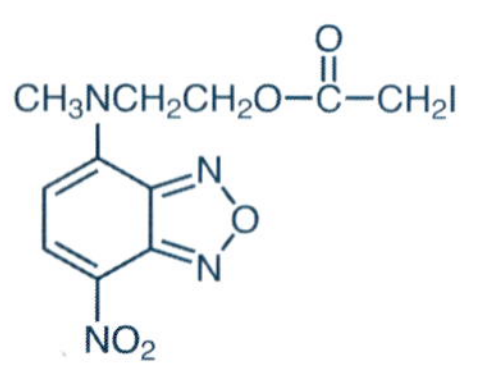

Figure 2.2.27 *N*-((2-(iodoacetoxy)ethyl)-*N*-methyl)amino-7-nitrobenz-2-oxa-1,3-diazole (IANBD ester, I9).

Figure 2.2.28 *N,N′*-dimethyl-*N*-(iodoacetyl)-*N′*-(7-nitrobenz-2-oxa-1,3-diazol-4-yl)ethylenediamine (IANBD amide, D2004).

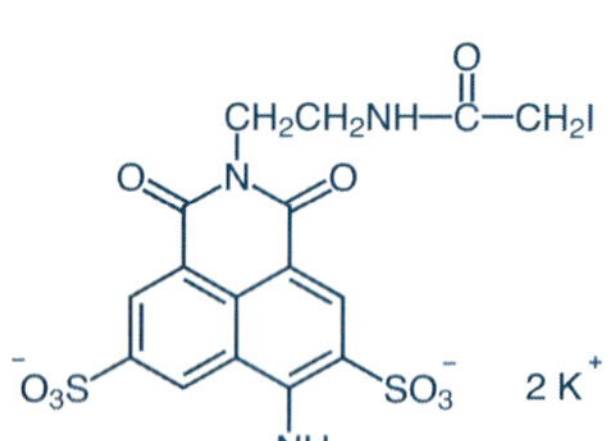

Figure 2.2.29 Lucifer yellow iodoacetamide, dipotassium salt (L1338).

PyMPO Maleimide

PyMPO maleimide (M6026, Figure 2.2.24) is an environment-sensitive thiol-reactive dye with a fluorescence excitation peak near 415 nm and an unusually long Stokes shift (fluorescence emission peak at ~560–580 nm). Its most widespread application is for labeling cysteine residues in the voltage-sensor domains of ion channels, where its fluorescence is exquisitely sensitive to structural rearrangements associated with channel gating.[40,49] This technique is commonly referred to as voltage-clamp fluorometry.[39]

Benzoxadiazole Derivatives, Including NBD Probes

NBD Chloride and NBD Fluoride

NBD chloride (C20260, Figure 2.2.25) and the more reactive NBD fluoride (F486) are common reagents for amine modification (Section 1.8). They also react with thiols[50–52] and cysteine in several proteins[53–57] to yield thioethers. NBD conjugates of thiols usually have much shorter-wavelength absorption and weaker fluorescence than do NBD conjugates of amines.[52] Selective modification of cysteines in the presence of reactive lysines and tyrosines is promoted by carrying out the reaction at pH <7;[58,59] however, NBD conjugates of thiols are often unstable, resulting in time-dependent label migration to adjacent lysine residues.[52,59]

ABD-F

Thiol conjugates of 7-fluorobenz-2-oxa-1,3-diazole-4-sulfonamide[60,61] (ABD-F, F6053; Figure 2.2.26) are much more stable in aqueous solution than are the thiol conjugates prepared from NBD chloride or NBD fluoride.[60] ABD-F is nonfluorescent until reacted with thiols and therefore can be used to quantitate thiols in solution,[62] as well as thiols separated by HPLC[63] or TLC.[64] ABD-F also reportedly reacts slowly with the hydroxy group of some tyrosine residues as well as α-amino groups in some proteins, forming products that are nonfluorescent but can be detected by absorbance at 385 nm.[65] ABD-F labeling is blocked by zinc binding to protein thiols and can therefore be used as an inverse proportionality indicator of bound Zn^{2+}.[66,67] In contrast, the fluorescent zinc indicators described in Section 19.7 primarily detect free Zn^{2+} ions. ABD–cysteine conjugates are very stable to acid hydrolysis, but labeling is partially reversed in basic solution containing DTT[68,69] (D1532; Section 2.1).

IANBD Ester and IANBD Amide

When conjugating the NBD fluorophore to thiols located in hydrophobic sites of proteins, we recommend using the NBD iodoacetate ester (IANBD ester, I9; Figure 2.2.27) or, preferably, the more hydrolytically stable NBD iodoacetamide (IANBD amide, D2004; Figure 2.2.28). These reactive reagents exhibit appreciable fluorescence only after reaction with thiols that are buried or unsolvated, and this fluorescence is highly sensitive to changes in protein conformation and assembly of molecular complexes.[70,71]

Lucifer Yellow Iodoacetamide

Lucifer yellow CH is a well-known polar tracer for neurons (Section 14.3). Its iodoacetamide derivative (L1338, Figure 2.2.29) similarly has high water solubility and visible absorption and emission spectra similar to those of lucifer yellow CH (Figure 2.2.30). As with the polar Alexa Fluor® maleimides and the stilbene iodoacetamide and maleimide (A484, A485; Section 2.3), a principal application of lucifer yellow iodoacetamide is the labeling of exposed thiols of proteins in solution, as well as in the outer membrane of live cells.[72] Lucifer yellow iodoacetamide has also been used as a fluorescence energy acceptor from aequorin in bioluminescence resonance energy transfer (BRET) assays.[73]

IMPORTANT NOTICE: The products described in this manual are covered by one or more Limited Use Label License(s). Please refer to the Appendix on page 971 and Master Product List on page 975. Products are For Research Use Only. Not intended for any animal or human therapeutic or diagnostic use.

molecular probes® | invitrogen™ by Life technologies™

TC-FlAsH™ and TC-ReAsH™ Detection of Tetracysteine-Tagged Proteins

TC-FlAsH™ and TC-ReAsH™ Detection Technology

TC-FlAsH™ and TC-ReAsH™ detection technology, based on the tetracysteine tag first described by Griffin, Adams and Tsien in 1998,[74,75] takes advantage of the high-affinity interaction of a biarsenical ligand (FlAsH-EDT_2 or ReAsH-EDT_2) with the thiols in a tetracysteine (TC) expression tag fused to the protein of interest. The FlAsH-EDT_2 ligand is essentially fluorescein that has been modified to contain two arsenic atoms at a set distance from each other, whereas the ReAsH-EDT_2 ligand is a similarly modified resorufin (Figure 2.2.31). Virtually nonfluorescent in the ethanedithiol (EDT)-bound state, these reagents become highly fluorescent when bound to the tetracysteine tag Cys-Cys-Xxx-Yyy-Cys-Cys, where Xxx-Yyy is typically Pro-Gly[76] (Figure 2.2.32). Modified tags with additional flanking sequences produce higher-affinity binding of the biarsenical ligand, resulting in improved signal-to-background characteristics.[77,78] Background due to off-target endogenous thiols can be diminished by washing with competitor dithiols such as 2,3-dimercaptopropanol (BAL). Although tetracysteine tag labeling is best suited to reducing intracellular environments, protocols involving co-administration of trialkylphosphine or dithiothreitol (DTT, D1532; Section 2.1) reducing agents have been devised for applications in oxidizing environments, including cell surfaces.[79,80] Photosensitized oxidation of diaminobenzidine (Fluorescent Probes for Photoconversion of Diaminobenzidine Reagents—Note 14.2) by ReAsH enables correlated fluorescence and electron microscopy of tetracysteine-tagged proteins.[80–82]

The six–amino acid tetracysteine tag is less likely to disrupt native protein structure and function than larger tags such as Green Fluorescent Protein[83] (GFP, 238 amino acids). Although the majority of TC-FlAsH™ and TC-ReAsH™ applications have been in mammalian cells (Figure 2.2.33), the reagents and associated methods are also particularly useful for nondisruptive labeling of viral coat proteins[84–88] and successful adaptations for labeling proteins in yeast,[83] bacteria,[89,90] *Dictyostelium discoideum*[91] and plants[92] have been described.

TC-FlAsH™ and TC-ReAsH™ Tetracysteine Tag Detection Kits

Transfecting the host cell line with an expression construct comprising the protein of interest fused to a tetracysteine tag (CCPGCC) is the first step into TC-FlAsH™ TC-ReAsH™ detection. The tagged protein is then detected by the addition of FlAsH-EDT_2 reagent or ReAsH-EDT_2 reagent, which generates green or red fluorescence, respectively, upon binding the tetracysteine motif. For detection of tetracysteine-tagged proteins expressed in cells, we offer the TC-FlAsH™ II and TC-ReAsH™ II In-Cell Tetracysteine Tag Detection Kits (T34561, T34562), which provide:

- FlAsH-EDT_2 or ReAsH-EDT_2 reagent (in Kit T34561 or T34562), respectively
- BAL wash buffer
- Detailed protocols

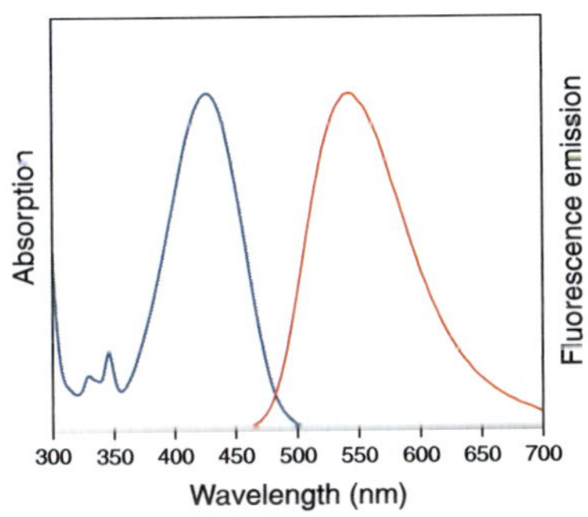

Figure 2.2.30 Absorption and fluorescence emission spectra of lucifer yellow CH in water.

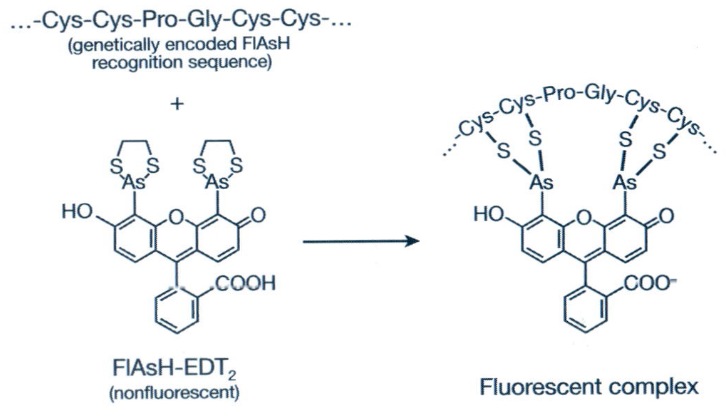

Figure 2.2.32 Binding of the nonfluorescent FlAsH-EDT_2 ligand to a recombinantly expressed tetracysteine sequence yields a highly fluorescent complex.

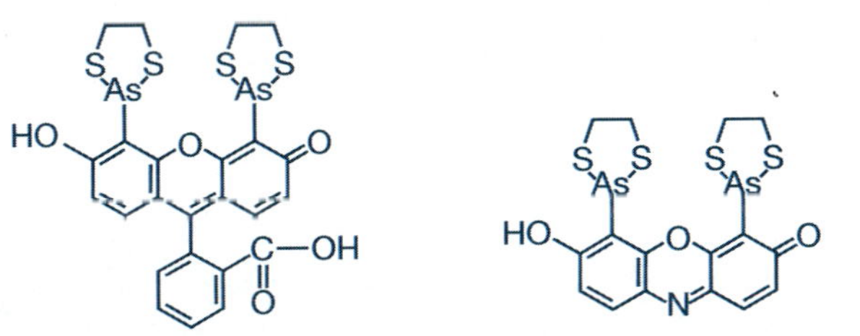

Figure 2.2.31 The structures of **A)** FlAsH-EDT_2 ligand and **B)** ReAsH-EDT_2 ligand, which are biarsenical labeling reagents provided in the TC-FlAsH™ II and TC-ReAsH™ II In-Cell Tetracysteine Tag Detection Kits (T34561, T34562), respectively.

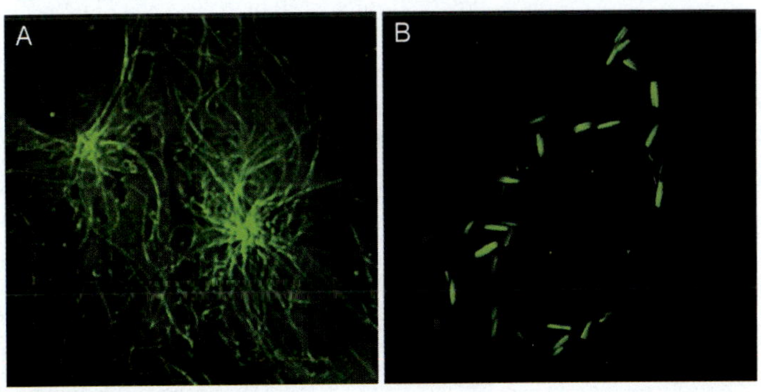

Figure 2.2.33 CHO-k1 cells expressing a tetracysteine-tagged version of β-tubulin labeled with FlAsH-EDT_2 reagent, provided in the TC-FlAsH™ II In Cell Tetracysteine Tag Detection Kit (T34561). Upon treatment with vinblastine, a compound known to perturb cytoskeletal structure, tubulin drastically rearranges from **A)** a reticular structure to **B)** rod-shaped structures.

IMPORTANT NOTICE: The products described in this manual are covered by one or more Limited Use Label License(s). Please refer to the Appendix on page 971 and Master Product List on page 975. Products are For Research Use Only. Not intended for any animal or human therapeutic or diagnostic use.

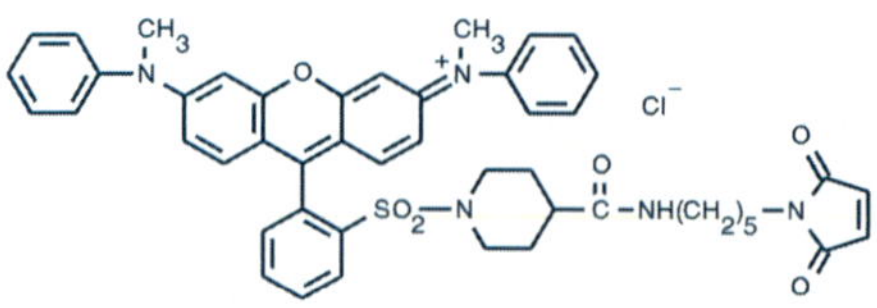

Figure 2.2.35 QSY® 7 C_5-maleimide (Q10257).

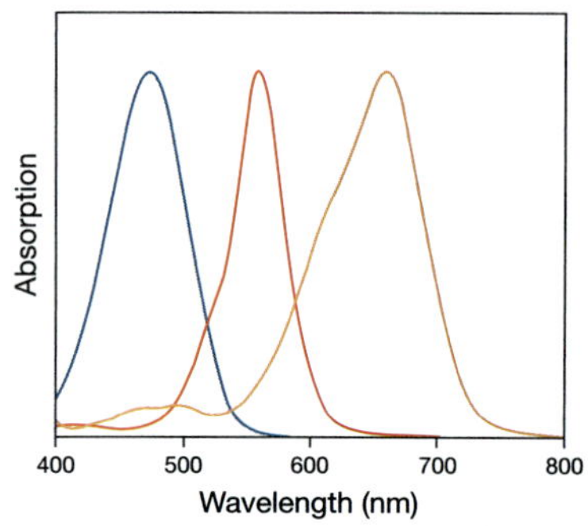

Figure 2.2.36 Normalized absorption spectra of the QSY® 35 (blue), QSY® 7 (red) and QSY® 21 (orange) dyes. The QSY® 7 and QSY® 9 dyes have essentially identical spectra.

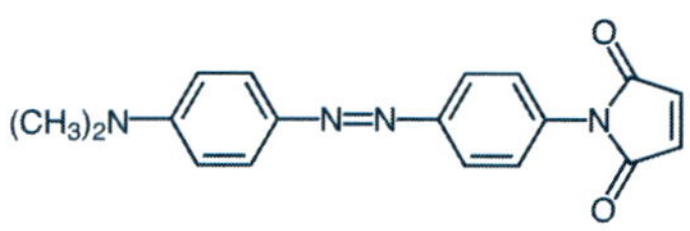

Figure 2.2.37 4-dimethylaminophenylazophenyl-4′-maleimide (DABMI, D1521).

We also offer these TC-FlAsH™ and TC-ReAsH™ detection reagents bundled with Gateway expression vectors for use in cloning the tetracysteine-tagged protein fusion. The TC-FlAsH™ II TC-ReAsH™ II In-Cell Tetracysteine Tag Detection Kit (with mammalian TC-Tag Gateway expression vectors) (T34563) provides:

- FlAsH-EDT_2 and ReAsH-EDT_2 reagents
- BAL wash buffer
- pcDNA 6.2/cTC-Tag-DEST
- pcDNA 6.2/nTC-Tag-DEST
- pcDNA 6.2/nTC-Tag-p64 control plasmid
- Detailed protocols

In addition to these kits for in-cell detection, we offer the TC-FlAsH™ Expression Analysis Detection Kits (A10067, A10068; Section 9.4), which are designed for detecting tetracysteine-tagged proteins in polyacrylamide gels (Figure 2.2.34).

Chromophoric Maleimides and Iodoacetamides

QSY® Maleimides and Iodoacetamide

QSY® 7 C_5-maleimide[93] (Q10257, Figure 2.2.35) and QSY® 9 C_5-maleimide (Q30457) are nonfluorescent, thiol-reactive diarylrhodamines with absorption spectra similar to those of our QSY® 7 and QSY® 9 succinimidyl esters (Q10193, Q20131; Section 1.6; Figure 2.2.36), respectively. Although the QSY® 7 and QSY® 9 chromophores are spectrally similar, QSY® 9 dye exhibits enhanced water solubility. QSY® 35 iodoacetamide (Q20348) is a nonfluorescent thiol-reactive analog of the amine-reactive nitrobenzoxadiazole (NBD) dye.

The principal applications of these thiol-reactive QSY® derivatives are as nonfluorescent acceptor dyes in fluorescence resonance energy transfer (FRET) assays[93–95] (Fluorescence Resonance Energy Transfer (FRET)—Note 1.2). The use of nonfluorescent acceptor dyes avoids the background fluorescence that often results from direct (i.e., nonsensitized) excitation of fluorescent acceptor dyes. The broad and strong absorption of QSY® 7 and QSY® 9 dyes (absorption maximum ~560 nm, EC ~90,000 $cm^{-1}M^{-1}$) yields extraordinarily efficient quenching of donors that have blue, green, orange or red fluorescence. QSY® 35 derivatives absorb light maximally near 470 nm (Figure 2.2.36), making their conjugates excellent FRET acceptors from UV light–excited donor dyes.

DABMI

DABMI (D1521, Figure 2.2.37) is the thiol-reactive analog of dabcyl succinimidyl ester (D2245, Section 1.8) and has similar properties and applications. Its principal application is as a nonfluorescent acceptor dye in fluorescence resonance energy transfer (FRET) assays[96–98] (Fluorescence Resonance Energy Transfer (FRET)—Note 1.2). The donor dyes in these assays typically include IAEDANS (I14) and other dyes described in Section 2.3. DABMI is also a useful derivatization reagent for MALDI-MS fragmentation analysis of cysteine-containing peptides.[99]

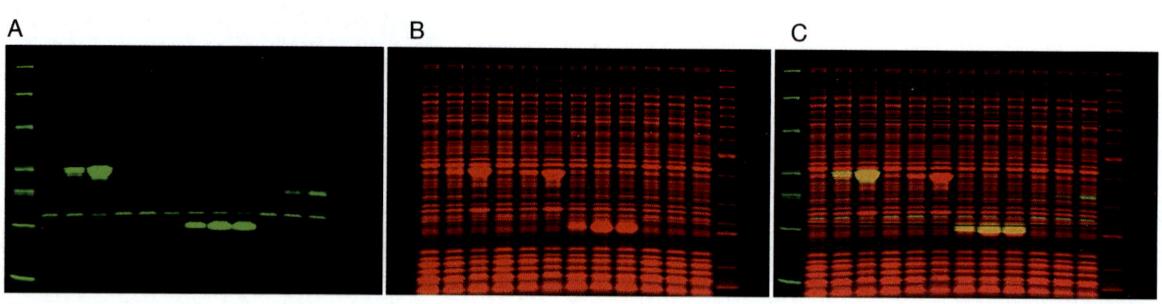

Figure 2.2.34 Protein gel staining using TC-FlAsH™ Expression Analysis Detection Kit (A10068). A) Tetracysteine-tagged proteins are labeled with FlAsH-EDT_2 reagent and fluoresce green. B) Total proteins are labeled with the Red total-protein stain provided in the kit and fluoresce red. C) an overlay of the two images reveals relative amounts of protein.

IMPORTANT NOTICE: The products described in this manual are covered by one or more Limited Use Label License(s). Please refer to the Appendix on page 971 and Master Product List on page 975. Products are For Research Use Only. Not intended for any animal or human therapeutic or diagnostic use.
www.invitrogen.com/probes
molecular probes® | invitrogen™ by life technologies™

NANOGOLD® Monomaleimide

In collaboration with Nanoprobes, Inc. (www.nanoprobes.com), we offer thiol-reactive NANOGOLD® monomaleimide (N20345). NANOGOLD® particles are small metal cluster complexes of gold particles for research applications in light or electron microscopy.[100,101] These cluster complexes are discrete chemical compounds, not gold colloids. NANOGOLD® monomaleimide (N20345) permits attachment of these very small (1.4 nm) yet uniformly sized gold particles to accessible thiol groups in biomolecules (Figure 2.2.38, Figure 2.2.39). NANOGOLD® monomaleimide, which is supplied as a set of five vials of a powder lyophilized from pH 7.5 HEPES buffer, is simply resuspended with the thiol-containing protein in deionized water at room temperature or below to form the conjugate, after which any excess NANOGOLD® monomaleimide is removed by gel filtration.[102,103]

In addition to its many uses for light and electron microscopy, NANOGOLD® monomaleimide has been shown to be an extremely efficient quencher for dyes in molecular beacons—probes that can be used for homogeneous fluorescence *in situ* hybridization assays.[104] NANOGOLD® conjugates of antibodies and streptavidin are described in Section 7.2 and Section 7.6, respectively, along with reagents and methods for silver enhancement to amplify electron microscopy detection.[104]

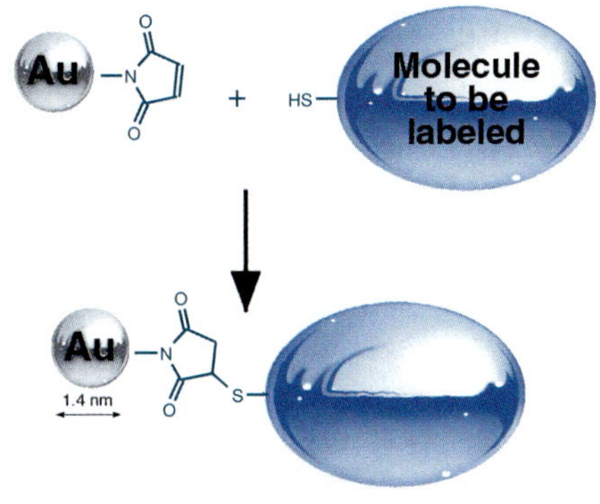

Figure 2.2.38 Reaction of NANOGOLD® monomaleimide (N20345) with a thiol. Image courtesy of Nanoprobes, Inc.

Figure 2.2.39 Scanning transmission electron microscope (STEM) image indicating that labeling with NANOGOLD® monomaleimide (N20345) occurs specifically at a hinge-thiol site on the IgG molecule. Image courtesy of Nanoprobes, Inc.

REFERENCES

1. J Histochem Cytochem (2003) 51:1699; **2**. PLoS One (2009) 4:e8115; **3**. Proc Natl Acad Sci U S A (2006) 103:12831; **4**. Proc Natl Acad Sci U S A (2003) 100:4001; **5**. Proteomics (2006) 6:1385; **6**. Lab Chip (2005) 5:1271; **7**. Electrophoresis (2005) 26:571; **8**. Bioconjug Chem (2009) 20:702; **9**. Clin Cancer Res (2008) 14:3840; **10**. Chemphyschem (2008) 9:2019; **11**. J Am Chem Soc (1994) 116:7801; **12**. Electrophoresis (2003) 24:2348; **13**. Biophys J (2007) 92:2162; **14**. J Thromb Haemost (2005) 3:2081; **15**. Blood (2009) 113:1332; **16**. Traffic (2007) 8:1179; **17**. Chem Commun (Camb) (2010) 46:2926; **18**. Anal Biochem (1997) 253:175; **19**. Electrophoresis (2002) 23:81; **20**. Anal Biochem (2002) 308:18; **21**. J Gen Microbiol (1993) 139:841; **22**. Adv Photochem (1993) 18:315; **23**. Photochem Photobiol (1983) 37:271; **24**. J Am Chem Soc (1977) 99:4306; **25**. J Cell Biol (1994) 126:901; **26**. Biochem Biophys Res Commun (2005) 330:454; **27**. Biochemistry (2002) 41:8630; **28**. Biochemistry (1990) 29:8283; **29**. Cytometry B Clin Cytom (2008) 74:244; **30**. Clin Lab Haematol (2003) 25:373; **31**. Br J Haematol (2004) 124:106; **32**. Biophys J (1993) 65:113; **33**. Biochemistry (1989) 28:2204; **34**. J Biol Chem (2006) 281:24036; **35**. Biophys J (1998) 74:3093; **36**. Biophys J (1998) 74:3083; **37**. Biochemistry (1992) 31:12431; **38**. ACS Chem Biol (2010) 5:415; **39**. Methods Mol Biol (2008) 491:213; **40**. J Gen Physiol (2008) 132:209; **41**. Biophys J (2007) 92:2403; **42**. Bioconjug Chem (1998) 9:160; **43**. Nat Struct Biol (2001) 8:540; **44**. Biophys J (2001) 81:2851; **45**. Nat Methods (2010) 7:377; **46**. Biochim Biophys Acta (2009) 1787:1499; **47**. BMC Cell Biol (2007) 8:43; **48**. Biochemistry (2005) 44:3694; **49**. Proc Natl Acad Sci U S A (2006) 103:12619; **50**. Anal Chem (1985) 57:1864; **51**. Anal Chem (1983) 55:1786; **52**. FEBS Lett (1970) 6:346; **53**. Arch Biochem Biophys (1990) 281:6; **54**. Biochemistry (1990) 29:10613; **55**. Biochemistry (1990) 29:7309; **56**. J Biochem (Tokyo) (1990) 107:563; **57**. Biochim Biophys Acta (1988) 956:217; **58**. J Biol Chem (1991) 266:13777; **59**. J Biol Chem (1983) 258:5419; **60**. Anal Chim Acta (1994) 290:3; **61**. Anal Chem (1984) 56:2461; **62**. Chem Pharm Bull (Tokyo) (1990) 38:2290; **63**. J Chromatogr (1990) 514:189; **64**. J Chromatogr (1990) 502:230; **65**. Anal Biochem (2003) 314:166; **66**. J Biol Inorg Chem (2006) 11:1049; **67**. Proc Natl Acad Sci U S A (2001) 98:5556; **68**. J Chromatogr A (1998) 798:47; **69**. Techniques in Protein Chemistry V, Crabb JW, Ed. 1994 p. 189; **70**. J Am Chem Soc (2009) 131:13286; **71**. Nat Methods (2010) 7:67; **72**. J Biol Chem (1999) 274:6626; **73**. Anal Bioanal Chem (2005) 381:1387; **74**. Methods Enzymol (2000) 327:565; **75**. Science (1998) 281:269; **76**. J Am Chem Soc (2002) 124:6063; **77**. J Am Chem Soc (2009) 131:4613; **78**. Nat Biotechnol (2005) 23:1308; **79**. Mol Biol Cell (2009) 20:233; **80**. Proc Natl Acad Sci U S A (2006) 103:17777; **81**. Nat Cell Biol (2007) 9:1057; **82**. Science (2002) 296:503; **83**. Mol Biol Cell (2004) 15:5616; **84**. Methods Mol Biol (2009) 485:151; **85**. PLoS Pathog (2008) 4:e1000015; **86**. J Virol (2009) 83:2611; **87**. Nat Methods (2008) 5:75; **88**. J Virol (2005) 79:4055; **89**. Appl Environ Microbiol (2010) 76:1241; **90**. Chem Biol (2008) 15:619; **91**. J Microsc (2009) 234:9; **92**. Biotechniques (2006) 41:569; **93**. Proc Natl Acad Sci U S A (2003) 100:13308; **94**. J Biol Chem (2008) 283:1501; **95**. Biochemistry (2005) 44:16835; **96**. J Biol Chem (2005) 280:2613; **97**. Biochemistry (2005) 44:4312; **98**. Biochemistry (2001) 40:1171; **99**. Protein Sci (2003) 12:1567; **100**. Microsc Res Tech (1998) 42:2; **101**. J Histochem Cytochem (2000) 48:471; **102**. J Biol Chem (2000) 275:30458; **103**. J Biol Chem (2000) 275:30465; **104**. Nat Biotechnol (2001) 19:365.

DATA TABLE 2.2 THIOL-REACTIVE PROBES EXCITED WITH VISIBLE LIGHT

Cat. No.	MW	Storage	Soluble	Abs	EC	Em	Solvent	Notes
A10254	720.66	F,DD,L	H_2O, DMSO	493	72,000	516	pH 7	1, 2, 3
A10255	812.88	F,DD,L	H_2O, DMSO	528	78,000	552	MeOH	1
A10256	908.97	F,DD,L	H_2O, DMSO	588	96,000	612	pH 7	1, 4
A10258	1034.37	F,DD,L	H_2O, DMSO	554	93,000	570	pH 7	1
A20341	880.92	F,DD,L	H_2O, DMSO	575	92,000	600	pH 7	1, 5
A20342	~1300	F,DD,L	H_2O, DMSO	622	143,000	640	MeOH	1
A20343	~900	F,DD,L	H_2O, DMSO	668	112,000	697	MeOH	1, 6
A20344	~1000	F,DD,L	H_2O, DMSO	684	175,000	714	MeOH	1, 7
A20346	~1250	F,DD,L	H_2O, DMSO	556	158,000	572	MeOH	1

continued on next page

IMPORTANT NOTICE: The products described in this manual are covered by one or more Limited Use Label License(s). Please refer to the Appendix on page 971 and Master Product List on page 975. Products are For Research Use Only. Not intended for any animal or human therapeutic or diagnostic use.

DATA TABLE 2.2 THIOL-REACTIVE PROBES EXCITED WITH VISIBLE LIGHT—*continued*

Cat. No.	MW	Storage	Soluble	Abs	EC	Em	Solvent	Notes
A20347	~1300	F,DD,L	H_2O, DMSO	651	265,000	671	MeOH	1, 8
A30459	~1350	F,DD,L	H_2O, DMSO	753	290,000	783	MeOH	1, 24
B1355	425.23	F,D,L	pH >6, DMF	492	81,000	515	pH 9	9
B2103	341.00	F,D,L	DMSO, MeCN	533	62,000	561	$CHCl_3$	10, 11
B10250	414.22	F,D,L	DMSO, MeCN	504	79,000	510	MeOH	11
B10621	840.47	F,D,L	DMSO	549	88,000	575	MeOH	12
B20340	788.44	F,D,L	DMSO	504	132,000	511	MeOH	13
B22802	449.14	F,D,L	DMSO, MeCN	658	73,000	678	$CHCl_3$	14
B30466	562.42	F,DD,L	DMSO, MeCN	544	60,000	570	MeOH	11
C20260	199.55	F,D,L	DMF, MeCN	336	9800	none	MeOH	15, 16
D1521	320.35	F,D,L	DMF, MeCN	419	34,000	none	MeOH	17
D2004	419.18	F,D,L	DMF, DMSO	478	25,000	541	MeOH	12, 17
D6003	417.00	F,D,L	DMSO, MeCN	502	76,000	510	MeOH	11, 12
D6004	431.03	F,D,L	DMSO, MeCN	508	69,000	543	MeOH	11, 12
D20350	419.24	F,D,L	DMSO	499	88,000	508	MeOH	18
D20351	575.38	F,D,L	DMSO	577	60,000	618	MeOH	18
E118	742.95	F,D,L	pH >6, DMF	524	103,000	545	MeOH	1, 19
F150	427.37	F,D,L	pH >6, DMF	492	83,000	515	pH 9	1, 9, 20
F486	183.10	F,D,L	MeCN, $CHCl_3$	328	8000	none	MeOH	15
F6053	217.17	F,D,L	DMF, DMSO	320	4800	none	MeOH	21
I9	406.14	F,D,L	DMF, MeCN	472	23,000	536	MeOH	12, 17
I30451	515.26	F,D,L	pH >6, DMF	492	78,000	515	pH 9	1, 9, 12
I30452	515.26	F,D,L	pH >6, DMF	491	82,000	516	pH 9	1, 9, 12
L1338	659.51	F,D,L	H_2O	426	11,000	531	pH 7	12
M6026	471.48	F,D,L	DMSO	412	23,000	561	MeOH	22
O6010	551.24	F,D,L	pH >6, DMF	491	68,000	516	pH 9	1, 12, 23
O6034	463.35	F,D,L	pH >6, DMF	491	81,000	515	pH 9	1, 23
Q10257	858.45	F,D,L	DMSO	560	92,000	none	MeOH	
Q20348	453.20	F,D,L	DMSO	475	24,000	none	MeOH	12
Q30457	1083.30	F,D,L	H_2O, DMSO	562	90,000	none	MeOH	1
R6029	680.79	F,D,L	DMSO	560	119,000	580	MeOH	
T6006	825.22	F,D,L	DMSO	543	87,000	567	MeOH	12
T6008	728.83	F,D,L	DMSO	582	112,000	600	MeOH	
T6009	811.80	F,D,L	DMSO	583	115,000	603	MeOH	
T6027	481.51	F,D,L	DMSO	541	95,000	567	MeOH	
T6028	481.51	F,D,L	DMSO	541	91,000	567	MeOH	
T30453	510.31	F,D,L	DMSO	503	80,000	510	MeOH	
T30454	658.52	F,D,L	DMSO	544	58,000	570	MeOH	
T30455	684.53	F,D,L	DMSO	589	63,000	617	MeOH	
T30456	710.57	F,D,L	DMSO	625	93,000	640	MeOH	
T34561	664.49	FF,D,L,AA	DMSO	508	70,000	530	pH 7.2	25, 26
T34562	545.37	FF,D,AA	DMSO	596	69,000	608	pH 7.2	25, 27

For definitions of the contents of this data table, see "Using *The Molecular Probes® Handbook*" in the introductory pages.

Notes

1. Aqueous stock solutions should be used within 24 hours; long-term storage is NOT recommended.
2. The fluorescence lifetime (τ) of the Alexa Fluor® 488 dye in pH 7.4 buffer at 20°C is 4.1 nanoseconds. Data provided by the SPEX Fluorescence Group, Horiba Jobin Yvon Inc.
3. Abs and Em of the Alexa Fluor® 488 dye are red-shifted by as much as 16 nm and 25 nm respectively on microarrays relative to aqueous solution values. The magnitude of the spectral shift depends on the array substrate material. (Biotechniques (2005) 38:127)
4. The fluorescence lifetime (τ) of the Alexa Fluor® 594 dye in pH 7.4 buffer at 20°C is 3.9 nanoseconds. Data provided by the SPEX Fluorescence Group, Horiba Jobin Yvon Inc.
5. The fluorescence lifetime (τ) of the Alexa Fluor® 568 dye in pH 7.4 buffer at 20°C is 3.6 nanoseconds. Data provided by the SPEX Fluorescence Group, Horiba Jobin Yvon Inc.
6. The fluorescence lifetime (τ) of the Alexa Fluor® 660 dye in pH 7.5 buffer at 20°C is 1.2 nanoseconds. Data provided by Pierre-Alain Muller, Max Planck Institute for Biophysical Chemistry, Göttingen.
7. The fluorescence lifetime (τ) of the Alexa Fluor® 680 dye in pH 7.5 buffer at 20°C is 1.2 nanoseconds. Data provided by Pierre-Alain Muller, Max Planck Institute for Biophysical Chemistry, Göttingen.
8. The fluorescence lifetime (τ) of the Alexa Fluor® 647 dye in H_2O at 20°C is 1.0 nanoseconds and 1.5 nanoseconds in EtOH. (Bioconjug Chem (2003) 14:195)
9. Absorption and fluorescence of fluorescein derivatives are pH dependent. Extinction coefficients and fluorescence quantum yields decrease markedly at pH <7.
10. B2103 spectra are for the unreacted reagent. The thiol adduct has Abs = 493 nm, Em = 503 nm in MeOH.
11. The absorption and fluorescence spectra of BODIPY® derivatives are relatively insensitive to the solvent.
12. Iodoacetamides in solution undergo rapid photodecomposition to unreactive products. Minimize exposure to light prior to reaction.
13. Fluorescence emission of B20340 is relatively weak until the disulfide linkage between its two BODIPY® FL fluorophores is reductively cleaved.
14. B22802 spectral data are for the unreacted reagent. The thiol adduct has Abs = 629 nm, Em = 647 nm in dichloromethane (CH_2Cl_2).
15. Spectra of 2-mercaptoethanol adduct of NBD chloride in MeOH: Abs = 425 nm (EC = 13,000 $cm^{-1}M^{-1}$), Em = 520 nm. NBD fluoride yields the same derivatives as NBD chloride but is more reactive.
16. This product is specified to equal or exceed 98% analytical purity by HPLC.
17. Spectral data of the 2-mercaptoethanol adduct.
18. Spectral data are for the unreacted reagent and are essentially unchanged upon reaction with thiols.
19. Eosin and erythrosin derivatives also exhibit phosphorescence with an emission maximum at ~680 nm. The phosphorescence lifetime is ~1 millisecond for eosin and 0.5 milliseconds for erythrosin. (Biochem J (1979) 183:561, Spectroscopy (1990) 5:20) Fluorescence lifetimes (τ) are 1.4 nanoseconds (QY = 0.2) for eosin and 0.1 nanoseconds (QY = 0.02) for erythrosin. (J Am Chem Soc (1977) 99:4306)
20. QY increases on reaction with thiols; Abs, EC and Em are essentially unchanged. (Anal Biochem (2001) 295:101)
21. F6053 reaction product with dimethylaminoethanethiol has Abs = 376 nm (EC ~8000 $cm^{-1}M^{-1}$), Em ~510 nm in MeOH.
22. Fluorescence emission spectrum shifts to shorter wavelengths in nonpolar solvents.
23. Absorption and fluorescence of Oregon Green® 488 derivatives are pH dependent only in moderately acidic solutions (pH <5).

IMPORTANT NOTICE: The products described in this manual are covered by one or more Limited Use Label License(s). Please refer to the Appendix on page 971 and Master Product List on page 975. Products are For Research Use Only. Not intended for any animal or human therapeutic or diagnostic use.

DATA TABLE 2.2 THIOL-REACTIVE PROBES EXCITED WITH VISIBLE LIGHT—*continued*

24. The fluorescence lifetime (τ) of the Alexa Fluor® 750 dye in H_2O at 22°C is 0.7 nanoseconds. Data provided by ISS Inc. (Champaign, IL).
25. This product is supplied as a ready-made solution in the solvent indicated under "Soluble."
26. Data for T34561 represents FlAsH complexed with the tetracysteine peptide FLNCCPGCCMEP. (Nat Biotechnol (2005) 23:1308, Nat Protoc (2008) 3:1527) The FlAsH-EDT_2 reagent is essentially nonfluorescent and has Abs = 496 nm (EC = 69,500 $cm^{-1}M^{-1}$) in 0.1 M NaOH. (Nat Protoc (2008) 3:1527)
27. Data for T34562 represents ReAsH complexed with the tetracysteine peptide FLNCCPGCCMEP. (Nat Biotechnol (2005) 23:1308, Nat Protoc (2008) 3:1527) The ReAsH-EDT_2 reagent is essentially nonfluorescent and has Abs = 579 nm (EC = 63,000 $cm^{-1}M^{-1}$) in 0.1 M NaOH. (Nat Protoc (2008) 3:1527)

PRODUCT LIST 2.2 THIOL-REACTIVE PROBES EXCITED WITH VISIBLE LIGHT

Cat. No.	Product	Quantity
A10254	Alexa Fluor® 488 C_5-maleimide	1 mg
A10255	Alexa Fluor® 532 C_5-maleimide	1 mg
A10258	Alexa Fluor® 546 C_5-maleimide	1 mg
A20346	Alexa Fluor® 555 C_2-maleimide	1 mg
A20341	Alexa Fluor® 568 C_5-maleimide	1 mg
A10256	Alexa Fluor® 594 C_5-maleimide	1 mg
A20342	Alexa Fluor® 633 C_5-maleimide	1 mg
A20347	Alexa Fluor® 647 C_2-maleimide	1 mg
A20343	Alexa Fluor® 660 C_2-maleimide	1 mg
A20344	Alexa Fluor® 680 C_2-maleimide	1 mg
A30459	Alexa Fluor® 750 C_5-maleimide	1 mg
B10621	bis-((*N*-iodoacetyl)piperazinyl)sulfonerhodamine	5 mg
B2103	BODIPY® 493/503 methyl bromide (8-bromomethyl-4,4-difluoro-1,3,5,7-tetramethyl-4-bora-3a,4a-diaza-*s*-indacene)	5 mg
B22802	BODIPY® 630/650 methyl bromide (8-bromomethyl-4,4-difluoro-3,5-bis-(2-thienyl)-4-bora-3a,4a-diaza-*s*-indacene)	1 mg
B10250	BODIPY® FL N-(2-aminoethyl)maleimide	5 mg
B20340	BODIPY® FL L-cystine	1 mg
B30466	BODIPY® TMR C_5-maleimide	1 mg
B1355	5-(bromomethyl)fluorescein	10 mg
C20260	4-chloro-7-nitrobenz-2-oxa-1,3-diazole (NBD chloride; 4-chloro-7-nitrobenzofurazan) *FluoroPure™ grade*	100 mg
D20351	4,4-difluoro-3,5-bis(4-methoxyphenyl)-8-(4-maleimidylphenyl)-4-bora-3a,4a-diaza-*s*-indacene (BODIPY® 577/618 maleimide)	5 mg
D6003	*N*-(4,4-difluoro-5,7-dimethyl-4-bora-3a,4a-diaza-*s*-indacene-3-yl)methyl)iodoacetamide (BODIPY® FL C_1-IA)	5 mg
D20350	4,4-difluoro-1,3,5,7-tetramethyl-8-(4-maleimidylphenyl)-4-bora-3a,4a-diaza-*s*-indacene (BODIPY® 499/508 maleimide)	5 mg
D6004	*N*-(4,4-difluoro-1,3,5,7-tetramethyl-4-bora-3a,4a-diaza-*s*-indacene-2-yl)iodoacetamide (BODIPY® 507/545 IA)	5 mg
D1521	4-dimethylaminophenylazophenyl-4′-maleimide (DABMI)	100 mg
D2004	*N,N′*-dimethyl-*N*-(iodoacetyl)-*N′*-(7-nitrobenz-2-oxa-1,3-diazol-4-yl)ethylenediamine (IANBD amide)	25 mg
E118	eosin-5-maleimide	25 mg
F150	fluorescein-5-maleimide	25 mg
F6053	7-fluorobenz-2-oxa-1,3-diazole-4-sulfonamide (ABD-F)	10 mg
F486	4-fluoro-7-nitrobenz-2-oxa-1,3-diazole (NBD fluoride; 4-fluoro-7-nitrobenzofurazan)	25 mg
I30451	5-iodoacetamidofluorescein (5-IAF)	25 mg
I30452	6-iodoacetamidofluorescein (6-IAF)	25 mg
I9	*N*-((2-(iodoacetoxy)ethyl)-*N*-methyl)amino-7-nitrobenz-2-oxa-1,3-diazole (IANBD ester)	100 mg
L1338	lucifer yellow iodoacetamide, dipotassium salt	25 mg
M6026	1-(2-maleimidylethyl)-4-(5-(4-methoxyphenyl)oxazol-2-yl)pyridinium methanesulfonate (PyMPO maleimide)	5 mg
N20345	NANOGOLD® monomaleimide *special packaging*	5 x 6 nmol
O6010	Oregon Green® 488 iodoacetamide *mixed isomers*	5 mg
O6034	Oregon Green® 488 maleimide	5 mg
Q10257	QSY® 7 C_5-maleimide	5 mg
Q30457	QSY® 9 C_5-maleimide	5 mg
Q20348	QSY® 35 iodoacetamide	5 mg
R6029	Rhodamine Red™ C_2-maleimide	5 mg
T34561	TC-FlAsH™ II In-cell Tetracysteine Tag Detection Kit *green fluorescence* *for live-cell imaging*	1 kit
T34562	TC-ReAsH™ II In-Cell Tetracysteine Tag Detection Kit *red fluorescence* *for live-cell imaging*	1 kit
T34563	TC-FlAsH™ TC-ReAsH™ II In-cell Tetracysteine Tag Detection Kit *with mammalian TC-Tag Gateway® expression vectors* *green fluorescence* *red fluorescence*	1 kit
T6006	tetramethylrhodamine-5-iodoacetamide dihydroiodide (5-TMRIA) *single isomer*	5 mg
T6027	tetramethylrhodamine-5-maleimide *single isomer*	5 mg
T6028	tetramethylrhodamine-6-maleimide *single isomer*	5 mg
T6008	Texas Red® C_2-maleimide	5 mg
T6009	Texas Red® C_5-bromoacetamide	5 mg
T30456	TS-Link™ BODIPY® 630/650 C_5-thiosulfate, sodium salt	5 mg
T30453	TS-Link™ BODIPY® FL C_2-thiosulfate, sodium salt	5 mg
T30454	TS-Link™ BODIPY® TMR C_5-thiosulfate, sodium salt	5 mg
T30455	TS-Link™ BODIPY® TR C_5-thiosulfate, sodium salt	5 mg

IMPORTANT NOTICE: The products described in this manual are covered by one or more Limited Use Label License(s). Please refer to the Appendix on page 971 and Master Product List on page 975. Products are For Research Use Only. Not intended for any animal or human therapeutic or diagnostic use.

2.3 Thiol-Reactive Probes Excited with Ultraviolet Light

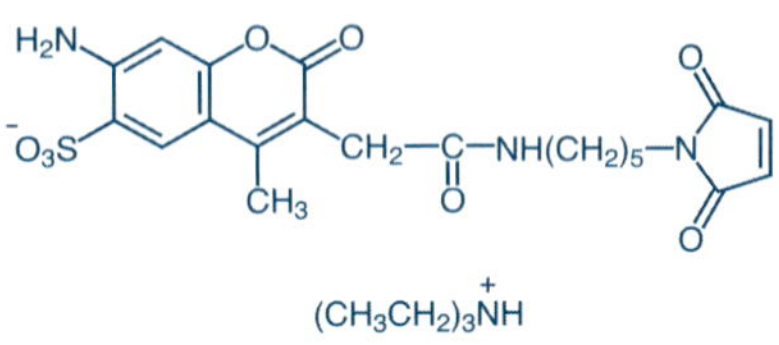

Figure 2.3.1 Alexa Fluor® 350 C_5-maleimide (A30505).

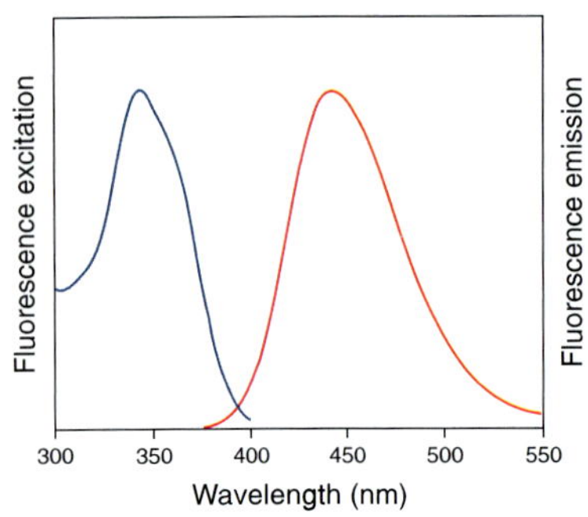

Figure 2.3.2 Fluorescence excitation and emission spectra of Alexa Fluor® 350 goat anti–mouse IgG antibody in pH 8.0 buffer.

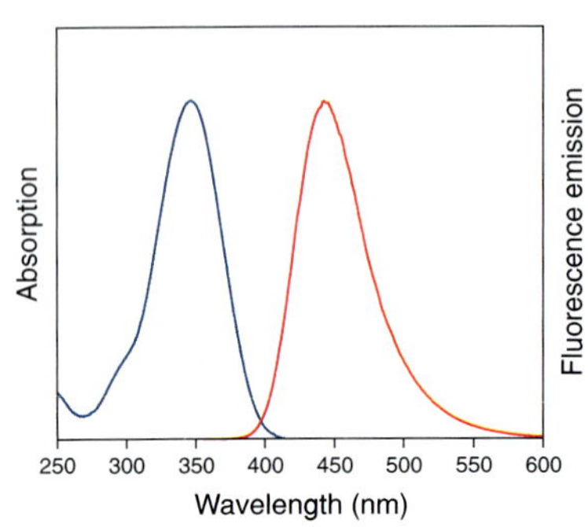

Figure 2.3.3 Absorption and fluorescence emission spectra of 7-amino-4-methylcoumarin in pH 7.0 buffer.

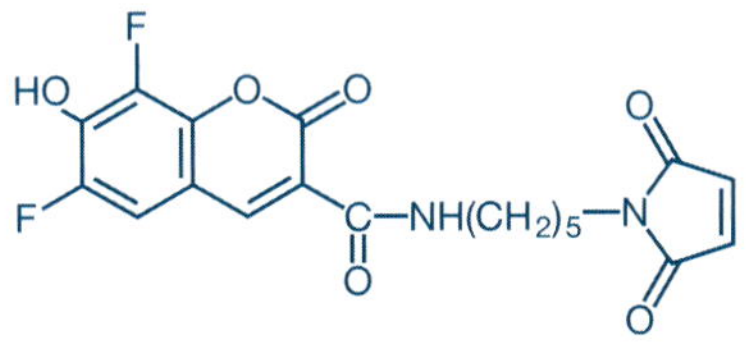

Figure 2.3.4 Pacific Blue™ C_5-maleimide (P30506).

The thiol-reactive dyes described in this section have their longest-wavelength absorption peaks at less than 410 nm (Table 2.2). Typically, these dyes exhibit blue fluorescence and have much weaker absorption than the dyes described in Section 2.2, with extinction coefficients often below 20,000 $cm^{-1}M^{-1}$.

The strong environmental dependence of the emission spectra and quantum yields of several of the dyes—especially the coumarin, benzoxadiazole (NBD probes and ABD-F, Section 2.2), aminonaphthalene (e.g., dansyl) and Dapoxyl® fluorophores—makes some of these thiol-reactive probes useful for investigating protein structure and assembly, following protein transport through membranes and studying ligand binding to receptors.[1]

Coumarin Derivatives

Alexa Fluor® 350 Maleimide

Alexa Fluor® 350 C_5-maleimide (A30505), a thiol-reactive, sulfonated coumarin derivative (Figure 2.3.1), produces protein conjugates that are optimally excited at 346 nm and have bright blue fluorescence emission (Figure 2.3.2) at wavelengths slightly shorter than AMCA or AMCA-X conjugates (Figure 2.3.3) (emission maximum ~442 nm versus 448 nm). The shorter emission maximum of Alexa Fluor® 350 conjugates reduces their spectral overlap with the emission of fluorescein and Oregon Green® 488 dyes.[2] Like our other Alexa Fluor® dyes, Alexa Fluor® 350 C_5-maleimide offers unrivaled brightness and pH-independent fluorescence, as well as water solubility and a low degree of quenching upon conjugation (Section 1.3).

Pacific Blue™ Maleimide

The Pacific Blue™ dye, which is based on the 6,8-difluoro-7-hydroxycoumarin fluorophore (Figure 2.3.4), exhibits bright blue fluorescence, with excitation/emission maxima of ~410/455 nm. Significantly, the pK_a value of this 6,8-difluoro-7-hydroxycoumarin derivative is 2–3 log units lower than that of the corresponding 7-hydroxycoumarin. Thus, the thiol-reactive

Table 2.2 Molecular Probes® thiol-reactive dyes excited with ultraviolet light.

Derivative	Abs *	Em *	Haloacetamide	Maleimide	Other
Alexa Fluor® 350	346	442		A30505	
Anilinonaphthalene †	326	462	IAANS, I7	MIANS, M8	
Benzophenone	282	NA		B1508	
Bimane	375	456	B30500	B30501	
Dibromobimane	394	490			bBBr, D1379 ‡
Diethylaminocoumarin	384	470	DCIA, D404	CPM, D346 ‡ MDCC, D10253	
Dimethylaminocoumarin	376	465	DACIA, D10252	DACM, D10251 ‡	
Dimethylaminonaphthalene †	391	500	badan, B6057		acrylodan, A433
Monobromobimane	394	490			mBBr, M1378 ‡ mBBr, M20381 ‡
Monochlorobimane	394	490			mBCl, M1381MP ‡
Naphthalene †	336	490	IAEDANS, I14		
Pacific Blue™	410	455		P30506	
Pacific Orange™	400	551		P30507	
Phenanthroline	270	NA	P6879		
Pyrene †	339	384	P29 P2007MP	P28	
Stilbene †	329	408	A484	A485	

* Approximate absorption (Abs) and emission (Em) maxima, in nm, for the reagent (if fluorescent) or the fluorescent thiol adduct. † Environment-sensitive fluorophore. ‡ Very weakly fluorescent until reacted with thiols. NA = Not applicable. More information on thiol-reactive dyes is available at www.invitrogen.com/handbook/labelingchemistry.

IMPORTANT NOTICE: The products described in this manual are covered by one or more Limited Use Label License(s). Please refer to the Appendix on page 971 and Master Product List on page 975. Products are For Research Use Only. Not intended for any animal or human therapeutic or diagnostic use.
www.invitrogen.com/probes
molecular probes® | invitrogen™ by life technologies™

Pacific Blue™ C_5-maleimide (P30506) yields conjugates that are strongly fluorescent, even at neutral pH. In addition, Pacific Blue™ conjugates are efficiently excited by the 405 nm spectral line of the violet diode laser developed for fluorescence microscopy and flow cytometry.[3]

Other Coumarin Maleimides and Iodoacetamides

We offer several blue-fluorescent thiol-reactive dialkylcoumarins (Table 2.2), including 7-diethylamino-3-(4′-maleimidylphenyl)-4-methylcoumarin (CPM, D346; Figure 2.3.5) and *N*-(7-dimethylamino-4-methylcoumarin-3-yl)maleimide (DACM, D10251) and the corresponding iodoacetamides DCIA (D404) and DACIA (D10252). The dialkylcoumarin fluorophore is an excellent fluorescence resonance energy acceptor from tryptophan and a good donor to fluorescein, NBD, Alexa Fluor® 488 dye, Green Fluorescent Protein (GFP) and the nonfluorescent QSY® 7, QSY® 9 and QSY® 35 quenchers, making these thiol-reactive coumarins especially valuable for studying protein structure and for detecting protein–membrane interactions.[4–6] Fluorescence emission of the dialkylcoumarin conjugates is moderately sensitive to environment.

Unlike MDCC (described below), which is intrinsically fluorescent, the maleimides CPM and DACM are essentially nonfluorescent until they react with thiols, permitting thiol quantitation without a separation step.[7,8] The environment-sensitive fluorescence of CPM is also a useful indicator of protein folding in thermal-shift assays of ligand- and mutagenesis-dependent protein stability[9] (Monitoring Protein-Folding Processes with Environment-Sensitive Dyes—Note 9.1).

We also offer 7-diethylamino-3-((((2-maleimidyl)ethyl)amino)carbonyl)coumarin (MDCC, D10253; Figure 2.3.6). When conjugated to a mutant phosphate-binding protein, MDCC has proven useful for direct, real-time measurement of inorganic phosphate release during enzymatic reactions.[10,11] Similarly, environment-sensitive fluorescence of MDCC site-specifically attached to *Escherichia coli* ParM nucleotide-binding protein provides the basis for a fluorogenic ADP biosensor.[12]

Pacific Orange™ Maleimide

The thiol-reactive Pacific Orange™ C_5-maleimide (P30507) yields conjugates with excitation/emission maxima of ~400/551 nm, making it ideal for use with violet diode laser–equipped flow cytometers and fluorescence microscopes. Moreover, Pacific Blue™ conjugates (described above with the other coumarins) and Pacific Orange™ conjugates are both excited with the 405 nm spectral line of the violet diode laser and emit at 455 and 551 nm, respectively, facilitating two-color analysis using 405 nm excitation and multiparameter analysis using the other flow cytometer channels.[3]

Pyrene Derivatives

Pyrene Maleimide

Not only is *N*-(1-pyrene)maleimide (pyrene maleimide, P28; Figure 2.3.7) essentially nonfluorescent until it has reacted with thiols, but once excited, pyrene–thiol conjugates can interact to form excited-state dimers (excimers) that emit at longer wavelengths than the excited monomeric fluorophore. Pyrene maleimide conjugates often have very long fluorescence lifetimes (>100 nanoseconds), giving proximal pyrene rings within 6–10 Å of each other ample time to form the spectrally altered excimer (Figure 2.3.8). The excimer-forming capacity of pyrene maleimide can be exploited for detection of conformational changes[13] and subunit assembly[14] of proteins and for analysis of protease activity.[15]

Pyrene Iodoacetamides

Fluorescence of the actin monomer labeled with pyrene iodoacetamide (P29, Figure 2.3.9) has been demonstrated to change upon polymerization, making this probe a widely utilized tool for following the kinetics

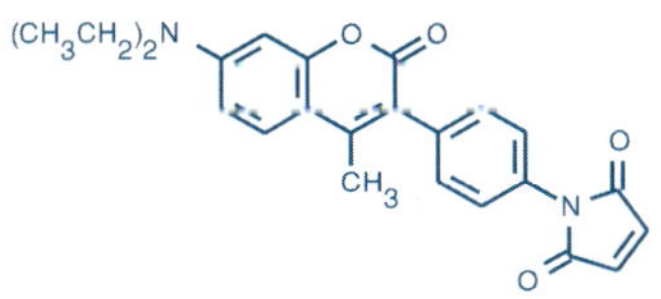

Figure 2.3.5 7-diethylamino-3-(4′-maleimidylphenyl)-4-methylcoumarin (CPM, D346).

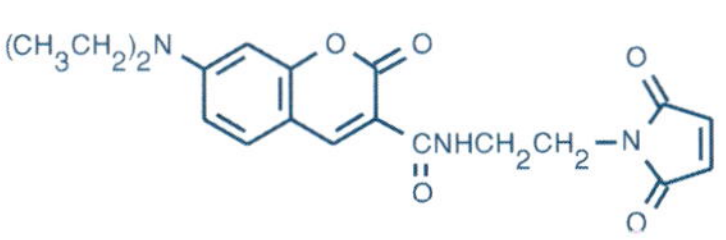

Figure 2.3.6 7-diethylamino-3-((((2-maleimidyl)ethyl)amino)carbonyl)coumarin (MDCC, D10253).

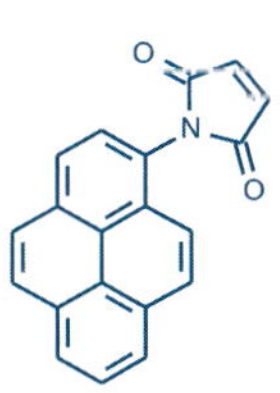

Figure 2.3.7 *N*-(1-pyrene)maleimide (P28).

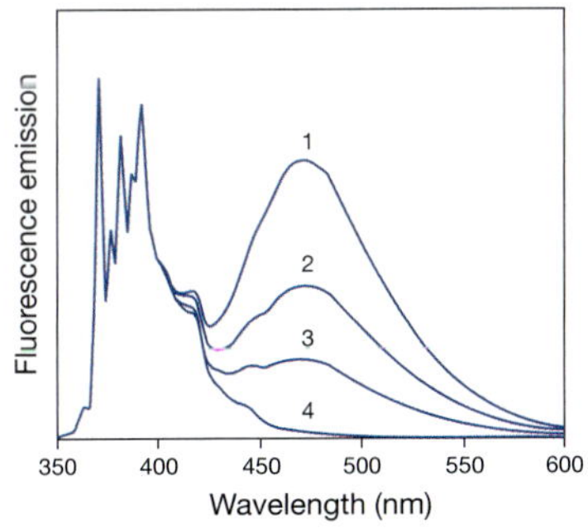

Figure 2.3.8 Excimer formation by pyrene in ethanol. Spectra are normalized to the 371.5 nm peak of the monomer. All spectra are essentially identical below 400 nm after normalization. Spectra are as follows: 1) 2 mM pyrene, purged with argon to remove oxygen; 2) 2 mM pyrene, air-equilibrated; 3) 0.5 mM pyrene (argon-purged); and 4) 2 µM pyrene (argon-purged). The monomer-to-excimer ratio (371.5 nm/470 nm) is dependent on both pyrene concentration and the excited-state lifetime, which is variable because of quenching by oxygen.

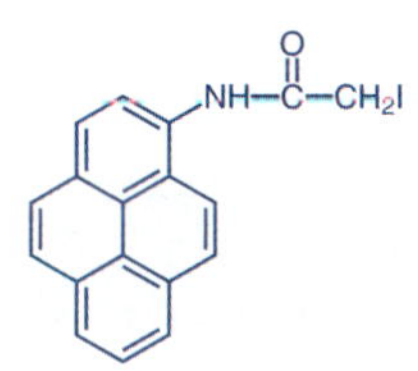

Figure 2.3.9 *N*-(1-pyrene)iodoacetamide (P29).

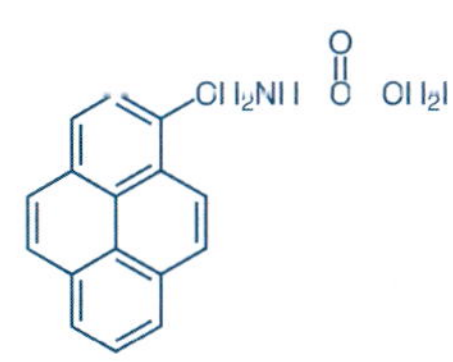

Figure 2.3.10 *N*-(1-pyrenemethyl)iodoacetamide (PMIA amide, P2007MP).

IMPORTANT NOTICE: The products described in this manual are covered by one or more Limited Use Label License(s). Please refer to the Appendix on page 971 and Master Product List on page 975. Products are For Research Use Only. Not intended for any animal or human therapeutic or diagnostic use.

of actin polymerization.[16–19] Conjugates of *N*-(1-pyrenemethyl)iodoacetamide (P2007MP, Figure 2.3.10) have the longest excited-state fluorescence lifetimes (>100 nanoseconds) of all reported thiol-reactive probes. Excimer formation can also be a useful indicator of protein folding.[20]

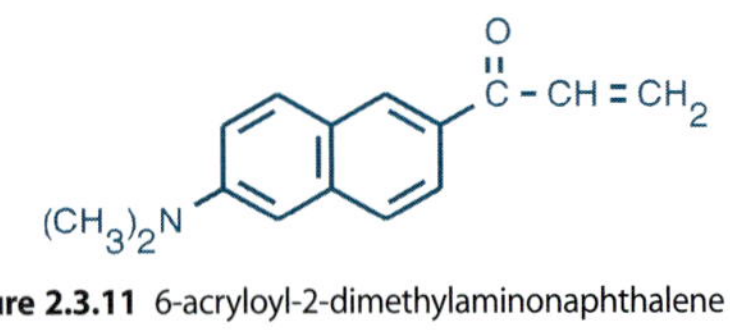

Figure 2.3.11 6-acryloyl-2-dimethylaminonaphthalene (acrylodan, A433).

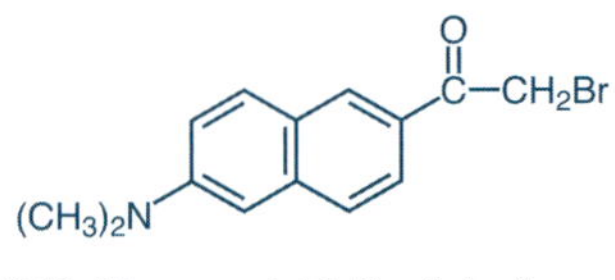

Figure 2.3.12 6-bromoacetyl-2-dimethylaminonaphthalene (badan, B6057).

Naphthalene Derivatives

Acrylodan and Badan

As compared with iodoacetamides or maleimides, acrylodan (A433, Figure 2.3.11) and 6-bromoacetyl-2-dimethylaminonaphthalene (badan, B6057, Figure 2.3.12) generally react with thiols more slowly but they form very strong thioether bonds that are expected to remain stable under conditions required for complete amino acid analysis. The fluorescence emission peak and intensity of these adducts (Figure 2.3.13) are particularly sensitive to conformational changes or ligand binding, making these dyes some of the most useful thiol-reactive probes for protein structure studies.[21–23] For example, the acrylodan conjugate of an intestinal fatty acid–binding protein, ADIFAB (A3880, Section 17.4), is a sensor for free fatty acids[24] (Figure 2.3.14).

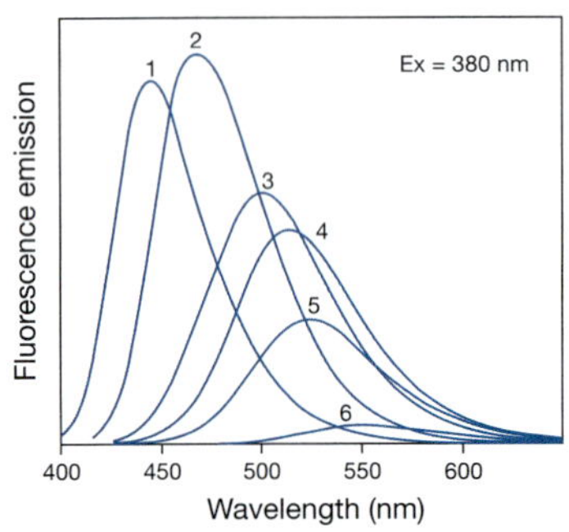

Figure 2.3.13 Fluorescence emission spectra of the 2-mercaptoethanol adduct of badan (B6057) in: 1) toluene, 2) chloroform, 3) acetonitrile, 4) ethanol, 5) methanol and 6) water. Each solution contains the same concentration of the adduct. Excitation of all samples is at 380 nm.

IAANS and MIANS

To develop appreciable fluorescence, both the reactive anilinonaphthalenesulfonate iodoacetamide (IAANS, I7; Figure 2.3.15) and maleimide (MIANS, also called Mal-ANS; M8, Figure 2.3.16) must be reacted with thiols that are located in hydrophobic sites. Often, however, buried unsolvated thiol residues are exceptionally reactive, allowing these sites to be selectively modified by these reagents. The environment-sensitive fluorescence properties of the protein conjugates of MIANS and IAANS are similar to those of the structurally related probes 1,8-ANS and 2,6-TNS (A47, T53; Section 13.5). The fluorescence intensity, and to a lesser extent, the emission wavelengths of the conjugates, tend to be very sensitive to substrate binding and folding and unfolding of the protein, as well as the association of the labeled protein with other proteins, membranes or nucleic acids.[25,26]

IAEDANS

The fluorescence of IAEDANS (I14, Figure 2.3.17) is quite dependent upon environment,[27] although less so than that of IAANS and MIANS conjugates. Its conjugates frequently respond to ligand binding by undergoing spectral shifts and changes in fluorescence intensity that are determined by the degree of aqueous solvation. Advantages of this reagent include high water solubility above pH 4 and a relatively long fluorescence lifetime (sometimes >20 nanoseconds, although commonly 10–15 nanoseconds), making the conjugates useful for fluorescence polarization assays[28,29] (Fluorescence Polarization (FP)—Note 1.4). The emission spectrum of IAEDANS overlaps well with the absorption of fluorescein, Alexa Fluor® 488 and Oregon Green® 488 dyes, as well as that of Green Fluorescent Protein (GFP). IAEDANS is an excellent reagent for fluorescence resonance energy transfer (FRET) measurements[30–32] (Fluorescence Resonance Energy Transfer (FRET)—Note 1.2). IAEDANS usually reacts with thiols; however, it has been reported to react with a lysine residue in tropomyosin.[33]

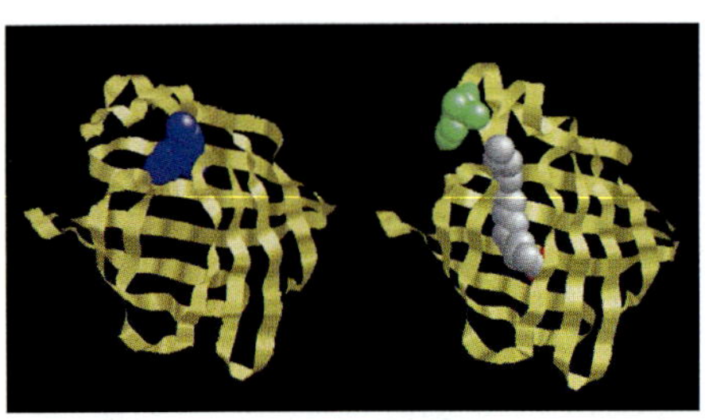

Figure 2.3.14 Ribbon representation of the ADIFAB free fatty acid indicator (A3880). In the left-hand image, the fatty acid binding site of intestinal fatty acid–binding protein (yellow) is occupied by a covalently attached acrylodan fluorophore (blue). In the right-hand image, a fatty acid molecule (gray) binds to the protein, displacing the fluorophore (green) and producing a shift of its fluorescence emission spectrum. Image contributed by Alan Kleinfeld, FFA Sciences LLC, San Diego.

Bimanes for Thiol Derivatization

Monobromobimane and Monochlorobimane

Monobromobimane (M1378, M20381; Figure 2.3.18), which is essentially nonfluorescent until conjugated, readily reacts with low molecular weight thiols,[34] including glutathione.[35] This reagent, originally described by Kosower and colleagues,[36,37] is also useful for detecting the distribution

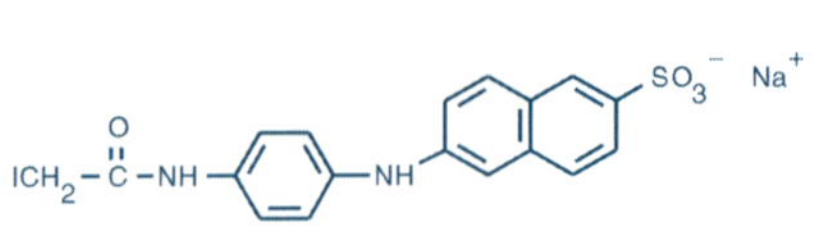

Figure 2.3.15 2-(4′-(iodoacetamido)anilino)naphthalene-6-sulfonic acid, sodium salt (IAANS, I7).

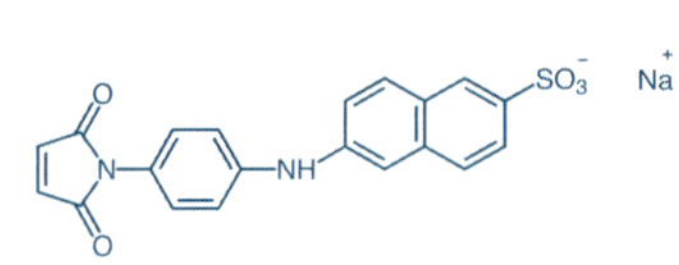

Figure 2.3.16 2-(4′-maleimidylanilino)naphthalene-6-sulfonic acid, sodium salt (MIANS, M8).

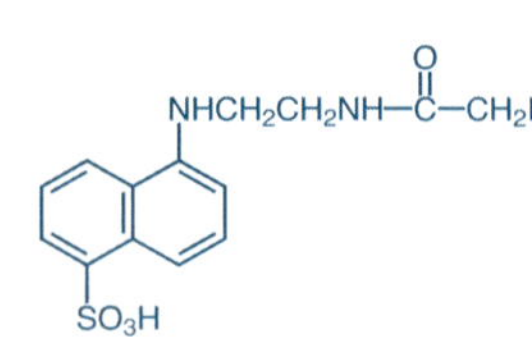

Figure 2.3.17 5-((((2-iodoacetyl)amino)ethyl)amino)naphthalene-1-sulfonic acid (1,5-IAEDANS), I14).

IMPORTANT NOTICE: The products described in this manual are covered by one or more Limited Use Label License(s). Please refer to the Appendix on page 971 and Master Product List on page 975. Products are For Research Use Only. Not intended for any animal or human therapeutic or diagnostic use.

of protein thiols in cells before and after chemical reduction of disulfides.[38] Monobromobimane is reportedly susceptible to inactivation by the disulfide reducing agent TCEP[39] (T2556, Section 2.1). Both monobromobimane and the more thiol-selective monochlorobimane (M1381MP) have been extensively used for detecting glutathione in live cells[40,41] (Section 15.6).

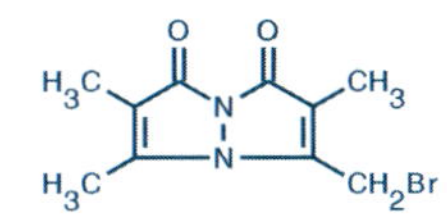

Figure 2.3.18 Monobromobimane (mBBr, M1378).

Dibromobimane

Dibromobimane (D1379, Figure 2.3.19) is an interesting homobifunctional crosslinking reagent for proteins[42–44] because it is unlikely to fluoresce until *both* of its alkylating groups have reacted. It has been used to crosslink thiols in myosin,[45] *Escherichia coli* lactose permease[46] and P-glycoprotein.[47] Despite its short length, dibromobimane is also an effective intermolecular crosslinker in some cases.[48]

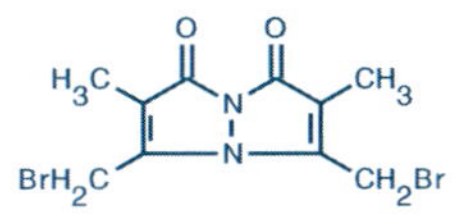

Figure 2.3.19 Dibromobimane (bBBr, D1379).

Bimane Iodoacetamide and Maleimide

Bimane iodoacetamide (B30500, Figure 2.3.20) and bimane C_3-maleimide (B30501, Figure 2.3.21) are blue-fluorescent thiol-reactive fluorophores with excitation/emission maxima of ~375/456 nm. The small size of the bimane fluorophore reduces the likelihood that the label will interfere with the function of the biomolecule, an important advantage for site-selective probes.

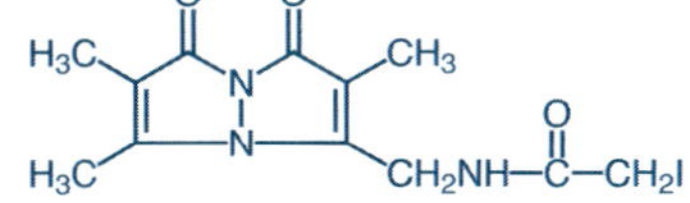

Figure 2.3.20 Bimane iodoacetamide (B30500).

Polar Reagents for Determining Thiol Accessibility

Like IAEDANS (I14), the iodoacetamide and maleimide derivatives of stilbene (A484, A485) have high water solubility and are readily conjugated to thiols. Their combination of high polarity and membrane impermeability makes these polysulfonated dyes useful for determining whether thiol-containing proteins and polypeptide chains are exposed at the extracellular or cytoplasmic membrane surface.

The sulfonated stilbene iodoacetamide (A484, Figure 2.3.22) was used to label single-cysteine mutants of staphylococcal α-hemolysin in order to determine structural changes that occur during oligomerization and pore formation[49] and of the lipid-binding region of *E. coli* pyruvate oxidase in order to detect conformational changes upon substrate binding.[50] Similarly, single-cysteine mutants of *Escherichia coli* Na^+-glutamate transporter GHS have been probed with the sulfonated stilbene maleimide (A485, Figure 2.3.23) to systematically study the topology of this membrane protein.[51]

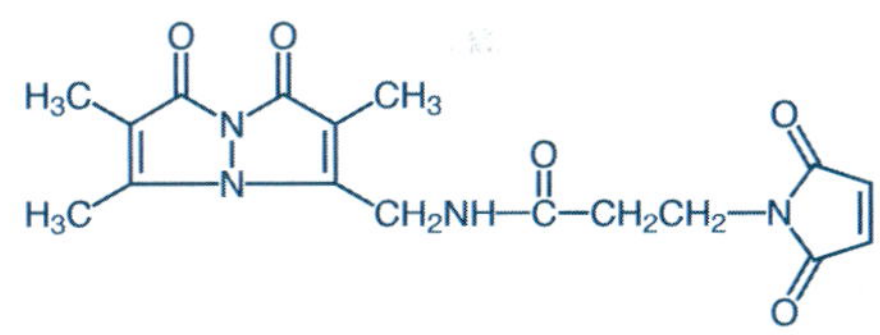

Figure 2.3.21 Bimane C_3-maleimide (B30501).

1,10-Phenanthroline Iodoacetamide for Preparing Metal-Binding Conjugates

Conjugation of *N*-(1,10-phenanthrolin-5-yl)iodoacetamide (P6879, Figure 2.3.24) to thiol-containing ligands confers the metal-binding properties of this important complexing agent on the ligand. For example, the covalent copper–phenanthroline complex of oligonucleotides or nucleic acid–binding molecules in combination with hydrogen peroxide acts as a chemical nuclease to selectively cleave DNA or RNA.[52,53]

Figure 2.3.22 4-acetamido-4′-((iodoacetyl)amino)stilbene-2,2′-disulfonic acid, disodium salt (A484).

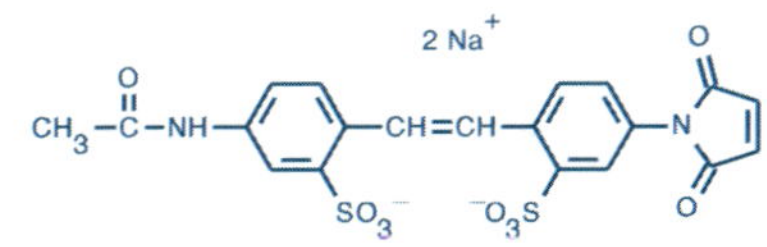

Figure 2.3.23 4-acetamido-4′-maleimidylstilbene-2,2′-disulfonic acid, disodium salt (A485).

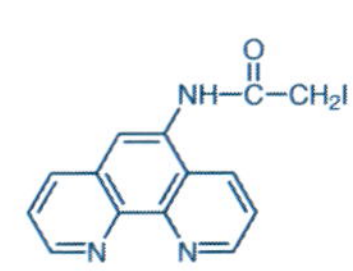

Figure 2.3.24 *N*-(1,10-phenanthrolin-5-yl)iodoacetamide (P6879).

REFERENCES

1. Neuron (2010) 66:170; **2.** Protein Sci (2009) 18:1033; **3.** Cytometry A (2006) 69:1153; **4.** J Biochem (2007) 141:889; **5.** Biophys J (2004) 87:344; **6.** Free Radic Biol Med (2008) 45:971; **7.** Assay Drug Dev Technol (2008) 6:361; **8.** Anal Biochem (1998) 265:8; **9.** Structure (2008) 16:351; **10.** Methods (2005) 37:183; **11.** Biochemistry (1994) 33:8262; **12.** J Biol Chem (2009) 284:33130; **13.** J Biol Chem (2004) 279:945; **14.** Biochemistry (2002) 41:3468; **15.** Anal Biochem (2002) 306:247; **16.** J Biol Chem (2008) 283:7135; **17.** Biochemistry (2007) 46:2707; **18.** Biophys J (2007) 92:2162; **19.** J Biol Chem (2006) 281:10635; **20.** FEBS Lett (1997) 420:63; **21.** J Am Chem Soc (2009) 131:13286; **22.** J Biol Chem (2008) 283:4124; **23.** Biochemistry (2007) 46:106; **24.** J Biol Chem (1992) 267:23495; **25.** J Biol Chem (2008) 283:9060; **26.** Anal Chem (2003) 75:3119; **27.** Biochemistry (1973) 12:4154; **28.** EMBO J (2007) 26:2786; **29.** Biochemistry (1973) 12:2250; **30.** Nat Struct Mol Biol (2008) 15:303; **31.** J Biol Chem (2007) 282:8265; **32.** J Biol Chem (2005) 280:2613; **33.** Eur J Biochem (1990) 187:155; **34.** J Chromatogr B Analyt Technol Biomed Life Sci (2004) 801:359; **35.** J Am Chem Soc (1986) 108:4527; **36.** Methods Enzymol (1987) 143:76; **37.** Proc Natl Acad Sci U S A (1979) 76:3382; **38.** Anal Chem (2006) 78:7959; **39.** Anal Biochem (2003) 318:325; **40.** Nat Protoc (2009) 4:1790; **41.** J Biol Chem (2006) 281:17420; **42.** Anal Biochem (2004) 331:27; **43.** Protein Sci (2001) 10:1293; **44.** Anal Biochem (1995) 225:174; **45.** Proc Natl Acad Sci U S A (2000) 97:1461; **46.** Proc Natl Acad Sci U S A (1996) 93:10123; **47.** J Biol Chem (1999) 274:35388; **48.** J Biol Chem (2006) 281:14163; **49.** FEBS Lett (1994) 356:66; **50.** Biochemistry (1997) 36:11564; **51.** Biochemistry (2007) 46:2326; **52.** Biochemistry (2000) 39:4068; **53.** Bioconjug Chem (1996) 7:413.

IMPORTANT NOTICE: The products described in this manual are covered by one or more Limited Use Label License(s). Please refer to the Appendix on page 971 and Master Product List on page 975. Products are For Research Use Only. Not intended for any animal or human therapeutic or diagnostic use.

DATA TABLE 2.3 THIOL-REACTIVE PROBES EXCITED WITH ULTRAVIOLET LIGHT

Cat. No.	MW	Storage	Soluble	Abs	EC	Em	Solvent	Notes
A433	225.29	L	DMF, MeCN	391	20,000	500	MeOH	1
A484	624.33	F,D,L	H_2O	329	39,000	408	pH 8	2, 3
A485	536.44	F,D	H_2O	322	35,000	411	pH 8	2
A30505	578.68	F,DD,L	H_2O, DMSO	345	17,000	444	pH 7	4
B6057	292.17	F,L	DMF, MeCN	387	21,000	520	MeOH	5
B30500	375.17	F,D,L	DMSO	375	5800	456	MeOH	3
B30501	358.35	F,D,L	DMSO	375	5700	458	MeOH	
D346	402.45	F,D,L	DMSO	384	33,000	469	MeOH	6
D404	490.34	F,D,L	DMSO	384	31,000	470	MeOH	2, 3
D1379	350.01	L	DMF, MeCN	391	6100	see Notes	MeOH	7
D10251	298.30	F,D,L	DMSO	383	27,000	463	MeOH	8
D10252	386.19	F,D,L	DMSO	376	24,000	465	MeOH	3
D10253	383.40	F,D,L	DMSO	419	50,000	466	MeOH	9
I7	504.27	F,D,L	DMF	326	27,000	462	MeOH	2, 3
I14	434.25	F,D,L	pH >6, DMF	336	5700	490	pH 8	3, 10
M8	416.38	F,D,L	DMSO, DMF	322	27,000	417	MeOH	11
M1378	271.11	F,L	DMF, MeCN	398	5000	see Notes	pH 7	7
M1381MP	226.66	F,L	DMSO	380	6000	see Notes	MeOH	7
M20381	271.11	F,L	DMF, MeCN	398	5000	see Notes	pH 7	7, 12
P28	297.31	F,D,L	DMF, DMSO	338	40,000	375	MeOH	13, 14
P29	385.20	F,D,L	DMF, DMSO	339	26,000	384	MeOH	2, 3
P2007MP	399.23	F,D,L	DMSO	341	41,000	377	MeOH	2, 3, 14
P6879	363.16	F,D,L	DMSO	270	28,000	none	$CHCl_3$	3
P30506	406.34	F,DD,L	DMSO	402	40,000	451	pH 9	15
P30507	~800	F,DD,L	DMSO	403	23,000	552	MeOH	

For definitions of the contents of this data table, see "Using *The Molecular Probes® Handbook*" in the introductory pages.

Notes

1. Fluorescence of unconjugated A433 is weak, increasing markedly upon reaction with thiols. Em (QY) for the 2-mercaptoethanol adduct are: 540 nm (0.18) in H_2O, 513 nm (0.57) in MeOH, 502 nm (0.79) in EtOH, 468 nm (0.78) in MeCN, 435 nm (0.83) in dioxane. (J Biol Chem (1983) 258:7541)
2. Spectral data of the 2-mercaptoethanol adduct.
3. Iodoacetamides in solution undergo rapid photodecomposition to unreactive products. Minimize exposure to light prior to reaction.
4. Aqueous stock solutions should be used within 24 hours; long-term storage is NOT recommended.
5. Em for 2-mercaptoethanol adduct of B6057: 550 nm in H_2O (pH 7), 523 nm in MeOH, 514 nm in EtOH, 502 nm in MeCN, 469 nm in $CHCl_3$, 457 nm in dioxane, 445 nm in toluene. Abs is relatively independent of solvent.
6. Spectral data are for the 2-mercaptoethanol adduct. The unreacted reagent is nonfluorescent, Abs = 384 nm (EC = 32,000 $cm^{-1}M^{-1}$) in MeOH.
7. Bimanes are almost nonfluorescent until reacted with thiols. For monobromobimane conjugated to glutathione, Abs = 394 nm, Em = 490 nm (QY ~0.1–0.3) in pH 8 buffer. (Methods Enzymol (1987) 143:76, Methods Enzymol (1995) 251:133)
8. Spectral data are for the 2-mercaptoethanol adduct. The unreacted reagent is nonfluorescent, Abs = 381 nm (EC = 27,000 $cm^{-1}M^{-1}$) in MeOH.
9. QY increases on reaction with thiols; Abs, EC and Em are unchanged. (J Chem Soc Perkin Trans I (1994) 2975)
10. The 2-mercaptoethanol adduct of I14 has essentially similar spectral characteristics in aqueous solution. (Biochemistry (1973) 12:4154) Fluorescence lifetime (τ) = 21 nsec when conjugated to myosin subfragment-1. (Biochemistry (1973) 12:2250)
11. Spectral data are for the 2-mercaptoethanol adduct. The unreacted reagent is nonfluorescent, Abs = 443 nm (EC = 13,000 $cm^{-1}M^{-1}$) in MeOH.
12. This product is specified to equal or exceed 98% analytical purity by HPLC.
13. Fluorescence of unreacted P28 is weak. Em data represent the 2-mercaptoethanol adduct.
14. Pyrene derivatives exhibit structured spectra. The absorption maximum is usually about 340 nm with a subsidiary peak at about 325 nm. There are also strong absorption peaks below 300 nm. The emission maximum is usually about 376 nm with a subsidiary peak at 396 nm. Excimer emission at about 470 nm may be observed at high concentrations.
15. The fluorescence quantum yield of Pacific Blue™ dye in 50 mM potassium phosphate, 150 mM NaCl, pH 7.2, at 22°C is 0.78.

IMPORTANT NOTICE: The products described in this manual are covered by one or more Limited Use Label License(s). Please refer to the Appendix on page 971 and Master Product List on page 975. Products are For Research Use Only. Not intended for any animal or human therapeutic or diagnostic use.

PRODUCT LIST 2.3 THIOL-REACTIVE PROBES EXCITED WITH ULTRAVIOLET LIGHT

Cat. No.	Product	Quantity
A484	4-acetamido-4′-((iodoacetyl)amino)stilbene-2,2′-disulfonic acid, disodium salt	25 mg
A485	4-acetamido-4′-maleimidylstilbene-2,2′-disulfonic acid, disodium salt	25 mg
A433	6-acryloyl-2-dimethylaminonaphthalene (acrylodan)	25 mg
A30505	Alexa Fluor® 350 C_5-maleimide	1 mg
B30501	bimane C_3-maleimide	5 mg
B30500	bimane iodoacetamide	5 mg
B6057	6-bromoacetyl-2-dimethylaminonaphthalene (badan)	10 mg
D1379	dibromobimane (bBBr)	25 mg
D404	7-diethylamino-3-((4′-(iodoacetyl)amino)phenyl)-4-methylcoumarin (DCIA)	25 mg
D10253	7-diethylamino-3-((((2-maleimidyl)ethyl)amino)carbonyl)coumarin (MDCC)	5 mg
D346	7-diethylamino-3-(4′-maleimidylphenyl)-4-methylcoumarin (CPM)	25 mg
D10252	*N*-(7-dimethylamino-4-methylcoumarin-3-yl)iodoacetamide (DACIA)	10 mg
D10251	*N*-(7-dimethylamino-4-methylcoumarin-3-yl)maleimide (DACM)	10 mg
I7	2-(4′-(iodoacetamido)anilino)naphthalene-6-sulfonic acid, sodium salt (IAANS)	100 mg
I14	5-((((2-iodoacetyl)amino)ethyl)amino)naphthalene-1-sulfonic acid (1,5-IAEDANS)	100 mg
M8	2-(4′-maleimidylanilino)naphthalene-6-sulfonic acid, sodium salt (MIANS)	100 mg
M1378	monobromobimane (mBBr)	25 mg
M20381	monobromobimane (mBBr) *FluoroPure™ grade*	25 mg
M1381MP	monochlorobimane (mBCl)	25 mg
P30506	Pacific Blue™ C_5-maleimide	1 mg
P30507	Pacific Orange™ C_5-maleimide	1 mg
P6879	*N*-(1,10-phenanthrolin-5-yl)iodoacetamide	5 mg
P29	*N*-(1-pyrene)iodoacetamide	100 mg
P28	*N*-(1-pyrene)maleimide	100 mg
P2007MP	*N*-(1-pyrenemethyl)iodoacetamide (PMIA amide)	25 mg

IMPORTANT NOTICE: The products described in this manual are covered by one or more Limited Use Label License(s). Please refer to the Appendix on page 971 and Master Product List on page 975. Products are For Research Use Only. Not intended for any animal or human therapeutic or diagnostic use.

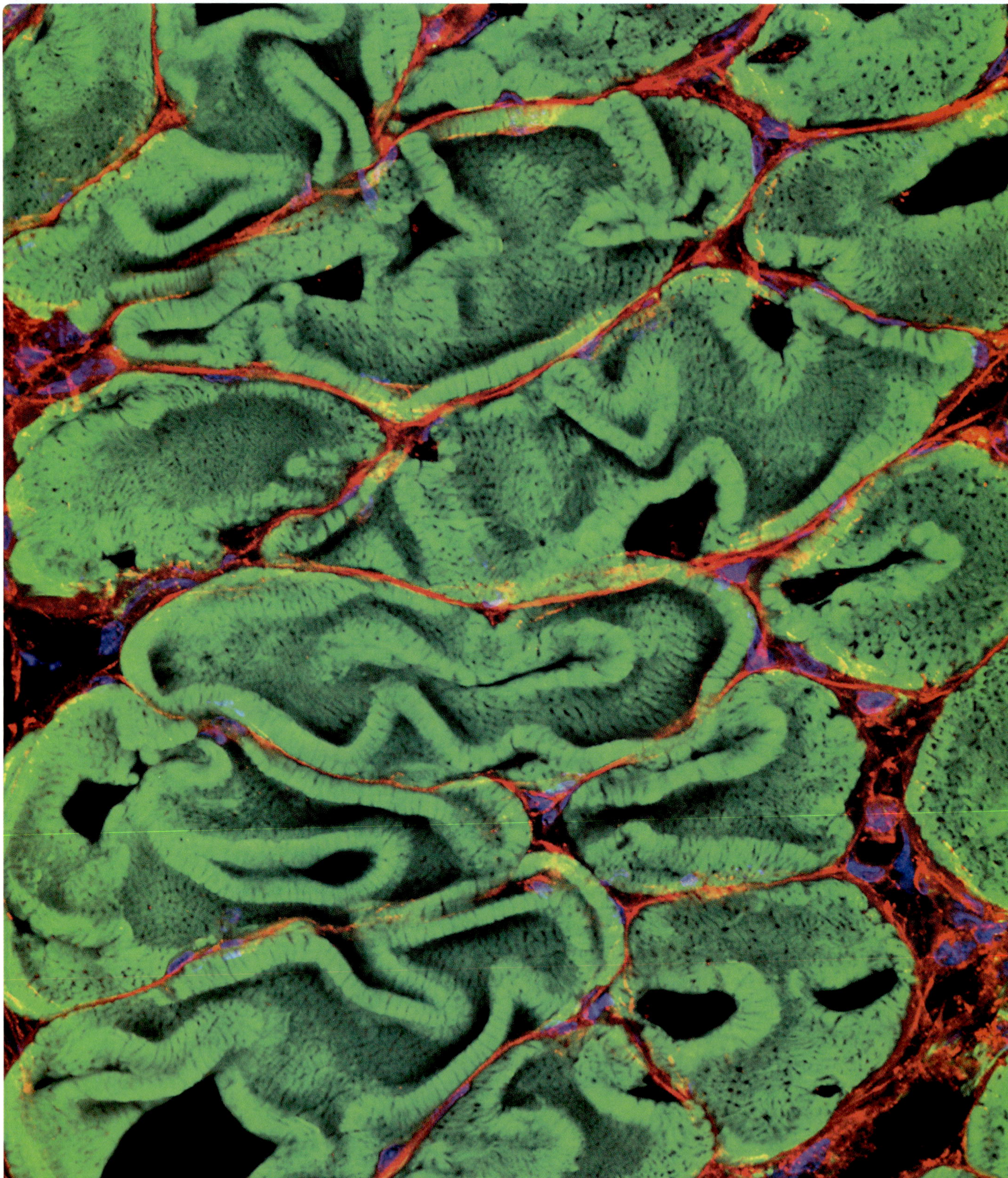

Texas Red® wheat germ agglutinin, Alexa Fluor® 488 phalloidin and Hoechst 33342.

IMPORTANT NOTICE: The products described in this manual are covered by one or more Limited Use Label License(s). Please refer to the Appendix on page 971 and Master Product List on page 975. Products are For Research Use Only. Not intended for any animal or human therapeutic or diagnostic use.

THREE

CHAPTER 3

Click Chemistry and Other Functional Group Modifications

IMPORTANT NOTICE: The products described in this manual are covered by one or more Limited Use Label License(s). Please refer to the Appendix on page 971 and Master Product List on page 975. Products are For Research Use Only. Not intended for any animal or human therapeutic or diagnostic use.

IMPORTANT NOTICE: The products described in this manual are covered by one or more Limited Use Label License(s). Please refer to the Appendix on page 971 and Master Product List on page 975. Products are For Research Use Only. Not intended for any animal or human therapeutic or diagnostic use.
www.invitrogen.com/probes

3.1 Click Chemistry

Principles of Bioorthogonal Labeling and Click Chemistry

The amine- and thiol-reactive labeling chemistries described in Chapter 1 and Chapter 2 are generally used in one of two ways: (1) labeling of purified proteins or other biopolymers yielding conjugates that are subsequently applied to cell or tissue specimens, or (2) nonselective *in situ* labeling of total cellular thiol or amine content. *In situ* labeling of specific molecular populations—such as proteins and nucleic acids that have been newly synthesized in some experimental time window of interest—is not feasible due to the ubiquitous distribution of amines and thiols in cells, as well as in the media in which they are maintained. Click-iT® labeling technology overcomes this obstacle by employing bioorthogonal reactive chemistry, in which the reaction partners have no endogenous representation in biological molecules, cells, tissues or model organisms.[1–3] In addition to reaction selectivity, *in situ* labeling methods should allow reactivity under mild conditions and in predominantly aqueous solvent conditions.

Although several known chemistries fulfill the requirements described above,[1,3] Click-iT® labeling technology is founded upon one of the most successful and versatile bioorthogonal labeling reactions currently available—the copper-catalyzed azide–alkyne cycloaddition[4,5] (Figure 3.1.1). Application of this reaction to *in situ* labeling of cells

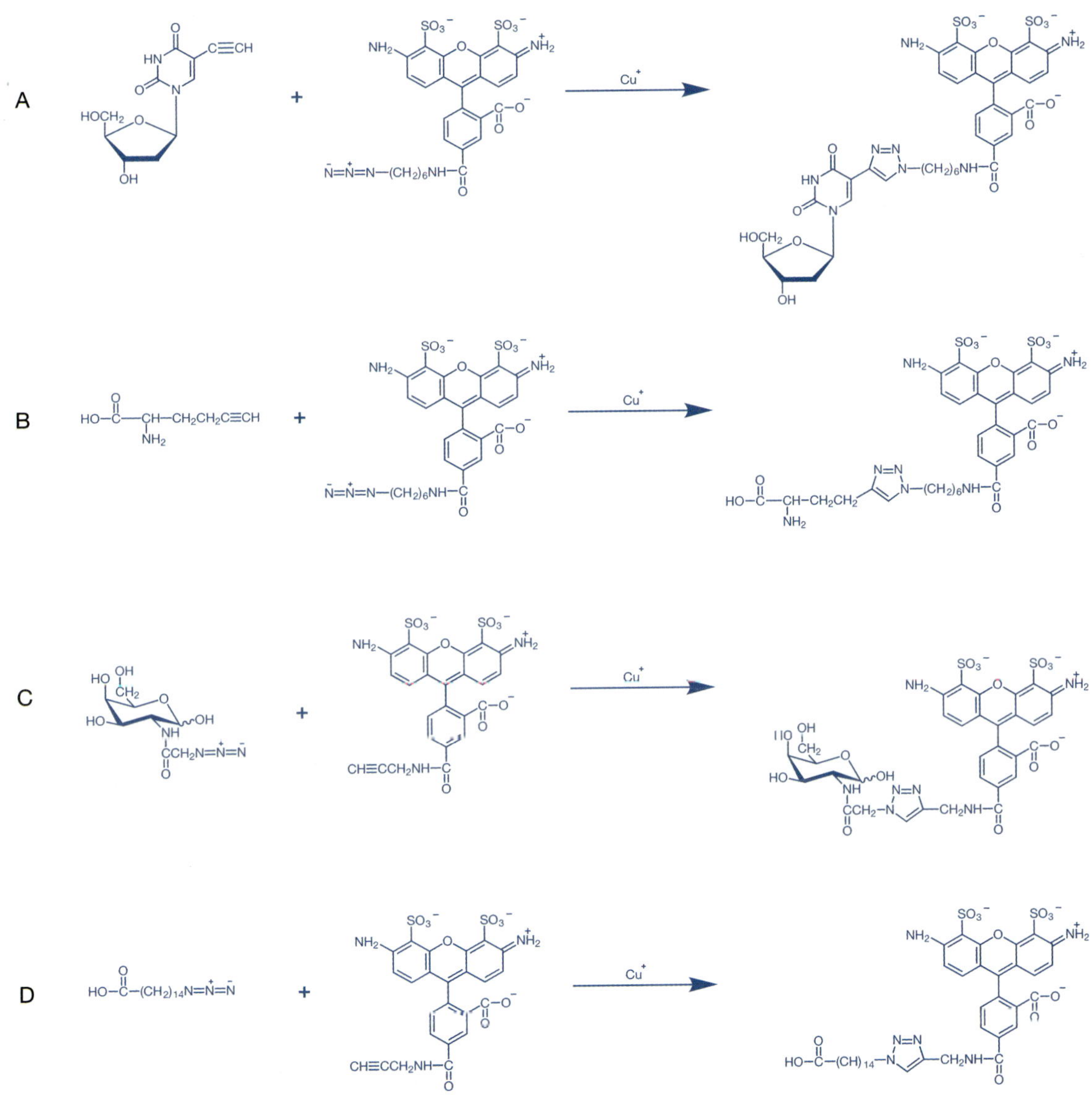

Figure 3.1.1 Click-iT® copper-catalyzed azide–alkyne cycloaddition chemistry applied to detection of **A**) nucleic acids, **B**) proteins, **C**) carbohydrates and **D**) lipids. The reaction partners are **A**) 5-ethynyl-2′-deoxyuridine (EdU) and Alexa Fluor® 488 azide, **B**) L-homopropargylglycine (HPG) and Alexa Fluor® 488 azide, **C**) *N*-azidoacetylgalactosamine and Alexa Fluor® 488 alkyne and **D**) 15-azidopentadecanoic acid and Alexa Fluor® 488 alkyne (D). In each case, the left-hand partner is a metabolic precursor that can be incorporated into proteins and nucleic acids via *de novo* synthesis or post-translational modification pathways.

is a two-step process. First, one reaction partner—either an azide or alkyne linked to a "building block" such as a nucleotide, nucleoside, amino acid, monosaccharide or fatty acid—is biosynthetically incorporated. Subsequently, the other reaction partner—the complementary alkyne or azide linked to a fluorescent dye, biotin or other detection reagent—is "clicked" into place in the presence of catalytic copper (I). One reaction partner must be an azide derivative and the other an alkyne derivative, but either functional moiety can serve as the biosynthetically incorporated molecule or the detection molecule[6] (e.g., L-azidohomoalanine (AHA) + Alexa Fluor® 488 alkyne is the inverse of the reaction scheme shown in Figure 3.1.1B).

Another important aspect of the azide and alkyne reaction partners is their small size (Figure 3.1.2). Expression tags such as Green Fluorescent Protein (GFP) provide the ultimate in labeling specificity because their linkage to proteins of interest is genetically prescribed. Once the GFP transgene has been inserted into a cell (BacMam Gene Delivery and Expression Technology—Note 11.1), *in situ* labeling is obtained without any outside intervention using the cellular transcription and translation machinery. However, the finite size of GFP (~27,000 daltons) sometimes causes functional perturbations and has spurred the development of alternative, smaller-sized expression tags such as the TC-FlAsH™ tetracysteine tag and biarsenical ligand system (T34561, T34562, T34563; Section 2.2). Furthermore, nucleic acids, lipids, glycans and post-translational protein modifications can only be detected indirectly by genetically encoded protein reporters. The small size of alkyne and azide tags allows the biosynthetic building blocks to which they are attached to be processed by enzymes, such as nucleotide polymerases and aminoacyl tRNA synthetases, that have poor tolerance for substrates with larger modifications such as fluorescent organic dyes.[7]

The 1,2,3-triazole linkage between the azide and alkyne reaction partners (Figure 3.1.1) is extremely stable. It is not susceptible to hydrolysis, oxidation or reduction, and it survives ionization in mass

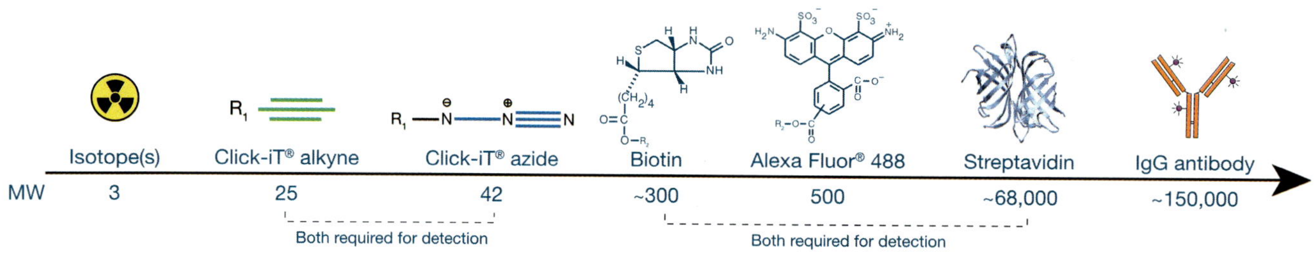

Figure 3.1.2 Relative size of detection molecules commonly used in cellular analysis. Because the azide and alkyne moieties can be used interchangeably to optimize labeling configurations, R_1 can be either the biomolecule of interest or the detection reagent. For biotin and Alexa Fluor® 488, R_2 represents the biomolecule of interest.

Table 3.1 Molecular Probes® azide and alkyne derivatives.

Compound or Fluorophore	Azide or Alkyne	Cat. No.	Application
Azide- or Alkyne-Modified Biomolecules for Metabolic or Enzymatic Labeling			
Click-iT® AHA (L-azidohomoalanine)	azide	C10102	Monitor nascent protein synthesis or inhibition
Click-iT® HPG (L-homopropargylglycine)	alkyne	C10186	
Click-iT® farnesyl alcohol, azide	azide	C10248	Identification of isoprenylated proteins
Click-iT® geranylgeranyl alcohol, azide	azide	C10249	
Click-iT® fucose alkyne (tetraacetylfucose alkyne)	alkyne	C10264	Identification of fucoslyated proteins
Click-iT® palmitic acid, azide	azide	C10265	Identification of protein fatty acylation
Click-iT® myristic acid, azide	azide	C10268	
Click-iT® GalNAz (tetraacetylated *N*-azidoacetylgalactosamine)	azide	C33365	Identification of *O*-linked glycoproteins, including *O*-GlcNAc
Click-iT® ManNAz (tetraacetylated *N*-azidoacetylmannosamine)	azide	C33366	Identification of sialic acid–modified glycoproteins
Click-iT® ManNAz (tetraacetylated *N*-azidoacetylglucosamine)	azide	C33367	Identification of *O*-GlcNAc–modified glycoproteins
Azide- or Alkyne-Modified Fluorophores and Haptens for Detection (Ex/Em) *			
Alexa Fluor® 488 (495/519)	azide alkyne	A10266 A10267	Fluorescent dye or hapten †
Alexa Fluor® 555 (555/565)	azide alkyne	A20012 A20013	Fluorescent dye
Alexa Fluor® 594 (590/617)	azide alkyne	A10270 A10275	Fluorescent dye
Alexa Fluor® 647 (650/668)	azide	A10277	Fluorescent dye
Biotin (NA)‡	azide alkyne	B10184 B10185	Hapten
Oregon Green® 488 (496/524)	azide alkyne	O10180 O10181	Fluorescent dye or hapten †
Tetramethylrhodamine (TAMRA) (555/580)	azide alkyne	T10182 T10183	Fluorescent dye or hapten †

* Absorption and fluorescence emission maxima, in nm. † As hapten, use with anti-dye antibody; see Section 7.4. ‡ Requires streptavidin for detection; see Section 7.6.

IMPORTANT NOTICE: The products described in this manual are covered by one or more Limited Use Label License(s). Please refer to the Appendix on page 971 and Master Product List on page 975. Products are For Research Use Only. Not intended for any animal or human therapeutic or diagnostic use.

molecular probes® | invitrogen™ by life technologies™

spectrometry (MS) analysis. The reaction is also regiospecific, yielding exclusively 1,4-disubstituted-1,2,3-triazole linkages (Figure 3.1.1). The copper (I) catalyst is both an essential feature of the reaction and its most problematic aspect in terms of applications.[1,3,8] Without the copper (I) catalyst, which accelerates the rate of the reaction by a factor of >10^6, the reaction is impractically slow. For convenience, copper (I) is usually prepared *in situ* by reduction of extraneously added copper (II) using ascorbate or TCEP (T2556, Section 2.1). Insufficient reductive capacity can result in attenuation of *in situ* reactions in highly oxidizing environments.[9] Copper is cytotoxic, due at least in part to its capacity for sensitizing oxidative damage to proteins and nucleic acids, and limiting applications of the azide–alkyne cyloaddition reaction in live cells. Copper/ascorbate treatment also causes extinction of R-phycoerythrin (R-PE) and GFP fluorescence.[10] An excellent analysis of these considerations, together with a list of practical recommendations for their management with respect to bioconjugation applications of copper-catalyzed azide–alkyne cycloaddition chemistry, has been published by Finn and co-workers.[8] Our Click-iT® product portfolio, consisting of individual azide and alkyne labeling reagents and application-specific kits, is described in detail below.

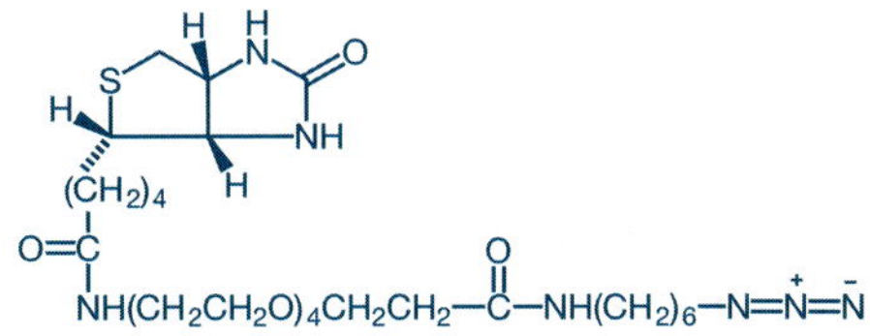

Figure 3.1.3 Biotin azide (PEG_4 carboxamide-6-azidohexanyl biotin, B10184).

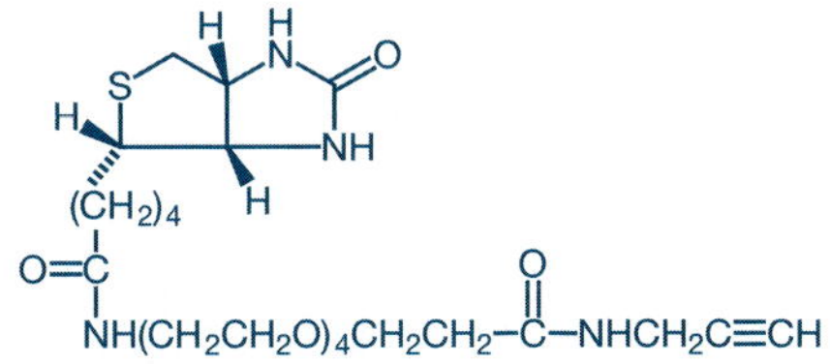

Figure 3.1.4 Biotin alkyne (PEG_4 carboxamide-propargyl biotin, B10185).

Click-iT® Azide and Alkyne Labeling Reagents

Although *in situ* labeling of biomolecules for cytochemical and proteomic analysis is perhaps the most notable application of Click-iT® technology, it is by no means the only one. This specific and direct labeling methodology can also be applied to bioconjugate preparation,[8] surface and particle functionalization and molecular ligations. Our Click-iT® azide and alkyne labeling reagents support these applications and also provide foundational tools for developing new *in situ* labeling applications (Table 3.1).

Azide- and Alkyne-Derivatized Dyes and Biotinylation Reagents

We offer a rich selection of azide- and alkyne-derivatized fluorescent dyes for coupling to complementary azide- and alkyne-functionalized biomolecules:

- Alexa Fluor® 488 azide (A10266, Figure 3.1.1) and alkyne (A10267, Figure 3.1.1)
- Alexa Fluor® 555 azide (A20012) and alkyne (A20013)
- Alexa Fluor® 594 azide (A10270) and alkyne (A10275)
- Alexa Fluor® 647 azide (A10277) and alkyne (A10278)
- Oregon Green® 488 azide (O10180) and alkyne (O10181)
- Tetramethylrhodamine (TAMRA) azide (T10182) and alkyne (T10183)
- Biotin azide (B10184, Figure 3.1.3) and alkyne (B10185, Figure 3.1.4)

Antibodies to Oregon Green® 488, tetramethylrhodamine and Alexa Fluor® 488 dyes (Section 7.4) and Tyramide Signal Amplification (TSA™) Kits (Section 6.2) are available to provide signal amplification if necessary.[11] The biotin azide and alkyne reagents facilitate western blotting applications and streptavidin enrichment in combination with our streptavidin or CaptAvidin™ agarose[12–14] (S951, C21386; Section 7.6).

Azide- and Alkyne-Modified Nucleosides and Amino Acids

We offer the following azide- and alkyne-modified nucleosides and amino acids for Click-iT® labeling protocols:

- 5-ethynyl uridine (EU, E10345)
- 5-ethynyl-2′-deoxyuridine (EdU; A10044, E10187, E10415; Figure 3.1.1A)
- Click-iT® AHA (L-azidohomoalanine for nascent protein synthesis, C10102)
- Click-iT® HPG (L-homopropargylglycine for nascent protein synthesis, C10186; Figure 3.1.1B)

The alkyne-modified nucleosides EdU and EU form the basis of our Click-iT® cell proliferation and nascent RNA assays described below. The individual packagings of these reagents provide the larger quantities required for *in vivo* labeling applications.[6,15] AHA and HPG are methionine surrogates providing nonradioactive alternatives to ^{35}S-methionine for pulse-chase detection of protein synthesis and degradation.[16–19]

IMPORTANT NOTICE: The products described in this manual are covered by one or more Limited Use Label License(s). Please refer to the Appendix on page 971 and Master Product List on page 975. Products are For Research Use Only. Not intended for any animal or human therapeutic or diagnostic use.

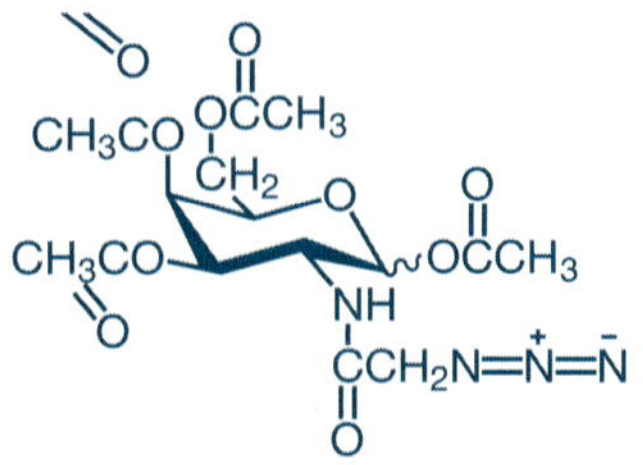

Figure 3.1.5 Click-iT® GalNAz metabolic glycoprotein labeling reagent (tetraacetylated N-azidoacetylgalactosamine, C33365).

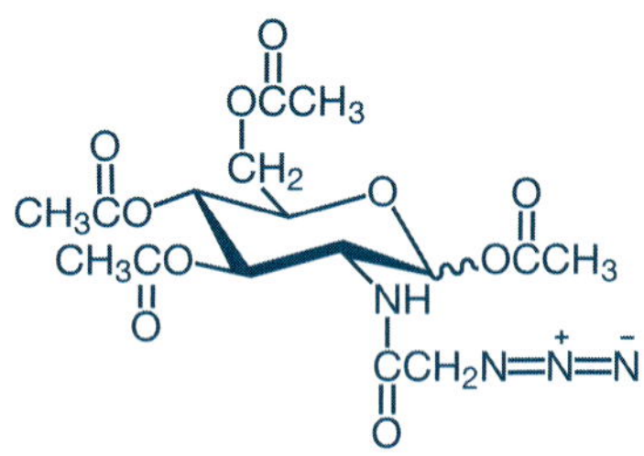

Figure 3.17 Click-iT® GlcNAz metabolic glycoprotein labeling reagent (tetraacetylated *N*-azidoacetylglucosamine for labeling *O*-linked *N*-acetylglucosamine (*O*-GlcNAc)–modified glycoproteins; C33367).

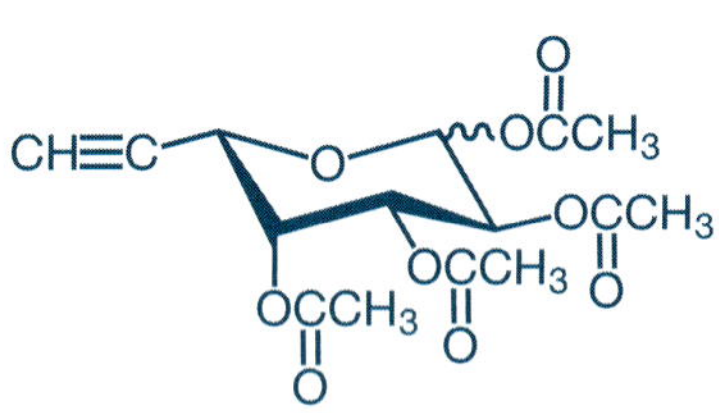

Figure 3.1.8 Click-iT® fucose alkyne (tetraacetylfucose alkyne, C10264).

Azide- and Alkyne-Modified Monosaccharides, Fatty Acids and Isoprenoids

The Click-iT® metabolic glycoprotein labeling reagents provide biosynthetic precursors for detecting and characterizing post-translational glycosylation of proteins:[14,20–22]

- Click-iT® GalNAz metabolic glycoprotein labeling reagent (tetraacetylated *N*-azidoacetylgalactosamine for labeling *O*-linked glycoproteins, C33365; Figure 3.1.1C, Figure 3.1.5)
- Click-iT® ManNAz metabolic glycoprotein labeling reagent (tetraacetylated *N*-azidoacetylmannosamine for labeling sialic acid–modified glycoproteins, C33366; Figure 3.1.6)
- Click-iT® GlcNAz metabolic glycoprotein labeling reagent (tetraacetylated *N*-azidoacetylglucosamine for labeling *O*-linked *N*-acetylglucosamine (*O*-GlcNAc)–modified glycoproteins, C33367; Figure 3.1.7)
- Click-iT® fucose alkyne (tetraacetylfucose alkyne, C10264; Figure 3.1.8)

Cultured cells are simply incubated with the modified sugars for 2–3 days or until cells reach the appropriate density. The acetyl groups improve cell permeability of the modified sugars and are removed by nonspecific intracellular esterases (Figure 3.1.8). The resulting azide- or alkyne-modified sugar is then metabolically incorporated through the permissive nature of the oligosaccharide biosynthesis pathway, yielding functionalized glycoproteins that can be chemoselectively coupled to complementary alkyne- or azide-functionalized fluorophores and biotinylation reagents for detection or affinity capture. We also offer the the Click-iT® *O*-GlcNAc Enzymatic Labeling System for *in vitro* enzyme-mediated *N*-azidoacetylgalactosamine labeling of *O*-GlcNAc–modified glycoproteins[26,27] (C33368, Section 9.4) and Click-iT® Protein Analysis Detection Kits (C33370, C33371, C33372; Section 9.4) for detection of azide-functionalized glycoproteins in 1D or 2D electrophoresis gels or western blots.

Similarly, our azide-functionalized isoprenoids and fatty acids enable detection of post-translational lipidation of proteins by in-gel fluorescence scanning, fluorescence microscopy and flow cytometry.[23–25] We offer the following azide-modified fatty acids and isoprenoids:

- Click-iT® farnesyl alcohol, azide (C10248, Figure 3.1.9)
- Click-iT® geranylgeranyl alcohol, azide (C10249, Figure 3.1.10)
- Click-iT® palmitic acid, azide (15-azidopentadecanoic acid, C10265; Figure 3.1.1D)
- Click-iT® myristic acid, azide (12-azidododecanoic acid, C10268)

Heterobifunctional Reagents

Our collection of heterobifunctional reagents provide a means for adapting the amine- and thiol-reactive labeling chemistries described in Chapters 1 and 2 with the azide–alkyne Click-iT® labeling protocols:

- Azide succinimidyl ester (A10280, Figure 3.1.11)
- Alkyne succinimidyl ester (A10279, Figure 3.1.12)
- Azide iodoacetamide (I10188, Figure 3.1.13)
- Alkyne iodoacetamide (I10189, Figure 3.1.14)

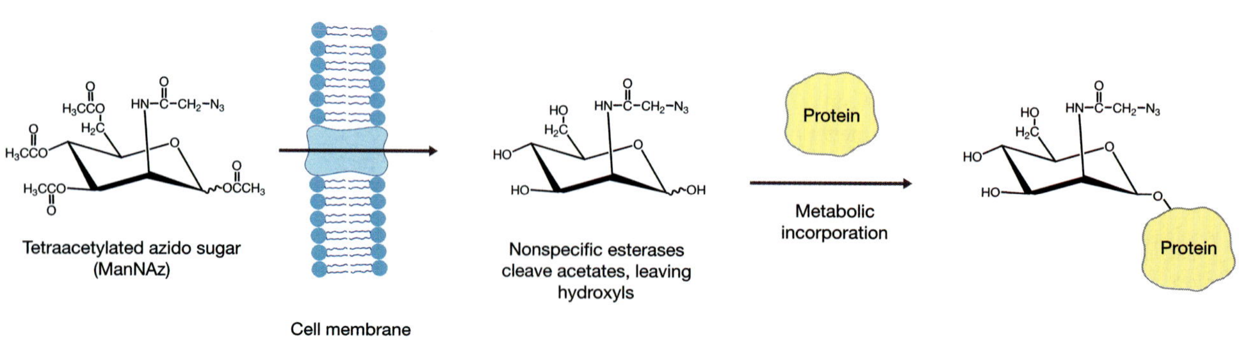

Figure 3.1.6 Metabolic incorporation of tetraacetylated azido sugars.

IMPORTANT NOTICE: The products described in this manual are covered by one or more Limited Use Label License(s). Please refer to the Appendix on page 971 and Master Product List on page 975. Products are For Research Use Only. Not intended for any animal or human therapeutic or diagnostic use.

The succinimidyl ester reagents can be used for azide or alkyne functionalization of amine-containing molecules and molecular assemblies including terminally or internally modified oligonucleotides and nanoparticles.[8,28]

Click-iT® Reaction Buffers

For added convenience, we offer Click-iT® Reaction Buffer Kits for protein or cell samples labeled with an azide- or alkyne-tagged biomolecule. The Click-iT® Cell Reaction Buffer Kit (C10269) includes sufficient reagents to perform 50 reactions based on a 0.5 mL reaction volume for subsequent analyses by flow cytometry, fluorescence microscopy or high-content screening (HCS). The Click-iT® Protein Reaction Buffer Kit (C10276) includes everything required for click coupling to functionalized proteins for subsequent standard protein biochemical analyses (e.g., western blots or mass spectrometry).

Click-iT® Tools for Cell Proliferation Analysis

The Click-iT® EdU cell proliferation assay provides a superior alternative to bromodeoxyuridine (BrdU) or ^{3}H-thymidine incorporation methods for measuring new DNA synthesis.[29,30] The alkynyl nucleoside analog EdU (5-ethynyl-2′-deoxyuridine; A10044, E10187, E10415) is incorporated into DNA during the synthesis phase (S phase) of the cell cycle and is subsequently detected by copper (I)–catalyzed click coupling to an azide-derivatized fluorophore[6] (Figure 3.1.2). The small size of the click-coupled fluorophore compared to that of antibodies required for immunodetection of BrdU (Figure 3.31) enables efficient penetration of complex samples without the need for harsh cell treatment, simplifying the assay considerably. The Click-iT® EdU assay protocol is compatible with both adherent cells and cell suspensions. From start to finish, the EdU detection assay is complete in as little as 90 minutes, as compared with the antibody-based BrdU method, which takes 6–24 hours to complete. In addition, the Click-iT® EdU cell proliferation assay can be multiplexed with surface and intracellular marker detection using Alexa Fluor® dye–labeled secondary antibodies[15,31–33] (Section 7.2) (Figure 3.1.15). Although the majority of applications are in cultured mammalian cells, Click-iT® EdU reagents and methods have also been successfully applied to a wide range of model organisms including:

- *Escherichia coli*[34]
- *Caenorhabditis elegans*[32]
- *Drosophila*[35]
- Zebrafish[36]
- Mouse[15]
- Plants[37,38] (alfalfa, *Arabidopsis*, grape, maize, rice and tobacco)

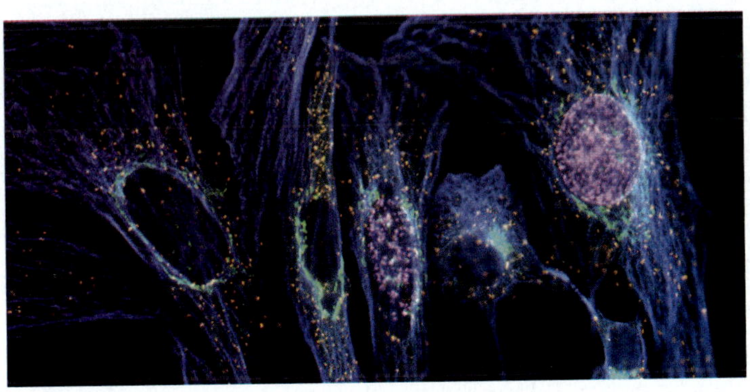

Figure 3.1.15 Multicolor imaging with the Click-iT® EdU Imaging Kits. Muntjac cells were treated with 10 µM EdU for 45 minutes. Cells were then fixed and permeabilized, and EdU that had been incorporated into newly synthesized DNA was detected by the far-red–fluorescent Click-iT® EdU Alexa Fluor® 647 HCS Assay Kit (C10356, C10357). Tubulin was labeled with an anti-tubulin antibody and visualized with an Alexa Fluor® 350 goat anti–mouse IgG antibody (A21049). The Golgi complex was stained with the green-fluorescent Alexa Fluor® 488 conjugate of lectin HPA from *Helix pomatia* (edible snail) (L11271), and peroxisomes were labeled with an anti-peroxisome antibody and visualized with an orange-fluorescent Alexa Fluor® 555 donkey anti–rabbit IgG antibody (A31572).

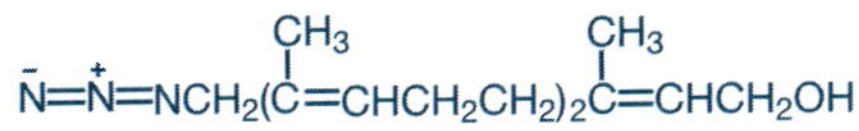

Figure 3.1.9 Click-iT® farnesyl alcohol, azide (C10248).

Figure 3.1.10 Click-iT® geranylgeranyl alcohol, azide (C10249).

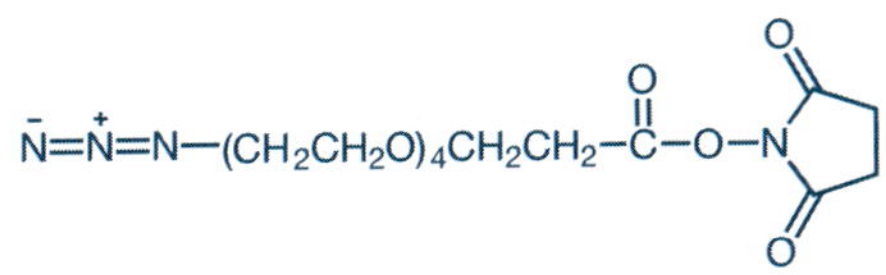

Figure 3.1.11 Azido (PEO)$_4$ propionic acid, succinimidyl ester (3-(azidotetra(ethyleneoxy))propionic acid, succinimidyl ester, A10280).

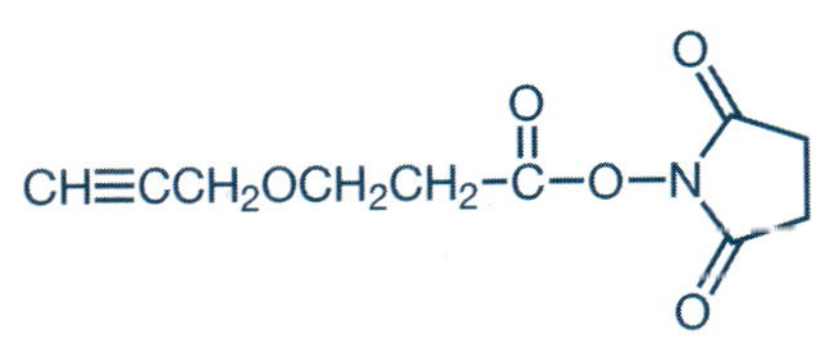

Figure 3.1.12 Alkyne, succinimidyl ester (3-propargyloxypropanoic acid, succinimidyl ester, A10279).

Figure 3.1.13 Iodoacetamide azide (I10188).

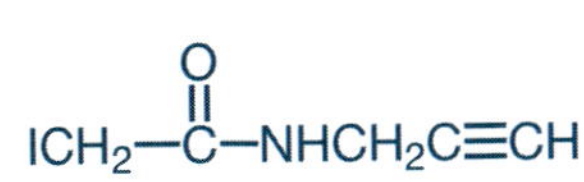

Figure 3.1.14 Iodoacetamide alkyne (I10189).

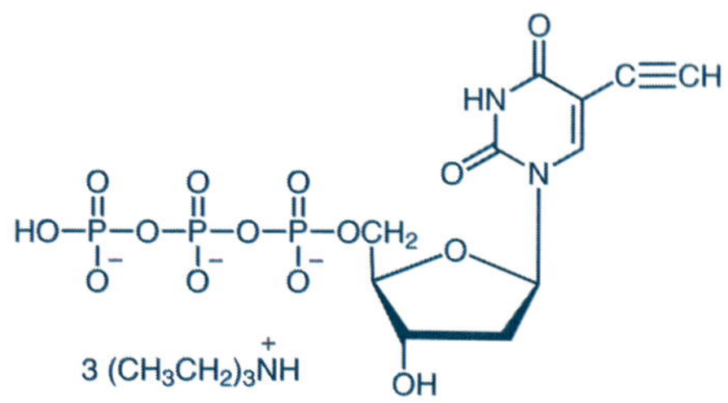

Figure 3.1.16 EdUTP nucleotide, provided in the Click-iT® TUNEL Imaging Assay Kits.

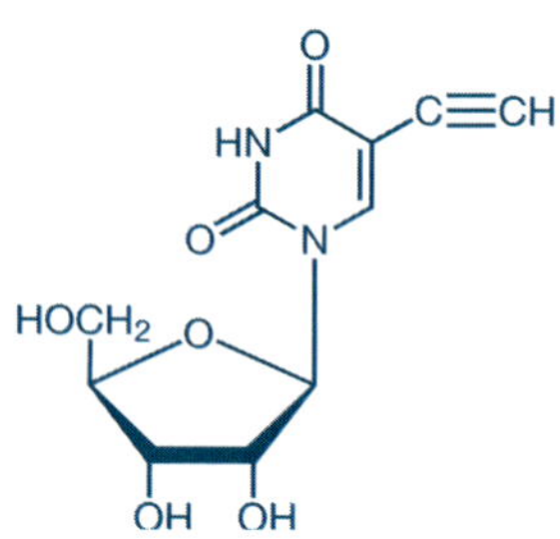

Figure 3.1.17 5-Ethynyl uridine (EU, E10345).

Click-iT® EdU Flow Cytometry Assay Kits

The Click-iT® EdU Flow Cytometry Assay Kits provide all the reagents needed to perform 50 assays using 0.5 mL reaction buffer per assay, including the nucleoside analog EdU and all components for fixation, permeabilization and labeling whole blood samples, adherent cells or suspension cells.:[30,39]

- Click-iT® EdU Alexa Fluor® 488 Flow Cytometry Assay Kit (C35002)
- Click-iT® EdU Alexa Fluor® 647 Flow Cytometry Assay Kit (A10202)
- Click-iT® EdU Pacific Blue™ Flow Cytometry Assay Kit (A10034)

Additionally, two cell-cycle stains compatible with the fluorescence excitation and emission characteristics of the fluorescent azide detection reagents are included.

Click-iT® EdU Imaging Kits

The Click-iT® EdU Imaging Kits contain all of the components needed to label and detect incorporated EdU on 50 coverslips using 0.5 mL reaction buffer per test, as well as the blue-fluorescent Hoechst 33342 nuclear stain for identification of cells irrespective of EdU incorporation status:

- Click-iT® EdU Alexa Fluor® 488 Imaging Kit (C10337)
- Click-iT® EdU Alexa Fluor® 555 Imaging Kit (C10338)
- Click-iT® EdU Alexa Fluor® 594 Imaging Kit (C10339)
- Click-iT® EdU Alexa Fluor® 647 Imaging Kit (C10340)

Click-iT® EdU HCS Assay Kits

The Click-iT® EdU HCS Assay Kits contain all of the materials needed to label and detect incorporated EdU in adherent cells in 96-well microplates and 100 µL reaction buffer per assay:

- Click-iT® EdU Alexa Fluor® 488 HCS Assay Kit (2-plate size, C10350; 10-plate size, C10351)
- Click-iT® EdU Alexa Fluor® 555 HCS Assay Kit (2-plate size, C10352; 10-plate size, C10353)
- Click-iT® EdU Alexa Fluor® 594 HCS Assay Kit (2-plate size, C10354; 10-plate size, C10355)
- Click-iT® EdU Alexa Fluor® 647 HCS Assay Kit (2-plate size, C10356; 10-plate size, C10357; Figure 3.1.15)

For cell registration or DNA profiling, these kits also include the blue-fluorescent HCS NuclearMask™ Blue stain.

Click-iT® EdU Microplate Assay Kit

The Click-iT® EdU Microplate Assay Kit provides a simple and rapid workflow with fewer wash steps resulting in a substantial time-savings advantage over traditional BrdU colorimetric or fluorescent cell proliferation assays. This assay uses Oregon Green® 488 azide for click coupling to synthetically incorporated EdU. The signal is amplified using immunodetection of the Oregon Green® 488 fluorophore by a rabbit anti–Oregon Green® horseradish peroxidase (HRP) conjugate followed by fluorogenic or chromogenic detection with our Amplex® UltraRed HRP substrate. The Click-iT® EdU microplate assay has been successfully tested in HeLa, A549, U2OS and A541 cells with a variety of reagents that modulate DNA synthesis, including the DNA synthesis inhibitor aphidicolin and the mitotic inhibitor paclitaxel. The Click-iT® EdU Microplate Assay Kit (C10214) contains sufficient reagents for performing 400 individual assays in a 96-well plate format.

Click-iT® TUNEL Assay

The terminal deoxynucleotidyl transferase-dUTP nick end labeling (TUNEL) assay—based on the incorporation of modified dUTPs by terminal deoxynucleotidyl transferase (TdT) at the 3′-OH ends of fragmented DNA—is probably the most widely used *in situ* test for studying apoptotic DNA fragmentation. For a sensitive and reliable TUNEL imaging assay, it is vital that the modified nucleotide is an efficient substrate for TdT. The minimally modified EdUTP nucleotide (Figure 3.1.16) used in the Click-iT® TUNEL imaging assay is rapidly incorporated by TdT, allowing samples to be rapidly fixed in order to preserve late-stage apoptotic cells, thereby lessening the possibility of false-negative results due to cell detachment and subsequent loss. The enzymatically incorporated nucleotide is detected by copper (I)–catalyzed click coupling to an azide-derivatized fluorophore. Compared with assays that use one-step incorporation of dye-modified nucleotides, the fast and sensitive Click-iT® TUNEL imaging assay can detect a higher percentage of apoptotic cells under identical conditions in 2 hours or less.

The Click-iT® TUNEL Imaging Assay Kits are available with a choice of azide-derivatized Alexa Fluor® dyes, providing flexibility for combination with other apoptosis detection reagents (Section 15.5):

- Click-iT® TUNEL Alexa Fluor® 488 Imaging Assay (C10245)
- Click-iT® TUNEL Alexa Fluor® 594 Imaging Assay (C10246)
- Click-iT® TUNEL Alexa Fluor® 647 Imaging Assay (C10247)

The Click-iT® TUNEL assay has been tested in HeLa, A549 and CHO K1 cells with a variety of reagents that induce apoptosis, including staurosporine, and multiplexed with antibody-based detection of other apoptosis biomarkers such as cleaved poly(ADP-ribose) polymerase (PARP), cleaved caspase-3 and phosphohistone 2B. It has also proven effective for detection of apoptosis induced by siRNA knockdown of the DEC2 transcription factor in human MCF-7 breast cancer cells.[40]

Click-iT® Tools for Detecting RNA and Protein Synthesis

Click-iT® RNA Kits

Click-iT® RNA Imaging and Click-iT® RNA HCS Assay Kits provide everything needed to detect newly synthesized RNA in adherent cells.:

- Click-iT® RNA Alexa Fluor® 488 Imaging Kit (C10329)
- Click-iT® RNA Alexa Fluor® 594 Imaging Kit (C10330)
- Click-iT® RNA Alexa Fluor® 488 HCS Assay (C10327)
- Click-iT® RNA Alexa Fluor® 594 HCS Assay (C10328)

Click-iT® RNA assays are ideal for imaging global RNA synthesis in multiplex analyses using traditional fluorescence microscopy or high-content screening (HCS). The Click-iT® RNA assays employ the alkyne-modified nucleoside EU (5-ethynyl uridine, E10345; Figure 3.1.17), which is supplied to cells and incorporated into nascent RNA.[41] The small size of the alkyne tag enables efficient incorporation by RNA polymerases without any apparent changes to the RNA levels of several housekeeping genes. Detection of incorporated EU is accomplished by copper (I)–catalyzed click coupling to an azide-derivatized fluorophore. The multiplexing capability of the assays makes them ideal for toxicological profiling or interrogation of disease models using high content imaging platforms.

The Click-iT® RNA HCS Assay Kits (C10329, C10330) contain sufficient reagents to label and detect newly synthesized RNA in whole cells using two 96-well microplates and 50 µL reaction volumes per well. These kits also supply the blue-fluorescent HCS NuclearMask™ blue stain as a nuclear counterstain for cell demarcation or for DNA profiling. The Click-iT® RNA Imaging Kits (C10327, C10328) contain sufficient reagents to label and detect newly synthesized RNA in whole cells using 25 coverslips and 500 µL reaction volume per well. These kits also supply the blue-fluorescent Hoechst 33342 dye as a nuclear counterstain or for DNA profiling.

The Click-iT® Nascent RNA Capture Kit (C10365) enables RNA synthesized during a time window defined by administration of EU to be selectively biotinylated via click coupling of EU to biotin azide. Biotinylated RNA is then captured using streptavidin-functionalized magnetic beads for reverse transcription and subsequent analysis by DNA sequencing, PCR or microarray hybridization.

Click-iT® Nascent Protein Synthesis Detection Kit

Detecting newly synthesized protein is key for researchers studying protein biosynthesis, trafficking and degradation. Click-iT® AHA (*L*-azidohomoalanine, Figure 3.1.18) incorporation provides a fast, sensitive and nonradioactive alternative to the traditional radioactive ^{35}S-methionine

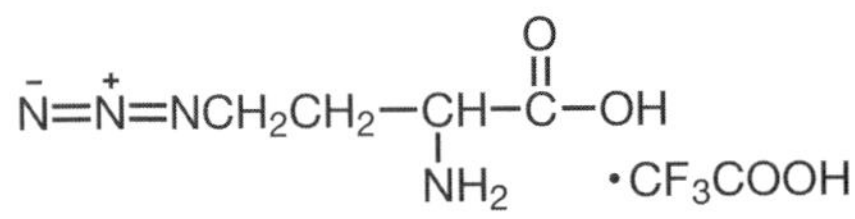

Figure 3.1.18 Click-iT® AHA (L-azidohomoalanine, C10102).

technique for the detection of nascent protein.[19,42] *L*-azidohomoalanine, an analog of *L*-methionine, is supplied to cultured cells and is biosynthetically incorporated into proteins. The incorporated amino acid is then detected by copper (I)–catalyzed click coupling to an alkyne-derivatized fluorophore. This two-step labeling and detection method provides detection sensitivity comparable with that obtained using the radioactive ^{35}S-methionine method and is compatible with downstream LC-MS/MS and MALDI-MS analysis. Click-iT® AHA is available as a stand-alone reagent (C10102) or in the Click-iT® AHA Alexa Fluor® 488 Protein Synthesis HCS Assay Kit (C10289), which contains Alexa Fluor® 488 alkyne for detection. Click-iT® AHA has proven to be a successful substitute for methionine in many cell types, including COS-7, 3T3-L1, HeLa, HEK 293 and Jurkat cells. Cellular incorporation of Click-iT® AHA should be carried out in methionine-free media, as methionine is the preferred substrate for methionyl tRNA transferase, and supplemented media (i.e., methionine-free DMEM) should be used in place of HBSS to achieve greater Click-iT® AHA incorporation at lower concentrations.

REFERENCES

1. Angew Chem Int Ed Engl (2009) 48:6974; **2.** Biochemistry (2009) 48:6571; **3.** ACS Chem Biol (2006) 1:644; **4.** Chem Rev (2009) 109:4207; **5.** Chem Rev (2008) 108:2952; **6.** Proc Natl Acad Sci U S A (2008) 105:2415; **7.** Protein Sci (2004) 13:2693; **8.** Angew Chem Int Ed Engl (2009) 48:9879; **9.** Bioconjug Chem (2009) 20:728; **10.** Methods Enzymol (1990) 186:161; **11.** J Histochem Cytochem (2010) 58:207; **12.** J Am Chem Soc (2010) 132:2504; **13.** Chem Res Toxicol (2008) 21:432; **14.** Nat Cell Biol (2008) 10:1224; **15.** J Neurosci Methods (2009) 177:122; **16.** Yeast (2008) 25:775; **17.** Bioorg Med Chem Lett (2008) 18:5995; **18.** Proc Natl Acad Sci U S A (2006) 103:9482; **19.** Nat Protoc (2007) 2:532; **20.** J Virol (2009) 83:13042; **21.** Methods Enzymol (2006) 415:230; **22.** Proc Natl Acad Sci U S A (2007) 104:2614; **23.** J Am Chem Soc (2009) 131:4967; **24.** J Am Chem Soc (2007) 129:2744; **25.** Proc Natl Acad Sci U S A (2004) 101:12479; **26.** J Biol Chem (2009) 284:21327; **27.** J Am Chem Soc (2008) 130:11576; **28.** Bioconjug Chem (2008) 19:1570; **29.** J Immunol Methods (2009) 350:29; **30.** Biotechniques (2008) 44:927; **31.** J Biol Chem (2010) 285:1544; **32.** Genetics (2009) 183:233; **33.** Nucleic Acids Res (2009) 37:e31; **34.** Methods (2009) 48:8; **35.** Science (2009) 325:340; **36.** Proc Natl Acad Sci U S A (2009) 106:14896; **37.** Plant Methods (2010) 6:5; **38.** Proc Natl Acad Sci U S A (2009) 106:11806; **39.** Cytometry A (2009) 75:862; **40.** Genes Cells (2010) 15:315; **41.** Proc Natl Acad Sci U S A (2008) 105:15779; **42.** J Neurosci (2009) 29:638.

DATA TABLE 3.1 CLICK CHEMISTRY

Cat. No.	MW	Storage	Soluble	Abs	EC	Em	Solvent	Notes
A10044	252.23	F,D	DMSO, H_2O	<300		none		1
A10266	861.04	F,D,L	DMSO	493	76,000	516	pH 7	
A10267	773.91	F,D,L	DMSO	494	76,000	520	pH 7	
A10270	948.16	F,D,L	DMSO	590	105,000	617	pH 7	
A10275	962.23	F,D,L	DMSO	588	100,000	616	pH 7	
A10277	~850	F,D,L	DMSO	646	270,000	668	MeOH	
A10278	~800	F,D,L	DMSO	646	250,000	661	pH 7	
A10279	225.20	F,D	DMSO	<300		none		
A10280	388.38	F,D,L	DMSO	<300		none		
A20012	~850	F,D,L	DMSO	554	151,000	568	pH 7	
A20013	~750	F,D,L	DMSO	554	150,000	567	pH 7	
B10184	615.79	F,D,L	DMSO	<300		none		
B10185	528.66	F,D	DMSO	<300		none		
C10102	258.16	F,DD	DMSO	<300		none		
C10186	127.14	F,D	DMSO	<300		none		
C10248	263.38	F,D,LL	DMSO	<300		none		
C10249	331.50	F,D,LL	DMSO	<300		none		
C10264	342.30	F,D	DMSO	<300		none		2
C10265	283.41	F,D,L	DMSO	<300		none		
C10268	241.33	F,D,L	DMSO	<300		none		
C33365	430.37	F,D	DMSO	<300		none		3
C33366	430.37	F,D	DMSO	<300		none		3
C33367	430.37	F,D	DMSO	<300		none		3
E10345	268.23	F,D	DMSO	<300		none		
I10188	310.14	F,D,L	DMSO	<300		none		4
I10189	223.01	F,D,L	DMSO	<300		none		4
O10180	637.68	F,D,L	DMSO	494	80,000	521	pH 9	5
O10181	449.37	F,D,L	DMSO	494	80,000	521	pH 9	5
T10182	554.65	F,D,L	DMSO	546	95,000	571	MeOH	6
T10183	467.52	F,D,L	DMSO	543	95,000	572	MeOH	6

For definitions of the contents of this data table, see "Using *The Molecular Probes® Handbook*" in the introductory pages.

Notes

1. E10187 and E10415 are an alternative packaging sof A10044 but are otherwise identical.
2. C10264 is deacetylated *in vivo* yielding the 6-ethynyl analog of fucose (MW = 174.15).
3. C33365, C33366 and C33367 are deacetylated *in vivo* yielding *N*-azidoacetylgalactosamine, *N*-azidoacetylmannosamine and *N*-azidoacetylglucosamine (all MW = 262.22), respectively.
4. Iodoacetamides in solution undergo rapid photodecomposition to unreactive products. Minimize exposure to light prior to reaction.
5. Absorption and fluorescence of Oregon Green® 488 dyes are pH dependent only in moderately acidic solutions (pH <5).
6. Abs and Em for tetramethylrhodamine dyes in pH 8 buffer are red-shifted approximately 8 nm compared to MeOH, with EC lower by ~10%.

IMPORTANT NOTICE: The products described in this manual are covered by one or more Limited Use Label License(s). Please refer to the Appendix on page 971 and Master Product List on page 975. Products are For Research Use Only. Not intended for any animal or human therapeutic or diagnostic use.
www.invitrogen.com/probes
molecular probes® | invitrogen™ by life technologies™

PRODUCT LIST 3.1 CLICK CHEMISTRY

Cat. No.	Product	Quantity
A10267	Alexa Fluor® 488 alkyne (Alexa Fluor® 488 5-carboxamido-(propargyl), bis(triethylammonium salt)) *5-isomer*	0.5 mg
A20013	Alexa Fluor® 555 alkyne, triethylammonium salt	0.5 mg
A10275	Alexa Fluor® 594 alkyne (Alexa Fluor® 594 carboxamido-(5-(and 6-)propargyl), bis(triethylammonium salt)) *mixed isomers*	0.5 mg
A10278	Alexa Fluor® 647 alkyne, triethylammonium salt	0.5 mg
A10266	Alexa Fluor® 488 azide (Alexa Fluor® 488 5-carboxamido-(6-azidohexanyl), bis(triethylammonium salt)) *5-isomer*	0.5 mg
A20012	Alexa Fluor® 555 azide, triethylammonium salt	0.5 mg
A10270	Alexa Fluor® 594 azide (Alexa Fluor® 594 carboxamido-(6-azidohexanyl), triethylammonium salt) *mixed isomers*	0.5 mg
A10277	Alexa Fluor® 647 azide, triethylammonium salt	0.5 mg
A10279	alkyne, succinimidyl ester (3-propargyloxypropanoic acid, succinimidyl ester)	1 mg
A10280	azido $(PEO)_4$ propionic acid, succinimidyl ester (3-(azidotetra(ethyleneoxy))propionic acid, succinimidyl ester)	1 mg
B10185	biotin alkyne (PEG_4 carboxamide-propargyl biotin)	1 mg
B10184	biotin azide (PEG_4 carboxamide-6-azidohexanyl biotin)	1 mg
C10102	Click-iT® AHA (L-azidohomoalanine) *for nascent protein synthesis*	5 mg
C10289	Click-iT® AHA Alexa Fluor® 488 Protein Synthesis HCS Assay *2-plate size*	1 kit
C10269	Click-iT® Cell Reaction Buffer Kit	1 kit
C35002	Click-iT® EdU Alexa Fluor® 488 Flow Cytometry Assay Kit *50 assays*	1 kit
A10202	Click-iT® EdU Alexa Fluor® 647 Flow Cytometry Assay Kit *50 assays*	1 kit
A10034	Click-iT® EdU Pacific Blue™ Flow Cytometry Assay Kit *50 assays*	1 kit
C10350	Click-iT® EdU Alexa Fluor® 488 HCS Assay *2-plate size*	1 kit
C10351	Click-iT® EdU Alexa Fluor® 488 HCS Assay *10-plate size*	1 kit
C10352	Click-iT® EdU Alexa Fluor® 555 HCS Assay *2-plate size*	1 kit
C10353	Click-iT® EdU Alexa Fluor® 555 HCS Assay *10-plate size*	1 kit
C10354	Click-iT® EdU Alexa Fluor® 594 HCS Assay *2-plate size*	1 kit
C10355	Click-iT® EdU Alexa Fluor® 594 HCS Assay *10-plate size*	1 kit
C10356	Click-iT® EdU Alexa Fluor® 647 HCS Assay *2-plate size*	1 kit
C10357	Click-iT® EdU Alexa Fluor® 647 HCS Assay *10-plate size*	1 kit
C10337	Click-iT® EdU Alexa Fluor® 488 Imaging Kit *for 50 coverslips*	1 kit
C10338	Click-iT® EdU Alexa Fluor® 555 Imaging Kit *for 50 coverslips*	1 kit
C10339	Click-iT® EdU Alexa Fluor® 594 Imaging Kit *for 50 coverslips*	1 kit
C10340	Click-iT® EdU Alexa Fluor® 647 Imaging Kit *for 50 coverslips*	1 kit
C10214	Click-iT® EdU Microplate Assay *400 assays*	1 kit
C10248	Click-iT® farnesyl alcohol, azide *mixed isomers*	1 mg
C10264	Click-iT® fucose alkyne (tetraacetylfucose alkyne)	5 mg
C33365	Click-iT® GalNAz metabolic glycoprotein labeling reagent (tetraacetylated *N*-azidoacetylgalactosamine) *for *O*-linked glycoproteins* *5.2 mg*	each
C10249	Click-iT® geranylgeranyl alcohol, azide *mixed isomers*	1 mg
C33367	Click-iT® GlcNAz metabolic glycoprotein labeling reagent (tetraacetylated *N*-azidoacetylglucosamine) *for *O*-GlcNAc modified proteins* *5.2 mg*	each
C10186	Click-iT® HPG (L-homopropargylglycine) *for nascent protein synthesis*	5 mg
C33366	Click-iT® ManNAz metabolic glycoprotein labeling reagent (tetraacetylated *N*-azidoacetyl-d-mannosamine) *for sialic acid glycoproteins* *5.2 mg*	each
C10268	Click-iT® myristic acid, azide (12-azidododecanoic acid)	1 mg
C10365	Click-iT® Nascent RNA Capture Kit *for gene expression analysis*	1 kit
C10265	Click-iT® palmitic acid, azide (15-azidopentadecanoic acid)	1 mg
C10276	Click-iT® Protein Reaction Buffer Kit	1 kit
C10327	Click-iT® RNA Alexa Fluor® 488 HCS Assay *2-plate size*	1 kit
C10328	Click-iT® RNA Alexa Fluor® 594 HCS Assay *2-plate size*	1 kit
C10329	Click-iT® RNA Alexa Fluor® 488 Imaging Kit *for 25 coverslips*	1 kit
C10330	Click-iT® RNA Alexa Fluor® 594 Imaging Kit *for 25 coverslips*	1 kit
C10245	Click-iT® TUNEL Alexa Fluor® 488 Imaging Assay *for microscopy and HCS* *50–100 assays*	1 kit
C10246	Click-iT® TUNEL Alexa Fluor® 594 Imaging Assay *for microscopy and HCS* *50–100 assays*	1 kit
C10247	Click-iT® TUNEL Alexa Fluor® 647 Imaging Assay *for microscopy and HCS* *50–100 assays*	1 kit
A10044	EdU (5-ethynyl-2′-deoxyuridine)	50 mg
E10187	EdU (5-ethynyl-2′-deoxyuridine)	500 mg
E10415	EdU (5-ethynyl-2′-deoxyuridine)	5 g
E10345	5-ethynyl uridine (EU)	5 mg
I10189	iodoacetamide alkyne	1 mg
I10188	iodoacetamide azide	1 mg
O10181	Oregon Green® 488 alkyne *6-isomer*	0.5 mg
O10180	Oregon Green® 488 azide (Oregon Green® 6-carboxamido-(6-azidohexanyl), triethylammonium salt) *6-isomer*	0.5 mg
T10183	tetramethylrhodamine (TAMRA) alkyne (5-carboxytetramethylrhodamine, propargylamide) *5-isomer*	0.5 mg
T10182	tetramethylrhodamine (TAMRA) azide (tetramethylrhodamine 5-carboxamido-(6-azidohexanyl)) *5-isomer*	0.5 mg

IMPORTANT NOTICE: The products described in this manual are covered by one or more Limited Use Label License(s). Please refer to the Appendix on page 971 and Master Product List on page 975. Products are For Research Use Only. Not intended for any animal or human therapeutic or diagnostic use.

3.2 Reagents for Modifying Alcohols

Alcohols in Proteins: Serine, Threonine and Tyrosine Residues

Although alcohols (including phenols such as tyrosine and the hydroxyl groups in serine, threonine, sterols and carbohydrates) are abundant in biomolecules, their chemical reactivity in aqueous solution is extremely low. Few reagents are selective for alcohols in aqueous solution, especially in the presence of more reactive nucleophiles such as thiols and amines. It is therefore difficult to selectively modify serine, threonine and tyrosine residues in proteins except when they exhibit unusual reactivity, such as by residing at an enzyme's active site.

Serine and Threonine Residues

Nonacylated N-terminal serine and threonine residues in peptides and proteins can be oxidized with periodate to yield aldehydes [1,2] (Figure 3.2.1) that can be subsequently modified with a variety of hydrazine, hydroxylamine or amine derivatives (Section 3.3, Table 3.2). In addition, peptides containing serine, threonine or tyrosine residues separated from a histidine residue by a single amino acid can be selectively acylated by the succinimidyl ester or sulfosuccinimidyl ester of biotin-X [3] (B1582, B6353; Section 4.2; Figure 3.2.2). This property may also permit selective modification of these sequences (Ser-x-His, Thr-x-His and Tyr-x-His, where "x" refers to any amino acid) in peptides and proteins with fluorescent succinimidyl esters (Chapter 1). *O*-acylation versus *N*-acylation can be detected by treatment with hydroxylamine, which cleaves esters but usually not amides.[4]

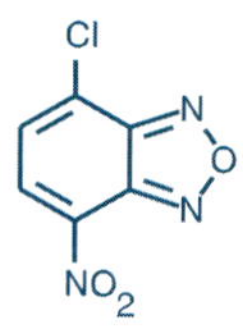

Figure 3.2.3 4-Chloro-7-nitrobenz-2-oxa-1,3-diazole (NBD chloride; 4-chloro-7-nitrobenzofurazan, C20260).

Figure 3.2.1 Sodium periodate oxidation of an N-terminal serine residue to an aldehyde, with the release of formaldehyde. The aldehyde thus formed from the protein can be subsequently modified with a variety of hydrazine, hydroxylamine or amine derivatives.

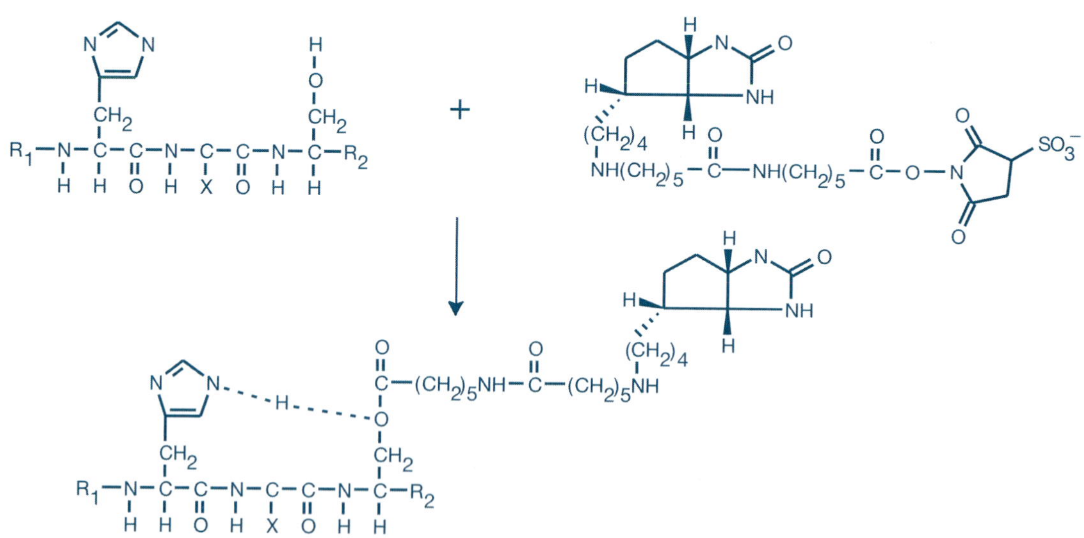

Figure 3.2.2 Nucleophilic attack of serine on the carbonyl group (C=O) of biotin-X, SSE (B6353) results in the stable *O*-acylated derivative. In addition to histidine-x-serine, this stable intermediate can be formed in the presence of linear sequences of histidine-x-tyrosine and histidine-x-threonine, where "x" refers to any amino acid.

IMPORTANT NOTICE: The products described in this manual are covered by one or more Limited Use Label License(s). Please refer to the Appendix on page 971 and Master Product List on page 975. Products are For Research Use Only. Not intended for any animal or human therapeutic or diagnostic use.
www.invitrogen.com/probes
molecular probes | invitrogen by life technologies

Tyrosine Residues

Modification of tyrosine residues is sometimes a side reaction when proteins are reacted with sulfonyl chlorides, iodoacetamides or other reactive dyes described in Chapter 1 and Chapter 2. For example, NBD chloride (C20260, Section 1.8, Figure 3.2.3) reacts with an active-site tyrosine in *Escherichia coli* F_1-ATPase, causing strong inhibition.[5]

Tyrosine residues in some proteins can be selectively modified by initial nitration of the *ortho* position of its phenol using tetranitromethane, and then reduction of the *o*-nitrotyrosine with sodium dithionite ($Na_2S_2O_4$) to form an *o*-aminotyrosine (Figure 3.2.4). Although much less reactive than aliphatic amines, the aromatic amine of *o*-aminotyrosine can react with most amine-reactive reagents (Chapter 1) between pH 5 and pH 7.[6] To obtain selective derivatization in complex samples, it is of course critical to block all aliphatic primary amines, typically by acetylation with acetic anhydride, prior to reduction of nitrotyrosine. Nitration of tyrosine residues by nitric oxide occurs naturally in cells through peroxynitrite radical intermediates (Section 18.3, Table 18.1), yielding derivatives that can be reduced to *o*-aminotyrosine and subsequently detected using amine-reactive reagents.[6,7] Tyrosine residues of peptides and proteins can be selectively coupled to aniline derivatives in the presence of formaldehyde.[8] Chemoselective derivatization of tyrosine has also been accomplished using a fluorescent diazodicarboxamide reagent prepared from carboxy-X-rhodamine (ROX) succinimidyl ester[9] (C1309, Section 1.6).

Another method for modifying tyrosine groups in peptides is to convert the phenol group in tyrosine residues to a salicylaldehyde derivative, and then to react the salicylaldehyde with 1,2-diamino-4,5-dimethoxybenzene (D1463, Section 3.3) to form a fluorescent benzimidazole.[10–12]

The tyramide signal amplification (TSA™) technology (Section 6.2), which was developed by NEN (now a part of PerkinElmer Corporation) and licensed for in-cell and in-tissue applications, permits significant amplification of the detectability of targets by a horseradish peroxidase–mediated scheme. In the TSA™ method, the labeled tyramide becomes covalently linked to tyrosine residues in or near the target. We have introduced an extensive selection of TSA™ Kits that utilize an Alexa Fluor® tyramide, Oregon Green® 488 tyramide or biotin-XX tyramide as the amplification reagents (Section 6.2).

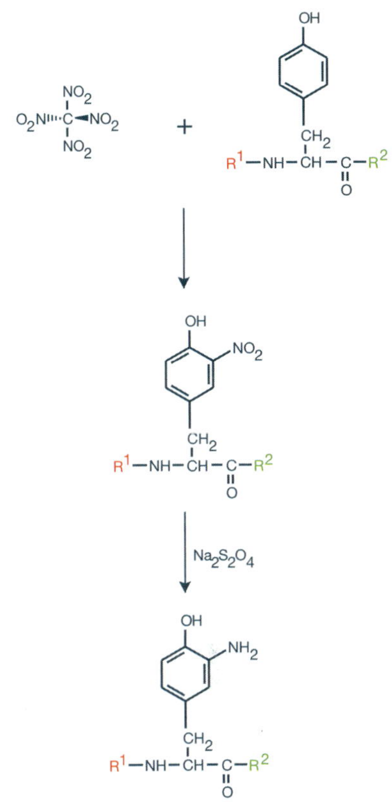

Figure 3.2.4 Reaction scheme for the conversion of tyrosine to *o*-aminotyrosine. Tyrosine undergoes nitration by reaction with tetranitromethane, followed by reduction with sodium dithionite, to yield an *o*-aminotyrosine.

Alcohols in Carbohydrates

As with derivatization of alcohols in proteins, it is difficult to selectively modify most carbohydrates in aqueous solution because of their low reactivity and the competing hydrolysis of the reactive reagents. However, several reagents are available for derivatizing reducing sugars (which contain a low equilibrium concentration of the reactive aldehyde function), as well as for modifying aldehydes and ketones obtained by periodate oxidation of various carbohydrates. To pursue this labeling approach, see Section 3.3 for a description of aldehyde- and ketone-reactive reagents.

Dichlorotriazines

Dichlorotriazines readily modify amines in proteins, and they are among the few reactive groups that are reported to react directly with polysaccharides and other alcohols in aqueous solution, provided that the pH is >9 and that other nucleophiles are absent. We offer the 5-isomer of fluorescein dichlorotriazine (5-DTAF, D16), with absorption/emission maxima of ~492/516 nm, as well as Texas Red® C_2-dichlorotriazine (T30200, Figure 3.2.5), with absorption/emission maxima of ~588/601 nm. 5-DTAF has been used to label a wide range of hydroxylated biopolymers including collagen,[13] cellulose,[14] cyclodextrins[15] and soluble beta-glucan,[16] as well as functionalized carbon nanotubes.[17]

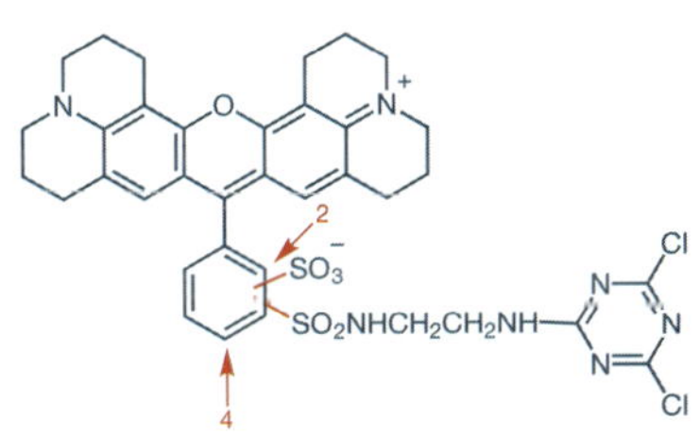

Figure 3.2.5 Texas Red® C_2-dichlorotriazine (T30200).

N-Methylisatoic Anhydride

In the absence of other reactive functional groups, *N*-methylisatoic anhydride (M25) will convert ribonucleotides and certain other carbohydrates[18,19] to fluorescent esters with excitation/emission maxima of ~350/446 nm in mildly basic aqueous solution.[20–24] The compactness and moderate environmental sensitivity of this fluorophore, which is a synthetic precursor to blue-fluorescent *N*-methylanthraniloyl (MANT) amides and esters,[23] may be advantageous for preparing site-selective probes. Low molecular weight alcohols are better derivatized by this reagent in aprotic organic solvents[25–27] (Figure 3.2.6).

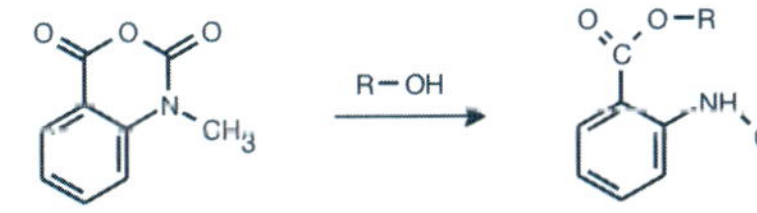

Figure 3.2.6 Reaction of *N*-methylisatoic anhydride (M25) with an alcohol to produce a blue-fluorescent (~350/446 nm) *N*-methylanthraniloyl (MANT) ester.

IMPORTANT NOTICE: The products described in this manual are covered by one or more Limited Use Label License(s). Please refer to the Appendix on page 971 and Master Product List on page 975. Products are For Research Use Only. Not intended for any animal or human therapeutic or diagnostic use.

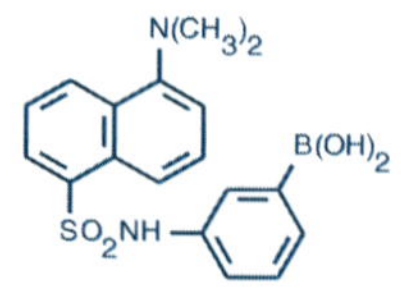

Figure 3.2.7 *m*-Dansylaminophenylboronic acid (D2281).

Dansyl Aminophenylboronic Acid

m-Dansylaminophenylboronic acid (D2281, Figure 3.2.7) reacts with vicinal diols (hydroxyl groups on adjacent carbon atoms) and certain amino alcohols[28] to form cyclic complexes[29] (Figure 3.2.8) that have a fluorescence intensity and peak emission dependent on the environment of the dansyl fluorophore. This interesting reagent binds reversibly to cell-wall carbohydrates,[30] as well as to glycosylated (but not deglycosylated) human serum albumin.[31] Dansylaminophenylboronic acid is also used as an HPLC derivatization reagent for vicinal diols[32] and as a detection reagent for glycolipids analyzed by thin-layer chromatography.[33]

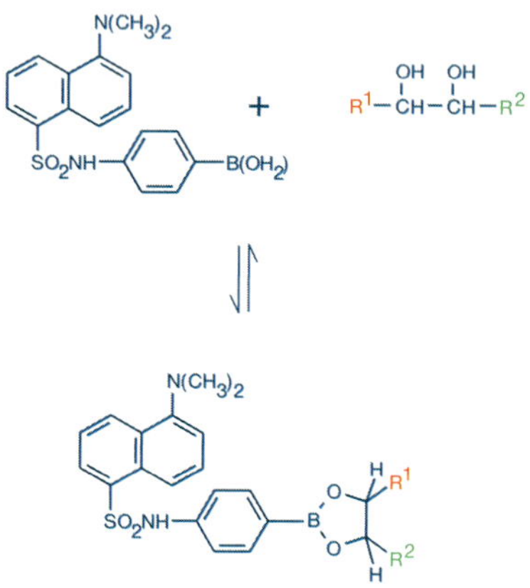

Figure 3.2.8 Reaction of *m*-dansylaminophenylboronic acid (D2281) with a vicinal diol to form a reversible fluorescent cyclic complex.

Alcohol Modification in Organic Solvents

Two functional groups—acyl azides and acyl nitriles—react directly with aliphatic amines to yield the same products as do the corresponding succinimidyl esters. When reacted in organic solvents, however, these reagents can also form derivatives of alcohols and phenols, making them extremely useful for sensitive analysis of alcohols by HPLC or capillary electrophoresis.

Isocyanates Prepared from Acyl Azides

Alcohols are much easier to modify in anhydrous organic solvents than in aqueous solution. Perhaps the most effective reagents are isocyanates, which are much more reactive with alcohols (and amines) than are isothiocyanates but are not sufficiently stable to permit their sale. Fortunately, isocyanates can often be prepared by Curtius rearrangement of acyl azides (Figure 3.2.9). When an acyl azide and alcohol are heated together in an organic solvent such as toluene, dioxane or DMF at 80°C, the acyl azide will rearrange to form an isocyanate that then reacts with the alcohol to form a stable urethane. As little as 50 femtograms of the urethane conjugates prepared from coumarin derivatives 7-methoxycoumarin-3-carbonyl azide (M1445) and 7-diethylaminocoumarin-3-carbonyl azide (D1446) can be detected using an HPLC fluorescence detector.[34,35] Alcohol conjugates (urethanes) prepared from the single-isomer carbonyl azides of fluorescein diacetate (F6218) and tetramethylrhodamine (T6219) may provide even higher sensitivity, particularly with instruments that employ the argon-ion laser. Following rearrangement and alcohol conjugation, the acetates of the fluorescein derivative can be removed by hydrolysis at pH 9–10. The diacetate of fluorescein-5-carbonyl azide has been used to synthesize a fluorogenic substrate for the anandamide transmembrane carrier.[36] Tetramethylrhodamine-5-carbonyl azide has been successfully conjugated to the hydrophobic poly(ε-caprolactone) (PCL) block of a diblock copolymer micelle in order to follow its cellular internalization[37,38] and has also been used to prepare riboflavin conjugates for the same purpose.[39,40]

Acyl Nitriles

9-Anthroylnitrile (A1440) reacts with alcohols, such as steroids and acylglycerols, in organic solvents to yield carboxylate esters that are useful for HPLC.[41,42] To optimize solid-phase organic synthesis, 9-anthroylnitrile has been used to quantitate the absolute amount of resin-bound hydroxyl groups directly on solid support.[43] In addition, 9-anthroylnitrile has been reported to be useful for the selective labeling of certain serine and threonine residues in myosin.[44–48] The lipophilicity of 9-anthroylnitrile may make it useful for modifying hydroxyl groups of proteins and hydroxylated fatty acids that are buried within cell membranes.

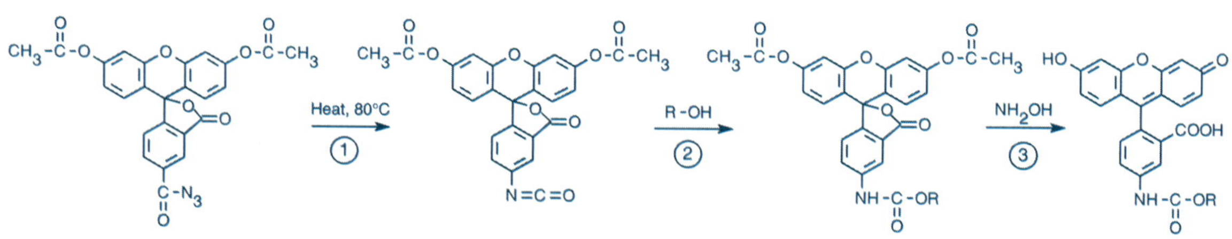

Figure 3.2.9 Derivatization of an alcohol using the diacetate of fluorescein-5-carbonyl azide (F6218). This process consists of three steps: 1) rearrangement of the acyl azide to an isocyanate, 2) reaction of the isocyanate with an alcohol to form a urethane and 3) deprotection of the nonfluorescent urethane derivative using hydroxylamine.

IMPORTANT NOTICE: The products described in this manual are covered by one or more Limited Use Label License(s). Please refer to the Appendix on page 971 and Master Product List on page 975. Products are For Research Use Only. Not intended for any animal or human therapeutic or diagnostic use.

REFERENCES

1. Bioconjug Chem (2003) 14:205; **2.** Bioconjug Chem (1992) 3:138; **3.** Anal Biochem (2010) 398:123; **4.** J Biol Chem (1992) 267:5060; **5.** J Biol Chem (2004) 279:31505; **6.** J Proteome Res (2007) 6:2257; **7.** Anal Biochem (1998) 259:127; **8.** Bioconjug Chem (2008) 19:153; **9.** J Am Chem Soc (2010) 132:1523; **10.** J Chromatogr (1988) 430:271; **11.** J Chromatogr (1986) 356:171; **12.** J Chromatogr (1985) 344:267; **13.** J Biomech (2009) 42:2279; **14.** Biomacromolecules (2003) 4:481; **15.** Biomacromolecules (2009) 10:505; **16.** J Immunol (2006) 177:1661; **17.** Langmuir (2009) 25:7573; **18.** Electrophoresis (2007) 28:1221; **19.** Anal Biochem (2000) 284:167; **20.** Biochemistry (1991) 30:422; **21.** Biochemistry (1990) 29:3309; **22.** Biochim Biophys Acta (1983) 742:496; **23.** J Biol Chem (1982) 257:13354; **24.** Arch Biochem Biophys (1973) 155:70; **25.** Anal Biochem (1996) 234:31; **26.** Synthesis (1982) 39:266; **27.** J Org Chem (1959) 24:1214; **28.** Anal Bioanal Chem (2006) 386:506; **29.** Med Res Rev (2010) 30:171; **30.** Biochem Biophys Res Commun (1980) 96:157; **31.** Clin Chim Acta (1985) 149:13; **32.** Anal Chim Acta (1990) 228:101; **33.** J Lipid Res (1995) 36:1848; **34.** Chem Pharm Bull (1985) 33:1164; **35.** J Lipid Res (1997) 38:429; **36.** J Pharmacol Exp Ther (2000) 293:289; **37.** Science (2003) 300:615; **38.** Bioconjug Chem (2002) 13:1259; **39.** Mol Pharm (2004) 1:257; **40.** J Pharmacol Exp Ther (2003) 306:681; **41.** Anal Chim Acta (1983) 147:397; **42.** J Chromatogr (1983) 276:289; **43.** Anal Chem (1999) 71:4564; **44.** Biochemistry (1996) 35:16061; **45.** Biochemistry (1994) 33:6867; **46.** J Biol Chem (1990) 265:18791; **47.** J Biol Chem (1990) 265:18786; **48.** J Biol Chem (1989) 264:18188.

DATA TABLE 3.2 REAGENTS FOR MODIFYING ALCOHOLS

Cat. No.	MW	Storage	Soluble	Abs	EC	Em	Solvent	Notes
A1440	231.25	F,D,L	DMF, MeCN	361	7500	470	MeOH	1
D16	495.28	F,D,L	pH >6, DMF	492	83,000	516	pH 9	2, 3
D1446	286.29	F,D,L	DMF, MeCN	436	57,000	478	MeOH	
D2281	370.23	D,L	DMF, DMSO	337	4600	517	MeOH	4
F6218	485.41	FF,D	DMF, MeCN	<300		none		
M25	177.16	D	DMF, DMSO	316	3500	386	MeOH	5
M1445	245.19	FF,D,L	DMF, MeCN	360	25,000	415	MeOH	
T6219	455.47	FF,D,L	DMF, MeCN	545	90,000	578	MeOH	
T30200	796.74	F,D,L	DMF, DMSO	583	87,000	604	MeOH	

For definitions of the contents of this data table, see "Using *The Molecular Probes® Handbook*" in the introductory pages.

Notes

1. The absorption spectrum of A1440 has subsidiary peaks at 380 nm and 344 nm. Emission spectrum is unstructured. Ester derivatives formed by reaction with alcohols have essentially similar spectra.
2. Unstable in water. Use immediately.
3. Absorption and fluorescence of fluorescein derivatives are pH dependent. Extinction coefficients and fluorescence quantum yields decrease markedly at pH <7.
4. Fluorescence of D2281 when bound to proteins is typically blue shifted (Em ~490 nm).
5. The amide reaction product of M25 with butylamine has Abs = 353 nm (EC = 5900 $cm^{-1}M^{-1}$), Em = 426 nm in MeOH. Ester reaction products with alcohols have Abs = 350 nm (EC = 5700 $cm^{-1}M^{-1}$), Em = 446 nm in water (pH 8).

PRODUCT LIST 3.2 REAGENTS FOR MODIFYING ALCOHOLS

Cat. No.	Product	Quantity
A1440	9-anthroylnitrile	25 mg
D2281	*m*-dansylaminophenylboronic acid	100 mg
D16	5-(4,6-dichlorotriazinyl)aminofluorescein (5-DTAF) *single isomer*	100 mg
D1446	7-diethylaminocoumarin-3-carbonyl azide	25 mg
F6218	fluorescein-5-carbonyl azide, diacetate	10 mg
M1445	7-methoxycoumarin-3-carbonyl azide	25 mg
M25	*N*-methylisatoic anhydride *high purity*	1 g
T6219	tetramethylrhodamine-5-carbonyl azide	5 mg
T30200	Texas Red® C_2-dichlorotriazine	5 mg

IMPORTANT NOTICE: The products described in this manual are covered by one or more Limited Use Label License(s). Please refer to the Appendix on page 971 and Master Product List on page 975. Products are For Research Use Only. Not intended for any animal or human therapeutic or diagnostic use.

3.3 Reagents for Modifying Aldehydes and Ketones

Aldehydes and ketones are present in a number of low molecular weight molecules such as drugs, steroid hormones, reducing sugars and metabolic intermediates (e.g., pyruvate and α-ketoglutarate). Except for polysaccharides containing free reducing sugars, however, biopolymers generally lack aldehyde and ketone groups. Even those aldehydes and ketones that are found in the open-ring form of simple carbohydrates are usually in equilibrium with the closed-ring form of the sugar.

The infrequent occurrence of aldehydes and ketones in biomolecules has stimulated the development of techniques to selectively introduce these functional groups, thus providing unique sites for chemical modification and greatly extending the applications of the probes found in this section. Fluorescent modification of aldehyde or carboxylic acid groups in carbohydrates is also frequently utilized for their analysis by HPLC, capillary electrophoresis [1] and other methods.

Introducing Aldehydes and Ketones into Biomolecules

Periodate Oxidation

The most common method for introducing aldehydes and ketones into polysaccharides and glycoproteins (including antibodies) is by periodate-mediated oxidation of vicinal diols. These introduced aldehydes and ketones can then be modified with fluorescent or biotinylated hydrazine, hydroxylamine or amine derivatives to label the polysaccharide or glycoprotein. For example, some of the hydrazine derivatives described in this section have been used to detect periodate-oxidized glycoproteins in gels.[2] The Pro-Q® Emerald 300 and Pro-Q® Emerald 488 Glycoprotein Gel and Blot Stain Kits (P21855, P21857, M33307; Section 9.4) are based on periodate oxidation of glycoproteins and subsequent labeling with a Pro-Q® Emerald dye.[3]

Periodate oxidation of the 3′-terminal ribose provides one of the few methods of selectively modifying RNA; periodate-oxidized ribonucleotides can subsequently be converted to fluorescent nucleic acid probes by reaction with fluorescent hydrazines, hydroxylamines and amines.[4–6] Alkenes from unsaturated fatty acids and ceramides can also be converted to glycols by osmium tetroxide and then oxidized by periodate to aldehydes, and periodate will oxidize certain β-aminoethanol derivatives such as the hydroxylysine residues in collagen, as well as methionine (to its sulfoxide) and certain thiols (usually to disulfides). These other reactions, however, usually occur at a slower rate than oxidation of vicinal diols.

In addition to vicinal diols, N-terminal serine and threonine residues of peptides and proteins can be selectively oxidized by periodate to aldehyde groups [7–9] (Figure 3.3.1). Moreover, because antibodies are glycosylated at sites distant from the antigen-binding region, modification of periodate-oxidized antibodies by hydrazines [10] and hydroxylamines usually does not inactivate the antibody, as sometimes occurs with amine-reactive labeling.

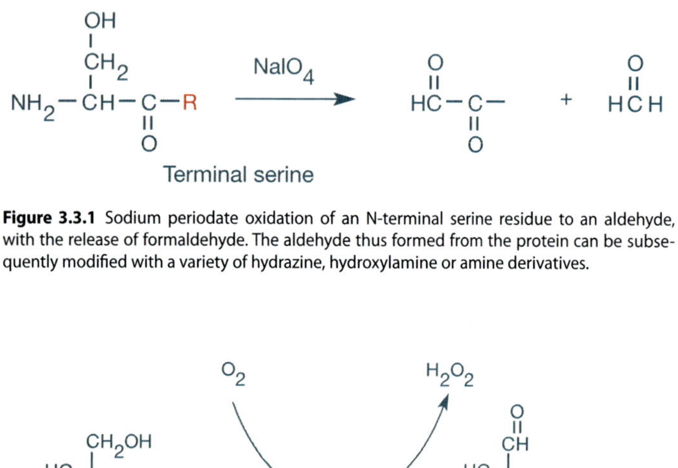

Figure 3.3.1 Sodium periodate oxidation of an N-terminal serine residue to an aldehyde, with the release of formaldehyde. The aldehyde thus formed from the protein can be subsequently modified with a variety of hydrazine, hydroxylamine or amine derivatives.

Galactose Oxidase–Mediated Oxidation and Other Methods

Galactose oxidase oxidizes terminal galactose residues to aldehydes, particularly in glycoproteins.[11–15] The introduction of galactose residues can be especially advantageous for structural studies because it provides a means of selectively labeling specific sites on biomolecules. For example, 2-keto-galactose has been specifically inserted into the Fc glycans of therapeutic antibodies, including Herceptin and Avastin, enabling site-specific labeling with Alexa Fluor® 488 hydroxylamine [16,17] (A30629). Galactose oxidase–modified lipopolysaccharides (LPS) have been modified with Alexa Fluor® 488 hydrazide (A10436) to probe for LPS-binding sites on cells.[18] Because galactose oxidase–mediated oxidation liberates a molecule of hydrogen peroxide for each molecule of aldehyde that is formed (Figure 3.3.2), horseradish peroxidase–catalyzed oxidation of the Amplex® Red reagent to red-fluorescent resorufin by hydrogen peroxide provides a ready means by which the number of aldehyde residues introduced into a biomolecule, including on a cell surface, can be quantitated. The Amplex® Red Galactose/Galactose Oxidase Assay Kit (A22179, Section 10.5) provides the reagents and a general protocol for this assay of introduced aldehyde residues. Other methods for aldehyde and ketone introduction include selective N-terminal transamination in the presence of pyridoxal-5′-phosphate,[19] ligation of a ketone analog of biotin to proteins with a biotin acceptor peptide (BAP) fusion tag by biotin ligase (BirA) [20] and co-translational modification of recombinantly tagged proteins by formylglycine-generating enzyme [21] (FGE).

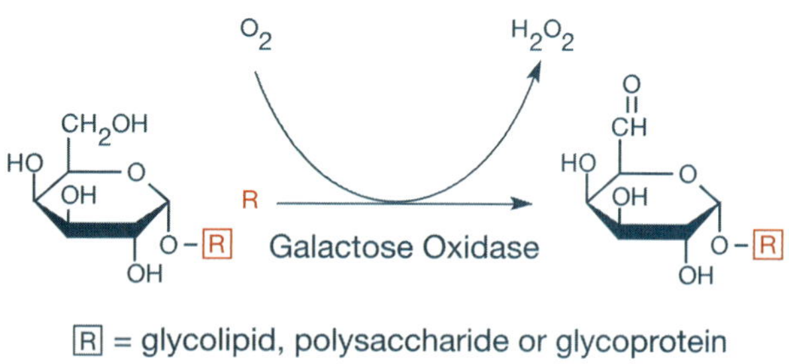

Figure 3.3.2 Oxidation of the terminal galactose residue of a glycoprotein, glycolipid or polysaccharide resultvs in the generation of an aldehyde, which can react with hydrazines, hydroxylamines or primary amine–containing compounds.

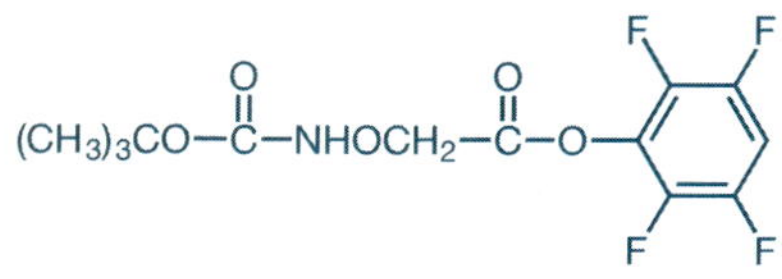

Figure 3.3.3 *N*-(*t*-BOC)-aminooxyacetic acid, tetrafluorophenyl ester (B30300).

IMPORTANT NOTICE: The products described in this manual are covered by one or more Limited Use Label License(s). Please refer to the Appendix on page 971 and Master Product List on page 975. Products are For Research Use Only. Not intended for any animal or human therapeutic or diagnostic use.

Coupling Hydrazines and Amines to Amine-Containing Biomolecules without Introducing Aldehydes and Ketones

Common tissue fixatives such as formaldehyde and glutaraldehyde can be used to couple hydrazine and amine derivatives to proteins and other amine-containing polymers. For example, lucifer yellow CH (L453) can be conjugated to surrounding biomolecules by common aldehyde-based fixatives in order to preserve the dye's staining pattern during subsequent tissue manipulations.[22]

Figure 3.3.4 8-Aminopyrene-1,3,6-trisulfonic acid, trisodium salt (APTS, A6257).

Introducing a Hydroxylamine into a Biomolecule

The tetrafluorophenyl (TFP) ester of *N*-(*t*-BOC)-aminooxyacetic acid (B30300, Figure 3.3.3) is an amine-reactive protected hydroxylamine that is useful for synthesizing new aldehyde- and ketone-reactive probes in an organic solvent. Following coupling to aliphatic amines, the *t*-BOC group can be quantitatively removed with trifluoroacetic acid. The resultant hydroxylamine probe can then spontaneously react with aldehydes, the reducing ends of saccharides and oligosaccharides, and abasic sites in oligonucleotides to form stable adducts.

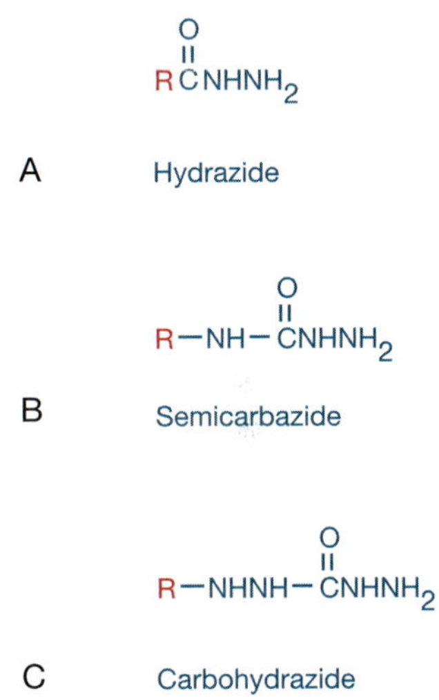

Figure 3.3.5 Structures of A) a hydrazide, B) a semicarbazide and C) a carbohydrazide.

Hydrazines and Hydroxylamines

Reactivity of Hydrazine and Hydroxylamine Derivatives

Although certain aromatic amines such as 8-aminonaphthalene-1,3,6-trisulfonic acid (ANTS, A350), 2-aminoacridone (A6289) and 8-aminopyrene-1,3,6-trisulfonic acid (APTS, A6257; Figure 3.3.4) have been extensively utilized to modify reducing sugars for analysis and sequencing, the most reactive reagents for forming stable conjugates of aldehydes and ketones are usually hydrazine derivatives, including hydrazides, semicarbazides and carbohydrazides (Figure 3.3.5), as well as hydroxylamine derivatives. Hydrazine derivatives react with ketones to yield relatively stable hydrazones (Figure 3.3.6), and with aldehydes to yield hydrazones that are somewhat less stable, though they may be formed faster. Hydroxylamine derivatives (aminooxy compounds) react with aldehydes and ketones to yield oximes. Oximes are superior to hydrazones with respect to hydrolytic stability.[23] Both hydrazones and oximes can be reduced with sodium borohydride ($NaBH_4$) to further increase the stability of the linkage. Rates and yields of aldehyde reactions with hydrazine and hydroxylamine derivatives are substantially enhanced by aniline catalysis.[24] This chemistry is sufficiently mild and efficient to be applicable for labeling periodate-oxidized sialylated glycoproteins on the surface of live cells.[25]

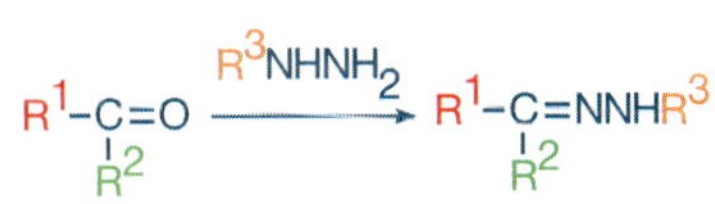

Figure 3.3.6 Modifying aldehydes and ketones with hydrazine derivatives.

Fluorescent Hydrazine and Hydroxylamine Derivatives Excited with Visible Light

We offer a large number of fluorescent hydrazine and hydroxylamine derivatives for reaction with aldehydes or ketones (Table 3.2). Because they are more photostable than the fluorescein derivatives, the Alexa Fluor®, BODIPY® and Texas Red® hydrazides should be among the most sensitive reagents for detecting aldehydes and ketones in laser-excited chromatographic methods.[26] However, with the exception of the Alexa Fluor® 555 and Alexa Fluor® 647 hydrazides and the Alexa Fluor® 647 hydroxylamine, the Alexa Fluor® reagents are mixed isomers and may resolve into multiple peaks when analyzed with high-resolution separation techniques. Fluorescent hydrazides and hydroxylamines are extensively used for labeling glycans via derivatization of aldehydes generated after periodate oxidation or via coupling to the reducing terminus.[27,28] Alexa Fluor® 488 hydroxylamine (A30629, Figure 3.3.7) is particularly useful for detecting aldehyde groups at abasic DNA lesions,[29,30] similar to the biotinylated hydroxylamine ARP described later in this section.

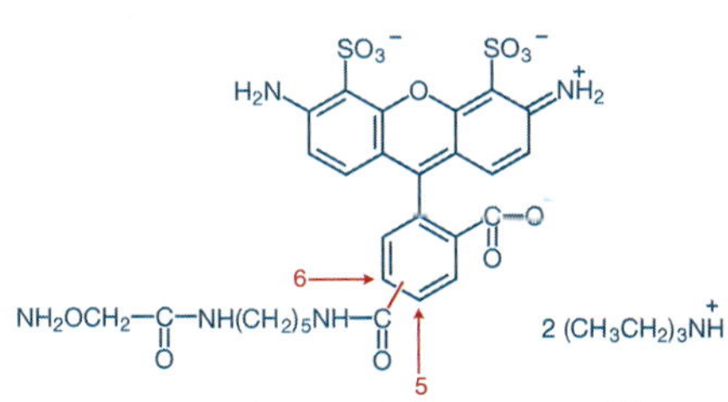

Figure 3.3.7 Alexa Fluor® 488 C_5-aminooxyacetamide, bis(triethylammonium) salt (Alexa Fluor® 488 hydroxylamine, A30629).

Fluorescent Hydrazine and Hydroxylamine Derivatives Excited with UV Light

Dansyl hydrazine (D100) has been by far the most widely used UV light–excitable hydrazine probe for derivatizing aldehydes and ketones for chromatographic analysis and mass spectrometry.[31,32] A unique application that has been reported for dansyl hydrazine, but that is likely a general reaction of hydrazine derivatives, is the detection of *N*-acetylated or *N*-formylated proteins through transfer of the acyl group to the fluorescent hydrazide.[33,34] Although dansyl hydrazine has been widely used as a UV light–excitable derivatization reagent, our 7-diethylaminocoumarin

IMPORTANT NOTICE: The products described in this manual are covered by one or more Limited Use Label License(s). Please refer to the Appendix on page 971 and Master Product List on page 975. Products are For Research Use Only. Not intended for any animal or human therapeutic or diagnostic use.

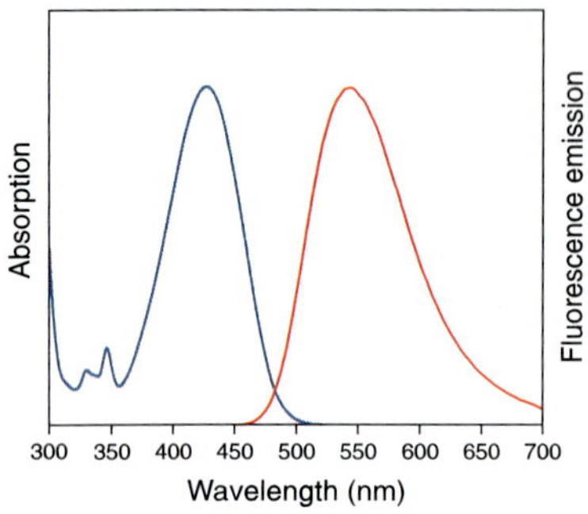

Figure 3.3.8 Absorption and fluorescence emission spectra of lucifer yellow CH in water.

and pyrene hydrazides (D355, P101) have much higher absorptivity and fluorescence, which should make their conjugates more detectable than those of dansyl hydrazine.

Polar Fluorescent Hydrazides and Hydroxylamines

Lucifer yellow CH (L453) is most commonly used as an aldehyde-fixable neuronal tracer with visible absorption and emission (Figure 3.3.8). This membrane-impermeant hydrazide also reacts with periodate-oxidized cell-surface glycoproteins,[35,36] oxidized ribonucleotides[37] and gangliosides.[38] Cascade Blue® hydrazide (C687) exhibits high absorptivity (EC >28,000 $cm^{-1}M^{-1}$), fluorescence quantum yield (0.54) and water solubility[39] (~1%). Like Cascade Blue® hydrazide, Alexa Fluor® 350 hydrazide (A10439) and Alexa Fluor® 350 hydroxylamine (A30627) also have high water solubility and bright blue fluorescence. These sulfonated pyrene and coumarin derivatives have applications similar to those of lucifer yellow CH, including as aldehyde-fixable polar tracers;[40] see Section 14.3 for a more complete discussion of this application.

Cell membrane–impermeant aldehyde- and ketone-reactive reagents are also important probes for assessing the topology of peptide and protein exposure on the surface of live cells. Periodate- or galactose oxidase–mediated oxidation of cell-surface glycoproteins and polysaccharides can be used to selectively introduce aldehyde residues on the cell surface, and these aldehydes can then be reacted with a membrane-impermeant hydrazide. The high polarity of our Alexa Fluor® hydrazides (A10436, A10437, A10438, A10439, A20501MP, A20502, A30634), Alexa Fluor® hydroxylamines

Table 3.2 Molecular Probes® hydrazine, hydroxylamine and amine derivatives.

Derivative	Hydrazines *	Hydroxylamines *	Cadaverines *	Other Amines *
Acridone				A6289 †
Alexa Fluor® 350	A10439	A30627	A30674	
Alexa Fluor® 405			A30675	
Alexa Fluor® 488	A10436	A30629	A30676	
Alexa Fluor® 555	A20501MP		A30677	
Alexa Fluor® 568	A10437		A30680	
Alexa Fluor® 594	A10438		A30678	
Alexa Fluor® 633	A30634			
Alexa Fluor® 647	A20502	A30632	A30679	
Bimane				B30633
Biotin and DSB-X™ biotin ‡	B1603, B2600, D20653	A10550 §	A1594, B1596	A1593, B1592, N6356
t-BOC		B30300 **		M6248
BODIPY® FL	D2371			D2390
BODIPY® TR			D6251	
Cascade Blue®	C687			C621
Coumarin	D355			A191 †
Dansyl	D100		D113	D112
Dapoxyl®				D10460
2,3-Diaminonaphthalene				D7918 †
Dimethoxybenzene				D1463 †
Dinitrophenyl				D1552 *
Fluorescein	C356, F121		A10466	A1351, A1353, A1363
Lucifer yellow	L453		A1340	A1339
Naphthalene				A91, A350 †
NBD	M20490			
Oregon Green® 488			O10465	
Pyrene	P101			A6257 †
QSY® 7				Q10464
Rhodamine			A1318	L2424
Texas Red®	T6256		T2425	

* Hydrazine, hydroxylamine and aromatic amine derivatives are discussed in Section 3.3, and aliphatic amine derivatives are discussed in Section 3.4, except for D1552, which appears in Section 12.3. † Aromatic amines used extensively for modifying aldehydes and ketones. ‡ Biotin and desthiobiotin (DSB-X™ biotin) derivatives are listed in Section 4.2. § ARP is the abbreviation for *N*-(aminooxyacetyl)-*N'*-(D-biotinoyl)hydrazine. ** B30300 requires *t*-BOC deprotection by trifluoroacetic acid to generate free hydroxylamine. More information on reactive dyes is available at www.invitrogen.com/handbook/labelingchemistry.

IMPORTANT NOTICE: The products described in this manual are covered by one or more Limited Use Label License(s). Please refer to the Appendix on page 971 and Master Product List on page 975. Products are For Research Use Only. Not intended for any animal or human therapeutic or diagnostic use.

(A30627, A30629, A30632), lucifer yellow CH (L453) and Cascade Blue® hydrazide (C687) make them the preferred labeling reagents.

NBD Methylhydrazine

NBD methylhydrazine (*N*-methyl-4-hydrazino-7-nitrobenzofurazan, M20490) has been used to monitor aldehydes and ketones in tobacco smoke [41] and automobile exhaust [42] and also to measure nitrite in water [43] (Section 21.2). NBD methylhydrazine reacts with carbonyl compounds in acidic media, forming the corresponding hydrazones (Figure 3.3.9). Following separation by HPLC, the hydrazones can be detected either by spectrophotometry (using wavelengths corresponding to the absorption maxima of the relevant hydrazone) or by fluorescence spectroscopy using excitation/emission at ~470/560 nm.

Biotin Hydrazides and Biotin Hydroxylamine

In addition to the fluorescent hydrazine and hydroxylamine derivatives, we offer several nonfluorescent biotin and DSB-X™ biotin hydrazides (B1603, B2600, D20653; Section 4.2) and the biotin hydroxylamine derivative ARP (A10550, Section 4.2), each of which can be detected using fluorescent dye– or enzyme-labeled avidin or streptavidin (Section 7.6, Table 7.9). DSB-X™ biotin hydrazide, which has moderate affinity for avidin and streptavidin that is rapidly reversed by low concentrations of free biotin, can be used to produce a DSB-X™ biotin–labeled molecule that reversibly binds avidin or streptavidin affinity matrices (Section 7.6).

We recommend the biotin hydroxylamine derivative ARP (aldehyde-reactive probe, A10550; Figure 3.3.10) as our most efficient reagent for incorporating biotins into aldehyde- or ketone-containing cell surfaces. ARP has been used extensively to modify the exposed aldehyde group at abasic lesions in DNA [44,45] (Figure 3.3.11). A quick and sensitive microplate assay for abasic sites can be performed using ARP.[46] In addition, ARP is membrane permeant, permitting detection of abasic sites in live cells.[47,48] Once the aldehyde groups in abasic sites are modified by ARP and the cells are fixed and permeabilized, the resulting biotinylated DNA can be detected with fluorescent dye–, Qdot® nanocrystal– or enzyme-conjugated streptavidin conjugates (Section 7.6, Table 7.9). Likewise, ARP can be used to detect and capture 4-hydroxynonenal (HNE)–modified proteins.[49] ARP has also been used to immobilize IgG antibodies on streptavidin-coated monolayer surfaces with their binding sites oriented toward the solution phase.[50] An alternative to ARP for detection of protein carbonyls is dinitrophenylhydrazine derivatization followed by immunolabeling with our Alexa Fluor® 488 dye–labeled anti-dinitrophenyl antibody [51] (A11097, Section 7.4).

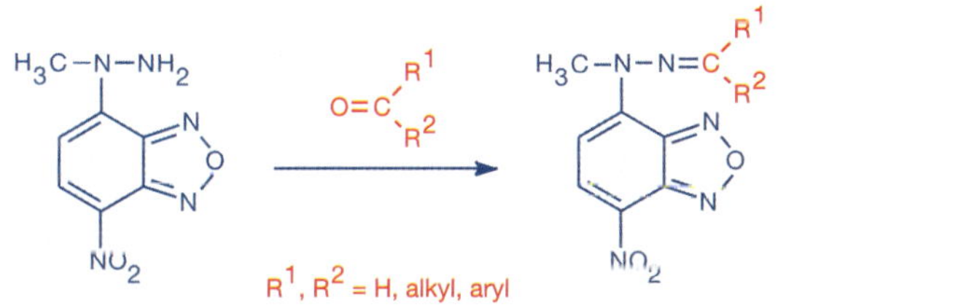

Figure 3.3.9 Reaction scheme illustrating the principle of ketone and aldehyde detection by NBD methylhydrazine (M20490).

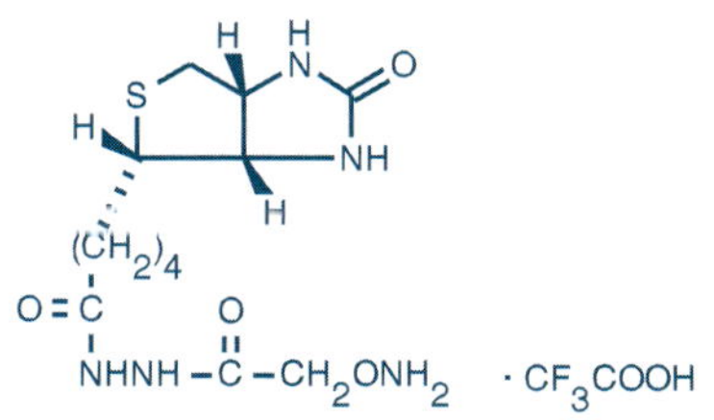

Figure 3.3.10 *N*-(aminooxyacetyl)-*N*′-(D-biotinoyl) hydrazine, trifluoroacetic acid salt (ARP, A10550).

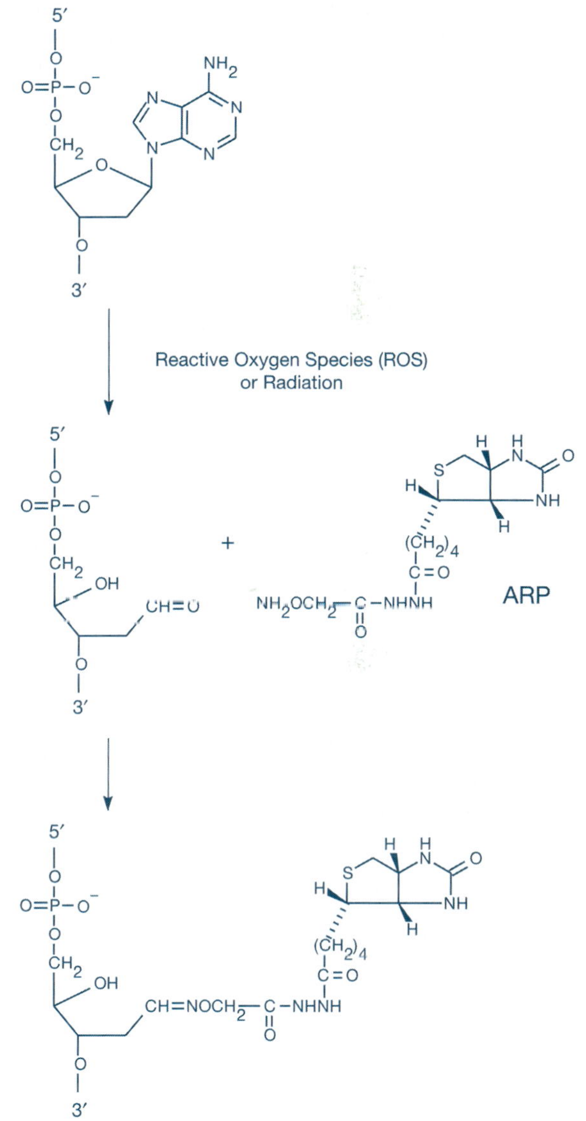

Figure 3.3.11 Aldehyde-reactive probe (ARP) used to detect DNA damage. The biotin hydroxylamine ARP (A10550) reacts with aldehyde groups formed when reactive oxygen species depurinate DNA. This reaction forms a covalent bond linking the DNA to biotin. The biotin can then be detected using fluorophore- or enzyme-linked streptavidin.

IMPORTANT NOTICE: The products described in this manual are covered by one or more Limited Use Label License(s). Please refer to the Appendix on page 971 and Master Product List on page 975. Products are For Research Use Only. Not intended for any animal or human therapeutic or diagnostic use.

Aliphatic and Aromatic Amines

Primary aliphatic and aromatic amines (Table 3.2) can be coupled reversibly to aldehydes and ketones to form hydrolytically unstable Schiff bases [11] (Figure 3.3.12). The reversibility of this modification makes reagents that contain amines less desirable unless the Schiff base is reduced by sodium borohydride [52,53] or sodium cyanoborohydride.[54] Chemical reduction also retains the amine's original charge. Sequencing of carbohydrate polymers using fluorescent derivatives has usually relied on derivatization of the reducing end of the polymer with a fluorescent amine.[55] Certain aromatic amines have been extensively utilized for coupling to aldehydes, ketones, monosaccharides and the reducing end of carbohydrate polymers:

- 2-Aminoacridone (A6289) forms conjugates that can be separated by HPLC [56,57] or, as their borate complexes, by polyacrylamide gel electrophoresis,[55,58,59] capillary electrophoresis [60–62] and micellar electrokinetic capillary chromatography [63,64] (MECC). Starting with as little as 25 µg of a glycoprotein, researchers have efficiently released and purified the carbohydrates, and then derivatized them with 2-aminoacridone for subsequent structural analysis.[65] 2-Aminoacridone derivatives of oligosaccharides have been directly analyzed by MALDI-TOF mass spectrometry.[66] 2-Aminoacridone is also used to prepare fluorogenic substrates for proteases.[67]
- 7-Amino-4-methylcoumarin (A191), which is a common base of protease substrates (Section 10.4), can be used for the reductive derivatization of oligosaccharides.[68,69]
- 8-Aminopyrene-1,3,6-trisulfonic acid (APTS, A6257) has been extensively used to derivatize carbohydrates prior to separation by gel or capillary electrophoresis.[70,71] Among the amines we offer, APTS is the aromatic amine that has the most favorable combination of strong absorbance, high quantum yield and ionic charge.
- ANTS (A350) has a high ionic charge, permitting electrophoretic separation of its products with complex oligosaccharides.[72]

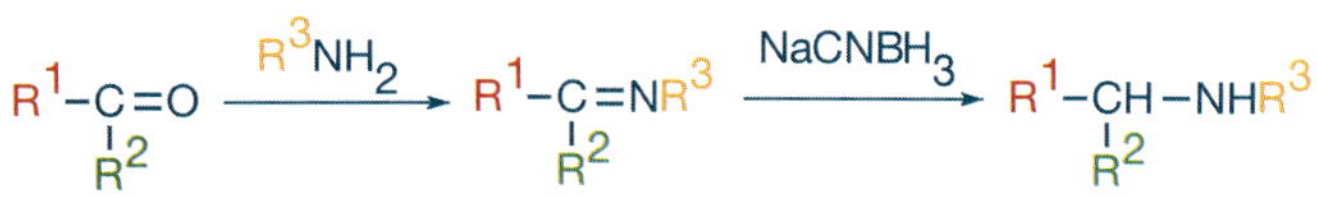

Figure 3.3.12 Modifying aldehydes and ketones with amine derivatives.

The aromatic diamine 1,2-diamino-4,5-dimethoxybenzene (DDB, D1463), which forms heterocyclic compounds with certain aldehydes and ketones, has been used to selectively detect aromatic aldehydes in the presence of aliphatic aldehydes, including carbohydrates.[73] DBB has proven to be a useful reagent for HPLC analysis of the cytotoxic metabolic by-product methylglyoxal in blood samples from diabetic patients.[74]

Alternatively, aldehydes and ketones can be transformed into primary aliphatic amines by reductive amination with ammonia, ethylenediamine or other nonfluorescent diamines.[75] This chemistry is particularly useful because the products can then be coupled with any of the amine-reactive reagents described in Chapter 1 such as the succinimidyl esters of TAMRA dye [76] (C1171, C6121, C6122; Section 1.6). Derivatization by succinimidyl esters has been extensively utilized for tagging oligosaccharides that are to be separated by capillary zone electrophoresis with laser-induced fluorescence detection.[76–79]

REFERENCES

1. J Biomed Mater Res A (2007) 83:1176; **2.** Anal Biochem (1987) 161:245; **3.** Electrophoresis (2003) 24:588; **4.** Nat Struct Mol Biol (2008) 15:494; **5.** Proc Natl Acad Sci U S A (2004) 101:7902; **6.** Bioconjug Chem (1994) 5:436; **7.** Bioconjug Chem (2003) 14:205; **8.** Bioconjug Chem (1992) 3:262; **9.** Bioconjug Chem (1994) 5:636; **10.** J Immunol (2009) 183:1551; **11.** Methods Enzymol (1994) 247:30; **12.** Anal Biochem (1988) 170:271; **13.** Methods Enzymol (1987) 138:429; **14.** Biochem Biophys Res Commun (1980) 92:1215; **15.** J Supramol Struct (1977) 6:291; **16.** Bioconjug Chem (2009) 20:1228; **17.** Bioconjug Chem (2009) 20:1383; **18.** Cytometry (2000) 41:316; **19.** ACS Chem Biol (2007) 2:247; **20.** Mol Ther (2008) 16:1467; **21.** Proc Natl Acad Sci U S A (2009) 106:3000; **22.** Nature (1981) 292:17; **23.** Angew Chem Int Ed Engl (2008) 47:7523; **24.** Bioconjug Chem (2008) 19:2543; **25.** Nat Methods (2009) 6:207; **26.** Electrophoresis (2007) 29:475; **27.** BMC Urol (2005) 5:4; **28.** J Am Chem Soc (2009) 131:10360; **29.** Anal Bioanal Chem (2007) 387:1883; **30.** Angew Chem Int Ed Engl (2007) 46:561; **31.** Anal Chem (2010) 82:2893; **32.** J Chromatogr A (2002) 979:409; **33.** J Cell Biol (1988) 106:1607; **34.** Anal Biochem (1969) 29:186; **35.** Biochemistry (1985) 24:322; **36.** Biochem Biophys Res Commun (1983) 112:872; **37.** Biochemistry (1988) 27:6039; **38.** J Cell Biol (1985) 100:721; **39.** Anal Biochem (1991) 198:119; **40.** Physiol Res (1997) 46:407; **41.** Fresenius J Anal Chem (2000) 366:396; **42.** Anal Chem (1999) 71:1893; **43.** Anal Chem (1999) 71:3003; **44.** Photochem Photobiol (2002) 76:123; **45.** Anal Chem (2001) 73:2229; **46.** Methods (2000) 22:164; **47.** Proc Natl Acad Sci U S A (2000) 97:686; **48.** J Biol Chem (2000) 275:6741; **49.** Anal Chem (2006) 78:6847; **50.** Anal Biochem (2003) 312:113; **51.** Anal Bioanal Chem (2008) 391:2591; **52.** Biochemistry (1987) 26:2162; **53.** Biochim Biophys Acta (1980) 597:285; **54.** Biochim Biophys Acta (1981) 670:181; **55.** Anal Biochem (1994) 222:270; **56.** Anal Chem (1998) 70:2530; **57.** Anal Chem (1997) 69:4985; **58.** Anal Biochem (1994) 216:243; **59.** Anal Biochem (1991) 196:238; **60.** Anal Biochem (1996) 240:68; **61.** Electrophoresis (1996) 17:406; **62.** Anal Biochem (1995) 230:115; **63.** Anal Chem (1996) 68:4424; **64.** J Chem Soc Chem Commun (1994) 14:1691; **65.** Anal Biochem (1998) 262:197; **66.** Rapid Commun Mass Spectrom (1997) 11:1635; **67.** Anal Biochem (1988) 171:393; **68.** Biophys J (1996) 71:2040; **69.** Anal Biochem (1983) 128:41; **70.** Nat Protoc (2006) 1:397; **71.** Electrophoresis (2005) 26:2034; **72.** Anal Biochem (2000) 283:136; **73.** J Chromatogr B Biomed Appl (1994) 659:85; **74.** Diabetes Metab (2006) 32:176; **75.** Proc Natl Acad Sci U S A (1991) 88:2302; **76.** J Chromatogr B Biomed Appl (1994) 657:307; **77.** Anal Biochem (1995) 227:368; **78.** Carbohydr Res (1996) 296:203; **79.** J Biomol Screen (1999) 4:239.

IMPORTANT NOTICE: The products described in this manual are covered by one or more Limited Use Label License(s). Please refer to the Appendix on page 971 and Master Product List on page 975. Products are For Research Use Only. Not intended for any animal or human therapeutic or diagnostic use.

molecular probes® | invitrogen™ by life technologies™

DATA TABLE 3.3 REAGENTS FOR MODIFYING ALDEHYDES AND KETONES

Cat. No.	MW	Storage	Soluble	Abs	EC	Em	Solvent	Notes
A191	175.19	L	DMF, DMSO	351	18,000	430	MeOH	
A350	427.33	L	H_2O	353	7200	520	H_2O	
A6257	523.39	D,L	H_2O	424	19,000	505	pH 7	
A6289	246.70	D,L	DMF, DMSO	425	5200	531	MeOH	1
A10436	570.48	D,L	H_2O	493	71,000	517	pH 7	
A10437	730.74	D,L	H_2O	576	86,000	599	pH 7	2
A10438	758.79	D,L	H_2O	588	97,000	613	pH 7	2
A10439	349.29	L	H_2O, DMSO	345	13,000	445	pH 7	
A20501MP	~1150	D,L	H_2O	554	150,000	567	pH 7	
A20502	~1200	D,L	H_2O	649	250,000	666	pH 7	
A30627	584.52	F,D,L	H_2O, DMSO	353	20,000	437	MeOH	3
A30629	895.07	F,D,L	H_2O, DMSO	494	77,000	518	pH 7	3, 4, 5
A30632	~1220	F,D,L	H_2O, DMSO	651	250,000	672	MeOH	3
A30634	~950	D,L	H_2O, DMSO	624	110,000	643	pH 7	
B30300	339.24	F,D	DMSO	<300		none		
C356	493.49	L	pH >7, DMF	492	78,000	516	pH 8	6
C687	596.44	L	H_2O	399	30,000	421	H_2O	7, 8
D100	265.33	L	EtOH	336	4400	534	MeOH	
D355	275.31	D,L	MeCN, DMF	420	46,000	468	MeOH	
D1463	241.12	D,L	EtOH	298	3100	359	MeOH	
D2371	306.12	F,D,L	MeOH, MeCN	503	71,000	510	MeOH	9
D7918	158.20	L	DMSO, MeOH	340	5100	377	MeOH	10
F121	421.43	D,L	pH >7, DMF	492	85,000	516	pH 9	6
L453	457.24	L	H_2O	428	12,000	536	H_2O	11, 12
M20490	209.16	F,L	MeCN	487	24,000	none	MeOH	13
P101	302.38	D,L	MeCN, DMF	341	43,000	376	MeOH	14
T6256	620.74	F,L	DMF	582	109,000	602	MeOH	

For definitions of the contents of this data table, see "Using *The Molecular Probes® Handbook*" in the introductory pages.

Notes

1. Spectra of this compound are in methanol containing a trace of KOH.
2. Maximum solubility in water is ~8% for A10437 and A10438.
3. Aqueous stock solutions should be used within 24 hours; long-term storage is NOT recommended.
4. The fluorescence lifetime (τ) of the Alexa Fluor® 488 dye in pH 7.4 buffer at 20°C is 4.1 nanoseconds. Data provided by the SPEX Fluorescence Group, Horiba Jobin Yvon Inc.
5. Abs and Em of the Alexa Fluor® 488 dye are red-shifted by as much as 16 nm and 25 nm respectively on microarrays relative to aqueous solution values. The magnitude of the spectral shift depends on the array substrate material. (Biotechniques (2005) 38:127)
6. Absorption and fluorescence of fluorescein derivatives are pH dependent. Extinction coefficients and fluorescence quantum yields decrease markedly at pH <7.
7. The Alexa Fluor® 405 and Cascade Blue® dyes have a second absorption peak at about 376 nm with EC ~80% of the 395–400 nm peak.
8. Maximum solubility in water is ~1% for C687, ~1% for C3221 and ~8% for C3239.
9. The absorption and fluorescence spectra of BODIPY® derivatives are relatively insensitive to the solvent.
10. Fluorescence of D7918 is weak. Reaction with α-ketoaldehydes yields fluorescent benzoquinoxaline derivatives (Abs = 365 nm, Em = 540 nm in H_2O at pH 8). (J Chromatogr B Biomed Sci Appl (1999) 729:237)
11. The fluorescence quantum yield of lucifer yellow CH in H_2O is 0.21. (J Am Chem Soc (1981) 103:7615)
12. Maximum solubility in water is ~8% for L453, ~6% for L682 and ~1% for L1177.
13. NBD methylhydrazine reacts with aldehydes and ketones in the presence of strong acid, yielding weakly fluorescent hydrazone products. (Anal Chem (1999) 71:1893) Abs = 493 nm, Em = 552 nm in MeOH for reaction product with acetone.
14. Pyrene derivatives exhibit structured spectra. The absorption maximum is usually about 340 nm with a subsidiary peak at about 325 nm. There are also strong absorption peaks below 300 nm. The emission maximum is usually about 376 nm with a subsidiary peak at 396 nm. Excimer emission at about 470 nm may be observed at high concentrations.

IMPORTANT NOTICE: The products described in this manual are covered by one or more Limited Use Label License(s). Please refer to the Appendix on page 971 and Master Product List on page 975. Products are For Research Use Only. Not intended for any animal or human therapeutic or diagnostic use.

PRODUCT LIST 3.3 REAGENTS FOR MODIFYING ALDEHYDES AND KETONES

Cat. No.	Product	Quantity
A30627	Alexa Fluor® 350 C_5-aminooxyacetamide, trifluoroacetate salt (Alexa Fluor® 350 hydroxylamine)	1 mg
A30629	Alexa Fluor® 488 C_5-aminooxyacetamide, bis(triethylammonium) salt (Alexa Fluor® 488 hydroxylamine)	1 mg
A30632	Alexa Fluor® 647 C_5-aminooxyacetamide, bis(triethylammonium) salt (Alexa Fluor® 647 hydroxylamine)	1 mg
A10439	Alexa Fluor® 350 hydrazide, sodium salt	5 mg
A10436	Alexa Fluor® 488 hydrazide, sodium salt	1 mg
A20501MP	Alexa Fluor® 555 hydrazide, tris(triethylammonium) salt	1 mg
A10437	Alexa Fluor® 568 hydrazide, sodium salt	1 mg
A10438	Alexa Fluor® 594 hydrazide, sodium salt	1 mg
A30634	Alexa Fluor® 633 hydrazide, bis(triethylammonium) salt	1 mg
A20502	Alexa Fluor® 647 hydrazide, tris(triethylammonium) salt	1 mg
A6289	2-aminoacridone, hydrochloride	25 mg
A191	7-amino-4-methylcoumarin *reference standard*	100 mg
A350	8-aminonaphthalene-1,3,6-trisulfonic acid, disodium salt (ANTS)	1 g
A6257	8-aminopyrene-1,3,6-trisulfonic acid, trisodium salt (APTS)	10 mg
B30300	*N*-(*t*-BOC)-aminooxyacetic acid, tetrafluorophenyl ester	25 mg
C356	5-(((2-(carbohydrazino)methyl)thio)acetyl)aminofluorescein	25 mg
C687	Cascade Blue® hydrazide, trisodium salt	10 mg
D1463	1,2-diamino-4,5-dimethoxybenzene, dihydrochloride (DDB)	100 mg
D7918	2,3-diaminonaphthalene	100 mg
D355	7-diethylaminocoumarin-3-carboxylic acid, hydrazide (DCCH)	25 mg
D2371	4,4-difluoro-5,7-dimethyl-4-bora-3a,4a-diaza-*s*-indacene-3-propionic acid, hydrazide (BODIPY® FL hydrazide)	5 mg
D100	5-dimethylaminonaphthalene-1-sulfonyl hydrazine (dansyl hydrazine)	100 mg
F121	fluorescein-5-thiosemicarbazide	100 mg
L453	lucifer yellow CH, lithium salt	25 mg
M20490	*N*-methyl-4-hydrazino-7-nitrobenzofurazan (NBD methylhydrazine)	25 mg
P101	1-pyrenebutanoic acid, hydrazide	100 mg
T6256	Texas Red® hydrazide *>90% single isomer*	5 mg

IMPORTANT NOTICE: The products described in this manual are covered by one or more Limited Use Label License(s). Please refer to the Appendix on page 971 and Master Product List on page 975. Products are For Research Use Only. Not intended for any animal or human therapeutic or diagnostic use.

3.4 Derivatization Reagents for Carboxylic Acids and Carboxamides

Carboxylic acids can be converted to esters, amides, acyl hydrazides or hydroxamic acids, all of which are discussed in this section. Alternatively, the half-protected *tert*-butyloxycarbonyl (*t*-BOC) propylenediamine derivative (M6248) is useful for converting organic solvent–soluble carboxylic acids into aliphatic amines. Following coupling of the half-protected aliphatic diamine to an activated carboxylic acid, the *t*-BOC group can be quantitatively removed with trifluoroacetic acid (Figure 3.4.1). The resultant aliphatic amine can then be modified with any of the amine-reactive reagents described in Chapter 1 or coupled to solid-phase matrices for affinity chromatography.

Coupling Hydrazines, Hydroxylamines and Amines to Carboxylic Acids

Modification in Aqueous Solutions

The carboxylic acids of water-soluble biopolymers such as proteins can be coupled to hydrazines, hydroxylamines and amines (Table 3.2) in aqueous solution using water-soluble carbodiimides such as 1-ethyl-3-(3-dimethylaminopropyl)carbodiimide (EDAC, E2247). Including *N*-hydroxysulfosuccinimide (H2249) in the reaction mixture has been shown to improve the coupling efficiency of EDAC-mediated protein–carboxylic acid conjugations [1,2] (Figure 3.4.2). To reduce intra- and inter-protein coupling to lysine residues,[3] which is a common side reaction, carbodiimide-mediated coupling should be performed in a concentrated protein solution at a low pH, using a large excess of the nucleophile. EDAC-mediated coupling has been used to functionalize Qdot® 605 ITK™ carboxyl quantum dots (Q21301MP, Section 6.6) with the GABA receptor agonist muscimol.[4] EDAC has been shown to be impermeable to membranes of live cells, which permits its use to distinguish between cytoplasmic and lumenal sites of reaction.[5]

Fluoresceinyl glycine amide (5-(aminoacetamido)fluorescein, A1363) and various hydrazines and hydroxylamines may be the best probes for this application because they are more likely to remain reactive at a lower pH than are aliphatic amines such as the cadaverines.[6] Fluoresceinyl glycine amide has been coupled to the carboxylic acid of a cyclosporin derivative by EDAC.[7]

ANTS (8-aminonaphthalene-1,3,6-trisulfonic acid, A350; Section 3.3) has a high ionic charge, which permits electrophoretic separation of its products with complex oligosaccharides.[8] Several of the fluorescent hydrazine and hydroxylamine derivatives described in Section 3.3 should have similar utility for carbodiimide-mediated derivatization of carboxylic acids.

Modification in Organic Solvents

Peptide synthesis research has led to the development of numerous methods for coupling carboxylic acids to amines in organic solution. One such method involves the conversion of carboxylic acids to succinimidyl esters or mixed anhydrides. Dicyclohexylcarbodiimide and diisopropylcarbodiimide are widely used to promote amide formation in organic solution. Another recommended derivatization method for coupling organic solvent–soluble carboxylic acids, including

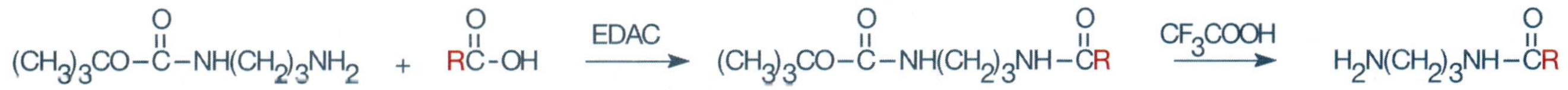

Figure 3.4.1 Conversion of a carboxylic acid group into an aliphatic amine. The activated carboxylic acid is derivatized with a half-protected aliphatic diamine (mono-*N*-(*t*-BOC)-propylenediamine, M6248), usually in an organic solvent, followed by removal of the *t*-BOC–protecting group with trifluoroacetic acid.

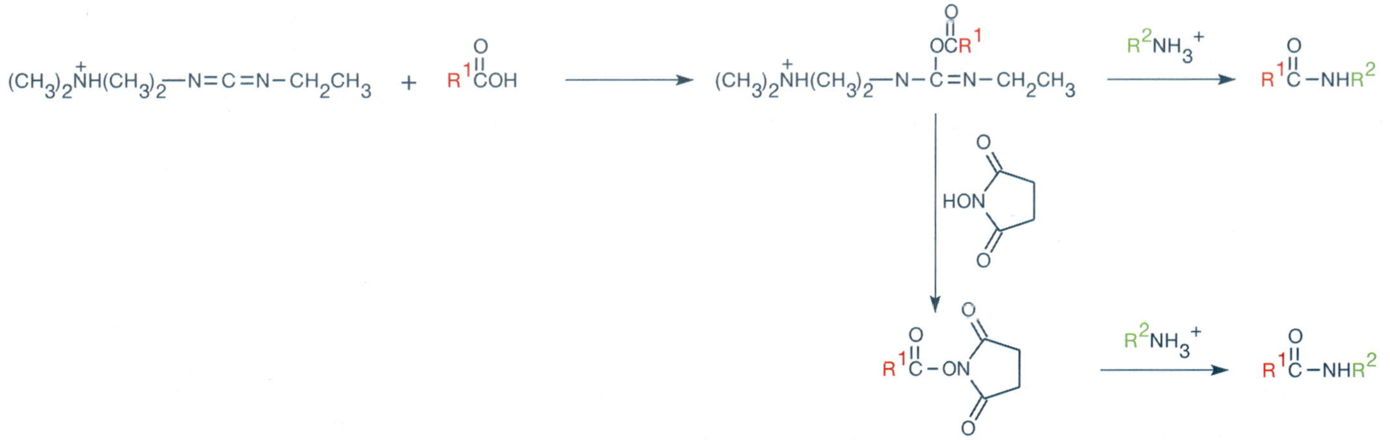

Figure 3.4.2 Stabilization of an unstable *O*-acylisourea intermediate by *N*-hydroxysuccinimide in a carbodiimide-mediated (EDAC, E2247) modification of a carboxylic acid with a primary amine.

peptides, to aliphatic amines without racemization is the combination of 2,2′-dipyridyldisulfide and triphenylphosphine.[9,10] Unlike fluorescent aliphatic amines, fluorescent aromatic amines such as those derived from 7-amino-4-methylcoumarin (A191) and 2-aminoacridone (A6289, Section 3.3) exhibit a shift in their absorption and emission (if any) to much shorter wavelengths upon forming carboxamides. This property makes these aromatic amines preferred reagents for preparing peptidase substrates (Section 10.4). Aromatic amines can generally be coupled to acid halides and anhydrides, with organic solvents usually required for efficient reaction.

Hydrazine, Hydroxylamine and Aliphatic Amine Derivatives

We provide a wide selection of carboxylic acid–reactive reagents (Table 3.2), including several different Dapoxyl®, Alexa Fluor®, BODIPY®, fluorescein, Oregon Green®, rhodamine, Texas Red® and QSY® hydrazine derivatives, hydroxylamine derivatives and amine derivatives, all of which are particularly useful for synthesizing drug analogs and as probes for fluorescence polarization immunoassays[11–13] (Fluorescence Polarization (FP)—Note 1.4). These probes all require a coupling agent such as a carbodiimide to react with carboxylic acids; they do not spontaneously react with carboxylic acids in solution. They do, however, react spontaneously with the common amine-reactive functional groups described in Section 1.1, including succinimidyl esters and isothiocyanates. Some of the more important probes and their applications include:

- Alexa Fluor® hydrazides (A10436, A10437, A10438, A10439, A30634, A20501MP, A20502; Section 3.3), Alexa Fluor® hydroxylamines (A30627, A30629, A30632; Section 3.3) and Alexa Fluor® cadaverines (A30674, A30675, A30676, A30677, A30678, A30679, A30680), our brightest and most photostable carboxylic acid–reactive probes
- BODIPY® aliphatic amines (D2390, D6251), for preparing pH-insensitive probes, such as BODIPY® FL etoposide,[14] from carboxylic acid derivatives
- Isomeric aminomethylfluoresceins (A1351, A1353), which are readily coupled to activated carboxylic acids[15]
- Dapoxyl® (2-aminoethyl)sulfonamide (D10460) for preparing conjugates with strong UV absorption and a Stokes shift of ~200 nm (Figure 3.4.3)
- Dansyl ethylenediamine (D112), dansyl cadaverine (D113), Dapoxyl® (2-aminoethyl)sulfonamide (D10460) and Lissamine rhodamine B ethylenediamine (L2424), for carboxylic acid derivatization[16] and glutamine transamidation reactions[17] (Figure 3.4.4)
- Bimane amine (B30633), a small blue-fluorescent dye for carboxylic acid derivatization
- EDANS (A91), for preparing radioactive IAEDANS,[13,18] energy transfer–quenched substrates for endopeptidases[19] (Section 10.4) and an ATP substrate analog for DNA-dependent RNA polymerase[20]
- QSY® 7 amine (Q10464, Figure 3.4.5), which is an essentially nonfluorescent dye with strong visible absorption for preparing highly efficient quenchers for bioassays based on fluorescence resonance energy transfer (FRET) (Fluorescence Resonance Energy Transfer (FRET)—Note 1.2)
- Hydrazine (Section 3.3) and amine derivatives of lucifer yellow (A1339), Alexa Fluor® 405 (A30675) and Cascade Blue® (C621) dyes, which are precursors of highly fluorescent, water-soluble probes
- Hydrazine and amine derivatives of biotin and desthiobiotin (Section 4.2), which are versatile intermediates for synthesizing biotin- and desthiobiotin-containing probes[21,22]

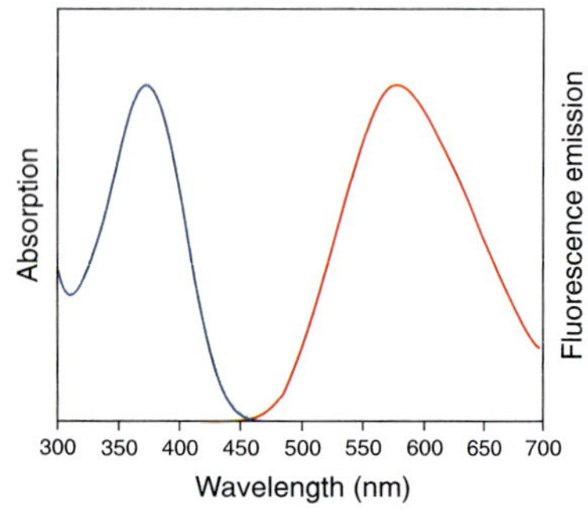

Figure 3.4.3 Absorption and fluorescence emission spectra of Dapoxyl® (2-aminoethyl) sulfonamide in methanol.

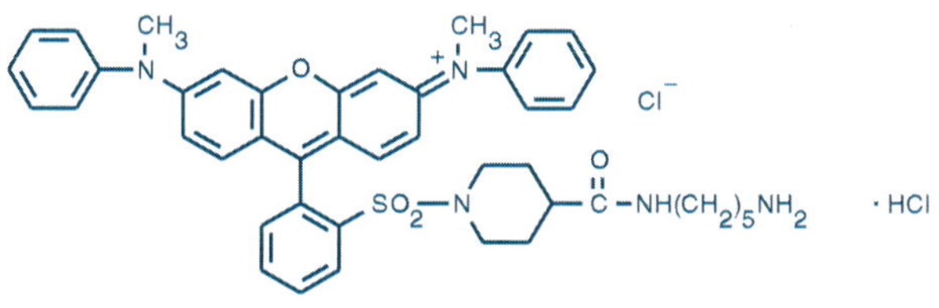

Figure 3.4.5 QSY® 7 amine, hydrochloride (Q10464).

Figure 3.4.4 Transglutaminase-mediated labeling of a protein using dansyl cadaverine (D113).

IMPORTANT NOTICE: The products described in this manual are covered by one or more Limited Use Label License(s). Please refer to the Appendix on page 971 and Master Product List on page 975. Products are For Research Use Only. Not intended for any animal or human therapeutic or diagnostic use.
www.invitrogen.com/probes
molecular probes® | invitrogen™ by life technologies™

Enzyme-Catalyzed Transamidation

A transglutaminase-catalyzed transamidation reaction of glutamine residues in some proteins and peptides enables their selective modification by amine-containing probes [23,24] (Figure 3.4.4). This unique method for selective protein modification requires formation of a complex consisting of the glutamine residue, the aliphatic amine probe and the enzyme. It has been found that a short aliphatic spacer in the amine probe enhances the reaction. The cadaverine ($-NH(CH_2)_5NH-$) spacer is usually optimal. Although dansyl cadaverine (D113) has been probably the most widely used reagent,[25–27] Alexa Fluor® cadaverines [24] (A30674, A30675, A30676, A30677, A30678, A30679, A30680), Oregon Green® 488 cadaverine (O10465), fluorescein cadaverine [28–33] (A10466), tetramethylrhodamine cadaverine [34] (A1318), Texas Red® cadaverine (T2425) and BODIPY® TR cadaverine (D6251) are among the most fluorescent transglutaminase substrates available. The intrinsic transglutaminase activity in sea urchin eggs has been used to covalently incorporate dansyl cadaverine during embryonic development.[35] Two biotin cadaverines (A1594, B1596; Section 4.2) are also available for transglutaminase-mediated reactions.[36–38] Amine-terminated peptides and fluorescent and biotin hydrazides, including Cascade Blue® hydrazide, have been successfully incorporated into protein fragments by transamidation during enzyme-catalyzed proteolysis.[39]

Transamidation of cell-surface glutamine residues by the combination of a transglutaminase enzyme and a fluorescent or biotinylated aliphatic amine can form stable amides.[27] Impermeability of the enzyme restricts this reaction to a limited number of proteins on the cell surface. This technique was used to selectively label erythrocyte band 3 protein with dansyl cadaverine (D113) and proteins of the extracellular matrix with fluorescein cadaverine [31,32] (A10466). Following protease treatment, the dansylated peptides were isolated using an anti-dansyl affinity column.[40]

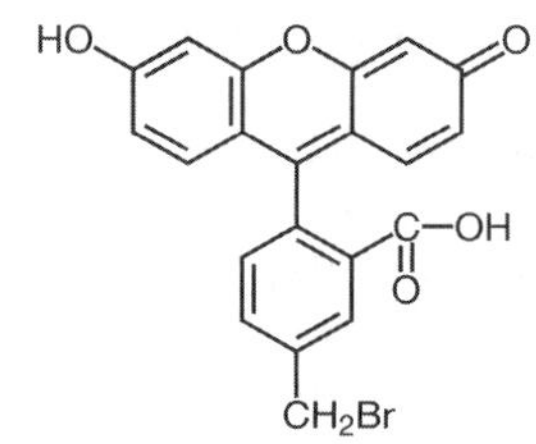

Figure 3.4.6 5-(Bromomethyl)fluorescein (B1355).

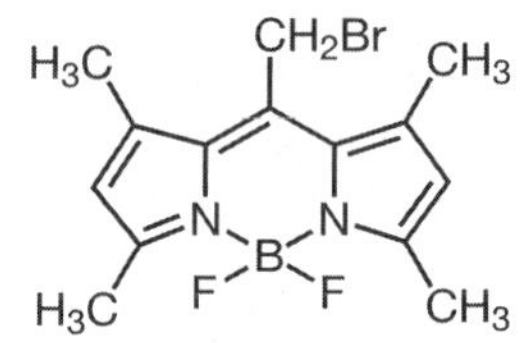

Figure 3.4.7 BODIPY® 493/503 methyl bromide (8-bromomethyl-4,4-difluoro-1,3,5,7-tetramethyl-4-bora-3a,4a-diaza-s-indacene, B2103).

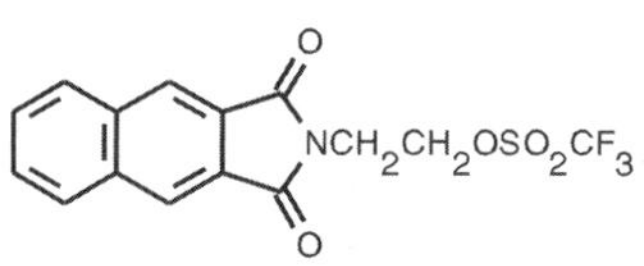

Figure 3.4.8 2-(2,3-Naphthalimino)ethyl trifluoromethanesulfonate (N2461).

Esterification of Carboxylic Acids with Fluorescent Diazoalkanes

Biologically important molecules, especially the nonchromophoric fatty acids, bile acids and prostaglandins, are typically esterified by carboxylic acid–reactive reagents in organic solvents. Esterification of carboxylic acids in aqueous solution is usually not possible, and esters tend to be unstable in water. Fluorescent derivatization reagents for biomedical chromatography have been extensively discussed in reviews.[41,42]

HPLC derivatization reagents for carboxylic acids include two fluorescent analogs of the common esterification reagent diazomethane. Diazoalkanes react without the addition of catalysts and may be useful for direct carboxylic acid modification of proteins and synthetic polymers. Fluorescent diazoalkanes also react with phosphates [43] and potentially with lipid-associated carboxylic acids in membrane-bound proteins or with free fatty acids.

The fluorescent diazomethyl derivative 9-anthryldiazomethane (ADAM, A1400) has been commonly used to derivatize biomolecules. Unfortunately, ADAM is not very stable and may decompose during storage. 1-Pyrenyldiazomethane [44–47] (PDAM, P1405) is recommended as a replacement for ADAM because it has much better chemical stability. Moreover, the detection limit for PDAM conjugates is reported to be about 20–30 femtomoles, which is five times better than reported for detection of ADAM conjugates.[45] In addition, fatty acids derivatized with these reagents have been used to measure phospholipase A_2 activity [48] (Section 17.4). It has been reported that photolysis of pyrenemethyl esters liberates the free carboxylic acid,[47] making PDAM a potential protecting group for carboxylic acids.

Fluorescent Alkyl Halides

The low nucleophilicity of carboxylic acids requires that they be converted to anions (typically cesium or quaternary ammonium are used as counterions) before they can be esterified with alkyl halides in organic solvents. Conjugates of 6-bromoacetyl-2-dimethylaminonaphthalene (badan, B6057) have a high Stokes shift, as well as spectral properties that are very sensitive

IMPORTANT NOTICE: The products described in this manual are covered by one or more Limited Use Label License(s). Please refer to the Appendix on page 971 and Master Product List on page 975. Products are For Research Use Only. Not intended for any animal or human therapeutic or diagnostic use.

to their environment. 5-(Bromomethyl)fluorescein (B1355, Figure 3.4.6) and BODIPY® 493/503 methyl bromide (B2103, Figure 3.4.7) have the strongest absorptivity and fluorescence of the currently available carboxylic acid–derivatization reagents.[49] An analytical method for estimating the degree of EDAC crosslinking of collagen has been developed based on derivatization of residual carboxyl groups by 5-bromomethylfluorescein.[50]

All of the alkyl halides in this section also react with thiol groups, including those in proteins.[51,52] Although more commonly used as thiol-reactive reagents, the monobromobimanes (M1378, M20381; Section 2.3) have been reported to react with carboxylic acids in organic solvents.[53] The coumarin iodoacetamide DCIA (D404, Section 2.3) has also been used to derivatize carboxylic acids;[54] other iodoacetamides described in Chapter 2 will probably react similarly.

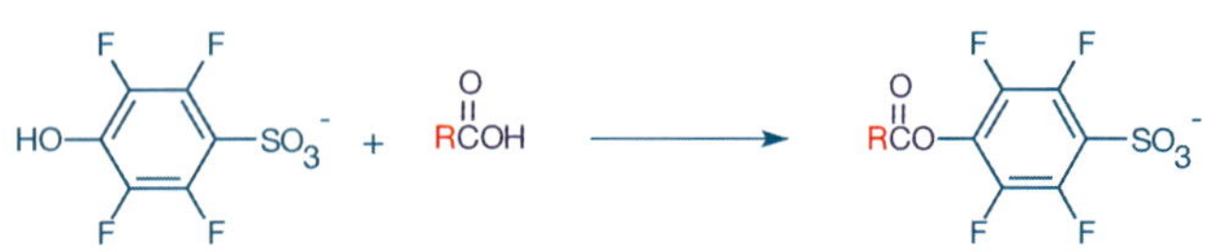

Figure 3.4.9 4-Sulfo-2,3,5,6-tetrafluorophenol (STP, S10490) can be used to prepare water-soluble activated esters from various carboxylic acids.

Fluorescent Trifluoromethanesulfonate

2-(2,3-Naphthalimino)ethyl trifluoromethanesulfonate (N2461, Figure 3.4.8) reacts rapidly with the anions of carboxylic acids in acetonitrile to give adducts that are reported to be detectable by absorption at 259 nm down to 100 femtomoles and by fluorescence at 394 nm down to 4 femtomoles.[55] This naphthalimide sulfonate ester will likely react with other nucleophiles too, including thiols, amines, phenols (e.g., tyrosine) and probably histidine. 2-(2,3-Naphthalimino)ethyl trifluoromethanesulfonate has been used for the sensitive reverse-phase HPLC detection of eicosanoids in brain tissue.[56]

4-Sulfo-2,3,5,6-Tetrafluorophenol (STP) and *N*-Hydroxysulfosuccinimide (NHSS)

4-Sulfo-2,3,5,6-tetrafluorophenol (STP, S10490) and *N*-hydroxysulfosuccinimide (NHSS, H2249) can be used to prepare water-soluble activated esters from various carboxylic acids (Figure 3.4.9). Coupling typically involves a carbodiimide such as EDAC (E2247) and is performed in an organic solvent. We have found that the resulting STP esters are much easier to purify and more stable than activated esters prepared from *N*-hydroxysulfosuccinimide.[57] NHSS esters of biotin and other derivatives considerably increase the aqueous solubility of the reagents.[58] We offer a variety of amine-reactive STP esters, which are discussed in Chapter 1.

REFERENCES

1. Bioconjug Chem (2008) 19:1880; **2.** Anal Biochem (1986) 156:220; **3.** Biophys J (2000) 78:1449; **4.** J Am Chem Soc (2006) 128:15701; **5.** J Biol Chem (2000) 275:977; **6.** Methods Enzymol (1987) 138:429; **7.** Bioconjug Chem (1992) 3:32; **8.** Anal Biochem (2000) 283:136; **9.** J Chromatogr (1993) 645:75; **10.** Tetrahedron Lett (1970) 22:1901; **11.** Anal Biochem (1987) 162:89; **12.** Clin Chem (1985) 31:1193; **13.** Biochemistry (1973) 12:4154; **14.** Biochem Pharmacol (1997) 53:715; **15.** Bioorg Med Chem Lett (2002) 12:1283; **16.** Biochim Biophys Acta (1991) 1085:223; **17.** Anal Biochem (1971) 44:221; **18.** J Labelled Compounds Radiopharmaceut (1983) 20:1265; **19.** J Biol Chem (2004) 279:20775; **20.** Arch Biochem Biophys (1986) 246:564; **21.** J Histochem Cytochem (1990) 38:377; **22.** Ann NY Acad Sci (1984) 463:214; **23.** J Am Chem Soc (2006) 128:4542; **24.** Protein Sci (2006) 15:640; **25.** Anal Biochem (1992) 201:270; **26.** FEBS Lett (1991) 278:51; **27.** Biochemistry (1978) 17:2163; **28.** Toxicol in Vitro (1999) 13:773; **29.** J Biol Chem (1999) 274:430; **30.** Biochem J (1998) 331:105; **31.** Exp Cell Res (1998) 239:119; **32.** Histochem J (1997) 29:593; **33.** Biochemistry (1988) 27:3483; **34.** Biochemistry (1988) 27:4512; **35.** Biochemistry (1990) 29:5103; **36.** J Biol Chem (1998) 273:11991; **37.** J Biol Chem (1994) 269:24596; **38.** Anal Biochem (1992) 205:166; **39.** J Biol Chem (1996) 271:28399; **40.** J Biol Chem (1994) 269:22907; **41.** J Chromatogr B Biomed Appl (1994) 659:139; **42.** J Chromatogr B Biomed Appl (1994) 659:85; **43.** J Chromatogr B Analyt Technol Biomed Life Sci (2006) 831:63; **44.** J Lipid Res (1997) 38:1913; **45.** Anal Chem (1988) 60:2067; **46.** J Chromatogr (1988) 456:421; **47.** Tetrahedron Lett (1987) 28:679; **48.** J Biol Chem (1988) 263:5724; **49.** Biomed Chromatogr (1996) 10:193; **50.** J Biomed Mater Res A (2007) 83:1176; **51.** J Biol Chem (1997) 272:17444; **52.** J Biol Chem (1996) 271:19964; **53.** J Org Chem (1981) 46:1666; **54.** Anal Chem (1987) 59:1203; **55.** J Chromatogr (1990) 508:133; **56.** J Chromatogr B Analyt Technol Biomed Life Sci (2004) 803:267; **57.** Tetrahedron Lett (1999) 40:1471; **58.** Biochemistry (1982) 21:3950.

IMPORTANT NOTICE: The products described in this manual are covered by one or more Limited Use Label License(s). Please refer to the Appendix on page 971 and Master Product List on page 975. Products are For Research Use Only. Not intended for any animal or human therapeutic or diagnostic use.

DATA TABLE 3.4 DERIVATIZATION REAGENTS FOR CARBOXYLIC ACIDS AND CARBOXAMIDES

Cat. No.	MW	Storage	Soluble	Abs	EC	Em	Solvent	Notes
A91	288.30	L	pH >10, DMF	335	5900	493	pH 8	
A191	175.19	L	DMF, DMSO	351	18,000	430	MeOH	1
A1318	514.62	F,D,L	DMF, EtOH	544	78,000	571	MeOH	
A1339	491.57	L	H_2O	425	12,000	532	H_2O	
A1340	533.65	L	H_2O	426	11,000	531	H_2O	
A1351	397.81	L	pH >6, DMF	492	80,000	516	pH 9	2
A1353	397.81	L	pH >6, DMF	492	68,000	516	pH 9	2
A1363	404.38	L	pH >6, DMF	491	80,000	515	pH 9	2
A1400	218.26	FF,D,L	DMF, MeCN	364	6100	411	MeOH	
A10466	653.38	D,L	pH >6, DMF	493	82,000	517	pH 9	2
A30674	397.45	F,D,L	H_2O	353	20,000	437	MeOH	
A30675	666.58	F,D,L	H_2O	399	29,000	422	H_2O	3
A30676	640.61	F,D,L	H_2O	493	73,000	516	pH 7	4, 5
A30677	~950	F,D,L	H_2O	555	155,000	572	MeOH	
A30678	806.94	F,D,L	H_2O	588	105,000	612	pH 7	
A30679	~1000	F,D,L	H_2O	651	245,000	672	MeOH	
A30680	812.95	F,D,L	H_2O	578	93,000	602	pH 7	
B1355	425.23	F,D,L	pH >6, DMF	492	81,000	515	pH 9	
B2103	341.00	F,D,L	DMSO, MeCN	533	62,000	561	$CHCl_3$	
B6057	292.17	F,L	DMF, MeCN	387	21,000	520	MeOH	
B30633	207.23	F,D,L	DMSO	375	6000	458	MeOH	
C621	624.49	L	H_2O	399	30,000	423	H_2O	3
D112	293.38	L	EtOH, DMF	335	4600	526	MeOH	
D113	335.46	L	EtOH, DMF	335	4600	518	MeOH	
D2390	370.64	F,D,L	DMSO, MeCN	503	76,000	510	MeOH	6
D6251	544.85	F,D,L	DMSO, MeCN	588	64,000	616	MeOH	6
D10460	386.47	L	DMF, DMSO	373	23,000	571	MeOH	7
E2247	191.70	F,D	H_2O	<300		none		
H2249	217.13	D	H_2O	<300		none		
L2424	600.75	L	DMF, DMSO	561	122,000	581	MeOH	
M6248	174.24	D,A	DMF, MeCN	<300		none		
N2461	373.30	FF,DD,L	DMF, $CHCl_3$	260	59,000	395	MeOH	
O10465	496.47	F,D,L	pH >6, DMF	494	75,000	521	pH 9	8
P1405	242.28	FF,L	DMF, MeCN	340	41,000	375	MeOH	
Q10464	814.87	L	DMSO	560	92,000	none	MeOH	
S10490	268.11	D	H_2O	<300		none		
T2425	690.87	L	DMF	591	85,000	612	pH 9	

For definitions of the contents of this data table, see "Using *The Molecular Probes® Handbook*" in the introductory pages.

Notes

1. A191 in aqueous solution (pH 7.0): Abs = 342 nm (EC = 16,000 $cm^{-1}M^{-1}$), Em = 441 nm.
2. Absorption and fluorescence of fluorescein derivatives are pH dependent. Extinction coefficients and fluorescence quantum yields decrease markedly at pH <7.
3. The Alexa Fluor® 405 and Cascade Blue® dyes have a second absorption peak at about 376 nm with EC ~80% of the 395–400 nm peak.
4. The fluorescence lifetime (τ) of the Alexa Fluor® 488 dye in pH 7.4 buffer at 20°C is 4.1 nanoseconds. Data provided by the SPEX Fluorescence Group, Horiba Jobin Yvon Inc.
5. Abs and Em of the Alexa Fluor® 488 dye are red-shifted by as much as 16 nm and 25 nm respectively on microarrays relative to aqueous solution values. The magnitude of the spectral shift depends on the array substrate material. (Biotechniques (2005) 38:127)
6. The absorption and fluorescence spectra of BODIPY® derivatives are relatively insensitive to the solvent.
7. Fluorescence emission spectrum shifts to shorter wavelengths in nonpolar solvents.
8. Absorption and fluorescence of Oregon Green® 488 derivatives are pH dependent only in moderately acidic solutions (pH <5).

IMPORTANT NOTICE: The products described in this manual are covered by one or more Limited Use Label License(s). Please refer to the Appendix on page 971 and Master Product List on page 975. Products are For Research Use Only. Not intended for any animal or human therapeutic or diagnostic use.

PRODUCT LIST 3.4 DERIVATIZATION REAGENTS FOR CARBOXYLIC ACIDS AND CARBOXAMIDES

Cat. No.	Product	Quantity
A30674	Alexa Fluor® 350 cadaverine	1 mg
A30675	Alexa Fluor® 405 cadaverine, trisodium salt4	1 mg
A30676	Alexa Fluor® 488 cadaverine, sodium salt	1 mg
A30677	Alexa Fluor® 555 cadaverine, disodium salt	1 mg
A30680	Alexa Fluor® 568 cadaverine, diammonium salt	1 mg
A30678	Alexa Fluor® 594 cadaverine	1 mg
A30679	Alexa Fluor® 647 cadaverine, disodium salt	1 mg
A1363	5-(aminoacetamido)fluorescein (fluoresceinyl glycine amide)	10 mg
A1339	*N*-(2-aminoethyl)-4-amino-3,6-disulfo-1,8-naphthalimide, dipotassium salt (lucifer yellow ethylenediamine)	25 mg
A91	5-((2-aminoethyl)amino)naphthalene-1-sulfonic acid, sodium salt (EDANS)	1 g
A191	7-amino-4-methylcoumarin *reference standard*	100 mg
A1351	4′-(aminomethyl)fluorescein, hydrochloride	25 mg
A1353	5-(aminomethyl)fluorescein, hydrochloride	10 mg
A1318	5-(and-6)-((*N*-(5-aminopentyl)amino)carbonyl)tetramethylrhodamine (tetramethylrhodamine cadaverine) *mixed isomers*	10 mg
A1340	*N*-(5-aminopentyl)-4-amino-3,6-disulfo-1,8-naphthalimide, dipotassium salt (lucifer yellow cadaverine)	25 mg
A10466	5-((5-aminopentyl)thioureidyl)fluorescein, dihydrobromide salt (fluorescein cadaverine)	25 mg
A1400	9-anthryldiazomethane (ADAM)	25 mg
B30633	bimane amine	5 mg
B2103	BODIPY® 493/503 methyl bromide (8-bromomethyl-4,4-difluoro-1,3,5,7-tetramethyl-4-bora-3a,4a-diaza-*s*-indacene)	5 mg
B6057	6-bromoacetyl-2-dimethylaminonaphthalene (badan)	10 mg
B1355	5-(bromomethyl)fluorescein	10 mg
C621	Cascade Blue® ethylenediamine, trisodium salt	10 mg
D10460	Dapoxyl® (2-aminoethyl)sulfonamide	10 mg
D2390	4,4-difluoro-5,7-dimethyl-4-bora-3a,4a-diaza-*s*-indacene-3-propionyl ethylenediamine, hydrochloride (BODIPY® FL EDA)	5 mg
D6251	5-(((4-(4,4-difluoro-5-(2-thienyl)-4-bora-3a,4a-diaza-*s*-indacene-3-yl)phenoxy)acetyl)amino)pentylamine, hydrochloride (BODIPY® TR cadaverine)	5 mg
D112	5-dimethylaminonaphthalene-1-(*N*-(2-aminoethyl))sulfonamide (dansyl ethylenediamine)	100 mg
D113	5-dimethylaminonaphthalene-1-(*N*-(5-aminopentyl))sulfonamide (dansyl cadaverine)	100 mg
E2247	1-ethyl-3-(3-dimethylaminopropyl)carbodiimide, hydrochloride (EDAC)	100 mg
H2249	*N*-hydroxysulfosuccinimide, sodium salt (NHSS)	100 mg
L2424	Lissamine rhodamine B ethylenediamine	10 mg
M6248	mono-*N*-(*t*-BOC)-propylenediamine	1 g
N2461	2-(2,3-naphthalimino)ethyl trifluoromethanesulfonate	100 mg
O10465	Oregon Green® 488 cadaverine *5-isomer*	5 mg
P1405	1-pyrenyldiazomethane (PDAM)	25 mg
Q10464	QSY® 7 amine, hydrochloride	5 mg
S10490	4-sulfo-2,3,5,6-tetrafluorophenol, sodium salt (STP)	100 mg
T2425	Texas Red® cadaverine (Texas Red® C_5)	5 mg

FOUR

CHAPTER 4

Biotin and Hapten Derivatives

IMPORTANT NOTICE: The products described in this manual are covered by one or more Limited Use Label License(s). Please refer to the Appendix on page 971 and Master Product List on page 975. Products are For Research Use Only. Not intended for any animal or human therapeutic or diagnostic use.

4.1 Introduction to Avidin–Biotin and Antibody–Hapten Techniques

This chapter is devoted to our biotinylation, desthiobiotinylation and haptenylation reagents (Section 4.2) and our biotin and desthiobiotin (DSB-X™ biotin) conjugates (Section 4.3). For the detection of biotin and hapten conjugates, we prepare a large assortment of labeled avidin and antibody probes, which are described in Section 7.6 and Section 7.4, respectively. Our avidin- and biotin-coated FluoSpheres® microspheres (Section 6.5) and Qdot® nanocrystals (Section 6.6) provide alternative detection technologies that offer a combination of fluorescence intensity and photostability far superior to that of any simple dye conjugate.

Avidin–Biotin and Antibody–Hapten Techniques and Their Applications

The high affinity and specificity of avidin–biotin and antibody–hapten interactions have been exploited for diverse applications in immunology, histochemistry, *in situ* hybridization, affinity chromatography and many other areas.[1–5] Biotinylation (Table 4.1) and haptenylation (Table 4.2) reagents provide the "tag" that transforms poorly detectable molecules into probes that can be recognized by a labeled detection reagent or an affinity-capture matrix. Once tagged with biotin or a hapten, a molecule of interest—such as an antibody, drug, oligonucleotide, polysaccharide or receptor ligand—can be used to probe cells and tissues, as well as protein and nucleic acid blots and arrays.[6,7] After finding its target, this tagged molecule can be

Table 4.1 Biotinylation and desthiobiotinylation reagents.

Reactive Moiety	Biotin Derivative	DSB-X™ Biotin Derivative
Aliphatic amine	A1593, A1594, B1596 (XX), N6356	
Alkyne	B10185	
Azide	B10184	
Carboxylic acid	B1595, B20656, B1592	D20657
DNP-X-biocytin-X, SE	B2604, F6348 (F)	
Hydrazide	B1603, B2600 (XX)	D20653 (X)
Hydroxylamine	A10550	
Iodoacetamide	B1591	D30753
Maleimide	M1602	
Succinimidyl ester (SE)	B1513, B1582 (X), B1606 (XX), B6353 (SSE)(X), B6352 (SSE) (XX), F2610 (F), F6347 (F), B30010 (Micro), B30756 (Micro)	D20655 (D)
TS-Link™ thiosulfate		T30754

(D) = DSB-X™ Biotin Protein Labeling Kit.
(F) = FluoReporter® Protein Labeling Kit.
(Micro) = Biotin-XX Microscale Protein Labeling Kit.
(SE) = Succinimidtyl ester. (SSE) = Sulfosuccinimidyl ester.
(X) = Aminohexanoyl (7-atom) spacer separating the biotin or desthiobiotin and the reactive moiety.
(XX) = Aminohexanoylaminohexanoyl (14-atom) spacer separating the biotin and the reactive moiety.

Table 4.2 Selected haptenylation reagents and their anti-hapten antibodies.

Cat. No.	Preferred Reactive Hapten(s)	Unlabeled and Labeled Anti-Hapten Antibodies (Cat. No.) *
A30000 A30100	Alexa Fluor® 405, SE	Anti–Alexa Fluor® 405/Cascade Blue® dye (A5760)
A20000 A20100 A30005 A30052	Alexa Fluor® 488, SE Alexa Fluor® 488, SE Alexa Fluor® 488, 5-TFP Alexa Fluor® 488, 5-SDP	Anti–Alexa Fluor® 488 dye (A11094)
A2952	3-Amino-3-deoxydigoxigenin hemisuccinamide, SE	Anti-digoxigenin (available from other suppliers)
B1582 B1606 B6353 B6352	Biotin-X, SE Biotin-XX, SE Biotin-X, SSE Biotin-XX, SSE	Anti-biotin (03-3700)
D6102 B10006	BODIPY® FL-X, SE BODIPY® FL, STP ester	Anti–BODIPY® FL dye (A5770) †
C2284	Cascade Blue® acetyl azide	Anti–Alexa Fluor® 405/Cascade Blue® dye (A5760)
D6104	Dansyl-X, SE	Anti-dansyl (A6398)
D2248 B2604	DNP-X, SE DNP-X-biocytin-X, SE	Anti-DNP (A6423, A6430, A6435, A11097, Q17421)
F2181 F6130	Fluorescein 5(6)-SFX Fluorescein-EX, SE	Anti–fluorescein/Oregon Green® dye (A889, A982, A6413, A6421, A11090, A11091, A11095, A21253, Q15421, Q15431) †
L1338	Lucifer yellow iodoacetamide	Anti–lucifer yellow dye (A5750, A5751)
O6185	Oregon Green® 488-X, SE	Anti–fluorescein/Oregon Green® dye (A889, A982, A6413, A6421, A11090, A11091, A11095, A21253, Q15421, Q15431) †
T6105	5(6)-TAMRA-X, SE	Anti-tetramethylrhodamine (A6397), anti–Texas Red® dye (A6399) †
R6160	Rhodamine Red™-X, SE	Anti-tetramethylrhodamine (A6397), anti–Texas Red® dye (A6399) †
T6134 T20175	Texas Red®-X, SE	Anti-tetramethylrhodamine (A6397), anti–Texas Red® dye (A6399) †

* See Section 7.4 for a description of these anti-hapten antibodies. † Both the anti-tetramethylrhodamine and the anti–Texas Red® dye antibodies cross-react with tetramethylrhodamine, Lissamine rhodamine, Rhodamine Red™ and Texas Red® fluorophores. Therefore, these fluorophores should not be used simultaneously to generate separate signals in a multicolor experiment. Similarly, the anti–BODIPY® FL dye antibody may cross-react with other BODIPY® dyes, and the anti–fluorescein/Oregon Green® dye antibody cross-reacts with both fluorescein and Oregon Green® dyes.

IMPORTANT NOTICE: The products described in this manual are covered by one or more Limited Use Label License(s). Please refer to the Appendix on page 971 and Master Product List on page 975. Products are For Research Use Only. Not intended for any animal or human therapeutic or diagnostic use.

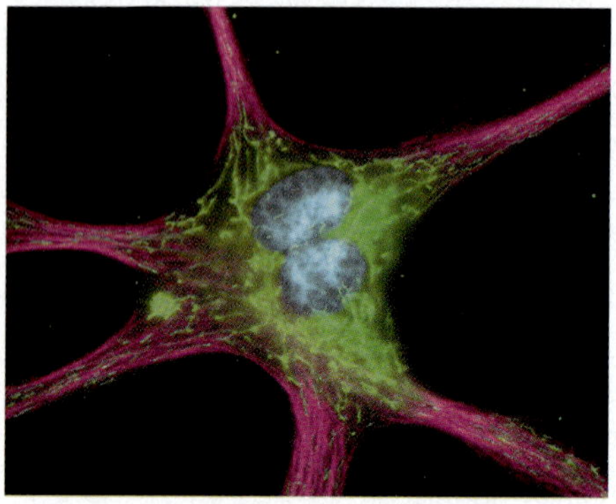

Figure 4.1.1 The cytoskeleton of a fixed and permeabilized bovine pulmonary artery endothelial cell detected using mouse monoclonal anti–α-tubulin antibody (A11126), visualized with Alexa Fluor® 647 goat anti–mouse IgG antibody (A21235) and pseudocolored magenta. Endogenous biotin in the mitochondria was labeled with green-fluorescent Alexa Fluor® 488 streptavidin (S11223, S32354) and DNA was stained with blue-fluorescent DAPI (D1306, D3571, D21490).

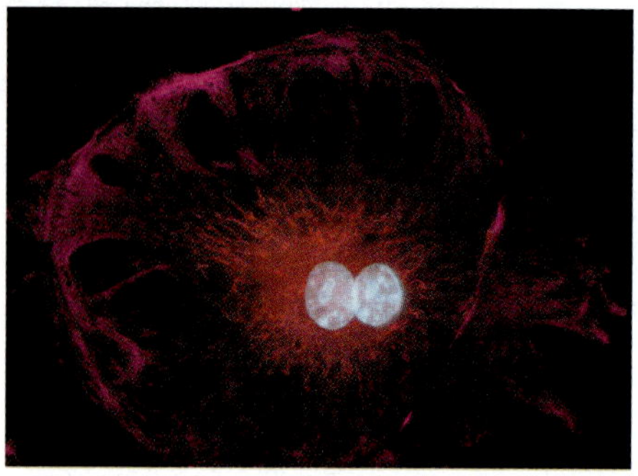

Figure 4.1.2 The intermediate filaments in bovine pulmonary artery endothelial cells, localized using our anti-desmin antibody (A21283), which was visualized with the Alexa Fluor® 647 goat anti-mouse IgG antibody (A21235). Endogenous biotin in the mitochondria was labeled with Alexa Fluor® 546 streptavidin (S11225) and DNA in the cell was stained with blue-fluorescent DAPI (D1306, D3571, D21490).

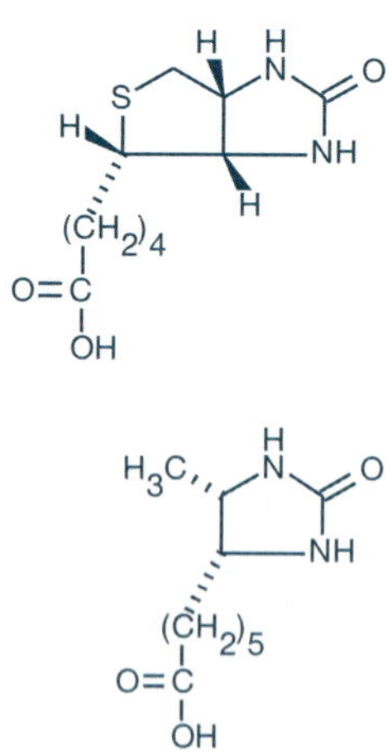

Figure 4.1.3 Comparison of the structures of D-biotin (top) and D-desthiobiotin (bottom).

detected with the appropriate avidin or anti-hapten antibody conjugate labeled with a fluorophore, fluorescent microsphere, enzyme, magnetic particle or colloidal gold. Biotinylated molecules can also be captured with various forms of immobilized streptavidin, such as streptavidin agarose (S951), CaptAvidin™ agarose (C21386) or streptavidin-coupled magnetic Dynabeads® (www.invitrogen.com/handbook/dynabeads). Biotinylated probes can be developed for electron microscopy with NANOGOLD® or Alexa Fluor® FluoroNanogold™ streptavidin (N24918, A24926, A24927) or streptavidin-coupled Qdot® nanocrystals [8–10] (Section 6.6). Our extensive array of avidin and streptavidin conjugates are described in Section 7.6.

Detection Methods Compatible with Avidin–Biotin Techniques

Avidin–biotin and antibody–hapten techniques are compatible with flow cytometry and light, electron and fluorescence microscopy, as well as with solution-based methods such as enzyme-linked immunosorbent assays (ELISAs). Moreover, avidin–biotin and antibody–hapten techniques are frequently combined for simultaneous, multicolor detection of multiple targets in complex tissue samples.[11] By judicious choice of detection reagents and sandwich protocols, these techniques can be employed to amplify signals from low-abundance analytes.[12] For example, the bridging method is a common immunohistochemical technique for signal amplification and improved tissue penetration in which avidin or streptavidin serves as a bridge between two biotinylated molecules.[13] Other amplification strategies include the tyramide signal amplification (TSA™) technology (Section 6.2).

Endogenous Biotin and Biotinidase

Mammalian cells and tissues contain biotin-dependent carboxylases, which are required for a variety of metabolic functions. These biotin-containing enzymes often produce substantial background signals when biotin–avidin or biotin–streptavidin detection systems are used to identify cellular targets [14,15] (Figure 4.1.1, Figure 4.1.2). Endogenous biotin is particularly prevalent in mitochondria [16,17] and in kidney, liver and brain tissues.[14,18] The reagents in the Endogenous Biotin-Blocking Kit (E21390), which is described in Section 7.6, can be used to minimize interference from endogenous biotin in these techniques. In mammalian serum and plasma, biotinylated proteins are susceptible to cleavage by endogenous biotinidases, producing free biotin and unlabeled protein.[19]

Modified Avidin and Biotin

CaptAvidin™ Biotin-Binding Protein

Although binding of biotin to native avidin or streptavidin is essentially irreversible, appropriately modified avidins can bind biotinylated probes reversibly, making them valuable reagents for isolating and purifying biotinylated molecules from complex mixtures. In the CaptAvidin™ biotin-binding protein (C21385, Section 7.6), selective nitration of tyrosine residues in the four biotin-binding sites of avidin considerably reduces the affinity of this protein for biotinylated molecules above pH 9.[20] Consequently, biotinylated probes can be adsorbed to CaptAvidin™ biotin-binding protein at neutral pH or below and released at ~pH 10.[20,21] CaptAvidin™ agarose (C21386, Section 7.6) is particularly useful for separating and purifying biotin conjugates from complex mixtures.[22,23]

DSB-X™ Biotin

In contrast to the modified avidin of the CaptAvidin™ products, our DSB-X™ biotin technology employs a modified biotin to provide a means of labeling and separating biomolecules, including live cells, under extremely gentle conditions.[24] The DSB-X™ biotin reagents, which are derivatives of desthiobiotin (Figure 4.1.3) with an additional seven-atom 'X' spacer, have moderate affinity for avidin and streptavidin that is rapidly reversed by low concentrations of free biotin or desthiobiotin at neutral pH and room temperature; the K_d for desthiobiotin binding to streptavidin has been reported to be 1.9 nM.[25] The DSB-X™ Biotin Protein Labeling Kit (D20655, Section 4.2) provides a convenient method for labeling proteins with the amine-reactive succinimidyl ester of DSB-X™ biotin and purifying the conjugate.

IMPORTANT NOTICE: The products described in this manual are covered by one or more Limited Use Label License(s). Please refer to the Appendix on page 971 and Master Product List on page 975. Products are For Research Use Only. Not intended for any animal or human therapeutic or diagnostic use.
www.invitrogen.com/probes
molecular probes® | invitrogen™ by life technologies™

REFERENCES

1. J Histochem Cytochem (2009) 57:701; 2. Bioconjug Chem (2009) 20:147; 3. Methods Mol Biol (2008) 418:187; 4. Nat Protoc (2008) 3:534; 5. Anal Biochem (1988) 171:1; 6. Glycoconj J (2008) 25:27; 7. Methods Mol Biol (2003) 224:55; 8. Nat Cell Biol (2008) 10:955; 9. Nat Methods (2005) 2:743; 10. J Histochem Cytochem (2004) 52:13; 11. Proteomics (2007) 7:1786; 12. J Immunol Methods (2004) 289:169; 13. J Neurosci Methods (2001) 112:43; 14. Methods Mol Biol (2008) 418:111; 15. Histochem J (2002) 34:567; 16. Methods Mol Biol (2008) 418:157; 17. Histochemistry (1993) 100:415; 18. J Comp Neurol (2004) 473:86; 19. Methods Mol Biol (2008) 418:209; 20. Anal Biochem (1996) 243:257; 21. Biochem J (1996) 316:193; 22. Endocrinology (2009) 150:1122; 23. FASEB J (2006) 20:518; 24. Anal Biochem (2002) 308:343; 25. Bioconjug Chem (2006) 17:366.

4.2 Biotinylation and Haptenylation Reagents

We are a primary manufacturer of a diverse array of biotinylation (Table 4.1) and haptenylation (Table 4.2) reagents for labeling biomolecules. Reviews of the methods that we use to prepare biotinylated[1] and fluorescent[2] conjugates of antibodies have been published. To make the labeling reactions particularly easy, we have developed some very useful kits for labeling proteins with biotin, DSB-X™ biotin, 2,4-dinitrophenyl (DNP) or a choice of several different fluorophores, as described below. Each of the protein labeling kits contains the preferred reactive dye or hapten—many of which have spacers to reduce interactions between the label and the biomolecule—along with a detailed protocol for preparing the conjugates. In most cases, these kits also provide the separation media for purifying labeled protein conjugates from the reaction mixture.

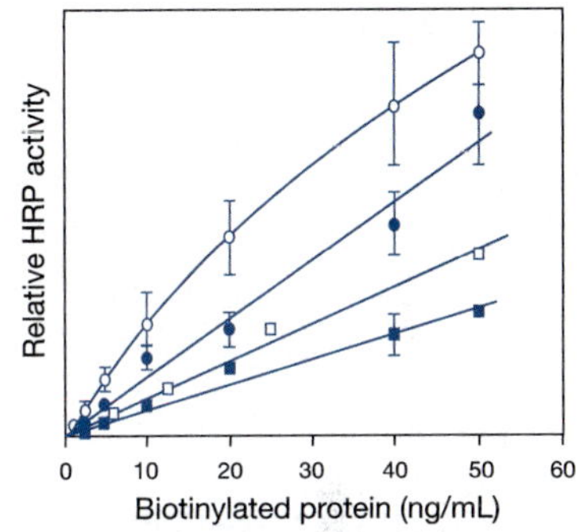

Figure 4.2.1 ELISA-type assay comparing the binding capacity of bovine serum albumin (BSA) and goat anti–mouse IgG antibody (GAM) biotinylated with either biotin-X succinimidyl ester (B1582) or biotin-XX succinimidyl ester (B1606). The assay was developed using horseradish peroxidase (HRP) streptavidin (S911, 0.2 µg/mL) and *o*-phenylenediamine dihydrochloride (OPD). The moles of biotin per mole of protein were: 4.0 biotin-X/GAM (●), 4.4 biotin-XX/GAM (○), 6.7 biotin-X/BSA (■) and 6.2 biotin-XX/GAM (□). Error bars on some data points have been omitted for clarity. Reprinted with permission from Methods Mol Biol (1995) 45:223.

Biotin and Biotinylation Reagents

The primary building blocks for preparing biotinylation reagents are biotin (B1595, B20656), biotin-X and biotin-XX, where "X" represents a seven-atom aminohexanoyl spacer between biotin and the reactive carboxylic acid. This spacer helps to separate the biotin moiety from its point of attachment, potentially reducing the interaction of biotin with the biomolecule to which it is conjugated and enhancing its ability to bind to the relatively deep biotin-binding sites in avidin[3] (Figure 4.2.1). D-Desthiobiotin (D20657) is the biological precursor to D-biotin[4,5] and a key reagent in our DSB-X™ biotin technology.

We also offer biocytin (ε-biotinoyl-L-lysine, B1592), biotin ethylenediamine (A1593), biotin cadaverine (A1594) and biotin-X cadaverine (B1596) each of which contains a primary amine that allows it to be fixed in cells with aldehyde-based fixatives, facilitating subsequent detection with conjugates of avidin and streptavidin (Table 7.9). Biocytin derivatives, including probes that contain both biotin and fluorophore moieties, are commonly employed as microinjectable cell tracers and are discussed in Section 4.3 and Section 14.3.

Amine-Reactive Biotinylation Reagents

Although biotin succinimidyl ester (B1513) can be used to biotinylate amines in peptides, proteins and other biomolecules,[6] we strongly recommend the biotin-X and biotin-XX succinimidyl esters (B1582, B1606), and especially the water-soluble biotin-X and biotin-XX sulfosuccinimidyl esters (B6353, B6352), because their additional 7- and 14-atom spacers greatly facilitate binding to avidin conjugates (Figure 4.2.1). We use biotin-X succinimidyl ester or the biotin-XX derivative to prepare all Molecular Probes® biotinylated protein and dextran conjugates. Red blood cells that were biotinylated with biotin-X succinimidyl ester—but surprisingly not those modified with biotin-X sulfosuccinimidyl ester—could be detected in circulation for almost 100 days after injection into dogs.[7]

The sulfosuccinimidyl esters of biotin-X and biotin-XX (B6353, B6352) have been extensively used as topological probes to label proteins in the outer membrane surface,[8] yielding conjugates that can be separated by electrophoresis, blotted onto membranes and then detected with a fluorophore- or enzyme-conjugated avidin derivative (Table 7.9). We utilize biotin-XX sulfosuccinimidyl ester as a component in our FluoReporter® Cell-Surface Biotinylation Kit (F20650).

Amine-Reactive Chromophoric Biotin Derivative

Determining a protein's degree of biotinylation is relatively difficult because of the lack of visible absorbance by the biotin molecule. To facilitate this determination, we offer an amine-reactive chromophoric derivative, biotin-X 2,4-dinitrophenyl-X-L-lysine succinimidyl ester (DNP-X–biocytin-X, SE; B2604; Figure 4.2.2). Following protein conjugation, the extent of biotinylation is easily determined from the absorbance of the DNP chromophore (EC_{360} = 15,000 $cm^{-1}M^{-1}$). Incorporation of the DNP moiety into the biotinylating reagent does not affect its complexation with avidin or with anti-biotin antibodies. Our FluoReporter® Biotin/DNP Protein Labeling Kit (F6348), described below, contains sufficient DNP-X–biocytin-X, SE for 5 to 10 protein labeling reactions of 0.2–2 mg each.

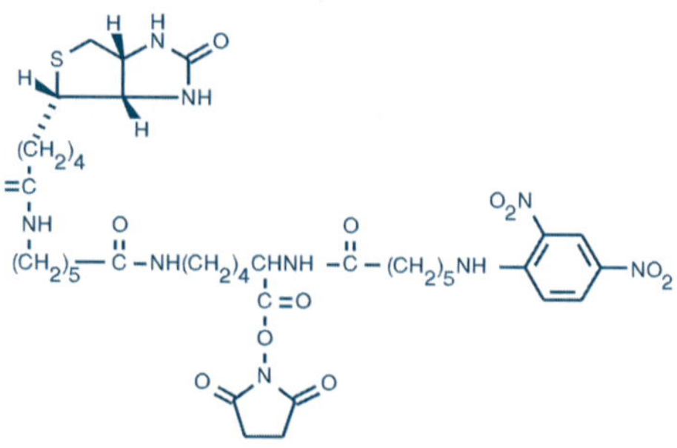

Figure 4.2.2 Biotin-X 2,4-dinitrophenyl-X-L-lysine, succinimidyl ester (DNP-X-biocytin-X, SE, B2604).

Thiol-Reactive Biotinylation Reagents

Although amine-reactive reagents are more commonly employed, the thiol-reactive biotin iodoacetamide, frequently identified in the literature by the acronym BIAM (B1591), biotin maleimide (M1602) and DSB-X™ biotin C_2-iodoacetamide (D30753) derivatives can also be used to label proteins and thiol-modified oligonucleotides.[9] Biotin iodoacetamide and biotin maleimide are primarily used for biotinylation of free protein thiols in relation to investigations of thiol–disulfide exchange, disulfide isomerization, *S*-glutathionylation and other posttranslational modification processes.[10–13]

We also offer TS-Link™ DSB-X™ biotin C_5-thiosulfate (TS-Link™ desthiobiotin-X C_5-thiosulfate, T30754), which is a water-soluble thiosulfate that reacts readily and selectively with a free thiol to form a disulfide bond (Figure 4.2.3). In contrast to the thioether bonds formed by maleimides and iodoacetamides, the disulfide bond formed by this TS-Link™ reagent is reversible—the TS-Link™ DSB-X™ hapten can easily be removed using a reducing agent such as dithiothreitol or tris-(2-carboxyethyl)phosphine (DTT, D1532; TCEP, T2556; Section 2.1), leaving the molecule of interest unchanged.

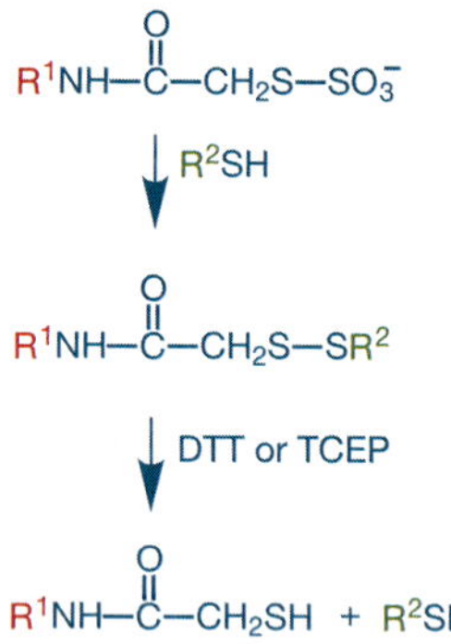

Figure 4.2.3 Reaction of a TS-Link™ reagent with a thiol, followed by removal of the label with a reducing agent.

Click-iT® Biotin Alkyne and Biotin Azide Reagents

Click-iT® labeling and detection technology uses bioorthogonal reactive chemistry in which the reaction partners have no endogenous representation in biological molecules, cells, tissues or model organisms.[14–16] The click reaction comprises a copper-catalyzed cycloaddition between an alkyne and an azide, forming a stable triazole conjugate (Section 3.1). The azide and alkyne moieties can be used interchangeably; either one can be used to tag the biomolecule of interest, whereas the other is used for subsequent detection. Moreover, the click chemistry labels—either the alkyne or the azide—provide a functional group that typically neither reacts with other cell components nor disrupts normal cell processes. The relative transparency to cell machinery of these labels means that tagged molecules are often acceptable substrates for enzymes that assemble these building blocks into biopolymers.

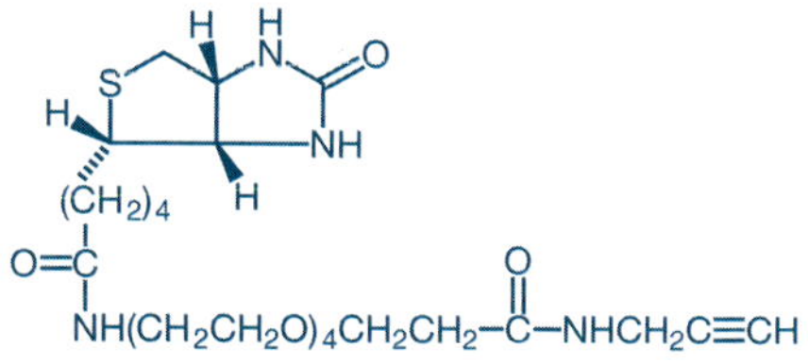

Figure 4.2.4 Biotin alkyne (PEG_4 carboxamide-propargyl biotin, B10185).

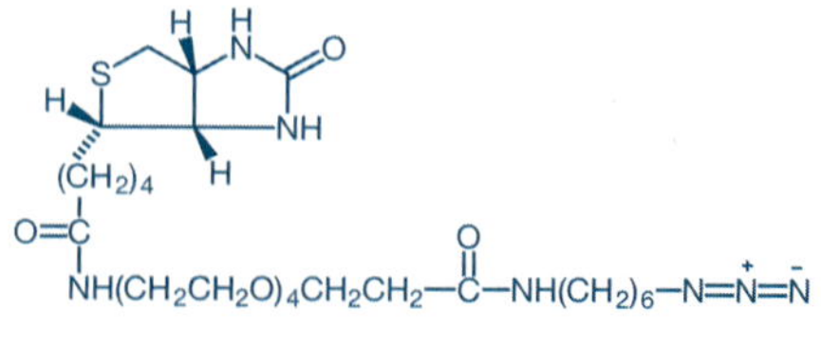

Figure 4.2.5 Biotin azide (PEG_4 carboxamide-6-azidohexanyl biotin, B10184).

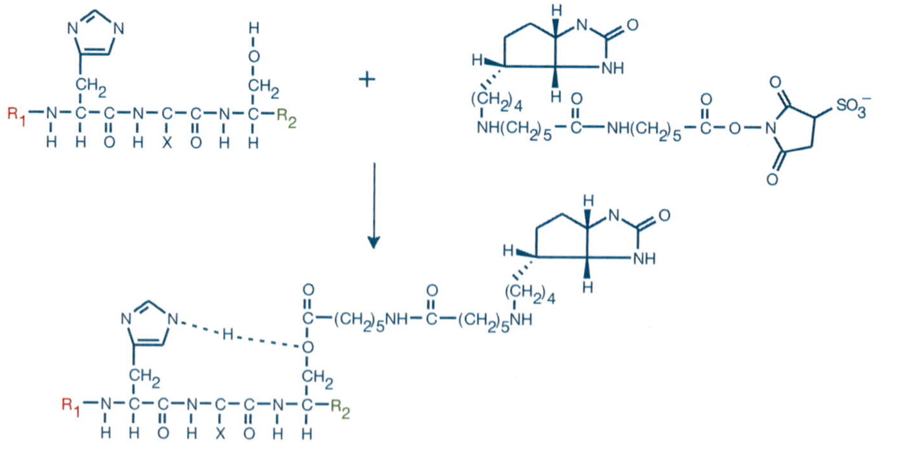

Figure 4.2.6 Nucleophilic attack of serine on the carbonyl group (C=O) of biotin-X, SSE (B6353) results in the stable *O*-acylated derivative. In addition to histidine-x-serine, this stable intermediate can be formed in the presence of linear sequences of histidine-x-tyrosine and histidine-x-threonine, where "x" refers to any amino acid.

IMPORTANT NOTICE: The products described in this manual are covered by one or more Limited Use Label License(s). Please refer to the Appendix on page 971 and Master Product List on page 975. Products are For Research Use Only. Not intended for any animal or human therapeutic or diagnostic use.

Each FluoReporter® Mini-Biotin-XX Labeling Kit contains:

- Biotin-XX sulfosuccinimidyl ester
- Reaction tubes, each containing a magnetic spin bar
- Purification resin, spin columns and collection tubes
- Dialysis tubing (molecular weight cut-off ~12,000–14,000 daltons)
- Protocols for preparing and purifying the biotinylated protein

The ready-to-use spin columns provide an extremely convenient method of purifying the biotinylated protein from excess biotinylation reagents. Alternatively, the researcher may choose to remove excess reagents by dialysis, thereby avoiding further dilution of the biotinylated protein. The FluoReporter® Mini-Biotin-XX Protein Labeling Kit contains sufficient reagents for five biotinylation reactions of 0.1–3 mg of protein each.

FluoReporter® Biotin-XX Protein Labeling Kit

We also offer the FluoReporter® Biotin-XX Protein Labeling Kit (F2610) for larger-scale biotinylation reactions. Once the labeled protein is purified from excess biotin reagent, its degree of biotinylation can be determined using an avidin–biotin displacement assay;[42,43] biotinylated goat IgG is provided as a control.

The FluoReporter® Biotin-XX Protein Labeling Kit supplies:

- Biotin-XX succinimidyl ester
- Dimethylsulfoxide (DMSO)
- Gel filtration column
- Avidin–HABA complex
- Biotinylated goat IgG
- Protocols for preparing and purifying the biotinylated protein, as well as for quantitating the degree of labeling

The FluoReporter® Biotin-XX Protein Labeling Kit provides sufficient reagents for five labeling reactions of 5–20 mg of protein each.

FluoReporter® Biotin/DNP Protein Labeling Kit

The FluoReporter® Biotin/DNP Protein Labeling Kit (F6348, Table 1.3) is similar to our other FluoReporter® Protein Labeling Kits, except that it contains DNP-X–biocytin-X succinimidyl ester as the reactive label. When proteins are labeled with this chromophoric biotin derivative, the degree of biotinylation can be readily assessed from the extinction coefficient of DNP (EC_{364} = 15,000 $cm^{-1}M^{-1}$). An additional feature of the conjugates labeled with DNP-X–biocytin-X succinimidyl ester is that they can be recognized by avidin derivatives (or anti-biotin antibodies) and by anti-DNP antibodies, enabling researchers to choose among several detection techniques suitable for fluorescence and electron microscopy.

Each FluoReporter® Biotin/DNP Protein Labeling Kit contains:

- DNP-X–biocytin-X succinimidyl ester
- Dimethylsulfoxide (DMSO)
- Reaction tubes, each containing a magnetic spin bar
- Spin columns plus collection tubes
- Protocols for preparing and purifying the protein conjugate, as well as for quantitating the degree of labeling

The FluoReporter® Biotin/DNP Protein Labeling Kit supplies sufficient reagents for 5 to 10 labeling reactions of 0.2–2 mg of protein each.

DSB-X™ Biotin Protein Labeling Kit

Our unique DSB-X™ biotin technology permits the facile reversal of the typically irreversible biotin–avidin interaction under extremely gentle conditions.[40] DSB-X™ biotin succinimidyl ester, a derivative of desthiobiotin (Figure 4.2.11) with an additional seven-atom spacer, reacts with amine groups of biomolecules to form stable amides. The DSB-X™ biotin conjugate can be detected with any of the avidin or streptavidin derivatives described in Table 7.9. Binding is almost totally reversed by addition of free biotin (B1595, B20656) at neutral pH and normal ionic strength. Significantly, DSB-X™ biotin–conjugated biopolymers can be separated from complex mixtures using agarose affinity matrices.

The DSB-X™ Biotin Protein Labeling Kit (D20655, Table 1.3) contains the reagents required for five protein conjugations of 0.5–3 mg each, including:

- DSB-X™ biotin succinimidyl ester
- Dimethylsulfoxide (DMSO)
- Reaction tubes
- Purification resin, spin columns and collection tubes for small-scale purifications
- Dialysis tubing for larger-scale purifications
- Detailed protocols for conjugations and purifications

FluoReporter® Cell-Surface Biotinylation Kit

Biotin-XX sulfosuccinimidyl ester is a membrane-impermeant, amine-reactive compound that may be used to label proteins exposed on the surface of live cells[8,44] (Figure 4.2.12). The sulfosuccinimidyl ester forms an extremely stable conjugate with cell surface proteins, and the biotin provides a convenient hapten for subsequent isolation or analysis with an avidin-based protein, including streptavidin, NeutrAvidin™ and CaptAvidin™ biotin-binding proteins (Table 7.9). Cell-surface biotinylation techniques have been employed to differentially label proteins in the apical and basolateral plasma membranes of epithelial cells.[45,46] These techniques are also well suited for studying internalization of membrane proteins and cell-surface targeting of proteins.[47–49]

The FluoReporter® Cell-Surface Biotinylation Kit (F20650) provides a convenient method to label proteins exposed on the cell surface including, but not limited to, membrane proteins. This kit includes:

- Biotin-XX sulfosuccinimidyl ester
- Anhydrous DMSO for preparing stable stock solutions
- Detailed protocols for cell-surface biotinylation

The supplied protocol for cell-surface biotinylation is easy to perform and can typically be completed in less than one hour. Biotinylated proteins can be subsequently identified using western blot techniques and labeled avidin conjugates[44] described in Section 6.6 and Section 7.6.

Zenon® Biotin Antibody Labeling Kits

Zenon® Antibody Labeling Kits provide a fast, versatile and reliable method for producing antibody conjugates that can be used in any application where a directly labeled primary antibody is suitable,

IMPORTANT NOTICE: The products described in this manual are covered by one or more Limited Use Label License(s). Please refer to the Appendix on page 971 and Master Product List on page 975. Products are For Research Use Only. Not intended for any animal or human therapeutic or diagnostic use.

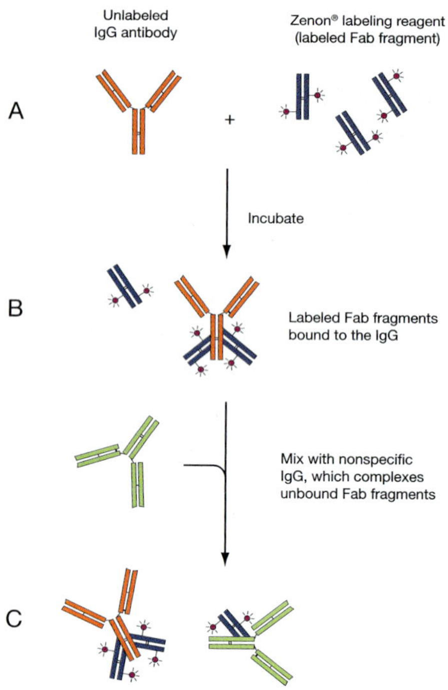

Figure 4.2.13 Labeling scheme utilized in the Zenon® Antibody Labeling Kits. A) An unlabeled IgG antibody is incubated with the Zenon® labeling reagent, which contains a fluorophore-labeled, Fc-specific anti-IgG Fab fragment. B) This labeled Fab fragment binds to the Fc portion of the IgG antibody. C) Excess Fab fragment is then neutralized by the addition of a nonspecific IgG, preventing cross-labeling by the Fab fragment in experiments where primary antibodies of the same type are present. Note that the Fab fragment used for labeling need not be coupled to a fluorophore, but could instead be coupled to an enzyme (such as HRP) or to biotin.

including microscopy, flow cytometry, high-throughput screening, and others. This enabling technology not only eliminates the need for secondary detection reagents in many protocols, but also simplifies immunolabeling protocols that previously were time consuming or impractical, such as use of multiple antibodies derived from the same species in the same protocol.[50,51] We have applied our exclusive Zenon® technology (Section 7.3) in several Zenon® Biotin-XX Antibody Labeling Kits, which permit the quantitative biotinylation of even submicrogram quantities of an antibody typically in less than 10 minutes.

The Zenon® Biotin-XX Antibody Labeling Kits include:

- Zenon® Biotin-XX Mouse IgG_1 Labeling Kit (Z25052)
- Zenon® Biotin XX Mouse IgG_{2a} Labeling Kit (Z25152)
- Zenon® Biotin-XX Mouse IgG_{2b} Labeling Kit (Z25252)
- Zenon® Biotin-XX Rabbit IgG Labeling Kit (Z25352)
- Zenon® Biotin-XX Human IgG Labeling Kit (Z25452)

Zenon® Antibody Labeling Kits are designed to label intact primary antibodies in amounts from less than 1 µg to as much as 50 µg, starting with a purified antibody fraction or with a crude antibody preparation such as ascites fluid or a hybridoma supernatant. Zenon® labeling technology takes advantage of the immunoselectivity of the antibody binding reaction by forming a complex between an intact primary IgG antibody and a biotin-labeled Fab fragment directed against the Fc portion of the IgG. Simple mixing of the labeled Fab fragment supplied in the Zenon® Antibody Labeling Kit with the corresponding primary antibody quantitatively produces the Fab–antibody complex typically in under 10 minutes, with no pre- or post-labeling purification required (Figure 4.2.13). This labeled Fab–antibody complex can be immediately used to stain cells, tissues and other targets in the same manner as a covalently labeled primary antibody.

Each Zenon® Biotin Antibody Labeling Kit provides:

- Zenon® antibody labeling reagent
- Zenon® blocking reagent
- Detailed protocols for antibody labeling

The Zenon® Biotin Antibody Labeling Kit contains sufficient reagents for 50 labelings, where one labeling is defined as the amount of Zenon® labeling reagent required to label 1 µg of an intact, affinity-purified IgG antibody at a Fab:antibody molar ratio of 3:1.

Biotin Quantitation Assay Kits

FluoReporter® Biotin Quantitation Assay Kit for Biotinylated Proteins

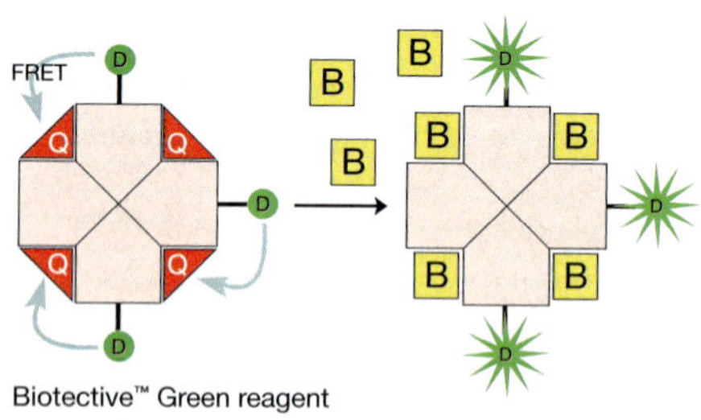

Figure 4.2.14 Schematic representation of the FluoReporter® biotin quantitation assay. This assay uses Biotective™ Green reagent, which consists of avidin labeled with a fluorescent dye (**D**) and with quencher dye ligands (**Q**) occupying the biotin-binding sites. Through fluorescence resonance energy transfer (FRET), the ligand quenches the fluorescence. Biotin (**B**) attached to a protein displaces the quencher dye from Biotective™ Green reagent, yielding fluorescence proportional to the amount of added biotin.

The FluoReporter® Biotin Quantitation Assay Kit for biotinylated proteins (F30751) is designed to provide a sensitive fluorometric assay for accurately determining the number of biotin labels on a protein. The assay is based on the displacement of a ligand tagged with a quencher dye from the biotin-binding sites of Biotective™ Green reagent[52] (Figure 4.2.14). The assay can detect from 4 to 80 picomoles of biotin in a sample (Figure 4.2.15), providing a 50-fold higher sensitivity than the traditional HABA biotin-binding assay.[53] Furthermore, unlike the HABA biotin-binding assay, which requires ~1 mg of protein sample, the FluoReporter® biotin quantitation assay requires only a minimum of 600 ng of a singly biotinylated IgG of MW 150,000. For proteins of lower molecular weight or multiple biotin labels, less protein can be used. To expose any biotin groups in a multiply labeled protein that are sterically restricted and inaccessible to the Biotective™ Green reagent, this kit includes protease and an optional protocol for digesting the protein. After this preliminary digestion, biotin assay values agree well with MALDI-TOF determinations. With excitation/emission maxima of 495/519 nm, this assay is compatible with any fluorescence-based microplate reader capable of detecting fluorescein (FITC) or Alexa Fluor® 488 dye; it can also be scaled up for fluorometer-based experiments.

IMPORTANT NOTICE: The products described in this manual are covered by one or more Limited Use Label License(s). Please refer to the Appendix on page 971 and Master Product List on page 975. Products are For Research Use Only. Not intended for any animal or human therapeutic or diagnostic use.
www.invitrogen.com/probes

Each FluoReporter® Biotin Quantitation Assay Kit for biotinylated proteins includes:

- Biotective™ Green reagent
- Biocytin
- Protease
- Concentrated phosphate-buffered saline (PBS)
- Biotinylated goat anti–mouse IgG antibody
- Detailed protocols

Sufficient reagents are provided for assaying 5 samples independently using eight wells in triplicate for the standard curve and three dilutions of the sample in triplicate (totaling 33 wells per assay). However, fewer wells may be used to conserve sample and a single standard curve can be used for multiple samples in the same experimental session. Biocytin (biotinylated lysine) is provided as a standard for the assay because it more closely represents the form of biotin present after proteolytic cleavage. Biotinylated goat anti–mouse IgG antibody is also provided as a positive control and biotinylated protein standard.

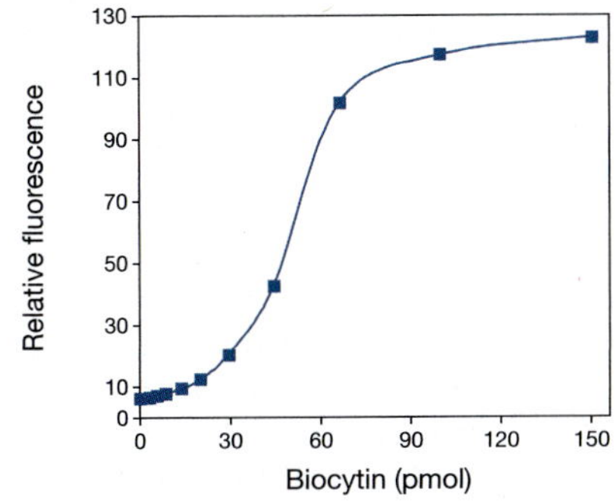

Figure 4.2.15 Standard curve showing dynamic range of the FluoReporter® Biotin Quantitation Assay Kit. Each reaction consisted of 1X PBS, 1X Biotective™ Green reagent and biocytin in a total volume of 100 µL. After a 5-minute incubation at room temperature in the dark, fluorescence was measured in a microplate reader using excitation at 485 ± 10 nm and fluorescence emission at 530 ± 12.5 nm.

FluoReporter® Biotin Quantitation Assay Kit for Biotinylated Nucleic Acids

The FluoReporter® Biotin Quantitation Assay Kit for biotinylated nucleic acids (F30755) provides a sensitive fluorometric assay for determining the number of biotin labels on a nucleic acid. The assay is based on the displacement of a quencher dye from the biotin-binding sites of Biotective™ Green reagent[52] (Figure 4.2.14). The assay can detect from 4 to 80 picomoles of biotin in a sample (Figure 4.2.16), providing a 50-fold higher sensitivity than the traditional HABA biotin-binding assay.[53] Analysis of multiply biotinylated nucleic acids requires a preliminary nuclease digestion step to avoid underestimation caused by steric restriction of avidin binding. This kit is ideal for determining the degree of biotinylation of cDNA samples used in Affymetrix or RLS microarray protocols, and it can be applied to as little as 13 ng of biotin-labeled nucleic acid (Table 4.3).

Each FluoReporter® Biotin Quantitation Assay Kit for biotinylated nucleic acids includes:

- Biotective™ Green reagent
- Biotin-dUMP
- Nuclease
- Concentrated phosphate-buffered saline (PBS)
- Biotinylated DNA postive control
- Concentrated nucleic acid digestion buffer
- Detailed protocols

Sufficient reagents are provided for assaying 10 samples independently using eight wells in triplicate for the standard curve and three dilutions of the sample in triplicate (totaling 33 wells per assay). However, fewer wells may be used to conserve sample and a single standard curve can be used for multiple samples in the same experimental session.

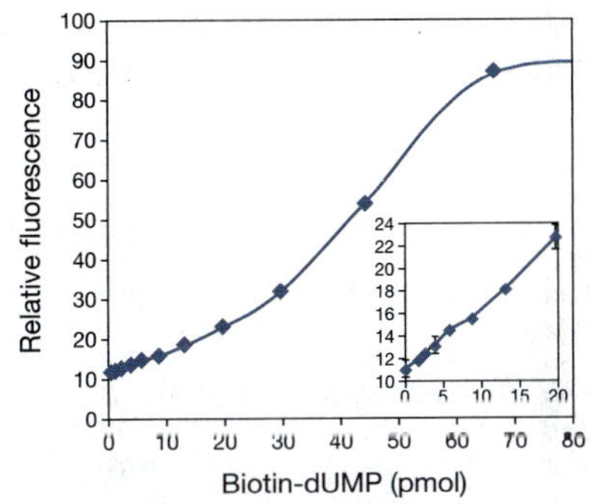

Figure 4.2.16 Standard curve showing dynamic range of the FluoReporter® Biotin Quantitation Assay Kit. Each reaction consisted of 0.5X PBS, 25 mM CHES buffer, pH 9.3, 1 mM $CaCl_2$, 1X Biotective™ Green reagent and biotin-dUMP in a total volume of 100 µL. After a 5-minute incubation at room temperature in the dark, fluorescence was measured in a microplate reader using excitation at 485 ± 7.5 nm and fluorescence emission at 535 ± 12.5 nm.

Haptenylation Reagents

A prerequisite for multicolor applications such as fluorescence *in situ* hybridization is the availability of multiple hapten molecules, along with their complementary binding proteins. The avidin–biotin system can provide only single-color detection, whereas antibody–hapten methods can generate a number of unique signals, limited only by the specificity of the antibody–hapten detection and the ability to distinguish the signals of different antibodies. The characteristics of a suitable hapten include a unique chemical structure that is not commonly found in cells (a bioorthogonal label), a high degree of antigenicity that elicits good antibody production, and a means for incorporating the hapten into the detection system. Our selection of haptenylation reagents enables researchers to covalently attach haptens to proteins, nucleotides and other biomolecules.

Table 4.3 Sensitivity of the FluoReporter® Biotin Quantitation Assay Kit for biotinylated nucleic acids with varying degrees of biotinylation.

Level of Labeling	Amount of DNA *Within Range of Assay (ng)
1 biotin: 10 bases	13–264
1 biotin: 50 bases	66–1300
1 biotin: 100 bases	132–2640
1 biotin: 200 bases	264–5280

* To convert ng of DNA to nanomoles or picomoles, use 330 g/mole for average molecular weight of a single base.

IMPORTANT NOTICE: The products described in this manual are covered by one or more Limited Use Label License(s). Please refer to the Appendix on page 971 and Master Product List on page 975. Products are For Research Use Only. Not intended for any animal or human therapeutic or diagnostic use.

In addition to our wide range of biotinylation reagents discussed above, we provide many unique haptenylation reagents, including an amine-reactive version of digoxigenin (A2952), dinitrophenyl-X[54] (DNP-X, SE; D2248) and several fluorophores (Table 4.2). We usually recommend haptenylation reagents that contain spacers between the hapten and the reactive groups to reduce potential interactions with the biomolecule to which it is conjugated and to make the hapten maximally available to secondary detection reagents. Most of the preferred haptenylation reagents in Table 4.2 possess this feature.

Fluorescein has been found to be an excellent hapten for *in situ* hybridization because it binds with high affinity to its anti-fluorescein antibody.[55–57] Anti-fluorescein antibodies cross-react with all of the Oregon Green® dyes (Section 1.5), permitting their use with conjugates prepared from these dyes. By adding antibodies that recognize the Alexa Fluor® 488, dansyl, tetramethylrhodamine and Texas Red® fluorophores to our line of detection reagents (Section 7.4), we have greatly expanded the number of potential haptens. Because the anti-tetramethylrhodamine and anti–Texas Red® dye antibodies cross-react with the tetramethylrhodamine, Lissamine rhodamine, Rhodamine Red™ and Texas Red® fluorophores, these antibody–fluorophore combinations should not be used simultaneously to generate separate signals in a multicolor experiment. Similarly, our antibody to the BODIPY® FL dye cross-reacts with some of the other BODIPY® dyes (Section 1.4).

REFERENCES

1. Methods Mol Biol (1998) 80:173; **2.** Methods Mol Biol (1995) 45:205; **3.** Biochemistry (1982) 21:978; **4.** Biochemistry (2001) 40:8343; **5.** Biochemistry (2001) 40:8352; **6.** Proc Natl Acad Sci U S A (1974) 71:3537; **7.** Ann Hematol (1997) 74:231; **8.** Proteomics (2008) 8:4012; **9.** Org Biomol Chem (2008) 6:908; **10.** Mol Cell Proteomics (2003) 2:242; **11.** J Biol Chem (2009) 284:22213; **12.** Methods Cell Biol (2007) 80:417; **13.** Proc Natl Acad Sci U S A (2006) 103:13932; **14.** Angew Chem Int Ed Engl (2009) 48:6974; **15.** Biochemistry (2009) 48:6571; **16.** ACS Chem Biol (2006) 1:644; **17.** J Am Chem Soc (2010) 132:2504; **18.** Chem Res Toxicol (2008) 21:432; **19.** Nat Cell Biol (2008) 10:1224; **20.** J Biol Chem (1992) 267:5060; **21.** Bioconjug Chem (2003) 14:205; **22.** Anal Biochem (2006) 354:54; **23.** Anal Biochem (1988) 170:271; **24.** Methods Mol Biol (2009) 536:457; **25.** Photochem Photobiol (2002) 76:123; **26.** Anal Chem (2001) 73:2229; **27.** Methods (2000) 22:164; **28.** Proc Natl Acad Sci U S A (2000) 97:686; **29.** J Biol Chem (2000) 275:6741; **30.** Anal Chem (2006) 78:6847; **31.** Anal Biochem (2003) 312:113; **32.** Anal Biochem (2002) 304:266; **33.** Proc Natl Acad Sci U S A (2009) 106:405; **34.** Bioconjug Chem (1996) 7:271; **35.** J Biol Chem (2003) 278:35184; **36.** J Biol Chem (2003) 278:4227; **37.** Anal Biochem (1994) 223:88; **38.** J Biol Chem (1994) 269:28309; **39.** Anal Biochem (1992) 205:166; **40.** Anal Biochem (2002) 308:343; **41.** Bioconjug Chem (2006) 17:366; **42.** FEBS Lett (1993) 328:165; **43.** Methods Enzymol (1970) 18:418; **44.** Methods Mol Biol (2005) 303:35; **45.** J Neurochem (2001) 77:1301; **46.** J Cell Sci (1996) 109:3025; **47.** Infect Immun (2006) 74:1148; **48.** J Virol (2001) 75:4744; **49.** J Biol Chem (1999) 274:36801; **50.** J Cell Biol (2009) 185:903; **51.** J Neurosci (2007) 27:1836; **52.** Biotechniques (2007) 43:503; **53.** Biochem J (1965) 94:23C; **54.** Eur J Cell Biol (1991) 56:223; **55.** Nucleic Acids Res (1991) 19:3237; **56.** J Histochem Cytochem (1990) 38:467; **57.** Nucleic Acids Res (2008) 36:4047.

DATA TABLE 4.2 BIOTINYLATION AND HAPTENYLATION REAGENTS

Cat. No.	MW	Storage	Soluble	Abs	EC	Em	Solvent	Notes
A1593	367.30	NC	DMF, DMSO	<300		none		
A1594	442.50	NC	DMF, DMSO	<300		none		
A2952	586.68	F,D	DMF, DMSO	<300		none		
A10550	445.41	F,D	H_2O, DMSO	<300		none		
A20000	643.41	F,DD,L	H_2O, DMSO	494	73,000	517	pH 7	1, 2, 3, 4
A20100	643.41	F,DD,L	H_2O, DMSO	494	73,000	517	pH 7	1, 2, 3, 4
A30000	1028.26	F,DD,L	H_2O, DMSO	400	35,000	424	pH 7	5, 6
A30005	884.91	F,DD,L	H_2O, DMSO	494	72,000	520	pH 7	2, 4, 7
A30052	825.46	F,DD,L	H_2O, DMSO	493	73,000	520	pH 7	2, 4, 7
A30100	1028.26	F,DD,L	H_2O, DMSO	400	35,000	424	pH 7	5, 6
B1513	341.38	F,D	DMF, DMSO	<300		none		
B1582	454.54	F,D	DMF, DMSO	<300		none		
B1591	454.33	F,D	DMF, DMSO	<300		none		8
B1592	372.48	NC	H_2O	<300		none		
B1595	244.31	NC	pH >6, DMF	<300		none		
B1596	555.65	NC	DMF, DMSO	<300		none		
B1603	386.51	D	pH >6, DMF	<300		none		
B1606	567.70	F,D	DMF, DMSO	<300		none		
B2600	484.66	D	DMF, DMSO	<300		none		
B2604	861.97	F,D,L	DMF	362	15,000	none	pH 8	
B6352	669.74	F,D	DMF, pH >6	<300		none		1
B6353	556.58	F,D	DMF, pH >6	<300		none		1
B10006	542.19	F,D,L	H_2O, DMSO	502	80,000	510	MeOH	11, 15
B10184	615.79	F,D,L		<300		none		
B10185	528.66	F,D		<300		none		
B20656	244.31	RO	pH >6	<300		none		9
C2284	607.42	F,D,LL	H_2O, MeOH	396	29,000	410	MeOH	5, 10
D2248	394.34	F,D,L	DMF, DMSO	348	18,000	none	MeOH	
D6102	502.32	F,D,L	DMSO, MeCN	504	85,000	510	MeOH	11
D6104	461.53	F,D,L	DMF, MeCN	335	4200	518	MeOH	
D20653	341.45	D	DMSO	<300		none		12
D20657	214.26	RO	pH >6	<300		none		9, 12
D30753	537.44	F,D	DMSO	<300		none		8, 12
F2181	586.55	F,D,L	DMF, DMSO	494	74,000	520	pH 9	13
F6130	590.56	F,D,L	DMF, DMSO	491	86,000	515	pH 9	13

IMPORTANT NOTICE: The products described in this manual are covered by one or more Limited Use Label License(s). Please refer to the Appendix on page 971 and Master Product List on page 975. Products are For Research Use Only. Not intended for any animal or human therapeutic or diagnostic use.
www.invitrogen.com/probes
molecular probes® | invitrogen™ by life technologies™

DATA TABLE 4.2 BIOTINYLATION AND HAPTENYLATION REAGENTS—*continued*

Cat. No.	MW	Storage	Soluble	Abs	EC	Em	Solvent	Notes
L1338	659.51	F,D,L	H_2O	426	11,000	531	pH 7	8
M1602	523.60	F,D	pH >6, DMF	<300		none		
N6356	251.77	D	DMF, pH <6	<300		none		
O6185	622.53	F,D,L	DMF, DMSO	494	84,000	517	pH 9	14
R6160	768.90	F,D,L	DMF, DMSO	560	129,000	580	MeOH	
T6105	640.69	F,D,L	DMF, DMSO	543	92,000	571	MeOH	
T6134	816.94	F,D,L	DMF, DMSO	583	112,000	603	MeOH	
T20175	816.94	F,D,L	DMF, DMSO	587	96,000	602	MeOH	
T30754	587.72	F,D	DMSO	<300		none		12

For definitions of the contents of this data table, see "Using *The Molecular Probes® Handbook*" in the introductory pages.

Notes

1. This sulfonated succinimidyl ester derivative is water soluble and may be dissolved in buffer at ~pH 8 for reaction with amines. Long-term storage in water is NOT recommended due to hydrolysis.
2. The fluorescence lifetime (τ) of the Alexa Fluor® 488 dye in pH 7.4 buffer at 20°C is 4.1 nanoseconds. Data provided by the SPEX Fluorescence Group, Horiba Jobin Yvon Inc.
3. A20100 is an alternative packaging of A20000 but is otherwise identical.
4. Abs and Em of the Alexa Fluor® 488 dye are red-shifted by as much as 16 nm and 25 nm respectively on microarrays relative to aqueous solution values. The magnitude of the spectral shift depends on the array substrate material. (Biotechniques (2005) 38:127)
5. The Alexa Fluor® 405 and Cascade Blue® dyes have a second absorption peak at about 376 nm with EC ~80% of the 395–400 nm peak.
6. A30100 is an alternative packaging of A30000 but is otherwise identical.
7. TFP and SDP ester derivatives are water-soluble and may be dissolved in buffer at ~pH 8 for reaction with amines. Long-term storage in water is NOT recommended due to hydrolysis.
8. Iodoacetamides in solution undergo rapid photodecomposition to unreactive products. Minimize exposure to light prior to reaction.
9. This product is supplied as a ready-made solution in the solvent indicated under "Soluble."
10. Unstable in water. Use immediately.
11. The absorption and fluorescence spectra of BODIPY® derivatives are relatively insensitive to the solvent.
12. The dissociation constant (K_d) for desthiobiotin binding to streptavidin is 1.9 nM. (Bioconjug Chem (2006) 17:366)
13. Absorption and fluorescence of fluorescein derivatives are pH dependent. Extinction coefficients and fluorescence quantum yields decrease markedly at pH <7.
14. Absorption and fluorescence of Oregon Green® 488 derivatives are pH dependent only in moderately acidic solutions (pH <5).
15. This sulfotetrafluorophenyl (STP) ester derivative is water soluble and may be dissolved in buffer at ~pH 8 for reaction with amines. Long-term storage in water is NOT recommended due to hydrolysis.

PRODUCT LIST 4.2 BIOTINYLATION AND HAPTENYLATION REAGENTS

Cat. No.	Description	Quantity
A30000	Alexa Fluor® 405 carboxylic acid, succinimidyl ester	1 mg
A30100	Alexa Fluor® 405 carboxylic acid, succinimidyl ester	5 mg
A20000	Alexa Fluor® 488 carboxylic acid, succinimidyl ester *mixed isomers*	1 mg
A20100	Alexa Fluor® 488 carboxylic acid, succinimidyl ester *mixed isomers*	5 mg
A30005	Alexa Fluor® 488 carboxylic acid, 2,3,5,6-tetrafluorophenyl ester (Alexa Fluor® 488 5-TFP) *5-isomer*	1 mg
A30052	Alexa Fluor® 488 5-SDP ester (Alexa Fluor® 488 sulfodichlorophenol ester)	1 mg
A2952	3-amino-3-deoxydigoxigenin hemisuccinamide, succinimidyl ester	5 mg
A1593	*N*-(2-aminoethyl)biotinamide, hydrobromide (biotin ethylenediamine)	25 mg
A10550	*N*-(aminooxyacetyl)-*N'*-(D-biotinoyl) hydrazine, trifluoroacetic acid salt (ARP)	10 mg
A1594	*N*-(5-aminopentyl)biotinamide, trifluoroacetic acid salt (biotin cadaverine)	25 mg
B1592	biocytin (ε-biotinoyl-L-lysine)	100 mg
B1603	biocytin hydrazide	25 mg
B1595	D-biotin	1 g
B20656	D-biotin *50 mM aqueous solution*	10 mL
B10185	biotin alkyne (PEG_4 carboxamide-propargyl biotin)	1 mg
B10184	biotin azide (PEG_4 carboxamide-6-azidohexanyl biotin)	1 mg
B1582	6-((biotinoyl)amino)hexanoic acid, succinimidyl ester (biotin-X, SE; biotinamidocaproate, *N*-hydroxysuccinimidyl ester)	100 mg
B6353	6-((biotinoyl)amino)hexanoic acid, sulfosuccinimidyl ester, sodium salt (Sulfo-NHS-LC-Biotin; biotin-X, SSE)	25 mg
B1606	6-((6-((biotinoyl)amino)hexanoyl)amino)hexanoic acid, succinimidyl ester (biotin-XX, SE)	100 mg
B6352	6-((6-((biotinoyl)amino)hexanoyl)amino)hexanoic acid, sulfosuccinimidyl ester, sodium salt (biotin-XX, SSE)	25 mg
B1591	*N*-(biotinoyl)-*N'*-(iodoacetyl)ethylenediamine	25 mg
B1513	D-biotin, succinimidyl ester (succinimidyl D-biotin)	100 mg
B1596	biotin-X cadaverine (5-(((*N*-(biotinoyl)amino)hexanoyl)amino)pentylamine, trifluoroacetic acid salt)	10 mg
B2604	biotin-X 2,4-dinitrophenyl-X-L-lysine, succinimidyl ester (DNP-X-biocytin-X, SE)	5 mg
B2600	biotin-XX hydrazide (6-((6-((biotinoyl)amino)hexanoyl)amino)hexanoic acid, hydrazide)	25 mg
B30010	Biotin-XX Microscale Protein Labeling Kit *for 20–100 µg protein* *3 labelings*	1 kit
B30756	Biotin-XX Microscale Protein Labeling Kit with FluoReporter® Biotin Quantitation Assay Kit *includes B30010 and F30751*	1 kit
B10006	BODIPY® FL, STP ester, sodium salt	5 mg
C2284	Cascade Blue® acetyl azide, trisodium salt	5 mg

continued on next page

PRODUCT LIST 4.2 BIOTINYLATION AND HAPTENYLATION REAGENTS—*continued*

Cat. No.	Description	Quantity
D20657	D-desthiobiotin *50 mM aqueous solution*	10 mL
D6102	6-((4,4-difluoro-5,7-dimethyl-4-bora-3a,4a-diaza-s-indacene-3-propionyl)amino)hexanoic acid, succinimidyl ester (BODIPY® FL-X, SE)	5 mg
D6104	6-((5-dimethylaminonaphthalene-1-sulfonyl)amino)hexanoic acid, succinimidyl ester (dansyl-X, SE)	25 mg
D2248	6-(2,4-dinitrophenyl)aminohexanoic acid, succinimidyl ester (DNP-X, SE)	25 mg
D30753	DSB-X™ biotin C_2-iodoacetamide (desthiobiotin-X C_2-iodoacetamide)	5 mg
D20653	DSB-X™ biotin hydrazide	5 mg
D20655	DSB-X™ Biotin Protein Labeling Kit *5 labelings*	1 kit
F30755	FluoReporter® Biotin Quantitation Assay Kit *for biotinylated nucleic acids* *10 determinations*	1 kit
F30751	FluoReporter® Biotin Quantitation Assay Kit *for biotinylated proteins* *5 determinations*	1 kit
F6348	FluoReporter® Biotin/DNP Protein Labeling Kit *5–10 labelings*	1 kit
F2610	FluoReporter® Biotin-XX Protein Labeling Kit *5 labelings of 5–20 mg protein each*	1 kit
F20650	FluoReporter® Cell-Surface Biotinylation Kit	1 kit
F6347	FluoReporter® Mini-biotin-XX Protein Labeling Kit *5 labelings of 0.1–3 mg protein each*	1 kit
F2181	6-(fluorescein-5-(and-6)-carboxamido)hexanoic acid, succinimidyl ester (5(6)-SFX) *mixed isomers*	10 mg
F6130	fluorescein-5-EX, succinimidyl ester	10 mg
L1338	lucifer yellow iodoacetamide, dipotassium salt	25 mg
M1602	Nα-(3-maleimidylpropionyl)biocytin	25 mg
N6356	norbiotinamine, hydrochloride	10 mg
O6185	Oregon Green® 488-X, succinimidyl ester *6-isomer*	5 mg
R6160	Rhodamine Red™-X, succinimidyl ester *5-isomer*	5 mg
T6105	6-(tetramethylrhodamine-5-(and-6)-carboxamido)hexanoic acid, succinimidyl ester (5(6)-TAMRA-X, SE) *mixed isomers*	10 mg
T6134	Texas Red®-X, succinimidyl ester *mixed isomers*	5 mg
T20175	Texas Red®-X, succinimidyl ester *single isomer*	2 mg
T30754	TS-Link™ DSB-X™ biotin C_5-thiosulfate (TS-Link™ desthiobiotin-X C_5-thiosulfate, sodium salt)	5 mg
Z25452	Zenon® Biotin-XX Human IgG Labeling Kit *50 labelings*	1 kit
Z25052	Zenon® Biotin-XX Mouse IgG_1 Labeling Kit *50 labelings*	1 kit
Z25152	Zenon® Biotin-XX Mouse IgG_{2a} Labeling Kit *50 labelings*	1 kit
Z25252	Zenon® Biotin-XX Mouse IgG_{2b} Labeling Kit *50 labelings*	1 kit
Z25352	Zenon® Biotin-XX Rabbit IgG Labeling Kit *50 labelings*	1 kit

4.3 Biotin and Desthiobiotin Conjugates

We prepare a wide array of biotin and desthiobiotin conjugates, all of which are included in this section's product list. We will also custom-conjugate biotin, desthiobiotin (DSB-X™), fluorophores or other haptens to proteins or other biomolecules of interest; contact Invitrogen Custom Services to request a quote.

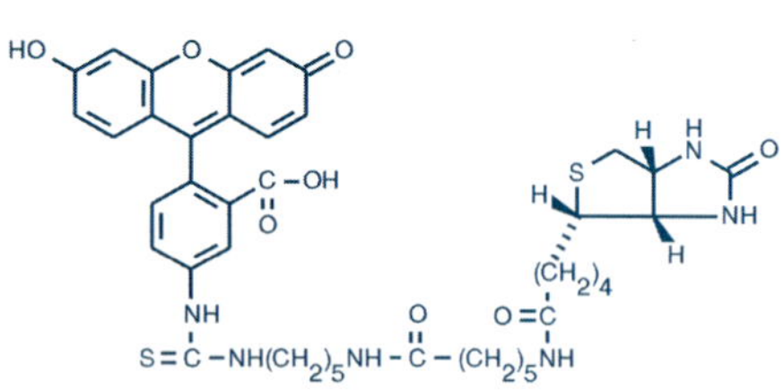

Figure 4.3.1 5-((N-(5-(N-(6-(biotinoyl)amino)hexanoyl)amino)pentyl)thioureidyl)fluorescein (fluorescein biotin, B1370).

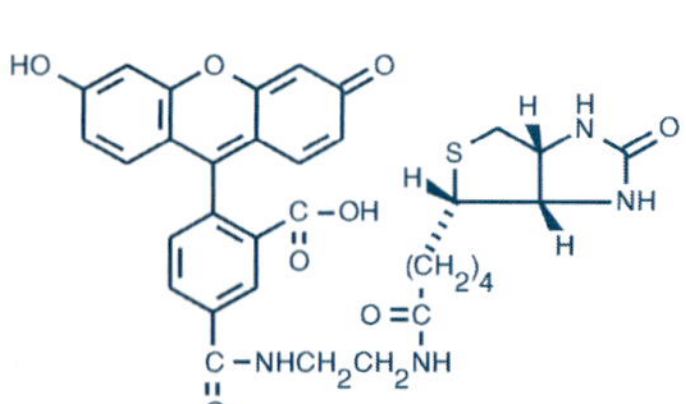

Figure 4.3.2 Biotin-4-fluorescein (B10570).

Fluorescent Biotin Derivatives

Fluorescein Biotin

Fluorescein biotin (B1370, Figure 4.3.1) was developed as an alternative to radioactive biotin for detecting and quantitating biotin-binding sites by either fluorescence or absorbance.[1] A fluorescence polarization–based assay that employs competitive binding of fluorescein biotin to assess the degree of protein biotinylation has been reported[2] (Fluorescence Polarization (FP)—Note 1.4). A similar derivative was used for determining avidin and biotin concentrations by fluorescence depolarization.[3]

Biotin-4-Fluorescein

Our biotin-4-fluorescein (B10570, Figure 4.3.2) offers a substantially improved method for quantitating biotin-binding sites. Biotin-4-fluorescein binds to avidin much faster than does conventional fluorescein biotin, allowing for rapid analysis.[4] The strong quenching associated with avidin binding to biotin-4-fluorescein can be used to accurately measure the concentration of avidin or streptavidin[5] (Figure 4.3.3). Engineered single-chain dimers created by circular

IMPORTANT NOTICE: The products described in this manual are covered by one or more Limited Use Label License(s). Please refer to the Appendix on page 971 and Master Product List on page 975. Products are For Research Use Only. Not intended for any animal or human therapeutic or diagnostic use.
www.invitrogen.com/probes

permutation of wild-type streptavidin exhibit substantial binding preference for biotin-4-fluorescein relative to biotin.[6]

Other Fluorescent Biotin Derivatives

In addition to supplying nonfluorescent biocytin (ε-biotinoyl-L-lysine, B1592, Section 4.2), we offer:

- Alexa Fluor® 488 biocytin (A12924, Figure 4.3.4)
- Alexa Fluor® 546 biocytin (A12923)
- Alexa Fluor® 594 biocytin (A12922, Figure 4.3.5)
- Lucifer yellow cadaverine biotin-X (L2601)
- Lucifer yellow biocytin (L6950)
- Oregon Green® 488 biocytin (O12920)
- Tetramethylrhodamine biocytin (T12921)

Each of these reagents contains both a fluorophore and biotin moiety in the same molecule. As with fluorescein biotin and biotin-4-fluorescein, which are described above, these reagents can be employed for detecting and quantitating biotin-binding proteins, but their principal application is as aldehyde-fixable polar cell tracers and as tracers for cell–cell communication[7,8] (Section 14.3). Our lucifer yellow cadaverine biotin-X is reportedly well retained in aldehyde-fixed tissues, even after sectioning, extraction with detergents and several washes.[9]

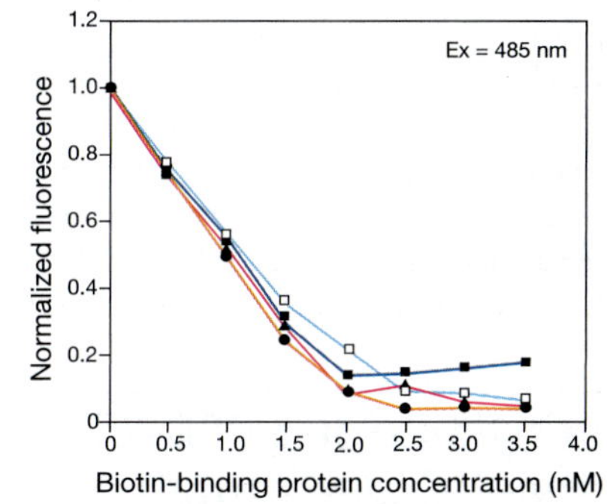

Figure 4.3.3 Quantitation of biotin-binding sites with 8 nM biotin-4-fluorescein (B10570). Both the fluorescence and absorbance of biotin-4-fluorescein are quenched upon binding to one of the four biotin-binding sites of streptavidin (S888, pink), avidin (A887, A2667; dark blue), or the streptavidin conjugates of the Alexa Fluor® 633 dye (S21375, orange) and alkaline phosphatase (S921, light blue). As a result, when a known concentration of biotin-4-fluorescein is added to a known amount of streptavidin, one can estimate the number of biotin-binding sites.

Biotinylated Dextrans

In addition to the low molecular weight biotinylated tracers described above, we prepare a variety of biotinylated dextrans (Figure 4.3.6, Table 14.4), including dextrans that are double-labeled with fluorophores and biotin moieties for correlated fluorescence and electron microscopy studies. We currently offer the following biotinylated dextrans:

- Dextran, biotin, 3000 MW, lysine fixable (BDA-3000) (D7135)
- Dextran, biotin, 10,000 MW, lysine fixable (BDA-10,000) (D1956)
- Dextran, biotin, 70,000 MW, lysine fixable (BDA-70,000) (D1957)
- Dextran, biotin, 500,000 MW, lysine fixable (BDA-500,000) (D7142)
- Dextran, fluorescein and biotin, 3000 MW, anionic, lysine fixable (micro-emerald) (D7156)
- Dextran, fluorescein and biotin, 10,000 MW, anionic, lysine fixable (mini-emerald) (D7178)
- Dextran, tetramethylrhodamine and biotin, 3000 MW, lysine fixable (micro-ruby) (D7162)
- Dextran, tetramethylrhodamine and biotin, 10,000 MW, lysine fixable (mini-ruby) (D3312)

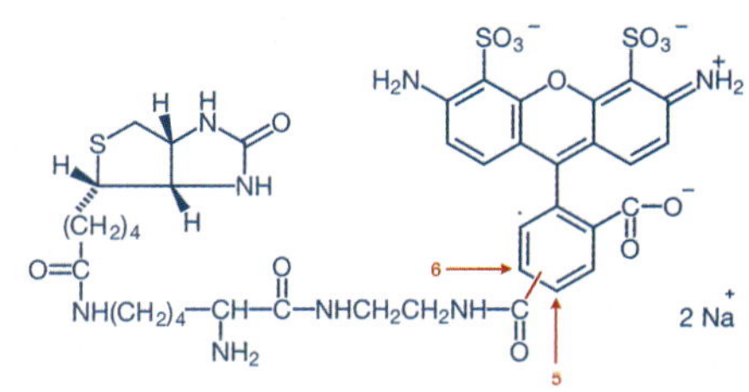

Figure 4.3.4 Alexa Fluor® 488 biocytin, disodium salt (biocytin Alexa Fluor® 488, A12924).

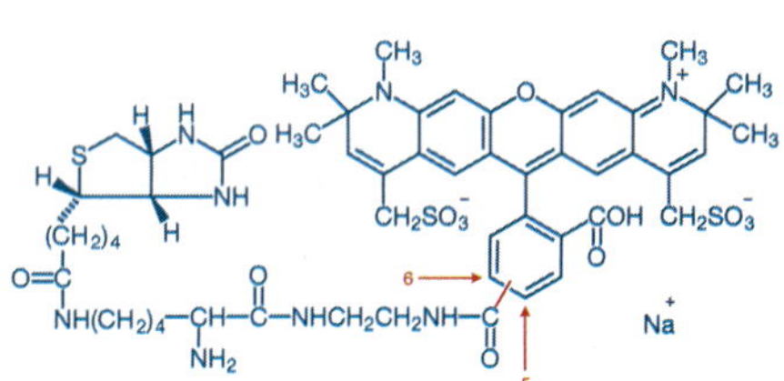

Figure 4.3.5 Alexa Fluor® 594 biocytin, sodium salt (biocytin Alexa Fluor® 594, A12922).

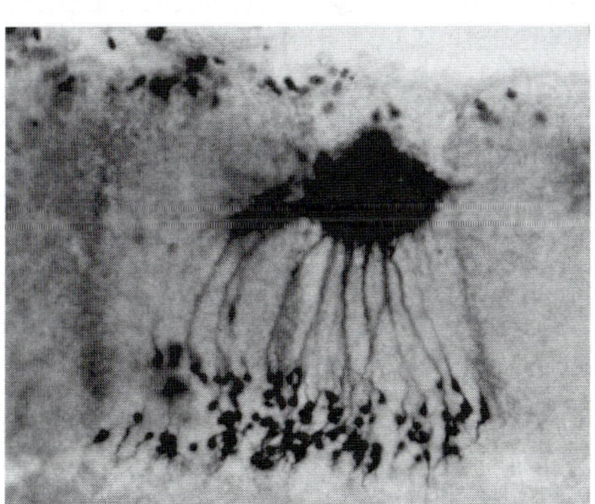

Figure 4.3.6 Motor neuron in a three-day chick embryo labeled with lysine-fixable, biotinylated 3000 MW dextran (BDA-3000, D7135). Filled neurons were detected with biotinylated horseradish peroxidase (P917) and diaminobenzidine using standard avidin/streptavidin bridging techniques. Reprinted with permission from J Neurosci Methods (1993) 50:95.

Dextrans are hydrophilic polysaccharides characterized by their moderate to high molecular weight, good water solubility and low toxicity. They are biologically inert due to their uncommon poly-(α-D-1,6-glucose) linkages, which render them resistant to cleavage by most endogenous cellular glycosidases. Dextrans are widely used as both anterograde and retrograde tracers in neurons and for many other diverse applications; see Section 14.5 for a discussion of the applications of these reagents, particularly as cell tracers.

Biotinylated and DSB-X™ Biotin–Labeled Proteins

Our biotinylated primary and secondary antibodies, F(ab′)₂ fragments, phycobiliproteins and enzymes are reliable detection reagents for a broad assortment of assays; for more information, see Chapter 6 and Chapter 7. Biotinylated R-phycoerythrin (P811) and biotinylated horseradish peroxidase (P917) can be used in combination with an avidin or streptavidin bridge to amplify the detection of biotinylated targets.

Biotinylated Primary Antibodies

We prepare biotin conjugates of several primary antibodies, including:

- Anti-bromodeoxyuridine, mouse IgG_1, monoclonal MoBU-1, biotin conjugate (biotinylated anti-BrdU, B35138)
- Anti–HuC/HuD neuronal protein (human), mouse IgG_{2b}, monoclonal 16A11, biotin-XX conjugate (A21272)
- Anti–α-tubulin (bovine), mouse IgG_1, monoclonal 236-10501, biotin-XX conjugate (A21371)
- Anti–dinitrophenyl-KLH, rabbit IgG fraction, biotin-XX conjugate (A6435)
- Anti–fluorescein/Oregon Green®, rabbit IgG fraction, biotin-XX conjugate (A982)
- Anti–lucifer yellow, rabbit IgG fraction, biotin-XX conjugate (A5751)
- Anti–Green Fluorescent Protein (GFP), rabbit IgG fraction, biotin-XX conjugate (A10259)
- Anti–Green Fluorescent Protein (GFP), chicken IgY fraction, biotin-XX conjugate (A10263)

For a complete list of primary antibodies available from Invitrogen, go to www.invitrogen.com/handbook/antibodies.

Biotinylated and DSB-X™ Biotin–Labeled Secondary Antibodies

We prepare biotin and DSB-X™ biotin conjugates of several commonly used secondary antibodies, including:

- Biotin-XX $F(ab')_2$ fragment of goat anti–mouse IgG (H+L) (B11027)
- Biotin-XX $F(ab')_2$ fragment of goat anti–rabbit IgG (H+L) (B21078)
- Biotin-XX goat anti–mouse IgG (H+L) (B2763)
- Biotin-XX goat anti–rabbit IgG (H+L) (B2770)
- DSB-X™ biotin goat anti–mouse IgG (H+L) (D20691)
- DSB-X™ biotin goat anti–mouse IgM (μ chain) (D20693)
- DSB-X™ biotin goat anti–rat IgG (H+L) (D20697)
- DSB-X™ biotin goat anti–chicken IgG (H+L) (D20701)
- DSB-X™ biotin donkey anti–goat IgG (H+L) (D20698)

Targets complexed with DSB-X™ biotin–labeled antibodies can be selectively detected with avidin or streptavidin conjugates or isolated on affinity matrices, including streptavidin agarose (S951, D20658; Section 7.6), and then rapidly released with D-biotin (B1595, B20656; Section 4.2) under extremely gentle conditions [10] (Figure 4.3.7). See Section 4.1 for a complete description of our unique DSB-X™ biotin technology.

BioGEE: A Biotinylated Glutathione Analog

Biotinylated glutathione ethyl ester (BioGEE, G36000; Figure 4.3.8) is a cell-permeant, biotinylated glutathione analog for detecting glutathiolation. Under conditions of oxidative stress, cells may transiently incorporate glutathione into proteins. Stressed cells incubated with BioGEE will also incorporate this biotinylated glutathione derivative into proteins, facilitating the identification of oxidation-sensitive proteins.[11,12] Once these cells are fixed and permeabilized, glutathiolation levels can be detected with a fluorescent streptavidin conjugate (Section 7.6, Table 7.9) using either flow cytometry or fluorescence microscopy. Proteins glutathiolated with BioGEE can be captured using streptavidin agarose (S951, Section 7.6) and analyzed by mass spectrometry or by western blotting methods.[13,14]

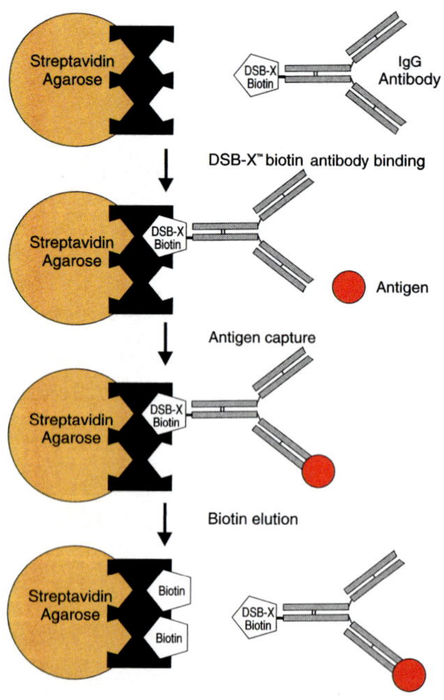

Figure 4.3.7 Diagram illustrating the use of streptavidin agarose and a DSB-X™ biotin bioconjugate in affinity chromatography. A DSB-X™ biotin–labeled IgG antibody and its target antigen are used as an example.

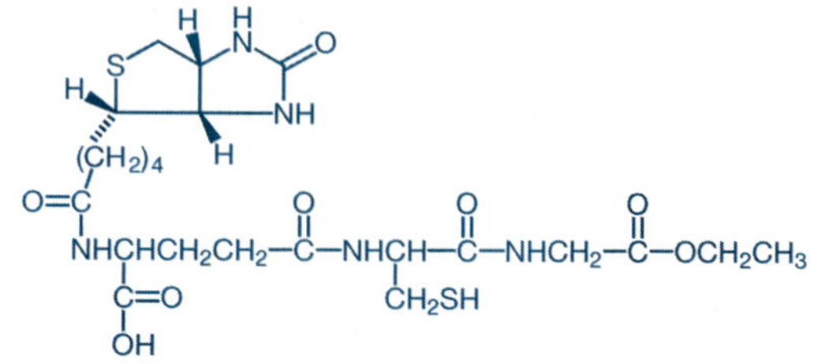

Figure 4.3.8 Glutathione ethyl ester, biotin amide (BioGEE, G36000).

IMPORTANT NOTICE: The products described in this manual are covered by one or more Limited Use Label License(s). Please refer to the Appendix on page 971 and Master Product List on page 975. Products are For Research Use Only. Not intended for any animal or human therapeutic or diagnostic use.

Biotinylated Microspheres

Biotinylated FluoSpheres® polystyrene microspheres have significant potential for signal amplification techniques, as described in Section 6.5. Like biotinylated R-phycoerythrin (P811), they can be used with bridging techniques to detect biotinylated targets. We currently offer the following biotinylated FluoSpheres® microspheres:

- FluoSpheres® biotin-labeled microspheres, 0.04 µm, yellow-green fluorescent (505/515), 1% solids (F8766)
- FluoSpheres® biotin-labeled microspheres, 0.2 µm, yellow-green fluorescent (505/515), 1% solids (F8767)
- FluoSpheres® biotin-labeled microspheres, 1.0 µm, yellow-green fluorescent (505/515), 1% solids (F8768)
- FluoSpheres® biotin-labeled microspheres, 1.0 µm, nonfluorescent, 1% solids (F8769)

FluoSpheres® polystyrene microspheres satisfy several prerequisites of ideal long-term biological tracers. Because the dyes in our microspheres are incorporated throughout the microsphere rather than just on its surface, the fluorescence output per microsphere is significantly greater than that obtained from protein or dextran conjugates and is relatively immune to photobleaching and other environment-dependent effects. FluoSpheres® microspheres are also biologically inert and physically durable, and are available with a large number of uniform sizes and surface properties.

Biotinylated Qdot® Nanocrystals

The Qdot® 605 and Qdot® 655 Biotin Conjugate Kits (Q10301MP, Q10321MP) provide Qdot® nanocrystals that have been functionalized with biotin on their surface via carbodiimide-mediated coupling. The biotinylated Qdot® 605 or biotinylated Qdot® 655 conjugate typically incorporates 5–7 biotin groups per Qdot® nanocrystal. In addition to biotinylated Qdot® nanocrystals, each kit also provides Qdot® Incubation Buffer, which is formulated specifically to achieve improved signal-to-noise ratios in immunohistochemical applications. Biotinylated Qdot® nanocrystals have been used alongside biotinylated microspheres for analysis of the effects of fluorescent label size on ligand–receptor binding dynamics and equilibrium by fluorescence correlation spectroscopy [15] (Fluorescence Correlation Spectroscopy (FCS)—Note 1.3).

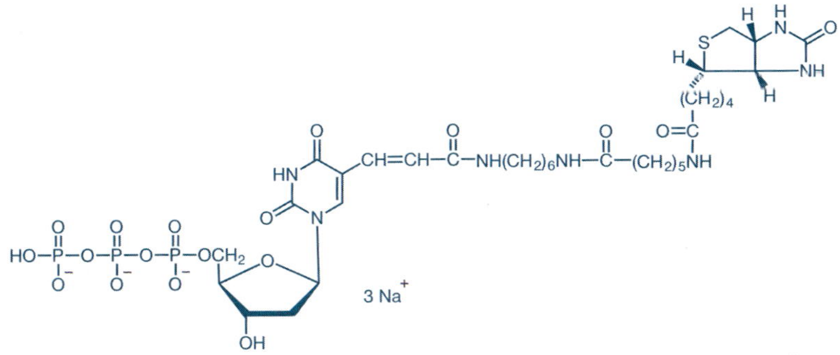

Figure 4.3.9 Biotin-aha-dUTP (B32766).

Biotinylated Nucleotides

Biotinylated nucleic acids are common nonisotopic probes used in hybridization techniques. Nucleoside triphosphate analogs such as our ChromaTide® UTP and dUTP nucleotides and aha-dUTP and aha-dCTP nucleotides (Section 8.2) are important reagents for preparing labeled nucleic acids for use as hybridization probes. The biotinylated aminohexylacrylamido-dUTP (biotin aha-dUTP) and biotinylated aminohexylacrylamido-dCTP (biotin aha-dCTP) derivatives (B32766, Figure 4.3.9; B32772) each contain a long 11-atom spacer between the biotin and its attachment point on the nucleic acid to facilitate its detection and signal amplification by fluorophore and enzyme conjugates of avidin and streptavidin (Section 7.6, Table 7.9).

Biotinylated Site-Selective Probes

Biotin conjugates of moderately low molecular weight ligands provide a means of amplifying the detection of ligand binding using fluorophore- or enzyme-labeled avidins or streptavidins. They may also be useful for immobilizing receptor ligands on streptavidin agarose (S951, D20658) or CaptAvidin™ agarose (C21386, Figure 4.3.10) for affinity isolation of receptors. See Section 7.6 for a description of these avidin conjugates, as well as other affinity isolation methods that use biotin- or DSB-X™ biotin–labeled reagents.

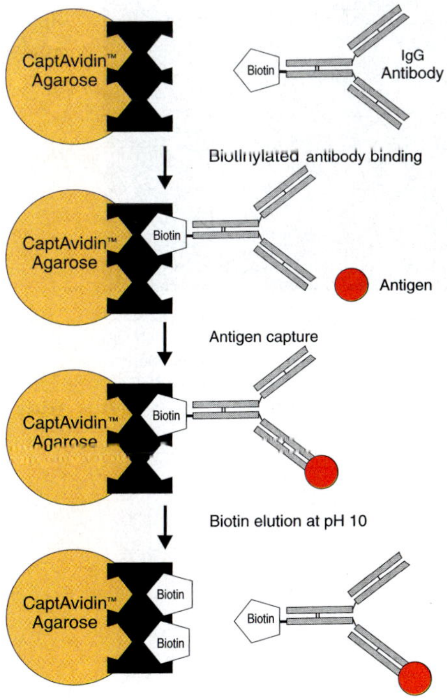

Figure 4.3.10 Diagram of the use of CaptAvidin™ agarose (C21386) in affinity chromatography. A biotinylated IgG molecule and target antigen are used as an example.

IMPORTANT NOTICE: The products described in this manual are covered by one or more Limited Use Label License(s). Please refer to the Appendix on page 971 and Master Product List on page 975. Products are For Research Use Only. Not intended for any animal or human therapeutic or diagnostic use.

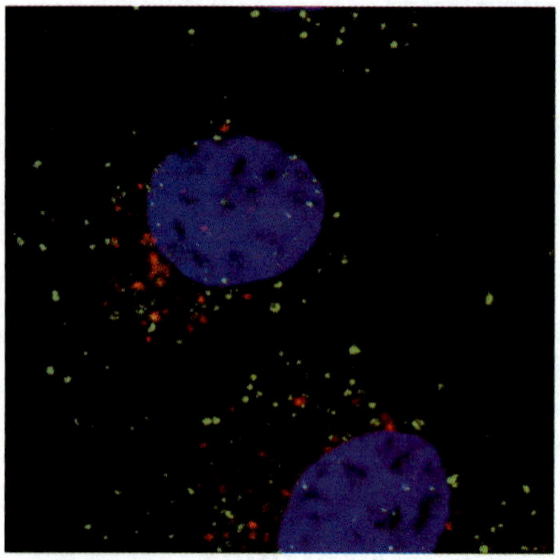

Figure 4.3.11 Early endosomes in live HeLa cells identified after a 10-minute incubation with green-fluorescent Alexa Fluor® 488 epidermal growth factor (E13345). The cells were subsequently fixed with formaldehyde and labeled with an antibody to the late endosomal protein, RhoB. That antibody was visualized with a red-orange–fluorescent secondary antibody. Nuclei were stained with TO-PRO®-3 iodide (T3605, pseudocolored blue). The image was contributed by Harry Mellor, University of Bristol.

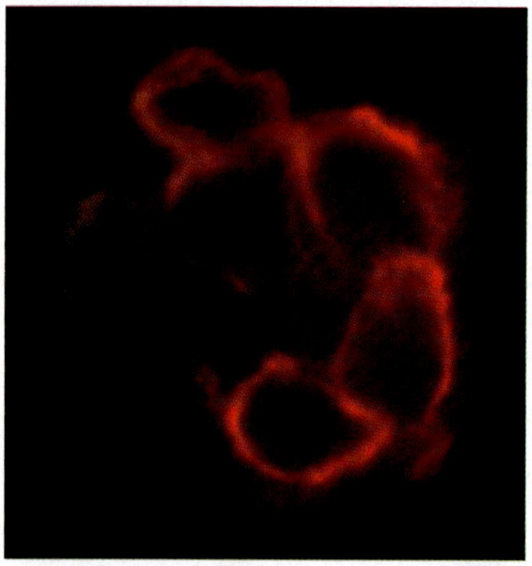

Figure 4.3.12 Lightly fixed human epidermoid carcinoma cells (A431) stained with biotinylated epidermal growth factor (EGF) complexed to Texas Red® streptavidin (E3480). An identical cell preparation stained in the presence of a 100-fold excess of unlabeled EGF (E3476) showed no fluorescent signal (data not shown).

Our biotinylated ligands include:

- Biotin-X conjugate of annexin V (A13204, Section 15.5), for detecting the externalization of phosphatidylserine, an early indicator of apoptosis
- Biotin-XX conjugate of α-bungarotoxin [16–19] (B1196, Section 16.2), for labeling the nicotinic acetylcholine receptor (nAChR)
- Biotin-XX cholera toxin subunit B (C34779, Section 7.7), which binds to cell-surface ganglioside G_{M1} receptors and is a marker of lipid rafts [20]
- Biotin-XX conjugate of epidermal growth factor [21–23] (EGF, E3477), as well as biotinylated EGF complexed with Alexa Fluor® 488 streptavidin (E13345, Figure 4.3.11), Alexa Fluor® 555 streptavidin (E35350), Alexa Fluor® 647 streptavidin (E35351) or Texas Red® streptavidin (E3480, Figure 4.3.12), for labeling EGF receptors (Section 16.1)
- Biotin-XX conjugate of isolectin IB_4 from *Griffonia simplicifolia* [24] (I21414, Section 7.7), for detecting α-galactosyl moieties in microglia and other glycoproteins
- Biotin-XX conjugate of phalloidin [25,26] (B7474, Section 11.1), for labeling F-actin
- Biotin-XX conjugate of transferrin [27] (T23363, Section 16.1), for following intracellular trafficking of this iron-carrying protein
- Biotin-XX tyramide, which is an amplification reagent used in several of our TSA™ Kits (Section 6.2, Table 6.1) for high-sensitivity detection of targets in cells and tissues

Biotinylated Lipids

Our extensive selection of labeled phospholipids includes two phospholipid derivatives of biotin and biotin-X:

- Biotin DHPE (*N*-(biotinoyl)-1,2-dihexadecanoyl-*sn*-glycero-3-phosphoethanolamine (B1550, Figure 4.3.13)
- Biotin-X DHPE (*N*-((6-(biotinoyl)amino)hexanoyl)-1,2-dihexadecanoyl-*sn*-glycero-3-phosphoethanolamine (B1616, Figure 4.3.14)

These biotinylated lipids, which are also described in Section 13.2, can be used to prepare liposomes that retain high affinity for avidin conjugates and can be captured for downstream analysis using streptavidin agarose [28] (S951, Section 7.6).

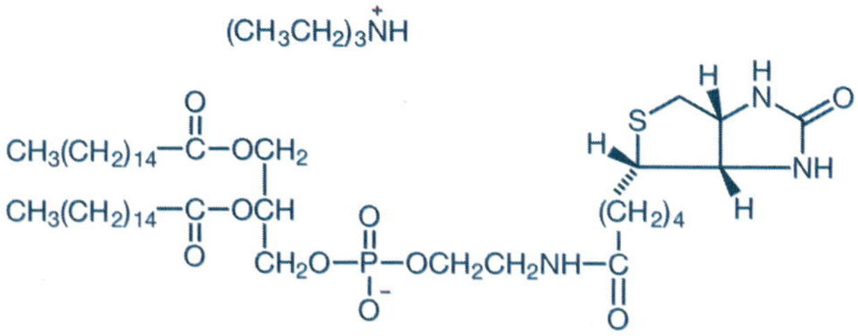

Figure 4.3.13 Biotin DHPE (*N*-(biotinoyl)-1,2-dihexadecanoyl-*sn*-glycero-3-phosphoethanolamine, triethylammonium salt, B1550).

IMPORTANT NOTICE: The products described in this manual are covered by one or more Limited Use Label License(s). Please refer to the Appendix on page 971 and Master Product List on page 975. Products are For Research Use Only. Not intended for any animal or human therapeutic or diagnostic use.
www.invitrogen.com/probes

REFERENCES

1. Biochim Biophys Acta (1998) 1381:203; **2.** Clin Chem (1994) 40:2112; **3.** Anal Chem (1988) 60:853; **4.** Biochim Biophys Acta (1999) 1427:33; **5.** Methods Mol Biol (2008) 418:73; **6.** Proc Natl Acad Sci U S A (2005) 102:8507; **7.** Science (2008) 321:417; **8.** Nat Neurosci (2005) 8:916; **9.** J Neurosci Methods (1993) 46:59; **10.** Anal Biochem (2002) 308:343; **11.** Amino Acids (2007) 33:51; **12.** Biochemistry (2000) 39:11121; **13.** J Biol Chem (2009) 284:22213; **14.** Nat Immunol (2008) 9:866; **15.** Biophys J (2008) 95:865; **16.** BMC Neurosci (2009) 10:80; **17.** Nano Lett (2008) 8:780; **18.** Mol Brain (2008) 1:18; **19.** J Neurosci (2005) 25:9949; **20.** Am J Pathol (2008) 172:1683; **21.** Methods Mol Biol (2007) 374:69; **22.** J Microsc (2006) 222:22; **23.** Nat Biotechnol (2004) 22:198; **24.** Nat Neurosci (2008) 11:429; **25.** J Cell Biol (1995) 130:591; **26.** Anal Biochem (1992) 200:199; **27.** J Cell Biol (2000) 149:901; **28.** Biotechniques (2005) 38:858.

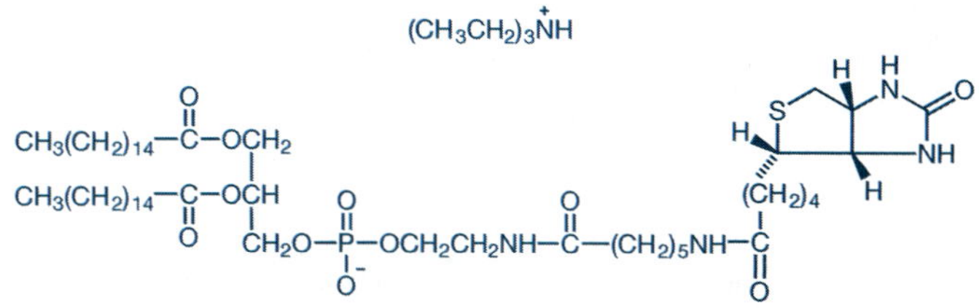

Figure 4.3.14 Biotin-X DHPE (*N*-((6-(biotinoyl)amino)hexanoyl)-1,2-dihexadecanoyl-*sn*-glycero-3-phosphoethanolamine, triethylammonium salt, B1616).

DATA TABLE 4.3 BIOTIN AND DESTHIOBIOTIN CONJUGATES

Cat. No.	MW	Storage	Soluble	Abs	EC	Em	Solvent	Notes
A12922	1141.31	D,L	DMSO, H_2O	591	80,000	618	pH 7	
A12923	1209.66	D,L	DMSO, H_2O	556	99,000	572	pH 7	
A12924	974.98	D,L	DMSO, H_2O	494	62,000	520	pH 7	
B1196	~8400	F,D	H_2O	<300		none		1
B1370	831.01	L	DMF, pH >6	494	75,000	518	pH 9	2
B1550	1019.45	FF,D	see Notes	<300		none		3
B1616	1132.61	FF,D	see Notes	<300		none		3
B7474	~1300	F	MeOH, H_2O	<300		none		1, 4
B10570	644.70	L	DMSO	494	68,000	523	pH 9	2
B32766	1041.78	FF	H_2O	<300		none		5
B32772	~1050	FF	H_2O	<300		none		5
E3477	~6600	FF,D	H_2O	<300		none		1
E3480	see Notes	FF,D,L	H_2O	596	ND	612	pH 7	6, 7
E13345	see Notes	FF,D,L	H_2O	497	ND	520	pH 8	6, 8
E35350	see Notes	FF,D,L	H_2O	554	ND	568	pH 7	6, 9
E35351	see Notes	FF,D,L	H_2O	653	ND	671	pH 7	6, 10
G36000	561.67	F,D	DMSO	<300		none		
L2601	873.10	D,L	H_2O	428	11,000	531	H_2O	
L6950	850.03	D,L	H_2O	428	11,000	532	pH 7	
O12920	887.39	L	DMSO, H_2O	495	66,000	522	pH 9	11
T12921	869.09	D,L	DMSO	554	103,000	581	pH 7	

For definitions of the contents of this data table, see "Using *The Molecular Probes® Handbook*" in the introductory pages.

Notes

1. α-Bungarotoxin, EGF and phallotoxin conjugates have approximately 1 label per peptide.
2. Absorption and fluorescence of fluorescein derivatives are pH dependent. Extinction coefficients and fluorescence quantum yields decrease markedly at pH <7.
3. Chloroform is the most generally useful solvent for preparing stock solutions of phospholipids (including sphingomyelins). Glycerophosphocholines are usually freely soluble in ethanol. Most other glycerophospholipids (phosphoethanolamines, phosphatidic acids and phosphoglycerols) are less soluble in ethanol, but solutions up to 1–2 mg/mL should be obtainable, using sonication to aid dispersion if necessary. Labeling of cells with fluorescent phospholipids can be enhanced by addition of cyclodextrins during incubation. (J Biol Chem (1999) 274:35359)
4. Although this phallotoxin is water soluble, storage in water is not recommended, particularly in dilute solution.
5. This product is supplied as a ready-made solution in the solvent indicated under "Soluble."
6. ND = not determined.
7. E3480 is a complex of E3477 with Texas Red® streptavidin, which typically incorporates 3 dyes/streptavidin (MW ~52,800).
8. E13345 is a complex of E3477 with Alexa Fluor® 488 streptavidin, which typically incorporates 5 dyes/streptavidin (MW ~52,800).
9. E35350 is a complex of E3477 with Alexa Fluor® 555 streptavidin, which typically incorporates 3 dyes/streptavidin (MW ~52,800).
10. E35351 is a complex of E3477 with Alexa Fluor® 647 streptavidin, which typically incorporates 3 dyes/streptavidin (MW ~52,800).
11. Absorption and fluorescence of Oregon Green® 488 derivatives are pH dependent only in moderately acidic solutions (pH <5).

IMPORTANT NOTICE: The products described in this manual are covered by one or more Limited Use Label License(s). Please refer to the Appendix on page 971 and Master Product List on page 975. Products are For Research Use Only. Not intended for any animal or human therapeutic or diagnostic use.

PRODUCT LIST 4.3 BIOTIN AND DESTHIOBIOTIN CONJUGATES

Cat. No.	Description	Quantity
A12924	Alexa Fluor® 488 biocytin, disodium salt (biocytin Alexa Fluor® 488)	250 µg
A12923	Alexa Fluor® 546 biocytin, sodium salt (biocytin Alexa Fluor® 546)	250 µg
A12922	Alexa Fluor® 594 biocytin, sodium salt (biocytin Alexa Fluor® 594)	250 µg
A13204	annexin V, biotin-X conjugate *100 assays*	500 µL
A6435	anti–dinitrophenyl-KLH, rabbit IgG fraction, biotin-XX conjugate *2 mg/mL*	0.5 mL
A982	anti–fluorescein/Oregon Green®, rabbit IgG fraction, biotin-XX conjugate *1 mg/mL*	0.5 mL
A10263	anti–Green Fluorescent Protein, chicken IgY fraction, biotin-XX conjugate (anti-GFP, IgY, biotin-XX conjugate) *2 mg/mL*	100 µL
A10259	anti–Green Fluorescent Protein, rabbit IgG fraction, biotin-XX conjugate (anti-GFP, IgG, biotin-XX conjugate) *2 mg/mL*	100 µL
A21272	anti–HuC/HuD neuronal protein (human), mouse IgG_{2b}, monoclonal 16A11, biotin-XX conjugate	100 µg
A5751	anti–lucifer yellow, rabbit IgG fraction, biotin-XX conjugate *3 mg/mL*	0.5 mL
A21371	anti–α-tubulin (bovine), mouse IgG_1, monoclonal 236-10501, biotin-XX conjugate	50 µg
B32772	biotin-aha-dCTP *1 mM in TE buffer*	25 µL
B32766	biotin-aha-dUTP *1 mM in TE buffer*	25 µL
B1550	biotin DHPE (*N*-(biotinoyl)-1,2-dihexadecanoyl-*sn*-glycero-3-phosphoethanolamine, triethylammonium salt)	10 mg
B10570	biotin-4-fluorescein	5 mg
B1370	5-((*N*-(5-(*N*-(6-(biotinoyl)amino)hexanoyl)amino)pentyl)thioureidyl)fluorescein (fluorescein biotin)	5 mg
B1616	biotin-X DHPE (*N*-((6-(biotinoyl)amino)hexanoyl)-1,2-dihexadecanoyl-*sn*-glycero-3-phosphoethanolamine, triethylammonium salt)	5 mg
B11027	biotin-XX $F(ab')_2$ fragment of goat anti-mouse IgG (H+L) *2 mg/mL*	250 µL
B21078	biotin-XX $F(ab')_2$ fragment of goat anti-rabbit IgG (H+L) *2 mg/mL*	250 µL
B2763	biotin-XX goat anti-mouse IgG (H+L) *2 mg/mL*	0.5 mL
B2770	biotin-XX goat anti-rabbit IgG (H+L) *2 mg/mL*	0.5 mL
B7474	biotin-XX phalloidin	50 U
B35138	BrdU, mouse monoclonal antibody (Clone MoBU-1), biotin conjugate (anti-BrdU, biotin conjugate) *0.2 mg/mL*	350 µL
B1196	α-bungarotoxin, biotin-XX conjugate	500 µg
C34779	cholera toxin subunit B (recombinant), biotin-XX conjugate	100 µg
D7135	dextran, biotin, 3000 MW, lysine fixable (BDA-3000)	10 mg
D1956	dextran, biotin, 10,000 MW, lysine fixable (BDA-10,000)	25 mg
D1957	dextran, biotin, 70,000 MW, lysine fixable (BDA-70,000)	25 mg
D7142	dextran, biotin, 500,000 MW, lysine fixable (BDA-500,000)	10 mg
D7156	dextran, fluorescein and biotin, 3000 MW, anionic, lysine fixable (micro-emerald)	5 mg
D7178	dextran, fluorescein and biotin, 10,000 MW, anionic, lysine fixable (mini-emerald)	10 mg
D7162	dextran, tetramethylrhodamine and biotin, 3000 MW, lysine fixable (micro-ruby)	5 mg
D3312	dextran, tetramethylrhodamine and biotin, 10,000 MW, lysine fixable (mini-ruby)	10 mg
D20658	DSB-X™ Bioconjugate Isolation Kit #1 *with streptavidin agarose* *5 isolations*	1 kit
D20698	DSB-X™ biotin donkey anti–goat IgG (H+L) *2 mg/mL*	0.5 mL
D20701	DSB-X™ biotin goat anti–chicken IgG (H+L) *2 mg/mL*	0.5 mL
D20691	DSB-X™ biotin goat anti–mouse IgG (H+L) *highly cross-adsorbed* *2 mg/mL*	0.5 mL
D20693	DSB-X™ biotin goat anti–mouse IgM (µ chain) *2 mg/mL*	250 µL
D20697	DSB-X™ biotin goat anti–rat IgG (H+L) *2 mg/mL*	0.5 mL
E3477	epidermal growth factor, biotin-XX conjugate (biotin EGF)	20 µg
E13345	epidermal growth factor, biotinylated, complexed to Alexa Fluor® 488 streptavidin (Alexa Fluor® 488 EGF complex)	100 µg
E35350	epidermal growth factor, biotinylated, complexed to Alexa Fluor® 555 streptavidin (Alexa Fluor® 555 EGF complex)	100 µg
E35351	epidermal growth factor, biotinylated, complexed to Alexa Fluor® 647 streptavidin (Alexa Fluor® 647 EGF complex)	100 µg
E3480	epidermal growth factor, biotinylated, complexed to Texas Red® streptavidin (Texas Red® EGF complex)	100 µg
F8769	FluoSpheres® biotin-labeled microspheres, 1.0 µm, nonfluorescent *1% solids*	0.4 mL
F8766	FluoSpheres® biotin-labeled microspheres, 0.04 µm, yellow-green fluorescent (505/515) *1% solids*	0.4 mL
F8767	FluoSpheres® biotin-labeled microspheres, 0.2 µm, yellow-green fluorescent (505/515) *1% solids*	0.4 mL
F8768	FluoSpheres® biotin-labeled microspheres, 1.0 µm, yellow-green fluorescent (505/515) *1% solids*	0.4 mL
G36000	glutathione ethyl ester, biotin amide (BioGEE) *glutathiolation detection reagent* *special packaging*	10 x 100 µg
I21414	isolectin GS-IB_4 from *Griffonia simplicifolia*, biotin-XX conjugate	500 µg
L6950	lucifer yellow biocytin, potassium salt (biocytin lucifer yellow)	5 mg
L2601	lucifer yellow cadaverine biotin-X, dipotassium salt	10 mg
O12920	Oregon Green® 488 biocytin (biocytin Oregon Green® 488)	5 mg
P917	peroxidase from horseradish, biotin-XX conjugate	10 mg
P811	R-phycoerythrin, biotin-XX conjugate *4 mg/mL*	0.5 mL
Q10301MP	Qdot® 605 Biotin Conjugate Kit	1 kit
Q10321MP	Qdot® 655 Biotin Conjugate Kit	1 kit
T12921	5-(and-6)-tetramethylrhodamine biocytin (biocytin TMR)	5 mg
T23363	transferrin from human serum, biotin-XX conjugate	5 mg

IMPORTANT NOTICE: The products described in this manual are covered by one or more Limited Use Label License(s). Please refer to the Appendix on page 971 and Master Product List on page 975. Products are For Research Use Only. Not intended for any animal or human therapeutic or diagnostic use.

molecular probes® | invitrogen™ by life technologies™

FIVE

CHAPTER 5
Crosslinking and Photoactivatable Reagents

IMPORTANT NOTICE: The products described in this manual are covered by one or more Limited Use Label License(s). Please refer to the Appendix on page 971 and Master Product List on page 975. Products are For Research Use Only. Not intended for any animal or human therapeutic or diagnostic use.

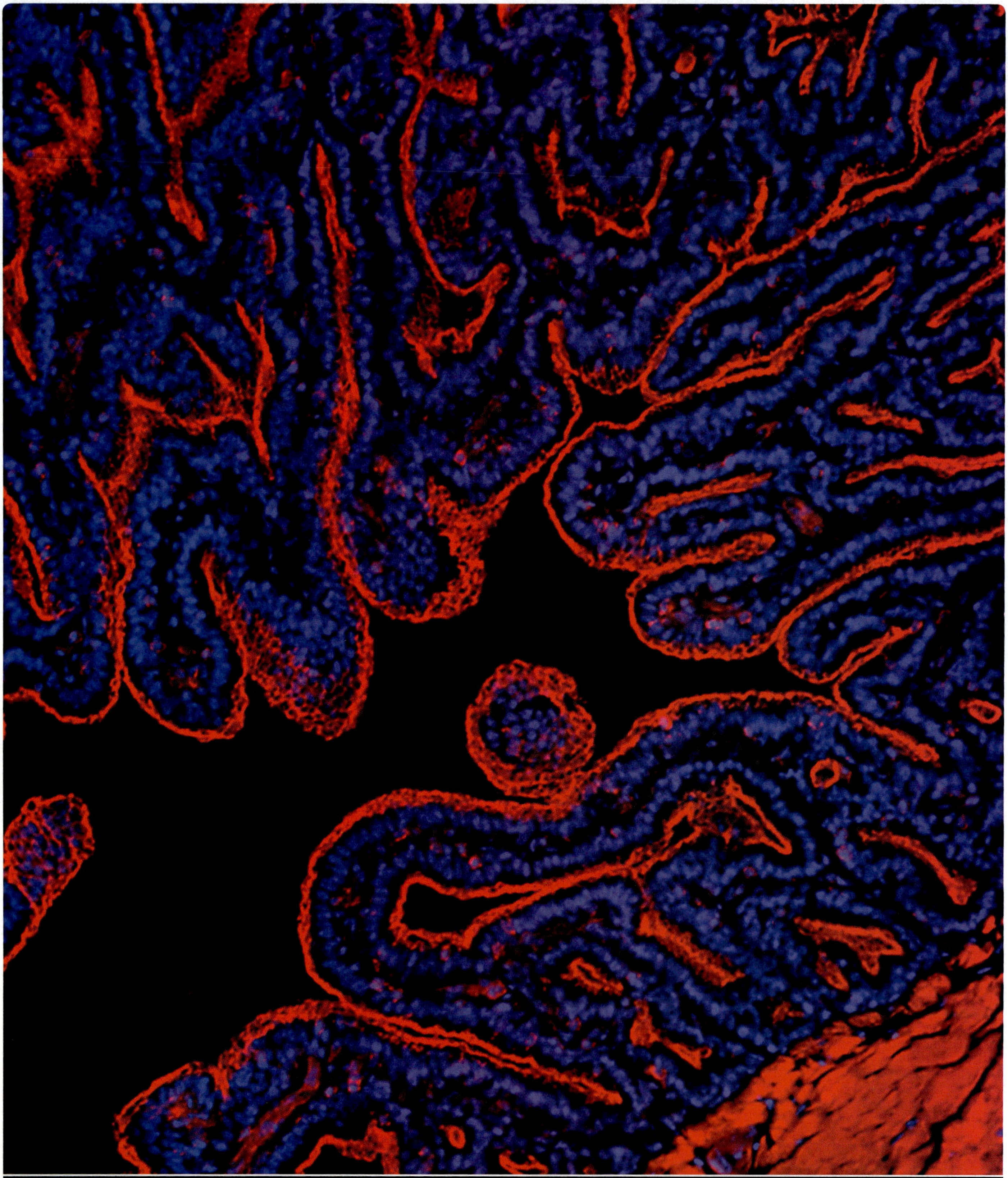

Alexa Fluor® 568 phalloidin and Hoechst 33342.

IMPORTANT NOTICE: The products described in this manual are covered by one or more Limited Use Label License(s). Please refer to the Appendix on page 971 and Master Product List on page 975. Products are For Research Use Only. Not intended for any animal or human therapeutic or diagnostic use.

5.1 Introduction to Crosslinking and Photoactivatable Reagents

Chemical and Photoreactive Crosslinkers

Bifunctional "crosslinking" reagents contain two reactive groups, thereby providing a means of covalently linking two target groups. The reactive groups in a chemical crosslinking reagent (Section 5.2) typically belong to the classes of functional groups—including succinimidyl esters, maleimides and iodoacetamides—described in Chapter 1 and Chapter 2.

In contrast, one of the reactive groups in each of our photoreactive crosslinking reagents (Section 5.3) requires light activation before reacting with a target group. Crosslinking of a biopolymer (such as an antibody, enzyme, avidin or nucleic acid) to a low molecular weight molecule (such as a drug, toxin, peptide or oligonucleotide) or to another biopolymer yields a stable heteroconjugate. This bioconjugate can serve as a detection reagent in a wide variety of research and diagnostic assays or as an immunogen designed to elicit antibody production. Crosslinking reagents are also useful for probing the spatial relationships and interactions within and between biomolecules.

In homobifunctional crosslinking reagents (Section 5.2), the reactive groups are identical. These reagents couple like functional groups—typically two thiols, two amines, two acids or two alcohols—and are predominantly used to form intramolecular crosslinks or to prepare polymers from monomers. When used to conjugate two different biomolecules, for example an enzyme to an antibody, these relatively nonspecific reagents tend to yield high molecular weight aggregates.

In heterobifunctional crosslinking reagents (Section 5.2, Table 5.1), the reactive groups have dissimilar chemistry, allowing the formation of crosslinks between unlike functional groups (Figure 5.2.2). As with homobifunctional crosslinking reagents, heterobifunctional crosslinking reagents can still form multiple intermolecular crosslinks to yield high molecular weight aggregates, but conjugations that use these reagents can be more easily controlled so as to optimize the stoichiometry of the target molecules. Thus, heterobifunctional crosslinking reagents are very useful for preparing conjugates between two different biomolecules.

The photoreactive crosslinking reagents (Section 5.3) are a special subset of the heterobifunctional crosslinking reagents. Upon UV illumination, these reagents react with nucleophiles or form C–H insertion products (Figure 5.1.1, Figure 5.1.2, Figure 5.1.3).

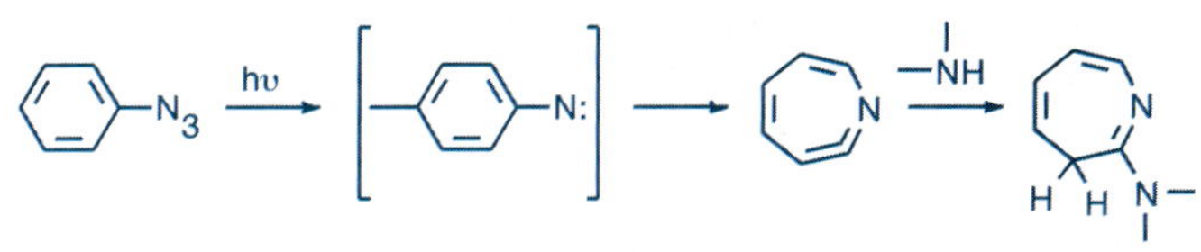

Figure 5.1.1 Photoreactive crosslinking reaction of a simple aryl azide.

Figure 5.1.2 Photoreactive crosslinking reaction of a fluorinated aryl azide.

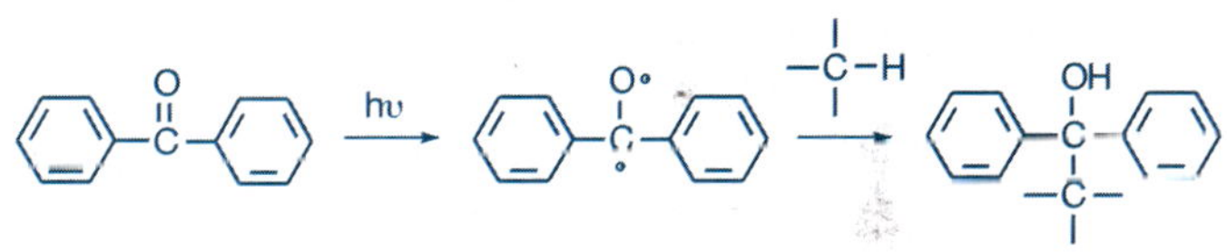

Figure 5.1.3 Photoreactive crosslinking reaction of a benzophenone derivative.

Table 5.1 Molecular Probes® heterobifunctional crosslinkers.

		Reactivity				
Cat. No.	**Crosslinker**	**Thiol (R–SH)**	**Amine ($R–NH_2$)**	**Azide * ($R–N_3$)**	**Alkyne ***	**Photoreactive †**
S1553	succinimidyl acetylthioacetate (SATA) ‡	●	●			
S1534	succinimidyl *trans*-4-(maleimidylmethyl) cyclohexane-1-carboxylate (SMCC) ‡	●	●			
S1531	succinimidyl 3-(2-pyridyldithio)propionate (SPDP)	●	●			
E2247	1-ethyl-3-(3-dimethylaminopropyl)carbodiimide (EDAC) §		●			
P6317	*N*-((2-pyridyldithio)ethyl)-4-azidosalicylamide (PEAS; AET) ‡	●				●
A2522	4-azido-2,3,5,6-tetrafluorobenzoic acid, succinimidyl ester (ATFB, SE)		●			●
B1508	benzophenone-4-maleimide	●				●
B1526	benzophenone-4-isothiocyanate		●			●
B1577	4-benzoylbenzoic acid, succinimidyl ester		●			●
I10188	iodoacetamide azide *	●			●	
I10189	iodoacetamide alkyne *	●		●		
A10280	azido $(PEO)_4$ propionic acid, succinimidyl ester *		●		●	
A10279	alkyne, succinimidyl ester *		●	●		

* Azides react with alkynes via the copper-catalyzed azide–alkyne cycloaddition reaction, as discussed in Section 3.1 (where these products are also described). † Reacts nonspecifically with available sites upon UV illumination. ‡ These reagents contain cryptic thiols that are exposed by disulfide reduction (SPDP or PEAS) or deacetylation (SATA) and can be subsequently disulfide-coupled to other thiolated molecules or thiolether-coupled to maleimides or iodoacetamides. § EDAC couples amines to carboyxlic acids.

IMPORTANT NOTICE: The products described in this manual are covered by one or more Limited Use Label License(s). Please refer to the Appendix on page 971 and Master Product List on page 975. Products are For Research Use Only. Not intended for any animal or human therapeutic or diagnostic use.

An additional variation is the "zero-length" crosslinking reagent—a reagent that forms a chemical bond between two groups without itself being incorporated into the product (Figure 5.1.4). The water-soluble carbodiimide EDAC (E2247, Section 5.2), which is used to couple carboxylic acids to amines, is an example of a zero-length crosslinking reagent.

A noncovalent interaction between two molecules that has very slow dissociation kinetics can also function as a crosslink. For example, reactive derivatives of phospholipids can be used to link the liposomes or cell membranes in which they are incorporated to antibodies or enzymes. Biotinylation and haptenylation reagents (Chapter 4) can also be thought of as heterobifunctional crosslinking reagents because they comprise a chemically reactive group as well as a biotin or hapten moiety that binds with high affinity to avidin or an anti-hapten antibody, respectively. Similarly, avidin, streptavidin, NeutrAvidin™ biotin-binding protein and CaptAvidin™ biotin-binding protein (Section 7.6) can tightly bind up to four molecules of a biotinylated target.

REFERENCES

1. ACS Chem Biol (2009) 4:409; **2.** Nat Methods (2007) 4:619; **3.** Nature (1984) 310:74; **4.** Biotechniques (1993) 15:848; **5.** Anal Chem (2003) 75:1387.

Photoactivatable (Caged) Probes

In addition to the photoreactive crosslinking reagents that are briefly mentioned above and described in detail in Section 5.3, we prepare photoactivatable probes. Flash photolysis of photoactivatable or "caged" probes provides a means of controlling the release—both spatially and temporally—of biologically active products or other reagents of interest.[1,2] The chemical caging process may also confer membrane permeability on the caged ligand, as is the case for caged cAMP[3] and caged luciferin.[4] Our selection of caged nucleotides, chelators, second messengers and neurotransmitters has tremendous potential for use with both live cells and isolated proteins (Section 5.3).

We prepare caged versions of biologically active molecules, as well as caged fluorescent dyes that are essentially nonfluorescent until after photolysis. These caged fluorophores have proven useful for photoactivation of fluorescence (PAF) experiments, which are analogous to fluorescence recovery after photobleaching (FRAP) experiments except that the fluorophore is activated upon illumination rather than bleached. Measuring the fluorescent signal of the photoactivated fluorophore against a dark background is intrinsically more sensitive than measuring a dark photobleached region against a bright field.[5]

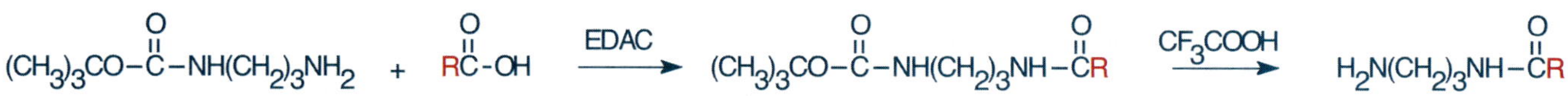

Figure 5.1.4 Conversion of a carboxylic acid group into an aliphatic amine. The activated carboxylic acid is derivatized with a half-protected aliphatic diamine (mono-*N*-(*t*-BOC)-propylenediamine, M6248), usually in an organic solvent, followed by removal of the *t*-BOC–protecting group with trifluoroacetic acid.

5.2 Chemical Crosslinking Reagents

The most common schemes for forming a well-defined heteroconjugate require the indirect coupling of an amine group on one biomolecule to a thiol group on a second biomolecule, usually by a two- or three-step reaction sequence. The high reactivity of thiols (Chapter 2) and—with the exception of a few proteins such as β-galactosidase—their relative rarity in most biomolecules make thiol groups ideal targets for controlled chemical crosslinking. If neither molecule contains a thiol group, then one or more can be introduced using one of several thiolation methods. The thiol-containing biomolecule is then reacted with an amine-containing biomolecule using a heterobifunctional crosslinking reagent such as one of those described in Amine–Thiol Crosslinking, below.

Thiolation of Biomolecules

Introducing Thiol Groups into Biomolecules

Several methods are available for introducing thiols into biomolecules, including the reduction of intrinsic disulfides, as well as the conversion of amine or carboxylic acid groups to thiol groups:

- Disulfide crosslinks of cystines in proteins can be reduced to cysteine residues by dithiothreitol[1] (DTT, D1532) or tris-(2-carboxyethyl)phosphine (TCEP, T2556; Figure 5.2.1). However, reduction may result in loss of protein activity or specificity. Excess DTT must be carefully removed under conditions that prevent reformation of the disulfide,[2] whereas excess TCEP usually does not need to be removed before carrying out the crosslinking reaction. TCEP is also more stable at higher pH values and at higher temperatures than is the air-sensitive DTT reagent.[3]
- Amines can be indirectly thiolated by reaction with succinimidyl acetylthioacetate[4] (SATA, S1553), followed by removal of the acetyl group with 50 mM hydroxylamine or hydrazine at near-neutral pH (Figure 5.2.2). This reagent is most useful when disulfides are essential for activity, as is the case for some peptide toxins.
- Amines can be indirectly thiolated by reaction with succinimidyl 3-(2-pyridyldithio)propionate[5] (SPDP, S1531), followed by reduction of the 3-(2-pyridyldithio)propionyl conjugate with DTT or TCEP (Figure 5.2.3). Reduction releases the 2-pyridinethione chromophore, which can be used to determine the degree of thiolation.
- Thiols can be incorporated at carboxylic acid groups by an EDAC-mediated reaction with cystamine, followed by reduction of the disulfide with DTT or TCEP;[6,7] see Amine–Carboxylic Acid Crosslinking below.
- Tryptophan residues in thiol-free proteins can be oxidized to mercaptotryptophan residues, which can then be modified by iodoacetamides or maleimides.[8–10]

IMPORTANT NOTICE: The products described in this manual are covered by one or more Limited Use Label License(s). Please refer to the Appendix on page 971 and Master Product List on page 975. Products are For Research Use Only. Not intended for any animal or human therapeutic or diagnostic use.
www.invitrogen.com/probes

Our preferred reagent combination for protein thiolation is SPDP/DTT or SPDP/TCEP.[11] We use SPDP to prepare a reactive R-phycoerythrin derivative (P806, Section 6.4), providing researchers with the optimal number of pyridyldisulfide groups for crosslinking the phycobiliprotein to thiolated antibodies, enzymes and other biomolecules through disulfide linkages.[12] More commonly, the pyridyldisulfide groups are first reduced to thiols, which are then reacted with maleimide- or iodoacetamide-derivatized proteins (Figure 5.2.3). SPDP can also be used to thiolate oligonucleotides [13] and—like all of the thiolation reagents in this section—to introduce the highly reactive thiol group into peptides, onto cell surfaces or onto affinity matrices for subsequent reaction with fluorescent, enzyme-coupled or other thiol-reactive reagents (Chapter 2). In addition, because the 3-(2-pyridyldithio)propionyl conjugate releases the 2-pyridinethione chromophore upon reduction, SPDP is useful for quantitating the number of reactive amines in an affinity matrix.[14]

Measuring Thiolation of Biomolecules

To ensure success in forming heterocrosslinks, it is important to know that a molecule has the proper degree of thiolation. We generally find that two to three thiol residues per protein are optimal. Following removal of excess reagents, the degree of thiolation in proteins or other molecules thiolated with SPDP can be directly determined by measuring release of the 2-pyridinethione chromophore [5] (EC ~8000 $cm^{-1}M^{-1}$ at 343 nm).

Alternatively, the degree of thiolation and presence of residual thiols in a solution can be assessed using 5,5′-dithiobis-(2-nitrobenzoic acid) (DTNB, Ellman's reagent; D8451), which stoichiometrically yields the 5-mercapto-2-nitrobenzoic acid chromophore (EC ~13,600 $cm^{-1}M^{-1}$ at 410 nm) upon reaction with a thiol group.[15,16] DTNB can also be used to quantitate residual phosphines in aqueous solutions, including TCEP; [17] in this case, two molecules of 5-mercapto-2-nitrobenzoic acid are formed per reaction with one molecule of a phosphine.

Measure-iT™ Thiol Assay Kit

The Measure-iT™ Thiol Assay Kit (M30550) provides easy and accurate quantitation of thiol. The kit supplies concentrated assay reagent, dilution buffer, and concentrated thiol standard. The assay has

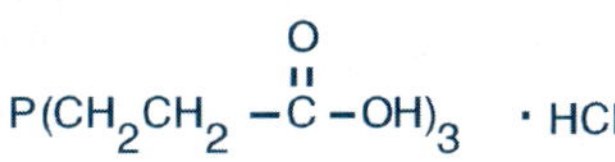

Figure 5.2.1 Tris-(2-carboxyethyl)phosphine, hydrochloride (TCEP, T2556).

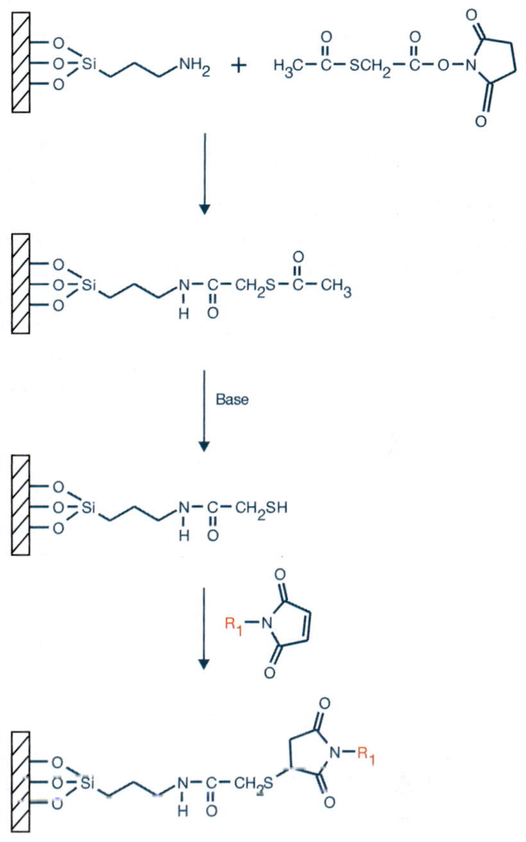

Figure 5.2.2 Schematic illustration of the heterobifunctional crosslinker succinimidyl acetylthioacetate (SATA, S1553): **A)** attachment to an aminosilane-modified surface, **B)** deprotection with base and **C)** reaction with a thiol-reactive biomolecule.

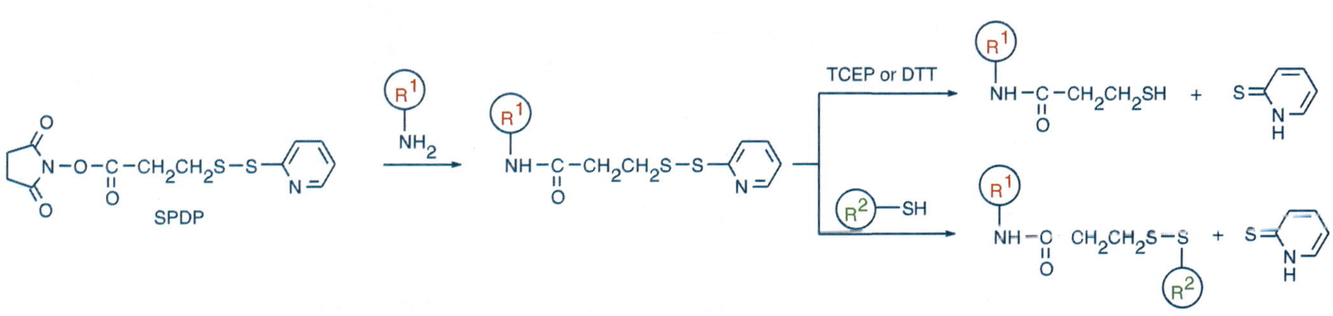

Figure 5.2.3 SPDP derivatization reactions. SPDP (S1531) reacts with an amine-containing biomolecule at pH 7 to 9, yielding a pyridyldithiopropionyl mixed disulfide. The mixed disulfide can then be reacted with a reducing agent such as DTT (D1532) or TCEP (T2556) to yield a 3-mercaptopropionyl conjugate or with a thiol-containing biomolecule to form a disulfide-linked tandem conjugate. Either reaction can be quantitated by measuring the amount of 2-pyridinethione chromophore released during the reaction.

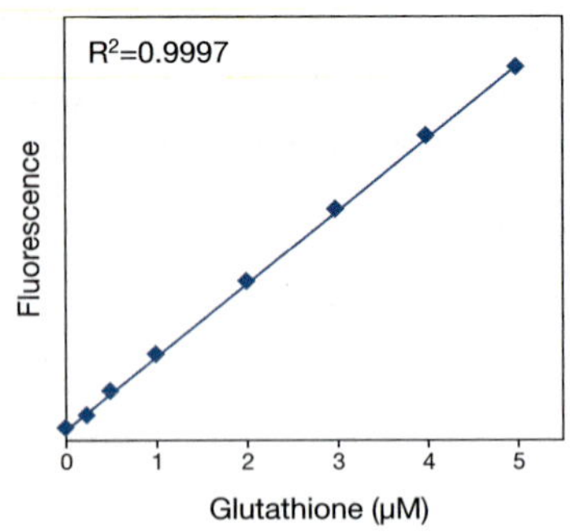

Figure 5.2.4 Linearity and sensitivity of the Measure-iT™ thiol assay. Triplicate 10 µL samples of glutathione were assayed using the Measure-iT™ Thiol Assay Kit (M30550). Fluorescence was measured using excitation/emission of 490/520 nm and plotted versus glutathione concentration. The variation (CV) of replicate samples was <2%.

a linear range of 0.05–5 µM thiol (Figure 5.2.4), making it up to 400 times more sensitive than colorimetric methods based on DTNB (Ellman's reagent).

Each Measure-iT™ Thiol Assay Kit contains:

- Measure-iT™ thiol quantitation reagent (100X concentrate in 1,2-propanediol)
- Measure-iT™ thiol quantitation buffer (50 mM potassium phosphate buffer)
- Measure-iT™ thiol quantitation standard (reduced glutathione)
- Detailed protocols

Simply dilute the reagent 1:100, load 100 µL into the wells of a microplate, add 1–10 µL sample volumes, mix, then read the fluorescence. Maximum fluorescence signal is attained within 5 minutes and is stable for at least 1 hour. The assay is performed at room temperature, and common contaminants are well tolerated in the assay. The Measure-iT™ Thiol Assay Kit provides sufficient materials for 500 assays, based on a 100 µL assay volume in a 96-well microplate format; this thiol assay can also be adapted for use in cuvettes or 384-well microplates.

Thiol and Sulfide Quantitation Kit

Ultrasensitive colorimetric quantitation of both protein and nonprotein thiols can also be achieved using the Thiol and Sulfide Quantitation Kit (T6060). In this assay, which is based on a method reported by Singh,[18,19] thiols reduce a disulfide-inhibited derivative of papain, stoichiometrically releasing the active enzyme. Activity of the enzyme is then measured using the chromogenic papain substrate L-BAPNA via spectrophotometric detection of *p*-nitroaniline release at 412 nm (Figure 5.2.5). Although thiols can also be quantitated using DTNB (Ellman's reagent), the enzymatic amplification step in this quantitation kit enables researchers to detect as little as 0.2 nanomoles of a thiol—a sensitivity that is about 100-fold better than that achieved with DTNB.[20,21] Thiols in proteins and potentially in other high molecular weight molecules can be detected indirectly by incorporating the disulfide cystamine into the solution. Cystamine undergoes an exchange reaction with protein thiols, yielding 2-mercaptoethylamine (cysteamine), which then releases active papain. Thiols that are alkylated by maleimides, iodoacetamides and other reagents are excluded from detection and can therefore be assayed subtractively.[22]

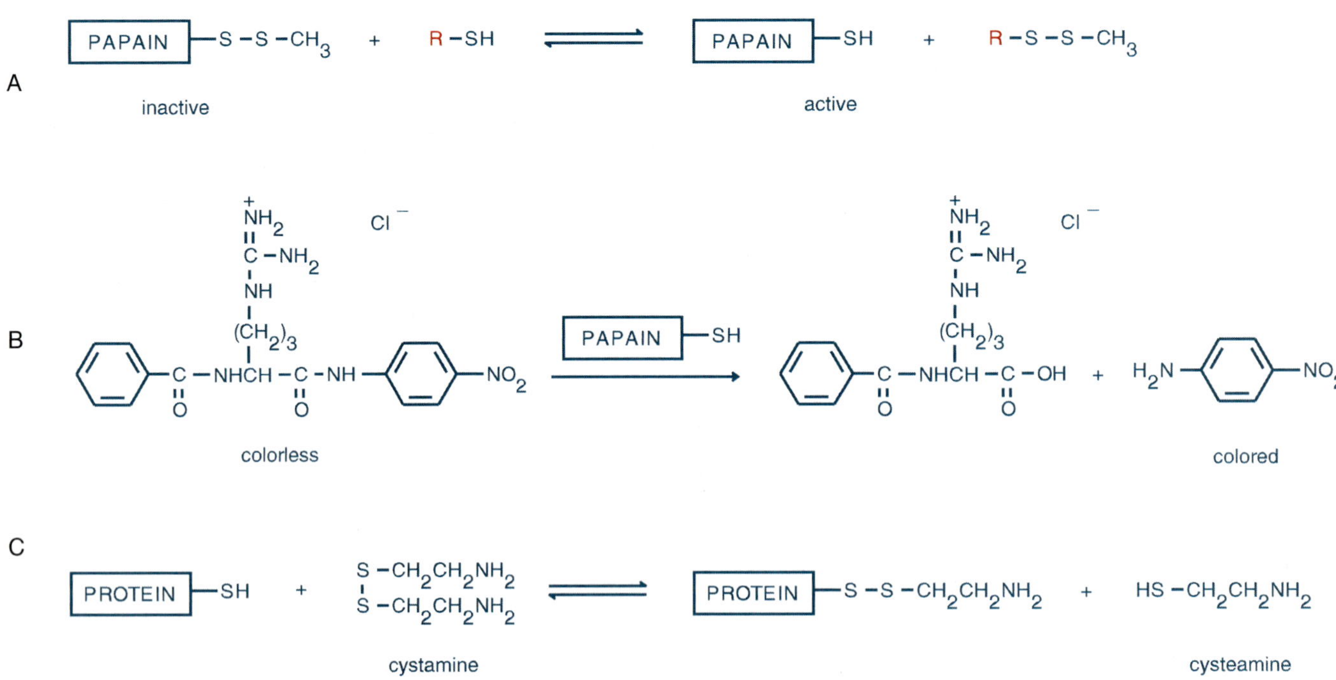

Figure 5.2.5 Chemical basis for thiol detection using the Thiol and Sulfide Quantitation Kit (T6060): **A)** the inactive disulfide derivative of papain, papain–$SSCH_3$, is activated in the presence of thiols; **B)** active papain cleaves the substrate L-BAPNA, releasing the *p*-nitroaniline chromophore; **C)** protein thiols, often poorly accessible, exchange with cystamine to generate 2-mercaptoethylamine (cysteamine), which is easily detected.

IMPORTANT NOTICE: The products described in this manual are covered by one or more Limited Use Label License(s). Please refer to the Appendix on page 971 and Master Product List on page 975. Products are For Research Use Only. Not intended for any animal or human therapeutic or diagnostic use.
www.invitrogen.com/probes

The Thiol and Sulfide Quantitation Kit contains:

- Papain–$SSCH_3$, the disulfide-inhibited papain derivative
- L-BAPNA, a chromogenic papain substrate
- DTNB (Ellman's reagent), for calibrating the assay
- Cystamine
- L-Cysteine, a thiol standard
- Buffer
- Detailed protocols for measuring thiols, inorganic sulfides and maleimides

Sufficient reagents are provided for approximately 50 assays using 1 mL assay volumes and standard cuvettes or 250 assays using a microplate format.

Thiol–Thiol Crosslinking

Oxidation

Thiol residues in close proximity can be oxidized to disulfides by either an intra- or intermolecular reaction. In many circumstances, however, this oxidation reaction is reversible and difficult to control.

Fluorescent Thiol–Thiol Crosslinkers

Dibromobimane (bBBr, D1379; Figure 5.2.6) is an interesting crosslinking reagent for proteins because it is unlikely to fluoresce until both of its alkylating groups have reacted.[23] It has been used to crosslink thiols in myosin,[24] actin,[25] hemoglobin,[26] *Escherichia coli* lactose permease[27] and mitochondrial ATPase.[28] It has also been shown to intramolecularly crosslink thiols in a complex of nebulin and calmodulin.[29] In addition, dibromobimane has been used to probe for the proximity of dual-cysteine mutagenesis sites in ArsA ATPase[30] and P-glycoprotein.[31–33] Dibromobimane, a stimulator of the ATPase activity of a cysteine-free P-glycoprotein, was used with cysteine-scanning mutagenesis to identify amino acid residues important for function.[34]

In addition to dibromobimane, we offer the thiol-reactive homobifunctional crosslinker bis-((*N*-iodoacetyl)piperazinyl)sulfonerhodamine (B10621), which is derived from a relatively rigid rhodamine dye (Figure 5.2.7). This crosslinker is similar to a thiol-reactive rhodamine-based crosslinking reagent that was used to label regulatory light-chains of chicken gizzard myosin for fluorescence polarization experiments.[35] Researchers have attached bis-((*N*-iodoacetyl)piperazinyl) sulfonerhodamine to the kinesin motor domain and determined the orientation of kinesin bound to microtubules in the presence of a nonhydrolyzable ATP analog by fluorescence polarization microscopy.[36,37] Images of single molecules of chicken calmodulin crosslinked between two engineered cysteines by bis-((*N*-iodoacetyl)piperazinyl)sulfonerhodamine have been used to generate comparisons of experimental and theoretical super-resolution point-spread functions[30] (PSF).

Amine–Amine Crosslinking

The scientific literature contains numerous references to reagents that form crosslinks between amines of biopolymers. Homobifunctional amine crosslinkers include glutaraldehyde, bis(imido esters), bis(succinimidyl esters), diisocyanates and diacid chlorides.[39] These reagents, however, tend to yield high molecular weight aggregates, making them unsuitable for reproducibly preparing well-defined conjugates between two different amine-containing biomolecules. For example, glutaraldehyde is still used by some companies and research laboratories to couple horseradish peroxidase, which has only six lysine residues,[40] to proteins with a larger number of lysine residues. Unfortunately, this practice can result in variable molecular weights and batch-to-batch inconsistency.

Well-defined conjugates between two amine-containing molecules are more reliably prepared by thiolating one or more amines on one of the biomolecules and converting one or more amines on the second biomolecule to a thiol-reactive functional group such as a maleimide or iodoacetamide, as described below in Amine–Thiol Crosslinking. For example, we prepare our horseradish peroxidase conjugates (Section 7.2, Section 7.6) using SPDP- and SMCC-mediated reactions (Figure 5.2.3, Figure 5.2.8).

Direct amine–amine crosslinking routinely occurs during fixation of proteins, cells and tissues with formaldehyde or glutaraldehyde. These common aldehyde-based fixatives are also used to crosslink amine and hydrazine derivatives to proteins and other amine-containing polymers. For example, lucifer yellow CH (L453, Section 14.3) is nonspecifically conjugated to surrounding biomolecules by aldehyde-based fixatives in order to preserve the dye's staining pattern during subsequent tissue manipulations.[41] Also, biotin hydrazides (Section 4.2) have been directly coupled to nucleic acids with glutaraldehyde,[42,43] a reaction that is potentially useful for conjugating fluorescent hydrazides and hydroxylamines to DNA.

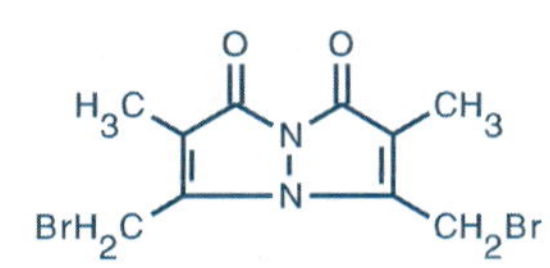

Figure 5.2.6 Dibromobimane (bBBr, D1379).

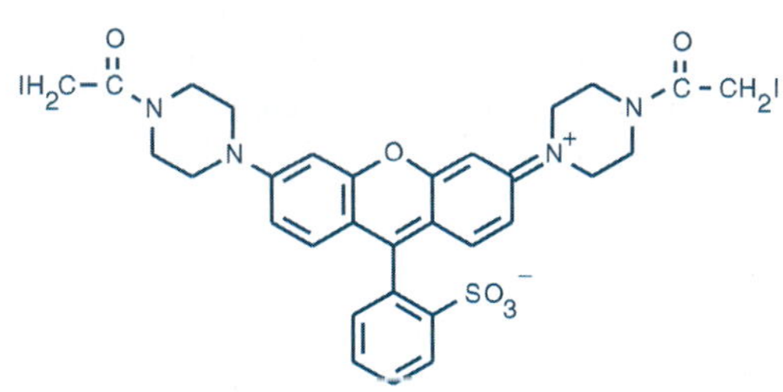

Figure 5.2.7 bis-((*N*-iodoacetyl)piperazinyl)sulfonerhodamine (B10621).

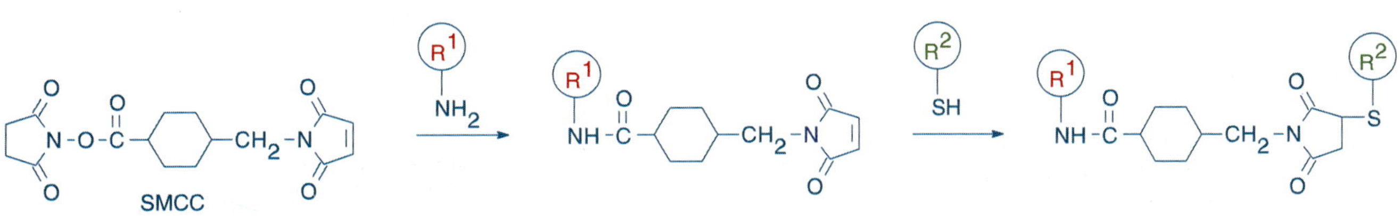

Figure 5.2.8 Two-step reaction sequence for crosslinking biomolecules using the heterobifunctional crosslinker SMCC (S1534).

IMPORTANT NOTICE: The products described in this manual are covered by one or more Limited Use Label License(s). Please refer to the Appendix on page 971 and Master Product List on page 975. Products are For Research Use Only. Not intended for any animal or human therapeutic or diagnostic use.

Amine–Thiol Crosslinking

Indirect crosslinking of the amines in one biomolecule to the thiols in a second biomolecule is the predominant method for forming a heteroconjugate. If one of the biomolecules does not already contain one or more thiol groups, it is necessary to introduce them using one of the thiolation procedures described above in Thiolation of Biomolecules. Thiol-reactive groups such as maleimides are typically introduced into the second biomolecule by modifying a one or more of its amines with a heterobifunctional crosslinker containing both a succinimidyl ester and a maleimide. The maleimide-modified biomolecule is then reacted with the thiol-containing biomolecule to form a stable thioether crosslink (Figure 5.2.8). Chromatographic methods are usually employed to separate the higher molecular weight heteroconjugate from the unconjugated biomolecules.

Introducing Maleimides at Amines

Succinimidyl *trans*-4-(maleimidylmethyl)cyclohexane-1-carboxylate[44] (SMCC, S1534) is our reagent of choice for introducing thiol-reactive groups at amine sites because of the superior chemical stability of its maleimide and its ease of use[45] (Figure 5.2.8).

Introducing Disulfides at Amines

Our preferred method for preparing heteroconjugates employs the thiolation reagent SPDP (S1531). The pyridyldisulfide intermediate that is initially formed by reaction of SPDP with amines can form an unsymmetrical disulfide through reaction with a second thiol-containing molecule[5,12] (Figure 5.2.3). The thiol-containing target can be a molecule such as β-galactosidase that contains intrinsic thiols or a molecule in which thiols have been introduced using one of the thiolation procedures described above in Thiolation of Biomolecules. In either case, it is essential that all reducing agents, such as DTT and TCEP, are absent. The heteroconjugate's disulfide bond is about as stable and resistant to reduction as disulfides found in proteins; it can be reduced with DTT or TCEP to generate two thiol-containing biomolecules.

Protein–Protein Crosslinking Kit

Our Protein–Protein Crosslinking Kit (P6305) provides all of the reagents and purification media required to perform three protein–protein conjugations in which neither protein contains thiol residues. The chemistry used to thiolate the amines of one of the proteins with SPDP and to convert the amines of the second protein to thiol-reactive maleimides with SMCC is shown in Figure 5.2.3 and Figure 5.2.8, respectively. Included in the kit are:

- SPDP, for thiolating amines
- SMCC, for converting amines to thiol-reactive maleimides
- TCEP, for reducing the pyridyldisulfide intermediate
- *N*-ethylmaleimide (NEM), for capping residual thiols
- Six reaction tubes, each containing a magnetic stir bar
- Spin columns plus collection tubes
- Dimethylsulfoxide (DMSO)
- Detailed crosslinking protocols

The Protein–Protein Crosslinking Kit was designed to prepare and purify protein–protein conjugates; however, it can be readily modified for generating peptide–protein or enzyme–nucleic acid conjugates or for conjugating biomolecules to affinity matrices.

We have considerable experience in preparing protein–protein conjugates and will apply this expertise to a researcher's particular application through our custom synthesis service. We provide custom conjugation services on an exclusive or nondisclosure basis when requested. For more information or a quote, please go to www.invitrogen.com/handbook/custom/bioconjugates.

Assaying Maleimide- and Iodoacetamide-Modified Biomolecules

The potential instability of maleimide derivatives and the photosensitivity of iodoacetamide derivatives may make it advisable to assay the modified biomolecule for thiol reactivity before conjugation with a thiol-containing biomolecule. SAMSA fluorescein (A685, Figure 5.2.9), which is currently our only fluorescent reagent that can generate a free thiol group, was designed for assaying whether or not a biomolecule is adequately labeled with a heterobifunctional maleimide or iodoacetamide crosslinker. Brief treatment of SAMSA fluorescein with NaOH at pH 10 liberates a free thiol. By adding base-treated SAMSA fluorescein to a small aliquot of the crosslinker-modified biomolecule, the researcher can check to see whether the biomolecule has been sufficiently labeled before proceeding to the next step. The degree of modification can be approximated from either the absorbance or the fluorescence of the conjugate following quick purification on a gel-filtration column.

Alternatively, thiol reactivity of the modified biomolecule can be assayed using the reagents provided in our Thiol and Sulfide Quantitation Kit (T6060), a product that is described above.[19,22] Once unconjugated reagents have been removed, a small aliquot of the maleimide- or iodoacetamide-modified biomolecule can be reacted with excess cysteine. Thiol-reactive groups can then be quantitated by determining the amount of cysteine consumed in this reaction with the Thiol and Sulfide Quantitation Kit.

Amine–Carboxylic Acid Crosslinking

1-Ethyl-3-(3-dimethylaminopropyl)carbodiimide (EDAC, E2247) can react with biomolecules to form "zero-length" crosslinks, usually within a molecule or between subunits of a protein complex. In this chemistry, the crosslinking reagent is not incorporated into the final product. The water-soluble carbodiimide EDAC crosslinks a specific amine and carboxylic acid between subunits of allophycocyanin, thereby stabilizing its assembly;[46] we use EDAC to stabilize allophycocyanin in its allophycocyanin conjugates (Section 6.4). EDAC has also been

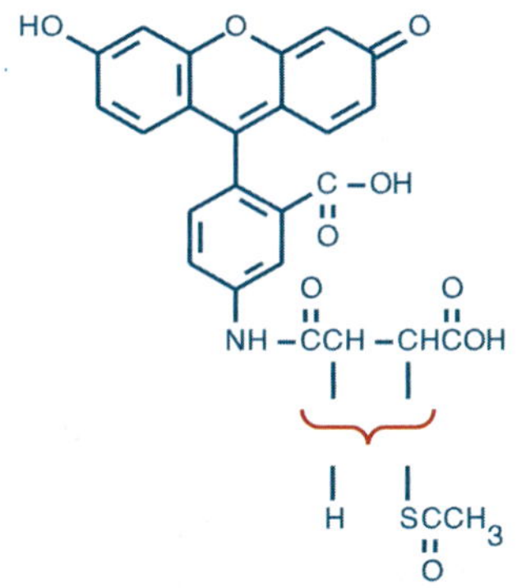

Figure 5.2.9 5-((2-(and-3)-*S*-(acetylmercapto)succinoyl)amino)fluorescein (SAMSA fluorescein, A685)

IMPORTANT NOTICE: The products described in this manual are covered by one or more Limited Use Label License(s). Please refer to the Appendix on page 971 and Master Product List on page 975. Products are For Research Use Only. Not intended for any animal or human therapeutic or diagnostic use.

used to form intramolecular crosslinks in myosin subfragment-1,[47] intermolecular crosslinks in actomyosin,[48] intersubunit crosslinks of chloroplast subunits [49] and DNA–protein crosslinks.[50] Addition of *N*-hydroxysuccinimide or *N*-hydroxysulfosuccinimide (NHSS, H2249) is reported to enhance the yield of carbodiimide-mediated conjugations,[51] indicating the *in situ* formation of a succinimidyl ester–activated protein (Figure 5.2.10). EDAC has been reported to be impermeant to cell membranes,[52] which should permit selective surface labeling of cellular carboxylic acids with fluorescent amines.

Reaction of carboxylic acids with cystamine ($H_2NCH_2CH_2S$–$SCH_2CH_2NH_2$) and EDAC followed by reduction with DTT results in thiolation at carboxylic acids.[7] This indirect route to amine–carboxylic acid coupling is particularly suited to acidic proteins with few amines, carbohydrate polymers,[6] heparin, poly(glutamic acid) and synthetic polymers lacking amines. The thiolated biomolecules can also be reacted with any of the probes described in Chapter 2.

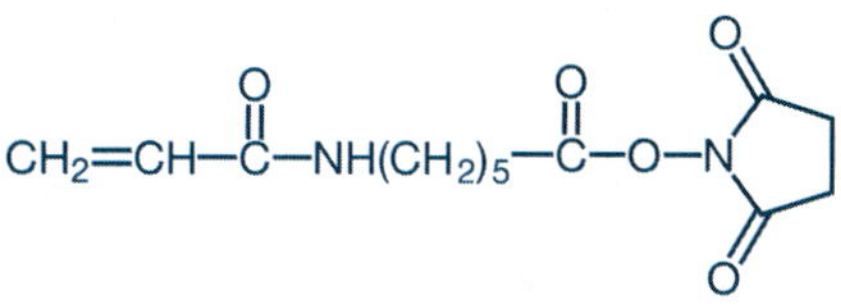

Figure 5.2.11 6-((acryloyl)amino)hexanoic acid, succinimidyl ester (acryloyl-X, SE, A20770).

Crosslinking Amines to Acrylamide Polymers

The succinimidyl ester of 6-((acryloyl)amino)hexanoic acid (acryloyl-X, SE; A20770; Figure 5.2.11) reacts with amines of proteins, amine-modified nucleic acids and other biomolecules to yield acrylamides that can be copolymerized into polyacrylamide matrices or onto surfaces, such as in microarrays and in biosensors. For example, streptavidin acrylamide (S21379, Section 7.6) copolymerizes with acrylamide on polymeric surfaces to create a uniform monolayer of the immobilized protein. The immobilized streptavidin can then bind biotinylated ligands, including biotinylated hybridization probes, enzymes, antibodies and drugs.[53]

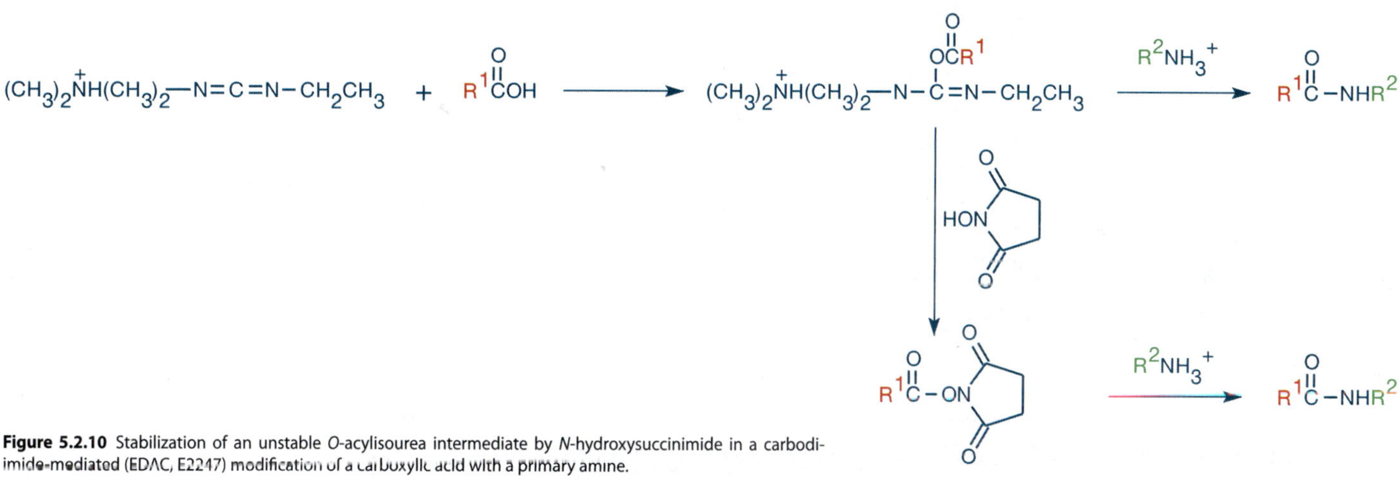

Figure 5.2.10 Stabilization of an unstable *O*-acylisourea intermediate by *N*-hydroxysuccinimide in a carbodiimide-mediated (EDAC, E2247) modification of a carboxylic acid with a primary amine.

REFERENCES

1. Bioconjug Chem (2001) 12:421; **2.** Methods Enzymol (1987) 143:246; **3.** Anal Biochem (2004) 325:137; **4.** Anal Biochem (1983) 132:68; **5.** Biochem J (1978) 173:723; **6.** Methods Md Biol (2008) 418:209; **7.** Biochim Biophys Acta (1990) 1038:382; **8.** Biochim Biophys Acta (1988) 971:307; **9.** Biochim Biophys Acta (1988) 971:298; **10.** J Biol Chem (1980) 255:10884; **11.** Methods Mol Biol (1995) 45:235; **12.** J Cell Biol (1982) 93:981; **13.** Nucleic Acids Res (1989) 17:4404; **14.** J Biochem Biophys Methods (1986) 12:349; **15.** Methods Enzymol (1994) 233:380; **16.** Methods Enzymol (1983) 91:49; **17.** Anal Biochem (1994) 220:5; **18.** Anal Biochem (1998) 265:8; **19.** Anal Biochem (1993) 213:49; **20.** Environ Sci Technol (2008) 42:8127; **21.** Anal Chem (2007) 79:1411; **22.** Bioconjug Chem (1994) 5:348; **23.** Anal Biochem (1995) 225:174; **24.** Proc Natl Acad Sci U S A (2000) 97:1461; **25.** J Mol Biol (2000) 299:421; **26.** Biochim Biophys Acta (1980) 622:201; **27.** Proc Natl Acad Sci U S A (1996) 93:10123; **28.** FEBS Lett (1982) 150:207; **29.** Biochemistry (2001) 40:7903; **30.** J Biol Chem (1996) 271:24465; **31.** J Biol Chem (2000) 275:39272; **32.** J Biol Chem (1999) 274:35388; **33.** Kidney Int (1997) 51:1797; **34.** J Biol Chem (1997) 272:31945; **35.** Bioconjug Chem (1998) 9:160; **36.** Nat Struct Biol (2001) 8:540; **37.** Biophys J (2001) 81:2851; **38.** Nat Methods (2010) 7:377; **39.** Methods Enzymol (1989) 172:584; **40.** Eur J Biochem (1979) 96:483; **41.** Nature (1981) 292:17; **42.** Nucleic Acids Res (1989) 17:4899; **43.** Chem Pharm Bull (Tokyo) (1989) 37:1831; **44.** Eur J Biochem (1979) 101:395; **45.** Anal Biochem (1991) 198:75; **46.** Cytometry (1987) 8:91; **47.** Biochemistry (1994) 33:6867; **48.** Biophys J (1995) 68:35S; **49.** Biochim Biophys Acta (1992) 1101:97; **50.** J Mol Biol (1978) 123:149; **51.** Anal Biochem (1986) 156:220; **52.** J Biol Chem (2000) 275:977; **53.** Anal Biochem (2000) 282:200.

IMPORTANT NOTICE: The products described in this manual are covered by one or more Limited Use Label License(s). Please refer to the Appendix on page 971 and Master Product List on page 975. Products are For Research Use Only. Not intended for any animal or human therapeutic or diagnostic use.

DATA TABLE 5.2 CHEMICAL CROSSLINKING REAGENTS

Cat. No.	MW	Storage	Soluble	Abs	EC	Em	Solvent	Notes
A685	521.50	F,D,L	pH >6, DMF	491	78,000	515	pH 9	
A20770	282.30	F,D,L	DMSO	<300		none		
B10621	840.47	F,D,L	DMSO	549	88,000	575	MeOH	1
D1379	350.01	L	DMF, MeCN	391	6100	see Notes	MeOH	2
D1532	154.24	D	H_2O	<300		none		
D8451	396.35	D	pH >6	324	18,000	none	pH 8	3
E2247	191.70	F,D	H_2O	<300		none		
H2249	217.13	D	H_2O	<300		none		
S1531	312.36	F,D	DMF, MeCN	282	4700	none	MeOH	4
S1534	334.33	F,D	DMF, MeCN	<300		none		
S1553	231.22	F,D	DMF, MeCN	<300		none		
T2556	286.65	D	pH >5	<300		none		

For definitions of the contents of this data table, see "Using *The Molecular Probes® Handbook*" in the introductory pages.

Notes

1. Iodoacetamides in solution undergo rapid photodecomposition to unreactive products. Minimize exposure to light prior to reaction.
2. Bimanes are almost nonfluorescent until reacted with thiols. For monobromobimane conjugated to glutathione, Abs = 394 nm, Em = 490 nm (QY ~0.1–0.3) in pH 8 buffer. (Methods Enzymol (1987) 143:76, Methods Enzymol (1995) 251:133)
3. D8451 reaction product with thiols has Abs = 410 nm (EC = 14,000 $cm^{-1}M^{-1}$). (Methods Enzymol (1994) 233:380)
4. After conjugation of S1531 the degree of substitution can be determined by measuring the amount of 2-pyridinethione formed by treatment with DTT (D1532) or TCEP (T2556) from its absorbance at 343 nm (EC = 8000 $cm^{-1}M^{-1}$). (Biochem J (1978) 173:723)

PRODUCT LIST 5.2 CHEMICAL CROSSLINKING REAGENTS

Cat. No.	Product	Quantity
A685	5-((2-(and-3)-S-(acetylmercapto)succinoyl)amino)fluorescein (SAMSA fluorescein) *mixed isomers*	25 mg
A20770	6-((acryloyl)amino)hexanoic acid, succinimidyl ester (acryloyl-X, SE)	5 mg
B10621	bis-((N-iodoacetyl)piperazinyl)sulfonerhodamine	5 mg
D1379	dibromobimane (bBBr)	25 mg
D8451	5,5′-dithiobis-(2-nitrobenzoic acid) (DTNB; Ellman's reagent)	10 g
D1532	dithiothreitol (DTT)	1 g
E2247	1-ethyl-3-(3-dimethylaminopropyl)carbodiimide, hydrochloride (EDAC)	100 mg
H2249	N-hydroxysulfosuccinimide, sodium salt (NHSS)	100 mg
M30550	Measure-iT™ Thiol Assay Kit *500 assays*	1 kit
P6305	Protein–Protein Crosslinking Kit *3 conjugations*	1 kit
S1553	succinimidyl acetylthioacetate (SATA)	100 mg
S1534	succinimidyl trans-4-(maleimidylmethyl)cyclohexane-1-carboxylate (SMCC)	100 mg
S1531	succinimidyl 3-(2-pyridyldithio)propionate (SPDP)	100 mg
T6060	Thiol and Sulfide Quantitation Kit *50–250 assays*	1 kit
T2556	tris-(2-carboxyethyl)phosphine, hydrochloride (TCEP)	1 g

5.3 Photoactivatable Reagents, Including Photoreactive Crosslinkers and Caged Probes

This section describes two types of photoactivatable probes: products that form short-lived, high-energy intermediates that can chemically couple to nearby residues, and "caged" probes that are designed to be biologically inactive until UV light–mediated photolysis releases a natural product. Photolysis of each of these photoactivatable probes can be accomplished with high spatial and temporal resolution, releasing active probe at the site of interest.

Nonfluorescent Photoreactive Crosslinking Reagents

In contrast to chemical crosslinking reagents (Section 5.2), which are often used to prepare bioconjugates, photoreactive crosslinking reagents are important tools for determining the proximity of two sites. These probes can be employed to define relationships between two reactive groups that are on a single protein, on a ligand and its receptor or on separate biomolecules within an assembly. In the latter case, photoreactive crosslinking reagents can potentially reveal interactions among proteins, nucleic acids and membranes in live cells. The general scheme for defining spatial relationships usually involves photoreactive crosslinking reagents that contain a chemically reactive group as well as a photoreactive group. These crosslinkers are first chemically reacted with one molecule, for example a receptor ligand, and then this modified molecule is coupled to a second molecule, for example the ligand's receptor, using UV illumination. Depending on the reactive properties of the chemical and photoreactive groups, these crosslinkers can be used to couple like or unlike functional groups.

IMPORTANT NOTICE: The products described in this manual are covered by one or more Limited Use Label License(s). Please refer to the Appendix on page 971 and Master Product List on page 975. Products are For Research Use Only. Not intended for any animal or human therapeutic or diagnostic use.
www.invitrogen.com/probes
molecular probes® | invitrogen™ by life technologies™

We offer three types of photoreactive reagents for covalent labeling:

- Simple aryl azides that upon illumination (usually at <360 nm) generate reactive intermediates that form bonds with nucleophilic groups (Figure 5.3.1)
- Fluorinated aryl azides that upon UV photolysis generate reactive nitrenes, thereby producing more C–H insertion products than the simple aryl azides (Figure 5.3.2)
- Benzophenone derivatives that can be repeatedly excited at <360 nm until they generate covalent adducts, without loss of reactivity (Figure 5.3.3)

Simple Aryl Azide Crosslinker

The "transferable" aryl azide *N*-((2-pyridyldithio)ethyl)-4-azidosalicylamide (PEAS; AET; P6317; Figure 5.3.4) is a unique reagent for assessing protein–protein or protein–nucleic acid interactions. This aryl azide undergoes disulfide–thiol interchange of its pyridyldisulfide groups with the thiol groups of biomolecules to form mixed disulfides in the same way as SPDP[1] (S1531, Section 5.2). UV photolysis induces covalent crosslinking to residues or biomolecules adjacent to the crosslinker. The mixed disulfide can then be cleaved with DTT or TCEP (D1532, T2556; Section 5.2). If the phenolic PEAS reagent is radioiodinated before the coupling and photolysis steps, then only the resulting target biomolecule will be radioactive at the conclusion of the reaction.

Fluorinated Aryl Azides: True Nitrene-Generating Reagents

Although the simple aryl azides may be initially photolyzed to electron-deficient aryl nitrenes, it has been shown that these rapidly ring-expand to form dehydroazepines—molecules that tend to react with nucleophiles rather than form C–H insertion products.[2,3] In contrast, Keana and Cai have shown that the photolysis products of the fluorinated aryl azides are clearly aryl nitrenes[4] and undergo characteristic nitrene reactions such as C–H bond insertion with high efficiency. Moreover, conjugates prepared from the amine-reactive succinimidyl ester of 4-azido-2,3,5,6-tetrafluorobenzoic acid (ATFB, SE; A2522) may have quantum yields for formation of photocrosslinked products that are superior to those of the nonfluorinated aryl azides. An important application of the succinimidyl ester of ATFB is the photofunctionalization of polymer surfaces[5,6] (Figure 5.3.5).

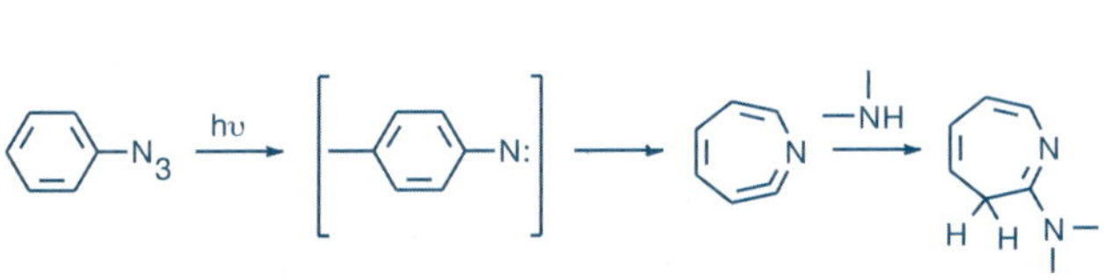

Figure 5.3.1 Photoreactive crosslinking reaction of a simple aryl azide.

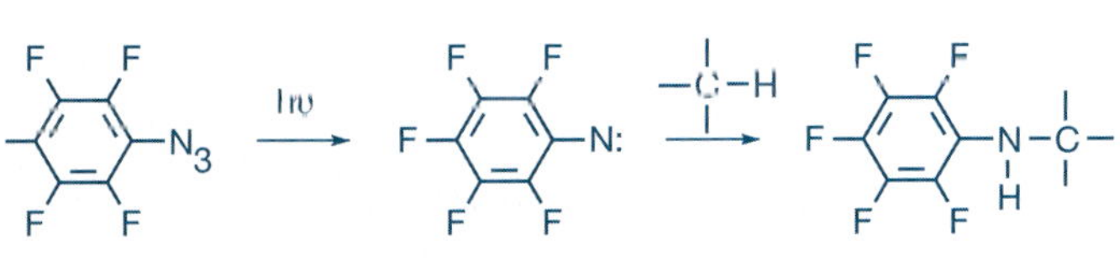

Figure 5.3.2 Photoreactive crosslinking reaction of a fluorinated aryl azide.

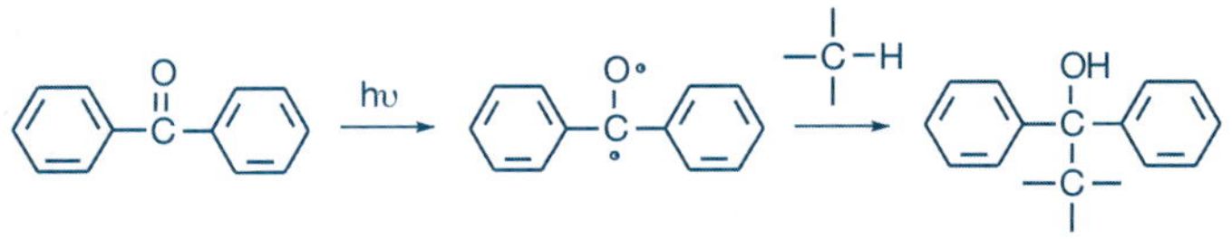

Figure 5.3.3 Photoreactive crosslinking reaction of a benzophenone derivative.

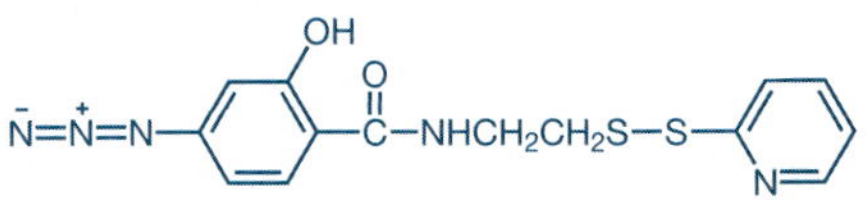

Figure 5.3.4 *N*-((2-pyridyldithio)ethyl)-4-azidosalicylamide (PEAS; AET, P6317).

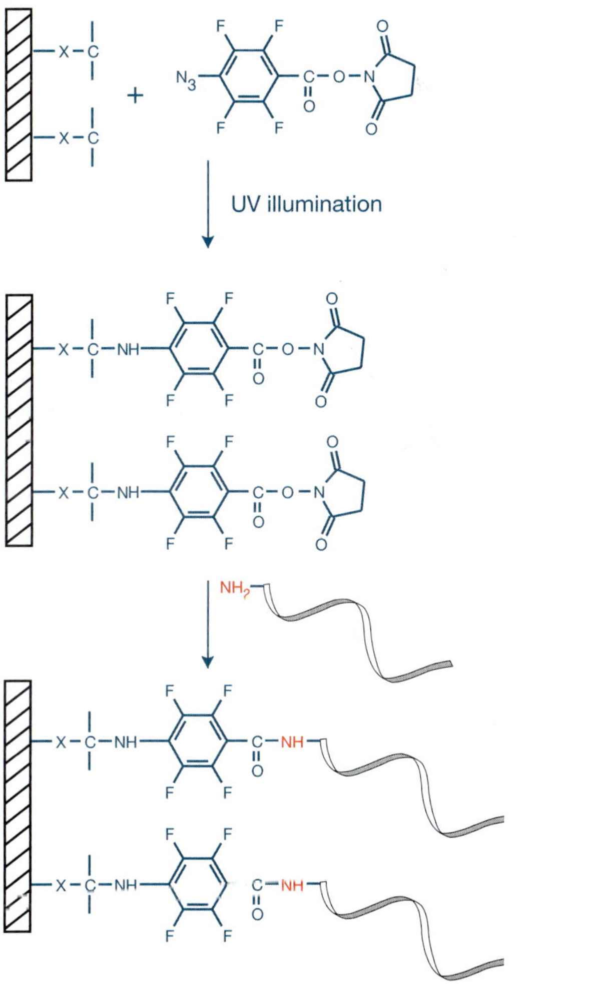

Figure 5.3.5 Schematic showing attachment of an amine-modified oligonucleotide to a surface using the photoreactive crosslinking reagent 4-azido-2,3,5,6-tetrafluorobenzoic acid, succinimidyl ester (ATFB, SE; A2522).

IMPORTANT NOTICE: The products described in this manual are covered by one or more Limited Use Label License(s). Please refer to the Appendix on page 971 and Master Product List on page 975. Products are For Research Use Only. Not intended for any animal or human therapeutic or diagnostic use.

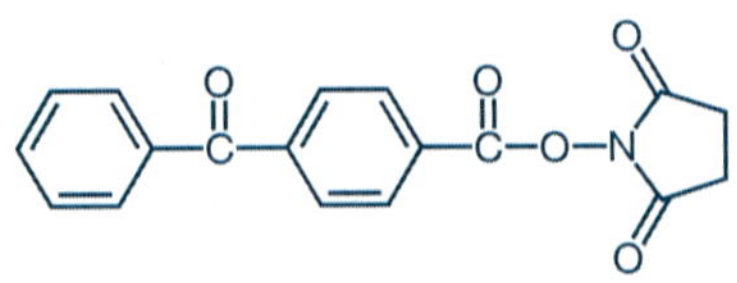

Figure 5.3.6 4-benzoylbenzoic acid, succinimidyl ester (B1577).

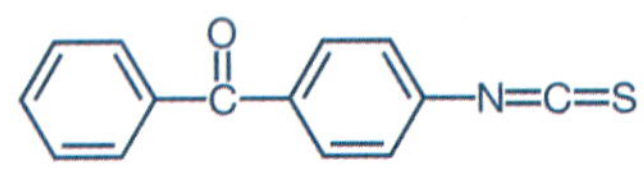

Figure 5.3.7 Benzophenone-4-isothiocyanate (B1526).

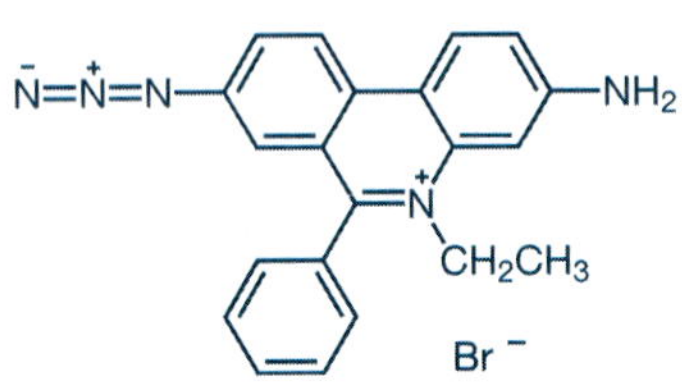

Figure 5.3.8 Ethidium monoazide bromide (EMA, E1374).

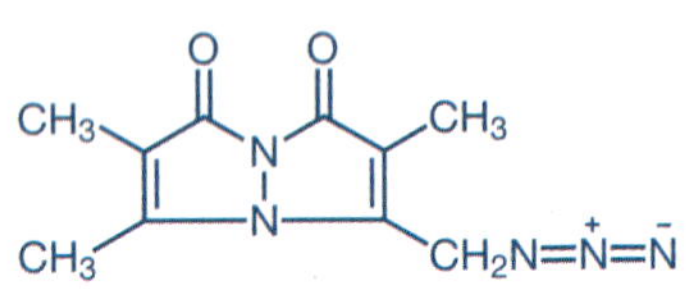

Figutre 5.3.9 Bimane azide (B30600).

Benzophenone-Based Photoreactive Reagents

Benzophenones generally have higher crosslinking yields than the aryl azide photoreactive reagents.[7] Benzophenone maleimide (B1508) has been used for efficient irreversible protein crosslinking of actin,[8] calmodulin,[9,10] myosin,[11,12] tropomyosin,[13] troponin,[14–17] ATP synthase[18,19] and other proteins. The succinimidyl ester of 4-benzoylbenzoic acid (B1577, Figure 5.3.6) and benzophenone isothiocyanate (B1526, Figure 5.3.7) have proven useful for synthesizing photoreactive peptides[20–23] and oligonucleotides.[24,25] A benzophenone-labeled ATP probe (BzBzATP, B22358) is described in Other Photoreactive Reagents below.

Other Photoreactive Reagents

Ethidium Monoazide for Photoreactive Fluorescent Labeling of Nucleic Acids

Ethidium monoazide (E1374, Figure 5.3.8) can be photolyzed in the presence of DNA or RNA to yield fluorescently labeled nucleic acids, both in solution and in cells.[26–29] The efficiency of the irreversible photolytic coupling of ethidium monoazide, which intercalates into nucleic acids like ethidium bromide, is unusually high[30] (>40%). The membrane-impermeant ethidium monoazide is reported to label only those cells with compromised membranes and can therefore serve as a fixable cell viability probe. This property, allied to the blocking of transcription caused by photoreaction of ethidium monoazide with DNA, provides a method for suppressing PCR amplification of dead-cell DNA.[31,32] Similarly, multiphoton-targeted photochemistry of vertebrate cells labeled with ethidium monoazide was used to selectively inactivate gene expression.[33] A mixed population of live and dead cells labeled with ethidium monoazide retains its staining pattern after aldehyde-based fixation, thereby reducing the investigator's exposure to potentially pathogenic cells during cell viability analysis.[34,35]

Bimane Azide for Photoaffinity Labeling of Proteins

Bimane azide (B30600, Figure 5.3.9) is a small blue-fluorescent photoreactive alkyl azide (excitation/emission maxima ~375/458 nm) for photoaffinity labeling of proteins. This reactive fluorophore's small size may reduce the likelihood that the label will interfere with the function of the biomolecule, an important advantage for site-selective probes.

Photoreactive ATP Derivative for Labeling Nucleotide-Binding Proteins

Functional ion channels can be assembled from both homomeric and heteromeric combinations of the seven P2X purinergic receptor subunits so far identified ($P2X_{1–7}$). Due to the lack of specific agonists or antagonists for P2X receptors, it is difficult to determine which receptor subtypes mediate particular cellular responses. We offer one of the most potent and widely used P2X receptor agonists,[36–39] BzBzATP (2′-(or 3′-)*O*-(4-benzoylbenzoyl)adenosine 5′-triphosphate, B22358; Figure 5.3.10). BzBzATP also has more general applications for site-directed irreversible modification of nucleotide-binding proteins via photoaffinity labeling.[40,41]

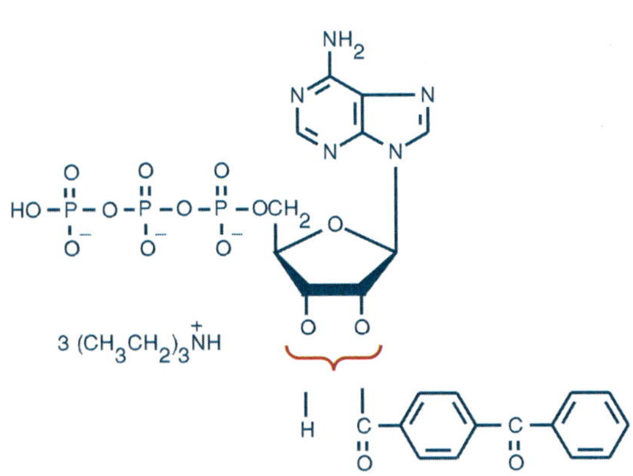

Figure 5.3.10 2′-(or-3′)-*O*-(4-benzoylbenzoyl)adenosine 5′-triphosphate, tris(triethylammonium) salt (BzBzATP, B22358).

Caged Probes and Their Photolysis

Flash photolysis of photoactivatable or "caged" probes provides a means of controlling the release—both spatially and temporally—of biologically active products or other reagents of interest.[42–48] The chemical caging process may also confer membrane permeability on the caged ligand, as is the case for caged cAMP[49] and caged luciferin.[50] Our extensive selection of caged nucleotides, second messengers (Figure 5.3.11), chelators and neurotransmitters has tremendous potential for use with both live cells and isolated proteins. These caged probes provide researchers with important tools for delivering physiological stimuli by naturally active biomolecules with spatial and temporal precision that far exceeds that of microinjection or perfusion. A recent review by Ellis-Davies describes the optical and chemical properties of many of our caged compounds, as well as of several common caging groups.[51]

Caging Groups

The caging moiety (Table 5.2) is designed to *maximally* interfere with the binding or activity of the molecule. It is detached in microseconds to milliseconds by flash photolysis at ≤360 nm, resulting in a pulse (concentration jump) of active product. Uncaging can easily be accomplished with UV illumination in a fluorescence microscope or with a UV laser or UV flashlamp. Low-cost light-emitting diodes[52] (LED) and 405 nm violet diode lasers[53] are providing increased access to experimentation using caged compounds. The effects of photolytic release are frequently monitored either with fluorescent probes that measure calcium, pH, other ions or membrane potential, or with electrophysiological techniques.

Most of the caged reagents described in the literature have been derivatives of *o*-nitrobenzylic compounds. The nitrobenzyl group is synthetically incorporated into the biologically active molecule by linkage to a heteroatom (usually *O, S* or *N*) as an ether, thioether, ester (including phosphate or thiophosphate esters), amine or similar functional group. Both the structure of the nitrobenzylic compound and the atom to which it is attached affect the efficiency and wavelength required for uncaging. We currently use six different photolabile protecting groups in our caged probes.[44] Their properties are summarized in Table 5.2.

- Probes caged with our α-carboxy-2-nitrobenzyl (CNB) caging group generally have the most advantageous properties. These include good water solubility, very fast uncaging rates in the microsecond range, high photolysis quantum yields (from 0.2–0.4) and biologically inert photolytic by-products. Although the absorption maximum of the CNB caging group is near 260 nm, its absorption spectrum tails out to approximately 360 nm, allowing successful photolysis using light with wavelengths ≤360 nm.
- The 1-(2-nitrophenyl)ethyl (NPE) caging group has properties similar to those of CNB and can also be photolyzed at ≤360 nm.
- As compared with CNB and NPE, the 4,5-dimethoxy-2-nitrobenzyl (DMNB) and 1-(4,5-dimethoxy-2-nitrophenyl)ethyl (DMNPE) caging groups have longer-wavelength absorption (absorption maximum ~355 nm) and therefore absorb 340–360 nm light more efficiently. However, photolysis rates and quantum yields of DMNB- and DMNPE-caged probes are generally lower than those obtained for CNB-caged probes.
- The 5-carboxymethoxy-2-nitrobenzyl (CMNB) caging group provides an absorption maximum of intermediate wavelength (absorption maximum ~310 nm), while imparting significant water solubility to the caged probe. Its photolysis rate and quantum yield are intermediate between those of CNB- and DMNB-caged probes.
- The nitrophenyl (NP) caging group is available on the caged calcium reagent NP-EGTA (N6802), a photolabile Ca^{2+} chelator that can be used to rapidly deliver a pulse of Ca^{2+} upon illumination with ultraviolet light, with a high photolysis quantum yield of 0.23.

Experiments utilizing probes caged with any of the above caging groups, except the CNB caging group, may require the addition of dithiothreitol (DTT, D1532; Section 2.1). This reducing reagent prevents the potentially cytotoxic reaction between amines and the 2-nitrosobenzoyl photolytic by-products.[54]

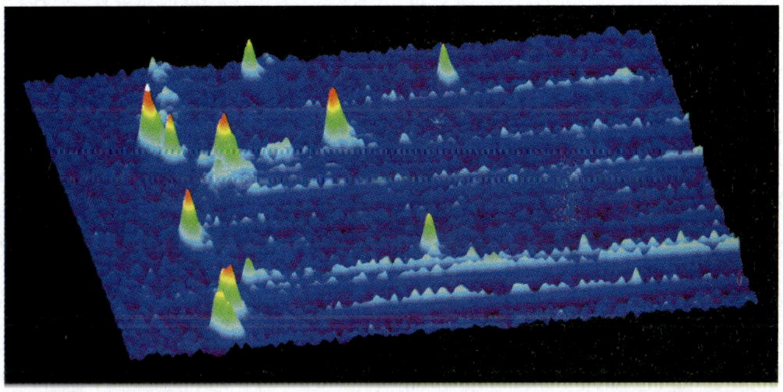

Figure 5.3.11 Confocal linescan image of calcium "puffs" in a *Xenopus* oocyte. Oregon Green® 488 BAPTA-1 (O6806) was used as the calcium indicator and Ca^{2+} liberation was evoked by flash photolysis of NPE-caged Ins 1,4,5-P_3 (I23580). Image contributed by Ian Parker and Nick Callamaras, University of California at Irvine.

Table 5.2 Properties of six different caging groups.

Caging Group	Uncaging Rate *	Photolysis Quantum Yield *	Inertness of Photolysis By-product	Confers Water Solubility	Long-Wavelength Absorption (≥360 nm)
CNB	++++	+++++	+++++	+++++	++
NPE	+++	+++	+++	+	++
DMNPE	+++	+++	+++	+	+++++
DMNB	+++	+++	++	+	+++++
CMNB	+++	+++	+	++++	+++
NP	+++	+++	+	+	++

+++++ = Optimal response. + = Poor response. * Both the structure of the nitrobenzyl moiety and the atom to which it is attached have some effect on the efficiency and wavelength required for uncaging.

IMPORTANT NOTICE: The products described in this manual are covered by one or more Limited Use Label License(s). Please refer to the Appendix on page 971 and Master Product List on page 975. Products are For Research Use Only. Not intended for any animal or human therapeutic or diagnostic use.

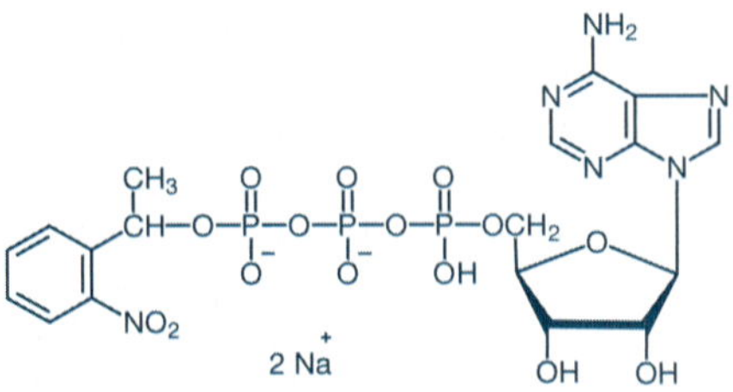

Figure 5.3.12 Adenosine 5′-triphosphate, P^3-(1-(2-nitrophenyl)ethyl) ester, disodium salt (NPE-caged ATP, A1048).

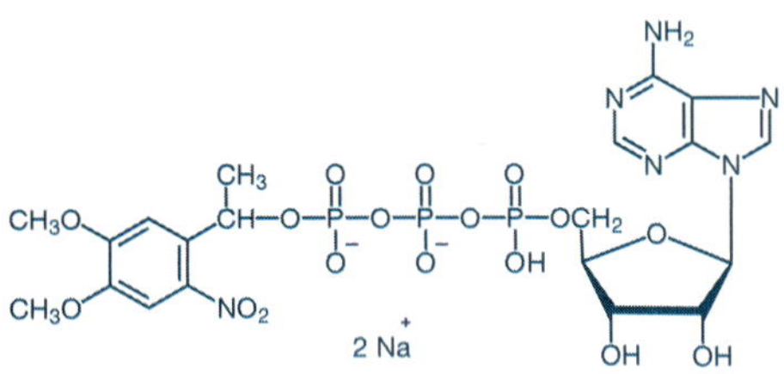

Figure 5.3.13 Adenosine 5′-triphosphate, P^3-(1-(4,5-dimethoxy-2-nitrophenyl)ethyl) ester, disodium salt (DMNPE-caged ATP, A1049).

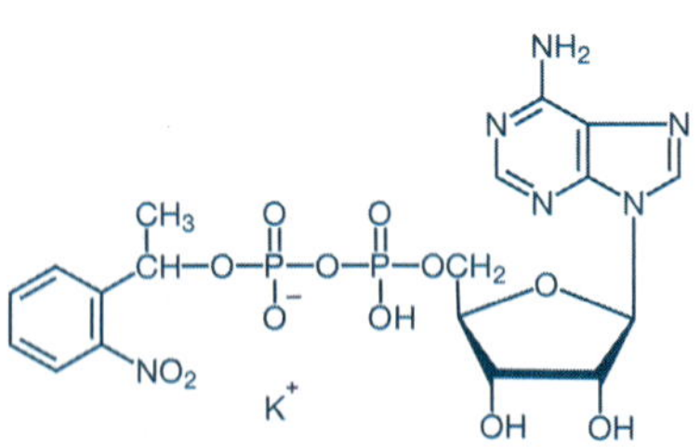

Figure 5.3.14 Adenosine 5′-diphosphate, P^2-(1-(2-nitrophenyl)ethyl) ester, monopotassium salt (NPE-caged ADP, A7056).

Caged Nucleotides

Photoactivatable nucleotides and phosphates have contributed significantly to our understanding of cytoskeleton dynamics, signal transduction pathways and other critical cellular processes.[46] Some of our caged nucleotides are available with a choice of caging group:

- Caged ATP (A1048, Figure 5.3.12; A1049, Figure 5.3.13), which has been shown to release ATP in skinned muscle fibers,[55] sarcoplasmic reticulum vesicles,[56] submitochondrial particles[57] and membrane fragments containing Na^+/K^+-ATPase[58]
- Caged ADP (A7056, Figure 5.3.14), which has been used to investigate the molecular basis of contraction of skeletal muscle fibers,[59,60] as well as transport by an ADP/ATP carrier[61]
- Caged cAMP (D1037, Figure 5.3.15), which is cell-permeant and rapidly photolyzed to cAMP[62]
- Caged inositol 1,4,5-triphosphate[63–67] (I23580, Figure 5.3.16) and caged cADP-ribose[68,69] (C7074, Figure 5.3.17), which are important probes for second messenger studies (Section 17.2)

NPE-caged Ins 1,4,5-P_3 can be used to generate rapid and precisely controlled release of Ins 1,4,5-P_3 in intact cells (Figure 5.3.11) and is widely employed in studies of Ins 1,4,5-P_3–mediated second messenger pathways.[63] Our NPE-caged Ins 1,4,5-P_3 (I23580) is a mixture of the physiologically inert, singly esterified P^4 and P^5 esters (Figure 5.3.16) and does not contain the somewhat physiologically active P^1 ester. NPE-caged Ins 1,4,5-P_3 exhibits essentially no biological activity prior to photolytic release of the biologically active Ins 1,4,5-P_3 (I3716, Section 17.2).

Cyclic ADP-ribose (cADP-ribose) is a potent intracellular Ca^{2+}–mobilizing agent that functions as a second messenger in an Ins 1,4,5-P_3–independent pathway.[70–75] Our NPE-caged cADP-ribose (C7074, Figure 5.3.17) induces Ca^{2+} mobilization in sea urchin egg homogenates only after photolysis, and this Ca^{2+} release is inhibited by the specific cADP-ribose antagonist 8-amino-cADP-ribose[69] (A7621, Section 17.2). Furthermore, when microinjected into live sea urchin eggs, NPE-caged cADP-ribose was shown to mobilize Ca^{2+} and activate cortical exocytosis after illumination with a mercury-arc lamp.[69]

Caged Ca^{2+} Reagents: NP-EGTA and DMNP-EDTA

Caged ions and caged chelators can be used to influence the ionic composition of both solutions and cells, particularly for ions such as Ca^{2+} that are present at low concentrations under normal physiological conditions. Developed by Ellis-Davies and Kaplan,[76] nitrophenyl EGTA (NP-EGTA) is a photolabile Ca^{2+} chelator that exhibits a high selectivity for Ca^{2+} ions, a dramatic increase in its K_d for Ca^{2+} upon illumination (from 80 nM to 1 mM) and a high photolysis quantum yield (0.23). NP-EGTA's affinity for Ca^{2+} *decreases* ~12,500-fold upon photolysis. Furthermore, its K_d for Mg^{2+} of 9 mM makes NP-EGTA essentially insensitive to

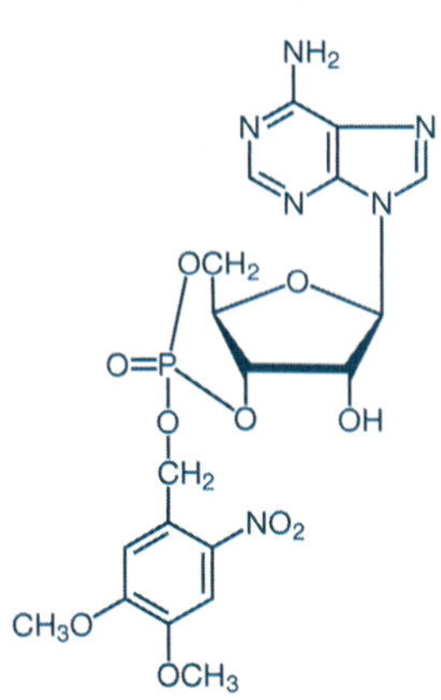

Figure 5.3.15 4,5-dimethoxy-2-nitrobenzyl adenosine 3′,5′-cyclicmonophosphate (DMNB-caged cAMP, D1037).

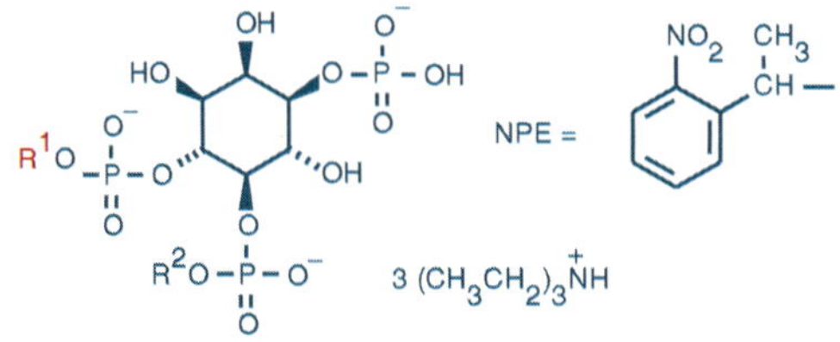

R^1 = NPE, R^2 = H or R^1 = H, R^2 = NPE

Figure 5.3.16 D-myo-inositol 1,4,5-triphosphate, $P_{4(5)}$-(1-(2-nitrophenyl)ethyl) ester, tris(triethylammonium) salt (NPE-caged Ins 1,4,5-P_3, I23580).

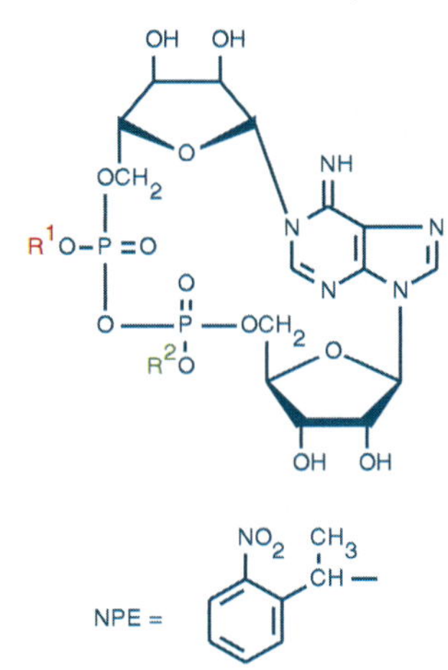

R^1 = NPE, R^2 = H or R^1 = H, R^2 = NPE

Figure 5.3.17 Cyclic adenosine 5′-diphosphate ribose, 1-(1-(2-nitrophenyl)ethyl) ester (NPE-caged cADP-ribose, C7074).

IMPORTANT NOTICE: The products described in this manual are covered by one or more Limited Use Label License(s). Please refer to the Appendix on page 971 and Master Product List on page 975. Products are For Research Use Only. Not intended for any animal or human therapeutic or diagnostic use.

physiological Mg^{2+} concentrations. We exclusively offer the tetrapotassium salt (N6802) and the acetoxymethyl (AM) ester (N6803) of NP-EGTA. The NP-EGTA salt can be complexed with Ca^{2+} to generate a caged Ca^{2+} reagent that will rapidly deliver Ca^{2+} upon photolysis[67,77,78] (Figure 5.3.18). The cell-permeant AM ester of NP-EGTA does not bind Ca^{2+} unless its AM ester groups are removed. This AM ester can serve as a photolabile chelator in cells because, once converted to NP-EGTA by intracellular esterases, it will bind free Ca^{2+} until photolyzed with UV light.

The first caged Ca^{2+} reagent described by Kaplan and Ellis-Davies was 1-(4,5-dimethoxy-2-nitrophenyl) EDTA (DMNP-EDTA, D6814), which they named DM-Nitrophen™[79,80] (now a trademark of Calbiochem-Novabiochem Corp.). Because its structure more resembles that of EDTA than EGTA, we named it as a caged EDTA derivative (Figure 5.3.19). Upon illumination, DMNP-EDTA's affinity for Ca^{2+} *decreases* ~600,000-fold and its K_d for Ca^{2+} rises from 5 nM to 3 mM. Thus, photolysis of DMNP-EDTA complexed with Ca^{2+} results in a pulse of free Ca^{2+}. DMNP-EDTA has a stronger absorbance at longer wavelengths than does NP-EGTA (Figure 5.3.20), which facilitates uncaging. Furthermore, DMNP-EDTA has significantly higher affinity for Mg^{2+} (K_d = 2.5 µM)[79] than does NP-EGTA (K_d = 9 mM),[76] making it a potentially useful caged Mg^{2+} reagent. Two reviews by Ellis-Davies discuss the uses and limitations of DMNP-EDTA.[51,81]

Diazo-2: A Photoactivatable Ca^{2+} Knockdown Reagent

In contrast to NP-EGTA and DMNP-EDTA, diazo-2 (D3034) is a photoactivatable Ca^{2+} scavenger. Diazo-2 (Figure 5.3.21), which was introduced by Adams, Kao and Tsien,[82,83] is a relatively weak chelator (K_d for Ca^{2+} = 2.2 µM). Following flash photolysis at ~360 nm, however, cytosolic free Ca^{2+} rapidly binds to the diazo-2 photolysis product, which has a high affinity for Ca^{2+} (K_d = 73 nM). Intracellular loading of NP-EGTA, DMNP-EDTA and diazo-2 is best accomplished by patch pipette infusion with the carboxylate salt form of the caged compound added to the internal pipette solution at 1–10 mM. These reagents are increasingly being applied *in vivo* for controlled intervention in calcium-regulated fundamental processes in neurobiology[84] and developmental biology.[85]

Caged Amino Acid Neurotransmitters

Once activated, caged amino acid neurotransmitters rapidly initiate neurotransmitter action (Figure 5.3.22), providing tools for kinetic studies of receptor binding or channel opening.[42,46] We offer caged carbamylcholine[86–94] (*N*-(CNB-caged) carbachol, C13654, Figure 5.3.23) and caged γ-aminobutyric acid[67,95–97] (*O*-(CNB-caged) GABA, A7110, Figure 5.3.24), as well as two caged versions of L-glutamic acid[96,98–104] (C7122, G7055), all of which are biologically inactive before photolysis.[44]

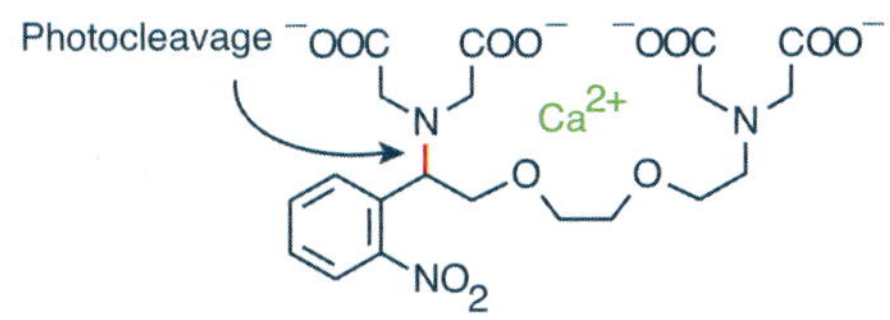

Figure 5.3.18 NP-EGTA (N6802) complexed with Ca^{2+}. Upon illumination, this complex is cleaved to yield free Ca^{2+} and two iminodiacetic acid photoproducts. The affinity of the photoproducts for Ca^{2+} is ~12,500-fold lower than that of NP-EGTA.

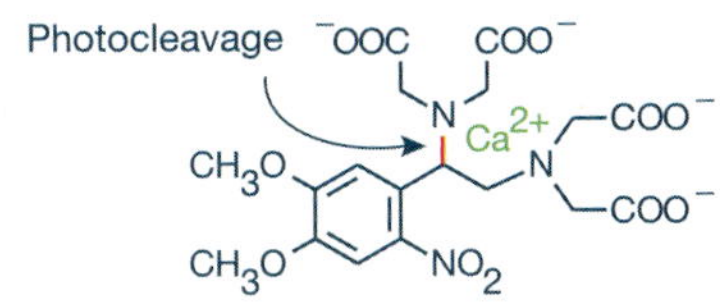

Figure 5.3.19 DMNP-EDTA (D6814) complexed with Ca^{2+}. Upon illumination, this complex is cleaved to yield free Ca^{2+} and two iminodiacetic acid photoproducts. The affinity of the photoproducts for Ca^{2+} is ~600,000-fold lower than that of DMNP-EDTA.

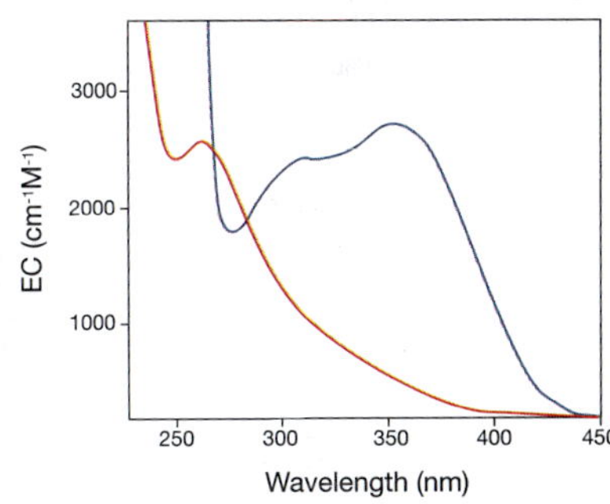

Figure 5.3.20 Spectral comparison of equimolar concentrations of the caged Ca^{2+} reagents NP-EGTA (N6802, red line) and DMNP-EDTA (D6814, blue line), illustrating the optimal wavelengths for photolysis and subsequent release of Ca^{2+} from these chelators. Spectra were taken in 100 mM KCl and 30 mM MOPS buffer containing 39.8 µM free Ca^{2+} at pH 7.2.

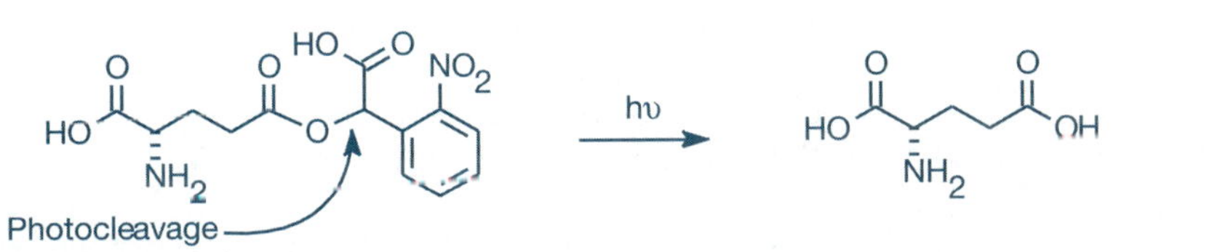

Figure 5.3.22 CNB-caged L-glutamic acid (G7055). The CNB-caging group is rapidly photocleaved with UV light to release L-glutamic acid.

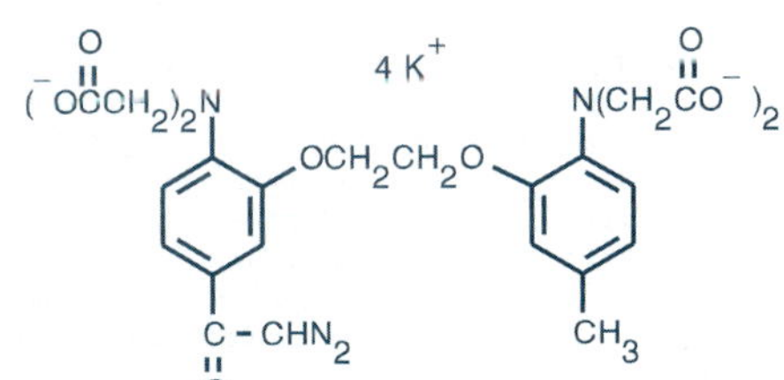

Figure 5.3.21 Diazo-2, tetrapotassium salt (D3034).

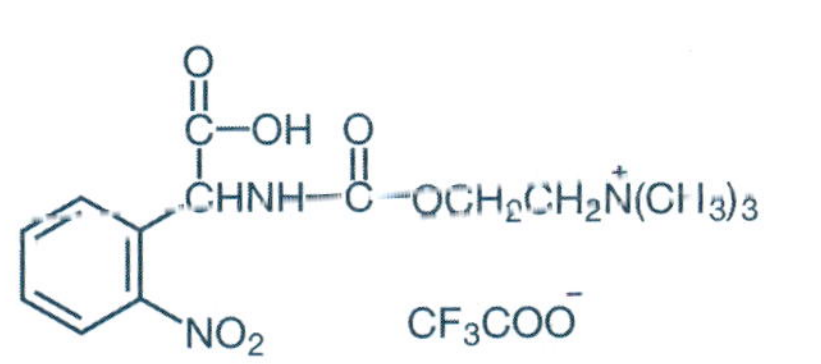

Figure 5.3.23 *N*-(CNB-caged) carbachol (N-(α-carboxy-2-nitrobenzyl)carbamylcholine, trifluoroacetic acid salt, C13654)

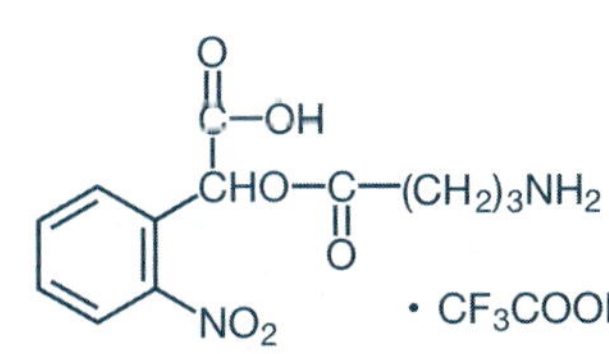

Figure 5.3.24 γ-aminobutyric acid, α-carboxy-2-nitrobenzyl ester, trifluoroacetic acid salt (O-(CNB-caged) GABA, A7110).

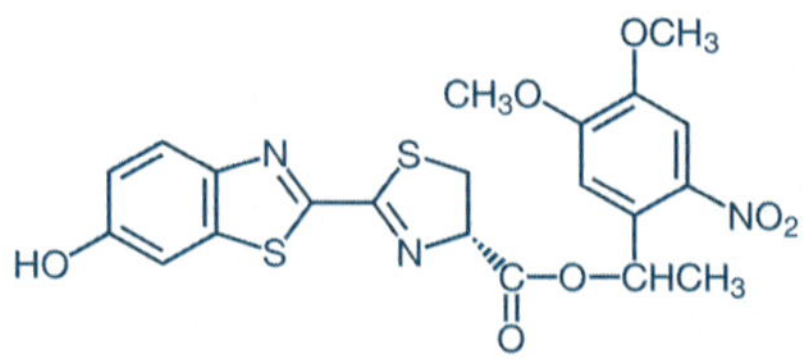

Figure 5.3.25 D-luciferin, 1-(4,5-dimethoxy-2-nitrophenyl)ethyl ester (DMNPE-caged luciferin, L7085).

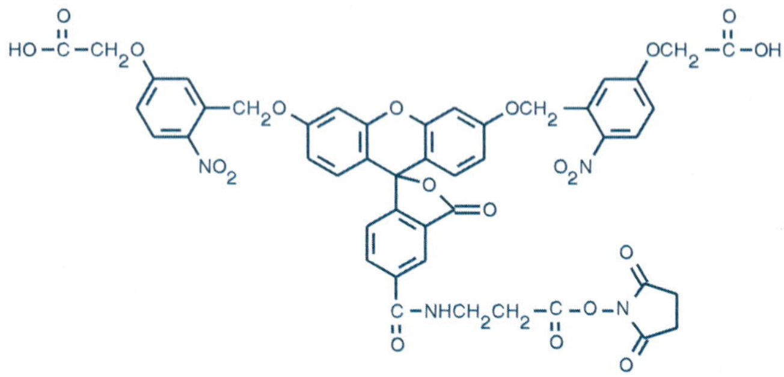

Figure 5.3.26 5-carboxyfluorescein-bis-(5-carboxymethoxy-2-nitrobenzyl) ether, β-alanine-carboxamide, succinimidyl ester (CMNB-caged carboxyfluorescein, SE, C20050).

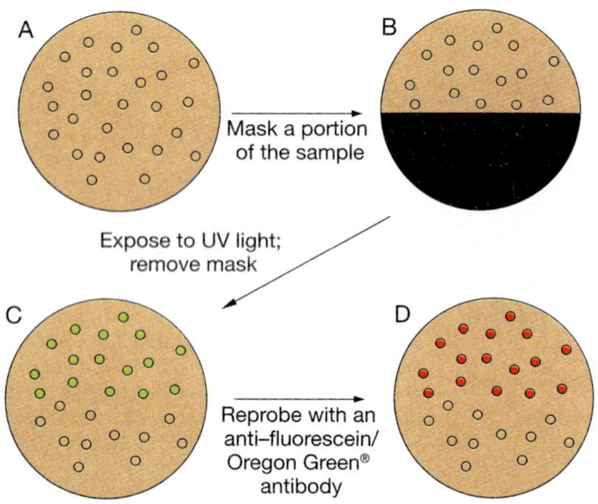

Figure 5.3.27 Schematic representation of photoactivated fluorescence combined with sample masking. Initially, no fluorescence is observed from samples stained with a CMNB-caged fluorescein-labeled secondary detection reagent (**A**). The desired mask is then placed over the sample (**B**), after which the sample is exposed to UV light. The mask is then removed; fluorescein molecules present in the unmasked portion of the sample are uncaged by the UV light and fluoresce brightly when viewed with the appropriate filters (**C**). Uncaged fluorescein may now also serve as a hapten for further signal amplification using our anti-fluorescein/Oregon Green® antibody. For example, probing with the anti–fluorescein/Oregon Green® antibody followed by staining with the Alexa Fluor® 594 goat anti–mouse IgG antibody can be used to change the color of the uncaged probe to red fluorescent (**D**).

Caged Luciferin

Luciferase produces light by the ATP-dependent oxidation of luciferin. The 560 nm chemiluminescence from this reaction peaks within seconds, with light output that is proportional to luciferase activity or ATP concentrations. DMNPE-caged luciferin (L7085, Figure 5.3.25) readily crosses cell membranes, allowing more efficient delivery of luciferin into intact cells.[50] Once the caged luciferin is inside the cell, active luciferin can be released either instantaneously by a flash of UV light, or continuously by the action of endogenous intracellular esterases found in many cell types.

Caged Fluorescent Dyes

Photoactivatable fluorescent dyes, which are generally colorless and nonfluorescent until photolyzed with UV light,[51] are particularly useful for investigating cell lineage[105] and for spatiotemporal interrogation of fluid flows.[106–108] In addition to CMNB-caged fluorescein[107] (F7103), we prepare the succinimidyl ester of CMNB-caged carboxyfluorescein (C20050, Figure 5.3.26), which can be used to attach the caged fluorophore to primary amine groups of a variety of biomolecules. CMNB-caged carboxyfluorescein succinimdyl ester is a key starting material in the preparation of probes for super-resolution photoactivation microscopy.[109,110] Furthermore, caged fluorescein probes are immunochemically cryptic; i.e., the probe is immunoreactive with anti–fluorescein/Oregon Green® antibodies (Section 7.4) after but not before photoactivation (Figure 5.3.27).

Kit for Caging Carboxylic Acids

Using organic synthesis methods, researchers can cage a diverse array of molecules. One of the preferred caging groups is the 1-(4,5-dimethoxy-2-nitrophenyl)ethyl (DMNPE) ester. Because the diazoethane precursor to DMNPE esters is unstable, we offer a kit (D2516) for the generation of 1-(4,5-dimethoxy-2-nitrophenyl)diazoethane and the subsequent preparation of DMNPE esters. This kit includes:

- 25 mg of the hydrazone precursor
- MnO_2 for oxidation
- Celite® for filtration of the reaction mixture
- Detailed protocols for caging carboxylic acids

A wide range of compounds containing a weak oxy acid (with a pK_a between 3 and 7), including carboxylic acids, phenols and phosphates, should react with the diazoethane to form the DMNPE-caged analogs[54] (Figure 5.3.28).

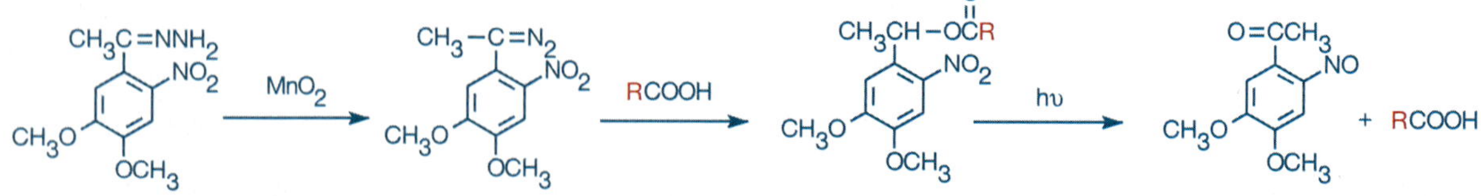

Figure 5.3.28 Caging of a carboxylic acid using the hydrazone precursor of DMNPE, a reagent that is provided in the 1-(4,5-Dimethoxy-2-nitrophenyl)diazoethane Generation Kit (D2516).

IMPORTANT NOTICE: The products described in this manual are covered by one or more Limited Use Label License(s). Please refer to the Appendix on page 971 and Master Product List on page 975. Products are For Research Use Only. Not intended for any animal or human therapeutic or diagnostic use.

REFERENCES

1. Bioconjug Chem (1996) 7:380; **2.** Annu Rev Biochem (1993) 62:483; **3.** Adv Photochem (1992) 17:69; **4.** J Org Chem (1990) 55:3640; **5.** Bioconjug Chem (1994) 5:151; **6.** J Am Chem Soc (1993) 115:814; **7.** Biochemistry (1994) 33:5661; **8.** Arch Biochem Biophys (1985) 240:627; **9.** Biochemistry (1994) 33:518; **10.** J Biol Chem (1988) 263:542; **11.** Arch Biochem Biophys (1991) 288:584; **12.** J Biol Chem (1991) 266:2272; **13.** Biochemistry (1986) 25:7633; **14.** J Muscle Res Cell Motil (1998) 19:479; **15.** Biochemistry (1996) 35:11026; **16.** Science (1990) 247:1339; **17.** Biochemistry (1987) 26:7042; **18.** J Biol Chem (1998) 273:15162; **19.** J Biol Chem (1996) 271:28341; **20.** J Virol (1981) 38:840; **21.** J Protein Chem (1985) 3:479; **22.** Proc Natl Acad Sci U S A (1986) 83:483; **23.** Biochemistry (1993) 32:2741; **24.** Nucleic Acids Res (1998) 26:1421; **25.** Bioconjug Chem (1999) 10:56; **26.** Biochemistry (1991) 30:5644; **27.** Photochem Photobiol (1986) 43:7; **28.** J Biol Chem (1984) 259:11090; **29.** Photochem Photobiol (1982) 36:31; **30.** Biochemistry (1981) 20:1887; **31.** Appl Environ Microbiol (2005) 71:1018; **32.** Appl Environ Microbiol (2006) 72:1997; **33.** Proc Natl Acad Sci U S A (2000) 97:9504; **34.** Methods Mol Biol (2009) 510:415; **35.** Cytometry (1995) 19:243; **36.** Biochemistry (1987) 26:7524; **37.** Proc Natl Acad Sci U S A (1993) 90:10449; **38.** J Physiol (1999) 519; **39.** Mol Pharmacol (1999) 56:1171; **40.** J Neurochem (1993) 61:1657; **41.** Biochemistry (1989) 28:3989; **42.** Photochem Photobiol Sci (2002) 1:441; **43.** Methods Enzymol (1998) 291:63; **44.** Methods Enzymol (1998) 291:30; **45.** Curr Opin Neurobiol (1996) 6:379; **46.** Biological Applications of Photochemical Switches, Morrison H, Ed. 1993; **47.** Optical Microscopy: Emerging Methods and Applications, Herman B, Lemasters JJ, Eds. 1993 p. 27; **48.** Annu Rev Physiol (1993) 55:755; **49.** Nature (1984) 310:74; **50.** Biotechniques (1993) 15:848; **51.** Nat Methods (2007) 4:619; **52.** Cell Calcium (2005) 37:565; **53.** J Neurosci Methods (2009) 180:9; **54.** Annu Rev Biophys Biophys Chem (1989) 18:239; **55.** Biochemistry (2004) 43:2804; **56.** Biophys J (2004) 86:815; **57.** J Biol Chem (1993) 268:25320; **58.** Biochim Biophys Acta (1988) 939:197; **59.** Biophys J (2001) 81:334; **60.** Biophys J (2001) 80:1905; **61.** Biochemistry (1997) 36:13865; **62.** Proc Natl Acad Sci U S A (2006) 103:12923; **63.** J Neurosci Methods (2004) 132:81; **64.** Biotechniques (1997) 23:268; **65.** J Biol Chem (2007) 282:13984t; **66.** J Physiol (1995) 487:343; **67.** Neuron (1995) 15:755; **68.** Methods Enzymol (1998) 291:403; **69.** J Biol Chem (1995) 270:7745; **70.** Cell Calcium (1997) 22:11; **71.** Physiol Rev (1997) 77:1133; **72.** Biochem J (1996) 315:721; **73.** EMBO J (1994) 13:2038; **74.** Mol Cell Biochem (1994) 138:229; **75.** Science (1993) 259:370; **76.** Proc Natl Acad Sci U S A (1994) 91:187; **77.** J Biol Chem (1995) 270:23966; **78.** Science (1995) 267:1997; **79.** Proc Natl Acad Sci U S A (1988) 85:6571; **80.** Science (1988) 241:842; **81.** Chem Rev (2008) 108:1603; **82.** Biochim Biophys Acta (1990) 1035:378; **83.** J Am Chem Soc (1989) 111:7957; **84.** Science (2009) 325:207; **85.** Dev Growth Differ (2009) 51:617; **86.** Proc Natl Acad Sci U S A (1996) 93:12964; **87.** J Neurosci Methods (1994) 54:151; **88.** Proc Natl Acad Sci U S A (1994) 91:6629; **89.** Biochemistry (1993) 32:3831; **90.** Biochemistry (1993) 32:989; **91.** Biochemistry (1992) 31:5507; **92.** Adv Exp Med Biol (1991) 287:75; **93.** Biochemistry (1989) 28:49; **94.** Biochemistry (1986) 25:1799; **95.** J Org Chem (1990) 55:1585; **96.** J Org Chem (1996) 61:1228; **97.** J Am Chem Soc (1994) 116:8366; **98.** Neuroscience (1998) 86:265; **99.** Science (1998) 279:1203; **100.** Proc Natl Acad Sci U S A (1994) 91:8752; **101.** J Am Chem Soc (2002) 124:7676; **102.** J Neurosci Methods (1994) 54:205; **103.** Science (1994) 265:255; **104.** Proc Natl Acad Sci U S A (1993) 90:7661; **105.** Methods Mol Biol (2000) 135:349; **106.** Anal Chem (2003) 75:1218; **107.** Anal Chem (2003) 75:1387; **108.** Anal Chem (2001) 73:3656; **109.** Methods Mol Biol (2009) 544:483; **110.** Angew Chem Int Ed Engl (2009) 48:6903.

DATA TABLE 5.3 PHOTOACTIVATABLE REAGENTS, INCLUDING PHOTOREACTIVE CROSSLINKERS AND CAGED PROBES

Cat #	MW	Storage	Soluble	Abs	EC	Em	Solvent	Notes
A1048	700.30	FF,D,LL	H_2O	259	18,000	none	MeOH	1, 2, 3
A1049	760.35	FF,D,LL	H_2O	351	4400	none	H_2O	1, 2
A2522	322.17	F,D,LL	DMF	273	23,000	none	EtOH	3
A7056	614.44	FF,D,LL	H_2O	259	15,000	none	MeOH	1, 2, 3
A7110	396.28	F,D,LL	H_2O	262	4500	none	pH 7	2, 3
B1508	277.28	F,D	DMF, MeCN	260	17,000	none	MeOH	3, 4
B1526	239.29	F,DD	DMF, MeCN	300	26,000	none	MeOH	3
B1577	323.30	F,D	DMF, MeCN	256	27,000	none	MeOH	3
B22358	1018.97	FF,L	H_2O	260	27,000	none	pH 7	3, 5, 6, 7
B30600	233.23	F,D,L	DMSO	375	6000	458	MeOH	
C7074	690.45	FF,D,LL	H_2O	259	16,000	none	H_2O	2, 3
C7122	326.26	F,D,LL	H_2O	266	4800	none	pH 7	2, 3
C13654	439.34	F,D,LL	H_2O	264	4200	none	H_2O	2, 3
C20050	962.79	F,D,LL	DMSO	289	9500	none	MeOH	2, 8
D1037	524.38	F,D,LL	DMSO	338	6100	none	MeOH	1, 2
D3034	710.86	F,D,LL	pH >6	369	18,000	none	pH 7.2	2, 9
D6814	473.39	D,LL	DMSO	348	4200	none	pH 7.2	2, 10
E1374	420.31	F,LL	DMF, EtOH	462	5400	625	pH 7	11
F7103	826.81	FF,D,LL	H_2O, DMSO	333	15,000	none	DMSO	2, 8, 12
G7055	440.29	F,D,LL	H_2O, DMSO	262	5100	none	pH 7	2, 3
I23580	872.82	FF,D,LL	H_2O	264	4200	none	H_2O	2, 3, 13
L7085	489.52	FF,D,LL	DMSO, DMF	334	22,000	none	MeOH	2, 14
N6802	653.81	FF,D,LL	pH >6	260	3500	none	pH 7.2	2, 3, 15
N6803	789.70	FF,D,LL	DMSO	250	4200	none	MeCN	16, 17
P6317	347.41	F,D,LL	DMSO	271	24,000	none	MeOH	18

For definitions of the contents of this data table, see "Using *The Molecular Probes® Handbook*" in the introductory pages.

Notes

1. Caged nucleotide esters are free of contaminating free nucleotides when initially prepared. However, some decomposition may occur during storage.
2. All photoactivatable probes are sensitive to light. They should be protected from illumination except when photolysis is intended.
3. This compound has weaker visible absorption at >300 nm but no discernible absorption peaks in this region.
4. Spectral data of the 2-mercaptoethanol adduct.
5. The molecular weight (MW) of this product is approximate because the degree of hydration and/or salt form has not been conclusively established.
6. This product is supplied as a ready-made solution in the solvent indicated under "Soluble."
7. This product can be activated by long-wavelength ultraviolet light (>300 nm) for photoaffinity labeling of proteins.
8. This product is colorless and nonfluorescent until it is activated by ultraviolet photolysis. Photoactivation generates a fluorescein derivative with spectral characteristics similar to C1359.

continued on next page

IMPORTANT NOTICE: The products described in this manual are covered by one or more Limited Use Label License(s). Please refer to the Appendix on page 971 and Master Product List on page 975. Products are For Research Use Only. Not intended for any animal or human therapeutic or diagnostic use.

DATA TABLE 5.3 PHOTOACTIVATABLE REAGENTS, INCLUDING PHOTOREACTIVE CROSSLINKERS AND CAGED PROBES—*continued*

9. The Ca^{2+} dissociation constant of diazo-2 is 2200 nM before photolysis and 73 nM after ultraviolet photolysis. The absorption spectrum of the photolysis product is similar to that of B1204. (J Am Chem Soc (1989) 111:7957)
10. $K_d(Ca^{2+})$ increases from 5 nM to 3 mM after ultraviolet photolysis. K_d values determined in 130 mM KCl, 10 mM HEPES, pH 7.1. (Proc Natl Acad Sci U S A (1988) 85:6571)
11. E1374 spectral data are for the free dye. Fluorescence is weak, but intensity increases ~15-fold on binding to DNA. After photocrosslinking to DNA, Abs = 504 nm (EC ~4000 $cm^{-1}M^{-1}$), Em = 600 nm. (Nucleic Acids Res (1978) 5:4891, Biochemistry (1980) 19:3221)
12. Unstable in water. Use immediately.
13. Ultraviolet photolysis of I23580 generates I3716 (Section 17.2).
14. L7085 is converted to bioluminescent luciferin (L2911, Section 10.6) upon ultraviolet photoactivation.
15. $K_d(Ca^{2+})$ increases from 80 nM to 1 mM after ultraviolet photolysis. K_d values determined in 100 mM KCl, 40 mM HEPES, pH 7.2. (Proc Natl Acad Sci U S A (1994) 91:187)
16. This product is intrinsically a liquid or an oil at room temperature.
17. N6803 is converted to N6802 via hydrolysis of its acetoxymethyl ester (AM) groups.
18. The absorption spectrum of P6317 includes an additional shoulder at 306 nm (EC = 10,000 $cm^{-1}M^{-1}$).

PRODUCT LIST 5.3 PHOTOACTIVATABLE REAGENTS, INCLUDING PHOTOREACTIVE CROSSLINKERS AND CAGED PROBES

Cat. No.	Product	Quantity
A7056	adenosine 5′-diphosphate, P^2-(1-(2-nitrophenyl)ethyl) ester, monopotassium salt (NPE-caged ADP)	5 mg
A1048	adenosine 5′-triphosphate, P^3-(1-(2-nitrophenyl)ethyl) ester, disodium salt (NPE-caged ATP)	5 mg
A1049	adenosine 5′-triphosphate, P^3-(1-(4,5-dimethoxy-2-nitrophenyl)ethyl) ester, disodium salt (DMNPE-caged ATP)	5 mg
A7110	γ-aminobutyric acid, α-carboxy-2-nitrobenzyl ester, trifluoroacetic acid salt (*O*-(CNB-caged) GABA)	5 mg
A2522	4-azido-2,3,5,6-tetrafluorobenzoic acid, succinimidyl ester (ATFB, SE)	25 mg
B1526	benzophenone-4-isothiocyanate	100 mg
B1508	benzophenone-4-maleimide	100 mg
B1577	4-benzoylbenzoic acid, succinimidyl ester	100 mg
B22358	2′-(or-3′)-*O*-(4-benzoylbenzoyl)adenosine 5′-triphosphate, tris(triethylammonium) salt (BzBzATP) *5 mM in buffer*	2 mL
B30600	bimane azide	5 mg
C20050	5-carboxyfluorescein-bis-(5-carboxymethoxy-2-nitrobenzyl) ether, β-alanine-carboxamide, succinimidyl ester (CMNB-caged carboxyfluorescein, SE)	1 mg
C13654	*N*-(CNB-caged) carbachol (*N*-(α-carboxy-2-nitrobenzyl)carbamylcholine, trifluoroacetic acid salt)	5 mg
C7122	*N*-(CNB-caged) L-glutamic acid (*N*-(α-carboxy-2-nitrobenzyl)-L-glutamic acid)	5 mg
C7074	cyclic adenosine 5′-diphosphate ribose, 1-(1-(2-nitrophenyl)ethyl) ester (NPE-caged cADP-ribose) *mixed isomers*	50 µg
D3034	diazo-2, tetrapotassium salt *cell impermeant*	1 mg
D6814	1-(4,5-dimethoxy-2-nitrophenyl)-1,2-diaminoethane-*N,N,N′,N′*-tetraacetic acid (DMNP-EDTA) *cell impermeant*	5 mg
D1037	4,5-dimethoxy-2-nitrobenzyl adenosine 3′,5′-cyclicmonophosphate (DMNB-caged cAMP)	5 mg
D2516	1-(4,5-Dimethoxy-2-nitrophenyl)diazoethane Generation Kit	1 kit
E1374	ethidium monoazide bromide (EMA)	5 mg
F7103	fluorescein bis-(5-carboxymethoxy-2-nitrobenzyl) ether, dipotassium salt (CMNB-caged fluorescein)	5 mg
G7055	L-glutamic acid, γ-(α-carboxy-2-nitrobenzyl) ester, trifluoroacetic acid salt (γ-(CNB-caged) L-glutamic acid)	5 mg
I23580	D-*myo*-inositol 1,4,5-triphosphate, $P_{4(5)}$-(1-(2-nitrophenyl)ethyl) ester, tris(triethylammonium) salt (NPE-caged Ins 1,4,5-P_3)	25 µg
L7085	D-luciferin, 1-(4,5-dimethoxy-2-nitrophenyl)ethyl ester (DMNPE-caged luciferin)	5 mg
N6803	*o*-nitrophenyl EGTA, AM (NP-EGTA, AM) *cell permeant* *special packaging*	20 x 50 µg
N6802	*o*-nitrophenyl EGTA, tetrapotassium salt (NP-EGTA) *cell impermeant*	1 mg
P6317	*N*-((2-pyridyldithio)ethyl)-4-azidosalicylamide (PEAS; AET)	10 mg

IMPORTANT NOTICE: The products described in this manual are covered by one or more Limited Use Label License(s). Please refer to the Appendix on page 971 and Master Product List on page 975. Products are For Research Use Only. Not intended for any animal or human therapeutic or diagnostic use.

www.invitrogen.com/probes

molecular probes® | invitrogen™ by life technologies™

PART II

Detecting Biomolecules: Signal Amplification and Secondary Detection

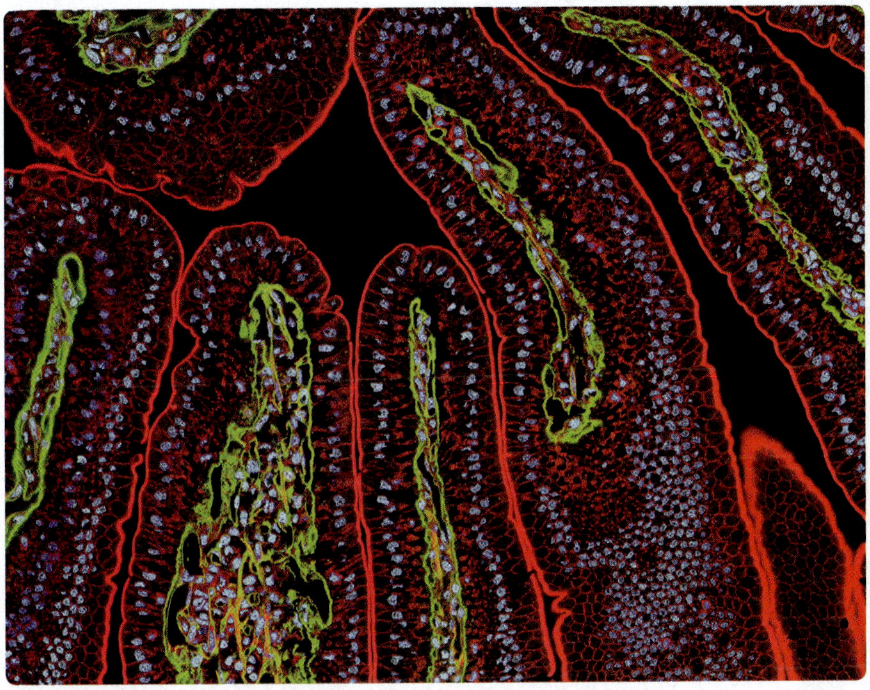
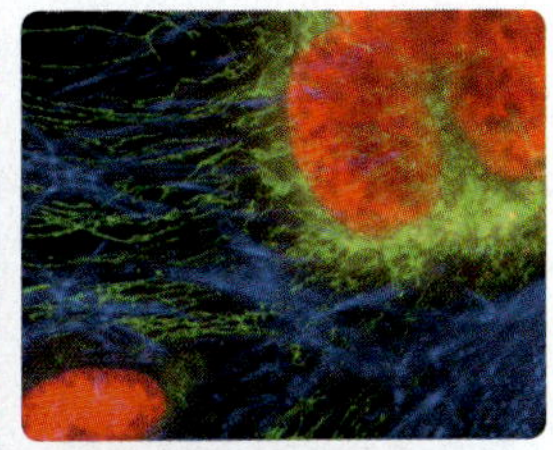
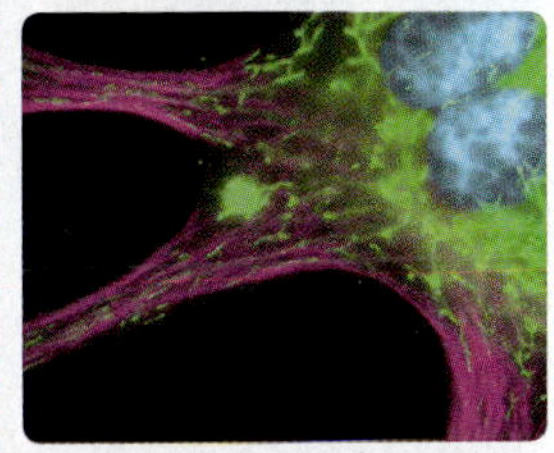
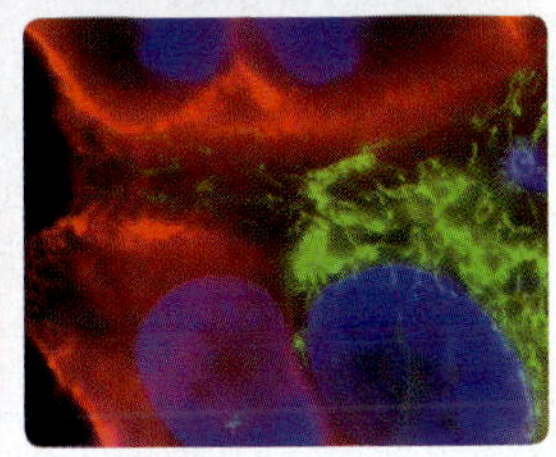

IMPORTANT NOTICE: The products described in this manual are covered by one or more Limited Use Label License(s). Please refer to the Appendix on page 971 and Master Product List on page 975. Products are For Research Use Only. Not intended for any animal or human therapeutic or diagnostic use.

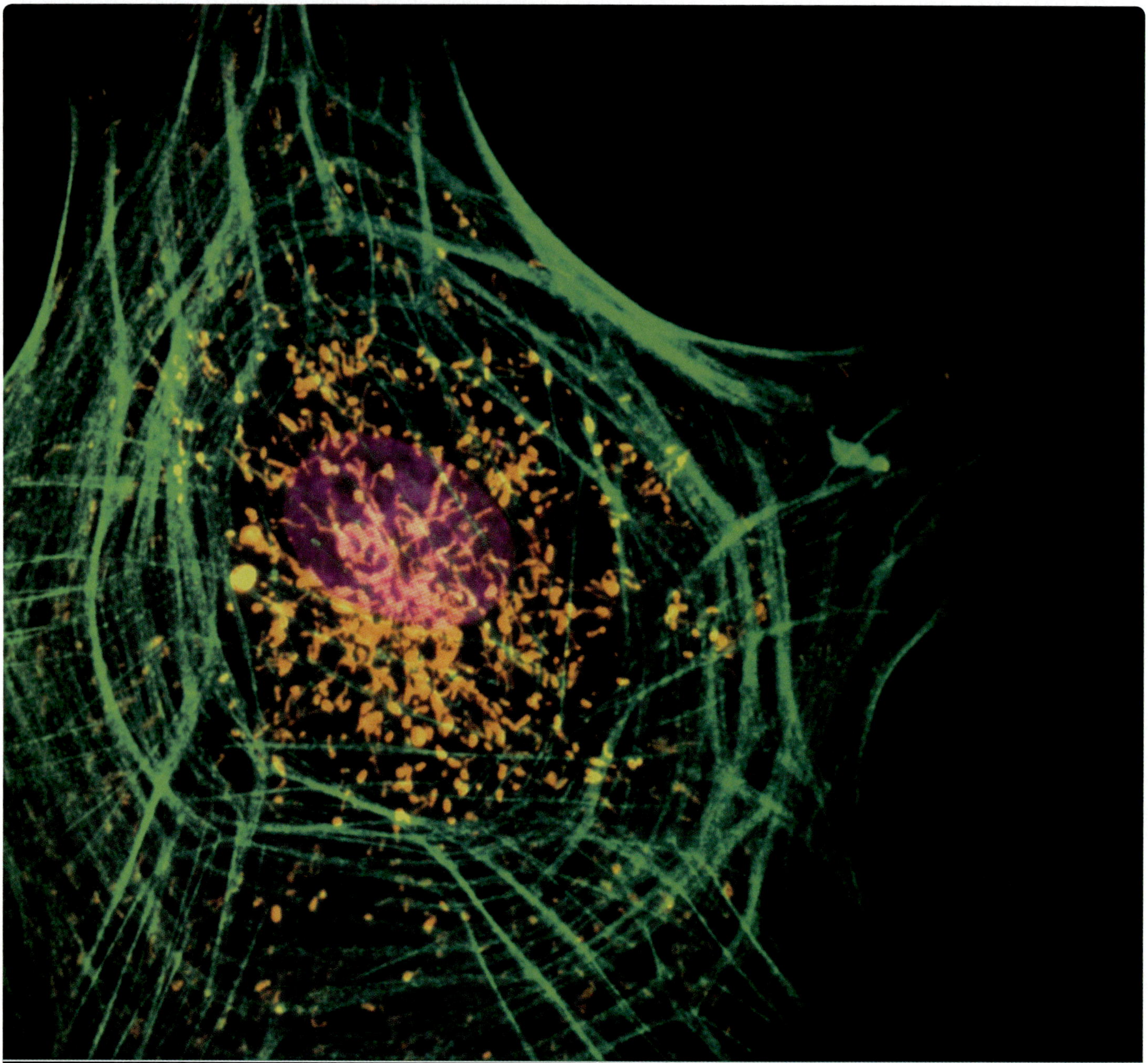

FluoCells® prepared slide #6 (F36925) showing a fixed, permeabilized and labeled muntjac skin fibroblast. Mitochondria were labeled with anti–OxPhos Complex V inhibitor protein mouse IgG1 and visualized using orange-fluorescent Alexa Fluor® 555 goat anti–mouse IgG (A21422). F-actin was labeled with green-fluorescent Alexa Fluor® 488 phalloidin (A12379), and the nucleus was stained with TO-PRO®-3 iodide (T3605, pseudocolored magenta). Image contributed by Michael S. Janes, Invitrogen.

This image appeared on the cover of *BioProbes 56*. *BioProbes*® newsletter is published several times each year. This award-winning publication is dedicated to furnishing researchers with the very latest information about cell biology products and their applications, and provides a great way to stay connected with the fluorescence community. You can subscribe and view the latest issues online at **www.invitrogen.com/handbook/bioprobes**.

SIX

CHAPTER 6
Ultrasensitive Detection Technology

IMPORTANT NOTICE: The products described in this manual are covered by one or more Limited Use Label License(s). Please refer to the Appendix on page 971 and Master Product List on page 975. Products are For Research Use Only. Not intended for any animal or human therapeutic or diagnostic use.

IMPORTANT NOTICE: The products described in this manual are covered by one or more Limited Use Label License(s). Please refer to the Appendix on page 971 and Master Product List on page 975. Products are For Research Use Only. Not intended for any animal or human therapeutic or diagnostic use.

6.1 Introduction to Signal Amplification

Signal Amplification: Why and How

The number of target molecules per unit volume of sample is a key variable in all biological detection applications. Although it is possible to control target abundance through strategies such as recombinant protein overexpression and siRNA knockdowns, there is an associated risk of fundamentally perturbing the finely balanced and intertwined molecular interaction networks that underlie cellular function. There is therefore always some degree of need to detect molecules at their native abundance levels, which can vary by many orders of magnitude. For example, proteins in mammalian cells have abundances varying by at least seven orders of magnitude (~10^1–10^8 copies per cell). Furthermore, the distribution of target molecules within the cell is neither spatially uniform nor temporally static—indeed, these spatial and temporal variations are often the subject for experimental investigation. Many functionally important proteins such as transcription factors and cell-surface cytokine receptors have native expression levels below the detection threshold of labeled primary and secondary antibodies and other affinity reagents. In this chapter, we describe a collection of signal amplification strategies that can be used facilitate the detection of low-abundance molecular targets either *in situ* or *ex situ*[1] (e.g., on microarrays). The signal amplification strategies described below are essentially of two types—enzyme labeling and macrofluorophore labeling. These two approaches are not necessarily singular and may be used in combination for additive effect.[2]

Enzyme labeling utilizes an enzyme linked to a target-specific affinity reagent, either by direct conjugation or through a secondary complex (Figure 6.1.1). The enzyme turns over multiple copies of a fluorogenic or chromogenic substrate (Chapter 10), resulting in much higher target-associated signal levels than are obtainable by dye-labeled affinity reagents. The two most widely used enzymes for this purpose are horseradish peroxidase (HRP, Section 6.2) and alkaline phosphatase (Section 6.3). Major applications of enzyme labeling include immunocytochemical and immunohistochemical detection and enzyme-linked immunosorbent assays (ELISAs). In immunocytochemical and immunohistochemical detection applications, it is essential that the product of the enzyme reaction is localized in the vicinity of the enzyme conjugate in order to convey information on the spatial distribution of the target. Tyramide substrates for HRP (Section 6.2, Figure 6.1.2) and our ELF® substrates for alkaline phosphatase (Section 6.3, Figure 6.1.3) fulfill this requirement. In ELISAs, the objective is macroscopic

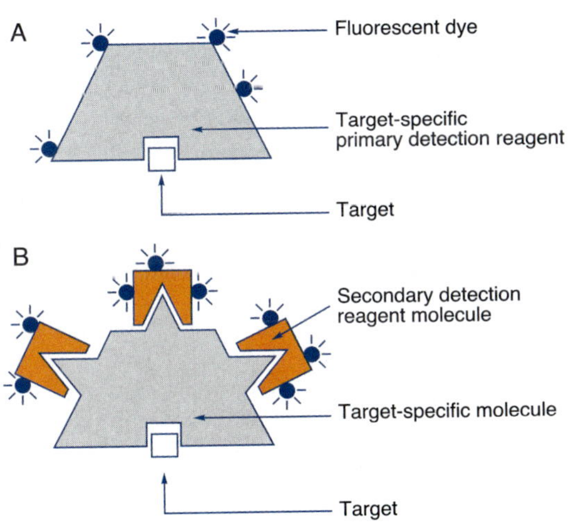

Figure 6.1.1 Schematic diagram of primary and secondary detection reagents. **A**) In primary detection methods, the target-specific molecule includes one or more detectable moieties, shown here as radiant orbs. **B**) In secondary detection methods, the target-specific molecule contains binding sites or haptens that can be selectively recognized by secondary detection reagents. For example, these sites might be antigenic epitopes that bind antibodies. Alternatively, the target-specific molecule might be conjugated to either biotin or fluorescent dyes, thereby creating a molecule that can be detected with any of our avidin and streptavidin conjugates or our anti-fluorescent dye antibodies. As shown here, the target-specific molecule may contain multiple sites for binding the secondary detection reagent, thereby providing a simple system for amplifying the signal.

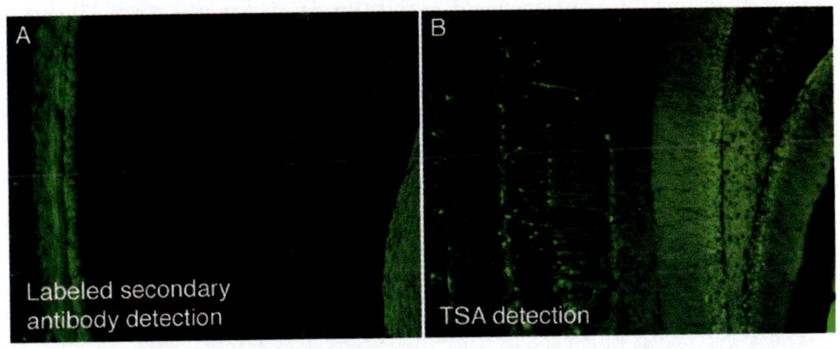

Figure 6.1.2 Tyramide signal amplification of immunofluorescent staining in mouse brain sections. Mice were transcardially perfused with phosphate-buffered saline followed by 4% formaldehyde in phosphate buffer. Serial sections (30 µm) were cut in a freezing microtome and transferred to phosphate-buffered saline. Free-floating sections were incubated with 1% hydrogen peroxide to quench endogenous peroxidase activity, blocked in 5% normal goat serum, then stained with a rabbit polyclonal antibody to calbindin D-28K (Chemicon) at a 1:1000 dilution. After washing, sections were incubated with Alexa Fluor® 488 goat anti–rabbit IgG antibody (A11008) at 5 µg/mL (**A**) or HRP–goat anti–rabbit IgG antibody at 1 µg/mL, followed by Alexa Fluor® 488 tyramide (in TSA™ Kit #12, T20922; **B**). Sections were washed, mounted on slides, coverslipped with ProLong® antifade reagent (in Kit P7481) and imaged under identical conditions (10× magnification, 250 millisecond exposure) using a bandpass filter set appropriate for fluorescein (FITC).

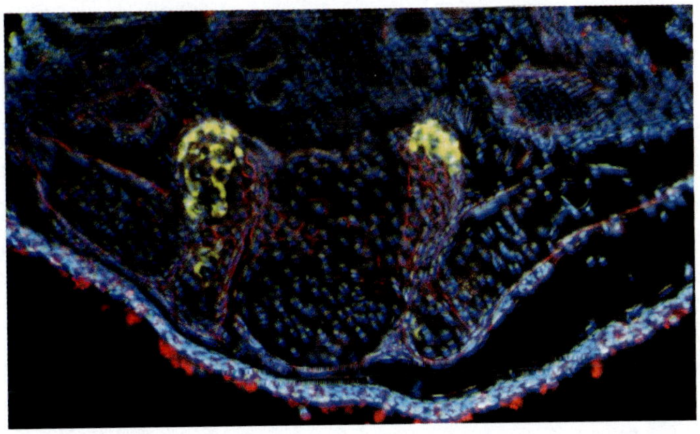

Figure 6.1.3 Endogenous phosphatase activity of osteoblast cells in a cartilaginous element of an adult zebrafish head was localized using the ELF® 97 Endogenous Phosphatase Detection Kit (E6601) to stain a cryosection. In addition to the yellow-green fluorescence of the ELF® 97 alcohol precipitate, the section was stained with red-fluorescent Texas Red®-X wheat germ agglutinin (W21405) and with the blue-fluorescent Hoechst 33342 nucleic acid stain (H1399, H3570, H21492). The triple-exposure image was acquired using bandpass filter sets appropriate for ELF® 97 alcohol, Texas Red® dye and DAPI.

quantitation rather than microscopic localization of the target, so substrates that yield diffusible products such as our fluorogenic Amplex® UltraRed substrate for HRP (Section 6.2) and our CSPD® and CDP-*Star*® chemiluminescent substrates for alkaline phosphatase (Section 6.3) are typically used. Detection signals that are amplified using enzyme reactions are necessarily time dependent. Therefore, in both immunocytochemical and immunoassay applications of enzyme labeling, careful control of timing is an essential prerequisite for obtaining quantitative and reproducible results.

Macrofluorophores are collections of fluorophores numbering in the tens (phycobiliproteins, Section 6.4) to millions (fluorescent microspheres, Section 6.5) attached to or incorporated in a common scaffold. The scaffold is coupled to a target-specific affinity reagent such as an antibody or streptavidin, and the incorporated fluorophores are thereby collectively associated with the target upon binding. From a physical perspective, quantum dot nanocrystals (Qdot® nanocrystals, Section 6.6) are single fluorophores, albeit ones with extraordinary photon output capacity. From a utilization standpoint however, they resemble macrofluorophores and are similar in size to our smallest fluorescent microspheres. Macrofluorophores are not subject to the time-dependent signal development considerations introduced by enzyme labeling but are more susceptible to nonspecific binding. Even phycobiliproteins, the smallest and most biocompatible of these macrofluorophores, are not immune to these effects.[3]

There are several other noteworthy approaches to the detection of low-abundance targets that may be applied either in combination with or as alternatives to the labeling technologies described in this chapter. In the case of nucleic acids, the capacity for self-replication allows the amount of target to be amplified through application of the polymerase chain reaction (PCR). Attempts to increase target-associated signals should generally be pursued in parallel with attempts to decrease off-target background signals. Blocking reagents such as our BlockAid™ blocking solution for use with fluorescent microspheres (B10710, Section 6.5) and our Image-iT® FX signal enhancer (I36922, Section 23.1) for use with dye-labeled antibodies can be employed in pursuit of this objective. For fluorescence detection in general, single-molecule detection techniques represent perhaps the ultimate in background reduction strategies.[4]

Primary and Secondary Detection Reagents

Both enzyme and macrofluorophore labels can be coupled directly to target-specific affinity reagents (primary detection) or to more generic affinity reagents that form stable complexes with unlabeled primary reagents, usually on the basis of immunorecognition (secondary detection). As indicated schematically in Figure 6.1.1, secondary detection inherently provides some degree of signal amplification, although sometimes at the expense of additional background due to nonspecific binding. These basic concepts of primary and secondary detection apply not only to the signal amplification techniques addressed in the current chapter but also to the dye-labeled affinity reagents described in Chapter 7.

Primary Detection Reagents

Any easily detectable molecule that binds directly to a specific target is a primary detection reagent. Such reagents are detected by fluorescence, chemiluminescence, absorption or electron diffraction conferred by stably attached labels. The conjugation and crosslinking chemistries used to create these stable attachments are discussed in detail in Chapter 1, Chapter 2 and Chapter 5. In addition to our fluorophore-labeled anti-dye antibodies (Section 7.4) and monoclonal antibodies (www.invitrogen.com/handbook/antibodies), many of the Molecular Probes® site-selective products can be considered primary detection reagents. These include our fluorescent lectins (Section 7.7), nucleic acid stains (Chapter 8), protein and glycoprotein stains (Section 9.3, Section 9.4), phallotoxins (Section 11.1), membrane probes (Chapter 13), annexin V conjugates for detecting apoptotic cells (Section 15.5) and various drug and toxin analogs (Section 16.2, Section 16.3). These primary detection reagents can typically be detected by fluorescence microscopy, fluorometry or flow cytometry methods.

Secondary Detection Reagents

Although many biomolecules, such as antibodies and lectins, bind selectively to a biological target, they usually need to be chemically modified before they can be detected. Often the biomolecule is conjugated to a fluorescent or chromophoric dye or to a heavy atom complex such as colloidal gold. However, the researcher may wish to avoid the time and expense required for these conjugations, choosing instead to use a more generic secondary detection reagent. Typically, secondary detection reagents recognize a particular class of molecules. For example, labeled goat anti–mouse IgG antibodies can be used to localize a tremendous variety of target-specific mouse monoclonal antibodies. Our extensive secondary antibody offering (Section 7.2) provides a wide selection of labels including our superior Alexa Fluor® dye series, phycobiliproteins, Alexa Fluor® dye–phycobiliprotein tandem fluorophores, Qdot® nanocrystals, biotin and enzyme labels (HRP and alkaline phosphatase). We also offer many options in terms of immunoreactivity, an essential consideration in avoiding confounding cross-reactivity when performing simultaneous secondary immunodetection of two or more targets. Our labeled secondary antibody portfolio contains antibodies against IgG and IgM from several mammalian species, including various isotypes of mouse IgG, as well as antibodies against avian (chicken) IgY. Our Zenon® antibody labeling technology (Section 7.3) uses conjugates of an Fc-specific anti-IgG Fab fragment for the rapid and quantitative labeling of the corresponding mouse, rabbit, goat or human antibody.

REFERENCES

1. Mol Cell Probes (2008) 22:294; **2.** J Histochem Cytochem (2003) 51:981; **3.** J Immunol Methods (1993) 162:269; **4.** Science (2007) 315:81.

6.2 TSA and Other Peroxidase-Based Signal Amplification Techniques

Principles of Tyramide Signal Amplification

We are committed to the extensive development of tyramide signal amplification (TSA) in combination with our Alexa Fluor® dyes to achieve high-resolution signal amplification in cell and tissue applications. TSA—sometimes called CARD, for Catalyzed Reporter Deposition—is an enzyme-mediated detection method that utilizes the catalytic activity of horseradish peroxidase (HRP) to generate high-density labeling of a target protein or nucleic acid sequence *in situ*.[1–5] TSA has been reported to increase detection sensitivity up to 100-fold, as compared with conventional avidin–biotinylated enzyme complex (ABC) procedures.[3,6–12] Moreover, for multiparameter detection of targets in either live or fixed cells or tissues, TSA can be combined with several other important technologies, including our nucleic acid labeling kits (Section 8.2), primary and secondary antibodies, avidin and lectin conjugates (Chapter 7), cytoskeletal stains (Chapter 11), organelle probes (Chapter 12) and cell tracers (Chapter 14). The Zenon® Horseradish Peroxidase Antibody Labeling Kits (Section 7.3, Table 7.7), which are described below, are of particular utility when used in combination with TSA technology.

TSA labeling is a combination of three (or four) elementary processes (Figure 6.2.1) that typically comprise:

- Binding of a probe to the target via immunoaffinity (proteins) or hybridization (nucleic acids) followed by secondary detection of the probe with an HRP-labeled antibody or streptavidin conjugate. Peroxidase conjugates of other targeting proteins such as lectins and receptor ligands are likely to be suitable for labeling targets, as is endogenous peroxidase activity.[13,14] Unconjugated HRP is also useful as a neuronal tracer; its use in combination with TSA is demonstrated in Figure 6.2.2.
- Activation of multiple copies of a labeled tyramide derivative by HRP. Most often a fluorescent or biotinylated tyramide has been used; however, labeling with other hapten-conjugated tyramides[15,16] or with polymeric reagents, including tyramide-conjugated gold particles, has also been reported.[17]
- Covalent coupling of the resulting highly reactive, short-lived tyramide radicals to residues (principally the phenol moiety of protein tyrosine residues) in the vicinity of the HRP–target interaction site, resulting in minimal diffusion-related loss of signal localization (Figure 6.2.3). In a unique application, fluorescein-labeled tyramine has been used to detect protein oxidation by reactive oxygen species (ROS, Section 18.2) in fibroblasts exposed to oxidative stress[18] and in the extracellular proteins of endothelial cells exposed to an oxidative burst from phorbol myristate acetate–activated neutrophils.[19]

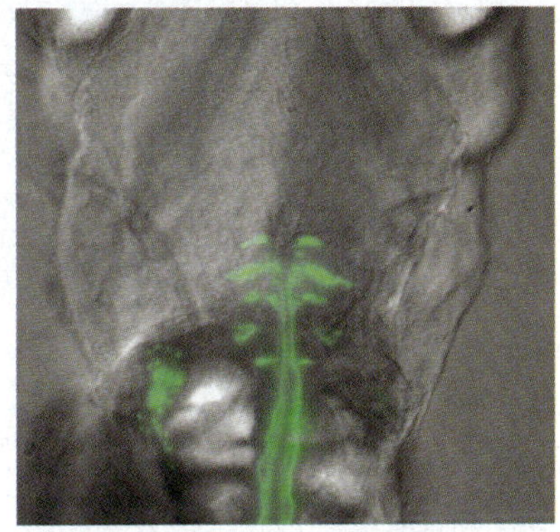

Figure 6.2.2 Horseradish peroxidase (HRP)–filled reticulospinal neurons of the hindbrain of a whole-mount zebrafish larva were detected using green-fluorescent Alexa Fluor® 488 tyramide (TSA™ Kit #2, T20912). The spinal cord was labeled by transection in the presence of HRP (1 mg/mL). After a 2-hour incubation, the specimen was fixed and the HRP was visualized using Alexa Fluor® 488 tyramide. This figure represents a projection (performed with AutoQuant Imaging, Inc., software) of a stack of 30 images.

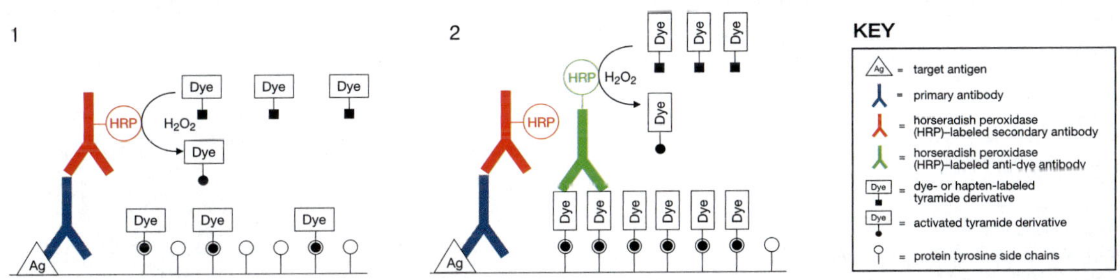

Figure 6.2.1 Schematic representation of TSA detection methods applied to immunolabeling of an antigen. The antigen is detected by a primary antibody, followed by a horseradish peroxidase (HRP)–labeled secondary antibody in conjunction with a dye-labeled tyramide, resulting in localized deposition of the activated tyramide derivative (Stage 1). Further dye deposition, and therefore higher levels of signal amplification, can be generated by detecting dye deposited in Stage 1 with a HRP-labeled anti-dye antibody in conjunction with a dye-labeled tyramide (Stage 2).

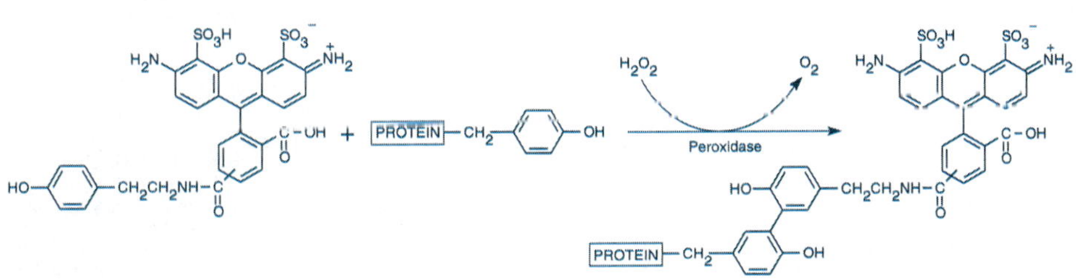

Figure 6.2.3 Coupling of Alexa Fluor® 488 tyramide to protein tyrosine side chains via peroxidase-mediated formation of an *O,O'*-dityrosine adduct.

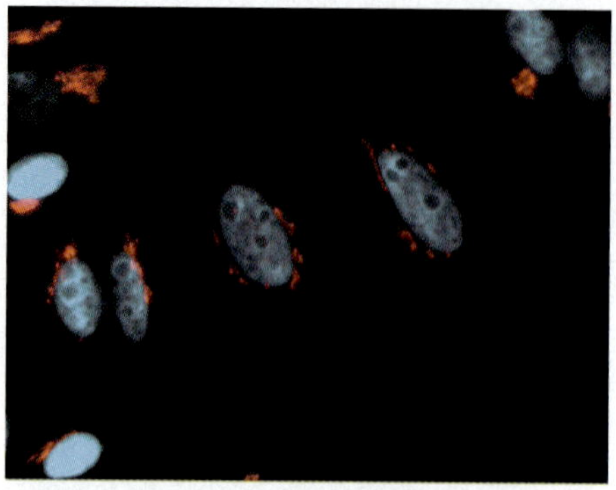

Figure 6.2.4 Golgi in HeLa cells detected with Alexa Fluor® 546 tyramide. Cells were fixed and permeabilized, then labeled with anti–human Golgin-97 antibody (A21270) and detected using HRP-conjugated goat anti–mouse IgG antibody and Alexa Fluor® 546 tyramide, which are components of the TSA™ Kit #3 (T20913). The nuclei were counterstained using DAPI (D1306, D3571, D21490). The images were acquired using filters appropriate for DAPI and Alexa Fluor® 546 and processed using MetaMorph® software from Universal Imaging Corp.

In direct TSA protocols, the fluorescent signal can be immediately detected, resulting in both excellent spatial resolution (Figure 6.2.4) and high signal intensity. When using a hapten-labeled tyramide such as biotin-XX tyramide, a subsequent detection step is required using a bioconjugate that recognizes the hapten, in this case a fluorescent streptavidin (Section 7.6). Alternatively, the hapten-labeled tyramide can be detected using an alkaline phosphate– or HRP-labeled hapten recognizer in conjunction with a fluorogenic or chromogenic substrate (Figure 6.2.5), resulting in another enzyme-amplified detection step. Chemiluminescent detection of an HRP-deposited biotin tyramide has also been reported.[20] The streptavidin conjugate of NANOGOLD® 1.4 nm gold clusters (N24918, Section 7.6) has been used to make biotin tyramide conjugates visible in light and electron microscopy.[21,22] The antibody and streptavidin conjugates of Alexa Fluor® FluoroNanogold™ 1.4 nm gold clusters (Section 7.2, Section 7.6) can also be used with hapten-labeled tyramides for correlated fluorescence, light and electron microscopy studies.

The signal amplification conferred by the turnover of multiple tyramide substrates per peroxidase label translates into practical benefits, namely ultrasensitive detection of low-abundance targets in fluorescence *in situ* hybridization,[3,23,24] immunohistochemistry,[6,25] neuroanatomical tracing[7,26] and other applications. For example, we have utilized TSA and Alexa Fluor® 488 tyramide to detect expression of low-abundance epidermal growth factor (EGF) and estrogen receptors by flow cytometry with far greater sensitivity than can be obtained using a directly labeled EGF probe (Figure 6.2.6) or fluorophore- or hapten-labeled antibodies to the estrogen receptor (Figure 6.2.7). Application of TSA resulted in significantly increased detectability of estrogen receptors in urinary bladder carcinomas, as compared with conventional immunohistochemical analysis.[27]

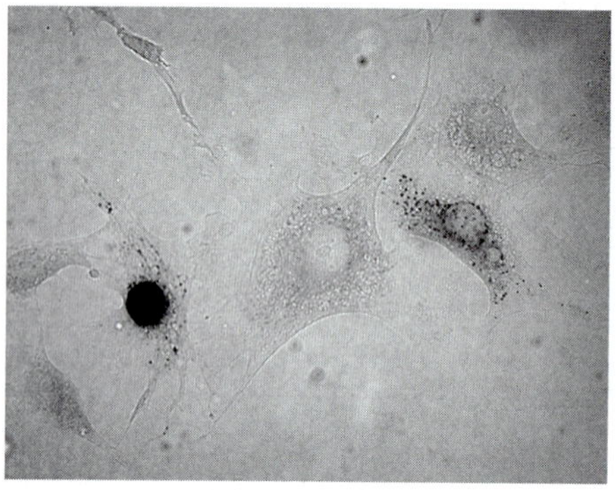

Figure 6.2.5 Nuclear and non-nuclear incorporation of 5-bromo-2′-deoxyuridine in live cells. Bovine pulmonary arterial endothelial (BPAE) cells were labeled with 5-bromo-2′-deoxyuridine (BrdU, B23151) applied at a concentration of 10 µM for 30 minutes. After fixation with 4% formaldehyde in phosphate-buffered saline for 30 minutes, chromatin was denatured by treatment with 2 M HCl for 20 minutes. Incorporated BrdU was detected with mouse monoclonal anti-bromodeoxyuridine antibody followed by HRP-conjugated goat anti–mouse IgG antibody and Oregon Green® 488 tyramide (TSA™ Kit #9, T20919). Tyramide labeling was further amplified and converted for visualization by bright-field microscopy by detection of the Oregon Green® 488 dye hapten using the HRP conjugate of anti–fluorescein/Oregon Green® antibody (A21253) and diaminobenzidine (DAB) staining. Both nuclear and non-nuclear (presumably mitochondrial) incorporation of BrdU is clearly visible in the resulting image.

A Variety of Kits for TSA Detection

TSA™ Kits

We have developed a number of TSA™ Kits that combine the versatile tyramide signal amplification technology with our high-performance Alexa Fluor® tyramides, Oregon Green® 488 tyramide or biotin-XX tyramide (Table 6.1). Each kit provides sufficient materials to stain 50–150 slide preparations and includes the following components:

- Tyramide labeled with an Alexa Fluor® dye, Oregon Green® 488 dye or biotin-XX
- HRP-conjugated anti–mouse IgG antibody, anti–rabbit IgG antibody or streptavidin
- Amplification reaction buffer
- H_2O_2 reaction additive
- TSA blocking reagent
- Detailed protocols for tyramide labeling

Our fluorescent and biotin-XX tyramides are not currently available as stand-alone reagents.

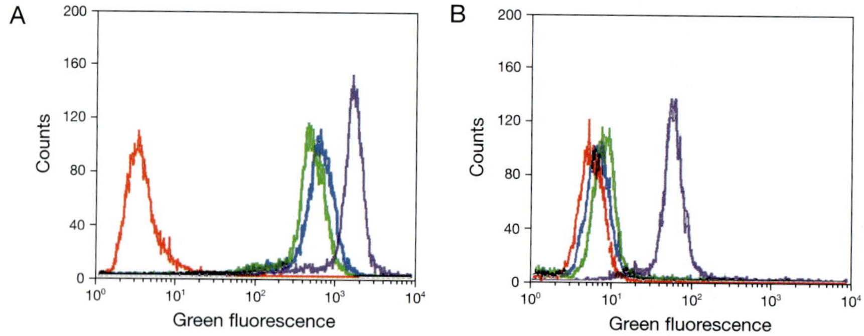

Figure 6.2.6 Detection of epidermal growth factor (EGF) receptors directly or with signal amplification. Cells expressing high (A431 cells, panel **A**) and low (NIH 3T3 cells, panel **B**) levels of EGF receptors were either directly labeled with the preformed Alexa Fluor® 488 complex of biotinylated epidermal growth factor (E13345, blue) or indirectly labeled with biotinylated EGF (E3477) followed by either Alexa Fluor® 488 streptavidin (S11223, green) or HRP-conjugated streptavidin and Alexa Fluor® 488 tyramide (purple), components of our TSA™ Kit #22 (T20932).

IMPORTANT NOTICE: The products described in this manual are covered by one or more Limited Use Label License(s). Please refer to the Appendix on page 971 and Master Product List on page 975. Products are For Research Use Only. Not intended for any animal or human therapeutic or diagnostic use.
www.invitrogen.com/probes

Zenon® Horseradish Peroxidase Antibody Labeling Kits

Our Zenon® Horseradish Peroxidase Antibody Labeling Kits, available for mouse IgG (Z25054, Z25154, Z25254), rabbit IgG (Z25354) and human IgG (Z25454), make it possible to quantitatively label even submicrogram quantities of a primary antibody with HRP immediately before it is applied to the sample (Section 7.3, Table 7.7, Figure 6.2.8). Antibodies labeled with HRP using these Zenon® Antibody Labeling Kits can replace the HRP-labeled goat anti–mouse IgG and goat anti–rabbit IgG antibody conjugates in any of the TSA™ Kits containing these secondary detection reagents.

Zenon® Antibody Labeling Kit Enhanced with TSA Technology

For mouse IgG_1 primary antibodies, we have developed the Zenon® Alexa Fluor® 488 Mouse IgG_1 Labeling Kit enhanced with TSA technology (Z25090), which provides the necessary reagents from both the Zenon® Horseradish Peroxidase Mouse IgG_1 Labeling Kit and the Alexa Fluor® 488 TSA™ Kit, for researchers who want both the ease of labeling mouse IgG_1 antibodies with Zenon® labeling reagents and the signal amplification afforded by TSA technology. Each kit provides sufficient reagents for 25 labelings, including:

- Zenon® HRP mouse IgG_1 labeling reagent
- Zenon® mouse IgG blocking reagent
- Alexa Fluor® 488 tyramide
- Dimethylsulfoxide (DMSO)
- TSA blocking reagent
- TSA amplification buffer
- Hydrogen peroxide (H_2O_2)
- Detailed protocols for Zenon® complex formation and tyramide labeling

The Zenon® HRP mouse IgG_1 labeling reagent contains Fab fragments of goat IgG antibodies directed against the Fc portion of intact mouse IgG_1 antibodies. These Fab fragments have been purified to ensure their selectivity for the Fc portion of the mouse IgG_1 antibody and then labeled with HRP. This Zenon® HRP mouse IgG_1 labeling reagent is simply mixed with any mouse IgG_1

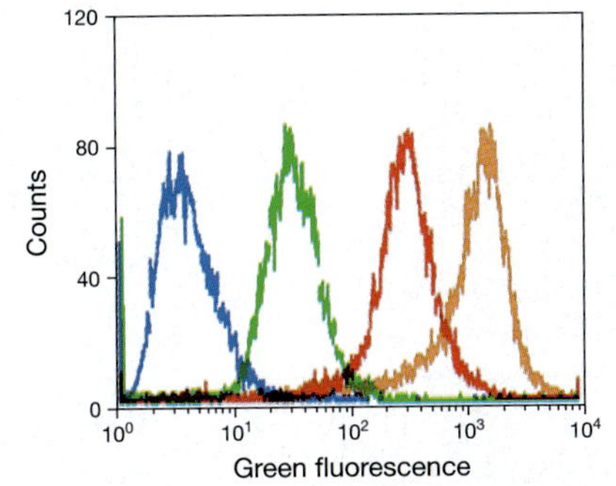

Figure 6.2.7 Enhancement of estrogen receptor detection sensitivity by tyramide signal amplification (TSA™). SKBR3 cells with characteristically low levels of estrogen receptor expression were fixed, permeabilized and treated with H_2O_2 to inhibit endogenous peroxidase activity. A mouse anti–human estrogen receptor monoclonal antibody (Chemicon) was labeled with the Alexa Fluor® 488 dye or with biotin using our Zenon® Alexa Fluor® 488 Mouse IgG_1 Labeling Kit (Z25002) or Zenon® Biotin-XX Mouse IgG_1 Labeling Kit (Z25052), respectively. Detection of estrogen receptors using the labeled antibodies was performed on a Becton Dickinson FACScan™ flow cytometer with excitation at 488 nm. The cellular fluorescence intensity histograms represent detection with Alexa Fluor® 488 dye–labeled antibodies (green), biotinylated antibodies coupled to Alexa Fluor® 488 streptavidin (S11223, red) and biotinylated antibodies coupled to HRP–streptavidin and Alexa Fluor® 488 tyramide (TSA™ Kit #22, T20932; orange). The blue histogram represents unstained cells.

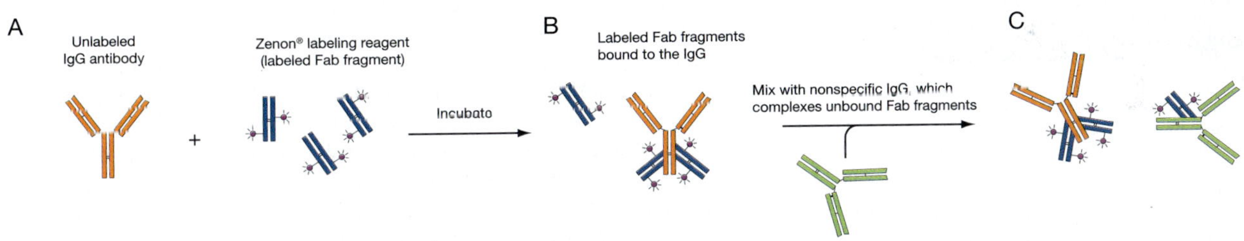

Figure 6.2.8 Labeling scheme utilized in the Zenon® Antibody Labeling Kits. An unlabeled IgG antibody is incubated with the Zenon® labeling reagent, which contains a fluorophore-labeled, Fc-specific anti-IgG Fab fragment (**A**). This labeled Fab fragment binds to the Fc portion of the IgG antibody (**B**). Excess Fab fragment is then neutralized by the addition of a nonspecific IgG (**C**), preventing cross-labeling by the Fab fragment in experiments where primary antibodies of the same type are present. Note that the Fab fragment used for labeling need not be coupled to a fluorophore, but could instead be coupled to an enzyme (such as HRP) or to biotin.

Table 6.1 Tyramide Signal Amplification (TSA) Kits.

	Peroxidase Conjugate		
Tyramide (Ex/Em) *	**Anti–Mouse IgG †**	**Anti–Rabbit IgG †**	**Streptavidin**
Alexa Fluor® 350 (346/442)	TSA™ Kit #7 (T20917)	TSA™ Kit #17 (T20927)	TSA™ Kit #27 (T20937)
Alexa Fluor® 488 (495/519)	TSA™ Kit #2 (T20912)	TSA™ Kit #12 (T20922)	TSA™ Kit #22 (T20932)
Alexa Fluor® 546 (556/573)	TSA™ Kit #3 (T20913)	TSA™ Kit #13 (T20923)	TSA™ Kit #23 (T20933)
Alexa Fluor® 555 (555/565)	TSA™ Kit #40 (T30953)	TSA™ Kit #41 (T30954)	TSA™ Kit #42 (T30955)
Alexa Fluor® 568 (578/603)	TSA™ Kit #4 (T20914)	TSA™ Kit #14 (T20924)	TSA™ Kit #24 (T20934)
Alexa Fluor® 594 (590/617)	TSA™ Kit #5 (T20915)	TSA™ Kit #15 (T20925)	TSA™ Kit #25 (T20935)
Alexa Fluor® 647 (650/668)	TSA™ Kit #6 (T20916)	TSA™ Kit #16 (T20926)	TSA™ Kit #26 (T20936)
Biotin-XX (NA)	TSA™ Kit #1 (T20911)	TSA™ Kit #11 (T20921)	TSA™ Kit #21 (T20931)

*Fluorescence excitation (Ex) and emission (Em) maxima, in nm. † Host = goat. NA = Not applicable.

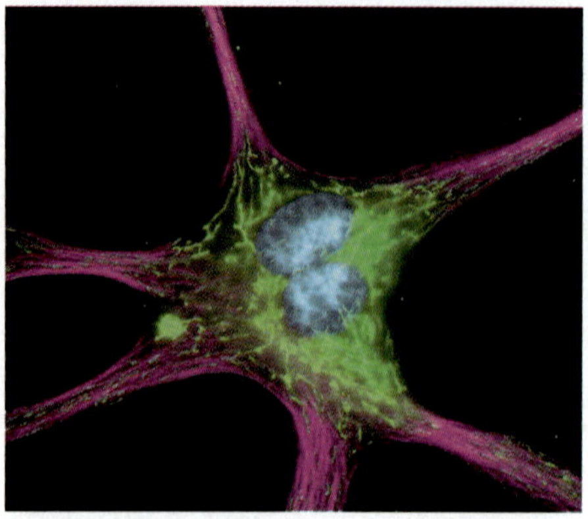

Figure 6.2.9 The cytoskeleton of a fixed and permeabilized bovine pulmonary artery endothelial cell detected using mouse monoclonal anti-α-tubulin antibody (A11126), visualized with Alexa Fluor® 647 goat anti-mouse IgG antibody (A21235) and pseudocolored magenta. Endogenous biotin in the mitochondria was labeled with green-fluorescent Alexa Fluor® 488 streptavidin (S11223) and DNA was stained with blue-fluorescent DAPI (D1306, D3571, D21490).

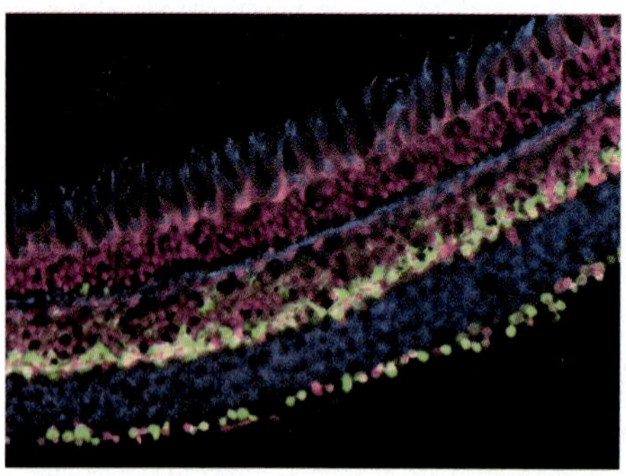

Figure 6.2.10 Immunohistochemical detection using tyramide signal amplification. A transverse section of fixed zebrafish retina was probed with mouse monoclonal FRet 34 antibody and subsequently developed for visualization using HRP-conjugated goat anti-mouse IgG antibody and Alexa Fluor® 488 tyramide, which are supplied in the TSA™ Kit #2 (T20912). The section was counterstained with the blue-fluorescent Alexa Fluor® 350 wheat germ agglutinin (W11263) and the far red–fluorescent TOTO®-3 nuclear stain (T3604).

primary antibody to form the Fab–mouse IgG_1 complexes, which can be used for immunolabeling similar to that of primary antibodies covalently labeled with HRP. TSA technology is then used to detect the target-bound Fab–mouse IgG_1 complex. Each HRP label on the Fab–mouse IgG_1 complexes can activate multiple copies of the Alexa Fluor® 488 tyramide to produce short-lived tyramide radicals that are highly reactive with residues near the interaction site, yielding an amplified green-fluorescent signal with minimal diffusion.

Other Horseradish Peroxidase Conjugates for Secondary Detection

For use in signal amplification of antibody- or biotin-labeled targets, we offer the horseradish peroxidase conjugates of:

- Streptavidin and NeutrAvidin™ biotin-binding protein (S911, A2664; Section 7.6)
- Goat anti–mouse IgG antibody (A10551, A10668, A10677, A10685, G21040)
- Goat anti–mouse $F(ab')_2$ fragment (F21453)
- Goat anti–rabbit IgG antibody (G21234)
- Goat anti–rabbit $F(ab')_2$ fragment (A10547)
- Goat anti–rat IgG antibody (A10549)
- Goat anti–rat $F(ab')_2$ fragment (A10548)
- Mouse anti–human IgG antibody (A10648, A10654)
- Rabbit anti–goat IgG antibody (R21459)
- Rabbit anti–mouse IgG antibody (R21455)

A more thorough discussion of our secondary antibodies can be found in Section 7.2 (Table 7.1).

Applying TSA Technology to Cells and Tissues

TSA technology has been used successfully for over two decades,[1,28] and there are many reports of the use of biotin tyramide for indirect target labeling and fluorescent tyramides for direct target labeling. Direct labeling methods have the considerable advantage of saving a second step in the detection scheme. Moreover, labeling targets with fluorescent tyramides instead of biotin tyramide has the further advantage of avoiding amplification of endogenous biotin in cells and tissues, such as we have observed in mitochondrial staining with streptavidin conjugates in the absence of a biotinylated probe (Figure 6.2.9). In most of the early reports, the biotin tyramide used did not have the additional 14-atom spacer that we utilize in our biotin-XX tyramide to make the probe more accessible to avidin conjugates, nor are the specific fluorescent dyes available in our TSA™ Kits. Therefore, the specific methods described in these references should be considered guides rather than definitive protocols, and results using our TSA reagents may differ from those reported. In our experience, the Alexa Fluor® 488 tyramide (Table 6.1) provides greater signal and significantly greater photostability than fluorescein tyramide, and the other Alexa Fluor® tyramides also yield intense staining of targets.

Immunohistochemical Detection Using TSA

TSA detection can be applied to a variety of immunohistochemical specimen preparations, including crytostat sections, formaldehyde-fixed paraffin-embedded sections, plastic-embedded sections and cultured cells. In immunohistochemical applications (Figure 6.2.10), sensitivity enhancements derived from TSA allow primary antibody dilutions to be increased—up to a 1:1,000,000 antibody dilution was possible in one reported case,[10] although a 5- to 50-fold increase over the normal dilution factor is more common[9]—in order to reduce nonspecific background signals.[7] Additionally, the strong signal amplification provided by the TSA method can overcome relatively high autofluorescence of cells and tissues.[6] Furthermore, because TSA and diaminobenzidine (DAB) oxidation are both peroxidase-mediated reactions, TSA is readily adaptable for correlated fluorescence and electron microscopy studies.[29,30] The significantly lower detection threshold of TSA as compared with fluorescent secondary antibodies allows detection of two targets by primary antibodies raised in the same host species, without substantial crosstalk between the signals.[31] The first target was detected using TSA and a primary antibody concentration that was so low that it was essentially undetectable by fluorescent secondary antibodies; the second target was then detected by conventional secondary immunofluorescence labeling.

IMPORTANT NOTICE: The products described in this manual are covered by one or more Limited Use Label License(s). Please refer to the Appendix on page 971 and Master Product List on page 975. Products are For Research Use Only. Not intended for any animal or human therapeutic or diagnostic use.

Fluorescence *In Situ* Hybridization Using TSA

The increased sensitivity afforded by TSA (Figure 6.2.11) can be critically important for detecting relatively short oligonucleotide probes and low-abundance mRNAs by fluorescence *in situ* hybridization[3,32] (FISH). Cosmid detection in formalin-fixed, paraffin-embedded sections is cumbersome, and the ability to use smaller cosmid probes of less than 1000 bases in conjunction with TSA detection technology is likely to be an important technique for FISH.[23] TSA is also faster than traditional FISH detection schemes, allowing definitive results to be obtained within a single day. In addition, a two-stage amplification method for ultrasensitive mRNA detection has been reported that combines TSA detection of biotinylated riboprobes with alkaline phosphatase–mediated fluorescence generation using Molecular Probes® ELF® 97 phosphatase substrate[33] (Section 6.3).

TSA, however, is not a panacea for FISH sensitivity problems. Because both specifically and nonspecifically bound probe signals are amplified, TSA will not compensate for suboptimal hybridization conditions. Optimal probe concentrations are typically 2- to 10-fold lower for TSA-detected FISH than for conventional immunocytochemical detection procedures, again saving on the cost of expensive hybridization probes.[24] Typically, FISH probes are labeled by indirect methods that use streptavidin- or antibody-conjugated HRP. Techniques for direct labeling of oligonucleotide probes have been developed to eliminate background signals due to nonspecific binding of peroxidase conjugates.[4,34,35]

As with some other detection systems, TSA technology allows several probes to be hybridized simultaneously to identify multiple targets. Signal development using multicolored fluorescent tyramides must then be carried out sequentially, with a peroxidase inactivation step between each TSA reaction to prevent crosstalk[24] (Figure 6.2.12). TSA amplification followed by peroxidase inactivation through mild acid treatment with 0.01 M HCl for 10 minutes at room temperature[36] and then reapplication of TSA using a fluorescent tyramide of a different fluorescent color has been used for triple-labeled *in situ* hybridization.[4,36]

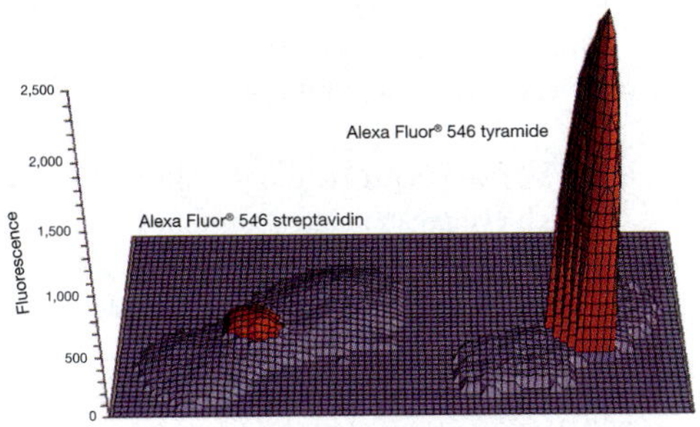

Figure 6.2.11 Digital image analysis comparison of *in situ*–hybridized biotinylated α-satellite probes detected using TSA™ Kit #23 (T20933) with HRP–streptavidin and Alexa Fluor® 546 tyramide (right) or Alexa Fluor® 546 streptavidin (S11225, left). Both images were converted to pixel intensity values using MetaMorph® software (Universal Imaging Corporation) and transferred to a Microsoft® Excel spreadsheet for plotting. Alexa Fluor® 546 dye and DAPI (counterstain) intensity values are shown in red and blue, respectively. Alexa Fluor® 546 dye intensity values below 35% of maximum were omitted for clarity.

Detection of Hapten-Labeled Tyramides

When a tyramide labeled with a hapten such as biotin-XX is used for TSA in an indirect labeling technique, a signal-generation reagent or scheme is necessary. A fluorescent tyramide such as our Oregon Green® 488 tyramide can also be utilized as a hapten for subsequent detection and amplification by an anti–fluorescein/Oregon Green® dye antibody (Section 7.4). Various reagents and reagent combinations have been reported for detecting enzyme-deposited biotin tyramide or fluorescein tyramide that should be equally suitable for use with our biotin-XX tyramide and Oregon Green® 488 tyramide, including:

- Streptavidin conjugate of alkaline phosphatase (S921, Section 7.6) in combination with NBT/BCIP[37,38] (N6495, N6547; Section 6.3)
- Streptavidin conjugate of HRP (S911, Section 7.6) or the rabbit anti–fluorescein/Oregon Green® antibody conjugate of HRP[32,39–42] (A21253, Section 7.4) in combination with a traditional chromogenic peroxidase substrate such as diaminobenzidine (DAB)[43,44] (Figure 6.2.5)
- Fluorescent conjugates of avidin or streptavidin (Section 7.6).
- Qdot® nanocrystal streptavidin conjugates[45] (Section 6.6)
- Diaminobenzidine (DAB) Histochemistry Kits (D22185, D22187; see below), for direct use with biotin-XX tyramide or conversion of fluorescent signals to permanent staining
- NANOGOLD® and Alexa Fluor® FluoroNanogold™ conjugates of antibodies (Section 7.2) and streptavidin (Section 7.6), for target localization using a combination of light and electron microscopy[21,46]

Anti–fluorescein/Oregon Green® antibody conjugates of HRP (Section 7.4, Table 7.8) have been used with fluorescein labeled probes and TSA to detect:

- Embryonic gene expression at the cellular level by FISH[47]
- mRNA in paraffin sections of organotypic multicellular spheroids[48]
- mRNA probe for a calcium transporter protein[49]
- Somatostatin receptor protein[41]
- Tissue antigens, with a 10- to 100-fold increase in sensitivity over conventional staining methods[39]

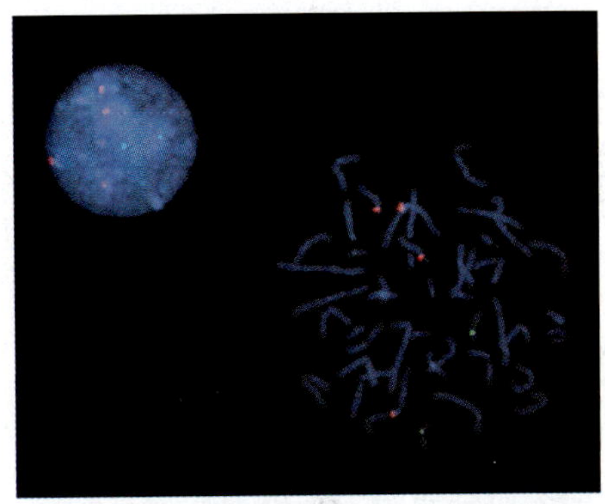

Figure 6.2.12 *In situ* hybridization of α-satellite probes to human chromosomes 1, 15 and 17 detected by tyramide signal amplification. α-Satellite probes to chromosomes 1, 15 and 17 were labeled by nick translation with biotin-11-dUTP, ChromaTide® Texas Red®-12-dUTP (C7631) and ChromaTide® Oregon Green® 488-5-dUTP, respectively. Following simultaneous hybridization of all three probes, the biotinylated chromosome 1 probe was detected with HRP–streptavidin conjugate and Alexa Fluor® 546 tyramide (TSA™ Kit #23, T20933). HRP activity from this first TSA detection step was then quenched by treatment with 1% hydrogen peroxide for 30 minutes. Lastly, the Oregon Green® 488 dye–labeled chromosome 17 probe was detected with anti–fluorescein/Oregon Green® antibody (A6421) followed by HRP-conjugated goat anti–mouse IgG antibody and Alexa Fluor® 594 tyramide (TSA™ Kit #5, T20915). HRP activity from this second TSA detection step was then quenched by treatment with 1% hydrogen peroxide for 30 minutes. The Texas Red® dye–labeled chromosome 15 probe was then detected with rabbit anti–Texas Red® antibody (A6399) followed by HRP-conjugated goat anti–rabbit IgG antibody and Alexa Fluor® 488 tyramide (TSA™ Kit #12, T20922). After counterstaining with Hoechst 33258 (H1398, H3569, H21491), the images were acquired using filters appropriate for DAPI, FITC, TRITC and Texas Red® dyes.

Similarly, our anti–Green Fluorescent Protein (anti-GFP) antibodies (Section 7.5) can be used in combination with TSA for ultrasensitive immunocytochemical detection of GFP in situations where the expression level is insufficient for direct fluorescence detection.[50]

Double and Sequential Amplification with TSA

To achieve greater signal amplification, sequential rounds of amplification can be achieved using TSA[51] (Figure 6.2.1). For example, in the first round biotin-XX tyramide can be deposited on a target using one of our biotin-XX tyramide TSA™ Kits (Table 6.1). In a subsequent step, the peroxidase conjugate of streptavidin that is used in TSA™ Kit #21 (T20931) is used again, but this time in combination with an Alexa Fluor® tyramide, Oregon Green® 488 tyramide or another round of biotin-XX tyramide. Presumably, this amplification can be continued for at least a third round, although some loss of spatial resolution may result. Biotin tyramide that has first been deposited at the binding site of a biotin-labeled riboprobe using the streptavidin conjugate of HRP has been further amplified with the streptavidin conjugate of alkaline phosphatase (S921, Section 7.6) in conjunction with ELF® 97 phosphate for the ultrasensitive detection of a scarce leptin receptor mRNA.[33] In another example demonstrating the versatility of the TSA technology, several labeling technologies were combined sequentially to detect the HIV-1 virus:[52]

1. *In situ* hybridization with a 15-base peptide nucleic acid probe labeled with a single fluorescein dye
2. Complexation with an HRP conjugate of anti-fluorescein antibody
3. Incubation with biotin tyramide
4. Incubation with the streptavidin conjugate of HRP
5. Re-incubation with biotin tyramide
6. Detection with the Alexa Fluor® 488 conjugate of streptavidin (S11223, Section 7.6)

Alternatively, for detection by light microscopy, the sample was incubated with the streptavidin conjugate of HRP in conjunction with DAB instead of Alexa Fluor® 488 streptavidin.

Additional Tips on Using TSA Technology

Use of the TSA technology is not without its precautions. Among these is the possibility of endogenous peroxidase activity in certain cells, especially eosinophils.[14] This activity can be at least partially blocked by incubation with 0.3–3% hydrogen peroxide for about 60 minutes. Second, when using biotin-XX tyramide, endogenous biotinylated proteins are a potential problem (Figure 6.2.9). Third, because of the significant signal amplification capability of TSA, nonspecific binding of labeled hybridization probes, antibodies and other targeting probes can lead to unacceptably high background staining. This nonspecific staining can be alleviated to some degree with appropriate blocking reagents.[53] Furthermore, the high sensitivity of TSA permits antibodies and nucleic acid probes to be highly diluted, far below the amount required for target saturation, thus reducing nonspecific background. Antibody and nucleic acid probe dilution can also substantially reduce the cost of an assay and the amount of a rare material required for staining.[22]

Mammalian cells and tissues contain biotin-dependent carboxylases, which are required for a variety of metabolic functions. These biotin-containing enzymes produce substantial background signals when biotin–streptavidin detection systems are used to identify cellular targets[54] (Figure 6.2.9). Because the TSA technology is so sensitive, we recommended preblocking endogenous biotin in cells with our Endogenous Biotin-Blocking Kit (E21390, Section 7.6) when using TSA™ Kits containing biotin-XX tyramide and the streptavidin conjugate of HRP. The Endogenous Biotin-Blocking Kit provides streptavidin and biotin solutions in convenient dropper bottles and an easy-to-follow protocol; sufficient material is provided for approximately one hundred 18 mm × 18 mm glass coverslips.

Improvement of TSA detection by post-incubation heating has been reported.[33] Addition of viscosity-increasing dextran sulfate, poly(vinyl alcohol), poly(ethylene glycol) or poly(vinyl pyrrolidone) to the medium is reported to decrease diffusion of the phenoxy radical intermediate, resulting in superior localization of the signal.[51,55] Hybridization probes that are directly labeled with HRP are reportedly useful for lowering nonspecific binding when working with labeled tyramides.[4,35,56] Endogenous peroxidase can be sufficient to yield labeling at the site of this activity in cells, as in the case of eosinophils.[14] The review by Speel, Hopman and Komminoth gives additional practical suggestions and references.[3]

Chromogenic and Chemiluminescent Peroxidase Substrates

DAB Histochemistry Kits

The use of HRP for enzyme-amplified immunodetection—commonly referred to as immunoperoxidase labeling—is a well-established standard histochemical technique.[57,58] The most widely used HRP substrate for these applications is diaminobenzidine (DAB), which generates a brown-colored polymeric oxidation product localized at HRP-labeled sites. The DAB reaction product can be visualized directly by bright-field light microscopy or, following osmication, by electron microscopy. We offer DAB Histochemistry Kits for detecting mouse IgG primary antibodies (D22185) and biotinylated antibodies and tracers (D22187). Each kit contains sufficient materials to stain approximately 200 slides, including:

- Diaminobenzidine (DAB)
- HRP-labeled goat anti–mouse IgG antibody (in Kit D22185) or streptavidin (in Kit D22187) conjugate
- H_2O_2 reaction additive
- Blocking reagent
- Staining buffer
- Detailed staining protocols

Luminol and MCLA: Chemiluminescent Peroxidase Substrates

Nonisotopic immunoassays utilizing peroxidase conjugates and the chemiluminescent horseradish peroxidase substrate luminol (L8455) have provided a rapid and sensitive method for quantitating a wide variety of analytes, including cholesterol,[59] digoxin[60] and acetylcholine.[61] Addition of trace amounts of luciferin (L2911, L2912, L2916; Section 10.6) has been shown to considerably enhance the sensitivity in the assay of thyroxine, digoxin, α-fetoprotein and other analytes.[62] A method that employs luminol has been developed for the quantitation of very limiting samples of human DNA from single hairs, saliva, small blood stains and paraffin-embedded and fixed tissue sections. Using a biotinylated oligodeoxynucleotide probe to membrane-immobilized DNA, horseradish peroxidase streptavidin and luminol, researchers have detected 150 pg of human DNA.[63]

IMPORTANT NOTICE: The products described in this manual are covered by one or more Limited Use Label License(s). Please refer to the Appendix on page 971 and Master Product List on page 975. Products are For Research Use Only. Not intended for any animal or human therapeutic or diagnostic use.
www.invitrogen.com/probes
molecular probes® | invitrogen™ by life technologies™

MCLA (M23800) is principally utilized as a superoxide-sensitive chemiluminescent probe (Section 18.2). MCLA has also been utilized for the determination of both horseradish peroxidase[64] and myeloperoxidase.[65,66,67]

Peroxidase-Based Amplex® ELISA Kits

The Amplex® ELISA Development Kits for Mouse IgG (A33851) and for Rabbit IgG (A33852) provide a comprehensive set of components for creating fluorescence-based ELISAs using mouse and rabbit detection antibodies, respectively.[68] This assay is based on the Amplex® UltraRed reagent, a fluorogenic substrate for horseradish peroxidase (HRP) that reacts with H_2O_2 in a 1:1 stoichiometric ratio to produce a brightly fluorescent and strongly absorbing reaction product (excitation/emission maxima ~568/581 nm). Because the Amplex® UltraRed peroxidation product has long-wavelength spectra, there is little interference from the blue or green autofluorescence found in most biological samples.

With a high extinction coefficient, good quantum efficiency and resistance to autooxidation, the fluorescence-based Amplex® UltraRed reagent delivers better sensitivity and a broader assay range than colorimetric reagents. In a sandwich ELISA format we can routinely detect 1 pg of C-reative protein using HRP-conjugated goat anti–rabbit IgG antibody (Figure 6.2.13); this detection limit is 25-fold lower than that obtained from the same sandwich ELISA format using the common colorimetric reagent TMB.

Each Amplex® ELISA Development Kit contains:

- Amplex® UltraRed reagent
- Dimethylsulfoxide (DMSO)
- Concentrated phosphate-buffered saline (PBS), pH 7.2
- Horseradish peroxidase conjugate of goat anti–mouse IgG antibody (in A33851) or goat anti–rabbit IgG antibody (in A33852)
- Amplex® stop reagent
- Hydrogen peroxide (H_2O_2)
- 0.1 M sodium bicarbonate buffer, pH ~9.3
- Bovine serum albumin (BSA)
- Tween® 20
- Nunc-Immuno™ MaxiSorp™ U96 plate
- Detailed protocols

Sufficient reagents are provided in each kit for 500 microplate assays in a 96-well fluorescence microplate format (100 µL per assay).

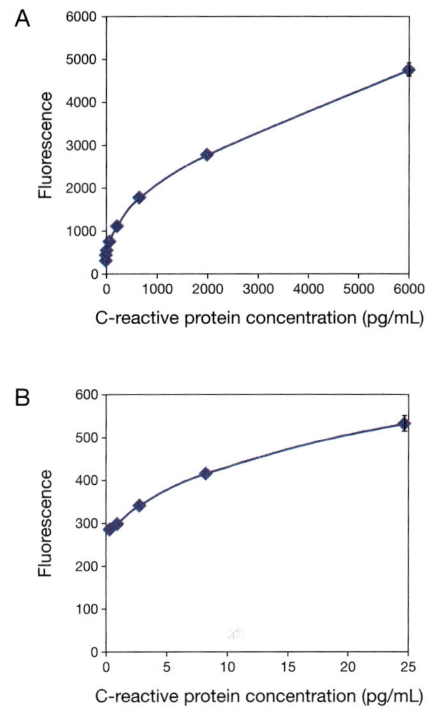

Figure 6.2.13 Detection range of C-reactive protein (CRP) using the Amplex® ELISA Development Kit for Rabbit IgG (A33852). The sandwich ELISA was carried out using a mouse anti-CRP capture antibody, CRP in a concentration range between 6000 pg/mL and 0.10 pg/mL and a rabbit polyclonal anti-CRP primary antibody (10 µL per well of a 50 ng/mL solution). The Z′ factor analysis of the data obtained gives a lower detection level for CRP in this assay of 1 pg/mL or 0.1 pg/well (based on a well volume of 100 µL in the sandwich ELISA).

REFERENCES

1. J Immunol Methods (1989) 125:279; **2.** Cytometry (1996) 23:48; **3.** J Histochem Cytochem (1999) 47:281; **4.** J Histochem Cytochem (1997) 45:375; **5.** US Patent No 5,196,306 (1993); **6.** J Histochem Cytochem (1997) 45:315; **7.** J Histochem Cytochem (1992) 40:1457; **8.** Histochem Cell Biol (2000) 114:447; **9.** J Histochem Cytochem (1997) 45:1455; **10.** Am J Clin Pathol (1996) 106:16; **11.** Lab Invest (1995) 73:149; **12.** J Histochem Cytochem (1994) 42:1635; **13.** J Histochem Cytochem (2001) 49:155; **14.** Histochem Cell Biol (1996) 106:447; **15.** Histochem Cell Biol (1998) 110:571; **16.** J Histochem Cytochem (1998) 46:771; **17.** J Histochem Cytochem (1999) 47:421; **18.** Biochemistry (2001) 40:7783; **19.** Med Sci Monit (2001) 7:606; **20.** Biotechniques (1997) 23:1076; **21.** J Histochem Cytochem (2000) 48:933; **22.** J Histochem Cytochem (1997) 45:359; **23.** Mutat Res (1998) 400:287; **24.** Current Protocols in Cytometry (2000) 8.9.1; **25.** Methods (1999) 18:459; **26.** J Histochem Cytochem (1998) 46:527; **27.** J Pathol (1998) 186:165; **28.** J Immunol Methods (1991) 137:103; **29.** J Histochem Cytochem (2000) 48:153; **30.** J Neurosci Methods (1999) 88:55; **31.** J Histochem Cytochem (2007) 55:545; **32.** J Histochem Cytochem (1999) 47:431; **33.** J Histochem Cytochem (2000) 48:1593; **34.** J Histochem Cytochem (1998) 46:1249; **35.** Histochem Cell Biol (2000) 113:175; **36.** J Histochem Cytochem (1997) 45:1439; **37.** J Virol Methods (1998) 70:119; **38.** J Histochem Cytochem (1997) 45:1629; **39.** J Histochem Cytochem (1996) 44:1353; **40.** J Histochem Cytochem (1998) 46:149; **41.** Endocrinology (1997) 138:2632; **42.** J Histochem Cytochem (1997) 45:1643; **43.** Brain Res (1999) 822:251; **44.** Brain Res (1998) 788:43; **45.** J Histochem Cytochem (2003) 51:981; **46.** J Histochem Cytochem (1999) 47:99; **47.** Histochem Cell Biol (1999) 111:435; **48.** Neuropathol Appl Neurobiol (1996) 22:548; **49.** J Biol Chem (2000) 275:28186; **50.** Exp Cell Res (2007) 313:1943; **51.** J Histochem Cytochem (1996) 44:389; **52.** J Pathol (2001) 194:130; **53.** J Histochem Cytochem (2003) 51:129; **54.** J Histochem Cytochem (1997) 45:1053; **55.** Histochem Cell Biol (1999) 111:89; **56.** Cytogenet Cell Genet (1996) 75:258; **57.** Arch Pathol Lab Med (1978) 102:113; **58.** J Histochem Cytochem (1988) 36:317; **59.** Biochim Biophys Acta (1994) 1210:151; **60.** Clin Chem (1985) 31:1335; **61.** J Neurochem (1982) 39:248; **62.** Nature (1983) 305:158; **63.** Nucleic Acids Res (1992) 20:5061; **64.** J Biolumin Chemilumin (1994) 9:355; **65.** Cell Mol Neurobiol (1998) 18:565; **66.** Methods Enzymol (1994) 233:495; **67.** Anal Biochem (1991) 199:191; **68.** J Biol Chem (2005) 280:37377.

IMPORTANT NOTICE: The products described in this manual are covered by one or more Limited Use Label License(s). Please refer to the Appendix on page 971 and Master Product List on page 975. Products are For Research Use Only. Not intended for any animal or human therapeutic or diagnostic use.

PRODUCT LIST 6.2 TSA AND OTHER PEROXIDASE-BASED SIGNAL AMPLIFICATION TECHNIQUES

Cat. No.	Product	Quantity
A33851	Amplex® ELISA Development Kit for Mouse IgG *with Amplex® UltraRed reagent* *500 assays*	1 kit
A33852	Amplex® ELISA Development Kit for Rabbit IgG *with Amplex® UltraRed reagent* *500 assays*	1 kit
D22185	Diaminobenzidine (DAB) Histochemistry Kit #1 *with goat anti-mouse IgG–HRP*	1 kit
D22187	Diaminobenzidine (DAB) Histochemistry Kit #3 *with streptavidin–HRP*	1 kit
L8455	luminol (3-aminophthalhydrazide)	25 g
M23800	2-methyl-6-(4-methoxyphenyl)-3,7-dihydroimidazo1,2-apyrazin-3-one, hydrochloride (MCLA)	5 mg
T20911	TSA™ Kit #1 *with HRP–goat anti-mouse IgG and biotin-XX tyramide* *50–150 slides*	1 kit
T20912	TSA™ Kit #2 *with HRP–goat anti-mouse IgG and Alexa Fluor® 488 tyramide* *50–150 slides*	1 kit
T20913	TSA™ Kit #3 *with HRP–goat anti-mouse IgG and Alexa Fluor® 546 tyramide* *50–150 slides*	1 kit
T20914	TSA™ Kit #4 *with HRP–goat anti-mouse IgG and Alexa Fluor® 568 tyramide* *50–150 slides*	1 kit
T20915	TSA™ Kit #5 *with HRP–goat anti-mouse IgG and Alexa Fluor® 594 tyramide* *50–150 slides*	1 kit
T20916	TSA™ Kit #6 *with HRP–goat anti-mouse IgG and Alexa Fluor® 647 tyramide* *50–150 slides*	1 kit
T20917	TSA™ Kit #7 *with HRP–goat anti-mouse IgG and Alexa Fluor® 350 tyramide* *50–150 slides*	1 kit
T20921	TSA™ Kit #11 *with HRP–goat anti-rabbit IgG and biotin-XX tyramide* *50–150 slides*	1 kit
T20922	TSA™ Kit #12 *with HRP–goat anti-rabbit IgG and Alexa Fluor® 488 tyramide* *50–150 slides*	1 kit
T20923	TSA™ Kit #13 *with HRP–goat anti-rabbit IgG and Alexa Fluor® 546 tyramide* *50–150 slides*	1 kit
T20924	TSA™ Kit #14 *with HRP–goat anti-rabbit IgG and Alexa Fluor® 568 tyramide* *50–150 slides*	1 kit
T20925	TSA™ Kit #15 *with HRP–goat anti-rabbit IgG and Alexa Fluor® 594 tyramide* *50–150 slides*	1 kit
T20926	TSA™ Kit #16 *with HRP–goat anti-rabbit IgG and Alexa Fluor® 647 tyramide* *50–150 slides*	1 kit
T20927	TSA™ Kit #17 *with HRP–goat anti-rabbit IgG and Alexa Fluor® 350 tyramide* *50–150 slides*	1 kit
T20931	TSA™ Kit #21 *with HRP–streptavidin and biotin-XX tyramide* *50–150 slides*	1 kit
T20932	TSA™ Kit #22 *with HRP–streptavidin and Alexa Fluor® 488 tyramide* *50–150 slides*	1 kit
T20933	TSA™ Kit #23 *with HRP–streptavidin and Alexa Fluor® 546 tyramide* *50–150 slides*	1 kit
T20934	TSA™ Kit #24 *with HRP–streptavidin and Alexa Fluor® 568 tyramide* *50–150 slides*	1 kit
T20935	TSA™ Kit #25 *with HRP–streptavidin and Alexa Fluor® 594 tyramide* *50–150 slides*	1 kit
T20936	TSA™ Kit #26 *with HRP–streptavidin and Alexa Fluor® 647 tyramide* *50–150 slides*	1 kit
T20937	TSA™ Kit #27 *with HRP–streptavidin and Alexa Fluor® 350 tyramide* *50–150 slides*	1 kit
T20939	TSA™ Kit #29 *with HRP–streptavidin and Oregon Green® 488 tyramide* *50–150 slides*	1 kit
T30953	TSA™ Kit #40 *with HRP–goat anti-mouse IgG and Alexa Fluor® 555 tyramide* *50–150 slides*	1 kit
T30954	TSA™ Kit #41 *with HRP–goat anti-rabbit IgG and Alexa Fluor® 555 tyramide* *50–150 slides*	1 kit
T30955	TSA™ Kit #42 *with HRP–streptavidin and Alexa Fluor® 555 tyramide* *50–150 slides*	1 kit
Z25090	Zenon® Alexa Fluor® 488 Mouse IgG_1 Labeling Kit *enhanced with TSA technology* *25 labelings*	1 kit
Z25454	Zenon® Horseradish Peroxidase Human IgG Labeling Kit *25 labelings*	1 kit
Z25054	Zenon® Horseradish Peroxidase Mouse IgG_1 Labeling Kit *25 labelings*	1 kit
Z25154	Zenon® Horseradish Peroxidase Mouse IgG_{2a} Labeling Kit *25 labelings*	1 kit
Z25254	Zenon® Horseradish Peroxidase Mouse IgG_{2b} Labeling Kit *25 labelings*	1 kit
Z25354	Zenon® Horseradish Peroxidase Rabbit IgG Labeling Kit *25 labelings*	1 kit

IMPORTANT NOTICE: The products described in this manual are covered by one or more Limited Use Label License(s). Please refer to the Appendix on page 971 and Master Product List on page 975. Products are For Research Use Only. Not intended for any animal or human therapeutic or diagnostic use.

6.3 Phosphatase-Based Signal Amplification Techniques

When detecting specific biomolecules in cells, in tissues and on microarrays, enzyme-mediated detection methods can provide significant signal amplification due to the catalytic turnover of fluorogenic, chemiluminescent or chromogenic substrates. Because fluorescence and chemiluminescence methods are potentially much more sensitive than colorimetric methods, we have been actively engaged in research to develop substrates for phosphatase-mediated detection and offer two complementary technologies—Enzyme-Labeled Fluorescence (ELF®) signal amplification and CSPD® and CDP-*Star*® chemiluminescent substrates.

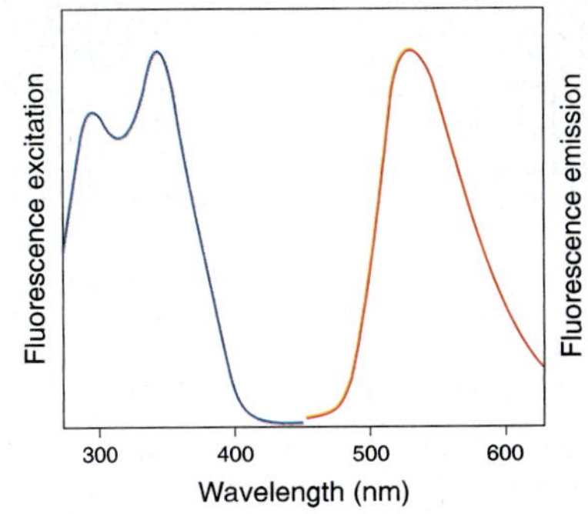

Figure 6.3.2 The normalized fluorescence excitation and emission spectra of the ELF® 97 alcohol precipitate (E6578), which is generated by enzymatic cleavage of the soluble ELF® 97 phosphatase substrate (E6588, E6589).

Principles of ELF® Signal Amplification

ELF® signal amplification is based on ELF® 97 phosphate, a substrate for both alkaline phosphatase and acid phosphatase.[1] Upon enzymatic cleavage (Figure 6.3.1), this weakly blue-fluorescent substrate yields a bright yellow-green–fluorescent precipitate that exhibits an unusually large Stokes shift (Figure 6.3.2) and excellent photostability (Figure 6.3.3). Water-soluble ELF® 97 phosphate (E6588, E6589) is converted to insoluble ELF® 97 alcohol (E6578), by action of a phosphatase (Figure 6.3.1). Like other crystalline fluorescent molecules containing an intramolecular hydrogen bond, the ELF® 97 alcohol precipitate provides a fluorescent signal that is not only extremely photostable but also has an exceptionally large Stokes shift.[2–4] The ELF® 97 signal is more photostable than signals achieved using either direct or indirect detection with fluorescein conjugates[1,5] (Figure 6.3.3). As a result of its extremely high photostability, the ELF® 97 phosphate-generated signal is well suited for immunolabeling targets to be viewed by UV-light–excited confocal laser-scanning microscopy[6] (Figure 6.3.4).

Furthermore, because the fluorescence emission of the ELF® 97 alcohol precipitate is separated from its excitation maximum by over 180 nm (Figure 6.3.2), the ELF® 97 signal can be clearly distinguished from most cell and tissue autofluorescence, which generally has a Stokes shift of

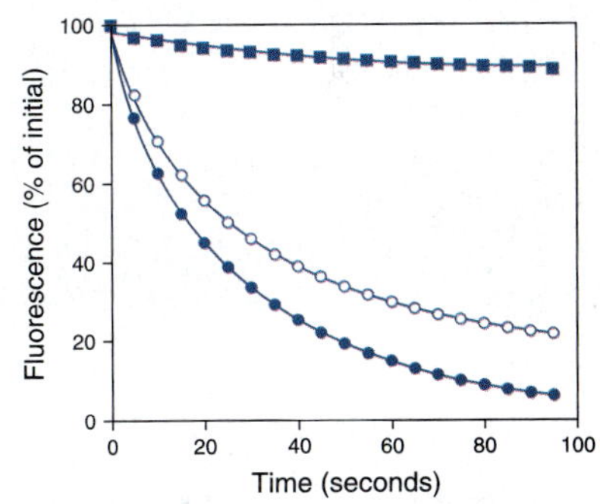

Figure 6.3.3 Photostability comparison for ELF® 97 alcohol- and fluorescein-labeled tubulin preparations. Tubulin in acetone-fixed CRE BAG 2 mouse fibroblasts was labeled with an anti–β-tubulin monoclonal antibody and then detected using biotin-XX goat anti–mouse IgG antibody (B2763) in conjunction with either our ELF® 97 Cytological Labeling Kit (E6603, ■) or fluorescein streptavidin (S869, ●). Alternatively, anti-tubulin labeling was detected directly using fluorescein goat anti–mouse IgG antibody (F2761, ○). The photostability of labeling produced by the three methods was compared by continuously illuminating stained samples on a fluorescence microscope using Omega® Optical longpass optical filter sets. Images were acquired every 5 seconds using a Star 1 CCD camera (Photometrics®); the average fluorescence intensity in the field of view was calculated with Image-1 software (Universal Imaging Corp.) and expressed as a fraction of the initial intensity. Three data sets, representing different fields of view, were averaged for each conjugate to obtain the plotted time courses.

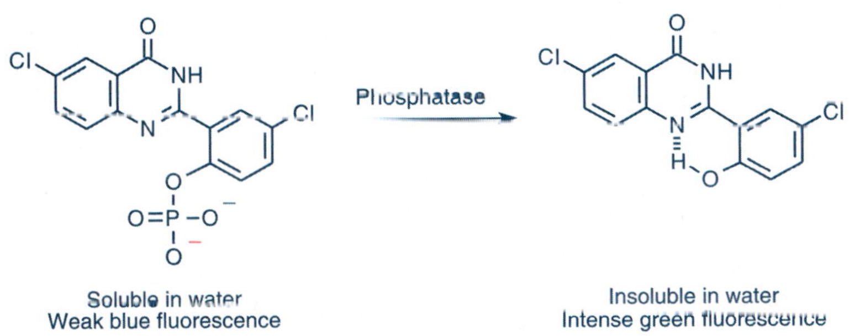

Figure 6.3.1 Principle of enzyme-mediated formation of the fluorescent ELF® 97 alcohol precipitate from the ELF® 97 phosphatase substrate.

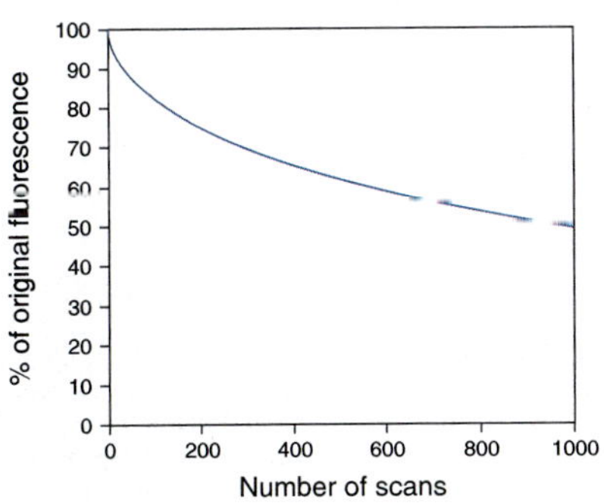

Figure 6.3.4 The ELF® 97 Cytological Labeling Kit (E6603) was used to label endogenous alkaline phosphatase activity of zebrafish kidney. The ELF® 97 fluorescence intensity was measured from each of 1000 consecutive scans on a confocal laser-scanning microscope (Bio-Rad® MRC 1024) with UV illumination (UV/Vis Coherent Innova Enterprise Model 622 argon-ion laser, 60 mW output) using the 363 nm spectral line at 100% power with a 40× objective. Fluorescence intensity, expressed as a percentage of initial value, is plotted against scan number. In this example, 970 scans were required to reduce the signal by 50%. This figure was contributed by J. Paul Robinson and Jennie Sturgis, Purdue University.

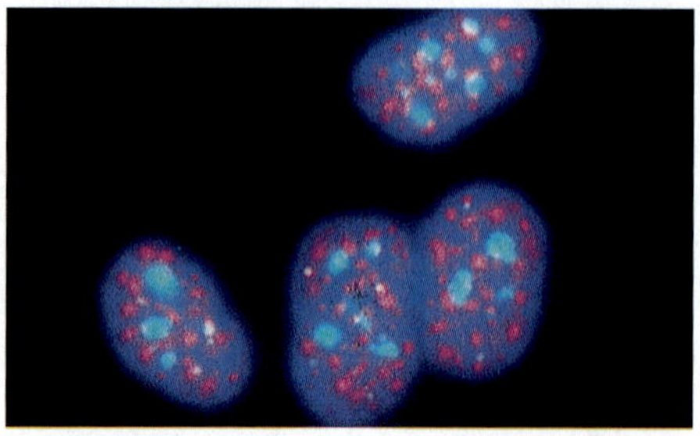

Figure 6.3.5 HeLa cell nuclei incubated with a mouse monoclonal antibody directed against nuclear-localized antigens (a gift from Benjamin Blencowe and Phil Sharp, Massachusetts Institute of Technology) in conjunction with Texas Red®-X streptavidin (S6370) and with a human anti-nuclear antibody in conjunction with biotin-XX goat anti–human IgG antibody. The reagents in the ELF® 97 Cytological Labeling Kit (E6603) were then used to detect the biotinylated secondary antibody. Cells were counterstained with Hoechst 33258 (H1398, H3569, H21491). Multiple exposures were taken using filters appropriate for the Texas Red® dye, fluorescein and DAPI.

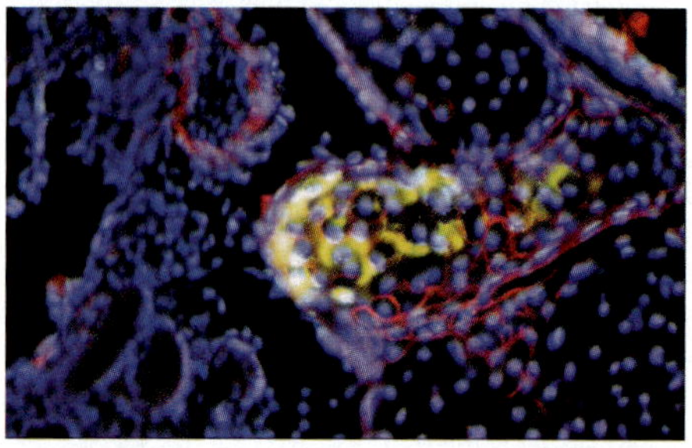

Figure 6.3.6 Endogenous alkaline phosphatase activity of osteoblast cells in a cartilaginous element of an adult zebrafish head cryosection. The activity was localized with the ELF® 97 Endogenous Phosphatase Detection Kit (E6601). In addition to the yellow-green fluorescence of the ELF® 97 precipitate, the section was stained with Texas Red®-X wheat germ agglutinin (W21405) and counterstained with the Hoechst 33342 nucleic acid stain (H1399, H3570, H21492). The triple-exposure image was acquired using bandpass filter sets appropriate for the ELF® precipitate, Texas Red® dye and AMCA.

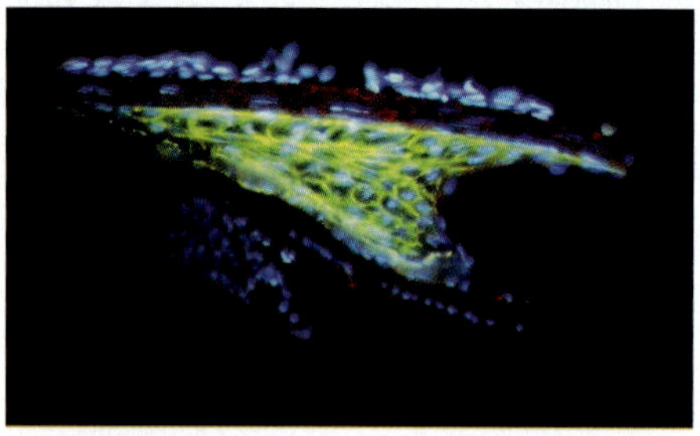

Figure 6.3.7 The yellow-green fluorescence of the ELF® 97 alcohol precipitate, demonstrating localization of endogenous alkaline phosphatase activity in the ciliary body of the zebrafish eye. Adult zebrafish cryosections were first stained with Texas Red®-X wheat germ agglutinin (W21405) followed by the ELF® 97 phosphate substrate in the ELF® 97 Endogenous Phosphatase Detection Kit (E6601), and finally the nuclei were counterstained with the blue-fluorescent Hoechst 33342 nucleic acid stain (H1399, H3570, H21492). The triple-exposure image was acquired using bandpass filters appropriate for the Texas Red® dye, ELF® 97 alcohol precipitate and AMCA.

much less than 100 nm.[5] This extremely high Stokes shift also makes the ELF® 97 phosphatase substrate ideal for use in multicolor applications (Figure 6.3.5, Figure 6.3.6, Figure 6.3.7). The yellow-green–fluorescent ELF® 97 signal can be visualized simultaneously with blue-fluorescent probes, such as Alexa Fluor® 350 and Alexa Fluor® 405 conjugates (Section 1.7), or with the DAPI and Hoechst counterstains (Section 12.5) using a fluorescence microscope fitted with a standard DAPI/Hoechst nuclear longpass optical filter set.[5] In addition, the excitation spectra of tetramethylrhodamine, Texas Red® and Alexa Fluor® dyes with absorption maxima beyond 530 nm are very well separated from that of the ELF® 97 alcohol precipitate, allowing sequential visualization with the appropriate optical filter sets without bleed-through.[5] Also, ELF® 97 alcohol signals can be distinguished from fluorescein, Oregon Green® 488 and Alexa Fluor® 488 dye signals because their excitation wavelengths do not significantly overlap.

Applications of ELF® Signal Amplification

Histochemical applications often require reliable, sensitive and stable detection of targets in complex samples that may have significant background from either natural sample autofluorescence or fluorescence created during sample preparation. Many histochemical procedures utilize enzyme-amplified detection methods in conjunction with chromophoric substrates such as 5-bromo-4-chloro-3-indolyl phosphate (BCIP), nitro blue tetrazolium (NBT) for phosphatase (N6495,N6547; see below) or diaminobenzidine (DAB; D22185,D22187; Section 6.2) for peroxidase to yield colored precipitates at the site of labeling. The ELF® technology yields staining that can be equal to or superior to NBT/BCIP staining and that forms an exceptionally persistent product. We have observed that ELF® 97 phosphate–based staining of fixed samples can persist for months to years with little if any loss of signal. ELF® 97 staining can also be combined with staining by other fluorophores to permit simultaneous analysis of multiple targets in the sample. Combination of the ELF® technology with TSA technology (Section 6.2) provides exceptional detection limits for low-abundance targets in cells and tissues that are not possible with existing histochemical methods.[7,8]

A general scheme for the secondary detection methods used to develop the ELF® 97 signal for *in situ* hybridization, cytological labeling, immunohistochemistry and microarrays[9–11] is shown in Figure 6.3.8. Please note that the components of these ELF® 97 Kits are not interchangeable. Through the course of our product development, we have discovered that to achieve optimal

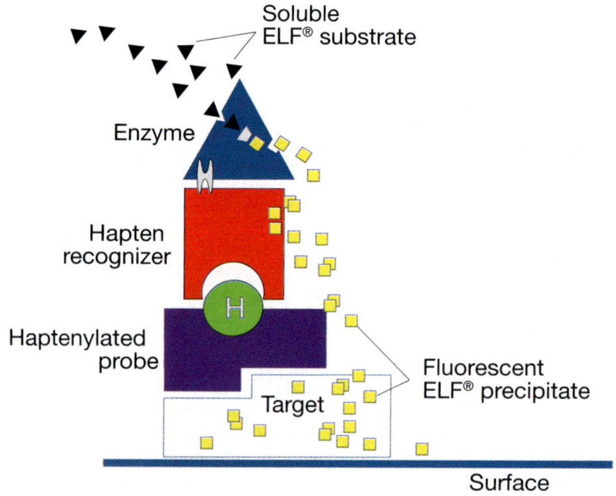

Figure 6.3.8 Schematic diagram of the method employed in our ELF® 97 mRNA *In Situ* Hybridization (E6604, E6605), Cytological Labeling (E6603) and Immunohistochemistry (E6600) Kits. Samples are probed with haptenylated or biotinylated target-specific probes such as antibodies or hybridization probes. Next, alkaline phosphatase conjugates of streptavidin or the hapten-specific probe are applied. Alternatively, a biotinylated antibody and biotinylated alkaline phosphatase can be used with standard bridging methods to increase the penetration in tissue, a method that is employed in our ELF® 97 Immunohistochemistry Kit. The sample is then incubated with the ELF® 97 phosphatase substrate, which forms an intense yellow-green–fluorescent ELF® 97 alcohol precipitate at the site of alkaline phosphatase activity.

IMPORTANT NOTICE: The products described in this manual are covered by one or more Limited Use Label License(s). Please refer to the Appendix on page 971 and Master Product List on page 975. Products are For Research Use Only. Not intended for any animal or human therapeutic or diagnostic use.

sensitivity in each type of biological application requires substrate formulations, buffers and protocols that are tailored to the application. For this reason, we strongly recommend that each kit be used only for the applications for which it was developed. However, we also offer the ELF® 97 phosphate (E6588, E6589) as well as the ELF® 97 alcohol cleavage product (E6578) separately, for those researchers who want to develop their own applications. We have found that addition of 1–5 µM ELF® 97 alcohol (E6578) to the enzyme detection medium usually improves the quality of precipitation by reducing the crystal size. Addition of other components to the buffers may also be required to maximize the signal intensity and localize the signal to the target.

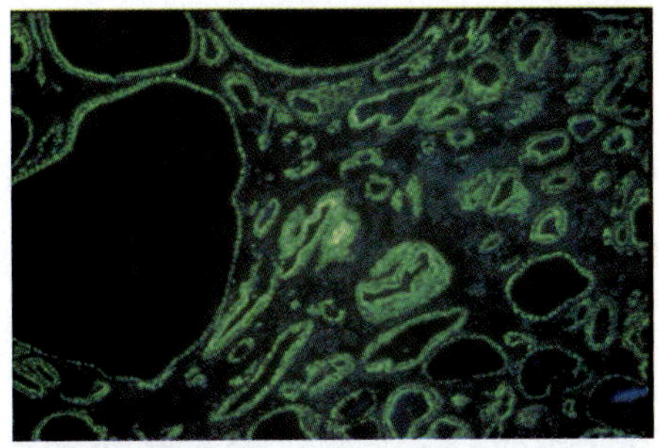

Figure 6.3.9 Tissue from a prostate carcinoma that has been fixed with formalin, embedded in paraffin, sectioned and hybridized with a biotinylated antisense RNA probe to gastrin-releasing peptide (GRP) receptor mRNA. Following *in situ* hybridization, the biotinylated probe was developed for visualization with alkaline phosphatase–mediated techniques using the ELF® 97 mRNA *In Situ* Hybridization Kit (E6604, E6605). Image contributed by Marty Bartholdi, Berlex Biosciences, Berlex Laboratories, Inc.

ELF® 97 mRNA *In Situ* Hybridization Kits

The optimized reagents and protocols in our ELF® 97 mRNA *In Situ* Hybridization Kits (Kit #1, E6604; Kit #2 with streptavidin, alkaline phosphatase conjugate, E6605) provide a rapid and sensitive assay for detecting mRNA *in situ* hybridization signals in cells and tissue sections[12] (Figure 6.3.9, Figure 6.3.10).

The ELF® 97 mRNA *In Situ* Hybridization Kits include:

- Alkaline phosphatase streptavidin (only in Kit #2, E6605)
- ELF® wash, blocking and developing buffers
- Application-specific ELF® 97 phosphatase substrate solution
- Hoechst 33342 nucleic acid counterstain
- ELF® mounting medium
- 50 plastic coverslips
- Detailed protocols

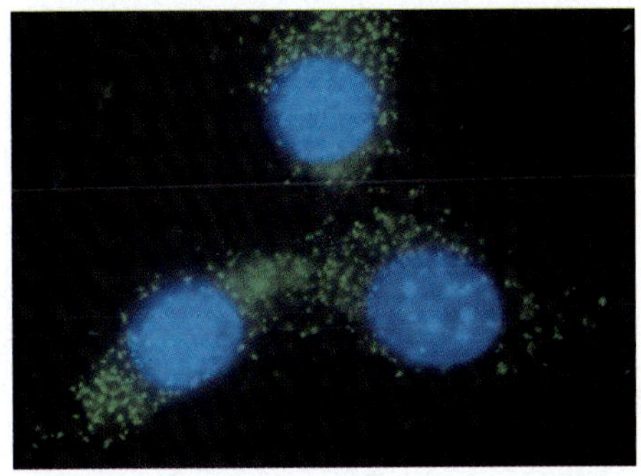

Figure 6.3.10 *lacZ* mRNA in transformed mouse fibroblasts (CRE BAG 2 cells) hybridized with a singly biotinylated, complementary oligonucleotide. Hybrids were then detected by incubation with a streptavidin–alkaline phosphatase conjugate in combination with the ELF® 97 alkaline phosphatase substrate, both of which are provided in our ELF® 97 mRNA *In Situ* Hybridization Kit (E6604, E6605). Cells were counterstained with DAPI (D1306, D3571, D21490) and photographed using a longpass optical filter appropriate for DAPI.

The ELF® 97 mRNA *In Situ* Hybridization Kit #2 (E6605), which contains alkaline phosphatase streptavidin, can be used to detect biotinylated DNA probes, biotinylated RNA probes or digoxigenin-labeled probes in conjunction with a biotinylated anti-digoxigenin antibody.[13] The ELF® 97 mRNA *In Situ* Hybridization Kit #1 (E6604), which does not include alkaline phosphatase streptavidin, is designed for use with other alkaline phosphatase conjugates that have been applied to detect DNA or RNA probes labeled with haptens other than biotin. Each kit contains sufficient reagents for 50 slides or coverslips.

ELF® 97 Cytological Labeling Kit

The ELF® 97 Cytological Labeling Kit (E6603) facilitates the detection of a broad range of cellular targets, including cell-surface sites, cytoplasmic organelles, nuclear antigens and cytoskeletal networks[14] (Figure 6.3.11). This versatile kit can potentially be used to detect any subcellular structure that can be selectively labeled with a biotinylated or haptenylated ligand (Figure 6.3.12).

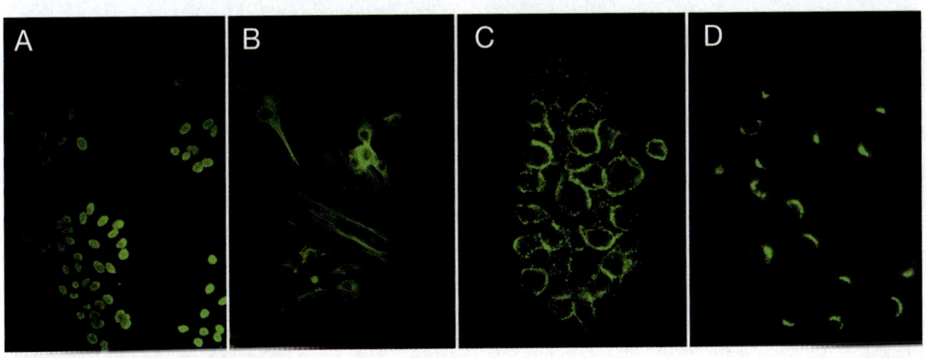

Figure 6.3.12 Cellular targets developed for visualization with the reagents in our ELF® 97 Cytological Labeling Kit (E6603), using the methods described in Figure 6.3.8. **A)** Nuclei in a commercial preparation of human epithelial cells (HEp-2) that have been labeled with human anti-nuclear antibodies and then incubated with biotin-XX goat anti–human IgG antibody. **B)** Acetone-fixed mouse fibroblast (CRE BAG 2) cells that have been treated with Triton X-100 and then incubated with biotin-XX phalloidin (B7474), a probe specific for actin stress fibers. **C)** Formaldehyde-fixed human carcinoma cells that have been incubated with biotin-XX epidermal growth factor (E3477). **D)** Mouse fibroblast cells that have been probed with antibodies directed against rat medial Golgi cisternae (a gift from Vivek Malhotra, University of California, San Diego) and then incubated with biotin-XX goat anti–rabbit IgG antibody (B2770). In each case, the biotinylated probe was detected with alkaline phosphatase streptavidin, followed by incubation with the ELF® 97 phosphatase substrate.

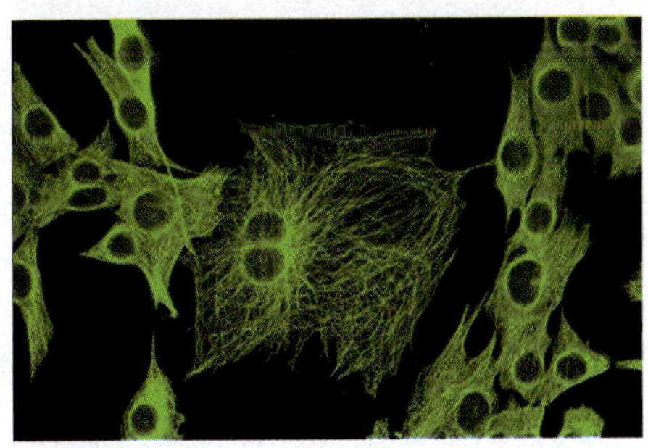

Figure 6.3.11 Mouse fibroblast microtubules labeled with a mouse monoclonal anti-ß-tubulin antibody in conjunction with biotin-XX goat anti–mouse IgG antibody (B2763) and then developed for visualization with alkaline phosphatase–mediated techniques using the ELF® 97 Cytological Labeling Kit (E6603). This kit's novel ELF® 97 phosphatase substrate yields a yellow-green–fluorescent precipitate at the site of alkaline phosphatase activity. Prior to antibody labeling, mouse fibroblasts were fixed and permeabilized in the presence of cytoskeletal stabilizing buffer and treated with paclitaxel (P3456) to stabilize microtubule structures. The image was deconvolved using Huygens software (Scientific Volume Imaging, http://www.svi.nl/).

IMPORTANT NOTICE: The products described in this manual are covered by one or more Limited Use Label License(s). Please refer to the Appendix on page 971 and Master Product List on page 975. Products are For Research Use Only. Not intended for any animal or human therapeutic or diagnostic use.

The ELF® 97 Cytological Labeling Kit includes:

- Alkaline phosphatase streptavidin
- Application-specific ELF® 97 phosphatase substrate solution plus additives
- ELF® wash, blocking and developing buffers
- ELF® mounting medium
- Detailed protocols

Each kit provides sufficient reagents for 50 slides or coverslips.

ELF® 97 Immunohistochemistry Kit

The extremely high Stokes shift for the ELF® 97 alcohol precipitate—greater than 180 nm—is an important advantage for immunohistochemistry applications. Although the ELF® 97 dye is excited in the ultraviolet range where autofluorescence from cells, tissues, paraffin sections and various fixatives can be quite high,[15] the high intensity of ELF® 97 staining overcomes most of this background. Because the ELF® 97 alcohol precipitate is extremely photostable, the background can be further reduced by pre-bleaching the entire sample with ultraviolet light before measuring the specific fluorescence of the ELF® 97 alcohol precipitate.[16,17]

The ELF® 97 phosphatase substrate in our ELF® 97 Immunohistochemistry Kit has been specially formulated to reduce nonspecific staining in immunohistochemical preparations and contains the key reagents for detecting antigens in tissue sections, including streptavidin and biotinylated alkaline phosphatase. The streptavidin provided in the kit is used to link the biotinylated alkaline phosphatase with a biotinylated secondary antibody—a common immunohistochemical technique for optimizing tissue penetration. We have used the ELF® 97 Immunohistochemistry Kit to characterize a number of antibodies generated against the zebrafish retina (Figure 6.3.13, Figure 6.3.14) and found that the ELF® 97 alcohol staining pattern was identical to that seen with fluorophore-conjugated secondary reagents.[5] The ELF® 97 signal could easily be visualized despite this tissue's considerable autofluorescence. Moreover, the staining was approximately 500 times more photostable than that produced by fluorescein-labeled secondary reagents.[5] ELF® 97 phosphate has been used to measure endogenous phosphatase activity and, in combination with alkaline phosphatase–conjugated immunoreagents, to detect intracellular cell cycle–associated proteins such as cyclin B1 by flow cytometry.[18–20]

The ELF® 97 Immunohistochemistry Kit (E6600) provides sufficient reagents for preparing 50 mL of detection solution, which is adequate for staining 250 to 1000 sections.

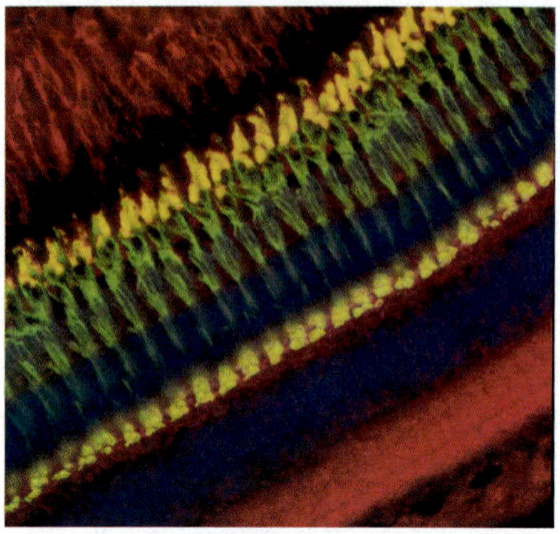

Figure 6.3.13 A transverse section of fixed zebrafish retina probed with FRet 43, a monoclonal antibody that binds to double cone cells, and developed for visualization with enzyme-mediated immunohistochemical techniques using our ELF® 97 Immunohistochemistry Kit (E6600). The yellow-green–fluorescent double cones stained with the ELF® 97 alcohol precipitate are oriented so that their outer segments are at the top of the stained configuration and the synaptic pedicles are at the bottom. This section has been counterstained with tetramethylrhodamine wheat germ agglutinin (W849), which makes the rod outer segments (top left) and the inner plexiform layer and ganglion cell axons (bottom right) appear bright red. Wheat germ agglutinin also binds in the region occupied by the cone outer segments and synaptic pedicles, which appear bright yellow because they are double-labeled with both the ELF® 97 alcohol precipitate and tetramethylrhodamine. Although this section has also been counterstained with Hoechst 33342 (H1399, H3570, H21492), the blue-stained nuclei are barely visible in this photomicrograph. However, the double cones' inner fibers traverse the region occupied by the Hoechst 33342 dye–stained rod nuclei, and thus appear light blue. The inner segments, myoids and nuclei of these double-cone cells are labeled only with the ELF® 97 alcohol precipitate, giving them a characteristic green appearance. The image was obtained by triple-exposure through optical filters appropriate for DAPI, tetramethylrhodamine and the ELF® precipitate. Used with permission from (J Histochem Cytochem (1995) 43:77).

Figure 6.3.14 A transverse section through adult zebrafish retina that has been probed with anti–glial fibrillary acidic protein antibody (anti-GFAP antibody, A21282) and then developed for visualization with alkaline phosphatase–mediated techniques using our ELF® 97 Immunohistochemistry Kit (E6600). This kit's ELF® 97 phosphatase substrate yields an extremely photostable yellow-green signal at the site of anti-GFAP binding, clearly showing that this antibody binds to Müller cells in the zebrafish retina. The retinal section has been counterstained with Hoechst 33342 (H1399, H3570, H21492), which stains all nuclei blue, and with tetramethylrhodamine wheat germ agglutinin (W849), which stains both the inner and outer plexiform layers as well as the photoreceptor outer segments red.

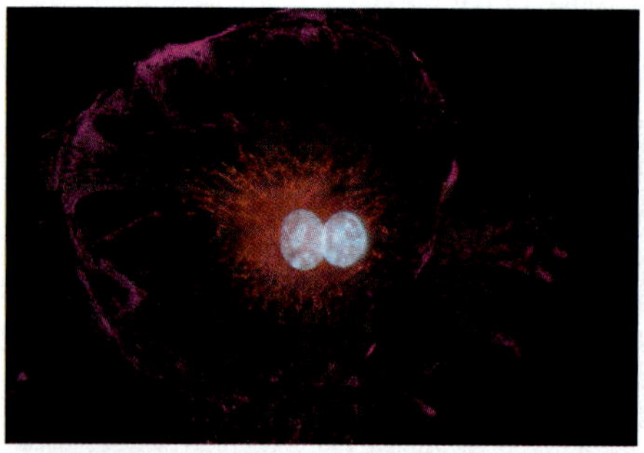

Figure 6.3.15 The intermediate filaments in bovine pulmonary artery endothelial cells, localized using our anti-desmin antibody (A21283), which was visualized with the Alexa Fluor® 647 goat anti-mouse IgG antibody (A21235). Endogenous biotin in the mitochondria was labeled with Alexa Fluor® 546 streptavidin (S11225) and DNA in the cell was stained with blue-fluorescent DAPI (D1306, D3571, D21490).

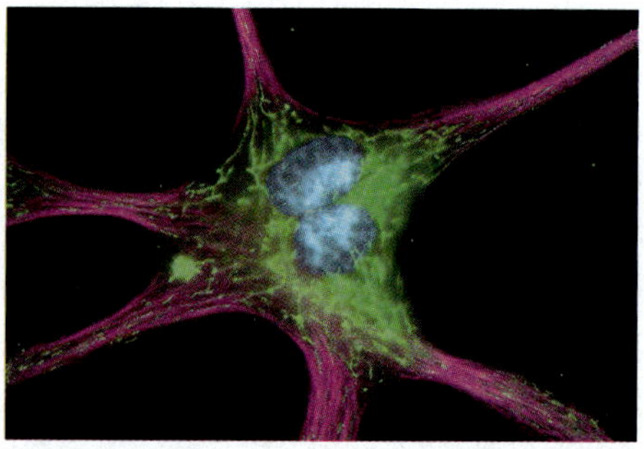

Figure 6.3.16 The cytoskeleton of a fixed and permeabilized bovine pulmonary artery endothelial cell detected using mouse monoclonal anti-α-tubulin antibody (A11126), visualized with Alexa Fluor® 647 goat anti-mouse IgG antibody (A21235) and pseudocolored magenta. Endogenous biotin in the mitochondria was labeled with green-fluorescent Alexa Fluor® 488 streptavidin (S11223) and DNA was stained with blue-fluorescent DAPI (D1306, D3571, D21490).

IMPORTANT NOTICE: The products described in this manual are covered by one or more Limited Use Label License(s). Please refer to the Appendix on page 971 and Master Product List on page 975. Products are For Research Use Only. Not intended for any animal or human therapeutic or diagnostic use.

The ELF® 97 Immunohistochemistry Kit includes:

- Application-specific ELF® 97 phosphatase substrate solution
- Streptavidin
- Biotinylated alkaline phosphatase
- ELF® reaction buffer
- ELF® mounting medium
- Detailed protocol

Mammalian cells and tissues contain biotin-dependent carboxylases, which are required for a variety of metabolic functions. These biotin-containing enzymes produce substantial background signals when biotin–streptavidin detection systems are used to identify cellular targets[21,22] (Figure 6.3.15, Figure 6.3.16). Because the ELF® technology is so sensitive, we recommend preblocking endogenous biotin in cells with our Endogenous Biotin-Blocking Kit (E21390, Section 7.6) when staining cells using our ELF® 97 Kits containing streptavidin conjugates.

ELF® 97 Endogenous Phosphatase Detection Kit

We have used the ELF® 97 phosphatase substrate to develop a novel fluorescence-based assay for detecting phosphatases in tissue sections and cells[23] (Figure 6.3.17, Figure 6.3.18, Figure 6.3.19). The ELF® 97 Endogenous Phosphatase Detection Kit (E6601) provides several advantages over traditional approaches to phosphatase histochemistry.[5,23]

The ELF® 97 endogenous phosphatase detection system uses a simple protocol and is compatible with many types of tissue preparations, including tissue cryosections, paraffin-embedded sections and cultured cells. The fluorescence excitation and emission maxima (~365/530 nm) of the ELF® 97 alcohol precipitate are well separated, allowing researchers to easily distinguish the signal from autofluorescence and from other fluorescent labels.

Phosphatases have been commonly used as enzyme markers, allowing researchers to identify primordial germ cells,[24] to distinguish subpopulations of bone marrow stromal cells[25] and to investigate *in vitro* differentiation in carcinoma cell lines.[26–28] A particularly prominent application of the ELF® 97 endogenous phosphatase detection system is the detection of TRAP (tartrate-resistant acid phosphatase) activity, a marker of osteoclast differentiation from hematopoietic stem cells[29–32] (HSC). Because ELF® 97 phosphate is intrinsically membrane impermeant, it can be utilized as a marker for membrane integrity in which the intracellular phosphatase activity of membrane-compromised cells is detected.[33] The patterns of ELF® 97 alcohol staining in intestine (Figure 6.3.17), kidney (Figure 6.3.20), ovary and gills (Figure 6.3.19) are essentially identical

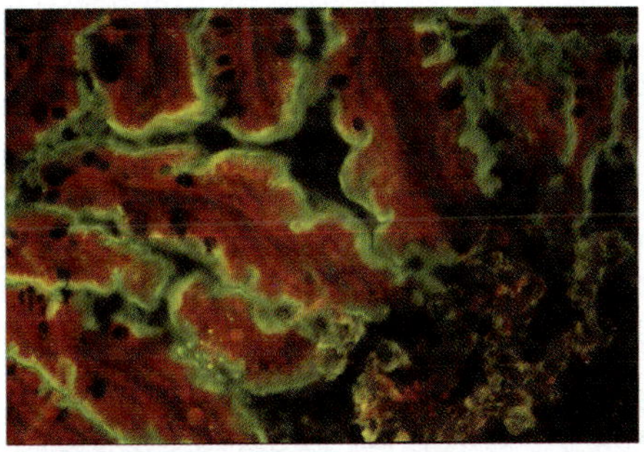

Figure 6.3.17 Lightly fixed adult zebrafish intestine stained with the substrate contained in the ELF® 97 Endogenous Phosphatase Detection Kit (E6601). This kit's ELF® substrate yields a green-fluorescent precipitate at the site of endogenous phosphatase activity. A partially digested brine shrimp, which also contains endogenous phosphatase activity, can be seen in the bottom right corner of this micrograph. The tissue was counterstained with propidium iodide (P1304MP, P3566, P21493).

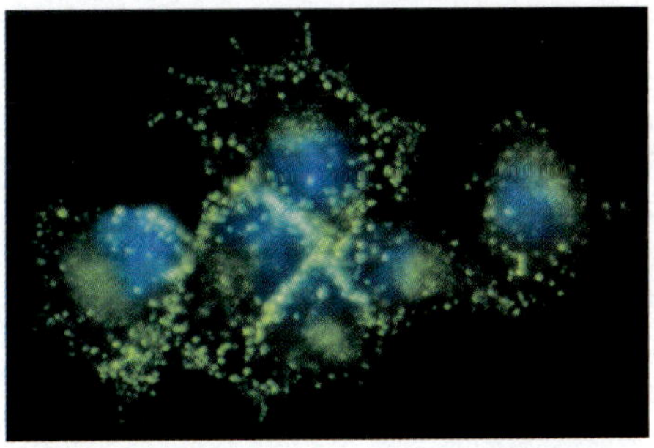

Figure 6.3.18 Endogenous alkaline phosphatase enzyme of osteosarcoma cells localized with the ELF® 97 Endogenous Phosphatase Detection Kit (E6601). Unlike other phosphatase substrates, the unique ELF® 97 phosphatase substrate forms a fluorescent precipitate at the site of enzymatic activity. The blue-fluorescent nucleic acid stain Hoechst 33342 (H1399, H3570, H21492) was used as a counterstain to the green fluorescence of the ELF® 97 alcohol precipitate. The double-exposure image was acquired using a bandpass filter set appropriate for ELF® 97 alcohol and a longpass filter set appropriate for DAPI.

Figure 6.3.19 A cryostat section of lightly fixed adult zebrafish gills that have been incubated with the ELF® 97 substrate in our ELF® 97 Endogenous Phosphatase Detection Kit (E6601). This kit's novel ELF® 97 phosphatase substrate yields a yellow green-fluorescent precipitate at the site of endogenous phosphatase activity. This staining pattern is identical to that seen when employing the conventional Gomori method for detecting phosphatase activity.

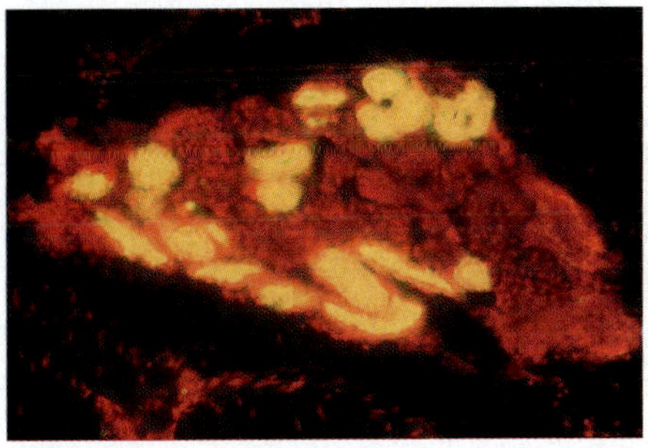

Figure 6.3.20 A cryostat section through adult zebrafish kidney that was incubated with the ELF® 97 phosphatase substrate in our ELF® 97 Endogenous Phosphatase Detection Kit (E6601) and counterstained with propidium iodide (P1304MP, P3566, P21493). The ELF® 97 alcohol signal, which is ordinarily green fluorescent, appears yellow in this section because the phosphatase activity is in close proximity to the nuclei, which are stained red by the propidium iodide. This double-exposure image was obtained using longpass filter sets appropriate for ELF® 97 alcohol product and tetramethylrhodamine.

IMPORTANT NOTICE: The products described in this manual are covered by one or more Limited Use Label License(s). Please refer to the Appendix on page 971 and Master Product List on page 975. Products are For Research Use Only. Not intended for any animal or human therapeutic or diagnostic use.

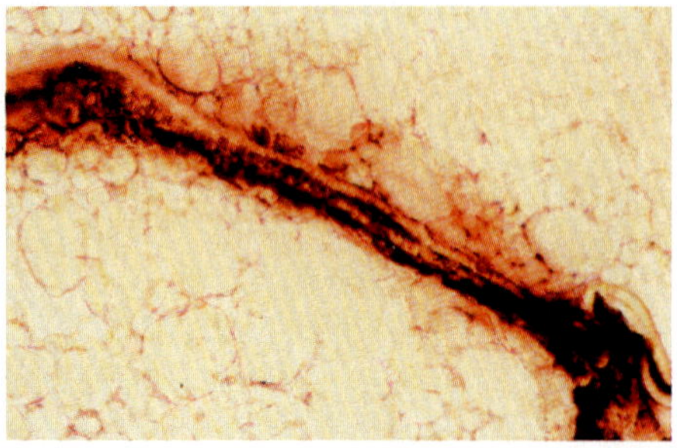

Figure 6.3.22 Endogenous phosphatase activity in a zebrafish ovary cross-section, stained with the BCIP (B6492) and NBT (N6495) reagents. These reagents are also available in the NBT/BCIP Reagent Kit (N6547).

to the patterns of black precipitate produced by the conventional Gomori technique, but show much higher spatial resolution. The ELF® 97 Endogenous Phosphatase Detection Kit has also been used to monitor and quantitate alkaline phosphatase activity in marine phytoplankton [34] and bacteria.[35]

Each ELF® 97 Endogenous Phosphatase Detection Kit (E6601) provides:

- ELF® 97 phosphatase substrate solution
- ELF® detection buffer
- ELF® mounting medium
- Easy-to-follow protocol

Sufficient materials are provided for staining 50–250 tissue sections.

ELF® Spin Filters

Like many enzyme substrates, the ELF® 97 developing solution should be filtered before use for optimal staining. The ELF® spin filters (E6606)—spin-filtration devices with a pore size of 0.2 µm—are both convenient and efficient, permitting a very small volume to be filtered without significant loss. These spin filters are recommended for use with all of our ELF® 97 Kits. These filters are equally suitable for rapid filtration of aqueous solutions of other probes.

Alkaline Phosphatase Conjugates for Secondary Detection

For use in signal amplification of antibody- or biotin-labeled targets, we offer the alkaline phosphatase conjugates of:

- Streptavidin (S921, Section 7.6)
- Goat anti–mouse IgG antibody (G21060)
- Goat anti–mouse $F(ab')_2$ fragment (F21452)
- Goat anti–rabbit IgG antibody (G21079)
- Goat anti–rabbit $F(ab')_2$ fragment (F21456)
- Goat anti–rat IgG antibody (A10546)
- Rabbit anti–goat IgG antibody (R21458)

A more thorough discussion of our secondary antibodies can be found in Section 7.2 (Table 7.1). We also offer the Zenon® Alkaline Phosphatase Rabbit IgG Labeling Kit (Z25350) for creating alkaline phosphatase conjugates of primary rabbit IgG antibodies. Our Zenon® antibody labeling technology (Section 7.3, Table 7.7) permits the rapid and quantitative formation of dye

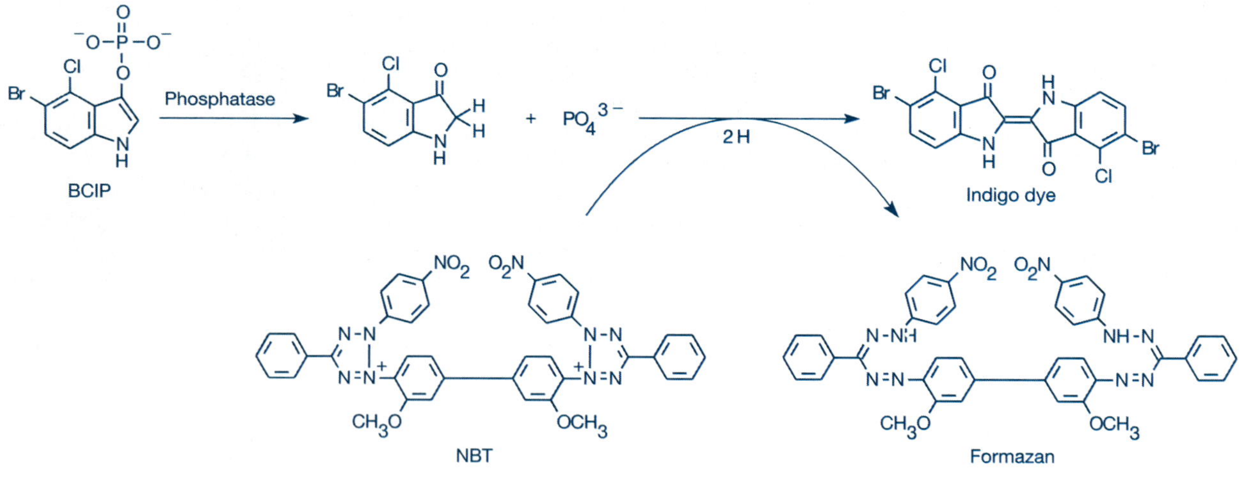

Figure 6.3.21 Principle of enzyme-linked detection using the reagents in our NBT/BCIP Reagent Kit (N6547). Phosphatase hydrolysis of BCIP is coupled to reduction of NBT, yielding a formazan and an indigo dye that together form a black-purple–colored precipitate.

and enzyme complexes with the Fc portion of an intact antibody. The alkaline phosphatase–antibody complexes formed with our Zenon® Alkaline Phosphatase Antibody Labeling Kit can be used in combination with the ELF® 97 Kits described above or with other applications for alkaline phosphatase–conjugated antibodies such as ELISAs.

Chromogenic Phosphatase Substrate

NBT/BCIP Reagent Kit

5-Bromo-4-chloro-3-indolyl phosphate (BCIP) is commonly used with a number of different chromogens in various histological and molecular biology techniques. Hydrolysis of this indolyl phosphate, followed by oxidation, produces a blue-colored precipitate at the site of enzymatic activity. For convenience, we offer the NBT/BCIP Reagent Kit (N6547), which provides 1 g samples of BCIP as well as its co-precipitant NBT.

NBT: A Co-Precipitant for the BCIP Reaction

Nitro blue tetrazolium (NBT, N6495) is the most commonly used electron-transfer agent and co-precipitant for the BCIP reaction, forming a dark blue, precisely localized precipitate in the presence of alkaline phosphatase [36,37] (Figure 6.3.21, Figure 6.3.22).

CSPD® and CDP-*Star*® Chemiluminescent ELISA Detection

The Chemiluminescent Alkaline Phosphatase ELISA Kits employ CSPD® or CDP-*Star*® 1,2-dioxetane substrates for alkaline phosphatase with Sapphire™-II or Emerald-II enhancer in a system designed for rapid and ultrasensitive analyte detection in enzyme-linked immunoassays.[38] Maximum light emission from alkaline phosphatase–activated CSPD® or CDP-*Star*® substrate is reached in 5 to 60 minutes, depending on the temperature and the substrate chosen. Enzymatic dephosphorylation of substrate occurs at a constant rate proportional to enzyme concentration; the resulting phenolate anion decomposes with a finite half-life (Figure 6.3.23). Light emission can be quantitated with a variety of luminometers without the need for solution injection.

Enzyme-linked immunosorbent assays (ELISAs) can be formatted in several configurations on a variety of solid supports including microplate wells, tubes, polystyrene beads or ferrite particles. The high sensitivity obtained with 1,2-dioxetane substrates is demonstrated in a sandwich immunoassay format that employs a biotinylated detector antibody and streptavidin alkaline phosphatase conjugate for quantitating recombinant human IL-6 (rhIL-6). The results obtained with CSPD® substrate/Sapphire™-II enhancer (Figure 6.3.24) show a significant improvement in signal-to-noise performance at all concentrations of rhIL-6 and a much wider assay dynamic range compared to those obtained with the fluorescent substrate 4-methylumbelliferyl phosphate (MUP, M6491; Section 10.3), and the colorimetric substrate *p*-nitrophenyl phosphate [38] (*p*NPP).

We offer four different Chemiluminescent Alkaline Phosphatase ELISA Kits, as well as a Chemiluminescent Alkaline Phosphatase ELISA Sampler Kit. Each kit provides concentrated assay buffer and blocking reagent, as well as one of the following substrate/enhancer solutions:

- CSPD® substrate/Sapphire™-II enhancer (in Kit #1, C10552)
- CSPD® substrate/Emerald-II™ enhancer (in Kit #2, C10553)
- CDP-*Star*® substrate/Sapphire™-II enhancer (in Kit #3, C10554)
- CDP-*Star*® substrate/Emerald-II™ enhancer (in Kit #4, C10555)

The Chemiluminescent Alkaline Phosphatase ELISA Sampler Kit (C10556) provides assay buffer, blocking reagent and sample sizes of all four substrate/enhancer solutions. All five kits provide sufficient reagents for 1000 assays based on the provided protocol.

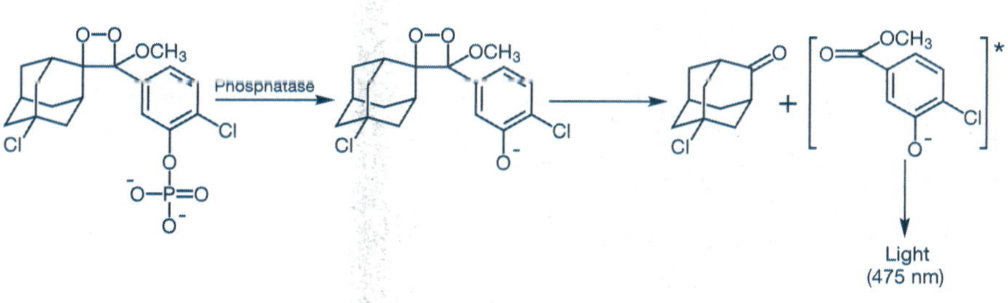

Figure 6.3.23 Mechanism of phosphatase-dependent chemiluminescence generation by CDP-*Star*® substrate. Luminescence emission is shifted to 461 nm or 542 nm respectively by the Sapphire™-II or Emerald-II™ enhancers.

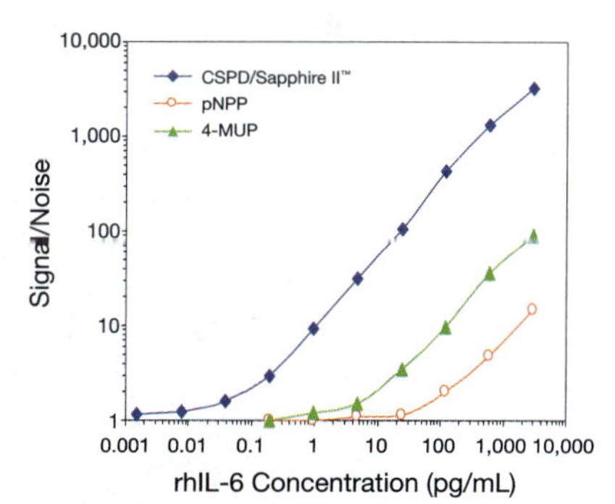

Figure 6.3.24 Comparison of detection sensitivities for the ELISA quantitation of recombinant human IL-6 (rhIL-6) using the chemiluminescent phosphatase substrate/enhancer in the Chemiluminescent Alkaline Phosphatase ELISA Kit #1 (CSPD® substrate/Sapphire™-II enhancer; C10552), the fluorescent phosphatase substrate 4-methylumbelliferyl phosphate (MUP, M6491) and the colorimetric phosphatase substrate *p*-nitrophenyl phosphate (*p*NPP).

IMPORTANT NOTICE: The products described in this manual are covered by one or more Limited Use Label License(s). Please refer to the Appendix on page 971 and Master Product List on page 975. Products are For Research Use Only. Not intended for any animal or human therapeutic or diagnostic use.

REFERENCES

1. Proc SPIE-Int Soc Opt Eng (1999) 3602:265; **2.** Optics Comm (1987) 64:457; **3.** J Phys Chem (1970) 74:4473; **4.** Anal Biochem (1992) 207:32; **5.** J Histochem Cytochem (1995) 43:77; **6.** Cytometry (2000) 40:42; **7.** Endocrinology (2002) 143:239; **8.** J Histochem Cytochem (2000) 48:1593; **9.** Appl Environ Microbiol (2003) 69:2950; **10.** Clin Chem (2001) 47:1451; **11.** Int J Food Microbiol (2001) 67:71; **12.** J Histochem Cytochem (1997) 45:345; **13.** Cell Growth Differ (2002) 13:227; **14.** J Microsc (2002) 206:106; **15.** J Pathol (2000) 191:452; **16.** Mol Biol Cell (2005) 16:881; **17.** Biotechniques (2001) 30:794; **18.** Cytometry (2003) 54A:48; **19.** Cytometry (2001) 43:117; **20.** Cytometry (1999) 37:314; **21.** Methods Mol Biol (2008) 418:111; **22.** Methods Mol Biol (2008) 418:157; **23.** J Histochem Cytochem (1999) 47:1443; **24.** Anatomical Record (1954) 118:135; **25.** J Histochem Cytochem (1992) 40:1059; **26.** Dev Biol (1981) 88:279; **27.** Cell (1975) 5:229; **28.** Proc Natl Acad Sci U S A (1973) 70:3899; **29.** Nature (2008) 454:221; **30.** Blood (2007) 109:3024; **31.** J Histochem Cytochem (2005) 53:1525; **32.** J Histochem Cytochem (2004) 52:411; **33.** J Biol Chem (2003) 278:36250; **34.** Marine & Freshwater Res (2005) 56:417; **35.** Cytometry A (2009) 75:163; **36.** Biotechniques (1992) 12:656; **37.** Histochemistry (1978) 58:203; **38.** J Immunol Methods (2001) 247:111.

DATA TABLE 6.3 PHOSPHATASE-BASED SIGNAL AMPLIFICATION TECHNIQUES

Cat. No.	MW	Storage	Soluble	Abs	EC	Em	Solvent	Notes
E6578	307.14	L	DMSO	345	ND	530	pH 8	1, 2, 3
E6588	431.08	F,L	H_2O	289	12,000	see Notes	pH 10	2, 4, 5
E6589	431.08	F,L	H_2O	289	12,000	see Notes	pH 10	2, 4, 5
N6495	817.65	D,L	H_2O, DMSO	256	64,000	see Notes	MeOH	6

For definitions of the contents of this data table, see "Using *The Molecular Probes® Handbook*" in the introductory pages.

Notes

1. ND = not determined.
2. This product is supplied as a ready-made solution in the solvent indicated under "Soluble."
3. ELF® 97 alcohol is insoluble in water. Spectral maxima listed are for an aqueous suspension; for this reason, the value of EC cannot be determined.
4. Enzymatic cleavage of this substrate yields E6578.
5. Fluorescence of the unhydrolyzed substrate is very weak.
6. Phosphatase hydrolysis of BCIP (B6492) is coupled to reduction of NBT (N6495), yielding a water-insoluble indigo dye (Abs ~615 nm) and a water-insoluble formazan (Abs ~605 nm), respectively.

PRODUCT LIST 6.3 PHOSPHATASE-BASED SIGNAL AMPLIFICATION TECHNIQUES

Cat. No.	Product	Quantity
C10552	Chemiluminescent Alkaline Phosphatase ELISA Kit #1 *with CSPD® Substrate/ Sapphire-II™ enhancer* *1000 assays*	1 kit
C10553	Chemiluminescent Alkaline Phosphatase ELISA Kit #2 *with CSPD® Substrate/ Emerald-II™ enhancer* *1000 assays*	1 kit
C10554	Chemiluminescent Alkaline Phosphatase ELISA Kit #3 *with CDP-*Star*® Substrate/ Sapphire-II™ enhancer* *1000 assays*	1 kit
C10555	Chemiluminescent Alkaline Phosphatase ELISA Kit #4 *with CDP-*Star*® Substrate/ Emerald-II™ enhancer* *1000 assays*	1 kit
C10556	Chemiluminescent Alkaline Phosphatase ELISA Sampler Kit *1000 assays*	1 kit
E6578	ELF® 97 alcohol *1 mM solution in DMSO*	1 mL
E6603	ELF® 97 Cytological Labeling Kit *with streptavidin, alkaline phosphatase conjugate* *50 assays*	1 kit
E6601	ELF® 97 Endogenous Phosphatase Detection Kit	1 kit
E6600	ELF® 97 Immunohistochemistry Kit	1 kit
E6604	ELF® 97 mRNA *In Situ* Hybridization Kit #1 *50 assays*	1 kit
E6605	ELF® 97 mRNA *In Situ* Hybridization Kit #2 *with streptavidin, alkaline phosphatase conjugate* *50 assays*	1 kit
E6588	ELF® 97 phosphatase substrate (ELF® 97 phosphate) *5 mM in water* *0.2 µm filtered*	1 mL
E6589	ELF® 97 phosphatase substrate (ELF® 97 phosphate) *5 mM in water* *contains 2 mM azide*	1 mL
E6606	ELF® spin filters *20 filters*	1 box
N6547	NBT/BCIP Reagent Kit	1 kit
N6495	nitro blue tetrazolium chloride (NBT)	1 g

IMPORTANT NOTICE: The products described in this manual are covered by one or more Limited Use Label License(s). Please refer to the Appendix on page 971 and Master Product List on page 975. Products are For Research Use Only. Not intended for any animal or human therapeutic or diagnostic use.
www.invitrogen.com/probes

6.4 Phycobiliproteins

Phycobiliproteins are a family of highly soluble and reasonably stable fluorescent proteins derived from cyanobacteria and eukaryotic algae. These proteins contain covalently linked tetrapyrrole groups that play a biological role in collecting light and, through fluorescence resonance energy transfer, conveying it to a special pair of chlorophyll molecules located in the photosynthetic reaction center.[1] Because of their role in light collection, phycobiliproteins have evolved to maximize both absorption and fluorescence and to minimize the quenching caused either by internal energy transfer or by external factors such as changes in pH or ionic composition.[2,3] Phycobiliproteins have several advantages when used as fluorescent probes, including:

- Intense long-wavelength excitation and emission to provide fluorescence that is relatively free of interference from other biological materials
- Relatively large Stokes shifts with extremely high emission quantum yields
- Fluorescence that is not quenched by external agents because the fluorophores are protected by covalent binding to the protein backbone
- Very high water solubility
- Homogeneous structure with defined molecular weights
- Multiple sites for stable conjugation to many biological and synthetic materials

Spectral Characteristics of Phycobiliproteins

B-Phycoerythrin, R-Phycoerythrin and Allophycocyanin

The phycobiliproteins B-phycoerythrin (B-PE), R-phycoerythrin (R-PE) and allophycocyanin (APC) are among the preferred dyes for applications that require either high sensitivity or simultaneous multicolor detection.[4–6] Quantum yields up to 0.98 and extinction coefficients up to 2.4 million $cm^{-1}M^{-1}$ have been reported for these fluorescent proteins (Table 6.2). On a molar basis, the fluorescence yield is equivalent to at least 30 unquenched fluorescein or 100 rhodamine molecules at comparable wavelengths. The fluorescence of a single molecule of B-PE has been detected.[7,8] B-PE is reportedly more photostable than R-PE, but photostability of R-PE conjugates can be improved by adding 1-propyl gallate.[9]

In practical applications such as flow cytometry and immunoassays,[10,11] the sensitivity of B-PE– and R-PE–conjugated antibodies is usually 5 to 10 times greater than that of the corresponding fluorescein conjugate.[12,13] Using R-PE–conjugated streptavidin, researchers have detected fewer than 100 receptor-bound biotinylated antibodies per cell by flow cytometry.[14] A multistep amplification method utilizing a fluoresceinated opioid, biotinylated anti–fluorescein/Oregon Green® antibody (A982, Section 7.4) and a phycoerythrin conjugate of avidin (A2660) was required to detect low-abundance κ-opioid receptors by flow cytometry. In imaging applications, APC and its conjugates are both brighter and more photostable than Cy®5 conjugates (Figure 6.4.1).

Tandem Conjugates of Phycobiliproteins

A phycoerythrin-labeled detection reagent can be used in combination with a green-fluorescent detection reagent to detect two different signals using simultaneous excitation with the 488 nm spectral line of the argon-ion laser.[15] By conjugating R-PE to longer-wavelength light–emitting fluorescence acceptors, an energy transfer cascade is established wherein excitation of the R-PE produces fluorescence of the acceptor dye by the process of fluorescence resonance energy transfer (FRET) (Fluorescence Resonance Energy Transfer (FRET)—Note 1.2). This process, which occurs naturally within single molecules and assemblies of phycobiliproteins (phycobilisomes), can be quite efficient, resulting in almost total transfer of energy from the phycobiliprotein to the acceptor dye of these "tandem conjugates." Thus, it is possible to combine a green-fluorescent antibody conjugate with an R-PE conjugate, as mentioned above, and then to add tandem conjugates of R-PE with either our Alexa Fluor® 610, Alexa Fluor® 647, Alexa Fluor® 680 or

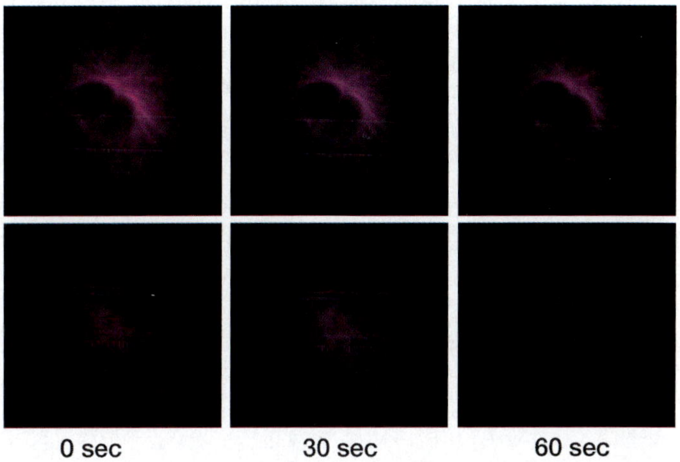

Figure 6.4.1 A comparison of the photobleaching rates of APC and Cy®5 conjugates. The microtubules of bovine pulmonary artery endothelial cells were stained with mouse anti-α-tubulin antibody (A11126) in combination with goat anti-mouse IgG labeled antibody with either crosslinked APC (A865, top series) or the Cy®5 dye (bottom series). The samples were exposed to continuous illumination, and the images were acquired at 30-second intervals with a Quantex cooled CCD camera (Photometrics®) using filter sets appropriate for both APC and Cy®5 dye.

Table 6.2 Spectral data for B-PE, R-PE and APC.

Cat. No.	Phycobiliprotein	Molecular Weight	Absorption Max (nm)	EC ($cm^{-1}M^{-1}$)	Emission Max (nm)	Fluorescence QY
P800	B-phycoerythrin	240,000	546, 565	2,410,000	575	0.98
P801	R-phycoerythrin	240,000	496, 546, 565	1,960,000	578	0.82
A803, A819	Allophycocyanin	104,000	650	700,000	660	0.68
EC = extinction coefficient. QY = quantum yield.						

IMPORTANT NOTICE: The products described in this manual are covered by one or more Limited Use Label License(s). Please refer to the Appendix on page 971 and Master Product List on page 975. Products are For Research Use Only. Not intended for any animal or human therapeutic or diagnostic use.

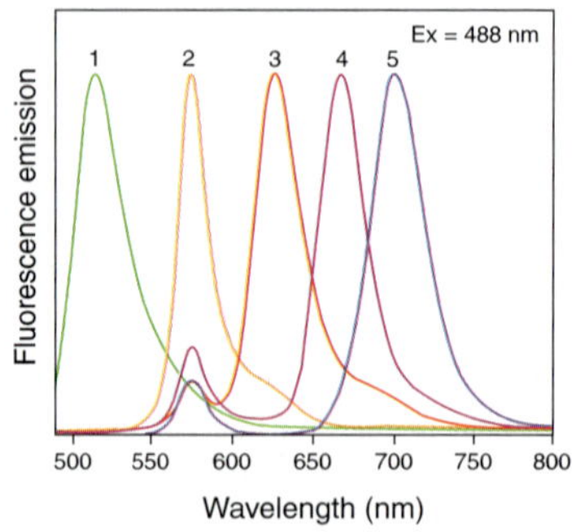

Figure 6.4.2 Normalized fluorescence emission spectra of **1**) Alexa Fluor® 488 goat anti–mouse IgG antibody (A11001), **2**) R-phycoerythrin goat anti–mouse IgG antibody (P852), **3**) Alexa Fluor® 610–R-phycoerythrin goat anti–mouse IgG antibody (A20980), **4**) Alexa Fluor® 647–R-phycoerythrin goat anti–mouse IgG antibody (A20990) and **5**) Alexa Fluor® 680–R-phycoerythrin goat anti–mouse IgG antibody (A20983). The tandem conjugates permit simultaneous multicolor labeling and detection of up to five targets with excitation by a single excitation source—the 488 nm spectral line of the argon-ion laser.

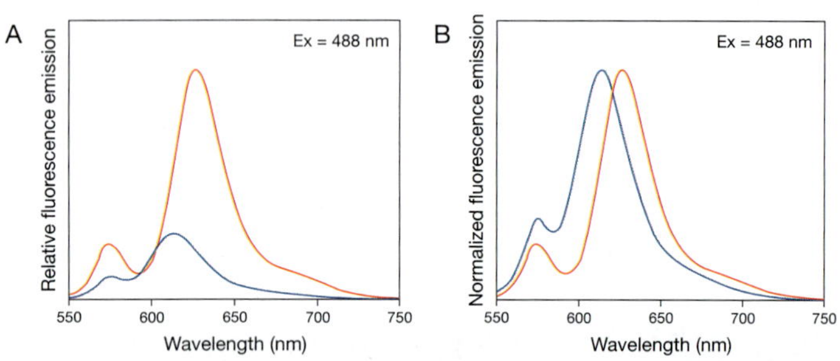

Figure 6.4.3 Fluorescence emission spectra of Alexa Fluor® 610–R-phycoerythrin streptavidin (S20982; red) and Texas Red®–R-phycoerythrin streptavidin (blue) tandem conjugates. Panel A shows a comparison of the spectra on a relative fluorescence intensity scale for samples prepared with equal absorbance at the excitation wavelength (488 nm). Panel B shows the same data normalized to the same peak intensity value to facilitate comparison of the spectral profiles.

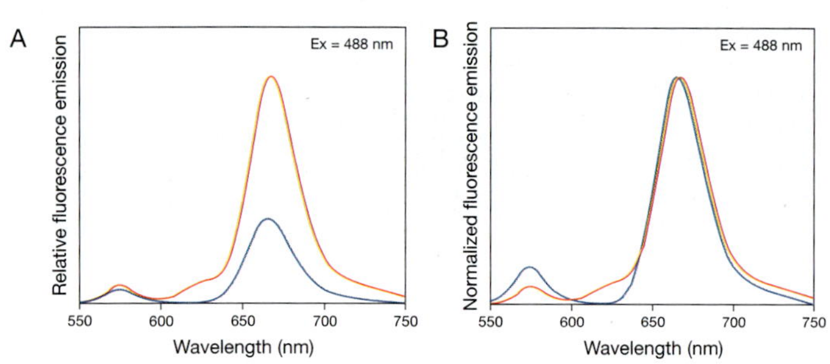

Figure 6.4.4 Fluorescence emission spectra of Alexa Fluor® 647–R-phycoerythrin streptavidin (S20992; red) and Cy®5–R-phycoerythrin streptavidin (blue) tandem conjugates. Panel A shows a comparison of the spectra on a relative fluorescence intensity scale for samples prepared with equal absorbance at the excitation wavelength (488 nm). Panel B shows the same data normalized to the same peak intensity value to facilitate comparison of the spectral profiles.

Alexa Fluor® 750 dyes for simultaneous detection of up to five targets using only 488 nm excitation (Figure 6.4.2).

Phycoerythrin has previously been conjugated to our Texas Red® dye to provide a third signal that is excitable at 488 nm; however, our Alexa Fluor® 610 tandem conjugates of R-PE (A20980, A20981, S20982) have emission properties superior to those of commercially available Texas Red® tandem conjugates of R-PE. Not only are the Alexa Fluor® 610–R-PE tandem conjugates more fluorescent than the commercially available Texas Red®–R-PE tandem conjugates, but also the fluorescence emission of Alexa Fluor® 610–R-PE tandem conjugates is shifted to somewhat longer wavelengths than is the emission of Texas Red®–R-PE conjugates, resulting in better separation from the emission of R-PE (Figure 6.4.3). Our Alexa Fluor® 647 (A20990, A20991, S20992) and Alexa Fluor® 680 (A20983, A20984, S20985) tandem conjugates of R-PE have emission spectra almost identical to those of Cy®5–R-PE and Cy®5.5–R-PE tandem conjugates but tend to have more intense long-wavelength emission and to require less compensation in the R-PE channel than the Cy® dye–R-PE tandem conjugates (Figure 6.4.4, Figure 6.4.5). Our Alexa Fluor® 750 tandem conjugate of R-PE (S32363) emits at 771 nm.

We have also conjugated APC to our Alexa Fluor® 680 (A21000, A21001MP, S21002), Alexa Fluor® 700 (A21005) and Alexa Fluor® 750 (A21006, S21008) dyes to provide tandem conjugates that can be excited by the He-Ne laser at 633 nm or by the krypton-ion laser at 647 nm. These Alexa Fluor® dye–APC tandem conjugates can potentially be combined with direct APC conjugates for simultaneous three- or four-color applications (Figure 6.4.6).

As the absorption and emission maxima of the acceptor dye move to longer wavelengths, the energy transfer efficiency from the R-PE to the

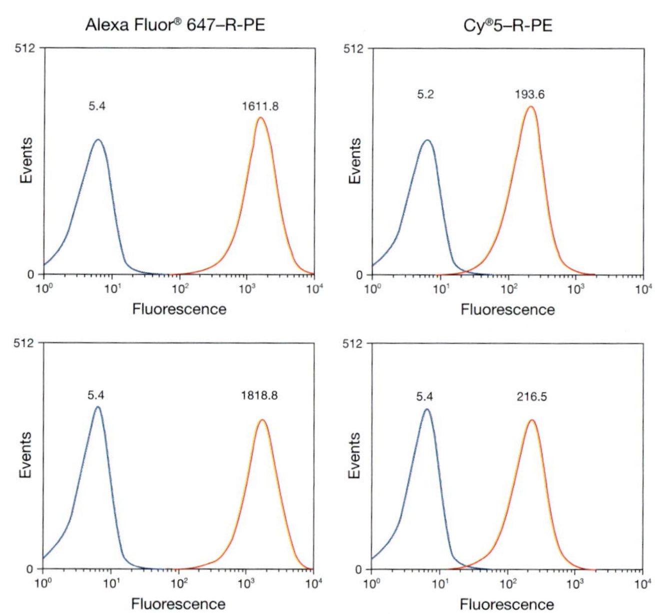

Figure 6.4.5 Comparison of immunofluorescent staining by R-phycoerythrin–dye tandem conjugates. EL4 cells labeled with a biotinylated anti-CD44 monoclonal antibody were detected with streptavidin conjugates of Alexa Fluor® 647–R-PE (S20992) or Cy®5–R-PE (Serotec). The cells were analyzed by flow cytometry on a Coulter® XL cytometer using excitation at 488 nm. Data were obtained using an bandpass emission filter (675 ± 20 nm; upper panels) or a longpass emission filter (>650 nm; lower panels). In each histogram, unstained and stained cells are represented by the blue and red lines, respectively. The numbers above each peak represent mean channel fluorescence intensities. Data provided by William Telford, NCI-NIH, Bethesda, MD.

IMPORTANT NOTICE: The products described in this manual are covered by one or more Limited Use Label License(s). Please refer to the Appendix on page 971 and Master Product List on page 975. Products are For Research Use Only. Not intended for any animal or human therapeutic or diagnostic use.

bound dyes tends to decrease; also, the quantum yields of the longer-wavelength acceptor dyes in the tandem conjugates tend to be lower than those of the shorter-wavelength dyes and to decrease further at high degrees of substitution. Consequently, the preparation of tandem conjugates necessarily involves careful optimization of both the energy transfer efficiency from the R-PE to the longer-wavelength–emitting acceptor dye and the total brightness of the tandem conjugate. For our Alexa Fluor® 647 and Alexa Fluor® 680 tandem conjugates of R-PE, the energy transfer efficiency from R-PE to the attached dye is about 99% and 98%, respectively, as determined from their fluorescence at 575 nm relative to unconjugated R-PE. The residual signal that overlaps the unquenched R-PE emission can be compensated by methods familiar to flow cytometrists.

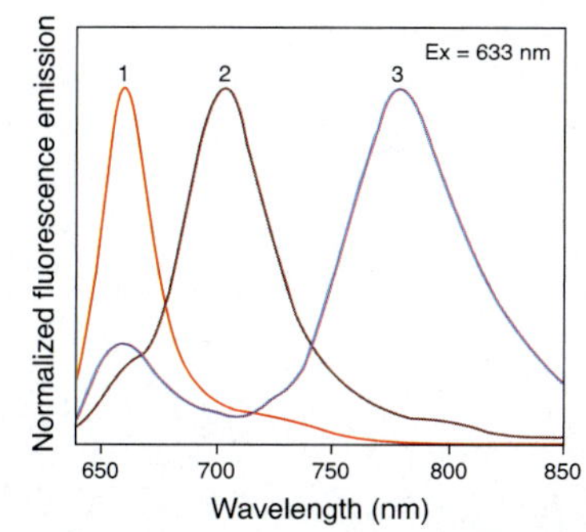

Figure 6.4.6 Normalized fluorescence emission spectra of **1)** allophycocyanin goat anti–mouse IgG antibody (A865), **2)** Alexa Fluor® 680–allophycocyanin goat anti–mouse IgG antibody (A21000) and **3)** Alexa Fluor® 750–allophycocyanin goat anti–mouse IgG antibody (A21006). The tandem conjugates permit simultaneous multicolor labeling and detection of up to three targets with excitation by a single excitation source—the 633 nm spectral line of the He-Ne laser.

Pure Phycobiliproteins

We were the first company to make the phycobiliproteins available for research, and we can supply bulk quantities of B-PE (P800), R-PE (P801), APC (A803) and chemically crosslinked APC (A819); more information is available at www.invitrogen.com/handbook/custombioconjugates.

Phycobiliproteins may undergo some loss of fluorescence upon freezing. The pure proteins are shipped in an ammonium sulfate suspension and are stable for at least one year when stored at 4°C. The conjugates and modified derivatives are shipped in solutions containing sodium azide to inhibit bacterial growth and typically have a useful life of more than six months. *All phycobiliproteins and their derivatives should be stored refrigerated, never frozen.*

Phycobiliprotein Conjugates

Reactive Phycobiliprotein Derivative

Conjugates of R-PE with other proteins are generally prepared from the pyridyldisulfide derivative of R-PE (P806). This derivative can be directly reacted with thiolated antibodies, enzymes and other biomolecules to form a disulfide linkage. More commonly, however, the pyridyldisulfide groups in this derivative are first reduced to thiols, which are then reacted with maleimide-derivatized proteins (Figure 6.4.7). Because the pyridyldisulfide derivative of R-PE is somewhat unstable, we recommend using it within three months of receipt. Phycobiliproteins can be conveniently crosslinked to other proteins using the reagents and protocol provided in our Protein–Protein Crosslinking Kit (P6305, Section 5.2).

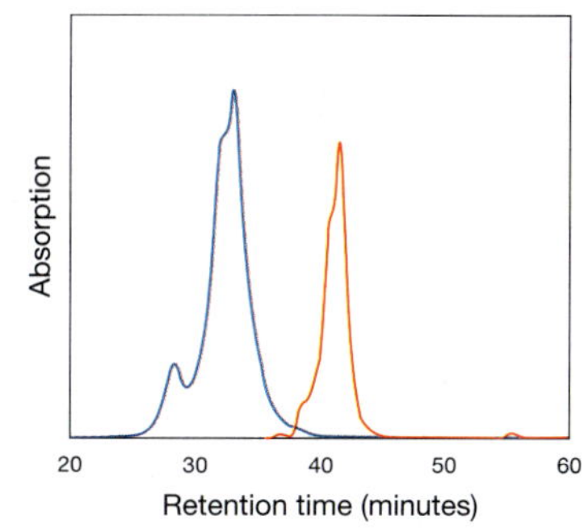

Figure 6.4.8 Analytical size-exclusion chromatograms of free streptavidin (S888; red curve, detected by absorption at 280 nm) and R-phycoerythrin streptavidin (SAPE; S866, S21388; blue curve, detected by absorption at 565 nm), demonstrating that the R-phycoerythrin conjugate is substantially free of unconjugated streptavidin.

Phycobiliprotein-Labeled Secondary Detection Reagents

We prepare R-PE conjugates of the goat anti–mouse IgG (P852) and goat anti–rabbit IgG (P2771MP) antibodies and NeutrAvidin™ biotin-binding protein (A2660), as well as both the R-PE (SAPE, S866) and B-PE (S32350) conjugates of streptavidin. R-PE conjugates of anti–mouse IgG_1, IgG_{2a} and IgG_{2b} antibodies are also available (P21129, P21139, P21149). Our streptavidin conjugates of R-PE and B-PE have been purified to help ensure that all unconjugated streptavidin has been removed (Figure 6.4.8), making them useful for multicolor flow cytometry and

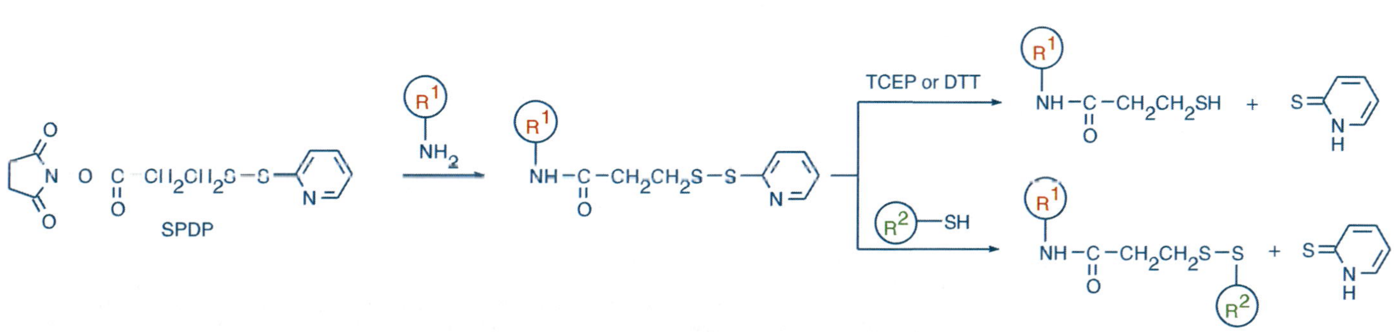

Figure 6.4.7 SPDP derivatization reactions. SPDP (S1531) reacts with an amine-containing biomolecule at pH 7 to 9, yielding a pyridyldithiopropionyl mixed disulfide. The mixed disulfide can then be reacted with a reducing agent such as DTT (D1532) or TCEP (T2556) to yield a 3-mercaptopropionyl conjugate or with a thiol-containing biomolecule to form a disulfide-linked tandem conjugate. Either reaction can be quantitated by measuring the amount of 2-pyridinethione chromophore released during the reaction.

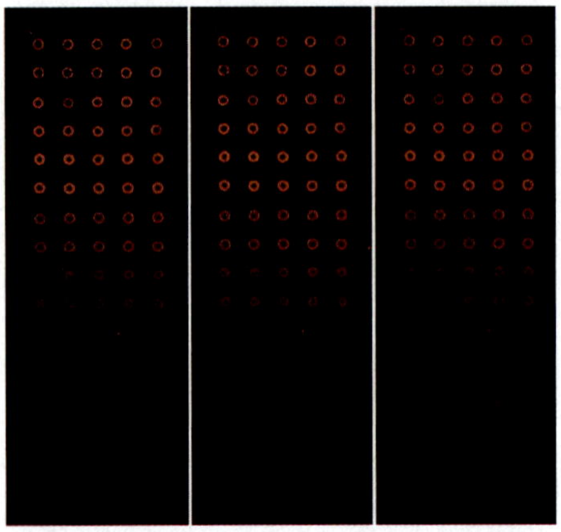

Figure 6.4.9 R-phycoerythrin used to detect DNA on a microarray. A DNA microarray containing a decreasing dilution of calf thymus DNA was hybridized with a biotinylated DNA probe and then incubated with R-phycoerythrin–streptavidin (SAPE; S866, S21388). After washing, the fluorescence signal was detected on a Packard ScanArray® 5000 using three different detection configurations: 488 nm excitation (argon-ion laser)/570 nm emission filter (left); 543.5 nm excitation (He-Ne laser)/570 nm emission filter (middle); 543.5 nm excitation (He-Ne laser)/592 nm emission filter (right).

microarray assays[16–23] (Figure 6.4.9). In addition, biotinylated R-PE (P811) can be used with standard avidin/streptavidin bridging techniques to detect biotinylated molecules.[24]

Because APC tends to dissociate into subunits when highly diluted or treated with chaotropic agents, we prepare all APC conjugates—including APC tandem conjugates, APC conjugates of the goat anti–mouse IgG (A865) and goat anti–rabbit IgG (A10931) antibodies and APC-labeled streptavidin (S868)—from chemically crosslinked APC (A819), a protein complex that does not dissociate even in strongly chaotropic salts.[25–27] We also prepare premium-grade R-PE and APC conjugates of streptavidin (S21388, S32362), which represent an even further fractionation of our R-PE streptavidin (S866) and APC streptavidin (S868), respectively.

Secondary Detection Reagents Labeled with Alexa Fluor® Dye–Phycobiliprotein Tandem Conjugates

We have conjugated R-PE with four of our Alexa Fluor® dyes—Alexa Fluor® 610, Alexa Fluor® 647, Alexa Fluor® 680 and Alexa Fluor® 750 dyes—and then conjugated these fluorescent proteins to antibodies or streptavidin to yield secondary detection reagents that can be excited with the 488 nm spectral line of the argon-ion laser (Table 6.3). The long-wavelength emission maxima are 628 nm for the Alexa Fluor® 610–R-PE conjugates, 668 nm for the Alexa Fluor® 647–R-PE conjugates, 702 nm for the Alexa Fluor® 680–R-PE conjugates (Figure 6.4.2) and 771 nm for the Alexa Fluor® 750–R-PE conjugates. Emission of the Alexa Fluor® 610–R-PE conjugates is shifted to longer wavelengths by about 13 nm relative to that of Texas Red® conjugates of R-PE (Figure 6.4.3). This slightly longer-wavelength emission maximum significantly improves the resolution that can be obtained when using the Alexa Fluor® 610–R-PE tandem conjugates in place of Texas Red®–R-PE tandem conjugates for multicolor flow cytometry. The Alexa Fluor® 647–R-PE tandem conjugates have spectra virtually identical to those of Cy®5 conjugates of R-PE but are about three-fold brighter (Figure 6.4.4). These tandem conjugates can potentially be used for simultaneous three-, four- or five-color labeling with a single excitation (Figure 6.4.2, Figure 6.4.10, Figure 6.4.11).

In addition, we have conjugated crosslinked APC (A819) to our Alexa Fluor® 680, Alexa Fluor® 700 and Alexa Fluor® 750 dyes, and then conjugated these fluorescent proteins to antibodies or streptavidin to yield secondary detection reagents that can be excited with the He-Ne laser

Table 6.3 Tandem conjugates of R-phycoerythrin (R-PE).

	Conjugate		
Acceptor Dye (Ex/Em) *	**Anti-Mouse IgG †**	**Anti-Rabbit IgG †**	**Streptavidin**
Alexa Fluor® 610 (565/628)	A20980	A20981	S20982
Alexa Fluor® 647 (565/668)	A20990	A20991	S20992
Alexa Fluor® 680 (565/702)	A20983	A20984	S20985

* Fluorescence excitation and emission maxima, in nm. † Host = goat.

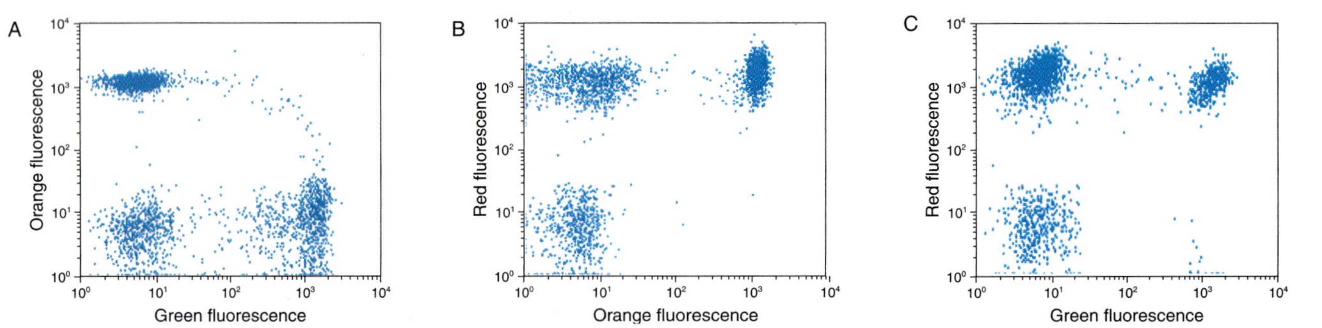

Figure 6.4.10 Simultaneous detection of three cell surface markers using an Alexa Fluor® 610–R-phycoerythrin tandem conjugate, Alexa Fluor® 488 dye and R-phycoerythrin labels. Lymphocytes from ammonium chloride RBC–lysed whole blood were labeled with a biotinylated mouse anti–human CD3 monoclonal antibody, washed with 1% BSA in PBS and then incubated with Alexa Fluor® 610–R-phycoerythrin tandem dye–labeled streptavidin (S20982). Cells were again washed and then labeled with directly conjugated primary antibodies against the CD8 and CD4 markers (Alexa Fluor® 488 dye–labeled mouse anti–human CD8 antibody and R-phycoerythrin–conjugated mouse anti–human CD4 antibody (A21337)). After a further wash in 1% BSA/PBS, labeling was analyzed on a Becton Dickinson FACScan™ flow cytometer using excitation at 488 nm. CD8 was detected in the green channel (525 ± 10 nm), CD4 in the orange channel (575 ± 10 nm) and CD3 in the red channel (>650 nm). The bivariate scatter plots show the expected mutually exclusive populations of CD4 and CD8 positive cells (**A**), together with co-positive CD3/CD4 (**B**) and CD3/CD8 (**C**) populations.

IMPORTANT NOTICE: The products described in this manual are covered by one or more Limited Use Label License(s). Please refer to the Appendix on page 971 and Master Product List on page 975. Products are For Research Use Only. Not intended for any animal or human therapeutic or diagnostic use.
www.invitrogen.com/probes

at 633 nm with emission beyond 700 nm (Table 6.4). The long-wavelength emission maxima are 702 nm for the Alexa Fluor® 680–APC conjugates, 719 nm for the Alexa Fluor® 700–APC conjugates and 779 nm for the Alexa Fluor® 750–APC conjugates (Figure 6.4.2). Our Alexa Fluor® dye–APC tandem conjugates can potentially be combined with direct APC conjugates for simultaneous three-color applications (Figure 6.4.6).

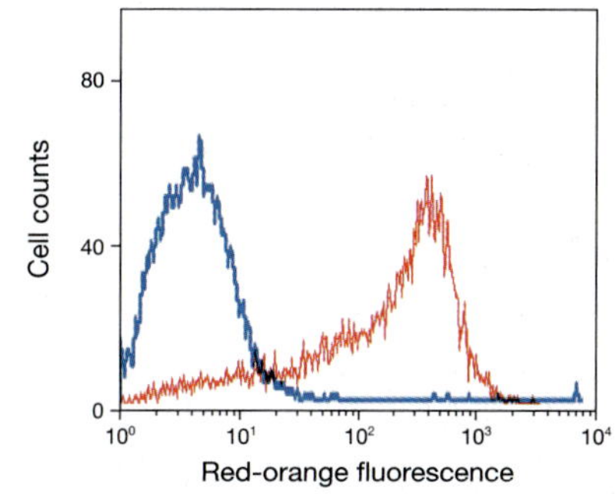

Figure 6.4.12 Color-shifting using a labeled anti–fluorescein/Oregon Green® antibody. Jurkat cells were first stained with a primary mouse anti–human CD3 antibody, followed by fluorescein goat anti–mouse IgG antibody (F2761), with the resultant fluorescence detected in the R-phycoerythrin (red-orange fluorescence) channel of a flow cytometer (blue curve). The weak signal was then shifted to better suit the R-phycoerythrin channel by the addition of an R-phycoerythrin conjugate of anti–fluorescein/Oregon Green® antibody (A21250). The resulting signal intensity is approximately two orders of magnitude greater (red curve) than the direct fluorescence from the first staining step (blue curve).

R-Phycoerythrin Anti–Fluorescein/Oregon Green® Antibody

The R-PE conjugate of the rabbit anti–fluorescein/Oregon Green® antibody (A21250) has the unique ability both to shift the green-fluorescent emission of fluorescein-labeled probes to longer wavelengths and to greatly intensify the signal (Figure 6.4.12). Anti-fluorescein antibodies strongly cross-react with our Oregon Green® dye conjugates, suggesting the possibility of amplifying the signal from nucleic acid probes labeled by our ULYSIS® Oregon Green® 488 Nucleic Acid Labeling Kit (U21659, Section 8.2) or for further amplifying the signal of Oregon Green® 488 tyramide, which is used in one of our TSA™ Kits (Section 6.2, Table 6.1).

Phycobiliprotein Conjugates of Annexin V

In collaboration with Nexins Research BV, we offer the highly fluorescent APC and R-PE conjugates of annexin V (A35110, A35111), in addition to several other fluorescent annexin V conjugates (Section 15.5). Highly fluorescent annexin V conjugates provide quick and reliable detection methods for studying the externalization of phosphatidylserine, an indicator of intermediate stages of apoptosis (Section 15.5).[28] Several of Molecular Probes® apoptosis assay kits (V35112, V35113, V35114; Section 15.5; Table 15.4) contain either R-PE–annexin V or APC–annexin V conjugates as well as SYTOX® Green nucleic acid stain to characterize mixed populations of apoptotic and non-apoptotic cells by flow cytometry.

Custom Phycobiliprotein Conjugates

We have carried out hundreds of successful conjugations with phycobiliproteins, beginning soon after their use was disclosed in 1982.[6] We are experts in doing custom conjugations of phycobiliproteins to antibodies and other proteins and welcome inquiries for specific conjugates (www.invitrogen.com/handbook/custombioconjugates).

Table 6.4 Tandem conjugates of allophycocyanin (APC).

	Conjugate		
Acceptor Dye (Ex/Em) *	**Anti-Mouse IgG †**	**Anti-Rabbit IgG †**	**Streptavidin**
Alexa Fluor® 680 (650/702)	A21000	A21001MP	S21002
Alexa Fluor® 700 (650/719)			S21005
Alexa Fluor® 750 (650/779)	A21006		S21008

* Fluorescence excitation and emission maxima, in nm. † Host = goat.

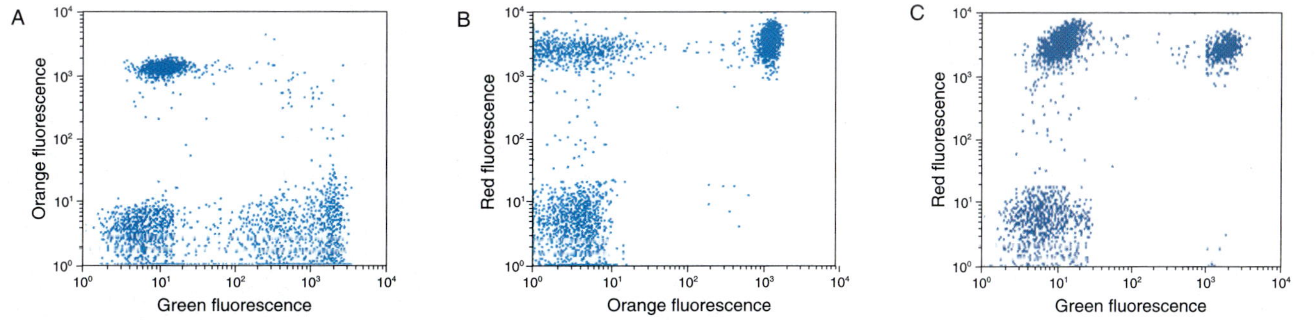

Figure 6.4.11 Simultaneous detection of three cell surface markers using an Alexa Fluor® 647–R-phycoerythrin tandem conjugate, Alexa Fluor® 488 dye and R-phycoerythrin labels. Lymphocytes from ammonium chloride RBC–lysed whole blood were labeled with a mouse anti–human CD3 monoclonal antibody, washed with 1% BSA in PBS and then incubated with a goat anti–mouse IgG antibody labeled with the Alexa Fluor® 647–R-phycoerythrin tandem dye (A20990). Cells were again washed and then labeled with directly conjugated primary antibodies against the CD8 and CD4 markers (Alexa Fluor® 488 dye–labeled mouse anti–human CD8 antibody and R-phycoerythrin–conjugated mouse anti–human CD4 antibody). After a further wash in 1% BSA/PBS, labeling was analyzed on a Becton Dickinson FACScan™ flow cytometer using excitation at 488 nm. CD8 was detected in the green channel (525 ± 10 nm), CD4 in the orange channel (575 ± 10 nm) and CD3 in the red channel (>650 nm). The bivariate scatter plots show the expected mutually exclusive populations of CD4 and CD8 positive cells (**A**), together with co-positive CD3/CD4 (**B**) and CD3/CD8 (**C**) populations.

IMPORTANT NOTICE: The products described in this manual are covered by one or more Limited Use Label License(s). Please refer to the Appendix on page 971 and Master Product List on page 975. Products are For Research Use Only. Not intended for any animal or human therapeutic or diagnostic use.

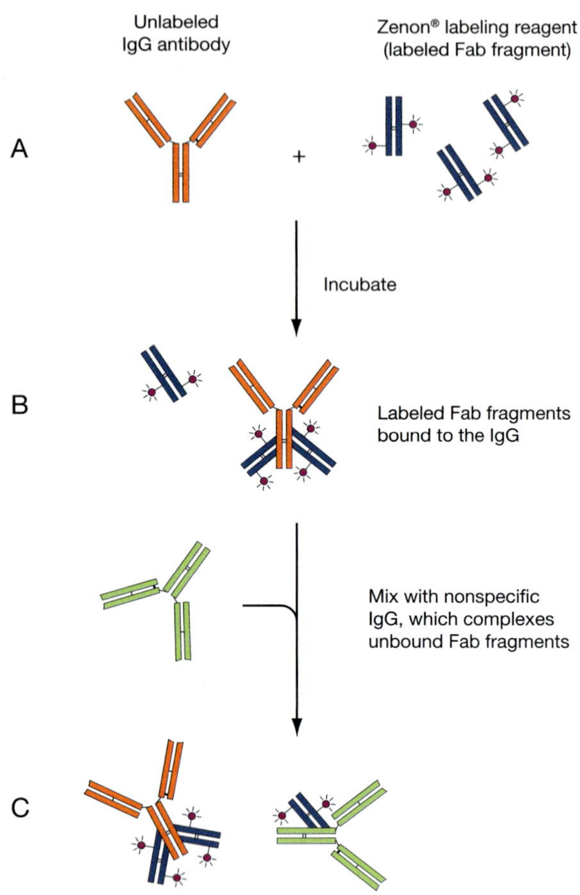

Figure 6.4.13 Labeling scheme utilized in the Zenon® Antibody Labeling Kits. An unlabeled IgG antibody is incubated with the Zenon® labeling reagent, which contains a fluorophore-labeled, Fc-specific anti-IgG Fab fragment (**A**). This labeled Fab fragment binds to the Fc portion of the IgG antibody (**B**). Excess Fab fragment is then neutralized by the addition of a nonspecific IgG (**C**), preventing crosslabeling by the Fab fragment in experiments where primary antivbodies of the same type are present. Note that the Fab fragment used for labeling need not be coupled to a fluorophore, but could instead be coupled to an enzyme (such as HRP) or to biotin.

Zenon® Antibody Labeling Technology

Chemical conjugation of phycobiliproteins to antibodies and other proteins is a moderately difficult and relatively low-yield process that cannot be done on very small quantities of proteins. The Protein–Protein Crosslinking Kit (P6305, Section 5.2) provides the reagents and a protocol for conjugating phycobiliproteins using our recommended procedure. However, instead of labeling each primary antibody, researchers typically use labeled secondary antibodies to detect their primary antibodies. Our exceptional Zenon® immunolabeling technology (Section 7.3) provides an easy, versatile and unique method of labeling antibodies with phycobiliproteins, as well as with many other premier dyes, haptens and enzymes. This enabling technology not only eliminates the need for secondary detection reagents in many applications, but also simplifies immunolabeling applications that were previously time consuming or impractical, including the use of multiple antibodies derived from the same species in the same protocol, as well as the detection of antibody binding in tissues when both the antibody and the tissue are derived from the same species.

Zenon® immunolabeling technology allows the rapid and quantitative preparation of antibody complexes from a purified antibody fraction or from a crude antibody preparation such as serum, ascites fluid or a hybridoma supernatant. The Zenon® antibody labeling procedure (Figure 6.4.13) has numerous advantages, particularly when preparing phycobiliprotein-labeled antibodies:

- Conjugations can be done on submicrogram quantities of a primary antibody.
- The reactions are usually quantitative with respect to the primary antibody.
- Labeling and purification of the complex can be completed in only minutes.
- Labeling is essentially irreversible under conditions of use.
- Multiple antibodies derived from the same species can be used in the same experiment.
- The fluorescence intensity of the cells can be adjusted by changing the ratio of labeling reagent to primary antibody, which even permits using identical dyes to detect multiple targets in cells by flow cytometry.
- Antibody complexes prepared from the Zenon® Antibody Labeling Kits can be combined with direct conjugates for multicolor labeling.
- Labeling is possible with a wide variety of fluorophores, including R-PE and APC as well as most of our Alexa Fluor® dyes (Table 7.7).
- Zenon® Antibody Labeling Kits with Alexa Fluor® dye–phycobiliprotein tandem conjugates are also available (Table 7.7), increasing the possible combinations of detection wavelengths in a multicolor experiment.

REFERENCES

1. J Fluorescence (1991) 1:135; **2.** J Biol Chem (1989) 264:1; **3.** Methods Enzymol (1988) 167:291; **4.** Methods Mol Biol (2003) 224:55; **5.** Proc Natl Acad Sci U S A (1988) 85:7312; **6.** J Cell Biol (1982) 93:981; **7.** Proc Natl Acad Sci U S A (1989) 86:4087; **8.** Anal Chem (1987) 59:2158; **9.** Anal Biochem (1987) 161:442; **10.** J Histochem Cytochem (1991) 39:921; **11.** Anal Lett (1991) 24:1075; **12.** Eur Biophys J (1987) 15:141; **13.** Clin Chem (1983) 29:1582; **14.** J Immunol Methods (1990) 135:247; **15.** J Immunol Methods (2000) 243:77; **16.** Biotechniques (2001) 31:490; **17.** Proc Natl Acad Sci U S A (2001) 98:8862; **18.** Proc Natl Acad Sci U S A (2000) 97:3260; **19.** Proc Natl Acad Sci U S A (1998) 95:3752; **20.** Proc Natl Acad Sci U S A (2000) 97:2680; **21.** Anal Biochem (1998) 255:188; **22.** J Biol Chem (2000) 275:11181; **23.** Nat Biotechnol (1996) 14:1675; **24.** J Biol Chem (1990) 265:15776; **25.** Arch Biochem Biophys (1983) 223:24; **26.** Biochemistry (1980) 19:2817; **27.** Cytometry (1987) 8:91; **28.** Nat Protoc (2009) 4:1383.

IMPORTANT NOTICE: The products described in this manual are covered by one or more Limited Use Label License(s). Please refer to the Appendix on page 971 and Master Product List on page 975. Products are For Research Use Only. Not intended for any animal or human therapeutic or diagnostic use.
www.invitrogen.com/probes

PRODUCT LIST 6.4 PHYCOBILIPROTEINS

Cat. No.	Product	Quantity
A20980	Alexa Fluor® 610–R-phycoerythrin goat anti–mouse IgG (H+L) *1 mg/mL*	100 µL
A20981	Alexa Fluor® 610–R-phycoerythrin goat anti–rabbit IgG (H+L) *1 mg/mL*	100 µL
A20990	Alexa Fluor® 647–R-phycoerythrin goat anti–mouse IgG (H+L) *1 mg/mL*	100 µL
A20991	Alexa Fluor® 647–R-phycoerythrin goat anti–rabbit IgG (H+L) *1 mg/mL*	100 µL
A21000	Alexa Fluor® 680–allophycocyanin goat anti–mouse IgG (H+L) *1 mg/mL*	100 µL
A21001MP	Alexa Fluor® 680–allophycocyanin goat anti–rabbit IgG (H+L) *1 mg/mL*	100 µL
A20983	Alexa Fluor® 680–R-phycoerythrin goat anti–mouse IgG (H+L) *1 mg/mL*	100 µL
A20984	Alexa Fluor® 680–R-phycoerythrin goat anti–rabbit IgG (H+L) *1 mg/mL*	100 µL
A21006	Alexa Fluor® 750–allophycocyanin goat anti–mouse IgG (H+L) *1 mg/mL*	100 µL
A803	allophycocyanin *4 mg/mL*	0.5 mL
A819	allophycocyanin, crosslinked (APC-XL) *4 mg/mL*	250 µL
A865	allophycocyanin, crosslinked, goat anti–mouse IgG (H+L) *1 mg/mL*	0.5 mL
A10931	allophycocyanin, crosslinked, goat anti–rabbit IgG (H+L) *1 mg/mL*	0.5 mL
A35110	annexin V, allophycocyanin conjugate (APC annexin V) *50 assays*	250 µL
A35111	annexin V, R-phycoerythrin conjugate (R-PE annexin V) *50 assays*	250 µL
A21250	anti–fluorescein/Oregon Green®, rabbit IgG fraction, R-phycoerythrin conjugate *2 mg/mL*	250 µL
A2660	avidin, NeutrAvidin™, R-phycoerythrin conjugate *1 mg/mL*	1 mL
P800	B-phycoerythrin *4 mg/mL*	0.5 mL
P801	R-phycoerythrin *4 mg/mL*	0.5 mL
P811	R-phycoerythrin, biotin-XX conjugate *4 mg/mL*	0.5 mL
P852	R-phycoerythrin goat anti–mouse IgG (H+L) *1 mg/mL*	1 mL
P21129	R-phycoerythrin goat anti–mouse IgG_1 (γ1) conjugate *1 mg/mL*	250 µL
P21139	R-phycoerythrin goat anti–mouse IgG_{2a} (γ2a) conjugate *1 mg/mL*	250 µL
P21149	R-phycoerythrin goat anti–mouse IgG_{2b} (γ2b) conjugate *1 mg/mL*	250 µL
P2771MP	R-phycoerythrin goat anti–rabbit IgG (H+L) *1 mg/mL*	0.5 mL
P806	R-phycoerythrin, pyridyldisulfide derivative *2 mg/mL*	1 mL
S20982	streptavidin, Alexa Fluor® 610–R-phycoerythrin conjugate (Alexa Fluor® 610–R-phycoerythrin streptavidin) *1 mg/mL*	100 µL
S20992	streptavidin, Alexa Fluor® 647–R-phycoerythrin conjugate (Alexa Fluor® 647–R-phycoerythrin streptavidin) *1 mg/mL*	100 µL
S21002	streptavidin, Alexa Fluor® 680–allophycocyanin conjugate (Alexa Fluor® 680–allophycocyanin streptavidin) *1 mg/mL*	100 µL
S20985	streptavidin, Alexa Fluor® 680–R-phycoerythrin conjugate (Alexa Fluor® 680–R-phycoerythrin streptavidin) *1 mg/mL*	100 µL
S21005	streptavidin, Alexa Fluor® 700–allophycocyanin conjugate (Alexa Fluor® 700–allophycocyanin streptavidin) *1 mg/mL*	100 µL
S21008	streptavidin, Alexa Fluor® 750–allophycocyanin conjugate (Alexa Fluor® 750–allophycocyanin streptavidin) *1 mg/mL*	100 µL
S32363	streptavidin, Alexa Fluor® 750–R-phycoerythrin conjugate (Alexa Fluor® 750–R-phycoerythrin streptavidin) *1 mg/mL*	100 µL
S32362	streptavidin, allophycocyanin conjugate *premium grade* *1 mg/mL*	250 µL
S868	streptavidin, allophycocyanin, crosslinked, conjugate *1 mg/mL*	0.5 mL
S32350	streptavidin, B-phycoerythrin conjugate *1 mg/mL*	1 mL
S866	streptavidin, R-phycoerythrin conjugate (SAPE) *1 mg/mL*	1 mL
S21388	streptavidin, R-phycoerythrin conjugate (SAPE) *premium grade* *1 mg/mL*	1 mL
Z25020	Zenon® Alexa Fluor® 610–R-Phycoerythrin Mouse IgG_1 Labeling Kit *10 labelings*	1 kit
Z25021	Zenon® Alexa Fluor® 647–R-Phycoerythrin Mouse IgG_1 Labeling Kit *10 labelings*	1 kit
Z25121	Zenon® Alexa Fluor® 647–R-Phycoerythrin Mouse IgG_{2a} Labeling Kit *10 labelings*	1 kit
Z25221	Zenon® Alexa Fluor® 647–R-Phycoerythrin Mouse IgG_{2b} Labeling Kit *10 labelings*	1 kit
Z25022	Zenon® Alexa Fluor® 680–R-Phycoerythrin Mouse IgG_1 Labeling Kit *10 labelings*	1 kit
Z25030	Zenon® Alexa Fluor® 700–Allophycocyanin Mouse IgG_1 Labeling Kit *10 labelings*	1 kit
Z25031	Zenon® Alexa Fluor® 750–Allophycocyanin Mouse IgG_1 Labeling Kit *10 labelings*	1 kit
Z25451	Zenon® Allophycocyanin Human IgG Labeling Kit *25 labelings*	1 kit
Z25051	Zenon® Allophycocyanin Mouse IgG_1 Labeling Kit *25 labelings*	1 kit
Z25151	Zenon® Allophycocyanin Mouse IgG_{2a} Labeling Kit *25 labelings*	1 kit
Z25251	Zenon® Allophycocyanin Mouse IgG_{2b} Labeling Kit *25 labelings*	1 kit
Z25351	Zenon® Allophycocyanin Rabbit IgG Labeling Kit *25 labelings*	1 kit
Z25455	Zenon® R-Phycoerythrin Human IgG Labeling Kit *25 labelings*	1 kit
Z25055	Zenon® R-Phycoerythrin Mouse IgG_1 Labeling Kit *25 labelings*	1 kit
Z25155	Zenon® R-Phycoerythrin Mouse IgG_{2a} Labeling Kit *25 labelings*	1 kit
Z25255	Zenon® R-Phycoerythrin Mouse IgG_{2b} Labeling Kit *25 labelings*	1 kit
Z25355	Zenon® R-Phycoerythrin Rabbit IgG Labeling Kit *25 labelings*	1 kit

IMPORTANT NOTICE: The products described in this manual are covered by one or more Limited Use Label License(s). Please refer to the Appendix on page 971 and Master Product List on page 975. Products are For Research Use Only. Not intended for any animal or human therapeutic or diagnostic use.

6.5 Microspheres

Figure 6.5.1 A photomicrograph of a multicolor mixture of Molecular Probes® FluoSpheres® fluorescent microspheres overlaid with a differential interference contrast (DIC) image of the same field. Molecular Probes® proprietary fluorescent dye technology produces a range of intensely fluorescent FluoSpheres® microspheres labeled with biotin, streptavidin and NeutrAvidin™ biotin-binding protein, providing important tools for improving the sensitivity of flow cytometry applications and immunodiagnostic assays.

Although low molecular weight reactive dyes are versatile and easy to use, they are not without limitations. For example, the fluorescence output of the dye–biomolecule conjugate is often limited by the number of dyes that can be attached to the biomolecule without disrupting its function. Our highly fluorescent microspheres—both the FluoSpheres® (Figure 6.5.1) and TransFluoSpheres® (Figure 6.5.2) beads—provide a means of overcoming this limitation. Moreover, our TransFluoSpheres® beads are designed to facilitate multicolor detection, particularly in applications that use lasers with their inherent limited number of excitation wavelengths. TransFluoSpheres® beads contain a series of two or more proprietary dyes that have been carefully chosen to ensure excited-state energy transfer between the dyes. This strategy enables fine-tuning of both the excitation and emission wavelengths of the microspheres so that they match a particular instruments, excitation source and detection sensitivity, and complement the spectra of other fluorophores in a multicolor experiment.

Properties of Molecular Probes® Fluorescent and Nonfluorescent Microspheres

Fluorescent FluoSpheres® and TransFluoSpheres® Microspheres

Molecular Probes® FluoSpheres® and TransFluoSpheres® beads are manufactured using high-quality, ultraclean polystyrene microspheres. These microspheres are internally labeled with Molecular Probes® dyes, making them among the brightest fluorescent microspheres available (Table 6.5). We employ methods to help ensure that each bead is heavily loaded with dye. The protective environment within the bead matrix shields the dyes from many of the environmental effects that cause photobleaching of exposed fluorophores. Not only are our yellow-green–fluorescent beads more photostable, but their emission spectra are not affected by changes in pH, as are conventional fluorescein-labeled microspheres. Fluorescent microspheres can be fixed in formalin and embedded in paraffin if care is used to avoid extraction of the noncovalently associated dyes from the microspheres.[1] The stability, uniformity and reproducibility of our fluorescent beads, as well as the extensive selection of colors available, make our microspheres the preferred tools for research and diagnostic assays that use fluorescence.

In addition to the microspheres described in this section, we have developed several important microsphere-based products for calibrating and aligning fluorescence microscopes (Section 23.1) and flow cytometers (Section 23.2). Custom preparation of microspheres with other colors, sizes and surface coatings is available; please go to www.invitrogen.com/handbook/customparticles for more information.

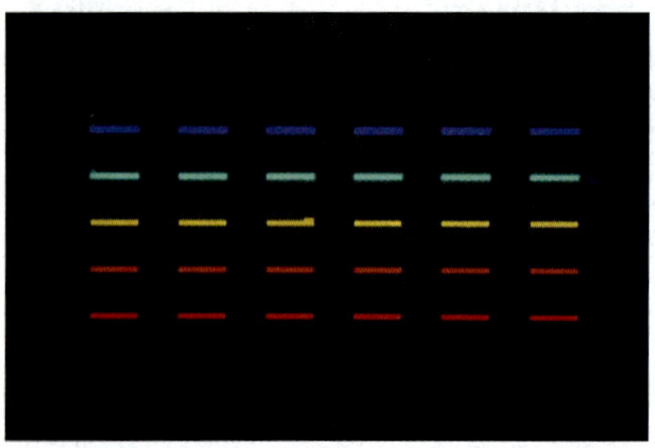

Figure 6.5.2 A positively charged membrane containing an approximately equal number of TransFluoSpheres® fluorescent microspheres per slot. Our proprietary TransFluoSpheres® polystyrene beads are designed to be excited with a common wavelength and then detected at a variety of longer wavelengths with minimal spectral overlap. This nylon membrane was excited with 365 nm epi-illumination and photographed through a 400 nm longpass optical filter.

Table 6.5 Molecular Probes® yellow-green–fluorescent FluoSpheres® beads compared with other commercially available yellow-green–fluorescent microspheres.

Brand/Supplier	Size (µm)	Fluorescence Intensity *	CV for Intensity †
Molecular Probes® F8852	1.02	1998	4.40%
Molecular Probes® F8853	2.07	8998	3.26%
Company A	0.84	3.7	30.28%
Company A	1.55	5.2	11.69%
Company B	1.01	12.6	2.49%
Company B	1.94	595	2.91%
Company C	0.93	116	4.62%
Company C	1.48	434	1.92%
Company D	0.85	17	5.19%
Company D	1.84	119	3.08%

* Median value for fluorescence intensity (in arbitrary units), measured for 10,000 individual beads per sample excited at 488 nm using flow cytometry. Values may vary slightly between batches of these products. † CV = coefficient of variation.

IMPORTANT NOTICE: The products described in this manual are covered by one or more Limited Use Label License(s). Please refer to the Appendix on page 971 and Master Product List on page 975. Products are For Research Use Only. Not intended for any animal or human therapeutic or diagnostic use.
www.invitrogen.com/probes

Colored and Unstained Microspheres

We also offer a wide selection of colored and unstained microspheres for research applications as well as for water- and air-flow testing and other bead-based applications. Through the acquisition of Interfacial Dynamics Corporation (IDC), we provide milliliter to 500-liter quantities of ultraclean microspheres with diameters from 20 nm to 10.0 µm and with more than 20 different surface functionalities; see www.invitrogen.com/handbook/latex for a complete listing of the microspheres available. IDC pioneered the commercial development of surfactant-free polymer particles used in bead-based assay systems and has been a key supplier of beads for Molecular Probes® fluorescent microspheres. Our outstanding capabilities in microsphere manufacturing allow a high level of control of the colloid engineering employed in the particle synthesis, providing excellent batch-to-batch consistency. Other microsphere manufacturers often use surfactants to prevent aggregation. However, standards of surfactant purity are generally not very high, leading to an undefined particle surface and variable protein attachment. In the manufacturing of our ultraclean microspheres, no surfactants are required to prevent aggregation, taking much of the guesswork out of stability and adsorption experiments. We can tailor-make colored and unstained microspheres of many sizes, surface chemistries, densities and volumes to meet the diverse needs of customers, including academic, industrial and government laboratories, as well as major global diagnostic companies; please go to www.invitrogen.com/handbook/customparticles for more information.

Applications for Fluorescent Microspheres

Fluorescent microspheres have been used as immunofluorescent reagents,[2–4] retrograde neuronal tracers,[5–7] microinjectable cell tracers[8,9] (Section 14.6) and standardization reagents for microscopy (Section 23.1) and flow cytometry[10] (Section 23.2). Suspension arrays of fluorescent microspheres that differ in intensity, size or excited-state lifetime are extensively used for simultaneous assays to determine multiple analytes in a single sample.[11–18] In addition, fluorescent microspheres are potentially more sensitive than colorimetric methods in most, if not all, of the major microsphere-based test systems presently in use, including latex-agglutination tests, filter-separation tests, particle-capture ELISA methods and two-particle sandwich techniques. Applications of our fluorescent microspheres for assessing tissue blood perfusion[19] are described in Section 14.6.

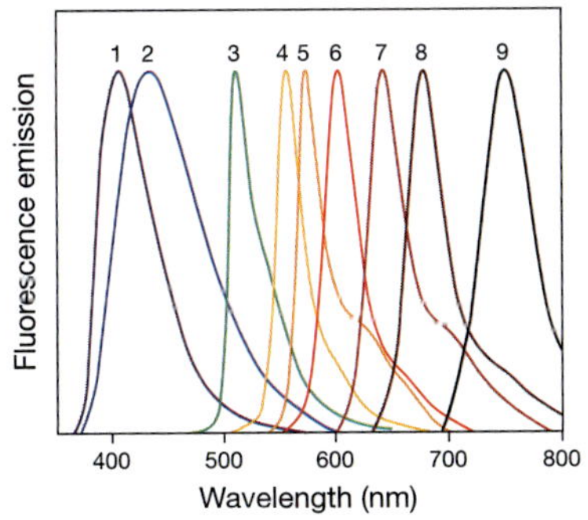

Figure 6.5.3 Normalized fluorescence emission spectra of our FluoSpheres® beads, named according to their excitation/emission maxima (nm): **1)** blue (365/415), **2)** blue (350/440), **3)** yellow-green (505/515), **4)** orange (540/560), **5)** red-orange (565/580), **6)** red (580/605), **7)** crimson (625/645), **8)** dark red (660/680) and **9)** infrared (715/755) FluoSpheres® beads.

FluoSpheres® Fluorescent Microspheres

A Wide Array of Fluorescent Colors

Molecular Probes® FluoSpheres® fluorescent microspheres contain dyes with excitation and emission wavelengths that cover the entire spectrum from the near UV to the near-infrared. Figure 6.5.3 shows the normalized emission spectra for 9 fluorescent colors of FluoSpheres® beads. Because long-wavelength (>680 nm) light can penetrate tissues, our infrared-fluorescent microspheres may allow researchers to conduct experiments that were not previously possible with beads that emit at shorter wavelengths. We would like to highlight the following FluoSpheres® products:

- Blue-fluorescent FluoSpheres® beads with excitation/emission maxima of 350/440 nm contain an improved blue-fluorescent dye that has exceptional brightness and a long shelf life. We also offer blue-fluorescent FluoSpheres® beads with slightly shorter-wavelength fluorescence spectra (excitation/emission maxima ~365/415 nm).
- Yellow-green–fluorescent FluoSpheres® beads have excitation/emission maxima of 505/515 nm and thus are excited very efficiently using the 488 nm spectral line of the argon-ion laser, resulting in exceptionally intense fluorescence (Table 6.5).
- Orange-, red-orange– and red-fluorescent FluoSpheres® beads have excitation maxima of 540 nm, 565 nm and 580 nm, respectively.
- Nile red–fluorescent FluoSpheres® beads have broad excitation/emission bandwidths at 535/575 nm, making them compatible with filter sets appropriate for fluorescein, rhodamine and Texas Red® dyes.
- Crimson- and dark red–fluorescent FluoSpheres® beads with excitation/emission maxima of 625/645 nm and 660/680 nm, respectively, are efficiently excited by the 633 nm spectral line of the He-Ne laser. Although the dark red–fluorescent beads are significantly less fluorescent than the crimson-fluorescent particles, they fluoresce at wavelengths that are longer than, and clearly distinguishable from, those of the crimson-fluorescent particles.
- Infrared-fluorescent FluoSpheres® beads with excitation/emission maxima of 715/755 nm are the longest-wavelength fluorescent microspheres currently available from any source. These beads absorb and emit at wavelengths at which most tissues are almost optically transparent.
- Europium luminescent and platinum luminescent FluoSpheres® beads have excitation/emission maxima of 340–370/610 nm (Figure 6.5.4) and ~390/650 nm (Figure 6.5.5), respectively, and decay times of >40 microseconds for the platinum microspheres and >100 microseconds for the europium microspheres, far longer than those of conventional fluorescent probes and autofluorescent samples. These beads should be useful as standards for time-resolved microscopy and for tracing applications in highly autofluorescent samples.

Table 6.6 Fluorescein equivalents in Molecular Probes® yellow-green–fluorescent FluoSpheres® beads.

Microsphere Diameter (µm)	Fluorescein Equivalents per Microsphere *
0.02	1.8×10^2
0.04	3.5×10^2
0.1	7.4×10^3
0.2	1.1×10^5
0.5	2.0×10^6
1.0	1.3×10^7
2.0	3.1×10^7
10	1.1×10^{10}
15	3.7×10^{10}

* Values may vary slightly between batches of these products. These data were obtained by taking known numbers of beads and totally dissolving the polystyrene to release the dye. The fluorescence of the released dye was then compared to a fluorescein standard curve. The relationship of fluorescence (fluorescein equivalents) to bead size scales as the cube of the bead radius, as it should for a spherical bead where the dye fills the full volume (as opposed to only being on the surface).

Our FluoSpheres® beads are many times brighter than fluorescent microspheres from other companies (Table 6.5). Table 6.6 shows the approximate number of unquenched fluorescein equivalents in our yellow-green–fluorescent FluoSpheres® beads. The intensity of the beads is sufficient to allow visualization of single particles, even for our smallest microspheres, which appear as point sources; see the description of the PS-Speck™ Microscope Point Source Kit (P7220) in Section 23.1. Our FluoSpheres® beads show little or no photobleaching, even when excited with the intense illumination required for fluorescence microscopy.

Although some of our FluoSpheres® beads are available in limited sizes, colors and surface functions, we will prepare custom orders upon request (www.invitrogen.com/handbook/customparticles). We also have considerable experience developing standards, including microsphere-based standards for companies selling fluorescence instrumentation. FluoSpheres® beads can also be prepared with intensities that are *lower* than those of our regular products—a desirable feature in some multicolor applications. FluoSpheres® beads with calibrated intensities are already available in our InSpeck™ Microscope Intensity Calibration Kits (Section 23.1) and LinearFlow™ Flow Cytometry Intensity Calibration Kits (Section 23.2).

A Wide Range of Sizes

To meet the diverse needs of academic and industry laboratories, we offer FluoSpheres® beads in a variety of sizes (Table 6.7). The smallest microspheres are currently about 0.02 µm in diameter, with a coefficient of variation (CV) of about 20% as determined by electron microscopy. The size uniformity improves with increasing size, with the CV decreasing from ~5% for 0.1 µm FluoSpheres® beads to ~1% for those with 10–15 µm diameters. The sizes

Table 6.7 Summary of FluoSpheres® fluorescent microspheres.

Microspheres (Ex/Em) *	0.02 µm	0.04 µm	0.1 µm	0.2 µm	0.5 µm	1.0 µm	2.0 µm	4.0 µm
Carboxylate-Modified Microspheres								
Europium luminescent (340–370/610)		F20880 (2 mL)		F20881 (2 mL)				
Platinum luminescent (390/650)		F20886 (2 mL)						
Blue (365/415)	F8781 (10 mL)			F8805 (10 mL)		F8814 (10 mL)	F8824 (2 mL)	
Blue (350/440)			F8797 (10 mL)			F8815 (10 mL)		
Blue-green (430/465)						F13080 ‡ (5 mL)		
Yellow-green (505/515)	F8787 (10 mL)	F8795 † (1 mL)	F8803 (10 mL)	F8811 (10 mL)	F8813 (10 mL)	F8823 (10 mL) F13081 ‡ (5 mL)	F8827 (2 mL)	
Nile red (535/575)	F8784 (10 mL)					F8819 (10 mL)	F8825 (2 mL)	
Orange (540/560)		F8792 † (1 mL)	F8800 (10 mL)	F8809 (10 mL)		F8820 (10 mL) F13082 ‡ (5 mL)		
Red-orange (565/580)		F8794 † (1 mL)						
Red (580/605)	F8786 (10 mL)	F8793 † (1 mL)	F8801 (10 mL)	F8810 (10 mL)	F8812 (10 mL)	F8821 (10 mL) F13083 ‡ (5 mL)	F8826 (2 mL)	
Crimson (625/645)	F8782 (2 mL)			F8806 (2 mL)		F8816 (2 mL)		
Dark red (660/680)	F8783 (2 mL)	F8789 † (1 mL)		F8807 (2 mL)				
Infrared (715/755)			F8799 (1 mL)					
Sulfate Microspheres								
Blue (365/415)								F8854 (2 mL)
Yellow-green (505/515)	F8845 (10 mL)			F8848 (10 mL)		F8852 (10 mL)	F8853 (2 mL)	F8859 (2 mL)
Red (580/605)						F8851 (10 mL)		F8858 (2 mL)
Aldehyde–Sulfate Microspheres								
Yellow-green (505/515)	F8760 (10 mL)					F8762 (10 mL)		
Amine-Modified Microspheres								
Yellow-green (505/515)				F8764 (5 mL)		F8765 (5 mL)		
Red (580/605)				F8763 (5 mL)				

* Excitation (Ex) and emission (Em) maxima in nm. FluoSpheres® beads are supplied as aqueous suspensions containing 2% solids and 2 mM sodium azide, except for the 0.04 µm microspheres (†), which are supplied as aqueous suspensions containing 5% solids without preservatives; the 1 µm microspheres for tracer studies (‡), which are supplied as aqueous suspensions containing 10^{10} microspheres per mL and 0.02% thimerosal; and the europium luminescent microspheres, which are supplied as aqueous suspensions containing 0.5% solids and 2 mM sodium azide. All sizes fall within a narrow range as discussed in Section 6.5. Sizes indicated in the above tables are nominal and may vary from batch to batch. Actual sizes, as determined by electron microscopy, are specified on the product labels.

IMPORTANT NOTICE: The products described in this manual are covered by one or more Limited Use Label License(s). Please refer to the Appendix on page 971 and Master Product List on page 975. Products are For Research Use Only. Not intended for any animal or human therapeutic or diagnostic use.
www.invitrogen.com/probes
molecular probes® | invitrogen™ by Life technologies™

specified in the product names are nominal bead diameters; because of batch-to-batch variation in the undyed microspheres, the actual mean diameters shown on the product labels may differ from the nominal diameters, especially for the smaller microspheres. Because of their small size, 0.02–0.04 µm microspheres are effectively transparent to light in aqueous suspensions and behave very much like true solutions.

Four Different Surface Functional Groups

We prepare FluoSpheres® beads with four different surface functional groups, making them compatible with a variety of conjugation strategies. Our fluorescent dyes have negligible effect on the surface properties of the polystyrene beads or on their protein adsorption. We caution, however, that the surface properties have an important role in the functional utility of the microspheres; we cannot guarantee the suitability of a particular bead type for all applications.

- Carboxylate-modified FluoSpheres® beads have pendent carboxylic acids, making them suitable for covalent coupling of proteins and other amine-containing biomolecules using water-soluble carbodiimide reagents such as EDAC (E2247, Section 3.4). In order to both decrease nonspecific binding and provide additional functional groups for conjugation, these FluoSpheres® beads have a high density of carboxylic acids on their surfaces.
- Sulfate FluoSpheres® beads are relatively hydrophobic particles that will passively and nearly irreversibly adsorb almost any protein, including albumin, IgG, avidin and streptavidin.
- Aldehyde–sulfate FluoSpheres® beads, which are sulfate microspheres that have been modified to add surface aldehyde groups, are designed to react with proteins and other amines under very mild conditions.
- Amine-modified FluoSpheres® beads can be coupled to a wide variety of amine-reactive molecules, including the succinimidyl esters and isothiocyanates of haptens and drugs or the carboxylic acids of proteins, using a water-soluble carbodiimide. The amine surface groups can also be reacted with SPDP (S1531, Section 5.2) to yield (after reduction) microspheres with pendent sulfhydryl groups.

Fluorescent Microspheres Conjugated to Biotin, Avidin and Streptavidin

We offer yellow-green–fluorescent FluoSpheres® microspheres conjugated to biotin or streptavidin, and yellow-green fluorescent, red-fluorescent, europium luminescent and nonfluorescent microspheres conjugated to NeutrAvidin™ biotin-binding protein (Table 6.8). NeutrAvidin™ biotin-binding protein has been specially processed to remove carbohydrates and lower the isoelectric point, resulting in a near-neutral protein that has significantly lower nonspecific binding than conventional avidin. These microsphere conjugates provide valuable tools for improving the sensitivity of flow cytometry applications and immunochemical assays.[2,20] They are also useful as tracers that can be detected with standard enzyme-mediated avidin/streptavidin methods.[21] Additional sizes and colors of these microspheres can be custom-ordered through Invitrogen Custom Services.

Fluorescent Microspheres Coated with Collagen

Fibroblasts phagocytose and then intracellularly digest collagen. These activities play an important role in the remodeling of the extracellular matrix during normal physiological turnover of connective tissues, in development, in wound repair and aging and in various disorders. A well-established procedure for observing collagen phagocytosis by either flow cytometry or fluorescence microscopy involves the use of collagen-coated fluorescent microspheres, which attach to the cell surface and become engulfed by fibroblasts.[22,23] We offer yellow-green–fluorescent FluoSpheres® collagen I–labeled microspheres in either 1.0 µm or 2.0 µm diameter (F20892, F20893) for use in these applications. These microspheres have collagen I from calf skin attached covalently to their surface.

Europium and Platinum Luminescent Microspheres for Time-Resolved Fluorometry

Detecting low levels of protein or DNA targets in a tissue sample or on a membrane using classic fluorochromes is sometimes difficult and prone to errors because specific fluorescence signals tend to be low and are usually mixed with nonspecific signals and autofluorescence. One approach to improve detectability is the use of time-resolved luminescence reagents, such as our FluoSpheres® europium luminescent microspheres and FluoSpheres® platinum luminescent microspheres. The FluoSpheres® europium luminescent beads contain Eu^{3+} coordination complexes with luminescence decay times of >100 microseconds[24]—much longer than the nanosecond decay times of conventional fluorophores and autofluorescence. The luminescence of the Pt^{2+} chelate in the FluoSpheres® platinum luminescent microspheres has a decay time of

Table 6.8 Summary of biotin-, streptavidin- and NeutrAvidin™ biotin-binding protein–labeled FluoSpheres® microspheres.

Microspheres (Ex/Em) *	0.04 µm	0.2 µm	1.0 µm
Biotin-Labeled Microspheres			
Yellow-green (505/515)	F8766 (0.4 mL)	F8767 (0.4 mL)	F8768 (0.4 mL)
Nonfluorescent			F8769 (0.4 mL)
Streptavidin-Labeled Microspheres			
Yellow-green (505/515)	F8780 (0.4 mL)		
NeutrAvidin™-Labeled Microspheres			
Europium luminescent (340–370/610)	F20883 (0.4 mL)	F20884 (0.4 mL)	
Yellow-green (505/515)	F8771 (0.4 mL)	F8774 (0.4 mL)	F8776 (0.4 mL)
Red (580/605)	F8770 (0.4 mL)		F8775 (0.4 mL)
Nonfluorescent	F8772 (0.4 mL)		F8777 (0.4 mL)

* Excitation (Ex) and emission (Em) maxima in nm. The streptavidin-labeled FluoSpheres® beads are supplied as aqueous suspensions containing 0.5% solids and 5 mM sodium azide. The NeutrAvidin™ biotin-binding protein–labeled europium luminescent FluoSpheres® are supplied as aqueous suspensions containing 0.5% solids and 5 mM azide. Other NeutrAvidin™ biotin-binding protein- and biotin-labeled FluoSpheres® beads are supplied as aqueous suspensions containing 1% solids, 5 mM sodium azide and 0.02% Tween®. All sizes fall within a narrow range as discussed in this section. Sizes indicated in the above tables are nominal and may vary from batch to batch. Actual sizes, as determined by electron microscopy, are specified on the product label.

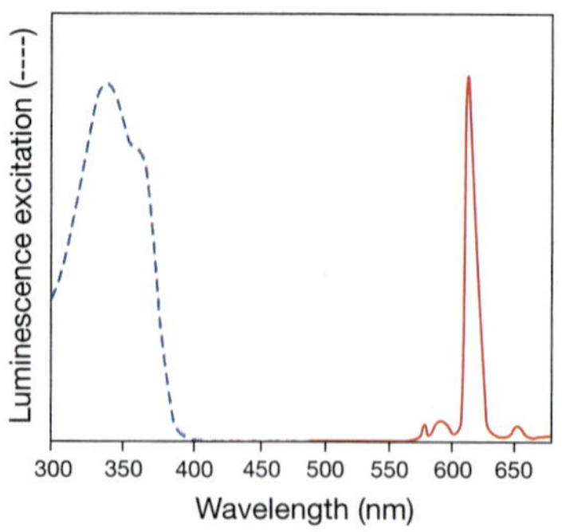

Figure 6.5.4 Luminescence excitation and emission spectra of the FluoSpheres® europium luminescent microspheres (F20880, F20881).

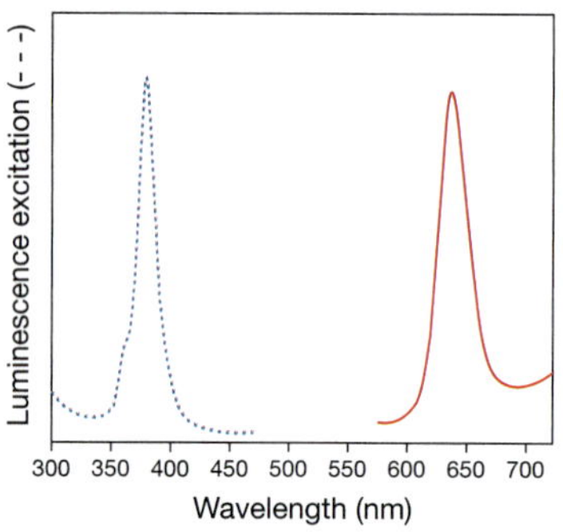

Figure 6.5.5 Luminescence excitation and emission spectra of the FluoSpheres® platinum luminescent microspheres (F20886).

Figure 6.5.6 Luminescent microsphere products provide an extensive range of sizes and fluorescent colors, illustrated by a sample of our Constellation™ microspheres for imaging (C14837).

>40 microseconds. Thus, time-gated fluorescence detection using these microspheres results in complete rejection of autofluorescence signals. In addition, both the europium luminescent and platinum luminescent microspheres feature long-wavelength emissions (610–650 nm) that are well separated from their excitation peaks (340–390 nm) (Figure 6.5.4, Figure 6.5.5). Because of this exceptionally large Stokes shift, filter combinations can be chosen that effectively isolate the desired luminescence signal.

These microspheres are available uncoated (F20880, F20881, F20886) or conjugated to NeutrAvidin™ biotin-binding protein (F20883, F20884), with nominal diameters of 0.04 µm or 0.2 µm. Beads labeled with NeutrAvidin™ biotin-binding protein can be used for the indirect detection of antigens and DNA targets in many biotin/avidin-based assays.

Fluorescent Microsphere Starter Kits

For applications requiring several different microsphere colors or sizes, we offer three types of fluorescent microsphere starter kits:

- The FluoSpheres® Fluorescent Color Kit (F10720) consists of 1 mL samples of yellow-green–, orange-, red- and dark red–fluorescent, carboxylate-modified 0.04 µm FluoSpheres® beads packaged as high-density, azide-free suspensions for microinjection.
- The FluoSpheres® Size Kits contain 1 mL samples of carboxylate-modified FluoSpheres® beads in 0.02, 0.1, 0.2, 0.5, 1.0 and 2.0 µm sizes. These beads are available in yellow-green– (F8888) or red- (F8887) fluorescent colors.
- The FluoSpheres® Blood Flow Determination Fluorescent Color Kits provide several different fluorescent colors of 10 µm (F8890) or 15 µm (F8891, F8892, F21015) FluoSpheres® microspheres; see Section 14.6 for a description of these kits.

Fluorescent Microspheres for Educational Purposes

Constellation™ microspheres for imaging (C14837) can be used to demonstrate hands-on techniques with a fluorescence microscope. Constellation™ microspheres consist of a selected mixture of beads in assorted sizes and colors (Figure 6.5.6) that can be used to practice adjusting the focus and switching filters on a fluorescence microscope. These microspheres are stable at room temperature, so they can be conveniently stored.

TransFluoSpheres® Fluorescent Microspheres: Tools for Multicolor Detection

Advantages of TransFluoSpheres® Fluorescent Microspheres

TransFluoSpheres® fluorescent microspheres (Table 6.9; Figure 6.5.7, Figure 6.5.2) are specially designed to overcome the limitations imposed by modern fluorescence instrumentation. Many flow cytometers, confocal laser-scanning microscopes and laser scanners incorporate the argon-ion laser as the excitation source, thereby limiting the available excitation wavelengths to the laser's 488 nm and 514 nm spectral lines and severely restricting simultaneous multicolor detection. Ideally, it would be useful to have a series of fluorescent dyes with absorption maxima close to the argon-ion laser's spectral lines, but with emission maxima at a variety of longer wavelengths. This approach would require that some of the dyes exhibit large Stokes shifts—defined as the separation of the absorption and emission maxima. Unfortunately, very few low molecular weight dyes have a combination of a large Stokes shift and a high molar absorptivity. For example, the Texas Red® fluorophore—often used in combination with fluorescein—has particularly weak absorption at 488 nm and 514 nm. In applications that employ the argon-ion laser as an excitation source, Texas Red® conjugates have a low fluorescence output that is easily obscured by the more intense fluorescein fluorescence, even when detected past 600 nm.

Our TransFluoSpheres® beads, which incorporate two or more fluorescent dyes that undergo excited-state energy transfer, exhibit Stokes shifts that can be extremely large. Each microsphere contains a dye with an excitation peak that maximally overlaps the spectral output

IMPORTANT NOTICE: The products described in this manual are covered by one or more Limited Use Label License(s). Please refer to the Appendix on page 971 and Master Product List on page 975. Products are For Research Use Only. Not intended for any animal or human therapeutic or diagnostic use.

of a commonly used excitation source (for example, the 488 nm spectral line of the argon-ion laser; Figure 6.5.8). In addition, each microsphere contains one or more longer-wavelength dyes that are carefully chosen to create a relay series that can efficiently transfer the energy from the initially excited dye to the longest-wavelength acceptor dye. The proprietary dyes used in the TransFluoSpheres® beads are optimally loaded to help ensure that the excitation energy is efficiently transferred from dye to dye so that essentially only the longest-wavelength dye in the series exhibits significant fluorescence. Because these TransFluoSpheres® beads fluoresce at a considerably longer wavelength than the excitation wavelength, they provide a signal that can be detected in samples with significant Rayleigh or Raman scattering or with endogenous fluorescent compounds such as bilins, flavins and certain drugs. Also, the large Stokes shifts exhibited by the TransFluoSpheres® beads allow the use of broadband filters, both to excite the sample and to detect the emission, resulting in a greater fluorescent signal (Figure 6.5.7).

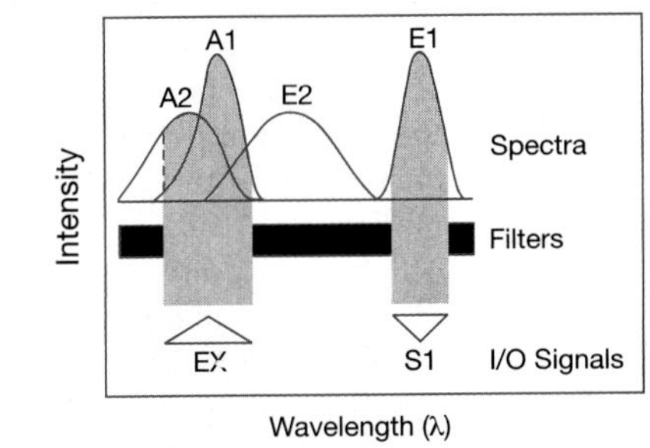

Figure 6.5.7 Schematic diagram of the advantages of the large Stokes shift exhibited by our TransFluoSpheres® beads. A1 and E1 represent the absorption and emission bands of a typical TransFluoSpheres® bead. The large separation of the absorption and emission maxima (Stokes shift) is characteristic of our TransFluoSpheres® beads. Unlike most low molecular weight fluorescent dyes, which show considerable overlap of their absorption and emission spectra, the TransFluoSpheres® beads can be excited (EX) across the entire absorption band A1 and the resulting fluorescence can be detected across the full emission band E1, thereby allowing the researcher to maximize the signal (S1). Moreover, because of the large Stokes shifts of the TransFluoSpheres® beads, researchers can often avoid problems associated with autofluorescence. The absorption and emission bands of a typical autofluorescent component are represented in this figure by A2 and E2. Although the endogenous fluorescent species will be excited simultaneously with the TransFluoSpheres® beads, the resulting emission (E2) does not coincide with E1 and is therefore readily rejected by suitably chosen optical filters.

TransFluoSpheres® Beads to Match Different Excitation Sources

We offer TransFluoSpheres® beads compatible with two different laser excitation sources. The argon-ion laser–excitable TransFluoSpheres® beads have an excitation maximum near 488 nm but emit at 560 nm (T8864, T8872, T8880) or 645 nm (T8883). The red He-Ne laser–excitable TransFluoSpheres® beads have excitation/emission maxima of 633/720 nm (T8870). TransFluoSpheres® beads can also be combined with our more traditional FluoSpheres® beads or with low molecular weight dyes for multicolor detection.[25]

Using carbodiimide reagents such as EDAC (E2247, Section 3.4), researchers can couple protein or other amine-containing molecules to our carboxylate-modified TransFluoSpheres® beads, making these microspheres suitable for a wide range of applications. TransFluoSpheres® beads can be used in the major microsphere-based test systems and in experiments that currently employ standard fluorescent microspheres to measure regional blood flow (Section 14.6), to study phagocytosis (Section 16.1), to detect cell-surface antigens and to trace neurons.[6,26–30]

In addition, we offer TransFluoSpheres® microspheres with excitation/emission maxima of 488/605 nm (T8860, T8861) conjugated to NeutrAvidin™ biotin- binding protein, as well as TransFluoSpheres® beads with excitation/emission maxima of 488/645 nm (T10711) conjugated to streptavidin. Flow cytometry studies demonstrate that the sensitivity of our 40 nm TransFluoSpheres® beads conjugated to streptavidin is superior to that of fluorescein streptavidin and comparable to that of R-phycoerythrin–streptavidin for detecting biotinylated epidermal growth factor (EGF) bound to EGF receptors.[25] In multicolor experiments, the long-wavelength fluorescence emission of these TransFluoSpheres® beads permits their use simultaneously with fluorescein- and R-phycoerythrin–labeled probes. For all applications requiring protein-coated microspheres, we strongly recommend using our BlockAid™ blocking solution (B10710) to reduce nonspecific binding.

If wavelength or bead-size requirements are not met by our current selection of products, we invite inquiries about custom synthesis by contacting Invitrogen Custom Services at www.invitrogen.com/handbook/customparticles. We can fine-tune the excitation and emission to match a particular excitation source or detection wavelength. In addition, we can covalently conjugate our TransFluoSpheres® beads to other target-specific proteins to provide detection reagents that have potentially greater sensitivity in flow cytometry applications and immunochemical assays.

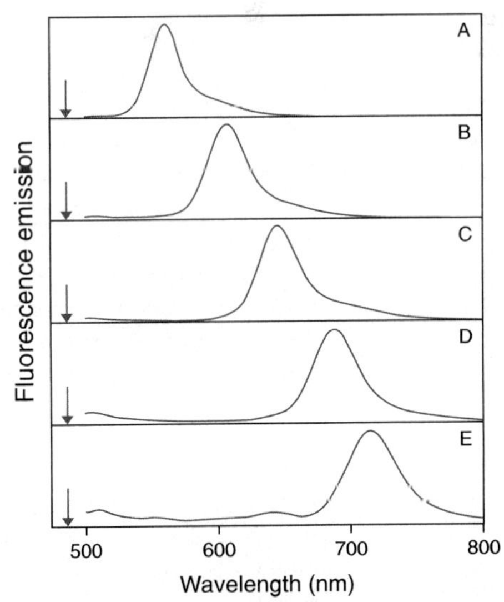

Figure 6.5.8 Fluorescence emission spectra of our five 488 nm light-excitable TransFluoSpheres® beads, named according to their excitation/emission maxima (nm): **A)** 488/560, **B)** 488/605, **C)** 488/645, **D)** 488/685 and **E)** 488/720. The arrow in each spectrum represents the 488 nm spectral line of the argon-ion laser. The beads shown in panels B, D and E are no longer routinely available; please inquire through Invitrogen Custom Services.

Table 6.9 Summary of TransFluoSpheres® fluorescent microspheres.

	Size		
Excitation/Emission (nm) *	**0.04 µm**	**0.1 µm**	**1.0 µm**
488/560	T8864	T8872	T8880
488/605 NeutrAvidin™-labeled	T8860	T8861	
488/645			T8883
488/645 Streptavidin-labeled	T10711		
633/720	T8870		

* All TransFluoSpheres® beads are based on carboxylate-modified polystyrene microspheres and are supplied as 0.5 mL of an aqueous suspension containing 2% solids and 2 mM sodium azide, except the TransFluoSpheres® beads labeled with NeutrAvidin™ biotin-binding protein, which are supplied as 0.4 mL of an aqueous suspension containing 1% solids and 2 mM sodium azide, and the streptavidin-labeled beads, which are supplied as 0.4 mL of an aqueous suspension containing 0.5% solids and 2 mM sodium azide.

BlockAid™ Blocking Solution

The intensely fluorescent and highly photostable FluoSpheres® and TransFluoSpheres® microspheres have significant potential for applications requiring probes that can deliver a strong signal. Unfortunately, microspheres conjugated to proteins have hydrophobic regions that may cause them to bind to nontarget surfaces in some experimental systems. This nonspecific binding can often be relieved by the use of a blocking solution. However, we have found that microspheres require a stronger blocking solution than those in common use, and therefore we have developed the BlockAid™ blocking solution (B10710).

The BlockAid™ reagent is a protein-based blocking solution designed for use with Fluospheres and TransFluoSpheres® microspheres conjugated to biotin, streptavidin, NeutrAvidin™ biotin-binding protein or other proteins. In our tests, BlockAid™ blocking solution has been found to be effective in flow cytometry applications with the NIH 3T3, A431, RAW and Jurkat cell lines; however, with the HMC-1 cell line, it did not appear to offer any advantages over standard blocking solutions. The BlockAid™ blocking solution has proven useful for reducing the nonspecific binding of protein-coated or other macromolecule-coated microspheres in a wide variety of flow cytometry, microscopy and microarray applications.[20,31,32]

REFERENCES

1. Biotech Histochem (1998) 73:291; **2.** Anal Biochem (1999) 272:165; **3.** Anal Biochem (1999) 271:143; **4.** Science (1980) 208:364; **5.** J Neurosci (1994) 14:6621; **6.** J Neurosci (1993) 13:5082; **7.** Nature (1984) 310:498; **8.** Cell Motil Cytoskeleton (1987) 8:293; **9.** Dev Growth Differ (1986) 28:461; **10.** Clin Immunol Immunopathol (1990) 55:173; **11.** J Immunol Methods (1999) 227:41; **12.** J Immunol Methods (2000) 243:243; **13.** Clin Chem (1998) 44:2054; **14.** Cytometry (1998) 33:318; **15.** Clin Chem (1998) 44:2057; **16.** Clin Chem (1997) 43:1799; **17.** Clin Chem (1997) 43:1749; **18.** Methods Cell Biol (1990) 33:613; **19.** Nat Protoc (2009) 4:1779; **20.** Mol Cell Probes (2008) 22:294; **21.** Eukaryot Cell (2009) 8:1665; **22.** PLoS One (2009) 4:e8402; **23.** Oncogene (2009) 28:1454; **24.** Anal Chem (2007) 79:122; **25.** J Immunol Methods (1998) 219:57; **26.** Biophys J (1993) 65:2396; **27.** Brain Res (1993) 630:115; **28.** J Membr Biol (1993) 135:83; **29.** Cytometry (1992) 13:423; **30.** Ophthalmic Surg Lasers (1997) 28:937; **31.** Proc Natl Acad Sci U S A (2005) 102:4225; **32.** Biophys J (2004) 86:1223.

PRODUCT LIST 6.5 MICROSPHERES

Cat. No.	Product	Quantity
B10710	BlockAid™ blocking solution *for use with microspheres*	50 mL
C14837	Constellation™ microspheres for imaging *mixture of assorted sizes and colors*	3 mL
F8760	FluoSpheres® aldehyde-sulfate microspheres, 0.02 µm, yellow-green fluorescent (505/515) *2% solids*	10 mL
F8762	FluoSpheres® aldehyde-sulfate microspheres, 1.0 µm, yellow-green fluorescent (505/515) *2% solids*	10 mL
F8763	FluoSpheres® amine-modified microspheres, 0.2 µm, red fluorescent (580/605) *2% solids*	5 mL
F8764	FluoSpheres® amine-modified microspheres, 0.2 µm, yellow-green fluorescent (505/515) *2% solids*	5 mL
F8765	FluoSpheres® amine-modified microspheres, 1.0 µm, yellow-green fluorescent (505/515) *2% solids*	5 mL
F8766	FluoSpheres® biotin-labeled microspheres, 0.04 µm, yellow-green fluorescent (505/515) *1% solids*	0.4 mL
F8767	FluoSpheres® biotin-labeled microspheres, 0.2 µm, yellow-green fluorescent (505/515) *1% solids*	0.4 mL
F8769	FluoSpheres® biotin-labeled microspheres, 1.0 µm, nonfluorescent *1% solids*	0.4 mL
F8768	FluoSpheres® biotin-labeled microspheres, 1.0 µm, yellow-green fluorescent (505/515) *1% solids*	0.4 mL
F8781	FluoSpheres® carboxylate-modified microspheres, 0.02 µm, blue fluorescent (365/415) *2% solids*	10 mL
F8782	FluoSpheres® carboxylate-modified microspheres, 0.02 µm, crimson fluorescent (625/645) *2% solids*	2 mL
F8783	FluoSpheres® carboxylate-modified microspheres, 0.02 µm, dark red fluorescent (660/680) *2% solids*	2 mL
F8784	FluoSpheres® carboxylate-modified microspheres, 0.02 µm, nile red fluorescent (535/575) *2% solids*	10 mL
F8786	FluoSpheres® carboxylate-modified microspheres, 0.02 µm, red fluorescent (580/605) *2% solids*	10 mL
F8787	FluoSpheres® carboxylate-modified microspheres, 0.02 µm, yellow-green fluorescent (505/515) *2% solids*	10 mL
F8789	FluoSpheres® carboxylate-modified microspheres, 0.04 µm, dark red fluorescent (660/680) *5% solids, azide free*	1 mL
F20880	FluoSpheres® carboxylate-modified microspheres, 0.04 µm, europium luminescent (365/610) *0.5% solids*	2 mL
F8792	FluoSpheres® carboxylate-modified microspheres, 0.04 µm, orange fluorescent (540/560) *5% solids, azide free*	1 mL
F20886	FluoSpheres® carboxylate-modified microspheres, 0.04 µm, platinum luminescent (390/650) *0.5% solids*	2 mL
F8793	FluoSpheres® carboxylate-modified microspheres, 0.04 µm, red fluorescent (580/605) *5% solids, azide free*	1 mL
F8794	FluoSpheres® carboxylate-modified microspheres, 0.04 µm, red-orange fluorescent (565/580) *5% solids, azide free*	1 mL
F8795	FluoSpheres® carboxylate-modified microspheres, 0.04 µm, yellow-green fluorescent (505/515) *5% solids, azide free*	1 mL
F8797	FluoSpheres® carboxylate-modified microspheres, 0.1 µm, blue fluorescent (350/440) *2% solids*	10 mL
F8799	FluoSpheres® carboxylate-modified microspheres, 0.1 µm, infrared fluorescent (715/755) *2% solids*	1 mL
F8800	FluoSpheres® carboxylate-modified microspheres, 0.1 µm, orange fluorescent (540/560) *2% solids*	10 mL
F8801	FluoSpheres® carboxylate-modified microspheres, 0.1 µm, red fluorescent (580/605) *2% solids*	10 mL
F8803	FluoSpheres® carboxylate-modified microspheres, 0.1 µm, yellow-green fluorescent (505/515) *2% solids*	10 mL
F8805	FluoSpheres® carboxylate-modified microspheres, 0.2 µm, blue fluorescent (365/415) *2% solids*	10 mL
F8806	FluoSpheres® carboxylate-modified microspheres, 0.2 µm, crimson fluorescent (625/645) *2% solids*	2 mL

IMPORTANT NOTICE: The products described in this manual are covered by one or more Limited Use Label License(s). Please refer to the Appendix on page 971 and Master Product List on page 975. Products are For Research Use Only. Not intended for any animal or human therapeutic or diagnostic use.
www.invitrogen.com/probes
molecular probes® | invitrogen™ by life technologies™

PRODUCT LIST 6.5 MICROSPHERES—*continued*

Cat. No.	Product	Quantity
F8807	FluoSpheres® carboxylate-modified microspheres, 0.2 µm, dark red fluorescent (660/680) *2% solids*	2 mL
F20881	FluoSpheres® carboxylate-modified microspheres, 0.2 µm, europium luminescent (365/610) *0.5% solids*	2 mL
F8809	FluoSpheres® carboxylate-modified microspheres, 0.2 µm, orange fluorescent (540/560) *2% solids*	10 mL
F8810	FluoSpheres® carboxylate-modified microspheres, 0.2 µm, red fluorescent (580/605) *2% solids*	10 mL
F8811	FluoSpheres® carboxylate-modified microspheres, 0.2 µm, yellow-green fluorescent (505/515) *2% solids*	10 mL
F8812	FluoSpheres® carboxylate-modified microspheres, 0.5 µm, red fluorescent (580/605) *2% solids*	10 mL
F8813	FluoSpheres® carboxylate-modified microspheres, 0.5 µm, yellow-green fluorescent (505/515) *2% solids*	10 mL
F8815	FluoSpheres® carboxylate-modified microspheres, 1.0 µm, blue fluorescent (350/440) *2% solids*	10 mL
F8814	FluoSpheres® carboxylate-modified microspheres, 1.0 µm, blue fluorescent (365/415) *2% solids*	10 mL
F8816	FluoSpheres® carboxylate-modified microspheres, 1.0 µm, crimson fluorescent (625/645) *2% solids*	2 mL
F8819	FluoSpheres® carboxylate-modified microspheres, 1.0 µm, nile red fluorescent (535/575) *2% solids*	10 mL
F8820	FluoSpheres® carboxylate-modified microspheres, 1.0 µm, orange fluorescent (540/560) *2% solids*	10 mL
F8821	FluoSpheres® carboxylate-modified microspheres, 1.0 µm, red fluorescent (580/605) *2% solids*	10 mL
F8823	FluoSpheres® carboxylate-modified microspheres, 1.0 µm, yellow-green fluorescent (505/515) *2% solids*	10 mL
F8824	FluoSpheres® carboxylate-modified microspheres, 2.0 µm, blue fluorescent (365/415) *2% solids*	2 mL
F8825	FluoSpheres® carboxylate-modified microspheres, 2.0 µm, nile red fluorescent (535/575) *2% solids*	2 mL
F8826	FluoSpheres® carboxylate-modified microspheres, 2.0 µm, red fluorescent (580/605) *2% solids*	2 mL
F8827	FluoSpheres® carboxylate-modified microspheres, 2.0 µm, yellow-green fluorescent (505/515) *2% solids*	2 mL
F20892	FluoSpheres® collagen I-labeled microspheres, 1.0 µm, yellow-green fluorescent (505/515) *0.5% solids*	0.4 mL
F20893	FluoSpheres® collagen I-labeled microspheres, 2.0 µm, yellow-green fluorescent (505/515) *0.5% solids*	0.4 mL
F10720	FluoSpheres® Fluorescent Color Kit, carboxylate-modified microspheres, 0.04 µm *four colors, 1 mL each* *5% solids, azide free*	1 kit
F20883	FluoSpheres® NeutrAvidin™ labeled microspheres, 0.04 µm, europium luminescent (365/610) *0.5% solids*	0.4 mL
F8772	FluoSpheres® NeutrAvidin™ labeled microspheres, 0.04 µm, nonfluorescent *1% solids*	0.4 mL
F8770	FluoSpheres® NeutrAvidin™ labeled microspheres, 0.04 µm, red fluorescent (580/605) *1% solids*	0.4 mL
F8771	FluoSpheres® NeutrAvidin™ labeled microspheres, 0.04 µm, yellow-green fluorescent (505/515) *1% solids*	0.4 mL
F20884	FluoSpheres® NeutrAvidin™ labeled microspheres, 0.2 µm, europium luminescent (365/610) *0.5% solids*	0.4 mL
F8774	FluoSpheres® NeutrAvidin™ labeled microspheres, 0.2 µm, yellow-green fluorescent (505/515) *1% solids*	0.4 mL
F8777	FluoSpheres® NeutrAvidin™ labeled microspheres, 1.0 µm, nonfluorescent *1% solids*	0.4 mL
F8775	FluoSpheres® NeutrAvidin™ labeled microspheres, 1.0 µm, red fluorescent (580/605) *1% solids*	0.4 mL
F8776	FluoSpheres® NeutrAvidin™ labeled microspheres, 1.0 µm, yellow-green fluorescent (505/515) *1% solids*	0.4 mL
F13080	FluoSpheres® polystyrene microspheres, 1.0 µm, blue-green fluorescent (430/465) *for tracer studies* *1.0x1010 beads/mL*	5 mL
F13082	FluoSpheres® polystyrene microspheres, 1.0 µm, orange fluorescent (540/560) *for tracer studies* *1.0x1010 beads/mL*	5 mL
F13083	FluoSpheres® polystyrene microspheres, 1.0 µm, red fluorescent (580/605) *for tracer studies* *1.0x1010 beads/mL*	5 mL
F13081	FluoSpheres® polystyrene microspheres, 1.0 µm, yellow-green fluorescent (505/515) *for tracer studies* *1.0x1010 beads/mL*	5 mL
F8887	FluoSpheres® Size Kit #1, carboxylate-modified microspheres, red fluorescent (580/605) *six sizes, 1 mL each* *2% solids*	1 kit
F8888	FluoSpheres® Size Kit #2, carboxylate-modified microspheres, yellow-green fluorescent (505/515) *six sizes, 1 mL each* *2% solids*	1 kit
F8780	FluoSpheres® streptavidin-labeled microspheres, 0.04 µm, yellow-green fluorescent (505/515) *0.5% solids*	0.4 mL
F8845	FluoSpheres® sulfate microspheres, 0.02 µm, yellow-green fluorescent (505/515) *2% solids*	10 mL
F8848	FluoSpheres® sulfate microspheres, 0.2 µm, yellow-green fluorescent (505/515) *2% solids*	10 mL
F8851	FluoSpheres® sulfate microspheres, 1.0 µm, red fluorescent (580/605) *2% solids*	10 mL
F8852	FluoSpheres® sulfate microspheres, 1.0 µm, yellow-green fluorescent (505/515) *2% solids*	10 mL
F8853	FluoSpheres® sulfate microspheres, 2.0 µm, yellow-green fluorescent (505/515) *2% solids*	2 mL
F8854	FluoSpheres® sulfate microspheres, 4.0 µm, blue fluorescent (365/415) *2% solids*	2 mL
F8858	FluoSpheres® sulfate microspheres, 4.0 µm, red fluorescent (580/605) *2% solids*	2 mL
F8859	FluoSpheres® sulfate microspheres, 4.0 µm, yellow-green fluorescent (505/515) *2% solids*	2 mL
T8864	TransFluoSpheres® carboxylate-modified microspheres, 0.04 µm (488/560) *2% solids*	0.5 mL
T8870	TransFluoSpheres® carboxylate-modified microspheres, 0.04 µm (633/720) *2% solids*	0.5 mL
T8872	TransFluoSpheres® carboxylate-modified microspheres, 0.1 µm (488/560) *2% solids*	0.5 mL
T8880	TransFluoSpheres® carboxylate-modified microspheres, 1.0 µm (488/560) *2% solids*	0.5 mL
T8883	TransFluoSpheres® carboxylate-modified microspheres, 1.0 µm (488/645) *2% solids*	0.5 mL
T8860	TransFluoSpheres® NeutrAvidin™ labeled microspheres, 0.04 µm (488/605) *1% solids*	0.4 mL
T8861	TransFluoSpheres® NeutrAvidin™ labeled microspheres, 0.1 µm (488/605) *1% solids*	0.4 mL
T10711	TransFluoSpheres® streptavidin-labeled microspheres, 0.04 µm (488/645) *0.5% solids*	0.4 mL

IMPORTANT NOTICE: The products described in this manual are covered by one or more Limited Use Label License(s). Please refer to the Appendix on page 971 and Master Product List on page 975. Products are For Research Use Only. Not intended for any animal or human therapeutic or diagnostic use.

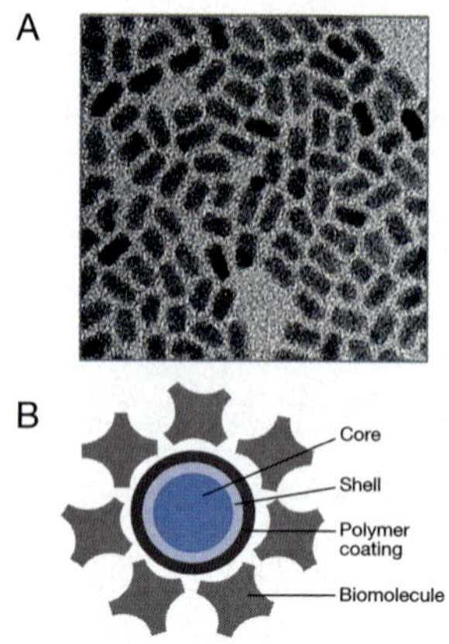

Figure 6.6.1 Structure of a Qdot® nanocrystal. **A)** Qdot® nanocrystals containing core and shell components only are shown in this transmission electron microscope image (200,000× magnification). **B)** In this schematic of the overall structure of a Qdot® nanocrystal conjugate, the layers represent the distinct structural elements and are roughly to scale.

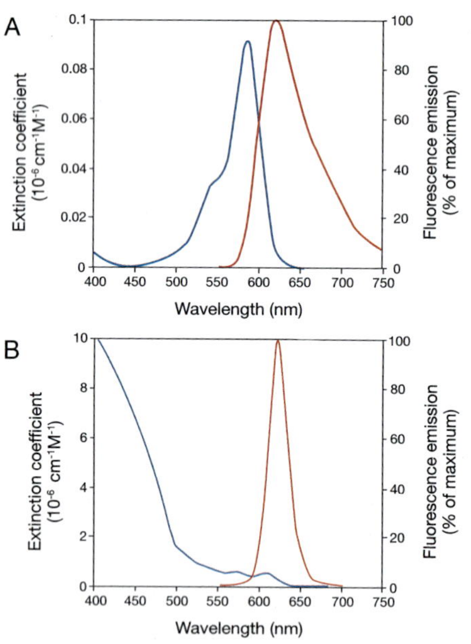

Figure 6.6.3 Absorption and fluorescence emission spectra of **A)** an organic dye (Alexa Fluor® 594 dye) and **B)** a quantum dot nanocrystal (Qdot® 625 nanocrystal). Absorption spectra are plotted in terms of molar extinction coefficient to illustrate the 100-fold larger values of this parameter typically exhibited by quantum dot nanocrystals relative to organic dyes. Comparison of the emission spectra illustrates the narrower bandwidths typical of quantum dot nanocrystals. In particular, Alexa Fluor® 594 dye exhibits significant emission intensity at wavelengths >700 nm whereas Qdot® 625 nanocrystal exhibits essentially none (note that the emission peak positions of the two fluorophores are almost identical). Overlay comparisons of Qdot® nanocrystal spectra with those of organic dyes and fluorescent proteins can be generated using our online Fluorescence SpectraViewer utility (www.invitrogen.com/handbook/spectraviewer). Note that molar extinction coefficients in cm^{-1} M^{-1} have been multiplied by a constant (1×10^{-6}) for scaling purposes.

6.6 Qdot® Nanocrystals

Properties of Qdot® Nanocrystals

Structural Properties

Qdot® nanocrystals are nanometer-scale atom clusters comprising a core, shell and surface coating (Figure 6.6.1). The core is made up of a few hundred to a few thousand atoms of a semiconductor material (cadmium selenide (CdSe) or cadmium telluride (CdTe)). A semiconductor shell (typically zinc sulfide (ZnS)) surrounds and stabilizes the core, improving both the optical and physical properties of the material.

The core–shell assembly as initially prepared is extremely hydrophobic. An amphiphilic polymer coating is then applied, serving two purposes. Firstly, it incorporates ionizable functional groups that confer water solubility essential for bioanalytical applications. Secondly, it provides a platform for covalent functionalization of the Qdot® nanocrystal with antibodies, streptavidin, oligonucleotides and other affinity reagents (denoted as "biomolecule" in Figure 6.6.1) that confer targeting specificity for biomolecular detection. As depicted in Figure 6.6.1, Qdot® nanocrystal conjugates typically incorporate multiple copies of the affinity reagent—the inverse of the situation encountered in dye-labeled conjugates in which multiple dyes are attached to a single affinity reagent.

In most of our Qdot® nanocrystal conjugates, the affinity reagent is coupled to the amphiphilic polymer coating via a functionalized polyethylene glycol (PEG) linker. PEG linkers have been shown to reduce nonspecific binding in flow cytometry [1] and imaging assays,[2] thereby improving signal-to-noise ratios. Qdot® primary and secondary antibody conjugates, Qdot® streptavidin conjugates, Qtracker® non-targeted quantum dots, and Qdot® ITK™ amino (PEG) quantum dots, as well as the reactive nanocrystals provided in the Qdot® Antibody Conjugation Kits, all utilize this PEG linker chemistry.

The overall size of functionalized Qdot® nanocrystals, in a spherical approximation, is around 20 nm in diameter (Figure 6.6.2). This is much larger than an organic dye and comparable to the size of a large (molecular weight >500 kDa) protein or our smallest FluoSpheres® fluorescent microspheres (Section 6.5). The novel physical properties of Qdot® nanocrystals coupled to the fact that they contain cadmium [3] gives rise to concerns regarding toxicity in applications involving live cells or *in vivo* administration. These concerns have prompted much investigation.[3–5] It is difficult to draw general conclusions from these studies due to the influence of the cellular, tissue or animal context on the results. Toxicity concerns, however, are generally mitigated by the extraordinarily high fluorescence output of Qdot® nanocrystals, which enables the use of very low (often nanomolar) working concentrations. The extent to which the size of Qdot® nanocrystals may cause perturbations of molecular interactions has also been subjected to detailed scrutiny.[6]

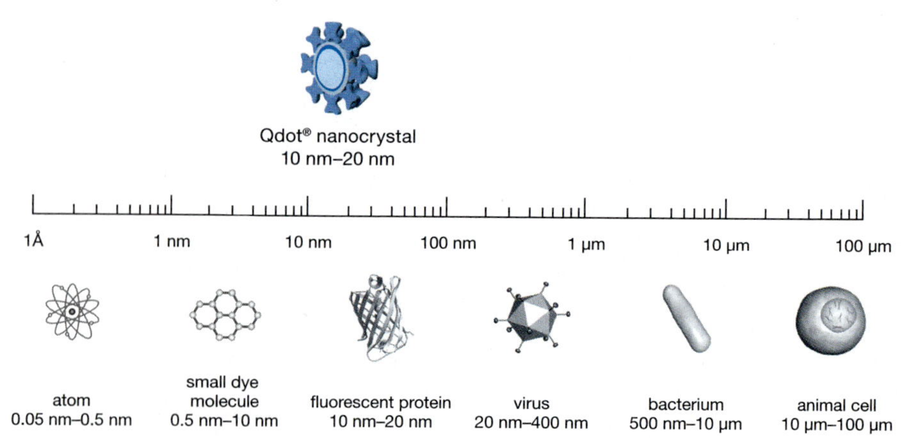

Figure 6.6.2 Relative size of Qdot® nanocrystals.

IMPORTANT NOTICE: The products described in this manual are covered by one or more Limited Use Label License(s). Please refer to the Appendix on page 971 and Master Product List on page 975. Products are For Research Use Only. Not intended for any animal or human therapeutic or diagnostic use.

Spectroscopic Properties

Although Qdot® nanocrystals are composed of semiconductor materials, their small size results in spectroscopic properties that are radically different from those of bulk semiconductors. Absorption of a photon causes an electron to move from the semiconductor valence band to the conductance band, creating an exciton (electron-hole pair). Absorption occurs as long as the energy of the incident photons is higher than the semiconductor bandgap energy; thus, excitons can be created over a wide range of energies within the nanocrystal core. The higher-energy excitons relax to the lowest bandgap energy before they recombine and emit a photon. Therefore, the wavelength range of the absorption spectrum is broad, whereas that of the emission spectrum is narrow (Figure 6.6.3, Figure 6.6.4, Figure 6.6.5).

Due to the small size of the nanocrystal, the electron-hole separation is confined to be smaller than the Bohr radius of the semiconductor. As the nanocrystal core is made progressively smaller, more energy is required to confine the exciton, and the energy of the emitted photons increases. Therefore, smaller nanocrystals exhibit shorter-wavelength emission. By controlled variation of the core size, we have developed a range of nanocrystals with distinctively different emission spectral characteristics (Figure 6.6.5) but largely overlapping absorption spectra (Figure 6.6.4) and generally similar physical properties. Currently, we offer eight Qdot® nanocrystal types identified by 3-digit numerals representing the peak emission wavelength in nanometers (e.g., Qdot® 655 nanocrystal, emission peak = 655 nm).

The spectral characteristics of Qdot® nanocrystals offer some distinctive advantages over those of organic dyes and fluorescent proteins,[7] particularly with regard to implementation of spectral multiplex detection schemes. The broad absorption spectra facilitate single-source (e.g., 405 nm) excitation of multiple Qdot® nanocrystal types. Furthermore, Qdot® nanocrystals lack the ubiquitous red-edge "tail" of organic dye and fluorescent protein emission spectra (Figure 6.6.3), reducing the likelihood of detection channel crosstalk. Qdot® nanocrystals have been extensively characterized and utilized as donors in fluorescence resonance energy transfer (FRET) assays (Fluorescence Resonance Energy Transfer (FRET)—Note 1.2) of molecular proximity.[8–11] Their utility as FRET acceptors is limited by their absorption spectral characteristics (Figure 6.6.4), making it difficult to avoid direct, proximity-independent fluorescence excitation. However, they work well as acceptors from lanthanide chelate donors in time-resolved FRET (TR-FRET) assays, in which fluorescence originating from direct excitation is eliminated by time-gating of the detection system.[12,13]

Another distinctive property of Qdot® nanocrystals relative to organic dyes and fluorescent proteins is their prodigious fluorescence output, which principally derives from their extraordinarily large extinction coefficients for absorption (Table 6.10, Figure 6.6.3). Organic dye and fluorescent protein extinction coefficients do not often exceed 10^5 $cm^{-1}M^{-1}$, whereas those of Qdot® nanocrystals are often >10^6 $cm^{-1}M^{-1}$, particularly in the blue visible and ultraviolet wavelength regions (Figure 6.6.3, Figure 6.6.4). Because their fluorescence quantum yields are high and comparable to those of organic dyes and fluorescent proteins,[14] the large capacity for photon absorption represented by the extinction coefficient is propagated into emission of fluorescence photons. The fluorescence lifetimes of Qdot® nanocrystals are typically in the 10–30 nanosecond range as compared to 1–10 nanoseconds for most organic dyes.[8] They also have very large two-photon absorption cross-sections,[15] resulting

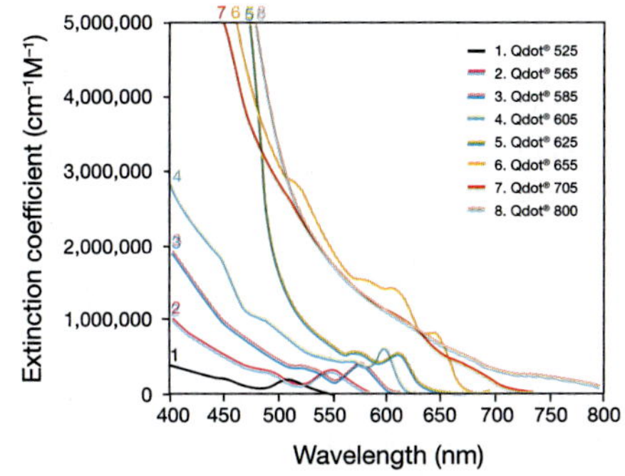

Figure 6.6.4 Absorption spectra of Qdot® nanocrystals plotted in terms of molar extinction coefficient.

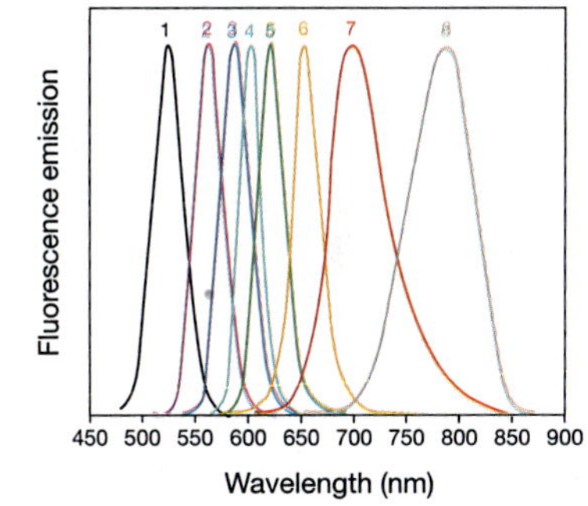

Figure 6.6.5 Normalized fluorescence emission spectra of Qdot® nanocrystals: **1)** Qdot® 525, **2)** Qdot® 565, **3)** Qdot® 585, **4)** Qdot® 605, **5)** Qdot® 625, **6)** Qdot® 655, **7)** Qdot® 705, **8)** Qdot® 800 nanocrystals.

Table 6.10 Extinction coefficients of Qdot® nanocrystal streptavidin conjugates at ultraviolet and visible wavelengths.

	Extinction Coefficient (EC) *			
Product	**350 nm**	**405 nm**	**488 nm**	**532 nm**
Qdot® 525 nanocrystals	710,000	360,000	130,000	NA
Qdot® 565 nanocrystals	1,900,000	1,100,000	290,000	139,000
Qdot® 585 nanocrystals	3,500,000	2,200,000	530,000	305,000
Qdot® 605 nanocrystals	4,400,000	2,800,000	1,100,000	580,000
Qdot® 625 nanocrystals	14,700,000	9,900,000	2,700,000	870,000
Qdot® 655 nanocrystals	9,100,000	5,700,000	2,900,000	2,400,000
Qdot® 705 nanocrystals	12,900,000	8,300,000	3,000,000	2,100,000
Qdot® 800 nanocrystals	15,800,000	10,600,000	4,100,000	2,100,000

* Molar extinction coefficient in $cm^{-1}M^{-1}$ determined at the corresponding wavelength. NA = Not applicable.

IMPORTANT NOTICE: The products described in this manual are covered by one or more Limited Use Label License(s). Please refer to the Appendix on page 971 and Master Product List on page 975. Products are For Research Use Only. Not intended for any animal or human therapeutic or diagnostic use.

in exceptional performance in two-photon excitation microscopy applications[8,16,17] (Fluorescent Probes for Two-Photon Microscopy—Note 1.5). Qdot® nanocrystals are much less susceptible to photobleaching than organic dyes and fluorescent proteins.[18,19] This feature is particularly beneficial in protocols involving repeated image captures on a single field of view such as confocal z-stack[20] or longitudinal time-lapse acquisitions.

Furthermore, Qdot® nanocrystals exhibit fluorescence intermittency, commonly referred to as blinking, whereby continuously excited single fluorophores will spontaneously alternate between emitting and non-emitting states across timescales from milliseconds to many seconds.[21–24] By its nature, blinking does not significantly impact ensemble measurements. Instead, it is most readily observable in single molecule detection trajectories where the binary nature of the "on" and "off" emission levels provides a useful confirmatory indication that the signals represent single fluorophores rather than clustered multiples.

Qdot® nanocrystals are intrinsically electron dense due to the tightly packed atomic lattice of the core–shell assembly, making them directly detectable by electron microscopy (EM).[25] They can therefore be used as dual-mode contrast agents for correlated fluorescence and EM without the diaminobenzidine photoconversion and subsequent osmification processing steps required for EM level visualization of organic dyes[26,27] (Fluorescent Probes for Photoconversion of Diaminobenzidine Reagents—Note 14.2). Furthermore, particle size and shape differences can be used as identifying characteristics in EM in the same way that emission spectra are used in fluorescence microscopy.[26]

Since the first reported applications of quantum dot nanocrystals for biomolecular detection in 1998,[28,29] many reviews have been published that may be consulted for further information on their properties and applications, both actual and prospective.[30–33]

Qdot® Nanocrystal Products and Applications

Qdot® Streptavidin Conjugates

Streptavidin conjugates provide access to a wide range of biomolecular functionality via high-affinity avidin–biotin coupling (Chapter 4), making them perhaps the most versatile subset in our range of Qdot® nanocrystal labeling reagents:

- Qdot® 525 streptavidin conjugate (Q10141MP)
- Qdot® 565 streptavidin conjugate (Q10131MP)
- Qdot® 585 streptavidin conjugate (Q10111MP)
- Qdot® 605 streptavidin conjugate (Q10101MP)
- Qdot® 625 streptavidin conjugate (A10196)
- Qdot® 655 streptavidin conjugate (Q10121MP)
- Qdot® 705 streptavidin conjugate (Q10161MP)
- Qdot® 800 streptavidin conjugate (Q10171MP)
- Qdot® Streptavidin Sampler Kit *1 µM solutions* (Q10151MP)

For example, nucleic acid hybridization can be detected by complexation of terminally biotinylated oligonucleotides with Qdot® streptavidin conjugates[34–37] (Figure 6.6.6). Qdot® streptavidin conjugates in combination with biotinylated antibodies provide increased multiplexing capacity for immunodetection of proteins on both imaging[38–42] and flow cytometry[1,43] platforms. Single-particle tracking of molecular motion is a particularly compelling application of Qdot® streptavidin conjugates that takes advantage of their extraordinary fluorescence output capacity.[44] The principal subjects for these measurements are cell-surface receptors[44–50] and molecular motors such as myosin.[51,52] Single-molecule detection uncovers details of molecular

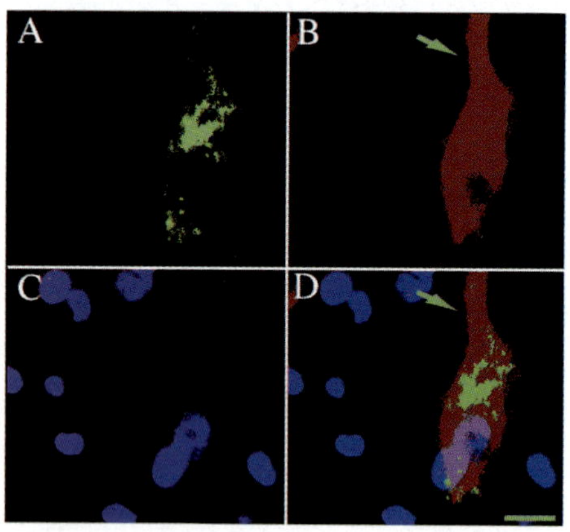

Figure 6.6.6 Multiplex mRNA *in situ* hybridization and immunofluorescence localization of tyrosine hydroxylase in mouse brain using Qdot® 525 and Qdot® 605 streptavidin conjugates (Q10141MP, Q10101MP). **A)** Vmat2 mRNA-positive neurons in substantia nigra were probed with a biotinylated oligonucleotide and labeled with Qdot® 525 streptavidin conjugate. **B)** The same cell was labeled with anti-tyrosine hydroxylase (TH) antibody in conjunction with biotinylated secondary antibody and Qdot® 605 streptavidin conjugate. The Vmat2 mRNA signal is restricted to the cytoplasm, whereas the labeling of TH is extended to the whole cell body and processes (arrow). **C)** Cell nuclei were labeled with DAPI. **D)** Overlay of all three stained images shows the different subcellular distributions of Vmat2 mRNA and TH immunoreactivity. Scale bar = 15 µm. Images contributed by Stuart Sealfon, Mount Sinai School of Medicine, and reprinted with permission from Nucleic Acids Res (2005) 33:e161.

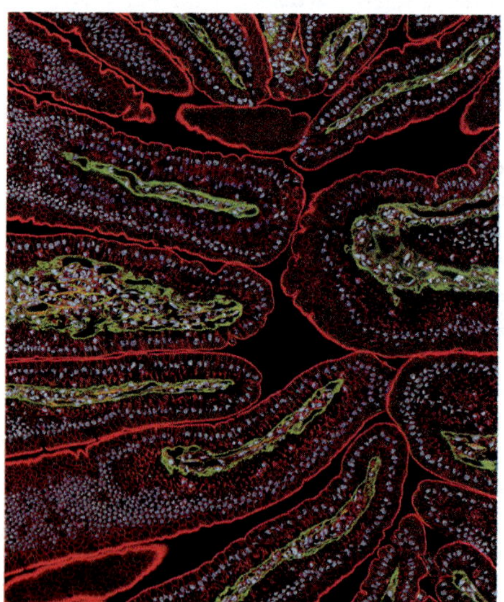

Figure 6.6.7 Multicolor immunofluorescence imaging with Qdot® secondary antibody conjugates. Actin in a mouse intestine section was detected with mouse anti-actin antibody and Qdot® 655 goat F(ab')$_2$ anti–mouse IgG antibody (red; Q11021MP, Q11022MP), laminin was detected with rabbit anti-laminin antibody and Qdot® 525 goat F(ab')$_2$ anti–rabbit IgG antibody (green, Q11441MP), and nuclei were stained with Hoechst 33342 (blue). Image contributed by Thomas Deerinck and Mark Ellisman, The National Center for Microscopy and Imaging Research, San Diego, California, USA.

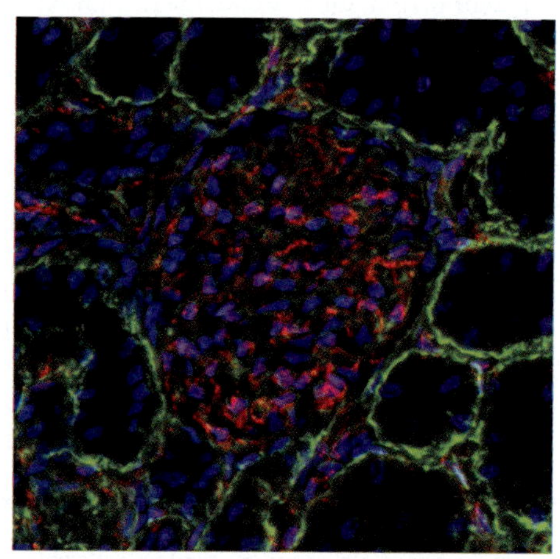

Figure 6.6.8 Multicolor immunofluorescence imaging with Qdot® secondary antibody conjugates. Laminin in a mouse kidney section was labeled with a rabbit anti-laminin primary antibody and visualized using green-fluorescent Qdot® 565 F(ab')$_2$ anti–rabbit IgG secondary antibody (Q11431MP, Q11432MP). PECAM-1 (platelet/endothelial cell adhesion molecule-1, CD31) was labeled with a rat anti–PECAM-1 primary antibody and visualized using red-fluorescent Qdot® 655 F(ab')$_2$ anti–rat IgG secondary antibody (Q11621MP). Nuclei were counterstained with blue-fluorescent Hoechst 33342. Image contributed by Stuart Shand, Center for Biologic Imaging, University of Pittsburgh.

IMPORTANT NOTICE: The products described in this manual are covered by one or more Limited Use Label License(s). Please refer to the Appendix on page 971 and Master Product List on page 975. Products are For Research Use Only. Not intended for any animal or human therapeutic or diagnostic use.

motion and assembly that are obscured by the averaging effects of bulk measurements. Qdot® nanocrystals are quite essential in such applications as fluorescence photon output ultimately propagates into the spatial and temporal resolution of the measurements.Typically, labeling is accomplished using biotinylated ligands such as biotin EGF (E3477, Section 4.3) and biotin-XX α-bungartoxin (B1196, Section 4.3) to couple Qdot® streptavidin conjugates to the receptor of interest. Because these conjugates incorporate multiple streptavidins (typically 5–10 per Qdot® nanocrystal), careful attention must be paid to the streptavidin:ligand stoichiometry to avoid receptor crosslinking mediated by the Qdot® streptavidin–biotinyated ligand complex.[49,53] A molar excess of Qdot® streptavidin conjugate over the biotinylated ligand biases the complex stoichiometry towards the desired outcome of one ligand per nanocrystal. Qdot® streptavidin conjugates are supplied as 1 µM solutions in 50 mM borate, pH 8.3 containing 0.05% sodium azide and 1 M betaine cryoprotectant in 200 µL units. The Qdot® Streptavidin Sampler Kit (Q10151MP) provides 50 µL units of six different Qdot® streptavidin conjugates in a single package.

We also offer Qdot® ITK™ streptavidin quantum dots, in which streptavidin is directly attached to the inner amphiphilic coating without a PEG linker:

- Qdot® 525 ITK™ Streptavidin Conjugate Kit (Q10041MP)
- Qdot® 565 ITK™ Streptavidin Conjugate Kit (Q10031MP)
- Qdot® 585 ITK™ Streptavidin Conjugate Kit (Q10011MP)
- Qdot® 605 ITK™ Streptavidin Conjugate Kit (Q10001MP)
- Qdot® 655 ITK™ Streptavidin Conjugate Kit (Q10021MP)
- Qdot® 705 ITK™ Streptavidin Conjugate Kit (Q10061MP)
- Qdot® 800 ITK™ Streptavidin Conjugate Kit (Q10071MP)

This formulation results in a smaller overall particle size and an increase in the number of streptavidins per nanocrystal relative to PEG-linked conjugates. These characteristics are advantageous in fluorescence resonance energy transfer (FRET) assays for detecting hybridization of target nucleic acid sequences to biotinylated oligonucleotides captured on the Qdot® nanocrystal surface.[11,54,55] Qdot® ITK™ streptavidin conjugates are supplied in 250 µL units of 2 µM solution in kits that also provide Qdot® incubation buffer (50 mM borate buffer, pH 8.3, containing 2% BSA and 0.05% sodium azide). Qdot® incubation buffer is also available in separately packaged 30 mL units (Q20001MP).

Biotinylated Qdot® Nanocrystals

The Qdot® 605 and Qdot® 655 Biotin Conjugate Kits (Q10301MP, Q10321MP) provide Qdot® nanocrystals that have been functionalized with biotin on their surface via carbodiimide-mediated coupling. The resulting conjugates typically incorporate 5–7 biotin groups per Qdot® nanocrystal. In addition to biotinylated Qdot® nanocrystals, these kits also provide Qdot® incubation buffer, which is formulated specifically to achieve improved signal-to-noise ratios in immunohistochemical applications. Biotinylated Qdot® nanocrystals have been used alongside biotinylated microspheres for analysis of the effects of fluorescent label size on ligand–receptor binding dynamics and equilibria by fluorescence correlation spectroscopy[6] (Fluorescence Correlation Spectroscopy (FCS)—Note 1.3).

WesternDot™ Western Blot Kits

The WesternDot™ 625 Goat Anti-Mouse and Goat Anti-Rabbit Western Blot Kits (W10132, W10142) provide Qdot® nanocrystal secondary reagents for the detection of subnanogram amounts of protein on western blots without the complications of time-dependent signal development imposed by enzyme-amplified chemiluminescence detection systems. These kits provide Qdot® 625 streptavidin conjugates and biotinylated secondary antibodies for use in combination with user-supplied mouse or rabbit primary antibodies against the protein of interest. In other respects, immunodetection of proteins immobilized on nitrocellulose (NC) or polyvinylidene difluoride (PVDF) membranes follows standard western blotting protocols. Detection of the fluorescence signal can be accomplished using standard UV- or blue light–based gel and blot imaging systems and does not require specialized emission filters.

Qdot® Secondary Antibody Conjugates

Qdot® secondary antibody conjugates (Table 6.11) combine the spectral characteristics of Qdot® nanocrystals with the selective binding of $F(ab')_2$ fragments from affinity-purified secondary antibodies, enabling a wide range of immunocytochemical and immunohistochemical applications[56,57] (Figure 6.6.7, Figure 6.6.8). Some examples include immunolabeling of $GABA_A$ receptor α1 and γ2 subunits for colocalization analysis in HEK 293 cells[58] and immunodetection of EGFR, E-cadherin and cytokeratin on formalin-fixed, paraffin-embedded (FFPE) tissue array slides.[59]

Table 6.11 Qdot® nanocrystal secondary antibody conjugates.

Antibody	Host	Qdot® 525	Qdot® 565	Qdot® 585	Qdot® 605	Qdot® 625	Qdot® 655	Qdot® 705	Qdot® 800
Anti-mouse IgG	Goat	Q11041MP	Q11031MP Q11032MP *	Q11011MP	Q11001MP Q11002MP *	A10195 *	Q11021MP Q11022MP *	Q11061MP Q11062MP *	Q11071MP
Anti-rabbit IgG	Goat	Q11441MP	Q11431MP Q11432MP *	Q11411MP	Q11401MP Q11402MP *	A10194 *	Q11421MP Q11422MP *	Q11461MP Q11462MP *	Q11471MP
Anti-rat IgG	Goat		Q11631MP		Q11601MP		Q11621MP		
Anti-human IgG	Goat				Q11201MP		Q11221MP		
Anti-chicken IgY	Goat						Q14421MP		
Anti-goat IgG	Rabbit						Q11821MP		

All Qdot® nanocrystal secondary antibody conjugates are prepared from affinity-purified $F(ab')_2$ fragments and are supplied as 1 µM solutions in 200 µL units unless specified otherwise (* denotes 100 µL units).

IMPORTANT NOTICE: The products described in this manual are covered by one or more Limited Use Label License(s). Please refer to the Appendix on page 971 and Master Product List on page 975. Products are For Research Use Only. Not intended for any animal or human therapeutic or diagnostic use.

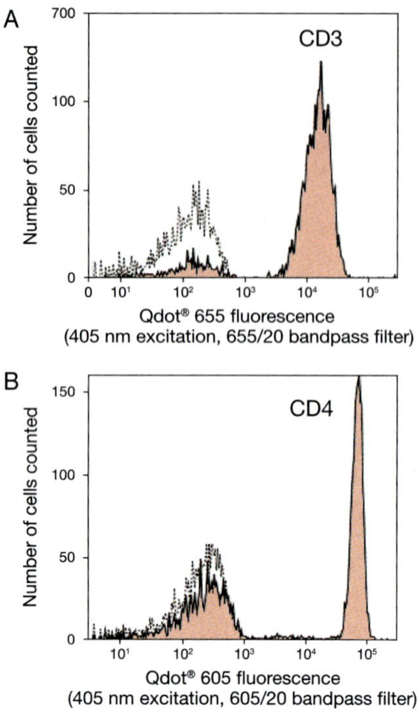

Figure 6.6.9 Antigen detection in human peripheral blood lymphocytes. Cells were incubated with **A)** Qdot® 655 mouse anti–human CD3 antibody (Q10012) or **B)** Qdot® 605 mouse anti–human CD4 antibody (Q10008). Samples were then subjected to flow cytometric analysis using LSR II flow cytometers and FACSDiva software (BD Biosciences) or FlowJo software (Tree Star, Inc.). The gray overlay peaks represent a sample of unstained cells.

Qmount® Qdot® Mounting Media

Qmount® Qdot® mounting media (Q10336) is a specialized mountant that preserves the fluorescence signal of Qdot® nanocrystals with little to no quenching of the initial signal intensity. The formulation cures within 12 hours and is provided in a convenient and easy-to-use dropper bottle. This mounting medium offers excellent compatibility with Qdot® secondary antibody and streptavidin conjugates in immunocytochemical and immunohistochemical detection applications, as well with Qnuclear™ Deep Red stain (Q10363, Section 12.5), which is the recommended nuclear counterstain for these applications.

Anti–Human CD Antibodies

Immunophenotyping of leukocytes based on expression of cell-surface CD (cluster of differentiation) markers is a mainstream flow cytometry application in both clinical and research settings.[60] Our Qdot® nanocrystal–conjugated mouse monoclonal anti–human CD antibodies (Table 6.12) bring extended multiplexing capacity and unmatched fluorescence output to enhance conventional leukocyte immunophenotyping protocols[61] (Figure 6.6.9). The large selection of Invitrogen™ anti–human CD antibodies labeled with Alexa Fluor® dyes, phycobiliproteins and phycobiliprotein tandem fluorophores[62] can be surveyed using our online primary antibody search utility at www.invitrogen.com/handbook/antibodies.

Qdot® Antibody Conjugation Kits

Qdot® Antibody Conjugation Kits provide reagents and protocols for researchers to prepare Qdot® nanocrystal conjugates of independently sourced antibodies:

- Qdot® 525 Antibody Conjugation Kit (Q22041MP)
- Qdot® 565 Antibody Conjugation Kit (Q22031MP)
- Qdot® 585 Antibody Conjugation Kit (Q22011MP)
- Qdot® 605 Antibody Conjugation Kit (Q22001MP)
- Qdot® 625 Antibody Conjugation Kit (A10197)
- Qdot® 655 Antibody Conjugation Kit (Q22021MP)
- Qdot® 705 Antibody Conjugation Kit (Q22061MP)
- Qdot® 800 Antibody Conjugation Kit (Q22071MP)

The conjugation reaction involves reduction of antibody interchain disulfides followed by coupling of the resulting free thiols to amine-derivatized, PEG-coated Qdot® nanocrystals via the heterobifunctional crosslinker SMCC (S1534, Section 5.2). Primary antibodies labeled using these methods have demonstrated utility in applications as diverse as multiplexed immunodetection of

Table 6.12 Qdot® nanocrystal anti–human CD antibody conjugates.

Target	Clone	Isotype	Qdot® 565	Qdot® 605	Qdot® 655	Qdot® 705	Qdot® 800
Human CD2	S5.5	Mouse IgG_{2a}		Q10172[1]			
Human CD3	UCHT1	Mouse IgG_{2a}		Q10054[1]			
Human CD3	S4.1	Mouse IgG_{2a}			Q10012[4]		
Human CD4	S3.5	Mouse IgG_{2a}		Q10008[1]	Q10007[4]	Q10060[5]	
Human CD8	3B5	Mouse IgG_{2a}	Q10152	Q10009[1]	Q10055[4]	Q10059[5]	
Human CD10	MEM-78	Mouse IgG_1		Q10153			Q10154[6]
Human CD14	TüK4	Mouse IgG_{2a}		Q10013[1]	Q10056[4]		Q10064[7]
Human CD19	SJ25-C1	Mouse IgG_1		Q10306[2]	Q10179		
Human CD20	HI47	Mouse IgG_3			Q10305		
Human CD27	CLB-27/1	Mouse IgG_{2a}		Q10065	Q10066[4]		
Human CD38	HIT2	Mouse IgG_1		Q10053[2]	Q10057		
Human CD45	HI30	Mouse IgG_1		Q10051[2]		Q10062	Q10156[6]
Human CD45RA	MEM-56	Mouse IgG_{2b}		Q10047[3]	Q10069		
Human CD56	MEM-188	Mouse IgG_{2b}		Q10307[3]			
Human HLA-DR	TÜ36	Mouse IgG_{2b}		Q10052[3]			Q10063

1. Isotope control Q10014; **2.** Isotope control Q10073; **3.** Isotype control Q10074; **4.** Isotype control Q10015; **5.** Isotype control Q10076; **6.** Isotype control Q10298; **7.** Isotype control Q10075. All mouse anti–human CD antibody conjugates and isotype controls are supplied as 1 µM solutions in 100 µL units.

IMPORTANT NOTICE: The products described in this manual are covered by one or more Limited Use Label License(s). Please refer to the Appendix on page 971 and Master Product List on page 975. Products are For Research Use Only. Not intended for any animal or human therapeutic or diagnostic use.

biomarkers in formalin-fixed, paraffin-embedded (FFPE) tissue sections[41,63] and *in vivo* imaging of HER2-positive tumors in mice.[64]

Other Qdot® Nanocrystal Conjugates

We offer Qdot® 565 and Qdot® 655 conjugates of goat anti-fluorescein antibody and a Qdot® 655 conjugate of rat anti-dinitrophenol (anti-DNP) antibody, which complement the Alexa Fluor® dye–labeled conjugates described in Section 7.4:

- Qdot® 565 goat anti-fluorescein conjugate (Q15431MP)
- Qdot® 655 goat anti-fluorescein conjugate (Q15421MP)
- Qdot® 655 rat anti-dinitrophenol conjugate (Q17421MP)

Fluorescein and DNP are excellent affinity tags that can be recognized by labeled primary antibodies, providing an alternative to the traditional biotin–streptavidin system in applications such as *in situ* hybridization,[65] enzyme-linked immunosorbent assays (ELISA) and western blot analysis. T

he Qdot® 655 wheat germ agglutinin (WGA) conjugate (Q12021MP) extends our selection of dye-labeled WGA conjugates (Section 7.7) for labeling of *N*-acetylglucosamine and *N*-acetylneuraminic acid (sialic acid) residues of glycoproteins on cell surfaces.[58] WGA conjugates are also useful for labeling of bacterial cell wall peptidoglycans, chitin and cartilage glycosaminoglycans.

Qtracker® Cell Labeling Kits

Qtracker® Cell Labeling Kits provide Qdot® nanocrystals that have been functionalized on their surface with polyarginine peptides to facilitate spontaneous uptake by live cells:[66]

- Qtracker® 525 Cell Labeling Kit (Q25041MP)
- Qtracker® 565 Cell Labeling Kit (Q25031MP)
- Qtracker® 585 Cell Labeling Kit (Q25011MP)
- Qtracker® 605 Cell Labeling Kit (Q25001MP)
- Qtracker® 625 Cell Labeling Kit (A10198)
- Qtracker® 655 Cell Labeling Kit (Q25021MP)
- Qtracker® 705 Cell Labeling Kit (Q25061MP)
- Qtracker® 800 Cell Labeling Kit (Q25071MP)

The mechanism whereby such cell-penetrating peptides induce the cellular uptake of cargoes—including oligonucleotides and proteins as well as Qdot® nanocrystals—remains a topic of active investigation and considerable debate.[67,68] However, it is reasonably well established that the uptake is passive and does not require specific receptors or active transport processes. Therefore, labeling of live cells is a simple matter of adding the Qtracker® cell labeling reagents supplied in the kits to adherent or suspension cells in complete growth medium, incubating at 37°C for 45–60 minutes, followed by a wash step to remove the unincorporated labeling reagents. The Qdot® nanocrystals are stably incorporated in cytoplasmic vesicles (Figure 6.6.10) and are passed to daughter cells upon cell division but are not transferred to adjacent cells in mixed cultures or host tissues. Fluorescence is not impacted by complex and varying cellular environments such as changes in intracellular pH, temperature and metabolic activity. A wide range of applications for Qtracker® Cell Labeling Kits have been reported, including fiducial marking of cell populations for identification during or after downstream analysis,[66,69] labeling of mouse cerebral vascular tissue for visualization of explantation,[70] imaging the assembly of cultured pulmonary artery adventitial fibroblasts and endothelial cells into three-dimensional structures[71] and tracking embryonic and mesenchymal stem cells.[72–74]

Qtracker® Non-Targeted Quantum Dots

Qtracker® non-targeted quantum dots are designed for small animal *in vivo* imaging, and especially for studying vascular structure after microinjection[75,76] (Figure 6.6.11). Our selection of Qtracker® non-targeted quantum dots includes:

- Qtracker® 565 non-targeted quantum dots (Q21031MP)
- Qtracker® 655 non-targeted quantum dots (Q21021MP)
- Qtracker® 705 non-targeted quantum dots (Q21061MP)
- Qtracker® 800 non-targeted quantum dots (Q21071MP)

These nanocrystals exhibit intense red or near-infrared fluorescence emission, enabling maximum transmission through tissues while avoiding interference from background autofluorescence. Qtracker®

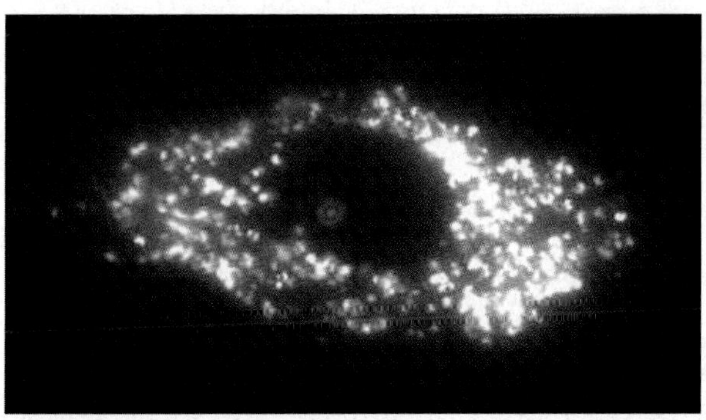

Figure 6.6.10 Distribution of Qdot® nanocrystals in cytoplasmic vesicles after labeling cells with the Qtracker® 655 Cell Labeling Kit (Q25021MP). HeLa cells were labeled with the Qtracker® 655 Cell Labeling Kit and then observed using a Leica TCS SP2 confocal microscope (excitation at 488 nm). This representative image shows the Qdot® nanocrystals distributed in vesicles throughout the cytoplasm.

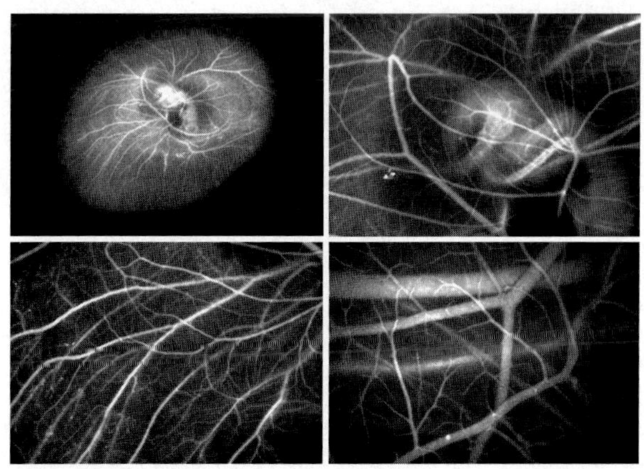

Figure 6.6.11 Chick embryo injected through the major vitelline vein with Qtracker® non-targeted quantum dots. Following a few minutes of circulation of the Qtracker® 705 non-targeted quantum dots (Q21061MP), fluorescence images of the embryo were captured at increasing magnification using 460 nm excitation and a digital imaging system equipped with appropriate emission filters. These Qtracker® reagents revealed highly detailed vascular structure at all levels of magnification. Images contributed by Greg Fisher, Carnegie Mellon University.

IMPORTANT NOTICE: The products described in this manual are covered by one or more Limited Use Label License(s). Please refer to the Appendix on page 971 and Master Product List on page 975. Products are For Research Use Only. Not intended for any animal or human therapeutic or diagnostic use.

non-targeted quantum dots have polyethylene glycol (PEG) surface coatings to minimize nonspecific binding interactions and associated inflammatory responses.[77,78] Because the PEG surface coating does not contain reactive functional groups, the Qtracker® non-targeted quantum dots are retained in circulation longer and can be imaged for up to 3 months without additional injections.[79] Qtracker® non-targeted quantum dots are supplied as 2 µM solutions in 50 mM borate buffer, pH 8.3 in units of 200 µL.

Qdot® ITK™ Quantum Dots

ITK™ (Innovator's Tool Kit) quantum dots enable researchers to equip Qdot® nanocrystals with customized functionalities by means of surface derivatization (Figure 6.6.12). Qdot® ITK™ quantum dots are available with three different surface chemistries—carboxyl groups, amino (PEG) groups and aliphatic hydrocarbon—and up to nine different fluorescent colors (525, 545, 565, 585, 605, 625, 655, 705 or 800 nm emission).

Qdot® ITK™ carboxyl quantum dots have a carboxyl-derivatized amphiphilic coating and can be coupled to amine groups of proteins and modified oligonucleotides in aqueous solution using EDAC-mediated condensation[55,80,81] (Section 3.4). Our Qdot® ITK™ carboxyl quantum dots are provided as 8 µM solutions and include:

- Qdot® 525 ITK™ carboxyl quantum dots (Q21341MP)
- Qdot® 545 ITK™ carboxyl quantum dots (Q21391MP)
- Qdot® 565 ITK™ carboxyl quantum dots (Q21331MP)
- Qdot® 585 ITK™ carboxyl quantum dots (Q21311MP)
- Qdot® 605 ITK™ carboxyl quantum dots (Q21301MP)
- Qdot® 625 ITK™ carboxyl quantum dots (A10200)
- Qdot® 655 ITK™ carboxyl quantum dots (Q21321MP)
- Qdot® 705 ITK™ carboxyl quantum dots (Q21361MP)
- Qdot® 800 ITK™ carboxyl quantum dots (Q21371MP)

Qdot® ITK™ amino (PEG) quantum dots have amine-derivatized PEG covalently attached to the amphiphilic inner coating and react efficiently with succinimidyl ester derivatives and other amine-reactive compounds.[68,82] Our Qdot® ITK™ amino (PEG) quantum dots are provided as 8 µM solutions and include:

- Qdot® 525 ITK™ amino (PEG) quantum dots (Q21541MP)
- Qdot® 545 ITK™ amino (PEG) quantum dots (Q21591MP)
- Qdot® 565 ITK™ amino (PEG) quantum dots (Q21531MP)
- Qdot® 585 ITK™ amino (PEG) quantum dots (Q21511MP)
- Qdot® 605 ITK™ amino (PEG) quantum dots (Q21501MP)
- Qdot® 655 ITK™ amino (PEG) quantum dots (Q21521MP)
- Qdot® 705 ITK™ amino (PEG) quantum dots (Q21561MP)
- Qdot® 800 ITK™ amino (PEG) quantum dots (Q21571MP)

Qdot® ITK™ organic quantum dots have an aliphatic hydrocarbon surface coating instead of an amphiphilic polymer coating. They are provided as a suspension in decane and are specifically designed for applications requiring organic solvents. Our Qdot® ITK™ organic quantum dots are provided as 1 µM solutions and include:

- Qdot® 545 ITK™ organic quantum dots (Q21791MP)
- Qdot® 565 ITK™ organic quantum dots (Q21731MP)
- Qdot® 585 ITK™ organic quantum dots (Q21711MP)
- Qdot® 605 ITK™ organic quantum dots (Q21701MP)
- Qdot® 655 ITK™ organic quantum dots (Q21721MP)
- Qdot® 705 ITK™ organic quantum dots (Q21761MP)
- Qdot® 800 ITK organic quantum dots (Q21771MP)

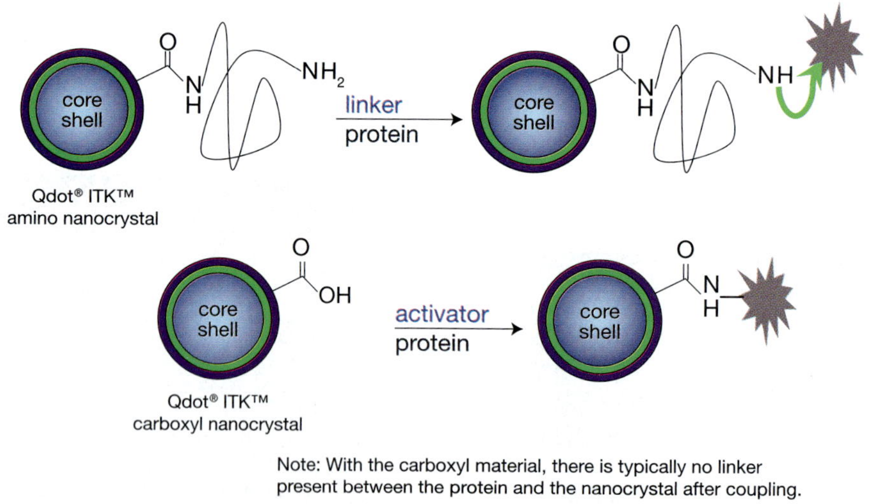

Figure 6.6.12 Coupling of Qdot® ITK™ quantum dots. Qdot® ITK™ amino (PEG) quantum dots can be coupled to biomolecules using amine-reactive crosslinking chemistries. Qdot® ITK™ carboxyl quantum dots can be coupled to amine-containing biomolecules by amide linkages generated by EDAC-mediated condensation.

IMPORTANT NOTICE: The products described in this manual are covered by one or more Limited Use Label License(s). Please refer to the Appendix on page 971 and Master Product List on page 975. Products are For Research Use Only. Not intended for any animal or human therapeutic or diagnostic use.
www.invitrogen.com/probes

REFERENCES

1. Nat Med (2006) 12:972; **2**. Bioconjug Chem (2005) 16:1488; **3**. Langmuir (2007) 23:1974; **4**. Toxicol Sci (2009) 110:138; **5**. Toxicol Appl Pharmacol (2008) 232:418; **6**. Biophys J (2008) 95:865; **7**. Nat Methods (2008) 5:763; **8**. Microsc Res Tech (2004) 65:169; **9**. Anal Biochem (2006) 357:68; **10**. J Am Chem Soc (2004) 126:301; **11**. Mol Ther (2008) 16:324; **12**. Angew Chem Int Ed Engl (2005) 44:7612; **13**. J Am Chem Soc (2006) 128:12800; **14**. Anal Biochem (2007) 364:193; **15**. Science (2003) 300:1434; **16**. Am J Physiol Regul Integr Comp Physiol (2006) 290:R114; **17**. J Neurosci Methods (2006) 151:276; **18**. J Histochem Cytochem (2003) 51:981; **19**. Nat Biotechnol (2003) 21:41; **20**. J Cell Sci (2005) 118:1091; **21**. J Phys Chem C Nanomater Interfaces (2009) 113:11541; **22**. Biophys J (2006) 91:3050; **23**. Proc Natl Acad Sci U S A (2005) 102:14284; **24**. J Am Chem Soc (2007) 129:8936; **25**. Biotechniques (2006) 41:295; **26**. Nat Methods (2005) 2:743; **27**. J Histochem Cytochem (2004) 52:13; **28**. Science (1998) 281:2013; **29**. Science (1998) 281:2016; **30**. Annu Rev Biomed Eng (2005) 7:55; **31**. Expert Rev Mol Diagn (2006) 6:231; **32**. Science (2005) 307:538; **33**. Nat Mater (2005) 4:435; **34**. Nano Lett (2005) 5:1693; **35**. Nucleic Acids Res (2005) 33:e17; **36**. Nucleic Acids Res (2005) 33:e161; **37**. J Mol Diagn (2007) 9:20; **38**. Clin Cancer Res (2009) 15:3568; **39**. Mod Pathol (2006) 19:1181; **40**. J Histochem Cytochem (2009) 57:701; **41**. Nat Protoc (2007) 2:1152; **42**. Biochem Biophys Res Commun (2008) 374:181; **43**. Nat Methods (2009) 6:520; **44**. Nat Methods (2008) 5:687; **45**. Proc Natl Acad Sci U S A (2007) 104:13666; **46**. Mol Biol Cell (2006) 17:4937; **47**. Proc Natl Acad Sci U S A (2007) 104:11251; **48**. BMC Neurosci (2009) 10:80; **49**. J Cell Biol (2005) 170:619; **50**. Methods Mol Biol (2007) 374:69; **51**. Biophys J (2009) 97:509; **52**. Proc Natl Acad Sci U S A (2007) 104:4332; **53**. Nat Cell Biol (2008) 10:955; **54**. Nat Mater (2005) 4:826; **55**. Mol Cell Probes (2007) 21:116; **56**. BMC Biotechnol (2009) 9:49; **57**. Appl Environ Microbiol (2007) 73:630; **58**. J Neurosci (2010) 30:4895; **59**. Nano Lett (2006) 6:2881; **60**. Nat Rev Microbiol (2008) 6:7; **61**. J Immunol Methods (2009) 344:6; **62**. Nat Protoc (2009) 4:372; **63**. Adv Mater (2007) 19:3146; **64**. Cancer Res (2007) 67:1138; **65**. Cytogenet Genome Res (2009) 124:351; **66**. Nano Lett (2004) 4:2019; **67**. Bioconjug Chem (2008) 19:1785; **68**. Bioconjug Chem (2009) 20:249; **69**. J Biomol Screen (2009) 14:845; **70**. Nat Neurosci (2008) 11:429; **71**. Am J Pathol (2006) 168:1793; **72**. Nano Lett (2006) 6:2826; **73**. Stem Cells (2007) 25:2128; **74**. BMC Biotechnol (2007) 7:67; **75**. Methods Mol Biol (2009) 544:393; **76**. Bioconjug Chem (2004) 15:79; **77**. J Invest Dermatol (2007) 127:143; **78**. Nano Lett (2007) 7:2513; **79**. Environ Health Perspect (2007) 115:1339; **80**. Nano Lett (2008) 8:2851; **81**. Nano Lett (2007) 7:2618; **82**. Clin Cancer Res (2008) 14:731.

PRODUCT LIST 6.6 QDOT® NANOCRYSTALS

Cat. No.	Product	Quantity
Q10172	CD2, mouse anti-human, Qdot® 605 conjugate (clone S5.5) *1 µM solution*	100 µL
Q10054	CD3, mouse anti-human, Qdot® 605 conjugate (clone UCHT1) *1 µM solution*	100 µL
Q10012	CD3, mouse anti-human, Qdot® 655 conjugate (clone S4.1) *1 µM solution*	100 µL
Q10008	CD4, mouse anti-human, Qdot® 605 conjugate (clone S3.5) *1 µM solution*	100 µL
Q10007	CD4, mouse anti-human, Qdot® 655 conjugate (clone S3.5) *1 µM solution*	100 µL
Q10060	CD4, mouse anti-human, Qdot® 705 conjugate (clone S3.5) *1 µM solution*	100 µL
Q10152	CD8, mouse anti-human, Qdot® 565 conjugate (clone 3B5) *1 µM solution*	100 µL
Q10009	CD8, mouse anti-human, Qdot® 605 conjugate (clone 3B5) *1 µM solution*	100 µL
Q10055	CD8, mouse anti-human, Qdot® 655 conjugate (clone 3B5) *1 µM solution*	100 µL
Q10059	CD8, mouse anti-human, Qdot® 705 conjugate (clone 3B5) *1 µM solution*	100 µL
Q10153	CD10, mouse anti-human, Qdot® 605 conjugate (clone MEM-78) *2 µM solution*	100 µL
Q10154	CD10, mouse anti-human, Qdot® 800 conjugate (clone MEM-78) *1 µM solution*	100 µL
Q10013	CD14, mouse anti-human, Qdot® 605 conjugate (clone TüK4) *2 µM solution*	100 µL
Q10056	CD14, mouse anti-human, Qdot® 655 conjugate (clone TüK4) *1 µM solution*	100 µL
Q10064	CD14, mouse anti-human, Qdot® 800 conjugate (clone TüK4) *1 µM solution*	100 µL
Q10306	CD19, mouse anti-human, Qdot® 605 conjugate (clone SJ25-C1) *1 µM solution*	100 µL
Q10179	CD19, mouse anti-human, Qdot® 655 conjugate (clone SJ25-C1) *1 µM solution*	100 µL
Q10305	CD20, mouse anti-human, Qdot® 655 conjugate (clone HI47) *1 µM solution*	100 µL
Q10065	CD27, mouse anti-human, Qdot® 605 conjugate (clone CLB-27/1) *2 µM solution*	100 µL
Q10066	CD27, mouse anti-human, Qdot® 655 conjugate (clone CLB-27/1) *2 µM solution*	100 µL
Q10053	CD38, mouse anti-human, Qdot® 605 conjugate (clone HIT2) *2 µM solution*	100 µL
Q10057	CD38, mouse anti-human, Qdot® 655 conjugate (clone HIT2) *2 µM solution*	100 µL
Q10051	CD45, mouse anti-human, Qdot® 605 conjugate (clone HI30) *1 µM solution*	100 µL
Q10062	CD45, mouse anti-human, Qdot® 705 conjugate (clone HI30) *1 µM solution*	100 µL
Q10156	CD45, mouse anti-human, Qdot® 800 conjugate (clone HI30) *1 µM solution*	100 µL
Q10047	CD45RA, mouse anti-human, Qdot® 605 conjugate (clone MEM-56) *1 µM solution*	100 µL
Q10069	CD45RA, mouse anti-human, Qdot® 655 conjugate (clone MEM-56) *1 µM solution*	100 µL
Q10307	CD56, mouse anti-human, Qdot® 605 conjugate (clone MEM-188) *2 µM solution*	100 µL
Q10052	HLA-DR, mouse anti-human, Qdot® 605 conjugate (clone Tu36) *1 µM solution*	100 µL
Q10063	HLA-DR, mouse anti-human, Qdot® 800 conjugate (clone Tu36) *1 µM solution*	100 µL
Q10073	Mouse IgG_1, Qdot® 605 conjugate (isotype control) *1 µM solution*	100 µL
Q10298	Mouse IgG_1, Qdot® 800 conjugate (isotype control) *1 µM solution*	100 µL
Q10014	Mouse IgG_{2a}, Qdot® 605 conjugate (isotype control) *1 µM solution*	100 µL
Q10015	Mouse IgG_{2a}, Qdot® 655 conjugate (isotype control) *1 µM solution*	100 µL

continued on next page

IMPORTANT NOTICE: The products described in this manual are covered by one or more Limited Use Label License(s). Please refer to the Appendix on page 971 and Master Product List on page 975. Products are For Research Use Only. Not intended for any animal or human therapeutic or diagnostic use.

PRODUCT LIST 6.6 QDOT® NANOCRYSTALS—*continued*

Cat. No.	Product	Quantity
Q10076	Mouse IgG_{2a}, Qdot® 705 conjugate (isotype control) *1 µM solution*	100 µL
Q10075	Mouse IgG_{2a}, Qdot® 800 conjugate (isotype control) *1 µM solution*	100 µL
Q10074	Mouse IgG_{2b}, Qdot® 605 conjugate (isotype control) *1 µM solution*	100 µL
Q22041MP	Qdot® 525 Antibody Conjugation Kit	1 kit
Q22031MP	Qdot® 565 Antibody Conjugation Kit	1 kit
Q22011MP	Qdot® 585 Antibody Conjugation Kit	1 kit
Q22001MP	Qdot® 605 Antibody Conjugation Kit	1 kit
A10197	Qdot® 625 Antibody Conjugation Kit	1 kit
Q22021MP	Qdot® 655 Antibody Conjugation Kit	1 kit
Q22061MP	Qdot® 705 Antibody Conjugation Kit	1 kit
Q22071MP	Qdot® 800 Antibody Conjugation Kit	1 kit
Q10301MP	Qdot® 605 Biotin Conjugate Kit	1 kit
Q10321MP	Qdot® 655 Biotin Conjugate Kit	1 kit
Q15431MP	Qdot® 565 goat anti-fluorescein conjugate *1 µM solution* *whole IgG*	200 µL
Q15421MP	Qdot® 655 goat anti-fluorescein conjugate *1 µM solution* *whole IgG*	200 µL
Q11201MP	Qdot® 605 goat $F(ab')_2$ anti-human IgG conjugate (H+L) *1 µM solution*	200 µL
Q11221MP	Qdot® 655 goat $F(ab')_2$ anti-human IgG conjugate (H+L) *1 µM solution*	200 µL
Q11041MP	Qdot® 525 goat $F(ab')_2$ anti-mouse IgG conjugate (H+L) *1 µM solution*	200 µL
Q11031MP	Qdot® 565 goat $F(ab')_2$ anti-mouse IgG conjugate (H+L) *1 µM solution*	200 µL
Q11032MP	Qdot® 565 goat $F(ab')_2$ anti-mouse IgG conjugate (H+L) *1 µM solution* *in 1 M betaine, 50 mM borate, pH 8.3 with 0.05% sodium azide*	100 µL
Q11011MP	Qdot® 585 goat $F(ab')_2$ anti-mouse IgG conjugate (H+L) *1 µM solution*	200 µL
Q11001MP	Qdot® 605 goat $F(ab')_2$ anti-mouse IgG conjugate (H+L) *1 µM solution*	200 µL
Q11002MP	Qdot® 605 goat $F(ab')_2$ anti-mouse IgG conjugate (H+L) *1 µM solution* *in 1 M betaine, 50 mM borate, pH 8.3 with 0.05% sodium azide*	100 µL
A10195	Qdot® 625 goat $F(ab')_2$ anti-mouse IgG conjugate (H+L) *1 µM solution*	100 µL
Q11021MP	Qdot® 655 goat $F(ab')_2$ anti-mouse IgG conjugate (H+L) *1 µM solution*	200 µL
Q11022MP	Qdot® 655 goat $F(ab')_2$ anti-mouse IgG conjugate (H+L) *1 µM solution* *in 1 M betaine, 50 mM borate, pH 8.3 with 0.05% sodium azide*	100 µL
Q11061MP	Qdot® 705 goat $F(ab')_2$ anti-mouse IgG conjugate (H+L) *1 µM solution*	200 µL
Q11062MP	Qdot® 705 goat $F(ab')_2$ anti-mouse IgG conjugate (H+L) *1 µM solution* *in 1 M betaine, 50 mM borate, pH 8.3 with 0.05% sodium azide*	100 µL
Q11071MP	Qdot® 800 goat $F(ab')_2$ anti-mouse IgG conjugate (H+L) *1 µM solution*	200 µL
Q11441MP	Qdot® 525 goat $F(ab')_2$ anti-rabbit IgG conjugate (H+L) *1 µM solution*	200 µL
Q11431MP	Qdot® 565 goat $F(ab')_2$ anti-rabbit IgG conjugate (H+L) *1 µM solution*	200 µL
Q11432MP	Qdot® 565 goat $F(ab')_2$ anti-rabbit IgG conjugate (H+L) *1 µM solution* *in 1 M betaine, 50 mM borate, pH 8.3 with 0.05% sodium azide*	100 µL
Q11411MP	Qdot® 585 goat $F(ab')_2$ anti-rabbit IgG conjugate (H+L) *1 µM solution*	200 µL
Q11401MP	Qdot® 605 goat $F(ab')_2$ anti-rabbit IgG conjugate (H+L) *1 µM solution*	200 µL
Q11402MP	Qdot® 605 goat $F(ab')_2$ anti-rabbit IgG conjugate (H+L) *1 µM solution* *in 1 M betaine, 50 mM borate, pH 8.3 with 0.05% sodium azide*	100 µL
A10194	Qdot® 625 goat $F(ab')_2$ anti-rabbit IgG conjugate (H+L) *1 µM solution*	100 µL
Q11421MP	Qdot® 655 goat $F(ab')_2$ anti-rabbit IgG conjugate (H+L) *1 µM solution*	200 µL
Q11422MP	Qdot® 655 goat $F(ab')_2$ anti-rabbit IgG conjugate (H+L) *1 µM solution* *in 1 M betaine, 50 mM borate, pH 8.3 with 0.05% sodium azide*	100 µL
Q11461MP	Qdot® 705 goat $F(ab')_2$ anti-rabbit IgG conjugate (H+L) *1 µM solution*	200 µL
Q11462MP	Qdot® 705 goat $F(ab')_2$ anti-rabbit IgG conjugate (H+L) *1 µM solution* *in 1 M betaine, 50 mM borate, pH 8.3 with 0.05% sodium azide*	100 µL
Q11471MP	Qdot® 800 goat $F(ab')_2$ anti-rabbit IgG conjugate (H+L) *1 µM solution*	200 µL
Q11631MP	Qdot® 565 goat $F(ab')_2$ anti-rat IgG conjugate (H+L) *1 µM solution*	200 µL
Q11601MP	Qdot® 605 goat $F(ab')_2$ anti-rat IgG conjugate (H+L) *1 µM solution*	200 µL
Q11621MP	Qdot® 655 goat $F(ab')_2$ anti-rat IgG conjugate (H+L) *1 µM solution*	200 µL
Q14421MP	Qdot® 655 goat whole IgG anti-chicken IgY (H+L) conjugate *1 µM solution* *in 1 M betaine, 50 mM borate, pH 8.3 with 0.05% sodium azide*	200 µL
Q20001MP	Qdot® incubation buffer	30 mL
Q21541MP	Qdot® 525 ITK™ amino (PEG) quantum dots *8 µM solution*	250 µL
Q21591MP	Qdot® 545 ITK™ amino (PEG) quantum dots *8 µM solution*	250 µL
Q21531MP	Qdot® 565 ITK™ amino (PEG) quantum dots *8 µM solution*	250 µL
Q21511MP	Qdot® 585 ITK™ amino (PEG) quantum dots *8 µM solution*	250 µL
Q21501MP	Qdot® 605 ITK™ amino (PEG) quantum dots *8 µM solution*	250 µL
Q21521MP	Qdot® 655 ITK™ amino (PEG) quantum dots *8 µM solution*	250 µL
Q21561MP	Qdot® 705 ITK™ amino (PEG) quantum dots *8 µM solution*	250 µL

IMPORTANT NOTICE: The products described in this manual are covered by one or more Limited Use Label License(s). Please refer to the Appendix on page 971 and Master Product List on page 975. Products are For Research Use Only. Not intended for any animal or human therapeutic or diagnostic use.

PRODUCT LIST 6.6 QDOT® NANOCRYSTALS—*continued*

Cat. No.	Product	Quantity
Q21571MP	Qdot® 800 ITK™ amino (PEG) quantum dots *8 µM solution*	250 µL
Q21341MP	Qdot® 525 ITK™ carboxyl quantum dots *8 µM solution*	250 µL
Q21391MP	Qdot® 545 ITK™ carboxyl quantum dots *8 µM solution*	250 µL
Q21331MP	Qdot® 565 ITK™ carboxyl quantum dots *8 µM solution*	250 µL
Q21311MP	Qdot® 585 ITK™ carboxyl quantum dots *8 µM solution*	250 µL
Q21301MP	Qdot® 605 ITK™ carboxyl quantum dots *8 µM solution*	250 µL
A10200	Qdot® 625 ITK™ carboxyl quantum dots *8 µM solution*	250 µL
Q21321MP	Qdot® 655 ITK™ carboxyl quantum dots *8 µM solution*	250 µL
Q21361MP	Qdot® 705 ITK™ carboxyl quantum dots *8 µM solution*	250 µL
Q21371MP	Qdot® 800 ITK™ carboxyl quantum dots *8 µM solution*	250 µL
Q21791MP	Qdot® 545 ITK™ organic quantum dots *1 µM solution*	4 mL
Q21731MP	Qdot® 565 ITK™ organic quantum dots *1 µM solution*	4 mL
Q21711MP	Qdot® 585 ITK™ organic quantum dots *1 µM solution*	4 mL
Q21701MP	Qdot® 605 ITK™ organic quantum dots *1 µM solution*	4 mL
Q21721MP	Qdot® 655 ITK™ organic quantum dots *1 µM solution*	4 mL
Q21761MP	Qdot® 705 ITK™ organic quantum dots *1 µM solution*	4 mL
Q21771MP	Qdot® 800 ITK™ organic quantum dots *1 µM solution*	4 mL
Q10041MP	Qdot® 525 ITK™ Streptavidin Conjugate Kit *2 µM solution*	1 kit
Q10091MP	Qdot® 545 ITK™ Streptavidin Conjugate Kit *2 µM solution*	1 kit
Q10031MP	Qdot® 565 ITK™ Streptavidin Conjugate Kit *2 µM solution*	1 kit
Q10011MP	Qdot® 585 ITK™ Streptavidin Conjugate Kit *2 µM solution*	1 kit
Q10001MP	Qdot® 605 ITK™ Streptavidin Conjugate Kit *2 µM solution*	1 kit
Q10021MP	Qdot® 655 ITK™ Streptavidin Conjugate Kit *2 µM solution*	1 kit
Q10061MP	Qdot® 705 ITK™ Streptavidin Conjugate Kit *2 µM solution*	1 kit
Q10071MP	Qdot® 800 ITK™ Streptavidin Conjugate Kit *2 µM solution*	1 kit
Q11821MP	Qdot® 655 rabbit F(ab')$_2$ anti-goat IgG conjugate (H+L) *1 µM solution* *in 1 M betaine, 50 mM borate, pH 8.3 with 0.05% sodium azide*	200 µL
Q17421MP	Qdot® 655 rat anti-dinitrophenol conjugate *1 µM solution* *whole monoclonal IgG*	200 µL
Q10141MP	Qdot® 525 streptavidin conjugate *1 µM solution*	200 µL
Q10131MP	Qdot® 565 streptavidin conjugate *1 µM solution*	200 µL
Q10111MP	Qdot® 585 streptavidin conjugate *1 µM solution*	200 µL
Q10101MP	Qdot® 605 streptavidin conjugate *1 µM solution*	200 µL
A10196	Qdot® 625 streptavidin conjugate *1 µM solution*	200 µL
Q10121MP	Qdot® 655 streptavidin conjugate *1 µM solution*	200 µL
Q10161MP	Qdot® 705 streptavidin conjugate *1 µM solution*	200 µL
Q10171MP	Qdot® 800 streptavidin conjugate *1 µM solution*	200 µL
Q10151MP	Qdot® Streptavidin Sampler Kit *1 µM solutions*	1 kit
Q12021MP	Qdot® 655 wheat germ agglutinin conjugate *1 µM solution*	200 µL
Q10336	Qmount® Qdot® mounting media	3 x 2 mL
Q25041MP	Qtracker® 525 Cell Labeling Kit	1 kit
Q25031MP	Qtracker® 565 Cell Labeling Kit	1 kit
Q25011MP	Qtracker® 585 Cell Labeling Kit	1 kit
Q25001MP	Qtracker® 605 Cell Labeling Kit	1 kit
A10198	Qtracker® 625 Cell Labeling Kit	1 kit
Q25021MP	Qtracker® 655 Cell Labeling Kit	1 kit
Q25061MP	Qtracker® 705 Cell Labeling Kit	1 kit
Q25071MP	Qtracker® 800 Cell Labeling Kit	1 kit
Q21031MP	Qtracker® 565 non-targeted quantum dots *2 µM solution*	200 µL
Q21021MP	Qtracker® 655 non-targeted quantum dots *2 µM solution*	200 µL
Q21061MP	Qtracker® 705 non-targeted quantum dots *2 µM solution*	200 µL
Q21071MP	Qtracker® 800 non-targeted quantum dots *2 µM solution*	200 µL
W10132	WesternDot™ 625 Goat Anti-Mouse IgG Western Blot Kit *20 minigel blots*	1 kit
W10142	WesternDot™ 625 Goat Anti-Rabbit IgG Western Blot Kit *20 minigel blots*	1 kit

IMPORTANT NOTICE: The products described in this manual are covered by one or more Limited Use Label License(s). Please refer to the Appendix on page 971 and Master Product List on page 975. Products are For Research Use Only. Not intended for any animal or human therapeutic or diagnostic use.

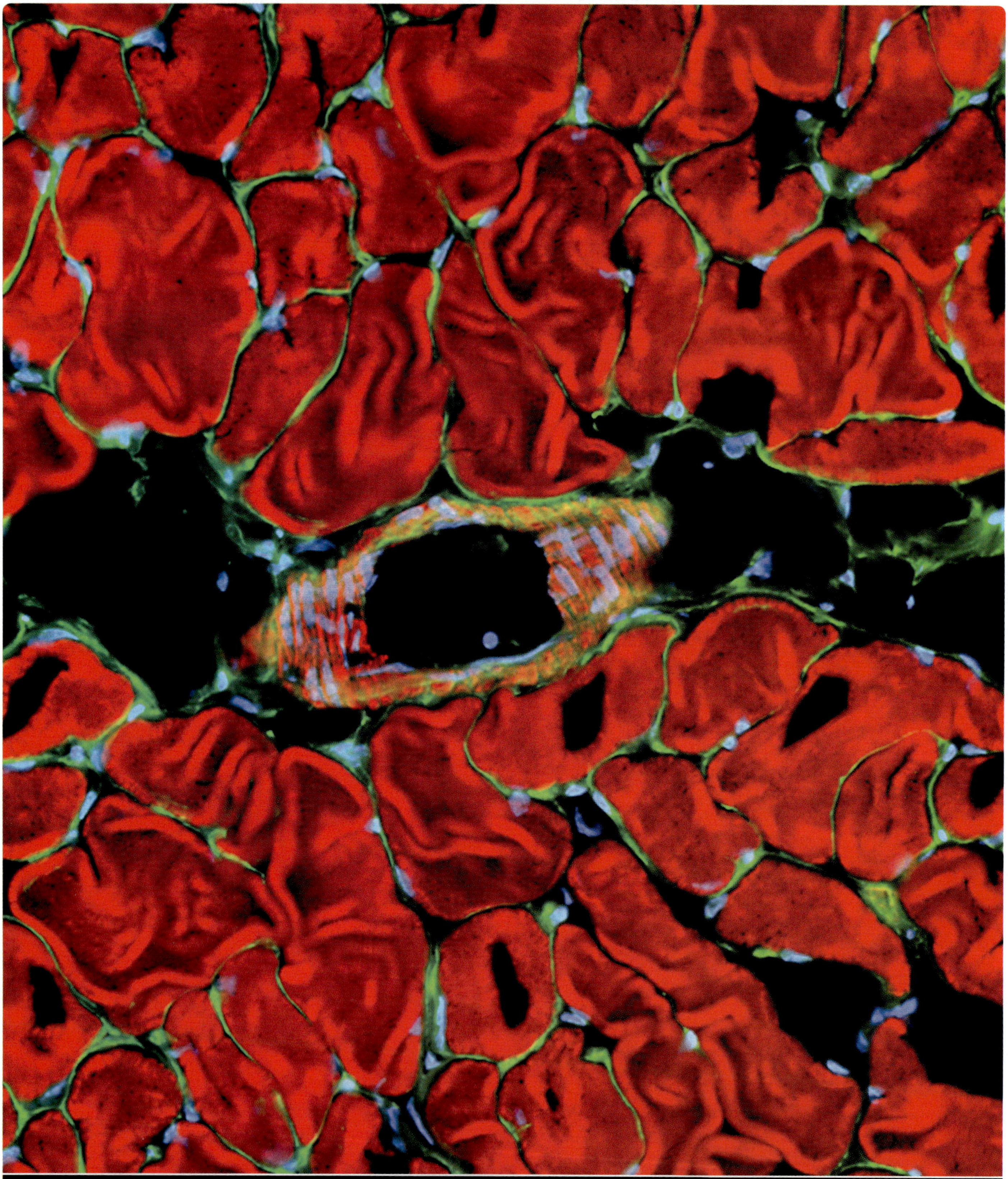

Alexa Fluor® 568 phalloidin, Oregon Green® wheat germ agglutinin and Hoechst 33342.

IMPORTANT NOTICE: The products described in this manual are covered by one or more Limited Use Label License(s). Please refer to the Appendix on page 971 and Master Product List on page 975. Products are For Research Use Only. Not intended for any animal or human therapeutic or diagnostic use.

SEVEN

CHAPTER 7
Antibodies, Avidins and Lectins

IMPORTANT NOTICE: The products described in this manual are covered by one or more Limited Use Label License(s). Please refer to the Appendix on page 971 and Master Product List on page 975. Products are For Research Use Only. Not intended for any animal or human therapeutic or diagnostic use.

IMPORTANT NOTICE: The products described in this manual are covered by one or more Limited Use Label License(s). Please refer to the Appendix on page 971 and Master Product List on page 975. Products are For Research Use Only. Not intended for any animal or human therapeutic or diagnostic use.

IMPORTANT NOTICE: The products described in this manual are covered by one or more Limited Use Label License(s). Please refer to the Appendix on page 971 and Master Product List on page 975. Products are For Research Use Only. Not intended for any animal or human therapeutic or diagnostic use.

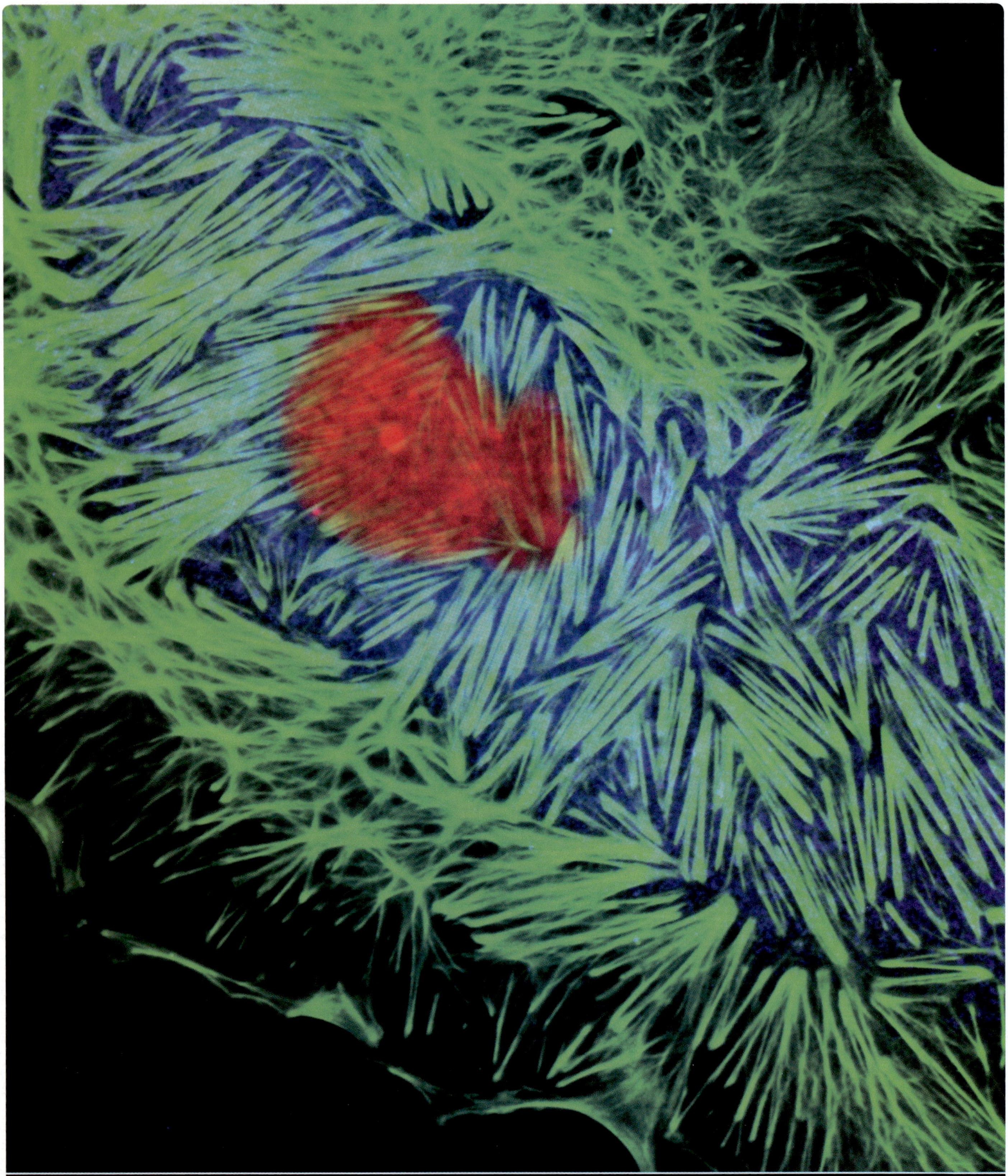

SYTOX® Orange, Alexa Fluor® 488 phalloidin and Alexa Fluor® 350 concanavalin A.

IMPORTANT NOTICE: The products described in this manual are covered by one or more Limited Use Label License(s). Please refer to the Appendix on page 971 and Master Product List on page 975. Products are For Research Use Only. Not intended for any animal or human therapeutic or diagnostic use.

7.1 Introduction to Antibodies, Avidins and Lectins

This chapter describes Molecular Probes® affinity reagents for protein detection. These reagents are primarily conjugated and unconjugated antibodies, augmented by avidin and streptavidin conjugates for detecting biotinylated proteins and lectin conjugates for detecting glycoproteins. Collectively, they represent the combination of our premium fluorophores with our bioconjugation expertise to produce an unsurpassed range of products for histochemical and cytochemical analysis.

Despite the advent of competing approaches such as recombinant peptide tagging and mass spectrometry, antibody-based detection remains the most broadly applicable means of localizing and quantitating specific proteins in complex samples.[1–3] Although the primary emphasis is on proteins, some of these reagents, particularly avidin and streptavidin conjugates, have important applications for the detection of nucleic acids[4,5] and other biomolecules. Non-antibody based approaches for protein detection are discussed in Chapter 9. Within this chapter, Molecular Probes® affinity reagents for protein detection are subdivided as described below.

Secondary Immunoreagents—Section 7.2

Labeled secondary antibodies form stable and specific complexes with unlabeled primary antibodies (Figure 7.1.1), providing the foundation for most immunofluorescence microscopy protocols. Molecular Probes® secondary antibody offering provides an extensive selection of labels that include our superior Alexa Fluor® dyes, phycobiliproteins, Alexa Fluor® dye–phycobiliprotein tandem fluorophores, Qdot® nanocrystals, biotin and enzyme labels (horseradish peroxidase and alkaline phosphatase). We also offer many options in terms of immunoreactivity, an essential consideration in avoiding confounding cross-reactivity when performing simultaneous secondary immunodetection of two or more targets.[6–8] Our labeled secondary antibody portfolio includes antibodies against IgG and IgM from several mammalian species, including various isotypes of mouse IgG, as well as against avian (chicken) IgY.

Zenon® Technology: Versatile Reagents for Immunolabeling—Section 7.3

Molecular Probes® Zenon® antibody labeling technology uses reagents that are technically secondary antibody conjugates in rapid immunolabeling protocols that would normally require preparation or purchase of directly labeled primary antibodies. Rapid processing enabled by elimination of the secondary reagent incubation step in conventional immunolabeling protocols is vital when identifying cells for mRNA extraction and downstream genomic analysis.[9] Elimination of the secondary reagent incubation step also enables the use of multiple antibodies derived from the same species in the same immunolabeling protocol.[10,11]

Anti-Dye and Anti-Hapten Antibodies—Section 7.4

Antibodies to fluorophores and other nonfluorescent labels provide unique opportunities for signal enhancement and for correlated fluorescence and electron microscopy studies.[12] For example, transplanted cells labeled with fluorescein-based tracers such as CFSE (C1157, V12883; Section 14.2) can be immunohistochemically detected using our anti–fluorescein/Oregon Green® antibodies in situations where the fluorescein fluorescence signal is undetectable due to cellular dissemination *in vivo*.[13–15] Dye/anti-dye antibody affinity pairs can also be used as alternatives to biotin/avidin detection methods in applications such as *in situ* hybridization and enzyme-linked immunosorbent assays (ELISAs).

Antibodies against Expression Tags—Section 7.5

This section presents a selected set of primary antibodies for detection of expression tags, including Green Fluorescent Protein (GFP) and Red Fluorescent Protein (RFP). The full range of Invitrogen™ primary antibodies can be surveyed using our online primary antibody search utility at www.invitrogen.com/handbook/antibodies. Direct labeling of primary antibodies can be accomplished via amine-reactive chemistry implemented in our comprehensive selection of convenient and easy-to-use protein labeling kits (Section 1.2). Alternatively, Zenon® immunolabeling technology (Section 7.3) can fulfill the same purposes and is less demanding in terms of antibody quantity and purity requirements.

Avidin and Streptavidin Conjugates—Section 7.6

Molecular Probes® fluorophore-conjugated avidins and streptavidins are extensively used in conjunction with biotinylated oligonucleotides or biotinylated antibodies for nucleic acid hybridization analysis,[4,16] immunohistochemistry[6,17] and multicolor flow cytometry.[18] Qdot® nanocrystal–streptavidin conjugates used for the same purposes are described in Section 6.6. Section 7.6 also features our enzyme-labeled avidins and streptavidins, NeutrAvidin™-coated microspheres and affinity matrices for capture and isolation of biotinylated proteins and nucleic acids. Biotin conjugates and the reagents and methods used for their preparation are described in Chapter 4.

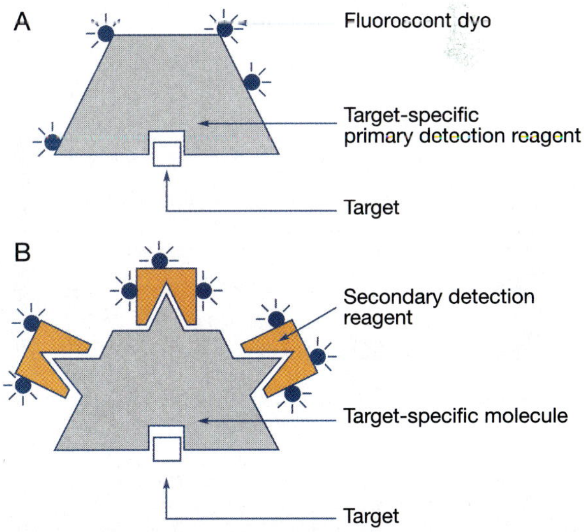

Figure 7.1.1 Schematic diagram of primary and secondary detection reagents. **A)** In primary detection methods, the target-specific molecule includes one or more detectable moieties, shown here as radiant orbs. **B)** In secondary detection methods, the target-specific molecule contains binding sites or haptens that can be selectively recognized by secondary detection reagents. For example, these sites might be antigenic epitopes that bind antibodies. Alternatively, the target-specific molecule might be conjugated to either biotin or fluorescent dyes, thereby creating a molecule that can be detected with any of our avidin and streptavidin conjugates or our anti–fluorescent dye antibodies. As shown here, the target-specific molecule may contain multiple sites for binding the secondary detection reagent, thereby providing a simple system for amplifying the signal.

Lectins and Other Carbohydrate-Binding Proteins—Section 7.7

Lectins and certain other carbohydrate-binding proteins that bind to specific glycans are versatile primary detection reagents for histochemical and cytochemical analysis of glycoproteins and glycolipids.[19–21] Complementary detection methods for glycoproteins based on Molecular Probes® Click-iT® labeling technology are described in Section 3.1 and Section 9.4.

REFERENCES

1. Biotechniques (2010) 48:197; 2. J Histochem Cytochem (2009) 57:1; 3. Mol Cell Proteomics (2008) 7:499; 4. Nat Protoc (2008) 3:563; 5. Methods Mol Biol (2003) 224:55; 6. Science (2010) 327:1380; 7. BMC Cell Biol (2008) 9:13; 8. J Neurosci (2007) 27:1836; 9. J Mol Diagn (2006) 8:246; 10. J Cell Biol (2009) 185:903; 11. Proc Natl Acad Sci U S A (2009) 106:6158; 12. J Histochem Cytochem (2006) 54:817; 13. Nat Biotechnol (2008) 26:215; 14. Am J Pathol (2006) 169:2223; 15. Development (1998) 125:201; 16. J Neurosci Methods (2007) 162:119; 17. J Histochem Cytochem (2009) 57:701; 18. Cytometry A (2008) 73:1001; 19. Development (2007) 134:211; 20. Am J Physiol Gastrointest Liver Physiol (2007) 293:G165; 21. J Histochem Cytochem (2007) 55:57.

7.2 Secondary Immunoreagents

We provide an extensive selection of Molecular Probes® secondary immunoreagents for use in fluorescence microscopy, flow cytometry and microplate assays, as well as protein and nucleic acid blots and microarrays. Because of their brightness and photostability, the Alexa Fluor® conjugates are superior to most conventional fluorescent secondary reagents (Figure 7.2.1, Figure 7.2.2, Figure 7.2.3) and are the detection reagents of choice for many immunofluorescence microscopy, *in situ* hybridization and bead-based applications. Properties of the organic fluorophores we use to prepare our conjugates are described in detail in Chapter 1. We also offer Qdot® secondary antibody conjugates (Table 6.11), which combine the fluorescence characteristics of Qdot® nanocrystals with the selective binding of $F(ab')_2$ fragments from affinity-purified secondary antibodies; the physical and spectroscopic properties of Qdot® nanocrystals are discussed in Section 6.6. In addition to our extensive line of species-specific antibodies, isotype-specific antibodies and $F(ab')_2$ fragments (Antibody Structure and Classification—Note 7.1), we prepare fluorescent conjugates of protein A and protein G—bacterial proteins that bind with high affinity to the Fc portion of various classes and subclasses of immunoglobulins from many species.

The next section of this chapter is devoted to our Zenon® antibody labeling technology (Section 7.3). The Zenon® antibody labeling technology uses affinity-purified dye- or enzyme-labeled Fab fragments of Fc-specific anti-IgG antibodies for the rapid and quantitative labeling

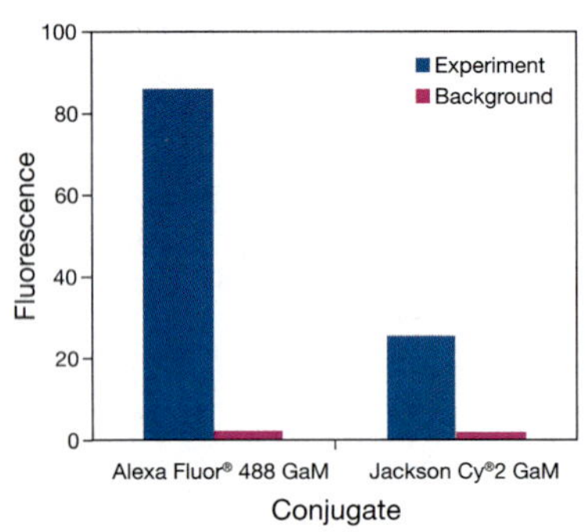

Figure 7.2.1 Brightness comparison of Molecular Probes® Alexa Fluor® 488 goat anti-mouse IgG antibody with Cy®2 goat anti-mouse IgG antibody from Jackson ImmunoResearch. Human blood was blocked with normal goat serum and incubated with an anti-CD3 mouse monoclonal antibody; cells were washed, resuspended and incubated with either Alexa Fluor® 488 or Cy®2 goat anti-mouse IgG antibody at equal concentrations. Red blood cells were lysed, and the samples were analyzed with a flow cytometer equipped with a 488 nm argon-ion laser and a 525 ± 10 nm bandpass emission filter.

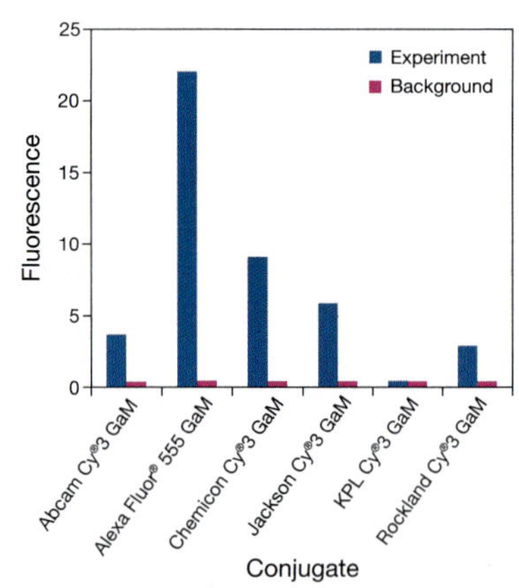

Figure 7.2.2 Brightness comparison of Molecular Probes® Alexa Fluor® 555 goat anti-mouse IgG antibody with Cy®3 goat anti-mouse IgG antibody conjugates commercially available from several other companies. Human blood was blocked with normal goat serum and incubated with an anti-CD3 mouse monoclonal antibody; cells were washed, resuspended and incubated with either the Alexa Fluor® 555 or Cy®3 goat anti-mouse IgG antibody at equal concentrations. Red blood cells were lysed and the samples were analyzed with a flow cytometer equipped with a 488 nm argon-ion laser and a 585 ± 21 nm bandpass emission filter.

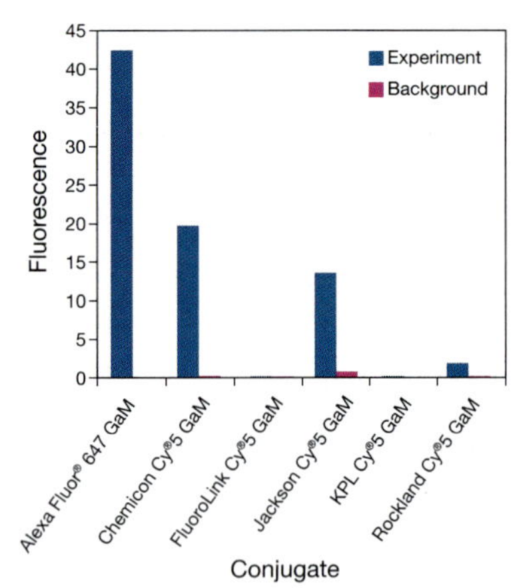

Figure 7.2.3 Brightness comparison of Molecular Probes® Alexa Fluor® 647 goat anti-mouse IgG antibody with Cy®5 goat anti-mouse IgG antibody conjugates commercially available from other companies. Human blood was blocked with normal goat serum and incubated with an anti-CD3 mouse monoclonal antibody; cells were washed, resuspended and incubated with either Alexa Fluor® 647 or Cy®5 goat anti-mouse IgG antibody at equal concentrations. Red blood cells were lysed and the samples were analyzed with a flow cytometer equipped with a 633 nm He-Ne laser and a longpass emission filter (>650 nm).

IMPORTANT NOTICE: The products described in this manual are covered by one or more Limited Use Label License(s). Please refer to the Appendix on page 971 and Master Product List on page 975. Products are For Research Use Only. Not intended for any animal or human therapeutic or diagnostic use.
www.invitrogen.com/probes

NOTE 7.1

Antibody Structure and Classification

The basic structural unit of most mammalian antibodies is a glycoprotein (MW ~150,000 daltons) comprising four polypeptide chains—two light chains and two heavy chains, which are connected by disulfide bonds (Figure 1). Each light chain has a molecular weight of ~25,000 daltons and is composed of two domains, one variable domain (V_L) and one constant domain (C_L). There are two types of light chains, lambda (λ) and kappa (κ). In humans, 60% of the light chains are κ, and 40% are λ, whereas in mice, 95% of the light chains are κ and only 5% are λ. A single antibody molecule contains either κ light chains or λ light chains, but never both.

Each heavy chain has a molecular weight of ~50,000 daltons and consists of a constant and variable region. The heavy and light chains contain a number of homologous sections consisting of similar but not identical groups of amino acid sequences. These homologous units consist of about 110 amino acids and are called immunoglobulin domains. The heavy chain contains one variable domain (V_H) and either three or four constant domains (C_H1, C_H2, C_H3 and C_H4, depending on the antibody class or isotype). The region between the C_H1 and C_H2 domains is called the hinge region and permits flexibility between the two Fab arms of the Y-shaped antibody molecule, allowing them to open and close to accommodate binding to two antigenic determinants separated by a fixed distance.

The heavy chain also serves to determine the functional activity of the antibody molecule. There are five antibody classes—IgG, IgA, IgM, IgE and IgD—which are distinguished by their heavy chains γ, α, μ, ε and δ, respectively (Table 1). The IgD, IgE and IgG antibody classes are each made up of a single structural unit, whereas IgA antibodies may contain either one or two units and IgM antibodies consist of five disulfide-linked structural units. IgG antibodies are further divided into four subclasses (often referred to as isotypes) although the nomenclature differs slightly depending on the species producing the antibody (Table 1).

Structure/function studies on IgG have been aided by the discovery that the proteolytic enzymes pepsin and papain cleave the molecule into specific fragments with specific biological properties. Treatment of an IgG molecule with pepsin generates the $F(ab')_2$ fragment, which broadly encompasses the two Fab regions linked by the hinge region. Because the $F(ab')_2$ molecule is bivalent, it is capable of precipitating an antigen. Papain cleaves the IgG molecule in the hinge region between the C_H1 and C_H2 domains to yield two identical Fab fragments, which retain their antigen-binding ability, and one non-antigen–binding fragment—the (Fc) region. The Fc region is glycosylated and has many effector functions (e.g., binding complement, binding to cell receptors on macrophages and monocytes), and serves to distinguish one class of antibody from another.

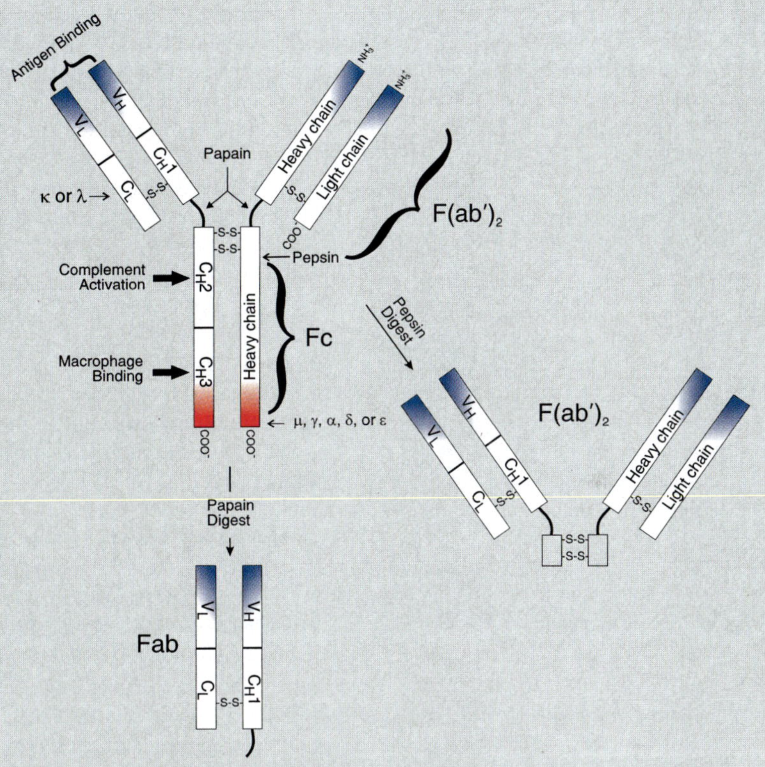

Figure 1 Schematic representation of an antibody molecule.

Table 1 Overview of antibody classes and subclasses.

Antibody	Human and Mouse					
	Light Chain	Subtype	Heavy Chain	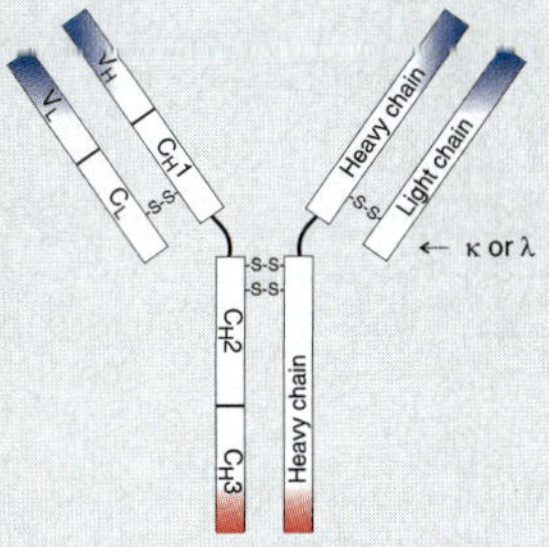		
IgA	κ or λ κ or λ	IgA_1 IgA_2	α_1 α_2			
IgE	κ or λ	None	ε			
IgD	κ or λ	None	δ			
IgM	κ or λ	None	μ			
	Human			Mouse		
	Light Chain	Subtype	Heavy Chain	Light Chain	Subtype	Heavy Chain
IgG	κ or λ κ or λ κ or λ κ or λ	IgG_1 IgG_2 IgG_3 IgG_4	γ_1 γ_2 γ_3 γ_4	κ or λ κ or λ κ or λ κ or λ	IgG_1 IgG_{2a} IgG_{2b} IgG_3	γ_1 γ_{2a} γ_{2b} γ_3

IMPORTANT NOTICE: The products described in this manual are covered by one or more Limited Use Label License(s). Please refer to the Appendix on page 971 and Master Product List on page 975. Products are For Research Use Only. Not intended for any animal or human therapeutic or diagnostic use.
www.invitrogen.com/probes

Table 7.1 Summary of Molecular Probes® secondary antibody conjugates.

Antibody	Host	Table No.
Anti-mouse IgG	Goat	Table 7.1A
Anti-mouse IgM	Goat	Table 7.1J
Anti-rabbit IgG	Goat	Table 7.1B
Anti-rat IgG	Goat	Table 7.1C
Anti-rat IgM	Goat	Table 7.1J
Anti-human IgG	Goat	Table 7.1D
Anti-human IgM	Goat	Table 7.1J
Anti-chicken IgG	Goat	Table 7.1G
Anti-guinea pig IgG	Goat	Table 7.1H
Anti-hamster IgG	Goat	Table 7.1I
Anti-mouse IgG	Rabbit	Table 7.1A
Anti-rat IgG	Rabbit	Table 7.1C
Anti-goat IgG	Rabbit	Table 7.1F
Anti-mouse IgG	Chicken	Table 7.1A
Anti-rabbit IgG	Chicken	Table 7.1B
Anti-rat IgG	Chicken	Table 7.1C
Anti-goat IgG	Chicken	Table 7.1F
Anti-mouse IgG	Donkey	Table 7.1A
Anti-rabbit IgG	Donkey	Table 7.1B
Anti-rat IgG	Donkey	Table 7.1C
Anti-sheep IgG	Donkey	Table 7.1E
Anti-goat IgG	Donkey	Table 7.1F

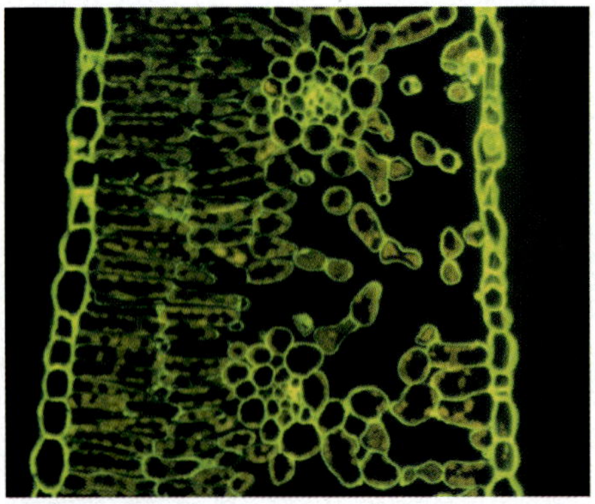

Figure 7.2.5 The primary cell walls in a 500 nm–thick apple leaf section identified with an antibody to the methyl-esterified regions of pectic polysaccharides or pectin, and visualized with green-fluorescent Alexa Fluor® 488 goat anti–rabbit IgG antibody (A11008). The orange regions inside the cells are due to the autofluorescent properties of chlorophyll localized in the chloroplasts. Image contributed by Paul Sutherland, EM Unit, Mount Albert Research Centre, Auckland, New Zealand.

of an intact mouse, rabbit, goat or human IgG antibody. The Zenon® labeling method has several advantages over use of the secondary antibodies discussed in this section, including:

- The Zenon® labeling method is suitable for quantitatively labeling submicrogram amounts of an antibody in a few minutes and is not affected by the presence of non-antibody proteins or amine-containing buffers in the sample.
- Multiple antibodies derived from the same species can be used in the same experiment.
- The conjugate's degree of substitution and brightness can be easily adjusted.
- The Zenon® labeled antibody complexes can be combined with labeled primary and secondary antibodies in most protocols.

The fluorescence intensity of the Zenon® labeled antibody complexes may be 2- to 3-fold less than that of labeled secondary antibodies.

Species-Specific Secondary Antibodies

Fluorescent Anti-IgG Antibodies

We offer fluorescent secondary antibody conjugates directed against IgG from a variety of species, including human, mouse, rabbit, rat, chicken, goat, guinea pig, hamster and sheep (Table 7.1). These anti-IgG antibodies are available with a wide selection of fluorophores, including our:

- Blue-fluorescent Alexa Fluor® 350 (Figure 7.2.4), Alexa Fluor® 405, Marina Blue®, Cascade Blue® and Pacific Blue™ dyes
- Green-fluorescent Alexa Fluor® 488 (Figure 7.2.5, Figure 7.2.6, Figure 7.2.7), Oregon Green® 488, Oregon Green® 514 (Figure 7.2.8), BODIPY® FL (Figure 7.2.9) and fluorescein dyes
- Yellow-green–fluorescent Alexa Fluor® 430 and Alexa Fluor® 514 dyes
- Yellow-fluorescent Alexa Fluor® 532 dye (Figure 7.2.10)
- Orange-fluorescent Pacific Orange™, Alexa Fluor® 546 (Figure 7.2.11), Alexa Fluor® 555, tetramethylrhodamine and R-phycoerythrin (R-PE) dyes
- Red-orange–fluorescent Alexa Fluor® 568 and Rhodamine Red™-X dyes
- Red-fluorescent Alexa Fluor® 594 (Figure 7.2.12), Alexa Fluor® 610, Texas Red® and Texas Red®-X dyes
- Far-red–fluorescent Alexa Fluor® 633, Alexa Fluor® 635, Alexa Fluor® 647 (Figure 7.2.13), Alexa Fluor® 660, Alexa Fluor® 680 and allophycocyanin dyes
- Infrared-fluorescent Alexa Fluor® 700 and Alexa Fluor® 750 dyes
- Alexa Fluor® dye–R-phycoerythrin (R-PE) tandem conjugates, which can each be excited with the 488 nm spectral line of the argon-ion laser, but exhibit long-wavelength emission maxima (627 nm for the Alexa Fluor® 610–R-PE conjugates, 667 nm for the Alexa Fluor® 647–R-PE conjugates and 702 nm for the Alexa Fluor® 680–R-PE conjugates)
- Alexa Fluor® dye–allophycocyanin (APC) tandem conjugates, which can each be excited by the He-Ne laser at 633 nm or 635 nm diode lasers with emission beyond 700 nm

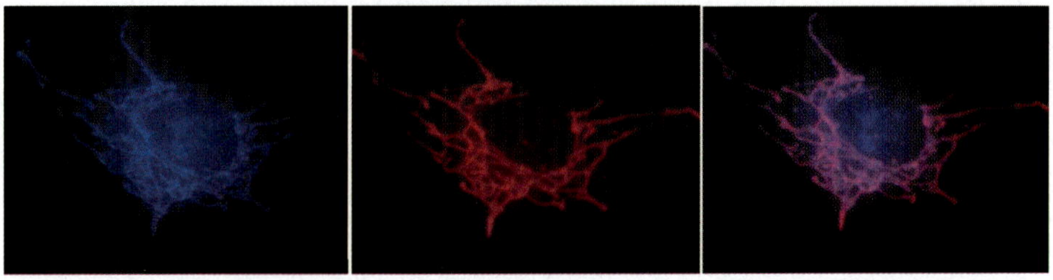

Figure 7.2.4 Live bovine pulmonary artery endothelial cells (BPAEC) loaded with red-fluorescent MitoTracker® Red CMXRos (M7512) (middle). After fixation and permeabilization, the cells were treated with a cocktail containing two mouse monoclonal antibodies to cytochrome oxidase (anti–OxPhos Complex IV subunit VIc and anti–OxPhos Complex IV subunit I antibodies), which were then detected using blue-fluorescent Alexa Fluor® 350 goat anti–mouse IgG antibody (A11045) (left). The image in the right panel is an overlay of the first two images.

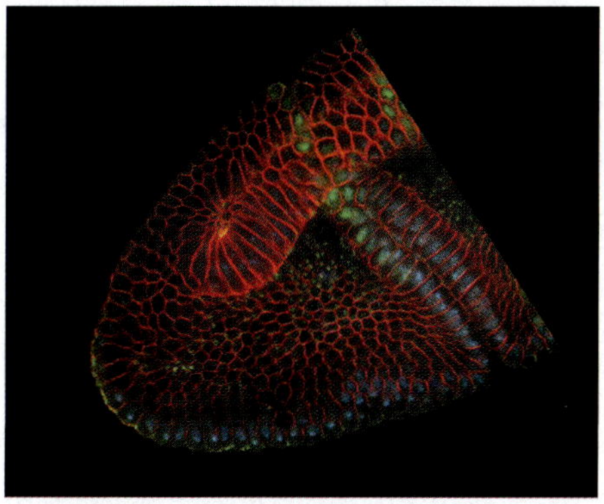

Figure 7.2.6 Formation of the cephalic furrow in the anterior end of a developing *Drosophila melanogaster* embryo visualized with the help of several fluorescent stains. A primary antibody to neurotactin was visualized using a red-fluorescent Cy®3 dye secondary antibody (Amersham Pharmacia® Biotech Ltd.). Primary antibodies to plasma membrane–bound myosin and to nuclear-localized even-skipped (Eve) protein were visualized with green-fluorescent Alexa Fluor® 488 goat anti–mouse IgG antibody (A11001) and Alexa Fluor® 488 goat anti–rabbit IgG antibody (A11008), respectively. The nuclei were stained with blue-fluorescent Hoechst 33342 (H1399, H3570, H21492). The sample was prepared by Eric Wieschaus, and the imaging was performed by Joe Goodhouse of Princeton University.

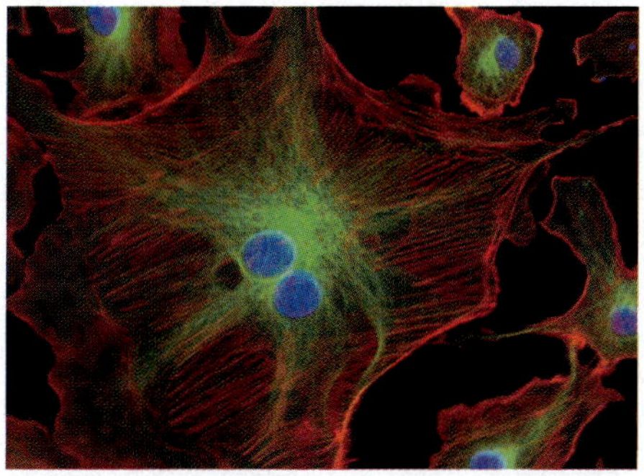

Figure 7.2.9 FluoCells® prepared slide #2 (F14781), which shows bovine pulmonary artery endothelial cells (BPAEC) that have been stained with a mouse monoclonal anti–β-tubulin antibody in conjunction with BODIPY® FL goat anti–mouse IgG antibody (B2752) for labeling microtubules, Texas Red®-X phalloidin (T7471) for labeling F-actin and DAPI (D1306, D3571, D21490) for labeling nuclei. This multiple-exposure image was acquired using bandpass optical filter sets appropriate for DAPI, fluorescein and Texas Red® dye.

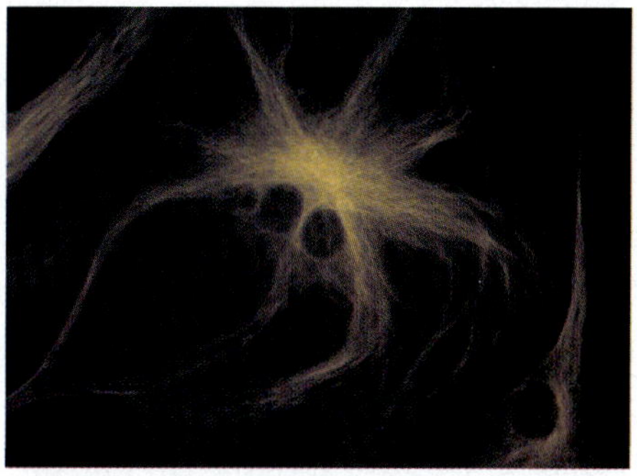

Figure 7.2.10 A bovine pulmonary artery endothelial cell (BPAEC) labeled with mouse monoclonal anti–α-tubulin antibody (A11126) in combination with Alexa Fluor® 532 goat anti–mouse IgG antibody (A11002) to stain microtubules. The image was acquired using a bandpass filter set (excitation/emission 535 ± 17.5/590 ± 17.5 nm).

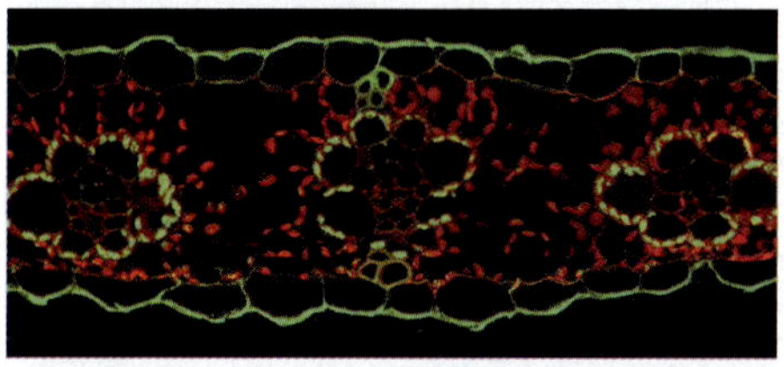

Figure 7.2.7 A 2.0 µm maize leaf section illustrating the immunolocalization of the enzyme ribulose bisphosphate carboxylase (rubisco) in the chloroplasts of the bundle sheath cells surrounding the vascular bundles. Maize is a C4 plant and, as a result, spatially segregates components of the photosynthetic process between the leaf mesophyll and the bundle sheath. Rubisco was localized using a rabbit anti-rubisco antibody and visualized using the highly cross-adsorbed Alexa Fluor® 488 goat anti-rabbit IgG antibody (A11034). The remaining fluorescence is due to the autofluorescence of chlorophyll, which appears red and is localized to the mesophyll plastids; lignin, which appears dull green and is localized to the xylem of the vascular bundle; and cutin, which appears bright green and is localized to the cuticle outside the epidermis. Image contributed by Todd Jones, DuPont.

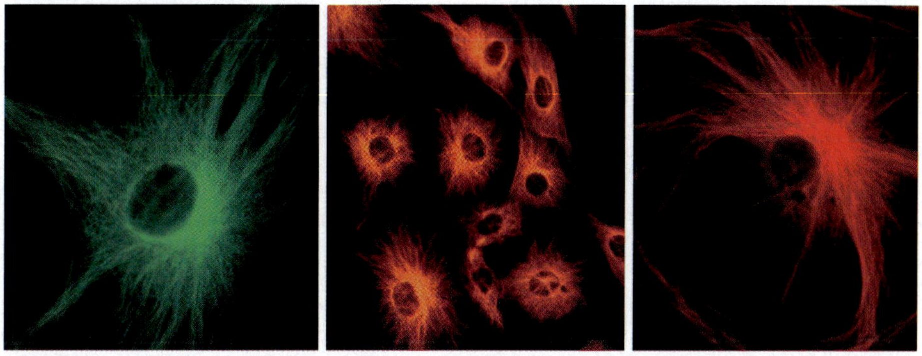

Figure 7.2.11 Microtubules of bovine pulmonary artery endothelial cells (BPAEC) tagged with mouse monoclonal anti–α-tubulin antibody (A11126) and subsequently probed with: Alexa Fluor® 488 goat anti–mouse IgG antibody (A11001, left), Alexa Fluor® 546 goat anti–mouse IgG antibody (A11003, middle) or Alexa Fluor® 594 goat anti–mouse IgG antibody (A11005, right). These images were acquired using a fluorescein bandpass optical filter set, a rhodamine bandpass optical filter set and a Texas Red® bandpass optical filter set, respectively.

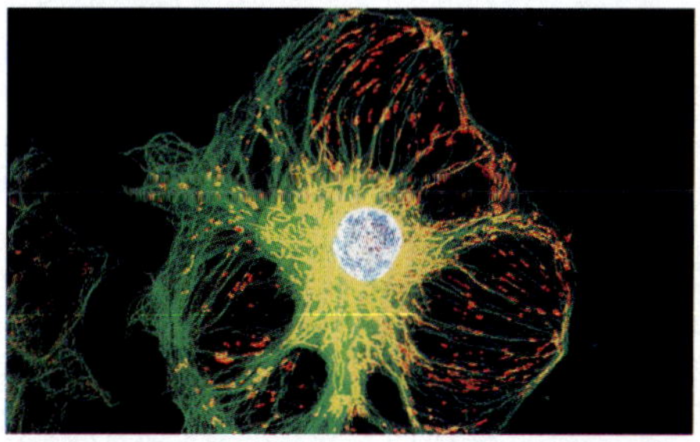

Figure 7.2.8 A bovine pulmonary artery endothelial cell (BPAEC) stained with mouse monoclonal anti–β-tubulin antibody in conjunction with Oregon Green® 514 goat anti–mouse IgG antibody (O6383), MitoTracker® Red CMXRos (M7512) and DAPI (D1306, D3571, D21490).

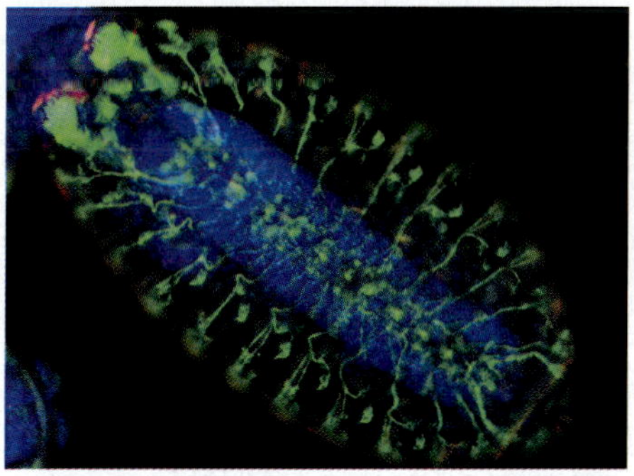

Figure 7.2.12 The peripheral nervous system of a wild type (Canton-S) *Drosophila melanogaster* embryo labeled with the monoclonal 22c10 antibody (which detects a microtubule-associated protein) and subsequently visualized using green-fluorescent Alexa Fluor® 488 rabbit anti–mouse IgG antibody (A11059). The actively dividing cells of the developing denticle bands were labeled with a rabbit anti-histone H3 antibody and visualized using red-fluorescent Alexa Fluor® 594 goat anti-rabbit IgG antibody (A11012). Finally, the nuclei, which are concentrated in the central nervous system, were counterstained with blue-fluorescent DAPI (D1306, D3571, D21490). Image contributed by Neville Cobbe, University of Edinburgh.

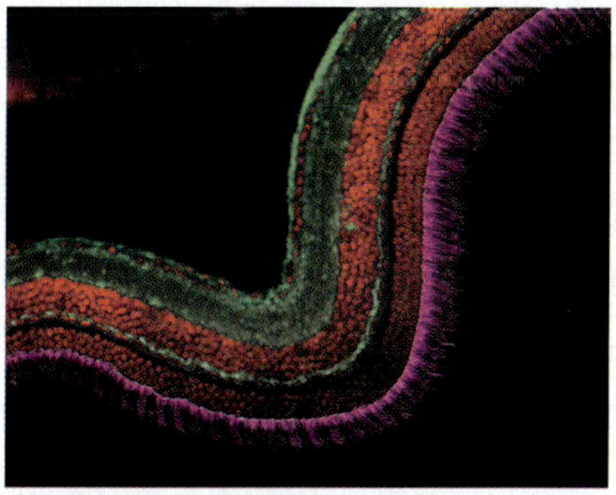

Figure 7.2.13 A zebrafish cryosection incubated with the biotin-XX conjugate of mouse monoclonal anti–α-tubulin antibody (A21371). The signal was amplified with TSA™ Kit #22, which includes a horseradish peroxidase–streptavidin conjugate and Alexa Fluor® 488 tyramide (T20932). The sample was then incubated with the mouse monoclonal FRET 6 antibody and was visualized with Alexa Fluor® 647 goat anti–mouse IgG antibody (A21235), which is pseudocolored magenta. Finally, the nuclei were counterstained with SYTOX® Orange nucleic acid stain (S11368).

Table 7.1A Molecular Probes® affinity-purified anti–mouse IgG antibody conjugates.

Cat. No.	Label	Emission Color	Abs *	Em *
Goat Anti–Mouse IgG				
A11045	Alexa Fluor® 350	Blue	346	442
M10991	Marina Blue®	Blue	365	460
C962	Cascade Blue®	Blue	400	420
A31553	Alexa Fluor® 405	Blue	402	421
P10993	Pacific Blue™	Blue	410	455
A11063	Alexa Fluor® 430	Yellow-green	434	539
F2761	Fluorescein	Green	494	518
A11001	Alexa Fluor® 488	Green	495	519
O6380	Oregon Green® 488	Green	496	524
B2752	BODIPY® FL	Green	505	513
O6383	Oregon Green® 514	Green	511	530
A31555	Alexa Fluor® 514	Yellow-green	518	540
A11002	Alexa Fluor® 532	Yellow	531	554
A10521	Cy®3	Orange	552	570
T2762	Tetramethylrhodamine	Orange	555	580
A11003	Alexa Fluor® 546	Orange	556	573
A21422	Alexa Fluor® 555	Orange	555	565
P852	R-phycoerythrin (R-PE)	Orange	496, 546, 565	578
A20980	Alexa Fluor® 610–R-PE	Red	496, 546, 565	628
A20990	Alexa Fluor® 647–R-PE	Far-red	496, 546, 565	668
A20983	Alexa Fluor® 680–R-PE	Far-red	496, 546, 565	702
R6393	Rhodamine Red™-X	Red-orange	570	590
A11004	Alexa Fluor® 568	Red-orange	578	603
A11005	Alexa Fluor® 594	Red	590	617
T862	Texas Red®	Red	595	615
T6390	Texas Red®-X	Red	595	615
A21050	Alexa Fluor® 633	Far-red	632	647
A31574	Alexa Fluor® 635	Far-red	633	647
A10524	Cy®5	Far-red	649	670
A865	Allophycocyanin	Far-red	650	660
A21000	Alexa Fluor® 680–allophycocyanin	Far-red	650	702
A21006	Alexa Fluor® 750–allophycocyanin	Infrared	650	775
A21235	Alexa Fluor® 647	Far-red	650	668
A21054	Alexa Fluor® 660	Far-red	663	690
A21057	Alexa Fluor® 680	Far-red	679	702
A21036	Alexa Fluor® 700	Infrared	702	723
A21037	Alexa Fluor® 750	Infrared	749	775
B2763	Biotin-XX	NA	<300	NA
G21060	Alkaline phosphatase	NA	NA	NA
G21040	Horseradish peroxidase	NA	NA	NA
G31567	β-Lactamase	NA	NA	NA
A10535	Unlabeled	NA	NA	NA
Goat Anti–Mouse IgG, Highly Cross-Adsorbed				
A21049	Alexa Fluor® 305	Blue	346	442
P31582	Pacific Blue™	Blue	410	455
A11029	Alexa Fluor® 488	Green	495	519
O11033	Oregon Green® 488	Green	496	524
A11030	Alexa Fluor® 546	Orange	556	573
A21424	Alexa Fluor® 555	Orange	555	565
A11031	Alexa Fluor® 568	Red-orange	578	603
A11032	Alexa Fluor® 594	Red	590	617
A21052	Alexa Fluor® 633	Far-red	632	647
A31575	Alexa Fluor® 635	Far-red	633	647
A21236	Alexa Fluor® 647	Far-red	650	668
A21055	Alexa Fluor® 660	Far-red	663	690
A21058	Alexa Fluor® 680	Far-red	679	702
D20691	DSB-X™ biotin	NA	<300	NA
Goat Anti–Mouse IgG, F(ab')$_2$ Fragments				
A11068	Alexa Fluor® 350	Blue	346	442
P31585	Pacific Orange™	Orange	400	551
P31581	Pacific Blue™	Blue	410	455
F11021	Fluorescein	Green	494	518
A11017	Alexa Fluor® 488	Green	495	519
A11018	Alexa Fluor® 546	Orange	556	573
A21425	Alexa Fluor® 555	Orange	555	565
A10543	R-phycoerythrin (R-PE)	Orange	496, 546, 565	578
A11019	Alexa Fluor® 568	Red-orange	578	603
A11020	Alexa Fluor® 594	Red	590	617
A21053	Alexa Fluor® 633	Far-red	632	647
A10539	Allophycocyanin	Far-red	650	660
A21237	Alexa Fluor® 647	Far-red	650	668
A21059	Alexa Fluor® 680	Far-red	679	702
B11027	Biotin-XX	NA	<300	NA
F21452	Alkaline phosphatase	NA	NA	NA
F21453	Horseradish peroxidase	NA	NA	NA
A10534	Unlabeled	NA	NA	NA
Goat Anti–Mouse IgG, IgM				
A10679	Fluorescein	Green	494	518
A10680	Alexa Fluor® 488	Green	495	519
A10677	Horseradish peroxidase	NA	NA	NA
Goat Anti–Mouse IgG, IgM; F(ab')$_2$ Fragments				
A10683	Fluorescein	Green	494	518
A10684	Alexa Fluor® 488	Green	495	519
Goat Anti–Mouse IgG, IgM, IgA				
A10667	Alexa Fluor® 488	Green	495	519
A10676	Biotin-XX	NA	<300	NA
A10668	Horseradish peroxidase	NA	NA	NA
A10666	Unlabeled	NA	NA	NA

continued on next page

More information on secondary antibodies is available at www.invitrogen.com/handbook/antibodies.

IMPORTANT NOTICE: The products described in this manual are covered by one or more Limited Use Label License(s). Please refer to the Appendix on page 971 and Master Product List on page 975. Products are For Research Use Only. Not intended for any animal or human therapeutic or diagnostic use.
www.invitrogen.com/probes
molecular probes® | invitrogen™ by life technologies™

Table 7.1A Molecular Probes® affinity-purified anti-mouse IgG antibody conjugates—*continued.*

Cat. No.	Label	Emission Color	Abs *	Em *
Rabbit Anti-Mouse IgG				
A21062	Alexa Fluor® 350	Blue	346	442
A11059	Alexa Fluor® 488	Green	495	519
A11060	Alexa Fluor® 546	Orange	556	573
A21427	Alexa Fluor® 555	Orange	555	565
A11061	Alexa Fluor® 568	Red-orange	578	603
A11062	Alexa Fluor® 594	Red	590	617
A21063	Alexa Fluor® 633	Far-red	632	647
A21239	Alexa Fluor® 647	Far-red	650	668
A21065	Alexa Fluor® 680	Far-red	679	702
R21455	Horseradish peroxidase	NA	NA	NA
Rabbit Anti-Mouse IgG, F(ab')$_2$ Fragments				
A21204	Alexa Fluor® 488	Green	495	519
A21205	Alexa Fluor® 594	Red	590	617
Chicken Anti-Mouse IgG				
A21200	Alexa Fluor® 488	Green	495	519
A21201	Alexa Fluor® 594	Red	590	617
A21463	Alexa Fluor® 647	Far-red	650	668
Donkey Anti-Mouse IgG				
A10035	Alexa Fluor® 350	Blue	346	442
A21202	Alexa Fluor® 488	Green	495	519
A10036	Alexa Fluor® 546	Orange	556	573
A31570	Alexa Fluor® 555	Orange	555	565
A10037	Alexa Fluor® 568	Red-orange	578	603
A21203	Alexa Fluor® 594	Red	590	617
A31571	Alexa Fluor® 647	Far-red	650	668
A10038	Alexa Fluor® 680	Far-red	679	702

* Absorption (Abs) and fluorescence emission (Em) maxima of conjugates, in nm. NA = Not applicable.

Table 7.1B Molecular Probes® affinity-purified anti-rabbit IgG antibody conjugates.

Cat. No.	Label	Emission Color	Abs *	Em *
Goat Anti-Rabbit IgG				
A11046	Alexa Fluor® 350	Blue	346	442
M10992	Marina Blue®	Blue	365	460
C2764	Cascade Blue®	Blue	400	420
P31584	Pacific Orange™	Orange	400	551
A31556	Alexa Fluor® 405	Blue	402	421
P10994	Pacific Blue™	Blue	410	455
A11064	Alexa Fluor® 430	Yellow-green	434	539
F2765	Fluorescein	Green	494	518
A11008	Alexa Fluor® 488	Green	495	519
O6381	Oregon Green® 488	Green	496	524
B2766	BODIPY® FL	Green	505	513
A31558	Alexa Fluor® 514	Yellow-green	518	540
A11009	Alexa Fluor® 532	Yellow	531	554
A10520	Cy®3	Orange	552	570
T2769	Tetramethylrhodamine	Orange	555	580
A11010	Alexa Fluor® 546	Orange	556	573
A21428	Alexa Fluor® 555	Orange	555	565
P2771MP	R-phycoerythrin (R-PE)	Orange	496, 546, 565	578

Table 7.1B Molecular Probes® affinity-purified anti-rabbit IgG antibody conjugates—*continued.*

Cat. No.	Label	Emission Color	Abs *	Em *
Goat Anti-Rabbit IgG—*continued*				
A20981	Alexa Fluor® 610–R-PE	Red	496, 546, 565	628
A20991	Alexa Fluor® 647–R-PE	Far-red	496, 546, 565	668
A20984	Alexa Fluor® 680–R-PE	Far-red	496, 546, 565	702
R6394	Rhodamine Red™-X	Red-orange	570	590
A11011	Alexa Fluor® 568	Red-orange	578	603
A11012	Alexa Fluor® 594	Red	590	617
T2767	Texas Red®	Red	595	615
T6391	Texas Red®-X	Red	595	615
A21070	Alexa Fluor® 633	Far-red	632	647
A31576	Alexa Fluor® 635	Far-red	633	647
A10523	Cy®5	Far-red	649	670
A10931	Allophycocyanin	Far-red	650	660
A21001MP	Alexa Fluor® 680–allophycocyanin	Far-red	650	702
A21244	Alexa Fluor® 647	Far-red	650	668
A21073	Alexa Fluor® 660	Far-red	663	690
A21076	Alexa Fluor® 680	Far-red	679	702
A21038	Alexa Fluor® 700	Infrared	702	723
A21039	Alexa Fluor® 750	Infrared	749	775
B2770	Biotin-XX	NA	<300	NA
G21079	Alkaline phosphatase	NA	NA	NA
G21234	Horseradish peroxidase	NA	NA	NA
G31568	β-Lactamase	NA	NA	NA
A10533	Unlabeled	NA	NA	NA
Goat Anti-Rabbit IgG, Highly Cross-Adsorbed				
A21068	Alexa Fluor® 350	Blue	346	442
A11034	Alexa Fluor® 488	Green	495	519
O11038	Oregon Green® 488	Green	496	524
A11035	Alexa Fluor® 546	Orange	556	573
A21429	Alexa Fluor® 555	Orange	555	565
A11036	Alexa Fluor® 568	Red-orange	578	603
A11037	Alexa Fluor® 594	Red	590	617
A21071	Alexa Fluor® 633	Far-red	632	647
A31577	Alexa Fluor® 635	Far-red	633	647
A21245	Alexa Fluor® 647	Far-red	650	668
A21074	Alexa Fluor® 660	Far-red	663	690
A21109	Alexa Fluor® 680	Far-red	679	702
Goat Anti-Rabbit IgG, F(ab')$_2$ Fragments				
A11069	Alexa Fluor® 350	Blue	346	442
A10526	Fluorescein	Green	494	518
A11070	Alexa Fluor® 488	Green	495	519
A11071	Alexa Fluor® 546	Orange	556	573
A21430	Alexa Fluor® 555	Orange	555	565
A10542	R-phycoerythrin (R-PE)	Orange	496, 546, 565	578
A21069	Alexa Fluor® 568	Red-orange	578	603
A11072	Alexa Fluor® 594	Red	590	617
A21072	Alexa Fluor® 633	Far-red	632	647
A21246	Alexa Fluor® 647	Far-red	650	668
A21077	Alexa Fluor® 680	Far-red	679	702

continued on next page

IMPORTANT NOTICE: The products described in this manual are covered by one or more Limited Use Label License(s). Please refer to the Appendix on page 971 and Master Product List on page 975. Products are For Research Use Only. Not intended for any animal or human therapeutic or diagnostic use.

Table 7.1B Molecular Probes® affinity-purified anti–rabbit IgG antibody conjugates—*continued.*

Cat. No.	Label	Emission Color	Abs *	Em *
Goat Anti–Rabbit IgG—*continued*				
B21078	Biotin-XX	NA	<300	NA
F21456	Alkaline phosphatase	NA	NA	NA
A10547	Horseradish peroxidase	NA	NA	NA
Chicken Anti–Rabbit IgG				
A21441	Alexa Fluor® 488	Green	495	519
A21442	Alexa Fluor® 594	Red	590	617
A21443	Alexa Fluor® 647	Far-red	650	668
Donkey Anti–Rabbit IgG				
A10039	Alexa Fluor® 350	Blue	346	442
A21206	Alexa Fluor® 488	Green	495	519
A10040	Alexa Fluor® 546	Orange	556	573
A31572	Alexa Fluor® 555	Orange	555	565
A10042	Alexa Fluor® 568	Red-orange	578	603
A21207	Alexa Fluor® 594	Red	590	617
A31573	Alexa Fluor® 647	Far-red	650	668
A10043	Alexa Fluor® 680	Far-red	679	702

* Absorption (Abs) and fluorescence emission (Em) maxima of conjugates, in nm. NA = Not applicable.

Table 7.1C Molecular Probes® affinity-purified anti–rat IgG antibody conjugates.

Cat. No.	Label	Emission Color	Abs *	Em *
Goat Anti–Rat IgG				
A21093	Alexa Fluor® 350	Blue	346	442
A10528	Fluorescein	Green	494	518
A11006	Alexa Fluor® 488	Green	495	519
O6382	Oregon Green® 488	Green	496	524
A10522	Cy®3	Orange	552	570
A10531	Tetramethylrhodamine	Orange	555	580
A11081	Alexa Fluor® 546	Orange	556	573
A21434	Alexa Fluor® 555	Orange	555	565
A10545	R-phycoerythrin (R-PE)	Orange	496, 546, 565	578
A11077	Alexa Fluor® 568	Red-orange	578	603
A11007	Alexa Fluor® 594	Red	590	617
T6392	Texas Red®-X	Red	595	615
A21094	Alexa Fluor® 633	Far-red	632	647
A10525	Cy®5	Far-red	649	670
A10540	Allophycocyanin	Far-red	650	660
A21247	Alexa Fluor® 647	Far-red	650	668
A21096	Alexa Fluor® 680	Far-red	679	702
A10517	Biotin-XX	NA	<300	NA
D20697	DSB-X™ biotin	NA	<300	NA
A10546	Alkaline phosphatase	NA	NA	NA
A10549	Horseradish peroxidase	NA	NA	NA
A10536	Unlabeled	NA	NA	NA
Goat Anti–Rat IgG, F(ab')$_2$ Fragments				
A10527	Fluorescein	Green	494	518
A10544	R-phycoerythrin (R-PE)	Orange	496, 546, 565	578
A10691	Cy®5–R-PE	Far-red	496, 546, 565	670
A10548	Horseradish peroxidase	NA	NA	NA

Table 7.1C Molecular Probes® affinity-purified anti–rat IgG antibody conjugates—*continued.*

Cat. No.	Label	Emission Color	Abs *	Em *
Rabbit Anti–Rat IgG				
A21210	Alexa Fluor® 488	Green	495	519
A21211	Alexa Fluor® 594	Red	590	617
Chicken Anti–Rat IgG				
A21470	Alexa Fluor® 488	Green	495	519
A21471	Alexa Fluor® 594	Red	590	617
A21472	Alexa Fluor® 647	Far-red	650	668
Donkey Anti–Rat IgG				
A21208	Alexa Fluor® 488	Green	495	519
A21209	Alexa Fluor® 594	Red	590	617

* Absorption (Abs) and fluorescence emission (Em) maxima of conjugates, in nm. NA = Not applicable.

Table 7.1D Molecular Probes® affinity-purified anti–human IgG antibody conjugates.

Cat. No.	Label	Emission Color	Abs *	Em *
Goat Anti–Human IgG				
A11013	Alexa Fluor® 488	Green	495	519
A21089	Alexa Fluor® 546	Orange	556	573
A21433	Alexa Fluor® 555	Orange	555	565
A21090	Alexa Fluor® 568	Red-orange	578	603
A11014	Alexa Fluor® 594	Red	590	617
A21091	Alexa Fluor® 633	Far-red	632	647
A21445	Alexa Fluor® 647	Far-red	650	668
Mouse Anti–Human IgG$_1$				
A10631	Alexa Fluor® 488	Green	495	519
A10650	Biotin-XX	NA	<300	NA
A10648	Horseradish peroxidase	NA	NA	NA
A10630	Unlabeled	NA	NA	NA
Mouse Anti–Human IgG$_4$				
A10663	Biotin-XX	NA	<300	NA
A10654	Horseradish peroxidase	NA	NA	NA
A10651	Unlabeled	NA	NA	NA

* Absorption (Abs) and fluorescence emission (Em) maxima of conjugates, in nm. NA = Not applicable.

Table 7.1E Molecular Probes® affinity-purified anti–sheep IgG antibody conjugates.

Cat. No.	Label	Emission Color	Abs *	Em *
Donkey Anti–Sheep IgG				
A21097	Alexa Fluor® 350	Blue	346	442
A11015	Alexa Fluor® 488	Green	495	519
A21098	Alexa Fluor® 546	Orange	556	573
A21436	Alexa Fluor® 555	Orange	555	565
A21099	Alexa Fluor® 568	Red-orange	578	603
A11016	Alexa Fluor® 594	Red	590	617
A21100	Alexa Fluor® 633	Far-red	632	647
A21448	Alexa Fluor® 647	Far-red	650	668
A21102	Alexa Fluor® 680	Far-red	679	702

* Absorption (Abs) and fluorescence emission (Em) maxima of conjugates, in nm.

IMPORTANT NOTICE: The products described in this manual are covered by one or more Limited Use Label License(s). Please refer to the Appendix on page 971 and Master Product List on page 975. Products are For Research Use Only. Not intended for any animal or human therapeutic or diagnostic use.

Table 7.1F Molecular Probes® affinity-purified anti–goat IgG antibody conjugates.

Cat. No.	Label	Emission Color	Abs *	Em *
Rabbit Anti–Goat IgG				
A10529	Fluorescein	Green	494	518
A11078	Alexa Fluor® 488	Green	495	519
A10532	Tetramethylrhodamine	Orange	555	580
A21085	Alexa Fluor® 546	Orange	556	573
A21431	Alexa Fluor® 555	Orange	555	565
A11079	Alexa Fluor® 568	Red-orange	578	603
A11080	Alexa Fluor® 594	Red	590	617
A21086	Alexa Fluor® 633	Far-red	632	647
A21446	Alexa Fluor® 647	Far-red	650	668
A21088	Alexa Fluor® 680	Far-red	679	702
A10518	Biotin-XX	NA	<300	NA
R21458	Alkaline phosphatase	NA	NA	NA
R21459	Horseradish peroxidase	NA	NA	NA
A10537	Unlabeled	NA	NA	NA
Rabbit Anti–Goat IgG, F(ab')$_2$ Fragments				
A21222	Alexa Fluor® 488	Green	495	519
A21223	Alexa Fluor® 594	Red	590	617
Chicken Anti–Goat IgG				
A21467	Alexa Fluor® 488	Green	495	519
A21468	Alexa Fluor® 594	Red	590	617
A21469	Alexa Fluor® 647	Far-red	650	668
Donkey Anti–Goat IgG				
A21081	Alexa Fluor® 350	Blue	346	442
A11055	Alexa Fluor® 488	Green	495	519
A11056	Alexa Fluor® 546	Orange	556	573
A21432	Alexa Fluor® 555	Orange	555	565
A11057	Alexa Fluor® 568	Red-orange	578	603
A11058	Alexa Fluor® 594	Red	590	617
A21082	Alexa Fluor® 633	Far-red	632	647
A21447	Alexa Fluor® 647	Far-red	650	668
A21083	Alexa Fluor® 660	Far-red	663	690
A21084	Alexa Fluor® 680	Far-red	679	702
D20698	DSB-X™ biotin	NA	<300	NA

* Absorption (Abs) and fluorescence emission (Em) maxima of conjugates, in nm. NA = Not applicable.

Table 7.1G Molecular Probes® affinity-purified anti–chicken IgG antibody conjugates.

Cat. No.	Label	Emission Color	Abs *	Em *
Goat Anti–Chicken IgG				
A11039	Alexa Fluor® 488	Green	495	519
A11040	Alexa Fluor® 546	Orange	556	573
A21437	Alexa Fluor® 555	Orange	555	565
A11041	Alexa Fluor® 568	Red-orange	578	603
A11042	Alexa Fluor® 594	Red	590	617
A21103	Alexa Fluor® 633	Far-red	632	647
A21449	Alexa Fluor® 647	Far-red	650	668
D20701	DSB-X™ biotin	NA	<300	NA

* Absorption (Abs) and fluorescence emission (Em) maxima of conjugates, in nm. NA = Not applicable.

Table 7.1H Molecular Probes® affinity-purified anti–guinea pig IgG antibody conjugates.

Cat. No.	Label	Emission Color	Abs *	Em *
Goat Anti–Guinea Pig IgG				
A11073	Alexa Fluor® 488	Green	495	519
A11074	Alexa Fluor® 546	Orange	556	573
A21435	Alexa Fluor® 555	Orange	555	565
A11075	Alexa Fluor® 568	Red-orange	578	603
A11076	Alexa Fluor® 594	Red	590	617
A21105	Alexa Fluor® 633	Far-red	632	647
A21450	Alexa Fluor® 647	Far-red	650	668

* Absorption (Abs) and fluorescence emission (Em) maxima of conjugates in nm.

Table 7.1I Molecular Probes® affinity-purified goat anti–hamster IgG antibody conjugates.

Cat. No.	Label	Emission Color	Abs *	Em *
Goat Anti–Hamster IgG				
A21110	Alexa Fluor® 488	Green	495	519
A21111	Alexa Fluor® 546	Orange	556	573
A21112	Alexa Fluor® 568	Red-orange	578	603
A21113	Alexa Fluor® 594	Red	590	617
A21451	Alexa Fluor® 647	Far-red	650	668

* Approximate absorption (Abs) and emission (Em) maxima of conjugates, in nm.

Table 7.1J Molecular Probes® affinity-purified goat anti–IgM (μ chain) antibody conjugates.

Fluorophore	Abs/Em *	Anti–Mouse IgM †	Anti–Rat IgM †	Anti–Human IgM †
Alexa Fluor® 350	346/442	A31552		
Alexa Fluor® 488	495/519	A21042	A21212	A21215
Alexa Fluor® 546	556/573	A21045		
Alexa Fluor® 555	555/565	A21426		
R-phycoerythrin	496, 546, 565/578	A10689 A10690 ‡		
Alexa Fluor® 568	578/603	A21043		
Alexa Fluor® 594	590/617	A21044	A21213	A21216
Alexa Fluor® 633	632/647	A21046		
Allophycocyanin	650/660	A10688		
Alexa Fluor® 647	650/668	A21238	A21248	A21249
Alexa Fluor® 680	679/702	A21048		
DSB-X™ biotin	NA	D20693		

* Absorption (Abs) and fluorescence emission (Em) maxima of conjugates, in nm. † May also react with IgM from other species. ‡ F(ab')$_2$ fragments. NA = Not applicable.

More information on secondary antibodies is available at www.invitrogen.com/handbook/antibodies.

IMPORTANT NOTICE: The products described in this manual are covered by one or more Limited Use Label License(s). Please refer to the Appendix on page 971 and Master Product List on page 975. Products are For Research Use Only. Not intended for any animal or human therapeutic or diagnostic use.

Our species-specific anti-IgG antibodies, which are raised against IgG heavy and light chains, are affinity purified and adsorbed against the sera of a number of species to minimize cross-reactivity. For multilabeling experiments in which cross-reactivity is critical, we offer highly cross-adsorbed goat anti–mouse IgG and goat anti–rabbit IgG antibodies. See the accompanying product information for a complete list of IgG and sera against which our anti-IgG antibodies have been cross-adsorbed.

Fluorescent Chicken IgY Antibodies

We also offer Alexa Fluor® conjugates of chicken IgY secondary antibodies, including anti–mouse IgG, –rabbit IgG, –rat IgG, –human IgG and –goat IgG antibodies. Chicken secondary antibodies have gained popularity because they demonstrate a lower level of nonspecific binding. Chicken IgY antibodies, which are functionally equivalent to mammalian IgG antibodies (and sometimes referred to as chicken IgG antibodies), lack a classical "Fc" domain and are not bound by protein A or protein G, nor do they bind to Fc receptors for mammalian IgG.

Fluorescent Anti-IgM Antibodies

Our anti-IgM conjugates are prepared from well-characterized antibodies that have been purified by IgM affinity chromatography. These anti-IgM conjugates react specifically with IgM heavy chains (μ chains) (Antibody Structure and Classification—Note 7.1). Due to their large size, IgM antibodies do not diffuse well into tissue. In addition, because the IgM μ chain is more highly conserved across different species than are IgG, IgA, or light chains, anti-IgM antibodies may react with IgM from other species.

Isotype-Specific Antibodies

We offer isotype-specific antibodies to aid in multilabeling experiments (Table 7.2). The Alexa Fluor® goat anti–mouse IgG isotype-specific antibodies have been cross-adsorbed against mouse IgM, IgA, pooled human sera, purified human paraproteins and other isotypes to minimize cross-reactivity.

$F(ab')_2$ Fragments

Our range of anti–mouse IgG, –rabbit IgG, –rat IgG and –goat IgG antibodies has been expanded to include fluorescent dye–, alkaline phosphatase–, horseradish peroxidase– and biotin-labeled $F(ab')_2$ fragments (Table 7.1). These $F(ab')_2$ fragments are often preferred to whole antibody conjugates because they lack the Fc region (Antibody Structure and Classification—Note 7.1), thereby eliminating nonspecific interactions with Fc receptor–bearing cell membranes and allowing for better penetration into tissue. Please note that the rabbit Fc region may bind nonspecifically to human tissue; consequently, we recommend the $F(ab')_2$ fragment when using rabbit-derived secondary antibodies on human tissues.

Table 7.2 Molecular Probes® goat anti-mouse isotype-specific antibodies.

Fluorophore	Abs/Em *	Goat Anti-Mouse Isotype-Specific Antibodies †			
		IgG_1 (γ_1)	IgG_{2a} (γ_{2a})	IgG_{2b} (γ_{2b})	IgG_3 (γ_3)
Alexa Fluor® 350	346/442	A21120	A21130	A21140	
Fluorescein	494/518	A10530			
Alexa Fluor® 488	495/519	A21121	A21131	A21141	A21151
Alexa Fluor® 546	556/573	A21123	A21133	A21143	
Alexa Fluor® 555	555/565	A21127	A21137	A21147	
R-phycoerythrin	496, 546, 565/578	P21129	P21139	P21149	
Alexa Fluor® 568	578/603	A21124	A21134	A21144	
Alexa Fluor® 594	590/617	A21125	A21135	A21145	A21155
Alexa Fluor® 633	632/647	A21126	A21136	A21146	
Allophycocyanin	650/660	A10541	A10686		
Alexa Fluor® 647	650/668	A21240	A21241	A21242	
Alexa Fluor® 680	679/702	A31562	A31563	A31564	
Horseradish peroxidase	NA	A10551	A10685		
Biotin-XX	NA	A10519			
Unlabeled	NA	A10538			

* Approximate absorption (Abs) and fluorescence emission (Em) maxima, in nm. † The Alexa Fluor® goat anti-mouse isotype-specific antibodies have been cross-adsorbed against mouse IgM, IgA, pooled sera, purified human paraproteins and the other isotypes to minimize cross-reactivity. NA = not applicable.

More information on secondary antibodies is available at www.invitrogen.com/handbook/antibodies.

Qdot® Secondary Antibody Conjugates

Because of their unique physical and spectroscopic properties, our Qdot® secondary antibody conjugates are discussed together with our complete line of Qdot® nanocrystal products in Section 6.6 (Table 6.11). The Qdot® nanocrystal secondary antibody conjugates exhibit the broad absorption spectra and intense fluorescence output characteristic of Qdot® nanocrystals along with the selective binding of $F(ab')_2$ fragments, enabling a wide range of immunocytochemical and immunohistochemical applications.[1,2] Some examples include immunolabeling of $GABA_A$ receptor α1 and γ2 subunits for colocalization analysis in HEK 293 cells[3] and immunodetection of EGFR, E-cadherin and cytokeratin on formalin-fixed, paraffin-embedded (FFPE) tissue array slides.[4]

Enzyme-Labeled and Biotinylated Secondary Antibodies

In addition to fluorescent secondary antibodies, we offer high-activity horseradish peroxidase and alkaline phosphatase conjugates of our species-specific secondary antibodies (Table 7.3), as well as several biotin conjugates, for enzyme-based signal amplification techniques. By using an avidin, streptavidin or NeutrAvidin™ biotin-binding protein bridge, the biotinylated secondary antibodies can be linked to a biotinylated enzyme—a method that is often preferred because it tends to reduce nonspecific staining. Enzyme and hapten conjugates of secondary antibodies are also commonly used in histochemical amplification schemes such as the tyramide signal amplification (TSA™) technology and Enzyme-Labeled Fluorescence (ELF®) technology, as well as in enzyme-linked immunosorbent assays (ELISAs) (Section 6.2, Section 6.3). The Zenon® Alkaline Phosphatase and Horseradish Peroxidase Antibody Labeling Kits (Section 7.3, Table 7.7) permit the formation of enzyme-labeled antibodies using submicrogram quantities of a primary antibody.

IMPORTANT NOTICE: The products described in this manual are covered by one or more Limited Use Label License(s). Please refer to the Appendix on page 971 and Master Product List on page 975. Products are For Research Use Only. Not intended for any animal or human therapeutic or diagnostic use.

molecular probes® | invitrogen™ by life technologies™

Image-iT® FX Signal Enhancer

By efficiently blocking nonspecific electrostatic interactions of anionic fluorescent dyes with cationic cell and tissue constituents, the Image-iT® FX signal enhancer (I36933) dramatically improves the signal-to-noise ratio of immunolabeled cells and tissues,[5–7] allowing clear visualization of targets that would normally be indistinguishable due to background fluorescence (Figure 7.2.14). Background staining seen with fluorescent conjugates of streptavidin (Table 7.4), goat anti–mouse IgG antibody or goat anti-rabbit IgG antibody is largely eliminated when Image-iT® FX signal enhancer is applied to fixed and permeabilized cells prior to staining. Image-iT® FX signal enhancer is particularly recommended for use with Alexa Fluor® secondary antibodies.

In addition to the stand-alone reagent, we offer the Image-iT® FX signal enhancer bundled with our Alexa Fluor® goat anti–mouse IgG or Alexa Fluor® goat anti-rabbit IgG secondary antibodies (either the standard or highly cross-adsorbed versions) in the Immunofluorescence SFX Detection Kits:

- Alexa Fluor® 488 Goat Anti-Mouse SFX Kit (standard, A31619; highly cross-adsorbed, A31620)
- Alexa Fluor® 555 Goat Anti-Mouse SFX Kit (standard, A31621; highly cross-adsorbed, A31622)
- Alexa Fluor® 594 Goat Anti-Mouse SFX Kit (standard, A31623; highly cross-adsorbed, A31624)
- Alexa Fluor® 647 Goat Anti-Mouse SFX Kit (standard, A31625; highly cross-adsorbed, A31626)
- Alexa Fluor® 488 Goat Anti-Rabbit SFX Kit (standard, A31627; highly cross-adsorbed, A31628)
- Alexa Fluor® 555 Goat Anti-Rabbit SFX Kit (standard, A31629; highly cross-adsorbed, A31630)
- Alexa Fluor® 594 Goat Anti-Rabbit SFX Kit (standard, A31631; highly cross-adsorbed, A31632)
- Alexa Fluor® 647 Goat Anti-Rabbit SFX Kit (standard, A31633; highly cross-adsorbed, A31634)

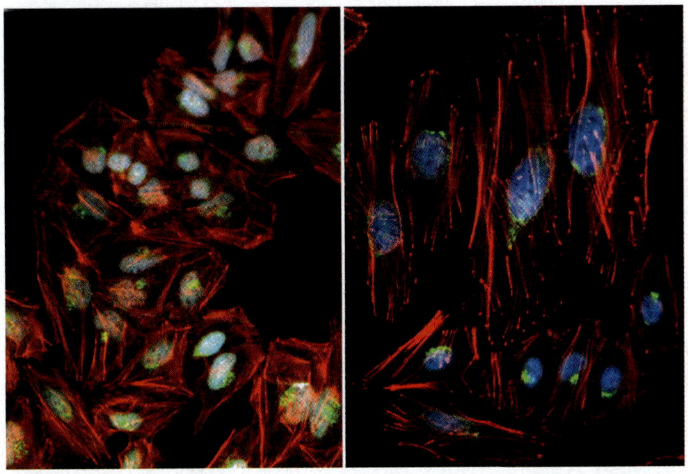

Figure 7.2.14 Golgi in fixed and permeabilized HeLa cells labeled with anti-golgin-97 antibody (A21270) and visualized with green-fluorescent Alexa Fluor® 488 goat anti–mouse IgG antibody (A11001). Actin was stained with red-fluorescent Alexa Fluor® 594 phalloidin (A12381); nuclei were stained with blue-fluorescent DAPI (D1306, D3571, D21490). Treatment with Image-iT® FX signal enhancer (I36933) largely eliminates nonspecific dye binding (right) as compared to untreated slide (left).

Table 7.3 Alkaline phosphatase and horseradish peroxidase enzyme conjugates.

Conjugate or Zenon® Antibody Labeling Kit	Enzyme Conjugate	
	Alkaline Phosphatase	Horseradish Peroxidase
Antibody (Host)		
Anti–mouse IgG (goat)	G21060	G21040
Anti–mouse IgG (γ1) (goat)		A10551
Anti–mouse IgG, IgM (goat)		A10677
Anti–mouse IgG, IgM, IgA (goat)		A10668
Anti–mouse IgG, F(ab')$_2$ fragment (goat)	F21452	F21453
Anti–mouse IgG (rabbit)		R21455
Anti–rabbit IgG (goat)	G21079	G21234
Anti–rabbit IgG, F(ab')$_2$ fragment (goat)	F21456	A10547
Anti–rat IgG (goat)		A10549
Anti–rat IgG, F(ab')$_2$ fragment (goat)		A10548
Anti–human IgG$_1$ (mouse)		A10648
Anti–human IgG$_4$ (mouse)		A10654
Anti–goat IgG (rabbit)	R21458	R21459
Anti–fluorescein/Oregon Green® (rabbit)		A21253
Anti–Green Fluorescent Protein (rabbit)		A10260
Zenon® Antibody Labeling Kit		
Zenon® Mouse IgG$_1$ Labeling Kit		Z25054
Zenon® Mouse IgG$_{2a}$ Labeling Kit		Z25154
Zenon® Mouse IgG$_{2b}$ Labeling Kit		Z25254
Zenon® Rabbit IgG Labeling Kit	Z25350	Z25354
Zenon® Human IgG Labeling Kit		Z25454
Avidin and Protein G Conjugates		
NeutrAvidin™ biotin-binding protein		A2664
Streptavidin	S921	S911
Protein G		P21041

Table 7.4 Fluorescent dyes successfully tested with the Image-iT® FX signal enhancer.

Dyes * with potentially strong background fluorescence that is reduced with the Image-iT® FX Signal Enhancer †		
Fluorescein	Atto 610	Alexa Fluor® 594
Oregon Green® 488	Cascade Blue®	Alexa Fluor® 610
Oregon Green® 514	Alexa Fluor® 405	Alexa Fluor® 633
Tetramethylrhodamine	Alexa Fluor® 430	Alexa Fluor® 635
Texas Red®	Alexa Fluor® 488	Alexa Fluor® 647
Cascade Yellow™		Alexa Fluor® 660
Dy 565	Alexa Fluor® 514	Alexa Fluor® 680
Dy 630	Alexa Fluor® 555	Alexa Fluor® 700
Atto 590	Alexa Fluor® 568	Alexa Fluor® 750
Dyes * with potentially weak background fluorescence that is reduced with the Image-iT® FX Signal Enhancer †		
Cy®5	Alexa Fluor® 532	Allophycocyanin
Dy 635	Alexa Fluor® 546	R-phycoerythrin
Marina Blue®		
Background-free fluorescent dyes * that are not affected by the Image-iT® FX Signal Enhancer ‡		
Alexa Fluor® 350	Pacific Blue™	Texas Red®-X
IRTM 790	Rhodamine B	Dy 550
Cy®3	Rhodamine Red™-X	Dy 610

* All dyes were conjugated to streptavidin and tested at 10 µg/mL. † Background staining was blocked by the Image-iT® FX signal enhancer. ‡ Staining was unaffected by the Image-iT® FX signal enhancer.

IMPORTANT NOTICE: The products described in this manual are covered by one or more Limited Use Label License(s). Please refer to the Appendix on page 971 and Master Product List on page 975. Products are For Research Use Only. Not intended for any animal or human therapeutic or diagnostic use.

Alexa Fluor® Signal Amplification Kits

Alexa Fluor® Signal Amplification Kits for Mouse Antibodies

The Alexa Fluor® Signal Amplification Kits for Mouse Antibodies provide a method for enhancing the detection of mouse primary antibodies using Alexa Fluor® 488, Alexa Fluor® 568 and Alexa Fluor® 594 dye conjugates, which yield green, red-orange and red fluorescence, respectively. These kits provide two Alexa Fluor® antibody conjugates to detect antibodies derived from mouse:

- The Alexa Fluor® 488 Signal Amplification Kit (A11054) provides Alexa Fluor® 488 rabbit anti–mouse IgG and Alexa Fluor® 488 goat anti–rabbit IgG antibodies and is compatible with fluorescein filter sets.
- The Alexa Fluor® 568 Signal Amplification Kit (A11066) provides Alexa Fluor® 568 rabbit anti–mouse IgG and Alexa Fluor® 568 goat anti–rabbit IgG antibodies and is compatible with tetramethylrhodamine filter sets.
- The Alexa Fluor® 594 Signal Amplification Kit (A11067) provides Alexa Fluor® 594 rabbit anti–mouse IgG and Alexa Fluor® 594 goat anti–rabbit IgG antibodies and it compatible with Texas Red® dye filter sets.

An Alexa Fluor® rabbit anti–mouse IgG antibody conjugate is first used to bind to the mouse-derived primary antibody. The fluorescence is then dramatically enhanced by the addition of an Alexa Fluor® goat anti–rabbit IgG antibody (Figure 7.2.15). The Alexa Fluor® 488, Alexa Fluor® 568 and Alexa Fluor® 594 Signal Amplification Kits for Mouse Antibodies can be used for both fluorescence microscopy and flow cytometry applications and contain sufficient materials for 60–300 assays.

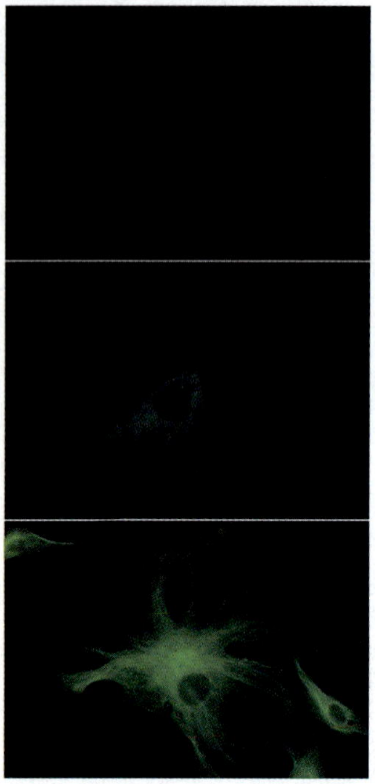

Figure 7.2.16 Demonstration of the amplification obtained with the Alexa Fluor® 488 Signal Amplification Kit for Fluorescein- and Oregon Green® Dye–Conjugated Probes (A11053). Bovine pulmonary artery endothelial cells were labeled with anti-α-tubulin antibody (A11126) in combination with fluorescein goat anti-mouse IgG antibody (F2761) (top). The center panel shows the cells after treatment with Alexa Fluor® 488 rabbit anti–fluorescein/Oregon Green® antibody (A11090), and the bottom panel show the cells after additional labeling with the Alexa Fluor® 488 goat anti–rabbit IgG antibody (A11008). The images were acquired using identical exposure times, and a bandpass filter set appropriate for fluorescein.

Alexa Fluor® 488 Signal Amplification Kit for Fluorescein- and Oregon Green® Dye–Conjugated Probes

The Alexa Fluor® 488 Signal Amplification Kit for Fluorescein-Conjugated Probes (A11053) is designed to simultaneously enhance the fluorescence and the photostability of virtually any fluorescein- or Oregon Green® dye–containing probe (Figure 7.2.16, Figure 7.2.17). This kit takes advantage of the superior spectral properties of Alexa Fluor® 488 conjugates. Alexa Fluor® 488 conjugates are considerably brighter and more photostable than fluorescein-labeled probes. In addition, the fluorescence of Alexa Fluor® 488 conjugates is not sensitive to pH over a wide pH range, unlike the fluorescence of fluorescein conjugates.

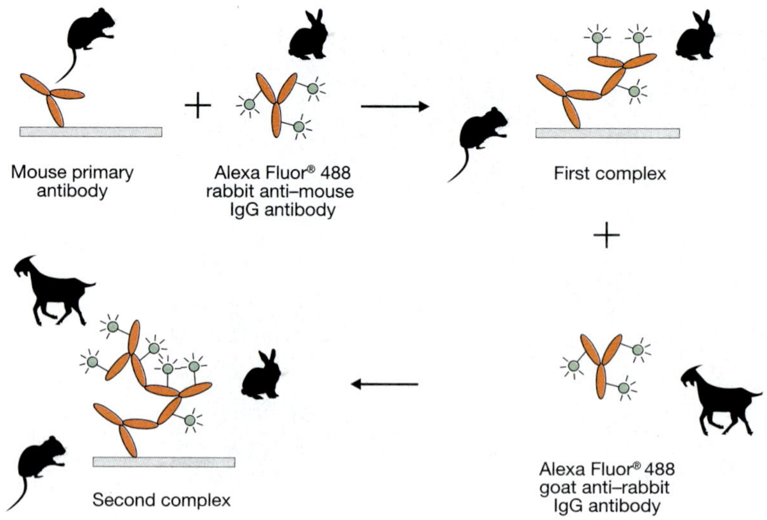

Figure 7.2.15 Antibody amplification scheme provided in the Alexa Fluor® 488 Signal Amplification Kits for Mouse Antibodies (A11054). The analogous amplification scheme is used in the Alexa Fluor® 568 (A11066) and Alexa Fluor® 594 (A11067) Signal Amplificaiton Kits for Mouse Antibodies. An Alexa Fluor® rabbit anti–mouse IgG antibody conjugate is first used to bind to the mouse primary antibody. The fluorescence signal is then dramatically enhanced by the addition of an Alexa Fluor® goat anti–rabbit IgG antibody.

IMPORTANT NOTICE: The products described in this manual are covered by one or more Limited Use Label License(s). Please refer to the Appendix on page 971 and Master Product List on page 975. Products are For Research Use Only. Not intended for any animal or human therapeutic or diagnostic use.
www.invitrogen.com/probes

The Alexa Fluor® 488 Signal Amplification Kit for Fluorescein- and Oregon Green® Dye–Conjugated Probes uses Alexa Fluor® 488 conjugates of two different antibodies to amplify the signals from fluorescein-labeled probes. Alexa Fluor® 488 anti–fluorescein/Oregon Green® antibody, which is prepared from a rabbit IgG fraction, is first used to bind to the fluorescein- or Oregon Green® dye–labeled target. The fluorescence signal is then dramatically enhanced by addition of the Alexa Fluor® 488 goat anti–rabbit IgG antibody. Because the spectra of Alexa Fluor® 488 conjugates are remarkably similar to those of fluorescein conjugates (Figure 7.2.18), this kit can be used with optical filters or instrument settings appropriate for fluorescein. The Alexa Fluor® 488 Signal Amplification Kit for Fluorescein-Conjugated Probes can be used for fluorescence microscopy, flow cytometry or other applications that employ fluorescein-conjugated probes; it contains sufficient reagents for 60–120 assays by microscopy or flow cytometry.

Fluorescence- and Chemiluminescence-Based ELISA Kits

Amplex® ELISA Development Kits for Mouse IgG and for Rabbit IgG

The Amplex® ELISA Development Kits for Mouse IgG and for Rabbit IgG (A33851, A33852; Section 6.2) provide a comprehensive set of components for creating fluorescence-based ELISAs using horseradish peroxidase conjugates of mouse and rabbit detection antibodies, respectively.[8] This assay is based on the Amplex® UltraRed reagent, a fluorogenic substrate for horseradish peroxidase (HRP) that reacts with H_2O_2 in a 1:1 stoichiometric ratio to produce a brightly fluorescent and strongly absorbing reaction product (excitation/emission maxima ~568/581 nm).

With a high extinction coefficient, good quantum efficiency and resistance to autooxidation, the fluorescence-based Amplex® UltraRed reagent allows for better sensitivity and a broader assay range than colorimetric reagents. In a sandwich ELISA format using C-reactive protein, we can routinely detect 75 pg of antigen using goat anti–mouse IgG antibody and 1 pg using goat anti–rabbit IgG antibody; these detection limits are 25-fold lower than those obtained from the same sandwich ELISA format using the common colorimetric reagent TMB. The Amplex® ELISA Development Kits for Mouse IgG and for Rabbit IgG are described in detail in Section 6.2.

Chemiluminescent Alkaline Phosphatase ELISA Kits

We offer four different Chemiluminescent Alkaline Phosphatase ELISA Kits, as well as a Chemiluminescent Alkaline Phosphatase ELISA Sampler Kit. These enzyme-linked immunoassays are based on the CSPD® or CDP-*Star*® 1,2-dioxetane substrates for alkaline phosphatase with Sapphire™-II or Emerald-II™ enhancer in a system designed for rapid and ultrasensitive analyte detection.[9] Each kit provides concentrated assay buffer and blocking reagent, as well as one of the following substrate/enhancer solutions:

- CSPD® substrate/Sapphire™-II enhancer (in Kit #1, C10552)
- CSPD® substrate/Emerald-II™ enhancer (in Kit #2, C10553)
- CDP-*Star*® substrate/Sapphire™-II enhancer (in Kit #3, C10554)
- CDP-*Star*® substrate/Emerald-II™ enhancer (in Kit #4, C10555)

The Chemiluminescent Alkaline Phosphatase ELISA Sampler Kit (C10556) provides assay buffer, blocking reagent and sample sizes of all four substrate/enhancer solutions. All five kits are described in detail in Section 6.3.

Fluorescence-Based Western Blot Kits

DyeChrome™ Double Western Blot Stain Kit

The DyeChrome™ Double Western Blot Stain Kit (D21887) provides a method for the simultaneous trichromatic detection of multiple

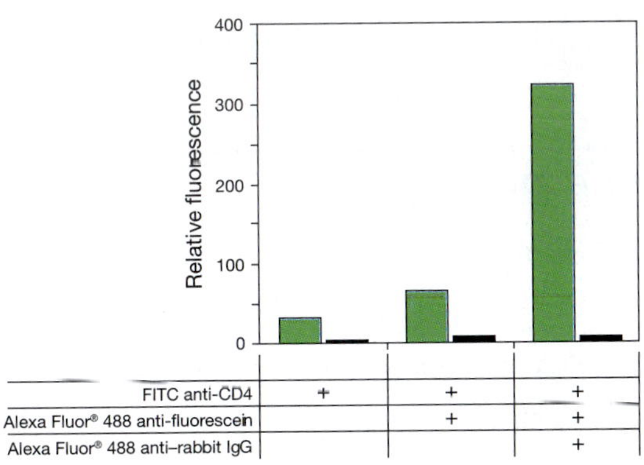

FITC anti-CD4	+	+	+
Alexa Fluor® 488 anti-fluorescein		+	+
Alexa Fluor® 488 anti-rabbit IgG			+

Figure 7.2.17 An example of flow cytometry results obtained using the Alexa Fluor® 488 Signal Amplification Kit for Fluorescein- and Oregon Green® Dye–Conjugated Probes (A11053). Human T-cell leukemia cells (Jurkat) were stained with fluorescein (FITC) mouse anti-CD4 antibody and, as indicated, with Alexa Fluor® 488 rabbit anti–fluorescein/Oregon Green® antibody (A11090) and Alexa Fluor® 488 goat anti–rabbit IgG antibody (A11008). The fluorescence values of the negative controls, in which the FITC anti-CD4 antibody was omitted, are shown (black) together with the fluorescence values of the experimental samples (green). The fluorescence values represent the average signals from the population of cells analyzed.

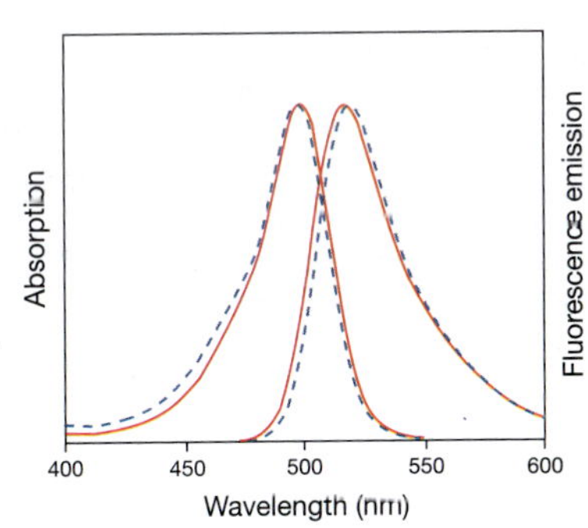

Figure 7.2.18 Absorption and fluorescence emission spectra of fluorescein goat anti–mouse IgG antibody (F2761, (–)) and Alexa Fluor® 488 goat anti–mouse IgG antibody (A11001, (- - -)). The fluorescence intensity of the Alexa Fluor® 488 conjugate was significantly higher than that of the fluorescein conjugate. The data are normalized to show the spectral similarity.

IMPORTANT NOTICE: The products described in this manual are covered by one or more Limited Use Label License(s). Please refer to the Appendix on page 971 and Master Product List on page 975. Products are For Research Use Only. Not intended for any animal or human therapeutic or diagnostic use.

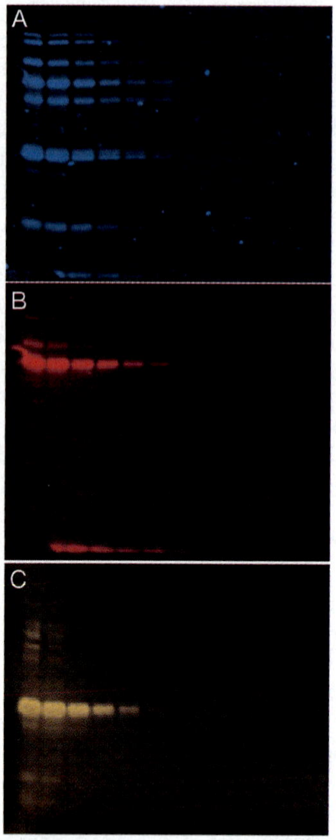

Figure 7.2.19 A total-protein profile and two specific protein bands visualized on a blot using the DyeChrome™ Double Western Blot Stain Kit. A 2-fold dilution series of a protein mixture containing bovine serum albumin (BSA), tubulin, ovalbumin, carbonic anhydrase and soybean trypsin inhibitor (from 1 µg to 0.24 ng each) was separated by electrophoresis through a 13% SDS-polyacrylamide gel and blotted onto a PVDF membrane. The DyeChrome™ Double Western Blot Stain Kit (D21887) was used, together with two primary antibodies, to stain all proteins and to visualize two specific proteins. A) The total-protein profile was stained with the blue-fluorescent dye MDPF. B) Tubulin was detected using mouse monoclonal anti–α-tubulin antibody followed by an alkaline phosphatase conjugate of goat anti–mouse IgG antibody, along with DDAO phosphate (red fluorescence). C) BSA was detected using a rabbit IgGt antibody against BSA followed by a horseradish peroxidase conjugate of goat anti–rabbit IgG antibody, along with the Amplex® Gold reagent (yellow fluorescence). The fluorescent signals were detected separately using appropriate excitation light and emission filters on either the Fluor-S MAX MultiImager documentation system (Bio-Rad Laboratories) or the FLA3000G laser scanner (Fuji Photo Film Co.).

targets on the same blot using two different enzyme-conjugated secondary antibodies and a general protein stain (Figure 7.2.19).

The components of the DyeChrome™ Double Western Blot Stain Kit are:

- Horseradish peroxidase (HRP) conjugate of goat anti–rabbit IgG antibody and the Amplex® Gold reagent, for yellow-fluorescent detection of a rabbit antibody to a specific protein or proteins
- Alkaline phosphatase conjugate of goat anti–mouse IgG antibody and DDAO phosphate, for far-red–fluorescent detection of a mouse antibody to a specific protein or proteins
- MDPF (2-methoxy-2,4-diphenyl-3(2H)-furanone) for blue-fluorescent detection of the total-protein profile
- Appropriate solvents and buffers for the enzymatic reactions
- Detailed protocols

Each DyeChrome™ Double Western Blot Stain Kit contains sufficient materials to stain approximately 20 minigel blots (6 cm × 9 cm). The two antigens are developed and detected simultaneously, and staining is stable indefinitely on dried blots. The fluorescent signals can be visualized using UV or visible-light illumination.

WesternDot™ Western Blot Kits

The WesternDot™ 625 Goat Anti-Mouse and Goat Anti-Rabbit Western Blot Kits (W10132, W10142) provide Qdot® nanocrystal secondary reagents for the detection of subnanogram amounts of protein on western blots without the complications of time-dependent signal development imposed by enzyme-amplified chemiluminescence detection systems. These kits provide Qdot® 625 streptavidin conjugates and biotinylated secondary antibodies for use in combination with user-supplied mouse or rabbit primary antibodies against the protein of interest. In other respects, immunodetection of proteins immobilized on nitrocellulose (NC) or polyvinylidene difluoride (PVDF) membranes follows standard western blotting protocols. The components of these kits are:

- Biotin-XX goat anti–mouse IgG antibody (in W10132) or biotin-XX goat anti–rabbit IgG antibody (in W10142)
- Qdot® 625 streptavidin
- WesternDot™ blocking buffer
- Wash buffer
- WesternDot™ staining dish
- Detailed protocols

Each WesternDot™ 625 Western Blot Kit contains sufficient materials to stain approximately 20 minigel blots (8 cm × 8 cm). Detection of the fluorescence signal can be accomplished using standard UV- or blue light–based gel and blot imaging systems and does not require specialized emission filters.

Gold-Labeled Immunoreagents

NANOGOLD® and Alexa Fluor® FluoroNanogold™ Conjugates

In collaboration with Nanoprobes, Inc. (www.nanoprobes.com), we offer NANOGOLD® and Alexa Fluor® FluoroNanogold™ conjugates of antibodies and streptavidin to facilitate immunoblotting, light microscopy and electron microscopy applications. These reagents include

Table 7.5 NANOGOLD®, Alexa Fluor® FluoroNanogold™ and colloidal gold conjugates.

	Label (Abs/Em) *			
Conjugate	NANOGOLD®	Alexa Fluor® 488 FluoroNanogold™ (495/519)	Alexa Fluor® 488 Colloidal Gold (495/519)	Alexa Fluor® 594 Fluoro-Nanogold™ (590/617)
Goat anti–mouse IgG	N24915 †	A24920 †	A31560 ‡, A31561 §	A24921 †
Goat anti–rabbit IgG	N24916 †	A24922 †	A31565 ‡, A31566 §	A24923 †
Streptavidin	N24918	A24926	A32360 ‡, A32361 §	A24927

* Approximate absorption (Abs) and fluorescence emission (Em) maxima, in nm, for conjugates. † Fab′ fragment. ‡ 5 nm colloidal gold. § 10 nm colloidal gold.

IMPORTANT NOTICE: The products described in this manual are covered by one or more Limited Use Label License(s). Please refer to the Appendix on page 971 and Master Product List on page 975. Products are For Research Use Only. Not intended for any animal or human therapeutic or diagnostic use.

molecular probes® | invitrogen™ by life technologies™

affinity-purified Fab′ fragments of the goat anti–mouse IgG and anti–rabbit IgG antibodies, as well as of streptavidin (Section 7.6, Table 7.5). Also available is NANOGOLD® monomaleimide (N20345, Section 2.2), which can be conjugated to thiols in the same way that dyes are conjugated to proteins and nucleic acids.[10]

NANOGOLD® antibody conjugates are covalently conjugated to the 1.4 nm NANOGOLD® gold cluster label, whereas Alexa Fluor® FluoroNanogold™ antibody conjugates are coupled to both a NANOGOLD® label and either the Alexa Fluor® 488 or Alexa Fluor® 594 fluorophore, resulting in gold clusters with green or red fluorescence, respectively. Alexa Fluor® FluoroNanogold™ conjugates have all the advantages of the NANOGOLD® cluster, with the additional benefit that they may be used for correlative fluorescence, light and electron microscopy[11,12] (Figure 7.2.20).

NANOGOLD® gold clusters have several advantages over colloidal gold. They develop better with silver than do most gold colloids and therefore provide higher sensitivity.[13] Silver enhancement, such as the system provided in the LI Silver Enhancement Kit (L24919), is described below. Additionally, NANOGOLD® particles do not have as high affinity for proteins as do gold colloids, thereby reducing background due to nonspecific binding.

Several additional features of NANOGOLD® and Alexa Fluor® FluoroNanogold™ conjugates include:

- NANOGOLD® gold clusters are an extremely uniform (1.4 nm ± 10% diameter) and stable compound, not a gold colloid.
- NANOGOLD® gold clusters are smaller than a complete IgG (H+L) antibody—approximately 1/15 the size of an Fab′ fragment—and therefore will be able to better penetrate cells and tissues, reaching antigens that are inaccessible to conjugates of larger gold particles.
- NANOGOLD® conjugates contain no aggregates, as they are chromatographically purified through gel filtration columns. This feature is in sharp contrast to colloidal gold conjugates, which are usually prepared by centrifugation to remove the largest aggregates and frequently contain significantly smaller aggregates.
- The ratio of NANOGOLD® particle to Fab′ fragment is nearly 1:1, making this product distinct from the 0.2–10 variable stoichiometry of most colloidal gold–antibody preparations.

NANOGOLD® and Alexa Fluor® FluoroNanogold™ products can be used in immunoblotting, light microscopy and electron microscopy (Figure 7.2.21). Standard immunostaining methodologies can be used successfully with NANOGOLD® and Alexa Fluor® FluoroNanogold™ immunoreagents. Also, because the concentration of antibody and gold is similar to most commercial preparations of colloidal gold antibodies, similar dilutions and blocking agents are appropriate.

Colloidal Gold Complexes

We also offer Alexa Fluor® 488 dye–labeled colloidal gold conjugates, including affinity-purified goat anti–mouse IgG and anti–rabbit IgG antibodies and streptavidin (Table 7.5). These conjugates, which have been adsorbed to 5 nm or 10 nm gold colloids, may be used as probes in immunoblotting, light microscopy, fluorescence microscopy or electron microscopy. The fluorescence of these conjugates can be easily detected by standard techniques, but visualization of colloidal gold can be greatly improved using silver-enhancement methods, such as those provided in the LI Silver Enhancement Kit (L24919) described below.

Combining fluorescent secondary detection reagents with colloidal gold to form functional complexes is difficult because the fluorescence of fluorophores such as fluorescein is significantly quenched by proximity to the colloidal gold.[14] We prepare fluorescent colloidal gold complexes with our Alexa Fluor® 488 dye, a dye that has superior brightness and photostability. Our Alexa Fluor® 488 dye–labeled colloidal gold complexes of anti-IgG antibody and of streptavidin may be used to perform correlated immunofluorescence and electron microscopy in a two-step labeling procedure, rather than in the three-step indirect labeling procedure that is required with conventional nonfluorescent colloidal gold complexes of anti-IgG antibodies or streptavidin.[15]

LI Silver Enhancement Kit

The LI Silver Enhancement Kit (L24919) provides a convenient, light-insensitive silver-enhancement system for use with the NANOGOLD®, Alexa Fluor® FluoroNanogold™ and colloidal gold reagents that can be used with electron or light microscopy or for visualization on blots. LI silver is nucleated quickly by NANOGOLD® gold clusters or colloidal gold, resulting in the precipitation of metallic silver and the formation of a dark brown to black signal. The system

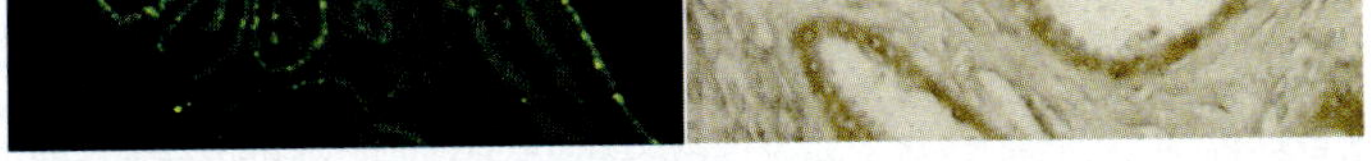

Figure 7.2.20 Human prostate tissue with adenocarcinoma imaged using an Alexa Fluor® 488 FluoroNanogold™ conjugate. Tissues were labeled with an anti-cytokeratin antibody, followed by staining with the Alexa Fluor® 488 FluoroNanogold™ Fab′ fragment of goat anti–mouse IgG antibody (A24920). The fluorescence image in the left panel was taken with optical filters appropriate for fluorescein. The right panel shows a bright-field image of the same system, processed with the LI Silver (LIS) Enhancement Kit (L24919) to visualize the gold distribution. Image courtesy of Nanoprobes, Inc.

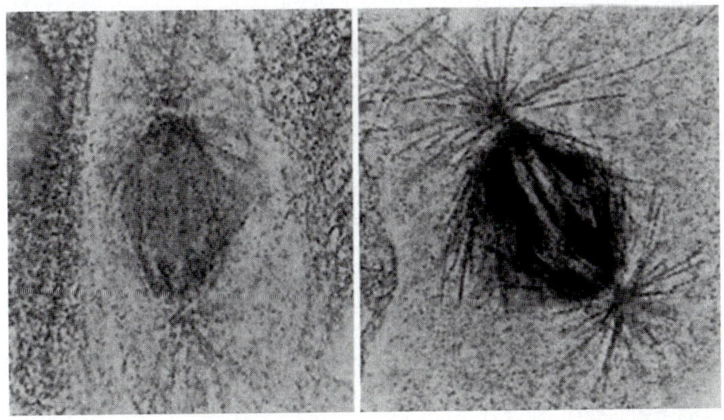

Figure 7.2.21 Spindle microtubules labeled with an anti-tubulin primary antibody followed by (left) goat anti–mouse IgG antibody colloidal gold or (right) NANOGOLD® Fab′ fragment of goat anti–mouse IgG antibody (N24915). Micrographs (magnification = 1300x) were generated by D. Vandre and R. Burry, Ohio State University. Image courtesy of Nanoprobes, Inc.

IMPORTANT NOTICE: The products described in this manual are covered by one or more Limited Use Label License(s). Please refer to the Appendix on page 971 and Master Product List on page 975. Products are For Research Use Only. Not intended for any animal or human therapeutic or diagnostic use.

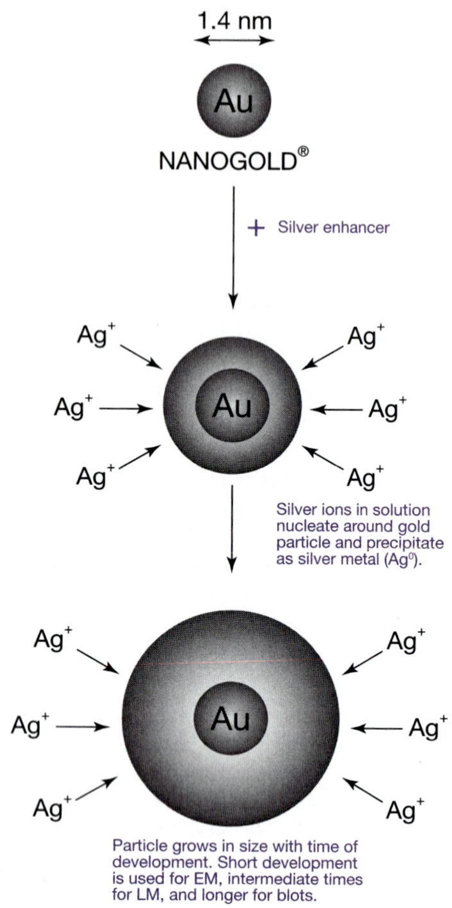

Figure 7.2.22 Mechanism of silver deposition utilized by the LI Silver Enhancement Kit (L24919). Image courtesy of Nanoprobes, Inc.

Table 7.6 Binding profiles of protein A and protein G.

Antibody	Protein A	Protein G
Bovine	+	++
Cat	++	–
Chicken	–	–
Dog	++	+
Goat	+	++
Guinea pig	+	++
Horse	–	++
Human IgG_1, IgG_2, IgG_4	++	++
Human IgG_3	–	++
Human IgM, IgA, IgE	++	–
Human IgD	–	–
Mouse IgG_1	–	++
Mouse (others)	++	++
Pig	++	++
Rabbit	++	++
Rat	–	+
Sheep	–	++

+ Moderate binding. ++ Strong binding. – Weak or no binding.

has markedly delayed self-nucleation, resulting in high contrast and very low backgrounds.

Gold particles in the presence of silver (I) ions and a reducing agent such as hydroquinone act as catalysts to reduce silver (I) ions to metallic silver (Figure 7.2.22). The silver is deposited onto the gold, enlarging the particles to between 30 and 100 nm in diameter. Tissues or blots stained with NANOGOLD® gold clusters or colloidal gold are "developed" by this autometallographic procedure to give black staining that can be seen in a light microscope. This immunogold silver staining (IGSS) method has been widely used with the NANOGOLD® cluster probe. It is one of the most sensitive immunodetection systems available, yielding highly visible, permanent staining with no fading and with detection limits rivaling that of chemiluminescence and radionuclide labeling. Silver-enhanced NANOGOLD® staining is compatible with double-labeling techniques, including enzyme-mediated staining. In blots, as little as 0.1 pg of a target IgG antibody can be detected using a NANOGOLD® gold cluster labeled with an Fab′ fragment of a secondary antibody. NANOGOLD® streptavidin (N24918, Section 7.6) has proven to be highly sensitive in detecting biotinylated nucleic acid probes in *in situ* hybridization studies.

The LI Silver Enhancement Kit (L24919), which provides both an initiator reagent and an enhancer reagent, is compatible with the NANOGOLD® and Alexa Fluor® FluoroNanogold™ reagents. Features of the LI silver enhancement method include:

- High-contrast signal for easy light microscope and immunoblot visibility
- Lower background than many other commercial developers
- High sensitivity
- Light-insensitive signal that can be observed under normal room lighting
- Relatively slow development (10–30 minutes) for precise monitoring
- Compatibility with all immunogold reagents

Protein A and Protein G Conjugates

Protein A and protein G are bacterial proteins that bind with high affinity to the Fc portion of various classes and subclasses of immunoglobulins from a variety of species (Table 7.6). We offer protein A conjugated to several different Alexa Fluor® fluorophores:

- Green-fluorescent Alexa Fluor® 488 dye (P11047)
- Orange-fluorescent Alexa Fluor® 546 dye (P11049)
- Far-red–fluorescent Alexa Fluor® 647 dye (P21462)

We also offer protein G conjugated to the Alexa Fluor® 488 dye (P11065) and to horseradish peroxidase (P21041).

REFERENCES

1. BMC Biotechnol (2009) 9:49; **2.** Appl Environ Microbiol (2007) 73:630; **3.** J Neurosci (2010) 30:4895; **4.** Nano Lett (2006) 6:2881; **5.** Mol Pharm (2009) 6:1170; **6.** J Immunol (2009) 182:4056; **7.** J Cell Sci (2007) 120:101; **8.** J Biol Chem (2005) 280:37377; **9.** J Immunol Methods (2001) 247:111; **10.** J Histochem Cytochem (2000) 48:471; **11.** J Histochem Cytochem (2003) 51:707; **12.** Placenta (2003) 24:557; **13.** J Microsc (2000) 199:163; **14.** Colloidal Gold: Principles. Methods and Applications, Vol. 1, Hayat MA, Ed. (1989) p.323; **15.** Proc Natl Acad Sci U S A (1985) 82:109.

PRODUCT LIST 7.2 SECONDARY IMMUNOREAGENTS

Cat. No.	Product	Quantity
A21081	Alexa Fluor® 350 donkey anti–goat IgG (H+L) *2 mg/mL*	0.5 mL
A10035	Alexa Fluor® 350 donkey anti–mouse IgG (H + L) *2 mg/mL*	0.5 mL
A10039	Alexa Fluor® 350 donkey anti–rabbit IgG (H + L) *2 mg/mL*	0.5 mL
A21097	Alexa Fluor® 350 donkey anti–sheep IgG (H+L) *2 mg/mL*	0.5 mL
A11068	Alexa Fluor® 350 F(ab')$_2$ fragment of goat anti–mouse IgG (H+L) *2 mg/mL*	250 µL
A11069	Alexa Fluor® 350 F(ab')$_2$ fragment of goat anti–rabbit IgG (H+L) *2 mg/mL*	250 µL
A11045	Alexa Fluor® 350 goat anti–mouse IgG (H+L) *2 mg/mL*	0.5 mL
A21049	Alexa Fluor® 350 goat anti–mouse IgG (H+L) *highly cross-adsorbed* *2 mg/mL*	0.5 mL
A21120	Alexa Fluor® 350 goat anti–mouse IgG$_1$ (γ1) *2 mg/mL*	250 µL
A21130	Alexa Fluor® 350 goat anti–mouse IgG$_{2a}$ (γ2a) *2 mg/mL*	250 µL
A21140	Alexa Fluor® 350 goat anti–mouse IgG$_{2b}$ (γ2b) *2 mg/mL*	250 µL
A31552	Alexa Fluor® 350 goat anti–mouse IgM (µ chain) *2 mg/mL*	250 µL
A11046	Alexa Fluor® 350 goat anti–rabbit IgG (H+L) *2 mg/mL*	0.5 mL
A21068	Alexa Fluor® 350 goat anti–rabbit IgG (H+L) *highly cross-adsorbed* *2 mg/mL*	0.5 mL
A21093	Alexa Fluor® 350 goat anti–rat IgG (H+L) *2 mg/mL*	0.5 mL
A21062	Alexa Fluor® 350 rabbit anti–mouse IgG (H+L) *2 mg/mL*	0.5 mL
A31553	Alexa Fluor® 405 goat anti–mouse IgG (H+L) *2 mg/mL*	0.5 mL
A31556	Alexa Fluor® 405 goat anti–rabbit IgG (H+L) *2 mg/mL*	0.5 mL
A11063	Alexa Fluor® 430 goat anti–mouse IgG (H+L) *2 mg/mL*	0.5 mL
A11064	Alexa Fluor® 430 goat anti–rabbit IgG (H+L) *2 mg/mL*	0.5 mL
A21467	Alexa Fluor® 488 chicken anti–goat IgG (H+L) *2 mg/mL*	0.5 mL
A21200	Alexa Fluor® 488 chicken anti–mouse IgG (H+L) *2 mg/mL*	0.5 mL
A21441	Alexa Fluor® 488 chicken anti–rabbit IgG (H+L) *2 mg/mL*	0.5 mL
A21470	Alexa Fluor® 488 chicken anti–rat IgG (H+L) *2 mg/mL*	0.5 mL
A11055	Alexa Fluor® 488 donkey anti–goat IgG (H+L) *2 mg/mL*	0.5 mL
A21202	Alexa Fluor® 488 donkey anti–mouse IgG (H+L) *2 mg/mL*	0.5 mL
A21206	Alexa Fluor® 488 donkey anti–rabbit IgG (H+L) *2 mg/mL*	0.5 mL
A21208	Alexa Fluor® 488 donkey anti–rat IgG (H+L) *2 mg/mL*	0.5 mL
A11015	Alexa Fluor® 488 donkey anti–sheep IgG (H+L) *2 mg/mL*	0.5 mL
A11017	Alexa Fluor® 488 F(ab')$_2$ fragment of goat anti–mouse IgG (H+L) *2 mg/mL*	250 µL
A10684	Alexa Fluor® 488 F(ab')$_2$ fragment of goat anti–mouse IgG, IgM (H+L) *2 mg/mL*	250 µL
A11070	Alexa Fluor® 488 F(ab')$_2$ fragment of goat anti–rabbit IgG (H+L) *2 mg/mL*	250 µL
A21222	Alexa Fluor® 488 F(ab')$_2$ fragment of rabbit anti–goat IgG (H+L) *2 mg/mL*	250 µL
A21204	Alexa Fluor® 488 F(ab')$_2$ fragment of rabbit anti–mouse IgG (H+L) *2 mg/mL*	250 µL
A24920	Alexa Fluor® 488 FluoroNanogold™ Fab' fragment of goat anti–mouse IgG *80 µg protein/mL*	1 mL
A24922	Alexa Fluor® 488 FluoroNanogold™ Fab' fragment of goat anti–rabbit IgG *80 µg protein/mL*	1 mL
A11039	Alexa Fluor® 488 goat anti–chicken IgG (H+L) *2 mg/mL*	0.5 mL
A11073	Alexa Fluor® 488 goat anti–guinea pig IgG (H+L) *highly cross-adsorbed* *2 mg/mL*	0.5 mL
A21110	Alexa Fluor® 488 goat anti–hamster IgG (H+L) *2 mg/mL*	0.5 mL
A11013	Alexa Fluor® 488 goat anti–human IgG (H+L) *2 mg/mL*	0.5 mL
A21215	Alexa Fluor® 488 goat anti–human IgM (µ chain) *2 mg/mL*	250 µL
A11001	Alexa Fluor® 488 goat anti–mouse IgG (H+L) *2 mg/mL*	0.5 mL
A11029	Alexa Fluor® 488 goat anti–mouse IgG (H+L) *highly cross-adsorbed* *2 mg/mL*	0.5 mL
A31561	Alexa Fluor® 488 goat anti–mouse IgG, 10 nm colloidal gold conjugate *30 µg protein/mL*	0.5 mL
A31560	Alexa Fluor® 488 goat anti–mouse IgG, 5 nm colloidal gold conjugate *30 µg protein/mL*	0.5 mL
A10667	Alexa Fluor® 488 goat anti–mouse IgG, IgA, IgM (H+L) *2 mg/mL*	250 µL
A10680	Alexa Fluor® 488 goat anti–mouse IgG, IgM (H+L) *2 mg/mL*	250 µL
A21121	Alexa Fluor® 488 goat anti–mouse IgG$_1$ (γ1) *2 mg/mL*	250 µL
A21131	Alexa Fluor® 488 goat anti–mouse IgG$_{2a}$ (γ2a) *2 mg/mL*	250 µL
A21141	Alexa Fluor® 488 goat anti–mouse IgG$_{2b}$ (γ2b) *2 mg/mL*	250 µL
A21151	Alexa Fluor® 488 goat anti–mouse IgG$_3$ (γ3) *2 mg/mL*	250 µL
A21042	Alexa Fluor® 488 goat anti–mouse IgM (µ chain) *2 mg/mL*	250 µL
A31619	Alexa Fluor® 488 Goat Anti-Mouse SFX Kit	1 kit
A31620	Alexa Fluor® 488 Goat Anti-Mouse SFX Kit *highly cross-adsorbed*	1 kit
A11008	Alexa Fluor® 488 goat anti–rabbit IgG (H+L) *2 mg/mL*	0.5 mL
A11034	Alexa Fluor® 488 goat anti–rabbit IgG (H+L) *highly cross-adsorbed* *2 mg/mL*	0.5 mL
A31566	Alexa Fluor® 488 goat anti–rabbit IgG, 10 nm colloidal gold conjugate *30 µg protein/mL*	0.5 mL
A31565	Alexa Fluor® 488 goat anti–rabbit IgG, 5 nm colloidal gold conjugate *30 µg protein/mL*	0.5 mL
A31627	Alexa Fluor® 488 Goat Anti-Rabbit SFX Kit	1 kit
A31628	Alexa Fluor® 488 Goat Anti-Rabbit SFX Kit *highly cross-adsorbed*	1 kit
A11006	Alexa Fluor® 488 goat anti–rat IgG (H+L) *2 mg/mL*	0.5 mL

continued on next page

IMPORTANT NOTICE: The products described in this manual are covered by one or more Limited Use Label License(s). Please refer to the Appendix on page 971 and Master Product List on page 975. Products are For Research Use Only. Not intended for any animal or human therapeutic or diagnostic use.

PRODUCT LIST 7.2 SECONDARY IMMUNOREAGENTS—*continued*

Cat. No.	Product	Quantity
A21212	Alexa Fluor® 488 goat anti–rat IgM (μ chain) *2 mg/mL*	250 μL
A10631	Alexa Fluor® 488 mouse anti–human IgG_1 *1 mg/mL*	250 μL
A11078	Alexa Fluor® 488 rabbit anti–goat IgG (H+L) *2 mg/mL*	0.5 mL
A11059	Alexa Fluor® 488 rabbit anti–mouse IgG (H+L) *2 mg/mL*	0.5 mL
A21210	Alexa Fluor® 488 rabbit anti–rat IgG (H+L) *2 mg/mL*	0.5 mL
A11053	Alexa Fluor® 488 Signal-Amplification Kit for Fluorescein- and Oregon Green® Dye–Conjugated Probes *60–120 assays*	1 kit
A11054	Alexa Fluor® 488 Signal-Amplification Kit for Mouse Antibodies *60–300 assays*	1 kit
A31555	Alexa Fluor® 514 goat anti–mouse IgG (H+L) *2 mg/mL*	0.5 mL
A31558	Alexa Fluor® 514 goat anti–rabbit IgG (H+L) *2 mg/mL*	0.5 mL
A11002	Alexa Fluor® 532 goat anti–mouse IgG (H+L) *2 mg/mL*	0.5 mL
A11009	Alexa Fluor® 532 goat anti–rabbit IgG (H+L) *2 mg/mL*	0.5 mL
A11056	Alexa Fluor® 546 donkey anti–goat IgG (H+L) *2 mg/mL*	0.5 mL
A10036	Alexa Fluor® 546 donkey anti–mouse IgG (H + L) *2 mg/mL*	0.5 mL
A10040	Alexa Fluor® 546 donkey anti–rabbit IgG (H + L) *2 mg/mL*	0.5 mL
A21098	Alexa Fluor® 546 donkey anti–sheep IgG (H+L) *2 mg/mL*	0.5 mL
A11018	Alexa Fluor® 546 $F(ab')_2$ fragment of goat anti–mouse IgG (H+L) *2 mg/mL*	250 μL
A11071	Alexa Fluor® 546 $F(ab')_2$ fragment of goat anti–rabbit IgG (H+L) *2 mg/mL*	250 μL
A11040	Alexa Fluor® 546 goat anti–chicken IgG (H+L) *2 mg/mL*	0.5 mL
A11074	Alexa Fluor® 546 goat anti–guinea pig IgG (H+L) *highly cross-adsorbed* *2 mg/mL*	0.5 mL
A21111	Alexa Fluor® 546 goat anti–hamster IgG (H+L) *2 mg/mL*	0.5 mL
A21089	Alexa Fluor® 546 goat anti–human IgG (H+L) *2 mg/mL*	0.5 mL
A11003	Alexa Fluor® 546 goat anti–mouse IgG (H+L) *2 mg/mL*	0.5 mL
A11030	Alexa Fluor® 546 goat anti–mouse IgG (H+L) *highly cross-adsorbed* *2 mg/mL*	0.5 mL
A21123	Alexa Fluor® 546 goat anti–mouse IgG_1 (γ1) *2 mg/mL*	250 μL
A21133	Alexa Fluor® 546 goat anti–mouse IgG_{2a} (γ2a) *2 mg/mL*	250 μL
A21143	Alexa Fluor® 546 goat anti–mouse IgG_{2b} (γ2b) *2 mg/mL*	250 μL
A21045	Alexa Fluor® 546 goat anti–mouse IgM (μ chain) *2 mg/mL*	250 μL
A11010	Alexa Fluor® 546 goat anti–rabbit IgG (H+L) *2 mg/mL*	0.5 mL
A11035	Alexa Fluor® 546 goat anti–rabbit IgG (H+L) *highly cross-adsorbed* *2 mg/mL*	0.5 mL
A11081	Alexa Fluor® 546 goat anti–rat IgG (H+L) *2 mg/mL*	0.5 mL
A21085	Alexa Fluor® 546 rabbit anti–goat IgG (H+L) *2 mg/mL*	0.5 mL
A11060	Alexa Fluor® 546 rabbit anti–mouse IgG (H+L) *2 mg/mL*	0.5 mL
A21432	Alexa Fluor® 555 donkey anti–goat IgG (H+L) *2 mg/mL*	0.5 mL
A31570	Alexa Fluor® 555 donkey anti–mouse IgG (H+L) *2 mg/mL*	0.5 mL
A31572	Alexa Fluor® 555 donkey anti–rabbit IgG (H+L) *2 mg/mL*	0.5 mL
A21436	Alexa Fluor® 555 donkey anti–sheep IgG (H+L) *2 mg/mL*	0.5 mL
A21425	Alexa Fluor® 555 $F(ab')_2$ fragment of goat anti–mouse IgG (H+L) *2 mg/mL*	250 μL
A21430	Alexa Fluor® 555 $F(ab')_2$ fragment of goat anti–rabbit IgG (H+L) *2 mg/mL*	250 μL
A21437	Alexa Fluor® 555 goat anti–chicken IgG (H+L) *2 mg/mL*	0.5 mL
A21435	Alexa Fluor® 555 goat anti–guinea pig IgG (H+L) *highly cross-adsorbed* *2 mg/mL*	0.5 mL
A21433	Alexa Fluor® 555 goat anti–human IgG (H+L) *2 mg/mL*	0.5 mL
A21422	Alexa Fluor® 555 goat anti–mouse IgG (H+L) *2 mg/mL*	0.5 mL
A21424	Alexa Fluor® 555 goat anti–mouse IgG (H+L) *highly cross-adsorbed* *2 mg/mL*	0.5 mL
A21127	Alexa Fluor® 555 goat anti–mouse IgG_1 (γ1) *2 mg/mL*	250 μL
A21137	Alexa Fluor® 555 goat anti–mouse IgG_{2a} (γ2a) *2 mg/mL*	250 μL
A21147	Alexa Fluor® 555 goat anti–mouse IgG_{2b} (γ2b) *2 mg/mL*	250 μL
A21426	Alexa Fluor® 555 goat anti–mouse IgM (μ chain) *2 mg/mL*	250 μL
A31621	Alexa Fluor® 555 Goat Anti-Mouse SFX Kit	1 kit
A31622	Alexa Fluor® 555 Goat Anti-Mouse SFX Kit *highly cross-adsorbed*	1 kit
A21428	Alexa Fluor® 555 goat anti–rabbit IgG (H+L) *2 mg/mL*	0.5 mL
A21429	Alexa Fluor® 555 goat anti–rabbit IgG (H+L) *highly cross-adsorbed* *2 mg/mL*	0.5 mL
A31629	Alexa Fluor® 555 Goat Anti-Rabbit SFX Kit	1 kit
A31630	Alexa Fluor® 555 Goat Anti-Rabbit SFX Kit *highly cross-adsorbed*	1 kit
A21434	Alexa Fluor® 555 goat anti–rat IgG (H+L) *2 mg/mL*	0.5 mL
A21431	Alexa Fluor® 555 rabbit anti–goat IgG (H+L) *2 mg/mL*	0.5 mL
A21427	Alexa Fluor® 555 rabbit anti–mouse IgG (H+L) *2 mg/mL*	0.5 mL
A11057	Alexa Fluor® 568 donkey anti–goat IgG (H+L) *2 mg/mL*	0.5 mL
A10037	Alexa Fluor® 568 donkey anti–mouse IgG (H + L) *2 mg/mL*	0.5 mL
A10042	Alexa Fluor® 568 donkey anti–rabbit IgG (H + L) *2 mg/mL*	0.5 mL
A21099	Alexa Fluor® 568 donkey anti–sheep IgG (H+L) *2 mg/mL*	0.5 mL
A11019	Alexa Fluor® 568 $F(ab')_2$ fragment of goat anti–mouse IgG (H+L) *2 mg/mL*	250 μL

IMPORTANT NOTICE: The products described in this manual are covered by one or more Limited Use Label License(s). Please refer to the Appendix on page 971 and Master Product List on page 975. Products are For Research Use Only. Not intended for any animal or human therapeutic or diagnostic use.

PRODUCT LIST 7.2 SECONDARY IMMUNOREAGENTS—*continued*

Cat. No.	Product	Quantity
A21069	Alexa Fluor® 568 F(ab')$_2$ fragment of goat anti–rabbit IgG (H+L) *2 mg/mL*	250 µL
A11041	Alexa Fluor® 568 goat anti–chicken IgG (H+L) *2 mg/mL*	0.5 mL
A11075	Alexa Fluor® 568 goat anti–guinea pig IgG (H+L) *highly cross-adsorbed* *2 mg/mL*	0.5 mL
A21112	Alexa Fluor® 568 goat anti–hamster IgG (H+L) *2 mg/mL*	0.5 mL
A21090	Alexa Fluor® 568 goat anti–human IgG (H+L) *2 mg/mL*	0.5 mL
A11004	Alexa Fluor® 568 goat anti–mouse IgG (H+L) *2 mg/mL*	0.5 mL
A11031	Alexa Fluor® 568 goat anti–mouse IgG (H+L) *highly cross-adsorbed* *2 mg/mL*	0.5 mL
A21124	Alexa Fluor® 568 goat anti–mouse IgG$_1$ (γ1) *2 mg/mL*	250 µL
A21134	Alexa Fluor® 568 goat anti–mouse IgG$_{2a}$ (γ2a) *2 mg/mL*	250 µL
A21144	Alexa Fluor® 568 goat anti–mouse IgG$_{2b}$ (γ2b) *2 mg/mL*	250 µL
A21043	Alexa Fluor® 568 goat anti–mouse IgM (µ chain) *2 mg/mL*	250 µL
A11011	Alexa Fluor® 568 goat anti–rabbit IgG (H+L) *2 mg/mL*	0.5 mL
A11036	Alexa Fluor® 568 goat anti–rabbit IgG (H+L) *highly cross-adsorbed* *2 mg/mL*	0.5 mL
A11077	Alexa Fluor® 568 goat anti–rat IgG (H+L) *2 mg/mL*	0.5 mL
A11079	Alexa Fluor® 568 rabbit anti–goat IgG (H+L) *2 mg/mL*	0.5 mL
A11061	Alexa Fluor® 568 rabbit anti–mouse IgG (H+L) *2 mg/mL*	0.5 mL
A11066	Alexa Fluor® 568 Signal-Amplification Kit for Mouse Antibodies *60–300 assays*	1 kit
A21468	Alexa Fluor® 594 chicken anti–goat IgG (H+L) *2 mg/mL*	0.5 mL
A21201	Alexa Fluor® 594 chicken anti–mouse IgG (H+L) *2 mg/mL*	0.5 mL
A21442	Alexa Fluor® 594 chicken anti–rabbit IgG (H+L) *2 mg/mL*	0.5 mL
A21471	Alexa Fluor® 594 chicken anti–rat IgG (H+L) *2 mg/mL*	0.5 mL
A11058	Alexa Fluor® 594 donkey anti–goat IgG (H+L) *2 mg/mL*	0.5 mL
A21203	Alexa Fluor® 594 donkey anti–mouse IgG (H+L) *2 mg/mL*	0.5 mL
A21207	Alexa Fluor® 594 donkey anti–rabbit IgG (H+L) *2 mg/mL*	0.5 mL
A21209	Alexa Fluor® 594 donkey anti–rat IgG (H+L) *2 mg/mL*	0.5 mL
A11016	Alexa Fluor® 594 donkey anti–sheep IgG (H+L) *2 mg/mL*	0.5 mL
A11020	Alexa Fluor® 594 F(ab')$_2$ fragment of goat anti–mouse IgG (H+L) *2 mg/mL*	250 µL
A11072	Alexa Fluor® 594 F(ab')$_2$ fragment of goat anti–rabbit IgG (H+L) *2 mg/mL*	250 µL
A21223	Alexa Fluor® 594 F(ab')$_2$ fragment of rabbit anti–goat IgG (H+L) *2 mg/mL*	250 µL
A21205	Alexa Fluor® 594 F(ab')$_2$ fragment of rabbit anti–mouse IgG (H+L) *2 mg/mL*	250 µL
A24921	Alexa Fluor® 594 FluoroNanogold™ Fab' fragment of goat anti–mouse IgG *80 µg protein/mL*	1 mL
A24923	Alexa Fluor® 594 FluoroNanogold™ Fab' fragment of goat anti–rabbit IgG *80 µg protein/mL*	1 mL
A11042	Alexa Fluor® 594 goat anti–chicken IgG (H+L) *2 mg/mL*	0.5 mL
A11076	Alexa Fluor® 594 goat anti–guinea pig IgG (H+L) *highly cross-adsorbed* *2 mg/mL*	0.5 mL
A21113	Alexa Fluor® 594 goat anti–hamster IgG (H+L) *2 mg/mL*	0.5 mL
A11014	Alexa Fluor® 594 goat anti–human IgG (H+L) *2 mg/mL*	0.5 mL
A21216	Alexa Fluor® 594 goat anti–human IgM (µ chain) *2 mg/mL*	250 µL
A11005	Alexa Fluor® 594 goat anti–mouse IgG (H+L) *2 mg/mL*	0.5 mL
A11032	Alexa Fluor® 594 goat anti–mouse IgG (H+L) *highly cross-adsorbed* *2 mg/mL*	0.5 mL
A21125	Alexa Fluor® 594 goat anti–mouse IgG$_1$ (γ1) *2 mg/mL*	250 µL
A21135	Alexa Fluor® 594 goat anti–mouse IgG$_{2a}$ (γ2a) *2 mg/mL*	250 µL
A21145	Alexa Fluor® 594 goat anti–mouse IgG$_{2b}$ (γ2b) *2 mg/mL*	250 µL
A21155	Alexa Fluor® 594 goat anti–mouse IgG$_3$ (γ3) *2 mg/mL*	250 µL
A21044	Alexa Fluor® 594 goat anti–mouse IgM (µ chain) *2 mg/mL*	250 µL
A31623	Alexa Fluor® 594 Goat Anti-Mouse SFX Kit	1 kit
A31624	Alexa Fluor® 594 Goat Anti-Mouse SFX Kit *highly cross-adsorbed*	1 kit
A11012	Alexa Fluor® 594 goat anti–rabbit IgG (H+L) *2 mg/mL*	0.5 mL
A11037	Alexa Fluor® 594 goat anti–rabbit IgG (H+L) *highly cross-adsorbed* *2 mg/mL*	0.5 mL
A31631	Alexa Fluor® 594 Goat Anti-Rabbit SFX Kit	1 kit
A31632	Alexa Fluor® 594 Goat Anti-Rabbit SFX Kit *highly cross-adsorbed*	1 kit
A11007	Alexa Fluor® 594 goat anti–rat IgG (H+L) *2 mg/mL*	0.5 mL
A21213	Alexa Fluor® 594 goat anti–rat IgM (µ chain) *2 mg/mL*	250 µL
A11080	Alexa Fluor® 594 rabbit anti–goat IgG (H+L) *2 mg/mL*	0.5 mL
A11062	Alexa Fluor® 594 rabbit anti–mouse IgG (H+L) *2 mg/mL*	0.5 mL
A21211	Alexa Fluor® 594 rabbit anti–rat IgG (H+L) *2 mg/mL*	0.5 mL
A11067	Alexa Fluor® 594 Signal-Amplification Kit for Mouse Antibodies *60–300 assays*	1 kit
A20980	Alexa Fluor® 610–R-phycoerythrin goat anti–mouse IgG (H+L) *1 mg/mL*	100 µL
A20981	Alexa Fluor® 610–R-phycoerythrin goat anti–rabbit IgG (H+L) *1 mg/mL*	100 µL
A21082	Alexa Fluor® 633 donkey anti–goat IgG (H+L) *2 mg/mL*	0.5 mL
A21100	Alexa Fluor® 633 donkey anti–sheep IgG (H+L) *2 mg/mL*	0.5 mL
A21053	Alexa Fluor® 633 F(ab')$_2$ fragment of goat anti–mouse IgG (H+L) *2 mg/mL*	250 µL

continued on next page

IMPORTANT NOTICE: The products described in this manual are covered by one or more Limited Use Label License(s). Please refer to the Appendix on page 971 and Master Product List on page 975. Products are For Research Use Only. Not intended for any animal or human therapeutic or diagnostic use.

PRODUCT LIST 7.2 SECONDARY IMMUNOREAGENTS—*continued*

Cat. No.	Product	Quantity
A21072	Alexa Fluor® 633 F(ab')$_2$ fragment of goat anti–rabbit IgG (H+L) *2 mg/mL*	250 µL
A21103	Alexa Fluor® 633 goat anti–chicken IgG (H+L) *2 mg/mL*	0.5 mL
A21105	Alexa Fluor® 633 goat anti–guinea pig IgG (H+L) *highly cross-adsorbed* *2 mg/mL*	0.5 mL
A21091	Alexa Fluor® 633 goat anti–human IgG (H+L) *2 mg/mL*	0.5 mL
A21050	Alexa Fluor® 633 goat anti–mouse IgG (H+L) *2 mg/mL*	0.5 mL
A21052	Alexa Fluor® 633 goat anti–mouse IgG (H+L) *highly cross-adsorbed* *2 mg/mL*	0.5 mL
A21126	Alexa Fluor® 633 goat anti–mouse IgG$_1$ (γ1) *2 mg/mL*	250 µL
A21136	Alexa Fluor® 633 goat anti–mouse IgG$_{2a}$ (γ2a) *2 mg/mL*	250 µL
A21146	Alexa Fluor® 633 goat anti–mouse IgG$_{2b}$ (γ2b) *2 mg/mL*	250 µL
A21046	Alexa Fluor® 633 goat anti–mouse IgM (µ chain) *2 mg/mL*	250 µL
A21070	Alexa Fluor® 633 goat anti–rabbit IgG (H+L) *2 mg/mL*	0.5 mL
A21071	Alexa Fluor® 633 goat anti–rabbit IgG (H+L) *highly cross-adsorbed* *2 mg/mL*	0.5 mL
A21094	Alexa Fluor® 633 goat anti–rat IgG (H+L) *2 mg/mL*	0.5 mL
A21086	Alexa Fluor® 633 rabbit anti–goat IgG (H+L) *2 mg/mL*	0.5 mL
A21063	Alexa Fluor® 633 rabbit anti–mouse IgG (H+L) *2 mg/mL*	0.5 mL
A31574	Alexa Fluor® 635 goat anti–mouse IgG (H+L) *2 mg/mL*	0.5 mL
A31575	Alexa Fluor® 635 goat anti–mouse IgG (H+L) *highly cross-adsorbed* *2 mg/mL*	0.5 mL
A31576	Alexa Fluor® 635 goat anti–rabbit IgG (H+L) *2 mg/mL*	0.5 mL
A31577	Alexa Fluor® 635 goat anti–rabbit IgG (H+L) *highly cross-adsorbed* *2 mg/mL*	0.5 mL
A21469	Alexa Fluor® 647 chicken anti–goat IgG (H+L) *2 mg/mL*	0.5 mL
A21463	Alexa Fluor® 647 chicken anti–mouse IgG (H+L) *2 mg/mL*	0.5 mL
A21443	Alexa Fluor® 647 chicken anti–rabbit IgG (H+L) *2 mg/mL*	0.5 mL
A21472	Alexa Fluor® 647 chicken anti–rat IgG (H+L) *2 mg/mL*	0.5 mL
A21447	Alexa Fluor® 647 donkey anti–goat IgG (H+L) *2 mg/mL*	0.5 mL
A31571	Alexa Fluor® 647 donkey anti–mouse IgG (H+L) *2 mg/mL*	0.5 mL
A31573	Alexa Fluor® 647 donkey anti–rabbit IgG (H+L) *2 mg/mL*	0.5 mL
A21448	Alexa Fluor® 647 donkey anti–sheep IgG (H+L) *2 mg/mL*	0.5 mL
A21237	Alexa Fluor® 647 F(ab')$_2$ fragment of goat anti–mouse IgG (H+L) *2 mg/mL*	250 µL
A21246	Alexa Fluor® 647 F(ab')$_2$ fragment of goat anti–rabbit IgG (H+L) *2 mg/mL*	250 µL
A21449	Alexa Fluor® 647 goat anti–chicken IgG (H+L) *2 mg/mL*	0.5 mL
A21450	Alexa Fluor® 647 goat anti–guinea pig IgG (H+L) *highly cross-adsorbed* *2 mg/mL*	0.5 mL
A21451	Alexa Fluor® 647 goat anti–hamster IgG (H+L) *2 mg/mL*	0.5 mL
A21445	Alexa Fluor® 647 goat anti–human IgG (H+L) *2 mg/mL*	0.5 mL
A21249	Alexa Fluor® 647 goat anti–human IgM (µ chain) *2 mg/mL*	250 µL
A21235	Alexa Fluor® 647 goat anti–mouse IgG (H+L) *2 mg/mL*	0.5 mL
A21236	Alexa Fluor® 647 goat anti–mouse IgG (H+L) *highly cross-adsorbed* *2 mg/mL*	0.5 mL
A21240	Alexa Fluor® 647 goat anti–mouse IgG$_1$ (γ1) *2 mg/mL*	250 µL
A21241	Alexa Fluor® 647 goat anti–mouse IgG$_{2a}$ (γ2a) *2 mg/mL*	250 µL
A21242	Alexa Fluor® 647 goat anti–mouse IgG$_{2b}$ (γ2b) *2 mg/mL*	250 µL
A21238	Alexa Fluor® 647 goat anti–mouse IgM (µ chain) *2 mg/mL*	250 µL
A31625	Alexa Fluor® 647 Goat Anti-Mouse SFX Kit	1 kit
A31626	Alexa Fluor® 647 Goat Anti-Mouse SFX Kit *highly cross-adsorbed*	1 kit
A21244	Alexa Fluor® 647 goat anti–rabbit IgG (H+L) *2 mg/mL*	0.5 mL
A21245	Alexa Fluor® 647 goat anti–rabbit IgG (H+L) *highly cross-adsorbed* *2 mg/mL*	0.5 mL
A31633	Alexa Fluor® 647 Goat Anti-Rabbit SFX Kit	1 kit
A31634	Alexa Fluor® 647 Goat Anti-Rabbit SFX Kit *highly cross-adsorbed*	1 kit
A21247	Alexa Fluor® 647 goat anti–rat IgG (H+L) *2 mg/mL*	0.5 mL
A21248	Alexa Fluor® 647 goat anti–rat IgM (µ chain) *2 mg/mL*	250 µL
A21446	Alexa Fluor® 647 rabbit anti–goat IgG (H+L) *2 mg/mL*	0.5 mL
A21239	Alexa Fluor® 647 rabbit anti–mouse IgG (H+L) *2 mg/mL*	0.5 mL
A20990	Alexa Fluor® 647–R-phycoerythrin goat anti–mouse IgG (H+L) *1 mg/mL*	100 µL
A20991	Alexa Fluor® 647–R-phycoerythrin goat anti–rabbit IgG (H+L) *1 mg/mL*	100 µL
A21083	Alexa Fluor® 660 donkey anti–goat IgG (H+L) *2 mg/mL*	0.5 mL
A21054	Alexa Fluor® 660 goat anti–mouse IgG (H+L) *2 mg/mL*	0.5 mL
A21055	Alexa Fluor® 660 goat anti–mouse IgG (H+L) *highly cross-adsorbed* *2 mg/mL*	0.5 mL
A21073	Alexa Fluor® 660 goat anti–rabbit IgG (H+L) *2 mg/mL*	0.5 mL
A21074	Alexa Fluor® 660 goat anti–rabbit IgG (H+L) *highly cross-adsorbed* *2 mg/mL*	0.5 mL
A21084	Alexa Fluor® 680 donkey anti–goat IgG (H+L) *2 mg/mL*	0.5 mL
A10038	Alexa Fluor® 680 donkey anti–mouse IgG (H + L) *2 mg/mL*	0.5 mL
A10043	Alexa Fluor® 680 donkey anti–rabbit IgG (H + L) *2 mg/mL*	0.5 mL
A21102	Alexa Fluor® 680 donkey anti–sheep IgG (H+L) *2 mg/mL*	0.5 mL

IMPORTANT NOTICE: The products described in this manual are covered by one or more Limited Use Label License(s). Please refer to the Appendix on page 971 and Master Product List on page 975. Products are For Research Use Only. Not intended for any animal or human therapeutic or diagnostic use.

PRODUCT LIST 7.2 SECONDARY IMMUNOREAGENTS—*continued*

Cat. No.	Product	Quantity
A21059	Alexa Fluor® 680 F(ab')$_2$ fragment of goat anti–mouse IgG (H+L) *2 mg/mL*	250 µL
A21077	Alexa Fluor® 680 F(ab')$_2$ fragment of goat anti–rabbit IgG (H+L) *2 mg/mL*	250 µL
A21057	Alexa Fluor® 680 goat anti–mouse IgG (H+L) *2 mg/mL*	0.5 mL
A21058	Alexa Fluor® 680 goat anti–mouse IgG (H+L) *highly cross-adsorbed* *2 mg/mL*	0.5 mL
A31562	Alexa Fluor® 680 goat anti–mouse IgG$_1$ (γ1) *2 mg/mL*	250 µL
A31563	Alexa Fluor® 680 goat anti–mouse IgG$_{2a}$ (γ2a) *2 mg/mL*	250 µL
A31564	Alexa Fluor® 680 goat anti–mouse IgG$_{2b}$ (γ2b) *2 mg/mL*	250 µL
A21048	Alexa Fluor® 680 goat anti–mouse IgM (µ chain) *2 mg/mL*	250 µL
A21076	Alexa Fluor® 680 goat anti–rabbit IgG (H+L) *2 mg/mL*	0.5 mL
A21109	Alexa Fluor® 680 goat anti–rabbit IgG (H+L) *highly cross-adsorbed* *2 mg/mL*	0.5 mL
A21096	Alexa Fluor® 680 goat anti–rat IgG (H+L) *2 mg/mL*	0.5 mL
A21088	Alexa Fluor® 680 rabbit anti–goat IgG (H+L) *2 mg/mL*	0.5 mL
A21065	Alexa Fluor® 680 rabbit anti–mouse IgG (H+L) *2 mg/mL*	0.5 mL
A21000	Alexa Fluor® 680–allophycocyanin goat anti–mouse IgG (H+L) *1 mg/mL*	100 µL
A21001MP	Alexa Fluor® 680–allophycocyanin goat anti–rabbit IgG (H+L) *1 mg/mL*	100 µL
A20983	Alexa Fluor® 680–R-phycoerythrin goat anti–mouse IgG (H+L) *1 mg/mL*	100 µL
A20984	Alexa Fluor® 680–R-phycoerythrin goat anti–rabbit IgG (H+L) *1 mg/mL*	100 µL
A21036	Alexa Fluor® 700 goat anti–mouse IgG (H+L) *2 mg/mL*	0.5 mL
A21038	Alexa Fluor® 700 goat anti–rabbit IgG (H+L) *2 mg/mL*	0.5 mL
A21037	Alexa Fluor® 750 goat anti–mouse IgG (H+L) *2 mg/mL*	0.5 mL
A21039	Alexa Fluor® 750 goat anti–rabbit IgG (H+L) *2 mg/mL*	0.5 mL
A21006	Alexa Fluor® 750–allophycocyanin goat anti–mouse IgG (H+L) *1 mg/mL*	100 µL
A10539	allophycocyanin, crosslinked, F(ab')$_2$ fragment of goat anti–mouse (H+L) *1 mg/mL*	250 µL
A865	allophycocyanin, crosslinked, goat anti–mouse IgG (H+L) *1 mg/mL*	0.5 mL
A10541	allophycocyanin, crosslinked, goat anti–mouse IgG$_1$ (γ1) *1 mg/mL*	250 µL
A10686	allophycocyanin, crosslinked, goat anti–mouse IgG$_{2a}$ (γ2a) *1 mg/mL*	250 µL
A10688	allophycocyanin, crosslinked, goat anti–mouse IgM (µ chain) *1 mg/mL*	250 µL
A10931	allophycocyanin, crosslinked, goat anti–rabbit IgG (H+L) *1 mg/mL*	0.5 mL
A10540	allophycocyanin, crosslinked, goat anti–rat IgG (H+L) *1 mg/mL*	250 µL
B11027	biotin-XX F(ab')$_2$ fragment of goat anti–mouse IgG (H+L) *2 mg/mL*	250 µL
B21078	biotin-XX F(ab')$_2$ fragment of goat anti–rabbit IgG (H+L) *2 mg/mL*	250 µL
B2763	biotin-XX goat anti–mouse IgG (H+L) *2 mg/mL*	0.5 mL
A10676	biotin-XX goat anti–mouse IgG, IgA, IgM (H+L) *2 mg/mL*	250 µL
A10519	biotin-XX goat anti–mouse IgG$_1$ (γ1) *2 mg/mL*	250 µL
B2770	biotin-XX goat anti–rabbit IgG (H+L) *2 mg/mL*	0.5 mL
A10517	biotin-XX goat anti–rat IgG (H+L) *2 mg/mL*	0.5 mL
A10650	biotin-XX mouse anti–human IgG$_1$ *1 mg/mL*	250 µL
A10663	biotin-XX mouse anti–human IgG$_4$ *1 mg/mL*	250 µL
A10518	biotin-XX rabbit anti–goat IgG (H+L) *2 mg/mL*	0.5 mL
B2752	BODIPY® FL goat anti–mouse IgG (H+L)	1 mg
B2766	BODIPY® FL goat anti–rabbit IgG (H+L)	1 mg
C962	Cascade Blue® goat anti–mouse IgG (H+L) *2 mg/mL*	0.5 mL
C2764	Cascade Blue® goat anti–rabbit IgG (H+L) *2 mg/mL*	0.5 mL
A10521	Cy®3 goat anti–mouse IgG (H+L) *2 mg/mL*	0.5 mL
A10520	Cy®3 goat anti–rabbit IgG (H+L) *2 mg/mL*	0.5 mL
A10522	Cy®3 goat anti–rat IgG (H+L) *2 mg/mL*	0.5 mL
A10524	Cy®5 goat anti–mouse IgG (H+L) *2 mg/mL*	0.5 mL
A10523	Cy®5 goat anti–rabbit IgG (H+L) *2 mg/mL*	0.5 mL
A10525	Cy®5 goat anti–rat IgG (H+L) *2 mg/mL*	0.5 mL
A10691	Cy®5–R-phycoerythrin, F(ab')$_2$ fragment of goat anti–rat IgG (H+L) *1 mg/mL*	100 µL
D22185	Diaminobenzidine (DAB) Histochemistry Kit #1 *with goat anti–mouse IgG–HRP*	1 kit
D20698	DSB-X™ biotin donkey anti–goat IgG (H+L) *2 mg/mL*	0.5 mL
D20701	DSB-X™ biotin goat anti–chicken IgG (H+L) *2 mg/mL*	0.5 mL
D20691	DSB-X™ biotin goat anti–mouse IgG (H+L) *highly cross-adsorbed* *2 mg/mL*	0.5 mL
D20693	DSB-X™ biotin goat anti–mouse IgM (µ chain) *2 mg/mL*	250 µL
D20697	DSB-X™ biotin goat anti–rat IgG (H+L) *2 mg/mL*	0.5 mL
D21887	DyeChrome™ Double Western Blot Stain Kit *for mouse IgG, rabbit IgG and total protein detection* *20 minigel blots*	1 kit
A10534	F(ab')$_2$ fragment of goat anti–mouse IgG (H+L) *2 mg/mL*	250 µL
F21452	F(ab')$_2$ fragment of goat anti–mouse IgG (H+L), alkaline phosphatase conjugate	0.5 mg
F21453	F(ab')$_2$ fragment of goat anti–mouse IgG (H+L), horseradish peroxidase conjugate	0.5 mg
F21456	F(ab')$_2$ fragment of goat anti–rabbit IgG (H+L), alkaline phosphatase conjugate	0.5 mg

continued on next page

PRODUCT LIST 7.2 SECONDARY IMMUNOREAGENTS—*continued*

Cat. No.	Product	Quantity
A10547	F(ab')$_2$ fragment of goat anti–rabbit IgG (H+L), horseradish peroxidase conjugate	0.5 mg
A10548	F(ab')$_2$ fragment of goat anti–rat IgG (H+L), horseradish peroxidase conjugate	0.5 mg
F11021	fluorescein F(ab')$_2$ fragment of goat anti–mouse IgG (H+L) *2 mg/mL*	250 µL
A10683	fluorescein F(ab')$_2$ fragment of goat anti–mouse IgG, IgM (H+L) *2 mg/mL*	250 µL
A10526	fluorescein F(ab')$_2$ fragment of goat anti–rabbit IgG (H+L) *2 mg/mL*	250 µL
A10527	fluorescein F(ab')$_2$ fragment of goat anti–rat IgG (H+L) *2 mg/mL*	250 µL
F2761	fluorescein goat anti–mouse IgG (H+L) *2 mg/mL*	0.5 mL
A10679	fluorescein goat anti–mouse IgG, IgM (H+L) *2 mg/mL*	250 µL
A10530	fluorescein goat anti–mouse IgG$_1$ (γ1) *2 mg/mL*	250 µL
F2765	fluorescein goat anti–rabbit IgG (H+L) *2 mg/mL*	0.5 mL
A10528	fluorescein goat anti–rat IgG (H+L) *2 mg/mL*	0.5 mL
A10529	fluorescein rabbit anti–goat (H+L) *2 mg/mL*	0.5 mL
A10535	goat anti–mouse IgG (H+L) *2 mg/mL*	0.5 mL
A10516	goat anti–mouse IgG (H+L) agarose *sedimented bead suspension*	2 mL
G21060	goat anti–mouse IgG (H+L), alkaline phosphatase conjugate	1 mg
G21040	goat anti–mouse IgG (H+L), horseradish peroxidase conjugate	1 mg
G31567	goat anti–mouse IgG (H+L), β-lactamase TEM-1 conjugate *0.5 mg net protein*	0.5 mg
A10538	goat anti–mouse IgG$_1$ (γ1) *2 mg/mL*	250 µL
A10551	goat anti–mouse IgG$_1$ (γ1), horseradish peroxidase conjugate	0.5 mg
A10685	goat anti–mouse IgG$_{2a}$ (γ2a), horseradish peroxidase conjugate	200 µg
A10666	goat anti–mouse IgG, IgA, IgM (H+L) *2 mg/mL*	250 µL
A10668	goat anti–mouse IgG, IgA, IgM (H+L), horseradish peroxidase conjugate	200 µg
A10677	goat anti–mouse IgG, IgM (H+L), horseradish peroxidase conjugate	200 µg
A10533	goat anti–rabbit IgG (H+L) *2 mg/mL*	0.5 mL
G21079	goat anti–rabbit IgG (H+L), alkaline phosphatase conjugate	1 mg
G21234	goat anti–rabbit IgG (H+L), horseradish peroxidase conjugate	1 mg
G31568	goat anti–rabbit IgG (H+L), β-lactamase TEM-1 conjugate *0.5 mg net protein*	0.5 mg
A10536	goat anti–rat IgG (H+L) *2 mg/mL*	0.5 mL
A10546	goat anti–rat IgG (H+L), alkaline phosphatase conjugate	0.5 mg
A10549	goat anti–rat IgG (H+L), horseradish peroxidase conjugate	0.5 mg
I36933	Image-iT® FX signal enhancer	10 mL
L24919	LI Silver (LIS) Enhancement Kit	1 kit
M10991	Marina Blue® goat anti–mouse IgG (H+L) *2 mg/mL*	0.5 mL
M10992	Marina Blue® goat anti–rabbit IgG (H+L) *2 mg/mL*	0.5 mL
A10630	mouse anti–human IgG$_1$ *1 mg/mL*	250 µL
A10648	mouse anti–human IgG$_1$, horseradish peroxidase conjugate	200 µg
A10651	mouse anti–human IgG$_4$ *1 mg/mL*	250 µL
A10654	mouse anti–human IgG$_4$, horseradish peroxidase conjugate	200 µg
N24915	NANOGOLD® Fab' fragment of goat anti–mouse IgG *80 µg protein/mL*	1 mL
N24916	NANOGOLD® Fab' fragment of goat anti–rabbit IgG *80 µg protein/mL*	1 mL
O6380	Oregon Green® 488 goat anti–mouse IgG (H+L) *2 mg/mL*	0.5 mL
O11033	Oregon Green® 488 goat anti–mouse IgG (H+L) *highly cross-adsorbed* *2 mg/mL*	0.5 mL
O6381	Oregon Green® 488 goat anti–rabbit IgG (H+L) *2 mg/mL*	0.5 mL
O11038	Oregon Green® 488 goat anti–rabbit IgG (H+L) *highly cross-adsorbed* *2 mg/mL*	0.5 mL
O6382	Oregon Green® 488 goat anti–rat IgG (H+L) *2 mg/mL*	0.5 mL
O6383	Oregon Green® 514 goat anti–mouse IgG (H+L) *2 mg/mL*	0.5 mL
P31581	Pacific Blue™ F(ab')$_2$ fragment of goat anti–mouse IgG (H+L) *2 mg/mL*	250 µL
P10993	Pacific Blue™ goat anti–mouse IgG (H+L) *2 mg/mL*	0.5 mL
P31582	Pacific Blue™ goat anti–mouse IgG (H+L) *highly cross-adsorbed* *2 mg/mL*	0.5 mL
P10994	Pacific Blue™ goat anti–rabbit IgG (H+L) *2 mg/mL*	0.5 mL
P31585	Pacific Orange™ F(ab')$_2$ fragment of goat anti–mouse IgG (H+L) *2 mg/mL*	250 µL
P31584	Pacific Orange™ goat anti–rabbit IgG (H+L) *2 mg/mL*	0.5 mL
P11047	protein A, Alexa Fluor® 488 conjugate	1 mg
P11049	protein A, Alexa Fluor® 546 conjugate	1 mg
P21462	protein A, Alexa Fluor® 647 conjugate	1 mg
P11065	protein G, Alexa Fluor® 488 conjugate	1 mg
P21041	protein G, horseradish peroxidase conjugate	1 mg
Q11041MP	Qdot® 525 goat F(ab')$_2$ anti–mouse IgG conjugate (H+L) *1 µM solution*	200 µL
Q11441MP	Qdot® 525 goat F(ab')$_2$ anti–rabbit IgG conjugate (H+L) *1 µM solution*	200 µL
Q11031MP	Qdot® 565 goat F(ab')$_2$ anti–mouse IgG conjugate (H+L) *1 µM solution*	200 µL
Q11032MP	Qdot® 565 goat F(ab')$_2$ anti–mouse IgG conjugate (H+L) *1 µM solution* *in 1 M betaine, 50 mM borate, pH 8.3 with 0.05% sodium azide*	100 µL

IMPORTANT NOTICE: The products described in this manual are covered by one or more Limited Use Label License(s). Please refer to the Appendix on page 971 and Master Product List on page 975. Products are For Research Use Only. Not intended for any animal or human therapeutic or diagnostic use.

www.invitrogen.com/probes

molecular probes® | invitrogen™ by life technologies™

PRODUCT LIST 7.2 SECONDARY IMMUNOREAGENTS—*continued*

Cat. No.	Product	Quantity
Q11431MP	Qdot® 565 goat F(ab')$_2$ anti-rabbit IgG conjugate (H+L) *1 µM solution*	200 µL
Q11432MP	Qdot® 565 goat F(ab')$_2$ anti-rabbit IgG conjugate (H+L) *1 µM solution* *in 1 M betaine, 50 mM borate, pH 8.3 with 0.05% sodium azide*	100 µL
Q11631MP	Qdot® 565 goat F(ab')$_2$ anti-rat IgG conjugate (H+L) *1 µM solution*	200 µL
Q11011MP	Qdot® 585 goat F(ab')$_2$ anti-mouse IgG conjugate (H+L) *1 µM solution*	200 µL
Q11411MP	Qdot® 585 goat F(ab')$_2$ anti-rabbit IgG conjugate (H+L) *1 µM solution*	200 µL
Q11201MP	Qdot® 605 goat F(ab')$_2$ anti-human IgG conjugate (H+L) *1 µM solution*	200 µL
Q11001MP	Qdot® 605 goat F(ab')$_2$ anti-mouse IgG conjugate (H+L) *1 µM solution*	200 µL
Q11002MP	Qdot® 605 goat F(ab')$_2$ anti-mouse IgG conjugate (H+L) *1 µM solution* *in 1 M betaine, 50 mM borate, pH 8.3 with 0.05% sodium azide*	100 µL
Q11401MP	Qdot® 605 goat F(ab')$_2$ anti-rabbit IgG conjugate (H+L) *1 µM solution*	200 µL
Q11402MP	Qdot® 605 goat F(ab')$_2$ anti-rabbit IgG conjugate (H+L) *1 µM solution* *in 1 M betaine, 50 mM borate, pH 8.3 with 0.05% sodium azide*	100 µL
Q11601MP	Qdot® 605 goat F(ab')$_2$ anti-rat IgG conjugate (H+L) *1 µM solution*	200 µL
A10195	Qdot® 625 goat F(ab')$_2$ anti-mouse IgG conjugate (H+L) *1 µM solution*	100 µL
A10194	Qdot® 625 goat F(ab')$_2$ anti-rabbit IgG conjugate (H+L) *1 µM solution*	100 µL
Q11221MP	Qdot® 655 goat F(ab')$_2$ anti-human IgG conjugate (H+L) *1 µM solution*	200 µL
Q11021MP	Qdot® 655 goat F(ab')$_2$ anti-mouse IgG conjugate (H+L) *1 µM solution*	200 µL
Q11022MP	Qdot® 655 goat F(ab')$_2$ anti-mouse IgG conjugate (H+L) *1 µM solution* *in 1 M betaine, 50 mM borate, pH 8.3 with 0.05% sodium azide*	100 µL
Q11421MP	Qdot® 655 goat F(ab')$_2$ anti-rabbit IgG conjugate (H+L) *1 µM solution*	200 µL
Q11422MP	Qdot® 655 goat F(ab')$_2$ anti-rabbit IgG conjugate (H+L) *1 µM solution* *in 1 M betaine, 50 mM borate, pH 8.3 with 0.05% sodium azide*	100 µL
Q11621MP	Qdot® 655 goat F(ab')$_2$ anti-rat IgG conjugate (H+L) *1 µM solution*	200 µL
Q14421MP	Qdot® 655 goat whole IgG anti-chicken IgY (H+L) conjugate *1 µM solution* *in 1 M betaine, 50 mM borate, pH 8.3 with 0.05% sodium azide*	200 µL
Q11821MP	Qdot® 655 rabbit F(ab')$_2$ anti-goat IgG conjugate (H+L) *1 µM solution* *in 1 M betaine, 50 mM borate, pH 8.3 with 0.05% sodium azide*	200 µL
Q11061MP	Qdot® 705 goat F(ab')$_2$ anti-mouse IgG conjugate (H+L) *1 µM solution*	200 µL
Q11062MP	Qdot® 705 goat F(ab')$_2$ anti-mouse IgG conjugate (H+L) *1 µM solution* *in 1 M betaine, 50 mM borate, pH 8.3 with 0.05% sodium azide*	100 µL
Q11461MP	Qdot® 705 goat F(ab')$_2$ anti-rabbit IgG conjugate (H+L) *1 µM solution*	200 µL
Q11462MP	Qdot® 705 goat F(ab')$_2$ anti-rabbit IgG conjugate (H+L) *1 µM solution* *in 1 M betaine, 50 mM borate, pH 8.3 with 0.05% sodium azide*	100 µL
Q11071MP	Qdot® 800 goat F(ab')$_2$ anti-mouse IgG conjugate (H+L) *1 µM solution*	200 µL
Q11471MP	Qdot® 800 goat F(ab')$_2$ anti-rabbit IgG conjugate (H+L) *1 µM solution*	200 µL
A10537	rabbit anti-goat IgG (H+L) *2 mg/mL*	0.5 mL
R21458	rabbit anti-goat IgG (H+L), alkaline phosphatase conjugate	1 mg
R21459	rabbit anti-goat IgG (H+L), horseradish peroxidase conjugate	1 mg
R21455	rabbit anti-mouse IgG (H+L), horseradish peroxidase conjugate	1 mg
R6393	Rhodamine Red™-X goat anti-mouse IgG (H+L) *2 mg/mL*	0.5 mL
R6394	Rhodamine Red™-X goat anti-rabbit IgG (H+L) *2 mg/mL*	0.5 mL
A10543	R-phycoerythrin F(ab')$_2$ fragment of goat anti-mouse IgG (H+L) *1 mg/mL*	250 µL
A10690	R-phycoerythrin F(ab')$_2$ fragment of goat anti-mouse IgM (µ chain) *1 mg/mL*	250 µL
A10542	R-phycoerythrin F(ab')$_2$ fragment of goat anti-rabbit IgG (H+L) *1 mg/mL*	250 µL
A10544	R-phycoerythrin F(ab')$_2$ fragment of goat anti-rat IgG (H+L) *1 mg/mL*	250 µL
P852	R-phycoerythrin goat anti-mouse IgG (H+L) *1 mg/mL*	1 mL
P21129	R-phycoerythrin goat anti-mouse IgG$_1$ (γ1) conjugate *1 mg/mL*	250 µL
P21139	R-phycoerythrin goat anti-mouse IgG$_{2a}$ (γ2a) conjugate *1 mg/mL*	250 µL
P21149	R-phycoerythrin goat anti-mouse IgG$_{2b}$ (γ2b) conjugate *1 mg/mL*	250 µL
A10689	R-phycoerythrin goat anti-mouse IgM (µ chain) *1 mg/mL*	250 µL
P2771MP	R-phycoerythrin goat anti-rabbit IgG (H+L) *1 mg/mL*	0.5 mL
A10545	R-phycoerythrin goat anti-rat IgG (H+L) *1 mg/mL*	250 µL
T2762	tetramethylrhodamine goat anti-mouse IgG (H+L) *2 mg/mL*	0.5 mL
T2769	tetramethylrhodamine goat anti-rabbit IgG (H+L) *2 mg/mL*	0.5 mL
A10531	tetramethylrhodamine goat anti-rat IgG (H+L) *2 mg/mL*	0.5 mL
A10532	tetramethylrhodamine rabbit anti-goat IgG (H+L) *2 mg/mL*	0.5 mL
T862	Texas Red® goat anti-mouse IgG (H+L) *2 mg/mL*	0.5 mL
T2767	Texas Red® goat anti-rabbit IgG (H+L) *2 mg/mL*	0.5 mL
T6390	Texas Red®-X goat anti-mouse IgG (H+L) *2 mg/mL*	0.5 mL
T6391	Texas Red®-X goat anti-rabbit IgG (H+L) *2 mg/mL*	0.5 mL
T6392	Texas Red®-X goat anti-rat IgG (H+L) *2 mg/mL*	0.5 mL
W10132	WesternDot™ 625 Goat Anti-Mouse IgG Western Blot Kit *20 minigel blots*	1 kit
W10142	WesternDot™ 625 Goat Anti-Rabbit IgG Western Blot Kit *20 minigel blots*	1 kit

IMPORTANT NOTICE: The products described in this manual are covered by one or more Limited Use Label License(s). Please refer to the Appendix on page 971 and Master Product List on page 975. Products are For Research Use Only. Not intended for any animal or human therapeutic or diagnostic use.

7.3 Zenon® Technology: Versatile Reagents for Immunolabeling

Our exceptional Zenon® immunolabeling technology provides an easy, versatile and truly unique method of labeling antibodies with Molecular Probes® premier dyes, haptens and enzymes. This enabling technology not only eliminates the need for secondary detection reagents in many applications, but also simplifies immunolabeling applications that previously were time consuming or impractical, including the use of multiple antibodies derived from the same species in the same protocol,[1–4] as well as the detection of antibody binding in tissues when both the antibody and the tissue are derived from the same species. Moreover, Zenon® immunolabeling technology permits the rapid and quantitative preparation of antibody complexes from a purified antibody fraction or from a crude antibody preparation such as serum, ascites fluid or a hybridoma supernatant.

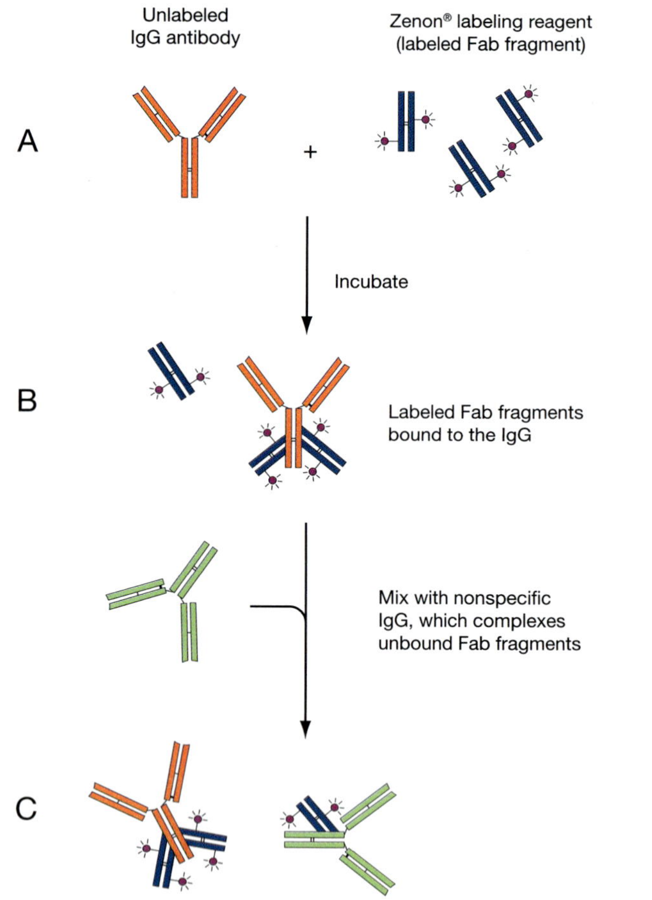

Figure 7.3.1 Labeling scheme utilized in the Zenon® Antibody Labeling Kits. **A)** An unlabeled IgG antibody is incubated with the Zenon® labeling reagent, which contains a fluorophore-labeled, Fc-specific anti-IgG Fab fragment. **B)** This labeled Fab fragment binds to the Fc portion of the IgG antibody. **C)** Excess Fab fragment is then neutralized by the addition of a nonspecific IgG, preventing crosslabeling by the Fab fragment in experiments where primary antibodies of the same type are present. Note that the Fab fragment used for labeling need not be coupled to a fluorophore, but could instead be coupled to an enzyme (such as HRP) or to biotin.

The Zenon® labeling method takes advantage of the immunoselectivity of the antibody binding reaction by forming a complex between an intact primary IgG antibody and a fluorophore-, biotin- or enzyme-labeled Fab fragment directed against the Fc portion of that IgG (Figure 7.3.1). Simple mixing of the labeled Fab fragment, which is supplied in the Zenon® Antibody Labeling Kits, with the corresponding primary antibody quantitatively produces the Fab–antibody complex in under 10 minutes, with no pre- or postlabeling purification required. This labeled Fab–antibody complex can be immediately used to stain cells, tissues[1,4] and other targets in the same manner as a covalently labeled primary antibody.

Molecular Probes® Zenon® immunolabeling technology has many outstanding features that open several avenues for research and development. These features, which are discussed in detail later in this section, include:

- **Rapid antibody labeling**. The Zenon® antibody labeling protocol permits rapid and reproducible labeling of submicrogram quantities of a primary antibody. The Zenon® complex is ready for use in cell-labeling protocols in less than 10 minutes. In the absence of competing antibodies of the same species, the Zenon® complex can also be stored for later use.
- **Quantitative antibody labeling**. Labeling is quantitative with respect to the primary antibody, and the extent of antibody labeling (and thus the intensity of the fluorophore or activity of a conjugated enzyme) can be adjusted by changing the molar ratio of the dye- or enzyme-labeled Fab fragment to the primary antibody. In flow cytometry, this unique property of the Zenon® reagents permits use of the same fluorescent color to label two or more cell populations in the same sample.[5]
- **Affinity-purified Fab fragments**. The Zenon® dye- and enzyme-labeled Fab fragments have been affinity purified during their preparation to help ensure their high affinity and selectivity for the Fc portion of the corresponding primary antibody. Furthermore, our procedure for chemical labeling of the Fab fragments protects the Fc-binding site, resulting in more active labeling reagents.
- **Function like covalently labeled primary antibodies**. The Zenon® complexes formed with the Zenon® Antibody Labeling Kits can be used to label cells and tissues without introducing nonspecific background, which can often arise from secondary antibody detection. Moreover, these Zenon® complexes are much easier to prepare and more versatile than covalently labeled primary antibodies, and they show higher activity because only the Fc portion of the primary antibody is labeled, leaving the antigen-binding sites available for cell and tissue labeling. Furthermore, the Zenon® Labeling Kits permit easy and efficient labeling of antibodies that are not commercially available as direct conjugates.
- **Many fluorophore and enzyme labels available**. Zenon® immunolabeling technology makes it very easy to change fluorescent color combinations or detection methodologies by simply using a different dye- or enzyme-labeled Fab fragment from our extensive selection of Zenon® Antibody Labeling Kits (Table 7.7).

IMPORTANT NOTICE: The products described in this manual are covered by one or more Limited Use Label License(s). Please refer to the Appendix on page 971 and Master Product List on page 975. Products are For Research Use Only. Not intended for any animal or human therapeutic or diagnostic use.
www.invitrogen.com/probes

Features of Zenon® Antibody Labeling Technology

Zenon® Immunolabeling Is Rapid and Quantitative

Labeling of a suitable primary antibody with the Zenon® Antibody Labeling Kits is *very* fast. The Zenon® labeling reagent contains dye- or enzyme-labeled Fab fragments that have been affinity purified to help ensure their high affinity for the Fc portion of the corresponding primary antibody. We recommend incubation of the Zenon® labeling reagent with the appropriate primary antibody for 5 minutes, followed by a 5-minute blocking reaction of any remaining labeled Fab fragments with excess nonspecific IgG; however, the actual rates of the complex formation and IgG blocking steps are likely to be even faster and may occur within the mixing time of the components (Figure 7.3.2).

Fab–antibody complex formation can be quantitative with respect to the primary antibody (Figure 7.3.3). Furthermore, the degree of labeling (and thus the intensity of the fluorophore or activity of a conjugated enzyme) can be adjusted to some extent by changing the molar ratio of the labeled Fab fragment to the primary antibody. We find that approximately equal *weights* of a dye-labeled Fab fragment generated from a goat anti-Fc antibody (MW ~50,000 daltons) and the intact mouse primary antibody (MW ~150,000 daltons)—a 3:1 *molar* ratio of the labeled Fab fragment to the primary antibody—yields an Fab–antibody complex that is suitable for most applications. Thus, the 50 µg of labeled Fab fragments in the Zenon® labeling reagent provided in the organic dye–based Zenon® Antibody Labeling Kits is sufficient for labeling approximately 50 µg of the corresponding intact IgG antibody; we have defined "one labeling" in all of our Zenon® Antibody Labeling Kits as the amount of Zenon® labeling reagent required for labeling 1 µg of an intact primary antibody. Use of a 5:1 molar ratio of the Zenon® labeling reagent gives an apparent 100% conversion of the primary antibody to the labeling complex (Figure 7.3.3) and a somewhat brighter total fluorescence when compared with the fluorescence of Zenon® complexes prepared using a 3:1 ratio.

Increasing the molar (or weight) ratio of the Zenon® labeling reagent to the antibody can yield a somewhat brighter conjugate and allows for quantitative utilization of the primary antibody, whereas decreasing the ratio can yield a somewhat less fluorescent complex. The ability to adjust the fluorescence intensity (or enzymatic activity) of a labeled complex is a feature of the Zenon® technology that is not at all practical with direct chemical labeling of antibodies.[5] This property also permits the researcher to rapidly optimize the *best* complex for their experiment, rather than depending on the quite variable degree of substitution that is typical of covalently labeled antibodies from different commercial sources or from chemical labeling in a research laboratory. A 3:1 molar ratio of Zenon® labeling reagent to primary antibody is also suitable for labeling with the phycobiliprotein- and enzyme-derived Zenon® labeling reagents. The Zenon® Antibody Labeling Kits containing a phycobiliprotein- or enzyme-labeled Fab fragment include sufficient reagents for labeling ~25 µg of an intact primary antibody; the Zenon® Antibody Labeling Kits containing an Alexa Fluor® dye–phycobiliprotein tandem conjugate include sufficient reagents for labeling ~10 µg of a primary antibody.

Zenon® Technology Simplifies the Use of Multiple Antibodies of the Same Isotype in the Same Protocol

Unlike detection with secondary antibodies, Zenon® immunolabeling technology allows staining of a cell or tissue sample with multiple antibodies of the same isotype. Our affinity purification of the Zenon® labeling reagent (which we perform subsequent to dye or enzyme conjugation) helps to ensure that the formation of the Fab–antibody complex is stable. Reversal of Zenon® complex formation by the excess of nonspecific IgG used to capture any uncomplexed Zenon® labeling reagent does occur, particularly at temperatures above room temperature. The stability of the Zenon® complex, however, is sufficient to allow sequential (or simultaneous) labeling of different targets in cells and tissues

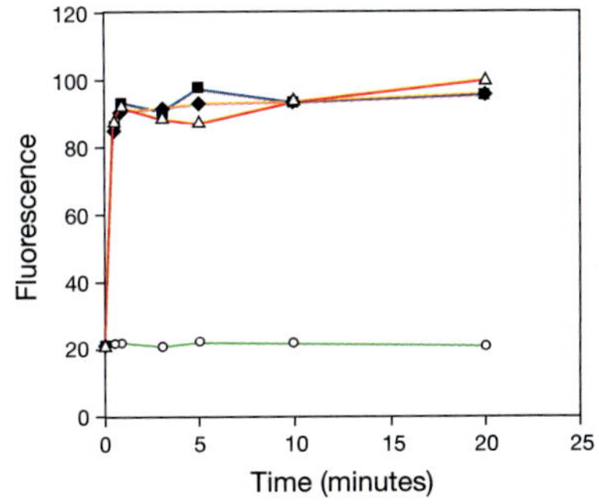

Figure 7.3.2 Formation of antibody–Fab complexes. An anti-biotin mouse IgG_1 monoclonal antibody was mixed with the Zenon® Alexa Fluor® 488 labeling reagent (a component of Kit Z25002) for varying time intervals before the reaction was quenched by the addition of excess mouse IgG blocking reagent. The quenched reactions were then added to a microplate well containing biotinylated bovine serum albumin and incubated for 20 minutes. After washing, the fluorescence of the remaining bound signal was measured. Binding was found to be essentially complete in less than 5 minutes. Three trials are shown, along with a control (O) where no labeling reagent was added.

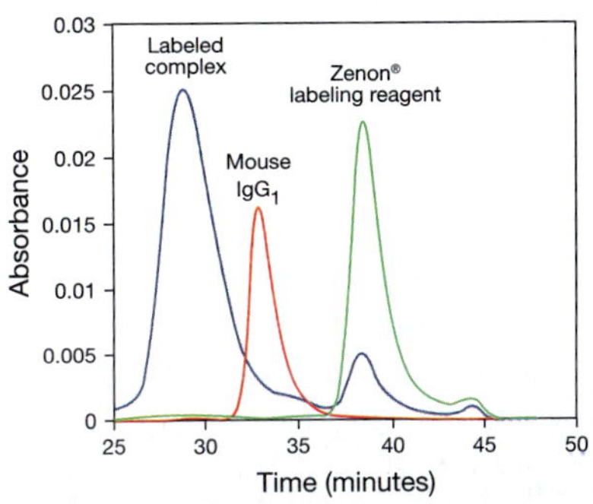

Figure 7.3.3 High-performance size-exclusion chromatographic analysis of the Zenon® Alexa Fluor® 488 labeling reagent (a component of Kit Z25002) binding to a mouse IgG_1 antibody. The Zenon® labeling reagent peak appears at 38 minutes; the mouse IgG_1 peak appears at 33 minutes. When combined at a molar ratio of ~5:1 (Zenon® labeling reagent: IgG_1), the IgG_1 antibody is quantitatively converted to a labeled complex, which appears as a peak at 29 minutes.

IMPORTANT NOTICE: The products described in this manual are covered by one or more Limited Use Label License(s). Please refer to the Appendix on page 971 and Master Product List on page 975. Products are For Research Use Only. Not intended for any animal or human therapeutic or diagnostic use.

with multiple antibody complexes (Figure 7.3.4, Figure 7.3.5, Figure 7.3.6). Subsequent to staining, an aldehyde-based fixation step can permanently block the transfer of Zenon® labels between different primary antibodies and will preserve the staining pattern. Zenon® complexes can be used together with other Zenon® complexes, with directly labeled primary antibodies, with labeled secondary antibodies to primary antibodies derived from other species, with antibodies that lack an Fc fragment and with avidin–biotin techniques (Section 7.6) for multiplexed immunolabeling. Some cross-reactivity of the Zenon® labeling reagent with the Fc portion of antibodies from other species may occur unless the excess Zenon® labeling reagent is captured with the soluble nonspecific IgG included in the Zenon® Antibody Labeling Kits.

Zenon® Immunolabeling Is Reliable, Even with Very Small Quantities of Reagents

Formation of the Fab–antibody complex with the Zenon® Antibody Labeling Kits is extremely reliable and reproducible, even with very small (submicrogram) amounts of primary antibody. Successful chemical labeling of submicrogram quantities of an antibody with a succinimidyl ester or an isothiocyanate of a dye is typically not possible; chemical labeling and purification of proteins usually requires at least 100 µg of the carrier-free protein. Because submicrogram amounts of an immunolabeling complex may be all that are required for an experiment, there is absolutely no waste of expensive or difficult-to-obtain antibodies when using the Zenon® Antibody Labeling Kits. Additionally, optimization of the degree of labeling by a dye or an enzyme is trivial, as compared with any chemical labeling method. Although we define "one labeling" in the Zenon® Antibody Labeling Kits as the amount of Zenon® labeling reagent required for labeling 1 µg of the primary antibody, we routinely label about 0.4 µg of the primary antibody dissolved in 2 µL of a buffer.

Unfortunately, many providers of mouse monoclonal antibodies do not indicate how much of the pure antibody they provide in their products. Furthermore, these antibodies are frequently supplied diluted with an albumin or other carrier protein, which precludes estimation of the antibody concentration from the optical density at 280 nm. However, the ease of preparing Zenon® complexes makes it practical to rapidly optimize the labeling reagent, even when the amount of primary antibody in the sample is unknown.

Zenon® Immunolabeling Does Not Require Antibody Pre-Purification

Unlike chemically modifying antibodies, labeling antibodies with the Zenon® Antibody Labeling Kits does not require removal of exogenous proteins such as serum albumin from the antibody. Serum albumin is frequently added to laboratory-derived or commercially supplied antibodies to help preserve the activity of dilute solutions of antibodies. Because the Zenon® labeling reagent selectively binds only to the Fc portion of the primary antibody, there should be limited or no effect of exogenous proteins that do not have an Fc fragment. The Zenon® labeling reagents function equally well when used to label a purified antibody or a crude antibody preparation such as serum, ascites fluid or a hybridoma supernatant. Zenon® labeling is usually quite successful with even dilute solutions of a primary antibody, as well as with antibodies that are dissolved in amine-containing buffers such as Tris.

Zenon® Immunolabeling Is Versatile

We currently offer Zenon® Antibody Labeling Kits containing a wide selection of different fluorescent labels (Table 7.7), with spectra that span ultraviolet, visible and near-infrared wavelengths. Zenon® immunolabeling technology makes it particularly easy to change fluorescent colors or detection methodologies by simply using a different Zenon® labeling reagent from our extensive selection of Zenon® Antibody Labeling Kits. There is no need to purchase multiple direct conjugates of the same mouse primary antibody (such as with the

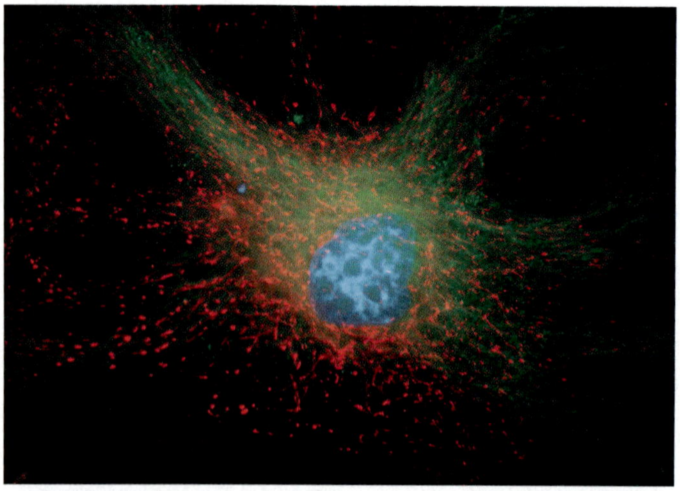

Figure 7.3.4 Bovine pulmonary artery endothelial cell labeled with probes for tubulin and mitochondria. Tubulin was detected with an anti–α-tubulin mouse IgG_{2b} monoclonal antibody prelabeled with the Zenon® Alexa Fluor® 488 Mouse IgG_{2b} Labeling Kit (Z25202), and mitochondria were labeled using an anti-OxPhos Complex V subunit a, mouse IgG_{2b} monoclonal antibody (A21350) prelabeled with the Zenon® Alexa Fluor® 555 Mouse IgG_{2b} Labeling Kit (Z25205). The nucleus was stained with DAPI (D1306, D3571, D21490).

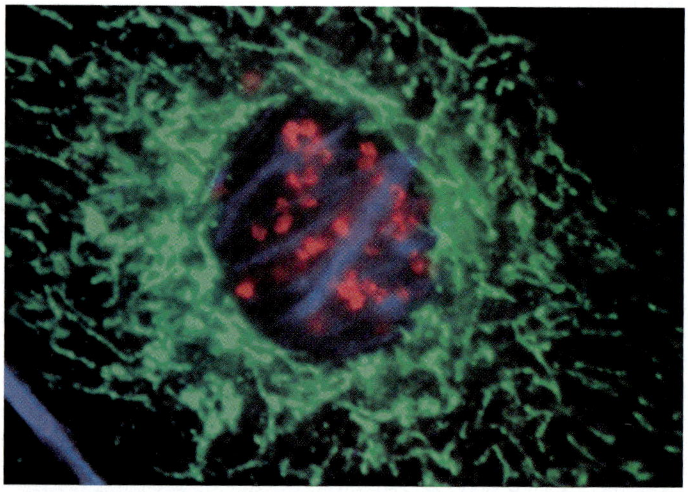

Figure 7.3.5 Bovine pulmonary artery endothelial cell labeled with probes for actin, mitochondria and the phosphorylated form of histone H3. Actin was labeled with blue-fluorescent Alexa Fluor® 350 phalloidin (A22281), and phosphorylated histone H3 was detected using a rabbit anti–phosphohistone H3 antibody prelabeled with the Zenon® Alexa Fluor® 555 Rabbit IgG Labeling Kit (red fluorescence, Z25305). Endogenous biotin associated with the mitochondria was labeled using streptavidin (S888) followed by a rabbit anti-streptavidin antibody prelabeled with the Zenon® Alexa Fluor® 488 Rabbit IgG Labeling Kit (green fluorescence, Z25302).

IMPORTANT NOTICE: The products described in this manual are covered by one or more Limited Use Label License(s). Please refer to the Appendix on page 971 and Master Product List on page 975. Products are For Research Use Only. Not intended for any animal or human therapeutic or diagnostic use.

Table 7.7 Zenon® Antibody Labeling Kits.

Label	Abs/Em *	Mouse IgG_1 Labeling Kit	Mouse IgG_{2a} Labeling Kit	Mouse IgG_{2b} Labeling Kit	Rabbit IgG Labeling Kit	Goat IgG Labeling Kit	Human IgG Labeling Kit
Alexa Fluor® Dyes							
Alexa Fluor® 350	346/442	Z25000	Z25100	Z25200	Z25300		Z25400
Alexa Fluor® 405	402/421	Z25013	Z25113	Z25213	Z25313		
Alexa Fluor® 430	434/539	Z25001			Z25301		
Alexa Fluor® 488	495/519	Z25002 Z25090 †	Z25102	Z25202	Z25302	Z25602	Z25402
Alexa Fluor® 532	531/554	Z25003			Z25303		
Alexa Fluor® 546	556/573	Z25004	Z25104	Z25204	Z25304		
Alexa Fluor® 555	555/565	Z25005	Z25105	Z25205	Z25305	Z25605	Z25405
Alexa Fluor® 568	578/603	Z25006	Z25106	Z25206	Z25306	Z25606	
Alexa Fluor® 594	590/617	Z25007	Z25107	Z25207	Z25307	Z25607	Z25407
Alexa Fluor® 647	650/668	Z25008	Z25108	Z25208	Z25308	Z25608	Z25408
Alexa Fluor® 660	663/690	Z25009					
Alexa Fluor® 680	679/702	Z25010	Z25110	Z25210	Z25310		
Alexa Fluor® 700	702/723	Z25011					
Alexa Fluor® 750	749/775				Z25312		
Classic Dyes							
Pacific Orange™	400/551	Z25256	Z25257				
Pacific Blue™	410/455	Z25041	Z25156		Z25341		
Fluorescein	494/518	Z25042			Z25342		
Texas Red®-X	595/615	Z25045					
Biotin							
Biotin-XX	NA	Z25052	Z25152	Z25252	Z25352		Z25452
Phycobiliproteins and Alexa Fluor® Dye–Phycobiliprotein Tandem Conjugates							
R-Phycoerythrin (R-PE)	496, 546, 565 ‡/578	Z25055	Z25155	Z25255	Z25355		Z25455
Alexa Fluor® 610–R-PE	496, 546, 565 ‡/628	Z25020					
Alexa Fluor® 647–R-PE	496, 546, 565 ‡/668	Z25021	Z25121	Z25221			
Alexa Fluor® 680–R-PE	496, 546, 565 ‡/702	Z25022					
Allophycocyanin (APC)	650/660	Z25051	Z25151	Z25251	Z25351		Z25451
Alexa Fluor® 700–APC	650/723	Z25030					
Alexa Fluor® 750–APC	650/775	Z25031					
Enzymes							
Horseradish peroxidase	NA	Z25054	Z25154	Z25254	Z25354		Z25454
Alkaline phosphatase	NA				Z25350		
Tricolor Labeling Kits							
Kit #1: Alexa Fluor® 488, Alexa Fluor® 555, Alexa Fluor® 647 dyes §		Z25060	Z25160	Z25260	Z25360		Z25460
Kit #2: Alexa Fluor® 350, Alexa Fluor® 488, Alexa Fluor® 594 dyes §		Z25070	Z25170	Z25270	Z25370		Z25470
Kit #3: Alexa Fluor® 488, R-PE, Alexa Fluor® 647–R-PE dyes §		Z25080	Z25180	Z25280	Z25380		

* Approximate absorption and emission maxima, in nm. † This Zenon® Antibody Labeling Kit is enhanced with TSA™ technology. ‡ Multiple absorption peaks. § See individual kit components for Abs/Em values. NA = Not applicable.

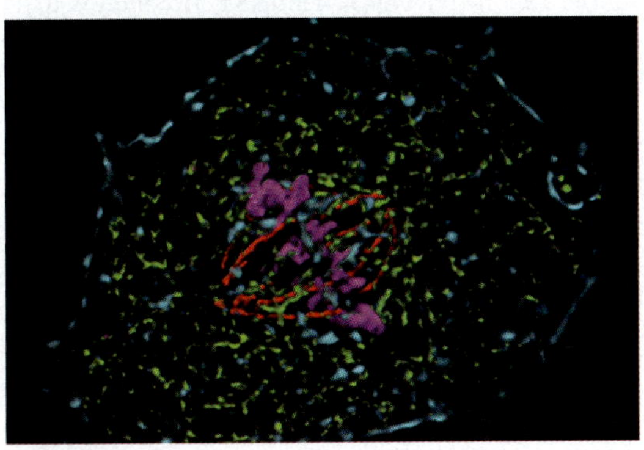

Figure 7.3.6 Four-color staining of muntjac skin fibroblast with probes for cytoskeletal, nuclear and mitochondrial proteins. Fixed and permeabilized cells were stained with Alexa Fluor® 350 phalloidin (A22281), an anti–α-tubulin antibody (A11126), an anti-cdc6 peptide antibody and an anti–OxPhos Complex V inhibitor protein antibody (A21355). The anti–OxPhos Complex V inhibitor protein antibody was prelabeled with the Zenon® Alexa Fluor® 488 Mouse IgG_1 Labeling Kit (Z25002), the anti–α-tubulin antibody was prelabeled with the Zenon® Alexa Fluor® 568 Mouse IgG_1 Labeling Kit (Z25002), and the anti-cdc6 peptide antibody was prelabeled with the Zenon® Alexa Fluor® 647 Mouse IgG_1 Labeling Kit (Z25008). The image was deconvolved using Huygens software (Scientific Volume Imaging, http://www.svi.nl/).

IMPORTANT NOTICE: The products described in this manual are covered by one or more Limited Use Label License(s). Please refer to the Appendix on page 971 and Master Product List on page 975. Products are For Research Use Only. Not intended for any animal or human therapeutic or diagnostic use.

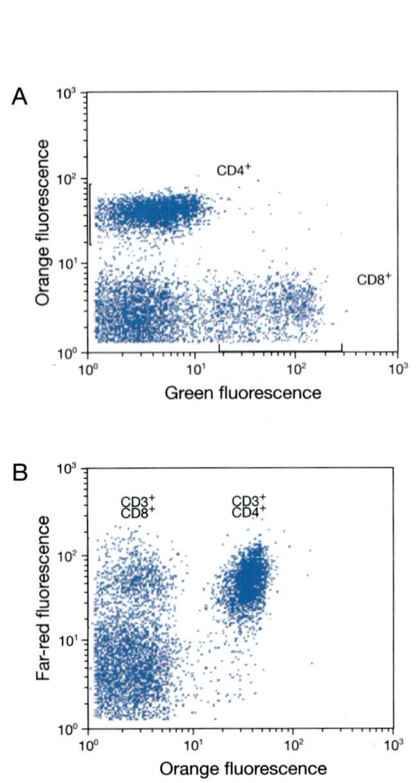

Figure 7.3.7 Human peripheral blood lymphocytes were stained with the following three antibodies: an anti-CD3 mouse IgG_1 antibody prelabeled with the Zenon® Alexa Fluor® 647 Mouse IgG_1 Labeling Kit (Z25008), an anti-CD4 mouse IgG_1 antibody prelabeled with the Zenon® R-Phycoerythrin Mouse IgG_1 Labeling Kit (Z25055) and an anti-CD8 mouse IgG_{2a} antibody prelabeled with the Zenon® Alexa Fluor® 488 Mouse IgG_{2a} Labeling Kit (Z25102). Panels A and B show that cells can be separated by plotting the orange-fluorescent versus green-fluorescent signal or red-fluorescent versus orange-fluorescent signal, respectively, demonstrating that the Zenon® label does not transfer to other antibodies in the same sample. The samples were analyzed on a Coulter® Elite flow cytometer using 488 nm excitation for R-phycoerythrin and the Alexa Fluor® 488 dye, and 633 nm excitation for the Alexa Fluor® 647 dye.

cluster of differentiation (CD) antibodies) in order to perform multicolor flow cytometry (Figure 7.3.7) and imaging protocols.

The fluorescence intensity or enzymatic activity of a Zenon® complex prepared using our protocol is usually similar to that of the corresponding directly labeled conjugate of the primary antibody, although it is typically somewhat lower than the intensity that can be obtained using a labeled secondary antibody. If necessary, the sensitivity of the assay can be increased through use of a Zenon® complex in combination with a signal amplification method such as TSA™ (Section 6.2); see below for a description of our Zenon® Alexa Fluor® 488 Mouse IgG_1 Labeling Kit enhanced with TSA™ technology (Z25090).

Zenon® Labeling Kits and Their Applications

Our wide selection of Zenon® Kits can be mixed and matched in the same experimental protocol and even in the same cell-labeling solution, providing the freedom to experiment with multiple dye–antibody combinations in flow cytometry and imaging applications. In addition, we offer Zenon® Mouse IgG_1 Labeling Kits enhanced with TSA™ technology (Z25090), which combine the advantages of Zenon® labeling technology with the sensitivity of tyramide signal amplification (TSA™).

We offer an extensive assortment of Zenon® Antibody Labeling Kits for use with mouse IgG_1, mouse IgG_{2a}, mouse IgG_{2b}, rabbit IgG, goat IgG and human IgG antibodies (Antibody Structure and Classification—Note 7.1). Each of these kits contain a Zenon® labeling reagent, comprising Fab fragments generated from goat secondary antibodies—or in the case of the Zenon® labeling reagents for goat IgG antibodies, from rabbit secondary antibodies—directed against the Fc portion of the corresponding primary antibody and covalently conjugated with a detectable label. These detectable labels include our outstanding Alexa Fluor® dyes (Section 1.3), as well as phycobiliproteins, tandem conjugates of phycobiliproteins with Alexa Fluor® dyes (Section 6.4), enzymes (horseradish peroxidase and alkaline phosphatase), biotin and some conventional dyes such as fluorescein. Three types of kits are available:

- Zenon® Antibody Labeling Kits that contain an affinity-purified Fab fragment of a goat anti-Fc antibody (or, in the case of the Zenon® Goat IgG Labeling Kits, a rabbit anti-Fc antibody) that has been conjugated to one of our premier Alexa Fluor® dyes or to Pacific Blue™, Pacific Orange™, fluorescein, Texas Red®-X dyes or biotin.
- Zenon® Antibody Labeling Kits that contain an Fab fragment of an anti-Fc antibody labeled with a R-phycoerythrin (R-PE), allophycocyanin (APC) or a tandem conjugate of R-PE or APC with one of our long-wavelength Alexa Fluor® dyes.
- Zenon® Antibody Labeling Kits that contain HRP or alkaline phosphatase conjugates for use in enzyme-amplified protocols (Section 6.2, Section 6.3); see below for a description of the Zenon® Alexa Fluor® 488 Mouse IgG_1 Labeling Kit enhanced with TSA™ technology (Z25090).

In addition to these Zenon® Antibody Labeling Kits, we offer three different types of Zenon® Tricolor Labeling Kits for mouse IgG antibodies,[5] rabbit IgG and human IgG antibodies (Table 7.7). The Zenon® Tricolor Labeling Kit #1 contains 10 µg each of the Alexa Fluor® 488, Alexa Fluor® 555 and Alexa Fluor® 647 Zenon® labeling reagents. This kit is designed for optimal triple-antibody staining with confocal laser-scanning microscopes but it can also be used with any suitably equipped fluorescence microscope. The Zenon® Tricolor Labeling Kit #2 contains 10 µg each of the Alexa Fluor® 350, Alexa Fluor® 488 and Alexa Fluor® 594 Zenon® labeling reagents, yielding simultaneous blue-, green- and red-fluorescent immunostaining that is useful with almost any fluorescence microscope. The Zenon® Tricolor Labeling Kit #3 is especially suitable for flow cytometry applications and contains 10 µg each of the Alexa Fluor® 488, R-phycoerythrin (R-PE) and Alexa Fluor® 647–R-PE Zenon® labeling reagents, which are each excited efficiently with an argon-ion laser and exhibit minimal spectral overlap.

Zenon® Antibody Labeling Kit Enhanced with TSA™ Technology

The Zenon® Alexa Fluor® 488 Mouse IgG_1 Labeling Kit enhanced with TSA™ technology (Z25090) provides exceptional target-identification capabilities. This kit provides the necessary reagents from both the Zenon® Horseradish Peroxidase Mouse IgG_1 Labeling Kit and the Alexa

IMPORTANT NOTICE: The products described in this manual are covered by one or more Limited Use Label License(s). Please refer to the Appendix on page 971 and Master Product List on page 975. Products are For Research Use Only. Not intended for any animal or human therapeutic or diagnostic use.

Fluor® 488 TSA™ Kit for researchers who want both the ease of labeling mouse IgG_1 antibodies with Zenon® labeling reagents and the signal amplification afforded by use of TSA™ technology. TSA™ technology is an enzyme-mediated detection method that utilizes the catalytic activity of horseradish peroxidase (HRP) to generate high-density labeling of a target protein (Section 6.2). Each HRP label on the target-bound Fab–mouse IgG_1 complexes can activate multiple copies of Alexa Fluor® 488 tyramide to produce short-lived tyramide radicals that are highly reactive with nucleophilic residues near the interaction site, yielding an amplified fluorescent signal with minimal diffusion. The Zenon® Alexa Fluor® 488 Mouse IgG_1 Labeling Kit enhanced with TSA™ technology provides sufficient reagents for 25 labelings, including:

- Zenon® HRP mouse IgG_1 labeling reagent
- Zenon® mouse IgG blocking reagent
- Alexa Fluor® 488 tyramide
- DMSO
- TSA™ blocking reagent
- TSA™ amplification buffer
- Hydrogen peroxide (H_2O_2)
- Detailed protocols for Zenon® complex formation and fluorescent tyramide labeling

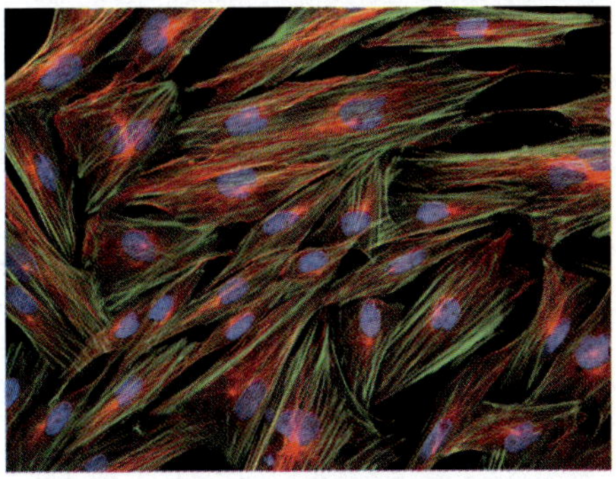

Figure 7.3.8 Fixed and permeabilized muntjac skin fibroblasts stained with Alexa Fluor® 488 phalloidin (A12379), an anti–α-tubulin antibody (A11126) and anti–cdc6 peptide antibody. The anti–α-tubulin antibody was prelabeled with the Zenon® Alexa Fluor® 568 Mouse IgG_1 Labeling Kit (Z25006), and the anti–cdc6 peptide antibody was prelabeled with the Zenon® Alexa Fluor® 350 Mouse IgG_1 Labeling Kit (Z25000).

Use and Applications of Zenon® Technology

With the exception of our Zenon® Tricolor Labeling Kits, our Zenon® Antibody Labeling Kits contain sufficient reagents for 50 labelings (with the low molecular weight dye–derived Zenon® labeling reagents), 25 labelings (with the R-PE–, APC-, HRP- and alkaline phosphatase–derived reagents) or 10 labelings (with the Zenon® labeling reagents containing one of five different Alexa Fluor® dye–phycobiliprotein tandem conjugates). The Zenon® Tricolor Labeling Kits #1, #2 and #3 contain sufficient reagents for 10 labelings of the same or different antibodies with each of three different Alexa Fluor® dyes. "One labeling" is defined as the amount of Zenon® labeling reagent required for labeling 1 µg of an intact mouse, rabbit, goat or human IgG antibody; however, smaller (or larger) quantities of a primary antibody can be reliably labeled with any of the Zenon® labeling reagents.

For single-color labeling, it is usually not necessary to block any residual Zenon® labeling reagent that has not complexed with the primary antibody; however, in applications that involve multiple antibodies of any type (including antibodies from other species or of other isotypes that may react to a small degree with the Zenon® labeling reagent), adsorption of residual Zenon® labeling reagent is essential to avoid cross-reactivity. Adsorption can be done with a solution of soluble nonspecific IgG, which is included in all of the Zenon® Antibody Labeling Kits. When the Zenon® complexes are in a solution containing the Zenon® labeling reagents adsorbed onto soluble nonspecific IgG, they should be used for staining within an hour to avoid possible transfer of the Zenon® labeling reagent to the excess nonspecific IgG. Zenon® complexes can be used with standard immunolabeling techniques; staining with Zenon® complexes can be performed sequentially or combined with each other or with additional dye- or enzyme-labeled primary antibody conjugates in a one-step, multiparameter labeling protocol. Fixation with aldehyde-based fixatives following staining is recommended to prevent transfer of the Zenon® label between antibodies.

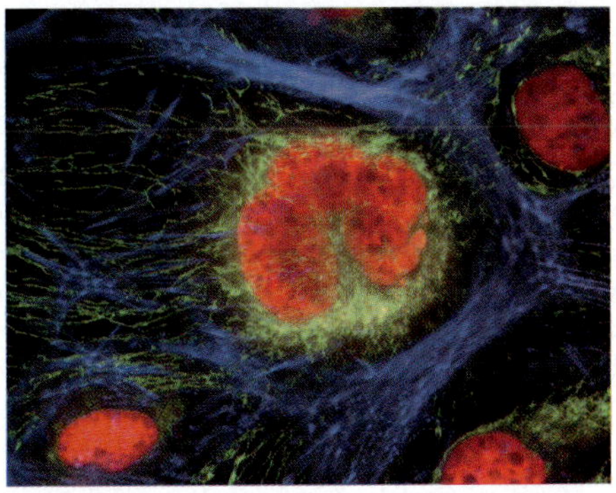

Figure 7.3.9 Fixed and permeabilized HeLa cells stained with Alexa Fluor® 350 phalloidin (A22281) and an anti–OxPhos Complex V inhibitor protein antibody (A21355) to label actin filaments and the mitochondria, respectively. An anti–cdc6 peptide antibody was used to label the nucleus. The anti–OxPhos Complex V inhibitor protein antibody was labeled with the Zenon® Alexa Fluor® 488 Mouse IgG_1 Labeling Kit (Z25002), and the anti–cdc6 peptide antibody was labeled with the Zenon® Alexa Fluor® 568 Mouse IgG_1 Labeling Kit (Z25006).

We have demonstrated the utility of Zenon® staining for imaging in an assortment of cells (Figure 7.3.4, Figure 7.3.8, Figure 7.3.9) and tissues (Figure 7.3.10). As with the use of any antibody conjugate in tissues, staining with Zenon® complexes requires good accessibility of the antibody to the target, which can be affected by the specific conditions of fixation and permeabilization. Zenon® complexes of intact primary antibodies have somewhat higher molecular weights than direct conjugates of antibodies, potentially reducing the accessibility of the complexes to tissue antigens in some cases.

Figure 7.3.10 A 14 µm coronal section of mouse hippocampus stained with an anti–α-tubulin antibody (A11126) labeled with the Zenon® Alexa Fluor® 488 Mouse IgG_1 Labeling Kit (Z25002). The section was counterstained with the NeuroTrace® 530/615 red fluorescent Nissl stain (N21482) to visualize neuronal cell bodies and Hoechst 33258 (H1398, H3569, H21491) to stain nuclei.

REFERENCES

1. Am J Physiol Renal Physiol (2004) 286:F161; **2.** Circulation (2004) 109:1401; **3.** Am J Pathol (2008) 172:1683; **4.** J Cell Biol (2009) 185:903; **5.** Cytometry A (2004) 61:142.

IMPORTANT NOTICE: The products described in this manual are covered by one or more Limited Use Label License(s). Please refer to the Appendix on page 971 and Master Product List on page 975. Products are For Research Use Only. Not intended for any animal or human therapeutic or diagnostic use.

PRODUCT LIST 7.3 ZENON® TECHNOLOGY: VERSATILE REAGENTS FOR IMMUNOLABELING

Cat. No.	Product	Quantity
Z25400	Zenon® Alexa Fluor® 350 Human IgG Labeling Kit *50 labelings*	1 kit
Z25000	Zenon® Alexa Fluor® 350 Mouse IgG_1 Labeling Kit *50 labelings*	1 kit
Z25100	Zenon® Alexa Fluor® 350 Mouse IgG_{2a} Labeling Kit *50 labelings*	1 kit
Z25200	Zenon® Alexa Fluor® 350 Mouse IgG_{2b} Labeling Kit *50 labelings*	1 kit
Z25300	Zenon® Alexa Fluor® 350 Rabbit IgG Labeling Kit *50 labelings*	1 kit
Z25013	Zenon® Alexa Fluor® 405 Mouse IgG_1 Labeling Kit *50 labelings*	1 kit
Z25113	Zenon® Alexa Fluor® 405 Mouse IgG_{2a} Labeling Kit *50 labelings*	1 kit
Z25213	Zenon® Alexa Fluor® 405 Mouse IgG_{2b} Labeling Kit *50 labelings*	1 kit
Z25313	Zenon® Alexa Fluor® 405 Rabbit IgG Labeling Kit *50 labelings*	1 kit
Z25001	Zenon® Alexa Fluor® 430 Mouse IgG_1 Labeling Kit *50 labelings*	1 kit
Z25301	Zenon® Alexa Fluor® 430 Rabbit IgG Labeling Kit *50 labelings*	1 kit
Z25602	Zenon® Alexa Fluor® 488 Goat IgG Labeling Kit *50 labelings*	1 kit
Z25402	Zenon® Alexa Fluor® 488 Human IgG Labeling Kit *50 labelings*	1 kit
Z25002	Zenon® Alexa Fluor® 488 Mouse IgG_1 Labeling Kit *50 labelings*	1 kit
Z25090	Zenon® Alexa Fluor® 488 Mouse IgG_1 Labeling Kit *enhanced with TSA™ technology* *25 labelings*	1 kit
Z25102	Zenon® Alexa Fluor® 488 Mouse IgG_{2a} Labeling Kit *50 labelings*	1 kit
Z25202	Zenon® Alexa Fluor® 488 Mouse IgG_{2b} Labeling Kit *50 labelings*	1 kit
Z25302	Zenon® Alexa Fluor® 488 Rabbit IgG Labeling Kit *50 labelings*	1 kit
Z25003	Zenon® Alexa Fluor® 532 Mouse IgG_1 Labeling Kit *50 labelings*	1 kit
Z25303	Zenon® Alexa Fluor® 532 Rabbit IgG Labeling Kit *50 labelings*	1 kit
Z25004	Zenon® Alexa Fluor® 546 Mouse IgG_1 Labeling Kit *50 labelings*	1 kit
Z25104	Zenon® Alexa Fluor® 546 Mouse IgG_{2a} Labeling Kit *50 labelings*	1 kit
Z25204	Zenon® Alexa Fluor® 546 Mouse IgG_{2b} Labeling Kit *50 labelings*	1 kit
Z25304	Zenon® Alexa Fluor® 546 Rabbit IgG Labeling Kit *50 labelings*	1 kit
Z25605	Zenon® Alexa Fluor® 555 Goat IgG Labeling Kit *50 labelings*	1 kit
Z25405	Zenon® Alexa Fluor® 555 Human IgG Labeling Kit *50 labelings*	1 kit
Z25005	Zenon® Alexa Fluor® 555 Mouse IgG_1 Labeling Kit *50 labelings*	1 kit
Z25105	Zenon® Alexa Fluor® 555 Mouse IgG_{2a} Labeling Kit *50 labelings*	1 kit
Z25205	Zenon® Alexa Fluor® 555 Mouse IgG_{2b} Labeling Kit *50 labelings*	1 kit
Z25305	Zenon® Alexa Fluor® 555 Rabbit IgG Labeling Kit *50 labelings*	1 kit
Z25606	Zenon® Alexa Fluor® 568 Goat IgG Labeling Kit *50 labelings*	1 kit
Z25006	Zenon® Alexa Fluor® 568 Mouse IgG_1 Labeling Kit *50 labelings*	1 kit
Z25106	Zenon® Alexa Fluor® 568 Mouse IgG_{2a} Labeling Kit *50 labelings*	1 kit
Z25206	Zenon® Alexa Fluor® 568 Mouse IgG_{2b} Labeling Kit *50 labelings*	1 kit
Z25306	Zenon® Alexa Fluor® 568 Rabbit IgG Labeling Kit *50 labelings*	1 kit
Z25607	Zenon® Alexa Fluor® 594 Goat IgG Labeling Kit *50 labelings*	1 kit
Z25407	Zenon® Alexa Fluor® 594 Human IgG Labeling Kit *50 labelings*	1 kit
Z25007	Zenon® Alexa Fluor® 594 Mouse IgG_1 Labeling Kit *50 labelings*	1 kit
Z25107	Zenon® Alexa Fluor® 594 Mouse IgG_{2a} Labeling Kit *50 labelings*	1 kit
Z25207	Zenon® Alexa Fluor® 594 Mouse IgG_{2b} Labeling Kit *50 labelings*	1 kit
Z25307	Zenon® Alexa Fluor® 594 Rabbit IgG Labeling Kit *50 labelings*	1 kit
Z25020	Zenon® Alexa Fluor® 610–R-Phycoerythrin Mouse IgG_1 Labeling Kit *10 labelings*	1 kit
Z25608	Zenon® Alexa Fluor® 647 Goat IgG Labeling Kit *50 labelings*	1 kit
Z25408	Zenon® Alexa Fluor® 647 Human IgG Labeling Kit *50 labelings*	1 kit
Z25008	Zenon® Alexa Fluor® 647 Mouse IgG_1 Labeling Kit *50 labelings*	1 kit
Z25108	Zenon® Alexa Fluor® 647 Mouse IgG_{2a} Labeling Kit *50 labelings*	1 kit
Z25208	Zenon® Alexa Fluor® 647 Mouse IgG_{2b} Labeling Kit *50 labelings*	1 kit
Z25308	Zenon® Alexa Fluor® 647 Rabbit IgG Labeling Kit *50 labelings*	1 kit
Z25021	Zenon® Alexa Fluor® 647–R-Phycoerythrin Mouse IgG_1 Labeling Kit *10 labelings*	1 kit
Z25121	Zenon® Alexa Fluor® 647–R-Phycoerythrin Mouse IgG_{2a} Labeling Kit *10 labelings*	1 kit
Z25221	Zenon® Alexa Fluor® 647–R-Phycoerythrin Mouse IgG_{2b} Labeling Kit *10 labelings*	1 kit
Z25009	Zenon® Alexa Fluor® 660 Mouse IgG_1 Labeling Kit *50 labelings*	1 kit
Z25010	Zenon® Alexa Fluor® 680 Mouse IgG_1 Labeling Kit *50 labelings*	1 kit
Z25110	Zenon® Alexa Fluor® 680 Mouse IgG_{2a} Labeling Kit *50 labelings*	1 kit
Z25210	Zenon® Alexa Fluor® 680 Mouse IgG_{2b} Labeling Kit *50 labelings*	1 kit
Z25310	Zenon® Alexa Fluor® 680 Rabbit IgG Labeling Kit *50 labelings*	1 kit
Z25022	Zenon® Alexa Fluor® 680–R-Phycoerythrin Mouse IgG_1 Labeling Kit *10 labelings*	1 kit
Z25011	Zenon® Alexa Fluor® 700 Mouse IgG_1 Labeling Kit *50 labelings*	1 kit
Z25030	Zenon® Alexa Fluor® 700–Allophycocyanin Mouse IgG_1 Labeling Kit *10 labelings*	1 kit

IMPORTANT NOTICE: The products described in this manual are covered by one or more Limited Use Label License(s). Please refer to the Appendix on page 971 and Master Product List on page 975. Products are For Research Use Only. Not intended for any animal or human therapeutic or diagnostic use.
www.invitrogen.com/probes

PRODUCT LIST 7.3 ZENON® TECHNOLOGY: VERSATILE REAGENTS FOR IMMUNOLABELING—*continued*

Cat. No.	Product	Quantity
Z25312	Zenon® Alexa Fluor® 750 Rabbit IgG Labeling Kit *50 labelings*	1 kit
Z25031	Zenon® Alexa Fluor® 750–Allophycocyanin Mouse IgG_1 Labeling Kit *10 labelings*	1 kit
Z25350	Zenon® Alkaline Phosphatase Rabbit IgG Labeling Kit *25 labelings*	1 kit
Z25451	Zenon® Allophycocyanin Human IgG Labeling Kit *25 labelings*	1 kit
Z25051	Zenon® Allophycocyanin Mouse IgG_1 Labeling Kit *25 labelings*	1 kit
Z25151	Zenon® Allophycocyanin Mouse IgG_{2a} Labeling Kit *25 labelings*	1 kit
Z25251	Zenon® Allophycocyanin Mouse IgG_{2b} Labeling Kit *25 labelings*	1 kit
Z25351	Zenon® Allophycocyanin Rabbit IgG Labeling Kit *25 labelings*	1 kit
Z25452	Zenon® Biotin-XX Human IgG Labeling Kit *50 labelings*	1 kit
Z25052	Zenon® Biotin-XX Mouse IgG_1 Labeling Kit *50 labelings*	1 kit
Z25152	Zenon® Biotin-XX Mouse IgG_{2a} Labeling Kit *50 labelings*	1 kit
Z25252	Zenon® Biotin-XX Mouse IgG_{2b} Labeling Kit *50 labelings*	1 kit
Z25352	Zenon® Biotin-XX Rabbit IgG Labeling Kit *50 labelings*	1 kit
Z25042	Zenon® Fluorescein Mouse IgG_1 Labeling Kit *50 labelings*	1 kit
Z25342	Zenon® Fluorescein Rabbit IgG Labeling Kit *50 labelings*	1 kit
Z25454	Zenon® Horseradish Peroxidase Human IgG Labeling Kit *25 labelings*	1 kit
Z25054	Zenon® Horseradish Peroxidase Mouse IgG_1 Labeling Kit *25 labelings*	1 kit
Z25154	Zenon® Horseradish Peroxidase Mouse IgG_{2a} Labeling Kit *25 labelings*	1 kit
Z25254	Zenon® Horseradish Peroxidase Mouse IgG_{2b} Labeling Kit *25 labelings*	1 kit
Z25354	Zenon® Horseradish Peroxidase Rabbit IgG Labeling Kit *25 labelings*	1 kit
Z25041	Zenon® Pacific Blue™ Mouse IgG_1 Labeling Kit *50 labelings*	1 kit
Z25156	Zenon® Pacific Blue™ Mouse IgG_{2a} Labeling Kit *50 labelings*	1 kit
Z25341	Zenon® Pacific Blue™ Rabbit IgG Labeling Kit *50 labelings*	1 kit
Z25256	Zenon® Pacific Orange™ Mouse IgG_1 Labeling Kit *50 labelings*	1 kit
Z25257	Zenon® Pacific Orange™ Mouse IgG_{2a} Labeling Kit *50 labelings*	1 kit
Z25455	Zenon® R-Phycoerythrin Human IgG Labeling Kit *25 labelings*	1 kit
Z25055	Zenon® R-Phycoerythrin Mouse IgG_1 Labeling Kit *25 labelings*	1 kit
Z25155	Zenon® R-Phycoerythrin Mouse IgG_{2a} Labeling Kit *25 labelings*	1 kit
Z25255	Zenon® R-Phycoerythrin Mouse IgG_{2b} Labeling Kit *25 labelings*	1 kit
Z25355	Zenon® R-Phycoerythrin Rabbit IgG Labeling Kit *25 labelings*	1 kit
Z25045	Zenon® Texas Red®-X Mouse IgG_1 Labeling Kit *50 labelings*	1 kit
Z25460	Zenon® Tricolor Human IgG Labeling Kit #1 *for green, orange and deep red fluorescence imaging* *3 x 10 labelings*	1 kit
Z25470	Zenon® Tricolor Human IgG Labeling Kit #2 *for blue, green and red fluorescence imaging* *3 x 10 labelings*	1 kit
Z25060	Zenon® Tricolor Mouse IgG_1 Labeling Kit #1 *for green, orange and deep red fluorescence imaging* *3 x 10 labelings*	1 kit
Z25070	Zenon® Tricolor Mouse IgG_1 Labeling Kit #2 *for blue, green and red fluorescence imaging* *3 x 10 labelings*	1 kit
Z25080	Zenon® Tricolor Mouse IgG_1 Labeling Kit #3 *for flow cytometry, 488 nm excitation* *3 x 10 labelings*	1 kit
Z25160	Zenon® Tricolor Mouse IgG_{2a} Labeling Kit #1 *for green, orange and deep red fluorescence imaging* *3 x 10 labelings*	1 kit
Z25170	Zenon® Tricolor Mouse IgG_{2a} Labeling Kit #2 *for blue, green and red fluorescence imaging* *3 x 10 labelings*	1 kit
Z25180	Zenon® Tricolor Mouse IgG_{2a} Labeling Kit #3 *for flow cytometry, 488 nm excitation* *3 x 10 labelings*	1 kit
Z25260	Zenon® Tricolor Mouse IgG_{2b} Labeling Kit #1 *for green, orange and deep red fluorescence imaging* *3 x 10 labelings*	1 kit
Z25270	Zenon® Tricolor Mouse IgG_{2b} Labeling Kit #2 *for blue, green and red fluorescence imaging* *3 x 10 labelings*	1 kit
Z25280	Zenon® Tricolor Mouse IgG_{2b} Labeling Kit #3 *for flow cytometry, 488 nm excitation* *3 x 10 labelings*	1 kit
Z25360	Zenon® Tricolor Rabbit IgG Labeling Kit #1 *for green, orange and deep red fluorescence imaging* *3 x 10 labelings*	1 kit
Z25370	Zenon® Tricolor Rabbit IgG Labeling Kit #2 *for blue, green and red fluorescence imaging* *3 x 10 labelings*	1 kit
Z25380	Zenon® Tricolor Rabbit IgG Labeling Kit #3 *for flow cytometry, 488 nm excitation* *3 x 10 labelings*	1 kit

IMPORTANT NOTICE: The products described in this manual are covered by one or more Limited Use Label License(s). Please refer to the Appendix on page 971 and Master Product List on page 975. Products are For Research Use Only. Not intended for any animal or human therapeutic or diagnostic use.

7.4 Anti-Dye and Anti-Hapten Antibodies

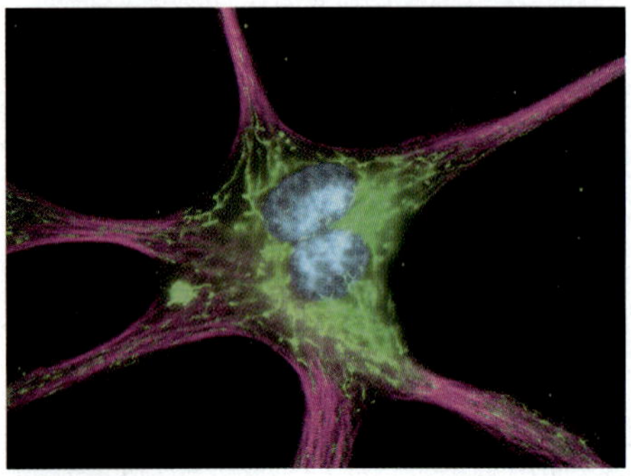

Figure 7.4.1 The cytoskeleton of a fixed and permeabilized bovine pulmonary artery endothelial cell detected using mouse monoclonal anti-α-tubulin antibody (A11126), visualized with Alexa Fluor® 647 goat anti–mouse IgG antibody (A21235) and pseudocolored magenta. Endogenous biotin in the mitochondria was labeled with green-fluorescent Alexa Fluor® 488 streptavidin (S11223) and DNA was stained with blue-fluorescent DAPI (D1306, D3571, D21490).

Anti-Dye and Anti-Hapten Antibodies

Fluorescent dyes and nonfluorescent ligands are excellent affinity tags that can be recognized by labeled primary antibodies, providing an alternative to the traditional avidin–biotin system in applications such as *in situ* hybridization,[1] enzyme-linked immunosorbent assays (ELISAs) and western blot analysis (Section 4.2, Table 4.2). Antibodies to dyes and other ligands provide unique opportunities both for signal enhancement and for correlated fluorescence and electron microscopy studies. Essentially all of the methods that use biotin and avidin reagents (Section 7.6) are also possible using dyes as haptens, as long as the corresponding anti-dye antibody is also available (Table 7.8).

One advantage of using fluorescent dyes as haptens instead of biotin is that the hapten signal is usually visible, or at least its concentration can be measured by its absorption in solution, preceding the secondary detection step. Unlike biotin, which is an endogenous ligand in mitochondria (Figure 7.4.1), dye-based haptens permit background-free staining of cells and tissues. We provide a large assortment of anti-dye antibodies (Table 7.8), including rabbit polyclonal IgG antibodies to the fluorescein, tetramethylrhodamine, Texas Red®, dansyl, Alexa Fluor® 488, BODIPY® FL, lucifer yellow and Alexa Fluor® 405/Cascade Blue® fluorophores, as well as a goat anti–fluorescein/Oregon Green® IgG antibody (A11095). We also provide antibodies against three nonfluorescent haptens: dinitrophenyl (DNP), biotin and nitrotyrosine (Table 7.8).

Table 7.8 Anti-fluorophore and anti-hapten antibodies.

Cat. No.	Anti-Fluorophore	Host	Label
Anti-Fluorophore Antibodies and Their Conjugates			
A5760	Alexa Fluor® 405/Cascade Blue® dye	Rabbit	None
A11094	Alexa Fluor® 488 dye	Rabbit	None
A5770	BODIPY® FL dye	Rabbit	None
A6398	Dansyl	Rabbit	None
A6421	Fluorescein/Oregon Green® dye	Mouse (clone 4-4-20)	None
A889	Fluorescein/Oregon Green® dye	Rabbit	None
A6413	Fluorescein/Oregon Green® dye	Rabbit, Fab fragment	None
A11095	Fluorescein/Oregon Green® dye	Goat	None
A11090	Fluorescein/Oregon Green® dye	Rabbit	Alexa Fluor® 488 dye
A11096	Fluorescein/Oregon Green® dye	Goat	Alexa Fluor® 488 dye
A11091	Fluorescein/Oregon Green® dye	Rabbit	Alexa Fluor® 594 dye
A21250	Fluorescein/Oregon Green® dye	Rabbit	R-phycoerythrin
Q15431MP	Fluorescein/Oregon Green® dye	Goat	Qdot® 565 nanocrystal *
Q15421MP	Fluorescein/Oregon Green® dye	Goat	Qdot® 655 nanocrystal *
A982	Fluorescein/Oregon Green® dye	Rabbit	Biotin-XX
A21253	Fluorescein/Oregon Green® dye	Rabbit	Horseradish peroxidase
A5750	Lucifer yellow dye	Rabbit	None
A5751	Lucifer yellow dye	Rabbit	Biotin-XX
A6397	Tetramethylrhodamine and Rhodamine Red™ dye	Rabbit	None
A6399	Texas Red® and Texas Red®-X dye	Rabbit	None
Anti-Hapten Antibodies and Their Conjugates			
03-3700	Biotin	Mouse	None
A6430	Dinitrophenyl	Rabbit	None
A6435	Dinitrophenyl	Rabbit	Biotin-XX
A11097	Dinitrophenyl	Rabbit	Alexa Fluor® 488
A6423	Dinitrophenyl	Rabbit	Fluorescein
Q17421MP	Dinitrophenyl	Rat	Qdot® 655 nanocrystal *
A21285	Nitrotyrosine	Rabbit	None

* Properties and applications of Qdot® nanocrystal conjugates are described in Section 6.6. More information is available at www.invitrogen.com/handbook/antibodies.

IMPORTANT NOTICE: The products described in this manual are covered by one or more Limited Use Label License(s). Please refer to the Appendix on page 971 and Master Product List on page 975. Products are For Research Use Only. Not intended for any animal or human therapeutic or diagnostic use.
www.invitrogen.com/probes
molecular probes® | invitrogen™ by life technologies™

Antibodies to Fluorescein and Oregon Green® Dyes

We have observed complete cross-reactivity of our anti-fluorescein antibodies with the Oregon Green® 488 and Oregon Green® 514 dyes (Section 1.5). These antibodies also quench the fluorescence of the structurally similar dye resorufin (R363, Section 10.1). The high affinity and specificity of anti–fluorescein/Oregon Green® antibodies (A889, A6413, A6421, A11095) makes fluorescein and Oregon Green® dyes ideal haptens for various detection schemes.[2,3] Researchers have found that fluorescein–anti-fluorescein ELISA techniques display low nonspecific binding and are similar in sensitivity to biotin–streptavidin methods.[4]

Our anti–fluorescein/Oregon Green® rabbit polyclonal antibody (A889) and anti–fluorescein/Oregon Green® goat polyclonal antibody (A11095) can be used in combination with any of our Zenon® Rabbit IgG and Zenon® Goat IgG Labeling Kits (Section 7.3, Table 7.7), respectively, to produce fluorophore-, biotin- or enzyme-labeled antibodies. In addition to these unlabeled anti–fluorescein/Oregon Green® antibodies, we offer an anti–fluorescein/Oregon Green® monoclonal antibody and an anti–fluorescein/Oregon Green® rabbit polyclonal Fab fragment (Antibody Structure and Classification—Note 7.1). The high-affinity anti–fluorescein/Oregon Green® mouse IgG_{2a} monoclonal 4-4-20 antibody (A6421) may reduce nonspecific binding in ELISAs and other second-step detection assays. The Fab fragment of our rabbit polyclonal anti–fluorescein/Oregon Green® antibody (A6413) provides researchers with a probe that more efficiently penetrates cell and tissue preparations. Furthermore, because the Fab fragment no longer contains the Fc portion, nonspecific interactions with Fc receptor–bearing cells are eliminated. As expected, none of our anti–fluorescein/Oregon Green® antibodies recognize the Alexa Fluor® or BODIPY® dyes.

We also offer the horseradish peroxidase (HRP) conjugate of the rabbit anti–fluorescein/Oregon Green® antibody (A21253). HRP conjugates are commonly used in histochemical amplification schemes such as tyramide signal amplification (TSA™) technology (Section 6.2). TSA™ technology provides a greater degree of resolution than many conventional enzyme-mediated fluorescent staining methods, and sequential TSA™ labeling followed by a second round of TSA™ labeling provides an extremely sensitive assay for detection of low-abundance targets in cells and tissues with high spatial resolution.[5]

Our Alexa Fluor® 488 dye–labeled rabbit or goat anti–fluorescein/Oregon Green® antibodies (A11090, A11096) can be used to enhance the intensity and photostability of the green-fluorescent signal of the fluorescein hapten without changing its fluorescence color.[6–8] Thus, this conjugate allows researchers to take advantage of the superior photostability of the Alexa Fluor® 488 dye, while utilizing existing fluorescein-labeled probes and fluorescein-compatible optics. This strategy has been exploited in our Alexa Fluor® 488 Signal Amplification Kit for Fluorescein Conjugated Probes (A11053), which is described in Section 7.2. The Alexa Fluor® 594 dye–labeled (A11091) anti–fluorescein/Oregon Green® antibody can be used to convert the green fluorescence of fluorescein conjugates to photostable red fluorescence, or potentially to amplify the signal from fluorescein conjugates (Figure 7.4.2).

The R-phycoerythrin conjugate of the rabbit anti–fluorescein/Oregon Green® IgG antibody (A21250), as well as the Qdot® 565 and Qdot® 655 conjugates of goat anti-fluorescein IgG antibody (Q15431MP, Q15421MP; Section 6.6) have the unique utility of both shifting the green-fluorescent emission of fluorescein-labeled probes to longer wavelengths and greatly intensifying the long-wavelength signal (Figure 7.4.3). Biotin-XX–labeled rabbit anti–fluorescein/Oregon Green® antibody (A982) is an excellent reagent for converting a fluorescence-based detection method into an enzyme-amplified light or electron microscopy technique. Biotin-XX anti–fluorescein/Oregon Green® can be combined with tyramide signal amplification (TSA™) technology (Section 6.2) in a variety of signal amplification schemes for cell and tissue labeling.

Some of the more important applications for anti–fluorescein/Oregon Green® antibodies—almost all of which could also be carried out with any of our other anti-dye antibodies and their complementary dyes—include:

- Amplification of the signal from a fluorescein tyramide in the TSA™ technology[9] (Section 6.2)
- Detection of fluorescein-labeled primary antibodies[10,11]
- Development of fluorescein-labeled cell preparations for electron microscopy[12]
- Investigation of the uptake of a fluorescein dextran in kidney proximal tubules[13]
- Localization of mRNA sequences in a double *in situ* hybridization experiment in which both fluorescein- and biotin-labeled oligonucleotides were used[2]

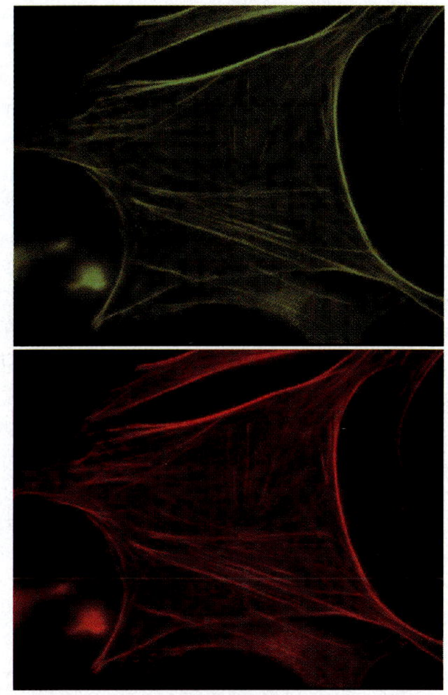

Figure 7.4.2 Fixed and permeabilized bovine pulmonary artery endothelial cells were labeled with the filamentous actin (F-actin) stain, fluorescein phalloidin (F432, top). An Alexa Fluor® 594 anti–fluorescein/Oregon Green® rabbit IgG antibody (A11091) converted the green fluorescence to red (bottom).

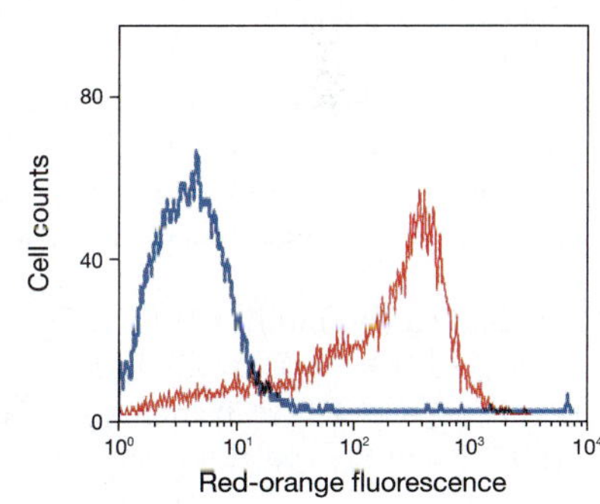

Figure 7.4.3 Color-shifting using a labeled anti–fluorescein/Oregon Green® antibody. Jurkat cells were first stained with a primary mouse anti-human CD3 antibody, followed by fluorescein goat anti-mouse IgG antibody (F2761), with the resultant fluorescence detected in the R-phycoerythrin (red-orange fluorescence) channel of a flow cytometer (blue curve). The weak signal was then shifted to better suit the R-phycoerythrin channel by the addition of an R-phycoerythrin conjugate of anti–fluorescein/Oregon Green® antibody (A21250). The resulting signal intensity is approximately two orders of magnitude greater (red curve) than the direct fluorescence from the first staining step (blue curve).

IMPORTANT NOTICE: The products described in this manual are covered by one or more Limited Use Label License(s). Please refer to the Appendix on page 971 and Master Product List on page 975. Products are For Research Use Only. Not intended for any animal or human therapeutic or diagnostic use.

- Preparation of an anti-fluorescein affinity matrix, which was used to immobilize a fluoresceinated protein in order to study protein–protein interactions *in vitro*[14,15]
- Separation of fluorescein antibody–labeled cell populations by immunoadsorption [16] or magnetic separation techniques
- Assessment of the accessibility of active site–bound fluorescein probes [17]
- Investigation of the internalization pathway of fluorescein transferrin [18,19] (T2871, Section 16.1)

Antibody to the Alexa Fluor® 488 Dye

We have prepared a rabbit polyclonal antibody to our green-fluorescent Alexa Fluor® 488 dye (A11094). In a manner analogous to the anti–fluorescein/Oregon Green® antibodies, the anti–Alexa Fluor® 488 dye antibody specifically recognizes and efficiently quenches most of the fluorescence of the Alexa Fluor® 488 dye. In contrast, the anti–Alexa Fluor® 488 dye antibody does not appreciably quench the fluorescence of fluorescein, carboxytetramethylrhodamine or the Alexa Fluor® 594 dye. The high affinity of the anti–Alexa Fluor® 488 dye antibody makes it potentially useful for various immunochemical applications. This antibody can also be used to further amplify the signals from our Alexa Fluor® 488 tyramide–containing TSA™ Kits (Section 6.2).

Antibodies to Tetramethylrhodamine and the Rhodamine Red™ and Texas Red® Dyes

As with the anti–fluorescein/Oregon Green® antibodies, the rabbit polyclonal antibodies to the tetramethylrhodamine and Texas Red® fluorophores (A6397, A6399) are effective reagents for binding these dye-based haptens and quenching their fluorescence. However, these antibodies strongly cross-react with some other rhodamines, including the Rhodamine Red™ and Lissamine rhodamine B dyes, and therefore cannot be used for simultaneous detection of more than one rhodamine-based dye. These anti-tetramethylrhodamine and anti–Texas Red® dye antibodies do not appear to cross-react with fluorescein or the Oregon Green® or Alexa Fluor® dyes, and our anti–fluorescein/Oregon Green® antibodies do not cross-react with tetramethylrhodamine or the Rhodamine Red™ or Texas Red® dyes. Anti-tetramethylrhodamine has been used to localize retrogradely transported tetramethylrhodamine dextrans by an immunoperoxidase-based amplification technique.[20]

Antibodies to the Lucifer Yellow Dye and the Cascade Blue® and Alexa Fluor® 405 Dyes

Lucifer yellow CH (L453) and Cascade Blue® hydrazide (C687) are frequently employed as polar tracers for neuronal cell labeling (Section 14.3). Our unconjugated (A5750, A5760) and biotinylated (A5751) rabbit polyclonal antibodies to these dyes are useful in standard enzyme-mediated immunohistochemical methods for permanently labeling neuronal tissue.[21–25] Anti–lucifer yellow dye antibody (A5750) has also been used to follow dye coupling in smooth muscle cells by electron microscopy.[26,27] The anti–Alexa Fluor® 405/Cascade Blue® dye antibody (A5760) has been employed in western blot analysis to identify cytoplasmic and luminal domains of the sarcoplasmic reticulum Ca^{2+}-ATPase, which had been photolabeled with Cascade Blue® aminoethyl 4-azidobenzamide.[28] The anti–Alexa Fluor® 405/Cascade Blue® dye antibody also recognizes conjugates of our Alexa Fluor® 405 succinimidyl ester (A30000, A30100; Section 1.7), which is a derivative of our Cascade Blue® dye with a 4-piperidinecarboxylic acid spacer that separates the fluorophore from its reactive moiety.

Antibody to the BODIPY® FL Dye

Our unlabeled rabbit polyclonal antibody to the BODIPY® FL fluorophore (A5770) cross-reacts with some other BODIPY® dyes but does not cross-react appreciably with any other fluorophores. This anti–BODIPY® FL dye antibody should therefore not be used for simultaneous detection of more than one dye based on the BODIPY® fluorophore. In solution assays, we have found that the anti–BODIPY® FL dye antibody effectively quenches most of the fluorescence of the BODIPY® FL dye, but quenches the BODIPY® TR dye to a lesser degree and does not significantly quench the BODIPY® TMR dye. The anti–BODIPY® FL dye antibody has been used in a fluorescence quenching assay to determine the accessibility of BODIPY® FL dye–labeled cysteine residues in the transmembrane domain of diphtheria toxin.[29–32]

Anti-Dansyl Antibody

In contrast to the other anti-fluorophore antibodies, which usually quench the fluorescence of the dye to which they bind, our rabbit polyclonal anti-dansyl antibody (A6398) typically *enhances* the fluorescence of dansyl amides by greater than 10-fold. Binding of the anti-dansyl antibody also blue shifts the emission spectrum of the fluorophore in water from ~520 nm to ~450 nm. These properties, combined with the unusually high Stokes shift of the dansyl dye (Figure 7.4.4), make this antibody particularly useful for determining the topography of dansyl-labeled probes, including that of dansyl-labeled phospholipids (Section 13.2) in cell and artificial membranes.[33] The dansyl hapten is preferably incorporated into biomolecules using the succinimidyl ester of dansyl-X (D6104, Section 4.2) because its aminohexanoyl spacer ("X") reduces the interaction of the fluorophore with the biomolecule to which it is conjugated and makes it more accessible to anti-dansyl antibodies.[33–35]

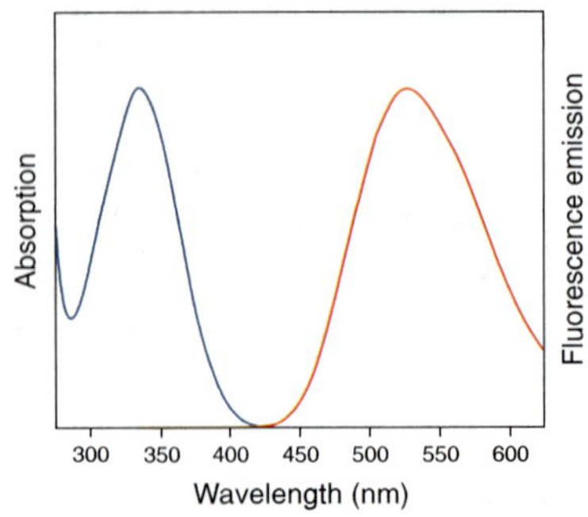

Figure 7.4.4 Absorption and fluorescence emission spectra of dansyl cadaverine in methanol.

Efficient Quenching by Anti-Fluorophore Antibodies

Quenching Efficiencies

Except for the anti-dansyl antibody, which enhances the fluorescence of the dansyl fluorophore, all of our anti-fluorophore antibodies strongly quench the fluorescence of their complementary dyes in free solution. For example, our anti–fluorescein/Oregon Green® antibodies typically effect up to 95% quenching of the fluorescence of both fluorescein and the Oregon Green® 488 dye. The anti–fluorescein/Oregon Green® antibody also quenches some other fluorescein derivatives, such as carboxyfluorescein, Calcium Green™-1 Ca^{2+} indicator and

IMPORTANT NOTICE: The products described in this manual are covered by one or more Limited Use Label License(s). Please refer to the Appendix on page 971 and Master Product List on page 975. Products are For Research Use Only. Not intended for any animal or human therapeutic or diagnostic use.
www.invitrogen.com/probes

BCECF pH indicator, making this antibody useful for reducing background fluorescence caused by leakage of these dyes from the cell.[36] However, quenching of our fluorescein-based Ca^{2+} indicators by our anti–fluorescein/Oregon Green® IgG antibody is apparently dependent on whether or not Ca^{2+} is bound; Calcium Green™-1 indicator is quenched by 89% in the presence of 5 µM Ca^{2+}, whereas it is quenched by only 46% in the presence of 10 mM EGTA. Maximal quenching efficiencies for fluorescein analogs (all at 5 nM dye using the rabbit anti–fluorescein/Oregon Green® IgG antibody, A889) are as follows (values may vary somewhat from batch to batch) and may be different using the goat anti–fluorescein/Oregon Green® IgG antibody, A11095, the mouse monoclonal anti–fluorescein/Oregon Green®, A6421 or the rabbit IgG Fab fragment of anti–fluorescein/Oregon Green®, A6413):

- Oregon Green® 488 dye, 95%
- Oregon Green® 514 dye, 92%
- Carboxyfluorescein, 93%
- Calcium Green™-1 indicator (in the presence of 5 µM Ca^{2+}), 89%
- Calcium Green™-1 indicator (in the presence of 10 mM EGTA), 46%
- BCECF indicator, 43%
- Fluo-3 indicator (in the presence of 5 µM Ca^{2+}), 32%
- Rhodamine Green™ dye, 9%
- Calcein, <5%
- Tetramethylrhodamine, <5%

Our preparations of the anti-tetramethylrhodamine and anti–Texas Red® dye antibodies are somewhat less effective as fluorescence quenchers of their complementary fluorophores, with maximal quenching efficiencies of ~75% and ~60%, respectively. Our rabbit anti–Alexa Fluor® 488 dye IgG antibody quenches the fluorescence of the free Alexa Fluor® 488 dye by >90%. Our antibody to the BODIPY® FL fluorophore typically quenches the dye's fluorescence by ~90%. It also quenches BODIPY® TR dye fluorescence by ~45%, but does not significantly quench BODIPY® TMR dye fluorescence. Antibodies to the lucifer yellow and Alexa Fluor® 405/Cascade Blue® fluorophores quench the fluorescence of their complementary dyes by ~85% and ~80%, respectively. In addition, anti-DNP antibodies have been reported to significantly quench the fluorescence of aminonitrobenzoxadiazoles[37] (NBD amines).

Quenching Assay

We use a sensitive fluorescence quenching–based assay to help ensure that the concentration of specific anti-dye antibody in its purified IgG fractions is provided at a consistently high titer. As supplied, 20 µL of the antibody solution is certified to produce ≥50% of the maximal fluorescence quenching (or enhancement, in the case of anti-dansyl antibody) of 1 mL of a 50 nM solution of the corresponding dye, assayed in 100 mM sodium phosphate, pH 8.0. All maximal quenching values are determined using the free fluorophore; the maximal quenching of a fluorophore covalently bound to a protein is often significantly less due to steric hindrance.

Applications for Fluorescence Quenching by Anti-Fluorophore Antibodies

Fluorescence quenching of dye haptens by anti-dye antibodies provides a useful measure of topography in cells, proteins and membranes. For example, researchers have used anti-fluorescein quenching assays to determine the accessibility of a fluorescein-labeled ATP-binding site in both Na^+/K^+-ATPase and Ca^{2+}/Mg^{2+}-ATPase.[17,38] Similarly, the anti–BODIPY® FL dye antibody has been employed to identify shallow- and deep-membrane–penetrating forms of diphtheria toxin T domain.[29–32] In addition, anti-fluorophore antibodies have been used as cell-impermeant probes for determining whether fluorescent dye–conjugated ligands, proteins, bacteria or other biomolecules have been internalized by endocytic or pinocytic processes[39–42] (Section 16.1). Anti-fluorophore antibodies also permit background-free observation of fusion events in an assay designed to monitor the fusion of membrane vesicles *in vitro*.[43] As noted above, however, these antibodies may quench dye-labeled proteins less effectively than they quench free dyes.

Anti-Dinitrophenyl Antibody

Because of its high affinity for the dinitrophenyl (DNP) hapten,[44,45] our anti-DNP polyclonal rabbit antibodies (Table 7.8) are excellent reagents for probing DNP-labeled molecules, including nucleic acid probes prepared using DNP-labeled nucleotides. Unlike assays that use biotin as the hapten, it is usually easy to determine the degree of substitution of the DNP hapten in bioconjugates from the dye's visible absorption near 350 nm[46] (EC ~18,000 $cm^{-1}M^{-1}$). It has been reported that human chromosomes can be probed with equal sensitivities using either biotinylated, DNP-modified or digoxigenin-modified cosmid probes.[47] Anti-DNP antibodies have been used to localize a DNP-labeled DNA probe in HIV-infected cells[48] and 2,4-dinitrophenylhydrazine-labeled proteins on blots.[49,50] Researchers have also reported using anti-DNP antibodies to probe for DNP-labeled IgG as a method for detecting sparse antigens[51] and, in conjunction with DNP-labeled bovine serum albumin (BSA), to study the Fc receptor–mediated endocytosis of IgG complexes.[52,53]

In addition to the unlabeled anti-DNP antibody (A6430), we offer anti-DNP antibody conjugates of biotin-XX (A6435), fluorescein (A6423), Alexa Fluor® 488 dye (A11097) and Qdot® 655 nanocrystal (Q17421MP, Section 6.6). Our anti-DNP antibody is prepared against DNP–keyhole limpet hemocyanin (DNP-KLH) and thus the antibody and its conjugates do not cross-react with BSA, a common blocking reagent in hybridization applications.

For use in conjunction with our anti-DNP antibody, we offer the DNP-X succinimidyl ester (D2248, Section 4.2) for labeling proteins and amine-modified DNA. The literature also describes methods for incorporating DNP into DNA using DNP-labeled primers.[54] In addition to recognizing DNP, our anti-DNP antibody cross-reacts with trinitrobenzenesulfonic acid–modified proteins, making this antibody useful both for localizing and for isolating cell-surface molecules labeled with either DNP or TNP.[54] Furthermore, anti-DNP antibodies have been reported to quench aminonitrobenzoxadiazoles (NBD amines),[37] indicating that NBD-X succinimidyl ester (S1167, Section 1.7) will also be a useful haptenylating reagent for use with this antibody.

Anti-Nitrotyrosine Antibody

Our antibody to nitrotyrosine (A21285) is raised in rabbits that have been immunized with nitrated KLH. Nitrotyrosine-modified proteins are the principal reaction products of nitric oxide (through the formation of peroxynitrite) in cells (Section 18.3). Because tyrosine residues are also conveniently converted to nitrotyrosine by reaction at near-neutral pH with tetranitromethane,[55,56] the nitrotyrosine hapten can be

IMPORTANT NOTICE: The products described in this manual are covered by one or more Limited Use Label License(s). Please refer to the Appendix on page 971 and Master Product List on page 975. Products are For Research Use Only. Not intended for any animal or human therapeutic or diagnostic use.

Figure 7.4.6 Specificity of our rabbit anti-nitrotyrosine antibody (A21285) to nitrated proteins. Equal amounts of avidin (A887, lane 1) and CaptAvidin™ biotin-binding protein (C21385, lane 2) were run on an SDS-polyacrylamide gel (4–20%) and blotted onto a PVDF membrane. CaptAvidin™ biotin-binding protein, a derivative of avidin, has nitrated tyrosine residues in the biotin-binding site. Nitrated proteins were identified with the anti-nitrotyrosine antibody, in combination with an alkaline phosphatase conjugate of goat anti–rabbit IgG antibody (G21079) and the red-fluorescent substrate, DDAO phosphate (D6487).

readily created in almost any peptide or protein that contains a tyrosine residue. A further advantage is that nitrotyrosine has pH-dependent visible absorbance (absorption maxima ~360 nm [55] and 428 nm [57]) that can be utilized to detect formation of the hapten in soluble biopolymers.

Our anti-nitrotyrosine antibody is useful for detecting nitrotyrosine-containing peptides and proteins both in immunohistochemical (Figure 7.4.5) and western blot applications (Figure 7.4.6). It can be used to identify nitrated proteins and to determine the level of protein nitrosylation in tissues.[58,59] This rabbit IgG antibody can be labeled with fluorophores, biotin or enzymes using any of our Zenon® Rabbit IgG Antibody Labeling Kits (Section 7.3, Table 7.7).

Anti-Biotin Antibody

The high affinity of avidin for biotin was first exploited in histochemical applications in the early 1970s.[60,61] The use of avidin–biotin techniques has since become standard for diverse detection schemes, although limitations of this method have also been recognized.[62,63] As an alternative to avidin-based reagents, we offer a high-affinity mouse monoclonal antibody to biotin (03-3700). This anti-biotin antibody can be used to detect biotinylated molecules in immunohistochemistry, *in situ* hybridization, ELISAs and flow cytometry applications.

It has been shown that certain monoclonal antibodies to biotin have biotin-binding motifs that are similar to those seen for avidin and streptavidin.[64] Anti-biotin antibody has been shown to selectively stain endogenous biotin-dependent carboxylase proteins used in fatty acid synthesis of the mitochondria.[65,66] Nonspecific staining of mitochondrial proteins by labeled avidins and by anti-biotin antibodies can be a complicating factor in using avidin–biotin techniques (Figure 7.4.1). This nonspecific binding can usually be blocked by pretreatment of the sample with the reagents in our Endogenous Biotin-Blocking Kit (E21390, Section 7.6).

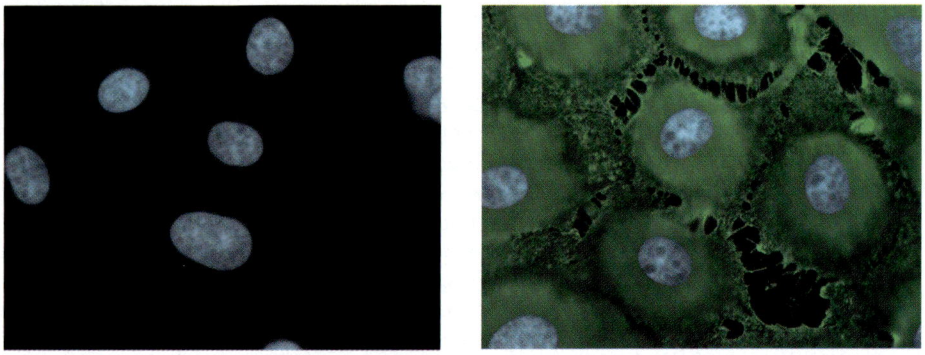

Figure 7.4.5 Fixed and permeabilized bovine pulmonary artery endothelial cells were treated with either degraded peroxynitrite (left) or 100 µM peroxynitrite (right) for 5 minutes at room temperature to induce protein nitration. Nitrated tyrosine residues were detected with our rabbit anti-nitrotyrosine antibody (A21285) and visualized with the green-fluorescent Alexa Fluor® 488 goat anti-rabbit IgG antibody (A11008). Nuclei were counterstained with blue-fluorescent DAPI (D1306, D3571, D21490).

REFERENCES

1. Cytogenet Genome Res (2009) 124:351; **2.** J Histochem Cytochem (1990) 38:467; **3.** Fluorescein Hapten: An Immunological Probe, Voss EW, Ed. (1984); **4.** J Immunol Methods (1989) 122:115; **5.** J Histochem Cytochem (2000) 48:1593; **6.** Appl Environ Microbiol (2000) 66:4258; **7.** J Biol Chem (2000) 275:11050; **8.** Mol Biol Cell (1999) 10:4385; **9.** J Immunol Methods (1991) 137:103; **10.** J Histochem Cytochem (1990) 38:325; **11.** J Immunol Methods (1989) 117:45; **12.** J Cell Biol (1990) 111:249; **13.** Am J Physiol (1990) 258:C309; **14.** J Clin Microbiol (1997) 35:578; **15.** J Biol Chem (1982) 257:13095; **16.** J Immunol Methods (1982) 53:321; **17.** FEBS Lett (1989) 253:273; **18.** J Cell Biol (1988) 106:1083; **19.** Proc Natl Acad Sci U S A (1982) 79:6186; **20.** J Neurosci Methods (1996) 65:157; **21.** Cell (1995) 81:631; **22.** J Neurosci (1995) 15:4851; **23.** Neuroscience Protocols, Wouterlood FG, Ed. (1993) p.93-050-06; **24.** J Neurosci Methods (1992) 41:45; **25.** Dev Biol (1982) 94:391; **26.** Circ Res (1992) 70:49; **27.** Front Neuroanat (2010) 4:24; **28.** Biochim Biophys Acta (1991) 1068:27; **29.** J Biol Chem (1998) 273:22950; **30.** Biochemistry (1999) 38:976; **31.** Science (1999) 284:955; **32.** J Biol Chem (1997) 272:25091; **33.** Biochim Biophys Acta (1992) 1104:9; **34.** Biochim Biophys Acta (1984) 776:217; **35.** Biochemistry (1982) 21:978; **36.** J Biol Chem (1991) 266:24540; **37.** Biochemistry (1977) 16:5150; **38.** Biochemistry (1991) 30:1692; **39.** Biochemistry (1991) 30:2888; **40.** Biochim Biophys Acta (1985) 817:238; **41.** Biochim Biophys Acta (1984) 778:612; **42.** J Biol Chem (1984) 259:5661; **43.** FEBS Lett (1986) 197:274; **44.** J Exp Med (1977) 145:931; **45.** Adv Immunol (1962) 2:1; **46.** J Immunol Methods (1992) 150:193; **47.** Science (1990) 247:64; **48.** Biotechniques (1990) 9:186; **49.** Free Radic Biol Med (1994) 17:429; **50.** Chem Res Toxicol (1993) 6:430; **51.** J Histochem Cytochem (1990) 38:69; **52.** Cell (1989) 58:317; **53.** J Cell Biol (1984) 98:1170; **54.** Nucleic Acids Res (1990) 18:3175; **55.** Electrophoresis (1999) 20:2519; **56.** Biochemistry (1995) 34:12524; **57.** J Biol Chem (1976) 251:308; **58.** Nat Clin Pract Cardiovasc Med (2008) 5:811; **59.** Methods Mol Biol (2008) 477:41; **60.** Proc Natl Acad Sci U S A (1974) 71:3537; **61.** Biochim Biophys Acta (1972) 264:165; **62.** Methods Mol Biol (2008) 418:111; **63.** Methods Mol Biol (2008) 418:157; **64.** FEBS Lett (1993) 322:47; **65.** J Histochem Cytochem (1997) 45:1053; **66.** Histochemistry (1993) 100:415.

IMPORTANT NOTICE: The products described in this manual are covered by one or more Limited Use Label License(s). Please refer to the Appendix on page 971 and Master Product List on page 975. Products are For Research Use Only. Not intended for any animal or human therapeutic or diagnostic use.

PRODUCT LIST 7.4 ANTI-DYE AND ANTI-HAPTEN ANTIBODIES

Cat. No.	Product	Quantity
A11053	Alexa Fluor® 488 Signal-Amplification Kit for Fluorescein- and Oregon Green® Dye–Conjugated Probes *60–120 assays*	1 kit
A5760	anti–Alexa Fluor® 405/Cascade Blue®, rabbit IgG fraction *3 mg/mL*	0.5 mL
A11094	anti–Alexa Fluor® 488, rabbit IgG fraction *1 mg/mL*	0.5 mL
03-3700	anti-Biotin, Mouse IgG monoclonal	0.5 mg
A5770	anti–BODIPY® FL, rabbit IgG fraction *3 mg/mL*	0.5 mL
A6398	anti-dansyl, rabbit IgG fraction *1 mg/mL*	0.5 mL
A6430	anti–dinitrophenyl-KLH, rabbit IgG fraction *2 mg/mL*	0.5 mL
A11097	anti–dinitrophenyl-KLH, rabbit IgG fraction, Alexa Fluor® 488 conjugate *2 mg/mL*	0.5 mL
A6435	anti–dinitrophenyl-KLH, rabbit IgG fraction, biotin-XX conjugate *2 mg/mL*	0.5 mL
A6423	anti–dinitrophenyl-KLH, rabbit IgG fraction, fluorescein conjugate *2 mg/mL*	0.5 mL
A11095	anti–fluorescein/Oregon Green®, goat IgG fraction *1 mg/mL*	0.5 mL
A11096	anti–fluorescein/Oregon Green®, goat IgG fraction, Alexa Fluor® 488 conjugate *1 mg/mL*	0.5 mL
A6421	anti–fluorescein/Oregon Green®, mouse IgG_{2a}, monoclonal 4-4-20	0.5 mg
A6413	anti–fluorescein/Oregon Green®, rabbit IgG Fab fragment *0.5 mg/mL*	0.5 mL
A889	anti–fluorescein/Oregon Green®, rabbit IgG fraction *1 mg/mL*	0.5 mL
A11090	anti–fluorescein/Oregon Green®, rabbit IgG fraction, Alexa Fluor® 488 conjugate *1 mg/mL*	0.5 mL
A11091	anti–fluorescein/Oregon Green®, rabbit IgG fraction, Alexa Fluor® 594 conjugate *1 mg/mL*	0.5 mL
A982	anti–fluorescein/Oregon Green®, rabbit IgG fraction, biotin-XX conjugate *1 mg/mL*	0.5 mL
A21253	anti–fluorescein/Oregon Green®, rabbit IgG fraction, horseradish peroxidase conjugate	0.5 mg
A21250	anti–fluorescein/Oregon Green®, rabbit IgG fraction, R-phycoerythrin conjugate *2 mg/mL*	250 µL
A5750	anti–lucifer yellow, rabbit IgG fraction *3 mg/mL*	0.5 mL
A5751	anti–lucifer yellow, rabbit IgG fraction, biotin-XX conjugate *3 mg/mL*	0.5 mL
A21285	anti-nitrotyrosine, rabbit IgG fraction *1 mg/mL*	0.5 mL
A6397	anti-tetramethylrhodamine, rabbit IgG fraction *1 mg/mL*	0.5 mL
A6399	anti-Texas Red®, rabbit IgG fraction *1 mg/mL*	0.5 mL
Q15431MP	Qdot® 565 goat anti-fluorescein conjugate *1 µM solution* *whole IgG*	200 µL
Q15421MP	Qdot® 655 goat anti-fluorescein conjugate *1 µM solution* *whole IgG*	200 µL
Q17421MP	Qdot® 655 rat anti-dinitrophenol conjugate *1 µM solution* *whole monoclonal IgG*	200 µL

7.5 Antibodies against Expression Tags

This section presents a selected set of primary antibodies for detecting expression tags, including Green Fluorescent Protein (GFP) and Red Fluorescent Protein (RFP) as well as oligohistidine, hemagglutinin and c-myc. The full range of Invitrogen™ primary antibodies can be surveyed using our online primary antibody search utility at www.invitrogen.com/handbook/antibodies.

Direct labeling of primary antibodies can be accomplished via amine-reactive chemistry implemented in our comprehensive selection of convenient and easy-to-use protein labeling kits (Section 1.2). Alternatively, Zenon® immunolabeling technology (Section 7.3) can be used to prepare fluorophore-, biotin- or enzyme-labeled primary antibodies but is less demanding than chemical labeling methods in terms of antibody quantity and purity requirements.

Antibodies Specific for Reporter Gene Products

Anti–Green Fluorescent Protein (GFP) Antibodies

Expression of the intrinsically fluorescent Green Fluorescent Protein (GFP) from the jellyfish *Aequorea victoria* has become a popular method for following gene expression and protein localization.[1–5] We offer a rabbit polyclonal antibody that is raised against GFP purified directly from *A. victoria*. This anti-GFP antibody, which is available as a complete serum (A6455) or as an IgG fraction (A11122) purified by ion-exchange chromatography, facilitates the detection of native GFP, recombinant GFP and GFP fusion proteins by immunofluorescence[6] (Figure 7.5.1), western blot analysis[7] and immunoprecipitation. Several direct conjugates derived from the IgG fraction are also available:

- Green-fluorescent Alexa Fluor® 488 rabbit anti-GFP antibody (A21311)
- Orange-fluorescent Alexa Fluor® 555 rabbit anti-GFP antibody (A31851)
- Red-fluorescent Alexa Fluor® 594 rabbit anti-GFP antibody (A21312)
- Far-red–fluorescent Alexa Fluor® 647 rabbit anti-GFP antibody (A31852)
- Biotin-XX rabbit anti-GFP antibody (A10259)
- Horseradish peroxidase (HRP) rabbit anti-GFP antibody (A10260)

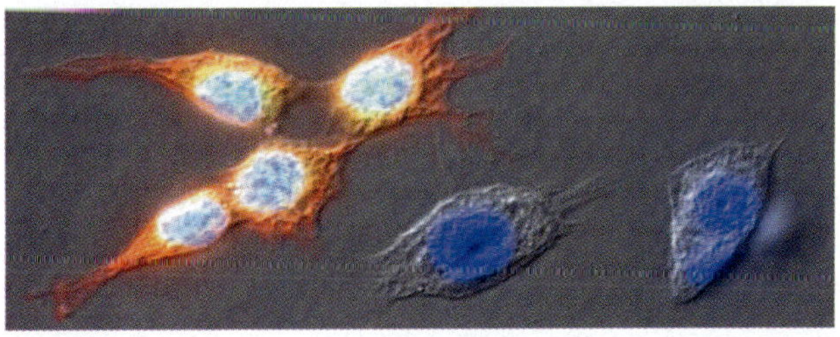

Figure 7.5.1 NIH 3T3 cells that were transiently transfected with a Green Fluorescent Protein (GFP) expression vector, then plated and allowed to attach and proliferate. The cells were fixed and labeled with our Alexa Fluor® 594 conjugate of the anti-GFP antibody (A21312). Cells expressing GFP show dual labeling of both GFP (green fluorescence) and the anti-GFP antibody (red fluorescence). In this overlay of fluorescence and differential interference contrast (DIC) micrographs, the GFP-transfected cells exhibit green and red signals that overlap to yield yellow, and DAPI (D1306, D3571, D21490) stains the nuclei with a light-blue fluorescence. In the cells that are not transfected, the DAPI-stained nuclei exhibit a bright blue fluorescence.

In addition, we offer an affinity-purified IgY fraction of chicken anti-GFP antibody, both unlabeled (A10262) and as the biotin-XX conjugate (A10263). Chicken IgY lacks a classic Fc domain and therefore does not bind to mammalian IgG Fc receptors, resulting in lower backgrounds during immunostaining protocols. The chicken IgY is also antigenically different from the mammalian IgG, allowing antibodies from both species to be used in a single immunostaining experiment.

Lastly, we offer a rabbit monoclonal anti-GFP antibody (G10362), as well as two mouse monoclonal anti-GFP antibodies. The rabbit monoclonal anti-GFP antibody is raised against full-length GFP and is suitable for applications such as ELISAs, immunohistochemistry, immunocytochemistry, immunoprecipitation and western blotting. Mouse monoclonal 11E5 (anti-GFP, clone 11E5; A11121) is optimized for western analysis, allowing colorimetric detection of as little as 10 ng of GFP or GFP-fusion proteins. Mouse monoclonal 3E6 (anti-GFP, clone 3E6; A11120) is useful for immunoprecipitation, immunocytochemical localization and immunosorbent assays (ELISAs).

Anti–Red Fluorescent Protein (RFP) Antibody

We offer an affinity-purified rabbit polyclonal anti–Red Fluorescent Protein antibody (anti-RFP, R10367) for western blot and immunocytochemistry applications. Full-length recombinant denatured and non-denatured TagRFP was used as the immunogen, making the antibody suitable for immunocytochemical detection of the expression products of our CellLight® RFP BacMam vectors (Table 11.1; Note 11.1—BacMam Gene Delivery and Expression Technology).

Anti–Glutathione *S*-Transferase Antibody

One common partner in protein fusions is glutathione *S*-transferase (GST), a protein with natural binding specificity for glutathione that can be exploited to facilitate its purification.[8] We prepare a highly purified rabbit polyclonal anti-GST antibody (A5800) that can be used for western blot analysis, immunodetection and immunoprecipitation of GST fusion proteins.[9,10] This highly specific and high-titer antibody, which was generated against a 260–amino acid N-terminal fragment of the *Schistosoma japonica* enzyme expressed in *Escherichia coli*, is particularly useful for detecting GST distribution in cells.[10–12] To facilitate the localization of GST and GST-fusion proteins using immunofluorescence techniques, we prepare the anti-GST antibody labeled with our green-fluorescent Alexa Fluor® 488 dye (A11131).

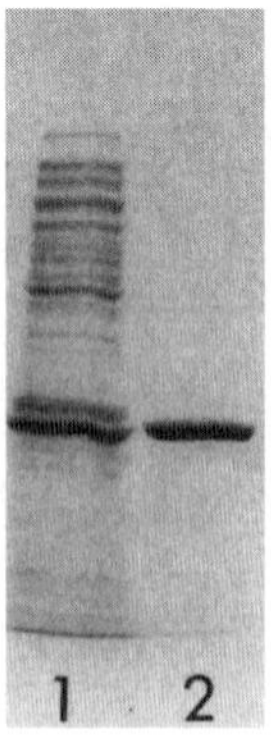

Figure 7.5.2 Coomassie brilliant blue–stained SDS-polyacrylamide gel, demonstrating the purification of a glutathione *S*-transferase (GST) fusion protein using glutathione agarose (G2879). Lane 1 contains crude supernatant from an *Escherichia coli* lysate and lane 2 contains the affinity-purified GST fusion protein.

Because the GST portion of the fusion protein retains its affinity and selectivity for glutathione, the fusion protein can be conveniently purified from the cell lysate in a single step by affinity chromatography on glutathione agarose[13–18] (Figure 7.5.2). For the purification of GST fusion proteins, we offer glutathione linked via the sulfur atom to cross-linked beaded agarose, which is available as 10 mL of a sedimented bead suspension (G2879, Section 9.4).

Anti–β-Glucuronidase Antibody

The *E. coli* β-glucuronidase (GUS) gene (*uidA*) is a popular reporter gene in plants.[19–22] For western blot and immunohistochemical analysis of transformed plant tissue[23,24] and transfected animal cells,[25] we offer unlabeled rabbit anti–β-glucuronidase antibody (A5790) raised against *E. coli* type X-A β-glucuronidase. Our fluorogenic and chromogenic β-glucuronidase substrates are described in Section 10.2.

Anti–β-Galactosidase Antibody

We offer a polyclonal antibody to the widely used reporter gene product β-galactosidase. Our rabbit anti–β-galactosidase antibody (A11132) is raised against *E. coli*–derived β-galactosidase and demonstrates high specificity for the enzyme. Whether the enzyme is used as a reporter or as a fusion protein, this antibody provides an easy tool for detecting the β-galactosidase protein. The antibody is suited to a variety of techniques, including immunoblotting, ELISA and immunoprecipitation. Our fluorogenic and chromogenic β-galactosidase substrates are described in Section 10.2.

Antibodies Specific for Epitope Tags

Epitope tagging is a powerful and versatile strategy for detecting and purifying proteins expressed by cloned genes. Protein expression vectors are often engineered with a nucleotide sequence that encodes a peptide epitope tag. Typically, a gene is cloned in-frame relative to the epitope tag such that, upon expression, a fusion protein containing the epitope tag is synthesized. Detection and purification of the epitope-tagged fusion protein can be mediated by antibodies to the engineered peptide, thus eliminating the need for antibodies to proteins from each newly cloned gene. Here we highlight a series of antibodies to common epitopes, including oligohistidine, hemagglutinin (HA) and c-myc, but a more comprehensive list can be found using our online primary antibody search utility at www.invitrogen.com/handbook/antibodies.

Penta·His Antibody

The oligohistidine domain is a Ni^{2+}-binding peptide sequence comprising a sequence of four to six histidine residues.[26,27] When the DNA sequence corresponding to an oligohistidine domain is fused in-frame with a gene of interest, the resulting recombinant fusion protein can be easily purified using a nickel-chelating resin.[27,28]

In collaboration with QIAGEN, we offer a highly selective antibody to oligohistidine fusion proteins (Penta·His mouse monoclonal antibody, P21315). The antibody is useful for detecting sequences of five or six histidine residues, whether present at the C-terminus, the N-terminus or an internal position in a protein, in both western blots and in immunohistochemistry.

The Penta·His antibody can detect as little as 1–2 ng in western blot applications using fluorescent or chemiluminescent development techniques (Figure 7.5.3); the supplied amount (100 µg) is sufficient for

IMPORTANT NOTICE: The products described in this manual are covered by one or more Limited Use Label License(s). Please refer to the Appendix on page 971 and Master Product List on page 975. Products are For Research Use Only. Not intended for any animal or human therapeutic or diagnostic use.

50–100 minigel blots. This antibody does not recognize tetrahistidine domains or domains in which the histidine string is interrupted by another amino acid. The antibody binds to the oligohistidine domain regardless of the surrounding amino acid context and even when the group is partially hidden, although subtle differences in the amino acid context may change the sensitivity limit for a particular fusion protein. The Penta·His antibody is directly useful for immunoprecipitation, ELISAs and immunohistochemistry.

Anti-HA Antibody

Our mouse monoclonal anti-hemagglutinin (HA) antibody was raised against the 12-amino acid peptide CYPYDVPDYASL. It recognizes the influenza hemagglutinin epitope YPYDVPDYA, which has been used extensively as a general epitope tag in expression vectors.[29] The extreme specificity of this antibody permits unambiguous identification and quantitative analysis of the tagged protein. This antibody is effective in immunoblotting, immunofluorescence and immunoprecipitation of tagged proteins. We offer Alexa Fluor® 488 (A21287) and Alexa Fluor® 594 (A21288) conjugates of the anti-HA antibody for those applications that require a directly conjugated antibody.

Anti–c-myc Antibody

We offer two antibodies to c-myc, which is commonly used in epitope tagging.[29] Our mouse monoclonal anti–c-myc antibody (IgG$_{1,\kappa}$, clone 238; A21280) was raised against the peptide AEEQKLISEEDLLRKRREQLKHKLEQLRNSCA, which corresponds to amino acids 408–439 of the human c-myc protein. The chicken anti–c-myc antibody (IgY, A21281) was raised against the peptide EQKLISEEDL. These antibodies specifically react with the C-terminal epitope (AEEQKLISEEDL)[30] of human c-myc protein encoded in many expression vectors.

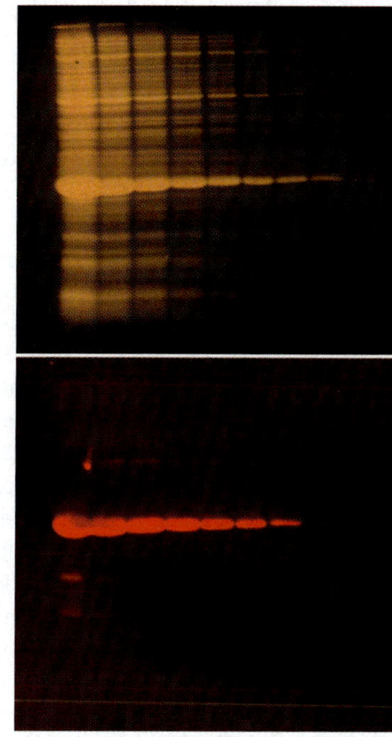

Figure 7.5.3 Detection of an oligohistidine fusion protein with the Penta·His mouse monoclonal antibody. Two-fold serial dilutions of an *Escherichia coli* lysate containing overexpressed oligomycin sensitivity–conferring protein (OSCP) fused with a hexahistidine domain were run an SDS-polyacrylamide gel and blotted onto a PVDF membrane. The blot was stained with the SYPRO® Ruby protein blot stain (S11791) to detect the entire protein profile (top). After imaging, the blot was incubated with Penta·His mouse IgG$_1$ monoclonal antibody (P21315), followed by immunodetection using an alkaline phosphatase conjugate of goat anti–mouse antibody in conjunction with DDAO phosphate (bottom).

REFERENCES

1. Nat Rev Mol Cell Biol (2002) 3:906; 2. Annu Rev Biochem (1998) 67:509; 3. Nat Biotechnol (1997) 15:961; 4. ACS Chem Biol (2009) 4:85; 5. Science (1994) 263:802; 6. J Cell Biol (2004) 167:851; 7. J Biol Chem (1998) 273:21054; 8. Proc Natl Acad Sci U S A (1986) 83:8703; 9. J Biol Chem (1997) 272:8133; 10. J Cell Biol (1996) 133:1403; 11. Mol Biol Cell (1996) 7:1209; 12. Cytometry (1995) 20:134; 13. Meth Mol Genet (1993) 1:267; 14. Biotechniques (1992) 13:856; 15. Biotechniques (1991) 10:178; 16. Nucleic Acids Res (1991) 19:4005; 17. Science (1991) 252:712; 18. Gene (1988) 67:31; 19. Mol Gen Genet (1989) 216:321; 20. Mol Gen Genet (1988) 215:38; 21. Plant Mol Biol (1988) 10:387; 22. Proc Natl Acad Sci U S A (1986) 83:8447; 23. J Biol Chem (1994) 269:17635; 24. Plant Mol Biol (1990) 15:821; 25. Biotechniques (1999) 27:896; 26. DNA Cell Biol (1993) 12:441; 27. Protein Expr Purif (1992) 3:263; 28. Methods Mol Biol (2000) 130:203; 29. Methods Enzymol (1991) 194:508; 30. Mol Cell Biol (1985) 5:3610.

PRODUCT LIST 7.5 ANTIBODIES AGAINST EXPRESSION TAGS

Cat. No.	Product	Quantity
A21281	anti–c-myc, chicken IgY fraction *1 mg/mL*	100 µL
A21280	anti–c-myc, mouse IgG$_1$, monoclonal 289-19510 *1 mg/mL*	100 µL
A11132	anti-β-galactosidase, rabbit IgG fraction *2 mg/mL*	0.5 mL
A5790	anti-β-glucuronidase, rabbit IgG fraction *2 mg/mL*	0.5 mL
A5800	anti-glutathione *S*-transferase, rabbit IgG fraction *3 mg/mL*	0.5 mL
A11131	anti-glutathione *S*-transferase, rabbit IgG fraction, Alexa Fluor® 488 conjugate *2 mg/mL*	0.5 mL
A10262	anti-green fluorescent protein, chicken IgY fraction (anti-GFP, IgY) *2 mg/mL*	100 µL
A10263	anti-green fluorescent protein, chicken IgY fraction, biotin-XX conjugate (anti-GFP, IgY, biotin-XX conjugate) *2 mg/mL*	100 µL
A11120	anti-green fluorescent protein, mouse IgG$_{2a}$, monoclonal 3E6 (anti-GFP, mAb 3E6)	100 µg
A11121	anti-green fluorescent protein, mouse IgG$_1$, monoclonal 11E5 (anti-GFP, mAb 11E5)	100 µg
A11122	anti-green fluorescent protein, rabbit IgG fraction (anti-GFP, IgG) *2 mg/mL*	100 µL
A21311	anti-green fluorescent protein, rabbit IgG fraction, Alexa Fluor® 488 conjugate (anti-GFP, IgG, Alexa Fluor® 488 conjugate) *2 mg/mL*	100 µL
A31851	anti-green fluorescent protein, rabbit IgG fraction, Alexa Fluor® 555 conjugate (anti-GFP, IgG, Alexa Fluor® 555 conjugate) *2 mg/mL*	100 µL
A21312	anti-green fluorescent protein, rabbit IgG fraction, Alexa Fluor® 594 conjugate (anti-GFP, IgG, Alexa Fluor® 594 conjugate) *2 mg/mL*	100 µL
A31852	anti-green fluorescent protein, rabbit IgG fraction, Alexa Fluor® 647 conjugate (anti-GFP, IgG, Alexa Fluor® 647 conjugate) *2 mg/mL*	100 µL
A10259	anti-green fluorescent protein, rabbit IgG fraction, biotin-XX conjugate (anti-GFP, IgG, biotin-XX conjugate) *2 mg/mL*	100 µL
A10260	anti-green fluorescent protein, rabbit IgG fraction, horseradish peroxidase conjugate (anti-GFP, IgG, HRP)	200 µg
A6455	anti-green fluorescent protein, rabbit serum (anti-GFP, serum)	100 µL
A21287	anti-hemagglutinin, mouse IgG$_1$, monoclonal 16B12, Alexa Fluor® 488 conjugate (anti-HA, Alexa Fluor® 488 conjugate) *1 mg/mL*	100 µL
A21288	anti-hemagglutinin, mouse IgG$_1$, monoclonal 16B12, Alexa Fluor® 594 conjugate (anti-HA, Alexa Fluor® 594 conjugate) *1 mg/mL*	100 µL
G10362	GFP, ABfinity™ recombinant rabbit monoclonal antibody, unconjugated (anti-GFP, rabbit mAb)	100 µg
P21315	Penta·His mouse IgG$_1$, monoclonal antibody (anti-pentahistidine) *BSA free*	100 µg
R10367	RFP, rabbit polyclonal antibody, unconjugated (anti-RFP)	100 µg

7.6 Avidin and Streptavidin Conjugates

The high affinity of avidin for biotin was first exploited in histochemical applications in the mid-1970s.[1,2] This egg-white protein and its bacterial counterpart, streptavidin, have since become standard reagents for diverse detection schemes.[3–6] In their simplest form, such avidin–biotin detection methods entail applying a biotinylated probe to a sample and then detecting the bound probe with a labeled avidin or streptavidin. These techniques are commonly used to localize antigens in cells and tissues[7,8] and to detect biomolecules in immunoassays and DNA hybridization procedures.[6,9–11] In some applications, immobilized avidins are used to capture and release biotinylated targets. In addition to our dye and enzyme conjugates of avidins and streptavidins, this section contains several products that can be used for the affinity isolation of biotin-conjugated molecules and their complexes in cell and tissues.

Our diverse set of biotinylation reagents and biotin conjugates are described in Section 4.3. Combining one of our biotinylated or DSB-X™ biotin–labeled secondary antibodies (Section 7.2, Table 7.1) with a fluorescent dye– or enzyme-labeled avidin provides an sensitive method for the indirect detection of antibodies from various animal sources.

Binding Characteristics of Biotin-Binding Proteins

Avidin, streptavidin and NeutrAvidin™ biotin-binding protein each bind four biotins per molecule with high affinity and selectivity. Dissociation of biotin from streptavidin (S888) is reported to be about 30 times faster than dissociation of biotin from avidin[12] (A887, A2667). Their multiple binding sites permit a number of techniques in which unlabeled avidin, streptavidin or NeutrAvidin™ biotin-binding protein can be used to bridge two biotinylated reagents. This bridging method, which is commonly used to link a biotinylated probe to a biotinylated enzyme in enzyme-linked immunohistochemical applications, often eliminates the background problems that can occur when using direct avidin– or streptavidin–enzyme conjugates. However, endogenously biotinylated proteins that have carboxylase activity are found in the mitochondria (Figure 7.6.1); therefore, sensitive detection of biotinylated targets in cells requires the use of biotin-blocking agents to reduce this background.[13,14] Our Endogenous Biotin-Blocking Kit (E21390) provides the reagents and a protocol for this application. Nonspecific binding of avidin conjugates of enzymes to nitrocellulose can be blocked more effectively by adding extra salts to buffers rather than by adding protein-based blocking reagents.[15]

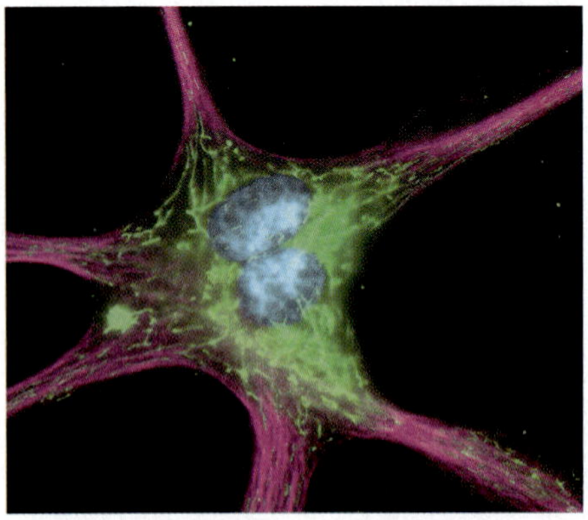

Figure 7.6.1 The cytoskeleton of a fixed and permeabilized bovine pulmonary artery endothelial cell detected using mouse monoclonal anti-α-tubulin antibody (A11126), visualized with Alexa Fluor® 647 goat anti–mouse IgG antibody (A21235) and pseudocolored magenta. Endogenous biotin in the mitochondria was labeled with green-fluorescent Alexa Fluor® 488 streptavidin (S11223) and DNA was stained with blue-fluorescent DAPI (D1306, D3571, D21490).

High-purity unlabeled avidin (A887), streptavidin (S888), NeutrAvidin™ biotin-binding protein (A2666) and CaptAvidin™ biotin-binding protein (C21385) are available in bulk. We also offer avidin specially packaged in a smaller unit size for extra convenience (A2667). Our avidin, streptavidin and deglycosylated NeutrAvidin™ biotin-binding protein typically bind greater than 12 µg of biotin per mg protein.

Avidin

Avidin (A887, A2667; Table 7.9) is a highly cationic 66,000-dalton glycoprotein[16,17] with an isoelectric point of about 10.5. It is thought that avidin's positively charged residues and its oligosaccharide component (heterogeneous structures composed largely of mannose and *N*-acetylglucosamine) can interact nonspecifically with negatively charged cell surfaces and nucleic acids, sometimes causing background problems in some histochemical applications and flow cytometry. Methods have been developed to suppress this nonspecific avidin binding.[14] In some cases, avidin's nonspecific binding can also be exploited. For example, avidin and its conjugates selectively bind to a component in rodent and human mast cell granules in fixed-cell preparations and can be used to identify mast cells in normal and diseased human tissue without requiring a biotinylated probe.[18,19]

Streptavidin

Streptavidin (S888, Table 7.9), a nonglycosylated 52,800-dalton protein with a near-neutral isoelectric point, reportedly exhibits less nonspecific binding than avidin. However, streptavidin contains the tripeptide sequence Arg–Tyr–Asp (RYD) that apparently mimics the Arg–Gly–Asp (RGD) binding sequence of fibronectin, a component of the extracellular matrix that specifically promotes cellular adhesion.[20] This universal recognition sequence binds integrins and related cell-surface molecules.[21,22] Background problems sometimes associated with streptavidin may be attributable to this tripeptide. We have particularly observed binding of streptavidin and anti-biotin[23] conjugates to mitochondria in some cells (Figure 7.6.1) that can be blocked with the reagents in our Endogenous Biotin-Blocking Kit (E21390).

NeutrAvidin™ Biotin-Binding Protein

We provide an alternative to the commonly used avidin and streptavidin. Our conjugates of NeutrAvidin™ biotin-binding protein (A2666, Table 7.9)—a protein that has been processed to remove the carbohydrate and lower its isoelectric point—can sometimes reduce background staining. The methods used to deglycosylate the avidin are reported to retain both its specific binding[24] and its complement of amine-conjugation sites. NeutrAvidin™ conjugates have been shown to provide improved detection of single-copy genes in metaphase chromosome spreads.[25]

Table 7.9 Molecular Probes® avidin, streptavidin, NeutrAvidin™ and CaptAvidin™ conjugates.

Label (Abs/Em) *	Streptavidin	Neutr-Avidin™	Avidin	Capt-Avidin™
Fluorophore Conjugates				
Alexa Fluor® 350 (346/442)	S11249	A11236		
Marina Blue® (365/460)	S11221			
Cascade Blue® (400/420)		A2663		
Alexa Fluor® 405 (402/421)	S32351			
Pacific Orange™ (400/551)	S32365			
Cascade Yellow™ (402/545)	S11228			
Pacific Blue™ (410/455)	S11222			
Alexa Fluor® 430 (434/539)	S11237			
Fluorescein (494/518)	S869	A2662	A821	
Alexa Fluor® 488 (495/519)	S11223, S32354 †		A21370	
Oregon Green® 488 (496/524)	S6368	A6374		
R-Phycoerythrin (496/578) ‡	S866, S21388 §	A2660		
Alexa Fluor® 610–R-PE (496/628) ‡	S20982			
Alexa Fluor® 647–R-PE (496/668) ‡	S20992			
Alexa Fluor® 680–R-PE (496/702) ‡	S20985			
Alexa Fluor® 750–R-PE (496/775) ‡	S32363			
Oregon Green® 514 (511/530)	S6369			
Alexa Fluor® 514 (518/540)	S32353			
Alexa Fluor® 532 (531/554)	S11224			
B-Phycoerythrin (546/575) ‡	S32350			
Tetramethylrhodamine (555/580)	S870	A6373		
Alexa Fluor® 546 (556/573)	S11225			
Alexa Fluor® 555 (555/565)	S21381, S32355 †			
Rhodamine B (570/590)	S871			
Rhodamine Red™-X (570/590)	S6366	A6378		
Alexa Fluor® 568 (578/603)	S11226			
Alexa Fluor® 594 (590/617)	S11227, S32356 †			
Texas Red® (595/615)	S872	A2665	A820	
Texas Red®-X (595/615)	S6370			
Alexa Fluor® 633 (632/647)	S21375			
Alexa Fluor® 635 (633/647)	S32364			
Alexa Fluor® 647 (650/668)	S21374, S32357 †			
Allophycocyanin (650/660)	S868, S32362 §			
Alexa Fluor® 680–allophycocyanin (650/702)	S21002			
Alexa Fluor® 700–allophycocyanin (650/723)	S21005			
Alexa Fluor® 750–allophycocyanin (650/775)	S21008			
Alexa Fluor® 660 (663/690)	S21377			
Alexa Fluor® 680 (679/702)	S21378, S32358 †			
Alexa Fluor® 700 (702/723)	S21383			
Alexa Fluor® 750 (749/775)	S21384			

Label (Abs/Em) *	Streptavidin	Neutr-Avidin™	Avidin	Capt-Avidin™
Qdot® Nanocrystal Conjugates				
Qdot® 525 (405/525)	Q10141MP **			
Qdot® 565 (405/565)	Q10131MP **			
Qdot® 585 (405/585)	Q10111MP **			
Qdot® 605 (405/605)	Q10101MP **			
Qdot® 625 (405/625)	A10196			
Qdot® 655 (405/655)	Q10121MP **			
Qdot® 705 (405/705)	Q10161MP **			
Qdot® 800 (405/800)	Q10171MP			
FluoSpheres® and TransFluoSpheres® Microsphere Conjugates				
Europium luminescent FluoSpheres® microspheres †† (365/610)		F20883, F20884		
Yellow-green–fluorescent FluoSpheres® microspheres †† (505/515)	F8780	F8771, F8774, F8776		
Red-fluorescent FluoSpheres® microspheres †† (580/605)		F8770, F8775		
TransFluoSpheres® microspheres †† (488/605)		T8860, T8861		
TransFluoSpheres® microspheres †† (488/645)	T10711			
Nonfluorescent FluoSpheres® microspheres ††		F8772, F8777		
Gold Conjugates				
NANOGOLD®	N24918			
Alexa Fluor® 488 FluoroNanogold™ (495/519)	A24926			
Alexa Fluor® 488 colloidal gold (495/519)	A32360 ‡‡, A32361 §§			
Alexa Fluor® 594 FluoroNanogold™ (590/617)	A24927			
Enzyme Conjugates				
Alkaline phosphatase	S921			
β-Galactosidase	S931			
Horseradish peroxidase	S911	A2664		
β-Lactamase	S31569			
Acrylamide and Agarose Conjugates				
Acrylamide	S21379			
Agarose	S951			C21386
Unlabeled Avidins				
Unlabeled	S888	A2666	A887, A2667	C21385

* Approximate absorption (Abs) and fluorescence emission (Em) maxima, in nm, for conjugates. † These Alexa Fluor® streptavidin conjugates are supplied in 0.5 mL units of a 2 mg/mL solution. ‡ Multiple absorption peaks. § Premium grade. ** Properties and applications of Qdot® nanocrystal conjugates are described in Section 6.6. The Qdot® Streptavidin Sampler Kit (Q10151MP) contains six different Qdot® streptavidin conjugates. †† FluoSpheres® and TransFluoSpheres® microspheres are available in different diameters, including 0.04 µm, 0.2 µm and 1.0 µm. ‡‡ 5 nm colloidal gold conjugate. §§ 10 nm colloidal gold conjugate.

IMPORTANT NOTICE: The products described in this manual are covered by one or more Limited Use Label License(s). Please refer to the Appendix on page 971 and Master Product List on page 975. Products are For Research Use Only. Not intended for any animal or human therapeutic or diagnostic use.

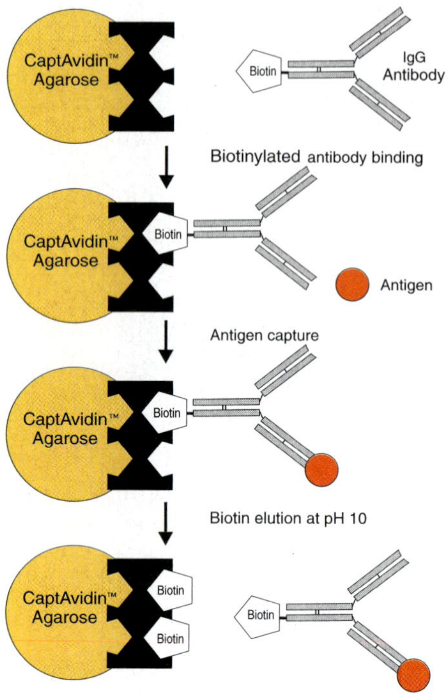

Figure 7.6.2 Diagram of the use of CaptAvidin™ agarose (C21386) in affinity chromatography. A biotinylated IgG molecule and target antigen are used as an example.

CaptAvidin™ Biotin-Binding Protein: Reversible Binding of Biotinylated Molecules

In addition to avidin, streptavidin and NeutrAvidin™ biotin-binding protein, we offer CaptAvidin™ biotin-binding protein (C21385, Table 7.9). Selective nitration of tyrosine residues in the four biotin-binding sites of avidin considerably reduces the affinity of the protein for biotinylated molecules above pH 9.[26] Consequently, biotinylated probes can be adsorbed at neutral pH and released at pH ~10 (Figure 7.6.2). We use free biotin to block any remaining high-affinity biotin-binding sites that have not been nitrated. CaptAvidin™ agarose (C21386) is particularly useful for separating and purifying biotin conjugates from complex mixtures. The biotin-binding capacity of CaptAvidin™ derivatives is typically at least 10 µg of free biotin per mg protein.

Secondary Detection with Avidins

Avidin, streptavidin and NeutrAvidin™ conjugates are extensively used as secondary detection reagents in histochemical applications (Figure 7.6.3, Figure 7.6.4), FISH (Figure 7.6.5), flow cytometry,[27,28] microarrays (Figure 7.6.6) and blot analysis. These reagents can also be employed to localize biocytin, biotin ethylenediamine or any of our fluorescent biocytins—all of which are biotin derivatives commonly used as neuroanatomical tracers[29,30] (Section 14.3).

The following are commonly used methods for employing avidin, streptavidin, NeutrAvidin™ biotin-binding protein and CaptAvidin™ biotin-binding protein as secondary detection reagents:

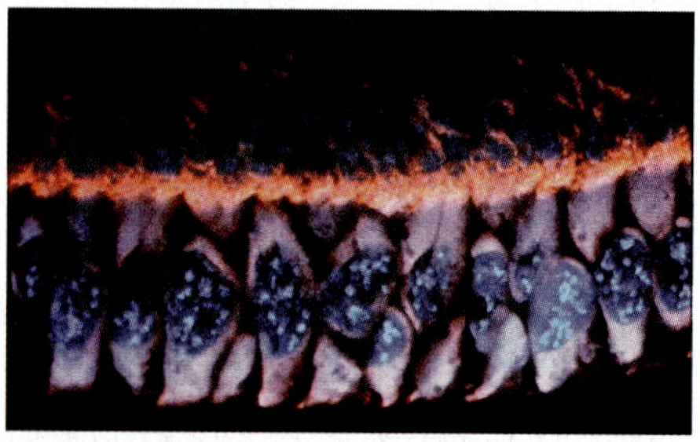

Figure 7.6.3 The cortical region of the developing follicle of the giant silkmoth *Antheraea polyphemus* stained with a monoclonal antibody against cytoskeletal actin. The primary antibody was visualized using biotin-XX goat anti–mouse IgG antibody (B2763), followed by incubation with Texas Red® streptavidin (S872). The orange to pink colors in this confocal laser-scanning micrograph show the distribution of cytoskeletal actin in the oocyte cortex and follicle cell cytoplasm. The blue color can be attributed to autofluorescence. The image was contributed by Ivo Sauman, Wesleyan University.

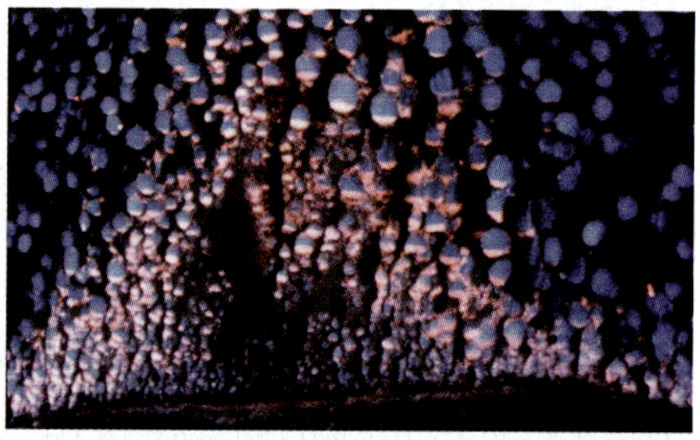

Figure 7.6.4 The "delta" region of a developing follicle of the giant silkmoth *Hyalophora cecropia* stained with an antibody against the largest subunit of the *Drosophila* RNA polymerase II (RNAp II). The primary antibody was visualized using a biotinylated secondary antibody followed by Texas Red® dye–conjugated streptavidin (S872). The distribution of the RNAp II appears violet because the Texas Red® staining colocalizes with the blue autofluorescence of the yolk granules. Image contributed by Ivo Sauman, Wesleyan University.

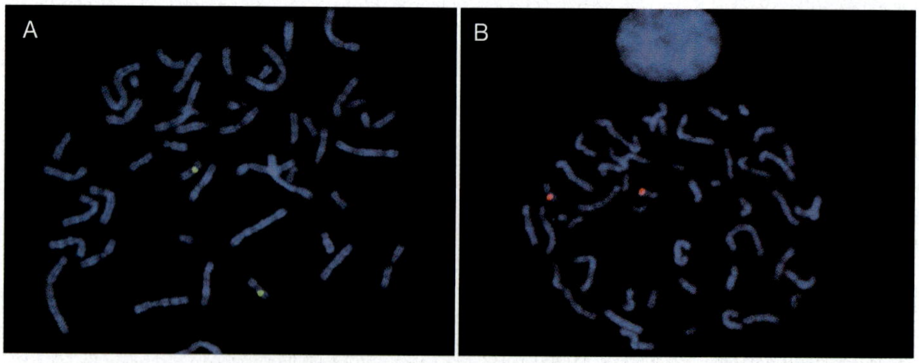

Figure 7.6.5 Fluorescence *in situ* hybridization detected by tyramide signal amplification. Chromosome spreads were prepared from the cultured fibroblast cell line MRC-5 and hybridized with a biotinylated α-satellite probe specific for chromosome 17. The probe was generated by nick translation in the presence of biotinylated dUTP. For detection by TSA™, hybridized chromosome spreads were labeled using TSA™ Kit #22 (T20932) with HRP-conjugated streptavidin and Alexa Fluor® 488 tyramide (**A**) or using TSA™ Kit #23 (T20933) with HRP-conjugated streptavidin and Alexa Fluor® 546 tyramide (**B**). After counterstaining with DAPI (D1306, D3571, D21490), images were obtained using filters appropriate for DAPI, FITC or TRITC.

IMPORTANT NOTICE: The products described in this manual are covered by one or more Limited Use Label License(s). Please refer to the Appendix on page 971 and Master Product List on page 975. Products are For Research Use Only. Not intended for any animal or human therapeutic or diagnostic use.

- **Direct procedure.** A biotinylated primary probe such as an antibody, single-stranded nucleic acid probe or lectin is bound to tissues, cells or other surfaces. Excess protein is removed by washing, and detection is mediated by reagents such as our fluorescent avidins, streptavidins or NeutrAvidin™ biotin-binding proteins (Figure 7.6.7) or our enzyme-conjugated streptavidins plus a fluorogenic, chromogenic or chemiluminescent substrate. Enzyme conjugates of streptavidin are key reagents in some of our Tyramide Signal Amplification (TSA™) Kits (Section 6.2, Figure 7.6.5).
- **Capture and release.** Our unique DSB-X™ biotin technology permits the fully reversible labeling of DSB-X™ biotin derivatives by avidin and streptavidin conjugates.[31] Consequently, targets in cells and tissues or on blots labeled with DSB-X™ biotin conjugates of antibodies or other DSB-X™ biotin reagents can initially be stained with fluorescent avidin or streptavidin conjugates, then the fluorescent staining can be reversed with D-biotin (B1595, B20656; Figure 7.6.8) and the sample restained with an enzyme-conjugated avidin or streptavidin derivative in conjunction with a permanent stain such as diaminobenzidine (DAB, D22187) or the combination of NBT and BCIP (N6495, B6492, N6547; Section 10.3).
- **Bridging methods.** A biotinylated antibody or oligonucleotide is used to probe a tissue, cell or other surface, and then this preparation is treated with unlabeled avidin, streptavidin or NeutrAvidin™ biotin-binding protein. Excess reagents are removed by washing, and detection is mediated by a biotinylated detection reagent such as a fluorescent biotin or biocytin dye (Section 4.3), biotinylated R-phycoerythrin[32] (P811, Section 6.4), biotinylated FluoSpheres® microspheres (Section 6.5) or biotinylated horseradish peroxidase (P917) plus a fluorogenic, chromogenic or chemiluminescent substrate.
- **Indirect procedure.** An unlabeled primary antibody is bound to a cell followed by a biotinylated species-specific secondary antibody. After washing, the complex is detected by one of the two procedures described above. Our Zenon® Biotin-XX Antibody Labeling Kits (Section 7.3, Table 7.7) permit the rapid and quantitative biotinylation of antibodies for combination with avidin–biotin detection methods.

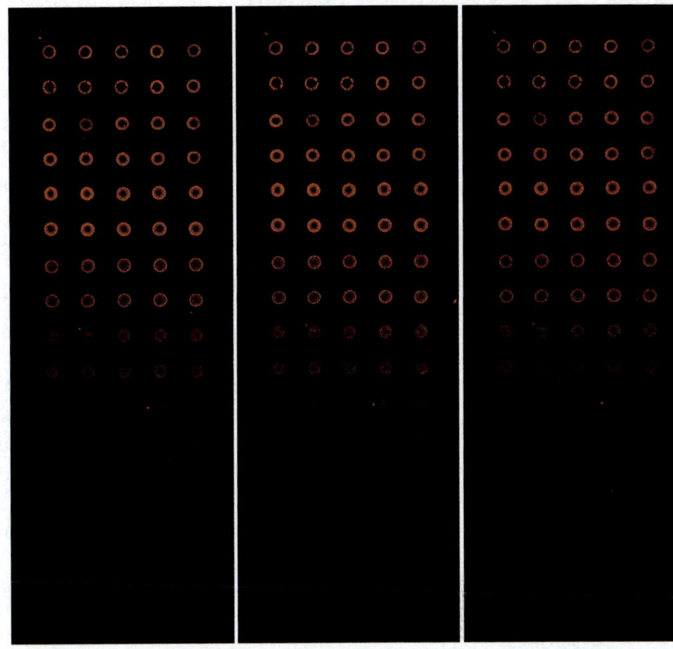

Figure 7.6.6 R-phycoerythrin used to detect DNA on a microarray. A DNA microarray containing a decreasing dilution of calf thymus DNA was hybridized with a biotinylated DNA probe and then incubated with R-phycoerythrin–streptavidin (SAPE; S866, S21388). After washing, the fluorescence signal was detected on a Packard ScanArray® 5000 using three different detection configurations: 488 nm excitation (argon-ion laser)/570 nm emission filter (left); 543.5 nm excitation (He-Ne laser)/570 nm emission filter (middle); 543.5 nm excitation (He-Ne laser)/592 nm emission filter (right).

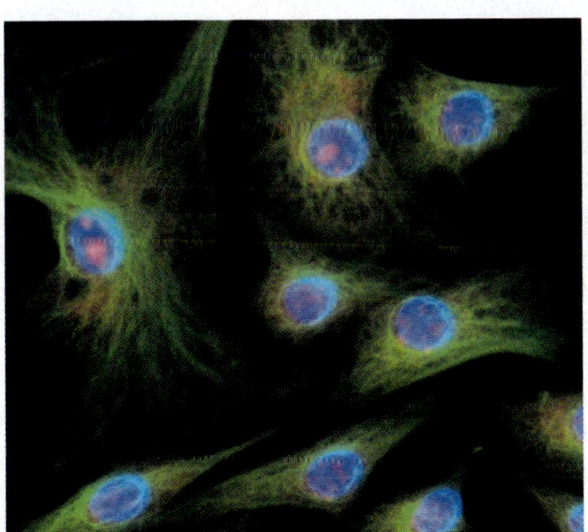

Figure 7.6.7 Microtubules of fixed bovine pulmonary artery endothelial cells (BPAEC) were localized with mouse monoclonal anti-α-tubulin antibody (A11126), followed by the biotin-XX–conjugated $F(ab')_2$ fragment of goat anti-mouse IgG antibody (B11027) and visualized with green-fluorescent Alexa Fluor® 488 streptavidin (S11223). Cells were counterstained with blue-fluorescent Hoechst 33342 (H1399, H3570, H21492) to image the DNA and red-fluorescent propidium iodide (P1304MP, P3566, P21493) to image nucleolar RNA. The multiple-exposure image was acquired using bandpass filter sets appropriate for Texas Red® dye, fluorescein and DAPI.

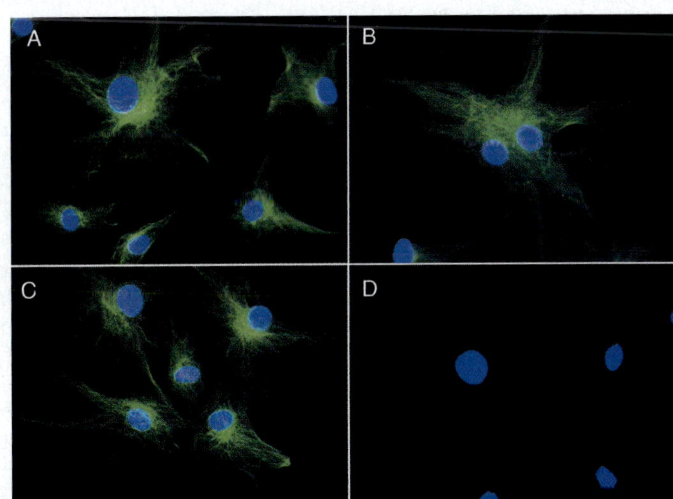

Figure 7.6.8 Reversible binding by DSB-X™ biotin. Microtubules of fixed bovine pulmonary artery endothelial cells were labeled with mouse monoclonal anti–α-tubulin antibody (A11126), detected with either biotin-XX goat anti–mouse IgG antibody (B2763, panel **A**) or DSB-X™ biotin goat anti-mouse IgG antibody (D20691, panel **B**) and visualized with green-fluorescent Alexa Fluor® 488 streptavidin (S11223). Nuclei were stained with blue-fluorescent DAPI (D1306, D3571, D21490). After incubating with 10 mM D-biotin (B1595, B20656), the binding between the biotinylated antibody is unaltered (panel **C**), whereas the streptavidin conjugate has been stripped from the DSB-X™ biotin–labeled antibody (panel **D**).

IMPORTANT NOTICE: The products described in this manual are covered by one or more Limited Use Label License(s). Please refer to the Appendix on page 971 and Master Product List on page 975. Products are For Research Use Only. Not intended for any animal or human therapeutic or diagnostic use.

Figure 7.6.9 A multinucleate HeLa cell in metaphase that was fixed and then stained with a combination of fluorescent dyes. The chromosomes were stained with DAPI (D1306, D3571, D21490). The cytoskeleton was detected with the biotin-XX conjugate of mouse monoclonal anti-α-tubulin antibody (A21371), which was then visualized with red-fluorescent Alexa Fluor® 568 streptavidin (S11226). The multiple-exposure image was acquired using filter sets appropriate for rhodamine and DAPI.

Endogenous Biotin and Biotinidase

Mammalian cells and tissues contain biotin-dependent carboxylases, which are required for a variety of metabolic functions. These biotin-containing enzymes often produce substantial background signals when avidin–biotin or streptavidin–biotin detection systems are used to identify cellular targets[33,34] (Figure 7.6.1). Endogenous biotin is particularly prevalent in mitochondria[35,36] and in kidney, liver and brain tissues.[33,37] In mammalian serum and plasma, biotinylated proteins are susceptible to cleavage by endogenous biotinidases, producing free biotin and unlabeled protein.[38]

The reagents in the Endogenous Biotin-Blocking Kit (E21390) can be used to minimize interference from endogenous biotin in these techniques. This kit provides streptavidin and biotin solutions in convenient dropper bottles and an easy-to-follow protocol. Sufficient material is provided for approximately one hundred 18 mm × 18 mm glass coverslips.

Fluorescent Avidin and Streptavidin Conjugates

Fluorophore-Labeled Avidin, Streptavidin and NeutrAvidin™ Biotin-Binding Protein

Fluorescent avidin and streptavidin are extensively used in DNA hybridization techniques,[39,40] immunohistochemistry (Figure 7.6.9), MHC tetramer technology[41] (MHC Tetramer Technology—Note 7.2) and multicolor flow cytometry.[42–44] Our selection of avidin, streptavidin and NeutrAvidin™ conjugates continues to expand as we introduce new and improved fluorophores and signal amplification technologies (Table 7.9). We continue to provide avidin, streptavidin and NeutrAvidin™ conjugates of fluorescein (A821, S869, A2662), tetramethylrhodamine

NOTE 7.2

MHC Tetramer Technology

Cytotoxic T lymphocytes (CTLs) or "killer" T cells perform a crucial function in the vertebrate immune system by destroying cells that display foreign antigens on their surfaces. CTLs are distinguished from other types of T lymphocytes (such as "helper" T cells) by their expression of CD8, a transmembrane protein with immunoglobulin-like domains that interact with class I Major Histocompatibility Complexes[1] (MHCs). MHC molecules have a dual role; their amino-terminal domains bind antigen for presentation to T cells and also serve as a marker identifying host cells to the immune system. Activation of CTLs to eliminate a target cell occurs only if the CD8 protein recognizes the proper MHC molecule and the α/β T cell receptor binds the MHC-presented antigen.[2]

Historically, assays for an antigen-specific T cell response made use of the limiting dilution analysis (LDA) method, which requires that single clones survive, divide and differentiate before they can be detected. However, this method likely underestimates the number of CTLs that respond to a particular antigen because it will not detect cells that are part of the expanded effector population that can no longer divide.[3] Alternate approaches attempt to directly measure CTL response using an antigenic peptide and MHC. However, these experiments are hampered by the low affinity of the interaction between the MHC/antigen complex and the corresponding CTL receptors.[4]

Joining multiple copies of the MHC/antigen complex into a single probe—tetramer technology—resolves the difficulties presented by the low affinity of the class I MHC molecule for the CD8 receptor.[5] CTL response assays using tetramer technology often detect a response rate that is 50–500 times higher than detected by the LDA method,[3] and they have proven indispensable for the study of the CTL response to HIV, cancers and transplants.6 Similar methods using class II MHC tetramers are also being employed to explore the response of helper (CD4 expressing) T cells.[7]

MHC tetramers are formed by first refolding MHCs in the presence of high concentrations of the desired antigenic peptide, followed by biotinylation of the carboxy-terminus of one chain of the MHC molecule. This MHC/peptide complex can then be bound to streptavidin. Because the latter has four biotin-binding sites, four MHC molecules can be linked together in a single complex (Figure 1). The use of fluorophore-labeled streptavidin or streptavidin-coated magnetic beads for tetramer formation allows for efficient detection by flow cytometry or immunomagnetic methods.[8–10] It is critical for MHC tetramer–based assays that the labeled streptavidin be free of unlabeled streptavidin because this lowers the apparent activity of the MHC complex by "blocking" the biotinylation site from binding to the labeled streptavidin.

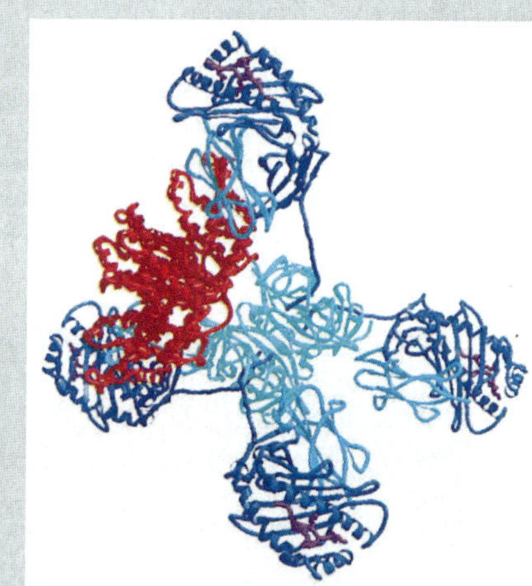

Figure 1 An MHC tetramer.

1. Annu Rev Immunol (1992) 10:645; **2.** Adv Immunol (1993) 53:59; **3.** J Exp Med (1998) 187:1367; **4.** Proc Natl Acad Sci U S A (1994) 91:12862; **5.** Immunol Rev (1996) 150:5; **6.** Immunology (2009) 126:147; **7.** J Clin Invest (1999) 104:1669; **8.** J Immunol Meth (2007) 319:13; **9.** Cytometry A (2008) 73:1001; **10.** Nat Methods (2009) 6:520.

IMPORTANT NOTICE: The products described in this manual are covered by one or more Limited Use Label License(s). Please refer to the Appendix on page 971 and Master Product List on page 975. Products are For Research Use Only. Not intended for any animal or human therapeutic or diagnostic use.
www.invitrogen.com/probes
molecular probes® | invitrogen™ by life technologies™

(S870, A6373), rhodamine B (S871) and Texas Red® (A820, S872, A2665, S6370) dyes. However, we strongly recommend that researchers evaluate our many newer fluorescent conjugates:

- The green-fluorescent Alexa Fluor® 488 and Oregon Green® conjugates are not only brighter than fluorescein conjugates, but also much more photostable (Figure 7.6.10) and less pH sensitive (Section 1.3).
- The orange- and red-orange–fluorescent Alexa Fluor® 546, Alexa Fluor® 555, Alexa Fluor® 568 and Rhodamine Red™-X dyes, and the red-fluorescent Alexa Fluor® 594 and Texas Red®-X dyes are more fluorescent than traditional Lissamine rhodamine B and Texas Red® conjugates, yet have similar excitation and emission maxima.
- The far-red– and infrared-fluorescent Alexa Fluor® 633, Alexa Fluor® 635, Alexa Fluor® 647, Alexa Fluor® 660, Alexa Fluor® 680, Alexa Fluor® 700 and Alexa Fluor® 750 conjugates of streptavidin have fluorescence that is not visible to the eye, but they are efficiently excited by laser light sources and their fluorescence is easily detected with infrared light–sensitive detectors. Conjugates of the Alexa Fluor® 555 and Alexa Fluor® 647 dyes, in particular, have fluorescence that is superior to that of the spectrally similar Cy®3 and Cy®5 dyes,[45] respectively, and their conjugates are more photostable than Cy®3 and Cy®5 conjugates (Figure 7.6.11).
- The blue-fluorescent Alexa Fluor® 350 streptavidin (S11249) displays significantly more fluorescence than AMCA streptavidin (Figure 7.6.12) in side-by-side testing. The blue-fluorescent Alexa Fluor® 405 streptavidin (S32351), Pacific Blue™ streptavidin (S11222), yellow-fluorescent Cascade Yellow™ streptavidin (S11228) and orange-fluorescent Pacific Orange™ streptavidin (S32365) absorb maximally between 400 and 410 nm, making them near-perfect matches to the 405 nm violet laser used for both fluorescence microscopy and flow cytometry.
- R-phycoerythrin (R-PE) conjugates of streptavidin (SAPE; S866, S21388) and NeutrAvidin™ biotin-binding protein (A2660) and the B-phycoerythrin (B-PE) conjugate of streptavidin (S32350) are significantly brighter than our organic dye–conjugated avidins. Our streptavidin conjugates of R-PE and B-PE have been purified to help ensure that all unconjugated streptavidin has been removed (Figure 7.6.13), making them particularly important labels for multicolor flow cytometry (Section 6.4).

Furthermore, we have conjugated R-PE with four of our Alexa Fluor® dyes—the Alexa Fluor® 610, Alexa Fluor® 647, Alexa Fluor® 680 and Alexa Fluor® 750 dyes—and then conjugated these tandem labels to streptavidin to yield labeled conjugates that can be excited with the 488 nm spectral line of the argon-ion laser. The long-wavelength emission maxima are 628 nm for the Alexa Fluor® 610–R-PE conjugate (S20982), 668 nm for the Alexa Fluor® 647–R-PE conjugate (S20992), 702 nm for the Alexa Fluor® 680–R-PE conjugate (S20985) and 775 nm for the Alexa Fluor® 750–R-PE conjugate (S32363). Emission of the Alexa Fluor® 610–R-PE conjugates is shifted to longer wavelengths by about 13 nm relative to that of Texas Red® conjugates of R-PE

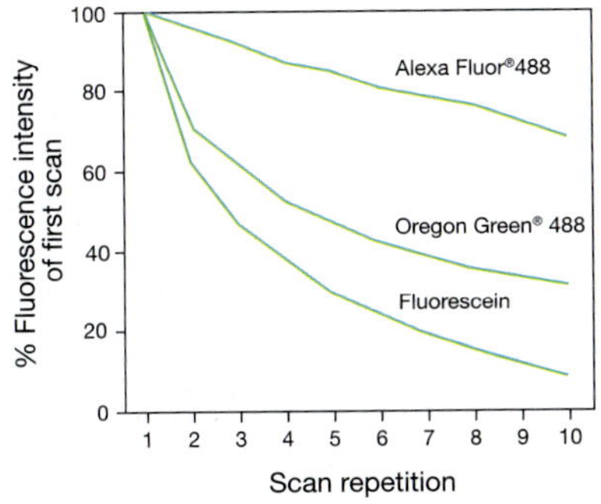

Figure 7.6.10 Photobleaching resistance of the green-fluorescent Alexa Fluor® 488, Oregon Green® 488 and fluorescein dyes, as determined by laser-scanning cytometry. EL4 cells were labeled with biotin-conjugated anti-CD44 antibody and detected by Alexa Fluor® 488 (S11223), Oregon Green® 488 (S6368) or fluorescein (S869) streptavidin. The cells were then fixed in 1% formaldehyde, washed and wet-mounted. After mounting, cells were scanned 10 times on a laser-scanning cytometer; laser power levels were 25 mW for the 488 nm spectral line of the argon-ion laser. Scan durations were approximately 5 minutes apiece, and each repetition was started immediately after completion of the previous scan. Data are expressed as percentages derived from the mean fluorescence intensity (MFI) of each scan divided by the MFI of the first scan. Data contributed by Bill Telford, Experimental Transplantation and Immunology Branch, National Cancer Institute.

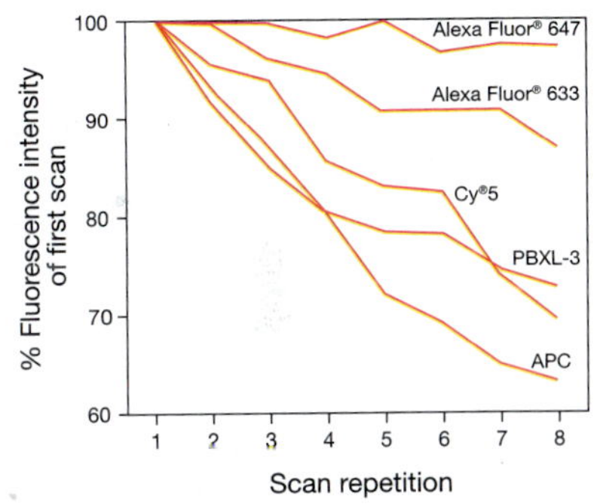

Figure 7.6.11 Photobleaching resistance of the red-fluorescent Alexa Fluor® 647, Alexa Fluor® 633, PBXL-3 and Cy®5 dyes and the allophycocyanin fluorescent protein, as determined by laser-scanning cytometry. EL4 cells were labeled with biotin-conjugated anti-CD44 antibody and detected by Alexa Fluor® 647 (S21374), Alexa Fluor® 633 (S21375), PBXL-3, Cy®5 or allophycocyanin (APC, S868) streptavidin. The cells were then fixed in 1% formaldehyde, washed and wet-mounted. After mounting, cells were scanned eight times on a laser-scanning cytometer; laser power levels were 18 mW for the 633 nm spectral line of the He-Ne laser. Scan durations were approximately 5 minutes apiece, and each repetition was started immediately after completion of the previous scan. Data are expressed as percentages derived from the mean fluorescence intensity (MFI) of each scan divided by the MFI of the first scan. Data contributed by Bill Telford, Experimental Transplantation and Immunology Branch, National Cancer Institute.

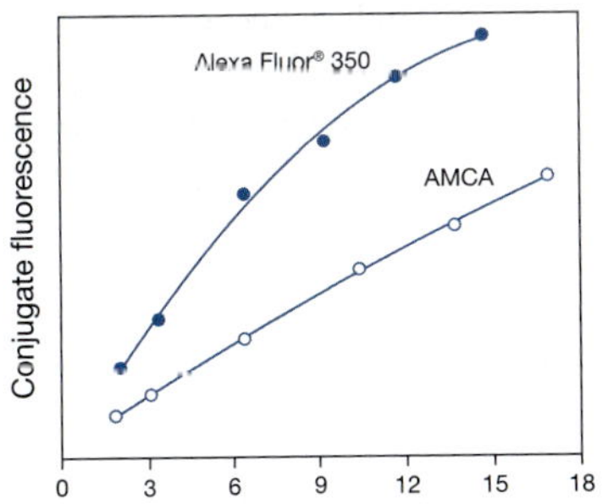

Figure 7.6.12 Comparison of the relative fluorescence of 7-amino-4-methylcoumarin-3-acetic acid (AMCA) streptavidin (○) and Alexa Fluor® 350 streptavidin, a sulfonated AMCA derivative (S11249, ●). Conjugate fluorescence is determined by measuring the fluorescence quantum yield of the conjugated dye relative to that of the free dye and multiplying by the number of fluorophores per protein.

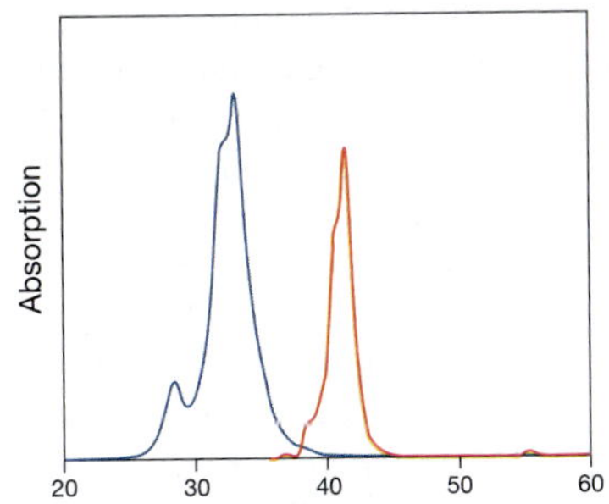

Figure 7.6.13 Analytical size-exclusion chromatograms of free streptavidin (S888; red curve, detected by absorption at 280 nm) and R-phycoerythrin streptavidin (SAPE; S866, S21388; blue curve, detected by absorption at 565 nm), demonstrating that the R-phycoerythrin conjugate is substantially free of unconjugated streptavidin.

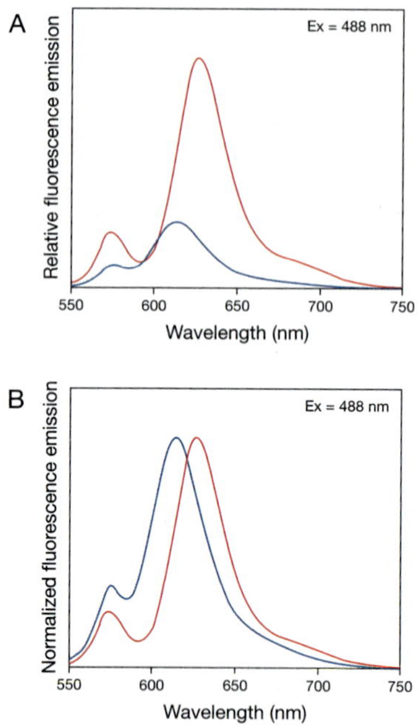

Figure 7.6.14 Fluorescence emission spectra of Alexa Fluor® 610–R-phycoerythrin streptavidin (S20982, red) and Texas Red® dye–R-phycoerythrin streptavidin (blue) tandem conjugates. **A**) Acomparison of the spectra on a relative fluorescence intensity scale for samples prepared with equal absorbance at the excitation wavelength (488 nm). **B**) The same data normalized to the same peak intensity value to facilitate comparison of the spectral profiles.

(Figure 7.6.14), significantly improving the resolution that can be obtained in multicolor flow cytometry. The Alexa Fluor® 647–R-PE tandem conjugates have spectra virtually identical to those of Cy®5 conjugates of R-PE but are about three times more fluorescent (Figure 7.6.15). We have also conjugated allophycocyanin (APC) with three of our Alexa Fluor® dyes—Alexa Fluor® 680, Alexa Fluor® 700 and Alexa Fluor® 750 dyes—and then conjugated these tandem labels to streptavidin (S21002, S21005, S21008). The resulting probes can all be excited by the He-Ne laser at 633 nm and have distinguishable emission spectra. A complete list of our current offerings of fluorophore-, enzyme- and gold-labeled avidins, streptavidins and NeutrAvidin™ biotin-binding proteins can be found in Table 7.9.

Streptavidin-, NeutrAvidin™- and Biotin-Labeled Fluorescent Microspheres

We offer streptavidin, NeutrAvidin™ and biotin conjugates of the intensely fluorescent FluoSpheres® and TransFluoSpheres® polystyrene microspheres in a variety of colors and sizes, including our europium and platinum luminescent beads labeled with the NeutrAvidin™ biotin-binding protein for time-resolved fluorometry (Table 14.8). Because single fluorescent microspheres can be detected, FluoSpheres® and TransFluoSpheres® beads have significant potential for ultrasensitive flow cytometry applications and immunoassays.[46,47] They may also be useful as tracers that can be detected with standard enzyme-mediated histochemical methods (Section 14.6).

BlockAid™ blocking solution (B10710) is a protein-based reagent designed principally for use with our streptavidin-, NeutrAvidin™- and biotin-labeled FluoSpheres® (Table 6.8) and our streptavidin- and NeutrAvidin™-labeled TransFluoSpheres® microspheres (Table 6.9). Protein- and other macromolecule-labeled microspheres have hydrophobic regions that may cause them to bind to nontarget surfaces in some experimental systems. Although this nonspecific binding can often be relieved by the use of a blocking solution, we have found that microspheres require a stronger blocking solution than those in common use. In our tests, the BlockAid™ blocking solution was mixed with streptavidin-labeled FluoSpheres® microspheres, which were then used to stain several different cell types for subsequent analysis by flow cytometry. We found the BlockAid™ blocking solution to be superior to other commercially available blocking solutions, as well as to several standard blocking solutions described in the scientific literature for reducing nonspecific binding of labeled microspheres. BlockAid™ blocking solution has been found to be effective in flow cytometry applications with the NIH 3T3, A431, RAW and Jurkat cell lines; however, with the HMC-1 cell line, it did not appear to offer any advantages over standard blocking solutions. We expect that the BlockAid™ blocking solution will be useful for reducing the nonspecific binding of protein-coated or other macromolecule-coated microspheres in a variety of flow cytometry and microscopy applications.[48–50]

Qdot® Streptavidin Conjugates

Qdot® streptavidin conjugates combine the highly specific binding properties of streptavidin with the exceptional photostability of Qdot® nanocrystals (Section 6.6). The large surface area afforded by the Qdot® nanocrystal allows simultaneous conjugation of multiple streptavidin molecules to a single fluorophore. Advantages conferred by this approach include increased avidity for targets, the potential for cooperative binding in some cases and the use of efficient signal amplification methodologies. For example, combining biotin-functionalized products with the streptavidin labels allows for successive enhancements in signal via "sandwiching" (streptavidin/biotin/streptavidin, etc.) following an initial labeling step.

These powerful fluorescence detection reagents offer unique performance advantages in a wide variety of tissue labeling and flow cytometry experiments; they are efficiently excited using the 405 nm violet laser, and the Qdot® nanocrystal fluorescence is extremely resistant to photobleaching. Not only can tissues stained with Qdot® nanocrystals be observed for hours, but these stained tissues can be archived permanently; re-analysis of properly stored archive samples remains as quantitative as it was during the initial assay.

Our selection of Qdot® streptavidin conjugates can all be excited by a single excitation source, enabling easy multicolor analysis of multiple targets or events in a single sample using color filtering to resolve the individual signals:

- Qdot® 525 streptavidin conjugate (Q10141MP)
- Qdot® 565 streptavidin conjugate (Q10131MP)
- Qdot® 585 streptavidin conjugate (Q10111MP)

IMPORTANT NOTICE: The products described in this manual are covered by one or more Limited Use Label License(s). Please refer to the Appendix on page 971 and Master Product List on page 975. Products are For Research Use Only. Not intended for any animal or human therapeutic or diagnostic use.
www.invitrogen.com/probes
molecular probes® | invitrogen™ by life technologies™

- Qdot® 605 streptavidin conjugate (Q10101MP)
- Qdot® 655 streptavidin conjugate (A10196)
- Qdot® 655 streptavidin conjugate (Q10121MP)
- Qdot® 705 streptavidin conjugate (Q10161MP)
- Qdot® 800 streptavidin conjugate (Q10171MP)
- Qdot® Streptavidin Sampler Kit (Q10151MP)

Nucleic acid hybridization can be detected by complexation of terminally biotinylated oligonucleotides with Qdot® streptavidin conjugates.[51–54] Qdot® streptavidin conjugates in combination with biotinylated antibodies provide increased multiplexing capacity for immunodetection of proteins on both imaging[55–59] and flow cytometry[60,61] platforms. Single-particle tracking of molecular motion is a particularly compelling application of Qdot® streptavidin conjugates that takes advantage of their extraordinary fluorescence capacity.[62] The principal subjects for these measurements are cell-surface receptors[62–68] and molecular motors such as myosin.[69,70] Single-molecule detection uncovers details of molecular motion and assembly that are obscured by the averaging effects of bulk measurements. Qdot® nanocrystals are quite essential in such applications, as fluorescence photon output ultimately propagates into the spatial and temporal resolution of the measurements.

Typically, labeling is accomplished using biotinylated ligands such as biotin EGF (E3477, Section 4.3) and biotin-XX α-bungartoxin (B1196, Section 4.3) to couple Qdot® streptavidin conjugates to the receptor of interest. Because these conjugates incorporate multiple streptavidins (typically 5–10) per Qdot® nanocrystal, careful attention must be paid to the streptavidin:ligand stoichiometry to avoid receptor crosslinking mediated by the Qdot® streptavidin–biotinyated ligand complex.[67,71] A molar excess of Qdot® streptavidin conjugate over the biotinylated ligand biases the complex stoichiometry towards the desired outcome of one ligand per nanocrystal. Qdot® streptavidin conjugates are supplied as 1 µM solutions in 50 mM borate, pH 8.3 containing 0.05% sodium azide and 1 M betaine cryoprotectant in 200 µL units. The Qdot® Streptavidin Sampler Kit (Q10151MP) provides 50 µL units of six different Qdot® streptavidin conjugates in a single package.

We also offer Qdot® ITK™ streptavidin quantum dots, in which streptavidin is directly attached to the inner amphiphilic coating without a PEG linker. This formulation results in a smaller overall particle size and an increase in the number of streptavidins per nanocrystal relative to PEG-linked conjugates. These characteristics are advantageous in fluorescence resonance energy transfer (FRET) assays for detecting hybridization of target nucleic acid sequences to biotinylated oligonucleotides captured on the Qdot® nanocrystal surface.[72–74] Qdot® ITK™ streptavidin conjugates are described in Section 6.6.

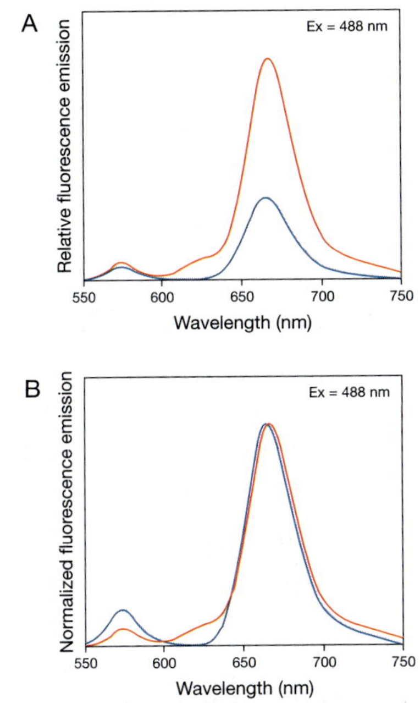

Figure 7.6.15 Fluorescence emission spectra of Alexa Fluor® 647–R-phycoerythrin streptavidin (S20992; red) and Cy®5–R-phycoerythrin streptavidin (blue) tandem conjugates. **A)** A comparison of the spectra on a relative fluorescence intensity scale for samples prepared with equal absorbance at the excitation wavelength (488 nm). **B)** The same data normalized to the same peak intensity value to facilitate comparison of the spectral profiles.

Enzyme Conjugates of Avidin and Streptavidin

Enzyme conjugates of avidin and streptavidin are extensively used in enzyme-linked immunosorbent assays (ELISAs),[75] blotting techniques,[76] *in situ* hybridization[77] and cytochemistry and histochemistry.[78] Enzyme-mediated *in situ* techniques using these conjugates provide better resolution and are safer, more sensitive and faster than radioactive methods. Most frequently, the enzymes of choice are horseradish peroxidase, alkaline phosphatase and *Escherichia coli* β-galactosidase because of their high turnover rate, stability, ease of conjugation and relatively low cost.

Our enzyme conjugates of streptavidin and NeutrAvidin™ biotin-binding protein are prepared by techniques that yield an approximate 1:1 ratio of enzyme to avidin analog, thus maximizing retention of both enzyme and carrier protein activity. We have prepared highly active streptavidin and NeutrAvidin™ biotin-binding protein conjugates of horseradish peroxidase[79] (S911, A2664), alkaline phosphatase (S921), β-galactosidase (S931) and β-lactamase TEM-1 (S31569). Fluorogenic, chromogenic and chemiluminescent substrates for these enzymes are described in Chapter 10. To decrease background problems, researchers often prefer to use the biotin-XX conjugate of horseradish peroxidase (P917) in conjunction with an avidin or streptavidin bridge for indirect detection of a wide array of biotinylated biomolecules.

A principal application of HRP and alkaline phosphatase conjugates of avidins and secondary antibodies is in enzyme-amplified histochemical staining of cells and tissues. Several of the Tyramide Signal Amplification (TSA™) Kits (Table 6.1) in Section 6.2 and Enzyme-Labeled Fluorescence (ELF®) Kits in Section 6.3 utilize enzyme conjugates of streptavidin to yield

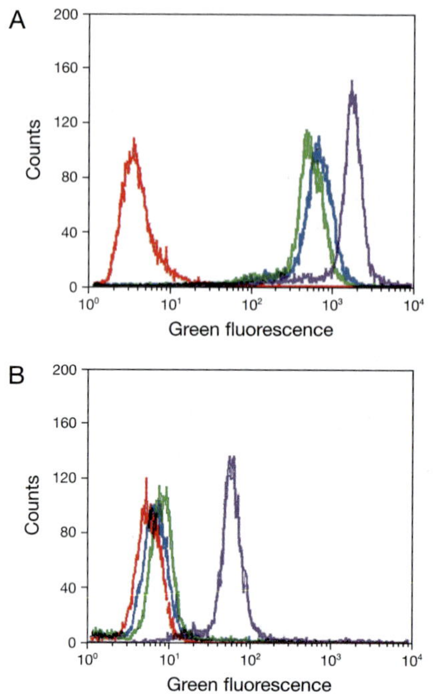

Figure 7.6.16 Detection of epidermal growth factor (EGF) receptors directly or with signal amplification. Cells expressing high (A431 cells, **A**) and low (NIH 3T3 cells, **B**) levels of EGF receptors were either directly labeled with the preformed Alexa Fluor® 488 complex of biotinylated epidermal growth factor (E13345, blue) or indirectly labeled with biotinylated EGF (E3477) followed by either Alexa Fluor® 488 streptavidin (S11223, green) or HRP-conjugated streptavidin and Alexa Fluor® 488 tyramide (purple), components of our TSA™ Kit #22 (T20932).

intensely fluorescent staining of cellular targets (Figure 7.6.5, Figure 7.6.16). These kits are very useful for immunofluorescence, *in situ* hybridization and flow cytometry. Use of a combination of the TSA™ and ELF® technologies or double application of TSA™ methods provides enhanced sensitivity for detection of low-abundance targets.[80]

Gold-Labeled Streptavidin

NANOGOLD® and Alexa Fluor® FluoroNanogold™ Streptavidin

In collaboration with Nanoprobes, Inc. (www.nanoprobes.com), we offer NANOGOLD® and Alexa Fluor® FluoroNanogold™ conjugates of streptavidin to facilitate immunoblotting, light microscopy and electron microscopy applications. NANOGOLD® conjugates are covalently conjugated to the 1.4 nm NANOGOLD® gold cluster label, whereas Alexa Fluor® FluoroNanogold™ conjugates are coupled to both a NANOGOLD® label and either the Alexa Fluor® 488 or Alexa Fluor® 594 fluorophore, resulting in gold clusters with green or red fluorescence, respectively. Alexa Fluor® FluoroNanogold™ streptavidin conjugates have all the advantages of the NANOGOLD® conjugates, with the additional benefit that they may be used for correlative fluorescence, light and electron microscopy[81,82] (Figure 7.6.17).

NANOGOLD® gold clusters have several advantages over colloidal gold. They develop better with silver than do most gold colloids and therefore provide higher sensitivity.[83] Silver enhancement, such as the system provided in the LI Silver Enhancement Kit (L24919), is described below. Additionally, NANOGOLD® particles do not have as high affinity for proteins as do gold colloids, thereby reducing background due to nonspecific binding. Several additional features of NANOGOLD® and Alexa Fluor® FluoroNanogold™ conjugates include:

- NANOGOLD® gold clusters are an extremely uniform (1.4 nm ± 10% diameter) and stable compound, not a gold colloid.
- NANOGOLD® gold clusters are smaller than a complete IgG (H+L) antibody—approximately 1/15 the size of an Fab fragment—and therefore will be able to better penetrate cells and tissues, reaching antigens that are inaccessible to conjugates of larger gold particles.
- NANOGOLD® conjugates contain absolutely no aggregates, as they are chromatographically purified through gel filtration columns. This feature is in sharp contrast to colloidal gold conjugates, which are usually prepared by centrifugation to remove the largest aggregates and frequently contain significantly smaller aggregates.
- The ratio of NANOGOLD® particle to Fab fragment is nearly 1:1, making this product distinct from the 0.2–10 variable stoichiometry of most colloidal gold–antibody preparations.

NANOGOLD® and Alexa Fluor® FluoroNanogold™ products can be used in immunoblotting, light microscopy and electron microscopy. Standard immunostaining methodologies can be used successfully with NANOGOLD® and Alexa Fluor® FluoroNanogold™ immunoreagents. Also, because the concentration of antibody and gold is similar to most commercial preparations of colloidal gold antibodies, similar dilutions and blocking agents are appropriate.

We offer several other NANOGOLD® and Alexa Fluor® FluoroNanogold™ reagents (Table 7.5), including the affinity-purified Fab fragments of the goat anti-mouse IgG and anti-rabbit IgG antibodies covalently conjugated to the 1.4 nm NANOGOLD® gold cluster label (Section 7.2). Also available is NANOGOLD® monomaleimide (N20345, Section 2.2), which can be conjugated to thiols in the same way that dyes are conjugated to proteins and nucleic acids.[84]

Colloidal Gold Complexes

We offer Alexa Fluor® 488 dye–labeled colloidal gold conjugates, including those of goat anti-mouse IgG (A31560, A31561; Section 7.2) and goat anti-rabbit IgG antibodies (A31565, A31566; Section 7.2) and streptavidin (A32360, A32361; Table 7.9). These conjugates, which have been adsorbed to 5 nm or 10 nm gold colloids, may be used as probes in immunoblotting, light microscopy, fluorescence microscopy or electron microscopy. The fluorescence of these conjugates can be easily detected by standard techniques, but visualization of colloidal gold can be greatly improved using silver-enhancement methods, such as those we provide in the LI Silver Enhancement Kit (L24919) described in Section 7.2.

IMPORTANT NOTICE: The products described in this manual are covered by one or more Limited Use Label License(s). Please refer to the Appendix on page 971 and Master Product List on page 975. Products are For Research Use Only. Not intended for any animal or human therapeutic or diagnostic use.
www.invitrogen.com/probes

Combining fluorescent secondary detection reagents with colloidal gold to form functional complexes is difficult because the fluorescence of fluorophores such as fluorescein is significantly quenched by proximity to the colloidal gold.[85] We prepare fluorescent colloidal gold complexes with our Alexa Fluor® 488 dye, a dye that has superior brightness and photostability. Our Alexa Fluor® 488 dye–labeled colloidal gold complexes of anti-IgG antibody and of streptavidin can potentially be used to perform correlated immunofluorescence and electron microscopy in a two-step labeling procedure, rather than in the three-step indirect labeling procedure that is required with conventional nonfluorescent colloidal gold complexes of anti-IgG antibodies or streptavidin.[86]

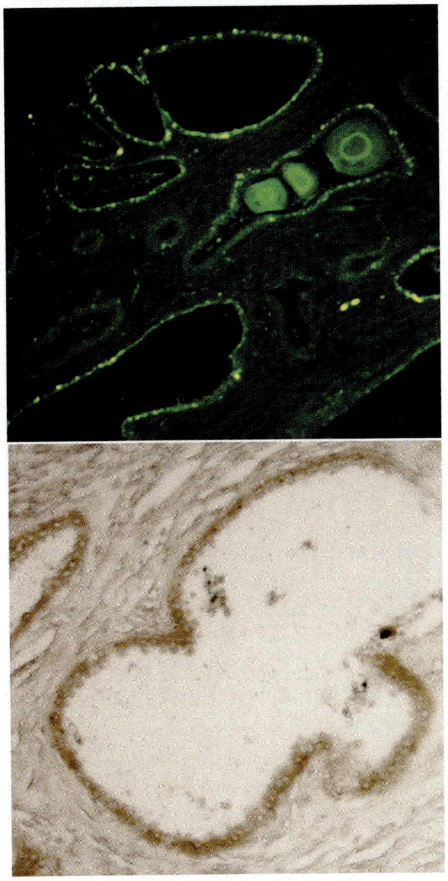

Figure 7.6.17 Human prostate tissue with adenocarcinoma imaged using an Alexa Fluor® 488 FluoroNanogold™ conjugate. Tissues were labeled with an anti-cytokeratin antibody, followed by staining with the Alexa Fluor® 488 FluoroNanogold™ Fab' fragment of goat anti-mouse IgG antibody (A24920). The fluorescence image in the top panel was taken with filters appropriate for fluorescein. The bottom panel shows a bright-field image of the same system, processed with the LI Silver (LIS) Enhancement Kit (L24919) to visualize the gold distribution. Image courtesy of Nanoprobes, Inc.

Affinity Chromatography

Streptavidin Agarose

We prepare streptavidin conjugated to 4% beaded crosslinked agarose (S951)—a matrix that can be used to isolate biotinylated peptides, proteins, hybridization probes, haptens and other molecules.[87] In addition, biotinylated antibodies can be bound to streptavidin agarose to generate an affinity matrix for the large-scale isolation of antigens.[87] For instance, staurosporine-treated myotubules have been incubated with biotinylated α-bungarotoxin (B1196, Section 16.2) in order to isolate the acetylcholine receptors (AChRs) on streptavidin agarose and assess staurosporine's effect on the degree of phosphorylation of this receptor.[88] Streptavidin agarose has also been used to investigate the turnover of cell-surface proteins that had previously been derivatized with an amine-reactive biotin[89] (B1582, Section 4.2).

CaptAvidin™ Agarose

CaptAvidin™ agarose (C21386) is another versatile form of a biotin-binding protein in that its affinity for biotinylated molecules can be completely reversed by raising the pH to 10, permitting the facile separation and isolation of biotin-labeled molecules from complex mixtures (Figure 7.6.2). This form of agarose-immobilized biotin-binding protein has been used to purify immunoglobulin from whole rabbit serum and to isolate anti-transferrin antibodies directly from rabbit antiserum.[26]

Acrylamide Conjugates for Immobilization of Avidins in Polymers

Streptavidin acrylamide (S21379), which is prepared from the succinimidyl ester of 6-((*N*-acryloyl)amino)hexanoic acid (acryloyl-X, SE; A20770, Section 5.2), is a reagent that may be useful for preparing biosensors.[90] A similar streptavidin acrylamide has been shown to copolymerize with acrylamide on a polymeric surface to create a uniform monolayer of the immobilized protein. The streptavidin can then bind biotinylated ligands, including biotinylated hybridization probes, enzymes, antibodies and drugs.

Anti-Biotin Antibody: An Alternative to Avidins

The high affinity of avidin for biotin was first exploited in histochemical applications in the early 1970s.[1,2] The use of avidin–biotin techniques has since become widely adopted for diverse detection schemes, although limitations of this method have also been recognized.[4,5] As an alternative to avidin-based reagents, we offer a high-affinity mouse monoclonal antibody to biotin (03-3700). This anti-biotin antibody can be used to detect biotinylated molecules in immunohistochemistry, *in situ* hybridization, ELISAs and flow cytometry applications.

It has been shown that certain monoclonal antibodies to biotin have biotin-binding motifs that are similar to those seen for avidin and streptavidin.[91] Anti-biotin antibody has been shown to selectively stain endogenous biotin-dependent carboxylase proteins used in fatty acid synthesis of the mitochondria.[23,36] Nonspecific staining of mitochondrial proteins by labeled avidins and by anti-biotin antibodies can be a complicating factor in using avidin–biotin techniques (Figure 7.6.1). This nonspecific binding can usually be blocked by pretreatment of the sample with the reagents in our Endogenous Biotin-Blocking Kit (E21390, Section 7.6).

IMPORTANT NOTICE: The products described in this manual are covered by one or more Limited Use Label License(s). Please refer to the Appendix on page 971 and Master Product List on page 975. Products are For Research Use Only. Not intended for any animal or human therapeutic or diagnostic use.

REFERENCES

1. Proc Natl Acad Sci U S A (1974) 71:3537; **2.** Biochim Biophys Acta (1972) 264:165; **3.** Clin Chem (1991) 37:625; **4.** Proteomics (2008) 8:4012; **5.** Nat Methods (2006) 3:267; **6.** Anal Biochem (1988) 171:1; **7.** J Cell Biol (1990) 111:1183; **8.** Physiol Plantarum (1990) 79:231; **9.** Cytometry (1990) 11:126; **10.** Proc Natl Acad Sci U S A (1990) 87:6223; **11.** Nat Protoc (2008) 3:563; **12.** J Immunol Methods (1990) 133:141; **13.** Biochemistry (1993) 32:8457; **14.** J Histochem Cytochem (1981) 29:1196; **15.** J Histochem Cytochem (1986) 34:1509; **16.** Adv Protein Chem (1975) 29:85; **17.** Proc Natl Acad Sci U S A (1993) 90:5076; **18.** J Histochem Cytochem (1985) 33:27; **19.** J Invest Dermatol (1984) 83:214; **20.** Biochem Biophys Res Commun (1990) 170:1236; **21.** Eur J Cell Biol (1992) 58:271; **22.** Eur J Cell Biol (1993) 60:1; **23.** J Histochem Cytochem (1997) 45:1053; **24.** Biochem J (1987) 248:167; **25.** Trends Genet (1993) 9:71; **26.** Anal Biochem (1996) 243:257; **27.** J Microbiol Methods (1990) 12:1; **28.** Biochemistry (1977) 16:5150; **29.** J Neurosci (1990) 10:3421; **30.** Brain Res (1989) 497:361; **31.** Anal Biochem (2002) 308:343; **32.** J Biol Chem (1990) 265:15776; **33.** Methods Mol Biol (2008) 418:111; **34.** Histochem J (2002) 34:567; **35.** Methods Mol Biol (2008) 418:157; **36.** Histochemistry (1993) 100:415; **37.** J Comp Neurol (2004) 473:86; **38.** Methods Mol Biol (2008) 418:209; **39.** Histochemistry (1986) 85:4; **40.** Proc Natl Acad Sci U S A (1986) 83:2934; **41.** Nat Med (2000) 6:707; **42.** J Immunol (1986) 137:1486; **43.** Methods Enzymol (1984) 108:197; **44.** J Immunol (1982) 129:532; **45.** J Histochem Cytochem (2003) 51:1699; **46.** Methods Mol Biol (2008) 418:187; **47.** J Immunol Methods (1998) 219:57; **48.** Mol Cell Probes (2008) 22:294; **49.** Proc Natl Acad Sci U S A (2005) 102:4225; **50.** Biophys J (2004) 86:1223; **51.** Nano Lett (2005) 5:1693; **52.** Nucleic Acids Res (2005) 33:e17; **53.** Nucleic Acids Res (2005) 33:e161; **54.** J Mol Diagn (2007) 9:20; **55.** Clin Cancer Res (2009) 15:3568; **56.** Mod Pathol (2006) 19:1181; **57.** J Histochem Cytochem (2009) 57:701; **58.** Nat Protoc (2007) 2:1152; **59.** Biochem Biophys Res Commun (2008) 374:181; **60.** Nat Med (2006) 12:972; **61.** Nat Methods (2009) 6:520; **62.** Nat Methods (2008) 5:687; **63.** Proc Natl Acad Sci U S A (2007) 104:13666; **64.** Mol Biol Cell (2006) 17:4937; **65.** Proc Natl Acad Sci U S A (2007) 104:11251; **66.** BMC Neurosci (2009) 10:80; **67.** J Cell Biol (2005) 170:619; **68.** Methods Mol Biol (2007) 374:69; **69.** Biophys J (2009) 97:509; **70.** Proc Natl Acad Sci U S A (2007) 104:4332; **71.** Nat Cell Biol (2008) 10:955; **72.** Mol Ther (2008) 16:324; **73.** Nat Mater (2005) 4:826; **74.** Mol Cell Probes (2007) 21:116; **75.** Antibodies: A Laboratory Manual, Harlow E, Lane D (1988) p.553; **76.** Short Protocols in Molecular Biology, 2nd Ed., Ausubel FM, et al., Eds. (1992); **77.** J Mol Diagn (2007) 9:604; **78.** J Histochem Cytochem (1979) 27:1131; **79.** Histochemistry (1986) 84:333; **80.** J Histochem Cytochem (2000) 48:1593; **81.** J Histochem Cytochem (2003) 51:707; **82.** Placenta (2003) 24:557; **83.** J Microsc (2000) 199:163; **84.** J Histochem Cytochem (2000) 48:471; **85.** Colloidal Gold: Principles. Methods and Applications, Vol. 1, Hayat MA, Ed. (1989) p.323; **86.** Proc Natl Acad Sci U S A (1985) 82:109; **87.** J Chromatogr (1990) 510:3; **88.** J Cell Biol (1994) 125:661; **89.** Biochemistry (1989) 28:574; **90.** Anal Biochem (2000) 282:200; **91.** FEBS Lett (1993) 322:47.

PRODUCT LIST 7.6 AVIDIN AND STREPTAVIDIN CONJUGATES

Cat. No.	Product	Quantity
A24926	Alexa Fluor® 488 FluoroNanogold™ streptavidin *80 µg protein/mL*	1 mL
A24927	Alexa Fluor® 594 FluoroNanogold™ streptavidin *80 µg protein/mL*	1 mL
A32360	Alexa Fluor® 488 streptavidin, 5 nm colloidal gold conjugate *30 µg protein/mL*	0.5 mL
A32361	Alexa Fluor® 488 streptavidin, 10 nm colloidal gold conjugate *30 µg protein/mL*	0.5 mL
A2667	avidin, egg white	5 mg
A887	avidin, egg white	100 mg
A21370	avidin, Alexa Fluor® 488 conjugate	1 mg
A821	avidin, fluorescein conjugate	5 mg
A820	avidin, Texas Red® conjugate	5 mg
A2666	avidin, NeutrAvidin™ biotin-binding protein	5 mg
A11236	avidin, NeutrAvidin™, Alexa Fluor® 350 conjugate	1 mg
A2663	avidin, NeutrAvidin™, Cascade Blue® conjugate	1 mg
A2662	avidin, NeutrAvidin™, fluorescein conjugate	1 mg
A2664	avidin, NeutrAvidin™, horseradish peroxidase conjugate	1 mg
A6374	avidin, NeutrAvidin™, Oregon Green® 488 conjugate	1 mg
A6378	avidin, NeutrAvidin™, Rhodamine Red™-X conjugate	1 mg
A2660	avidin, NeutrAvidin™, R-phycoerythrin conjugate *1 mg/mL*	1 mL
A6373	avidin, NeutrAvidin™, tetramethylrhodamine conjugate	1 mg
A2665	avidin, NeutrAvidin™, Texas Red® conjugate	1 mg
B1595	D-biotin	1 g
B20656	D-biotin *50 mM aqueous solution*	10 mL
B10710	BlockAid™ blocking solution *for use with microspheres*	50 mL
C21386	CaptAvidin™ agarose *sedimented bead suspension*	5 mL
C21385	CaptAvidin™ biotin-binding protein	1 mg
D20657	D-desthiobiotin *50 mM aqueous solution*	10 mL
D22187	Diaminobenzidine (DAB) Histochemistry Kit #3 *with streptavidin–HRP*	1 kit
E21390	Endogenous Biotin-Blocking Kit *100 assays*	1 kit
F8769	FluoSpheres® biotin-labeled microspheres, 1.0 µm, nonfluorescent *1% solids*	0.4 mL
F8766	FluoSpheres® biotin-labeled microspheres, 0.04 µm, yellow-green fluorescent (505/515) *1% solids*	0.4 mL
F8767	FluoSpheres® biotin-labeled microspheres, 0.2 µm, yellow-green fluorescent (505/515) *1% solids*	0.4 mL
F8768	FluoSpheres® biotin-labeled microspheres, 1.0 µm, yellow-green fluorescent (505/515) *1% solids*	0.4 mL
F20883	FluoSpheres® NeutrAvidin™ labeled microspheres, 0.04 µm, europium luminescent (365/610) *0.5% solids*	0.4 mL
F20884	FluoSpheres® NeutrAvidin™ labeled microspheres, 0.2 µm, europium luminescent (365/610) *0.5% solids*	0.4 mL

IMPORTANT NOTICE: The products described in this manual are covered by one or more Limited Use Label License(s). Please refer to the Appendix on page 971 and Master Product List on page 975. Products are For Research Use Only. Not intended for any animal or human therapeutic or diagnostic use.

PRODUCT LIST 7.6 AVIDIN AND STREPTAVIDIN CONJUGATES—*continued*

Cat. No.	Product	Quantity
F8772	FluoSpheres® NeutrAvidin™ labeled microspheres, 0.04 µm, nonfluorescent *1% solids*	0.4 mL
F8777	FluoSpheres® NeutrAvidin™ labeled microspheres, 1.0 µm, nonfluorescent *1% solids*	0.4 mL
F8770	FluoSpheres® NeutrAvidin™ labeled microspheres, 0.04 µm, red fluorescent (580/605) *1% solids*	0.4 mL
F8775	FluoSpheres® NeutrAvidin™ labeled microspheres, 1.0 µm, red fluorescent (580/605) *1% solids*	0.4 mL
F8771	FluoSpheres® NeutrAvidin™ labeled microspheres, 0.04 µm, yellow-green fluorescent (505/515) *1% solids*	0.4 mL
F8774	FluoSpheres® NeutrAvidin™ labeled microspheres, 0.2 µm, yellow-green fluorescent (505/515) *1% solids*	0.4 mL
F8776	FluoSpheres® NeutrAvidin™ labeled microspheres, 1.0 µm, yellow-green fluorescent (505/515) *1% solids*	0.4 mL
F8780	FluoSpheres® streptavidin-labeled microspheres, 0.04 µm, yellow-green fluorescent (505/515) *0.5% solids*	0.4 mL
I36933	Image-iT® FX signal enhancer	10 mL
L24919	LI Silver (LIS) Enhancement Kit	1 kit
N24918	NANOGOLD® streptavidin (streptavidin, NANOGOLD® conjugate) *80 µg protein/mL*	1 mL
P917	peroxidase from horseradish, biotin-XX conjugate	10 mg
Q10141MP	Qdot® 525 streptavidin conjugate *1 µM solution*	200 µL
Q10131MP	Qdot® 565 streptavidin conjugate *1 µM solution*	200 µL
Q10111MP	Qdot® 585 streptavidin conjugate *1 µM solution*	200 µL
Q10101MP	Qdot® 605 streptavidin conjugate *1 µM solution*	200 µL
A10196	Qdot® 625 streptavidin conjugate *1 µM solution*	200 µL
Q10121MP	Qdot® 655 streptavidin conjugate *1 µM solution*	200 µL
Q10161MP	Qdot® 705 streptavidin conjugate *1 µM solution*	200 µL
Q10171MP	Qdot® 800 streptavidin conjugate *1 µM solution*	200 µL
Q10151MP	Qdot® Streptavidin Sampler Kit *1 µM solutions*	1 kit
S888	streptavidin	5 mg
S21379	streptavidin acrylamide	1 mg
S951	streptavidin agarose *sedimented bead suspension*	5 mL
S11249	streptavidin, Alexa Fluor® 350 conjugate	1 mg
S32351	streptavidin, Alexa Fluor® 405 conjugate	1 mg
S11237	streptavidin, Alexa Fluor® 430 conjugate	1 mg
S11223	streptavidin, Alexa Fluor® 488 conjugate	1 mg
S32354	streptavidin, Alexa Fluor® 488 conjugate *2 mg/mL*	0.5 mL
S32353	streptavidin, Alexa Fluor® 514 conjugate	1 mg
S11224	streptavidin, Alexa Fluor® 532 conjugate	1 mg
S11225	streptavidin, Alexa Fluor® 546 conjugate	1 mg
S21381	streptavidin, Alexa Fluor® 555 conjugate	1 mg
S32355	streptavidin, Alexa Fluor® 555 conjugate *2 mg/mL*	0.5 mL
S11226	streptavidin, Alexa Fluor® 568 conjugate	1 mg
S11227	streptavidin, Alexa Fluor® 594 conjugate	1 mg
S32356	streptavidin, Alexa Fluor® 594 conjugate *2 mg/mL*	0.5 mL
S20982	streptavidin, Alexa Fluor® 610–R-phycoerythrin conjugate (Alexa Fluor® 610–R-phycoerythrin streptavidin) *1 mg/mL*	100 µL
S21375	streptavidin, Alexa Fluor® 633 conjugate	1 mg
S32364	streptavidin, Alexa Fluor® 635 conjugate	1 mg
S21374	streptavidin, Alexa Fluor® 647 conjugate	1 mg
S32357	streptavidin, Alexa Fluor® 647 conjugate *2 mg/mL*	0.5 mL
S20992	streptavidin, Alexa Fluor® 647–R-phycoerythrin conjugate (Alexa Fluor® 647–R-phycoerythrin streptavidin) *1 mg/mL*	100 µL
S21377	streptavidin, Alexa Fluor® 660 conjugate	1 mg
S21378	streptavidin, Alexa Fluor® 680 conjugate	1 mg
S32358	streptavidin, Alexa Fluor® 680 conjugate *2 mg/mL*	0.5 mL
S21002	streptavidin, Alexa Fluor® 680–allophycocyanin conjugate (Alexa Fluor® 680–allophycocyanin streptavidin) *1 mg/mL*	100 µL
S20985	streptavidin, Alexa Fluor® 680–R-phycoerythrin conjugate (Alexa Fluor® 680–R-phycoerythrin streptavidin) *1 mg/mL*	100 µL
S21383	streptavidin, Alexa Fluor® 700 conjugate	1 mg
S21005	streptavidin, Alexa Fluor® 700–allophycocyanin conjugate (Alexa Fluor® 700–allophycocyanin streptavidin) *1 mg/mL*	100 µL
S21384	streptavidin, Alexa Fluor® 750 conjugate	1 mg
S21008	streptavidin, Alexa Fluor® 750–allophycocyanin conjugate (Alexa Fluor® 750–allophycocyanin streptavidin) *1 mg/mL*	100 µL
S32363	streptavidin, Alexa Fluor® 750–R-phycoerythrin conjugate (Alexa Fluor® 750–R-phycoerythrin streptavidin) *1 mg/mL*	100 µL
S921	streptavidin, alkaline phosphatase conjugate *2 mg/mL*	0.5 mL
S32362	streptavidin, allophycocyanin conjugate *premium grade* *1 mg/mL*	250 µL
S868	streptavidin, allophycocyanin, crosslinked, conjugate *1 mg/mL*	0.5 mL
S32350	streptavidin, B-phycoerythrin conjugate *1 mg/mL*	1 mL
S11228	streptavidin, Cascade Yellow™ conjugate	1 mg

continued on next page

IMPORTANT NOTICE: The products described in this manual are covered by one or more Limited Use Label License(s). Please refer to the Appendix on page 971 and Master Product List on page 975. Products are For Research Use Only. Not intended for any animal or human therapeutic or diagnostic use.

PRODUCT LIST 7.6 AVIDIN AND STREPTAVIDIN CONJUGATES—*continued*

Cat. No.	Product	Quantity
S869	streptavidin, fluorescein conjugate	1 mg
S931	streptavidin, β-galactosidase conjugate	1 mg
S911	streptavidin, horseradish peroxidase conjugate	1 mg
S31569	streptavidin, β-lactamase TEM-1 conjugate *0.5 mg net protein*	0.5 mg
S11221	streptavidin, Marina Blue® conjugate	1 mg
S6368	streptavidin, Oregon Green® 488 conjugate	1 mg
S6369	streptavidin, Oregon Green® 514 conjugate	1 mg
S11222	streptavidin, Pacific Blue™ conjugate	1 mg
S32365	streptavidin, Pacific Orange™ conjugate	1 mg
S871	streptavidin, rhodamine B conjugate	1 mg
S6366	streptavidin, Rhodamine Red™-X conjugate	1 mg
S866	streptavidin, R-phycoerythrin conjugate (SAPE) *1 mg/mL*	1 mL
S21388	streptavidin, R-phycoerythrin conjugate (SAPE) *premium grade* *1 mg/mL*	1 mL
S870	streptavidin, tetramethylrhodamine conjugate	1 mg
S872	streptavidin, Texas Red® conjugate	1 mg
S6370	streptavidin, Texas Red®-X conjugate	1 mg
T8860	TransFluoSpheres® NeutrAvidin™ labeled microspheres, 0.04 µm (488/605) *1% solids*	0.4 mL
T8861	TransFluoSpheres® NeutrAvidin™ labeled microspheres, 0.1 µm (488/605) *1% solids*	0.4 mL
T10711	TransFluoSpheres® streptavidin-labeled microspheres, 0.04 µm (488/645) *0.5% solids*	0.4 mL

7.7 Lectins and Other Carbohydrate-Binding Proteins

Cellular proteoglycans, glycoproteins and glycolipids may contain any of a wide variety of oligosaccharides. Although most abundant on the cell surface, oligosaccharide residues are sometimes also found covalently attached to constituents within the cell.[1] Often, specific oligosaccharides are associated with a certain cell type or organelle. Lectins and certain other carbohydrate-binding proteins that bind to specific configurations of sugar molecules can thus serve to identify cell types or cellular components, making them versatile primary detection reagents in histochemical applications and flow cytometry[2] (Figure 7.7.1, Figure 7.7.2, Figure 7.7.3). Fluorescent derivatives of carbohydrate-binding proteins and other reagents have been used to detect cell-surface and intracellular glycoconjugates by microscopy[3,4] and flow cytometry,[5] to localize glycoproteins in gels and on protein blots,[6–9] to precipitate glycoproteins from solution and to cause agglutination of specific cell types.[10,11] In addition, lectins are also useful markers of certain cancers because these cells often display altered surface glycoproteins.[12,13]

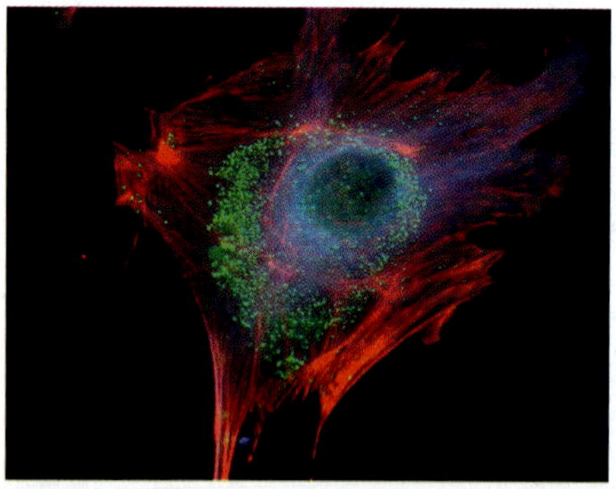

Figure 7.7.1 Microtubules of fixed bovine pulmonary artery endothelial cells localized with mouse monoclonal anti–α-tubulin antibody (A11126), which was subsequently visualized with Alexa Fluor® 350 goat anti–mouse IgG antibody (A11045). Next, the F-actin was labeled with Alexa Fluor® 594 phalloidin (A12381). Finally, the cells were incubated with Alexa Fluor® 488 wheat germ agglutinin (W11261) to stain components of endosomal pathways. The superimposed and pseudocolored images were acquired sequentially using bandpass filter sets appropriate for DAPI, the Texas Red® dye and fluorescein, respectively.

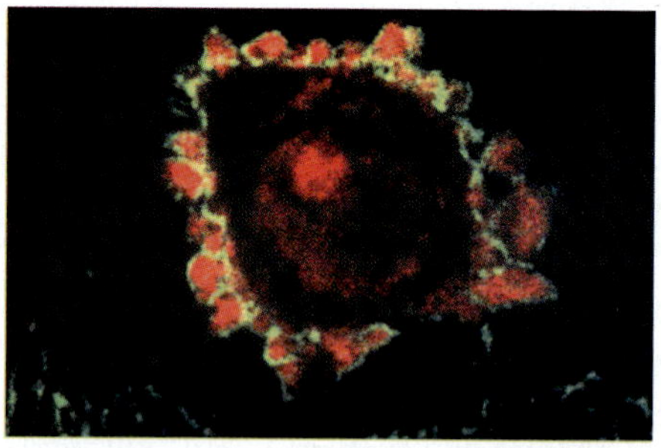

Figure 7.7.2 Apoptotic human keratinocytes that were fixed with formaldehyde and stained with fluorescein concanavalin A (C827). The cells were subsequently permeabilized with acetone, stained with propidium iodide (P1304MP, P3566, P21493) and visualized by confocal laser-scanning microscopy. This photomicrograph clearly shows that the green-fluorescent lectin staining outlines the apoptotic surface blebs, whereas the red-fluorescent propidium iodide stains the interior of the blebs. Image contributed by Livia Casciola-Rosen and Antony Rose, Johns Hopkins University. Reproduced from J Exp Med (1994) 179:1317 by copyright permission of the Rockefeller University Press.

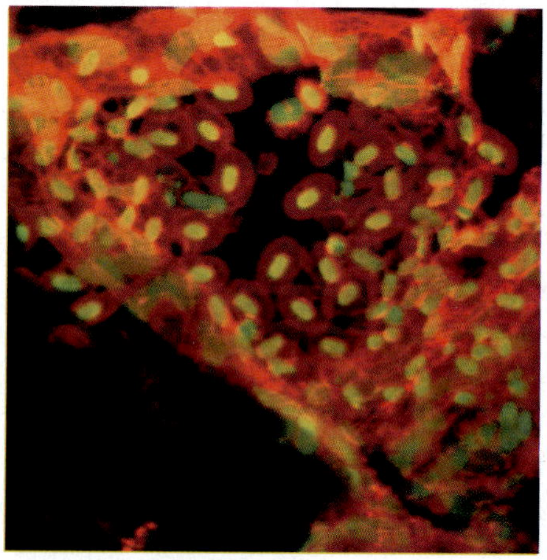

Figure 7.7.3 Zebrafish tissue section labeled with Texas Red®-X wheat germ agglutinin (W21405) and SYTOX® Green nucleic acid stain (S7020).

Concanavalin A and Wheat Germ Agglutinin and Their Conjugates

Properties of Concanavalin A and Wheat Germ Agglutinin

Concanavalin A (Con A) selectively binds to α-mannopyranosyl and α-glucopyranosyl residues. In moderately acidic solutions (pH 4.5–5.6), Con A exists as a dimer with a molecular weight of approximately 52,000 daltons; above pH 7, it is primarily a tetramer with a molecular weight of 104,000 daltons. Con A is a metalloprotein that requires one Ca^{2+} and one Mn^{2+} per subunit for carbohydrate binding.[14,15] The 36,000-dalton dimeric wheat germ agglutinin (WGA), which is normally cationic, binds to *N*-acetylneuraminic acid (sialic acid) and *N*-acetylglucosaminyl residues.

When Con A is succinylated with succinic anhydride, as in the Alexa Fluor® 488 conjugate of succinylated Con A (C21401), it is irreversibly converted to a dimer that retains the same sugar-binding specificity as the parent lectin. Succinylated Con A, however, has a profile of biological activities quite different from that of the tetrameric form. In contrast to the tetramer, succinylated Con A does not induce capping of cell-surface glycoprotein receptors, nor does it inhibit capping of cell-surface immunoglobulin receptors or strongly agglutinate erythrocytes or spleen cells. The mitogenic effect of succinylated Con A is similar to that of the native lectin, although it is mitogenic over a significantly wider range of concentrations than is the tetramer.[16,17]

Con A and WGA Conjugates

We offer several fluorescent conjugates of Con A and WGA—two of the most commonly used lectins in cell biology (Table 7.10). Although we continue to provide fluorescein and tetramethylrhodamine conjugates of these lectins, we strongly recommend that researchers also evaluate our other fluorescent conjugates, which exhibit excitation and emission maxima nearly identical to these traditional reagents but exhibit superior brightness and photostability. In particular, our green-fluorescent Alexa Fluor® 488 conjugates of Con A, succinylated Con A and WGA (C11252, C21401, W11261) are not only brighter than fluorescein conjugates, but also much more photostable and pH insensitive in the physiological range (Section 1.3). Moreover, the red-fluorescent Alexa Fluor® 594 conjugates of Con A and WGA (C11253, W11262) are even more fluorescent than their Texas Red® counterparts. The Alexa Fluor® 633 and Alexa Fluor® 647 conjugates of Con A (C21402, C21421) and WGA (W21404, W32466) and the Alexa Fluor® 680 conjugate of WGA (W32465) emit light at long wavelengths and are exceptionally useful for multicolor fluorescent labeling of cells and tissues, including those that have high intrinsic autofluorescence. For blue-fluorescent labeling, we recommend the Alexa Fluor® 350 conjugates of Con A and WGA (C11254, W11263; Figure 7.7.4). For researchers interested in testing our fluorescent WGA conjugates in their various applications, we offer a Wheat Germ Agglutinin Sampler Kit (W7024), which

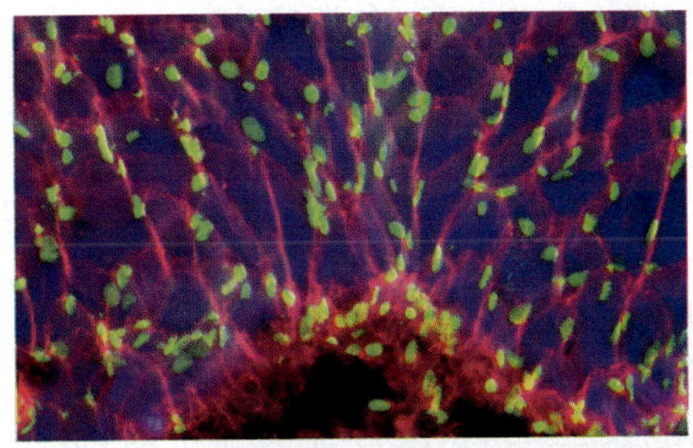

Figure 7.7.4 A zebrafish tissue section labeled with FRET 43 mouse monoclonal IgG in conjunction with Texas Red®-X goat anti-mouse IgG (T6390), Alexa Fluor® 350 wheat germ agglutinin (W11263) and SYTOX® Green nucleic acid stain (S7020). The multiple-exposure image was acquired using bandpass filter sets appropriate for DAPI, fluorescein and Texas Red® dye.

Table 7.10 Molecular Probes® lectin conjugates.

Label (Abs/Em) *	Lectin Conjugate †							
	Con A	WGA	IB_4	GS-II	PHA-L	HPA	SBA	PNA
Alexa Fluor® 350 (346/442)	C11254	W11263						
Fluorescein (494/518)	C827	W834						
Alexa Fluor® 488 (495/519)	C11252 ‡	W11261	I21411	L21415	L11270	L11271	L11272	L21409
Oregon Green® 488 (496/524)		W6748						
Alexa Fluor® 555 (555/565)		W32464						
Tetramethylrhodamine (555/580)	C860	W849						
Alexa Fluor® 568 (578/603)			I21412					L32458
Alexa Fluor® 594 (590/617)	C11253	W11262	I21413	L21416	L32456		L32462	L32459
Texas Red® (595/615)	C825							
Texas Red®-X (595/615)		W21405						
Alexa Fluor® 633 (632/647)	C21402	W21404						
Alexa Fluor® 647 (650/668)	C21421	W32466	I32450	L32451	L32457	L32454	L32463	L32460
Alexa Fluor® 680 (679/702)		W32465						
Qdot® 655 (405/655) §		Q12021MP						
Biotin-XX (NA)			I21414					

* Approximate absorption (Abs) and fluorescence emission (Em) maxima, in nm, for conjugates. † Lectin sources: Concanavalin A (Con A) isolated from *Canavalia ensiformis* (jack bean); Wheat Germ Agglutinin (WGA) isolated from *Triticum vulgaris*; IB_4 isolated from *Griffonia simplicifolia* (an African legume); GS-II isolated from *Griffonia simplicifolia*; PHA-L isolated from *Phaseolus vulgaris* (red kidney bean); HPA isolated from *Helix pomatia* (edible snail); SBA isolated from *Glycine max* (soybean); PNA isolated from *Arachis hypogaea* (peanut). ‡ Succinylated Alexa Fluor® 488 concanavalin A is also available, C21401. § The properties and applications of Qdot® nanocrystal conjugates are described in Section 6.6. NA = Not applicable.

For researchers interested in testing our fluorescent WGA conjugates in their application, we also offer a Wheat Germ Agglutinin Stain Sampler Kit (W7024), which contains 1 mg quantities each of WGA conjugates of the Alexa Fluor® 350, Oregon Green® 488, tetramethylrhodamine and Texas Red®-X dyes.

IMPORTANT NOTICE: The products described in this manual are covered by one or more Limited Use Label License(s). Please refer to the Appendix on page 971 and Master Product List on page 975. Products are For Research Use Only. Not intended for any animal or human therapeutic or diagnostic use.

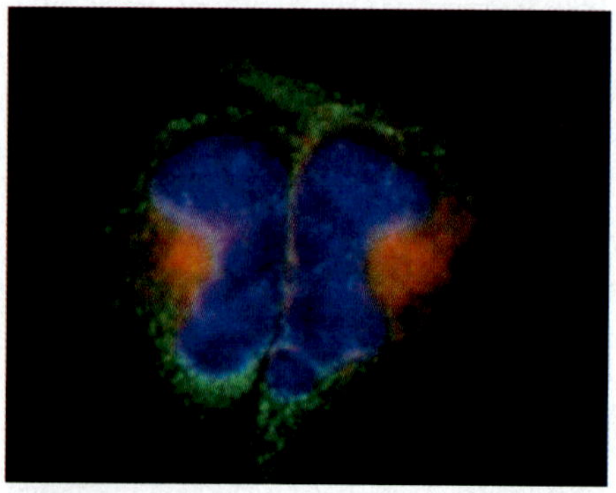

Figure 7.7.5 Fixed and permeabilized osteosarcoma cells simultaneously stained with the fluorescent lectins Alexa Fluor® 488 concanavalin A (Con A) (C11252) and Alexa Fluor® 594 wheat germ agglutinin (WGA) (W11262). Con A selectively binds α-glucopyranosyl residues, whereas WGA selectively binds sialic acid and *N*-acetylglucosaminyl residues. The nuclei were counterstained with blue-fluorescent Hoechst 33342 nucleic acid stain (H1399, H3570, H21492). The image was acquired using bandpass filter sets appropriate for the Texas Red® dye, fluorescein and DAPI.

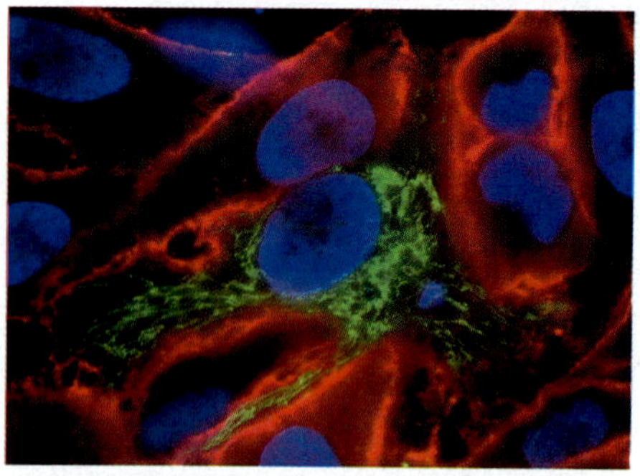

Figure 7.7.6 Live, highly confluent HeLa cells were transfected using pShooter™ vector pCMV/myc/mito/GFP and Lipofectamine® 2000 transfection reagent and stained with the reagents in the Image-iT® LIVE Plasma Membrane and Nuclear Labeling Kit (I34406). Plasma membranes were labeled with Alexa Fluor® 594 wheat germ agglutinin and nuclei were stained with Hoechst 33342. Cells were visualized using epifluorescence microscopy.

contains 1 mg quantities each of WGA conjugates of the Alexa Fluor® 350, Oregon Green® 488, tetramethylrhodamine and Texas Red®-X dyes.

The Qdot® 655 WGA conjugate (Q12021MP) extends our selection of dye-labeled WGA conjugates for labeling of *N*-acetylglucosaminyl and sialic acid residues of glycoproteins on cell surfaces;[18] the physical and spectroscopic properties of Qdot® nanocrystals are described in Section 6.6.

Applications for Fluorescent Con A and WGA

Although the distribution of oligosaccharides that may be bound by Con A and WGA varies widely among cell types, these two lectins have proven to be useful reagents for a number of applications, including immunolocalization of oncogene products,[19] specific intracellular enzymes,[20,21] viral proteins[22] and components of the cytoskeleton[23] (Figure 7.7.5). Con A also reportedly binds specifically to isolated Golgi fractions from rat liver, which enabled researchers to use fluorescein-labeled Con A to examine the effect of chronic ethanol intake on carbohydrate content in these organelles using flow cytometry.[24] Fluorescent Con A has been used to:

- Determine if human sperm cells have undergone the progesterone-induced acrosome reaction[25]
- Investigate receptor capping in leukocytes[26]
- Measure lateral diffusion of glycoproteins, glycolipids and viruses in membranes[27,28]
- Show the redistribution of cell-surface glycoproteins in murine fibroblasts that had been induced to migrate by exposure to an electric field[29]

WGA conjugates are useful for labeling of bacterial cell wall peptidoglycans, chitin and cartilage glycosaminoglycans. In addition, nuclear core complexes have been found to contain several proteins with *O*-linked *N*-acetylglucosaminyl residues.[30–34] In a study of nuclear protein transport, nuclei isolated from monkey kidney epithelial cells were demonstrated to be intact by their bright staining with fluorescein WGA (W834); fluorescein Con A (C827), which binds to residues accessible only in nuclei with compromised membranes, was used as a negative control for intact nuclei.[35] Fluorescent WGA has also been employed to monitor reconstitution of the nuclear core complex in *Xenopus* egg extracts.[36] WGA conjugates undergo axonal transport[37,38] and have been shown to cross from axonal nerve endings into adjacent neurons.[39]

Fluorescent lectins are also useful in microbiology applications. Fluorescent WGA conjugates stain chitin in fungal cell walls[40] and have been reported to stain gram-positive but not gram-negative bacteria for subsequent analysis by either imaging[41] or flow cytometry.[42] Fluorescent WGA conjugates are utilized in our ViaGram™ Red+ Bacterial Gram Stain and Viability Kit (V7023, Section 15.3) to differentiate gram-positive and gram-negative bacteria. Fluorescent WGA also binds to sheathed microfilariae and has been used to detect filarial infection in blood smears.[43]

In addition to these nuclear core and microbiology studies, fluorescent WGA has been used to:

- Bind the sarcolemma of rat and dog cardiac myocytes, even within the intercalated discs and transverse tubules, allowing researchers to map the distribution of gap junctions in these cell types[44]
- Determine the intracellular distribution of altered lysosomal proteins, enabling the definition of the sequence requirements for proper cell sorting[45]
- Identify the differentiation state of Madin–Darby canine kidney (MDCK) cells[46]
- Investigate plant hemicelluloses[47]
- Measure cell membrane potential in combination with potential-sensitive membrane probes[48]

Image-iT® LIVE Plasma Membrane and Nuclear Labeling Kit

The Image-iT® LIVE Plasma Membrane and Nuclear Labeling Kit (I34406) provides two stains—red-fluorescent Alexa Fluor® 594 wheat germ agglutinin (WGA) (excitation/emission maxima ~590/617 nm) and blue-fluorescent Hoechst 33342 dye (excitation/emission maxima when bound to DNA ~350/461 nm)—for highly selective staining of the plasma membrane and nucleus, respectively, of live cells expressing Green Fluorescent Protein (GFP, Figure 7.7.6). These dyes can be combined into one staining solution using the protocol provided, saving labeling time and wash steps while still providing optimal staining. Cell-impermeant Alexa Fluor® 594 WGA binds selectively to *N*-acetylglucosamine and *N*-acetylneuraminic acid residues.[49]

IMPORTANT NOTICE: The products described in this manual are covered by one or more Limited Use Label License(s). Please refer to the Appendix on page 971 and Master Product List on page 975. Products are For Research Use Only. Not intended for any animal or human therapeutic or diagnostic use.
www.invitrogen.com/probes
molecular probes® | invitrogen™ by life technologies™

When used according to the protocol, Alexa Fluor® 594 WGA provides highly selective labeling of the plasma membrane with minimal background, although labeling may not be as distinct for flat cell types when viewed using standard epifluorescence microscopy or low magnification. Alexa Fluor® 594 WGA is retained after formaldehyde fixation and permeabilization with Triton® X-100. This fluorescent lectin conjugate can also be used to label fixed cells; however, to avoid labeling intracellular components, formaldehyde-fixed cells should not be permeabilized before labeling. It is important to note that Alexa Fluor® 594 WGA can stimulate biological activity, including clustering of glycosylated cell-surface proteins.

The kit also includes Hoechst 33342 dye, a cell-permeant nucleic acid stain that is selective for DNA and is spectrally similar to DAPI. This dye should not interfere with GFP fluorescence and is retained after fixation and permeabilization. The Image-iT® LIVE Plasma Membrane and Nuclear Labeling Kit contains:

- Alexa Fluor® 594 wheat germ agglutinin (WGA)
- Hoechst 33342
- Detailed experimental protocols

Each kit provides enough staining solution for 360 assays using the protocol provided for labeling live, cultured cells that are adhering to coverslips.

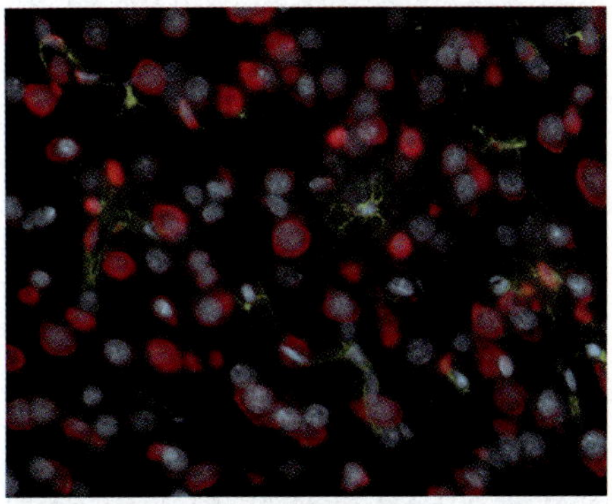

Figure 7.7.7 Capillary endothelial cells and microglia in a rat brain cryosection labeled with the green-fluorescent Alexa Fluor® 488 isolectin IB_4 (I21411). Neurons were identified using the NeuroTrace® 530/615 red-fluorescent Nissl stain (N21482) and nuclei were counterstained with blue-fluorescent DAPI (D1306, D3571, D21490). The digital image was obtained using filters appropriate for fluorescein, rhodamine and DAPI.

Other Carbohydrate-Binding Proteins and Their Conjugates

Lectins from *Griffonia simplicifolia*

Isolectin GS-IB_4 is a 114,000-dalton glycoprotein that is part of a family of five tetrameric type I isolectins (IA_4, IA_3B, IA_2B_2, IAB_3 and IB_4) isolated from the seeds of the tropical African legume *Griffonia simplicifolia* (formerly *Bandeiraea simplicifolia*). The A and B subunits are very similar, differing in amino acid sequence only at the N-terminus. The subunits, however, display remarkably different binding specificities. The A subunit prefers *N*-acetyl-D-galactosamine end groups, whereas the B subunit is selective for terminal α-D-galactosyl residues.[50]

We offer Alexa Fluor® 488, Alexa Fluor® 568, Alexa Fluor® 594 and Alexa Fluor® 647 dye conjugates of this versatile isolectin (I21411, I21412, I21413, I32450), as well as its biotinylated[51–53] (I21414) derivative. Isolectin GS-IB_4 specifically agglutinates blood group B erythrocytes and was originally employed for this purpose.[54] Subsequent work has shown that isolectin GS-IB_4 is cytotoxic to several normal and tumor cell types[55] and has particularly strong affinity for brain microglial and perivascular cells[56] (Figure 7.7.7). Conjugates of isolectin GS-IB_4 have been particularly valuable as histochemical and flow cytometric probes for specifically labeling endothelial cells in a number of species.[57,58] They have also been used effectively for tracing central and peripheral neuronal pathways following local injections.[59–61]

Lectin GS-II from *G. simplicifolia* differs from isolectin S-IB_4 in that it is the only known lectin that binds with high selectivity to terminal, nonreducing α- and β-*N*-acetyl-D-glucosaminyl (GlcNAc) residues of glycoproteins. Lectin GS-II is a tetrameric protein with each site binding a single carbohydrate and an aggregate molecular weight of 113,000 daltons. Because of its affinity for GlcNAc, lectin GS-II conjugates are useful for staining intermediate-to-trans Golgi—the site of *N*-acetylglucosaminyltransferase activity. The Golgi apparatus of oligodendrocytes and ganglion neurons are also stained by fluorescent GS-II conjugates. We have prepared the green-fluorescent Alexa Fluor® 488 (L21415, Figure 7.7.8), red-fluorescent Alexa Fluor® 594 (L21416) and far-red–fluorescent Alexa Fluor® 647 (L32451) conjugates of lectin GS-II for use in cell staining.

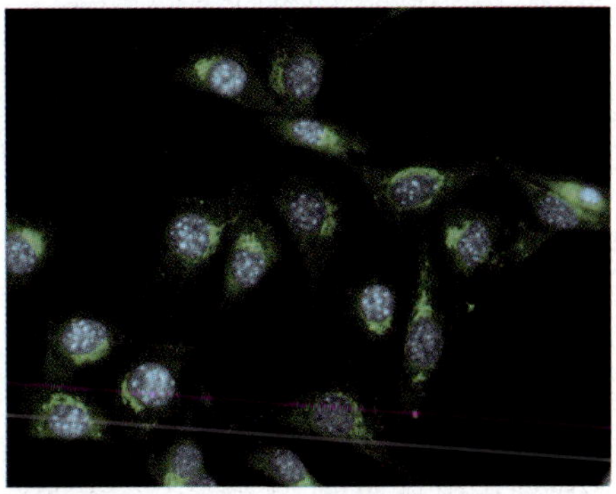

Figure 7.7.8 Fixed and permeabilized NIH 3T3 cells were labeled with the Alexa Fluor® 488 conjugate of lectin GS-II from *Griffonia simplicifolia* (L21415) and counterstained with DAPI (D1306, D3571, D21490). The fluorescent images are shown overlaid with the differential interference contrast (DIC) image.

Phaseolus vulgaris (Red Kidney Bean) Lectin

Phaseolus vulgaris lectin (PHA-L) is a tetrameric protein with a molecular weight of 120,000 daltons. Its binding to glycoproteins is strongly inhibited by *N*-acetylglucosaminyl (1-2) mannopyranosyl residues.[11] Like WGA, PHA-L is widely used as an injectable neuronal tracer.[62] Iontophoretically injected PHA-L clearly demonstrates the morphological features of the filled neurons at the injection site and the labeled axons and axon terminals. Furthermore, PHA-L that has been transported is not degraded over long periods. PHA-L has been used in combination with our biotinylated lysine-fixable dextrans (BDA dextrans; Section 14.5, Table 14.4) and the fluorescent dextrans fluoro-ruby and mini-ruby (D1817, D3312; Section 14.5) to demonstrate

Figure 7.7.9 Capillaries in the hippocampal region of a mouse brain cryosection were visualized with the green-fluorescent Alexa Fluor® 488 conjugate of lectin HPA from *Helix pomatia* (L11271), which specifically binds to type A erythrocytes and α-*N*-acetylgalactosaminyl residues. The nuclei were counterstained with nuclear yellow (N21485). The multiple-exposure image was acquired using a DAPI longpass filter set and a filter set appropriate for fluorescein.

their similar transport properties and to show the distribution of anterograde-labeled fibers.[63] We have prepared Alexa Fluor® 488, Alexa Fluor® 594 and Alexa Fluor® 647 conjugates of PHA-L (L11270, L32456, L32457).

Arachis hypogaea (Peanut) Lectin

Arachis hypogaea lectin (PNA) is a tetrameric protein with a molecular weight of ~110,000 daltons that has specificity for terminal β-galactose residues of glycoproteins. Lactose strongly inhibits binding of PNA to these glycoproteins, with D-galactose somewhat less effective. PNA binds to a broad range of receptors in human tissues, with some preference for glycoproteins that have been treated with neuraminidase (sialidase) to remove terminal sialic acids.[64] We offer Alexa Fluor® 488, Alexa Fluor® 568, Alexa Fluor® 594 and Alexa Fluor® 647 conjugates of PNA (L21409, L32458, L32459, L32460).

PNA-binding sites are widespread in human tissues, with staining patterns varying by tissue type.[65] Research has shown PNA to be selective for acrosomes in rat and human sperm,[66,67] and PNA serves as a marker for certain melanomas.[13,68] PNA has also been used to label the synaptic extracellular matrix in the study of developing neuromuscular junctions.[69]

Applications for fluorescent PNA conjugates include:

- Binding to desialylated CD44, the PNA receptor in keratinocytes, as a marker of terminal differentiation[70]
- Detecting the acrosome reaction of sperm[71]
- Labeling of photoreceptors in chick embryos and in intact retina and dissociated retinal cells[72]
- Staining colonic mucins in cultured human tumor cells, transitional cell carcinomas in the bladder and leukemic cells[73,74]

Helix pomatia (Edible Snail) Agglutinin

Helix pomatia agglutinin (HPA), a hexameric protein with a molecular weight of 70,000 daltons, selectively binds to type A erythrocytes and α-*N*-acetylgalactosaminyl residues[11] (Figure 7.7.9). In some cell types, HPA conjugates may be used as markers for the Golgi complex.[75,76] Furthermore, HPA conjugates have been shown to be as effective as monoclonal antibodies for the detection and differentiation of herpes simplex virus type 1 and type 2 (HSV-1 and HSV-2) in cultured cells.[75] Our Alexa Fluor® 488 and Alexa Fluor® 647 conjugates of HPA (L11271, L32454) should be particularly useful for cell staining.[77]

Glycine max (Soybean) Agglutinin

Soybean agglutinin (SBA) from *Glycine max* is a tetrameric protein that consists of a mixture of isolectins that have an aggregate molecular weight of ~120,000 daltons.[11] SBA has a stronger reaction with type A1 blood cells than with type A2 blood cells and is known to selectively bind to terminal α- and β-*N*-acetylgalactosaminyl and galactopyranosyl residues of glycoproteins. Neuraminidase-treated cells react more strongly with SBA conjugates than do untreated cells.[67] SBA has selective affinity for human $CD34^+$ hematopoietic stem cells and lymphocytes, and its immobilized conjugates are important reagents for the depletion of T cells in bone marrow transplantation.[78] In addition, a fluorescein conjugate of SBA was shown to be useful for assessing the stage of lymphoid cell differentiation in human leukemic–lymphoma cell lines.[79] Like many lectins, fluorescent SBA conjugates have been shown to bind with high affinity to several types of tumor cells.[80,81] SBA conjugates are also reported to be selective stains for glial cells[82] and the zona pellucida of the mammalian egg.[67] For detecting SBA-binding cells, we have prepared the Alexa Fluor® 488, Alexa Fluor® 594 and Alexa Fluor® 647 conjugates of SBA (L11272, L32462, L32463).

Cholera Toxin Subunit B

Cholera toxin comprises two subunits, A and B, arranged in an AB_5 conformation. Subunit A is an ADP-ribosyltransferase, which disrupts the proper signaling of G proteins and eventually leads to dehydration of the cell.[83] The 12,000-dalton nontoxic subunit B ("choleragenoid"), which is assembled into a 60,000-dalton pentamer above pH 2,[84] allows the protein complex to bind to cellular surfaces via the pentasaccharide chain of ganglioside G_{M1}.[85]

IMPORTANT NOTICE: The products described in this manual are covered by one or more Limited Use Label License(s). Please refer to the Appendix on page 971 and Master Product List on page 975. Products are For Research Use Only. Not intended for any animal or human therapeutic or diagnostic use.

Fluorescent cholera toxin subunit B, which binds to galactosyl moieties, is a marker of lipid rafts—regions of cell membranes high in ganglioside G_{M1} that are thought to be important in cell signaling.[86,87] Lipid rafts are detergent-insoluble, sphingolipid- and cholesterol-rich membrane microdomains that form lateral assemblies in the plasma membrane.[88–94] Lipid rafts also sequester glycophosphatidylinositol (GPI)-linked proteins and other signaling proteins and receptors, which may be regulated by their selective interactions with these membrane microdomains.[95–100] Research has demonstrated that lipid rafts play a role in a variety of cellular processes—including the compartmentalization of cell-signaling events,[101–108] the regulation of apoptosis[109–111] and the intracellular trafficking of certain membrane proteins and lipids[112–114]—as well as in the infectious cycles of several viruses and bacterial pathogens.[115–120]

The Vybrant® Lipid Raft Labeling Kits (V34403, V34404, V34405; Section 14.4) provide the key reagents for fluorescently labeling lipid rafts *in vivo* with our bright and extremely photostable Alexa Fluor® dyes (Figure 7.7.10). Live cells are first labeled with the green-fluorescent Alexa Fluor® 488, orange-fluorescent Alexa Fluor® 555 or red-fluorescent Alexa Fluor® 594 conjugate of cholera toxin subunit B (CT-B). This CT-B conjugate binds to the pentasaccharide chain of plasma membrane ganglioside G_{M1}, which selectively partitions into lipid rafts.[84,99] An antibody that specifically recognizes CT-B is then used to crosslink the CT-B–labeled lipid rafts into distinct patches on the plasma membrane, which are easily visualized by fluorescence microscopy.[121,122] Each Vybrant® Lipid Raft Labeling Kit contains sufficient reagents to label 50 live-cell samples, including:

- Recombinant cholera toxin subunit B (CT-B) labeled with the Alexa Fluor® 488 (in Kit V34403), Alexa Fluor® 555 (in Kit V34404) or Alexa Fluor® 594 (in Kit V34405) dye
- Anti–cholera toxin subunit B antibody (anti–CT-B)
- Concentrated phosphate-buffered saline (PBS)
- Detailed labeling protocol

Cholera toxin subunit B and its conjugates are also established as superior tracers for retrograde labeling of neurons.[123,124] Cholera toxin subunit B conjugates bind to the pentasaccharide chain of ganglioside G_{M1} on neuronal cell surfaces and are actively taken up and transported; alternatively, they can be injected by iontophoresis. Unlike the carbocyanine-based neuronal tracers such as DiI (D282, D3911, V22885; Section 14.4), cholera toxin subunit B conjugates can be used on tissue sections that will be fixed and frozen.[125]

All of our cholera toxin subunit B conjugates are prepared from recombinant cholera toxin subunit B, which is completely free of the toxic subunit A, thus eliminating any concern for toxicity or ADP-ribosylating activity. The Alexa Fluor® 488 (C22841, C34775), Alexa Fluor® 555 (C22843, C34776), Alexa Fluor® 594 (C22842, C34777) and Alexa Fluor® 647 (C34778) conjugates of cholera toxin subunit B combine this versatile tracer with the superior brightness of our Alexa Fluor® dyes to provide sensitive and selective receptor labeling and neuronal tracing. We also offer biotin-XX (C34779) and horseradish peroxidase (C34780) conjugates of cholera toxin subunit B for use in combination with diaminobenzidine (DAB) oxidation,[126] tyramide signal amplification (TSA™) and Qdot® nanocrystal–streptavidin conjugates.[127]

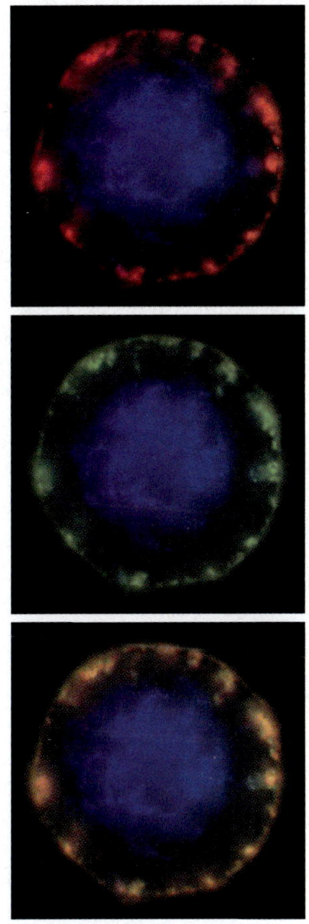

Figure 7.7.10 A J774 mouse macrophage cell sequentially stained with BODIPY® FL ganglioside G_{M1} (B13950) and then with Alexa Fluor® 555 dye–labeled cholera toxin subunit B (C22843, C34776; also available as a component of V34404). The cell was then treated with an anti–CT-B antibody (a component of V34404) to induce crosslinking. Alexa Fluor® 555 dye fluorescence (top, red) and BODIPY® FL dye fluorescence (middle, green) were imaged separately and overlaid to emphasize the coincident staining (bottom, yellow). Nuclei were stained with blue-fluorescent Hoechst 33342 (H1399, H3570, H21492).

REFERENCES

1. J Biol Chem (2000) 275:29179; **2.** Science (1989) 246:227; **3.** J Parasitol (1990) 76:130; **4.** Histochemistry (1978) 56:265; **5.** Cytometry (1995) 19:112; **6.** Electrophoresis (2003) 24:588; **7.** The Protein Protocols Handbook, 2nd Ed., Walker JM, Ed. (2002) p. 761; **8.** Proteomics (2001) 1:841; **9.** Anal Biochem (1979) 96:208; **10.** Mol Biochem Parasitol (1987) 23:165; **11.** Adv Immunol (1983) 34:213; **12.** J Histochem Cytochem (1998) 46:793; **13.** Hum Pathol (1999) 30:556; **14.** Biochemistry (1964) 4:876; **15.** J Biol Chem (1936) 115:583; **16.** Biochemistry (1993) 32:5116; **17.** Proc Natl Acad Sci U S A (1973) 70:1012; **18.** J Neurosci (2010) 30:4895; **19.** J Cell Biol (1990) 111:3097; **20.** J Biol Chem (1994) 269:1727; **21.** J Cell Biol (1990) 111:2851; **22.** J Cell Biol (1990) 110:625; **23.** J Cell Biol (1990) 111:1929; **24.** Exp Cell Res (1993) 207:136; **25.** Mol Cell Endocrinol (1994) 101:221; **26.** J Virol Methods (1990) 29:257; **27.** FEBS Lett (1989) 246:65; **28.** Cell (1981) 23:423; **29.** J Cell Biol (1994) 127:117; **30.** Annu Rev Cell Biol (1992) 8:495; **31.** Science (1992) 258:942; **32.** Biochim Biophys Acta (1991) 1071:83; **33.** Cell (1991) 64:489; **34.** Physiol Rev (1991) 71:909; **35.** J Biol Chem (1994) 269:4910; **36.** J Biol Chem (1994) 269:9289; **37.** J Neurosci Methods (1983) 9:185; **38.** J Neurosci (1982) 2:647; **39.** Brain Res (1985) 344:41; **40.** Invest Ophthalmol Vis Sci (1986) 27:500; **41.** Appl Environ Microbiol (1990) 56:2245; **42.** Appl Environ Microbiol (2003) 69:2857; **43.** Int J Parasitol (1990) 20:1099; **44.** J Mol Cell Cardiol (1992) 24:1443; **45.** J Cell Biol (1990) 111:955; **46.** Cell Physiol Biochem (1993) 3:42; **47.** Protoplasma (1990) 156:67;

continued on next page

IMPORTANT NOTICE: The products described in this manual are covered by one or more Limited Use Label License(s). Please refer to the Appendix on page 971 and Master Product List on page 975. Products are For Research Use Only. Not intended for any animal or human therapeutic or diagnostic use.

REFERENCES—*continued*

48. Biophys J (1995) 69:1272; **49**. J Mol Biol (1984) 178:91; **50**. J Biol Chem (1977) 252:4739; **51**. J Immunol (2000) 164:5446; **52**. Somatosens Mot Res (1997) 14:17; **53**. Histochem Cell Biol (1996) 106:331; **54**. Subcell Biochem (1999) 32:127; **55**. Cancer Res (1982) 42:2977; **56**. Histochemistry (1994) 102:483; **57**. Histochem J (1987) 19:225; **58**. Am J Pathol (1989) 134:1227; **59**. Biol Bull (1999) 197:115; **60**. Neurosci Lett (1997) 222:53; **61**. Brain Res (1998) 811:34; **62**. Cell Vision (1995) 2:184; **63**. Histochem Cell Biol (1997) 107:391; **64**. J Biol Chem (1975) 250:8518; **65**. Hum Pathol (1984) 15:904; **66**. Mol Reprod Dev (2000) 55:289; **67**. Histochem J (1997) 29:583; **68**. Oncol Res (1993) 5:235; **69**. J Neurosci (1994) 14:796; **70**. J Cell Sci (1995) 108:1959; **71**. J Androl (1998) 19:542; **72**. Invest Ophthalmol Vis Sci (1984) 25:546; **73**. Cancer (1984) 53:272; **74**. Immunology (1983) 49:147; **75**. Proc Natl Acad Sci U S A (1993) 90:2798; **76**. Histochemistry (1990) 94:397; **77**. Curr Biol (2001) 11:1847; **78**. Mol Biotechnol (1999) 11:181; **79**. Leuk Res (1987) 11:589; **80**. Surg Today (1997) 27:293; **81**. Cancer Res (1992) 52:5235; **82**. J Neurocytol (1992) 21:211; **83**. J Biol Chem (1980) 255:1252; **84**. Biochemistry (1996) 35:16069; **85**. Mol Microbiol (1994) 13:745; **86**. Science (2010) 327:46; **87**. FEMS Microbiol Lett (2007) 266:129; **88**. J Cell Biol (2003) 162:365; **89**. J Lipid Res (2003) 44:655; **90**. Eur J Biochem (2002) 269:737; **91**. Science (2000) 290:1721; **92**. Mol Membr Biol (1999) 16:145; **93**. Trends Cell Biol (1999) 9:87; **94**. Annu Rev Cell Dev Biol (1998) 14:111; **95**. Proc Natl Acad Sci U S A (2003) 100:5813; **96**. J Immunol (2003) 170:1329; **97**. J Membr Biol (2002) 189:35; **98**. Proc Natl Acad Sci U S A (2001) 98:9098; **99**. J Cell Biol (1999) 147:447; **100**. Mol Biol Cell (1999) 10:3187; **101**. Biochim Biophys Acta (2003) 1610:247; **102**. Annu Rev Immunol (2003) 21:457; **103**. Mol Immunol (2002) 38:1247; **104**. Nat Rev Immunol (2002) 2:96; **105**. Biol Res (2002) 35:127; **106**. Nat Rev Mol Cell Biol (2000) 1:31; **107**. J Exp Med (1999) 190:1549; **108**. J Cell Biol (1998) 143:637; **109**. Immunity (2003) 18:655; **110**. J Biol Chem (2002) 277:39541; **111**. Biochem Biophys Res Commun (2002) 297:876; **112**. Biol Chem (2002) 383:1475; **113**. J Cell Biol (2001) 153:529; **114**. J Cell Sci (2001) 114:3957; **115**. J Virol (2003) 77:9542; **116**. Exp Cell Res (2003) 287:67; **117**. Traffic (2002) 3:705; **118**. J Clin Virol (2001) 22:217; **119**. Curr Biol (2000) 10:R823; **120**. J Virol (2000) 74:3264; **121**. J Cell Biol (1998) 141:929; **122**. J Biol Chem (1994) 269:30745; **123**. Nat Protoc (2009) 4:1157; **124**. Brain Struct Funct (2009) 213:367; **125**. J Neurosci (2002) 22:9419; **126**. J Neurosci Methods (2005) 149:101; **127**. BMC Neurosci (2009) 10:80.

PRODUCT LIST 7.7 LECTINS AND OTHER CARBOHYDRATE-BINDING PROTEINS

Cat. No.	Product	Quantity
C34775	cholera toxin subunit B (recombinant), Alexa Fluor® 488 conjugate	100 µg
C22841	cholera toxin subunit B (recombinant), Alexa Fluor® 488 conjugate	500 µg
C34776	cholera toxin subunit B (recombinant), Alexa Fluor® 555 conjugate	100 µg
C22843	cholera toxin subunit B (recombinant), Alexa Fluor® 555 conjugate	500 µg
C34777	cholera toxin subunit B (recombinant), Alexa Fluor® 594 conjugate	100 µg
C22842	cholera toxin subunit B (recombinant), Alexa Fluor® 594 conjugate	500 µg
C34778	cholera toxin subunit B (recombinant), Alexa Fluor® 647 conjugate	100 µg
C34779	cholera toxin subunit B (recombinant), biotin-XX conjugate	100 µg
C34780	cholera toxin subunit B (recombinant), horseradish peroxidase conjugate	100 µg
C11254	concanavalin A, Alexa Fluor® 350 conjugate	5 mg
C11252	concanavalin A, Alexa Fluor® 488 conjugate	5 mg
C11253	concanavalin A, Alexa Fluor® 594 conjugate	5 mg
C21402	concanavalin A, Alexa Fluor® 633 conjugate	5 mg
C21421	concanavalin A, Alexa Fluor® 647 conjugate	5 mg
C827	concanavalin A, fluorescein conjugate	10 mg
C21401	concanavalin A, succinylated, Alexa Fluor® 488 conjugate	5 mg
C860	concanavalin A, tetramethylrhodamine conjugate	10 mg
C825	concanavalin A, Texas Red® conjugate	10 mg
I34406	Image-iT® LIVE Plasma Membrane and Nuclear Labeling Kit *counterstains for GFP-expressing cells*	1 kit
I21411	isolectin GS-IB4 from *Griffonia simplicifolia*, Alexa Fluor® 488 conjugate	500 µg
I21412	isolectin GS-IB4 from *Griffonia simplicifolia*, Alexa Fluor® 568 conjugate	500 µg
I21413	isolectin GS-IB4 from *Griffonia simplicifolia*, Alexa Fluor® 594 conjugate	500 µg
I32450	isolectin GS-IB4 from *Griffonia simplicifolia*, Alexa Fluor® 647 conjugate	500 µg
I21414	isolectin GS-IB4 from *Griffonia simplicifolia*, biotin-XX conjugate	500 µg
L21415	lectin GS-II from *Griffonia simplicifolia*, Alexa Fluor® 488 conjugate	500 µg
L21416	lectin GS-II from *Griffonia simplicifolia*, Alexa Fluor® 594 conjugate	500 µg
L32451	lectin GS-II from *Griffonia simplicifolia*, Alexa Fluor® 647 conjugate	500 µg
L11271	lectin HPA from *Helix pomatia* (edible snail), Alexa Fluor® 488 conjugate	1 mg
L32454	lectin HPA from *Helix pomatia* (edible snail), Alexa Fluor® 647 conjugate	1 mg
L11270	lectin PHA-L from *Phaseolus vulgaris* (red kidney bean), Alexa Fluor® 488 conjugate	1 mg
L32456	lectin PHA-L from *Phaseolus vulgaris* (red kidney bean), Alexa Fluor® 594 conjugate	1 mg
L32457	lectin PHA-L from *Phaseolus vulgaris* (red kidney bean), Alexa Fluor® 647 conjugate	1 mg
L21409	lectin PNA from *Arachis hypogaea* (peanut), Alexa Fluor® 488 conjugate	1 mg
L32458	lectin PNA from *Arachis hypogaea* (peanut), Alexa Fluor® 568 conjugate	1 mg
L32459	lectin PNA from *Arachis hypogaea* (peanut), Alexa Fluor® 594 conjugate	1 mg
L32460	lectin PNA from *Arachis hypogaea* (peanut), Alexa Fluor® 647 conjugate	1 mg

IMPORTANT NOTICE: The products described in this manual are covered by one or more Limited Use Label License(s). Please refer to the Appendix on page 971 and Master Product List on page 975. Products are For Research Use Only. Not intended for any animal or human therapeutic or diagnostic use.

PRODUCT LIST 7.7 LECTINS AND OTHER CARBOHYDRATE-BINDING PROTEINS—*continued*

Cat. No.	Product	Quantity
L11272	lectin SBA from *Glycine max* (soybean), Alexa Fluor® 488 conjugate	1 mg
L32462	lectin SBA from *Glycine max* (soybean), Alexa Fluor® 594 conjugate	1 mg
L32463	lectin SBA from *Glycine max* (soybean), Alexa Fluor® 647 conjugate	1 mg
Q12021MP	Qdot® 655 wheat germ agglutinin conjugate *1 µM solution*	200 µL
W11263	wheat germ agglutinin, Alexa Fluor® 350 conjugate	5 mg
W11261	wheat germ agglutinin, Alexa Fluor® 488 conjugate	5 mg
W32464	wheat germ agglutinin, Alexa Fluor® 555 conjugate	5 mg
W11262	wheat germ agglutinin, Alexa Fluor® 594 conjugate	5 mg
W21404	wheat germ agglutinin, Alexa Fluor® 633 conjugate	5 mg
W32466	wheat germ agglutinin, Alexa Fluor® 647 conjugate	5 mg
W32465	wheat germ agglutinin, Alexa Fluor® 680 conjugate	5 mg
W834	wheat germ agglutinin, fluorescein conjugate	5 mg
W6748	wheat germ agglutinin, Oregon Green® 488 conjugate	5 mg
W849	wheat germ agglutinin, tetramethylrhodamine conjugate	5 mg
W21405	wheat germ agglutinin, Texas Red®-X conjugate	1 mg
W7024	Wheat Germ Agglutinin Sampler Kit *four fluorescent conjugates, 1 mg each*	1 kit

IMPORTANT NOTICE: The products described in this manual are covered by one or more Limited Use Label License(s). Please refer to the Appendix on page 971 and Master Product List on page 975. Products are For Research Use Only. Not intended for any animal or human therapeutic or diagnostic use.

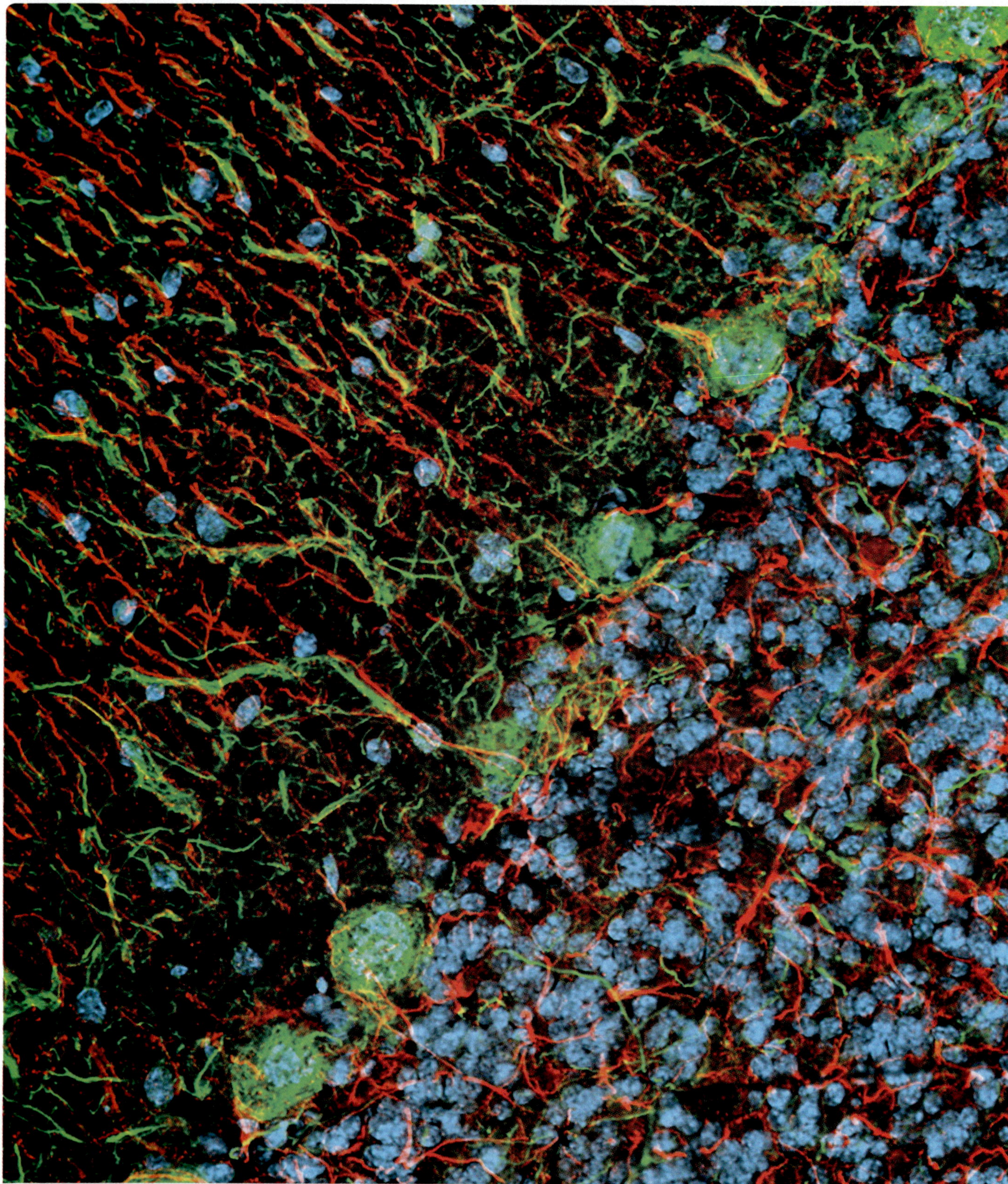

Alexa Fluor® 568 goat anti-rabbit IgG, Alexa Fluor® 488 goat anti-mouse IgG and Hoechst 33342.

IMPORTANT NOTICE: The products described in this manual are covered by one or more Limited Use Label License(s). Please refer to the Appendix on page 971 and Master Product List on page 975. Products are For Research Use Only. Not intended for any animal or human therapeutic or diagnostic use.

PART III

Detecting Cell Components: Nucleic Acid and Protein Analysis

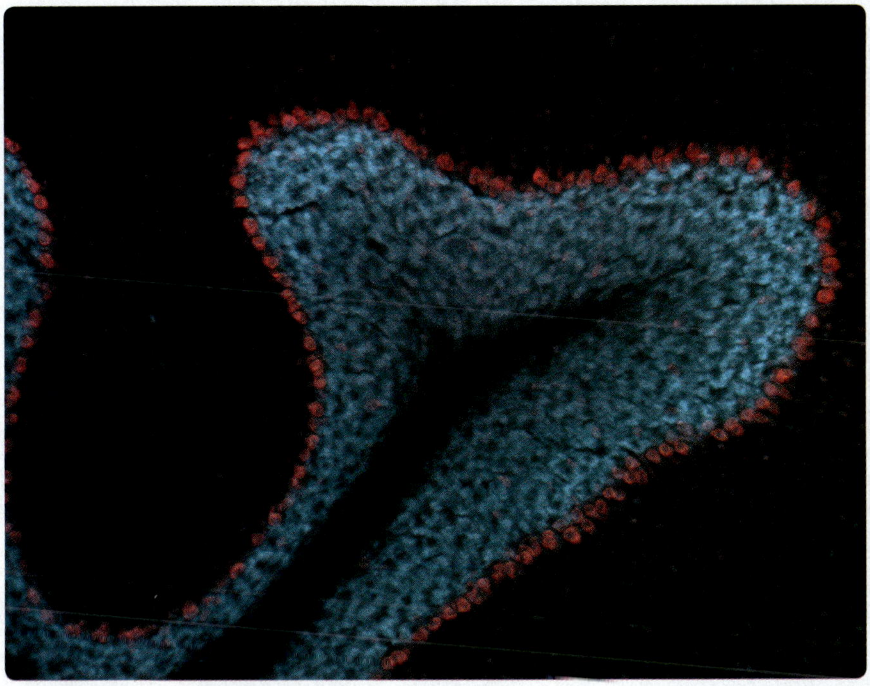
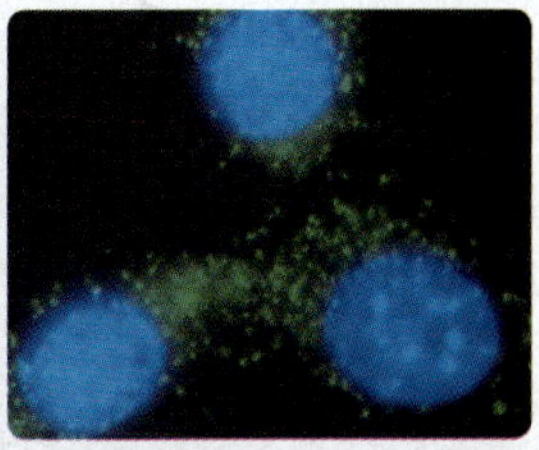
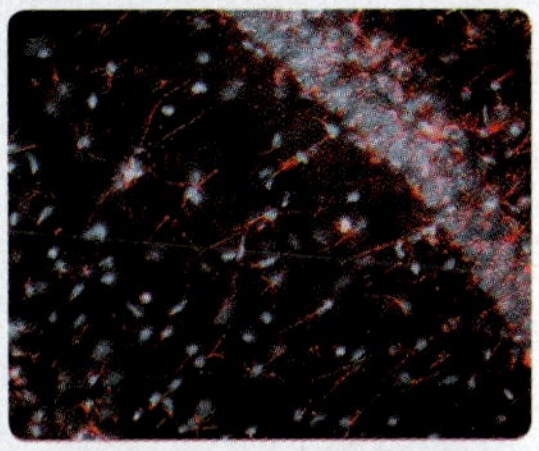
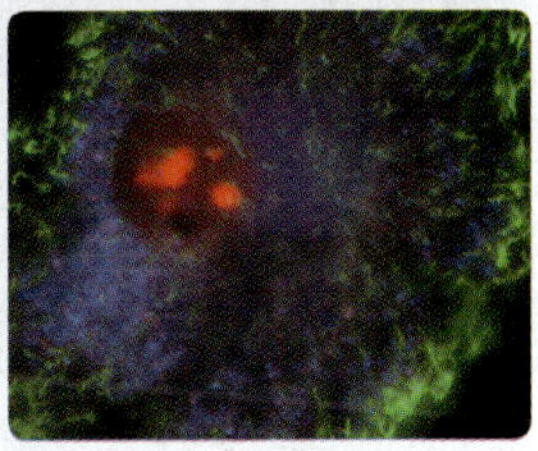

IMPORTANT NOTICE: The products described in this manual are covered by one or more Limited Use Label License(s). Please refer to the Appendix on page 971 and Master Product List on page 975. Products are For Research Use Only. Not intended for any animal or human therapeutic or diagnostic use.

Schematic illustration of a biarsenical version of fluorescein (FlAsH) bound to a tetracysteine (TC) tag appended to a protein of interest. The FlAsH molecule is presented in bright green. The side chains of the four cysteines that make up the TC tag are shown in grey and yellow spheres with black bonds. The protein of interest is represented as a blue ribbon. The TC-FlAsH™ technology is useful for live-cell imaging, easy protein detection in SDS-PAGE gels, and affinity purification of tagged proteins. Illustration by Lydia Jablonski.

This image appeared on the cover of *BioProbes 54*. *BioProbes*® newsletter is published several times each year. This award-winning publication is dedicated to furnishing researchers with the very latest information about cell biology products and their applications, and provides a great way to stay connected with the fluorescence community. You can subscribe and view the latest issues online at **www.invitrogen.com/handbook/bioprobes**.

IMPORTANT NOTICE: The products described in this manual are covered by one or more Limited Use Label License(s). Please refer to the Appendix on page 971 and Master Product List on page 975. Products are For Research Use Only. Not intended for any animal or human therapeutic or diagnostic use.

EIGHT

CHAPTER 8

Nucleic Acid Detection and Analysis

IMPORTANT NOTICE: The products described in this manual are covered by one or more Limited Use Label License(s). Please refer to the Appendix on page 971 and Master Product List on page 975. Products are For Research Use Only. Not intended for any animal or human therapeutic or diagnostic use.

IMPORTANT NOTICE: The products described in this manual are covered by one or more Limited Use Label License(s). Please refer to the Appendix on page 971 and Master Product List on page 975. Products are For Research Use Only. Not intended for any animal or human therapeutic or diagnostic use.

IMPORTANT NOTICE: The products described in this manual are covered by one or more Limited Use Label License(s). Please refer to the Appendix on page 971 and Master Product List on page 975. Products are For Research Use Only. Not intended for any animal or human therapeutic or diagnostic use.

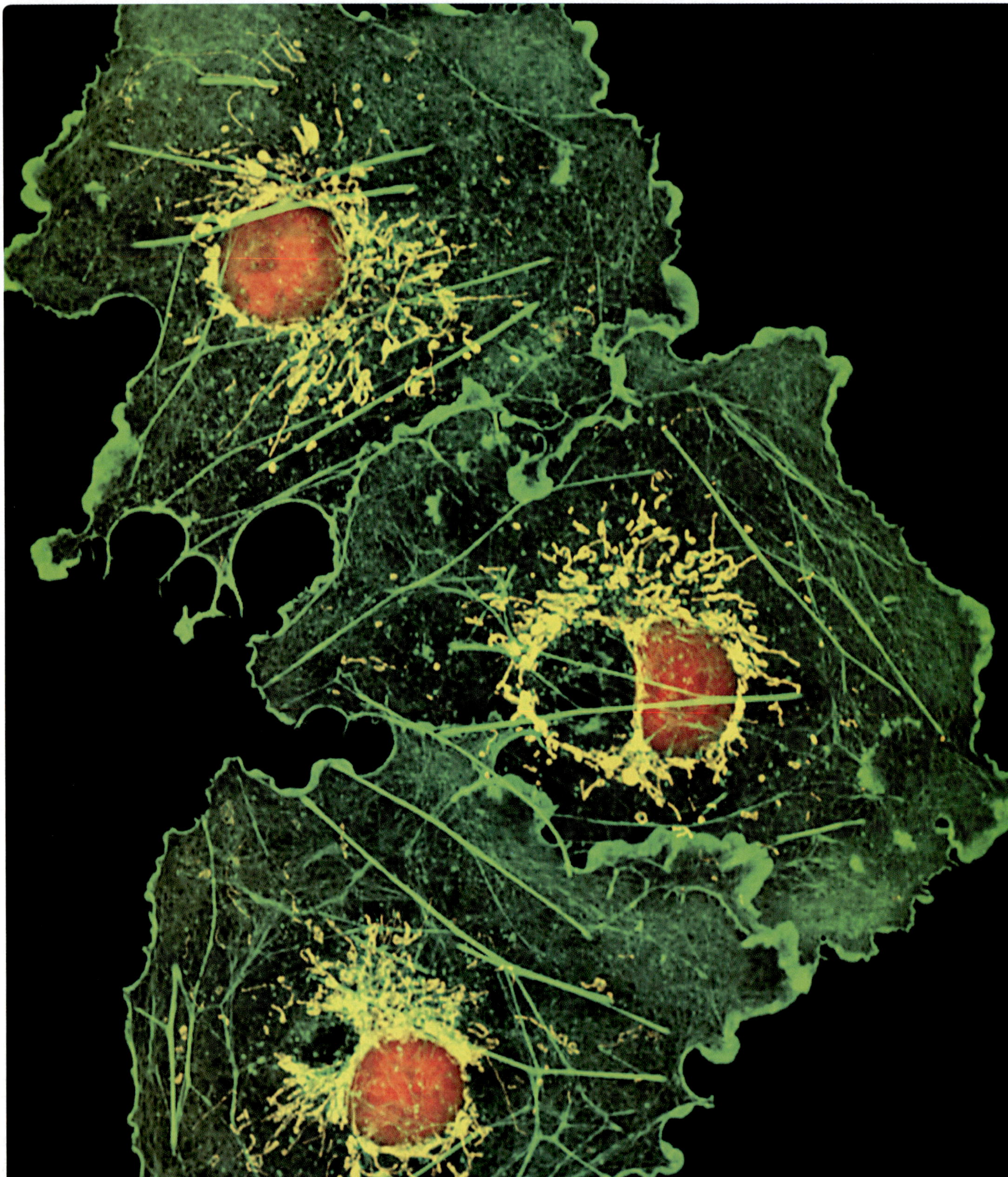

TO-PRO®-3 iodide, MitoTracker® Orange CMTMRos and BODIPY® FL phallacidin.

8.1 Nucleic Acid Stains

We offer an extensive assortment of nucleic acid stains, many of which have been developed in our research laboratories. This section discusses the physical properties of the various classes of dyes listed below. The sections that follow describe applications of these nucleic acid stains for genomics research. The four classes of Molecular Probes® cyanine dyes include:

- Premier dyes for ultrasensitive nucleic acid quantitation and gel staining (Table 8.1)
- Cell-impermeant TOTO®, TO-PRO® and SYTOX® families of dyes (Table 8.2)
- Cell-permeant SYTO® family of dyes (Table 8.3)
- Amine-reactive SYBR® dye that can be used to form bioconjugates

The three classes of classic nucleic acid stains (Table 8.4) include:

- Intercalating dyes (Figure 8.1.1), such as ethidium bromide and propidium iodide
- Minor-groove binders (Figure 8.1.1), such as DAPI and the Hoechst dyes
- Other types of nucleic acid binders (Figure 8.1.1) such as acridine orange, 7-AAD, LDS 751 and hydroxystilbamidine

Properties of Cyanine Dyes

Molecular Probes® nucleic acid–binding cyanine dyes share several important spectroscopic and physical properties:

- High molar absorptivity, with extinction coefficients typically greater than 50,000 $cm^{-1}M^{-1}$ at visible wavelengths
- Very low intrinsic fluorescence, with quantum yields usually less than 0.01 when not bound to nucleic acids
- Large fluorescence enhancements (often over 1000-fold) upon binding to nucleic acids, with increases in quantum yields to as high as 0.9
- Moderate to very high affinity for nucleic acids, with little or no staining of other biopolymers

Their fluorescence aborption and emission spectra span the visible-light spectrum from blue to near-infrared (Figure 8.1.2) with additional absorption peaks in the UV, making them compatible with many different types of instrumentation. The cyanine dyes also show important differences in some physical characteristics—particularly in cell membrane permeability and nucleic acid specificity—that allow their distribution into distinct classes, and these classes are discussed in detail below and in the following sections of this chapter.

Premier Cyanine Dyes for Ultrasensitive Nucleic Acid Detection and Quantitation

Several of our cyanine dyes provide excellent sensitivity in specific nucleic acid assays (Table 8.1). For these dyes, we have developed detailed and extensively tested protocols to facilitate reproducible, high-sensitivity results in these assays.

- PicoGreen®, OliGreen® and RiboGreen® quantitation reagents and their Quant-iT™ reagent counterparts in Section 8.3 set a benchmark for the detection and quantitation of DNA, oligonucleotides and RNA in solution. These reagents offer extremely simple and rapid protocols as well as linear ranges that span up to four orders of magnitude in nucleic acid concentration.
- SYBR® Gold, SYBR® Green I and SYBR® Green II nucleic acid gel stains in Section 8.4 are ultrasensitive gel stains that surpass the sensitivity of ethidium bromide by more than an order of magnitude in nucleic acid detection.
- SYBR® Safe DNA gel stain (Section 8.4) is not only significantly less mutagenic than ethidium bromide, but SYBR® Safe stain's detection sensitivity is comparable to that of ethidium bromide. SYBR® Safe stain showed no or very low mutagenic activity when tested by an independent, licensed testing laboratory, and it is not classified as hazardous waste under U.S. Federal regulations.
- CyQUANT® GR dye (C7026), discussed in Section 15.4, is a cyanine dye designed to quantitate cell proliferation and can reliably detect the nucleic acids in as few as 50 cells.

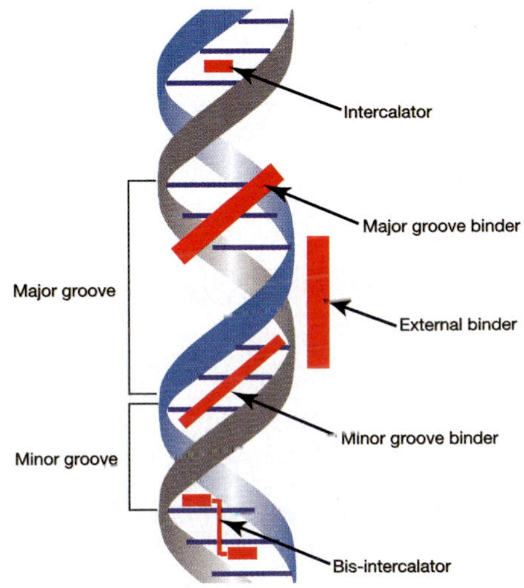

Figure 8.1.1 Schematic diagram showing the different binding modes of dyes (and other ligands) to DNA.

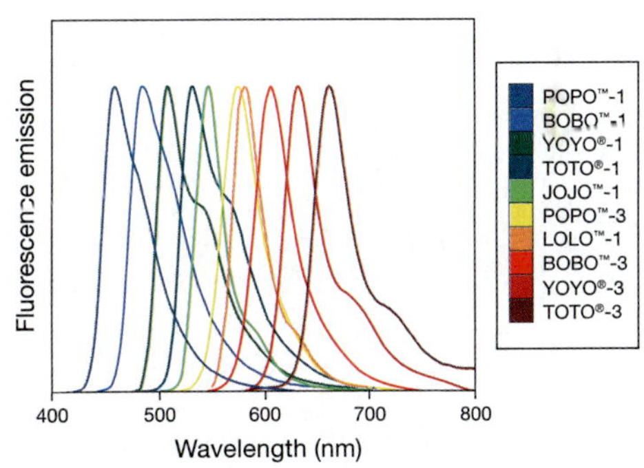

Figure 8.1.2 Normalized fluorescence emission spectra of DNA-bound cyanine dimers, identified by the color key on the sidebar.

Table 8.1 Specialty nucleic acid reagents for molecular biology.

Cat. No.	Dye	Ex/Em *	Application
Dyes for Ultrasensitive Solution Quantitation †			
P7581, P11495, P7589, P11496	Quant-iT™ PicoGreen® dsDNA reagent and Kits	502/523	Ultrasensitive reagent for solution quantitation of dsDNA
O7582, O11492	Quant-iT™ OliGreen® ssDNA reagent and Kit	498/518	Ultrasensitive reagent for solution quantitation of ssDNA and oligonucleotides
R11491, R11490, R32700	Quant-iT™ RiboGreen® RNA reagent and Kits	500/525	Ultrasensitive reagent for solution quantitation of RNA
Dyes for Sensitive Detection of Nucleic Acids in Gels and on Blots			
S11494	SYBR® Gold nucleic acid gel stain	495/537	Ultrasensitive gel stain for single- or double-stranded DNA or RNA post-electrophoresis
S7563, S7567, S7585	SYBR® Green I nucleic acid gel stain	494/521	Ultrasensitive gel stain for double-stranded DNA and oligonucleotides post-electrophoresis; also useful for real-time PCR assays
S7564, S7568, S7586	SYBR® Green II RNA gel stain	492/513	Sensitive stain for RNA and single-stranded DNA post-electrophoresis
S33100, S33101, S33102, S33110, S33111, S33112	SYBR® Safe DNA stain	502/530	Sensitive DNA gel stain with significantly reduced mutagenicity

* All excitation (Ex) and emission (Em) maxima were determined for dyes bound to double-stranded calf thymus DNA in aqueous solution. † In addition to these dyes, we offer a wide selection of Quant-iT™ Kits for use with the Qubit® fluorometer and with microplate readers; these are listed in Table 8.8.

Table 8.2 Cell membrane–impermeant cyanine nucleic acid stains.

Cat. No.	Dye	Ex/Em *
SYTOX® Dyes: Dead-Cell Stains †		
S11348 S34857	SYTOX® Blue	445/470
S7020	SYTOX® Green	504/523
S11368	SYTOX® Orange	547/570
S34859	SYTOX® Red	640/658
Cyanine Dimers: High-Affinity Stains ‡		
P3580	POPO™-1	434/456
B3582	BOBO™-1	462/481
Y3601	YOYO®-1	491/509
T3600	TOTO®-1	514/533
J11372	JOJO™-1	529/545
P3584	POPO™-3	534/570
L11376	LOLO™-1	565/579
B3586	BOBO™-3	570/602
Y3606	YOYO®-3	612/631
T3604	TOTO®-3	642/660
N7565	Dimer Sampler Kit §	Various
Cyanine Monomers: Nuclear Counterstains **		
P3581	PO-PRO™-1	435/455
Y3603	YO-PRO®-1	491/509
T3602	TO-PRO®-1	515/531
J11373	JO-PRO™-1	530/546
P3585	PO-PRO™-3	539/567
Y3607	YO-PRO®-3	612/631
T3605	TO-PRO®-3	642/661
T7596	TO-PRO®-5	747/770

* Wavelengths of excitation (Ex) and emission (Em) maxima, in nm. All excitation and emission maxima were determined for dyes bound to double-stranded calf thymus DNA in aqueous solution. † Products supplied as 250 µL of a 5 mM solution. ‡ Products (except N7565) supplied as 200 µL of a 1 mM solution. § Includes 10 µL each of a 1 mM solution of the TOTO®-1, TOTO®-3, YOYO®-1, YOYO®-3, BOBO™-1, BOBO™-3, POPO™-1 and POPO™-3 dyes. ** Products supplied as 1 mL of a 1 mM solution.

Cell-Impermeant Cyanine Dimers: The TOTO® Family of Dyes

The cyanine dimer dyes listed in Table 8.2—sometimes referred to as the TOTO® family of dyes—are symmetric dimers of cyanine dyes with exceptional sensitivity for nucleic acids. This sensitivity is due to a high affinity for nucleic acids, in combination with a very high fluorescence enhancement and quantum yield upon binding. The unique physical characteristics of these dyes and some illustrative applications are discussed below; specific applications are discussed in later sections of this chapter.

Each of the cyanine dimer dyes is available separately. For researchers designing new applications, the Nucleic Acid Stains Dimer Sampler Kit (N7565) provides samples of eight spectrally distinct analogs of the dimeric cyanine dyes for testing (Table 8.2).

High Affinity for Nucleic Acids

Appropriately designed dimers of nucleic acid–binding dyes have nucleic acid–binding affinities that are several orders of magnitude greater than those of their parent monomer dyes.[1–3] For example, the intrinsic DNA binding affinity constants of ethidium bromide and ethidium homodimer-1 (E1169) are reported to be 1.5×10^5 and 2×10^8 M^{-1}, respectively, in 0.2 M Na^+.[4] As a result, the dimeric cyanine dyes are among the highest-affinity fluorescent probes available for nucleic acid staining.

For example, in the TOTO®-1 dimeric cyanine dye (T3600), the positively charged side chains of the TO-PRO®-1 monomeric cyanine dye (T3602, Figure 8.1.3) are covalently linked to form the TOTO®-1 molecule, with four positive charges (Figure 8.1.4). This linkage gives TOTO®-1 dye a greatly enhanced affinity for nucleic acids—more than 100 times greater than that of the TO-PRO®-1 monomer. TOTO®-1 dye

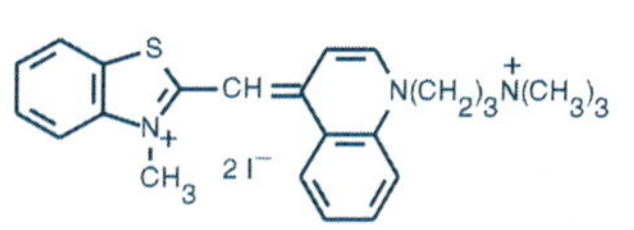

Figure 8.1.3 TO-PRO®-1 iodide (515/531, T3602).

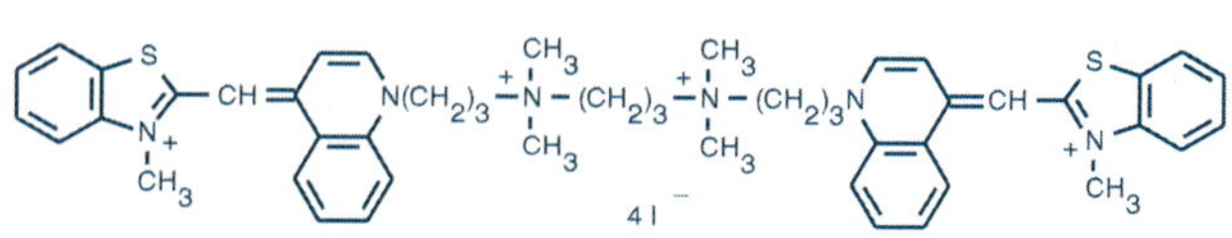

Figure 8.1.4 TOTO®-1 iodide (514/533, T3600).

IMPORTANT NOTICE: The products described in this manual are covered by one or more Limited Use Label License(s). Please refer to the Appendix on page 971 and Master Product List on page 975. Products are For Research Use Only. Not intended for any animal or human therapeutic or diagnostic use.

exhibits a higher affinity for double-stranded DNA (dsDNA) than even the ethidium homodimers and also binds to both single-stranded DNA (ssDNA) and RNA. The extraordinary stability of TOTO®-1–nucleic acid complexes[2,5,6] allows the dye–DNA association remains stable, even during electrophoresis (Figure 8.1.5); thus, samples can be stained with nanomolar dye concentrations *prior to* electrophoresis,[7,8] thereby reducing the hazards inherent in handling large volumes of ethidium bromide staining solutions.[2,6,9] In contrast, the binding of thiazole orange—the parent compound of TOTO®-1 and TO-PRO®-1—is rapidly reversible, limiting the dye's sensitivity and rendering its nucleic acid complex unstable to electrophoresis.[9]

High Fluorescence Enhancements and High Quantum Yields upon Binding to Nucleic Acids

In addition to their superior binding properties, TOTO®-1 dye and the other cyanine dimers are essentially nonfluorescent in the absence of nucleic acids and exhibit fluorescence enhancements upon DNA binding of 100- to 1000-fold,[5,10] which compares favorably with the fluorescence enhancement of thiazole orange upon DNA binding[11] (~3000-fold). Furthermore, the fluorescence quantum yields of the cyanine dimers bound to DNA are high (generally between 0.2 and 0.6), and their extinction coefficients are an order of magnitude greater than those of the ethidium homodimers.[5] This sensitivity is sufficient for detecting single molecules of labeled nucleic acids by optical imaging (Figure 8.1.6) and flow cytometry and for tracking dye-labeled virus particles in microbial communities and aquatic systems by fluorescence microscopy.[12,13] These dyes are generally considered to be cell impermeant, although their use to stain reticulocytes permeabilized by 5% DMSO has been reported.[14]

Modifying the Dimers Creates Compounds with Different Spectral Characteristics

Simply by changing the aromatic rings and the number of carbon atoms linking the cyanine monomers, we were able to synthesize an extended series of these dyes with different spectral characteristics (Table 8.2). Chemical modifications produce dramatic shifts in the absorption and emission spectra and reduce the quantum yields of the bound dyes but cause little or no change in their high affinity for DNA. The names of the dyes reflect their basic structure and spectral characteristics. For example, YOYO®-1 iodide (491/509) has one carbon atom bridging the aromatic rings of the oxacyanine dye and exhibits absorption/emission maxima of 491/509 nm when bound to dsDNA. YOYO®-3 dye (612/631)—which differs from YOYO®-1 dye only in the number of bridging carbon atoms—has absorption/emission maxima of 612/631 nm when bound to dsDNA. Fluorescence spectra for the POPO™, BOBO™, YOYO®, TOTO®, JOJO™ and LOLO™ dyes bound to dsDNA are shown in Figure 8.1.2. The spectra of these dyes at dye:base ratios of less than 1:1 are essentially the same for the corresponding dye–ssDNA and dye–RNA complexes. At higher dye:base ratios, however, ssDNA and RNA complexes of all of the monomethine ("-1") dyes of the TOTO® series and TO-PRO® series have red-shifted emissions, whereas corresponding complexes of the trimethine ("-3") analogs do not. Thus, the cyanine dimer family provides dyes with a broad range of spectral characteristics to match the output of almost any available excitation source.

Binding Modes of the Cyanine Dimers

The studies on cyanine dimer binding modes have focused on the YOYO®-1 and TOTO®-1 dyes. YOYO®-1 dye was found to exhibit at least two distinct binding modes. At low dye:base pair ratios, the binding mode appears to consist primarily of bis-intercalation[15–19] (Figure 8.1.1). Each monomer unit intercalates between bases, with the benzazolium ring system sandwiched between the pyrimidines and the quinolinium ring between the purine rings, causing the helix to unwind. The distortion in the local DNA structure caused by YOYO®-1 bis-intercalation has been observed by two-dimensional NMR spectroscopy. At high dye:base pair ratios, a second, less well characterized mode of external binding begins to contribute.[15,16] Circular dichroism measurements also indicate a possible difference in the binding modes of YOYO®-1 dye to ssDNA and dsDNA. These data are consistent with our own results, including the observation that the fluorescence emission of the YOYO®-1 dye complex with nucleic acids shifts to longer wavelengths at high dye:base ratios upon binding to single-stranded nucleic acids and that the salt, ethanol and sodium dodecyl sulfate (SDS) sensitivity of YOYO®-1 dye binding to DNA is a function of the dye:base pair ratio.[19]

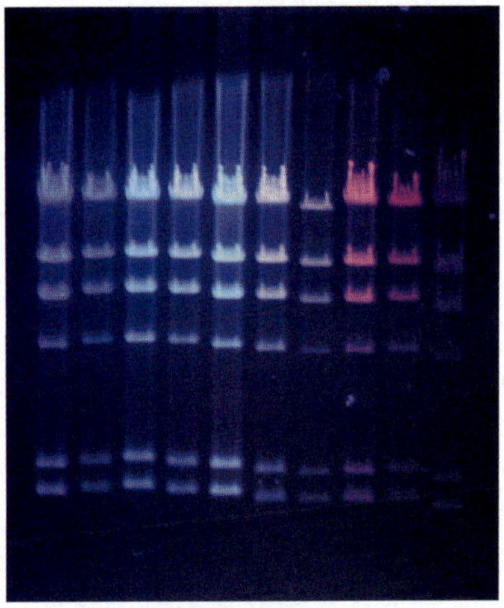

Figure 8.1.5 Lambda bacteriophage *Hin*dIII fragments were prestained with various nucleic acid dyes, run on a 0.7% agarose gel and visualized using a standard 300 nm UV transilluminator. From left to right, the dyes used were: POPO™-1 (P3580), BOBO™-1 (B3582), YOYO®-1 (Y3601), TOTO®-1 (T3600), JOJO™-1 (J11372), POPO™-3 (P3584), LOLO™-1 (L11376), BOBO™-3 (B3586), YOYO®-3 (Y3606) and TOTO®-3 (T3604) nucleic acid stains. The longest-wavelength stains are barely visible to the eye but can be detected with infrared-enhanced films and imaging equipment.

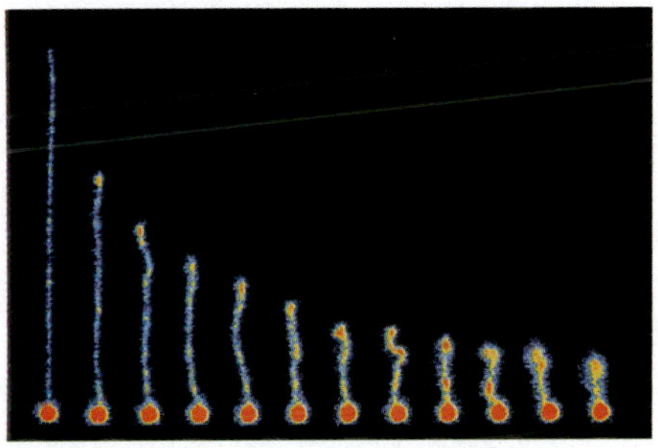

Figure 8.1.6 The relaxation of a single, 39 µm–long DNA molecule stained with YOYO®-1 iodide (Y3601) imaged at 4.5 second intervals. After the 1 µm polystyrene sphere was trapped with optical tweezers, the attached DNA was stretched to its full extension in a fluid flow and then allowed to relax upon stoppage of fluid flow due to its entropic elasticity (Science (1994) 264:822). The YOYO®-1 iodide–DNA complex is excited with the 488 nm spectral line of the argon-ion laser and visualized through a 515 nm longpass optical filter using a Hamamatsu SIT camera with image processing. Image contributed by Thomas Perkins, Stanford University.

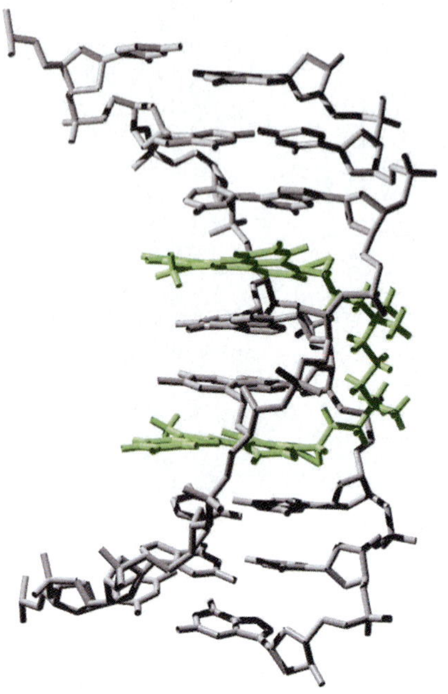

Figure 8.1.7 NMR solution structure of the TOTO®-1 dye (T3600) bound to DNA; the image was derived from data submitted to the Protein Data Bank (number PDB 108D, www.rcsb.org/pdb/) (Nucleic Acids Res (2000) 28:235). The NMR structure shows that TOTO®-1 binds to DNA through bis-intercalation (Biochemistry (1995) 34:8542).

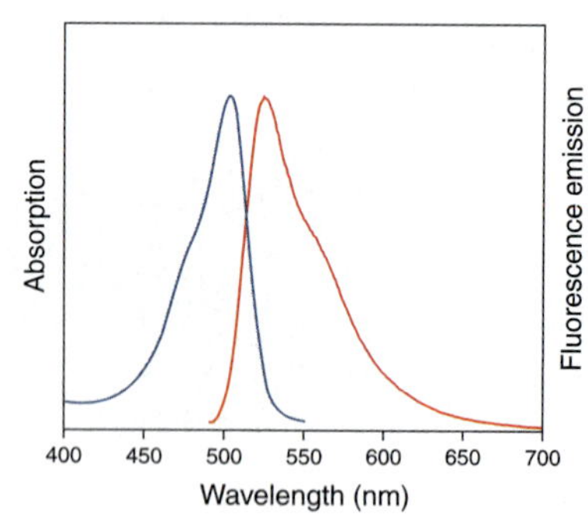

Figure 8.1.8 Absorption and fluorescence emission spectra of the SYTOX® Green nucleic acid stain bound to DNA.

TOTO®-1 dye is also capable of bis-intercalation,[20] although it reportedly interacts with dsDNA and ssDNA with similarly high affinity.[1] NMR studies of TOTO®-1 dye interactions with a double-stranded 8-mer indicate that TOTO®-1 dye is a bis-intercalator, with the fluorophores intercalating between the bases and the linker region having interactions in the minor groove[21] (Figure 8.1.7). Binding of the dye partially unwinds the DNA,[21] distorting and elongating the helix.[22] However, another study using fluorescence polarization measurements suggests that an external binding mode, where the dipole of the dye molecule is aligned with the DNA grooves, may be more important.[23] TOTO®-1 dye reportedly exhibits some sequence selectivity for the site 5′-CTAG-3′, although it will bind to almost any sequence in dsDNA.[24–27] TOTO®-1 dye does not exhibit cooperative binding to DNA, suggesting that it may be a suitable dye for detecting nucleic acids in gels.[25]

The binding modes of the other members of the TOTO® dye series have also been partially characterized. Electrophoresis and fluorescence lifetime measurements have shown that YOYO®-3 dye also appears to intercalate into DNA.[28] During application development, we have determined that staining of nucleic acids by BOBO™-1 and POPO™-1 dyes is much faster (occurring within minutes) than staining by YOYO®-1 or TOTO®-1 dyes (which can take several hours to reach equilibrium under the same experimental conditions),[20] indicating possible differences in their binding mechanisms. Fluorescence yield and lifetime measurements have been used to assess the base selectivity of an extensive series of these dyes.[10] Circular dichroism measurements have shown that bis-intercalation is the predominant binding mode for the POPO™-1 dye.[29]

Working with Cyanine Dimers

All of the dyes in the TOTO® series (Table 8.2) are supplied as 1 mM solutions in dimethylsulfoxide (DMSO), except for POPO™-3 (P3584), which is supplied as a 1 mM solution in dimethylformamide (DMF). These cationic dyes appear to be readily adsorbed out of aqueous solutions onto surfaces (particularly glass) but are very stable once complexed to nucleic acids.

Cell-Impermeant Cyanine Monomers: The TO-PRO® Family of Dyes

Our TO-PRO® family of dyes, all of which are listed in Table 8.2, each comprise a single cyanine dye and a cationic side chain (Figure 8.1.3). The monomeric dyes in the TO-PRO® series are spectrally analogous to the corresponding dimeric cyanine dyes; however, with only two positive charges and one intercalating unit, the TO-PRO® dyes exhibit somewhat reduced affinity for nucleic acids relative to the dyes in the TOTO® series. Like their dimeric counterparts, these monomeric cyanine dyes are typically impermeant to cells,[30] although YO-PRO®-1 (Y3603) dye has been shown to be permeant to apoptotic cells, providing a convenient indicator of apoptosis[31–34] (Section 15.5). YO-PRO®-1 has also been observed to pass through $P2X_7$ receptor channels of live cells.[35–37]

Spectral Characteristics of the Cyanine Dye Monomers

The TO-PRO® family of dyes retains all of the exceptional spectral properties of the dimeric cyanine dyes discussed above. The absorption and emission spectra of these monomeric cyanine dyes cover the visible and near-infrared spectrum (Table 8.2). They also have relatively narrow emission bandwidths, thus facilitating multicolor applications in imaging and flow cytometry. YO-PRO®-1 (491/509) and TO-PRO®-1 (515/531) dyes are optimally excited by the 488 nm and 514 nm spectral lines of the argon-ion laser, respectively. In flow cytometric analysis, the TO-PRO®-3 (642/661) complex with nucleic acids has been excited directly by the red He-Ne laser[38] and indirectly by the argon-ion laser by using fluorescence resonance energy transfer (FRET) from co-bound propidium iodide[39] (Fluorescence Resonance Energy Transfer (FRET)—Note 1.2). The TO-PRO®-3 complex with nucleic acids has also been detected in a flow cytometer equipped with an inexpensive 3 mW visible-wavelength diode laser that provides excitation at 635 nm.[40] Although the DNA-induced fluorescence enhancement of TO-PRO®-5 dye (T7596) is not as large as that observed with our other cyanine dyes, its spectral characteristics (excitation/emission maxima ~745/770 nm) provide a unique alternative for multicolor applications.

IMPORTANT NOTICE: The products described in this manual are covered by one or more Limited Use Label License(s). Please refer to the Appendix on page 971 and Master Product List on page 975. Products are For Research Use Only. Not intended for any animal or human therapeutic or diagnostic use.

Working with Cyanine Monomers

The binding affinity of the TO-PRO® series of dyes to dsDNA is lower than that of the TOTO® series of dyes but is still very high, with dissociation constants in the micromolar range.[41] TO-PRO® dyes also bind to RNA and ssDNA, although typically with somewhat lower fluorescence quantum yields. Fluorescence polarization studies indicate that TO-PRO®-1 and PO-PRO™-1 dyes bind by intercalation, with unwinding angles of 2° and 31°, respectively.[29] Binding of these dyes to dsDNA is not sequence selective.[42] All dyes of the TO-PRO® series (Table 8.2) are supplied as 1 mM solutions in DMSO.

Cell-Impermeant SYTOX® Dyes for Dead-Cell Staining

SYTOX® Green Stain

Our SYTOX® nucleic acid stains (Table 8.2) are cell-impermeant cyanine dyes that are particularly useful as dead-cell stains. SYTOX® Green nucleic acid stain (S7020) is a high-affinity nucleic acid stain that easily penetrates cells with compromised plasma membranes and yet will not cross the membranes of live cells. It is especially useful for staining both gram-positive and gram-negative bacteria, in which an exceptionally bright signal is required. Following brief incubation with the SYTOX® Green stain, dead cells fluoresce bright green when excited with the 488 nm spectral line of the argon-ion laser or with any other 450–500 nm source (Figure 8.1.8). Because all of the SYTOX® dyes are essentially nonfluorescent in aqueous medium, no wash steps are required. Unlike the DAPI or Hoechst dyes, the SYTOX® Green nucleic acid stain shows little base selectivity. These properties, combined with its ~1000-fold fluorescence enhancement upon nucleic acid binding and high quantum yield, make our SYTOX® Green stain a simple and quantitative single-step dead-cell indicator for use with epifluorescence and confocal laser-scanning microscopes, fluorometers, fluorescence microplate readers and flow cytometers (Figure 8.1.9).

The SYTOX® Green nucleic acid stain can be used with blue- and red-fluorescent labels for multiparameter analyses (Figure 8.1.10). It is also possible to combine the SYTOX® Green nucleic acid stain with the SYTO® 17 red-fluorescent nucleic acid stain (S7579) for two-color visualization of dead and live cells. Because the SYTOX® Green nucleic acid stain is an excellent DNA counterstain for chromosome labeling and for fixed cells and tissues (Figure 8.1.11), we have incorporated it into our SelectFX® Nuclear Labeling Kit (S33025), which is discussed in Section 12.5.

SYTOX® Blue Stain

SYTOX® Blue stain (S11348, S34857) is a high-affinity nucleic acid stain that typically penetrates only cells with compromised plasma membranes (Figure 8.1.12). The SYTOX® Blue stain

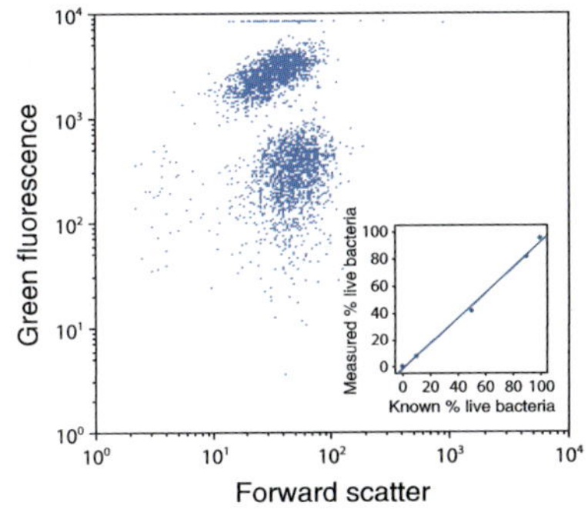

Figure 8.1.9 Quantitative flow cytometric analysis of *Escherichia coli* viability using the SYTOX® Green nucleic acid stain (S7020). A bacterial suspension containing an equal number of live and isopropyl alcohol–killed *E. coli* was stained with SYTOX® Green and analyzed using excitation at 488 nm. A bivariate frequency distribution for forward light scatter versus log fluorescence intensity (collected with a 510 nm longpass optical filter) shows two clearly distinct populations. When live and dead bacteria were mixed in varying proportions, a linear relationship between the population numbers and the actual percentage of live cells in the sample was obtained (see inset).

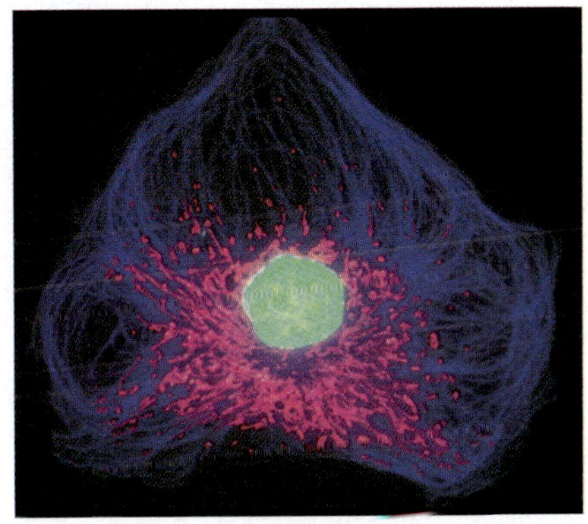

Figure 8.1.10 Bovine pulmonary artery endothelial cells (BPAEC) incubated with the fixable, mitochondrion-selective MitoTracker® Red CMXRos (M7512). After staining, the cells were formaldehyde-fixed, acetone-permeabilized, treated with DNase-free RNase and counterstained using SYTOX® Green nucleic acid stain (S7020). Microtubules were labeled with a mouse monoclonal anti–ß-tubulin antibody, biotin-XX goat anti–mouse IgG antibody (B2763) and Cascade Blue® NeutrAvidin™ biotin-binding protein (A2663). This photograph was taken using multiple exposures through bandpass optical filters appropriate for Texas Red® dye, fluorescein and DAPI using a Nikon® Labophot 2 microscope equipped with a Quadfluor epi-illumination system.

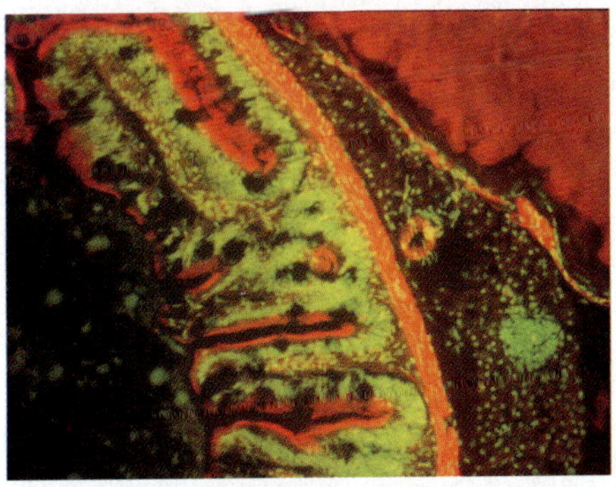

Figure 8.1.11 Adult zebrafish gut cryosections that have been incubated with BODIPY® TR-X phallacidin (B7464), followed by the SYTOX® Green nucleic acid stain (S7020), and then dehydrated and mounted. The image was obtained by taking multiple exposures through bandpass optical filter sets appropriate for fluorescein and the Texas Red® dye.

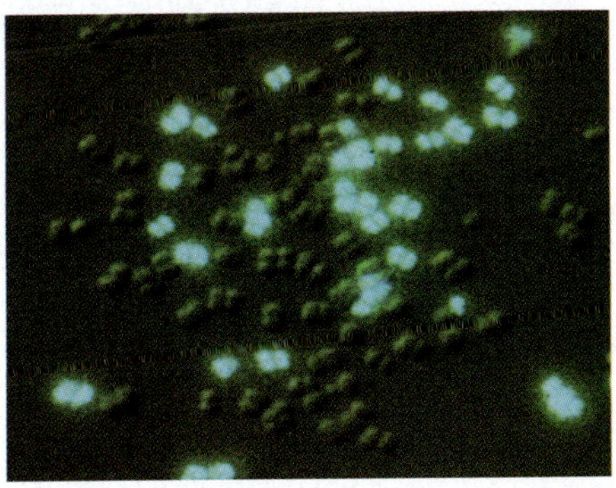

Figure 8.1.12 A mixed population of live and isopropyl alcohol–killed *Micrococcus luteus* stained with SYTOX® Blue nucleic acid stain (S11348), which does not penetrate intact plasma membranes. Dead cells exhibit bright blue-fluorescent staining. The image was acquired using a longpass optical filter set appropriate for the Cascade Blue® dye.

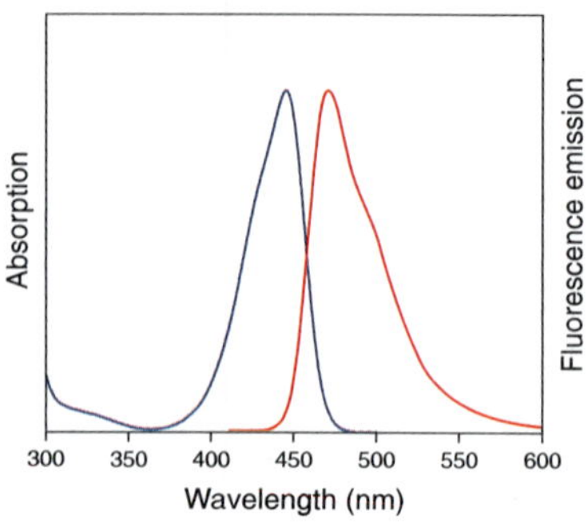

Figure 8.1.13 Absorption and fluorescence emission spectra of SYTOX® Blue nucleic acid stain bound to DNA.

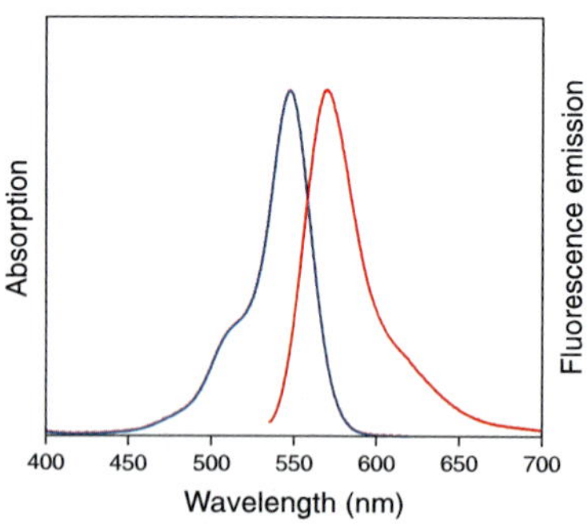

Figure 8.1.14 Absorption and fluorescence emission spectra of SYTOX® Orange nucleic acid stain bound to DNA.

labels both DNA and RNA with extremely bright fluorescence centered near 470 nm (Figure 8.1.13). The absorption maximum of the nucleic acid–bound SYTOX® Blue stain (~445 nm) permits very efficient fluorescence excitation by the 436 nm spectral line of the mercury-arc lamp. Unlike many blue-fluorescent dyes, the SYTOX® Blue stain is also efficiently excited by tungsten–halogen lamps and other sources that have relatively poor emission in the UV portion of the spectrum. The brightness of the SYTOX® Blue stain allows sensitive detection with fluorometers, microplate readers, arc-lamp–equipped flow cytometers and epifluorescence microscopes, including those not equipped with UV-pass optics.

Table 8.3 Cell-permeant cyanine nucleic acid stains.

Cat. No.	Dye *	Ex/Em †
Blue-Fluorescent SYTO® dyes		
S11351	SYTO® 40 blue-fluorescent nucleic acid stain	419/445
S11352	SYTO® 41 blue-fluorescent nucleic acid stain	426/455
S11353	SYTO® 42 blue-fluorescent nucleic acid stain	430/460
S11356	SYTO® 45 blue-fluorescent nucleic acid stain	452/484
S11350	SYTO® Blue-Fluorescent Nucleic Acid Stain Sampler Kit (SYTO® dyes 40–45) ‡	Various
Green-Fluorescent SYTO® Dyes		
S32703	SYTO® RNASelect™ green-fluorescent cell stain §	490/530 †
S34854	SYTO® 9 green-fluorescent nucleic acid stain §	483/503
S32704	SYTO® 10 green-fluorescent nucleic acid stain §	484/505
S34855	SYTO® BC green-fluorescent nucleic acid stain §	485/500
S7575	SYTO® 13 green-fluorescent nucleic acid stain	488/509
S7578	SYTO® 16 green-fluorescent nucleic acid stain **	488/518
S7559	SYTO® 24 green-fluorescent nucleic acid stain	490/515
S7556	SYTO® 21 green-fluorescent nucleic acid stain	494/517
S7574	SYTO® 12 green-fluorescent nucleic acid stain	500/522
S7573	SYTO® 11 green-fluorescent nucleic acid stain	508/527
S7576	SYTO® 14 green-fluorescent nucleic acid stain	517/549
S7560	SYTO® 25 green-fluorescent nucleic acid stain	521/556
S7572	SYTO® Green-Fluorescent Nucleic Acid Stain Sampler Kit (SYTO® dyes 11–14, 16, 21, 24 and 25) ‡	Various
Orange-Fluorescent SYTO® dyes		
S11362	SYTO® 81 orange-fluorescent nucleic acid stain	530/544
S11361	SYTO® 80 orange-fluorescent nucleic acid stain	531/545
S11363	SYTO® 82 orange-fluorescent nucleic acid stain	541/560
S11364	SYTO® 83 orange-fluorescent nucleic acid stain	543/559
S11365	SYTO® 84 orange-fluorescent nucleic acid stain	567/582
S11366	SYTO® 85 orange-fluorescent nucleic acid stain	567/583
S11360	SYTO® Orange-Fluorescent Nucleic Acid Stain Sampler Kit (SYTO® dyes 80–85) ‡	Various
Red-Fluorescent SYTO® dyes		
S11346	SYTO® 64 red-fluorescent nucleic acid stain §	598/620
S11343	SYTO® 61 red-fluorescent nucleic acid stain	620/647
S7579	SYTO® 17 red-fluorescent nucleic acid stain	621/634
S11341	SYTO® 59 red-fluorescent nucleic acid stain §	622/645
S11344	SYTO® 62 red-fluorescent nucleic acid stain	649/680
S11342	SYTO® 60 red-fluorescent nucleic acid stain	652/678
S11345	SYTO® 63 red-fluorescent nucleic acid stain	654/675
S11340	SYTO® Red-Fluorescent Nucleic Acid Stain Sampler Kit (SYTO® dyes 17, 59–64) ‡	Various

* All products supplied as 250 µL of a 5 mM solution, with exceptions noted. † Wavelengths of excitation (Ex) and emission (Em) maxima, in nm. All excitation and emission maxima were determined for dyes bound to double-stranded calf thymus DNA in aqueous solution, except for the SYTO® RNASelect™ green-fluorescent cell stain, which was determined for the dye bound to *Escherichia coli* RNA. ‡ Supplied as individual 50 µL vials. § Unit size = 100 µL. ** Supplied as 250 µL of a 1 mM solution.

IMPORTANT NOTICE: The products described in this manual are covered by one or more Limited Use Label License(s). Please refer to the Appendix on page 971 and Master Product List on page 975. Products are For Research Use Only. Not intended for any animal or human therapeutic or diagnostic use.

In a side-by-side comparison with the SYTOX® Green stain, the SYTOX® Blue stain yielded identical results when quantitating membrane-compromised bacterial cells. Furthermore, like the SYTOX® Green stain, the SYTOX® Blue stain does not interfere with bacterial cell growth. Because their emission spectra overlap somewhat, we have found that it is not ideal to use the SYTOX® Blue stain and green-fluorescent dyes together; however, fluorescence emission of the SYTOX® Blue stain permits clear discrimination from orange- or red-fluorescent probes, facilitating the development of multicolor assays with minimal spectral overlap between signals.

SYTOX® Orange Stain

SYTOX® Orange nucleic acid stain (S11368) is designed to clearly distinguish dead bacteria, yeast or mammalian cells from live cells. As compared with propidium iodide, SYTOX® Orange stain has shorter-wavelength emission and its spectra more closely matches the rhodamine filter set (Figure 8.1.14). In addition, the SYTOX® Orange stain has a much higher molar absorptivity (extinction coefficient) than propidium iodide and a far greater fluorescence enhancement upon binding DNA, suggesting that it may have a higher sensitivity as a dead-cell stain or as a nuclear counterstain. The SYTOX® Orange stain was shown to be extremely useful for DNA fragment sizing by single-molecule flow cytometry when using a Nd:YAG excitation source, with a 450-fold enhancement upon binding to dsDNA.[43]

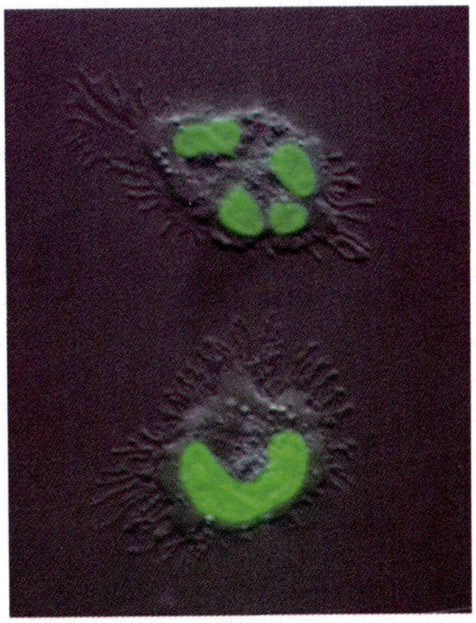

Figure 8.1.15 Human neutrophil nuclei stained with SYTO® 13 live-cell nucleic acid stain (S7575). The photo was acquired using an optical filter appropriate for fluorescein, and differential interference contrast (DIC) sequentially in a Nikon® Eclipse E800 microscope.

Cell-Permeant Cyanine Dyes: The SYTO® Nucleic Acid Stains

SYTO® Nucleic Acid Stains for DNA and RNA

The SYTO® dyes are somewhat lower-affinity nucleic acid stains that passively diffuse through the membranes of most cells. These UV- or visible light–excitable dyes can be used to stain RNA and DNA in both live and dead eukaryotic cells, as well as in gram-positive and gram-negative bacteria. We have synthesized a large number of SYTO® dyes (Table 8.3) that share several important characteristics:

- Permeability to virtually all cell membranes, including mammalian cells and bacteria (Chapter 15)
- High molar absorptivity, with extinction coefficients greater than 50,000 $cm^{-1}M^{-1}$ at visible absorption maxima
- Extremely low intrinsic fluorescence, with quantum yields typically less than 0.01 when not bound to nucleic acids
- Quantum yields typically greater than 0.4 when bound to nucleic acids

Available as blue-, green-, orange- or red-fluorescent dyes, these novel SYTO® stains provide researchers with visible light–excitable dyes for labeling DNA and RNA in live cells (Figure 8.1.15). SYTO® dyes differ from each other in one or more characteristics, including cell permeability, fluorescence enhancement upon binding nucleic acids, excitation and emission spectra (Table 8.3), DNA/RNA selectivity and binding affinity. The SYTO® dyes are compatible with a variety of fluorescence-based instruments that use either laser excitation or a conventional broadband illumination source (e.g., mercury- and xenon-arc lamps).

The SYTO® dyes can stain both DNA and RNA. In most cases, the fluorescence wavelengths and emission intensities are similar for solution measurements of DNA or RNA binding. Exceptions include the SYTO® 12 and SYTO® 14 dyes, which are about twice as fluorescent when complexed with RNA as with DNA, and SYTO® 16, which is about twice as fluorescent on DNA than RNA. Consequently, the SYTO® dyes do not act exclusively as nuclear stains in live cells and should not be equated in this regard with DNA-selective compounds such as DAPI or the Hoechst 33258 and Hoechst 33342 dyes, which readily stain cell nuclei at low concentrations in most cells. SYTO® dye–stained eukaryotic cells will generally show diffuse cytoplasmic staining, as well as nuclear staining. The SYTO® 14 dye (S7576) has been used to visualize the translocation of endogenous RNA found in polyribosome complexes in living cells.[44,45] Particularly intense staining of intranuclear bodies is frequently observed. Because these dyes are generally cell permeant and most of the SYTO® dyes contain a net positive charge at neutral pH, they may also stain mitochondria. In addition, the SYTO® dyes will stain most gram-positive and

IMPORTANT NOTICE: The products described in this manual are covered by one or more Limited Use Label License(s). Please refer to the Appendix on page 971 and Master Product List on page 975. Products are For Research Use Only. Not intended for any animal or human therapeutic or diagnostic use.

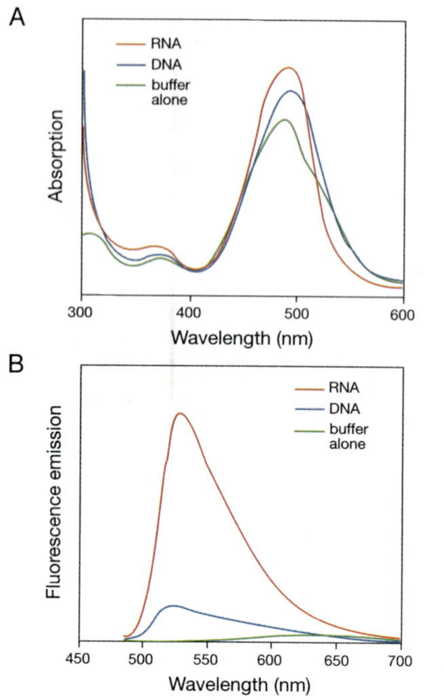

Figure 8.1.16 Relative absorption (**A**) and fluorescence emission (**B**) spectra of SYTO® RNASelect™ green-fluorescent cell stain (S32703) in the presence of *Escherichia coli* DNA or in buffer alone.

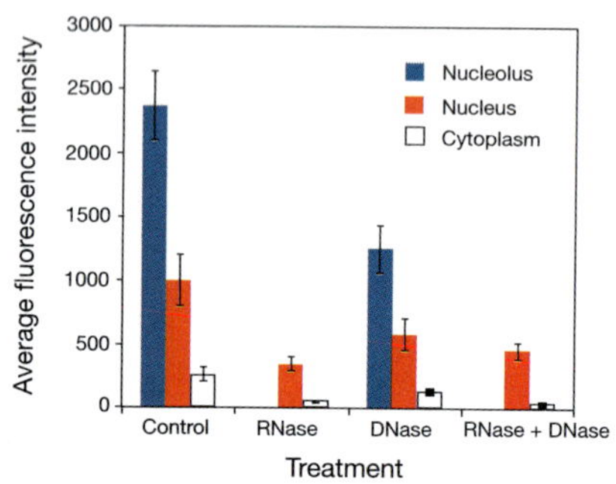

Figure 8.1.17 Methanol-fixed bovine pulmonary artery endothelial cells treated with RNase, DNase or both, and then labeled with SYTO® RNASelect™ Green cell stain (S32703). Removal of RNA with RNase prevented nucleolar labeling and greatly decreased nuclear and cytoplasmic labeling. Use of DNase resulted in less of a loss of label intensity in these cell compartments, reflecting the RNA-selective nature of this dye.

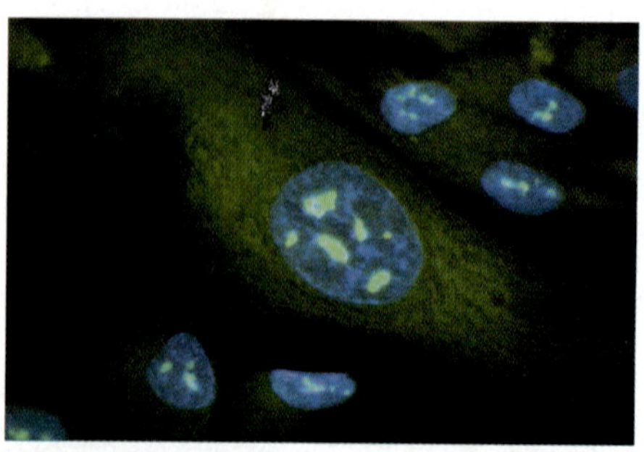

Figure 8.1.18 Methanol-fixed MRC-5 cells stained with SYTO® RNASelect™ green-fluorescent cell stain (S32703). Nuclei were stained with DAPI (D1306, D3571, D21490); the densely stained areas are nucleoli.

gram-negative bacterial cells. Dead yeast cells are brightly stained with the SYTO® dyes, and live yeast cells typically exhibit staining of both the mitochondria and the nucleus. Some of the SYTO® dyes have been reported to be useful for detecting apoptosis[46,47] (Section 15.5), and dyes structurally similar to the SYTO® dyes have been used to detect multidrug-resistant cells[48] (Section 15.6). The red-fluorescent SYTO® dyes are proving useful as counterstains (Section 12.5) when combined with green-fluorescent antibodies, lectins or the cell-impermeant SYTOX® Green nucleic acid stain.

All of the SYTO® dyes are available separately (Table 8.3), and several SYTO® dyes are included in our LIVE/DEAD® Viability Kits (Section 15.3, Table 15.2). The green-fluorescent SYBR® 14 dye, a component of our LIVE/DEAD® Sperm Viability Kit (L7011, Section 15.3) is also in the SYTO® family of dyes. To facilitate testing the SYTO® dyes in new applications, we offer several sampler kits containing sample sizes of SYTO® dyes in each color set (Table 8.3). The recommended dye concentration for cell staining depends on the assay and may vary widely but is typically 1–20 µM for bacteria, 1–100 µM for yeast and 10 nM–5 µM for other eukaryotes.

SYTO® RNASelect™ Green-Fluorescent Cell Stain

SYTO® RNASelect™ green-fluorescent cell stain (S32703, Section 15.2) is a cell-permeant nucleic acid stain that selectively stains RNA (Figure 8.1.16, Figure 8.1.17). Although virtually nonfluorescent in the absence of nucleic acids, the SYTO® RNASelect™ stain exhibits bright green fluorescence when bound to RNA (absorption/emission maxima ~490/530 nm), but only a weak fluorescent signal when bound to DNA (Figure 8.1.16). Filter sets that are suitable for imaging cells labeled with fluorescein (FITC) will work well for imaging cells stained with SYTO® RNASelect™ stain (Figure 8.1.18).

Amine-Reactive Cyanine Dye

The amine-reactive succinimidyl ester of SYBR® 101 dye (S21500) can be conjugated to peptides, proteins, drugs, polymeric matrices and biomolecules with primary amine groups. The conjugates are expected to be essentially nonfluorescent until they complex with nucleic acids, resulting in strong green fluorescence. Thus, they may be useful for studies of nucleic acid binding to various biomolecules, such as DNA-binding proteins. It is also possible that the fluorescence enhancement upon nucleic acid binding of SYBR® 101 dye conjugates will be useful for monitoring their transport into the nucleus. SYBR® 101 dye conjugates of solid or semisolid matrices (such as microspheres, magnetic particles or various resins) may be useful for detection or affinity isolation of nucleic acids.

The reactive SYBR® 101 dye may also be conjugated to amine-modified nucleic acids. Although it is possible that the SYBR® 101 dye may show some fluorescence when conjugated to amine groups on nucleic acids, they may be useful for developing homogeneous hybridization assays in which a specific sequence can be quantitated in solution without the need to separate bound and free probes. For example, a similar reactive nucleic acid stain has been used to label peptide–nucleic acid conjugates (PNA) for use as probes in real-time PCR. The labeled PNA probes exhibited a fluorescence increase upon hybridization to their complementary sequence and have been used to identify a single-base mismatch in a 10-base target sequence.[49,50]

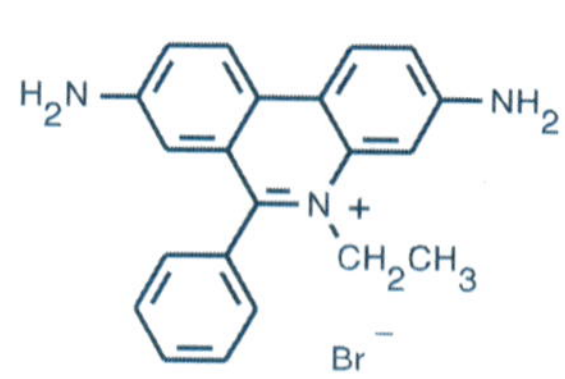

Figure 8.1.19 Ethidium bromide (15585-011).

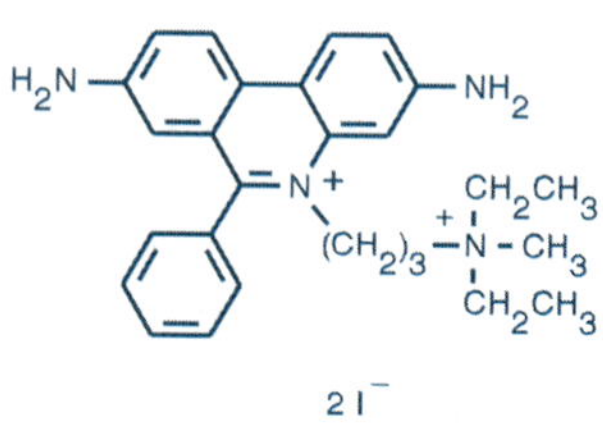

Figure 8.1.20 Propidium iodide (P1304MP).

IMPORTANT NOTICE: The products described in this manual are covered by one or more Limited Use Label License(s). Please refer to the Appendix on page 971 and Master Product List on page 975. Products are For Research Use Only. Not intended for any animal or human therapeutic or diagnostic use.

Phenanthridines and Acridines: Classic Intercalating Dyes

Cell-Impermeant Ethidium Bromide and Propidium Iodide

Ethidium bromide (EtBr, 15585-011; Figure 8.1.19) and propidium iodide (PI, P1304MP; P3566, P21493; Figure 8.1.20) are structurally similar phenanthridinium intercalators. PI is more soluble in water and less membrane-permeant than EtBr, although both dyes are generally excluded from viable cells. EtBr and PI can be excited with mercury- or xenon-arc lamps or with the argon-ion laser, making them suitable for fluorescence microscopy, confocal laser-scanning microscopy (Figure 8.1.21), flow cytometry and fluorometry. These dyes bind with little or no sequence preference at a stoichiometry of one dye per 4–5 base pairs of DNA.[51] Excitation of the EtBr–DNA complex may result in photobleaching of the dye and single-strand breaks.[52] Both EtBr and PI also bind to RNA, necessitating treatment with nucleases to distinguish between RNA and DNA. Once these dyes are bound to nucleic acids, their fluorescence is enhanced ~10-fold, their excitation maxima are shifted ~30–40 nm to the red and their emission maxima are shifted ~15 nm to the blue[53] (Figure 8.1.22, Table 8.4). Although their molar absorptivities (extinction coefficients) are relatively low, EtBr and PI exhibit sufficiently large Stokes shifts to allow simultaneous detection of nuclear DNA and fluorescein-labeled antibodies, provided that the proper optical filters are used.

PI is commonly used as a nuclear or chromosome counterstain (Section 12.5, Figure 8.1.21) and as a stain for dead cells (Section 15.2, Figure 8.1.23). EtBr is the conventional dye used for nucleic acid gel staining (Section 8.4). However, our SYBR® Gold and SYBR® Green nucleic acid gel stains are far more sensitive than EtBr, and the SYBR® Green I stain has been shown to be significantly less mutagenic than EtBr by Ames testing[54] (Section 8.4). Furthermore, our SYBR® Safe DNA gel stain, which is as sensitive as EtBr and less mutagenic in the standard Ames test, has tested negative in three mammalian cell–based assays for genotoxicity and is not classified as hazardous waste under U.S. Federal regulations (Section 8.4).

EtBr and PI are potent mutagens and must be handled with extreme care. Solutions containing EtBr or PI can be decontaminated by filtration through activated charcoal, which is then incinerated, thus providing an economical decontamination procedure.[55] Alternatively, the dyes can be completely degraded in buffer by reaction with sodium nitrite and hypophosphorous acid.[56] PI is offered as a solid (P1304MP, P21493), and both EtBr and PI are available as aqueous solutions (15585-011, P3566).

Cell-Permeant Hexidium Iodide

Hexidium iodide (H7593) is a moderately lipophilic phenanthridinium dye (Figure 8.1.24) that is permeant to mammalian cells and selectively stains almost all gram-positive bacteria in the presence of gram-negative bacteria. Our LIVE *Bac*Light™ Bacterial Gram Stain Kit and ViaGram™ Red⁺ Bacterial Gram Stain and Viability Kit (L7005, V7023; Section 15.3) use

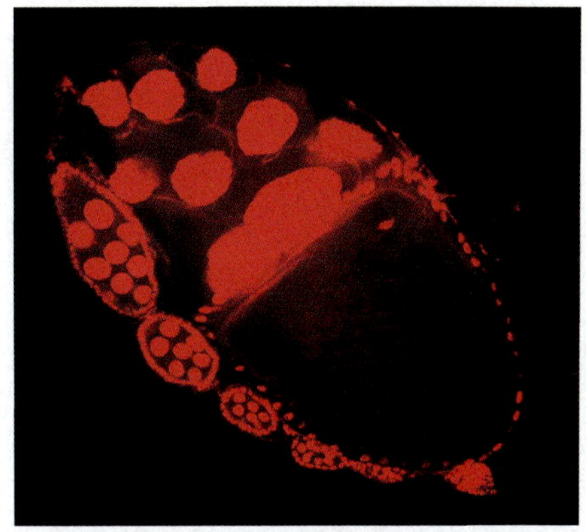

Figure 8.1.21 Day 10 of development of a *Drosophila* ovarian egg chamber assembly line. The nuclei of follicle and nurse cells were labeled with propidium iodide (P1304MP, P3566, P21493) and visualized by confocal laser-scanning microscopy using excitation by the 568 nm spectral line of an Ar-Kr laser. Image contributed by Sandra Orsulic, University of North Carolina at Chapel Hill.

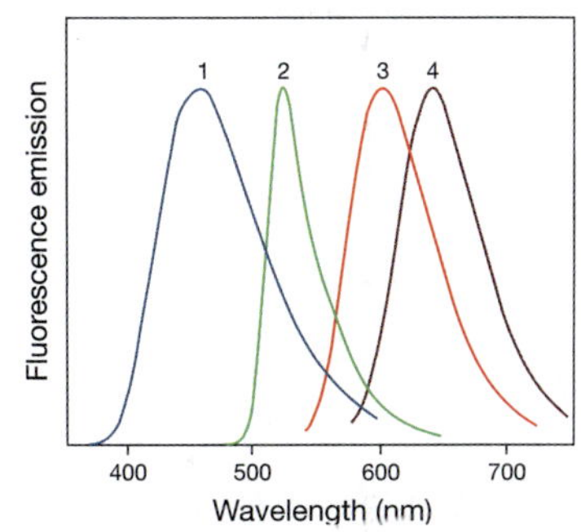

Figure 8.1.22 Normalized fluorescence emission spectra of DNA-bound **1)** Hoechst 33258 (H1398, H3569, H21491), **2)** acridine orange (A1301, A3568), **3)** ethidium bromide (15585-011) and **4)** 7-aminoactinomycin D (A1310).

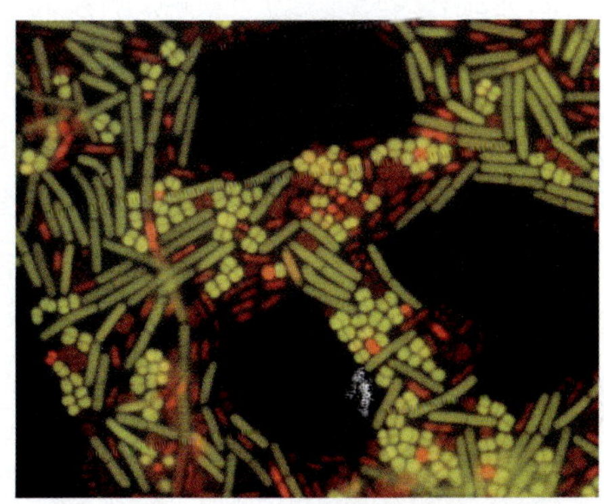

Figure 8.1.23 A mixed population of live and isopropyl alcohol–killed *Micrococcus luteus* and *Bacillus cereus* stained with the LIVE/DEAD® *Bac*Light™ Bacterial Viability Kit (L7007, L7012). Bacteria with intact cell membranes exhibit green fluorescence, whereas bacteria with damaged membranes exhibit red fluorescence. Prior to imaging, the bacteria were placed onto a polycarbonate filter and immersed in *Bac*Light™ mounting oil. This multiple exposure image was acquired with a triple-bandpass optical filter set appropriate for simultaneous imaging of DAPI, fluorescein and Texas Red® dyes.

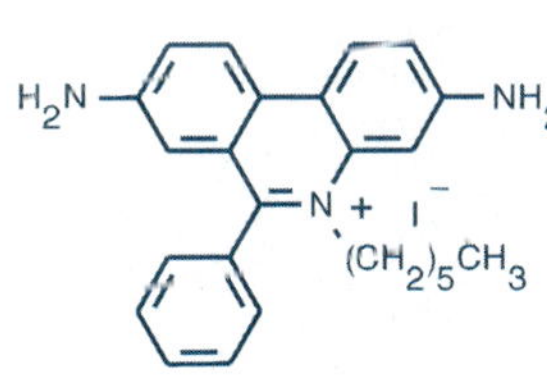

Figure 8.1.24 Hexidium iodide (H7593).

Figure 8.1.25 A mixed population of *Bacillus cereus* and *Pseudomonas aeruginosa* stained with the dye mixture provided in our LIVE *Bac*Light™ Gram Stain Kit (L7005). When live bacteria are incubated with this kit's cell-permeant nucleic acid stains, gram-positive organisms fluoresce orange and gram-negative organisms fluoresce green. The bacteria were photographed through an Omega® Optical triple bandpass filter set.

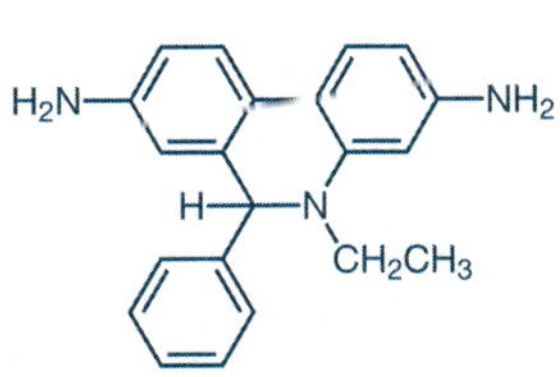

Figure 8.1.26 Dihydroethidium (hydroethidine, D1168).

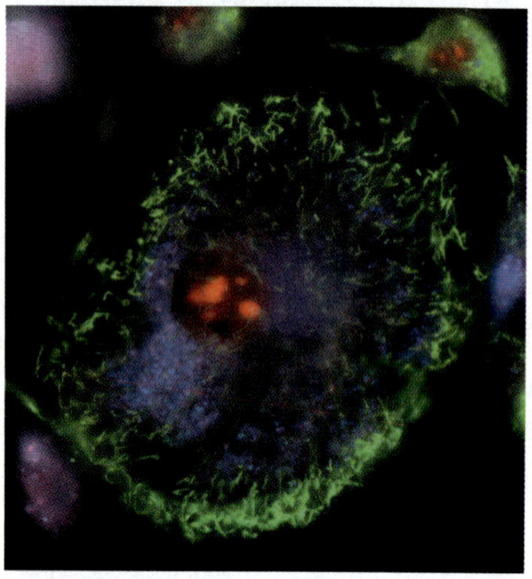

Figure 8.1.27 Live bovine pulmonary artery endothelial cells (BPAEC) were incubated with the cell-permeant, weakly blue-fluorescent dihydroethidium (D1168, D11347, D23107) and the green-fluorescent mitochondrial stain, MitoTracker® Green FM® (M7514). Upon oxidation, red-fluorescent ethidium accumulated in the nucleus.

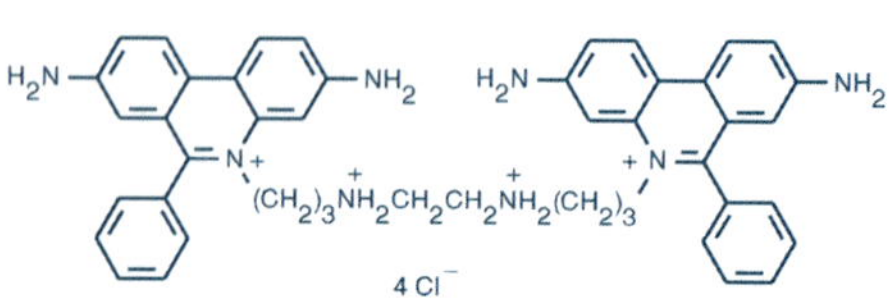

Figure 8.1.28 Ethidium homodimer-1 (EthD-1, E1169).

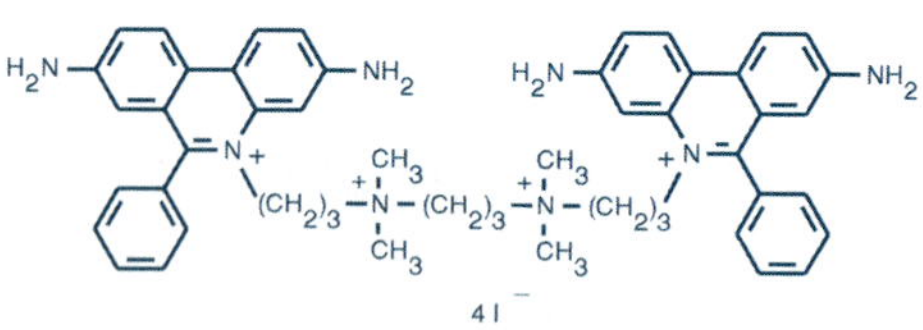

Figure 8.1.29 Ethidium homodimer-2 (EthD-2, E3599).

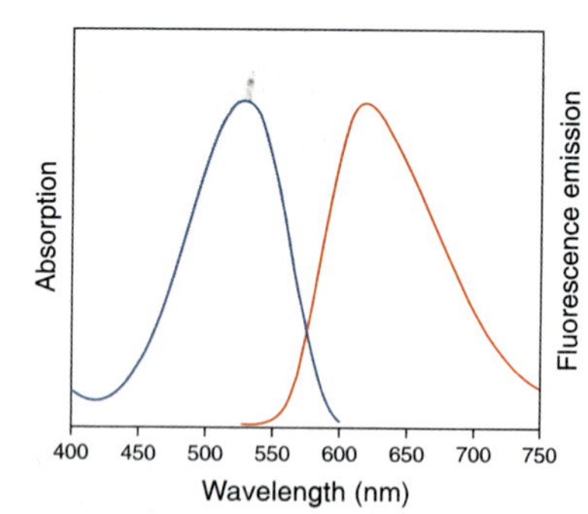

Figure 8.1.30 Absorption and fluorescence emission spectra of ethidium homodimer-1 bound to DNA.

hexidium iodide for the discrimination of bacterial gram sign (Figure 8.1.25). Hexidium iodide yields slightly shorter-wavelength spectra upon DNA binding than our ethidium or propidium dyes. Generally, both the cytoplasm and nuclei of eukaryotic cells show staining with hexidium iodide; however, mitochondria and nucleoli may also be stained.

Cell-Permeant Dihydroethidium (Hydroethidine)

Dihydroethidium (also known as hydroethidine) is a chemically reduced ethidium derivative (Figure 8.1.26) that is permeant to live cells and exhibits blue fluorescence in the cytoplasm. Many viable cells oxidize the probe to ethidium, which then fluoresces red upon DNA intercalation[57–59] (Figure 8.1.27). Dihydroethidium, which is somewhat air sensitive, is available in a 25 mg vial (D1168) or specially packaged in 10 vials of 1 mg each (D11347); the special packaging is strongly recommended when small quantities of the dye will be used at a time. Dihydroethidium is also available as a 5 mM stabilized solution in dimethylsulfoxide (D23107).

Table 8.4 Properties of classic nucleic acid stains.

Cat. No.	Dye	Ex/Em *	Fluorescence Emission Color	Applications †
A666	Acridine homodimer	431/498	Green	• Impermeant • AT-selective • High-affinity DNA binding
A1301 A3568 ‡	Acridine orange	500/526 (DNA) 460/650 (RNA)	Green/Red	• Permeant • RNA/DNA discrimination measurements • Lysosome labeling • Flow cytometry • Cell-cycle studies
A1310	7-AAD (7-amino-actinomycin D)	546/647	Red	• Weakly permeant • GC-selective • Flow cytometry • Chromosome banding
A7592	Actinomycin D	442	None	• Chromosome banding
A1324	ACMA	419/483	Blue	• AT-selective • Alternative to quinacrine for chromosome Q banding • Membrane phenomena
D1306 D3571 D21490	DAPI	358/461	Blue	• Semi-permeant • AT-selective • Cell-cycle studies • Mycoplasma detection • Chromosome and nuclei counterstain • Chromosome banding
D1168 D11347 D23107	Dihydroethidium	518/605	Red §	• Permeant • Blue fluorescent until oxidized to ethidium
15585-011	Ethidium bromide	518/605	Red	• Impermeant • dsDNA intercalator • Dead-cell stain • Chromosome counterstain • Electrophoresis • Flow cytometry • Argon-ion laser excitable
E1169	Ethidium homodimer-1 (EthD-1)	528/617	Red	• Impermeant • High-affinity DNA labeling • Dead-cell stain • Electrophoresis prestain • Argon-ion and green He-Ne laser excitable
E3599	Ethidium homodimer-2 (EthD-2)	535/624	Red	• Impermeant • Very high-affinity DNA labeling • Electrophoresis prestain

* Excitation (Ex) and emission (Em) maxima in nm. All excitation and emission maxima were determined for dyes bound to double-stranded calf thymus DNA in aqueous solution, unless otherwise indicated. † Indication of dyes as "permeant" or "impermeant" are for the most common applications; permeability to cell membranes may vary considerably with the cell type, dye concentrations and other staining conditions. ‡ Available in aqueous solution for those wishing to avoid potentially hazardous and mutagenic powders. § After oxidation to ethidium. ** Prior to photolysis; after photolysis the spectra of the dye/DNA complexes are similar to those of ethidium bromide–DNA complexes.

IMPORTANT NOTICE: The products described in this manual are covered by one or more Limited Use Label License(s). Please refer to the Appendix on page 971 and Master Product List on page 975. Products are For Research Use Only. Not intended for any animal or human therapeutic or diagnostic use.
www.invitrogen.com/probes
molecular probes® | invitrogen™ by life technologies™

High-Affinity Ethidium Homodimers

Ethidium homodimer-1 (EthD-1, E1169; Figure 8.1.28) and ethidium homodimer-2 (EthD-2, E3599; Figure 8.1.29) strongly bind to dsDNA, ssDNA, RNA and oligonucleotides with a significant fluorescence enhancement (>40-fold). EthD-1 also binds with high affinity to triplex nucleic acid structures.[60] One molecule of EthD-1 binds per four base pairs in dsDNA,[4] and the dye's intercalation is not sequence selective.[61] It was originally reported that only one of the two phenanthridinium rings of EthD-1 is bound at a time;[4] subsequent reports indicate that bis-intercalation appears to be involved in staining both double-stranded and triplex nucleic acids.[20,60]

The spectra and other properties of the EthD-1 and EthD-2 dimers are almost identical (Figure 8.1.30). However, the DNA affinity of EthD-2 is about twice that of EthD-1. EthD-2 is also about twice as fluorescent bound to dsDNA than to RNA. Because both EthD-1 and EthD-2 can be excited with UV light or by the 488 nm spectral line of the argon-ion laser, either dye can be used in combination with the TOTO®-1, YOYO®-1 or SYTOX® Green nucleic acid stains for multicolor experiments (Figure 8.1.31). The ethidium homodimer dyes are impermeant to cells with intact membranes, a property that makes EthD-1 useful as a dead-cell indicator in our

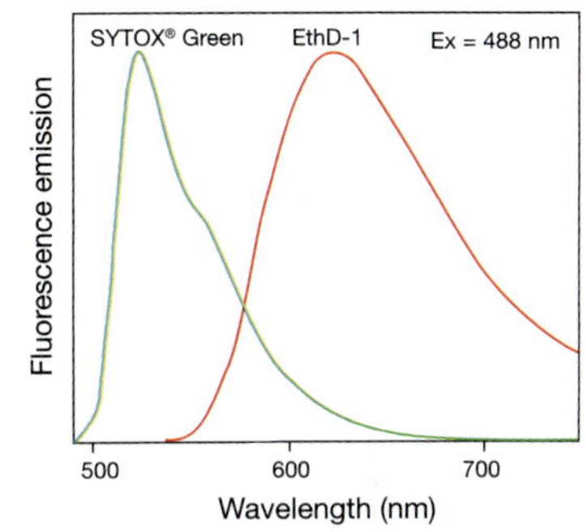

Figure 8.1.31 Normalized fluorescence emission spectra of DNA-bound SYTOX® Green nucleic acid stain (S7020) and ethidium homodimer-1 (EthD-1, E1169). Both spectra were obtained using excitation at 488 nm.

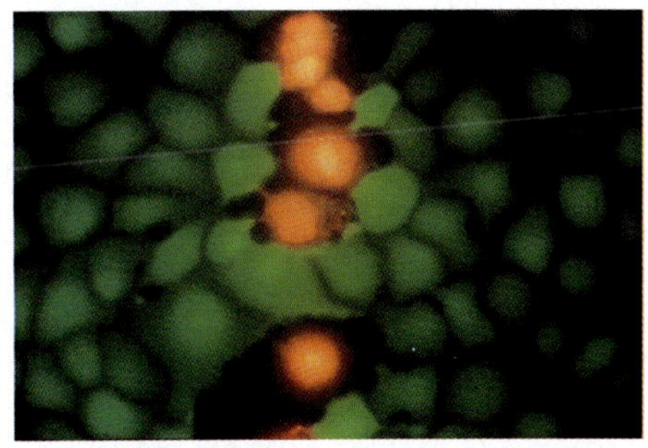

Figure 8.1.32 Live and dead kangaroo rat (PtK2) cells stained with ethidium homodimer-1 and the esterase substrate calcein AM, both of which are provided in our LIVE/DEAD® Viability/Cytotoxicity Kit (L3224). Live cells fluoresce a bright green, whereas dead cells with compromised membranes fluoresce red-orange.

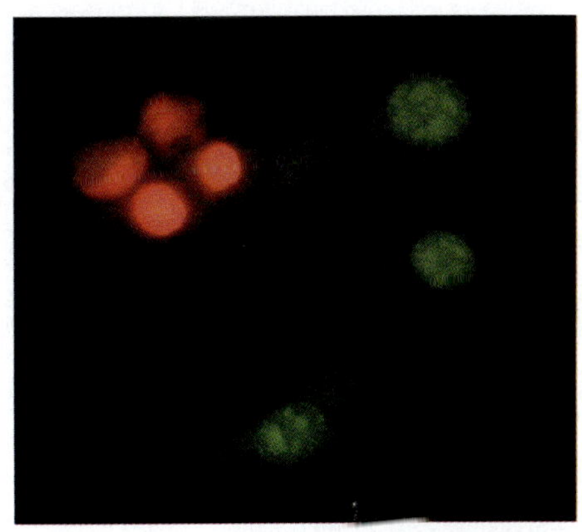

Figure 8.1.33 A mixed population of trinitrophenyl-sensitized and nonsensitized bovine pulmonary artery endothelial cells that have been probed with anti–dinitrophenyl-KLH antibody (A6430), incubated with rabbit complement and then stained using the LIVE/DEAD® Reduced Biohazard Viability/Cytotoxicity Kit #1 (L7013). With this kit, the membrane-permeant SYTO® 10 dye labels the nucleic acids of live cells with green fluorescence, and the membrane-impermeant DEAD Red™ dye labels nucleic acids of membrane-compromised cells with red fluorescence. Subsequent aldehyde-based fixation inactivates pathogens without distorting the original staining pattern. This multiple-exposure image was acquired with bandpass optical filters appropriate for fluorescein and Texas Red® dyes.

Table 8.4 Properties of classic nucleic acid stains—*continued.*

Cat. No.	Dye	Ex/Em *	Fluorescence Emission Color	Applications †
E1374	Ethidium monoazide	464/625 (unbound)**	Red	• Impermeant • Photocrosslinkable • Compatible with fixation procedures
H7593	Hexidium iodide	518/600	Red	• Permeant, except gram-negative bacteria • Stains nuclei and cytoplasm of eukaryotes and some bacteria
H1398 H3569 ‡ H21491	Hoechst 33258 (bis-benzimide)	352/461	Blue	• Permeant • AT-selective • Minor groove–binding • dsDNA-selective binding • Cell-cycle studies • Chromosome and nuclear counterstain
H1399 H3570 ‡ H21492	Hoechst 33342	350/461	Blue	• Permeant • AT-selective • Minor groove–binding • dsDNA-selective binding • Cell-cycle studies • Chromosome and nuclear counterstain
H21486	Hoechst 34580	392/498	Blue	• Permeant • AT-selective • Minor groove–binding • dsDNA-selective binding • Cell-cycle studies • Chromosome and nuclear counterstain
H22845	Hydroxy-stilbamidine	385/emission varies with nucleic acid	Varies	• AT-selective • Spectra dependent on secondary structure and sequence • RNA/DNA discrimination • Nuclear stain in tissue
L7595	LDS 751	543/712 (DNA) 590/607 (RNA)	Red/infrared	• Permeant • High Stokes shift • Long-wavelength spectra • Flow cytometry
N21485	Nuclear yellow	355/495	Yellow	• Impermeant • Nuclear counterstain
P1304MP P3566 ‡ P21493	Propidium iodide (PI)	530/625	Red	• Impermeant • Dead-cell stain • Chromosome and nuclear counterstain

* Excitation (Ex) and emission (Em) maxima in nm. All excitation and emission maxima were determined for dyes bound to double-stranded calf thymus DNA in aqueous solution, unless otherwise indicated. † Indication of dyes as "permeant" or "impermeant" are for the most common applications; permeability to cell membranes may vary considerably with the cell type, dye concentrations and other staining conditions. ‡ Available in aqueous solution for those wishing to avoid potentially hazardous and mutagenic powders. § After oxidation to ethidium. ** Prior to photolysis; after photolysis the spectra of the dye/DNA complexes are similar to those of ethidium bromide–DNA complexes.

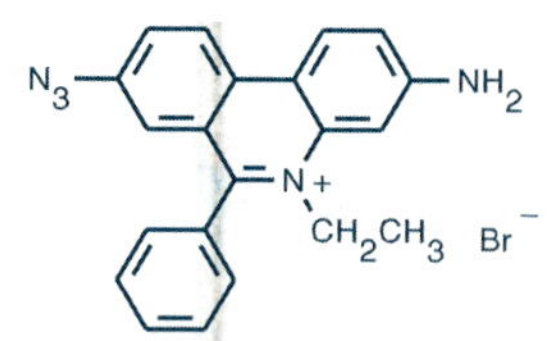

Figure 8.1.34 Ethidium monoazide bromide (EMA, E1374).

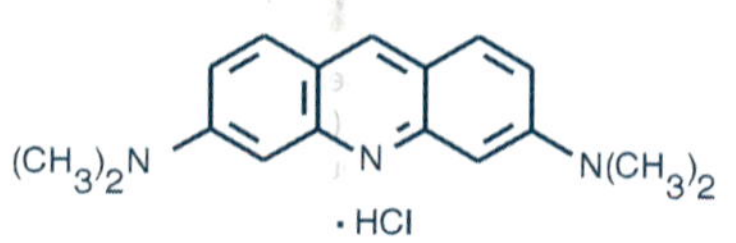

Figure 8.1.35 Acridine orange (A1301).

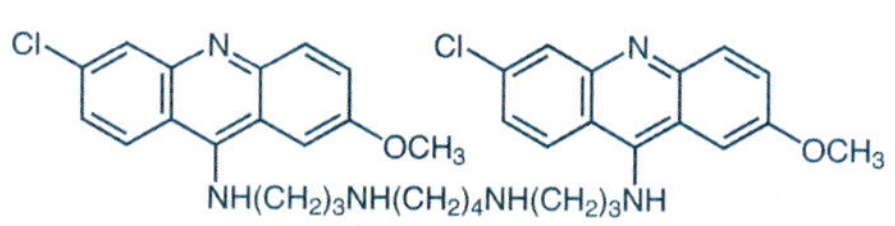

Figure 8.1.36 Acridine homodimer (bis-(6-chloro-2-methoxy-9-acridinyl)spermine, A666).

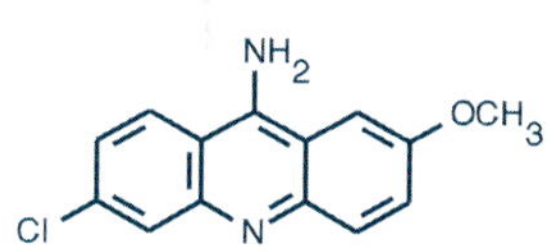

Figure 8.1.37 9-Amino-6-chloro-2-methoxyacridine (ACMA, A1324).

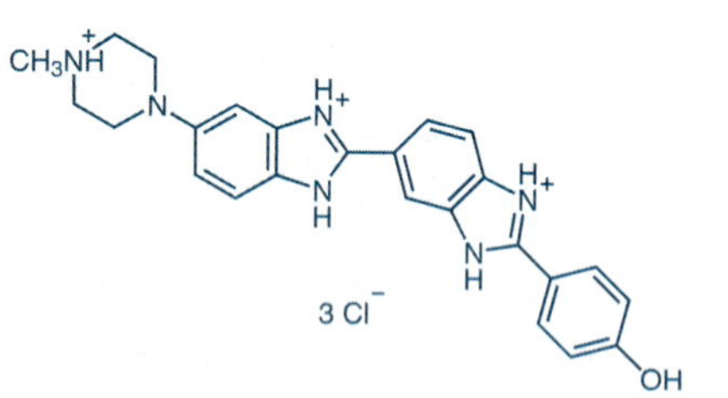

Figure 8.1.38 Hoechst 33258 (bis-benzimide, H1398).

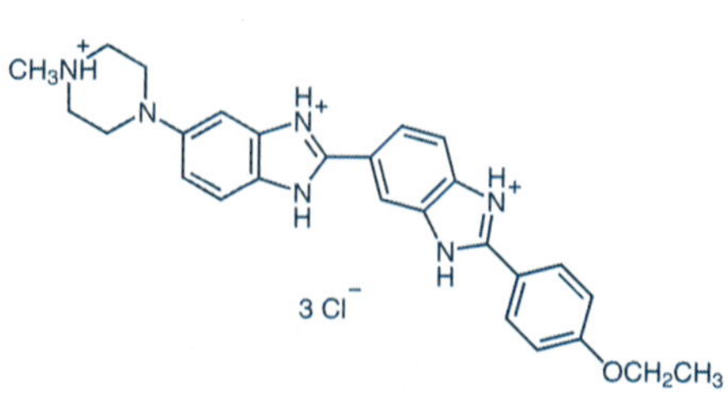

Figure 8.1.39 Hoechst 33342 (H1399).

LIVE/DEAD® Viability/Cytotoxicity Kit (L3224, Section 15.3, Figure 8.1.32) and EthD-2 (as Dead Red™ nucleic acid stain) a suitable dead-cell indicator in our LIVE/DEAD® Reduced Biohazard Cell Viability Kit #1 (L7013, Section 15.3, Figure 8.1.33). These dyes have also been used to detect DNA in solution,[61] although they are not as sensitive or as easy to use as our Quant-iT™ PicoGreen® dsDNA reagent (Section 8.3).

Ethidium Monoazide: A Photocrosslinking Reagent

Nucleic acids can be covalently photolabeled by various DNA intercalators. Ethidium monoazide (E1374, Figure 8.1.34) is a fluorescent photoaffinity label that, after photolysis, binds covalently to nucleic acids both in solution and in cells that have compromised membranes.[62–66] The quantum yield for covalent photolabeling by ethidium monoazide is unusually high (>0.4).

The membrane-impermeant ethidium monoazide is reported to only label dead cells and is therefore particularly useful for assaying the viability of pathogenic cells (Section 15.2). A mixed population of live and dead cells incubated with this reagent can be illuminated with a visible-light source, washed, fixed and then analyzed in order to determine the viability of the cells at the time of photolysis.[67] This method not only reduces some of the hazards inherent in working with pathogenic cells, but also is compatible with immunocytochemical analyses requiring fixation. We have developed alternative assays for determining the original viability of fixed samples and provide these in the LIVE/DEAD® Reduced Biohazard Cell Viability Kit #1 (L7013) and the LIVE/DEAD® Fixable Dead Cell Stain Kits, which are described in Section 15.3.

In addition to its utility as a viability indicator, ethidium monoazide has been used to irreversibly label the DNA of *Candida albicans* in order to investigate phagocytic capacity of leukocytes.[68] Ethidium monoazide has also been employed to "footprint" drug-binding sites on DNA,[69] to probe for ethidium-binding sites in DNA[70] and transfer RNA (tRNA)[65] and to selectively photoinactivate the expression of genes in vertebrate cells.[71]

Acridine Orange: A Dual-Fluorescence Nucleic Acid Stain

We offer highly purified, flow cytometry–grade acridine orange, a dye that interacts with DNA and RNA by intercalation or electrostatic attractions. In condensed chromatin, however, the bulk of DNA is packed in a way that does not allow efficient acridine orange intercalation.[72] This cationic dye (Figure 8.1.35) has green fluorescence with an emission maximum at 525 nm when bound to DNA. Upon association with RNA, its emission is shifted to ~650 nm (red fluorescence). Acridine orange is available as a solid (A1301) and, for ease of handling, as a 10 mg/mL aqueous solution (A3568).

AT-Selective Acridine Homodimer

The water-soluble acridine homodimer bis-(6-chloro-2-methoxy-9-acridinyl)spermine (A666, Figure 8.1.36) is one of several acridine dimers that have been described in the literature. This dye has extremely high affinity for AT-rich regions of nucleic acids, making it particularly useful for chromosome banding.[73,74] Acridine homodimer emits a blue-green fluorescence when bound to DNA, yielding fluorescence that is proportional to the fourth power of the AT base-pair content.[75] Because of its greater brightness and photostability, acridine homodimer has been recommended as an alternative to quinacrine for Q banding.[73]

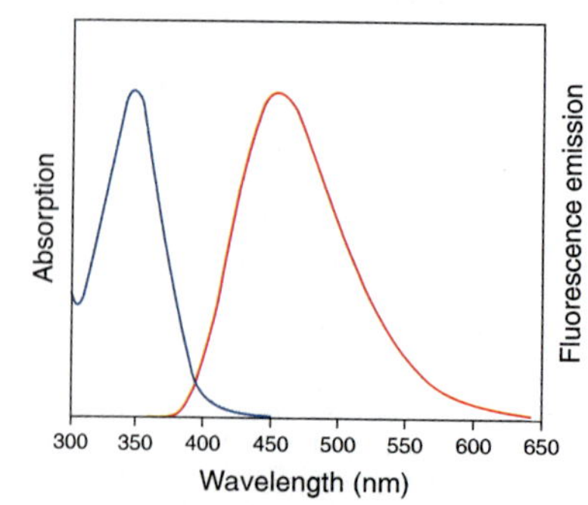

Figure 8.1.40 Absorption and fluorescence emission spectra of Hoechst 33258 bound to DNA.

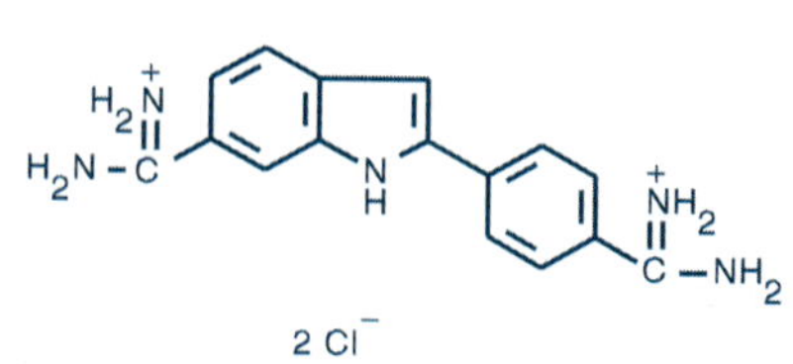

Figure 8.1.41 4′,6-Diamidino-2-phenylindole, dihydrochloride (DAPI, D1306).

IMPORTANT NOTICE: The products described in this manual are covered by one or more Limited Use Label License(s). Please refer to the Appendix on page 971 and Master Product List on page 975. Products are For Research Use Only. Not intended for any animal or human therapeutic or diagnostic use.
www.invitrogen.com/probes

AT-Selective ACMA

ACMA (9-amino-6-chloro-2-methoxyacridine, A1324, Figure 8.1.37) is a DNA intercalator that selectively binds to poly(d(AT)) with a binding affinity constant of $2 \times 10^5\ M^{-1}$ at pH 7.4.[76,77] Excitation of the ACMA–DNA complex (excitation/emission maxima ~419/483 nm) is possible with most UV-light sources, making it compatible for use with both shorter- and longer-wavelength dyes. ACMA also apparently binds to membranes in the energized state and becomes quenched if a pH gradient forms.[78] It has been extensively employed to follow cation and anion movement across membranes[78–81] and to study the proton-pumping activity of various membrane-bound ATPases[82,83] (Section 20.3).

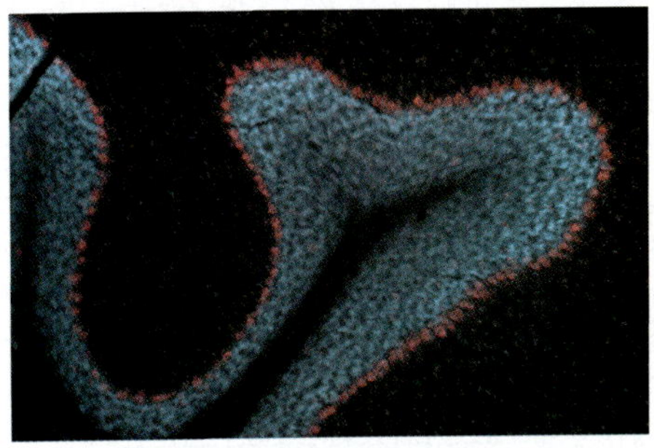

Figure 8.1.42 Neural somata labeled with NeuroTrace® 530/615 red-fluorescent Nissl stain (N21482). Nuclei in this mouse cerebellum section were counterstained with DAPI (D1306, D3571, D21490). The image is a composite of two micrographs acquired using filters appropriate for tetramethylrhodamine and DAPI.

Indoles and Imidazoles: Classic Minor Groove–Binding Dyes

DNA-Selective Hoechst Dyes

The bisbenzimide dyes—Hoechst 33258 (Figure 8.1.38), Hoechst 33342 (Figure 8.1.39) and Hoechst 34580—are cell membrane–permeant, minor groove–binding DNA stains (Figure 8.1.1) that fluoresce bright blue upon binding to DNA. Hoechst 33342 has slightly higher membrane permeability than Hoechst 33258,[53] but both dyes are quite soluble in water (up to 2% solutions can be prepared) and relatively nontoxic. These Hoechst dyes, which can be excited with the UV spectral lines of the argon-ion laser and by most conventional fluorescence excitation sources, exhibit relatively large Stokes shifts (Figure 8.1.40) (excitation/emission maxima ~350/460 nm), making them suitable for multicolor labeling experiments. Hoechst 34580[84] has somewhat longer-wavelength spectra than the other Hoechst dyes when bound to nucleic acids.

The Hoechst 33258 and Hoechst 33342 dyes have complex, pH-dependent spectra when not bound to nucleic acids, with a much higher fluorescence quantum yield at pH 5 than at pH 8. Their fluorescence is also enhanced by surfactants such as sodium dodecyl sulfate[85] (SDS). These dyes appear to show a wide spectrum of sequence-dependent DNA affinities and bind with sufficient strength to poly(d(AT)) sequences that they can displace several known DNA intercalators.[86] They also exhibit multiple binding modes and distinct fluorescence emission spectra that are dependent on dye:base pair ratios.[87] Hoechst dyes are used in many cellular applications, including cell-cycle and apoptosis studies (Section 15.4, Section 15.5), and they are common nuclear counterstains (Section 12.5). Hoechst 33258, which is selectively toxic to malaria parasites,[88] is also useful for flow cytometric screening of blood samples for malaria parasites and for assessing their susceptibility to drugs.[89–91]

The Hoechst 33258 and Hoechst 33342 dyes are available as solids (H1398, H1399), as high-purity solids (H21491, H21492) and, for ease of handling, as 10 mg/mL aqueous solutions (H3569, H3570). The Hoechst 34580 dye is available as a solid (H21486).

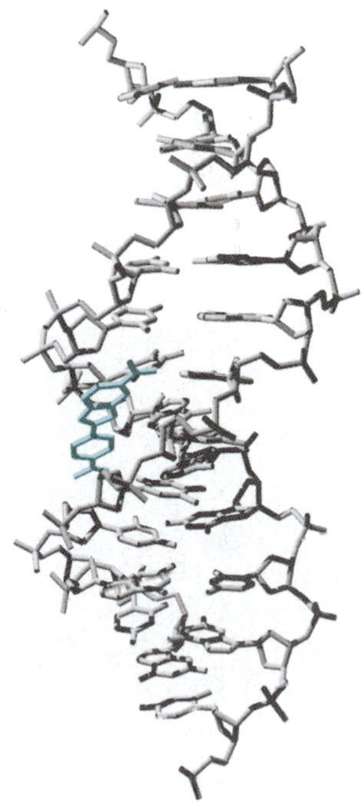

Figure 8.1.43 X-ray crystal structure of DAPI (D1306, D3571, D21490) bound to DNA; the image was derived from data submitted to the Protein Data Bank (number PDB 1D30, www.rcsb.org/pdb) (Nucleic Acids Res (2000) 28:235). X-ray crystallography shows that DAPI binds to DNA in the minor groove (J Biomol Struct Dyn (1989) 7:477).

AT-Selective DAPI

DAPI (4′,6-diamidino-2-phenylindole; D1306, D3571, D21490; Figure 8.1.41) shows blue fluorescence (Figure 8.1.42) upon binding DNA and can be excited with a mercury-arc lamp or with the UV lines of the argon-ion laser. Like the Hoechst dyes, the blue-fluorescent DAPI stain apparently associates with the minor groove of dsDNA (Figure 8.1.43), preferentially binding to AT clusters;[92] there is evidence that DAPI also binds to DNA sequences that contain as few as two consecutive AT base pairs, perhaps employing a different binding mode.[93–95] DAPI is thought to employ an intercalating binding mode with RNA that is AU selective.[96]

The selectivity of DAPI for DNA over RNA is reported to be greater than that displayed by ethidium bromide and propidium iodide.[97] Furthermore, the DAPI–RNA complex exhibits a longer-wavelength fluorescence emission maximum than the DAPI–dsDNA complex (~500 nm versus ~460 nm) but a quantum yield that is only about 20% as high.[98]

Binding of DAPI to dsDNA produces an ~20-fold fluorescence enhancement, apparently due to the displacement of water molecules from both DAPI and the minor groove.[99] Although the Hoechst dyes may be somewhat brighter in some applications, their photostability when bound to dsDNA is less than that of DAPI. In the presence of appropriate salt concentrations, DAPI usually does not exhibit fluorescence enhancement upon binding to ssDNA or GC base pairs.[100] However, the fluorescence of DAPI does increase significantly upon binding to detergents,[101] dextran sulfate,[102] polyphosphates and other polyanions.[103] A review by Kapuscinski discusses the mechanisms of DAPI binding to nucleic acids, its spectral properties and its uses

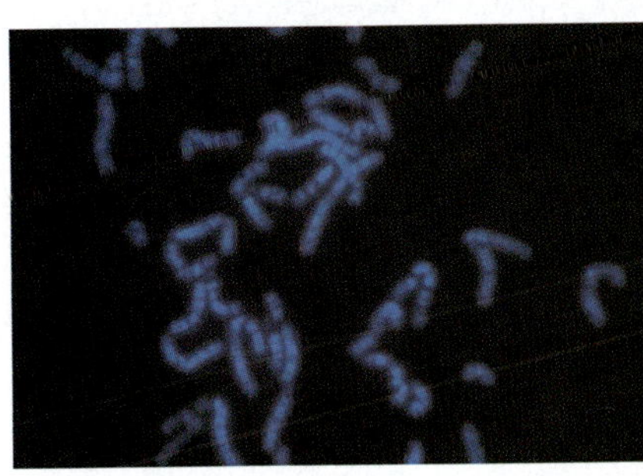

Figure 8.1.44 Human metaphase chromosomes stained with DAPI (D1306, D3571, D21490).

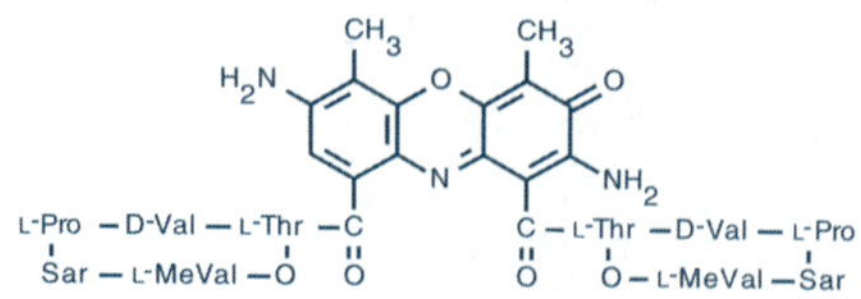

Figure 8.1.45 7-Aminoactinomycin D (7-AAD, A1310).

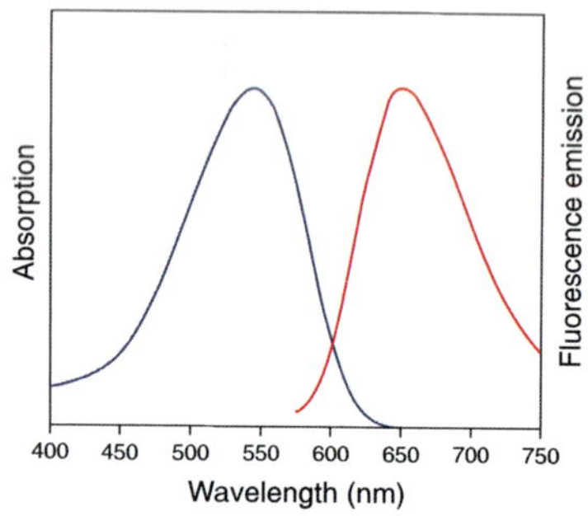

Figure 8.1.46 Absorption and fluorescence emission spectra of 7-aminoactinomycin D (7-AAD) bound to DNA.

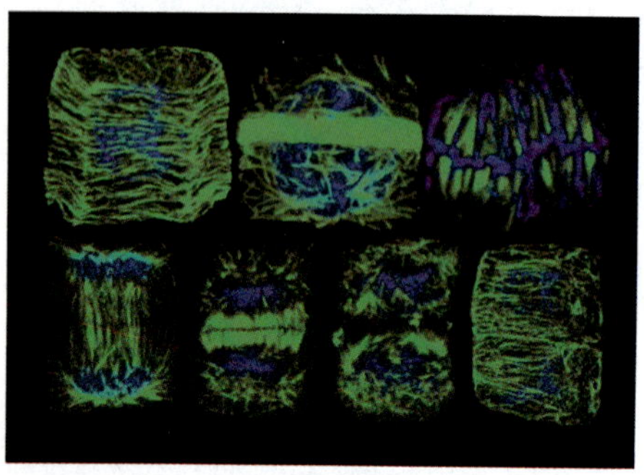

Figure 8.1.47 Panel of confocal micrographs showing cells from wheat root tips in seven stages of the cell cycle. DNA was stained with 7-aminoactinomycin D (A1310), and microtubules were labeled with an anti–ß-tubulin antibody in conjunction with a fluorescein-labeled secondary antibody. Cells vary in width from about 15 µm to about 25 µm. The stages are (from top left): interphase cortical microtubule array; preprophase band of microtubules (predicts future plane of division); metaphase mitotic spindle; telophase, showing early phragmoplast and cell plate; fully developed phragmoplast during cytokinesis; late cytokinesis (plane of division matching plane of earlier pre-prophase band); restoration of cortical arrays in daughter cells. Image contributed by B.E.S. Gunning, Plant Cell Biology Group, Research School of Biological Sciences, Australian National University. Used with permission from Gunning, B.E.S. and Steer, M.W., Plant Cell Biology: Structure and Function, Jones and Bartlett Publishers (1995).

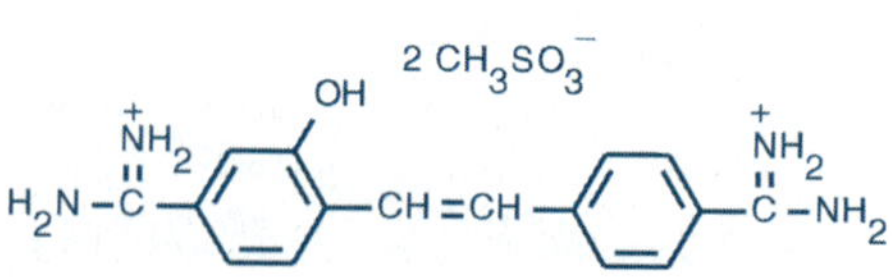

Figure 8.1.48 Hydroxystilbamidine, methanesulfonate (H22845).

in flow cytometry and for chromosome staining.[104] DAPI is an excellent nuclear counterstain, showing a distinct banding pattern in chromosomes (Figure 8.1.44), and we have included it in our SelectFX® Nuclear Labeling Kit (S33025, Section 12.5). DAPI is quite soluble in water but has limited solubility in phosphate-buffered saline.

We also offer DAPI premixed with our ProLong® Gold and *SlowFade*® Gold antifade reagents (P36931, P36935, S36938, S36939; Section 23.1). This combination of nucleic acid dye and antifade reagent permits simultaneous staining and protection of the stained sample from photobleaching.

Other Nucleic Acid Stains

7-Aminoactinomycin D and Actinomycin D: Fluorescent Intercalators

7-AAD (7-aminoactinomycin D, A1310; Figure 8.1.45) is a fluorescent intercalator that undergoes a spectral shift upon association with DNA. 7-AAD–DNA complexes can be excited by the argon-ion laser and emit beyond 610 nm (Table 8.4, Figure 8.1.22, Figure 8.1.46), making this nucleic acid stain useful for multicolor fluorescence microscopy (Figure 8.1.47), confocal laser-scanning microscopy and immunophenotyping by flow cytometry.[105–110] 7-AAD appears to be generally excluded from live cells, although it has been reported to label the nuclear region of live cultured mouse L cells and salivary gland polytene chromosomes of *Chironomus thummi* larvae.[111] 7-AAD binds selectively to GC regions of DNA,[112] yielding a distinct banding pattern in polytene chromosomes and chromatin.[111,113] This sequence selectivity has been exploited for chromosome banding studies.[114]

Actinomycin D (A7592) is a nonfluorescent intercalator that exhibits high GC selectivity and causes distortion at its binding site.[115] Binding of the nonfluorescent actinomycin D to nucleic acids changes the absorbance of the dye.[116] Like 7-AAD, actinomycin D has been used for chromosome banding studies.[117] Binding of actinomycin D to ssDNA is reported to inhibit reverse transcriptase and other polymerases.[118]

Hydroxystilbamidine

The trypanocidal drug hydroxystilbamidine (H22845, Figure 8.1.48) is an interesting probe of nucleic acid conformation; its nucleic acid staining properties were first described in 1973.[119] Hydroxystilbamidine, a nonintercalating dye, exhibits AT-selective binding that is reported to favor regions of nucleic acids that have secondary structure. The interaction between hydroxystilbamidine and DNA has been investigated using binding isotherms[120] and temperature-jump relaxation studies.[121]

Hydroxystilbamidine has some unique spectral properties upon binding nucleic acids. At pH 5, the free dye exhibits UV excitation maxima at ~330 nm and ~390 nm, with dual emission at ~450 nm and ~600 nm (Figure 8.1.49). Although the red-fluorescent component remains present when bound to DNA, it is never observed when the dye is bound to RNA, permitting potential discrimination to be made between these two types of nucleic acids. The enhancement of its metachromatic fluorescence upon binding to DNA is proportional to the square of the AT base-pair content. Hydroxystilbamidine is reported to exhibit red fluorescence when bound to calf thymus DNA and T5 DNA, orange fluorescence with *Micrococcus lysodeikticus* DNA and blue-violet fluorescence on poly(d(AT)).[119] It has been used for the treatment of myeloma, binding selectively to myeloma cells in the bone marrow.[122]

Murgatroyd described the use of hydroxystilbamidine's metachromatic fluorescence properties for the selective permanent staining of DNA (with yellow fluorescence), mucosubstances and elastic fibers in paraffin sections.[123] He also reported that hydroxystilbamidine (as its isethionate salt, which we do not currently offer) is nonmutagenic in *Salmonella typhimurium* by the Ames test.[123]

Long-Wavelength LDS 751 Dye

LDS 751 (L7595, Figure 8.1.50) is a cell-permeant nucleic acid stain that has been used to discriminate intact nucleated cells from nonnucleated and damaged nucleated cells,[124,125] as well as to identify distinct cell types in mixed populations of neutrophils, leukocytes and monocytes by flow cytometry.[126] LDS 751, which has its peak excitation at ~543 nm on dsDNA, can be excited by the argon-ion laser at 488 nm and is particularly useful in multicolor analyses due to its long-wavelength emission maximum (~712 nm). Binding of LDS 751 to dsDNA results in an ~20-fold fluorescence enhancement. When LDS 751 binds to RNA, we have observed a significant red shift

IMPORTANT NOTICE: The products described in this manual are covered by one or more Limited Use Label License(s). Please refer to the Appendix on page 971 and Master Product List on page 975. Products are For Research Use Only. Not intended for any animal or human therapeutic or diagnostic use.
www.invitrogen.com/probes
molecular probes® | invitrogen™ by life technologies™

in its excitation maximum to 590 nm and blue shift in its emission maxima to 607 nm, which may permit its use to discriminate DNA and RNA in cells. A report has ascribed the name LDS 751 to a dye called styryl 8;[127] however, their chemical structures are not the same.

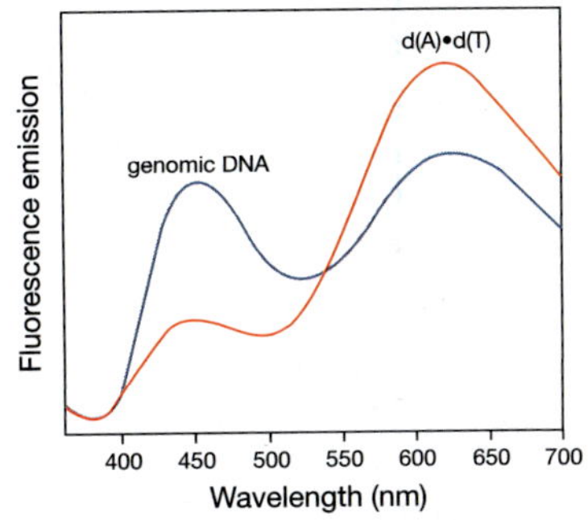

Figure 8.1.49 Fluorescence spectra of hydroxystilbamidine bound to different forms of DNA. Hydroxystilbamidine (H22845) was incubated with either calf thymus DNA (red) or a hybrid of poly(d(A)) and poly(d(T)) homopolymers (blue) in 50 mM sodium acetate, pH 5.0. The fluorescence emission spectra changes when the dye is bound to AT-rich DNA versus calf-thymus genomic DNA.

NeuroTrace® Fluorescent Nissl Stains

The Nissl substance, described by Franz Nissl more than 100 years ago, is unique to neuronal cells.[128] Composed of an extraordinary amount of rough endoplasmic reticulum, the Nissl substance reflects the unusually high protein synthesis capacity of neuronal cells. Various fluorescent or chromophoric "Nissl stains" have been used for this counterstaining, including acridine orange,[129] ethidium bromide,[129] neutral red (N3246, Section 15.2), cresyl violet,[130] methylene blue, safranin-O and toluidine blue-O.[131] We have developed five fluorescent Nissl stains (Table 14.2) that not only provide a wide spectrum of fluorescent colors for staining neurons, but also are far more sensitive than the conventional dyes:

- NeuroTrace® 435/455 blue-fluorescent Nissl stain (N21479)
- NeuroTrace® 500/525 green-fluorescent Nissl stain (N21480)
- NeuroTrace® 515/535 yellow-fluorescent Nissl stain (N21481)
- NeuroTrace® 530/615 red-fluorescent Nissl stain (Figure 8.1.42) (N21482)
- NeuroTrace® 640/660 deep red–fluorescent Nissl stain (N21483)

In addition, the Nissl substance redistributes within the cell body in injured or regenerating neurons. Therefore, these Nissl stains can also act as markers for the physiological state of the neuron. Staining by the Nissl stains is completely eliminated by pretreatment of tissue specimens with RNase; however, these dyes are not specific stains for RNA in solutions. The strong fluorescence (emission maximum ~515–520 nm) of NeuroTrace® 500/525 green-fluorescent Nissl stain (N21480) makes it a preferred dye for use as a counterstain in combination with orange- or red-fluorescent neuroanatomical tracers such as DiI[132] (D282, D3911, V22885; Section 14.4).

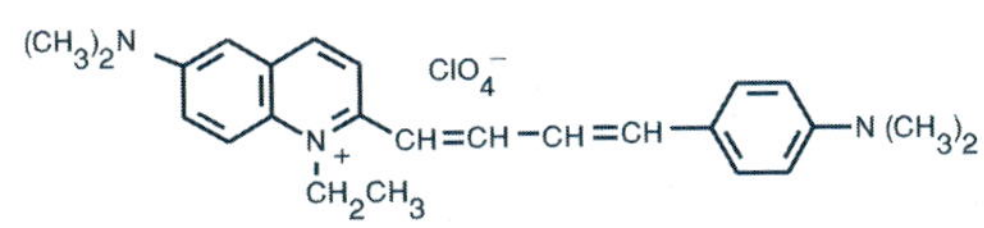

Figure 8.1.50 LDS 751 (L7595).

REFERENCES

1. Nucleic Acids Res (1995) 23:1215; **2.** Nucleic Acids Res (1991) 19:327; **3.** Biochemistry (1978) 17:5071; **4.** Biochemistry (1978) 17:5078; **5.** Nucleic Acids Res (1992) 20:2803; **6.** Proc Natl Acad Sci U S A (1990) 87:3851; **7.** Nucleic Acids Res (1993) 21:5720; **8.** Nature (1992) 359:859; **9.** Biotechniques (1991) 10:616; **10.** J Phys Chem (1995) 99:17936; **11.** Cytometry (1986) 7:508; **12.** Appl Environ Microbiol (2000) 66:2283; **13.** Appl Environ Microbiol (1995) 61:3623; **14.** Anal Biochem (1994) 221:78; **15.** J Am Chem Soc (1994) 116:8459; **16.** J Phys Chem (1994) 98:10313; **17.** J Biomol Struct Dyn (1998) 16:205; **18.** J Am Chem Soc (2006) 128:7661; **19.** Biophys J (2009)97:835; **20.** Nucleic Acids Res (1995) 23:2413; **21.** Biochemistry (1995) 34:8542; **22.** Biochemistry (1998) 37:16863; **23.** Cytometry (1999) 37:230; **24.** Bioconjug Chem (1999) 10:824; **25.** Nucleic Acids Res (1996) 24:859; **26.** Nucleic Acids Res (1995) 23:753; **27.** Acta Chem Scand (1998) 52:641; **28.** Biochem Mol Biol Int (1994) 34:1189; **29.** Biopolymers (1997) 41:481; **30.** Appl Environ Microbiol (1994) 60:3284; **31.** Cancer Res (1997) 57:3804; **32.** Blood (1996) 87:4959; **33.** J Immunol Methods (1995) 185:249; **34.** J Exp Med (1995) 182:1759; **35.** Br J Pharmacol (2000) 130:513; **36.** Br J Pharmacol (1998) 125:1194; **37.** J Biol Chem (2001) 276:125; **38.** Cytometry (1994) 17:185; **39.** Cytometry (1994) 17:310; **40.** Cytometry (1994) 15:267; **41.** Nucleic Acids Res (2003) 31:2561; **42.** J Phys Chem B (2000) 104:7221; **43.** Anal Biochem (2000) 286:138; **44.** Proc Natl Acad Sci U S A (1997) 94:14804; **45.** J Neurosci (1996) 16:7812; **46.** Cytometry (1995) 21:265; **47.** Neuron (1995) 15:961; **48.** Cytometry (1995) 20:218; **49.** Anal Biochem (2000) 287:179; **50.** Anal Biochem (2000) 281:26; **51.** J Mol Biol (1965) 13:269; **52.** Biochemistry (1990) 29:981; **53.** Methods Cell Biol (1989) 30:417; **54.** Mutat Res (1999) 439:37; **55.** Chromatographia (1990) 29:167; **56.** Anal Biochem (1987) 162:453; **57.** J Immunol Methods (1994) 170:117; **58.** FEMS Microbiol Lett (1992) 101:173; **59.** J Histochem Cytochem (1986) 34:1109; **60.** Bioorg Med Chem (1995) 3:701; **61.** Anal Biochem (1979) 94:259; **62.** Biochemistry (1991) 30:5644; **63.** Photochem Photobiol (1986) 43:7; **64.** J Biol Chem (1984) 259:11090; **65.** Photochem Photobiol (1982) 36:31; **66.** J Mol Biol (1975) 92:319; **67.** Cytometry (1991) 12:133; **68.** Cytometry (1990) 11:610; **69.** Eur J Biochem (1989) 182:437; **70.** Biochemistry (1980) 19:3221; **71.** Proc Natl Acad Sci U S A (2000) 97:9504; **72.** Exp Cell Res (1991) 194:147; **73.** Methods Mol Biol (1994) 29:83; **74.** Biochemistry (1979) 18:3354; **75.** Proc Natl Acad Sci U S A (1975) 72:2915; **76.** Eur J Biochem (1989) 180:359; **77.** J Biomol Struct Dyn (1987) 5:361; **78.** Biochim Biophys Acta (1983) 722:107; **79.** Biochim Biophys Acta (1993) 1143:215; **80.** Eur Biophys J (1991) 19:189; **81.** Biochemistry (1980) 19:1922; **82.** J Biol Chem (1994) 269:10221; **83.** Biochim Biophys Acta (1993) 1183:161; **84.** Cytometry (2001) 44:133; **85.** Photochem Photobiol (2001) 73:339; **86.** Biochemistry (1990) 29:9029; **87.** J Histochem Cytochem (1985) 33:333; **88.** Mol Biochem Parasitol (1993) 58:7; **89.** Am J Trop Med Hyg (1990) 43:602; **90.** Cytometry (1993) 14:276; **91.** Methods Cell Biol (1994) 42:295; **92.** Biochemistry (1987) 26:4545; **93.** Biochemistry (1993) 32:2987; **94.** J Biol Chem (1993) 268:3944; **95.** Biochemistry (1990) 29:8452; **96.** Biochemistry (1992) 31:3103; **97.** Nucleic Acids Res (1979) 6:3535; **98.** J Histochem Cytochem (1990) 38:1323; **99.** Biochem Biophys Res Commun (1990) 170:270; **100.** Nucleic Acids Res (1979) 6:3519; **101.** Nucleic Acids Res (1978) 5:3775; **102.** Can J Microbiol (1980) 26:912; **103.** Biochim Biophys Acta (1982) 721:394; **104.** Biotech Histochem (1995) 70:220; **105.** Exp Parasitol (2001) 97:141; **106.** Br J Haematol (1999) 104:530; **107.** Cytometry (1991) 12:221; **108.** Cytometry (1991) 12:172; **109.** J Immunol (1986) 136:2769; **110.** J Histochem Cytochem (1975) 23:793; **111.** Histochem J (1985) 17:131; **112.** Biopolymers (1979) 18:1749; **113.** Cytometry (1995) 20:296; **114.** Chromosoma (1978) 68:287; **115.** J Mol Biol (1992) 225:445; **116.** Biochemistry (1993) 32:5554; **117.** Cancer Genet Cytogenet (1980) 1:187; **118.** Biochemistry (1996) 35:3525; **119.** Biochemistry (1973) 12:4827; **120.** Biochim Biophys Acta (1975) 407:24; **121.** Biochim Biophys Acta (1975) 407:43; **122.** J Lab Clin Med (1951) 37:562; **123.** Histochemistry (1982) 74:107; **124.** J Immunol Methods (1989) 123:103; **125.** Cytometry (1988) 9:477; **126.** J Immunol Methods (1993) 163:155; **127.** J Photochem Photobiol A (1994) 84:45; **128.** Neuroscience Protocols, Wouterlood FG, Ed. (1993) p. 93-050-12; **129.** Proc Natl Acad Sci U S A (1980) 77:2260; **130.** J Neurosci Methods (1990) 33:129; **131.** J Neurosci Methods (1997) 72:49; **132.** Neurosci Lett (1995) 184:169.

IMPORTANT NOTICE: The products described in this manual are covered by one or more Limited Use Label License(s). Please refer to the Appendix on page 971 and Master Product List on page 975. Products are For Research Use Only. Not intended for any animal or human therapeutic or diagnostic use.

DATA TABLE 8.1 NUCLEIC ACID STAINS

Cat. No.	MW	Storage	Soluble	Abs	EC	Em	Solvent	Notes
A666	685.69	L	DMSO, DMF	431	ND	498	H_2O/DNA	1, 2
A1301	301.82	L	H_2O, EtOH	500	53,000	526	H_2O/DNA	3, 4
A1310	1270.45	F,L	DMF, DMSO	546	25,000	647	H_2O/DNA	3
A1324	258.71	L	DMF, DMSO	412	8200	471	MeOH	5
A3568	301.82	RR,L	H_2O	500	53,000	526	H_2O/DNA	3, 4, 6
A7592	1255.43	F,L	DMF, DMSO	442	23,000	none	MeOH	
B3582	1202.66	F,D,L	DMSO	462	114,000	481	H_2O/DNA	3, 6, 7, 8
B3586	1254.73	F,D,L	DMSO	570	148,000	604	H_2O/DNA	3, 6, 7, 8
D1168	315.42	FF,L,AA	DMF, DMSO	355	14,000	see Notes	MeCN	9, 10
D1306	350.25	L	H_2O, DMF	358	24,000	461	H_2O/DNA	3, 11
D3571	457.49	L	H_2O, MeOH	358	24,000	461	H_2O/DNA	3, 11
D11347	315.42	FF,L,AA	DMF, DMSO	355	14,000	see Notes	MeCN	9, 10
D21490	350.25	L	H_2O, DMF	358	24,000	461	H_2O/DNA	3, 11, 12
D23107	315.42	FF,D,L,AA	DMSO	355	14,000	see Notes	MeCN	10, 13
E1169	856.77	F,D,L	DMSO	528	7000	617	H_2O/DNA	3, 7, 14
E1374	420.31	F,LL	DMF, EtOH	462	5400	625	pH 7	15
E3599	1292.71	F,D,L	DMSO	535	8000	624	H_2O/DNA	3, 6, 7, 14
H1398	623.96	L	H_2O, DMF	352	40,000	461	H_2O/DNA	3, 16, 17
H1399	615.99	L	H_2O, DMF	350	45,000	461	H_2O/DNA	3, 16, 18
H3569	623.96	RR,L	H_2O	352	40,000	461	H_2O/DNA	3, 6, 16, 17
H3570	615.99	RR,L	H_2O	350	45,000	461	H_2O/DNA	3, 6, 16, 18
H7593	497.42	L	DMSO	518	3900	600	H_2O/DNA	3, 19
H21486	560.96	L	DMSO	392	47,000	440	H_2O/DNA	3
H21491	623.96	L	H_2O, DMF	352	40,000	461	H_2O/DNA	3, 12, 16, 17
H21492	615.99	L	H_2O, DMF	350	45,000	461	H_2O/DNA	3, 12, 16, 18
H22845	472.53	F,D,L	H_2O, DMSO	360	27,000	625	H_2O/DNA	3, 20
J11372	1272.63	F,D,L	DMSO	530	171,000	545	H_2O/DNA	3, 6, 7, 8
J11373	630.31	F,D,L	DMSO	532	94,000	544	H_2O/DNA	3, 6, 7, 8
L7595	471.98	L	DMSO, EtOH	543	46,000	712	H_2O/DNA	3
L11376	1462.54	F,D,L	DMSO	566	108,000	580	H_2O/DNA	3, 6, 7, 8
N21479	see Notes	F,D,L	DMSO	435	see Notes	457	H_2O/RNA	6, 8, 21
N21480	see Notes	F,D,L	DMSO	497	see Notes	524	H_2O/RNA	6, 8, 21
N21481	see Notes	F,D,L	DMSO	515	see Notes	535	H_2O/RNA	6, 8, 21
N21482	see Notes	F,D,L	DMSO	530	see Notes	619	H_2O/RNA	6, 8, 21
N21483	see Notes	F,D,L	DMSO	644	see Notes	663	H_2O/RNA	6, 8, 21
N21485	651.01	L	DMSO	355	36,000	495	H_2O/DNA	3
O7582	see Notes	F,D,L	DMSO	498	see Notes	518	H_2O/DNA	6, 8, 21
P1304MP	668.40	L	H_2O, DMSO	535	5400	617	H_2O/DNA	3, 22
P3566	668.40	RR,L	H_2O	535	5400	617	H_2O/DNA	3, 6, 22
P3580	1170.53	F,D,L	DMSO	434	92,000	456	H_2O/DNA	3, 6, 7, 8
P3581	579.26	F,D,L	DMSO	435	50,000	455	H_2O/DNA	3, 6, 7, 8
P3584	1222.61	F,D,L	DMF	534	146,000	570	H_2O/DNA	3, 6, 7, 8
P3585	605.30	F,D,L	DMSO	539	88,000	567	H_2O/DNA	3, 6, 7, 8
P7581	see Notes	F,D,L	DMSO	502	see Notes	523	H_2O/DNA	6, 8, 21
P11495	see Notes	F,D,L	DMSO	502	see Notes	523	H_2O/DNA	6, 8, 21
P21493	668.40	L	H_2O, DMSO	535	5400	617	H_2O/DNA	3, 12, 22
R11491	see Notes	F,D,L	DMSO	500	see Notes	525	H_2O/RNA	6, 8, 21
S7020	~600	F,D,L	DMSO	504	67,000	523	H_2O/DNA	3, 6, 8, 23
S7556	~500	F,D,L	DMSO	494	43,000	517	H_2O/DNA	3, 6, 8, 23
S7559	~550	F,D,L	DMSO	490	58,000	515	H_2O/DNA	3, 6, 8, 23
S7560	~450	F,D,L	DMSO	521	57,000	556	H_2O/DNA	3, 6, 8, 23
S7573	~400	F,D,L	DMSO	508	75,000	527	H_2O/DNA	3, 6, 8, 23
S7574	~300	F,D,L	DMSO	500	54,000	522	H_2O/DNA	3, 6, 8, 23
S7575	~400	F,D,L	DMSO	488	74,000	509	H_2O/DNA	3, 6, 8, 23, 24
S7576	~500	F,D,L	DMSO	517	60,000	549	H_2O/DNA	3, 6, 8, 23
S7578	~450	F,D,L	DMSO	488	42,000	518	H_2O/DNA	3, 6, 8, 23
S7579	~650	F,D,L	DMSO	621	88,000	634	H_2O/DNA	3, 6, 8, 23
S11341	~550	F,D,L	DMSO	622	112,000	645	H_2O/DNA	3, 6, 8, 23
S11342	~500	F,D,L	DMSO	652	83,000	678	H_2O/DNA	3, 6, 8, 23
S11343	~500	F,D,L	DMSO	620	85,000	647	H_2O/DNA	3, 6, 8, 23
S11344	~550	F,D,L	DMSO	649	76,000	680	H_2O/DNA	3, 6, 8, 23
S11345	~550	F,D,L	DMSO	654	119,000	675	H_2O/DNA	3, 6, 8, 23
S11346	~400	F,D,L	DMSO	598	84,000	620	H_2O/DNA	3, 6, 8, 23
S11348	~400	F,D,L	DMSO	445	38,000	470	H_2O/DNA	3, 6, 8, 23
S11351	~250	F,D,L	DMSO	419	33,000	445	H_2O/DNA	3, 6, 8, 23
S11352	~450	F,D,L	DMSO	426	34,000	455	H_2O/DNA	3, 6, 8, 23
S11353	~350	F,D,L	DMSO	430	31,000	460	H_2O/DNA	3, 6, 8, 23
S11356	~300	F,D,L	DMSO	452	43,000	484	H_2O/DNA	3, 6, 8, 23
S11361	~400	F,D,L	DMSO	531	89,000	545	H_2O/DNA	3, 6, 8, 23
S11362	~300	F,D,L	DMSO	530	82,000	544	H_2O/DNA	3, 6, 8, 23
S11363	~350	F,D,L	DMSO	541	76,000	560	H_2O/DNA	3, 6, 8, 23
S11364	~350	F,D,L	DMSO	543	68,000	559	H_2O/DNA	3, 6, 8, 23
S11365	~500	F,D,L	DMSO	567	95,000	582	H_2O/DNA	3, 6, 8, 23
S11366	~350	F,D,L	DMSO	567	86,000	583	H_2O/DNA	3, 6, 8, 23
S11368	~500	F,D,L	DMSO	547	79,000	570	H_2O/DNA	3, 6, 8, 23

IMPORTANT NOTICE: The products described in this manual are covered by one or more Limited Use Label License(s). Please refer to the Appendix on page 971 and Master Product List on page 975. Products are For Research Use Only. Not intended for any animal or human therapeutic or diagnostic use.

www.invitrogen.com/probes

molecular probes® | invitrogen™ by life technologies™

DATA TABLE 8.1 NUCLEIC ACID STAINS—*continued*

Cat. No.	MW	Storage	Soluble	Abs	EC	Em	Solvent	Notes
S21500	~600	F,D,L	DMSO	494	57,000	519	H_2O/DNA	3, 8, 23
S32703	~800	F,D,L	DMSO	491	107,000	532	H_2O/RNA	3, 6, 8, 23
S32704	~350	F,D,L	DMSO	484	67,000	505	H_2O/DNA	3, 6, 8, 23
S34854	~400	F,D,L	DMSO	483	65,000	503	H_2O/DNA	3, 6, 8, 23
S34855	~400	F,D,L	DMSO	480	66,000	502	H_2O/DNA	3, 6, 8, 23
S34857	~400	F,D,L	DMSO	445	38,000	470	H_2O/DNA	3, 6, 8, 23
S34859	~450	F,D,L	DMSO	640	92,000	658	H_2O/DNA	3, 6, 8, 23
T3600	1302.78	F,D,L	DMSO	514	117,000	533	H_2O/DNA	3, 6, 7, 8
T3602	645.38	F,D,L	DMSO	515	63,000	531	H_2O/DNA	3, 6, 7, 8
T3604	1354.85	F,D,L	DMSO	642	154,000	660	H_2O/DNA	3, 6, 7, 8
T3605	671.42	F,D,L	DMSO	642	102,000	661	H_2O/DNA	3, 6, 7, 8
T7596	697.46	F,D,L	DMSO	747	108,000	770	H_2O/DNA	3, 6, 7, 8
Y3601	1270.65	F,D,L	DMSO	491	99,000	509	H_2O/DNA	3, 6, 7, 8
Y3603	629.32	F,D,L	DMSO	491	52,000	509	H_2O/DNA	3, 6, 7, 8
Y3606	1322.73	F,D,L	DMSO	612	167,000	631	H_2O/DNA	3, 6, 7, 8
Y3607	655.36	F,D,L	DMSO	612	100,000	631	H_2O/DNA	3, 6, 7, 8
15585-011(EtBr)	394.31	RR,L	H_2O	518	5200	605	H_2O/DNA	3, 6, 25

For definitions of the contents of this data table, see "Using *The Molecular Probes® Handbook*" in the introductory pages.

Notes

1. ND = not determined.
2. A666 in MeOH: Abs = 418 nm (EC = 12,000 $cm^{-1}M^{-1}$), Em = 500 nm.
3. Spectra represent aqueous solutions of nucleic acid–bound dye. EC values are derived by comparing the absorbance of the nucleic acid–bound dye with that of free dye in a reference solvent (H_2O or MeOH).
4. Acridine orange bound to RNA: Abs ~460 nm, Em ~650 nm. (Methods Cell Biol (1994) 41:401, Cytometry (1982) 2:201)
5. Spectra of this compound are in methanol acidified with a trace of HCl.
6. This product is supplied as a ready-made solution in the solvent indicated under "Soluble."
7. Although this compound is soluble in water, preparation of stock solutions in water is not recommended because of possible adsorption onto glass or plastic.
8. This product is essentially nonfluorescent except when bound to DNA or RNA.
9. This compound is susceptible to oxidation, especially in solution. Store solutions under argon or nitrogen. Oxidation may be induced by illumination.
10. Dihydroethidium has blue fluorescence (Em ~420 nm) until oxidized to ethidium. The reduced dye does not bind to nucleic acids. (FEBS Lett (1972) 26:169)
11. DAPI in H_2O: Abs = 342 nm (EC = 28,000 $cm^{-1}M^{-1}$), Em = 450 nm. The fluorescence quantum yield of DAPI bound to dsDNA is 0.34, representing an ~20-fold increase relative to the free dye in H_2O. (Photochem Photobiol (2001) 73:585)
12. This product is specified to equal or exceed 98% analytical purity by HPLC.
13. This product is supplied as a ready-made solution in DMSO with sodium borohydride added to inhibit oxidation.
14. E1169 in H_2O: Abs = 493 nm (EC = 9100 $cm^{-1}M^{-1}$). E3599 in H_2O: Abs = 498 nm (EC = 10,800 $cm^{-1}M^{-1}$). Both compounds are very weakly fluorescent in H_2O. QY increases >40-fold on binding to dsDNA.
15. E1374 spectral data are for the free dye. Fluorescence is weak, but intensity increases ~15-fold on binding to DNA. After photocrosslinking to DNA, Abs = 504 nm (EC ~4000 $cm^{-1}M^{-1}$), Em = 600 nm. (Nucleic Acids Res (1978) 5:4891, Biochemistry (1980) 19:3221)
16. MW is for the hydrated form of this product.
17. The fluorescence quantum yield of Hoechst 33258 bound to dsDNA is 0.42, representing an ~30-fold increase relative to the free dye in H_2O. (Photochem Photobiol (2001) 73:585)
18. The fluorescence quantum yield of Hoechst 33342 bound to dsDNA is 0.38, representing an ~10-fold increase relative to the free dye in H_2O. (Photochem Photobiol (2001) 73:585)
19. H7593 in H_2O: Abs = 482 nm (EC = 5500 $cm^{-1}M^{-1}$), Em = 625 nm (weakly fluorescent).
20. Nucleic acid–bound hydroxystilbamidine has a second fluorescence emission peak at ~450 nm. The relative amplitudes of the two emission peaks are dependent on the nucleotide content of the nucleic acid. (Biochemistry (1973) 12:4827)
21. The active ingredient of this product is an organic dye with MW <1000. The exact MW and extinction coefficient values for this dye are proprietary.
22. Propidium iodide in H_2O: Abs = 493 nm (EC = 5900 $cm^{-1}M^{-1}$), Em = 636 nm (weakly fluorescent). Fluorescence is enhanced >10-fold on binding to dsDNA.
23. MW: The preceding ~ symbol indicates an approximate value, not including counterions.
24. The fluorescence quantum yield (QY) of SYTO® 13 dye bound to dsDNA is 0.56 (measured at 22°C).
25. Ethidium bromide in H_2O: Abs = 480 nm (EC = 5600 $cm^{-1}M^{-1}$), Em = 620 nm (weakly fluorescent). Fluorescence is enhanced >10-fold on binding to dsDNA.

IMPORTANT NOTICE: The products described in this manual are covered by one or more Limited Use Label License(s). Please refer to the Appendix on page 971 and Master Product List on page 975. Products are For Research Use Only. Not intended for any animal or human therapeutic or diagnostic use.

PRODUCT LIST 8.1 NUCLEIC ACID STAINS

Cat. No.	Product	Quantity
A666	acridine homodimer (bis-(6-chloro-2-methoxy-9-acridinyl)spermine)	10 mg
A1301	acridine orange	1 g
A3568	acridine orange *10 mg/mL solution in water*	10 mL
A7592	actinomycin D	10 mg
A1310	7-aminoactinomycin D (7-AAD)	1 mg
A1324	9-amino-6-chloro-2-methoxyacridine (ACMA)	100 mg
B3582	BOBO™-1 iodide (462/481) *1 mM solution in DMSO*	200 µL
B3586	BOBO™-3 iodide (570/602) *1 mM solution in DMSO*	200 µL
D1306	4',6-diamidino-2-phenylindole, dihydrochloride (DAPI)	10 mg
D21490	4',6-diamidino-2-phenylindole, dihydrochloride (DAPI) *FluoroPure™ grade*	10 mg
D3571	4',6-diamidino-2-phenylindole, dilactate (DAPI, dilactate)	10 mg
D1168	dihydroethidium (hydroethidine)	25 mg
D23107	dihydroethidium (hydroethidine) *5 mM stabilized solution in DMSO*	1 mL
D11347	dihydroethidium (hydroethidine) *special packaging*	10 x 1 mg
E1169	ethidium homodimer-1 (EthD-1)	1 mg
E3599	ethidium homodimer-2 (EthD-2) *1 mM solution in DMSO*	200 µL
E1374	ethidium monoazide bromide (EMA)	5 mg
H7593	hexidium iodide	5 mg
H1398	Hoechst 33258, pentahydrate (bis-benzimide)	100 mg
H3569	Hoechst 33258, pentahydrate (bis-benzimide) *10 mg/mL solution in water*	10 mL
H21491	Hoechst 33258, pentahydrate (bis-benzimide) *FluoroPure™ grade*	100 mg
H1399	Hoechst 33342, trihydrochloride, trihydrate	100 mg
H3570	Hoechst 33342, trihydrochloride, trihydrate *10 mg/mL solution in water*	10 mL
H21492	Hoechst 33342, trihydrochloride, trihydrate *FluoroPure™ grade*	100 mg
H21486	Hoechst 34580	5 mg
H22845	hydroxystilbamidine, methanesulfonate	10 mg
J11372	JOJO™-1 iodide (529/545) *1 mM solution in DMSO*	200 µL
J11373	JO-PRO™-1 iodide (530/546) *1 mM solution in DMSO*	1 mL
L7595	LDS 751	10 mg
L11376	LOLO™-1 iodide (565/579) *1 mM solution in DMSO*	200 µL
N21479	NeuroTrace® 435/455 blue fluorescent Nissl stain *solution in DMSO*	1 mL
N21480	NeuroTrace® 500/525 green fluorescent Nissl stain *solution in DMSO*	1 mL
N21481	NeuroTrace® 515/535 yellow fluorescent Nissl stain *solution in DMSO*	1 mL
N21482	NeuroTrace® 530/615 red fluorescent Nissl stain *solution in DMSO*	1 mL
N21483	NeuroTrace® 640/660 deep-red fluorescent Nissl stain *solution in DMSO*	1 mL
N21485	nuclear yellow (Hoechst S769121, trihydrochloride, trihydrate)	10 mg
N7565	Nucleic Acid Stains Dimer Sampler Kit	1 kit
P3580	POPO™-1 iodide (434/456) *1 mM solution in DMSO*	200 µL
P3584	POPO™-3 iodide (534/570) *1 mM solution in DMF*	200 µL
P3581	PO-PRO™-1 iodide (435/455) *1 mM solution in DMSO*	1 mL
P3585	PO-PRO™-3 iodide (539/567) *1 mM solution in DMSO*	1 mL
P1304MP	propidium iodide	100 mg
P3566	propidium iodide *1.0 mg/mL solution in water*	10 mL
P21493	propidium iodide *FluoroPure™ grade*	100 mg
O11492	Quant-iT™ OliGreen® ssDNA Assay Kit *2000 assays*	1 kit
O7582	Quant-iT™ OliGreen® ssDNA reagent *2000 assays*	1 mL
P7589	Quant-iT™ PicoGreen® dsDNA Assay Kit *2000 assays*	1 kit
P11496	Quant-iT™ PicoGreen® dsDNA Assay Kit *2000 assays* *10 x 100 µL*	1 kit
P7581	Quant-iT™ PicoGreen® dsDNA reagent *2000 assays*	1 mL
P11495	Quant-iT™ PicoGreen® dsDNA reagent *2000 assays* *10 x 100 µL*	10 x 100 µL
R11490	Quant-iT™ RiboGreen® RNA Assay Kit *2000 assays*	1 kit
R11491	Quant-iT™ RiboGreen® RNA reagent *2000 assays*	1 mL
R32700	RediPlate™ 96 RiboGreen® RNA Quantitation Kit *one 96-well microplate*	1 kit
S21500	SYBR® 101, succinimidyl ester	1 mg
S11494	SYBR® Gold nucleic acid gel stain *10,000X concentrate in DMSO*	500 µL
S7563	SYBR® Green I nucleic acid gel stain *10,000X concentrate in DMSO*	500 µL
S7567	SYBR® Green I nucleic acid gel stain *10,000X concentrate in DMSO*	1 mL
S7585	SYBR® Green I nucleic acid gel stain *10,000X concentrate in DMSO* *special packaging*	20 x 50 µL

IMPORTANT NOTICE: The products described in this manual are covered by one or more Limited Use Label License(s). Please refer to the Appendix on page 971 and Master Product List on page 975. Products are For Research Use Only. Not intended for any animal or human therapeutic or diagnostic use.
www.invitrogen.com/probes
molecular probes® | invitrogen™ by life technologies™

PRODUCT LIST 8.1 NUCLEIC ACID STAINS—*continued*

Cat. No.	Product	Quantity
S7564	SYBR® Green II RNA gel stain *10,000X concentrate in DMSO*	500 µL
S7568	SYBR® Green II RNA gel stain *10,000X concentrate in DMSO*	1 mL
S7586	SYBR® Green II RNA gel stain *10,000X concentrate in DMSO* *special packaging*	20 x 50 µL
S7580	SYBR® Green Nucleic Acid Gel Stain Starter Kit	1 kit
S33102	SYBR® Safe DNA gel stain *10,000X concentrate in DMSO*	400 µL
S33100	SYBR® Safe DNA gel stain in 0.5X TBE	1 L
S33101	SYBR® Safe DNA gel stain in 0.5X TBE	4 L
S33111	SYBR® Safe DNA gel stain in 1X TAE	1 L
S33112	SYBR® Safe DNA gel stain in 1X TAE	4 L
S33110	SYBR® Safe DNA Gel Stain Starter Kit *with 1 L of SYBR® Safe DNA gel stain in 0.5X TBE (S33100) and one photographic filter (S37100)*	1 kit
S11351	SYTO® 40 blue fluorescent nucleic acid stain *5 mM solution in DMSO*	250 µL
S11352	SYTO® 41 blue fluorescent nucleic acid stain *5 mM solution in DMSO*	250 µL
S11353	SYTO® 42 blue fluorescent nucleic acid stain *5 mM solution in DMSO*	250 µL
S11356	SYTO® 45 blue fluorescent nucleic acid stain *5 mM solution in DMSO*	250 µL
S11350	SYTO® Blue Fluorescent Nucleic Acid Stain Sampler Kit *SYTO® dyes 40–45* *50 µL each*	1 kit
S34854	SYTO® 9 green fluorescent nucleic acid stain *5 mM solution in DMSO*	100 µL
S32704	SYTO® 10 green fluorescent nucleic acid stain *5 mM solution in DMSO*	100 µL
S7573	SYTO® 11 green fluorescent nucleic acid stain *5 mM solution in DMSO*	250 µL
S7574	SYTO® 12 green fluorescent nucleic acid stain *5 mM solution in DMSO*	250 µL
S7575	SYTO® 13 green fluorescent nucleic acid stain *5 mM solution in DMSO*	250 µL
S7576	SYTO® 14 green fluorescent nucleic acid stain *5 mM solution in DMSO*	250 µL
S7578	SYTO® 16 green fluorescent nucleic acid stain *1 mM solution in DMSO*	250 µL
S7556	SYTO® 21 green fluorescent nucleic acid stain *5 mM solution in DMSO*	250 µL
S7559	SYTO® 24 green fluorescent nucleic acid stain *5 mM solution in DMSO*	250 µL
S7560	SYTO® 25 green fluorescent nucleic acid stain *5 mM solution in DMSO*	250 µL
S34855	SYTO® BC green fluorescent nucleic acid stain *5 mM solution in DMSO*	100 µL
S7572	SYTO® Green Fluorescent Nucleic Acid Stain Sampler Kit *SYTO® dyes 11–14,16, 21, 24, and 25* *50 µL each*	1 kit
S11361	SYTO® 80 orange fluorescent nucleic acid stain *5 mM solution in DMSO*	250 µL
S11362	SYTO® 81 orange fluorescent nucleic acid stain *5 mM solution in DMSO*	250 µL
S11363	SYTO® 82 orange fluorescent nucleic acid stain *5 mM solution in DMSO*	250 µL
S11364	SYTO® 83 orange fluorescent nucleic acid stain *5 mM solution in DMSO*	250 µL
S11365	SYTO® 84 orange fluorescent nucleic acid stain *5 mM solution in DMSO*	250 µL
S11366	SYTO® 85 orange fluorescent nucleic acid stain *5 mM solution in DMSO*	250 µL
S11360	SYTO® Orange Fluorescent Nucleic Acid Stain Sampler Kit *SYTO® dyes 80–85* *50 µL each*	1 kit
S7579	SYTO® 17 red fluorescent nucleic acid stain *5 mM solution in DMSO*	250 µL
S11341	SYTO® 59 red fluorescent nucleic acid stain *5 mM solution in DMSO*	100 µL
S11342	SYTO® 60 red fluorescent nucleic acid stain *5 mM solution in DMSO*	250 µL
S11343	SYTO® 61 red fluorescent nucleic acid stain *5 mM solution in DMSO*	250 µL
S11344	SYTO® 62 red fluorescent nucleic acid stain *5 mM solution in DMSO*	250 µL
S11345	SYTO® 63 red fluorescent nucleic acid stain *5 mM solution in DMSO*	250 µL
S11346	SYTO® 64 red fluorescent nucleic acid stain *5 mM solution in DMSO*	100 µL
S11340	SYTO® Red Fluorescent Nucleic Acid Stain Sampler Kit *SYTO® dyes 17 and 59–64* *50 µL each*	1 kit
S32703	SYTO® RNASelect™ green fluorescent cell stain *5 mM solution in DMSO*	100 µL
S34857	SYTOX® Blue dead cell stain *for flow cytometry* *1000 assays* *1 mM solution in DMSO*	1 mL
S11348	SYTOX® Blue nucleic acid stain *5 mM solution in DMSO*	250 µL
S7020	SYTOX® Green nucleic acid stain *5 mM solution in DMSO*	250 µL
S11368	SYTOX® Orange nucleic acid stain *5 mM solution in DMSO*	250 µL
S34859	SYTOX® Red dead cell stain *for 633 or 635 nm excitation* *5 µM solution in DMSO*	1 mL
T3602	TO-PRO®-1 iodide (515/531) *1 mM solution in DMSO*	1 mL
T3605	TO-PRO®-3 iodide (642/661) *1 mM solution in DMSO*	1 mL
T7596	TO-PRO®-5 iodide (745/770) *1 mM solution in DMSO*	1 mL
T3600	TOTO®-1 iodide (514/533) *1 mM solution in DMSO*	200 µL
T3604	TOTO®-3 iodide (642/660) *1 mM solution in DMSO*	200 µL
15585-011	UltraPure™ Ethidium Bromide *10 mg/ml*	10 mL
Y3603	YO-PRO®-1 iodide (491/509) *1 mM solution in DMSO*	1 mL
Y3607	YO-PRO®-3 iodide (612/631) *1 mM solution in DMSO*	1 mL
Y3601	YOYO®-1 iodide (491/509) *1 mM solution in DMSO*	200 µL
Y3606	YOYO®-3 iodide (612/631) *1 mM solution in DMSO*	200 µL

IMPORTANT NOTICE: The products described in this manual are covered by one or more Limited Use Label License(s). Please refer to the Appendix on page 971 and Master Product List on page 975. Products are For Research Use Only. Not intended for any animal or human therapeutic or diagnostic use.

8.2 Labeling Oligonucleotides and Nucleic Acids

To facilitate the preparation of optimally labeled nucleic acids, we offer several modified nucleotides, as well as a wide selection of nucleic acid labeling kits that include reactive versions of Molecular Probes® fluorophores. Our available technologies include:

- ChromaTide® dUTP, ChromaTide® OBEA-dCTP and ChromaTide® UTP nucleotides, which provide researchers with a large selection of fluorophore- and hapten-labeled nucleotides that can be enzymatically incorporated into DNA or RNA probes for FISH (fluorescence *in situ* hybridization), DNA arrays/microarrays and other hybridization techniques.
- Unlabeled aminoallyl derivatives of dUTP and UTP, as well as unlabeled and labeled aminohexylacrylamide (aha) derivatives of dUTP and dCTP, that are easy to incorporate into nucleic acids for subsequent conjugation with any of our amine-reactive probes (Chapter 1).
- ARES™ DNA Labeling Kits, which employ a versatile, two-step method for labeling DNA with fluorescent dyes to achieve a uniformity and consistency of labeling that is difficult to obtain with conventional enzymatic incorporation of labeled nucleotides.[1]
- FISH Tag™ DNA and RNA Kits, which provide a complete workflow solution for creating labeled DNA and RNA probes for use in fluorescence *in situ* hybridization (FISH) applications.
- ULYSIS® Nucleic Acid Labeling Kits, which employ a fast, simple and reliable chemical method for labeling nucleic acids without enzymatic incorporation of labeled nucleotides.
- Alexa Fluor® Oligonucleotide Amine Labeling Kits, which use familiar chemical labeling of amine-terminated oligonucleotides to prepare the best singly labeled fluorescent oligonucleotide conjugates.

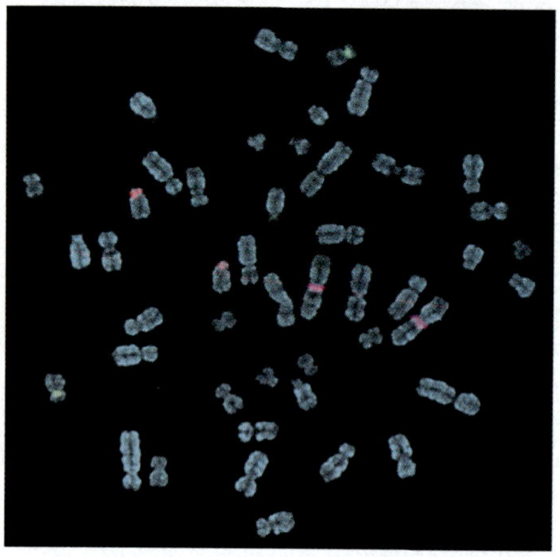

Figure 8.2.1 Human metaphase chromosomes hybridized to centromere probes labeled with ChromaTide® OBEA-dCTP nucleotides. Probes specific to the α satellite sequences from human chromosome 17, chromosome 15 and chromosome 1 were labeled using nick translation with ChromaTide® Alexa Fluor® 488-7-OBEA-dCTP (green, C21555), ChromaTide® Alexa Fluor® 594-7-OBEA-dCTP (red) and ChromaTide® Alexa Fluor® 647-12-OBEA-dCTP (pink, C21559), respectively. The probes were hybridized to metaphase spreads from peripheral blood lymphocyte cultures. The chromosomes were counterstained with Hoechst 33342 (blue) (H1399, H3570, H21492).

ChromaTide® Nucleotides

We offer a series of uridine triphosphates (UTP), deoxyuridine triphosphates (dUTP) and deoxycytidine triphosphates (OBEA-dCTP) conjugated to an extensive selection of fluorophores and haptens, including our intensely fluorescent Alexa Fluor® dyes (Table 8.5). These ChromaTide® nucleotides are useful for generating labeled nucleic acids for molecular biology and molecular cytogenetics applications, including chromosome and mRNA fluorescence *in situ* hybridization (FISH) experiments[2–5] (Figure 8.2.1), gene expression and mutation detection on arrays and microarrays[6–16] (Figure 8.2.2), and *in situ* PCR and RT-PCR. Our diverse fluorescent labels provide the ideal tools for multicolor techniques such as spectral karyotyping,[17–23] multilocus FISH analysis,[24] chromosome painting[25] and comparative genome hybridization.[26,27]

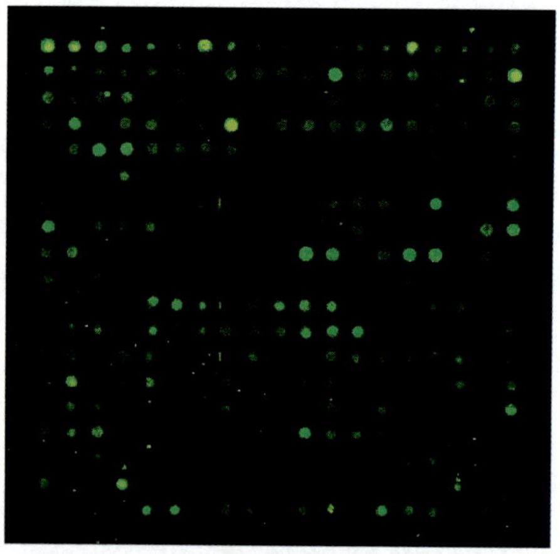

Figure 8.2.2 A microarray hybridized to Alexa Fluor® 546 dye–labeled cDNA. cDNA was labeled by reverse transcription from *Vibrio cholerae* total RNA using ChromaTide® Alexa Fluor® 546-14-dUTP (C11404). Labeled cDNA was then hybridized to a *V. cholerae* O1 El Tor microarray. The array was imaged with ScanArray® 5000XL scanner (PerkinElmer LAS, Inc.). The image was contributed by Kimberly Chong and Gary Schoolnik, Stanford University.

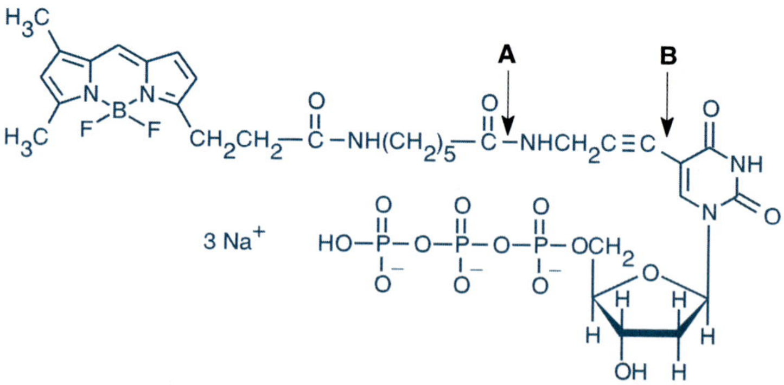

Figure 8.2.3 Structure of ChromaTide® BODIPY® FL-14-dUTP (C7614). This structure is representative of our other labeled ChromaTide® aminoalkynyl dUTP nucleotides. Fluorophore labels are attached via a four-atom aminoalkynyl spacer (between arrows **A** and **B**) to either deoxyuridine triphosphate (dUTP) or uridine triphosphate (UTP).

IMPORTANT NOTICE: The products described in this manual are covered by one or more Limited Use Label License(s). Please refer to the Appendix on page 971 and Master Product List on page 975. Products are For Research Use Only. Not intended for any animal or human therapeutic or diagnostic use.
www.invitrogen.com/probes

Structures of the ChromaTide® Nucleotides

The ChromaTide® UTP and dUTP nucleotides are modified at the *C*-5 position of UTP or dUTP via a unique aminoalkynyl linker (Figure 8.2.3). The *C*-5 position of UTP and dUTP is not involved in Watson–Crick base-pairing and so interferes little with probe hybridization. The aminoalkynyl linker between the fluorophore and the nucleotide in the ChromaTide® UTP and dUTP nucleotides is designed to reduce the fluorophore's interaction with enzymes or target binding sites. In addition to this four-atom bridge, several of these nucleotides contain a seven- to 10-atom spacer that further separates the dye from the base. The number in the product's name (e.g., the "12" in ChromaTide® fluorescein-12-dUTP) indicates the net length of the spacer in atoms. Longer spacers typically result in brighter conjugates and increased hapten accessibility for secondary detection reagents.

The ChromaTide® OBEA-deoxycytidine triphosphates (OBEA-dCTP) are modified at the *N*-4 position of cytosine using a 2-aminoethoxyethyl (OBEA) linker (Figure 8.2.4). The Alexa Fluor® 546 and Alexa Fluor® 647 OBEA-dCTP conjugates (C21555, C21559) also have a built-in spacer that reduces possible steric interference caused by the presence of the dye.

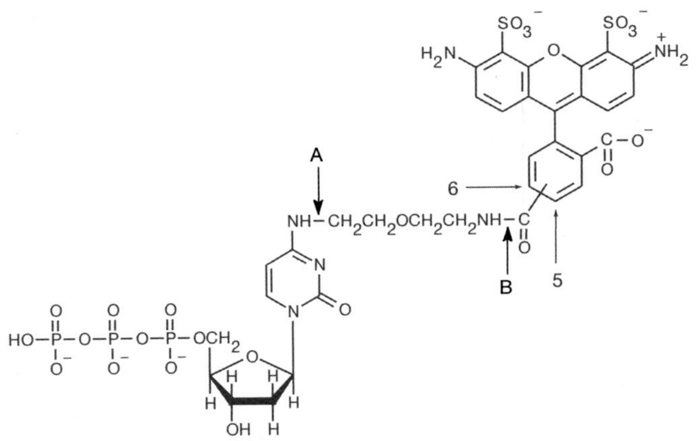

Figure 8.2.4 Structure of ChromaTide® Alexa Fluor® 488-7-OBEA-dCTP (C21555). This structure is representative of our labeled ChromaTide® OBEA-dCTP nucleotides. Fluorophore labels are attached via the OBEA spacer (between arrows **A** and **B**) to deoxycytidine triphosphate (dCTP).

Table 8.5 Molecular Probes® ChromaTide® nucleotides and aha-dUTP and aha-dCTP nucleotides.

Cat. No.	dUTP, aha-dUTP, aha-dCTP, OBEA-dCTP or UTP Nucleotide *	Ex/Em †	Applications ‡
Nonfluorescent			
B32766	Biotin-aha-dUTP	NA	RT
B32772	Biotin-aha-dCTP	NA	RT
Green Fluorescence			
C11397	Alexa Fluor® 488-5-dUTP	490/520	PCR, TDT, RP, RT, NT
C7604	Fluorescein-12-dUTP	495/525	PCR, TDT, RP, NT
C7614	BODIPY® FL-14-dUTP	505/515	PCR, TDT, RP
F32767	Fluorescein-aha-dUTP	495/525	RT
C21555	Alexa Fluor® 488-7-OBEA-dCTP	490/520	RT, NT
C11403	Alexa Fluor® 488-5-UTP	490/520	SP6, T3, T7
Yellow Fluorescence			
C11398	Alexa Fluor® 532-5-dUTP	525/550	PCR, TDT, RP, RT, NT
Orange Fluorescence			
C7606MP	Tetramethylrhodamine-6-dUTP	550/570	PCR, TDT, RP
C11401	Alexa Fluor® 546-14-dUTP	555/570	PCR, TDT, RP, RT, NT
A32762	Alexa Fluor® 555-aha-dUTP	555/570	RT
A32770	Alexa Fluor® 555-aha-dCTP	555/570	RT
C21556	Alexa Fluor® 546-16-OBEA-dCTP	555/570	RT, NT
C11404	Alexa Fluor® 546-14-UTP	555/570	SP6, T3, T7
Red Fluorescence			
C11399	Alexa Fluor® 568-5-dUTP	575/600	PCR, TDT, RP, NT
C7631	Texas Red®-12-dUTP	595/610	PCR, TDT, RP
C11400	Alexa Fluor® 594-5-dUTP	590/615	PCR, TDT, RP, NT
Far-Red Fluorescence			
A32763	Alexa Fluor® 647-aha-dUTP	650/670	RT
A32771	Alexa Fluor® 647-aha-dCTP	650/670	RT
C21559	Alexa Fluor® 647-12-OBEA-dCTP	650/670	RT, NT

* All products are supplied as either 25 µL (dUTPs and UTPs) or 50 µL (dCTPs) of a 1 mM solution in TE buffer. † Excitation (Ex) and emission (Em) maxima, in nm, for the labeled nucleotide. ‡ Except where otherwise noted, the following applications were tested: PCR, Taq polymerase for DNA amplification; TDT, terminal deoxynucleotidyl transferase for 3′-end labeling of dsDNA; RP, Klenow polymerase for labeling DNA by random hexamer priming; RT, murine leukemia virus (MLV) reverse transcriptase for synthesizing DNA from an RNA template; NT, nick translation using DNase I and DNA polymerase I. ChromaTide® UTP nucleotides were incorporated into RNA by standard transcription reactions using the indicated RNA polymerases (SP6, T3, T7). We also offer unlabeled aha-dUTP (50 µL of a 50 mM solution in TE buffer, A32761) and unlabeled aha-dCTP (50 µL of a 50 mM solution in TE buffer, A32769; 500 µL of a 2 mM solution in TE buffer, A32768).

Fluorescent ChromaTide® Nucleotides

The spectral diversity of the ChromaTide® UTP, ChromaTide® dUTP and ChromaTide® OBEA-dCTP nucleotides (Table 8.5) gives researchers significant flexibility in choosing a label that is compatible with a particular optical detection system or multicolor experiment. Probes made from the fluorescent ChromaTide® nucleotides can be imaged directly; alternatively, some fluorophores can be used as a hapten for signal amplification (Section 7.4). In many cases, the TSA™ (Section 6.2) or ELF® technologies (Section 6.3) can be used inidividually or together[28] to significantly amplify the signal of dye-labeled hybridization probes in cells and tissues and on microarrays.

The Alexa Fluor® conjugates of UTP, OBEA-dCTP and dUTP provide fluorophore labels with demonstrably superior fluorescence properties, as compared with conventional dyes. The Alexa Fluor® 488, Alexa Fluor® 568 and Alexa Fluor® 594 nucleotides are spectrally similar to fluorescein, Lissamine rhodamine B and Texas Red® conjugates, respectively, but the Alexa Fluor® conjugates exhibit superior spectral and chemical properties. ChromaTide® OBEA-dCTP nucleotides have been prepared from four of our brightest and most photostable dyes—Alexa Fluor® 488, Alexa Fluor® 546, Alexa Fluor® 594 and Alexa Fluor® 647 dyes—with spectra virtually identical to those of fluorescein, Cy®3, Texas Red® and Cy®5 dyes, respectively.

The ChromaTide® Alexa Fluor® dUTP and ChromaTide® Alexa Fluor® OBEA-dCTP nucleotides are highly water soluble, as are the DNA probes that contain them. Thus, Alexa Fluor® dye–labeled DNA probes do not aggregate or precipitate, even in high-salt hybridization solutions. Fluorescence of the Alexa Fluor® conjugates is not pH sensitive in the range used for hybridization or microscopy mounting media. Additionally, the enhanced photostability of these conjugates makes them ideal for imaging applications.

Using ChromaTide® Nucleotides in Enzymatic Labeling Methods

The ChromaTide® nucleotides can be incorporated into DNA and RNA using conventional enzymatic labeling techniques (Table 8.5). Protocols for many of these techniques are provided with the ChromaTide® nucleotides; please note that not all ChromaTide® nucleotides have been tested in all applications. Enzymes that we have used successfully include:

- Taq polymerase in polymerase chain reaction (PCR) assays
- DNA polymerase I in nick-translation and primer-extension assays
- Klenow polymerase in random-primer labeling
- Terminal deoxynucleotidyl transferase (TdT) for 3′-end labeling
- Reverse transcriptase for synthesizing DNA from RNA templates
- SP6 RNA polymerase, T3 RNA polymerase and T7 RNA polymerase for *in vitro* transcription

ChromaTide® nucleotides have also been used in the TUNEL assay for detecting DNA fragmentation in apoptotic cells[29–32] (Section 15.5, Figure 8.2.5). Microinjected fluorescent nucleotides have been utilized to follow the dynamics of chromosome formation and cell proliferation in live cells.[33–35]

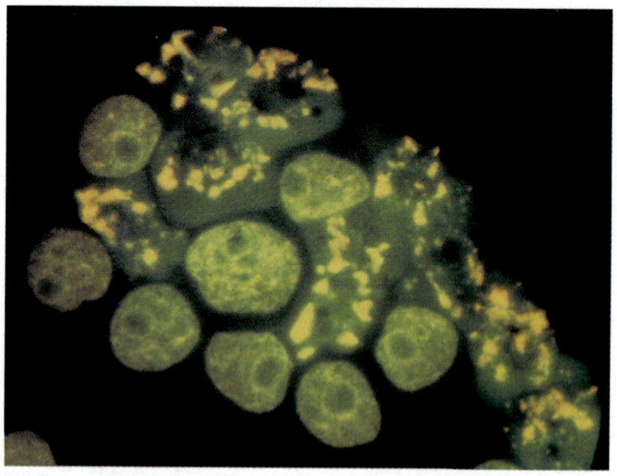

Figure 8.2.5 HL-60 cells treated with camptothecin for three hours. The DNA strand nicks characteristic of apoptosis were detected with the TUNEL (terminal deoxynucleotidyl transferase–mediated dUTP nick end-labeling) assay using the fluorescently labeled nucleotide ChromaTide® BODIPY® FL-14-dUTP (C7614). Image contributed by Zbigniew Darzynkiewicz, Cancer Research Institute, New York Medical College.

Amine-Modified Nucleotides

Unlabeled and Labeled aha-dUTP and aha-dCTP

5-Aminohexylacrylamido-dUTP (aha-dUTP) and 5-aminohexylacrylamido-dCTP (aha-dCTP) can be used to produce amine-modified DNA by conventional enzymatic incorporation methods such as reverse transcription, nick translation, random primed labeling or PCR. The amine-modified DNA can then be labeled with any amine-reactive dye or hapten. This two-step technique permits a uniform and high degree of DNA labeling that is difficult to obtain by other methods. The protocol provided with the aha-dUTP and aha-dCTP nucleotides yields a labeling efficiency of ~5–8 dyes per 100 bases, which we have found to be optimal for fluorescence *in situ* hybridization (FISH), dot blot hybridization and especially microarray applications, in which the consistency of labeling between samples is critical for accurate interpretation of results. The aha-dUTP and aha-dCTP nucleotides are available as 50 µL of a 50 mM solution (A32761, A32769) in 10 mM Tris, 1 mM EDTA, pH 7.5 (TE); aha-CTP is also available as 500 µL of a 2 mM solution (A32768). We provide a wide variety of amine-reactive reagents for labeling amine-modified DNA, including succinimidyl esters of our Alexa Fluor® dyes, conventional fluorophores, biotin and dinitrophenyl (DNP) (Chapter 1).

The labeled aha-dUTP and aha-dCTP nucleotides can be used to generate labeled nucleic acid hybridization probes for many molecular biology and molecular cytogenetics applications, including two-color microarray assays, northern and Southern blots, colony and plaque

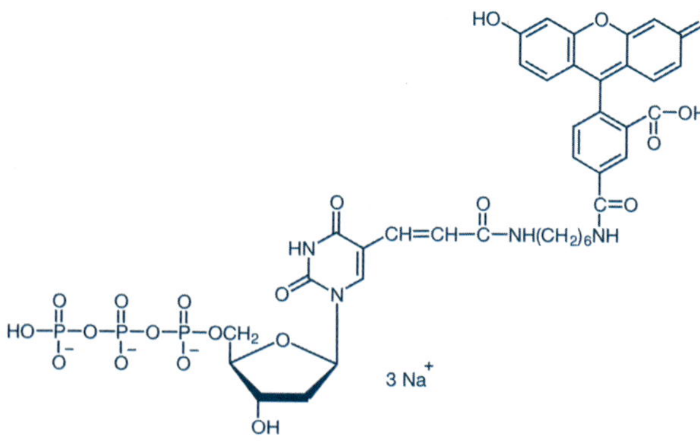

Figure 8.2.6 Fluorescein-aha-dUTP (F32767).

hybridizations, DNA sequencing, primer extension, DNA and RNA amplification and bead-based separation techniques. In these applications, the labeled samples can be detected by the intrinsic fluorescence of the labeled probe or with enzyme-conjugated secondary detection reagents in conjunction with fluorescent, chemiluminescent or colorimetric substrates. These nucleotides are modified at the *C*-5 position of uridine and cytosine, respectively, with a unique hexylacrylamide linker, which serves as a spacer between the nucleotide and the dye or hapten (Figure 8.2.6). This spacer reduces interactions between the nucleotide and the dye or hapten, producing brighter conjugates and increased hapten accessibility for secondary detection reagents. The Alexa Fluor® 555 aha-dUTP and aha-dCTP nucleotides (A32762, A32770), with excitation/emission maxima of 555/570 nm, and the Alexa Fluor® 647 aha-dUTP and aha-dCTP nucleotides (A32763, A32771), with excitation/emission maxima of 650/670 nm, respectively, are compatible with commonly used microarray scanners. These fluorescent nucleotides provide greater signal correlation (R^2) values than do the spectrally similar Cy®3 and Cy®5 dye pair, thereby improving the resolution of two-color microarray gene expression assays.[36] The exceptionally bright and photostable Alexa Fluor® dyes are also essentially insensitive to pH and are highly water soluble.

We also offer biotin aha-dUTP and aha-dCTP (B32766, B32772) and fluorescein aha-dUTP (F32767) nucleotides, which can be used to generate nucleic acid probes that can be detected with streptavidin conjugates (Section 7.6, Table 7.9) or labeled anti-fluorescein antibodies (Section 7.4), respectively. Nucleic acid probes labeled with biotin have generally been the most common nonisotopic probes used in hybridization techniques. Biotinylated probes are readily detected with fluorophore or enzyme conjugates of avidins or streptavidins, providing amplification of the signal (Figure 8.2.7). The signal from biotin-labeled hybridization probes can be considerably amplified, while retaining excellent spatial resolution, using tyramide signal amplification technology (Section 6.2, Figure 8.2.8) or Enzyme-Labeled Fluorescence (ELF®) technology (Section 6.3, Figure 8.2.9). In addition, biotinylated nucleic acids can be adsorbed onto streptavidin agarose or CaptAvidin™ agarose (S951, C21386; Section 7.6) or detected with NANOGOLD® or Alexa Fluor® FluoroNanogold™ streptavidin[37–39] (N24918, A24926, A24927; Section 7.6).

Aminoallyl dUTP

Aminoallyl dUTP[40] (5-aminoallyl-2′-deoxyuridine 5′-triphosphate, A21664) can be incorporated into DNA using conventional enzymatic incorporation techniques, as described above for the ChromaTide® dUTP nucleotides.[41] The resulting amine-modified nucleic acid can then be labeled using the amine-reactive dyes[1] described in Chapter 1. Lacking bulky dye groups, the aminoallyl-modified nucleotides can be incorporated to extremely high and consistent levels. Subsequent reaction of the amine-modified nucleic acid with an excess of amine-reactive reagent achieves correspondingly high and consistent labeling efficiency, regardless of the labeling reagent chosen. We typically obtain labeling efficiencies of ~5–8 dyes per 100 bases. This two-step labeling method also eliminates the need to optimize an enzymatic reaction to accommodate different dye modified nucleotides, which may incorporate at very different rates. This

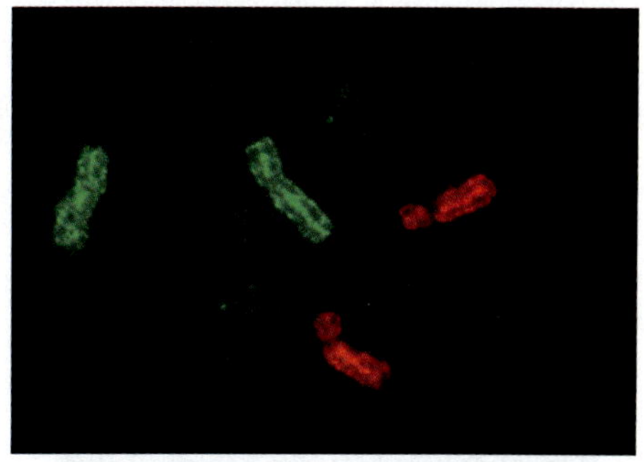

Figure 8.2.7 Labeled paint probes hybridized to human metaphase chromosomes. A biotinylated chromosome 5 probe was detected with Alexa Fluor® 594 streptavidin (S11227), and a digoxigenin-labeled chromosome 2 probe detected with mouse anti-digoxigenin in combination with Alexa Fluor® 488 goat anti–mouse IgG antibody (A11001). Image contributed by Joop Wiegant, Leiden University Medical Center, Leiden, The Netherlands.

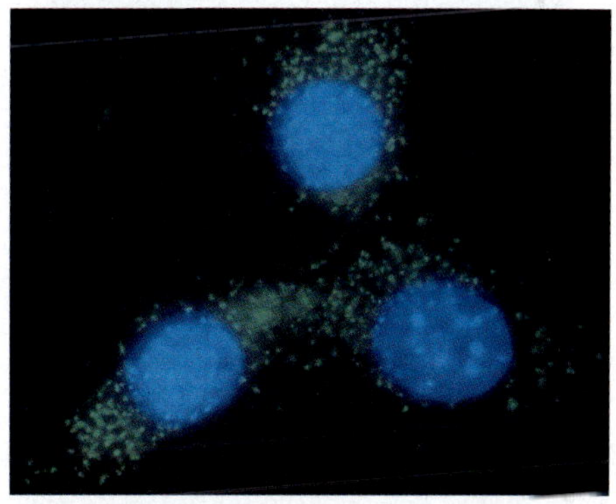

Figure 8.2.9 *lacZ* mRNA in transformed mouse fibroblasts (CRE BAG 2 cells) hybridized with a singly biotinylated, complementary oligonucleotide. Hybrids were then detected by incubation with a streptavidin–alkaline phosphatase conjugate in combination with the ELF® 97 alkaline phosphatase substrate, both of which are provided in our ELF® 97 mRNA *In Situ* Hybridization Kit (E6604, E6605). Cells were counterstained with DAPI (D1306, D3571, D21490) and photographed using a longpass optical filter appropriate for DAPI.

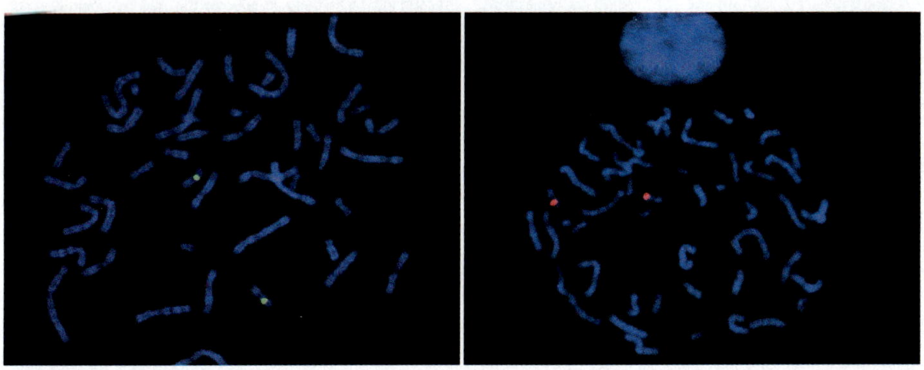

Figure 8.2.8 Fluorescence *in situ* hybridization detected by tyramide signal amplification. Chromosome spreads were prepared from the cultured fibroblast cell line MRC-5 and hybridized with a biotinylated α-satellite probe specific for chromosome 17. The probe was generated by nick translation in the presence of biotinylated dUTP. For detection by TSA, hybridized chromosome spreads were labeled using TSA™ Kit #22 (T20932) with HRP-conjugated streptavidin and Alexa Fluor® 488 tyramide (left) or using TSA™ Kit #23 (T20933) with HRP-conjugated streptavidin and Alexa Fluor® 546 tyramide (right). After counterstaining with DAPI (D1306, D3571, D21490), images were obtained using filters appropriate for DAPI, FITC or TRITC.

IMPORTANT NOTICE: The products described in this manual are covered by one or more Limited Use Label License(s). Please refer to the Appendix on page 971 and Master Product List on page 975. Products are For Research Use Only. Not intended for any animal or human therapeutic or diagnostic use.

labeling method is ideal for both FISH probes (Figure 8.2.10, Figure 8.2.11) and microarray-based experiments (Figure 8.2.12). Aminoallyl dUTP labeling is the basis of both our ARES™ DNA Labeling Kits and FISH Tag™ DNA Kits (see below), which provide aminoallyl dUTP, amine-reactive dyes and carefully tested protocols.

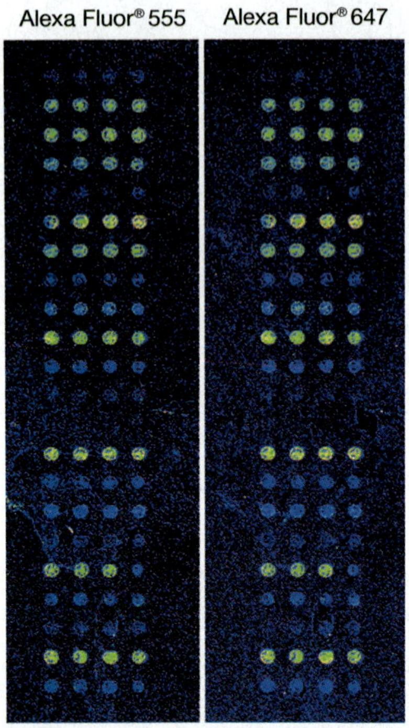

Figure 8.2.12 DNA microarray hybridized to DNA labeled using the ARES™ DNA Labeling Kits. Total human RNA was labeled by reverse transcription using either ARES™ Alexa Fluor® 555 DNA Labeling Kit (A21677) or ARES™ Alexa Fluor® 647 DNA Labeling Kit (A21676). Labeled DNA was hybridized to a microarray containing human housekeeping genes. After hybridization, the array was imaged using a ScanArray®5000XL microarray scanner (PerkinElmer LAS, Inc.) using the appropriate lasers and filter sets. The image was pseudocolored so that white areas show the most intense signal and blue areas show the least intense signal.

Alexa Fluor® Amine-Reactive Dye Decapacks for Labeling Amine-Modified DNA and RNA

For labeling amine-modified DNA or RNA probes in microarray-based experiments, we offer three amine-reactive Alexa Fluor® dyes conveniently packaged in 10 single-use vials and rigorously tested for the ability to efficiently label amine-modified DNA:

- Alexa Fluor® 488 reactive dye decapack (A32750)
- Alexa Fluor® 555 reactive dye decapack (A32756)
- Alexa Fluor® 647 reactive dye decapack (A32757)
- A set containing both the Alexa Fluor® 555 and Alexa Fluor® 647 reactive dye decapacks (A32755), for two-color experiments

These specially packaged amine-reactive dyes can be used in conjunction with our aminohexylacrylamido-dUTP (aha-dUTP, A32760) or aminoallyl dUTP (A21664) nucleotides or with commercially available aminoallyl nucleotide–based nucleic acid labeling kits. With excitation/emission maxima of 495/519 nm, 555/565 nm and 650/668 nm, respectively, the Alexa Fluor® 488, Alexa Fluor® 555 and Alexa Fluor® 647 succinimidyl esters match the most popular wavelength channels used to scan microarrays. We have found that when single-stranded DNA is labeled with equivalent levels of Alexa Fluor® 647 dye or Cy®5 dye using aminoallyl

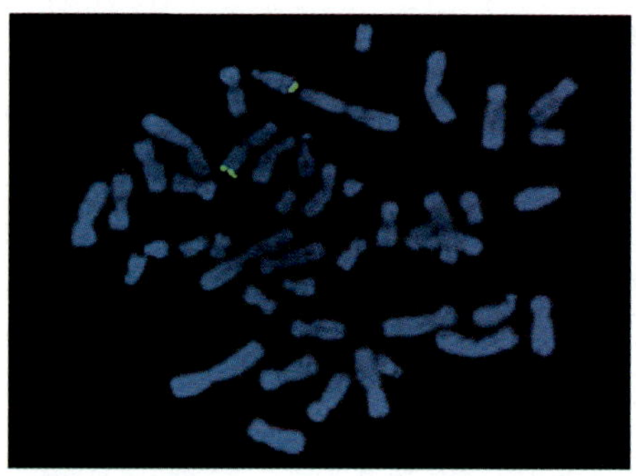

Figure 8.2.10 Fluorescence *in situ* hybridization (FISH) mapping of a BAC clone on human metaphase chromosomes. FISH was performed using a BAC clone labeled using the ARES™ Alexa Fluor® 488 DNA Labeling Kit (A21665). The chromosomes were counterstained with DAPI (D1306, D3571, D21490). Image contributed by Nallasivam Palanisamy, Cancer Genetics Inc.

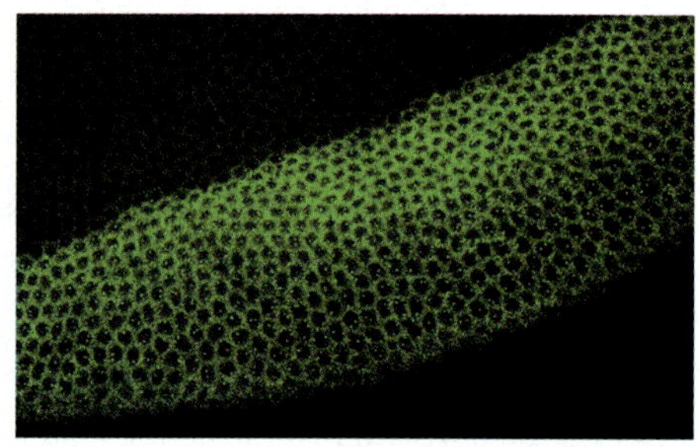

Figure 8.2.11 Expression of snail RNA in an early-stage fruit fly embryo visualized by FISH. A 1.7 kb RNA probe corresponding to the snail gene was labeled by *in vitro* transcription using aminoallyl UTP (A21663) followed by reaction with Alexa Fluor® 488 carboxylic acid, succinimidyl ester (A20000, A20100). The probe was hybridized to *Drosophila melanogaster* embryos and imaged using confocal laser-scanning microscopy. Image contributed by David Kosman and Ethan Bier, University of California, San Diego.

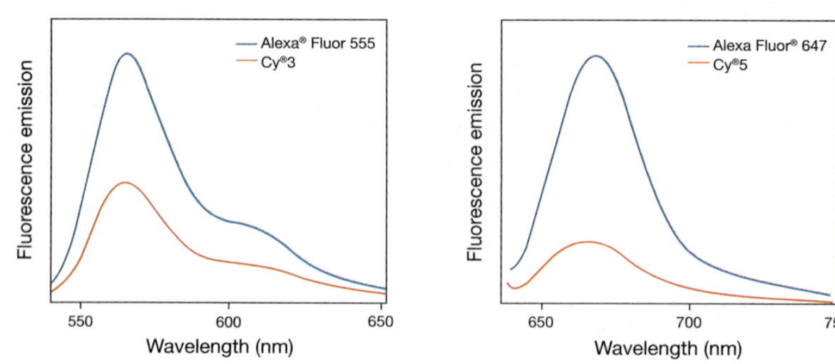

Figure 8.2.13 Fluorescence emission of DNA labeled with Alexa Fluor® dyes versus Cy® dyes. DNA was amine modified by reverse transcription in the presence of aminoallyl dUTP (A21664). The modified DNA was then labeled with Alexa Fluor® 555 or Alexa Fluor® 647 succinimidyl ester (A21677, A21676) or with Cy®3 or Cy®5 reactive dye to a level optimal for hybridization. The fluorescence emission spectra for DNA samples labeled to the same degree show that the Alexa Fluor® dye–labeled DNA was consistently brighter than the spectrally similar Cy® dye–labeled DNA.

dUTP incorporation, the fluorescence emission from the Alexa Fluor® 647 dye–labeled DNA is several times more intense that that of Cy® 5 dye–labeled DNA; similarly, Alexa Fluor® 555 dye–labeled DNA is consistently brighter than spectrally similar Cy®3 dye–labeled DNA (Figure 8.2.13). Furthermore, the Alexa Fluor® 555/Alexa Fluor® 647 dye pair has been shown to display higher signal correlation coefficients than the Cy®3/Cy®5 dye pair in two-color DNA microarray assays.[36]

ARES™ DNA Labeling Kits

ARES™ DNA Labeling Kits (Table 8.6) provide a versatile two-step method for labeling DNA with our Alexa Fluor® dyes[1] (Figure 8.2.14). This method achieves uniformity and consistency of labeling that is difficult to obtain with conventional enzymatic incorporation of labeled nucleotides. In the first step, an amine-modified nucleotide, 5-(3-aminoallyl)-dUTP, is enzymatically incorporated into DNA. This step permits relatively uniform labeling of the probe with primary amine groups. The aminoallyl dUTP substrate used in this reaction is taken up efficiently by reverse transcription or nick translation, for which we provide the protocols; other enzymatic methods are also likely to be compatible. In the second step, the amine-modified DNA is chemically labeled using an amine-reactive Alexa Fluor® dye. This chemical reaction varies little in its efficiency from dye to dye, so that it is possible to use any combination of the ARES™ Kits and obtain consistent labeling for every DNA sample. The labeling protocols provided generally result in about one dye per 12–20 bases, which we have determined to be optimal for fluorescence *in situ* hybridization (FISH) and dot-blot hybridization. Nucleic acids labeled using this method are ideal for FISH (Figure 8.2.15) or microarray experiments (Figure 8.2.12).

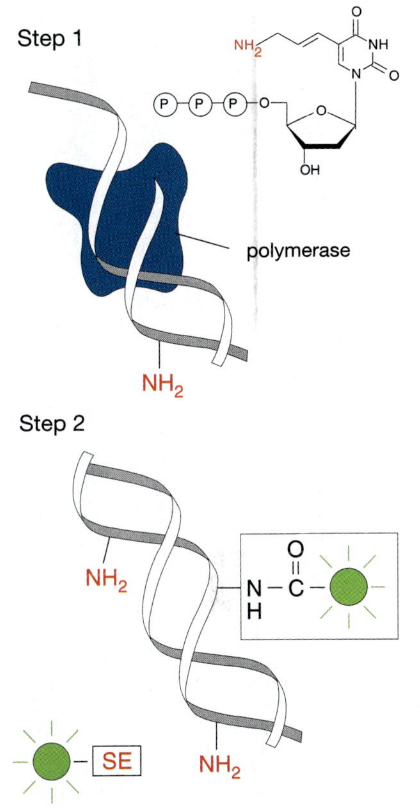

Figure 8.2.14 Schematic diagram of the labeling method provided in our ARES™ DNA Labeling Kits. The ARES™ DNA Labeling Kits use a two-step method to label DNA. Step 1) The aminoallyl dUTP is enzymatically incorporated. Step 2) A reactive fluorophore is used to label the incorporated aminoallyl group.

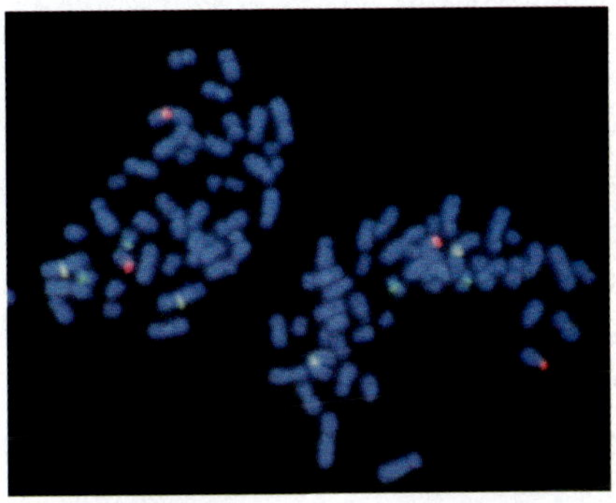

Figure 8.2.15 Fluorescent probes generated with ARES™ DNA Labeling Kits hybridized to human metaphase chromosome spreads. Centromere probes specific for chromosomes 17, 1 and 15 were prepared by nick translation and labeled with kits containing green-fluorescent Alexa Fluor® 488 (A21665), red-orange–fluorescent Alexa Fluor® 546 (A21667) and red-fluorescent Alexa Fluor® 594 (A21669) dyes. DNA was counterstained with the blue-fluorescent Hoechst 33342 dye (H1399, H3570, H21491), and the slides were mounted using the ProLong® Antifade Kit (P7481). This multiple-exposure image was obtained with bandpass filter sets appropriate for fluorescein, rhodamine, Texas Red® dye and DAPI.

Table 8.6 Molecular Probes® nucleic acid labeling kits.

Fluorophore	Ex/Em *	Similar Dyes	ULYSIS® Nucleic Acid Labeling Kits	ARES™ DNA Labeling Kits	FISH Tag™ DNA Kits †	FISH Tag™ RNA Kits †	FluoReporter® Oligonucleotide Amine Labeling Kits
Alexa Fluor® 488	490/520	Fluorescein (FITC or FAM), SpectrumGreen	U21650	A21665	F32947	F32952	A20191
Oregon Green® 488	495/520	Fluorescein (FITC or FAM), SpectrumGreen	U21659				
Alexa Fluor® 532	525/550	Rhodamine 6G	U21651				
Alexa Fluor® 546	555/570	Cy®3 dye, tetramethylrhodamine (TRITC), SpectrumOrange	U21652	A21667			
Alexa Fluor® 555	555/570	Cy®3 dye, tetramethylrhodamine (TRITC), SpectrumOrange		A21677	F32948	F32953	
Alexa Fluor® 568	575/600	Lissamine rhodamine B dye	U21653				
Alexa Fluor® 594	590/615	Texas Red® dye, SpectrumRed	U21654	A21669	F32949	F32954	
Alexa Fluor® 647	650/670	Cy®5 dye	U21660	A21676	F32950	F32955	A20196

* Excitation (Ex) and Emission (Em) maxima, in nm. † Also available are the FISH Tag™ DNA Multicolor Kit (F32951) and the FISH Tag™ RNA Multicolor Kit (F32956), each containing four different amine-reactive Alexa Fluor® dyes.

IMPORTANT NOTICE: The products described in this manual are covered by one or more Limited Use Label License(s). Please refer to the Appendix on page 971 and Master Product List on page 975. Products are For Research Use Only. Not intended for any animal or human therapeutic or diagnostic use.

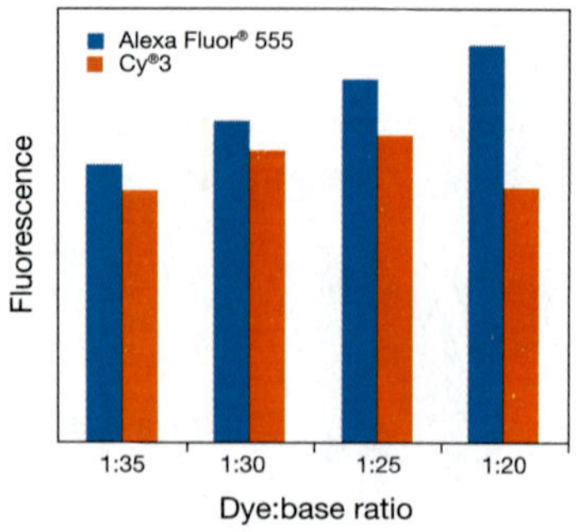

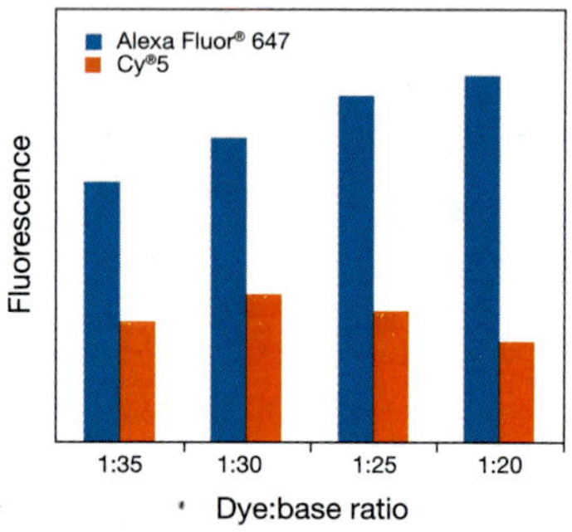

Figure 8.2.16 Change in signal brightness with level of labeling for Alexa Fluor® versus Cy® dye–labeled DNA. Chromosome 17 α-satellite DNA was amine-modified by nick translation in the presence of varying ratios of aminoallyl dUTP (A21664) to dTTP. The pools of modified DNA were split into equal parts and labeled with Alexa Fluor® 555 or Alexa Fluor® 647 succinimidyl ester (A21677, A21676) or with Cy®3 or Cy®5 reactive dye. The degree of labeling was calculated for each reaction; each probe was then hybridized to human metaphase chromosomes and the brightness of each signal measured. The brightness of the signal was plotted against the degree of labeling for each dye. At higher levels of labeling, the Alexa Fluor® dyes become brighter, whereas the corresponding Cy® dyes become quenched.

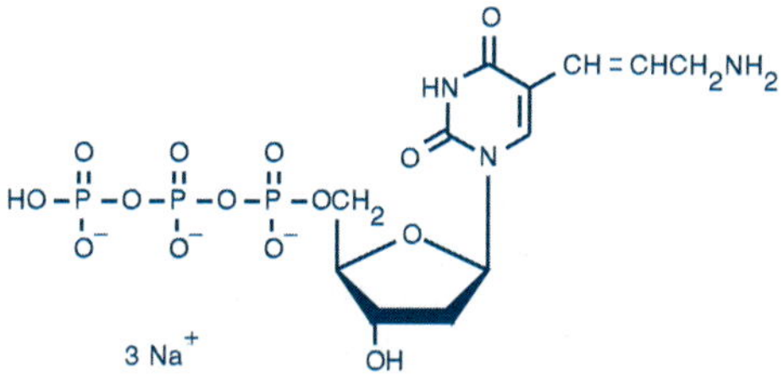

Figure 8.2.17 Aminoallyl dUTP (5-(3-aminoallyl)-2′-deoxyuridine 5′-triphosphate, trisodium salt; A21664).

In addition to 5-(3-aminoallyl)-dUTP, the ARES™ DNA Labeling Kits provide amine-reactive Alexa Fluor® dyes, which exhibit spectral properties that are superior to conventional dyes for labeling nucleic acids (Figure 8.2.16). Each kit contains sufficient reagents for 5–10 labelings of 1–5 µg DNA each, including:

- 5-(3-Aminoallyl)-dUTP
- Amine-reactive fluorescent dye and an appropriate solvent
- Sodium bicarbonate
- Nuclease-free H_2O
- Detailed protocols for labeling DNA using reverse transcriptase or nick translation

The 5-(3-aminoallyl)-dUTP (A21664, Figure 8.2.17) and amine-reactive dyes are also available as stand-alone reagents. Enzymatic incorporation of 5-(3-aminoallyl)-dUTP permits incorporation of almost any amine-reactive dye in Chapter 1 into nucleic acids.[40]

FISH Tag™ DNA and FISH Tag™ RNA Kits

The FISH Tag™ DNA and FISH Tag™ RNA Kits (Table 8.6) employ the same aminoallyl nucleotide labeling method as the ARES™ DNA labeling Kit but provide a complete workflow solution for fluorescence *in situ* hybridization (FISH) applications. Each FISH Tag™ Kit provides all of the reagents needed for enzymatically incorporating the amine-modified nucleotide (aminoallyl dUTP or aminoallyl UTP) into DNA or RNA, followed by fluorescent labeling with an amine-reactive Alexa Fluor® dye and purification of the labeled probe using PureLink™ nucleic acid purification technology.

The FISH Tag™ DNA Kits are supplied with one of four spectrally distinct Alexa Fluor® dyes or with all four of these dyes in the FISH Tag™ DNA Multicolor Kit:

- FISH Tag™ DNA Green Kit with Alexa Fluor® 488 dye (F32947)
- FISH Tag™ DNA Orange Kit with Alexa Fluor® 555 dye (F32948)
- FISH Tag™ DNA Red Kit with Alexa Fluor® 594 dye (F32949)
- FISH Tag™ DNA Far-Red Kit with Alexa Fluor® 647 dye (F32950)
- FISH Tag™ DNA Multicolor Kit (F32951)

Each kit provides sufficient reagents for 10 labelings, as well as detailed instructions for synthesizing amine-modified DNA by nick translation of template DNA, labeling with an amine-reactive Alexa Fluor® dye and purifying the fluorescent DNA probe. Kit contents include:

- Alexa Fluor® amine-reactive dye
- Dimethylsulfoxide (DMSO)
- Dithiothreitol (DTT), glycogen, sodium bicarbonate and nuclease-free water
- DNA nucleotide mix, containing the optimal ratio of aminoallyl dUTP:dTTP
- DNA polymerase 1
- DNase 1
- Nick translation buffer
- Binding, wash and elution buffers
- Spin columns and collection tubes, for purifying both the amine-modified DNA and the fluorescent DNA probe
- *SlowFade*® Gold antifade reagent, for use in the hybridization/imaging protocol
- Detailed protocols

The FISH Tag™ RNA Kits are also supplied with one of four spectrally distinct Alexa Fluor® dyes or with all four of these dyes in the FISH Tag™ RNA Multicolor Kit:

- FISH Tag™ RNA Green Kit with Alexa Fluor® 488 dye (F32952)
- FISH Tag™ RNA Orange Kit with Alexa Fluor® 555 dye (F32953)

IMPORTANT NOTICE: The products described in this manual are covered by one or more Limited Use Label License(s). Please refer to the Appendix on page 971 and Master Product List on page 975. Products are For Research Use Only. Not intended for any animal or human therapeutic or diagnostic use.

- FISH Tag™ RNA Red Kit with Alexa Fluor® 594 dye (F32954)
- FISH Tag™ RNA Far-Red Kit with Alexa Fluor® 647 dye (F32955)
- FISH Tag™ RNA Multicolor Kit (F32956)

Each kit provides sufficient reagents for 10 labelings, as well as detailed instructions for synthesizing amine-modified RNA by *in vitro* transcription of template DNA (containing a T3, T7 or SP6 promoter), labeling with an amine-reactive Alexa Fluor® dye and purifying the fluorescent RNA probe. Kit contents include:

- Alexa Fluor® amine-reactive dye
- Dimethylsulfoxide (DMSO)
- Dithiothreitol (DTT), glycogen, sodium bicarbonate and nuclease-free water
- RNA nucleotide mix, containing the optimal ratio of aminoallyl UTP:UTP
- T7, T3 and SP6 RNA polymerases
- DNase 1
- T3/T7 and SP6 transcription buffers
- RNaseOUT™ ribonuclease inhibitor
- Binding, wash and elution buffers
- Spin columns and collection tubes, for purifying both the amine-modified RNA and the fluorescent RNA probe
- *SlowFade*® Gold antifade reagent, for use in the hybridization/imaging protocol
- Detailed protocols

ULYSIS® Nucleic Acid Labeling Kits

The ULYSIS® Alexa Fluor® Nucleic Acid Labeling Kits (Table 8.6) provide a simple, reliable method for producing fluorescent hybridization probes by combining our Alexa Fluor® fluorophores with the versatile, patented Universal Linkage System (ULS®) platinum-based chemistry developed by KREATECH Biotechnology BV. The ULS® technology is based on the use of a platinum dye reagent that forms a stable adduct with the *N*-7 position of guanine and, to a lesser extent, adenine bases in DNA, RNA, peptide–nucleic acid conjugates (PNA) and oligonucleotides (Figure 8.2.18). In protein contexts, ULS® reagents are reactive with cysteine residues and other thiols.[42] The labeling reaction typically takes only 15 minutes, and separation of the labeled nucleic acids from the unreacted ULS® reagent can be accomplished through use of a simple spin-column procedure (Figure 8.2.19). DNA longer than ~1000 base pairs requires a 10-minute DNase digestion before labeling, which both optimizes labeling and fragments the probe for efficient hybridization.

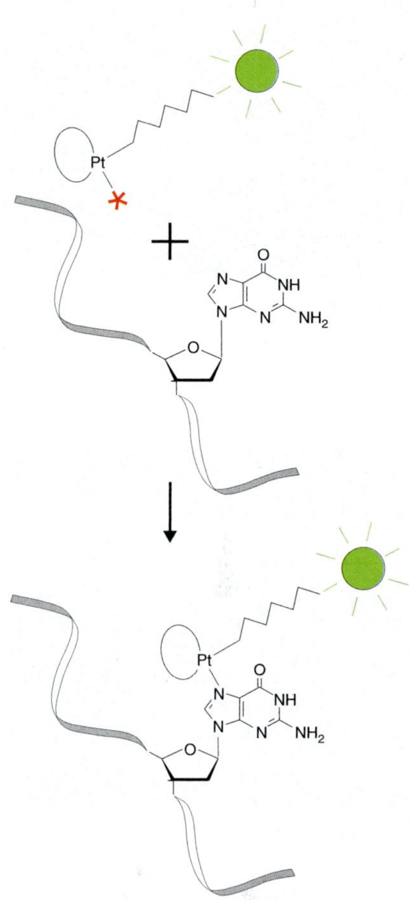

Figure 8.2.18 Schematic diagram of the labeling method provided in our ULYSIS® Nucleic Acid Labeling Kits. The ULS® reagent in the ULYSIS® Nucleic Acid Labeling Kits reacts with the *N*-7 position of guanine residues to provide a stable coordination complex between the nucleic acid and the fluorophore label.

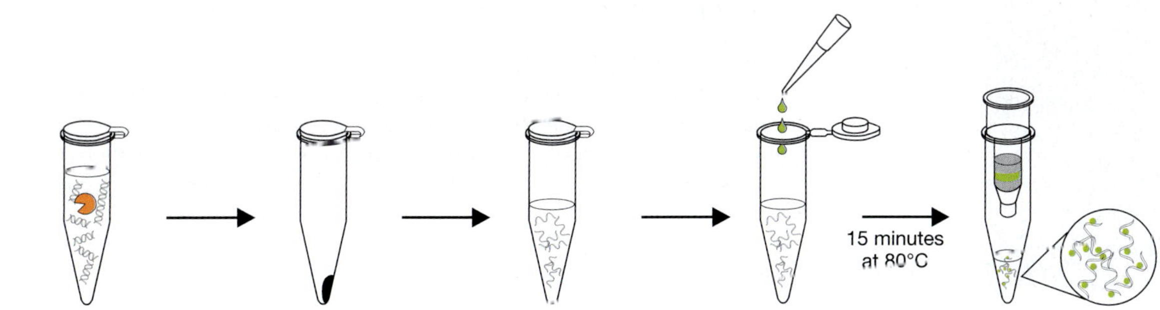

Figure 8.2.19 Nucleic acid labeling method provided in our ULYSIS® Nucleic Acid Labeling Kits.

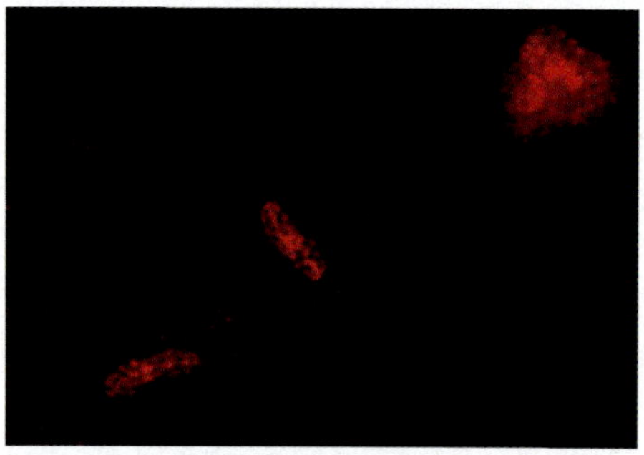

Figure 8.2.20 A paint probe for chromosome 2 was labeled with the ULYSIS® Alexa Fluor® 546 Nucleic Acid Labeling Kit (U21652) and hybridized to human metaphase chromosomes. Image contributed by Joop Wiegant, Leiden University Medical Center, Leiden, The Netherlands.

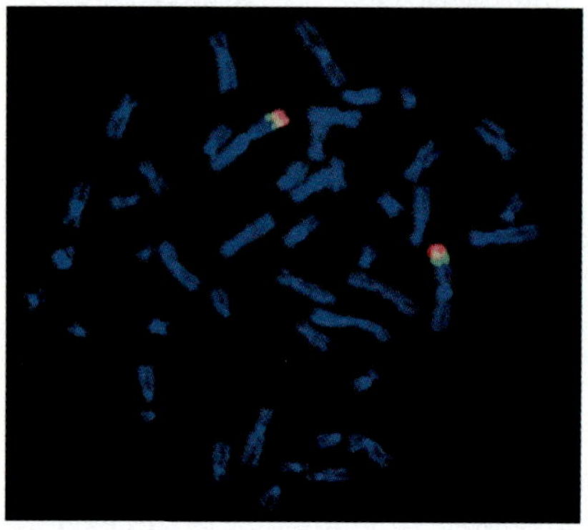

Figure 8.2.21 Human metaphase chromosomes hybridized to fluorescent probes from two overlapping microdissection libraries. Probes specific to chromosome regions 1p34–35 and 1p36 were labeled using the ULYSIS® Oregon Green® 488 (U21659) and Alexa Fluor® 594 (U21654) Nucleic Acid Labeling Kits, respectively. The chromosomes were counterstained with DAPI (D1306, D3571, D21490). Image contributed by Jingwei Yu, Colorado Genetics Laboratory.

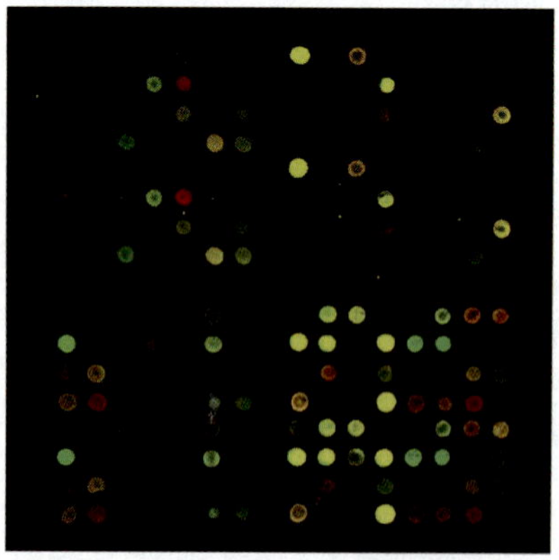

Figure 8.2.22 ULYSIS® reagent–labeled RNA hybridized to a microarray. Poly(A)+ RNA samples from the spleen of an irradiated or unirradiated mouse were labeled using the Alexa Fluor® 594 ULS® reagent or Alexa Fluor® 546 ULS® reagent, respectively. The labeled samples were mixed and hybridized to a cDNA microarray. The image was contributed by Rahul Mitra (Baylor College of Medicine) and Mini Kapoor, Thomas H. Burrows, and Rachel Grier (MD Anderson Cancer Center).

In addition to ULYSIS® Alexa Fluor® Nucleic Acid Labeling Kits, we offer the ULYSIS® Oregon Green® 488 Nucleic Acid Labeling Kit. Each of these ULYSIS® Kits provides sufficient reagents for 20 labelings of 1 µg DNA each, including:

- ULS® labeling reagent and appropriate solvent
- Labeling buffer
- Deoxyribonuclease I (DNase I), for digesting DNA longer than 1000 base-pairs prior to labeling
- DNase I storage and reaction buffers
- Control DNA from calf thymus
- Nuclease-free H_2O
- Detailed protocols for preparing fluorescent DNA hybridization probes for chromosome *in situ* hybridization and dot-blot hybridization

Probes labeled using the ULYSIS® Kits are stable indefinitely and hybridize effectively to target DNA. The ULS® method has been used to prepare labeled probes for dot, northern and Southern blot analysis, RNA and DNA *in situ* hybridization, multicolor fluorescence *in situ* hybridization (FISH; Figure 8.2.20, Figure 8.2.21), comparative genome hybridization (CGH) and microarray analysis (Figure 8.2.22).

Labeled Oligonucleotides

Panomer™ 9 Random Oligodeoxynucleotides

RNA samples can be labeled by reverse transcription using fluorophore-labeled random oligonucleotides as primers in combination with unlabeled deoxynucleotide triphosphates. Panomer™ 9 random-sequence oligodeoxynucleotides are covalently labeled on the 5′-terminus with one of four different Alexa Fluor® dyes:

- Panomer™ 9 random oligodeoxynucleotide, Alexa Fluor® 488 conjugate (P21680)
- Panomer™ 9 random oligodeoxynucleotide, Alexa Fluor® 546 conjugate (P21681)
- Panomer™ 9 random oligodeoxynucleotide, Alexa Fluor® 555 conjugate (P21687)
- Panomer™ 9 random oligodeoxynucleotide, Alexa Fluor® 647 conjugate (P21686)

The Panomer™ 9 oligonucleotides are also useful as primers for synthesizing labeled DNA via Klenow DNA polymerase or reverse transcriptase. In these reactions, the primer provides the fluorescent label, whereas unlabeled nucleotides are incorporated by the enzyme. This labeling strategy allows for efficient and unbiased incorporation of nucleotides because the bulky dye molecule does not interfere with nucleotide incorporation. However, because the resulting DNA fragments contain only a single label, the detection sensitivity will typically be lower than that achieved by incorporating fluorophore- or hapten-labeled nucleotides.

Alexa Fluor® Oligonucleotide Amine Labeling Kits

The Alexa Fluor® Oligonucleotide Amine Labeling Kits (Table 8.6) provide the reagents required for labeling synthetic oligonucleotides that have amine groups incorporated at their 5′- or 3′-terminus. Following purification by standard chromatographic or electrophoretic procedures, these singly labeled oligonucleotides can serve as hybridization or ligation probes for a variety of applications.[43–45]

Each Alexa Fluor® Oligonucleotide Amine Labeling Kit contains sufficient reagents for three labelings of 50 µg each of an amine-modified oligonucleotide, including:

- Three vials of amine-reactive dye
- Dimethylsulfoxide (DMSO)
- Three vials of labeling buffer
- Labeling protocol

Labeling Amine-, Thiol- or Phosphate-Modified Oligonucleotides

Amine or thiol groups can be incorporated into a chemically synthesized oligonucleotide. These groups can then be directly conjugated to an amine-reactive (Chapter 1) or thiol-reactive (Chapter 2) fluorophore or hapten. For labeling amine-modified DNA or RNA probes in microarray-based experiments, we offer three amine-reactive Alexa Fluor® dyes conveniently packaged in 10 single-use vials and rigorously tested for the ability to efficiently label amine-modified DNA:

- Alexa Fluor® 488 reactive dye decapack (A32750)
- Alexa Fluor® 555 reactive dye decapack (A32756)
- Alexa Fluor® 647 reactive dye decapack (A32757)
- A set containing both the Alexa Fluor® 555 and Alexa Fluor® 647 reactive dye decapacks (A32755), for two-color experiments

For use in fluorescence resonance energy transfer (FRET) applications (Fluorescence Resonance Energy Transfer (FRET)—Note 1.2), amine-modified oligonucleotides can be labeled with our nonfluorescent, amine-reactive QSY® 7, QSY® 9 and QSY® 21 dyes (Section 1.6). These QSY® quencher dyes have absorption in the visible and near-infrared spectrum, making them excellent energy transfer acceptors from a wide variety of dyes that emit in the visible range, including fluoresceins, Oregon Green® dyes, rhodamines, Texas Red®, Cy®3 and several of the Alexa Fluor® dyes.[46] Conjugates of the QSY® 35 dye (Section 1.8) have somewhat shorter-wavelength absorption and are useful as quenchers of UV light–excited fluorescent dyes. Oligonucleotide conjugates of the nonfluorescent dabcyl succinimidyl ester (D2245) have also been extensively used for FRET-based and quencher-based assays.[47–50]

A fluorophore or hapten containing an aliphatic amine may be conjugated to the 3′- or 5′-phosphate group of an oligonucleotide by using the zero-length crosslinker EDAC (E2247, Section 3.4) in an *N*-methylimidazole buffer at pH 9. This reaction results in a phosphoramidate bond that is stable in most molecular biology assays. The method can be used in combination with T4 polynucleotide kinase to fluorescently label oligonucleotides lacking a 5′-phosphate, or to double-label radioactively labeled oligonucleotides. We have found that this reaction is very efficient—labeling over 90% of the oligonucleotides that contain a phosphate group—and much easier than conventional methods for modifying terminal phosphate groups, which typically require multistep synthesis.[51–53] For this reaction we recommend cadaverine-conjugated fluorophores (Section 3.4, Table 3.2).

It has also been reported that DNA can be reacted quantitatively with carbonyl diimidazole and a diamine (such as ethylenediamine) or a carbohydrazide to yield a phosphoramidate that has a free primary amine; the amine can then be modified with amine-reactive reagents[51–55] of the type described in Chapter 1. Fluorescent or biotinylated amines have been coupled to the 5′-phosphate of tRNA using dithiodipyridine and triphenylphosphine.[56] Wang and Giese[57] have reported a general method that employs an imidazole derivative prepared from our BODIPY® FL hydrazide (D2371, Section 3.3) to label phosphates, including nucleotides, for capillary electrophoresis applications.

Amine-Reactive SYBR® Dye

The amine-reactive succinimidyl ester of the SYBR® 101 dye (S21500) can be coupled with amine-derivatized oligonucleotides. The conjugates may fluoresce green as the result of intramolecular or intermolecular association of the dye with the nucleic acid backbone; however, changes in intensity and fluorescence polarization may occur during hybridization reactions. Similar amine-reactive versions of the cyanine dyes have been utilized to label peptide–nucleic acid conjugates (PNA) and to detect their hybridization to target nucleic acids in solution.[58–61]

Dyes for Sequencing Applications

We are a major manufacturer of several dyes that are used directly or indirectly in nucleic acid sequencing and provide these dyes in reactive forms for preparing conjugates (Table 8.7). Because the electrophoretic separation step during sequencing is highly sensitive to the chemical structure of the fragments, the use of single-isomer labels is essential. In addition to providing high-purity reactive succinimidyl esters of the common FAM™, JOE™, TAMRA™ and ROX™ dyes, we prepare amine-reactive single isomers of carboxyrhodamine 6G (CR 6G). The 6-isomer of the CR 6G dye has spectroscopic and electrophoretic properties that are superior to the JOE™ dye often used for automated DNA sequencing. Please go to www.invitrogen.com/handbook/customorganics for information about availability of any of our reactive dyes in bulk.

Certain BODIPY® dyes (Section 1.4) have been shown to be very useful for DNA sequencing,[62–64] in part because the dyes have a minimal effect on the mobility of the fragment during electrophoresis and also exhibit well-resolved spectra with narrow bandwidths. BODIPY® dyes, which are all high-purity single isomers, exhibit high extinction

Table 8.7 Amine-reactive dyes for nucleic acid sequencing.

Cat. No.	Reactive Dye *	Ex/Em §	Handbook Location
C2210	5-FAM™, SE †	494/518	Section 1.5
C6164	6-FAM™, SE	494/518	Section 1.5
C6127	5-CR 6G, SE	525/555	Section 1.6
C6171MP	6-JOE™, SE †	522/550	Section 1.5
C2211	5-TAMRA™, SE	555/580	Section 1.6
C6123	6-TAMRA™, SE †	555/580	Section 1.6
C6125	5-ROX™, SE	580/605	Section 1.6
C6126	6-ROX™, SE †	580/605	Section 1.6
D2184	BODIPY® FL, SE ‡	505/513	Section 1.4
D6140	BODIPY® FL, SSE	505/513	Section 1.4
D6102	BODIPY® FL-X, SE	505/513	Section 1.4
D6180	BODIPY® R6G, SE ‡	528/550	Section 1.4
D6117	BODIPY® TMR-X, SE	542/574	Section 1.4
D2222	BODIPY® 564/570, SE ‡	565/571	Section 1.4
D2228	BODIPY® 581/591, SE ‡	584/592	Section 1.4
D6116	BODIPY® TR-X, SE	589/617	Section 1.4

* FAM™ = carboxyfluorescein; CR 6G = carboxyrhodamine 6G, JOE™ = 6-carboxy-4′,5′-dichloro-2′,7′-dimethoxyfluorescein; TAMRA™ = carboxytetramethylrhodamine; ROX™ = carboxy-X-rhodamine; BODIPY® = a substituted 4,4-difluoro-4-bora-3a,4a-diaza-*s*-indacene derivative; SE = succinimidyl ester; SSE = water-soluble sulfosuccinimidyl ester. † These are the most widely used isomers for DNA sequencing.[1–4] ‡ These BODIPY® derivatives were reported to be useful for automated DNA sequencing, in part because the dyes have a minimal effect on the mobility of the fragment during electrophoresis and also exhibit well-resolved spectra with narrow bandwidths.[5] § Excitation (Ex) and emission (Em) maxima in nm.

1. Anal Biochem (1994) 223:39; **2.** Nucleic Acids Res (1992) 20:2471; **3.** Proc Natl Acad Sci U S A (1989) 86:9178; **4.** Genome Res (1996) 6:995; **5.** Science (1996) 271:1420.

IMPORTANT NOTICE: The products described in this manual are covered by one or more Limited Use Label License(s). Please refer to the Appendix on page 971 and Master Product List on page 975. Products are For Research Use Only. Not intended for any animal or human therapeutic or diagnostic use.

coefficients, excellent quantum yields and fluorescence emission that is quite photostable and insensitive to pH. BODIPY® FL-X, BODIPY® TMR-X and BODIPY® TR-X succinimidyl esters are reactive versions of the BODIPY® fluorophores with emission properties similar to those of fluorescein, (Section 1. 4) tetramethylrhodamine and Texas Red® dyes, respectively; BODIPY® 630/650-X and BODIPY® 650/665-X succinimidyl esters provide long-wavelength amine-reactive fluorophores that match filter sets optimized for the Alexa Fluor® 647 and Cy®5 dyes. Oligonucleotide conjugates of the BODIPY® FL, BODIPY® R6G, BODIPY® 564/570 and BODIPY® FL, BODIPY® R6G, BODIPY® 564/570 and BODIPY® 581/591 dyes have been found to be particularly useful for automated DNA sequencing.[62–64]

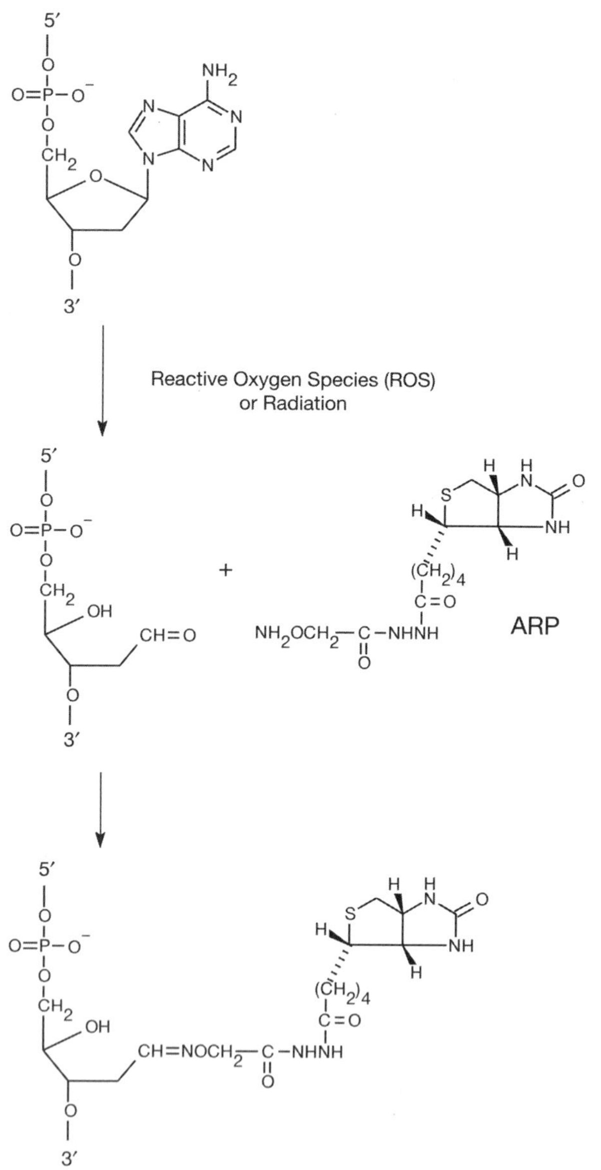

Figure 8.2.23 Aldehyde-reactive probe (ARP) used to detect DNA damage. The biotin hydroxylamine ARP (A10550) reacts with aldehyde groups formed when reactive oxygen species depurinate DNA. This reaction forms a covalent bond linking the DNA to biotin. The biotin can then be detected using fluorophore- or enzyme-linked streptavidin.

Other Labeling Methods for Nucleic Acids

Labeling Nascent DNA and RNA for Cell Proliferation Studies

Cells can naturally incorporate the thymidine analog 5-bromo-2′-deoxyuridine (BrdU, B23151; Section 15.4) into their DNA during cell division, making this nucleoside analog an excellent marker of both cell cycle and cell proliferation.[65] Analysis of incorporated BrdU can be either by detection with an antibody to BrdU-modified DNA or by modification of the fluorescence of a nucleic acid stain. For instance, the fluorescence of TO-PRO®-3 and LDS 751 is considerably enhanced by the presence of BrdU in DNA,[66] whereas that of the Hoechst dyes is specifically quenched.[67] 5-Bromo-2′-deoxyuridine 5′-triphosphate (BrdUTP, B21550; Section 15.4) is commonly used in TUNEL-based methods to detect proliferating or apoptotic cells.[30,31] BrdUTP is a substrate for reverse transcriptase[68,69] and Klenow polymerase,[70,71] and has been used in a sensitive nonisotopic assay for detecting HIV-1–associated reverse transcriptase activity.[68,72] Similarly, the corresponding brominated ribonucleotide, 5-Bromouridine 5′-triphosphate (BrUTP, B21551; Section 15.4) is an excellent substrate for RNA polymerase[73] and has been used to monitor nucleolar transcription *in situ*.[74,75]

The Click-iT® EdU cell proliferation assay (Section 15.4) provides a superior alternative to bromodeoxyuridine (BrdU) or ^{3}H-thymidine incorporation methods for measuring new DNA synthesis.[76,77] The alkynyl nucleoside analog EdU (5-ethynyl-2′-deoxyuridine; A10044, E10187; Section 3.1) is incorporated into DNA during the synthesis phase (S phase) of the cell cycle and is subsequently detected by copper (I)–catalyzed click coupling to an azide-derivatized fluorophore[78] (Section 3.1). The small size of the click-coupled fluorophore compared to that of antibodies required for immunodetection of BrdU enables efficient penetration of complex samples without the need for harsh cell treatment, simplifying the assay considerably.

Similarly, the Click-iT® RNA assay employs the alkyne-modified nucleoside EU (5-ethynyl uridine, E10345; Section 3.1), which is supplied to cells and incorporated into nascent RNA.[79] The small size of the alkyne tag enables efficient incorporation by RNA polymerases without any apparent changes to the RNA levels of several housekeeping genes. Detection of incorporated EU is accomplished by copper (I)–catalyzed click coupling to an azide-derivatized fluorophore (Section 3.1). The multiplexing capability of the assays makes them ideal for toxicological profiling or interrogation of disease models using high-content imaging platforms.

Labeling Abasic Sites with ARP

Abasic sites in DNA can be generated spontaneously or through the action of free radicals, ionizing radiation or mutagens like MMS (methyl methanesulfonate). These apurinic and apyrimidinic sites are very common lesions in DNA and are thought to be important intermediates in mutagenesis. A quick and sensitive microplate assay for abasic sites can be performed using ARP (A10550, Figure 8.2.23), a biotin hydroxylamine that reacts with the exposed aldehyde group at abasic sites. Biotins bound to the abasic sites can be quantitated with our fluorescent– or enzyme-conjugated streptavidin complexes[80–83] (Section 7.6, Table 7.9). ARP is permeant to cell membranes, permitting detection of abasic sites in live cells.[84,85] Alexa Fluor® hydroxylamine derivatives (Section 3.3) may have utility similar to that of ARP but are not membrane permeant.

IMPORTANT NOTICE: The products described in this manual are covered by one or more Limited Use Label License(s). Please refer to the Appendix on page 971 and Master Product List on page 975. Products are For Research Use Only. Not intended for any animal or human therapeutic or diagnostic use.
www.invitrogen.com/probes
molecular probes® | invitrogen™ by life technologies™

Labeling Cytidine Residues

DNA and RNA can be modified by reacting their cytidine residues with sodium bisulfite to form sulfonate intermediates that can then be directly coupled to hydrazides or aliphatic amines.[86–88] For example, biotin hydrazides (Section 4.2) have been used in a bisulfite-mediated reaction to couple biotin to cytidine residues in oligonucleotides.[89] The fluorescent hydrazides, hydroxylamines and aliphatic amines listed in Table 3.2 might be useful in this reaction. The bisulfite-activated cytidylic acid can also be coupled to aliphatic diamines such as ethylenediamine.[90] The amine-modified DNA or RNA can then be modified with any of the amine-reactive dyes described in this section or in Chapter 1.

Specialized Methods for Nucleic Acid Modification

A few other specialized methods have been developed for nucleic acid modification. These include:

- Use of a fluorescent iodoacetamide or maleimide, along with T4 polynucleotide kinase and ATP-γ-*S* (ATP with a sulfur in the terminal phosphate), to introduce a thiophosphate at the 5′-terminus of 5′-dephosphorylated RNA [91,92] or DNA
- Introduction of 4-thiouridine at the 3′-terminus of DNA using calf thymus terminal deoxynucleotidyl transferase followed by treatment with ribonuclease and reaction with thiol-reactive probes [93,94]
- Direct reaction of thiol-reactive reagents with 4-thiouridine residues in nucleic acids [56,95–98]
- Selective oxidation of the 3′-terminus of RNA by sodium metaperiodate to a dialdehyde, which can be coupled with a fluorescent or biotin hydrazide or hydroxylamine reagent [55,92,99–103]
- Direct reaction of amine- or thiol-reactive reagents with aminoacyl tRNA or thioacetylated aminoacyl tRNA [56,104,105]
- Reaction of the X-base of tRNA with isothiocyanates [101] or replacement of other uncommon bases in tRNA by fluorophores [106–108]
- Photolabeling of plasmid DNA with fluorescent 4-azido-2,3,5,6-tetrafluorobenzyl derivatives [109]
- Coupling of labeled diazonium salts to nucleic acids [110]

REFERENCES

1. Biotechniques (2004) 36:114; **2.** Trends Genet (1997) 13:475; **3.** Bioessays (1997) 19:75; **4.** Mol Pathol (1998) 51:62; **5.** Cell Vis (1998) 5:49; **6.** Nat Genet (1999) 21:48; **7.** Nat Genet (1999) 21:42; **8.** Nat Genet (1999) 21:33; **9.** Nat Genet (1999) 21:25; **10.** Nat Genet (1999) 21:20; **11.** Nat Genet (1999) 21:15; **12.** Nat Genet (1999) 21:10; **13.** Nat Genet (1999) 21:5; **14.** Mol Psychiatry (1998) 3:483; **15.** Nat Biotechnol (1998) 16:45; **16.** Biotechniques (1995) 19:442; **17.** Nat Genet (1996) 14:312; **18.** Histochem Cell Biol (1997) 108:299; **19.** Genes Chromosomes Cancer (2000) 28:318; **20.** Cytometry (1999) 35:214; **21.** Genes Chromosomes Cancer (2000) 27:418; **22.** Eur J Hum Genet (1999) 7:2; **23.** Science (1996) 273:494; **24.** Am J Hum Genet (1997) 61:16; **25.** Cytobios (1997) 90:7; **26.** J Mol Med (1997) 75:801; **27.** J Biolol Tech (2005) 16:104; **28.** J Histochem Cytochem (2000) 48:1593; **29.** Cytometry (1997) 27:1; **30.** Exp Cell Res (1996) 222:28; **31.** Cell Prolif (1995) 28:571; **32.** Cytometry (1995) 20:172; **33.** J Cell Biol (1999) 144:813; **34.** Biophys J (1999) 77:2871; **35.** Dev Biol (1999) 206:232; **36.** Anal Biochem (2004) 331:243; **37.** Eur J Histochem (1998) 42:111; **38.** Cell Vis (1998) 5:83; **39.** Am J Pathol (1997) 150:1553; **40.** Biotechniques (2000) 28:518; **41.** Proc Natl Acad Sci U S A (1993) 90:4206; **42.** Clin Chem (2002) 48:1352; **43.** J Am Chem Soc (2006) 128:11423; **44.** Biochemistry (2006) 45:4164; **45.** Anal Chem (2003) 75:1664; **46.** J Forensic Sci (2003) 48:282; **47.** Anal Biochem (1999) 276:177; **48.** Biotechniques (1999) 27:1116; **49.** Biotechniques (1999) 26:552; **50.** J Am Chem Soc (1999) 121:2921; **51.** Anal Biochem (1994) 218:444; **52.** Biochem Biophys Res Commun (1994) 200:1239; **53.** Methods Mol Biol (1994) 26:145; **54.** J Chromatogr (1992) 608:171; **55.** J Mol Biol (1991) 221:441; **56.** Biochemistry (1990) 29:10734; **57.** Anal Chem (1993) 65:3518; **58.** Biochemistry (2000) 39:4327; **59.** Anal Biochem (2000) 281:26; **60.** Anal Biochem (2000) 287:179; **61.** J Am Chem Soc (2001) 123:803; **62.** Biotechniques (1998) 25:446; **63.** Science (1996) 271:1420; **64.** Nucleic Acids Res (1992) 20:2471; **65.** Methods Cell Biol (1994) 41:297; **66.** Cytometry (1994) 17:310; **67.** Exp Cell Res (1987) 173:256; **68.** Biotechnol Appl Biochem (1999) 29:241; **69.** Biotechnol Appl Biochem (1996) 23:95; **70.** Exp Cell Res (1997) 234:498; **71.** Biochem J (1988) 253:637; **72.** J Virol Methods (1991) 31:181; **73.** J Biochem (Tokyo) (1984) 96:1501; **74.** Histochem Cell Biol (2000) 113:181; **75.** Mol Biol Cell (1999) 10:211; **76.** J Immunol Methods (2009) 350:29; **77.** Biotechniques (2008) 44:927; **78.** Proc Natl Acad Sci U S A (2008) 105:2415; **79.** Proc Natl Acad Sci U S A (2008) 105:15779; **80.** Anal Biochem (1999) 267:331; **81.** Biochemistry (1972) 11:3610; **82.** Biochemistry (1992) 31:3703; **83.** Biochemistry (1993) 32:8276; **84.** Proc Natl Acad Sci U S A (2000) 97:686; **85.** J Biol Chem (2000) 275:6741; **86.** J Clin Microbiol (1986) 23:311; **87.** Biochemistry (1980) 19:1774; **88.** Biochemistry (1976) 15:2677; **89.** Biochem Biophys Res Commun (1987) 142:519; **90.** Biochem J (1968) 108:883; **91.** Anal Biochem (2004) 325:137; **92.** Biochemistry (1991) 30:4821; **93.** Anal Biochem (1988) 170:271; **94.** Nucleic Acids Res (1979) 7:1485; **95.** Nucleic Acids Res (1988) 16:2203; **96.** Anal Biochem (1983) 131:419; **97.** Biochemistry (1985) 24:692; **98.** J Mol Biol (1982) 156:113; **99.** Bioconjug Chem (1994) 5:436; **100.** Biochemistry (1980) 19:5947; **101.** Eur Biophys J (1988) 16:45; **102.** Biochemistry (1986) 25:5298; **103.** Eur J Biochem (1984) 142:261; **104.** J Am Chem Soc (1991) 113:2722; **105.** Eur J Biochem (1988) 172:663; **106.** Eur J Biochem (1979) 98:465; **107.** Methods Enzymol (1974) 29:667; **108.** FEBS Lett (1971) 18:214; **109.** Bioconjug Chem (2000) 11:51; **110.** Nucleic Acids Res (1988) 16:7197.

PRODUCT LIST 8.2 LABELING OLIGONUCLEOTIDES AND NUCLEIC ACIDS

Cat. No.	Product	Quantity
A32768	aha-dCTP (5-aminohexylacrylamido-dCTP) *2 mM in TE buffer*	500 µL
A32769	aha-dCTP (5-aminohexylacrylamido-dCTP) *50 mM in TE buffer*	50 µL
A32761	aha-dUTP (5-aminohexylacrylamido-dUTP) *50 mM in TE buffer*	50 µL
A32770	Alexa Fluor® 555-aha-dCTP *1 mM in TE buffer*	50 µL
A32762	Alexa Fluor® 555-aha-dUTP *1 mM in TE buffer*	50 µL
A32771	Alexa Fluor® 647-aha-dCTP *1 mM in TE buffer*	50 µL
A32763	Alexa Fluor® 647-aha-dUTP *1 mM in TE buffer*	50 µL
A20191	Alexa Fluor® 488 Oligonucleotide Amine Labeling Kit *3 labelings*	1 kit
A20196	Alexa Fluor® 647 Oligonucleotide Amine Labeling Kit *3 labelings*	1 kit
A32750	Alexa Fluor® 488 reactive dye decapack *for microarrays* *set of 10 vials*	1 set
A32756	Alexa Fluor® 555 reactive dye decapack *for microarrays* *set of 10 vials*	1 set

continued on next page

IMPORTANT NOTICE: The products described in this manual are covered by one or more Limited Use Label License(s). Please refer to the Appendix on page 971 and Master Product List on page 975. Products are For Research Use Only. Not intended for any animal or human therapeutic or diagnostic use.

PRODUCT LIST 8.2 LABELING OLIGONUCLEOTIDES AND NUCLEIC ACIDS—*continued*

Cat. No.	Product	Quantity
A32757	Alexa Fluor® 647 reactive dye decapack *for microarrays* *set of 10 vials*	1 set
A32755	Alexa Fluor® 555 and Alexa Fluor® 647 reactive dye decapacks *for microarrays* *set of 2 x 10 vials* *includes A32756 and A32757 decapacks*	1 set
A21664	aminoallyl dUTP (5-(3-aminoallyl)-2′-deoxyuridine 5′-triphosphate, trisodium salt) *2 mM in TE buffer*	500 µL
A10550	*N*-(aminooxyacetyl)-*N*′-(D-biotinoyl) hydrazine, trifluoroacetic acid salt (ARP)	10 mg
A21665	ARES™ Alexa Fluor® 488 DNA Labeling Kit *10 labelings*	1 kit
A21667	ARES™ Alexa Fluor® 546 DNA Labeling Kit *10 labelings*	1 kit
A21677	ARES™ Alexa Fluor® 555 DNA Labeling Kit *10 labelings*	1 kit
A21669	ARES™ Alexa Fluor® 594 DNA Labeling Kit *10 labelings*	1 kit
A21676	ARES™ Alexa Fluor® 647 DNA Labeling Kit *10 labelings*	1 kit
B32772	biotin-aha-dCTP *1 mM in TE buffer*	25 µL
B32766	biotin-aha-dUTP *1 mM in TE buffer*	25 µL
C11397	ChromaTide® Alexa Fluor® 488-5-dUTP *1 mM in TE buffer*	25 µL
C11403	ChromaTide® Alexa Fluor® 488-5-UTP *1 mM in TE buffer*	25 µL
C21555	ChromaTide® Alexa Fluor® 488-7-OBEA-dCTP *1 mM in TE buffer*	50 µL
C11398	ChromaTide® Alexa Fluor® 532-5-dUTP *1 mM in TE buffer*	25 µL
C11401	ChromaTide® Alexa Fluor® 546-14-dUTP *1 mM in TE buffer*	25 µL
C11404	ChromaTide® Alexa Fluor® 546-14-UTP *1 mM in TE buffer*	25 µL
C21556	ChromaTide® Alexa Fluor® 546-16-OBEA-dCTP *1 mM in TE buffer*	50 µL
C11399	ChromaTide® Alexa Fluor® 568-5-dUTP *1 mM in TE buffer*	25 µL
C11400	ChromaTide® Alexa Fluor® 594-5-dUTP *1 mM in TE buffer*	25 µL
C21559	ChromaTide® Alexa Fluor® 647-12-OBEA-dCTP *1 mM in TE buffer*	50 µL
C7614	ChromaTide® BODIPY® FL-14-dUTP *1 mM in TE buffer*	25 µL
C7604	ChromaTide® fluorescein-12-dUTP *1 mM in TE buffer*	25 µL
C7606MP	ChromaTide® tetramethylrhodamine-6-dUTP *1 mM in TE buffer*	25 µL
C7631	ChromaTide® Texas Red®-12-dUTP *1 mM in TE buffer*	25 µL
F32950	FISH Tag™ DNA Far Red Kit *with Alexa Fluor® 647 dye* *10 reactions*	1 kit
F32947	FISH Tag™ DNA Green Kit *with Alexa Fluor® 488 dye* *10 reactions*	1 kit
F32951	FISH Tag™ DNA Multicolor Kit *Alexa Fluor® dye combination* *10 reactions*	1 kit
F32948	FISH Tag™ DNA Orange Kit *with Alexa Fluor® 555 dye* *10 reactions*	1 kit
F32949	FISH Tag™ DNA Red Kit *with Alexa Fluor® 594 dye* *10 reactions*	1 kit
F32955	FISH Tag™ RNA Far Red Kit *with Alexa Fluor® 647 dye* *10 reactions*	1 kit
F32952	FISH Tag™ RNA Green Kit *with Alexa Fluor® 488 dye* *10 reactions*	1 kit
F32956	FISH Tag™ RNA Multicolor Kit *Alexa Fluor® dye combination* *10 reactions*	1 kit
F32953	FISH Tag™ RNA Orange Kit *with Alexa Fluor® 555 dye* *10 reactions*	1 kit
F32954	FISH Tag™ RNA Red Kit *with Alexa Fluor® 594 dye* *10 reactions*	1 kit
F30755	FluoReporter® Biotin Quantitation Assay Kit *for biotinylated nucleic acids* *10 determinations*	1 kit
F32767	fluorescein-aha-dUTP *1 mM in TE buffer*	25 µL
P21680	Panomer™ 9 random oligodeoxynucleotide, Alexa Fluor® 488 conjugate	10 nmol
P21681	Panomer™ 9 random oligodeoxynucleotide, Alexa Fluor® 546 conjugate	10 nmol
P21687	Panomer™ 9 random oligodeoxynucleotide, Alexa Fluor® 555 conjugate	10 nmol
P21686	Panomer™ 9 random oligodeoxynucleotide, Alexa Fluor® 647 conjugate	10 nmol
S21500	SYBR® 101, succinimidyl ester	1 mg
U21650	ULYSIS® Alexa Fluor® 488 Nucleic Acid Labeling Kit *20 labelings*	1 kit
U21651	ULYSIS® Alexa Fluor® 532 Nucleic Acid Labeling Kit *20 labelings*	1 kit
U21652	ULYSIS® Alexa Fluor® 546 Nucleic Acid Labeling Kit *20 labelings*	1 kit
U21653	ULYSIS® Alexa Fluor® 568 Nucleic Acid Labeling Kit *20 labelings*	1 kit
U21654	ULYSIS® Alexa Fluor® 594 Nucleic Acid Labeling Kit *20 labelings*	1 kit
U21660	ULYSIS® Alexa Fluor® 647 Nucleic Acid Labeling Kit *20 labelings*	1 kit
U21659	ULYSIS® Oregon Green® 488 Nucleic Acid Labeling Kit *20 labelings*	1 kit

IMPORTANT NOTICE: The products described in this manual are covered by one or more Limited Use Label License(s). Please refer to the Appendix on page 971 and Master Product List on page 975. Products are For Research Use Only. Not intended for any animal or human therapeutic or diagnostic use.

8.3 Nucleic Acid Quantitation in Solution

Through intensive research efforts in both chemical synthesis and bioassay development, we have developed rapid and exceptionally sensitive fluorescence-based assays for quantitation of nucleic acids in solution. The Quant-iT™ Assay Kits represent advanced quantitation systems for DNA, RNA or protein samples. These state-of-the-art assays are designed to deliver high sensitivity and specificity together with a streamlined protocol, prediluted standards and a ready-to-use buffer. The Quant-iT™ PicoGreen® dsDNA reagent, Quant-iT™ OliGreen® ssDNA reagent and Quant-iT™ RiboGreen® RNA reagent—optimized for double-stranded DNA, oligonucleotides and RNA, respectively—have a high affinity for nucleic acids and an extremely large fluorescence enhancement upon binding, making possible the direct detection of minute amounts of nucleic acids in complex solutions within minutes, usually without interference from other biomolecules. These reagents and quantitation assays provide the following advantages:

- Sensitivity. The PicoGreen® dye–, OliGreen® dye– and RiboGreen® dye–based fluorescence assays are up to 10,000-fold more sensitive than UV absorbance measurements and at least 400-fold more sensitive than assays that use the Hoechst 33258 dye [1] (H1398, H3569, H21491), requiring much less sample for quantitation.
- Accuracy. Unlike measurements of UV absorbance, these assays are not affected by the presence of proteins, free nucleotides or very short oligonucleotides, making quantitation of intact oligonucleotides and nucleic acids much more accurate in complex mixtures such as serum or whole blood.
- Precision. The average standard deviations of triplicate assays using these reagents are typically less than 5%.
- Simplicity. These assays have a very simple protocol that requires no separation steps, making them ideal for automated, high-throughput measurements.
- Broad dynamic range. Quantitation is accurate over four orders of magnitude for the PicoGreen® and OliGreen® assays, with a single dye concentration. The RiboGreen® assay is accurate over three orders of magnitude.
- Instrument compatibility. Quantitation assays can be performed using a fluorescence microplate reader with standard filters optimized for fluorescein-like dyes, a relatively inexpensive filter-based spectrofluorometer or a standard spectrofluorometer. Additionally, we offer a set of Quant-iT™ Assay Kits specifically designed for the Qubit® fluorometer.
- Convenience. Each reagent is available separately or in a kit that additionally contains a nuclease-free buffer and standards. The RiboGreen® RNA quantitation assay is also available in the RediPlate™ 96 format—a prepared 96-well microplate with standards included—for high-throughput applications.

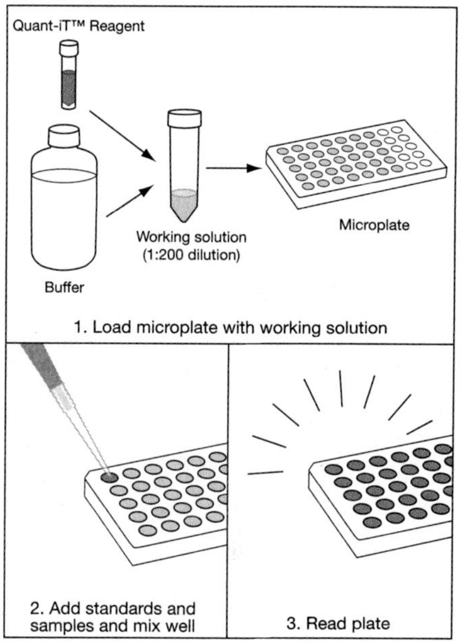

Figure 8.3.1 DNA, RNA or protein quantitation with the Quant-iT™ Assay Kits (Q33120, Q33130, Q33140, Q33210).

Quant-iT™ Assay Kits for DNA and RNA

The Quant-iT™ family of assay kits provides state-of-the-art reagents designed for sensitive and selective quantitation of DNA, RNA or protein samples using a standard fluorescence microplate reader or the Qubit® fluorometer (Table 8.8). These kits have been specially formulated with ready-to-use buffers, prediluted standards and easy-to-follow instructions, allowing quantitation to be both accurate and extremely easy (Figure 8.3.1). Each Quant-iT™ assay is:

- Ready to use. Only the dye is diluted in the supplied buffer; dilution of standards or buffer is not required.
- Easy to perform. Just add the sample to the diluted dye and read the fluorescence.
- Highly sensitive. The Quant-iT™ protein assay is orders of magnitude more sensitive than UV absorbance measurements.
- Highly selective. Separate kits are available for quantitating DNA, RNA (see below) or protein (Section 9.2), with minimal interference from common contaminants.
- Precise. CVs are generally less than 5%.

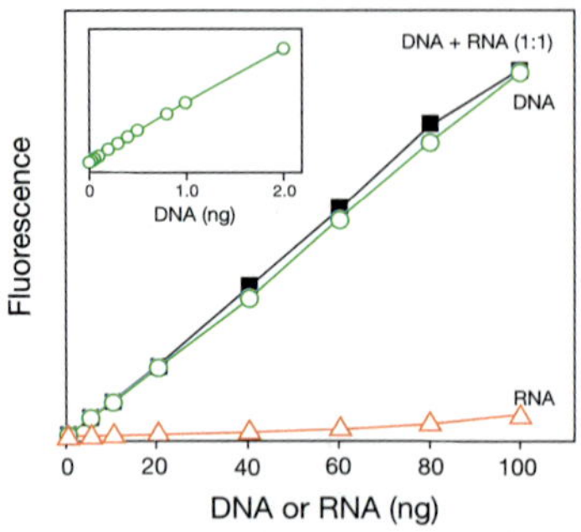

Figure 8.3.2 DNA selectivity and sensitivity of the Quant-iT™ dsDNA high-sensitivity assay. Triplicate 10 µL samples of DNA (O), *Escherichia coli* rRNA (Δ) or a 1:1 mixture of DNA and RNA (■) were assayed with the Quant-iT™ dsDNA High-Sensitivity Assay Kit (Q33120). Fluorescence was measured at 485/530 nm and plotted versus the mass of nucleic acid for the DNA alone, the mass of nucleic acid for the RNA alone or the mass of the DNA component in the 1:1 mixture. The coefficient of variation (CV) of replicate DNA determinations was ≤2%. The inset, a separate experiment with octuplicate determinations, shows the extreme sensitivity of the assay for DNA. Background fluorescence has not been subtracted.

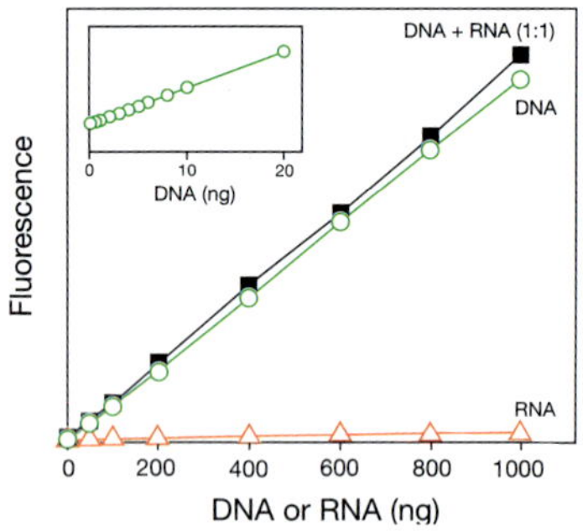

Figure 8.3.3 DNA selectivity and sensitivity of the Quant-iT™ dsDNA broad-range assay. Triplicate 10 µL samples of DNA (O), *Escherichia coli* rRNA (Δ)or a 1:1 mixture of DNA and RNA (■) were assayed with the Quant-iT™ dsDNA Broad-Range Assay Kit (Q33130). Fluorescence was measured at 485/530 nm and plotted versus the mass of nucleic acid for the DNA alone, the mass of nucleic acid for the RNA alone or the mass of the DNA component in the 1:1 mixture. The coefficient of variation (CV) of replicate DNA determinations was ≤3%. The inset, a separate experiment with octuplicate determinations, shows the sensitivity of the assay for DNA. Background fluorescence has not been subtracted.

Because the fluorescent dye in each Quant-iT™ Kit matches common fluorescence excitation and emission filter sets in microplate readers, these assay kits are ideal for high-throughput environments, as well as for small numbers of samples.

Quant-iT™ DNA Assay Kits

The Quant-iT™ DNA Assay Kits simplify DNA quantitation without sacrificing sensitivity. The Quant-iT™ dsDNA High-Sensitivity Assay Kit (Q33120) provides a linear detection range between 0.2 ng and 100 ng double-stranded DNA (dsDNA) (Figure 8.3.2), corresponding to initial experimental sample concentrations between 10 pg/µL and 100 ng/µL. This high-sensitivity DNA assay is ideal for quantitating PCR products, viral DNA, DNA fragments for subcloning and other applications requiring small amounts of DNA. The Quant-iT™ dsDNA Broad-Range Assay Kit (Q33130) provides a linear detection range between 2 ng and 1000 ng dsDNA (Figure 8.3.3), corresponding to initial experimental sample concentrations between 100 pg/µL and 1000 ng/µL. This broad-range DNA assay minimizes the need to dilute concentrated samples, such as genomic DNA and miniprep DNA, prior to high-throughput procedures. Both Quant-iT™ DNA assays are highly selective for dsDNA over RNA, and have shown to be accurate even in the presence of an equal mass of RNA. Moreover, their fluorescence signals are unaffected by many common contaminants, including free nucleotides, salts, solvents and proteins. Each Quant-iT™ DNA Assay Kit contains:

- Quant-iT™ DNA HS reagent (in Kit Q33120) or Quant-iT™ DNA BR reagent (in Kit Q33130)
- Quant-iT™ DNA HS buffer (in Kit Q33120) or Quant-iT™ DNA BR buffer (in Kit Q33130)
- A set of eight prediluted λ DNA standards between 0 and 10 ng/µL (in Kit Q33120) or between 0 and 100 ng/µL (in Kit Q33130)
- Easy-to-follow instructions for the high-sensitivity DNA assay or the broad-range DNA assay

Table 8.8 Selection guide for the Quant-iT™ Assay Kits.

Cat. No.	Kit	Target	Useful Range *	No. of Assays
Quant-iT™ Assay Kits for Use with the Qubit® Fluorometer †				
Q32851 Q32854	Quant-iT™ dsDNA High-Sensitivity (HS) Assay Kit	dsDNA	0.2–100 ng	100 500
Q32850 Q32853	Quant-iT™ dsDNA Broad-Range (BR) Assay Kit	dsDNA	2–1000 ng	100 500
Q10212	Quant-iT™ ssDNA Assay Kit	dsDNA	1–200 ng	100
Q32852 Q32855	Quant-iT™ RNA Assay Kit	RNA	5–100 ng	100 500
Q10210 Q10211	Quant-iT™ BR RNA Assay Kit	RNA	20–1000 ng	100 500
Q33211 Q33212	Quant-iT™ Protein Assay Kit	Protein	0.25–5 µg	100 500
Quant-iT™ Assay Kits for Use with Microplate Readers ‡				
Q33120	Quant-iT™ dsDNA High-Sensitivity (HS) Assay Kit (Ex/Em = 510/527 nm)	dsDNA	0.2–100 ng	1000
Q33130	Quant-iT™ dsDNA Broad-Range (BR) Assay Kit (Ex/Em = 510/527 nm)	dsDNA	2–1000 ng	1000
Q33140	Quant-iT™ RNA Assay Kit (Ex/Em = 644/673 nm)	RNA	5–100 ng	1000
Q10213	Quant-iT™ RNA BR Assay Kit (Ex/Em = 644/673 nm)	RNA	20–1000 ng	1000
Q33210	Quant-iT™ Protein Assay Kit (Ex/Em = 470/570 nm)	Protein	0.25–5 µg	1000
Quant-iT™ Bulk Reagents and Kits ‡				
P7589 P11496 P7581 P11495	Quant-iT™ PicoGreen® dsDNA Assay Kit and Reagent (Ex/Em = 502/523 nm)	dsDNA	0.2–100 ng	2000
O11492 O7582	Quant-iT™ OliGreen® ssDNA Assay Kit and Reagent (Ex/Em = 498/518 nm)	ssDNA	1–200 ng	2000
R11490 R11491	Quant-iT™ RiboGreen® RNA Assay Kit and Reagent (Ex/Em = 500/525 nm)	RNA	20–1000 ng	2000

* The useful range assumes a 1–20 µL sample volume in a 96-well microplate assay. † These kits are formulated for use with the Qubit® fluorometer (Q32857). Also available is the Qubit® Quantitation Starter Kit (Q32860), which contains the Qubit® fluorometer and four Quant-iT™ Assay Kits. ‡ Excitation (Ex) and emission (Em), in nm, of the Quant-iT™ detection reagent is indicated in parentheses.

IMPORTANT NOTICE: The products described in this manual are covered by one or more Limited Use Label License(s). Please refer to the Appendix on page 971 and Master Product List on page 975. Products are For Research Use Only. Not intended for any animal or human therapeutic or diagnostic use.
www.invitrogen.com/probes

Sufficient reagents are provided to perform 1000 assays, based on a 200 µL assay volume in a 96-well microplate format; this assay can also be adapted for use in cuvettes or 384-well microplates. Both the high-sensitivity assay and the broad-range assay can be detected using standard fluorescein filters, and the fluorescence signal is stable for three hours at room temperature.

Quant-iT™ RNA Assay Kit

The Quant-iT™ RNA Assay Kit (Q33140) provides the first homogeneous assay ever developed for quantitating RNA in the presence of DNA. This RNA assay exhibits a linear detection range between 5 ng and 100 ng RNA (Figure 8.3.4), corresponding to initial experimental sample concentrations between 250 pg/µL and 100 ng/µL. We also offer the Quant-iT™ RNA Broad-Range Assay Kit (Q10213), with a linear detection range between 20 ng and 1000 ng. Because of the high selectivity of the Quant-iT™ RNA reagent for RNA over dsDNA, this assay enables accurate RNA quantitation even in the presence of an equal mass of DNA. The fluorescence signal is unaffected by many common contaminants, including free nucleotides, salts, solvents and proteins, making this assay ideal for measuring samples for microarray, RT-PCR and northern blot procedures. Each Quant-iT™ RNA Assay Kit contains:

- Quant-iT™ RNA reagent
- Quant-iT™ RNA buffer
- A set of eight prediluted *Escherichia coli* rRNA standards between 0 and 10 ng/µL
- Easy-to-follow instructions

Sufficient reagents are provided to perform 1000 assays, based on a 200 µL assay volume in a 96-well microplate format; this assay can also be adapted for use in cuvettes or 384-well microplates. The fluorescence signal exhibits excitation/emission maxima of 644/673 nm and is stable for three hours at room temperature.

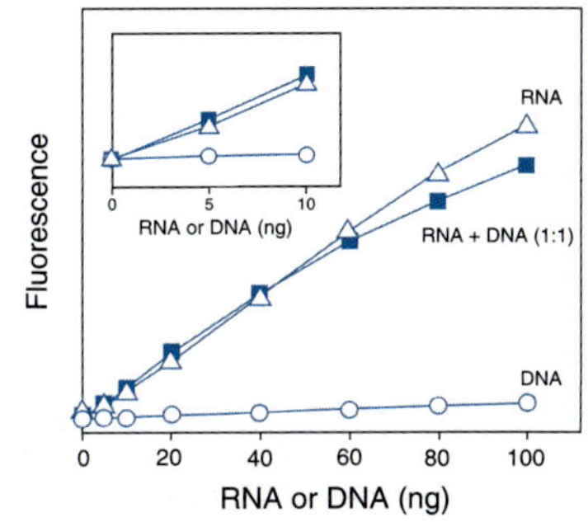

Figure 8.3.4 RNA selectivity and sensitivity of the Quant-iT™ RNA assay. Triplicate 10 µL samples of *Escherichia coli* rRNA (△), DNA (○) or a 1:1 mixture of RNA and DNA (■) were assayed with the Quant-iT™ RNA Assay Kit (Q33140). Fluorescence was measured at 630/680 nm and plotted versus the mass of nucleic acid for the RNA alone, the mass of nucleic acid for the DNA alone or the mass of the RNA component in the 1:1 mixture. The coefficient of variation (CV) of replicate RNA determinations was ≤10%. The inset is an enlargement of the graph to show the sensitivity of the assay for RNA. Background fluorescence has not been subtracted.

Quant-iT™ Assay Kits for Use with the Qubit® Fluorometer

The Qubit® fluorometer (Figure 8.3.5) is designed to work seamlessly with the Quant-iT™ Assay Kits. Together, they form the Qubit® Quantitation Platform, an efficient combination of sophisticated, accurate and highly sensitive fluorescence-based quantitation assays for DNA, RNA and protein, along with an extremely user-friendly fluorometer. This powerful pairing offers:

- Selective quantitation—more accurate than UV absorbance readings
- High sensitivity—use as little as 1 µL of sample for quantitation
- Intuitive integrated platform—sophisticated quantitation in 5 minutes or less

In addition to providing quantitation results that are fast, easy and reliable, the Qubit® fluorometer features software that is fully upgradable. The Qubit® fluorometer incorporates sophisticated data analysis algorithms designed to produce accurate measurements and an intuitive user interface for seamless integration with the full range of Quant-iT™ Assay Kits. We offer the Qubit® fluorometer separately (Q32857; USB cable, Q32858; replacement international power cord, Q32859) or bundled in a starter kit that includes four different Quant-iT™ assay kits as well as a supply of 500 Qubit® assay tubes (Q32860); a set of 500 Qubit® assay tubes is also available separately (Q32856).

In addition to the Quant-iT™ Assay Kits described above for microplate readers, we offer a set of Quant-iT™ Assay Kits specifically formulated for use with the Qubit® fluorometer (Table 8.8). These Quant-iT™ Assay Kits are up to 1000 times as sensitive as UV absorbance readings, and as little as 1 µL of sample is all that's needed to get accurate, reliable quantitation. There are Quant-iT™ Assay Kits for a variety of quantitation needs:

- Quant-iT™ dsDNA Broad-Range (BR) and Quant-iT™ High-Sensitivity (HS) Assay Kits (Q32850, Q32853; Q32851, Q32854)—for sequencing samples, genomic DNA samples and routine cloning experiments
- Quant-iT™ ssDNA Assay Kit (Q10212)—for hybridization experiments, site-directed mutagenesis and antisense gene suppression
- Quant-iT™ RNA and Quant-iT™ RNA BR Assay Kits (Q32852, Q32855; Q10210, Q10211)—for microarray experiments, real-time PCR samples and northern blots
- Quant-iT™ Protein Assay Kits—for western blotting, activity assays and SDS-PAGE or 2D gel electrophoresis (Q33211, Q33212; Section 9.2)

Figure 8.3.5 The Qubit® fluorometer. The Qubit® fluorometer (Q32857) is designed to work seamlessly with the Quant-iT™ assays for DNA, RNA and protein quantitation. It features state-of-the-art optical components, sophisticated data analysis and an intuitive user interface. The result is an easier method to obtain much more accurate and sensitive quantitation than can be obtained with UV absorbance assays, increasing sample performance in downstream applications.

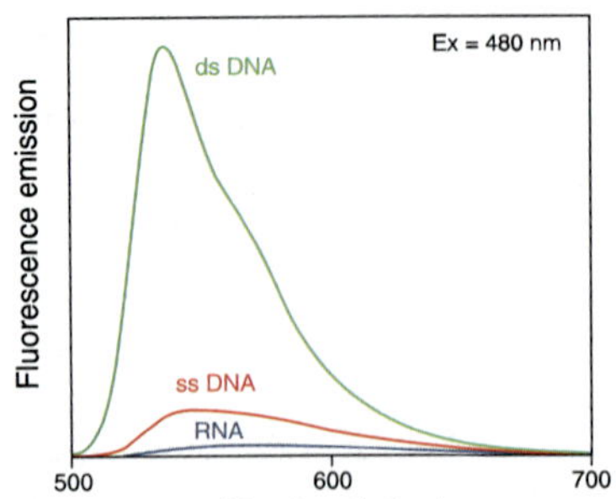

Figure 8.3.6 Fluorescence enhancement of the Quant-iT™ PicoGreen® dsDNA reagent upon to binding dsDNA, ssDNA and RNA. Samples containing 500 ng/mL calf thymus DNA, M13 ssDNA or *Escherichia coli* ribosomal RNA were added to cuvettes containing Quant-iT™ PicoGreen® dsDNA reagent (P7581, P7589, P11495, P11496) in TE buffer. Samples were excited at 480 nm, and the fluorescence emission spectra were collected using a spectrofluorometer. Emission spectra for samples containing dye and nucleic acids, as well as for dye alone (baseline), are shown.

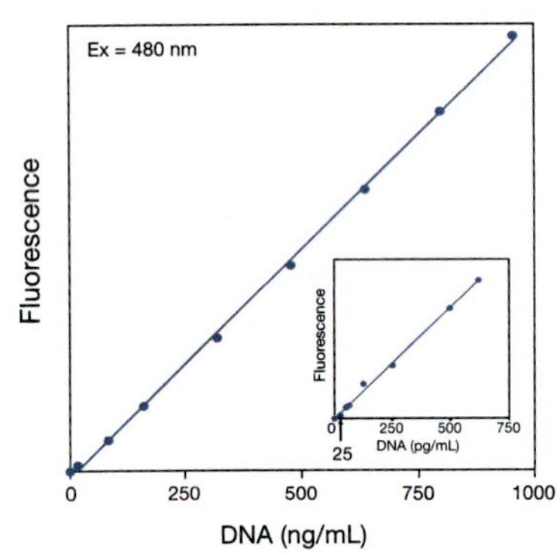

Figure 8.3.7 Linear quantitation of calf thymus DNA from 25 pg/mL to 1000 ng/mL using the Quant-iT™ PicoGreen® dsDNA reagent (P7581, P7589, P11495, P11496). Samples in 10 mm × 10 mm cuvettes were excited at 480 nm. The fluorescence emission intensity was measured at 520 nm using a spectrofluorometer and plotted as a function of DNA concentration. The inset shows an enlargement of the results obtained with DNA concentrations between 0 and 750 pg/mL.

Quant-iT™ PicoGreen® dsDNA Quantitation Assay

The Quant-iT™ PicoGreen® dsDNA reagent [2] (P7581, P11495) and Kits (P7589, P11496) are designed to accurately quantitate as little as 25 pg/mL of dsDNA in a fluorometer or 250 pg/mL (typically 50 pg in a 200 µL volume) in a fluorescence microplate reader. The PicoGreen® dsDNA quantitation assay is more than 10,000 times as sensitive as conventional UV absorbance measurements at 260 nm (an A_{260} of 0.1 corresponds to an ~5 µg/mL dsDNA solution) and at least 400 times more sensitive than the Hoechst 33258 dye–based assay.[3] It is even more sensitive than assays based on our YO-PRO®-1 and YOYO®-1 dyes, which have reported detection limits of approximately 2.5 ng/mL and 0.5 ng/mL,[4] respectively.

Although the PicoGreen® reagent is not specific for dsDNA, it shows a >1000-fold fluorescence enhancement upon binding to dsDNA, and much less fluorescence enhancement upon binding to single-stranded DNA (ssDNA) or RNA, making it possible to quantitate dsDNA in the presence of equimolar amounts of ssDNA, RNA or proteins [5–7] (Figure 8.3.6). The PicoGreen® reagent also selectively detects DNA–RNA hybrids in the presence of ssDNA and RNA. Differences in the emission lifetimes of the PicoGreen® complexes with dsDNA and ssDNA make it possible to quantitate the relative amounts of each species in solution by time-resolved measurements. By contrast, UV absorbance measurements cannot distinguish between dsDNA, ssDNA and RNA or proteins. Thus, the PicoGreen® reagent allows direct quantitation of PCR amplicons without purification from the reaction mixture and makes it possible to detect low levels of DNA contamination in recombinant protein products.[6,8] In comparison to the Hoechst 33258 dye, which shows significant AT selectivity, the PicoGreen® reagent shows little if any AT- or GC-selectivity, enabling accruate DNA quantitation from many sources.

The protocol for the PicoGreen® dsDNA quantitation assay is simple and requires very few steps—the dye is simply added to the sample and incubated for five minutes, then the fluorescence is measured. In addition, the fluorescence signal from binding of the PicoGreen® reagent to dsDNA is linear over at least four orders of magnitude (Figure 8.3.7) with a single dye concentration, whereas assays using ethidium bromide, Hoechst 33258 or the YOYO®-1 dye exhibit a much more limited linear range.[4,9,10] We have found that this linearity is maintained in the presence of several compounds commonly found in nucleic acid preparations, including salts, urea, ethanol, chloroform, detergents, proteins and agarose.[5] The PicoGreen® reagent can be excited at 488 nm with an argon-ion laser, and is reported to be a superior nucleic acid stain for analysis of single DNA molecules in a flow cytometer.[11] A method that utilizes the PicoGreen® reagent for the absolute quantitation of cDNA using real-time PCR has been reported.[12]

The PicoGreen® assay is useful for quantitating DNA templates for PCR,[13] labeling reactions, electrophoretic mobility-shift (bandshift) assays, DNA-footprinting assays and filter-binding assays, and for measuring yields from PCR reactions,[14] DNA minipreps and maxipreps, cDNA synthesis and nuclease protection assays. The simplicity and selectivity of the assay also make it ideal for high-throughput automated quantitation assays used in forensic and genomics research. Furthermore, the PicoGreen® reagent has been used for:

- Genotyping by allele-specific PCR [15]
- Quantitating dsDNA samples before and after PCR amplification or after agarose gel electrophoresis [8,16]
- Determining PCR amplification yields before sequencing [14]
- Automating quantitation of DNA isolated from biological samples or obtained from PCR reactions, prior to running DNA typing gels for high-throughput genotyping [17]
- Quantitating DNA from buccal scrapes prior to DNA profiling by short tandem repeat (STR) analysis [18]
- Identifying contaminating DNA in recombinant protein products [19] or purified monoclonal antibody preparations [20,21]
- Monitoring DNA strand breaks in plasmids and bacterial artificial chromosomes [22]
- Quantitating the efficiency of DNA extraction from frozen and formalin-fixed tissue sections [23]
- Detecting mammalian telomerase activity in tumor cells using the PCR-based TRAP assay [24–26]
- Developing assays for DNA polymerase

IMPORTANT NOTICE: The products described in this manual are covered by one or more Limited Use Label License(s). Please refer to the Appendix on page 971 and Master Product List on page 975. Products are For Research Use Only. Not intended for any animal or human therapeutic or diagnostic use.

- Measuring the activity of reverse transcriptase or DNase I[7]
- Measuring supercoiled DNA forms in solution, based on their renaturation properties[28]
- Monitoring plasmid production during fermentation and downstream processing[29]
- Quantitating DNA denaturation as a measure for DNA damage in purified DNA preparations, cell lysates or homogenized solid tissues[30–32]

Each vial of the Quant-iT™ PicoGreen® dsDNA reagent (P7581) contains a sufficient amount of dye for at least 200 assays using a 2 mL assay volume and a standard fluorometer, or 2000 assays using a 200 µL assay volume and a fluorescence microplate reader. The product is accompanied by a simple protocol that enables linear and reproducible quantitation of dsDNA. We also provide the PicoGreen® reagent in the Quant-iT™ PicoGreen® dsDNA Assay Kit (P7589), which contains:

- Quant-iT™ PicoGreen® dsDNA quantitation reagent
- Low-fluorescence, nucleic acid– and nuclease-free assay buffer concentrate, essential for the accurate measurement of dsDNA
- dsDNA standard solution for assay calibration
- Detailed protocols for dsDNA quantitation

This kit provides sufficient reagents for 200 assays using a 2 mL assay volume and a standard fluorometer or 2000 assays using a 200 µL assay volume and a fluorescence microplate reader. Both the stand-alone reagent and the kit are also available in special packaging, in which the Quant-iT™ PicoGreen® reagent is supplied as 10 vials of 100 µL aliquots for added convenience (P11495; Kit, P11496). The special packaging reduces thawing times, provides individual aliquots for each person performing the assay, and allows smaller amounts of dye to be taken into the field for analysis of water or other samples.

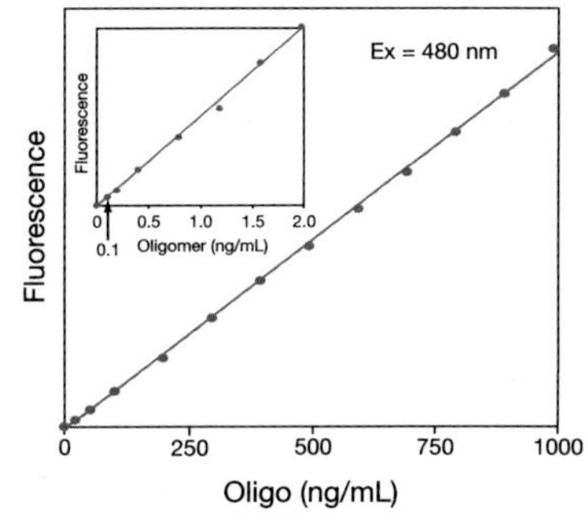

Figure 8.3.8 Linear quantitation of a synthetic 24-mer (an M13 sequencing primer) from 0.1 to 1000 ng/mL using the Quant-iT™ OliGreen® ssDNA reagent (O7582, O11492). Samples in 10 mm × 10 mm cuvettes were excited at 480 nm. The fluorescence emission intensity was measured at 520 nm using a spectrofluorometer and plotted as a function of oligonucleotide concentration. The inset shows an enlargement of the results obtained with oligonucleotide concentrations between zero and 2.0 ng/mL.

Quant-iT™ OliGreen® ssDNA Quantitation Assay

For researchers working with ssDNA and oligonucleotides, as well as for companies that synthesize oligonucleotides, we offer the Quant-iT™ OliGreen® ssDNA reagent (O7582) and Quant-iT™ OliGreen® ssDNA Assay Kit (O11492). Short, synthetic oligonucleotides are used in a number of molecular biology techniques, including DNA sequencing, site-directed mutagenesis, DNA amplification, antisense gene suppression and *in situ* and blot hybridization. The conventional methods for quantitating oligonucleotides are not very sensitive, often requiring highly concentrated samples, and are quite subject to interference from sample contaminants. The most commonly used technique for measuring oligonucleotide and ssDNA concentrations is the determination of absorbance at 260 nm (an A_{260} of 0.1 corresponds to an ~3 µg/mL solution of a synthetic 24-mer M13 sequencing primer).

The Quant-iT™ OliGreen® ssDNA reagent enables researchers to routinely quantitate as little as 100 pg/mL of ssDNA or oligonucleotide (200 pg in a 2 mL assay volume with a standard fluorometer) or 200 pg in a 200 µL assay volume using a fluorescence microplate reader (Figure 8.3.8). Thus, quantitation with the OliGreen® reagent is about 10,000 times more sensitive than quantitation with UV absorbance methods and at least 500 times more sensitive (and far faster, with a greater throughput) than detecting oligonucleotides on electrophoretic gels stained with ethidium bromide. Using an easy-to-follow protocol and fluorescein excitation and emission wavelengths, we have quantitated oligonucleotides that range from 10 to 50 nucleotides in length, as well as several sources of ssDNA, such as M13 and φX174 viral DNA and denatured calf thymus DNA, and obtained similar sensitivity. The Quant-iT™ OliGreen® reagent has also been used to detect phosphodiester and phosphorothioate oligonucleotides.[33]

Significant disadvantages of the UV absorbance method for oligonucleotide quantitation include the large relative contribution of free nucleotides to the signal and the interference caused by contaminants commonly found in nucleic acid preparations. By contrast, nucleotides and short oligonucleotides of six bases or less do not interfere with the OliGreen® ssDNA quantitation assay. However, the OliGreen® reagent does exhibit fluorescence enhancement when bound to dsDNA and RNA. Like the PicoGreen® assay, the linear detection range of the OliGreen® ssDNA

IMPORTANT NOTICE: The products described in this manual are covered by one or more Limited Use Label License(s). Please refer to the Appendix on page 971 and Master Product List on page 975. Products are For Research Use Only. Not intended for any animal or human therapeutic or diagnostic use.

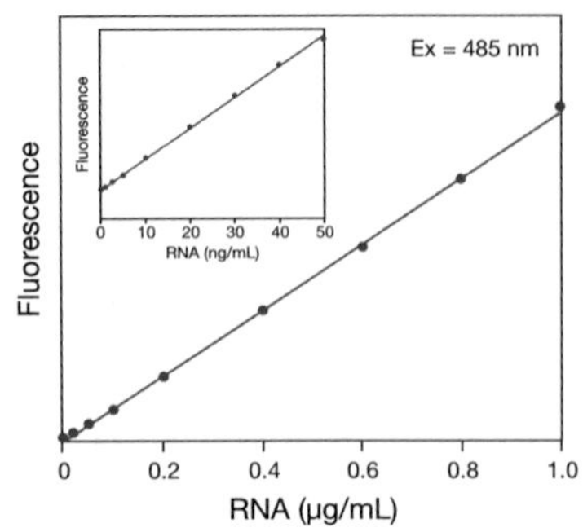

Figure 8.3.9 Linear quantitation of ribosomal RNA using the Quant-iT™ RiboGreen® RNA reagent (R11491, R11490). For the high-range assay, the RiboGreen® reagent was diluted 200-fold into 10 mM Tris-HCl, 1 mM EDTA, pH 7.5 (TE), and 100 µL of the reagent solution was added to microplate wells containing 100 µL ribosomal RNA in TE. For the low-range assay (see inset), the RiboGreen® reagent was diluted 2000-fold into TE, and 100 µL of the reagent solution was added to 100 µL of ribosomal RNA in TE. Samples were excited at 485 ± 10 nm, and the fluorescence emission intensity was measured at 530 ± 12.5 nm using a fluorescence microplate reader. Background fluorescence was not subtracted.

quantitation assay in a standard fluorometer extends over four orders of magnitude—from 100 pg/mL to 1 µg/mL—with a single dye concentration (Figure 8.3.8). The linearity of the Quant-iT™ OliGreen® assay is maintained in the presence of several compounds commonly found to contaminate nucleic acid preparations, including salts, urea, ethanol, chloroform, detergents, proteins, ATP and agarose; however, many of these compounds do affect the signal intensity to some extent, so standard curves should be generated using solutions that closely mimic those of the samples. The OliGreen® assay can even be performed using samples as complex as whole blood or serum.[33]

Our experiments with homopolymers have demonstrated that the OliGreen® reagent may exhibit significant base selectivity. The Quant-iT™ OliGreen® reagent shows a large fluorescence enhancement when bound to poly(dT) but only a relatively small fluorescence enhancement when bound to poly(dG) and little signal with poly(dA) and poly(dC). Thus, it is important to use an oligonucleotide with similar base composition when generating the standard curve.

The remarkable properties of our Quant-iT™ OliGreen® reagent make it ideal for the fast and accurate detection and quantitation of:

- Antisense oligonucleotides
- Aptamers
- Genomic DNA isolated under denaturing conditions
- PCR primers
- Phosphorothioate and phosphodiester oligodeoxynucleotides[33]
- Sequencing primers
- Single-stranded phage DNA

Each vial of Quant-iT™ OliGreen® ssDNA reagent (O7582) contains a sufficient amount of dye for at least 200 assays using a 2 mL assay volume and a standard fluorometer, or 2000 assays using a 200 µL assay volume and a fluorescence microplate reader. The product is accompanied by a simple protocol that ensures linear and reproducible quantitation of ssDNA. We also provide the Quant-iT™ OliGreen® reagent in the Quant-iT™ OliGreen® ssDNA Assay Kit (O11492), which contains:

- Quant-iT™ OliGreen® ssDNA quantitation reagent
- Low-fluorescence, nucleic acid– and nuclease-free assay buffer concentrate, essential for the accurate measurement of ssDNA
- Oligonucleotide standard (M13 sequencing primer) solution for assay calibration
- Detailed protocols for ssDNA quantitation

This kit provides sufficient reagents for 200 assays using a 2 mL assay volume and a standard fluorometer or 2000 assays using a 200 µL assay volume and a fluorescence microplate reader.

Quant-iT™ RiboGreen® RNA Quantitation Assay

Quant-iT™ RiboGreen® RNA Reagent and Kit

The Quant-iT™ RiboGreen® RNA reagent (R11491) is our premier stain for quantitating RNA in solution. Like the Quant-iT™ PicoGreen® dsDNA and Quant-iT™ OliGreen® ssDNA quantitation assays, the RiboGreen® RNA quantitation assay relies on a proprietary dye that exhibits a large fluorescence enhancement upon binding to nucleic acids. The extinction coefficient (EC) of the Quant-iT™ RiboGreen® reagent, as well as its quantum yield (QY) and fluorescence enhancement upon binding RNA, are all significantly greater than those of ethidium bromide.

- Quant-iT™ RiboGreen® reagent: EC_{482} = 67,000 $cm^{-1}M^{-1}$, QY = 0.65, fluorescence enhancement >1000-fold.
- Ethidium bromide: EC_{482} = 5,500 $cm^{-1}M^{-1}$, QY <0.3, fluorescence enhancement <30-fold.

The Quant-iT™ RiboGreen® assay allows detection of as little as 1 ng/mL RNA in a standard fluorometer, fluorescence microplate reader or filter-based fluorometer using standard fluorescein excitation and emission settings (Figure 8.3.9). This sensitivity is at least 200-fold better than that achieved with ethidium bromide[34] and at least 1000-fold better than that achieved using

IMPORTANT NOTICE: The products described in this manual are covered by one or more Limited Use Label License(s). Please refer to the Appendix on page 971 and Master Product List on page 975. Products are For Research Use Only. Not intended for any animal or human therapeutic or diagnostic use.
www.invitrogen.com/probes
molecular probes® | invitrogen™ by life technologies™

conventional absorbance measurements at 260 nm (an A_{260} of 0.1 corresponds to an ~4 µg/mL RNA solution). Unlike UV absorbance measurements at 260 nm, the Quant-iT™ RiboGreen® reagent does not detect significant sample contamination by free nucleotides.[35] Thus, the Quant-iT™ RiboGreen® reagent more accurately measures the amount of intact RNA polymers in potentially degraded samples.

Using two different dye concentrations to cover its full dynamic range of three orders of magnitude, we have observed a linear correlation between the RNA concentration and fluorescence for 1.0 ng/mL to 50 ng/mL RNA using a 4000-fold dilution of the RiboGreen® reagent, and for 20 ng/mL to 1 µg/mL using a 400-fold dilution of the dye (Figure 8.3.9). Assay linearity is maintained in the presence of several compounds commonly found in nucleic acid preparations, including salts, urea, ethanol, chloroform, detergents, proteins and agarose.[35]

The Quant-iT™ RiboGreen® reagent is not appreciably selective for RNA—the dye also shows significant fluorescence enhancement upon binding to DNA. However, a simple DNase pretreatment of samples removes the contribution of DNA to the signal. The Quant-iT™ RiboGreen® reagent may also have some base selectivity; it exhibits about 60% less fluorescence when bound to poly(G) homopolymers and virtually no fluorescence when bound to poly(U) or poly(C) homopolymers compared with the fluorescence when bound to poly(A) homopolymers or to rRNA.[35]

Using the Quant-iT™ RiboGreen® RNA reagent, we have reproducibly quantitated RNA from a wide variety of sources, including ribosomal RNA (rRNA), transfer RNA (tRNA), viral RNA, polyA+ fractions and total cellular RNA.[35] The RiboGreen® reagent is useful for:

- Fast and accurate measurements of RNA yields before generating cDNA
- Determination of RNA yields from *in vitro* transcription
- Accurate measurements of RNA before performing northern blot analysis,[36] S1 nuclease assays, RNase-protection assays, reverse-transcription PCR and differential-display PCR
- Assay of DNA-dependent RNA polymerase activity[37]
- Detection of capillary electrophoresis–separated viral RNA that has been stained *in vitro*[38]

Each vial of the Quant-iT™ RiboGreen® RNA reagent (R11491) contains a sufficient amount of dye for at least 200 high-range assays or 2000 low-range assays using a 2 mL assay volume and a standard fluorometer. With a fluorescence microplate reader and a 96-well microplate, the assay volume is reduced to 200 µL, allowing 2000 high-range assays or 20,000 low-range assays. Included with each vial of the Quant-iT™ RiboGreen® reagent is a simple protocol that permits linear and reproducible quantitation of RNA. We also provide the RiboGreen® reagent in the Quant-iT™ RiboGreen® RNA Assay Kit (R11490), which contains:

- Quant-iT™ RiboGreen® RNA reagent
- Low-fluorescence, nucleic acid– and nuclease-free assay buffer concentrate, essential for the accurate measurement of RNA
- Ribosomal RNA standard (16S and 23S rRNA from *Escherichia coli*) solution for assay calibration
- Detailed protocols

This kit provides sufficient reagents for at least 200 assays using a 2 mL assay volume and a standard fluorometer or at least 2000 assays using a 200 µL assay volume and a fluorescence microplate reader. The RNase-free TE buffer concentrate, which is essential to the success of the assay, is also available separately (T11493) and can be used to extend the number of low-concentration assays possible with the kit.

RediPlate™ 96 RiboGreen® RNA Quantitation Kit

The RiboGreen® assay is also available in a convenient RediPlate™ 96 RiboGreen® RNA Quantitation Kit in which the RiboGreen® reagent is predispensed into a 96-well microplate (R32700). The buffer and sample are simply added to the microplate wells—there is no need to handle the RiboGreen® reagent. After a 10-minute incubation, the microplate is ready to read in a fluorescence microplate reader. The RediPlate™ RiboGreen® assay has a linear range of ~15–1000 ng/mL (~3–200 ng in a 200 µL assay volume) with a single dye concentration (Figure 8.3.10). The microplate used in the RediPlate™ 96 RiboGreen® RNA Quantitation Kit is provided in a resealable foil packet, and it snaps apart into twelve strips to permit assays in any multiple

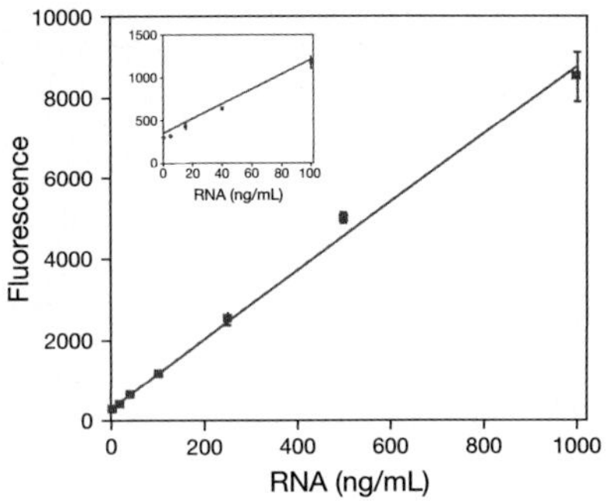

Figure 8.3.10 Dynamic range and sensitivity of the RediPlate™ 96 RiboGreen® RNA quantitation assay. The RNA standards provided in the RediPlate™ 96 RiboGreen® RNA Quantitation Kit (R32700) were added in quadruplicate to assay wells as described in the accompanying protocol, and fluorescence was measured in a fluorescence microplate reader using excitation at 485 ± 12.5 nm and fluorescence detection at 530 ± 15 nm. Fluorescence was plotted against the RNA concentration with no background subtraction. The inset shows the sensitivity of the assay at very low levels of RNA.

IMPORTANT NOTICE: The products described in this manual are covered by one or more Limited Use Label License(s). Please refer to the Appendix on page 971 and Master Product List on page 975. Products are For Research Use Only. Not intended for any animal or human therapeutic or diagnostic use.

Figure 8.3.11 A RediPlate™ 96 microplate.

Figure 8.3.12 An environmental sample containing marine viruses (smallest dots), bacteria (larger, brighter dots) and a diatom (long thin cell with prominent nucleus) stained with SYBR® Green I nucleic acid stain (S7563, S7567, S7585). Image contributed by Jed Fuhrman, University of Southern California.

of eight (Figure 8.3.11). Eleven of the strips are preloaded with the RiboGreen® reagent; the remaining strip, marked with blackened tabs, contains a series of RNA standards for generating a standard curve. In addition to the 96-well microplate, each RediPlate™ RiboGreen® 96 RNA Quantitation Kit includes RNase-free reaction buffer and detailed instructions.

Other Stains for Nucleic Acid Quantitation in Solution

Cyanine Dyes and Phenanthridine Dyes for Nucleic Acid Quantitation in Solution

The dimeric cyanine dyes TOTO®-1 and YOYO®-1 are useful for sensitive fluorometric measurement of dsDNA, ssDNA and RNA in solution,[4] although the PicoGreen®, OliGreen® and RiboGreen® reagents generally are faster, have greater sensitivity and show a linear response over a broader range of nucleic acid concentrations. The linear range of assays that use the TOTO®-1 and YOYO®-1 dyes for DNA quantitation encompasses about two orders of magnitude, with a sensitivity limit of about 0.5 ng/mL. The TOTO®-1, YOYO®-1 and YO-PRO®-1 nucleic acid stains have been used to quantitate PCR amplification products in a homogeneous human leukocyte antigen (HLA) typing method that requires no transfer or washing steps, thus minimizing the risk of sample contamination.[39,40] Other dyes for nucleic acid quantitation in solution and their applications include:

- YOYO®-1 dye (Y3601, Section 8.1) for solution quantitation of oligonucleotides,[41] PCR products[42] and nuclear run-on assays[43]
- YO-PRO®-1 dye (Y3603, Section 8.1) for quantitating dsDNA in a fluorescence microplate reader, with a reported sensitivity limit of about 2.5 ng/mL[3]
- YO-PRO®-1 dye for quantitating RNA isolated from *Xenopus* embryos[44]
- YO-PRO®-1 dye and SYBR® Green I dye for direct counting of viruses in marine and freshwater environments[45–50] (Figure 8.3.12)
- PO-PRO™-3 dye (P3585, Section 8.1) for quantitating DNA in a fluorescence microplate reader,[51] with fluorescence measurements reported to be independent of base-pair composition
- Ethidium bromide for measuring the yield of PCR products[52]
- Ethidium bromide and acridine homodimer (A666, Section 8.1) for quantitating covalently closed, circular DNA and for measuring the activity of polymerases, deoxynucleotidyl transferases, ligases, gyrases, topoisomerases and nucleases[9,53,54]
- YOYO®-1 dye for measuring the activity of DNases[55]
- SYBR® Gold nucleic acid gel stain (S11494, Section 8.4), for detecting DNA mutations in a simple PCR-based assay[56]

Dyes such as the ethidium homodimers and our dimeric cyanine dyes—the TOTO®, YOYO®, BOBO™, POPO™, JOJO™ and LOLO™ dyes (Table 8.2)—exhibit a high affinity for double-stranded nucleic acids but label small single-stranded oligonucleotides less well. This characteristic of ethidium homodimer-1 (E1169, Section 8.1) was exploited to analyze short self-annealing oligonucleotides for their ability to hybridize.[57] Because our dimeric cyanine dyes and the Quant-iT™ PicoGreen® dsDNA quantitation reagent have extremely low intrinsic fluorescence in the absence of DNA, high fluorescence enhancements upon binding, higher quantum yields and much larger extinction coefficients than ethidium homodimer-1,[3,58] they should prove superior in this application.

Hoechst 33258 Dye for Quantitating DNA in Solution

The Hoechst 33258 dye (H1398, H3569, H21491) has been extensively used to quantitate dsDNA in solution. Hoechst 33258 shows a fluorescence increase upon binding nucleic acids and a preference for binding to AT regions. Hoechst 33258 is selective (but not specific) for dsDNA over RNA in high-salt buffers and for dsDNA over ssDNA in low-salt buffers. The Hoechst 33258 dye can quantitatively detect from 10 ng/mL to ~10 µg/mL dsDNA when two different dye concentrations are used.[10] While this assay uses principles that are similar to other fluorescent assays, newer dyes such as the Quant-iT™ PicoGreen® reagent provide much higher sensitivity, better selectivity and a much broader dynamic range with a single dye concentration.

IMPORTANT NOTICE: The products described in this manual are covered by one or more Limited Use Label License(s). Please refer to the Appendix on page 971 and Master Product List on page 975. Products are For Research Use Only. Not intended for any animal or human therapeutic or diagnostic use.

The FluoReporter® Blue Fluorometric dsDNA Quantitation Kit (F2962) provides the protocols developed by Rago and colleagues[1] for analyzing cellular DNA with the blue-fluorescent Hoechst 33258 nucleic acid stain. The kit enables researchers to detect ~10 ng of isolated calf thymus DNA or ~1000 mouse NIH 3T3 cells in a 200 µL sample (substantially lower levels are detectable using our CyQUANT® Cell Proliferation Assay Kit described in Section 15.4). With this kit, quantitation of cellular DNA is rapid, and all manipulations can be carried out in microplate wells. The cells are lysed by freezing them in distilled water, which circumvents the requirement for extraction procedures used in other Hoechst 33258 dye–based protocols.[10,59–62] The diluted dye solution is then added to lysed cells and the fluorescence is measured. Kit components include:

- Hoechst 33258 in dimethylsulfoxide (DMSO)/H_2O
- TNE buffer
- Detailed protocols

Each kit provides sufficient reagents for assaying 2000 samples using a fluorescence microplate reader.

Real-Time Quantitative PCR Using SYBR® Green I Nucleic Acid Gel Stain

Measurements of PCR products can be taken during the linear portion of the amplification reactions, allowing accurate quantitation of templates. Several methods exist for real-time quantitation of PCR products, including fluorescence resonance energy transfer techniques using fluorescently labeled primers or molecular beacon. Identification of PCR products during the reaction can also be monitored using SYBR® Green I nucleic acid gel stain (S7563, S7567, S7585); this method has been shown to be more precise than TaqMan® assays using labeled oligonucleotide probes.[63] In addition, individual DNA molecules have been detected with on-line capillary PCR coupled with laser-induced fluorescence detection by adding SYBR® Green I stain to the reaction mixture.[64]

SYBR® Green I stain binds preferentially to dsDNA, allowing accurate quantitation of double-stranded product in the presence of single-stranded oligonucleotide primers.[65] SYBR® Green I stain is stable to the extremes of temperature required for PCR reactions and does not interfere with Taq polymerase. Improved specificity for quantitating desired products can be achieved by using SYBR® Green I stain after the assay to measure the melting temperature of the products.[66–74] Double-stranded DNA with no base mismatches will show a higher melting temperature than the nonspecific templates that contain mismatches. Real-time quantitative PCR experiments can be carried out using instruments specialized for the application[75,76] or by quantitating amplification products manually at different time points.[77] Real-time quantitative PCR with SYBR® Green I stain has been used to develop reliable and simple assays for detecting genetic mutations, including duplications and deletions in mosquito drug-resistance genes,[78] chromosomal translocations in human disease genes,[70,79] and base substitutions.[67,80,81] It has also been used for the unequivocal identification of viral, bacterial or fungal pathogens.[66,68,71,82–84] In addition, this method has been used successfully for quantitative reverse-transcription PCR.[63,75]

Invitrogen has introduced the next generation of real-time PCR detection reagents, including the SYBR® GreenER qPCR reagent system, which produces the extremely reliable gene expression data from real-time quantitative PCR. SYBR® GreenER reagent is a novel dsDNA-binding dye that exhibits a brighter signal and significantly reduced PCR inhibition compared to the original SYBR® Green dye, but nearly identical spectral characteristics. More information about the SYBR® GreenER qPCR reagent system and the latest developments in real-time PCR detection can be found at www.invitrogen.com/handbook/qrtpcr.

REFERENCES

1. Anal Biochem (1990) 191:31; **2.** Nucleic Acids Res (2004) 32:e103; **3.** Anal Chem (2006) 78:4630; **4.** Anal Biochem (1993) 208:144; **5.** Anal Biochem (1997) 249:228; **6.** J Pharm Biomed Anal (2009) 49:997; **7.** Nucleic Acids Res (2003) 31:e111; **8.** Anal Biochem (2000) 279:111; **9.** Anal Biochem (1995) 230:353; **10.** Anal Biochem (1980) 102:344; **11.** Anal Chem (1999) 71:5470; **12.** J Immunol Methods (2003) 278:261; **13.** Mol Cell Probes (1995) 9:145; **14.** Biotechniques (1996) 20:676; **15.** Biotechniques (1998) 24:206, 210, 212; **16.** Biotechniques (1996) 21:372; **17.** Genome Res (1996) 6:781; **18.** Biotechniques (1997) 23:18; **19.** Biotechniques (1997) 23:532; **20.** Biotechol Tech (1999) 13:681; **21.** Bioseparation (1999) 8:281; **22.** Nucleic Acids Res (2003) 31:e65; **23.** Am J Pathol (2000) 156:1189; **24.** Anal Biochem (2003) 323:65; **25.** Clin Chem (1998) 44:2133; **26.** Proc Natl Acad Sci U S A (1996) 93:6091; **27.** Anal Biochem (2000) 281:95; **28.** Nucleic Acids Res (2000) 28:E57; **29.** Biotechnol Bioeng (1999) 66:195; **30.** Anal Biochem (1999) 270:195; **31.** Anal Chem (1999) 71:4423; **32.** Eur J Biochem (1985) 153:105; **33.** Antisense Nucleic Acid Drug Dev (1997) 7:133; **34.** Anal Biochem (1966) 17:100; **35.** Anal Biochem (1998) 265:368; **36.** J Immunol (1998) 161:4332; **37.** Anal Biochem (2004) 324:183; **38.** Anal Chem (2004) 76:882; **39.** Hum Immunol (1994) 39:1; **40.** Anal Biochem (1994) 221:340; **41.** Biotechniques (1994) 16:1032; **42.** Anal Biochem (1994) 218:458; **43.** Anal Biochem (1994) 221:202; **44.** Neuron (1995) 14:865; **45.** Nature (1999) 399:541; **46.** Aquat Microbial Ecol (1998) 14:113; **47.** Appl Environ Microbiol (1999) 65:45; **48.** Appl Environ Microbiol (1998) 64:1725; **49.** Appl Environ Microbiol (1997) 63:186; **50.** Limnol Oceanogr (1995) 40:1050; **51.** Biotechniques (1995) 18:136; **52.** Biotechniques (1990) 9:310; **53.** Nucleic Acids Res (1979) 7:571; **54.** Nucleic Acids Res (1979) 7:547; **55.** Biotechniques (1995) 18:231; **56.** Nucleic Acids Res (2000) 28:E36; **57.** Biotechniques (1993) 15:1060; **58.** Nucleic Acids Res (1992) 20:2803; **59.** In Vitro Toxicol (1990) 3:219; **60.** J Immunol Methods (1993) 162:41; **61.** Cancer Res (1989) 49:565; **62.** Anal Biochem (1983) 131:538; **63.** Anal Biochem (2000) 285:194; **64.** Anal Chem (2001) 73:1537; **65.** Biotechniques (1997) 22:130, 134; **66.** J Clin Microbiol (2000) 38:2756; **67.** Clin Chem (2000) 46:1540; **68.** J Clin Microbiol (2000) 38:795; **69.** J Microbiol Methods (1999) 35:23; **70.** Lab Invest (1999) 79:337; **71.** Syst Appl Microbiol (1998) 21:89; **72.** Anal Biochem (1998) 259:112; **73.** Biotechniques (1998) 24:954; **74.** Anal Biochem (1997) 245:154; **75.** Nat Protoc (2006) 1:1559; **76.** Methods Mol Biol (2007) 353:237; **77.** PCR Methods Appl (1995) 4:234; **78.** Biochem J (2000) 346:17; **79.** Am J Pathol (1999) 154:97; **80.** Mol Genet Metab (1999) 68:357; **81.** Anal Biochem (1998) 260:142; **82.** Biotechniques (2001) 31:278; **83.** J Clin Microbiol (1999) 37:987; **84.** J Clin Microbiol (2000) 38:586.

IMPORTANT NOTICE: The products described in this manual are covered by one or more Limited Use Label License(s). Please refer to the Appendix on page 971 and Master Product List on page 975. Products are For Research Use Only. Not intended for any animal or human therapeutic or diagnostic use.

PRODUCT LIST 8.3 NUCLEIC ACID QUANTITATION IN SOLUTION

Cat. No.	Product	Quantity
F2962	FluoReporter® Blue Fluorometric dsDNA Quantitation Kit *200–2000 assays*	1 kit
H1398	Hoechst 33258, pentahydrate (bis-benzimide)	100 mg
H3569	Hoechst 33258, pentahydrate (bis-benzimide) *10 mg/mL solution in water*	10 mL
H21491	Hoechst 33258, pentahydrate (bis-benzimide) *FluoroPure™ grade*	100 mg
Q33130	Quant-iT™ dsDNA Assay Kit, Broad Range, 1000 assays *2–1000 ng*	1 kit
Q33120	Quant-iT™ dsDNA Assay Kit, High Sensitivity, 1000 assays *0.2–100 ng*	1 kit
Q32850	Quant-iT™ dsDNA BR Assay Kit, 100 assays *for use with the Qubit® fluorometer* *2–1000 ng*	1 kit
Q32853	Quant-iT™ dsDNA BR Assay Kit, 500 assays *for use with the Qubit® fluorometer* *2–1000 ng*	1 kit
Q32851	Quant-iT™ dsDNA HS Assay Kit, 100 assays *0.2–100 ng* *for use with the Qubit® fluorometer*	1 kit
Q32854	Quant-iT™ dsDNA HS Assay Kit, 500 assays *0.2–100 ng* *for use with the Qubit® fluorometer*	1 kit
O11492	Quant-iT™ OliGreen® ssDNA Assay Kit *2000 assays*	1 kit
O7582	Quant-iT™ OliGreen® ssDNA reagent *2000 assays*	1 mL
P7589	Quant-iT™ PicoGreen® dsDNA Assay Kit *2000 assays*	1 kit
P11496	Quant-iT™ PicoGreen® dsDNA Assay Kit *2000 assays* *10 x 100 µL*	1 kit
P7581	Quant-iT™ PicoGreen® dsDNA reagent *2000 assays*	1 mL
P11495	Quant-iT™ PicoGreen® dsDNA reagent *2000 assays* *10 x 100 µL*	10 x 100 µL
R11490	Quant-iT™ RiboGreen® RNA Assay Kit *2000 assays*	1 kit
R11491	Quant-iT™ RiboGreen® RNA reagent *2000 assays*	1 mL
Q33140	Quant-iT™ RNA Assay Kit, 1000 assays *5–100 ng*	1 kit
Q32852	Quant-iT™ RNA Assay Kit, 100 assays *5–100 ng* *for use with the Qubit® fluorometer*	1 kit
Q32855	Quant-iT™ RNA Assay Kit, 500 assays *5–100 ng* *for use with the Qubit® fluorometer*	1 kit
Q10213	Quant-iT™ RNA Assay Kit, Broad Range, 1000 assays *20–1000 ng*	1 kit
Q10210	Quant-iT™ RNA BR Assay Kit, 100 assays *20–1000 ng* *for use with the Qubit® fluorometer*	1 kit
Q10211	Quant-iT™ RNA BR Assay Kit, 500 assays *20–1000 ng* *for use with the Qubit® fluorometer*	1 kit
Q10212	Quant-iT™ ssDNA Assay Kit, 100 assays *1–200 ng* *for use with the Qubit® fluorometer*	1 kit
Q32856	Qubit™ assay tubes *set of 500*	1 set
Q32857	Qubit® fluorometer	each
Q32859	Qubit™ fluorometer international power cord *replacement*	each
Q32858	Qubit™ fluorometer USB cable	each
Q32861	Qubit™ Quantitation Lab Starter Kit	1 kit
Q32860	Qubit™ Quantitation Starter Kit	1 kit
R32700	RediPlate™ 96 RiboGreen® RNA Quantitation Kit *one 96-well microplate*	1 kit
S7563	SYBR® Green I nucleic acid gel stain *10,000X concentrate in DMSO*	500 µL
S7567	SYBR® Green I nucleic acid gel stain *10,000X concentrate in DMSO*	1 mL
S7585	SYBR® Green I nucleic acid gel stain *10,000X concentrate in DMSO* *special packaging*	20 x 50 µL
T11493	20X TE buffer *RNase free*	100 mL

IMPORTANT NOTICE: The products described in this manual are covered by one or more Limited Use Label License(s). Please refer to the Appendix on page 971 and Master Product List on page 975. Products are For Research Use Only. Not intended for any animal or human therapeutic or diagnostic use.
www.invitrogen.com/probes
molecular probes® | invitrogen™ by life technologies™

8.4 Nucleic Acid Detection on Gels, Blots and Arrays

Nucleic Acid Detection in Gels

Molecular Probes® fluorescent nucleic acid gel stains—SYBR® Gold, SYBR® Green I, SYBR® Green II and SYBR® Safe dyes—are high-sensitivity reagents for staining DNA and RNA in electrophoretic gels.[1] These gel stains provide greater sensitivity with lower background fluorescence than the conventional ethidium bromide gel stain. In addition, SYBR® Safe DNA gel stain showed very low mutagenic activity when tested by an independent, licensed testing laboratory, and it is not classified as hazardous waste or as a pollutant under U.S. Federal regulations.

SYBR® Gold Nucleic Acid Gel Stain

SYBR® Gold nucleic acid gel stain (S11494) is our most sensitive stain for detecting DNA or RNA in gels using a standard 300 nm UV transilluminator and Polaroid® 667 black-and-white print film. Although SYBR® Green I and SYBR® Green II gel stains are preferred for specific applications, several characteristics of SYBR® Gold stain represent a further improvement over our SYBR® Green I and SYBR® Green II gel stains for routine gel analysis. SYBR® Gold nucleic acid gel stain provides:

- **Maximum sensitivity.** Upon binding to nucleic acids, SYBR® Gold stain exhibits a >1000-fold fluorescence enhancement and a quantum yield of ~0.6.[2] By comparison, ethidium bromide exhibits <30-fold fluorescence enhancement upon binding nucleic acids[3] and a quantum yield of ~0.15.[4] Because of its superior fluorescence characteristics, SYBR® Gold stain is greater than 10-fold more sensitive than ethidium bromide for detecting DNA and RNA in gels using a 300 nm UV transilluminator and black-and-white photography (Figure 8.4.1). We routinely detect as little as 25 pg of dsDNA or 1 ng of RNA per band using a 300 nm UV transilluminator or a blue-light transilluminator—sensitivity levels even higher than those of silver staining.[2] For detecting glyoxalated RNA with 300 nm transillumination, SYBR® Gold stain is 25–100 times more sensitive than ethidium bromide (Figure 8.4.2) and thus represents a significant advance for protocols requiring sensitive RNA detection. SYBR® Gold stain has also proven to be more sensitive than SYBR® Green II RNA gel stain for detecting single-strand conformation polymorphism (SSCP) products.
- **Rapid gel penetration.** Staining gels with SYBR® Gold stain after electrophoresis followed by gel photography provides the optimal sensitivity. SYBR® Gold stain penetrates agarose gels faster and stains thick and high-percentage gels better than other post-electrophoresis stains.
- **Versatility.** SYBR® Gold stain is a universal nucleic acid gel stain that provides high detection sensitivity for dsDNA, ssDNA and RNA detection in many gel types, including high-percentage agarose, glyoxal/agarose, formaldehyde/agarose, native polyacrylamide–and urea–polyacrylamide gels. No wash step is required in order to achieve maximal sensitivities.[2]
- **Ease of use.** As a result of the low intrinsic fluorescence of the unbound dye, gel staining with SYBR® Gold dye shows extremely low background fluorescence and does not require a destaining step, even when staining agarose/formaldehyde gels. After incubating the gel in SYBR® Gold staining solution for 10–40 minutes (depending on the thickness and percentage of the agarose or polyacrylamide gel) the golden-yellow–fluorescent DNA or RNA bands are ready to be photographed.
- **Compatibility with other molecular biology techniques.** The presence of unbound SYBR® Gold dye in stained gels at standard staining concentrations does not interfere with restriction endonuclease or ligase activity or with subsequent PCR reactions. SYBR® Gold nucleic acid staining is compatible with both northern and Southern blotting—the stain transfers with the DNA or RNA to the blot and is washed off during incubation in the prehybridization mix.[2] SYBR® Gold stain is also easily removed from dsDNA by simple ethanol precipitation, leaving templates ready for subsequent manipulation or analysis.

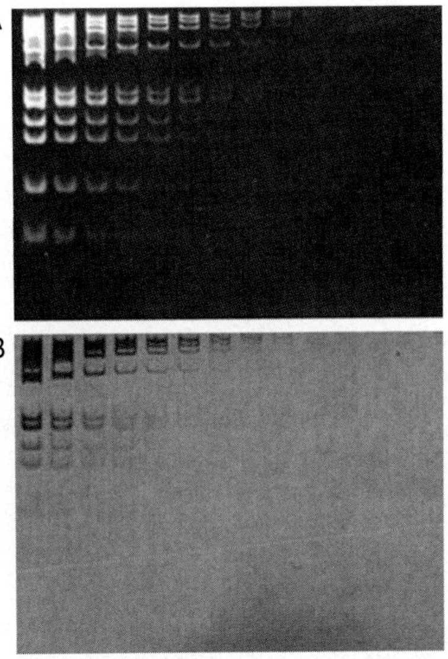

Figure 8.4.1 Comparison of the sensitivity achieved using SYBR® Gold stain (S11494) with 300 nm transillumination (**A**) or silver stain to detect double-stranded DNA separated on native polyacrylamide gels (**B**).

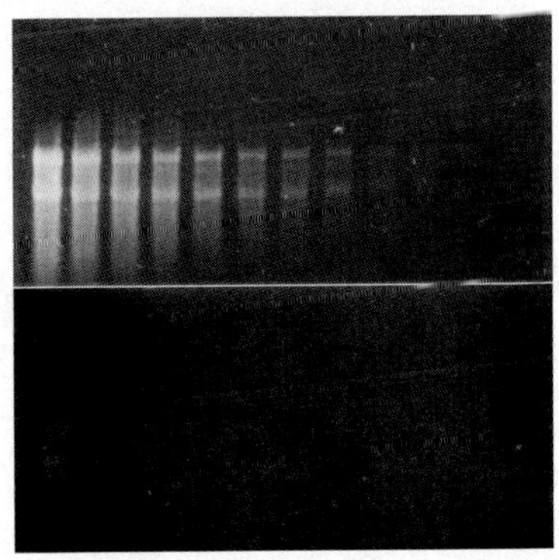

Figure 8.4.2 Comparison of glyoxylated RNA stained with ethidium bromide and with SYBR® Gold nucleic acid gel stain (S11494). Identical 2-fold dilutions of glyoxylated *Escherichia coli* 16S and 23S ribosomal RNA were separated on 1% agarose minigels using standard methods and stained for 30 minutes with SYBR® Gold stain in TBE buffer (**A**) or 0.5 µg/mL ethidium bromide in 0.1 M ammonium acetate (**B**). Both gels were subjected to 300 nm transillumination and photographed using Polaroid® 667 black-and-white print film and a SYBR® photographic filter (S7569, **A**) or an ethidium bromide photographic filter (**B**).

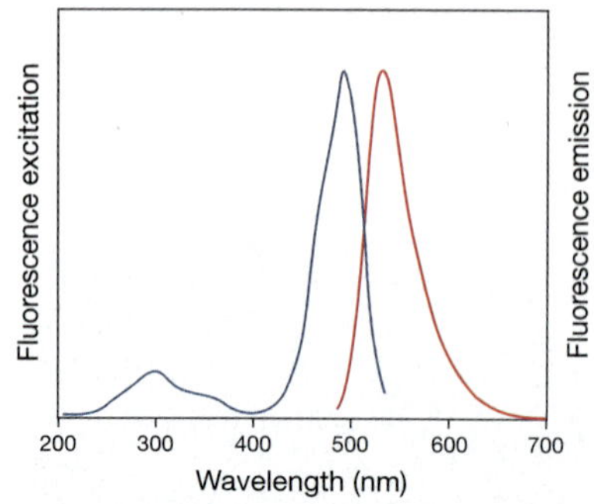

Figure 8.4.3 Fluorescence excitation and emission spectra of SYBR® Gold nucleic acid gel stain (S11494) bound to double-stranded DNA.

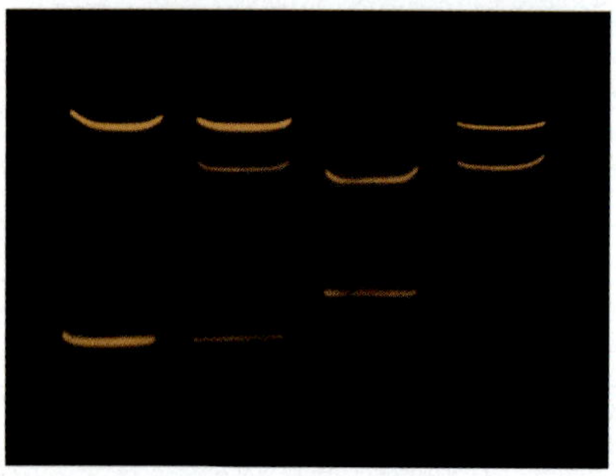

Figure 8.4.4 Sensitive and direct visualization of single-strand conformation polymorphism (SSCP) in exon 1 of human K-ras using SYBR® Gold nucleic acid gel stain (S11494). Lane 1 contains wild-type DNA and lanes 2–4 contain DNA from various adenocarcinoma samples with mutant alleles. Image contributed by Valerie DeGroff and Chris Weghorst, Ohio State University.

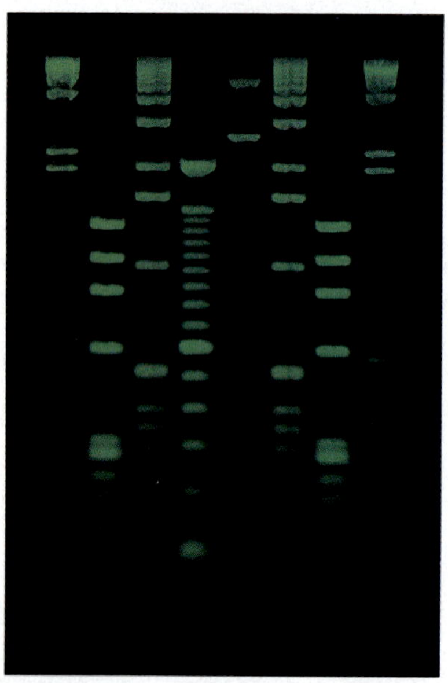

Figure 8.4.5 DNA molecular weight ladders that have been electrophoresed on a 1% agarose gel and then stained for 30 minutes with a 1:10,000 dilution of SYBR® Green I nucleic acid gel stain (S7563, S7567, S7585). Lanes 1 and 8 contain *Hind*III-cut DNA; lanes 2 and 7, HaeIII-cut fX174 RF DNA; lanes 3 and 6, 1 kilobase pair DNA ladder (GIBCO®); lane 4, 100 base pair DNA ladder (GIBCO®); lane 5, EcoR-I-cut pUC19 DNA mixed with PstI-cut fX174 RF DNA. Gel staining was visualized using 254 nm epi-illumination.

- **Instrument compatibility.** Because the nucleic acid–bound SYBR® Gold dye exhibits excitation maxima at both ~495 nm and ~300 nm (the emission maximum is ~537 nm) (Figure 8.4.3), it is compatible with a wide variety of instrumentation, ranging from UV epi- and transilluminators and blue-light transilluminators, to mercury-arc lamp– and argon-ion laser–based gel scanners. Short-wavelength (254 nm) epi-illumination is not required to obtain high sensitivity with SYBR® Gold stain. For optimal sensitivity with black-and-white print film and UV illumination, SYBR® Gold dye–stained gels should be photographed through a SYBR® photographic filter (S7569, see below).

SYBR® Gold dye should prove invaluable in applications such as agarose/formaldehyde gel electrophoresis prior to northern blot analysis, denaturing gradient gel electrophoresis (DGGE) and single-strand conformation polymorphism (SSCP) studies (Figure 8.4.4), as well as routine gel analysis. The high signals of the dye–DNA complex and the remarkable photostability observed with 300 nm transilluminators will make it easier to cut out low-abundance bands for subsequent manipulation, including subcloning, bandshift assays or other dsDNA template–based reactions. SYBR® Gold nucleic acid stain is useful for methylation-sensitive single-strand conformation analysis [5] (MS-SSCA). It is also reportedly more sensitive and easier to use than silver staining in the telomeric repeat amplification protocol [6] (TRAP). In the electrophoretic analysis of DNA forms in liposomes, the dye showed 40-fold greater sensitivity and more consistent staining between different isoforms of DNA compared with ethidium bromide.[7] In addition, it has been used to monitor the formation of crosslinked peptide–DNA complexes.[8]

Each 500 µL vial of SYBR® Gold nucleic acid gel stain (S11494) contains sufficient reagent to stain at least 100 agarose or polyacrylamide minigels. SYBR® Gold nucleic acid stain is accompanied by detailed instructions for use of the dye in staining nucleic acids. Answers to frequently asked questions about all our SYBR® dyes are available in a separate information sheet.

SYBR® Green I Nucleic Acid Gel Stain

SYBR® Green I nucleic acid gel stain [9] (S7563, S7567, S7585) is an extremely sensitive fluorescent stain for detecting nucleic acids in agarose and polyacrylamide gels (Figure 8.4.5). As with SYBR® Gold stain, this remarkable sensitivity can be attributed to a combination of unique dye characteristics. SYBR® Green I stain exhibits exceptional affinity for DNA and a large fluorescence enhancement upon binding to DNA—at least an order of magnitude greater than that of ethidium bromide when detected by photography. Also, the fluorescence quantum yield of the SYBR® Green I dye–DNA complex (~0.8) is over five times greater than that of the ethidium complex of DNA (~0.15). Furthermore, SYBR® Green I stain has been shown to be significantly less mutagenic than EtBr by Ames testing.[10] SYBR® Green I stain is somewhat less sensitive than our SYBR® Gold stain, but has some important characteristics that make it the preferred reagent for certain applications:

- **Preferential DNA staining.** SYBR® Green I nucleic acid stain has a much greater fluorescence enhancement when bound to dsDNA and oligonucleotides than when bound to RNA. With a standard 300 nm UV transilluminator and photographic detection, as little as 60 pg dsDNA per band can be detected with SYBR® Green I stain [11–13] (Figure 8.4.6), whereas SYBR® Green I

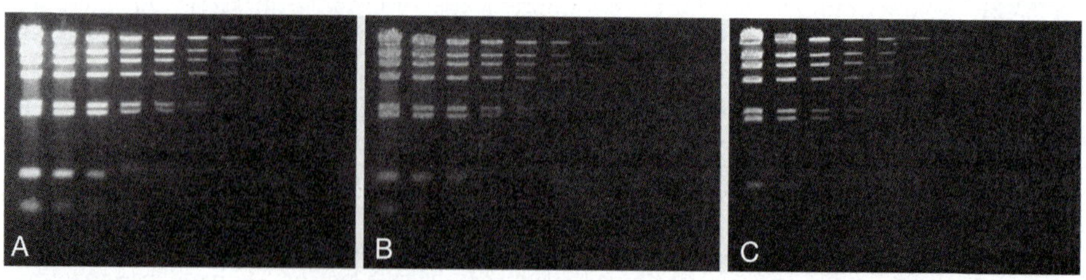

Figure 8.4.6 Comparison of dsDNA detection in native gels using SYBR® Green I nucleic acid gel stain and ethidium bromide. Identical 3-fold dilutions of bacteriophage lambda DNA digested with *Hind*III restriction endonuclease were electrophoresed on 1% agarose gels. Gels were stained for 30 minutes with a 1:10,000 dilution of SYBR® Green I nucleic acid gel stain (S7563, S7567, S7585) and not destained (**A** and **B**) or with 5 µg/mL ethidium bromide for 30 minutes and destained for a further 30 minutes in water (**C**). Gel staining was visualized using 254 nm epi-illumination (**A**) or 300 nm transillumination (**B** and **C**) and then photographed using Polaroid® 667 black-and-white print film and a SYBR® photographic filter (**A** and **B**, S7569) or an ethidium bromide gel stain photographic filter (**C**).

IMPORTANT NOTICE: The products described in this manual are covered by one or more Limited Use Label License(s). Please refer to the Appendix on page 971 and Master Product List on page 975. Products are For Research Use Only. Not intended for any animal or human therapeutic or diagnostic use.
www.invitrogen.com/probes

stain is not much more sensitive than ethidium bromide for staining RNA. This quality makes SYBR® Green I dye the ideal gel stain for applications in which RNA in the sample may obscure the results, such as when visualizing DNA fragmentation ladders from apoptotic cells [14,15] (Figure 8.4.7). Fluorescence of nucleic acid–bound SYBR® Green I dye is also of sufficient sensitivity to allow detection and discrimination of viruses by flow cytometry.[16]

- **Sensitivity for oligonucleotide detection.** We have determined that SYBR® Green I nucleic acid gel stain is nearly two orders of magnitude more sensitive than ethidium bromide for staining oligonucleotides in gels, provided that the gel is photographed according to the provided protocol. Using 254 nm epi-illumination, it is possible to detect 1–2 ng of a synthetic 24-mer on 5% polyacrylamide gels (Figure 8.4.8).
- **Exceptionally low background.** SYBR® Green I stain shows very low background fluorescence in the gel, making it the preferred dye for some laser-scanning instruments, in which background fluorescence can produce unacceptable noise levels.
- **Spectral compatibility with lasers and filter sets.** SYBR® Green I stain has a UV-excitation peak of ~250 nm (Figure 8.4.9). Thus, higher sensitivity can be achieved with SYBR® Green I stain using 254 nm transillumination, as compared with the more common 300 nm transillumination. However, the visible excitation peak of SYBR® Green I dye–stained nucleic acids near 497 nm is very close to the principal emission lines of many laser-scanning instruments. Because nucleic acid–bound SYBR® Green I dye exhibits spectral characteristics (excitation/emission maxima ~497/520 nm) very similar to those of fluorescein, it is compatible with most common filter sets used in laser scanners.

Like SYBR® Gold stain, SYBR® Green I dye is very easy to use—the staining procedure can be completed in 10–40 minutes (somewhat longer for thicker gels), with no destaining step required prior to photography. Presence of typical staining concentrations of SYBR® Green I dye does not significantly inhibit the ability of several restriction endonucleases to cleave DNA.[11] This property makes staining with SYBR® Green I dye compatible with in-gel subcloning protocols.[17] SYBR® Green I stain is also easily removed from dsDNA by simple ethanol precipitation. For optimal sensitivity with black-and-white print film and UV illumination, gels stained with SYBR® Green dye should be photographed through a SYBR® photographic filter (S7569); a CCD camera also provides good detection sensitivity.

The ultrasensitivity of SYBR® Green I dye makes it useful for detecting the products of DNA and RNA amplification reactions by gel electrophoresis,[18] restriction mapping small amounts of DNA and detecting the products of bandshift [19] and nuclease-protection assays. PCR amplification products that are at the limit of detection using ethidium bromide are easily detected using SYBR® Green I dye. Reverse-transcription PCR (RT-PCR) reaction products have been detected with high sensitivity following gel electrophoresis and staining with SYBR® Green I dye, allowing the cycle number to be lowered, which reduces heteroduplex formation during amplification.[20] SYBR® Green I stain was used to detect RT-PCR products amplified from B cells,[21] *Xenopus laevis* embryos [22] and smooth muscle cells.[23] Using a laser scanner and SYBR® Green I stain, researchers have developed a high-throughput RT-PCR DNA profiling assay in multiwell agarose gels.[24] SYBR® Green I dye was used to stain DNA in high-resolution gels capable of resolving 100–200 base-pair DNA fragments and differing by as few as two base pairs.[25] Use of SYBR® Green I stain also eliminates the need to label PCR products with radioisotopes in a kinetic PCR assay.[26–28] The high chemical stability of SYBR® Green I nucleic acid stain and the dye's selective sensitivity for detecting double-stranded products made in the presence of single-stranded oligonucleotide

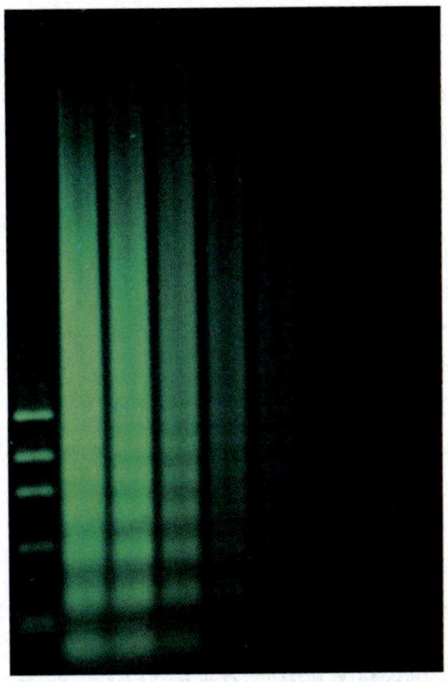

Figure 8.4.7 DNA extracts from camptothecin-treated HL-60 cells separated on an agarose gel and stained with SYBR® Green I nucleic acid gel stain (S7563, S7567, S7585). The 200 to 5000 bp DNA fragments characteristic of apoptotic cells (which appear as "ladders") are clearly visualized with this sensitive nucleic acid stain. Cell preparations were gifts of Zbigniew Darzynkiewicz, Cancer Research Institute, New York Medical College.

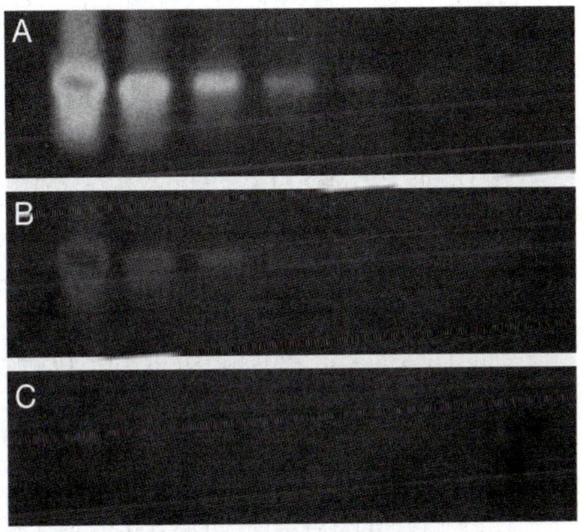

Figure 8.4.8 Comparison of single-stranded oligonucleotide detection using SYBR® Green I nucleic acid gel stain and ethidium bromide. Identical 3-fold dilutions of a synthetic, single-stranded 24-mer were electrophoresed on 10% polyacrylamide gels. Gels were stained for 30 minutes with a 1:10,000 dilution of SYBR® Green I nucleic acid gel stain (S7563, S7567, S7585) and not destained (**A** and **B**), or with 5 µg/mL ethidium bromide for 30 minutes and destained for a further 30 minutes in water (**C**). Gel staining was visualized using 254 nm epi-illumination (**A**) or 300 nm transillumination (**B** and **C**), and then photographed using Polaroid® 667 black-and-white print film and a SYBR® photographic filter (S7569, **A** and **B**) or an ethidium bromide gel stain photographic filter (**C**).

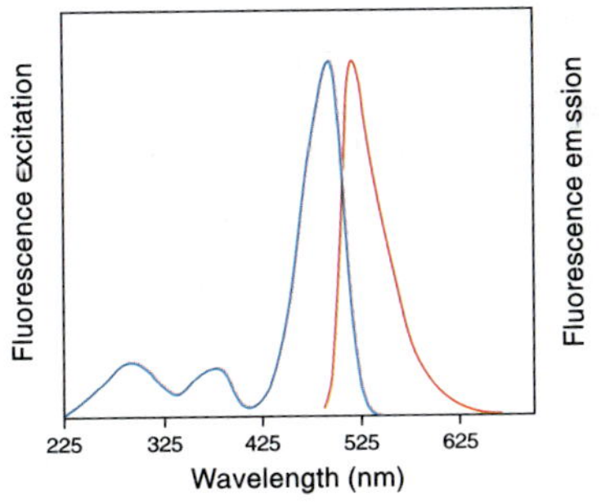

Figure 8.4.9 Fluorescence excitation and emission spectra of dsDNA-bound SYBR® Green I nucleic acid gel stain (S7563, S7567, S7585).

primers make SYBR® Green I stain the preferred dye for real-time quantitative analysis of PCR products in a solution assay (Section 8.3).

In other gel-based techniques, SYBR® Green I nucleic acid gel stain has enabled researchers to eliminate silver staining and radioactivity from their protocols. SYBR® Green I dye staining was shown to be as sensitive as silver staining—as well as being more rapid, less laborious and less expensive—in a nonradioactive method for detecting hypervariable simple sequence repeats in electrophoretic gels.[29] SYBR® Green I dye staining has replaced conventional silver staining techniques for routine identity testing in some forensics laboratories. In addition, SYBR® Green I stain is as sensitive as silver staining, but less expensive, for detecting STR (short tandem repeat) polymorphisms,[30,31] and mitochondrial DNA deletions.[32]

Likewise, in a gel assay for detection of telomerase activity (telomeric repeat amplification protocol or TRAP) in human cells and tumors, SYBR® Green I dye staining was found to be more sensitive than silver staining and gave results comparable to those achieved with a radioisotope-based TRAP assay.[33,34] Moreover, unlike silver stains, SYBR® Green I stain does not label proteins carried over from the reaction mixture. SYBR® Green I dye staining has also been shown to be as sensitive as ^{3}H-labeled thymidine for detecting double-strand breaks in mammalian cells.[35] It may be possible to further increase the sensitivity of some of these reported applications of SYBR® Green I stain by using SYBR® Gold nucleic acid gel stain (see above).

Each milliliter of our concentrated SYBR® Green I nucleic acid gel stain (S7563, S7567, S7585) contains sufficient reagent to stain at least 100 agarose or polyacrylamide minigels. Reuse of the staining solution can significantly increase the number of gels stained per vial. In some applications, such as preparative agarose gel electrophoresis, the amount of SYBR® Green I dye used per gel can be significantly reduced if the dye is added directly to the loading buffer. However, because the dye affects DNA mobility and dissociates from the smaller DNA fragments during electrophoresis, this method should not be used for size determination or for DNA fragments less than ~100 base pairs in length.[36]

SYBR® Green II RNA Gel Stain

SYBR® Green II RNA gel stain (S7564, S7568, S7586) is a highly sensitive dye for detecting RNA or ssDNA in agarose or polyacrylamide gels (Figure 8.4.10, Figure 8.4.11). Some outstanding features of SYBR® Green II RNA gel stain include its high binding affinity for RNA and its large fluorescence enhancement and exceptionally high quantum yield upon binding to RNA. Although it is not a specific stain for RNA, SYBR® Green II dye exhibits a larger fluorescence quantum yield when bound to RNA (~0.54) than to dsDNA (~0.36). This property is unusual among nucleic acid stains; most show far greater quantum yields and fluorescence enhancements when bound to double-stranded nucleic acids. Moreover, the fluorescence quantum yield of the SYBR® Green II complex of RNA is over seven times greater than that of the ethidium bromide–RNA complex[37] (~0.07). The affinity of SYBR® Green II RNA gel stain for RNA is also higher than that of ethidium bromide, and its fluorescence enhancement upon binding to RNA is well over an order of magnitude greater. Like SYBR® Green I stain, our SYBR® Green II stain gives the greatest sensitivity on 254 nm transillumination or laser scanners. However, the best sensitivity for RNA detection using 300 nm transillumination is achieved with SYBR® Gold dye (see above). Other important properties of SYBR® Green II RNA gel stain include:

- **Sensitivity.** Using 254 nm epi-illumination, Polaroid® 667 black-and-white print film and a SYBR® photographic filter (S7569), we have been able to detect as little as 100 pg of ribosomal RNA (rRNA) per band on native 1% agarose gels and <1 ng rRNA per band on 5% polyacrylamide gels stained with SYBR® Green II RNA gel stain. The detection limit of SYBR® Green II dye–stained native gels excited with 300 nm transillumination is approximately 500 pg per band, as compared with about 1.5 ng for ethidium bromide–stained gels[13] (Figure 8.4.10).

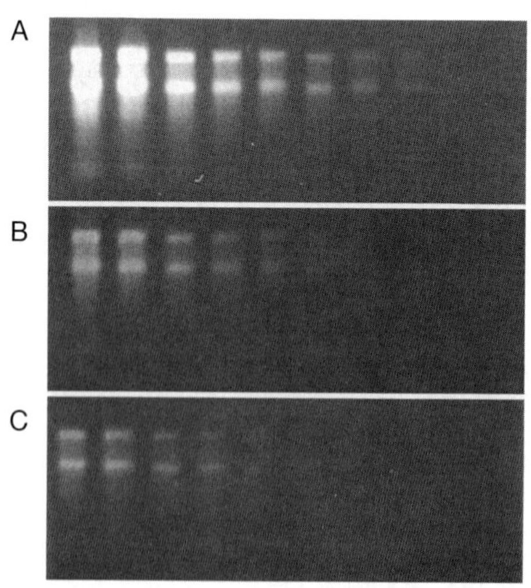

Figure 8.4.10 Comparison of RNA detection in nondenaturing gels using SYBR® Green II RNA gel stain and ethidium bromide. Identical 2-fold dilutions of *Escherichia coli* ribosomal RNA were electrophoresed on 1% agarose gels using Tris-borate buffer. Gels were stained for 20 minutes with a 1:10,000 dilution of SYBR® Green II RNA gel stain (S7564, S7568, S7586) and not destained (**A** and **B**), or with 5 µg/mL ethidium bromide for 20 minutes and destained for a further 20 minutes in water (**C**). Gel staining was visualized using 254 nm epi-illumination (**A**) or 300 nm transillumination (**B** and **C**), and then photographed using Polaroid® 667 black-and-white print film and a SYBR® photographic filter (**A** and **B**, S7569) or an ethidium bromide gel stain photographic filter (**C**).

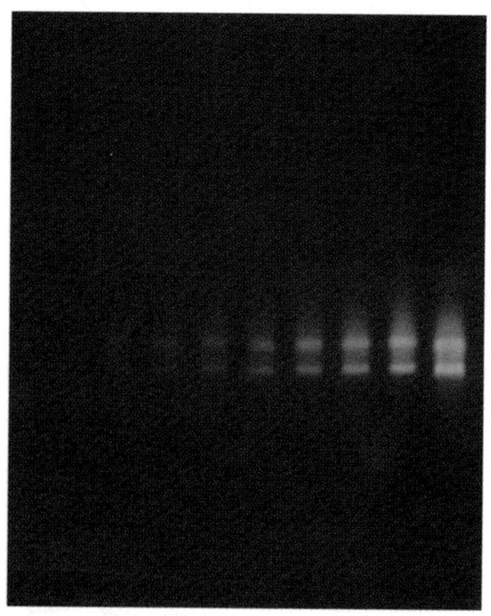

Figure 8.4.11 A 2-fold dilution series of *Escherichia coli* 16S and 23S ribosomal RNA (rRNA) that has been electrophoresed on a nondenaturing 1% agarose gel and then stained with our SYBR® Green II RNA gel stain (S7564, S7568, S7586). Gel staining was visualized with 254 nm epi-illumination.

IMPORTANT NOTICE: The products described in this manual are covered by one or more Limited Use Label License(s). Please refer to the Appendix on page 971 and Master Product List on page 975. Products are For Research Use Only. Not intended for any animal or human therapeutic or diagnostic use.

- **Ease of use.** Like SYBR® Green I and SYBR® Gold stains, SYBR® Green II RNA gel stain has a very low intrinsic fluorescence, eliminating the need to destain gels.
- **Compatibility with urea and formaldehyde gels.** Fluorescence of SYBR® Green II dye–RNA complexes does not appear to be quenched in the presence of urea or formaldehyde, so that denaturing gels do not have to be washed free of the denaturant prior to staining.
- **Broad linear dynamic range.** When used on a laser scanner, SYBR® Green II stain shows a dynamic range of over five orders of magnitude—far greater than the linear dynamic range of ethidium bromide—allowing more accurate quantitation of bands in the gel.[38]
- **Compatibility with northern blots.** SYBR® Green II dye staining is compatible with agarose/formaldehyde gels. The formaldehyde does not have to be removed prior to staining, and the sensitivity of SYBR® Green II dye staining is 5–10 times better than that of ethidium bromide on these gels. In addition, staining agarose/formaldehyde gels with SYBR® Green II dye does not interfere with transfer of the RNA to filters or subsequent hybridization in northern blot analysis, provided that 0.1% to 0.3% SDS is included in the prehybridization and hybridization buffers.[39] Thus, SYBR® Green II stain can be used to normalize the hybridization signal to the amount of RNA loaded on the gel.[38,40,41]

SYBR® Green II RNA gel stain facilitates the detection of viroid RNAs and other multicopy cellular RNA species. This gel stain has been used to visualize the migration behavior of 5S rRNA species after electrophoresis through a denaturing gradient gel, a method that was used to discriminate among different acidophile species in a mixed culture.[42] SYBR® Green II RNA gel stain should also improve the analysis of small aliquots from an RNA preparation, leaving the researcher with more material to carry out the primary experiment, be it northern blotting, start-site mapping or cDNA preparation.

In addition to its use for detecting RNA, SYBR® Green II RNA gel stain is useful for single-strand conformation polymorphism (SSCP) analysis,[43,44] which demands extremely sensitive detection techniques.[45] Many of the nonradioisotopic SSCP methods currently in use, such as silver staining or chemiluminescence-mediated signal amplification, require long, complex procedures.[46–49] An SSCP assay using precast polyacrylamide minigels and SYBR® Green II stain not only provides the precise temperature control required for the assay,[50] but it is more rapid and less labor-intensive than assays that use silver staining for detection.[51] In another report, SYBR® Green II RNA gel stain was used to detect Ki-*ras* mutants by SSCP analysis and was reported to yield 10-fold better sensitivity than standard silver-staining techniques.[44] SYBR® Green II stain is compatible with amplification by PCR; after SSCP analysis, SYBR® Green II dye–stained bands can be excised out of the gel and used in cycle-sequencing.[52] SYBR® Green II nucleic acid stain also provides high-sensitivity staining for rRNA separated by high-resolution denaturing gradient electrophoresis (DGGE), making it possible to discriminate between closely related species of bacteria.[42]

SYBR® Green Nucleic Acid Gel Stains: Special Packaging and a Starter Kit

In addition to providing SYBR® Green nucleic acid gel stains packaged as 500 µL or 1 mL stock solutions in DMSO (S7563, S7564, S7567, S7568), we make both SYBR® Green I and SYBR® Green II available as a set of 20 individual vials, each containing 50 µL of the DMSO stock solution (S7585, S7586). This convenient packaging makes it easy to supply members of the laboratory with an aliquot of stock solution, or to share stock with other laboratories. Special packaging also minimizes potential losses due to contamination, spills and light exposure. Each milliliter of the concentrated gel stain provides sufficient reagent to prepare 10 liters of a staining solution. Although best results are obtained with freshly diluted dye, properly prepared staining solution can be stored for up to a week, if kept refrigerated and protected from light, and can be reused 2–3 times with little loss of signal. SYBR® Green nucleic acid stains are accompanied by detailed instructions for their use in staining gels; answers to frequently asked questions about all SYBR® dyes are available in a separate information sheet.

Our SYBR® Green Nucleic Acid Gel Stain Starter Kit (S7580) is designed for laboratories that want to sample these products. The kit includes single 50 µL vials of both SYBR® Green I and SYBR® Green II stains and a SYBR® gel stain photographic filter, along with complete directions for their use.

SYBR® Safe DNA Gel Stain

SYBR® Safe DNA gel stain (S33100, S33101, S33102, S33111, S33112) provides sensitive DNA detection with significantly reduced mutagenicity, making it safer than ethidium bromide for staining DNA in agarose or acrylamide gels. Not only is SYBR® Safe stain less mutagenic than ethidium bromide, but the detection sensitivity of SYBR® Safe gel stain is comparable to that of ethidium bromide and 400 times greater than that of colorimetric stains for detecting DNA in electrophoretic gels. SYBR® Safe stain is provided as a premixed solution that can directly replace ethidium bromide in standard protocols; SYBR® Safe stain can either be cast into the gel or be used as a post-electrophoresis stain. DNA bands stained with SYBR® Safe DNA gel stain can be detected using a standard UV transilluminator, a visible-light transilluminator or a laser scanner. SYBR® Safe stain is also suitable for detecting RNA in gels. Bound to nucleic acids, SYBR® Safe stain has fluorescence excitation maxima at 280 nm and 502 nm, and an emission maximum at 530 nm (Figure 8.4.12). SYBR® Safe DNA gel stain offers:

- **Increased safety.** SYBR® Safe DNA gel stain has tested negative in three mammalian cell–based assays for genotoxicity, is less mutagenic than ethidium bromide in standard Ames tests and is not classified as hazardous waste under U.S. Federal regulations (SYBR® Safe DNA Gel Stain—Note 8.1).

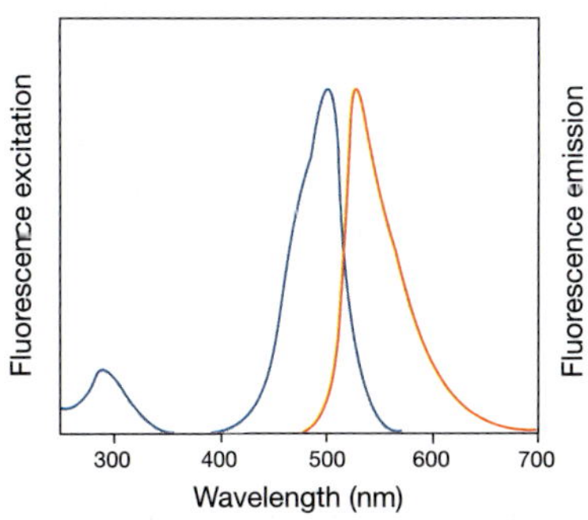

Figure 8.4.12 Normalized fluorescence excitation and emission spectra of SYBR® Safe DNA gel stain (S33100, S33101, S33110), determined in the presence of double-stranded DNA.

- **Better performance.** SYBR® Safe DNA gel stain is as sensitive as ethidium bromide and 400 times as sensitive as colorimetric stains for detecting DNA in electrophoretic gels.
- **Convenience.** SYBR® Safe stain is provided as a ready-to-use solution in 0.5X TBE or 1X TAE; it can be cast directly in the gel or used as a post-electrophoresis stain (Figure 8.4.13).
- **Quick staining protocol.** Simply incubate the gel in staining solution for 30 minutes; no destaining is required.

Furthermore, we have demonstrated a vast improvement in cloning efficiency with DNA fragments isolated from agarose gels using SYBR® Safe stain and blue light versus the same DNA fragments isolated using ethidium bromide and UV light. In the experiment, a 1.25 kilobase gene was amplified by PCR. Seven equal amounts of the PCR product were electrophoresed on duplicate agarose gels; one gel was visualized with SYBR® Safe stain and blue-light illumination, while the other gel was visualized with ethidium bromide and UV illumination. Bands were excised after defined exposure times, then purified using the Invitrogen™ PureLink™ Gel Extraction Kit. The purified product was then used in a Gateway® BxP cloning reaction. A portion of each reaction product was transformed into One Shot® TOP10 chemically competent bacteria; three serial dilutions were plated, and colonies were counted using an Alpha Innotech imaging system. The results showed an 80% reduction in transformation efficiency after only 30 seconds of exposure to ethidium bromide and UV light; after only 2 minutes of exposure, the number of transformants was nearly zero. In contrast, the transformation

NOTE 8.1

SYBR® Safe DNA Gel Stain

SYBR® Safe stain tests negative in standardized mammalian cell tests for genotoxicity. Certified testing has shown that SYBR® Safe DNA gel stain does not induce transformations in primary cultures of Syrian hamster embryo (SHE) cells when compared with solvent alone (Table 1). This test has a high concordance (>80%) with rodent carcinogenesis assays;[1] thus, a negative test strongly indicates that the SYBR® Safe stain is noncarcinogenic. In contrast, ethidium bromide tests positive in the SHE cell assay,[2] consistent with its known activity as a strong mutagen.

In addition, SYBR® Safe stain does not cause mutations in mouse lymphoma cells at the TK locus, nor does it induce chromosomal aberrations in cultured human peripheral blood lymphocytes, with or without S9 metabolic activation, using standardized tests with appropriate controls (Table 1).

- SYBR® Safe stain is less mutagenic than ethidium bromide in the standard Ames test.
- Compared with ethidium bromide, SYBR® Safe DNA gel stain causes fewer mutations in the Ames test, as measured in several different strains of *Salmonella typhimurium*. Weakly positive results for SYBR® Safe stain in this test occurred in four out of seven strains and only with activation by a mammalian S9 fraction (Figure 1).
- SYBR® Safe stain is not classified as hazardous waste.
- SYBR® Safe DNA gel stain has been extensively tested for environmental safety. According to the test results (Table 2), this stain is not classified as hazardous waste under U.S. Federal regulations (Resource Conservation and Recovery Act (RCRA)). Contact your safety officer for proper disposal procedures.

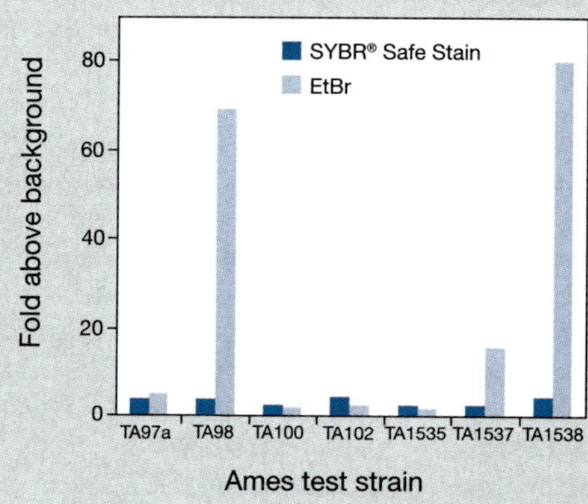

Figure 1 Summary of Ames test results for mutagenicity of SYBR® Safe DNA gel stain and ethidium bromide. Samples were pretreated with a mammalian S9 fraction and then tested using the indicated test strain. With strains TA97a, TA98, TA100 and TA102, an increase in revertants of more than 2-fold over background is a positive result for mutagenicity in this test. With strains TA1535, TA1537 and TA1538, an increase in revertants of more than 3-fold over background is a positive result. All tests were performed by Covance Laboratories, Inc., Vienna, VA, an independent testing laboratory.

1. Toxicol Sci (2001) 60:28; 2. Mutat Res (1999) 439:37.

Table 1 Summary of mammalian cell test results for SYBR® Safe DNA gel stain.

Test *	Cell Type	With S9 Activation †	Without S9 Activation
Transformation	Syrian hamster embryo (SHE) cells	NA	Negative
Chromosomal aberrations	Cultured human peripheral blood lymphocytes	Negative	Negative
Forward mutation	L5178YTK+/- mouse lymphoma cells	Negative	Negative

* All tests were performed by Covance Laboratories Inc., Vienna, VA, an independent testing laboratory. † Mammalian S9 fraction obtained from Aroclor 1254–induced rat liver. NA = Not applicable.

Table 2 Environmental safety test results.

Analysis *	Method	Results
Ignitability	EPA 1010	Not ignitable (>212°F)
Corrosivity	EPA 150.1	Not corrosive (pH = 8.25)
Reactivity	EPA 9010B/9030A	No reactivity detected
Corrosivity (by Corrositex®)	DOT-E 10904	Category 2 noncorrosive
Aquatic toxicity	Fathead minnow CA Title 22 acute screening	Not classified as hazardous or toxic to aquatic life

* All tests were independently confirmed by AMEC Earth and Environmental San Diego Bioassay Laboratory, San Diego, CA.

IMPORTANT NOTICE: The products described in this manual are covered by one or more Limited Use Label License(s). Please refer to the Appendix on page 971 and Master Product List on page 975. Products are For Research Use Only. Not intended for any animal or human therapeutic or diagnostic use.
www.invitrogen.com/probes
molecular probes® | invitrogen™ by Life technologies™

efficiency attained using SYBR® Safe stain and blue light remained at virtually 100% of the control value throughout the entire 16-minute time course of the experiment.

SYBR® Safe DNA gel stain is supplied ready-to-use in two different sizes and in two different buffers. The 1 L unit size in 0.5X TBE or 1X TAE (S33100, S33111) provides sufficient reagent to stain ~20 minigels; the 4 L unit size in 0.5X TBE or 1X TAE (S33101, S33112) provides sufficient reagent to stain ~80 minigels and is supplied in a cube-shaped container with a removable spigot. We also offer a 400 µL unit size of 10,000X concentrate in DMSO (S33102). SYBR® Safe DNA Gel Stain Starter Kit (S33110) is a convenient packaging of the 1 L unit size of SYBR® Safe stain in 0.5X TBE plus one SYBR® Safe photographic filter (S37100).

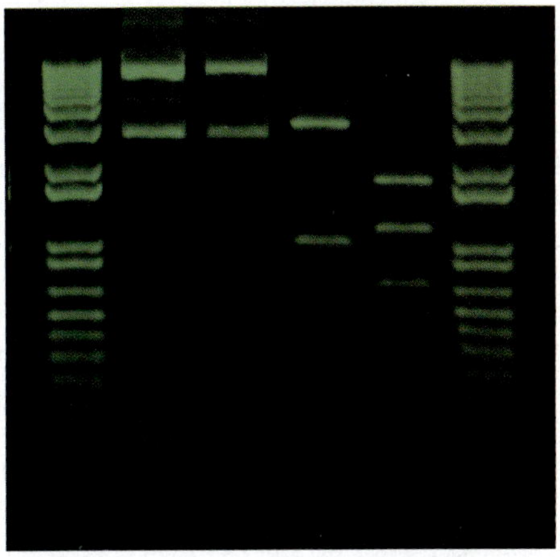

Figure 8.4.13 DNA fragments were electrophoresed through an agarose gel, then stained with SYBR® Safe DNA gel stain (S33100, S33101, S33110).

Ethidium Bromide

Ethidium bromide (EtBr, 15585-011) is the most commonly used dye for DNA and RNA detection in gels.[3] It binds to single-, double- and triple-stranded DNA.[53–55] Ethidium bromide has also been used to detect protein–DNA complex formation in bandshift assays[56] and to observe single DNA molecules undergoing gel electrophoresis.[57,58] We offer a 10 mg/mL aqueous solution of ethidium bromide (15585-011), which can be used as provided or diluted to the desired concentration.

Cyanine Monomers for Staining DNA in Electrophoretic Gels

Although SYBR® dyes are now the preferred gel stains, at least six of the monomeric cyanine dyes—TO-PRO®-1, YO-PRO®-1, BO-PRO™-1, PO-PRO™-1, JO-PRO™-1 and LO-PRO™-1 (Section 8.1, Table 8.2)—are also sensitive reagents for staining gels after electrophoresis and are compatible with UV trans- or epi-illumination or with laser-excited gel scanners. Their range of absorption maxima may make them superior to SYBR® dyes when using some lasers as excitation sources. We have determined that the limit of detection of dsDNA with some of these dyes is about 60 pg/band, using 254 nm epi-illumination and Polaroid® 667 black-and-white print film photography.

Our TO-PRO®-3 dye (T3605) can detect less than 0.1 ng/band DNA in an ultrathin-layer agarose gel–based electrophoretogram when excited by an inexpensive 640 nm red diode laser.[59] Preloading of the gel buffer with the TO-PRO®-3 dye has been recommended for this application when analyzing migrating allele-specific PCR fragments.[60,61]

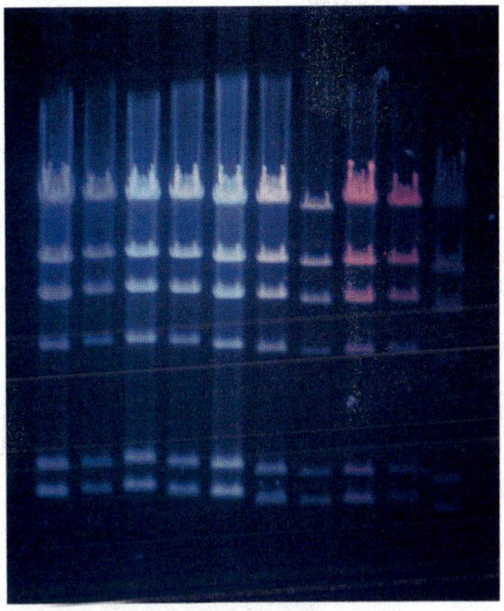

Figure 8.4.14 Lambda bacteriophage *Hin*dIII fragments were prestained with various nucleic acid dyes, run on a 0.7% agarose gel and visualized using a standard 300 nm UV transilluminator. From left to right, the dyes used were: POPO™-1 (P3580), BOBO™-1 (B3582), YOYO®-1 (Y3601), TOTO®-1 (T3600), JOJO™-1 (J11372), POPO™-3 (P3584), LOLO™-1 (L11376), BOBO™-3 (B3586), YOYO®-3 (Y3606) and TOTO®-3 (T3604) nucleic acid stains. The longest-wavelength stains are barely visible to the eye but can be detected with infrared-enhanced films and imaging equipment.

Cyanine and Ethidium Dimers for Staining DNA Prior to Electrophoresis

The extraordinary stability of the nucleic acid complexes formed with our dimeric cyanine dyes or ethidium homodimers[62–64] (Table 8.2) allows the dye–DNA association to remain stable during electrophoresis. Thus, samples can be prestained with subsaturating nanomolar dye concentrations before electrophoresis[65,66] (Figure 8.4.14), thereby reducing the hazards inherent in handling large volumes of ethidium bromide staining solutions.[63,67] The fluorescence intensities of both the EthD-1–DNA and TOTO®-1–DNA complexes are directly proportional to the amount of DNA in a band; however, TOTO®-1 dye staining has less effect on the electrophoretic mobility of DNA fragments than does EthD-1. Furthermore, unlike EthD-1-labeled DNA, in which up to two-thirds of the bound dye can be transferred to excess unlabeled DNA, the extent of transfer of TOTO®-1 dye to unlabeled DNA is reported to be only about 15–20%, even when the TOTO®-1–DNA complexes are incubated for up to 10 hours with a 100-fold excess of uncomplexed dsDNA.[62,68,69] This property is valuable for multiplexed electrophoretic separations,[62] especially because our cyanine nucleic acid stains are available in so many visually distinct colors. If two DNA populations are stained with spectrally distinct cyanine dimer dyes and run in the same lane, simultaneous two-color detection can potentially eliminate errors caused by lane-to-lane variations in electrophoretic mobility. Binding of the TOTO®-1 dye (T3600), YOYO®-1 dye (Y3601) and ethidium homodimer-1 (E1169) to DNA initially results in inhomogeneous binding that yields double bands in DNA gel electrophoresis.[70] These double bands can be avoided by incubating complexes for times long enough to allow binding to come to equilibrium or by heating samples to 50°C for at least two hours. Binding of our other dimeric nucleic acid stains (Table 8.2) to DNA does not seem to give this problem.

An extremely sensitive confocal laser–based gel scanner has been exploited in multiplexed electrophoretic separations to detect as little as four picograms per band of TOTO®-1 dye– and YOYO®-1 dye–stained dsDNA;[62,66,69,71] although sophisticated equipment is required for achieving these low detection limits, such equipment is not essential for detecting somewhat larger quantities of these nucleic acid–dye complexes. TOTO®-1 dye has been used to label DNA prior to electrophoresis in order to detect cystic fibrosis mutant alleles with a laser-excited fluorescence gel scanner,[72] as well as to detect DNA amplification products on agarose gels with standard UV transillumination.[73] TOTO®-1 dye has also been used to label nine DNA fragments of the dystrophin gene that were simultaneously generated using the polymerase chain reaction.[74] The resolution obtained by gel electrophoresis of these labeled fragments compared favorably to that observed using fluorophore-labeled primers. TOTO®-3 and POPO™-1 dyes (T3604, P3580; Section 8.1) have been similarly used to analyze DNA with a xenon lamp–based luminescence analyzer.[75]

Ethidium homodimer-1 (EthD-1, E1169) has been used for fluorescence detection of 30–60 picograms DNA per band on polyacrylamide gels using a confocal laser–based scanning system.[63,67] Ethidium homodimer-2 (EthD-2, E3599), which has a higher affinity for nucleic acids than does ethidium homodimer-1, may also be useful for this application.

Electrophoretic Mobility-Shift (Bandshift) Assay (EMSA) Kit

The Electrophoretic Mobility-Shift Assay (EMSA) Kit (E33075) provides a fast and quantitative fluorescence-based method to detect both nucleic acids and proteins in the same gel (Figure 8.4.15), doubling the information that can be obtained from bandshift assays.[19] This kit uses two fluorescent dyes for detection—SYBR® Green EMSA nucleic acid gel stain for RNA or DNA and SYPRO® Ruby EMSA protein gel stain for proteins. Because the nucleic acids and proteins are stained in the gel after electrophoresis, there is no need to prelabel the DNA or RNA with a radioisotope, biotin or a fluorescent dye before the binding reaction, and therefore there is no possibility that the label will interfere with protein binding. Staining takes only about 20 minutes for the nucleic acid stain, and about 4 hours for the subsequent protein stain, yielding results much faster than radioisotope labeling (which may require multiple exposure times) or chemiluminescence-based detection (which requires blotting and multiple incubation steps).

This kit also makes it possible to perform ratiometric measurements of nucleic acid and protein in the same band, providing more detailed information on the binding interaction. The signals from the two stains are linear over a broad range, allowing accurate determination of the amount of nucleic acid and protein, even in a single band, with detection limits of ~1 ng for nucleic acids and ~20 ng for proteins. Both stains can be detected using a standard 300 nm UV illuminator, a 254 nm epi-illuminator or a laser scanner (Figure 8.4.15). Digital images can easily be overlaid for a two-color representation of nucleic acid and protein in the gel. The EMSA Kit contains sufficient reagents for 10 nondenaturing polyacrylamide minigel assays, including:

- SYBR® Green EMSA nucleic acid gel stain
- SYPRO® Ruby EMSA protein gel stain
- Trichloroacetic acid, for preparing the working solution of SYPRO® Ruby EMSA protein gel stain
- Concentrated EMSA gel-loading solution
- Concentrated binding buffer
- Detailed protocols

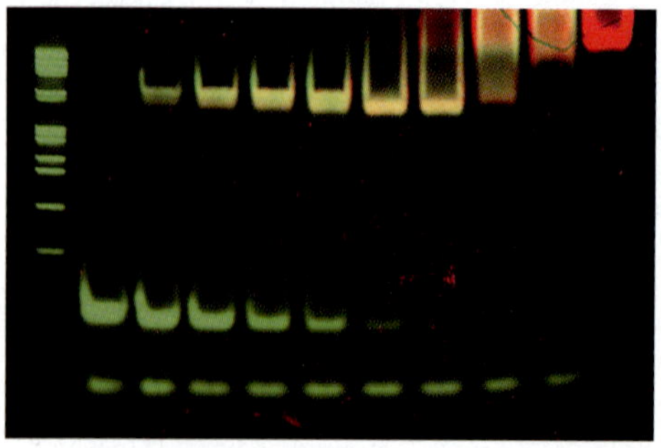

Figure 8.4.15 Titration of lac operator DNA with lac repressor protein. Increasing amounts of lac repressor protein were mixed with 40 ng of lac operator DNA, incubated for 20 minutes and then separated on a 6% nondenaturing polyacrylamide gel. The gel was stained with SYBR® Green EMSA stain (green) followed by SYPRO® Ruby EMSA stain (red), components of the Electrophoretic Mobility-Shift Assay Kit (E33075). After each staining, the image was documented using an FLA-3000 laser-based scanner (Fuji®) and the digital images pseudocolored and overlaid. Yellow bands indicate areas stained with both stains.

Other Nucleic Acid Stains for Gel-Staining Applications

DAPI (D1306, D3571, D21490; Section 8.1) reportedly provides a significantly more sensitive means of detecting dsDNA in agarose gels than ethidium bromide.[76] Selective detection of dsDNA in the presence of dsRNA in gels with DAPI has been reported.[77] Likewise, the Hoechst 33258 and Hoechst 33342 dyes (Section 8.1) have been used to detect DNA in the presence of RNA in agarose gels.[78] DNA conformational changes during gel electrophoresis have been investigated with acridine orange[79,80] (A1301, A3568; Section 8.1).

Safe Imager™ Blue-Light Transilluminator

The Safe Imager™ 2.0 blue-light transilluminator (G6600) is a blue-light transilluminator designed for viewing stained gels on the laboratory bench top. Light from the LED source inside the transilluminator passes through a blue filter, producing light with a narrow emission peak centered at approximately 470 nm. This 470 nm transillumination is effective for the excitation of the SYBR® Safe DNA gel stain, as well as many of our other nucleic acid and protein stains such as the SYBR® Gold, SYBR® Green I and SYBR® Green II nucleic acid stains and the SYPRO® Ruby, SYPRO® Orange, Coomassie Fluor™ Orange and Pro-Q® Diamond protein stains. Sensitivity obtained using this instrument is comparable to that obtained with a standard UV transilluminator. Unlike UV transilluminators, however, the Safe Imager™ blue-light transilluminator does not produce UV light and therefore does not require UV-protective equipment during use. Moreover, as compared to UV transillumination, blue-light transillumination results in dramatically increased cloning efficiencies.

The Safe Imager™ 2.0 blue-light transilluminator is supplied with viewing glasses (S37103), amber filter (G6601) and an international power cord (G6602), each of which is also available separately. The Safe Imager™ viewing glasses allow the user to visualize gel bands on the Safe Imager™ transilluminator without the use of the Safe Imager™ amber filter, thereby facilitating band excision.

SYBR® Photographic Filters

We offer a number of fluorescent reagents for staining nucleic acids and proteins in gels and on blots. Preeminent among these stains are our SYBR® Gold, SYBR® Green and SYBR® Safe nucleic acid gel stains (Section 8.4) and our SYPRO® protein stains for gels and blots (Section 9.3). To achieve optimal sensitivity with these exceptional fluorescent dyes, it is *essential* to photograph the gel or blot because the camera's

IMPORTANT NOTICE: The products described in this manual are covered by one or more Limited Use Label License(s). Please refer to the Appendix on page 971 and Master Product List on page 975. Products are For Research Use Only. Not intended for any animal or human therapeutic or diagnostic use.

integrating capability can make bands visible that are not detected by eye. Photographs should be taken using a photographic filter with spectral properties closely matched to those of the fluorescent dye used. We offer 75 mm × 75 mm gelatin filters (Figure 8.4.16) optimized for photographing stained gels or blots with a Polaroid® camera and Polaroid® 667 black-and-white print film. Note that these gelatin filters are generally not suitable for use with portable or stationary gel-documentation systems or with CCD cameras.

SYBR® Photographic Filter

To achieve optimal sensitivity using Polaroid® 667 black-and-white print film and UV illumination, DNA or RNA gels stained with SYBR® Gold, SYBR® Green I and SYBR® Green II nucleic acid gel stains should be photographed through the SYBR® photographic filter (S7569, Figure 8.4.17).

SYBR® Safe Photographic Filter

The SYBR® Safe photographic filter (S37100) is ideal for black-and-white photography of gels stained with the SYBR® Safe DNA gel stain. Note that the SYBR® Safe photographic filter is identical to the SYPRO® photographic filter (Figure 8.4.18).

Figure 8.4.16 Molecular Probes® 75 mm × 75 mm gelatin photographic filters for use with Polaroid® black-and-white print film photography.

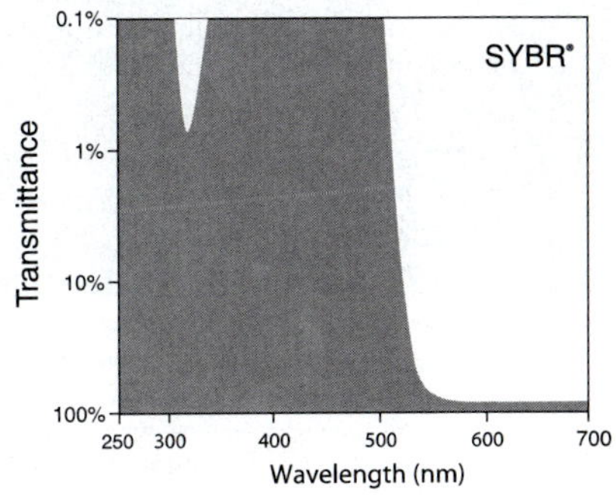

Figure 8.4.17 Transmittance profile of the SYBR® photographic filter (S7569).

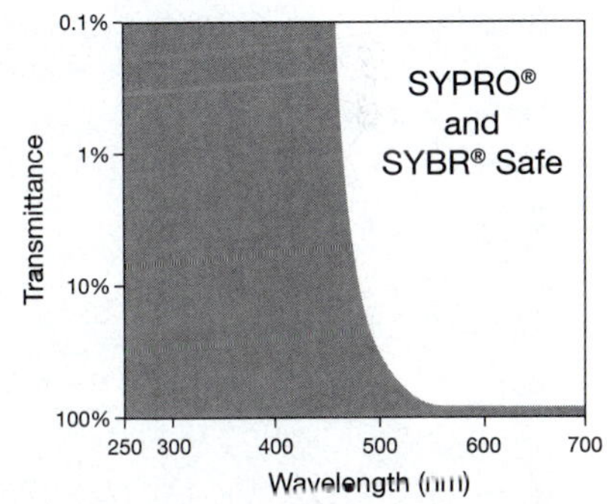

Figure 8.4.18 Transmittance profile of the SYBR® Safe photographic filter (S37100) and the SYPRO® photographic filter (S6656), which are identical.

Capillary Electrophoresis and Channel Electrophoresis

Capillary Electrophoresis

Capillary gel electrophoresis (CGE) performs separations of nucleic acids in a manner analogous to standard slab-gel electrophoresis, but with the advantages of faster run times, higher resolution and greater sensitivity. The use of on-line detection by laser-induced fluorescence (LIF) increases the sensitivity by several orders of magnitude over UV detection, eliminates the time spent staining and photographing the gel and allows for the possibility of automated sample processing. CGE-LIF is widely used for the separation and identification of DNA fragments and has increased the efficiency of DNA typing and forensics analyses.[81,82] Researchers are using several of our high-sensitivity nucleic acid stains with CGE for resolving similar-length DNA fragments.[83]

In CGE applications, the nucleic acid stain can be chosen to match available laser excitation sources;[84–89] furthermore, multiple dyes can be used to prestain samples, which can then be used for multiplexed capillary electrophoresis.[90]

- SYBR® Green I nucleic acid gel stain exhibits a large linear detection range and high resolution of DNA fragments from 100 to 1000 base pairs in length. A clinically applicable high-throughput screen was developed using SYBR® Green I stain to detect mutations in the methylenetetrahydrofolate reductase gene.[91,92]
- SYBR® Gold stain has sufficient sensitivity to detect electrophoretically separated nucleic acids from single cells.[93]
- Quant-iT™ OliGreen® ssDNA reagent (O7582, O11492; Section 8.3) has been employed to detect short single-stranded oligonucleotides using CGE with laser-induced fluorescence detection; formation of the fluorescent oligonucleotide complexes is accomplished on the column.[94]
- Both TOTO®-1 dye (T3600) and ethidium bromide (15585-011) have been used with capillary array electrophoresis for high-speed, high-throughput parallel separation of DNA fragments.[95,96]
- YOYO®-1 dye (Y3601) has been used wtih CGE to quantitate DNA complexes in polymerase chain reaction (PCR) mixtures.[84,97]
- Hepatitis B viral fragments have been detected by incorporation of submicromolar concentrations of either POPO™-3 dye (P3584) or ethidium homodimer-2 (E3599) in the detection buffer.[98] Sensitivity was as great as 3.9×10^{-16} M (390 attomolar) and increased with fragment length.
- YO-PRO®-1 dye (Y3603) has been used to develop a more rapid screening technique for identifying hypervariable regions in mitochondrial DNA. RFLP fragments were generated after PCR amplification and detected in CGE-LIF.[99] Automated CGE-LIF with the YO-PRO®-1 dye made it possible to replace time-consuming slab gel methods of analyzing variable number of tandem repeats (VNTR) in DNA typing labs.[100–102] YO-PRO®-3 dye (T3605) has proven useful for identifying single-sequence-repeat polymorphisms with high accuracy, using as little as 80 zeptomoles of sample DNA.[103]

IMPORTANT NOTICE: The products described in this manual are covered by one or more Limited Use Label License(s). Please refer to the Appendix on page 971 and Master Product List on page 975. Products are For Research Use Only. Not intended for any animal or human therapeutic or diagnostic use.

- CGE using ethidium bromide, SYBR® Green I or SYBR® Gold stain has been used in single-nucleotide polymorphism (SNP) analysis, making it possible to analyze as many as 96 samples in parallel.[104,105]
- CGE-LIF has been used for short tandem repeat (STR) genotyping using nucleic acids stained by an on-column labeling technique with either TO-PRO®-1, YO-PRO®-1, TOTO®-1 or YOYO®-1.[106]
- YO-PRO®-1 dye and ethidium bromide have both been used in heteroduplex analysis (HDA) by CGE.[107]
- The use of CGE-LIF with the YO-PRO®-1 dye makes it possible to accurately quantitate RNA transcripts from competitive RT-PCR.[108] CGE-LIF with the YO-PRO®-1 dye can also detect fragmented DNA from apoptotic cells, making it possible to use 1000–2000-fold fewer cells than are needed for ladder detection on conventional slab gels.[109]

Figure 8.4.19 DNA microarrays stained with nucleic acid stains for quality control. DNA microarrays were stained with dilutions of SYBR® Green II dye (left, green; S7568, S7564, S7586), POPO™-3 dye (center, orange, P3584) or SYTO® 59 dye (right, magenta, S11341) in aqueous buffer. The microarrays were imaged on a ScanArray® 5000XL microarray scanner (PerkinElmer LAS, Inc.) using the appropriate lasers and filters. Staining reveals the variable amounts of DNA spotted onto the different microarrays.

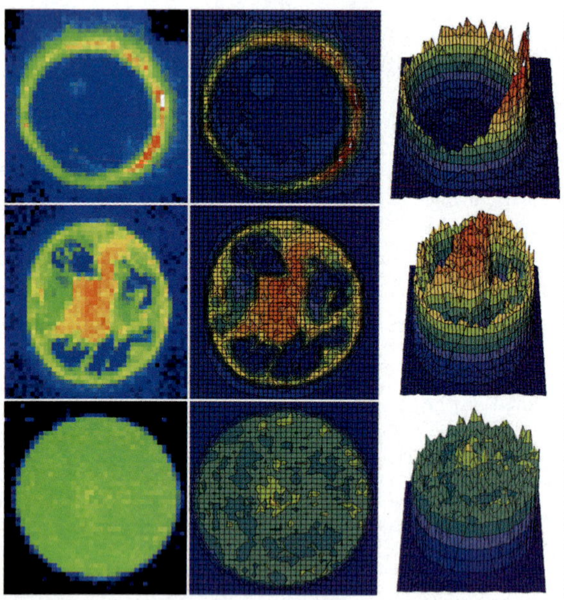

Figure 8.4.20 Panomer™ 9 oligodeoxynucleotides for quality control of microarray spotting. Three microarrays were made using three different spotting protocols. Each microarray was then hybridized to a Panomer™ 9 oligonucleotide, conjugated with either the Alexa Fluor® 488 dye (top) or the Alexa Fluor® 546 dye (middle and bottom), washed and imaged using a ScanArray® 5000XL microarray scanner (PerkinElmer LAS, Inc.). One representative spot from each slide was selected for comparison purposes. The spots were analyzed using Metamorph software (Universal Imaging, Inc.), and the data is presented as a three-dimensional graph with high-intensity areas as peaks and low-intensity areas as valleys. For further clarification, the graphs are color coded so that the highest intensity areas are red and the lowest intensity areas are blue. The comparison shows that the Panomer™ 9 oligodeoxynucleotides are ideal for microanalysis of spot morphology on DNA microarrays.

- Using POPO™-3 dye (P3584) or ethidium homodimer-2, researchers have been able to detect as little as 3.9×10^{-13} M duck hepatitis B virus.[110]

As an alternative to using nucleic acid stains for CGE, amine- or thiol-derivatized oligonucleoctides can be chemically derivatized pre- or post-separation with many of the dyes described in Chapter 1 and Chapter 2. The thiol-reactive Alexa Fluor®, BODIPY®, fluorescein and Oregon Green® dyes are particularly suitable for labeling thiolated oligonucleotides and for applications that use ultrasensitive laser-scanning techniques.[111] Several papers have been published on the separation of fluorescent oligonucleotides by capillary electrophoresis.[112]

Channel Electrophoresis

Similar in concept to capillary electrophoresis, channel electrophoresis on microchips has the potential to provide even higher throughput by using completely automated nucleic acid analysis. Our intensely fluorescent nucleic acid dyes make sensitive on-line detection possible. SYBR® Green I dye was used to detect amplified DNA on a nanoliter device that mixes DNA samples, amplifies DNA fragments and separates the products in a channel for on-line detection.[113] TO-PRO®-1 dye (T3602) was used to detect DNA fragments from bacterial DNA that had been extracted, amplified and separated in channels on the same microchip.[114] YOYO®-1 dye (Y3601) has been used to detect as little as a few zeptomoles (10^{-21} mole) of DNA fragments on a chip device,[115] and YO-PRO®-1 dye (Y3603) made it possible to distinguish triplet repeat DNA fragments in a 6 mm channel in only 12 seconds.[116] A novel radial microchip device simultaneously separates 96 DNA samples prestained with YOYO®-1 dye.[117] The TO-PRO®-3 dye has been used to detect DNA in a polycarbonate channel electrophoresis device.[118] Fluorescence-based sequencing using dye-labeled primers (see above) in capillary electrophoresis chips allowed sequencing of ~200 bases in as little as 10 minutes.[119]

Quality Control Testing on Microarrays and Blots

Whether using northern blots, macroarrays on membranes or microarrays on glass slides, it is important to normalize the signal to the amount of nucleic acid on the solid support. Differences in purification, loading and transfer may create differences in RNA levels on blots and, even with sophisticated robotics, direct spotting techniques for creating arrays of nucleic acids on solid supports vary widely in reproducibility (Figure 8.4.19, Figure 8.4.20). A method to qualitatively determine the amount of nucleic acid on a support is desirable for quality control purposes. Quantitative data that can be used for signal normalization is even more useful, making it possible to validate perceived changes in RNA expression between samples, despite differences in the nucleic acid levels on the solid support. Often, hybridization of a "constitutively expressed" RNA sequence is used for normalization on northern blots. However, it is sometimes discovered that the level of such RNA sequences is not constant through changing physiological states of a cell or tissue. Direct measurement of the levels of nucleic acid spotted on the slide provides a reliable method for normalization and can also be used to assess the amount of nucleic acid remaining on a support after stripping off a probe and before reusing the blot or array.

In addition to providing a means of normalizing hybridization signals, staining denatured DNA or RNA directly on filter membranes after blotting protocols provides for more accurate comparison of the sample

IMPORTANT NOTICE: The products described in this manual are covered by one or more Limited Use Label License(s). Please refer to the Appendix on page 971 and Master Product List on page 975. Products are For Research Use Only. Not intended for any animal or human therapeutic or diagnostic use.
www.invitrogen.com/probes

to molecular weight markers and eliminates guesswork about transfer efficiency. However, direct staining on blotting membranes has not been widely used because the most common methods for detecting denatured DNA or RNA—ethidium bromide[120] or methylene blue staining—give rise to high background fluorescence. Silver staining or gold staining followed by silver enhancement provides 10- to 100-fold better sensitivity than ethidium bromide but is expensive, time-consuming and tedious. Also, because of the affinity of gold for sulfur, only agarose gels containing less than 0.1% sulfate are suitable for use with gold staining; higher amounts of sulfate invariably result in unacceptably high background signals.

Nucleic Acid Stains for Standardizing Microarrays

The simplest technique for comparing the amounts of DNA spotted onto arrays is to use a fluorescent nucleic acid stain. Molecular Probes® nucleic acid stains exhibit a strong fluorescence signal when bound to nucleic acids, providing an easy and effective method for measuring the amount of nucleic acids on solid supports (Figure 8.4.19). Nucleic acid stains that have been used effectively for microarrays include:

- POPO™-3 dye (P3584)
- SYBR® Green II RNA stain (S7568, S7564, S7586)
- SYTO® 59 dye (S11341, Section 8.1)

Other dyes in these families (Section 8.1) should also prove useful in this application. When staining nucleic acids on solid supports with nucleic acid stains, it is important to choose a dye that matches the light sources and filter sets available in the image analysis system or the array reader. For example, the POPO™-3 dye has a maximum fluorescence excitation at 534 nm and a maximum fluorescence emission at 570 nm, which is compatible with filter sets for the Alexa Fluor® 546 dye, Alexa Fluor® 555 dye or Cy®3 dyes. In one case in which POPO™-3 staining was used to determine the number of spots on the microarray that contained PCR products, 1281 of 9216 spots did not display a significant POPO™-3 signal due to poor growth of the bacterial clones, failure of the PCR amplification or improper printing of the spots due to irregularities on the array.[121] When testing dyes to use for normalization of signals, it is important to carefully test several dilutions of the dye to determine the one that gives the best linear response over the range of DNA in the spot. It is also important to determine any effects of the dye on subsequent hybridization.

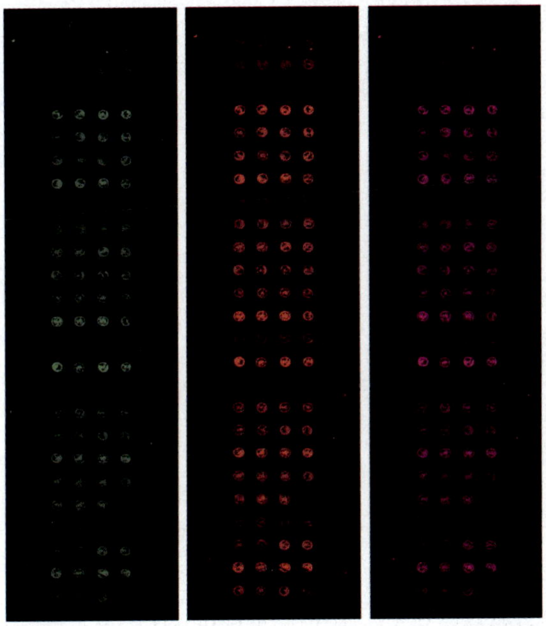

Figure 8.4.21 DNA microarray hybridized to Panomer™ 9 random oligodeoxynucleotides. A DNA microarray slide was hybridized sequentially with one of three different Panomer™ 9 random oligodeoxynucleotides at room temperature for two minutes. After each hybridization, the slide was washed first in 2X SSC and 0.2% SDS and then in 1X SSC. After drying, the slide was imaged on a ScanArray® 5000XL microarray reader (PerkinElmer LAS, Inc.), using appropriate lasers and filter sets. After imaging, the Panomer™ 9 oligodeoxynucleotide was stripped from the microarray by incubation in deionized water for one minute at room temperature and the microarray was hybridized to another Panomer™ 9 oligodeoxynucleotide. From left to right, the images show the array hybridized to Panomer™ 9 oligodeoxynucleotides labeled with Alexa Fluor® 488 dye (P21680), Alexa Fluor® 546 dye (P21681) and Alexa Fluor® 647 dye (P21686). Each image has been pseudocolored to indicate the different dyes. The hybridization results reveal the variable amounts of DNA spotted onto this microarray.

Panomer™ Random-Sequence Oligonucleotides

Fluorescently labeled random-sequence oligonucleotides—such as the Panomer™ 9 random oligodeoxynucleotides—provide an alternative method for assessing the level of nucleic acids immobilized on solid supports:

- Panomer™ 9 random oligodeoxynucleotide, Alexa Fluor® 488 conjugate (P21680)
- Panomer™ 9 random oligodeoxynucleotide, Alexa Fluor® 546 conjugate (P21681)
- Panomer™ 9 random oligodeoxynucleotide, Alexa Fluor® 555 conjugate (P21687)
- Panomer™ 9 random oligodeoxynucleotide, Alexa Fluor® 647 conjugate (P21686)

This method assays the capability of spotted DNA to hybridize, making it possible to determine if hybridization efficiency varies across the array[122] (Figure 8.4.21). These nine-base, random-sequence Panomer™ oligodeoxynucleotides are covalently labeled on the 5′-end with a fluorescent dye. The variety of available fluorescent dyes makes it possible to use any fluorescence channel of interest and to compare relative signal intensities per spot in several different channels (Figure 8.4.22). It is also possible to use Panomer™ 9 oligodeoxynucleotides for quality control of spotting techniques or to assay the stability of DNA spots after the array is subjected to washing, boiling, hybridization or other conditions (Figure 8.4.21).

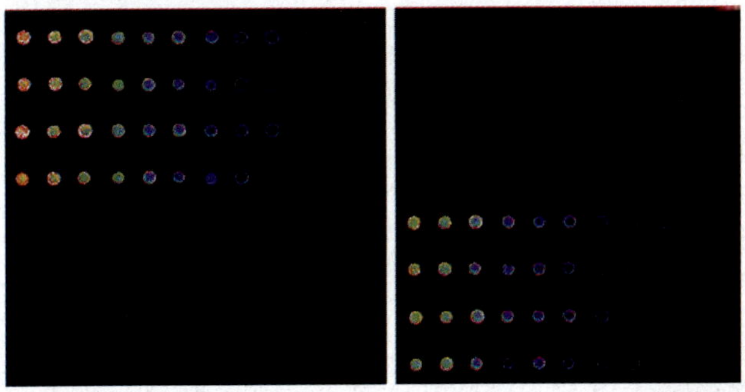

Figure 8.4.22 Microarray spotted with Panomer™ 9 oligodeoxynucleotides labeled with Alexa Fluor® 546 dye (P21681) and Alexa Fluor® 647 dye (P21686). Equal amounts of two Panomer™ 9 random oligodeoxyribonucleotides were spotted directly onto microscope slides and the fluorescence documented using two different channels (left and right panels) of a ScanArray® 5000XL microarray reader (PerkinElmer LAS, Inc.). The Alexa Fluor® 546 Panomer™ 9 random oligodeoxyribonucleotide signal (left panel) was completely separated from the Alexa Fluor® 647 Panomer™ 9 random oligodeoxyribonucleotide signal (right panel).

IMPORTANT NOTICE: The products described in this manual are covered by one or more Limited Use Label License(s). Please refer to the Appendix on page 971 and Master Product List on page 975. Products are For Research Use Only. Not intended for any animal or human therapeutic or diagnostic use.

REFERENCES

1. Fluorescent and Luminescent Probes for Biological Activity, 2nd Ed., Mason WT, Ed., (1999) p. 51; **2.** Anal Biochem (1999) 268:278; **3.** Anal Biochem (1966) 17:100; **4.** Methods Biochem Anal (1971) 20:41; **5.** Hum Mutat (1999) 14:289; **6.** Mol Pathol (1998) 51:342; **7.** Biochim Biophys Acta (2000) 1509:176; **8.** J Biol Chem (2000) 275:9970; **9.** Nucleic Acids Res (2004) 32:e103; **10.** Mutat Res (1999) 439:37; **11.** Electrophoresis (2007) 28:749; **12.** Clin Chem Lab Med (2005) 43:841; **13.** Electrophoresis (1998) 19:2416; **14.** J Biol Chem (2000) 275:288; **15.** Proc Natl Acad Sci U S A (1997) 94:12419; **16.** J Virol Methods (2000) 85:175; **17.** Biotechniques (1985) 3:452; **18.** Mol Cell Probes (1995) 9:145; **19.** Electrophoresis (2004) 25:2439; **20.** PCR Methods Appl (1995) 4:234; **21.** Immunity (1996) 5:377; **22.** Cell (1997) 88:757; **23.** J Biol Chem (1998) 273:7643; **24.** Biotechniques (1997) 22:1107; **25.** Anal Biochem (1999) 275:116; **26.** Anal Biochem (1996) 237:204; **27.** J Microbiol Methods (1999) 35:23; **28.** J Clin Microbiol (1999) 37:987; **29.** Biotechniques (1995) 19:223; **30.** Biotechniques (1997) 22:976; **31.** Nat Biotechnol (1998) 16:91; **32.** Biochim Biophys Acta (1997) 1360:193; **33.** Cell Res (2000) 10:71; **34.** Methods Cell Sci (1995) 17:1; **35.** Nucleic Acids Res (1997) 25:2945; **36.** Biotechniques (1999) 27:34; **37.** Cytometry (1986) 7:508; **38.** Biotechniques (1999) 26:46; **39.** J Chinese Biochem Soc (1995) 32:1; **40.** J Biol Chem (2000) 275:6945; **41.** J Biol Chem (2000) 275:32846; **42.** Appl Environ Microbiol (1996) 62:1969; **43.** Cancer Res (1998) 58:4227; **44.** Diagnostic Mol Pathol (1996) 5:260; **45.** Genomics (1989) 5:874; **46.** Trends Genet (1992) 8:49; **47.** Nucleic Acids Res (1991) 19:3154; **48.** Nucleic Acids Res (1991) 19:2500; **49.** Nucleic Acids Res (1991) 19:405; **50.** Nucleic Acids Res (1993) 21:3637; **51.** Anal Biochem (1996) 236:373; **52.** Proc Natl Acad Sci U S A (1997) 94:10745; **53.** J Biol Chem (1991) 266:5417; **54.** Nucleic Acids Res (1991) 19:1521; **55.** J Mol Biol (1965) 13:269; **56.** Anal Biochem (1990) 190:331; **57.** Biopolymers (1989) 28:1491; **58.** Science (1989) 243:203; **59.** J Chromatogr A (2000) 871:289; **60.** J Chromatogr A (1998) 828:481; **61.** Electrophoresis (1999) 20:497; **62.** Nucleic Acids Res (1992) 20:2803; **63.** Nucleic Acids Res (1991) 19:327; **64.** Proc Natl Acad Sci U S A (1990) 87:3851; **65.** Nucleic Acids Res (1993) 21:5720; **66.** Nature (1992) 359:859; **67.** Biotechniques (1991) 10:616; **68.** Nucleic Acids Res (1995) 23:1215; **69.** Methods Enzymol (1993) 217:414; **70.** Nucleic Acids Res (1995) 23:2413; **71.** Rev Sci Instrum (1994) 65:807; **72.** Mol Cell Probes (1994) 8:245; **73.** Mod Pathol (1994) 7:385; **74.** Biotechniques (1993) 15:274; **75.** Biotechniques (1996) 20:708; **76.** J Biochem Biophys Methods (1982) 6:95; **77.** Nucleic Acids Res (1979) 6:3535; **78.** Nucleic Acids Res (1987) 15:10589; **79.** Annu Rev Biophys Biophys Chem (1991) 20:415; **80.** Biochemistry (1990) 29:3396; **81.** Forensic Sci Int (1998) 92:89; **82.** Electrophoresis (1998) 19:2695; **83.** Anal Chem (1992) 64:1737; **84.** J Chromatogr B Biomed Appl (1994) 658:271; **85.** Anal Biochem (1995) 231:359; **86.** Anal Chem (2007) 79:7691; **87.** J Chromatogr A (1994) 669:205; **88.** Appl Theor Electrophor (1993) 3:235; **89.** J Microcolumn Separation (1993) 5:275; **90.** Anal Chem (1997) 69:1355; **91.** Clin Chem (1997) 43:267; **92.** Biotechniques (1997) (23):58; **93.** J Chromatogr A (2001) 911:269; **94.** J Chromatogr A (1996) 755:271; **95.** Anal Chem (1994) 66:1424; **96.** Anal Biochem (1993) 215:163; **97.** Anal Biochem (1995) 224:140; **98.** Vis Neurosci (1992) 8:295; **99.** Electrophoresis (1998) 19:119; **100.** Electrophoresis (1998) 19:80; **101.** Chem Pharm Bull (Tokyo) (1998) 46:294; **102.** Electrophoresis (1995) 16:974; **103.** Genome Res (1996) 6:893; **104.** Biotechniques (2001) 30:334, 338; **105.** Anal Chem (2000) 72:2499; **106.** J Chromatogr B Biomed Sci Appl (1999) 732:365; **107.** Anal Chem (2000) 72:5483; **108.** Biotechniques (1998) 25:130; **109.** J Chromatogr A (1995) 700:151; **110.** J Chromatogr A (1999) 853:309; **111.** J Am Chem Soc (1994) 116:7801; **112.** Anal Chem (2000) 72:5583; **113.** Science (1998) 282:484; **114.** Anal Chem (1998) 70:5172; **115.** Anal Chem (1997) 69:3451; **116.** Electrophoresis (2000) 21:176; **117.** Anal Chem (1999) 71:5354; **118.** Anal Chem (2001) 73:4196; **119.** Anal Chem (1995) 67:3676; **120.** J Neurosci Methods (1992) 42:211; **121.** Nucleic Acids Res (2002) 30:e116; **122.** Nucleic Acids Res (2003) 31:e18.

PRODUCT LIST 8.4 NUCLEIC ACID DETECTION ON GELS, BLOTS AND ARRAYS

Cat. No.	Product	Quantity
E33075	Electrophoretic Mobility-Shift Assay (EMSA) Kit *with SYBR® Green and SYPRO® Ruby EMSA stains* *10 minigel assays*	1 kit
E1169	ethidium homodimer-1 (EthD-1)	1 mg
E3599	ethidium homodimer-2 (EthD-2) *1 mM solution in DMSO*	200 µL
P21680	Panomer™ 9 random oligodeoxynucleotide, Alexa Fluor® 488 conjugate	10 nmol
P21681	Panomer™ 9 random oligodeoxynucleotide, Alexa Fluor® 546 conjugate	10 nmol
P21687	Panomer™ 9 random oligodeoxynucleotide, Alexa Fluor® 555 conjugate	10 nmol
P21686	Panomer™ 9 random oligodeoxynucleotide, Alexa Fluor® 647 conjugate	10 nmol
P3584	POPO™-3 iodide (534/570) *1 mM solution in DMF*	200 µL
S37102	Safe Imager™ blue-light transilluminator	each
S37104	Safe Imager™ international power cord *replacement*	each
S37103	Safe Imager™ viewing glasses	each
S11494	SYBR® Gold nucleic acid gel stain *10,000X concentrate in DMSO*	500 µL
S7563	SYBR® Green I nucleic acid gel stain *10,000X concentrate in DMSO*	500 µL
S7567	SYBR® Green I nucleic acid gel stain *10,000X concentrate in DMSO*	1 mL
S7585	SYBR® Green I nucleic acid gel stain *10,000X concentrate in DMSO* *special packaging*	20 x 50 µL
S7564	SYBR® Green II RNA gel stain *10,000X concentrate in DMSO*	500 µL
S7568	SYBR® Green II RNA gel stain *10,000X concentrate in DMSO*	1 mL
S7586	SYBR® Green II RNA gel stain *10,000X concentrate in DMSO* *special packaging*	20 x 50 µL
S7580	SYBR® Green Nucleic Acid Gel Stain Starter Kit	1 kit
S7569	SYBR® photographic filter	each
S33102	SYBR® Safe DNA gel stain *10,000X concentrate in DMSO*	400 µL
S33100	SYBR® Safe DNA gel stain in 0.5X TBE	1 L
S33101	SYBR® Safe DNA gel stain in 0.5X TBE	4 L
S33111	SYBR® Safe DNA gel stain in 1X TAE	1 L
S33112	SYBR® Safe DNA gel stain in 1X TAE	4 L
S33110	SYBR® Safe DNA Gel Stain Starter Kit *with 1 L of SYBR® Safe DNA gel stain in 0.5X TBE (S33100) and one photographic filter (S37100)*	1 kit
S37100	SYBR® Safe photographic filter	each
T3602	TO-PRO®-1 iodide (515/531) *1 mM solution in DMSO*	1 mL
T3605	TO-PRO®-3 iodide (642/661) *1 mM solution in DMSO*	1 mL
T3600	TOTO®-1 iodide (514/533) *1 mM solution in DMSO*	200 µL
T3604	TOTO®-3 iodide (642/660) *1 mM solution in DMSO*	200 µL
15585-011	UltraPure™ Ethidium Bromide *10 mg/ml*	10 mL
Y3603	YO-PRO®-1 iodide (491/509) *1 mM solution in DMSO*	1 mL
Y3601	YOYO®-1 iodide (491/509) *1 mM solution in DMSO*	200 µL

IMPORTANT NOTICE: The products described in this manual are covered by one or more Limited Use Label License(s). Please refer to the Appendix on page 971 and Master Product List on page 975. Products are For Research Use Only. Not intended for any animal or human therapeutic or diagnostic use.

molecular probes® | invitrogen™ by life technologies™

NINE

CHAPTER 9
Protein Detection and Analysis

IMPORTANT NOTICE: The products described in this manual are covered by one or more Limited Use Label License(s). Please refer to the Appendix on page 971 and Master Product List on page 975. Products are For Research Use Only. Not intended for any animal or human therapeutic or diagnostic use.

IMPORTANT NOTICE: The products described in this manual are covered by one or more Limited Use Label License(s). Please refer to the Appendix on page 971 and Master Product List on page 975. Products are For Research Use Only. Not intended for any animal or human therapeutic or diagnostic use.

molecular probes® | invitrogen™ by life technologies™

9.1 Introduction to Protein Detection

Proteomics: A Rapidly Developing Field

For decades, polyacrylamide gel electrophoresis and related blotting techniques have formed the core technologies for protein analysis. Traditionally, these technologies have been paired with chromogenic dye–based protein detection techniques, such as silver or Coomassie brilliant blue staining. With the rapid growth of proteomics,[1,2] however, the limitations and experimental disadvantages of absorption-based detection technologies and labor-intensive silver staining techniques have become glaringly apparent. The field of proteomics requires new, highly quantitative electrophoresis and blotting techniques that can interface seamlessly with improved microanalysis methods and that can perform in an increasingly high-throughput environment. These requirements are particularly important for quantitative proteomics and multiplexing techniques.

Fluorescence- or luminescence-based detection technologies offer the opportunity for multicolor labeling, making multiplexed analysis possible (Figure 9.1.1). In particular, Molecular Probes® fluorescent and luminescent protein stains enable the identification of specific protein modifications (for example, phosphorylation, glycosylation and epitope tags) within the context of the entire protein profile. Simultaneous measurement of several variables greatly increases the amount of data that can be collected in a single experiment. In addition, directly comparing multiple measurements leads to more controlled experiments, more accurate data and fewer ambiguities. The detection characteristics of our protein stains greatly streamline protocols for proteome analysis and promise to bring to proteomics the same capability for rapid, large-scale data acquisition that fluorescence has brought to genomics and other fields.

Molecular Probes® Detection Technology for Proteomics

We are meeting the demands of the rapidly expanding field of proteomics through the development of fluorescence- and luminescence-based detection methods for proteins in solutions and on gels, blots and microarrays. We are continuing to develop new reagents and detection methods for proteins and their modifications, such as phosphorylation, glycosylation and epitope tags, as well as improved methods of separating and analyzing peptides and proteins. Our advanced technologies are compatible with modern needs for sensitivity, specificity, sequencing compatibility and automatability. Application of our unique detection reagents requires minimal investment in labor, as compared with older technologies, while significantly increasing throughput, reducing total cost and improving accuracy. Furthermore, the greater sensitivity and linearity of most of our premier reagents makes it possible to do quantitative proteomics and perform comparative protein expression measurements on very small samples.

In this chapter, Section 9.2 includes Molecular Probes® reagents and kits (including the Quant-iT™, NanoOrange®, CBQCA and EZQ® reagents) for quantitating total protein in solution. Section 9.3 includes the important SYPRO® stains for detecting and quantitating total protein on gels and blots. Reagents for the qualitative and quantiative detection of phosphorylation, glycosylation and other post-translational modifications are described in Section 9.4 and include:

- Click-iT® reagents for detecting nascent protein synthesis, as well as post-translational modifications
- Pro-Q® Diamond phosphoprotein gel and blot stains—a breakthrough technology for selectively detecting phosphoproteins in gels and on blots
- Pro-Q® Emerald 300 and Pro-Q® Emerald 488 glycoprotein gel and blot stains—effective and easy-to-use stains for detecting periodate-oxidized glycoproteins in gels and on blots

Section 9.5 describes reagents used in the synthesis of fluorophore- or hapten-labeled peptides and fluorogenic protease substrates, as well as in peptide analysis and sequencing.

REFERENCES

1. J Proteome Res (2007) 6:1418; **2.** Proteomics (2006) 6:5385.

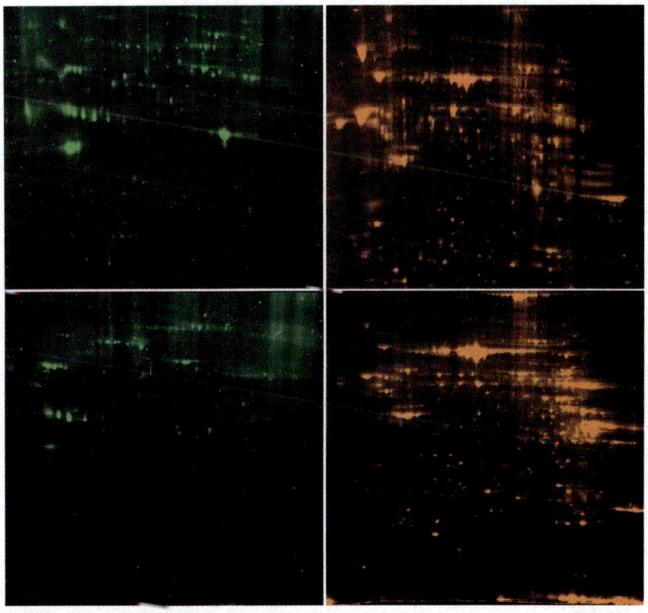

Figure 9.1.1 Two-dimensional protein gels of tumor versus normal cells stained for glycoproteins and total proteins. Lysates from rat liver tumor cells (top panels) or rat normal liver cells (bottom panels) were run on identical 2D gels. Following electrophoresis, the gels were stained with the Pro-Q® Emerald 300 glycoprotein detection reagent (left panels) (available in the Pro-Q® Emerald 300 Kits P21855 and P21857). After documentation of the fluorescence signal, the gel was stained with the SYPRO® Ruby protein gel stain (right panels) (S12000, S12001, S21900).

IMPORTANT NOTICE: The products described in this manual are covered by one or more Limited Use Label License(s). Please refer to the Appendix on page 9/1 and Master Product List on page 975. Products are For Research Use Only. Not intended for any animal or human therapeutic or diagnostic use.

9.2 Protein Quantitation in Solution

Several colorimetric methods have been described for quantitating proteins in solution, including the widely used Bradford, Lowry and BCA (bicinchoninic acid) assays (Table 9.1). Because they rely on absorption-based measurements, however, these methods are inherently limited in both sensitivity and effective range. We have developed four unique fluorometric methods for quantitating proteins in solution—the Quant-iT™ Protein Assay Kit (Q33210), the NanoOrange® Protein Quantitation Kit (N6666), the CBQCA Protein Quantitation Kit (C6667) and the EZQ® Protein Quantitation Kit (R33200)—that outperform these conventional methods (Table 9.1). We also offer the Qubit® fluorometer (Q32857), which is designed to work seamlessly with the Quant-iT™ Assay Kits, as well as several other fluorescent reagents useful for protein detection in solution.

Table 9.1 A comparison of reagents for detecting and quantitating proteins in solution.

Assay	Detection Wavelength(s) (nm) *	Sensitivity and Effective Range	Mechanism of Action	Notes
Quant-iT™ protein quantitation assay (Q33210, Q33211, Q33212)	470/570	Quasi-linear from 0.5 to 4 µg in a 200 µL assay volume, with a sample volume of 1–20 µL	Binds to detergent coating on proteins and hydrophobic regions of proteins; the unbound dye is nonfluorescent	• Extremely fast and easy—just add sample to diluted dye and read fluorescence • High sensitivity • Little protein-to-protein variation • Compatible with salts, solvents, 2-mercaptoethanol, amino acids and DNA, but not detergents
NanoOrange® protein quantitation assay (N6666)	470/570	10 ng/mL to 10 µg/mL	Binds to detergent coating on proteins and hydrophobic regions of proteins; the unbound dye is nonfluorescent	• High sensitivity • Little protein-to-protein variation • Rapid assay with a simple procedure • Compatible with reducing agents, but not detergents
CBQCA protein quantitation assay (C6667)	450/550	10 ng/mL to 150 µg/mL	Reacts with primary amine groups on proteins in the presence of cyanide or thiols; the unreacted dye is nonfluorescent	• High sensitivity • Sensitivity depends on the number of amines present • Linear over an extended range of protein concentration • Compatible with detergents and lipophilic proteins • Not compatible with buffers containing amines or thiols
EZQ® protein quantitation assay (R33200)	280 and 450/618	50 µg/mL to 5 mg/mL, with a sample volume of 1 µL	Binds electrostatically to basic amino acids, supplemented by additional hydrophobic interactions	• Ideal for determining protein concentration prior to electrophoresis • Solid-phase format designed for high-throughput analysis • Little protein-to-protein variation • Compatible with detergents, reducing agents, urea and tracking dyes
Bradford assay [1] (Coomassie brilliant blue)	595	1 µg/mL to 1.5 mg/mL	Directly binds specific amino acids and protein tertiary structures; the dye's color changes from brown to blue	• High protein-to-protein variation • Not compatible with detergents • Rapid assay • Useful when accuracy is not crucial
BCA method [2] (bicinchoninic acid)	562	0.5 µg/mL to 1.2 mg/mL	Cu^{2+} is reduced to Cu^{+} in the presence of proteins at high pH; the BCA reagent chelates Cu^{+} ions, forming purple-colored complexes	• Compatible with detergents, chaotropes and organic solvents • Not compatible with reducing agents • The sample must be read within 10 minutes
Lowry assay [3] (biuret reagent plus Folin–Ciocalteu reagent)	750	1 µg/mL to 1.5 mg/mL	Cu^{2+} is reduced to Cu^{+} in the presence of proteins at high pH; the biuret reagent chelates the Cu^{+} ion, then the Folin–Ciocalteu reagent enhances the blue color	• Lengthy procedure with carefully timed steps • Not compatible with detergents or reducing agents
Fluorescamine [4–7] (F2332, F20261)	390/475	0.3 µg/mL to 13 µg/mL	Reacts with primary amine groups on proteins; unbound dye is nonfluorescent	• Sensitivity depends on the number of amines present • Reagent is unstable • Not compatible with amine-containing buffers
OPA [8–10] (o-phthaldialdehyde) (P2331MP)	340/455	0.2 µg/mL to 25 µg/mL	Reacts with primary amine groups on proteins in the presence of 2-mercaptoethanol; unbound dye is nonfluorescent	• Sensitivity depends on the number of amines present • Not compatible with amine-containing buffers • Low cost
UV absorption [11]	280	10 µg/mL to 50 µg/mL or 50 µg/mL to 2 mg/mL	Peptide bond absorption; tryptophan and tyrosine absorption	• Sensitivity depends on the number of aromatic amino acid residues present • Nondestructive • Low cost

* Excitation and emission wavelength maxima or absorption wavelength maximum, in nm.

1. Anal Biochem (1976) 72:248; **2.** Anal Biochem (1985) 150:76; **3.** J Biol Chem (1951) 193:265; **4.** Science (1972) 178:871; **5.** Clin Chim Acta (1986) 157:73; **6.** J Lipid Res (1986) 27:792; **7.** Anal Biochem (1993) 214:346; **8.** Anal Biochem (1981) 115:203; **9.** Biotechniques (1986) 4:130; **10.** J Immunol Methods (1994) 172:141; **11.** Methods Enzymol (2009) 463:73.

IMPORTANT NOTICE: The products described in this manual are covered by one or more Limited Use Label License(s). Please refer to the Appendix on page 971 and Master Product List on page 975. Products are For Research Use Only. Not intended for any animal or human therapeutic or diagnostic use.

Quant-iT™ Protein Assay Kits

Overview of the Quant-iT™ Assay Kits

The Quant-iT™ family of assay kits provides state-of-the-art reagents for sensitive and selective quantitation of protein, DNA or RNA (Table 8.8) samples using a standard fluorescence microplate reader (Figure 9.2.1) or the Qubit® flurometer. These kits have been specially formulated with ready-to-use buffers, prediluted standards and easy-to-follow instructions, to help make quantitation both accurate and extremely easy. Each Quant-iT™ assay is:

- **Ready to use.** Only the dye is diluted in the supplied buffer; no dilution of standards or buffer required.
- **Easy to perform.** Just add the sample to the diluted dye and read the fluorescence.
- **Highly sensitive.** The Quant-iT™ assays are orders of magnitude more sensitive than UV absorbance measurements.
- **Highly selective.** Separate kits are available for quantitating protein, DNA or RNA (Section 8.3), with minimal interference from common contaminants.
- **Precise.** CVs are generally less than 5% for typical users.

The user-friendly Qubit® fluorometer has been designed to work seamlessly with the Quant-iT™ Assay Kits. The combination of the Quant-iT™ Assay Kits and the Qubit® fluorometer produces a completely integrated quantitation platform. Moreover, because the fluorescent dye in each Quant-iT™ Kit matches common fluorescence excitation and emission filter sets in microplate readers, these assay kits are ideal for high-throughput environments, as well as for small numbers of samples.

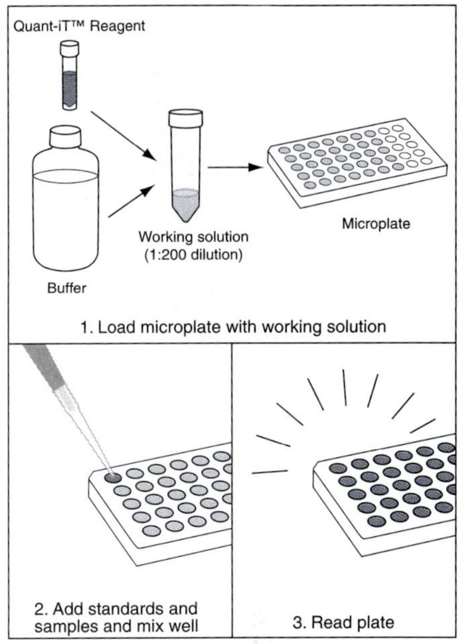

Figure 9.2.1 Microplate assay implementation of DNA, RNA or protein quantitation with the Quant-iT™ Assay Kits.

Quant-iT™ Protein Assay Kits

The Quant-iT™ Protein Assay Kit (Q33210) simplifies protein quantitation without sacrificing sensitivity.[1–3] This protein assay exhibits a detection range between 0.25 and 5 µg protein, and the response curve is sigmoidal (quasi-linear from 0.5 to 4 µg) with little protein-to-protein difference in signal intensity (Figure 9.2.2). Common contaminants, including salts, solvents, reducing agents (dithiothreitol, 2-mercaptoethanol), amino acids, nucleotides and DNA, are well tolerated in this assay; however, it is not compatible with detergents; slight modifications in the procedure are required for other contaminants. Each Quant-iT™ Protein Assay Kit contains:

- Quant-iT™ protein reagent
- Quant-iT™ protein buffer
- A set of eight prediluted bovine serum albumin (BSA) standards between 0 and 500 ng/µL
- Easy-to-follow instructions

Sufficient reagents are provided to perform 1000 assays, based on a 200 µL assay volume in a 96-well microplate format; this assay can also be adapted for use in cuvettes or 384-well microplates. The fluorescence signal exhibits excitation/emission maxima of 470/570 nm and is stable for three hours at room temperature. The Quant-iT™ protein reagent is an improved formulation of Molecular Probes® NanoOrange® reagent, which is described below.

We also offer Quant-iT™ Protein Assay Kits specifically designed for use with the Qubit® fluorometer (100-assay size, Q33211; 500-assay size, Q33212). Similar to the general Quant-iT™ Protein Assay Kit described above, these kits provide:

- Quant-iT™ protein reagent
- Quant-iT™ protein buffer
- Three pre-diluted protein standards
- Easy-to-follow instructions

The assay steps are simple: dilute the reagent using the buffer provided, add the sample (any volume between 1 µL and 20 µL is acceptable) and read the concentration using the Qubit® fluorometer. The assay is performed at room temperature, and the signal is stable for 3 hours. This assay is accurate for initial sample concentrations from 12.5 µg/mL to 5 mg/mL.

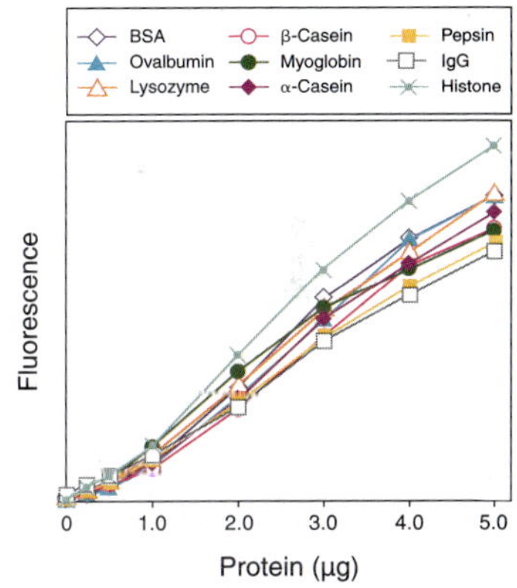

Figure 9.2.2 Low protein-to-protein variation in the Quant-iT™ protein assay. Solutions of the following proteins were prepared, diluted and assayed with the Quant-iT™ Protein Assay Kit (Q33210): bovine serum albumin (BSA), chicken-egg ovalbumin, chicken-egg lysozyme, bovine-milk β-casein, equine myoglobin, bovine-milk α-casein, porcine pepsin, mouse immunoglobulin G (IgG) and calf-thymus histone. Fluorescence was measured at 485/590 nm and plotted versus the mass of protein sample. Background fluorescence has not been subtracted.

Qubit® Fluorometer

Figure 9.2.3 The Qubit® fluorometer. The Qubit® fluorometer (Q32857) is designed to work seamlessly with the Quant-iT™ assays for DNA, RNA and protein quantitation. It features state-of-the-art optical components, sophisticated data analysis and an intuitive user interface. The result is an easier method to obtain much more accurate and sensitive quantitation than can be obtained with UV absorbance assays, increasing sample performance in downstream applications.

The Qubit® fluorometer (Figure 9.2.3) is designed to work seamlessly with the Quant-iT™ Assay Kits. Together, they form the Qubit™ Quantitation Platform, an efficient combination of sophisticated, accurate, and highly sensitive fluorescence-based quantitation assays for DNA, RNA and protein, along with an extremely user-friendly fluorometer. This powerful pairing offers:

- Selective quantitation—more accurate than UV absorbance readings
- High sensitivity—use as little as 1 µL of sample for quantitation
- Intuitive integrated platform—sophisticated quantitation in 5 minutes or less

In addition to providing quantitation results that are fast, easy and reliable, the Qubit® fluorometer features software that is fully upgradable. The Qubit® fluorometer incorporates sophisticated data analysis algorithms designed to produce accurate measurements and an intuitive user interface for seamless integration with the full range of Quant-iT™ Assay Kits.. We offer the Qubit® fluorometer separately (Q32857; USB cable, Q32858; replacement international power cord, Q32859) or bundled in a starter kit that includes four different Quant-iT™ assay kits as well as a supply of 500 Qubit® assay tubes (Q32860); a set of 500 Qubit® assay tubes is also available separately (Q32856).

Quant-iT™ Assay Kits use advanced Molecular Probes® fluorophores that become fluorescent upon binding to DNA, RNA or protein. The selectivity of these interactions ensures more accurate results than can be obtained with UV absorbance readings; Quant-iT™ Assay Kits only report the concentration of the molecule of interest, not sample contaminants. Quant-iT™ Assay Kits are up to 1000 times as sensitive as UV absorbance readings, and as little as 1 µL of sample is all that's needed for accurate, reliable quantitation. We offer Quant-iT™ Assay Kits for a variety of quantitation needs:

- Quant-iT™ dsDNA Broad-Range (BR) and High-Sensitivity (HS) Assay Kits—for sequencing samples, genomic DNA samples and routine cloning experiments (Section 8.3)
- Quant-iT™ RNA Assay Kits—for microarray experiments, real-time PCR samples and northern blots (Section 8.3)
- Quant-iT™ Protein Assay Kits—for analysis of samples destined for western blotting and SDS-PAGE or 2D gel electrophoresis

More information is available at www.invitrogen.com/handbook/qubit.

NanoOrange® Protein Quantitation Kit

The NanoOrange® Protein Quantitation Kit (N6666) provides an ultrasensitive assay for measuring the concentration of proteins in solution.[2,4] The NanoOrange® Protein Quantitation Kit has several important features:

- **Ease of use**. The NanoOrange® assay protocol is much easier to perform than the Lowry method (Figure 9.2.4). Protein samples are simply added to the diluted NanoOrange® reagent in a lipid-containing medium, and the mixtures are heated at 95°C for 10 minutes. After cooling the mixtures to room temperature, their fluorescence emissions are measured directly. The interaction of the lipid-coated proteins with the NanoOrange® reagent produces a large fluorescence enhancement that can be used to generate a standard curve for protein determination; fluorescence of the reagent in aqueous solutions in the absence of proteins is negligible.
- **Sensitivity and effective range**. The NanoOrange® assay can detect proteins at a final concentration as low as 10 ng/mL when a standard spectrofluorometer or minifluorometer is used. A single protocol is suitable for quantitating protein concentrations between 10 ng/mL and 10 µg/mL—an effective range of three orders of magnitude (Figure 9.2.5).

IMPORTANT NOTICE: The products described in this manual are covered by one or more Limited Use Label License(s). Please refer to the Appendix on page 971 and Master Product List on page 975. Products are For Research Use Only. Not intended for any animal or human therapeutic or diagnostic use.

molecular probes® | invitrogen™ by life technologies™

- **Stability.** The NanoOrange® reagent and its protein complex have high chemical stability. In contrast to the Bradford and BCA assays, readings can be taken for up to six hours after sample preparation with no loss in signal, provided that samples are protected from light.
- **Little protein-to-protein variability** (Figure 9.2.6). The NanoOrange® assay is not only more sensitive, but shows less protein-to-protein variability than Bradford assays.
- **Insensitivity to sample contaminants.** Unlike the Lowry and BCA assays, the NanoOrange® assay is compatible with the presence of reducing agents. Furthermore, the high sensitivity of the assay and stability of the protein–dye complex make it possible to dilute out most potential contaminants, including detergents and salts. Nucleic acids do not interfere with protein quantitation using the NanoOrange® reagent. Although unusually high concentrations of lipids in the sample can interfere with the NanoOrange® assay, this interference can be eliminated by acetone precipitation of the protein, followed by delipidation with diethyl ether.[5]

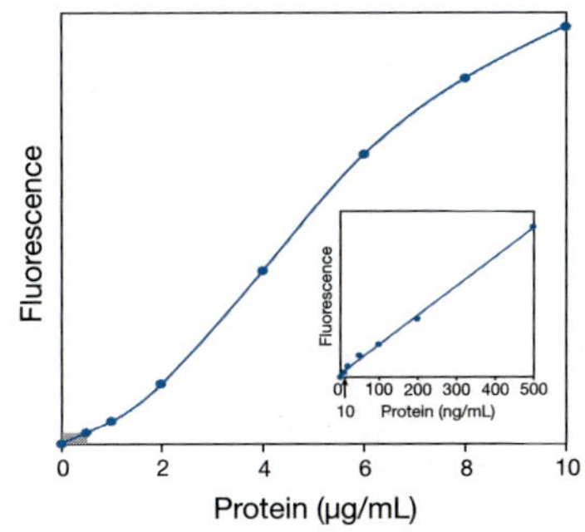

Figure 9.2.5 Quantitative analysis of bovine serum albumin (BSA) using the NanoOrange® Protein Quantitation Kit (N6666). Fluorescence measurements were carried out on an SLM SPF-500C fluorometer using excitation/emission wavelengths of 485/590 nm. The inset shows an enlargement of the results obtained (0–500 ng protein per mL) and illustrates the detection limit of ~10 ng/mL.

The NanoOrange® protein quantitation reagent, with an excitation/emission maxima of 470/570 nm when bound to proteins, is suitable for use with a variety of instrumentation. Fluorescence is typically measured using instrument settings or filters that provide excitation/emission at ~485/590 nm, which are commonly available for both spectrofluorometers and microplate readers. A spectrofluorometer—either a standard fluorometer or a minifluorometer—offers the greatest effective range and lowest detection limits for this assay. With fluorescence microplate readers, the NanoOrange® assay is useful over a somewhat narrower range—from 100 ng/mL to 10 μg/mL in final protein concentration.

The NanoOrange® Protein Quantitation Kit (N6666) supplies:

- Concentrated NanoOrange® reagent in dimethylsulfoxide (DMSO)
- Concentrated NanoOrange® diluent
- Bovine serum albumin (BSA) as a protein reference standard
- Detailed protocols for protein quantitation

The amount of dye supplied in this kit is sufficient for ~200 assays using a 2 mL assay volume and a standard fluorometer or minifluorometer, or ~2000 assays using a 200 μL assay volume and a fluorescence microplate reader.

The NanoOrange® reagent is ideal for quantitating protein samples before gel electrophoresis[5] and western blot analysis.[6] It has also been used to measure bound versus free protein levels in protein binding assays, and was even able to detect protein trapped in filters during a separation step.[7] The NanoOrange® reagent is also an optimal reagent for detecting proteins that have been separated by microchip capillary electrophoresis.[8,9] A high-throughput assay that may be suitable for clinical samples has been developed for quantitating human serum albumin using a fluorescence microplate reader and using capillary electrophoresis laser-induced fluorescence[10] (CE-LIF). Additionally, the NanoOrange® reagent has been shown to be useful in cell-based assays, including an assay designed to measure total protein content of cell cultures[11] and a rapid method for demonstrating flagellar movement of bacteria.[12]

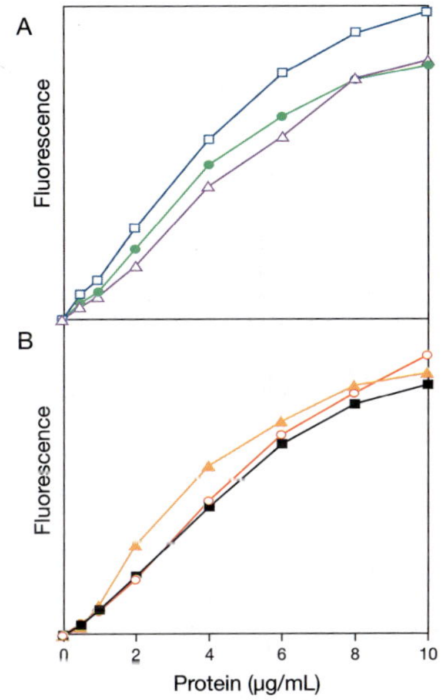

Figure 9.2.6 Quantitative analysis of six different proteins using the NanoOrange® Protein Quantitation Kit (N6666): A) bovine serum albumin (BSA, □), trypsin (●) and carbonic anhydrase (△); B) IgG (■), streptavidin (O) and RNase A (▲). The y-axis fluorescence intensity scale is the same in both panels, illustrating the minimal protein-to-protein staining variation of the NanoOrange® assay. Data were collected using a microplate reader with excitation/emission wavelengths set at 485 ± 20 nm/590 ± 35 nm.

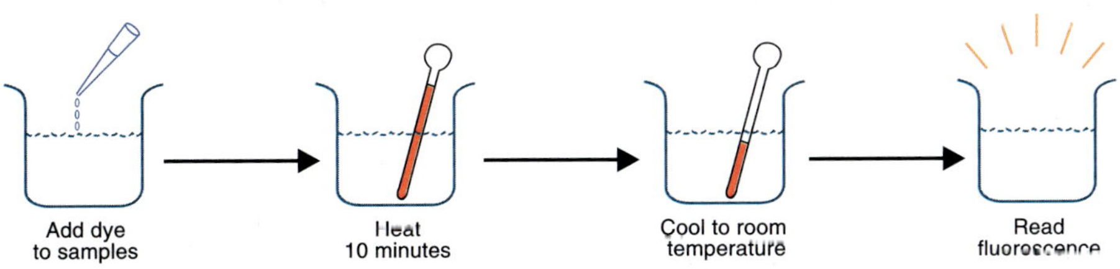

Figure 9.2.4 Protein quantitation with the NanoOrange® Protein Quantitation Kit. The NanoOrange® assay (N6666) is simple to perform: after adding diluted dye, the samples are heated to denature the proteins, cooled to room temperature and the fluorescence read in either a microplate reader or a fluorometer.

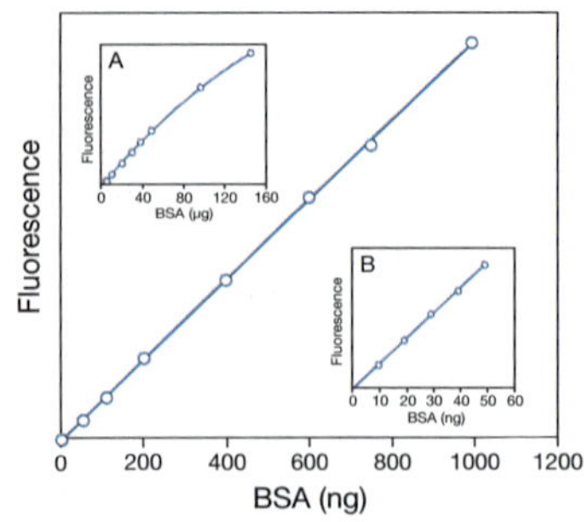

Figure 9.2.8 Detection of bovine serum albumin (BSA) using the CBQCA Protein Quantitation Kit (C6667). The primary plot shows detection of BSA from 50 ng to 1000 ng. Inset A shows that the detection range can extend up to 150 µg. Inset B shows that the lower detection limit can extend down to 10 ng. Each point is the average of four determinations.

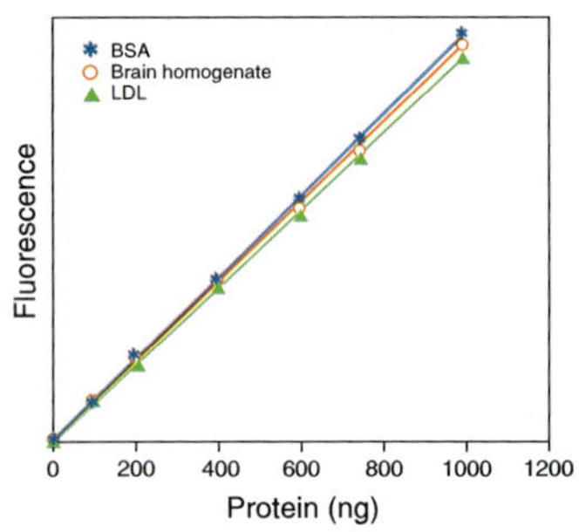

Figure 9.2.9 Quantitation of proteins in a lipid environment using the CBQCA Protein Quantitation Kit (C6667). The protein concentrations of an LDL preparation and a bovine brain homogenate were first determined by the modified Lowry method using BSA as a standard. Assays were then performed using the CBQCA Protein Quantitation Kit on samples containing from 100 ng to 1000 ng protein in 0.1 M sodium borate buffer, pH 9.3, containing 0.1% Triton X-100. Similar results were obtained without the addition of detergent (data not shown). Fluorescence was measured using a fluorescence microplate reader with excitation at 485 ± 10 nm and emission detection at 530 ± 12.5 nm. Each point is the average of three determinations.

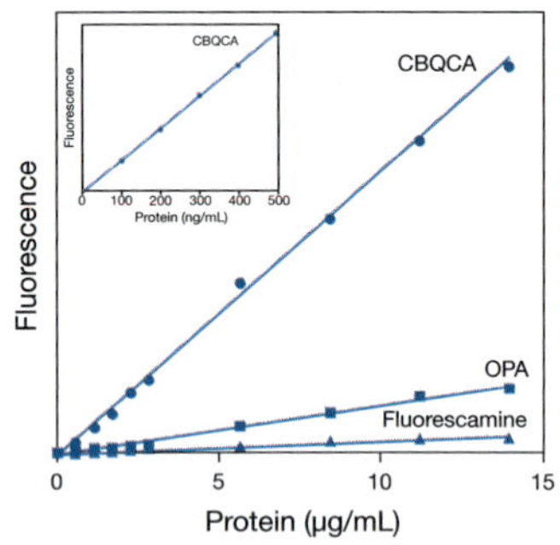

Figure 9.2.10 Comparison of the fluorometric quantitation of bovine serum albumin (BSA) using ATTO-TAG™ CBQCA (which is supplied in the CBQCA Protein Quantitation Kit, C6667), OPA (P2331MP) or fluorescamine (F2332, F20261). BSA samples were derivatized using large molar excesses of the fluorogenic reagents and were analyzed using a fluorescence microplate reader. Excitation/emission wavelengths were 360/460 nm for OPA and fluorescamine and 485/530 nm for ATTO-TAG™ CBQCA. The inset shows an enlargement of the results obtained using CBQCA to assay protein concentrations between 0 and 500 ng/mL.

CBQCA Protein Quantitation Kit

The ATTO-TAG™ CBQCA reagent (A6222) was originally developed as a chromatographic derivatization reagent for amines[13–15] (Section 1.8), but this reagent is also useful for quantitating proteins by virtue of its rapid and quantitative reaction with their accessible amines. We have developed the CBQCA Protein Quantitation Kit (C6667, Figure 9.2.7, Figure 9.2.8), which employs the ATTO-TAG™ CBQCA reagent for rapid and sensitive protein quantitation in solution[1,2,16] (Table 9.1). The CBQCA protein quantitation assay functions well in the presence of lipids and detergents,[16,17] substances that interfere with many other protein determination methods.[16] For example, the CBQCA-based assay can be used directly to determine the protein content of lipoprotein samples or lipid–protein mixtures (Figure 9.2.9). The CBQCA assay has been shown to give faster and more sensitive detection of both free amino acids in human plasma[18] and both low and high molecular weight primary amines in clinical samples from hemodialysis.[19] ATTO-TAG™ CBQCA is more water soluble than either fluorescamine or *o*-phthaldialdehyde and much more stable in aqueous solution than fluorescamine. Moreover, ATTO-TAG™ CBQCA provides greater sensitivity for protein quantitation in solution than either fluorescamine or *o*-phthaldialdehyde (Figure 9.2.10). As little as 10 ng of BSA can be detected in a 100–200 µL assay volume using a fluorescence microplate reader, and the effective range extends up to 150 µg (Figure 9.2.8). Alternatively, the reaction mixtures can be diluted to 1–2 mL for fluorescence measurement in a standard fluorometer or minifluorometer.

Each CBQCA Protein Quantitation Kit (C6667) contains:

- ATTO-TAG™ CBQCA detection reagent
- Potassium cyanide
- Dimethylsulfoxide (DMSO)
- Bovine serum albumin (BSA) protein reference standard
- Detailed protocols for protein quantitation

The CBQCA Protein Quantitation Kit provides sufficient reagents for 300–800 assays using a standard fluorometer, minifluorometer or fluorescence microplate reader.

EZQ® Protein Quantitation Kit

The EZQ® Protein Quantitation Kit (R33200) provides a fast and easy high-throughput assay for proteins. Because detergents, reducing agents, urea and tracking dyes do not interfere, this fluorescence-based protein quantitation assay is ideal for determining the protein concentration of samples prior to polyacrylamide gel electrophoresis.[20] This convenient kit can also provide a quick assessment of protein content during protein purification schemes and fractionation procedures.

The EZQ® assay requires only 1 µL of a sample per spot, and up to 96 samples, including standards, can be assayed in one session. The protein samples are simply spotted onto one of the provided assay papers, fixed with methanol and then stained with the EZQ® protein quantitation reagent. This assay paper is then clamped to the specially designed 96-well microplate

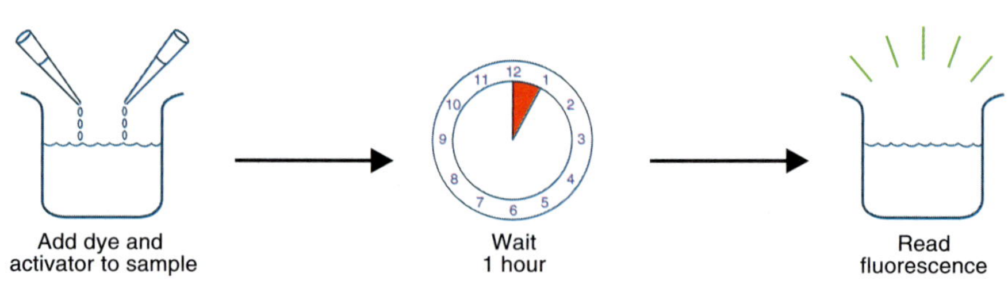

Figure 9.2.7 Protein quantitation with the CBQCA Protein Quantitation Kit. The CBQCA assay (C6667) is simple to perform: after the dye and an activator are added, the sample is incubated for 1 hour and the fluorescence is read in either a microplate reader or a fluorometer.

IMPORTANT NOTICE: The products described in this manual are covered by one or more Limited Use Label License(s). Please refer to the Appendix on page 971 and Master Product List on page 975. Products are For Research Use Only. Not intended for any animal or human therapeutic or diagnostic use.
www.invitrogen.com/probes

supersctructure for quick analysis in a top- or bottom-reading fluorescence microplate reader. For added versatility, the solid-phase assay format and provided 96-well microplate are also compatible with laser scanners equipped with 450, 473 or 488 nm lasers and with UV illuminators in combination with photographic or CCD cameras for image documentation and analysis. Once the samples are spotted, the assay protocol can be completed in about 1 hour. The protein concentration is determined from a standard curve, and the effective range for the assay is generally 0.05–5 mg/mL or 0.05–5 µg per spot (Figure 9.2.11). Protein-to-protein sensitivity differences in the assay are minimal—the observed coefficient of variation is typically ~16% (Figure 9.2.12).

Each EZQ® Protein Quantitation Kit contains:

- EZQ® protein quantitation reagent
- EZQ® 96-well microplate cassette
- Assay paper
- Ovalbumin, for preparing protein standards
- Detailed protocols for protein quantitation using a variety of fluorescence-detection instruments

Sufficient reagent and assay paper are provided for ~2000 protein quantitation assays.

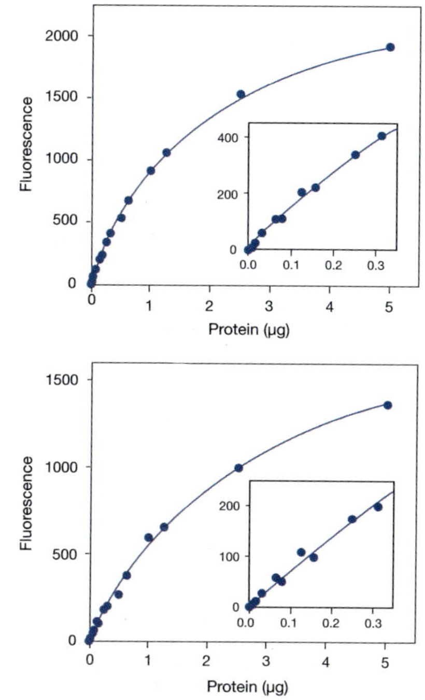

Figure 9.2.11 EZQ® protein quantitation assay of ovalbumin. A dilution series of ovalbumin was prepared, assayed with the EZQ® Protein Quantitation Kit (R33200) and then quantitated using both a 473 nm laser-based scanning instrument (top panel) and a fluorescence microplate reader (bottom panel). The assays were performed over a broad range; the insets show the low range in greater detail. The assays were performed in triplicate, and the mean values, in arbitrary fluorescence units, were plotted after subtracting background values of 86 (upper panel) or 18 (lower panel).

Other Reagents for Protein Quantitation in Solution

Most traditional fluorogenic reagents for general protein quantitation in solution detect accessible primary amines. The sensitivity of assays based on these reagents therefore depends on the number of amines available—a function of both the protein's three-dimensional structure and its amino acid composition. For example, horseradish peroxidase (MW ~40,000 daltons), which has only six lysine residues,[21] will be detected less efficiently than egg white avidin (MW ~66,000 daltons), which has 36 lysine residues,[22] and bovine serum albumin (MW ~66,000 daltons), which has 59 lysine residues.[23] However, the assays are generally rapid and easy to conduct, particularly in minifluorometer and fluorescence microplate reader formats.

Certain dyes that detect primary aliphatic amines, including ATTO-TAG™ CBQCA (A6222), fluorescamine (F2332, F20261) and *o*-phthaldialdehyde (OPA, P2331MP), have been the predominant reagents for fluorometric determination of proteins in solution (Table 9.1). These same reagents, and others such as naphthalene-2,3-dicarboxaldehyde[24,25] (NDA, N1138; Section 1.8), have frequently been used for amino acid analysis of hydrolyzed proteins.

Fluorescamine

Fluorescamine (F2332, F20261) is intrinsically nonfluorescent but reacts in milliseconds with primary aliphatic amines, including peptides and proteins, to yield a fluorescent derivative[26] (Figure 9.2.13). This amine-reactive reagent has been shown to be useful for determining protein concentrations of aqueous solutions[27–29] and for measuring the number of accessible lysine residues in proteins.[23] Protein quantitation with fluorescamine is particularly well suited to a minifluorometer or fluorescence microplate reader.[30] Fluorescamine can also be used to detect proteins in gels and to analyze low molecular weight amines by TLC, HPLC and capillary electrophoresis.[31]

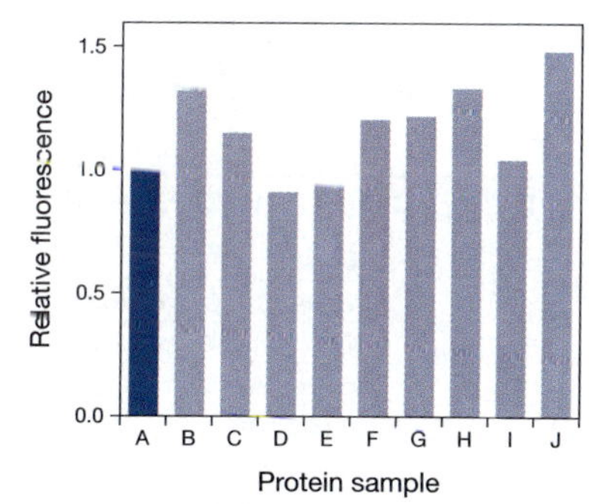

Figure 9.2.12 Protein-to-protein variation in the EZQ® protein quantitation assay. Triplicate 1 µg samples of various proteins were assayed using the EZQ® Protein Quantitation Kit (R33200) and a fluorescence microplate reader. The mean fluorescence values, after correcting for background fluorescence, are expressed relative to that of ovalbumin. The coefficient of variation is ~16%. The protein samples are: A, ovalbumin; B, bovine serum albumin (BSA); C, myoglobin; D, soybean trypsin inhibitor; E, β-casein; F, carbonic anhydrase; G, transferrin; H, mouse IgG; I, lysozyme; and J, histones.

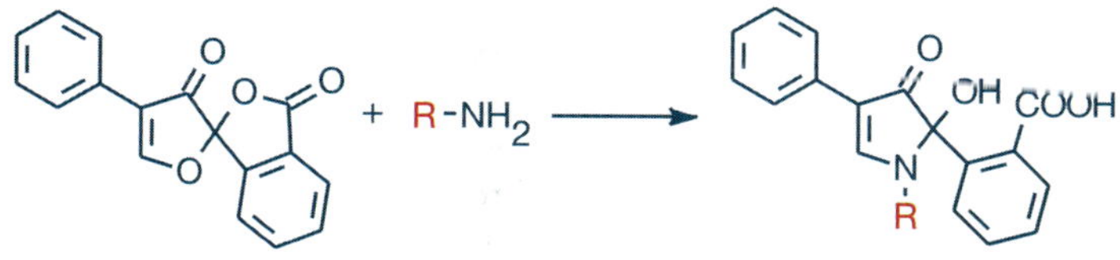

Fluorescamine

Figure 9.2.13 Fluorogenic amine-derivatization reaction of fluorescamine (F2332, F20261).

o-Phthaldialdehyde

The combination of *o*-phthaldialdehyde (OPA, P2331MP) and 2-mercaptoethanol provides a rapid and simple method of determining protein concentrations in the range of 0.2 µg/mL to 25 µg/mL[32] (Figure 9.2.14). As compared with fluorescamine, OPA is both more soluble and stable in aqueous buffers and its sensitivity for detection of peptides is reported to be 5–10 times better.[33] The OPA assay for lysine content is reasonably reliable over a broad range of proteins.[23] OPA can also be used to detect *increases* in the concentration of free amines that result from protease-catalyzed protein hydrolysis.[34]

SYPRO® Red and SYPRO® Orange Protein Gel Stains

An assay has been reported that uses the SYPRO® Red protein gel stain (S6653, S6654; Section 9.3) for quantitating total protein content of bacterial cells by flow cytometry.[35] This assay provides an accurate method to measure planktonic bacterial biomass in marine samples. Fluorescence of the SYPRO® Orange protein gel stain (S6650, S6651; Section 9.3) has been used to follow isothermal protein denaturation[36] (Note 9.1—Monitoring Protein-Folding Processes with Environment-Sensitive Dyes) and to selectivly stain proteins in biofilms prior to two-photon laser-scanning microscopy.[37]

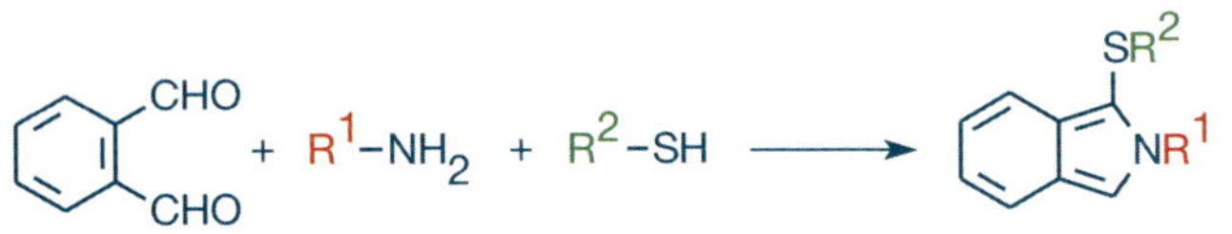

Figure 9.2.14 Fluorogenic amine-derivatization reaction of *o*-phthaldialdehyde (OPA) (P2331MP).

REFERENCES

1. Mol Biotechnol (2007) 39:99; **2.**Methods Enzymol (2009)463:71; **3.** Anal Chem (2010) 82:336; **4.** Biotechniques (2003) 34:850; **5.** J Biol Chem (1997) 272:12762; **6.** J Biol Chem (2000) 275:3256; **7.** J Biol Chem (1999) 274:35367; **8.** Anal Chem (2001) 73:4994; **9.** Anal Chem (2000) 72:4608; **10.** J Chromatogr B Biomed Sci Appl (2001) 754:345; **11.** J Cell Biol (1998) 142:1313; **12.** Appl Environ Microbiol (2000) 66:3632; **13.** Anal Chem (1991) 63:408; **14.** Anal Chem (1991) 63:413; **15.** J Chromatogr (1990) 499:579; **16.** Anal Biochem (1997) 244:277; **17.** J Biol Chem (1999) 274:25461; **18.** J Chromatogr B Biomed Sci Appl (2001) 754:217; **19.** Clin Chim Acta (2001) 308:147; **20.** Electrophoresis (2004) 25:2478; **21.** Eur J Biochem (1979) 96:483; **22.** Adv Protein Chem (1975) 29:85; **23.** Anal Biochem (1981) 115:203; **24.** Anal Chem (1990) 62:1580; **25.** Anal Chem (1990) 62:1577; **26.** Science (1972) 178:871; **27.** Clin Chim Acta (1986) 157:73; **28.** J Lipid Res (1986) 27:792; **29.** Anal Biochem (1997) 248:195; **30.** Anal Biochem (1993) 214:346; **31.** J Chromatogr (1990) 502:247; **32.** J Immunol Methods (1994) 172:141; **33.** Proc Natl Acad Sci U S A (1975) 72:619; **34.** J Biochem Biophys Methods (2008) 70:878; **35.** Appl Environ Microbiol (1999) 65:3251; **36.** Anal Biochem (2001) 292:40; **37.** Appl Environ Microbiol (2002) 68:901.

DATA TABLE 9.2 PROTEIN QUANTITATION IN SOLUTION

Cat. No.	MW	Storage	Soluble	Abs	EC	Em	Solvent	Notes
A6222	305.29	F,D,L	MeOH	465	ND	560	MeOH	1, 2, 3
F2332	278.26	F,D,L	MeCN	380	7800	464	MeCN	4
F20261	278.26	F,D,L	MeCN	380	8400	464	MeCN	4, 5
P2331MP	134.13	L	EtOH	334	5700	455	pH 9	6

For definitions of the contents of this data table, see "Using *The Molecular Probes® Handbook*" in the introductory pages.

Notes

1. Spectral data are for the reaction product with glycine in the presence of cyanide. Unreacted reagent in MeOH: Abs = 254 nm (EC = 46,000 $cm^{-1}M^{-1}$), nonfluorescent.
2. ND = not determined.
3. Solubility in methanol is improved by addition of base (e.g., 1–5% (v/v) 0.2 M KOH).
4. Fluorescamine spectra are for the reaction product with butylamine. The fluorescence quantum yield and lifetime of the butylamine adduct in EtOH are 0.23 and 7.5 nanoseconds, respectively. (Arch Biochem Biophys (1974) 163:390) The unreacted reagent is nonfluorescent (Abs = 234 nm, EC = 28,000 $cm^{-1}M^{-1}$ in MeCN).
5. This product is specified to equal or exceed 98% analytical purity by HPLC.
6. Spectral data are for the reaction product of P2331MP with alanine and 2-mercaptoethanol. The spectra and stability of the adduct depend on the amine and thiol reactants. (Biochim Biophys Acta (1979) 576:440) Unreacted reagent in H_2O: Abs = 257 nm (EC = 1000 $cm^{-1}M^{-1}$).

PRODUCT LIST 9.2 PROTEIN QUANTITATION IN SOLUTION

Cat. No.	Product	Quantity
A6222	ATTO-TAG™ CBQCA derivatization reagent (CBQCA; 3-(4-carboxybenzoyl)quinoline-2-carboxaldehyde)	10 mg
C6667	CBQCA Protein Quantitation Kit *300–800 assays*	1 kit
R33200	EZQ® Protein Quantitation Kit *2000 assays*	1 kit
F2332	fluorescamine	100 mg
F20261	fluorescamine *FluoroPure™ grade*	100 mg
N6666	NanoOrange® Protein Quantitation Kit *200–2000 assays*	1 kit
P2331MP	o-phthaldialdehyde (OPA) *high purity*	1 g
Q33211	Quant-iT™ Protein Assay Kit, 100 assays *0.25–5 µg* *for use with the Qubit® fluorometer*	1 kit
Q33212	Quant-iT™ Protein Assay Kit, 500 Assays *0.25–5 µg* *for use with the Qubit® fluorometer*	1 kit
Q33210	Quant-iT™ Protein Assay Kit, 1000 assays *0.25–5 µg*	1 kit
Q32856	Qubit™ assay tubes *set of 500*	1 set
Q32857	Qubit® fluorometer	each
Q32859	Qubit™ fluorometer international power cord *replacement*	each
Q32858	Qubit™ fluorometer USB cable	each
Q32860	Qubit™ Quantitation Starter Kit	1 kit

IMPORTANT NOTICE: The products described in this manual are covered by one or more Limited Use Label License(s). Please refer to the Appendix on page 971 and Master Product List on page 975. Products are For Research Use Only. Not intended for any animal or human therapeutic or diagnostic use.
www.invitrogen.com/probes
molecular probes® | invitrogen™ by life technologies™

9.3 Protein Detection on Gels, Blots and Arrays

SYPRO® Protein Gel Stains

The luminescent SYPRO® protein gel stains are revolutionizing the detection of the total-protein profile in polyacrylamide gels. SYPRO® protein gel stains exhibit several important characteristics that together make them far superior to traditional staining methods, including:

- Fast and easy staining protocols
- High sensitivity
- Minimal protein-to-protein variation in staining
- Broad linear quantitation range
- Compatibility with subsequent microanalysis and a variety of instrumentation

Conventional methods for universal profiling of proteins in gels include Coomassie brilliant blue staining [1] and silver staining.[2] Although Coomassie brilliant blue is an inexpensive reagent, its staining is relatively insensitive and time consuming. Silver staining may be up to 100 times more sensitive than Coomassie brilliant blue staining, but it is relatively expensive and entails several labor-intensive and time-sensitive steps. Silver staining also exhibits a high degree of protein-to-protein variability; staining intensity and color are very dependent on each polypeptide's sequence and degree of glycosylation, and some proteins are detectable only as negatively stained patches. Moreover, silver staining shows very poor linearity with protein concentration (Figure 9.3.1) and poor reproducibility in staining from gel to gel, making it inadequate for comparative studies of protein expression in cells.

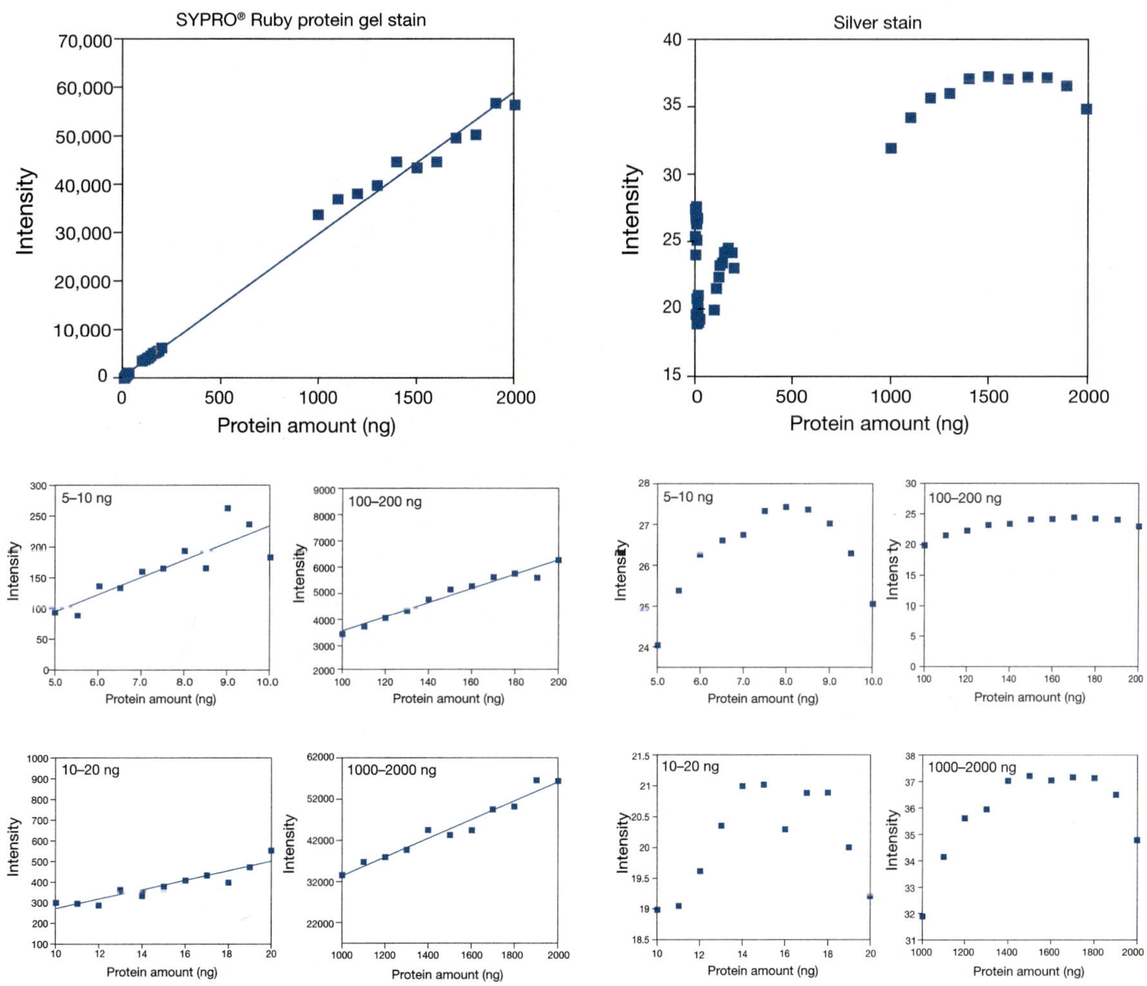

Figure 9.3.1 Quantitation of proteins in gels using SYPRO® Ruby protein gel stain versus silver stain. Dilutions of proteins were electrophoresed on eight different SDS-polyacrylamide gels, two gels for each of four dilution ranges. The gels were stained with either SYPRO® Ruby protein gel stain (S12000, S12001, S21900) or a silver stain. Staining intensities were quantitated using either the Fluor-S MAX MultiImager documentation system (Bio-Rad® Laboratories) or the FLA3000G laser scanner (Fuji® Photo Film Co.) and plotted against the protein amount for bovine serum albumin. SYPRO® Ruby protein gel stain shows a linear quantitation range over three orders of magnitude, as well as consistent staining intensities from gel to gel. In contrast, the silver stain shows linear quantitation over only a small range, a very shallow slope and inconsistent staining intensities from gel to gel, even when corrected for background differences.

Many of the drawbacks of these traditional stains can be overcome by using one of the SYPRO® protein gel stains, without sacrificing detection sensitivity.[3] We have developed a SYPRO® dye optimized for protein profiling in nearly every type of gel (Table 9.2) or blot application (Table 9.3). The characteristics and applications of the individual SYPRO® protein gel and blot stains for detecting the total-protein profile of a sample are described in this section. Section 9.4 discusses the use of these SYPRO® stains in conjunction with phosphoprotein and glycoprotein detection reagents for multiparameter staining.

SYPRO® Ruby Protein Gel Stain: Ultrasensitive Protein Detection in 1D, 2D and IEF Gels

SYPRO® Ruby protein gel stain (S12000, S12001, S21900) is a ready-to-use protein stain that has sensitivity equal to or exceeding that of the best silver staining techniques, is compatible with Edman sequencing and mass spectrometry[4] and can be visualized with a simple UV transilluminator or a laser scanner.[3,5–17] SYPRO® Ruby protein gel stain offers several advantages over conventional staining techniques:

- **High-sensitivity staining**. SYPRO® Ruby protein gel stain provides at least the same subnanogram sensitivity as the best silver staining techniques in 1D, 2D[11,13] or IEF gels (Figure 9.3.2).
- **Simple staining protocol**. After fixation, the gel is incubated in the staining solution. No stop solutions or destaining steps are required and, unlike silver staining, gels can be left in the dye solution for indefinite periods without overstaining, vastly simplifying the simultaneous processing of multiple gels and making it possible to perform high-throughput staining without investing in robotic staining devices.
- **Accurate peptide and protein detection**. SYPRO® Ruby protein gel stain shows little protein-to-protein variability in staining[18] and detects some proteins that are completely missed by silver staining (Figure 9.3.3), such as heavily glycosylated proteins. Unlike silver staining, SYPRO® Ruby dye does not stain extraneous nucleic acids, lipids or carbohydrates in the sample.[13]
- **Excellent performance in comparative protein expression studies**. SYPRO® Ruby stain shows a greater linear quantitation range than either silver or Coomassie brilliant blue staining—extending over three orders of magnitude—making it possible to accurately compare protein expression levels[13] (Figure 9.3.1, Figure 9.3.2). Gel-to-gel staining is extremely consistent; same-spot intensity comparisons between identical 2D gels show a correlation coefficient of 0.9.[12] Multiple gels can easily be compared using available software (Figure 9.3.4). Other protein quantitation methods, including running multiple prestained samples on the same gel, generally do not produce results that approach this level of discrimination.[19]

Table 9.2 SYPRO® and Coomassie Fluor™ luminescent and fluorescent protein gel stains.

Dye (Cat. No.)	Ex/Em *	Major Applications	Features
SYPRO® Ruby protein gel stain (S12000, S12001, S21900)	280, 450/610	2D gels, IEF gels, 1D SDS-PAGE	• High sensitivity (1–2 ng/band; comparable to silver staining) • Linear quantitation range over three orders of magnitude
SYPRO® Orange protein gel stain (S6650, S6651)	300, 470/570	1D SDS-PAGE	• Very good sensitivity (4–8 ng/band; higher than Coomassie brilliant blue staining) • Little protein-to-protein variability • Linear quantitation range over three orders of magnitude
SYPRO® Red protein gel stain (S6653, S6654)	300, 550/630	1D SDS-PAGE	• Very good sensitivity (4–8 ng/band; higher than Coomassie brilliant blue staining) • Little protein-to-protein variability • Linear quantitation range over three orders of magnitude
SYPRO® Tangerine protein gel stain (S12010)	300, 490/640	1D SDS-PAGE, blotting applications, zymography, electroelution	• Very good sensitivity (4–8 ng/band; higher than Coomassie brilliant blue staining) • Little protein-to-protein variability • Linear quantitation range over three orders of magnitude • No fixation required
Coomassie Fluor™ Orange protein gel stain (C33250, C33251)	300, 470/570	1D SDS-PAGE	• Premixed and ready-to-use solution • Very good sensitivity (8 ng/band; higher than Coomassie brilliant blue staining) • Little protein-to-protein variability • Linear quantitation range over at least two orders of magnitude

All SYPRO® gel and blot stains are compatible with Edman sequencing and mass spectrometry. *Excitation (Ex) and emission (Em) maxima, in nm. For maximum sensitivity, use excitation sources and optical filters matched to these values.

Table 9.3 Molecular Probes® fluorescent and luminescent protein blot stains.

Dye (Cat. No.)	Ex/Em *	Major Applications	Features
SYPRO® Ruby protein blot stain (S11791)	280, 450/618	Mass spectrometry, microsequencing, counterstain for blot-based detection techniques (nitrocellulose or PVDF membranes)	• High sensitivity (1–2 ng/band; comparable to colloidal gold staining) • Reversible
BODIPY® FL-X succinimidyl ester (D6102) †	365, 505/575	Counterstain for blot-based detection techniques (PVDF membranes)	• Very good sensitivity (4–8 ng/band) • Permanent, covalent staining for multicolor techniques
BODIPY® TR-X succinimidyl ester (D6116) ‡	300, 590/615	Counterstain for blot-based detection techniques (PVDF membranes)	• Very good sensitivity (4–8 ng/band) • Permanent, covalent staining for multicolor techniques

* Excitation (Ex) and emission (Em) maxima, in nm. For maximum sensitivity, use excitation sources and optical filters matched to these values. † Available as a stand-alone reagent (D6102, Section 1.4). ‡ Available as a stand-alone reagent (D6116, Section 1.4).

IMPORTANT NOTICE: The products described in this manual are covered by one or more Limited Use Label License(s). Please refer to the Appendix on page 971 and Master Product List on page 975. Products are For Research Use Only. Not intended for any animal or human therapeutic or diagnostic use.

- **Compatibility with microsequencing and mass spectrometry.**[4] Unlike silver staining techniques, which use glutaraldehyde- or formaldehyde-based fixatives, SYPRO® Ruby dye is a gentle stain that interacts noncovalently with proteins. Thus, high-quality Edman sequencing or mass spectrometry data[4,12,20,21] (Figure 9.3.5) can be obtained immediately after staining, without modification steps that may compromise sensitivity. Automated in-gel digestion methods have been used in the analysis of femtomole levels of SYPRO® Ruby dye–stained proteins.[22]
- **Utility for isoelectric focusing (IEF).** SYPRO® Ruby protein gel stain also provides a reliable, high-sensitivity staining method for isoelectric focusing (IEF) gels (Figure 9.3.6) without the problems typically encountered with silver staining, such as ampholyte staining or mirroring effects on the plastic gel backing.
- **Minimal hazardous waste.** As compared with silver stains, SYPRO® Ruby protein gel stain generates much less hazardous waste, minimizing the time and expense associated with waste disposal.

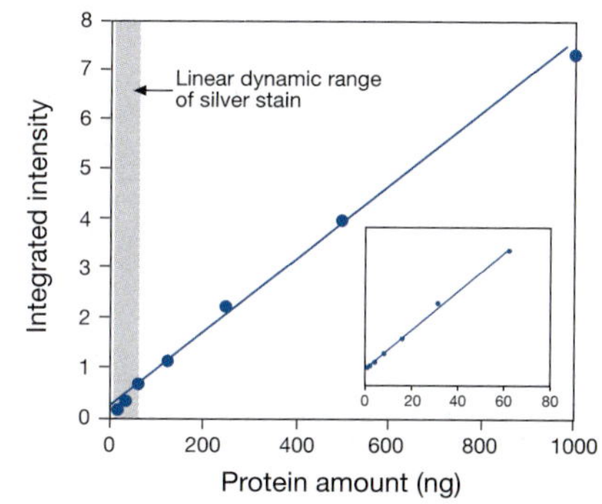

Figure 9.3.2 Amounts of carbonic anhydrase ranging from 1 ng to 1000 ng were separated on an SDS-polyacrylamide gel and stained with SYPRO® Ruby protein gel stain (S12000, S12001, S21900). The inset shows the excellent linearity in the lower part of the range from 1 ng to 60 ng protein. Staining intensities were quantitated using the Molecular Imager® FX documentation system (Bio-Rad® Laboratories). For comparison, the gray band shows the linear range for the same protein detected with silver staining.

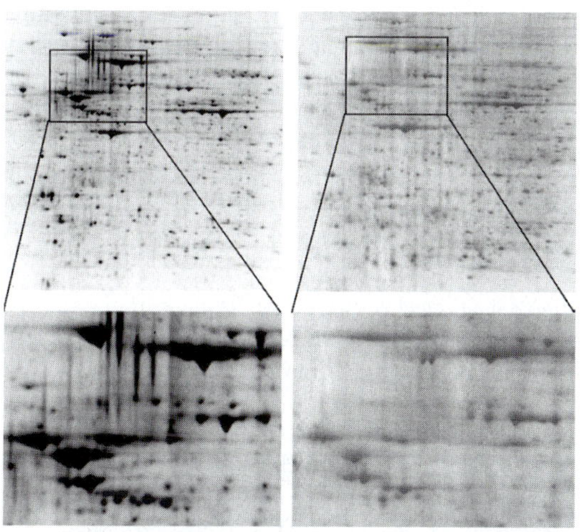

Figure 9.3.3 SYPRO® Ruby protein gel stain (S12000, S12001, S21900) shows less protein-to-protein variation than silver staining. Proteins from a cell lysate were run on a 2D gel and stained with SYPRO® Ruby protein gel stain (left) or silver stain (right). The grayscale values of the SYPRO® Ruby dye–stained gel have been inverted for easier comparison with the silver-stained gel.

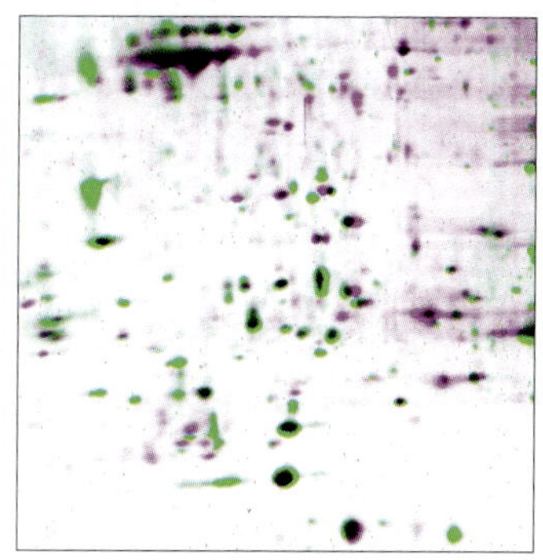

Figure 9.3.4 Comparison of two protein samples run on 2D gels. Proteins from either a normal liver tissue sample or a liver tumor sample were run on two 2D gels and stained with SYPRO® Ruby protein gel stain (S12000, S12001, S21900). Images of the gels were captured using the FLA-3000 scanner (Fuji®). Images from a portion of the two gels were then pseudocolored either pink or green, overlaid and matched spot-for-spot using Z3 software (Compugen). Green spots represent proteins expressed in the liver tumor samples; pink spots represent proteins expressed in the normal liver tissue sample. Black spots represent proteins expressed in both tissues.

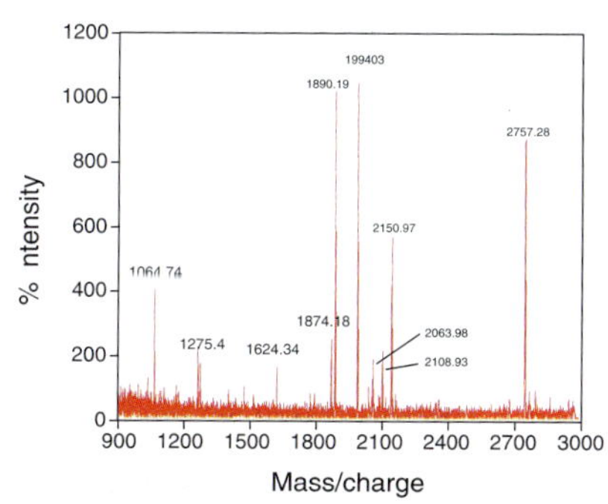

Figure 9.3.5 Mass spectrum profile of NADH:ubiquinone reductase precursor (75,000-dalton subunit) obtained after 2D gel electrophoresis of bovine heart mitochondria and staining with SYPRO® Ruby protein gel stain (S12000, S12001, S21900). Bovine heart mitochondria were a gift of Roderick Capaldi, University of Oregon.

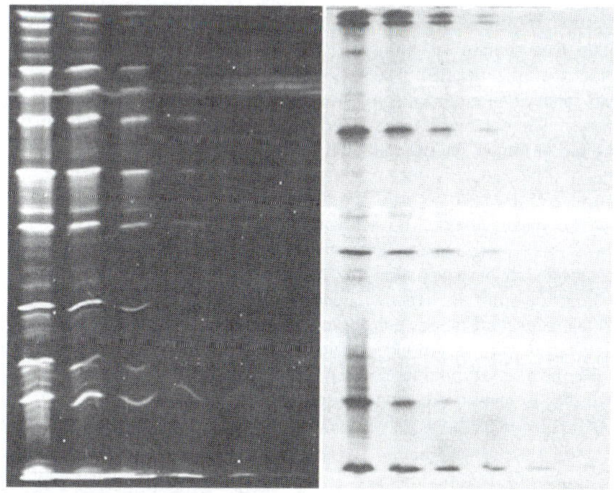

Figure 9.3.6 SYPRO® Ruby protein gel stain versus silver stain for IEF gels. Serial dilutions of isoelectric focusing protein standards were electrophoresed on two identical polyacrylamide gels. One gel was stained with SYPRO® Ruby protein gel stain (S12000, S12001, S21900) (left) and the other with silver stain (right). Both stains show a similar limit of sensitivity for all proteins.

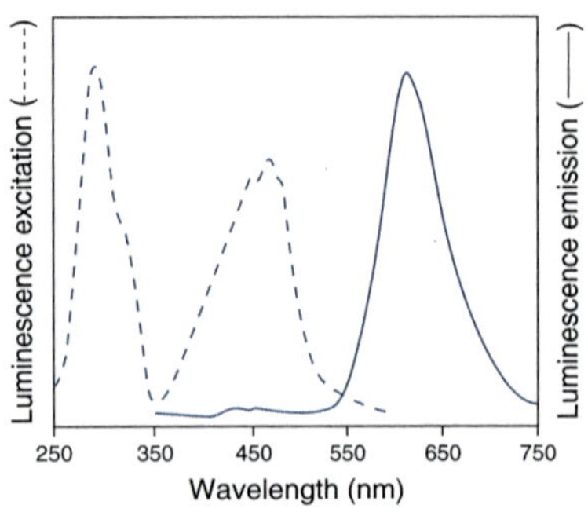

Figure 9.3.7 Luminescence excitation (dashed line) and emission (solid line) spectra of the SYPRO® Ruby protein gel stain (S12000, S12001, S21900) and SYPRO® Ruby protein blot stain (S11791).

SYPRO® Ruby protein gel stain is based on an organometallic ruthenium complex, which shows an extremely bright and photostable red-orange luminescence when excited with either UV or blue light (Figure 9.3.7). Stained proteins can be visualized using a UV transilluminator, a blue-light transilluminator or a laser-scanning instrument. Gels can then be documented using Polaroid® 667 black-and-white print film, a CCD camera with an image documentation system or a laser-scanning instrument.[13] For optimal sensitivity using a UV transilluminator and Polaroid® 667 black-and-white print film, the SYPRO® photographic filter (S6656) is recommended.

SYPRO® Ruby protein gel stain is supplied as 200 mL of a 1X staining solution (S12001), sufficient for staining about four minigels, or 1 L of a 1X staining solution (S12000), sufficient for staining about 20 minigels or two standard 2D gels. Additionally, we offer SYPRO® Ruby protein gel stain in a 5 L box (S21900), sufficient for staining about 100 minigels or 10 standard 2D gels. These boxes are easy to stack and store, and the convenient spigot makes it easy to dispense just the right amount of stain.

SYPRO® Orange and SYPRO® Red Protein Gel Stains: For Routine Detection of Proteins in 1D SDS-Polyacrylamide Gels

SYPRO® Orange (S6650, S6651) and SYPRO® Red (S6653, S6654) protein gel stains provide a fluorescence-based alternative for protein detection in SDS-polyacrylamide gels that is not only faster and more sensitive than Coomassie brilliant blue staining, but can be as sensitive as short-protocol silver staining methods (Figure 9.3.8) at a fraction of the time, effort and cost.[23–28]

In the presence of excess SDS, nonpolar regions of polypeptides are coated with detergent molecules, forming a micelle-like structure with a nearly constant SDS/protein ratio (1.4 g SDS:1.0 g protein); this constant charge-per-mass ratio is the basis of molecular weight determination by SDS-polyacrylamide gel electrophoresis. SYPRO® Orange and SYPRO® Red dyes bind to the SDS coat that surrounds proteins in SDS-polyacrylamide gels. Thus, the staining observed with these dyes exhibits relatively little protein-to-protein variation and is linearly related to protein mass (Figure 9.3.9). Some important features of SYPRO® Orange and SYPRO® Red protein gel stains include:

- **Ease of use**. Following electrophoresis, the gel is stained for 10–60 minutes and then briefly rinsed—no separate fixation, stop or destaining steps are required. After staining, the gel is immediately ready for photography, or it can be stored, in or out of the staining solution, for days.

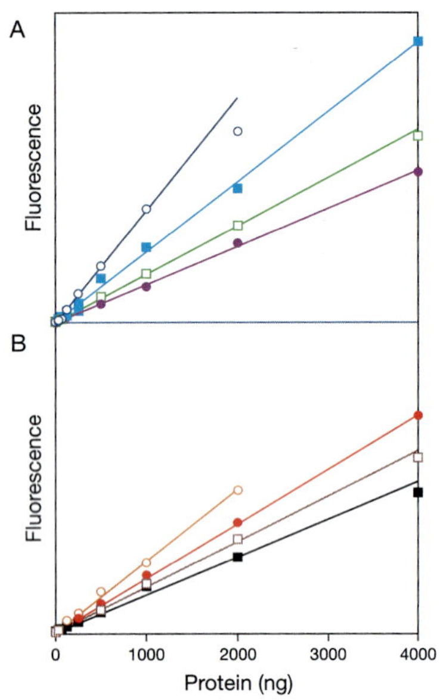

Figure 9.3.9 Quantitation of proteins in a gel using SYPRO® Orange protein gel stain (S6650, S6651). A protein mixture was serially diluted and electrophoresed on a 15% SDS-polyacrylamide gel and then stained with SYPRO® Orange protein gel stain. The gel was then scanned using a Storm gel and blot analysis system (excitation/emission 488/>520 nm) and analyzed to yield the fluorescence intensities of the stained bands. The fluorescence intensity scale is the same in both panels, illustrating the minimal protein-to-protein staining variation of SYPRO® Orange gel stain. Detection limits are between 2 and 16 ng of protein; the linear detection ranges are approximately 1000-fold. Proteins represented are: **A)** β-galactosidase (O), lysozyme (■), bovine serum albumin (BSA, □) and phosphorylase B (●); **B)** myosin (O), soybean trypsin inhibitor (■), ovalbumin (□) and carbonic anhydrase (●).

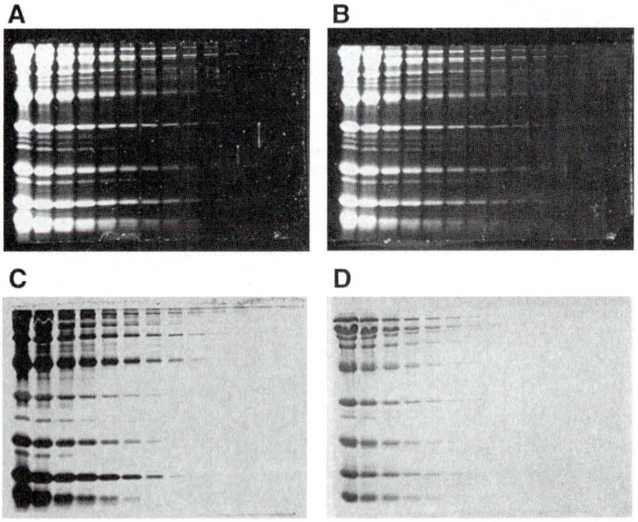

Figure 9.3.8 Comparison of the sensitivity achieved with SYPRO® Orange, SYPRO® Red, silver and Coomassie brilliant blue stains. Identical SDS-polyacrylamide gels were stained with **A)** SYPRO® Orange protein gel stain (S6650, S6651), **B)** SYPRO® Red protein gel stain (S6653, S6654), **C)** silver stain and **D)** Coomassie brilliant blue stain, according to standard protocols. SYPRO® Orange and SYPRO® Red dye–stained gels were photographed using 300 nm transillumination, a SYPRO® photographic filter (S6656) and Polaroid® 667 black-and-white print film. The silver- and Coomassie brilliant blue–stained gels were photographed with transmitted white light and Polaroid® 667 black-and-white print film; no photographic filter was used to photograph these gels.

IMPORTANT NOTICE: The products described in this manual are covered by one or more Limited Use Label License(s). Please refer to the Appendix on page 971 and Master Product List on page 975. Products are For Research Use Only. Not intended for any animal or human therapeutic or diagnostic use.
www.invitrogen.com/probes

- **High sensitivity**. SYPRO® Orange and SYPRO® Red protein gel stains routinely provide a sensitivity level of at least 8–16 ng per band in SDS-polyacrylamide minigels when visualized with standard 300 nm transillumination (Figure 9.3.8). Photography using Polaroid® 667 black-and-white print film and a SYPRO® photographic filter (S6656) enhances the detection of staining with SYPRO® Orange or SYPRO® Red dye by several-fold over visible detection because the film integrates the signal throughout the duration of the exposure. Laser scanners also detect nanogram quantities of SYPRO® dye–stained proteins in gels.
- **Broad linear quantitation range**. Protein detection in gels stained with either SYPRO® Orange or SYPRO® Red stain is linear over three orders of magnitude in protein quantity[26] (Figure 9.3.9).
- **Uniform protein staining**. Unlike silver staining,[29] SYPRO® Orange and SYPRO® Red dyes exhibit relatively low protein-to-protein variability in SDS-polyacrylamide gels[27] (Figure 9.3.9) and do not stain nucleic acids, which are sometimes found in protein mixtures from cell or tissue extracts.[26] In addition, SYPRO® Orange and SYPRO® Red dyes only weakly stain lipopolysaccharides in bacterial lysates, whereas these biopolymers are strongly detected by some types of silver staining.[26] Glycoproteins (such as the IgG variable subunit) and proteins with prosthetic groups (such as bovine cytochrome oxidase) are also efficiently stained with SYPRO® Orange and SYPRO® Red dyes.[27]
- **Photostability.** Proteins stained with SYPRO® Orange or SYPRO® Red dye are relatively photostable, enabling the researcher to acquire multiple photographic images and to use long film-exposure times (2–8 seconds). Gels that are illuminated for long periods of time may partially photobleach but can be restained with little loss of sensitivity.[30]
- **Compatibility with many types of instruments**. Although their excitation maxima are in the visible range (Figure 9.3.10), SYPRO® dye–stained gels are readily visualized using standard 300 nm transilluminators.[23] SYPRO® Orange protein gel stain also exhibits good sensitivity when viewed with a blue-light transilluminator, and both SYPRO® Orange and SYPRO® Red protein gel stains are suitable for use with many laser-scanning instruments.[27]
- **Chemical stability**. SYPRO® Orange and SYPRO® Red gel stains are chemically stable; if protected from light, fluorescence of the stained gel is stable for several days, and staining solutions can be stored for months.
- **Economy**. SYPRO® Orange and SYPRO® Red gel stains are not only less expensive than silver-staining kits but faster and less laborious to use. Additionally, use of SYPRO® Orange or SYPRO® Red dye avoids the costs of purchase and disposal of organic solvents that are required for Coomassie brilliant blue staining.
- **Compatibility with mass spectroscopy**. Unlike silver stains, SYPRO® Orange and SYPRO® Red dyes do not covalently bind to proteins, allowing subsequent analysis of stained proteins by microsequencing or mass spectrometry.[4,25,31]

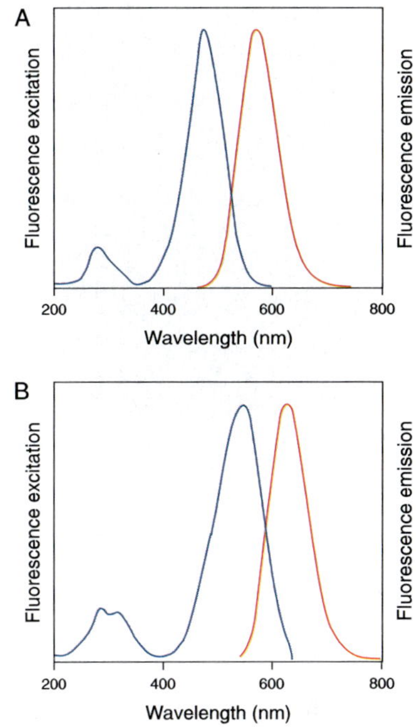

Figure 9.3.10 Fluorescence excitation and emission spectra of **A)** SYPRO® Orange (S6650, S6651) and **B)** SYPRO® Red (S6653, S6654) protein gel stains diluted 1:10,000 in water containing 0.05% SDS and 150 µg/mL bovine serum albumin (BSA).

SYPRO® Orange and SYPRO® Red stains have very similar staining properties, although we have observed that proteins stained with SYPRO® Orange dye are slightly brighter, whereas gels stained with SYPRO® Red dye tend to have lower background fluorescence. For maximum sensitivity with UV transilluminators, we recommend documenting the signal using Polaroid® 667 black-and-white print film and the SYPRO® photographic filter (S6656). For maximum sensitivity with laser scanners, we recommend matching the appropriate SYPRO® dye with the excitation light source of the instrument. SYPRO® Orange protein gel stain is most suitable for gel scanners that employ argon-ion lasers with output at 488 nm, whereas SYPRO® Red protein gel stain is best matched to laser-scanning instruments that employ Nd:YAG lasers with output at 532 nm. SYPRO® Red protein gel stain has been used as a prestain for protein analysis in an automated ultrathin-layer gel electrophoretic technique.[32,33] SYPRO® Orange protein gel stain has been used for protein sizing on a microchip[34] and for analyzing the kinetics of isothermal protein denaturation (Note 9.1— Monitoring Protein-Folding Processes with Environment-Sensitive Dyes).[35]

SYPRO® Orange and SYPRO® Red protein gel stains are compatible with SDS or urea/SDS gels. Staining native proteins in gels in the absence of SDS results in more protein-to-protein variation and lower sensitivity than staining SDS-denatured proteins, due to variations in

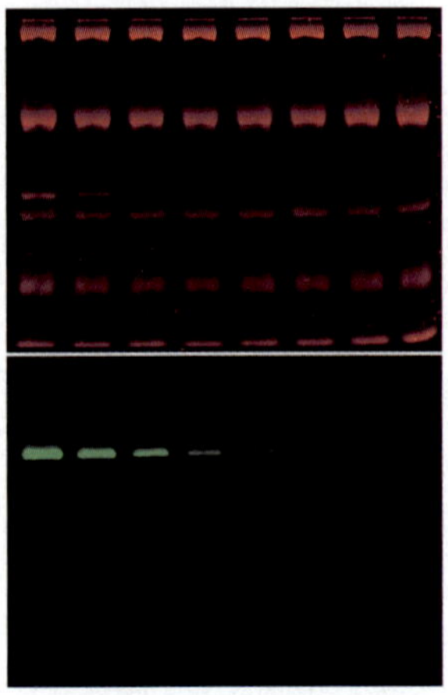

Figure 9.3.11 Analysis of β-glucuronidase activity and total protein content using SYPRO® Tangerine protein gel stain and ELF® β-D-glucuronide. Molecular weight standards containing decreasing amounts of *Escherichia coli* β-glucuronidase were run on an SDS-polyacrylamide gel and stained with SYPRO® Tangerine protein gel stain (orange, S12010) followed by incubation with the ELF® 97 β-D-glucuronidase substrate (green) to detect the activity of β-glucuronidase.

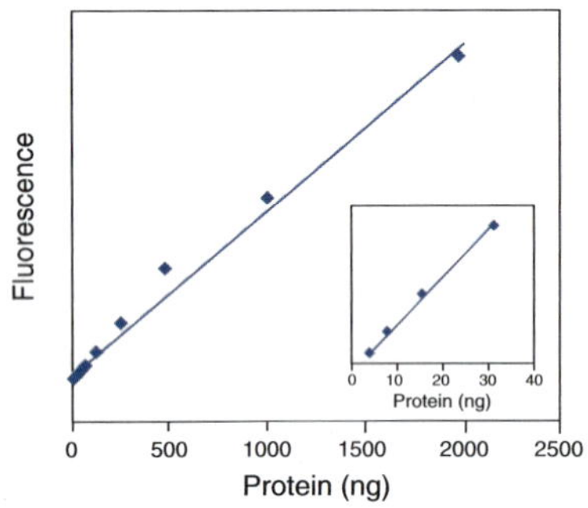

Figure 9.3.12 Linearity of SYPRO® Tangerine protein gel stain (S12010). A dilution series of carbonic anhydrase was electrophoresed through a 13% SDS-polyacrylamide gel and stained with SYPRO® Tangerine protein gel stain in 7% acetic acid. The fluorescence signal was quantified and plotted versus the protein amount, revealing a sensitivity limit of about 4 ng/band and a linear quantitation range over at least three orders of magnitude.

Figure 9.3.13 Protein molecular weight standards (P6649) separated on a 15% SDS-polyacrylamide gel and stained with SYPRO® Orange protein gel stain (S6650, S6651).

hydrophobicity of the target polypeptides. However, sensitivity of SYPRO® dye staining in native gels can be improved if gels are soaked in 0.05% SDS solution after electrophoresis but prior to staining. For optimal staining of proteins in 2D gels and IEF gels, we recommend SYPRO® Ruby protein gel stain (S12000, S12001, S21900).

Because SYPRO® Orange and SYPRO® Red dyes do not covalently bind to proteins, stained proteins can be subsequently analyzed by microsequencing or mass spectrometry.[4,25,31] These dyes, however, are not recommended for staining gels prior to blotting, as there is a significant loss of sensitivity when proteins are stained with SYPRO® Orange or SYPRO® Red dyes in typical western blotting buffers. To obtain maximum sensitivity when staining proteins in western blotting protocols, we recommend SYPRO® Tangerine protein gel stain (S12010) for staining proteins on the gel *before* blotting or SYPRO® Ruby protein blot stain (S11791) for staining proteins on nitrocellulose or PVDF membranes *after* blotting.

SYPRO® Orange and SYPRO® Red protein gel stains are available as 500 µL stock solutions in dimethylsulfoxide (DMSO), either in a single vial (S6650, S6653) or specially packaged as a set of 10 vials, each containing 50 µL (S6651, S6654). The reagents are supplied as 5000X concentrates; thus, 500 µL of either stain yields 2.5 L of staining solution. Photography of proteins in gels, which is essential for obtaining the maximum sensitivity, requires use of the SYPRO® photographic filter (S6656).

SYPRO® Tangerine Protein Gel Stain: Sensitive Protein Detection without Fixation for Electroelution, Zymography and Classroom Use

SYPRO® Tangerine protein gel stain[3,36] (S12010), which stains proteins in gels without the need for either acids or organic solvents, serves as an alternative to conventional protein stains that fix proteins in the gel.[36] Whereas SYPRO® Orange and SYPRO® Red stains require a staining solution containing acetic acid for maximum performance, staining with SYPRO® Tangerine protein gel stain is possible in almost any buffer that contains NaCl. Because proteins are not fixed during the staining procedure, they can be readily eluted from gels, used for zymography (in-gel enzyme activity assays, Figure 9.3.11) or analyzed by mass spectrometry.[36] SYPRO® Tangerine stain can also be used to stain gels before transferring the proteins to nitrocellulose or PVDF membranes for immunostaining[36] (western blotting). Like SYPRO® Orange and SYPRO® Red protein gel stains, SYPRO® Tangerine protein gel stain shows high sensitivity (down to ~4 ng/band) and a broad linear quantitation range (Figure 9.3.12). Easy to use, SYPRO® Tangerine protein gel stain is also ideal for use in educational settings, especially when used with UV-free blue-light transilluminators.

SYPRO® Tangerine stain is compatible with conventional SDS-polyacrylamide gel electrophoresis, but it is not recommended for 2D or IEF gels. Stained proteins can be visualized using a UV transilluminator, a blue-light transilluminator or a laser scanner. Photography of stained gels using Polaroid® 667 black-and-white print film requires use of the SYPRO® photographic filter (S6656) for optimal sensitivity. SYPRO® Tangerine protein gel stain[3,36] (S12010) is available as 500 µL of a 5000X concentrate in dimethylsulfoxide (DMSO), an amount sufficient to stain about 100 minigels.

SYPRO® Protein Gel Stain Starter Kit

SYPRO® Orange, SYPRO® Red and SYPRO® Tangerine protein gel stains are all available in a SYPRO® Protein Gel Stain Starter Kit (S12012) for first-time users. Each kit includes:

- 50 µL of SYPRO® Orange protein gel stain, sufficient for 5–20 minigels
- 50 µL of SYPRO® Red protein gel stain, sufficient for 5–20 minigels
- 50 µL of SYPRO® Tangerine protein gel stain, sufficient for 5–20 minigels
- SYPRO® protein gel stain photographic filter
- Detailed protocols

Protein Molecular Weight Standards

We offer a protein mixture for use as molecular weight markers in SDS-polyacrylamide gel electrophoresis (Figure 9.3.13). This broad-range marker mixture (P6649) contains a balanced formulation of 11 polypeptides with molecular weights from 6500 to 205,000 daltons. These protein molecular weight standards give rise to sharp, well-separated bands when the gel is stained with any of our protein gel or blot stains. The mixture provides amounts of marker

IMPORTANT NOTICE: The products described in this manual are covered by one or more Limited Use Label License(s). Please refer to the Appendix on page 971 and Master Product List on page 975. Products are For Research Use Only. Not intended for any animal or human therapeutic or diagnostic use.

proteins sufficient for loading about 200 gel lanes. We also have available PeppermintStick™ phosphoprotein molecular weight standards (P33350) and CandyCane™ glycoprotein molecular weight standards (C21852), which are described in Section 9.4.

Electrophoretic Mobility-Shift (Bandshift) Assay (EMSA) Kit

To make bandshift assays easier, we offer the Electrophoretic Mobility-Shift Assay (EMSA) Kit (E33075), which provides a fast and quantitative fluorescence-based method for detecting both nucleic acid and protein in the same gel [37] (Figure 9.3.14). This kit uses two different stains for detection: the fluorescent SYBR® Green EMSA nucleic acid gel stain for RNA or DNA and the luminescent SYPRO® Ruby EMSA protein gel stain for proteins. Because the nucleic acids and proteins are stained in the gel after electrophoresis, there is no need to prelabel the DNA or RNA with a radioisotope, biotin or a fluorescent dye before the binding reaction, and therefore there is no possibility that the label will interfere with protein binding. Staining takes only about 20 minutes for the nucleic acid stain, and about 4 hours for the subsequent protein stain, yielding results much faster than radioisotope labeling (which may require multiple exposure times) or chemiluminescence-based detection (which requires blotting and multiple incubation steps).

This kit also makes it possible to perform ratiometric measurements of nucleic acid and protein in the same band, providing more detailed information on the binding interaction.[37] The signals from the two stains are linear over a broad range, allowing accurate determination of the amount of nucleic acid and protein, even in a single band, with detection limits of ~1 ng for nucleic acids and ~20 ng for protein. Both stains can be visualized using a standard 300 nm UV transilluminator, a 254 nm epi-illuminator or a laser scanner (Figure 9.3.14). For optimal sensitivity using a UV transilluminator and Polaroid® 667 black-and-white print film, the SYPRO® photographic filter (S6656) and the SYBR® photographic filter (S7569, Section 8.4) are recommended. Digital images can easily be overlaid for a two-color representation of nucleic acid and protein in the gel.

The EMSA Kit contains sufficient reagents for 10 nondenaturing polyacrylamide minigel assays, including:

- SYBR® Green EMSA nucleic acid gel stain
- SYPRO® Ruby EMSA protein gel stain
- Trichloroacetic acid, for preparing the working solution of SYPRO® Ruby EMSA protein gel stain
- Concentrated EMSA gel-loading solution
- Concentrated binding buffer
- Detailed protocols

SYPRO® Photographic Filter

To achieve optimal sensitivity with the SYPRO® stains, it is essential to photograph the gel or blot because the camera's integrating capability can make bands visible that are not detected by eye. Photographs should be taken using a photographic filter with spectral properties closely matched to those of the fluorescent or luminescent dye used.

When using Polaroid® 667 black-and-white print film and UV illumination, protein gels or blots stained with any of the SYPRO® protein stains (including SYPRO® Ruby, SYPRO® Orange, SYPRO® Red and SYPRO® Tangerine gel stains as well as SYPRO® Ruby protein blot stain) should be photographed through the SYPRO® photographic filter (Figure 9.3.15, Figure 9.3.16), a 75 mm × 75 mm gelatin filter. Note that these gelatin filters are generally not suitable for use with portable or stationary gel-documentation systems or with CCD cameras.

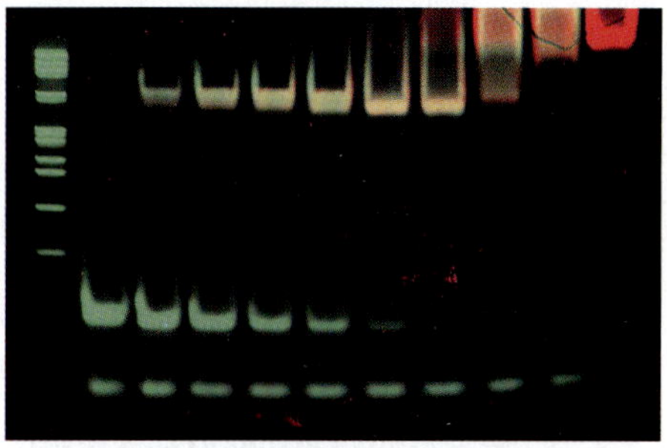

Figure 9.3.14 Titration of *lac* operator DNA with lac repressor protein. Increasing amounts of lac repressor protein were mixed with 40 ng of *lac* operator DNA, incubated for 20 minutes and then separated on a 6% nondenaturing polyacrylamide gel. The gel was stained with SYBR® Green EMSA stain (green) followed by SYPRO® Ruby EMSA stain (red), components of the Electrophoretic Mobility-Shift Assay Kit (E33075). After each staining, the image was documented using an FLA-3000 laser-based scanner (Fuji®) and the digital images pseudocolored and overlaid. Yellow bands indicate areas stained with both stains.

Figure 9.3.15 Molecular Probes® 75 mm × 75 mm gelatin photographic filter for use with Polaroid® black-and-white print film photography.

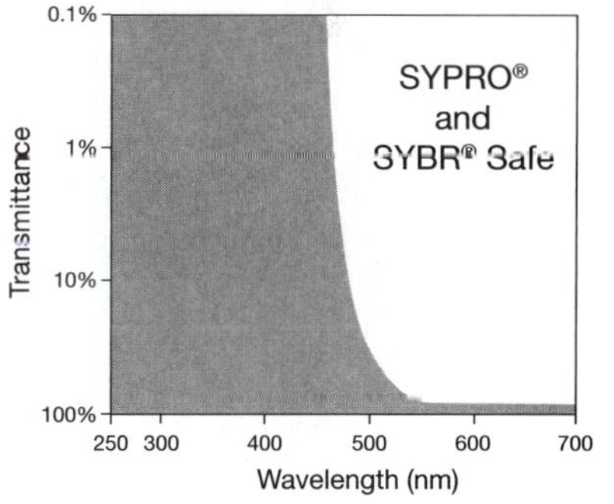

Figure 9.3.16 Transmission profile of the SYBR® Safe photographic filter (S37100) and the SYPRO® photographic filter (S6656), which are identical.

IMPORTANT NOTICE: The products described in this manual are covered by one or more Limited Use Label License(s). Please refer to the Appendix on page 971 and Master Product List on page 975. Products are For Research Use Only. Not intended for any animal or human therapeutic or diagnostic use.

NOTE 9.1

Monitoring Protein-Folding Processes with Environment-Sensitive Dyes

1,8-ANS and bis-ANS (A47, B153; Section 13.5) have proven to be sensitive probes for partially folded intermediates in protein-folding pathways. These applications take advantage of the strong fluorescence enhancement exhibited by these amphiphilic dyes when their exposure to water is lowered (Figure 1, Figure 2). Consequently, fluorescence of ANS increases substantially when proteins to which it is bound undergo transitions from unfolded to fully or partially folded states that provide shielding from water. Molten globule intermediates are characterized by particularly high ANS fluorescence intensities due to the exposure of hydrophobic core regions that are inaccessible to the dye in the native structure.[1] Binding of 1,8-ANS and bis-ANS to proteins is noncovalent and involves a combination of electrostatic and hydrophobic modes.[2–6] Some investigators have noted that the dye-binding event itself may induce protein conformational changes, indicating the advisability of correlating ANS fluorescence measurements with data obtained using other physical techniques.[5,7–9] In particular, high-resolution structural analysis of an ANS–protein complex by X-ray crystallography has demonstrated the occurrence of local rearrangements of the protein structure to accommodate the dye.[10]

The general mechanism of protein-folding detection by ANS has been developed as the basis of fluorescence thermal shift (a.k.a. differential scanning fluorometry) assays for high-throughput analysis of protein stability.[11–14] The assay readout is a profile of protein–dye complex fluorescence intensity as a function of temperature. Profiles are obtained and compared for multiple samples in which environmental or structural factors influencing protein stability are systematically varied. Thermocycler instruments designed for real-time PCR monitoring provide a readily adaptable instrument platform for these measurements.[12,13,15] Most high-throughput fluorescence thermal shift assays use SYPRO® Orange dye (S6650, S6651; Section 9.3) instead of ANS. Other environment-sensitive dyes with demonstrated utility include SYPRO® Red dye (S6653, S6654; Section 9.3), Dapoxyl® sulfonic acid (D12800, Section 13.5), nile red (N1142, Section 13.5) and CPM (D346, Section 2.3). Within the broad context of protein stability optimization, fluorescence thermal shift assays have many applications including:

- Analysis of ligand binding to proteins of unknown function[16]
- Identification of protein–protein interaction inhibitors[17]
- Analysis of protein stabilization by peptide aptamers[18] and amino acid ligands[19]
- Characterization of engineered protein variants[20]
- Optimization of protein crystallization conditions[21,22]
- Enhancement of recombinant protein quality and yield[23]

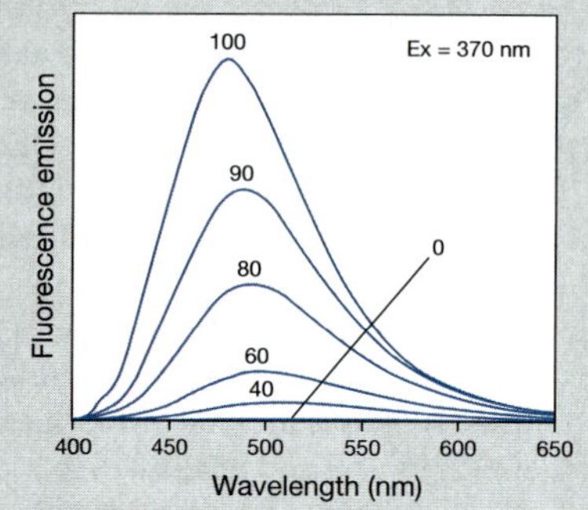

Figure 1 Fluorescence emission spectra of equal concentrations of 1,8-ANS (A47) in ethanol:water mixtures. The labels adjacent to each curve indicate the percentage of ethanol in the solvent mixture.

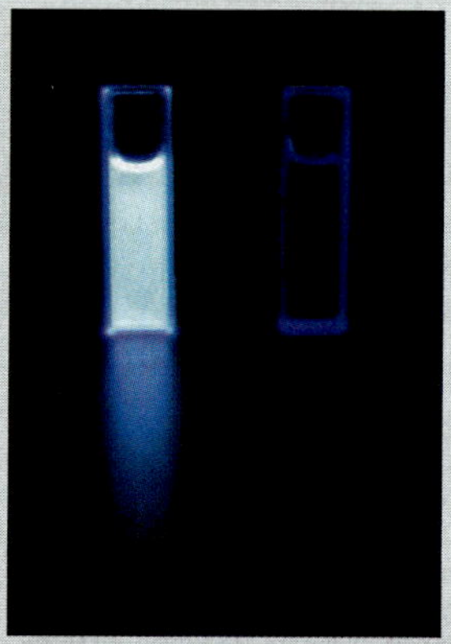

Figure 2 Fluorescence enhancement of 1,8-ANS (1-anilinonaphthalene-8-sulfonic acid, A47) upon binding to protein. The image shows aqueous solutions of 1,8-ANS excited by ultraviolet light. Addition of protein (bovine serum albumin) to the solution in the cuvette on the left results in intense blue fluorescence. In comparison, the fluorescence of uncomplexed free dye in the cuvette on the right is negligible.

1. Biopolymers (1991) 31:119; **2.** Biophys J (1998) 74:422; **3.** Biophys J (1996) 70:69; **4.** Biochemistry (1994) 33:7536; **5.** Biopolymers (1999) 49:451; **6.** Biochem J (1992) 282:589; **7.** Biochemistry (1999) 38:13635; **8.** Biophys J (1998) 75:2195; **9.** Biochemistry (2001) 40:4484; **10.** Proc Natl Acad Sci U S A (2000) 97:6345; **11.** Mol Biosyst (2009) 5:217; **12.** Methods Mol Biol (2008) 426:299; **13.** Nat Protoc (2007) 2:2212; **14.** J Biomol Screen (2001) 6:429; **15.** Anal Biochem (2004) 332:153; **16.** J Biol Chem (2005) 280:11704; **17.** J Gen Virol (2009) 90:1294; **18.** J Mol Biol (2010) 395:871; **19.** Biochemistry (2008) 47:13974; **20.** J Am Chem Soc (2009) 131:3794; **21.** Anal Biochem (2006) 357:289; **22.** Proc Natl Acad Sci U S A (2006) 103:15835; **23.** J Biomol Screen (2007) 12:418.

IMPORTANT NOTICE: The products described in this manual are covered by one or more Limited Use Label License(s). Please refer to the Appendix on page 971 and Master Product List on page 975. Products are For Research Use Only. Not intended for any animal or human therapeutic or diagnostic use.

Coomassie Fluor™ Orange Protein Gel Stain

Coomassie Fluor™ Orange protein gel stain (C33250, C33251) provides a fast, simple and sensitive method for staining of proteins in SDS-polyacrylamide electrophoretic gels. Gels do not need to be washed before staining with the Coomassie Fluor™ Orange dye or destained after staining. After electrophoresis, the gel is simply stained, rinsed and photographed on a standard UV transilluminator. Our premixed and ready-to-use Coomassie Fluor™ Orange protein gel stain offers the following advantages over conventional colorimetric stains:

- **High sensitivity**. Coomassie Fluor™ Orange protein gel stain detects as little as 8 ng of protein per minigel band (Figure 9.3.17). This sensitivity matches the best colloidal Coomassie stains and exceeds standard Coomassie brilliant blue stains.
- **Broad linear range of detection**. The fluorescence intensity of Coomassie Fluor™ Orange dye–stained bands is linear with protein quantity over at least two orders of magnitude, permitting accurate quantitation.
- **Rapid staining**. Staining is complete in less than an hour, and there is no risk of overstaining the gel.
- **Compatibility with standard laboratory equipment.** Stained proteins can be visualized using a standard 300 nm UV transilluminator or a laser scanner (Figure 9.3.18). For optimal sensitivity using a UV transilluminator and Polaroid® 667 black-and-white print film, the SYPRO® photographic filter (S6656) is recommended.
- **Low protein-to-protein variability.** Because Coomassie Fluor™ Orange dye interacts with the SDS coat around proteins in the gel, it gives more consistent staining between different types of proteins, as compared with Coomassie brilliant blue staining, and it never exhibits negative staining. Coomassie Fluor™ Orange dye also stains glycoproteins.
- **High selectivity for proteins.** Coomassie Fluor™ Orange protein gel stain detects a variety of proteins down to ~6500 daltons without staining nucleic acid or lipopolysaccharide contaminants that are sometimes found in protein preparations derived from cell or tissue extracts.

Coomassie Fluor™ Orange protein gel stain is not recommended for staining proteins in 2D, IEF or nondenaturing gels; for these applications we recommend our SYPRO® Ruby protein gel stain (S12000, S12001, S21900).

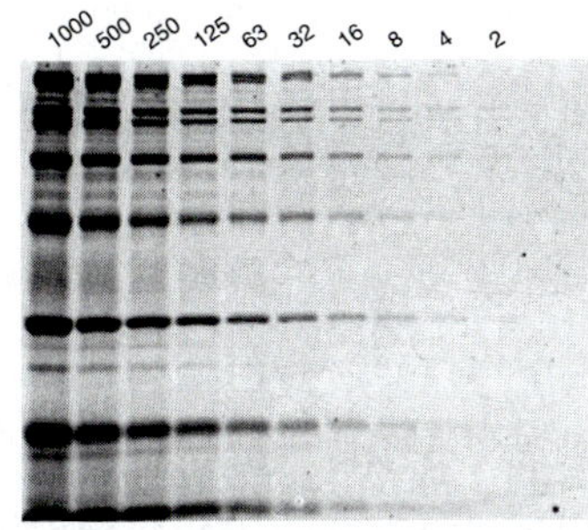

Figure 9.3.17 Protein molecular weight standards (in nanograms) were separated on a 12% SDS-polyacrylamide gel and then stained with Coomassie Fluor™ Orange protein gel stain (C33250, C33251). The colors have been digitally reversed.

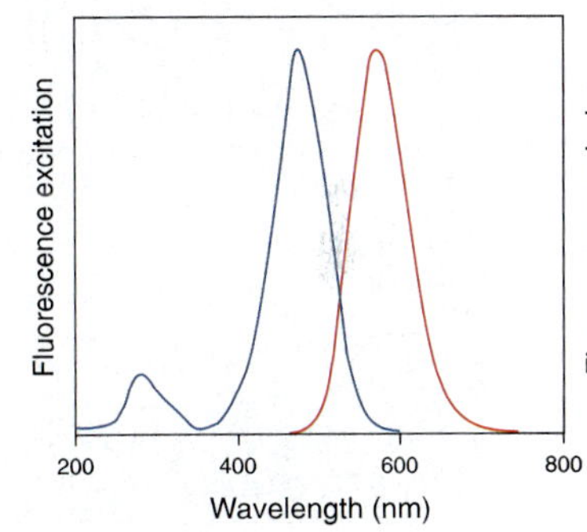

Figure 9.3.18 Fluorescence excitation and emission spectra of the Coomassie Fluor™ Orange protein gel stain (C33250, C33251) in a solution of 150 µg/mL bovine serum albumin (BSA) with 0.05% SDS.

Rhinohide™ Polyacrylamide Gel Strengthener

Rhinohide™ polyacrylamide gel strengthener improves upon classic polyacrylamide gel technology by making gels much stronger, providing easier handling and much greater resistance to tearing without adverse side effects (Figure 9.3.19). Rhinohide™ polyacrylamide gel strengthener is highly recommended for low-percentage gels, large-format gels and gels subject to multiple staining and handling steps, and it is compatible with fluorescent protein stains as well as silver and Coomassie stains.

SDS-polyacrylamide gels supplemented with Rhinohide™ polyacrylamide gel strengthener exhibit resolution capabilities comparable to traditional SDS-polyacrylamide gels, with clear, focused bands and without the undesirable side effects common with other gel strengtheners. For example, film-backed gels and polyester fabric–reinforced gels interfere with blotting techniques and can negatively affect protein staining. Alternatively, strengthening gels by the addition of pre-formed polymers causes turbidity and can produce serious spot-morphology artifacts, such as the distortion of high molecular weight bands or doubling of protein spots in the molecular weight dimension of 2D gels.[18] Rhinohide™ polyacrylamide gel strengthener produces gels with excellent transparency, providing exceptional image viewing and scanning of fluorescently stained gels with minimal background staining. We offer a concentrated form of the Rhinohide™ polyacrylamide gel strengthener (R33400) for adding to existing stock solutions of acrylamide/bis-acrylamide (37.5:1). Because prestained proteins, such as prestained molecular weight markers, will not migrate correctly in acrylamide gels containing the Rhinohide™ polyacrylamide gel strengthener, we recommend using only unstained proteins as markers.

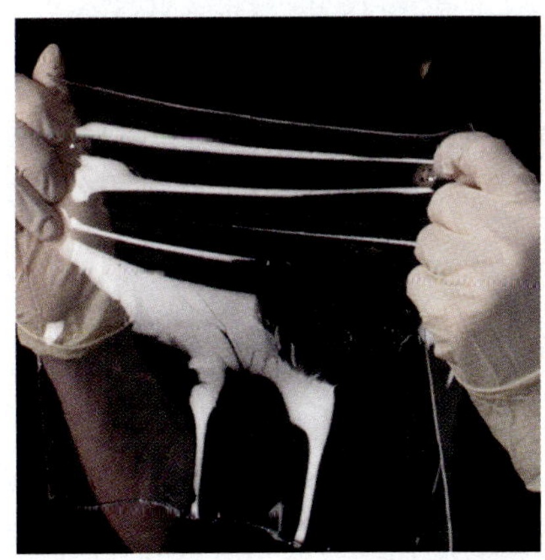

Figure 9.3.19 Demonstration of the strength and durability of gels made with the Rhinohide™ polyacrylamide gel strengthener (R33400).

IMPORTANT NOTICE: The products described in this manual are covered by one or more Limited Use Label License(s). Please refer to the Appendix on page 971 and Master Product List on page 975. Products are For Research Use Only. Not intended for any animal or human therapeutic or diagnostic use.

SYPRO® Ruby Protein Blot Stain

To characterize specific proteins in complex mixtures, proteins are frequently separated by electrophoresis, then blotted onto nitrocellulose or PVDF (poly(vinylidene difluoride)) membranes (blots) for immunostaining[38] (western blotting), glycoprotein staining, sequencing or mass spectrometry. Total-protein stains provide valuable information about the protein samples on blots, making it possible to assess the efficiency of protein transfer to the blot, detect contaminating proteins in the sample and compare the sample with molecular weight standards. For blots of 2D gels, staining of the entire protein profile makes it easier to localize a protein to a particular spot in the complex protein pattern. The superior properties of our fluorescent and luminescent protein stains, compared with conventional colorimetric stains, make it possible to quickly and easily obtain this information without running duplicate gels. Our luminescent and fluorescent protein staining technology can also be combined with fluorescent reagents for glycoprotein and phosphoprotein detection, which are described in detail in Section 9.4, to create multiparameter staining for multiplex analyses.

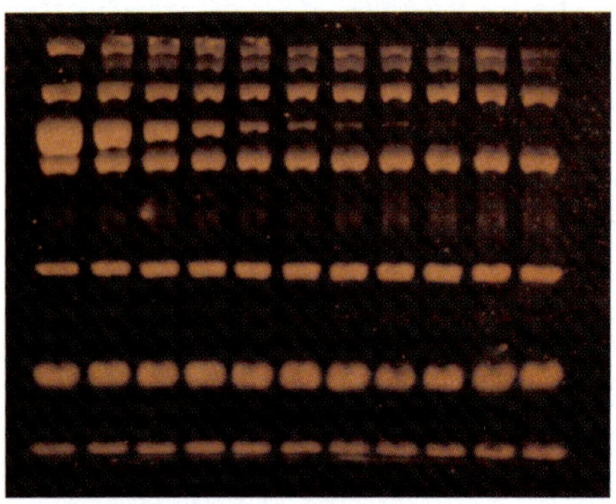

Figure 9.3.20 Protein detection with SYPRO® Ruby protein blot stain (S11791). Samples of protein molecular weight standards containing a twofold dilution series of α-tubulin, starting with 1 µg of tubulin in the left-most lane, were run on an SDS-polyacrylamide gel, blotted onto a PVDF membrane and stained with SYPRO® Ruby protein blot stain.

SYPRO® Ruby protein blot stain[3,16,39,40] (Table 9.3) is designed to provide fast and highly sensitive detection of proteins blotted onto membranes, making it easy to assess the efficiency of protein transfer to the blot and to determine if lanes are loaded equally[41] (Figure 9.3.20). Because the stain does not covalently alter the proteins, it can be used to locate proteins on blots before sequencing or mass spectrometry.[4,39] It can also be used before chromogenic, fluorogenic or chemiluminescent immunostaining procedures to locate molecular weight markers and visualize the total-protein profile in the sample.[39] Furthermore, the stain does not interfere with subsequent identification techniques, minimizing the need to run duplicate gels, vastly simplifying the comparison of total protein and target protein on western blots, and allowing precise localization of the target protein relative to other proteins on electroblots of 2D gels. SYPRO® Ruby protein blot stain is also compatible with our Pro-Q® Emerald glycoprotein blot stains for glycoproteins[42] (Section 9.4). The superior properties of SYPRO® Ruby protein blot stain, as compared with conventional protein blot stains, make it possible to routinely stain blots for total protein before continuing with specific protein detection techniques. SYPRO® Ruby protein blot stain[3,10,16,39,40] (S11791) combines the following superior staining characteristics:

- **High sensitivity**. SYPRO® Ruby protein blot stain can detect as little as 0.25–1 ng protein/mm^2 (~2–8 ng/band) blotted onto PVDF or nitrocellulose membranes after only 15 minutes of staining[39] (Figure 9.3.20). This sensitivity on blots is far better than that of colorimetric stains, such as Ponceau S, amido black or Coomassie brilliant blue, and rivals the best colloidal gold blot-staining techniques (Figure 9.3.21).

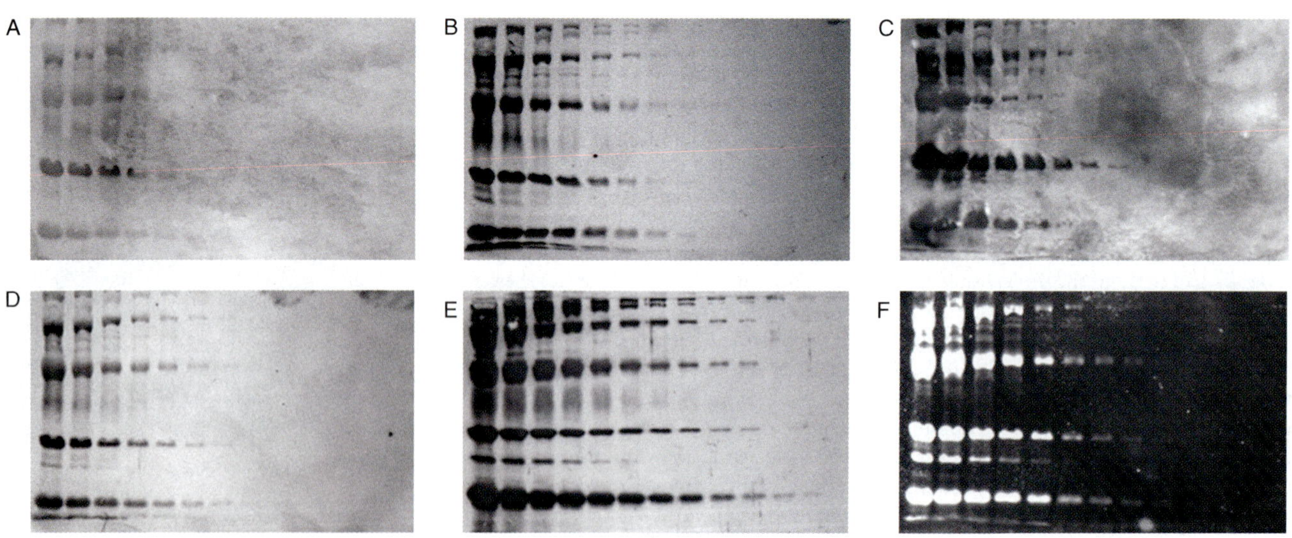

Figure 9.3.21 Comparison of commonly used stains for proteins on blots. Twofold serial dilutions of protein molecular weight standards ranging from 2000 to 1 ng/band were run on six identical SDS-polyacrylamide gels and blotted to PVDF membrane. The membranes were then stained with **A**) Ponceau S stain, **B**) Coomassie brilliant blue stain, **C**) colloidal silver stain, **D**) amido black stain, **E**) colloidal gold stain or **F**) SYPRO® Ruby protein blot stain (S11791).

IMPORTANT NOTICE: The products described in this manual are covered by one or more Limited Use Label License(s). Please refer to the Appendix on page 971 and Master Product List on page 975. Products are For Research Use Only. Not intended for any animal or human therapeutic or diagnostic use.
www.invitrogen.com/probes

- **Rapid total-protein staining procedure.** The SYPRO® Ruby protein blot-staining protocol takes less than an hour—including fixation and wash steps—and maximum sensitivity is achieved after only 15 minutes of dye staining, even for some peptides as small as seven amino acids. In contrast, gold or silver staining procedures may require overnight incubations to achieve maximum sensitivity and usually include extensive wash procedures that must be carefully timed.
- **Compatibility with western blot protocols.** Staining the total-protein profile on the blot minimizes guesswork about transfer efficiency and removes the need to run two gels for comparison of total and target protein.[41] SYPRO® Ruby protein blot stain is gentle and, unlike colorimetric or colloidal gold blot stains, does not interfere with subsequent colorimetric or chemiluminescent immunodetection of proteins on western blots.[39]
- **Compatibility with standard microsequencing and mass spectrometry protocols.** Whereas colloidal gold, Coomassie brilliant blue and amido black staining can interfere with post-staining analysis,[43] SYPRO® Ruby protein blot stain binds noncovalently to proteins and is thus fully compatible with Edman sequencing or mass spectrometry.[4,39]

SYPRO® Ruby protein blot stain is based on an organometallic ruthenium complex. Because the ruthenium complex has dual-excitation maxima (Figure 9.3.7), the dye exhibits luminescence upon excitation with either UV or visible light. This property makes it possible to visualize the luminescence with many types of instruments, including UV epi-illumination sources, UV or blue-light transilluminators and a variety of laser-scanning instruments, including those with excitation light at 450 nm, 473 nm, 488 nm or 532 nm. Also, SYPRO® Ruby dye–stained proteins can be indirectly excited by the chemiluminescence of the high-energy intermediate produced in the reaction of bis-(2,4,6-trichlorophenyl) oxalate (TCPO) with H_2O_2.[44] The red luminescence of the ruthenium complex has a peak at ~618 nm that is well separated from these excitation peaks, minimizing the amount of background signal seen from the excitation source. The staining signal can be documented using Polaroid® 667 black-and-white print film and a SYPRO® photographic filter (S6656; Figure 9.3.15), using a CCD-based imaging system equipped with a 600 nm bandpass or 490 nm longpass filter, or by using the appropriate filter set and software in a laser scanner. SYPRO® Ruby protein blot stain has exceptional photostability, allowing long exposure times for maximum sensitivity.

SYPRO® Ruby protein blot stain (S11791) is supplied as a 1X staining solution, which is sufficient for staining ~1600 cm^2 of blotting membrane, and is accompanied by a detailed protocol for its use.

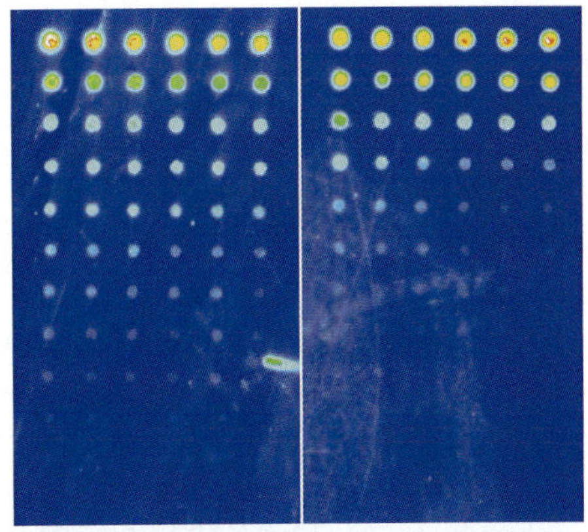

Figure 9.3.22 Protein detection on microarrays using BODIPY® FL-X SE or SYPRO® Ruby protein blot stain. Bovine serum albumin was arrayed onto a PVDF membrane using a Packard BioScience PiezoTip dispenser. The array contains 72 spots with 12 dilutions of the dye, in replicates of 6, ranging from 666–0.325 pg per spot. The proteins were stained with either BODIPY® FL-X SE (D6102, left) in sodium borate buffer, pH 9.5, or with SYPRO® Ruby protein blot stain (S11791, right). Arrays were imaged on a ScanArray® 5000XL microarray analysis system (Packard BioScience) and pseudocolored such that the different colors indicate different signal intensities. The sensitivity limit was 1.3–2.6 pg of protein in a 175 µm spot for the BODIPY® FL-X SE detection technique and 104 pg of protein per 175 µm spot for the SYPRO® Ruby protein blot stain technique.

Protein Detection on Microarrays

SYPRO® Ruby Protein Blot Stain for Reversible Protein Detection on Microarrays

We have found that SYPRO® Ruby protein blot stain (S11791) performa particularly well when staining proteins on PVDF microarrays for quality control and normalization purposes.[45] SYPRO® Ruby protein blot stain shows good sensitivity on protein microarrays (Figure 9.3.22) and should be very useful for staining proteins before exposing the microarray to the sample. The stain washes off PVDF membranes very easily under conditions used with typical western blot blocking buffers.

Reactive Fluorescent Dyes for Permanent Protein Detection on Microarrays

The BODIPY® reactive dyes described in Section 1.4 label amine groups (predominantly lysine residues) on proteins. As compared to the reversible SYPRO® blot stains described above, the amine-reactive BODIPY® FL-X succinimidyl ester (D6102) shows even greater sensitivity for microarray staining (Figure 9.3.22), making it useful for quality control testing or as an internal normalization standard. Both BODIPY® FL-X succinimidyl ester and and the red-fluorescent BODIPY® TR-X succinimidyl ester (D6116) are particularly effective general stains for proteins on PVDF membranes[10,46] (Table 9.3).

This unique method of staining total protein on blots with the reactive BODIPY® dyes has an approximately 30-fold linear dynamic range (Figure 9.3.23), although the absolute intensity between proteins may vary somewhat with the nature of the protein. Reactive BODIPY® dye–based staining is designed to be rapid, simple and highly sensitive, permitting detection of as little as 4 ng of a protein per band in about an hour. Because the reactive dyes form a

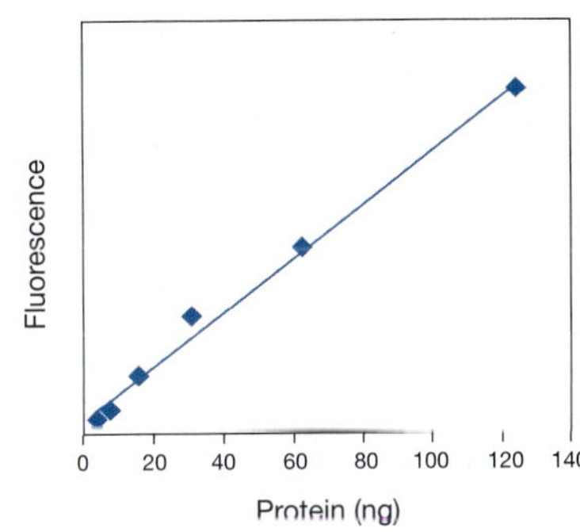

Figure 9.3.23 Linear dynamic range of detection for BODIPY® FL-X succinimidyl ester, used as a blot stain. A two-fold dilution series of molecular weight markers (P6649) was loaded onto a gel, electrophoresed and electroblotted to a PVDF membrane. The proteins on the blot were then stained with BODIPY® FL-X succinimidyl ester (D6102). The fluorescence intensity for one of the proteins (carbonic anhydrase) was measured and plotted against the amount of protein loaded in the lane. The result shows an approximately linear dynamic range from 4 ng to 125 ng.

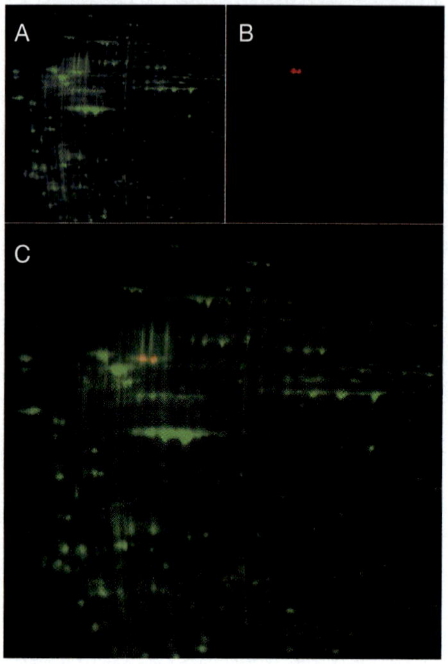

Figure 9.3.24 Dual labeling of proteins on a western blot. Proteins from a rat fibroblast lysate were separated by 2D gel electrophoresis and blotted onto a PVDF membrane. The proteins are acidic to basic (left to right) and high to low molecular weight (top to bottom). After electrophoresis, the blot was stained with BODIPY® FL-X succinimidyl ester (green, D6102) to detect total protein. The blot was then incubated with an anti-α-tubulin antibody (A11126), followed by the alkaline phosphatase conjugate of goat anti-mouse IgG antibody (G21060). Finally, the blot was stained with DDAO phosphate (red, D6487), which produces a red-fluorescent product in the presence of alkaline phosphatase. The fluorescent signals were visualized using UV epi-illumination. The signals were documented separately using appropriate photographic filters (**A**, **B**), and the resulting images overlaid (**C**).

covalent bond with the protein, the staining is permanent and lasts through any subsequent blot manipulations. The covalent modification appears to minimally interfere with subsequent immunostaining, as we have successfully performed two-color labeling with reactive BODIPY® dyes and either fluorogenic immunostains or glycoprotein stains and visualized both stains simultaneously (Figure 9.3.24). Simultaneous dual labeling of a sample simplifies localization of a specific protein with respect to other proteins in the sample, particularly on electroblots of 2D gels, pairs of which are difficult to align. Note that because reaction of the dye covalently modifies the protein at random locations, staining by the amine-reactive BODIPY® dyes may complicate or preclude subsequent analysis by mass spectrometry or microsequencing.

Protein Detection in Capillary Electrophoresis

Capillary electrophoresis (CE) is an exceptionally powerful tool for the resolution of biomolecules.[47,48] Fluorescent detection of proteins that are separated by capillary electrophoresis can occur either during the run—the more common procedure—or subsequent to the separation on isolated fractions. When detected during the electrophoretic separation, the protein is either derivatized with a fluorescent reagent prior to the separation or labeled with a fluorescent dye that is incorporated into the running medium. In general, the same reagents may be useful for fluorometric detection of peptides and proteins that are separated by high-performance liquid chromatography (HPLC). In addition to the total-protein staining techniques described below, many selective staining techniques, such as the use of BODIPY® FL GTP-γ-*S* (G22183, Section 17.3) to detect GTP-binding proteins,[49] can be applied to proteins separated by capillary electrophoresis.

Use of SYPRO® Dyes for Capillary Electrophoresis

SDS–capillary gel electrophoresis (SDS-CGE) separates proteins based on principles similar to those of standard slab-gel electrophoresis, but with the advantages of faster run times, higher resolution and greater sensitivity. The use of online detection by laser-induced fluorescence (LIF) increases the sensitivity several orders of magnitude over UV detection, eliminates the time spent staining and photographing the gel and allows for the possibility of automated sample processing. SDS-CGE analyzed by LIF is widely used for the separation and identification of DNA fragments and has increased the efficiency of genomics, DNA typing and forensics laboratories.[47,48] SDS-CGE promises to be just as useful for proteomics laboratories and other laboratories that require characterization of a large number of protein samples.

For SDS-CGE of protein samples, the amine groups of lysine residues and the N-terminus of the proteins are typically derivatized with a fluorescent or fluorogenic dye such as the ATTO-TAG™ CBQCA[50,51] (A2333, A6222) or ATTO-TAG™ FQ[52–55] (A2334, A10192) reagents before separation in the capillary. The derivatized proteins are then coated with SDS and travel through the capillary gel towards the positive electrode based on their size, with smaller proteins traveling faster. The derivatized proteins are detected by fluorescence emitted as they pass a laser that excites the fluorophores.

One disadvantage of this labeling technique is that the proteins generally contain multiple amine groups, each of which can react with the derivatization reagent. Typically, only a few of the amine groups on each protein molecule react, and the result is an enormous number of different derivatives, creating broad peaks that may be difficult to correlate with the original protein's structure or abundance.[55] In addition, variations between runs make it difficult to reproducibly estimate molecular weights. In contrast, use of SYPRO® Red protein gel stain (S6653, S6654) to prestain SDS-coated proteins allows more accurate determination of molecular weights because the proteins are relatively uniformly coated with SDS and the dye. Staining the SDS-coated proteins with SYPRO® Red protein gel stain leads to molecular weight determinations similar in accuracy to those achieved with polyacrylamide slab gels, with a limit of detection estimated to exceed the detection limit of silver staining in slab gels.[56] The NanoOrange® reagent (N6666, Section 9.2) is also reported to be an effective reagent for detecting proteins that have been separated by microchip capillary electrophoresis.[57–59]

IMPORTANT NOTICE: The products described in this manual are covered by one or more Limited Use Label License(s). Please refer to the Appendix on page 971 and Master Product List on page 975. Products are For Research Use Only. Not intended for any animal or human therapeutic or diagnostic use.
www.invitrogen.com/probes

Protein Derivatization Reagents

Several of the same reagents that were described in Section 9.2 for protein quantitation in solution are also useful for peptide and protein derivatization, either prior to or following separation by capillary electrophoresis. However, chemical derivatization prior to separation is likely to change the electronic charge and always changes the mass of the protein. Furthermore, incomplete derivatization of amines or thiols on the protein can lead to a pure protein resolving into multiple species in the electrophoretogram.

In an improved procedure for fluorescent analysis of peptides by capillary electrophoresis, Zhou and colleagues [60] modified all α- and ε-amino groups of the peptide with phenyl isothiocyanate. Following one cycle of Edman degradation, the single free α-amino group was modified with fluorescent reagents to give a homogeneous, dye-labeled peptide.

The preferred reagents for derivatizing amine residues in proteins either prior to or following electrophoretic separation are those that are essentially nonfluorescent until reacted with the protein. These include:

- ATTO-TAG™ CBQCA, which is available in the ATTO-TAG™ CBQCA Amine-Derivatization Kit (A2333) and as a stand-alone reagent (A6222). ATTO-TAG™ CBQCA reacts with primary amines to form highly fluorescent isoindoles [50,61–69] and has been extensively used for the derivatization of amino acids,[51,61,70–72] peptides [63,73–76] and carbohydrates [62,64,67,77–81] prior to capillary electrophoretic separation. ATTO-TAG™ CBQCA has been used to derivatize a fusion protein expressed in the bacterium *Escherichia coli* before purification by capillary zone electrophoresis. After purification, the fluorescent isoindole can be removed by acid treatment to allow sequencing of the purified protein.[50]
- ATTO-TAG™ FQ (3-(2-furoyl)quinoline-2-carboxaldehyde), which is available in our ATTO-TAG™ FQ Amine-Derivatization Kit (A2334) and as a stand-alone reagent (A10192). ATTO-TAG™ FQ has been used as a protein detection reagent in capillary electrophoresis.[82–85] It has been reported that ATTO-TAG™ FQ can detect as little as 200 attomoles of a protein by capillary electrophoresis.[84] Excitation of amine derivatives of ATTO-TAG™ FQ by the 488 nm spectral line of the argon-ion laser is more efficient than that of ATTO-TAG™ CBQCA derivatives. A report [85] describes the solid-phase derivatization of dilute peptide solutions (10^{-8} M) that have been immobilized on Immobilon® CD membranes. This technique permits the quantitative derivatization and analysis by capillary electrophoresis of only a few picomoles of the analyte.
- Fluorescamine (F2332, F20261), a nonfluorescent reagent that rapidly reacts with amines to give a fluorescent product. Fluorescamine has been used for solution quantitation of proteins and peptides (Section 9.2). It is also useful as a peptide and protein detection reagent for capillary electrophoresis.[76,86,87] Use of fluorescamine to derivatize a standard protein of known molecular weight together with use of the ATTO-TAG™ FQ reagent to derivatize the sample protein allows the sample to be run simultaneously with the standard, improving the accuracy of molecular weight determination.[83] Chiral separation of fluorescamine-labeled amino acids has been optimized using capillary electrophoresis in the presence of hydroxypropyl-β-cyclodextran, a method designed for use in extraterrestrial exploration on Mars.[88]
- Dialdehydes OPA and NDA (P2331MP, N1138), which react with amines in the presence of a nucleophile to give fluorescent products. These inexpensive reagents have been used for capillary electrophoresis of peptides and proteins.[89–96]

Chapter 1 describes a variety of other amine-reactive reagents that have been used or may be useful for peptide and protein detection in capillary electrophoresis, including dansyl chloride [97] (D21), NBD chloride (C20260), NBD fluoride (F486), FITC [60,98] (F143), Pacific Blue™ succinimidyl ester (P10163)[99] and other common reagents described in Section 1.8.

Derivatization reagents that react with thiols or other functional groups have also been used for protein detection in capillary electrophoresis. Thiol-reactive reagents that are essentially nonfluorescent until conjugated to thiols, such as the coumarin maleimides CPM and DACM (D346, D10251), monobromobimane (M1378, M20381) and *N*-(1-pyrene)maleimide (P28), should work well in this application and are described in Section 2.3. Although intrinsically fluorescent, BODIPY® iodoacetamides and maleimides (Section 2.2) have also been used to detect thiol-containing proteins in SDS gels and by reverse phase HPLC,[100] and 5-iodoacetamidofluorescein (I30451, Section 2.2) has been used to label proteins for analysis by capillary electrophoresis.[101]

IMPORTANT NOTICE: The products described in this manual are covered by one or more Limited Use Label License(s). Please refer to the Appendix on page 971 and Master Product List on page 975. Products are For Research Use Only. Not intended for any animal or human therapeutic or diagnostic use.

REFERENCES

1. Biochim Biophys Acta (1963) 71:377; **2.** Methods Enzymol (1990) 182:477; **3.** Biotechniques (2000) 28:944, 950; **4.** Electrophoresis (2001) 22:906; **5.** Electrophoresis (2002) 23:2203; **6.** Proteomics (2001) 1:54; **7.** Biotechniques (2001) 31:146; **8.** Electrophoresis (2001) 22:829; **9.** Proteomics (2006) 6:5385; **10.** Electrophoresis (2001) 22:950; **11.** Electrophoresis (2000) 21:3657; **12.** Electrophoresis (2000) 21:3673; **13.** Electrophoresis (2000) 21:2509; **14.** Electrophoresis (2000) 21:1082; **15.** Electrophoresis (2000) 21:1037; **16.** Electrophoresis (2000) 21:1123; **17.** Scientist (2000) 14:26; **18.** Electrophoresis (2000) 21:486; **19.** Proteomics (2001) 1:377; **20.** J Proteome Res (2007) 6:1418; **21.** J Biol Chem (2006) 281:23686; **22.** Electrophoresis (2003) 24:3508; **23.** Anal Biochem (1997) 248:168; **24.** Biotechnol Intl (1997) 1:339; **25.** J Mass Spectrom (2000) 35:672; **26.** Anal Biochem (1996) 239:238; **27.** Anal Biochem (1996) 239:223; **28.** Electrophoresis (2007) 28:749; **29.** Anal Biochem (1987) 165:33; **30.** J NIH Res (1995) 7:82; **31.** IDrugs (1998) 1:299; **32.** J Chromatogr A (2000) 894:329; **33.** Anal Chem (2000) 72:2519; **34.** Anal Chem (2001) 73:1207; **35.** Anal Biochem (2001) 292:40; **36.** Electrophoresis (2000) 21:497; **37.** Electrophoresis (2004) 25:2439; **38.** Proc Natl Acad Sci U S A (1979) 76:4350; **39.** Anal Biochem (1999) 276:129; **40.** Electrophoresis (2000) 21:2196; **41.** J Biol Chem (2000) 275:32846; **42.** J Chromatogr B Analyt Technol Biomed Life Sci (2003) 793:127; **43.** Electrophoresis (1998) 19:752; **44.** J Chromatogr B Analyt Technol Biomed Life Sci (2003) 793:75; **45.** Mol Cell Proteomics (2008) 7:1902; **46.** Electrophoresis (2001) 22:896; **47.** Anal Chem (2007) 79:345; **48.** Electrophoresis (1998) 19:2695; **49.** Anal Chem (2003) 75:4297; **50.** J Chromatogr B Biomed Sci Appl (1997) 695:67; **51.** Anal Chem (1994) 66:3512; **52.** Anal Chem (2003) 75:3163; **53.** Anal Chem (2003) 75:3502; **54.** Mol Cell Proteomics (2002) 1:69; **55.** Anal Chem (1998) 70:2493; **56.** Electrophoresis (1998) 19:2169; **57.** Biotechniques (2003) 34:850; **58.** Anal Chem (2001) 73:4994; **59.** Anal Chem (2000) 72:4608; **60.** J Chromatogr (1992) 608:239; **61.** J Neurosci Methods (1996) 65:33; **62.** Anal Chem (1994) 66:3477; **63.** Anal Chem (1993) 65:563; **64.** Electrophoresis (1993) 14:373; **65.** Anal Chem (1991) 63:413; **66.** Anal Chem (1991) 63:408; **67.** J Chromatogr (1991) 559:223; **68.** Proc Natl Acad Sci U S A (1991) 88:2302; **69.** J Chromatogr (1990) 499:579; **70.** Anal Chim Acta (1995) 299:319; **71.** Chromatographia (1994) 39:7; **72.** J Chromatogr B Biomed Appl (1994) 659:185; **73.** J Chromatogr A (1995) 716:389; **74.** Biopolymers (1993) 33:1299; **75.** J Chromatogr B Analyt Technol Biomed Life Sci (2006) 843:240; **76.** J Chromatogr (1990) 519:189; **77.** J Chromatogr A (1995) 716:221; **78.** Anal Chem (1994) 66:3466; **79.** Glycobiology (1994) 4:397; **80.** Proc Natl Acad Sci U S A (1993) 90:9451; **81.** Anal Chem (1992) 64:973; **82.** J Chromatogr B Analyt Technol Biomed Life Sci (2003) 793:141; **83.** Electrophoresis (1998) 19:2175; **84.** Anal Chem (1997) 69:3015; **85.** Electrophoresis (1995) 16:534; **86.** J Chromatogr A (1998) 814:213; **87.** Trends Anal Chem (1992) 11:114; **88.** J Chromatogr A (2003) 1021:191; **89.** Anal Chem (1995) 67:4261; **90.** Anal Chem (1991) 63:417; **91.** J Chromatogr (1991) 540:343; **92.** Anal Chem (1990) 62:2189; **93.** Science (1988) 242:224; **94.** Anal Chem (1995) 67:58; **95.** Science (1989) 246:57; **96.** Anal Chem (1999) 71:28; **97.** Anal Biochem (1998) 258:38; **98.** Bioconjug Chem (1995) 6:447; **99.** Anal Chem (2009) 81:2537; **100.** Electrophoresis (2003) 24:2348; **101.** Electrophoresis (2003) 24:2796.

DATA TABLE 9.3 PROTEIN DETECTION ON GELS, BLOTS AND ARRAYS

Cat. No.	MW	Storage	Soluble	Abs	EC	Em	Solvent	Notes
A2333	305.29	F,D,L	MeOH	465	ND	560	MeOH	1, 2, 3, 4
A2334	251.24	F,D,L	EtOH	486	ND	591	MeOH	2, 4, 5
A6222	305.29	F,D,L	MeOH	465	ND	560	MeOH	1, 2, 3
A10192	251.24	F,L	EtOH	486	ND	591	MeOH	2, 5
D6102	502.32	F,D,L	DMSO, MeCN	504	85,000	510	MeOH	
D6116	634.46	F,D,L	DMSO, MeCN	588	68,000	616	MeOH	
F2332	278.26	F,D,L	MeCN	380	7800	464	MeCN	6
F20261	278.26	F,D,L	MeCN	380	8400	464	MeCN	6, 7
N1138	184.19	L	DMF, MeCN	419	9400	493	see Notes	8
P2331MP	134.13	L	EtOH	334	5700	455	pH 9	9
S6650	see Notes	D,L	DMSO	467	see Notes	570	H_2O/BSA	10, 11, 12
S6651	see Notes	D,L	DMSO	467	see Notes	570	H_2O/BSA	10, 11, 12
S6653	see Notes	D,L	DMSO	542	see Notes	630	H_2O/BSA	10, 11, 12
S6654	see Notes	D,L	DMSO	542	see Notes	630	H_2O/BSA	10, 11, 12
S12000	see Notes	L	see Notes	462	see Notes	610	MeOH	13, 14, 15
S12001	see Notes	L	see Notes	462	see Notes	610	MeOH	13, 14, 15
S12010	see Notes	D,L	DMSO	492	see Notes	639	H_2O/BSA	10, 11, 12
S21900	see Notes	L	see Notes	462	see Notes	610	MeOH	13, 14, 15

For definitions of the contents of this data table, see "Using *The Molecular Probes® Handbook*" in the introductory pages.

Notes

1. Spectral data are for the reaction product with glycine in the presence of cyanide. Unreacted reagent in MeOH: Abs = 254 nm (EC = 46,000 $cm^{-1}M^{-1}$), nonfluorescent.
2. ND = not determined.
3. Solubility in methanol is improved by addition of base (e.g., 1–5% (v/v) 0.2 M KOH).
4. Data represent the reactive dye component of this labeling kit.
5. Spectral data are for the reaction product with glycine in the presence of cyanide. Unreacted reagent in MeOH: Abs = 282 nm (EC = 21,000 $cm^{-1}M^{-1}$), nonfluorescent.
6. Fluorescamine spectra are for the reaction product with butylamine. The fluorescence quantum yield and lifetime of the butylamine adduct in EtOH are 0.23 and 7.5 nanoseconds, respectively. (Arch Biochem Biophys (1974) 163:390) The unreacted reagent is nonfluorescent (Abs = 234 nm, EC = 28,000 $cm^{-1}M^{-1}$ in MeCN).
7. This product is specified to equal or exceed 98% analytical purity by HPLC.
8. Spectral data are for the reaction product with glycine in the presence of cyanide, measured in pH 7.0 buffer/MeCN (40:60). (Anal Chem (1987) 59:1102) Unreacted reagent in MeOH: Abs = 279 nm (EC = 5500 $cm^{-1}M^{-1}$), Em = 330 nm.
9. Spectral data are for the reaction product of P2331MP with alanine and 2-mercaptoethanol. The spectra and stability of the adduct depend on the amine and thiol reactants. (Biochim Biophys Acta (1979) 576:440) Unreacted reagent in H_2O: Abs = 257 nm (EC = 1000 $cm^{-1}M^{-1}$).
10. This product is supplied as a ready-made solution in the solvent indicated under "Soluble."
11. The active ingredient of this product is an organic dye with MW <1000. The exact MW and extinction coefficient values for this dye are proprietary.
12. Abs and Em values are for the dye complexed with bovine serum albumin (H_2O/BSA).
13. This product is supplied as a ready-made aqueous staining solution.
14. The active ingredient of this product is an organometallic complex with MW <1500. The exact MW value and extinction coefficient of the complex are proprietary.
15. SYPRO® Ruby protein gel stain also has an absorption peak at 278 nm with about 4-fold higher EC than the 462 nm peak.

IMPORTANT NOTICE: The products described in this manual are covered by one or more Limited Use Label License(s). Please refer to the Appendix on page 971 and Master Product List on page 975. Products are For Research Use Only. Not intended for any animal or human therapeutic or diagnostic use.
www.invitrogen.com/probes
molecular probes® | invitrogen™ by life technologies™

PRODUCT LIST 9.3 PROTEIN DETECTION ON GELS, BLOTS AND ARRAYS

Cat. No.	Product	Quantity
A2333	ATTO-TAG™ CBQCA Amine-Derivatization Kit	1 kit
A6222	ATTO-TAG™ CBQCA derivatization reagent (CBQCA; 3-(4-carboxybenzoyl)quinoline-2-carboxaldehyde)	10 mg
A2334	ATTO-TAG™ FQ Amine-Derivatization Kit	1 kit
A10192	ATTO-TAG™ FQ derivatization reagent (FQ; 3-(2-furoyl)quinoline-2-carboxaldehyde)	10 mg
C33250	Coomassie Fluor™ Orange protein gel stain *ready-to-use solution*	1 L
C33251	Coomassie Fluor™ Orange protein gel stain *ready-to-use solution* *bulk packaging*	5 L
D6102	6-((4,4-difluoro-5,7-dimethyl-4-bora-3a,4a-diaza-s-indacene-3-propionyl)amino)hexanoic acid, succinimidyl ester (BODIPY® FL-X, SE)	5 mg
D6116	6-(((4-(4,4-difluoro-5-(2-thienyl)-4-bora-3a,4a-diaza-s-indacene-3-yl)phenoxy)acetyl)amino)hexanoic acid, succinimidyl ester (BODIPY® TR-X, SE)	5 mg
E33075	Electrophoretic Mobility-Shift Assay (EMSA) Kit *with SYBR® Green and SYPRO® Ruby EMSA stains* *10 minigel assays*	1 kit
F2332	fluorescamine	100 mg
F20261	fluorescamine *FluoroPure™ grade*	100 mg
N1138	naphthalene-2,3-dicarboxaldehyde (NDA)	100 mg
P2331MP	*o*-phthaldialdehyde (OPA) *high purity*	1 g
P6649	protein molecular weight standards *broad range* *200 gel lanes*	400 µL
R33400	Rhinohide™ polyacrylamide gel strengthener concentrate *sufficient additive for 1 L of 30% acrylamide/bis-acrylamide (37.5:1)*	200 mL
S6650	SYPRO® Orange protein gel stain *5000X concentrate in DMSO*	500 µL
S6651	SYPRO® Orange protein gel stain *5000X concentrate in DMSO* *special packaging*	10 x 50 µL
S6656	SYPRO® photographic filter	each
S12012	SYPRO® Protein Gel Stain Starter Kit	1 kit
S6653	SYPRO® Red protein gel stain *5000X concentrate in DMSO*	500 µL
S6654	SYPRO® Red protein gel stain *5000X concentrate in DMSO* *special packaging*	10 x 50 µL
S11791	SYPRO® Ruby protein blot stain *10–40 blots*	200 mL
S12001	SYPRO® Ruby protein gel stain	200 mL
S12000	SYPRO® Ruby protein gel stain	1 L
S21900	SYPRO® Ruby protein gel stain *bulk packaging*	5 L
S12010	SYPRO® Tangerine protein gel stain *5000X concentrate in DMSO*	500 µL

IMPORTANT NOTICE: The products described in this manual are covered by one or more Limited Use Label License(s). Please refer to the Appendix on page 971 and Master Product List on page 975. Products are For Research Use Only. Not intended for any animal or human therapeutic or diagnostic use.

9.4 Detecting Protein Modifications

Click-iT® Reagents for Detecting Protein Synthesis and Modifications

Click-iT® labeling technology employs a bioorthogonal reactive chemistry for the *in situ* labeling of specific molecular populations, such as proteins that have been newly synthesized or post-translationally modified in some experimental time window of interest. The Click-iT® labeling reaction is based on a copper-catalyzed azide–alkyne cyloaddition [1,2] and derives its high degree of specificity from the fact that the azide and alkyne reaction partners have no endogenous representation in biological molecules, cells, tissues or model organisms.[3–5]

Application of this reaction to *in situ* labeling of cells is a two-step process. First, one reaction partner—either an azide or alkyne linked to a "building block" such as an amino acid, monosaccharide, fatty acid, nucleotide or nucleoside—is biosynthetically incorporated. Subsequently, the other reaction partner—the complementary alkyne or azide linked to a fluorescent dye, biotin or other detection reagent—is "clicked" into place in the presence of catalytic copper (I), providing a detection moiety (Figure 9.4.1). One reaction partner must be an azide derivative and the other an alkyne derivative, but either functional moiety can serve as the biosynthetically incorporated molecule or the detection molecule [6] (e.g., L-azidohomoalanine (AHA) + Alexa Fluor® 488 alkyne is the inverse of the reaction scheme shown in Figure 9.4.1A).

The small size of alkyne and azide tags allows the biosynthetic building blocks to which they are attached to be processed by enzymes, such as aminoacyl tRNA synthetases and nucleotide polymerases, that have poor tolerance for substrates with larger modifications such as fluorescent organic dyes.[7] Furthermore, the 1,2,3-triazole linkage between the azide and alkyne reaction partners (Figure 9.4.1) is extremely stable. It is not susceptible to hydrolysis, oxidation or reduction, and it survives ionization in mass spectrometry (MS) analysis. Click-iT® labeling technology and the details of the click reaction are discussed in Section 3.1. For a complete list of azide and alkyne derivatives compatible with Click-iT® labeling technology, see Table 3.1. Here we highlight the azide and alkyne derivatives that can be used for labeling newly synthesized proteins and detecting post-translational protein modifications.

Azide- and Alkyne-Modified Amino Acids

The Click-iT® AHA (L-azidohomoalanine, C10102) and Click-iT® HPG (L-homopropargylglycine, C10186; Figure 9.4.1) reagents are methionine surrogates that provide nonradioactive alternatives to ^{35}S-methionine for pulse-chase detection of protein synthesis and degradation.[8–11] These amino acid analogs are fed to cultured cells and incorporated into proteins during active protein synthesis. The enzymatically incorporated Click-iT® AHA or Click-iT® HPG is then detected with a fluorescent alkyne or fluorescent azide, respectively, using a Cu(I)-catalyzed click reaction. These Click-iT® reagents provide detection sensitivity comparable to that obtained using the radioactive ^{35}S-methionine method and are compatible with downstream LC-MS/MS and MALDI MS analysis, as well as with total-protein, glycoprotein and phosphoprotein gel stains for differential analyses of newly synthesized protein together with post-translational modifications.

Click-iT® AHA is also available in the Click-iT® AHA Alexa Fluor® 488 Protein Synthesis HCS Assay Kit (C10289), which provides Alexa Fluor® 488 alkyne for detection. Click-iT® AHA has proven to be a successful substitute for methionine in many cell types, including COS-7,

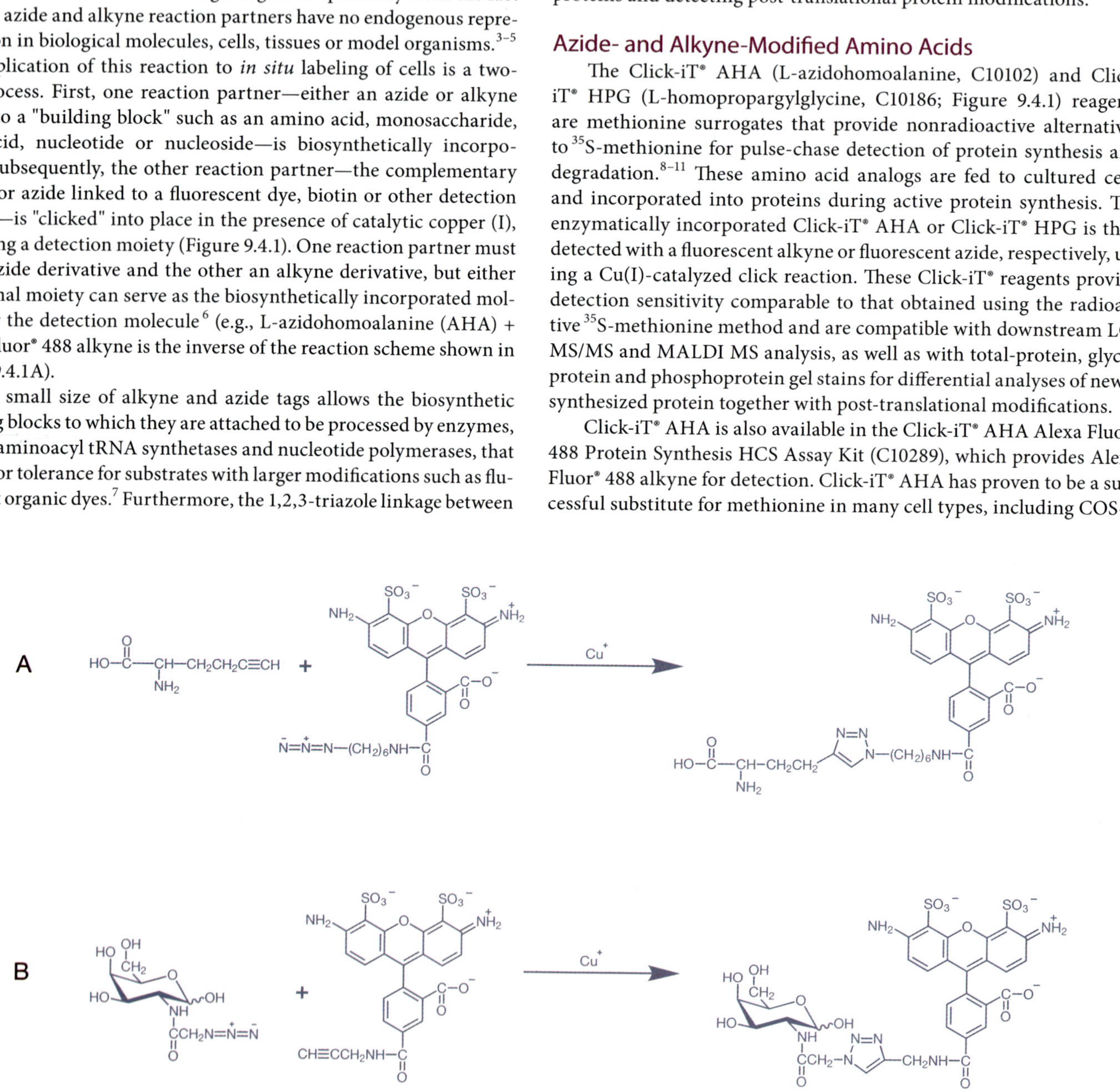

Figure 9.4.1 Click-iT® copper-catalyzed azide–alkyne cycloaddition chemistry applied to detection of **A**) proteins and **B**) carbohydrates. The reaction partners are **A**) L-homopropargylglycine (HPG) and Alexa Fluor® 488 azide and **B**) *N*-azidoacetylgalactosamine and Alexa Fluor® 488 alkyne. In both cases, the left-hand partner is a metabolic precursor that can be incorporated into proteins via *de novo* synthesis or post-translational modification pathways.

IMPORTANT NOTICE: The products described in this manual are covered by one or more Limited Use Label License(s). Please refer to the Appendix on page 971 and Master Product List on page 975. Products are For Research Use Only. Not intended for any animal or human therapeutic or diagnostic use.

3T3-L1, HeLa, HEK 293 and Jurkat cells. Note that cells should be labeled in methionine-free media, as methionine is the preferred substrate for methionyl tRNA transferase, and supplemented media (i.e., methionine-free DMEM) should be used in place of HBSS to achieve greater Click-iT® AHA incorporation at lower concentrations.

Azide- and Alkyne-Modified Monosaccharides

The Click-iT® metabolic glycoprotein labeling reagents provide biosynthetic precursors for detecting and characterizing post-translational glycosylation of proteins.[12–15] Four azide- or alkyne-modified monosaccharides are available for metabolic incorporation into a specific subclass of protein glycan structures:

- Click-iT® GalNAz metabolic glycoprotein labeling reagent (tetraacetylated *N*-azidoacetylgalactosamine, C33365; Figure 9.4.1), for labeling *O*-linked glycoproteins
- Click-iT® ManNAz metabolic glycoprotein labeling reagent (tetraacetylated *N*-azidoacetylmannosamine, C33366; Figure 9.4.2), for labeling sialic acid–modified glycoproteins
- Click-iT® GlcNAz metabolic glycoprotein labeling reagent (tetraacetylated *N*-azidoacetylglucosamine, C33367; Figure 9.4.3), for labeling *O*-linked *N*-acetylglucosamine (*O*-GlcNAc)–modified glycoproteins
- Click-iT® fucose alkyne (tetraacetylfucose alkyne, C10264; Figure 9.4.4), for labeling fucosylated proteins

Cultured cells are simply incubated with the modified sugars for 2–3 days or until cells reach the appropriate density. The acetyl groups improve cell permeability of the modified sugars and are removed by nonspecific intracellular esterases (Figure 9.4.2). The resulting azide- or alkyne-modified sugar is then metabolically incorporated through the permissive nature of the oligosaccharide biosynthesis pathway, yielding functionalized glycoproteins that can be chemoselectively coupled to complementary alkyne- or azide-functionalized fluorophores and biotinylation reagents for detection or affinity capture. We offer three Click-iT® Protein Analysis Detection Kits (C33370, C33371, C33372) described below for the detection of azide-functionalized glycoproteins in 1D or 2D electrophoresis gels or western blots. These labeled glycoproteins are compatible with total-protein, glycoprotein and phosphoprotein gel stains and provide a detection sensitivity of a few hundred femtomoles, allowing an in-depth analysis of low-abundance glycoproteins as well as glycoproteins with a small degree of glycosylation.

Glycoproteins labeled with the Click-iT® labeling and detection reagents are also compatible with downstream LC-MS/MS and MALDI-MS analyses for further identification and characterization. For added convenience, we offer an *O*-GlcNAc peptide LC/MS standard (C33374) from the transcription factor CREB for LC-MS/MS and MALDI-MS analyses of the *O*-GlcNAc posttranslational modification. This peptide is also available together with its phosphorylated counterpart for use as LC/MS standards (C33373) in differential mass spectrometry–based studies of the corresponding modifications, as well as for characterizing differential β-elimination/addition conditions.

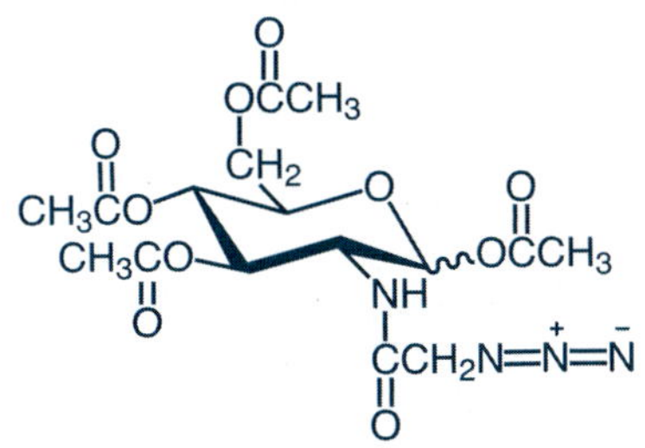

Figure 9.4.3 Click-iT® GlcNAz metabolic glycoprotein labeling reagent (tetraacetylated *N*-azidoacetylglucosamine for labeling *O*-linked *N*-acetylglucosamine *O*-GlcNAc)–modified glycoproteins; C33367).

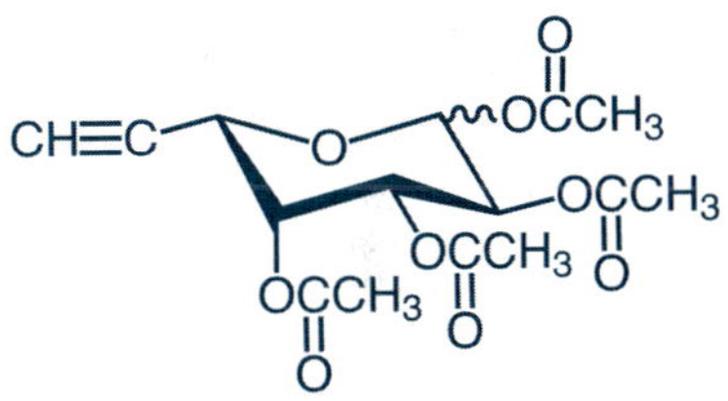

Figure 9.4.4 Click-iT® fucose alkyne (tetraacetylfucose alkyne) (C10264).

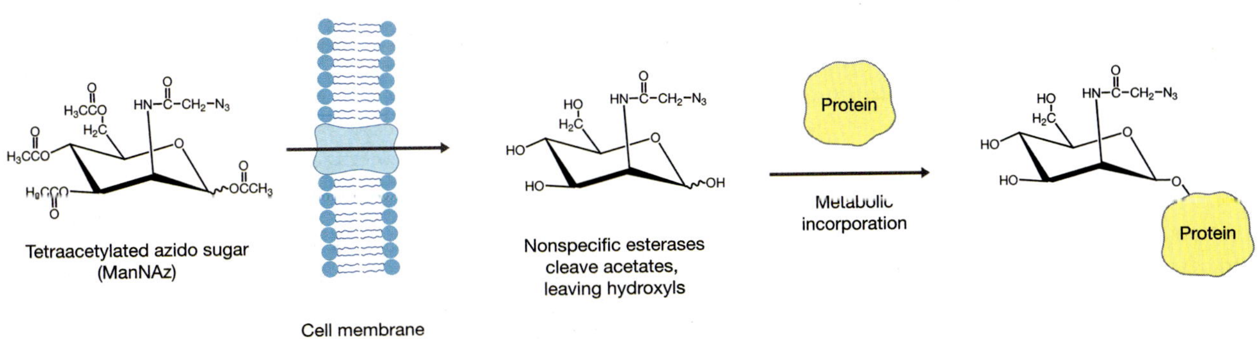

Figure 9.4.2 Metabolic incorporation of tetraacetylated azido sugars.

Click-iT® O-GlcNAc Enzymatic Labeling System

We also offer the the Click-iT® *O*-GlcNAc Enzymatic Labeling System for *in vitro* enzyme-mediated *N*-azidoacetylgalactosamine labeling of *O*-GlcNAc–modified glycoproteins [16,17] (C33368). Proteins are enzymatically labeled using the permissive mutant β-1,4-galactosyltransferase (Gal-T1, Y289L), which transfers azido-modified galactose (GalNAz) from UDP-GalNAz to *O*-GlcNAc residues on the target proteins. Target proteins can then be detected using an alkyne-derivatized fluorophore or one of the Click-iT® Protein Analysis Detection Kits described below. Using the Click-iT® *O*-GlcNAc Enzymatic Labeling System in conjunction with the Click-iT® Tetramethylrhodamine (TAMRA) Protein Analysis Detection Kit, we have detected as little as 1 picomole of α-crystallin, a protein which is only 2–10% *O*-GlcNAc modified.

Each Click-iT® *O*-GlcNAc Enzymatic Labeling System provides:

- UDP-GalNAz
- Gal-T1 (Y289L)
- Click-iT® *O*-GlcNAc enzymatic labeling buffer
- $MnCl_2$
- α-crystallin, for use as a positive control
- Detailed labeling protocols

Azide-Modified Isoprenoids and Fatty Acids

We offer several azide-modified isoprenoids and fatty acids, including:

- Click-iT® farnesyl alcohol, azide (C10248), for identifying farnesylated proteins
- Click-iT® geranylgeranyl alcohol, azide (C10249), for identifying geranylgeranylated proteins
- Click-iT® palmitic acid, azide (15-azidopentadecanoic acid, C10265), for identifying protein fatty acylation
- Click-iT® myristic acid, azide (12-azidododecanoic acid, C10268), for identifying protein fatty acylation

These azide-functionalized isoprenoids and fatty acids enable detection of post-translational lipidation of proteins by in-gel fluorescence scanning, fluorescence microscopy and flow cytometry.[18–20]

Azide- and Alkyne-Derivatized Dyes and Biotinylation Reagents

We offer a rich selection of azide- and alkyne-derivatized fluorescent dyes for coupling to complementary azide- and alkyne-functionalized biomolecules (Table 3.1, Section 3.1), including:

- Alexa Fluor® 488 azide (A10266, Figure 9.4.1) and alkyne (A10267, Figure 9.4.1)
- Alexa Fluor® 555 azide (A20012) and alkyne (A20013)
- Alexa Fluor® 594 azide (A10270) and alkyne (A10275)
- Alexa Fluor® 647 azide (A10277) and alkyne (A10278)
- Oregon Green® 488 azide (O10180) and alkyne (O10181)
- Tetramethylrhodamine (TAMRA) azide (T10182) and alkyne (T10183)
- Biotin azide (B10184) and alkyne (B10185)

Antibodies to Oregon Green® 488, tetramethylrhodamine and Alexa Fluor® 488 dyes (Section 7.4) and Tyramide Signal Amplification (TSA™) Kits (Section 6.2) are available to provide signal amplification if necessary.[21] The biotin azide and alkyne reagents facilitate western blotting applications and streptavidin enrichment in combination with our streptavidin and CaptAvidin™ agarose affinity matrices [14,22,23] (S951, C21386; Section 7.6).

Click-iT® Protein Analysis Detection Kits

In addition to azide- and alkyne-derivatized dyes and biotinylation reagents, we offer three Click-iT® Protein Analysis Detection Kits (C33370, C33371, C33372) that provide labeled alkynes for the detection of azide-labeled biomolecules. These Click-iT® Protein Analysis Detection Kits provide sufficient reagents for 10 labeling reactions based on the provided protocol and include:

- TAMRA alkyne (in (C33370), Dapoxyl® alkyne (in C33371) or biotin alkyne (in C33372)
- Click-iT® reaction buffer
- $CuSO_4$
- Click-iT® reaction buffer additives
- Detailed labeling protocols

Click-iT® Reaction Buffers

For added convenience, we offer Click-iT® Reaction Buffer Kits for protein or cell samples labeled with an azide- or alkyne-tagged biomolecule. The Click-iT® Cell Reaction Buffer Kit (C10269) includes sufficient reagents to perform 50 reactions based on a 0.5 mL reaction volume for subsequent analyses by flow cytometry, fluorescence microscopy or high-content screening (HCS). The Click-iT® Protein Reaction Buffer Kit (C10276) includes everything required for click coupling to functionalized proteins for subsequent standard protein biochemical analyses (e.g., western blots or mass spectrometry).

Pro-Q® Diamond Phosphoprotein Stain for Gels and Blots

We have developed a suite of compatible methodologies for the differential staining of specific proteins (phosphoproteins, glycoproteins or membrane proteins) and the total-protein profile in two or more visually distinguishable colors, producing a more complete picture of the proteome. This set of protein stains not only offer the capacity for the simultaneous detection of multiple protein targets in a single sample, but also provides a combination of high sensitivity and simplicity that can streamline protocols in 1D and 2D polyacrylamide gels or on blots.

The Pro-Q® Diamond phosphoprotein gel stain, Pro-Q® Emerald glycoprotein gel stains and SYPRO® Ruby protein gel stain—which we have optimized to complement each other in selectivity, sensitivity and staining protocols—can be used in serial detection of phosphoproteins, glycoproteins and total proteins on a single protein sample separated by 1D or 2D gel electrophoresis (Figure 9.4.5). Our Rhinohide™ polyacrylamide gel strengthener (R33400, Section 9.3) greatly improves the strength of any polyacrylamide gel, making it

IMPORTANT NOTICE: The products described in this manual are covered by one or more Limited Use Label License(s). Please refer to the Appendix on page 971 and Master Product List on page 975. Products are For Research Use Only. Not intended for any animal or human therapeutic or diagnostic use.

molecular probes® invitrogen™ by life technologies™

easy to perform these multiple staining procedures without special gel handling. After each staining step, an image of the gel is collected. Once collected, the three images can be overlaid in any combination for analysis of phosphorylation, glycosylation and total-protein expression. Because all three stains are used on the same gel, unambiguous spot matching of phosphoproteins and glycoproteins is made simple by direct comparison with the total-protein profile provided by the SYPRO® Ruby protein gel stain. This simultaneous measurement of several variables allows for perfect spatial registration of signals and increases the amount of data that can be collected in a single experiment, leading to more controlled experiments, more accurate data comparisons and fewer ambiguities.

Pro-Q® Diamond Phosphoprotein Gel Stain and Destain

Pro-Q® Diamond phosphoprotein gel stain [24–27] is a breakthrough technology that provides a simple, direct method for selectively staining *O*-linked phosphoproteins in polyacrylamide gels (Figure 9.4.6). It is ideal for the identification of kinase targets in signal transduction pathways and for phosphoproteomic studies. This proprietary fluorescent stain allows direct, in-gel detection of phosphate groups attached to tyrosine, serine or threonine residues. The Pro-Q® Diamond phosphoprotein gel stain can be used with standard SDS-polyacrylamide

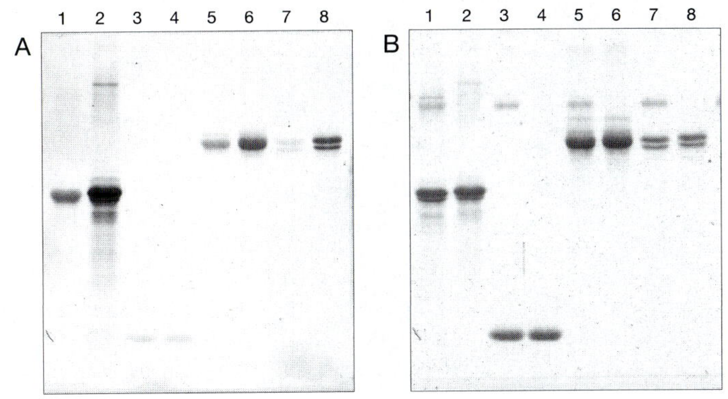

Figure 9.4.6 Selectivity of the Pro-Q® Diamond phosphoprotein gel stain (P33300, P33301, P33302) for phosphoproteins. A polyacrylamide gel containing various proteins was stained with **A)** Pro-Q® Diamond phosphoprotein gel stain, followed by **B)** SYPRO® Ruby protein gel stain. This gel shows a nonphosphorylated protein, lysozyme (lanes 3 and 4), as well as several phosphoproteins, α-casein (lanes 1 and 2), ovalbumin (lanes 5 and 6) and pepsin (lanes 7 and 8), before (even lanes) and after (odd lanes) treatment with phosphatases. Loss of Pro-Q® Diamond staining indicates loss of all phosphates from pepsin, partial loss of phosphates from α-casein and ovalbumin and no change in the nonphosphorylated protein lysozyme.

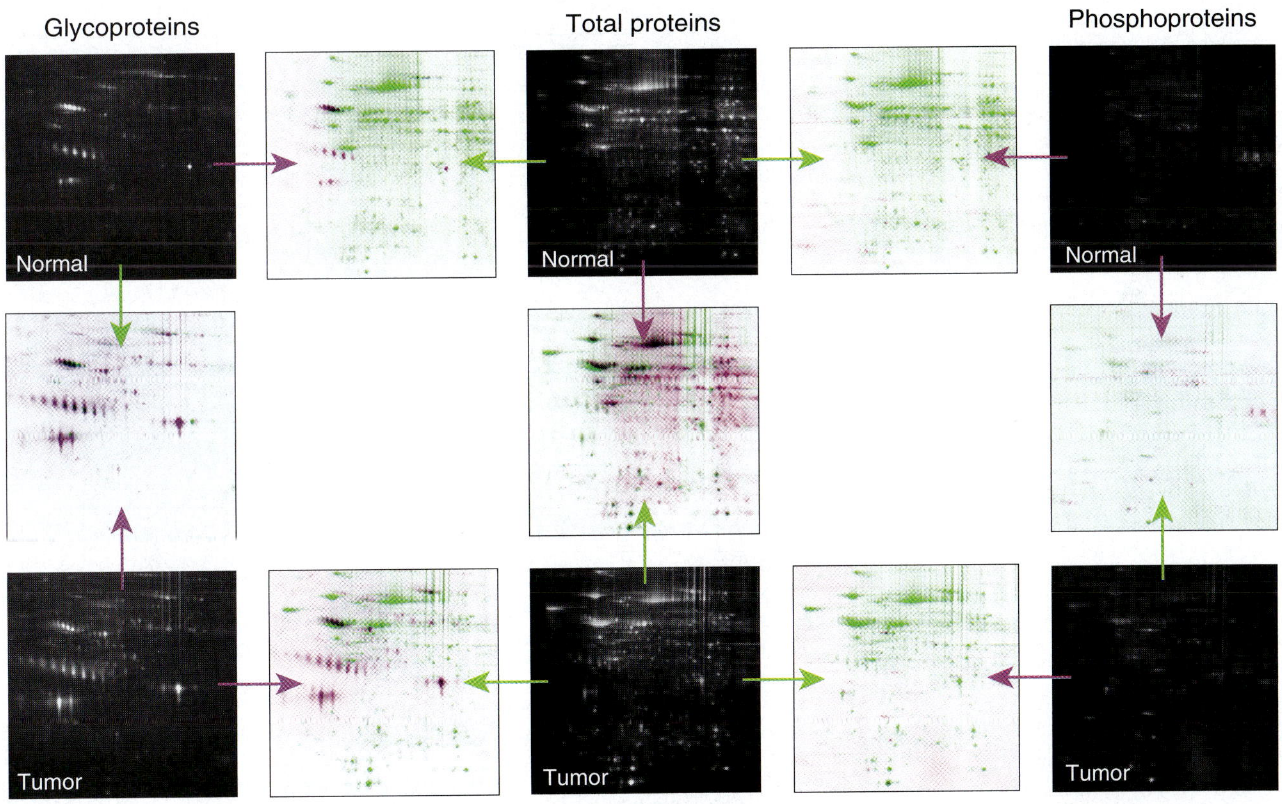

Figure 9.4.5 An overview of the Multiplexed Proteomics® approach. Images collected after each staining step can be overlaid in any combination for analysis of protein expression, phosphorylation and glycosylation between samples.

IMPORTANT NOTICE: The products described in this manual are covered by one or more Limited Use Label License(s). Please refer to the Appendix on page 971 and Master Product List on page 975. Products are For Research Use Only. Not intended for any animal or human therapeutic or diagnostic use.

gels (Figure 9.4.6) or with 2D gels (Figure 9.4.7)—blotting is not required and there is no need for radioisotopes, phosphoprotein-specific antibodies or western blot detection reagents. The simple and reliable staining protocol delivers results in as little as 4 to 5 hours. The stain is also compatible with mass spectrometry, allowing analysis of the phosphorylation state of entire proteomes. The Pro-Q® Diamond phosphoprotein gel stain provides:

- **Simple in-gel detection**. Proteins containing phosphate groups attached to tyrosine, serine or threonine residues can be detected directly in either 1D or 2D polyacrylamide gels after the gel is fixed, stained and destained; no antibodies are required and no blotting is necessary.
- **Selectivity without radioactivity**. The Pro-Q® Diamond phosphoprotein gel stain is a fluorescent stain that selectively detects phosphoproteins; radioisotopes are not used and therefore no radioactive waste is generated.
- **Sensitivity**. The Pro-Q® Diamond phosphoprotein gel stain allows the detection of as little as 1–16 ng of phosphoprotein per band, depending on the phosphorylation level of the protein.

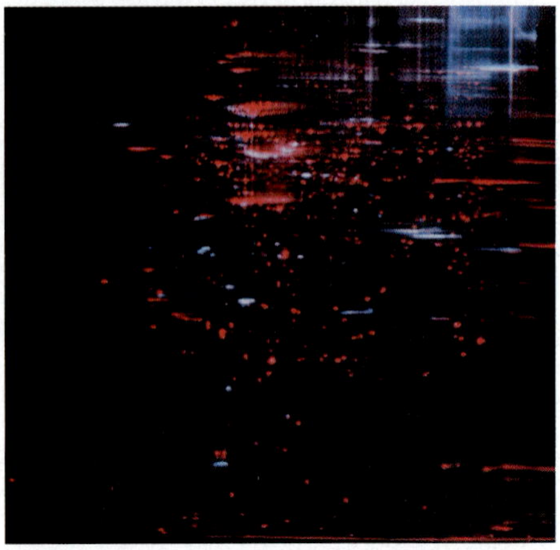

Figure 9.4.7 Visualization of total protein and phosphoproteins in a 2D gel. Proteins from a Jurkat T-cell lymphoma line cell lysate were separated by 2D gel electrophoresis and stained with Pro-Q® Diamond phosphoprotein gel stain (P33300, P33301, P33302, blue) followed by SYPRO® Ruby protein gel stain (S12000, S12001, S21900, red). After each dye staining, the gel was imaged on an FLA-3000 scanner (Fuji®). The digital images were acquired using Z3 software (Compugen), and the resulting composite image was digitally pseudocolored and overlaid.

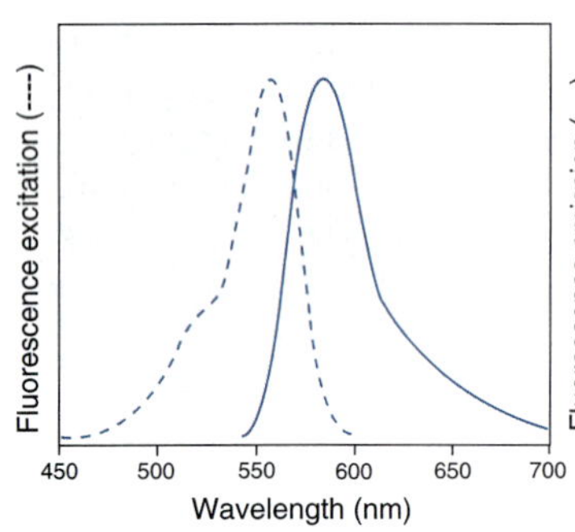

Figure 9.4.9 Fluorescence excitation and emission spectra of the Pro-Q® Diamond phosphoprotein gel stain (P33300, P33301, P33302).

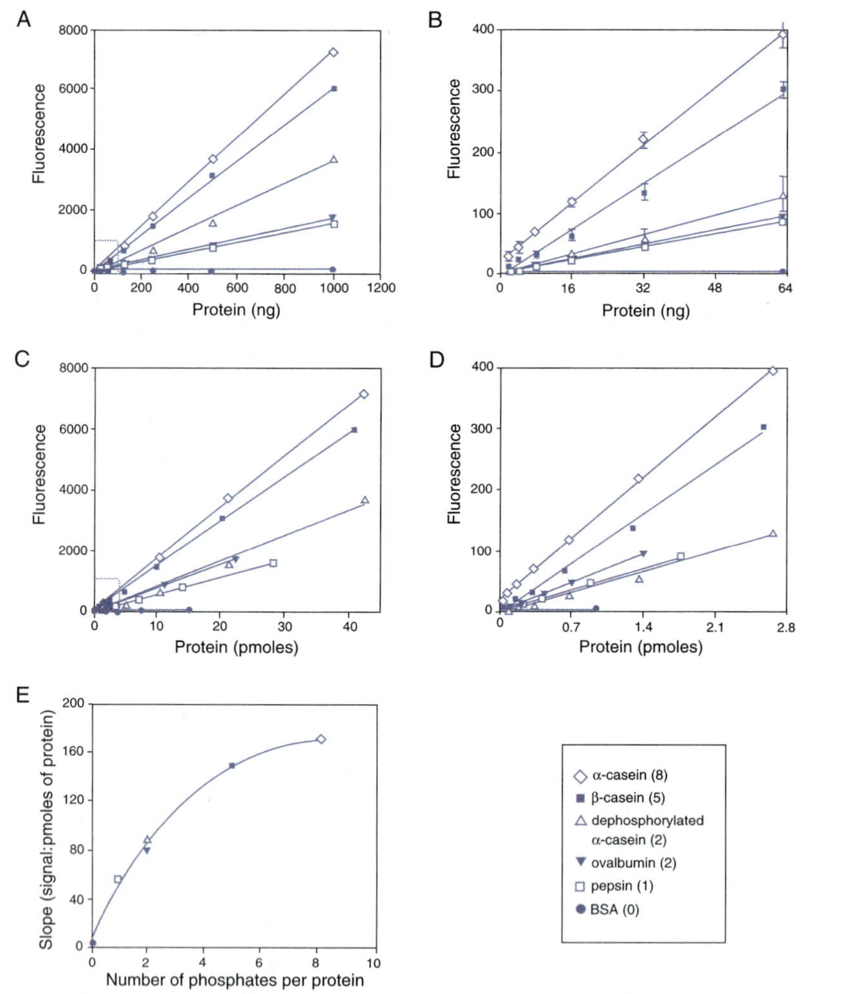

Figure 9.4.8 Sensitivity and linear range of Pro-Q® Diamond phosphoprotein gel stain (P33300, P33301, P33302). Six different proteins were serially diluted and run on separate SDS-polyacrylamide gels, which were then stained with Pro-Q® Diamond phosphoprotein gel stain. The images were documented on a fluorescence imager, and the fluorescence emission from each band was quantitated. The number of known phosphate groups on each protein is indicated in the figure legend. **A)** Fluorescence emission of the band, plotted as a function of protein amount, in nanograms. **B)** Magnification of data points in the highlighted box in panel A. **C)** The fluorescence emission of the band, plotted as a function of picomoles of protein. **D)** Magnification of the data points in the highlighted box in panel C. **E)** The slope of the line for each protein in panel C, plotted against the known number of phosphates per protein.

IMPORTANT NOTICE: The products described in this manual are covered by one or more Limited Use Label License(s). Please refer to the Appendix on page 971 and Master Product List on page 975. Products are For Research Use Only. Not intended for any animal or human therapeutic or diagnostic use.

- **Quantitation.** The Pro-Q® Diamond signal for individual phosphoproteins is linear over three orders of magnitude and correlates with the number of phosphate groups (Figure 9.4.8).
- **Compatibility.** Pro-Q® Diamond gel stain (excitation/emission maxima ~555/580 nm; Figure 9.4.9) is compatible with a visible-light–scanning instrument, a visible-light transilluminator or (with reduced sensitivity) a 300 nm transilluminator, as well as with mass spectrometry analysis.
- **Multiplexing capability.** Pro-Q® Diamond gel stain can be used with SYPRO® Ruby protein gel stain (Figure 9.4.6) and Pro-Q® Emerald glycoprotein gel stain on the same gel for multiparameter staining.

The Pro-Q® Diamond phosphoprotein gel stain (Table 9.4) is supplied ready-to-use in three different sizes: a 200 mL size (P33301) suitable for staining approximately four minigels; a 1 L size (P33300) suitable for staining approximately 20 minigels or two large-format gels, e.g., 2D gels; and a 5 L bulk-packaging size (P33302). In addition, we offer Pro-Q® Diamond Phosphoprotein Gel Staining Kits (MPP33300, MPP33301, MPP33302) that include both the Pro-Q® Diamond gel stain and the PeppermintStick™ phosphoprotein molecular weight standards. All products are accompanied by a simple and reliable staining and destaining protocol that delivers results in as little as four to five hours. For convenient destaining, we also offer the Pro-Q® Diamond phosphoprotein gel destaining solution as a ready-to-use solution in either a 1 L (P33310) or 5 L (P33311) size.

Multiplexed Proteomics® Kits for Phosphoprotein and Total-Protein Gel Staining

When used together, the Pro-Q® Diamond phosphoprotein gel stain and the SYPRO® Ruby protein gel stain (S12000, S12001, S21900; Section 9.3) make a powerful combination for proteome analysis. The SYPRO® Ruby dye is a total-protein stain that, like the Pro-Q® Diamond gel stain, is quantitative over three orders of magnitude. Determining the ratio of the Pro-Q® Diamond dye to SYPRO® Ruby dye signal intensities for each band or spot thus provides a measure of the phosphorylation level normalized to the total amount of protein. Using both stains in combination makes it possible to distinguish a low amount of a highly phosphorylated protein from a higher amount of a less phosphorylated protein. To make this staining more convenient and economical, we offer the Multiplexed Proteomics® Kit #2 with 200 mL of the Pro-Q® Diamond phosphoprotein gel stain and 200 mL of the SYPRO® Ruby protein gel stain (M33306), the Multiplexed Proteomics® Kit #1 with 1 L of each stain (M33305) and the Multiplexed Proteomics® Phosphoprotein Gel Stain Kits (MPM33305, MPM33306), which include the Pro-Q® Diamond phosphoprotein gel stain, SYPRO® Ruby protein gel stain and PeppermintStick™ phosphoprotein molecular weight standards; Table 9.4 summarizes all of our Pro-Q® Diamond gel stain reagents and kits.

Pro-Q® Diamond Phosphoprotein Blot Stain Kit

The Pro-Q® Diamond Phosphoprotein Blot Stain Kit (P33356) provides a simple and quick method for directly detecting phosphoproteins on poly(vinylidene difluoride) (PVDF) or nitrocellulose membranes without the use of radioactivity or antibodies. As with the gel stain, the Pro-Q® Diamond phosphoprotein blot stain detects phosphoserine-, phosphothreonine- and phosphotyrosine-containing proteins, independent of the sequence context of the phosphorylated amino acid residue. Thus, the native phosphorylation levels of proteins from a variety of sources, including tissue specimens and body fluids, can be analyzed. Protein samples are separated by 1D or 2D gel electrophoresis, electroblotted to the membrane, stained and destained using a protocol similar to that typically performed with amido black or Ponceau S staining of total-protein profiles on membranes. After staining, gels are simply imaged using any of a variety of laser scanners, xenon-arc lamp–based scanners or CCD-based imaging devices employing UV transilluminators; the excitation/emission maxima of the Pro-Q® Diamond phosphoprotein blot stain are ~555/580 nm (Figure 9.4.9). The limits of detection for the stain on PVDF membrane blots are typically 8–16 ng of phosphoprotein, with a linear dynamic range of approximately 15-fold. The sensitivity of the Pro-Q® Diamond phosphoprotein blot stain is decreased when using nitrocellulose blots. Each Pro-Q® Diamond Phosphoprotein Blot Stain Kit provides sufficient reagents for staining ~20 minigel electroblots, including:

- Pro-Q® Diamond phosphoprotein blot stain reagent
- Pro-Q® Diamond blot stain buffer
- Detailed protocols for staining and photographing the blot

Table 9.4 Pro-Q® Diamond gel stain reagents and kits.

Product	Cat. No.	Includes: Pro-Q® Diamond Phosphoprotein Gel Stain	SYPRO® Ruby Protein Gel Stain	Phosphoprotein MW Standard	Number of Gels Stained
Pro-Q® Diamond phosphoprotein gel stain	P33300	1 L			~20 minigels or two large-format gels (e.g., 2D gels)
Pro-Q® Diamond phosphoprotein gel stain	P33301	200 mL			~4 minigels
Pro-Q® Diamond phosphoprotein gel stain	P33302	5 L			~100 minigels or 10 large-format gels
Multiplexed Proteomics® Phosphoprotein Gel Stain Kit #1	M33305	1 L	1 L		~20 minigels or two large-format gels
Multiplexed Proteomics® Phosphoprotein Gel Stain Kit #2	M33306	200 mL	200 mL		~4 minigels
Pro-Q® Diamond Phosphoprotein Gel Staining Kit	MPP33300	1 L		40 µL	~20 minigels or two large-format gels
Pro-Q® Diamond Phosphoprotein Gel Staining Kit	MPP33301	200 mL		40 µL	~4 minigels
Pro-Q® Diamond Phosphoprotein Gel Staining Kit	MPP33302	5 L		40 µL	~100 minigels or 10 large-format gels
Multiplexed Proteomics® Phosphoprotein Gel Stain Kit	MPM33305	1 L	1 L	40 µL	~20 minigels or two large-format gels
Multiplexed Proteomics® Phosphoprotein Gel Stain Kit	MPM33306	200 mL	200 mL	40 µL	~4 minigels

IMPORTANT NOTICE: The products described in this manual are covered by one or more Limited Use Label License(s). Please refer to the Appendix on page 971 and Master Product List on page 975. Products are For Research Use Only. Not intended for any animal or human therapeutic or diagnostic use.

The Pro-Q® Diamond phosphoprotein blot stain binds noncovalently to phosphoproteins and is thus fully compatible with matrix-assisted laser desorption ionization time-of-flight mass spectrometry (MALDI-TOF MS) and Edman sequencing. Furthermore, the Pro-Q® Diamond phosphoprotein blot stain is compatible with the standard colorimetric, fluorometric and chemiluminescent detection techniques employed in immunoblotting. This phosphoprotein blot stain may be used in conjunction with the SYPRO® Ruby protein blot stain, a total-protein stain that is quantitative over two orders of magnitude on blots. Using the SYPRO® and Pro-Q® Diamond blot stains in combination makes it possible to distinguish a low amount of a highly phosphorylated protein from a higher amount of a less phosphorylated protein.

PeppermintStick™ Phosphoprotein Molecular Weight Standards

PeppermintStick™ phosphoprotein molecular weight standards are a mixture of phosphorylated and nonphosphorylated proteins with molecular weights from 14,400 to 116,250 daltons. Separation by polyacrylamide gel electrophoresis resolves this mixture into two phosphorylated and four nonphosphorylated protein bands (Figure 9.4.10). These standards serve both as molecular weight markers and as positive and negative controls for our Pro-Q® Diamond phosphoprotein gel stain and other methods that detect phosphorylated proteins. We offer two different unit sizes of the PeppermintStick™ phosphoprotein molecular weight standards: a 40 µL unit size sufficient for 20–40 gel lanes (P27167) and a 400 µL unit size sufficient for 200–400 gel lanes (P33350).

Phosphopeptide Standard Mixture

Formulated especially for MALDI-MS, the phosphopeptide standard mixture (P33357) contains equimolar amounts of three unphosphorylated and four phosphorylated peptides, ranging in mass between 1047 and 2192 and representing phosphoserine (pS), phosphothreonine (pT) and phosphotyrosine (pY) monophosphopeptides, as well as a peptide containing both pT and pY. This mixture is ideal for use as an internal or external control for LC/MS, MALDI analysis or β-elimination reactions.

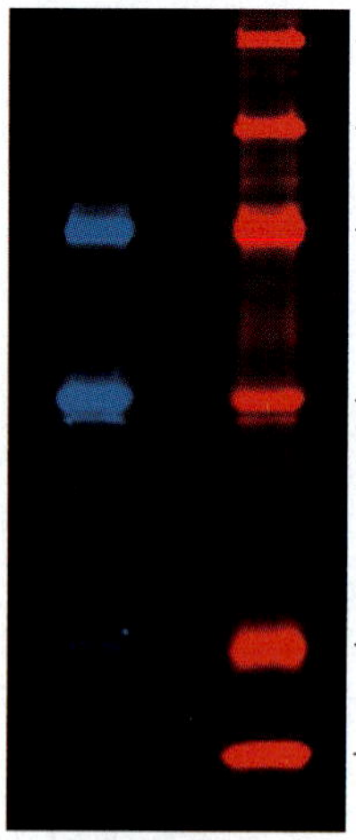

Figure 9.4.10 PeppermintStick™ phosphoprotein molecular weight standards (P33350, P27167) separated on a 13% SDS polyacrylamide gel. The markers contain (from largest to smallest) ß-galactosidase, bovine serum albumin (BSA), ovalbumin, ß-casein, avidin and lysozyme. Ovalbumin and ß-casein are phosphorylated. The gel was stained with Pro-Q® Diamond phosphoprotein gel stain (blue) followed by SYPRO® Ruby protein gel stain (red). The digital images were pseudocolored.

Pro-Q® Diamond Phosphoprotein Enrichment Kits

The Pro-Q® Diamond Phosphoprotein Enrichment Kit (P33358) enables efficient, nonradioactive isolation of phosphoproteins from complex cellular extracts. This kit provides resin, reagents and columns designed to isolate phosphoproteins from 0.5–1.0 mg of total cellular protein per column. The column bed volume can be easily scaled up or down depending on the amount of available starting material. The phosphoprotein-binding properties of the resin allow efficient capture of both native and denatured proteins. Therefore, cell or tissue samples can be denatured in lysis buffers and stored in the freezer prior to the phosphoprotein enrichment procedure. Each Pro-Q® Diamond Phosphoprotein Enrichment Kit contains:

- Phosphoprotein Enrichment Module
- Resin (50% v/v slurry)
- Disposable 2 mL columns, 10 columns
- Lysis buffer
- Wash buffer
- Elution buffer
- Vivaspin® filtration concentrators with 10 kDa cut-off polyethersulfone membrane, 10 concentrators

  Protease Inhibitor and Endonuclease Module
- Protease inhibitor
- Endonuclease

Protocols for both undenatured and denatured lysates are provided, and these procedures can be completed in approximately three hours. For added convenience, the Pro-Q® Diamond Phosphoprotein Enrichment and Detection Kit (P33359) provides all the reagents in the Pro-Q® Diamond Phosphoprotein Enrichment Kit, as well as Pro-Q® Diamond phosphoprotein gel stain and PeppermintStick™ phosphoprotein molecular weight markers for detecting phosphoproteins on SDS-polyacrylamide gels.

Pro-Q® Diamond Phosphoprotein/ Phosphopeptide Microarray Stain Kit

The Pro-Q® Diamond Phosphoprotein/Phosphopeptide Microarray Stain Kit (P33706) provides a method for selective staining of phosphoproteins or phosphopeptides on microarrays, without the use of antibodies or radioactivity. This kit permits direct detection of phosphate groups attached to tyrosine, serine or threonine residues in a microarray environment and has been optimized for microarrays with acrylamide gel surfaces. Each Pro-Q® Diamond Phosphoprotein/Phosphopeptide Microarray Stain Kit provides:

- Pro-Q® Diamond phosphoprotein/phosphopeptide microarray stain
- Pro-Q® Diamond microarray destain solution
- Microarray staining gasket with seal tabs, 10 chambers
- Slide holder tube, 20 tubes
- Detailed protocols

The Pro-Q® Diamond Phosphoprotein/Phosphopeptide Microarray Stain Kit is useful for identifying kinase targets in signal transduction pathways and for phosphoproteomics studies.

IMPORTANT NOTICE: The products described in this manual are covered by one or more Limited Use Label License(s). Please refer to the Appendix on page 971 and Master Product List on page 975. Products are For Research Use Only. Not intended for any animal or human therapeutic or diagnostic use.

Pro-Q® Diamond LC Phosphopeptide Detection Kit

The Pro-Q® Diamond LC Phosphopeptide Detection Kit (P33203) is designed to provide sensitive and selective fluorescence-based detection of phosphorylated peptides during liquid chromatography separations. The Pro-Q® Diamond LC phosphopeptide detection reagent interacts selectively with phosphoserine-, phosphothreonine- and phosphotyrosine-containing peptides to form highly fluorescent dye–phosphopeptide complexes that elute from an HPLC column with altered retention times, allowing identification and purification of phosphopeptides prior to analysis by mass spectrometry. This kit is ideal for isolating phosphopeptides from chromatographic fractions or from complex peptide mixtures such as the tryptic digest of a phosphoprotein. The Pro-Q® Diamond LC Phosphopeptide Detection Kit provides:

- Pro-Q® Diamond LC phosphopeptide detection reagent
- Concentrated activation buffer
- Positive control phosphopeptide RII
- Kemptide, a negative control peptide
- Detailed protocols

Sufficient reagents are provided for 20 HPLC separations; a single separation will selectively detect 20 picomoles or less of a monophosphorylated peptide using a standard microbore C_{18} HPLC column.

Other Reagents for Phosphoproteomics

The Antibody Beacon™ Tyrosine Kinase Assay Kit (A35725), described in detail in Section 10.3, provides a simple yet robust solution assay for real-time monitoring of tyrosine kinase activity and the effectiveness of potential inhibitors and modulators. The key to this tyrosine kinase assay is a small-molecule tracer ligand labeled with our bright green-fluorescent Oregon Green® 488 dye. When an anti-phosphotyrosine antibody binds this tracer ligand to form the Antibody Beacon™ detection complex, the fluorescence of the Oregon Green® 488 dye is efficiently quenched. In the presence of a phosphotyrosine-containing peptide, however, this Antibody Beacon™ detection complex is rapidly disrupted, releasing the tracer ligand and relieving its antibody-induced quenching. Upon its displacement by a phosphotyrosine residue, the Oregon Green® 488 dye–labeled tracer ligand exhibits an approximately 4-fold enhancement in its fluorescence, enabling the detection of as little as 50 nM phosphotyrosine-containing peptide with excellent signal-to-background discrimination.

Pro-Q® Glycoprotein Stain Kits for Gels and Blots

Glycoproteins play important roles as cell-surface markers, as well as in cell adhesion, immune recognition and inflammation reactions.[28] To facilitate research on glycoproteins, we offer the Pro-Q® Glycoprotein Stain Kits for Gels and for Blots, which provide extraordinary sensitivity, linearity and ease of use for selective detection of glycoproteins.

Pro-Q® Emerald Glycoprotein Stain Kits for Gels and for Blots

Pro-Q® Emerald 300 and Pro-Q® Emerald 488 Glycoprotein Stain Kits (Table 9.5) provide advanced reagents for detecting glycoproteins in gels and on blots.[29] The Pro-Q® Emerald glycoprotein stains react with periodate-oxidized carbohydrate groups, creating a bright green-fluorescent signal on glycoproteins (Figure 9.4.11). The staining procedure requires only three steps: fixation, oxidation and staining—no reduction step is required. Depending on the nature and degree of glycosylation, the Pro-Q® Emerald 300 stain allows the detection of as little as 1 ng of a glycoprotein per band in gels (4 ng/band with the Pro-Q® Emerald 488 stain), making these stains about 50 times more sensitive than the standard periodic acid–Schiff base method using acidic fuchsin dye. Blot staining is not quite as sensitive (2–18 ng of a glycoprotein per band can be detected) and is more time consuming, but provides an opportunity to combine glycoprotein staining with immunostaining or other blot-based detection techniques. The Pro-Q® Emerald 300 stain is best visualized using 300 nm UV illumination, whereas the Pro-Q® Emerald 488 stain is best visualized using visible light with wavelengths near its 510 nm excitation maximum. The Pro-Q® Emerald dye is also used as the detection reagent in our Pro-Q® Emerald 300 Lipopolysaccharide Gel Stain Kit (P20495), which is described in Section 13.3.

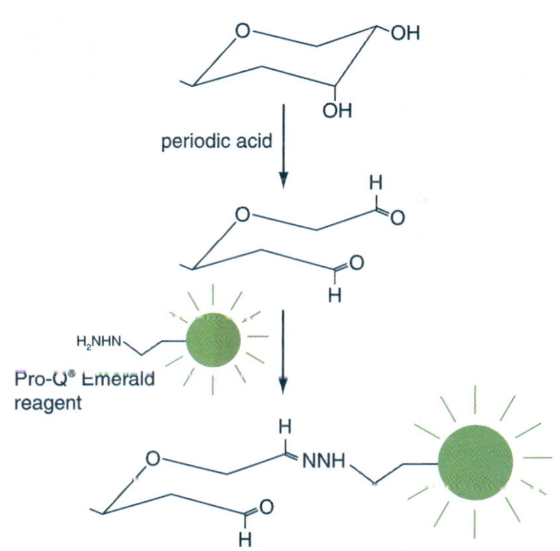

Figure 9.4.11 Detecting glycoproteins with Pro-Q® Emerald glycoprotein detection reagents. Oxidation with periodic acid converts *cis*-glycols to dialdehydes, which can then react with the hydrazide-based Pro-Q® Emerald reagents to form a covalent bond.

Table 9.5 Pro-Q® Emerald glycoprotein stain kits for gels and for blots.w

Product	Glycoprotein Stain	Kit Type	Cat. No.
Pro-Q® Emerald 300 Glycoprotein Stain Kits	Pro-Q® Emerald 300 stain, Ex/Em = 280/530 nm	Gel Stain Kit (includes SYPRO® Ruby protein gel stain *)	P21855
		Gel and Blot Stain Kit (does not include a total-protein stain)	P21857
Pro-Q® Emerald 488 Glycoprotein Stain Kit	Pro-Q® Emerald 488 stain, Ex/Em = 510/520 nm	Gel and Blot Stain Kit (does not include a total-protein stain)	P21875

* See Section 9.3 for a description of the SYPRO® Ruby protein gel stain.

IMPORTANT NOTICE: The products described in this manual are covered by one or more Limited Use Label License(s). Please refer to the Appendix on page 971 and Master Product List on page 975. Products are For Research Use Only. Not intended for any animal or human therapeutic or diagnostic use.

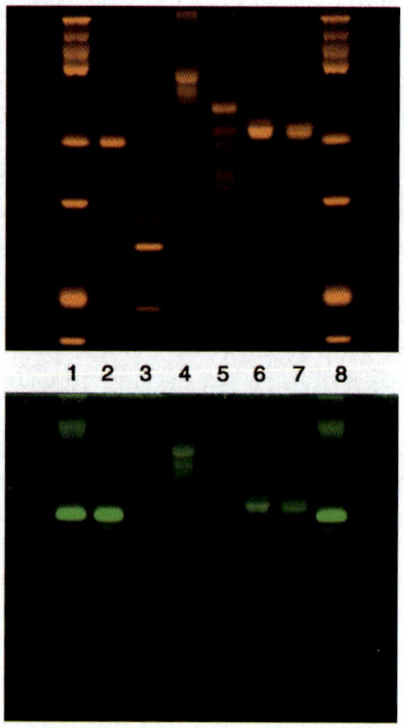

Figure 9.4.13 Mobility-shift gel assay using deglycosylating enzymes, stained with the SYPRO® Ruby protein gel stain (top; S12000, S12001, S21900) and Pro-Q® Emerald 300 glycoprotein stain (bottom). The glycoproteins α1-acidic glycoprotein, fetuin and horseradish peroxidase (HRP) are shown before (lanes 2, 4 and 6, respectively) and after (lanes 3, 5 and 7, respectively) glycosidase treatment. Glycosidase treatment resulted in a mobility shift and loss of green-fluorescent Pro-Q® Emerald 300 staining for α1-acidic glycoprotein and fetuin, indicating that the carbohydrate groups had been cleaved off. HRP, which contains an α-(1,3)-fucosylated asparagine GlcNac-linkage that is resistant to many glycosidases, showed neither a mobility shift nor a loss of green-fluorescent Pro-Q® Emerald 300 staining. Thus, use of the Pro-Q® Emerald 300 Glycoprotein Stain Kits (P21855, P21857, M33307) identifies glycoproteins that are not susceptible to the glycosidases used in the assay, providing important structural information about the glycoprotein's carbohydrate moiety.

The Pro-Q® Emerald glycoprotein stains can be combined with general protein stains for dichromatic detection of glycoproteins and total proteins in gels [29] and on blots, making it much easier to identify the location of the glycoproteins in the total-protein profile (Figure 9.4.12, Figure 9.4.13, Figure 9.4.14, Figure 9.4.15). The SYPRO® Ruby protein gel and blot stains (described in Section 9.3) provide the same sensitivity as silver staining (gels) or colloidal gold staining (blots) but, unlike these chromogenic techniques, do not require formaldehyde or glutaraldehyde, which can produce false-positive responses when glycoproteins are stained. These total-protein stains make it possible to visualize the entire protein complement of a sample and to thus identify contaminating proteins, to compare stained proteins to molecular weight standards and to provide a control for protease contamination in glycosidase mobility-shift experiments. The SYPRO® Ruby protein blot stain is additionally useful for assessing the efficiency of protein transfer to a blot (Figure 9.4.14), which is especially important when working with glycoproteins, because they often transfer poorly to blotting membranes. Proteins labeled with the SYPRO® Ruby total-protein stains exhibit red-orange fluorescence when excited with either a 300 nm UV light source or a laser scanner with a 473, 488 or 532 nm laser light source.

The Pro-Q® Emerald Glycoprotein Stain Kits also include our CandyCane™ molecular weight standards, a mixture of glycosylated and nonglycosylated proteins that, when separated by electrophoresis, provide alternating positive and negative controls (Figure 9.4.16). The CandyCane™ molecular weight standards are also available separately (C21852). In addition, we offer a Pro-Q®

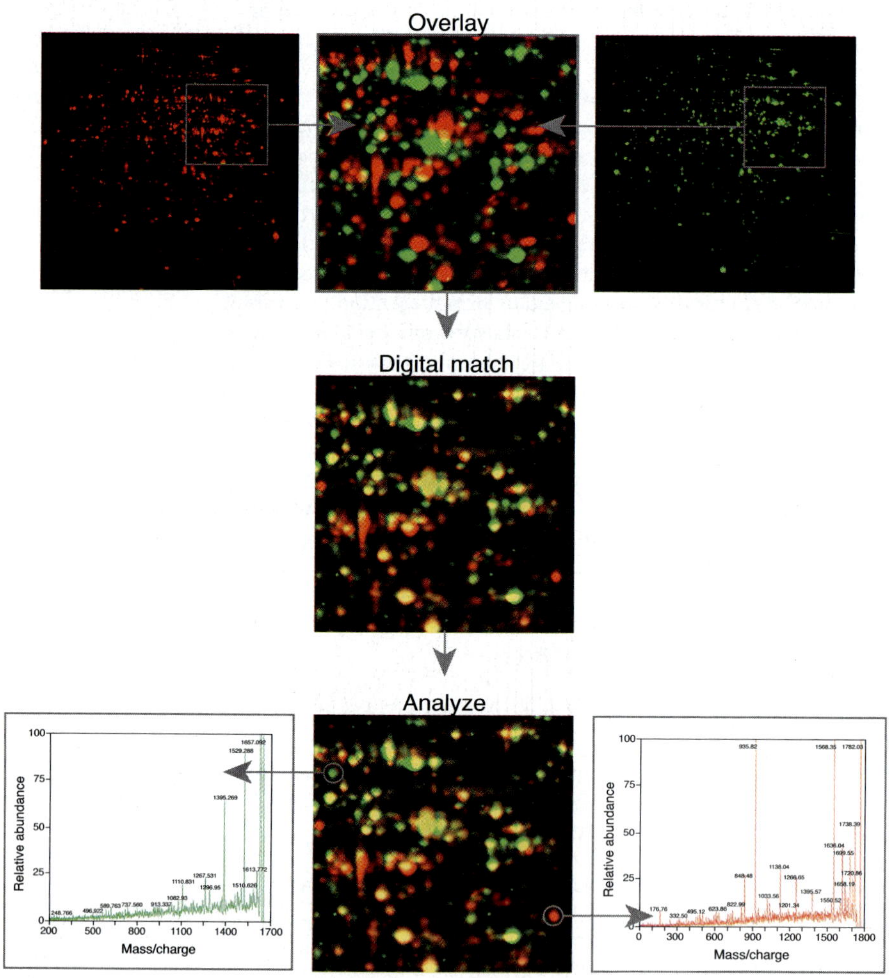

Figure 9.4.12 Multiplexed analysis of normal liver cells (left) and liver tumor cells (right). Gels are stained for total protein and glycoproteins using SYPRO® Ruby protein gel stain (S12000, S12001, S21900) and a Pro-Q® Emerald glycoprotein stain (P21855, P21857, P21875), respectively. Individual gel images can then be digitally overlaid, matched and analyzed for changes in protein expression and glycosylation.

IMPORTANT NOTICE: The products described in this manual are covered by one or more Limited Use Label License(s). Please refer to the Appendix on page 971 and Master Product List on page 975. Products are For Research Use Only. Not intended for any animal or human therapeutic or diagnostic use.

Emerald 300 Glycoprotein Gel Stain Kit (P21855) that includes our SYPRO® Ruby protein gel stain for detecting total proteins. The Pro-Q® Emerald Glycoprotein Stain Kits for gels and blots (P21855, P21857, P21875) contain sufficient materials to stain approximately ten 8 cm × 10 cm gels or blots, including:

- Pro-Q® Emerald 300 glycoprotein stain (in Kits P21855 and P21857) or Pro-Q® Emerald 488 glycoprotein stain (in Kit P21875)
- Pro-Q® Emerald 300 staining buffer (in Kits P21855 and P21857 or Pro-Q® Emerald 488 staining buffer (in Kit P21875)
- Oxidizing reagent (periodic acid)
- SYPRO® Ruby protein gel stain (in Kit P21855 only)
- CandyCane™ glycoprotein molecular weight standards
- Detailed protocols

Multiplexed Proteomics® Kit for Glycoprotein and Total-Protein Gel Staining

When used together, the Pro-Q® Emerald 300 glycoprotein gel stain and the SYPRO® Ruby protein gel stain (S12000, S12001, S21900; Section 9.3) make a powerful combination for proteome analysis (Figure 9.4.12). Determining the ratio of the Pro-Q® Emerald dye to SYPRO® Ruby dye signal intensities for each band or spot provides a measure of the glycosylation level normalized to the total amount of protein. Using both stains in combination makes it possible to distinguish a lightly glycosylated, high-abundance protein from a heavily glycosylated, low-abundance protein. To make this staining more convenient and economical, we offer the Multiplexed Proteomics® Glycoprotein Gel Stain Kit with 1 L of our Pro-Q® Emerald 300 glycoprotein gel stain and 1 L of our SYPRO® Ruby protein gel stain (M33307).

CandyCane™ Glycoprotein Molecular Weight Standards

CandyCane™ glycoprotein molecular weight standards (C21852) contain a mixture of glycosylated and nonglycosylated proteins with molecular weights from 14,000 to 180,000 daltons. When separated by polyacrylamide gel electrophoresis, the standards appear as alternating bands corresponding to glycosylated and nonglycosylated proteins (Figure 9.4.16). Thus, these standards serve both as molecular weight markers and as positive and negative controls for methods that detect glycosylated proteins, such as those provided in our Pro-Q® Emerald Glycoprotein Stain Kits.

Other Specialized Techniques for Detecting Specific Proteins on Gels and Blots

Detecting Calcium-Binding Proteins

The luminescent lanthanide terbium, which is available as its chloride salt (Tb^{3+} from $TbCl_3$, T1247), selectively stains calcium-binding proteins in SDS-polyacrylamide gels.[30] With some modifications to the staining protocol, these lanthanides can also be used to detect all protein bands.[30] Terbium chloride has also been used as a rapid negative stain for proteins in SDS-polyacrylamide gels, in which the background is green fluorescent and the proteins are unstained.[31]

Detecting Penicillin-Binding Proteins

BOCILLIN™ FL penicillin and BOCILLIN™ 650/665 penicillin (B13233, B13234) are green- and infrared-fluorescent penicillin analogs, respectively, that bind selectively and with high affinity to

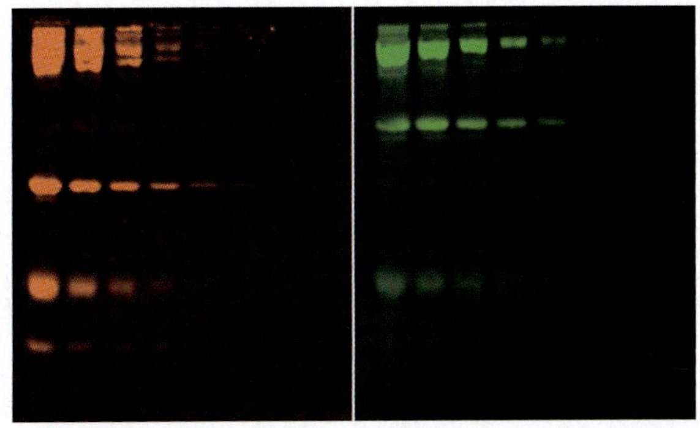

Figure 9.4.14 Staining glycoproteins and the total protein profile on blots using the Pro-Q® Emerald 300 Glycoprotein Gel and Blot Stain Kit (P21857). A 2-fold dilution series of the CandyCane™ glycoprotein molecular weight standards (C21852) was run an SDS-polyacrylamide gel and blotted onto a PVDF membrane. The blot was first stained with the SYPRO® Ruby protein blot stain (S11791) to detect the total protein profile (left). After documentation of the signal, the blot was stained with the Pro-Q® Emerald 300 glycoprotein stain (right) provided in the Pro-Q® Emerald 300 Glycoprotein Gel and Blot Stain Kit.

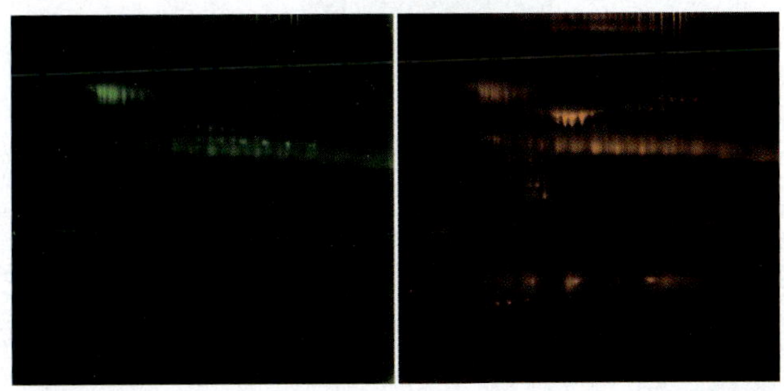

Figure 9.4.15 2D gel stained with the SYPRO® Ruby protein gel stain and the Pro-Q® Emerald 300 reagent. Combined Cohn fractions II and III from cow plasma, containing primarily ß- and γ-globulins, were run on a 2D gel and stained first with the Pro-Q® Emerald 300 reagent (P21855, P21857; left) and then with the SYPRO® Ruby protein gel stain (S12000, S12001, S21900, and in P21855; right).

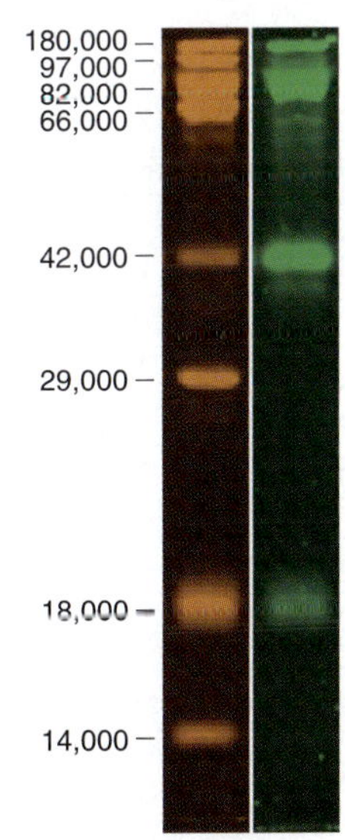

Figure 9.4.16 Glycosylated and nonglycosylated proteins in the CandyCane™ glycoprotein molecular weight standards (C21852). The standards were electrophoresed through two identical 13% polyacrylamide gels. Both lanes contain ~0.5 µg of protein in each band. The left lane was stained with our SYPRO® Ruby protein gel stain (S12000, S12001, S21900) to detect all eight marker proteins. The right lane was stained using the reagents in the Pro-Q® Emerald 300 Glycoprotein Gel Stain Kit (P21855).

IMPORTANT NOTICE: The products described in this manual are covered by one or more Limited Use Label License(s). Please refer to the Appendix on page 971 and Master Product List on page 975. Products are For Research Use Only. Not intended for any animal or human therapeutic or diagnostic use.

penicillin-binding proteins present on the cytoplasmic membranes of eubacteria.[32,33] When electrophoresed under nonreducing conditions, the dye-labeled penicillin-binding proteins are easily visible in the gel with sensitivity in the low nanogram range[34] (Figure 9.4.17). BOCILLIN™ FL penicillin, synthesized from penicillin V and the BODIPY® FL dye (spectrally similar to fluorescein), has been used to determine the penicillin-binding protein profiles of *Escherichia coli, Pseudomonas aeruginosa* and *Streptococcus pneumoniae*, and these binding profiles are found to be similar to those reported by researchers using radioactively labeled penicillin V.[32] Fluorescently labeled penicillin has also been used for direct labeling and rapid detection of whole *E. coli* and *Bacillus licheniformis*[35] and of *Enterobacter pneumoniae*.[36]

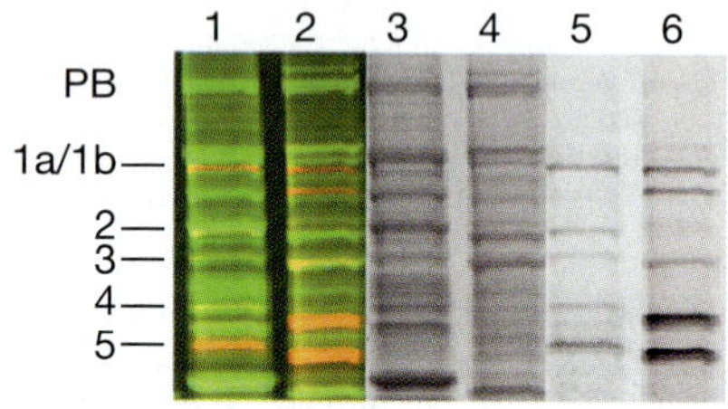

Figure 9.4.17 Detection of penicillin-binding proteins (PBPs) from *Escherichia coli* and *Pseudomonas aeruginosa*. The membrane fractions from *E. coli* and *P. aeruginosa* were prepared as previously described and labeled with BOCILLIN™ 650/665 penicillin (B13234). The labeled membranes were separated on an SDS-polyacrylamide gel, stained with SYPRO® Ruby and visualized using a Typhoon imager. The location of PBPs from *E. coli* are labeled to the left of the gels. Lanes 1, 3, and 5 are *E. coli* membrane preparations; lanes 2, 4, and 6 are *P. aeruginosa* membrane preparations; lanes 1 and 2 are overlays of images obtained from total protein (green) and PBP (red) scans; lanes 3 and 4 are total protein visualized with the SYPRO® Ruby protein gel stain; lanes 5 and 6 are PBPs as detected by BOCILLIN™ 650/665. Image used with permission from Wiley VCH publishers.

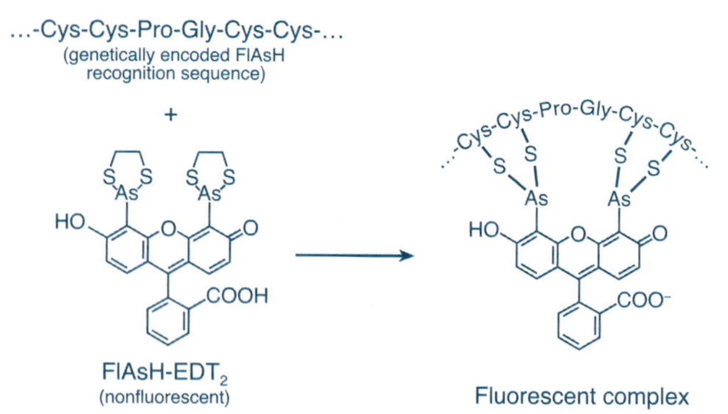

Figure 9.4.18 Binding of the nonfluorescent FlAsH-EDT$_2$ ligand to a recombinantly expressed tetracysteine sequence yields a highly fluorescent complex.

The β-lactam sensor-transducer (BlaR), an integral membrane protein from *Staphylococcus aureus*, covalently and stoichiometrically reacts with β-lactam antibiotics, including BOCILLIN™ FL penicillin, by acylation of its active-site serine residue.[37]

Detecting TC-Tagged Fusion Proteins with TC-FlAsH™ and TC-ReAsH™ Reagents

TC-FlAsH™ and TC-ReAsH™ detection technology, based on the tetracysteine tag first described by Griffin, Adams and Tsien in 1998,[38,39] takes advantage of the high-affinity interaction of a biarsenical ligand (FlAsH-EDT$_2$ or ReAsH-EDT$_2$) with the thiols in a tetracysteine (TC) expression tag fused to the protein of interest. The FlAsH-EDT$_2$ ligand is essentially fluorescein that has been modified to contain two arsenic atoms at a set distance from each other, whereas the ReAsH-EDT$_2$ ligand is a similarly modified resorufin. Virtually nonfluorescent in the ethanedithiol (EDT)-bound state, these reagents become highly fluorescent when bound to the thiol-containing tetracysteine tag Cys-Cys-Xxx-Yyy-Cys-Cys, where Xxx-Yyy is typically Pro-Gly[40] (Figure 9.4.18). Modified tags with additional flanking sequences produce higher affinity binding of the biarsenical ligand, resulting in improved signal-to-background characteristics.[41,42]

Transduction of the host cell line with an expression construct comprising the protein of interest fused to a tetracysteine tag is followed by addition FlAsH-EDT$_2$ reagent (Figure 9.4.18) or ReAsH-EDT$_2$ reagent, generating green or red fluorescence, respectively, upon binding the tetracysteine motif. For detection of tetracysteine-tagged proteins expressed in cells, we offer the TC-FlAsH™ II and TC-ReAsH™ II In-Cell Tetracysteine Tag Detection Kits (T34561, T34562, T34563), which are described in Section 2.2.

As an alternative to in-cell detection, we offer the TC-FlAsH™ Expression Analysis Detection Kits (A10067, A10068) for detecting tetracysteine-tagged proteins in polyacrylamide gels (Figure 9.4.19). These kits provide:

- FlAsH loading buffer
- Orange total-protein stain (in A10067) or Red total-protein stain (in A10068)
- Dimethylsulfoxide (DMSO)
- BenchMark™ Protein Ladder
- Detailed protocols

Sufficient reagents are provided for ten 17-well minigels, based on a 12 µL reaction volume. When bound to TC-tagged protein, FlAsH dye exhibits excitation/emission maxima of 505/530 nm. The Orange and Red total-protein gel stains supplied in these detection kits exhibit emission maxima of 585/620 nm and 650/660 nm, respectively.

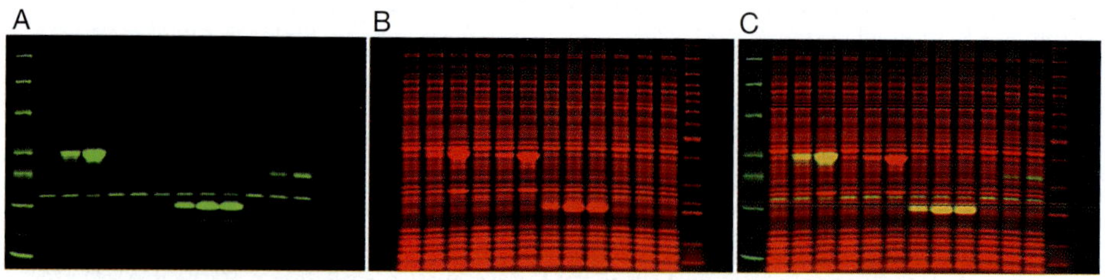

Figure 9.4.19 Protein gel staining using TC-FlAsH™ Expression Analysis Detection Kit (A10068). **A)** Tetracysteine-tagged proteins are labeled with FlAsH-EDT$_2$ reagent and fluoresce green. **B)** Total proteins are labeled with the Red total-protein stain provided in the kit and fluoresce red. **C)** An overlay of the two images reveals relative amounts of protein.

IMPORTANT NOTICE: The products described in this manual are covered by one or more Limited Use Label License(s). Please refer to the Appendix on page 971 and Master Product List on page 975. Products are For Research Use Only. Not intended for any animal or human therapeutic or diagnostic use.
www.invitrogen.com/probes

Detecting Oligohistidine Fusion Proteins

The oligohistidine domain is a Ni^{2+}-binding peptide sequence comprising a string of four to six histidine residues. When the DNA sequence corresponding to the oligohistidine domain is fused in frame with a gene of interest, the resulting recombinant protein can be easily purified using a nickel-chelating resin.[43,44]

Developed by QIAGEN, the Penta·His mouse IgG_1 monoclonal antibody (P21315) provides a sensitive method for specific detection of fusion proteins that have an oligohistidine domain comprising five or six consecutive histidine residues. The antibody does not recognize tetrahistidine domains or domains in which the histidine string is interrupted by another amino acid. The Penta·His antibody binds to the oligohistidine domain regardless of the surrounding amino acid context and even when the group is partially hidden, although subtle differences in the amino acid context may change the sensitivity limit for a particular fusion protein. The antibody is ideal for detecting oligohistidine fusion proteins on western blots (Figure 9.4.20). The Penta·His antibody is also useful for immunoprecipitation, ELISA assays and immunohistochemistry.

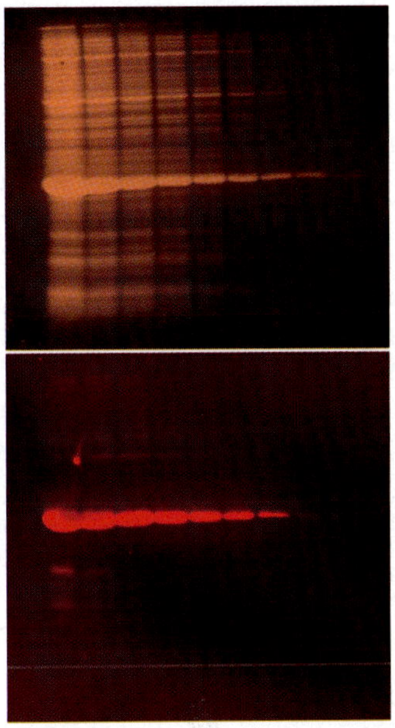

Figure 9.4.20 Detection of an oligohistidine fusion protein with the Penta·His mouse monoclonal antibody. Two-fold serial dilutions of an *Escherichia coli* lysate containing overexpressed oligomycin sensitivity–conferring protein (OSCP) fused with a hexahistidine domain were run an SDS-polyacrylamide gel and blotted onto a PVDF membrane. The blot was stained with the SYPRO® Ruby protein blot stain (S11791) to detect the entire protein profile (top). After imaging, the blot was incubated with Penta·His mouse IgG_1 monoclonal antibody (P21315), followed by immunodetection using an alkaline phosphatase conjugate of goat anti–mouse antibody in conjunction with DDAO phosphate (bottom).

Anti-Dinitrophenyl Antibody for Measuring Protein Carbonyls

Oxidative injury can be monitored by following the formation of protein-derived aldehydes and ketones. Traditionally, protein-derived aldehydes and ketones have been quantitated using a colorimetric assay based on their reaction with 2,4-dinitrophenylhydrazine to yield protein-bound dinitrophenyl (DNP) moieties. A much more sensitive ELISA method has been developed that detects the protein-bound DNP using unlabeled or biotin-labeled anti-DNP antibodies[45] (A6430, A6435; Section 7.4). The bound anti-DNP antibody is subsequently detected with horseradish peroxidase–conjugated secondary detection reagents (Section 7.2). Our Alexa Fluor® 488 and fluorescein conjugates of the anti-DNP antibody (A11097, A6423; Section 7.2) can also be applied to this detection scheme.[46] Our polyclonal antibody to nitrotyrosine (A21285, Section 7.4) can be used similarly to separate and detect proteins of cell extracts that have been naturally nitrated by nitric oxide (Section 18.3, Figure 9.4.21).

Detecting Glutathiolation with BioGEE

Biotinylated glutathione ethyl ester (BioGEE, G36000; Figure 9.4.22) is a cell-permeant, biotinylated glutathione analog for detecting glutathiolation. Under conditions of oxidative stress, cells may transiently incorporate glutathione into proteins. Stressed cells incubated with BioGEE will also incorporate this biotinylated glutathione derivative into proteins, facilitating the identification of oxidation-sensitive proteins.[47,48] Once these cells are fixed and permeabilized, glutathiolation levels can be detected with a fluorescent streptavidin conjugate (Section 7.6, Table 7.9) using either flow cytometry or fluorescence microscopy. Proteins glutathiolated with BioGEE can be captured using streptavidin or CaptAvidin™ agarose (S951, C21386; Section 7.6) and analyzed by mass spectrometry or by western blotting methods.[49,50]

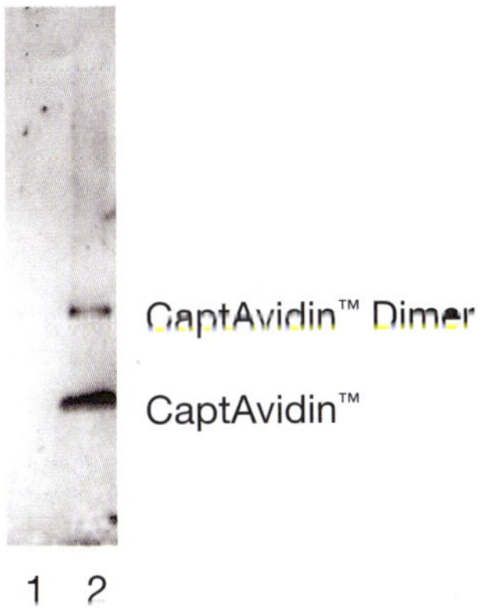

Figure 9.4.21 Specificity of our rabbit anti-nitrotyrosine antibody (A21285) to nitrated proteins. Equal amounts of avidin (A887, lane 1) and CaptAvidin™ biotin-binding protein (C21385, lane 2) were run on an SDS-polyacrylamide gel (4–20%) and blotted onto a PVDF membrane. CaptAvidin™ biotin-binding protein, a derivative of avidin, has nitrated tyrosine residues in the biotin-binding site. Nitrated proteins were identified with the anti-nitrotyrosine antibody in combination with an alkaline phosphatase conjugate of goat anti–rabbit IgG antibody (G21079) and the red-fluorescent substrate, DDAO phosphate (D6487).

Glutathione Agarose and Anti–Glutathione *S*-Transferase Antibody for GST Fusion Protein Identification and Purification

In protein fusion techniques, the coding sequence of one protein is fused in-frame with another so that the expressed hybrid protein possesses desirable properties of both parent proteins. One common partner in these engineered products is glutathione *S*-transferase (GST),

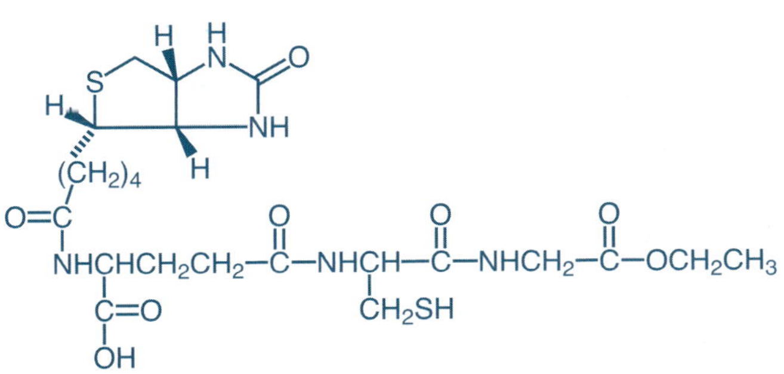

Figure 9.4.22 Glutathione ethyl ester, biotin amide (BioGEE, G36000).

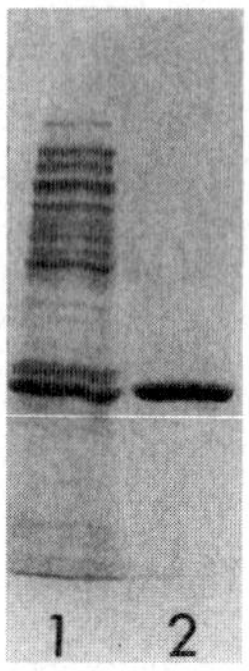

Figure 9.4.23 Coomassie brilliant blue–stained SDS-polyacrylamide gel, demonstrating the purification of a glutathione *S*-transferase (GST) fusion protein using glutathione agarose (G2879). Lane 1 contains crude supernatant from an *Escherichia coli* lysate and lane 2 contains the affinity-purified GST fusion protein.

a protein with natural binding specificity that can be exploited to facilitate its purification.[51] Because the GST portion of the fusion protein retains its affinity and selectivity for glutathione, the fusion protein can be conveniently purified from the cell lysate in a single step by affinity chromatography on glutathione agarose[51–57] (Figure 9.4.23). For purification of GST fusion proteins, we offer glutathione linked via the sulfur atom to crosslinked beaded agarose (10 mL of sedimented bead suspension, G2879). Each milliliter of gel can bind approximately 5–6 mg of bovine-liver GST. Adding excess free glutathione liberates the GST fragment from the matrix, which can then be regenerated by washing with a high-salt buffer.

We also offer a highly purified rabbit polyclonal anti-GST antibody (A5800) that can be used to purify GST fusion proteins by immunoprecipitation.[58] This highly specific antibody, which was generated against a 260–amino acid N-terminal fragment of the *Schistosoma japonica* enzyme expressed in *Escherichia coli*, is also useful for detecting GST fusion proteins on western blots and for detecting GST distribution in cells. The intensely green-fluorescent Alexa Fluor® 488 conjugate of anti–glutathione *S*-transferase (A11131) is also available for direct detection of GST fusion proteins.

Following purification, the fusion protein can serve as an immunogen for antibody production[59,60] or its properties can be compared with those of the native polypeptide to provide insights on the normal function of the polypeptide of interest. Such methods have been used to investigate biological properties of many proteins. Examples include cleavage of the capsid assembly protein ICP35 by the herpes simplex virus type 1 protease,[61] the role of the Rho GTP-binding protein in *lbc* oncogene function[62] and the association of v-Src with cortactin in Rous sarcoma virus–transformed cells.[63] In fact, the Ca^{2+}-binding properties of a protein kinase C–GST fusion protein were examined while the GST fusion protein was still bound to the glutathione agarose.[64] Likewise, interactions of a DNA-binding protein–GST fusion protein have been assessed using an affinity column consisting of the fusion protein bound to glutathione agarose.[52] Alternatively, the GST fusion expression vector can be engineered to encode a recognition sequence for a site-specific protease, such as thrombin or factor Xa, between the GST structural gene and gene of interest.[65–68] Once the fusion protein is bound to the affinity matrix, the site-specific enzyme can be added to release the protein.

Streptavidin Acrylamide and Reactive Acrylamide Derivatives

Streptavidin acrylamide (S21379), which is prepared from the succinimidyl ester of 6-((acryloyl)amino)hexanoic acid (acryloyl-X, SE, A20770), may be useful for the preparation of biosensors.[69] A similar streptavidin acrylamide has been shown to copolymerize with acrylamide on a polymeric surface to create a uniform monolayer of the immobilized protein. The streptavidin can then bind biotinylated ligands, including biotinylated hybridization probes, enzymes, antibodies and drugs.

Like streptavidin and CaptAvidin™ biotin-binding protein, other amine-containing biomolecules can be crosslinked to acrylamides using acryloyl-X, SE. Acryloyl-X, SE reacts with amines of proteins, amine-modified nucleic acids and other biomolecules to yield acrylamides that can be copolymerized into polyacrylamide matrices or on surfaces, such as in microarrays and in biosensors. We prepare both streptavidin and CaptAvidin™ biotin-binding protein conjugated to 4% beaded crosslinked agarose (S951, C21386; Section 7.6)—matrices that can be used to isolate biotinylated peptides, proteins, hybridization probes, haptens and other molecules.

REFERENCES

1. Chem Rev (2009) 109:4207; **2.** Chem Rev (2008) 108:2952; **3.** Angew Chem Int Ed Engl (2009) 48:6974; **4.** Biochemistry (2009) 48:6571; **5.** ACS Chem Biol (2006) 1:644; **6.** Proc Natl Acad Sci U S A (2008) 105:2415; **7.** Protein Sci (2004) 13:2693; **8.** Yeast (2008) 25:775; **9.** Bioorg Med Chem Lett (2008) 18:5995; **10.** Proc Natl Acad Sci U S A (2006) 103:9482; **11.** Nat Protoc (2007) 2:532; **12.** J Virol (2009) 83:13042; **13.** Methods Enzymol (2006) 415:230; **14.** Nat Cell Biol (2008) 10:1224; **15.** Proc Natl Acad Sci U S A (2007) 104:2614; **16.** J Biol Chem (2009) 284:21327; **17.** J Am Chem Soc (2008) 130:11576; **18.** J Am Chem Soc (2009) 131:4967; **19.** J Am Chem Soc (2007) 129:2744; **20.** Proc Natl Acad Sci U S A (2004) 101:12479; **21.** J Histochem Cytochem (2010) 58:207; **22.** J Am Chem Soc (2010) 132:2504; **23.** Chem Res Toxicol (2008) 21:432; **24.** J Biol Chem (2003) 278:27251; **25.** Methods Mol Biol (2008) 446:21; **26.** Proteomics (2003) 3:1128; **27.** Comb Chem High Throughput Screen (2003) 6:331; **28.** Chem Rev (2008) 108:1708; **29.** J Proteomics (2010) 73:879; **30.** Anal Biochem (1994) 216:439; **31.** Anal Biochem (1994) 220:218; **32.** Antimicrob Agents Chemother (1999) 43:1124; **33.** J Biol Chem (2000) 275:17693; **34.** Electrophoresis (2001) 22:960; **35.** Biochem J (1993) 291:19; **36.** Biochem J (1994) 300:141; **37.** J Biol Chem (2003) 278:18419; **38.** Methods Enzymol (2000) 327:565; **39.** Science (1998) 281:269; **40.** J Am Chem Soc (2002) 124:6063; **41.** J Am Chem Soc (2009) 131:4613; **42.** Nat Biotechnol (2005) 23:1308; **43.** J Chromatogr (1987) 411:117; **44.** J Chromatogr (1987) 411:177; **45.** Methods Enzymol (1999) 300:106; **46.** Anal Bioanal Chem (2008) 391:2591; **47.** Amino Acids (2007) 33:51; **48.** Biochemistry (2000) 39:11121; **49.** J Biol Chem (2009) 284:22213; **50.** Nat Immunol (2008) 9:866; **51.** Methods Mol Biol (2004) 261:175; **52.** Meth Mol Genet (1993) 1:267; **53.** Biotechniques (1992) 13:856; **54.** Biotechniques (1991) 10:178; **55.** Nucleic Acids Res (1991) 19:4005; **56.** Science (1991) 252:712; **57.** Gene (1988) 67:31; **58.** J Biol Chem (1997) 272:8133; **59.** J Cell Biol (1995) 131:1003; **60.** J Cell Biol (1995) 130:651; **61.** J Biol Chem (1995) 270:30168; **62.** J Biol Chem (1995) 270:9031; **63.** J Biol Chem (1995) 270:26613; **64.** J Biol Chem (1993) 268:3715; **65.** J Biol Chem (1995) 270:24525; **66.** J Cell Biol (1995) 129:189; **67.** Mol Biol Cell (1995) 6:247; **68.** Biochemistry (1992) 31:5841; **69.** Anal Biochem (2000) 282:200.

IMPORTANT NOTICE: The products described in this manual are covered by one or more Limited Use Label License(s). Please refer to the Appendix on page 971 and Master Product List on page 975. Products are For Research Use Only. Not intended for any animal or human therapeutic or diagnostic use.
www.invitrogen.com/probes

PRODUCT LIST 9.4 DETECTING PROTEIN MODIFICATIONS

Cat. No.	Product	Quantity
A20770	6-((acryloyl)amino)hexanoic acid, succinimidyl ester (acryloyl-X, SE)	5 mg
A5800	anti–glutathione *S*-transferase, rabbit IgG fraction *3 mg/mL*	0.5 mL
A11131	anti–glutathione *S*-transferase, rabbit IgG fraction, Alexa Fluor® 488 conjugate *2 mg/mL*	0.5 mL
B13234	BOCILLIN™ 650/665 penicillin, sodium salt	1 mg
B13233	BOCILLIN™ FL penicillin, sodium salt	1 mg
C21852	CandyCane™ glycoprotein molecular weight standards *200 gel lanes*	400 µL
C10102	Click-iT® AHA (L-azidohomoalanine) *for nascent protein synthesis*	5 mg
C10289	Click-iT® AHA Alexa Fluor® 488 Protein Synthesis HCS Assay *2-plate size*	1 kit
C33372	Click-iT® Biotin Protein Analysis Detection Kit *10 reactions*	1 kit
C10269	Click-iT® Cell Reaction Buffer Kit	1 kit
C33371	Click-iT® Dapoxyl® Protein Analysis Detection Kit *for UV excitation* *10 reactions*	1 kit
C10248	Click-iT® farnesyl alcohol, azide *mixed isomers*	1 mg
C10264	Click-iT® fucose alkyne (tetraacetylfucose alkyne)	5 mg
C33365	Click-iT® GalNAz metabolic glycoprotein labeling reagent (tetraacetylated N-azidoacetylgalactosamine) *for O-linked glycoproteins* *5.2 mg*	each
C10249	Click-iT® geranylgeranyl alcohol, azide *mixed isomers*	1 mg
C33368	Click-iT® O-GlcNAc Enzymatic Labeling System *for O-linked GlcNAc glycoproteins* *10 labelings*	1 kit
C33373	Click-iT® O-GlcNAc peptide and phosphopeptide LC/MS standards *5 nmol each*	1 set
C33374	Click-iT® O-GlcNAc peptide LC/MS standard (H-Thr-Ala-Pro-Thr-(O-GlcNAc)Ser-Thr-Ile-Ala-Pro-Gly-OH) *Theoretical Mass (M+H): 1118.50*	5 nmol
C33367	Click-iT® GlcNAz metabolic glycoprotein labeling reagent (tetraacetylated N-azidoacetylglucosamine) *for O-GlcNAc-modified proteins* *5.2 mg*	each
C10186	Click-iT® HPG (L-homopropargylglycine) *for nascent protein synthesis*	5 mg
C33366	Click-iT® ManNAz metabolic glycoprotein labeling reagent (tetraacetylated N-azidoacetyl-d-mannosamine) *for sialic acid glycoproteins* *5.2 mg*	each
C10268	Click-iT® myristic acid, azide (12-azidododecanoic acid)	1 mg
C10265	Click-iT® palmitic acid, azide (15-azidopentadecanoic acid)	1 mg
C10276	Click-iT® Protein Reaction Buffer Kit	1 kit
C33370	Click-iT® Tetramethylrhodamine (TAMRA) Protein Analysis Detection Kit *UV/532 nm excitation* *10 reactions*	1 kit
G2879	glutathione agarose, linked through sulfur *sedimented bead suspension*	10 mL
G36000	glutathione ethyl ester, biotin amide (BioGEE) *glutathiolation detection reagent* *special packaging*	10 x 100 µg
M33307	Multiplexed Proteomics® Glycoprotein Gel Stain Kit *with 1 L each of Pro-Q® Emerald 300 and SYPRO® Ruby (S12000) gel stains*	1 kit
MPM33305	Multiplexed Proteomics® Phosphoprotein Gel Stain Kit *includes MPP33300 and S12000*	1 kit
MPM33306	Multiplexed Proteomics® Phosphoprotein Gel Stain Kit *includes MPP33301 and S12001*	1 kit
M33305	Multiplexed Proteomics® Phosphoprotein Gel Stain Kit #1 *with 1 L each of Pro-Q® Diamond (P33300) and SYPRO® Ruby (S12000) gel stains*	1 set
M33306	Multiplexed Proteomics® Phosphoprotein Gel Stain Kit #2 *with 200 mL each of Pro-Q® Diamond (P33301) and SYPRO® Ruby (S12001) gel stains*	1 set
P21315	Penta·His mouse IgG_1, monoclonal antibody (anti-pentahistidine) *BSA free*	100 µg
P27167	PeppermintStick™ phosphoprotein molecular weight standards	40 µL
P33350	PeppermintStick™ phosphoprotein molecular weight standards *200 gel lanes*	400 µL
P33357	phosphopeptide standard mixture *400 pmol of each peptide*	2800 pmol
P33203	Pro-Q® Diamond LC Phosphopeptide Detection Kit	1 kit
P33356	Pro-Q® Diamond Phosphoprotein Blot Stain Kit *20 minigel blots*	1 kit
P33359	Pro-Q® Diamond Phosphoprotein Enrichment and Detection Kit	1 kit
P33358	Pro-Q® Diamond Phosphoprotein Enrichment Kit	1 kit
P33310	Pro-Q® Diamond phosphoprotein gel destaining solution	1 L
P33311	Pro-Q® Diamond phosphoprotein gel destaining solution *bulk packaging*	5 L
P33301	Pro-Q® Diamond phosphoprotein gel stain	200 mL
P33300	Pro-Q® Diamond phosphoprotein gel stain	1 L
P33302	Pro-Q® Diamond phosphoprotein gel stain *bulk packaging*	5 L
MPP33301	Pro-Q® Diamond Phosphoprotein Gel Staining Kit *includes 200 mL stain and 40 µL standard*	1 kit
MPP33300	Pro-Q® Diamond Phosphoprotein Gel Staining Kit *includes 1 L stain and 40 µL standard*	1 kit
MPP33302	Pro-Q® Diamond Phosphoprotein Gel Staining Kit *includes 5 L stain and 400 µL standard*	1 kit
P33706	Pro-Q® Diamond Phosphoprotein/Phosphopeptide Microarray Stain Kit	1 kit
P21857	Pro-Q® Emerald 300 Glycoprotein Gel and Blot Stain Kit *10 minigels or minigel blots*	1 kit
P21855	Pro-Q® Emerald 300 Glycoprotein Gel Stain Kit *with SYPRO® Ruby protein gel stain* *10 minigels*	1 kit
P21875	Pro-Q® Emerald 488 Glycoprotein Gel and Blot Stain Kit *10 minigels or minigel blots*	1 kit
S21379	streptavidin acrylamide	1 mg
A10067	TC-FlAsH™ Expression Analysis Detection Kit - Orange *fluorescent in-gel detection of TC-tagged and total protein*	1 kit
A10068	TC-FlAsH™ Expression Analysis Detection Kit - Red *fluorescent in-gel detection of TC-tagged and total protein*	1 kit
T1247	terbium(III) chloride, hexahydrate	1 g

IMPORTANT NOTICE: The products described in this manual are covered by one or more Limited Use Label License(s). Please refer to the Appendix on page 971 and Master Product List on page 975. Products are For Research Use Only. Not intended for any animal or human therapeutic or diagnostic use.

9.5 Peptide Analysis, Sequencing and Synthesis

This section describes Molecular Probes® reagents used in the synthesis of fluorophore- or hapten-labeled peptides and fluorogenic protease substrates, as well as in peptide and protein sequencing. Some of our fluorescent probes and research chemicals have also been used for N-terminal amino acid analysis and peptide sequencing, as well as for protein fragment modification prior to Edman sequencing.

N-Terminal Amino Acid Analysis

Except when it is already blocked by formylation, acetylation, pyroglutamic acid formation or other chemistry, the N-terminal amino acid of proteins can be labeled with a variety of fluorescent and chromophoric reagents from Chapter 1. However, only those functional groups that survive complete protein hydrolysis, such as sulfonamides, are useful for N-terminal amino acid analysis. Dansyl chloride (D21) is the most commonly employed reagent for such analyses.[1]

Nonacylated N-terminal serine and threonine residues of proteins can be periodate-oxidized to aldehydes[2–4] (Figure 9.5.1) that can then be modified by a variety of hydrazine and hydroxylamine derivatives listed in Section 3.3. Only peptides and proteins that contain these two terminal amino acids become fluorescent, although oxidation of the carbohydrate portion of glycoproteins to aldehydes may cause interference in this analysis.

N-Acetylated or *N*-formylated proteins have been detected by transfer of the acyl group to dansyl hydrazine (D100) and subsequent chromatographic separation of the fluorescent product.[5,6] The sensitivity of this method can likely be improved by the use of other fluorescent hydrazine and hydroxylamine derivatives described in Section 3.3.

Peptide Sequencing

The dominant chemistry for sequencing peptides employs the nonfluorescent reagent phenyl isothiocyanate, which forms phenylthiohydantoins (PTH) in the sequencing reaction. As analogs of phenyl isothiocyanate, fluorescein-5-isothiocyanate (FITC; F143, F1906, F1907; Section 1.5) and other fluorescent isothiocyanates yield peptide conjugates that are also susceptible to Edman degradation via their thiohydantoins. Thus, these fluorescent reagents are potentially useful for ultrasensitive amino acid sequencing.[7–10]

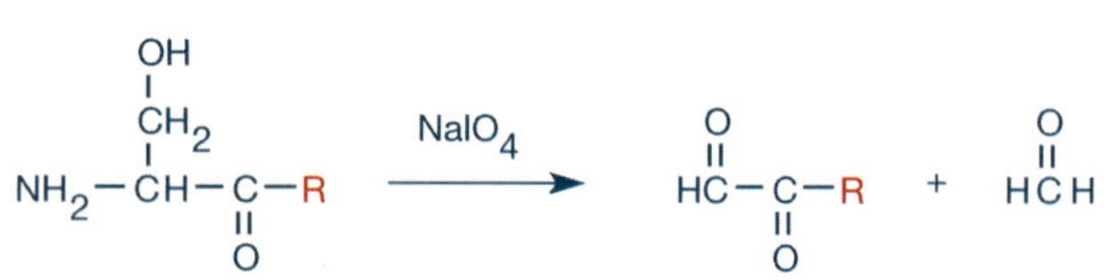

Figure 9.5.1 Sodium periodate oxidation of an N-terminal serine residue to an aldehyde, with the release of formaldehyde. The aldehyde thus formed from the protein can be subsequently modified with a variety of hydrazine, hydroxylamine or amine derivatives.

Peptide Synthesis

Peptides specifically labeled with fluorescent dyes, haptens, photoactive groups or radioisotopes are important both as probes for receptors and as substrates for enzymes (Section 10.4). Labeled peptides can be prepared by modifying isolated peptides or by incorporating the label during solid-phase synthesis. We offer several fluorescent neuropeptides, most of which are described in Section 16.2.

Labeling Peptides in Solution

Appropriately substituted synthetic peptides can be labeled in solution by almost any of the reactive probes described in Chapters 1–5 (Labeling Small Peptides with Amine-Reactive Dyes in Organic Solvents—Note 9.2). Many peptides contain multiple residues that can be modified, potentially leading to complex mixtures of products, some of which may be biologically inactive. Modification of a peptide's thiol group by one of the thiol-reactive reagents described in Chapter 2 is usually easy, selective and very efficient. If the peptide is synthetic, or can be modified by site-directed mutagenesis, incorporation of a cysteine residue at the desired site of labeling is recommended. The N-terminal α-amine of peptides, which has a lower pK_a than the ε-amino group of lysine residues, can sometimes be labeled in the presence of other amines if the pH is kept near neutral. Conversion of tyrosine residues to *o*-aminotyrosines (Section 3.2) can be used to provide selective sites for peptide modification, unless the tyrosine residues are essential for the biological activity of the peptide.

Solid-Phase Synthesis of Labeled Peptides

If specific labeling of peptides in solution is problematic, it may be more convenient to conjugate the fluorophore to the N-terminus of a resin-bound peptide *before* removal of other protecting groups and release of the labeled peptide from the resin. About five equivalents of an amine-reactive fluorophore are usually used per amine of the immobilized peptide. The fluorescein, Alexa Fluor®, Oregon Green®, Rhodamine Green™, tetramethylrhodamine, Rhodamine Red™, Texas Red®, coumarin and NBD fluorophores, as well as the QSY®, dabcyl and dabsyl chromophores and the biotin hapten, are all expected to be reasonably stable to hydrogen fluoride (HF) as well as to most other acids.[11–16] These fluorophores, chromophores and biotin are also expected to be stable to reagents used for deprotection of peptides synthesized using FMOC chemistry.[17] In contrast, the BODIPY® fluorophore may be unstable to the conditions used to remove some protecting groups.

We prepare unique reagents for use in the automated synthesis of peptides that are specifically labeled with fluorophores, chromophores and haptens. These precursors permit the incorporation of a specific label at a specific site in the peptide's sequence. The α-FMOC derivative of ε-dabcyl-L-lysine (D6216) can be used to incorporate the dabcyl chromophore at selected sites in the peptide sequence. The dabcyl chromophore, which has broad visible absorption (Figure 9.5.2), has been extensively used as a quenching group in the automated synthesis of HIV protease and renin (H2930, R2931; Section 10.4) and other fluorogenic

IMPORTANT NOTICE: The products described in this manual are covered by one or more Limited Use Label License(s). Please refer to the Appendix on page 971 and Master Product List on page 975. Products are For Research Use Only. Not intended for any animal or human therapeutic or diagnostic use.

peptidase substrates.[18–21] The dabcyl group can also be incorporated at the N-terminus by using dabcyl succinimidyl ester [16,22] (D2245). The aminonaphthalene derivative EDANS (A91) has been the most common fluorophore for pairing with the dabcyl quencher in fluorescence resonance energy transfer (FRET) experiments because its fluorescence emission spectrum overlaps the absorption spectrum of dabcyl (Figure 9.5.2) (Fluorescence Resonance Energy Transfer (FRET)—Note 1.2). This fluorophore is conveniently introduced during automated synthesis of peptides by using γ-EDANS-α-FMOC-L-glutamic acid (F11831). The tetramethylrhodamine fluorophore can be incorporated during automated FMOC synthesis of peptides using our single-isomer α-(FMOC)-ε-TMR-L-lysine building block (F11830).

QSY® dyes (Section 1.6, Section 1.8) have broad visible to near-infrared absorption (Figure 9.5.3). These dyes, which are essentially nonfluorescent, are particularly useful as energy acceptors from blue-, green-, orange- or red-fluorescent donor dyes (Note 1.2). The QSY® 7, QSY® 9, QSY® 21 and QSY® 35 chromophores can be conjugated to amines via their succinimidyl esters (Q10193, Q20131, Q20132, Q20133). The QSY® 7 and QSY® 35 chromophores can also be conjugated to peptide thiols or thiol-modified oligonucleotides via its maleimide [23] (Q10257) or iodoacetamide (Q20348); additionally, peptide amides can be prepared from QSY® 7 amine (Q10464). We have also prepared α-(FMOC)-ε-QSY® 7-L-lysine and α-FMOC-β-QSY® 35-L-alanine (Q21930, Q21931), which can be used in the automated synthesis of QSY® 7 quencher– or QSY® 35 quencher–containing peptides.

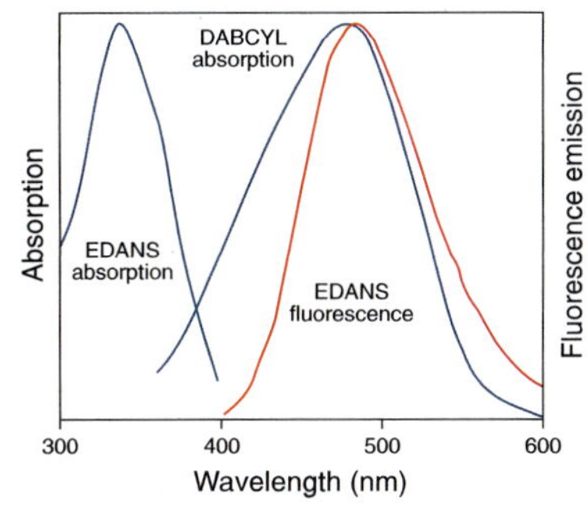

Figure 9.5.2 Spectral overlap between EDANS fluorescence and dabcyl absorption, which is required for efficient quenching of EDANS fluorescence by resonance energy transfer to the nonfluorescent dabcyl chromophore. Spectra are normalized to the same intensities.

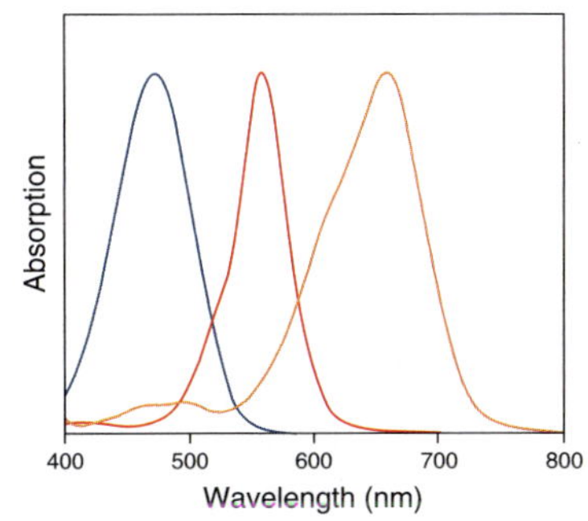

Figure 9.5.3 Normalized absorption spectra of the QSY® 35 (blue), QSY® 7 (red) and QSY® 21 (orange) dyes. The QSY® 7 and QSY® 9 dyes have essentially identical spectra.

REFERENCES

1. Anal Biochem (1988) 174:38; **2.** Biochem J (1968) 108:883; **3.** Biochem J (1965) 95:180; **4.** Biochem J (1965) 94:17; **5.** J Cell Biol (1988) 106:1607; **6.** Anal Biochem (1969) 29:186; **7.** Biosci Biotechnol Biochem (1994) 58:300; **8.** Biol Chem Hoppe Seyler (1986) 367:1259; **9.** FEBS Lett (1986) 198:150; **10.** Anal Biochem (1984) 141:446; **11.** Biochemistry (1994) 33:7211; **12.** Biochemistry (1994) 33:6966; **13.** J Biol Chem (1994) 269:15124; **14.** Bioorg Med Chem Lett (2008) 18:2555; **15.** Anal Biochem (1992) 202:68; **16.** J Med Chem (1992) 35:3727; **17.** Biochemistry (1994) 33:10951; **18.** Bioorg Med Chem Lett (1992) 2:1665; **19.** J Protein Chem (1990) 9:663; **20.** Science (1990) 247:954; **21.** Tetrahedron Lett (1990) 31:6493; **22.** FEBS Lett (1992) 297:100; **23.** Proc Natl Acad Sci U S A (2003) 100:13308.

NOTE 9.2

Labeling Small Peptides with Amine-Reactive Dyes in Organic Solvents

Most of the product literature associated with our amine-reactive dyes provides protocols for labeling proteins, typically IgG antibodies in aqueous buffers. The following protocol is a starting point for labeling peptides in organic solvents. Please note that the reaction conditions, including concentrations of the reactants and the reaction times, may require optimization. Furthermore, many peptides are not soluble in a 100% organic solution. It is very important to test the solubility of the peptide in dimethylsulfoxide (DMSO) or dimethylformamide (DMF) before attempting this procedure.

1. Dissolve the peptide to be labeled in DMSO or DMF at 0.1–1 mM.
2. Add 100 mM triethylamine to the reaction solution. This will ensure that the amines to be derivatized are deprotonated.
3. Add the amine-reactive dye to the reaction solution. The reactive dye should be in a 1:1 to 3:1 molar ratio to the peptide.
4. React at room temperature or at 4°C for at least 4 hours with continuous stirring, protected from light. The reaction can proceed overnight. Thin-layer chromatography may be useful for monitoring the reaction's progress.
5. Purify the conjugate by an appropriate method, such as HPLC-based separation.

IMPORTANT NOTICE: The products described in this manual are covered by one or more Limited Use Label License(s). Please refer to the Appendix on page 971 and Master Product List on page 975. Products are For Research Use Only. Not intended for any animal or human therapeutic or diagnostic use.

DATA TABLE 9.5 PEPTIDE ANALYSIS, SEQUENCING AND SYNTHESIS

Cat. No.	MW	Storage	Soluble	Abs	EC	Em	Solvent	Notes
A91	288.30	L	pH >10, DMF	335	5900	493	pH 8	
D21	269.75	F,DD,L	DMF, MeCN	372	3900	none	$CHCl_3$	1, 2
D100	265.33	L	EtOH	336	4400	534	MeOH	
D2245	366.38	F,D,L	DMF, DMSO	453	32,000	none	MeOH	3
D6216	619.72	F,D,L	DMF, MeCN	427	30,000	none	MeOH	
F11830	780.88	F,D,L	DMF, MeCN	543	92,000	570	MeOH	
F11831	617.67	F,D,L	DMF, MeCN	341	5200	471	MeOH	
Q10193	791.32	F,D,L	DMSO	560	90,000	none	MeOH	
Q10257	858.45	F,D,L	DMSO	560	92,000	none	MeOH	
Q10464	814.87	L	DMSO	560	92,000	none	MeOH	
Q20131	951.43	F,D,L	H_2O, DMSO	562	88,000	none	MeOH	4
Q20132	815.34	F,D,L	DMSO	661	90,000	none	MeOH	
Q20133	411.33	F,D,L	DMSO	475	23,000	none	MeOH	
Q20348	453.20	F,D,L	DMSO	475	24,000	none	MeOH	5
Q21930	1044.66	F,D,L	DMF, MeCN	560	90,000	none	MeOH	
Q21931	565.54	F,D,L	DMF, MeCN	475	23,000	none	MeOH	

For definitions of the contents of this data table, see "Using *The Molecular Probes® Handbook*" in the introductory pages.

Notes

1. D21 butylamine derivative has Abs = 337 nm (EC = 5300 $cm^{-1}M^{-1}$), Em = 492 nm in $CHCl_3$. Em and QY are highly solvent dependent: Em = 496 nm (QY = 0.45) in dioxane, 536 nm (QY = 0.28) in MeOH and 557 nm (QY = 0.03) in H_2O. (Biochemistry (1967) 6:3408) EC typically decreases upon conjugation to proteins (EC = 3400 $cm^{-1}M^{-1}$ at 340 nm). (Biochemistry (1986) 25:513) Fluorescence lifetimes (τ) of protein conjugates are typically 12–20 nanoseconds. (Arch Biochem Biophys (1969) 133:263, Arch Biochem Biophys (1968) 128:163)
2. Do NOT dissolve in DMSO.
3. D2245 is nonfluorescent both before and after reaction with amines. Reaction product with butylamine has Abs = 428 nm (EC = 32,000 $cm^{-1}M^{-1}$) in MeOH.
4. This sulfonated succinimidyl ester derivative is water soluble and may be dissolved in buffer at ~pH 8 for reaction with amines. Long-term storage in water is NOT recommended due to hydrolysis.
5. Iodoacetamides in solution undergo rapid photodecomposition to unreactive products. Minimize exposure to light prior to reaction.

PRODUCT LIST 9.5 PEPTIDE ANALYSIS, SEQUENCING AND SYNTHESIS

Cat. No.	Product	Quantity
A91	5-((2-aminoethyl)amino)naphthalene-1-sulfonic acid, sodium salt (EDANS)	1 g
D21	5-dimethylaminonaphthalene-1-sulfonyl chloride (dansyl chloride)	1 g
D100	5-dimethylaminonaphthalene-1-sulfonyl hydrazine (dansyl hydrazine)	100 mg
D2245	4-((4-(dimethylamino)phenyl)azo)benzoic acid, succinimidyl ester (dabcyl, SE)	100 mg
D6216	ε-(4-((4-(dimethylamino)phenyl)azo)benzoyl)-α-9-fluorenylmethoxycarbonyl-L-lysine (ε-dabcyl-α-FMOC-L-lysine)	100 mg
F11830	N^{α}-(9-fluorenylmethoxycarbonyl)-N^{ϵ}-tetramethylrhodamine-(5-carbonyl)-L-lysine (α-FMOC-ε-TMR-L-lysine)	25 mg
F11831	5-((2-(FMOC)-γ-L-glutamylaminoethyl)amino)naphthalene-1-sulfonic acid (γ-EDANS-α-FMOC-L-glutamic acid)	100 mg
Q10464	QSY® 7 amine, hydrochloride	5 mg
Q10193	QSY® 7 carboxylic acid, succinimidyl ester	5 mg
Q10257	QSY® 7 C_5-maleimide	5 mg
Q21930	N^{ϵ}-(QSY® 7)-N^{α}-(9-fluorenylmethoxycarbonyl)-L-lysine (α-FMOC-ε-QSY® 7-L-lysine)	5 mg
Q20131	QSY® 9 carboxylic acid, succinimidyl ester	5 mg
Q20132	QSY® 21 carboxylic acid, succinimidyl ester	5 mg
Q20133	QSY® 35 acetic acid, succinimidyl ester	5 mg
Q21931	N^{β}-(QSY® 35)-N^{α}-(9-fluorenylmethoxycarbonyl)-L-alanine (α-FMOC-β-QSY® 35-L-alanine)	5 mg
Q20348	QSY® 35 iodoacetamide	5 mg

IMPORTANT NOTICE: The products described in this manual are covered by one or more Limited Use Label License(s). Please refer to the Appendix on page 971 and Master Product List on page 975. Products are For Research Use Only. Not intended for any animal or human therapeutic or diagnostic use.

TEN

CHAPTER 10

Enzyme Substrates and Assays

IMPORTANT NOTICE: The products described in this manual are covered by one or more Limited Use Label License(s). Please refer to the Appendix on page 971 and Master Product List on page 975. Products are For Research Use Only. Not intended for any animal or human therapeutic or diagnostic use.

IMPORTANT NOTICE: The products described in this manual are covered by one or more Limited Use Label License(s). Please refer to the Appendix on page 971 and Master Product List on page 975. Products are For Research Use Only. Not intended for any animal or human therapeutic or diagnostic use.

IMPORTANT NOTICE: The products described in this manual are covered by one or more Limited Use Label License(s). Please refer to the Appendix on page 971 and Master Product List on page 975. Products are For Research Use Only. Not intended for any animal or human therapeutic or diagnostic use.

IMPORTANT NOTICE: The products described in this manual are covered by one or more Limited Use Label License(s). Please refer to the Appendix on page 971 and Master Product List on page 975. Products are For Research Use Only. Not intended for any animal or human therapeutic or diagnostic use.

10.1 Introduction to Enzyme Substrates and Their Reference Standards

We offer a large assortment of fluorogenic and chromogenic enzyme substrates, which are described in the following sections. We prepare substrates for detecting very low levels of enzymatic activity in fixed cells, tissues, cell extracts and purified preparations, as well as substrates for enzyme-linked immunosorbent assays (ELISAs). Our RediPlate™ product line includes enzyme substrates predispensed in 96-well plates for high-throughput applications, along with the appropriate reference standards and other reaction components. We have also developed effective methods for detecting some enzymes in live cells.

In this section, we describe the characteristics of our enzyme substrates and the fluorophores and chromophores from which they are derived, focusing primarily on the suitability of these substrates for different types of enzyme assays. The fluorophores that are available as reference standards—including a NIST-traceable fluorescein standard (F36915)—can be found in the data table and product list associated with this section. Substrates for specific enzymes are described in subsequent sections of this chapter.

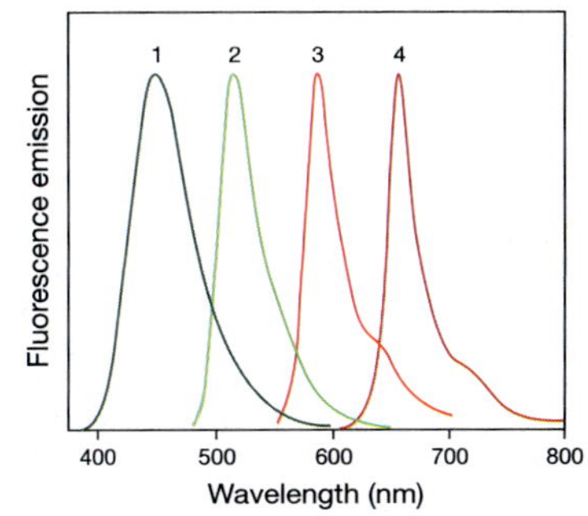

Figure 10.1.1 Normalized emission spectra of **1)** 7-hydroxy-4-methylcoumarin (H189) and 6,8-difluoro-7-hydroxy-4-methylcoumarin (DiFMU, D6566), **2)** fluorescein (F1300, F36915), **3)** resorufin (R363) and **4)** DDAO (H6482) in aqueous solution at pH 9. These fluorophores correspond to the hydrolysis, oxidation or reduction products of several of our fluorogenic enzyme substrates.

Substrates Yielding Soluble Fluorescent Products

Solution assays designed to quantitate enzymatic activity in cell extracts or other biological fluids typically employ substrates that yield highly fluorescent or intensely absorbing water-soluble products. ELISAs also rely on these substrates for indirect quantitation of analytes.[1–3] An ideal fluorogenic substrate for fluorescence-based solution assays yields a highly fluorescent, water-soluble product with optical properties significantly different from those of the substrate. If the fluorescence spectra of the substrate and product overlap significantly, analysis will likely require a separation step, especially when using excess substrate to obtain pseudo–first-order kinetics. Fortunately, many substrates have low intrinsic fluorescence or are metabolized to products that have longer-wavelength excitation or emission spectra (Figure 10.1.1). These fluorescent products can typically be quantitated in the presence of the unreacted substrate using a fluorometer or a fluorescence microplate reader. Microplate readers facilitate high-throughput analysis and require relatively small assay volumes, which usually reduces reagent costs. Moreover, the front-face optics in many microplate readers allows researchers to use more concentrated solutions, which may both improve the linearity of the kinetics and reduce inner-filter effects.

When the spectral characteristics of the substrate and its metabolic product are similar, techniques such as thin-layer chromatography (TLC), high-performance liquid chromatography (HPLC), capillary electrophoresis, solvent extraction or ion exchange can be used to separate the product from unconsumed substrate prior to analysis. For example, our *FAST* CAT® Chloramphenicol Acetyltransferase Assay Kits (F2900, F6616, F6617; Section 10.6) utilize chromatography to separate the intrinsically fluorescent substrates from their fluorescent products.

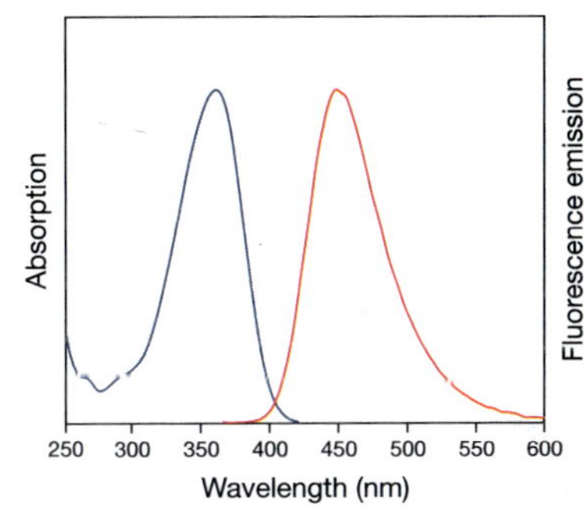

Figure 10.1.2 Absorption and fluorescence emission spectra of 7-hydroxy-4-methylcoumarin (H189) in pH 9.0 buffer. The spectra of 6,8-difluoro-7-hydroxy-4-methylcoumarin (D6566) are essentially identical.

Substrates Derived from Water-Soluble Coumarins

Hydroxy- and amino-substituted coumarins have been the most widely used fluorophores for preparing fluorogenic substrates. Phosphate, ester and ether derivatives of 7-hydroxy-4-methylcoumarin (β-methylumbelliferone, H189), typified by 4-methylumbelliferyl phosphate (MUP, M6491; Section 10.3), are only weakly fluorescent in the optimal configuration for 7-hydroxy-4-methylcoumarin detection (excitation/emission ~ 360/460 nm, Figure 10.1.2). Thus, enzymatic cleavage of the phosphate, ester or ether results in an increase in fluorescence under these conditions. Because the fluorescence of 7-hydroxycoumarin dyes is pH dependent, this detection configuration is only optimal when the phenolate anion form is the predominant species (i.e., at pH >8 for 7-hydroxy-4-methylcoumarin). At lower pH values, the excitation wavelength must be switched to about 320 nm, which is optimal for the protonated phenol form. 6,8-Difluoro-7-hydroxy-4-methylcoumarin (DiFMU, D6566; Figure 10.1.2, Figure 10.1.3) has a much lower

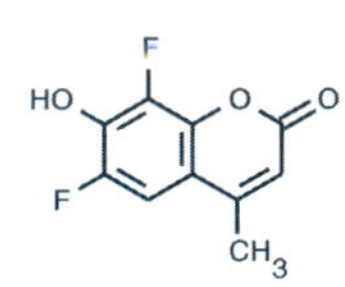

Figure 10.1.3 6,8-Difluoro-7-hydroxy-4-methylcoumarin (DiFMU, D6566).

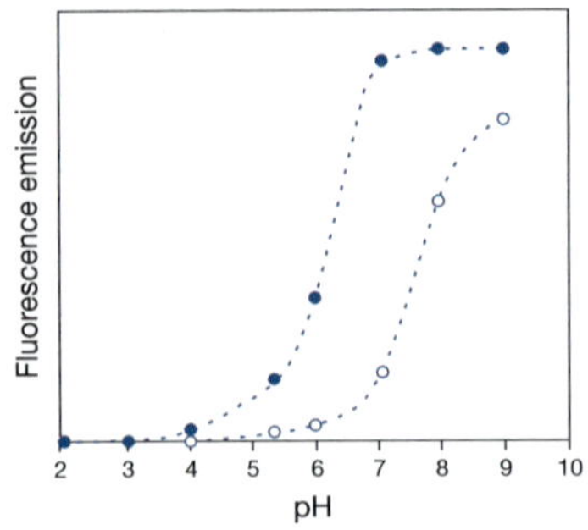

Figure 10.1.4 Comparison of the pH-dependent fluorescence changes produced by attachment of electron-withdrawing fluorine atoms to a hydroxycoumarin. 7-Hydroxy-4-methylcoumarin-3-acetic acid (○, H1428) and 6,8-difluoro-7-hydroxy-4-methylcoumarin (●, D6566) demonstrate the decrease of the pK_a from ~7.4 to ~6.2. Fluorescence intensities were measured for equal concentrations of the two dyes using excitation/emission at 360/450 nm.

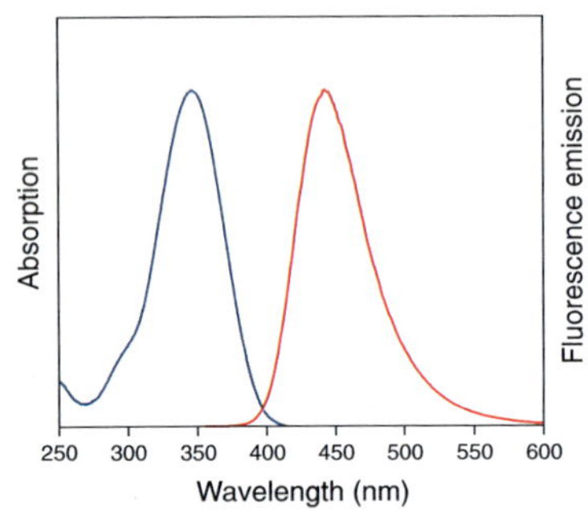

Figure 10.1.5 Absorption and fluorescence emission spectra of 7-amino-4-methylcoumarin in pH 7.0 buffer.

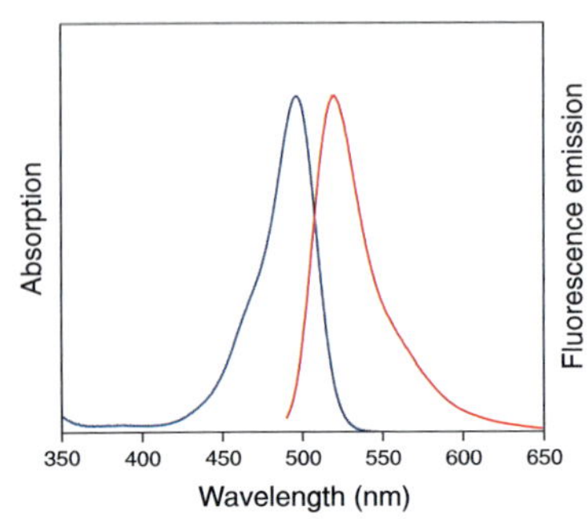

Figure 10.1.6 Absorption and fluorescence emission spectra of rhodamine 110 in pH 7.0 buffer.

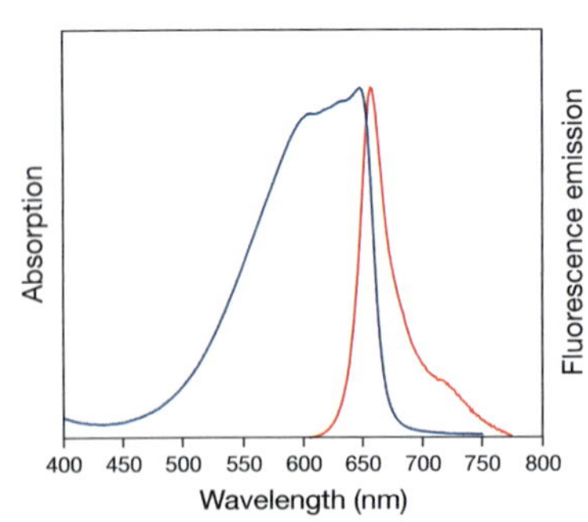

Figure 10.1.8 Normalized absorption and fluorescence emission spectra of DDAO, which is formed by alkaline phosphatase–mediated hydrolysis of DDAO phosphate (D6487).

phenol/phenolate[4] pK_a (Figure 10.1.4), allowing measurements to be made over a broader pH range without adjusting the excitation/emission wavelength settings to accomodate the pH-dependent spectral shift of the 7-hydroxycoumarin fluorophore. Furthermore, the electron-withdrawing effect of the fluorine substituents manifested in the lower pK_a also facilitates cleavage of the phosphate, ester or ether substituent, resulting in accelerated enzyme kinetics. Consequently, the phosphate ester of 6,8-difluoro-7-hydroxy-4-methylcoumarin (DiFMUP, D6567, D22065, E12020; Section 10.3) is one of the most sensitive fluorogenic substrates for continuous high-throughput assay of alkaline phosphatase and its bioconjugates.

Aromatic amines, including the commonly used 7-amino-4-methylcoumarin (AMC, A191; Figure 10.1.5), are partially protonated at low pH (less than ~5) but fully deprotonated at physiological pH. Thus, their fluorescence spectra are not subject to variability due to pH-dependent protonation/deprotonation when assayed near or above physiological pH. AMC is widely used to prepare peptidase substrates in which the amide has shorter-wavelength absorption and emission spectra than the amine hydrolysis product.

Substrates Derived from Water-Soluble Green to Yellow Fluorophores

As compared with coumarin-based substrates, substrates derived from fluoresceins, rhodamines and resorufins often provide significantly greater sensitivity in fluorescence-based enzyme assays. In addition, most of these longer-wavelength dyes have extinction coefficients that are 5 to 25 times that of coumarins, nitrophenols or nitroanilines, making them additionally useful as sensitive chromogenic substrates.

Hydrolytic substrates based on the derivatives of fluorescein (fluorescein reference standard, F1300; fluorescein NIST-traceable standard, F36915) or rhodamine 110 (R110, R6479; Figure 10.1.6) usually incorporate two moieties, each of which serves as a substrate for the enzyme. Consequently, they are cleaved first to the monosubstituted analog and then to the free fluorophore. Because the monosubstituted analog often absorbs and emits light at the same wavelengths as the ultimate hydrolysis product, this initial hydrolysis complicates the interpretation of hydrolysis kinetics.[5] When highly purified, however, the disubstituted fluorescein- and rhodamine 110–based substrates have virtually no visible-wavelength absorbance or background fluorescence, making them extremely sensitive detection reagents. For example, researchers have reported that the activity of as few as 1.6 *molecules* of β-galactosidase can be detected with fluorescein di-β-D-galactopyranoside (FDG) and capillary electrophoresis.[6] Fluorogenic substrates based on either the AMC and R110 fluorophore are used in our EnzChek® Caspase Assay Kits (Section 10.4, Section 15.5) to detect apoptotic cells.

Chemical reduction of fluorescein- and rhodamine-based dyes yields colorless and nonfluorescent dihydrofluoresceins and dihydrorhodamines. Although extremely useful for detection of reactive oxygen species (ROS) in phagocytic and other cells (Section 18.2), these dyes tend to be insufficiently stable for solution assays.

Substrates Derived from Water-Soluble Red Fluorophores

Long-wavelength fluorophores are often preferred because background absorbance and autofluorescence are generally lower when longer excitation wavelengths are used. Substrates derived from the red-fluorescent resorufin (R363, Figure 10.1.7) and the dimethylacridinone derivative 7-hydroxy-9*H*-(1,3-dichloro-9,9-dimethylacridin-2-one) (DDAO, H6482; Figure 10.1.8; Figure 10.1.9) contain only a single hydrolysis-sensitive moiety, thereby avoiding the biphasic kinetics of both fluorescein- and rhodamine-based substrates.[7]

Resorufin is used to prepare several substrates for glycosidases, hydrolytic enzymes and dealkylases. In most cases, the relatively low pK_a of resorufin (~6.0) permits continuous measurement of enzymatic activity. Thiols such as DTT or 2-mercaptoethanol should be avoided in

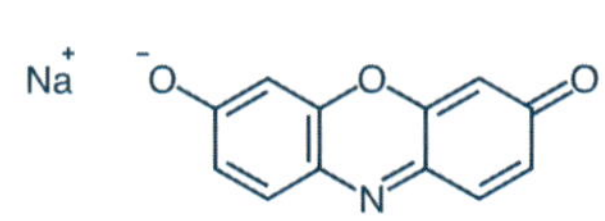

Figure 10.1.7 Resorufin, sodium salt (R363).

Figure 10.1.9 7-Hydroxy-9*H*-(1,3-dichloro-9,9-dimethylacridin-2-one; DDAO, H6482).

IMPORTANT NOTICE: The products described in this manual are covered by one or more Limited Use Label License(s). Please refer to the Appendix on page 971 and Master Product List on page 975. Products are For Research Use Only. Not intended for any animal or human therapeutic or diagnostic use.
www.invitrogen.com/probes
molecular probes® | invitrogen™ by life technologies™

assays utilizing resorufin-based substrates. Our Amplex® Red peroxidase substrate (A12222, A22177; Section 10.5) is a chemically reduced, colorless form of resorufin that is oxidized to resorufin by horseradish peroxidase (HRP) in combination with hydrogen peroxide. Resorufin is also the product of enzyme-catalyzed reduction of resazurin (R12204; Section 10.6, Section 15.2)—also known as alamarBlue® dye. Our Amplex® UltraRed reagent (A36006, Section 10.5) improves upon the performance of the Amplex® Red reagent, offering brighter fluorescence and enhanced sensitivity on a per-mole basis in peroxidase or peroxidase-coupled enzyme assays.

Substrates derived from DDAO, a red He-Ne laser–excitable fluorophore, generally exhibit good water solubility, low K_M and high turnover rates. In addition, the difference between the excitation maximum of the DDAO-based substrates and that of the phenolic DDAO product is greater than 150 nm (Figure 10.1.8), which allows the two species to be easily distinguished.

Substrates for Live-Cell Enzyme Assays

We have developed a number of innovative strategies for investigating enzymatic activity in live cells.[8,9] For example, we offer a diverse set of probes that can passively enter the cell; once inside, they are processed by intracellular enzymes to generate products with improved cellular retention (Section 14.2). We also offer kits and reagents for detecting the expression of several common reporter genes in cells and cell extracts. These include substrates for β-galactosidase and β-glucuronidase (Section 10.2), secreted alkaline phosphatase (SEAP, Section 10.3), chloramphenicol acetyltransferase (CAT) and luciferase (Section 10.6). Some of our EnzChek® and DQ™ Kits (Section 10.4) are useful for study of the uptake and metabolism of proteins during phagocytosis (Section 16.1). Substrates for specific proteases are also useful for the detection of apoptosis (Section 15.5).

Thiol-Reactive Fluorogenic Substrates

We prepare a number of enzyme substrates for live-cell assays that incorporate a mildly thiol-reactive chloromethyl moiety. Once inside the cell, this chloromethyl group undergoes what is believed to be a glutathione *S*-transferase–mediated reaction to produce a membrane-impermeant, glutathione–fluorescent dye adduct, although our experiments suggest that they may also react with other intracellular components. Regardless of the mechanism, many cell types loaded with these chloromethylated substrates are both fluorescent and viable for at least 24 hours after loading and often through several cell divisions. Furthermore, unlike the free dye, the peptide–fluorescent dye adducts contain amino groups and can therefore be covalently linked to surrounding biomolecules by fixation with formaldehyde or glutaraldehyde. This property permits long-term storage of the labeled cells or tissue and, in cases where the anti-dye antibody is available (Section 7.4), amplification of the conjugate by standard immunochemical techniques, including tyramide signal amplification (TSA, Section 6.2). Chloromethyl analogs of fluorogenic substrates for glycosidases (for example, in the DetectaGene™ Green CMFDG Kit, (D2920); Section 10.2) and peptidases are available. Our CellTracker™ Blue CMAC and CellTracker™ Blue CMF2HC dyes (C2110, Figure 10.1.10; C12881, Figure 10.1.11) are precursors to peptidase and glycosidase substrates, respectively, and are also used for long-term cell tracing (Section 14.2).

Lipophilic Fluorophores

Lipophilic analogs of fluorescein and resorufin exhibit many of the same properties as the water-soluble fluorophores, including relatively high extinction coefficients and good quantum yields. In most cases, however, substrates based on these lipophilic analogs load more readily into cells, permitting use of much lower substrate concentrations in the loading medium, and their fluorescent products are better retained after cleavage than their water-soluble counterparts. Lipophilic substrates and their products probably also distribute differently in cells and likely associate with lipid regions of the cell. When passive cell loading or enhanced dye retention are critical parameters of the experiment, we recommend using our lipophilic substrates for glycosidases (such as our ImaGene Green™ and ImaGene Red™ products, Section 10.2). Like resazurin (R12204, Section 15.2), dodecylresazurin—the substrate in our LIVE/DEAD® Cell Vitality Assay Kit, Vybrant® Cell Metabolic Assay Kit and Metabolic Activity/Annexin V/Dead Cell Apoptosis Kit (L34951, V23110, V35114; Section 15.3, Section 15.5)—is reduced to dodecylresorufin by metabolically active cells; however, this lipophilic substrate is more useful than resazurin for microplate assays of all metabolic activity and permits single-cell analysis of cell metabolism by flow cytometry and cell counting. Dodecylresorufin is also the hydrolysis product of the β-galactosidase substrate used in our ImaGene Red™ C_{12}RG *lacZ* Gene Expression Kit (I2906, Section 10.2).

Substrates Yielding Insoluble Fluorescent or Chromophoric Products

Alkaline phosphatase, β-galactosidase and horseradish peroxidase (HRP) conjugates are widely used as secondary detection reagents for immunohistochemical analysis and *in situ* hybridization, as well as for protein and nucleic acid detection by Southern, northern and western blots. In order to precisely localize enzymatic activity in a tissue, cell, gel or blot, the substrate must yield a product that immediately precipitates or reacts at the site of enzymatic activity. In addition, various methods such as chromatography, isoelectric focusing and gel electrophoresis are commonly employed to separate enzymes preceding their detection. A review by Weder and Kaiser discusses the use of a wide variety of fluorogenic substrates for the detection of electrophoretically separated hydrolases.[10]

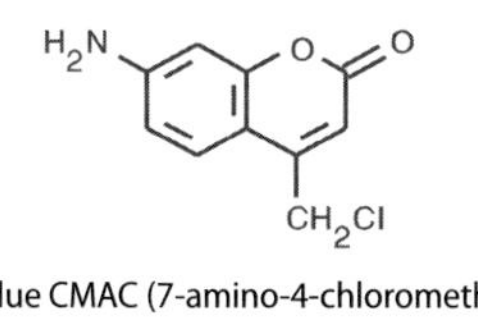

Figure 10.1.10 CellTracker™ Blue CMAC (7-amino-4-chloromethylcoumarin, C2110).

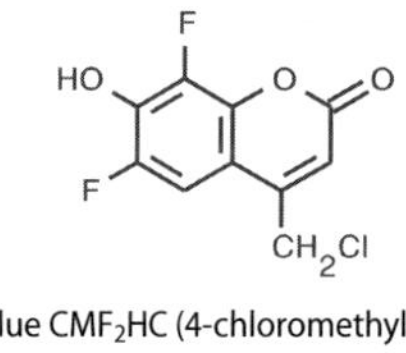

Figure 10.1.11 CellTracker™ Blue CMF$_2$HC (4-chloromethyl-6,8-difluoro-7-hydroxycoumarin, C12881).

IMPORTANT NOTICE: The products described in this manual are covered by one or more Limited Use Label License(s). Please refer to the Appendix on page 971 and Master Product List on page 975. Products are For Research Use Only. Not intended for any animal or human therapeutic or diagnostic use.

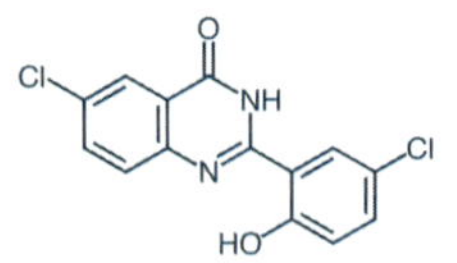

Figure 10.1.12 ELF® 97 alcohol (E6578).

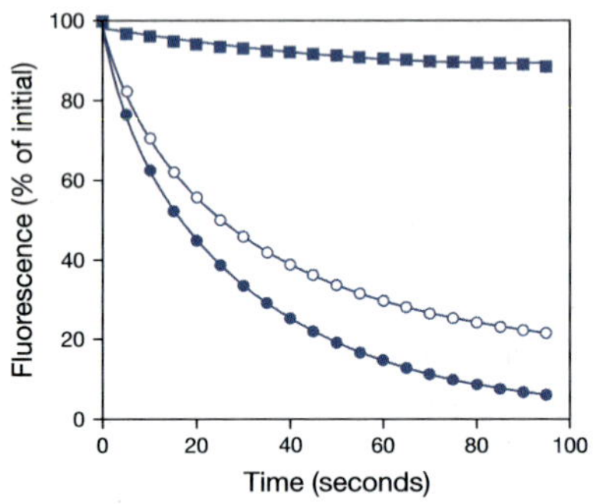

Figure 10.1.13 Photostability comparison for ELF® 97 alcohol– and fluorescein-labeled tubulin preparations. Tubulin in acetone-fixed CRE BAG 2 mouse fibroblasts was labeled with an anti–β-tubulin monoclonal antibody and then detected using biotin-XX goat anti–mouse IgG antibody (B2763) in conjunction with either our ELF® 97 Cytological Labeling Kit (E6603, ■) or fluorescein streptavidin (S869, ●). Alternatively, anti-tubulin labeling was detected directly using fluorescein goat anti–mouse IgG antibody (F2761, ○). The photostability of labeling produced by the three methods was compared by continuously illuminating stained samples on a fluorescence microscope using Omega® Optical longpass optical filter sets. Images were acquired every 5 seconds using a Star 1 CCD camera (Photometrics®); the average fluorescence intensity in the field of view was calculated with Image-1 software (Universal Imaging Corp.) and expressed as a fraction of the initial intensity. Three data sets, representing different fields of view, were averaged for each conjugate to obtain the plotted time courses.

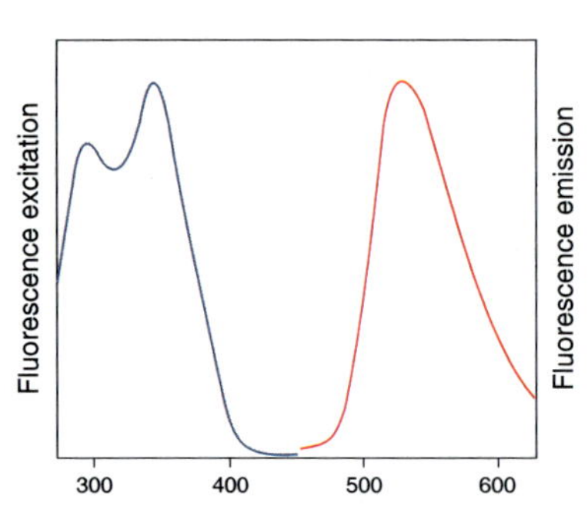

Figure 10.1.14 The normalized fluorescence excitation and emission spectra of the ELF® 97 alcohol precipitate (E6578), which is generated by enzymatic cleavage of the soluble ELF® 97 phosphatase substrate (E6588, E6589).

Substrates Yielding Insoluble Fluorescent Products

In addition to the commonly used chromogenic substrates, including BCIP, NBT and X-gal, we have developed a fluorogenic ELF® 97 phosphatase substrate that fluoresces only weakly in the blue range. Upon enzymatic cleavage, however, this substrate forms the intensely yellow-green–fluorescent ELF® 97 alcohol (E6578, Figure 10.1.12), which precipitates immediately at the site of enzymatic activity. The fluorescent ELF® alcohol precipitate is exceptionally photostable (Figure 10.1.13) and has a high Stokes shift (Figure 10.1.14). We offer several ELF® Kits based on the ELF® 97 phosphatase substrate; see Section 6.3 for a complete discussion of ELF® technology.

Tyramide signal amplification (TSA) technology utilizes fluorescent or biotin-labeled tyramide substrates that are activated by HRP to phenoxyl radicals. These phenoxy radicals are then trapped by reaction with nearby tyrosine residues, resulting in localized tyramide deposition. We offer a wide assortment of TSA™ Kits based on this detection technology; see Section 6.2 for a complete discussion of TSA technology.

Substrates Yielding Insoluble Chromophoric Products

A number of chromogenic substrates for hydrolytic enzymes are derived from indolyl chromophores. These initially form a colorless—and sometimes blue-fluorescent—3-hydroxyindole ("indoxyl"), which spontaneously, or through mediation of an oxidizing agent such as nitro blue tetrazolium (NBT, N6495; Section 10.3) or potassium ferricyanide,[11] is converted to an intensely colored indigo dye that typically precipitates from the medium (Figure 10.1.15). Halogenated indolyl derivatives, including 5-bromo-4-chloro-3-indolyl β-D-galactopyranoside (X-Gal, B1690; Section 10.2) and 5-bromo-4-chloro-3-indolyl phosphate (BCIP, N6547) are generally preferred because they produce finer precipitates that are less likely to diffuse from the site of formation, making them especially useful for detecting enzymatic activity in cells, tissues, gels and blots.

Substrates Based on Excited-State Energy Transfer

The principle of excited-state energy transfer can also be used to generate fluorogenic substrates (Fluorescence Resonance Energy Transfer (FRET)—Note 1.2). For example, the EDANS fluorophore in our HIV protease and renin substrates is effectively quenched by a nearby dabcyl acceptor chromophore (Figure 10.1.16). This chromophore has been carefully chosen for maximal overlap of its absorbance with the fluorophore's fluorescence, thus ensuring that the fluorescence is quenched through excited-state energy transfer. Proteolytic cleavage of the substrate results in spatial separation of the fluorophore and the acceptor chromophore, thereby restoring the fluorophore's fluorescence.[12–15]

Many of the dyes described in Chapter 1 have been used to form energy-transfer pairs, some of which can be introduced during automated synthesis of peptides using modified amino acids described in Section 9.5. Table 1.10 (Section 1.6) lists our nonfluorescent quenching dyes and their spectral properties.

The protease substrates in three of our EnzChek® Protease Assay Kits (Section 10.4) are heavily labeled casein conjugates; the close proximity of dye molecules results in self-quenching. Hydrolysis of the protein to smaller fragments is accompanied by a dramatic increase in fluorescence, which forms the basis of a simple and sensitive continuous assay for a variety of proteases. In addition, we offer a phospholipase A substrate (bis-BODIPY® FL C_{11}-PC, B7701; Section 17.4) that contains a BODIPY® FL fluorophore on each phospholipid acyl chain. Proximity of the BODIPY® FL fluorophores on adjacent phospholipid acyl chains causes fluorescence self-quenching that is relieved only when the fluorophores are separated by phospholipase A–mediated cleavage. PED6, a phospholipid with a green-fluorescent BODIPY® fatty acid on the lipid portion of the molecule and a 2,4-dinitrophenyl quencher on the polar head group (PED6, D23739; Section 17.4) is useful as a specific phospholipase-A_2 substrate.[16]

IMPORTANT NOTICE: The products described in this manual are covered by one or more Limited Use Label License(s). Please refer to the Appendix on page 971 and Master Product List on page 975. Products are For Research Use Only. Not intended for any animal or human therapeutic or diagnostic use.
www.invitrogen.com/probes

Fluorescent Derivatization Reagents for Discontinuous Enzyme Assays

The mechanism of some enzymes makes it difficult to obtain a continuous optical change during reaction with an enzyme substrate. However, a discontinuous assay can often be developed by derivatizing the reaction products with one of the reagents described in Chapter 1, Chapter 2 and Chapter 3, usually followed by a separation step in order to generate a product-specific fluorescent signal.

For example, fluorescamine or *o*-phthaldialdehyde (OPA) (F2332, F20261, P2331MP; Section 1.8) can detect the rate of *any* peptidase reaction by measuring the increase in the concentration of free amines in solution.[17,18] The activity of enzymes that produce free coenzyme A from its esters can be detected using thiol-reactive reagents such as 5,5′-dithiobis-(2-nitrobenzoic acid) (DTNB, D8451; Section 5.2) or 7-fluorobenz-2-oxa-1,3-diazole-4-sulfonamide[19] (ABD-F, F6053; Section 2.2). The products of enzymes that metabolize low molecular weight substrates can frequently be detected by chromatographic or electrophoretic analysis. HPLC or capillary zone electrophoresis can also be used to enhance the sensitivity of reactions that yield fluorescent products.[20]

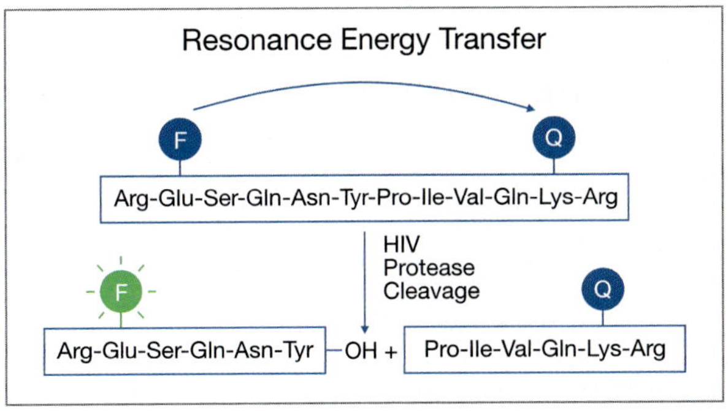

Figure 10.1.16 Principle of the fluorogenic response to protease cleavage exhibited by HIV protease substrate 1 (H2930). Quenching of the EDANS fluorophore (**F**) by distance-dependent resonance energy transfer to the dabcyl quencher (**Q**) is eliminated upon cleavage of the intervening peptide linker.

REFERENCES

1. Methods Mol Biol (1994) 32:461; **2.** J Immunol Methods (1992) 150:5; **3.** Anal Biochem (2005) 345:227; **4.** Bioorg Med Chem Lett (1998) 8:3107; **5.** Biochemistry (1991) 30:8535; **6.** Anal Biochem (1995) 226:147; **7.** Angew Chem Int Ed Engl (1991) 30:1646; **8.** Biotech Histochem (1995) 70:243; **9.** J Fluorescence (1993) 3:119; **10.** J Chromatogr A (1995) 698:181; **11.** Histochemie (1970) 23:266; **12.** Bioorg Med Chem Lett (1992) 2:1665; **13.** J Protein Chem (1990) 9:663; **14.** Science (1990) 247:954; **15.** Tetrahedron Lett (1990) 31:6493; **16.** Anal Biochem (1999) 276:27; **17.** Biochemistry (1990) 29:6670; **18.** Anal Biochem (1982) 123:41; **19.** Chem Pharm Bull (Tokyo) (1990) 38:2290; **20.** Anal Biochem (1981) 115:177.

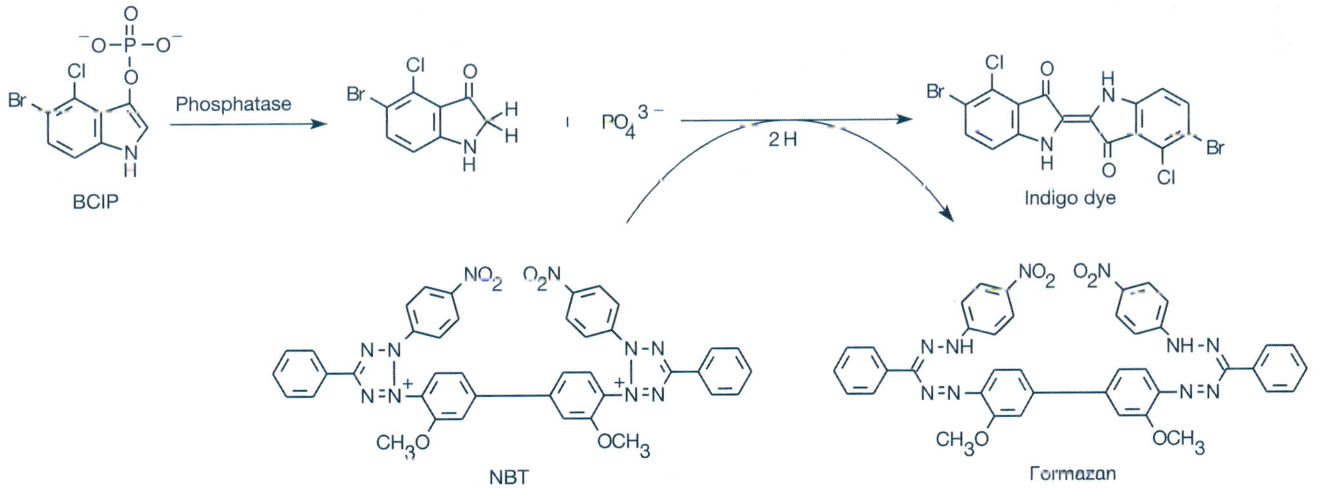

Figure 10.1.15 Principle of enzyme-linked detection using the reagents in our NBT/BCIP Reagent Kit (N6547). Phosphatase hydrolysis of BCIP is coupled to reduction of NBT, yielding a formazan and an indigo dye that together form a black-purple–colored precipitate.

IMPORTANT NOTICE: The products described in this manual are covered by one or more Limited Use Label License(s). Please refer to the Appendix on page 971 and Master Product List on page 975. Products are For Research Use Only. Not intended for any animal or human therapeutic or diagnostic use.

DATA TABLE 10.1 INTRODUCTION TO ENZYME SUBSTRATES AND THEIR REFERENCE STANDARDS

Cat. No.	MW	Storage	Soluble	Abs	EC	Em	Solvent	Notes
A191	175.19	L	DMF, DMSO	351	18,000	430	MeOH	1
C183	187.15	L	pH >8, DMF	408	43,000	450	pH 9	
C2110	209.63	F,D,L	DMSO	353	14,000	466	pH 9	2
C12881	246.60	F,D,L	DMSO	371	16,000	464	pH 9	3
D6566	212.15	L	DMSO, DMF	358	18,000	452	pH 9	3, 4
E6578	307.14	L	DMSO	345	ND	530	pH 8	5, 6, 7
F1300	332.31	L	pH >6, DMF	490	93,000	514	pH 9	8
F36915	332.31	RO,L	see Notes	490	93,000	514	pH 9.5	8, 9
H189	176.17	L	DMSO, MeOH	360	19,000	449	pH 9	3
H6482	308.16	L	DMF	646	41,000	659	pH 10	
R363	235.17	L	pH >7, DMF	571	62,000	585	pH 9	10, 11, 12
R6479	366.80	L	DMSO	499	92,000	521	MeOH	13

For definitions of the contents of this data table, see "Using *The Molecular Probes® Handbook*" in the introductory pages.

Notes

1. A191 in aqueous solution (pH 7.0): Abs = 342 nm (EC = 16,000 $cm^{-1}M^{-1}$), Em = 441 nm.
2. C2110 in MeOH: Abs = 364 nm (EC = 16,000 $cm^{-1}M^{-1}$), Em = 454 nm.
3. Spectra of hydroxycoumarins are pH dependent. Below the pK_a, Abs shifts to shorter wavelengths (325–340 nm) and fluorescence intensity decreases. Approximate pK_a values are 7.8 (H189, C2111), 7.5 (H185) and 4.9 (D6566, C12881).
4. The fluorescence quantum yield of DiFMU (D6566) is 0.89 measured in 0.1 M phosphate buffer, pH 10. (Bioorg Med Chem Lett (1998) 8:3107)
5. ND = not determined.
6. This product is supplied as a ready-made solution in the solvent indicated under "Soluble."
7. ELF® 97 alcohol is insoluble in water. Spectral maxima listed are for an aqueous suspension; for this reason, the value of EC cannot be determined.
8. Absorption and fluorescence of fluorescein derivatives are pH dependent. Extinction coefficients and fluorescence quantum yields decrease markedly at pH <7.
9. F36915 consists of a fluorescein solution in 100 mM sodium borate buffer pH 9.5. The concentration of fluorescein is set spectrophotometrically to be equivalent to that of NIST Standard Reference Material (SRM®) 1932.
10. The fluorescence quantum yield of resorufin in pH 9.5 buffer is 0.74 and the fluorescence lifetime (τ) is 2.9 nanoseconds. (Photochem Photobiol (2002) 76:385)
11. Absorption and fluorescence of resorufin are pH dependent. Below the pK_a (~6.0), Abs shifts to ~480 nm and both EC and the fluorescence quantum yield are markedly lower.
12. Resorufin is unstable in the presence of thiols such as dithiothreitol (DTT) and 2-mercaptoethanol. (Bioorg Chem (1998) 26:63)
13. R6479 in aqueous solution (pH 7.0): Abs = 496 nm (EC = 83,000 $cm^{-1}M^{-1}$), Em = 520 nm.

PRODUCT LIST 10.1 INTRODUCTION TO ENZYME SUBSTRATES AND THEIR REFERENCE STANDARDS

Cat. No.	Product	Quantity
A191	7-amino-4-methylcoumarin *reference standard*	100 mg
C2110	CellTracker™ Blue CMAC (7-amino-4-chloromethylcoumarin)	5 mg
C12881	CellTracker™ Blue CMF_2HC (4-chloromethyl-6,8-difluoro-7-hydroxycoumarin)	5 mg
C183	3-cyano-7-hydroxycoumarin	100 mg
D6566	6,8-difluoro-7-hydroxy-4-methylcoumarin (DiFMU) *reference standard*	10 mg
E6578	ELF® 97 alcohol *1 mM solution in DMSO*	1 mL
F36915	fluorescein *NIST-traceable standard* *nominal concentration 50 µM* *special packaging*	5 x 1 mL
F1300	fluorescein *reference standard*	1 g
H6482	7-hydroxy-*9H*-(1,3-dichloro-9,9-dimethylacridin-2-one) (DDAO) *reference standard*	10 mg
H189	7-hydroxy-4-methylcoumarin *reference standard*	1 g
R363	resorufin, sodium salt *reference standard*	100 mg
R6479	rhodamine 110 (R110) *reference standard*	25 mg

IMPORTANT NOTICE: The products described in this manual are covered by one or more Limited Use Label License(s). Please refer to the Appendix on page 971 and Master Product List on page 975. Products are For Research Use Only. Not intended for any animal or human therapeutic or diagnostic use.
www.invitrogen.com/probes
molecular probes® | invitrogen™ by life technologies™

10.2 Detecting Glycosidases

Glycosidase enzymes exhibit very high selectivity for hydrolysis of their preferred sugars. For example, β-galactosidase rapidly hydrolyzes β-D-galactopyranosides but usually does not hydrolyze either the anomeric α-D-galactopyranosides or the isomeric β-D-glucopyranosides. Endogenous glycosidase activity is frequently used to characterize strains of microorganisms [1–3] and to selectively label organelles of mammalian cells; [4–6] defects in glycosidase activity are characteristic of several diseases.[7–9]

In addition, glycosidases are important reporter gene markers. Specifically, *lacZ*, which encodes β-galactosidase, is extensively used as a reporter gene in animals and yeast, whereas the β-glucuronidase (GUS) gene is a popular reporter gene in plants.[10–14] Glycosidase substrates are also used in conjunction with glycosidase-conjugated secondary detection reagents in immunohistochemical techniques and enzyme-linked immunosorbent assays [15] (ELISAs). Molecular Probes® fluorogenic and chromogenic glycosidase substrates are listed in Table 10.1.

Fluorogenic β-Galactosidase Substrates

Fluorescein Digalactoside

Probably the most sensitive fluorogenic substrate for detecting β-galactosidase is fluorescein di-β-D-galactopyranoside [16] (FDG, F1179; Figure 10.2.1). Nonfluorescent FDG is sequentially hydrolyzed by β-galactosidase, first to fluorescein monogalactoside (FMG) and then to highly fluorescent fluorescein (F1300, F36915; Section 10.1). Enzyme-mediated hydrolysis of FDG can be followed by the increase in either absorbance or fluorescence. Although the turnover rates of FDG and its analogs are considerably slower than that of the common spectrophotometric galactosidase substrate, *o*-nitrophenyl β-D-galactopyranoside (ONPG),[17,18] the absorbance of fluorescein is about 5-fold greater than that of *o*-nitrophenol. Moreover, fluorescence-based measurements can be several orders of magnitude more sensitive than absorption-based measurements. Fluorescence-based assays employing FDG are also reported to be 100- to 1000-fold more sensitive than radioisotope-based ELISAs.[19]

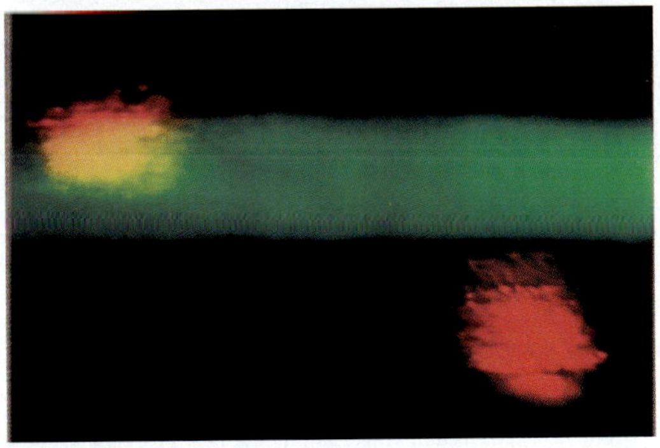

Figure 10.2.1 Sternomastoid muscle fibers of a living mouse that have been transfected with YOYO®-1 dye–stained DNA (red) containing the *lacZ* reporter gene and then stained with the β-galactosidase substrate fluorescein di-β-D-galactopyranoside (FDG, F1179). DNA stained with YOYO®-1 (Y3601) prior to implantation could still be localized 5 days after application. Fluorescence signals were visualized *in situ* by epifluorescence microscopy with a low–light level SIT camera and a computer imaging system. Image contributed by Peter van Mier, Washington University School of Medicine.

In addition to its use in ELISAs, the FDG substrate has proven very effective for identifying *lacZ*-positive cells with fluorescence microscopy [20–23] and flow cytometry.[24–29] FDG has been employed to identify cells infected with recombinant herpesvirus,[30] to detect unique patterns of β-galactosidase expression in live transgenic zebrafish embryos [20] and to monitor β-galactosidase expression in bacteria.[21,31,32] The purity of FDG and its analogs is very important because a reagent with extremely low fluorescence background is essential for most applications. Our stringent quality control helps ensure that the fluorescent contamination of FDG is less than 50 ppm.

The FluoReporter® *lacZ* Flow Cytometry Kit (F1930) provides materials and protocols for quantitating β-galactosidase activity with FDG in single cells using flow cytometry.

The FluoReporter® *lacZ* Flow Cytometry Kit contains:

- FDG
- Phenylethyl β-D-thiogalactopyranoside (PETG; also available separately as a solid, P1692), a broad-spectrum β-galactosidase inhibitor for stopping the reaction [33]
- Chloroquine diphosphate for inhibiting hydrolysis of the substrate in acidic organelles by endogenous galactosidase enzymes
- Propidium iodide for detecting dead cells
- Detailed protocol

Each kit provides sufficient reagents for 50 flow cytometry assays. This assay enables researchers to detect heterogeneous expression patterns and to sort and clone individual cells expressing known quantities of β-galactosidase. Practical reviews on using FDG for flow cytometric analysis and sorting of *lacZ*-positive cells are available.[34,35]

The fluorescent hydrolysis product of FDG is fluorescein (F1300, F36915; Section 10.1), which rapidly leaks from cells under physiological conditions, making the use of FDG problematic for prolonged studies. Our DetectaGene™ Green, ImaGene Green™ and ImaGene Red™ substrates (described below) have been specifically designed to improve retention of the fluorescent hydrolysis products in cells.

Resorufin Galactoside

Unlike FDG, resorufin β-D-galactopyranoside (R1159) requires only a single-step hydrolysis reaction to attain full fluorescence.[36] This substrate is especially useful for sensitive enzyme measurements in ELISAs.[15,37] The relatively low pK_a (~6.0) of its hydrolysis product, resorufin (R363, Section 10.1), permits its use for continuous measurement of enzymatic activity at physiological pH. Resorufin galactoside has also been used to quantitate β-galactosidase activity in single yeast cells by flow cytometry [38] and to detect immobilized β-galactosidase activity in bioreactors.[39,40]

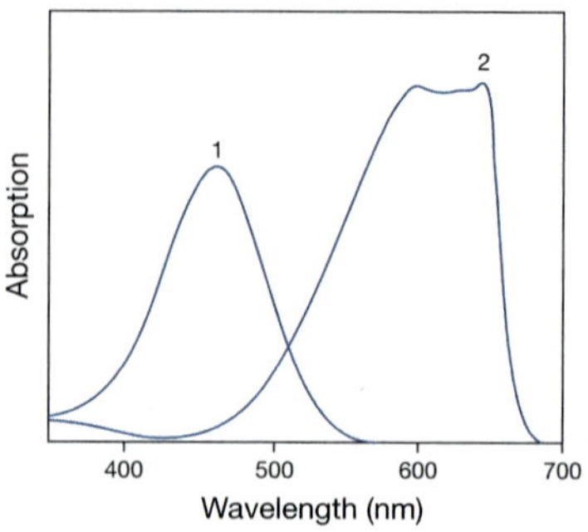

Figure 10.2.2 Absorption spectra of **1**) DDAO galactoside (D6488) and **2**) DDAO (H6482) at equal concentrations in pH 9 aqueous buffer. These spectra show the large spectral shift accompanying enzymatic cleavage of DDAO-based substrates.

DDAO Galactoside

Although substrates based on DDAO (7-hydroxy-9*H*-(1,3-dichloro-9,9-dimethylacridin-2-one)) are intrinsically fluorescent (excitation/emission ~460/610 nm), β-galactosidase–catalyzed hydrolysis of DDAO galactoside (D6488) liberates the DDAO fluorophore, which absorbs and emits light at much longer wavelengths (excitation/emission ~645/660 nm) (Figure 10.2.2). Not only can DDAO (H6482, Section 10.1) be excited without interference from the unhydrolyzed substrate, but its fluorescence emission is detected at wavelengths that are well beyond the autofluorescence exhibited by most biological samples. The relatively low pK_a of DDAO (~5.5) permits continuous monitoring of β-galactosidase activity at physiological pH.

Methylumbelliferyl Galactoside

The fluorogenic β-galactosidase substrate β-methylumbelliferyl β-D-galactopyranoside (MUG, M1489MP) is commonly used to detect β-galactosidase activity in cell extracts,[41–43] lysosomes[44] and human blood serum.[45] However, the hydrolysis product, 7-hydroxy-4-methylcoumarin (β-methylumbelliferone, H189; Section 10.1; Figure 10.2.3), has a relatively high pK_a (~7.8), precluding its use for continuous measurement of enzymatic activity.

Carboxyumbelliferyl Galactoside and the FluoReporter® *lacZ*/Galactosidase Quantitation Kit

Hydrolysis of 3-carboxyumbelliferyl β-D-galactopyranoside (CUG) by β-galactosidase yields 7-hydroxycoumarin-3-carboxylic acid (H185, Section 10.1). 7-Hydroxycoumarin has a pK_a below the pH at which the turnover rate is optimal, facilitating the use of CUG for continuous measurements of β-galactosidase activity. Unlike most substrates for β-galactosidase, CUG is quite water-soluble and can be used over a wide range of concentrations in enzymatic activity measurements.[46–48] Our

Table 10.1 Glycosidase enzymes and their fluorogenic and chromogenic substrates.

Carbohydrate (Enzyme)	Notes on Enzyme Activity	Labeled Substrate (Abs/Em of the products) *		Cat. No.	Ref
β-D-Galactopyranoside (β-Galactosidase, E.C. 3.2.1.23)	• Useful as a reporter gene marker[1–4] • Useful for ELISAs[5–8] • Useful for enumerating coliforms from the family *Enterobacteriaceae*[9–11] • Useful for classifying mycobacteria[12]	Blue-fluorescent product	4-Methylumbelliferyl (360/449)	M1489MP	13, 14
		Green-fluorescent product	Fluorescein (490/514)	F1179	1–3
			5-(Pentafluorobenzoylamino)-fluorescein (490/514)	P11948	
			C_2-Fluorescein (490/514)	A22010	15
			C_{12}-Fluorescein (490/514)	D2893, I2904‡	16–18
			5-Chloromethylfluorescein (490/514)	D2920†	19
		Red-fluorescent product	C_{12}-Resorufin (571/585)	I2906‡	20, 21
			DDAO (646/659)	D6488	
			Resorufin (571/585)	R1159	22
		Chromogenic substrate	5-Bromo-4-chloro-3-indoyl (615/NA)	X-Gal; B1690	23
β-D-Glucopyranoside (β-Glucosidase, E.C. 3.2.1.21)	• Deficiency in acid β-glucosidase, which leads to abnormal lysosomal storage, characterizes Gaucher disease[24] • Useful as a marker for the endoplasmic reticulum[25]	Green-fluorescent product	Fluorescein (490/514)	F2881	24, 26
			5-(Pentafluorobenzoylamino)fluorescein (490/514)	P11947	27
β-D-Glucuronide (β-Glucuronidase, GUS; E.C. 3.2.1.31)	• Useful as a reporter gene marker[1–4] • Useful for ELISAs[5–8] • Useful for enumerating coliforms from the family *Enterobacteriaceae*[9–11] • Useful for classifying mycobacteria[12]	Blue-fluorescent product	4-Methylumbelliferyl (360/449)	M1490	13, 28, 30–32
		Green-fluorescent product	Fluorescein (490/514)	F2915	33
			5-(Pentafluorobenzoylamino)fluorescein (490/514)	P11949	33
			C_{12}-Fluorescein (490/514)	I2908‡	34, 35
		Chromogenic substrate	5-Bromo-4-chloro-3-indoyl (615/NA)	X-GlcU, B1691	28, 32

* Approximate absorption (Abs) and fluorescence emission (Em) maxima of enzymatic hydrolysis product, in nm. † DetectaGene™ Green Gene Expression Kit. ‡ ImaGene Green™ or ImaGene Red™ Gene Expression Kit. NA = Not applicable.

1. Cytometry (1994) 17:216; **2.** Dev Biol (1994) 161:77; **3.** Proc Natl Acad Sci U S A (1988) 85:2603; **4.** Biophys J (1998) 74:11; **5.** Anal Biochem (1985) 146:211; **6.** Exp Parasitol (1991) 73:440; **7.** J Immunol Methods (1982) 54:297; **8.** J Virol Methods (1981) 3:155; **9.** Microbiol Rev (1991) 55:335; **10.** J Appl Bacteriol (1988) 64:65; **11.** Appl Environ Microbiol (1978) 35:136; **12.** Zentralbl Bakteriol (1994) 280:476; **13.** Anal Biochem (1980) 104:182; **14.** Anal Biochem (1993) 215:24; **15.** FEMS Microbiol Lett (1999) 179:317; **16.** Nat Protoc (2009) 4:1798; **17.** Proc Natl Acad Sci U S A (1992) 89:10681; **18.** FASEB J (1991) 5:3108; **19.** J Neurosci (1995) 15:1025; **20.** Biotechnol Bioeng (1993) 42:1113; **21.** US Patent No 5,242,805 (1993); **22.** Anal Chim Acta (1984) 163:67; **23.** Biotechniques (1989) 7:576; **24.** Cell Biochem Funct (1993) 11:167; **25.** Nature (1994) 369:113; **26.** Anal Biochem (1997) 247:268; **27.** Blood (1997) 89:3412; **28.** J Appl Bacteriol (1993) 74:223; **29.** Arch Biochem Biophys (1991) 286:394; **30.** Cell Signal (1991) 3:625; **31.** Plant Sci (1991) 78:73; **32.** J Immunol Methods (1987) 100:211; **33.** J Biol Chem (1999) 274:657; **34.** Microbiology (1997) 143:267; **35.** Plant J (1996) 10:745.

IMPORTANT NOTICE: The products described in this manual are covered by one or more Limited Use Label License(s). Please refer to the Appendix on page 971 and Master Product List on page 975. Products are For Research Use Only. Not intended for any animal or human therapeutic or diagnostic use.

FluoReporter® *lacZ*/Galactosidase Quantitation Kit (F2905) provides a CUG-based method for quantitating β-galactosidase activity in ELISAs or *lacZ*-positive cell extracts. Each kit contains:

- CUG
- 7-Hydroxycoumarin-3-carboxylic acid, a reference standard
- Detailed protocols suitable for use with any fluorescence microplate reader

Sufficient reagents are provided for approximately 1000 β-galactosidase assays. We have demonstrated a practical detection limit of ~0.5 pg of β-galactosidase using this kit and a fluorescence microplate reader.

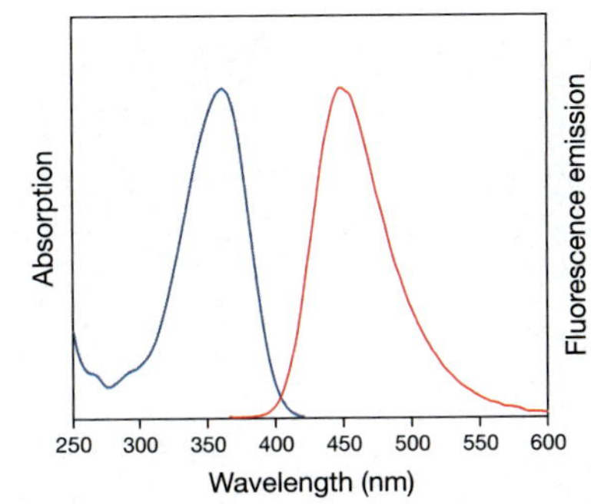

Figure 10.2.3 Absorption and fluorescence emission spectra of 7-hydroxy-4-methylcoumarin (H189) in pH 9.0 buffer. The spectra of 6,8-difluoro-7-hydroxy-4-methylcoumarin (D6566) are essentially identical.

Fluorescent Glycosphingolipids

β-Galactosidase enzymes that act on the lipophilic sphingosyl galactosides, including galactosylceramidase (EC 3.2.1.46) and G_{M1} ganglioside β-galactosidase (EC 3.2.1.23), are particularly important in neurochemistry. The preferred substrates for these enzymes are sphingolipids derived from galactose (Section 13.3). Galactosylceramidase converts substrates such as our BODIPY® FL C_{12}-galactosylceramide (D7519) back to the ceramide. Purified G_{M1} ganglioside galactosidase removes the terminal galactose residue from lactosylceramides such as our BODIPY® FL C_5-lactosylceramide[49] (D13951), yielding the corresponding glucosylceramide.[50] BODIPY® FL C_5-lactosylceramide complexed with bovine serum albumin (BSA) (B34402) may also be useful for assaying G_{M1} ganglioside galactosidase. Complexing fluorescent lipids with BSA facilitates the preparation of aqueous solutions by eliminating the need for organic solvents to dissolve the lipophilic probe. The lack of a spectral shift between the substrates and the hydrolysis products in these reactions means that extraction and chromatographic separation of the products is necessary for assessment of enzyme activity.

Modified Fluorogenic β-Galactosidase Substrates with Improved Cellular Retention

The primary problems associated with detecting *lacZ* expression in live cells using fluorogenic substrates are:

- Difficulty in loading the substrates under physiological conditions
- Leakage of the fluorescent product from live cells
- High levels of endogenous β-galactosidase activity in many cells

Our DetectaGene™ Green, ImaGene Green™ and ImaGene Red™ Kits are designed to improve the sensitivity of β-galactosidase assays by yielding products that are better retained in viable cells and, in the case of the ImaGene Green™ and ImaGene Red™ Kits, by providing substrates that can be passively loaded into live cells. The high level of endogenous β-galactosidase activity remains an obstacle when detecting low levels of *lacZ* expression.

DetectaGene™ Green *lacZ* Gene Expression Kit

The substrate in our DetectaGene™ Green *lacZ* Gene Expression Kit (D2920)—5-chloromethylfluorescein di-β-D-galactopyranoside (CMFDG)—is a galactose derivative that has been chemically modified to include a mildly thiol-reactive chloromethyl group (Figure 10.2.4). Once loaded into the cell using the Influx™ pinocytic cell-loading reagent (I14402,

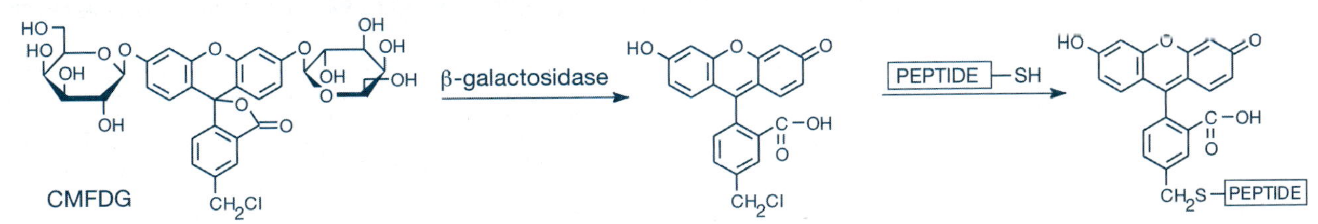

Figure 10.2.4 Sequential β-galactosidase hydrolysis and peptide conjugate formation of CMFDG, a component of the DetectaGene™ Green CMFDG *lacZ* Gene Expression Kit (D2920).

IMPORTANT NOTICE: The products described in this manual are covered by one or more Limited Use Label License(s). Please refer to the Appendix on page 971 and Master Product List on page 975. Products are For Research Use Only. Not intended for any animal or human therapeutic or diagnostic use.

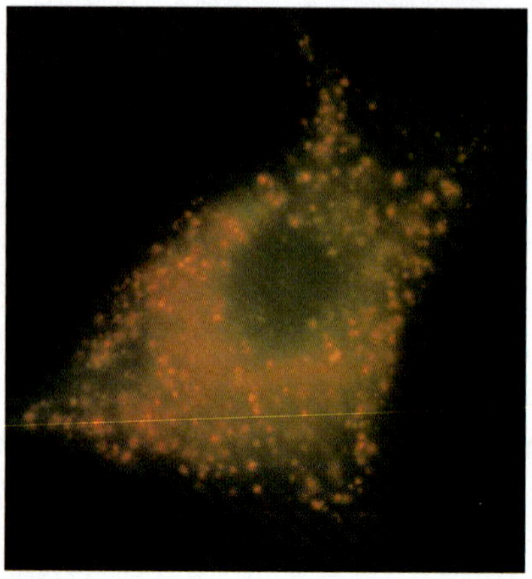

Figure 10.2.5 Live CRE BAG 2 cells loaded with the β-galactosidase substrate from the DetectaGene™ Green CMFDG *lacZ* Gene Expression Kit (D2920). The substrate was loaded into cells with the aid of the Influx™ pinocytic cell-loading reagent (I14402), a component of the DetectaGene™ Green Kit. The cells were then stained with LysoTracker® Red DND-99 (L7528) to differentiate lysosomal from cytoplasmic localization of the green-fluorescent CMFDG product, as well as to illustrate the selective rupture of only the Influx™ reagent–induced pinocytic vesicles. The double-exposure image was acquired using bandpass filters appropriate for fluorescein and Texas Red® dye.

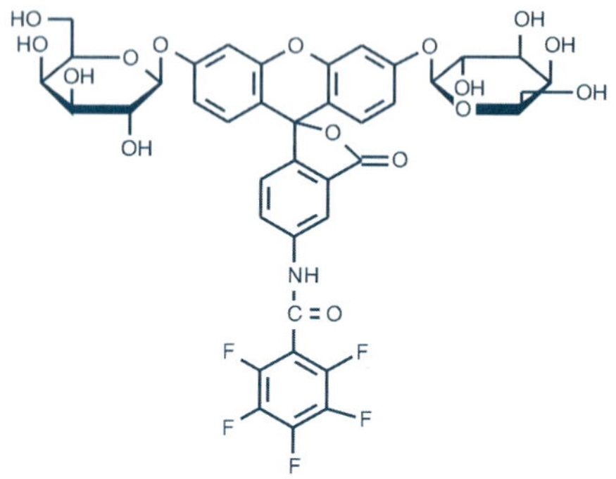

Figure 10.2.6 5-(Pentafluorobenzoylamino)fluorescein di-β-D-galactopyranoside (PFB-FDG, P11948).

included in Kit D2920; Figure 10.2.5) or by microinjection, hypotonic shock or another technique (Table 14.1, Section 1), the DetectaGene™ substrate undergoes two reactions: 1) its two galactose moieties are cleaved by intracellular β-galactosidase and 2) either simultaneously or sequentially, its chloromethyl moiety reacts with glutathione and possibly other intracellular thiols to form a membrane-impermeant, peptide–fluorescent dye adduct[51] (Figure 10.2.4). Because peptides do not readily cross the plasma membrane, the resulting fluorescent adduct is much better retained than is the free dye, even in cells that have been kept at 37°C. We have found that *lacZ*-positive cells loaded from medium containing 1 mM CMFDG are as fluorescent as cells loaded with 40-fold higher concentrations of FDG. Furthermore, unlike the free dye, the peptide–fluorescent dye adducts contain amine groups and can therefore be covalently linked to surrounding biomolecules by fixation with formaldehyde or glutaraldehyde. This property permits long-term storage of the labeled cells or tissue, as well as amplification of the conjugate by standard immunohistochemical techniques using our anti–fluorescein/Oregon Green® antibody (Section 7.4, Table 7.8).

The CMFDG substrate in our DetectaGene™ Green *lacZ* Gene Expression Kit was used to stain *lacZ*-expressing floor plate cells in tissue dissected from a developing mouse embryo,[52] to identify *lacZ*-enhancer–trapped *Drosophila* neurons in culture and to detect β-galactosidase activity in hippocampus slices.[53] In the latter study, the fluorescence of the neurons could still be visualized 24 hours after dye loading, and the fluorescent CMFDG-loaded neurons exhibited a normal pattern and time course of axonal outgrowth and branching.[54] CMFDG also has been microinjected into primary hepatocytes, fibroblasts and glioma cells to detect β-galactosidase activity[55] and has been incorporated into an electrophysiological recording pipette to confirm the identity of neurons cotransfected with the *lacZ* gene and a second gene encoding Ca^{2+}/calmodulin-dependent protein kinase II[56] (CaM kinase II).

The DetectaGene™ Green CMFDG *lacZ* Gene Expression Kit (D2920) contains:

- DetectaGene™ Green CMFDG substrate (Figure 10.2.4)
- Phenylethyl β-D-thiogalactopyranoside (PETG; also available separately as a solid, P1692), a broad-spectrum β-galactosidase inhibitor for stopping the reaction[33]
- Verapamil for inhibiting product efflux[57,58]
- Chloroquine diphosphate for inhibiting acidic hydrolysis of the substrate
- Propidium iodide for detecting dead cells
- Influx™ pinocytic cell-loading reagent for introducing CMFDG into cells
- Detailed protocols for detecting β-galactosidase activity

When used at the recommended dilutions, sufficient reagents are provided for approximately 200 flow cytometry tests with the DetectaGene™ Green CMFDG Kit. Verapamil has been added to the DetectaGene™ Green CMFDG *lacZ* Gene Expression Kit because we have observed that cell retention of the fluorescent dye–peptide adduct can be considerably improved in many cell types by adding verapamil to the medium.[57]

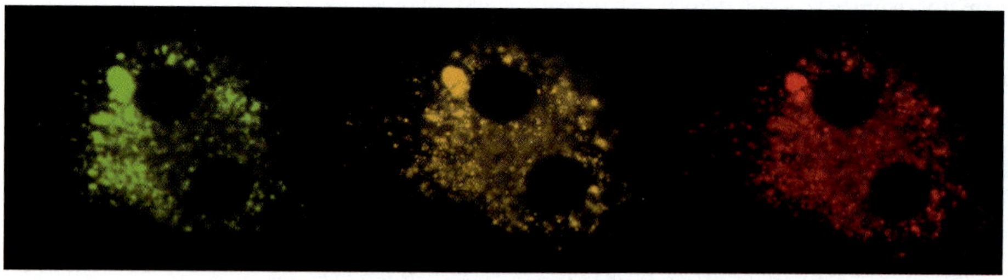

Figure 10.2.7 Bovine pulmonary artery endothelial cells simultaneously stained with LysoTracker® Red DND-99 (L7528), a cell-permeant, fixable lysosomal stain, and with 5-(pentafluorobenzoylamino)fluorescein di-β-D-galactopyranoside (PFB-FDG, P11948), a fluorogenic substrate for β-galactosidase. PFB-FDG is nonfluorescent until enzymatically hydrolyzed to green-fluorescent PFB-F. The center image demonstrates colocalization of the LysoTracker® Red DND-99 dye and PFB-F to the lysosomes. The left image was acquired with a bandpass filter set appropriate for fluorescein, the right image was acquired with a bandpass filter set appropriate for Texas Red® dye, and the center image was acquired with a double bandpass optical filter set appropriate for fluorescein and the Texas Red® dye.

IMPORTANT NOTICE: The products described in this manual are covered by one or more Limited Use Label License(s). Please refer to the Appendix on page 971 and Master Product List on page 975. Products are For Research Use Only. Not intended for any animal or human therapeutic or diagnostic use.

PFB Aminofluorescein Digalactoside

5-(Pentafluorobenzoylamino)fluorescein di-β-D-galactopyranoside (PFB-FDG, P11948; Figure 10.2.6) yields the relatively lipophilic, green-fluorescent 5-(pentafluorobenzoylamino) fluorescein (PFB-F), which appears to localize to endosomal and lysosomal compartments when loaded into cells by pinocytosis (Figure 10.2.7), similar to our PFB aminofluorescein diglucoside (PFB-FDGlu, P11947). Thus, PFB-FDG is potentially useful for studying lysosomal storage diseases, including Krabbe disease, G_{M1} gangliosidosis, galactosialidosis and Morquio syndrome, which are all associated with deficient lysosomal β-galactosidase activity.[59]

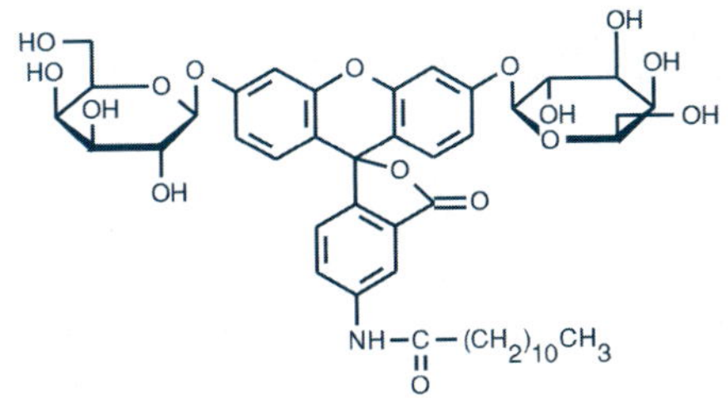

Figure 10.2.8 ImaGene Green™ C_{12}FDG *lacZ* substrate (I2904).

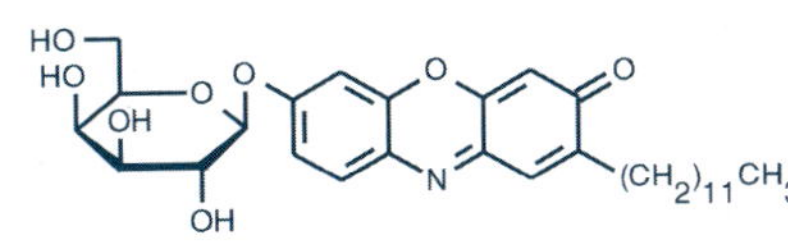

Figure 10.2.9 ImaGene Red™ C_{12}RG *lacZ* substrate (I2906).

ImaGene Green™ and ImaGene Red™ *lacZ* Reagents and Gene Expression Kits

The fluorescein- and resorufin-based galactosidase substrates in our ImaGene Green™ and ImaGene Red™ *lacZ* Gene Expression Kits (I2904, I2906) have been covalently modified to include a 12-carbon lipophilic moiety. Unlike FDG or CMFDG, these lipophilic fluorescein- and resorufin-based substrates—abbreviated C_{12}FDG (Figure 10.2.8) and C_{12}RG (Figure 10.2.9) for the ImaGene Green™ and ImaGene Red™ substrates, respectively—can be loaded simply by adding the substrate to the aqueous medium in which the cells or organisms are growing, either at ambient temperatures or at 37°C. Once inside the cell, the substrates are cleaved by β-galactosidase, producing fluorescent products that are well retained by the cells, probably by incorporation of their lipophilic tails within the cellular membranes. Mammalian NIH 3T3 *lacZ*-positive cells grown for several days in medium containing 60 µM C_{12}FDG appear morphologically normal, continue to undergo cell division and remain fluorescent for up to three cell divisions after replacement with substrate-free medium.[51,60]

The C_{12}FDG substrate in our ImaGene Green™ *lacZ* Expression Kit (I2904) is superior to FDG for flow cytometric detection of β-galactosidase activity in live mammalian cells.[61] Using C_{12}FDG with flow cytometric methods, researchers have:

- Analyzed senescence-associated β-galactosidase activity[62]
- Identified endocrine cell precursors in dissociated fetal pancreatic tissue based on their high levels of endogenous acid β-galactosidase[63]
- Measured β-galactosidase activity in single recombinant *Escherichia coli* bacteria[64]
- Detected the activity of β-galactosidase fusion proteins in yeast[65]
- Sorted β-galactosidase–expressing mouse sperm cells[66] and insect cells that harbor recombinant baculovirus[67,68]

The C_{12}FDG substrate was also useful in a fluorescence microscopy study of zebrafish expressing a *lacZ* reporter gene that was under the control of a mammalian homeobox gene promoter.[69] In some cell types, C_{12}FDG produces high levels of background fluorescence that may prohibit its use in assaying low β-galactosidase expression.[22]

The ImaGene Green™ C_{12}FDG or ImaGene Red™ C_{12}RG *lacZ* Gene Expression Kits contain:

- ImaGene Green™ C_{12}FDG (in Kit I2904) or ImaGene Red™ C_{12}RG (in Kit I2906)
- Phenylethyl β-D-thiogalactopyranoside (PETG; also available separately as a solid, P1692), a broad-spectrum β-galactosidase inhibitor for stopping the reaction[33]
- Chloroquine diphosphate for inhibiting acidic hydrolysis of the substrate
- Detailed protocols for detecting β-galactosidase activity

Each kit provides sufficient reagents for 100–200 assays, depending on the volume used for each experiment. 5-Dodecanoylaminofluorescein di-β-D-galactopyranoside (C_{12}FDG) is available separately (D2893). We also offer 5-acetylaminofluorescein di-β-D-galactopyranoside (C_2FDG, A22010), which has been shown to be useful for detecting *lacZ* reporter gene expression in slow-growing mycobacteria, including *Mycobacterium tuberculosis*, using a fluorescence microplate reader or a flow cytometer.[70]

NovaBright™ β-Galactosidase Chemiluminescent Detection Kits

The NovaBright™ β-Galactosidase Enzyme Reporter Gene Chemiluminescent Detection Kits for mammalian cells (N10563, N10564) and for yeast cells (N10565, N10566) are designed for the rapid, simple and sensitive detection of the β-galactosidase reporter enzyme directly in microwell cultures of mammalian and yeast cells, respectively. This chemiluminescence-based assay uses the Galacton-*Star*® 1,2-dioxetane substrate (Figure 10.2.10) with a luminescence enhancer to generate glow light emission kinetics. A single reagent, which provides cell lysis components, luminescent substrate and enhancer, is added directly to cells in culture medium. Light emission typically reaches maximum intensity within 60–90 minutes and remains constant for 45–90 minutes or longer, depending upon assay temperature. Light emission is measured in a variety of luminometers without the need for solution injection.

The simple assay format and glow light emission achieved with the NovaBright™ β-Galactosidase Enzyme Reporter Gene Chemiluminescent Detection Kit provide an ideal assay system for automated high-throughput screening applications, enabling simple processing and measurement of multiple microplates. The dynamic range of the assay spans five orders of magnitude, from picogram to nanogram levels, enabling detection of a wide range of reporter enzyme concentration in cells.

Each NovaBright™ β-Galactosidase Enzyme Reporter Gene Chemiluminescent Detection Kit provides:

- Galacton-*Star*® substrate concentrate
- Assay buffer for mammalian cells (200 assays, N10563; 1000 assays, N10564) or for yeast cells (200 assays, N10565; 1000 assays, N10566)
- Detailed protocols

Chemiluminescent 1,2-dioxetane substrates for β-galactosidase, including the Galacton-*Star*® substrate described here and the Galacton-Plus substrate provided in the NovaBright™ β-Galactosidase and Firefly Luciferase Dual Enzyme Reporter Gene Chemiluminescent Detection Kits (N10561, N10562; Section 10.6) provide highly sensitive enzyme detection and have been utilized extensively in reporter assays in mammalian cell and tissue extracts.

Amplex® Red Galactose/ Galactose Oxidase Assay Kit

Molecular Probes® Amplex® Red reagent (10-acetyl-3,7-dihydroxyphenoxazine, A12222, A22177) is an unusually stable peroxidase substrate that we have used in coupled reactions to detect a wide variety of analytes, including both enzymes and their substrates (see Section 10.5 for a list of all of our Amplex® Red Kits and reagents). Most of the assays can be performed as continuous assays at neutral or slightly acidic pH and are particularly suitable for automation and high-throughput screening using either an absorption- or fluorescence-based microplate reader.

Rather than requiring an unnatural fluorogenic or chromogenic substrate for β-galactosidase (or α-galactosidase), our Amplex® Red reagent–based technology permits the direct quantitation of free galactose, which is produced by a wide variety of enzymes. Even enzymes that act on polysaccharides and glycolipids but are difficult to assay using known chromogenic substrates can, in some cases, be detected and their activity quantitated using the Amplex® Red reagent in combination with galactose oxidase and horseradish peroxidase. Unlike glucose oxidase, galactose oxidase produces H_2O_2 from either free galactose or from polysaccharides—including glycoproteins in solution and on cell surfaces—and from certain glycolipids in which galactose is the terminal residue (Figure 10.2.11). Because the galactose oxidase–catalyzed reaction does not require prior cleavage of the glycoside to free galactose by a galactosidase, appropriate control reactions must be used to ascertain whether the rate-limiting step is the galactosidase- or galactose oxidase–mediated reaction.

The Amplex® Red Galactose/Galactose Oxidase Assay Kit (A22179) provides an ultrasensitive fluorescence-based method for detecting galactose (Figure 10.2.12) and galactose oxidase (Figure 10.2.13) activity. This assay utilizes the Amplex® Red reagent to detect H_2O_2 generated by galactose oxidase–mediated oxidation of desialated galactose moieties. In the presence of horseradish peroxidase (HRP), the H_2O_2 thus produced reacts with the Amplex® Red reagent in a 1:1 stoichiometry to generate the red-fluorescent oxidation product resorufin.[71] Resorufin has absorption and fluorescence emission maxima of approximately 571 nm and 585 nm, respectively, and because its extinction coefficient is high (54,000 $cm^{-1}M^{-1}$), the assay can be performed either fluorometrically or spectrophotometrically. The Amplex® Red Galactose/Galactose Oxidase Assay Kit provides all the reagents and a general protocol for the assay of galactose-producing enzymes or for the assay of galactose oxidase, including:

- Amplex® Red reagent
- Dimethylsulfoxide (DMSO)
- Horseradish peroxidase (HRP)
- Hydrogen peroxide (H_2O_2)
- Concentrated reaction buffer
- Galactose oxidase from *Dactylium dendroides*
- D-Galactose
- Detailed protocols

Sufficient reagents are provided for approximately 400 assays using either a fluorescence- or absorption-based microplate reader and reaction volumes of 100 µL per assay. The Amplex® Red galactose/galactose oxidase assay is designed to accurately measure as little as 4 µM galactose and 2 mU/mL galactose oxidase activity (Figure 10.2.12, Figure 10.2.13). Because of the high absorbance of resorufin, the spectrophotometric assay has only slightly lower sensitivity than the fluorometric assay.

Fluorogenic β-Glucuronidase Substrates

The substrate 4-methylumbelliferyl β-D-glucuronide (MUGlcU, M1490) is probably the most commonly used fluorogenic reagent for identifying *E. coli* contamination and for detecting GUS reporter gene expression in plants and plant extracts.[72,73] However, β-glucuronidase substrates based on fluorescein may be much more sensitive and yield products that are fluorescent at physiological pH, making them useful for continuous monitoring of enzymatic activity.

IMPORTANT NOTICE: The products described in this manual are covered by one or more Limited Use Label License(s). Please refer to the Appendix on page 971 and Master Product List on page 975. Products are For Research Use Only. Not intended for any animal or human therapeutic or diagnostic use.

Fluorescein Diglucuronide

Fluorescein di-β-D-glucuronide (FDGlcU, F2915) is colorless and nonfluorescent until it is hydrolyzed to the monoglucuronide and then to highly fluorescent fluorescein (F1300, F36915; Section 10.1). FDGlcU has been used to detect β-glucuronidase activity in plants and animals containing the GUS reporter gene[73,74] and may also be useful for assaying lysosomal enzyme release from neutrophils.[75,76] FDGlcU has also been used in the flow cytometry assay of individual mammalian cells expressing the *E. coli* β-glucuronidase gene.[77]

PFB Aminofluorescein Diglucuronide

5-(Pentafluorobenzoylamino)fluorescein di-β-D-glucuronide (PFB-FDGlcU, P11949) yields the green-fluorescent 5-(pentafluorobenzoylamino)fluorescein (PFB-F), which appears to localize in endosomal and lysosomal compartments when PFB-FDGlcU is loaded into cells by pinocytosis, similar to PFB aminofluorescein diglucoside (PFB-FDGlu, P11947) described below. PFB-FDGlcU has been used for the quantitative analysis of β-glucuronidase activity in viable cells and for sorting high-expressing cells by flow cytometry.[78]

Coumarin Glucuronides

4-Methylumbelliferyl β-D-glucuronide (MUGlcU, M1490) has been used extensively to detect *E. coli* in food,[79,80] water,[81] urine[82] and environmental samples.[83] MUGlcU is stable to the conditions required for sterilization of media. A fluorogenic bioassay using MUGlcU has been developed to assess the detrimental effects of Li^+, Al^{3+}, Cr^{6+} and Hg^{2+} on the proliferation of *E. coli*.[84] MUGlcU is also commonly used to identify plant tissue expressing the GUS reporter gene,[85–87] including nondestructive assays that allow propagation of the transformed plant lines.[88,89] In addition, MUGlcU has served as a sensitive substrate for lysosomal enzyme release from neutrophils.[75,76]

ImaGene Green™ β-D-Glucuronidase Substrate

We also offer a lipophilic analog of fluorescein di-β-D-glucuronide in the ImaGene Green™ C_{12}FDGlcU GUS Gene Expression Kit (I2908). As with our similar ImaGene Green™ and ImaGene Red™ substrates for β-galactosidase, we have shown that this lipophilic β-glucuronidase substrate freely diffuses across the membranes of viable cultured tobacco leaf cells or protoplasts under physiological conditions. Furthermore, the fluorescent cleavage product is retained in the plant cell for hours to days, facilitating long-term measurements of GUS gene expression. In thin sections of transgenic tomato leaf, the ImaGene Green™ C_{12}FDGlcU GUS Gene Expression Kit provided a simple and reliable GUS assay that, coupled with confocal laser-scanning microscopy,

Figure 10.2.10 Galacton-*Star*® chemiluminescent β-galactosidase substrate.

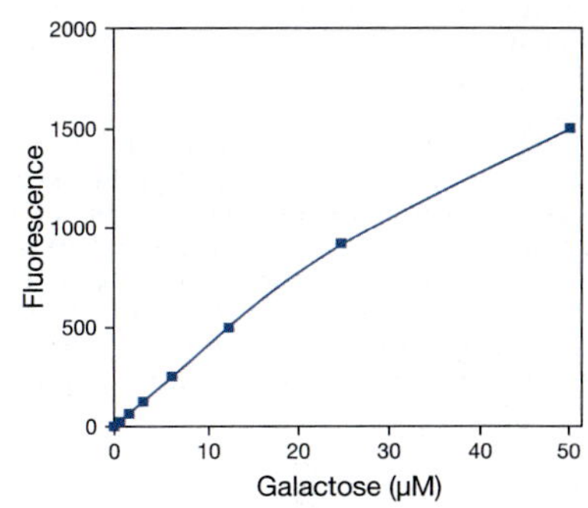

Figure 10.2.12 Detection of galactose using the Amplex® Red Galactose/Galactose Oxidase Assay Kit (A22179). Each reaction contained 50 µM Amplex® Red reagent, 0.1 U/mL HRP, 2 U/mL of galactose oxidase and the indicated amount of galactose in 1X reaction buffer. Reactions were incubated at 37°C. After 30 minutes, fluorescence was measured in a fluorescence microplate reader using excitation at 530 ± 12.5 nm and fluorescence detection at 590 ± 17.5 nm. A background fluorescence of 93 units was subtracted from each data point.

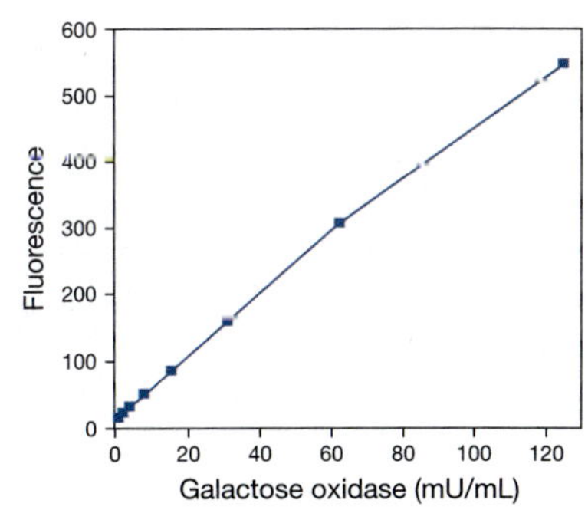

Figure 10.2.13 Detection of galactose oxidase activity using the Amplex® Red Galactose/Galactose Oxidase Assay Kit (A22179). Each reaction contained 50 µM Amplex® Red reagent, 0.1 U/mL HRP, 100 µM galactose and the indicated amount of galactose oxidase in 1X reaction buffer. Reactions were incubated at 37°C. After 20 minutes, fluorescence was measured in a fluorescence microplate reader using excitation at 530 ± 12.5 nm with fluorescence detection at 590 ± 17.5 nm.

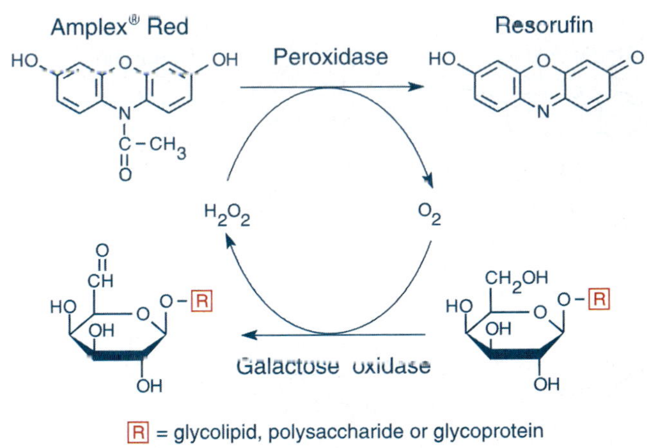

Figure 10.2.11 Detection scheme utilized in the Amplex® Red Galactose/Galactose Oxidase Assay Kit (A22179). Oxidation of the terminal galactose residue of a glycoprotein, glycolipid or polysaccharide results in the generation of H_2O_2, which, in the presence of horseradish peroxidase (HRP), reacts stoichiometrically with the Amplex® Red reagent to generate the red-fluorescent oxidation product, resorufin.

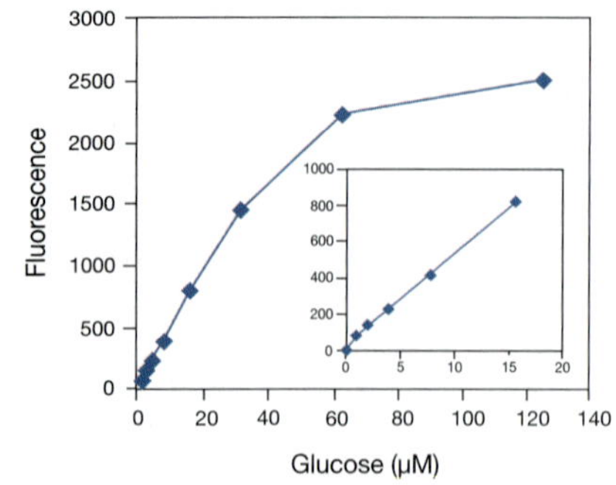

Figure 10.2.14 Detection of glucose using the Amplex® Red Glucose/Glucose Oxidase Assay Kit (A22189). Reactions containing 50 μM Amplex® Red reagent, 0.1 U/mL HRP, 1 U/mL glucose oxidase and the indicated amount of glucose in 50 mM sodium phosphate buffer, pH 7.4, were incubated for 1 hour at room temperature. Fluorescence was then measured with a fluorescence microplate reader using excitation at 530 ± 12.5 nm and fluorescence detection at 590 ± 17.5 nm. Background fluorescence (arbitrary units), determined for a no-glucose control reaction, has been subtracted from each value. The inset shows the sensitivity and linearity of the assay at low levels of glucose (0–15 μM).

yielded good cellular resolution.[90] The substrate has also been used to detect β-glucuronidase activity in an *Acremonium* transformant containing the GUS reporter gene.[91]

The ImaGene Green™ C_{12}FDGlcU GUS Gene Expression Kit contains:

- ImaGene Green™ C_{12}FDGlcU
- D-Glucaric acid-1,4-lactone, a β-glucuronidase inhibitor for stopping the reaction
- Detailed protocols for detecting β-glucuronidase activity

Each kit provides sufficient reagents for approximately 100 tests, depending on the volume used for each experiment.

Fluorogenic β-Glucosidase Substrates

β-Glucosidase, which is a marker for the endoplasmic reticulum (Section 12.4), is present in nearly all species. Its natural substrate is a glucosylceramide (Section 13.3). People with Gaucher disease have mutations in the acid β-glucosidase gene that result in abnormal lysosomal storage.[9,92] Enzyme replacement therapy in Gaucher disease patients[93] requires sensitive and selective methods for measuring β-glucosidase activity (Table 10.1).

Fluorescein Diglucoside

As with the other fluorescein diglycosides, our fluorogenic fluorescein di-β-D-glucopyranoside (FDGlu, F2881) is likely to yield the greatest sensitivity for detecting β-glucosidase activity in both cells[92] and emulsion droplets.[94] This substrate has been used to demonstrate the utility of *Saccharomyces cerevisiae* and *Candida albicans* exo-1,3-β-glucanase genes as reporter genes.[95] Because these reporter genes encode secreted proteins, assays for reporter gene expression do not require cell permeabilization. FDGlu has been reported to be a selective substrate for the flow cytometric assay of lysosomal glucocerebrosidase activity in a variety of cells.[96] The assay demonstrated the inordinately low glucocerebrosidase activity present in fibroblasts of Gaucher disease patients.

PFB Aminofluorescein Diglucoside

Through a collaboration with Matthew Lorincz and Leonard A. Herzenberg at Stanford University Medical School,[97] our PFB aminofluorescein diglucoside (PFB-FDGlu, P11947) has shown to be an excellent substrate for the flow cytometric discrimination of normal peripheral blood mononuclear cells (PBMC) from the PBMC of patients with Gaucher disease, a genetic deficiency in lysosomal β-glucocerebrosidase activity.[59]

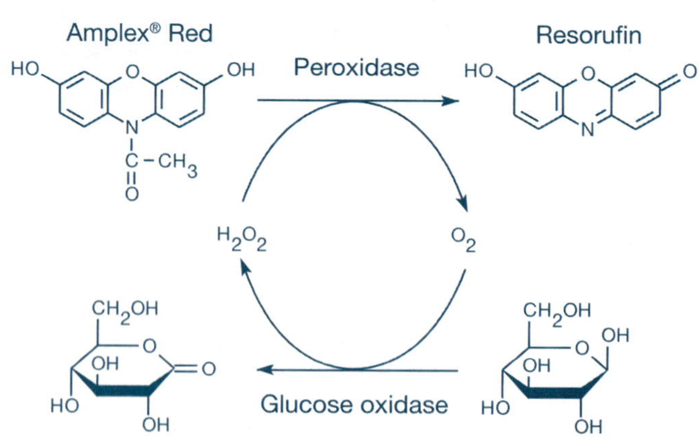

Figure 10.2.15 Principle of coupled enzymatic assays using our Amplex® Red reagent. Oxidation of glucose by glucose oxidase results in generation of H_2O_2, which is coupled to conversion of the Amplex® Red reagent to fluorescent resorufin by HRP. The detection scheme shown here is used in our Amplex® Red Glucose/Glucose Oxidase Assay Kit (A22189).

Amplex® Red Glucose/Glucose Oxidase Assay Kit

The Amplex® Red reagent (10-acetyl-3,7-dihydroxyphenoxazine, A12222, A22177; Section 10.5) is a colorless, stable and extremely versatile peroxidase substrate.[98] In an application similar to our use of the Amplex® Red reagent to detect galactose-producing enzymes, we have shown that it is practical to detect free glucose with high specificity at levels as low as 0.5 µg/mL using the Amplex® Red reagent in combination with excess glucose oxidase (Figure 10.2.14). Because the peroxidase- and glucose oxidase–mediated reactions can be coupled, it is potentially possible to measure the release of glucose by *any* glucosidase enzyme—for instance, α-glucosidase, β-glucosidase and glucocerebrosidase—in either a continuous or discontinuous assay (Figure 10.2.15). This assay should also be very useful for quantitation of glucose levels in foods, fermentation media and bodily fluids. The long-wavelength spectral properties of resorufin and high sensitivity of the assay result in little interference from colored components in the samples.

Our Amplex® Red Glucose/Glucose Oxidase Assay Kit (A22189) provides all the reagents required for the assay of glucose and enzymes that produce glucose. This kit is also useful for the assay of glucose oxidase activity from cell extracts. We have even shown that the Amplex® Red reagent can detect glucose liberated from native dextrans by dextranase[99] and from carboxymethylcellulose by cellulase.[100] The Amplex® Red Glucose/Glucose Oxidase Assay Kit contains:

- Amplex® Red reagent
- DMSO
- Horseradish peroxidase (HRP)
- Hydrogen peroxide (H_2O_2) for use as a positive control
- Concentrated reaction buffer
- Glucose oxidase
- D-glucose
- Detailed protocols

The kit provides sufficient reagents for approximately 500 assays using either a fluorescence- or absorption-based microplate reader and a reaction volume of 100 µL per assay.

Amplex® Red Neuraminidase/Sialidase Assay Kit

Neuraminidase (NA, also known as sialidase) is a very common enzyme that hydrolyzes terminal sialic acid residues on polysaccharide chains, most often exposing a galactose residue. Although NA is found in mammals, it is predominantly expressed in microorganisms such as bacteria and viruses,[101] including the negative-stranded RNA influenza virus. NA located on the surface of the influenza virus is thought to play a key role in its invasion of target cells and subsequent replication through cleavage of target cell receptor sialic acid moieties. Anti-influenza drug design has therefore focused on the inhibition of NA.[102] Various methods using chemiluminescence, absorption and fluorescence have been developed to quantitate NA in biological fluids for detecting influenza virus[103,104] and for screening inhibitors of NA activity in drug development.[104,105] The ultimate goal has been to develop a rapid, single-step assay that is sensitive and adaptable for a high-throughput screening format.

The Amplex® Red Neuraminidase (Sialidase) Assay Kit (A22178) provides an ultrasensitive fluorescence-based method for detecting NA activity. This assay utilizes the Amplex® Red reagent to detect H_2O_2 generated by galactose oxidase–mediated oxidation of desialated galactose, the end result of NA action. In the presence of horseradish peroxidase (HRP), the H_2O_2 thus produced reacts with a 1:1 stoichiometry with the Amplex® Red reagent to generate the red-fluorescent oxidation product, resorufin.[71] Resorufin has absorption and fluorescence emission maxima of approximately 571 nm and 585 nm, respectively, and because the extinction coefficient is high (54,000 $cm^{-1}M^{-1}$), the assay can be performed either fluorometrically or spectrophotometrically. In a purified system with fetuin as the substrate, NA levels as low as 0.2 mU/mL have been detected with the Amplex® Red Neuraminidase (Sialidase) Assay Kit (Figure 10.2.16). NA activity can also be detected in biological samples such as serum (Figure 10.2.17).

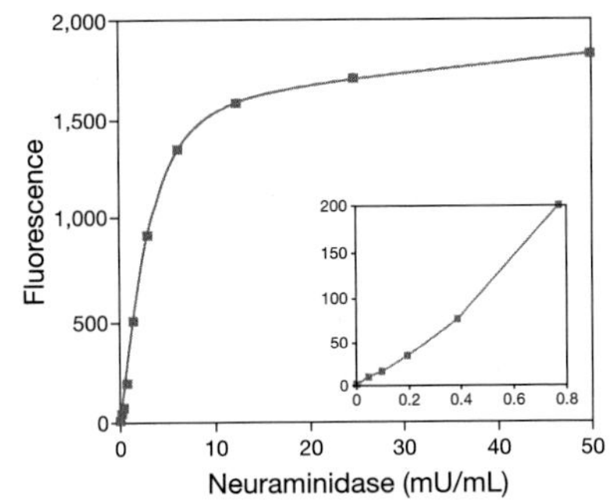

Figure 10.2.16 Detection of neuraminidase activity using the Amplex® Red Neuraminidase (Sialidase) Assay Kit (A22178). Each reaction contained 50 µM Amplex® Red reagent, 0.1 U/mL HRP, 2 U/mL galactose oxidase, 250 µg/mL fetuin and the indicated amount of neuraminidase in 1X reaction buffer. Reactions were incubated at 37°C. After 30 minutes, fluorescence was measured in a fluorescence microplate reader using excitation at 530 ± 12.5 nm and fluorescence detection at 590 ± 17.5 nm. A background fluorescence of 70 fluorescence units was subtracted from each data point. The inset shows the sensitivity and linearity of the assay at low levels of neuraminidase (0–0.8 mU/mL).

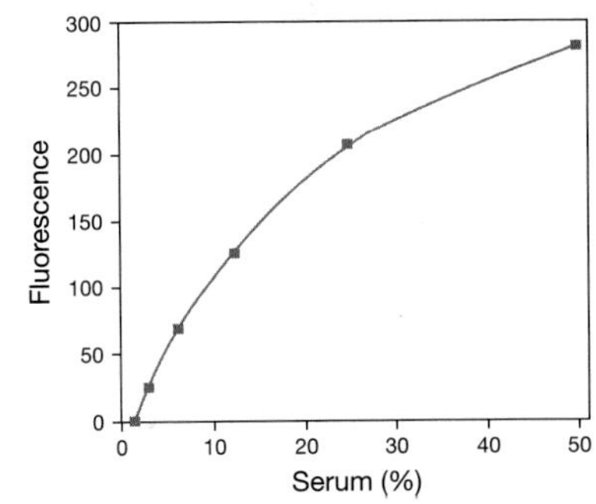

Figure 10.2.17 Detection of neuraminidase activity in serum using the Amplex® Red Neuraminidase (Sialidase) Assay Kit (A22178). Each reaction contained 50 µM Amplex® Red reagent, 0.1 U/mL HRP, 2 U/mL galactose oxidase, 250 µg/mL fetuin and the indicated amount of serum in 1X reaction buffer. Reactions were incubated at 37°C. After 60 minutes, fluorescence was measured in a fluorescence microplate reader using excitation at 530 ± 12.5 nm and fluorescence detection at 590 ± 17.5 nm. A background fluorescence of 112 units was subtracted from each data point.

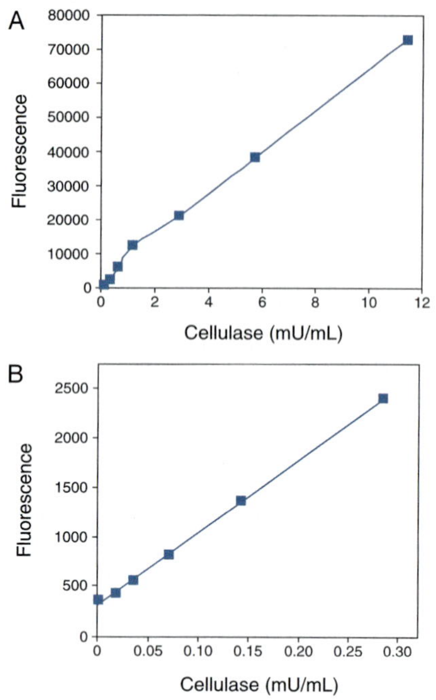

Figure 10.2.18 Sample standard curves obtained with the EnzChek® cellulase substrate. **A)** Cellulase from *Trichoderma reesei* was assayed with EnzChek® cellulase substrate (E33953) in 100 mM sodium acetate (pH 5.0) digestion buffer using cellulase activities from 0.018 to 11.4 mU/mL. **B)** This standard curve shows the same experiment at cellulase activities from 0.018 to 0.285 mU/mL. Samples were incubated for 30 minutes at room temperature and fluorescence was measured in a fluorescence microplate reader using excitation/emission wavelengths of 360/460 nm. Background fluorescence has not been subtracted.

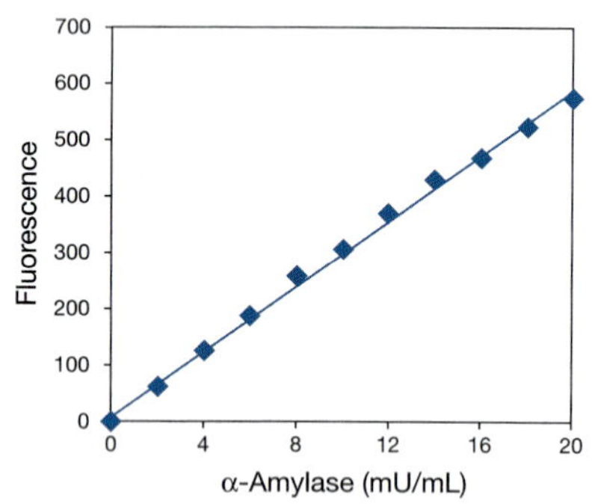

Figure 10.2.19 Assay of α-amylase from *Bacillus* sp. using the EnzChek® *Ultra* Amylase Assay Kit. Reactions contained 200 µg/mL DQ™ starch and the indicated amount of α-amylase in 100 mM MOPS (pH 6.9). After incubating at room temperature for 30 minutes, the fluorescence was measured in a microplate reader using excitation at 485 ± 12 nm and fluorescence detection at 520 ± 12 nm. The fluorescence of the blank sample (without enzyme) was subtracted from each value.

The Amplex® Red Neuraminidase/Sialidase Assay Kit contents include:

- Amplex® Red reagent
- Dimethylsulfoxide (DMSO)
- Horseradish peroxidase (HRP)
- Hydrogen peroxide (H_2O_2)
- Concentrated reaction buffer
- Galactose oxidase from *Dactylium dendroides*
- Fetuin from fetal calf serum
- Neuraminidase from *Clostridium perfringens*
- Detailed protocols

Each kit provides sufficient reagents for approximately 400 assays using either a fluorescence- or absorption-based microplate reader and reaction volumes of 100 µL per assay.

EnzChek® Cellulase Substrate

The EnzChek® cellulase substrate (E33953) was developed for simple and rapid fluorescence-based quantitation of cellulase (EC 3.2.1.4). This cellulase assay can be completed in 30 minutes or less using a simple mix-incubate-read format. In contrast to other more complex, multistep assays, the EnzChek® cellulase assay is ideal for high-throughput screening, with a detection limit as low as 40 µU/mL using a fluorescence microplate reader (Figure 10.2.18).

In the presence of cellulase, the EnzChek® cellulase substrate exhibits an increase in absorption at 360 nm, which is proportional to the amount of enzyme present and can be used to effectively quantitate enzyme activity using an absorption-based microplate reader or spectrophotometer. In addition, celluase activity can be assayed using the Amplex® Red Glucose/Glucose Oxidase Assay Kit (A22189), as described above.[100]

EnzChek® *Ultra* Amylase Assay

α-Amylase is a hydrolytic enzyme that catalyzes the conversiovn of starch to a mixture of glucose, maltose, maltotriose and dextrins. The levels of α-amylase in various fluids of the human body are of clinical importance in the diagnosis of disease states, including pancreatitis and diabetes.[106–109] Plant and microbial α-amylases are important enzymes for industrial applications ranging from the manufacture of baked goods and dairy products to the production of ethanol and paper.

The EnzChek® *Ultra* Amylase Assay Kit (E33651) provides a fast, sensitive and convenient solution-based assay for measuring amylase activity[110] or for screening amylase inhibitors in a high-throughput format. This EnzChek® Kit contains DQ™ starch, a starch derivative that is labeled with BODIPY® FL dye to such a degree that the fluorescence is quenched. Amylase-catalyzed hydrolysis of this substrate relieves the quenching, yielding brightly fluorescent BODIPY® FL dye–labeled fragments. The accompanying increase in fluorescence is proportional to amylase activity and can be monitored with a fluorescence microplate reader or fluorometer, using standard fluorescein filters. In tests using α-amylase from *Bacillus* sp. and 200 µg/mL of the DQ™ starch substrate (30-minute incubation at room temperature), the EnzChek® *Ultra* Amylase Assay Kit could be used to detect α-amylase activity down to a final concentration of 0.002 U/mL, where one unit is defined as the amount of enzyme required to liberate 1 mg of maltose from starch in 3 minutes at 20°C, pH 6.9 (Figure 10.2.19).

The EnzChek® *Ultra* Amylase Assay Kit provides:

- DQ™ starch—a heavily labeled, highly quenched BODIPY® FL conjugate of starch from corn
- Concentrated reaction buffer
- Substrate solvent
- BODIPY® FL propionic acid in dimethylsulfoxide (DMSO), for use as a fluorescence standard
- Detailed protocols

IMPORTANT NOTICE: The products described in this manual are covered by one or more Limited Use Label License(s). Please refer to the Appendix on page 971 and Master Product List on page 975. Products are For Research Use Only. Not intended for any animal or human therapeutic or diagnostic use.
www.invitrogen.com/probes
molecular probes® invitrogen™ by life technologies™

Each kit provides sufficient reagents for 500 assays using 100 µL assay volumes in a 96-well microplate assay format.

EnzChek® Xylanase Assay Kit

The hydrolysis of xylosidic linkages in hemicellulose polysaccharides by xylanase (EC 3.2.1.8) is important in a wide range of industrial processes, including baking, pulp and paper manufacturing and animal feed production. Xylanases occur in a variety of bacteria and fungi and are classified into two families—glycosyl hydrolase families 10 and 11.[111] Family 11 xylanases are more specific for xylans; family 10 xylanases also exhibit cellulase activity. Existing xylanase assay methods typically require separation or heating steps. The chromogenic substrate *o*-nitrophenyl-β-D-xylobioside can be used in a simpler homogeneous assay format; however, *o*-nitrophenol is pH sensitive ($pK_a = 7.2$) and therefore less than ideal for spectrophotometric assays of xylanases, which exhibit a wide range of pH optima (from pH 2 to 9).

The EnzChek® *Ultra* Xylanase Assay Kit (E33650) provides a quick and convenient mix-and-read assay for measuring xylanase activity or for screening xylanase inhibitors in a high-throughput format using a fluorescence microplate reader or a standard fluorometer (excitation/emission maxima ~358/455 nm).[112] This kit can be used for continuous detection of xylanase activity and offers broad dynamic and pH ranges (1.5 to 200 mU/mL and pH 4–10, respectively; Figure 10.2.20), high sensitivity (as low as 1.5 mU/mL) and excellent temperature tolerance. It has been tested with xylanases from *Trichoderma viride*, *T. longibrachiatum*, *Thermomyces lanuginosus*, *Aspergillus niger* and other bacterial and fungal species.

The EnzChek® *Ultra* Xylanase Assay Kit contains:

- Xylanase substrate
- Concentrated reaction buffer
- Fluorescence standard
- Detailed protocols

Each kit provides sufficient reagents for 500 assays using 100 µL assay volumes in a 96-well microplate assay format.

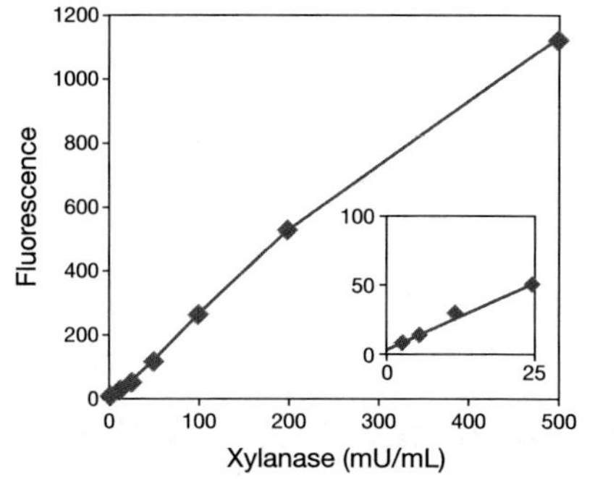

Figure 10.2.20 The EnzChek® xylanase assay was performed at pH 4.6 under standard conditions (30-minute incubation) as described in the experimental protocol, using *Aspergillus niger* xylanase and the EnzChek® xylanase substrate. The fluorescence signal for 500 mU/mL of xylanase is equivalent to 49 µM fluorescence standard, indicating 94% turnover of the xylanase substrate (initially 52 µM).

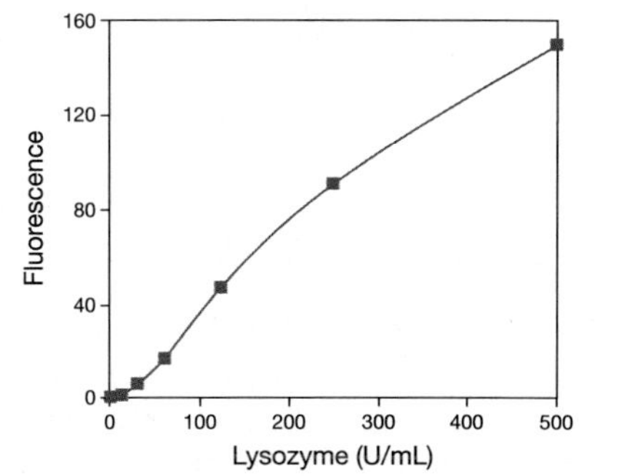

Figure 10.2.21 Detection of lysozyme activity using the EnzChek® Lysozyme Assay Kit (E22013). Increasing amounts of lysozyme were incubated with the DQ™ lysozyme substrate for 60 minutes at 37°C. The fluorescence was measured in a fluorescence microplate reader using excitation/emission wavelengths of ~485/530 nm. Background fluorescence, determined for a no-enzyme control, was subtracted from each value.

EnzChek® Lysozyme Assay Kit

Lysozyme (muramidase) hydrolyzes β-1-4-glycosidic linkages between *N*-acetylmuramic acid and *N*-acetyl-D-glucosamine residues present in the mucopolysaccharide cell wall of a variety of microorganisms. Lysozyme is present in human serum, urine, tears, seminal fluid and milk. Serum and urine lysozyme levels may be elevated in acute myelomonocytic leukemia (FAB-M4), chronic myelomonocytic leukemia (CMML) and chronic myelocytic leukemia[113] (CML). Increased serum lysozyme activity is also present in tuberculosis,[114] sarcoidosis, megaloblastic anemias,[115] acute bacterial infections, ulcerative colitis[116] and Crohn disease.[114,116] Elevated levels of urine and serum lysozyme occur during severe renal insufficiency, renal transplant rejection,[117] urinary tract infections,[118] glomerulonephritis and nephrosis.[119]

We have developed a simple and sensitive fluorescence-based EnzChek® assay that can continuously measure lysozyme activity in solution. This assay is based on a DQ™ lysozyme substrate comprising *M. lysodeikticus* cell walls that are labeled with fluorescein to such a degree that fluorescence is quenched. Lysozyme action relieves the quenching, yielding a dramatic increase in fluorescence that is proportional to lysozyme activity. This increase in fluorescence can be measured with any spectrofluorometer, mini-fluorometer or fluorescence microplate reader that can detect fluorescein (excitation/emission maxima ~490/525 nm). Our fluorescence-based EnzChek® Lysozyme Assay Kit (E22013) permits the detection of as little as 30 U/mL of lysozyme (Figure 10.2.21). One unit of lysozyme is the quantity of enzyme that produces a decrease in turbidity of 0.001 optical density units per minute at 450 nm measured at pH 7.0 (25°C) using a 0.3 mg/mL suspension of *Micrococcus lysodeikticus* cells as substrate.[120]

IMPORTANT NOTICE: The products described in this manual are covered by one or more Limited Use Label License(s). Please refer to the Appendix on page 971 and Master Product List on page 975. Products are For Research Use Only. Not intended for any animal or human therapeutic or diagnostic use.

The EnzChek® Lysozyme Assay Kit (E22013) contains:

- DQ™ lysozyme substrate—a heavily labeled, highly quenched fluorescein conjugate of *Micrococcus lysodeikticus*
- Reaction buffer
- Lysozyme from chicken egg white, for use as a positive control
- Detailed protocols

Each kit contains sufficient materials for approximately 400 assays of 100 µL in a fluorescence microplate reader.

Chromogenic Glycosidase Substrates

The widely used β-galactosidase substrate—5-bromo-4-chloro-3-indolyl β-D-galactopyranoside (X-Gal, B1690)—yields a dark blue precipitate at the site of enzymatic activity. X-Gal is useful for numerous histochemical and molecular biology applications, including detection of *lacZ* activity in cells and tissues. In contrast to β-glucuronidase as a gene marker, β-galactosidase can be fixed in cells and tissues with glutaraldehyde without loss of activity and detected with high resolution with X-Gal.[121]

The chromogenic substrate 5-bromo-4-chloro-3-indolyl β-D-glucuronic acid (X-GlcU, B1691) forms a dark blue precipitate. X-GlcU is routinely used to detect GUS expression in transformed plant cells and tissues.[10,122–128] X-GlcU can also be used to detect *E. coli* contamination in food and water.[129,130]

Auxiliary Products for Glycosidase Research

Phenylethyl β-D-Thiogalactopyranoside (PETG)

Phenylethyl β-D-thiogalactopyranoside (PETG, P1692) is a cell-permeant inhibitor of β-galactosidase activity.[33,131] We provide PETG in our FluoReporter®, DetectaGene™ Green, ImaGene Green™ and ImaGene Red™ *lacZ* Gene Expression Kits for stopping the enzymatic reaction.

Streptavidin Conjugate of β-Galactosidase

We also offer the streptavidin conjugate of β-galactosidase (S931), a reagent used in a variety of ELISAs.[132] β-D-Galactosidase streptavidin reportedly provided enhanced sensitivity over that obtained with the avidin conjugate of horseradish peroxidase in the detection of a variety of mammalian interleukins and their receptors by ELISA.[133] This reagent has also been used in fluorometric-reverse (IgE-capture)[134] and fluorescence-sandwich[135] ELISAs.

Rabbit Anti–β-Galactosidase Antibody

We offer a polyclonal antibody to the widely used reporter gene product, β-galactosidase. Our rabbit anti–β-galactosidase antibody (A11132) is raised against *E. coli*–derived β-galactosidase and demonstrates high selectivity for the enzyme. Whether it is being used as a reporter gene or to generate fusion proteins, anti–β-galactosidase provides an easy tool for detecting the enzyme. The antibody is suited to a variety of techniques, including immunoblotting, ELISA, immunoprecipitation and most immunological methods. β-Galactosidase has been used as a tag for quantitative detection of molecules expressed on a cell surface in unfixed, live cells, using anti–β-galactosidase and a β-galactosidase substrate for detection.[136] This novel "cell-ELISA" technique is reported to be applicable to adherent cells and nonadherent cells and to have utility for large-scale screening for expression of cell-surface molecules and of hybridomas for production of antibodies to cell-surface epitopes.

Rabbit Anti–β-Glucuronidase Antibody

In combination with a fluorophore- or enzyme-labeled anti–rabbit IgG secondary antibody (Section 7.2, Table 7.1), our anti–β-glucuronidase antibody (A5790) can be used to detect the GUS enzyme in transformed plant tissue[10,137,138] and in transfected animal cells[139] using western blotting or immunohistochemical techniques. Furthermore, this antibody, which is raised in rabbits against *E. coli*–type X-A β-glucuronidase, can be immobilized in microplate wells in order to capture the GUS enzyme from cell lysates.[140] The enzymatic activity can subsequently be determined using any of our fluorogenic or chromogenic β-glucuronidase substrates.[141]

Related Products for Carbohydrate Research

We offer an extensive assortment of reagents for detection and analysis of carbohydrates that are described in other sections of this *Handbook*. These products include:

- Hydrazine, hydroxylamine and aromatic amine reagents for derivatization and analysis of reducing sugars and periodate-oxidized carbohydrates by electrophoretic methods (Section 3.3, Table 3. 2)
- Lectins and fluorescent lectin conjugates (Section 7.7, Table 7.10)
- Pro-Q® Glycoprotein Blot and Gel Stain Kits (Section 9.4)
- Pro-Q® Emerald 300 Lipopolysaccharide Gel Stain Kit (P20495, Section 3.3)
- Fluorescent lipopolysaccharides (Section 16.1, Table 16.1)
- Fluorescent glycolipids, including BODIPY® FL C_5-ganglioside G_{M1} (B13950, Section 13.3)
- Fluorescent and biotinylated dextrans (Section 14.5, Table 14.4)
- NBD-glucosamine derivatives for glucose-transport studies (N13195, N23106; Section 15.2)
- Fluorescein insulin (I13269, Section 16.1)

REFERENCES

1. J Clin Microbiol (1992) 30:1402; **2.** Microbiol Rev (1991) 55:335; **3.** Meth Microbiol (1987) 19:105; **4.** Nature (1994) 369:113; **5.** J Cell Biol (1990) 110:309; **6.** Anal Biochem (1985) 148:50; **7.** Traffic (2000) 1:836; **8.** Clin Chim Acta (1992) 205:87; **9.** Crit Rev Biochem Mol Biol (1990) 25:385; **10.** Plant Physiol(2002) 129:333; **11.** Mol Gen Genet (1989) 216:321; **12.** Mol Gen Genet (1988) 215:38; **13.** Plant Mol Biol (1988) 10:387; **14.** Proc Natl Acad Sci U S A (1986) 83:8447; **15.** J Immunol Methods (1992) 150:5; **16.** Anal Biochem (1998) 257:234; **17.** J Immunol Methods (1992) 150:23; **18.** J Immunol Methods (1982) 48:133; **19.** Exp Parasitol (1991) 73:440; **20.** Dev Biol (1994) 161:77; **21.** Mol Microbiol (1994) 13:655; **22.** J Neurosci (1993) 13:1418; **23.** Eur J Immunol (1989) 19:1619; **24.** Cytometry (1994) 17:216; **25.** Eur J Cell Biol (1993) 62:324; **26.** J Biol Chem (1993) 268:9762; **27.** Neuron (1992) 9:1117; **28.** Science (1991) 251:81; **29.** Proc Natl Acad Sci U S A (1988) 85:2603; **30.** J Virol Methods (1993) 44:99; **31.** Proc Natl Acad Sci U S A (1993) 90:8194; **32.** Appl Environ Microbiol (1990) 56:3861; **33.** Carbohydr Res (1977) 56:153; **34.** Methods (1991) 2:261; **35.** Methods

continued on next page

IMPORTANT NOTICE: The products described in this manual are covered by one or more Limited Use Label License(s). Please refer to the Appendix on page 971 and Master Product List on page 975. Products are For Research Use Only. Not intended for any animal or human therapeutic or diagnostic use.

REFERENCES—*continued*

(1991) 2:248; **36.** Anal Chim Acta (1984) 163:67; **37.** Oncogene (1995) 10:2323; **38.** Cytometry (1988) 9:394; **39.** Ann N Y Acad Sci (1990) 613:333; **40.** Anal Chim Acta (1988) 213:245; **41.** Anal Biochem (2009) 389:59; **42.** Neuron (1993) 10:427; **43.** Proc Natl Acad Sci U S A (1987) 84:156; **44.** J Histochem Cytochem (1985) 33:965; **45.** Clin Chim Acta (1965) 12:647; **46.** Biotechniques (2001) 30:776; **47.** Infect Immun (1993) 61:5231; **48.** Anal Biochem (1985) 146:211; **49.** Proc Natl Acad Sci U S A (1998) 95:6373; **50.** J Biochem (Tokyo) (1986) 100:707; **51.** J Fluorescence (1993) 3:119; **52.** Development (1993) 119:1217; **53.** Neuron (1995) 14:685; **54.** J Neurosci (1995) 15:1025; **55.** Exp Cell Res (1995) 219:372; **56.** Science (1994) 266:1881; **57.** Cytometry (1997) 28:36; **58.** J Histochem Cytochem (1996) 44:1363; **59.** Proc Natl Acad Sci U S A (2006) 103:13813; **60.** FASEB J (1991) 5:3108; **61.** Appl Environ Microbiol (1994) 60:4638; **62.** Nat Protoc (2009) 4: 1798; **63.** J Clin Endocrinol Metab (1994) 78:1232; **64.** Biotechnol Bioeng (1993) 42:708; **65.** J Biol Chem (1996) 271:29312; **66.** Proc Natl Acad Sci U S A (1992) 89:10681; **67.** Methods Cell Biol (1994) 42:563; **68.** Biotechniques (1993) 14:274; **69.** Genes Dev (1992) 6:591; **70.** FEMS Microbiol Lett (1999) 179:317; **71.** J Immunol Methods (1997) 202:133; **72.** Biotechniques (1990) 8:38; **73.** Plant Mol Biol Rep (1988) 5:387; **74.** Gene Ther (2007) 14:565; **75.** Cell Signal (1991) 3:625; **76.** J Immunol Methods (1987) 100:211; **77.** Cytometry (1996) 24:321; **78.** J Biol Chem (1999) 274:657; **79.** J Assoc Off Anal Chem (1988) 71:589; **80.** Appl Environ Microbiol (1985) 50:1383; **81.** Can J Microbiol (1993) 39:1066; **82.** J Microbiol Methods (1990) 12:51; **83.** J Microbiol Methods (1990) 12:235; **84.** Biotechniques (1994) 16:888; **85.** J Biol Chem (2001) 276:9855; **86.** Plant Sci (1991) 78:73; **87.** Plant Mol Biol (1990) 15:527; **88.** Plant Mol Biol Rep (1992) 10:37; **89.** J Biochem Biophys Methods (1996) 33:197; **90.** Plant J (1996) 10:745; **91.** Microbiology (1997) 143:267; **92.** Cell Biochem Funct (1993) 11:167; **93.** Neurochem Res (1999) 24:301; **94.** Appl Biochem Biotechnol (2010) 161:301; **95.** Yeast (1994) 10:747; **96.** Anal Biochem (1997) 247:268; **97.** Blood (1997) 89:3412; **98.** Anal Biochem (1997) 253:162; **99.** Anal Biochem (1998) 260:257; **100.** J Microbiol Methods (2009) 79:174; **101.** J Biochem Biophys Methods (1991) 22:23; **102.** Nat Biotechnol (2000) 18:835; **103.** Antiviral Res (2000) 47:1; **104.** J Med Chem (1970) 13:697; **105.** Anal Biochem (2000) 280:291; **106.** Clin Chem (1976) 22:57; **107.** Medicine (Baltimore) (1976) 55:269; **108.** Clin Chem (1976) 21:57; **109.** Clin Chem (1984) 30:387; **110.** J Lipid Res (2008) 49:1588; **111.** Curr Opin Biotechnol (1996) 7:337; **112.** Anal Biochem (2007) 362:63; **113.** Mod Pathol (1994) 7:771; **114.** Acta Pathol Jpn (1978) 28:689; **115.** N Engl J Med (1967) 277:10; **116.** J Clin Pathol (1983) 36:1312; **117.** Clin Chem (1986) 32:1807; **118.** Toxicology (1983) 28:347; **119.** Nephron (1993) 63:423; **120.** Biochim Biophys Acta (1952) 8:302; **121.** EMBO J (1989) 8:343; **122.** Biotechniques (1995) 19:106; **123.** Biotechnology (N Y) (1990) 8:833; **124.** Plant Cell Physiol (1990) 31:805; **125.** Science (1990) 249:1285; **126.** Science (1990) 248:471; **127.** Nature (1989) 342:837; **128.** EMBO J (1987) 6:3901; **129.** J Appl Bacteriol (1993) 74:223; **130.** Lett Appl Microbiol (1991) 13:212; **131.** Anal Biochem (1991) 199:119; **132.** J Immunol Methods (1989) 125:279; **133.** Biochemistry (1987) 26:4922; **134.** J Immunol Methods (1989) 116:181; **135.** J Immunol Methods (1988) 110:129; **136.** J Immunol Methods (2000) 234:153; **137.** J Biol Chem (1994) 269:17635; **138.** Plant Mol Biol (1990) 15:821; **139.** Biotechniques (1999) 27:896; **140.** J Clin Microbiol (1994) 32:1444; **141.** Appl Environ Microbiol (1987) 53:1073.

DATA TABLE 10.2 DETECTING GLYCOSIDASES

Cat. No.	MW	Storage	Soluble	Abs	EC	Em	Solvent	Product	Notes
A22010	713.65	F,D	DMSO	289	5500	none	MeOH	see Notes	1
B1690	408.63	F,D	DMSO	290	4900	none	H_2O	see Notes	2
B1691	521.79	F,D	pH >6	292	4800	none	MeOH	see Notes	2
B34402	~66,000	F,D,L	H_2O	505	80,000	511	MeOH	see Notes	3, 4, 5
D2893	853.92	F,D	DMSO	289	6000	none	MeOH	D109	
D2920	705.07	F,L	see Notes	273	4800	none	MeOH	see Notes	6, 7, 8
D6488	470.31	F,D,L	DMSO	465	24,000	608	pH 7	H6482	
D7519	861.96	FF,D,L	DMSO, EtOH	505	85,000	511	MeOH	see Notes	3, 4
D13951	925.91	FF,D,L	DMSO, EtOH	505	80,000	511	MeOH	see Notes	3, 4
E6587	483.26	F,D,L	DMSO, H_2O	302	14,000	see Notes	MeOH	E6578	9
F1179	656.60	F,D	DMSO	273	6500	none	MeOH	F1300	10
F2881	656.60	F,D	DMSO	272	6200	none	MeOH	F1300	
F2905	368.30	F,D,L	pH 7	330	16,000	396	pH 8	H185	8, 9, 11
F2915	684.56	F,D	pH >6, DMSO	272	5700	none	MeOH	F1300	
I2904	853.92	F,D,L	DMSO	289	6000	none	MeOH	D109	8, 11
I2906	543.66	F,D,L	see Notes	448	20,000	none	MeOH	see Notes	8, 12, 13
I2908	881.88	F,D	see Notes	290	5400	none	MeOH	D109	6, 8
M1489MP	338.31	D	DMSO, H_2O	316	14,000	376	pH 9	H189	9
M1490	352.30	F,D	pH >6	316	12,000	375	pH 9	H189	9
P1692	300.37	F,D	DMSO, H_2O	<300		none			
P11947	865.67	F,D	DMSO	260	26,000	none	MeOH	P12925	
P11948	865.67	F,D	DMSO	260	25,000	none	MeOH	P12925	
P11949	893.64	F,D	pH >6, DMSO	260	21,000	none	MeOH	P12925	
R1159	375.33	F,D,L	DMSO	469	19,000	none	pH 9	R363	

For definitions of the contents of this data table, see "Using *The Molecular Probes® Handbook*" in the introductory pages.

Notes

1. Enzymatic cleavage of this substrate yields a 5-acylaminofluorescein derivative with spectroscopic properties similar to D109 .
2. Enzymatic cleavage of this substrate yields a water-insoluble, blue-colored indigo dye (Abs ~615 nm).
3. Em for BODIPY® FL sphingolipids shifts to ~620 nm when high concentrations of the probe (>5 mol %) are incorporated in lipid mixtures. (J Cell Biol (1991) 113:1267)
4. Enzymatic cleavage of this product yields a fluorescent ceramide or glycosylceramide with unchanged spectral properties.
5. This product is a lipid complexed with bovine serum albumin (BSA). Spectroscopic data are for the free lipid in MeOH.
6. This product is packaged as a solution in 1:1 (v/v) DMSO/H_2O.
7. Enzymatic cleavage of this substrate yields 5-chloromethylfluorescein, with spectroscopic properties similar to C1904.
8. Data represent the substrate component of this kit.
9. Fluorescence of the unhydrolyzed substrate is very weak.
10. F1179 is soluble at 1 mM in water, but it is best to prepare a stock solution in DMSO.
11. This product is supplied as a ready-made solution in the solvent indicated under "Soluble."
12. This product is packaged as a solution in 7:3 (v/v) DMSO/EtOH.
13. Enzymatic cleavage of this substrate yields 2-dodecylresorufin, Abs = 578 nm (EC = 69,000 $cm^{-1}M^{-1}$), Em = 597 nm in MeOH.

IMPORTANT NOTICE: The products described in this manual are covered by one or more Limited Use Label License(s). Please refer to the Appendix on page 971 and Master Product List on page 975. Products are For Research Use Only. Not intended for any animal or human therapeutic or diagnostic use.

PRODUCT LIST 10.2 DETECTING GLYCOSIDASES

Cat. No.	Product	Quantity
A22010	5-acetylaminofluorescein di-β-D-galactopyranoside (C_2FDG)	5 mg
A22179	Amplex® Red Galactose/Galactose Oxidase Assay Kit *400 assays*	1 kit
A22189	Amplex® Red Glucose/Glucose Oxidase Assay Kit *500 assays*	1 kit
A22178	Amplex® Red Neuraminidase (Sialidase) Assay Kit *400 assays*	1 kit
A11132	anti-β-galactosidase, rabbit IgG fraction *2 mg/mL*	0.5 mL
A5790	anti-β-glucuronidase, rabbit IgG fraction *2 mg/mL*	0.5 mL
B34402	BODIPY® FL C_5-lactosylceramide complexed to BSA	1 mg
B1690	5-bromo-4-chloro-3-indolyl β-D-galactopyranoside (X-Gal)	1 g
B1691	5-bromo-4-chloro-3-indolyl β-D-glucuronide, cyclohexylammonium salt (X-GlcU, CHA)	100 mg
D2920	DetectaGene™ Green CMFDG *lacZ* Gene Expression Kit	1 kit
D6488	*9H*-(1,3-dichloro-9,9-dimethylacridin-2-one-7-yl) β-D-galactopyranoside (DDAO galactoside)	5 mg
D7519	*N*-(4,4-difluoro-5,7-dimethyl-4-bora-3a,4a-diaza-*s*-indacene-3-dodecanoyl)sphingosyl 1-β-D-galactopyranoside (BODIPY® FL C_{12}-galactocerebroside)	25 µg
D13951	*N*-(4,4-difluoro-5,7-dimethyl-4-bora-3a,4a-diaza-*s*-indacene-3-pentanoyl)sphingosyl 1-β-D-lactoside (BODIPY® FL C_5-lactosylceramide)	25 µg
D2893	5-dodecanoylaminofluorescein di-β-D-galactopyranoside ($C_{12}FDG$)	5 mg
E33953	EnzChek® cellulase substrate *blue fluorescent, 339/452*	1 mg
E22013	EnzChek® Lysozyme Assay Kit *400 assays*	1 kit
E33651	EnzChek® Ultra Amylase Assay Kit *500 assays*	1 kit
E33650	EnzChek® Ultra Xylanase Assay Kit *500 assays*	1 kit
F1930	FluoReporter® *lacZ* Flow Cytometry Kit *50 assays*	1 kit
F2905	FluoReporter® *lacZ*/Galactosidase Quantitation Kit *1000 assays*	1 kit
F1179	fluorescein di-β-D-galactopyranoside (FDG)	5 mg
F2881	fluorescein di-β-D-glucopyranoside (FDGlu)	5 mg
F2915	fluorescein di-β-D-glucuronide (FDGlcU)	5 mg
I2904	ImaGene Green™ $C_{12}FDG$ *lacZ* Gene Expression Kit	1 kit
I2908	ImaGene Green™ $C_{12}FDGlcU$ GUS Gene Expression Kit	1 kit
I2906	ImaGene Red™ $C_{12}RG$ *lacZ* Gene Expression Kit	1 kit
M1489MP	4-methylumbelliferyl β-D-galactopyranoside (MUG)	1 g
M1490	4-methylumbelliferyl β-D-glucuronide (MUGlcU)	100 mg
N10563	NovaBright™ β-galactosidase Enzyme Reporter Gene Chemiluminescent Detection Kit *for mammalian cells* *200 assays*	1 kit
N10564	NovaBright™ β-galactosidase Enzyme Reporter Gene Chemiluminescent Detection Kit *for mammalian cells* *1000 assays*	1 kit
N10565	NovaBright™ β-galactosidase Enzyme Reporter Gene Chemiluminescent Detection Kit *for yeast cells* *200 assays*	1 kit
N10566	NovaBright™ β-galactosidase Enzyme Reporter Gene Chemiluminescent Detection Kit *for yeast cells* *1000 assays*	1 kit
P11948	5-(pentafluorobenzoylamino)fluorescein di-β-D-galactopyranoside (PFB-FDG)	5 mg
P11947	5-(pentafluorobenzoylamino)fluorescein di-β-D-glucopyranoside (PFB-FDGlu)	5 mg
P11949	5-(pentafluorobenzoylamino)fluorescein di-β-D-glucuronide (PFB-FDGlcU)	5 mg
P1692	phenylethyl β-D-thiogalactopyranoside (PETG)	10 mg
R1159	resorufin β-D-galactopyranoside	25 mg
S931	streptavidin, β-galactosidase conjugate	1 mg

10.3 Detecting Enzymes That Metabolize Phosphates and Polyphosphates

Cells utilize a wide variety of phosphate and polyphosphate esters as enzyme substrates, second messengers, membrane structural components and vital energy reservoirs. This section includes an assortment of reagents and methods for detecting the metabolism of phosphate esters. Our diverse array of fluorogenic, chemiluminescent and chromogenic substrates include substrates for phosphatases, as well as reagents to measure the activity of enzymes such as ATPases, GTPases, DNA and RNA polymerases and paraoxonase. Unlike phosphatases, which hydrolyze phosphate esters, kinases use ATP to phosphorylate their target. Our Antibody Beacon™ Tyrosine Kinase Assay Kit (A35725, described below) provides a general and high-throughput solution assay for measuring the activity of tyrosine kinases using unlabeled peptides. In addition to these reagents, we offer several nucleotide analogs and substrates for phosphodiesterases and phospholipases that are described in Section 17.3 and Section 17.4, respectively.

By far the largest group of chromogenic and fluorogenic substrates for phosphate-ester metabolizing enzymes are those for simple phosphatases such as alkaline and acid phosphatase, both of which hydrolyze phosphate monoesters to an alcohol and inorganic phosphate. Conjugates of calf intestinal alkaline phosphatase are extensively used as secondary detection reagents in ELISAs,[1] immunohistochemical techniques[2] and Southern, northern and western blot analyses. In addition, phosphatases serve as enzyme markers, allowing researchers to identify primordial germ cells,[3] to distinguish subpopulations of bone marrow stromal cells[4] and to investigate *in vitro* differentiation in carcinoma cell lines.[5–7] *PALP-1*, the gene for human placental alkaline phosphatase, has been used as a eukaryotic reporter gene that is superior to *lacZ* for lineage studies in murine retina.[8,9] This gene has also been engineered to produce a secreted alkaline phosphatase (SEAP), allowing quantitation of gene expression without disrupting the cells.[10]

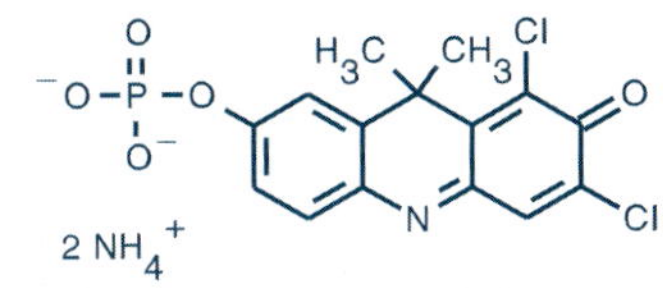

Figure 10.3.1 *9H*-(1,3-dichloro-9,9-dimethylacridin-2-one-7-yl) phosphate, diammonium salt (DDAO phosphate, D6487).

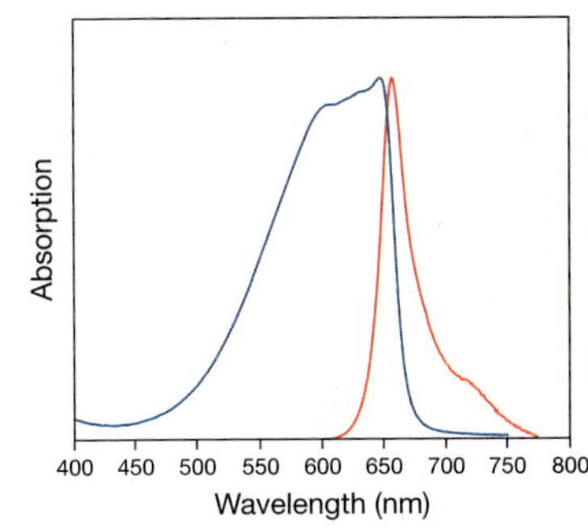

Figure 10.3.2 Normalized absorption and fluorescence emission spectra of DDAO, which is formed by alkaline phosphatase–mediated hydrolysis of DDAO phosphate (D6487).

Phosphatase Substrates Yielding Soluble Fluorescent Products

Fluorescein Diphosphate

First described in 1963,[11] fluorescein diphosphate (FDP, F2999) is perhaps the most sensitive fluorogenic phosphatase substrate available. The colorless and nonfluorescent FDP reagent is hydrolyzed to fluorescein (F1300, F36915; Section 10.1), which exhibits superior spectral properties (EC ~90,000 $cm^{-1}M^{-1}$, quantum yield ~0.92). We have succeeded in preparing a highly purified FDP and find it to be an excellent substrate for alkaline phosphatase in ELISAs, providing detection limits at least 50 times lower than those obtained with the chromogenic 4-nitrophenyl phosphate.[12] The relatively high pH required to monitor alkaline phosphatase activity is advantageous because it also enhances fluorescein's fluorescence. FDP has been used for a diverse set of applications, including:

- Assaying alkaline phosphatase in a microfluidic device[13]
- Assaying diarrheic shellfish toxins in a microplate assay[14]
- Detecting alkaline phosphatase immobilized on microspheres in an optical sensor array[15]
- Measuring endogenous phosphatase in an assay for cell adhesion and migration that is reported to be as sensitive as ^{51}Cr-release assays[16]
- Quantitating the effect of inhibitors on tyrosine phosphatases[17–19]

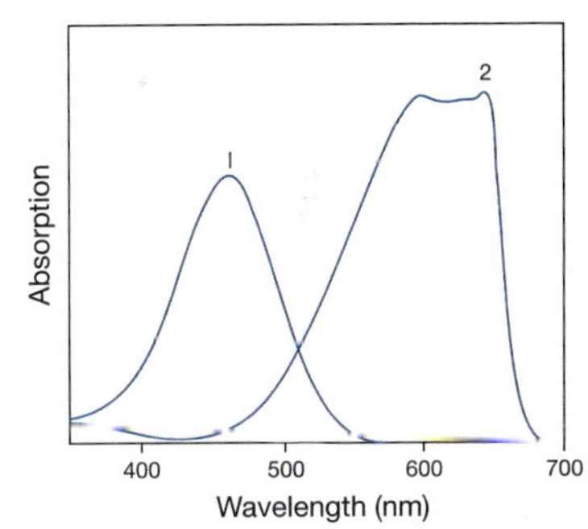

Figure 10.3.3 Absorption spectra of 1) DDAO galactoside (D6488) and 2) DDAO (H6482) at equal concentrations in pH 9 aqueous buffer. These spectra show the large spectral shift accompanying enzymatic cleavage of DDAO-based substrates. DDAO phosphate (D6487) has very similar spectra to DDAO galactoside.

Dimethylacridinone (DDAO) Phosphate

Our phosphatase substrate DDAO phosphate (9H-(1,3-dichloro-9,9-dimethylacridin-2-one-7-yl) phosphate, D6487; Figure 10.3.1) yields a hydrolysis product that is efficiently excited by the 633 nm spectral line of the He-Ne laser to produce bright red fluorescence with absorption/emission maxima of ~646/659 nm (Figure 10.3.2). Although the substrate itself is fluorescent, the difference between the substrate's excitation maximum and that of the phenolic hydrolysis product is over 200 nm (Figure 10.3.3), allowing the two species to be easily distinguished. Like other DDAO-based substrates, DDAO phosphate has good water solubility, a low K_M and a high turnover rate, making it particularly useful for both fluorescence- and absorption-based microplate assays.[20]

Methylumbelliferyl Phosphate (MUP) and Difluorinated Methylumbelliferyl Phosphate (DiFMUP)

We offer 4-methylumbelliferyl phosphate (MUP, M6491), a widely used fluorogenic substrate for alkaline phosphatase detection. MUP has been used for a variety of ELISA protocols[21] in which the relatively high pH optimum of alkaline phosphatase permits continuous detection of the rate of formation of 4-methylumbelliferone (7-hydroxy-4-methylcoumarin, H189, Section 10.1). MUP has also been used to count cells based on their alkaline phosphatase activity,[22] to detect PCR amplification products[23,24] and to identify and characterize bacteria.[25,26]

We also offer 6,8-difluoro-4-methylumbelliferyl phosphate (DiFMUP, D6567, D22065; Figure 10.3.4), which exhibits extraordinary spectral properties that are advantageous for the assay of both acid and alkaline phosphatase activity and is probably the best general substrate available for measuring the activity of the protein phosphatases that are important for high-throughput screening applications under physiological conditions.[27] The hydrolysis product of DiFMUP—6,8-difluoro-4-methylumbelliferone (6,8-difluoro-7-hydroxy-4-methylcoumarin, D6566; Section 10.1)—exhibits both a lower pK_a (4.9 versus 7.8, Figure 10.3.5) and a higher fluorescence quantum yield (0.89 versus 0.63) than the hydrolysis product of MUP.[28] The low pK_a of its hydrolysis product makes DiFMUP a sensitive substrate for the continuous assay of acid phosphatases (Figure 10.3.6, Figure 10.3.7), which is not possible with MUP because its fluorescence must be measured at alkaline pH (pH >9) if the conventional detection configuration (excitation/emission = 360/460 nm) is used. Furthermore, with its high fluorescence quantum yield, DiFMUP increases the sensitivity of both acid and alkaline phosphatase measurements. As with our fluorinated fluorescein derivatives (Oregon Green® dyes, Section 1.5), fluorination renders the methylumbelliferone fluorophore much less susceptible to photobleaching, yet does not significantly affect the extinction coefficient or excitation/emission maxima. DiFMUP is available as a single 5 mg vial (D6567) or as a set of 10 vials, each containing 10 mg of the substrate (D22065) for high-throughput screening applications. DiFMUP is also used in our EnzChek® Phosphatase Detection Kit (E12020), our RediPlate™ 96 EnzChek® Tyrosine Phosphatase Assay Kits (R22067) and our RediPlate™ 96 EnzChek® Serine/Threonine Phosphatase Assay Kit (R33700), which are described below with our other phosphatase assay kits.

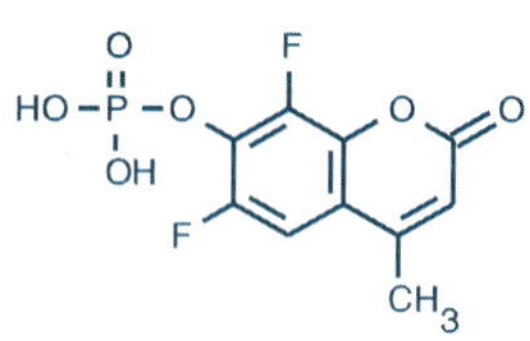

Figure 10.3.4 6,8-Difluoro-4-methylumbelliferyl phosphate (DiFMUP, D6567).

Coumarin-Based Probe for Detection of Inorganic Phosphate

When conjugated to a mutant phosphate-binding protein, 7-diethylamino-3-((((2-maleimidyl)ethyl)amino)carbonyl)coumarin (MDCC, D10253; Section 2.3) has proven useful for direct, real-time measurement of inorganic phosphate release during enzymatic reactions.[29–33] A similar (7-diethylamino-3-((((2-iodoacetamido)ethyl)amino)carbonyl) coumarin (IDCC) conjugate of a mutant nucleoside diphosphate kinase has been used as a fluorescent sensor of the phosphorylation state of the kinase and to monitor purine nucleoside diphosphate concentrations in real time.[34]

ELF® 97 Phosphate: A Phosphatase Substrate That Yields a Fluorescent Precipitate

ELF® 97 phosphate (E6588, E6589) yields a fluorescent precipitate at the site of phosphatase activity (Figure 10.3.8)—a process we call Enzyme-Labeled Fluorescence (ELF®). Upon enzymatic cleavage (Figure 10.3.9), this weakly blue-fluorescent substrate yields an extremely photostable green-fluorescent precipitate that is up to 40 times brighter than the signal achieved when using either directly labeled fluorescent hybridization probes or fluorescent secondary detection methods in comparable applications.[35–37] A particularly prominent application of ELF® 97 phosphate is the detection of tartrate-resistant acid

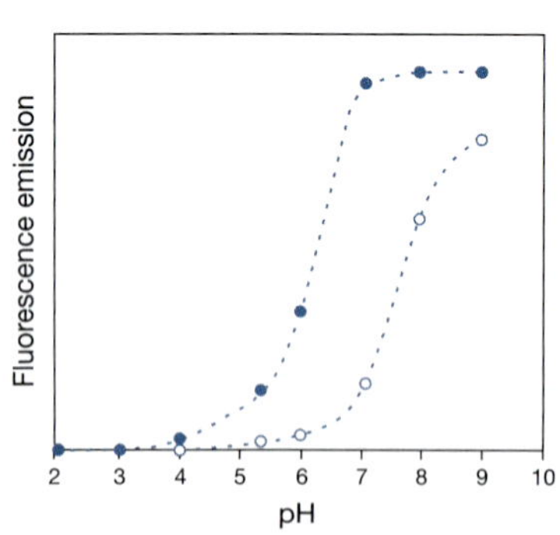

Figure 10.3.5 Comparison of the pH-dependent fluorescence changes produced by attachment of electron-withdrawing fluorine atoms to a hydroxycoumarin. 7-Hydroxy-4-methylcoumarin-3-acetic acid (○, H1428) and 6,8-difluoro-7-hydroxy-4-methylcoumarin (●, D6566) demonstrate the decrease of the pK_a from ~7.4 to ~6.2. Fluorescence intensities were measured for equal concentrations of the two dyes using excitation/emission at 360/450 nm.

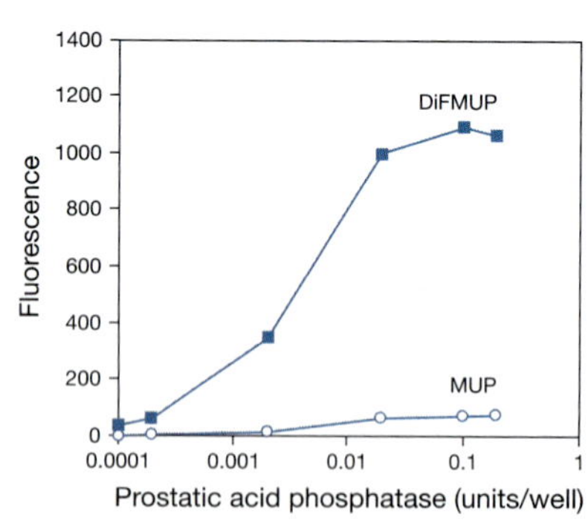

Figure 10.3.6 Comparison of DiFMUP (D6567, D22065) with MUP (M6491) for the detection of acid phosphatase activity at pH 5.5. Increasing amounts of prostatic acid phosphatase from human semen were reacted with 100 µM DiFMUP, the substrate in the EnzChek® Acid Phosphatase Assay Kit (E12020), or 100 µM MUP, in 100 mM sodium acetate, pH 5.5, at room temperature. Fluorescence was measured after 60 minutes in a fluorescence microplate reader using excitation at 360 ± 20 nm and emission detection at 460 ± 20 nm.

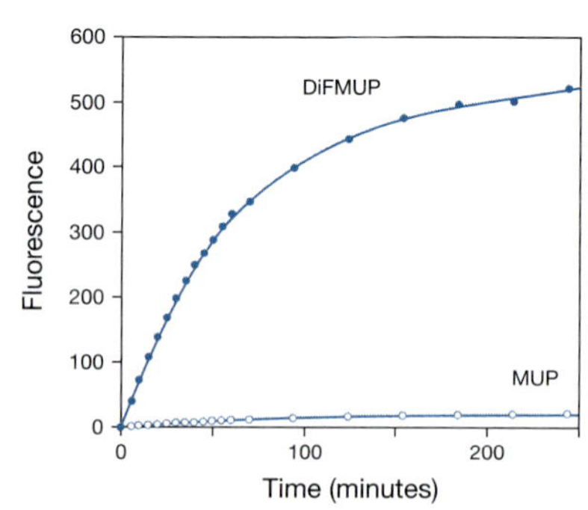

Figure 10.3.7 Time course of the reaction of prostatic acid phosphatase with DiFMUP (D6567, D22065) and MUP (M6491). Prostatic acid phosphatase from human semen (0.002 units) was reacted with 100 µM DiFMUP or 100 µM MUP in 100 mM sodium acetate, pH 5.5, at room temperature. Fluorescence was measured at the indicated times in a fluorescence microplate reader using excitation at 360 ± 20 nm and emission detection at 460 ± 20 nm.

IMPORTANT NOTICE: The products described in this manual are covered by one or more Limited Use Label License(s). Please refer to the Appendix on page 971 and Master Product List on page 975. Products are For Research Use Only. Not intended for any animal or human therapeutic or diagnostic use.

phosphatase (TRAP) activity, a marker of osteoclast differentiation from hematopoietic stem cells[38–41] (HSC). Using ELF® 97 phosphate and confocal laser-scanning microscopy, researchers have developed a semiautomated method for analyzing the position within a regenerating newt limb of transfected cells expressing the secreted alkaline phosphatase reporter gene.[42]

Kits based on our ELF® 97 phosphate include:

- ELF® 97 mRNA *In Situ* Hybridization Kits (E6604, E6605)
- ELF® 97 Cytological Labeling Kit (E6603)
- ELF® 97 Immunohistochemistry Kit (E6600)
- ELF® 97 Endogenous Phosphatase Detection Kit (E6601)

These kits and their contents are described in detail in Section 6.3. ELF® 97 phosphate is also available separately as a 5 mM solution in water containing sodium azide (E6589) or in a 0.2 µm–filtered solution (E6588). Filtration of ELF® 97 phosphate through an ELF® spin filter (E6606) is recommended before use. The ELF® 97 alcohol (Figure 10.3.10) is available as a concentrated solution in DMSO (E6578, Section 10.1).

BODIPY® FL ATP-γ-*S* and BODIPY® FL GTP-γ-*S* Thioesters: Substrates for the Fhit Nucleotide-Binding Protein

BODIPY® FL ATP-γ-*S* (A22184) and BODIPY® FL GTP-γ-*S* (G22183) thioesters are important substrates for Fhit (Figure 10.3.11), a member of the histidine triad superfamily of nucleotide-binding proteins that binds and cleaves diadenosine polyphosphates.[43–45] Fhit, one of the most

Figure 10.3.8 A transverse section of fixed zebrafish retina probed with FRet 43, a monoclonal antibody that binds to double cone cells, and developed for visualization with enzyme-mediated immunohistochemical techniques using our ELF® 97 Immunohistochemistry Kit (E6600). The yellow-green–fluorescent double cones stained with the ELF® 97 alcohol precipitate are oriented so that their outer segments are at the left of the stained configuration and the synaptic pedicles are at the right. This section has been counterstained with tetramethylrhodamine wheat germ agglutinin (W849), which makes the rod outer segments (bottom left) and the inner plexiform layer and ganglion cell axons (top right) appear bright red. Wheat germ agglutinin also binds in the region occupied by the cone outer segments and synaptic pedicles, which appear bright yellow because they are double-labeled with both the ELF® 97 alcohol precipitate and tetramethylrhodamine. Although this section has also been counterstained with Hoechst 33342 (H1399, H3570, H21492), the blue-stained nuclei are barely visible in this photomicrograph. However, the double cones' inner fibers traverse the region occupied by the Hoechst 33342 dye–stained rod nuclei, and thus appear light blue. The inner segments, myoids and nuclei of these double-cone cells are labeled only with the ELF® 97 alcohol precipitate, giving them a characteristic green appearance. The image was obtained by triple-exposure through optical filters appropriate for DAPI, tetramethylrhodamine and the ELF® precipitate. Used with permission from (J Histochem Cytochem (1995) 43:77.).

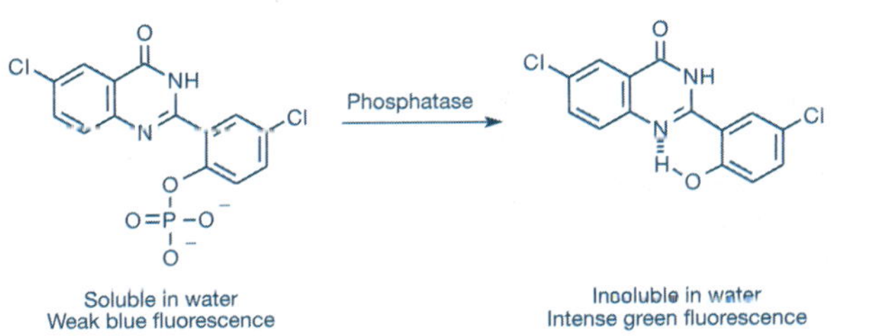

Figure 10.3.9 Principle of enzyme mediated formation of the fluorescent ELF® 97 alcohol precipitate from the ELF® 97 phosphatase substrate.

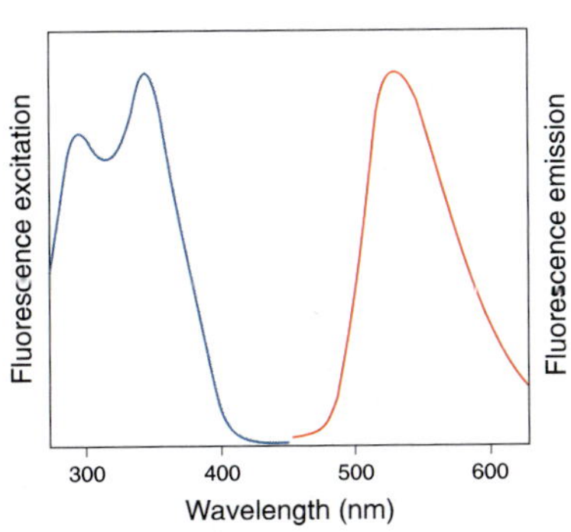

Figure 10.3.10 The normalized fluorescence excitation and emission spectra of the ELF® 97 alcohol precipitate (E6578), which is generated by enzymatic cleavage of the soluble ELF® 97 phosphatase substrate (E6588, E6589).

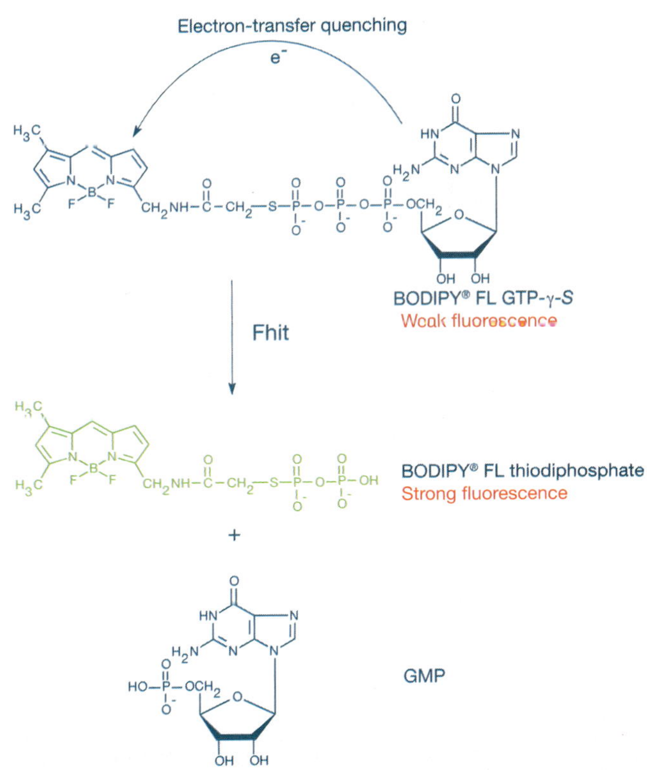

Figure 10.3.11 Principle of fluorescence-based detection of the diadenosine triphosphate hydrolase activity of Fhit using BODIPY® FL GTP-γ-*S* thioester (G22183) as a substrate analog.

IMPORTANT NOTICE: The products described in this manual are covered by one or more Limited Use Label License(s). Please refer to the Appendix on page 971 and Master Product List on page 975. Products are For Research Use Only. Not intended for any animal or human therapeutic or diagnostic use.

frequently inactivated proteins in lung cancer, functions as a tumor suppressor by inducing apoptosis.[44,46,47] In addition to their use for screening potential Fhit inhibitors, these BODIPY® FL nucleotides should also be important for studying other nucleotide-binding proteins because they bind to ATP- or GTP-binding sites but are not metabolized (Section 17.3). As with other fluorescent nucleotides, the fluorescence polarization properties of the bound nucleotide should differ from that of the free nucleotide. The fluorescence of BODIPY® FL GTP-γ-*S* thioester is quenched ~90% relative to that of the free dye (Figure 10.3.12) but is recovered upon binding to at least some G-proteins.[48,49]

Chemiluminescent Phosphatase Substrates

NovaBright™ Secreted Placental Alkaline Phosphatase (SEAP) Enzyme Reporter Gene Chemiluminescent Detection System

The NovaBright™ Secreted Placental Alkaline Phosphatase (SEAP) Enzyme Reporter Gene Chemiluminescent Detection System is a chemiluminescent reporter gene assay designed for the rapid and sensitive detection of secreted placental alkaline phosphatase (SEAP) in cell culture media. SEAP, a truncated form of human placental alkaline phosphatase (PLAP), is commonly used as a reporter protein that is secreted into the cell culture media and detected by testing aliquots of media, leaving cells intact for further experimentation. Detection of nonsecreted placental alkaline phosphatase is also possible using the NovaBright™ SEAP Enzyme Reporter Gene Chemiluminescent Detection System.

The NovaBright™ SEAP Enzyme Reporter Gene Chemiluminescent Detection System employs the CSPD® chemiluminescent substrate and Emerald luminescence enhancer to achieve sensitive SEAP detection with a wide dynamic range.[50,51] In this NovaBright™ assay, SEAP activity is measured 48–72 hours after cell transfection by first incubating the cell culture medium or cell lysate with a buffer that selectively inhibits nonplacental alkaline phosphatase (serum and endogeneous cellular alkaline phosphatases), and then with CSPD®-containing reaction buffer until maximum light emission is reached (approximately 20 minutes). The light emission kinetics provide a persistent glow signal that enables measurement over a wide time interval. Light signal output is measured in a luminometer, without the need for automated injection capability. The NovaBright™ SEAP Enzyme Reporter Gene Chemiluminescent Detection System has been used to detect SEAP activity in cell culture media and to quantitate nonsecreted placental alkaline phosphatase in both cell and tissue extracts.

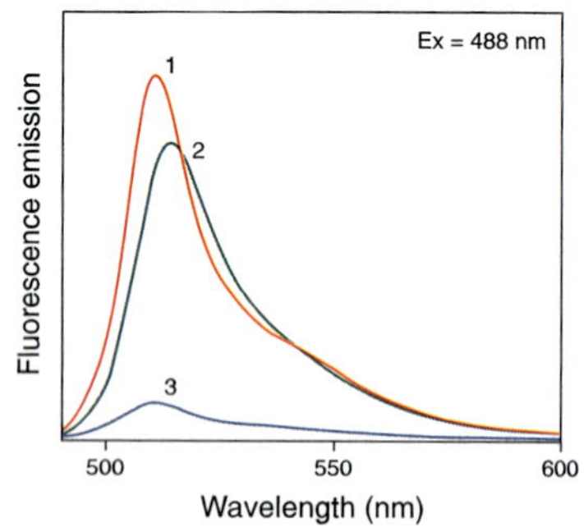

Figure 10.3.12 Fluorescence emission spectra of **1)** free BODIPY® FL dye in phosphate-buffered saline, pH 7.2; **2)** BODIPY® FL ATP (A12410); and **3)** BODIPY® FL GTP (G12411). Samples were prepared with equal absorbance at the excitation wavelength (488 nm). The areas under the curves are therefore proportional to the relative fluorescence quantum yields, clearly showing the quenching effect caused by interaction of the BODIPY® FL fluorophore with the guanine base of GTP.

Chemiluminescent reporter assays for secreted placental alkaline phosphatase may be conducted in cells that have endogenous nonplacental alkaline phosphatase activity. Endogenous nonplacental enzyme activity is significantly reduced with a combination of heat inactivation and differential inhibitors that do not significantly inhibit the transfected placental alkaline phosphatase. It is important to determine the level of endogenous enzyme in media of nontransfected cells in order to establish assay background. Certain cell lines, such as HeLa and others derived from cervical cancers, may express placental alkaline phosphatase, which will likely produce high assay backgrounds when shed into the media, and therefore we do not recommend the use of secreted alkaline phosphatase as a reporter system in these cell lines.

Each NovaBright™ SEAP Enzyme Reporter Gene Chemiluminescent Detection System provides:

- CSPD® substrate
- Purified human placental alkaline phosphatase
- Dilution buffer
- Assay buffer, containing a mixture of nonplacental alkaline phosphatase inhibitors
- Reaction buffer, containing the Emerald luminescence enhancer
- Detailed protocols

Sufficient reagents are supplied for 400 (N10559) or 1200 (N10560) microplate assays based on the protocol provided.

NovaBright™ Secreted Placental Alkaline Phosphatase (SEAP) Enzyme Reporter Gene Chemiluminescent Detection System 2.0

The next-generation NovaBright™ SEAP Enzyme Reporter Gene Chemiluminescent Detection System 2.0 employs CSPD® chemiluminescent substrate and an advanced Emerald-III™ luminescence enhancer that produces higher signal intensity and higher assay signal/noise performance than the original assay system. In addition, all the reagents in the new assay system are provided ready-to-use without the need for any reagent preparation steps, allowing fewer assay steps and a shorter assay duration. Depending on the expression system, secreted placental alkaline phosphatase can be measured in as little as 6 hours after cell transfection.

In this assay, cell culture medium is incubated first with an assay buffer that contains selective inhibitors of nonplacental alkaline phosphatase (serum and endogenous cellular alkaline phosphatases) and then with the reaction buffer containing CSPD® substrate and Emerald-III™ light emission enhancer until maximum light signal is reached in approximately 20 minutes. The light emission kinetics provide a persistent glow signal, with light emission half-life of approximately 2.5 hours, enabling measurement over a wide time interval. Light emission is measured in a luminometer without the need for automated injection capability. As with the original system, we do not recommend the use of this assay in cell lines such as HeLa and others derived from cervical cancers, that may express placental alkaline phosphatase.

IMPORTANT NOTICE: The products described in this manual are covered by one or more Limited Use Label License(s). Please refer to the Appendix on page 971 and Master Product List on page 975. Products are For Research Use Only. Not intended for any animal or human therapeutic or diagnostic use.

Each NovaBright™ SEAP Enzyme Reporter Gene Chemiluminescent Detection System 2.0 provides:

- Reaction buffer, containing CSPD® substrate and Emerald-III™ luminescence enhancer
- Purified human placental alkaline phosphatase
- Assay buffer, containing a mixture of nonplacental alkaline phosphatase inhibitors
- Detailed protocols

Sufficient reagents are supplied for 192 (N10577) or 960 (N10578) microplate assays based on the protocol provided.

Chemiluminescent Alkaline Phosphatase ELISA Sampler Kit

The Chemiluminescent Alkaline Phosphatase ELISA Kits employ CSPD® or CDP-*Star*® 1,2-dioxetane substrates for alkaline phosphatase with Sapphire™-II or Emerald-II™ enhancer in a system designed for rapid and ultrasensitive analyte detection in enzyme-linked immunoassays.[52] Maximum light emission from alkaline phosphatase–activated CSPD® or CDP-*Star*® substrate is reached in 5 to 60 minutes, depending on the temperature and the substrate chosen. Enzymatic dephosphorylation of substrate occurs at a constant rate proportional to enzyme concentration; the resulting phenolate anion decomposes with a finite half-life (Figure 10.3.13). Light emission can be quantitated with a variety of luminometers without the need for solution injection.

Enzyme-linked immunosorbent assays (ELISAs) can be formatted in several configurations on a variety of solid supports including microplate wells, tubes, polystyrene beads or ferrite particles. The high sensitivity obtained with 1,2-dioxetane substrates is demonstrated in a sandwich immunoassay format that employs a biotinylated detector antibody and streptavidin alkaline phosphatase conjugate for quantitating recombinant human IL-6 (rhIL-6). The results obtained with CSPD® substrate/Sapphire™-II enhancer (Figure 10.3.14) show a significant improvement in signal-to-noise performance at all concentrations of rhIL-6 and a much wider assay dynamic range compared to those obtained with the fluorescent substrate 4-methylumbelliferyl phosphate (MUP, M6491), and the colorimetric substrate *p*-nitrophenyl phosphate[52] (*p*NPP).

We offer four different Chemiluminescent Alkaline Phosphatase ELISA Kits, as well as a Chemiluminescent Alkaline Phosphatase ELISA Sampler Kit. Each kit provides concentrated assay buffer and blocking reagent, as well as one of the following substrate/enhancer solutions:

- CSPD® substrate/Sapphire™-II enhancer (in Kit #1, C10552)
- CSPD® substrate/Emerald-II™ enhancer (in Kit #2, C10553)
- CDP-*Star*® substrate/Sapphire™-II enhancer (in Kit #3, C10554)
- CDP-*Star*® substrate/Emerald-II™ enhancer (in Kit #4, C10555)

The Chemiluminescent Alkaline Phosphatase ELISA Sampler Kit (C10556) provides assay buffer, blocking reagent and sample sizes of all four substrate/enhancer solutions. All five kits provide sufficient reagents for 1000 assays based on the provided protocol.

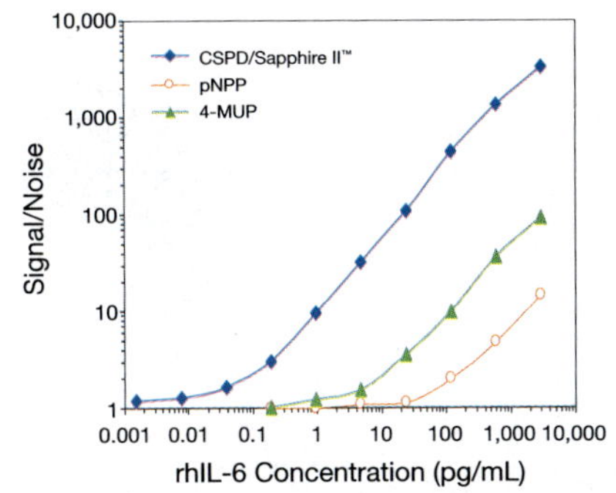

Figure 10.3.14 Comparison of detection sensitivities for the ELISA quantitation of recombinant human IL-6 (rhIL-6) using the chemiluminescent phosphatase substrate/enhancer in the Chemiluminescent Alkaline Phosphatase ELISA Kit #1 (CSPD® substrate/Sapphire™-II enhancer in C10552), the fluorescent phosphatase substrate 4-methylumbelliferyl phosphate (MUP, M6491) and the colorimetric phosphatase substrate *p*-nitrophenyl phosphate (*p*NPP).

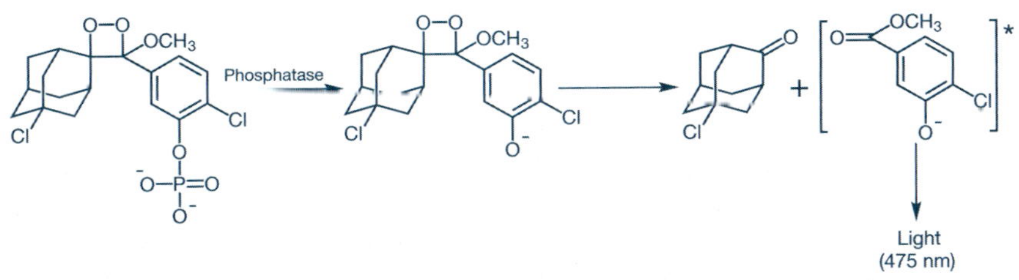

Figure 10.3.13 Mechanism of phosphatase-dependent chemiluminescence generation by CDP-*Star*® substrate. Luminescence emission is shifted to 461 nm or 542 nm, respectively, by the Sapphire™-II and Emerald-II™ enhancers.

Chromogenic Phosphatase Substrate

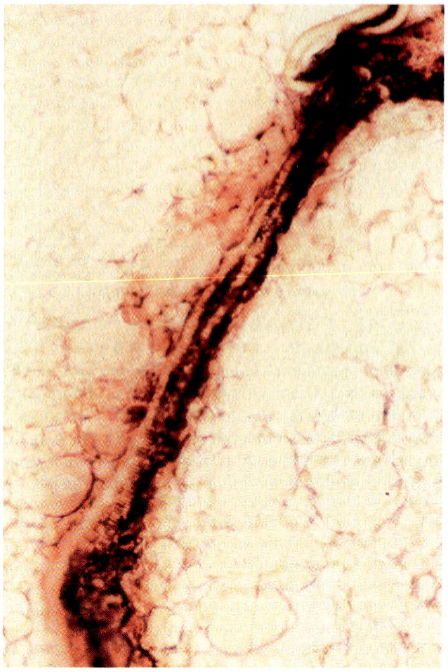

Figure 10.3.16 Endogenous phosphatase activity in a zebrafish ovary cross-section, stained with the BCIP (B6492) and NBT (N6495) reagents. These reagents are also available in the NBT/BCIP Reagent Kit (N6547).

NBT/BCIP Reagent Kit

5-Bromo-4-chloro-3-indolyl phosphate (BCIP) is commonly used with a number of different chromogens in various histological and molecular biology techniques. Hydrolysis of this indolyl phosphate, followed by oxidation, produces a blue-colored precipitate at the site of enzymatic activity. For convenience, we offer the NBT/BCIP Reagent Kit (N6547), which provides 1 g samples of BCIP as well as its co-precipitant NBT.

NBT: A Co-Precipitant for the BCIP Reaction

Nitro blue tetrazolium (NBT, N6495) is the most commonly used electron-transfer agent and co-precipitant for the BCIP reaction, forming a dark blue, precisely localized precipitate in the presence of alkaline phosphatase[53,54] (Figure 10.3.15, Figure 10.3.16).

Kits for Detecting Phosphatases, Polymerases and Nucleases

We have developed some unique products for following the activity of phosphatases, polymerases and nucleases. Our P_iPer™ Phosphate and P_iPer™ Pyrophosphate Assay Kits provide ultrasensitive assays for free phosphate and pyrophosphate, respectively, through the formation of resorufin, which can be detected either fluorometrically or spectrophotometrically. Our EnzChek® Phosphate and EnzChek® Pyrophosphate Assay Kits provide spectrophotometric assays for inorganic phosphate and pyrophosphate, respectively. Both the P_iPer™ and the EnzChek® Phosphate Assay Kits permit continuous measurement of the activity of ATPases, GTPases, phosphatases, nucleotidases and a number of enzymes that produce or consume inorganic phosphate. Moreover, these enzyme assay kits utilize the natural substrate for the enzyme rather than a synthetic analog and detect the inorganic phosphate produced using enzyme-coupled reactions. The P_iPer™ Kit is more sensitive and better suited for high-throughput studies in microplates, but requires an additional enzymatic step, whereas the EnzChek® Kit is less sensitive but has a simpler protocol. In both the P_iPer™ and the EnzChek® Pyrophosphate Assay Kits, the phosphate assay is coupled with the enzyme, pyrophosphatase, in order to monitor the activity of pyrophosphate-producing enzymes such as DNA and RNA polymerases, adenylate cyclase and guanylate cyclase.[55,56]

In addition to these kits, our RediPlate™ 96 EnzChek® Tyrosine Phosphatase Assay Kit (R22067) and RediPlate™ 96 EnzChek® Serine/Threonine Phosphatase Assay Kit (R33700) utilize DiFMUP for assaying tyrosine phosphatase activity and serine/threonine phosphatase

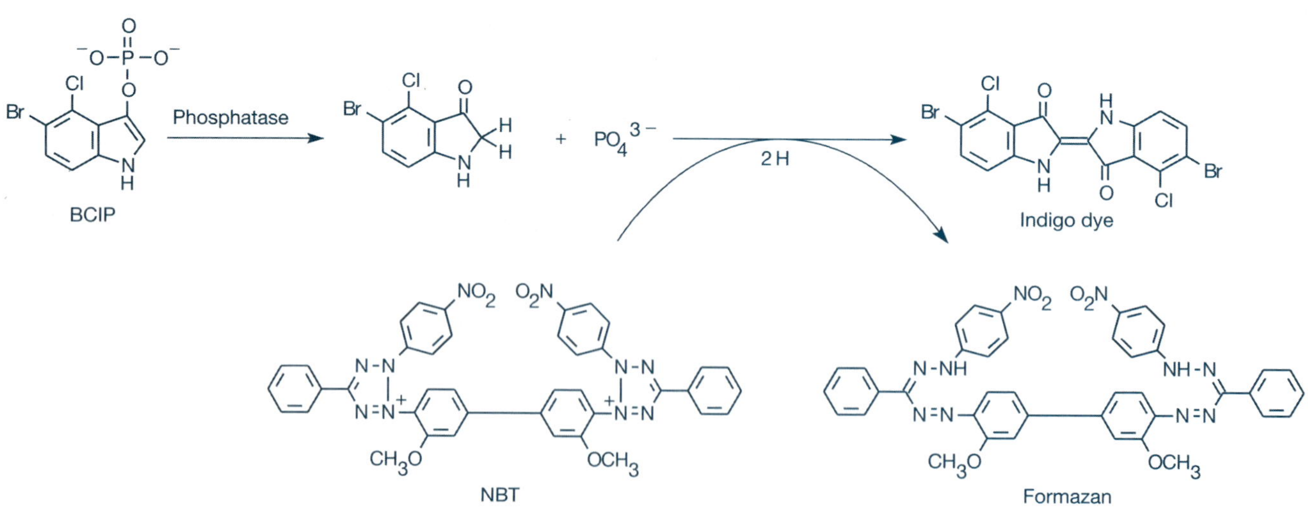

Figure 10.3.15 Principle of enzyme-linked detection using the reagents in our NBT/BCIP Reagent Kit (N6547). Phosphatase hydrolysis of BCIP is coupled to reduction of NBT, yielding a formazan and an indigo dye that together form a black-purple–colored precipitate.

IMPORTANT NOTICE: The products described in this manual are covered by one or more Limited Use Label License(s). Please refer to the Appendix on page 971 and Master Product List on page 975. Products are For Research Use Only. Not intended for any animal or human therapeutic or diagnostic use.
www.invitrogen.com/probes

activity, respectively, as well as for screening phosphatase inhibitors in a convenient high-throughput format.

P_iPer™ Phosphate Assay Kit

The P_iPer™ Phosphate Assay Kit (P22061) provides an ultrasensitive assay that detects free phosphate in solution through formation of the fluorescent product resorufin. Because resorufin also has strong absorption, the assay can be performed either fluorometrically or spectrophotometrically. This kit can be used to detect inorganic phosphate (P_i) in a variety of samples or to monitor the kinetics of phosphate release by a variety of enzymes, including ATPases, GTPases, 5′-nucleotidase, protein phosphatases, acid and alkaline phosphatases and phosphorylase kinase. Furthermore, the assay can be modified to detect virtually any naturally occurring organic phosphate molecule by including an enzyme that can specifically digest the organic phosphate to liberate inorganic phosphate.

In the P_iPer™ phosphate assay (Figure 10.3.17), maltose phosphorylase converts maltose (in the presence of P_i) to glucose 1-phosphate and glucose. Then, glucose oxidase converts the glucose to gluconolactone and H_2O_2. Finally, with horseradish peroxidase as a catalyst, the H_2O_2 reacts with the Amplex® Red reagent (10-acetyl-3,7-dihydroxyphenoxazine) to generate resorufin, which has absorption/emission maxima of ~571/585 nm.[57,58] The resulting increase in fluorescence or absorption is proportional to the amount of P_i in the sample. This kit can be used to detect as little as 0.2 µM P_i by fluorescence (Figure 10.3.18) or 0.4 µM P_i by absorption.

Each P_iPer™ Phosphate Assay Kit contains:

- Amplex® Red reagent
- Dimethylsulfoxide (DMSO)
- Concentrated reaction buffer
- Recombinant maltose phosphorylase from *Escherichia coli*
- Maltose
- Glucose oxidase from *Aspergillus niger*
- Horseradish peroxidase
- Phosphate standard
- Hydrogen peroxide
- Detailed protocols for detecting phosphatase activity

Sufficient reagents are provided for approximately 1000 assays using a reaction volume of 100 µL per assay and either a fluorescence or absorbance microplate reader.

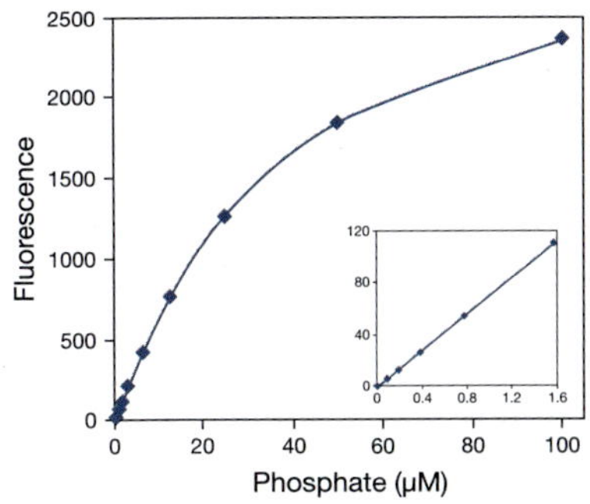

Figure 10.3.18 Detection of inorganic phosphate using the P_iPer™ Phosphate Assay Kit (P22061). Each reaction contained 50 µM Amplex® Red reagent, 2 U/mL maltose phosphorylase, 1 mM maltose, 1 U/mL glucose oxidase and 0.2 U/mL HRP in 1X reaction buffer. Reactions were incubated at 37°C. After 60 minutes, fluorescence was measured in a fluorescence microplate reader using excitation at 530 ± 12.5 nm and fluorescence detection at 590 ± 17.5 nm. Data points represent the average of duplicate reactions, and a background value of 43 (arbitrary units) was subtracted from each reading.

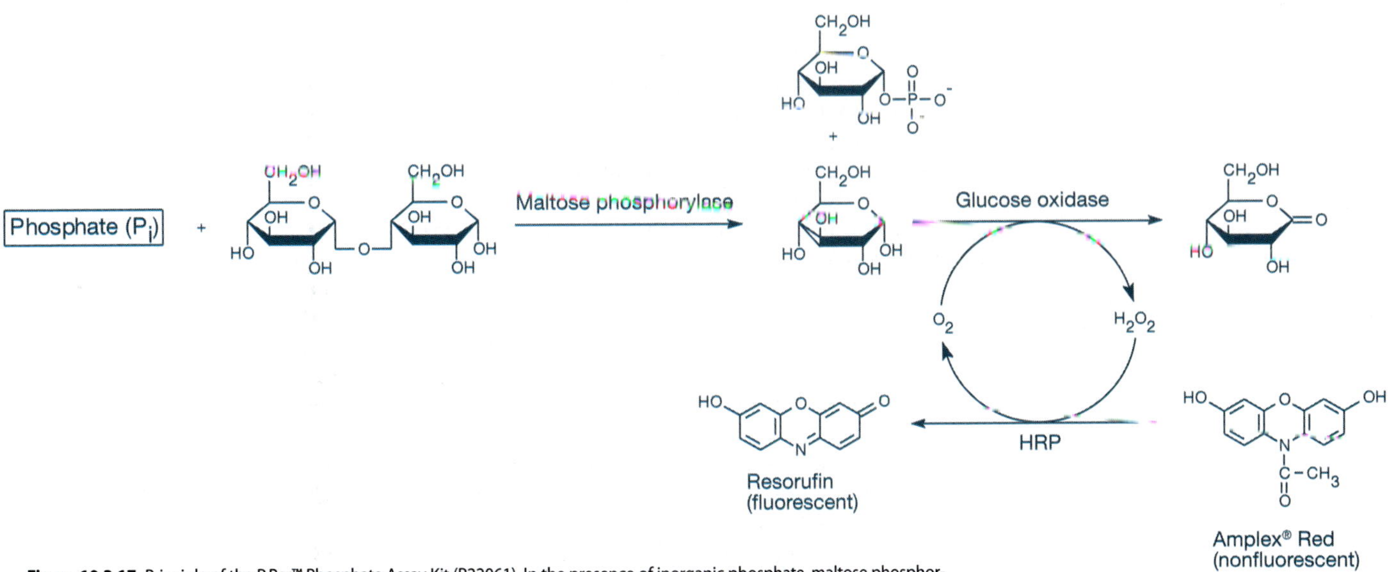

Figure 10.3.17 Principle of the P_iPer™ Phosphate Assay Kit (P22061). In the presence of inorganic phosphate, maltose phosphorylase converts maltose to glucose 1-phosphate and glucose. Then, glucose oxidase converts the glucose to gluconolactone and H_2O_2. Finally, with horseradish peroxidase (HRP) as a catalyst, the H_2O_2 reacts with the Amplex® Red reagent to generate the highly fluorescent resorufin. The resulting increase in fluorescence or absorption is proportional to the amount of P_i in the sample.

IMPORTANT NOTICE: The products described in this manual are covered by one or more Limited Use Label License(s). Please refer to the Appendix on page 971 and Master Product List on page 975. Products are For Research Use Only. Not intended for any animal or human therapeutic or diagnostic use.

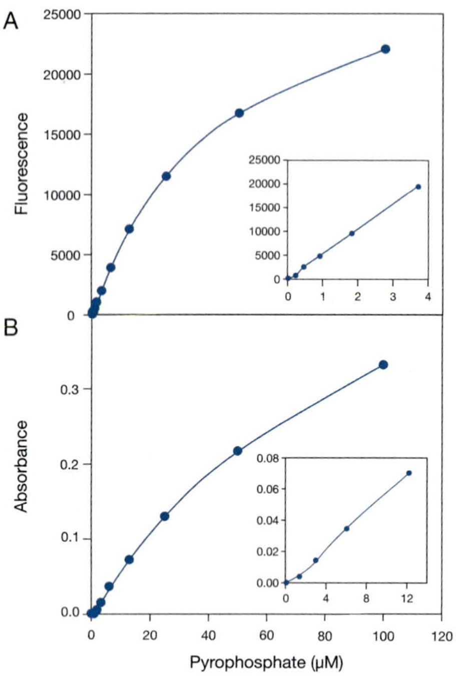

Figure 10.3.19 Detection of pyrophosphate using the P_iPer™ Pyrophosphate Assay Kit (P22062). Each reaction contained 50 µM Amplex Red reagent, 0.01 U/mL inorganic pyrophosphatase, 2 U/mL maltose phosphorylase, 0.2 mM maltose, 1 U/mL glucose oxidase and 0.2 U/mL HRP in 1X reaction buffer. Reactions were incubated at 37°C. After 60 minutes, **A**) fluorescence was measured in a fluorescence-based microplate reader using excitation at 530 ± 12.5 nm and fluorescence detection at 590 ± 17.5 nm or **B**) absorbance was measured in an absorption-based microplate reader at 576 ± 5 nm. Data points represent the average of duplicate reactions. In panel A, a background value of 78 (arbitrary units) was subtracted from each reading; in panel B, a background absorbance of 0.011 was subtracted from each reading.

P_iPer™ Pyrophosphate Assay Kit

The P_iPer™ Pyrophosphate Assay Kit (P22062) provides a sensitive fluorometric or colorimetric method for measuring the inorganic pyrophosphate (PP_i) in experimental samples or for monitoring the kinetics of PP_i release by a variety of enzymes, including DNA and RNA polymerases, adenylate cyclase and *S*-acetyl coenzyme A synthetase. In the P_iPer™ pyrophosphate assay, inorganic pyrophosphatase hydrolyzes PP_i to two molecules of inorganic phosphate (P_i). The P_i then enters into the same cascade of reactions as it does in the P_iPer™ Phosphate Assay Kit (Figure 10.3.17). In this case, the resulting increase in fluorescence or absorption is proportional to the amount of PP_i in the sample. This kit can be used to detect as little as 0.1 µM PP_i by fluorescence or 0.2 µM PP_i by absorption (Figure 10.3.19). Each P_iPer™ Pyrophosphate Assay Kit contains:

- Amplex® Red reagent
- Dimethylsulfoxide (DMSO)
- Concentrated reaction buffer
- Recombinant maltose phosphorylase from *Escherichia coli*
- Maltose
- Glucose oxidase from *Aspergillus niger*
- Horseradish peroxidase
- Inorganic pyrophosphatase from baker's yeast
- Pyrophosphate standard
- Detailed protocols for detecting pyrophosphatase activity

Sufficient reagents are provided for approximately 1000 assays using a reaction volume of 100 µL per assay and either a fluorescence or absorbance microplate reader.

EnzChek® Phosphate Assay Kit

The EnzChek® Phosphate Assay Kit (E6646), which is based on a method originally described by Webb,[59,60] provides an easy enzymatic assay for detecting P_i from multiple sources through formation of a chromophoric product (Figure 10.3.20). Although this kit is usually used to determine the P_i produced by a wide variety of enzymes such as ATPases, kinases and phosphatases, it can also be used to specifically quantitate P_i with a sensitivity of ~2 µM P_i (~0.2 µg/mL) (Figure 10.3.21). Moreover, this spectrophotometric assay has proven useful for determining the level of P_i contamination in the presence of high concentrations of acid-labile phosphates using a microplate reader.[61] Because the sulfate anion competes with P_i for binding to purine nucleoside phosphorylase (PNP), this kit can be adapted for measurement of sulfate concentrations between 0.1 and 10 mM in the presence of a low (<100 µM) fixed P_i concentration.[62] Each EnzChek® Phosphate Assay Kit contains:

- 2-Amino-6-mercapto-7-methylpurine riboside (MESG)
- Purine nucleoside phosphorylase (PNP)
- Concentrated reaction buffer
- KH_2PO_4 standard
- Detailed protocols for detecting and quantitating P_i

In the presence of P_i, MESG is enzymatically converted by PNP to ribose 1-phosphate and the chromophoric product 2-amino-6-mercapto-7-methylpurine (Figure 10.3.20). This kit contains sufficient reagents for about 100 phosphate assays using 1 mL assay volumes and standard cuvettes. The substrate MESG and the enzyme PNP included in our EnzChek® Phosphate Assay Kit have been adapted for monitoring the kinetics of phosphate release by:

- Actin-activated myosin ATPase [60,63]
- Actin polymerization [64]
- Aminoacyl-tRNA synthetase [65]
- Aspartate transcarbamylase [59,66]
- ArsA ATPase, the catalytic subunit of the arsenite pump [67]
- Dethiobiotin synthetase [68]
- Glycerol kinase [60]
- Glycogen phosphorylase [69]
- GTPases [70–75]

IMPORTANT NOTICE: The products described in this manual are covered by one or more Limited Use Label License(s). Please refer to the Appendix on page 971 and Master Product List on page 975. Products are For Research Use Only. Not intended for any animal or human therapeutic or diagnostic use.

- ATPases[66,76,77]
- *myo*-Inositol monophosphatase[78]
- Phospholysine and phosphohistidine phosphatases[79,80]
- Phosphorylase *a* phosphatase[81]
- Phosphorylase kinase[82]
- Serine phosphatase[83]
- Self-assembly of actin and tubulin[84]
- Autodephosphorylation of CheY, a response regulator that mediates bacterial chemotaxis[85]

Moreover, the EnzChek® phosphate assay is sufficiently fast and quantitative to permit stopped-flow kinetic experiments on enzymes that produce phosphate, an important development for mechanistic enzyme studies.[60] Although this kit is usually used to determine the inorganic phosphate produced by enzymes such as ATPases and GTPases, it can also be used to specifically quantitate inorganic phosphate contamination of reagents and solution, with a detection limit of ~2 µM P_i (~0.2 µg/mL) and an effective range between 2 and 150 µM inorganic phosphate (between 2 and 150 nanomoles phosphate in a 1 mL volume). For example, the assay has been used for the rapid assay of inorganic phosphate in the presence of high concentrations of acid-labile phosphates using a microplate reader.[61] The reagents in this kit can also be used as a phosphate "mop" to remove almost all inorganic phosphate from a protein solution.[33]

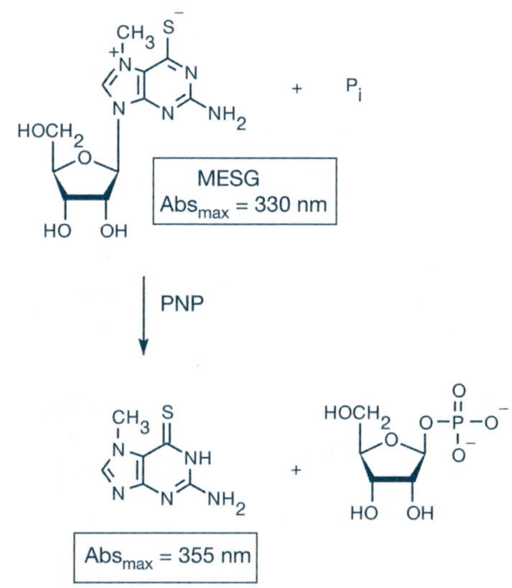

Figure 10.3.20 Enzymatic conversion of 2-amino-6-mercapto-7-methylpurine riboside (MESG) to ribose 1-phosphate and 2-amino-6-mercapto-7-methylpurine by purine nucleoside phosphorylase (PNP), reagents supplied in the EnzChek® Phosphate Assay Kit (E6646). The accompanying change in the absorption maximum (Abs) allows quantitation of inorganic phosphate (P_i) consumed in the reaction.

EnzChek® Pyrophosphate Assay Kit

In the EnzChek® Pyrophosphate Assay Kit (E6645), we have adapted the method provided in the EnzChek® Phosphate Assay Kit to permit the sensitive spectrophotometric detection of PP_i, which is converted by the enzyme pyrophosphatase to P_i.[56] Because two moles of P_i are released per mole of PP_i consumed, the sensitivity limit of the EnzChek® Pyrophosphate Assay Kit is 1 µM PP_i (~0.2 µg/mL). This assay has been modified to continuously detect several enzymes that liberate PP_i[86–89] such as aminoacyl-tRNA synthetase,[65] luciferase, cytidylyl transferase[90] and *S*-acetyl coenzyme A synthetase[56] adenylate cyclase and guanylyl cyclase.[55] Each EnzChek® Pyrophosphate Assay Kit contains:

- Inorganic pyrophosphatase
- 2-Amino-6-mercapto-7-methylpurine riboside (MESG)
- Purine nucleoside phosphorylase (PNP)
- Concentrated reaction buffer
- $Na_2P_2O_7$ standard
- Detailed protocols for detecting and quantitating PP_i

This kit contains sufficient reagents for about 100 PP_i assays using standard 1 mL assay volumes and standard cuvettes.

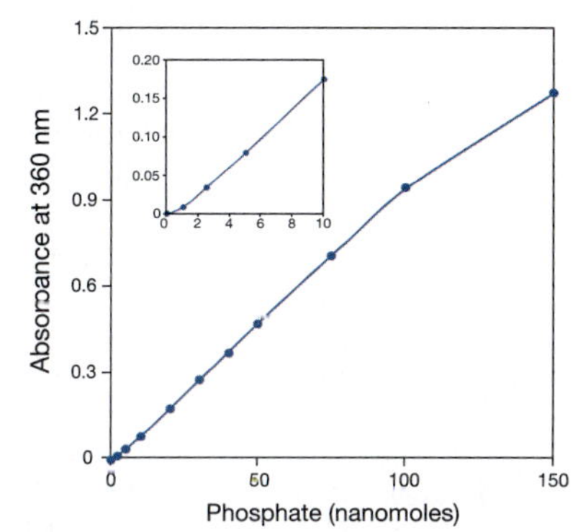

Figure 10.3.21 Quantitative analysis of inorganic phosphate using the EnzChek® Phosphate Assay Kit (E6646). KH_2PO_4 was used as the source for the inorganic phosphate, and the absorbance at 360 nm was corrected for background absorbance. The inset shows an enlargement of the standard curve, demonstrating the lower range of the assay; the units are the same.

EnzChek® Phosphatase Assay Kit

The EnzChek® Phosphatase Assay Kit[91] (E12020), which includes the DiFMUP substrate, can be used to continuously detect phosphatase activity at neutral, basic or moderately acidic pH. Using the assay outlined in the kit protocol, we have found that DiFMUP is about 100 times more sensitive than MUP for the detection of prostatic acid phosphatase at pH 5.5 (Figure 10.3.6) and can measure this activity at a pH that is optimal for the enzyme (Figure 10.3.7). The EnzChek® Phosphatase Assay Kit is perfect for the continuous assay of prostatic acid phosphatase, protein phosphatase 1 or almost any other phosphatase that can be assayed with nonprotein-based substrates such as MUP or 4-nitrophenyl phosphate (*p*NPP). Each EnzChek® Phosphatase Assay Kit contains:

- DiFMUP substrate
- Concentrated reaction buffer
- Acid phosphatase from potato for use as a positive control
- 6,8-Difluoro-7-hydroxy-4-methylcoumarin for use as a reference standard
- Detailed protocols for detecting phosphatase activity

The kit provides sufficient reagents for performing approximately 1000 assays, using a reaction volume of 100 µL per assay.

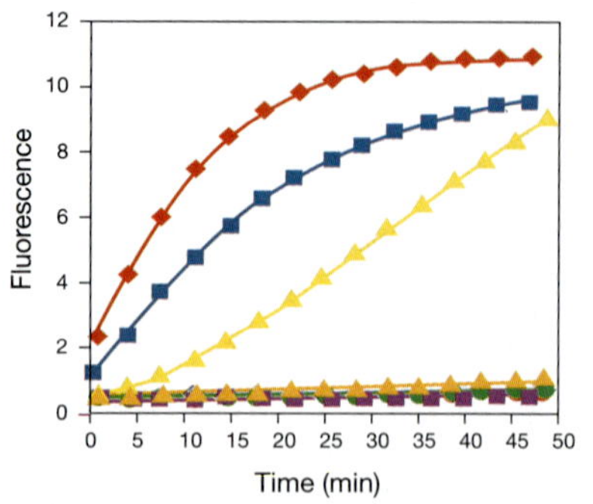

Symbol	Enzyme (Class)	Enzyme Units*
◆ (red)	CD-45 (tyrosine phosphatase)	1 U/mL
■ (blue)	PTP-1B (tyrosine phosphatase)	1 mU/mL
▲ (yellow)	PTPase (tyrosine phosphatase)	1 U/mL
▲ (orange)	Acid phosphatase	1 U/mL
● (green)	Alkaline phosphatase	1 U/mL
● (red)	PP2A (ser/thr phosphatase)†	1 U/mL
■ (purple)	PP1 (ser/thr phosphatase)†	1 U/mL
◆ (gray)	PP-2B (ser/thr phosphatase)†	500 U/mL

* Enzyme unit (U) definitions are standard definitions for each enzyme.
† Serine theonine phosphatase.

Figure 10.3.23 Specificity of the RediPlate™ 96 EnzChek® Tyrosine Phosphatase Assay Kit (R22067). The phosphatases listed in the table were applied to a RediPlate™ 96 EnzChek® tyrosine phosphatase assay microplate. At the indicated time points, the fluorescence was measured in a fluorescence microplate reader using excitation at 355 ± 20 nm and emission at 460 ± 12.5 nm.

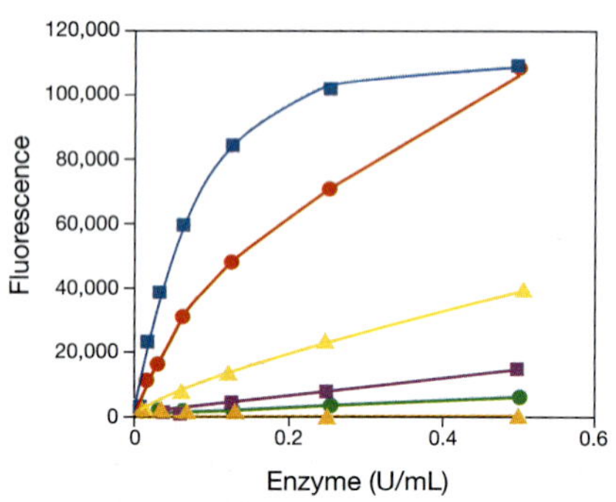

Symbol	Enzyme (Class)
■ (blue)	PP-2A (Ser/Thr phosphatase)
● (red)	PP-1 (Ser/Thr phosphatase)
▲ (yellow)	PP-2B (Ser/Thr phosphatase)
■ (purple)	Alkaline phosphatase
● (green)	Acid phosphatase
▲ (orange)	LAR (tyrosine phosphatase)

Figure 10.3.24 Specificity of the RediPlate™ 96 EnzChek® Serine/Threonine Phosphatase Assay Kit (R33700) for serine/threonine phosphatases. The phosphatases listed in the table were applied at the indicated concentrations to a RediPlate™ 96 EnzChek® serine/threonine phosphatase assay microplate. Reactions were incubated at 37°C. After 1 hour, fluorescence was measured in a fluorescence microplate reader using excitation at 355 ± 20 nm and emission at 460 ± 12.5 nm.

RediPlate™ 96 EnzChek® Tyrosine Phosphatase Assay Kits

Protein tyrosine phosphatases (PTP) represent a large family of enzymes that play a very important role in intra- and intercellular signaling. PTPs work antagonistically with protein tyrosine kinases to regulate signal transduction pathways in response to a variety of signals, including hormones and mitogens.[92] Our RediPlate™ 96 EnzChek® Tyrosine Phosphatase Assay Kit (R22067) provides researchers with a sensitive and convenient method to monitor PTP and screen PTP inhibitors in a variety of research areas[93–96] (Figure 10.3.22).

The EnzChek® tyrosine phosphatase assay is based on 6,8-difluoro-4-methylumbelliferyl phosphate[97] (DiFMUP; D6567, D22065). The EnzChek® tyrosine phosphatase assay is continuous, allowing researchers to easily measure fluorescence at various time points in order to follow the kinetics of the reaction. Furthermore, the assay is not affected by free phosphate and is compatible with most nonionic detergents, resulting in minimal sample processing before analysis. Most importantly, each assay well contains inhibitors to help ensure that the assay is selective for tyrosine phosphatases; other phosphatases, including serine/threonine phosphatases, will not hydrolyze DiFMUP under our assay conditions (Figure 10.3.23). Unlike phosphopeptide-based assays, this DiFMUP-based assay can be used to monitor a variety of tyrosine phosphatases, including PTP-1B and CD-45 (Figure 10.3.23). Tyrosine phosphatase inhibitors can be evaluated quantitatively in the assay for their effect on tyrosine phosphatase activity.

Each RediPlate™ 96 EnzChek® Tyrosine Phosphatase Assay Kit (R22067) includes:

- One RediPlate™ 96 EnzChek® tyrosine phosphatase assay 96-well microplate
- Reaction buffer
- Detailed assay protocols

RediPlate™ 96 EnzChek® Serine/Threonine Phosphatase Assay Kit

The majority of protein phosphorylation occurs on serine and threonine residues, with <0.01–0.05% on tyrosine residues. Serine/threonine phosphatases represent a large family of enzymes that have been implicated in the regulation of metabolism, transcription, translation, differentiation, cell cycle, cytoskeletal dynamics, oncogenesis and signal transduction. The RediPlate™ 96 EnzChek® Serine/Threonine Phosphatase Assay Kit (R33700) provides a fast, simple and direct fluorescence-based assay for detecting serine/threonine phosphatases and their corresponding modulators and inhibitors[97] (Figure 10.3.22).

As with the RediPlate™ 96 EnzChek® Tyrosine Phosphatase Kit, the substrate incorporated in the RediPlate™ 96 EnzChek® Serine/Threonine Phosphatase Assay Kit is DiFMUP. Inhibitors are included in each assay well to help ensure that the assay is selective for serine/threonine phosphatases; under the prescribed assay conditions, other phosphatases, including tyrosine phosphatases, do not significantly react with the substrate (Figure 10.3.24). Furthermore, unlike phosphopeptide-based assays, this DiFMUP-based assay can be used to monitor a variety of serine/threonine phosphatases including PP-1, PP-2A and PP-2B (Figure 10.3.24). Serine/threonine phosphatase inhibitors can be evaluated quantitatively in the assay for their effect on

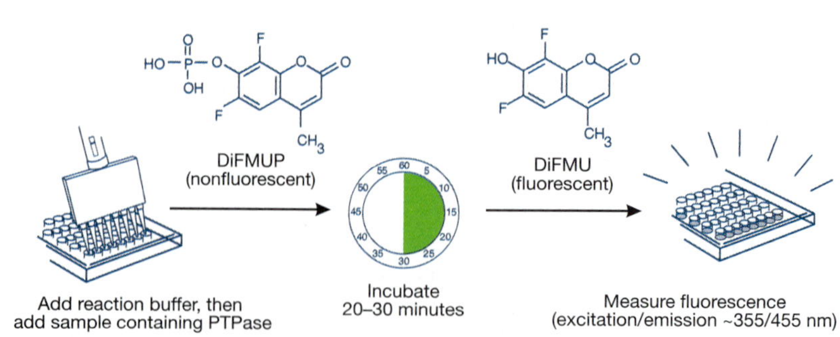

Figure 10.3.22 Schematic diagram of the method used in the RediPlate™ EnzChek® Phosphatase Assay Kits (R22067, R33700).

IMPORTANT NOTICE: The products described in this manual are covered by one or more Limited Use Label License(s). Please refer to the Appendix on page 971 and Master Product List on page 975. Products are For Research Use Only. Not intended for any animal or human therapeutic or diagnostic use.

serine/threonine phosphatase activity (Figure 10.3.25). Additional advantages of this RediPlate™ assay include compatibility with nonionic detergents and insensitivity to free phosphate, resulting in minimal sample processing before analysis.

Each RediPlate™ 96 EnzChek® Serine/Threonine Phosphatase Assay Kit includes:

- One RediPlate™ 96 EnzChek® serine/threonine phosphatase assay 96-well microplate
- Concentrated reaction buffer
- $NiCl_2$
- $MnCl_2$
- Dithiothreitol
- Detailed assay protocols

To help ensure the integrity of the pre-dispensed reagents, the 96-well microplate provided in both RediPlate™ Protein Phosphatase Assay Kits is packaged in a resealable foil pouch and consists of twelve removable strips, each with eight wells (Figure 10.3.26). Eleven of the strips (88 wells) are preloaded with the fluorogenic substrate DiFMUP; the remaining strip, marked with black tabs, contains a dilution series of the DiFMU reference standard for generating a standard curve.

EnzChek® *Ultra* Phytase Assay Kit

Phytases catalyze the sequential hydrolysis of phytate (myoinositol hexakis phosphate; phytin; phytic acid) to less phosphorylated myo-inositol compounds and inorganic phosphate. The EnzChek® *Ultra* Phytase Assay Kit (E33701) utilizes a series of linked enzymatic reactions for the detection of phytase activity based on the measurement of phosphate released from the substrate phytic acid (Figure 10.3.27). In the initial step, phytase catalyzes the release of inorganic phosphate from phytic acid; in the presence of P_i, maltose phosphorylase converts maltose to glucose 1-phosphate and glucose. Glucose oxidase then converts the glucose to gluconolactone and H_2O_2. Finally, with horseradish peroxidase (HRP) as a catalyst, the H_2O_2 reacts with the Amplex® UltraRed reagent to generate a brightly fluorescent and strongly absorbing reaction product (excitation/emission maxima ~568/581 nm). The resulting increase in fluorescence is proportional to the amount of P_i in the sample. This relative measure of phytase activity can be used as the basis of a standard curve of phytase enzyme standards with known activity. The assay takes 1 hour at 37°C; under these conditions, the limit of detection is 0.001 FTU/mL and the intra-assay CV is ≤10%. The signal is not linear, but can cover a two-log concentration range for phytase (Figure 10.3.28).

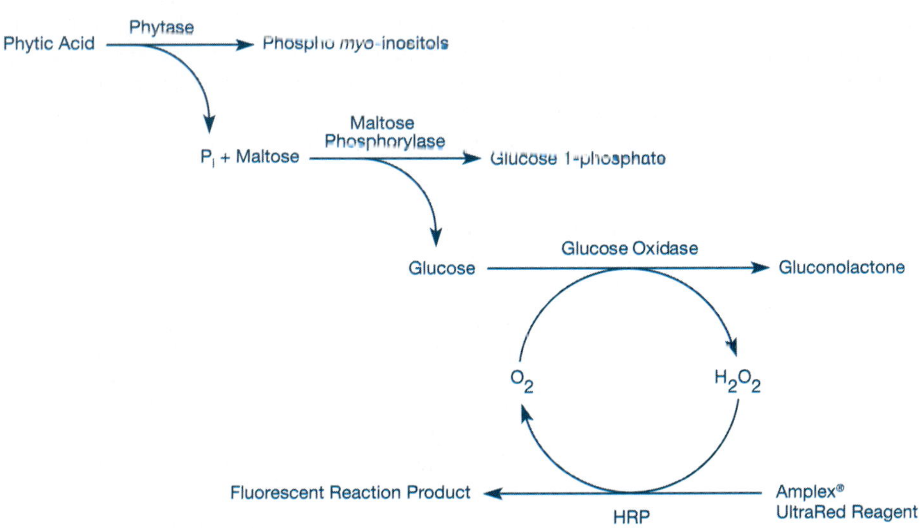

Figure 10.3.27 Principle of the EnzChek® *Ultra* Phytase Assay Kit (E33701).

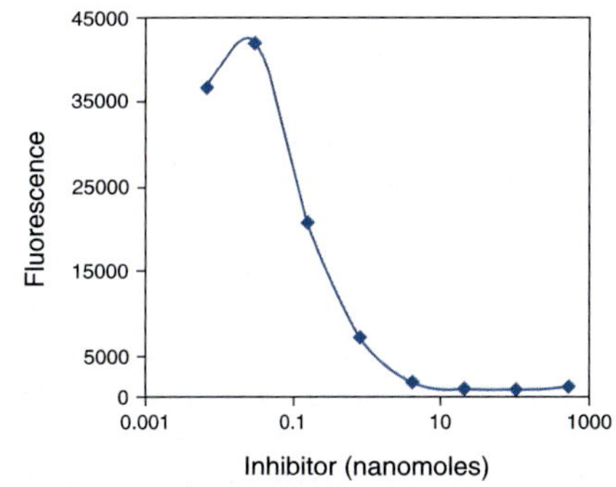

Figure 10.3.25 Detection of PP-2A inhibition by okadaic acid using the RediPlate™ 96 EnzChek® Serine/Threonine Phosphatase Assay Kit (R33700). Each reaction contained 50 µM DiFMUP, 10 mU/mL PP-2A and the indicated concentration (log scale) of okadaic acid in reaction buffer containing 50 mM Tris-HCl, 0.1 mM $CaCl_2$, 1 mM $NiCl_2$, 125 µg/mL bovine serum albumin (BSA) and 0.05% Tween® 20. Reactions were incubated at 37°C. After 30 minutes, fluorescence was measured in a fluorescence microplate reader using excitation at 355 ± 20 nm and emission at 460 ± 12.5 nm.

Figure 10.3.26 A RediPlate™ 96 microplate.

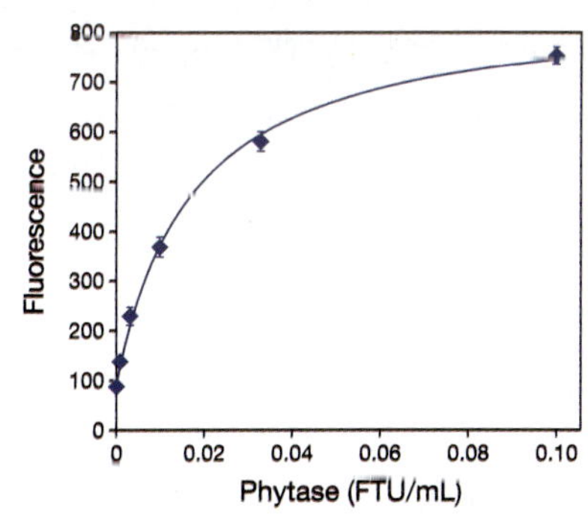

Figure 10.3.28 Assay of a standard phytase preparation using the EnzChek® *Ultra* phytase assay. A dilution series of Natuphos 10000 L phytase (BASF) in 0.1 M acetate buffer (pH 5.5) was assayed with the Enzchek® *Ultra* Phytase Assay Kit (E33701). Incubation time was 60 minutes; phytase activity is shown in FTU/mL (FTU= phytase unit, as defined by the manufacturer).

Each EnzChek® *Ultra* Phytase Assay Kit provides:

- Amplex® UltraRed reagent
- Dimethylsulfoxide (DMSO)
- Concentrated reaction buffer
- Sodium acetate, pH 5.5
- Phytic acid
- Maltose phosphorylase, recombinant from *Escherichia coli* (1 unit = the amount of enzyme that will convert maltose, in the presence of inorganic phosphate, to 1.0 mole of D-glucose and D-glucose 1-phosphate per minute at 37°C, pH 7.0)
- Glucose oxidase from *Aspergillus niger* (1 unit = the amount of enzyme that will oxidize 1.0 mole of -D-glucose to D-gluconolactone and H_2O_2 per minute at 35°C, pH 5.1)
- Detailed protocols

Sufficient reagents are provided for five 96-well microplate assay with 6 mL per plate and 20% excess for pipetting considerations or partial-plate assays.

EnzChek® Paraoxonase Assay Kit

Paraoxonase is a mammalian enzyme associated with high-density lipoprotein ("good cholesterol") in serum. Low serum paraoxonase levels are positively correlated with risk of cardiovascular disease,[98] and paraoxonase activity is a better marker than the PON1 genotype for predicting susceptibility to vascular disease.[99] Paraoxonase has multiple activities including organophosphatase, phosphotriesterase, arylesterase and thiolactonase. The organophosphatase activity confers protection against toxic organophosphates such as insecticides, which are a common source of chemical intolerance, as well as nerve agents such as sarin and VX. Current methods for measuring paraoxonase in serum, including the colorimetric paraoxon assay,[100] are relatively insensitive and often toxic themselves, and are compromised by high background and low signal as well.

The EnzChek® Paraoxonase Assay Kit (E33702) is a highly sensitive, homogeneous fluorometric assay (excitation/emission maxima ~360/450 nm) for the organophosphatase activity of paraoxonase and is based on the hydrolysis of a fluorogenic organophosphate analog. Under standard conditions, this assay requires only 5 µL of serum, yields a signal in as little as 15 minutes and is linear for up to 60 minutes. The EnzChek® paraoxonase assay is >10-fold more sensitive than the colorimetric paraoxon assay and, unlike the colorimetric assay, can distinguish samples of very similar paraoxonase activity. Furthermore, the EnzChek® paraoxonase assay requires only a single homogeneous reaction, which may be either continuously monitored or terminated using a stop solution. Correlation between the inhibition curves of the fluorogenic paraoxonase assay versus the colorimetric assay is excellent. The Z′ factor (a statistical parameter for evaluating the signal window of an assay)[101] is 0.95 when the assay is performed as described in the 96-well format.

Each EnzChek® Paraoxonase Assay Kit provides:

- Paraoxonase substrate
- Fluorescent reference standard
- Concentrated reaction buffer
- Stop reagent
- 96-well microplates
- Adhesive microplate covers
- Organophosphatase positive control
- Dimethylsulfoxide (DMSO)
- Detailed protocols

Sufficient reagents are provided for 100 assays using a reaction volume of 110 µL per well and a fluorescence microplate reader.

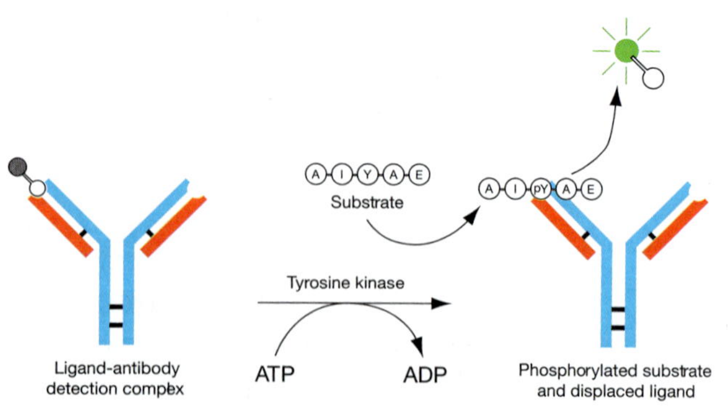

Figure 10.3.29 Reaction scheme for the tyrosine kinase assay used in the Antibody Beacon™ Tyrosine Kinase Assay Kit (A35725). The unlabeled natural substrate (AIYAE) is phosphorylated by the tyrosine kinase to AIY(P)AE, which displaces the quenched Oregon Green® 488 dye–labeled peptide from the anti-phosphotyrosine antibody, resulting in a large increase in its fluorescence that is proportional to the amount of AIY(P)AE formed in the reaction.

Antibody Beacon™ Tyrosine Kinase Assay Kit

The Antibody Beacon™ Tyrosine Kinase Assay Kit (A35725) provides a homogeneous solution assay for measuring the activity of tyrosine kinases and the effectiveness of potential inhibitors and modulators.[92] The key to this tyrosine kinase assay is a small-molecule tracer ligand labeled with our bright green-fluorescent Oregon Green® 488 dye. When an anti-phosphotyrosine antibody binds this tracer ligand to form the Antibody Beacon™ detection complex, the fluorescence of the Oregon Green® 488 dye is efficiently quenched. In the presence of a phosphotyrosine-containing peptide, however, this Antibody Beacon™ detection complex is rapidly disrupted, releasing the tracer ligand and relieving its antibody-induced quenching (Figure 10.3.29). Upon its displacement by a phosphotyrosine residue, the Oregon Green® 488 dye–labeled tracer ligand exhibits an approximately 4-fold fluorescence enhancement, enabling the detection of as little as 50 nM phosphotyrosine-containing peptide with excellent signal-to-background discrimination.

Key benefits of the Antibody Beacon™ Tyrosine Kinase Assay Kit include:

- **Real-time measurements.** Unlike many other commercially available tyrosine kinase assays, the Antibody Beacon™ Tyrosine Kinase Assay Kit permits real-time monitoring of kinase activity (Figure 10.3.30). Not only is the Antibody Beacon™ detection complex rapidly dissociated in the presence of phosphotyrosine residues, but the assay components have been designed to be simultaneously combined, eliminating any delay in the measurements.
- **Simple detection protocol.** Tyrosine kinase activity is measured by a simple increase in fluorescence intensity; no special equipment, additional reagents, or extra steps are required. This assay is readily compatible with any fluorescence microplate reader.
- **Use of natural substrates.** The Antibody Beacon™ tyrosine kinase assay utilizes unlabeled peptide or protein substrates, is compatible

IMPORTANT NOTICE: The products described in this manual are covered by one or more Limited Use Label License(s). Please refer to the Appendix on page 971 and Master Product List on page 975. Products are For Research Use Only. Not intended for any animal or human therapeutic or diagnostic use.

with substrates that are pre-phosphorylated at serine or threonine (but *not* at tyrosine) residues and is applicable to the assay of a wide variety of kinases.

- **Compatibility.** The anti-phosphotyrosine antibody provided in the Antibody Beacon™ Tyrosine Kinase Assay Kit is specific for phosphotyrosine residues; assay components such as ATP (up to 1 mM) and reducing agents such as dithiothreitol (DTT, up to 1 mM) do not interfere with this assay. This anti-phosphotyrosine antibody was selected from among several clones to produce the greatest fluorescence enhancement by the kinase-phosphorylated product.
- **Reliability.** This tyrosine kinase assay has a broad signal window, indicated by a Z′ factor of >0.85.

The Antibody Beacon™ Tyrosine Kinase Assay Kit comes with all the reagents needed to perform this assay, including:

- Oregon Green® 488 dye–labeled tracer ligand
- Anti-phosphotyrosine antibody
- Concentrated tyrosine kinase reaction buffer
- Two generic tyrosine kinase substrate solutions: a poly(Glu:Tyr) solution and a poly(Glu:Ala:Tyr) solution
- Dithiothreitol (DTT)
- Adenosine triphosphate (ATP)
- Phosphotyrosine-containing peptide, phospho-pp60 c-src (521–533), for use as a reference
- Detailed protocols

Each kit provides sufficient reagents to perform ~400 assays using a 50 µL assay volume in a fluorescence microplate reader. We provide an extensive range of assays for protein kinases that utilize a variety of strategies to detect phosphorylation of peptide and protein substrates (Table 17.1).

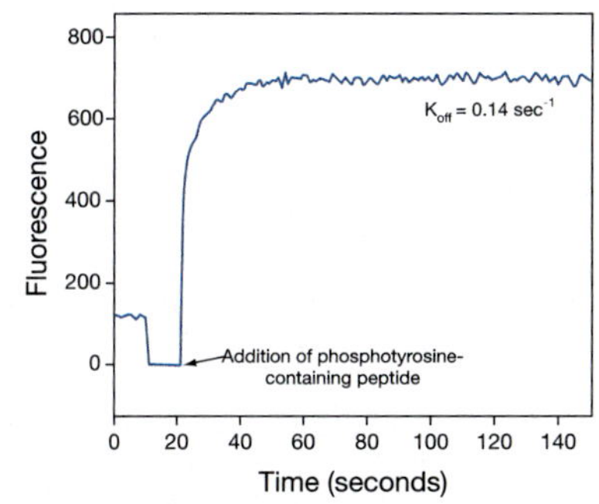

Figure 10.3.30 Real-time detection capability of the Antibody Beacon™ Tyrosine Kinase Assay Kit (A35725). Fluorescence of the Antibody Beacon™ detection complex in tyrosine kinase assay buffer was monitored over time. After ~15 seconds, an excess of phosphotyrosine-containing peptide was added to the Antibody Beacon™ detection complex and the off-rate was calculated.

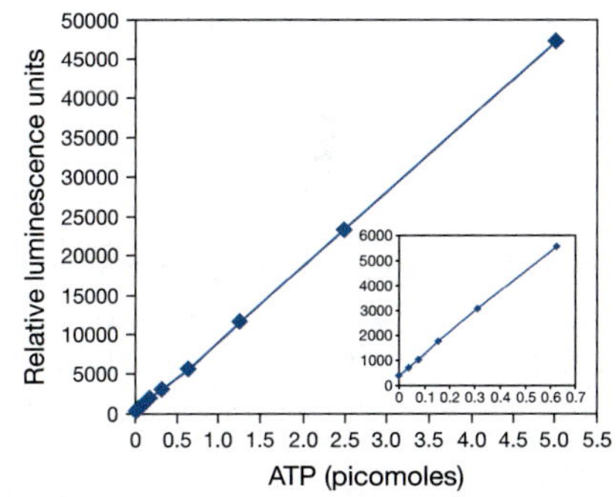

Figure 10.3.31 Detection of ATP using the ATP Determination Kit (A22066). Each reaction contained 1.25 µg/mL of firefly luciferase, 50 µM D-luciferin and 1 mM DTT in 1X reaction buffer. Luminescence was measured after a 15-minute incubation.

ATP Determination Kit

The ATP Determination Kit (A22066) is designed to provide a convenient method for the sensitive bioluminescence-based detection of ATP with recombinant firefly luciferase and its substrate luciferin. This assay is based on luciferase's absolute requirement for ATP to produce light. In the presence of Mg^{2+}, luciferase catalyzes the reaction of luciferin, ATP and O_2 to form oxyluciferin, AMP, CO_2, pyrophosphate and ~560 nm light (Section 10.6).

The luciferin–luciferase bioluminescence assay is extremely sensitive; most luminometers can detect as little as 1 picomole of pre-existing ATP or ATP as it is generated in kinetic systems (Figure 10.3.31). This sensitivity has led to its widespread use for detecting ATP in various enzymatic reactions, as well as for measuring viable cell number [102] (Section 15.2, Section 15.3) and for detecting low-level bacterial contamination in samples such as blood, milk, urine, soil and sludge.[103–107] The luciferin–luciferase bioluminescence assay has also been used to determine cell proliferation and cytotoxicity in both bacterial [108,109] and mammalian cells,[110,111] and to distinguish cytostatic versus cytocidal potential of anticancer drugs on malignant cell growth.[112]

Each ATP Determination Kit (A22066) contains:

- Luciferin (5 vials, each containing 3.0 mg)
- Luciferase
- Dithiothreitol (DTT)
- ATP
- Concentrated reaction buffer
- Detailed protocols for ATP quantitation

Unlike most other commercially available ATP detection kits, our ATP Determination Kit provides the luciferase and luciferin packaged separately, which enables researchers to optimize the reaction conditions for their particular instruments and samples. The ATP Determination Kit provides sufficient reagents for performing 200 ATP assays using 0.5 mL sample volumes or 500 ATP assays using 0.2 mL sample volumes.

REFERENCES

1. Methods Mol Biol (1994) 32:461; **2.** J Clin Microbiol (1984) 19:230; **3.** Anatomical Record (1954) 118:135; **4.** J Histochem Cytochem (1992) 40:1059; **5.** Dev Biol (1981) 88:279; **6.** Cell (1975) 5:229; **7.** Proc Natl Acad Sci U S A (1973) 70:3899; **8.** Biotechniques (1993) 14:818; **9.** Proc Natl Acad Sci U S A (1992) 89:693; **10.** Methods Enzymol (1992) 216:362; **11.** Proc Natl Acad Sci U S A (1963) 50:1; **12.** J Immunol Methods (1992) 149:261; **13.** Proc Natl Acad Sci U S A (2008) 105:4099; **14.** Anal Biochem (1997) 248:258; **15.** Anal Chem (1998) 70:1242; **16.** J Immunol Methods (1996) 192:165; **17.** J Biol Chem (1997) 272:22472; **18.** Biochem Pharmacol (1997) 54:721; **19.** Biochem Pharmacol (1997) 54:703; **20.** Toxicon (2000) 38:1833; **21.** J Immunol Methods (1992) 150:23; **22.** In Vitro Cell Dev Biol (1989) 25:105; **23.** Anal Biochem (1992) 205:1; **24.** Mol Cell Probes (1992) 6:489; **25.** J Microbiol Methods (2000) 40:147; **26.** Microbiol Rev (1991) 55:335; **27.** Anal Biochem (1999) 273:41; **28.** Bioorg Med Chem Lett (1998) 8:3107; **29.** Biochemistry (1998) 37:10381; **30.** Biophys J (1998) 74:3120; **31.** J Physiol (1997) 501:125; **32.** FEBS Lett (1995) 364:59; **33.** Biochemistry (1994) 33:8262; **34.** Biochemistry (2001) 40:5087; **35.** Cytometry (2001) 43:117; **36.** J Histochem Cytochem (1999) 47:1443; **37.** J Histochem Cytochem (1997) 45:345; **38.** Nature (2008) 454:221; **39.** Blood (2007) 109:3024; **40.** J Histochem Cytochem (2005) 53:1525; **41.** J Histochem Cytochem (2004) 52:411; **42.** J Histochem Cytochem (1996) 44:559; **43.** Proc Natl Acad Sci U S A (2003) 100:1592; **44.** Curr Biol (2000) 10:907; **45.** J Biol Chem (2000) 275:4555; **46.** Am J Pathol (2000) 156:419; **47.** J Natl Cancer Inst (2000) 92:338; **48.** J Biol Chem (2001) 276:29275; **49.** Anal Biochem (2001) 291:109; **50.** Clin Chem (1996) 42:1542; **51.** Biotechniques (1994) 17:172; **52.** J Immunol Methods (2001) 247:111; **53.** Biotechniques (1992) 12:656; **54.** Histochemistry (1978) 58:203; **55.** Biochemistry (1996) 35:11013; **56.** Anal Biochem (1996) 243:41; **57.** Anal Biochem (1997) 253:162; **58.** J Immunol Methods (1997) 202:133; **59.** Anal Biochem (1994) 218:449; **60.** Proc Natl Acad Sci U S A (1992) 89:4884; **61.** Anal Biochem (1995) 230:173; **62.** Biophys Chem (1997) 63:107; **63.** Biochemistry (1997) 36:11837; **64.** Biosci Rep (1994) 14:309; **65.** Nucleic Acids Res (1995) 23:2886; **66.** Anal Biochem (1997) 246:86; **67.** J Biol Chem (1999) 274:16153; **68.** Biochemistry (1995) 34:10976; **69.** Anal Biochem (1994) 221:348; **70.** J Biol Chem (2000) 275:25299; **71.** Biochemistry (1998) 37:5249; **72.** J Biol Chem (1997) 272:32830; **73.** Biochemistry (1995) 34:15592; **74.** Biochem J (1992) 287:555; **75.** J Biol Chem (1997) 272:21999; **76.** Biochemistry (1996) 35:10922; **77.** Biochemistry (1998) 37:17209; **78.** Biochem J (1995) 307:585; **79.** Anal Biochem (1994) 222:14; **80.** Biochem J (1993) 296:293; **81.** Anal Biochem (1995) 226:68; **82.** Anal Biochem (1995) 230:55; **83.** Biochemistry (1994) 33:2380; **84.** Biochemistry (1996) 35:12038; **85.** J Biol Chem (2001) 276:18478; **86.** J Biol Chem (2000) 275:17962; **87.** Biochemistry (2000) 39:2297; **88.** J Biol Chem (1998) 273:16555; **89.** J Biol Chem (1998) 273:22151; **90.** Biochemistry (2001) 40:5041; **91.** J Virol (2002) 76:11505; **92.** Free Radic Biol Med (2009) 47:983; **93.** J Biol Chem (2008) 283:3401; **94.** Nat Med (2006) 12:549; **95.** Nature (2005) 437:911; **96.** Nat Cell Biol (2005) 7:78; **97.** Comb Chem High Throughput Screen (2003) 6:341; **98.** Circulation (2003) 107:2775; **99.** Arterioscler Thromb Vasc Biol (2003) 23:1465; **100.** Anal Biochem (1989) 180:242; **101.** J Biomol Screen (1999) 4:67; **102.** J Biolumin Chemilumin (1995) 10:29; **103.** Anal Biochem (1988) 175:14; **104.** Anal Biochem (2002) 306:323; **105.** J Clin Microbiol (1984) 20:644; **106.** J Clin Microbiol (1983) 18:521; **107.** Methods Enzymol (1978) 57:3; **108.** Biotechnol Bioeng (1993) 42:30; **109.** J Biolumin Chemilumin (1991) 6:193; **110.** Biochem J (1993) 295:165; **111.** J Immunol Methods (1993) 160:81; **112.** J Natl Cancer Inst (1986) 77:1039-1045.

DATA TABLE 10.3 DETECTING ENZYMES THAT METABOLIZE PHOSPHATES AND POLYPHOSPHATES

Cat. No.	MW	Storage	Soluble	Abs	EC	Em	Solvent	Product	Notes
A22184	878.28	FF,L	H_2O	504	68,000	514	pH 7	see Notes	1, 2, 3
D6487	422.20	F,D,L	DMSO, H_2O	478	26,000	628	pH 7	H6482	
D6567	292.13	F,D	DMSO, H_2O	320	14,000	385	pH 9	D6566	4
D22065	292.13	F,D	DMSO, H_2O	320	14,000	385	pH 9	D6566	4
E6588	431.08	F,L	H_2O	289	12,000	see Notes	pH 10	E6578	2, 4
E6589	431.08	F,L	H_2O	289	12,000	see Notes	pH 10	E6578	2, 4
E6645	313.33	FF,D	H_2O	332	16,000	none	pH 7	see Notes	5, 6
E6646	313.33	F,D	H_2O	332	16,000	none	pH 7	see Notes	5, 6
F2999	560.39	F,D	H_2O	272	5300	none	MeOH	F1300	7
G22183	894.28	FF,L	H_2O	504	68,000	510	pH 7	see Notes	1, 2, 3, 8
M6491	256.15	F,D	DMSO, H_2O	319	15,000	383	pH 9	H189	4
N6495	817.65	D,L	H_2O, DMSO	256	64,000	none	MeOH	see Notes	9

For definitions of the contents of this data table, see "Using *The Molecular Probes® Handbook*" in the introductory pages.

Notes

1. The molecular weight (MW) of this product is approximate because the degree of hydration and/or salt form has not been conclusively established.
2. This product is supplied as a ready-made solution in the solvent indicated under "Soluble."
3. This compound is a nonhydrolyzable substrate analog.
4. Fluorescence of the unhydrolyzed substrate is very weak.
5. Data represent the substrate component of this kit.
6. Enzymatic phosphorylation of this substrate yields 2-amino-6-mercapto-7-methylpurine (Abs = 355 nm). (Proc Natl Acad Sci U S A (1992) 89:4884)
7. Unstable in water. Use immediately.
8. Fluorescence of BODIPY® dye–labeled guanosine derivatives is generally weak due to base-specific intramolecular quenching. (Anal Biochem (2001) 291:109)
9. Phosphatase hydrolysis of BCIP (B6492) is coupled to reduction of NBT (N6495), yielding a water-insoluble indigo dye (Abs ~615 nm) and a water-insoluble formazan (Abs ~605 nm), respectively.

IMPORTANT NOTICE: The products described in this manual are covered by one or more Limited Use Label License(s). Please refer to the Appendix on page 971 and Master Product List on page 975. Products are For Research Use Only. Not intended for any animal or human therapeutic or diagnostic use.

PRODUCT LIST 10.3 DETECTING ENZYMES THAT METABOLIZE PHOSPHATES AND POLYPHOSPHATES

Cat. No.	Product	Quantity
A22184	adenosine 5′-*O*-(3-thiotriphosphate), BODIPY® FL thioester, sodium salt (BODIPY® FL ATP-γ-*S*, thioester) *5 mM in buffer*	50 µL
A35725	Antibody Beacon™ Tyrosine Kinase Assay Kit *400 assays*	1 kit
A22066	ATP Determination Kit *special packaging* *200–1000 assays*	1 kit
C10552	Chemiluminescent Alkaline Phosphatase ELISA Kit #1 *with CSPD® Substrate/ Sapphire-II™ enhancer* *1000 assays*	1 kit
C10553	Chemiluminescent Alkaline Phosphatase ELISA Kit #2 *with CSPD® Substrate/ Emerald-II™ enhancer* *1000 assays*	1 kit
C10554	Chemiluminescent Alkaline Phosphatase ELISA Kit #3 *with CDP-*Star*® Substrate/ Sapphire-II™ enhancer* *1000 assays*	1 kit
C10555	Chemiluminescent Alkaline Phosphatase ELISA Kit #4 *with CDP-*Star*® Substrate/ Emerald-II™ enhancer* *1000 assays*	1 kit
C10556	Chemiluminescent Alkaline Phosphatase ELISA Sampler Kit *1000 assays*	1 kit
D6487	9*H*-(1,3-dichloro-9,9-dimethylacridin-2-one-7-yl) phosphate, diammonium salt (DDAO phosphate)	5 mg
D6567	6,8-difluoro-4-methylumbelliferyl phosphate (DiFMUP)	5 mg
D22065	6,8-difluoro-4-methylumbelliferyl phosphate (DiFMUP) *packaged for high-throughput screening*	10 x 10 mg
E6603	ELF® 97 Cytological Labeling Kit *with streptavidin, alkaline phosphatase conjugate* *50 assays*	1 kit
E6601	ELF® 97 Endogenous Phosphatase Detection Kit	1 kit
E6600	ELF® 97 Immunohistochemistry Kit	1 kit
E6604	ELF® 97 mRNA *In Situ* Hybridization Kit #1 *50 assays*	1 kit
E6605	ELF® 97 mRNA *In Situ* Hybridization Kit #2 *with streptavidin, alkaline phosphatase conjugate* *50 assays*	1 kit
E6588	ELF® 97 phosphatase substrate (ELF® 97 phosphate) *5 mM in water* *0.2 µm filtered*	1 mL
E6589	ELF® 97 phosphatase substrate (ELF® 97 phosphate) *5 mM in water* *contains 2 mM azide*	1 mL
E6606	ELF® spin filters *20 filters*	1 box
E33702	EnzChek® Paraoxonase Assay Kit *100 assays*	1 kit
E12020	EnzChek® Phosphatase Assay Kit *1000 assays*	1 kit
E6646	EnzChek® Phosphate Assay Kit *100 assays*	1 kit
E6645	EnzChek® Pyrophosphate Assay Kit *100 assays*	1 kit
E33701	EnzChek® *Ultra* Phytase Assay Kit *500 assays*	1 kit
F2999	fluorescein diphosphate, tetraammonium salt (FDP)	5 mg
G22183	guanosine 5′-*O*-(3-thiotriphosphate), BODIPY® FL thioester, sodium salt (BODIPY® FL GTP-γ-*S*, thioester) *5 mM in buffer*	50 µL
M6491	4-methylumbelliferyl phosphate, free acid (MUP)	1 g
N6547	NBT/BCIP Reagent Kit	1 kit
N6495	nitro blue tetrazolium chloride (NBT)	1 g
N10559	NovaBright™ Secreted Placental Alkaline Phosphatase (SEAP) Enzyme Reporter Gene Chemiluminescent Detection Kit *400 assays*	1 kit
N10560	NovaBright™ Secreted Placental Alkaline Phosphatase (SEAP) Enzyme Reporter Gene Chemiluminescent Detection Kit *1200 assays*	1 kit
N10577	NovaBright™ Secreted Placental Alkaline Phosphatase (SEAP) Enzyme Reporter Gene Chemiluminescent Detection Kit 2.0 *192 assays*	1 kit
N10578	NovaBright™ Secreted Placental Alkaline Phosphatase (SEAP) Enzyme Reporter Gene Chemiluminescent Detection Kit 2.0 *960 assays*	1 kit
P22061	P_iPer™ Phosphate Assay Kit *1000 assays*	1 kit
P22062	P_iPer™ Pyrophosphate Assay Kit *1000 assays*	1 kit
R33700	RediPlate™ 96 EnzChek® Serine/Threonine Phosphatase Assay Kit *one 96-well microplate*	1 kit
R22067	RediPlate™ 96 EnzChek® Tyrosine Phosphatase Assay Kit *one 96-well microplate*	1 kit

IMPORTANT NOTICE: The products described in this manual are covered by one or more Limited Use Label License(s). Please refer to the Appendix on page 971 and Master Product List on page 975. Products are For Research Use Only. Not intended for any animal or human therapeutic or diagnostic use.

10.4 Detecting Peptidases and Proteases

Peptidases and proteases play essential roles in protein activation, cell regulation and signaling, as well as in the generation of amino acids for protein synthesis or utilization in other metabolic pathways. In general, peptidases cleave shorter peptides, and proteases cleave longer peptides and proteins. Depending on their site of cleavage, peptidases can be classified as exopeptidases if they preferentially hydrolyze amino acid residues from the terminus of a peptide, or endopeptidases if they cleave internal peptide bonds.[1] Exopeptidases are further divided into aminopeptidases and carboxypeptidases depending on whether they hydrolyze residues from the amine or the carboxy terminus.

Although the spectral properties of fluorogenic peptidase and protease substrates and their hydrolysis products are easily predictable, the utility of a given substrate for an enzyme depends on the kinetics of hydrolysis by the enzyme, which, in turn, depends on the substrate's concentration and amino acid sequence, as well as on the pH, temperature and presence of cofactors in the medium. For measurements in live cells, the suitability of a particular substrate also hinges on its accessibility to the enzyme and the cellular retention of the hydrolysis product.[2] In addition to these factors, the chromophore or fluorophore conjugated to the substrate can influence its hydrolysis rate and specificity, as well as the permeability of the substrate and its hydrolysis product.

We prepare a variety of protease substrates, including selective protease substrates for caspase-3 and caspase-8—enzymes that are activated during apoptosis (Section 15.5)—and for HIV protease and renin. Our EnzChek® and DQ™ protease substrates include:

- EnzChek® Protease Assay Kits (E6638, E6639, R22132), which use a heavily BODIPY® dye–labeled DQ™ casein derivative for the assay of a wide variety of proteases (Table 10.4).
- EnzChek® Peptidase/Protease Assay Kit (E33758), which uses a fluorescence resonance energy transfer (FRET)–based substrate to detect protease activity.
- The EnzChek® Polarization Assay Kit for Proteases (E6658), which uses a green-fluorescent BODIPY® FL casein conjugate with an optimal degree of labeling for fluorescence polarization-based general protease assays (Fluorescence Polarization (FP)—Note 1.4).
- EnzChek® Gelatinase/Collagenase Assay Kit (E12055) and EnzChek® Elastase Assay Kit (E12056), which use DQ™ gelatin or DQ™ elastin as substrates, provide the speed, high sensitivity and convenience required for measuring gelatinase (collagenase) or elastase activity[3] and for screening protease inhibitors in a high-throughput format.
- DQ™ collagen (D12052, D12060), DQ™ BSA (D12050, D12051) and DQ™ ovalbumin (D12053), which are useful for screening protease activity and inhibitors (including *in situ* zymography[4]) and for studying antigen processing.

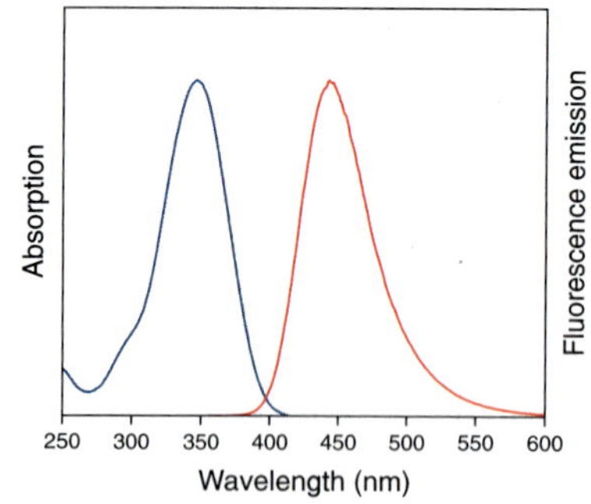

Figure 10.4.1 Absorption and fluorescence emission spectra of 7-amino-4-methylcoumarin in pH 7.0 buffer.

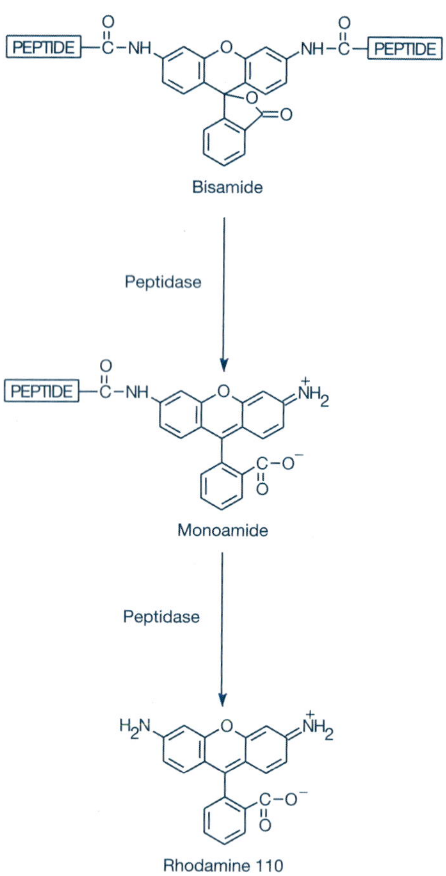

Figure 10.4.2 Sequential peptidase cleavage of a rhodamine 110–based substrate. The nonfluorescent bisamide substrate is first converted to the fluorescent monoamide and then to the highly fluorescent rhodamine 110.

Peptidase Substrates

The carboxy terminus of single amino acids and short peptides can be conjugated to certain amine-containing fluorophores to create fluorogenic peptidase substrates. The dyes used to make these substrates are fluorescent at physiological pH; however, when the dyes are coupled in an amide linkage to peptides, their absorption maxima are usually shortened significantly. The resulting substrates are sometimes fluorescent but with relatively short-wavelength emission spectra. In an extreme case such as that of rhodamine 110–based substrates, detectable long-wavelength absorbance and fluorescence are completely eliminated by amide formation. Peptidase activity releases the fluorophore, restoring its free-dye fluorescence.

UV Light–Excitable Substrates Based on 7-Aminocoumarins

7-Amino-4-methylcoumarin (AMC, A191; Section 10.1) is a blue-fluorescent dye (Figure 10.4.1) whose peptide amides are used extensively as substrates for detecting enzymatic activity in cells, homogenates and solutions. The CBZ-L-phenylalanyl-L-arginine amide of AMC (A6521) is a substrate for a variety of serine proteases, including cathepsins, kallikrein and plasmin.[5–14] AMC and 7-amino-4-trifluoromethylcoumarin (AFC)—a dye with somewhat longer-wavelength

IMPORTANT NOTICE: The products described in this manual are covered by one or more Limited Use Label License(s). Please refer to the Appendix on page 971 and Master Product List on page 975. Products are For Research Use Only. Not intended for any animal or human therapeutic or diagnostic use.

spectra than AMC (excitation/emission maxima of ~380/500 nm) at pH 7—are released from the caspase-3, caspase-7 and caspase-8 substrates listed in Table 15.5; these caspases are activated during early stages of apoptosis (Section 15.5). The Z-DEVD-AMC substrate is also a component of the EnzChek® Caspase-3 Assay Kit #1 (E13183).

7-Amino-4-chloromethylcoumarin (CMAC, C2110; Section 10.1) is a mildly thiol-reactive analog of AMC; CMAC-based substrates yield fluorescent peptidase products with improved retention in live cells. The fluorogenic *t*-BOC-Leu-Met-CMAC substrate [15,16] (A6520) has been used to measure calpain activity in hepatocytes following the addition of extracellular ATP.[17,18]

Visible Light–Excitable Substrates Based on Rhodamine 110

Rhodamine 110 (R110, R6479; Section 10.1) is a visible light–excitable dye with much stronger absorption than AMC; R110-based substrates usually comprise two identical amino acids or peptides attached to a single fluorophore. Molecular Probes® bisamide derivatives of rhodamine 110 are sensitive and selective substrates for assaying protease activity in solution or inside live cells. Originally developed by Walter Mangel and colleagues, these substrates comprise an amino acid or peptide covalently linked to each of R110's amino groups, thereby suppressing both its visible absorption and fluorescence.[19,20] Upon enzymatic cleavage, the nonfluorescent bisamide substrate is converted in a two-step process first to the fluorescent monoamide and then to the even more fluorescent R110 (Figure 10.4.2). The fluorescence intensities of the monoamide and of R110 are constant from pH 3–9. Both of these hydrolysis products exhibit spectral properties similar to those of fluorescein, with peak excitation and emission wavelengths of 496 nm and 520 nm (Figure 10.4.3), respectively, making them compatible with flow cytometers [21] and other instrumentation based on the argon-ion laser. Substrates based on R110 may also be useful for sensitive spectrophotometric assays because the R110 dye has intense visible absorption (extinction coefficient at 496 nm = ~80,000 $cm^{-1}M^{-1}$ in pH 6 solution).

We prepare a variety of substrates based on the rhodamine 110 fluorophore (Table 10.2). Bis-(CBZ-Arg)-R110 (BZAR, R6501) is a general substrate for serine proteases that has proven to be 50- to 300-fold more sensitive than the analogous AMC-based substrate.[19,22] This enhanced sensitivity can be attributed both to the greater fluorescence of the enzymatic product and to the enhanced reactivity of the cleavage site. In addition, BZAR inhibits guanidinobenzoatase activity in tumor cells.[22,23] The tripeptide derivative bis-(CBZ-Ile-Pro-Arg)-R110 (BZiPAR, R6505) allows direct and continuous monitoring of enzyme turnover, making it useful for determining individual kinetic constants of fast-acting, irreversible trypsin inhibitors.[24] BZiPAR has been shown to enter intact cells where it is cleaved by lysosomal proteases.[25] Simultaneous

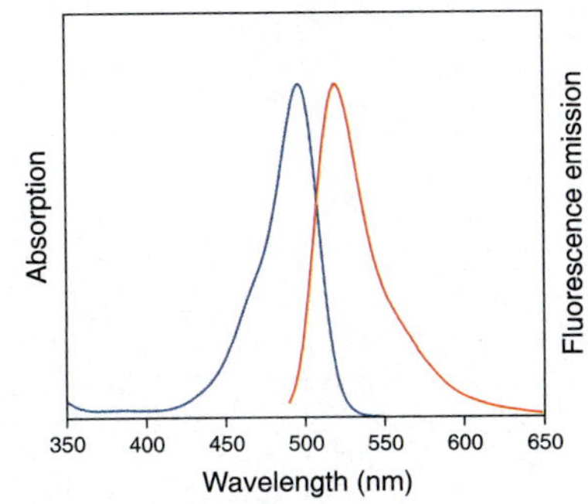

Figure 10.4.3 Absorption and fluorescence emission spectra of rhodamine 110 in pH 7.0 buffer.

Table 10.2 Rhodamine 110–based bis-peptide substrates.

Cat. No.	Proteinase Substrate *	Enzymes †
R22122	$(Asp)_2$-R110	caspase-3 and other apoptosis-related proteases [1]
R6506	$(CBZ\text{-}Ala\text{-}Ala\text{-}Ala\text{-}Ala)_2$-R110	elastase [2]
R33752	$(CBZ\text{-}Ala\text{-}Ala\text{-}Asp)_2$-R110	granzyme B
R6508	$(CBZ\text{-}Ala\text{-}Arg)_2$-R110	trypsin [3]
R6501	$(CBZ\text{-}Arg)_2$-R110	trypsin [3,4]
R22120	$(CBZ\text{-}Asp\text{-}Glu\text{-}Val\text{-}Asp)_2$-R110	caspase-3, -6, -7, -8, -10 [1]
R6505	$(CBZ\text{-}Ile\text{-}Pro\text{-}Arg)_2$-R110	trypsin [5,6]
R22125, R22126	$(CBZ\text{-}Ile\text{-}Glu\text{-}Thr\text{-}Asp)_2$-R110	caspase-8
R33753	$(CBZ\text{-}Leu\text{-}Glu\text{-}Glu\text{-}Asp)_2$-R110	caspase-13
R6502	$(CBZ\text{-}Phe\text{-}Arg)_2$-R110	plasmin, cathepsin L [7–9]
R33750	$(CBZ\text{-}Tyr\text{-}Val\text{-}Ala\text{-}Asp)_2$-R110	caspase-1, -4
R33755	$(CBZ\text{-}Val\text{-}Asp\text{-}Val\text{-}Ala\text{-}Asp)_2$-R110	caspase-2
R33754	$(CBZ\text{-}Val\text{-}Glu\text{-}Ile\text{-}Asp)_2$-R110	caspase-6
R22124	$(p\text{-}tosyl\text{-}Gly\text{-}Pro\text{-}Arg)_2$-R110	thrombin [10]

* CBZ = benzyloxycarbonyl; R110 = rhodamine 110; SC = succinoyl; *p*-tosyl = *p*-toluenesulfonyl. † Caspase substrates can often be cleaved by multiple enzymes. The caspase most often associated with the given peptide sequence is listed first.
1. Biochemistry (1999) 38:13906; **2.** Anal Chem (1993) 65:2352; **3.** Biochem J (1983) 215:253; **4.** Biochem J (1983) 209:299; **5.** Biochim Biophys Acta (1984) 788:74; **6.** Photochem Photobiol (1986) 44:461; **7.** Biol Chem Hoppe Seyler (1992) 373:547; **8.** Glia (1993) 7:183; **9.** Biol Chem Hoppe Seyler (1992) 373:433; **10.** Biomed Instrum Technol (1996) 30:245.

measurement of enzymatic activity with BZiPAR and Ca^{2+} transients with fura-2 (F1201, F1221, F1225, F14185; Section 19.2) has been reported.[26] Bis-(CBZ-Phe-Arg)-R110 (R6502) has been employed for flow cytometric analysis of the cysteine proteases cathepsin B and L in human monocytes and rat macrophages.[27–29] Bis-(CBZ-Ala-Ala-Ala-Ala)-R110 (R6506), an elastase substrate, has been used in a novel DNA detection assay.[30] Bis-(CBZ-Ala-Arg)-R110 (R6508) is a fluorogenic substrate for both elastase and trypsin.[19,28] The bis-(tosyl-Gly-Pro-Arg) amide of rhodamine 110 (R22124) is a selective substrate for thrombin.[31] Turnover of this substrate by thrombin on a membrane in the presence of thromboplastin produces both color and fluorescence that has been reported to model coagulation and blood clot formation. We also offer the human renin substrate 1 (R2931) for measuring the activity of this important blood-pressure–regulating enzyme.

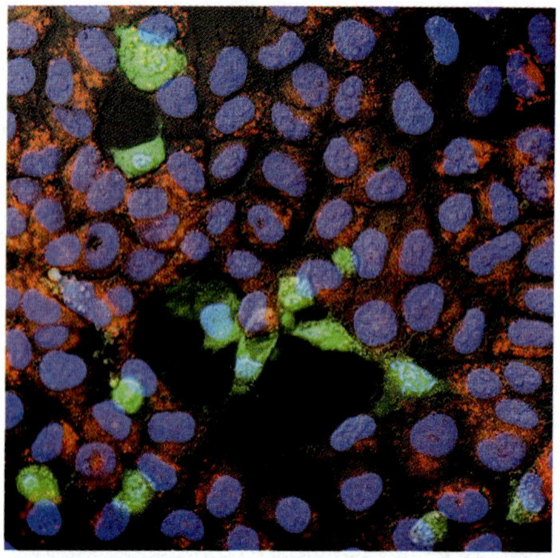

Figure 10.4.4 Detection of apoptosis in SK-N-MC neuroblastoma cells. Following a 6-hour exposure to hydrogen peroxide, cells were labeled with Hoechst 33342 (H1399, H3570, H21492), tetramethylrhodamine ethyl ester (TMRE, T669) and rhodamine 110, bis-L-aspartic acid amide (R22122) for 15 minutes. Apoptotic cells show green cytosolic fluorescence resulting from cleavage of the rhodamine 110, bis-L-aspartic acid amide substrate by active caspase-3. The staining pattern of the Hoechst 33342 dye reveals that the majority of the rhodamine 110–positive cells also contain condensed or fragmented nuclei characteristic of apoptosis. Furthermore, the rhodamine 110–positive cells are also characterized by an absence of polarized mitochondria, as indicated by their failure to load the positively charged mitochondrial indicator TMRE. The image was contributed by A.K. Stout and J.T. Greenamyre, Emory University.

Caspase Substrates and Assay Kits

Members of the caspase (CED-3/ICE) family of cysteine–aspartic acid specific proteases have been identified as crucial mediators of the complex biochemical events associated with apoptosis.[32–35] The recognition site for caspases is marked by three to four amino acids followed by an aspartic acid residue, with the cleavage occurring after the aspartate.[36] The caspase proteases are typically synthesized as inactive precursors. Inhibitor release or cofactor binding activates the caspase through cleavage at internal aspartates, either by autocatalysis or by the action of another protease.[33]

Caspase-3 Substrates

Caspase-3 (CPP32/apopain) is a key effector in the apoptosis pathway, amplifying the signal from initiator caspases (such as caspase-8) and signifying full commitment to cellular disassembly. In addition to cleaving other caspases in the enzyme cascade, caspase-3 has been shown to cleave poly(ADP-ribose) polymerase (PARP), DNA-dependent protein kinase, protein kinase C and actin.[34,37] We offer a selection of fluorogenic caspase substrates (Table 15.5). The Z-DEVD-R110 substrate,[38,39] which contains the caspase-3 recognition site Asp-Glu-Val-Asp (DEVD), is a component of our EnzChek® Caspase-3 Assay Kit #2 (E13184) and RediPlate™ 96 EnzChek® Caspase-3 Assay Kit (R35100; Table 10.3) described below and is available separately in a 20 mg unit size for high-throughput screening applications (R22120, Table 10.2). This nonfluorescent bisamide is first converted by caspase-3 (or a closely related protease) to the fluorescent monoamide and then to the even more fluorescent rhodamine 110 (excitation/emission maxima ~496/520 nm). In addition, the bis-L-aspartic acid amide of R110 (D_2-R110, R22122), which contains rhodamine 110 (R110) flanked by aspartic acid residues, may serve as a substrate for a variety of apoptosis-related proteases, including caspase-3 and caspase-7,[38] and does not appear to require any invasive techniques such as osmotic shock to gain entrance into the cytoplasm (Figure 10.4.4).

Table 10.3 RediPlate™ Assay Kits.

RediPlate™ Assay Kit	Cat. No.	Assay Details
EnzChek® Caspase-3 Assay Kit (Section 10.4)	R35100	A direct fluorescence-based assay for detecting caspase-3 and other DEVD-selective protease activities (e.g., caspase-7) in cell extracts and purified enzyme preparations. Caspase activity is indicated by increasing green fluorescence (excitation/emission maxima ~496/520 nm).
EnzChek® Protease Assay Kit *red fluorescence* (Section 10.4)	R22132	A direct fluorescence-based assay for detecting metallo-, serine, acid and sulfhydryl proteases and their inhibitors. Protease activity is indicated by increasing red fluorescence (excitation/emission maxima ~589/617 nm).
EnzChek® Tyrosine Phosphatase Assay Kit (Section 10.3)	R22067	An assay for monitoring protein tyrosine phosphatases (PTP) and for screening PTP inhibitors and modulators. PTP activity is indicated by increasing blue fluorescence (excitation/emission maxima ~358/452 nm).
EnzChek® Serine/Threonine Phosphatase Assay Kit (Section 10.3)	R33700	An assay for monitoring serine/threonine (Ser/Thr) phosphatases and for screening Ser/Thr inhibitors and modulators. Ser/Thr phosphatase activity is indicated by increasing blue fluorescence (excitation/emission maxima ~358/452 nm).
RiboGreen® RNA Quantitation Kit (Section 8.3)	R32700	An assay for accurately quantitating low levels of RNA (1 ng/mL) with a broad dynamic range—over three orders of magnitude using two different dye concentrations. When bound to RNA, the RiboGreen® reagent exhibits excitation/emission maxima of ~500/520 nm.

IMPORTANT NOTICE: The products described in this manual are covered by one or more Limited Use Label License(s). Please refer to the Appendix on page 971 and Master Product List on page 975. Products are For Research Use Only. Not intended for any animal or human therapeutic or diagnostic use.

Caspase-8 Substrates

Caspase-8 plays a critical role in the early cascade of apoptosis, acting as an initiator of the caspase activation cascade. Activation of the enzyme itself is accomplished through direct interaction with the death domains of cell-surface receptors for apoptosis-inducing ligands.[40,41] The activated protease has been shown to be involved in a pathway that mediates the release of cytochrome *c* from the mitochondria[42] and is also known to activate downstream caspases, such as caspase-3.[43] The fluorogenic substrate Z-IETD-R110 (R22125, R22126; green fluorescent after cleavage) contains the caspase-8 recognition sequence Ile-Glu-Thr-Asp (IETD) (Table 15.5).

Other Caspase and Granzyme B Substrates

In addition to our R110-derived caspase-3 and -8 substrates, we offer R110-based substrates for caspase-1, -2, -6, -9 and -13, as well as substrates for granzyme B (Table 15.5). Granzyme B, a serine protease contained within cytotoxic T lymphocytes and natural killer cells, is thought to induce apoptosis in target cells by activating caspases and causing mitochondrial cytochrome *c* release.[44]

EnzChek® Caspase-3 Assay Kits

The EnzChek® Caspase-3 Assay Kits permit the detection of apoptosis by assaying for increases in caspase-3 and caspase-3–like protease activities (Figure 10.4.5, Figure 10.4.6). Our EnzChek® Caspase-3 Assay Kit #1 (E13183) contains the 7-amino-4-methylcoumarin (AMC)–derived substrate Z-DEVD-AMC (Figure 10.4.7) (where Z represents a benzyloxycarbonyl group). This substrate, which is weakly fluorescent in the UV spectral range (excitation/emission maxima ~330/390 nm), yields the blue–fluorescent product AMC (A191, Section 10.1, Figure 10.4.1), which has excitation/emission maxima of 342/441 nm upon proteolytic cleavage.

The EnzChek® Caspase-3 Assay Kit #2 (E13184) contains the R110-derived substrate, Z-DEVD-R110[38,39] (Figure 10.4.8). This substrate is a bisamide derivative of R110, containing DEVD peptides covalently linked to each of R110's amino groups, thereby suppressing both the dye's visible absorption and fluorescence. Upon enzymatic cleavage by caspase-3 (or a closely related protease), the nonfluorescent bisamide substrate is converted in a two-step process first to the fluorescent monoamide and then to the even more fluorescent R110 (R6479, Section 10.1, Figure 10.4.2, Figure 10.4.3). Both of these hydrolysis products exhibit spectral properties similar to those of fluorescein, with excitation/emission maxima of 496/520 nm. The Z-DEVD-R110 substrate (R22120) is also available separately in a 20 mg unit size for high-throughput screening applications.

Either kit can be used to continuously measure the activity of caspase-3 and closely related proteases in cell extracts and purified enzyme preparations using a fluorescence microplate reader or fluorometer. AMC-based DEVD substrates, which yield blue fluorescence upon proteolytic cleavage, are widely used to monitor caspase-3 activity.[37,45] The longer-wavelength spectra and higher extinction coefficient of the green-fluorescent products of the R110-based substrate in Kit #2 (E13184) should provide even greater sensitivity.[38,39] The reversible aldehyde-based inhibitor Ac-DEVD-CHO can be used to confirm that the observed fluorescence signal in both induced and control cell populations is due to the activity of caspase-3–like proteases.[37]

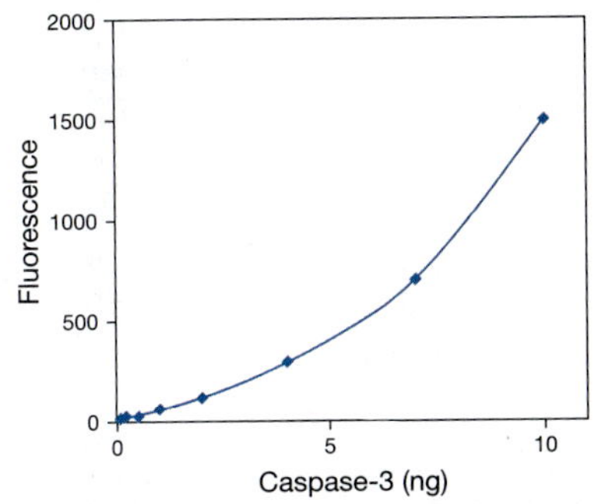

Figure 10.4.5 Detection of caspase-3 activity using the EnzChek® Caspase-3 Assay Kit #1 (E13183). Increasing amounts of purified active human (recombinant) caspase-3 (PharMingen) were allowed to react with 100 µM Z-DEVD–AMC in 1X reaction buffer for ~45 minutes at room temperature. Fluorescence was measured in a fluorescence microplate reader using excitation at 360 ± 17.5 nm and emission detection at 465 ± 17.5 nm. Background fluorescence (386 arbitrary units), determined for a no-enzyme control, was subtracted from each value.

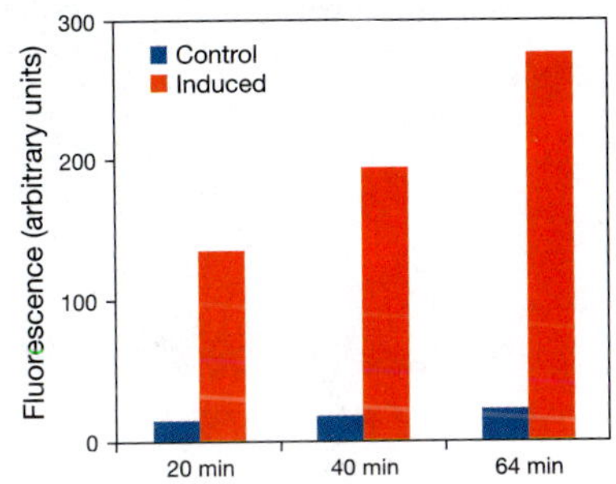

Figure 10.4.6 Detection of protease activity in Jurkat cells using the EnzChek® Caspase-3 Assay Kit #1 with Z-DEVD-AMC substrate (E13183). Cells were either treated with 10 µM camptothecin for 4 hours at 37°C to induce apoptosis (induced) or left untreated (control). Both induced and control cells were then harvested, lysed and assayed. Reactions were carried out at room temperature, and fluorescence was measured in a fluorescence microplate reader using excitation at 360 ± 20 nm with emission detection at 460 ± 20 nm after the indicated amount of time.

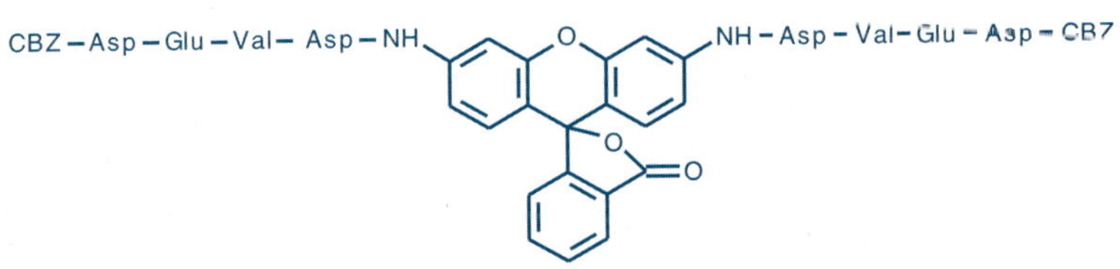

Figure 10.4.8 Rhodamine 110, bis-(*N*-CBZ-L-aspartyl-L-glutamyl- L-valyl-L-aspartic acid amide, Z-DEVD-R110, E13184).

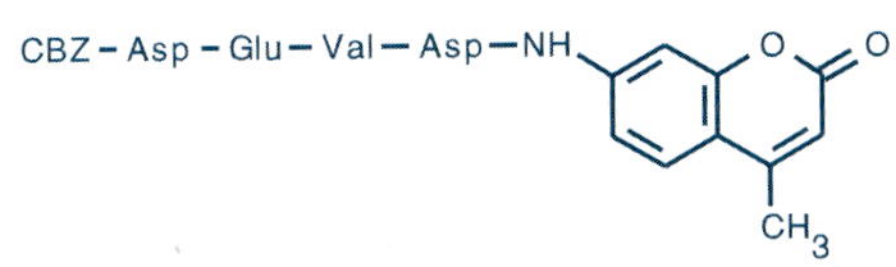

Figure 10.4.7 7-amino-4-methylcoumarin, *N*-CBZ-L-aspartyl-L-glutamyl-L-valyl-L-aspartic acid amide (Z-DEVD-AMC, E13183).

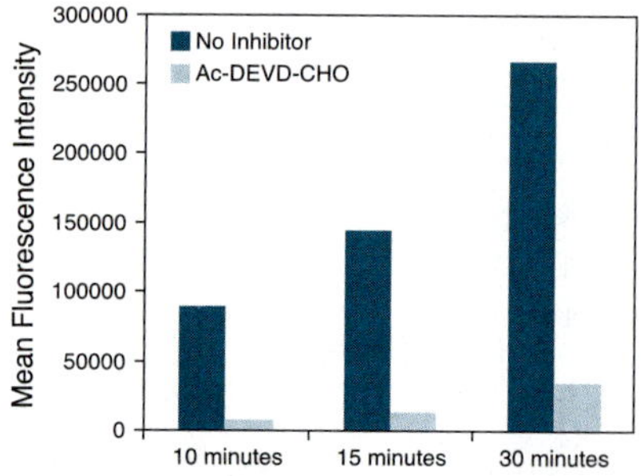

Figure 10.4.9 Detection of protease activity in Jurkat cells using the RediPlate™ 96 EnzChek® Caspase-3 Assay Kit (R35100). Jurkat human T-cell leukemia cells were first exposed to 10 µM camptothecin at 37°C to induce apoptosis, and then harvested and lysed according to the kit protocol. The cell lysate was separated into two samples, one of which was treated with the Ac-DEVD-CHO inhibitor (provided in the RediPlate™ 96 EnzChek® Caspase-3 Assay Kit). Assay reactions on both the inhibited and the uninhibited samples were carried out at 37°C, and fluorescence was measured in a fluorescence microplate reader (excitation/emission = 485/535 nm).

Figure 10.4.10 A RediPlate™ 96 microplate.

The EnzChek® Caspase-3 Assay Kits contain:

- Z-DEVD-AMC[46,47] (in Kit #1, E13183) or Z-DEVD-R110[38,48] (in Kit #2, E13184)
- Dimethylsulfoxide (DMSO)
- Concentrated cell-lysis buffer
- Concentrated reaction buffer
- Dithiothreitol (DTT)
- Ac-DEVD-CHO, a reversible aldehyde-based inhibitor
- 7-Amino-4-methylcoumarin (AMC) (in Kit E13183) or rhodamine 110 (in Kit E13184) reference standard to quantitate the amount of fluorophore released in the reaction
- Detailed protocols

Each kit provides sufficient reagents for performing ~500 assays using a volume of 100 µL per assay. For information about additional kits and reagents for studying apoptosis, see Section 15.5.

RediPlate™ 96 EnzChek® Caspase-3 Assay Kit

Our EnzChek® Caspase-3 Assay Kit #2 is also available as a convenient RediPlate™ 96 EnzChek® Caspase-3 Assay Kit (R35100), which includes one 96-well microplate, contained in a resealable foil packet to ensure the integrity of the fluorogenic components, plus all necessary buffers and reagents for performing the assay (Figure 10.4.9). The enzyme sample to be assayed is added to the microplate in a suitable buffer, along with any compounds to be tested. Then, after incubation, the resultant fluorescence is quantitated on a fluorescence microplate reader equipped with filters appropriate for the green-fluorescent R110, with excitation/emission maxima of 496/520 nm. The microplate consists of twelve removable strips, each with eight wells, allowing researchers to perform only as many assay as required for the experiment (Figure 10.4.10). Eleven of the strips (88 wells) are preloaded with the Z-DEVD-R110 substrate. The remaining strip, marked with blackened tabs, contains a dilution series of free R110 that may be used as a fluorescence reference standard. The reversible aldehyde-based inhibitor Ac-DEVD-CHO, which is supplied in a separate vial, can be used to confirm that the observed fluorescence signal in both induced and control cell populations is due to the activity of caspase-3-like proteases.[37] Table 10.3 summarizes our other RediPlate™ 96 Assay Kits for protease activity, phosphatase activity (Section 10.3) and RNA quantitation (Section 8.3).

Image-iT® LIVE Green Caspase Detection Kits for Fluorescence Microscopy

The Image-iT® LIVE Green Caspase-3 and -7 Detection Kit, Image-iT® LIVE Green Caspase-8 Detection Kit and Image-iT® LIVE Green Poly Caspases Detection Kit (I35106, I35105, I35104) employ a novel approach to detect active caspases that is based on a fluorescent inhibitor of caspases (FLICA® methodology). The FLICA® inhibitor comprises a fluoromethyl ketone (FMK) moiety, which can react covalently with a cysteine, a caspase-selective amino acid sequence and a fluorescent carboxyfluorescein reporter group. Essentially an affinity label, the FLICA® inhibitor is thought to interact with the enzymatic reactive center of an activated caspase via the recognition sequence, and then to attach covalently to a cysteine through the reactive FMK moiety.[49] The FLICA® inhibitor's recognition sequence is aspartic acid–glutamic acid–valine–aspartic acid (DEVD) for caspase-3 and-7 detection, leucine–glutamic acid–threonine–aspartic acid (LETD) for caspase-8 detection and valine–alanine–aspartic acid (VAD) for detection of most caspases (including caspase-1, -3, -4, -5, -6, -7, -8 and -9). Importantly, the FLICA® inhibitor is cell permeant and not cytotoxic; unbound FLICA® molecules diffuse out of the cell and are washed away. The remaining green-fluorescent signal (excitation/emission maxima ~488/530 nm) can be used as a direct measure of the amount of active caspase that was present at the time the inhibitor was added. FLICA® reagents have been used widely to study apoptosis with flow cytometry and microscopy.[50–54] Recent work indicates that cellular fluorescence from the bound FLICA® reagent is strongly linked to caspase activity in apoptotic cells; however, the interaction of the FLICA® reagent with other cellular sites may contribute to signal intensity in nonapoptotic cells.[55] Appropriate controls should be included in any experimental design.

IMPORTANT NOTICE: The products described in this manual are covered by one or more Limited Use Label License(s). Please refer to the Appendix on page 971 and Master Product List on page 975. Products are For Research Use Only. Not intended for any animal or human therapeutic or diagnostic use.
www.invitrogen.com/probes
molecular probes® | invitrogen™ by life technologies™

The Image-iT® LIVE Green Caspase Detection Kit includes:

- FAM-DEVD-FMK caspase-3 and -7 reagent (in Kit I35106), FAM-LETD-FMK caspase-8 reagent (in Kit I35105) or FAM-VAD-FMK poly caspases reagent (in Kit I35104)
- Hoechst 33342
- Propidium iodide
- Dimethylsulfoxide (DMSO)
- Apoptosis fixative solution
- Concentrated apoptosis wash buffer
- Detailed protocols for fluorescence microcscopy assays

In addition to a specific FLICA® reagent, each kit provides Hoechst 33342 and propidium iodide stains, which allow the simultaneous evaluation of caspase activation, nuclear morphology and plasma membrane integrity. Sufficient reagents are provided for 25 assays, based on labeling volumes of 300 µL. These Image-iT® LIVE Green Caspase Detection Kits can also be used in combination with other reagents for multiparametric study of apoptosis.

Image-IT® LIVE Red Caspase Detection Kits for Fluorescence Microscopy

The Image-iT® LIVE Red Caspase-3 and -7 Detection Kit and Image-iT® LIVE Red Poly Caspases Detection Kit (I35102, I35101) are analogous to the Image-iT® LIVE Green Caspase Detection Kits except that the FLICA® reagent contains a red-fluorescent sulforhodamine (SR) reporter group instead of a green-fluorescent carboxyfluorescein (FAM) reporter group. This assay's red-fluorescent signal (excitation/emission maxima ~550/595 nm) can be used as a direct measure of the amount of active caspase that was present at the time the inhibitor was added.

The Image-iT® LIVE Red Caspase Detection Kit includes:

- SR-DEVD-FMK caspase-3 and -7 reagent (in Kit I35102) or SR-VAD-FMK poly caspases reagent (in Kit I35101)
- Hoechst 33342
- SYTOX® Green nucleic acid stain
- Dimethylsulfoxide (DMSO)
- Apoptosis fixative solution
- Concentrated apoptosis wash buffer
- Detailed protocols for fluorescence microcscopy assays

In addition to a specific FLICA® reagent, each kit provides Hoechst 33342 and SYTOX® Green nucleic acid stains, which allow the simultaneous evaluation of caspase activation, nuclear morphology and plasma membrane integrity. Sufficient reagents are provided for 25 assays, based on labeling volumes of 300 µL.

Vybrant® FAM Caspase Assay Kits for Flow Cytometry

Like the Image-iT® Kits described above, the Vybrant® FAM Caspase Assay Kits for flow cytometry are based on a fluorescent caspase inhibitor (FLICA® methodology). We offer three different Vybrant® FAM Caspase Assay Kits designed to target different caspases. The Vybrant® FAM Caspase-3 and -7 Assay Kit (V35118) provides a FLICA® inhibitor containing the caspase-3 and -7 recognition sequence DEVD; the Vybrant® FAM Caspase-8 Assay Kit (V35119) provides a FLICA® inhibitor containing the caspase-8 recognition sequence Leu-Glu-Thr-Asp (LETD); and the Vybrant® FAM Poly Caspases Assay Kit (V35117) provides a FLICA® inhibitor containing the caspase recognition sequence Val-Ala-Asp (VAD), which is recognized by caspase-1, -3, -4, -5, -6, -7, -8 and -9. In addition to the selective FLICA® reagent, these kits contain the Hoechst 33342 and propidium iodide nucleic acid stains to permit simultaneous evaluation of caspase activation, membrane permeability and cell cycle.

Each Vybrant® FAM Caspase Assay Kit includes:

- FAM-DEVD-FMK caspase-3 and -7 reagent (in Kit V35118), FAM-LETD-FMK caspase-8 reagent (in Kit V35119) or FAM-VAD-FMK poly caspases reagent (in Kit V35117)
- Hoechst 33342
- Propidium iodide
- Dimethylsulfoxide (DMSO)
- Apoptosis fixative solution
- Concentrated apoptosis wash buffer
- Detailed protocols for flow cytometry assays

Sufficient reagents are provided for 25 assays, based on labeling volumes of 300 µL. These Vybrant® FAM Caspase Assay Kits can be used in combination with other fluorescent probes, such as the far-red–fluorescent allophycocyanin annexin V (A35110), for a multiparameter study of apoptosis.

Substrates for HIV Protease and Renin

Alternative strategies have been employed to create substrates specifically for some endopeptidases. Our HIV protease and renin substrates (H2930, R2931) utilize fluorescence resonance energy transfer (FRET) to generate a spectroscopic response to protease cleavage (Fluorescence Resonance Energy Transfer (FRET)—Note 1.2). In this type of substrate, both an acceptor molecule and a fluorescent molecule are attached to the peptide or protein. The acceptor molecule is carefully chosen so that its absorbance overlaps with the fluorophore's excited-state fluorescence (Figure 10.4.11), thus ensuring that the fluorescence is quenched through resonance energy transfer.[56] Enzyme hydrolysis of the substrate results in spatial separation of the fluorophore and the acceptor molecule, thereby restoring the fluorophore's fluorescence (Figure 10.4.12). See Section 9.5 for a discussion of our reagents

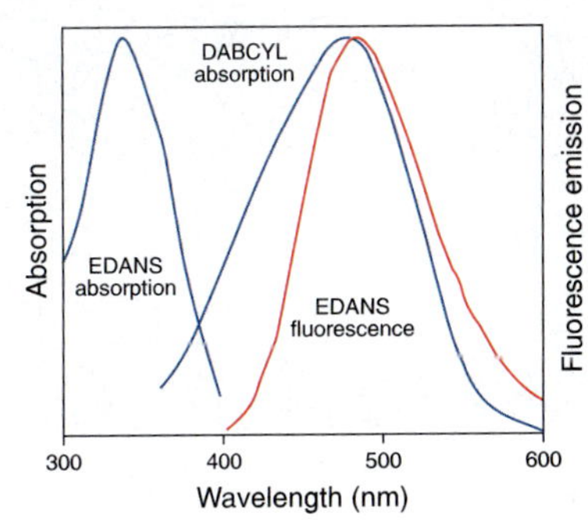

Figure 10.4.11 Spectral overlap between EDANS fluorescence and dabcyl absorption, which is required for efficient quenching of EDANS fluorescence by resonance energy transfer to the nonfluorescent dabcyl chromophore. Spectra are normalized to the same intensities.

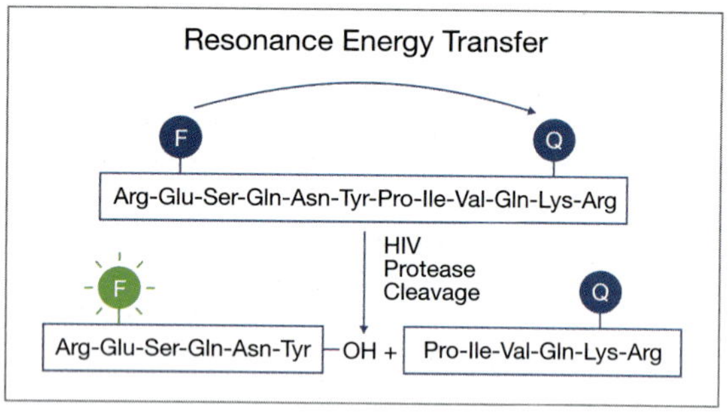

Figure 10.4.12 Principle of the fluorogenic response to protease cleavage exhibited by HIV protease substrate 1 (H2930). Quenching of the EDANS fluorophore (**F**) by distance-dependent resonance energy transfer to the dabcyl quencher (**Q**) is eliminated upon cleavage of the intervening peptide linker.

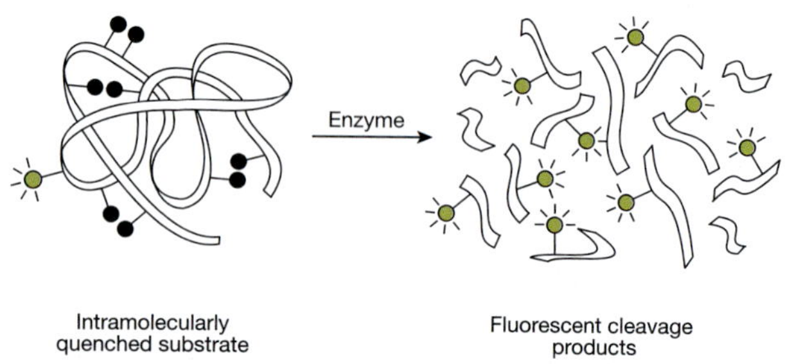

Figure 10.4.13 Principle of enzyme detection via the disruption of intramolecular self-quenching. Enzyme-catalyzed hydrolysis of the heavily labeled and almost totally quenched substrates provided in the EnzChek® Protease Assay Kits (E6638, E6639), EnzChek® *Ultra* Amylase Assay Kit (E33651), EnzChek® Gelatinase/Collagenase Assay Kit (E12055), EnzChek® Elastase Kit (E12056), EnzChek® Lysozyme Assay Kit (E22013)—as well as the stand-alone quenched substrates DQ™ BSA (D12050, D12051), DQ™ collagen (D12052, D12060), DQ™ ovalbumin (D12053) and DQ™ gelatin (D12054)—relieves the intramolecular self-quenching, yielding brightly fluorescent reaction products.

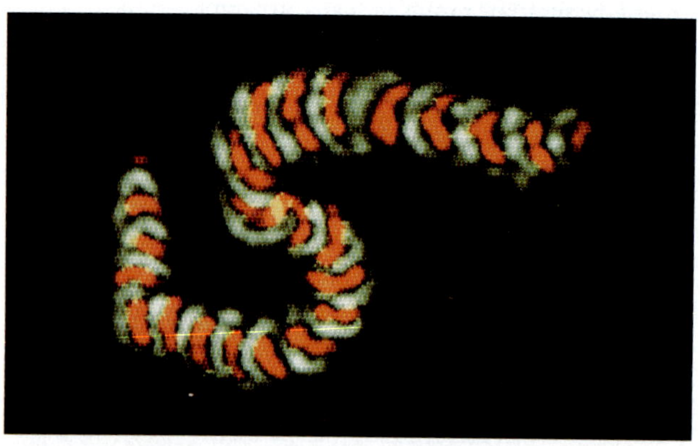

Figure 10.4.14 Unlabeled neutrophils that have been allowed to migrate spontaneously in gelatin matrices containing a prototype of DQ™ Green BSA (D12050) and dihydrotetramethylrosamine. Extracellular release of reactive oxygen metabolites oxidizes dihydrotetramethylrosamine to tetramethylrosamine, yielding orange fluorescence. Extracellular protease activity cleaves highly quenched DQ™ Green BSA, yielding green-fluorescent peptides. The micrograph illustrates an alternating pattern of orange and green fluorescence and shows that extracellular release of oxidants and proteolytic activity oscillate 180° out of phase during cell locomotion. Image contributed by Howard R. Petty, Wayne State University.

for synthesizing labeled peptides and peptidase substrates, including our QSY® series of nonfluorescent dyes, which have broad visible or near-infrared absorption spectra and serve as almost universal quenchers of most fluorescent donors that emit in the visible, with unusually high efficiency (Table 1.1).

Substrate for Detecting HIV Protease Activity

HIV protease substrate 1 (H2930) is a peptide that includes the HIV protease cleavage site, along with two covalently modified amino acid residues—one that has been linked to EDANS and the other to dabcyl.[57–59] Proteolytic cleavage releases a fragment containing only the EDANS fluorophore, thus liberating it from the quenching effect of the nearby dabcyl chromophore (Figure 10.4.12). HIV protease activity can be measured by exciting the sample at ~340 nm and measuring the resulting fluorescence at 490 nm. HIV protease substrate 1 has been used to analyze the effects of solvent composition, incubation time and enzyme concentration on HIV-1 protease activity[60] and to investigate a newly designed inhibitor of the enzyme.[61] HIV protease substrate 1 has also been employed to follow the inhibition of HIV-1 protease activity after the enzyme's two cysteine residues are reversibly modified by nitric oxide.[62] One milligram of HIV protease substrate 1 is sufficient for approximately 120 enzyme assays using 2 mL assay volumes and standard fluorescence cuvettes or ~1600 assays using 150 µL assay volumes and microcuvettes.

Human Renin Substrate 1

Assaying renin activity with human renin substrate 1 (R2931) is analogous to assaying HIV protease activity with the HIV protease substrate described above. Renin, an aspartic protease, plays an important role in blood-pressure regulation and is therefore a target for antihypertension therapeutics. Using renin substrate 1, researchers have discovered a stable, partially active conformational variant of recombinant human prorenin.[63] This substrate has also been used to investigate the kinetics and pH stability of recombinant human renin.[64] A fluorogenic substrate similar to our renin substrate 1 was used to develop a microplate assay for screening renin inhibitors.[65] One milligram of the renin substrate 1 is sufficient for approximately 100 enzyme assays using 2 mL volumes and standard fluorescence cuvettes or ~1400 assays using 150 µL assay volumes and microcuvettes. The short-wavelength excitation maximum (335 nm) of the EDANS fluorophore precludes use of this substrate in most fluorescence microplate readers. We also have a fluorogenic substrate for thrombin (rhodamine 110, bis-(tosyl-Gly-Pro-Arg) amide, R22124) that is useful for measuring the activity of this enzyme, which is important for coagulation and blood clot formation.[31]

EnzChek® Protease Assay Kits and Fluorescein Casein

Often it is necessary to have fluorogenic substrates for the assay of purified enzymes with unknown specificity or for which there are no known useful substrates. Assay for contamination of biological preparations by unknown proteases requires substrates that can detect a variety of enzymes. We have developed a method of relieving the fluorescence quenching of BODIPY® dye–labeled biopolymers by enzymatic hydrolysis (Figure 10.4.13), which we have applied to several of the general or selective protease assay kits and DQ™ reagents described in this section. We have also described the method's use in

IMPORTANT NOTICE: The products described in this manual are covered by one or more Limited Use Label License(s). Please refer to the Appendix on page 971 and Master Product List on page 975. Products are For Research Use Only. Not intended for any animal or human therapeutic or diagnostic use.

an assay for dextranase,[66] and others have applied it to the assay of O-sialoglycoprotein endopeptidase[67] and enzymes that process vesicular stomatitis virus[68] (VSV). The DQ™ and EnzChek® protease assay reagents may have significant potential for detecting matrix metalloproteinase (MMP) activity in living tissues by simple incubation of the tissue with the protein-based fluorogenic substrates.[2] Fluorescent products have been shown to accumulate on the cell's surface where proteases are active, including in living human breast cancer cells.[69] These quenched protease substrates are particularly useful for following cell migration through matrices[70–72] (Figure 10.4.14).

The EnzChek® Protease Assay Kits provide exceptionally fast, simple and direct fluorescence assays for detecting metallo-, serine, acid and thiol proteases. The EnzChek® Protease and EnzChek® Peptidase/Protease Assay Kits (E6638, E6639, E33758) measure the increase in fluorescence intensity that results from protease hydrolysis of a quenched substrate, whereas our EnzChek® Polarization Assay Kit for Proteases (E6658) monitors fluorescence polarization changes that occur during protease hydrolysis of a lightly labeled fluorescent casein derivative. Although the detection principles of these protease assays are quite different, no separation steps are required for either, and both types of assays are rapid, sensitive and versatile.

EnzChek® Protease Assay Kits for Fluorescence Intensity Measurements

The EnzChek® Protease Assay Kits contain a casein derivative that is heavily labeled with either the green-fluorescent BODIPY® FL or red-fluorescent BODIPY® TR-X dye, resulting in almost total quenching of the conjugate's fluorescence; they typically exhibit <3% of the fluorescence of the corresponding free dyes.[73–75] Protease-catalyzed hydrolysis relieves this quenching, yielding brightly fluorescent BODIPY® FL dye– or BODIPY® TR-X dye–labeled peptides[74,75] (Figure 10.4.13). The increase in fluorescence, which can be measured with a spectrofluorometer, minifluorometer or fluorescence microplate reader, is directly proportional to protease activity.

In contrast to the conventional fluorescein thiocarbamoyl (FTC)–casein protease assay, these EnzChek® assays do not involve any separation steps and, consequently, can be used to continuously measure the kinetics of a variety of exopeptidases and endopeptidases over a wide pH range. They can also be used to measure the total substrate turnover at a fixed time following addition of the enzyme. We have found that these protease assays are over 100-times more sensitive and much easier to perform than the labor-intensive FTC–casein assay. Detection limits for fluorescence intensity measurements with these kits are given in Table 10.4.

Hydrolysis of the fluorogenic substrates by proteases provides a sensitive assay of cell proliferation and a means of detecting the sterility of a sample. In addition to their utility for detecting protease contamination in culture medium and other experimental samples, BODIPY® FL casein and BODIPY® TR-X casein have significant potential as general, nontoxic, pH-insensitive markers for phagocytic cells in culture (Section 16.1). We have shown that uptake of these quenched conjugates by neutrophils is accompanied by hydrolysis of the labeled proteins by intracellular proteases and generation of fluorescent products that are well retained in cells. This phagocytosis assay is readily performed in a fluorescence microplate reader or a flow cytometer; localization of the fluorescent products can be determined by fluorescence microscopy. The same substrates can readily detect secretion of proteases from live cells.

BODIPY® FL casein and BODIPY® TR-X casein can be used interchangeably, depending on whether green or red fluorescence is desired. The peptide hydrolysis products of BODIPY® FL casein exhibit green fluorescence that is optimally excited by the argon-ion laser, permitting flow sorting of the cells. The red-fluorescent BODIPY® TR-X–labeled peptides, with excitation and emission spectra similar to those of the Texas Red® fluorophore, should be useful for multilabeling experiments or measurements in the presence of green autofluorescence.

The EnzChek® Protease Assay Kit includes:

- BODIPY® FL casein (in Kit E6638) or BODIPY® TR-X casein (in Kit E6639)
- Concentrated digestion buffer
- Detailed protocols

Each kit provides sufficient reagents for ~100 assays using 2 mL assay volumes and standard fluorescence cuvettes or ~1000 assays using 200 µL assay volumes and 96-well microplates.

The EnzChek® Protease Assay Kit containing red-fluorescent BODIPY® TR-X casein is also available as a convenient RediPlate™ 96 Protease Assay Kit (R22132). Each RediPlate™ 96 EnzChek® Protease Assay Kit includes one 96-well microplate, with all of the necessary reagents predispensed into the wells, where 88 wells (11 lanes) are intended for assays and 8 wells (1 lane) include a dilution series of an appropriate reference standard for generation of standard curves. The enzyme sample to be assayed is added to the microplate in a suitable buffer, along with any compounds to be tested. Then, after incubation, the resultant fluorescence is quantitated on a fluorescence microplate reader equipped with filters appropriate for the green- or red-fluorescent dye. Each RediPlate™ 96 microplate has removable lanes that allow researchers to perform only as many assays as required for the experiment

Table 10.4 Detection limits of the EnzChek® Protease Assay Kits.

Enzyme (Source)	Class	Detection Limit (Units) *	Buffer Conditions
Elastase, Type IV (porcine pancreas)	Serine protease	2.2×10^{-3}	10 mM Tris-HCl, pH 8.8
Chymotrypsin, Type II (bovine pancreas)	Serine protease	5.0×10^{-5}	10 mM Tris-HCl, pH 7.8
Thermolysin (*Bacillus thermoproteolyticus* Rokko)	Acid protease	4.4×10^{-5}	10 mM Tris-HCl, pH 7.8
Trypsin, Type IX (porcine pancreas)	Serine protease	1.3×10^{-2}	10 mM Tris-HCl, pH 7.8
Papain (papaya latex)	Cysteine protease	2.1×10^{-4}	10 mM MES, pH 6.2
Pepsin (porcine stomach mucosa)	Acid protease	2.1×10^{-3}	10 mM HCl, pH 2.0
Elastase (*Pseudomonas aeruginosa*)	Metallo-protease	1.0×10^{-3}	20 mM sodium phosphate, pH 8.0
Cathepsin D	Acid protease	2.0×10^{-4}	20 mM sodium citrate, pH 5.0
Elastase (human leukocyte)	Serine protease	1.0×10^{-3}	10 mM Tris-HCl, pH 7.5

* The detection limit is defined as the amount of enzyme required to cause a 10–20% change in fluorescence compared with the control sample at 22°C. Enzyme unit definitions are standard definitions for each individual enzyme. Detection limits were determined with BODIPY® FL casein and with BODIPY® TR-X casein; both substrates yielded similar results. Detection limits may vary with instrumentation.

IMPORTANT NOTICE: The products described in this manual are covered by one or more Limited Use Label License(s). Please refer to the Appendix on page 971 and Master Product List on page 975. Products are For Research Use Only. Not intended for any animal or human therapeutic or diagnostic use.

(Figure 10.4.10). Resealable packaging ensures plate and well integrity between experiments. Table 10.3 summarizes our other RediPlate™ Assay Kits for caspase-3 activity, phosphatase activity (Section 10.3) and RNA quantitation (Section 8.3).

EnzChek® Peptidase/Protease Assay Kit

The EnzChek® Peptidase/Protease Assay Kit (E33758) provides a FRET (fluorescence resonance energy transfer)–based method for the simple and accurate quantitation of a wide range of protease activities. The EnzChek® peptidase/protease substrate comprises a fluorophore and a quencher moiety separated by an amino acid sequence. Upon protease cleavage, the fluorophore separates from the quencher and is free to emit a detectable fluorescent signal (excitation/emission maxima ~502/528 nm). The magnitude of the resultant signal is proportional to the degree of substrate cleavage and can therefore be used to quantitate the enzyme activity present (Figure 10.4.15). The assay is performed in a simple mix–incubate–read format and can be completed in 60 minutes or less.

The EnzChek® Peptidase/Protease Assay Kit contains:

- EnzChek® peptidase/protease substrate
- Concentrated digestion buffer (200 mM Tris-HCl, pH 7.8)
- Substrate solvent (50% DMSO in 10 mM Tris-HCl, pH 7.8)
- Detailed protocols

Sufficient materials are supplied for 100 assays, based on a 100 µL assay volume in a 96-well microplate format. The EnzChek® peptidase/protease assay can be adapted for use in cuvettes or 384-well microplates.

EnzChek® Protease Assay Kit for Fluorescence Polarization Measurements

When a fluorescent molecule tethered to a protein is excited by polarized fluorescent light, the polarization of fluorescence emission is dependent on the rate of molecular tumbling. Upon proteolytic cleavage of the fluorescently labeled protein, the resultant smaller peptides tumble faster, and the emitted light is depolarized relative to the light measured from the intact conjugate. The EnzChek® Polarization Assay Kit for Proteases[76] (E6658) contains green-fluorescent BODIPY® FL casein with an optimal degree of labeling for fluorescence polarization–based protease assays.[77,78] Fluorescence polarization technology is more sensitive than many nonradioactive protease assays and allows measurements to be taken in real time, permitting the collection of kinetics data (Fluorescence Polarization (FP)—Note 1.4). Our BODIPY® FL dye has an adequate fluorescence lifetime and pH-insensitive fluorescence—two prerequisites for successful measurement of protease activity by fluorescence polarization.

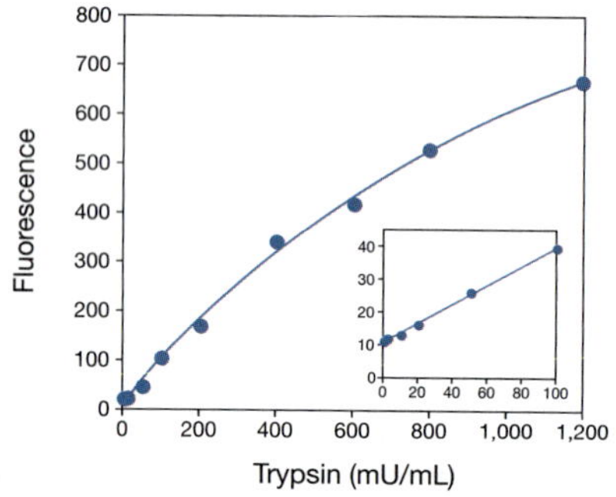

Figure 10.4.15 Sample standard curves obtained with the EnzChek® Peptidase/Protease Assay Kit. Trypsin (EC 3.4.21.4) was assayed in 10 mM Tris-HCl (pH 7.8) digestion buffer from 10 to 1200 mU/mL using the EnzChek® peptidase/protease substrate. The inset shows a separate experiment using the same enzyme, but at activities from 2 to 100 mU/mL. Samples were incubated for 60 minutes at room temperature. Fluorescence was measured at 490/520 nm; background fluorescence was subtracted from the inset data.

The EnzChek® Polarization Assay Kit for Proteases contains:

- BODIPY® FL casein
- Concentrated digestion buffer
- Detailed protocols

Each kit provides sufficient reagents for ~100 assays using 2 mL assay volumes and standard fluorescence cuvettes or ~1000 assays using 200 µL assay volumes and 96-well microplates. With the advent of high-capacity automated instrumentation, this kit provides an important tool for high-throughput screening of proteases and their inhibitors in research laboratories.

Fluorescein Casein

We also offer fluorescein casein (C2990) for assaying protease activity. In this assay, unhydrolyzed fluorescein casein must be precipitated with trichloroacetic acid, separated by centrifugation, transferred for measurement and then pH-adjusted for fluorescein signal enhancement.[79–83] Fluorescein casein may be useful for a continuous assay if monitored by fluorescence polarization.[78] Fluorescein casein is rapidly degraded by *Bacteroides gingivalis* but only slowly by streptococci.[84]

EnzChek® Gelatinase/Collagenase Assay Kit

Collagen is a major component of the extracellular matrix, which not only serves as scaffolding to stabilize tissue structure, but also influences the development, migration, proliferation and metabolism of cells that contact it. Gelatinases and collagenases—matrix metalloproteinases (MMPs) that digest collagen or gelatin (denatured collagen)—are increasingly important to our understanding of both normal development and carcinogenesis.[85] For example, gelatinase A (20,000-dalton MMP-2) is primarily responsible for degrading the helical domains of type IV collagen, the principal collagen of basement membranes.[86] Thus, gelatinase A likely plays a major role in the turnover of basement membrane during fetal tissue development, wound healing, angiogenesis and tumor invasion.

The EnzChek® Gelatinase/Collagenase Assay Kit (E12055) provides a fast, sensitive and convenient assay for measuring gelatinase or collagenase activity or for screening inhibitors[87] in a high-throughput format. This kit contains:

- DQ™ gelatin, a gelatin conjugate so heavily labeled with fluorescein that its fluorescence is quenched; this substrate is also available separately (D12054)
- 1,10-Phenanthroline, a general metalloproteinase inhibitor
- Type IV collagenase from *Clostridium histolyticum* for use as a positive control
- Concentrated reaction buffer
- Detailed protocols

IMPORTANT NOTICE: The products described in this manual are covered by one or more Limited Use Label License(s). Please refer to the Appendix on page 971 and Master Product List on page 975. Products are For Research Use Only. Not intended for any animal or human therapeutic or diagnostic use.
www.invitrogen.com/probes

DQ™ gelatin, the highly quenched gelatinase/collagenase substrate provided, is efficiently digested by most, if not all, gelatinases and collagenases, releasing brightly fluorescent peptides (Figure 10.4.13). The increase in fluorescence upon digestion is proportional to proteolytic activity and can be monitored with a fluorescence microplate reader, minifluorometer or spectrofluorometer. Depending on the substrate concentration used in each reaction, sufficient reagents are supplied for 250–1000 assays using 200 µL assay volumes and 96-well microplates. Using 100 µg/mL DQ™ gelatin and a 2-hour incubation period, we have detected as little as 2×10^{-3} U/mL (7 ng protein/mL) of *C. histolyticum* collagenase, where one unit is defined as the amount of enzyme required to liberate 1 µmole L-leucine equivalents from collagen in 5 hours at 37°C, pH 7.5. Longer incubation times increase the sensitivity, whereas higher enzyme concentrations decrease the incubation times. Using human gelatinase A (not provided), 100 µg/mL DQ™ gelatin and a 24-hour incubation period, we have detected concentrations of gelatinase as low as 3×10^{-4} U/mL, where one unit is defined as the amount of enzyme that can hydrolyze 1 mg of type IV collagen in 1 hour at 37°C, pH 7.5. DQ™ gelatin and some of our other highly quenched protease substrates have been utilized for *in situ* detection of matrix metalloproteinase and other protease activity in cell preparations, tissue sections and SDS gels.[3,88–91]

EnzChek® Elastase Assay Kit

The EnzChek® Elastase Assay Kit (E12056) provides a fast, sensitive and convenient assay for measuring elastase activity or for screening inhibitors in a high-throughput format. This kit contains DQ™ elastin—soluble bovine neck ligament elastin that has been labeled with our BODIPY® FL dye such that the conjugate's fluorescence is quenched. Upon digestion by elastase or other proteases, the fluorescence is revealed (Figure 10.4.13). The resulting increase in fluorescence can be monitored with a fluorescence microplate reader, minifluorometer or spectrofluorometer. Digestion products from the DQ™ elastin substrate have absorption maxima at ~505 nm and fluorescence emission maxima at ~515 nm. Because the assay is continuous, kinetics data can be obtained easily. Furthermore, because fluorescence of the BODIPY® FL dye is pH insensitive between pH 3 and 9, the assay can be performed under a variety of buffer conditions. Please note that DQ™ elastin is also digested by proteases other than elastase.

The EnzChek® Elastase Assay Kit (A10256) contains:

- DQ™ elastin substrate
- Concentrated reaction buffer
- Elastase from pig pancreas for use as a positive control
- *N*-Methoxysuccinyl-Ala-Ala-Pro-Val-chloromethyl ketone, a selective elastase inhibitor
- Detailed protocols

Each kit provides sufficient reagents for approximately 600 assays using a fluorescence microplate reader and reaction volumes of 200 µL. The *N*-methoxysuccinyl-Ala-Ala-Pro-Val-chloromethyl ketone inhibitor can be used to confirm the identity of the protease responsible for substrate digestion or, alternatively, as a control inhibitor for use when screening for elastase inhibitors.

DQ™ Substrates

DQ™ Collagens

Our DQ™ collagens, type I (D12060) and type IV (D12052), are complementary reagents to the DQ™ gelatin provided in the EnzChek® Gelatinase/Collagenase Assay Kit. Like DQ™ gelatin, these highly quenched substrates are heavily labeled with fluorescein and release fluorescent peptides when enzymatically cleaved. DQ™ collagen, type I should be useful in assays detecting matrix metalloprotease (MMP-1) activity.[92] DQ™ collagen, type IV may prove particularly useful in the development of assays for gelatinase A (MMP-2) and gelatinase A inhibitors, as well as for other gelatinases and collagenases that specifically degrade type IV collagen. DQ™ collagens may be used with the EnzChek® Gelatinase/Collagenase Assay Kit, and a sample protocol is included. Because of its more complex structure, DQ™ collagen generally requires either greater amounts of the enzyme or longer incubation periods than does DQ™ gelatin. Please note that both the DQ™ collagens and DQ™ gelatin can be digested by proteases other than gelatinases and collagenases. The fluorescence generated by hydrolysis of DQ™ collagen (D12052) by cellular collagenase has been used to visualize the migratory pathway followed by tumor cells during invasion of a gelatin matrix and to image proteolysis by living breast cancer cells.[69,70]

In addition to the DQ™ substrates, we have prepared gelatin and collagen conjugates that have been labeled to maximize probe fluorescence and minimize dye quenching. We offer two green-fluorescent conjugates of gelatin, one in which gelatin is coupled to our photostable Oregon Green® 488 dye (G13186, Section 15.6) and the other to fluorescein (G13187, Section 15.6).

DQ™ BSA

DQ™ Green BSA (D12050, excitation/emission maxima ~500/506 nm) and DQ™ Red BSA (D12051, excitation/emission maxima ~589/617 nm) are bovine serum albumin (BSA) conjugates that have been labeled to such a high degree that the BODIPY® dyes used to label them are strongly self-quenched. Proteolysis of the DQ™ BSA can easily be monitored as proteolytic fragments containing the fluorophores are released from the larger conjugate and become brightly fluorescent[93] (Figure 10.4.13). An unlabeled neutrophil making its way through a gelatin matrix containing a DQ™ Green BSA prototype, as well as dihydrotetramethylrosamine, a nonfluorescent probe that fluoresces bright orange upon oxidation, leaves behind a fluorescent trail[72] (Figure 10.4.14). The alternating green- and orange-fluorescent bands dramatically demonstrate that the proteolytic and oxidative activities of the migrating neutrophil are oscillatory and are 180° out of phase with each other. Intracellular processing of a similar BODIPY® BSA conjugate by J774 macrophages can be completely inhibited by protease inhibitors.[93] DQ™ Green BSA has been embedded in a gelatin matrix and used to image both extracellular and intracellular proteolysis in living cells. DQ™ BSA can be complexed with our rabbit IgG fraction anti-BSA (A11133, Section 7.5) to form an immune complex that is internalized through the Fc receptor and processed in the phagovacuole to highly fluorescent peptides (Section 16.1). DQ™ Green BSA has been used for imaging proteolysis in living breast cancer cells.[69]

IMPORTANT NOTICE: The products described in this manual are covered by one or more Limited Use Label License(s). Please refer to the Appendix on page 971 and Master Product List on page 975. Products are For Research Use Only. Not intended for any animal or human therapeutic or diagnostic use.

DQ™ Ovalbumin

DQ™ ovalbumin[94,95] (D12053) is a self-quenched ovalbumin conjugate designed specifically for the study of antigen processing. Ovalbumin is efficiently processed through mannose receptor–mediated endocytosis by antigen-presenting cells and is widely used for studying antigen processing.[96–98] DQ™ ovalbumin is labeled with our pH-insensitive, green-fluorescent BODIPY® FL dye such that the fluorescence is almost completely quenched. Upon endocytosis and proteolysis, highly fluorescent peptides are released. DQ™ ovalbumin appears to be an excellent indicator of macrophage-mediated antigen processing in flow cytometry and microscopy assays.

Alternative Methods for Detecting Protease Activity

Peptidases typically liberate a free amine for each hydrolysis step. Thus, fluorogenic amine detection reagents such as fluorescamine (F2332, F20261; Section 1.8) and *o*-phthaldialdehyde (P2331MP, Section 1.8) have been used to detect the rate of amine production by peptidases.[99–102]

Peptidases that liberate single free amino acids for which specific oxidases exist can be analyzed by coupling the hydrolytic reaction to oxidation of our Amplex® Red reagent (Section 10.5) to the red-fluorescent dye resorufin. For example, glutamic acid production can be monitored using glutamate oxidase[103] and D-amino acid liberation monitored using a D-amino acid oxidase.

Protease assays conducted in highly autofluorescent or strongly light-scattering solutions (such as crude cell and tissue extracts) often can be improved by extracting the fluorescent hydrolysis product from the assay mixture with an organic solvent such as toluene, chloroform or ethyl acetate.[104] Most unhydrolyzed peptidase substrates will remain in the aqueous layer.

Endopeptidase substrates that are singly labeled at the amine terminus with a fluorophore usually do not undergo a fluorescence change upon hydrolysis of internal peptide bonds; however, fluorescence (or absorbance) of the fluorophore that remains attached to the cleaved peptide can be used to detect the hydrolysis product following separation by TLC, HPLC or capillary electrophoresis.

Protease Inhibitors

Alexa Fluor® 488 Soybean Trypsin Inhibitor

Trypsin inhibitor from soybean (SBTI) is a 21,000-dalton protein that inhibits the catalytic activity of serine proteases.[105–107] SBTI binds to acrosin, an acrosomal serine protease that is associated with binding of spermatozoa and penetration of the zona pellucida,[108,109] and SBTI binding patterns in non-fixed human sperm have demonstrated its usefulness for detecting acrosome-reacted sperm.[110] In particular, an Alexa Fluor® 488 conjugate of the protein has been used to measure the acrosomal status of macaque sperm and to determine the localization of acrosin during the reaction.[111] We provide a fluorescent SBTI conjugate with one of our best fluorophores, the Alexa Fluor® 488 dye (T23011). Alexa Fluor® 488 dye (Section 1.3) has spectral characteristics similar to fluorescein (excitation at 495 nm and emission at 519 nm) but produces conjugates that are brighter and more photostable. Furthermore, the fluorescence of Alexa Fluor® 488 conjugates is insensitive to pH from 4 to 10.

BODIPY® FL Pepstatin

Pepstatin A is an inhibitor of carboxyl (acid) proteases that contain aspartate residues at their active sites, including cathepsin D, pepsin and renin.[112,113] We have prepared the green-fluorescent BODIPY® FL pepstatin A analog (P12271), which we have shown binds to cathepsin D in live cells[114,115] (Section 15.5).

Leupeptin Hemisulfate

Leupeptin hemisulfate (L6543) can be used to inhibit serine and cysteine proteases. It does not inhibit chymotrypsin and thrombin.

IMPORTANT NOTICE: The products described in this manual are covered by one or more Limited Use Label License(s). Please refer to the Appendix on page 971 and Master Product List on page 975. Products are For Research Use Only. Not intended for any animal or human therapeutic or diagnostic use.
www.invitrogen.com/probes

REFERENCES

1. Methods Enzymol (1994) 244:1; **2.** J Histochem Cytochem (2001) 49:1473; **3.** Am J Physiol Lung Cell Mol Physiol (2004) 287:L184; **4.** J Histochem (2004) 52:711; **5.** J Biol Chem (1995) 270:558; **6.** Arch Biochem Biophys (1994) 314:171; **7.** FEBS Lett (1994) 341:197; **8.** J Biochem (Tokyo) (1993) 113:441; **9.** FEBS Lett (1989) 257:388; **10.** Arch Biochem Biophys (1987) 259:131; **11.** Biochem J (1982) 201:367; **12.** Biochem J (1981) 193:187; **13.** Methods Enzymol (1981) 80:341; **14.** J Biochem (Tokyo) (1977) 82:1495; **15.** J Biol Chem (2000) 275:2390; **16.** J Biol Chem (2000) 275:9452; **17.** J Biol Chem (1993) 268:23593; **18.** J Biol Chem (1999) 274:787; **19.** Biochem J (1983) 215:253; **20.** Biochem J (1983) 209:299; **21.** Cytometry (1995) 20:334; **22.** Anticancer Res (1988) 8:1179; **23.** J Enzyme Inhib (1988) 2:209; **24.** Biochim Biophys Acta (1984) 788:74; **25.** Photochem Photobiol (1986) 44:461; **26.** Proc Natl Acad Sci U S A (2000) 97:13126; **27.** Glia (1993) 7:183; **28.** Biol Chem Hoppe Seyler (1992) 373:547; **29.** Biol Chem Hoppe Seyler (1992) 373:433; **30.** Anal Chem (1993) 65:2352; **31.** Biomed Instrum Technol (1996) 30:245; **32.** Immunol Cell Biol (1998) 76:1; **33.** Science (1998) 281:1312; **34.** Trends Biochem Sci (1997) 22:388; **35.** Cell Death Differ (1999) 6:1067; **36.** J Biol Chem (1997) 272:17907; **37.** Nature (1995) 376:37; **38.** Biochemistry (1999) 38:13906; **39.** Bioorg Med Chem Lett (1999) 9:3231; **40.** Cell Death Differ (1999) 6:821; **41.** Science (1998) 281:1305; **42.** J Biol Chem (1999) 274:17484; **43.** Exp Cell Res (1999) 250:203; **44.** J Biol Chem (2001) 276:6974; **45.** J Neurosci Res (1998) 52:334; **46.** Biochemistry (2000) 39:16056; **47.** Oncogene (1998) 17:1295; **48.** J Biol Chem (2000) 275:288; **49.** Cell Death Differ (1999) 6:1081; **50.** Exp Cell Res (2000) 259:308; **51.** Biotechniques (2001) 31:608; **52.** Leukemia (2002) 16:1589; **53.** Cytometry (2002) 47:143; **54.** J Immunol Methods (2002) 265:111; **55.** Cytometry A (2003) 55:50; **56.** J Biochem Biophys Methods (1996) 33:135; **57.** Science (1990) 247:954; **58.** Tetrahedron Lett (1990) 31:6493; **59.** J Biol Chem (1997) 272:15603; **60.** J Biol Chem (1992) 267:20028; **61.** Science (1990) 249:527; **62.** Biochemistry (1999) 38:13407; **63.** J Protein Chem (1990) 9:663; **64.** J Protein Chem (1991) 10:553; **65.** Anal Biochem (1993) 210:351; **66.** Anal Biochem (1998) 260:257; **67.** Anal Biochem (1998) 259:8; **68.** J Virol Methods (1998) 70:45; **69.** Neoplasia (2000) 2:496; **70.** FASEB J (2001) 15:932; **71.** Exp Cell Res (2000) 260:292; **72.** Biophys J (1998) 74:90; **73.** Anal Biochem (2000) 279:170; **74.** Anal Biochem (1997) 251:144; **75.** Anal Biochem (1997) 254:144; **76.** Anal Biochem (1996) 243:1; **77.** J Biomol Screen (1996) 1:33; **78.** Biotechniques (1994) 17:585; **79.** Cytometry (1994) 15:213; **80.** Anal Biochem (1991) 197:347; **81.** Anal Biochem (1990) 191:133; **82.** Anal Biochem (1984) 143:30; **83.** Anal Biochem (1982) 121:290; **84.** FEMS Microbiol Lett (1990) 55:257; **85.** Chem Biol (1996) 3:895; **86.** J Biol Chem (1995) 270:5872; **87.** J Biol Chem (2001) 276:11347; **88.** J Histochem Cytochem (2004) 52:711; **89.** Biol Reprod (2002) 66:685; **90.** Biol Reprod (2002) 66:1083; **91.** J Neurosci (1999) 19:8464; **92.** Lab Invest (1998) 78:687; **93.** Eur J Cell Biol (1998) 75:192; **94.** J Immunol (2000) 165:49; **95.** Proc Natl Acad Sci U S A (1999) 96:15056; **96.** Eur J Immunol (1995) 25:1823; **97.** J Immunol (1992) 149:2894; **98.** J Immunol (1990) 145:417; **99.** Biochemistry (1990) 29:6670; **100.** Anal Biochem (1982) 123:41; **101.** Anal Biochem (1978) 87:556; **102.** Biochem Biophys Res Commun (1973) 53:75; **103.** Anal Biochem (2000) 284:382; **104.** Anal Biochem (1978) 87:257; **105.** J Cell Biol (2003) 161:79; **106.** J Biol Chem (1966) 241:3955; **107.** Methods Enzymol (1976) 45:700; **108.** Development (1990) 109:41; **109.** Zygote (1999) 7:143; **110.** Fertil Steril (1994) 62:1044; **111.** Zygote (2000) 8:127; **112.** J Biol Chem (1992) 267:24725; **113.** Biochemistry (1985) 24:3165; **114.** J Neurosci (2008) 26:6926; **115.** J Biochem Biophys Methods (2000) 42:137.

DATA TABLE 10.4 DETECTING PEPTIDASES AND PROTEASES

Cat. No.	MW	Storage	Soluble	Abs	EC	Em	Solvent	Product	Notes
A6520	554.10	F,D	DMSO	330	13,000	403	MeOH	C2110	1
A6521	649.14	F,D	DMSO	326	19,000	384	MeOH	A191	1
E13183	767.74	F,D,L	DMSO	325	16,000	395	pH 7	A191	1, 2
E13184	1515.46	F,D,L	DMSO	232	52,000	none	MeOH	R6479	2
H2930	~2016	F,D,L	DMF, H_2O	430	23,000	none	MeOH	see Notes	3
L6543	~475	F,D	H_2O	<300		none			
P12271	1044.14	F,D,L	DMSO	504	86,000	511	MeOH		
R2931	~2281	F,D,L	DMF, H_2O	460	13,000	none	H_2O	see Notes	3
R6501	983.91	F,D	DMSO, DMF	232	55,000	none	MeOH	R6479	
R6502	1278.26	F,D	DMSO, DMF	232	60,000	none	MeOH	R6479	
R6505	1404.46	F,D	DMSO, DMF	231	44,000	none	MeOH	R6479	
R6506	1167.24	F,D	DMSO, DMF	232	56,000	none	MeOH	R6479	
R6508	1126.06	F,D	DMSO, DMF	232	57,000	none	MeOH	R6479	
R22120	1515.46	F,D	DMSO, DMF	232	52,000	none	MeOH	R6479	
R22122	[illegible]	F,D	DMSO, DMF	232	55,000	none	MeOH	R6479	
R22124	1259.42	F,D	DMSO, DMF	232	[illegible]	none	MeOH	R6479	
R22125	1515.55	F,D	DMSO, DMF	232	52,000	none	MeOH	R6479	
R22126	1515.55	F,D	DMSO, DMF	232	52,000	none	MeOH	R6479	
R33750	1495.56	F,D	DMSO, DMF	230	76,000	none	MeOH	R6479	
R33752	1113.10	F,D	DMSO, DMF	232	57,000	none	MeOH	R6479	
R33753	1571.57	F,D	DMSO, DMF	232	57,000	none	MeOH	R6479	
R33754	1511.60	F,D	DMSO, DMF	232	57,000	none	MeOH	R6479	
R33755	1597.65	F,D	DMSO, DMF	232	57,000	none	MeOH	R6479	

For definitions of the contents of this data table, see "Using *The Molecular Probes® Handbook*" in the introductory pages.

Notes

1. Fluorescence of the unhydrolyzed substrate is very weak.
2. Data represent the substrate component of this kit.
3. Fluorescence of this substrate is >99% quenched. The proteolytic cleavage products fluoresce at 500 nm (excitation at 335 nm)

PRODUCT LIST 10.4 DETECTING PEPTIDASES AND PROTEASES

Cat. No.	Product	Quantity
A6520	7-amino-4-chloromethylcoumarin, *t*-BOC-L-leucyl-L-methionine amide (CMAC, *t*-BOC-Leu-Met)	5 mg
A6521	7-amino-4-methylcoumarin, CBZ-L-phenylalanyl-L-arginine amide, hydrochloride	25 mg
C2990	casein, fluorescein conjugate	25 mg
D12060	DQ™ collagen, type I from bovine skin, fluorescein conjugate	1 mg
D12052	DQ™ collagen, type IV from human placenta, fluorescein conjugate	1 mg
D12054	DQ™ gelatin from pig skin, fluorescein conjugate *special packaging*	5 x 1 mg
D12050	DQ™ Green BSA *special packaging*	5 x 1 mg
D12053	DQ™ ovalbumin *special packaging*	5 x 1 mg
D12051	DQ™ Red BSA *special packaging*	5 x 1 mg
E13183	EnzChek® Caspase-3 Assay Kit #1 *Z-DEVD-AMC substrate* *500 assays*	1 kit
E13184	EnzChek® Caspase-3 Assay Kit #2 *Z-DEVD-R110 substrate* *500 assays*	1 kit
E12056	EnzChek® Elastase Assay Kit *600 assays*	1 kit
E12055	EnzChek® Gelatinase/Collagenase Assay Kit *250–2000 assays*	1 kit
E33758	EnzChek® Peptidase/Protease Assay Kit *100 assays*	1 kit
E6658	EnzChek® Polarization Assay Kit for Proteases *green fluorescence* *100–1000 assays*	1 kit
E6638	EnzChek® Protease Assay Kit *green fluorescence* *100–1000 assays*	1 kit
E6639	EnzChek® Protease Assay Kit *red fluorescence* *100–1000 assays*	1 kit
H2930	HIV Protease Substrate 1 (Arg-Glu(EDANS)-Ser-Gln-Asn-Tyr-Pro-Ile-Val-Gln-Lys(dabcyl)-Arg)	1 mg
I35106	Image-iT® LIVE Green Caspase-3 and -7 Detection Kit *for microscopy* *25 tests*	1 kit
I35105	Image-iT® LIVE Green Caspase-8 Detection Kit *for microscopy* *25 tests*	1 kit
I35104	Image-iT® LIVE Green Poly Caspases Detection Kit *for microscopy* *25 tests*	1 kit
I35102	Image-iT® LIVE Red Caspase-3 and -7 Detection Kit *for microscopy*	1 kit
I35101	Image-iT® LIVE Red Poly Caspases Detection Kit *for microscopy*	1 kit
L6543	leupeptin hemisulfate	10 mg
P12271	pepstatin A, BODIPY® FL conjugate	25 µg
R35100	RediPlate™ 96 EnzChek® Caspase-3 Assay Kit *Z-DEVD-R110 substrate* *one 96-well microplate*	1 kit
R22132	RediPlate™ 96 EnzChek® Protease Assay Kit *red fluorescence* *one 96-well microplate*	1 kit
R2931	Renin Substrate 1 (Arg-Glu(EDANS)-Ile-His-Pro-Phe-His-Leu-Val-Ile-His-Thr-Lys(dabcyl)-Arg)	1 mg
R22122	rhodamine 110, bis-(L-aspartic acid amide), trifluoroacetic acid salt	1 mg
R6506	rhodamine 110, bis-(CBZ-L-alanyl-L-alanyl-L-alanyl-L-alanine amide)	5 mg
R33752	rhodamine 110, bis-(*N*-CBZ-L-alanyl-L-alanyl-L-aspartic acid amide) (Z-AAD-R110)	2 mg
R6508	rhodamine 110, bis-(CBZ-L-alanyl-L-arginine amide), dihydrochloride	5 mg
R6501	rhodamine 110, bis-(CBZ-L-arginine amide), dihydrochloride (BZAR)	5 mg
R22120	rhodamine 110, bis-(*N*-CBZ-L-aspartyl-L-glutamyl-L-valyl-L-aspartic acid amide) (Z-DEVD-R110) *bulk packaging*	20 mg
R22125	rhodamine 110, bis-(*N*-CBZ-L-isoleucyl-L-glutamyl-L-threonyl-L-aspartic acid amide) (Z-IETD-R110)	2 mg
R22126	rhodamine 110, bis-(*N*-CBZ-L-isoleucyl-L-glutamyl-L-threonyl-L-aspartic acid amide) (Z-IETD-R110) *bulk packaging*	20 mg
R6505	rhodamine 110, bis-(CBZ-L-isoleucyl-L-prolyl-L-arginine amide), dihydrochloride (BZiPAR)	5 mg
R33753	rhodamine 110, bis-(*N*-CBZ-L-leucyl-L-glutamyl-L-glutamyl-L-aspartic acid amide) (Z-LEED-R110)	2 mg
R6502	rhodamine 110, bis-(CBZ-L-phenylalanyl-L-arginine amide), dihydrochloride	5 mg
R33750	rhodamine 110, bis-(*N*-CBZ-L-tyrosinyl-L-valyl-L-alanyl-L-aspartic acid amide) (Z-YVAD-R110)	2 mg
R33755	rhodamine 110, bis-(*N*-CBZ-L-valyl-L-aspartyl-L-valyl-L-alanyl-L-aspartic acid amide) (Z-VDVAD-R110)	2 mg
R33754	rhodamine 110, bis-(*N*-CBZ-L-valyl-L-glutamyl-L-isoleucyl-L-aspartic acid amide) (Z-VEID-R110)	2 mg
R22124	rhodamine 110, bis-(*p*-tosyl-L-glycyl-L-prolyl-L-arginine amide)	2 mg
V35118	Vybrant® FAM Caspase-3 and -7 Assay Kit *for flow cytometry* *25 assays*	1 kit
V35119	Vybrant® FAM Caspase-8 Assay Kit *for flow cytometry* *25 assays*	1 kit
V35117	Vybrant® FAM Poly Caspases Assay Kit *for flow cytometry* *25 assays*	1 kit

IMPORTANT NOTICE: The products described in this manual are covered by one or more Limited Use Label License(s). Please refer to the Appendix on page 971 and Master Product List on page 975. Products are For Research Use Only. Not intended for any animal or human therapeutic or diagnostic use.

10.5 Substrates for Oxidases, Including Amplex® Red Kits

Oxidases, the most useful of which is undoubtedly horseradish peroxidase (HRP), are important enzymes that are used in a wide variety of bioassays. Peroxidase activity is also present in many cells. Reagents for quantitating peroxidase and the activity of a variety of other oxidases are described in this section; reagents for detecting the activity of cellular peroxidases and the oxygen radicals produced by these peroxidases are described in Section 16.1 and Section 18.2. Antibody and streptavidin conjugates of horseradish peroxidase are listed in the product list for this section and described in Section 7.2 and Section 7.6. Tyramide signal amplification (TSA) technology, described in Section 6.2, makes extensive use of peroxidase-conjugated reagents and fluorescent dye- or hapten-labeled tyramides to deposit a detectable product at the site of enzymatic activity.

We have used our extremely versatile Amplex® Red reagent—one of the most stable and sensitive fluorogenic substrate known for horseradish peroxidase—to develop a variety of novel fluorogenic and chromogenic assays for enzymes that produce hydrogen peroxide. Furthermore, these coupled assays permit the ultrasensitive quantitation of a diverse assortment of analytes, including glucose, galactose, cholesterol, glutamic acid, xanthine (or hypoxanthine), uric acid, choline and acetylcholine, as well as hydrogen peroxide. The P_iPer™ Phosphate and P_iPer™ Pyrophosphate Assay Kits (P22061, P22062), described in Section 10.3, also utilize the Amplex® Red reagent for the continuous assays of enzymes that produce either inorganic phosphate or pyrophosphate.

Figure 10.5.1 Amplex® Red reagent (A12222).

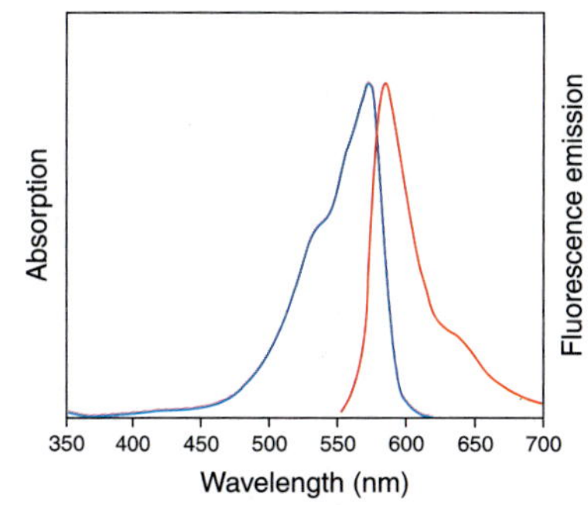

Figure 10.5.3 Absorption and fluorescence emission spectra of resorufin in pH 9.0 buffer.

Peroxidases

Amplex® Red Reagent: Stable Substrate for Peroxidase Detection

In the presence of horseradish peroxidase (HRP), Amplex® Red reagent (10-acetyl-3,7-dihydroxyphenoxazine, A12222, A22177; Figure 10.5.1) reacts with H_2O_2 in a 1:1 stoichiometry to produce highly fluorescent resorufin [1] (R363, Section 10.1, Figure 10.5.2). Amplex® Red reagent has greater stability, yields less background and produces a red-fluorescent product (excitation/emission maxima ~571/585 nm, Figure 10.5.3) that is more readily detected than the similar reduced methylene blue derivatives commonly used for colorimetric determination of lipid peroxides in plasma, sera, cell extracts and a variety of membrane systems.[2–4]

Amplex® Red reagent has been used to detect the release of H_2O_2 from activated human leukocytes,[1,5] to measure the activity of monoamine oxidase in bovine brain tissue,[6] to demonstrate the extracellular production of H_2O_2 produced by UV light stimulation of human keratinocytes [7–9] and for microplate assays of H_2O_2 and lipid hydroperoxide generation by isolated mitochondria.[10–12] Amplex® Red reagent is available in a single 5 mg vial (A12222) or packaged as a set of 10 vials, each containing 10 mg of the substrate, for high-throughput screening applications (A22177).

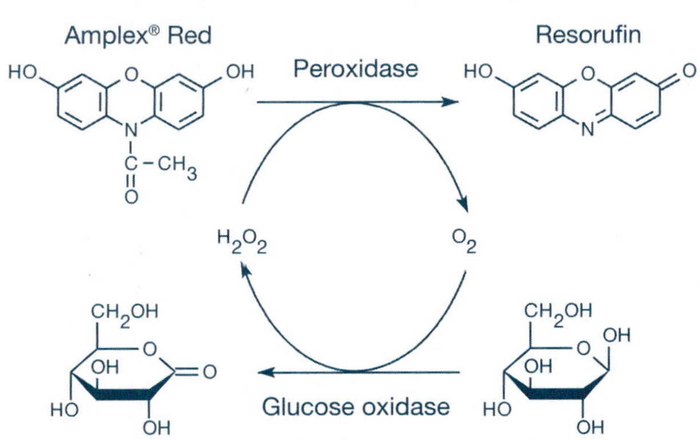

Figure 10.5.2 Principle of coupled enzymatic assays using Amplex® Red reagent. Oxidation of glucose by glucose oxidase results in generation of H_2O_2, which is coupled to conversion of the Amplex® Red reagent to fluorescent resorufin by HRP. The detection scheme shown here is used in the Amplex® Red Glucose/Glucose Oxidase Assay Kit (A22189).

IMPORTANT NOTICE: The products described in this manual are covered by one or more Limited Use Label License(s). Please refer to the Appendix on page 971 and Master Product List on page 975. Products are For Research Use Only. Not intended for any animal or human therapeutic or diagnostic use.

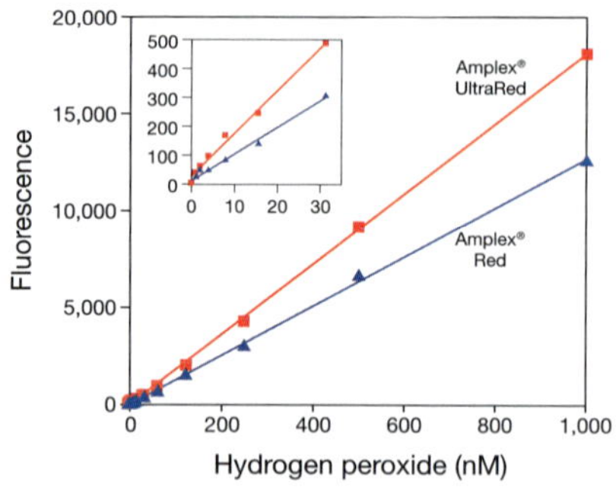

Figure 10.5.4 Detection of H_2O_2 using Amplex® UltraRed reagent (■) or Amplex® Red reagent (▲). Reactions containing 50 µM Amplex® UltraRed or Amplex® Red reagent, 1 U/mL HRP and the indicated amount of H_2O_2 in 50 mM sodium phosphate buffer, pH 7.4, were incubated for 30 minutes at room temperature. The inset shows the sensitivity and linearity of the Amplex® UltraRed assay at low levels of H_2O_2.

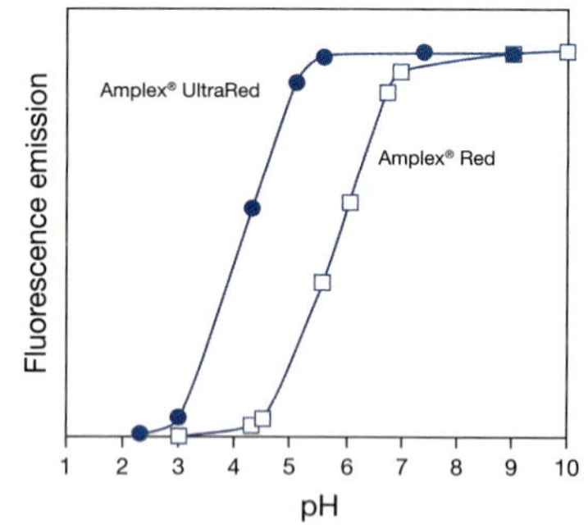

Figure 10.5.5 Comparison of pH-dependent fluorescence of Amplex® UltraRed reagent (●) and Amplex® Red reagent (□). Fluorescence intensities were measured using excitation/emission of ~570/585 nm.

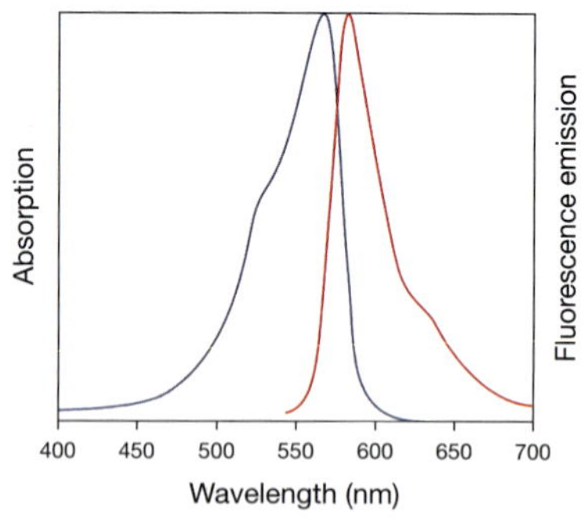

Figure 10.5.6 Absorption and fluorescence emission spectra of the product generated by horseradish peroxidase–mediated oxidation of the Amplex® UltraRed reagent in pH 7.5 buffer.

Amplex® UltraRed Reagent: Brighter and More Sensitive than the Amplex® Red Reagent

Amplex® UltraRed reagent (A36006) improves upon the performance of Amplex® Red reagent, offering brighter fluorescence and enhanced sensitivity on a per-mole basis in horseradish peroxidase or horseradish peroxidase–coupled enzyme assays [13] (Table 10.5, Figure 10.5.4). Fluorescence of oxidized Amplex® UltraRed reagent is also less sensitive to pH (Figure 10.5.5), and the substrate and its oxidation product exhibit greater stability than Amplex® Red reagent in the presence of H_2O_2 or thiols such as dithiothreitol (DTT). Like Amplex® Red reagent, nonfluorescent Amplex® UltraRed reagent reacts with H_2O_2 in a 1:1 stoichiometric ratio to produce a brightly fluorescent and strongly absorbing reaction product (excitation/emission maxima ~568/581 nm) (Figure 10.5.6). Although the primary applications of the Amplex® UltraRed reagent are enzyme-linked immunosorbent assays (ELISAs) and HRP-coupled solution assays,[13] it is also frequently used (in combination with HRP) to detect H_2O_2 production by isolated mitochondria [14] and cell cultures.[15,16]

Amplex® Red/UltraRed Stop Reagent

Amplex® Red and Amplex® UltraRed Assay Kits provide sensitive biomolecular assays based on hydrogen peroxide–generating enzyme systems linked to peroxidase-mediated oxidation of the fluorogenic Amplex® Red or Amplex® UltraRed reagent. Typically, detection reactions are performed in microplate wells and are initiated by adding the fluorogenic Amplex® or Amplex® UltraRed reagent, resulting in a continuous fluorescence increase that proceeds for 30 minutes or more. Ultimately, unknown analyte concentrations are determined by referencing fluorescence intensities measured at a certain time point during the reaction to parallel measurements at the same time point on standard samples of known concentration. Clearly, it is quite critical to ensure that the timing of the standard and unknown sample measurements is the same. The Amplex® Red/UltraRed stop reagent (A33855) provides convenience and control by allowing the fluorescence signal-generating reaction to be terminated at a user-determined time point. After addition of the stop reagent, the fluorescence signal remains stable for at least 3 hours. The Amplex® Red/UltraRed stop reagent is compatible with all Amplex® Red and Amplex® UltraRed Assay Kits and will terminate reactions containing up to 0.1 units/mL HRP and 5 µM H_2O_2.

Coupled Enzymatic Reactions with the Amplex® Red and Amplex® UltraRed Reagents

Because H_2O_2 is produced in many different enzymatic reactions, the Amplex® Red and Amplex® UltraRed reagents can be used in coupled enzymatic reactions to detect the activity of many different enzymes such as NADPH oxidase [17] and lysyl oxidase,[18] or, when the substrate concentration is limited, to assay solutions for metabolically active constituents such as glucose, pyruvate,[13] acetylcholine and cholesterol.

Table 10.5 Advantages of the Amplex® UltraRed reagent over chromogenic reagents.

Peroxidase Detection Reagent	Abs *	Em *	Notes
Amplex® UltraRed reagent	568	581	• More resistant to oxidation • More sensitive • Fast fluorescence (and color) development
Amplex® Red reagent	570	585	• Resistant to oxidation • Sensitive • Fast fluorescence (and color) development
ABTS (2′-azino-*bis*-(3-ethylbenzothiazoline-6-sulfonic acid))	405	NA	• Readily oxidized • Slow color development
OPD (o-phenylenediamine)	492	NA	• Light sensitive
TMB (3,3′,5,5′-tetramethylbenzidine)	450 (stopped reaction) 653 (kinetic assay)	NA	• Readily oxidizes • Precipitates easily due to low water solubility

* Absorption (Abs) and fluorescence emission (Em) maxima, in nm. NA = Not applicable.

IMPORTANT NOTICE: The products described in this manual are covered by one or more Limited Use Label License(s). Please refer to the Appendix on page 971 and Master Product List on page 975. Products are For Research Use Only. Not intended for any animal or human therapeutic or diagnostic use.
www.invitrogen.com/probes
molecular probes® | invitrogen™ by life technologies™

Advantages of Amplex® Red reagent– and Amplex® UltraRed reagent–based assays include the following:

- The Amplex® Red and Amplex® UltraRed reagents are fluorogenic substrates with extremely low background color or fluorescence.
- Stock solutions of the Amplex® Red and Amplex® UltraRed reagents are chemically stable.
- The fluorescent peroxidase reaction products are also stable.
- The long-wavelength spectra of the peroxidase reaction products (Figure 10.5.3, Figure 10.5.6) result in little interference from blue or green autofluorescence in biological samples.
- The peroxidase reaction products can be detected by either fluorescence- or absorption-based methods.
- In most cases, Amplex® Red and Amplex® UltraRed assays can detect either unlabeled natural biomolecules, including amino acids, sugars or lipids, or the activity of enzymes that metabolize these substrates.

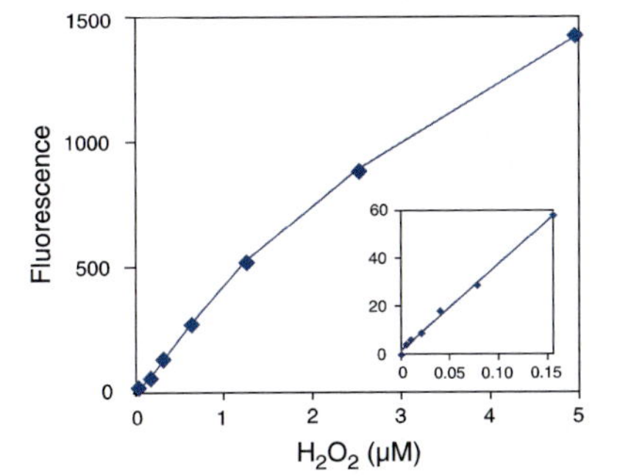

Figure 10.5.7 Detection of H_2O_2 using the Amplex® Red Hydrogen Peroxide/Peroxidase Assay Kit (A22188). Reactions containing 50 µM Amplex® Red reagent, 1 U/mL HRP and the indicated amount of H_2O_2 in 50 mM sodium phosphate buffer, pH 7.4, were incubated for 30 minutes at room temperature. Fluorescence was measured with a fluorescence microplate reader using excitation at 530 ± 12.5 nm and fluorescence detection at 580 ± 25 nm. Background fluorescence (24 units), determined for a no-H_2O_2 control reaction, was subtracted from each value. The inset shows the sensitivity and linearity of the assay at low levels of H_2O_2.

The Amplex® Red reagent is also utilized as the detection reagent in our many Amplex® Red assay kits. Substituting the Amplex® UltraRed reagent in these kits should offer even greater sensitivity.

The Amplex® Red assay kits include:

- Amplex® Red Hydrogen Peroxide/Peroxidase Assay Kit (A22188)
- Amplex® Red Catalase Assay Kit (A22180)
- Amplex® Red Monoamine Oxidase Assay Kit (A12214)
- Amplex® Red Glutamic Acid/Glutamate Oxidase Assay Kit [19] (A12221)
- Amplex® Red Glucose/Glucose Oxidase Assay Kit (A22189)
- Amplex® Red Galactose/Galactose Oxidase Assay Kit (A22179)
- Amplex® Red Neuraminidase (Sialidase) Assay Kit (A22178)
- Amplex® Red Cholesterol Assay Kit (A12216)
- Amplex® Red Acetylcholine/Acetylcholinesterase Assay Kit (A12217)
- Amplex® Red Phosphatidylcholine-Specific Phospholipase C Assay Kit [20] (A12218, Section 17.4)
- Amplex® Red Phospholipase D Assay Kit [20,21] (A12219, Section 17.4)
- Amplex® Red Sphingomyelinase Assay Kit (A12220, Section 13.3)
- Amplex® Red Uric Acid/Uricase Assay Kit (A22181)
- Amplex® Red Xanthine/Xanthine Oxidase Assay Kit (A22182)

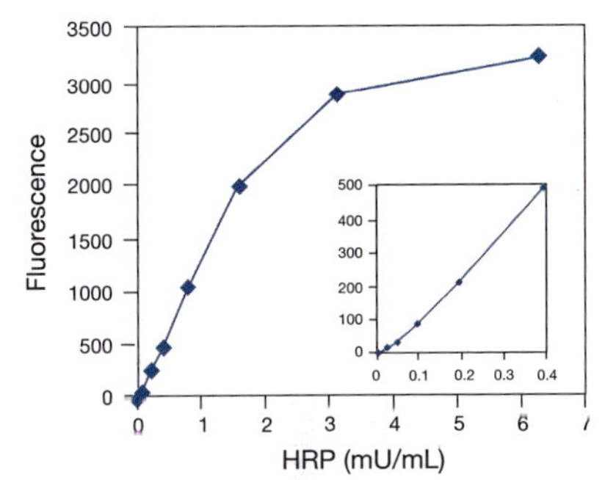

Figure 10.5.8 Detection of HRP using the Amplex® Red Hydrogen Peroxide/Peroxidase Assay Kit (A22188). Reactions containing 50 µM Amplex® Red reagent, 1 mM H_2O_2 and the indicated amount of HRP in 50 mM sodium phosphate buffer, pH 7.4, were incubated for 30 minutes at room temperature. Fluorescence was measured with a fluorescence microplate reader using excitation at 530 ± 12.5 nm and fluorescence detection at 590 ± 17.5 nm. Background fluorescence (3 units), determined for a no-HRP control reaction, was subtracted from each value. The inset shows the sensitivity of the assay at very low levels of HRP.

The Amplex® UltraRed assay kits include:

- Amplex® ELISA Development Kits for Mouse IgG and for Rabbit IgG (A33851, A33852; Section 6.2)
- EnzChek® Myeloperoxidase (MPO) Activity Assay Kit (E33856)
- Zen™ Myeloperoxidase (MPO) ELISA Kit (Z33857)
- EnzChek® *Ultra* Phytase Assay Kit (E33701)

Most of these Amplex® Red and Amplex® UltraRed kits are further discussed in this section; however, some are only presented in the sections listed above.

Amplex® Red Hydrogen Peroxide/Peroxidase Assay Kit

The Amplex® Red Hydrogen Peroxide/Peroxidase Assay Kit (A22188) provides a simple, sensitive, one-step assay for detecting H_2O_2 (Figure 10.5.7) or the activity of horseradish peroxidase either by measuring fluorescence with a fluorescence-based microplate reader or a fluorometer (Figure 10.5.8) or by measuring absorption with an absorption-based microplate reader or a spectrophotometer. The Amplex® Red peroxidase substrate can detect the presence of active peroxidases and the release of H_2O_2 from biological samples, including cells and cell extracts.[1,10,22,23]

The Amplex® Red Hydrogen Peroxide/Peroxidase Assay Kit contains:

- Amplex® Red reagent
- Dimethylsulfoxide (DMSO)
- Horseradish peroxide (HRP)
- H_2O_2 for use as a positive control
- Concentrated reaction buffer
- Detailed protocols

Each kit provides sufficient reagents for approximately 500 assays using a fluorescence- or absorption-based microplate reader and a reaction volume of 100 µL per assay.

Other Substrates for Peroxidase Assays

Although HRP is an important enzyme for both histochemistry and ELISAs, fluorogenic peroxidase substrates have not been extensively used for its detection. Fluorogenic peroxidase substrates such as the dihydrofluoresceins (also known as fluorescins) (D399, C400, C13293), dihydrocalcein AM (D23805, Figure 10.5.9), dihydrorhodamines (D632, D633, D23806; Section 18.3) and dihydroethidium (hydroethidine; D1168, D11347, D23107; Section 18.2) are converted to highly fluorescent products in the presence of the enzyme and hydrogen peroxide. Because these substrates are insufficiently stable for routine use in ELISA assays, we have converted the dihydrofluoresceins to diacetates. When used in intracellular applications, the acetates are cleaved by endogenous esterases, releasing the intact substrate. However, when used for *in vitro* assays, an esterase or a mild base must first be added to cleave the acetates, releasing the substrate. The dihydrofluoresceins have been used to measure peroxidase activity[24] and to detect hydroperoxide formation.[25–28]

In addition to being a reagent for derivatization of aldehydes and ketones (Section 3.3) and detection of nitric oxide (Section 18.3), NBD methylhydrazine (*N*-methyl-4-hydrazino-7-nitrobenzofurazan, M20490) has been reported to be useful as a fluorogenic peroxidase substrate, with a sensitivity limit for detection of H_2O_2 of about 75 nM.[29]

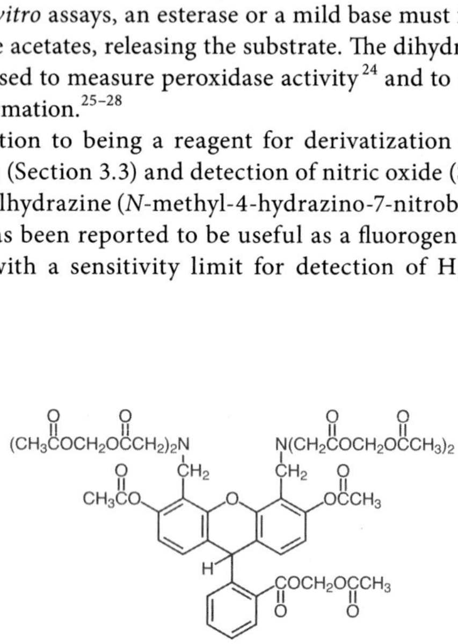

Figure 10.5.9 Dihydrocalcein, AM (D23805).

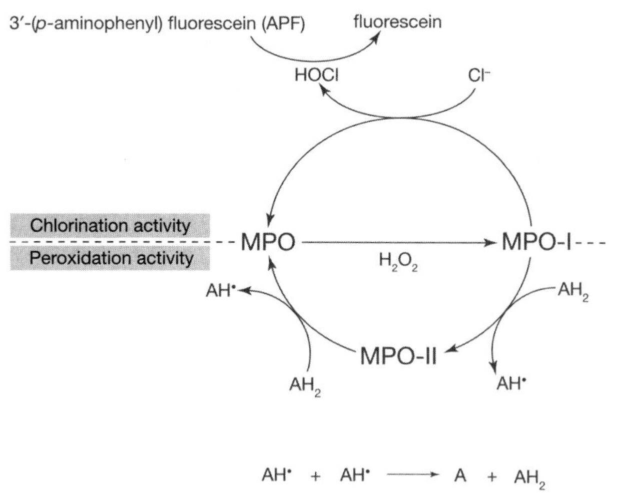

Figure 10.5.10 Schematic diagram for detection of chlorination and peroxidation activity of MPO using the EnzChek® Myeloperoxidase (MPO) Activity Assay Kit (E33856). AH_2 represents the nonfluorescent Amplex® UltraRed substrate, and A represents its fluorescent oxidation product.

Luminol and MCLA: Chemiluminescent Peroxidase Substrates

Nonisotopic immunoassays utilizing peroxidase conjugates and the chemiluminescent horseradish peroxidase substrate luminol (L8455) have provided a rapid and sensitive method for quantitating a wide variety of analytes, including cholesterol,[30] digoxin[31] and acetylcholine.[32] Addition of trace amounts of luciferin (L2911, L2912, L2916; Section 10.6) has been shown to considerably enhance the sensitivity in the assay of thyroxine, digoxin, α-fetoprotein and other analytes.[33] A method that employs luminol has been developed for the quantitation of very limiting samples of human DNA from single hairs, saliva, small blood stains and paraffin-embedded and fixed tissue sections. Using a biotinylated oligodeoxynucleotide probe to membrane-immobilized DNA, horseradish peroxidase streptavidin and luminol, researchers have detected 150 pg of human DNA.[34]

MCLA (M23800) is principally utilized as a superoxide-sensitive chemiluminescent probe (Section 18.2). MCLA has also been utilized for the determination of both horseradish peroxidase[35] and myeloperoxidase.[36,37,38]

DAB Histochemistry Kits

The use of horseradish peroxidase (HRP) for enzyme-amplified immunodetection—commonly referred to as immunoperoxidase labeling—is a well-established standard histochemical technique.[39,40] The most widely used HRP substrate for these applications is diaminobenzidine (DAB), which generates a brown-colored polymeric oxidation product localized at HRP-labeled sites. The DAB reaction product can be visualized directly by bright-field light microscopy or, following osmication, by electron microscopy. We offer DAB Histochemistry Kits for detecting mouse IgG primary antibodies (D22185; Section 7.2) and biotinylated antibodies and tracers (D22187; Section 7.6). Each kit contains sufficient materials to stain approximately 200 slides, including:

- Diaminobenzidine (DAB)
- HRP-labeled goat anti–mouse IgG antibody (in Kit D22185) or streptavidin (in Kit D22187) conjugate
- Hydrogen peroxide (H_2O_2) reaction additive
- Blocking reagent
- Staining buffer
- Detailed staining protocols

Myeloperoxidase

Myeloperoxidase (MPO, EC 1.11.1.7) is a lysosomal hemoprotein located in the azurophilic granules of polymorphonuclear (PMN) leukocytes and monocytes. It is a dimeric protein composed of two 59 kD and two 13.5 kD subunits. MPO is a unique peroxidase that catalyzes the conversion of hydrogen peroxide (H_2O_2) and chloride to hypochlorous acid, a strong oxidant with powerful antimicrobial activity and broad-spectrum reactivity with biomolecules. MPO is considered an important marker for inflammatory diseases, autoimmune diseases and cancer. MPO is also experimentally and clinically important for distinguishing myeloid from lymphoid leukemia and, due to its role in the pathology of atherogenesis, has been advocated as a prognostic marker of cardiovascular disease.

The ferric, or native, MPO reacts with hydrogen peroxide (H_2O_2) to form the active intermediate MPO-I, which oxidizes chloride (Cl^-) to HOCl; these reactions make up the chlorination cycle (Figure 10.5.10). MPO also oxidizes a variety of substrates, including phenols and

IMPORTANT NOTICE: The products described in this manual are covered by one or more Limited Use Label License(s). Please refer to the Appendix on page 971 and Master Product List on page 975. Products are For Research Use Only. Not intended for any animal or human therapeutic or diagnostic use.

anilines, via the classic peroxidation cycle. The relative concentrations of chloride and the reducing substrate determine whether MPO uses hydrogen peroxide for chlorination or peroxidation. Assays based on measurement of chlorination activity are more specific for MPO than those based on peroxidase substrates such as tetramethylbenzidine (TMB).

EnzChek® Myeloperoxidase (MPO) Activity Assay Kit

The EnzChek® Myeloperoxidase (MPO) Activity Assay Kit (E33856) provides assays that enable rapid and sensitive determination of both chlorination and peroxidation activities of MPO in solution and in cell lysates[41–43] (Figure 10.5.10). For detection of chlorination, the kit provides nonfluorescent 3′-(*p*-aminophenyl) fluorescein (APF), which is selectively cleaved by hypochlorite (^{-}OCl) to yield fluorescein. Peroxidation is detected using nonfluorescent Amplex® UltraRed reagent, which is oxidized by the H_2O_2-generated redox intermediates MPO-I and MPO-II to form a fluorescent product. The EnzChek® Myeloperoxidase Activity Assay Kit can be used to continuously detect these activities at room temperature over a broad dynamic range (1.5 to 200 ng/mL) (Figure 10.5.11). The speed (30 minutes), sensitivity, and mix-and-read convenience make this kit ideal for measuring MPO activities and for high-throughput screening for MPO-specific inhibitors.

The EnzChek® Myeloperoxidase (MPO) Activity Assay Kit contains:

- 3′-(*p*-aminophenyl) fluorescein (APF)
- Amplex® UltraRed reagent
- Human myeloperoxidase (MPO) standard
- Chlorination inhibitor
- Peroxidation inhibitor
- Hydrogen peroxide (H_2O_2)
- Phosphate-buffered saline (PBS)
- Dimethylsulfoxide (DMSO)
- Detailed protocols

Sufficient reagents are provided to perform 200 assays for chlorination and 200 assays for peroxidation activity in a 96-well fluorescence microplate format (100 µL per assay).

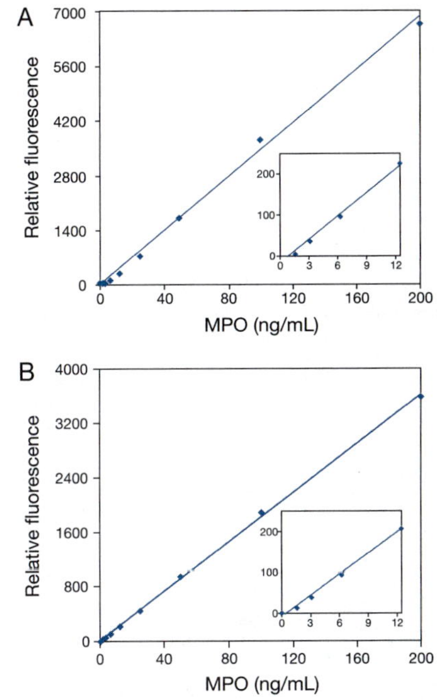

Figure 10.5.11 Typical standard curves for detection of MPO using the APF-based chlorination assay (**A**) and Amplex® UltraRed–based peroxidation assay (**B**) provided in the EnzChek® Myeloperoxidase (MPO) Activity Assay Kit (E33856). Reactions were incubated at room temperature for 30 minutes. Values on the x-axes are concentrations of MPO in the standards prior to adding the detection reagent. Fluorescence was measured with a fluorescence microplate reader using fluorescence excitation and emission at 485 and 530 nm, respectively, for the APF assay, or excitation and emission at 530 and 590 nm, respectively, for the Amplex® UltraRed assay. The background fluorescence measured for each zero-MPO control reaction was subtracted from each fluorescence measurement before plotting.

Zen™ Myeloperoxidase (MPO) ELISA Kit

The Zen™ Myeloperoxidase (MPO) ELISA Kit (Z33857) provides a comprehensive set of components is designed for accurate and sensitive quantitation of human MPO in a variety of biological samples, including human serum. This sandwich immunoassay utilizes the Amplex® UltraRed reagent, a fluorogenic substrate for horseradish peroxidase (HRP) that reacts with H_2O_2 in a 1:1 stoichiometric ratio to produce the brightly fluorescent and strongly absorbing Amplex® UltraRed oxidation product (excitation/emission maxima ~568/581 nm). Because the Amplex® UltraRed product has long-wavelength emission, there is little interference from the blue or green autofluorescence found in most biological samples. With a high extinction coefficient, good quantum efficiency, and resistance to autooxidation, the fluorescence-based Amplex® UltraRed reagent delivers better sensitivity and a broader assay range than colorimetric reagents.

The Zen™ Myeloperoxidase (MPO) ELISA Kit contains:

- Amplex® UltraRed reagent
- Dimethylsulfoxide (DMSO)
- Concentrated phosphate-buffered saline (PBS)
- Horseradish peroxidase (HRP)–labeled goat anti–rabbit IgG antibody
- Amplex® stop reagent
- Hydrogen peroxide (H_2O_2)
- MPO standard
- Bovine serum albumin (BSA)
- Tween® 20
- Mouse anti-MPO antibody (capture antibody)
- Rabbit anti-MPO antibody (detection antibody)
- Zen™ microplates for oriented capture antibody coating
- Detailed protocols

IMPORTANT NOTICE: The products described in this manual are covered by one or more Limited Use Label License(s). Please refer to the Appendix on page 971 and Master Product List on page 975. Products are For Research Use Only. Not intended for any animal or human therapeutic or diagnostic use.

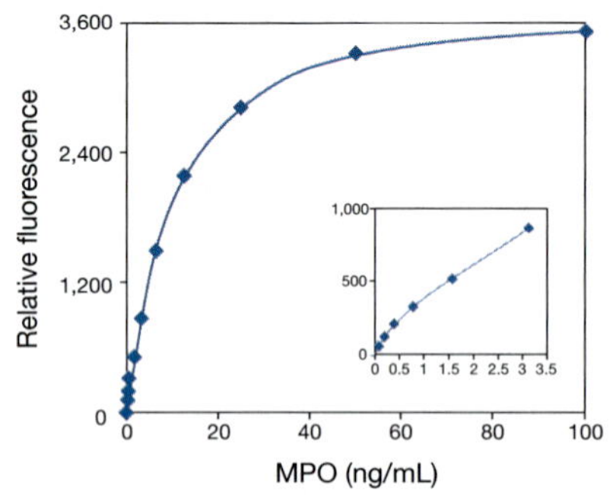

Figure 10.5.12 Typical standard curve for detection of MPO using the Zen™ Myeloperoxidase (MPO) ELISA Kit (Z33857). The sandwich ELISA was carried out as described in the protocol using a mouse anti-MPO primary capture antibody, MPO standards ranging from 0.2 ng/mL to 100 ng/mL, and a rabbit anti-MPO detection antibody.

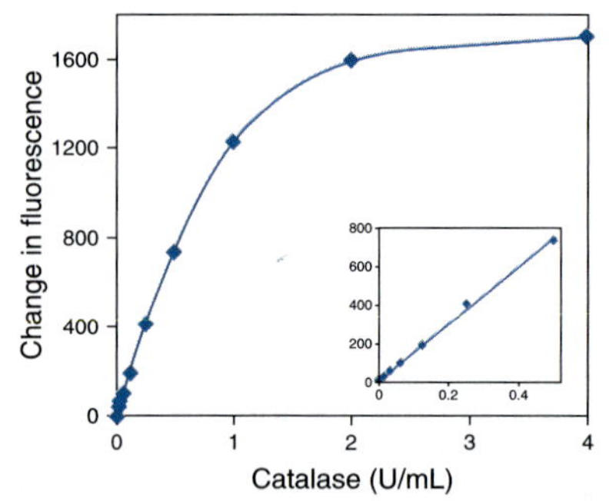

Figure 10.5.13 Subtractive detection of catalase using the Amplex® Red Catalase Assay Kit (A22180). Reactions containing the indicated amount of catalase and 20 µM H_2O_2 in 1X reaction buffer were incubated for 30 minutes. The final reactions containing 50 µM Amplex® Red reagent and 0.2 U/mL HRP and was incubated at 37°C. After 30 minutes, fluorescence was measured in a fluorescence microplate reader using excitation at 530 ± 12.5 nm and fluorescence detection at 590 ± 17.5 nm. The change in fluorescence is reported as the observed fluorescence intensity subtracted from that of a no-catalase control.

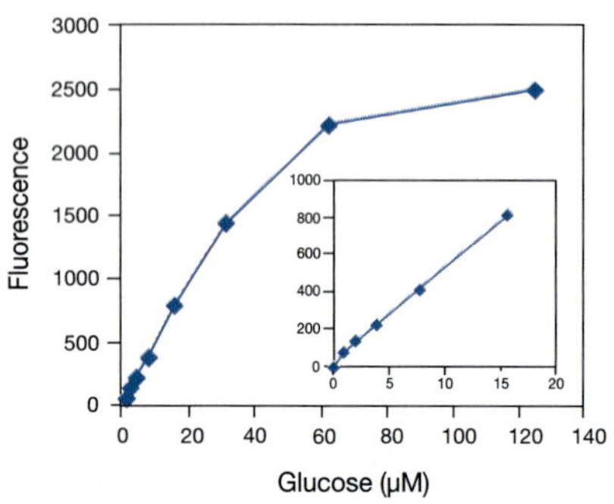

Figure 10.5.14 Detection of glucose using the Amplex® Red Glucose/Glucose Oxidase Assay Kit (A22189). Reactions containing 50 µM Amplex® Red reagent, 0.1 U/mL HRP, 1 U/mL glucose oxidase and the indicated amount of glucose in 50 mM sodium phosphate buffer, pH 7.4, were incubated for 1 hour at room temperature. Fluorescence was then measured with a fluorescence microplate reader using excitation at 530 ± 12.5 nm and fluorescence detection at 590 ± 17.5 nm. Background fluorescence (arbitrary units), determined for a no-glucose control reaction, has been subtracted from each value. The inset shows the sensitivity and linearity of the assay at low levels of glucose (0–15 µM).

Sufficient reagents are provided for 200 assays in a microplate format, using a 100 µL per well reaction volume. The Zen™ Myeloperoxidase (MPO) ELISA Kit can be used to detect from 0.2 to 100 ng/mL MPO at room temperature (Figure 10.5.12).

Catalase

The Amplex® Red Catalase Assay Kit (A22180) provides a sensitive subtractive assay for measuring catalase activity. Catalase is a heme-containing redox protein found in nearly all animal and plant cells, as well as in aerobic microorganisms. In eukaryotic cells, it is concentrated in the peroxisomes. Catalase is an important enzyme because H_2O_2 is a powerful oxidizing agent that is potentially damaging to cells. By preventing excessive buildup of H_2O_2, catalase allows important cellular processes that produce H_2O_2 as a by-product to take place safely.

In the assay, catalase first reacts with H_2O_2 to produce water and oxygen (O_2). Next, the Amplex® Red reagent reacts with a 1:1 stoichiometry with any unreacted H_2O_2 in the presence of horseradish peroxidase to produce the highly fluorescent oxidation product, resorufin. Therefore, as catalase activity increases, the signal from resorufin decreases (Figure 10.5.13). The results are typically plotted by subtracting the observed fluorescence from that of a no-catalase control. Using this kit, it is possible to detect catalase activity in a purified system at levels as low as 50 mU/mL.

The Amplex® Red Catalase Assay Kit contains:

- Amplex® Red reagent
- Dimethylsulfoxide (DMSO)
- Horseradish peroxidase (HRP)
- Hydrogen peroxide (H_2O_2)
- Concentrated reaction buffer
- Catalase
- Detailed protocols

Each kit provides sufficient reagents for approximately 400 assays using either a fluorescence- or absorption-based microplate reader and a reaction volume of 100 µL per assay.

Glucose and Glucose Oxidase

Glucose oxidase is widely used for glucose determination and, when conjugated to antibodies, for use in enzyme immunoassays (EIAs). Amplex® Red reagent can be utilized for the ultrasensitive detection of both glucose and glucose oxidase. In this enzyme-coupled assay, glucose oxidase reacts with glucose to form gluconolactone and H_2O_2. The H_2O_2 is then detected using the Amplex® Red reagent peroxidase substrate (Figure 10.5.2). The Amplex® Red Glucose/Glucose Oxidase Assay Kit (A22189) can be used to detect glucose levels as low as 3 µM or 0.5 µg/mL (Figure 10.5.14) and is at least 10-fold more sensitive than assays using *o*-dianisidine as the peroxidase substrate. This kit can also be used to detect glucose oxidase levels as low as 0.05 mU/mL (Figure 10.5.15). We have even shown that the Amplex® Red reagent can detect glucose liberated from native dextrans by dextranase[44] and from carboxymethylcellulose by cellulase.[45]

The Amplex® Red Glucose/Glucose Oxidase Assay Kit contains:

- Amplex® Red reagent
- DMSO
- Horseradish peroxidase (HRP)
- Hydrogen peroxide (H_2O_2) for use as a positive control
- Concentrated reaction buffer
- Glucose oxidase
- D-glucose
- Detailed protocols

Each kit provides sufficient reagents for approximately 500 assays using either a fluorescence- or absorption-based microplate reader and a reaction volume of 100 µL per assay.

Galactose and Galactose Oxidase

The Amplex® Red Galactose/Galactose Oxidase Assay Kit (A22179) provides the reagents and a general protocol for the assay of terminal galactosylated proteins, galactose-producing enzymes and for the assay of galactose oxidase. Unlike glucose oxidase, galactose oxidase can produce H_2O_2 from either free galactose or from polysaccharides—including glycoproteins in solution or on cell surfaces—in which galactose is the terminal residue, producing an aldehyde moiety on the 6-position of the galactose (Figure 10.5.16). We have used our Amplex® Red galactose oxidase assay for the quantitative assay of mucin-type glycoproteins by using a method similar to one described by Kinosita and collaborators.[46]

The Amplex® Red Galactose/Galactose Oxidase Assay Kit (A22179) contains:

- Amplex® Red reagent
- Dimethylsulfoxide (DMSO)
- Horseradish peroxidase (HRP)
- Hydrogen peroxide (H_2O_2)
- Concentrated reaction buffer
- Galactose oxidase from *Dactylium dendroides*
- D-Galactose
- Detailed protocols

Sufficient reagents are provided for approximately 400 assays using either a fluorescence- or absorption-based microplate reader and a reaction volume of 100 µL per assay. The Amplex® Red galactose/galactose oxidase assay is designed to accurately measure as low as 4 µM galactose and 2 mU/mL galactose oxidase activity (Figure 10.5.17, Figure 10.5.18). Because of the high absorption of resorufin, the spectrophotometric assay has only slightly lower sensitivity than the fluorometric assay. The Amplex® Red Neuraminidase (Sialidase) Assay Kit (A22178) utilizes this galactose oxidase–coupled chemistry for continuous assay of neuraminidase-catalyzed hydrolysis of fetuin, a sialoglyconjugate. This product is described in detail in Section 10.2.

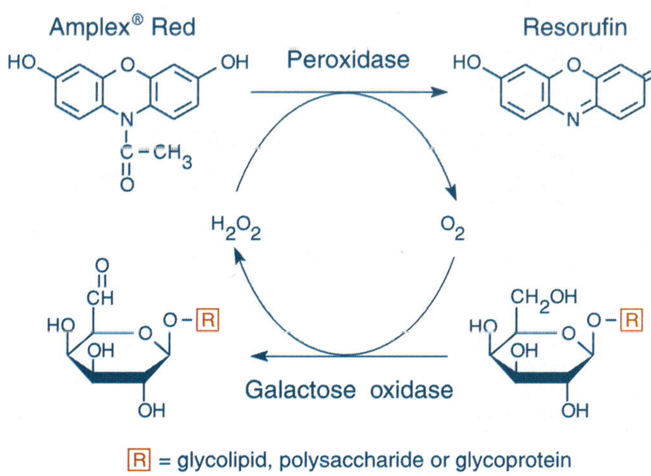

Figure 10.5.16 Detection scheme utilized in the Amplex® Red Galactose/Galactose Oxidase Assay Kit (A22179). Oxidation of the terminal galactose residue of a glycoprotein, glycolipid or polysaccharide results in the generation of H_2O_2, which, in the presence of horseradish peroxidase (HRP), reacts stoichiometrically with the Amplex® Red reagent to generate the red-fluorescent oxidation product, resorufin.

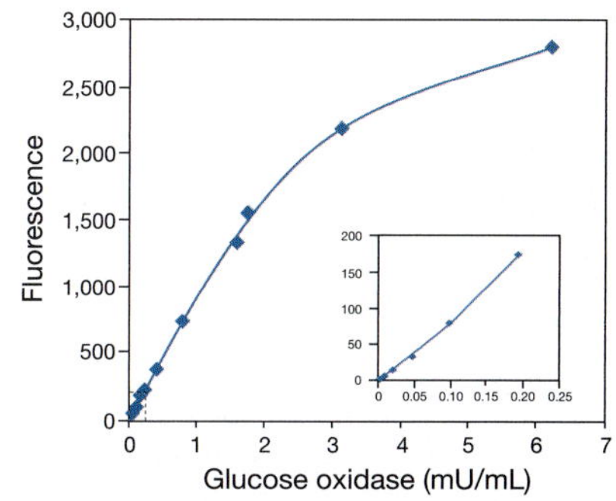

Figure 10.5.15 Detection of glucose oxidase using the Amplex® Red Glucose/Glucose Oxidase Assay Kit (A22189). Reactions containing 50 µM Amplex® Red reagent, 1 U/mL HRP, 50 mM glucose and the indicated amount of glucose oxidase in 50 mM sodium phosphate buffer, pH 7.4, were incubated for 30 minutes at room temperature. Fluorescence was measured with a fluorescence microplate reader using excitation at 530 ± 12.5 nm and fluorescence detection at 590 ± 17.5 nm. Background fluorescence (19 units) determined for a no-glucose oxidase control reaction was subtracted from each value. The inset shows the assay's sensitivity at low levels of glucose oxidase (0–0.2 mU/mL).

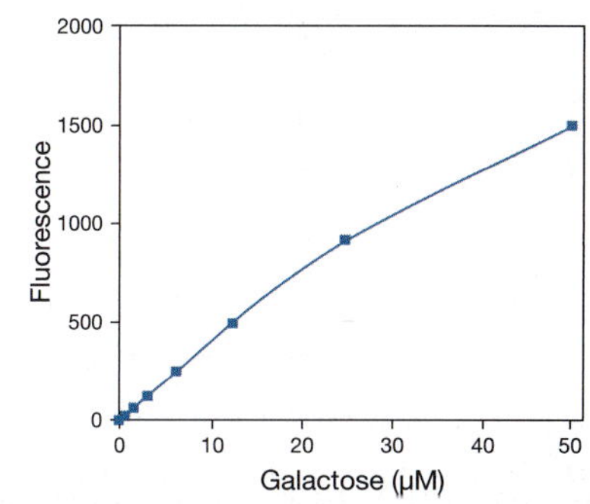

Figure 10.5.17 Detection of galactose using the Amplex® Red Galactose/Galactose Oxidase Assay Kit (A22179). Each reaction contained 50 µM Amplex® Red reagent, 0.1 U/mL HRP, 2 U/mL of galactose oxidase and the indicated amount of galactose in 1X reaction buffer. Reactions were incubated at 37°C. After 30 minutes, fluorescence was measured in a fluorescence microplate reader using excitation at 530 ± 12.5 nm and fluorescence detection at 590 ± 17.5 nm. A background fluorescence of 93 units was subtracted from each data point.

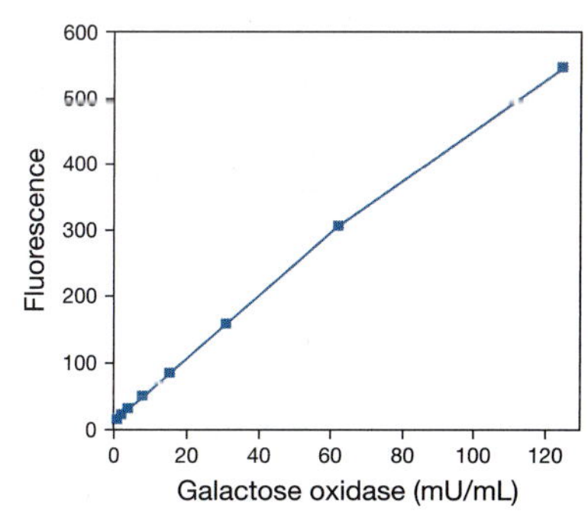

Figure 10.5.18 Detection of galactose oxidase activity using the Amplex® Red Galactose/Galactose Oxidase Assay Kit (A22179). Each reaction contained 50 µM Amplex® Red reagent, 0.1 U/mL HRP, 100 µM galactose and the indicated amount of galactose oxidase in 1X reaction buffer. Reactions were incubated at 37°C. After 20 minutes, fluorescence was measured in a fluorescence microplate reader using excitation at 530 ± 12.5 nm with fluorescence detection at 590 ± 17.5 nm.

IMPORTANT NOTICE: The products described in this manual are covered by one or more Limited Use Label License(s). Please refer to the Appendix on page 971 and Master Product List on page 975. Products are For Research Use Only. Not intended for any animal or human therapeutic or diagnostic use.

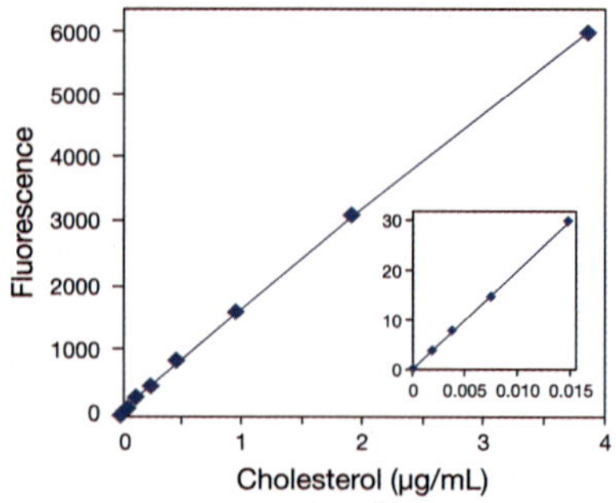

Figure 10.5.19 Detection of cholesterol using the Amplex® Red reagent–based assay. Each reaction contained 150 µM Amplex® Red reagent, 1 U/mL HRP, 1 U/mL cholesterol oxidase, 0.1 µM cholesterol esterase and the indicated amount of the cholesterol in 1X Reaction Buffer. Reactions were incubated at 37°C for 30 minutes. Fluorescence was measured with a fluorescence microplate reader using excitation at 560 ± 10 nm and fluorescence detection at 590 ± 10 nm. Background fluorescence (340 arbitrary units), determined for the no-cholesterol control reaction, has been subtracted from each value.

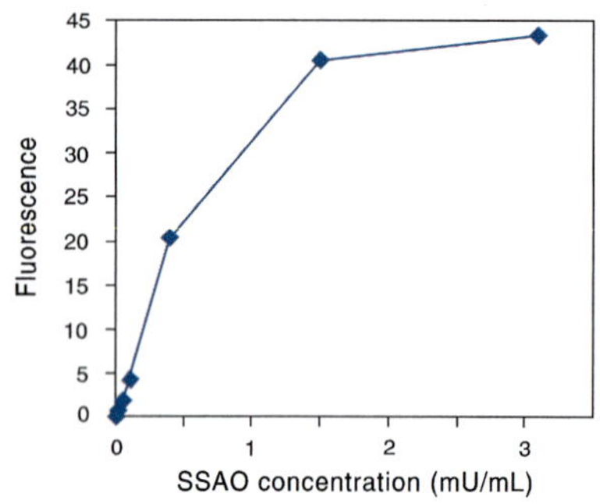

Figure 10.5.21 Detection of semicarbazide-sensitive amine oxidase (SSAO) activity using the Amplex® Red Monoamine Oxidase Assay Kit (A12214) and benzylamine as the substrate. Each reaction contained 1 mM benzylamine, 1 U/mL HRP, 200 µM Amplex® Red reagent and the indicated amount of SSAO in 50 mM potassium phosphate, pH 7.4. Reactions were incubated at room temperature for 15 minutes. Fluorescence was measured with a fluorescence microplate reader using excitation at 560 ± 10 nm and fluorescence detection at 590 ± 10 nm.

Cholesterol and Cholesterol Oxidase

The Amplex® Red Cholesterol Assay Kit (A12216) provides an exceptionally sensitive assay for both cholesterol and cholesteryl esters in complex mixtures that is suitable for use with either fluorescence microplate readers or fluorometers. The assay provided in this kit can detect as little as 5 ng/mL (5×10^{-4} mg/dL) cholesterol (Figure 10.5.19) and is designed to accurately measure the cholesterol or cholesteryl ester content in the equivalent of 0.01 µL of human serum.[47] The assay uses an enzyme-coupled reaction scheme in which cholesteryl esters are hydrolyzed by cholesterol esterase into cholesterol, which is then oxidized by cholesterol oxidase to yield H_2O_2 and the corresponding ketone steroidal product (Figure 10.5.20). The H_2O_2 is then detected using the Amplex® Red reagent in combination with HRP. The Amplex® Red cholesterol assay is continuous and requires no separation or wash steps. These characteristics make the assay particularly well suited for the rapid and direct analysis of cholesterol in blood and food samples using automated instruments. By performing two separate measurements in the presence and absence of cholesterol esterase, this assay is also useful for determining the fraction of cholesterol that is in the form of cholesteryl esters within a sample.[48] In addition, by adding an excess of cholesterol to the reaction, this assay can be used to sensitively detect the activity of cholesterol oxidase.

The Amplex® Red Cholesterol Assay Kit contains:

- Amplex® Red reagent
- Dimethylsulfoxide (DMSO)
- Horseradish peroxidase (HRP)
- Hydrogen peroxide (H_2O_2) for use as a positive control
- Concentrated reaction buffer
- Cholesterol oxidase from *Streptomyces*
- Cholesterol esterase from *Pseudomonas*
- Cholesterol for preparation of a standard curve
- Detailed protocols

Each kit provides sufficient reagents for approximately 500 assays using a fluorescence microplate reader and a reaction volume of 100 µL per assay.

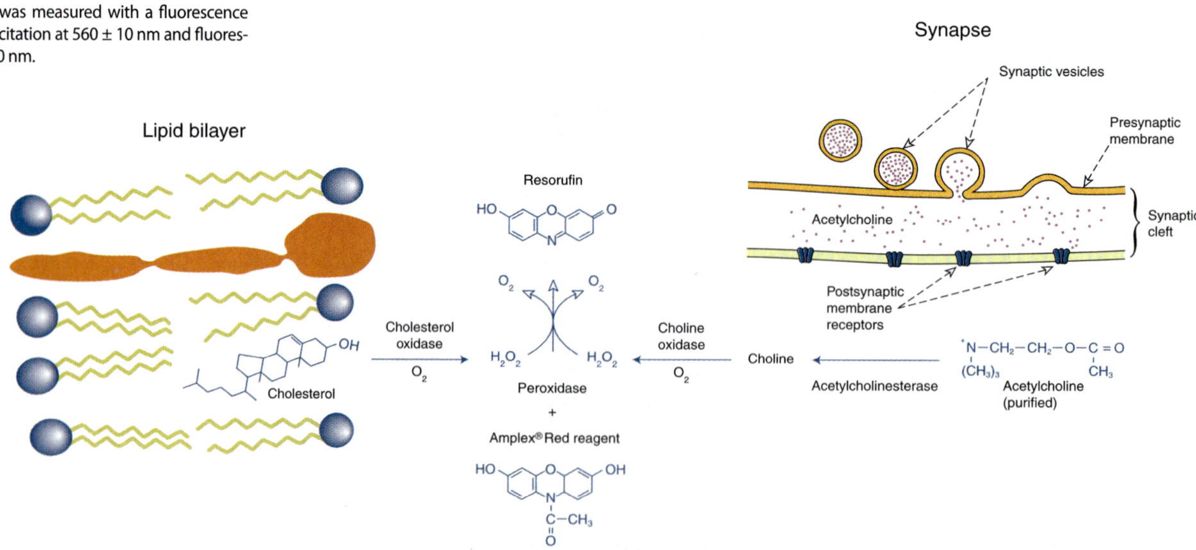

Figure 10.5.20 Enzyme-coupled Amplex® Red assays. Enzyme reactions that produce H_2O_2 can be made into Amplex® Red assays. The Amplex® Red Cholesterol Assay Kit (A12216) uses cholesterol oxidase to produce H_2O_2, which is then detected by the Amplex® Red reagent in the presence of horseradish peroxidase (HRP). Similarly, the Amplex® Red Acetylcholine/Acetylcholinesterase Assay Kit (A12217) uses choline oxidase to produce H_2O_2.

IMPORTANT NOTICE: The products described in this manual are covered by one or more Limited Use Label License(s). Please refer to the Appendix on page 971 and Master Product List on page 975. Products are For Research Use Only. Not intended for any animal or human therapeutic or diagnostic use.

Monoamine Oxidase

Monoamine oxidase, which inactivates several primary, secondary and tertiary amines via oxidative transamination, serves to regulate tissue levels of amine neurotransmitters and dietary amines. The Amplex® Red Monoamine Oxidase Assay Kit (A12214) provides a simple fluorometric method for the continuous measurement of amine oxidase activity in tissue homogenates or purified preparations. We have found that the assay is able to sensitively detect both monoamine oxidase (MAO) activity and semicarbazide-sensitive amine oxidase (SSAO) activity and is useful for performing both end-point and continuous measurements of amine oxidase activity. The assay is able to detect both MAO-A and MAO-B from cow brain tissue using as little as 200 µg of total protein per sample[6] and has been used to measure plasma amine oxidase (an SSAO) activity levels as low as 1.2×10^{-5} U/mL using a commercially available enzyme (Figure 10.5.21).

To facilitate discrimination of MAO-A and MAO-B activity, two MAO substrates and two MAO inhibitors are included in the kit. *p*-Tyramine is a substrate for both MAO-A and MAO-B, whereas benzylamine is a substrate for MAO-B.[49] Both *p*-tyramine and benzylamine are also substrates for SSAO enzymes. Clorgyline is a specific inhibitor of MAO-A activity, and pargyline is a specific inhibitor of MAO-B activity.[50,51] The potential applications of this kit include the measurement of amine oxidase activity in normal and diseased tissues, blood samples and other biological fluids, the screening of drugs as possible MAO inhibitors or substrates and the determination of kinetic constants for different amine oxidase substrates.

The Amplex® Red Monoamine Oxidase Assay Kit contains:

- Amplex® Red reagent
- Dimethylsulfoxide (DMSO)
- Horseradish peroxidase (HRP)
- Hydrogen peroxide (H_2O_2) for use as a positive control
- Concentrated reaction buffer
- Benzylamine, a substrate for MAO-B and SSAO enzymes
- *p*-Tyramine, a substrate for MAO-A, MAO-B and SSAO enzymes
- Clorgyline, a specific inhibitor of MAO-A activity
- Pargyline, a specific inhibitor of MAO-B activity
- Resorufin, for use as a reference standard
- Detailed protocols

Each kit provides sufficient reagents for approximately 500 assays using a reaction volume of 200 µL per assay.

Glutamic Acid and Glutamate Oxidase

The Amplex® Red Glutamic Acid/Glutamate Oxidase Assay Kit (A12221) provides an ultrasensitive method for continuously detecting glutamic acid[52] or for monitoring glutamate oxidase activity in a fluorescence microplate reader or a fluorometer.[19] In this assay, L-glutamic acid is oxidized by glutamate oxidase to produce α-ketoglutarate, NH_3 and H_2O_2. L-Alanine and L-glutamate–pyruvate transaminase are also included in the reaction. Thus, the L-glutamic acid is regenerated by transamination of α-ketoglutarate, resulting in multiple cycles of the initial reaction and a significant amplification of the H_2O_2 produced. Hydrogen peroxide reacts with the Amplex® Red reagent in a 1:1 stoichiometry in a reaction catalyzed by horseradish peroxidase (HRP) to generate the highly fluorescent product resorufin[1,5] (R363, Section 10.1). Because resorufin has absorption/emission maxima of ~571/ 585 nm, there is little interference from autofluorescence in most biological samples.

If the concentration of L-glutamic acid is limiting in this assay, then the fluorescence increase is proportional to the initial L-glutamic acid concentration. The Amplex® Red Glutamic Acid/Glutamate Oxidase Assay Kit allows detection of as little as 10 nM L-glutamic acid in purified systems using a 30-minute reaction time (Figure 10.5.22). If the reaction is modified to include an excess of L-glutamic acid, then this kit can be used to continuously monitor glutamate oxidase activity. For example, purified L-glutamate oxidase from *Streptomyces* can be detected at levels as low as 0.04 mU/mL (Figure 10.5.23). The Amplex® Red reagent has been used to quantitate the activity of glutamate-producing enzymes in a high-throughput assay for drug discovery.[52]

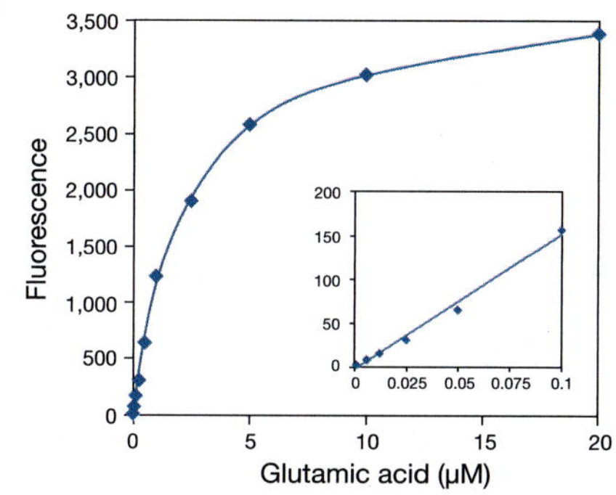

Figure 10.5.22 Detection of L-glutamic acid using the Amplex® Red Glutamic Acid/Glutamate Oxidase Assay Kit (A12221). Each reaction contained 50 µM Amplex® Red reagent, 0.125 U/mL HRP, 0.04 U/mL L-glutamate oxidase, 0.25 U/mL L-glutamate–pyruvate transaminase, 100 µM L-alanine and the indicated amount of L-glutamic acid in 1X reaction buffer. Reactions were incubated at 37°C. After 30 minutes, fluorescence was measured in a fluorescence microplate reader using excitation at 530 ± 12.5 nm and fluorescence detection at 590 ± 17.5 nm. The inset shows the assay sensitivity at low levels of glutamic acid (0–0.1 µM).

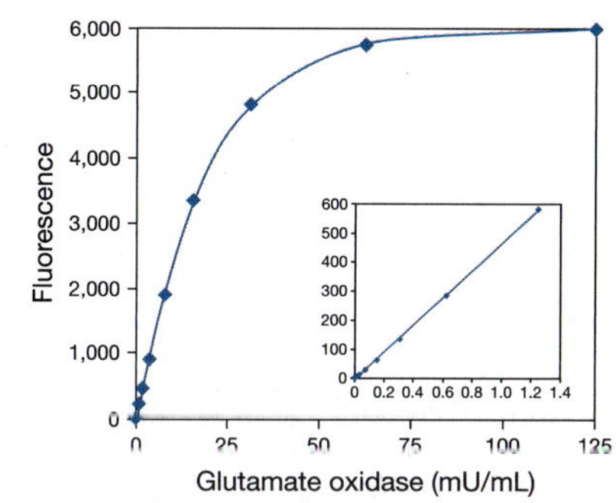

Figure 10.5.23 Detection of L-glutamate oxidase using the Amplex® Red Glutamic Acid/Glutamate Oxidase Assay Kit (A12221). Each reaction contained 50 µM Amplex® Red reagent, 0.125 U/mL HRP, 0.25 U/mL L-glutamate–pyruvate transaminase, 20 µM L-glutamic acid, 100 µM L-alanine and the indicated amount of *Streptomyces* L-glutamate oxidase in 1X reaction buffer. Reactions were incubated at 37°C. After 60 minutes, fluorescence was measured in a fluorescence microplate reader using excitation at 530 ± 12.5 nm and fluorescence detection at 590 ± 17.5 nm. The inset represents data from a separate experiment for lower L-glutamate oxidase concentrations and incubation time of 60 minutes (0–1.25 mU/mL).

IMPORTANT NOTICE: The products described in this manual are covered by one or more Limited Use Label License(s). Please refer to the Appendix on page 971 and Master Product List on page 975. Products are For Research Use Only. Not intended for any animal or human therapeutic or diagnostic use.

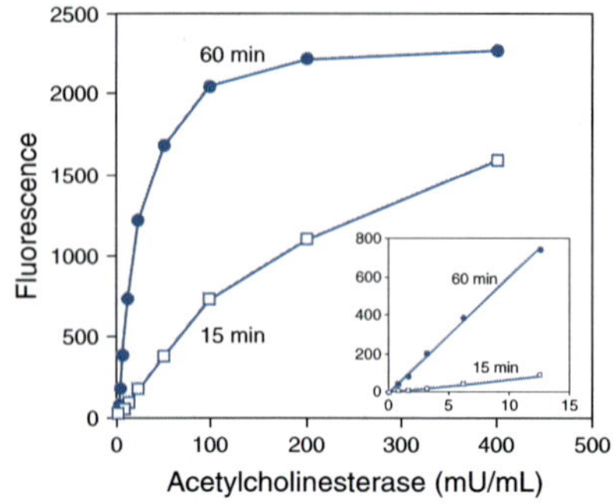

Figure 10.5.24 Detection of electric eel acetylcholinesterase activity using the Amplex® Red Acetylcholine/Acetylcholinesterase Assay Kit (A12217). Each reaction contained 50 µM acetylcholine, 200 µM Amplex® Red reagent, 1 U/mL HRP, 0.1 U/mL choline oxidase and the indicated amount of acetylcholinesterase in 1X reaction buffer. Reactions were incubated at room temperature. After 15 and 60 minutes, fluorescence was measured in a fluorescence microplate reader using excitation at 560 ± 10 nm and fluorescence detection at 590 ± 10 nm. The inset shows the sensitivity of the 15 min (□) and 60 min (●) assays at low levels of acetylcholinesterase activity (0–13 mU/mL).

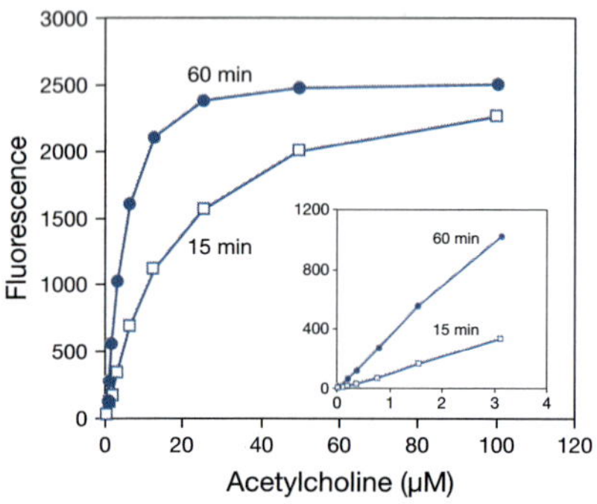

Figure 10.5.25 Detection of acetylcholine using the Amplex® Red Acetylcholine/Acetylcholinesterase Assay Kit (A12217). Each reaction contained 200 µM Amplex® Red reagent, 1 U/mL HRP, 0.1 U/mL choline oxidase, 0.5 U/mL acetylcholinesterase and the indicated amount of acetylcholine in 1X reaction buffer. Reactions were incubated at room temperature. After 15 and 60 minutes, fluorescence was measured with a fluorescence microplate reader using excitation at 560 ± 10 nm and fluorescence detection at 590 ± 10 nm. The inset shows the sensitivity of the 15 min (□) and 60 min (●) assays at low levels of acetylcholine (0–3 µM).

The Amplex® Red Glutamic Acid/Glutamate Oxidase Assay Kit contains:

- Amplex® Red reagent
- Dimethylsulfoxide (DMSO)
- Horseradish peroxidase (HRP)
- Hydrogen peroxide (H_2O_2)
- Concentrated reaction buffer
- L-Glutamate oxidase from *Streptomyces* sp.
- L-Glutamate–pyruvate transaminase from pig heart
- L-Glutamic acid
- L-Alanine
- Detailed protocols

Each kit provides sufficient reagents for approximately 200 assays using a fluorescence microplate reader and a reaction volume of 100 µL per assay.

Acetylcholine, Acetylcholinesterase and Choline Oxidase

Acetylcholine, the neurotransmitter released from the nerve terminal at neuromuscular junctions, binds to the acetylcholine receptor and opens its transmitter-gated ion channel (Section 16.3). The action of acetylcholine (ACh) is regulated by acetylcholinesterase (AChE), which hydrolyzes ACh to choline and acetate. The Amplex® Red Acetylcholine/Acetylcholinesterase Assay Kit (A12217) provides an ultrasensitive method for continuously monitoring AChE activity or for detecting ACh in a fluorescence microplate reader or a fluorometer. Other uses for this kit include screening for AChE inhibitors and measuring the release of ACh from synaptosomes.[53,54]

In the assay, AChE activity is monitored indirectly using the Amplex® Red reagent (Figure 10.5.20). First, AChE converts the acetylcholine substrate to choline. Choline is in turn oxidized by choline oxidase to betaine and H_2O_2, the latter of which, in the presence of HRP, reacts with the Amplex® Red reagent to generate the red-fluorescent product resorufin. Experiments with purified AChE from electric eel indicate that the Amplex® Red Acetylcholine/Acetylcholinesterase Assay Kit can detect AChE levels as low as 0.002 U/mL using a reaction time of 1 hour (Figure 10.5.24). We have been able to detect acetylcholinesterase activity from a tissue sample with total protein content as low as 200 ng/mL or 20 ng/well in a microplate assay. By providing an excess of AChE in the assay, the kit can also be used to detect acetylcholine levels as low as 0.3 µM, with a range of detection from 0.3 µM to ~100 µM acetylcholine (Figure 10.5.25).

The Amplex® Red Acetylcholine/Acetylcholinesterase Assay Kit contains:

- Amplex® Red reagent
- Dimethylsulfoxide (DMSO)
- Horseradish peroxidase (HRP)
- Hydrogen peroxide (H_2O_2) for use as a positive control
- Concentrated reaction buffer
- Choline oxidase from *Alcaligenes* sp.
- Acetylcholine (ACh)
- Acetylcholinesterase (AChE) from electric eel
- Detailed protocols

Each kit provides sufficient reagents for approximately 500 assays using a fluorescence microplate reader and a reaction volume of 200 µL per assay.

Xanthine and Xanthine Oxidase

Xanthine oxidase (E.C. 1.2.3.2) plays a key role in the production of free radicals, including superoxide, in the body. The Amplex® Red Xanthine/Xanthine Oxidase Assay Kit (A22182) provides an ultrasensitive method for detecting xanthine or hypoxanthine or for monitoring xanthine oxidase activity. In the assay, xanthine oxidase catalyzes the oxidation of purine nucleotides,

IMPORTANT NOTICE: The products described in this manual are covered by one or more Limited Use Label License(s). Please refer to the Appendix on page 971 and Master Product List on page 975. Products are For Research Use Only. Not intended for any animal or human therapeutic or diagnostic use.
www.invitrogen.com/probes

hypoxanthine or xanthine, to uric acid and superoxide. In the reaction mixture, the superoxide spontaneously degrades to H_2O_2, which in the presence of HRP reacts stoichiometrically with Amplex® Red reagent to generate the red-fluorescent oxidation product, resorufin. Resorufin has absorption and fluorescence emission maxima of approximately 571 nm and 585 nm, respectively, and because the extinction coefficient is high (54,000 $cm^{-1}M^{-1}$), the assay can be performed either fluorometrically or spectrophotometrically.

The Amplex® Red Xanthine/Xanthine Oxidase Assay Kit (A22182) contains:

- Amplex® Red reagent
- Dimethylsulfoxide (DMSO)
- Horseradish peroxidase (HRP)
- H_2O_2
- Concentrated reaction buffer
- Xanthine oxidase from buttermilk
- Hypoxanthine
- Xanthine
- Detailed protocols

Each kit provides sufficient reagents for approximately 400 assays using either a fluorescence- or absorption-based microplate reader and a reaction volume of 100 µL per assay.

In healthy individuals, xanthine oxidase is present in appreciable amounts only in the liver and jejunum. In various liver disorders, however, the enzyme is released into circulation. Therefore, determination of serum xanthine oxidase levels serves as a sensitive indicator of acute liver damage such as jaundice. The Amplex® Red xanthine/xanthine oxidase assay has been used as a marker of recovery from exercise stress.[55] Previously, researchers have utilized chemiluminescence or absorbance to monitor xanthine oxidase activity. The Amplex® Red Xanthine/Xanthine Oxidase Assay Kit permits the detection of xanthine oxidase in a purified system at levels as low as 0.1 mU/mL by fluorescence (Figure 10.5.26). This kit can also be used to detect as little as 200 nM hypoxanthine or xanthine (Figure 10.5.27), and, when coupled to the purine nucleotide phosphorylase enzyme, to detect inorganic phosphate.[56]

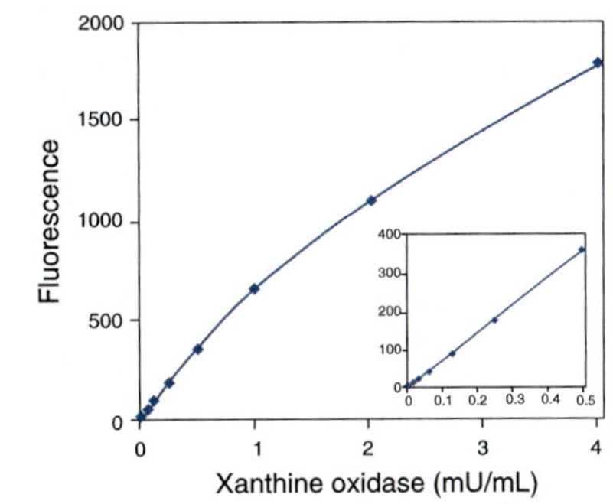

Figure 10.5.26 Detection of xanthine oxidase using the Amplex® Red Xanthine/Xanthine Oxidase Assay Kit (A22182). Each reaction contained 50 µM Amplex® Red reagent, 0.2 U/mL horseradish peroxidase, 0.1 mM hypoxanthine and the indicated amount of xanthine oxidase in 1X reaction buffer. After 30 minutes, fluorescence was measured in a fluorescence microplate reader using excitation at 530 ± 12.5 nm and detection at 590 ± 17.5 nm. A background of 65 fluorescence units was subtracted from each data point. The inset shows the assay's sensitivity and linearity at low hypoxanthine concentrations.

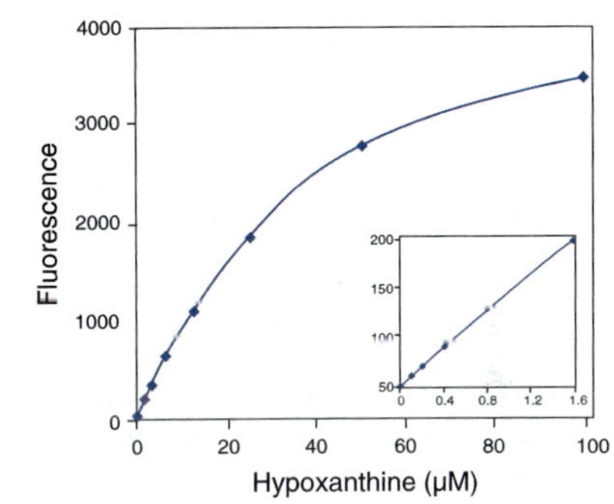

Figure 10.5.27 Detection of hypoxanthine using the Amplex® Red Xanthine/Xanthine Oxidase Assay Kit (A22182). Each reaction contained 50 µM Amplex® Red reagent, 0.2 U/mL horseradish peroxidase, 20 mU/mL xanthine oxidase and the indicated amount of hypoxanthine in 1X reaction buffer. Reactions were incubated at 37°C. After 30 minutes, fluorescence was measured in a fluorescence microplate reader using excitation at 530 ± 12.5 nm and detection at 590 ± 17.5 nm. A background of 54 fluorescence units was subtracted from each data point. The inset shows the assay's sensitivity and linearity at low enzyme concentrations.

Uric Acid and Uricase

Serum uric acid is the end product of purine metabolism in the body tissues and is cleared through the kidneys by glomerular filtration. Most animals can metabolize uric acid to more readily excreted products, but humans lack the necessary enzyme, urate oxidase (uricase), as a result of the presence of two "nonsense mutations" in the human gene for uricase. Increased uric acid levels may result from leukemia, polycythemia, ingestion of foods high in nucleoproteins (e.g., liver and kidney) or impaired renal function. Gout results from the deposit of uric acid in body joints.

The Amplex® Red Uric Acid/Uricase Assay Kit (A22181) provides an ultrasensitive method for detecting uric acid or for monitoring uricase activity.[57] In the assay, uricase catalyzes the conversion of uric acid to allantoin, H_2O_2 and carbon dioxide. In the presence of HRP, the H_2O_2 reacts stoichiometrically with Amplex® Red reagent to generate the red-fluorescent oxidation product, resorufin. Resorufin has absorption and fluorescence emission maxima of approximately 571 nm and 585 nm, respectively, and because the extinction coefficient is high (54,000 $cm^{-1}M^{-1}$), the assay can be performed either fluorometrically or spectrophotometrically. Previous literature reports colorimetric detection limits at 3.6 µM, whereas the Amplex® Red Uric Acid/Uricase Assay Kit can be used to detect as little as 100 nM uric acid in a purified system. A biosensor containing an encapsulated uricase–peroxidase system and the Amplex® Red reagent exhibits a high specificity for uric acid in the presence of interfering species and a linear response from 20 nM to 1 µM uric acid.[57] The Amplex® Red Uric Acid/Uricase Assay Kit can also be used to detect as little as 0.2 mU/mL uricase in a purified system.

The Amplex® Red Uric Acid/Uricase Assay Kit (A22181) contains:

- Amplex® Red reagent
- Dimethylsulfoxide (DMSO)
- Horseradish peroxidase (HRP)
- Hydrogen peroxide (H_2O_2)
- Concentrated reaction buffer
- Uricase
- Uric acid
- Detailed protocols

Each kit provides sufficient reagents for approximately 400 assays using either a fluorescence- or absorption-based microplate reader and a reaction volume of 100 µL per assay.

REFERENCES

1. Anal Biochem (1997) 253:162; **2.** Proc Soc Exp Biol Med (1994) 206:53; **3.** Free Radic Biol Med (1992) 12:389; **4.** Biochem Int (1985) 10:205; **5.** J Immunol Methods (1997) 202:133; **6.** Anal Biochem (1997) 253:169; **7.** J Invest Dermatol (1999) 112:751; **8.** Free Radic Biol Med (1999) 27:1197; **9.** J Invest Dermatol (1998) 110:966; **10.** Methods Mol Biol (2009) 476:28; **11.** J Biol Chem (2006) 281:39766; **12.** J Biol Chem (2009) 284:46; **13.** Anal Biochem (2010) 403:123; **14.** Methods Enzymol (2009) 456:381; **15.** Am J Physiol Lung Cell Mol Physiol (2007) 292:L1289; **16.** J Biol Chem (2007) 282:14186; **17.** J Biol Chem (2000) 275:15749; **18.** Anal Biochem (2002) 300:245; **19.** Anal Chim Acta (1999) 402:47; **20.** Proc Natl Acad Sci U S A (2004) 101:9745; **21.** J Biol Chem (2002) 277:45592; **22.** J Immunol (2010) 184:582; **23.** J Neurochem (2001) 79:266; **24.** Anal Biochem (1965) 11:6; **25.** J Clin Invest (1991) 87:711; **26.** J Lab Clin Med (1991) 117:291; **27.** Anal Biochem (1990) 187:129; **28.** Anal Biochem (1983) 134:111; **29.** Angew Chem Int Ed Engl (2000) 39:1453; **30.** Biochim Biophys Acta (1994) 1210:151; **31.** Clin Chem (1985) 31:1335; **32.** J Neurochem (1982) 39:248; **33.** Nature (1983) 305:158; **34.** Nucleic Acids Res (1992) 20:5061; **35.** J Biolumin Chemilumin (1994) 9:355; **36.** Cell Mol Neurobiol (1998) 18:565; **37.** Methods Enzymol (1994) 233:495; **38.** Anal Biochem (1991) 199:191; **39.** J Histochem Cytochem (1988) 36:317; **40.** Arch Pathol Lab Med (1978) 102:113; **41.** Exp Gerontol (2008) 43:563; **42.** Eur J Immunol (2007) 37:467; **43.** Blood (2007) 109:4716; **44.** Anal Biochem (1998) 260:257; **45.** J Microbiol Methods (2009) 79:174; **46.** Anal Biochem (2000) 284:87; **47.** J Biochem Biophys Methods (1999) 38:43; **48.** J Lipid Res (2004) 45:396; **49.** Methods Enzymol (1987) 142:617; **50.** Anal Biochem (1997) 244:384; **51.** Biochem Pharmacol (1969) 18:1447; **52.** Anal Biochem (2000) 284:382; **53.** Toxicol Sci (2006) 94:342; **54.** J Exp Med (2007) 204:1273; **55.** Am J Physiol Endocrinol Metab (2002) 282:E474; **56.** Anal Biochem (2003) 320:292; **57.** Anal Biochem (2003) 322:238.

DATA TABLE 10.5 SUBSTRATES FOR OXIDASES, INCLUDING AMPLEX® RED KITS

Cat. No.	MW	Storage	Soluble	Abs	EC	Em	Solvent	Product	Notes
A12222	257.25	FF,D,A	DMSO	280	6000	none	pH 8	R363	1
A22177	257.25	FF,D,A	DMSO	280	6000	none	pH 8	R363	1
A36006	~300	FF,D,A	DMSO	293	11,000	none	pH 8	see Notes	2
C400	531.30	F,D	DMSO, EtOH	290	5600	none	MeCN	see Notes	3
C13293	498.39	F,D	DMSO, EtOH	290	5500	none	MeCN	see Notes	4
D399	487.29	F,D	DMSO, EtOH	258	11,000	none	MeOH	see Notes	3
D23805	1068.95	F,D	DMSO	285	5800	none	MeCN	C481	
L8455	177.16	D,L	DMF	355	7500	411	MeOH	see Notes	5
M20490	209.16	F,L	MeCN	487	24,000	none	MeOH	see Notes	6
M23800	291.74	FF,D,LL,AA	DMSO	430	8400	546	MeOH	see Notes	7

For definitions of the contents of this data table, see "Using *The Molecular Probes® Handbook*" in the introductory pages.

Notes

1. This substrate is used for peroxidase-coupled detection in our Amplex® Red Assay Kits.
2. Peroxidase-catalyzed reaction of the Amplex® UltraRed reagent (A36006) with H_2O_2 yields a fluorescent product with Abs = 568 nm (EC = 57,000 $cm^{-1}M^{-1}$), Em = 581 nm in pH 7.5 buffer.
3. Dihydrofluorescein diacetates are colorless and nonfluorescent until both of the acetate groups are hydrolyzed and the products are subsequently oxidized to fluorescein derivatives. The materials contain less than 0.1% of oxidized derivative when initially prepared. The oxidation products of C400, C2938, C6827, D399 and D2935 are 2′,7′-dichlorofluorescein derivatives with spectra similar to C368.
4. Difluorodihydrofluorescein diacetates are colorless and nonfluorescent. Acetate hydrolysis and subsequent oxidation generate a fluorescent 2′,7′-difluorofluorescein derivative with spectra similar to O6146.
5. This compound emits chemiluminescence upon oxidation in basic aqueous solutions. Emission peaks are at 425 nm (L8455) and 470 nm (L6868).
6. Peroxidase-catalyzed oxidation of NBD methylhydrazine generates fluorescent *N*-methyl-4-amino-7-nitrobenzofurazan, Abs = 470 nm, Em = 547 nm in aqueous buffer (pH 5.8). (Angew Chem Int Ed Engl (2000) 39:1453)
7. Generates chemiluminescence (Em = 455 nm) upon reaction with superoxide.

IMPORTANT NOTICE: The products described in this manual are covered by one or more Limited Use Label License(s). Please refer to the Appendix on page 971 and Master Product List on page 975. Products are For Research Use Only. Not intended for any animal or human therapeutic or diagnostic use.
www.invitrogen.com/probes

PRODUCT LIST 10.5 SUBSTRATES FOR OXIDASES, INCLUDING AMPLEX® RED KITS

Cat. No.	Product	Quantity
A33851	Amplex® ELISA Development Kit for Mouse IgG *with Amplex® UltraRed reagent* *500 assays*	1 kit
A33852	Amplex® ELISA Development Kit for Rabbit IgG *with Amplex® UltraRed reagent* *500 assays*	1 kit
A12217	Amplex® Red Acetylcholine/Acetylcholinesterase Assay Kit *500 assays*	1 kit
A22180	Amplex® Red Catalase Assay Kit *400 assays*	1 kit
A12216	Amplex® Red Cholesterol Assay Kit *500 assays*	1 kit
A22179	Amplex® Red Galactose/Galactose Oxidase Assay Kit *400 assays*	1 kit
A22189	Amplex® Red Glucose/Glucose Oxidase Assay Kit *500 assays*	1 kit
A12221	Amplex® Red Glutamic Acid/Glutamate Oxidase Assay Kit *200 assays*	1 kit
A22188	Amplex® Red Hydrogen Peroxide/Peroxidase Assay Kit *500 assays*	1 kit
A12214	Amplex® Red Monoamine Oxidase Assay Kit *500 assays*	1 kit
A22178	Amplex® Red Neuraminidase (Sialidase) Assay Kit *400 assays*	1 kit
A12218	Amplex® Red Phosphatidylcholine-Specific Phospholipase C Assay Kit *500 assays*	1 kit
A12219	Amplex® Red Phospholipase D Assay Kit *500 assays*	1 kit
A12222	Amplex® Red reagent	5 mg
A22177	Amplex® Red reagent *packaged for high-throughput screening*	10 x 10 mg
A12220	Amplex® Red Sphingomyelinase Assay Kit *500 assays*	1 kit
A22181	Amplex® Red Uric Acid/Uricase Assay Kit *400 assays*	1 kit
A22182	Amplex® Red Xanthine/Xanthine Oxidase Assay Kit *400 assays*	1 kit
A33855	Amplex® Red/UltraRed stop reagent *500 tests* *set of 5 vials*	1 set
A36006	Amplex® UltraRed reagent	5 x 1 mg
C13293	5-(and-6)-carboxy-2',7'-difluorodihydrofluorescein diacetate (carboxy-H_2DFFDA) *mixed isomers*	5 mg
C400	5-(and-6)-carboxy-2',7'-dichlorodihydrofluorescein diacetate (carboxy-H_2DCFDA) *mixed isomers*	25 mg
D22185	Diaminobenzidine (DAB) Histochemistry Kit #1 *with goat anti-mouse IgG–HRP*	1 kit
D22187	Diaminobenzidine (DAB) Histochemistry Kit #3 *with streptavidin–HRP*	1 kit
D399	2',7'-dichlorodihydrofluorescein diacetate (2',7'-dichlorofluorescin diacetate; H_2DCFDA)	100 mg
D23805	dihydrocalcein, AM *special packaging*	20 x 50 µg
E33856	EnzChek® Myeloperoxidase (MPO) Activity Assay Kit *400 assays* *for myeloperoxidase chlorination and peroxidation activity*	1 kit
E33701	EnzChek® Ultra Phytase Assay Kit *500 assays*	1 kit
G21040	goat anti-mouse IgG (H+L), horseradish peroxidase conjugate	1 mg
L8455	luminol (3-aminophthalhydrazide)	25 g
M20490	*N*-methyl-4-hydrazino-7-nitrobenzofurazan (NBD methylhydrazine)	25 mg
M23800	2-methyl-6-(4-methoxyphenyl)-3,7-dihydroimidazo1,2-apyrazin-3-one, hydrochloride (MCLA)	5 mg
P21041	protein G, horseradish peroxidase conjugate	1 mg
Z33857	Zen™ Myeloperoxidase (MPO) ELISA Kit *200 assays*	1 kit

IMPORTANT NOTICE: The products described in this manual are covered by one or more Limited Use Label License(s). Please refer to the Appendix on page 971 and Master Product List on page 975. Products are For Research Use Only. Not intended for any animal or human therapeutic or diagnostic use.

10.6 Substrates for Microsomal Dealkylases, Acetyltransferases, Luciferases and Other Enzymes

Fluorogenic substrates that detect glycosidases (Section 10.2) and phosphatases (Section 10.3) have been by far the dominant probes for measuring enzymatic activity. Exactly the same fluorophores and chromophores—fluoresceins, resorufins and umbelliferones (7-hydroxycoumarins)—can be used to prepare substrates for other hydrolytic enzymes and ether-metabolizing microsomal dealkylase enzymes or peroxidases. In addition, we offer substrates for chloramphenicol acetyltransferase (CAT), luciferase and β-lactamase, which are usually not widely expressed in cells. These substrates are important tools for detecting cells transfected with reporter genes that encode these enzymes. We also have available several reagents that are substrates for detecting enzyme-catalyzed chemical reduction associated with cells, including the tetrazolium salts MTT and XTT [1] (M6494, X6493; Section 15.2) and resazurin (R12204), which is useful for quantitatively measuring cell-mediated cytotoxicity,[2] cell proliferation [3,4] and mitochondrial metabolic activity in isolated neural tissue.[5]

Microsomal Dealkylases

Metabolic oxidation of chemical compounds, including many pollutants, is the function of the cytochrome-mediated monooxygenase or mixed-function oxidase system. Several enzymes are involved, including cytochrome P448 monooxygenase (aryl hydrocarbon hydroxylase), which is induced by carcinogenic polyaromatic hydrocarbons. Cytochrome P450 (CYP) is a useful marker of endoplasmic reticulum membranes.[6] The very low turnover rate of these enzymes can be followed using various fluorogenic alkyl ether derivatives of coumarin,[7] resorufin [8] and fluorescein,[9] all of which yield cleavage products with longer-wavelength spectral properties than the parent substrates.

Resorufin-Based Microsomal Dealkylase Substrates

Resorufin ether–based substrates (R352, R441, R1147), which all yield red-fluorescent resorufin (R363, Section 10.1; excitation/emission maxima ~571/585 nm), have been extensively used to differentiate isozymes of cytochrome P450.[8,10–16] Ethoxyresorufin *O*-deethylase (EROD) and total protein concentration have been simultaneously assayed in a fluorescence microplate reader using ethoxyresorufin (resorufin ethyl ether, R352) and fluorescamine [17] (F2332, F20261; Section 9.2).

Coumarin-Based Microsomal Dealkylase Substrates

Fluorescence detection of the deethylation of 3-cyano-7-ethoxycoumarin [10,18] (C684) is reported to be 50–100 times more sensitive than that of ethoxyresorufin, primarily because of the faster turnover rate of 3-cyano-7-ethoxycoumarin;[19,20] however, ethoxyresorufin exhibits lower fluorescence background due to its more favorable spectral shifts. The deethylase product of 3-cyano-7-ethoxycoumarin, 3-cyano-7-hydroxycoumarin (C183, Section 10.1), has a lower pK_a than that of 7-ethoxycoumarin,[21,22] allowing continuous measurements of enzyme activity at pH 7.

The cytochrome P450 substrate 7-ethoxy-4-trifluoromethylcoumarin (E2882) yields a product with a fluorescence emission that is distinct from that of the substrate and of NADPH, making this substrate useful for the direct measurement of enzymatic activity.[23–26] Researchers have shown that this substrate is cleaved by at least the 1A2, 2E1 and 2B1 isozymes of cytochrome P450.[24,27]

EnzChek® Epoxide Hydrolase Substrate

The EnzChek® epoxide hydrolase substrate (E33956) is ideal for studying the epoxide hydrolase family of enzymes, including the arachidonic epoxide hydrolases (implicated in the regulation of inflammation and blood pressure) and microsomal epoxide hydrolases (reported to detoxify epoxides into diols), and their respective inhibitors. This substrate permits accurate detection of epoxide hydrolase activity in solution, with better sensitivity than colorimetric epoxide hydrolase substrates. In the presence of epoxide hydrolases, the nonfluorescent EnzChek® epoxide hydrolase substrate produces a bright blue-fluorescent product with excitation and emission maxima of ~358 nm and 452 nm, respectively. Furthermore, the blue-fluorescent product of the EnzChek® epoxide hydrolase substrate exhibits pH-insensitive spectra in the physiological pH range and is compatible with optics used for coumarin detection in fluorometers.

Lipases

Lipases play an essential role in the transfer of lipids in cell signaling and metabolism and generally include glycerol ester hydrolases and cholesterol esterases. Phospholipase A selectively hydrolyzes lipophilic esters of phospholipids. Because of their importance to the process of signal transduction in cells, our extensive selection of substrates and other probes for phospholipases is discussed in Section 17.4.

EnzChek® Lipase Substrate

The triacylglycerol-based EnzChek® lipase substrate (E33955) offers higher throughput and better sensitivity than chromogenic (TLC or HPLC) assays, and a visible wavelength–detection alternative to pyrene-based fluorescent substrates. In the presence of lipases, the nonfluorescent EnzChek® lipase substrate produces a bright, green-fluorescent product (excitation/emission maxima of ~505/515 nm) that enables the accurate and sensitive detection of lipase activity in solution. Furthermore, the green-fluorescent product of the EnzChek® lipase substrate exhibits pH-insensitive spectra in the physiological pH range and is compatible with optics used for fluorescein detection in fluorometers.

Coumarin-Based Lipase Substrates

The fluorogenic lipase substrate *O*-pivaloyloxymethyl umbelliferone (C-POM, P35901) was developed to deliver optimal performance in assays of lipase activity. Standard lipase substrates may exhibit high levels of undesirable nonspecific reactivity, either through spontaneous hydrolysis or direct reaction of the substrate with noncatalytic proteins such as BSA.[28] C-POM is much less prone to these unwanted side reactions, and the resulting low level of background fluorescence yields a better signal-to-noise ratio, providing a more accurate measure of lipase catalysis. Enzymatic conversion of the essentially nonfluorescent C-POM yields a bright blue-fluorescent reaction product (excitation/emission ~360/460 nm). C-POM has been shown to serve as a substrate for a variety of lipases and displays excellent stability in solution, making it an ideal substrate for specific lipases or for general or high-throughput screening.

Unlike lipase substrates that are esters of 7-hydroxy-4-methylcoumarin (β-methylumbelliferone, H189; Section 10.1) 6,8-difluoro-4-methylumbelliferyl octanoate (DiFMU octanoate, D12200) can be used for the continuous *in vitro* assay of lipases at a pH greater than or equal to 6;[29] the blue-fluorescent hydrolysis product of DiFMU octanoate, 6,8-difluoro-7-hydroxy-4-methylcoumarin (DiFMU, D6566; Section 10.1), has a pK_a of 4.9.[30]

Cholesterol Esterase Assay

The cholesterol produced by cholesterol esterases is readily quantitated using the Amplex® Red Cholesterol Assay Kit (A12216), which is discussed in Section 10.5 and Section 13.3. The Amplex® Red cholesterol assay can be adapted to continuously measure cholesterol generated by the activity of cholesterol esterases.

Chloramphenicol Acetyltransferase (CAT)

Because of the close correlation between its transcript levels and enzymatic activity and the excellent sensitivity of the enzyme assay, the chloramphenicol acetyltransferase (CAT) reporter gene system has proven to be a powerful tool for investigating transcriptional elements in animal[31,32] and plant cells.[33,34] Most conventional CAT assays require incubation of cell extracts with radioactive substrates, typically ^{14}C chloramphenicol or ^{14}C acetyl CoA, followed by organic extraction and autoradiography or scintillation counting.[35–37] The *FAST* CAT® Chloramphenicol Acetyltransferase Assay Kits contain unique BODIPY® chloramphenicol fluorescent substrates that take advantage of the exquisite sensitivity of fluorescence techniques, thus eliminating the need for hazardous radiochemicals, film, scintillation counters and expensive radioactive waste disposal.[38,39] The original *FAST* CAT® Kit and our *FAST* CAT® Green and Yellow (deoxy) Kits provide detection limits similar to those achieved with conventional radioactive methods and yield results that are easily visualized using a hand-held UV lamp.

FAST CAT® Chloramphenicol Acetyltransferase Assay Kit

The green-fluorescent BODIPY® FL chloramphenicol substrate in our original *FAST* CAT® Chloramphenicol Acetyltransferase Assay Kit (F2900) has a K_M for purified CAT of 7.4 µM and a V_{max} of 375 picomoles/unit/minute, values that are similar to those of ^{14}C-labeled chloramphenicol[39,40] (K_M = 12 µM and V_{max} = 120 picomole/unit/minute). To perform the assay, cell extracts are simply incubated with BODIPY® FL chloramphenicol and acetyl coenzyme A. After a suitable incubation period, the products and remaining substrate are extracted and separated by thin-layer chromatography (TLC). The brightly fluorescent, well-resolved spots can be immediately visualized with a hand-held UV lamp or quantitated with a laser scanner or CCD camera. Alternatively, quantitation can be accomplished using a fluorometer or spectrophotometer after a simple extraction. HPLC analysis of the fluorescent products has also been used to further enhance the assay's sensitivity.[41,42]

These attributes have enabled researchers to use this *FAST* CAT® substrate to measure CAT activity in crude cellular extracts of transfected ovarian granulosa cells.[40] The *FAST* CAT® Chloramphenicol Acetyltransferase Assay Kit has also been employed to study hormonal regulation of prodynorphin gene expression[43,44] and to measure the rate of hair growth in single follicles of transgenic mice.[42]

Each *FAST* CAT® Chloramphenicol Acetyltransferase Assay Kit (F2900) is available in a 100-test size and includes:

- BODIPY® FL chloramphenicol substrate
- Mixture of the 1- and 3-acetyl and 1,3-diacetyl BODIPY® FL derivatives, which serve as a reference standard for the fluorescent products
- Detailed protocols

FAST CAT® (Deoxy) Chloramphenicol Acetyltransferase Assay Kits

The *FAST* CAT® (deoxy) Chloramphenicol Acetyltransferase Assay Kits (F6616, F6617; Figure 10.6.1) contain substrates that greatly simplify the quantitation of chloramphenicol acetyltransferase (CAT) activity and extend the linear detection range of the original Molecular Probes® *FAST* CAT® assay.[45] The BODIPY® FL chloramphenicol substrate in our original *FAST* CAT® kit contains two acetylation sites, only one of which is acetylated by the CAT enzyme. Once the CAT enzyme adds an acetyl group to this position, the acetyl group can be nonenzymatically transferred to the second site, leaving the original position open for another enzymatic acetylation.[46–48] Therefore, enzyme

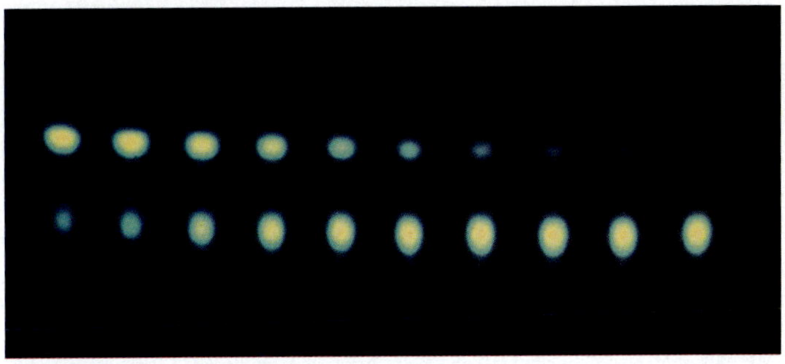

Figure 10.6.1 Chloramphenicol acetyltransferase (CAT) assays using our *FAST* CAT® Yellow (deoxy) Chloramphenicol Acetyltransferase Assay Kit (F6617). Decreasing amounts of purified CAT enzyme (2-fold dilutions) were incubated with the corresponding deoxy substrate in the presence of acetyl CoA; the reaction mixture was then separated with standard thin-layer chromatography (TLC) methods and visualized with 366 nm epi-illumination. The bottom row of fluorescent spots in each TLC represents the substrate; the top, the monoacetylated reaction product.

IMPORTANT NOTICE: The products described in this manual are covered by one or more Limited Use Label License(s). Please refer to the Appendix on page 971 and Master Product List on page 975. Products are For Research Use Only. Not intended for any animal or human therapeutic or diagnostic use.

acetylation of our original BODIPY® FL *FAST* CAT® substrate produces three products—one diacetylated and two monoacetylated chloramphenicols—thus complicating the quantitative analysis of CAT gene activity. More importantly, because the nonenzymatic transacetylation is the rate-limiting step, the rate of product accumulation may not accurately reflect CAT activity.[46,49,50]

To overcome this limitation, we have modified the original *FAST* CAT® substrate, producing reagents that undergo a single acetylation reaction (Figure 10.6.2). The green-fluorescent BODIPY® FL *deoxy*-chloramphenicol and yellow-fluorescent BODIPY® 543/569 *deoxy*-chloramphenicol substrates in the *FAST* CAT® Green and *FAST* CAT® Yellow (deoxy) Chloramphenicol Acetyltransferase Assay Kits (F6616, F6617) are acetylated at a single position, yielding only one fluorescent product[45,50] (Figure 10.6.2). This simplified reaction scheme is designed to provide a straightforward and reliable measure of CAT activity and to extend the linear detection range of our original *FAST* CAT® assay.

The *FAST* CAT® Green and *FAST* CAT® Yellow (deoxy) Chloramphenicol Acetyltransferase Assay Kits are available in a 100-test size and include:

- BODIPY® FL 1-deoxychloramphenicol substrate (in Kit F6616) or BODIPY® TMR 1-deoxychloramphenicol substrate (in Kit F6617)
- 3-Acetyl BODIPY® FL derivative (in Kit F6616) or BODIPY® TMR derivative (in Kit F6617), which serves as a reference standard for the fluorescent product
- Detailed protocols

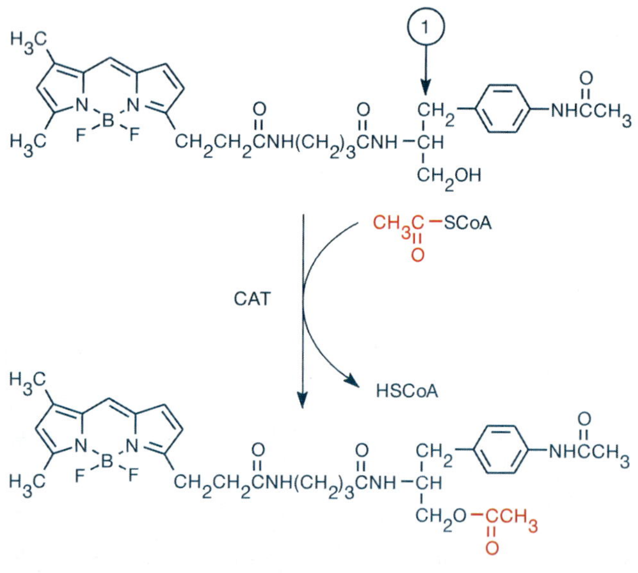

Figure 10.6.2 The green-fluorescent BODIPY® FL 1-deoxychloramphenicol substrate in our *FAST* CAT® Green (deoxy) Chloramphenicol Acetyltransferase Assay Kit (F6616). CAT-mediated acetylation of this substrate and of the BODIPY® TMR 1-deoxychloramphenicol in our *FAST* CAT® Yellow (deoxy) Chloramphenicol Acetyltransferase Assay Kit (F6617) results in single fluorescent products because these substrates contain only one hydroxyl group that can be acetylated. In contrast, the BODIPY® FL chloramphenicol substrate in our original *FAST* CAT® Kit (F2900) contains a second hydroxyl group at the 1-position (indicated by the labeled arrow). This hydroxyl group undergoes a nonenzymatic transacetylation step, restoring the original hydroxyl for a second acetylation. CAT-mediated acetylation of this chloramphenicol substrate produces three fluorescent products, thus complicating the analysis.

The CAT substrate in our *FAST* CAT® Green (deoxy) Chloramphenicol Acetyltransferase Assay Kit (F6616) is spectrally identical to the green-fluorescent BODIPY® FL chloramphenicol substrate in our original *FAST* CAT® Kit. The *FAST* CAT® Yellow (deoxy) Chloramphenicol Acetyltransferase Assay Kit (F6617) contains a red-orange–fluorescent BODIPY® TMR derivative. The availability of two spectrally distinct CAT substrates allows researchers to choose the optimal fluorophore for a particular excitation source or multicolor labeling experiment.

Luciferases

Firefly luciferase (*Photinus*-luciferin:oxygen 4-oxidoreductase or luciferin 4-monooxygenase, EC 1.13.12.7) produces light by the ATP-dependent oxidation of luciferin (Figure 10.6.3). The 560 nm chemiluminescence from this reaction peaks within seconds, with light output that is proportional to luciferase concentration when substrates are present in excess.[51] The *luc* gene, which encodes the 62,000-dalton firefly luciferase, is a popular reporter gene for plants,[33,52–55] bacteria[56,57] and mammalian cells[58,59] and for monitoring baculovirus gene expression in insects.[60,61] Chemiluminescent techniques are virtually background-free, making the *luc* reporter gene ideal for detecting low-level gene expression.[62]

Luciferin

The substrate for firefly luciferase, D-(–)-2-(6′-hydroxy-2′-benzothiazolyl)thiazoline-4-carboxylic acid, commonly known as luciferin, was first isolated by Bitler and McElroy.[63] In the firefly, spent luciferin (oxyluciferin) is recycled back to luciferin.[64] We prepare synthetic luciferin (L2911) and its water-soluble sodium (L2912) and potassium (L2916) salts. The physical properties of these derivatives are identical to those of the natural compound. Typically, luciferase expression is measured by adding the substrates ATP and luciferin to cell lysates and then analyzing light production with a luminometer. As little as 0.02 pg (250,000 molecules) of luciferase can be reliably measured using a standard scintillation counter.[65] Moreover, a CCD-based imaging method of detecting *luc* gene expression in single cells has been developed.[66]

NovaBright™ β-Galactosidase and Firefly Luciferase Dual Enzyme Reporter Gene Chemiluminescent Detection Kit

The NovaBright™ β-Galactosidase and Firefly Luciferase Dual Enzyme Reporter Gene Chemiluminescent Detection Kit (200 assays, N10561; 600 assays, N10562) allows rapid and sensitive sequential detection of firefly luciferase and β-galactosidase, enabling experimental and control reporter gene enzymes to be measured in the same cell extract sample.[67]

This kit employs the chemiluminescent luciferase substrate luciferin and the chemiluminescent β-galactosidase substrate Galacton-Plus for the detection of 1 fg to 20 ng and 10 fg to 20 ng of purified luciferase and β-galactosidase, respectively. Cell lysate is mixed with assay buffer for the luciferase reaction, and the luciferase signal is measured immediately after the injection of substrate dilution buffer, which contains both luciferin and Galacton-Plus substrates. The enhanced luciferase reaction produces a light signal that decays with a half-life of approximately 1 minute. Light signal from the β-galactosidase reaction is negligible due to lack of enzyme turnover time, low pH (7.8) and absence

IMPORTANT NOTICE: The products described in this manual are covered by one or more Limited Use Label License(s). Please refer to the Appendix on page 971 and Master Product List on page 975. Products are For Research Use Only. Not intended for any animal or human therapeutic or diagnostic use.
www.invitrogen.com/probes

of enhancer. After a 30–60 minute incubation, light signal from the accumulated product of the β-galactosidase/Galacton-Plus reaction is initiated by adding the accelerator, which raises the pH and provides the Sapphire™-II luminescence enhancer to increase light intensity. Light emission from the β-galactosidase reaction exhibits glow kinetics with a half-life of 180 minutes. Residual light from the luciferase reaction is minimal, due to rapid kinetic signal decay and quenching by accelerator. Generally, only very high luciferase concentrations (ng levels of enzyme) interfere with detection of β-galactosidase. A longer delay after the addition of accelerator prior to measurement results in decreased residual luciferase signal when extremely high levels are present. It is important, however, to maintain consistent timing for the addition of substrate dilution buffer and the measurement of the β-galactosidase signal after adding accelerator.

Each NovaBright™ β-Galactosidase and Firefly Luciferase Dual Enzyme Reporter Gene Chemiluminescent Detection Kit provides:

- Lysis buffer
- Assay buffer
- Substrate dilution buffer
- Galacton-Plus substrate
- Accelerator
- Detailed protocols

Chemiluminescent 1,2-dioxetane substrates for β-galactosidase, including the Galacton-Plus substrate described here and the Galacton-*Star*® substrate provided in the NovaBright™ β-Galactosidase Enzyme Reporter Gene Chemiluminescent Detection Kits (N10563, N10564, N10565, N10566; Section 10.2) permit highly sensitive enzyme detection and have been utilized extensively in reporter assays in mammalian cell and tissue extracts.

Luciferin–Luciferase Assays for ATP, Anesthetics and Hormones

Luciferin has been used in an exquisitely sensitive and specific ATP assay,[68,69] which allows the detection of femtomolar quantities of ATP.[70] This bioluminescent ATP assay has been employed to determine cell proliferation and cytotoxicity in both bacteria[71,72] and mammalian cells.[73,74] We provide all the reagents needed for this important assay in the ATP Determination Kit (A22066, Section 10.3). Researchers have also adapted the luciferin–luciferase ATP assay system for detecting single base changes in a solid-phase DNA sequencing method.[75] In addition, amphipathic and hydrophobic substances, including certain anesthetics and hormones, compete with luciferin for the hydrophobic site on the luciferase molecule, providing a convenient method to assay subnanomolar concentrations of these substances.[76] A protein A–luciferase fusion protein has been developed that can be used in bioluminescence-based immunoassays.[77,78]

Caged Luciferin

Although luciferase activity is sometimes measured in live cells,[66,79] *in vivo* quantitation appears to be limited by the difficulty in delivering luciferin into intact cells.[80] Molecular Probes® DMNPE-caged luciferin (L7085) readily crosses cell membranes. Once the caged luciferin is inside the cell, active luciferin can be released either instantaneously by a flash of UV light or continuously by the action of endogenous intracellular esterases, which are found in many cell types. This probe facilitates *in vivo* luciferase assays in two important ways. First, caged luciferin improves the sensitivity and quantitative analysis of these assays by allowing more efficient delivery of luciferin into intact cells. Second, hydrolysis by intracellular esterases provides a continuous supply of active luciferin, permitting long-term measurements and reducing the need for rapid mixing protocols and costly injection devices. Moreover, DMNPE-caged luciferin may make it easier to follow dynamic changes in gene expression in live cells. We also offer DMNPE-caged ATP (A1049, Section 5.3), which can be used in conjunction with DMNPE-caged luciferin for *in vivo* luciferase assays.[81]

Coelenterazines for *Renilla* Luciferase

Coelenterazine and its analogs are substrates for the bioluminescent *Renilla* luciferase.[82,83] We offer coelenterazine (C2944) and several synthetic coelenterazine analogs, including coelenterazine *cp*, *f* and *h* (C14260, C6779, C6780; Section 19.5; Table 19.4). Luciferin and coelenterazine have been used together for dual detection of firefly and *Renilla* luciferases in live mice.[84] Coelenterazine analogs have been characterized for their effectiveness in measuring *Renilla* luciferase in both live cells and live animals.[85]

β-Lactamases

Fluorocillin™ Green 495/525 β-Lactamase Substrate

Fluorocillin™ Green 495/525 substrate (F33952) is a robust reporter for ELISA protocols that employ TEM-1 β-lactamase conjugates. Upon cleavage, Fluorocillin™ Green 495/525 reagent is converted to a green-fluorescent soluble product (excitation/emission maxima ~495/525 nm). Fluorocillin™ Green 495/525 reagent has a broad dynamic range of fluorescence signal, is more sensitive than common colorimetric substrates such as nitrocefin and displays only modest hydrolysis when incubated for extended periods from pH 5.5 to 8.0. Additionally, Fluorocillin™ Green 495/525 reagent consistently reports β-lactamase activity in the presence of EDTA, many detergents, salts and sodium azide. Fluorocillin™ Green 495/525 reagent is available as a dry powder, packaged in five vials each containing 100 µg (F33952), and as a component of the SensiFlex™ ELISA Development Kits (S33853, S33854; Section 7.2).

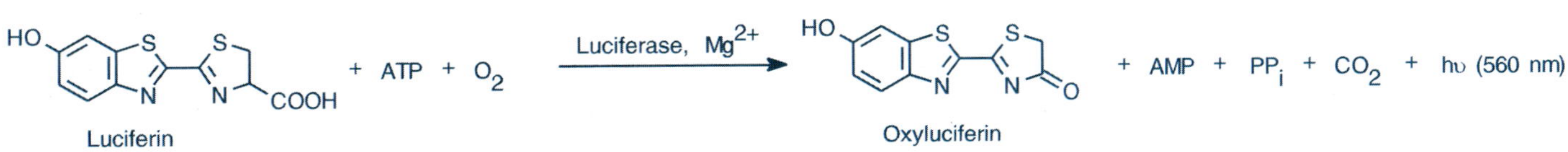

Figure 10.6.3 Reaction scheme for bioluminescence generation via luciferase-catalyzed conversion of luciferin (L2911, L2912, L2916) to oxyluciferin.

IMPORTANT NOTICE: The products described in this manual are covered by one or more Limited Use Label License(s). Please refer to the Appendix on page 971 and Master Product List on page 975. Products are For Research Use Only. Not intended for any animal or human therapeutic or diagnostic use.

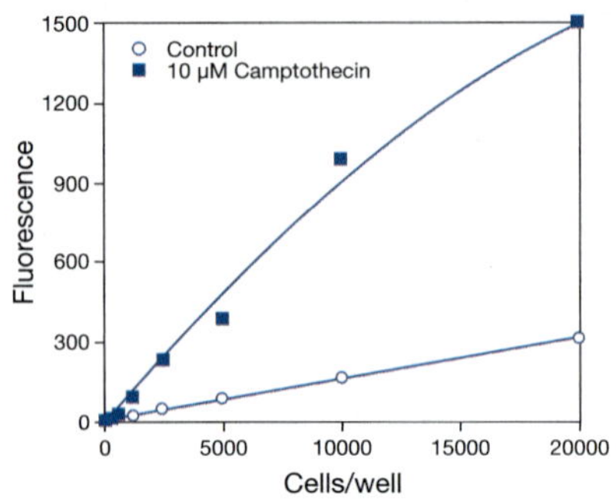

Figure 10.6.4 Detection of dead and dying cells using the Vybrant® Cytotoxicity Assay Kit (V23111). Jurkat cells were treated with 10 µM camptothecin for 6 hours, then assayed for glucose 6-phosphate dehydrogenase release. An untreated control sample is shown for comparison. The fluorescence was measured in a microplate reader (excitation/emission ~530/590 nm). A background of 55 fluorescence units was subtracted from each value.

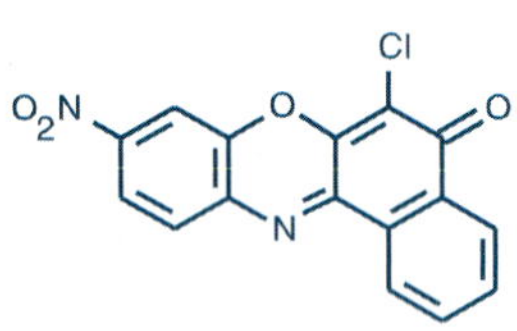

Figure 10.6.5 6-Chloro-9-nitro-5-oxo-5*H*-benzoaphenoxazine (CNOB, C22220).

Fluorocillin™ Green 345/530 β-Lactamase Substrate

Fluorocillin™ Green 345/530 β-lactamase substrate (F33951) was developed as a precipitating dye generally compatible with a variety of β-lactamase enzymes and corresponding antibody conjugates. We have demonstrated the utility of this reagent in immunohistochemistry applications, and it may also prove useful as a marker for endogenous β-lactamase activity in prokaryotes, possibly detected by microscopy or even flow cytometry. It is important to note, however, that in its precipitated form, Fluorocillin™ Green 345/530 β-lactamase substrate produces a crystal size incompatible with immunocytochemical analysis. This form of β-lactamase substrate is also incompatible with TEM-1 β-lactamase and its conjugates, but can, for example, be cleaved by P99 β-lactamase.

Dehydrogenases

Resazurin (R12204), which under the name alamarBlue® dye has been reported to be useful for quantitatively measuring cell-mediated cytotoxicity,[2] cell proliferation [3,4] and mitochondrial metabolic activity in isolated neural tissue, is also a useful substrate for measuring the dehydrogenase activity or a wide variety of dehydrogenase enzymes *in vitro*.

Among the assays reported are the use of resazurin to detect:

- Argininosuccinate lyase and NAPDH by a coupled diaphorase–resazurin reaction sequence [86]
- Bile acids in human urine, feces and serum using NAD^+ 3α-hydroxysteroid dehydrogenase [87–91]
- Glucose 6-phosphate dehydrogenase (G6PD) activity [92]
- NADH and bile acids with NADH oxidoreductase [93,94]
- Serum formate using formate dehydrogenase and NAD^+ [95]
- Triacylglycerols with glycerol dehydrogenase [96]
- Urinary acylcarnitines in an immobilized enzyme reactor [97,98]

Our Vybrant® Cytotoxicity Assay Kit (V23111, Section 15.3) employs the resazurin dehydrogenase substrate to monitor the release of G6PD from the cytosol of damaged cells into the surrounding medium (Figure 10.6.4); this method, however, also provides an extremely sensitive and specific assay for G6PD in cell-free extracts. The dehydrogenase substrate in our Vybrant® Cell Metabolic Assay Kit and LIVE/DEAD® Cell Vitality Assay Kit (V23110, L34951; Section 15.3) is dodecylresazurin, a more lipophilic version of resazurin. Because this substrate readily penetrates the membranes of live cells and its fluorescent reduction product (dodecylresorufin) is better retained in cells, it is preferred for both microscopy and flow cytometry assays.

Nitroreductase and Nitrate Reductase

We have developed a unique fluorogenic substrate that can detect the enzymatic activity of certain enzymes that reduce nitro compounds to amines or inorganic nitrate to nitrite. 6-Chloro-9-nitro-5-oxo-5*H*-benzoaphenoxazine (CNOB, C22220; Figure 10.6.5) is reduced to an aminophenoxazine dye that absorbs maximally at ~620 nm and has an emission maximum near 630 nm. We have shown that CNOB is a good substrate for at least some bacterial nitroreductases but apparently is not a good substrate for a mammalian nitroreductase. The utility of CNOB for detection of nitroreductase activity or detection of hypoxia in tumor cells has not yet been tested; however, it is known that some nitroimidazoles and other nitroaromatic compounds are reduced to amines under highly reducing conditions.[99]

REFERENCES

1. J Immunol Methods (1991) 142:257; **2.** J Immunol Methods (1998) 213:157; **3.** J Clin Lab Anal (1995) 9:89; **4.** J Immunol Methods (1994) 175:181; **5.** J Neurosci Res (1996) 45:216; **6.** J Biol Chem (1995) 270:24327; **7.** Anal Biochem (1990) 191:354; **8.** Methods Mol Biol (2006) 320:85; **9.** Anal Biochem (1983) 133:46; **10.** Anal Biochem (1997) 248:188; **11.** Biochem Pharmacol (1994) 47:893; **12.** Biochem Pharmacol (1993) 46:933; **13.** Anal Biochem (1990) 188:317; **14.** Biochem Pharmacol (1990) 40:2145; **15.** Eur J Immunol (1986) 16:829; **16.** Biochem Pharmacol (1985) 34:3337; **17.** Anal Biochem (1994) 222:217; **18.** Biopharm Drug Dispos (2003) 24:375; **19.** Anal Biochem (1988) 172:304; **20.** Biochem J (1987) 247:23; **21.** Anal Biochem (1981) 115:177; **22.** Toxicol Sci (2010) 113:293; **23.** Biochemistry (1998) 37:13184; **24.** Biochem Pharmacol (1993) 46:1577; **25.** Biochem Pharmacol (1988) 37:1731; **26.** Biochemistry (1997) 36:11707; **27.** Arch Biochem Biophys (1995) 323:303; **28.** Bioorg Med Chem Lett (2003) 13:2105; **29.** Microbiology (2004) 150:1947; **30.** Bioorg Med Chem Lett (1998) 8:3107; **31.** Development (1990) 109:577; **32.** Proc Natl Acad Sci U S A (1990) 87:6848; **33.** Methods Cell Biol (1995) 50:425; **34.** Methods Mol Biol (1995) 55:147; **35.** Gene (1988) 67:271; **36.** J Virol (1988) 62:297; **37.** Anal Biochem (1986) 156:251; **38.** Methods Enzymol (1992) 216:369; **39.** Biotechniques (1990) 8:170; **40.** Anal Biochem (1991) 197:401; **41.** J Biol Chem (1995) 270:28392; **42.** Eur J Clin Chem Clin Biochem (1993) 31:41; **43.** Mol Cell Neurosci (1992) 3:278; **44.** Mol Endocrinol (1992) 6:2244; **45.** Biotechniques (1995) 19:488; **46.** Biochemistry (1991) 30:3763; **47.** Biochemistry (1991) 30:3758; **48.** Biochemistry (1990) 29:2075; **49.** Annu Rev Biophys Biophys Chem (1991) 20:363; **50.** Nucleic Acids Res (1991) 19:6648; **51.** Mol Cell Biol (1987) 7:725; **52.** J Biolumin Chemilumin (1993) 8:267; **53.** Methods Enzymol (1992) 216:397; **54.** Dev Genet (1990) 11:224; **55.** J Biolumin Chemilumin (1990) 5:141; **56.** J Gen Microbiol (1992) 138:1289; **57.** Methods Mol Cell Biol (1989) 1:107; **58.** Methods Mol Biol (1991) 7:237; **59.** Biotechniques (1989) 7:1116; **60.** FEBS Lett (1990) 274:23; **61.** Gene (1990) 91:135; **62.** Anal Biochem (1989) 176:28; **63.** Arch Biochem Biophys (1957) 72:358; **64.** J Biol Chem (2001) 276:36508; **65.** Anal Biochem (1988) 171:404; **66.** J Biolumin Chemilumin (1990) 5:123; **67.** Biotechniques (1996) 21:520; **68.** J Appl Biochem (1981) 3:473; **69.** Anal Biochem (1969) 29:381; **70.** Lett Appl Microbiol (1990) 1:208; **71.** Biotechnol Bioeng (1993) 42:30; **72.** J Biolumin Chemilumin (1991) 6:193; **73.** Biochem J (1993) 295:165; **74.** J Immunol Methods (1993) 160:81; **75.** Anal Biochem (1993) 208:171; **76.** Anal Biochem (1990) 190:304; **77.** Anal Biochem (1993) 208:300; **78.** J Immunol Methods (1991) 137:199; **79.** Biotechnology (N Y) (1992) 10:565; **80.** Biochem J (1991) 276:637; **81.** Biotechniques (1993) 15:848; **82.** Biotechniques (1998) 24:185; **83.** Biochem Biophys Res Commun (1997) 233:349; **84.** Proc Natl Acad Sci U S A (2002) 99:377; **85.** Mol Imaging (2004) 3:43; **86.** Anal Biochem (1987) 164:482; **87.** Clin Chem (1983) 29:171; **88.** Arch Biochem Biophys (1997) 337:121; **89.** J Membr Biol (1997) 159:197; **90.** Clin Chim Acta (1976) 70:79; **91.** Clin Chem (1978) 24:1150; **92.** Biochim Biophys Acta (1977) 484:249; **93.** Clin Chim Acta (1980) 107:149; **94.** Clin Chim Acta (1980) 102:241; **95.** J Anal Toxicol (1984) 8:273; **96.** Clin Chem (1980) 26:613; **97.** Clin Chim Acta (1993) 216:135; **98.** Clin Chem (1990) 36:2072; **99.** Anticancer Drug Des (1998) 13:687.

DATA TABLE 10.6 SUBSTRATES FOR MICROSOMAL DEALKYLASES, ACETYLTRANSFERASES, LUCIFERASES AND OTHER ENZYMES

Cat. No.	MW	Storage	Soluble	Abs	EC	Em	Solvent	Product	Notes
C684	215.21	L	DMSO	356	20,000	411	pH 7	C183	
C2944	423.47	FF,D,LL,AA	MeOH	429	7500	see Notes	pH 7		
C6779	425.46	FF,D,LL,AA	MeOH	437	8700	see Notes	MeOH		
C6780	407.47	FF,D,LL,AA	MeOH	437	9500	see Notes	MeOH		
C14260	415.49	FF,D,LL,AA	MeOH	430	7000	see Notes	MeOH		
C22220	326.70	F,D,L	DMSO	448	13,000	none	MeOH	see Notes	1
D12200	338.35	F,D	MeCN	312	5000	none	MeCN	D6566	
E2882	258.20	L	DMSO, DMF	333	14,000	415	MeOH	T659	
F2900	583.44	F,D,L	MeOH	504	80,000	511	MeOH	see Notes	2, 3
F6616	567.44	F,D,L	MeOH	504	81,000	510	MeOH	see Notes	3, 4
F6617	673.57	F,D,L	MeOH	545	60,000	570	MeOH	see Notes	3, 4
L2911	280.32	F,D,L,A	pH >6, DMSO	328	18,000	532	pH 7	see Notes	5
L2912	302.30	F,D,L,A	pH >6	328	17,000	533	pH 7	see Notes	5
L2916	318.41	F,D,L,A	pH >6	328	18,000	533	pH 7	see Notes	5
L7085	489.52	FF,D,LL	DMSO, DMF	334	22,000	none	MeOH	see Notes	6, 7
P35901	276.29	F,D,L	DMSO	316	14,000	380	MeOH	see Notes	8
R352	241.25	L	DMSO	464	23,000	none	MeOH	R363	
R441	303.32	L	DMSO	463	21,000	none	MeOH	R363	
R1147	283.33	L	DMSO	465	21,000	none	MeOH	R363	
R12204	251.17	L	H_2O, MeOH	604	60,000	none	MeOH	R363	

For definitions of the contents of this data table, see "Using *The Molecular Probes® Handbook*" in the introductory pages.

Notes

1. Enzymatic reduction of C22220 yields a fluorescent aminobenzophenoxazine derivative (Abs = 617 nm, Em = 625 nm).
2. Acetylation by chloramphenicol acetyltransferase (CAT) yields a mixture of 1-acetyl, 3-acetyl and 1,3-diacetyl chloramphenicol derivatives. Spectroscopic properties of these products are similar to the substrate.
3. Data represent the substrate component of this kit.
4. Acetylation by chloramphenicol acetyltransferase (CAT) yields a 3-acetyl-1-deoxychloramphenicol derivative with similar spectroscopic properties to the substrate.
5. ATP-dependent oxidation of luciferin by luciferase results in bioluminescence (Em = 560 nm) at neutral and alkaline pH. Bioluminescence is red-shifted (Em = 617 nm) under acidic conditions. (J Am Chem Soc (1966) 88:2015)
6. All photoactivatable probes are sensitive to light. They should be protected from illumination except when photolysis is intended.
7. L7085 is converted to bioluminescent luciferin (L2911) upon ultraviolet photoactivation.
8. Enzymatic cleavage of this substrate yields 7-hydroxycoumarin (umbelliferone), which has similar spectroscopic properties to H189.

IMPORTANT NOTICE: The products described in this manual are covered by one or more Limited Use Label License(s). Please refer to the Appendix on page 971 and Master Product List on page 975. Products are For Research Use Only. Not intended for any animal or human therapeutic or diagnostic use.

PRODUCT LIST 10.6 SUBSTRATES FOR MICROSOMAL DEALKYLASES, ACETYLTRANSFERASES, LUCIFERASES AND OTHER ENZYMES

Cat. No.	Product	Quantity
R441	benzyloxyresorufin (resorufin benzyl ether)	10 mg
C22220	6-chloro-9-nitro-5-oxo-*5H*-benzoaphenoxazine (CNOB)	1 mg
C2944	coelenterazine	250 µg
C14260	coelenterazine *cp*	250 µg
C6779	coelenterazine *f*	250 µg
C6780	coelenterazine *h*	250 µg
C684	3-cyano-7-ethoxycoumarin	10 mg
D12200	6,8-difluoro-4-methylumbelliferyl octanoate (DiFMU octanoate)	10 mg
E33956	EnzChek® epoxide hydrolase substrate	100 µg
E33955	EnzChek® lipase substrate *green fluorescent, 505/515*	100 µg
R352	ethoxyresorufin (resorufin ethyl ether)	5 mg
E2882	7-ethoxy-4-trifluoromethylcoumarin	25 mg
F2900	*FAST* CAT® Chloramphenicol Acetyltransferase Assay Kit *100 assays*	1 kit
F6616	*FAST* CAT® Green (deoxy) Chloramphenicol Acetyltransferase Assay Kit *100 assays*	1 kit
F6617	*FAST* CAT® Yellow (deoxy) Chloramphenicol Acetyltransferase Assay Kit *100 assays*	1 kit
F33951	Fluorocillin™ Green 345/530 β-lactamase substrate *precipitating product*	5 mg
F33952	Fluorocillin™ Green 495/525 β-lactamase substrate *soluble product*	5 x 100 µg
L7085	D-luciferin, 1-(4,5-dimethoxy-2-nitrophenyl)ethyl ester (DMNPE-caged luciferin)	5 mg
L2911	D-luciferin, free acid	25 mg
L2916	D-luciferin, potassium salt	25 mg
L2912	D-luciferin, sodium salt	25 mg
N10561	NovaBright™ β-galactosidase and Firefly Luciferase Dual Enzyme Reporter Gene Chemiluminescent Detection Kit *200 assays*	1 kit
N10562	NovaBright™ β-galactosidase and Firefly Luciferase Dual Enzyme Reporter Gene Chemiluminescent Detection Kit *600 assays*	1 kit
R1147	pentoxyresorufin (resorufin pentyl ether)	5 mg
P35901	*O*-pivaloyloxymethyl umbelliferone (C-POM) *lipase substrate* *special packaging*	5 x 100 µg
R12204	resazurin, sodium salt	10 mg

IMPORTANT NOTICE: The products described in this manual are covered by one or more Limited Use Label License(s). Please refer to the Appendix on page 971 and Master Product List on page 975. Products are For Research Use Only. Not intended for any animal or human therapeutic or diagnostic use.
www.invitrogen.com/probes
molecular probes® | invitrogen™ by life technologies™

PART IV
Probing Cell Structure: From Organelles and Membranes to Whole Cells

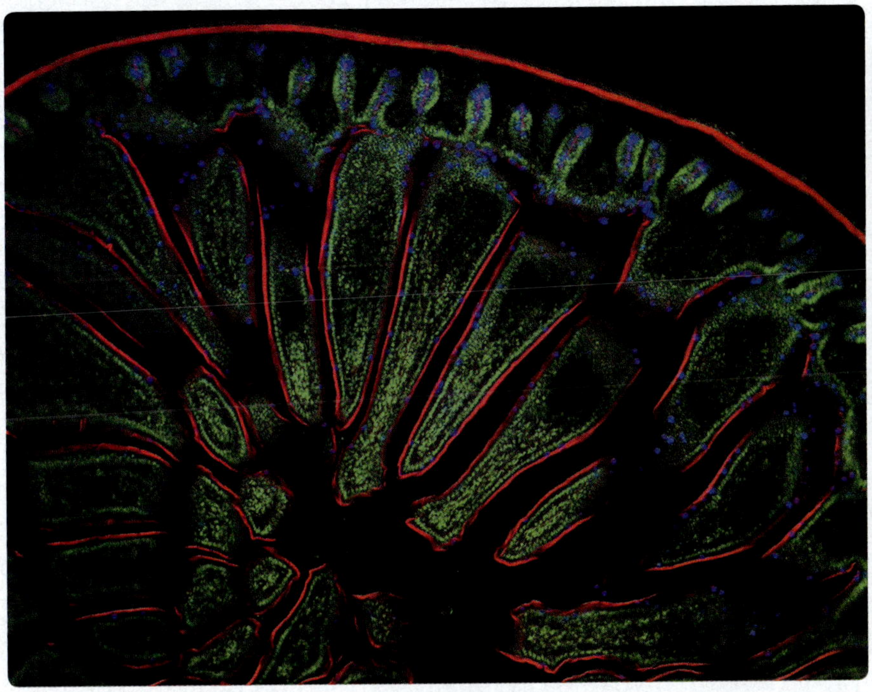
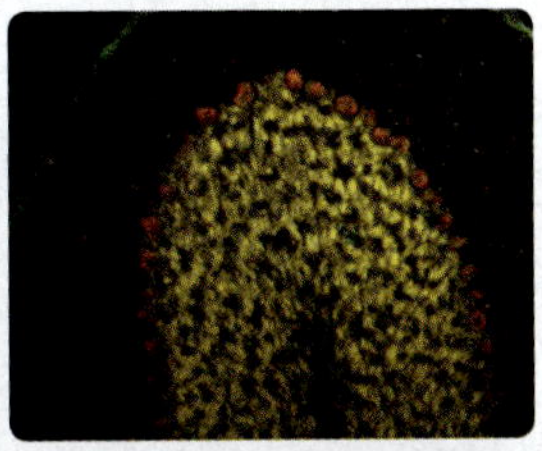
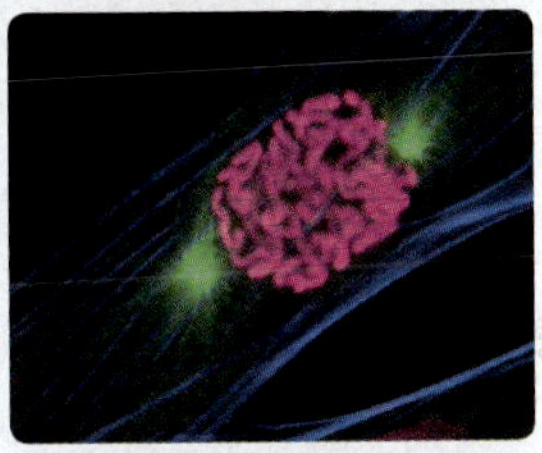
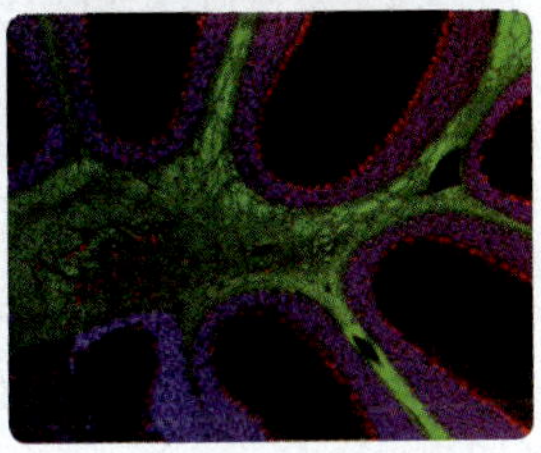

IMPORTANT NOTICE: The products described in this manual are covered by one or more Limited Use Label License(s). Please refer to the Appendix on page 971 and Master Product List on page 975. Products are For Research Use Only. Not intended for any animal or human therapeutic or diagnostic use.

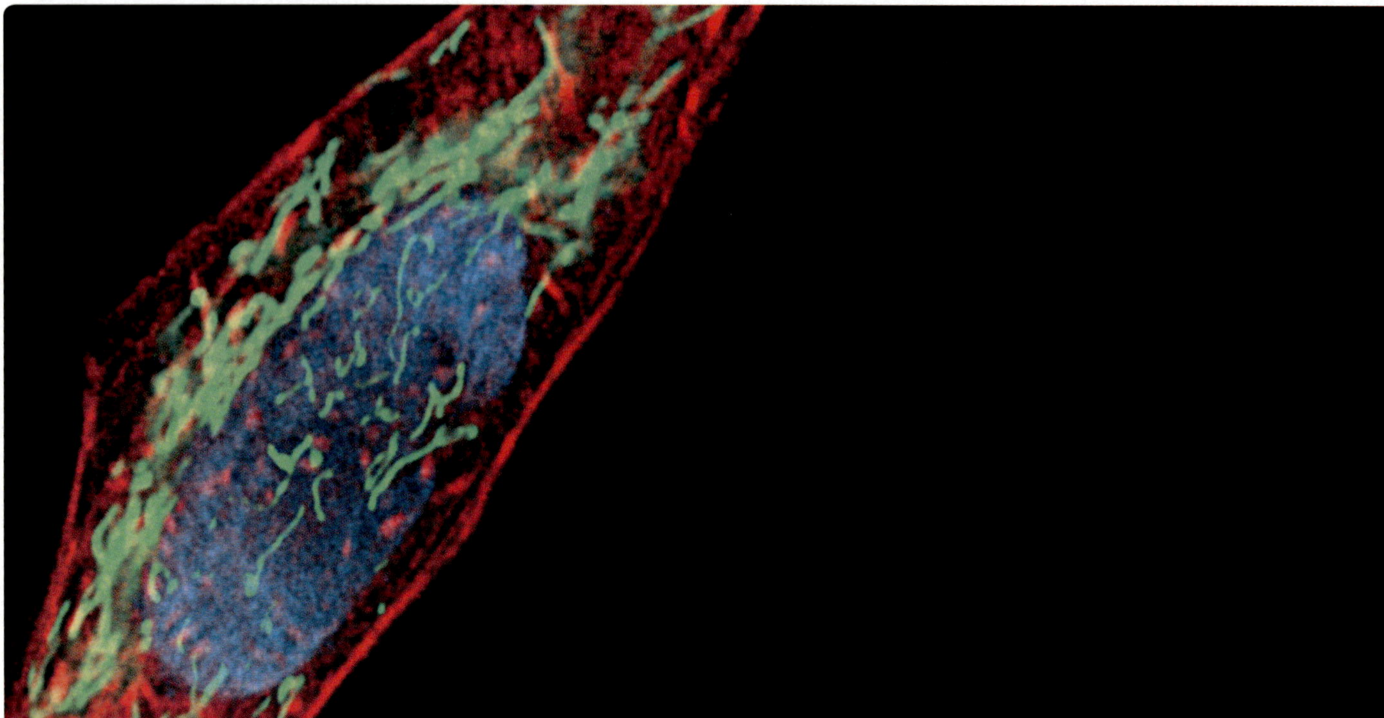

Visualize cytoskeletal and mitochondrial dynamics and organization with CellLight® Talin-RFP and CellLight® Mitochondria-GFP. CellLight® reagents use fluorescent protein–signal peptide fusions for accurate and specific targeting to subcellular structures in living cells. Their high spatial and temporal resolution enable unique and powerful insights into biological systems without modifying cell function. To create this image, HeLa cells were incubated with CellLight® Talin-RFP (C10612) and CellLight® Mitochondria-GFP (C10600) for ~2 hr. CellLight® Talin-RFP facilitates the visualization of focal adhesions in live cells via talin, a protein that links actin to the extracellular matrix; CellLight® Mitochondria-GFP enables the visualization of mitochondria independent of mitochondrial membrane potential. Nuclei were stained with Hoechst 33342 (H3570). Imaging was performed on live cells using a DeltaVision® Core microscope and standard DAPI/FITC/TRITC filter sets. Image contributed by Nicholas Dolman, Life Technologies Corporation.

This image appeared on the cover of *BioProbes 59*. *BioProbes®* newsletter is published several times each year. This award-winning publication is dedicated to furnishing researchers with the very latest information about cell biology products and their applications, and provides a great way to stay connected with the fluorescence community. You can subscribe and view the latest issues online at **www.invitrogen.com/handbook/bioprobes**.

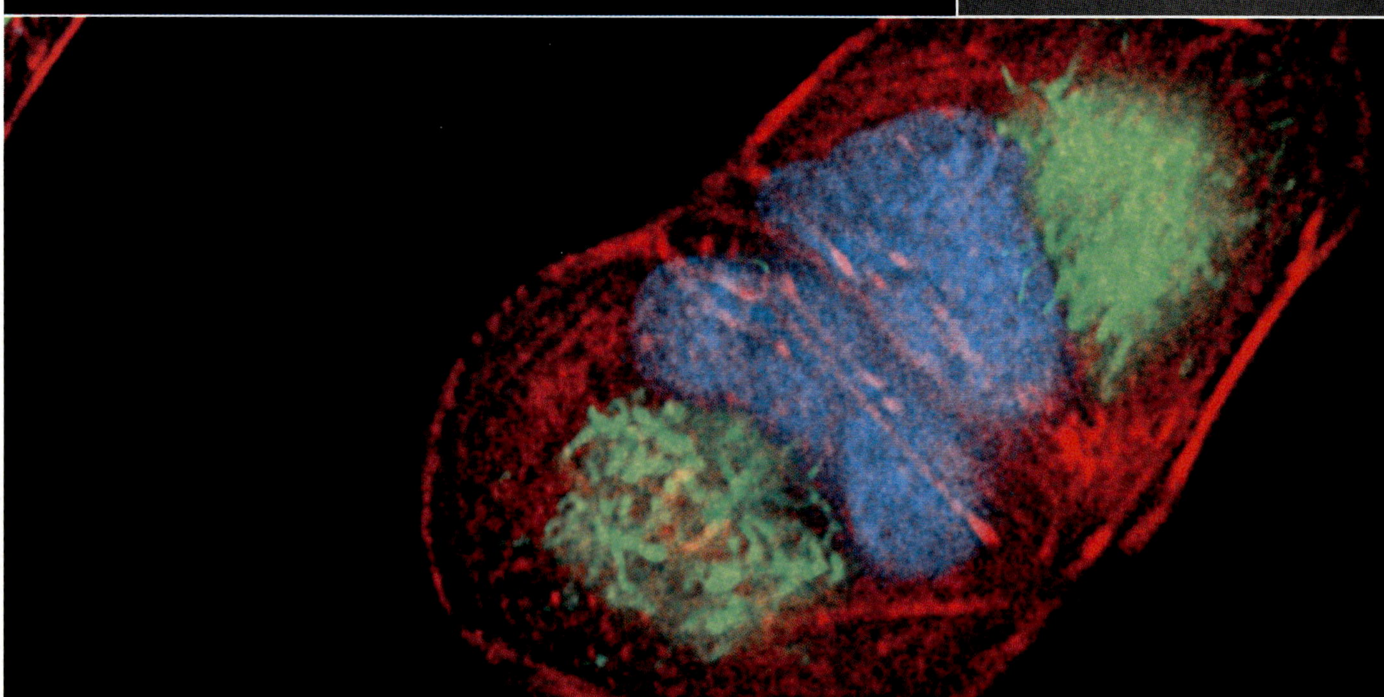

IMPORTANT NOTICE: The products described in this manual are covered by one or more Limited Use Label License(s). Please refer to the Appendix on page 971 and Master Product List on page 975. Products are For Research Use Only. Not intended for any animal or human therapeutic or diagnostic use.

molecular probes® | invitrogen™ by life technologies™

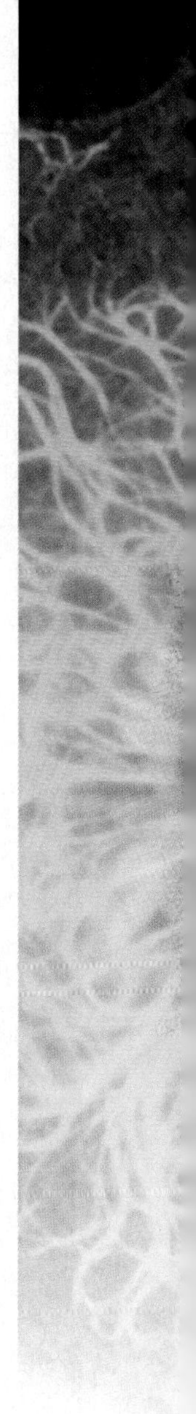

CHAPTER 11
Probes for Cytoskeletal Proteins

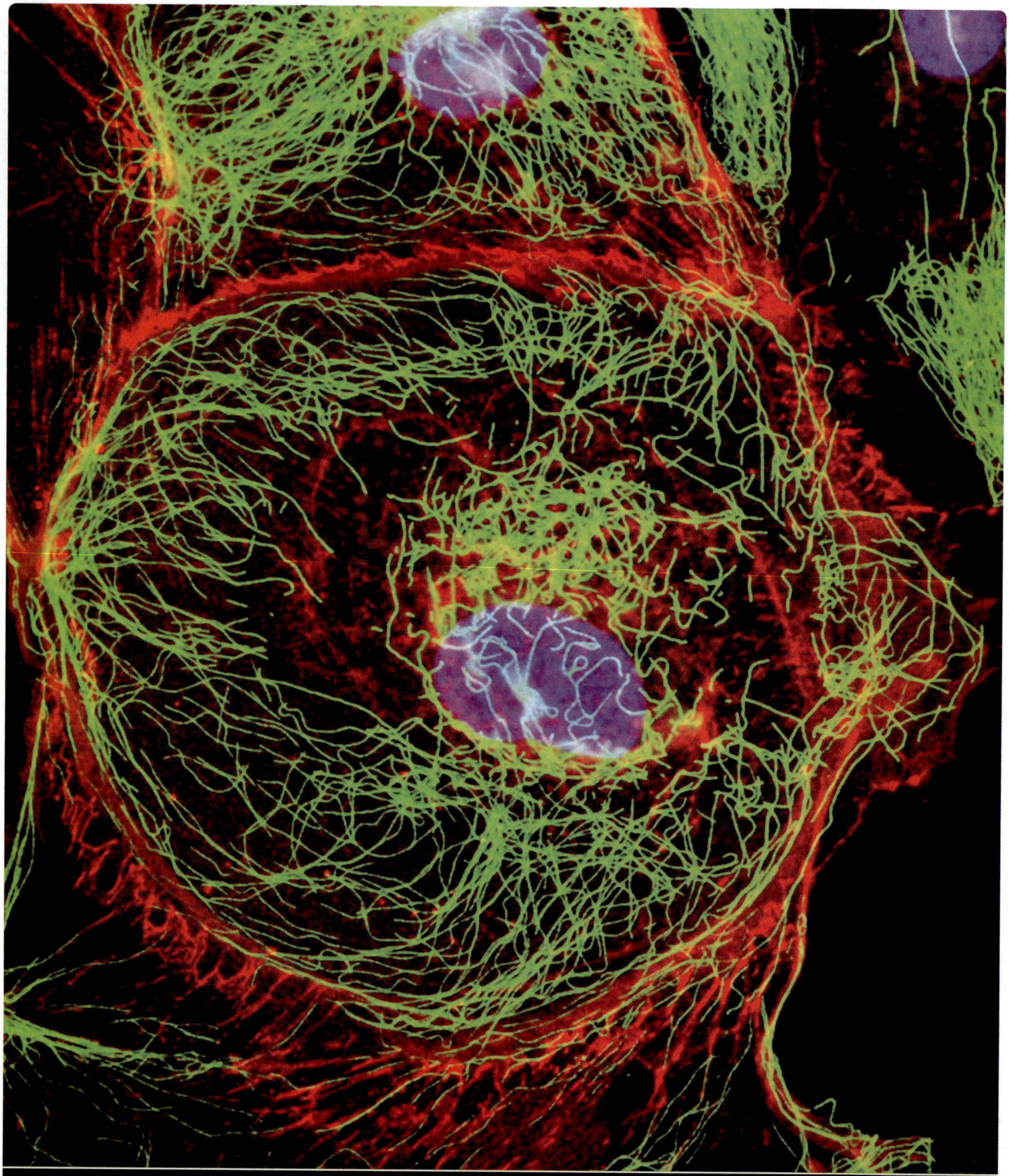

Rhodamine Red™ goat anti-rabbit IgG, Alexa Fluor® 488 goat anti-mouse IgG and Hoechst 33258.

IMPORTANT NOTICE: The products described in this manual are covered by one or more Limited Use Label License(s). Please refer to the Appendix on page 971 and Master Product List on page 975. Products are For Research Use Only. Not intended for any animal or human therapeutic or diagnostic use.

molecular probes® | invitrogen™ by life technologies™

11.1 Probes for Actin

The cytoskeleton is an essential component of a cell's structure and one of the easiest to label with fluorescent reagents. This section describes Molecular Probes® labeling reagents for both monomeric actin (G-actin) and filamentous actin (F-actin); reagents for staining tubulin and other cytoskeletal proteins are described in Section 11.2.

Fluorescent Actin

Alexa Fluor® Actin and Unlabeled Actin

Fluorescently labeled actin (Figure 11.1.1) is an important tool for investigating the structural dynamics of the cytoskeleton.[1–3] We offer highly purified actin from rabbit muscle (A12375), as well as fluorescent actin conjugates labeled with four of our brightest and most photostable dyes. The green-fluorescent Alexa Fluor® 488 actin conjugate (A12373) has excitation and emission maxima similar to fluorescein actin, but it is brighter and more photostable, and its spectra are much less pH dependent. The red-orange–fluorescent Alexa Fluor® 568 (A12374, Figure 11.1.2), red-fluorescent Alexa Fluor® 594 (A34050) and far-red–fluorescent Alexa Fluor® 647 (A34051) actin conjugates are more fluorescent than the spectrally similar Lissamine rhodamine B, Texas Red® and Cy®5 conjugates, respectively.

Our fluorescent actin conjugates are prepared by reacting amine residues of polymerized F-actin with the succinimidyl ester of the appropriate dye using a modification of the method described by Alberts and co-workers.[4] After labeling, the conjugates are subjected to depolymerization and subsequent polymerization to help ensure that the actin conjugates are able to assemble properly. The labeled actin that polymerizes is then separated from remaining monomeric actin by centrifugation, depolymerized and packaged in monomeric form.

GFP- and RFP Labeled Actin

The requirement for intracellular delivery of Alexa Fluor® dye–labeled actin conjugates by microinjection typically limits their applications for live-cell imaging to experiments involving no more than a few (<10) cells. For applications such as high-content screening (HCS) assays requiring larger sample sizes, GFP–actin fusions are well-established probes for imaging cytoskeletal structure and dynamics.[5] CellLight® Actin-GFP (C10582) and CellLight® Actin-RFP (C10583, Figure 11.1.3)

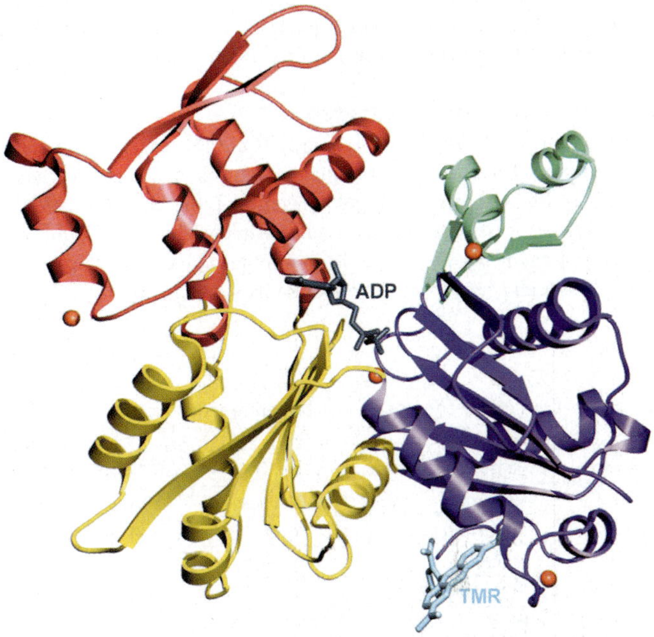

Figure 11.1.1 Ribbon diagram of the structure of uncomplexed actin in the ADP state. The four subdomains are represented in different colors, and ADP is bound at the center where the four subdomains meet. Four Ca^{2+} ions bound to the actin monomer are represented as gold spheres. In this structure, tetramethylrhodamine-5-maleimide (T6027) has been used to covalently attach the dye to a specific cysteine residue (Cys 374). Image provided by Roberto Dominguez, Boston Biomedical Research Institute, Watertown, Massachusetts. Reprinted with permission from Science (2001) 293:708. Copyright 2001 American Association for the Advancement of Science.

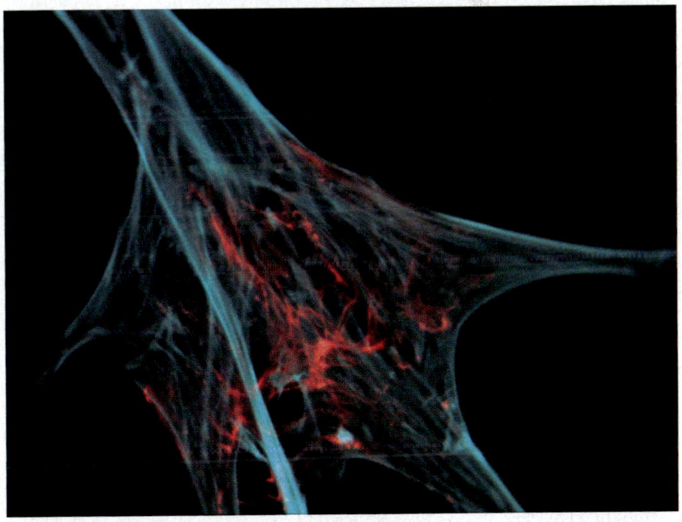

Figure 11.1.2 Chick embryo fibroblasts injected with the Alexa Fluor® 568 conjugate of actin from rabbit muscle (A12374). The cells were then fixed and permeabilized, and the filamentous actin was stained with coumarin phallacidin (C606). The double-exposure image was acquired using longpass filter sets appropriate for rhodamine and DAPI. Image contributed by Heiti Paves, Laboratory of Molecular Genetics, National Institute of Chemical Physics and Biophysics, Estonia.

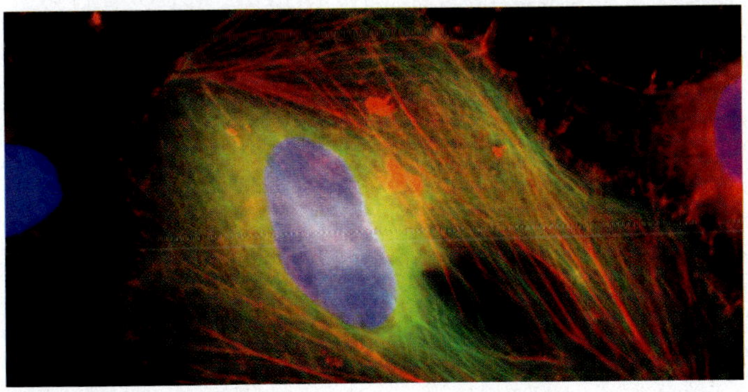

Figure 11.1.3 HeLa cell labeled with CellLight® Actin-RFP (C10583) and CellLight® MAP4-GFP (C10598) reagents and with Hoechst 33342 nucleic acid stain.

IMPORTANT NOTICE: The products described in this manual are covered by one or more Limited Use Label License(s). Please refer to the Appendix on page 971 and Master Product List on page 975. Products are For Research Use Only. Not intended for any animal or human therapeutic or diagnostic use.

expression vectors (Table 11.1) generate autofluorescent proteins fused to the N-terminus of human β-actin and incorporate all the generic advantages of BacMam 2.0 delivery technology (BacMam Gene Delivery and Expression Technology—Note 11.1). In particular, the viral dose can be readily adjusted to modulate expression levels if GFP- or RFP-dependent perturbation of cellular structural or functional properties is a concern.

CellLight® Null Control Reagent

The CellLight® Null (control) reagent (C10615), a suspension of baculovirus particles lacking mammalian genetic elements, is designed for use in parallel with our CellLight® reagents (Table 11.1). For example, microarray expression analysis on cells treated with the CellLight® Null (control) reagent can be used to assess down-regulation or up-regulation of host cell genes elicited by baculovirus infection.

Table 11.1 CellLight® reagents and their targeting sequences.

Target	Targeting Sequence	Ref	GFP (489/508 nm)*	RFP (555/584 nm)*	Handbook Section
Actin	Human actin	1	C10582	C10583	11.1
Tubulin	Human tubulin	2	C10613	C10614	11.2
MAP4	MAP4	3	C10598	C10599	11.2
Talin	Human talin $^{2341-2541}$	4	C10611	C10612	11.2
Chromatin	Histone 2B (H2B)	5	C10594	C10595	12.5
Mitochondria	Leader sequence of E1α pyruvate dehydrogenase	6	C10600	C10601	12.2
Lysosomes	Lamp1 (lysosomal-associated membrane protein 1)	7	C10596	C10597	12.3
Peroxisomes	Peroxisomal C-terminal targeting sequence	8	C10604		12.3
Endosomes	Rab5a	9	C10586	C10587	12.3, 16.1
Synaptosomes	Synaptophysin	10	C10609	C10610	16.1
Endoplasmic reticulum	ER signal sequence of calreticulin and KDEL (ER retention signal)	11	C10590	C10591	12.4
Golgi apparatus	Human golgi-resident enzyme *N*-acetylgalactosaminyltransferase 2	12	C10592	C10593	12.4
Nucleus	LSV40 nuclear localization sequence	13	C10602	C10603	12.5
Plasma membrane	Myristoylation/palmitoylation sequence from Lck tyrosine kinase	14	C10607†	C10608†	14.4
Cytoplasm	No targeting sequence		B10383		14.7

* Approximate absorption (Abs) and fluorescence (Em) maxima, in nm; GFP (Green Fluorescent Protein) and RFP (Red Fluorescent Protein, Nat Methods (2007) 4:555) can be imaged using optical filters for fluorescein (FITC) and tetramethylrhodamine (TRITC) dyes, respectively. † Also available is CellLight® Plasma Membrane-CFP (C10606), which generates a cyan-autofluorescent protein fused to the plasma membrane targeting sequence from Lck tyrosine kinase.

1. Curr Biol (1997) 7:176; **2.** PLoS One (2009) 4:e8171; **3.** J Cell Biol (1995) 130:639; **4.** Plant J (2003) 33:775; **5.** Curr Biol (1998) 8:377; **6.** J Biol Chem (2004) 279:13044; **7.** J Cell Sci (2005) 118:5243; **8.** J Cell Biol (1989) 108:1657; **9.** J Biol Chem (2009) 284:29218; **10.** J Neurosci (2006) 26:3604; **11.** FEBS Lett (1997) 405:18; **12.** J Cell Biol (1998) 143:1505; **13.** Trends Biochem Sci (1991) 16:478; **14.** EMBO J (1997) 16:4983.

Phallotoxins for Labeling F-Actin

We prepare a number of fluorescent and biotinylated derivatives of phalloidin and phallacidin for selectively labeling F-actin. Phallotoxins are bicyclic peptides isolated from the deadly *Amanita phalloides* mushroom [6] (www.grzyby.pl/gatunki/Amanita_phalloides.htm). They can be used interchangeably in most applications and bind competitively to the same sites on F-actin. Table 11.2 lists the available phallotoxin derivatives, along with their spectral properties.

A detailed staining protocol is included with each phallotoxin derivative. One vial of the fluorescent phallotoxin contains sufficient reagent for staining ~300 microscope slide preparations; one vial of biotin-XX phalloidin, which must be used at a higher concentration, contains sufficient reagent for ~50 microscope slide preparations. We also offer unlabeled phalloidin (P3457) for blocking F-actin staining by labeled phallotoxins and for promoting actin polymerization.

Table 11.2 Spectral characteristics of Molecular Probes® actin-selective probes.

Cat. No.	Actin-Selective Probe	Ex/Em *	Approximate MW
F-Actin–Selective Probes			
A22281	Alexa Fluor® 350 phalloidin	346/446 †	1100
C606	Coumarin phallacidin	355/443	1100
N354	NBD phallacidin	465/536	1040
A12379	Alexa Fluor® 488 phalloidin	495/517 †	1320
F432	Fluorescein phalloidin	496/516 †	1175
O7466	Oregon Green® 488 phalloidin	496/520 †	1180
B607	BODIPY® FL phallacidin	505/512	1125
O7465	Oregon Green® 514 phalloidin	511/528 †	1280
A22282	Alexa Fluor® 532 phalloidin	528/555 †	1350
R415	Rhodamine phalloidin	540/565 †	1250
A22283	Alexa Fluor® 546 phalloidin	554/570 †	1800
A34055	Alexa Fluor® 555 phalloidin	555/565 †	1800
B3475	BODIPY® 558/568 phalloidin	558/569	1115
A12380	Alexa Fluor® 568 phalloidin	578/600 †	1590
A12381	Alexa Fluor® 594 phalloidin	593/617 †	1620
T7471	Texas Red®-X phalloidin	591/608 †	1490
A22284	Alexa Fluor® 633 phalloidin	625/645 †	1900
A34054	Alexa Fluor® 635 phalloidin	633/648 †	1900
B12382	BODIPY® 650/665 phalloidin	647/661	1200
A22287	Alexa Fluor® 647 phalloidin	649/666 †	1950
A22285	Alexa Fluor® 660 phalloidin	661/689 †	1750
A22286	Alexa Fluor® 680 phalloidin	677/699 †	1850
B7474	Biotin-XX phalloidin	NA	1300
P3457	Phalloidin	NA	790
G-Actin–Selective Probes			
D12371	Alexa Fluor® 488 DNase I	495/519	>31,000
D12372	Alexa Fluor® 594 DNase I	590/617	>31,000

* Excitation (Ex) and emission (Em) maxima, in nm. Spectra of phallotoxins are either in aqueous buffer, pH 7–9 (denoted †) or in methanol. Spectra of DNase I conjugates are in aqueous buffer, pH 7–8. NA = Not applicable.

IMPORTANT NOTICE: The products described in this manual are covered by one or more Limited Use Label License(s). Please refer to the Appendix on page 971 and Master Product List on page 975. Products are For Research Use Only. Not intended for any animal or human therapeutic or diagnostic use.

NOTE 11.1

BacMam Gene Delivery and Expression Technology

Baculovirus-Mediated Transduction of Mammalian Cells

BacMam technology uses a modified insect cell baculovirus as a vehicle to efficiently deliver and express genes in mammalian cells with minimum effort and toxicity.[1–4] We have combined the BacMam gene delivery and expression system with genetically encoded Premo™ sensors as well as with genetically encoded CellLight® targeted fluorescent proteins to yield robust and easy-to-use cell-based assays (Figure 1).

BacMam particles carrying the biosensor or targeted fluorescent protein cDNA under the control of the CMV promoter are taken up by endocytosis. The viral DNA traffics to the nucleus where only the CMV promoter–driven gene is transcribed; baculovirus promoters are not recognized by the mammalian transcriptional machinery. Following transcription, the biosensor or targeted fluorescent protein mRNA is expressed in the cytosol and cells are soon ready to assay. This process begins within 4–6 hours after transduction and in many cell types is completed after an overnight period.

BacMam 2.0 vectors incorporated in our CellLight® reagents extend the applicability of BacMam-mediated transgene delivery and expression. Cells such as primary neurons that were not amenable to BacMam transduction with version 1.0 (used in the corresponding Organelle Lights™ and Cellular Lights™ reagents) can now be transduced quantitatively in a simple, one-step process. The improved performance is due to inclusion of a pseudotyped capsid protein for more efficient cell entry as well as genetic elements (enhanced CMV promoter and Woodchuck Post-Transcriptional Regulatory Element) that boost expression levels.

Inducible, division-arrested or transient expression systems such as the BacMam system are increasingly methods of choice to decrease variability of expression in cell-based assays. Constitutively expressed ion channels and other cell-surface proteins have been shown to contribute to cell toxicity in some systems, and they may also be subject to clonal drift and other inconsistencies that hamper successful experimentation and screening. Moreover, the BacMam gene delivery and expression system provides a method for simultaneously delivering multiple genes per cell, an important feature when expressing multisubunit proteins.[1]

Advantages of the BacMam Delivery and Expression System

Baculoviruses have been used extensively for protein production in insect cells for over two decades; however, its use with mammalian cells is relatively new. BacMam technology has opened up new avenues for mammalian cell–based assays in drug discovery applications.[3,5] In addition to producing ready-to-use viral stocks, BacMam delivery and expression technology has many advantages when compared with lipids and other viral delivery methods:

- High transduction efficiency across a broad range of cell types, including primary and stem cells
- Minimal microscopically observable cytopathic effects
- Highly reproducible and titratable transient expression
- Biosafety level 1 rating (baculovirus is not pathogenic to vertebrates and does not replicate in mammalian cells)
- Ability to simultaneously deliver multiple genes

Furthermore, it is possible to divide the BacMam-transduced, homogeneous cell population into aliquots that can be stored frozen for use at a later time, approximating the consistency of a stable cell line in a transient expression format. More information is available at www.invitrogen.com/handbook/bacmam2.0.

1. Nat Biotechnol (2004) 22:1583; 2. Br J Pharmacol (2008) 153:544; 3. Drug Discov Today (2007) 12:396; 4. Nat Biotechnol (2005) 23:567; 5. Adv Virus Res (2006) 68:255.

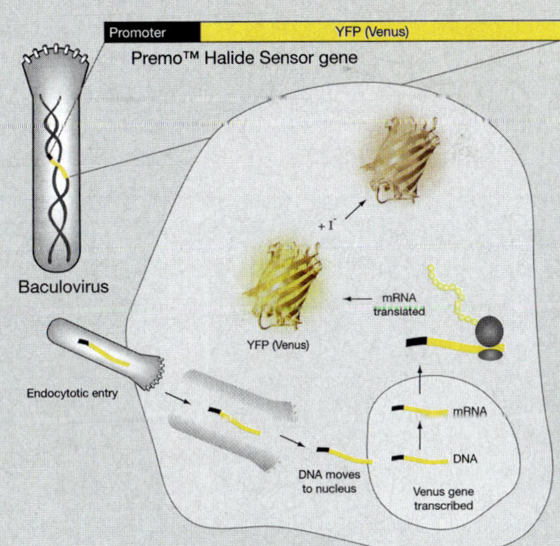

Figure 1 Schematic representation of BacMam transgene delivery and expression as exemplified by Premo™ Halide Sensor (P10229).

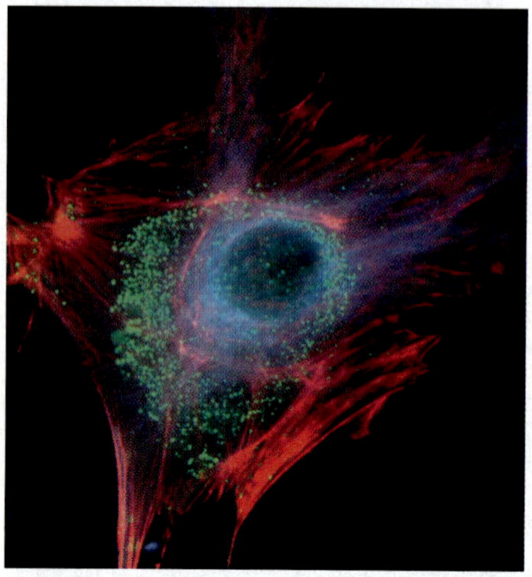

Figure 11.1.4 Microtubules of fixed bovine pulmonary artery endothelial cells localized with mouse monoclonal anti–α-tubulin antibody (A11126), which was subsequently visualized with Alexa Fluor® 350 goat anti–mouse IgG antibody (A11045). Next, the F-actin was labeled with Alexa Fluor® 594 phalloidin (A12381). Finally, the cells were incubated with Alexa Fluor® 488 wheat germ agglutinin (W11261) to stain components of endosomal pathways. The superimposed and pseudocolored images were acquired sequentially using bandpass filter sets appropriate for DAPI, the Texas Red® dye and fluorescein, respectively.

Properties of Phallotoxin Derivatives

The fluorescent and biotinylated phallotoxin derivatives stain F-actin selectively at nanomolar concentrations and are readily water soluble, thus providing convenient labels for identifying and quantitating actin in tissue sections, cell cultures or cell-free preparations.[7–11] F-actin in live neurons can be efficiently labeled using cationic liposomes containing fluorescent phallotoxins, such as BODIPY® FL phallacidin [12] (B607). This procedure permits the labeling of entire cell cultures with minimum disruption. Because fluorescent phalloidin conjugates are not permeant to most live cells, they can be used to detect cells that have compromised membranes. However, it has been reported that unlabeled phalloidin, and potentially dye-labeled phalloidins, can penetrate the membranes of certain hypoxic cells.[13] An extensive study on visualizing the actin cytoskeleton with various fluorescent probes in cell preparations, as well as in live cells, has been published.[7]

Labeled phallotoxins have similar affinity for both large and small filaments and bind in a stoichiometric ratio of about one phallotoxin per actin subunit in both muscle and nonmuscle cells; they reportedly do not bind to monomeric G-actin, unlike some antibodies against actin.[9,14] Phallotoxins have further advantages over antibodies for actin labeling, in that 1) their binding properties do not change appreciably with actin from different species, including plants and animals; and 2) their nonspecific staining is negligible; thus, the contrast between stained and unstained areas is high.

Phallotoxins shift actin's monomer/polymer equilibrium toward the polymer, lowering the critical concentration for polymerization as much as 30-fold.[15,16] Furthermore, depolymerization of F-actin by cytochalasins, potassium iodide and elevated temperatures is inhibited by phallotoxin binding. Because the phallotoxin derivatives are relatively small, with approximate diameters of 12–15 Å and molecular weights below 2000 daltons, a wide variety of actin-binding proteins—including myosin, tropomyosin, troponin and DNase I—can still bind to actin after treatment with fluorescent phallotoxins. Even more significantly, phallotoxin-labeled actin filaments retain certain functional characteristics; labeled glycerinated muscle fibers still contract, and labeled actin filaments still move on solid-phase myosin substrates.[17–19]

Alexa Fluor® Phalloidins

We have taken advantage of the outstanding fluorescence characteristics of our Alexa Fluor® dyes (Section 1.3) to create a series of Alexa Fluor® dye–labeled phalloidins (Figure 11.1.4, Figure 11.1.5, Figure 11.1.6, Figure 11.1.7), which are widely used F-actin stains for many applications

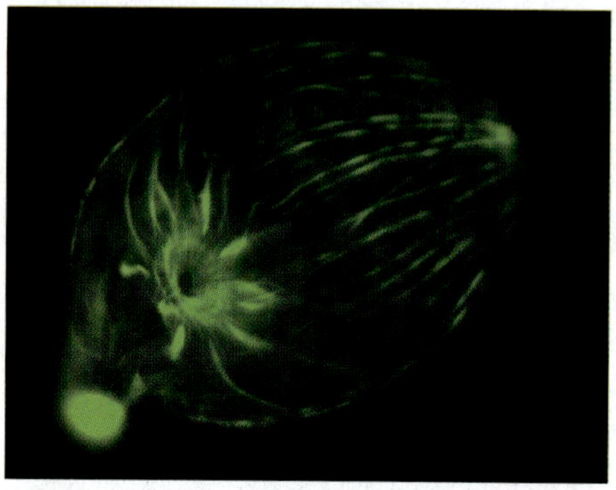

Figure 11.1.5 Actin filaments of the turbellarian flatworm *Archimonotresis* sp. stained with Alexa Fluor® 488 phalloidin (A12379) to reveal a meshwork of longitudinal, circular and diagonal muscles. The large, bright ring with muscle fibers radiating outward is the muscular pharynx, and the small, bright ring at the posterior is part of the reproductive system. This epifluorescence image was contributed by Matthew D. Hooge and Seth Tyler, University of Maine, Orono.

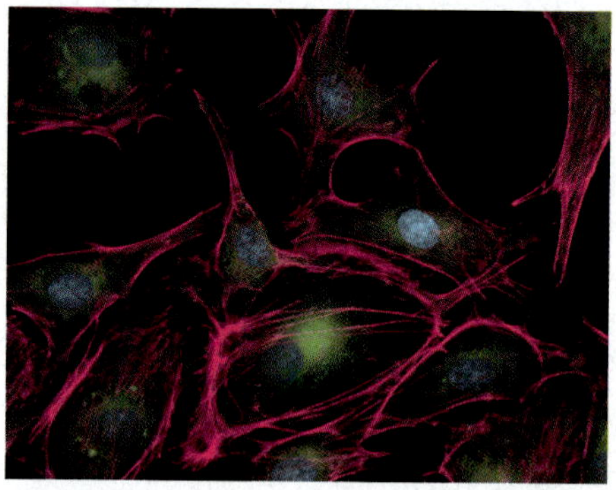

Figure 11.1.6 Subcellular structures in fixed and permeabilized bovine pulmonary artery endothelial cells visualized with several fluorescent dyes. Filamentous actin (F-actin) was identified with Alexa Fluor® 633 phalloidin (A22284), which is pseudocolored magenta. Intracellular membranes were stained with green-fluorescent $DiOC_6(3)$ (D273). Finally, nuclei were counterstained with blue-fluorescent DAPI (D1306, D3571, D21490). The image was acquired using filters appropriate for fluorescein and DAPI and a special filter (courtesy of Omega® Optical) for the Alexa Fluor® 633 dye, consisting of a narrow band exciter (630DF10), dichroic (640DRLP) and emitter (660DF10).

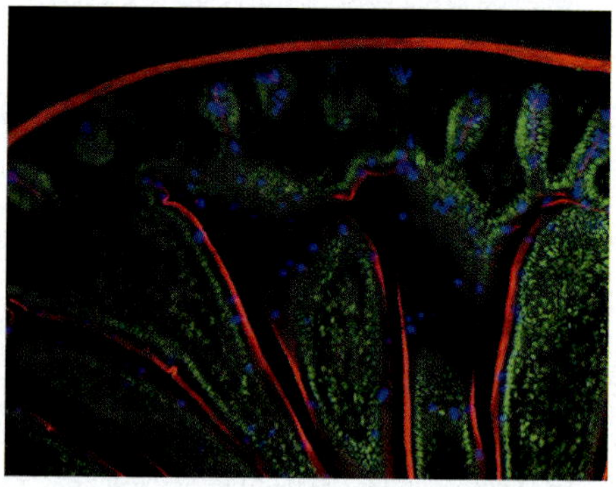

Figure 11.1.7 FluoCells® prepared slide #4 (F24631) contains a section of mouse intestine stained with a combination of fluorescent stains. Alexa Fluor® 350 wheat germ agglutinin (W11263) is a blue-fluorescent lectin that was used to stain the mucus of goblet cells. The filamentous actin prevalent in the brush border was stained with red-orange–fluorescent Alexa Fluor® 568 phalloidin (A12380). Finally, the nuclei were stained with SYTOX® Green nucleic acid stain (S7020). This image is a composite of three digitized images obtained with filter sets appropriate for fluorescein, DAPI and tetramethylrhodamine.

IMPORTANT NOTICE: The products described in this manual are covered by one or more Limited Use Label License(s). Please refer to the Appendix on page 971 and Master Product List on page 975. Products are For Research Use Only. Not intended for any animal or human therapeutic or diagnostic use.
www.invitrogen.com/probes

across the full spectral range. The Alexa Fluor® phalloidin conjugates (Figure 11.1.8) provide researchers with fluorescent probes that are superior in brightness and photostability to other spectrally similar conjugates tested (Figure 11.1.9). For improved fluorescence detection of F-actin in fixed and permeabilized cells, we encourage researchers to try these fluorescent phalloidins in their actin-labeling protocols. A series of videos showing Alexa Fluor® 488 phalloidin–stained actin[20] is available at the *Journal of Cell Biology* web site (www.jcb.org/cgi/content/full/150/2/361/DC1).

Oregon Green® Phalloidins

Green-fluorescent actin stains are popular reagents for labeling F-actin in fixed and permeabilized cells. Unfortunately, the green-fluorescent fluorescein phalloidin and NBD phallacidin photobleach rapidly, making their photography difficult. We have used two of our Oregon Green® dyes (Section 1.5) to prepare Oregon Green® 488 phalloidin (O7466, Figure 11.1.10) and the slightly longer-wavelength Oregon Green® 514 phalloidin (O7465). The excitation and emission spectra of the Oregon Green® 488 dye are virtually superimposable on those of fluorescein, and both the Oregon Green® 488 and Oregon Green® 514 dyes may be viewed with standard fluorescein optical filter sets. As shown in Figure 11.1.11, Oregon Green® 514 phalloidin is more photostable than fluorescein phalloidin, making it easier to visualize and photograph.

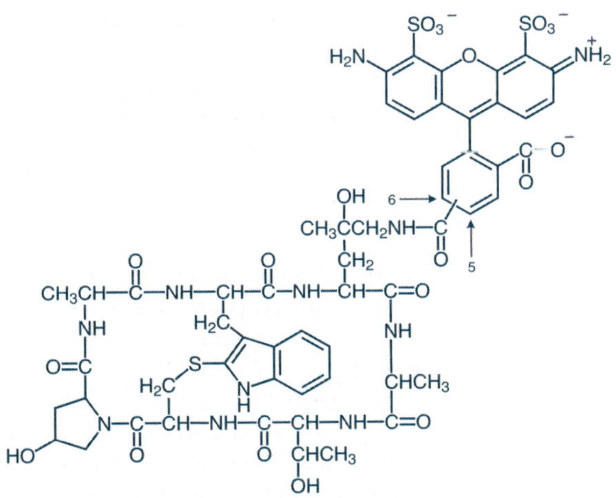

Figure 11.1.8 Alexa Fluor® 488 phalloidin (A12379).

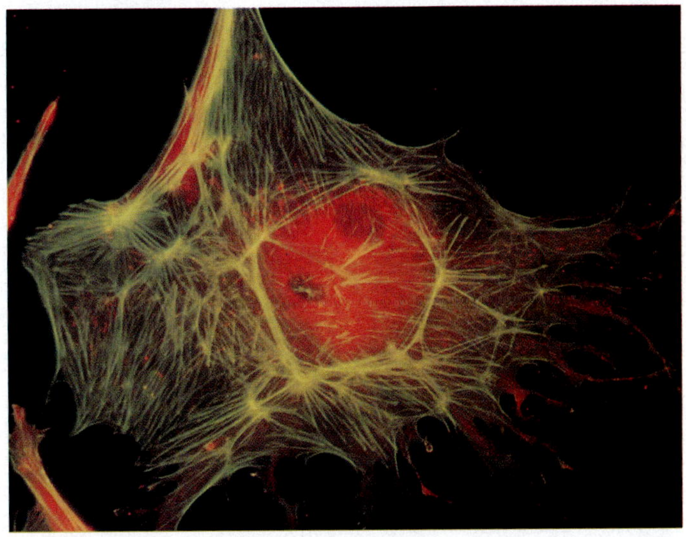

Figure 11.1.10 Simultaneous visualization of F- and G-actin in a bovine pulmonary artery endothelial cell (BPAEC) using F-actin–specific Oregon Green® 488 phalloidin (O7466) and G-actin–specific Texas Red® deoxyribonuclease I. The G-actin appears as diffuse red fluorescence that is more intense in the nuclear region where the cell thickness is greater and stress fibers are less dense. The image was obtained by taking multiple exposures through bandpass optical filter sets appropriate for fluorescein and the Texas Red® dye.

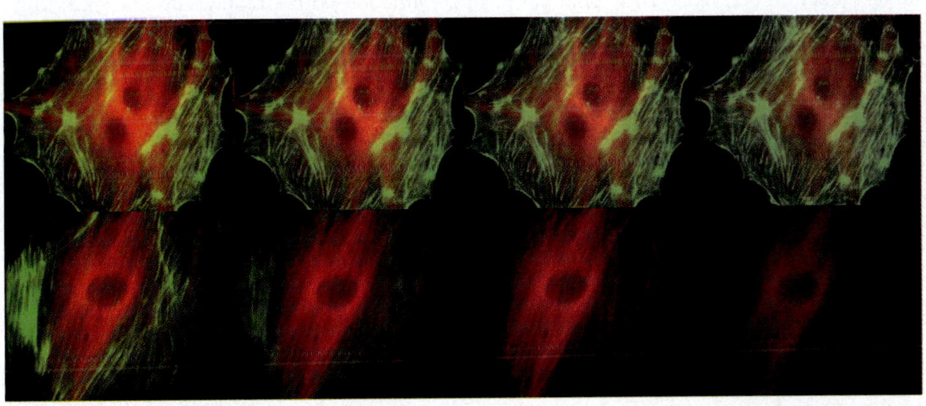

Figure 11.1.9 Comparison of the photobleaching rates of the Alexa Fluor® 488 and Alexa Fluor® 546 dyes and the well-known fluorescein and Cy®3 fluorophores. The cytoskeleton of bovine pulmonary artery endothelial cells (BPAEC) was labeled with (top series) Alexa Fluor® 488 phalloidin (A12379) and mouse monoclonal anti-α-tubulin antibody (A11126) in combination with Alexa Fluor® 546 goat anti–mouse IgG antibody (A11003) or (bottom series) fluorescein phalloidin (F432) and the anti-α-tubulin antibody in combination with a commercially available Cy®3 goat anti–mouse IgG antibody. The pseudocolored images were taken at 30-second intervals (0, 30, 90 and 210 seconds of exposure). The images were acquired with bandpass filter sets appropriate for fluorescein and rhodamine.

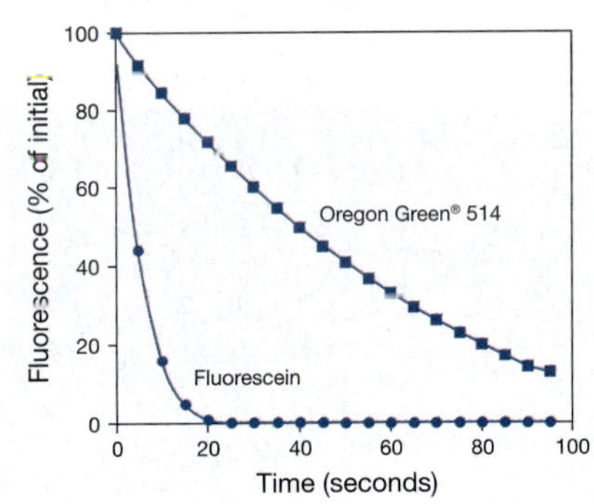

Figure 11.1.11 Photostability comparison for Oregon Green® 514 phalloidin (O7465) and fluorescein phalloidin (F432). CRE BAG 2 fibroblasts were fixed with formaldehyde, permeabilized with acetone and then stained with the fluorescent phallotoxins. Samples were continuously illuminated and images were acquired every 5 seconds using a Star 1 CCD camera (Photometrics); the average fluorescence intensity in the field of view was calculated with Image-1 software (Universal Imaging Corp.) and expressed as a fraction of the initial intensity. Three data sets, representing different fields of view, were averaged for each labeled phalloidin to obtain the plotted time courses.

IMPORTANT NOTICE: The products described in this manual are covered by one or more Limited Use Label License(s). Please refer to the Appendix on page 971 and Master Product List on page 975. Products are For Research Use Only. Not intended for any animal or human therapeutic or diagnostic use.

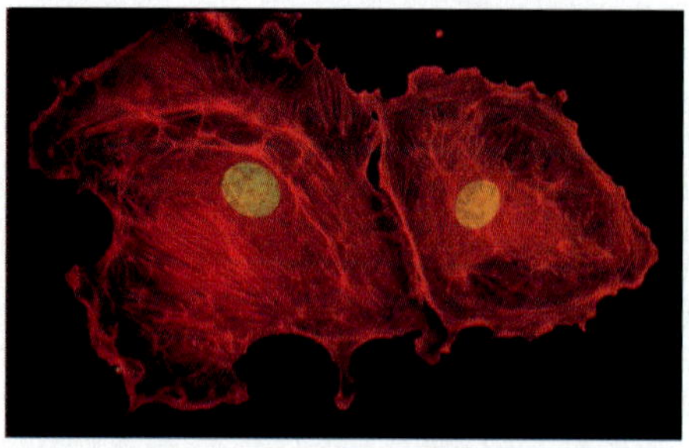

Figure 11.1.12 Permeabilized bovine pulmonary artery endothelial cells stained with SYTOX® Green nucleic acid stain (S7020) to label the nuclei and with BODIPY® TR-X phallacidin (B7464) to label the F-actin. The image was acquired by taking sequential exposures through bandpass optical filter sets appropriate for fluorescein and the Texas Red® dye.

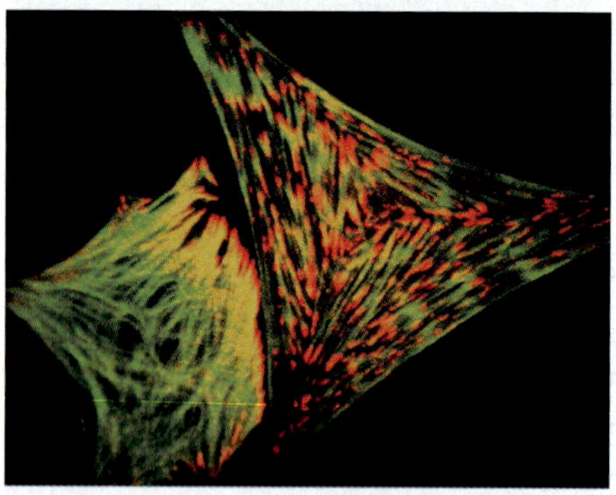

Figure 11.1.13 Actin labeled with BODIPY® FL phallacidin (B607) and vinculin, a cytoskeletal focal adhesion protein, tagged with a monoclonal anti-vinculin antibody that was subsequently probed with Texas Red® goat anti–mouse IgG antibody (T862). The large triangular cell is a fibroblast containing green actin stress fibers terminating in red focal adhesions. The neighboring polygonal cell, a rat neonatal cardiomyocyte, contains green striated actin in the myofibrils terminating in the focal adhesions. The close apposition of the two stains results in a yellowish-orange color. Image contributed by Mark B. Snuggs and W. Barry VanWinkle, University of Texas, Houston.

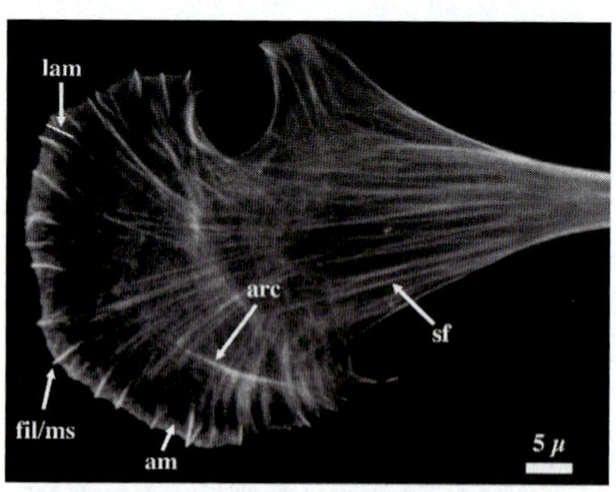

Figure 11.1.14 Actin filaments of chick heart fibroblasts stained with rhodamine phalloidin (R415). The subcompartments in the cytoskeleton are readily apparent and labeled as follows: **sf**, stress fiber; **lam**, lamellipodium; **fil/ms**, filipodium/microspike; **am**, actin meshwork; **arc**, dorsal arc. Figure reprinted from "Visualizing the Actin Cytoskeleton." J. Small *et al.* Microscopy Research and Technique (1999) 47:3. Reprinted by permission of Wiley-Liss, Inc., a subsidiary of John Wiley & Sons, Inc., and J. Victor Small.

BODIPY® Phallotoxins

BODIPY® phallotoxin conjugates (B607, B3475, B12382; Figure 11.1.12, Figure 11.1.13) have some important advantages over the conventional NBD, fluorescein and rhodamine phallotoxins. BODIPY® dyes are more photostable than these traditional fluorophores [21] and have narrower emission bandwidths (Section 1.4), making them especially useful for double- and triple-labeling experiments. BODIPY® FL phallacidin (B607), which reportedly gives a signal superior to that of fluorescein phalloidin,[22] has been used for quantitating F-actin and determining its distribution in cells.[23,24]

The BODIPY® FL and BODIPY® 558/568 phallotoxins (B607, B3475) exhibit excitation and emission spectra similar to those of fluorescein and rhodamine B, respectively, and can be used with standard optical filter sets. BODIPY® 650/665 phalloidin (B12382) is the longest-wavelength BODIPY® phallotoxin conjugate available, increasing the options for multicolor analysis. BODIPY® 650/665 phalloidin, Alexa Fluor® 647 phalloidin (A22287) and Alexa Fluor® 660 phalloidin (A22285) are among the few probes available that can be excited by the 647 nm spectral line of the Ar-Kr laser.

Rhodamine Phalloidin and Other Red-Fluorescent Phalloidins

Rhodamine phalloidin (R415, Figure 11.1.14) has been the standard for red-fluorescent phallotoxins.[25–27] Rhodamine phalloidin is excited efficiently by the mercury-arc lamp in most fluorescence microscopes. However, our Alexa Fluor® 546, Alexa Fluor® 568, Alexa Fluor® 594 and Texas Red®-X phalloidins [28] (A22283, A12380, A12381, T7471; Figure 11.1.15, Figure 11.1.16) will be welcome replacements for rhodamine phalloidin in many multicolor applications because their emission spectra are better separated from those of the green-fluorescent Alexa Fluor® 488, Oregon Green® and fluorescein dyes.

Other Labeled Phallotoxins

The original yellow-green–fluorescent NBD phallacidin (N354) and green-fluorescent fluorescein phalloidin (F432) remain in use despite their relatively poor photostability (Figure 11.1.11). Photostability of fluorescein phalloidin and some other fluorescent phallotoxins can be considerably improved (Figure 11.1.17) by mounting the stained samples with our ProLong® Antifade Kit or ProLong® Gold antifade reagent (P7481, P36930, P36934; Section 23.1). We recommend the Alexa Fluor® 488, Oregon Green® 488, Oregon Green® 514 and BODIPY® FL phallotoxins for photostable, green-fluorescent actin staining. Alexa Fluor® 350 phalloidin (A22281) and coumarin phallacidin (C606, Figure 11.1.2) are the only blue-fluorescent phallotoxin conjugates currently available for staining actin.[29]

Biotin-XX phalloidin (B7474) permits detection of F-actin by electron microscopy and light microscopy techniques.[30] This biotin conjugate can be visualized with fluorophore- or enzyme-labeled avidin and streptavidin (Section 7.6) or with tyramide signal amplification (TSA™) technology (Section 6.2). Biotin-XX phalloidin, in conjunction with streptavidin or CaptAvidin™ agarose (S951, C21386; Section 7.6), can be used to precipitate F-actin from the cytosolic anti-phosphotyrosine–reactive fraction in macrophages stimulated with colony-stimulating factor-1.[31]

DNase I Conjugates for Staining G-Actin

Bovine pancreatic deoxyribonuclease (DNase I, ~31,000 daltons) binds much more strongly to monomeric G-actin than to filamentous F-actin, with binding constants of $5 \times 10^8\ M^{-1}$ and $1.2 \times 10^4\ M^{-1}$, respectively.[32–35] Because of this strong, selective binding to G-actin, fluorescent DNase I conjugates have proven very useful for detecting and quantitating the proportion of unpolymerized actin in a cell. We have triple-labeled endothelial cells with fluorescein DNase I, BODIPY® 581/591 phalloidin and a monoclonal anti-actin antibody detected with a Cascade Blue® dye–labeled secondary antibody [36] (C962, Section 7.2). We found that the monoclonal antibody, which binds to both G-actin and F-actin, colocalized with the DNase I and phalloidin conjugates, suggesting that these three probes recognize unique binding sites on the actin molecule. Researchers can choose DNase I conjugates labeled with either the green-fluorescent Alexa Fluor® 488 (D12371) or red-fluorescent Alexa Fluor® 594 (D12372) dyes, depending on their multicolor application and their detection instrumentation (Table 11.2).

IMPORTANT NOTICE: The products described in this manual are covered by one or more Limited Use Label License(s). Please refer to the Appendix on page 971 and Master Product List on page 975. Products are For Research Use Only. Not intended for any animal or human therapeutic or diagnostic use.

Alexa Fluor® 488 and Alexa Fluor® 594 DNase I conjugates have been used in combination with fluorescently labeled phallotoxins to simultaneously visualize G-actin pools and filamentous F-actin[37,38] and to study the disruption of microfilament organization in live nonmuscle cells.[39] Rhodamine phalloidin (R415) has been used in conjunction with Oregon Green® 488 DNase I to determine the F-actin:G-actin ratio in *Dictyostelium* using confocal laser-scanning microscopy.[40] A mouse fibroblast labeled with both Texas Red® DNase I and Oregon Green® 488 phalloidin (O7466) permitted visualization of the G-actin and the complex network of F-actin throughout the cytoplasm, as well as at the cell periphery (Figure 11.1.10). The influence of cytochalasins on actin structure in monocytes has been quantitated by flow cytometry using Texas Red® DNase I and BODIPY® FL phallacidin (B607) to stain the G-actin and F-actin pools, respectively.[41] Fluorescent DNase I has also been used as a model system to study the interactions of nucleotides, cations and cytochalasin D with monomeric actin.[42]

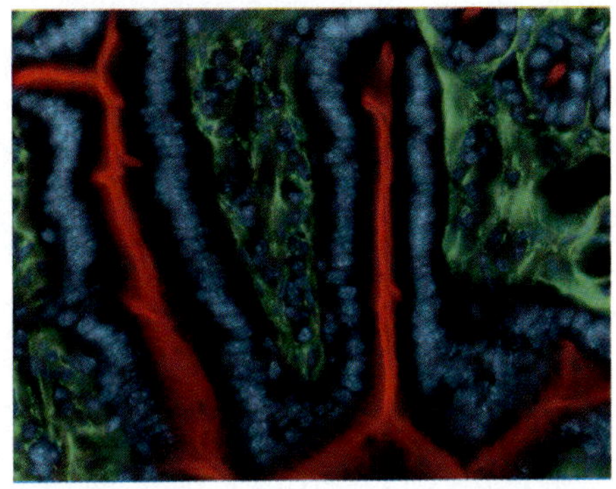

Figure 11.1.15 A section of mouse intestine stained with a combination of fluorescent stains. Fibronectin, an extracellular matrix adhesion molecule, was labeled using a chicken primary antibody against fibronectin and visualized using green-fluorescent Alexa Fluor® 488 goat anti–chicken IgG antibody (A11039). The filamentous actin (F-actin) prevalent in the brush border was stained with red-fluorescent Alexa Fluor® 568 phalloidin (A12380). Finally, the nuclei were stained with DAPI (D1306, D3571, D21490).

Probes for Actin Quantitation, Actin Polymerization and Actin-Binding Proteins

Assays for Quantitating F-Actin and G-Actin Polymerization

Quantitative assays for F-actin have employed fluorescein phalloidin,[43,44] rhodamine phalloidin,[45] BODIPY® FL phallacidin[24] and NBD phallacidin.[46] An F-actin assay based on fluorescein phalloidin was used to demonstrate the loss of F-actin from cells during apoptosis.[47] The addition of propidium iodide (P1304MP, P3566, P21493; Section 8.1) to the cell suspensions enabled these researchers to estimate the cell-cycle distributions of both the apoptotic and nonapoptotic cell populations. The change in F-actin content in proliferating adherent cells has been quantitated using the ratio of rhodamine phalloidin fluorescence to ethidium bromide fluorescence.[48] The spectral separation of the signals in this assay may be improved by using a green-fluorescent stain for F-actin and a high-affinity red-fluorescent nucleic acid stain, such as the combination of Alexa Fluor® 488 phalloidin (A12379) and ethidium homodimer-1 (E1169, Section 8.1).

The fluorescence of actin monomers labeled with pyrene iodoacetamide (P29) has been demonstrated to change upon polymerization, making this probe an excellent tool for following the kinetics of actin polymerization and the effects of actin-binding proteins on polymerization.[49–51]

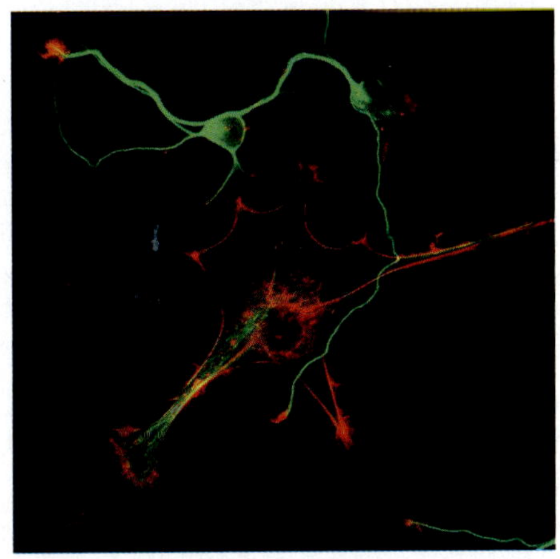

Figure 11.1.16 Confocal micrograph of the cytoskeleton of a mixed population of granule neurons and glial cells. The F-actin was stained with red-fluorescent Texas Red®-X phalloidin (T7471). The microtubules were detected with a mouse monoclonal anti–ß-tubulin primary antibody and subsequently visualized with the green-fluorescent Alexa Fluor® 488 goat anti–mouse IgG antibody (A11001). The image was contributed by Jonathan Zmuda, Immunomatrix, Inc.

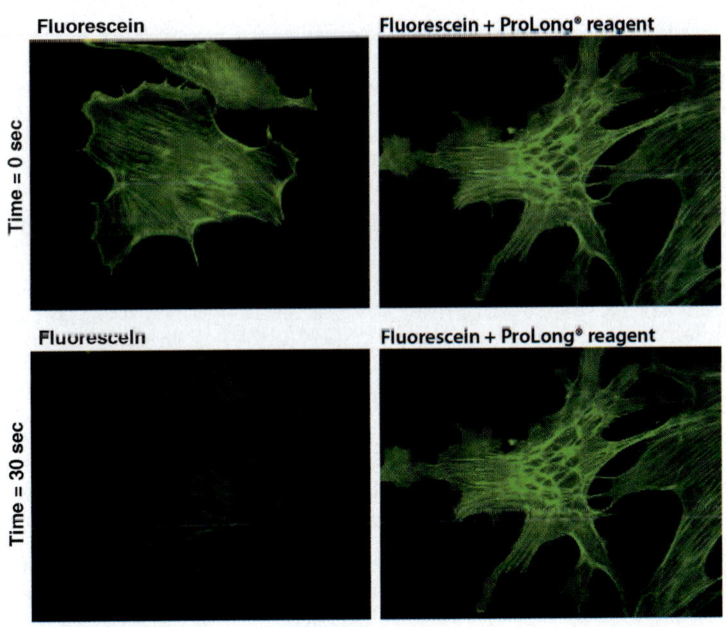

Figure 11.1.17 Bovine pulmonary artery endothelial cells were labeled with fluorescein phalloidin (F432), which labels filamentous actin, and placed under constant illumination on the microscope with a FITC filter set using a 60× objective. Images were acquired at one-second intervals for 30 seconds. Under these illumination conditions, fluorescein photobleached to about 12% of its initial value in 30 seconds in PBS (left), but stayed at the initial value under the same illumination conditions when mounted using the reagents in the ProLong® Antifade Kit (right, P7481).

IMPORTANT NOTICE: The products described in this manual are covered by one or more Limited Use Label License(s). Please refer to the Appendix on page 971 and Master Product List on page 975. Products are For Research Use Only. Not intended for any animal or human therapeutic or diagnostic use.

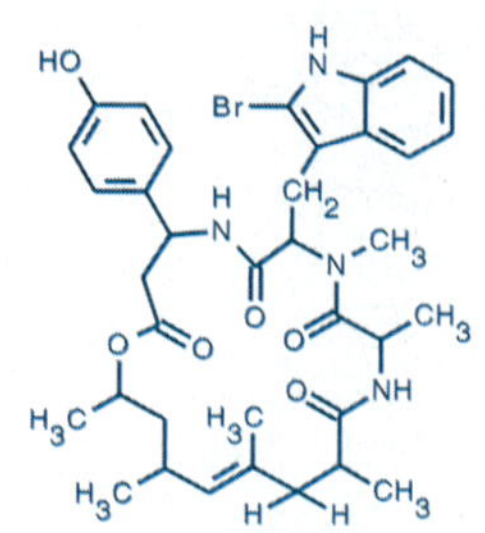

Figure 11.1.18 Jasplakinolide (J7473).

Jasplakinolide: A Cell-Permeant F-Actin Probe

We offer jasplakinolide (J7473, Figure 11.1.18), a macrocyclic peptide isolated from the marine sponge *Jaspis johnstoni*.[52–54] Jasplakinolide is a potent inducer of actin polymerization *in vitro* by stimulating actin filament nucleation[55,56] and competes with phalloidin for actin binding[57] (K_d = 15 nM). Moreover, unlike other known actin stabilizers such as phalloidins and virotoxins, jasplakinolide appears to be somewhat cell permeant and therefore can potentially be used to manipulate actin polymerization in live cells. This peptide, which also exhibits fungicidal, insecticidal and antiproliferative activity,[53,58–60] is particularly useful for investigating cell processes mediated by actin polymerization and depolymerization, including cell adhesion, locomotion, endocytosis and vesicle sorting and release. Jasplakinolide has been reported to enhance apoptosis induced by cytokine deprivation.[61]

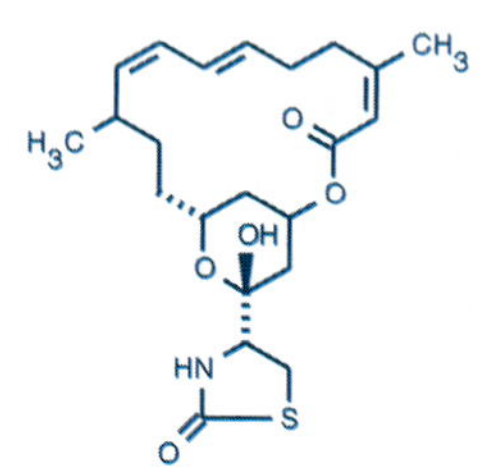

Figure 11.1.19 Latrunculin A (L12370).

Latrunculin A and Latrunculin B: Cell-Permeant Actin Antagonists

Latrunculins are powerful disruptors of microfilament organization. Isolated from a Red Sea sponge, these G-actin binding compounds inhibit fertilization and early embryological development,[62] alter the shape of cells[63,64] and inhibit receptor-mediated endocytosis.[65] Latrunculin A[61,63,66] (L12370, Figure 11.1.19) binds to monomeric G-actin in a 1:1 ratio at submicromolar concentrations (Howard Petty, Wayne State University, personal communication) and is frequently used to establish the effects of F-actin disassembly on particular physiological functions such as ion transport[67] and protein localization.[68] The activity of latrunculin B (L22290) mimics that of latrunculin A in most applications.[63,65,69–71]

Assays for Actin-Binding Proteins

Enhancement of the fluorescence of certain phallotoxins upon binding to F-actin can be a useful tool for following the kinetics and extent of binding of specific actin-binding proteins. We have used the change in fluorescence of rhodamine phalloidin (R415) to determine the dissociation constant of various phallotoxins.[72] The enhancement of rhodamine phalloidin's fluorescence upon actin binding has also been used to measure the kinetics and extent of gelsolin severing of actin filaments.[73] The affinity and rate constants for rhodamine phalloidin binding to actin are not affected by saturation of actin with either myosin subfragment-1 or tropomyosin, indicating that these two actin-binding proteins do not bind to the same sites as the phalloidin.[12]

REFERENCES

1. J Cell Biol (2009) 185:323; **2.** J Am Chem Soc (2008) 130:16840; **3.** Biophys J (2007) 92:1081; **4.** Development (1988) 103:675; **5.** Mol Biotechnol (2002) 21:241; **6.** Proc Natl Acad Sci U S A (1974) 71:2803; **7.** Microsc Res Tech (1999) 47:3; **8.** Biophys J (1998) 74:2451; **9.** Biophys J (2005) 88:2727; **10.** Methods Enzymol (1991) 194:729; **11.** J Muscle Res Cell Motil (1988) 9:370; **12.** Neurosci Lett (1996) 207:17; **13.** J Lab Clin Med (1994) 123:357; **14.** Biochemistry (1994) 33:14387; **15.** Eur J Biochem (1987) 165:125; **16.** J Cell Biol (1987) 105:1473; **17.** J Cell Biol (1991) 115:67; **18.** Nature (1987) 326:805; **19.** Proc Natl Acad Sci U S A (1986) 83:6272; **20.** J Cell Biol (2000) 150:361; **21.** J Cell Biol (1991) 114:1179; **22.** J Cell Biol (1994) 127:1637; **23.** J Cell Biol (1992) 116:197; **24.** Histochem J (1990) 22:624; **25.** Biochemistry (2008) 47:6460; **26.** BMC Cell Biol (2007) 8:43; **27.** Biotechniques (2006) 40:745; **28.** J Histochem Cytochem (2001) 49:1351; **29.** J Muscle Res Cell Motil (1993) 14:594; **30.** J Cell Biol (1995) 130:591; **31.** Biol Chem (1998) 273:17128; **32.** Anal Biochem (1983) 135:22; **33.** Exp Cell Res (1983) 147:240; **34.** Eur J Biochem (1980) 104:367; **35.** J Biol Chem (1980) 255:5668; **36.** J Histochem Cytochem (1994) 42:345; **37.** Stem Cells (2005) 23:507; **38.** Am J Physiol Heart Circ Physiol (2005) 288:H660; **39.** Proc Natl Acad Sci U S A (1990) 87:5474; **40.** J Cell Biol (1998) 142:1325; **41.** J Biol Chem (1994) 269:3159; **42.** Eur J Biochem (1989) 182:267; **43.** Proc Natl Acad Sci U S A (1980) 77:6624; **44.** J Cell Sci (1991) 100:187; **45.** J Cell Biol (1995) 130:613; **46.** J Cell Biol (1984) 98:1265; **47.** Cytometry (1995) 20:162; **48.** J Cell Biol (1995) 129:1589; **49.** Curr Biol (2006) 16:1924; **50.** J Biol Chem (2008) 283:7135; **51.** Biophys J (2007) 92:2162; **52.** J Cell Biol (1997) 137:399; **53.** J Am Chem Soc (1986) 108:3123; **54.** Tetrahedron Lett (1986) 27:2797; **55.** Methods Mol Biol (2001) 161:109; **56.** J Biol Chem (2000) 275:5163; **57.** J Biol Chem (1994) 269:14869; **58.** J Natl Cancer Inst (1995) 87:46; **59.** Cancer Chemother Pharmacol (1992) 30:401; **60.** Antimicrob Agents Chemother (1988) 32:1154; **61.** J Biol Chem (1999) 274:4259; **62.** Science (1983) 219:493; **63.** J Biol Chem (2000) 275:28120; **64.** FEBS Lett (1987) 213:316; **65.** Exp Cell Res (1986) 166:191; **66.** Cell Motil Cytoskeleton (1989) 13:127; **67.** J Biol Chem (1997) 272:20332; **68.** Am J Physiol (1997) 272:C254; **69.** J Biol Chem (2001) 276:23056; **70.** J Cell Sci (2001) 114:1025; **71.** Cell Motil Cytoskeleton (2001) 48:96; **72.** Anal Biochem (1992) 200:199; **73.** J Biol Chem (1994) 269:32916.

IMPORTANT NOTICE: The products described in this manual are covered by one or more Limited Use Label License(s). Please refer to the Appendix on page 971 and Master Product List on page 975. Products are For Research Use Only. Not intended for any animal or human therapeutic or diagnostic use.

DATA TABLE 11.1 PROBES FOR ACTIN

Cat. No.	MW	Storage	Soluble	Abs	EC	Em	Solvent	Notes
A12379	~1320	F,L	MeOH, H_2O	494	78,000	517	pH 7	1, 2, 3
A12380	~1590	F,L	MeOH, H_2O	578	88,000	600	pH 7	1, 2, 3
A12381	~1620	F,L	MeOH, H_2O	593	92,000	617	pH 7	1, 2, 3
A22281	~1100	F,L	MeOH, H_2O	346	17,000	446	pH 7	1, 2, 3
A22282	~1350	F,L	MeOH, H_2O	528	81,000	555	pH 7	1, 2, 3
A22283	~1800	F,L	MeOH, H_2O	554	112,000	570	pH 7	1, 2, 3
A22284	~1900	F,L	MeOH, H_2O	621	159,000	639	MeOH	1, 2, 3, 4
A22285	~1650	F,L	MeOH, H_2O	668	132,000	697	MeOH	1, 2, 3, 4
A22286	~1850	F,L	MeOH, H_2O	684	183,000	707	MeOH	1, 2, 3, 4
A22287	~1950	F,L	MeOH, H_2O	650	275,000	672	MeOH	1, 2, 3, 4
A34054	~1800	F,L	MeOH, H_2O	622	145,000	640	MeOH	1, 2, 3, 4
A34055	~1900	F,L	MeOH, H_2O	555	155,000	572	MeOH	1, 2, 3
B607	~1160	F,L	MeOH, H_2O	505	83,000	512	MeOH	1, 2, 3
B3475	~1115	F,L	MeOH, H_2O	558	85,000	569	MeOH	1, 2, 3
B7474	~1300	F	MeOH, H_2O	<300		none		1, 2
B12382	~1200	F,L	MeOH	647	102,000	661	MeOH	1, 3, 5
C606	~1100	F,L	MeOH, H_2O	355	16,000	443	MeOH	1, 2, 3
F432	~1175	F,L	MeOH, H_2O	496	84,000	516	pH 8	1, 2, 3
J7473	709.68	F,D	MeOH	278	8000	none	MeOH	
L12370	421.55	F,D	DMSO	<300		none		
L22290	395.51	F,D	DMSO	<300		none		
N354	~1040	F,L	MeOH, H_2O	465	24,000	536	MeOH	1, 2, 3
O7465	~1280	F,L	MeOH, H_2O	511	85,000	528	pH 9	1, 2, 3
O7466	~1180	F,L	MeOH, H_2O	496	86,000	520	pH 9	1, 2, 3
P29	385.20	F,D,L	DMF, DMSO	339	26,000	384	MeOH	6, 7
P3457	~790	F	MeOH, H_2O	<300		see Notes		2, 8
R415	~1250	F,L	MeOH, H_2O	542	85,000	565	MeOH	1, 2, 3, 9
T7471	~1490	F,L	MeOH, H_2O	583	95,000	603	MeOH	1, 2, 3, 9

For definitions of the contents of this data table, see "Using *The Molecular Probes® Handbook*" in the introductory pages.

Notes

1. α-Bungarotoxin, EGF and phallotoxin conjugates have approximately 1 label per peptide.
2. Although this phallotoxin is water-soluble, storage in water is not recommended, particularly in dilute solution.
3. The value of EC listed for this phallotoxin conjugate is for the labeling dye in free solution. Use of this value for the conjugate assumes a 1:1 dye:peptide labeling ratio and no change of EC due to dye–peptide interactions.
4. In aqueous solutions (pH 7.0), Abs/Em = 625/645 nm for A22284, 633/648 nm for A34054, 649/666 nm for A22287, 661/689 nm for A22285 and 677/699 nm for A22286.
5. B7464 and B12382 are not directly soluble in H_2O. Aqueous dispersions can be prepared by dilution of a stock solution in MeOH.
6. Spectral data of the 2-mercaptoethanol adduct.
7. Iodoacetamides in solution undergo rapid photodecomposition to unreactive products. Minimize exposure to light prior to reaction.
8. This bicyclic peptide is very weakly fluorescent in aqueous solution (Em ~380 nm). (Biochim Biophys Acta (1983) 760:411)
9. In aqueous solutions (pH 7.0), Abs/Em = 554/573 nm for R415 and 591/608 nm for T7471.

IMPORTANT NOTICE: The products described in this manual are covered by one or more Limited Use Label License(s). Please refer to the Appendix on page 971 and Master Product List on page 975. Products are For Research Use Only. Not intended for any animal or human therapeutic or diagnostic use.

PRODUCT LIST 11.1 PROBES FOR ACTIN

Cat. No.	Product	Quantity
A12375	actin from rabbit muscle	1 mg
A12373	actin from rabbit muscle, Alexa Fluor® 488 conjugate *in solution*	200 µg
A12374	actin from rabbit muscle, Alexa Fluor® 568 conjugate *in solution*	200 µg
A34050	actin from rabbit muscle, Alexa Fluor® 594 conjugate *in solution*	200 µg
A34051	actin from rabbit muscle, Alexa Fluor® 647 conjugate *in solution*	200 µg
A22281	Alexa Fluor® 350 phalloidin	300 U
A12379	Alexa Fluor® 488 phalloidin	300 U
A22282	Alexa Fluor® 532 phalloidin	300 U
A22283	Alexa Fluor® 546 phalloidin	300 U
A34055	Alexa Fluor® 555 phalloidin	300 U
A12380	Alexa Fluor® 568 phalloidin	300 U
A12381	Alexa Fluor® 594 phalloidin	300 U
A22284	Alexa Fluor® 633 phalloidin	300 U
A34054	Alexa Fluor® 635 phalloidin	300 U
A22287	Alexa Fluor® 647 phalloidin	300 U
A22285	Alexa Fluor® 660 phalloidin	300 U
A22286	Alexa Fluor® 680 phalloidin	300 U
B7474	biotin-XX phalloidin	50 U
B3475	BODIPY® 558/568 phalloidin	300 U
B12382	BODIPY® 650/665 phalloidin	300 U
B607	BODIPY® FL phallacidin	300 U
C10582	CellLight® Actin-GFP *BacMam 2.0*	1 mL
C10583	CellLight® Actin-RFP *BacMam 2.0*	1 mL
C10615	CellLight® Null (control) *BacMam 2.0*	1 mL
C606	coumarin phallacidin	300 U
D12371	deoxyribonuclease I, Alexa Fluor® 488 conjugate	5 mg
D12372	deoxyribonuclease I, Alexa Fluor® 594 conjugate	5 mg
F432	fluorescein phalloidin	300 U
J7473	jasplakinolide	100 µg
L12370	latrunculin A	100 µg
L22290	latrunculin B	100 µg
N354	*N*-(7-nitrobenz-2-oxa-1,3-diazol-4-yl)phallacidin (NBD phallacidin)	300 U
O7466	Oregon Green® 488 phalloidin	300 U
O7465	Oregon Green® 514 phalloidin	300 U
P3457	phalloidin	1 mg
P29	*N*-(1-pyrene)iodoacetamide	100 mg
R415	rhodamine phalloidin	300 U
T7471	Texas Red®-X phalloidin	300 U

IMPORTANT NOTICE: The products described in this manual are covered by one or more Limited Use Label License(s). Please refer to the Appendix on page 971 and Master Product List on page 975. Products are For Research Use Only. Not intended for any animal or human therapeutic or diagnostic use.

www.invitrogen.com/probes

molecular probes® | invitrogen™ by life technologies™

11.2 Probes for Tubulin and Other Cytoskeletal Proteins

Paclitaxel Probes

Paclitaxel

We offer paclitaxel (P3456) for *research purposes only* at a purity of >98% by HPLC. Paclitaxel, formerly referred to as taxol in some scientific literature, is the approved generic name for the anticancer pharmaceutical Taxol® (Bristol-Myers Squibb Co.). The diterpenoid paclitaxel is a potent anti-neoplastic agent [1,2] originally isolated from the bark and needles of the western yew tree, *Taxus brevifolia*.[3,4] The anti-mitotic and cytotoxic action of paclitaxel is related to its ability to promote tubulin assembly into stable aggregated structures that cannot be depolymerized by dilution, calcium ions, cold or a number of microtubule-disrupting drugs; [5–7] paclitaxel also decreases the critical concentration of tubulin required for microtubule assembly. Cultured cells treated with paclitaxel are blocked in the G_2 (the "gap" between DNA synthesis and mitosis) and M (mitosis) phases of the cell cycle.[8]

TubulinTracker™ Green Reagent

TubulinTracker™ Green reagent (T34075) provides green-fluorescent staining of polymerized tubulin in live cells.[9–11] Also known as Oregon Green® 488 paclitaxel bis-acetate (a bi-acetylated version of Oregon Green® 488 paclitaxel (P22310), see below), TubulinTracker™ Green reagent is an uncharged, nonfluorescent compound (Figure 11.2.1) that easily passes through the plasma membrane of live cells. Once inside the cell, the lipophilic blocking group is cleaved by nonspecific esterases, resulting in a green-fluorescent, charged paclitaxel.

TubulinTracker™ Green reagent is provided as a set of two components: lyophilized TubulinTracker™ Green reagent and a 20% Pluronic® F-127 solution in dimethylsulfoxide (DMSO), a solubilizing agent for making stock solutions and facilitating cell loading. Please note that because paclitaxel binds polymerized tubulin, TubulinTracker™ Green reagent will inhibit cell division and possibly other functions utilizing polymerized tubulin in live cells.

Fluorescent Paclitaxel Conjugates

In addition to unlabeled paclitaxel and TubulinTracker™ Green reagent, we provide three fluorescent derivatives of paclitaxel: Oregon Green® 488 paclitaxel (Flutax-2, P22310), BODIPY® FL paclitaxel (P7500) and BODIPY® 564/570 paclitaxel (P7501). These fluorescent paclitaxel derivatives are promising tools for imaging microtubule formation and motility. Their fluorescent attributes should also make these conjugates useful reagents for screening compounds that affect microtubule assembly.

Oregon Green® 488 paclitaxel [12–16] is an important probe for labeling tubulin filaments in live cells. The fluorescent label on this probe is attached by derivatizing the 7β-hydroxy group of native paclitaxel (Figure 11.2.2), a strategy that permits selective binding of the probe to microtubules with high affinity at 37°C [16] (K_d ~10^{-7} M). Oregon Green® 488 paclitaxel has been utilized in a high-throughput fluorescence polarization–based assay to screen for paclitaxel biomimetics.[14] We have successfully used Oregon Green® 488 paclitaxel to label microtubules

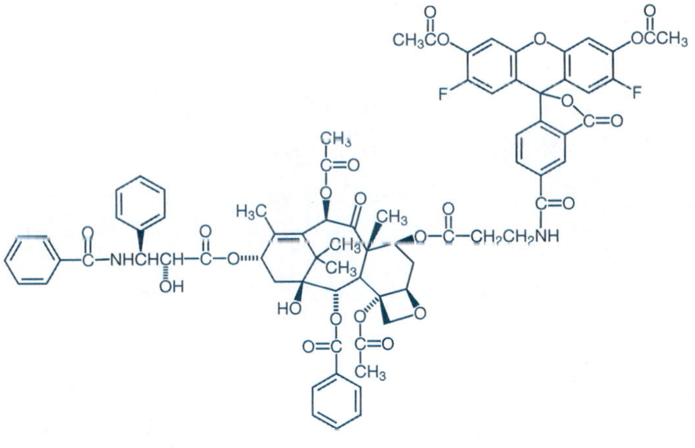

Figure 11.2.1 TubulinTracker™ Green (Oregon Green® 488 Taxol®, bis-acetate; T34075).

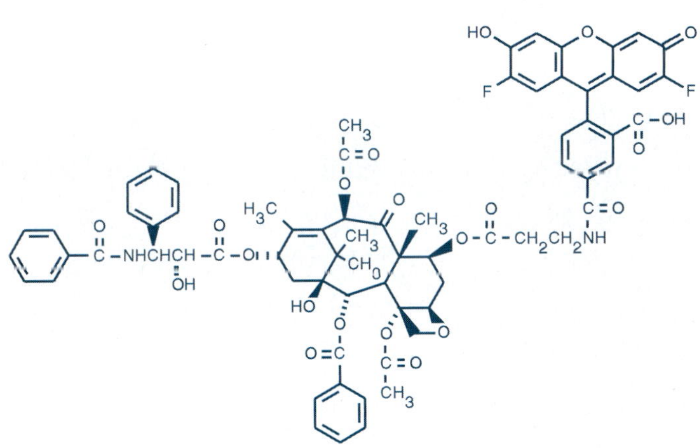

Figure 11.2.2 Paclitaxel, Oregon Green® 488 conjugate (Oregon Green® 488 Taxol®, Flutax-2; P22310).

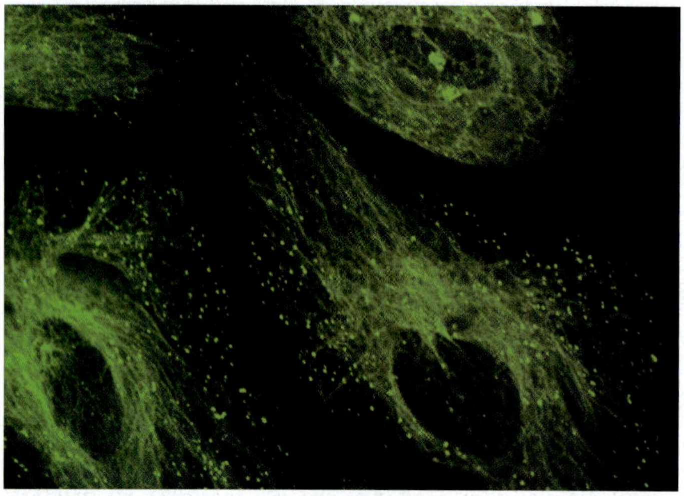

Figure 11.2.3 Microtubules were assembled, stabilized and visualized with the aid of green-fluorescent Oregon Green® 488 paclitaxel (P22310). Viable HeLa cells were incubated with the conjugate for 1 hour, followed by several washes with phosphate-buffered saline containing 2% bovine serum albumin. The image was acquired using a confocal laser-scanning microscope and a filter set appropriate for fluorescein.

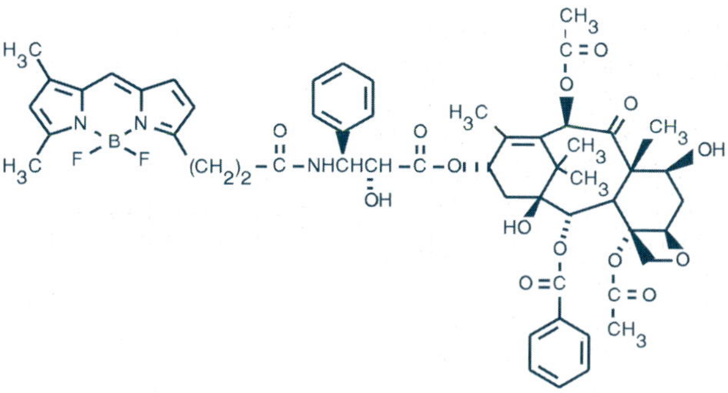

Figure 11.2.4 Paclitaxel, BODIPY® FL conjugate (BODIPY® FL Taxol®, P7500).

of live HeLa (Figure 11.2.3), NIH 3T3, A-10 and BC3H1 cells. *Xenopus laevis*[17] and bovine brain[18] microtubules have also been stained with Oregon Green® 488 paclitaxel.

In the BODIPY® FL and BODIPY® 564/570 paclitaxel derivatives, the *N*-benzoyl substituent of the 3-phenylisoserine portion of native paclitaxel is replaced by a BODIPY® propionyl substituent (Figure 11.2.4). As an alternative to chemically modifying tubulin with a reactive fluorophore, a published method describes the use of these BODIPY® paclitaxel derivatives to generate fluorescent microtubules that are stable at room temperature for one week or longer.[19] In contrast to the Oregon Green® 488 derivative, the BODIPY® FL and BODIPY® 564/570 paclitaxel derivatives do not appear to be suitable for labeling intracellular tubulin in most cases.

Tubulin-Selective Probes

GFP- and RFP-Labeled Tubulin and MAP4

GFP–tubulin fusions are well-established probes for imaging cytokinesis and other dynamic rearrangements of microtubules in live cells.[20] CellLight® Tubulin-GFP and CellLight® Tubulin-RFP expression vectors (C10613, C10614; Table 11.1) generate autofluorescent proteins fused to the N-terminus of human β-tubulin and incorporate all the generic advantages of BacMam 2.0 delivery technology (BacMam Gene Delivery and Expression Technology—Note 11.1).

In context-specific instances where GFP–tubulin fusion protein incorporation into microtubules is inefficient, CellLight® expression vectors encoding GFP (C10598; Figure 11.2.5) or RFP (C10599) fused to the N-terminus of the mammalian microtubule-associated protein MAP4 provide a second option for microtubule visualization. However, because MAP4 stabilizes polymerized tubulin, CellLight® Tubulin-GFP and CellLight® Tubulin-RFP are generally preferable for molecular-level investigations of microtubule dynamic instability.

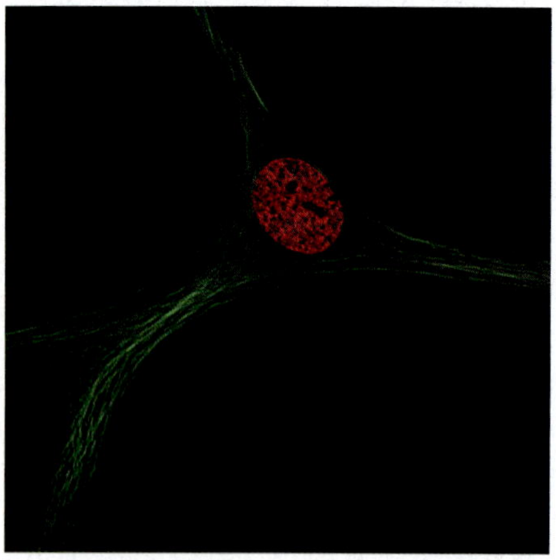

Figure 11.2.5 Human mesenchymal stem cell labeled with CellLight® MAP4-GFP (C10598) and CellLight® Histone 2B-RFP (C10595) reagents.

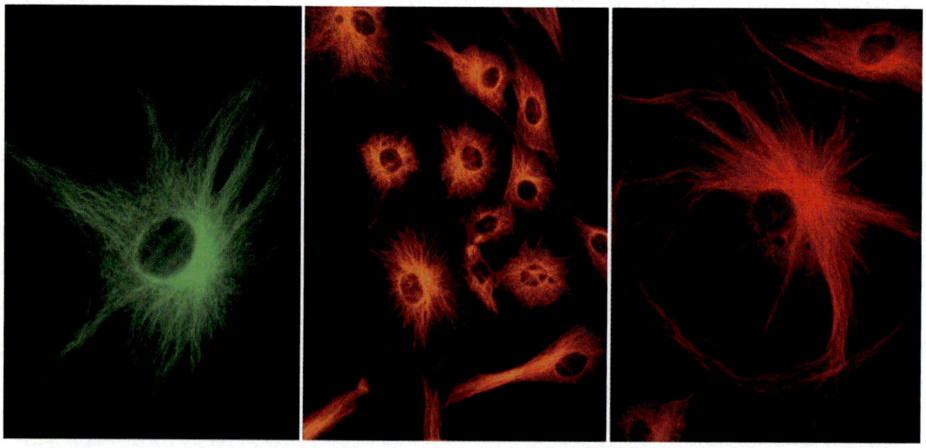

Figure 11.2.6 Microtubules of bovine pulmonary artery endothelial cells tagged with mouse monoclonal anti-α-tubulin antibody (A11126) and subsequently probed with: Alexa Fluor® 488 goat anti-mouse IgG antibody (A11001, left), Alexa Fluor® 546 goat anti-mouse IgG antibody (A11003, middle) or Alexa Fluor® 594 goat anti-mouse IgG antibody (A11005, right). These images were acquired using a fluorescein bandpass optical filter set, a rhodamine bandpass optical filter set and a Texas Red® bandpass optical filter set, respectively.

IMPORTANT NOTICE: The products described in this manual are covered by one or more Limited Use Label License(s). Please refer to the Appendix on page 971 and Master Product List on page 975. Products are For Research Use Only. Not intended for any animal or human therapeutic or diagnostic use.

Anti–α-Tubulin Monoclonal Antibody

When used in conjunction with an anti–mouse IgG secondary immunoreagent (Section 7.2, Table 7.1), our anti–α-tubulin monoclonal antibody (A11126) enables researchers to visualize microtubules in fixed cells (Figure 11.2.6, Figure 11.2.7, Figure 11.2.8, Figure 11.2.9) and in fixed or frozen tissue sections from various species. This mouse monoclonal antibody, which recognizes amino acid residues 69–97 of the N-terminal structural domain, is also useful for detecting tubulin by ELISA or western blotting, for screening expression libraries and as a probe for the N-terminal domain of α-tubulin.

The anti–α-tubulin monoclonal antibody is available either unlabeled (A11126) or as a biotin-XX conjugate (A21371). For detecting the biotinylated antibody, we carry a wide variety of fluorophore- and enzyme-labeled avidin, streptavidin and NeutrAvidin™ biotin-binding protein conjugates and NANOGOLD® and Alexa Fluor® FluoroNanogold™ streptavidin (Section 7.6, Table 7.9).

We have extensively utilized the mouse IgG_1 monoclonal anti–α-tubulin antibody during development and evaluation of our Zenon® technology (Section 7.3, Table 7.7), which allows labeling of submicrogram quantities of primary antibodies in minutes (Figure 11.2.10, Figure 11.2.11). A comprehensive listing of our primary antibodies for cytoskeletal proteins can be found at www.invitrogen.com/handbook/antibodies.

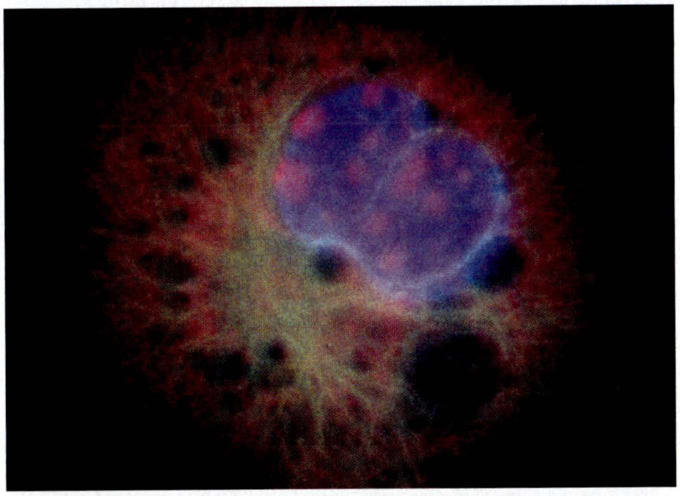

Figure 11.2.7 Microtubules of fixed bovine pulmonary artery endothelial cells were labeled with our mouse monoclonal anti–α-tubulin antibody (A11126), detected with the biotin-XX–conjugated $F(ab')_2$ fragment of goat anti–mouse IgG antibody (B11027) and visualized with Alexa Fluor® 488 streptavidin (S11223). The actin filaments were then labeled with orange-fluorescent Alexa Fluor® 568 phalloidin (A12380), and the cell was counterstained with blue-fluorescent Hoechst 33342 (H1399, H3570, H21492) to image the DNA, and red-fluorescent propidium iodide (P1304MP, P3566, P21493) to image the nucleolar RNA. The multiple-exposure image was acquired using bandpass filters appropriate for the Texas Red® dye, fluorescein and DAPI.

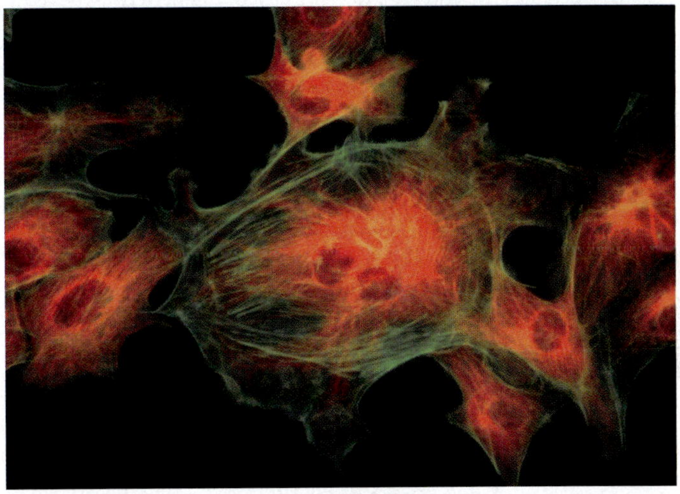

Figure 11.2.8 Bovine pulmonary artery endothelial cells were labeled with Alexa Fluor® 488 phalloidin (A12379) to stain F-actin and our mouse monoclonal anti–α-tubulin antibody (A11126) in combination with Alexa Fluor® 594 dye–conjugated $F(ab')_2$ fragment of goat anti–mouse IgG antibody (A11020) to stain microtubules. The multiple-exposure image was acquired using bandpass filter sets appropriate for Texas Red® dye and fluorescein.

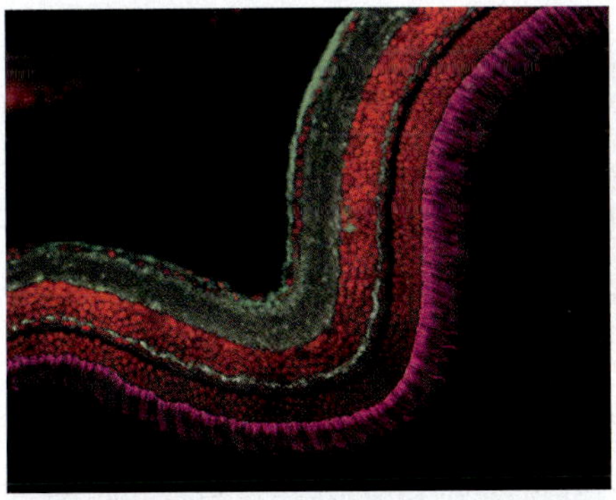

Figure 11.2.9 A zebrafish cryosection incubated with the biotin-XX conjugate of mouse monoclonal anti–α-tubulin antibody (A21371). The signal was amplified with TSA™ Kit #22, which includes HRP–streptavidin and Alexa Fluor® 488 tyramide (T20932). The sample was then incubated with the mouse monoclonal FRet 6 antibody and was visualized with Alexa Fluor® 647 goat anti–mouse IgG (A21235), which is pseudocolored magenta. Finally, the nuclei were counterstained with SYTOX® Orange nucleic acid stain (S11368).

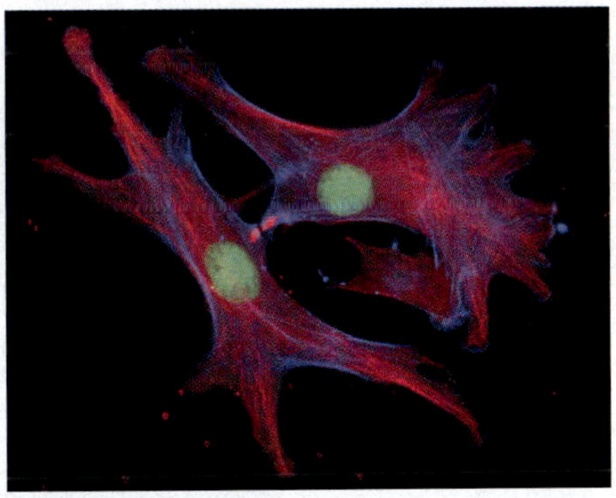

Figure 11.2.10 Fixed and permeabilized bovine pulmonary artery endothelial cells stained with Alexa Fluor® 350 phalloidin (A22281), an anti–α-tubulin antibody (A11126) and the anti-cdc6 peptide antibody (A21286). The anti–α-tubulin antibody was labeled with the Zenon® Alexa Fluor® 568 Mouse IgG_1 Labeling Kit (Z25006) and the anti-cdc6 peptide antibody was labeled with the Zenon® Alexa Fluor® 488 Mouse IgG_1 Labeling Kit (Z25002).

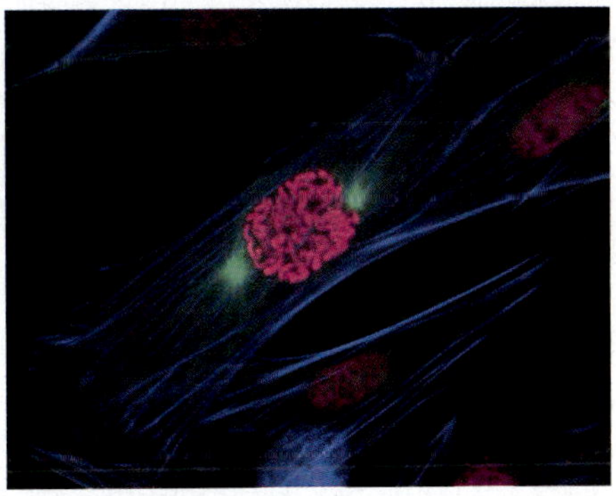

Figure 11.2.11 A prometaphase muntjac skin fibroblast stained with Alexa Fluor® 350 phalloidin (A22281), an anti–α-tubulin antibody (A11126) and an anti-cdc6 peptide antibody (A21286). The anti–α-tubulin antibody was prelabeled with the Zenon® Alexa Fluor® 488 Mouse IgG_1 Labeling Kit (Z25002) and the anti-cdc6 peptide antibody was prelabeled with the Zenon® Alexa Fluor® 647 Mouse IgG_1 Labeling Kit (Z25008).

IMPORTANT NOTICE: The products described in this manual are covered by one or more Limited Use Label License(s). Please refer to the Appendix on page 971 and Master Product List on page 975. Products are For Research Use Only. Not intended for any animal or human therapeutic or diagnostic use.

BODIPY® FL Vinblastine

BODIPY® FL vinblastine (V12390, Figure 11.2.12), a fluorescent analog of the anticancer drug vinblastine, is a useful probe for labeling β-tubulin and for investigating drug-transport mechanisms.[21,22] Vinblastine inhibits cell proliferation by capping microtubule ends, thereby suppressing mitotic spindle microtubule dynamics.[23] Another fluorescent vinblastine derivative, vinblastine 4′-anthranilate, reportedly binds to the central portion of the primary sequence of β-tubulin and inhibits polymerization.[21,24–26]

In addition, intracellular accumulation of vinblastine has been associated with a vinblastine-specific modulating site on P-glycoprotein, a drug-efflux pump that is overexpressed in multidrug-resistant (MDR) cells[27] (Section 15.6). This highly lipophilic P-glycoprotein substrate has also been used to study the role of P-glycoprotein in drug penetration through the blood-brain barrier.[28] Fluorescently labeled vinblastine analogs, including BODIPY® FL vinblastine, have been employed to measure drug-transport kinetics in MDR cells.[29]

Other Probes for Tubulin

The nuclear stain DAPI (D1306, D3571, D21490) binds tightly to purified tubulin *in vitro* without interfering with microtubule assembly or GTP hydrolysis. DAPI binds to tubulin at sites different from those of paclitaxel, colchicine and vinblastine, and its binding is accompanied by shifts in the absorption spectra and fluorescence enhancement. The affinity of DAPI for polymeric tubulin is 7-fold greater than for dimeric tubulin, making DAPI a sensitive tool for investigating microtubule assembly kinetics.[30–33] DAPI has been used to screen for potential antimicrotubule drugs in a high-throughput assay.[34]

Bis-ANS (B153) is a potent inhibitor of *in vitro* microtubule assembly.[35] This fluorescent probe binds to the hydrophobic clefts of proteins with an affinity approximately 10–100 times higher than that of 1,8-ANS (A47, Section 13.5) and exhibits a significant fluorescence enhancement upon binding. The bis-ANS binding site on tubulin lies near the critical contact region for microtubule assembly, but it is distinct from the binding sites for colchicine, vinblastine, podophyllotoxin and maytansine.[36–38] Bis-ANS was used to investigate structural changes in tubulin monomers and dimers during time- and temperature-dependent decay.[39,40]

DCVJ (4-(dicyanovinyl)julolidine; D3923), which binds to a specific site on the tubulin dimer,[41] has been reported to be a useful probe for following polymerization of tubulin in live cells.[42] DCVJ staining in live cells is mostly blocked by cytochalasin D.[43] Additionally, DCVJ emits strong green fluorescence upon binding to bovine brain calmodulin.[44] The hydrophobic surfaces of tubulin have also been investigated with the environment-sensitive probes nile red[45] (N1142) and prodan[46] (P248).

Probes for Other Cytoskeletal Proteins

GFP- and RFP-Labeled Talin

Talin is a cytoskeletal protein that is concentrated in focal adhesions, linking integrins to the actin cytoskeleton either directly or indirectly by interacting with vinculin and α-actinin. CellLight® Talin-GFP and CellLight® Talin-RFP expression vectors (C10611, C10612; Table 11.1; Figure 11.2.13) generate autofluorescent proteins fused to the C-terminal actin-binding domain of human talin and incorporate all the generic advantages of BacMam 2.0 delivery technology (BacMam Gene Delivery and Expression Technology—Note 11.1). These CellLight® reagents have potential applications in image-based high-content screening (HCS) assays of integrin-mediated cell adhesion, as well as for general-purpose labeling of cytoskeletal actin in live cells.

Anti–Glial Fibrillary Acidic Protein (GFAP) Antibody

The 50,000-dalton type-III intermediate filament protein known as glial fibrillary acidic protein (GFAP) is a major structural component of astrocytes and some ependymal cells.[47] GFAP associates with the calcium-binding protein annexin II2-p11(2) and S-100.[48,49] Association with these proteins together with phosphorylation regulates GFAP polymerization. Astrocytes respond to brain injury by proliferation (astrogliosis); one of the first events to occur during astrocyte proliferation is increased GFAP expression. Our anti-GFAP antibody (A21282) and its Alexa Fluor® 488 and Alexa Fluor® 594 conjugates (A21294, A21295; Figure 11.2.14) can be used to aid in the identification of cells of glial lineage. Interestingly, antibodies to GFAP have been detected in

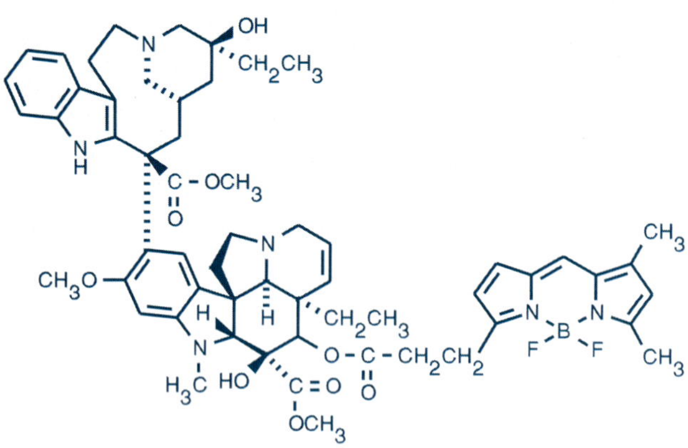

Figure 11.2.12 Vinblastine, BODIPY® FL conjugate (BODIPY® FL vinblastine, V12390).

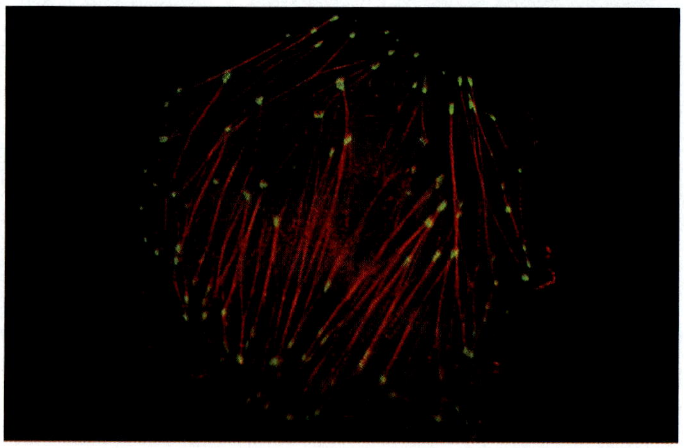

Figure 11.2.13 HeLa cell labeled with CellLight® Talin-GFP (C10611) and CellLight® Actin-RFP (C10583) reagents.

IMPORTANT NOTICE: The products described in this manual are covered by one or more Limited Use Label License(s). Please refer to the Appendix on page 971 and Master Product List on page 975. Products are For Research Use Only. Not intended for any animal or human therapeutic or diagnostic use.

individuals with dementia.[50] In the central nervous system, anti-GFAP antibody stains both astrocytes and ependymal cells. In the peripheral nervous system, Schwann cells, satellite cells and enteric glial cells are stained; tumors of glial origin contain high amounts of GFAP. No positive staining is observed in skin, connective tissue, adipose tissue, lymphatic tissue, muscle, kidney, ureter, bladder or gastrointestinal tract, including liver and pancreas. Our anti-GFAP antibody does not cross-react with vimentin, which is frequently co-expressed in glioma cells and some astrocytes, nor does it cross-react with Bergmann glia cells, gliomas or other glial cell–derived tumors.

Anti-Desmin Antibody

Desmin, encoded by a gene belonging to the intermediate filament protein gene family,[51–53] is the main intermediate filament in mature skeletal, cardiac and smooth muscle cells. Both striated and smooth muscle cells can be labeled by an anti-desmin antibody, although not all muscle tissue contains desmin (e.g., aorta smooth muscle). Identification of desmin is useful in distinguishing habdomyosarcomas and leiomyosarcomas from other vimentin-positive sarcomas. We offer a mouse IgG_1 monoclonal anti-desmin antibody (A21283), which can be used with our fluorescent secondary antibodies (Section 7.2, Figure 11.2.15) as a marker for typing soft tissue sarcomas. Anti-desmin immunohistochemical staining in cell-block preparations may also be helpful in distinguishing mesothelial cells from carcinoma.[54]

Anti-Synapsin I Antibody

Synapsin I, an actin-binding protein, is localized exclusively to synaptic vesicles and thus serves as an excellent marker for synapses in brain and other neuronal tissues.[55,56] Synapsin I inhibits neurotransmitter release, an effect that is abolished upon its phosphorylation by Ca^{2+}/calmodulin–dependent protein kinase II. For assaying the localization and abundance of synapsin I by western blot analysis, immunohistochemistry (Figure 11.2.16), enzyme-linked immunosorbent assay (ELISA) or immunoprecipitation, we offer a polyclonal rabbit anti–synapsin I antibody as an affinity-purified IgG fraction (A6442). Although raised against bovine synapsin I, this antibody also recognizes human, rat and mouse synapsin I; it has little or no activity against synapsin II.

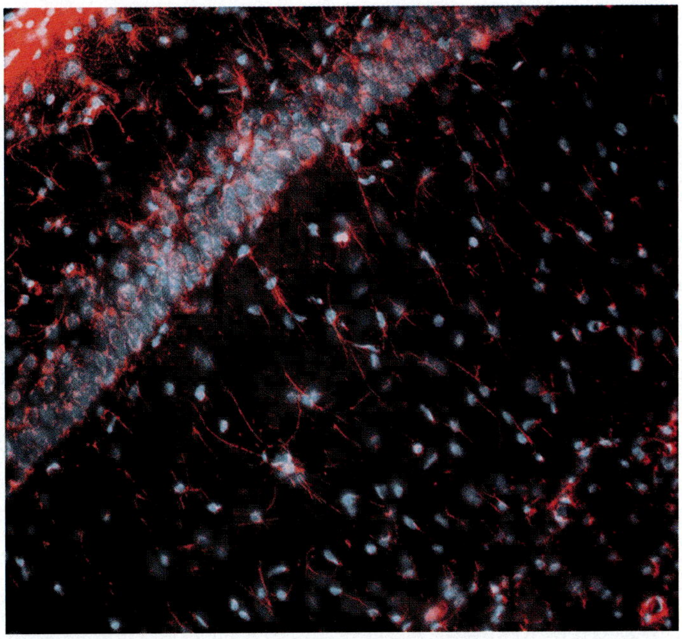

Figure 11.2.14 Rat brain cryosections were labeled with the red-fluorescent Alexa Fluor® 594 conjugate of anti–glial fibrillary acidic protein antibody (A21295). Nuclei were counterstained with TOTO®-3 iodide (T3604, pseudocolored blue).

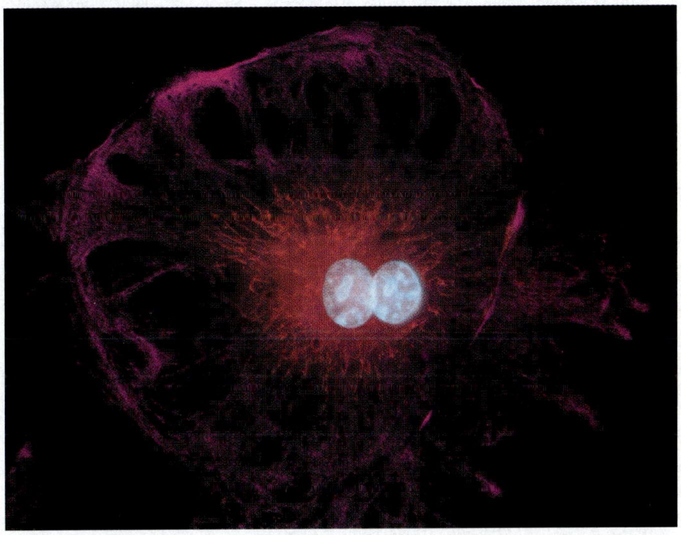

Figure 11.2.15 The intermediate filaments in bovine pulmonary artery endothelial cells, localized using our anti-desmin antibody (A21283), which was visualized with the Alexa Fluor® 647 goat anti-mouse IgG antibody (A21235). Endogenous biotin in the mitochondria was labeled with Alexa Fluor® 546 streptavidin (S11225) and DNA in the cell was stained with blue-fluorescent DAPI (D1306, D3571, D21490).

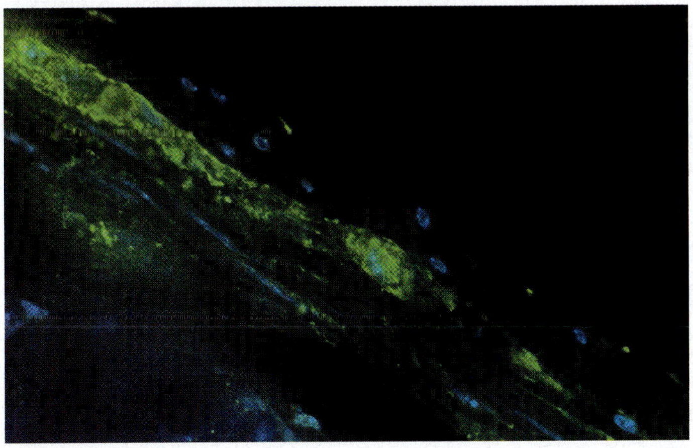

Figure 11.2.16 Peripheral neurons in mouse intestinal cryosections were labeled with rabbit anti–synapsin I antibody (A6442) and detected using Alexa Fluor® 488 goat anti-rabbit IgG antibody (A11008). This tissue was counterstained with DAPI (D1306, D3571, D21490).

IMPORTANT NOTICE: The products described in this manual are covered by one or more Limited Use Label License(s). Please refer to the Appendix on page 971 and Master Product List on page 975. Products are For Research Use Only. Not intended for any animal or human therapeutic or diagnostic use.

REFERENCES

1. BMC Cancer (2006) 6:86; **2.** J Am Chem Soc (1971) 93:2325; **3.** J Am Chem Soc (1988) 110:5917; **4.** Tetrahedron (1986) 42:4451; **5.** J Biol Chem (1994) 269:23399; **6.** J Cell Biol (1991) 112:1177; **7.** Pharmacol Ther (1984) 25:83; **8.** Cancer Treat Rep (1978) 62:1219; **9.** PLoS Biol (2008) 6:e209; **10.** J Neurosci (2008) 28:2601; **11.** J Neurochem (2007) 102:1009; **12.** J Biol Chem (2003) 278:8407; **13.** Biochemistry (2002) 41:12436; **14.** Biochemistry (2001) 40:11975; **15.** Cell Motil Cytoskeleton (2001) 49:1; **16.** J Biol Chem (2000) 275:26265; **17.** J Cell Biol (2000) 148:883; **18.** Chem Biol (2000) 7:275; **19.** Biotechniques (1998) 25:188; **20.** Mol Biotechnol (2002) 21:241; **21.** Mol Pharmacol (2002) 62:1238; **22.** Cancer Res (2002) 62:6864; **23.** Mol Biol Cell (1995) 6:1215; **24.** Mol Pharmacol (2002) 62:1; **25.** FEBS Lett (1997) 416:251; **26.** J Biol Chem (1996) 271:14707; **27.** Eur J Biochem (1997) 244:664; **28.** J Neurochem (1996) 67:1688; **29.** Biochemistry (1994) 33:12665; **30.** Acta Histochem (1993) 94:54; **31.** Arch Biochem Biophys (1993) 303:159; **32.** Eur J Biochem (1987) 165:613; **33.** J Biol Chem (1985) 260:2819; **34.** Anal Biochem (2003) 315:49; **35.** J Biol Chem (1984) 259:14647; **36.** Biochemistry (1994) 33:11900; **37.** Biochemistry (1994) 33:11891; **38.** Biochemistry (1986) 25:3536; **39.** Biochemistry (1998) 37:4687; **40.** Biochemistry (1995) 34:13367; **41.** Cell (1990) 62:579; **42.** Biochemistry (1989) 28:6678; **43.** Immunol Lett (1992) 33:285; **44.** J Biochem (Tokyo) (1991) 109:499; **45.** J Biol Chem (1990) 265:14899; **46.** Eur J Biochem (1992) 204:127; **47.** Neurochem Res (2000) 25:1439; **48.** Biochem Biophys Res Commun (1995) 208:910; **49.** Biochim Biophys Acta (1996) 1313:268; **50.** J Neurol Sci (1997) 151:41; **51.** Proc Natl Acad Sci U S A (1976) 73:4344; **52.** J Cell Sci (1977) 23:243; **53.** EMBO J (1982) 1:1649; **54.** Acta Cytol (2000) 44:976; **55.** Science (1984) 226:1209; **56.** J Cell Biol (1983) 96:1337.

DATA TABLE 11.2 PROBES FOR TUBULIN AND OTHER CYTOSKELETAL PROTEINS

Cat. No.	MW	Storage	Soluble	Abs	EC	Em	Solvent	Notes
B153	672.85	L	pH >6	395	23,000	500	MeOH	1, 2
D1306	350.25	L	H_2O, DMF	342	28,000	450	pH 7	3
D3571	457.49	L	H_2O, MeOH	342	28,000	450	pH 7	3
D3923	249.31	L	DMF, DMSO	456	61,000	493	MeOH	4
D21490	350.25	L	H_2O, DMF	342	28,000	450	pH 7	3, 5
N1142	318.37	L	DMF, DMSO	552	45,000	636	MeOH	6
P248	227.31	L	DMF, MeCN	363	19,000	497	MeOH	7
P3456	853.92	F,D	MeOH, DMSO	228	30,000	none	MeOH	
P7500	1023.89	FF,D,L	DMSO	504	66,000	511	MeOH	
P7501	1098.98	FF,D,L	DMSO	565	121,000	571	MeOH	
P22310	1319.28	FF,D,L	DMSO, EtOH	494	80,000	522	pH 9	
V12390	1043.02	F,D,L	DMSO, DMF	503	83,000	510	MeOH	

For definitions of the contents of this data table, see "Using *The Molecular Probes® Handbook*" in the introductory pages.

Notes

1. B153 is soluble in water at 0.1–1.0 mM after heating.
2. Bis-ANS (B153) bound to tubulin has Abs = 392 nm, Em = 490 nm and a fluorescence quantum yield of 0.23. (Biochemistry (1994) 33:11900)
3. DAPI undergoes an approximately 9-fold fluorescence enhancement on binding to polymerized tubulin. Abs = 345 nm, Em = 446 nm. (J Biol Chem (1985) 260:2819)
4. The absorption and fluorescence emission maxima of DCVJ (D3923) bound to tubulin are essentially the same as in methanol. (Biochemistry (1989) 28:6678)
5. This product is specified to equal or exceed 98% analytical purity by HPLC.
6. The fluorescence emission maximum of nile red (N1142) bound to tubulin is 623 nm. (J Biol Chem (1990) 265:14899)
7. The fluorescence emission maximum of prodan (P248) bound to tubulin is ~450 nm. (Eur J Biochem (1992) 204:127)

PRODUCT LIST 11.2 PROBES FOR TUBULIN AND OTHER CYTOSKELETAL PROTEINS

Cat. No.	Product	Quantity
A21283	anti-desmin, mouse IgG_1, monoclonal 131-15014 *1 mg/mL*	100 µL
A21282	anti-GFAP (anti–glial fibrillary acidic protein, mouse IgG_1, monoclonal 131-17719) *1 mg/mL*	100 µL
A21294	anti-GFAP, Alexa Fluor® 488 conjugate (anti–glial fibrillary acidic protein, mouse IgG_1, monoclonal 131-17719, Alexa Fluor® 488 conjugate) *1 mg/mL*	50 µL
A21295	anti-GFAP, Alexa Fluor® 594 conjugate (anti–glial fibrillary acidic protein, mouse IgG_1, monoclonal 131-17719, Alexa Fluor® 594 conjugate) *1 mg/mL*	50 µL
A6442	anti-synapsin I (bovine), rabbit IgG fraction *affinity purified*	10 µg
A11126	anti-α-tubulin (bovine), mouse IgG_1, monoclonal 236-10501	50 µg
A21371	anti-α-tubulin (bovine), mouse IgG_1, monoclonal 236-10501, biotin-XX conjugate	50 µg
B153	bis-ANS (4,4'-dianilino-1,1'-binaphthyl-5,5'-disulfonic acid, dipotassium salt)	10 mg
C10598	CellLight® MAP4-GFP *BacMam 2.0*	1 mL
C10599	CellLight® MAP4-RFP *BacMam 2.0*	1 mL
C10611	CellLight® Talin-GFP *BacMam 2.0*	1 mL
C10612	CellLight® Talin-RFP *BacMam 2.0*	1 mL
C10613	CellLight® Tubulin-GFP *BacMam 2.0*	1 mL
C10614	CellLight® Tubulin-RFP *BacMam 2.0*	1 mL
D3923	DCVJ (4-(dicyanovinyl)julolidine)	25 mg
D1306	4',6-diamidino-2-phenylindole, dihydrochloride (DAPI)	10 mg
D21490	4',6-diamidino-2-phenylindole, dihydrochloride (DAPI) *FluoroPure™ grade*	10 mg
D3571	4',6-diamidino-2-phenylindole, dilactate (DAPI, dilactate)	10 mg
N1142	nile red	25 mg
P3456	paclitaxel (Taxol® equivalent) *for use in research only*	5 mg
P7501	paclitaxel, BODIPY® 564/570 conjugate (BODIPY® 564/570 Taxol®)	10 µg
P7500	paclitaxel, BODIPY® FL conjugate (BODIPY® FL Taxol®)	10 µg
P22310	paclitaxel, Oregon Green® 488 conjugate (Oregon Green® 488 Taxol®; Flutax-2)	100 µg
P248	prodan (6-propionyl-2-dimethylaminonaphthalene)	100 mg
T34075	TubulinTracker™ Green (Oregon Green® 488 Taxol®, bis-acetate) *for live-cell imaging*	1 set
V12390	vinblastine, BODIPY® FL conjugate (BODIPY® FL vinblastine)	100 µg

IMPORTANT NOTICE: The products described in this manual are covered by one or more Limited Use Label License(s). Please refer to the Appendix on page 971 and Master Product List on page 975. Products are For Research Use Only. Not intended for any animal or human therapeutic or diagnostic use.

www.invitrogen.com/probes

molecular probes | invitrogen by life technologies

TWELVE

CHAPTER 12
Probes for Organelles

IMPORTANT NOTICE: The products described in this manual are covered by one or more Limited Use Label License(s). Please refer to the Appendix on page 971 and Master Product List on page 975. Products are For Research Use Only. Not intended for any animal or human therapeutic or diagnostic use.

molecular probes® | invitrogen™ by life technologies™

IMPORTANT NOTICE: The products described in this manual are covered by one or more Limited Use Label License(s). Please refer to the Appendix on page 971 and Master Product List on page 975. Products are For Research Use Only. Not intended for any animal or human therapeutic or diagnostic use.

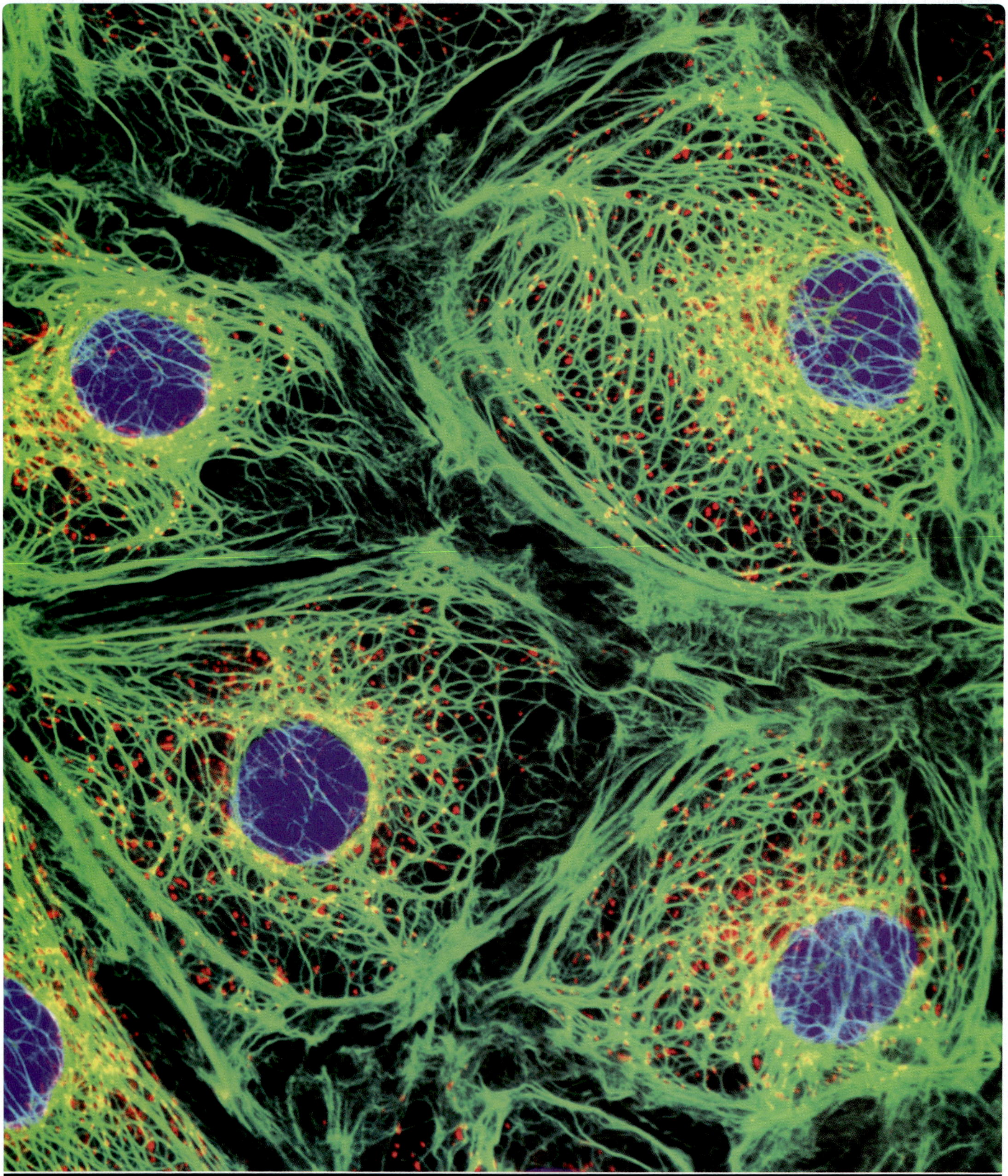

MitoTracker® Red CMXRos and DAPI.

IMPORTANT NOTICE: The products described in this manual are covered by one or more Limited Use Label License(s). Please refer to the Appendix on page 971 and Master Product List on page 975. Products are For Research Use Only. Not intended for any animal or human therapeutic or diagnostic use.

12.1 A Diverse Selection of Organelle Probes

We offer a diverse array of cell-permeant fluorescent stains that selectively associate with the mitochondria, lysosomes, endoplasmic reticulum, Golgi apparatus and nucleus in live cells (Figure 12.1.1). These probes, which include our MitoTracker®, LysoTracker®, LysoSensor™, RedoxSensor™ and ER-Tracker™ organelle stains (Table 12.1), are compatible with most fluorescence instrumentation and provide researchers with powerful tools for investigating respiration, mitosis, apoptosis, multidrug resistance, substrate degradation and detoxification, intracellular transport and sorting and more. Importantly, unlike antibodies, these fluorescent probes can be used to investigate organelle structure and activity in live cells with minimal disruption of cellular function (Figure 12.1.2). The red-fluorescent organelle stains are particularly useful for demonstrating colocalization with Green Fluorescent Protein (GFP) expression (Using Organic Fluorescent Probes in Combination with GFP—Note 12.1).

In addition to these organelle-selective organic dyes, we offer CellLight® targeted fluorescent protein–based markers, comprising BacMam expression vectors encoding an autofluorescent protein fused to a site-selective targeting sequence. These reagents are available for labeling a variety of organelles—including mitochondria, lysosomes, endoplasmic reticulum, Golgi apparatus and nucleus, as well as several subcellular structures such as actin filaments and microtubules—in live mammalian cells (Table 11.1). In this chapter, we discuss the CellLight® reagents targeted specifically to organelles. CellLight® reagents incorporate all the customary advantages of BacMam delivery technology, including high transduction efficiency and long-lasting and titratable expression (BacMam Gene Delivery and Expression Technology—Note 11.1).

Fluorescent cytoskeleton probes, including CellLight® targeted fluorescent proteins, are discussed in Chapter 11. Fluorescent plasma membrane stains, as well as CellLight® membrane-targeted fluorescent proteins, are discussed in Section 14.4. A variety of probes for phagovacuoles, endosomes and lysosomes—such as membrane markers, ligands for studying receptor-mediated endocytosis and CellLight® endosome-targeted fluorescent proteins—are discussed in Section 16.1. Our online Cell Staining Tool allows you to design your own multicolor experiments using these organelle probes; more information is available at www.invitrogen.com/handbook/cellstainingtool.

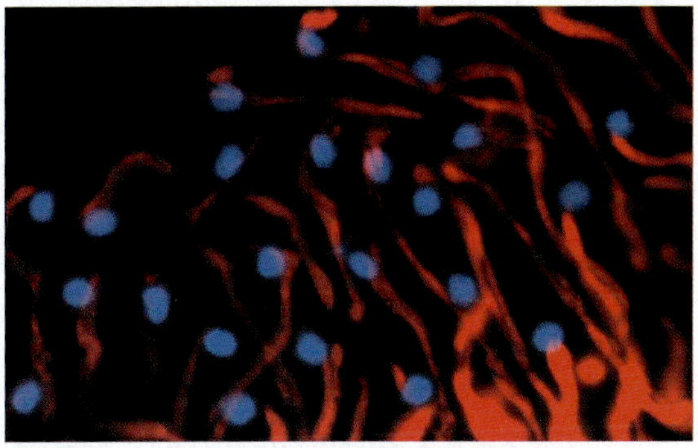

Figure 12.1.2 Euphyrene spermatozoa from the pupal testis of the gypsy moth (*Lymantria dispar*) that were incubated in a solution of MitoTracker® Red CMXRos (M7512). The sample was subsequently fixed and the nuclear material was counterstained with the blue-fluorescent dsDNA dye, DAPI (D1306, D3571, D21490). Image contributed by Laura K. Garvey, University of Connecticut.

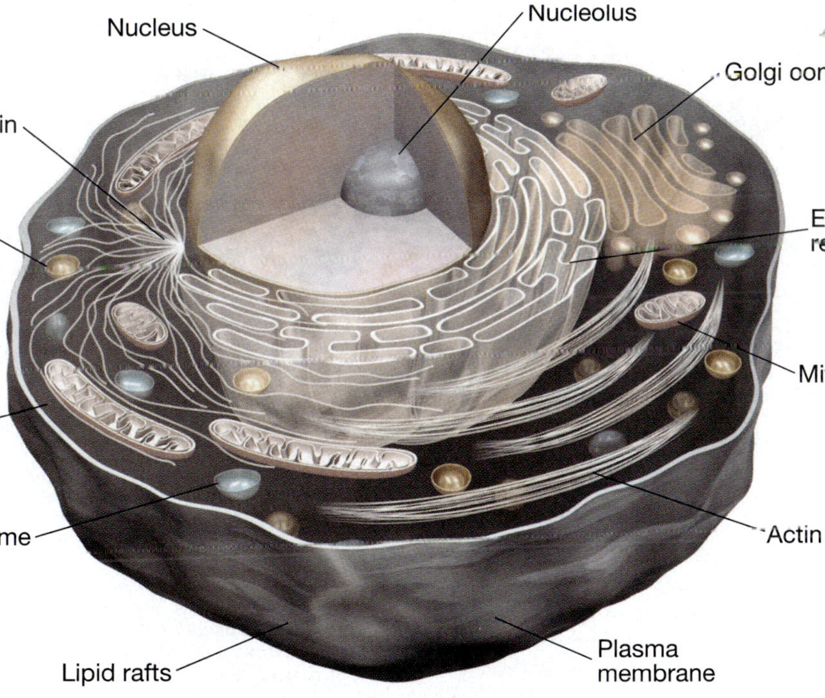

Figure 12.1.1 Diagram of an animal cell.

Table 12.1 Molecular Probes® fluorescent organelle stains.

Green-Fluorescent Probes		Yellow- and Orange-Fluorescent Probes		Red-Fluorescent Probes		Blue-Fluorescent and Other Detectable Probes	
Cat. No.	**Probe**	**Cat. No.**	**Probe**	**Cat. No.**	**Probe**	**Cat. No.**	**Probe**
Probes for Mitochondria: Section 12.2							
A1372	Nonyl acridine orange	D288	4-Di-1-ASP (DASPMI)	M7512	MitoTracker® Red CMXRos *	L6868	Lucigenin †
D273	$DiOC_6(3)$ ‡	D426	DASPEI	T3168	JC-1 §		
D378	$DiOC_7(3)$ ‡	M7510	MitoTracker® Orange CMTMRos *	D22421	JC-9 §		
M7514	MitoTracker® Green FM® *	R634	Rhodamine 6G	M22425	MitoTracker® Red FM® *		
R302 R22420	Rhodamine 123	T639	Tetramethylrosamine	M22426	MitoTracker® Deep Red FM® *		
Y7530	SYTO® 18 yeast mitochondrial stain	T668	Tetramethylrhodamine, methyl ester (TMRM)				
T3168	JC-1 §	T669	Tetramethylrhodamine, ethyl ester (TMRE)				
D22421	JC-9 §						
Oxidation-Sensitive Probes for Mitochondria: Section 12.2							
D632	Dihydrorhodamine 123	D633	Dihydrorhodamine 6G	M7513	MitoTracker® Red CM-H_2XRos *		
		M7511	MitoTracker® Orange CM-H_2TMRos *	R14060	RedoxSensor™ Red CC-1 **		
Probes for Acidic Organelles Including Lysosomes: Section 12.3							
L7526	LysoTracker® Green DND-26	D113	Dansyl cadaverine	L7528	LysoTracker® Red DND-99	D1552	DAMP ††
		L12491	LysoTracker® Yellow HCK-123	R14060	RedoxSensor™ Red CC-1 **	H22845	Hydroxystilbamidine
						L7525	LysoTracker® Blue DND-22
pH-Sensitive Probes for Acidic Organelles: Section 12.3							
L7534	LysoSensor™ Green DND-153	A1301	Acridine orange	N3246	Neutral red	L7533	LysoSensor™ Blue DND-167
L7535	LysoSensor™ Green DND-189	L7545	LysoSensor™ Yellow/Blue DND-160 §			L7545	LysoSensor™ Yellow/Blue DND-160 §
		L22460	LysoSensor™ Yellow/Blue 10,000 MW dextran §			L22460	LysoSensor™ Yellow/Blue 10,000 MW dextran §
Probes for the Endoplasmic Reticulum: Section 12.4							
E34251	ER-Tracker™ Green	D282	$DiIC_{18}(3)$	E34250	ER-Tracker™ Red	E12353	ER-Tracker™ Blue-White DPX
D272	$DiOC_5(3)$	D384	$DiIC_{16}(3)$				
D273	$DiOC_6(3)$	R648MP	Rhodamine B, hexyl ester				
Probes for the Golgi Apparatus: Section 12.4							
D3521 B22650	BODIPY® FL C_5-ceramide §			D3521 B22650	BODIPY® FL C_5-ceramide §		
N1154 N22651	NBD C_6-ceramide			D7540 B34400	BODIPY® TR ceramide		
Probes for the Nucleus: Section 12.5 includes a complete listing of SYTOX®, HCS NuclearMask™, Hoechst, DAPI and other nuclear stains.							

* Aldehyde-fixable probe. † Chemiluminescent probe. ‡ Selective for mitochondria only at low applied concentrations (<100 nM). § Dual-emission spectrum. ** The differential distribution of the oxidized product between mitochondria and lysosomes appears to depend on the oxidation–reduction (redox) potential of the cytosol. †† Detect using anti-dinitrophenyl antibody (Section 7.4).

NOTE 12.1

Using Organic Fluorescent Probes in Combination with GFP

Probes for Multiplexed Detection of GFP-Expressing Cells

The Green Fluorescent Protein (GFP) reporter has added a new dimension to the analysis of protein localization, allowing real-time examination in live cells of processes that have conventionally been observed through immunocytochemical "snapshots" in fixed specimens.[1] Using spectrally distinct, organic fluorescent probes and markers (Table 1) adds extra data dimensions and reference points to these experiments (Figure 1).

The majority of the applications summarized in Table 1 involve live cells, tissues and organisms. There are many other instances where research objectives call for complementary use of immunochemical and GFP-based protein localization techniques. These experiments demand

continued on page 502

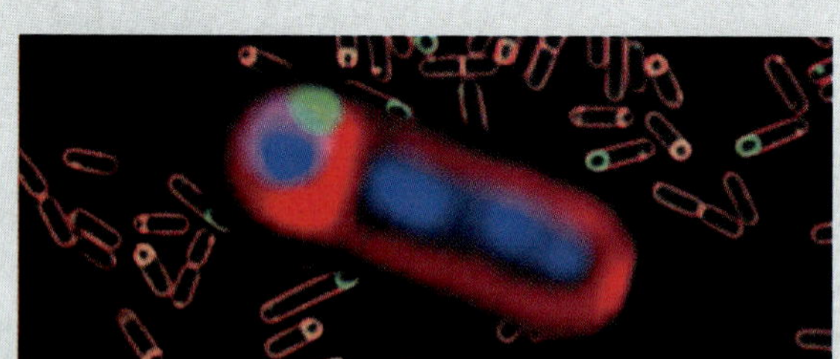

Figure 1 The morphology of sporulating *Bacillus subtilis* in the early stages of forespore engulfment. The membranes and chromosomes of both the forespore and the larger mother cell are stained with FM® 4-64 (red; T3166, T13320) and DAPI (blue; D1306, D3571, D21490), respectively. The small green-fluorescent patch indicates the localization of a GFP fusion to SpoIIIE, a protein essential for translocation of the forespore chromosome that may also regulate membrane fusion events (see Proc Natl Acad Sci U S A (1999) 96: 14553). The background contains sporangia at various stages in the engulfment process stained with MitoTracker® Green FM® (green, M7514) and FM® 4-64 (red).

IMPORTANT NOTICE: The products described in this manual are covered by one or more Limited Use Label License(s). Please refer to the Appendix on page 971 and Master Product List on page 975. Products are For Research Use Only. Not intended for any animal or human therapeutic or diagnostic use.

Table 1 Probes for multiplexed detection of GFP-expressing* cells.

Target	Probe	Cat. No.	Ex/Em †	GFP Fusion Partner	Specimen	Reference
Physiological Indicators						
Intracellular Ca^{2+}	Fura-2 AM	F1201, F1221, F1225, F14185	335/505 ‡	Protein kinase C (PKC)	BHK cells	Biochem J (1999) 337:211
Intracellular Ca^{2+}	X-Rhod-1 AM	X14210	580/602	Trpm5 (melastatin-related cation channel)	CHO cells	Nat Neurosci (2002) 5:1169
Intracellular Ca^{2+}	Fura Red™ AM	F3020, F3021	488/650	GFP expressed specifically in pancreatic β-cells	Mouse pancreatic islets	Am J Physiol Endocrinol Metab (2003) 284:E177
Intracellular pH	5-(and 6-)Carboxy SNARF®-1 AM ester acetate	C1271	568/635	Human growth hormone (hGH)	RIN1046-38 insulinoma cells	Am J Physiol Cell Physiol (2002) 283:C429
Mitochondrial membrane potential	TMRM	T668	555/580	Cytochrome *c*	MCF-7 human breast carcinoma, HeLa	J Cell Sci (2003) 116:525
Superoxide (O_2^-)	Dihydroethidium	D1168	518/605	Cytochrome *c*	MCF-7 human breast carcinoma	J Biol Chem (2003) 278:12645
Synaptic activity	FM® 4-64	T3166, T13320	506/750 §	VAMP (vesicle-associated membrane protein)	Rat hippocampal neurons	Nat Neurosci (2000) 3:445
Receptors and Endocytosis						
Acetylcholine receptor	Tetramethylrhodamine α-bungarotoxin	T1175	553/577	Rapsyn (receptor-aggregating protein)	Zebrafish	J Neurosci (2001) 21:5439
Epidermal growth factor (EGF)	Rhodamine EGF	E3481	555/581	EGF receptor	MTLn3 rat mammary adenocarcinoma	Mol Biol Cell (2000) 11:3873
Endosomes	Transferrin from human serum, Alexa Fluor® 546 conjugate	T23364	556/573	β2-adrenergic receptor (β2AR)	HEK 293, rat hippocampal neurons	Brain Res (2003) 984:21
Endosomes	Transferrin from human serum, Alexa Fluor® 568 conjugate	T23365	578/603	PrP^C (cellular prion protein)	SN56 cells	J Biol Chem (2002) 277:33311
Endosomes	FM® 4-64	T3166, T13320	506/750 §	PrP^C (cellular prion protein)	SN56 cells	J Biol Chem (2002) 277:33311
Organelles						
Endoplasmic reticulum	ER-Tracker™ Blue-White DPX	E12353	375/520 ‡	HSD17B7 gene product (3-ketosteroid reductase)	HeLa, NIH 3T3	Mol Endocrinol (2003) 17:1715
Golgi complex	BODIPY® TR ceramide	D7540	589/617	PrP^C (cellular prion protein)	SN56 cells	J Biol Chem (2002) 277:33311
Lysosomes	LysoTracker® Red	L7528	577/590	Heparanase	Primary human fibroblasts, MDA-231 (human breast carcinoma)	Exp Cell Res (2002) 281:50
Mitochondria	MitoTracker® Red	M7512	578/599	Sam5p (mitochondrial carrier for *S*-adenosylmethionine)	Yeast (*Saccharomyces cerevisiae*)	EMBO J (2003) 22:5975
Nuclear DNA	DAPI	D1306, D3571, D21490	358/461	Histone H2B	HeLa	Methods (2003) 29:42
Nuclear DNA	Hoechst 33342	H1399, H3570, H21492	350/461	Histone H1	BALB/c 3T3 fibroblasts	Nature (2000) 408:877
Nuclear DNA	SYTO® 17	S7579	621/634	HIV-1 integrase	HeLa	J Biol Chem (2003) 278:33528
Nuclear DNA	SYTO® 59	S11341	622/645	Microtubule plus-end binding protein	Porcine kidney epithelial cells (LLCPK)	Mol Biol Cell (2003) 14:916
Nuclear DNA	TO-PRO®-3	T3605	642/661	Citron kinase	HeLa	J Cell Sci (2001) 114:3273
Plasma membrane	DiI	D282, D3911, N22880	549/565	Synaptobrevin	*Xenopus* optic neurons	Nat Neurosci (2001) 4:1093
Other Subcellular Structures						
F-actin	Rhodamine phalloidin	R415	554/573	ERM (ezrin-radixin-moesin) proteins	Human peripheral blood T cells (PBT)	Nat Immunol (2004) 5:272
F-actin	Alexa Fluor® 568 phalloidin	A12380	578/603	Calponin	NIH 3T3	J Cell Sci (2000) 113:3725
Lipid rafts	Cholera toxin subunit B (recombinant), Alexa Fluor® 594 conjugate	C22842	590/617	Histocompatibility leukocyte antigen (HLA)-Cw4	NK cell–B-cell immunological synapse	Proc Natl Acad Sci U S A (2001) 98:14547

* This list covers only *Aequoria victoria* GFP, optimized mutants (e.g., EGFP) and green-fluorescent proteins from other species (e.g., *Renilla reniformis*). Fluorescent proteins with distinctly different excitation and emission characteristics (CFP, YFP, dsRed, etc.) are not included. † Fluorescence excitation (Ex) and emission (Em) maxima, in nm. ‡ Simultaneous imaging of GFP with fura-2 or ER-Tracker™ Blue-White DPX requires excitation wavelength–switching capability, because the fluorescence emission spectra overlap extensively. Even under these conditions, signal bleedthrough from one detection channel to the other may still be problematic, depending on the expression level and localization of the GFP chimera. See Biochem J (2001) 356:345 for further discussion. § The fluorescence emission spectra of styryl dyes such as FM® 1-43 and FM® 4-64 are broad and extend into the green emission range of GFP. In some cases, FM® dye emission can overspill into the GFP detection channel, causing degraded resolution of image features. The excitation and emission spectra of FM® 1-43 overlap those of GFP more extensively than those of FM® 4-64. Therefore, using FM® 4-64 instead of FM® 1-43 is recommended to minimize this problem.

continued on next page

IMPORTANT NOTICE: The products described in this manual are covered by one or more Limited Use Label License(s). Please refer to the Appendix on page 971 and Master Product List on page 975. Products are For Research Use Only. Not intended for any animal or human therapeutic or diagnostic use.

continued from previous page

Table 1 Probes for multiplexed detection of GFP-expressing* cells—*continued*.

Target	Probe	Cat. No.	Ex/Em †	GFP Fusion Partner	Specimen	Reference
Cell Classification Markers						
Apoptotic cells	Annexin V, Alexa Fluor® 594 conjugate	A13203	590/617	GRASP65 (Golgi stacking protein)	HeLa	J Cell Biol (2002) 156:495
Transformed B lymphocytes (Raji cells)	CellTracker™ Orange CMTMR	C2927	550/575	ICAM-3 (intercellular adhesion molecule-3)	T-lymphocytes and antigen-presenting B cells	Nat Immunol (2002) 3:159
Cell-surface antigens	R-Phycoerythrin (streptavidin conjugate)	S866, S21388	565/575	GFP gene expression	NIH 3T3	Cytometry (1996) 25:211
Neurons	NeuroTrace® 530/615 red-fluorescent Nissl stain	N21482	530/620	Tau microtubule-binding protein (Purkinje cell marker)	Mouse brain slice	J Neurosci (2003) 23:6392
Neurons	Alexa Fluor® 594 hydrazide	A10438, A10442	588/613	Synaptophysin	*Aplysia californica* sensory neurons	Neuron (2003) 40:151

* This list covers only *Aequoria victoria* GFP, optimized mutants (e.g., EGFP) and green-fluorescent proteins from other species (e.g., *Renilla reniformis*). Fluorescent proteins with distinctly different excitation and emission characteristics (CFP, YFP, dsRed, etc.) are not included. † Fluorescence excitation (Ex) and emission (Em) maxima, in nm. ‡ Simultaneous imaging of GFP with fura-2 or ER-Tracker™ Blue-White DPX requires excitation wavelength–switching capability, because the fluorescence emission spectra overlap extensively. Even under these conditions, signal bleedthrough from one detection channel to the other may still be problematic, depending on the expression level and localization of the GFP chimera. See Biochem J (2001) 356:345 for further discussion. § The fluorescence emission spectra of styryl dyes such as FM® 1-43 and FM® 4-64 are broad and extend into the green emission range of GFP. In some cases, FM® dye emission can overspill into the GFP detection channel, causing degraded resolution of image features. The excitation and emission spectra of FM® 1-43 overlap those of GFP more extensively than those of FM® 4-64. Therefore, using FM® 4-64 instead of FM® 1-43 is recommended to minimize this problem.

continued from page 500

the combination of brightness, photostability and spectral separation provided by our Alexa Fluor® dye–labeled secondary detection reagents. For two-color combinations with GFP, we recommend the Alexa Fluor® 555, Alexa Fluor® 568 or Alexa Fluor® 594 dye–labeled secondary antibodies (Section 7.2, Table 7.1). The addition of Alexa Fluor® 635 or Alexa Fluor® 647 dye–labeled antibodies allows three-color detection. Some immunohistochemical procedures such as paraffin embedding of fixed tissue result in loss of the intrinsic fluorescence of GFP. In other cases, GFP expression levels may simply be too low for detection above background autofluorescence.[2] Antibodies to GFP provide remedies for these problems (Figure 2). We offer unlabeled mouse and rabbit monoclonal and rabbit and chicken polyclonal antibodies to GFP as well as Alexa Fluor® dye–labeled rabbit polyclonal antibodies to GFP (Section 7.5).

Alexa Fluor® Dyes: Highly Fluorescent FRET Acceptors

Proximity-dependent fluorescence resonance energy transfer (FRET) allows detection of protein–protein interactions with much higher spatial resolution than conventional diffraction-limited microscopy.[3] Alexa Fluor® dyes with strong absorption in the 500–600 nm wavelength range are excellent FRET acceptors from GFP (Table 2). An assay to detect activation of GFP–GTPase fusions developed by researchers at Scripps Research Institute[4] utilizes the GTPase-binding domain (PBD) of PAK1, a protein that binds to GTPases only in their activated GTP-bound form. GTPase activation is indicated by FRET from GFP to PDB labeled with Alexa Fluor® 546 C_5-maleimide at a single N-terminal cysteine residue. This assay has been used to determine the location and dynamics of rac and Cdc42 GTPase activation in live cells.[4–6]

Normalizing Expression and Translation Signals

In 2002, researchers in Scott Fraser's laboratory at the California Institute of Technology reported a method of coinjecting Texas Red® dye–labeled 10,000 MW dextran and GFP vectors into sea urchin embryos. This method overcomes a multitude of problems inherent in making intra- and inter-embryo comparisons of gene expression levels using confocal microscopy. In particular, laser excitation and fluorescence collection efficiencies vary with the depth of the fluorescent protein in the embryo, and the orientation of different embryos on the coverslip varies relative to the microscope objective. Measuring the ratio of GFP fluorescence to Texas Red® dextran fluorescence corrects for these spatial factors, providing a gene expression readout that is 2–50 times more accurate than conventional confocal microscopy procedures depending on the localization of GFP within an embryo.[7] A similar strategy was previously used to determine translation efficiencies of GFP-encoding mRNAs.[8]

1. Nat Rev Mol Cell Biol (2002) 3:906; 2. Anal Biochem (2001) 291:175; 3. J Cell Biol (2003) 160:629; 4. Science (2000) 290:333; 5. J Biol Chem (2003) 278:31020; 6. Nat Cell Biol (2002) 4:32; 7. Proc Natl Acad Sci U S A (2002) 99:12895; 8. J Cell Biol (1999) 147:247.

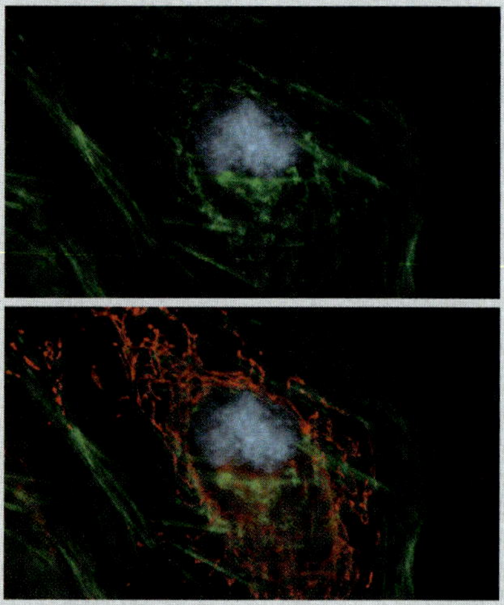

Figure 2 HeLa cell transfected with pShooter™ pCMV/myc/mito/GFP, then fixed and permeabilized. Green Fluorescent Protein (GFP) localized in the mitochondria was labeled with mouse IgG_{2a} anti-GFP antibody (A11120) and detected with orange-fluorescent Alexa Fluor® 555 goat anti–mouse IgG antibody (A21422), which colocalized with the dim GFP fluorescence. F-actin was labeled with green-fluorescent Alexa Fluor® 488 phalloidin (A12379), and the nucleus was stained with blue-fluorescent DAPI (D1306, D3571, D21490). The sample was mounted using ProLong® Gold antifade reagent (P36930). Some GFP fluorescence is retained in the mitochondria after fixation (top), but immunolabeling and detection greatly improve visualization (bottom).

Table 2 R_0 values for FRET from EGFP to Alexa Fluor® dyes.

Acceptor Dye	R_0 (Å)*
Alexa Fluor® 546 dye	57
Alexa Fluor® 555 dye	63
Alexa Fluor® 568 dye	54
Alexa Fluor® 594 dye	53

* R_0 values in angstroms (Å) represent the distance at which fluorescence resonance energy transfer from the donor dye to the acceptor dye is 50% efficient. Values were calculated from spectroscopic data as outlined (Fluorescence Resonance Energy Transfer (FRET)—Note 1.2).

IMPORTANT NOTICE: The products described in this manual are covered by one or more Limited Use Label License(s). Please refer to the Appendix on page 971 and Master Product List on page 975. Products are For Research Use Only. Not intended for any animal or human therapeutic or diagnostic use.
www.invitrogen.com/probes

12.2 Probes for Mitochondria

Mitochondria are found in eukaryotic cells, where they make up as much as 10% of the cell volume. They are pleomorphic organelles with structural variations depending on cell type, cell-cycle stage and intracellular metabolic state. The key function of mitochondria is energy production through oxidative phosphorylation (OxPhos) and lipid oxidation.[1] Several other metabolic functions are performed by mitochondria, including urea production and heme, non-heme iron and steroid biogenesis, as well as intracellular Ca^{2+} homeostasis. Mitochondria also play a pivotal role in apoptosis—the genetically controlled ablation of cells during normal development[2–4] (Section 15.5). For many of these mitochondrial functions, there is only a partial understanding of the components involved, with even less information on mechanism and regulation.

The morphology of mitochondria is highly variable. In dividing cells, the organelle can switch between a fragmented morphology with many ovoid-shaped mitochondria, as is often shown in textbooks, and a reticulum in which the organelle is a single, many-branched structure. The cell cycle– and metabolic state–dependent changes in mitochondrial morphology are controlled by a set of proteins that cause fission and fusion of the organelle mass. Mutations in these proteins are the cause of several human diseases, indicating the importance of overall morphology for cell functioning (Mitochondria in Diseases—Note 12.2). Organelle morphology is also controlled by cytoskeletal elements, including actin filaments and microtubules. In nondividing tissue, overall mitochondrial morphology is very cell-type dependent, with mitochondria spiraling around the axoneme in spermatozoa, and ovoid bands of mitochondria intercalating between actomyosin filaments. There is evidence of functionally significant heterogeneity of mitochondrial forms within individual cells.

The abundance of mitochondria varies with cellular energy level and is a function of cell type, cell-cycle stage and proliferative state. For example, brown adipose tissue cells,[5] hepatocytes[6] and certain renal epithelial cells[7] tend to be rich in active mitochondria, whereas quiescent immune-system progenitor or precursor cells show little staining with mitochondrion-selective dyes.[8] The number of mitochondria is reduced in Alzheimer disease and their proteins and nucleic acids are susceptible to damage by reactive oxygen species, including nitric oxide[9–11] (Chapter 18).

We have a range of mitochondrion-selective dyes with which to monitor mitochondrial morphology and organelle functioning. The uptake of most mitochondrion-selective dyes is dependent on the mitochondrial membrane potential. These dyes thereby enable researchers to probe mitochondrial activity, localization and abundance,[9,12,13] as well as to monitor the effects of some pharmacological agents that alter mitochondrial function.[14]

MitoTracker® Probes: Fixable Mitochondrion-Selective Probes

Although conventional fluorescent stains for mitochondria, such as rhodamine 123 and tetramethylrosamine, are readily sequestered by functioning mitochondria, they are subsequently washed out of the cells once the mitochondrion's membrane potential is lost. This characteristic limits their use in experiments in which cells must be treated with aldehyde-based fixatives or other agents that affect the energetic state of the mitochondria. To overcome this limitation, we have developed MitoTracker® probes—a series of mitochondrion-selective stains that are concentrated by active mitochondria and well retained during cell fixation.[15] Because the MitoTracker® Orange, MitoTracker® Red and MitoTracker® Deep Red probes are also retained following permeabilization, the sample retains the fluorescent staining pattern characteristic of live cells during subsequent processing steps for immunocytochemistry, *in situ* hybridization or electron microscopy. In addition, MitoTracker® reagents eliminate some of the difficulties of working with pathogenic cells because, once the mitochondria are stained, the cells can be treated with fixatives before the sample is analyzed.

NOTE 12.2

Mitochondria in Diseases

Given the multiple functions and numerous proteins present in the mitochondria, it is not surprising that genetically inherited defects of mitochondrial function are a major cause of morbidity and mortality in humans.[1] In particular, there are several human diseases that have known defects in the proteins responsible for oxidative phosphorylation (OxPhos) in cells. Typically, such defects produce lactic acidemia, exercise intolerance or neurological disorders.

Diseases of OxPhos are notoriously difficult to diagnose, and it is even more difficult to correlate their phenotype–genotype relationships. A subset of OxPhos defects is maternally inherited. These defects result from mutations in mitochondrial DNA (mtDNA), a small, 16-kb genome present in hundreds to thousands of copies per cell.[2,3] mtDNA, which encodes 13 polypeptides of the OxPhos machinery, differs from the nuclear genome in its absence of histones, poor repair mechanisms and very limited recombination frequencies. As a result, mtDNA in somatic cells builds up mutations over time due to errors in replication that are not repaired and physical insult from a variety of toxins.[4] Such accumulated mutations are implicated in a number of neurodegenerative diseases[5]—notably Parkinson disease and Alzheimer disease—where the mutation load triggers premature apoptotic or necrotic cell death. For example, a strong link has been established between exposure to the pesticide rotenone, a well-defined and specific inhibitor of OxPhos, and Parkinson disease. mtDNA mutations function by reducing energy production within the cell and are thought to contribute to cancer and aging. Likewise, mutations in the nuclear-encoded subunits of OxPhos have been found to regulate the life span in flies and worms. Many of the products listed in this section are useful tools for studying degenerative conditions.[6–9]

1. Annu Rev Physiol (2010) 72:61; **2.** Anal Chem (2007) 79:7691; **3.** Exp Cell Res (2005) 303:432; **4.** Biochem Biophys Res Commun (2009) 378:450; **5.** Brain (2010) 133:797; **6.** J Neurosci (2009) 29:9090; **7.** J Biol Chem (2009) 284:18754; **8.** Mol Cell (2009) 33:627; **9.** Methods Enzymol (2009) 453:217.

IMPORTANT NOTICE: The products described in this manual are covered by one or more Limited Use Label License(s). Please refer to the Appendix on page 971 and Master Product List on page 975. Products are For Research Use Only. Not intended for any animal or human therapeutic or diagnostic use.

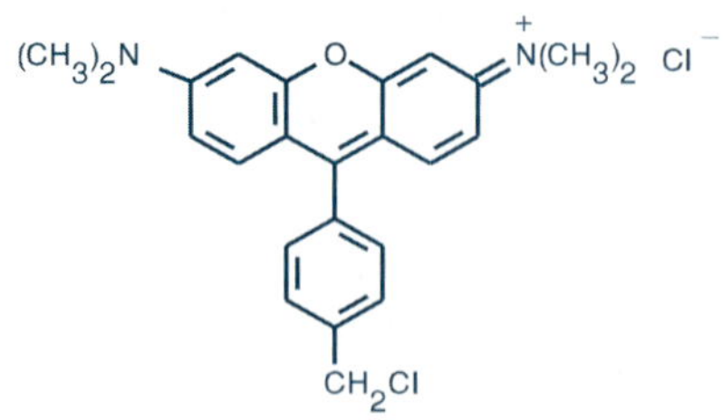

Figure 12.2.1 MitoTracker® Orange CMTMRos (M7510).

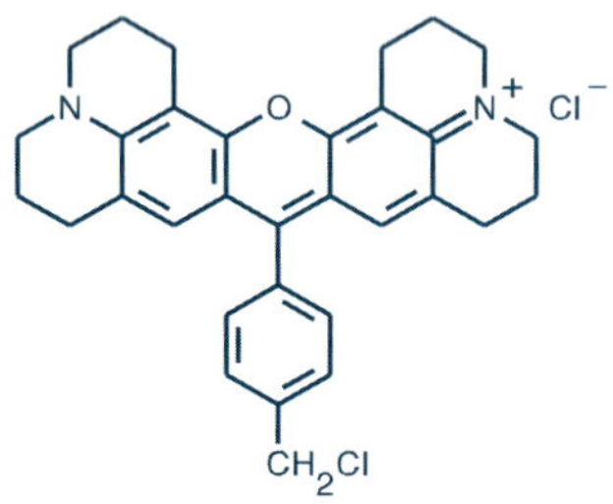

Figure 12.2.2 MitoTracker® Red CMXRos (M7512).

Properties of MitoTracker® Probes

MitoTracker® probes are cell-permeant mitochondrion-selective dyes that contain a mildly thiol-reactive chloromethyl moiety. The chloromethyl group appears to be responsible for keeping the dye associated with the mitochondria after fixation.[16] To label mitochondria, cells are simply incubated in submicromolar concentrations of the MitoTracker® probe, which passively diffuses across the plasma membrane and accumulates in active mitochondria. Once their mitochondria are labeled, the cells can be treated with aldehyde-based fixatives to allow further processing of the sample; with the exception of MitoTracker® Green FM®, subsequent permeabilization with cold acetone does not appear to disturb the staining pattern of the MitoTracker® dyes.

We offer seven MitoTracker® reagents that differ in spectral characteristics, oxidation state and fixability (Table 12.2). MitoTracker® probes are provided in specially packaged sets of 20 vials, each containing 50 µg for reconstitution as required.

Orange-, Red- and Infrared-Fluorescent MitoTracker® Dyes

We offer MitoTracker® derivatives of the orange-fluorescent tetramethylrosamine (MitoTracker® Orange CMTMRos, M7510; Figure 12.2.1) and the red-fluorescent X-rosamine (MitoTracker® Red CMXRos, M7512; Figure 12.2.2), as well as MitoTracker® Red FM® and MitoTracker® Deep Red FM® probes (M22425, M22426; Figure 12.2.3, Figure 12.2.4). Because the MitoTracker® Red CMXRos, MitoTracker® Red FM® and MitoTracker® Deep Red FM® probes produce longer-wavelength fluorescence that is well resolved from the fluorescence of green-fluorescent dyes, they are suitable for multicolor labeling experiments[17,18] (Figure 12.2.5, Figure 12.2.6, Figure 12.2.7). Also available are chemically reduced forms of the tetramethylrosamine (MitoTracker® Orange CM-H_2TMRos, M7511; Figure 12.2.8) and X-rosamine (MitoTracker® Red CM-H_2XRos, M7513; Figure 12.2.9) MitoTracker® probes. Unlike MitoTracker® Orange CMTMRos and MitoTracker® Red CMXRos, the reduced versions of these probes do not fluoresce until they enter an actively respiring cell, where they are oxidized to the fluorescent mitochondrion-selective probe and then sequestered in the mitochondria[19] (Figure 12.2.10, Figure 12.2.11, Figure 12.2.12).

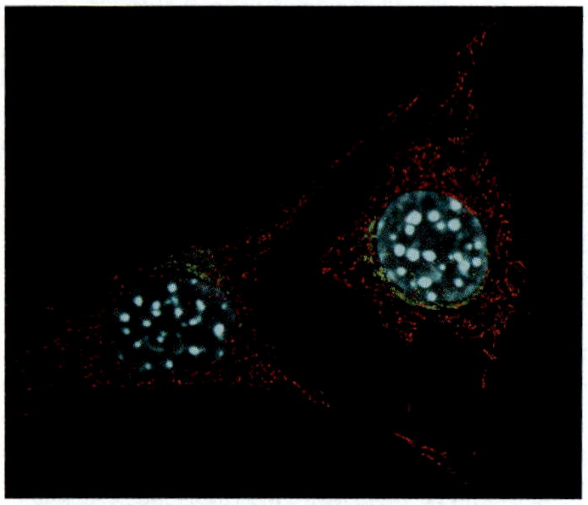

Figure 12.2.3 Live NIH 3T3 cells labeled with probes for mitochondria, Golgi and the nucleus. Mitochondria were labeled with MitoTracker® Red FM® (M22425), Golgi with BODIPY® FL ceramide (D3521, B22650), and the nucleus with Hoechst 33342 (H1399, H3570, H21492).

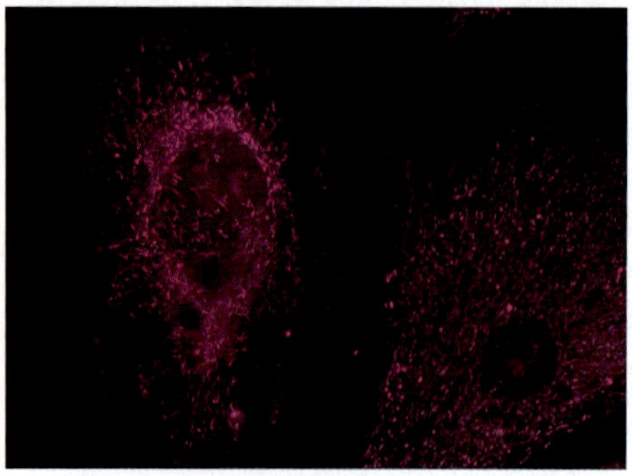

Figure 12.2.4 Mitochondria of live bovine pulmonary artery endothelial cells stained with the MitoTracker® Deep Red FM® dye (M22426).

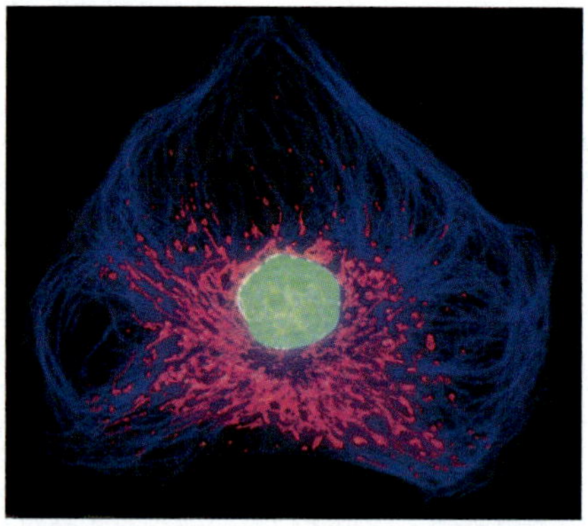

Figure 12.2.5 Bovine pulmonary artery endothelial cells (BPAEC) incubated with the fixable, mitochondrion-selective MitoTracker® Red CMXRos (M7512). After staining, the cells were formaldehyde-fixed, acetone-permeabilized, treated with DNase-free RNase and counterstained using SYTOX® Green nucleic acid stain (S7020). Microtubules were labeled with a mouse monoclonal anti–ß-tubulin antibody, biotin-XX goat anti-mouse IgG antibody (B2763) and Cascade Blue® NeutrAvidin™ biotin-binding protein (A2663). This photograph was taken using multiple exposures through bandpass optical filters appropriate for Texas Red® dye, fluorescein and DAPI using a Nikon® Labophot 2 microscope equipped with a Quadfluor epi-illumination system.

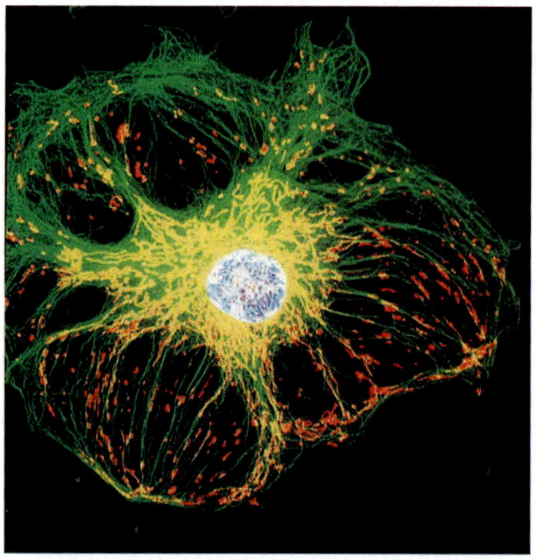

Figure 12.2.6 A bovine pulmonary artery endothelial cell (BPAEC) stained with mouse monoclonal anti–ß-tubulin in conjunction with Oregon Green® 514 goat anti–mouse IgG antibody (O6383), MitoTracker® Red CMXRos (M7512) and DAPI (D1306, D3571, D21490).

IMPORTANT NOTICE: The products described in this manual are covered by one or more Limited Use Label License(s). Please refer to the Appendix on page 971 and Master Product List on page 975. Products are For Research Use Only. Not intended for any animal or human therapeutic or diagnostic use.
www.invitrogen.com/probes
molecular probes® | invitrogen™ by life technologies™

Table 12.2 Spectral characteristics of the MitoTracker® probes.

Cat. No.	MitoTracker® Probe	Abs * (nm)	Em * (nm)	Oxidation State
M7514	MitoTracker® Green FM® †	490	516	NA
M7510	MitoTracker® Orange CMTMRos	551	576	Oxidized
M7511	MitoTracker® Orange CM-H_2TMRos	551 ‡	576 ‡	Reduced
M7512	MitoTracker® Red CMXRos	578	599	Oxidized
M7513	MitoTracker® Red CM-H_2XRos	578 ‡	599 ‡	Reduced
M22425	MitoTracker® Red FM®	581	644	NA
M22426	MitoTracker® Deep Red FM®	644	665	NA

* Absorption (Abs) and fluorescence emission (Em) maxima, determined in methanol; values may vary somewhat in cellular environments. † MitoTracker® Green FM® is nonfluorescent in aqueous environments. ‡ These reduced MitoTracker® probes are not fluorescent until oxidized. NA = Not applicable.

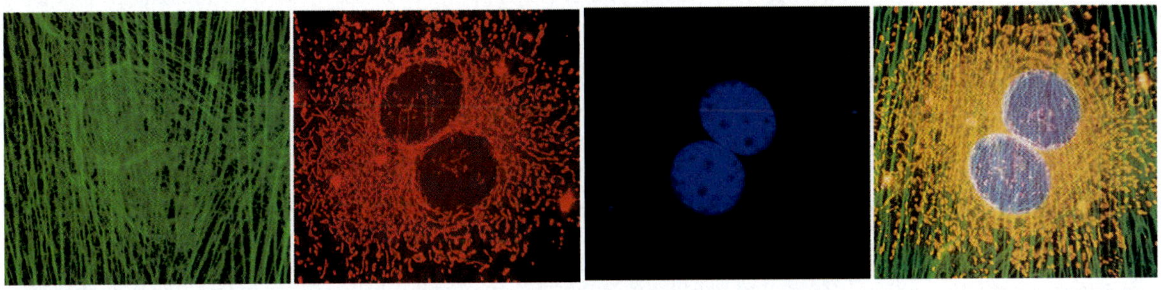

Figure 12.2.7 Four-panel composite image of mouse fibroblasts that were incubated with MitoTracker® Red CMXRos (M7512), and then formaldehyde-fixed, acetone-permeabilized and stained with the F-actin–specific probe, BODIPY® FL phallacidin (B607) and with DAPI (D1306, D3571, D21490). Images were obtained by taking single and multiple exposures through bandpass optical filter sets appropriate for fluorescein, the Texas Red® dye and DAPI.

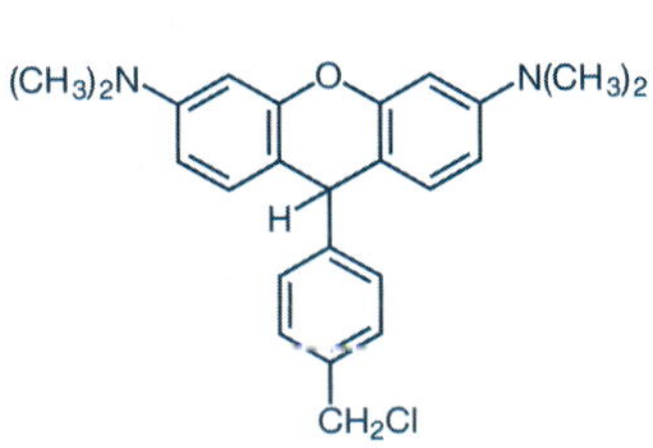

Figure 12.2.8 MitoTracker® Orange CM-H_2TMRos (M7511).

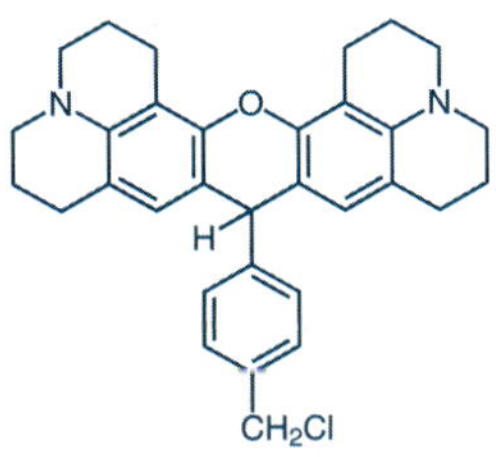

Figure 12.2.9 MitoTracker® Red CM-H_2XRos (M7513).

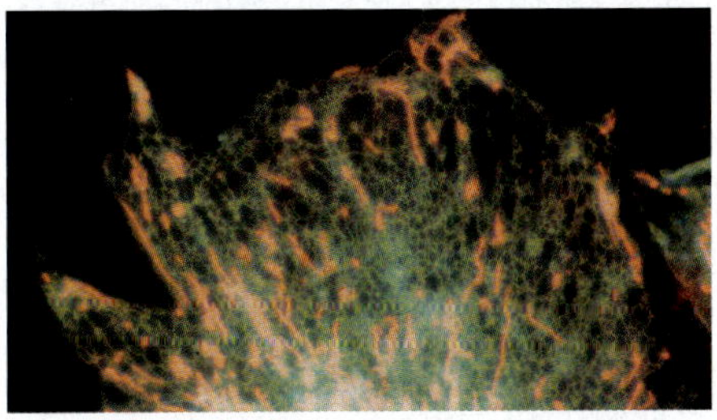

Figure 12.2.11 Live bovine pulmonary artery endothelial cells stained with ER-Tracker™ Blue-White DPX (E12353) and MitoTracker® Red CM-H_2XRos (M7513). The endoplasmic reticulum appears green and the mitochondria appear orange. The image was acquired using a fluorescence microscope equipped with a triple-bandpass filter set appropriate for DAPI, fluorescein and Texas Red® dyes.

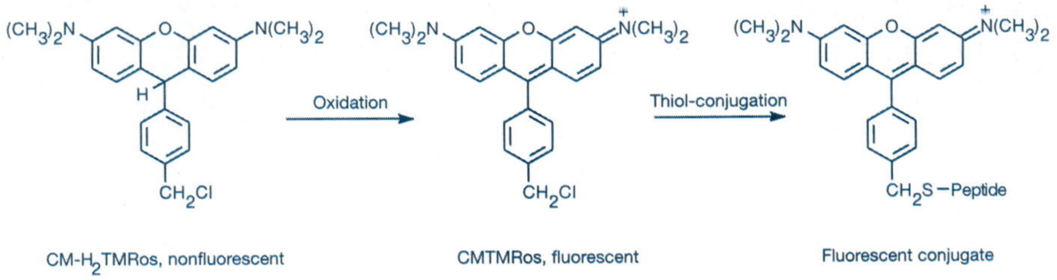

Figure 12.2.10 Intracellular reactions of our fixable, mitochondrion-selective MitoTracker® Orange CM-H_2TMRos (M7511). When this cell-permeant probe enters an actively respiring cell, it is oxidized to MitoTracker® Orange CMTMRos and sequestered in the mitochondria, where it can react with thiols on proteins and peptides to form aldehyde-fixable conjugates.

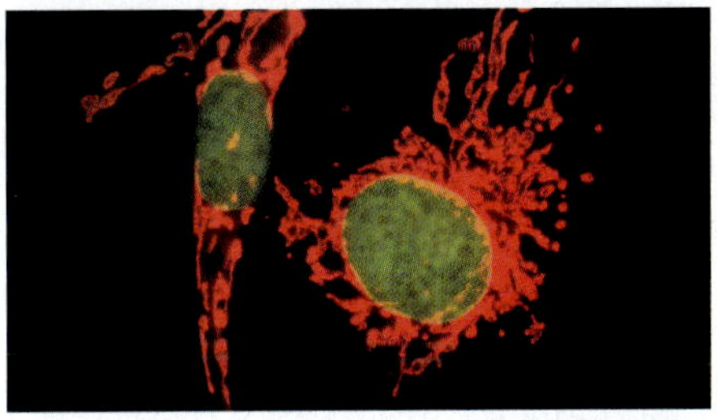

Figure 12.2.12 The mitochondria of bovine pulmonary artery endothelial cells stained with MitoTracker® Red CM-H_2XRos (M7513). The cells were subsequently fixed, permeabilized and treated with RNase. Then the nuclei were stained with SYTOX® Green nucleic acid stain (S7020). The multiple-exposure photomicrograph was acquired using a fluorescence microscope equipped with bandpass filter sets appropriate for fluorescein and Texas Red® dyes.

IMPORTANT NOTICE: The products described in this manual are covered by one or more Limited Use Label License(s). Please refer to the Appendix on page 971 and Master Product List on page 975. Products are For Research Use Only. Not intended for any animal or human therapeutic or diagnostic use.

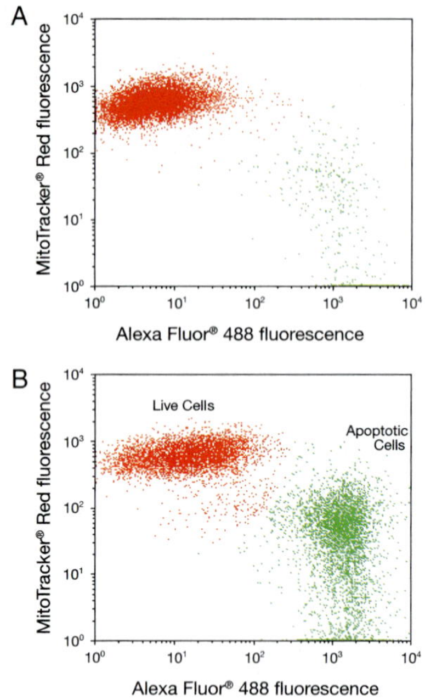

Figure 12.2.13 Flow cytometric analysis of Jurkat cells using the Mitochondrial Membrane Potential/Annexin V Apoptosis Kit (V35116). Jurkat human T-cell leukemia cells in complete medium were **B**) first exposed to 10 µM camptothecin for 4 hours or **A**) left untreated. Both cell populations were then treated with the reagents in the Mitochondrial Membrane Potential/Annexin V Apoptosis Kit and analyzed by flow cytometry. Note that the apoptotic cells show higher reactivity for annexin V and lower MitoTracker® Red dye fluorescence than do live cells.

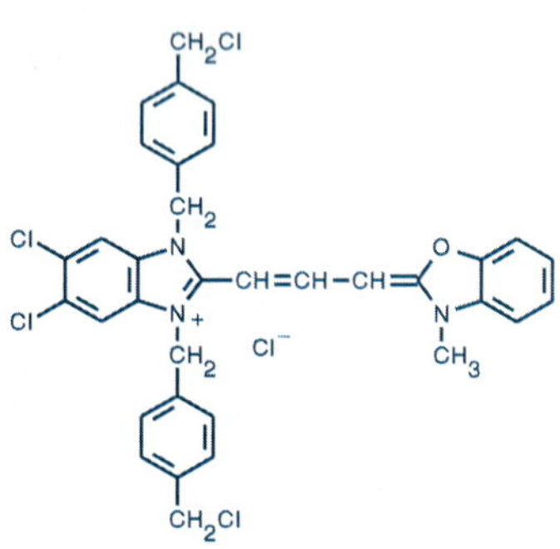

Figure 12.2.14 MitoTracker® Green FM® (M7514).

Our Mitochondrial Membrane Potential/Annexin V Apoptosis Kit (V35116, Section 15.5) utilizes MitoTracker® CMXRos in combination with Alexa Fluor® 488 annexin V in a two-color assay of apoptotic cells (Figure 12.2.13). Following fixation, the oxidized forms of the tetramethylrosamine and X-rosamine MitoTracker® dyes can be detected directly by fluorescence or indirectly with either anti-tetramethylrhodamine or anti–Texas Red® dye antibodies (A6397, A6399; Section 7.4).

MitoTracker® Green FM® Dye

Mitochondria in cells stained with nanomolar concentrations of MitoTracker® Green FM® dye (M7514, Figure 12.2.14) exhibit bright green, fluorescein-like fluorescence (Figure 12.2.15, Figure 12.2.16, Figure 12.2.17). The MitoTracker® Green FM® probe has the added advantage that it is essentially nonfluorescent in aqueous solutions and only becomes fluorescent once it accumulates in the lipid environment of mitochondria. Hence, background fluorescence is negligible, enabling researchers to clearly visualize mitochondria in live cells immediately following addition of the stain, without a wash step.

Unlike MitoTracker® Orange CMTMRos and MitoTracker® Red CMXRos, the MitoTracker® Green FM® probe appears to preferentially accumulate in mitochondria regardless of mitochondrial membrane potential in certain cell types, making it a possible tool for determining mitochondrial mass.[20] Furthermore, the MitoTracker® Green FM® dye is substantially more photostable than the widely used rhodamine 123 fluorescent dye and produces a brighter, more mitochondrion-selective signal at lower concentrations. Because its emission maximum is blue-shifted approximately 10 nm relative to the emission maximum of rhodamine 123, the MitoTracker® Green FM® dye produces a fluorescent staining pattern that should be better resolved from that of red-fluorescent probes in double-labeling experiments. The mitochondrial proteins that are selectively labeled by the MitoTracker® Green FM® reagent have been separated by capillary electrophoresis.[16]

Image-iT® LIVE Mitochondrial and Nuclear Labeling Kit

The Image-iT® LIVE Mitochondrial and Nuclear Labeling Kit (I34154) provides two stains—red-fluorescent MitoTracker® Red CMXRos dye (excitation/emission maxima ~578/599 nm) and blue-fluorescent Hoechst 33342 dye (excitation/emission maxima when bound to DNA ~350/461 nm)—for highly selective mitochondrial and nuclear staining, respectively, in live, Green Fluorescent Protein (GFP)–transfected cells. These dyes can be combined into one staining solution using the protocol provided, saving labeling time and wash steps while still providing optimal staining. Both dyes are retained after formaldehyde fixation and permeabilization. The Image-iT® LIVE Mitochondrial and Nuclear Labeling Kit contains:

- MitoTracker® CMXRos dye
- Hoechst 33342 dye
- Dimethylsulfoxide (DMSO)
- Labeling protocols

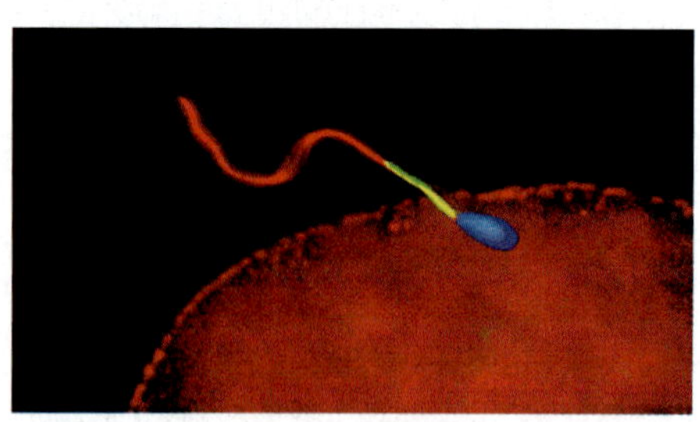

Figure 12.2.15 Bull sperm prelabeled with MitoTracker® Green FM® (M7514) and used for *in vitro* fertilization of bovine oocytes. After fertilization, eggs with bound or incorporated sperm were fixed in 2% formaldehyde, made permeable with Triton X-100 and labeled with an anti-tubulin antibody followed by a tetramethylrhodamine-labeled secondary antibody and counterstained with DAPI (D1306, D3571, D21490). Image contributed by Peter Sutovsky, University of Wisconsin.

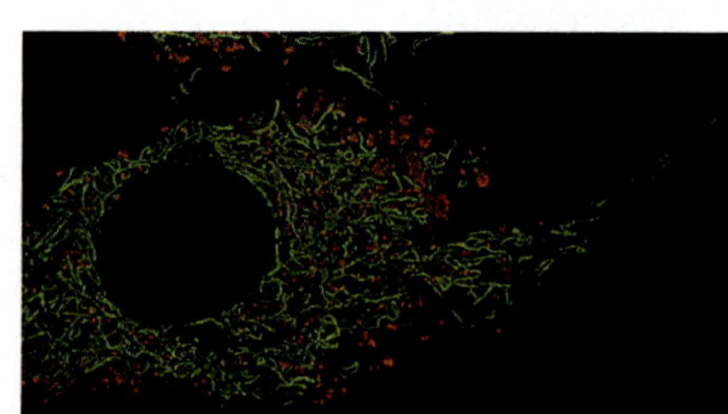

Figure 12.2.16 Bovine pulmonary artery endothelial cells (BPAEC) incubated simultaneously with 50 nM LysoTracker® Red DND-99 (L7528) and 75 nM MitoTracker® Green FM® (M7514) at 37°C for 30 minutes. Both dyes showed excellent cellular retention, even after cells were fixed in 3% glutaraldehyde for 30 minutes. The image was deconvolved using Huygens software (Scientific Volume Imaging, http://www.svi.nl/).

IMPORTANT NOTICE: The products described in this manual are covered by one or more Limited Use Label License(s). Please refer to the Appendix on page 971 and Master Product List on page 975. Products are For Research Use Only. Not intended for any animal or human therapeutic or diagnostic use.
www.invitrogen.com/probes
molecular probes | invitrogen by life technologies™

Each kit provides enough staining solution for 500 assays using the protocol provided for labeling live, cultured cells that are adhering to coverslips.

Fluorescent Protein–Based Markers for Mitochondria

CellLight® Mitochondria-GFP (C10600, Figure 12.2.18) and CellLight® Mitochondria-RFP (C10601) are BacMam expression vectors encoding GFP or RFP [21] fused to the leader sequence of E1α pyruvate dehydrogenase. These CellLight® reagents (Table 11.1) incorporate all the customary advantages of BacMam delivery technology including high transduction efficiency and long-lasting and titratable expression (BacMam Gene Delivery and Expression Technology—Note 11.1). In contrast to MitoTracker® Red CMXRos, TMRE, rhodamine 123 and other cationic dyes, mitochondrial localization of fluorescent protein–based markers is not driven by membrane potential.[22] They can therefore be used in combination with cationic dye probes to investigate relationships between mitochondrial morphology and membrane potential.

MitoSOX™ Red Mitochondrial Superoxide Indicator

Mitochondrial superoxide is generated as a by-product of oxidative phosphorylation. In an otherwise tightly coupled electron transport chain, approximately 1–3% of mitochondrial oxygen consumed is incompletely reduced; these "leaky" electrons can quickly interact with molecular oxygen to form superoxide anion, the predominant reactive oxygen species in mitochondria.[23,24] Increases in cellular superoxide production have been implicated in cardiovascular diseases, including hypertension, atherosclerosis and diabetes-associated vascular injuries,[25] as well as in neurodegenerative diseases such as Parkinson disease, Alzheimer disease and amyotrophic lateral sclerosis (ALS).[24]

MitoSOX™ Red mitochondrial superoxide indicator (M36008) is a cationic derivative of dihydroethidum (also known as hydroethidine; see below) designed for highly selective detection of superoxide in the mitochondria of live cells (Figure 12.2.19). The cationic triphenylphosphonium substituent of MitoSOX™ Red indicator is responsible for the electrophoretically driven uptake of the probe in actively respiring mitochondria. Oxidation of MitoSOX™ Red indicator (or dihydroethidium) by superoxide results in hydroxylation at the 2-position (Figure 12.2.20). 2-Hydroxyethidium (and the corresponding derivative of MitoSOX™ Red indicator) exhibit a fluorescence excitation

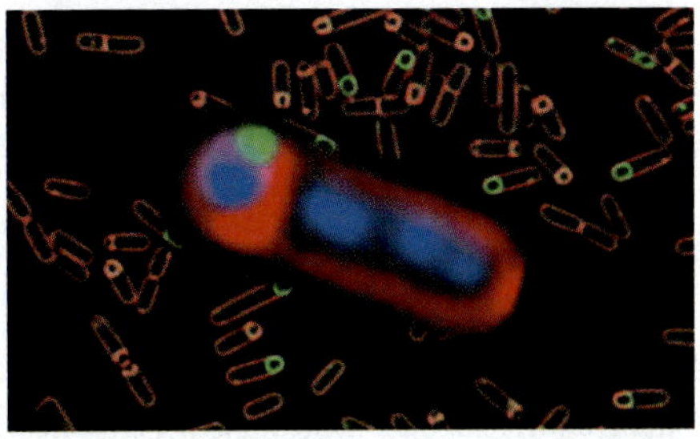

Figure 12.2.17 The morphology of sporulating *Bacillus subtilis* in the early stages of forespore engulfment. The membranes and chromosomes of both the forespore and the larger mother cell are stained with FM® 4-64 (red; T3166, T13320) and DAPI (blue, D1306, D3571, D21490), respectively. The small green-fluorescent patch indicates the localization of a GFP fusion to SPoIIIE, a protein essential for translocation of the forespore chromosome that may also regulate membrane fusion events (Proc Natl Acad Sci U S A (1999) 96:14553). The background contains sporangia at various stages in the engulfment process stained with MitoTracker® Green FM® (green, M7514) and FM® 4-64 (red).

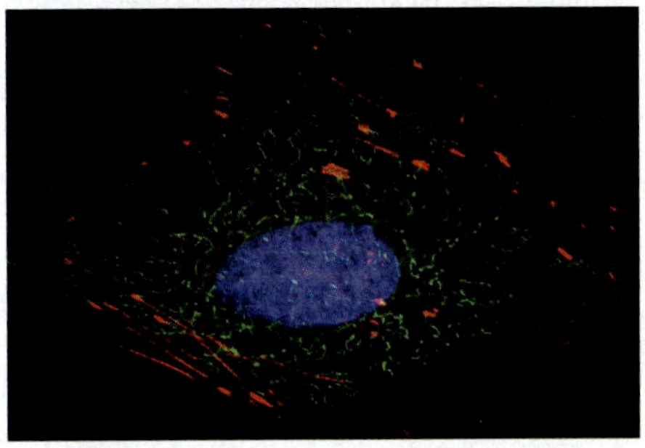

Figure 12.2.18 HeLa cell labeled with CellLight® Mitochondria-GFP (C10600) and CellLight® Talin-RFP (C10612) reagents and with Hoechst 33342 nucleic acid stain.

Figure 12.2.19 Detection of superoxide in live cells using MitoSOX™ Red superoxide indicator (M36008). Live 3T3 cells were treated with FeTCPP, a superoxide scavenger, (right) or left untreated (left). Cells were then labeled with MitoSOX™ Red reagent, which fluoresces when oxidized by superoxide, and nuclei were stained with blue-fluorescent Hoechst 33342. The mitochondria of untreated cells exhibited red fluorescence, indicating the presence of superoxide, whereas the mitochondria of treated cells showed minimal fluorescence.

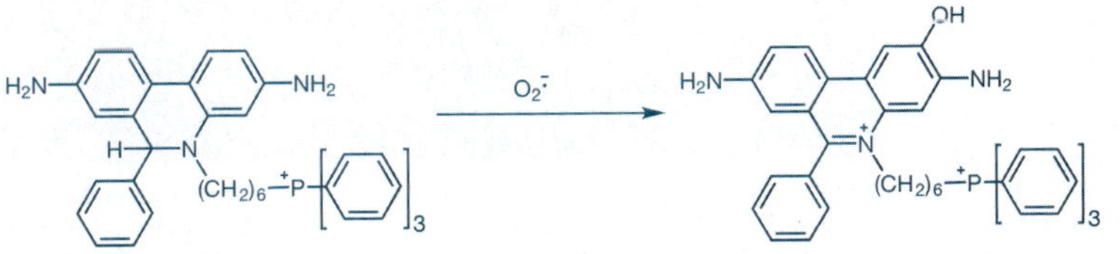

Figure 12.2.20 Oxidation of MitoSOX™ Red mitochondrial superoxide indicator to 2-hydroxy-5-(triphenylphosphonium)hexylethidium by superoxide ($\cdot O_2^-$).

peak at ~400 nm[26] that is absent in the excitation spectrum of the ethidium oxidation product generated by reactive oxygen species other than superoxide. Thus, fluorescence excitation at 400 nm with emission detection at ~590 nm provides optimum discrimination of superoxide from other reactive oxygen species[26–28] (Figure 12.2.21).

Measurements of mitochondrial superoxide generation using MitoSOX™ Red indicator in mouse cortical neurons expressing caspase-cleaved tau microtubule-associated protein have been correlated with readouts from fluorescent indicators of cytosolic and mitochondrial calcium and mitochondrial membrane potential.[29] The relationship of mitochondrial superoxide generation to dopamine transporter activity, measured using the aminostyryl dye substrate 4-Di-1-ASP (D288, see below), has been investigated in mouse brain astrocytes.[30] MitoSOX™ Red indicator has been used for confocal microscopy analysis of reactive oxygen species (ROS) production by mitochondrial NO synthase (mtNOS) in permeabilized cat ventricular myocytes[31] and, in combination with Amplex® Red reagent, for measurement of mitochondrial superoxide and hydrogen peroxide production in rat vascular endothelial cells.[32] In addition to imaging and microscope photometry measurements, several flow cytometry applications of MitoSOX™ Red have also been reported. Detailed protocols for simultaneous measurements of mitochondrial superoxide generation and apoptotic markers APC annexin V (A35110, Section 15.5) and SYTOX® Green (S7020, Section 8.1) in human coronary artery endothelial cells by flow cytometry have been published by Mukhopadhyay and co-workers.[33]

RedoxSensor™ Red CC-1 Stain

RedoxSensor™ Red CC-1 stain (2,3,4,5,6-pentafluorotetramethyldihydrorosamine, R14060; Figure 12.2.22) passively enters live cells and is subsequently oxidized in the cytosol to a red-fluorescent product (excitation/emission maxima ~540/600 nm), which then accumulates in the mitochondria. Alternatively, this nonfluorescent probe may be transported to the lysosomes where it is oxidized.

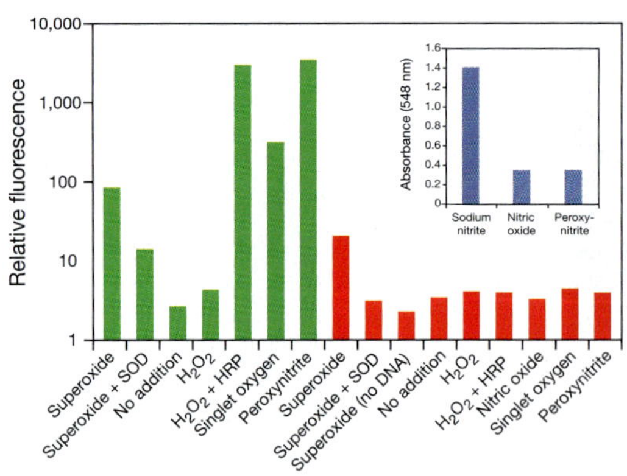

Figure 12.2.21 Selectivity of the MitoSOX™ Red mitochondrial superoxide indicator (M36008). Cell-free systems were used to generate a variety of reactive oxygen species (ROS) and reactive nitrogen species (RNS); each oxidant was then added to a separate 10 µM solution of MitoSOX™ Red reagent and incubated at 37°C for 10 minutes. Excess DNA was added (unless otherwise noted) and the samples were incubated for an additional 15 minutes at 37°C before fluorescence was measured. The Griess Reagent Kit (G7921) (for nitric oxide, peroxynitrite, and nitrite standards only; blue bars) and dihydrorhodamine 123 (DHR 123, D632); green bars) were employed as positive controls for oxidant generation. Superoxide dismutase (SOD), a superoxide scavenger, was used as a negative control for superoxide. The results show that the MitoSOX™ Red probe (red bars) is readily oxidized by superoxide but not by the other oxidants.

The differential distribution of the oxidized product between mitochondria and lysosomes appears to depend on the redox potential of the cytosol.[34–36] In proliferating cells, mitochondrial staining predominates; whereas in contact-inhibited cells, the staining is primarily lysosomal (Figure 12.2.23).

JC-1 and JC-9: Dual-Emission Potential-Sensitive Probes

The green-fluorescent JC-1 probe (5,5',6,6'-tetrachloro-1,1',3,3'-tetraethylbenzimidazolylcarbocyanine iodide, T3168; Figure 12.2.24) exists as a monomer at low concentrations or at low membrane potential. However, at higher concentrations (aqueous solutions above 0.1 µM) or higher potentials, JC-1 forms red-fluorescent "J-aggregates" that exhibit a broad excitation spectrum and an emission maximum at ~590 nm (Figure 12.2.25, Figure 12.2.26, Figure 12.2.27). Thus, the emission of this cyanine dye can be used as a sensitive measure of mitochondrial membrane potential. Various types of ratio measurements are possible by combining signals from the green-fluorescent JC-1 monomer (absorption/emission maxima ~514/529 nm in water) and the J-aggregate (emission maximum 590 nm), which can be effectively excited anywhere between 485 nm and its absorption maximum at 585 nm (Figure 12.2.28). The ratio of red-to-green JC-1 fluorescence is dependent only on the membrane potential and not on other factors that may influence single-component fluorescence signals, such as mitochondrial size, shape and density. Optical filters designed for fluorescein and

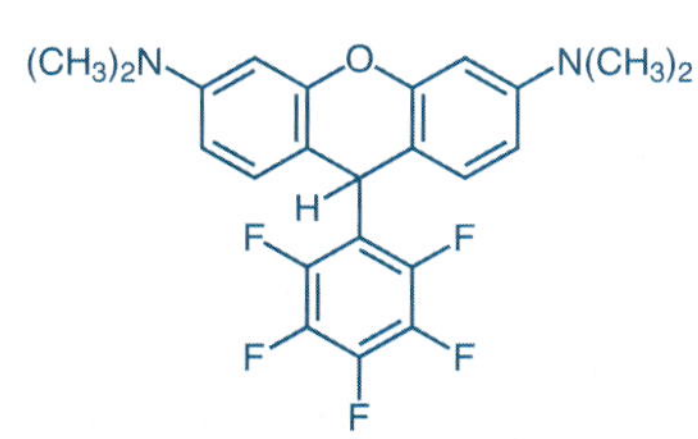

Figure 12.2.22 RedoxSensor™ Red CC-1 (R14060).

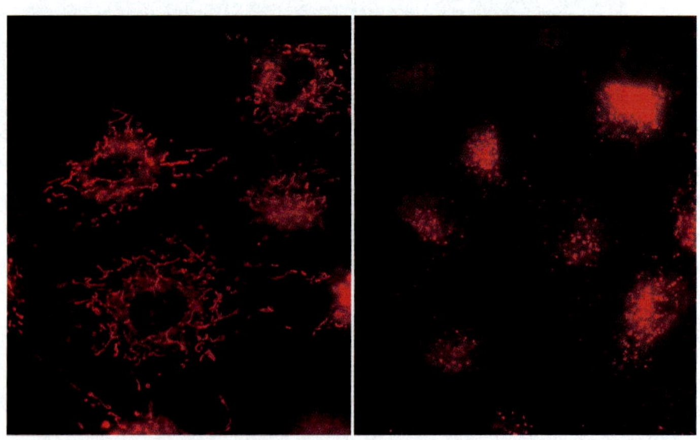

Figure 12.2.23 Cellular proliferation state determines the distribution of the oxidized product of RedoxSensor™ Red CC-1 (R14060). Normal rat kidney (NRK) cells in different growth states were stained with RedoxSensor™ Red CC-1. In proliferating cells (left), the oxidized dye accumulates in mitochondria. In quiescent cells (right), the oxidized product localizes in the lysosomes.

IMPORTANT NOTICE: The products described in this manual are covered by one or more Limited Use Label License(s). Please refer to the Appendix on page 971 and Master Product List on page 975. Products are For Research Use Only. Not intended for any animal or human therapeutic or diagnostic use.
www.invitrogen.com/probes
molecular probes® | invitrogen™ by life technologies™

tetramethylrhodamine can be used to separately visualize the monomer and J-aggregate forms, respectively. Alternatively, both forms can be observed simultaneously using a standard fluorescein longpass optical filter set. Chen and colleagues have used JC-1 to investigate mitochondrial potentials in live cells by ratiometric techniques[37–39] (Figure 12.2.29).

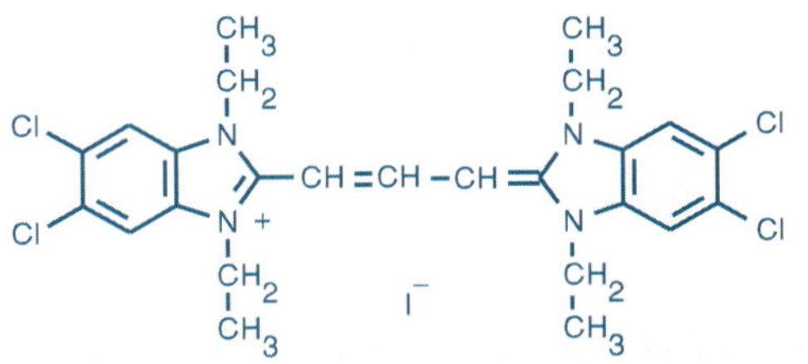

Figure 12.2.24 5,5′,6,6′-Tetrachloro-1,1′,3,3′-tetraethylbenzimidazolylcarbocyanine iodide (JC-1, T3168).

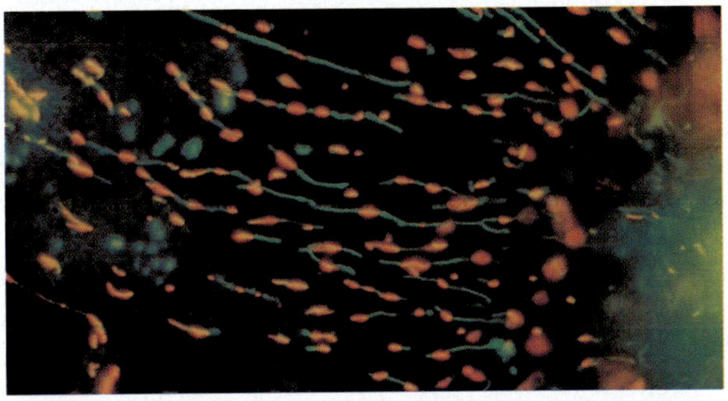

Figure 12.2.25 Potential-dependent staining of mitochondria in CCL64 fibroblasts by JC-1 (T3168). The mitochondria were visualized by epifluorescence microscopy using a 520 nm longpass optical filter. Regions of high mitochondrial polarization are indicated by red fluorescence due to J-aggregate formation by the concentrated dye. Depolarized regions are indicated by the green fluorescence of the JC-1 monomers. The image was contributed by Lan Bo Chen, Dana Farber Cancer Institute, Harvard Medical School.

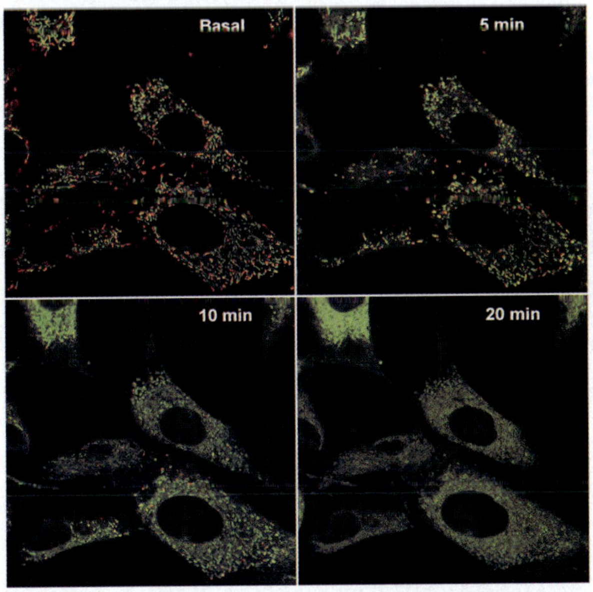

Figure 12.2.26 NIH 3T3 fibroblasts stained with JC-1 (T3168), showing the progressive loss of red J-aggregate fluorescence and cytoplasmic diffusion of green monomer fluorescence following exposure to hydrogen peroxide. Images show the same field of cells viewed before H_2O_2 treatment and 5, 10 and 20 minutes after treatment. The images were contributed by Ildo Nicoletti, Perugia University Medical School.

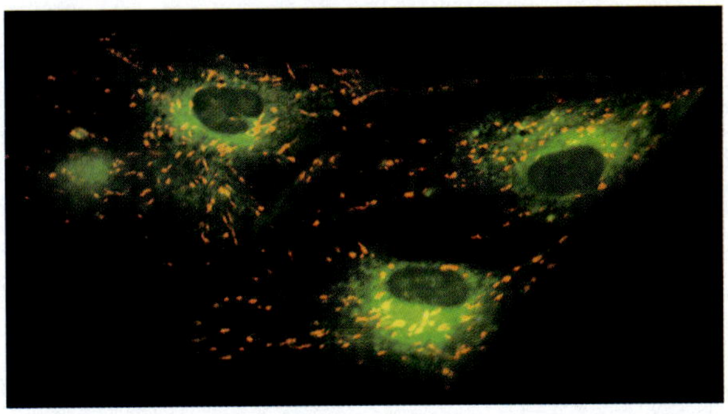

Figure 12.2.27 Cultured human pre-adipocytes loaded with the ratiometric mitochondrial potential indicator JC-1 (T3168) at 5 µM for 30 minutes at 37°C. In live cells, JC-1 exists either as a green-fluorescent monomer at depolarized membrane potentials or as an orange-fluorescent J-aggregate at hyperpolarized membrane potentials. Cells were then treated with 50 nM FCCP, a protonophore, to depolarize the mitochondrial membrane. Approximately 10 minutes after the addition of the uncoupler, the cells were illuminated at 488 nm and the emission was collected between 515–545 nm and 575–625 nm. Image contributed by Bob Terry, BioImage A/S, Denmark.

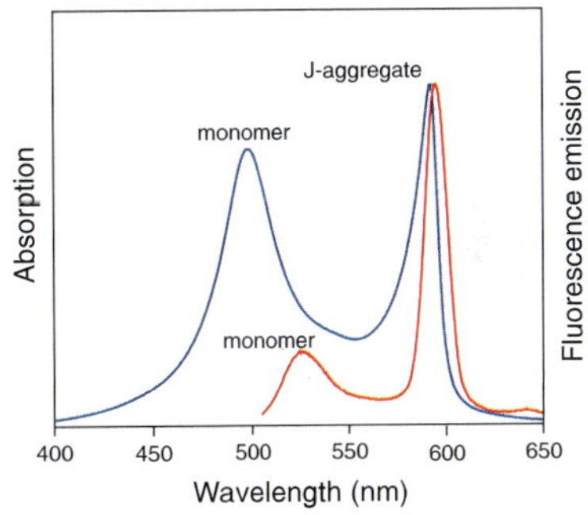

Figure 12.2.28 Absorption and fluorescence emission (excited at 488 nm) spectra of JC-1 in pH 8.2 buffer containing 1% (v/v) DMSO.

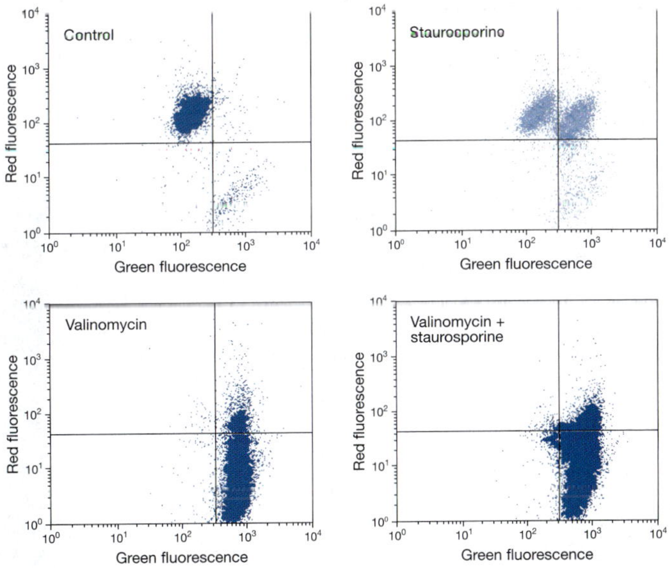

Figure 12.2.29 Bivariate JC-1 (T3168) analysis of mitochondrial membrane potential in HL60 cells by flow cytometry. The sensitivity of this technique is demonstrated by the response to depolarization using K+/valinomycin (V1644) (bottom two panels). Distinct populations of cells with different extents of mitochondrial depolarization are detectable following apoptosis-inducing treatment with 5 µM staurosporine for 2 hours (top right panel). Figure courtesy of Andrea Cossarizza, University of Modena and Reggio Emilia, Italy.

IMPORTANT NOTICE: The products described in this manual are covered by one or more Limited Use Label License(s). Please refer to the Appendix on page 971 and Master Product List on page 975. Products are For Research Use Only. Not intended for any animal or human therapeutic or diagnostic use.

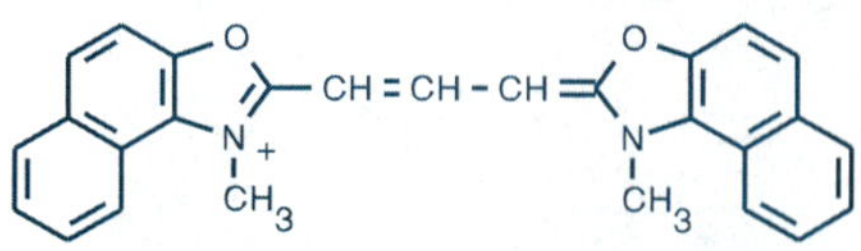

Figure 12.2.31 3,3′-Dimethyl-α-naphthoxacarbocyanine iodide (JC-9; $DiNOC_1(3)$, D22421).

JC-1 has been combined with Alexa Fluor® 647 annexin V (A23204, Section 15.5) to permit simultaneous assessment of phosphatidylserine externalization and mitochondrial function by flow cytometry.[40] We also offer JC-1 as part of the MitoProbe™ JC-1 Assay Kit for flow cytometry (M34152, Section 22.3). We have discovered another mitochondrial marker, JC-9 (3,3'-dimethyl-β-naphthoxazolium iodide, D22421; Figure 12.2.30), with a different chemical structure (Figure 12.2.31) but similar potential-dependent spectroscopic properties. However, the green fluorescence of JC-9 is essentially invariant with membrane potential, whereas the red fluorescence is significantly increased at hyperpolarized membrane potentials.

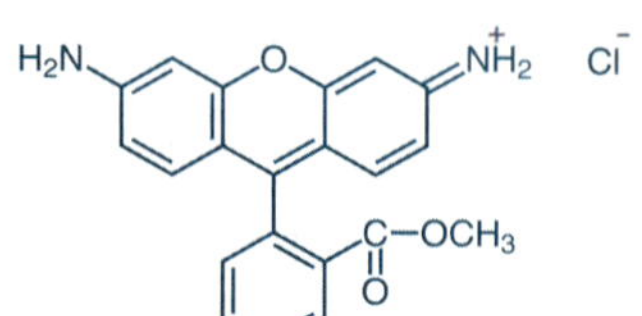

Figure 12.2.32 Rhodamine 123 (R302).

Mitochondrion-Selective Rhodamines and Rosamines

Rhodamine 123

Rhodamine 123 (R302, R22420; Figure 12.2.32) is a cell-permeant, cationic, fluorescent dye that is readily sequestered by active mitochondria without inducing cytotoxic effects.[41] Uptake and equilibration of rhodamine 123 is rapid (a few minutes) compared with dyes such as DASPMI (4-Di-1-ASP, D288), which may take 30 minutes or longer.[42] Viewed through a fluorescein longpass optical filter, fluorescence of the mitochondria of cells stained by rhodamine 123 appears yellow-green. Viewed through a tetramethylrhodamine longpass optical filter, however, these same mitochondria appear red. Unlike the lipophilic rhodamine and carbocyanine dyes, rhodamine 123 apparently does not stain the endoplasmic reticulum.

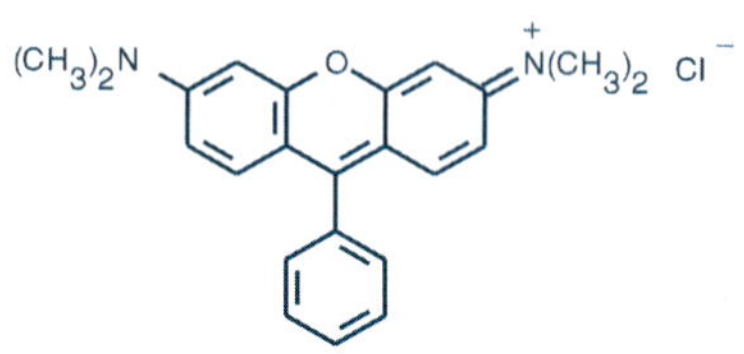

Figure 12.2.33 Tetramethylrosamine chloride (T639).

Rhodamine 123 has been used with a variety of cell types such as astrocytes, neurons,[43,44] live bacteria,[45] plants[46,47] and human spermatozoa.[48] Using flow cytometry, researchers employed rhodamine 123 in combination with Hoechst 33342 (H1399, H3570, H21492; Section 12.5) for the characterization of hematopoietic stem cells.[49] Rhodamine 123 is widely used as a substrate for functional assays of ATP-binding cassette (ABC) drug transporters[50] (Section 15.6).

Rosamines and Other Rhodamine Derivatives, Including TMRM and TMRE

Other mitochondrion-selective dyes include tetramethylrosamine (T639, Figure 12.2.33), whose fluorescence contrasts well with that of fluorescein for multicolor applications, and rhodamine 6G[51–54] (R634, Figure 12.2.34), which has an absorption maximum between that of rhodamine 123 and tetramethylrosamine. Tetramethylrosamine and rhodamine 6G have both been used to examine the efficiency of P-glycoprotein–mediated exclusion from multidrug-resistant cells[55] (Section 15.6). Rhodamine 6G has been employed to study microvascular reperfusion injury[56] and the stimulation and inhibition of F_1-ATPase from the thermophilic bacterium PS3.[57]

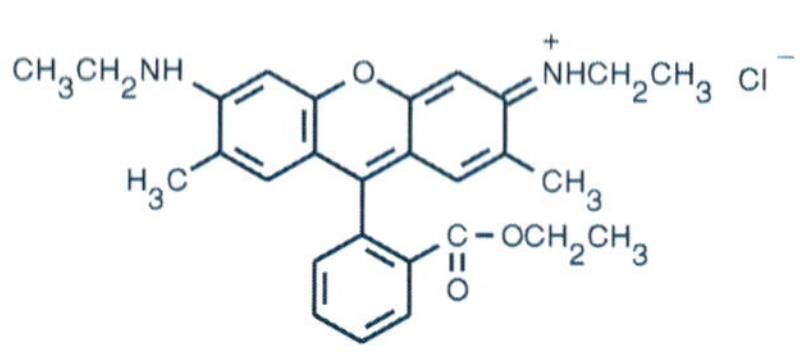

Figure 12.2.34 Rhodamine 6G chloride (R634).

At low concentrations, certain lipophilic rhodamine dyes selectively stain mitochondria in live cells.[58] We have observed that low concentrations of the hexyl ester of rhodamine B (R648MP) accumulate selectively in mitochondria (Figure 12.2.35) and appear to be relatively

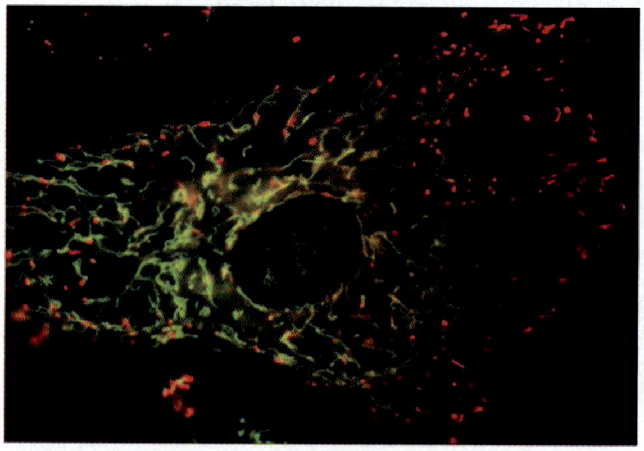

Figure 12.2.30 A viable bovine pulmonary artery endothelial cell incubated with the ratiometric mitochondrial potential indicator, JC-9 (D22421). In live cells, JC-9 exists either as a green-fluorescent monomer at depolarized membrane potentials, or as a red-fluorescent J-aggregate at hyperpolarized membrane potentials.

IMPORTANT NOTICE: The products described in this manual are covered by one or more Limited Use Label License(s). Please refer to the Appendix on page 971 and Master Product List on page 975. Products are For Research Use Only. Not intended for any animal or human therapeutic or diagnostic use.

nontoxic. We have included this probe in our Yeast Mitochondrial Stain Sampler Kit (Y7530, see below for description). At higher concentrations, rhodamine B hexyl ester and rhodamine 6G stain the endoplasmic reticulum of animal cells[58] (Section 12.4).

The accumulation of tetramethylrhodamine methyl and ethyl esters (TMRM, T668; TMRE, T669) in mitochondria and the endoplasmic reticulum has also been shown to be driven by their membrane potential[59,60] (Section 22.3). Moreover, because of their reduced hydrophobic character, these probes exhibit potential-independent binding to cells that is 10 to 20 times lower than that seen with rhodamine 6G.[61] Tetramethylrhodamine ethyl ester has been described as one of the best fluorescent dyes for dynamic and *in situ* quantitative measurements—better than rhodamine 123—because it is rapidly and reversibly taken up by live cells.[62–64] TMRM and TMRE have been used to measure mitochondrial depolarization related to cytosolic Ca^{2+} transients[65] and to image time-dependent mitochondrial membrane potentials.[63] A high-throughput assay utilizes TMRE and our low-affinity Ca^{2+} indicator fluo-5N AM (F14204, Section 19.3) to screen inhibitors of the opening of the mitochondrial transition pore.[66] Researchers have also taken advantage of the red shift exhibited by TMRM, TMRE and rhodamine 123 upon membrane potential–driven mitochondrial uptake to develop a ratiometric method for quantitating membrane potential.[67]

Reduced Rhodamines and Rosamines

Inside live cells, the colorless dihydrorhodamines and dihydrotetramethylrosamine are oxidized to fluorescent products that stain mitochondria.[68] However, the oxidation may occur in organelles other than the mitochondria. Dihydrorhodamine 123 (D632, D23806; Figure 12.2.36) reacts with hydrogen peroxide in the presence of peroxidases,[69] iron or cytochrome *c*[70] to form rhodamine 123. This reduced rhodamine has been used to monitor reactive oxygen intermediates in rat mast cells[71] and to measure hydrogen peroxide in endothelial cells.[70] Dihydrorhodamine 6G (D633, Figure 12.2.37) is another reduced rhodamine that has been shown to be taken up and oxidized by live cells.[72–74] Chloromethyl derivatives of reduced rosamines (MitoTracker® Orange CM-H_2TMRos, M7511; MitoTracker® Red CM-H_2XRos, M7513), which can be fixed in cells by aldehyde-based fixatives, have been described above.

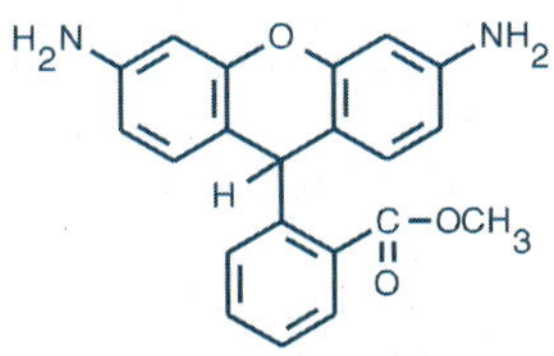

Figure 12.2.36 Dihydrorhodamine 123 (D632).

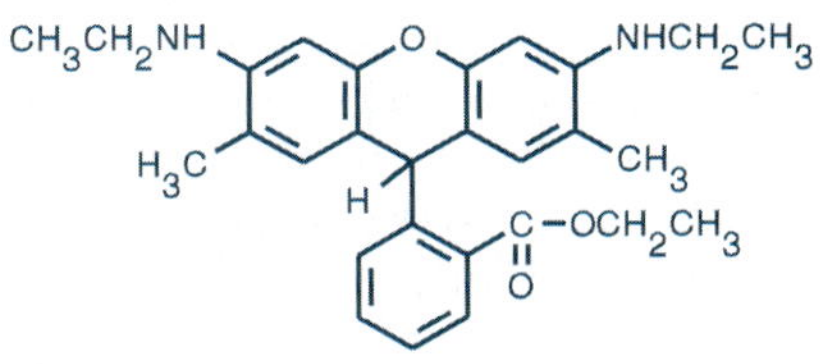

Figure 12.2.37 Dihydrorhodamine 6G (D633).

Other Mitochondrion-Selective Probes

Carbocyanines

Most carbocyanine dyes with short (C_1–C_6) alkyl chains stain mitochondria of live cells when used at low concentrations (<100 nM); those with pentyl or hexyl substituents also stain the endoplasmic reticulum when used at higher concentrations (>1 µM). DiOC_6(3) (D273) stains mitochondria in live yeast[75–78] and other eukaryotic cells,[54,79] as well as sarcoplasmic reticulum in beating heart cells.[80] It has also been used to demonstrate mitochondria moving along microtubules.[81] Photolysis of mitochondrion- or endoplasmic reticulum–bound

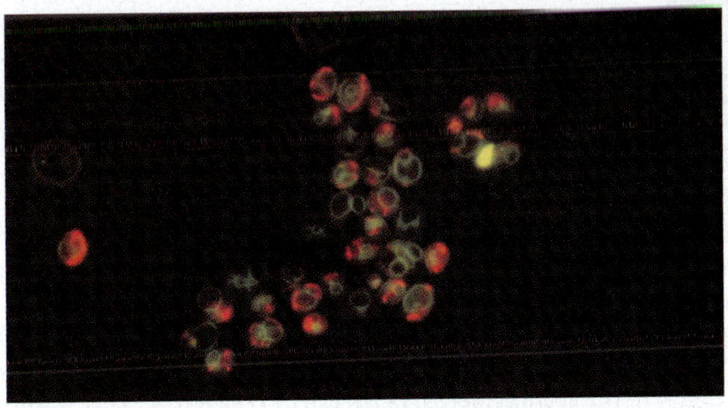

Figure 12.2.35 *Saccharomyces cerevisiae* stained sequentially with the red-fluorescent rhodamine B hexyl ester (R648MP), which selectively labels yeast mitochondria under these conditions, and the green-fluorescent yeast vacuole membrane marker MDY-64 (Y7536). These probes are also provided in our Yeast Mitochondrial Stain Sampler Kit (Y7530) and Yeast Vacuole Marker Sampler Kit (Y7531), respectively. Stained yeast were photographed in a single exposure through an Omega® Optical triple-bandpass filter set.

$DiOC_6(3)$ specifically destroys the microtubules of cells without affecting actin stress fibers, producing a highly localized inhibition of intracellular organelle motility.[82] We have included $DiIC_1(5)$ and $DiOC_2(3)$ in two of our MitoProbe™ Assay Kits for flow cytometry (M34151, M34150; Section 22.3). Several other potential-sensitive carbocyanine probes described in Section 22.3 also stain mitochondria in live cultured cells.[54] The carbocyanine $DiOC_7(3)$ (D378), which exhibits spectra similar to those of fluorescein, is a versatile dye that has been reported to be a sensitive probe for mitochondria in plant cells.[83]

Styryl Dyes

The styryl dyes DASPMI (4-Di-1-ASP, D288) and DASPEI (D426) can be used to stain mitochondria in live cells and tissues.[84,85] These dyes have large fluorescence Stokes shifts and are taken up relatively slowly as a function of membrane potential. The kinetics of mitochondrial staining with styrylpyridinium dyes has been investigated using the concentration jump method.[86]

Nonyl Acridine Orange

Nonyl acridine orange (A1372) is well retained in the mitochondria of live HeLa cells for up to 10 days, making it a useful probe for following mitochondria during isolation and after cell fusion.[87–89] The mitochondrial uptake of this metachromatic dye is reported not to depend on membrane potential. It is toxic at high concentrations[90] and apparently binds to cardiolipin in all mitochondria, regardless of their energetic state.[91–94] This derivative has been used to analyze mitochondria by flow cytometry,[95] to characterize multidrug resistance[96] (Section 15.6) and to measure changes in mitochondrial mass during apoptosis in rat thymocytes.[97]

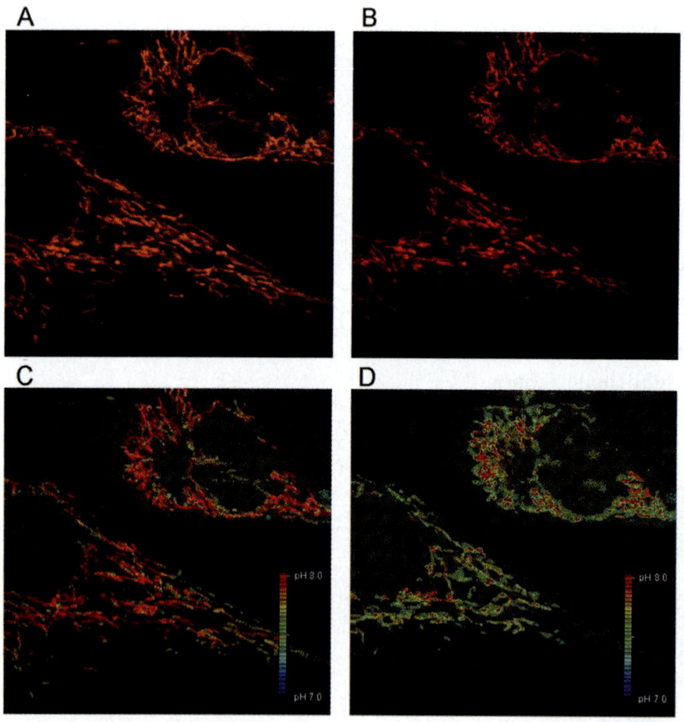

Figure 12.2.38 Selective loading of carboxy SNARF®-1 into mitochondria. BHK cells were loaded with 10 µM carboxy SNARF®-1, AM, acetate (C1271, C1272) for 10 minutes, followed by incubation for 4 hours at room temperature. **A)** Confocal image (488 nm excitation) of mitochondrial-selective loading of carboxy SNARF®-1 visualized through a 560–600 nm bandpass filter. **B)** Confocal image of the same cells as in **A**, but using a 605 nm dichroic mirror and a 610 nm longpass filter. **C)** Ratio image (**A** and **B**) of mitochondria in cells pseudocolored to represent different pH levels. **D)** Change in mitochondrial pH following the addition of 10 µM carbonyl cyanide m-chlorophenylhydrazone (CCCP), resulting in a decrease (acidification) of mitochondrial pH. Image contributed by Brian Herman, University of Texas Health Science Center, San Antonio, and reprinted with permission from Biotechniques (2001) 30:804.

Carboxy SNARF®-1 pH Indicator

A special cell-loading technique permits ratiometric measurement of intramitochondrial pH with our SNARF® dyes. Cell loading with 10 µM 5-(and 6-)carboxy SNARF®-1, acetoxymethyl ester, acetate (C1271, C1272; Section 20.2), followed by 4 hours of incubation at room temperature leads to highly selective localization of the carboxy SNARF®-1 dye in mitochondria (Figure 12.2.38), where it responds to changes in mitochondrial pH.[98]

Lucigenin

The well-known chemiluminescent probe lucigenin (L6868) accumulates in mitochondria of alveolar macrophages.[99] Relatively high concentrations of the dye (~100 µM) are required to obtain fluorescent staining; however, low concentrations reportedly yield a chemiluminescent response to stimulated superoxide generation within the mitochondria.[99] Molecular Probes® lucigenin has been highly purified to remove a bright blue-fluorescent contaminant that is found in some commercial samples.

Mitochondrial Transition Pore Assays

Image-iT® LIVE Mitochondrial Transition Pore Assay Kit for Fluorescence Microscopy

The mitochondrial permeability transition pore, a nonspecific channel formed by components from the inner and outer mitochondrial membranes, appears to be involved in the release of mitochondrial components during apoptotic and necrotic cell death. In a healthy cell, the inner mitochondrial membrane is responsible for maintaining the electrochemical gradient that is essential for respiration and energy production. As Ca^{2+} is taken up and released by mitochondria, a low-conductance permeability transition pore appears to flicker between open and closed states.[100] During cell death, the opening of the mitochondrial permeability transition pore dramatically alters the permeability of mitochondria. Continuous pore activation results from mitochondrial Ca^{2+} overload, oxidation of mitochondrial glutathione, increased levels of reactive oxygen species in mitochondria and other pro-apoptotic conditions.[101] Cytochrome *c* release from mitochondria and loss of mitochondrial membrane potential are observed subsequent to continuous pore activation.

The Image-iT® LIVE Mitochondrial Transition Pore Assay Kit (I35103), based on published experimentation for mitochondrial transition pore opening,[102,103] permits a more direct method of measuring mitochondrial permeability transition pore opening than assays relying on mitochondrial membrane potential alone. This assay employs the acetoxymethyl (AM) ester of calcein, a colorless and nonfluorescent esterase substrate, and $CoCl_2$, a quencher of calcein fluorescence, to selectively label mitochondria. Cells are loaded with calcein AM, which passively diffuses into the cells and accumulates in cytosolic compartments, including the mitochondria. Once inside cells, calcein AM is cleaved by intracellular esterases to liberate the very polar fluorescent dye calcein, which does not cross the mitochondrial or plasma membranes in appreciable amounts over relatively short periods of time. The fluorescence from cytosolic calcein is quenched by the addition of $CoCl_2$, while the fluorescence from the mitochondrial calcein is maintained. As a control, cells that have been loaded with calcein AM and $CoCl_2$ can also be treated with a Ca^{2+} ionophore such as ionomycin (I24222, Section 19.8) to allow entry of excess Ca^{2+} into the cells, which triggers mitochondrial pore activation and subsequent loss of mitochondrial calcein fluorescence. This ionomycin response can be blocked with cyclosporine A, a compound reported to prevent mitochondrial transition pore formation by binding cyclophilin D.

The Image-iT® LIVE Mitochondrial Transition Pore Assay Kit has been tested with HeLa cells and bovine pulmonary artery endothelial cells (BPAEC). Each Image-iT® LIVE Mitochondrial Transition Pore Assay Kit provides sufficient reagents for 100 assays (based on 1 mL labeling volumes), including:

- Calcein AM
- MitoTracker® Red CMXRos, a red-fluorescent mitochondrial stain (excitation/emission maxima ~579/599 nm)
- Hoechst 33342, a blue-fluorescent nuclear stain (excitation/emission maxima ~350/461 nm)
- Ionomycin
- $CoCl_2$
- Dimethylsulfoxide (DMSO)
- Detailed protocols

MitoProbe™ Transition Pore Assay Kit for Flow Cytometry

The MitoProbe™ Transition Pore Assay Kit (M34153), based on published experimentation for mitochondrial transition pore opening,[102,103] provides a more direct method of measuring mitochondrial permeability transition pore opening than assays relying on mitochondrial membrane potential alone (Figure 12.2.39). As with the Image-iT® LIVE mitochondrial transition pore assay described above, this assay employs the acetoxymethyl (AM) ester of calcein, a colorless and nonfluorescent esterase substrate, and $CoCl_2$, a quencher of calcein fluorescence, to selectively label mitochondria. Cells are loaded with calcein AM, which passively diffuses into the cells and accumulates in cytosolic compartments, including the mitochondria. Once inside cells, calcein AM is cleaved by intracellular esterases to liberate the polar fluorescent dye calcein, which does not cross the mitochondrial or plasma membranes in appreciable amounts over relatively short periods of time. The fluorescence from cytosolic calcein is quenched by the addition of $CoCl_2$, while the fluorescence from the mitochondrial calcein is maintained. As a control, cells that have been loaded with calcein AM and $CoCl_2$ can also be treated with a Ca^{2+} ionophore such as ionomycin (I24222, Section 19.8) to allow entry of excess Ca^{2+} into the cells, which triggers mitochondrial pore activation and subsequent loss of mitochondrial calcein fluorescence. This ionomycin response can be blocked with cyclosporine A, a compound reported to prevent mitochondrial transition pore formation by binding cyclophilin D.

The MitoProbe™ Transition Pore Assay Kit has been tested with Jurkat cells, MH1C1 cells and bovine pulmonary artery endothelial cells (BPAEC). Each MitoProbe™ Transition Pore Assay Kit provides sufficient reagents for 100 assays (based on 1 mL labeling volumes), including:

- Calcein AM
- $CoCl_2$
- Ionomycin
- Dimethylsulfoxide (DMSO)
- Detailed protocols

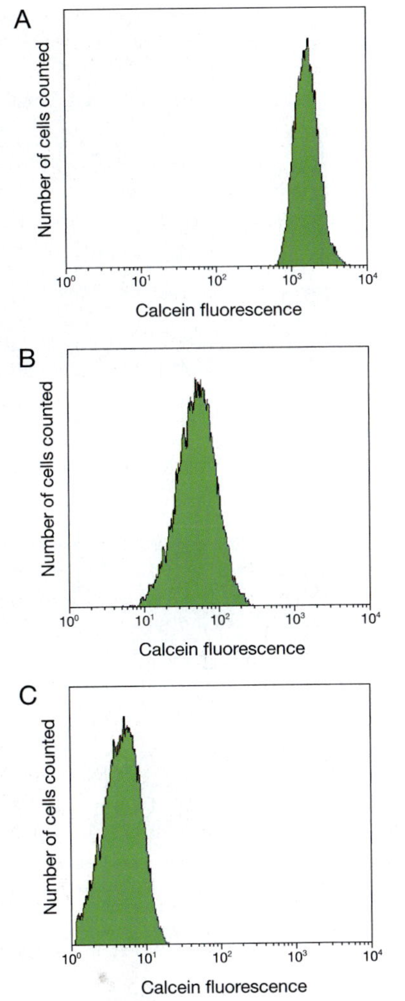

Figure 12.2.39 Flow cytometric analysis of Jurkat cells using the MitoProbe™ Transition Pore Assay Kit (M34153). Jurkat cells were incubated with the reagents in the MitoProbe™ Transition Pore Assay Kit and analyzed by flow cytometry. In the absence of $CoCl_2$ and ionomycin, fluorescent calcein is present in the cytosol as well as the mitochondria, resulting in a bright signal (**A**). In the presence of $CoCl_2$, calcein in the mitochondria emits a signal, but the cytosolic calcein fluorescence is quenched; the overall fluorescence is reduced, as compared with calcein alone (**B**). When ionomycin, a Ca^{2+} ionophore, and $CoCl_2$ are added to the cells at the same time that calcein AM is added, the fluorescent signals from both the cytosol and mitochondria are largely abolished (**C**). The change in fluorescence between panels B and C indicates the continuous activation of mitochondrial permeability transition pores.

Yeast Mitochondrial Stain Sampler Kit

Because fluorescence microscopy has been extensively used to study yeast,[104] we offer a Yeast Mitochondrial Stain Sampler Kit (Y7530). This kit contains sample quantities of five

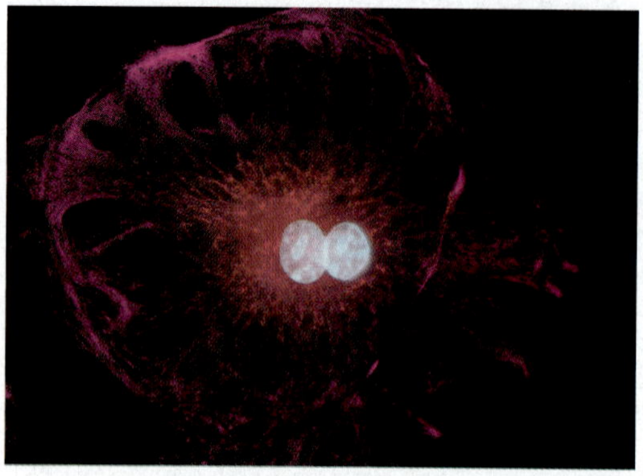

Figure 12.2.40 The intermediate filaments in bovine pulmonary artery endothelial cells, localized using our anti-desmin antibody (A21283), which was visualized with the Alexa Fluor® 647 goat anti–mouse IgG antibody (A21235). Endogenous biotin in the mitochondria was labeled with Alexa Fluor® 546 streptavidin (S11225) and DNA in the cell was stained with blue-fluorescent DAPI (D1306, D3571, D21490).

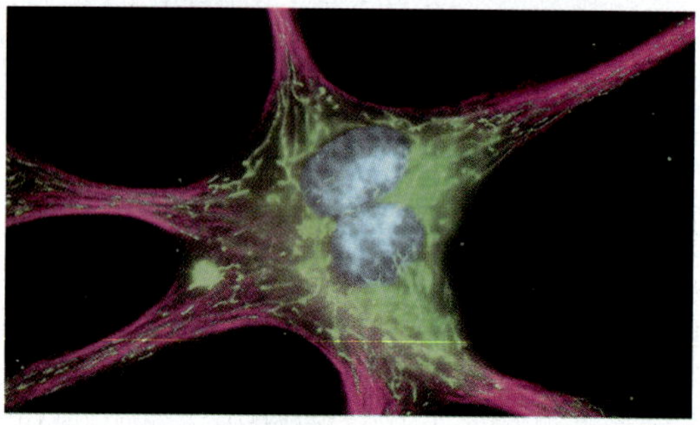

Figure 12.2.41 The cytoskeleton of a fixed and permeabilized bovine pulmonary artery endothelial cell detected using mouse monoclonal anti–α-tubulin antibody (A11126), visualized with Alexa Fluor® 647 goat anti–mouse IgG antibody (A21235) and pseudocolored magenta. Endogenous biotin in the mitochondria was labeled with green-fluorescent Alexa Fluor® 488 streptavidin (S11223) and DNA was stained with blue-fluorescent DAPI (D1306, D3571, D21490).

different probes that have been found to selectively label yeast mitochondria. Both well-characterized and proprietary mitochondrion-selective probes are provided:

- Rhodamine 123 [105–108]
- Rhodamine B hexyl ester [58] (Figure 12.2.35)
- MitoTracker® Green FM®
- SYTO® 18 yeast mitochondrial stain [109]
- $DiOC_6(3)$ [76–78,110–116]

The mitochondrion-selective nucleic acid stain included in this kit—SYTO® 18 yeast mitochondrial stain—exhibits a pronounced fluorescence enhancement upon binding to nucleic acids, resulting in very low background fluorescence even in the presence of dye. SYTO® 18 is an effective mitochondrial stain in live yeast but neither penetrates nor stains the mitochondria of higher eukaryotic cells. Several of the components of the Yeast Mitochondrial Stain Sampler Kit are also available separately.

Avidin Conjugates for Staining Mitochondria

Endogenously biotinylated proteins in mammalian cells, bacteria, yeast and plants—biotin carboxylase enzymes—are present almost exclusively in mitochondria, where biotin synthesis occurs; consequently, mitochondria can be selectively stained by almost any fluorophore- or enzyme-labeled avidin or streptavidin derivative (Section 7.6; Table 7.9; Figure 12.2.40, Figure 12.2.41) without applying any biotinylated ligand.[117,118] This staining, which can complicate the use of avidin–biotin techniques in sensitive cell-based assays, can be blocked by the reagents in our Endogenous Biotin-Blocking Kit (E21390, Section 7.6).

REFERENCES

1. Nat Biotechnol (2008) 26:343; **2.** Eur J Neurosci (2009) 29:114; **3.** Nat Cell Biol (2007) 9:1057; **4.** Apoptosis (2007) 12:803; **5.** FEBS Lett (1984) 170:181; **6.** Arch Biochem Biophys (1990) 282:358; **7.** J Microsc (1983) 132:143; **8.** Cytometry (1991) 12:179; **9.** Brain (2010) 133:797; **10.** Front Biosci (2009) 14:4809; **11.** J Neurosci (2009) 29:9090; **12.** J Neurosci Methods (2009) 178:378; **13.** Am J Physiol Heart Circ Physiol (2009) 297:H13; **14.** Toxicol Sci (2008) 103:335; **15.** J Histochem Cytochem (1996) 44:1363; **16.** J Chromatogr B Analyt Technol Biomed Life Sci (2003) 793:141; **17.** J Histochem Cytochem (2009) 57:687; **18.** Anal Chim Acta (2006) 575:223; **19.** J Immunol Methods (2005) 306:68; **20.** Cytometry A (2004) 61:162; **21.** Nat Methods (2007) 4:555; **22.** Cytometry A (2008) 73:129; **23.** J Biol Chem (2004) 279:4127; **24.** J Cell Mol Med (2002) 6:175; **25.** Am J Physiol Heart Circ Physiol (2003) 284:H605; **26.** Nat Protoc (2008) 3:8; **27.** Nat Protoc (2008) 3:941; **28.** Proc Natl Acad Sci U S A (2006) 103:15038; **29.** J Biol Chem (2009) 284:18754; **30.** PLoS ONE (2008) 3:e1616; **31.** J Physiol (2009) 587:851; **32.** Am J Physiol Heart Circ Physiol (2008) 294:H2121; **33.** Nat Protoc (2007) 2:2295; **34.** Am J Pathol (2009) 174:101; **35.** Am J Physiol Renal Physiol (2007) 292:F523; **36.** Free Radic Biol Med (2000) 28:1266; **37.** Fluorescent and Luminescent Probes for Biological Activity, Mason WT, Ed. (1993) p. 124; **38.** Biochemistry (1991) 30:4480; **39.** Proc Natl Acad Sci U S A (1991) 88:3671; **40.** Nat Protoc (2007) 2:2719; **41.** Proc Natl Acad Sci U S A (1980) 77:990; **42.** Int Rev Cytol (1990) 122:1; **43.** Proc Natl Acad Sci U S A (2009) 106:2007; **44.** J Neurosci Methods (2008) 171:87; **45.** Antonie Van Leeuwenhoek (2009) 96:227; **46.** Plant Physiol (1992) 98:279; **47.** Planta (1987) 17:346; **48.** J Histochem Cytochem (1993) 41:1247; **49.** Methods Mol Biol (2004) 263:181; **50.** J Pharmacol Exp Ther (2009) 331:1118; **51.** Biophys J (1989) 56:979; **52.** Histochemistry (1990) 94:303; **53.** Exp Pathol (1987) 31:47; **54.** J Cell Biol (1981) 88:526; **55.** Eur J Biochem (1997) 248:104; **56.** Transplantation (1994) 58:403; **57.** J Bioenerg Biomembr (1993) 25:679; **58.** J Cell Sci (1992) 101:315; **59.** Biophys J (1989) 56:1053; **60.** Biophys J (1988) 53:785; **61.** J Fluorescence (1993) 3:265; **62.** J Neurosci (2007) 27:8238; **63.** Biophys J (1993) 65:2396; **64.** Optical Microscopy for Biology, Herman B, Jacobson K, Eds. (1990) p. 131; **65.** Proc Natl Acad Sci U S A (1994) 91:12579; **66.** Anal Biochem (2001) 295:220; **67.** Biophys J (1999) 76:469; **68.** Methods Cell Biol (1989) 29:103; **69.** Eur J Biochem (1993) 217:973; **70.** Arch Biochem Biophys (1993) 302:348; **71.** APMIS (1994) 102:474; **72.** Nat Med (2009) 15:300; **73.** Proc Natl Acad Sci U S A (1996) 93:1167; **74.** Methods Enzymol (1989) 172:102; **75.** Mol Biol Cell (1998) 9:917; **76.** J Cell Biol (1994) 126:1375; **77.** Cell Motil Cytoskeleton (1993) 25:111; **78.** Methods Cell Biol (1989) 31:357; **79.** Methods Cell Biol (1989) 29:125; **80.** Exp Cell Res (1980) 125:514; **81.** Cell (1994) 79:1209; **82.** Cancer Res (1995) 55:2063; **83.** Plant Physiol (1987) 84:1385; **84.** Am J Respir Cell Mol Biol (2008) 39:180; **85.** J Biomed Opt (2004) 9:385; **86.** Histochemistry (1993) 99:75; **87.** Histochemistry (1985) 82:51; **88.** Histochemistry (1984) 80:385; **89.** Histochemistry (1983) 79:443; **90.** FEBS Lett (1990) 260:236; **91.** J Dent Res (1995) 74:1295; **92.** Eur J Biochem (1995) 228:113; **93.** Eur J Biochem (1990) 194:389; **94.** Biochem Biophys Res Commun (1989) 164:185; **95.** Basic Appl Histochem (1989) 33:71; **96.** Cancer Res (1991) 51:4665; **97.** Exp Cell Res (1994) 214:323; **98.** Biotechniques (2001) 30:804; **99.** Free Radic Biol Med (1994) 17:117; **100.** Am J Physiol Cell Physiol (2000) 279:C852; **101.** Biochem J (1999) 341:233; **102.** Biofactors (1998) 8:263; **103.** Biophys J (1999) 76:725; **104.** BMC Bioinformatics (2010) 11:263; **105.** J Biol Chem (1999) 274:543; **106.** Mol Biol Cell (1998) 9:523; **107.** Yeast (1998) 14:147; **108.** Curr Genet (1990) 18:265; **109.** Biochim Biophys Acta (1998) 1366:177; **110.** J Cell Biol (1998) 143:359; **111.** J Cell Biol (1998) 143:333; **112.** J Cell Biol (1998) 141:1371; **113.** Cytometry (1996) 23:28; **114.** J Cell Biol (1995) 130:345; **115.** Mol Biol Cell (1995) 6:1381; **116.** Biochem Int (1981) 2:503; **117.** Am J Physiol Endocrinol Metab (2004) 287:E574; **118.** Histochemistry (1993) 100:415.

IMPORTANT NOTICE: The products described in this manual are covered by one or more Limited Use Label License(s). Please refer to the Appendix on page 971 and Master Product List on page 975. Products are For Research Use Only. Not intended for any animal or human therapeutic or diagnostic use.

DATA TABLE 12.2 PROBES FOR MITOCHONDRIA

Cat. No.	MW	Storage	Soluble	Abs	EC	Em	Solvent	Notes
A1372	472.51	L	DMSO, EtOH	495	84,000	519	MeOH	
D273	572.53	D,L	DMSO	484	154,000	501	MeOH	
D288	366.24	L	DMF	475	45,000	605	MeOH	1
D378	600.58	D,L	DMSO	482	148,000	504	MeOH	
D426	380.27	L	DMF	461	39,000	589	MeOH	1
D632	346.38	F,D,L,AA	DMF, DMSO	289	7100	none	MeOH	2, 3
D633	444.57	F,D,L,AA	DMF, DMSO	296	11,000	none	MeOH	2, 3
D22421	532.38	D,L	DMSO, DMF	522	143,000	535	$CHCl_3$	4
D23806	346.38	F,D,L,AA	DMSO	289	7100	none	MeOH	3, 5
L6868	510.50	L	H_2O	455	7400	505	H_2O	6, 7
M7510	427.37	F,D,L	DMSO	551	102,000	576	MeOH	
M7511	392.93	F,D,L,AA	DMSO	235	57,000	none	MeOH	2, 3
M7512	531.52	F,D,L	DMSO	578	116,000	599	MeOH	
M7513	497.08	F,D,L,AA	DMSO	245	45,000	none	MeOH	2, 3
M7514	671.88	F,D,L	DMSO	490	119,000	516	MeOH	
M22425	724.00	F,D,L	DMSO	588	81,000	644	MeOH	
M22426	543.58	F,D,L	DMSO	640	194,000	662	MeOH	
M36008	759.71	FF,L,AA	DMSO	356	10,000	410	MeCN	2, 8
R302	380.83	F,D,L	MeOH, DMF	507	101,000	529	MeOH	
R634	479.02	F,D,L	EtOH	528	105,000	551	MeOH	
R648MP	627.18	F,D,L	DMF, DMSO	556	123,000	578	MeOH	
R14060	434.41	F,D,L,AA	DMSO	239	52,000	none	MeOH	2, 9
R22420	380.83	F,D,L	MeOH, DMF	507	101,000	529	MeOH	10
T639	378.90	L	DMF, DMSO	550	87,000	574	MeOH	
T668	500.93	F,D,L	DMSO, MeOH	549	115,000	573	MeOH	
T669	514.96	F,D,L	DMSO, EtOH	549	109,000	574	MeOH	
T3168	652.23	D,L	DMSO, DMF	514	195,000	529	MeOH	11

For definitions of the contents of this data table, see "Using *The Molecular Probes® Handbook*" in the introductory pages.

Notes

1. Abs and Em of styryl dyes are at shorter wavelengths in membrane environments than in reference solvents such as methanol. The difference is typically 20 nm for absorption and 80 nm for emission, but varies considerably from one dye to another. Styryl dyes are generally nonfluorescent in water.
2. This compound is susceptible to oxidation, especially in solution. Store solutions under argon or nitrogen. Oxidation may be induced by illumination.
3. These compounds are essentially colorless and nonfluorescent until oxidized. Oxidation products (in parentheses) are as follows: D632 and D23806 (R302); D633 (R634); M7511 (M7510); M7513 (M7512).
4. JC-9 exhibits long-wavelength J-aggregate emission at ~635 nm in aqueous solutions and polarized mitochondria.
5. This product is supplied as a ready-made solution in DMSO with sodium borohydride added to inhibit oxidation.
6. L6868 has much stronger absorption at shorter wavelengths (Abs = 368 nm (EC = 36,000 $cm^{-1}M^{-1}$)).
7. This compound emits chemiluminescence at 470 nm upon oxidation in basic aqueous solutions.
8. The product generated by reaction of M36008 with superoxide has similar spectroscopic properties to E1305.
9. R14060 is colorless and nonfluorescent until oxidized. The spectral characteristics of the oxidation product (2,3,4,5,6-pentafluorotetramethylrosamine) are similar to those of T639.
10. This product is specified to equal or exceed 98% analytical purity by HPLC.
11. JC-1 forms J-aggregates with Abs/Em = 585/590 nm at concentrations above 0.1 µM in aqueous solutions (pH 8.0). (Biochemistry (1991) 30:4480)

PRODUCT LIST 12.2 PROBES FOR MITOCHONDRIA

Cat. No.	Product	Quantity
A1372	acridine orange 10-nonyl bromide (nonyl acridine orange)	100 mg
C10600	CellLight® Mitochondria-GFP *BacMam 2.0*	1 mL
C10601	CellLight® Mitochondria-RFP *BacMam 2.0*	1 mL
D378	3,3′-diheptyloxacarbocyanine iodide (DiOC7(3))	100 mg
D273	3,3′-dihexyloxacarbocyanine iodide (DiOC6(3))	100 mg
D632	dihydrorhodamine 123	10 mg
D23806	dihydrorhodamine 123 *5 mM stabilized solution in DMSO*	1 mL
D633	dihydrorhodamine 6G	25 mg
D426	2-(4-(dimethylamino)styryl) -N-ethylpyridinium iodide (DASPEI)	1 g
D288	4-(4-(dimethylamino)styryl) -N-methylpyridinium iodide (4-Di-1-ASP)	1 g
D22421	3,3′-dimethyl-α-naphthoxacarbocyanine iodide (JC-9; DiNOC1(3))	5 mg
I34154	Image-iT® LIVE Mitochondrial and Nuclear Labeling Kit *counterstains for GFP-expressing cells*	1 kit
I35103	Image-iT® LIVE Mitochondrial Transition Pore Assay Kit *for microscopy*	1 kit
L6868	lucigenin (bis-N-methylacridinium nitrate) *high purity*	10 mg
M34153	MitoProbe™ Transition Pore Assay Kit *for flow cytometry* *100 assays*	1 kit
M36008	MitoSOX™ Red mitochondrial superoxide indicator *for live-cell imaging*	10 x 50 µg
M22426	MitoTracker® Deep Red FM® *special packaging*	20 x 50 µg
M7514	MitoTracker® Green FM® *special packaging*	20 x 50 µg
M7511	MitoTracker® Orange CM-H2TMRos *special packaging*	20 x 50 µg
M7510	MitoTracker® Orange CMTMRos *special packaging*	20 x 50 µg
M7513	MitoTracker® Red CM-H2XRos *special packaging*	20 x 50 µg

continued on next page

IMPORTANT NOTICE: The products described in this manual are covered by one or more Limited Use Label License(s). Please refer to the Appendix on page 971 and Master Product List on page 975. Products are For Research Use Only. Not intended for any animal or human therapeutic or diagnostic use.

PRODUCT LIST 12.2 PROBES FOR MITOCHONDRIA—*continued*

M7512	MitoTracker® Red CMXRos *special packaging*	20 x 50 µg
M22425	MitoTracker® Red FM® *special packaging*	20 x 50 µg
R14060	RedoxSensor™ Red CC-1 *special packaging*	10 x 50 µg
R302	rhodamine 123	25 mg
R22420	rhodamine 123 *FluoroPure™ grade*	25 mg
R634	rhodamine 6G chloride	1 g
R648MP	rhodamine B, hexyl ester, perchlorate (R 6)	10 mg
T3168	5,5',6,6'-tetrachloro- 1,1',3,3'-tetraethylbenzimidazolylcarboc yanine iodide (JC-1; CBIC2(3))	5 mg
T669	tetramethylrhodamine, ethyl ester, perchlorate (TMRE)	25 mg
T668	tetramethylrhodamine, methyl ester, perchlorate (TMRM)	25 mg
T639	tetramethylrosamine chloride	25 mg
Y7530	Yeast Mitochondrial Stain Sampler Kit	1 kit

12.3 Probes for Lysosomes, Peroxisomes and Yeast Vacuoles

Molecular Probes® acidotropic reagents can be used to stain lysosomes and yeast vacuoles, as well as several other types of acidic compartments such as trans-Golgi vesicles, endosomes and subpopulations of coated vesicles in fibroblasts, secretory vesicles in insulin-secreting pancreatic β-cells, acrosomes of spermatozoa and plant vacuoles.[1] Lysosomes contain glycosidases, acid phosphatases, elastase, cathepsins, carboxypeptidases and a variety of other proteases. Chapter 10 describes a number of substrates for detecting the activity of these hydrolytic enzymes. An excellent compendium of human diseases that affect intracellular transport processes through lysosomes, Golgi and endoplasmic reticulum (ER) has been published.[2]

Like lysosomes, peroxisomes are single membrane–bound vesicles that contain digestive enzymes. The chief function of these basic organelles is to enzymatically oxidize fatty acids and to subsequently catalyze the breakdown of H_2O_2, a by-product of fatty acid degradation. Recently, interest in peroxisomes has increased, especially studies related to peroxisomal origin and maintenance.[3,4] Morphological abnormalities in peroxisomes related to disease states and diet have also been the subject of current research.[5,6] The SelectFX® Alexa Fluor® 488 Peroxisome Labeling Kit (S34201), described below, provides an antibody-based method for labeling peroxisomes in fixed cells.

LysoTracker® Probes: Acidic Organelle–Selective Cell-Permeant Probes

LysoTracker® Probes

Weakly basic amines selectively accumulate in cellular compartments with low internal pH and can be used to investigate the biosynthesis and pathogenesis of lysosomes.[7,8] One frequently used probe for acidic organelles, DAMP (D1552), is not fluorescent and therefore must be used in conjunction with anti-DNP antibodies (Section 7.4) directly or indirectly conjugated to a fluorophore or enzyme in order to visualize the staining pattern.[9] The fluorescent probes neutral red (N3246) and acridine orange (A1301, A3568) are also commonly used for staining acidic organelles, though they lack specificity.[1,10]

These limitations have motivated us to search for alternative acidic organelle–selective probes, both for short-term and long-term tracking studies. The LysoTracker® probes are fluorescent acidotropic probes for labeling and tracing acidic organelles in live cells. These probes have several important features, including high selectivity for acidic organelles and effective labeling of live cells at nanomolar concentrations. Furthermore, the LysoTracker® probes are available in several fluorescent colors (Table 12.3, Figure 12.3.1), making them especially suitable for multicolor applications.

The LysoTracker® probes, which comprise a fluorophore linked to a weak base that is only partially protonated at neutral pH, are freely permeant to cell membranes and typically concentrate in spherical organelles (Figure 12.3.2). We have found that the fluorescent LysoTracker® probes must be used at very low concentrations—usually about 50 nM—to achieve optimal selectivity. Their mechanism of retention has not been firmly established but is likely to involve protonation and retention in the organelles' membranes, although staining is generally not reversed by subsequent treatment of the cells with weakly basic cell-permeant compounds. Unfortunately, these lysosomal probes can exhibit an alkalinizing effect on the lysosomes, such that longer incubation with these probes can induce an increase in lysosomal pH. Therefore, we recommend incubating cells with these probes for only one to five minutes before imaging.

The larger acidic compartments of cells stained with LysoTracker® Red DND-99 (L7528; Figure 12.3.3, Figure 12.3.4) usually retain their staining pattern following fixation with aldehydes. Simultaneous staining of lysosomes by two LysoTracker® dyes—LysoTracker® Yellow HCK-123 (L12491) and LysoTracker® Red DND-99 (L7528)—yields identical staining patterns when viewed through either the bandpass filter set appropriate for fluorescein or a longpass filter set appropriate for rhodamine (Figure 12.3.5). The LysoTracker® probes were principally developed for fluorescence microscopy applications. The lysosomal fluorescence in LysoTracker® dye–stained cells may constitute only a portion of total cellular fluorescence due to cellular autofluorescence or nonspecific staining. Consequently, successful application of these probes for quantitating the number of lysosomes by flow cytometry or fluorometry will likely depend on the particular cell lines and staining protocols used.

LysoTracker® Green DND-26 (L7526) was used to identify acidic compartments in a study of a membrane protein that facilitates vesicular sequestration of zinc,[11] to visualize acidic organelles labeled with rhodamine B in denervated skeletal muscle[12] and to assess acrosomal

IMPORTANT NOTICE: The products described in this manual are covered by one or more Limited Use Label License(s). Please refer to the Appendix on page 971 and Master Product List on page 975. Products are For Research Use Only. Not intended for any animal or human therapeutic or diagnostic use.
www.invitrogen.com/probes
molecular probes® | invitrogen™ by life technologies™

integrity in cryopreserved bovine spermatozoa.[13] This LysoTracker® probe also proved useful in a continuous assay for the secretion of pulmonary surfactant by exocytosis of lamellar bodies.[14] LysoTracker® Red DND-99 provided researchers with a probe for examining lysosome damage in *Trypanosoma brucei* after specific uptake of cytokine tumor necrosis factor-α,[15] for studying apoptosis in organogenesis-stage mouse embryos [16] and for determining the subcellular localization of receptor and channel proteins.[17–19]

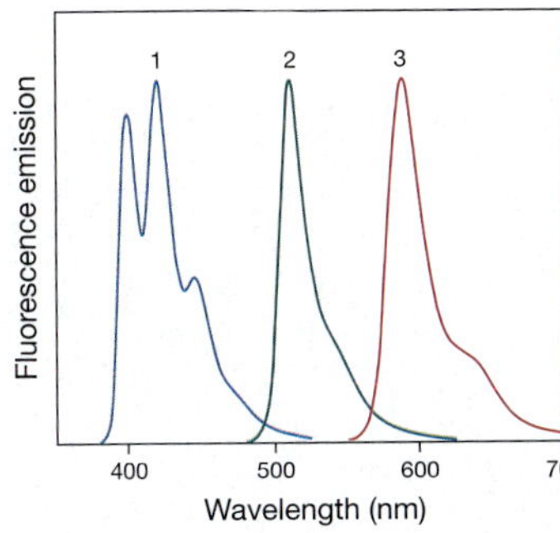

Figure 12.3.1 Normalized fluorescence emission spectra of **1**) LysoTracker® Blue DND-22 (L7525), **2**) LysoTracker® Green DND-26 (L7526) and **3**) LysoTracker® Red DND-99 (L7528) in aqueous solutions, pH 6.0.

Table 12.3 Summary of our LysoTracker® and LysoSensor™ probes.

Cat. No.	Probe	Abs * (nm)	Em * (nm)	pK_a
LysoTracker® Probes				
L7525	LysoTracker® Blue DND-22	373	422	ND
L12491	LysoTracker® Yellow HCK-123	465	535	ND
L7526	LysoTracker® Green DND-26	504	511	ND
L7528	LysoTracker® Red DND-99	577	590	ND
LysoSensor™ Probes				
L7533	LysoSensor™ Blue DND-167	373	425	5.1
L7535	LysoSensor™ Green DND-189	443	505	5.2
L7534	LysoSensor™ Green DND-153	442	505	7.5
L7545	LysoSensor™ Yellow/Blue DND-160	384 329	540 † 440 ‡	3.9
L22460	LysoSensor™ Yellow/Blue 10,000 MW dextran	381 335	521 † 452 ‡	4.2

* Absorption (Abs) and fluorescence emission (Em) maxima, determined in aqueous buffer or methanol; values may vary somewhat in cellular environments. † At pH 3. ‡ At pH 7; this dye has pH-dependent dual-excitation and dual-emission peaks. ND = Not determined.

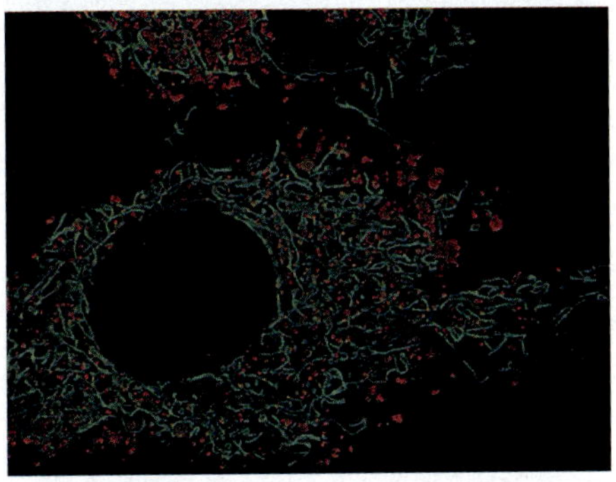

Figure 12.3.3 Bovine pulmonary artery endothelial cells (BPAEC) incubated simultaneously with 50 nM LysoTracker® Red DND-99 (L7528) and 75 nM MitoTracker® Green FM® (M7514) at 37°C for 30 minutes. Both dyes showed excellent cellular retention, even after cells were fixed in 3% glutaraldehyde for 30 minutes. The image was deconvolved using Huygens software (Scientific Volume Imaging, http://www.svi.nl/).

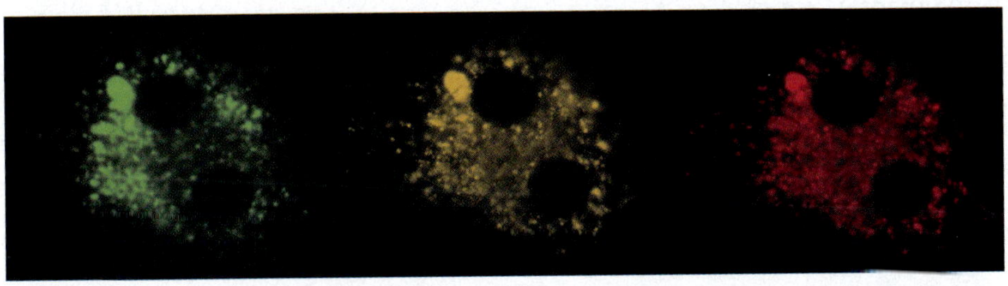

Figure 12.3.2 Bovine pulmonary artery endothelial cells simultaneously stained with LysoTracker® Red DND-99 (L7528), a cell-permeant, fixable lysosomal stain, and with 5-(pentafluorobenzoylamino)fluorescein di-β-D-galactopyranoside (PFB-FDG, P11948), a fluorogenic substrate for β-galactosidase. PFB-FDG is nonfluorescent until enzymatically hydrolyzed to green-fluorescent PFB-F. The center image demonstrates colocalization of the LysoTracker® Red DND-99 dye and PFB-F to the lysosomes. The left image was acquired with a bandpass filter set appropriate for fluorescein, the right image was acquired with a bandpass filter set appropriate for Texas Red® dye, and the center image was acquired with a double bandpass optical filter set appropriate for fluorescein and the Texas Red® dye.

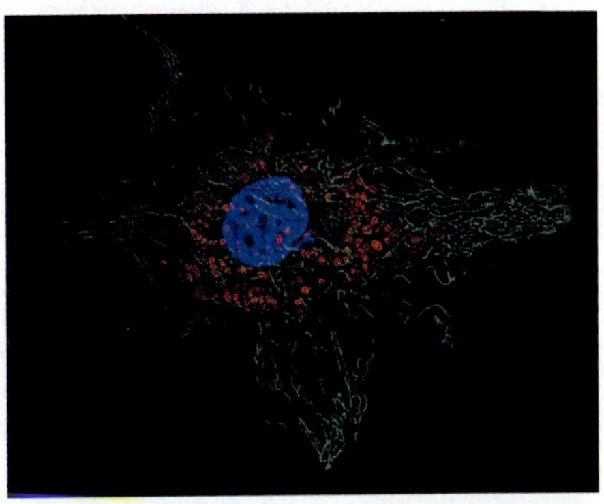

Figure 12.3.4 Live bovine pulmonary artery endothelial cells (BPAEC) were first stained with LysoTracker® Red DND-99 (L7528). Then, a solution of dihydrorhodamine 123 (D632, D23806) and Hoechst 33258 (H1398, H3569, H21491) was added and allowed to incubate with the cells for an additional 10 minutes before the cells were subsequently washed and visualized. The green-fluorescent oxidation product (rhodamine 123, R302) localized primarily to the mitochondria. The red-fluorescent LysoTracker® Red DND-99 stain accumulated in the lysosomes, and the blue-fluorescent Hoechst 33258 dye stained the nuclei. The image was acquired with filters appropriate for DAPI, fluorescein and the Texas Red® dye. The image was deconvolved using Huygens software (Scientific Volume Imaging, http://www.svi.nl/). 3D reconstruction was performed using Imaris software (Bitplane AG, http://www.bitplane.com/).

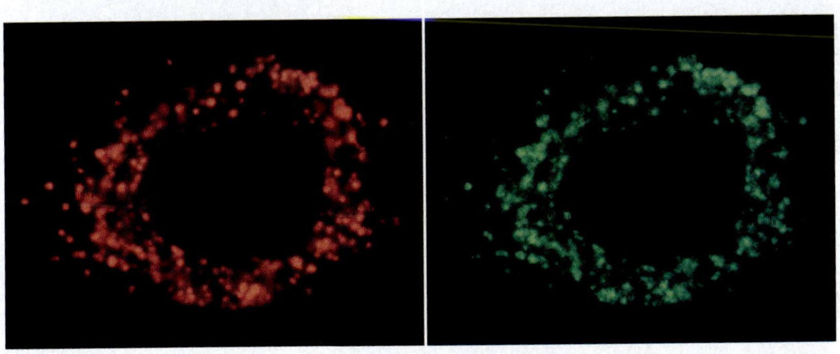

Figure 12.3.5 Rat fibroblasts stained with LysoTracker® Yellow HCK-123 (L12491) and LysoTracker® Red DND-99 (L7528). The lysosomes fluoresce green when visualized through a bandpass filter set appropriate for fluorescein (right image) and red when visualized through a longpass filter set appropriate for rhodamine (left image).

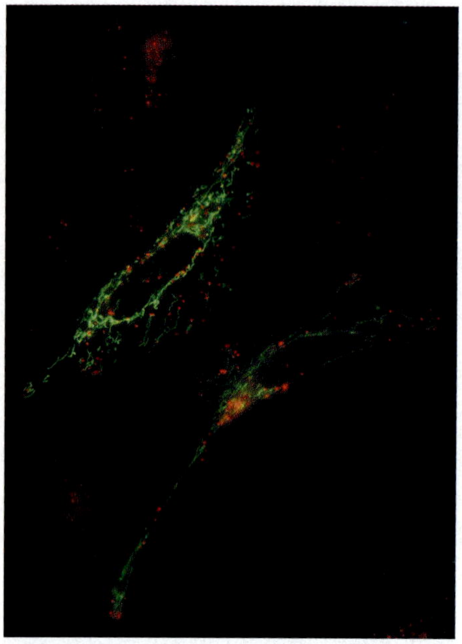

Figure 12.3.6 Live HeLa cells were transfected using pShooter™ vector pCMV/myc/mito/GFP and Lipofectamine® 2000 transfection reagent and stained with the reagents in the Image-iT® LIVE Lysosomal and Nuclear Labeling Kit (I34202). Lysosomes were stained with LysoTracker® Red DND-99. Cells were visualized using epifluorescence microscopy.

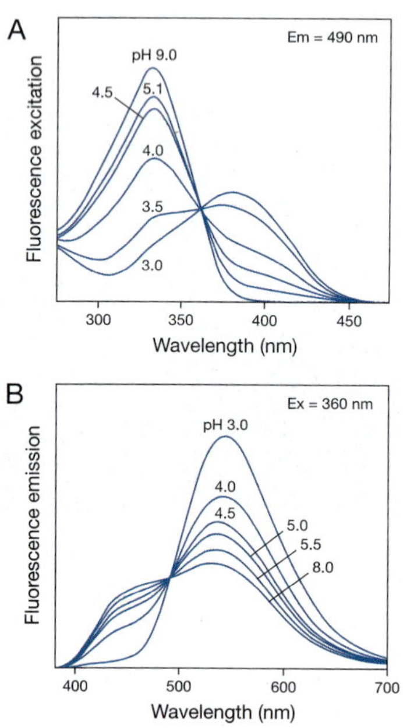

Figure 12.3.7 The pH-dependent spectral response of LysoSensor™ Yellow/Blue DND-160 (L7545): **A**) fluorescence excitation spectra and **B**) fluorescence emission spectra.

Image-iT® LIVE Lysosomal and Nuclear Labeling Kit

The Image-iT® LIVE Lysosomal and Nuclear Labeling Kit (I34202) provides two stains—red-fluorescent LysoTracker® Red DND-99 dye (excitation/emission maxima ~577/590 nm) and blue-fluorescent Hoechst 33342 dye (excitation/emission maxima when bound to DNA ~350/461 nm)—for highly selective staining of lysosomes and the nucleus, respectively, in live, Green Fluorescent Protein (GFP)–transfected cells (Figure 12.3.6). When used according to the sample protocol, cell-permeant LysoTracker® Red DND-99 dye provides highly selective lysosomal staining with minimal background. A significant amount of specific staining is retained after formaldehyde fixation, although some cytoplasmic background staining may be seen. Hoechst 33342 dye, a cell-permeant nucleic acid stain that is selective for DNA and spectrally similar to DAPI, is UV excitable and emits blue fluorescence when bound to DNA. This dye does not interfere with GFP fluorescence and is retained after fixation and permeabilization. It is not recommended that the dyes be combined into one staining solution; they should instead be used in separate labeling steps, with Hoechst 33342 staining first.

The Image-iT® LIVE Lysosomal and Nuclear Labeling Kit contains:

- LysoTracker® Red DND-99 dye
- Hoechst 33342 dye
- Labeling protocols

Each kit provides enough staining solution for 500 assays using the protocol provided for labeling live, cultured cells that are adhering to coverslips.

LysoSensor™ Probes: Acidic Organelle–Selective pH Indicators

LysoSensor™ Probes

For researchers studying the dynamic aspects of lysosome biogenesis and function in live cells, we have developed the LysoSensor™ probes—fluorescent pH indicators that partition into acidic organelles. The LysoSensor™ dyes are acidotropic probes that appear to accumulate in acidic organelles as the result of protonation. This protonation also relieves the fluorescence quenching of the dye by its weakly basic side chain, resulting in an increase in fluorescence intensity. Thus, the LysoSensor™ reagents exhibit a pH-dependent increase in fluorescence intensity upon acidification, in contrast to the LysoTracker® probes, which exhibit fluorescence that is not substantially enhanced at acidic pH.

We offer four LysoSensor™ reagents that differ in color and pK_a (Table 12.3). Because these probes may localize in the membranes of organelles, it is probable that the pK_a values listed in Table 12.3 will not be equivalent to those measured in cellular environments and that only qualitative and semiquantitative comparisons of organelle pH will be possible. The green-fluorescent LysoSensor™ probes are available with optimal pH sensitivity in either the acidic or neutral range

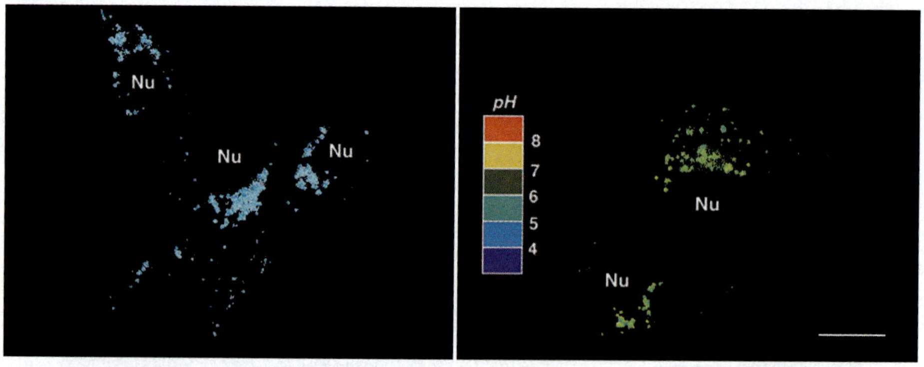

Figure 12.3.8 Dual-emission ratiometric measurement of lysosomal pH using LysoSensor™ Yellow/Blue DND-160 (L7545). Madin-Darby canine kidney cells were exposed to pH-calibration buffers (pH 4.5 or 7.0) in the presence of nigericin (N1495) and monensin. These pseudocolored images were constructed from two emission images at 450 ± 33 nm and 510 ± 20 nm, both excited at 365 ± 8 nm.

IMPORTANT NOTICE: The products described in this manual are covered by one or more Limited Use Label License(s). Please refer to the Appendix on page 971 and Master Product List on page 975. Products are For Research Use Only. Not intended for any animal or human therapeutic or diagnostic use.
www.invitrogen.com/probes

(pK_a ~5.2 or ~7.5 in aqueous buffers). With their low pK_a values, LysoSensor™ Blue DND-167 (L7533) and LysoSensor™ Green DND-189 (L7535) are almost nonfluorescent except when inside acidic compartments, whereas LysoSensor™ Green DND-153 (L7534) is brightly fluorescent at neutral pH. LysoSensor™ Yellow/Blue DND-160 (PDMPO, L7545) is unique in that it exhibits both dual-excitation and dual-emission spectral peaks that are pH dependent (Figure 12.3.7, Figure 12.3.8).

LysoSensor™ Yellow/Blue DND-160 exhibits predominantly yellow fluorescence in acidic organelles, and in less acidic organelles it exhibits blue fluorescence. Dual-emission measurements facilitate ratio imaging of the pH in acidic organelles such as lysosomes,[20] myeloid leukemic cells[21] and acidic vacuoles of plant cells.[22] LysoSensor™ Yellow/Blue DND-160, frequently referred to by the acronym PDMPO, has been widely utilized as a tracer of silica deposition and transport in marine diatoms.[23–26] Kinetic studies on the internalization of LysoSensor™ Yellow/Blue DND-160 indicate that the probe is taken up by live cells within seconds. Unfortunately, this lysosomal probe can exhibit an alkalinizing effect on the lysosomes, such that longer incubation with this probe can induce an increase in lysosomal pH. Therefore, it is a useful pH indicator only when incubation times are kept short; we recommend incubating cells for only one to five minutes before imaging.

The cell-permeant LysoSensor™ probes can be used singly or in combination to investigate the acidification of lysosomes and alterations of lysosomal function or trafficking that occur in cells. For example, lysosomes in some tumor cells have a lower pH than normal lysosomes,[27] whereas other tumor cells contain lysosomes with higher pH.[28] In addition, cystic fibrosis and other diseases result in defects in the acidification of some intracellular organelles, and the LysoSensor™ probes are useful in studying these aberrations.[29,30] LysoSensor™ Green DND-189 has been used to selectively label acidic compartments within granule cell neurites[31] and, along with LysoSensor™ Green DND-153, to examine the acidification of endosomes and lysosomes in a mutant CHO cell line.[32] LysoSensor™ Yellow/Blue DND-160 was employed in a study that demonstrated the involvement of lysosomes in the acquired drug-resistance phenotype of a doxorubicin-selected variant of human U-937 myeloid leukemia cells.[21]

As with the LysoTracker® probes, the cell-permeant LysoSensor™ probes were originally developed for fluorescence microscopy applications. The lysosomal fluorescence in LysoSensor™ dye–stained cells may constitute only a portion of total cellular fluorescence due to cellular autofluorescence or nonspecific staining. Therefore, the successful application of these probes for quantitating the number of lysosomes or their pH by flow cytometry or fluorometry will likely depend on the particular cell lines and staining protocols used.

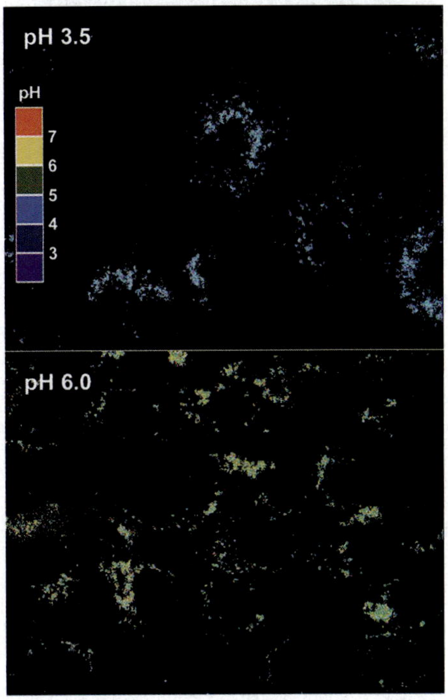

Figure 12.3.9 Dual-emission ratiometric measurement of lysosomal pH using LysoSensor™ Yellow/Blue dextran (L22460). MDCK cells labeled with the fluorescent dextran were exposed to pH-calibration buffers (pH 3.5 or pH 6.0) in the presence of nigericin (N1495) and monensin. Pseudocolored images were constructed from two emission images at 450 ± 33 nm and 510 ± 20 nm, both excited at 365 ± 8 nm (Chem Biol (1999) 6:411).

LysoSensor™ Yellow/Blue Dextran

We have prepared a 10,000 MW dextran conjugate of the LysoSensor™ Yellow/Blue dye (L22460). As this labeled dextran is taken up by the cells and moves through the endocytic pathway, the fluorescence of the LysoSensor™ dye changes from blue fluorescent in the near-neutral endosomes to longer-wavelength yellow fluorescent in the acidic lysosomes.[33] The greatest change in fluorescence emission occurs near the pK_a of the dye at pH ~3.9. Unlike the cell-permeant LysoSensor™ dyes, LysoSensor™ Yellow/Blue dextran allows measurement of pH in lysosomes using either fluorescence microscopy (Figure 12.3.9) or flow cytometry.

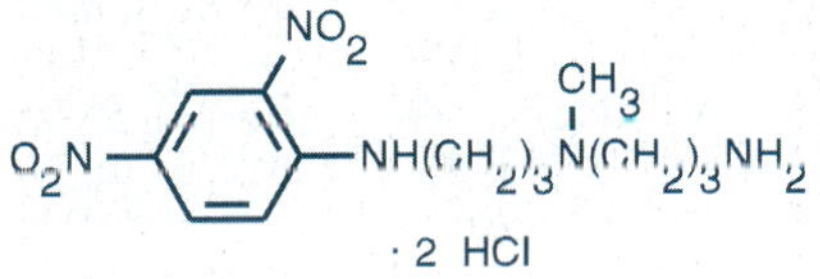

Figure 12.3.10 *N*-(3-((2,4-dinitrophenyl)amino)propyl)-*N*-(3-aminopropyl)methylamine, dihydrochloride (DAMP, D1552).

DAMP and Other Lysosomotropic Probes

DAMP

The reagent DAMP (*N*-(3-((2,4-dinitrophenyl)amino)propyl)-*N*-(3-aminopropyl)methylamine, dihydrochloride; D1552; Figure 12.3.10) is a weakly basic amine that is taken up in acidic organelles of live cells. This cell-permeant acidotropic reagent can be detected with anti-DNP antibodies (Section 7.4), including those labeled with Alexa Fluor® 488 dye, biotin, Qdot® 655 nanocrystal or enzymes,[9] making DAMP broadly applicable for detecting acidic organelles by electron and light microscopy. For example, DAMP has been used to investigate:

- Endocytic and secretory pathways[34–36]
- Defective acidification of intracellular organelles in cells from cystic fibrosis patients[37]
- Dependence on pH of the conversion of proinsulin to insulin in beta cells[38]
- Development of autophagic vacuoles[39,40]
- Location of intracellular acidic compartments during viral infection[41]

IMPORTANT NOTICE: The products described in this manual are covered by one or more Limited Use Label License(s). Please refer to the Appendix on page 971 and Master Product List on page 975. Products are For Research Use Only. Not intended for any animal or human therapeutic or diagnostic use.

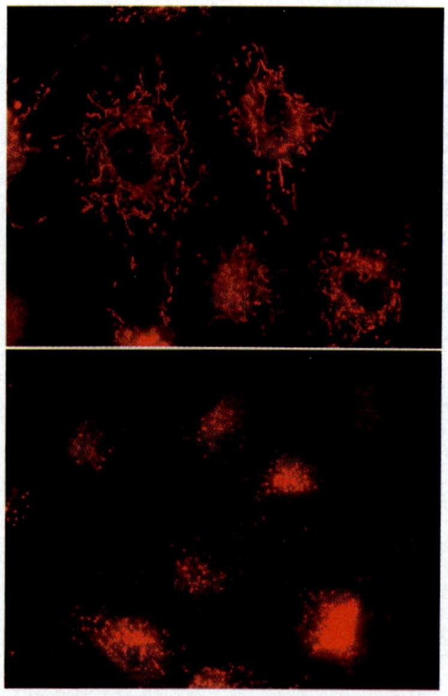

Figure 12.3.11 Cellular proliferation state determines the distribution of the oxidized product of RedoxSensor™ Red CC-1 (R14060). Normal rat kidney (NRK) cells in different growth states were stained with RedoxSensor™ Red CC-1. In proliferating cells (top panel), the oxidized dye accumulates in mitochondria. In quiescent cells (bottom panel), the oxidized product localizes in the lysosomes.

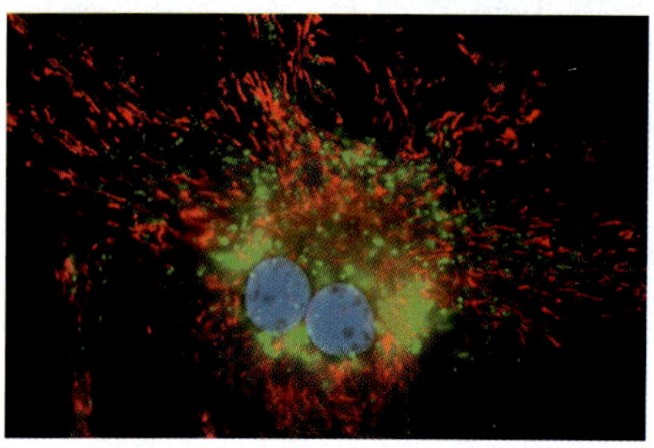

Figure 12.3.12 Viable bovine pulmonary artery endothelial cells simultaneously stained with BODIPY® FL histamine (B22461), MitoTracker® Red CMXRos (M7512) and Hoechst 33342 (H1399, H3570, H21492). Green-fluorescent BODIPY® FL histamine localized to lysosomes, red-fluorescent MitoTracker® Red CMXRos accumulated in the mitochondria, and the blue-fluorescent Hoechst 33342 dye stained the nuclei. This multiple-exposure image was acquired with bandpass filters appropriate for fluorescein, the Texas Red® dye and DAPI.

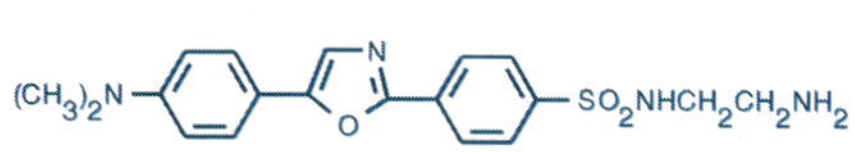

Figure 12.3.13 Dapoxyl® (2-aminoethyl)sulfonamide (D10460).

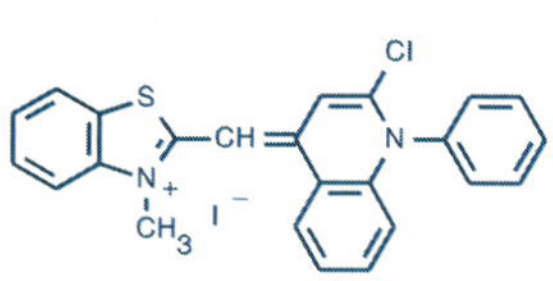

Figure 12.3.14 FUN® 1 cell stain (F7030).

As alternatives to DAMP, our cell-permeant fluorescent LysoTracker® and LysoSensor™ probes described above have significant potential in many of these applications. Because they can be visualized directly without any secondary detection reagents, the LysoTracker® and LysoSensor™ reagents enable researchers to study acidic organelles and follow their dynamic processes in live cells.

RedoxSensor™ Red CC-1 Stain

RedoxSensor™ Red CC-1 stain (2,3,4,5,6-pentafluorotetramethyldihydrorosamine, R14060) passively enters live cells and is subsequently oxidized in the cytosol to a red-fluorescent product (excitation/emission maxima ~540/600 nm), which then accumulates in the mitochondria. Alternatively, this nonfluorescent probe may be transported to the lysosomes where it is oxidized. The differential distribution of the oxidized product between mitochondria and lysosomes appears to depend on the redox potential of the cytosol.[42–44] In proliferating cells, mitochondrial staining predominates; whereas in contact-inhibited cells, the staining is primarily lysosomal (Figure 12.3.11). The best method we have found to quantitate the distribution of the oxidized product is to use the mitochondrion-selective MitoTracker® Green FM® stain (M7514) in conjunction with the RedoxSensor™ Red CC-1 stain.[44]

Other Lysosomotropic Probes

BODIPY® FL histamine (B22461) combines the pH-insensitive, bright green-fluorescent BODIPY® FL dye with the weakly basic imidazole moiety of histamine. When used at low concentrations, this probe selectively stains lysosomes (Figure 12.3.12).

As with the LysoTracker® and LysoSensor™ probes, the weak basicity of the amine group in Dapoxyl® (2-aminoethyl)sulfonamide (D10460) leads to its accumulation in acidic organelles. Dapoxyl® (2-aminoethyl)sulfonamide[45] (Figure 12.3.13) uptake by the acidic lumen of the intact acrosome of mouse sperm is accompanied by significant enhancement of this probe's fluorescence.[46] The fluorescence of Dapoxyl® (2-aminoethyl)sulfonamide is considerably reduced upon loss of the pH gradient at the onset of the acrosome reaction.[46]

Our high-purity neutral red (N3246) is a common lysosomal probe that stains lysosomes a fluorescent red.[10,47] It has also been used to determine the number of adherent and nonadherent cells in a microplate assay[48] and to stain cells in brain tissue.[49,50]

In addition, dansyl cadaverine[51,52] (D113) and the DNA intercalator acridine orange[10,53] (A1301, A3568) have been reported to be useful lysosomotropic reagents. Dansyl cadaverine has been shown to selectively label autophagic vacuoles, at least some of which had already fused with lysosomes; it did not, however, accumulate in early or late endosomes.[54]

Cell-Permeant Probes for Yeast Vacuoles

Biogenesis of the yeast vacuole has been extensively studied as a model system for eukaryotic organelle assembly.[55–58] Using a combination of genetic and biochemical approaches, researchers have isolated a large collection of yeast vacuolar protein sorting (*vps*) mutants[59] and characterized the vacuolar H^+-ATPase (V-ATPase) responsible for compartment acidification.[60] To facilitate the investigation of yeast vacuole structure and function, we offer membrane-permeant reagents and a Yeast Vacuole Marker Sampler Kit (Y7531).

FUN® 1 Vital Cell Stain for Yeast

The FUN® 1 (Figure 12.3.14) vital cell stain (F7030) exploits endogenous biochemical processing mechanisms that appear to be well conserved among different species of yeast and other fungi.[61,62] When used at micromolar concentrations, the FUN® 1 cell stain is freely taken up by several species of yeast and fungi and converted from a diffusely distributed pool of yellow-green–fluorescent intracellular stain into compact red-orange–fluorescent intravacuolar structures (Figure 12.3.15). This conversion requires both plasma membrane integrity and metabolic capability. Only metabolically active cells are marked clearly with fluorescent intravacuolar structures, while dead cells exhibit extremely bright, diffuse, yellow-green fluorescence[63,64] (Figure 12.3.16, Figure 12.3.17). FUN® 1 staining has been used to detect antifungal activity against *Candida* species[65] and to measure susceptibility of fungi to fungicides by flow cytometry.[66,67] The FUN® 1 cell stain is also available as a component in the LIVE/DEAD® Yeast Viability Kit (L7009, Section 15.3).

IMPORTANT NOTICE: The products described in this manual are covered by one or more Limited Use Label License(s). Please refer to the Appendix on page 971 and Master Product List on page 975. Products are For Research Use Only. Not intended for any animal or human therapeutic or diagnostic use.

FM® 4-64 and FM® 5-95

One of our FM® styryl dyes, FM® 4-64, has been reported to selectively stain yeast vacuolar membranes with red fluorescence[68] (excitation/emission maxima ~515/640 nm). This styryl dye is proving to be an important tool for visualizing vacuolar organelle morphology and dynamics, for studying the endocytic pathway and for screening and characterizing yeast endocytosis mutants.[69–74] We offer FM® 4-64 in 1 mg vials (T3166) or specially packaged in 10 vials of 100 µg each (T13320). The increasing number of successful applications for our FM® dyes has prompted us to synthesize FM® 5-95 (T23360), a slightly less lipophilic analog of FM® 4-64 with essentially identical spectroscopic properties.

Yeast Vacuole Marker Sampler Kit

The Yeast Vacuole Marker Sampler Kit (Y7531) contains sample quantities of a series of both novel and well-established vacuole marker probes that show promise for the study of yeast cell biology:

- 5-(and 6-)Carboxy-2',7'-dichlorofluorescein diacetate (carboxy-DCFDA)[55,68,75–79]
- CellTracker™ Blue CMAC[80]
- Aminopeptidase substrate Arg-CMAC (Figure 12.3.18)
- Dipeptidyl peptidase substrate Ala-Pro-CMAC
- Yeast vacuole membrane marker MDY-64[81] (Figure 12.3.19)

Our experiments have demonstrated that several cell-permeant derivatives of 7-amino-4-chloromethylcoumarin (CMAC) are largely sequestered within yeast vacuoles. The corresponding 7-amino-4-methylcoumarin derivatives are known to be substrates for yeast vacuolar enzymes.[82–84] This sampler kit's three coumarin-based vacuole markers selectively stain the lumen of the yeast vacuole. To complement the blue-fluorescent staining of the lumen, we provide a novel green-fluorescent membrane marker MDY-64 for staining the yeast vacuole membrane. Membrane staining can also be accomplished using the red-fluorescent probe FM® 4-64, as described above. The commonly used vacuole marker 5-(and 6-)carboxy-2',7'-dichlorofluorescein diacetate (carboxy-DCFDA) is supplied for use as a standard.[55,76] Three of the components in the Yeast Vacuole Marker Sampler Kit—CellTracker™ Blue CMAC (C2110, Section 14.2), the proprietary yeast vacuole membrane marker MDY-64[81,85] (Y7536) and carboxy-DCFDA (C369, Section 15.2)—are also available separately for those researchers who find that one of these dyes is well suited for their application.

Fluorescent Protein–Based Markers for Lysosomes, Peroxisomes and Endosomes

CellLight® reagents are BacMam expression vectors encoding site-selective autofluorescent protein fusions. These reagents incorporate all the customary advantages of BacMam delivery technology, including high efficiency transduction of mammalian cells and long-lasting, titratable expression (BacMam Gene Delivery and Expression Technology—Note 11.1). A complete list of our CellLight® reagents and their targeting sequences can be found in Table 11.1.

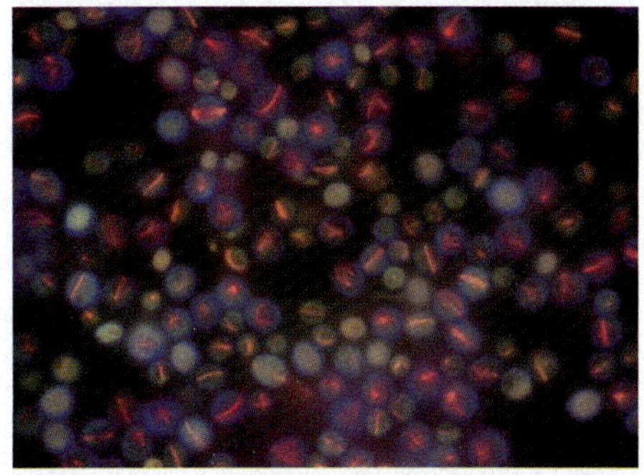

Figure 12.3.15 A culture of *Saccharomyces cerevisiae* incubated in medium containing the FUN® 1 viability indicator (F7030) and the counterstain Calcofluor White M2R, both of which are provided in our LIVE/DEAD® Yeast Viability Kit (L7009). Metabolically active yeast process the FUN® 1 dye, forming numerous red-fluorescent cylindrical structures within their vacuoles. Calcofluor stains the cell walls fluorescent blue, regardless of the yeast's metabolic state. The yeast were photographed in a single exposure through an Omega® Optical triple bandpass filter set.

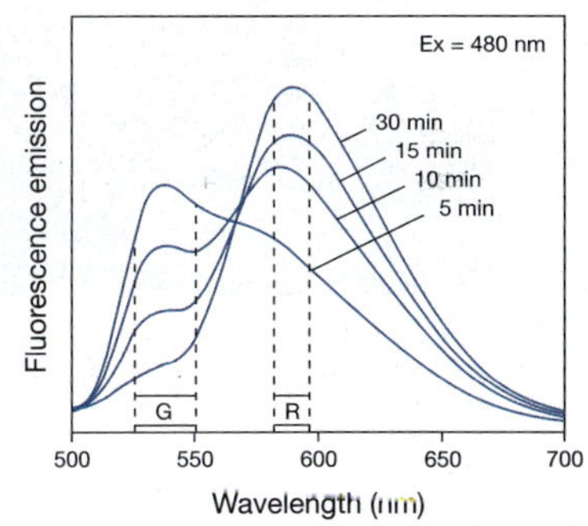

Figure 12.3.16 Fluorescence emission spectra of a *Saccharomyces cerevisiae* suspension that has been stained with the FUN® 1 cell stain, which is available separately (F7030) or in the LIVE/DEAD® Yeast Viability Kit (L7009). After the FUN® 1 reagent was added to the medium, the fluorescence emission spectrum (excited at 480 nm) was recorded in a spectrofluorometer at the indicated times during a 30-minute incubation period. The shift from green (G) to red (R) fluorescence reflects the processing of FUN® 1 by metabolically active yeast cells.

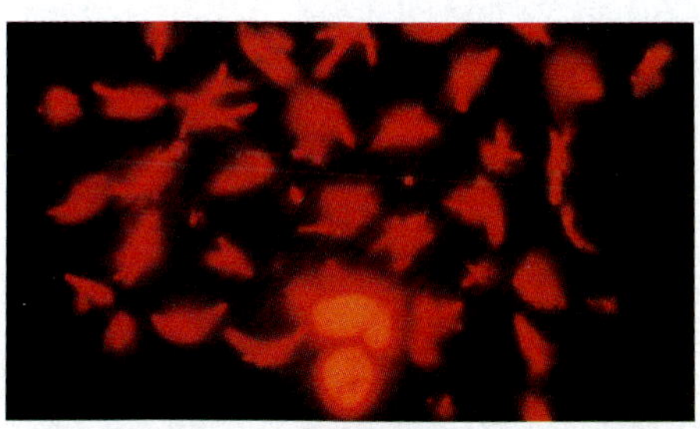

Figure 12.3.17 *Saccharomyces cerevisiae* stained with the FUN® 1 dye, available separately (F7030) or in our LIVE/DEAD® Yeast Viability Kit (L7009). Metabolically active yeast process the FUN® 1 dye, forming numerous red fluorescent cylindrical structures within their vacuoles.

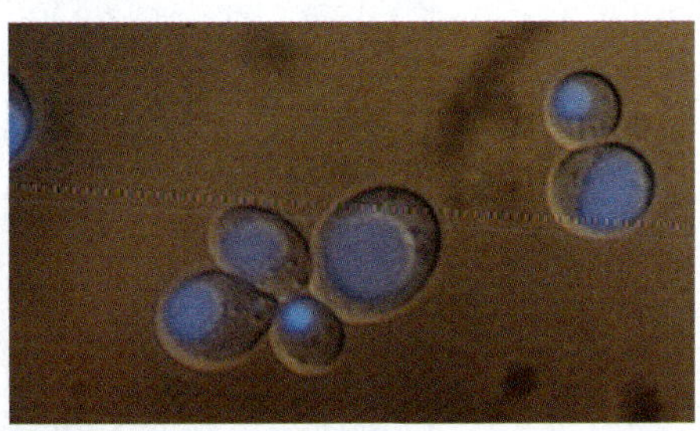

Figure 12.3.18 *Saccharomyces cerevisiae* vacuolar lumen stained with 7-amino-4-chloromethylcoumarin, L-arginine amide (Arg-CMAC), available in our Yeast Vacuole Marker Sampler Kit (Y7531). This kit contains three additional fluorescent vacuolar lumen stains plus a green-fluorescent vacuolar membrane stain. After staining, yeast were viewed by epifluorescence and DIC microscopy.

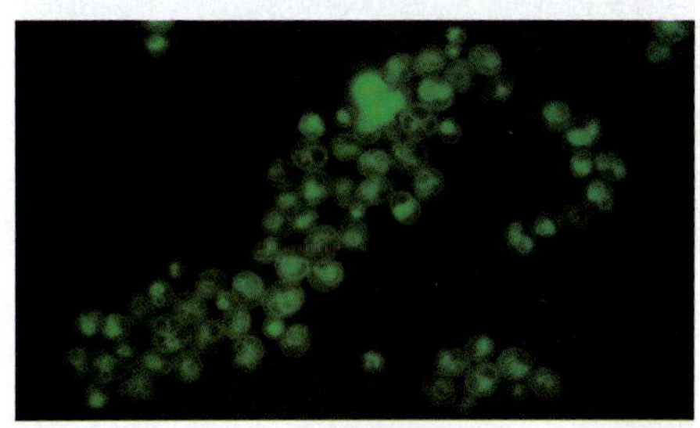

Figure 12.3.19 *Saccharomyces cerevisiae* vacuolar membranes stained with the yeast vacuole membrane marker MDY-64, which is available separately (Y7536) or in our Yeast Vacuole Marker Sampler Kit (Y7531). The kit contains four additional fluorescent vacuolar lumen stains.

IMPORTANT NOTICE: The products described in this manual are covered by one or more Limited Use Label License(s). Please refer to the Appendix on page 971 and Master Product List on page 975. Products are For Research Use Only. Not intended for any animal or human therapeutic or diagnostic use.

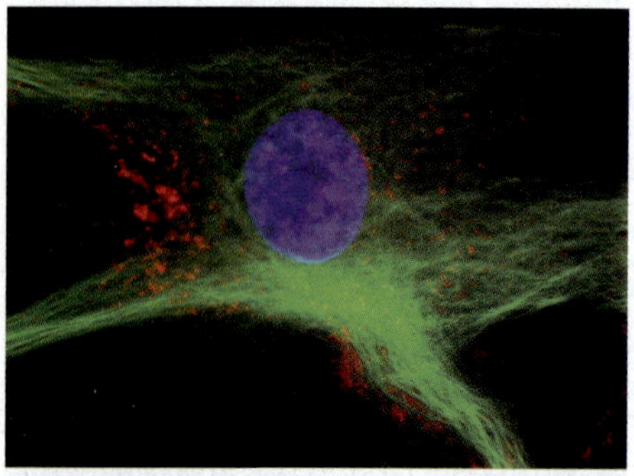

Figure 12.3.20 Human aortic smooth muscle cells (HASMC) labeled with CellLight® Lysosomes-RFP (C10597) and CellLight® MAP4-GFP (C10598) reagents and with Hoechst 33342 nucleic acid stain.

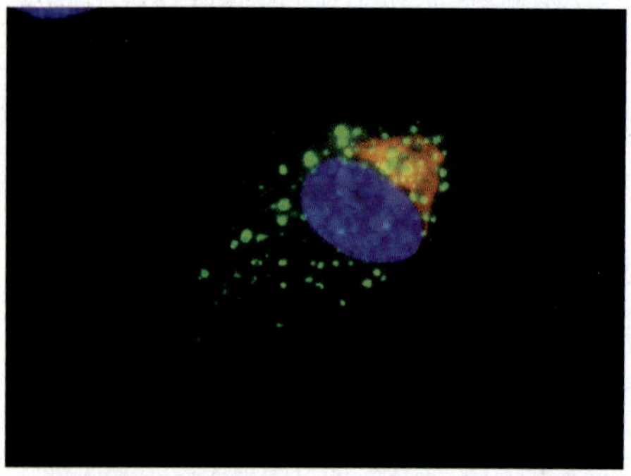

Figure 12.3.21 Human aortic smooth muscle cells (HASMC) labeled with CellLight® Early Endosomes-GFP (C10586) and Organelle Lights™ Golgi-OFP reagents and with Hoechst 33342 nucleic acid stain.

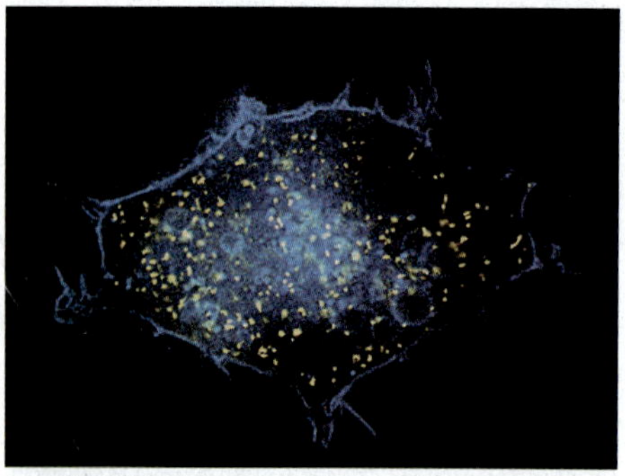

Figure 12.3.22 HEK 293 cell labeled with CellLight® Peroxisomes-GFP (C10604) and CellLight® Plasma Membrane-CFP (C10606) reagents.

CellLight® Lysosomes-GFP (C10596) and CellLight® Lysosomes-RFP (C10597, Figure 12.3.20) are BacMam expression vectors encoding fusions of Green Fluorescent Protein (GFP) or Red Fluorescent Protein [86] (RFP) with the targeting sequence from Lamp1 (lysosomal-associated membrane protein 1). These CellLight® reagents generate lysosomally localized fluorescent labeling in live cells that is retained after fixation and permeabilization procedures—procedures that will dissipate LysoTracker® Red DND-99 staining.[87] The titratable expression capacity of BacMam vectors is a particularly useful feature in the context of the Lamp1–GFP fusion, as high levels of overexpression have sometimes been found to induce aberrant aggregation of late-endocytic organelles.[88]

CellLight® Early Endosomes-GFP (C10586, Figure 12.3.21) and CellLight® Early Endosomes-RFP (C10587) reagents provide BacMam expression vectors encoding fusions of GFP or RFP with the small GTPase Rab5a. Rab5a fusions with autofluorescent proteins are sensitive and precise early endosome markers for real-time imaging of clathrin-mediated endocytosis in live cells.[87,89,90]

CellLight® Peroxisome-GFP (C10604, Figure 12.3.22) is a BacMam expression vector encoding GFP C-terminally linked to a peroxisomal targeting sequence [91] (GFP–PTS1). Live-cell imaging with the GFP–PTS1 fusion has provided many insights into normal and pathologically abnormal biogenesis and degradation of peroxisomes and the controlling influence of peroxisome proliferator–activated receptors (PPARs).

SelectFX® Alexa Fluor® 488 Peroxisome Labeling Kit

Peroxisomes, single membrane–bound vesicles found in most eukaryotic cells, function to enzymatically oxidize fatty acids and to subsequently catalyze the breakdown of H_2O_2, a by-product of fatty acid degradation. Peroxisomes are similar in size to lysosomes (0.5–1.5 µm). The SelectFX® Alexa Fluor® 488 Peroxisome Labeling Kit (S34201) provides all the reagents required for labeling peroxisomes in fixed cells, including cell fixation and permeabilization reagents. To specifically detect peroxisomes, this kit uses an antibody directed against peroxisomal membrane protein 70 (PMP 70), which is a high-abundance integral membrane protein in peroxisomes,[6] and an Alexa Fluor® 488 dye–labeled secondary antibody (Figure 12.3.23). The Alexa Fluor® 488 dye exhibits bright green fluorescence that is compatible with filters and instrument settings appropriate for fluorescein. PMP 70 is significantly induced by administration of hypolipidemic agents, in parallel with peroxisome proliferation and the induction of peroxisomal fatty acid β-oxidation enzymes.[6]

Each SelectFX® Alexa Fluor® 488 Peroxisome Labeling Kit contains:

- Rabbit IgG anti–peroxisomal membrane protein 70 (PMP 70) antibody
- Highly cross-adsorbed Alexa Fluor® 488 goat anti–rabbit IgG antibody
- Concentrated fixative solution
- Concentrated phosphate-buffered saline (PBS)
- Concentrated permeabilization solution
- Concentrated blocking solution
- Detailed protocols for mammalian cell preparation and staining

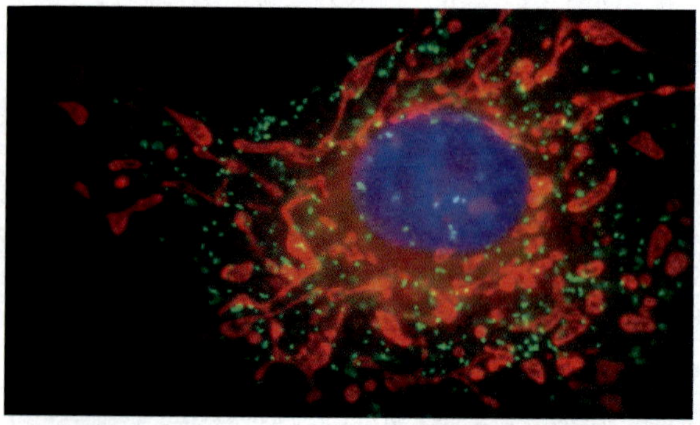

Figure 12.3.23 Peroxisome labeling in fixed and permeabilized bovine pulmonary artery endothelial cells. Peroxisomes were labeled using an antibody directed at peroxisomal membrane protein 70 (PMP-70) and detected with Alexa Fluor® 488 dye–labeled goat anti-mouse IgG secondary antibody. Mitochondria were stained with MitoTracker® Red CMXRos (M7512) prior to fixation; nuclei were stained with blue-fluorescent DAPI (D1306, D3571, D21490).

IMPORTANT NOTICE: The products described in this manual are covered by one or more Limited Use Label License(s). Please refer to the Appendix on page 971 and Master Product List on page 975. Products are For Research Use Only. Not intended for any animal or human therapeutic or diagnostic use.

molecular probes® | invitrogen™ by life technologies™

REFERENCES

1. J Cell Biol (1988) 106:539; **2.** Traffic (2000) 1:836; **3.** Am J Hum Genet (2003) 73:233; **4.** Mol Biol Cell (2003) 14:2900; **5.** J Biol Chem (2008) 283:2246; **6.** Cell Biochem Biophys (2000) 32:131; **7.** Cell (1988) 52:329; **8.** Lysosomes in Biology and Pathology, Dingle JT, Fell HB, Eds. (1969); **9.** Proc Natl Acad Sci U S A (1984) 81:4838; **10.** Lysosomes in Biology and Pathology, Vol. 2, Dingle JT, Fell HB, Eds. (1969) p. 600; **11.** EMBO J (1996) 15:1784; **12.** J Histochem Cytochem (1996) 44:267; **13.** Biol Reprod (1997) 56:991; **14.** Proc Natl Acad Sci U S A (1998) 95:1579; **15.** J Cell Biol (1997) 137:715; **16.** Cytometry (1998) 33:348; **17.** J Biol Chem (1998) 273:22466; **18.** J Biol Chem (1997) 272:14817; **19.** J Neurosci (1997) 17:1582; **20.** Chem Biol (1999) 6:411; **21.** Blood (1997) 89:3745; **22.** Plant Cell (1998) 10:685; **23.** Proc Natl Acad Sci U S A (2008) 105:1579; **24.** Eukaryot Cell (2007) 6:271; **25.** Limnol Oceanogr Meth (2005) 3:462; **26.** Chem Biol (2001) 8:1051; **27.** Br J Cancer (2003) 88:1327; **28.** J Biol Chem (1990) 265:4775; **29.** Methods Enzymol (2009) 453:417; **30.** J Biol Chem (2009) 284:7681; **31.** J Neurochem (1997) 69:1927; **32.** J Cell Biol (1997) 139:1183; **33.** Nucleic Acids Res (2002) 30:1338; **34.** Mol Biol Cell (2004) 15:3132; **35.** J Biol Chem (2003) 278:27180; **36.** J Cell Biol (2001) 152:809; **37.** Nature (1991) 352:70; **38.** J Cell Biol (1994) 126:1149; **39.** Methods Enzymol (2009) 453:111; **40.** J Histochem Cytochem (2006) 54:85; **41.** Nat Methods (2006) 3:817; **42.** Am J Pathol (2009) 174:101; **43.** Am J Physiol Renal Physiol (2007) 292:F523; **44.** Free Radic Biol Med (2000) 28:1266; **45.** Photochem Photobiol (1997) 66:424; **46.** Mol Reprod Dev (2000) 55:335; **47.** In Vitro Toxicol (1990) 3:219; **48.** Anal Biochem (1993) 213:426; **49.** Jpn J Physiol (1993) 43:161; **50.** Brain Res (1992) 573:1; **51.** Immunology (1984) 51:319; **52.** J Immunol (1983) 131:125; **53.** Anal Biochem (1991) 192:316; **54.** Eur J Cell Biol (1995) 66:3; **55.** Methods Enzymol (1991) 194:644; **56.** Trends Biochem Sci (1989) 14:347; **57.** J Biol Chem (2007) 282:16295; **58.** J Biol Chem (2006) 281:27158; **60.** J Biol Chem (1989) 264:19236; **61.** J Microsc (2007) 225:100; **62.** Appl Environ Microbiol (1997) 63:2897; **63.** Biotechnol Intl (1997) 1:291; **64.** J Cell Biol (1994) 126:1375; **65.** Nat Protoc (2008) 3:1909; **66.** Cytometry (1998) 31:307; **67.** J Clin Microbiol (1997) 35:5; **68.** J Cell Biol (1995) 128:779; **69.** J Cell Biol (1999) 146:85; **70.** Science (1999) 285:1084; **71.** J Cell Biol (1996) 135:1535; **72.** J Cell Biol (1996) 135:1485; **73.** Mol Biol Cell (1996) 7:1375; **74.** Mol Biol Cell (1996) 7:985; **75.** Eur J Cell Biol (1994) 65:305; **76.** Methods Cell Biol (1989) 31:357; **77.** Mol Biol Cell (1995) 6:525; **78.** J Cell Biol (1994) 125:283; **79.** J Cell Biol (1990) 111:877; **80.** J Biol Chem (1999) 274:1835; **81.** J Biol Chem (2006) 281:29916; **82.** Methods Enzymol (1991) 194:428; **83.** Arch Biochem Biophys (1983) 226:292; **84.** FEBS Lett (1981) 131:296; **85.** Fungal Genet Biol (1998) 24:86; **86.** Nat Methods (2007) 4:555; **87.** J Histochem Cytochem (2009) 57:687; **88.** J Cell Sci (2005) 118:5243; **89.** Biomaterials (2010) 31:1757; **90.** J Biol Chem (2009) 284:34296; **91.** J Cell Biol (1997) 136:71.

DATA TABLE 12.3 PROBES FOR LYSOSOMES, PEROXISOMES AND YEAST VACUOLES

Cat. No.	MW	Storage	Soluble	Abs	EC	Em	Solvent	Notes
A1301	301.82	L	H_2O, EtOH	489	65,000	520	MeOH	
A3568	301.82	RR,L	H_2O	489	65,000	520	MeOH	1
B22461	385.22	F,D,L	DMSO	503	82,000	511	MeOH	
D113	335.46	L	EtOH, DMF	335	4600	518	MeOH	
D1552	384.26	F,D,L	pH <7, DMF	349	16,000	none	MeOH	
D10460	386.47	L	DMF, DMSO	373	23,000	571	MeOH	
F7030	528.84	F,D,L	DMSO	508	71,000	none	pH 7	1, 2
L7525	524.40	F,D,L	DMSO	373	9600	422	pH 7	1, 3
L7526	398.69	F,D,L	DMSO	504	80,000	511	MeOH	1
L7528	399.25	F,D,L	DMSO	577	78,000	590	MeOH	1, 4
L7533	376.50	F,D,L	DMSO	373	11,000	425	pH 5	1, 5
L7534	356.43	F,D,L	DMSO	442	17,000	505	pH 5	1, 5
L7535	398.46	F,D,L	DMSO	443	16,000	505	pH 5	1, 5
L7545	366.42	F,D,L	DMSO	384	21,000	540	pH 3	1, 6
L12491	364.4	F,D,L	DMSO	466	22,000	536	MeOH	1
L22460	see Notes	F,D,L	H_2O	384	ND	540	pH 3	6, 7, 8
N3246	288.78	D,L	H_2O, EtOH	541	39,000	640	see Notes	9
R14060	434.41	F,D,L,AA	DMSO	239	52,000	none	MeOH	
T3166	607.51	D,L	H_2O, DMSO	505	47,000	725	see Notes	10, 11
T13320	607.51	D,L	H_2O, DMSO	505	47,000	725	see Notes	10, 11
T23360	565.43	D,L	H_2O, DMSO	560	43,000	734	$CHCl_3$	10
Y7536	384.48	F,L	DMSO, DMF	456	27,000	505	MeOH	

For definitions of the contents of this data table, see "Using *The Molecular Probes® Handbook*" in the introductory pages.

Notes

1. This product is supplied as a ready-made solution in the solvent indicated under "Soluble."
2. F7030 is fluorescent when bound to DNA (Em = 538 nm). Uptake and processing of the dye by live yeast results in red-shifted fluorescence (Em ~590 nm).
3. L7525 has structured absorption and fluorescence spectra with additional peaks at Abs = 394 nm and Em = 401 nm.
4. The pK_a of the dimethylamino substituent of LysoTracker® Red DND-99 is 7.5. (J Biol Chem (2004) 279:32367) The absorption and fluorescence spectra of the dye are insensitive to protonation of this substituent.
5. This LysoSensor™ dye exhibits increasing fluorescence as pH decreases with no spectral shift. L7533 has additional absorption and fluorescence emission peaks at Abs = 394 nm and Em = 401 nm.
6. LysoSensor™ Yellow/Blue spectra are pH dependent. Abs and Em shift to shorter wavelengths at pH >5.
7. The molecular weight is nominally as specified in the product name but may have a broad distribution.
8. ND = not determined.
9. Spectra of N3246 are pH dependent (pK_a ~6.7). Data reported are for 1:1 (v/v) EtOH/1% acetic acid.
10. FM® 4-64 and FM® 5-95 are nonfluorescent in water. For two-color imaging in GFP-expressing cells, these dyes can be excited at 560 nm with emission detection at 690–730 nm. (Am J Physiol Cell Physiol (2001) 281:C624)
11. Abs, EC and Em determined for dye bound to detergent micelles (20 mg/mL CHAPS in H_2O). These dyes are essentially nonfluorescent in pure water.

IMPORTANT NOTICE: The products described in this manual are covered by one or more Limited Use Label License(s). Please refer to the Appendix on page 971 and Master Product List on page 975. Products are For Research Use Only. Not intended for any animal or human therapeutic or diagnostic use.

PRODUCT LIST 12.3 PROBES FOR LYSOSOMES, PEROXISOMES AND YEAST VACUOLES

Cat. No.	Product	Quantity
A1301	acridine orange	1 g
A3568	acridine orange *10 mg/mL solution in water*	10 mL
B22461	BODIPY® FL histamine	1 mg
C10586	CellLight® Early Endosomes-GFP *BacMam 2.0*	1 mL
C10587	CellLight® Early Endosomes-RFP *BacMam 2.0*	1 mL
C10596	CellLight® Lysosomes-GFP *BacMam 2.0*	1 mL
C10597	CellLight® Lysosomes-RFP *BacMam 2.0*	1 mL
C10604	CellLight® Peroxisome-GFP *BacMam 2.0*	1 mL
D10460	Dapoxyl® (2-aminoethyl)sulfonamide	10 mg
D113	5-dimethylaminonaphthalene-1-(*N*-(5-aminopentyl))sulfonamide (dansyl cadaverine)	100 mg
D1552	*N*-(3-((2,4-dinitrophenyl)amino)propyl)-*N*-(3-aminopropyl)methylamine, dihydrochloride (DAMP)	100 mg
F7030	FUN® 1 cell stain *10 mM solution in DMSO*	100 µL
I34202	Image-iT® LIVE Lysosomal and Nuclear Labeling Kit *counterstains for GFP-expressing cells*	1 kit
L7533	LysoSensor™ Blue DND-167 *1 mM solution in DMSO* *special packaging*	20 x 50 µL
L7534	LysoSensor™ Green DND-153 *1 mM solution in DMSO* *special packaging*	20 x 50 µL
L7535	LysoSensor™ Green DND-189 *1 mM solution in DMSO* *special packaging*	20 x 50 µL
L22460	LysoSensor™ Yellow/Blue dextran, 10,000 MW, anionic, fixable	5 mg
L7545	LysoSensor™ Yellow/Blue DND-160 (PDMPO) *1 mM solution in DMSO*	20 x 50 µL
L7525	LysoTracker® Blue DND-22 *1 mM solution in DMSO* *special packaging*	20 x 50 µL
L7526	LysoTracker® Green DND-26 *1 mM solution in DMSO* *special packaging*	20 x 50 µL
L7528	LysoTracker® Red DND-99 *1 mM solution in DMSO* *special packaging*	20 x 50 µL
L12491	LysoTracker® Yellow HCK-123 *1 mM solution in DMSO* *special packaging*	20 x 50 µL
N3246	neutral red *high purity*	25 mg
R14060	RedoxSensor™ Red CC-1 *special packaging*	10 x 50 µg
S34201	SelectFX® Alexa Fluor® 488 Peroxisome Labeling Kit *for fixed cells*	1 kit
T3166	*N*-(3-triethylammoniumpropyl)-4-(6-(4-(diethylamino)phenyl)hexatrienyl)pyridinium dibromide (FM® 4-64)	1 mg
T13320	*N*-(3-triethylammoniumpropyl)-4-(6-(4-(diethylamino)phenyl)hexatrienyl)pyridinium dibromide (FM® 4-64) *special packaging*	10 x 100 µg
T23360	*N*-(3-trimethylammoniumpropyl)-4-(6-(4-(diethylamino)phenyl)hexatrienyl)pyridinium dibromide (FM® 5-95)	1 mg
Y7531	Yeast Vacuole Marker Sampler Kit	1 kit
Y7536	yeast vacuole membrane marker MDY-64	1 mg

12.4 Probes for the Endoplasmic Reticulum and Golgi Apparatus

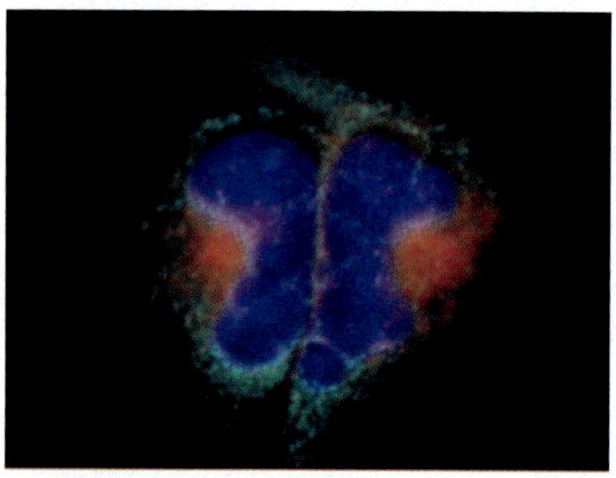

Figure 12.4.1 Fixed and permeabilized osteosarcoma cells simultaneously stained with the fluorescent lectins Alexa Fluor® 488 concanavalin A (Con A) (C11252) and Alexa Fluor® 594 wheat germ agglutinin (WGA) (W11262). Con A selectively binds α-glucopyranosyl residues, whereas WGA selectively binds sialic acid and *N*-acetylglucosaminyl residues. The nuclei were counterstained with blue-fluorescent Hoechst 33342 nucleic acid stain (H1399, H3570, H21492). The image was acquired using bandpass filter sets appropriate for the Texas Red® dye, fluorescein and AMCA.

The endoplasmic reticulum (ER) and Golgi apparatus are primarily responsible for the proper sorting of lipids and proteins in cells.[1] Consequently, most of the cell-permeant probes for these organelles are either lipids or chemicals that affect protein movement. Several of the most effective probes for the Golgi apparatus are fluorescent ceramides and sphingolipids, which are discussed below and in Section 13.3. Certain aspects of lipid trafficking through the ER and Golgi apparatus related to signal transduction are described in Section 17.4. In both live and fixed cells, the flattened membranous sacs of the ER and the Golgi apparatus can be stained with a variety of lipophilic probes and then distinguished by their morphology.

In addition to these fluorescent organelle stains, we offer several CellLight® reagents that comprise BacMam expression vectors encoding targeted autofluorescent proteins for visualizing the endoplasmic reticulum and Golgi apparatus in live mammalian cells. For labeling fixed-cell preparations, we offer the SelectFX® Alexa Fluor® 488 Endoplasmic Reticulum Labeling Kit (S34200), which contains an antibody directed against the ER-associated protein disulfide isomerase (PDI). Enzymes in the ER are also involved in synthesis of cholesterol and in the detoxification of hydrophobic drugs through the cytochrome P450 system (Section 10.6). Furthermore, some fluorescent lectins are useful markers for the Golgi apparatus because several enzymes in this organelle function to glycosylate lipids and proteins (Figure 12.4.1). Nissl bodies principally comprise ordered structures of alternate lamellae of rough endoplasmic reticulum and polyribosome arrays. Our NeuroTrace® fluorescent Nissl stains are described in Section 12.5. An excellent compendium of human diseases that affect intracellular transport processes through lysosomes, Golgi and ER has been published.[2]

IMPORTANT NOTICE: The products described in this manual are covered by one or more Limited Use Label License(s). Please refer to the Appendix on page 971 and Master Product List on page 975. Products are For Research Use Only. Not intended for any animal or human therapeutic or diagnostic use.
www.invitrogen.com/probes

ER-Tracker™ Dyes for Live-Cell Endoplasmic Reticulum Labeling

ER-Tracker™ dyes are cell-permeant, live-cell stains that are highly selective for the ER. These dyes rarely stain mitochondria, unlike the conventional ER stain $DiOC_6(3)$ (D273), and staining at low concentrations does not appear to be toxic to cells. When cells are stained using the optimized protocol provided, staining patterns are retained after treatment with formaldehyde, although at reduced intensities.

ER-Tracker™ Blue-White DPX Dye

ER-Tracker™ Blue-White DPX (E12353, Figure 12.4.2) is a highly selective and photostable stain for the ER in live cells[3-5] (Figure 12.4.3, Figure 12.4.4). ER-Tracker™ Blue-White DPX is a member of our Dapoxyl® dye family[6] and thus exhibits an unusually large Stokes shift and long-wavelength emission with a high extinction coefficient and high quantum yield when in a hydrophobic environment. Its fluorescence is highly environment sensitive—with increasing solvent polarity, the fluorescence maximum shifts to longer wavelengths (Figure 12.4.5) and the quantum yield decreases—and peak fluorescence emission ranges from 430 nm to 640 nm; we recommend visualizing its ER staining with a standard DAPI or UV longpass optical filter set. The ER-Tracker™ Blue-White DPX dye is also readily visualized by two-photon microscopy.[7]

ER-Tracker™ Green and Red Dyes

ER-Tracker™ Green and ER-Tracker™ Red endoplasmic reticulum stains (E34251, E34250) are fluorescent sulfonylureas—BODIPY® FL glibenclamide (Figure 12.4.6) and BODIPY® TR glibenclamide—which exhibit excitation/emission maxima of ~504/511 nm and 587/615 nm, respectively. Glibenclamide binds to sulfonylurea (SUR) receptors of ATP-sensitive K^+ channels, which are prominent on ER but may have more disseminated tissue- and cell type–dependent distributions.[8] BODIPY® FL glibenclamide also generates SUR-independent labeling in some cases.[9] Despite these mechanistic nuances, ER-Tracker™ Green (Figure 12.4.7) and ER-Tracker™ Red are effective and widely used endoplasmic reticulum markers in live-cell imaging applications.[10–12]

Carbocyanine Dyes

Short-Chain Carbocyanine Dyes

Terasaki and co-workers used the short-chain carbocyanine $DiOC_6(3)$ (D273) to visualize the ER in both live and aldehyde-fixed cells.[13–15] This dye and the similar $DiOC_5(3)$ (D272) have since been used extensively to study structural interactions and dynamics of the ER in neurons,[16,17] yeast[18] and onion epidermis,[19] and to examine the morphological relationships between the ER, mitochondria, intermediate filaments and microtubules in various cell types.[20–22] $DiOC_6(3)$ and $DiOC_5(3)$ pass through the plasma membrane and stain intracellular membranes

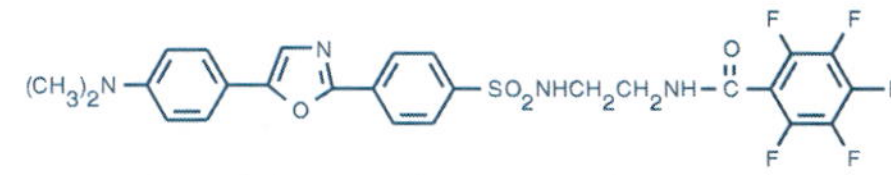

Figure 12.4.2 ER-Tracker™ Blue-White DPX (E12353).

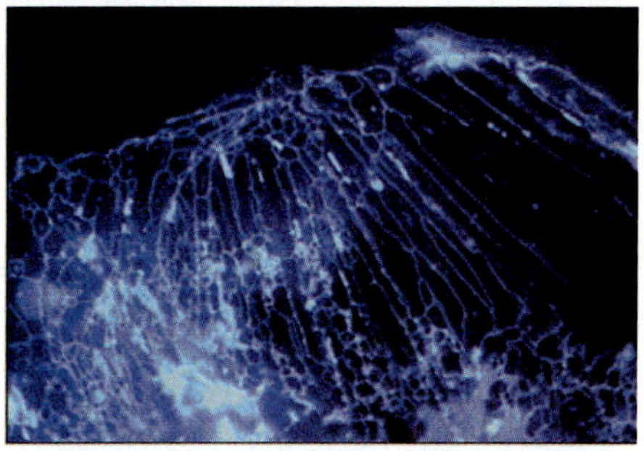

Figure 12.4.3 Live bovine pulmonary artery endothelial cells (BPAEC) stained with ER-Tracker™ Blue-White DPX (E12353), a Dapoxyl® derivative. This image was acquired using a DAPI bandpass optical filter.

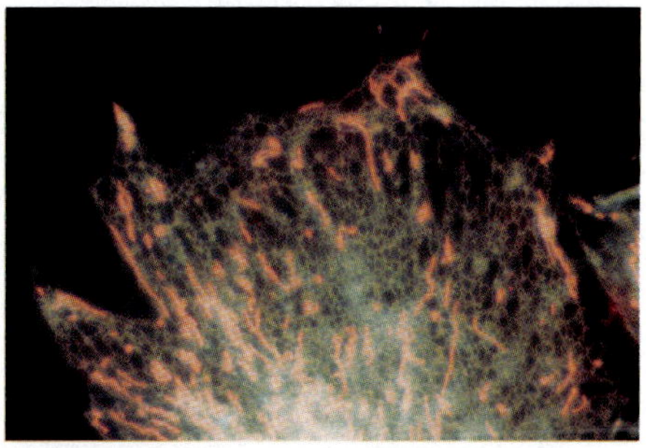

Figure 12.4.4 Live bovine pulmonary artery endothelial cells stained with ER-Tracker™ Blue-White DPX (E12353) and MitoTracker® Red CM-H_2XRos (M7513). The endoplasmic reticulum appears green and the mitochondria appear orange. The image was acquired using a fluorescence microscope equipped with a triple-bandpass filter set appropriate for DAPI, fluorescein and Texas Red® dyes.

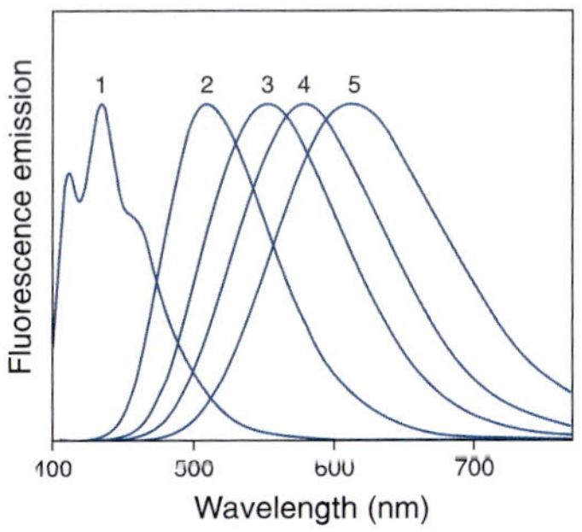

Figure 12.4.5 Normalized fluorescence emission spectra of Dapoxyl® (2-aminoethyl)sulfonamide (D10460) in 1) hexane, 2) chloroform, 3) acetone, 4) acetonitrile and 5) 1:1 acetonitrile:water.

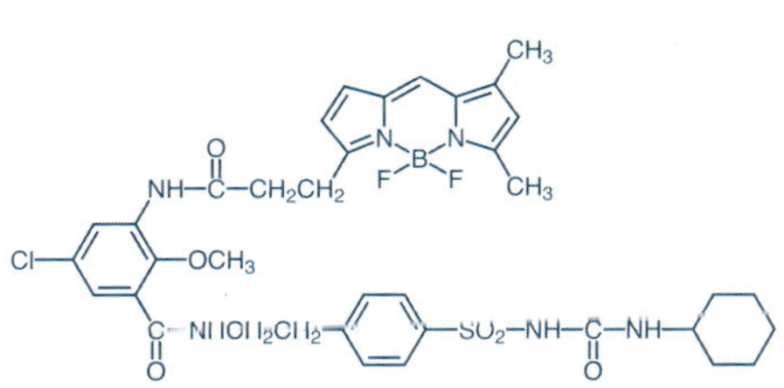

Figure 12.4.6 ER-Tracker™ Green (BODIPY® FL glibenclamide, E34251).

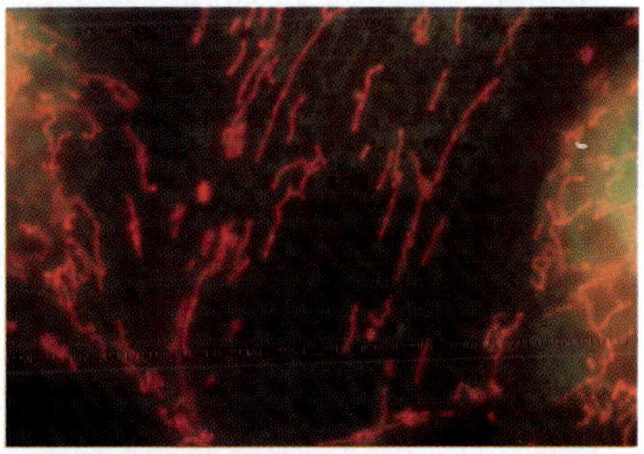

Figure 12.4.7 Organelle staining of live bovine pulmonary artery endothelial cells. Endoplasmic reticulum was labeled with ER-Tracker™ Green (E34251); mitochondria were visualized with MitoTracker® Red CMXRos (M7512).

IMPORTANT NOTICE: The products described in this manual are covered by one or more Limited Use Label License(s). Please refer to the Appendix on page 971 and Master Product List on page 975. Products are For Research Use Only. Not intended for any animal or human therapeutic or diagnostic use.

with a fluorescein-like fluorescence; ER membranes can easily be distinguished by their characteristic morphology.[23] Caution must be exercised, however, in using the carbocyanines as probes for the ER. It has been reported that ER staining with $DiOC_6(3)$ does not occur until the mitochondria round up and lose the fluorochrome.[24] Rhodamine 6G and the hexyl ester of rhodamine B (R634, R648MP; Section 12.2) appear to stain like $DiOC_6(3)$, except they are apparently less toxic and they fluoresce orange, providing possibilities for multicolor labeling.[23,25] When used at very low concentrations, these slightly lipophilic rhodamine dyes tend to stain only mitochondria of live cells.[26]

Long-Chain Carbocyanine Dyes

Terasaki and Jaffe have used the long-chain carbocyanines $DiIC_{16}(3)$ and $DiIC_{18}(3)$ (D384, D282) to label ER membranes. They achieved selective labeling of the ER by microinjecting a saturated solution of DiI in oil into sea urchin eggs.[27,28] This method has been successful in several other egg types but was not effective in molluscan or arthropod axons. As noted in the discussion of dialkylcarbocyanine and dialkylaminostyryl probes in Section 13.4, DiI diffuses only in continuous membranes.

Fluorescent Ceramide Analogs

NBD C_6-ceramide and BODIPY® FL C_5-ceramide (N1154, D3521), both of which can be used with fluorescein optical filter sets, are selective stains for the Golgi apparatus.[29–31] With spectral properties similar to those of Texas Red® dye, BODIPY® TR ceramide[32,33] (D7540) is especially useful for double-labeling in combination with Green Fluorescent Protein (GFP) fusion proteins[34] (Using Organic Fluorescent Probes in Combination with GFP—Note 12.1), as well as for staining cells and tissues that have substantial amounts of green autofluorescence. In addition, the BODIPY® TR fluorophore is ideal for imaging microscopy with CCD cameras or other red-sensitive detectors. Uptake of fluorescent ceramides, at least in *Paramecium* cells, appears to be an ATP-dependent process.[35]

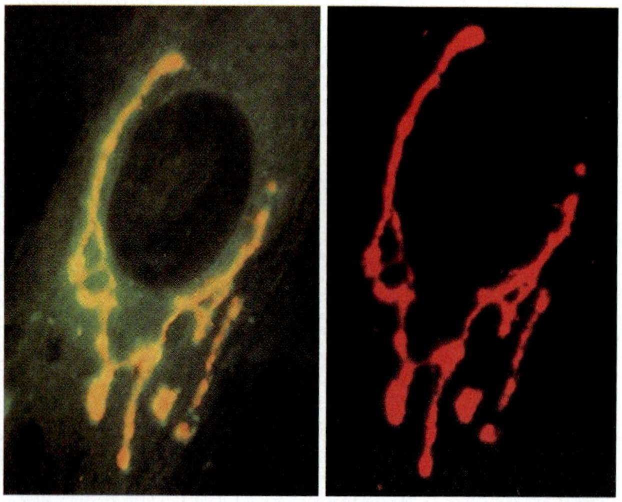

Figure 12.4.8 Selective staining of the Golgi apparatus using the green-fluorescent BODIPY® FL C_5-ceramide (D3521) (left). At high concentrations, the BODIPY® FL fluorophore forms excimers that can be visualized using a red longpass optical filter (right). The BODIPY® FL C_5-ceramide accumulation in the trans-Golgi is sufficient for excimer formation (J Cell Biol (1991) 113:1267). Images contributed by Richard Pagano, Mayo Foundation.

NBD Ceramide and NBD Sphingomyelin

NBD C_6-ceramide (N1154) and NBD C_6-ceramide complexed with defatted BSA (N22651) have been used extensively as a selective stain of the trans-Golgi in both live and fixed cells.[36–43] Complexing fluorescent ceramides with bovine serum albumin (BSA) facilitates cell labeling without requiring the use of organic solvents to dissolve the probe.[29] Furthermore, the fluorescence of NBD C_6-ceramide is apparently sensitive to the cholesterol content of the Golgi apparatus, a phenomenon that is not observed with BODIPY® FL C_5-ceramide.[44] If NBD C_6-ceramide–containing cells are starved for cholesterol, the NBD C_6-ceramide that accumulates within the Golgi apparatus appears to be severely photolabile. However, this NBD photobleaching can be reduced by stimulation of cholesterol synthesis. Thus, NBD C_6-ceramide may be useful in monitoring the cholesterol content of the Golgi apparatus in live cells.[44]

NBD C_6-ceramide's conversion to the NBD C_6-glycosyl ceramide and NBD C_6-sphingomyelin (N3524) has been observed *in vivo*.[45–47] Metabolism of the probe in live Chinese hamster ovary (CHO) fibroblasts has been used to define lipid-transport pathways.[45,48] Like NBD C_6-ceramide, NBD C_6-sphingomyelin has been used for the study of lipid trafficking between organelles.[49] Normal fibroblasts hydrolyze NBD C_6-sphingomyelin and process it through the Golgi apparatus.[50] However, in human skin fibroblasts from patients with Niemann–Pick disease, which is characterized by a lack of lysosomal sphingomyelinase activity, NBD C_6-sphingomyelin accumulates in the lysosomes.

BODIPY® Ceramides, BODIPY® Sphingomyelin and Related Derivatives

The green-fluorescent BODIPY® FL C_5-ceramide (D3521) is more fade-resistant and brighter than the NBD derivative and can likely be substituted for the NBD C_6-ceramide in many of its applications. The red-fluorescent BODIPY® TR ceramide (D7540) has proven useful for two-color immunofluorescence using a fluorescein-labeled antibody.[51] As with NBD C_6-ceramide, we also offer BODIPY® FL C_5-ceramide and BODIPY® TR ceramide complexed with defatted BSA (B22650, B34400) to facilitate cell labeling without the use of organic solvents to dissolve the probe.[29]

During normal resting intracellular transport, the kinetics of dye loading and transport may differ somewhat between the BODIPY® and NBD analogs.[52] BODIPY® FL C_5-ceramide has proven to be an excellent structural marker for the Golgi apparatus, visualized either by fluorescence microscopy[53,54] or, following diaminobenzidine (DAB) conversion, electron microscopy.[55–57] BODIPY® FL C_5-ceramide has also been used to:

- Delineate the Golgi apparatus in the cytoarchitecture of size-excluding compartments in live cells[58]
- Investigate both the inhibition of glycoprotein transport by ceramides[59] and the possible link between protein secretory pathways and sphingolipid biosynthesis[60]
- Isolate mammalian secretion mutants[61]
- Study sphingolipid distribution during human keratinocyte differentiation[62]
- Visualize tubovesicular membranes induced by *Plasmodium falciparum*[63,64]

BODIPY® FL C_5-ceramide exhibits concentration-dependent fluorescence properties that provide additional benefits for imaging the Golgi apparatus. At high concentrations, the nonpolar BODIPY® FL

IMPORTANT NOTICE: The products described in this manual are covered by one or more Limited Use Label License(s). Please refer to the Appendix on page 971 and Master Product List on page 975. Products are For Research Use Only. Not intended for any animal or human therapeutic or diagnostic use.

molecular probes® | invitrogen™ by life technologies™

fluorophore forms excimers, resulting in a shift of the fluorophore's emission maximum from 515 nm (green) to ~620 nm (red). BODIPY® FL C_5-ceramide accumulation is sufficient for excimer formation in the trans-Golgi but not in the surrounding cytoplasm. Longpass optical filters that isolate the red emission can thus be used to selectively visualize the Golgi apparatus (Figure 12.4.8, Figure 12.4.9). Moreover, this two-color property can be used to quantitate BODIPY® FL C_5-ceramide accumulation by ratio imaging.[30,60,65] Like BODIPY® FL C_5-ceramide, the red-fluorescent BODIPY® TR ceramide appears to form long-wavelength excimers when concentrated in the Golgi apparatus; in this case, however, the excimers exhibit infrared fluorescence. In an unexpected application, it has been shown that cells infected with some intracellular bacteria, including *Chlamydia psittaci*, accumulate BODIPY® FL C_5-ceramide (D3521) in their inclusion membranes rather than in the Golgi of the host cells.[66,67] Certain CellTracker™ reagents (Section 14.2) that were used in combination with BODIPY® FL C_5-ceramide were also found to selectively label intracellular bacteria and parasites.[66]

We also offer BODIPY® FL C_5-sphingomyelin (D3522)—the likely metabolic product of BODIPY® FL C_5-ceramide[30]—as well as BODIPY® FL C_{12}-sphingomyelin[68] (D7711) and BODIPY® FL C_5-lactosylceramide (D13951, B34402). The concentration-dependent fluorescence shift of BODIPY® FL C_5-sphingomyelin from green to red has been used to follow the initial steps of lipid uptake and transport by early endosomes through the cytoplasm.[69] BODIPY® FL C_5-glucocerebroside is reportedly internalized by endocytic and nonendocytic pathways that are quite different from those governing the internalization of BODIPY® FL C_5-sphingomyelin[70] (D3522). Addition of BODIPY® FL C_5-lactosylceramide to the culture medium of cells from patients with sphingolipid-storage diseases (sphingolipidosis) results in fluorescent product accumulation in lysosomes, whereas this probe accumulates in the Golgi apparatus of normal cells and cells from patients with other storage diseases.[71,72] Pagano and collaborators have published reviews of the use of BODIPY® ceramides and BODIPY® sphingolipids to study the endocytic pathway in mammalian cells.[29,73]

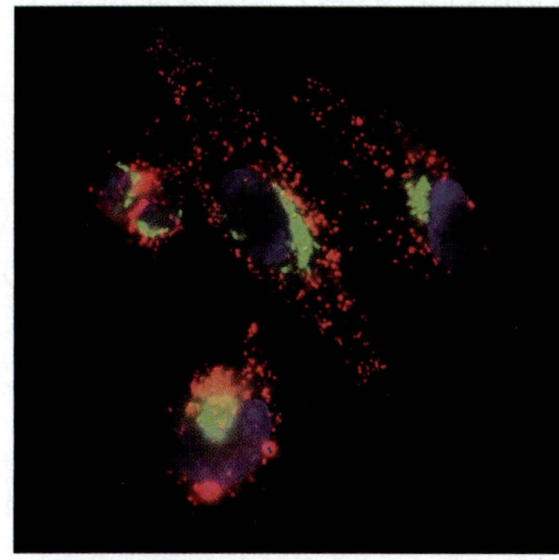

Figure 12.4.9 Viable Madin-Darby canine kidney (MDCK) cells sequentially stained with BODIPY® FL C_5-ceramide (D3521, B22650), LysoTracker® Red DND-99 (L7528) and Hoechst 33258 (H1398, H3569, H21491). Green-fluorescent BODIPY® FL C_5-ceramide localized to the Golgi apparatus, red-fluorescent LysoTracker® Red stain accumulated in the lysosomes, and the blue-fluorescent Hoechst 33258 dye stained the nuclei. The multiple-exposure image was acquired with bandpass filters appropriate for fluorescein, Texas Red® dye and DAPI.

Fluorescent Protein–Based Markers for the Endoplasmic Reticulum and Golgi Apparatus

CellLight® reagents are BacMam expression vectors encoding site-selective autofluorescent protein fusions. These CellLight® reagents incorporate all the customary advantages of BacMam delivery technology including high efficiency transduction of mammalian cells and long-lasting, titratable expression (BacMam Gene Delivery and Expression Technology—Note 11.1). A complete list of our CellLight® reagents and their targeting sequences can be found in Table 11.1.

The CellLight® ER and Golgi markers are generally useful for identification and demarcation of their respective target organelles in live-cell imaging investigations of protein trafficking. CellLight® ER-GFP (C10590, Figure 12.4.10) and CellLight® ER-RFP (C10591) are BacMam expression vectors encoding fusions of Green Fluorescent Protein (GFP) or Red Fluorescent Protein[74] (RFP) with the calreticulin ER insertion sequence and the KDEL tetrapeptide retention sequence. Because the localization of CellLight® ER-GFP and CellLight® ER-RFP is directed by cellular protein trafficking infrastructure, it is more specific than that of dyes such as $DiOC_6(3)$, which is largely driven by simple hydrophobic partition.

CellLight® Golgi-GFP (C10592, Figure 12.4.11) and CellLight® Golgi-RFP (C10593) are BacMam expression vectors encoding fusions of GFP or RFP with the human Golgi-resident enzyme *N*-acetylgalactosaminyltransferase 2.[75]

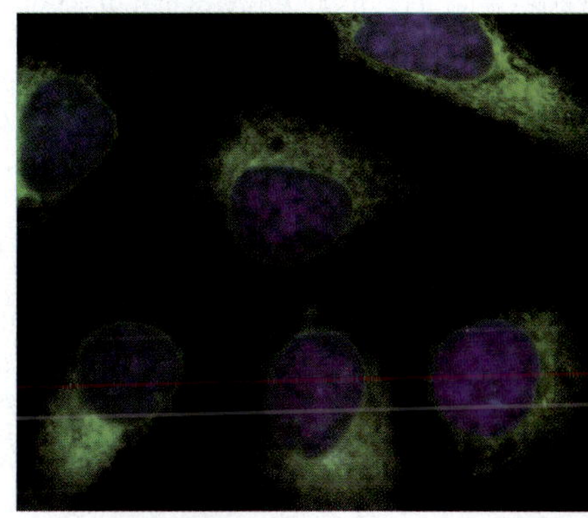

Figure 12.4.10 U2OS osteosarcoma cells labeled with CellLight® ER-GFP reagent (C10590) and with Hoechst 33342 nucleic acid stain.

SelectFX® Alexa Fluor® 488 Endoplasmic Reticulum Labeling Kit

The SelectFX® Alexa Fluor® 488 Endoplasmic Reticulum Labeling Kit (S34200) provides all the reagents required to fix and permeabilize mammalian cells and then specifically label the ER. To achieve ER labeling, this kit employs a primary antibody directed against an ER-associated protein, protein disulfide isomerase (PDI), and an Alexa Fluor® 488 dye–labeled secondary antibody. The Alexa Fluor® 488 dye exhibits bright green fluorescence that is compatible with filters and instrument settings appropriate for fluorescein.

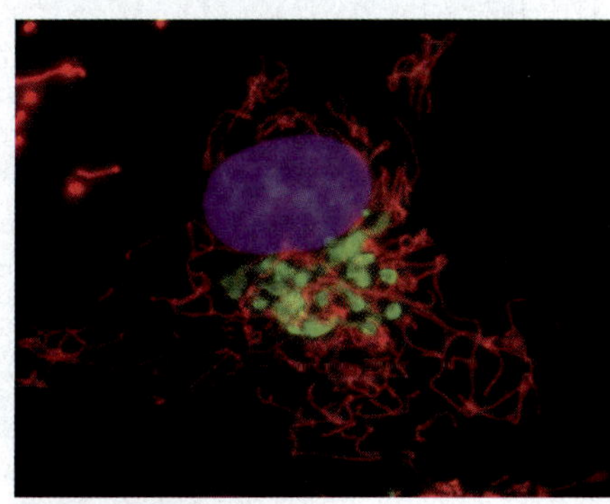

Figure 12.4.11 Human aortic smooth muscle cells (HASMC) labeled with CellLight® Golgi-GFP (C10592) and CellLight® Mitochondria-RFP (C10601) reagents and with Hoechst 33342 nucleic acid stain.

IMPORTANT NOTICE: The products described in this manual are covered by one or more Limited Use Label License(s). Please refer to the Appendix on page 971 and Master Product List on page 975. Products are For Research Use Only. Not intended for any animal or human therapeutic or diagnostic use.

Each SelectFX® Alexa Fluor® 488 Endoplasmic Reticulum Labeling Kit contains:

- Mouse IgG_{2b} anti–protein disulfide isomerase (PDI) antibody
- Highly cross-adsorbed Alexa Fluor® 488 goat anti–mouse IgG antibody
- Concentrated fixative solution
- Concentrated phosphate-buffered saline (PBS)
- Concentrated permeabilization solution
- Concentrated blocking solution
- Detailed protocols for mammalian cell preparation and staining

Lectins for Staining the Golgi Apparatus

Wheat Germ Agglutinin and Concanavalin A

Various proteins and lipids found in the Golgi apparatus are glycosylated; consequently, lectin conjugates (Section 7.7) have been found to be particularly useful for staining Golgi structures in fixed-cell preparations (Figure 12.4.12). Wheat germ agglutinin (WGA) conjugates are commonly used as markers of the trans-Golgi.[76–78] Fluorescent conjugates of concanavalin A (Con A) also stain the Golgi but with reduced specificity.[79] We prepare WGA and Con A conjugates whose fluorescence spans the entire visible and near-infrared spectrum (Table 7.10). Our Alexa Fluor® conjugates of these important lectins are particularly recommended for their enhanced brightness and photostability. We also offer a Wheat Germ Agglutinin Sampler Kit (W7024), which contains 1 mg quantities each of WGA conjugates of the Alexa Fluor® 350, Oregon Green® 488, tetramethylrhodamine and Texas Red®-X dyes.

Griffonia simplicifolia Lectin GS-II

Lectin GS-II from *Griffonia simplicifolia* is the only known lectin that binds with high selectivity to terminal, nonreducing α- and β-*N*-acetyl-D-glucosaminyl (GlcNAc) residues of glycoproteins. Because of the affinity of lectin GS-II for GlcNAc, conjugates of this lectin are useful for staining intermediate-to-trans Golgi—the site of *N*-acetylglucosaminyltransferase activity.[80] The Golgi apparatus of oligodendrocytes and ganglion neurons are readily stained by fluorescent GS-II conjugates. We have prepared the green-fluorescent Alexa Fluor® 488 (L21415, Figure 12.4.13), red-fluorescent Alexa Fluor® 594 (L21416) and far-red–fluorescent Alexa Fluor® 647 (L32451) conjugates of lectin GS-II for use in Golgi staining.[34]

Helix pomatia (Edible Snail) Agglutinin

Helix pomatia agglutinin (HPA) selectively binds to terminal α-*N*-acetylgalactosaminyl residues—an intermediate sugar added in *O*-linkage to serine and threonine residues in *cis*-Golgi cisternae and then substituted with galactose and sialic acid in the *trans*-Golgi.[81] HPA conjugates are principally used as markers for the Golgi. Our fluorescent Alexa Fluor® 488 and Alexa Fluor® 647 conjugates of HPA (L11271, L32454) should be particularly useful for Golgi staining.

Brefeldin A

Isolated from from *Penicillium brefeldianum*, brefeldin A (BFA, B7450) has multiple targets in cells.[82] Exposing cells to BFA causes a distortion in intracellular protein traffic from the ER to the Golgi apparatus and the eventual loss of Golgi apparatus morphology; removal of BFA completely reverses these effects.[83–88] BFA also alters the morphology of endosomes and lysosomes.[89] BFA has been used to prevent retinoic acid potentiation of immunotoxins,[90] to study translocation of proteins in polarized epithelial cells[91] and to investigate the regulation of ADP-ribosylation factor binding to the Golgi apparatus.[92] BFA action can be monitored using fluorescent endosomal markers such as lucifer yellow CH[89] (L453, L682, L1177, L12926; Section 14.3) and tetramethylrhodamine-labeled transferrin[93] (T2872, Section 16.1). Researchers have also used BFA to detect the intracellular expression of cytokines.[94,95] BFA disrupts Golgi-mediated intracellular transport and allows cytokines to accumulate, producing an enhanced cytokine signal that can be detected by flow cytometry.

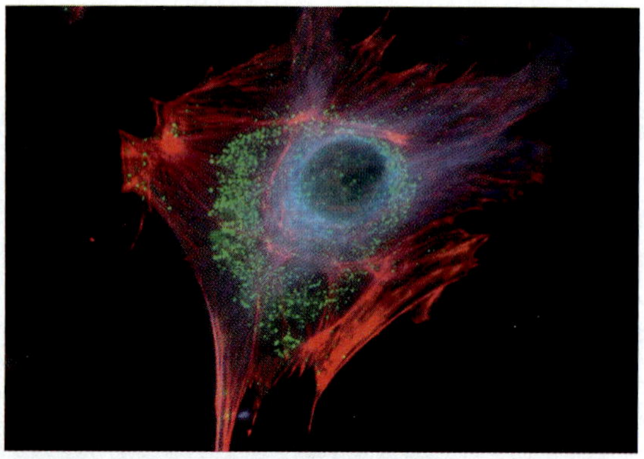

Figure 12.4.12 Microtubules of fixed bovine pulmonary artery endothelial cells localized with mouse monoclonal anti–α-tubulin antibody (A11126), which was subsequently visualized with Alexa Fluor® 350 goat anti–mouse IgG antibody (A11045). Next, the F-actin was labeled with Alexa Fluor® 594 phalloidin (A12381). Finally, the cells were incubated with Alexa Fluor® 488 wheat germ agglutinin (W11261) to stain components of endosomal pathways. The superimposed and pseudocolored images were acquired sequentially using bandpass filter sets appropriate for DAPI, the Texas Red® dye and fluorescein, respectively.

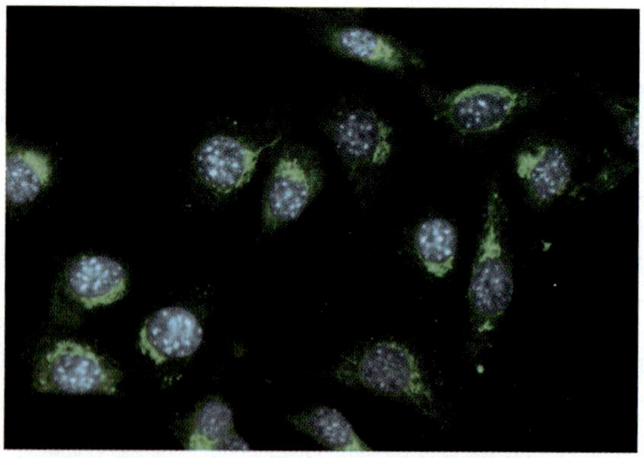

Figure 12.4.13 Fixed and permeabilized NIH 3T3 cells were labeled with the Alexa Fluor® 488 conjugate of lectin GS-II from *Griffonia simplicifolia* (L21415) and counterstained with DAPI (D1306, D3571, D21490). The fluorescent images are shown overlaid with the differential interference contrast (DIC) image.

IMPORTANT NOTICE: The products described in this manual are covered by one or more Limited Use Label License(s). Please refer to the Appendix on page 971 and Master Product List on page 975. Products are For Research Use Only. Not intended for any animal or human therapeutic or diagnostic use.

REFERENCES

1. Science (1998) 282:2172; **2.** Traffic (2000) 1:836; **3.** Fungal Genet Biol (2000) 29:95; **4.** Mol Biol Cell (2000) 11:1815; **5.** J Microsc (2000) 197:239; **6.** Photochem Photobiol (1997) 66:424; **7.** J Biol Chem (2000) 275:22487; **8.** Diabetologia (2007) 50:1889; **9.** Biochem Pharmacol (2004) 67:1437; **10.** Biomaterials (2010) 31:1757; **11.** J Histochem Cytochem (2009) 57:687; **12.** Apoptosis (2007) 12:1155; **13.** Cell Biology: A Laboratory Handbook, 2nd Ed., Vol. 2, Celis JE, Ed. (1998) p. 501; **14.** J Cell Biol (1986) 103:1557; **15.** Cell (1984) 38:101; **16.** J Cell Biol (1994) 127:1021; **17.** Nature (1984) 310:53; **18.** Cell Motil Cytoskeleton (1993) 25:111; **19.** Eur J Cell Biol (1990) 52:328; **20.** J Cell Biol (1997) 137:1199; **21.** Biochem Cell Biol (1992) 70:1174; **22.** Cell Motil Cytoskeleton (1990) 15:71; **23.** J Cell Sci (1992) 101:315; **24.** Microsc Res Tech (1994) 27:198; **25.** J Cell Biol (1998) 143:861; **26.** Histochemistry (1990) 94:303; **27.** J Cell Biol (1993) 120:1337; **28.** J Cell Biol (1991) 114:929; **29.** Cell Biology: A Laboratory Handbook, 2nd Ed., Vol. 2, Celis JE, Ed. (1998) p. 507; **30.** J Cell Biol (1991) 113:1267; **31.** J Cell Biol (1989) 109:2067; **32.** Infect Immun (2000) 68:5960; **33.** Mol Biochem Parasitol (2000) 106:21; **34.** Nat Cell Biol (2006) 8:238; **35.** J Histochem Cytochem (2004) 52:557; **36.** J Biol Chem (1993) 268:18390; **37.** Am J Physiol (1991) 260:G119; **38.** Neurochem Res (1991) 16:551; **39.** Biochem Soc Trans (1990) 18:361; **40.** Eur J Cell Biol (1990) 53:173; **41.** Science (1985) 229:1051; **42.** Science (1985) 228:745; **43.** Proc Natl Acad Sci U S A (1983) 80:2608; **44.** Proc Natl Acad Sci U S A (1993) 90:2661; **45.** J Cell Biol (1990) 111:977; **46.** Biochemistry (1988) 27:6197; **47.** J Cell Biol (1987) 105:1623; **48.** J Cell Biol (1989) 108:2169; **49.** J Biol Chem (2005) 280:15794; **50.** J Cell Biol (1990) 111:429; **51.** Proc Natl Acad Sci U S A (1996) 93:10217; **52.** Methods Cell Biol (1993) 38:221; **53.** Cytometry (1993) 14:251; **54.** J Cell Biol (1993) 120:399; **55.** J Cell Biol (1994) 127:29; **56.** Cell (1993) 73:1079; **57.** Eur J Cell Biol (1992) 58:214; **58.** J Cell Sci (1993) 106:565; **59.** J Biol Chem (1993) 268:4577; **60.** Biochemistry (1992) 31:3581; **61.** Mol Biol Cell (1995) 6:135; **62.** J Biol Chem (1998) 273:9651; **63.** Science (1997) 276:1122; **64.** J Cell Biol (1994) 124:449; **65.** FASEB J (1994) 8:573; **66.** J Microbiol Methods (2000) 40:265; **67.** EMBO J (1996) 15:964; **68.** J Cell Biol (1998) 140:39; **69.** Biophys J (1997) 72:37; **70.** J Cell Biol (1994) 125:769; **71.** Nat Cell Biol (1999) 1:386; **72.** Lancet (1999) 354:901; **73.** Ann N Y Acad Sci (1998) 845:152; **74.** Nat Methods (2007) 4:555; **75.** J Biol Chem (2009) 284:1636; **76.** Cytometry (1995) 19:112; **77.** J Cell Biol (1980) 85:429; **78.** J Cell Biochem (1997) 66:165; **79.** Exp Cell Res (1993) 207:136; **80.** J Struct Biol (1999) 128:131; **81.** J Biol Chem (2006) 281:20171; **82.** Cell (1991) 67:449; **83.** J Biol Chem (1986) 261:11398; **84.** J Biol Chem (1993) 268:2341; **85.** J Cell Biol (1992) 116:1071; **86.** Proc Natl Acad Sci U S A (1991) 88:9818; **87.** Cell (1990) 60:821; **88.** J Cell Biol (1990) 111:2295; **89.** J Cell Biol (1992) 119:273; **90.** J Cell Biol (1994) 125:743; **91.** J Cell Biol (1994) 124:83; **92.** Nature (1993) 364:818; **93.** J Cell Biol (1992) 118:267; **94.** Blood (1995) 86:1357; **95.** J Immunol Methods (1993) 159:197.

DATA TABLE 12.4 PROBES FOR THE ENDOPLASMIC RETICULUM AND GOLGI APPARATUS

Cat. No.	MW	Storage	Soluble	Abs	EC	Em	Solvent	Notes
B7447	554.44	F,D,L	DMSO, EtOH	503	83,000	510	MeOH	
B7449	608.51	F,D,L	DMSO, EtOH	559	80,000	568	MeOH	
B7450	280.36	F,D	DMSO, EtOH	<300		none		
B22650	~66,000	F,D,L	H_2O	505	91,000	511	MeOH	1, 2
B34400	~66,000	F,D,L	H_2O	589	65,000	616	MeOH	2
B34402	~66,000	F,D,L	H_2O	505	80,000	511	MeOH	1, 2
D272	544.47	D,L	DMSO	484	155,000	500	MeOH	
D273	572.53	D,L	DMSO	484	154,000	501	MeOH	
D282	933.88	L	DMSO, EtOH	549	148,000	565	MeOH	
D384	877.77	L	DMSO, EtOH	549	148,000	565	MeOH	
D3521	601.63	FF,D,L	$CHCl_3$, DMSO	505	91,000	511	MeOH	1
D3522	766.75	FF,D,L	see Notes	505	77,000	512	MeOH	1, 3
D7540	705.71	FF,D,L	$CHCl_3$, DMSO	589	65,000	616	MeOH	
D7711	864.94	FF,D,L	DMSO	505	75,000	513	MeOH	1, 4
D13951	925.91	FF,D,L	DMSO, EtOH	505	80,000	511	MeOH	
E12353	580.53	F,D,L	DMSO	374	25,000	575	MeOH	4, 5
N1154	575.75	FF,D,L	$CHCl_3$, DMSO	466	22,000	536	MeOH	6
N3524	740.88	FF,D,L	see Notes	466	22,000	536	MeOH	3, 6
N22651	~66,000	F,D,L	H_2O	466	22,000	536	MeOH	2, 6

For definitions of the contents of this data table, see "Using *The Molecular Probes® Handbook*" in the introductory pages.

Notes

1. Em for BODIPY® FL sphingolipids shifts to ~620 nm when high concentrations of the probe (>5 mol %) are incorporated in lipid mixtures. (J Cell Biol (1991) 113:1267)
2. This product is a lipid complexed with bovine serum albumin (BSA). Spectroscopic data are for the free lipid in MeOH.
3. Chloroform is the most generally useful solvent for preparing stock solutions of phospholipids (including sphingomyelins). Glycerophosphocholines are usually freely soluble in ethanol. Most other glycerophospholipids (phosphoethanolamines, phosphatidic acids and phosphoglycerols) are less soluble in ethanol, but solutions up to 1–2 mg/mL should be obtainable, using sonication to aid dispersion if necessary. Labeling of cells with fluorescent phospholipids can be enhanced by addition of cyclodextrins during incubation. (J Biol Chem (1999) 274:35359)
4. This product is supplied as a ready-made solution in the solvent indicated under "Soluble."
5. ER-Tracker™ Blue-White DPX Abs = 372 nm, Em = 556 nm bound to phospholipid bilayer membranes. The emission spectrum is extremely broad (~200 nm at half-maximum). Fluorescence in water is very weak.
6. Fluorescence of NBD and its derivatives in water is relatively weak. QY and τ increase and Em decreases in aprotic solvents and other nonpolar environments relative to water. (Biochemistry (1977) 16:5150, Photochem Photobiol (1991) 54:361)

PRODUCT LIST 12.4 PROBES FOR THE ENDOPLASMIC RETICULUM AND GOLGI APPARATUS

Cat. No.	Product	Quantity
B22650	BODIPY® FL C_5-ceramide complexed to BSA	5 mg
B34402	BODIPY® FL C_5-lactosylceramide complexed to BSA	1 mg
D7540	BODIPY® TR ceramide (*N*-((4-(4,4-difluoro-5-(2-thienyl)-4-bora-3a,4a-diaza-*s*-indacene-3-yl)phenoxy)acetyl)sphingosine)	250 µg
B34400	BODIPY® TR ceramide complexed to BSA	5 mg
B7450	brefeldin A *from *Penicillium brefeldianum**	5 mg
C10590	CellLight® ER-GFP *BacMam 2.0*	1 mL
C10591	CellLight® ER-RFP *BacMam 2.0*	1 mL
C10592	CellLight® Golgi-GFP *BacMam 2.0*	1 mL
C10593	CellLight® Golgi-RFP *BacMam 2.0*	1 mL
D7711	*N*-(4,4-difluoro-5,7-dimethyl-4-bora-3a,4a-diaza-*s*-indacene-3-dodecanoyl)sphingosyl phosphocholine (BODIPY® FL C_{12}-sphingomyelin) *1 mg/mL in DMSO*	250 µL
D3521	*N*-(4,4-difluoro-5,7-dimethyl-4-bora-3a,4a-diaza-*s*-indacene-3-pentanoyl)sphingosine (BODIPY® FL C_5-ceramide)	250 µg
D13951	*N*-(4,4-difluoro-5,7-dimethyl-4-bora-3a,4a-diaza-*s*-indacene-3-pentanoyl)sphingosyl 1-β-D-lactoside (BODIPY® FL C_5-lactosylceramide)	25 µg
D3522	*N*-(4,4-difluoro-5,7-dimethyl-4-bora-3a,4a-diaza-*s*-indacene-3-pentanoyl)sphingosyl phosphocholine (BODIPY® FL C_5-sphingomyelin)	250 µg
D384	1,1′-dihexadecyl-3,3,3′,3′-tetramethylindocarbocyanine perchlorate (DiIC_{16}(3))	100 mg
D273	3,3′-dihexyloxacarbocyanine iodide (DiOC_6(3))	100 mg
D272	3,3′-dipentyloxacarbocyanine iodide (DiOC_5(3))	100 mg
D282	1,1′-dioctadecyl-3,3,3′,3′-tetramethylindocarbocyanine perchlorate ('DiI'; DiIC_{18}(3))	100 mg
E12353	ER-Tracker™ Blue-White DPX *for live-cell imaging* *1 mM solution in DMSO*	20 x 50 µL
E34251	ER-Tracker™ Green (BODIPY® FL glibenclamide) *for live-cell imaging*	100 µg
E34250	ER-Tracker™ Red (BODIPY® TR glibenclamide) *for live-cell imaging*	100 µg
L21415	lectin GS-II from *Griffonia simplicifolia*, Alexa Fluor® 488 conjugate	500 µg
L21416	lectin GS-II from *Griffonia simplicifolia*, Alexa Fluor® 594 conjugate	500 µg
L32451	lectin GS-II from *Griffonia simplicifolia*, Alexa Fluor® 647 conjugate	500 µg
L11271	lectin HPA from *Helix pomatia* (edible snail), Alexa Fluor® 488 conjugate	1 mg
L32454	lectin HPA from *Helix pomatia* (edible snail), Alexa Fluor® 647 conjugate	1 mg
N1154	NBD C_6-ceramide (6-((*N*-(7-nitrobenz-2-oxa-1,3-diazol-4-yl)amino)hexanoyl)sphingosine)	1 mg
N22651	NBD C_6-ceramide complexed to BSA	5 mg
N3524	NBD C_6-sphingomyelin (6-((*N*-(7-nitrobenz-2-oxa-1,3-diazol-4-yl)amino)hexanoyl)sphingosyl phosphocholine)	1 mg
S34200	SelectFX® Alexa Fluor® 488 Endoplasmic Reticulum Labeling Kit *for fixed cells*	1 kit
W11263	wheat germ agglutinin, Alexa Fluor® 350 conjugate	5 mg
W11261	wheat germ agglutinin, Alexa Fluor® 488 conjugate	5 mg
W32464	wheat germ agglutinin, Alexa Fluor® 555 conjugate	5 mg
W11262	wheat germ agglutinin, Alexa Fluor® 594 conjugate	5 mg
W21404	wheat germ agglutinin, Alexa Fluor® 633 conjugate	5 mg
W32466	wheat germ agglutinin, Alexa Fluor® 647 conjugate	5 mg
W32465	wheat germ agglutinin, Alexa Fluor® 680 conjugate	5 mg
W834	wheat germ agglutinin, fluorescein conjugate	5 mg
W6748	wheat germ agglutinin, Oregon Green® 488 conjugate	5 mg
W849	wheat germ agglutinin, tetramethylrhodamine conjugate	5 mg
W21405	wheat germ agglutinin, Texas Red®-X conjugate	1 mg
W7024	Wheat Germ Agglutinin Sampler Kit *four fluorescent conjugates, 1 mg each*	1 kit

IMPORTANT NOTICE: The products described in this manual are covered by one or more Limited Use Label License(s). Please refer to the Appendix on page 971 and Master Product List on page 975. Products are For Research Use Only. Not intended for any animal or human therapeutic or diagnostic use.

12.5 Probes for the Nucleus

This section describes the use of Molecular Probes® nucleic acid stains for visualizing nuclei and chromosomes, as well as for analyzing chromosome banding patterns. The general chemical and spectroscopic properties of these nucleic acid stains are described in Section 8.1. The application of nucleic acid stains to the study of cell viability, cell proliferation and apoptosis is discussed in Chapter 15.

The counterstains described in this section are compatible with a wide range of cytological labeling techniques, including direct or indirect antibody-based detection methods, *in situ* hybridization and the detection of specific subcellular structures with fluorescent probes such as the mitochondrion-selective MitoTracker® reagents (Section 12.2), F-actin–selective phalloidin (Section 11.1) and autofluorescent proteins (Using Organic Fluorescent Probes in Combination with GFP—Note 12.1).

Nuclear Counterstains for Live Cells and Unfixed Tissues

Cell-permeant nucleic acid stains make it possible to stain live cells or tissues that have been minimally processed. Nuclear staining can reveal the natural location of cells in tissues and can provide a means to follow nuclear changes throughout processes such as mitosis and apoptosis (Section 15.5). Most of these dyes have little effect on cell function, allowing live cells to be traced as they move during development or as they infect other cells.

Cell-Permeant Blue-Fluorescent Counterstains

The membrane-permeant Hoechst 33342 dye (H1399, H3570, H21492) has been extensively used for staining the nuclei of live cells. Hoechst 33342 dye shows AT-selective staining, and Hoechst dye–stained cells and tissues show virtually no cytoplasmic staining. Hoechst 33342 is commonly used to distinguish the compact chromatin of apoptotic nuclei, in combination with BrdU labeling to identify replicating cells and to sort cells based on DNA content (Section 15.5).

While not all of the blue-fluorescent cell-permeant SYTO® dyes in Section 8.1 show selective nuclear staining, SYTO® 40 (S11351) shows excellent staining of the nuclei in a freshwater snail embryo (Figure 12.5.1). All of the blue-fluorescent SYTO® dyes listed in Table 8.3 are available individually as solutions in DMSO (Section 8.1) or in a sampler kit (S11350) to facilitate finding the best counterstain for a particular cell or tissue type.

Cell-Permeant Green-Fluorescent Counterstains

Some of the green-fluorescent cell-permeant SYTO® dyes (Table 8.3, Section 8.1) are excellent nuclear stains for live cells in culture (Figure 12.5.2) and for unfixed tissue sections. The green-fluorescent SYTO® 11 dye (S7573) has shown selective nuclear staining in heart tissue, vascular endothelium and cultured myocytes[1] and in cultured aortic vascular smooth muscle cells,[2] showing promise for broad use in noninvasive confocal laser-scanning microscope investigations. Staining with SYTO® 11 and SYTO® 13 (S7575) dyes facilitated counting cells in brain slices without disrupting the three-dimensional environment.[3] Staining with SYTO® 11 dye was used to follow the movement of cells during development in whole-mount zebrafish embryos.[4] SYTO® 11 dye has also been used to identify meiotic cells in developing brain tissue.[5] *Trypanosoma cruzi* stained with SYTO® 11 dye can easily be detected within the cells they have infected.[6] SYTO® 13 dye was used in a double-labeling experiment with BODIPY® 558/568 phalloidin (B3475, Section 11.1) to stain actin fibers, making it possible to look at nuclear changes and cytoskeletal changes concurrently in apoptotic cells.[7] SYTO® 12 dye (S7574) was used to follow chromosome movement during meiosis in live maize myocytes,[8] and SYTO® 14 dye (S7576) allowed researchers to follow RNA localization within live cells.[9,10] SYTO® 16 dye (S7578) served as an effective nuclear counterstain in cultured cells[11] and has been used to stain nuclei in whole maize roots.[12]

The green-fluorescent SYTO® dyes listed in Table 8.3 are available individually as solutions in DMSO (Section 8.1) or as components in the SYTO® Green-Fluorescent Nucleic Acid Stain Sampler Kit (S7572). This SYTO® Stain Sampler Kit contains 50 µL each of eight different green-fluorescent SYTO® dyes to facilitate finding the best counterstain for a particular cell or tissue type.

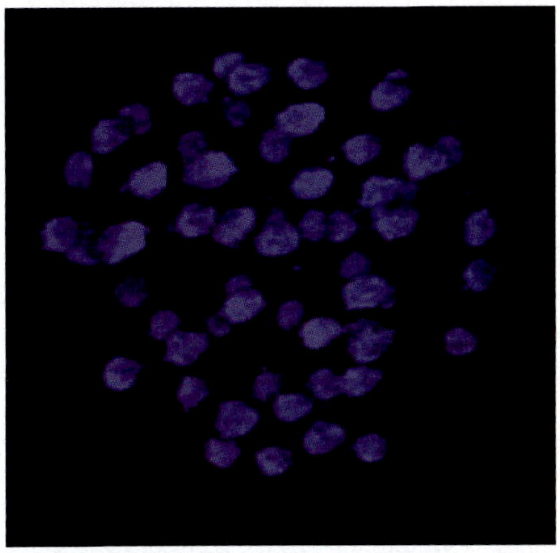

Figure 12.5.1 The developing embryo of a freshwater snail stained with SYTO® 40 blue-fluorescent nucleic acid stain (S11351), a component of the SYTO® Blue Fluorescent Stain Sampler Kit (S11350). A series of z-plane images was acquired with a wide-field optical sectioning confocal microscope. A three-dimensional volume rendering was generated from the deconvolved image series. Image contributed by Bruce Roth and Paul Millard, Molecular Probes®, Inc., and Hillary MacDonald, Applied Precision, Inc.

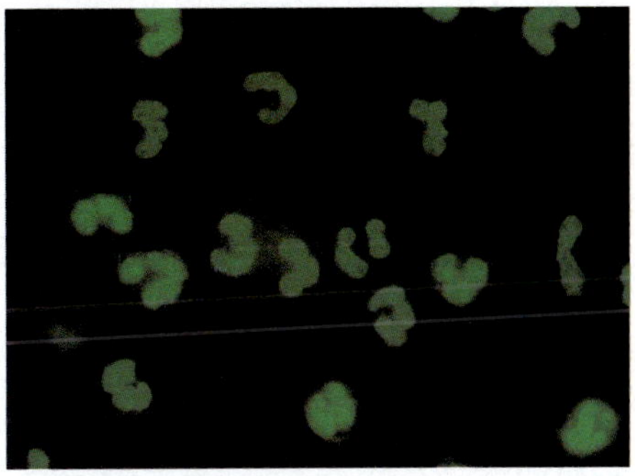

Figure 12.5.2 Adherent cells from human peripheral blood stained with the SYTO® 13 dye (S7575), one of the six visible light-excitable cell-permeant nucleic acid stains in our SYTO® Green Fluorescent Nucleic Acid Stain Sampler Kit #1 (S7572). The multilobed nuclei of these polymorphonuclear leukocytes are particularly striking in this field of view.

Cell-Permeant Orange- and Red-Fluorescent Counterstains

The orange- and red-fluorescent cell-permeant SYTO® nucleic acid stains listed in Table 8.3 may also prove useful as nuclear counterstains for live cells. The red-fluorescent SYTO® 17 dye (S7579) was used as a nuclear counterstain for the green-fluorescent membrane stain $DiOC_6(3)$ (D273, Section 12.4) and with fluorescein immunostaining, as well as with the TUNEL apoptosis assay using ChromaTide® fluorescein-12-dUTP (C7604, Section 15.4) to investigate chromatin degradation and denucleation of lens tissue.[13,14] *Leishmania* cells stained with SYTO® 17 dye could later be identified in cells they had infected.[15] SYTO® 59 dye (S11341) has been used as a red-fluorescent nuclear counterstain in combination with the Green Fluorescent Protein (GFP) expressed in lymphoid cells[16] and human embryonic kidney cells[17] (Using Organic Fluorescent Probes in Combination with GFP—Note 12.1). SYTO® 59 dye has also proven very useful in the study of *Cryptosporidium* oocytes because the intensity of staining appears to be related to the infectivity of the oocytes.[18]

All of these orange- or red-fluorescent SYTO® dyes are available individually as solutions in DMSO (Section 8.1) or as components in the SYTO® Orange Fluorescent Stain Sampler Kit (S11360) or the SYTO® Red Fluorescent Stain Sampler Kit (S11340), which contain 50 µL each of six different orange-fluorescent or seven different red-fluorescent SYTO® dyes to facilitate finding the best counterstain for a particular cell or tissue type.

HCS NuclearMask™ and HCS CellMask™ Stains

In image-based high-content screening (HCS) assays, cell or object identification is the first step of automated image acquisition and analysis. For many image anlaysis algorithms, the cell identification process begins with the detection of fluorescently stained nuclei.[19–21] Using the position of the stained nucleus as a guide, the software then extrapolates to build a mask that marks the probable position of the cytoplasmic region.

The versatile HCS NuclearMask™ stains, which survive standard formaldehyde-based fixation and detergent-based permeabilization methods, can be applied to live or fixed cells. For additional flexibility in multiplexing experiments, HCS NuclearMask™ reagents are available in three fluorescent colors (Figure 12.5.3):

- HCS NuclearMask™ Blue stain (excitation/emission maxima ~350/461 nm, H10325)
- HCS NuclearMask™ Red stain (excitation/emission maxima ~622/645 nm, H10326)
- HCS NuclearMask™ Deep Red stain (excitation/emission maxima ~638/686 nm, H10294)

These three HCS NuclearMask™ stains leave the wavelength region from 500–600 nm clear for multiplexing with green- or orange-fluorescent probes. Sufficient quantities are provided to stain ten 96-well plates using an assay volume of 100 µL per well.

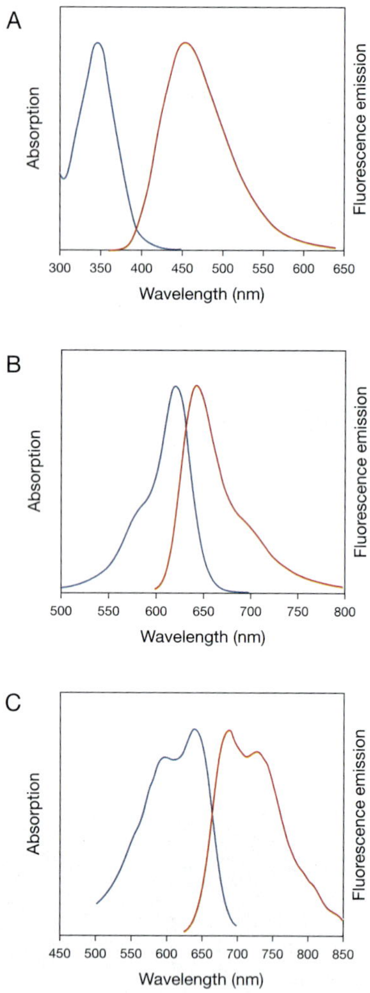

Figure 12.5.3 Absorption and fluorescence emission spectra of **A)** HCS NuclearMask™ Blue (H10325), **B)** HCS NuclearMask™ Red (H10326) and **C)** HCS NuclearMask™ Deep Red (H10294) stains.

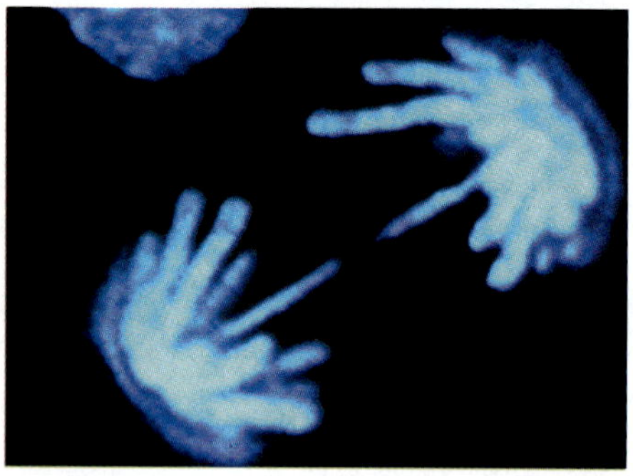

Figure 12.5.4 DAPI-stained condensed chromatin in PtK2 cells during the later stages of mitosis. DAPI (D1306, D3571, D21490) binds to the minor groove of DNA with significant fluorescence enhancement.

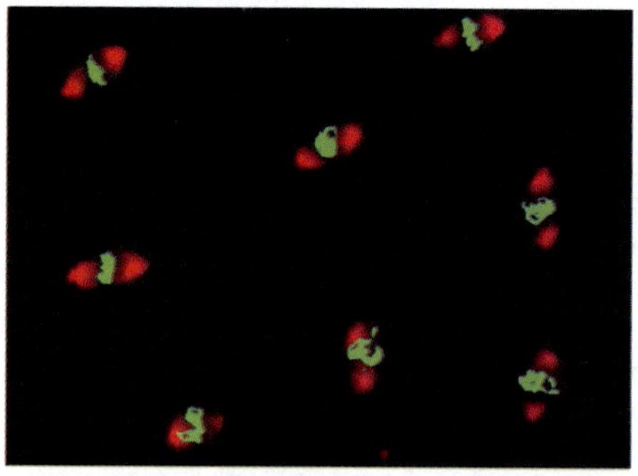

Figure 12.5.5 Mitotic divisions in early *Drosophila* embryos. The spindles were labeled with an anti-α-tubulin primary antibody and probed with a secondary antibody labeled with Lissamine rhodamine B sulfonyl chloride (L20, L1908). After a brief RNase treatment to reduce background fluorescence, the chromosomes were counterstained with SYTOX® Green nucleic acid stain (S7020). The image was contributed by Tulle Hazelrigg and Amy MacQueen, Columbia University.

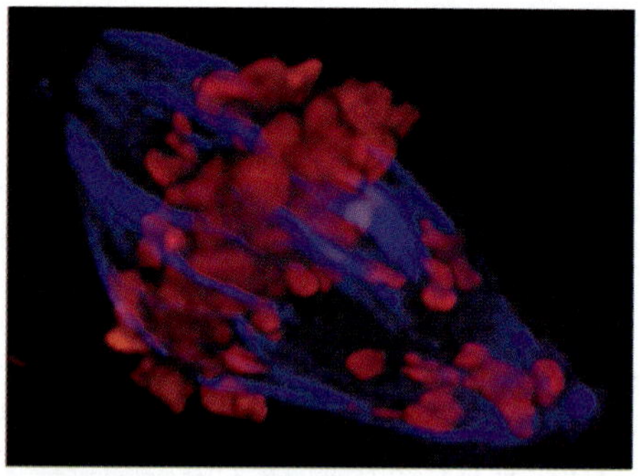

Figure 12.5.6 Microtubules of bovine pulmonary artery endothelial cells that have been labeled with mouse monoclonal anti-α-tubulin antibody (A11126), followed by biotin-XX goat anti-mouse IgG antibody (B2763), and then visualized with Marina Blue® streptavidin (S11221). The cells were next treated with RNase, and the chromosomes were labeled with TO-PRO®-3 iodide (T3605). A series of Z-plane images was acquired with a wide-field optical sectioning confocal laser-scanning microscope. A three-dimensional volume rendering was generated from the deconvolved image series.

In some HCS applications, cell identification based on nuclear staining alone is not adequate because the cytoplasmic region assigned by some image analysis algorithms does not accurately identify the actual cell boundaries. In addition to these HCS NuclearMask™ stains, we offer a series of HCS CellMask™ reagents that label the entire cell (i.e., cytoplasm and the nucleus), designed to provide an accurate backdrop against which to assess the features of interest:

- HCS CellMask™ Blue stain (excitation/emission maxima ~346/442 nm, H32720)
- HCS CellMask™ Green stain (excitation/emission maxima ~493/516 nm, H32714)
- HCS CellMask™ Orange stain (excitation/emission maxima ~556/572 nm, H32713)
- HCS CellMask™ Red stain (excitation/emission maxima ~588/612 nm, H32712)
- HCS CellMask™ Deep Red stain (excitation/emission maxima ~650/655 nm, H32721)

HCS CellMask™ stains are applied to cells immediately after fixation and permeabilization or after labeling with antibodies. Sufficient quantities are provided to stain ten 96-well plates using assay volumes of 100 µL per well.

Tracking Chromosomes through Mitosis

Many nucleic acid stains can be used to observe chromosomes caught in the act of cell division in fixed cells and tissues (Figure 12.5.4, Figure 12.5.5, Figure 12.5.6, Figure 12.5.7). Dimeric cyanine dyes (Table 8.2, Section 8.1) have been used to observe mitotic chromosome movement in live cells. For example, YOYO®-1 dye (Y3601) has been microinjected into cells in order to follow mitotic chromosomes through at least six cell cycles in fertilized sea urchin eggs (M. Terasaki and L. Jaffe, personal communication) (Figure 12.5.8).

Another useful technique for tracking chromosomes through mitosis involves metabolic incorporation of microinjected fluorescent nucleotides, including our fluorescein-12-dUTP (C7604, Section 8.2) by endogenous cellular enzymes into DNA. Incorporation of the fluorescent tracer does not interfere with subsequent progress through the cell cycle, and fluorescent strands of DNA can be followed as they assemble into chromosomes and segregate into daughters and granddaughters.[22–24]

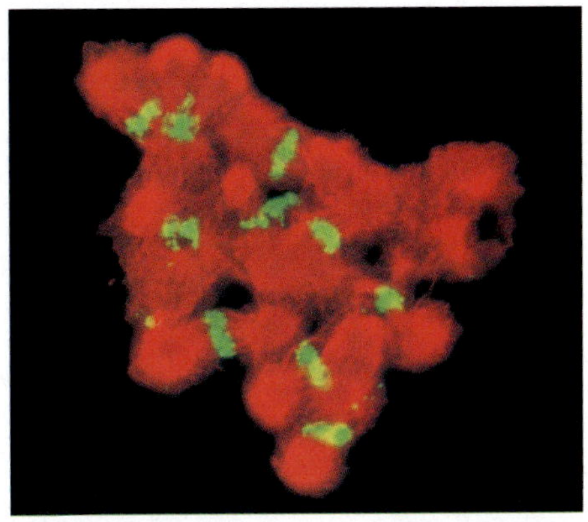

Figure 12.5.7 Mitotic spindles isolated from sea urchin eggs that are labeled with YOYO®-1 iodide (Y3601) and a monoclonal anti–ß-tubulin antibody in conjunction with Texas Red® goat anti–rat IgG antibody (T6392). This image was generated by epifluorescence microscopy. Image contributed by John Murray, University of Pennsylvania.

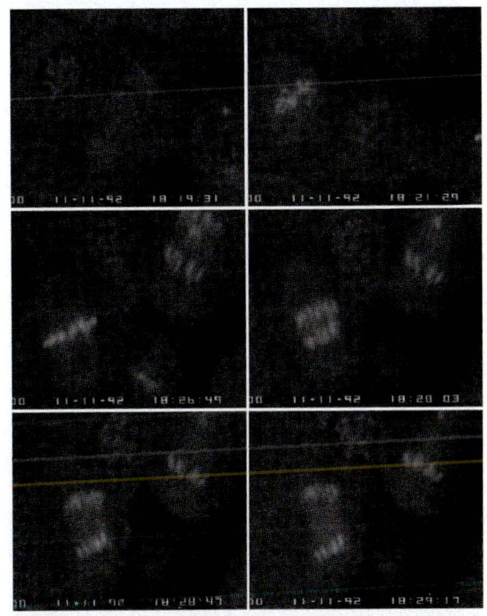

Figure 12.5.8 Using the YOYO®-1 dye to follow cell division in a sea urchin egg. The YOYO®-1 dye (Y3601) was injected into an unfertilized sea urchin egg. The egg was fertilized and then observed by confocal laser-scanning microscopy. Images were obtained every 15 sec in this sequence. Every fourth image is shown in the first part, then every image is shown during chromosome separation. The image was contributed by Mark Terasaki, University of Connecticut Health Center.

Fluorescent Protein–Based Markers for the Nucleus

GFP- and RFP-Labeled Nuclear Markers

CellLight® reagents are BacMam expression vectors encoding site-selective autofluorescent protein fusions. These reagents incorporate all the customary advantages of BacMam delivery technology, including high efficiency transduction of mammalian cells and long-lasting, titratable expression (BacMam Gene Delivery and Expression Technology—Note 11.1). A complete list of our CellLight® reagents and their targeting sequences can be found in Table 11.1.

CellLight® Nucleus-GFP (C10602) and CellLight® Nucleus-RFP (C10603) are BacMam expression vectors encoding fusions of Green Fluorescent Protein (GFP) or Red Fluorescent Protein[25] (RFP) with the SV40 nuclear localization sequence (NLS). In addition to general purpose identification and demarcation of the nucleus in live-cell imaging experiments, SV40 NLS-GFP is extensively used for analysis of nucleocytoplasmic transport and nuclear envelope integrity.[26,27]

CellLight® Histone 2B–GFP (C10594) and CellLight® Histone 2B–RFP (C10595, Figure 12.5.9) are BacMam expression vectors encoding fusions of GFP or RFP with histone 2B. Labeling with histone 2B-GFP is a well established and minimally invasive approach for visualization of chromatin in live cells[28] and is particularly useful for real-time imaging of mitotic cell division.

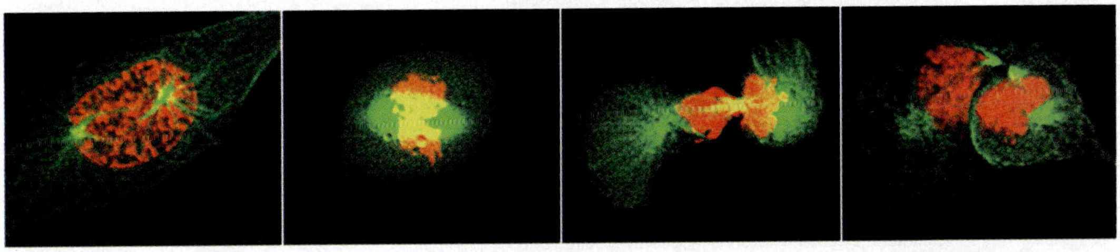

Figure 12.5.9 Image montage sampled from a continuous 16-hour time-lapse illustrating cytoskeletal and histone dynamics during mitosis using U2OS osteosarcoma cells transduced with CellLight® Histone 2B-RFP (C10595) and CellLight® MAP4-GFP (C10598) reagents.

IMPORTANT NOTICE: The products described in this manual are covered by one or more Limited Use Label License(s). Please refer to the Appendix on page 971 and Master Product List on page 975. Products are For Research Use Only. Not intended for any animal or human therapeutic or diagnostic use.

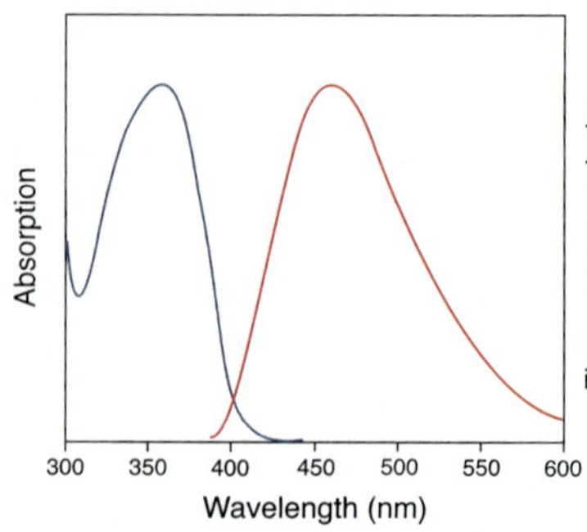

Figure 12.5.10 Absorption and fluorescence emission spectra of DAPI bound to DNA.

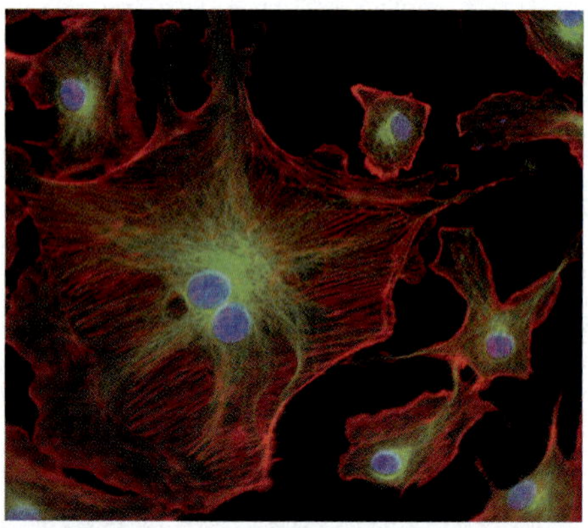

Figure 12.5.11 FluoCells® prepared slide #2 (F14781), which shows bovine pulmonary artery endothelial cells (BPAEC) that have been stained with an anti–ß-tubulin mouse monoclonal antibody in conjunction with BODIPY® FL goat anti–mouse IgG (B2752) for labeling microtubules, Texas Red®-X phalloidin (T7471) for labeling F-actin and DAPI (D1306, D3571, D21490) for labeling nuclei. This multiple-exposure image was acquired using bandpass optical filter sets appropriate for DAPI, fluorescein and Texas Red® dye.

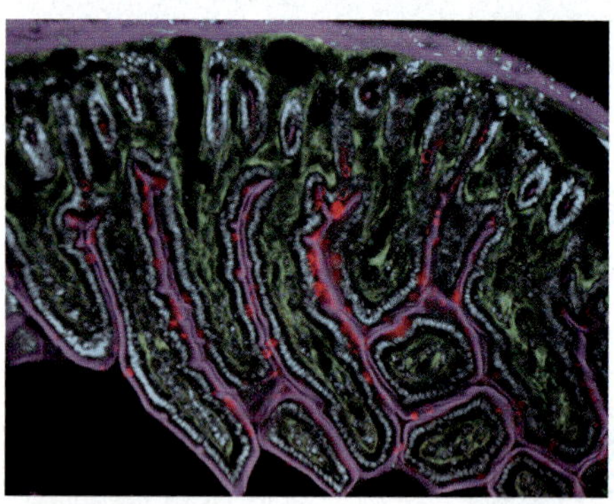

Figure 12.5.12 Mouse intestine cryosection showing basement membranes labeled with chicken IgY anti-fibronectin antibody and Alexa Fluor® 488 goat anti–chicken IgG antibody (A11039, green). Goblet cells and crypt cells were labeled with Alexa Fluor® 594 wheat germ agglutinin (W11262, red). The microvillar brush border and smooth muscle layers were visualized with Alexa Fluor® 680 phalloidin (A22286, pseudocolored purple). The section was counterstained with DAPI (D1306, D3571, D21490, blue).

Alexa Fluor® 488 Histone H1

The Alexa Fluor® 488 conjugate of the lysine-rich calf thymus histone H1 (H13188) is a useful probe for nuclear protein transport assays.[29] Nuclear-to-mitochondrial translocation of histone H1 is indicative of dsDNA strand breaks. Fluorescent histone H1 conjugates can also be used to detect membrane-surface exposure of acidic phospholipids such as phosphatidylserine.[30]

Nuclear Counterstaining of Fixed Cells and Tissues

Blue-Fluorescent Counterstains

DAPI (D1306, D3571, D21490; Figure 12.5.10) is the classic nuclear and chromosome counterstain for identifying nuclei and observing chromosome-banding patterns. DAPI binds selectively to dsDNA and thus shows little to no background staining of the cytoplasm. Its relatively low-level fluorescence emission does not overwhelm signals from green- or red-fluorescent secondary antibodies or FISH probes. DAPI is semipermeant to live cells and can be used on unfixed cells or tissue sections (Figure 12.5.11, Figure 12.5.12). We also offer DAPI premixed with our *SlowFade*®, *SlowFade*® Gold and ProLong® Gold antifade reagents (S36938, S36939, P36931, P36935; Section 23.1) for simultaneous nuclear staining and antifade protection.

The Hoechst 33342 dye (H1399, H3570, H21492) has been used widely for staining the nuclei of live cells. Hoechst dyes preferentially bind to AT regions, making them quite selective (but not specific) for DNA; Hoechst dye–stained cells and tissues show virtually no cytoplasmic staining (Figure 12.5.13). The Hoechst 33342 dye is commonly used in combination with labeling by 5-bromo-2'-deoxyuridine (BrdU, B23151; Section 8.2) to distinguish the compact chromatin of apoptotic nuclei, to identify replicating cells and to sort cells based on their DNA content (Section 15.4, Section 15.5).

The blue-fluorescent BOBO™-1 nucleic acid stain (B3582) emits a brighter fluorescent signal than does DAPI. BOBO™-1 has been used effectively as a counterstain for *Drosophila* chromosomes in combination with Cy®3 dye– or fluorescein-labeled antibodies.[31] SYTOX® Blue nucleic acid stain (S11348, S34857) also emits bright blue fluorescence upon binding to nucleic acids and

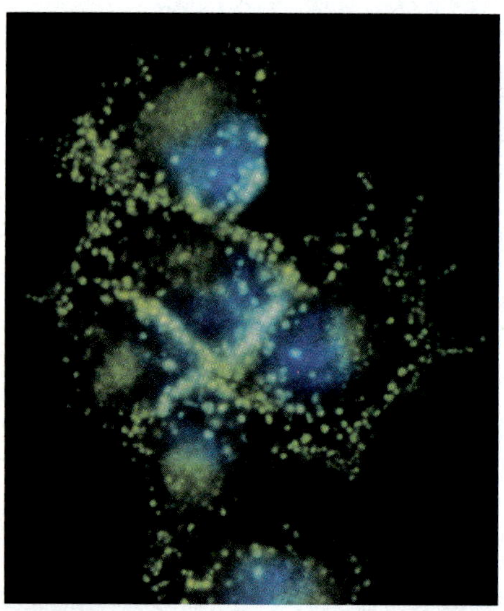

Figure 12.5.13 Endogenous alkaline phosphatase enzyme of osteosarcoma cells localized with the ELF® 97 Endogenous Phosphatase Detection Kit (E6601). Unlike other phosphatase substrates, the unique ELF® 97 phosphatase substrate forms a fluorescent precipitate at the site of enzymatic activity. The blue-fluorescent nucleic acid stain Hoechst 33342 (H1399, H3570, H21492) was used as a counterstain to the green fluorescence of the ELF® 97 alcohol precipitate. The double-exposure image was acquired using a bandpass filter set appropriate for ELF® 97 alcohol and a longpass filter set appropriate for DAPI.

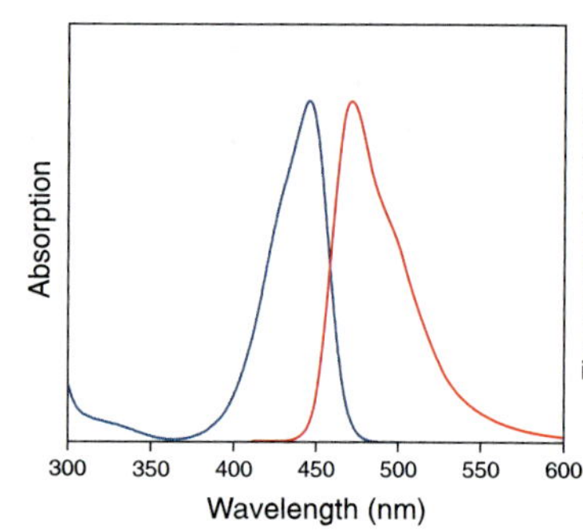

Figure 12.5.14 Absorption and fluorescence emission spectra of SYTOX® Blue nucleic acid stain bound to DNA.

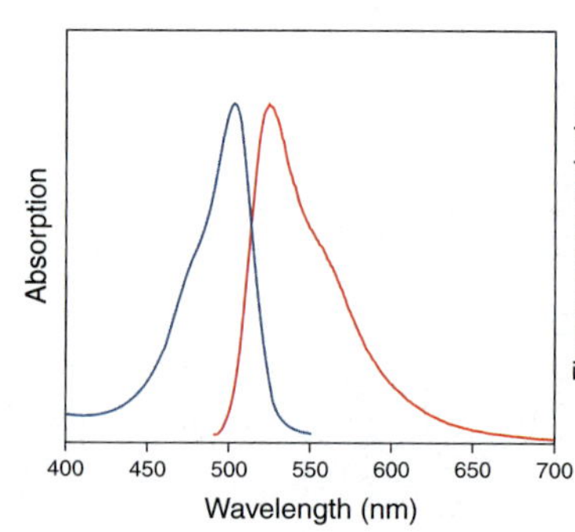

Figure 12.5.15 Absorption and fluorescence emission spectra of the SYTOX® Green nucleic acid stain bound to DNA.

IMPORTANT NOTICE: The products described in this manual are covered by one or more Limited Use Label License(s). Please refer to the Appendix on page 971 and Master Product List on page 975. Products are For Research Use Only. Not intended for any animal or human therapeutic or diagnostic use.

www.invitrogen.com/probes

molecular probes® | invitrogen™ by life technologies™

is a very good nuclear counterstain. Fluorescence emission of the SYTOX® Blue complex with nucleic acids (Figure 12.5.14) somewhat overlaps the emission of fluorescein, Alexa Fluor® 488 and Oregon Green® 488 dyes and thus we recommend SYTOX® Blue dye only as a counterstain for orange- or red-fluorescent dyes.

Green-Fluorescent Counterstains

Some of our cyanine dyes (Table 8.1, Table 8.2, Table 8.3; Section 8.1) have been found to be useful as green-fluorescent nuclear counterstains. YO-PRO®-1 dye (Y3603) and SYTOX® Green stain (S7020, Figure 12.5.15) are excellent nuclear counterstains for cells in culture or for whole-mount tissues[32–34] (Figure 12.5.16, Figure 12.5.17, Figure 12.5.18, Figure 12.5.19) and are useful counterstains for tissue sections as well. SYTOX® Green dye shows highly selective nuclear staining; YO-PRO®-1 dye shows more intense staining but also weakly stains the cytoplasm and nucleolus.[32,33] SYTOX® Green dye has been used to follow changes in nuclear morphology in apoptotic cells (Figure 12.5.20) and is a component in some Molecular Probes® apoptosis assay kits (Section 15.5). SYTOX® Green stain has been used as a specific nuclear counterstain for multicolor labeling in *Drosophila* imaginal disc cells.[35] YO-PRO®-1, also a component in some of Molecular Probes® apoptosis assay kits (Section 15.5), is selectively permeant to apoptotic cells, enabling facile identification of this cell population by flow cytometry[36,37] (Figure 12.5.21). Nuclear staining by YO-PRO®-1 dye has provided a method to identify individual cells within single live, perfused mesentery microvessels.[38]

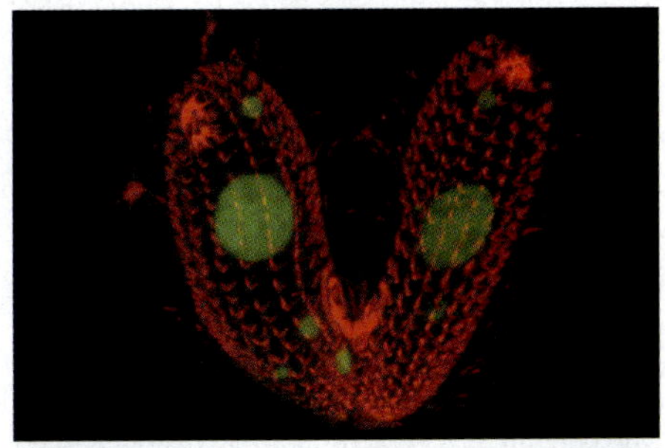

Figure 12.5.16 Confocal micrograph illustrating sexual reproduction of the ciliate protist, *Tetrahymena thermophila*, 6 hours after mating. After fixation and permeabilization, the cytoskeleton was labeled with an anti-tubulin antibody and subsequently visualized with Texas Red®-X goat anti–mouse IgG antibody (T6390). The macro- and micronuclei were stained with SYTOX® Green nucleic acid stain (S7020). Image contributed by David Asai and Amy Walanski, Purdue University.

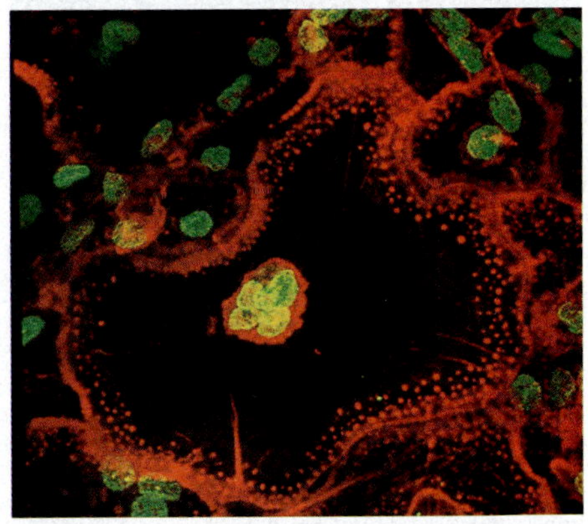

Figure 12.5.17 Macrophages cultured on a polymer surface that have fused to form a foreign-body giant cell following treatment with interleukin-4. Cells were fixed with 3.7% formaldehyde, treated with RNase A and stained with rhodamine phalloidin (R415) to visualize F-actin, and with YO-PRO®-1 iodide (Y3603) to visualize cell nuclei. Cells were imaged with a Bio-Rad® MRC600 confocal laser-scanning microscope. The image is a composite of optical sections taken through the Z-axis of the cell. F-actin (red) is restricted to the periphery of the multinucleated cell and surrounds a cluster of nuclei (green). Image contributed by Kristin DeFife and James M. Anderson, Institute of Pathology, Case Western Reserve University.

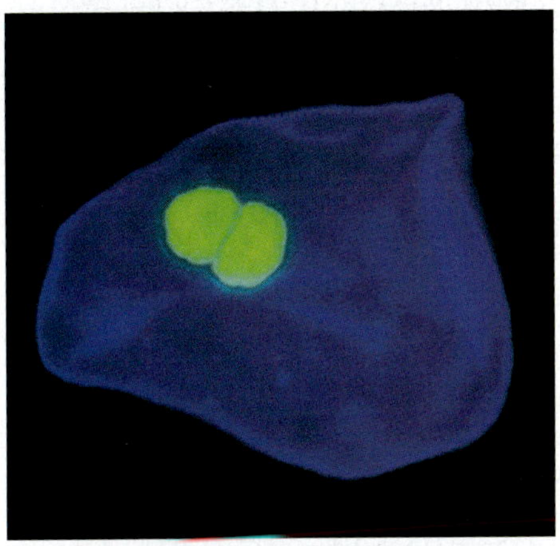

Figure 12.5.19 Human cheek epithelial cells labeled with Alexa Fluor® 350 wheat germ agglutinin (W11263) and stained with SYTOX® Green nucleic acid stain (S7020). This multiple-exposure image was acquired using bandpass filter sets appropriate for DAPI and fluorescein.

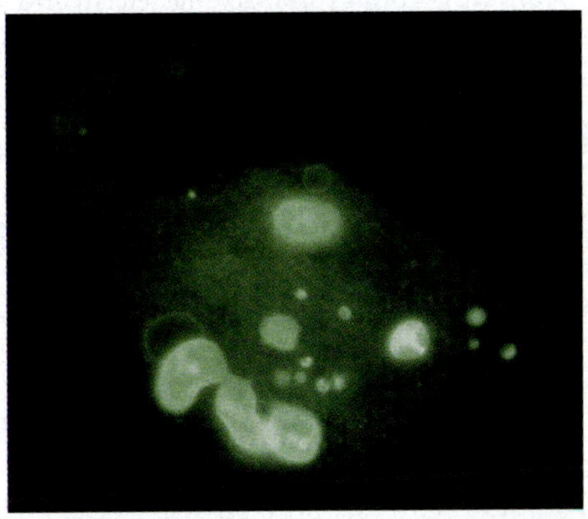

Figure 12.5.20 Nuclear deformation of an apoptotic cell visualized with the SYTOX® Green dye (S7020). Bovine pulmonary artery endothelial cells were treated with camptothecin for 24 hours, stained with the SYTOX® Green nucleic acid stain and photographed using a fluorescence microscope equipped with a bandpass filter set designed for fluorescein like dyes.

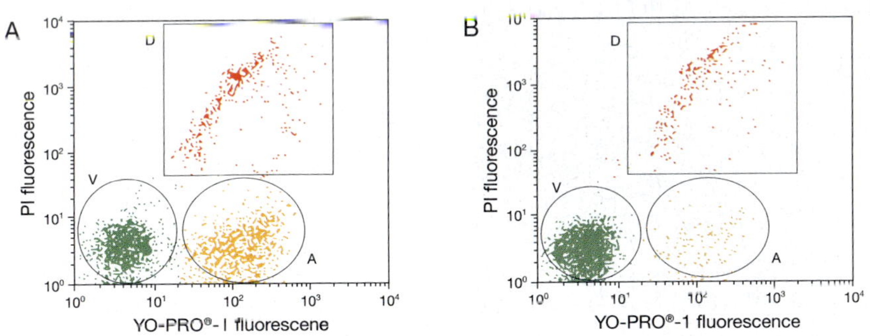

Figure 12.5.21 Flow cytometric analysis of Jurkat cells using the Membrane Permeability/Dead Cell Apoptosis Kit (V13243), which contains YO-PRO®-1 and propidium iodide. Jurkat human T-cell leukemia cells were first exposed to 10 µM camptothecin for 4 hours (**A**) or left untreated (as control, **B**). Cells were then treated with the YO-PRO®-1 and propidium iodide (PI) dyes provided in the Membrane Permeability/Dead Cell Apoptosis Kit and analyzed by flow cytometry. Note that the camptothecin-treated cells (**A**) have a significantly higher percentage of apoptotic cells (indicated by an "A") than the basal level of apoptosis seen in the control cells (**B**). V = viable cells, D = dead cells.

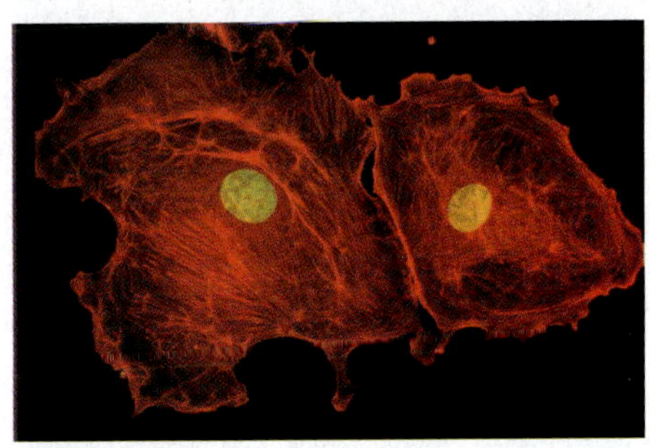

Figure 12.5.18 Permeabilized bovine pulmonary artery endothelial cells stained with SYTOX® Green nucleic acid stain (S7020) to label the nuclei and with BODIPY® TR-X phallacidin (B7464) to label the F-actin. The image was acquired by taking sequential exposures through bandpass optical filter sets appropriate for fluorescein and the Texas Red® dye.

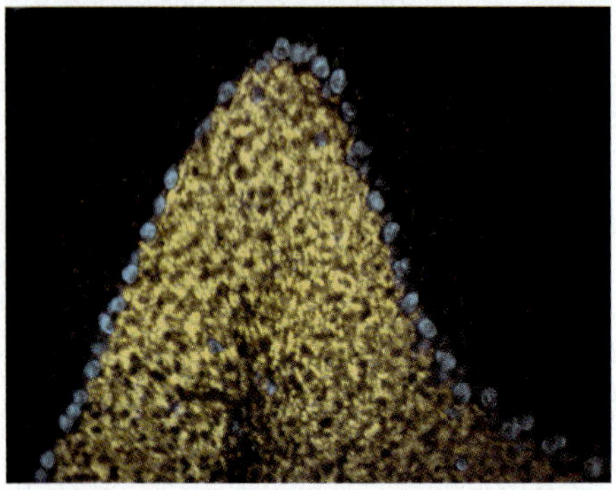

Figure 12.5.22 Mouse brain section stained with NeuroTrace® 435/455 blue-fluorescent Nissl stain (N21479) and counterstained with nuclear yellow (N21485).

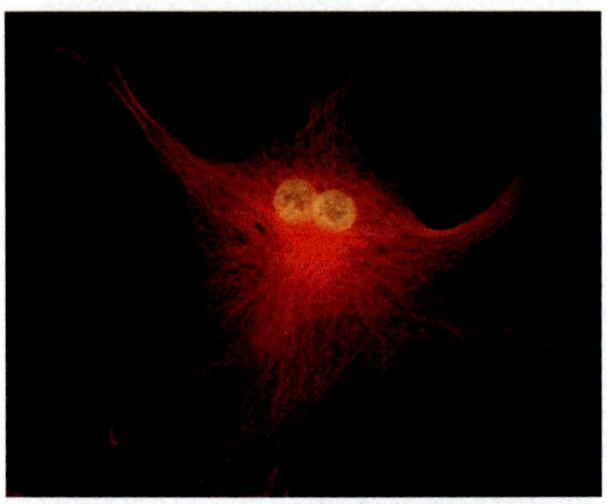

Figure 12.5.23 A binucleate bovine pulmonary artery endothelial cell labeled with the biotin-XX conjugate of anti–α-tubulin antibody (A21371) and Alexa Fluor® 568 streptavidin (S11226), then counterstained with nuclear yellow (N21485).

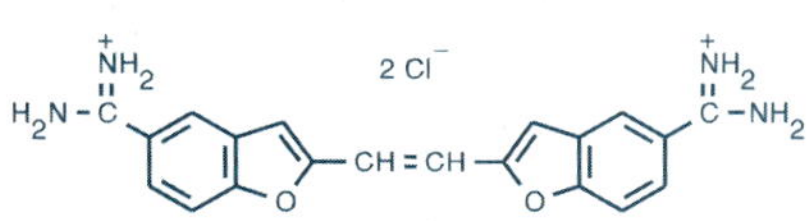

Figure 12.5.24 True blue chloride (T1323).

Staining with the TOTO®-1 (T3600) and YOYO®-1 (Y3601) nucleic acid stains enables extremely sensitive flow cytometric analysis of nuclei and isolated human chromosomes.[39] In this study, YOYO®-1 dye staining produced more than 1000 times the fluorescence signal obtained with mithramycin; furthermore, histograms of both TOTO®-1 and YOYO®-1 on RNase-treated nuclei provided coefficients of variation that were at least as low as those found with propidium iodide or mithramycin. These researchers also found that when nuclei were simultaneously stained with the YOYO®-1 and Hoechst 33258 (H1398, H3569, H21491) dyes, the ratio of the fluorescence of these two dyes varied as a function of cell cycle. This observation suggests that our cyanine dyes might be useful for examining cell cycle–dependent changes that occur in chromatin structure. YOYO®-1 dye staining also permitted the detection of discrete ribosome-containing domains within the cytoplasm of mature cell axons, which are traditionally thought to contain no transcriptional activity.[40] In addition, YOYO®-1 dye has been used as a counterstain for immunofluorescent staining of chromatin in the nuclei of developing *Drosophila* embryos.[31]

Yellow-Fluorescent Counterstain

The long-wavelength tracer nuclear yellow (Hoechst S769121, N21485; Figure 12.5.22, Figure 12.5.23) is often combined with the popular retrograde tracer true blue (T1323, Section 14.3) for two-color neuronal mapping. In neuronal cells, nuclear yellow primarily stains the nucleus with yellow fluorescence,[41–44] whereas true blue is a UV light–excitable, divalent cationic dye (Figure 12.5.24) that stains the cytoplasm with blue fluorescence.[41,45–48] Both nuclear yellow and true blue are stable when subjected to immunohistochemical processing and can be used to photoconvert DAB into an insoluble, electron-dense reaction product[49–52] (Fluorescent Probes for Photoconversion of Diaminobenzidine Reagents—Note 14.2).

Orange-Fluorescent Counterstains

BOBO™-3 (B3586) and SYTOX® Orange (S11368, Figure 12.5.25) cyanine dyes have fluorescence emission that is similar to that of PI, but show greater fluorescence enhancement upon binding to DNA and so should provide brighter nuclear staining. BOBO™-3 dye has been used as a nuclear stain in whole-mount *Xenopus laevis* embryos.[53] YOYO®-3 (Y3606) and YO-PRO®-3 (Y3607) dyes show strong and specific staining of the nucleus in most cultured cells.[32]

Red-Fluorescent Counterstains

Propidium iodide (PI; P1304MP, P3566, P21493) has long been a preferred dye for red-fluorescent counterstaining of nuclei and chromosomes (Figure 12.5.26). Under some fixation conditions, PI shows highly selective nuclear staining. Other preparations of cells and tissues require a simple treatment with a ribonuclease (RNase) to achieve specific nuclear staining. PI provides an excellent counterstain for cells stained with green-fluorescent probes or secondary antibodies conjugated to Alexa Fluor® 488, Oregon Green®, BODIPY® FL or fluorescein dyes.

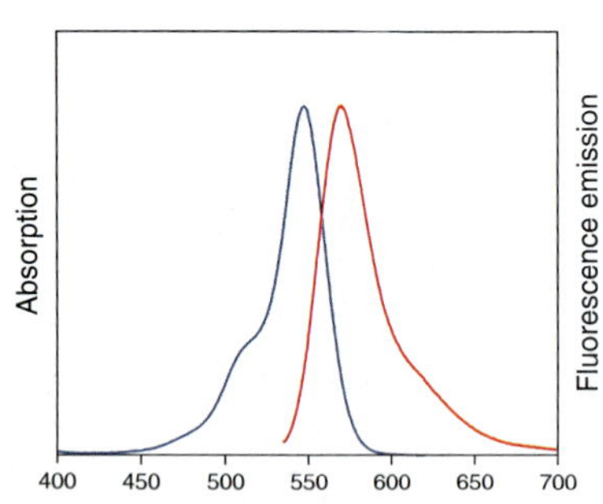

Figure 12.5.25 Absorption and fluorescence emission spectra of SYTOX® Orange nucleic acid stain bound to DNA.

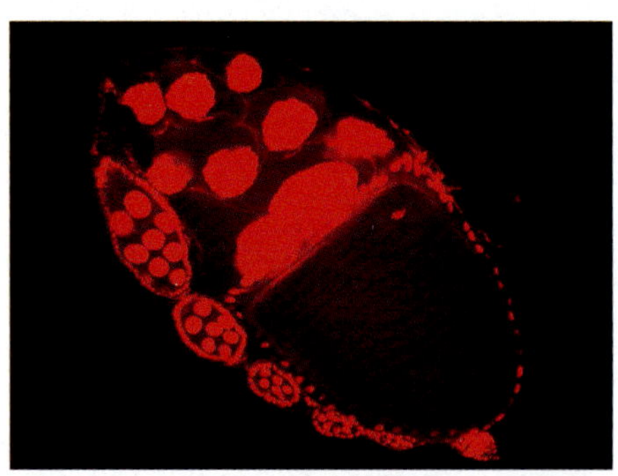

Figure 12.5.26 Day 10 of development of a *Drosophila* ovarian egg chamber assembly line. The nuclei of follicle and nurse cells were labeled with propidium iodide (P1304MP, P3566, P21493) and visualized by confocal laser-scanning microscopy using excitation by the 568 nm spectral line of an Ar-Kr laser. Image contributed by Sandra Orsulic, University of North Carolina at Chapel Hill.

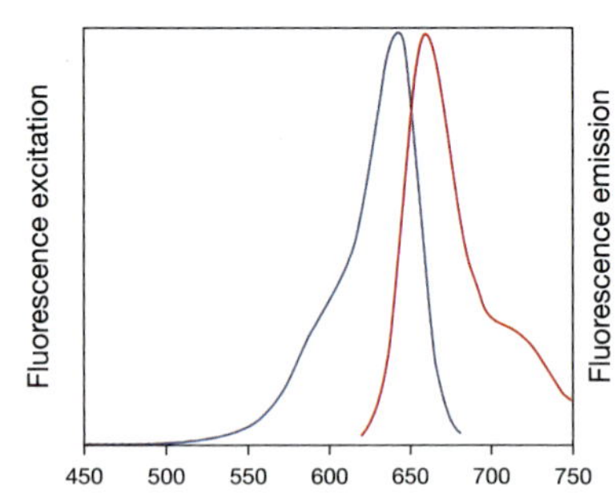

Figure 12.5.27 Fluorescence excitation and emission spectra of SYTOX® Red nucleic acid stain bound to DNA.

IMPORTANT NOTICE: The products described in this manual are covered by one or more Limited Use Label License(s). Please refer to the Appendix on page 971 and Master Product List on page 975. Products are For Research Use Only. Not intended for any animal or human therapeutic or diagnostic use.

SYTOX® Red dead cell stain (excitation/emission maxima ~640/658 nm, S34859) is a high-affinity nucleic acid stain that easily penetrates cells with compromised plasma membranes but will not cross uncompromised cell membranes. After brief incubation with SYTOX® Red stain, the nucleic acids of dead cells fluoresce bright red when excited at 633 nm or 635 nm (Figure 12.5.27). These properties, combined with its >500-fold fluorescence enhancement upon nucleic acid binding, make the SYTOX® Red stain a simple and quantitative single-step dead-cell indicator for use with red laser–equipped flow cytometers. Using 633 nm or 635 nm excitation, SYTOX® Red dead cell stain is distinct from other dead cell probes like 7-AAD and PI, which are excited using 488 nm. The emission of SYTOX® Red stain is limited to one channel with minimal spectral overlap, effectively freeing the channels of the 488 nm laser line.

The long-wavelength TOTO®-3 (T3604) and TO-PRO®-3 (T3605) dyes provide nuclear counterstains whose fluorescence is well separated from that of commonly used fluorophores, such as our popular Alexa Fluor® dyes, Oregon Green®, fluorescein (FITC), rhodamine (TRITC), Texas Red®, coumarin (AMCA), Marina Blue® and Pacific Blue™ dyes. Their long-wavelength spectra make these red-fluorescent nucleic acid stains particularly useful for three- or even four-color labeling using confocal laser-scanning or standard epifluorescence microscopes (Figure 12.5.28, Figure 12.5.29). The absorbance peaks of the TOTO®-3 (Figure 12.5.30) and TO-PRO®-3 (Figure 12.5.31) dyes closely match the 633 nm and 635 nm laser lines of many confocal laser-scanning microscopes and the spectra match filter sets typically used for the Alexa Fluor® 647 and Cy®5 dyes.

Long-wavelength light–absorbing dyes have the advantage that their fluorescence is usually not obscured by the autofluorescence of tissues. For example, analysis of fluorescently stained whole-mount *Xenopus laevis* embryos has traditionally been difficult due to the large amount of autofluorescence from the yolk. Two reports have shown that the TO-PRO®-3 dye can be used as a fluorescent nuclear stain in these embryos, allowing them to be analyzed by confocal laser-scanning microscopy. When either the 633 nm or 635 nm spectral lines of a confocal laser-scanning microscope is used with long-wavelength filter sets, the autofluorescence from the yolk is almost completely eliminated.[53,54]

The TOTO®-3 and TO-PRO®-3 dyes were tested as counterstains for aldehyde-fixed frozen rat tissue sections. The TO-PRO®-3 dye showed less cytoplasmic staining and little overlap with signals from fluorescein- or tetramethylrhodamine-labeled secondary antibodies in the same section.[55] The TO-PRO®-3 dye gives strong and selective staining of the nucleus in cultured cells.[32] A high selectivity for nuclear staining over cytoplasmic staining made TO-PRO®-3 the preferred dye for detecting amplification of the *Her-2/neu* gene by dual-color FISH in paraffin sections.[56] Although its nucleic acid complex reportedly bleaches relatively rapidly,[32] photobleaching can be slowed with antifade reagents such as are provided in our *SlowFade®*, *SlowFade®* Gold, ProLong® and ProLong® Gold antifade reagents (Section 23.1). Nuclear staining by TO-PRO®-3 dye has been used to study the structure of the nucleus in interphase cells[57] and to demonstrate segregation of chromosomes during meiosis in mouse oocytes.[58] TO-PRO®-3 dye was also used to counterstain the chromatin in nuclei of developing *Drosophila* embryos that were immunostained with Cy®3 dye– or fluorescein-labeled antibodies.[31] TOTO®-3 dye has been used as a counterstain for TUNEL assays[59] and for annexin V–based apoptosis assays[60] (Section 15.5). TOTO®-3 dye has also been used in combination with Cy®3 dye–labeled anti-BrdU antibody staining to show that replicons labeled with BrdU form clusters in the nucleus that are stable through several cell cycles.[61]

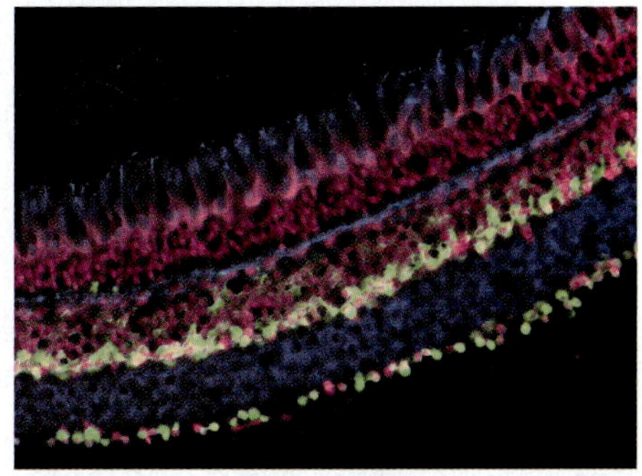

Figure 12.5.28 Immunohistochemical detection using tyramide signal amplification. A transverse section of fixed zebrafish retina was probed with mouse monoclonal FRet 34 antibody and subsequently developed for visualization using HRP-conjugated goat anti-mouse IgG antibody and Alexa Fluor® 488 tyramide, which are supplied in the TSA™ Kit #2 (T20912). The section was counterstained with the blue-fluorescent Alexa Fluor® 350 wheat germ agglutinin (W11263) and the far red–fluorescent TOTO®-3 nuclear stain (T3604).

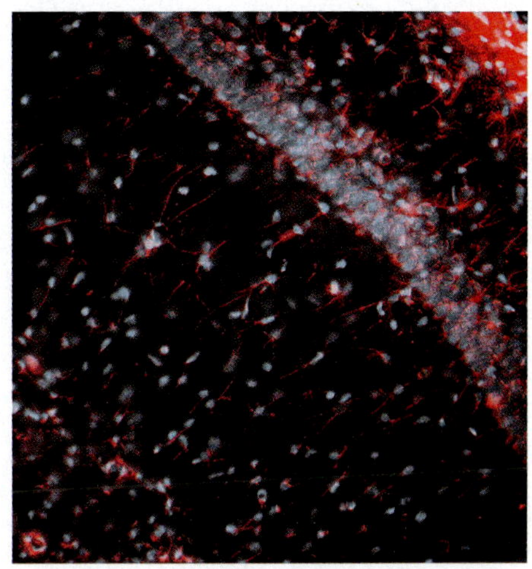

Figure 12.5.29 Rat brain cryosections were labeled with the red-fluorescent Alexa Fluor® 594 conjugate of anti-glial fibrillary acidic protein antibody (A21295). Nuclei were counterstained with TOTO®-3 iodide (T3604, pseudocolored blue).

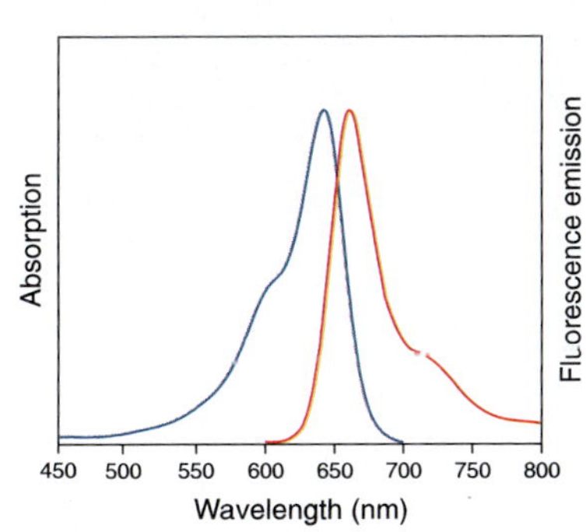

Figure 12.5.30 Absorption and fluorescence emission spectra of TOTO®-3 bound to DNA.

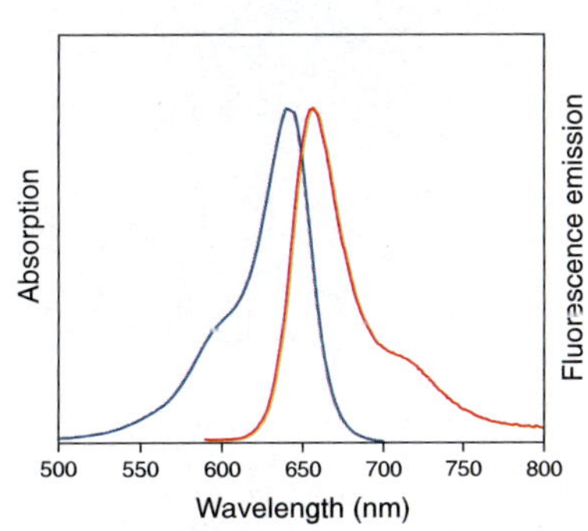

Figure 12.5.31 Absorption and fluorescence emission spectra of TO-PRO®-3 bound to DNA.

IMPORTANT NOTICE: The products described in this manual are covered by one or more Limited Use Label License(s). Please refer to the Appendix on page 971 and Master Product List on page 975. Products are For Research Use Only. Not intended for any animal or human therapeutic or diagnostic use.

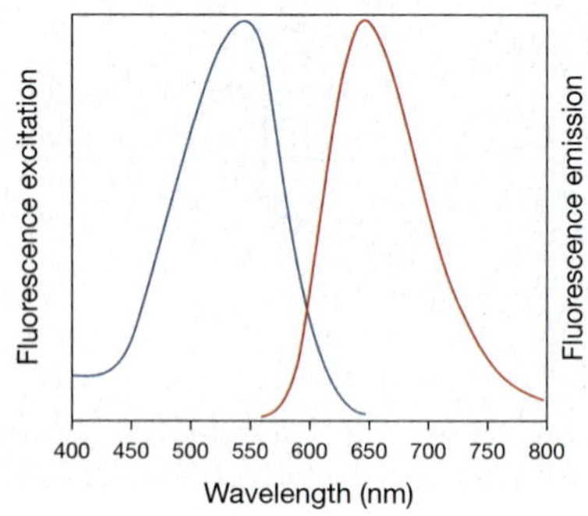

Figure 12.5.32 Fluorescence excitation and emission spectra of SYTOX® AADvanced™ dead cell stain bound to DNA.

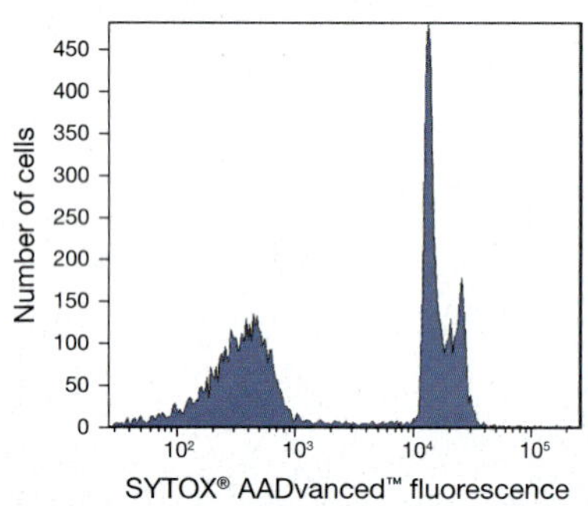

Figure 12.5.33 A mixture of heat-killed and untreated Jurkat cells were stained with 1 µM SYTOX® AADvanced™ dead cell stain for 5 minutes. Cells were analyzed on a flow cytometer equipped with a 488 nm laser and a 695/40 nm bandpass filter. Live cells are easily distinguished from the dead cell population.

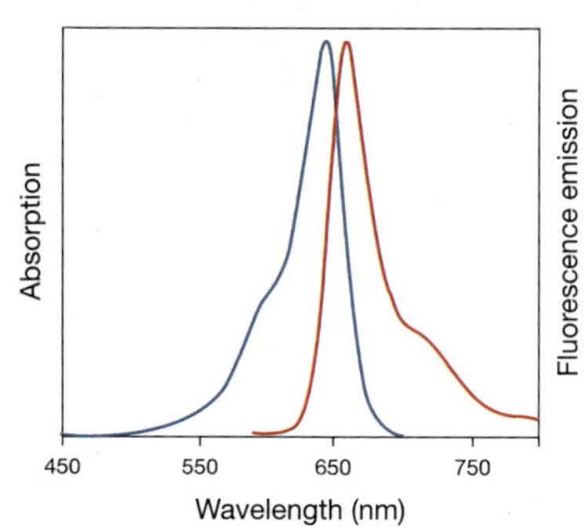

Figure 12.5.35 Absorption and fluorescence emission spectra for Qnuclear™ Deep Red stain (Q10363).

SYTOX® AADvanced™ Dead Cell Stain

Especially formulated for flow cytometry applications, SYTOX® AADvanced™ dead cell stain (excitation/emission maxima ~546/647 nm; S10274, S10349) labels the nucleic acids of dead cells with a bright red fluorescence when excited with the 488 nm spectral line of the argon-ion laser, with minimal compensation in the green, orange and near-infrared channels (Figure 12.5.32). This high-affinity red-fluorescent nucleic acid stain easily penetrates cells with compromised plasma membranes, but will not cross healthy cell membranes. These properties, combined with its >500-fold fluorescence enhancement upon nucleic acid binding, make the SYTOX® AADvanced™ dead cell stain a simple and quantitative single-step dead-cell indicator (Figure 12.5.33). Labeling of dead cells is achieved very rapidly, typically within 5 minutes. SYTOX® AADvanced™ dead cell stain may also be useful for DNA content cell-cycle analysis with the addition of RNase A in fixed cells.

Qnuclear™ Deep Red Stain

Qnuclear™ Deep Red stain (Q10363) is a bright and photostable nuclear counterstain designed for use with fixed and permeabilized cells that have been labeled with Qdot® nanocrystals (Section 6.6) and mounted in Qmount® Qdot® mounting media (Q10336, Section 23.1) (Figure 12.5.34). With excitation and emission maxima of 640 and 663 nm, respectively (Figure 12.5.35), this counterstain can be visualized with standard fluorescence microscopy filter sets. Qnuclear™ Deep Red stain is compatible with Qdot® 525, 565, 585, 605 and 625 nanocrystals; it can also be used with Qdot® 655 nanocrystals and its conjugates, though care must be taken to use appropriate excitation wavelengths and filter sets given the fluorescence emission overlap.

Qnuclear™ Deep Red stain is provided as a convenient, concentrated dimethylsulfoxide (DMSO) solution with a labeling protocol optimized for cells that have been formaldehyde fixed and and permeabilized with Triton X-100; other fixation techniques may result in nonspecific staining or abnormal cellular morphology. Nuclear labeling with Qnuclear™ Deep Red stain should be the last step of the cell-staining protocol.

SelectFX® Nuclear Labeling Kit

The SelectFX® Nuclear Labeling Kit (S33025) provides four spectrally distinct fluorescent dyes—blue-fluorescent DAPI, green-fluorescent SYTOX® Green stain, red-fluorescent 7-aminoactinomycin D (7-AAD) and far-red–fluorescent TO-PRO®-3 dye—for staining nuclei in fixed and permeabilized cells and tissues with essentially no cytoplasmic background staining. When used according to the protocol provided, the dyes in the SelectFX® Nuclear Labeling Kit provide highly selective nuclear staining with little or no cytoplasmic labeling; they are ideal for use as counterstains in multicolor applications. The stained nuclei stand out in vivid contrast to other fluorescently labeled cell structures when observed by fluorescence microscopy. These dyes have excitation wavelengths that match the common laser lines for confocal microscopy and flow cytometry and can be used with standard filter sets on fluorescence microscopes and microplate readers. The staining protocol provided is compatible with a wide range of cytological

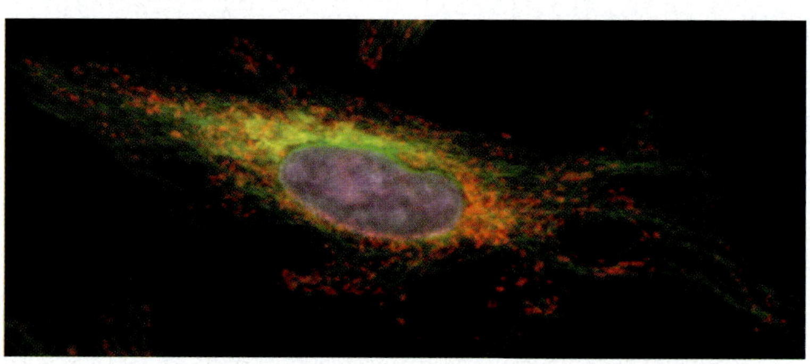

Figure 12.5.34 Human carcinoma (HeLa) cell labeled with Qdot® nanocrystals and mounted with Qmount® Qdot® mounting media. Mitochondria were detected with anti–OxPhos complex V inhibitor protein IgG (A21355) and labeled with Qdot® 625 goat F(ab')$_2$ anti–mouse IgG antibody (A10195, red fluorescence); the Golgi apparatus was detected with rabbit anti-giantin antibody and labeled with Qdot® 585 goat F(ab')$_2$ anti–rabbit IgG antibody (Q11411MP, yellow fluorescence); tubulin was detected with rat anti-tubulin antibody and labeled with DSB-X™ biotin goat anti–rat IgG antibody (D20697) and Qdot® 525 streptavidin (Q10141MP, green fluorescence). The nucleus was labeled with Qnuclear™ Deep Red Stain (Q10363, purple fluorescence), and the slide was mounted with Qmount® Qdot® mounting media (Q10336).

IMPORTANT NOTICE: The products described in this manual are covered by one or more Limited Use Label License(s). Please refer to the Appendix on page 971 and Master Product List on page 975. Products are For Research Use Only. Not intended for any animal or human therapeutic or diagnostic use.

molecular probes® | invitrogen™ by life technologies™

labeling techniques, including direct or indirect antibody-based detection methods, mRNA *in situ* hybridization and staining methods that incorporate organelle- and cytoskeleton-selective fluorescent probes (including MitoTracker® mitochondrion-selective probes and Alexa Fluor® dye–conjugated phalloidins). The dyes can also be used to fluorescently stain cells for analysis in multicolor flow cytometry experiments. All dyes are provided as stock solutions, convenient for diluting and staining, and each dye is also available separately.

The SelectFX® Nuclear Labeling Kit contains:

- DAPI, a blue-fluorescent counterstain (excitation/emission maxima ~358/451 nm)
- SYTOX® Green, a green-fluorescent counterstain (excitation/emission maxima ~504/523 nm)
- 7-Aminoactinomycin D (7-AAD), a red-fluorescent counterstain (excitation/emission maxima ~546/647 nm)
- TO-PRO®-3 iodide, a far-red–fluorescent counterstain (excitation/emission maxima ~642/661 nm)
- Detailed staining protocols

The SelectFX® Nuclear Labeling Kit contains sufficient reagents to prepare ~100 assays with each stain at 300 µL per assay.

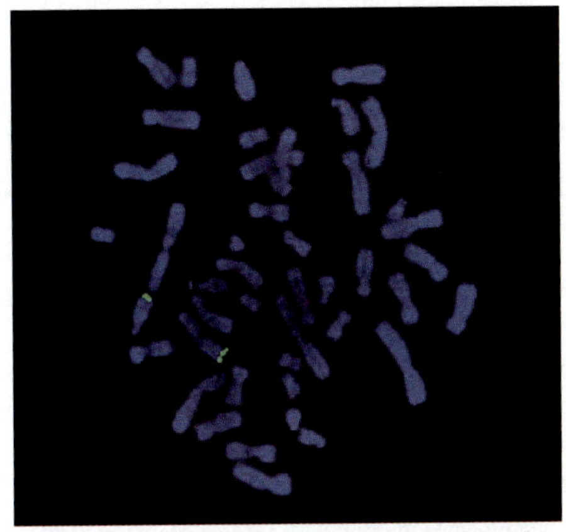

Figure 12.5.36 Fluorescence *in situ* hybridization (FISH) mapping of a BAC clone on human metaphase chromosomes. FISH was performed using a BAC clone labeled using the ARES™ Alexa Fluor® 488 DNA Labeling Kit (A21665). The chromosomes were counterstained with DAPI (D1306, D3571, D21490). Image contributed by Nallasivam Palanisamy, Cancer Genetics Inc.

Chromosome Counterstaining

Blue-Fluorescent Chromosome Counterstains

DAPI (D1306, D3571, D21490) is the classic blue-fluorescent nuclear and chromosome counterstain. DAPI binds selectively to dsDNA and thus shows little to no cytoplasmic background staining.[62–64] DAPI's relatively low-level fluorescence emission does not overwhelm signals from green- or red-fluorescent secondary antibodies or FISH probes (Figure 12.5.36, Figure 12.5.37). We also offer DAPI premixed with our *SlowFade*®, *SlowFade*® Gold and ProLong® Gold antifade reagents (S36938, S36939, P36931, P36935). Hoechst 33342 dye (H1399, H3570, H21492) is also commonly used for chromosome counterstaining; SYTOX® Blue nucleic acid stain (S11348, S34857), which is essentially nonfluorescent except when bound to nucleic acids, may be similarly useful.

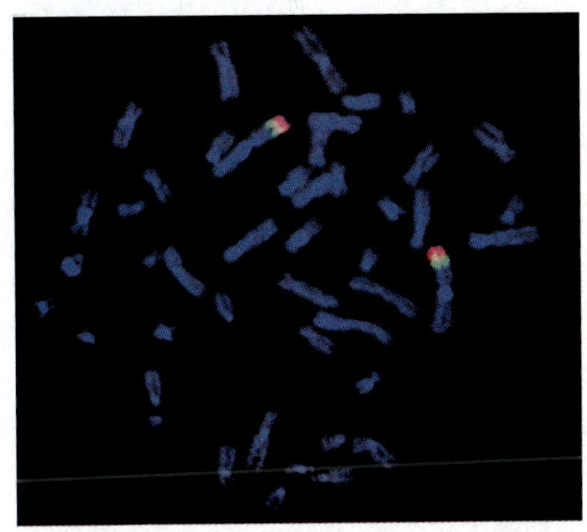

Figure 12.5.37 Human metaphase chromosomes hybridized to fluorescent probes from two overlapping microdissection libraries. Probes specific to chromosome regions 1p34–35 and 1p36 were labeled using the ULYSIS® Oregon Green® 488 (U21659) and Alexa Fluor® 594 (U21654) Nucleic Acid Labeling Kits, respectively. The chromosomes were counterstained with DAPI (D1306, D3571, D21490). Image contributed by Jingwei Yu, Colorado Genetics Laboratory.

Green-Fluorescent Chromosome Counterstains

SYTOX® Green (S7020) and YOYO®-1 (Y3601) nucleic acid stains are useful green-fluorescent nuclear counterstains because of their bright nuclear signal and low cytoplasmic background staining. Both dyes show intense green fluorescence upon binding to nucleic acids, and a wash step is not required because the dyes are essentially nonfluorescent in aqueous medium. We have found that both SYTOX® Green and YOYO®-1 dyes provide simple and reliable green-fluorescent counterstains for FISH analysis, though they differ somewhat in their properties and applications. Optimal staining by the YOYO®-1 dye requires RNase treatment for background reduction, whereas SYTOX® Green dye staining does not. In addition, counterstaining with the SYTOX® Green dye is more rapid than YOYO®-1 dye counterstaining. Although the spectral properties of the two green-fluorescent dyes differ slightly, nucleic acids counterstained with either of these green-fluorescent dyes can be efficiently excited with the mercury-arc lamp or argon-ion laser and can be visualized using standard fluorescein optical filter sets.

Red-Fluorescent Chromosome Counterstains

Propidium iodide (PI; P1304MP, P3566, P21493) is the traditional red-fluorescent chromosome counterstain and can be excited with the same excitation filters used for the green-fluorescent probes. Some of our longer-wavelength cyanine dyes, including the YO-PRO®-3, TO-PRO®-3, YOYO®-3 and TOTO®-3 dyes yield red-fluorescent nuclear staining that can be excited without also exciting the fluorescence of green-fluorescent dyes. TO-PRO®-3 (T3605) and TOTO®-3 (T3604) dyes exhibit long-wavelength fluorescence emissions (maxima at ~660 nm, Figure 12.5.30) that are well separated from the emissions of other commonly used fluorophores, such as Texas Red® dye, fluorescein or the Alexa Fluor® dyes that absorb maximally below 600 nm (Section 1.3), making three- or even four-color labeling possible. *Drosophila* polytene chromosomes and nuclei of cultured mammalian cells stained with the TO-PRO®-3 dye have also been detected with two-photon scanning near-field optical microscopy.[65]

Chromosome Banding Dyes

SYTOX® Green Nucleic Acid Stain

Chromosomes stained with SYTOX® Green dye (S7020) in combination with methyl green—a major-groove–binding dye that binds selectively to AT sequences along the chromosome—exhibit a banding pattern that indicates the location of AT-rich regions (Figure 12.5.38), representing an extremely simple, rapid, fluorescence-based banding method that may prove useful for general karyotype analysis. This observation has been exploited to examine metaphase chromatin structure.[66] The green-fluorescent SYTOX® Green dye is efficiently excited by the argon-ion laser, permitting analysis of chromosome structure by confocal laser-scanning microscopy. In addition, use of SYTOX® Green dye eliminates the need for RNase treatment of slides.

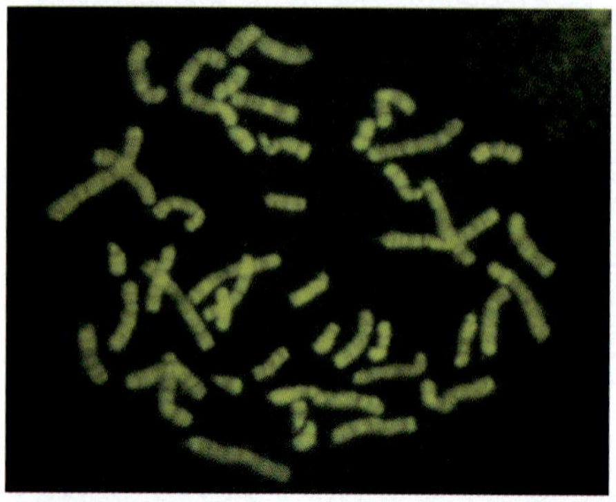

Figure 12.5.38 Human metaphase chromosomes stained with SYTOX® Green nucleic acid stain (S7020) and methyl green, and then mounted in Cytoseal 60 mounting medium.

Acridine Homodimer

The water-soluble acridine homodimer (A666) has extremely high affinity for AT-rich regions of nucleic acids, making it particularly useful for chromosome banding.[67,68] Acridine homodimer emits a blue-green fluorescence when bound to DNA, yielding fluorescence that is proportional to the fourth power of the AT base-pair content.[69] Acridine homodimer has been recommended as an alternative to quinacrine for Q banding because of its greater brightness and higher photostability.[67]

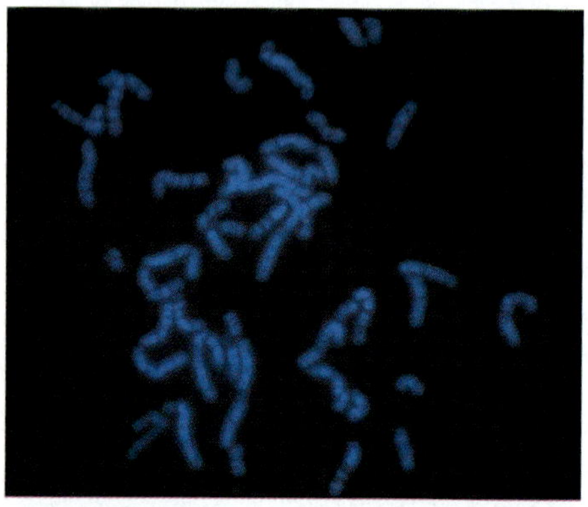

Figure 12.5.39 Human metaphase chromosomes stained with DAPI (D1306, D3571, D21490).

Other Chromosome Banding Dyes

A wide variety of fluorescent nucleic acid stains have been used for chromosome banding:[67,70–72]

- 7-Aminoactinomycin D (7-AAD, A1310) binds selectively to GC regions of DNA,[73] yielding a distinct banding pattern in polytene chromosomes and chromatin.[74,75]
- 9-Amino-6-chloro-2-methoxyacridine (ACMA, A1324) fluoresces with greater intensity in AT-rich regions on chromosomes,[76] yielding a staining pattern similar to the Q-banding pattern produced with quinacrine.
- DAPI (Figure 12.5.39) or combinations of DAPI or Hoechst 33258 (H1398, H3569, H21491) with nonfluorescent DNA-binding drugs have been used for chromosome-banding studies.[77]
- High-resolution flow karyotyping has also been carried out with DAPI[78,79] (D1306, D3571, D21490).
- Hoechst 33342 dye (H1399, H3570, H21492) has been used in chromosome sorting, multivariate analysis and karyotyping.[80]
- Hoechst dyes have been employed in combination with chromomycin and a high-resolution, dual-laser method to sort 21 unique human chromosome types onto nitrocellulose filters, followed by hybridization to gene-specific probes.[81]

Figure 12.5.40 Pyramidal cells of the hippocampus and dentate gyrus in a transverse cryosection of formaldehyde-fixed mouse brain. NeuroTrace® green fluorescent Nissl stain (N21480) is localized to neuronal somata, while non-neuronal cells can be identified by the presence of DAPI-stained nuclei. This image is a composite of images taken using a 10× objective and filters appropriate for fluorescein and DAPI.

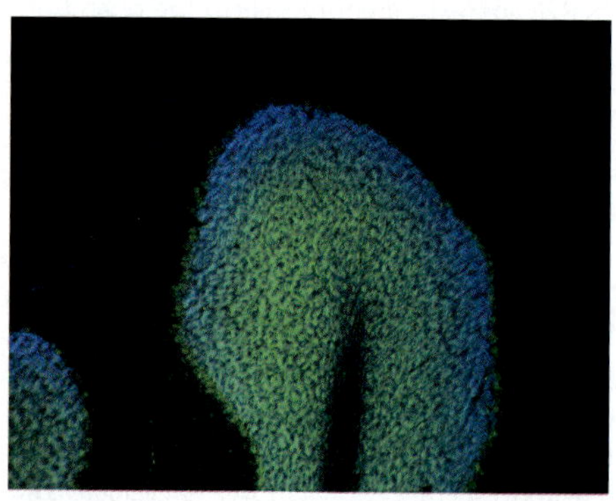

Figure 12.5.41 A mouse brain cryosection stained with the neuron-selective NeuroTrace® 500/525 green-fluorescent Nissl stain (N21480). The nuclei of the non-neuronal cells appear blue after incubation with the cell-permeant DNA counterstain, DAPI (D1306, D3571, D21490). The image is a composite of two micrographs acquired using filter sets appropriate for fluorescein and DAPI.

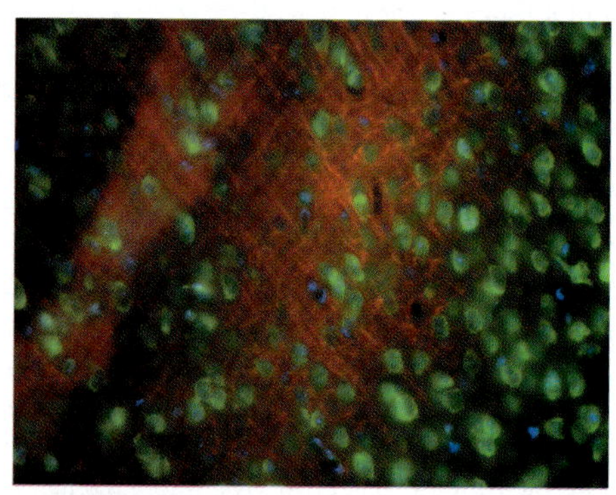

Figure 12.5.42 Neural somata from a mouse brain section labeled with NeuroTrace® 500/525 green-fluorescent Nissl stain (N21480). Nuclei are labeled with DAPI (D1306, D3571, D21490). Myelin and other lipophilic areas are stained with the the red-orange fluorescence from CellTracker™ CM-DiI (C7000, C7001). The image is a composite of three micrographs acquired using filters appropriate for fluorescein, tetramethylrhodamine and DAPI.

IMPORTANT NOTICE: The products described in this manual are covered by one or more Limited Use Label License(s). Please refer to the Appendix on page 971 and Master Product List on page 975. Products are For Research Use Only. Not intended for any animal or human therapeutic or diagnostic use.
www.invitrogen.com/probes

NeuroTrace® Fluorescent Nissl Stains

The Nissl substance, described by Franz Nissl more than 100 years ago, is unique to neuronal cells.[82] Composed of an extraordinary amount of rough endoplasmic reticulum, the Nissl substance reflects the unusually high protein synthesis capacity of neurons. Various fluorescent or chromophoric "Nissl stains" have been used for this counterstaining, including acridine orange,[83] ethidium bromide,[83] neutral red (N3246, Section 15.2), cresyl violet,[84] methylene blue, safranin-O and toluidine blue-O.[85] We have developed five fluorescent Nissl stains (Table 14.2) that not only provide a wide spectrum of fluorescent colors for staining neurons, but also are far more sensitive than the conventional dyes:

- NeuroTrace® 435/455 blue-fluorescent Nissl stain (N21479, Figure 12.5.22)
- NeuroTrace® 500/525 green-fluorescent Nissl stain (N21480; Figure 12.5.40, Figure 12.5.41, Figure 12.5.42)
- NeuroTrace® 515/535 yellow-fluorescent Nissl stain (N21481, Figure 12.5.43)
- NeuroTrace® 530/615 red-fluorescent Nissl stain (N21482; Figure 12.5.44, Figure 12.5.45)
- NeuroTrace® 640/660 deep red–fluorescent Nissl stain (N21483)

In addition, the Nissl substance redistributes within the cell body in injured or regenerating neurons. Therefore, these Nissl stains can also act as markers for physically or chemically induced neurostructural damage.[86,87] Staining by the Nissl stains is completely eliminated by pretreatment of tissue specimens with RNase; however, these dyes are not specific stains for RNA in solutions. The strong fluorescence (emission maximum ~515–520 nm) of NeuroTrace® 500/525 green-fluorescent Nissl stain (N21480) makes it a good choice for use as a counterstain in combination with orange- or red-fluorescent neuroanatomical tracers such as DiI[88] (D282, D3911, V22885; Section 14.4).

Figure 12.5.43 Mouse brain section stained with NeuroTrace® 515/535 yellow-fluorescent Nissl stain (N21481) and counterstained with nuclear yellow (N21485).

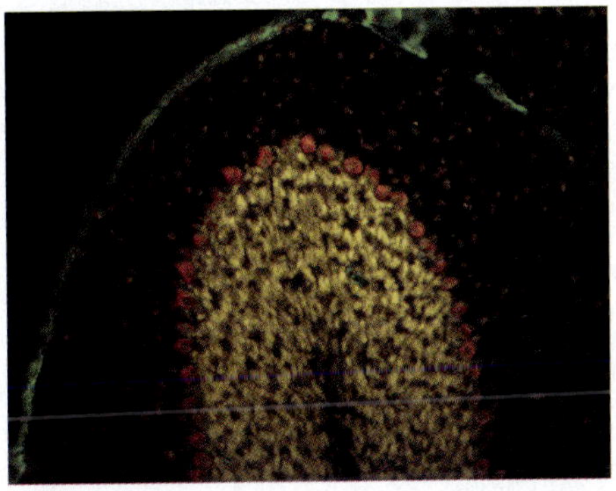

Figure 12.5.44 Neurons in a mouse cerebellum section labeled with NeuroTrace® 530/615 red-fluorescent Nissl stain (N21482). The Nissl substance, ribosomal RNA associated with the rough endoplasmic reticulum, is specific to neuronal cells. Other cells in the sample are identified with the contrasting nuclear counterstain nuclear yellow (N21485). Smaller neurons appear to be labeled primarily with the nuclear yellow stain. This image is a composite of two micrographs acquired using a DAPI longpass filter set and a filter set appropriate for tetramethylrhodamine.

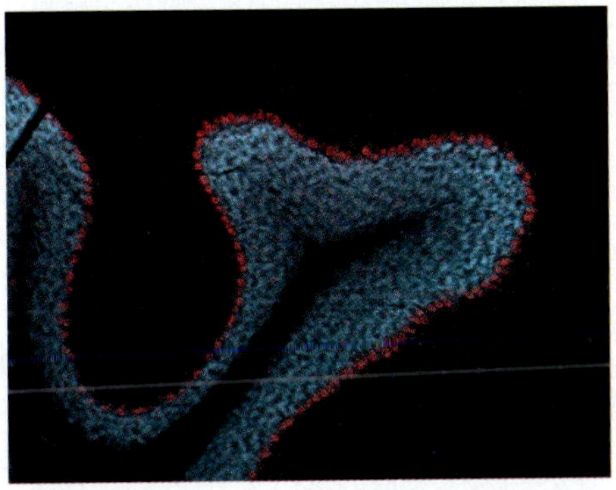

Figure 12.5.45 Neural somata labeled with NeuroTrace® 530/615 red-fluorescent Nissl stain (N21482). Nuclei in this mouse cerebellum section were counterstained with DAPI (D1306, D3571, D21490). The image is a composite of two micrographs acquired using filters appropriate for tetramethylrhodamine and DAPI.

REFERENCES

1. Mol Cell Biochem (1997) 172:171; 2. Can J Physiol Pharmacol (1997) 75:652; 3. Cytometry (1998) 32:66; 4. Dev Biol (1996) 180:184; 5. Cell (1995) 82:631; 6. J Biol Chem (1997) 272:12482; 7. J Immunol (1998) 160:2626; 8. J Cell Biol (1997) 139:831; 9. J Neurosci (1996) 16:7812; 10. Proc Natl Acad Sci U S A (1997) 94:14804; 11. J Neurochem (1996) 67:2484; 12. Protoplasma (1996) 192:70; 13. Invest Ophthalmol Vis Sci (1997) 38:301; 14. Invest Ophthalmol Vis Sci (1997) 38:1678; 15. Am J Trop Med Hyg (1998) 59:182; 16. J Biol Chem (1998) 273:28040; 17. Invest Ophthalmol Vis Sci (2000) 41:2849; 18. Int J Parasitol (1997) 27:787; 19. PLoS One (2009) 4:e7124; 20. Assay Drug Dev Technol (2008) 6:693; 21. J Biomol Screen (2008) 13:527; 22. J Cell Biol (1999) 144:813; 23. Biophys J (1999)

continued on next page

IMPORTANT NOTICE: The products described in this manual are covered by one or more Limited Use Label License(s). Please refer to the Appendix on page 971 and Master Product List on page 975. Products are For Research Use Only. Not intended for any animal or human therapeutic or diagnostic use.

REFERENCES—*continued*

77:2871; **24.** Dev Biol (1999) 206:232; **25.** Nat Methods (2007) 4:555; **26.** J Cell Biochem (2009) 107:1160; **27.** Methods Mol Biol (2002) 183:181; **28.** Methods (2003) 29:42; **29.** Plant J (2009) 57:680; **30.** Biochemistry (2004) 43:10192; **31.** J Cell Biol (1998) 141:469; **32.** Acta Histochem Cytochem (1997) 30:309; **33.** Acta Histochem Cytochem (1998) 31:297; **34.** J Cell Biol (1998) 143:1329; **35.** Genes Dev (1998) 12:435; **36.** Cancer Res (1997) 57:3804; **37.** J Immunol Methods (1995) 185:249; **38.** Microcirculation (1995) 2:267; **39.** Cytometry (1994) 15:129; **40.** J Neurosci (1996) 16:1400; **41.** Neuroscience (1994) 60:125; **42.** Neuroscience (1989) 28:725; **43.** Biotech Histochem (2000) 75:132; **44.** Neurosci Lett (1980) 18:25; **45.** Neurosci Lett (1991) 128:137; **46.** J Neurosci Methods (1990) 35:175; **47.** J Neurosci Methods (1990) 32:15; **48.** Meth Neurosci (1990) 3:275; **49.** Neuroscience Protocols, Wouterlood FG, Ed. (1993) p. 93-050-06; **50.** Microsc Res Tech (1993) 24:2; **51.** J Comp Neurol (1987) 258:230; **52.** J Neurosci Methods (1985) 14:273; **53.** J Histochem Cytochem (1996) 44:399; **54.** Trends Genet (1995) 11:9; **55.** J Histochem Cytochem (1997) 45:49; **56.** Histochem Cell Biol (2001) 115:293; **57.** J Cell Biol (1997) 136:531; **58.** Science (1995) 270:1595; **59.** Nat Med (1996) 2:1361; **60.** J Biomol Screen (1999) 4:193; **61.** J Cell Biol (1998) 140:1285; **62.** J Microsc (1990) 157:73; **63.** Proc Natl Acad Sci U S A (1990) 87:9358; **64.** Proc Natl Acad Sci U S A (1990) 87:6634; **65.** Biophys J (1998) 75:1513; **66.** Cell (1994) 76:609; **67.** Methods Mol Biol (1994) 29:83; **68.** Biochemistry (1979) 18:3354; **69.** Proc Natl Acad Sci U S A (1975) 72:2915; **70.** Hum Genet (1981) 57:1; **71.** Bioessays (1993) 15:349; **72.** Am J Hum Genet (1992) 51:17; **73.** Biopolymers (1979) 18:1749; **74.** Cytometry (1995) 20:296; **75.** Histochem J (1985) 17:131; **76.** Exp Cell Res (1978) 117:451; **77.** Eur J Cell Biol (1980) 20:290; **78.** Cytometry (1990) 11:184; **79.** Cancer Res (1999) 59:141; **80.** Cytometry (1990) 11:80; **81.** Science (1984) 225:57; **82.** Neuroscience Protocols, Wouterlood FG, Ed. (1993) p. 93-050-12; **83.** Proc Natl Acad Sci U S A (1980) 77:2260; **84.** J Neurosci Methods (1990) 33:129; **85.** J Neurosci Methods (1997) 72:49; **86.** J Neurosci (2004) 24:5549; **87.** Nat Med (2005) 11:1355; **88.** Neurosci Lett (1995) 184:169.

DATA TABLE 12.5 PROBES FOR THE NUCLEUS

Cat. No.	MW	Storage	Soluble	Abs	EC	Em	Solvent	Notes
A1310	1270.45	F,L	DMF, DMSO	546	25,000	647	H_2O/DNA	1
A1324	258.71	L	DMF, DMSO	412	8200	471	MeOH	2
B3582	1202.66	F,D,L	DMSO	462	114,000	481	H_2O/DNA	1, 3, 4, 5
B3586	1254.73	F,D,L	DMSO	570	148,000	604	H_2O/DNA	1, 3, 4, 5
D1306	350.25	L	H_2O, DMF	358	24,000	461	H_2O/DNA	1, 6
D3571	457.49	L	H_2O, MeOH	358	24,000	461	H_2O/DNA	1, 6
D21490	350.25	L	H_2O, DMF	358	24,000	461	H_2O/DNA	1, 6, 7
H1398	623.96	L	H_2O, DMF	352	40,000	461	H_2O/DNA	1, 8, 9
H1399	615.99	L	H_2O, DMF	350	45,000	461	H_2O/DNA	1, 8, 10
H3569	623.96	RR,L	H_2O	352	40,000	461	H_2O/DNA	1, 3, 8, 9
H3570	615.99	RR,L	H_2O	350	45,000	461	H_2O/DNA	1, 3, 8, 10
H21491	623.96	L	H_2O, DMF	352	40,000	461	H_2O/DNA	1, 7, 8, 9
H21492	615.99	L	H_2O, DMF	350	45,000	461	H_2O/DNA	1, 7, 8, 10
N21479	see Notes	F,D,L	DMSO	435	see Notes	457	H_2O/RNA	3, 5, 11
N21480	see Notes	F,D,L	DMSO	497	see Notes	524	H_2O/RNA	3, 5, 11
N21481	see Notes	F,D,L	DMSO	515	see Notes	535	H_2O/RNA	3, 5, 11
N21482	see Notes	F,D,L	DMSO	530	see Notes	619	H_2O/RNA	3, 5, 11
N21483	see Notes	F,D,L	DMSO	644	see Notes	663	H_2O/RNA	3, 5, 11
N21485	651.01	L	DMSO	355	36,000	495	H_2O/DNA	1
P1304MP	668.40	L	H_2O, DMSO	535	5400	617	H_2O/DNA	1, 12
P3566	668.40	RR,L	H_2O	535	5400	617	H_2O/DNA	1, 3, 12
P21493	668.40	L	H_2O, DMSO	535	5400	617	H_2O/DNA	1, 7, 12
S7020	~600	F,D,L	DMSO	504	67,000	523	H_2O/DNA	1, 3, 5, 13
S11348	~400	F,D,L	DMSO	445	38,000	470	H_2O/DNA	1, 3, 5, 13
S11368	~500	F,D,L	DMSO	547	79,000	570	H_2O/DNA	1, 3, 5, 13
S34857	~400	F,D,L	DMSO	445	38,000	470	H_2O/DNA	1, 3, 5, 13
S34859	~450	F,D,L	DMSO	640	92,000	658	H_2O/DNA	1, 3, 5, 13
T3600	1302.78	F,D,L	DMSO	514	117,000	533	H_2O/DNA	1, 3, 4, 5
T3604	1354.85	F,D,L	DMSO	642	154,000	660	H_2O/DNA	1, 3, 4, 5
T3605	671.42	F,D,L	DMSO	642	102,000	661	H_2O/DNA	1, 3, 4, 5
Y3601	1270.65	F,D,L	DMSO	491	99,000	509	H_2O/DNA	1, 3, 4, 5
Y3603	629.32	F,D,L	DMSO	491	52,000	509	H_2O/DNA	1, 3, 4, 5
Y3606	1322.73	F,D,L	DMSO	612	167,000	631	H_2O/DNA	1, 3, 4, 5
Y3607	655.36	F,D,L	DMSO	612	100,000	631	H_2O/DNA	1, 3, 4, 5

For definitions of the contents of this data table, see "Using *The Molecular Probes® Handbook*" in the introductory pages.

Notes

1. Spectra represent aqueous solutions of nucleic acid–bound dye. EC values are derived by comparing the absorbance of the nucleic acid–bound dye with that of free dye in a reference solvent (H_2O or MeOH).
2. Spectra of this compound are in methanol acidified with a trace of HCl.
3. This product is supplied as a ready-made solution in the solvent indicated under "Soluble."
4. Although this compound is soluble in water, preparation of stock solutions in water is not recommended because of possible adsorption onto glass or plastic.
5. This product is essentially nonfluorescent except when bound to DNA or RNA.
6. DAPI in H_2O: Abs = 342 nm (EC = 28,000 $cm^{-1}M^{-1}$), Em = 450 nm. The fluorescence quantum yield of DAPI bound to dsDNA is 0.34, representing an ~20-fold increase relative to the free dye in H_2O. (Photochem Photobiol (2001) 73:585)
7. This product is specified to equal or exceed 98% analytical purity by HPLC.
8. MW is for the hydrated form of this product.
9. The fluorescence quantum yield of Hoechst 33258 bound to dsDNA is 0.42, representing an ~30-fold increase relative to the free dye in H_2O. (Photochem Photobiol (2001) 73:585)
10. The fluorescence quantum yield of Hoechst 33342 bound to dsDNA is 0.38, representing an ~10-fold increase relative to the free dye in H_2O. (Photochem Photobiol (2001) 73:585)
11. The active ingredient of this product is an organic dye with MW <1000. The exact MW and extinction coefficient values for this dye are proprietary.
12. Propidium iodide in H_2O: Abs = 493 nm (EC = 5900 $cm^{-1}M^{-1}$), Em = 636 nm (weakly fluorescent). Fluorescence is enhanced >10-fold on binding to dsDNA.
13. MW: The preceding ~ symbol indicates an approximate value, not including counterions.

IMPORTANT NOTICE: The products described in this manual are covered by one or more Limited Use Label License(s). Please refer to the Appendix on page 971 and Master Product List on page 975. Products are For Research Use Only. Not intended for any animal or human therapeutic or diagnostic use.

PRODUCT LIST 12.5 PROBES FOR THE NUCLEUS

Cat. No.	Product	Quantity
A1310	7-aminoactinomycin D (7-AAD)	1 mg
A1324	9-amino-6-chloro-2-methoxyacridine (ACMA)	100 mg
B3582	BOBO™-1 iodide (462/481) *1 mM solution in DMSO*	200 µL
B3586	BOBO™-3 iodide (570/602) *1 mM solution in DMSO*	200 µL
C10594	CellLight® Histone 2B-GFP *BacMam 2.0*	1 mL
C10595	CellLight® Histone 2B-RFP *BacMam 2.0*	1 mL
C10602	CellLight® Nucleus-GFP *BacMam 2.0*	1 mL
C10603	CellLight® Nucleus-RFP *BacMam 2.0*	1 mL
D1306	4',6-diamidino-2-phenylindole, dihydrochloride (DAPI)	10 mg
D21490	4',6-diamidino-2-phenylindole, dihydrochloride (DAPI) *FluoroPure™ grade*	10 mg
D3571	4',6-diamidino-2-phenylindole, dilactate (DAPI, dilactate)	10 mg
H32720	HCS CellMask™ Blue stain *for 10 X 96-well plates*	1 set
H32721	HCS CellMask™ Deep Red stain *for 10 X 96-well plates*	1 set
H32714	HCS CellMask™ Green stain *for 10 X 96-well plates*	1 set
H32713	HCS CellMask™ Orange stain *for 10 X 96-well plates*	1 set
H32712	HCS CellMask™ Red stain *for 10 X 96-well plates*	1 set
H10325	HCS NuclearMask™ Blue stain *for 10 X 96-well plates* *2000X concentrate*	65 µL
H10294	HCS NuclearMask™ Deep Red stain *250X concentrate in DMSO*	400 µL
H10326	HCS NuclearMask™ Red stain *for 10 X 96-well plates* *1000X concentrate*	125 µL
H13188	histone H1 from calf thymus, Alexa Fluor® 488 conjugate	1 mg
H1398	Hoechst 33258, pentahydrate (bis-benzimide)	100 mg
H3569	Hoechst 33258, pentahydrate (bis-benzimide) *10 mg/mL solution in water*	10 mL
H21491	Hoechst 33258, pentahydrate (bis-benzimide) *FluoroPure™ grade*	100 mg
H1399	Hoechst 33342, trihydrochloride, trihydrate	100 mg
H3570	Hoechst 33342, trihydrochloride, trihydrate *10 mg/mL solution in water*	10 mL
H21492	Hoechst 33342, trihydrochloride, trihydrate *FluoroPure™ grade*	100 mg
N21479	NeuroTrace® 435/455 blue fluorescent Nissl stain *solution in DMSO*	1 mL
N21480	NeuroTrace® 500/525 green fluorescent Nissl stain *solution in DMSO*	1 mL
N21481	NeuroTrace® 515/535 yellow fluorescent Nissl stain *solution in DMSO*	1 mL
N21482	NeuroTrace® 530/615 red fluorescent Nissl stain *solution in DMSO*	1 mL
N21483	NeuroTrace® 640/660 deep-red fluorescent Nissl stain *solution in DMSO*	1 mL
N21485	nuclear yellow (Hoechst S769121, trihydrochloride, trihydrate)	10 mg
P1304MP	propidium iodide	100 mg
P3566	propidium iodide *1.0 mg/mL solution in water*	10 mL
P21493	propidium iodide *FluoroPure™ grade*	100 mg
Q10363	Qnuclear™ Deep Red stain	100 µL
S33025	SelectFX® Nuclear Labeling Kit *DAPI, SYTOX® Green, 7-AAD, TO-PRO®-3 iodide* *for fixed cells*	1 kit
S11350	SYTO® Blue Fluorescent Nucleic Acid Stain Sampler Kit *SYTO® dyes 40–45* *50 µL each*	1 kit
S7572	SYTO® Green Fluorescent Nucleic Acid Stain Sampler Kit *SYTO® dyes 11–14,16,21,24, and 25* *50 µL each*	1 kit
S11360	SYTO® Orange Fluorescent Nucleic Acid Stain Sampler Kit *SYTO® dyes 80–85* *50 µL each*	1 kit
S11340	SYTO® Red Fluorescent Nucleic Acid Stain Sampler Kit *SYTO® dyes 17 and 59–64* *50 µL each*	1 kit
S10349	SYTOX® AADvanced™ Dead Cell Stain Kit *for flow cytometry* *for 488 nm excitation* *100 tests*	1 kit
S10274	SYTOX® AADvanced™ Dead Cell Stain Kit *for flow cytometry* *for 488 nm excitation* *500 tests*	1 kit
S34857	SYTOX® Blue dead cell stain *for flow cytometry* *1000 assays* *1 mM solution in DMSO*	1 mL
S11348	SYTOX® Blue nucleic acid stain *5 mM solution in DMSO*	250 µL
S7020	SYTOX® Green nucleic acid stain *5 mM solution in DMSO*	250 µL
S11368	SYTOX® Orange nucleic acid stain *5 mM solution in DMSO*	250 µL
S34859	SYTOX® Red dead cell stain *for 633 or 635 nm excitation* *5 µM solution in DMSO*	1 mL
T3605	TO-PRO®-3 iodide (642/661) *1 mM solution in DMSO*	1 mL
T3600	TOTO®-1 iodide (514/533) *1 mM solution in DMSO*	200 µL
T3604	TOTO®-3 iodide (642/660) *1 mM solution in DMSO*	200 µL
Y3603	YO-PRO®-1 iodide (491/509) *1 mM solution in DMSO*	1 mL
Y3607	YO-PRO®-3 iodide (612/631) *1 mM solution in DMSO*	1 mL
Y3601	YOYO®-1 iodide (491/509) *1 mM solution in DMSO*	200 µL
Y3606	YOYO®-3 iodide (612/631) *1 mM solution in DMSO*	200 µL

IMPORTANT NOTICE: The products described in this manual are covered by one or more Limited Use Label License(s). Please refer to the Appendix on page 971 and Master Product List on page 975. Products are For Research Use Only. Not intended for any animal or human therapeutic or diagnostic use.

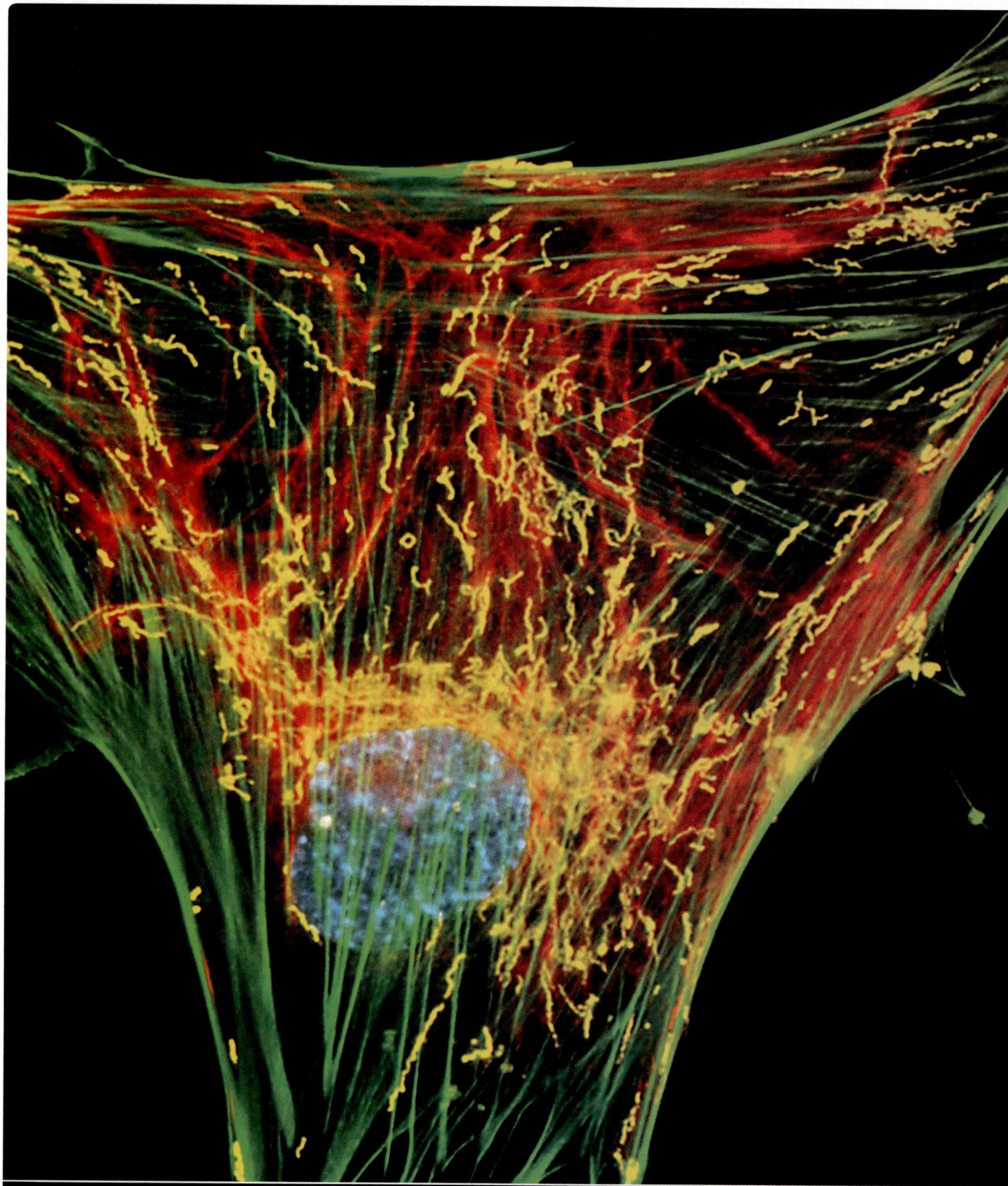

MitoTracker® Deep Red 633, BODIPY® FL phallacidin and Hoechst 33258.

IMPORTANT NOTICE: The products described in this manual are covered by one or more Limited Use Label License(s). Please refer to the Appendix on page 971 and Master Product List on page 975. Products are For Research Use Only. Not intended for any animal or human therapeutic or diagnostic use.

THIRTEEN

CHAPTER 13

Probes for Lipids and Membranes

IMPORTANT NOTICE: The products described in this manual are covered by one or more Limited Use Label License(s). Please refer to the Appendix on page 971 and Master Product List on page 975. Products are For Research Use Only. Not intended for any animal or human therapeutic or diagnostic use.

IMPORTANT NOTICE: The products described in this manual are covered by one or more Limited Use Label License(s). Please refer to the Appendix on page 971 and Master Product List on page 975. Products are For Research Use Only. Not intended for any animal or human therapeutic or diagnostic use.

13.1 Introduction to Membrane Probes

Fluorescent and Biotinylated Membrane Probes

The plasma membranes and intracellular membranes of live cells and the artificial membranes of liposomes represent a significant area of application for fluorescent probes. Membrane probes include fluorescent analogs of natural lipids, as well as lipophilic organic dyes that have little structural resemblance to natural biomolecules. We offer a wide range of both types of membrane probes. These probes are used for structural and biophysical analysis of membranes, for following lipid transport and metabolism in live cells (Figure 13.1.1) and for investigating synaptosome recycling (Section 16.1) and lipid-mediated signal transduction processes (Chapter 17). Due to their low toxicity and stable retention, some lipid probes are particularly useful for long-term cell tracing (Section 14.4). Other, slightly less lipophilic probes are used as membrane markers of endocytosis and exocytosis (Section 16.1).

Fluorescent Analogs of Natural Lipids

We offer fluorescent and, in a few cases, biotinylated analogs of five naturally occurring lipid classes—phospholipids, sphingolipids (including ceramides), fatty acids, triglycerides and steroids. Phospholipids are the principal building blocks of cell membranes. Most phospholipids are derivatives of glycerol comprising two fatty acyl residues (nonpolar tails) and a single phosphate ester substituent (polar head group). Despite their overall structural similarity (Figure 13.1.2), natural phospholipids exhibit subtle differences in their fatty acid compositions, degree of acyl

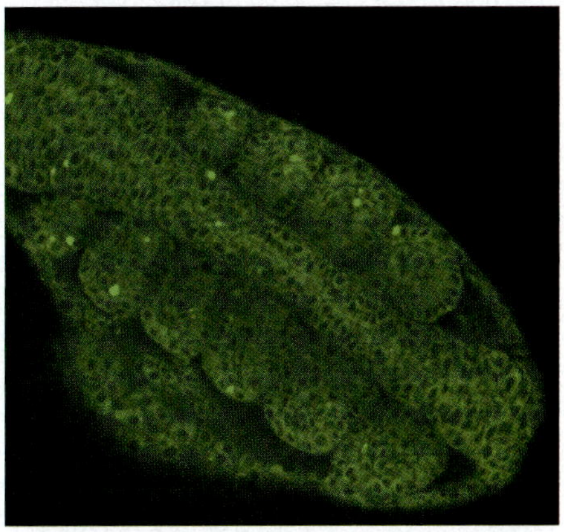

Figure 13.1.1 The cytoplasm of a live zebrafish embryo labeled with the green-fluorescent lipophilic tracer BODIPY® 505/515 (D3921). The image was contributed by Arantza Barrios, University College, London.

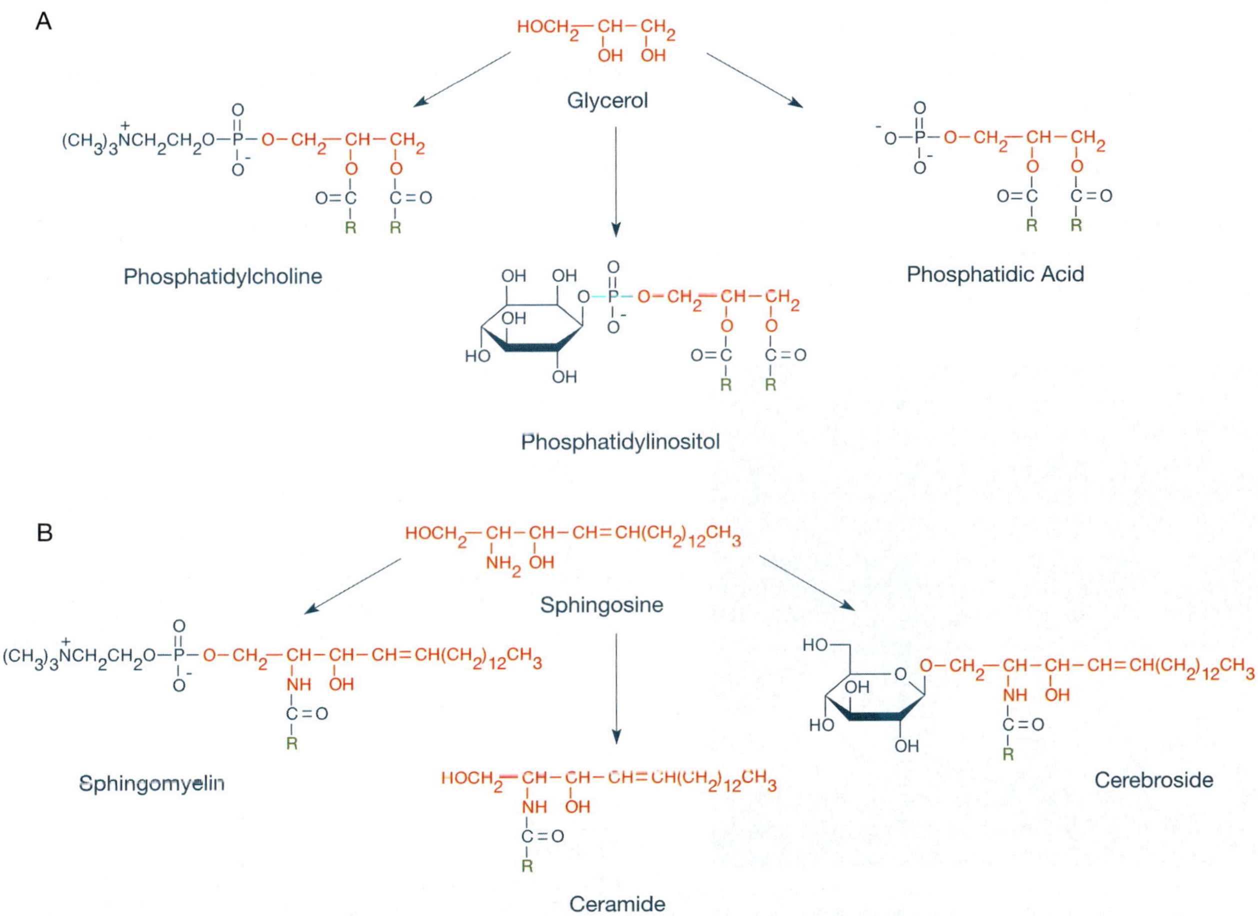

Figure 13.1.2 A) Phosphatidylcholines, phosphatidylinositols and phosphatidic acids are examples of glycerolipids derived from glycerol. B) Sphingomyelins, ceramides and cerebrosides are examples of sphingolipids derived from sphingosine. In all the structures shown, R represents the hydrocarbon tail portion of a fatty acid residue.

chain unsaturation and type of polar head group.[1] These differences can produce significant variations in membrane physical properties, in the location of phospholipids in a lipid bilayer and in their biological activity. Fluorescent phospholipid analogs (Section 13.2) can be classified according to where the fluorophore is attached. The fluorophore can be attached to one (or both) of the fatty acyl chains or to the polar head group. The attachment position of the fluorophore determines whether it is located in the nonpolar interior or at the water/lipid interface when the phospholipid analog is incorporated into a lipid bilayer membrane.

Fatty acids are the building blocks for a diverse set of biomolecules. Some fatty acids (e.g., arachidonic acid) are important in cell signaling.[2] Fatty acids are liberated by the enzymatic action of phospholipase A on phospholipids (Section 17.4) and also by various other lipases. Fluorescent fatty acids can often be used interchangeably with the corresponding phospholipids as membrane probes; however, fatty acids transfer more readily between aqueous and lipid phases.[3] Although fatty acids are ionized at neutral pH in water (pK_a ~5), their pK_a is typically ~7 in membranes, and thus a significant fraction of membrane-bound fatty acids are neutral species.[3] Certain fluorescent fatty acids (Section 13.2) are readily metabolized by live cells to phospholipids, mono-, di- and triacylglycerols, cholesteryl esters and other lipid derivatives.[4]

Sphingolipids play critical roles in processes such as proliferation, apoptosis, signal transduction and molecular recognition at cell surfaces.[1,5,6] Defects in the lysosomal breakdown of sphingolipids are the underlying cause of lipid storage disorders such as Niemann–Pick, Tay–Sachs, Krabbe and Gaucher diseases. The sphingolipids described in Section 13.3 include ceramides, sphingomyelins, glycosylceramides (cerebrosides) and gangliosides. The structural backbone of sphingolipids is the lipophilic amino–dialcohol sphingosine (2-amino-4-octadecen-1,3-diol, Figure 13.1.2) to which a single fatty acid residue is attached via an amide linkage. Our fluorescent analogs of sphingolipids are prepared by replacing the natural amide-linked fatty acid with a fluorescent analog. Sphingolipids with an unmodified hydroxyl group in the 1-position are classified as ceramides. As part of the lipid-sorting process in cells, ceramides are glycosylated to cerebrosides (Figure 13.1.2) or converted to sphingomyelins (Figure 13.1.2) in the Golgi complex. Glycosylated sphingolipids (cerebrosides and gangliosides) occur in the plasma membranes of all eukaryotic cells and are involved in cell recognition, signal transduction and modulation of receptor function.[7] Gangliosides have complex oligosaccharide head groups containing at least one sialic acid residue in place of the single galactose or glucose residues of cerebrosides.

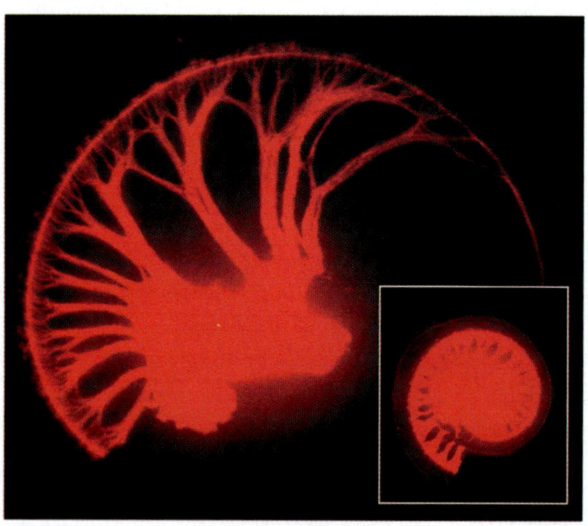

Figure 13.1.3 The neuronal tracer DiI (D282, D3911) used as a diagnostic tool to evaluate patterns of innervation in newborn mouse cochlea. The larger image is of a mutant cochlea and the inset is of a wild-type cochlea. Image contributed by Bernd Fritzsch, Creighton University, and L. Reichardt and I. Farinas, Howard Hughes Medical Institute, San Francisco.

Fluorescent cholesteryl esters and triglycerides (Section 13.3) can be used as structural probes and transport markers for these important lipid constituents of membranes and lipoproteins.[8] They may also serve as fluorescent substrates for lipases and lipid-transfer proteins and can be incorporated into low-density lipoproteins (LDL, Section 16.1).

Other Lipophilic and Amphiphilic Fluorescent Probes

The probes described in Section 13.4 and Section 13.5 are not analogs of any particular biological lipid class, but they have a general structural resemblance that facilitates labeling of membranes, lipoproteins and other lipid-based molecular assemblies. Particularly notable members of this group are the lipophilic carbocyanines DiI (Figure 13.1.3), DiO, DiD and DiR, the lipid fluidity probes DPH and TMA-DPH and the membrane-surface probes ANS and laurdan. These probes generally have limited water solubility and exhibit substantially enhanced fluorescence upon partition into lipid environments. They can be classified as either amphiphilic (having both polar and nonpolar structural elements) or neutral (lacking charges and most soluble in very nonpolar environments). We use similar neutral lipophilic dyes for internal staining of our fluorescent polystyrene microspheres (Section 6.5).

Other Probes for Studying Cell Membranes

In addition to the lipophilic probes described in this chapter, we have available the following products for studying the properties and functions of cell membranes:

- Moderately lipophilic stains for the endoplasmic reticulum and Golgi apparatus (Section 12.4)
- FM® dyes—amphiphilic probes for cell membrane labeling (Section 14.4, Section 16.1)
- CellLight® Plasma Membrane-CFP, CellLight® Plasma Membrane-GFP and CellLight® Plasma Membrane-RFP, which are BacMam 2.0 vectors encoding fluorescent proteins targeted to the plasma membrane (C10606, C10607, C10608; Section 14.4)
- Alexa Fluor® dye–labeled cholera toxin subunit B conjugates for labeling lipid rafts (Section 14.7)
- Annexin V conjugates for detection of phosphatidylserine exposure in apoptotic cell membranes (Section 15.5)
- Fluorescent and fluorogenic phospholipase A substrates (Section 17.4)
- Amplex® Red Phosphatidylcholine-Specific Phospholipase C Assay Kit and Amplex® Red Phospholipase D Assay Kit (A12218, A12219; Section 17.4)
- Antibodies to phosphatidylinositol phosphates (Section 17.4)
- Lipophilic pH indicators (Section 20.4)
- Membrane potential–sensitive probes (Section 22.2, Section 22.3)

REFERENCES

1. Nat Rev Mol Cell Biol (2008) 9:112; **2.** Biochim Biophys Acta (1994) 1212:26; **3.** J Lipid Res (1998) 39:467; **4.** Chem Phys Lipids (1991) 58:111; **5.** Annu Rev Biochem (1998) 67:27; **6.** Biochim Biophys Acta (1991) 1082:113; **7.** Ann N Y Acad Sci (1998) 845:57; **8.** Nat Rev Mol Cell Biol (2008) 9:125.

IMPORTANT NOTICE: The products described in this manual are covered by one or more Limited Use Label License(s). Please refer to the Appendix on page 971 and Master Product List on page 975. Products are For Research Use Only. Not intended for any animal or human therapeutic or diagnostic use.
www.invitrogen.com/probes

13.2 Fatty Acid Analogs and Phospholipids

The probes in this section and in Section 13.3 bear some structural resemblance to natural lipids. Included in this section are fluorescent fatty acid analogs, as well as phospholipids wherein one or both fatty acid esters are replaced by fluorescent fatty acid esters. The fluorophores in these probes tend to remain buried in the hydrophobic interior of lipid bilayer membranes.[1,2] In this location, they are sensitive to membrane properties such as lipid fluidity, lateral domain formation and structural perturbation by proteins, drugs and other additives. Also included in this section are several head group–modified phospholipid analogs incorporating a fluorophore or biotin (Table 13.1).

Sphingolipids, steroids and lipopolysaccharides are discussed in Section 13.3. Important applications of the fluorescent phosphatidylinositol derivatives as probes for signal transduction and various fluorescent phospholipids as phospholipase substrates are further described in Section 17.4. A review of fluorescent lipid probes and their use in biological and biophysical research has been published.[3]

Fluorescent Fatty Acid Analogs

Our fluorescent fatty acid analogs have a fluorophore linked within the fatty acid chain or, more commonly, at the terminal (omega) carbon atom that is furthest from the carboxylate moiety. Although fluorescent fatty acid analogs are sometimes used as direct probes for membranes and liposomes, their most common applications have been for synthesis of fluorescent phospholipids and for metabolic incorporation by live cells. Our fluorescent fatty acids currently include derivatives based on the BODIPY®, nitrobenzoxadiazole (NBD), pyrene and dansyl fluorophores, as well as the naturally fluorescent polyunsaturated fatty acid, *cis*-parinaric acid.

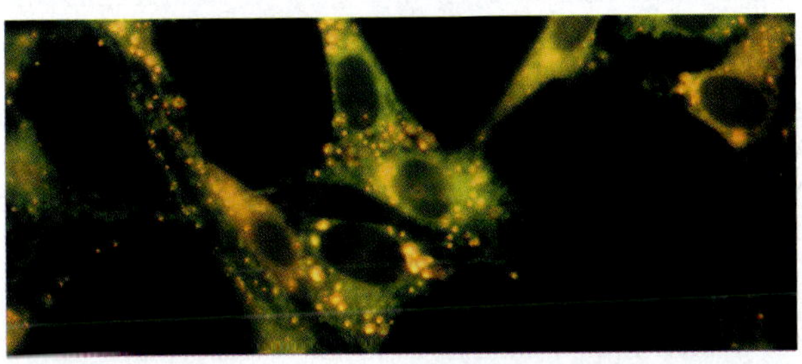

Figure 13.2.2 BHK cells incubated in medium containing the fluorescent fatty acid analog BODIPY® 500/510 C_1,C_{12} (D3823). This photomicrograph, obtained through a standard fluorescein longpass filter set, reveals reticular green-fluorescent staining as well as yellow-orange–fluorescent spherical structures. These fluorescent structures are indicative of the metabolic accumulation of BODIPY® dye–labeled neutral lipids in cytoplasmic droplets. Image contributed by Juha Kasurinen, University of Helsinki, Finland.

Table 13.1 Phospholipids with labeled head groups.

Label (Ex/Em) *	Cat. No.
Dansyl (336/517)	D57
Marina Blue® (365/460)	M12652
Pacific Blue™ (410/455)	P22652
NBD (463/536)	N360
Fluorescein (496/519)	F362
Oregon Green® 488 (501/526)	O12650
BODIPY® FL (505/511)	D3800
BODIPY® 530/550	D3815
Tetramethylrhodamine (540/566)	T1391
Lissamine rhodamine (560/581)	L1392
Texas Red® (582/601)	T1395MP
Biotin (<250/none)	B1550, B1616

* Fluorescence excitation (Ex) and emission (Em) spectral maxima, in nm, are in methanol. The spectra may be different in membranes.

BODIPY® Fatty Acids

BODIPY® fatty acids are, by far, the most fluorescent fatty acid analogs that we have available.[4] The lack of ionic charge on the BODIPY® fluorophore is unusual for long-wavelength fluorescent dyes and results in exclusive localization of the fluorophore within the membrane (Figure 13.2.1). BODIPY® derivatives typically have extinction coefficients greater than 90,000 $cm^{-1}M^{-1}$ with absorption maxima beyond 500 nm. A useful spectroscopic property of BODIPY® dyes is the concentration-dependent formation of excited-state dimers ("excimers") with red-shifted emission. We have observed this phenomenon particularly with our green-fluorescent BODIPY® fatty acid derivatives, which undergo a considerable red shift in their emission when metabolically incorporated into lipophilic products[5] (Figure 13.2.2). Pyrene

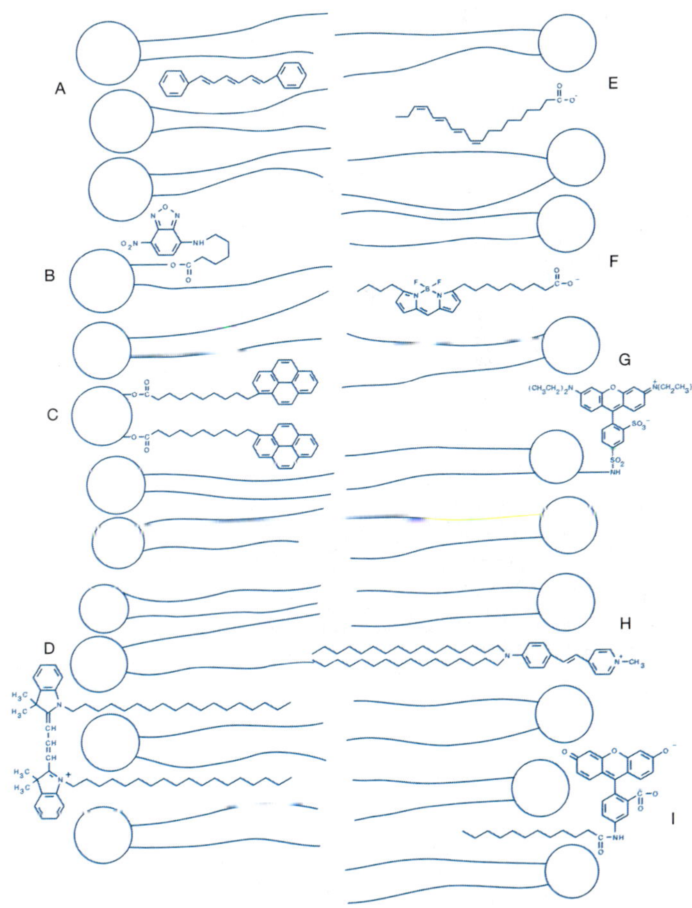

Figure 13.2.1 Location and orientation of representative fluorescent membrane probes in a phospholipid bilayer: **A)** DPH (D202), **B)** NBD-C_6-HPC (N3786), **C)** bis-pyrene-PC (B3782), **D)** DiI (D282), **E)** *cis*-parinaric acid (P36005), **F)** BODIPY® 500/510 C_4, C_9 (B3824), **G)** *N*-Rh-PE (L1392), **H)** DiA (D3883) and **I)** C_{12}-fluorescein (D109).

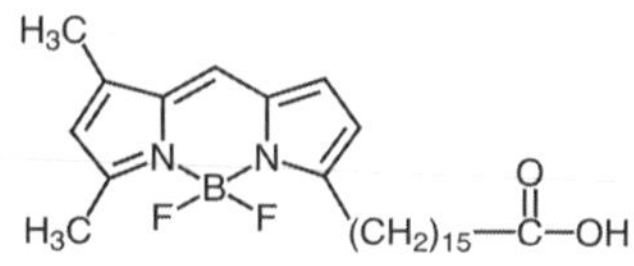

Figure 13.2.3 4,4-Difluoro-5,7-dimethyl-4-bora-3a,4a-diaza-*s*-indacene-3-hexadecanoic acid (BODIPY® FL C_{16}, D3821).

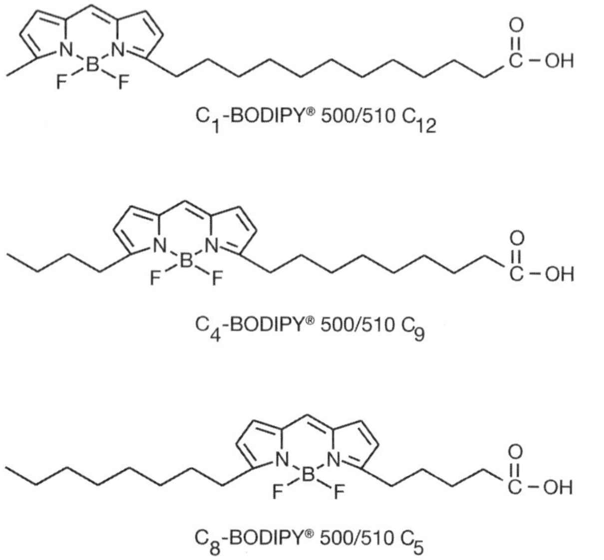

Figure 13.2.4 Structural representations showing the positional shift of the fluorophore with respect to the terminal carboxyl group in a homologous series of BODIPY® 500/510 fatty acids (D3823, B3824, D3825).

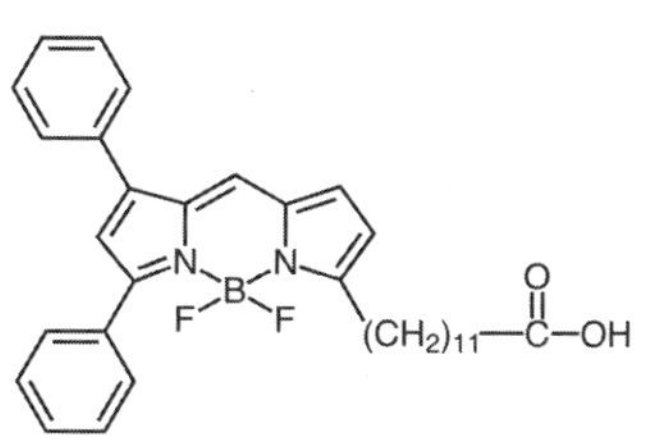

Figure 13.2.5 4,4-Difluoro-5,7-diphenyl-4-bora-3a,4a-diaza-*s*-indacene-3-dodecanoic acid (BODIPY® 530/550 C_{12}, D3832).

fatty acids also exhibit excimer formation but their emission is at much shorter wavelengths than that of the BODIPY® dyes and they are therefore less suitable for the study of live cells.

The fluorophores in our current selection of BODIPY® fatty acids and their approximate absorption/emission maxima (in nm) are:

- BODIPY® 503/512 (BODIPY® FL; D3821, Figure 13.2.3; D3822; D3834; D3862)
- BODIPY® 500/510 (D3823, B3824, D3825; Figure 13.2.4)
- BODIPY® 530/550 (D3832, Figure 13.2.5)
- BODIPY® 558/568 (D3835, Figure 13.2.6)
- BODIPY® 581/591 (D3861)

BODIPY® fatty acids are synthetic precursors to a wide variety of fluorescent phospholipids (described below), as well as several important sphingolipid probes described in Section 13.3. Some BODIPY® fatty acids are readily metabolized by live cells to phospholipids, di- and triacylglycerols, cholesteryl esters and other natural lipids.[6–9] Analysis of cellular lipid extracts by HPLC has shown that glycerophosphocholines constitute more than 90% of the products of biosynthetic incorporation of BODIPY® 500/510 dodecanoic acid (D3823) by BHK cells.[5]

The three BODIPY® 500/510 probes form a unique series in which the green-fluorescent fluorophore is located within the fatty acid chain at different distances from the terminal carboxylate group.[4] The overall length of the probe is constant and, including the fluorophore, is about equivalent to that of an 18-carbon fatty acid (Figure 13.2.4).

BODIPY® 581/591 undecanoic acid (D3861) is particularly useful for detecting reactive oxygen species in cells and membranes.[10–13] Oxidation of the polyunsaturated butadienyl portion of the BODIPY® 581/591 dye (Figure 13.2.7) truncates the conjugated π-electron system, resulting in a shift of the fluorescence emission peak from ~590 nm to ~510 nm.[10,13,14] This oxidation response mechanism is similar to that of the naturally occurring polyunsaturated fatty acid *cis*-parinaric acid. In comparison to *cis*-parinaric acid, advantages of BODIPY® 581/591 undecanoic acid include:

- Long-wavelength excitation, compatible with confocal laser-scanning microscopes and flow cytometers
- Avoidance of photooxidation effects induced by ultraviolet excitation
- Less interference by colored oxidant and antioxidant additives when detecting probe fluorescence[12]
- Greater resistance to spontaneous oxidation
- Red-to-green fluorescence shift, allowing the use of fluorescence ratio detection methods[10,13]

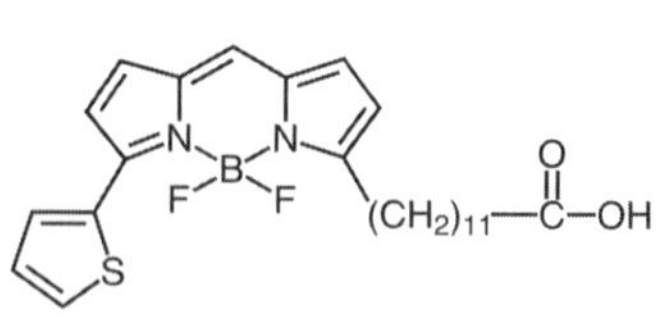

Figure 13.2.6 4,4-Difluoro-5-(2-thienyl)-4-bora-3a,4a-diaza-*s*-indacene-3-dodecanoic acid (BODIPY® 558/568 C_{12}, D3835).

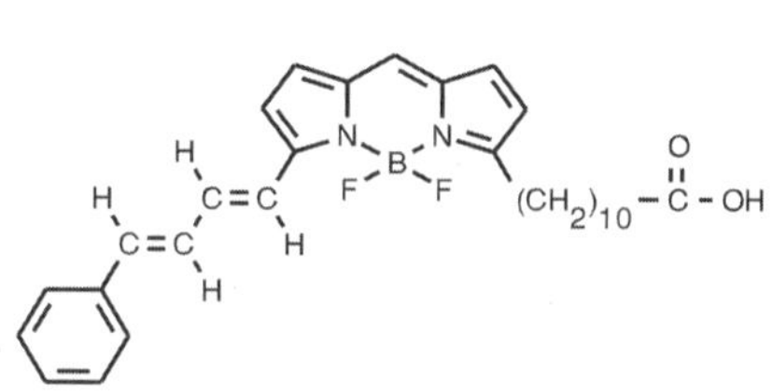

Figure 13.2.7 4,4-Difluoro-5-(4-phenyl-1,3-butadienyl)-4-bora-3a,4a-diaza-*s*-indacene-3-undecanoic acid (BODIPY® 581/591 C_{11}, D3861).

IMPORTANT NOTICE: The products described in this manual are covered by one or more Limited Use Label License(s). Please refer to the Appendix on page 971 and Master Product List on page 975. Products are For Research Use Only. Not intended for any animal or human therapeutic or diagnostic use.

An alternative technique for detecting lipid peroxidation utilizes the oxidation-induced decrease of concentration-dependent excimer formation by BODIPY® FL dye–labeled fatty acids.[15]

NBD Fatty Acids

Fluorescence of the nitrobenzoxadiazole (NBD) fluorophore is highly sensitive to its environment. Although it is moderately fluorescent in aprotic solvents, in aqueous solvents it is almost nonfluorescent.[16] The NBD fluorophore is moderately polar and both its homologous 6-carbon and 12-carbon fatty acid analogs (N316, Figure 13.2.8; N678) and the phospholipids derived from these probes (N3786, N3787) tend to sense the lipid–water interface region of membranes instead of the hydrophobic interior[17] (see part B of Figure 13.2.1). The environmental sensitivity of NBD fatty acids can be usefully exploited to probe the ligand-binding sites of fatty acid and sterol carrier proteins.[18] NBD fatty acids are not well metabolized by live cells.[9,19]

Figure 13.2.8 NBD-X (6-(*N*-(7-nitrobenz-2-oxa-1,3-diazol-4-yl)amino)hexanoic acid; N316).

Pyrene Fatty Acids

The hydrophobic pyrene fluorophore is readily accommodated within the membrane.[20] ω-Pyrene derivatives of longer-chain fatty acids (Figure 13.2.9) were first described by Galla and Sackmann in 1975.[21] We offer pyrene derivatives of the 4-, 10-, 12- and 16-carbon fatty acids (P1903MP, P31, P96, P243, respectively). Pyrenebutanoic acid—frequently called pyrenebutyric acid (P1903MP)—has rarely been used as a membrane probe; however, its conjugates have exceptionally long excited-state lifetimes (τ >100 nanoseconds) and are consequently useful for time-resolved fluorescence immunoassays and nucleic acid detection.[22,23] The long excited-state lifetime of pyrenebutyric acid also makes it useful as a probe for oxygen in cells[24–27] and lipid vesicles.[28]

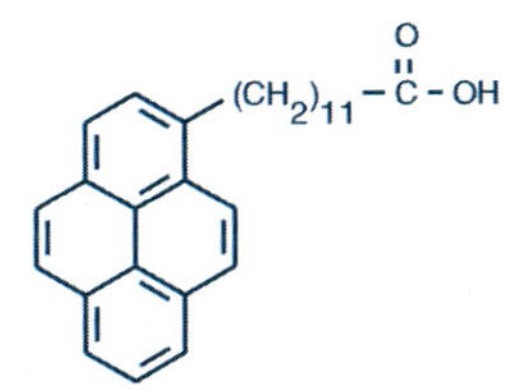

Figure 13.2.9 1-Pyrenedodecanoic acid (P96).

Pyrene derivatives form excited-state dimers (excimers) with red-shifted fluorescence emission[29–31] (Figure 13.2.10). Pyrene excimers can even form when two pyrenes are tethered by a short trimethine spacer, as in 1,3-bis-(1-pyrenyl)propane (B311, Section 13.5). Pyrene excimer formation is commonly exploited for assaying membrane fusion[32,33] (Lipid-Mixing Assays of Membrane Fusion—Note 13.1) and for detecting lipid domain formation.[34–36] Pyrene fatty acids are metabolically incorporated into phospholipids, di- and tri-acylglycerols and cholesteryl esters by live cells.[19,37,38]

Other uses of pyrene fatty acids include:

- Detecting lipid–protein interactions[9,40]
- Inducing photodynamic damage[41,42]
- Investigating phospholipase A_2 action on lipid assemblies[43–45]
- Studying lipid transport mechanisms and transfer proteins[46–48]
- Synthesizing fluorescent sphingolipid probes[49–52]

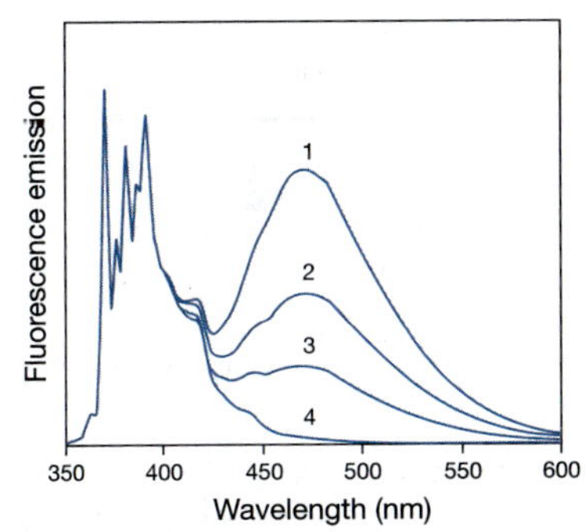

Figure 13.2.10 Excimer formation by pyrene in ethanol. Spectra are normalized to the 371.5 nm peak of the monomer. All spectra are essentially identical below 400 nm after normalization. Spectra are as follows: **1)** 2 mM pyrene, purged with argon to remove oxygen; **2)** 2 mM pyrene, air-equilibrated; **3)** 0.5 mM pyrene (argon-purged); and **4)** 2 µM pyrene (argon-purged). The monomer-to-excimer ratio (371.5 nm/470 nm) is dependent on both pyrene concentration and the excited-state lifetime, which is variable because of quenching by oxygen.

Dansyl Undecanoic Acid

Dansyl undecanoic acid (DAUDA, D94; Figure 13.2.11) incorporates a polar, environment-sensitive dansyl fluorophore that preferentially locates in the polar headgroup region of lipid bilayer membranes.[53] DAUDA exhibits a 60-fold fluorescence enhancement and a large emission spectral shift to shorter wavelengths on binding to certain proteins.[54] This property has been exploited to analyze fatty acid–binding proteins[54–57] and also to develop a fluorometric phospholipase A_2 assay (Section 17.4) based on competitive fatty acid displacement.[58–61]

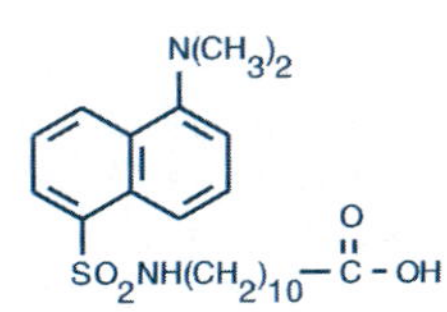

Figure 13.2.11 11-((5-Dimethylaminonaphthalene-1-sulfonyl)amino)undecanoic acid (DAUDA, D94).

cis-Parinaric Acid

The naturally occurring polyunsaturated fatty acid *cis*-parinaric acid (P36005, Figure 13.2.12) was initially developed as a membrane probe by Hudson and co-workers and published in 1975.[62] *cis*-Parinaric acid is the closest structural analog of intrinsic membrane lipids among currently available fluorescent probes (Figure 13.2.1). The chemical and physical properties of *cis*-parinaric acid have been well characterized. The lowest absorption band of *cis*-parinaric acid has two main peaks around 300 nm and 320 nm, with a high extinction coefficient. *cis*-Parinaric acid offers several experimentally advantageous optical properties, including a very large fluorescence Stokes shift (~100 nm) and an almost complete lack of fluorescence in water. In addition, the fluorescence decay lifetime of *cis*-parinaric acid varies from 1 to ~40 nanoseconds, depending on the molecular packing density in phospholipid bilayers. Consequently, minutely detailed information on lipid-bilayer dynamics can be obtained.

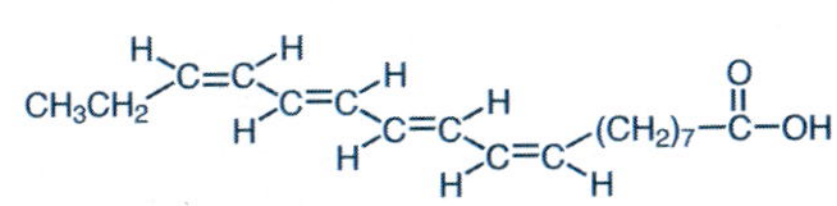

Figure 13.2.12 *cis*-Parinaric acid (P36005).

IMPORTANT NOTICE: The products described in this manual are covered by one or more Limited Use Label License(s). Please refer to the Appendix on page 971 and Master Product List on page 975. Products are For Research Use Only. Not intended for any animal or human therapeutic or diagnostic use.

NOTE 13.1

Lipid-Mixing Assays of Membrane Fusion

Fluorometric methods for assaying membrane fusion exploit processes, such as nonradiative energy transfer, fluorescence quenching and pyrene excimer formation, that are dependent on probe concentration.[1–8] Assays of membrane fusion report either the mixing of membrane lipids (described here) or the mixing of the aqueous contents of the fused entities (Assays of Volume Change, Membrane Fusion and Membrane Permeability—Note 14.3). Chapter 13 describes additional methods for detecting membrane fusion based on image analysis.

NBD–Rhodamine Energy Transfer

Principle: Struck, Hoekstra and Pagano introduced lipid-mixing assays based on NBD–rhodamine energy transfer.[9] In this method (Figure 1), membranes labeled with a combination of fluorescence energy transfer donor and acceptor lipid probes—typically NBD-PE and *N*-Rh-PE (N360, L1392; Section 13.2), respectively—are mixed with unlabeled membranes. Fluorescence resonance energy transfer (FRET), detected as rhodamine emission at ~585 nm resulting from NBD excitation at ~470 nm, decreases when the average spatial separation of the probes is increased upon fusion of labeled membranes with unlabeled membranes. The reverse detection scheme, in which FRET increases upon fusion of membranes that have been separately labeled with donor and acceptor probes, has also proven to be a useful lipid-mixing assay.[10]

Applications: Applications of the NBD–rhodamine assay are described in footnoted references.[11–20]

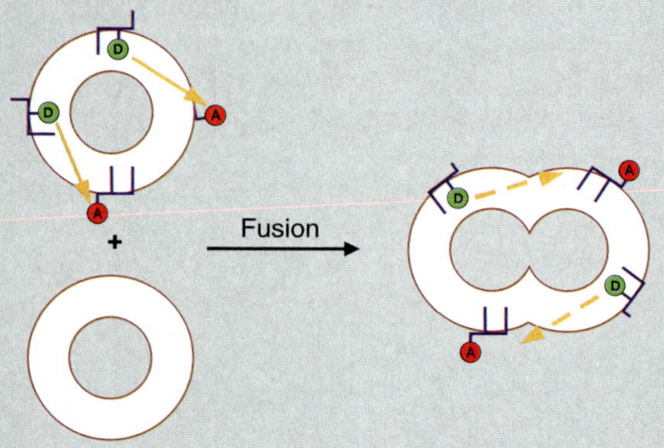

Figure 1 Pictorial representation of a lipid-mixing assay based on fluorescence resonance energy transfer (FRET). The average spatial separation of the donor (**D**) and acceptor (**A**) lipid probes increases upon fusion of labeled membranes with unlabeled membranes, resulting in decreased efficiency of proximity-dependent FRET (represented by yellow arrows). Decreased FRET efficiency is registered by increased donor fluorescence intensity and decreased acceptor fluorescence intensity.

Octadecyl Rhodamine B Self-Quenching

Principle: Lipid-mixing assays based on self-quenching of octadecyl rhodamine B (R18, O246; Section 13.5) were originally described by Hoekstra and co-workers.[21] Octadecyl rhodamine B self-quenching occurs when the probe is incorporated into membrane lipids at concentrations of 1–10 mole percent.[22] Unlike phospholipid analogs, octadecyl rhodamine B can readily be introduced into existing membranes in large amounts. Fusion with unlabeled membranes results in dilution of the probe, which is accompanied by increasing fluorescence[23,24] (excitation/emission maxima 560/590 nm) (Figure 2). The assay may be compromised by effects such as spontaneous transfer of the probe to unlabeled membranes, quenching of fluorescence by proteins and probe-related inactivation of viruses; the prevalence of these effects is currently debated.[25–27]

Applications: The octadecyl rhodamine B self-quenching assay is extensively used for detecting virus–cell fusion.[28–39]

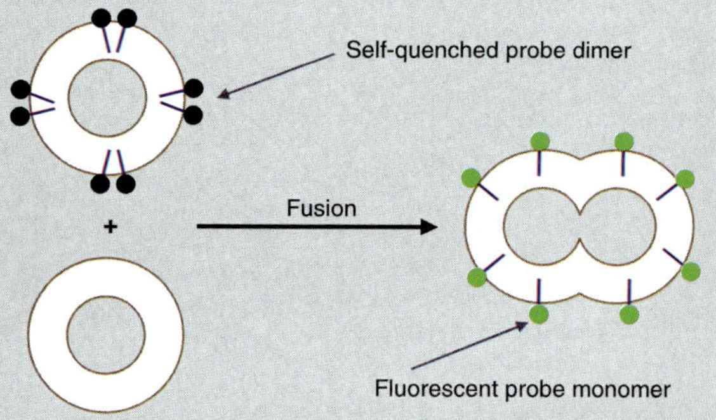

Figure 2 Pictorial representation of a lipid-mixing assay based on fluorescence self-quenching. Fluorescence of octadecyl rhodamine B (O246), incorporated at >1:100 with respect to host membrane lipids, is quenched due to dye–dye interactions. Fusion with unlabeled membranes causes dispersion of the probe, resulting in a fluorescence increase that is represented here by a color change from black to green.

Pyrene Excimer Formation

Principle: Pyrene-labeled fatty acids (e.g., P31, P96, P243; Section 13.2) can be biosynthetically incorporated into viruses and cells in sufficient quantities to produce the degree of labeling required for long-wavelength pyrene excimer fluorescence (Figure 3). This excimer fluorescence is diminished upon fusion of labeled membranes with unlabeled membranes (Figure 4). Fusion can be monitored by following the increase in the ratio of monomer (~400 nm) to excimer (~470 nm) emission, with excitation at about 340 nm. This method appears to circumvent some of the potential artifacts of the octadecyl rhodamine B self-quenching technique[26] and, therefore, provides a useful alternative for virus–cell fusion applications.

Applications: Applications of pyrene excimer assays for membrane fusion are described in the footnoted references.[26,28,40–43]

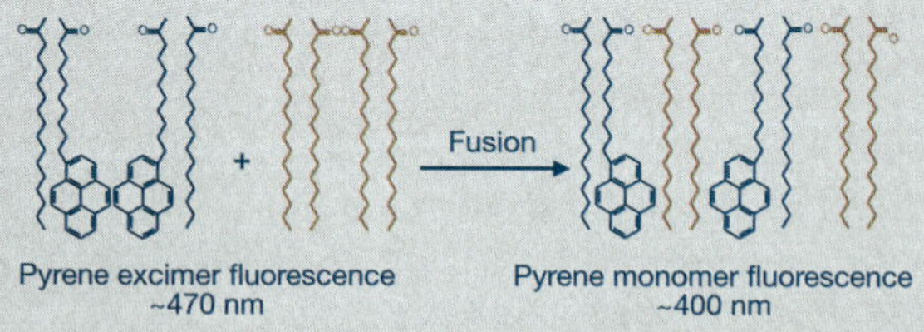

Figure 4 Pictorial representation of a lipid-mixing assay based on pyrene excimer formation. Locally concentrated pyrene-labeled lipid probes emit red-shifted fluorescence due to formation of excimers (excited-state dimers). Probe dilution by unlabeled lipids as a result of membrane fusion is registered by the replacement of excimer emission by blue-shifted monomer fluorescence.

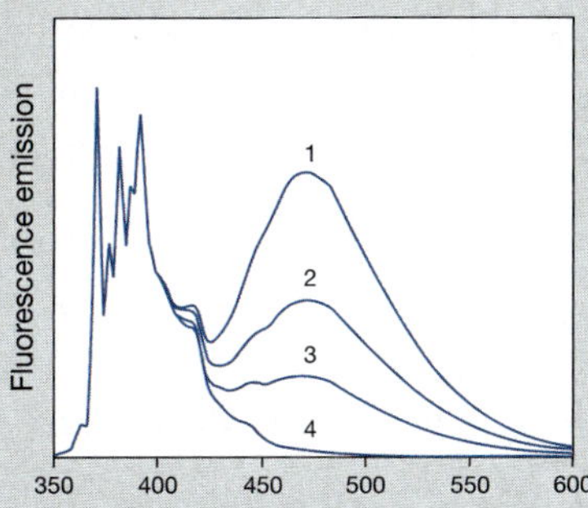

Figure 3 Excimer formation by pyrene in ethanol. Spectra are normalized to the 371.5 nm peak of the monomer. All spectra are essentially identical below 400 nm after normalization. Spectra are as follows: 1) 2 mM pyrene, purged with argon to remove oxygen; 2) 2 mM pyrene, air-equilibrated; 3) 0.5 mM pyrene (argon-purged); and 4) 2 µM pyrene (argon-purged). The monomer-to-excimer ratio (371.5 nm/470 nm) is dependent on both pyrene concentration and the excited-state lifetime, which is variable because of quenching by oxygen.

1. Chem Phys Lipids (2002) 116:39; **2.** Anal Biochem (2009) 386:91; **3.** Methods Enzymol (1993) 220:3; **4.** Methods Enzymol (1993) 220:15; **5.** Proc Natl Acad Sci U S A (2009) 106:979; **6.** Annu Rev Biophys Biophys Chem (1989) 18:187; **7.** Hepatology (1990) 12:61S-66S; **8.** Biochemistry (1987) 26:8435; **9.** Biochemistry (1981) 20:4093; **10.** Methods Enzymol (1993) 221:239; **11.** Biochemistry (1994) 33:12615; **12.** Biochemistry (1994) 33:5805; **13.** Biochemistry (1994) 33:3201; **14.** Biophys J (1994) 67:1117; **15.** J Biol Chem (1994) 269:15124; **16.** J Biol Chem (1994) 269:4050; **17.** J Biol Chem (1993) 268:1716; **18.** Biochemistry (1992) 31:2629; **19.** Biochemistry (1991) 30:5319; **20.** J Biol Chem (1991) 266:3252; **21.** Biochemistry (1984) 23:5675; **22.** J Biol Chem (1990) 265:13533; **23.** Biophys J (1993) 65:325; **24.** Biophys J (1990) 58:1157; **25.** Biochim Biophys Acta (1994) 1190:360; **26.** Biochemistry (1993) 32:11330; **27.** Biochemistry (1993) 32:900; **28.** Biochemistry (1994) 33:9110; **29.** Biochemistry (1994) 33:1977; **30.** Biochim Biophys Acta (1994) 1191:375; **31.** J Biol Chem (1994) 269:5467; **32.** Biochem J (1993) 294:325; **33.** J Biol Chem (1993) 268:25764; **34.** J Biol Chem (1993) 268:9267; **35.** Virology (1993) 195:855; **36.** Biochemistry (1992) 31:10108; **37.** Exp Cell Res (1991) 195:137; **38.** J Virol (1991) 65:4063; **39.** Biochemistry (1990) 29:4054; **40.** EMBO J (1993) 12:693; **41.** J Virol (1992) 66:7309; **42.** Biochemistry (1988) 27:30; **43.** Biochim Biophys Acta (1986) 860:301.

Selected applications of *cis*-parinaric acid include:

- Measurement of peroxidation in lipoproteins [63–65] and the relationship of peroxidation to cytotoxicity [66,67] and apoptosis [68–71]
- Evaluation of antioxidants [72–75]
- Detection of lipoproteins following chromatographic separation [76] and structural characterization of lipoproteins [77]
- Detection of lipid–protein interactions [78–80] and lipid clustering [81]
- High-affinity binding to a hydrophobic pocket between the heavy chain of myosin subfragment-1 and its essential light chain [82]
- Investigation of the mechanism of fatty acid–binding proteins [83–85] and phospholipid-transfer proteins [86,87]

The extensive unsaturation of *cis*-parinaric acid makes it quite susceptible to oxidation. Consequently, we offer *cis*-parinaric acid in a 10 mL unit size of a 3 mM solution in deoxygenated ethanol (P36005); if stored protected from light under an inert argon atmosphere at –20°C, this stock solution should be stable for at least six months. During experiments, we strongly advise handling *cis*-parinaric acid samples under inert gas and preparing solutions using degassed buffers and solvents. *cis*-Parinaric acid is also somewhat photolabile, undergoing photodimerization under intense illumination, resulting in loss of fluorescence.[88]

ADIFAB Fatty Acid Indicator

Fatty acid–binding proteins are small cytosolic proteins found in a variety of mammalian tissues, and studies of their physiological function frequently involve fluorescent fatty acid probes.[89] To facilitate these studies, we offer ADIFAB reagent (A3880), a dual-wavelength fluorescent indicator of free fatty acids [90–92] (Figure 13.2.13, Figure 13.2.14). ADIFAB reagent is a conjugate of I-FABP, a rat intestinal fatty acid–binding protein with a low molecular weight (15,000 daltons) and a high binding affinity for free fatty acids,[93] and the polarity-sensitive acrylodan fluorophore (A433, Section 2.3). It is designed to provide quantitative monitoring of free fatty acids without resorting to separative biochemical methods.[44,94,95] With appropriate precautions, which are described in the product information sheet accompanying this product, ADIFAB can be used to determine free fatty acid concentrations between 1 nM and >20 µM.

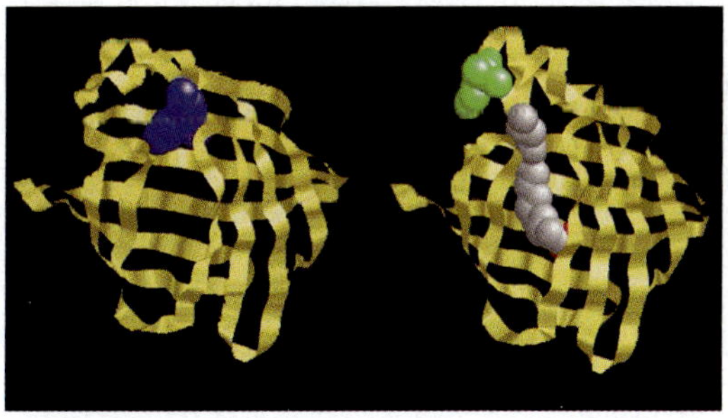

Figure 13.2.13 Ribbon representation of the ADIFAB free fatty acid indicator (A3880). In the left-hand image, the fatty acid binding site of intestinal fatty acid–binding protein (yellow) is occupied by a covalently attached acrylodan fluorophore (blue). In the right-hand image, a fatty acid molecule (gray) binds to the protein, displacing the fluorophore (green) and producing a shift of its fluorescence emission spectrum. Image contributed by Alan Kleinfeld, FFA Sciences LLC, San Diego.

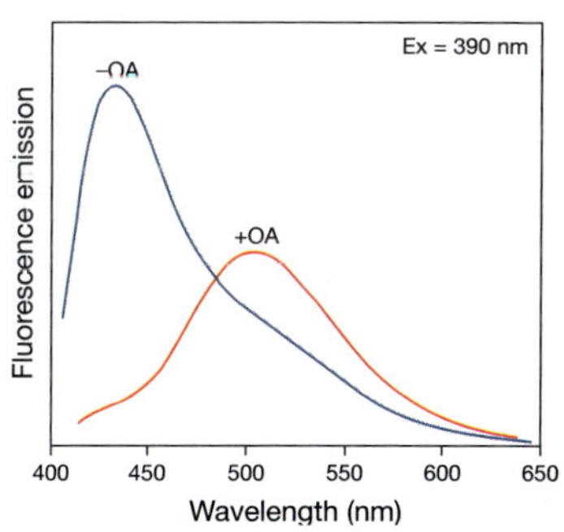

Figure 13.2.14 The free fatty acid–dependent spectral shift of ADIFAB (A3880). Spectra shown represent 0.2 µM ADIFAB in pH 8.0 buffer with (+OA) and without (–OA) addition of 4.7 µM *cis*-9-octadecenoic (oleic) acid (OA). The ratio of fluorescence emission intensities at 505 nm and 432 nm can be quantitatively related to free fatty acid concentrations.

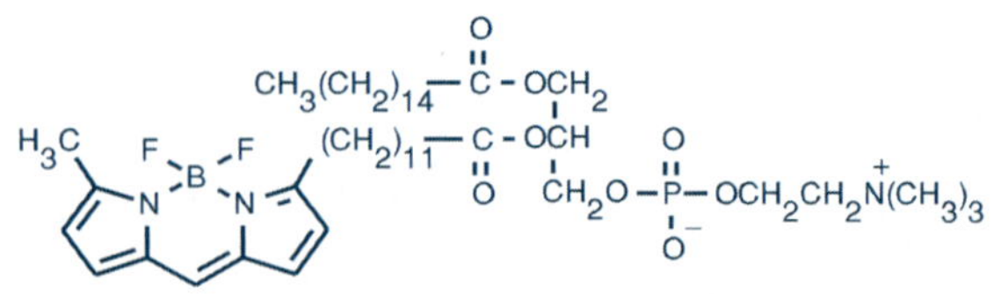

Figure 13.2.15 2-(4,4-Difluoro-5-methyl-4-bora-3a,4a-diaza-*s*-indacene-3-dodecanoyl)-1-hexadecanoyl-*sn*-glycero-3-phosphocholine (β-BODIPY® 500/510 C_{12}-HPC, D3793).

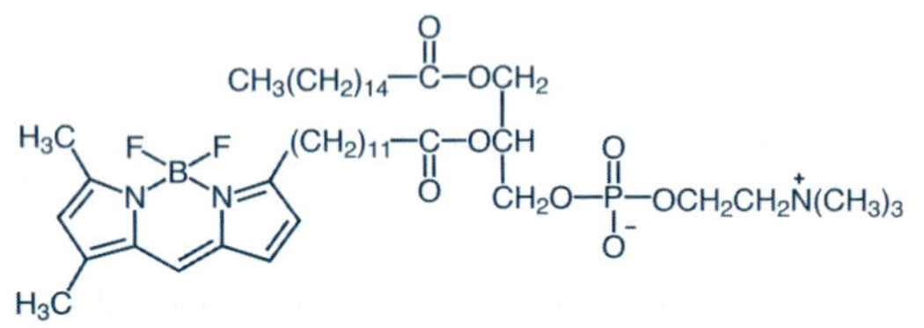

Figure 13.2.16 2-(4,4-Difluoro-5,7-dimethyl-4-bora-3a,4a-diaza-*s*-indacene-3-dodecanoyl)-1-hexadecanoyl-*sn*-glycero-3-phosphocholine (β-BODIPY® FL C_{12}-HPC, D3792).

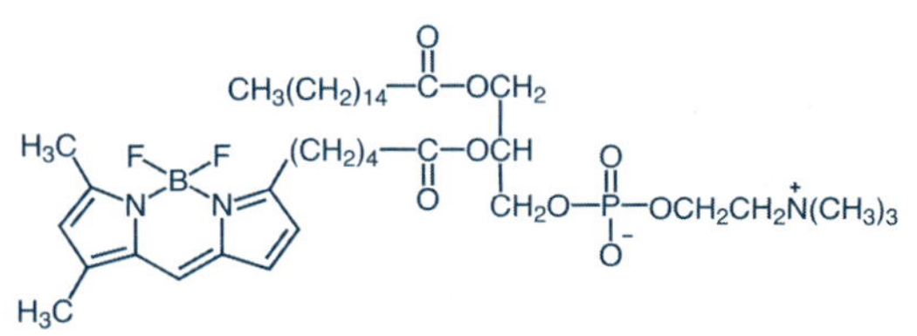

Figure 13.2.17 2-(4,4-Difluoro-5,7-dimethyl-4-bora-3a,4a-diaza-*s*-indacene-3-pentanoyl)-1-hexadecanoyl-*sn*-glycero-3-phosphocholine (β-BODIPY® FL C_5-HPC, D3803).

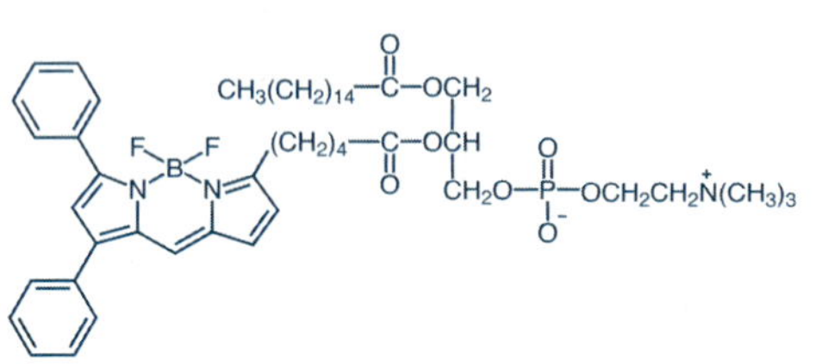

Figure 13.2.18 2-(4,4-Difluoro-5,7-diphenyl-4-bora-3a,4a-diaza-*s*-indacene-3-pentanoyl)-1-hexadecanoyl-*sn*-glycero-3-phosphocholine (β-BODIPY® 530/550 C_5-HPC, D3815).

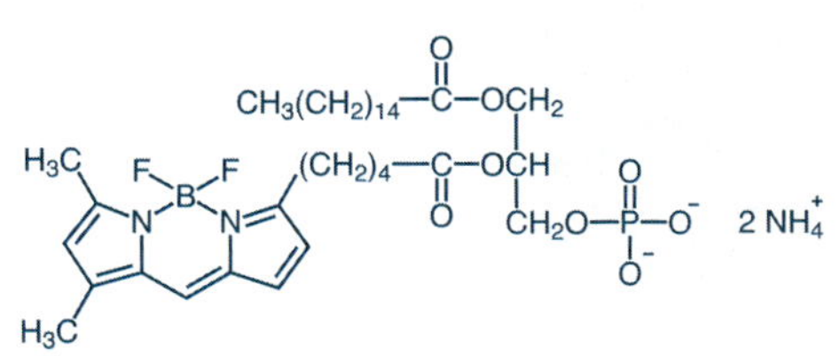

Figure 13.2.19 2-(4,4-Difluoro-5,7-dimethyl-4-bora-3a,4a-diaza-*s*-indacene-3-pentanoyl)-1-hexadecanoyl-*sn*-glycero-3-phosphate, diammonium salt (β-BODIPY® FL C_5-HPA, D3805).

Phospholipids with BODIPY® Dye–Labeled Acyl Chains

BODIPY® Glycerophospholipids

We offer several glycerophospholipid analogs labeled with a single green-fluorescent BODIPY® 500/510 or a BODIPY® FL fluorophore or red-orange–fluorescent BODIPY® 530/550 fluorophore on the *sn*-2 acyl chain, including:

- BODIPY® 500/510 dye–labeled glycerophosphocholine (D3793, Figure 13.2.15)
- BODIPY® FL dye–labeled glycerophosphocholine (D3792, Figure 13.2.16; D3803, Figure 13.2.17)
- BODIPY® 530/550 dye–labeled glycerophosphocholine (D3815, Figure 13.2.18)
- BODIPY® FL dye–labeled phosphatidic acid (D3805, Figure 13.2.19)

In addition, we prepare a glycerophosphocholine analog with a single nonhydrolyzable ether-linked BODIPY® FL fluorophore on the *sn*-1 position (D3771, Figure 13.2.20), as well as several doubly labeled glycerophospholipids. These doubly labeled glycerophospholipids, which are discussed in greater detail in Section 17.4, are designed primarily for detection of phospholipase A_1 and phospholipase A_2 and include:

- Glycerophosphoethanolamine with a BODIPY® FL dye–labeled *sn*-1 acyl chain and a dinitrophenyl quencher–modified headgroup (PED-A1, A10070; Figure 13.2.21)
- Glycerophosphoethanolamine with a BODIPY® FL dye–labeled *sn*-2 acyl chain and a dinitrophenyl quencher–modified headgroup [96] (PED6, D23739; Figure 13.2.22)
- Glycerophosphocholine with two BODIPY® FL dye–labeled acyl chains (bis-BODIPY® FL C_{11}-PC, B7701; Figure 13.2.22)
- Glycerophosphocholine with a BODIPY® 558/568 dye–labeled *sn*-1 alkyl chain and a BODIPY® FL dye–labeled *sn*-2 acyl chain (Red/Green BODIPY® PC-A2, A10072; Figure 13.2.23)

The spectral properties of BODIPY® FL dye–labeled phospholipids are summarized in Table 13.2. Unlike the nitrobenzoxadiazole (NBD) fluorophore, the BODIPY® FL and BODIPY® 500/510 fluorophores are intrinsically lipophilic and readily localize in the membrane's interior.[1] The fluorophore is completely inaccessible to the membrane-impermeant anti–BODIPY® FL antibody (A5770, Section 7.4), which also recognizes the BODIPY® 500/510 derivative. As shown in Figure 13.2.24, the emission spectrum of the BODIPY® 500/510 fluorophore is much narrower than that of the NBD fluorophore. Because both the extinction coefficient of the BODIPY® 500/510 fluorophore and its quantum yield in a lipophilic environment (EC ~90,000 $cm^{-1}M^{-1}$ and QY ~0.9) are much higher than those of the NBD fluorophore (EC ~20,000 $cm^{-1}M^{-1}$ and QY ~0.3), much less BODIPY® 500/510 dye–labeled phospholipid is required for labeling membranes.[4]

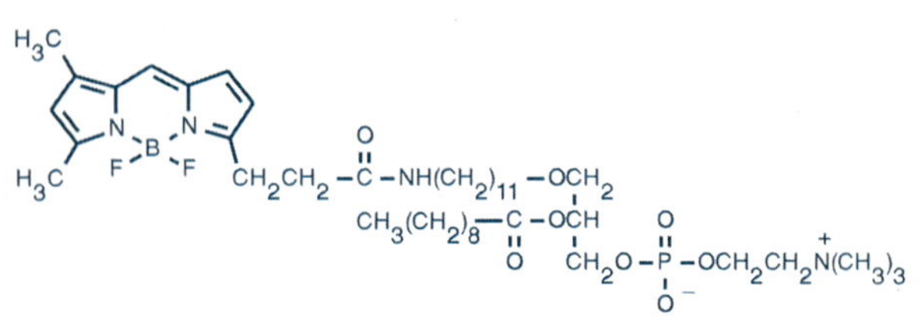

Figure 13.2.20 2-Decanoyl-1-(*O*-(11-(4,4-difluoro-5,7-dimethyl-4-bora-3a,4a-diaza-*s*-indacene-3-propionyl)amino)undecyl)-*sn*-glycero-3-phosphocholine (D3771).

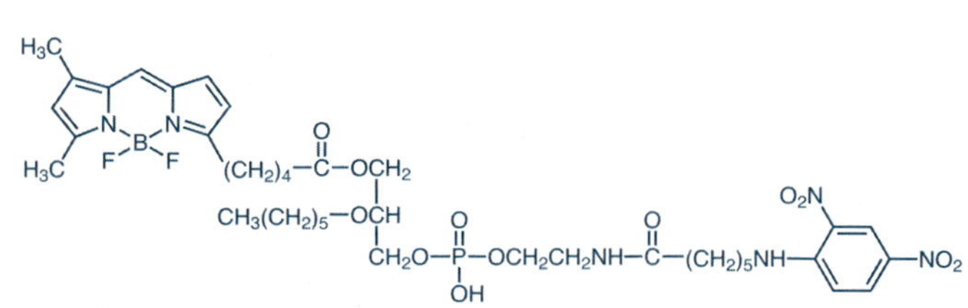

Figure 13.2.21 PED-A1 (*N*-((6-(2,4-DNP)amino)hexanoyl)-1-(BODIPY® FL C_5)-2-hexyl-*sn*-glycero-3-phosphoethanolamine; A10070).

IMPORTANT NOTICE: The products described in this manual are covered by one or more Limited Use Label License(s). Please refer to the Appendix on page 971 and Master Product List on page 975. Products are For Research Use Only. Not intended for any animal or human therapeutic or diagnostic use.

Phospholipids with a Fluorescent or Biotinylated Head Group

Phospholipid with a Dansyl-Labeled Head Group

The phospholipid analog incorporating the environment-sensitive[200] dansyl fluorophore (dansyl DHPE, D57; Figure 13.2.35) is a useful probe of lipid–water interfaces.[53,201] It is sensitive to the interactions of a number of proteins, including protein kinase C,[202,203] annexins[204,205] and phospholipase A_2,[206–208] with membrane surfaces. Dansyl DHPE has also been used to examine the effects of cholesterol on the accessibility of the dansyl hapten to antibodies[209] (Antibodies for Detecting Membrane-Surface Labels—Note 13.2).

Phospholipid with a Marina Blue® Dye–Labeled Head Group

Marina Blue® DHPE (M12652, Figure 13.2.36) is optimally excited by the intense 365 nm spectral line of the mercury-arc lamp and exhibits bright blue fluorescence emission near 460 nm. Significantly, the pK_a value of this 6,8-difluoro-7-hydroxycoumarin derivative is 2–3 log units lower than that of nonfluorinated 7-hydroxycoumarin analogs; consequently, Marina Blue® DHPE is expected to be strongly fluorescent in membranes, even at neutral pH.

Phospholipid with a Pacific Blue™ Dye–Labeled Head Group

The Pacific Blue™ dye–labeled phospholipid (Pacific Blue™ DMPE, P22652; Figure 13.2.37) is our only head group–labeled phospholipid with tetradecanoyl (myristoyl) esters rather than hexadecanoyl (palmitoyl) esters. This blue-fluorescent phospholipid is structurally similar to a phospholipid described by Gonzalez and Tsien for use in a FRET-based measurement of membrane potential.[210]

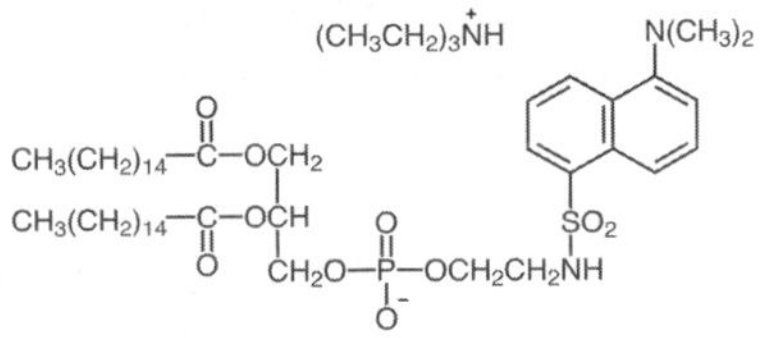

Figure 13.2.35 *N*-(5-dimethylaminonaphthalene-1-sulfonyl)-1,2-dihexadecanoyl-*sn*-glycero-3-phosphoethanolamine, triethylammonium salt (dansyl DHPE, D57).

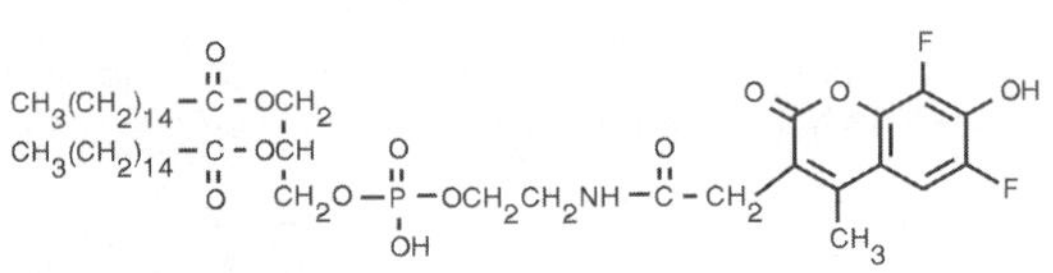

Figure 13.2.36 Marina Blue® 1,2-dihexadecanoyl-*sn*-glycero-3-phosphoethanolamine (Marina Blue® DHPE, M12652).

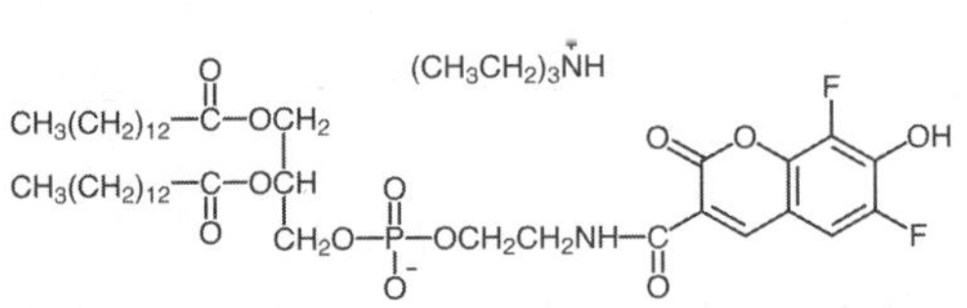

Figure 13.2.37 Pacific Blue™ DMPE (Pacific Blue™ 1,2-ditetradecanoyl-*sn*-glycero-3-phosphoethanolamine, triethylammonium salt; P22652).

Phospholipid with an NBD-Labeled Head Group

The widely used membrane probe nitrobenzoxadiazolyldihexadecanoylglycerophosphoethanolamine[17] (NBD-PE, N360; Figure 13.2.38) has three important optical properties: photolability, which makes it suitable for photobleaching recovery measurements; concentration-dependent self-quenching; and fluorescence resonance energy transfer to rhodamine acceptors (usually rhodamine DHPE, L1392). Spectroscopic characteristics of NBD-PE are generally similar to those described for our phospholipids with NBD-labeled acyl chains (N3786, N3787). NBD-PE is frequently used in NBD–rhodamine fluorescence energy transfer experiments to monitor membrane fusion (Lipid-Mixing Assays of Membrane Fusion—Note 13.1). In addition, this method can be used to detect lipid domain formation[165] and intermembrane lipid transfer[211–214] and to determine the transbilayer distribution of phospholipids.[151] Attachment of the NBD fluorophore to the head group makes NBD-PE resistant to transfer between vesicles.[142] NBD-PE has been used in combination with either rhodamine DHPE (L1392) or Texas Red® DHPE (T1395MP) for visualizing the spatial relationships of lipid populations by fluorescence resonance energy transfer microscopy.[215] The nitro group of NBD can be reduced with sodium dithionite, irreversibly eliminating the dye's fluorescence (Figure 13.2.31). This technique can be employed to determine whether the probe is localized on the outer or inner leaflet of the cell membrane.[171,216–218] The argon-ion laser–excitable NBD-PE is also a frequent choice for fluorescence recovery after photobleaching (FRAP) measurements of lateral diffusion in membranes.[219–222] In addition, NBD-PE is of particular value for monitoring bilayer-to-hexagonal phase transitions, because these transitions cause an increase in NBD-PE's fluorescence intensity.[223–225]

Phospholipid with a Fluorescein-Labeled Head Group

Fluorescein-derivatized dihexadecanoylglycerophosphoethanolamine (fluorescein DHPE, F362; Figure 13.2.39) is a membrane-surface probe that is sensitive to both the local electrostatic potential and pH.[226–228] An anti-fluorescein/Oregon Green® antibody (A889, Section 7.4) has been employed in combination with fluorescein DHPE

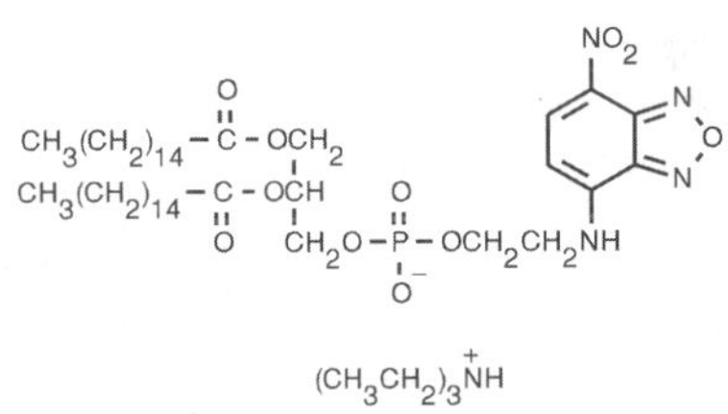

Figure 13.2.38 NBD-PE (*N*-(7-nitrobenz-2-oxa-1,3-diazol-4-yl)-1,2-dihexadecanoyl-*sn*-glycero-3-phosphoethanolamine, triethylammonium salt; N360).

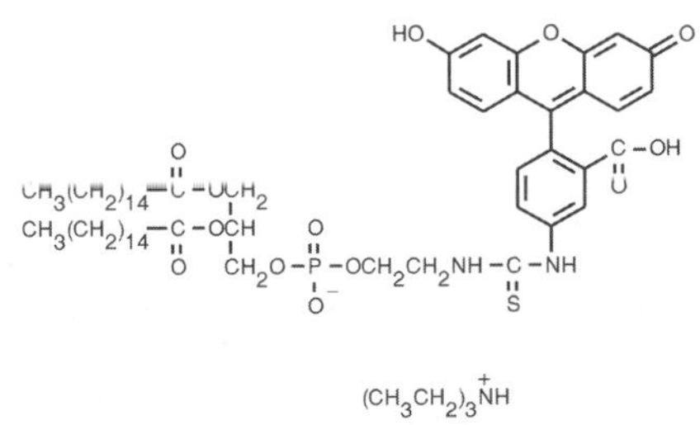

Figure 13.2.39 Fluorescein DHPE (*N*-(fluorescein-5-thiocarbamoyl)-1,2-dihexadecanoyl-*sn*-glycero-3-phosphoethanolamine, triethylammonium salt; F362).

to investigate specific recognition interactions at membrane surfaces[229,230] (Antibodies for Detecting Membrane-Surface Labels—Note 13.2).

Because of fluorescein's photolability, fluorescein DHPE is a useful reagent for measuring lateral diffusion in membranes using fluorescence photobleaching recovery methods.[231,232] Another technique, single-particle tracking (SPT), provides direct measurements of diffusion rates by calculating the trajectories of fluorescent polystyrene beads or colloidal gold particles from time-sequential images.[233,234] FluoSpheres® fluorescent microspheres (Section 6.5) were labeled with streptavidin and then coupled to fluorescein DHPE using a biotinylated conjugate of anti–fluorescein/Oregon Green® monoclonal 4-4-20 (A6421, Section 7.4). Diffusion rates measured with this bridged conjugate in glass-supported phospholipid bilayers were the same as those determined with streptavidin beads coupled directly to biotin-X DHPE (B1616). Fluorescein DHPE has also been used in conjunction with polyclonal anti–fluorescein/Oregon Green® antibody (A889, Section 7.4) to prepare colloidal gold probes for SPT diffusion measurements in supported phospholipid bilayers and in keratocyte plasma membranes.[235]

Phospholipid with an Oregon Green® 488 Dye–Labeled Head Group

With absorption and emission spectra that are virtually superimposable on those of fluorescein, our Oregon Green® 488 DHPE (O12650, Figure 13.2.40) provides an important alternative to fluorescein DHPE in its many applications. When compared with the fluorescein derivative, Oregon Green® 488 DHPE exhibits greater photostability and a lower pK_a (pK_a = 4.7 versus 6.4 for fluorescein); however, these pK_a values may differ when the probes are bound to membranes.

Phospholipid with a BODIPY® FL Dye–Labeled Head Group

Our phospholipid with the green-fluorescent BODIPY® FL dye attached to the head group (BODIPY® FL DHPE, D3800; Figure 13.2.41) has significant potential for studies of molecular recognition interactions at membrane surfaces (Antibodies for Detecting Membrane-Surface Labels—Note 13.2). Spectral properties of this BODIPY® probe is generally the same as those described above for phospholipids with BODIPY® FL dye–labeled acyl chains.

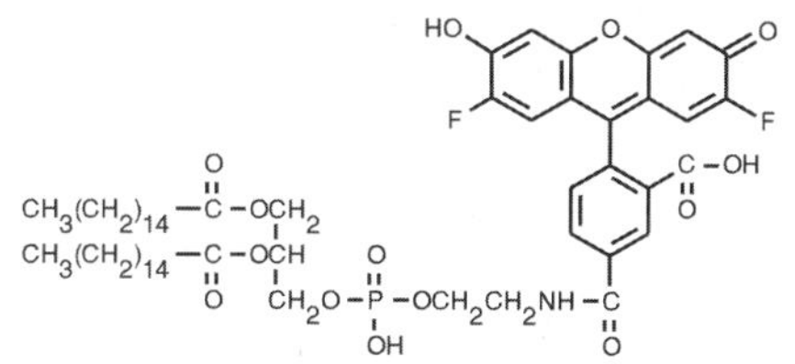

Figure 13.2.40 Oregon Green® 488 DHPE (Oregon Green® 488 1,2-dihexadecanoyl-*sn*-glycero-3-phosphoethanolamine; O12650).

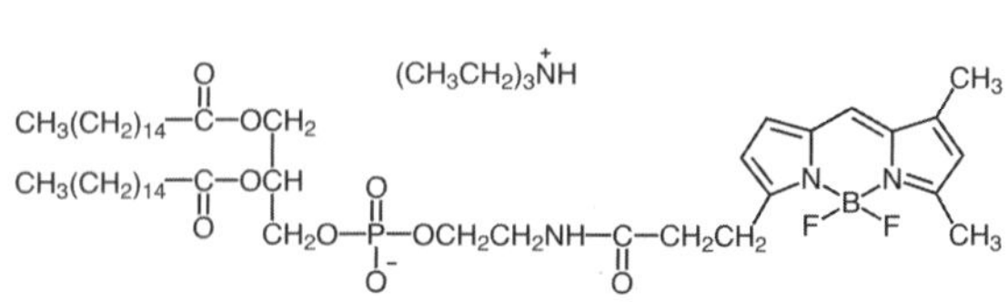

Figure 13.2.41 *N*-(4,4-difluoro-5,7-dimethyl-4-bora-3a,4a-diaza-s-indacene-3-propionyl)-1,2-dihexadecanoyl-*sn*-glycero-3-phosphoethanolamine, triethylammonium salt (BODIPY® FL DHPE, D3800).

Phospholipids with a Rhodamine or Texas Red® Dye–Labeled Head Group

The rhodamine-labeled phospholipids TRITC DHPE (T1391, Figure 13.2.42) and rhodamine DHPE (often referred to as *N*-Rh-PE, L1392; Figure 13.2.1) do not readily transfer between separated lipid bilayers.[140,236] This property has led to the extensive use of rhodamine DHPE for membrane fusion assays based on fluorescence resonance energy transfer from NBD-PE (Lipid-Mixing Assays of Membrane Fusion—Note 13.1). In addition, these probes are good resonance energy transfer acceptors from fluorescent lipid analogs such as the BODIPY® and NBD phospholipids[237] and from protein labels such as 5-iodoacetamidofluorescein 5-IAF, I30451; Section 2.2) and IAEDANS[238,239] (I14, Section 2.3). Rhodamine-labeled phospholipids have also been used as tracers for membrane traffic during endocytosis[240] and for lipid processing in hepatocytes.[241] Texas Red® DHPE (T1395MP) is principally employed as an energy transfer acceptor from NBD, BODIPY® and fluorescein lipid probes. The longer emission wavelength of the Texas Red® dye provides superior separation of the donor and acceptor emission signals in resonance energy transfer microscopy.[216,242] This technique has enabled visualization of ATP-dependent fusion of liposomes with the Golgi apparatus.[243] Membrane flux during hemagglutinin-mediated cell–cell fusion has been visualized using Texas Red® DHPE and the lipophilic carbocyanine DiI (D282, D3911; Section 13.4) as membrane labels.[244]

Phospholipids with a Biotinylated Head Group

We offer phospholipids labeled with a biotin at the head group to facilitate binding of labeled membranes to other biomolecules. The biotinylated phospholipids (biotin DHPE, B1550; biotin-X DHPE, B1616, Figure 13.2.43) can be used to couple avidin or streptavidin (Table 7.9) to cell membranes, liposomes and lipid monolayers.[245–248] Avidin can then be employed as a bridge for antibody coupling or for assembling liposomes into multiplex structures.[249,250] Liposomes incorporating biotinylated phospholipids can also be used to immobilize membrane-bound proteins for analysis by affinity chromatography.[251] Interactions of biotinylated lipids with streptavidin provide a model for molecular

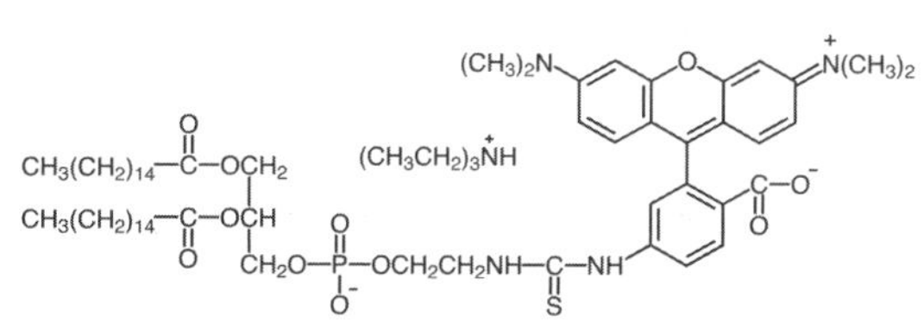

Figure 13.2.42 TRITC DHPE (*N*-(6-tetramethylrhodaminethiocarbamoyl)-1,2-dihexadecanoyl-*sn*-glycero-3-phosphoethanolamine, triethylammonium salt; T1391).

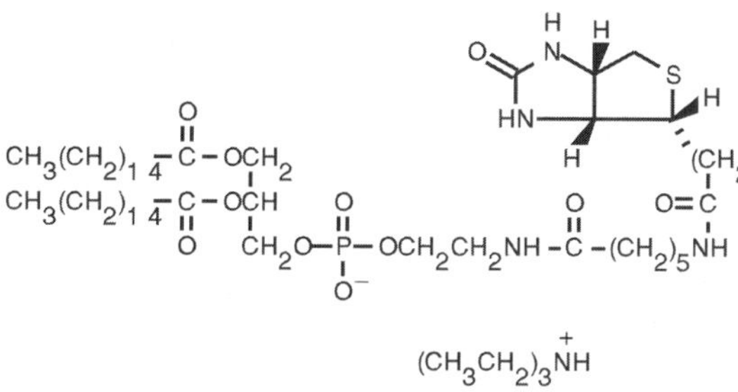

Figure 13.2.43 Biotin-X DHPE (*N*-((6-(biotinoyl)amino)hexanoyl)-1,2-dihexadecanoyl-*sn*-glycero-3-phosphoethanolamine, triethylammonium salt; B1616).

IMPORTANT NOTICE: The products described in this manual are covered by one or more Limited Use Label License(s). Please refer to the Appendix on page 971 and Master Product List on page 975. Products are For Research Use Only. Not intended for any animal or human therapeutic or diagnostic use.

recognition processes at membrane surfaces.[252–254] The phase structure of lipid assemblies incorporating biotinylated phospholipids has been studied by X-ray diffraction,[218] ^{31}P NMR and differential scanning calorimetry.[255,256]

LipidTOX™ Phospholipid and Neutral Lipid Stains for High-Content Screening

With the resolution inherent in an image-based methodology and the productivity of high-throughput assays, high-content screening (HCS) or automated imaging provides a powerful tool for studying biology in a spatial and temporal context. Using HCS technology, researchers can examine multiple cellular targets and parameters in a large number of individually imaged cells and quantitatively assess the data. While many Molecular Probes® products can be directly applied to HCS protocols, we have developed validated tools and assays specifically for HCS platforms. These HCS products are:

- Validated on multiple imaging platforms
- Packaged in automation-compatible formulations
- Compatible with multiplex applications

Although designed for HCS platforms, HCS products and kits can also be used with conventional fluorescence microscopes equipped with standard optical filter sets.

HCS LipidTOX™ Phospholipidosis Detection Reagents

Phospholipidosis is often triggered by cationic amphiphilic drugs, which become enriched in lysosomes to high concentrations and inhibit normal metabolism of phospholipids. The subsequent intracellular accumulation of phospholipids and formation of lamellar bodies—phospholipidosis—can be detected in cells incubated in the presence of phospholipids conjugated to fluorescent dyes.

HCS LipidTOX™ Green and HCS LipidTOX™ Red phospholipidosis detection reagents (H34350, H34351), also called LipidTOX™ phospholipid stains, were specifically developed to characterize the potentially toxic side effects of compounds on lipid metabolism in mammalian cell lines using image-based HCS assays.[257] Key advantages of this series of phospholipidosis detection reagents over conventional stains such as NBD-PE (N360) include their ready-to-use aqueous formulation, their narrow emission profiles (excitation/emission maxima ~495/525 nm for HCS LipidTOX™ Green phospholipidosis detection reagent and ~595/615 nm for HCS LipidTOX™ Red phospholipidosis detection reagent) and their compatibility with HCS LipidTOX™ neutral lipid stains.

HCS LipidTOX™ phospholipidosis detection reagents have not been observed to affect the normal growth of cells, and their live-cell staining patterns are maintained after formaldehyde fixation. These reagents are designed for fixed–end point workflows in which formaldehyde-fixed cells in microplates are processed, imaged and analyzed. HCS LipidTOX™ phospholipidosis detection reagents can easily be detected with fluorescence microscopes or HCS readers equipped with standard filter sets.

HCS LipidTOX™ Neutral Lipid Stains

As with phospholipidosis, steatosis or the intracellular accumulation of neutral lipids as lipid droplets or globules is often triggered by drugs that affect the metabolism of fatty acids or neutral lipids. HCS LipidTOX™ neutral lipid stains were developed to characterize the effects of drugs and other compounds on lipid metabolism in mammalian cell lines. HCS LipidTOX™ neutral lipid stains have an extremely high affinity for neutral lipid droplets. These reagents are added after cell fixation and do not require subsequent wash steps after incubation with the sample. Key advantages of this series of neutral lipid stains over conventional stains such as nile red (N1142; Section 13.5) include their ready-to-use formulations, their flexibility for multiplexing protocols and their compatibility with HCS LipidTOX™ phospholipidosis detection reagents. HCS LipidTOX™ neutral lipid stains can also be used to monitor the formation and differentiation of adipocytes, a process called adipogenesis. Adipogenesis is of acute interest to the biomedical and drug discovery community as it plays an important role in diseases such as obesity, diabetes and atherosclerosis.

Described more thoroughly in Section 13.5, HCS LipidTOX™ neutral lipid stains are available with green, red and deep red fluorescence emission:

- HCS LipidTOX™ Green neutral lipid stain (H34475), with excitation/emission maxima ~495/505 nm (Figure 13.2.44)
- HCS LipidTOX™ Red neutral lipid stain (H34476), with excitation/emission maxima ~577/609 nm
- HCS LipidTOX™ Deep Red neutral lipid stain (H34477), with excitation/emission maxima ~637/655 nm

HCS LipidTOX™ neutral lipid stains are designed for fixed–end point workflows in which formaldehyde-fixed cells in microplates are processed, imaged and analyzed. These stains can easily be detected with fluorescence microscopes or HCS readers equipped with standard filter sets.

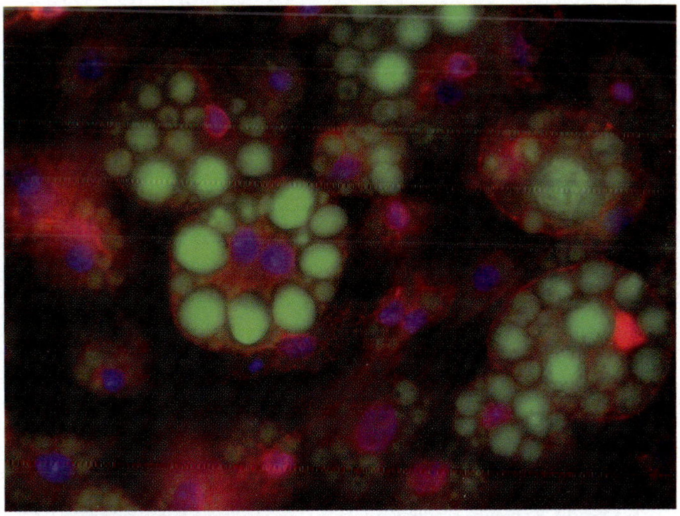

Figure 13.2.44 LipidTOX™ Green neutral lipid stain and fatty acid–binding protein (FABP4) antibody labeling in adipocytes. Adipocytes differentiated from 3T3-L1 mouse fibroblasts were fixed with formaldehyde and permeabilized with saponin before labeling with rabbit anti–fatty acid binding protein (FABP4) IgG (red). These cells were then stained with LipidTOX™ Green neutral lipid stain (H34475, green), counterstained with DAPI (D1306, D21490; blue) and mounted in ProLong® Gold antifade reagent (P36930).

HCS LipidTOX™ Phospholipidosis and Steatosis Detection Kit

The detection and analysis of prelethal mechanisms in toxicological profiling and compound screening are extremely important components of the drug discovery process. The cationic amphiphilic drugs are among the most prominent examples of compounds that impact lipid metabolism of cells. These drugs tend to become enriched in lysosomes to high concentrations and inhibit the normal metabolism of phospholipids, which in turn causes the intracellular accumulation of phospholipids and the formation of lamellar bodies. Other drug classes more adversely affect various aspects of fatty acid or neutral lipid metabolism, leading to the cytoplasmic accumulation of neutral lipid as lipid droplets or globules.

The HCS LipidTOX™ Phospholipidosis and Steatosis Detection Kit (H34157, H34158) provides a complete set of reagents for performing validated HCS assays to detect and distinguish these two facets of cytotoxicity—the intracellular accumulation of phospholipids (phospholipidosis) and of neutral lipids (steatosis)—in mammalian cell lines after exposure to test compounds.[258] This kit includes an aqueous, red-fluorescent formulation of labeled phospholipids (LipidTOX™ Red phospholipid stain, excitation/emission ~595/615 nm) and a ready-to-use, highly selective green-fluorescent stain for neutral lipids (LipidTOX™ Green neutral lipid stain, excitation/emission ~495/505 nm), which can be used sequentially for the analysis of phospholipidosis and steatosis, respectively, or can be used separately for single-parameter analysis.

After incubation with LipidTOX™ Red phospholipid stain and a test compound, the cells are fixed with formaldehyde and labeled with LipidTOX™ Green neutral lipid stain (Figure 13.2.45). Neither LipidTOX™ Red phospholipid stain, nor LipidTOX™ Green phospholipid stain described above, requires sonication or organic solvents. Furthermore, LipidTOX™ Green neutral lipid stain (as well as the other LipidTOX™ neutral lipid stains described above) is more selective than nile red, allowing you to easily distinguish neutral lipids (such as those in adipocytes and cells undergoing steatosis) from other types of lipids.

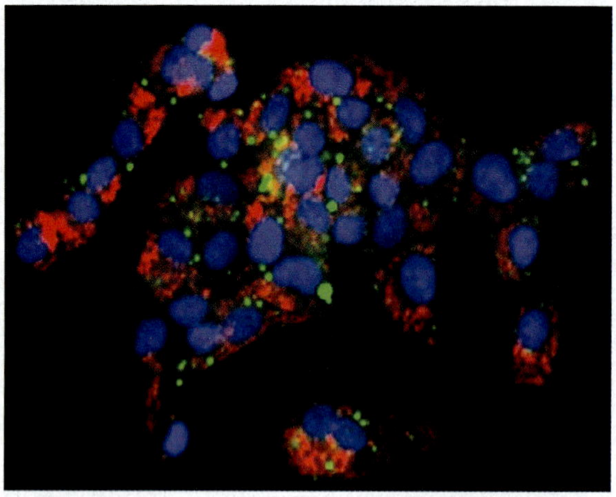

Figure 13.2.45 Multiplex detection of phospholipidosis and steatosis in HepG2 cells using the HCS LipidTOX™ Phospholipidosis and Steatosis Detection Kit (H34157, H34158). HepG2 cells were co-incubated with tamoxifen and LipidTOX™ Red phospholipid stain, followed by fixation with formaldehyde and labeling with HCS LipidTOX™ Green neutral lipid stain and Hoechst 33342 (H1399, H3570, H21492).

Each HCS LipidTOX™ Phospholipidosis and Steatosis Detection Kit provides:

- LipidTOX™ Red phospholipid stain
- LipidTOX™ Green neutral lipid stain
- Hoechst 33342 for nuclear labeling
- Propranolol, a positive-control compound for inducing phospholipidosis
- Cyclosporin A, a positive-control compound for inducing steatosis
- Dimethylsulfoxide (DMSO)
- Detailed protocols

Sufficient reagents are provided for 240 assays (H34157, 2-plate size) or 1200 assays (H34158, 10-plate size), based on assay volumes of 100 µL per well. These kits are designed for fixed–end point workflows in which formaldehyde-fixed cells in microplates are processed, imaged and analyzed. The fluorescent stains used for the analysis of phospholipidosis and steatosis can easily be detected with fluorescence microscopes or HCS readers equipped with standard filter sets.

REFERENCES

1. Biochim Biophys Acta (1998) 1375:13; **2.** Biochemistry (1992) 31:5312; **3.** Chem Phys Lipids (2002) 116:3; **4.** Anal Biochem (1991) 198:228; **5.** Biochem Biophys Res Commun (1992) 187:1594; **6.** J Biol Chem (2001) 276:1391; **7.** J Biol Chem (1997) 272:8531; **8.** Exp Parasitol (1997) 86:133; **9.** Am J Physiol (1995) 269:G842; **10.** Methods Enzymol (2000) 319:603; **11.** Biochim Biophys Acta (2000) 1487:61; **12.** Anal Biochem (1998) 265:290; **13.** FEBS Lett (1999) 453:278; **14.** Free Radic Biol Med (2002) 33:473; **15.** J Biochem Biophys Methods (1997) 35:23; **16.** Photochem Photobiol (1991) 54:361; **17.** Chem Phys Lipids (1990) 53:1; **18.** Biochemistry (1995) 34:11919; **19.** Chem Phys Lipids (1991) 58:111; **20.** Biophys J (2001) 80:832; **21.** J Am Chem Soc (1975) 97:4114; **22.** Anal Biochem (1988) 174:101; **23.** Anal Biochem (1989) 183:231; **24.** Arch Biochem Biophys (1997) 341:34; **25.** Exp Cell Res (1974) 89:105; **26.** Adv Exp Med Biol (1975) 75:47; **27.** Biochim Biophys Acta (1980) 591:187; **28.** Photochem Photobiol (1977) 26:221; **29.** Chem Phys Lipids (2002) 116:57; **30.** Chem Phys Lipids (1980) 27:199; **31.** Biophys J (2001) 80:832; **32.** J Biol Chem (1997) 272:3369; **33.** Biochemistry (1993) 32:11330; **34.** J Biol Chem (1996) 271:3085; **35.** Biophys J (1991) 60:110; **36.** Biochim Biophys Acta (1992) 1106:178; **37.** Chem Phys Lipids (1989) 50:191; **38.** J Biol Cm (1992) 267:6563; **39.** Biochemistry (1995) 34:7271; **40.** Biochemistry (1995) 34:9913; **41.** Biochim Biophys Acta (1998) 1402:61; **42.** J Biol Chem (2000) 275:21715; **43.** J Neurosci Methods (2000) 100:127; **44.** Biochemistry (1998) 37:10709; **45.** Biochemistry (1998) 37:14128; **46.** Biochemistry (1994) 33:15382; **47.** J Biol Chem (1998) 273:27800; **48.** Biophys J (1997) 72:1732; **49.** Chem Phys Lipids (1993) 65:43; **50.** Biochemistry (1990) 29:697; **51.** Biochemistry (1987) 26:5943; **52.** Biochim Biophys Acta (1987) 918:250; **53.** Biochemistry (1998) 37:4603; **54.** Biochem J (1986) 238:419; **55.** Biochem J (2001) 356:387; **56.** Biochemistry (1999) 38:16932; **57.** Analyst (1992) 117:1859; **58.** Biochim Biophys Acta (2000) 1484:195; **59.** Anal Biochem (1993) 212:65; **60.** Biochem J (1991) 278:843; **61.** Biochem J (1990) 266:435; **62.** Proc Natl Acad Sci U S A (1975) 72:1649; **63.** Biochem Pharmacol (2003) 66:947; **64.** Proc Natl Acad Sci U S A (1994) 91:1183; **65.** Arch Biochem Biophys (1992) 297:147; **66.** Biochim Biophys Acta (1997) 1330:127; **67.** Neurochem Res (1997) 22:1187; **68.** Biochemistry (2000) 39:127; **69.** Biochemistry (1998) 37:13781; **70.** Nature (1995) 378:776; **71.** Cell (1993) 75:241; **72.** Biochem Pharmacol (1997) 54:937; **73.** Biochemistry (1993) 32:10692; **74.** J Biol Chem (1993) 268:10906; **75.** Anal Biochem (1991) 196:443; **76.** Clin Chem (1998) 44:2148; **77.** Chem Phys Lipids (1991) 60:1; **78.** Biochemistry (1993) 32:12420; **79.** Biochemistry (1991) 30:6195; **80.** Biochemistry (1990) 29:6714; **81.** Biochemistry (1992) 31:1816; **82.** Biophys J (1997) 72:2268; **83.** Biochem J (2000) 349:377; **84.** Biochemistry (1994) 33:3327; **85.** J Biol Chem (1993) 268:7885; **86.** Lipids (1997) 32:1201; **87.** Biochemistry (1990) 29:8548; **88.** Proc Natl Acad Sci USA (1980) 77:26; **89.** Mol Cell Biochem (1993) 123:45; **90.** Mol Cell Biochem (1999) 192:87; **91.** J Biol Chem (1994) 269:23918; **92.** J Biol Chem (1992) 267:23495; **93.** J Biol Chem

IMPORTANT NOTICE: The products described in this manual are covered by one or more Limited Use Label License(s). Please refer to the Appendix on page 971 and Master Product List on page 975. Products are For Research Use Only. Not intended for any animal or human therapeutic or diagnostic use.
www.invitrogen.com/probes
molecular probes® | invitrogen™ by Life technologies™

REFERENCES—*continued*

(1992) 267:23534; **94**. Biochemistry (1998) 37:8011; **95**. J Lipid Res (1995) 36:229; **96**. Anal Biochem (1999) 276:27; **97**. Biochemistry (2001) 40:8292; **98**. J Fluorescence (1994) 4:295; **99**. Biochem Biophys Res Commun (1995) 207:508; **100**. Biochemistry (1997) 36:8840; **101**. Biophys J (1998) 74:2184; **102**. Biophys J (1996) 71:2656; **103**. J Cell Sci (1993) 106:485; **104**. J Biochem (Tokyo) (2001) 129:19; **105**. Biochim Biophys Acta (1999) 1438:185; **106**. Biochemistry (1996) 35:3063; **107**. Biophys J (2002) 83:1511; **108**. Proc Natl Acad Sci U S A (1999) 96:8461; **109**. Biophys J (2001) 80:2775; **110**. J Cell Biol (1995) 130:781; **111**. J Biol Chem (1999) 274:19338; **112**. J Biol Chem (1992) 267:21465; **113**. Plant Physiol (1996) 110:979; **114**. Br J Pharmacol (1998) 124:1675; **115**. J Biol Chem (1997) 272:2542; **116**. Biochim Biophys Acta (1999) 1448:390; **117**. Biochim Biophys Acta (1998) 1390:73; **118**. Biochim Biophys Acta (1982) 692:196; **119**. Biochemistry (1998) 37:8180; **120**. J Biol Chem (1996) 271:11627; **121**. Biochemistry (2000) 39:10928; **122**. Biochemistry (1999) 38:4604; **123**. Biochemistry (1997) 36:4675; **124**. Biochemistry (1997) 36:10630; **125**. Biochim Biophys Acta (1988) 946:85; **126**. Biophys J (1997) 72:2660; **127**. Biophys J (1996) 71:892; **128**. Biophys J (1996) 71:878; **129**. Methods Enzymol (1994) 233:459; **130**. Biochemistry (1997) 36:5827; **131**. Biochemistry (1997) 36:6251; **132**. J Fluorescence (1994) 4:153; **133**. J Biol Chem (1996) 271:31878; **134**. Biochemistry (1998) 37:16653; **135**. Biochemistry (1991) 30:4193; **136**. Biophys J (2001) 80:822; **137**. Biochemistry (1993) 32:10826; **138**. Biochim Biophys Acta (1988) 938:24; **139**. Biochemistry (1987) 26:39; **140**. Biochemistry (1982) 21:1720; **141**. Biochemistry (1981) 20:2783; **142**. J Biol Chem (1980) 255:5404; **143**. J Biol Chem (1994) 269:22517; **144**. Biochemistry (1993) 32:3714; **145**. Proc Natl Acad Sci U S A (1989) 86:9896; **146**. J Biol Chem (2000) 275:23065; **147**. Biophys J (2000) 78:2628; **148**. Biochemistry (1998) 37:14833; **149**. Biochemistry (1994) 33:6721; **150**. Biochim Biophys Acta (2007) 1768:502; **151**. Biochemistry (1992) 31:2865; **152**. J Cell Biol (1993) 123:1403; **153**. J Cell Biol (1991) 113:235; **154**. J Cell Biol (1991) 112:267; **155**. J Biol Chem (1990) 265:5337; **156**. Biochim Biophys Acta (1991) 1082:255; **157**. Biochemistry (1988) 27:1889; **158**. J Biol Chem (1987) 262:14172; **159**. Biochemistry (2001) 40:6475; **160**. Am J Physiol (1994) 267:G80; **161**. J Biol Chem (1983) 258:5368; **162**. Biochemistry (1994) 33:8225; **163**. J Cell Biol (1991) 112:1143; **164**. J Cell Biol (1987) 105:755; **165**. Biophys J (2001) 80:1819; **166**. Biochemistry (1994) 33:4483; **167**. Biochemistry (1993) 32:12591; **168**. Chem Phys Lipids (1991) 57:227; **169**. Proc Natl Acad Sci U S A (1991) 88:1364; **170**. Chem Phys Lipids (1991) 57:29; **171**. Biochemistry (1998) 37:15114; **172**. J Fluoresc (2007) 17:97; **173**. Biophys J (2007) 92:126; **174**. Biochim Biophys Acta (2000) 1487:82; **175**. Biochim Biophys Acta (2000) 1467:281; **176**. Biochemistry (1993) 32:11074; **177**. Biochemistry (1992) 31:5912; **178**. Biochemistry (1990) 29:1593; **179**. Biochemistry (2001) 40:4181; **180**. Biochemistry (1998) 37:17562; **181**. Biophys J (1994) 66:1981; **182**. Biophys J (1992) 63:903; **183**. Biochemistry (1988) 27:3433; **184**. Chem Phys Lipids (1990) 53:129; **185**. J Biol Chem (1993) 268:1074; **186**. Biochemistry (1993) 32:11711; **187**. Biochemistry (1993) 32:5411; **188**. Biochemistry (1993) 32:5373; **189**. Biophys J (1994) 66:729; **190**. Biochemistry (1997) 36:14325; **191**. Biochemistry (1999) 38:7803; **192**. Anal Chem (1987) 59:279; **193**. Biophys J (1985) 47:613; **194**. J Cell Physiol (1981) 107:329; **195**. Biochim Biophys Acta (1972) 279:393; **196**. Anal Biochem (1981) 116:553; **197**. Biochim Biophys Acta (1994) 1192:132; **198**. Anal Biochem (1994) 219:1; **199**. Anal Biochem (1995) 232:7; **200**. Biochim Biophys Acta (1985) 815:351; **201**. Biochim Biophys Acta (1996) 1284:191; **202**. J Biol Chem (1997) 272:6167; **203**. Biochemistry (1993) 32:66; **204**. Biochemistry (1997) 36:8189; **205**. Biochemistry (1994) 33:13231; **206**. Biochemistry (2000) 39:9623; **207**. Biochemistry (1998) 37:6697; **208**. Biochemistry (1998) 37:8516; **209**. Biochim Biophys Acta (1992) 1104:9; **210**. Chem Biol (1997) 4:269; **211**. J Biol Chem (1994) 269:10517; **212**. Biochemistry (1990) 29:879; **213**. Biochim Biophys Acta (1989) 981:178; **214**. Biochemistry (1988) 27:3925; **215**. Methods Enzymol (1989) 171:850; **216**. Biochemistry (1994) 33:9968; **217**. Chem Phys Lipids (1994) 70:205; **218**. Biochemistry (1993) 32:14194; **219**. Biochemistry (1977) 16:3836; **220**. J Cell Biol (1993) 122:1253; **221**. J Cell Biol (1991) 115:1585; **222**. J Cell Biol (1991) 115:245; **223**. Biophys J (1992) 63:309; **224**. Biochemistry (1990) 29:2976; **225**. Biochemistry (1988) 27:3947; **226**. Z Naturforsch C (2000) 55:418; **227**. Biochim Biophys Acta (1998) 1374:63; **228**. J Biol Chem (1999) 274:29951; **229**. Biophys J (1992) 63:823; **230**. J Am Chem Soc (1981) 103:6797; **231**. Biophys J (1994) 66:25; **232**. J Cell Biol (1986) 103:807; **233**. J Membr Biol (1993) 135:83; **234**. Proc Natl Acad Sci U S A (1991) 88:6274; **235**. J Cell Biol (1993) 120:25; **236**. Biochemistry (1985) 24:6390; **237**. Biochemistry (1981) 20:4093; **238**. J Biol Chem (1991) 266:12082; **239**. Biochemistry (1990) 29:1607; **240**. Eur J Cell Biol (1990) 53:173; **241**. Biochem J (1992) 284:259; **242**. J Cell Biol (1986) 103:1221; **243**. Cell (1988) 55:797; **244**. J Cell Biol (1993) 121:543; **245**. J Immunol Methods (1993) 158:183; **246**. Anal Biochem (1992) 207:341; **247**. Biophys J (1991) 59:387; **248**. Biochim Biophys Acta (1990) 1028:73; **249**. Anal Chem (2001) 73:91; **250**. Science (1994) 264:1753; **251**. Anal Biochem (2000) 280:94; **252**. Angew Chem Int Ed Engl (1990) 29:1269; **253**. Anal Biochem (1994) 217:128; **254**. Biophys J (1993) 65:2160; **255**. Biophys J (1994) 66:31; **256**. Biochemistry (1993) 32:9960; **257**. Toxicol Sci (2007) 99:162; **258**. Cytometry A (2009) 77:231.

DATA TABLE 13.2 FATTY ACID ANALOGS AND PHOSPHOLIPIDS

Cat. No.	MW	Storage	Soluble	Abs	EC	Em	Solvent	Notes
A3880	~15,350	FF,L,AA	H_2O	365	10,500	432	H_2O	1
A10070	880.68	FF,D,L	DMSO	505	92,000	512	MeOH	16
A10072	986.67	FF,D,L	DMSO	505	85,000	567	MeOH	17, 18
B1550	1019.45	FF,D	see Notes	<300		none		2
B1616	1132.61	FF,D	see Notes	<300		none		2
B3781	797.88	FF,D,L	see Notes	342	75,000	471	EtOH	3
B3782	966.20	FF,D,L	see Notes	340	62,000	473	EtOH	4
B3824	404.31	F,L	DMSO	509	101,000	515	MeOH	5
B7701	1029.80	FF,D,L	see Notes	505	123,000	512	MeOH	2, 6
D57	1026.44	FF,D,L	see Notes	336	4500	517	MeOH	2
D94	434.59	F,L	DMSO, EtOH	335	4800	519	MeOH	
D476	782.01	FF,D,L	see Notes	354	81,000	428	MeOH	2, 7
D3771	854.86	FF,D,L	see Notes	506	71,000	512	EtOH	2
D3792	895.95	FF,D,L	see Notes	506	86,000	513	EtOH	2, 5
D3793	881.93	FF,D,L	see Notes	509	86,000	513	EtOH	2, 5
D3800	1067.23	FF,D,L	see Notes	505	87,000	511	MeOH	2, 5
D3803	797.77	FF,D,L	see Notes	503	80,000	512	MeOH	2, 5
D3805	746.68	FF,D,L	see Notes	504	79,000	511	MeOH	2, 5
D3815	921.91	FF,D,L	see Notes	534	64,000	552	MeOH	2, 5
D3821	474.44	F,L	DMSO	505	90,000	512	MeOH	5
D3822	418.33	F,L	DMSO	505	87,000	511	MeOH	5
D3823	404.31	F,L	DMSO	508	97,000	514	MeOH	5
D3825	404.31	F,L	DMSO	509	100,000	515	MeOH	5
D3832	542.47	F,L	DMSO	534	76,000	552	MeOH	5
D3834	320.15	F,L	DMSO, MeCN	505	96,000	511	MeOH	5

continued on next page

IMPORTANT NOTICE: The products described in this manual are covered by one or more Limited Use Label License(s). Please refer to the Appendix on page 971 and Master Product List on page 975. Products are For Research Use Only. Not intended for any animal or human therapeutic or diagnostic use.

DATA TABLE 13.2 FATTY ACID ANALOGS AND PHOSPHOLIPIDS—*continued*

Cat. No.	MW	Storage	Soluble	Abs	EC	Em	Solvent	Notes
D3835	472.40	F,L	DMSO	559	91,000	568	MeOH	5
D3861	504.43	F,L	DMSO	582	140,000	591	MeOH	8
D3862	404.31	F,L	DMSO	505	92,000	510	MeOH	5
D23739	1136.13	FF,D,L	DMSO	505	92,000	511	MeOH	2, 9
F362	1182.54	FF,D,L	see Notes	496	88,000	519	MeOH	2, 10
H361	850.13	FF,D,L	see Notes	342	37,000	376	MeOH	2, 11, 12
H3809	856.09	FF,D,L	see Notes	341	38,000	376	MeOH	2, 11, 12
H34350	~1100	F,L	H_2O	495	84,000	525	MeOH	15
H34351	~1400	F,L	H_2O	595	112,000	615	MeOH	15
L1392	1333.81	FF,D,L	see Notes	560	75,000	581	MeOH	2
M12652	944.14	FF,D,L	see Notes	365	18,000	460	MeOH	2, 10
N316	294.27	L	DMSO	467	23,000	539	MeOH	13
N360	956.25	FF,D,L	see Notes	463	21,000	536	MeOH	2, 13
N678	378.43	L	DMSO	467	24,000	536	MeOH	13
N3786	771.89	FF,D,L	see Notes	465	21,000	533	EtOH	2, 13
N3787	856.05	FF,D,L	see Notes	465	22,000	534	EtOH	2, 13
O12650	1086.25	FF,D,L	see Notes	501	85,000	526	MeOH	2, 10
P31	372.51	L	DMF, DMSO	341	43,000	377	MeOH	11, 12
P96	400.56	L	DMF, DMSO	341	44,000	377	MeOH	11, 12
P243	456.67	L	DMF, DMSO	341	43,000	377	MeOH	11, 12
P1903MP	288.35	L	DMF, DMSO	341	43,000	376	MeOH	11, 12
P22652	961.17	FF,D,L	see Notes	411	40,000	454	MeOH	2
P36005	276.42	FF,LL,AA	EtOH	304	77,000	416	MeOH	14, 15
T1391	1236.68	FF,D,L	see Notes	540	93,000	566	MeOH	2
T1395MP	1381.85	FF,D,L	see Notes	583	115,000	601	MeOH	2

For definitions of the contents of this data table, see "Using *The Molecular Probes® Handbook*" in the introductory pages.

Notes

1. ADIFAB fatty acid indicator is a protein conjugate with a molecular weight of approximately 15,350. Em shifts from about 432 nm to 505 nm upon binding of fatty acids. (Mol Cell Biochem (1999) 192:87)
2. Chloroform is the most generally useful solvent for preparing stock solutions of phospholipids (including sphingomyelins). Glycerophosphocholines are usually freely soluble in ethanol. Most other glycerophospholipids (phosphoethanolamines, phosphatidic acids and phosphoglycerols) are less soluble in ethanol, but solutions up to 1–2 mg/mL should be obtainable, using sonication to aid dispersion if necessary. Labeling of cells with fluorescent phospholipids can be enhanced by addition of cyclodextrins during incubation. (J Biol Chem (1999) 274:35359)
3. Phospholipase A cleavage generates a fluorescent fatty acid (P1903MP) and a fluorescent lysophospholipid.
4. Phospholipase A cleavage generates a fluorescent fatty acid (P31) and a fluorescent lysophospholipid.
5. The absorption and fluorescence spectra of BODIPY® derivatives are relatively insensitive to the solvent.
6. Phospholipase A cleavage results in increased fluorescence with essentially no wavelength shift. The cleavage products are D3862 and a fluorescent lysophospholipid.
7. Diphenylhexatriene (DPH) and its derivatives are essentially nonfluorescent in water. Absorption and emission spectra have multiple peaks. The wavelength, resolution and relative intensity of these peaks are environment dependent. Abs and Em values are for the most intense peak in the solvent specified.
8. Oxidation of the polyunsaturated butadienyl portion of the BODIPY® 581/591 dye results in a shift of the fluorescence emission peak from ~590 nm to ~510 nm. (Methods Enzymol (2000) 319:603, FEBS Lett (1999) 453:278)
9. Phospholipase A_2 cleavage results in increased fluorescence with essentially no wavelength shift. The cleavage products are D3834 and a dinitrophenylated lysophospholipid.
10. Spectra of this compound are in methanol containing a trace of KOH.
11. Alkylpyrene fluorescence lifetimes are up to 110 nanoseconds and are very sensitive to oxygen.
12. Pyrene derivatives exhibit structured spectra. The absorption maximum is usually about 340 nm with a subsidiary peak at about 325 nm. There are also strong absorption peaks below 300 nm. The emission maximum is usually about 376 nm with a subsidiary peak at 396 nm. Excimer emission at about 470 nm may be observed at high concentrations.
13. Fluorescence of NBD and its derivatives in water is relatively weak. QY and τ increase and Em decreases in aprotic solvents and other nonpolar environments relative to water. (Biochemistry (1977) 16:5150, Photochem Photobiol (1991) 54:361)
14. *Cis*-parinaric acid is highly oxygen sensitive. Use under N_2 or Ar. *Cis*-parinaric acid is essentially nonfluorescent in water.
15. This product is supplied as a ready-made solution in the solvent indicated under "Soluble."
16. Phospholipase A_1 cleavage results in increased fluorescence with essentially no wavelength shift. The cleavage products are D3834 and a dinitrophenylated lysophospholipid.
17. A10072 exhibits dual emission (Em = 510 nm and 567 nm in MeOH, 513 nm and 575 nm when incorporated in phospholipid bilayer membranes). Phospholipase A_2 cleavage results in increased 510–513 nm emission and reciprocally diminshed 567–575 nm emission.
18. A10072 is also soluble at 2 mM in 2-methoxyethanol.

IMPORTANT NOTICE: The products described in this manual are covered by one or more Limited Use Label License(s). Please refer to the Appendix on page 971 and Master Product List on page 975. Products are For Research Use Only. Not intended for any animal or human therapeutic or diagnostic use.

molecular probes® | invitrogen™ by life technologies™

PRODUCT LIST 13.2 FATTY ACID ANALOGS AND PHOSPHOLIPIDS

Cat. No.	Product	Quantity
A3880	ADIFAB fatty acid indicator	200 µg
B1550	biotin DHPE (*N*-(biotinoyl)-1,2-dihexadecanoyl-*sn*-glycero-3-phosphoethanolamine, triethylammonium salt)	10 mg
B1616	biotin-X DHPE (*N*-((6-(biotinoyl)amino)hexanoyl)-1,2-dihexadecanoyl-*sn*-glycero-3-phosphoethanolamine, triethylammonium salt)	5 mg
B7701	1,2-bis-(4,4-difluoro-5,7-dimethyl-4-bora-3a,4a-diaza-*s*-indacene-3-undecanoyl)-*sn*-glycero-3-phosphocholine (bis-BODIPY® FL C_{11}-PC)	100 µg
B3781	1,2-bis-(1-pyrenebutanoyl)-*sn*-glycero-3-phosphocholine	1 mg
B3782	1,2-bis-(1-pyrenedecanoyl)-*sn*-glycero-3-phosphocholine	1 mg
B3824	BODIPY® 500/510 C_4, C_9 (5-butyl-4,4-difluoro-4-bora-3a,4a-diaza-*s*-indacene-3-nonanoic acid)	1 mg
D3771	2-decanoyl-1-(*O*-(11-(4,4-difluoro-5,7-dimethyl-4-bora-3a,4a-diaza-*s*-indacene-3-propionyl)amino)undecyl)-*sn*-glycero-3-phosphocholine	1 mg
D3822	4,4-difluoro-5,7-dimethyl-4-bora-3a,4a-diaza-*s*-indacene-3-dodecanoic acid (BODIPY® FL C_{12})	1 mg
D3792	2-(4,4-difluoro-5,7-dimethyl-4-bora-3a,4a-diaza-*s*-indacene-3-dodecanoyl)-1-hexadecanoyl-*sn*-glycero-3-phosphocholine (β-BODIPY® FL C_{12}-HPC)	100 µg
D3821	4,4-difluoro-5,7-dimethyl-4-bora-3a,4a-diaza-*s*-indacene-3-hexadecanoic acid (BODIPY® FL C_{16})	1 mg
D3834	4,4-difluoro-5,7-dimethyl-4-bora-3a,4a-diaza-*s*-indacene-3-pentanoic acid (BODIPY® FL C_5)	1 mg
D3805	2-(4,4-difluoro-5,7-dimethyl-4-bora-3a,4a-diaza-*s*-indacene-3-pentanoyl)-1-hexadecanoyl-*sn*-glycero-3-phosphate, diammonium salt (β-BODIPY® FL C_5-HPA)	100 µg
D3803	2-(4,4-difluoro-5,7-dimethyl-4-bora-3a,4a-diaza-*s*-indacene-3-pentanoyl)-1-hexadecanoyl-*sn*-glycero-3-phosphocholine (β-BODIPY® FL C_5-HPC)	100 µg
D3800	*N*-(4,4-difluoro-5,7-dimethyl-4-bora-3a,4a-diaza-*s*-indacene-3-propionyl)-1,2-dihexadecanoyl-*sn*-glycero-3-phosphoethanolamine, triethylammonium salt (BODIPY® FL DHPE)	100 µg
D3862	4,4-difluoro-5,7-dimethyl-4-bora-3a,4a-diaza-*s*-indacene-3-undecanoic acid (BODIPY® FL C_{11})	1 mg
D3832	4,4-difluoro-5,7-diphenyl-4-bora-3a,4a-diaza-*s*-indacene-3-dodecanoic acid (BODIPY® 530/550 C_{12})	1 mg
D3815	2-(4,4-difluoro-5,7-diphenyl-4-bora-3a,4a-diaza-*s*-indacene-3-pentanoyl)-1-hexadecanoyl-*sn*-glycero-3-phosphocholine (β-BODIPY® 530/550 C_5-HPC)	100 µg
D3823	4,4-difluoro-5-methyl-4-bora-3a,4a-diaza-*s*-indacene-3-dodecanoic acid (BODIPY® 500/510 C_1, C_{12})	1 mg
D3793	2-(4,4-difluoro-5-methyl-4-bora-3a,4a-diaza-*s*-indacene-3-dodecanoyl)-1-hexadecanoyl-*sn*-glycero-3-phosphocholine (β-BODIPY® 500/510 C_{12}-HPC)	100 µg
D3825	4,4-difluoro-5-octyl-4-bora-3a,4a-diaza-*s*-indacene-3-pentanoic acid (BODIPY® 500/510 C_8, C_5)	1 mg
D3861	4,4-difluoro-5-(4-phenyl-1,3-butadienyl)-4-bora-3a,4a-diaza-*s*-indacene-3-undecanoic acid (BODIPY® 581/591 C_{11})	1 mg
D3835	4,4-difluoro-5-(2-thienyl)-4-bora-3a,4a-diaza-*s*-indacene-3-dodecanoic acid (BODIPY® 558/568 C_{12})	1 mg
D94	11-((5-dimethylaminonaphthalene-1-sulfonyl)amino)undecanoic acid (DAUDA)	100 mg
D57	*N*-(5-dimethylaminonaphthalene-1-sulfonyl)-1,2-dihexadecanoyl-*sn*-glycero-3-phosphoethanolamine, triethylammonium salt (dansyl DHPE)	25 mg
D23739	*N*-((6-(2,4-dinitrophenyl)amino)hexanoyl)-2-(4,4-difluoro-5,7-dimethyl-4-bora-3a,4a-diaza-*s*-indacene-3-pentanoyl)-1-hexadecanoyl-*sn*-glycero-3-phosphoethanolamine, triethylammonium salt (PED6)	1 mg
D476	β-DPH HPC (2-(3-(diphenylhexatrienyl)propanoyl)-1-hexadecanoyl-*sn*-glycero-3-phosphocholine)	1 mg
F362	fluorescein DHPE (*N*-(fluorescein-5-thiocarbamoyl)-1,2-dihexadecanoyl-*sn*-glycero-3-phosphoethanolamine, triethylammonium salt)	5 mg
H34350	HCS LipidTOX™ Green phospholipidosis detection reagent *1000X aqueous solution* *for cellular imaging* *10-plate size*	each
H34157	HCS LipidTOX™ Phospholipidosis and Steatosis Detection Kit *for high-content screening* *for cellular imaging* *2-plate size*	1 kit
H34158	HCS LipidTOX™ Phospholipidosis and Steatosis Detection Kit *for high-content screening* *for cellular imaging* *10-plate size*	1 kit
H34351	HCS LipidTOX™ Red phospholipidosis detection reagent *1000X aqueous solution* *for cellular imaging* *10-plate size*	each
H361	1-hexadecanoyl-2-(1-pyrenedecanoyl)-*sn*-glycero-3-phosphocholine (β-py-C10-HPC)	1 mg
H3809	1-hexadecanoyl-2-(1-pyrenedecanoyl)-*sn*-glycero-3-phosphoglycerol, ammonium salt (β-py-C10-PG)	1 mg
L1392	Lissamine rhodamine B 1,2-dihexadecanoyl-*sn*-glycero-3-phosphoethanolamine, triethylammonium salt (rhodamine DHPE)	5 mg
M12652	Marina Blue® 1,2-dihexadecanoyl-*sn*-glycero-3-phosphoethanolamine (Marina Blue® DHPE)	1 mg
N360	NBD-PE (*N*-(7-nitrobenz-2-oxa-1,3-diazol-4-yl)-1,2-dihexadecanoyl-*sn*-glycero-3-phosphoethanolamine, triethylammonium salt)	10 mg
N316	NBD-X (6-(*N*-(7-nitrobenz-2-oxa-1,3-diazol-4-yl)amino)hexanoic acid)	100 mg
N678	12-(*N*-(7-nitrobenz-2-oxa-1,3-diazol-4-yl)amino)dodecanoic acid	100 mg
N3787	2-(12-(7-nitrobenz-2-oxa-1,3-diazol-4-yl)amino)dodecanoyl-1-hexadecanoyl-*sn*-glycero-3-phosphocholine (NBD C_{12}-HPC)	5 mg
N3786	2-(6-(7-nitrobenz-2-oxa-1,3-diazol-4-yl)amino)hexanoyl-1-hexadecanoyl-*sn*-glycero-3-phosphocholine (NBD C_6-HPC)	5 mg
O12650	Oregon Green® 488 DHPE (Oregon Green® 488 1,2-dihexadecanoyl-*sn*-glycero-3-phosphoethanolamine)	1 mg
P22652	Pacific Blue™ DMPE (Pacific Blue™ 1,2-ditetradecanoyl-*sn*-glycero-3-phosphoethanolamine, triethylammonium salt)	1 mg
P36005	*cis*-parinaric acid *3 mM in ethanol*	10 mL
A10070	PED-A1 (*N*-((6-(2,4-DNP)amino)hexanoyl)-1-(BODIPY® FL C_5)-2-hexyl-*sn*-glycero-3-phosphoethanolamine) *phospholipase A_1 selective substrate*	100 µg
P1903MP	1-pyrenebutanoic acid *high purity*	100 mg
P31	1-pyrenedecanoic acid	25 mg
P96	1-pyrenedodecanoic acid	25 mg
P243	1-pyrenehexadecanoic acid	5 mg
A10072	Red/Green BODIPY® PC-A2 (1-*O*-(6-BODIPY® 558/568-aminohexyl)-2-BODIPY® FL C_5-*sn*-glycero-3-phosphocholine) *ratiometric phospholipase A_2 substrate*	100 µg
T1395MP	Texas Red® DHPE (Texas Red® 1,2-dihexadecanoyl-*sn*-glycero-3-phosphoethanolamine, triethylammonium salt)	1 mg
T1391	TRITC DHPE (*N*-(6-tetramethylrhodaminethiocarbamoyl)-1,2-dihexadecanoyl-*sn*-glycero-3-phosphoethanolamine, triethylammonium salt)	1 mg

IMPORTANT NOTICE: The products described in this manual are covered by one or more Limited Use Label License(s). Please refer to the Appendix on page 971 and Master Product List on page 975. Products are For Research Use Only. Not intended for any animal or human therapeutic or diagnostic use.

13.3 Sphingolipids, Steroids, Lipopolysaccharides and Related Probes

Sphingolipids

Structure and Activity

Sphingolipids are essential components of the plasma membrane of eukaryotic cells, where they are typically found in the outer leaflet. Although particularly abundant in mammalian cells, sphingolipids are also present in *Saccharomyces cerevisiae*,[1] other fungi and plants. Sphingolipids differ from phospholipids in that they are based on a lipophilic amino alcohol (sphingosine, Figure 13.3.1) rather than glycerol. Sphingolipids play important roles in signal transduction processes[2,3] (Chapter 17). Genetic defects in enzymes in the metabolic pathways of sphingolipid synthesis and degradation, including those involved in type I Gaucher (Ashkenazi) disease, type A Niemann–Pick disease, Krabbe disease,[4–8] and other lysosomal storage diseases, can be detected at the cellular level using our fluorescent analogs of sphingolipids.

Ceramides are the biological building blocks of more complex sphingolipids. Metabolism of ceramides typically occurs in Golgi and endoplasmic reticulum membranes, and fluorescent ceramide analogs (Section 12.4) are important probes for measuring the intracellular distribution and transport of the labeled molecules in live cells.[9]

If the hydroxyl group of the ceramide is esterified to phosphocholine, the sphingolipid is a sphingomyelin (Figure 13.3.1). The main pathway of sphingomyelin biosynthesis in mammalian cells is based on the transfer of phosphocholine from glycerophosphocholine to ceramide, catalyzed by sphingomyelin synthase in the Golgi membrane. Synthesis is followed by exocytosis of the sphingomyelin to the plasma membrane, apparently via a vesicular pathway and flip-flop to the outer membrane.[2] Sphingomyelinases, which are functionally analogous to phospholipase C in their chemistry, hydrolyze sphingomyelins back to ceramides. Generation of ceramides by hydrolysis of sphingomyelins appears to play a role in mediating the effects of exposure to tumor necrosis factor-α[10] (TNF-α), γ-interferon and several other agents, all of which induce an apoptosis-like cell death.[11–15] Section15.5 describes our extensive selection of reagents for following the diverse morphological and biochemical changes that occur during apoptosis. Sensitive fluorescence-based measurements of sphingomyelinase activity using natural, unlabeled sphingomyelin as the substrate can be carried out using our Amplex® Red Sphingomyelinase Assay Kit (A12220), described below.

In glycosylsphingolipids, the free hydroxyl group of the ceramide is glycosylated to give a sphingosyl glycoside (cerebroside, Figure 13.3.2) or a ganglioside (Figure 13.3.3). These glycosphingolipids form cell-type–specific patterns at the cell surface that change with cell growth, differentiation, viral transformation and oncogenesis.[16]

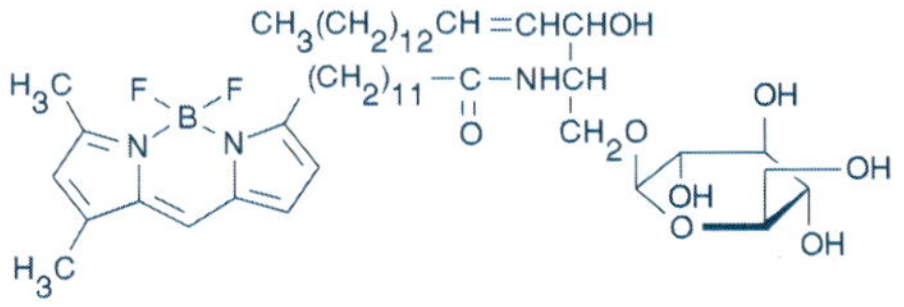

Figure 13.3.2 BODIPY® FL C_{12}-glucocerebroside.

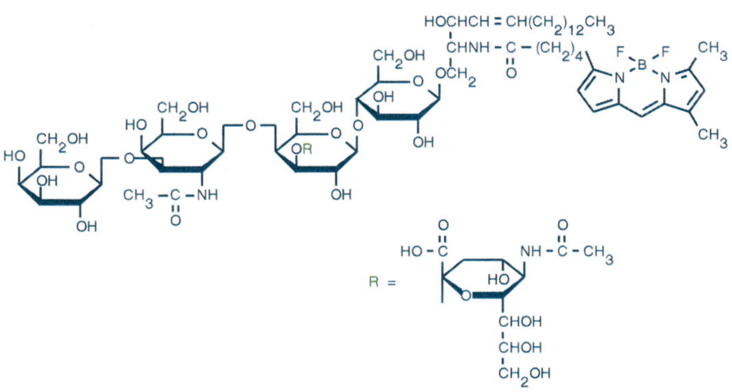

Figure 13.3.3 BODIPY® FL C_5-ganglioside G_{M1} (B13950).

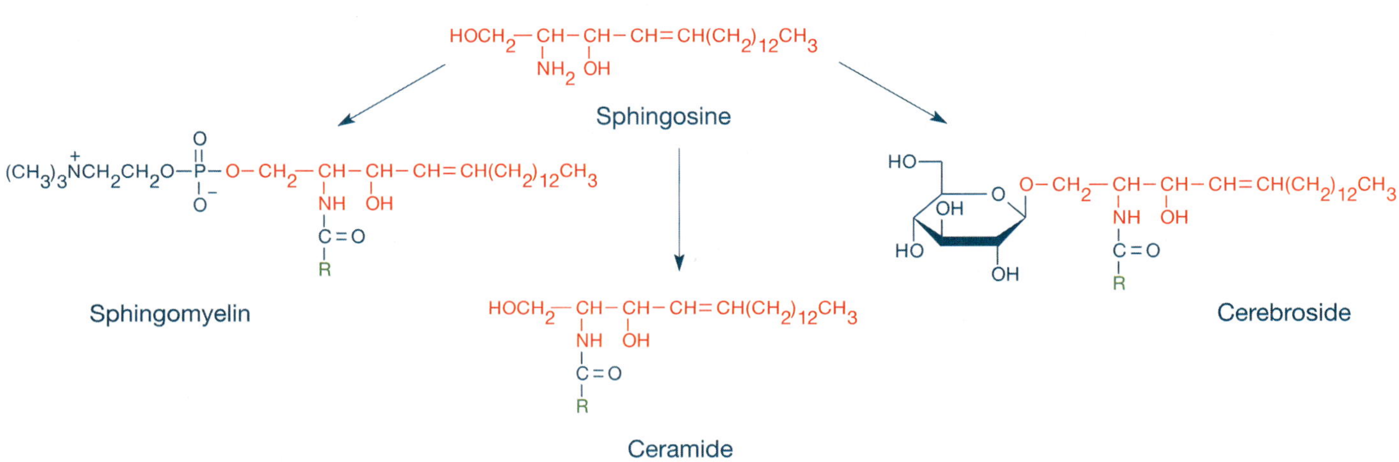

Figure 13.3.1 Sphingomyelins, ceramides and cerebrosides are examples of sphingolipids derived from sphingosine. R represents the hydrocarbon tail portion of a fatty acid residue.

Glycosphingolipids interact at the cell surface with toxins, viruses and bacteria, as well as with receptors and enzymes[17] and are involved in cell-type–specific adhesion processes.[16] Gangliosides modulate the trophic factor–stimulated dimerization, tyrosine phosphorylation and subsequent signal transduction events of several tyrosine kinase receptors.[17] Ganglioside G_{M1} has anti-neurotoxic, neuroprotective and neurorestorative effects on various central neurotransmitter systems.[18] Gangliosides, including ganglioside G_{M1}, partition into lipid rafts—detergent-insoluble, sphingolipid- and cholesterol-rich membrane microdomains that form lateral assemblies in the plasma membrane.[19–25] We offer Vybrant® Lipid Raft Labeling Kits (V34403, V34404, V34405), as well as Alexa Fluor® dye conjugates of subunit B of cholera toxin (Section 7.7), a protein that selectively binds to ganglioside G_{M1} in lipid rafts.

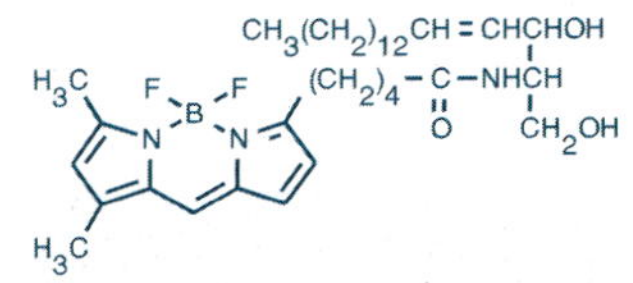

Figure 13.3.4 *N*-(4,4-difluoro-5,7-dimethyl-4-bora-3a,4a-diaza-*s*-indacene-3-pentanoyl)sphingosine (BODIPY® FL C_5-ceramide, D3521).

BODIPY® Sphingolipids

Ceramides (*N*-acylsphingosines), like diacylglycerols, are lipid second messengers that function in signal transduction processes.[26–28] The concentration-dependent spectral properties of BODIPY® FL C_5-ceramide (D3521, B22650; Figure 13.3.4), BODIPY® FL C_5-sphingomyelin[29–31] (D3522, Figure 13.3.5) and BODIPY® FL C_{12}-sphingomyelin[32] (D7711) make them particularly suitable for investigating sphingolipid transport, metabolism and microdomains,[31,33–37] in addition to their well-documented use as structural markers for the Golgi complex[38] (Section 12.4, Figure 13.3.6). BODIPY® FL C_5-ceramide can be visualized by fluorescence microscopy[39,40] (Figure 13.3.7, Figure 13.3.8) or by electron microscopy following diaminobenzidine (DAB) photoconversion to an electron-dense product[41] (Fluorescent Probes for Photoconversion of Diaminobenzidine Reagents—Note 14.2).

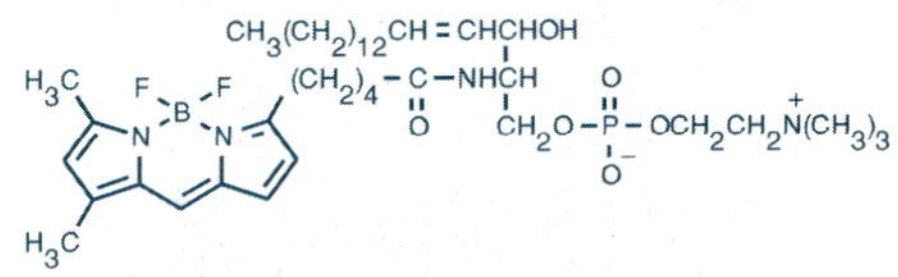

Figure 13.3.5 *N*-(4,4-difluoro-5,7-dimethyl-4-bora-3a,4a-diaza-*s*-indacene-3-pentanoyl)sphingosyl phosphocholine (BODIPY® FL C_5-sphingomyelin, D3522).

Our range of BODIPY® sphingolipids also includes the long-wavelength light–excitable BODIPY® TR ceramide[42,43] (D7540, Figure 13.3.9), as well as BODIPY® FL C_5-lactosylceramide[44–49] (D13951), BODIPY® FL C_5-ganglioside G_{M1}[50] (B13950, Figure 13.3.3) and BODIPY® FL C_{12}-galactocerebroside[51] (D7519). All Molecular Probes® sphingolipids are prepared from D-*erythro*-sphingosine and therefore have the same stereochemical conformation as natural biologically active sphingolipids.[52]

Complexing fluorescent lipids with bovine serum albumin (BSA) facilitates cell labeling by eliminating the need for organic solvents to dissolve the lipophilic probe—the BSA-complexed probe can be directly dissolved in water. We offer four BODIPY® sphingolipid–BSA complexes for the study of lipid metabolism and trafficking, including BODIPY® FL C_5-ceramide, BODIPY® TR ceramide, BODIPY® FL C_5-ganglioside G_{M1} and BODIPY® FL C_5-lactosylceramide, each complexed with defatted BSA (B22650, B34400, B34401, B34402, respectively).

BODIPY® FL C_5-ceramide has been used to investigate the linkage of sphingolipid metabolism to protein secretory pathways[53–56] and neuronal growth.[47,57] Internalization of BODIPY® FL C_5-sphingomyelin from the plasma membrane of human skin fibroblasts results in a mixed population of labeled endosomes that can be distinguished based on the concentration-dependent green

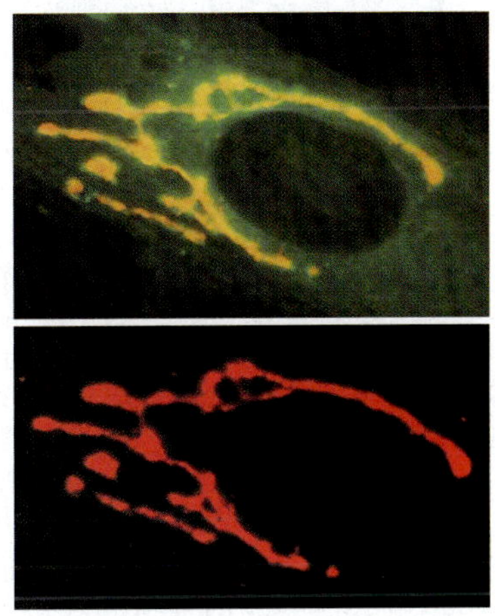

Figure 13.3.6 Selective staining of the Golgi apparatus using the green-fluorescent BODIPY® FL C_5-ceramide (D3521) (top). At high concentrations, the BODIPY® FL fluorophore forms excimers that can be visualized using a red longpass optical filter (bottom). The BODIPY® FL C_5-ceramide accumulation in the trans-Golgi is sufficient for excimer formation (J Cell Biol (1991) 113:1267). Images contributed by Richard Pagano, Mayo Foundation.

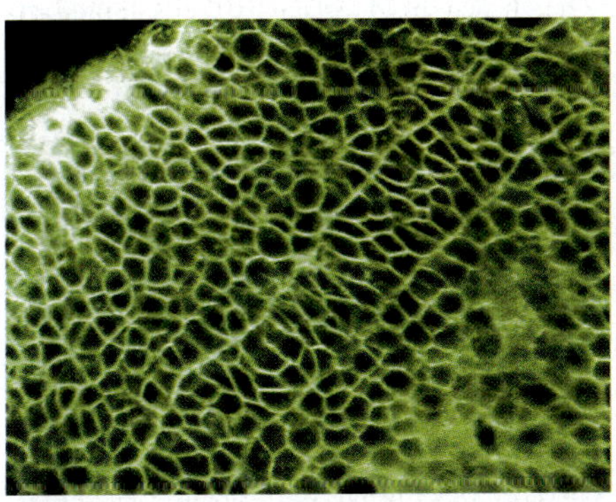

Figure 13.3.7 Cells in the notochord rudiment of a zebrafish embryo undergoing mediolateral intercalation to lengthen the forming notochord. BODIPY® FL C_5-ceramide (D3521) localizes in the interstitial fluid of the zebrafish embryo and freely diffuses between cells, illuminating cell boundaries. This confocal image was obtained using a Bio-Rad® MRC-600 microscope. Image contributed by Mark Cooper, University of Washington.

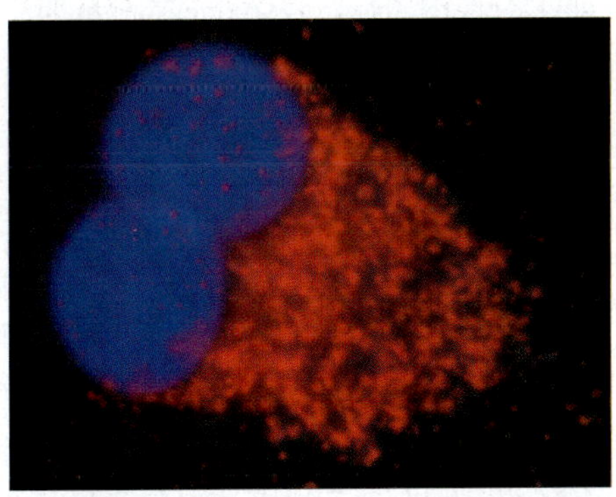

Figure 13.3.8 Nucleus and Golgi apparatus of a bovine pulmonary artery endothelial cell (BPAEC) labeled with Hoechst 33342 (H1399, H3569, H21492) and the BSA complex of BODIPY® FL C_5-ceramide (D3521, B22650), respectively.

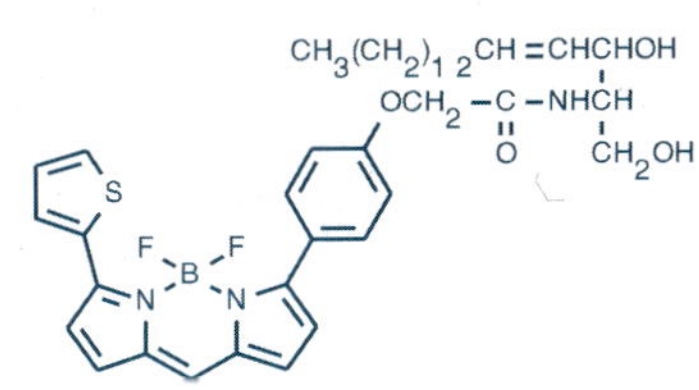

Figure 13.3.9 BODIPY® TR ceramide (*N*-((4-(4,4-difluoro-5-(2-thienyl)-4-bora-3a,4a-diaza-*s*-indacene-3-yl)phenoxy)acetyl)sphingosine; D7540).

IMPORTANT NOTICE: The products described in this manual are covered by one or more Limited Use Label License(s). Please refer to the Appendix on page 971 and Master Product List on page 975. Products are For Research Use Only. Not intended for any animal or human therapeutic or diagnostic use.

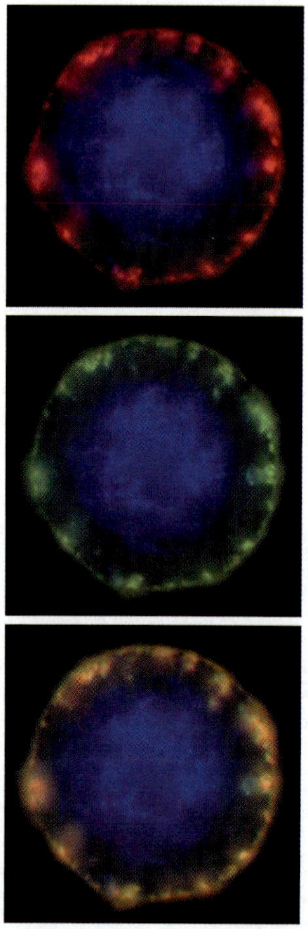

Figure 13.3.10 A J774 mouse macrophage cell sequentially stained with BODIPY® FL ganglioside G_{M1} (B13950) and then with Alexa Fluor® 555 dye–labeled cholera toxin subunit B (C22843, C34776; also available as a component of V34404). The cell was then treated with an anti–CT-B antibody (a component of V34404) to induce crosslinking. Alexa Fluor® 555 dye fluorescence (top panel, red) and BODIPY® FL dye fluorescence (middle panel, green) were imaged separately and overlaid to emphasize the coincident staining (bottom panel, yellow). Nuclei were stained with blue-fluorescent Hoechst 33342 (H1399, H3570, H21492).

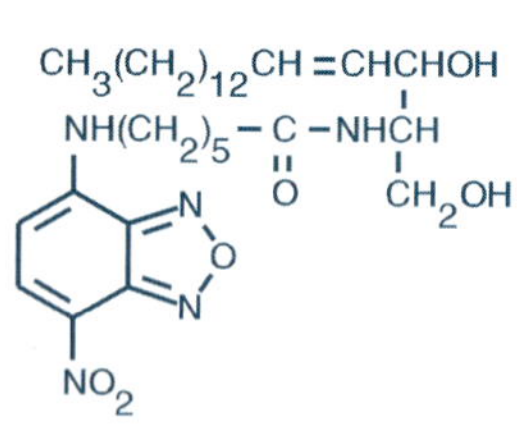

Figure 13.3.11 NBD C_6-ceramide (6-((*N*-(7-nitrobenz-2-oxa-1,3-diazol-4-yl)amino)hexanoyl)sphingosine, N1154).

(~515 nm) or red (~620 nm) emission of the probe[31] (Figure 13.3.6). BODIPY® C_5-sphingomyelin has also been used to assess sphingomyelinase gene transfer and expression in hematopoietic stem and progenitor cells.[4] Studies by Martin and Pagano have shown that the internalization routes for BODIPY® FL C_5-glucocerebroside follow both endocytic and nonendocytic pathways and are quite different from those for BODIPY® FL C_5-sphingomyelin.[58]

BODIPY® FL C_5-lactosylceramide, BODIPY® FL C_5-ganglioside G_{M1} and BODIPY® FL cerebrosides are useful tools for the study of glycosphingolipid transport and signaling pathways in cells[59,60] and for diagnosis of lipid-storage disorders such as Niemann–Pick disease,[61] Gaucher disease, G_{M1} gangliosidosis, Morquio syndrome and type IV mucolipidosis[6,49,62–67] (ML-IV). Addition of BODIPY® FL C_5-lactosylceramide to the culture medium of cells from patients with sphingolipid-storage diseases (sphingolipidosis) results in fluorescent product accumulation in lysosomes, whereas this probe accumulates in the Golgi apparatus of normal cells and cells from patients with other storage diseases.[46,48] BODIPY® FL C_5-ganglioside G_{M1} has been shown to form cholesterol-enhanced clusters in membrane complexes with amyloid β-protein in a model of Alzheimer disease amyolid fibrils.[68] As observed by fluorescence microscopy, the colocalization of BODIPY® FL C_5-ganglioside G_{M1} and fluorescent cholera toxin B conjugates (Section 7.7) provides a direct indication of the association of these molecules in lipid rafts[50] (Figure 13.3.10).

NBD Sphingolipids

NBD C_6-ceramide (N1154, Figure 13.3.11) and NBD C_6-sphingomyelin (N3524) analogs predate their BODIPY® counterparts and have been extensively used for following sphingolipid metabolism in cells[9,59,69,70] and in multicellular organisms.[71] As with BODIPY® FL C_5-ceramide, we also offer NBD C_6-ceramide complexed with defatted BSA (N22651) to facilitate cell loading without the use of organic solvents to dissolve the probe. Koval and Pagano have prepared NBD analogs of both the naturally occurring D-*erythro* and the nonnatural L-*threo* stereoisomers of sphingomyelin and have compared their intracellular transport behavior in Chinese hamster ovary (CHO) fibroblasts.[72]

NBD C_6-ceramide lacks the useful concentration-dependent optical properties of the BODIPY® FL analog and is less photostable; however, the fluorescence of NBD C_6-ceramide is apparently sensitive to the cholesterol content of the Golgi apparatus, a phenomenon that is not observed with BODIPY® FL C_5-ceramide. If NBD C_6-ceramide–containing cells are starved for cholesterol, the NBD C_6-ceramide that accumulates within the Golgi apparatus appears to be severely photolabile but this NBD photobleaching can be reduced by stimulation of cholesterol synthesis. Thus, NBD C_6-ceramide may be useful in monitoring the cholesterol content of the Golgi apparatus in live cells.[73]

Vybrant® Lipid Raft Labeling Kits

The Vybrant® Lipid Raft Labeling Kits (V34403, V34404, V34405) are designed to provide convenient, reliable and extremely bright fluorescent labeling of lipid rafts in live cells. Lipid rafts are detergent-insoluble, sphingolipid- and cholesterol-rich membrane microdomains that form lateral assemblies in the plasma membrane.[19–25] Lipid rafts also sequester glycophosphatidylinositol (GPI)-linked proteins and other signaling proteins and receptors, which may be regulated by their selective interactions with these membrane microdomains.[50,74–78] Lipid rafts play a role in a variety of cellular processes—including the compartmentalization of cell-signaling events,[79–86] the regulation of apoptosis[87–89] and the intracellular trafficking of certain membrane proteins and lipids[90–92]—as well as in the infectious cycles of several viruses and bacterial pathogens.[93–98] Examining the formation and regulation of lipid rafts is a critical step in understanding these aspects of eukaryotic cell function.

The Vybrant® Lipid Raft Labeling Kits provide the key reagents for fluorescently labeling lipid rafts *in vivo* with our bright and extremely photostable Alexa Fluor® dyes (Figure 13.3.10). Live cells are first labeled with the green-fluorescent Alexa Fluor® 488, orange-fluorescent Alexa Fluor® 555 or red-fluorescent Alexa Fluor® 594 conjugate of cholera toxin subunit B (CT-B). This CT-B conjugate binds to the pentasaccharide chain of plasma membrane ganglioside G_{M1}, which selectively partitions into lipid rafts.[50,99,100] All Molecular Probes® CT-B conjugates are prepared from recombinant CT-B and are completely free of the toxic subunit A, thus eliminating any concern for toxicity or ADP-ribosylating activity. An antibody that specifically recognizes CT-B is then

IMPORTANT NOTICE: The products described in this manual are covered by one or more Limited Use Label License(s). Please refer to the Appendix on page 971 and Master Product List on page 975. Products are For Research Use Only. Not intended for any animal or human therapeutic or diagnostic use.

used to crosslink the CT-B–labeled lipid rafts into distinct patches on the plasma membrane, which are easily visualized by fluorescence microscopy.[101,102]

Each Vybrant® Lipid Raft Labeling Kit contains sufficient reagents to label 50 live-cell samples in a 2 mL assay, including:

- Recombinant cholera toxin subunit B (CT-B) labeled with the Alexa Fluor® 488 (in Kit V34403), Alexa Fluor® 555 (in Kit V34404) or Alexa Fluor® 594 (in Kit V34405) dye
- Anti–cholera toxin subunit B antibody (anti–CT-B)
- Concentrated phosphate-buffered saline (PBS)
- Detailed labeling protocols

Because they are compatible with various multilabeling schemes, the Vybrant® Lipid Raft Labeling Kits can also serve as important tools for identifying physiologically significant membrane proteins that associate with lipid rafts. Cells can be labeled with other live-cell probes during the lipid raft labeling protocol or immediately following the antibody crosslinking step, depending on the specific labeling requirements of the other probes. Alternatively, once the lipid rafts have been labeled and crosslinked, the cells can be fixed for long-term storage or fixed and permeabilized for subsequent labeling with antibodies or other probes that are impermeant to live cells.

Amplex® Red Sphingomyelinase Assay Kit

The Amplex® Red Sphingomyelinase Assay Kit (A12220) is designed for measuring sphingomyelinase activity in solution using a fluorescence microplate reader or fluorometer[103] (Figure 13.3.12). This assay should be useful for screening sphingomyelinase activators or inhibitors or for detecting sphingomyelinase activity in cell and tissue extracts. The assay, which uses natural sphingomyelin as the principal substrate, employs an enzyme-coupled detection scheme in which phosphocholine liberated by the action of sphingomyelinase is cleaved by alkaline phosphatase to generate choline. Choline is, in turn, oxidized by choline oxidase, generating H_2O_2, which drives the conversion of the Amplex® Red reagent (A12222, A22177; Section 10.5) to red-fluorescent resorufin. This sensitive assay technique has been employed to detect activation of acid sphingomyelinase associated with ultraviolet radiation–induced apoptosis[104] and to characterize an insecticidal sphingomyelinase C produced by *Bacillus cereus*.[105]

The Amplex® Red Sphingomyelinase Assay Kit contains:

- Amplex® Red reagent
- Dimethylsulfoxide (DMSO)
- Horseradish peroxidase (HRP)
- H_2O_2 for use as a positive control
- Concentrated reaction buffer
- Choline oxidase from *Alcaligenes* sp.
- Alkaline phosphatase from calf intestine
- Sphingomyelin
- Triton X-100
- Sphingomyelinase from *Bacillus* sp.
- Detailed protocols

Each kit provides sufficient reagents for approximately 500 assays using a fluorescence microplate reader and a reaction volume of 200 µL per assay.

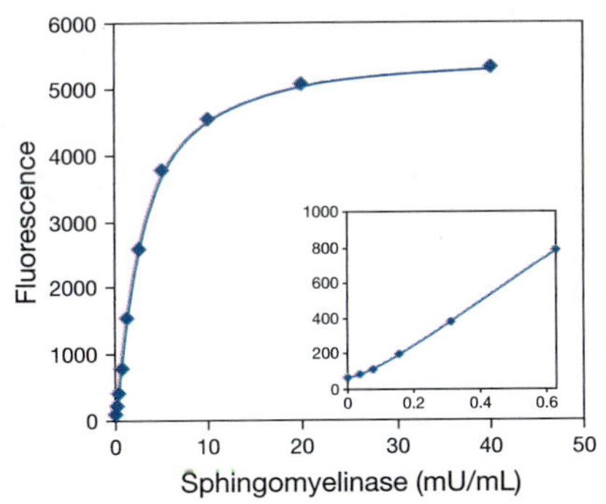

Figure 13.3.12 Measurement of sphingomyelinase activity using the Amplex® Red Sphingomyelinase Assay Kit (A12220). Each reaction contained 50 µM Amplex® Red reagent, 1 U/mL horseradish peroxidase (HRP), 0.1 U/mL choline oxidase, 4 U/mL of alkaline phosphatase, 0.25 mM sphingomyelin and the indicated amount of *Staphylococcus aureus* sphingomyelinase in 1X reaction buffer. Reactions were incubated at 37°C for one hour. Fluorescence was measured with a fluorescence microplate reader using excitation at 530 ± 12.5 nm and fluorescence detection at 590 ± 17.5 nm.

Steroids

Most steroids are neutral lipids and, as such, localize primarily within the cell's membranes, in lipid vacuoles and bound to certain lipoproteins. Fluorescent analogs of these biomolecules, most of which are derived from BODIPY® and NBD dyes, are highly lipophilic probes. One application of these probes is to detect enzymatic activity—either *in vitro* or *in vivo*—through hydrolysis of the fatty acid esters to fluorescent fatty acids.[106] Although the substrates and products in these enzyme assays typically have similar fluorescence properties, they are readily extracted by an organic solvent and separated by chromatography.

We have also developed sensitive fluorometric assays for cholesterol, cholesteryl esters and enzymes that metabolize natural cholesterol derivatives; the assay reagents and protocols are available in our Amplex® Red Cholesterol Assay Kit (A12216) described below. A review of the cellular organization, functions and transport of cholesterol has recently been published.[107]

BODIPY® Cholesteryl Esters

Cholesteryl esters consist of a fatty acid esterified to the 3β-hydroxyl group of cholesterol (Figure 13.3.13). These very nonpolar species are the predominant lipid components of atherosclerotic plaque and low- and high-density lipoprotein (LDL and HDL) cores. We offer cholesteryl esters of three of our BODIPY® fatty acids—BODIPY® FL C_{12} (C3927MP), BODIPY® 542/563 C_{11} (C12680) and BODIPY® 576/589 C_{11} (C12681)—all of which have long-wavelength visible emission. BODIPY®

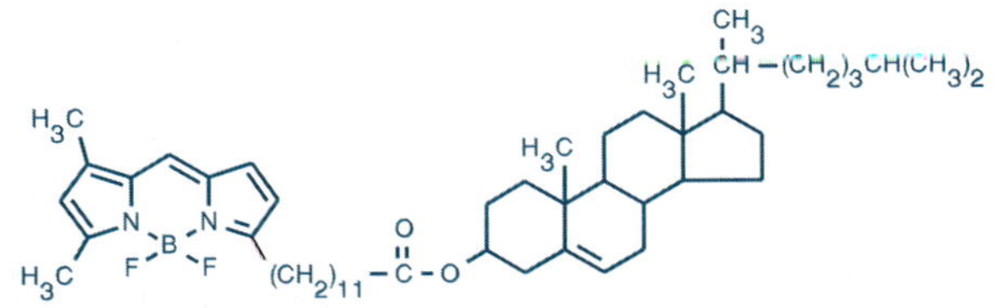

Figure 13.3.13 Cholesteryl BODIPY® FL C_{12} (cholesteryl 4,4-difluoro-5,7-dimethyl-4-bora-3a,4a-diaza-*s*-indacene-3-dodecanoate; C3927MP).

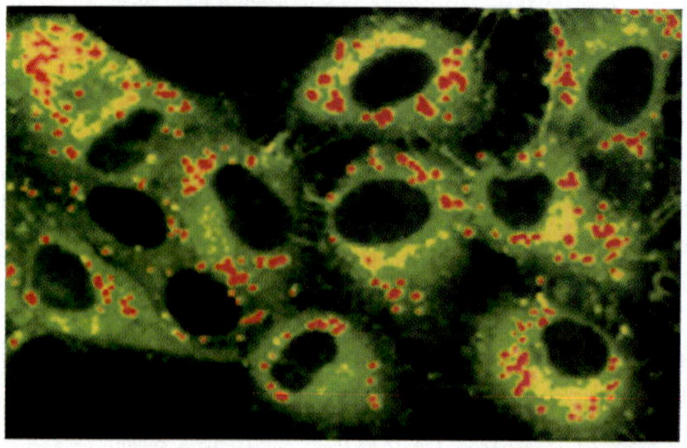

Figure 13.3.14 Selective uptake of cholesteryl esters (CE) in rat ovarian granulosa cells as monitored with cholesteryl BODIPY® FL C_{12} (C3927MP). The hormone-stimulated cells internalized and stored CEs derived from reconstituted high-density lipoprotein (HDL)–BODIPY® CE complexes (J Biol Chem (1996) 271:16208). A low-light (<100 µW beam power) computerized imaging system minimized any photobleaching of the fluorophore. This pseudocolored image uses yellow-green to illustrate the low-level fluorescence of the cytoplasmic membranes, yellow to illustrate the medium-level fluorescence of the Golgi, and red to illustrate the high-level fluorescence of the lipid droplets. Image contributed by Eve Reaven, VA Medical Center, Palo Alto, California.

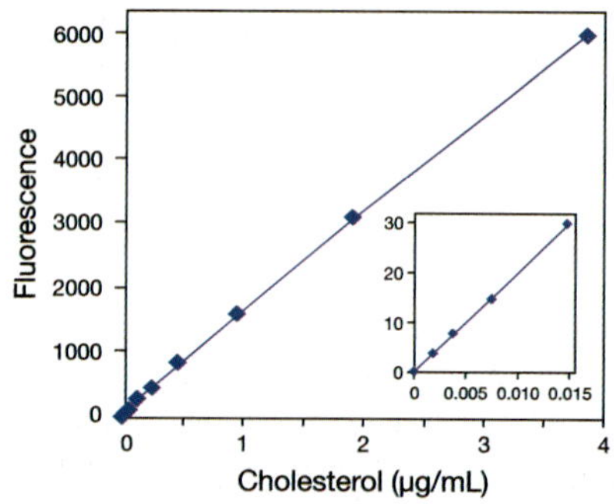

Figure 13.3.15 Detection of cholesterol using the Amplex® Red Cholesterol Assay Kit (A12220). Each reaction contained 150 µM Amplex® Red reagent, 1 U/mL horseradish peroxidase (HRP), 1 U/mL cholesterol oxidase, 1 U/mL cholesterol esterase and the indicated amount of cholesterol in 1X reaction buffer. Reactions were incubated at 37°C for 30 minutes. Fluorescence was measured with a fluorescence microplate reader using excitation at 560 ± 10 nm and fluorescence detection at 590 ± 10 nm. The insert above shows the high sensitivity and excellent linearity of the assay at low cholesterol levels (0–10 ng/mL).

FL cholesteryl ester can be used as a tracer of cholesterol transport and receptor-mediated endocytosis of lipoproteins by fluorescence microscopy[108–110] (Figure 13.3.14) and as a general nonexchangeable membrane marker. Addition of methyl β-cyclodextrin to BODIPY® FL cholesteryl ester is reported to facilitate its uptake by cells and tissues.[111] Researchers have extensively used BODIPY® FL cholesteryl ester to measure cholesteryl ester–transfer protein (CETP) activity using fluorescence microplate readers.[112–115] The longer-wavelength BODIPY® 542/563 and BODIPY® 576/589 cholesteryl esters likely have similar applications.

Side Chain–Modified Cholesterol Analog

We offer an NBD-labeled cholesterol analog in which the fluorophore replaces the terminal segment of cholesterol's flexible alkyl tail. The environment-sensitive NBD fluorophore of the NBD cholesterol analog (N1148) localizes in the membrane's interior, unlike the anomalous positioning of NBD-labeled phospholipid acyl chains.[116] As with other NBD lipid analogs, this probe is useful for investigating lipid transport processes[117,118] and lipid–protein interactions.[119,120] NBD cholesterol is selectively taken up by high-density lipoproteins via the scavenger receptor B1.[117] A lipid droplet–specific protein binds unesterified NBD cholesterol with extremely high affinity[117] (K_d = 2 nM).

Amplex® Red Cholesterol Assay Kit

The Amplex® Red Cholesterol Assay Kit (A12216) provides an exceptionally sensitive assay for both cholesterol and cholesteryl esters in complex mixtures and is suitable for use with either fluorescence microplate readers or fluorometers. The assay provided in this kit is designed to detect as little as 5 ng/mL (5×10^{-4} mg/dL) cholesterol (Figure 13.3.15) and to accurately measure the cholesterol or cholesteryl ester content in the equivalent of 0.01 µL of human serum.[121] The assay uses an enzyme-coupled reaction scheme in which cholesteryl esters are hydrolyzed by cholesterol esterase into cholesterol, which is then oxidized by cholesterol oxidase to yield H_2O_2 and the corresponding ketone steroidal product (Figure 13.3.16). The H_2O_2 is then detected using the Amplex® Red reagent in combination with horseradish peroxidase (HRP).

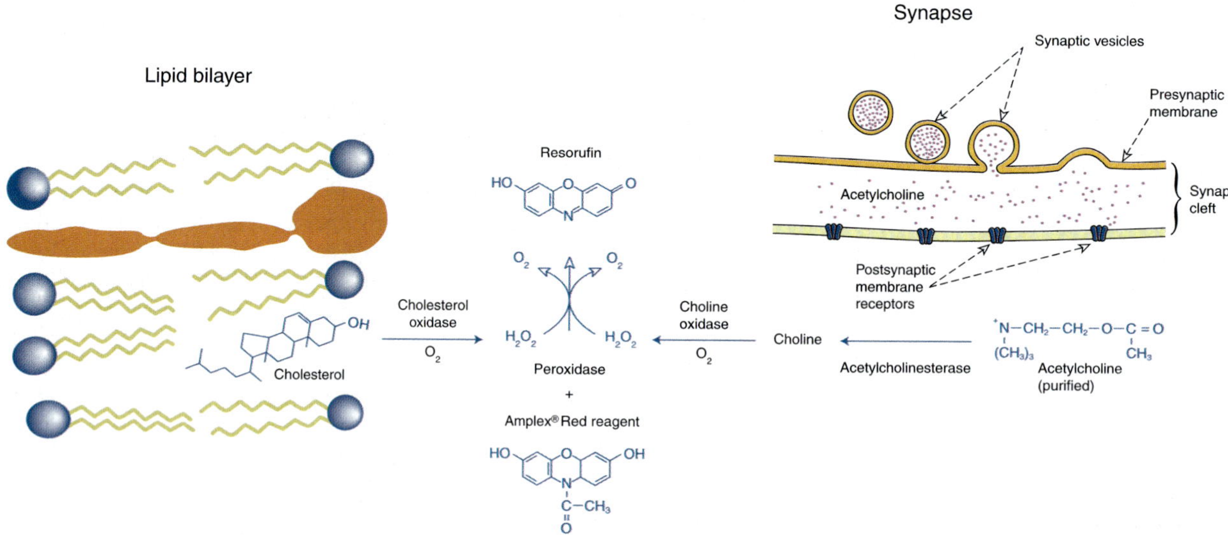

Figure 13.3.16 Enzyme-coupled Amplex® Red assays. Enzyme reactions that produce H_2O_2 can be made into Amplex® Red assays. The Amplex® Red Cholesterol Assay Kit (A12216) uses cholesterol oxidase to produce H_2O_2, which is then detected by the Amplex® Red reagent in the presence of horseradish peroxidase (HRP). Similarly, the Amplex® Red Acetylcholine/Acetylcholinesterase Assay Kit (A12217) uses choline oxidase to produce H_2O_2.

IMPORTANT NOTICE: The products described in this manual are covered by one or more Limited Use Label License(s). Please refer to the Appendix on page 971 and Master Product List on page 975. Products are For Research Use Only. Not intended for any animal or human therapeutic or diagnostic use.

The Amplex® Red cholesterol assay is continuous and requires no separation or wash steps. These characteristics make the assay particularly well suited for the rapid and direct analysis of cholesterol in blood and food samples using automated instruments. By performing two separate measurements in the presence and absence of cholesterol esterase, this assay is also potentially useful for determining the fraction of cholesterol that is in the form of cholesteryl esters within a sample. In addition, by adding an excess of cholesterol to the reaction, this assay can be used to sensitively detect the activity of cholesterol oxidase. The Amplex® Red Cholesterol Assay Kit contains:

- Amplex® Red reagent
- Dimethylsulfoxide (DMSO)
- Horseradish peroxidase (HRP)
- H_2O_2 for use as a positive control
- Concentrated reaction buffer
- Cholesterol oxidase from *Streptomyces*
- Cholesterol esterase from *Pseudomonas*
- Cholesterol for preparation of a standard curve
- Detailed protocols

Each kit provides sufficient reagents for approximately 500 assays using a fluorescence microplate reader and a reaction volume of 100 µL per assay.

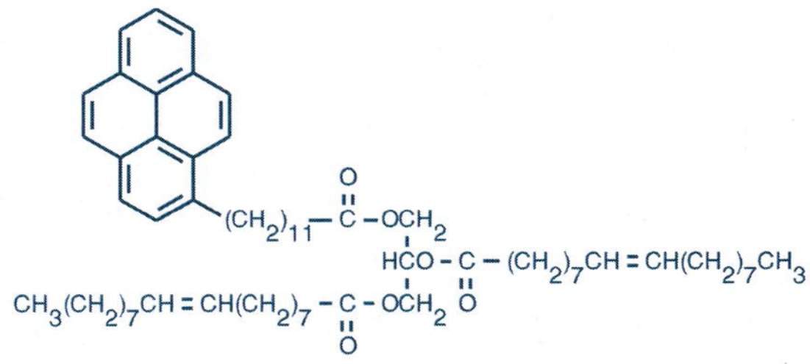

Figure 13.3.17 1,2-Dioleoyl-3-(1-pyrenedodecanoyl)-*rac*-glycerol (D6562).

Fluorescent Triacylglycerol

The fluorescent triacylglycerol 1,2-dioleoyl-3-(1-pyrenedodecanoyl)-*rac*-glycerol (D6562) has a pyrene fatty acid ester replacing one of the three fatty acyl residues of a natural triacylglycerol (Figure 13.3.17). Pyrene has the important spectral property of forming excimers (Figure 13.3.18) when two fluorophores are in close proximity during the excited state. Pyrene triacylglycerols are useful for measuring cholesteryl ester transfer protein–mediated triacylglycerol transport between plasma lipoproteins.[122] They are also excellent substrates for lipoprotein lipase and hepatic triacylglycerol lipase.[123]

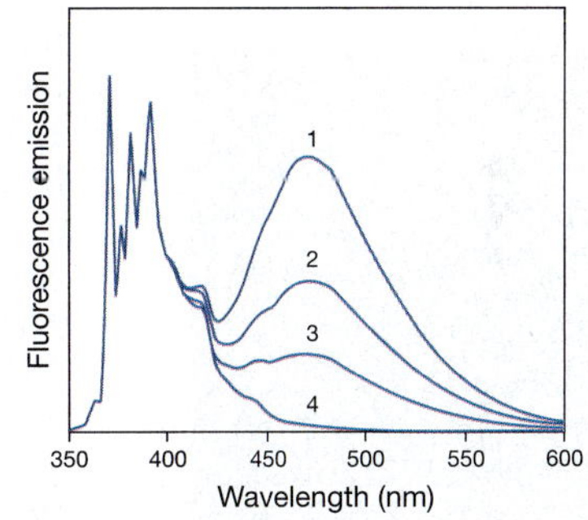

Figure 13.3.18 Excimer formation by pyrene in ethanol. Spectra are normalized to the 371.5 nm peak of the monomer. All spectra are essentially identical below 400 nm after normalization. Spectra are as follows: 1) 2 mM pyrene, purged with argon to remove oxygen; 2) 2 mM pyrene, air-equilibrated; 3) 0.5 mM pyrene (argon-purged); and 4) 2 µM pyrene (argon-purged). The monomer-to-excimer ratio (371.5 nm/470 nm) is dependent on both pyrene concentration and the excited-state lifetime, which is variable because of quenching by oxygen.

Lipopolysaccharides

Fluorescent Lipopolysaccharides

We offer fluorescent conjugates of lipopolysaccharides (LPS) from *Escherichia coli* and *Salmonella minnesota* (Section 16.1, Table 16.1), including:

- Alexa Fluor® 488 LPS from *E. coli* serotype 055:B5 (L23351)
- Alexa Fluor® 488 LPS from *S. minnesota* (L23356)
- Alexa Fluor® 568 LPS from *E. coli* serotype 055:B5 (L23352)
- Alexa Fluor® 594 LPS from *E. coli* serotype 055:B5 (L23353)
- BODIPY® FL LPS from *E. coli* serotype 055:B5 (L23350)

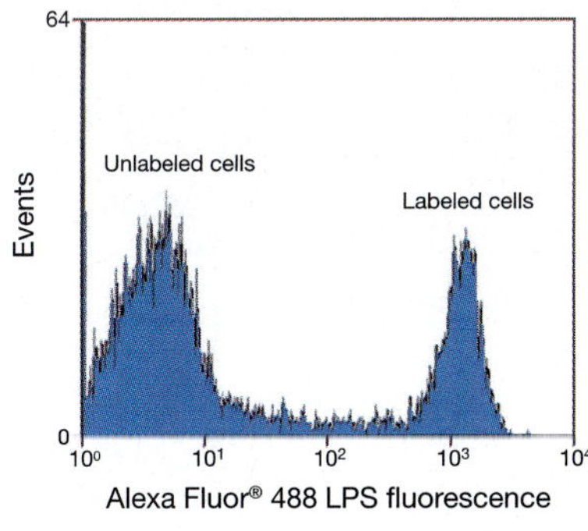

Figure 13.3.19 Flow cytometry analysis of blood using an Alexa Fluor® 488 lipopolysaccharide (LPS). Human blood was incubated with Alexa Fluor® 488 LPS from *Escherichia coli* (L23351) and anti-CD14 antibody on ice for 20 minutes. The red blood cells were lysed and the sample was analyzed on a flow cytometer equipped with a 488 nm Ar-Kr excitation source and a 525 ± 12 nm bandpass emission filter. Monocytes were identified based on their light scatter and CD14 expression.

LPS, also known as endotoxins, are a family of complex glycolipid molecules located on the surface of gram-negative bacteria. LPS play a large role in protecting the bacterium from host defense mechanisms and antibiotics. Binding of LPS to the CD14 cell-surface receptor of phagocytes (Figure 13.3.19) is the key initiation step in the mammalian immune response to infection by gram-negative bacteria. The structural

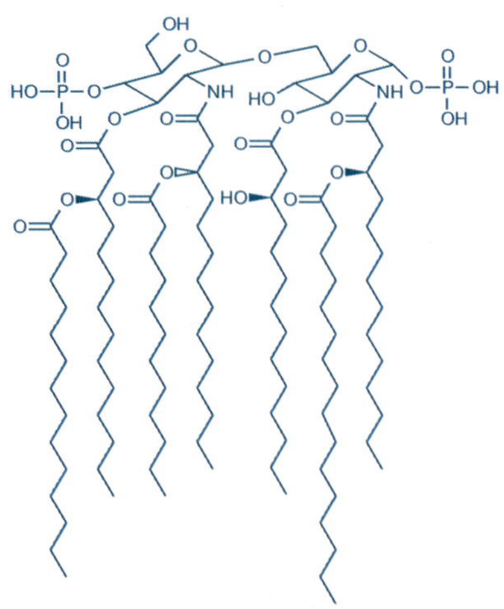

Figure 13.3.20 Structure of the lipid A component of *Salmonella minnesota* lipopolysaccharide.

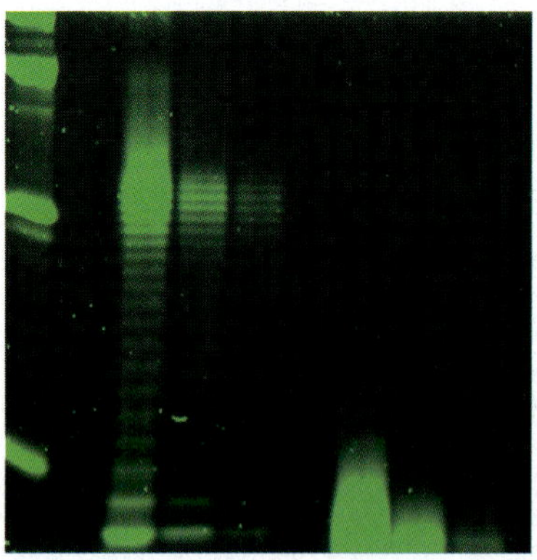

Figure 13.3.21 Lipopolysaccharide staining with the Pro-Q® Emerald 300 Lipopolysaccharide Gel Stain Kit. Lipopolysaccharides (LPS) were electrophoresed through a 13% acrylamide gel and stained using the Pro-Q® Emerald 300 Lipopolysaccharide Gel Stain Kit (P20495). From left to right, the lanes contain: CandyCane™ glycoprotein molecular weight standards (~250 ng/band), blank, 4, 1 and 0.25 µg of LPS from *Escherichia coli* smooth serotype 055:B5 and 4, 1 and 0.25 µg of LPS from *E. coli* rough mutant EH100 (Ra mutant).

core of LPS, and the primary determinant of its biological activity, is an *N*-acetylglucosamine derivative, lipid A (Figure 13.3.20). Two plasma proteins, LPS-binding protein (LBP) and soluble CD14 (sCD14), play primary roles in transporting LPS and mediating cellular responses.[124–129] If the fatty acid residues are removed from the lipid A component, the toxicity of the LPS can be reduced significantly; however, the mono- or diphosphoryl forms of lipid A are inherently toxic. In many gram-negative bacterial infections, LPS are responsible for clinically significant symptoms like fever, low blood pressure and tissue edema, which can lead to disseminated intravascular coagulation, organ failure and death. Studies also clearly indicate that LPS induce various signal transduction pathways, including those involving protein kinase C[130,131] and protein myristylation,[132] and stimulate a variety of immunochemical responses, including B lymphocyte[133] and G-protein activation.[134]

The fluorescent BODIPY® FL and Alexa Fluor® LPS conjugates, which are labeled with succinimidyl esters of these dyes, allow researchers to follow LPS binding, transport and cell internalization processes. Lipopolysaccharide internalization activates endotoxin-dependent signal transduction in cardiomyocytes.[135] The Alexa Fluor® 488 LPS conjugates (L23351, L23356) selectively label microglia in a mixed culture containing oligodendrocyte precursors, astrocytes and microglia.[136] A biologically active conjugate of galactose oxidase–oxidized *S. minnesota* LPS and our Alexa Fluor® 488 hydrazide (A10436, Section 3.3; A10440) has been used to elucidate molecular mechanisms of septic shock.[137]

The BODIPY® FL derivative of LPS from *E. coli* strain LCD25 (L23350) was used to measure the transfer rate of LPS from monocytes to high-density lipoprotein[138] (HDL). Another study utilized a BODIPY® FL derivative of LPS from *S. minnesota* to demonstrate transport to the Golgi apparatus in neutrophils,[124,125] although this could have been due to probe metabolism. It has been reported that organelles other than the Golgi are labeled by some fluorescent or nonfluorescent LPS.[139,140] Cationic lipids are reported to assist the translocation of fluorescent lipopolysaccharides into live cells;[141] cell surface–bound LPS can be quenched by trypan blue.[138] Molecular Probes® fluorescent LPS can potentially be combined with other fluorescent indicators, such as Ca^{2+}-, pH- or organelle-specific stains, for monitoring intracellular localization and real-time changes in cellular response to LPS.

Pro-Q® Emerald 300 Lipopolysaccharide Gel Stain Kit

The Pro-Q® Emerald 300 Lipopolysaccharide Gel Stain Kit (P20495) provides a simple, rapid and highly sensitive method for staining lipopolysaccharides (LPS) in gels (Figure 13.3.21, Figure 13.3.22, Figure 13.3.23). The structure of this important class of molecules can be analyzed by SDS-polyacrylamide gel electrophoresis, during which the heterogeneous mixture of polymers separates into a characteristic ladder pattern. This ladder has conventionally been detected using silver staining.[142–144] However, despite the long and complex procedures required, silver staining provides poor sensitivity and cannot differentiate LPS from proteins in the sample. An alternative

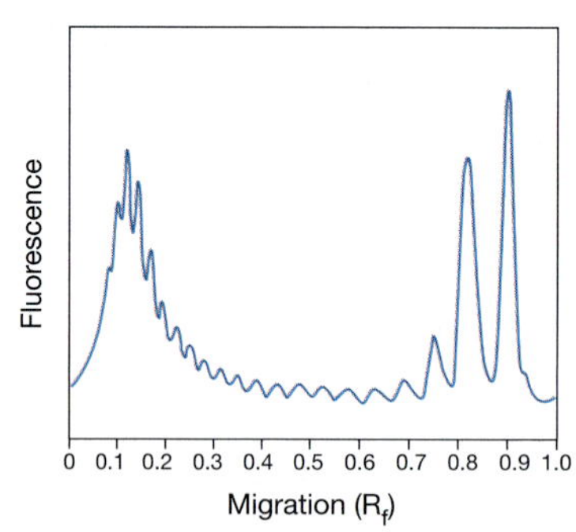

Figure 13.3.22 Characterization of lipopolysaccharides. Lipopolysaccharides (LPS) from *Escherichia coli* smooth serotype 055:B5 were loaded onto a 13% polyacrylamide gel. Following electrophoresis, the gel was stained using the Pro-Q® Emerald 300 Lipopolysaccharide Gel Stain Kit (P20495), and the fluorescence was measured for the lane. A plot of fluorescence signal versus the relative distance from the dye front shows a characteristic laddering profile for smooth-type LPS.

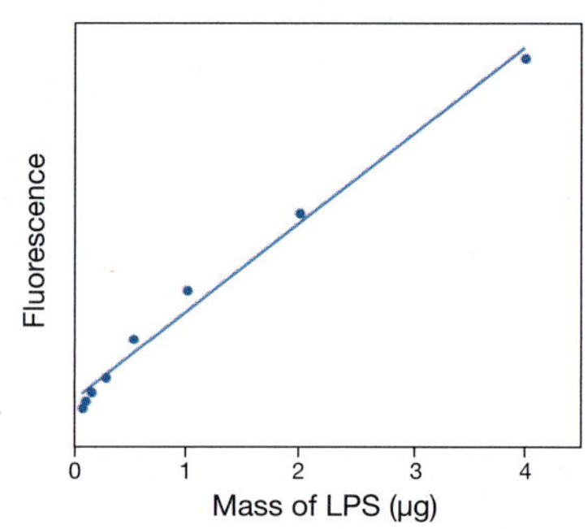

Figure 13.3.23 Linearity of the Pro-Q® Emerald 300 stain for lipopolysaccharide (LPS) detection. A dilution series of lipopolysaccharides from *Escherichia coli* smooth serotype 055:B5 was loaded onto a 13% polyacrylamide gel. Following electrophoresis, the gel was stained using the Pro-Q® Emerald 300 Lipopolysaccharide Gel Stain Kit (P20495) and the same band from each lane was quantitated using a CCD camera. A plot of the fluorescence intensity versus the mass of LPS loaded shows a linear range over two orders of magnitude.

IMPORTANT NOTICE: The products described in this manual are covered by one or more Limited Use Label License(s). Please refer to the Appendix on page 971 and Master Product List on page 975. Products are For Research Use Only. Not intended for any animal or human therapeutic or diagnostic use.

staining method that makes use of the reaction of the carbohydrates with detectable hydrazides obtains higher sensitivity, but requires blotting to a membrane and time- and labor-intensive procedures.[145–149]

By comparison, the staining technology used in the Pro-Q® Emerald 300 Lipopolysaccharide Gel Stain Kit vastly simplifies detection of LPS in SDS-polyacrylamide gels. The key to this novel methodology is our bright green-fluorescent Pro-Q® Emerald 300 dye, which covalently binds to periodate-oxidized carbohydrates of LPS. This dye allows the detection of as little as 200 pg of LPS in just a few hours using a simple UV transilluminator. The sensitivity is at least 50–100 times that of silver staining and requires much less hands-on time. This dye is also used in our Pro-Q® Emerald 300 and Multiplexed Proteomics® Glycoprotein Stain Kits (P21855, P21857, M33307; Section 9.4) and may be useful for detection of other molecules containing carbohydrates or aldehydes.

The Pro-Q® Emerald 300 Lipopolysaccharide Gel Stain Kit contains:

- Pro-Q® Emerald 300 reagent
- Pro-Q® Emerald 300 staining buffer
- Oxidizing reagent (periodic acid)
- Smooth LPS standard from *Escherichia coli* serotype 055-B5
- Detailed protocols

Sufficient materials are supplied to stain ten 8 cm × 10 cm gels, 0.5–0.75 mm thick.

REFERENCES

1. Annu Rev Biochem (1998) 67:27; **2.** Prog Lipid Res (1997) 36:153; **3.** Biochim Biophys Acta (1998) 1436:233; **4.** Blood (1999) 93:80; **5.** Biochim Biophys Acta (1984) 793:169; **6.** Clin Chim Acta (1982) 124:123; **7.** Clin Chim Acta (1984) 142:313; **8.** J Cell Biol (1990) 111:429; **9.** Methods Cell Biol (1993) 38:221; **10.** J Biol Chem (1993) 268:17762; **11.** Cell Signal (1998) 10:685; **12.** Trends Biochem Sci (1995) 20:73; **13.** Curr Opin Oncol (1998) 10:552; **14.** J Inherit Metab Dis (1998) 21:472; **15.** Science (1993) 259:1769; **16.** Ann N Y Acad Sci (1998) 845:139; **17.** Ann N Y Acad Sci (1998) 845:57; **18.** J Neurochem (1998) 70:1335; **19.** J Cell Biol (2003) 162:365; **20.** J Lipid Res (2003) 44:655; **21.** Eur J Biochem (2002) 269:737; **22.** Science (2000) 290:1721; **23.** Mol Membr Biol (1999) 16:145; **24.** Trends Cell Biol (1999) 9:87; **25.** Annu Rev Cell Dev Biol (1998) 14:111; **26.** Biochemistry (2001) 40:4893; **27.** Trends Cell Biol (2000) 10:408; **28.** J Biol Chem (1994) 269:3125; **29.** Chem Phys Lipids (1999) 102:55; **30.** Ann N Y Acad Sci (1998) 845:152; **31.** Biophys J (1997) 72:37; **32.** J Cell Biol (1998) 140:39; **33.** Histochem Cell Biol (2008) 130:819; **34.** Methods Enzymol (2000) 312:293; **35.** Methods Enzymol (2000) 312:523; **36.** Methods (2005) 36:186; **37.** J Cell Biol (1996) 134:1031; **38.** J Cell Biol (1991) 113:1267; **39.** Cytometry (1993) 14:251; **40.** J Cell Biol (1993) 120:399; **41.** Eur J Cell Biol (1992) 58:214; **42.** Mol Biochem Parasitol (2000) 106:21; **43.** Infect Immun (2000) 68:5960; **44.** J Cell Biol (2001) 154:535; **45.** Am J Physiol Lung Cell Mol Physiol (2001) 280:L938; **46.** Nat Cell Biol (1999) 1:386; **47.** J Neurochem (1999) 73:1375; **48.** Lancet (1999) 354:901; **49.** Proc Natl Acad Sci U S A (1998) 95:6373; **50.** J Cell Biol (1999) 147:447; **51.** J Cell Biol (2002) 157:327; **52.** Biophys J (1999) 77:1498; **53.** J Cell Sci (2006)119:2084; **54.** Mol Biol Cell (1995) 6:135; **55.** J Biol Chem (1993) 268:4577; **56.** Biochemistry (1992) 31:3581; **57.** J Biol Chem (1993) 268:14476; **58.** J Cell Biol (1994) 125:769; **59.** Biochim Biophys Acta (1992) 1113:277; **60.** Brain Res (1992) 597:108; **61.** Anal Biochem (2001) 293:204; **62.** Biochim Biophys Acta (1999) 1455:85; **63.** Traffic (2000) 1:807; **64.** J Biol Chem (1993) 268:14861; **65.** Biochim Biophys Acta (1987) 915:87; **66.** Biochem Biophys Res Comm (1965) 18:221; **67.** Anal Biochem (1984) 136:223; **68.** J Biol Chem (2001) 276:24985; **69.** Adv Cell Mol Biol Membranes (1993) 1:199; **70.** Biochim Biophys Acta (1991) 1082:113; **71.** Parasitology (1992) 105:81; **72.** J Cell Biol (1989) 108:2169; **73.** Proc Natl Acad Sci U S A (1993) 90:2661; **74.** Proc Natl Acad Sci U S A (2003) 100:5813; **75.** J Immunol (2003) 170:1329; **76.** J Membr Biol (2002) 189:35; **77.** Proc Natl Acad Sci U S A (2001) 98:9098; **78.** Mol Biol Cell (1999) 10:3187; **79.** Biochim Biophys Acta (2003) 1610:247; **80.** Annu Rev Immunol (2003) 21:457; **81.** Mol Immunol (2002) 38:1247; **82.** Nat Rev Immunol (2002) 2:96; **83.** Biol Res (2002) 35:127; **84.** Nat Rev Mol Cell Biol (2000) 1:31; **85.** J Exp Med (1999) 190:1549; **86.** J Cell Biol (1998) 143:637; **87.** Immunity (2003) 18:655; **88.** J Biol Chem (2002) 277:39541; **89.** Biochem Biophys Res Commun (2002) 297:876; **90.** Biol Chem (2002) 383:1475; **91.** J Cell Biol (2001) 153:529; **92.** J Cell Sci (2001) 114:3957; **93.** J Virol (2003) 77:9542; **94.** Exp Cell Res (2003) 287:67; **95.** Traffic (2002) 3:705; **96.** J Clin Virol (2001) 22:217; **97.** Curr Biol (2000) 10:R823; **98.** J Virol (2000) 74:3264; **99.** Biochemistry (1996) 35:16069; **100.** Mol Microbiol (1994) 13:745; **101.** J Cell Biol (1998) 141:929; **102.** J Biol Chem (1994) 269:30745; **103.** Am J Pathol (2002) 161:1061; **104.** J Biol Chem (2001) 276:11775; **105.** Eur J Biochem (2004) 271:601; **106.** J Lipid Res (1995) 36:1602; **107.** Nat Rev Mol Cell Biol (2008) 9:125; **108.** Proc Natl Acad Sci U S A (2001) 98:1613; **109.** J Biol Chem (1997) 272:25283; **110.** PLoS ONE (2007) 2:e511; **111.** Am J Physiol (1999) 277:G1017; **112.** Biochemistry (1995) 34:12560; **113.** Chem Phys Lipids (1995) 77:51; **114.** Lipids (1994) 29:811; **115.** J Lipid Res (1993) 34:1625; **116.** J Phys Chem (1999) 103:8180; **117.** J Biol Chem (2000) 275:12769; **118.** J Lipid Res (1999) 40:1747; **119.** Biochim Biophys Acta (1999) 1437:37; **120.** J Biol Chem (1999) 274:35425; **121.** J Biochem Biophys Methods (1999) 38:43; **122.** J Biochem (Tokyo) (1998) 124:237; **123.** Lipids (1988) 23:605; **124.** J Exp Med (1999) 190:523; **125.** J Exp Med (1999) 190:509; **126.** J Biol Chem (1996) 271:4100; **127.** J Exp Med (1995) 181:1743; **128.** J Exp Med (1994) 180:1025; **129.** J Exp Med (1994) 179:269; **130.** J Biol Chem (1984) 259:10048; **131.** J Exp Med (1996) 183:1899; **132.** Proc Natl Acad Sci U S A (1986) 83:5817; **133.** Adv Immunol (1979) 28:293; **134.** Eur J Immunol (1989) 19:125; **135.** Circ Res (2001) 88:491; **136.** J Neurosci (2002) 22:2478; **137.** Cytometry (2000) 41:316; **138.** J Biol Chem (1999) 274:34116; **139.** Electron Microsc Rev (1992) 5:381; **140.** J Periodontol (1985) 56:553; **141.** Biotechniques (2000) 28:510; **142.** J Clin Microbiol (1990) 28:2627; **143.** Microbiol Immunol (1991) 35:331; **144.** J Biochem Biophys Methods (1993) 26:81; **145.** Electrophoresis (1998) 19:2398; **146.** Appl Environ Microbiol (1995) 61:2845; **147.** Electrophoresis (1999) 20:462; **148.** Electrophoresis (2000) 21:526; **149.** Anal Biochem (1990) 188:285.

DATA TABLE 13.3 SPHINGOLIPIDS, STEROIDS, LIPOPOLYSACCHARIDES AND RELATED PROBES

Cat. No.	MW	Storage	Soluble	Abs	EC	Em	Solvent	Notes
B13950	1582.50	F,D,L	DMSO, EtOH	505	80,000	512	MeOH	1
B22650	~66,000	F,D,L	H_2O	505	91,000	511	MeOH	1, 2
B34400	~66,000	F,D,L	H_2O	589	65,000	616	MeOH	2
B34401	~66,000	F,D,L	H_2O	505	80,000	512	MeOH	1, 2
B34402	~66,000	F,D,L	H_2O	505	80,000	511	MeOH	1, 2
C3927MP	786.98	F,D,L	$CHCl_3$	505	86,000	511	MeOH	3
C12680	851.02	F,D,L	$CHCl_3$	543	57,000	563	MeOH	3
C12681	809.97	F,D,L	$CHCl_3$	579	98,000	590	MeOH	3
D3521	601.63	FF,D,L	$CHCl_3$, DMSO	505	91,000	511	MeOH	1
D3522	766.75	FF,D,L	see Notes	505	77,000	512	MeOH	1, 4
D6562	1003.54	FF,D,L,A	$CHCl_3$	341	40,000	376	MeOH	5, 6
D7519	861.96	FF,D,L	DMSO, EtOH	505	85,000	511	MeOH	1
D7540	705.71	FF,D,L	$CHCl_3$, DMSO	589	65,000	616	MeOH	

continued on next page

IMPORTANT NOTICE: The products described in this manual are covered by one or more Limited Use Label License(s). Please refer to the Appendix on page 971 and Master Product List on page 975. Products are For Research Use Only. Not intended for any animal or human therapeutic or diagnostic use.

DATA TABLE 13.3 SPHINGOLIPIDS, STEROIDS, LIPOPOLYSACCHARIDES AND RELATED PROBES—*continued*

Cat. No.	MW	Storage	Soluble	Abs	EC	Em	Solvent	Notes
D7711	864.94	FF,D,L	DMSO	505	75,000	513	MeOH	1, 7
D13951	925.91	FF,D,L	DMSO, EtOH	505	80,000	511	MeOH	1
N1148	494.63	L	$CHCl_3$, MeCN	469	21,000	537	MeOH	8
N1154	575.75	FF,D,L	$CHCl_3$, DMSO	466	22,000	536	MeOH	8
N3524	740.88	FF,D,L	see Notes	466	22,000	536	MeOH	4, 8
N22651	~66,000	F,D,L	H_2O	466	22,000	536	MeOH	2, 8

For definitions of the contents of this data table, see "Using *The Molecular Probes® Handbook*" in the introductory pages.

Notes

1. Em for BODIPY® FL sphingolipids shifts to ~620 nm when high concentrations of the probe (>5 mol %) are incorporated in lipid mixtures. (J Cell Biol (1991) 113:1267)
2. This product is a lipid complexed with bovine serum albumin (BSA). Spectroscopic data are for the free lipid in MeOH.
3. The absorption and fluorescence spectra of BODIPY® derivatives are relatively insensitive to the solvent.
4. Chloroform is the most generally useful solvent for preparing stock solutions of phospholipids (including sphingomyelins). Glycerophosphocholines are usually freely soluble in ethanol. Most other glycerophospholipids (phosphoethanolamines, phosphatidic acids and phosphoglycerols) are less soluble in ethanol, but solutions up to 1–2 mg/mL should be obtainable, using sonication to aid dispersion if necessary. Labeling of cells with fluorescent phospholipids can be enhanced by addition of cyclodextrins during incubation. (J Biol Chem (1999) 274:35359)
5. Alkylpyrene fluorescence lifetimes are up to 110 nanoseconds and are very sensitive to oxygen.
6. Pyrene derivatives exhibit structured spectra. The absorption maximum is usually about 340 nm with a subsidiary peak at about 325 nm. There are also strong absorption peaks below 300 nm. The emission maximum is usually about 376 nm with a subsidiary peak at 396 nm. Excimer emission at about 470 nm may be observed at high concentrations.
7. This product is supplied as a ready-made solution in the solvent indicated under "Soluble."
8. Fluorescence of NBD and its derivatives in water is relatively weak. QY and τ increase and Em decreases in aprotic solvents and other nonpolar environments relative to water. (Biochemistry (1977) 16:5150, Photochem Photobiol (1991) 54:361)

PRODUCT LIST 13.3 SPHINGOLIPIDS, STEROIDS, LIPOPOLYSACCHARIDES AND RELATED PROBES

Cat. No.	Product	Quantity
A12216	Amplex® Red Cholesterol Assay Kit *500 assays*	1 kit
A12220	Amplex® Red Sphingomyelinase Assay Kit *500 assays*	1 kit
B22650	BODIPY® FL C_5-ceramide complexed to BSA	5 mg
B13950	BODIPY® FL C_5-ganglioside GM1	25 µg
B34401	BODIPY® FL C_5-ganglioside GM1 complexed to BSA	1 mg
B34402	BODIPY® FL C_5-lactosylceramide complexed to BSA	1 mg
D7540	BODIPY® TR ceramide (*N*-((4-(4,4-difluoro-5-(2-thienyl)-4-bora-3a,4a-diaza-*s*-indacene-3-yl)phenoxy)acetyl)sphingosine)	250 µg
B34400	BODIPY® TR ceramide complexed to BSA	5 mg
C12680	cholesteryl BODIPY® 542/563 C_{11} (cholesteryl 4,4-difluoro-5-(4-methoxyphenyl)-4-bora-3a,4a-diaza-*s*-indacene-3-undecanoate)	1 mg
C12681	cholesteryl BODIPY® 576/589 C_{11} (cholesteryl 4,4-difluoro-5-(2-pyrrolyl)-4-bora-3a,4a-diaza-*s*-indacene-3-undecanoate)	1 mg
C3927MP	cholesteryl BODIPY® FL C_{12} (cholesteryl 4,4-difluoro-5,7-dimethyl-4-bora-3a,4a-diaza-*s*-indacene-3-dodecanoate)	1 mg
D7519	*N*-(4,4-difluoro-5,7-dimethyl-4-bora-3a,4a-diaza-*s*-indacene-3-dodecanoyl)sphingosyl 1-β-D-galactopyranoside (BODIPY® FL C_{12}-galactocerebroside)	25 µg
D7711	*N*-(4,4-difluoro-5,7-dimethyl-4-bora-3a,4a-diaza-*s*-indacene-3-dodecanoyl)sphingosyl phosphocholine (BODIPY® FL C_{12}-sphingomyelin) *1 mg/mL in DMSO*	250 µL
D3521	*N*-(4,4-difluoro-5,7-dimethyl-4-bora-3a,4a-diaza-*s*-indacene-3-pentanoyl)sphingosine (BODIPY® FL C_5-ceramide)	250 µg
D13951	*N*-(4,4-difluoro-5,7-dimethyl-4-bora-3a,4a-diaza-*s*-indacene-3-pentanoyl)sphingosyl 1-β-D-lactoside (BODIPY® FL C_5-lactosylceramide)	25 µg
D3522	*N*-(4,4-difluoro-5,7-dimethyl-4-bora-3a,4a-diaza-*s*-indacene-3-pentanoyl)sphingosyl phosphocholine (BODIPY® FL C_5-sphingomyelin)	250 µg
D6562	1,2-dioleoyl-3-(1-pyrenedodecanoyl)-*rac*-glycerol	1 mg
L23351	lipopolysaccharides from *Escherichia coli* serotype 055:B5, Alexa Fluor® 488 conjugate	100 µg
L23352	lipopolysaccharides from *Escherichia coli* serotype 055:B5, Alexa Fluor® 568 conjugate	100 µg
L23353	lipopolysaccharides from *Escherichia coli* serotype 055:B5, Alexa Fluor® 594 conjugate	100 µg
L23350	lipopolysaccharides from *Escherichia coli* serotype 055:B5, BODIPY® FL conjugate	100 µg
L23356	lipopolysaccharides from *Salmonella minnesota*, Alexa Fluor® 488 conjugate	100 µg
N1154	NBD C_6-ceramide (6-((*N*-(7-nitrobenz-2-oxa-1,3-diazol-4-yl)amino)hexanoyl)sphingosine)	1 mg
N22651	NBD C_6-ceramide complexed to BSA	5 mg
N3524	NBD C_6-sphingomyelin (6-((*N*-(7-nitrobenz-2-oxa-1,3-diazol-4-yl)amino)hexanoyl)sphingosyl phosphocholine)	1 mg
N1148	NBD cholesterol (22-(*N*-(7-nitrobenz-2-oxa-1,3-diazol-4-yl)amino)-23,24-bisnor-5-cholen-3β-ol)	10 mg
P20495	Pro-Q® Emerald 300 Lipopolysaccharide Gel Stain Kit *10 minigels*	1 kit
V34403	Vybrant® Alexa Fluor® 488 Lipid Raft Labeling Kit *50 labelings*	1 kit
V34404	Vybrant® Alexa Fluor® 555 Lipid Raft Labeling Kit *50 labelings*	1 kit
V34405	Vybrant® Alexa Fluor® 594 Lipid Raft Labeling Kit *50 labelings*	1 kit

IMPORTANT NOTICE: The products described in this manual are covered by one or more Limited Use Label License(s). Please refer to the Appendix on page 971 and Master Product List on page 975. Products are For Research Use Only. Not intended for any animal or human therapeutic or diagnostic use.

13.4 Dialkylcarbocyanine and Dialkylaminostyryl Probes

The dyes in this section are all amphiphilic probes—molecules that comprise a charged fluorophore that localizes the probe at the membrane's surface and lipophilic aliphatic "tails" that insert into the membrane and thus anchor the probe to the membrane. In addition to labeling model membranes, most of these probes are very useful for cell tracing applications (Section 14.4). Table 14.3 lists all of our lipophilic carbocyanine and aminostyryl tracers and compares their properties and uses. Our FM® dyes, which are also amphiphilic styryl dyes but with less lipophilic character than the dyes in this section, are particularly useful for labeling membranes of live cells and for following synaptosome recycling (Section 16.1).

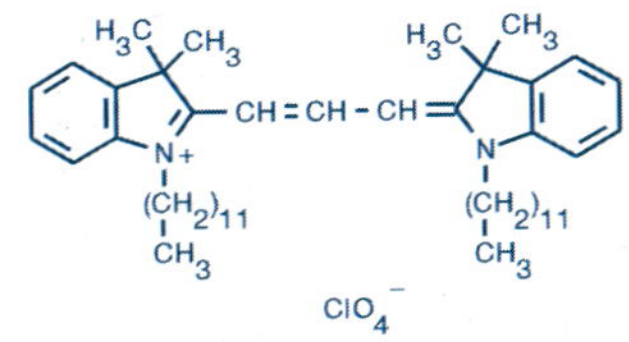

Figure 13.4.3 $DiIC_{12}(3)$ (1,1'-didodecyl-3,3,3',3'-tetramethylindocarbocyanine perchlorate; D383)

Dialkylcarbocyanine Probes

Carbocyanines are among the most strongly light-absorbing dyes known and have proven to be useful tools in several different areas of research. Carbocyanines with short alkyl tails attached to the imine nitrogens are employed both as membrane-potential sensors (Section 22.3) and as organelle stains for mitochondria and the endoplasmic reticulum (Section 12.2, Section 12.4). Those with longer alkyl tails (≥12 carbons) have an overall lipophilic character that makes them useful for neuronal tracing[1] and long-term labeling of cells in culture[2,3] (Section 14.4), as well as for noncovalent labeling of lipoproteins (Section 16.1). This section describes the use and properties of dialkylcarbocyanines as general-purpose probes of membrane structure and dynamics.

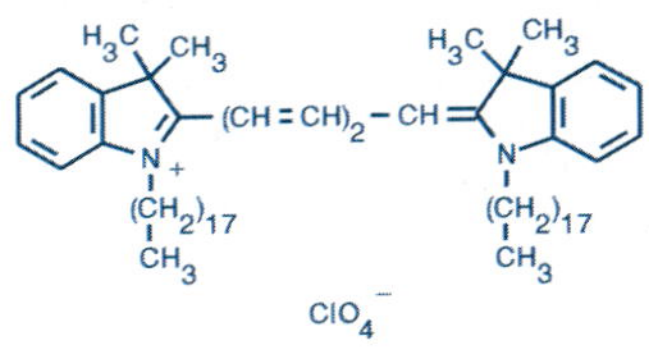

Figure 13.4.4 'DiD'; $DiIC_{18}(5)$ (1,1'-dioctadecyl-3,3,3',3'-tetramethylindodicarbocyanine perchlorate; D307).

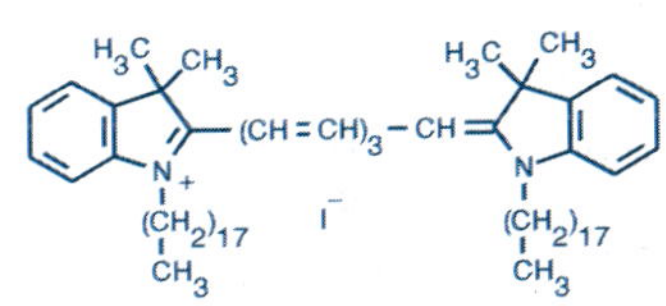

Figure 13.4.5 'DiR'; $DiIC_{18}(7)$ (1,1'-dioctadecyl-3,3,3',3'-tetramethylindotricarbocyanine iodide; D12731)

DiI, DiO, DiD, DiR and Analogs

The most widely used carbocyanine membrane probes have been the octadecyl (C_{18}) indocarbocyanines (D282, D3911; Figure 13.4.1) and oxacarbocyanines (D275, Figure 13.4.2) often referred to by the generic acronyms DiI and DiO, or more specifically as $DiIC_{18}(3)$ and $DiOC_{18}(3)$, where the subscript is the number of carbon atoms in each alkyl tail and the bracketed numeral is the number of carbon atoms in the bridge between the indoline or benzoxazole ring systems. We also offer several variations on these basic structures (Section 14.4, Table 14.3):

- DiI and DiO analogs with unsaturated alkyl tails (Δ^9-DiI, D3886; *FAST* DiO™, D3898; *FAST* DiI™, D3899, D7756)
- DiI and DiO analogs with shorter alkyl tails ($DiIC_{12}(3)$, D383; Figure 13.4.3; $DiIC_{16}(3)$, D384; $DiOC_{16}(3)$, D1125)
- Long-wavelength light–excitable carbocyanines (DiD, D307, D7757; Figure 13.4.4)
- Infrared light–excitable carbocyanine (DiR, D12731; Figure 13.4.5)
- Chloromethylbenzamido DiI and sulfonated DiI and DiO derivatives

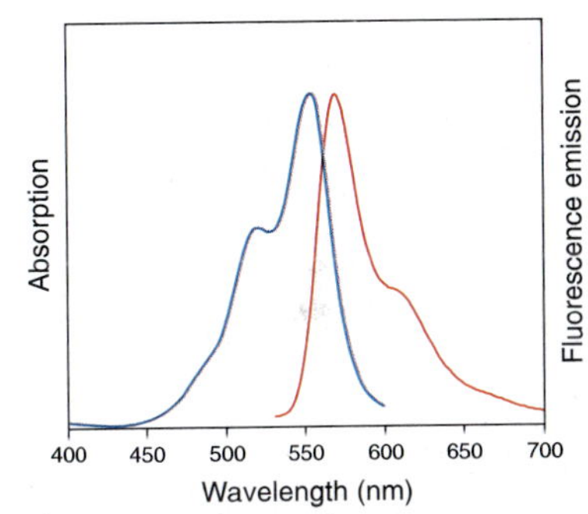

Figure 13.4.6 Absorption and fluorescence emission spectra of $DiIC_{18}(3)$ ("DiI") bound to phospholipid bilayer membranes.

Spectral Properties of Dialkylcarbocyanines

The spectral properties of dialkylcarbocyanines are largely independent of the lengths of the alkyl chains, and are instead determined by the heteroatoms in the terminal ring systems and the length of the connecting bridge. The $DiIC_n(3)$ probes have absorption and fluorescence spectra compatible with rhodamine (TRITC) optical filter sets (Figure 13.4.6), whereas $DiOC_n(3)$ analogs can be used with fluorescein (FITC) optical filter sets (Figure 13.4.7). The emission maxima of

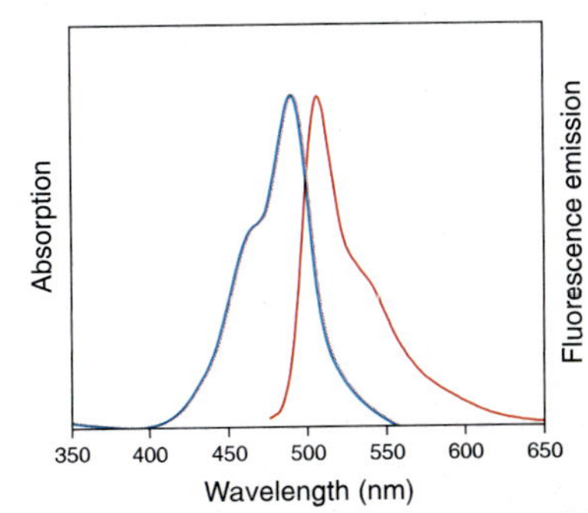

Figure 13.4.7 Absorption and fluorescence emission spectra of $DiOC_{18}(3)$ ("DiO") bound to phospholipid bilayer membranes.

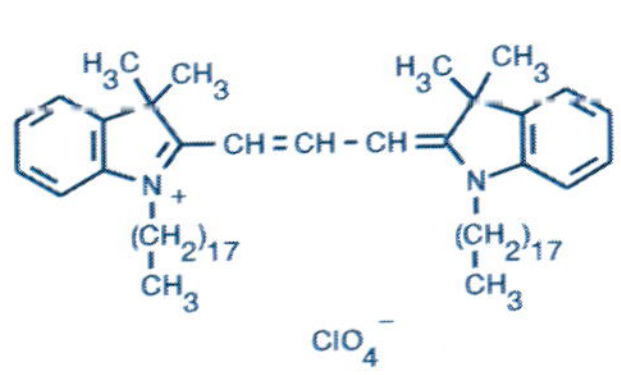

Figure 13.4.1 'DiI'; $DiIC_{18}(3)$ (1,1'-dioctadecyl-3,3,3',3'-tetramethylindocarbocyanine perchlorate; D282).

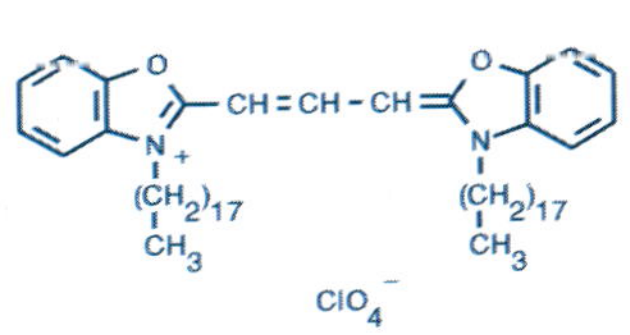

Figure 13.4.2 'DiO'; $DiOC_{18}(3)$ (3,3'-dioctadecyloxacarbocyanine perchlorate; D275).

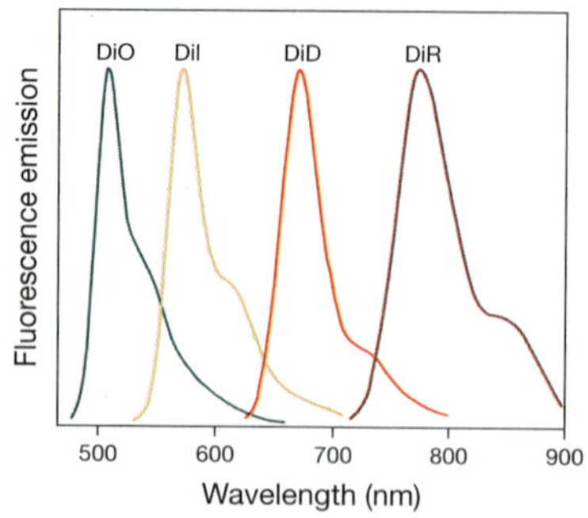

Figure 13.4.8 Normalized fluorescence emission spectra of DiO (D275), DiI (D282), DiD (D307) and DiR (D12731) bound to phospholipid bilayer membranes.

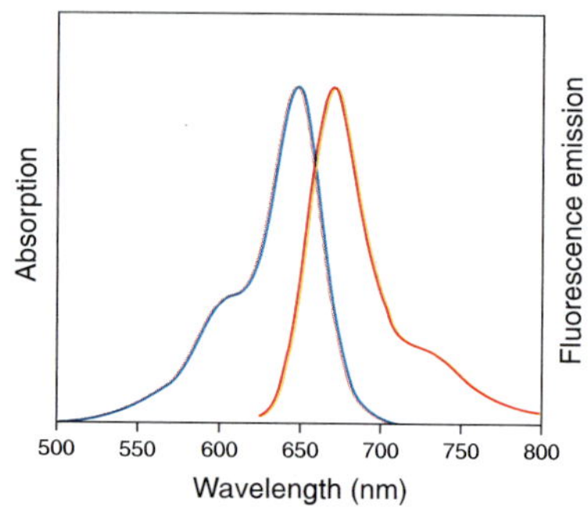

Figure 13.4.9 Absorption and fluorescence emission spectra of $DiIC_{18}(5)$ ("DiD") bound to phospholipid bilayer membranes.

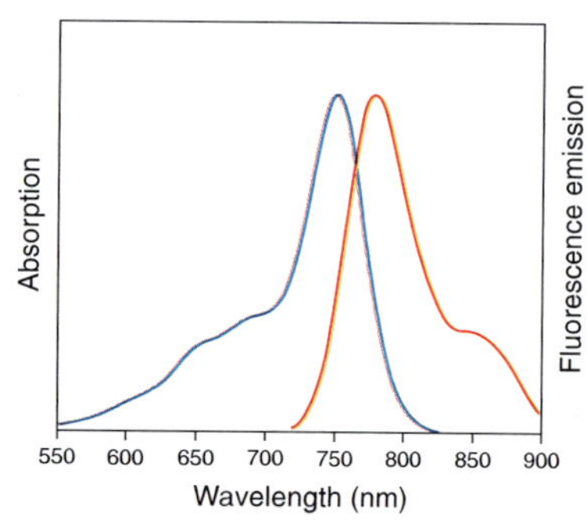

Figure 13.4.10 Fluorescence excitation and emission spectra of $DiIC_{18}(7)$ ("DiR") bound to phospholipid bilayer membranes.

$DiIC_{18}(3)$ and $DiOC_{18}(3)$ incorporated in dioctadecenoylphosphocholine (dioleoyl PC or DOPC) liposomes (Figure 13.4.8) are similar to those of the dyes in methanol.

The very large molar extinction coefficients of carbocyanine fluorophores are their most outstanding spectral property. Their fluorescence quantum yields are only modest—about 0.07 for DiI in methanol and about three-times greater in amphiphilic solvents such as octanol.[4,5] Their fluorescence in water is quite weak.[6] The excited-state lifetimes of carbocyanine fluorophores in lipid environments are short (~1 nanosecond), which is an advantage for flow cytometry applications because it allows more excitation/de-excitation cycles during flow transit; the overall decay is multi-exponential.[7] Dialkylcarbocyanines are also exceptionally photostable.[8]

The red He-Ne laser–excitable indodicarbocyanines such as DiD ($DiIC_{18}(5)$; D307, D7757) have long-wavelength absorption and red emission (Figure 13.4.8, Figure 13.4.9). Their extinction coefficients are somewhat larger and fluorescence quantum yields much larger than those of carbocyanines such as DiI.[5] Moreover, photoexcitation of DiD seems to cause less collateral damage than photoexcitation of DiI in live cells.[9] The $DiIC_{18}(7)$ tricarbocyanine probe (DiR, D12731) has excitation and emission in the infrared (Figure 13.4.10), which may make the dye useful as an *in vivo* tracer for labeled cells and liposomes in live organisms.[4,10]

Substituted DiI and DiO Derivatives

We have synthesized various derivatives of DiI, DiO and DiD. All of these derivatives have octadecyl (C_{18}) tails identical to those of DiI (D282, D3911) and DiO (D275), thereby preserving the excellent membrane retention characteristics of the parent molecules. A variety of substitutions have been made on the indoline or benzoxazole ring systems:

- Chloromethylbenzamido DiI derivatives (CellTracker™ CM-DiI; C7000, C7001; Figure 13.4.11)
- Anionic sulfophenyl derivatives[11] of DiI and DiO (SP-$DiIC_{18}(3)$, D7777, Figure 13.4.12; SP-$DiOC_{18}(3)$, D7778, Figure 13.4.13)
- Sulfonate derivatives of DiI and DiD ($DiIC_{18}(3)$-DS, D7776, Figure 13.4.14; $DiIC_{18}(5)$-DS, D12730, Figure 13.4.15)

Although these derivatives have primarily been developed to provide improved fixation and labeling in long-term cell tracing applications (Section 14.4), they also offer several features that can potentially be exploited for investigating membrane structure and dynamics. For researchers wishing to carry out comparative evaluations, our Lipophilic Tracer Sampler Kit (L7781) provides 1 mg samples of each of nine different carbocyanine derivatives, including several of the newer substituted derivatives:

- DiI ($DiIC_{18}(3)$)
- DiD ($DiIC_{18}(5)$)
- DiR ($DiIC_{18}(7)$)
- DiO ($DiOC_{18}(3)$)
- DiA (4-Di-16-ASP)
- $DiIC_{18}(3)$-DS
- SP-$DiIC_{18}(3)$
- SP-$DiOC_{18}(3)$
- 5,5′-Ph_2-$DiIC_{18}(3)$

The fluorescence quantum yields of the sulfophenyl and phenyl derivatives (measured in methanol) are generally 2- to 3-fold greater than those of DiI and DiO. In particular, we have found that the sulfophenyl derivatives (SP-$DiIC_{18}(3)$, D7777; SP-$DiOC_{18}(3)$, D7778) bound to phospholipid model membranes have approximately 5-fold higher quantum yields than DiI and DiO. $DiIC_{18}(5)$-DS (D12730) has been used in combination with an NBD-labeled glycerophosphoserine probe in a novel resonance energy transfer assay that detects inner monolayer membrane hemifusion, avoiding erroneous indications of membrane fusion due to lipid mixing and other environmental effects in the outer monolayer.[12] The negative charge and greater water solubility of the sulfonated carbocyanines results in modified lateral and transverse distributions of these probes in lipid bilayers relative to those of DiI and DiO. This characteristic has been exploited to identify plasma membrane lipid domains that are responsive to electrical stimulation of outer hair cells in the inner ear.[13]

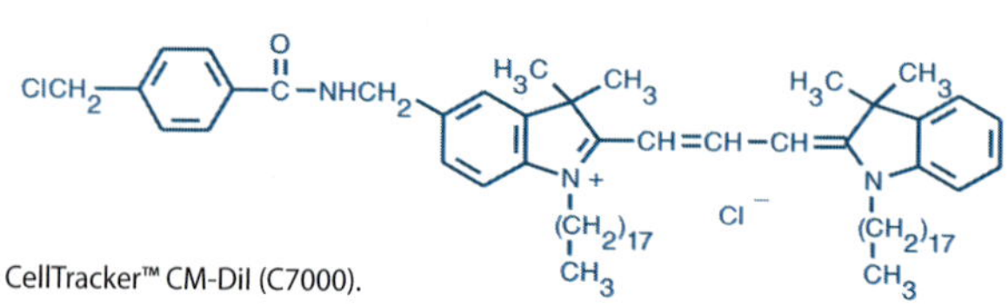

Figure 13.4.11 CellTracker™ CM-DiI (C7000).

IMPORTANT NOTICE: The products described in this manual are covered by one or more Limited Use Label License(s). Please refer to the Appendix on page 971 and Master Product List on page 975. Products are For Research Use Only. Not intended for any animal or human therapeutic or diagnostic use.

DiI and DiO as Probes of Membrane Structure

The orientation of $DiIC_{18}(3)$ in membranes has been determined by fluorescence polarization microscopy.[14] The long axis of the fluorophore is parallel to the membrane surface, and the two alkyl chains protrude perpendicularly into the lipid interior. There are conflicting reports in the literature regarding the ease of transbilayer migration ("flip-flop") of lipophilic indocarbocyanines.[15–17] The lateral partitioning behavior of dialkylindocarbocyanines in membranes has been investigated by fluorescence recovery after photobleaching (FRAP),[18] calorimetry,[19] lifetime measurements[8] and fluorescence resonance energy transfer techniques[20] (Fluorescence Resonance Energy Transfer (FRET)—Note 1.2). These studies demonstrate that the probe distribution between coexisting fluid and gel phases depends on the similarity of the alkyl chain lengths of the probe and the lipid. In general, the more dissimilar the lengths, the greater the preference for fluid-phase over gel-phase lipids. For example, the shorter-chain $DiIC_{12}(3)$ has a substantial preference for the fluid phase (~6:1) in DOPC, whereas $DiIC_{18}(3)$ is predominantly distributed in the gel phase[21] (~1:10). Consequently, long-chain dialkylcarbocyanines are among the best probes for detecting particularly rigid gel phases.

Lipophilic carbocyanines have been used to visualize membrane fusion and cell permeabilization that occurs in response to electric fields,[22–24] as well as fusion of liposomes with planar bilayers.[25] Membrane fusion can also be measured by fluorescence resonance energy transfer to $DiIC_{18}(3)$ from dansyl- or NBD-labeled phospholipid donors[26] or by direct imaging.[27] In Langmuir–Blodgett films, excited-state energy transfer from $DiIC_{18}(3)$ to $DiIC_{18}(5)$ is exceptionally efficient because of the favorable orientations of the fluorophores.[28] Energy transfer from $DiIC_{18}(5)$ to $DiIC_{18}(7)$ should be similarly effective. Lipophilic carbocyanines have also been used to elicit photosensitized destabilization of liposomes,[29] to sensitize photoaffinity labeling of the viral glycoprotein hemagglutinin,[30] to image membrane domains in lipid monolayers[31] and to develop a fiber-optic potassium sensor.[32]

DiI and DiO as Probes of Membrane Dynamics

Despite their reasonably good photostability, dialkylcarbocyanines are widely employed to measure lateral diffusion processes using fluorescence recovery after photobleaching (FRAP) techniques.[33–36] Their lateral diffusion coefficients in isolated fluid- and gel-phase bilayers are independent of the carbocyanine alkyl chain length.[18] Phase-separated populations of lipophilic carbocyanine dyes can be distinguished by their diffusion rates and can therefore be used to define lateral domains in cell membranes.[37,38] Combined lateral diffusion measurements of labeled proteins and lipids have demonstrated that transformed[39] and permeabilized[40] cells show marked changes in protein diffusion, whereas lipid diffusion rates remain unchanged. In other cases, coupling of lipid and protein mobility has been identified in the form of relatively immobilized lipid domains in yeast plasma membranes[41] and around IgE receptor complexes.[42] A different photobleaching technique, which depends on the *absence* of diffusional fluorescence recovery, was employed to determine lipid flow direction in locomoting cells by following the movement of a photobleached stripe of $DiIC_{16}(3)$[43] (D384).

Dialkylaminostyryl Probes

The lipophilic aminostyryl probes 4-Di-10-ASP (D291, Figure 13.4.16), DiA (4-Di-16-ASP, D3883; Figure 13.4.17) and *FAST* DiA™ (D7758, Figure 13.4.18) insert in membranes with their two alkyl

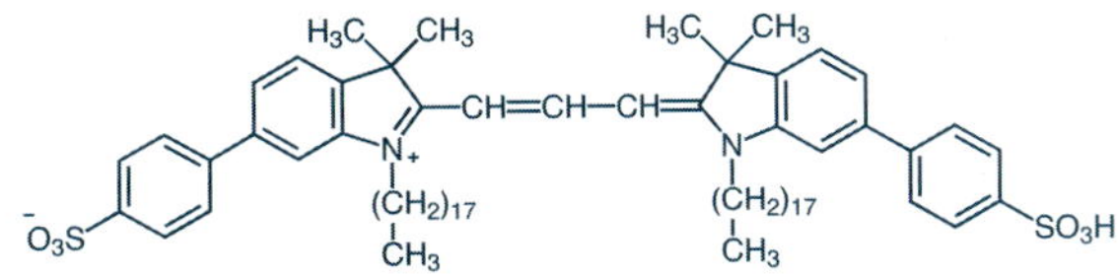

Figure 13.4.12 1,1′-Dioctadecyl-6,6′-di(4-sulfophenyl)-3,3,3′,3′-tetramethylindocarbocyanine (SP-$DiIC_{18}(3)$, D7777).

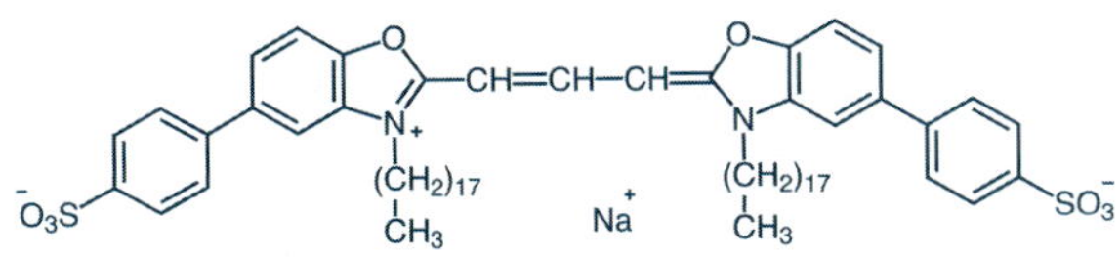

Figure 13.4.13 3,3′-Dioctadecyl-5,5′-di(4-sulfophenyl)oxacarbocyanine, sodium salt (SP-$DiOC_{18}(3)$, D7778).

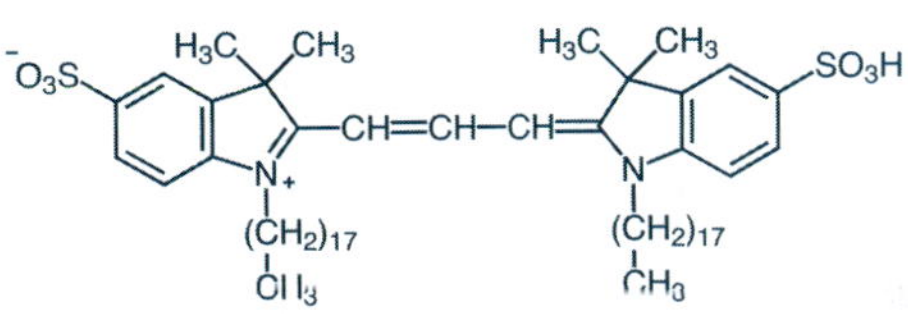

Figure 13.4.14 $DiIC_{18}(3)$-DS (1,1′-dioctadecyl-3,3,3′,3′-tetramethylindocarbocyanine-5,5′-disulfonic acid; D7776).

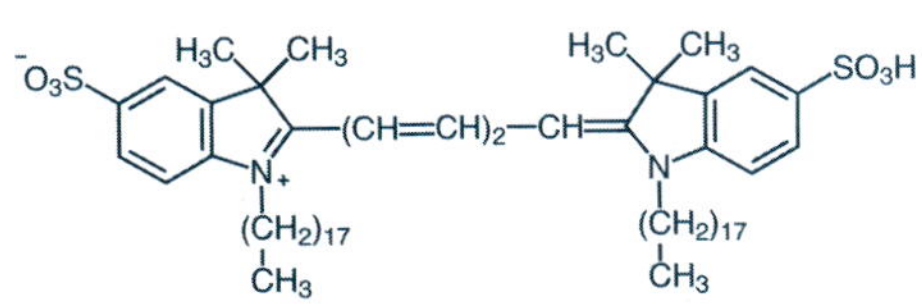

Figure 13.4.15 $DiIC_{18}(5)$-DS (1,1′-dioctadecyl-3,3,3′,3′-tetramethylindodicarbocyanine-5,5′-disulfonic acid; D12730).

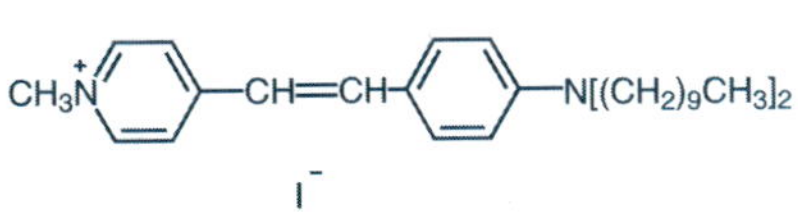

Figure 13.4.16 4-(4-(Didecylamino)styryl)-*N*-methylpyridinium iodide (4-Di-10-ASP, D291).

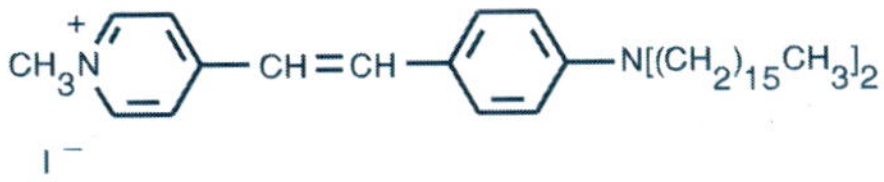

Figure 13.4.17 DiA; 4-Di-16-ASP (4-(4-(dihexadecylamino)styryl)-*N*-methylpyridinium iodide; D3883).

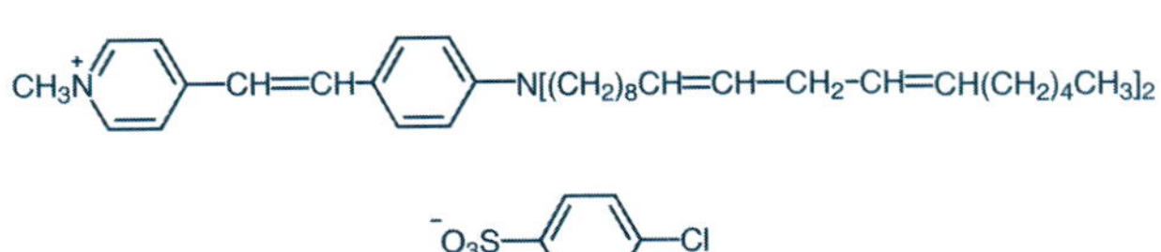

Figure 13.4.18 4-(4-(Dilinoleylamino) styryl)-*N*-methylpyridinium 4-chlorobenzenesulfonate (*FAST* DiA™ solid; $DiΔ^{9,12}$-C_{18}ASP, CBS; D7758).

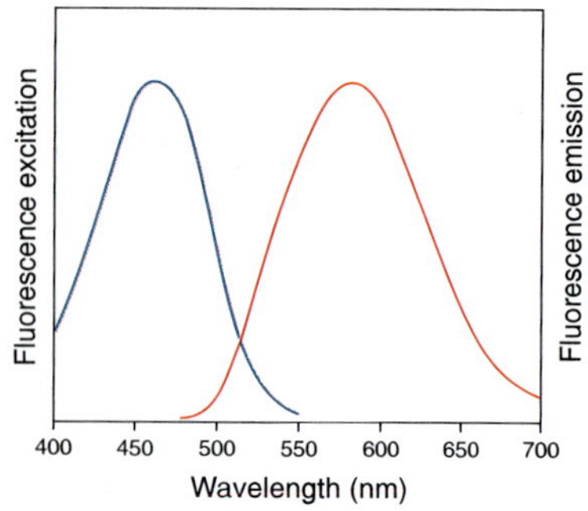

Figure 13.4.19 Fluorescence excitation and emission spectra of DiA bound to phospholipid bilayer membranes.

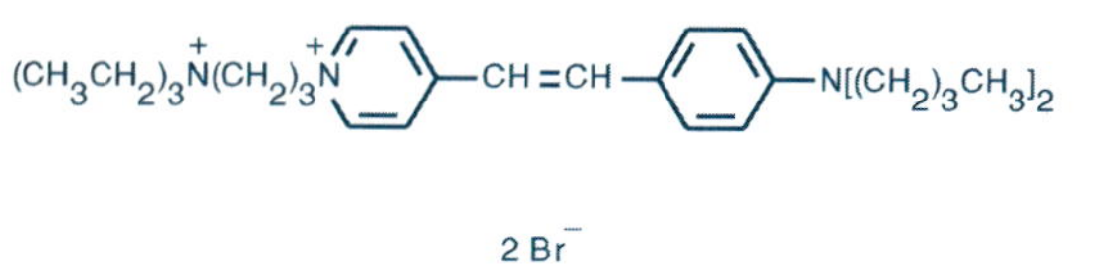

Figure 13.4.20 *N*-(3-triethylammoniumpropyl)-4-(4-(dibutylamino)styryl)pyridinium dibromide (FM® 1-43, T3163).

tails and their fluorophore oriented parallel to the phospholipid acyl chains.[44] When these dialkylaminostyryl probes bind to membranes, they exhibit a strong fluorescence enhancement; their fluorescence in water is minimal. The interfacial solvation of the aminostyryl fluorophore causes a large blue shift of the absorption spectrum of the membrane-bound probe.[44] For example, the absorption maximum of DiA is 456 nm when incorporated into DOPC liposomes and 490 nm when in methanol. The fluorescence emission maximum of DiA in the membrane environment is 590 nm, which is quite close to that observed for probes with shorter alkyl tails such as 4-Di-10-ASP;[44] however, the fluorescence spectrum of DiA is very broad, with appreciable intensity from about 510 nm to 690 nm (Figure 13.4.19). Consequently, DiA can be detected as green, orange or even red fluorescence, depending on the optical filter employed. Like the lipophilic carbocyanines, DiA is commonly used for neuronal membrane tracing (Section 14.4). *FAST* DiA™ (D7758), the diunsaturated analog of DiA, is intended to facilitate these studies by accelerating dye diffusion within the membrane.

The FM® 1-43 (Figure 13.4.20), FM® 1-43FX, FM® 4-64 and FM® 5-95 dyes, which are discussed in detail in Section 16.1, are styryl dyes that also exhibit high Stokes shifts and broad fluorescence emission but have less lipophilic character than the 4-Di-10-ASP and DiA probes. The FM® dyes are commonly used to define the outer membranes of liposomes and live cells and to detect synaptosome recycling.

REFERENCES

1. Trends Neurosci (1989) 12:333, 340; **2.** Histochemistry (1992) 97:329; **3.** Brain Res (2008) 1215:11; **4.** J Biomed Opt (2009) 14:054005; **5.** Biochemistry (1974) 13:3315; **6.** Chem Phys Lipids (2001) 109:175; **7.** Biochemistry (1985) 24:5176; **8.** J Cell Biol (1985) 100:1309; **9.** J Histochem Cytochem (1984) 32:608; **10.** J Am Chem Soc (2007) 129:5798; **11.** Bioorg Med Chem Lett (1996) 6:1479; **12.** Biochim Biophys Acta (2000) 1467:227; **13.** J Assoc Res Otolaryngol (2002) 3:289; **14.** Biophys J (1979) 26:557; **15.** J Cell Biol (1986) 103:807; **16.** Biochemistry (1985) 24:582; **17.** Nature (1981) 294:718; **18.** Biochemistry (1980) 19:6199; **19.** Biochim Biophys Acta (1994) 1191:164; **20.** Biochim Biophys Acta (2000) 1467:101; **21.** Biochim Biophys Acta (1990) 1023:25; **22.** Biophys J (1994) 67:427; **23.** Biophys J (1993) 65:568; **24.** Biochemistry (1990) 29:8337; **25.** J Membr Biol (1989) 109:221; **26.** Biochim Biophys Acta (1983) 735:243; **27.** J Cell Biol (1993) 121:543; **28.** Chem Phys Lett (1989) 159:231; **29.** FEBS Lett (2000) 467:52; **30.** J Biol Chem (1994) 269:14614; **31.** Biophys J (1993) 65:1019; **32.** Analyst (1990) 115:353; **33.** Biochemistry (1977) 16:3836; **34.** Biophys J (1998) 75:1131; **35.** Biophys J (1995) 68:766; **36.** Bioessays (1987) 6:117; **37.** Chem Phys Lipids (1994) 73:139; **38.** J Cell Biol (1991) 112:1143; **39.** Biochim Biophys Acta (1992) 1107:193; **40.** J Cell Physiol (1994) 158:7; **41.** J Membr Biol (1993) 131:115; **42.** J Cell Biol (1994) 125:795; **43.** Science (1990) 247:1229; **44.** Biophys J (1981) 34:353.

DATA TABLE 13.4 DIALKYLCARBOCYANINE AND DIALKYLAMINOSTYRYL PROBES

Cat. No.	MW	Storage	Soluble	Abs	EC	Em	Solvent	Notes
C7000	1051.50	F,D,L	DMSO, EtOH	553	134,000	570	MeOH	
C7001	1051.50	F,D,L	DMSO, EtOH	553	134,000	570	MeOH	
D275	881.72	L	DMSO, DMF	484	154,000	501	MeOH	
D282	933.88	L	DMSO, EtOH	549	148,000	565	MeOH	
D291	618.73	L	DMSO, EtOH	492	53,000	612	MeOH	1
D307	959.92	L	DMSO, EtOH	644	260,000	665	MeOH	2
D383	765.56	L	DMSO, EtOH	549	144,000	565	MeOH	3
D384	877.77	L	DMSO, EtOH	549	148,000	565	MeOH	
D1125	825.61	L	DMSO, DMF	484	156,000	501	MeOH	
D3883	787.05	L	DMSO, EtOH	491	52,000	613	MeOH	1
D3886	925.49	F,L,AA	DMSO, EtOH	549	144,000	564	MeOH	2
D3898	873.65	F,L,AA	DMSO, DMF	484	138,000	499	MeOH	
D3899	925.82	F,L,AA	DMSO, EtOH	549	143,000	564	MeOH	2
D3911	933.88	L	DMSO, EtOH	549	148,000	565	MeOH	
D7756	1017.97	F,L,AA	DMSO, EtOH	549	148,000	564	MeOH	
D7757	1052.08	L	DMSO, EtOH	644	193,000	663	MeOH	
D7758	899.80	F,L,AA	DMSO, EtOH	492	41,000	612	MeOH	1
D7776	993.54	L	DMSO, EtOH	555	144,000	570	MeOH	
D7777	1145.73	L	DMSO, EtOH	556	164,000	573	MeOH	
D7778	1115.55	L	DMSO, EtOH	497	175,000	513	MeOH	
D12730	1019.58	L	DMSO, EtOH	650	247,000	670	MeOH	
D12731	1013.41	L	DMSO, EtOH	748	270,000	780	MeOH	

For definitions of the contents of this data table, see "Using *The Molecular Probes® Handbook*" in the introductory pages.

Notes

1. Abs and Em of styryl dyes are at shorter wavelengths in membrane environments than in reference solvents such as methanol. The difference is typically 20 nm for absorption and 80 nm for emission, but varies considerably from one dye to another. Styryl dyes are generally nonfluorescent in water.
2. This product is intrinsically a liquid or an oil at room temperature.
3. This product is intrinsically a sticky gum at room temperature.

IMPORTANT NOTICE: The products described in this manual are covered by one or more Limited Use Label License(s). Please refer to the Appendix on page 971 and Master Product List on page 975. Products are For Research Use Only. Not intended for any animal or human therapeutic or diagnostic use.
www.invitrogen.com/probes

PRODUCT LIST 13.4 DIALKYLCARBOCYANINE AND DIALKYLAMINOSTYRYL PROBES

Cat. No.	Product	Quantity
C7001	CellTracker™ CM-DiI	1 mg
C7000	CellTracker™ CM-DiI *special packaging*	20 x 50 µg
D291	4-(4-(didecylamino)styryl)-*N*-methylpyridinium iodide (4-Di-10-ASP)	25 mg
D383	1,1′-didodecyl-3,3,3′,3′-tetramethylindocarbocyanine perchlorate (DiIC_{12}(3))	100 mg
D3883	4-(4-(dihexadecylamino)styryl)-*N*-methylpyridinium iodide (DiA; 4-Di-16-ASP)	25 mg
D1125	3,3′-dihexadecyloxacarbocyanine perchlorate (DiOC_{16}(3))	25 mg
D384	1,1′-dihexadecyl-3,3,3′,3′-tetramethylindocarbocyanine perchlorate (DiIC_{16}(3))	100 mg
D7758	4-(4-(dilinoleylamino)styryl)-*N*-methylpyridinium 4-chlorobenzenesulfonate (FAST DiA™ solid; DiA$\Delta^{9,12}$-C_{18}ASP, CBS)	5 mg
D3898	3,3′-dilinoleyloxacarbocyanine perchlorate (FAST DiO™ solid; DiO$\Delta^{9,12}$-C_{18}(3), ClO_4)	5 mg
D7756	1,1′-dilinoleyl-3,3,3′,3′-tetramethylindocarbocyanine, 4-chlorobenzenesulfonate (FAST DiI™ solid; DiI$\Delta^{9,12}$-C_{18}(3), CBS)	5 mg
D3899	1,1′-dilinoleyl-3,3,3′,3′-tetramethylindocarbocyanine perchlorate (FAST DiI™ oil; DiI$\Delta^{9,12}$-C_{18}(3), ClO_4)	5 mg
D7778	3,3′-dioctadecyl-5,5′-di(4-sulfophenyl)oxacarbocyanine, sodium salt (SP-DiOC_{18}(3))	5 mg
D7777	1,1′-dioctadecyl-6,6′-di(4-sulfophenyl)-3,3,3′,3′-tetramethylindocarbocyanine (SP-DiIC_{18}(3))	5 mg
D275	3,3′-dioctadecyloxacarbocyanine perchlorate ('DiO'; DiOC_{18}(3))	100 mg
D7776	1,1′-dioctadecyl-3,3,3′,3′-tetramethylindocarbocyanine-5,5′-disulfonic acid (DiIC_{18}(3)-DS)	5 mg
D282	1,1′-dioctadecyl-3,3,3′,3′-tetramethylindocarbocyanine perchlorate ('DiI'; DiIC_{18}(3))	100 mg
D3911	1,1′-dioctadecyl-3,3,3′,3′-tetramethylindocarbocyanine perchlorate *crystalline* ('DiI'; DiIC_{18}(3))	25 mg
D7757	1,1′-dioctadecyl-3,3,3′,3′-tetramethylindodicarbocyanine, 4-chlorobenzenesulfonate salt ('DiD' solid; DiIC_{18}(5) solid)	10 mg
D12730	1,1′-dioctadecyl-3,3,3′,3′-tetramethylindodicarbocyanine-5,5′-disulfonic acid (DiIC_{18}(5)-DS)	5 mg
D307	1,1′-dioctadecyl-3,3,3′,3′-tetramethylindodicarbocyanine perchlorate ('DiD' oil; DiIC_{18}(5) oil)	25 mg
D12731	1,1′-dioctadecyl-3,3,3′,3′-tetramethylindotricarbocyanine iodide ('DiR'; DiIC_{18}(7))	10 mg
D3886	1,1′-dioleyl-3,3,3′,3′-tetramethylindocarbocyanine methanesulfonate (Δ^9-DiI)	25 mg
L7781	Lipophilic Tracer Sampler Kit	1 kit

13.5 Other Nonpolar and Amphiphilic Probes

Amphiphilic Rhodamine, Fluorescein and Coumarin Derivatives

Each of our amphiphilic probes comprises a moderately polar fluorescent dye with a lipophilic "tail." When used to stain membranes, including liposomes, the lipophilic portion of the probe tends to insert in the membrane and the polar fluorophore resides on the surface, where it senses the membrane's surface environment and the surrounding medium.[1] Our lipophilic carbocyanines and styryl dyes (Section 13.4) are also amphiphilic molecules with a similar binding mode.

This section includes the classic membrane probes DPH, TMA-DPH, ANS, bis-ANS, TNS, prodan, laurdan and nile red, and also some lipophilic BODIPY® and Dapoxyl® dyes developed in our laboratories. Although they bear little resemblance to natural products, these probes tend to localize within cell membranes or liposomes or at their aqueous interfaces, where they are often used to report on characteristics of their local environment, such as viscosity, polarity and lipid order.

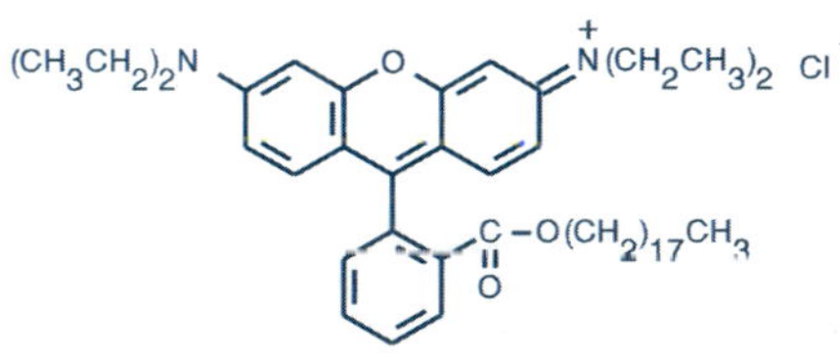

Figure 13.5.1 Octadecyl rhodamine B chloride (O246).

Octadecyl Rhodamine B

The relief of the fluorescence self-quenching of octadecyl rhodamine B (O246, Figure 13.5.1) can be used to monitor membrane fusion[2–7]—one of several experimental approaches developed for this application (Lipid-Mixing Assays of Membrane Fusion—Note 13.1). Octadecyl rhodamine B has been reported to undergo a potential-dependent "flip-flop" from one monolayer of a fluid-state phospholipid bilayer membrane to the other, with partial relief of its fluorescence quenching.[8,9] Investigators have used octadecyl rhodamine B in conjunction with video microscopy[10–12] or digital imaging techniques[13] to monitor viral fusion processes. Membrane fusion can

IMPORTANT NOTICE: The products described in this manual are covered by one or more Limited Use Label License(s). Please refer to the Appendix on page 971 and Master Product List on page 975. Products are For Research Use Only. Not intended for any animal or human therapeutic or diagnostic use.

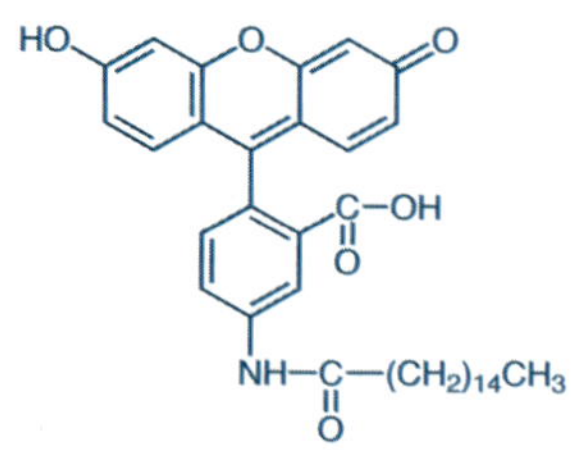

Figure 13.5.2 5-Hexadecanoylaminofluorescein (H110).

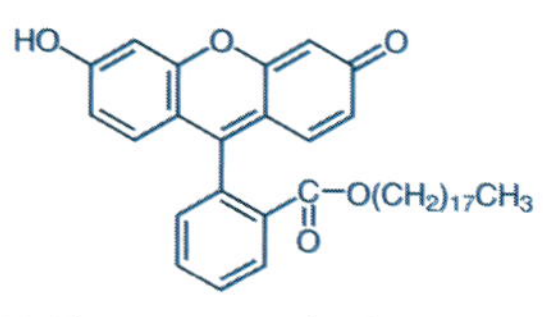

Figure 13.5.3 Fluorescein octadecyl ester (F3857).

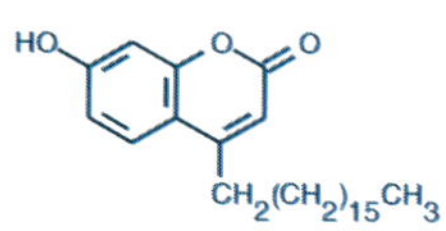

Figure 13.5.4 4-Heptadecyl-7-hydroxycoumarin (H22730).

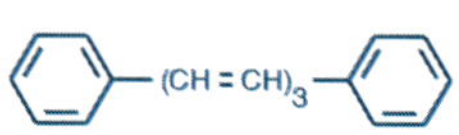

Figure 13.5.5 DPH (1,6-diphenyl-1,3,5-hexatriene; D202).

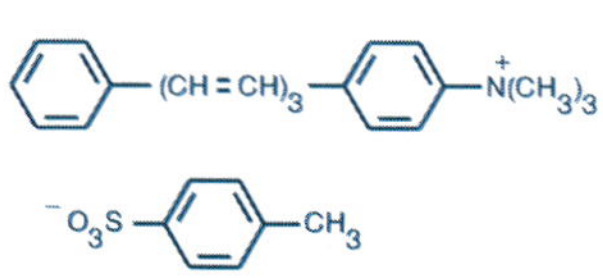

Figure 13.5.6 TMA-DPH (1-(4-trimethylammoniumphenyl)-6-phenyl-1,3,5-hexatriene *p*-toluenesulfonate; T204).

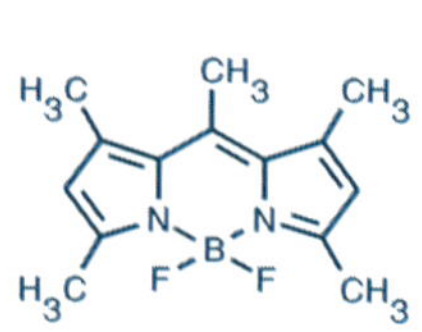

Figure 13.5.7 4,4-Difluoro-1,3,5,7,8-pentamethyl-4-bora-3a,4a-diaza-*s*-indacene (BODIPY® 493/503, D3922).

also be followed by monitoring fluorescence resonance energy transfer to octadecyl rhodamine B from an acylaminofluorescein donor such as 5-hexadecanoylaminofluorescein [5,7,14,15] (H110, Figure 13.5.2).

Fluorescence resonance energy transfer from fluorescein or dansyl labels to octadecyl rhodamine B has been used for structural studies of the blood coagulation factor IXa, EGF receptor and receptor-bound IgE.[16–18] Octadecyl rhodamine B has also been used to stain kinesin-generated membrane tubules,[19] to characterize detergent micelles,[20] to assay for lysosomal degradation of lipoproteins [21] and to investigate the influence of proteins on lipid dynamics using time-resolved fluorescence anisotropy.[22]

Amphiphilic Fluoresceins

The amphiphilic fluorescein probes bind to membranes with the fluorophore at the aqueous interface and the alkyl tail protruding into the lipid interior. 5-Dodecanoylaminofluorescein (D109) is the hydrolysis product of our ImaGene Green™ C_{12}-FDG β-galactosidase substrate (D2893, Section 10.2). We also offer the homologous membrane probe 5-hexadecanoylaminofluorescein [1,15,23] (H110, Figure 13.5.2) and the octadecyl ester of fluorescein [24,25] (F3857, Figure 13.5.3).

Amphiphilic fluorescein probes are commonly used for fluorescence recovery after photobleaching (FRAP) measurements of lipid lateral diffusion.[26] Some researchers have reported that 5-hexadecanoylaminofluorescein stays predominantly in the outer membrane leaflet of epithelia and does not pass through tight junctions, whereas the dodecanoyl derivative can "flip-flop" to the inner leaflet at 20°C (but not at <10°C) and may also pass through tight junctions.[27,28] More recent studies have indicated that the lack of tight junction penetration of 5-hexadecanoylaminofluorescein is due to probe aggregation rather than a significant difference in its transport properties.[29]

Amphiphilic Coumarin

4-Heptadecyl-7-hydroxycoumarin (H22730, Figure 13.5.4) is an alkyl derivative of the pH-sensitive blue-fluorescent 7-hydroxycoumarin (umbelliferone) fluorophore. As with other amphiphilic coumarins,[30] 4-heptadecyl-7-hydroxycoumarin is primarily useful as a probe of membrane surfaces. Deprotonation of the 7-hydroxyl group is expected to be strongly dependent on membrane-surface electrostatic potential. The pK_a of 4-heptadecyl-7-hydroxycoumarin varies from 6.35 in the cationic detergent CTAB to 11.15 in the anionic detergent sodium dodecyl sulfate (SDS), as measured by its fluorescence response.[31] However, its pK_a in lipid assemblies is strongly dependent on the ionic composition of the membrane surface,[31,32] making it a sensitive probe of membrane-surface electrostatic potential.[33] 4-Heptadecyl-7-hydroxycoumarin has been used to measure pH differences at membrane interfaces in isolated plasma membranes of normal and multidrug-resistant murine leukemia cells.[34,35] 4-Heptadecyl-7-hydroxycoumarin has also been employed as a structural probe for the head-group region of phospholipid bilayers.[36]

DPH and DPH Derivatives

Diphenylhexatriene (DPH)

1,6-Diphenyl-1,3,5-hexatriene (DPH, D202; Figure 13.5.5) continues to be a popular fluorescent probe of membrane interiors. We also offer the cationic DPH derivative TMA-DPH, as well as the phospholipid analog (D476, Section 13.2). The orientation of DPH within lipid bilayers is loosely constrained. It is generally assumed to be oriented parallel to the lipid acyl chain axis, but it can also reside in the center of the lipid bilayer parallel to the surface, as demonstrated by time-resolved fluorescence anisotropy and polarized fluorescence measurements of oriented samples.[37–40] DPH shows no partition preference between coexisting gel- and fluid-phase phospholipids.[41] Intercalation of DPH and its derivatives into membranes is accompanied by strong enhancement of their fluorescence; their fluorescence is practically negligible in water. The fluorescence decay of DPH in lipid bilayers is complex.[42–44] Fluorescence decay data are often analyzed in terms of continuous lifetime distributions,[45–48] which are in turn interpreted as being indicative of lipid environment heterogeneity.

DPH and its derivatives are cylindrically shaped molecules with absorption and fluorescence emission transition dipoles aligned approximately parallel to their long molecular

IMPORTANT NOTICE: The products described in this manual are covered by one or more Limited Use Label License(s). Please refer to the Appendix on page 971 and Master Product List on page 975. Products are For Research Use Only. Not intended for any animal or human therapeutic or diagnostic use.

molecular probes® | invitrogen™ by life technologies™

axis. Consequently, their fluorescence polarization is high in the absence of rotational motion and is very sensitive to reorientation of the long axis resulting from interactions with surrounding lipids. These properties have led to their extensive use for membrane fluidity measurements.[49] The exact physical interpretation of these measurements has some contentious aspects. For instance, the probes are largely sensitive to only the angular reorientation of lipid acyl chains—a motion that does not necessarily correlate with other dynamic processes such as lateral diffusion.[50] Reviews on this subject[39,49,51,52] should be consulted for further discussion. Time-resolved fluorescence polarization measurements of lipid order are more physically rigorous because they allow the angular range of acyl chain reorientation ("lipid order") to be resolved from its rate, and considerable research has been devoted to the interpretation of these measurements.[37,45,53,54]

TMA-DPH

Designed to improve the localization of DPH in the membrane, TMA-DPH (T204, Figure 13.5.6) contains a cationic trimethylammonium substituent that acts as a surface anchor.[55–57] Like DPH, this derivative readily partitions from aqueous dispersions into membranes and other lipid assemblies, accompanied by strong fluorescence enhancement. The lipid–water partition coefficient (K_P) for TMA-DPH ($K_P = 2.4 \times 10^5$) is lower than for DPH ($K_P = 1.3 \times 10^6$), reflecting the increased water solubility caused by the polar substituents.[58] The fluorescence decay lifetime of TMA-DPH is more sensitive to changes in lipid composition and temperature than is the fluorescence decay lifetime of DPH.[59–61]

Staining of cell membranes by TMA-DPH is much more rapid than staining by DPH; however, the duration of plasma membrane surface staining by TMA-DPH before internalization into the cytoplasm is quite prolonged.[62,63] As a consequence, TMA-DPH introduced into Madin–Darby canine kidney (MDCK) cell plasma membranes does not diffuse through tight junctions and remains in the apical domain, whereas the anionic DPH propionic acid accumulates rapidly in intracellular membranes.[64] TMA-DPH residing in the plasma membrane can be extracted by washing with medium, thus providing a method for isolating internalized probe and monitoring endocytosis[65] (Section 16.1). Furthermore, because TMA-DPH is virtually nonfluorescent in water and binds in proportion to the available membrane surface,[66] its fluorescence intensity is sensitive to increases in plasma membrane surface area resulting from exocytosis.[65,67,68]

TMA-DPH fluorescence polarization measurements can be combined with video microscopy to provide spatially resolved images of phospholipid order in large liposomes and single cells.[69–72] Information regarding lipid order heterogeneity among cell populations can be obtained in a similar way using flow cytometry.[73–75]

Nonpolar BODIPY® Probes

BODIPY® Fluorophores

BODIPY® fluorophore derivatives offer an unusual combination of nonpolar structure (Figure 13.5.7) and long-wavelength absorption and fluorescence.[76] BODIPY® dyes have small fluorescence Stokes shifts, extinction coefficients that are typically greater than 80,000 $cm^{-1}M^{-1}$ and high fluorescence quantum yields that are not diminished in water.[77] These dyes have applications as stains for neutral lipids and as tracers for oils and other nonpolar liquids. In addition, their photostability is generally high; this, together with other favorable characteristics (very low triplet–triplet absorption), make the BODIPY® 493/503 and BODIPY® 505/515 fluorophores excellent choices for flashlamp-pumped laser dyes.[78,79]

Staining with the BODIPY® 493/503 dye (D3922, Figure 13.5.7) has been shown by flow cytometry to be more specific for cellular lipid droplets than staining with nile red[80] (N1142). The low molecular weight of the BODIPY® 493/503 dye (262 daltons) results in the probe having a relatively fast diffusion rate in membranes.[81] The BODIPY® 493/503 dye has also been used to detect neutral compounds in a microchip channel separation device.[82]

BODIPY® 505/515 (D3921, Figure 13.5.8) rapidly permeates cell membranes of live zebrafish embryos,[83,84] selectively staining cytoplasmic yolk platelets. This staining provides dramatic contrast enhancement of cytoplasm relative to nucleoplasm and interstitial spaces, allowing individual cell boundaries and cell nuclei to be imaged clearly with a confocal laser-scanning microscope (Figure 13.5.9).

The very long-wavelength BODIPY® 665/676 dye (B3932, Figure 13.5.10) has fluorescence that is not visible to the human eye; however, it has found use as a probe for reactive oxygen species[85] (Section 18.2).

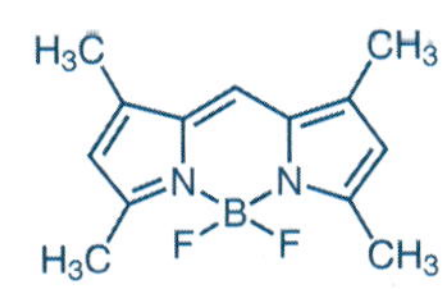

Figure 13.5.8 4,4-Difluoro-1,3,5,7-tetramethyl-4-bora-3a,4a-diaza-s-indacene (BODIPY® 505/515, D3921).

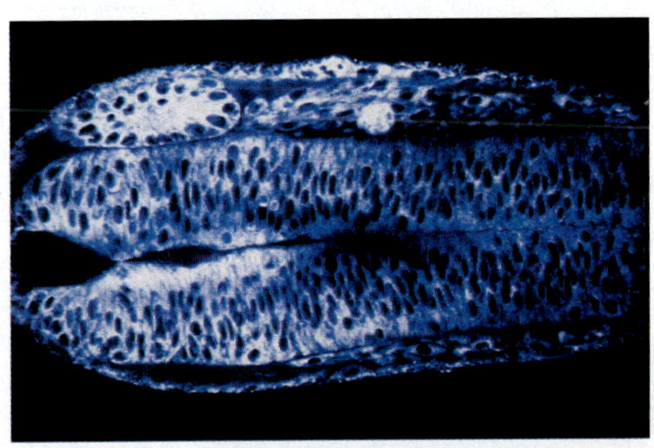

Figure 13.5.9 Dorsal view of the midbrain/hindbrain region of a 15-somite stage zebrafish embryo labeled with BODIPY® 505/515 (D3921). BODIPY® 505/515 localizes in lipidic yolk platelets, producing selective cytoplasmic staining. This pseudocolored confocal image was obtained using a Bio-Rad® MRC-600 microscope. Image contributed by Mark Cooper, University of Washington.

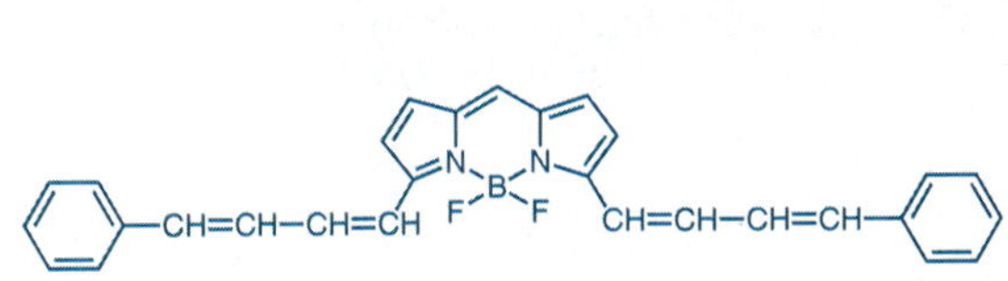

Figure 13.5.10 (E,E)-3,5-bis-(4-phenyl-1,3-butadienyl)-4,4-difluoro-4-bora-3a,4a-diaza-s-indacene (BODIPY® 665/676, B3932).

IMPORTANT NOTICE: The products described in this manual are covered by one or more Limited Use Label License(s). Please refer to the Appendix on page 971 and Master Product List on page 975. Products are For Research Use Only. Not intended for any animal or human therapeutic or diagnostic use.

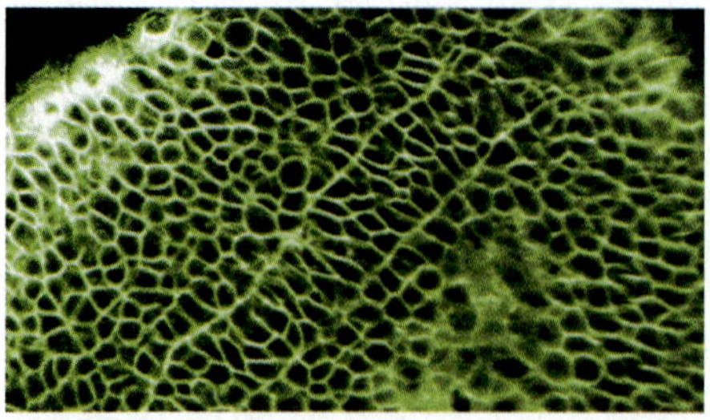

Figure 13.5.11 Cells in the notochord rudiment of a zebrafish embryo undergoing mediolateral intercalation to lengthen the forming notochord. BODIPY® FL C_5-ceramide (D3521) localizes in the interstitial fluid of the zebrafish embryo and freely diffuses between cells, illuminating cell boundaries. This confocal image was obtained using a Bio-Rad® MRC-600 microscope. Image contributed by Mark Cooper, University of Washington.

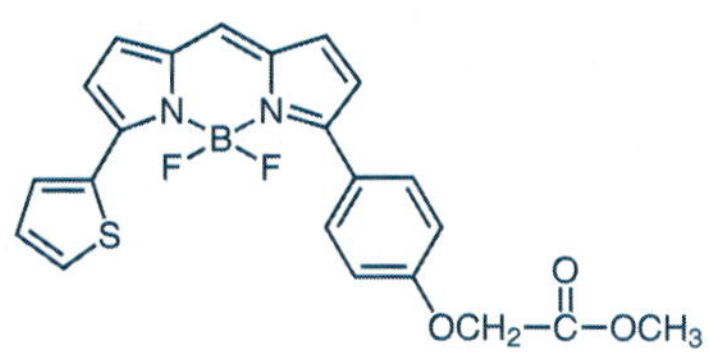

Figure 13.5.12 CellTrace™ BODIPY® TR methyl ester (C34556).

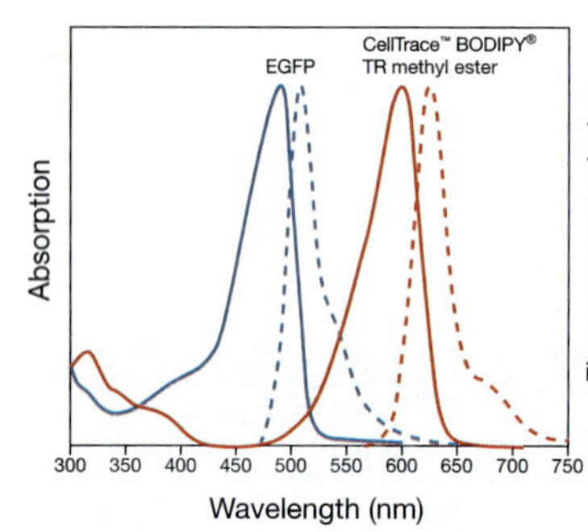

Figure 13.5.13 Normalized absorption (—) and fluorescence emission (- - -) spectra of enhanced Green Fluorescent Protein (EGFP) and CellTrace™ BODIPY® TR methyl ester (C34556).

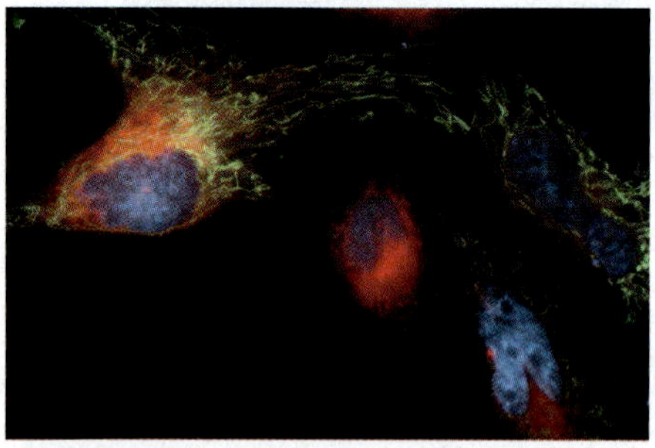

Figure 13.5.14 Live HeLa cells were transfected using pShooter™ vector pCMV/myc/mito/GFP and Lipofectamine® 2000 transfection reagent and stained with the reagents in the Image-iT® LIVE Intracellular Membrane and Nuclear Labeling Kit (I34407). Intracellular membranes were stained with CellTrace™ BODIPY® TR methyl ester, and nuclei were stained with Hoechst 33342. Cells were visualized using epifluorescence microscopy.

BODIPY® FL C_5-Ceramide

BODIPY® FL C_5-ceramide (D3521, B22650; Section 13.3) stains the plasma membrane, Golgi apparatus and cytoplasmic particles within the superficial enveloping layer (EVL) of embryos. Once the fluorescent lipid percolates through the EVL epithelium, however, it remains localized within the interstitial fluid of the embryo and freely diffuses between cells (Figure 13.5.11). Vital staining with BODIPY® FL C_5-ceramide thus allows hundreds of cells to be imaged *en masse* during morphogenetic movements.[83,86]

CellTrace™ BODIPY® TR Methyl Ester

Many research and biotechnological applications require detailed three- and four-dimensional visualization of embryonic cells labeled with Green Fluorescent Protein (GFP) within their native tissue environments. Fluorescent counterstains that label all the cells in a living embryo provide a histological context for the GFP-expressing cells in the specimen. The red-fluorescent CellTrace™ BODIPY® TR methyl ester (C34556, Figure 13.5.12) is an excellent counterstain for cells and tissues that are expressing GFP.[87] This dye readily permeates cell membranes and selectively stains mitochondria and endomembranous organelles such as endoplasmic reticulum and the Golgi apparatus, but does not appear to localize in the plasma membrane. These localization properties make the dye an ideal vital stain that can be used to reveal: (1) the location and shapes of cell nuclei, (2) the shapes of cells within embryonic tissues and (3) the boundaries of organ-forming tissues within the whole embryo.[87] Furthermore, CellTrace™ BODIPY® TR methyl ester staining is retained after formaldehyde fixation and permeabilization with Triton X-100, and the dye does not appear to produce any teratogenic effects on embryonic development. The emission spectra of enhanced GFP (EGFP) and CellTrace™ BODIPY® TR methyl ester are well separated, with peaks at 508 nm and 625 nm, respectively (Figure 13.5.13), allowing simultaneous dual-channel confocal imaging without significant overspill of GFP fluorescence into the CellTrace™ BODIPY® TR methyl ester detection channel.

The Image-iT® LIVE Intracellular Membrane and Nuclear Labeling Kit (I34407, Section 14.4) provides the red-fluorescent CellTrace™ BODIPY® TR methyl ester along with the blue-fluorescent Hoechst 33342 dye for highly selective staining of the intracellular membranes and nuclei, respectively, of live or fixed cells or tissues (Figure 13.5.14). These two fluorescent stains were especially chosen for their compatibility with live GFP-expressing cells, and they can be combined into one staining solution to save labeling time and wash steps while still providing optimal staining.

Pyrene, Nile Red and Bimane Probes

Nonpolar Pyrene Probe

1,3-Bis-(1-pyrene)propane (B311, Figure 13.5.15) has two pyrene moieties linked by a three-carbon alkylene spacer. This probe is somewhat analogous to the bis-pyrenyl phospholipids (Section 13.2) in that excimer formation (and, consequently, the fluorescence emission wavelength) is controlled by intramolecular rather than bimolecular interactions. Thus, this probe is highly sensitive to constraints imposed by its environment, and can therefore be used as a viscosity sensor for interior regions of lipoproteins, membranes, micelles, liquid crystals and synthetic polymers.[88] Because excimer formation results in a spectral shift (Figure 13.5.16), the probe may be useful for ratio imaging of molecular mobility.[89] However, pyrene fatty acids (Section 13.2) appear to be preferable for this purpose because the uptake of 1,3-bis-(1-pyrene)propane by cells is limited.

Nile Red

The phenoxazine dye nile red (N1142, Figure 13.5.17) is used to localize and quantitate lipids, particularly neutral lipid droplets within cells.[80,90–92] It is selective for neutral lipids such as cholesteryl esters[93,94] (and also, therefore, for lipoproteins) and is suitable for staining lysosomal phospholipid inclusions.[95] Nile red is almost nonfluorescent in water and other polar solvents but undergoes fluorescence enhancement and large absorption and emission blue shifts in nonpolar environments.[96,97] Its fluorescence enhancement upon binding to proteins is weaker than that produced by its association with lipids[97] (Figure 13.5.18). Ligand-binding studies on tubulin and tryptophan synthase[98] have exploited the environmental sensitivity of nile red's fluorescence. Nile red has also been used to detect sphingolipids on thin-layer chromatograms[99] and to stain proteins after SDS-polyacrylamide gel electrophoresis.[100]

IMPORTANT NOTICE: The products described in this manual are covered by one or more Limited Use Label License(s). Please refer to the Appendix on page 971 and Master Product List on page 975. Products are For Research Use Only. Not intended for any animal or human therapeutic or diagnostic use.
www.invitrogen.com/probes
molecular probes® | invitrogen™ by life technologies™

Bimane Azide

Bimane azide (B30600, Figure 13.5.19) is a small blue-fluorescent photoreactive alkyl azide (excitation/emission maxima ~375/458 nm) for photoaffinity labeling of proteins, potentially including membrane proteins from within the cell membrane. This reactive fluorophore's small size may reduce the likelihood that the label will interfere with the function of the biomolecule, an important advantage for site-selective probes.

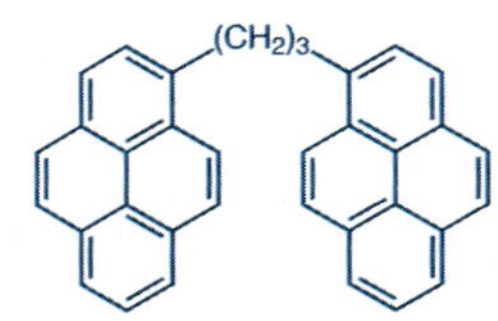

Figure 13.5.15 1,3-Bis-(1-pyrenyl)propane (B311).

LipidTOX™ Neutral Lipid Stains

Steatosis, the intracellular accumulation of neutral lipids as lipid droplets or globules, is often triggered by drugs that affect the metabolism of fatty acids or neutral lipids. LipidTOX™ neutral lipid stains were developed to characterize the effects of drugs and other compounds on lipid metabolism in mammalian cell lines. LipidTOX™ neutral lipid stains have an extremely high affinity for neutral lipid droplets. These reagents are added after cell fixation and do not require subsequent wash steps after incubation with the sample. Key advantages of this series of neutral lipid stains over conventional stains such as nile red include their ready-to-use formulations, their flexibility for multiplexing protocols and their compatibility with LipidTOX™ phospholipid stains (H34350, H34351; Section 13.2).

LipidTOX™ neutral lipid stains are available with green, red and deep red fluorescence emission:

- HCS LipidTOX™ Green neutral lipid stain (H34475), with excitation/emission maxima ~495/505 nm (Figure 13.5.20)
- HCS LipidTOX™ Red neutral lipid stain (H34476), with excitation/emission maxima ~577/609 nm
- HCS LipidTOX™ Deep Red neutral lipid stain (H34477), with excitation/emission maxima ~637/655 nm

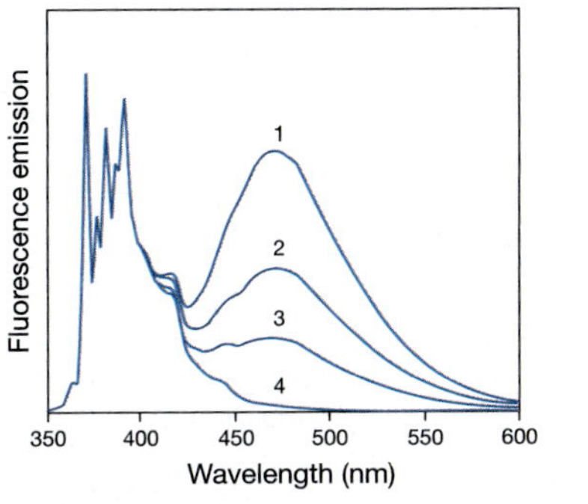

Figure 13.5.16 Excimer formation by pyrene in ethanol. Spectra are normalized to the 371.5 nm peak of the monomer. All spectra are essentially identical below 400 nm after normalization. Spectra are as follows: 1) 2 mM pyrene, purged with argon to remove oxygen; 2) 2 mM pyrene, air-equilibrated; 3) 0.5 mM pyrene (argon-purged); and 4) 2 µM pyrene (argon-purged). The monomer-to-excimer ratio (371.5 nm/470 nm) is dependent on both pyrene concentration and the excited-state lifetime, which is variable because of quenching by oxygen.

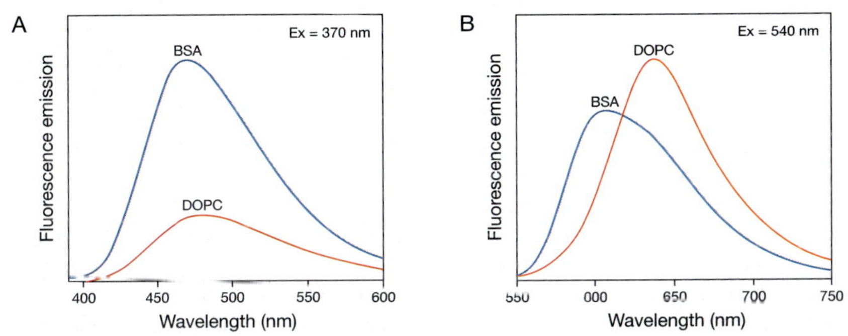

Figure 13.5.18 Fluorescence emission spectra of **A)** 1,8-ANS (A47) and **B)** nile red (N1142) bound to protein and phospholipid vesicles. Samples comprised 1 µM dye added to 20 µM bovine serum albumin (BSA) or 100 µM dioctadecenoylglycerophosphocholine (DOPC).

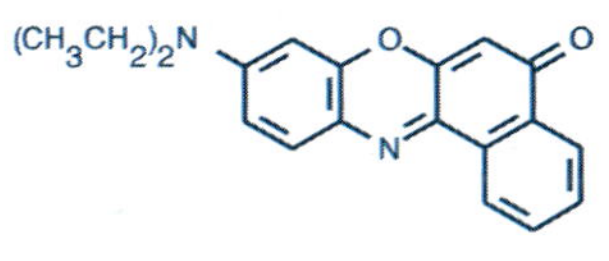

Figure 13.5.17 Nile red (N1142).

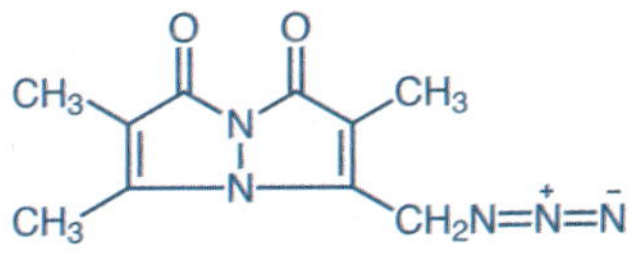

Figure 13.5.19 Bimane azide (B30600).

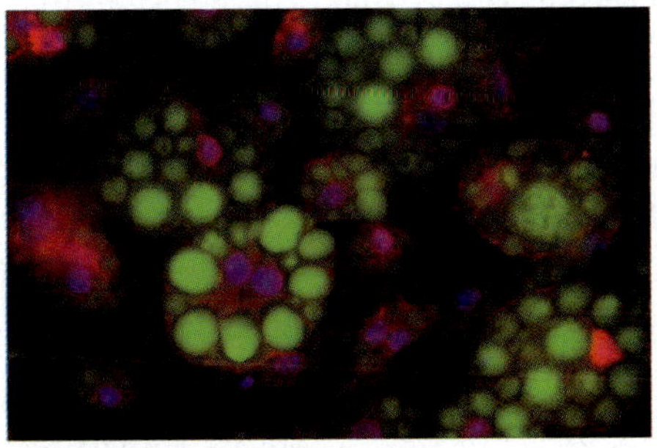

Figure 13.5.20 FABP4 antibody labeling in adipocytes. Adipocytes differentiated from 3T3-L1 mouse fibroblasts were fixed with formaldehyde and permeabilized with saponin before labeling with rabbit anti–fatty acid binding protein (FABP4) IgG (red). These cells were then stained with LipidTOX™ Green neutral lipid stain (H34475, green), counterstained with DAPI (D1306, D21490; blue) and mounted in ProLong® Gold antifade reagent (P36930).

IMPORTANT NOTICE: The products described in this manual are covered by one or more Limited Use Label License(s). Please refer to the Appendix on page 971 and Master Product List on page 975. Products are For Research Use Only. Not intended for any animal or human therapeutic or diagnostic use.

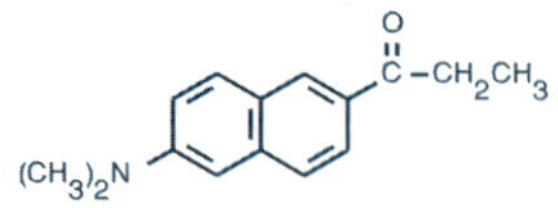

Figure 13.5.21 Prodan (6-propionyl-2-dimethylaminonaphthalene; P248).

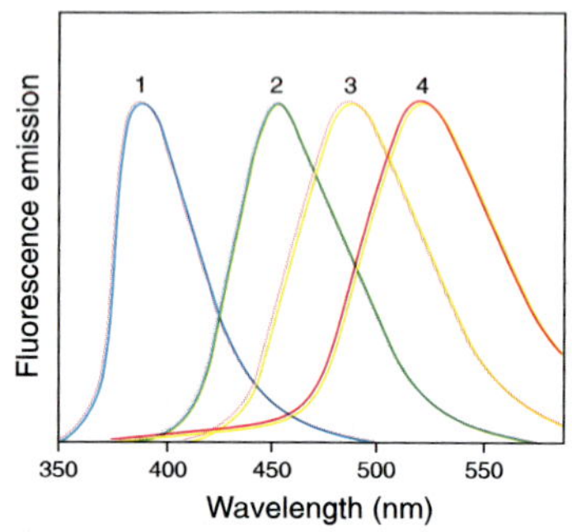

Figure 13.5.22 Normalized emission spectra of prodan (P248) excited at 345 nm in **1**) cyclohexane, **2**) dimethylformamide, **3**) ethanol and **4**) water.

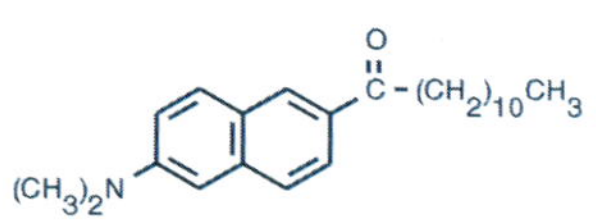

Figure 13.5.23 6-Dodecanoyl-2-dimethylaminonaphthalene (laurdan, D250).

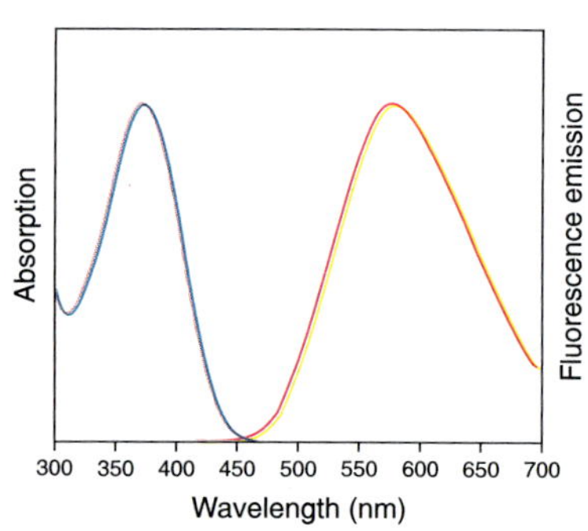

Figure 13.5.24 Absorption and fluorescence emission spectra of Dapoxyl® (2-aminoethyl)sulfonamide in methanol.

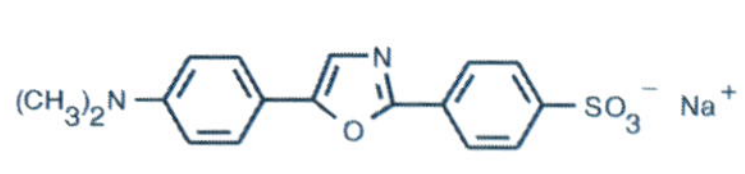

Figure 13.5.25 Dapoxyl® sulfonic acid, sodium salt (D12800).

These HCS LipidTOX™ neutral lipid stains have been used to image intracellular lipid accumulation in rat cortical neurons, COS-7 cells and hepatitis C virus (HCV)–infected FT3-7 human hepatoma cells.[101–103] HCS LipidTOX™ Red neutral lipid stain was used to detect RNAi knockdown of acyl-coenzyme A:cholesterol acyl transferase, isoform 1 (ACAT-1), an endoplasmic reticulum enzyme that regulates the equilibrium between free cholesterol and cholesteryl esters in cells.[104]

LipidTOX™ Green neutral lipid stain is also a component of the HCS LipidTOX™ Phospholipidosis and Steatosis Detection Kit (H34157, H34158; Section 13.2), which provides a complete set of reagents for performing high-content screening (HCS) assays to detect and distinguish the intracellular accumulation of phospholipids (phospholipidosis) and of neutral lipids (steatosis) in mammalian cell lines after exposure to test compounds. In addition, HCS LipidTOX™ neutral lipid stains can be used to monitor the formation and differentiation of adipocytes, a process called adipogenesis. Adipogenesis is of acute interest to the biomedical and drug discovery community as it plays an important role in diseases such as obesity, diabetes and atherosclerosis.

HCS LipidTOX™ neutral lipid stains are designed for fixed–end point workflows in which formaldehyde-fixed cells in microplates are processed, imaged and analyzed. These stains can easily be detected with fluorescence microscopes or HCS readers equipped with standard filter sets.

Membrane Probes with Environment-Sensitive Spectral Shifts

Prodan and Laurdan

Prodan (P248, Figure 13.5.21), introduced by Weber and Farris in 1979, has both electron-donor and electron-acceptor substituent, resulting in a large excited-state dipole moment and extensive solvent polarity–dependent fluorescence shifts[105] (Figure 13.5.22). Several variants of the original probe have since been prepared, including the lipophilic derivative laurdan (D250, Figure 13.5.23) and thiol-reactive derivatives acrylodan and badan (A433, B6057; Section 2.3), which can be used to confer the environment-sensitive properties of this fluorophore on bioconjugates.

When prodan or its derivatives are incorporated into membranes, their fluorescence spectra are sensitive to the physical state of the surrounding phospholipids.[106] In membranes, prodan appears to localize at the surface,[107] although Fourier transform infrared (FTIR) measurements indicate some degree of penetration into the lipid interior.[108] Excited-state relaxation of prodan is sensitive to the nature of the linkage (ester or ether) between phospholipid hydrocarbon tails and the glycerol backbone.[109] In contrast, laurdan's excited-state relaxation is independent of head-group type, and is instead determined by water penetration into the lipid bilayer.[110,111] Two-photon infrared excitation techniques have been successfully applied to both prodan and laurdan, although both probes nominally require ultraviolet excitation[112–115] (~360 nm).

Much experimental work using these probes has sought to characterize coexisting lipid domains based on their distinctive fluorescence spectra,[113,116–120] an approach that is intrinsically amenable to dual-wavelength ratio measurements.[111,121] Other applications include detecting non-bilayer lipid phases,[122,123] mapping changes in membrane structure induced by cholesterol and alcohols[124–127] and assessing the polarity of lipid/water interfaces.[128,129] Like ANS, prodan is also useful as a noncovalently interacting probe for proteins.[130–133]

Dapoxyl® Derivative

We have developed a variety of probes based on our Dapoxyl® fluorophore.[134] Dapoxyl® sulfonamide derivatives exhibit UV absorption with maxima near 370 nm, extinction coefficients >24,000 $cm^{-1}M^{-1}$ and Stokes shifts in excess of 200 nm (Figure 13.5.24). Dapoxyl® sulfonic acid (D12800, Figure 13.5.25) is an amphiphilic Dapoxyl® derivative with generally similar properties and applications to anilinonaphthalene sulfonate (ANS) (Monitoring Protein-Folding Processes with Environment-Sensitive Dyes—Note 9.1). Both ANS and Dapoxyl® sulfonic acid have been used in a drug-discovery assay based on the detection of protein thermal denaturation shifts.[135] Reactive versions of the Dapoxyl® fluorophore are described in Section 1.7 and Section 3.4.

Anilinonaphthalenesulfonate (ANS) and Related Derivatives

The use of anilinonaphthalene sulfonates (ANS) as fluorescent probes dates back to the pioneering work of Weber in the 1950s, and this class of probes remains valuable for studying both membrane

IMPORTANT NOTICE: The products described in this manual are covered by one or more Limited Use Label License(s). Please refer to the Appendix on page 971 and Master Product List on page 975. Products are For Research Use Only. Not intended for any animal or human therapeutic or diagnostic use.
www.invitrogen.com/probes
molecular probes® | invitrogen™ by life technologies™

surfaces and proteins. Slavik's 1982 review of its properties is recommended reading, especially for the extensive compilation of spectral data.[136] The primary member of this class, 1,8-ANS (A47, Figure 13.5.26), and its analogs 2,6-ANS (A50) and 2,6-TNS (T53) are all essentially nonfluorescent in water, only becoming appreciably fluorescent when bound to membranes (quantum yields ~0.25) or proteins (quantum yields ~0.7)[136–138] (Figure 13.5.18). This property makes them sensitive indicators of protein folding, conformational changes[139–142] and other processes that modify the exposure of the probe to water (Monitoring Protein-Folding Processes with Environment-Sensitive Dyes—Note 9.1). Fluorescence of 2,6-ANS is also enhanced by cyclodextrins, permitting a sensitive method for separating and analyzing cyclodextrins with capillary electrophoresis.[143]

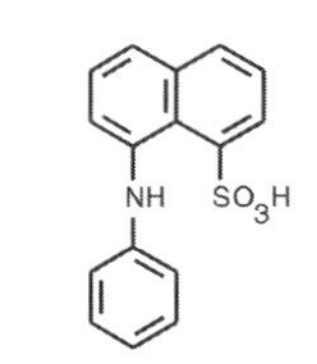

Figure 13.5.26 1,8-ANS (1-anilinonaphthalene-8-sulfonic acid, A47).

Bis-ANS

Bis-ANS (B153, Figure 13.5.27) is superior to 1,8-ANS as a probe for nonpolar cavities in proteins, often binding with an affinity that is orders-of-magnitude higher.[144–147] Bis-ANS has particularly high affinity for nucleotide-binding sites of some proteins.[148–150] It is also useful as a structural probe for tubulin[151,152] and as an inhibitor of microtubule assembly.[153–155] Covalent photoincorporation of bis-ANS into proteins has been reported.[156]

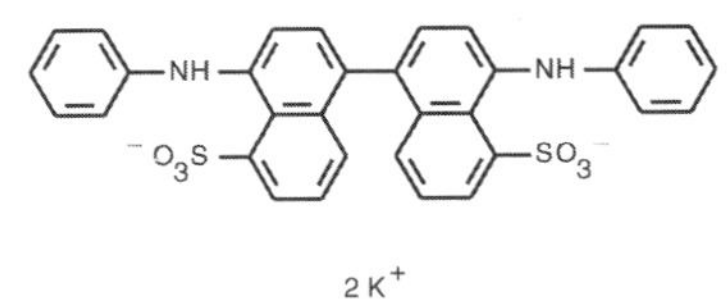

Figure 13.5.27 bis-ANS (4,4′-dianilino-1,1′-binaphthyl-5,5′-disulfonic acid, dipotassium salt; B153).

DCVJ

The styrene derivative DCVJ (D3923, Figure 13.5.28) is a sensitive indicator of tubulin assembly and actin polymerization.[157,158] The fluorescence quantum yield of DCVJ is strongly dependent on environmental rigidity, resulting in large fluorescence increases when the dye binds to antibodies[159] and when it is compressed in synthetic polymers or phospholipid membrane interiors.[160,161] DCVJ has been used for microviscosity measurements of phospholipid bilayers.[161]

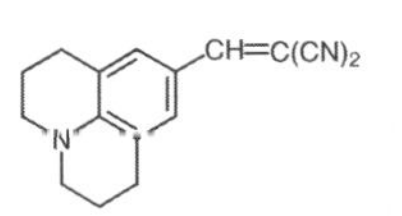

Figure 13.5.28 DCVJ (4-(dicyanovinyl)julolidine, D3923).

REFERENCES

1. Biochim Biophys Acta (1998) 1374:63; **2**. Biophys J (1999) 77:943; **3**. Photochem Photobiol (1994) 60:563; **4**. Biophys Chem (1989) 34:283; **5**. Chem Phys Lipids (2002) 116:39; **6**. Biochemistry (1984) 23:5675; **7**. J Cell Sci (1977) 28:167; **8**. Biophys J (1996) 71:2680; **9**. Biochim Biophys Acta (1995) 1237:121; **10**. Biophys J (1992) 63:710; **11**. J Gen Physiol (1991) 97:1101; **12**. Proc Natl Acad Sci U S A (1990) 87:1850; **13**. FEBS Lett (1989) 250:487; **14**. Biophys J (1999) 76:1812; **15**. Biochim Biophys Acta (1994) 1189:175; **16**. J Biol Chem (1992) 267:17012; **17**. Biochemistry (1991) 30:9125; **18**. Biochemistry (1990) 29:8741; **19**. J Cell Biol (1988) 107:2233; **20**. Langmuir (1994) 10:658; **21**. Eur J Cell Biol (1992) 59:116; **22**. Eur Biophys J (1990) 18:277; **23**. FEBS Lett (1989) 257:10; **24**. Cytometry (1996) 24:368; **25**. Biophys J (1996) 70:988; **26**. J Cell Sci (1991) 100:473; **27**. Exp Cell Res (1989) 181:375; **28**. Nature (1981) 294:718; **29**. Biochem Biophys Res Commun (1992) 184:160; **30**. Biochim Biophys Acta (1996) 1284:191; **31**. J Phys Chem (1977) 81:1755; **32**. Biochim Biophys Acta (1973) 323:326; **33**. Methods Enzymol (1989) 171:376; **34**. Biochemistry (1990) 29:7275; **35**. Biochim Biophys Acta (1983) 729:185; **36**. Biochemistry (1985) 24:573; **37**. Biochemistry (1991) 30:5565; **38**. Chem Phys Lipids (1991) 57:39; **39**. Biochimie (1989) 71:23; **40**. Biochim Biophys Acta (1986) 859:209; **41**. Biochim Biophys Acta (1988) 941:102; **42**. Biophys Chem (1993) 48:205; **43**. Biophys J (1991) 59:466; **44**. Biophys J (1989) 56:723; **45**. Biophys Chem (1994) 48:337; **46**. Biochim Biophys Acta (1992) 1104:273; **47**. Biochemistry (1990) 29:3248; **48**. Chem Phys Lipids (1989) 50:1; **49**. Chem Phys Lipids (1993) 64:99; **50**. Biochim Biophys Acta (1981) 649:471; **51**. Chem Phys Lipids (1993) 64:117; **52**. Biochim Biophys Acta (1986) 854:38; **53**. Chem Phys (1994) 185:393; **54**. Chem Phys Lett (1993) 216:559; **55**. Biochemistry (1998) 37:8180; **56**. Biochemistry (1988) 27:7723; **57**. Biochemistry (1981) 20:7333; **58**. Biochem Biophys Res Commun (1991) 181:166; **59**. Chem Phys Lipids (1990) 55:29; **60**. Biochemistry (1987) 26:5121; **61**. Biochemistry (1987) 26:5113; **62**. Biochim Biophys Acta (1985) 845:60; **63**. Cell Biophys (1983) 5:129; **64**. Am J Physiol (1988) 255:F22; **65**. Biochim Biophys Acta (1995); **66**. Biochemistry (1986) 25:2149; **67**. J Cell Biol (1996) 135:1741; **68**. Biochim Biophys Acta (1993) 1147:194; **69**. Fluorescent and Luminescent Probes for Biological Activity, Mason WT, Ed. (1993) p. 420; **70**. Am J Physiol (1991) 260:C1; **71**. FASEB J (1991) 5:2078; **72**. Biophys J (1990) 57:1199; **73**. Cytometry (2000) 39:151; **74**. Biochim Biophys Acta (1991) 1067:71; **75**. Plant Physiol (1990) 94:729; **76**. J Photochem Photobiol A (1999) 121:177; **77**. Anal Biochem (1991) 198:228; **78**. Heteroatomic Chem (1990) 1:389; **79**. Optics Comm (1989) 70:425; **80**. Cytometry (1994) 17:151; **81**. Biophys J (1996) 71:2656; **82**. Electrophoresis (2003) 24:3253; **83**. Methods Mol Biol (1999) 122:185; **84**. Neuron (1998) 20:1081; **85**. J Agric Food Chem (2000) 48:1150; **86**. Methods Cell Biol (1999) 59:179; **87**. Dev Dyn (2007) 232:359; **88**. Biochim Biophys Acta (1993) 1149:86; **89**. Eur J Cell Biol (1994) 65:172; **90**. J Histochem Cytochem (1997) 45:743; **91**. J Cell Biol (1993) 123:1567; **92**. Exp Cell Res (1992) 199:29; **93**. J Cell Biol (1989) 108:2201; **94**. J Chromatogr (1987) 421:136; **95**. Histochemistry (1992) 97:349; **96**. Anal Chem (1990) 62:615; **97**. Anal Biochem (1987) 167:228; **98**. J Biol Chem (1995) 270:6357; **99**. Anal Biochem (1993) 208:121; **100**. Biotechniques (1996) 21:625; **101**. J Biol Chem (2009) 284:3049; **102**. J Virol (2008) 82:7624; **103**. J Biol Chem (2009) 284:2383; **104**. FEBS Lett (2007) 581:1688; **105**. Photochem Photobiol (1993) 58:499; **106**. Photochem Photobiol (1999) 70:557; **107**. Biochemistry (1988) 27:399; **108**. Biochemistry (1989) 28:8358; **109**. Biochemistry (1990) 29:11134; **110**. Biophys J (1994) 66:763; **111**. Biophys J (1991) 60:179; **112**. Anal Chem (2001) 73:2302; **113**. Biophys J (2000) 78:290; **114**. Biophys J (1999) 77:2090; **115**. Biophys J (1997) 72:2413; **116**. Biophys J (2001) 80:1417; **117**. Biophys J (1994) 66:120; **118**. Photochem Photobiol (1993) 57:420; **119**. Biophys J (1990) 57:1179; **120**. J Biol Chem (1990) 265:20044; **121**. Photochem Photobiol (1993) 57:403; **122**. Biochemistry (1992) 31:1550; **123**. Biophys J (1990) 57:925; **124**. Biochim Biophys Acta (2001) 1511:330; **125**. Biophys J (1995) 68:1895; **126**. Biophys J (1993) 65:1404; **127**. Biochemistry (1992) 31:9473; **128**. J Biol Chem (1994) 269:10298; **129**. J Biol Chem (1994) 269:7429; **130**. Biochemistry (1998) 37:7167; **131**. Biochem J (1993) 290:411; **132**. Eur J Biochem (1992) 204:127; **133**. Nature (1986) 319:70; **134**. J Photochem Photobiol A (2000) 131:95; **135**. J Biomol Screen (2001) 6:429; **136**. Biochim Biophys Acta (1982) 694:1; **137**. Biophys J (1998) 74:422; **138**. Biochemistry (1968) 7:3381; **139**. Biochemistry (1999) 38:2110; **140**. Biochemistry (1998) 37:4621; **141**. Biochemistry (1998) 37:13862; **142**. Biochemistry (1998) 37:16802; **143**. J Chromatogr A (1994) 680:233; **144**. Arch Biochem Biophys (1989) 268:239; **145**. Biochemistry (1985) 24:3852; **146**. Biochemistry (1985) 24:2034; **147**. Biochemistry (1969) 8:3915; **148**. Biochemistry (1992) 31:2982; **149**. Biochim Biophys Acta (1990) 1040:66; **150**. Proc Natl Acad Sci U S A (1977) 74:2334; **151**. Biochemistry (1998) 37:4687; **152**. Biochemistry (1994) 33:11891; **153**. Biochemistry (1998) 37:17571; **154**. Biochemistry (1992) 31:6470; **155**. J Biol Chem (1984) 259:14647; **156**. Biochemistry (1995) 34:7443; **157**. Anal Biochem (1992) 204:110; **158**. Biochemistry (1989) 28:6678; **159**. Biochemistry (1993) 32:7589; **160**. Chem Phys (1993) 169:351; **161**. Biochemistry (1986) 25:6114.

IMPORTANT NOTICE: The products described in this manual are covered by one or more Limited Use Label License(s). Please refer to the Appendix on page 971 and Master Product List on page 975. Products are For Research Use Only. Not intended for any animal or human therapeutic or diagnostic use.

DATA TABLE 13.5 OTHER NONPOLAR AND AMPHIPHILIC PROBES

Cat. No.	MW	Storage	Soluble	Abs	EC	Em	Solvent	Notes
A47	299.34	L	pH >6, DMF	372	7800	480	MeOH	1
A50	299.34	L	DMF	319	27,000	422	MeOH	1
B153	672.85	L	pH >6	395	23,000	500	MeOH	1, 2
B311	444.57	L	MeCN, $CHCl_3$	344	80,000	378	MeOH	3
B3932	448.32	F,L	DMSO, $CHCl_3$	665	161,000	676	MeOH	4
B30600	233.23	F,D,L	DMSO	375	6000	458	MeOH	
C34556	438.25	F,D,L	DMSO	588	68,000	616	MeOH	5
D109	529.63	L	DMSO, EtOH	495	85,000	518	MeOH	6
D202	232.32	L	DMF, MeCN	350	88,000	452	MeOH	7, 8
D250	353.55	L	DMF, MeCN	364	20,000	497	MeOH	9
D3921	248.08	F,L	EtOH, DMSO	502	98,000	510	MeOH	4
D3922	262.11	F,L	EtOH, DMSO	493	89,000	504	MeOH	4
D3923	249.31	L	DMF, DMSO	456	61,000	493	MeOH	
D12800	366.37	L	DMSO, H_2O	358	25,000	517	MeOH	10
F3857	584.79	L	DMSO, EtOH	504	95,000	525	MeOH	6
H110	585.74	L	DMSO, EtOH	497	92,000	519	MeOH	6
H22730	400.60	L	DMSO, EtOH	366	20,000	453	MeOH	6
H34475	~300	F,L	DMSO	495	94,000	505	MeOH	5, 13
H34476	~400	F,L	DMSO	574	62,000	609	MeOH	5, 13
H34477	~350	F,L	DMSO	626	68,000	648	MeOH	5, 13
N1142	318.37	L	DMF, DMSO	552	45,000	636	MeOH	11
O246	731.50	F,DD,L	DMSO, EtOH	556	125,000	578	MeOH	12
P248	227.31	L	DMF, MeCN	363	19,000	497	MeOH	9
T53	335.35	L	DMF	318	26,000	443	MeOH	1
T204	461.62	D,L	DMF, DMSO	355	75,000	430	MeOH	7

For definitions of the contents of this data table, see "Using *The Molecular Probes® Handbook*" in the introductory pages.

Notes

1. Fluorescence quantum yields of ANS and its derivatives are environment dependent and are particularly sensitive to the presence of water. QY of A47 is about 0.4 in EtOH, 0.2 in MeOH and 0.004 in water. Em is also somewhat solvent dependent. (Biochim Biophys Acta (1982) 694:1)
2. B153 is soluble in water at 0.1–1.0 mM after heating.
3. Absorption spectra of bis-pyrenyl alkanes have additional peaks at ~325 nm and <300 nm. Emission spectra include both monomer (~380 nm and ~400 nm) and excimer (~470 nm) peaks.
4. The absorption and fluorescence spectra of BODIPY® derivatives are relatively insensitive to the solvent.
5. This product is supplied as a ready-made solution in the solvent indicated under "Soluble."
6. Spectra of this product are pH dependent. Data listed are for basic solutions prepared in methanol containing a trace of KOH.
7. Diphenylhexatriene (DPH) and its derivatives are essentially nonfluorescent in water. Absorption and emission spectra have multiple peaks. The wavelength, resolution and relative intensity of these peaks are environment dependent. Abs and Em values are for the most intense peak in the solvent specified.
8. Stock solutions of DPH (D202) are often prepared in in tetrahydrofuran (THF). Long-term storage of THF solutions is not recommended because of possible peroxide formation in that solvent.
9. The emission spectrum of P248 is solvent dependent. Em = 401 nm in cyclohexane, 440 nm in $CHCl_3$, 462 nm in MeCN, 496 nm in EtOH and 531 nm in H_2O. (Biochemistry (1979) 18:3075) Abs is only slightly solvent dependent. The emission spectra of D250 in these solvents are similar to those of P248.
10. Em = 520 nm when bound to phospholipid bilayer membranes. Fluorescence in H_2O is weak (Em ~600 nm).
11. The absorption and fluorescence spectra and fluorescence quantum yield of N1142 are highly solvent dependent. (J Lipid Res (1985) 26:781, Anal Biochem (1987) 167:228)
12. This product is intrinsically a sticky gum at room temperature.
13. Abs/Em in trioctanylglycerol = 498/507 nm, 582/616 nm and 635/652 nm for H34475, H34476 and H34477 respectively.

IMPORTANT NOTICE: The products described in this manual are covered by one or more Limited Use Label License(s). Please refer to the Appendix on page 971 and Master Product List on page 975. Products are For Research Use Only. Not intended for any animal or human therapeutic or diagnostic use.

www.invitrogen.com/probes

molecular probes® | invitrogen™ by life technologies™

PRODUCT LIST 13.5 OTHER NONPOLAR AND AMPHIPHILIC PROBES

Cat. No.	Product	Quantity
A47	1,8-ANS (1-anilinonaphthalene-8-sulfonic acid) *high purity*	100 mg
A50	2,6-ANS (2-anilinonaphthalene-6-sulfonic acid)	100 mg
B30600	Bimane azide	5 mg
B153	bis-ANS (4,4′-dianilino-1,1′-binaphthyl-5,5′-disulfonic acid, dipotassium salt)	10 mg
B3932	(E,E)-3,5-bis-(4-phenyl-1,3-butadienyl)-4,4-difluoro-4-bora-3a,4a-diaza-*s*-indacene (BODIPY® 665/676)	5 mg
B311	1,3-bis-(1-pyrenyl)propane	25 mg
C34556	CellTrace™ BODIPY® TR methyl ester *lipophilic counterstain for GFP* *solution in DMSO*	1 mL
D12800	Dapoxyl® sulfonic acid, sodium salt	10 mg
D3923	DCVJ (4-(dicyanovinyl)julolidine)	25 mg
D3922	4,4-difluoro-1,3,5,7,8-pentamethyl-4-bora-3a,4a-diaza-*s*-indacene (BODIPY® 493/503)	10 mg
D3921	4,4-difluoro-1,3,5,7-tetramethyl-4-bora-3a,4a-diaza-*s*-indacene (BODIPY® 505/515)	10 mg
D109	5-dodecanoylaminofluorescein	100 mg
D250	6-dodecanoyl-2-dimethylaminonaphthalene (laurdan)	100 mg
D202	DPH (1,6-diphenyl-1,3,5-hexatriene)	100 mg
F3857	Fluorescein octadecyl ester	10 mg
H34477	HCS LipidTOX™ Deep Red neutral lipid stain *solution in DMSO* *for cellular imaging*	125 µL
H34475	HCS LipidTOX™ Green neutral lipid stain *solution in DMSO* *for cellular imaging*	125 µL
H34476	HCS LipidTOX™ Red neutral lipid stain *solution in DMSO* *for cellular imaging*	125 µL
H22730	4-heptadecyl-7-hydroxycoumarin	10 mg
H110	5-hexadecanoylaminofluorescein	100 mg
N1142	Nile red	25 mg
O246	Octadecyl rhodamine B chloride (R18)	10 mg
P248	Prodan (6-propionyl-2-dimethylaminonaphthalene)	100 mg
T204	TMA-DPH (1-(4-trimethylammoniumphenyl)-6-phenyl-1,3,5-hexatriene p-toluenesulfonate)	25 mg
T53	2,6-TNS (2-(p-toluidinyl)naphthalene-6-sulfonic acid, sodium salt)	100 mg

IMPORTANT NOTICE: The products described in this manual are covered by one or more Limited Use Label License(s). Please refer to the Appendix on page 971 and Master Product List on page 975. Products are For Research Use Only. Not intended for any animal or human therapeutic or diagnostic use.

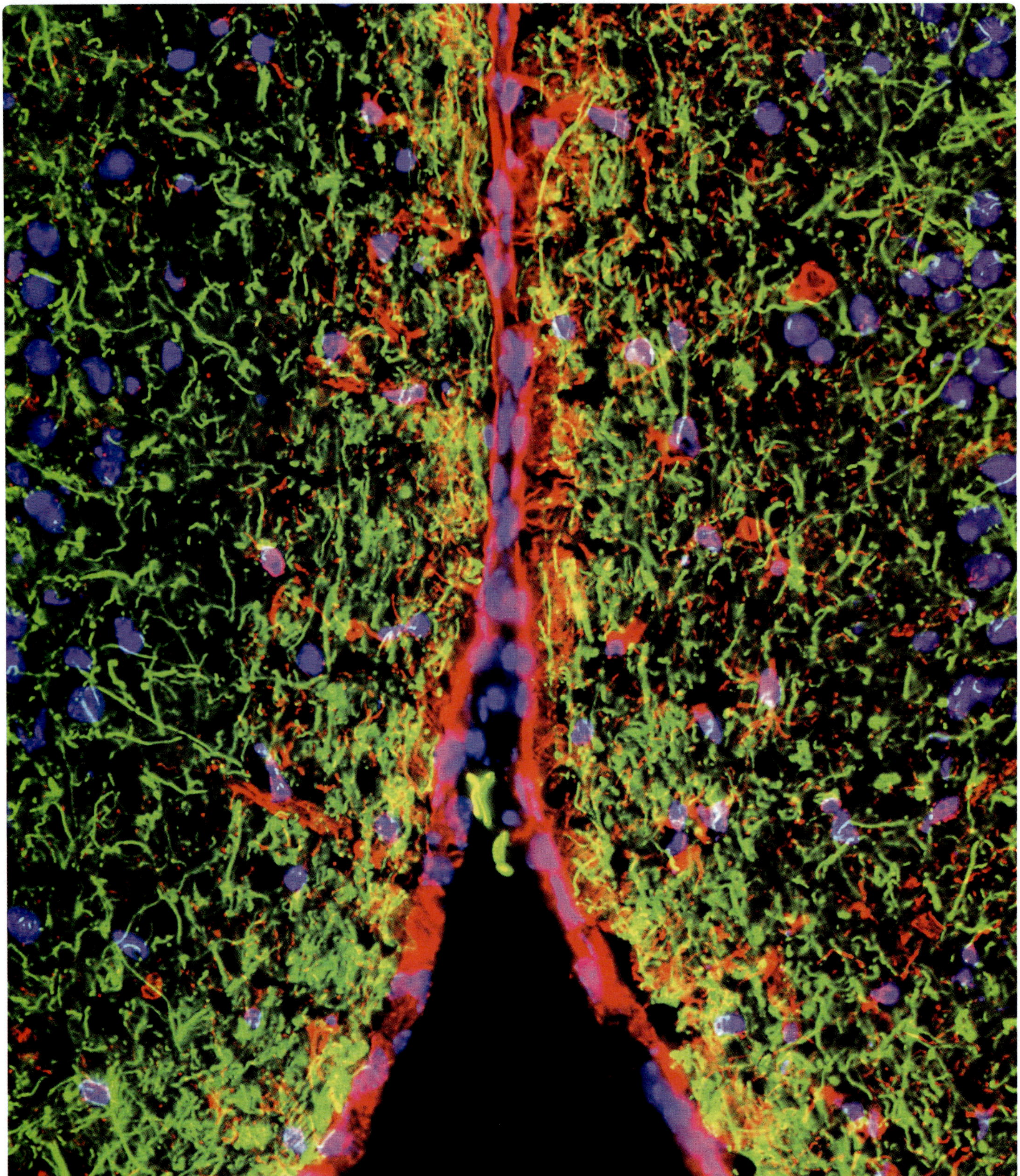

Alexa Fluor® 568 goat anti-rabbit IgG, Alexa Fluor® 488 anti-chicken IgG, and Hoechst 33342.

IMPORTANT NOTICE: The products described in this manual are covered by one or more Limited Use Label License(s). Please refer to the Appendix on page 971 and Master Product List on page 975. Products are For Research Use Only. Not intended for any animal or human therapeutic or diagnostic use.

FOURTEEN

CHAPTER 14

Fluorescent Tracers of Cell Morphology and Fluid Flow

IMPORTANT NOTICE: The products described in this manual are covered by one or more Limited Use Label License(s). Please refer to the Appendix on page 971 and Master Product List on page 975. Products are For Research Use Only. Not intended for any animal or human therapeutic or diagnostic use.

IMPORTANT NOTICE: The products described in this manual are covered by one or more Limited Use Label License(s). Please refer to the Appendix on page 971 and Master Product List on page 975. Products are For Research Use Only. Not intended for any animal or human therapeutic or diagnostic use.

IMPORTANT NOTICE: The products described in this manual are covered by one or more Limited Use Label License(s). Please refer to the Appendix on page 971 and Master Product List on page 975. Products are For Research Use Only. Not intended for any animal or human therapeutic or diagnostic use.

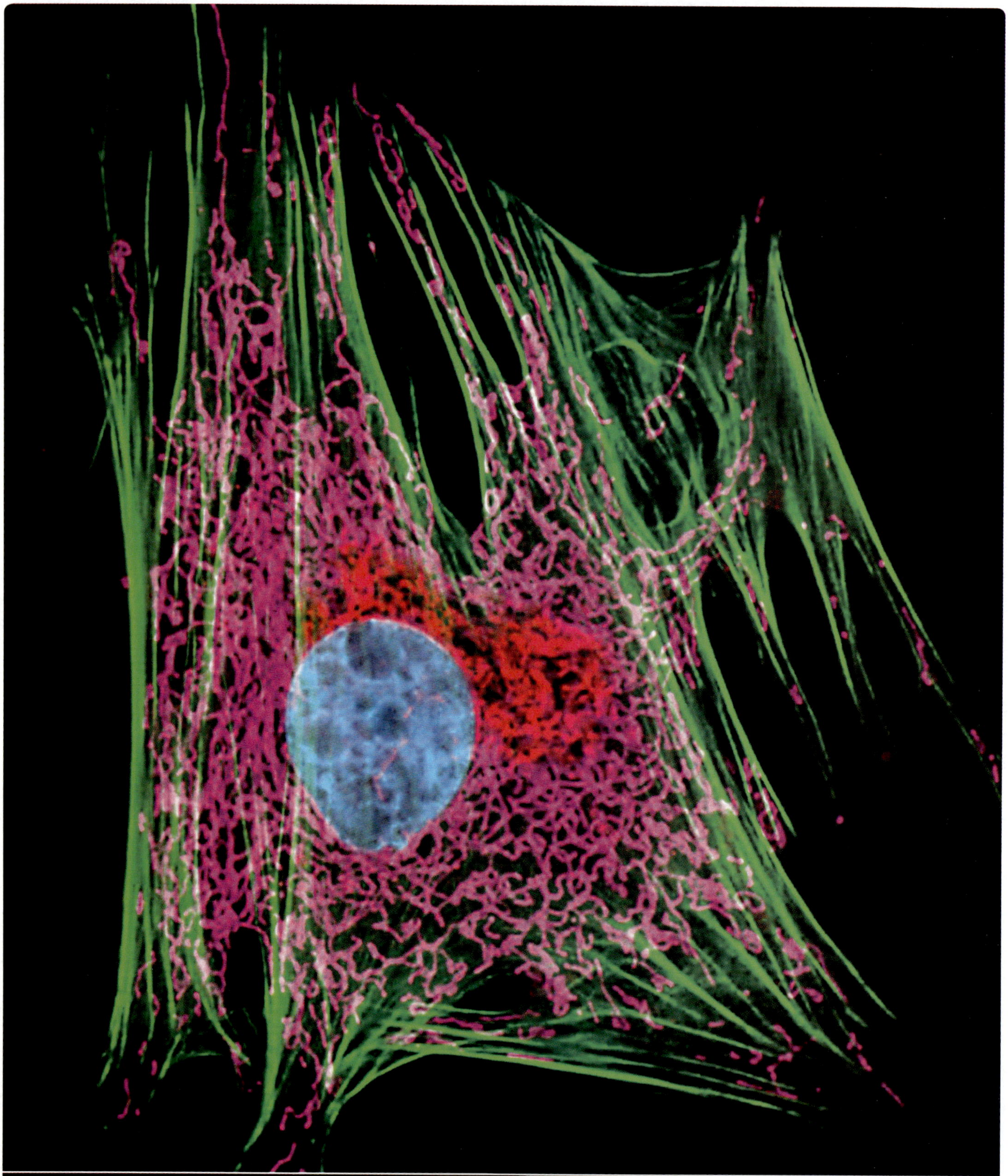

MitoTracker® Deep Red 633, Alexa Fluor® 568 goat anti–rabbit IgG, Alexa Fluor® 488 phalloidin and Hoechst 33258.

IMPORTANT NOTICE: The products described in this manual are covered by one or more Limited Use Label License(s). Please refer to the Appendix on page 971 and Master Product List on page 975. Products are For Research Use Only. Not intended for any animal or human therapeutic or diagnostic use.

14.1 Choosing a Tracer

To serve as an effective tracer of cell morphology, a fluorescent probe or other detectable molecule must have the capacity for localized introduction into a cell or organelle, as well as for long-term retention within that structure. If used with live cells and tissues, then the tracer should also be biologically inert and nontoxic. When these conditions are satisfied, the fluorescence or other detectable properties of the tracer can be used to track the position of the tracer over time. Fluorescent tracers can be employed to investigate flow in capillaries, to define neuronal cell connectivity and to study dye translocation through gap junctions, as well as to follow cell division, cell lysis or liposome fusion. Furthermore, they can be used to track the movements of labeled cells in culture, tissues or intact organisms. The review of techniques for tracing neuronal pathways by Bohland and co-workers [1] is particularly recommended.

Although the predominant tracers have been fluorescent, not all of the useful tracers are intrinsically detectable. For example, biotin derivatives are widely used as polar tracers, especially in neurons. However, when a biotinylated or haptenylated tracer is used in live cells, detection requires cell fixation and permeabilization to allow access to fluorescent dye– or enzyme-labeled conjugates of avidin and streptavidin (Section 7.6) or of antibodies (Section 7.4).

In many of these tracing applications, the physical dimensions of the tracer molecule are an important consideration. We offer fluorescent tracers ranging in size from small molecules about 1 nm in diameter to polystyrene microspheres up to 15 µm in diameter. This chapter discusses our diverse selection of fluorescent tracers, as well as biotin derivatives and other tracers:

- **Cell-permeant cytoplasmic labels (Section 14.2).** We have developed several thiol-reactive CellTracker™ probes (Figure 14.1.1), which yield fluorescent products that are retained in many live cells through several generations and are not transferred to adjacent cells in a population, except possibly by transport through gap junctions. These probes represent a significant breakthrough in the cellular retention of fluorescent dyes and are ideal long-term tracers for transplanted cells or tissues.
- **Microinjectable cytoplasmic labels (Section 14.3).** Polar tracers such as lucifer yellow CH, Cascade Blue® hydrazide, the Alexa Fluor® hydrazides and biocytin are membrane-impermeant probes that are usually introduced into cells by whole-cell patch clamping, iontophoresis, osmotic lysis of pinocytic vesicles or comparable methods [2,3] (Table 14.1). These tracers are commonly used to investigate cell–cell and cell–liposome fusion, as well as membrane permeability [4] and transport through gap junctions [5] or cell uptake during pinocytosis (Section 16.1).
- **Nissl stains for retrograde tracing in neurons (Section 14.3).** We have developed five fluorescent Nissl stains that not only provide a wide spectrum of fluorescent colors for staining neurons, but are also more sensitive than the conventional dyes.

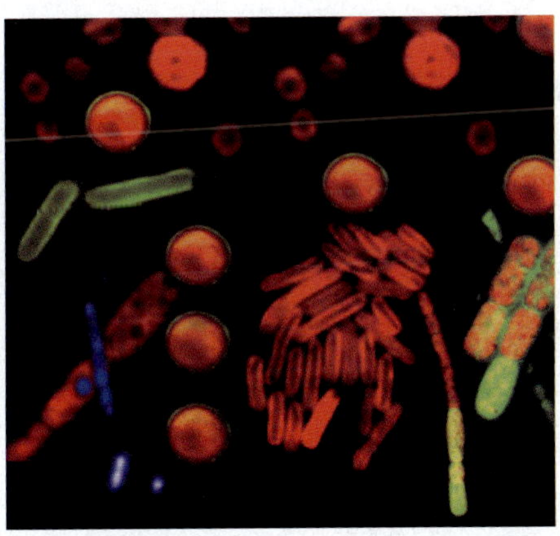

Figure 14.1.1 Collage of images of cyanobacteria stained with various blue- or green-fluorescent probes to complement the natural red autofluorescence from chlorophyll and phycobilisomes. The round cells are *Synechocystis* sp. (strain PCC 6803), and their membranes were labeled with green-fluorescent BODIPY® FL propionic acid (D2183). The cylindrical cells are *Synechococcus* sp. (strain PCC 7942), stained with blue-fluorescent DAPI (D1306, D3571, D21490). The filamentous cyanobacteria, *Anabaena cylindrica*, were labeled with either the green-fluorescent cytosolic stain, CellTracker™ Green BODIPY® (C2102), or with the lipophilic membrane stain BODIPY® FL (D2183). The image was contributed by Mary Sarcina, University College, London.

Table 14.1 Techniques for loading molecules into the cytoplasm.

Method of Plasma Membrane Breach	Size of Molecules Loaded (MW) *
Chemical	
ATP [1]	1000
Cell-penetrating peptides [2–5] (CPP)	150,000; (DNA)
Influx™ pinocytic cell-loading reagent [6–8] (I14402)	150,000
α-Toxin of *Staphylococcus aureus* [9,10]	1000
Vehicle	
Red blood cell fusion [11,12]	300,000
Cationic lipids [13–15] (Lipofectamine® reagents)	(DNA)
Mechanical	
Microinjection [16,17]	150,000
Whole-cell patch clamping [18]	Very high
Carbon nanotubes and nanowires [19–21]	150,000; (DNA)
Scrape loading [8]	500,000, (DNA)
Ultrasonication [22]	70,000; (DNA)
Microprojectile bombardment [23,24]	10,000; (DNA)
Glass beads [25]	150,000
Electrical	
Electroporation [26–29] †	150,000; (DNA)

* Molecular weight (MW) of largest molecules reported loaded (DNA is listed separately if it has been successfully introduced by a technique). † We recommend the Neon™ Transfection System (www.invitrogen.com/handbook/neon) for efficient electroporation of cultured cell lines, primary cells and stem cells.

1. J Neurosci (2003) 23:4054; **2.** Chembiochem (2010) 11:325; **3.** Bioconjug Chem (2009) 20:249; **4.** Methods Mol Biol (2009) 480:101; **5.** Org Biomol Chem (2008) 6:4516; **6.** Methods Enzymol (2006) 414:211; **7.** Infect Immun (2006) 74:3673; **8.** Antisense Nucleic Acid Drug Dev (2000) 10:263; **9.** Proc Natl Acad Sci U S A (2001) 98:3185; **10.** Methods (2003) 30:191; **11.** Cell (1975) 5:371; **12.** J Cell Biol (1985) 101:19; **13.** J Biol Chem (2007) 282:22953; **14.** Mol Ther (2004) 9:443; **15.** Biotechniques (2000) 28:510; **16.** Dev Cell (2007) 13:523; **17.** J Biomed Sci (2003) 10:328; **18.** Neuron (2001) 30:737; **19.** Nano Lett (2009) 9:2193; **20.** Proc Natl Acad Sci U S A (2010) 107:1870; **21.** Proc Natl Acad Sci U S A (2007) 104:8218; **22.** BMC Cancer (2002) 2:20; **23.** J Neurosci Methods (2009) 184:332; **24.** J Neurosci Methods (2005) 141:41; **25.** J Cell Sci (1987) 88:669; **26.** PLoS One (2009) 4:e7966; **27.** Nat Protoc (2009) 4:862; **28.** J Neurosci Meth (2009) 177:273; **29.** J Neurosci Methods (2009) 178:80.

IMPORTANT NOTICE: The products described in this manual are covered by one or more Limited Use Label License(s). Please refer to the Appendix on page 971 and Master Product List on page 975. Products are For Research Use Only. Not intended for any animal or human therapeutic or diagnostic use.

- **Membrane tracers—DiI, DiO, DiD, DiR, DiA, R18, FM® 1-43, FM® 4-64 and their analogs (Section 14.4).** Lipophilic carbocyanine, aminostyryl and rhodamine dyes can be introduced into membranes by direct application of a dye crystal onto a cell, by bulk loading from aqueous dispersions prepared from our Vybrant® DiI, DiO and DiD cell-labeling solutions or by application of the NeuroTrace® DiI, DiO and DiD tissue-labeling pastes. Lateral diffusion of the dye within the membrane eventually stains the entire cell. These probes are widely used for neuroanatomical tracing and long-term assays of cell–cell association. Some of our DiI and DiO analogs exhibit superior solubility and brightness and, in some cases, produce a cell-staining pattern that persists through fixation by aldehyde-based reagents and through acetone permeabilization (Figure 14.1.2).
- **Fluorescent and biotinylated dextran conjugates (Section 14.5).** Dextran conjugates are ideal cell-lineage tracers because they are relatively inert, exhibit low toxicity and are retained in cells for long periods. These membrane-impermeant probes are usually loaded into cells by invasive techniques such as microinjection, whole-cell patch clamping, scrape loading, microprojectile bombardment, electroporation or osmotic shock (Table 14.1). Availability of dextrans in a range of molecular weights makes them useful as size-exclusion probes for determining pore sizes in membranes.
- **Fluorescent microspheres (Section 14.6).** Molecular Probes® FluoSpheres® and TransFluoSpheres® fluorescent microspheres—which contain ~10^2 to ~10^{10} fluorescent dyes per bead are intensely fluorescent tracers (Figure 14.1.3). Although other multiply labeled particles such as our BioParticles® fluorescent bacteria (Section 16.1) may be used as tracers, they are often not biologically inert nor are they as physically durable as fluorescent microspheres. These properties make fluorescent beads particularly useful as long-term markers for transplantation studies. Submicron microspheres can be injected into cells or taken up by phagocytosis. Much larger (10–15 µm) beads provide an alternative to radioactive microspheres for determination of organ blood flow, and intermediate-sized (1–5 µm) microspheres are useful for studies that trace inhaled particles.
- **Qdot® nanocrystal tracers (Section 14.6).** Qtracker® Cell Labeling Kits provide spectrally distinct Qdot® nanocrystals that have been functionalized on their surface with polyarginine peptides to facilitate spontaneous uptake by live cells. Qtracker® non-targeted quantum dots are designed for small animal *in vivo* imaging, and especially for studying vascular structure after microinjection. These nanocrystals exhibit intense red or near-infrared fluorescence emission for maximum transmission through tissues and avoidance of background autofluorescence
- **Proteins and protein conjugates (Section 14.7).** Our fluorescent protein tracers have molecular weights between ~12,000 (cholera toxin subunit B conjugates) and ~240,000 daltons (B- and R-phycoerythrin). Their applications are sometimes similar to those of the fluorescent dextrans; however, unlike the polydisperse dextrans, fluorescent protein tracers have molecular weights that are reasonably well defined.

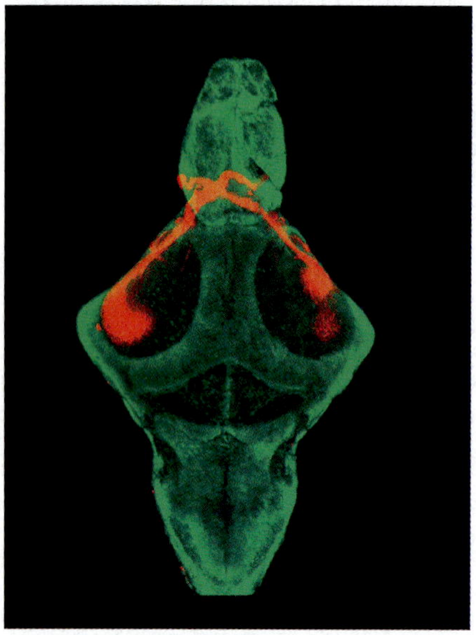

Figure 14.1.2 The tracer DiI (D282, D3911) used as a diagnostic tool to evaluate retinal and tectal patterns in a chimeric 5-day-old zebrafish brain. *Acerebellar* (*Fgf8*) gene expression is required for development of midbrain polarity and correct retinotopic mapping of retinal ganglion cell axons. One eye of this brain was wild type, whereas the other was mutant for *fgf8* (derived from homozygous acerebellar donors). The chimera was obtained by transplanting eye primordia, and the eyes were subsequently removed for photography. Retinal ganglion cell axons in the dorsonasal retina were labeled with DiI and nuclei were counterstained with SYTOX® Green nucleic acid stain (S7020). Axons from the wild-type eye terminate correctly in the posterior tectum only, whereas axons from the mutant eye terminate inappropriately in the anterior and posterior tectum. This digital image, previously published on the cover of Development 125 (13) (1999), was created by Alexander Picker and Michael Brand, Neurobiology, University of Heidelberg. The image is used with the permission of The Company of Biologists, Ltd.

REFERENCES

1. PLoS Comput Biol (2009) 5:e1000334; **2.** J Neurosci Methods (2002) 117:159; **3.** Nat Protoc (2009) 4:862; **4.** Cytometry (1995) 21:230; **5.** Biotechniques (2008) 45:33.

Figure 14.1.3 A photomicrograph of a multicolor mixture of Molecular Probes® FluoSpheres® fluorescent microspheres overlaid with a differential interference contrast (DIC) image of the same field. Molecular Probes® proprietary fluorescent dye technology is used to produce a range of intensely fluorescent FluoSpheres® microspheres labeled with biotin, streptavidin and NeutrAvidin™ biotin-binding protein, providing important tools for improving the sensitivity of flow cytometry applications and immunodiagnostic assays.

14.2 Membrane-Permeant Reactive Tracers

Thiol-Reactive CellTracker™ Probes

Molecular Probes® CellTracker™ reagents are fluorescent chloromethyl derivatives that freely diffuse through the membranes of live cells (Figure 14.2.1, Figure 14.2.2, Figure 14.2.3). Once inside the cell, these mildly thiol-reactive probes undergo what is believed to be a glutathione *S*-transferase–mediated reaction to produce membrane-impermeant glutathione–fluorescent dye adducts, although our experiments suggest that they may also react with other intracellular components (Figure 14.2.4). Regardless of the mechanism, many cell types loaded with the CellTracker™ probes are both fluorescent and viable for at least 24 hours after loading and often through several cell divisions. Most other cell-permeant fluorescent dyes, including the acetoxymethyl (AM) esters of calcein and BCECF (Section 15.2), are retained in viable cells for no more than a few hours at physiological temperatures (Figure 14.2.5). Furthermore, unlike the free dye, the peptide–fluorescent dye adducts contain amino groups and can therefore be covalently linked to surrounding biomolecules by fixation with formaldehyde or glutaraldehyde. This property permits long-term storage of the labeled cells or tissue and, in cases where the cognate anti-dye antibody is available (see below), amplification of the dye hapten by standard immunohistochemical techniques. Fixation without loss of the tracer also facilitates the safe handling and analysis of cells containing pathogens.

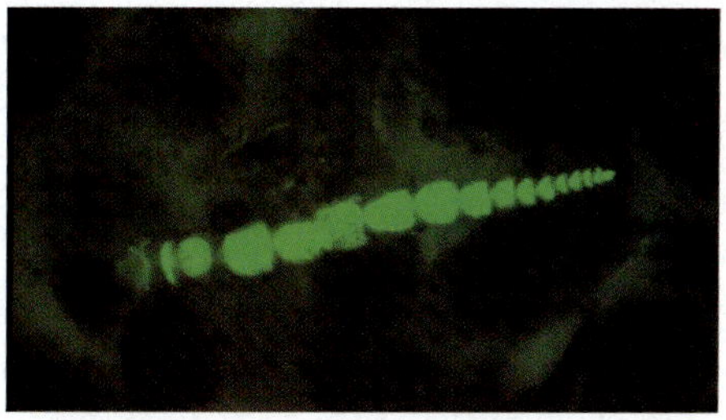

Figure 14.2.1 Detection of organisms in marine sediments by incubating an intact sediment core sample with the fixable, cell-permeant CellTracker™ Green CMFDA (C2925, C7025). The core sample was subsequently embedded, sectioned and examined for fluorescently labeled organisms. The micrograph reveals the microorganism *Leptohalysis scotti*, a marine benthic foraminifera. Image contributed by Joan M. Bernhard, Wadsworth Center, New York State Department of Health.

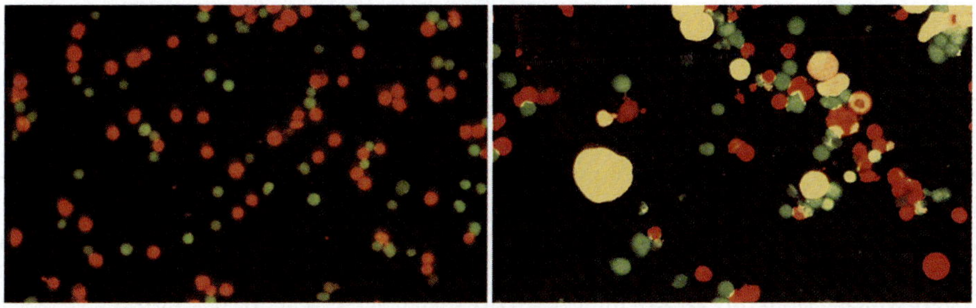

Figure 14.2.2 HL60 cells that have been stained with CellTracker™ Orange CMTMR (C2927) and then mixed with WEHI 7.1 cells stained with CellTracker™ Green CMFDA (C2925, C7025) (left). Several minutes after initiating cell–cell electrofusion, a CMTMR-stained HL60 cell is observed fusing with a CMFDA-stained WEHI 7.1 cell; cytoplasmic mixing is evident by the appearance of yellow fluorescence. After electrofusion is complete, dual-fluorescing (yellow) hybrids can be easily distinguished (right). Images contributed by Mark J. Jaroszeski, University of South Florida.

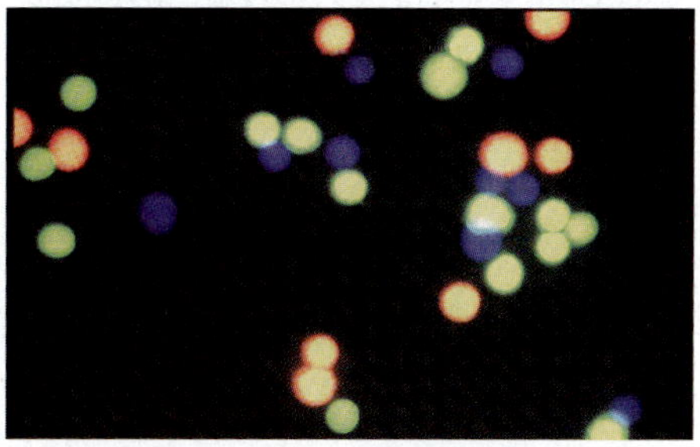

Figure 14.2.3 Individual populations of mouse myeloma (P3X) cells stained with our reactive tracers—CellTracker™ Orange CMTMR (C2927), CellTracker™ Green CMFDA (C2925) and CellTracker™ Blue CMAC (C2110). Each of three cell populations was stained with a different tracer and then the populations were mixed, demonstrating that these tracers allow simultaneous long-term monitoring of different groups of cells in transplantation and other assays.

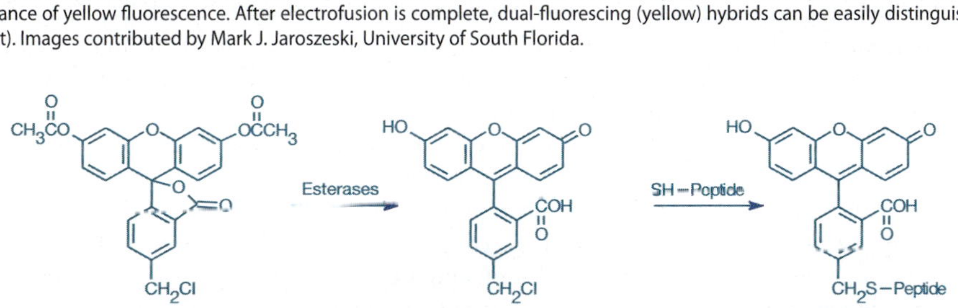

Figure 14.2.4 Intracellular reactions of our fixable CellTracker™ Green CMFDA (5-chloromethylfluorescein diacetate; C2925, C7025). Once this membrane-permeant probe enters a cell, esterase hydrolysis converts nonfluorescent CMFDA to fluorescent 5-chloromethylfluorescein, which can then react with thiols on proteins and peptides to form aldehyde-fixable conjugates. This probe may also react with intracellular thiol-containing biomolecules first, but the conjugate is nonfluorescent until its acetates are removed.

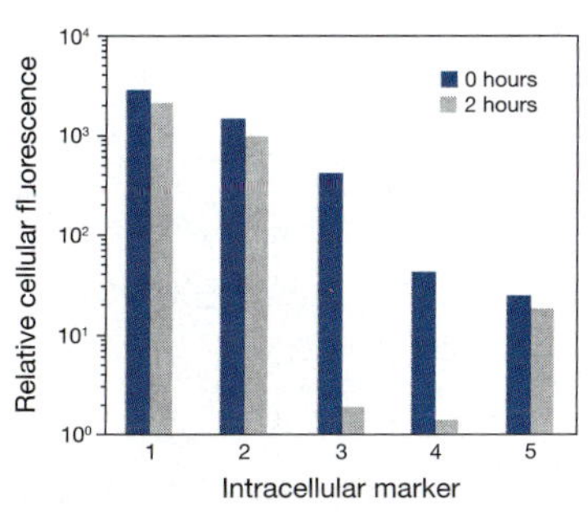

Figure 14.2.5 Loading and retention characteristics of intracellular marker dyes. Cells of a human lymphoid line (GePa) were loaded with the following cell-permeant acetoxymethyl ester (AM) or acetate derivatives of fluorescein: **1**) calcein AM (C1430, C3099, C3100MP), **2**) BCECF AM (B1150), **3**) fluorescein diacetate (FDA, F1303), **4**) carboxyfluorescein diacetate (CFDA, C1354) and **5**) CellTracker™ Green CMFDA (5-chloromethylfluorescein diacetate, C2925, C7025). Cells were incubated in 4 µM staining solutions in Dulbecco's modified eagle medium containing 10% fetal bovine serum (DMEM+) at 37°C. After incubation for 30 minutes, cell samples were immediately analyzed by flow cytometry to determine the average fluorescence per cell at time zero (0 hours). Retained cell samples were subsequently washed twice by centrifugation, resuspended in DMEM+, maintained at 37°C for 2 hours and then analyzed by flow cytometry. The decrease in the average fluorescence intensity per cell in these samples relative to the time zero samples indicates the extent of intracellular dye leakage during the 2-hour incubation period.

IMPORTANT NOTICE: The products described in this manual are covered by one or more Limited Use Label License(s). Please refer to the Appendix on page 971 and Master Product List on page 975. Products are For Research Use Only. Not intended for any animal or human therapeutic or diagnostic use.

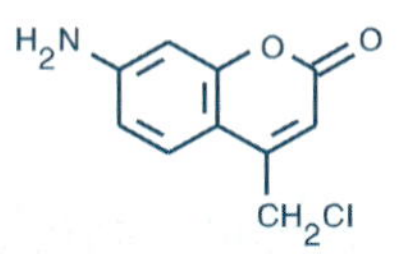

Figure 14.2.6 CellTracker™ Blue CMAC (7-amino-4-chloromethylcoumarin, C2110).

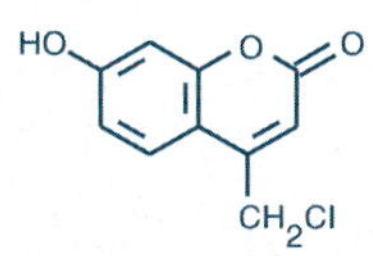

Figure 14.2.7 CellTracker™ Blue CMHC (4-chloromethyl-7-hydroxycoumarin, C2111).

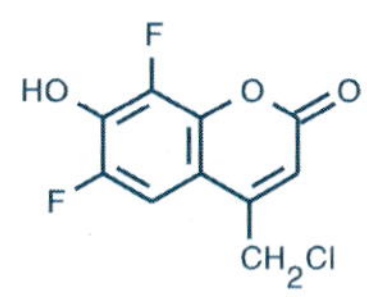

Figure 14.2.8 CellTracker™ Blue CMF_2HC (4-chloromethyl-6,8-difluoro-7-hydroxycoumarin; C12881).

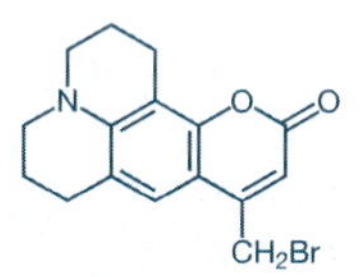

Figure 14.2.9 CellTracker™ Violet BMQC (2,3,6,7-tetrahydro-9-bromomethyl-1*H*,5*H*-quinolizino(9,1-*gh*)coumarin; C10094).

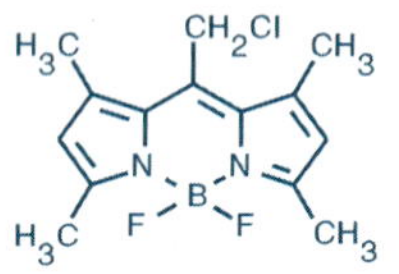

Figure 14.2.10 CellTracker™ Green BODIPY® (8-chloromethyl-4,4-difluoro-1,3,5,7-tetramethyl-4-bora-3a,4a-diaza-*s*-indacene; C2102).

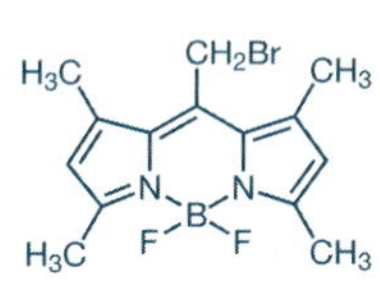

Figure 14.2.14 BODIPY® 493/503 methyl bromide (8-bromomethyl-4,4-difluoro-1,3,5,7-tetramethyl-4-bora-3a,4a-diaza-*s*-indacene; B2103).

CellTracker™ Probes in a Variety of Fluorescent Colors

Molecular Probes® CellTracker™ product line includes reactive chloromethyl derivatives of:

- Blue-fluorescent 7-aminocoumarin (CellTracker™ Blue CMAC, C2110, Figure 14.2.6)
- Blue-fluorescent 7-hydroxycoumarin (CellTracker™ Blue CMHC, C2111, Figure 14.2.7)
- Blue-fluorescent 6,8-difluoro-7-hydroxycoumarin (CellTracker™ Blue CMF_2HC, C12881, Figure 14.2.8)
- Violet-fluorescent 2,3,6,7-tetrahydro-9- bromomethyl-1*H*,5*H*-quinolizino(9,1-*gh*)coumarin (CellTracker™ Violet BMQC, C10094, Figure 14.2.9)
- Green-fluorescent fluorescein diacetate (CellTracker™ Green CMFDA, C2925 and C7025, Figure 14.2.4)
- Green-fluorescent BODIPY® derivative (CellTracker™ Green BODIPY®, C2102, Figure 14.2.10)
- Orange-fluorescent tetramethylrhodamine (CellTracker™ Orange CMTMR, C2927, Figure 14.2.11)
- Orange-fluorescent CellTracker™ Orange CMRA (C34551, Figure 14.2.12)
- Red-fluorescent CellTracker™ Red CMTPX (C34552, Figure 14.2.13)

CellTracker™ Green CMFDA freely diffuses into the cell, where cytosolic esterases cleave the acetate groups, releasing the fluorescent product; CellTracker™ Orange CMRA also requires esterase cleavage to activate its fluorescence. The CellTracker™ Blue CMAC, CMHC and CMF_2HC, CellTracker™ Violet BMQC, CellTracker™ Green BODIPY®, CellTracker™ Orange CMTMR and CellTracker™ Red CMTPX probes do not require enzymatic cleavage to activate their fluorescence. CellTracker™ Orange CMRA is a rhodol-based fluorophore with an overall net charge at neutral pH that is expected to be negative, in contrast with the overall net positive charge of the tetramethylrhodamine-based CellTracker™ Orange CMTMR. Thus, unlike CellTracker™ Orange CMTMR, CellTracker™ Orange CMRA should remain primarily in the cytoplasm instead of being sequestered inside actively respiring mitochondria. The long-wavelength CellTracker™ Red CMTPX exhibits bright red fluorescence that is easily distinguished from that of blue-, green- and far-red–fluorescent probes, including CellTracker™ Green CMFDA and CFSE [1–3] (see below). We also offer the green-fluorescent BODIPY® 493/503 methyl bromide (B2103, Figure 14.2.14), which has slightly greater thiol-reactivity than do chloromethyl derivatives.

Applications for CellTracker™ Probes

The thiol-reactive CellTracker™ Green CMFDA is suitable for long-term cell labeling and has been frequently used for *in vitro* labeling of cells prior to adoptive transfer to animals and tissues.[2,4,5] Other prominent applications include the analysis of cell–cell fusion,[6,7] cell adhesion [8,9] and multidrug resistance transporter function.[10,11]

CellTracker™ Green CMFDA was used to track wild-type and myosin II mutant *Dictyostelium discoideum* cells within aggregation streams during early multicellular morphogenesis; differentiation and morphogenesis pathways were reportedly unaffected in labeled cells imaged over several days by confocal laser-scanning microscopy.[12,13] CellTracker™ Green BODIPY® was used to label $CD4^+$ T cells and follow their activation and proliferation in mice with the immunodeficiency syndrome MAIDS.[14] CellTracker™ Green CMFDA and some other CellTracker™ dyes have been utilized to selectively label intracellular *Chlamydia psittaci* bacteria in infected cells.[15]

CellTracker™ Orange CMTMR has been used to stain the cytoplasm of *engrailed*-expressing *Drosophila* cells in an *in vitro* reconstruction experiment,[16] to follow T-cell differentiation and maturation [17] and to trace implanted tumor cells for at least two weeks.[18,19] Chlamydia labeled

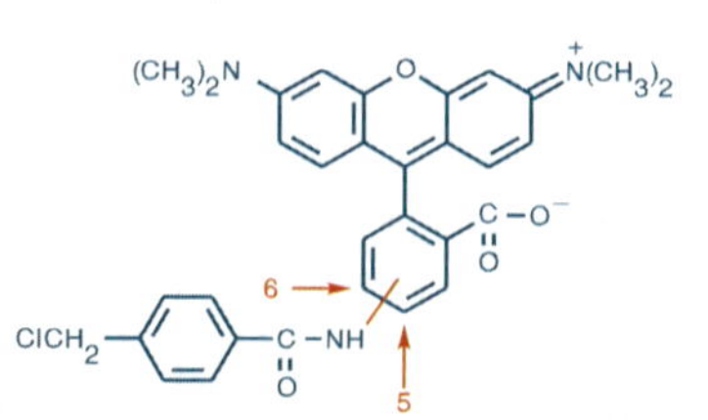

Figure 14.2.11 CellTracker™ Orange CMTMR (5-(and-6)-(((4-chloromethyl)benzoyl)amino)tetramethylrhodamine; C2927).

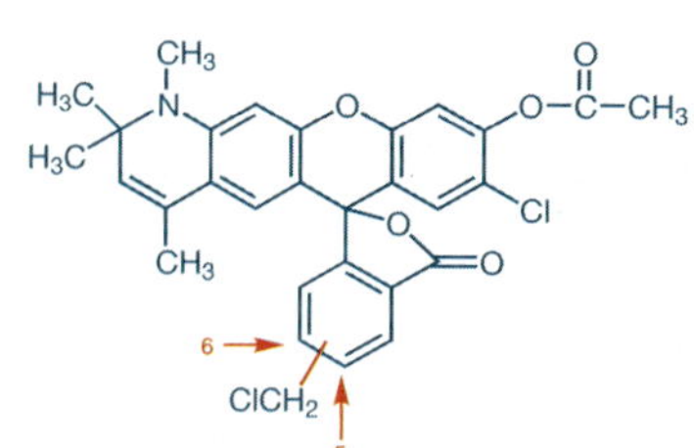

Figure 14.2.12 CellTracker™ Orange CMRA (C34551).

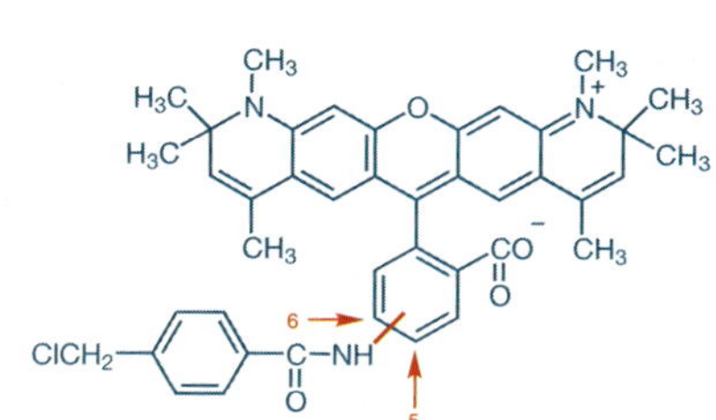

Figure 14.2.13 CellTracker™ Red CMTPX (C34552).

IMPORTANT NOTICE: The products described in this manual are covered by one or more Limited Use Label License(s). Please refer to the Appendix on page 971 and Master Product List on page 975. Products are For Research Use Only. Not intended for any animal or human therapeutic or diagnostic use.

www.invitrogen.com/probes

with CellTracker™ Orange CMTMR continued to grow and differentiate, and labeled chlamydia isolated from infected cells remained infectious.[15]

Metabolic activity and drug-induced cytotoxicity were measured with CellTracker™ Blue CMAC in a fluorescence-based microplate assay.[20] Peptidase substrates derived from CellTracker™ Blue CMAC yield blue-fluorescent products that are well retained in live cells[21] (Section 10.4). Our CellTracker™ Blue CMF_2HC (C12881) has a low pK_a that helps ensure that the dye's conjugates will be ionized and have bright blue fluorescence in the cytoplasm.[1,22] CellTracker™ Blue CMAC was used in combination with calcein AM (C1430, C3099, C3100MP; Section 15.6) and the lipophilic tracer DiI (D282, D3911, V22885; Section 14.4) to identify lipid mixing and cytoplasm mixing between labeled effector cells and labeled target cells.[23]

The ability to fix the intracellular products of CellTracker™ Green CMFDA, CellTracker™ Green BODIPY® and CellTracker™ Orange CMTMR in permeabilized cells permits the stained cells to be probed with our anti–fluorescein/Oregon Green®, anti–BODIPY® FL and anti-tetramethylrhodamine antibodies, respectively (Section 7.4, Table 7.8).

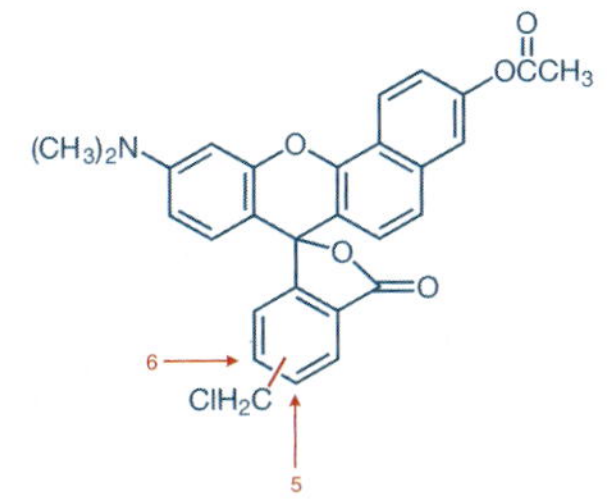

Figure 14.2.15 5-(and-6)-Chloromethyl SNARF®-1, acetate (C6826).

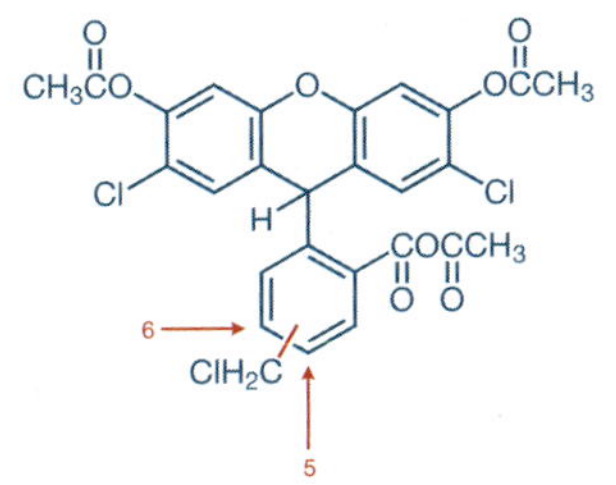

Figure 14.2.16 5-(and-6)-Chloromethyl-2′,7′-dichlorodihydrofluorescein diacetate, acetyl ester (CM-H_2DCFDA, C6827).

Other Thiol-Reactive Tracers

Chloromethyl Derivatives of SNARF®-1 and H_2DCFDA

Although they were designed for other purposes, the chloromethyl derivatives of our SNARF®-1 pH indicator (C6826, Figure 14.2.15) and of 2′,7′-dichlorodihydrofluorescein diacetate, acetate ester (CM-H_2DCFDA, C6827, Figure 14.2.16) possess some unique properties for tracking cells. As with the CellTracker™ probes, cytoplasmic enzymes hydrolytically remove the acetate groups from the membrane-permeant probes, and the chloromethyl moieties become conjugated to intracellular thiols.[24] With its long Stokes shift (Figure 14.2.17), the SNARF®-1 derivative has easily distinguished, red-orange fluorescence in the cytoplasm when excited at the same wavelengths used for the green-fluorescent CellTracker™ Green CMFDA (Figure 14.2.18).

As with other dihydrofluorescein derivatives (Section 18.2), CM-H_2DCFDA requires an additional oxidation step before becoming fluorescent. This probe is useful for following stimulation of oxidative activity by external agents or natural killer (NK) cells over extended periods, as well as for passively labeling cells that lack appropriate oxidative activity and then following their ingestion by scavengers such as neutrophils. CM-H_2DCFDA has been used to measure intracellular reactive oxygen species (ROS) in cardiac myocytes[25] and in human embryonic kidney 293 (HEK 293) cells stably transfected with the human vanilloid receptor 1 (VR1) cation channel.[26]

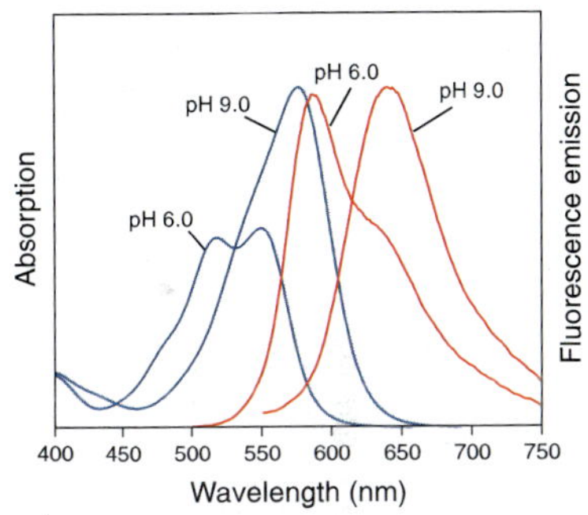

Figure 14.2.17 Absorption and fluorescence emission (excited at 488 nm) spectra of carboxy SNARF®-1 in pH 9.0 and pH 6.0 buffers.

Bimanes: Blue-Fluorescent Reactive Tracers

The bimane derivatives, monobromobimane (mBBr; M1378, M20381) and monochlorobimane (mBCl, M1381MP), are important thiol-derivatization reagents (Section 2.3). The essentially nonfluorescent mBBr and mBCl dyes are known to passively diffuse across the plasma membrane into the cytoplasm, where they form blue-fluorescent adducts with intracellular glutathione and thiol-containing proteins[27] (Section 15.6).

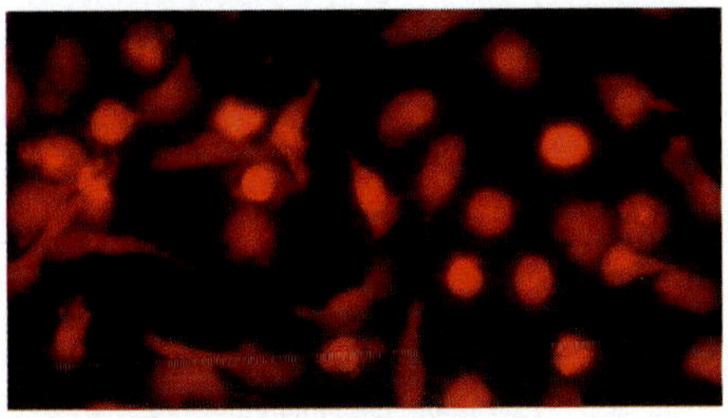

Figure 14.2.18 Human neutrophils loaded with 5-(and-6)-chloromethyl SNARF®-1 acetate (C6826).

Amine-Reactive Tracers

CFSE and Its Derivatives

Carboxyfluorescein diacetate succinimidyl ester (5(6)-CFDA, SE; C1157) is commonly referred to as CFSE. Among its many applications, the two most prominent are *ex vivo* labeling of cells for adoptive transfer and flow cytometric tracking of cell division by label partition analysis. CFSE is colorless and nonfluorescent until its acetate groups are cleaved by intracellular esterases to yield highly fluorescent, amine-reactive 5(6)-carboxyfluorescein succinimidyl ester (equivalent to C1311, Section 1.5). Upon reaction with amine-containing residues of intracellular proteins, the resulting dye–protein adducts are well retained in cells as they move and divide during embryonic development.[28,29] In addition, these adducts survive subsequent fixation with formaldehyde or glutaraldehyde. Because it is intrinsically more reactive, the succinimidyl ester of CFDA is more likely to react at sites on the extracellular surface than is CMFDA.

Once incorporated into cells, CFSE remains there—even through cell division—thus preventing transfer to unlabeled cells.[30] Labeled cells can subsequently be immunohistochemically

IMPORTANT NOTICE: The products described in this manual are covered by one or more Limited Use Label License(s). Please refer to the Appendix on page 971 and Master Product List on page 975. Products are For Research Use Only. Not intended for any animal or human therapeutic or diagnostic use.

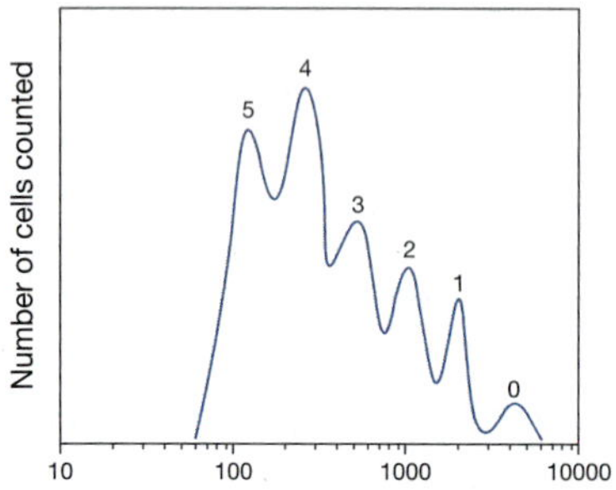

Figure 14.2.19 Tracking of asynchronous cell division using 5-(and 6-)carboxyfluorescein diacetate, succinimidyl ester (5(6)-CFDA SE or CFSE; C1157; V12883) labeling and flow cytometry. Cell division results in sequential halving of the initial fluorescence, resulting in a multipartite cellular fluorescence histogram. The peaks labeled 0, 1, 2, 3, 4 and 5 represent successive generations.

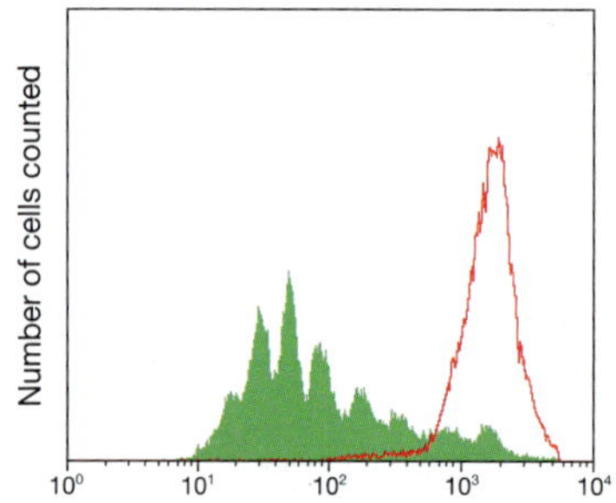

Figure 14.2.20 Following cell proliferation in human peripheral blood lymphocytes using the CellTrace™ CFSE Cell Proliferation Kit C34554). Human peripheral blood lymphocytes were harvested and stained with CellTrace™ CFSE (carboxyfluorescein diacetate, succinimidyl ester; 5(6)-CFDA, SE) on Day 0. A portion of the population was arrested at the parent generation using mitomycin C (red peak). The remainder of the sample was stimulated with phytohemagglutinin and allowed to proliferate for 5 days. Solid green peaks represent successive generations.

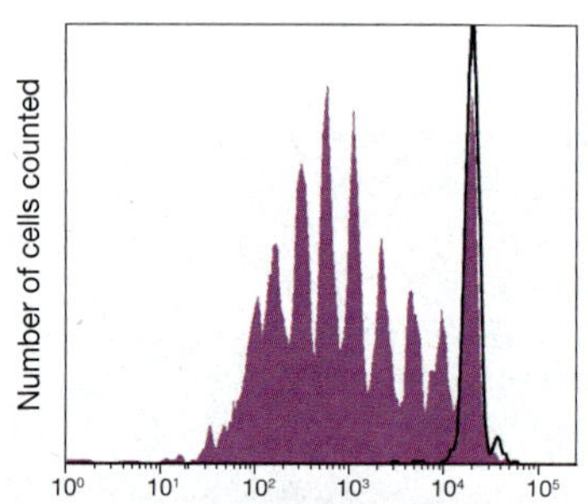

Figure 14.2.21 Human peripheral blood lymphocytes were harvested and stained with CellTrace™ Violet stain (C34557). The violet peaks represent successive generations of cells stimulated with mouse anti–human CD3 and Interleukin-2 and grown in culture for 7 days. The peak outlined in black represents cells that were grown in culture for 7 days with no stimulus.

detected post-mortem using our anti–fluorescein/Oregon Green® antibodies [31–33] (Section 7.4). The feasibility of using cell-permeant fluorescent tracers to follow asynchronous cell division of natural killer (NK) cells, B cells, T cells, thymocytes, lymphocytes, fibroblasts and hematopoietic cells has been demonstrated with CFSE [34–37] (Figure 14.2.19). Lymphocytes labeled with CFSE were detected up to eight weeks after injection into mice during lymphocyte migration studies,[38–40] and similarly labeled hepatocytes were easily located by fluorescence microscopy even 20 days after intrahepatic transplantation.[41] CFSE has also been successfully employed to quantitate adhesion of neutrophils [42] and leukocytes,[43] to assay T-cadherin–mediated cell aggregation,[44] to follow neurite growth in an *in vitro* bioassay [45] and to trace fetal cells in culture.[46] Different loading concentrations of CFSE can be used to produce cells that can be distinguished by their relative brightness.[47] CFSE has been utilized for tracing the transport of viable bacterial cells in groundwater under no-growth conditions for periods of at least 28 days.[48] CFSE is also available conveniently packaged for cell-tracing applications in our Vybrant® CFDA SE Cell Tracer Kit (V12883) and for cell proliferation studies in our CellTrace™ CFSE Cell Proliferation Kit (C34554, Figure 14.2.20). The fluorescent CFSE product has excitation/emission maxima of ~492/517 nm and can be detected using a fluorescence microscope, flow cytometer or fluorescence microplate reader.

Carboxyeosin diacetate succinimidyl ester (C22803) has applications that are expected to be similar to CFSE; however, its fluorescence can be excited and detected at longer wavelengths, possibly permitting two-color tracing experiments of mixed-cell populations. Eosin derivatives are also effective reagents for photoablation of cells. Unlike CFSE, fluorescence of the intracellular hydrolysis products of the succinimidyl ester of 5-(and 6-)carboxy-2′,7′-dichlorofluorescein diacetate (C1165) is relatively insensitive to fluctuations in pH. This amine-reactive tracer was reported to be more useful than the lipophilic marker DiI (D282, Section 14.4) in an investigation of palatal fusion in rodent embryos.[49]

CellTrace™ Violet Cell Proliferation Kit

CellTrace™ Violet stain is an esterase-activated phenolic fluorophore with a succinimidyl ester substituent for coupling to cell surface and intracellular amines. It is functionally analogous to CFSE, equally partitioning between daughter cells during division resulting in successive 2-fold reductions in cell-associated fluorescence intensity. When analyzed by flow cytometry, this progressive label partitioning provides a direct indication of cell proliferation status (Figure 14.2.21). In contrast to CFSE, CellTrace™ Violet stain is optimally excited by 405 nm violet diode lasers and generates blue fluorescence (emission peak ~455 nm). Consequently, it can be used in combination with CFSE to track cells from different origins after mixing or to analyze proliferation of GFP-expressing cells. The CellTrace™ Violet Cell Proliferation Kit (C34557) includes the CellTrace™ Violet stain together with dimethylsulfoxide (DMSO) for preparation of a stock solution.

CellTrace™ Oregon Green® 488 Carboxylic Acid Diacetate Succinimidyl Ester

The succinimidyl ester of Oregon Green® 488 carboxylic acid diacetate (carboxy-DFFDA SE) offers several important advantages over CFSE as a fluorescent cell tracer. This Oregon Green® 488 probe passively diffuses into cells, where it is colorless and nonfluorescent until its acetate groups are removed by intracellular esterases to yield a highly green-fluorescent, amine-reactive dye. Upon reaction with intracellular amines, the probe forms Oregon Green® 488 conjugates that are well-retained by cells. Unlike fluorescein derivatives, however, Oregon Green® 488 derivatives exhibit bright green fluorescence that is not pH dependent at typical cellular pH values. Moreover, Oregon Green® 488 probes are usually brighter and more photostable than fluorescein probes. We offer carboxy-DFFDA SE in a 1 mg unit size (O34550) and specially packaged in a set of 20 vials, each containing 50 µg (CellTrace™ Oregon Green® 488 carboxylic acid diacetate succinimidyl ester, C34555).

CellTrace™ Far Red DDAO-SE

CellTrace™ Far Red DDAO-SE (C34553) is a fixable, far-red–fluorescent tracer for long-term cell labeling. The succinimidyl ester (SE) reactive group forms a strong covalent attachment to primary amines that occur in proteins and other biomolecules inside and outside of cells. With its far-red fluorescence, CellTrace™ Far Red DDAO-SE has minimal spectral overlap with most other fluorophores (Figure 14.2.22) and thus can be used simultaneously with almost any blue, green or orange fluorophores including Green Fluorescent Protein [50–52] (GFP).

IMPORTANT NOTICE: The products described in this manual are covered by one or more Limited Use Label License(s). Please refer to the Appendix on page 971 and Master Product List on page 975. Products are For Research Use Only. Not intended for any animal or human therapeutic or diagnostic use.

Amine-Reactive SNARF®-1 Carboxylic Acid Acetate

To permit simultaneous long-term tracing of mixed-cell populations using different fluorescent colors, we have developed an amine-reactive probe whose applications are similar to those of CFSE.[53] Cells labeled with our succinimidyl ester of SNARF®-1 carboxylic acid acetate (S22801, Figure 14.2.23) have a red-orange fluorescence that can easily be distinguished from that of cells loaded with green-fluorescent tracers such as CFSE. However, the fluorescence intensity of cells loaded with this SNARF®-1 derivative will not be as high as that of cells loaded with the same concentration of CFSE. Thus, it is necessary to adjust the reagent concentration and/or select optical filters to appropriately balance the fluorescence intensities when doing two-color experiments.

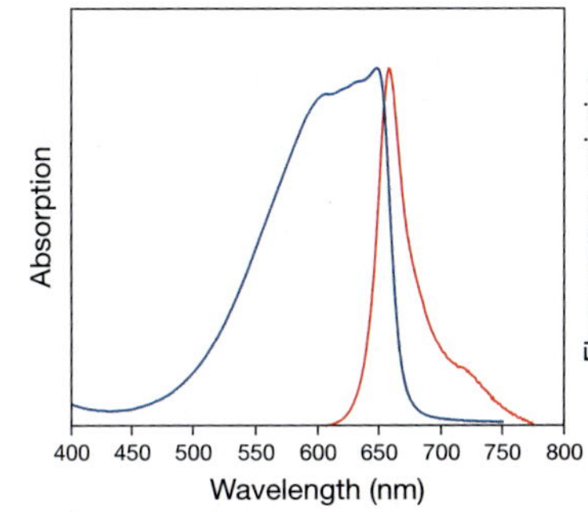

Figure 14.2.22 Normalized absorption and fluorescence emission spectra of DDAO.

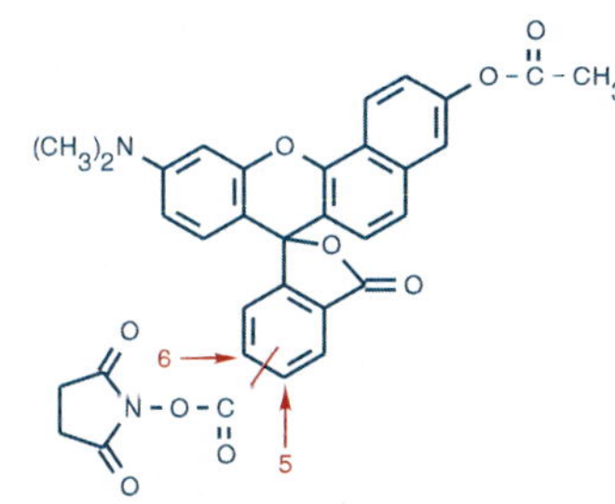

Figure 14.2.23 SNARF®-1 carboxylic acid, acetate, succinimidyl ester (S22801).

REFERENCES

1. Nat Immunol (2008) 9:155; **2.** J Immunol (2008) 181:3947; **3.** Cancer Res (2008) 68:6341; **4.** J Neurosci (2008) 28:5965; **5.** Nat Med (2008) 14:213; **6.** Nat Methods (2009) 6:147; **7.** Proc Natl Acad Sci U S A (2003) 100:2397; **8.** Anal Chem (2008) 80:7543; **9.** Nature (2008) 451:465; **10.** Eur J Pharm Biopharm (2008) 69:396; **11.** Mol Pharm (2008) 5:787; **12.** Dev Biol (1995) 170:434; **13.** J Cell Sci (1995) 108:1105; **14.** J Immunol (2002) 169:722; **15.** J Microbiol Methods (2000) 40:265; **16.** Nature (1993) 363:549; **17.** Stem Cells (1996) 14:132; **18.** Anticancer Res (1995) 15:719; **19.** Prostate (1998) 35:1; **20.** Clin Chem (1995) 41:1906; **21.** J Biol Chem (1993) 268:23593; **22.** Science (2009) 323:1743; **23.** J Cell Biol (1998) 140:315; **24.** Cytometry A (2007) 71:709; **25.** J Biol Chem (1999) 274:19323; **26.** J Pharmacol Exp Ther (2002) 300:9; **27.** Glia (2004) 45:59; **28.** J Cell Biol (1985) 101:610; **29.** J Cell Biol (1986) 103:2649; **30.** J Cell Sci (1992) 102:789; **31.** Nat Biotechnol (2008) 26:215; **32.** Am J Pathol (2006) 169:2223; **33.** Development (1998) 125:201; **34.** Methods Mol Biol (2009) 510:415; **35.** Clin Chem Lab Med (2005) 43:841; **36.** Nat Protoc (2007) 2:2049; **37.** Nat Protoc (2007) 2:2057; **38.** J Immunol Methods (2004) 286:69; **39.** J Immunol Methods (1990) 133:87; **40.** Cytometry (1992) 13:739; **41.** Transplant Proc (1992) 24:2820; **42.** J Immunol Methods (1994) 172:25; **43.** J Immunol Methods (1994) 172:115; **44.** J Cell Biol (1992) 119:451; **45.** J Neurosci Methods (1991) 39:193; **46.** J Neurosci Methods (1992) 44:7; **47.** J Exp Med (1996) 183:2313; **48.** Appl Environ Microbiol (2000) 66:4486; **49.** Development (1992) 116:1087; **50.** Cytometry A (2009) 75:440; **51.** J Immunol (2008) 181:4043; **52.** J Immunol (2007) 179:6808; **53.** Cytometry B Clin Cytom (2007) 72:458.

DATA TABLE 14.2 MEMBRANE-PERMEANT REACTIVE TRACERS

Cat. No.	MW	Storage	Soluble	Abs	EC	Em	Solvent	Notes
B2103	341.00	F,D,L	DMSO, MeCN	533	62,000	561	$CHCl_3$	1
C1157	557.47	F,D	DMF, DMSO	<300		none		2
C1165	626.36	F,D	DMF, MeCN	<300		none		3
C2102	296.55	F,D,L	DMSO	522	72,000	529	MeOH	
C2110	209.63	F,D,L	DMSO	353	14,000	466	pH 9	4
C2111	210.62	F,D,L	DMSO	372	16,000	470	pH 9	5
C2925	464.86	F,D	DMSO	<300		none		2
C2927	554.04	F,D,L	DMSO	541	91,000	565	MeOH	
C6826	499.95	F,D	DMSO	<350		none		6
C6827	577.80	F,D,AA	DMSO	287	9100	none	MeOH	7
C7025	464.86	F,D	DMSO	<300		none		2
C10094	334.21	F,D,L,A	DMSO	416	12,000	526	pH 7	
C12881	246.60	F,D,L	DMSO	371	16,000	464	pH 9	5
C22803	873.05	F,D	DMSO	<300		none		8
C34551	550.44	F,D,L	DMSO	348	6300	none	MeCN	9
C34552	686.25	F,D,L	DMSO	577	118,000	602	MeOH	
C34553	505.35	F,D,L	DMSO	647	35,000	657	pH 10	
C34555	593.45	F,D,L	DMSO	<300		none		10
M1378	271.11	F,L	DMF, MeCN	398	5000	see Notes	pH 7	11
M1381MP	226.66	F,L	DMSO	380	6000	see Notes	MeOH	11
M20381	271.11	F,L	DMF, MeCN	398	5000	see Notes	pH 7	11, 12
O34550	593.45	F,D,L	DMSO	<300		none		10
S22801	592.56	F,D	DMSO	<350		none		6

For definitions of the contents of this data table, see "Using *The Molecular Probes® Handbook*" in the introductory pages.

Notes

1. B2103 spectra are for the unreacted reagent. The thiol adduct has Abs = 493 nm, Em = 503 nm in MeOH.
2. Acetate hydrolysis of this compound yields a fluorescent product with similar pH-dependent spectral characteristics to C1904.
3. C1165 is converted to a fluorescent product with spectra similar to C368 after acetate hydrolysis.
4. C2110 in MeOH: Abs = 364 nm (EC = 16,000 $cm^{-1}M^{-1}$), Em = 454 nm.
5. Spectra of hydroxycoumarins are pH dependent. Below the pK_a (~7.8), Abs shifts to shorter wavelengths (325–340 nm) and fluorescence intensity decreases.
6. C6826 and S22801 are converted to fluorescent products with spectra similar to C1270 after acetate hydrolysis.
7. Acetate hydrolysis and subsequent oxidation of C6827 generate fluorescent 5-(and-6)-chloromethyl-2′,7′-dichlorofluorescein with spectra similar to C368.
8. Acetate hydrolysis of C22803 yields a fluorescent product with spectra similar to 5-carboxyeosin, Abs = 519 nm (EC = 100,000 $cm^{-1}M^{-1}$), Em = 542 nm in pH 9 buffer.
9. The fluorescent phenolic dye generated by acetate hydrolysis of C34551 has Abs = 550 nm (EC = 72,000 $cm^{-1}M^{-1}$), Em = 572 nm in pH 7 buffer.
10. Acetate hydrolysis of this compound yields a fluorescent product with similar spectral characteristics to O6146.
11. Bimanes are almost nonfluorescent until reacted with thiols. For monobromobimane conjugated to glutathione, Abs = 394 nm, Em = 490 nm (QY ~0.1–0.3) in pH 8 buffer. (Methods Enzymol (1987) 143:76, Methods Enzymol (1995) 251:133)
12. This product is specified to equal or exceed 98% analytical purity by HPLC.

IMPORTANT NOTICE: The products described in this manual are covered by one or more Limited Use Label License(s). Please refer to the Appendix on page 971 and Master Product List on page 975. Products are For Research Use Only. Not intended for any animal or human therapeutic or diagnostic use.

PRODUCT LIST 14.2 MEMBRANE-PERMEANT REACTIVE TRACERS

Cat. No.	Product	Quantity
B2103	BODIPY® 493/503 methyl bromide (8-bromomethyl-4,4-difluoro-1,3,5,7-tetramethyl-4-bora-3a,4a-diaza-*s*-indacene)	5 mg
C1165	CDCFDA, SE (5-(and-6)-carboxy-2',7'-dichlorofluorescein diacetate, succinimidyl ester) *mixed isomers*	25 mg
C22803	CEDA, SE (5-(and-6)-carboxyeosin diacetate, succinimidyl ester) *mixed isomers*	5 mg
C34554	CellTrace™ CFSE Cell Proliferation Kit *for flow cytometry*	1 kit
C34553	CellTrace™ Far Red DDAO-SE *special packaging*	20 x 50 µg
C34555	CellTrace™ Oregon Green® 488 carboxylic acid diacetate, succinimidyl ester (carboxy-DFFDA, SE) *cell permeant* *mixed isomers*	20 x 50 µg
C34557	CellTrace™ Violet Cell Proliferation Kit *for flow cytometry*	1 kit
C2110	CellTracker™ Blue CMAC (7-amino-4-chloromethylcoumarin)	5 mg
C12881	CellTracker™ Blue CMF$_2$HC (4-chloromethyl-6,8-difluoro-7-hydroxycoumarin)	5 mg
C2111	CellTracker™ Blue CMHC (4-chloromethyl-7-hydroxycoumarin)	5 mg
C2102	CellTracker™ Green BODIPY® (8-chloromethyl-4,4-difluoro-1,3,5,7-tetramethyl-4-bora-3a,4a-diaza-*s*-indacene)	5 mg
C2925	CellTracker™ Green CMFDA (5-chloromethylfluorescein diacetate)	1 mg
C7025	CellTracker™ Green CMFDA (5-chloromethylfluorescein diacetate) *special packaging*	20 x 50 µg
C34551	CellTracker™ Orange CMRA *special packaging*	20 x 50 µg
C2927	CellTracker™ Orange CMTMR (5-(and-6)-(((4-chloromethyl)benzoyl)amino)tetramethylrhodamine) *mixed isomers*	1 mg
C34552	CellTracker™ Red CMTPX *special packaging*	20 x 50 µg
C10094	CellTracker™ Violet BMQC (2,3,6,7-tetrahydro-9-bromomethyl-1H,5H-quinolizino(9,1-gh)coumarin)	5 x 100 µg
C1157	5(6)-CFDA, SE; CFSE (5-(and-6)-carboxyfluorescein diacetate, succinimidyl ester) *mixed isomers*	25 mg
C6827	5-(and-6)-chloromethyl-2',7'-dichlorodihydrofluorescein diacetate, acetyl ester (CM-H$_2$DCFDA) *mixed isomers* *special packaging*	20 x 50 µg
C6826	5-(and-6)-chloromethyl SNARF®-1, acetate *mixed isomers* *special packaging*	20 x 50 µg
M1378	monobromobimane (mBBr)	25 mg
M20381	monobromobimane (mBBr) *FluoroPure™ grade*	25 mg
M1381MP	monochlorobimane (mBCl)	25 mg
O34550	Oregon Green® 488 carboxylic acid diacetate, succinimidyl ester (carboxy-DFFDA, SE) *mixed isomers*	1 mg
S22801	SNARF®-1 carboxylic acid, acetate, succinimidyl ester *special packaging*	10 x 50 µg
V12883	Vybrant® CFDA SE Cell Tracer Kit	1 kit

14.3 Polar Tracers

Fixable Polar Tracers

We prepare a wide variety of highly water-soluble dyes and other detectable probes that can be used as cell tracers. In most cases, the polarity of these water-soluble probes is too high to permit them to passively diffuse through cell membranes. Consequently, special methods for loading the dyes into cells must be employed, including microinjection, pinocytosis or techniques that temporarily permeabilize the cell's membrane [1,2] (Table 14.1). Our Influx™ pinocytic cell-loading reagent (I14402, see below and Section 19.8) is particularly useful for loading many of the polar tracers in this section—as well as the dextrans and fluorescent proteins described in Section 14.5 and Section 14.7—into many types of cells. Permeabilization of cells with staphylococcal α-toxin [3] or the saponin ester β-escin is reported to make the membrane of smooth muscle cells permeable to low molecular weight (<1000 daltons) molecules, while retaining high molecular weight compounds.[4–6] Electroporation has been used to transport several of the polar tracers through the skin [7] and into cells.[8,9] Many of these tracers can also be loaded into cells noninvasively as their cell-permeant acetoxymethyl (AM) esters, which are discussed in more detail in Section 15.2.

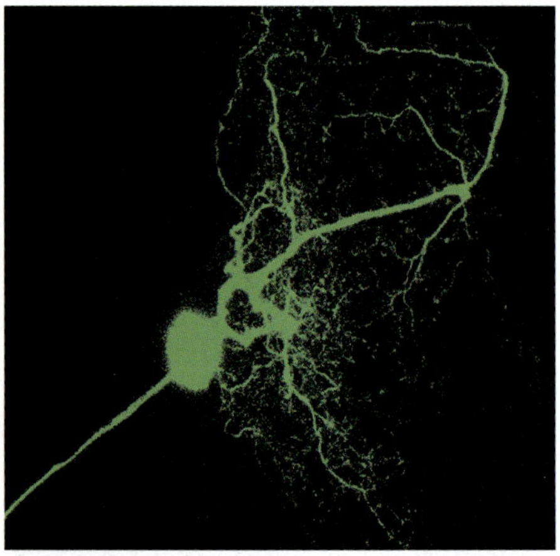

Figure 14.3.1 The APR motor neuron of a larval moth, *Manduca sexta*, labeled by the intracellular injection of Alexa Fluor® 488 hydrazide, sodium salt (A10436). This pseudocolored image was created by combining 21 optical sections obtained using a confocal laser-scanning microscope equipped with a bandpass filter appropriate for fluorescein. The image was contributed by Jack Gray and Walter Metcalfe, Institute of Neuroscience, University of Oregon.

Alexa Fluor® Hydrazides and Hydroxylamines

Our selection of fluorescent hydrazide and hydroxylamine derivatives continues to expand (Section 3.3, Table 3.2). The blue-fluorescent Alexa Fluor® 350 hydrazide and Alexa Fluor® 350 hydroxylamine (A10439, A30627), green-fluorescent Alexa Fluor® 488 hydrazide and Alexa Fluor® 488 hydroxylamine (A10436, A30629; Figure 14.3.1), orange-fluorescent Alexa Fluor® 555

IMPORTANT NOTICE: The products described in this manual are covered by one or more Limited Use Label License(s). Please refer to the Appendix on page 971 and Master Product List on page 975. Products are For Research Use Only. Not intended for any animal or human therapeutic or diagnostic use.

and Alexa Fluor® 568 hydrazides (A20501MP, A10437, A10441; Figure 14.3.2), red-fluorescent Alexa Fluor® 594 hydrazide [10] (A10438, A10442) and far-red–fluorescent Alexa Fluor® 633 hydrazide, Alexa Fluor® 647 hydrazide and Alexa Fluor® 647 hydroxylamine (A30634, A20502, A30632) are likely among the best overall polar tracers in each of their various spectral ranges.[11] These low molecular weight, cell membrane–impermeant molecules (Alexa Fluor® 350 hydrazide, 349 daltons; Alexa Fluor® 350 hydroxylamine, 585 daltons; ~570–760 daltons for the Alexa Fluor® 488, 568 and 594 hydrazides and hydroxylamine; and about 1200 daltons for the Alexa Fluor® 555 and 647 hydrazides and hydroxylamine) possess several properties that are superior to those of the widely used neuronal tracer lucifer yellow CH (L453, L682, L1177, L12926). Like lucifer yellow CH, the hydrazide moiety of the Alexa Fluor® derivatives makes these tracers fixable by common aldehyde-based fixatives. We have determined that Alexa Fluor® 594 hydrazide has a water solubility of ~84 mg/mL and the other Alexa Fluor® hydrazides are likely to have comparable or higher water solubility.

Our rabbit polyclonal antibody to the Alexa Fluor® 488 fluorophore (A11094, Section 7.4) quenches the fluorescence of the Alexa Fluor® 488 dye (Anti–Lucifer Yellow Dye, Anti–Alexa Fluor® 405/Cascade Blue® Dye and Anti–Alexa Fluor® 488 Dye Antibodies—Note 14.1) and, following cell fixation and permeabilization, can be used in conjunction with the reagents in our Tyramide Signal Amplification (TSA™) Kits (Section 6.2) to amplify the signal or with the anti–rabbit IgG conjugate of NANOGOLD® or Alexa Fluor® FluoroNanogold™ 1.4 nm gold clusters (N24916, A24926, A24927; Section 7.2) and the associated LI Silver Enhancement Kit (L24919, Section 7.2) for correlated fluorescence and light microscopy studies.

Although lucifer yellow CH can be used for confocal laser-scanning microscopy, its extinction coefficient at the 488 nm spectral line of the argon-ion laser (~700 $cm^{-1}M^{-1}$) is only about 1% of that of Alexa Fluor® 488 hydrazide and Alexa Fluor® 488 hydroxylamine (≥71,000 $cm^{-1}M^{-1}$) (Figure 14.3.3). Furthermore, the high photostability of the Alexa Fluor® dyes permits their detection in very fine structures that cannot be seen with lucifer yellow CH staining. All of these Alexa Fluor® dyes are remarkably bright and photostable. In addition, the Alexa Fluor® hydrazide salts have high water solubility (typically greater than 8%). We offer the Alexa Fluor® 568 and Alexa Fluor® 594 hydrazides either as solids (A10437, A10438) or as 10 mM solutions in 200 mM KCl (A10441, A10442). The 10 mM solutions have been filtered through a 0.2 µm filter to remove any insoluble material prior to packaging. Alexa Fluor® 488 hydrazide, Alexa Fluor® 555 hydrazide, Alexa Fluor® 633 hydrazide and Alexa Fluor® 647 hydrazide (A10436, A20501MP, A30634, A20502) and Alexa Fluor® 350, Alexa Fluor® 488 and Alexa Fluor® 647 hydroxylamines (A30627, A30629, A30632) are only available as solids. Our Alexa Fluor® 350 hydrazide and Alexa Fluor® 350 hydroxylamine, which

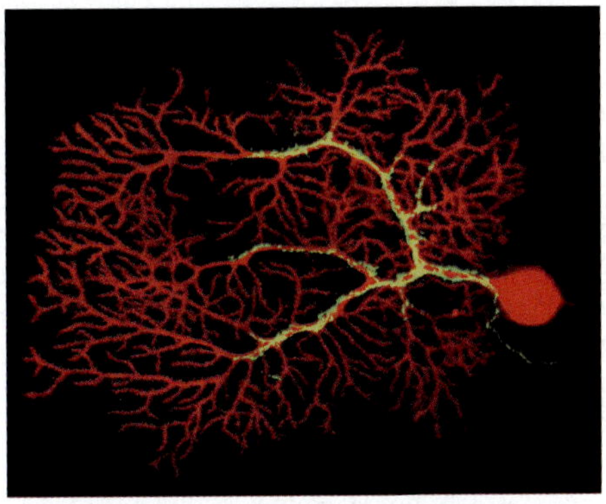

Figure 14.3.2 Confocal image stack of a 10,000 MW Calcium Green™ dextran–labeled (C3713) climbing fiber in a sagittal cerebellar slice, showing incoming axon and terminal arborization (in yellow). The Purkinje cell innervated by this climbing fiber was labeled with Alexa Fluor® 568 hydrazide (A10437, A10441) via a patch pipette and visually identified using bright-field microscopy. Image contributed by Anatol Kreitzer, Department of Neurobiology, Harvard Medical School.

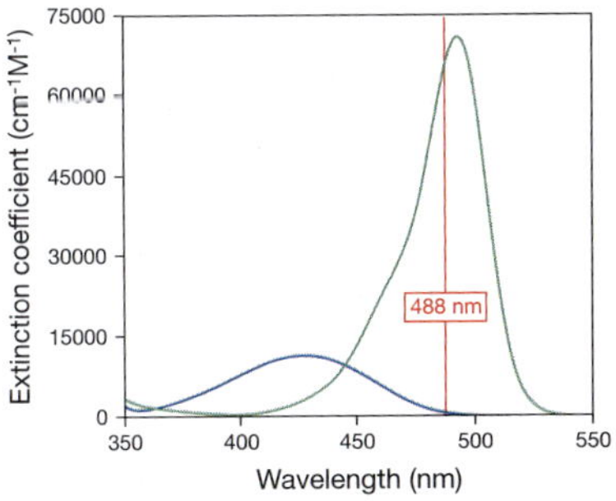

Figure 14.3.3 Absorption spectra showing that the molar extinction coefficient (EC) at 488 nm of Alexa Fluor® 488 hydrazide (A10436) in water (green line) is approximately 100-fold greater than that of lucifer yellow CH (L453, L682, L1177, L12926) in water (blue line).

NOTE 14.1

Anti–Lucifer Yellow Dye, Anti–Alexa Fluor® 405/Cascade Blue® Dye and Anti–Alexa Fluor® 488 Dye Antibodies

Molecular Probes® anti–lucifer yellow dye antibodies were specifically developed to overcome certain limitations of lucifer yellow CH (L453, L682, L12926, L1177). Lucifer yellow CH is an aldehyde-fixable fluorescent cell tracer that has long been used by neuroscientists to identify patterns of gap junctional communication,[1] to assay the outgrowth of developing neurons [2] and to characterize the morphology of neurons from which electrical recordings have been made.[3] Even though the cell soma of a lucifer yellow CH–filled neuron may be brightly stained, its finer processes can sometimes be faint and may fade rapidly or be obscured by the more intensely stained portions of the neuron. Investigators have been able to overcome these limitations by using anti–lucifer yellow dye antibodies in conjunction with standard enzyme-mediated immunohistochemical methods to develop a more permanent, fade-free signal for light microscopy.[4–7] Anti–lucifer yellow dye antibodies have also been used to develop tissue for electron microscopy [8] and to distinguish neurons filled with lucifer yellow CH from those injected with the lectin *Phaseolus vulgaris* leucoagglutinin [9] (PHA-L). Molecular Probes® Cascade Blue® hydrazide (C3239), which can also be fixed in place with aldehyde-based fixatives, can potentially be used as a second label with lucifer yellow CH to characterize the morphology of interacting neurons.

For these applications, we offer unconjugated and biotinylated rabbit polyclonal anti–lucifer yellow dye (A5750, A5751) and anti–Alexa Fluor® 405/Cascade Blue® dye (A5760) antibodies. Similarly, our rabbit polyclonal anti–Alexa Fluor® 488 dye antibody (A11094) can be used to detect Alexa Fluor® 488 hydrazide (A10436), Alexa Fluor® 488 hydroxylamine (A30629), Alexa Fluor® 488 cadaverine (A30676) and Alexa Fluor® 488 biocytin (A12924) in fixed-cell preparations. See Section 7.4 (Table 7.8) for a complete description of these and other anti-fluorophore and anti-biotin antibodies.

1. Biotechniques (2008) 45:33; **2.** Science (1988) 242:700; **3.** J Neurosci (2003) 23:4700; **4.** J Neurosci (1994) 14:5267; **5.** J Neurosci Methods (1992) 41:45; **6.** J Comp Neurol (1990) 296:598; **7.** Dev Biol (1982) 94:391; **8.** Circ Res (1992) 70:49; **9.** J Neurosci Methods (1990) 33:207.

IMPORTANT NOTICE: The products described in this manual are covered by one or more Limited Use Label License(s). Please refer to the Appendix on page 971 and Master Product List on page 975. Products are For Research Use Only. Not intended for any animal or human therapeutic or diagnostic use.

are sulfonated coumarin derivatives (Figure 14.3.4), are some of the few polar tracers that exhibit bright blue fluorescence.

Other Alexa Fluor® Derivatives

To allow amplification of signals, especially in the finer processes of dye-filled neurons, we also offer Alexa Fluor® 488 biocytin (A12924), Alexa Fluor® 546 biocytin (A12923) and Alexa Fluor® 594 biocytin (A12922). These unique probes combine our Alexa Fluor® 488, Alexa Fluor® 546 and Alexa Fluor® 594 fluorophores with biotin and an aldehyde-fixable primary amine (see "Fluorescent Biotin Derivatives," below). In addition, we offer the bright blue-fluorescent Alexa Fluor® 405 cadaverine (A30675, see below) as well as several other Alexa Fluor® cadaverines (Section 3.3, Table 3.2), all of which should be useful as tracing molecules because they are exceptionally bright, small and water soluble, and they each contain an aldehyde-fixable functional group. Alexa Fluor® 546 biocytin has been used to label streptavidin-coated particles in order to quantitate fluorescence signals in an automated imaging system designed for analyzing immobilized particle arrays.[12]

Lucifer Yellow CH

Lucifer yellow CH (LY-CH or LY, Figure 14.3.5) has long been a favorite tool for studying neuronal morphology because it contains a carbohydrazide (CH) group that allows it to be covalently linked to surrounding biomolecules during aldehyde-based fixation.[13,14] Loading of this polar tracer and other similar impermeant dyes is usually accomplished by microinjection,[15] pinocytosis,[16] scrape loading,[17] ATP-induced permeabilization[18] or osmotic shock[19] (Table 14.1), but can also be accomplished in cell suspensions or with adherent cells by using our Influx™ pinocytic cell-loading reagent (I14402, see below). Lucifer yellow CH localizes in the plant vacuole when taken up either through what is thought to be anion-transport channels[20–22] or by fluid-phase endocytosis.[23] Upon injection into the epidermal cells of *Egeria densa* leaves, lucifer yellow CH reportedly moved into the cytoplasm of adjacent cells, localized in the plant vacuole or moved in and out of the nucleus.[24] The lithium salt of lucifer yellow CH is widely used for microinjection because of its relatively high water solubility (~8%). In addition to the solid (L453), we offer the lithium salt of lucifer yellow CH as a filtered 100 mM solution (L12926), ready for microinjection. The potassium salt (L1177, solubility ~1%) or the ammonium salt of lucifer yellow CH (L682, solubility ~6%) may be preferred in applications where lithium ions interfere with biological function.

Although its weak absorption at 488 nm (EC ~700 $cm^{-1}M^{-1}$) (Figure 14.3.3) makes it inefficiently excited with the argon-ion laser, lucifer yellow CH has been used as a neuronal tracer in some confocal laser-scanning microscopy studies.[25–27] For electron microscopy studies, lucifer yellow CH can be used to photoconvert diaminobenzidine (DAB) into an insoluble, electron-dense reaction product[28–30] (Fluorescent Probes for Photoconversion of Diaminobenzidine Reagents—Note 14.2). Alternatively, rabbit anti–lucifer yellow dye antibodies (Section 7.4) can be used in conjunction with our goat anti–rabbit IgG antibody conjugated to either NANOGOLD® or Alexa Fluor® FluoroNanogold™ 1.4 nm gold clusters (Section 7.2) and the LI Silver (LIS) Enhancement Kit (L24919, Section 7.2) to develop a more permanent, fade-free colorimetric or electron-dense signal from dye-filled neurons that is suitable for light or electron microscopy[31–36] (Anti–Lucifer Yellow Dye, Anti–Alexa Fluor® 405/Cascade Blue® Dye and Anti–Alexa Fluor® 488 Dye Antibodies—Note 14.1).

Intracellular injection of lucifer yellow CH has been extensively employed to delineate neuronal morphology in live neurons[37–39] (Figure 14.3.6) and in fixed brain slices,[31] as well as to investigate intercellular communication through gap junctions.[40–42] Lucifer yellow CH can also be used to label neurons by using dye-filled electrodes during electrophysiological recording in order to correlate neuronal function with structure and connectivity.

Other Lucifer Yellow Derivatives

Like lucifer yellow CH, lucifer yellow ethylenediamine (A1339) is fixable with standard aldehyde-based fixatives and can be used as a building block for new lucifer yellow derivatives.[43] The thiol-reactive lucifer yellow iodoacetamide (L1338) can also be used as a microinjectable polar tracer, as well as for preparing fluorescent liposomes[44] and for detecting the accessibility of thiols in membrane-bound proteins.[45–47] In addition to these lucifer yellow derivatives, we offer a lucifer yellow–conjugated 10,000 MW dextran (D1825, Section 14.5).

Cascade Blue® Hydrazide

Molecular Probes® Cascade Blue® hydrazide is a fixable analog of the nonfixable, bright blue-fluorescent tracer methoxypyrenetrisulfonic acid[48] (MPTS). All of the Cascade Blue® hydrazide derivatives have

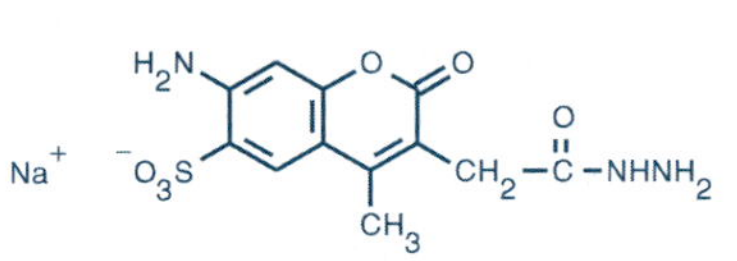

Figure 14.3.4 Alexa Fluor® 350 hydrazide, sodium salt (A10439).

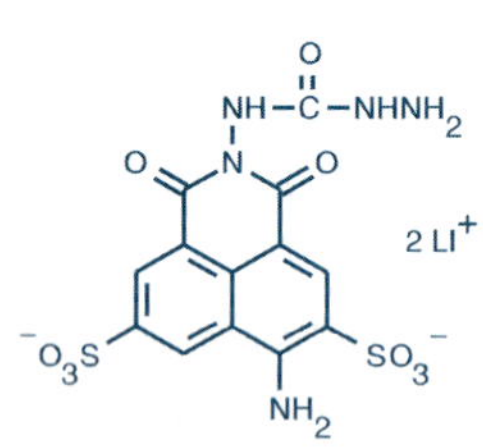

Figure 14.3.5 Lucifer yellow CH, lithium salt (L453).

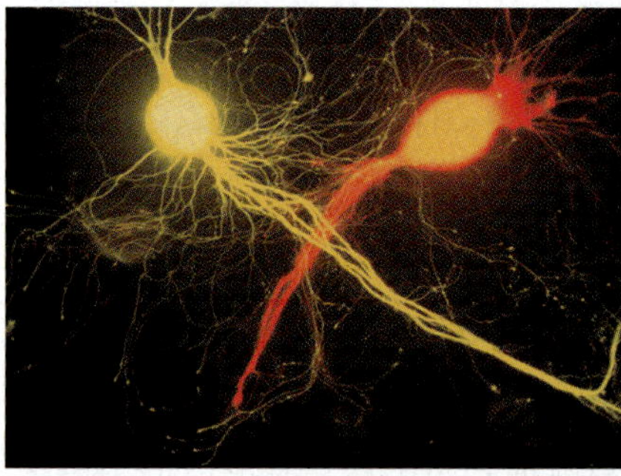

Figure 14.3.6 Cultured left-upper quadrant neurons from *Aplysia californica* that have been microinjected with either lucifer yellow CH (L453, L682, L1177, L12926) or sulforhodamine 101 (S359). These neurons display an extensive array of overlapping processes (J Neurophysiol (1991) 66:316). Image contributed by David Kleinfeld, AT&T Bell Laboratories, and Brian Salzberg, University of Pennsylvania School of Medicine.

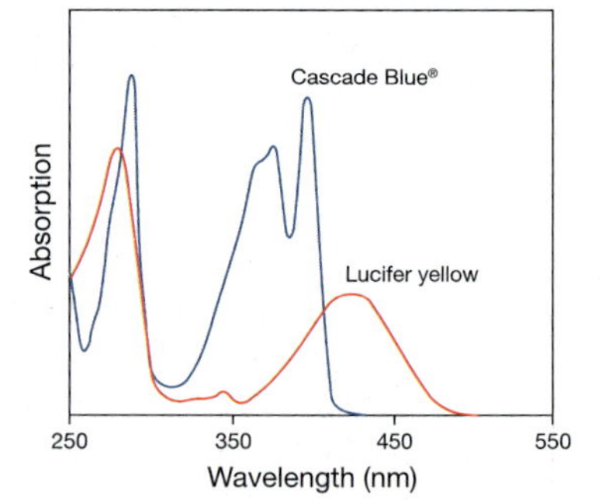

Figure 14.3.7 Absorption spectra for equal concentrations of Cascade Blue® hydrazide (C687, C3221, C3239) and lucifer yellow CH (L453, L682, L1177, L12926) in water.

IMPORTANT NOTICE: The products described in this manual are covered by one or more Limited Use Label License(s). Please refer to the Appendix on page 971 and Master Product List on page 975. Products are For Research Use Only. Not intended for any animal or human therapeutic or diagnostic use.

reasonable water solubility, ~1% for the sodium and potassium salts (C687, C3221) and ~8% for the lithium salt (C3239). They also exhibit a stronger absorption (EC_{400} > 28,000 $cm^{-1}M^{-1}$) and quantum yield (~0.54 in water) than lucifer yellow CH. In addition, Cascade Blue® derivatives have good photostability and emissions that are well resolved from those of fluorescein and lucifer yellow CH.[49] Cascade Blue® hydrazide, which readily passes through gap junctions, and lucifer yellow derivatives can be simultaneously excited at 405 nm (Figure 14.3.7) for two-color detection at about 430 and 530 nm.[50] Cascade Blue® dyes, lucifer yellow CH and sulforhodamine 101 can be used in combination for three-color mapping of neuronal processes (Figure 14.3.8). We also offer anti–Alexa Fluor® 405/Cascade Blue® dye antibodies (Section 7.4) for localizing Cascade Blue® dye–filled cells following fixation (Anti–Lucifer Yellow Dye, Anti–Alexa Fluor® 405/Cascade Blue® Dye and Anti–Alexa Fluor® 488 Dye Antibodies—Note 14.1). Like lucifer yellow CH, Cascade Blue® hydrazide and some other polar tracers are taken up by plants and sequestered into their central vacuoles. In onion epidermal cells, this uptake of Cascade Blue® hydrazide is blocked by probenecid (P36400; Section 19.8), indicating that transfer may be through anion-transport channels.[21,22]

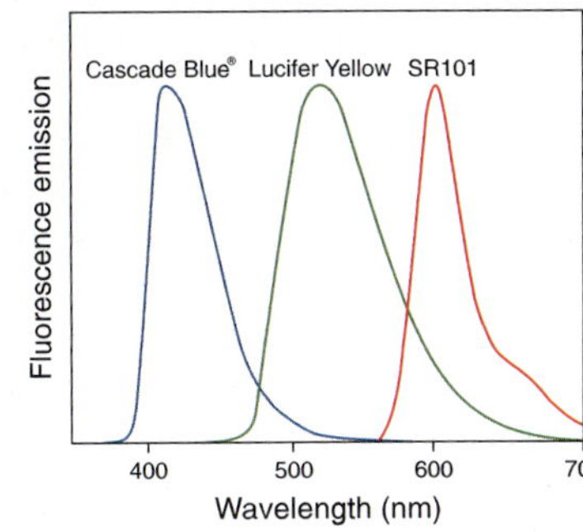

Figure 14.3.8 Normalized fluorescence emission spectra for Cascade Blue® hydrazide (C687, C3221, C3239), lucifer yellow CH (L453, L682, L1177, L12926) and sulforhodamine 101 (S359) in water.

NOTE 14.2

Fluorescent Probes for Photoconversion of Diaminobenzidine Reagents

Photoconversion of Diaminobenzidine

We offer a variety of fluorescent probes for photoconverting diaminobenzidine (DAB), enabling researchers to take advantage of an important development in correlated fluorescence, transmitted and electron microscopy. In 1982, Maranto first described the use of the fluorophore lucifer yellow for DAB photoconversion.[1] When a fluorophore is exposed to light of an appropriate wavelength, excitation from the electronic ground state to a higher singlet state occurs. Instead of emitting a photon, the excited state of the fluorophore may undergo intersystem crossing to the triplet state. Transfer of energy to ground state triplet oxygen ($^3O_2$) generates toxic and highly reactive singlet oxygen ($^1O_2$), which is capable of causing damage to lipids, proteins and nucleic acids.[2] However, the reactive potential of $^1O_2$ can also be harnessed to oxidize diaminobenzidine (DAB) into an electron-opaque osmiophilic precipitate within cells (Figure 1). The resulting DAB reaction product exhibits exceptionally uniform, nondiffusible staining properties, making it extremely useful for subsequent electron microscopy investigation of cellular ultrastructure.[1] Punctate background staining can reportedly be eliminated without affecting the DAB photoconversion signal by treating the sample with the peroxisomal catalase inhibitor aminotriazole.[3]

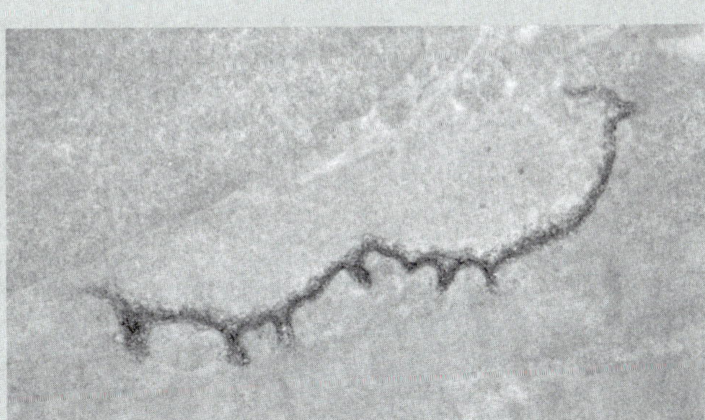

Figure 1 Electron micrograph of an 80 nm–thick section of formaldehyde-fixed rat soleus muscle, which was first stained with eosin α-bungarotoxin and then used to photoconvert DAB into an insoluble osmiophilic polymer. Photo contributed by Thomas J. Deerinck, University of California, San Diego.

Eosin Probes

In 1994, Deerinck and colleagues reported a simple method for eosin-mediated photoconversion of DAB.[4,5] Halogenated derivatives of fluorescein dyes are known to be effective photosensitizers and singlet oxygen generators.[6] Eosin is a brominated analog of fluorescein that has a $^1O_2$ yield 19 times greater than fluorescein and is an excellent dye for photoconverting DAB.[4,7] Furthermore, the small size of eosin promotes exceptional penetration into tissues resulting in increased resolution for electron microscopy.[8] We offer amine- and thiol-reactive eosin derivatives for preparing eosin-labeled bioconjugates (Section 1.5, Section 2.2). Some other fluorescent tracers that have been used to photoconvert DAB include:[9]

- BODIPY® FL C_5-ceramide [10] (D3521, B22650; Section 12.4)
- DiI [11,12] (D282, Section 14.4)
- Fluorescent polystyrene microspheres [13] (Section 6.5)
- Fluoro-ruby dextran [14] (D1817, Section 14.5)
- Lucifer yellow [15] (L453, L12926; Section 14.3)
- Propidium iodide [16] (P1304MP, P3566, P21493; Section 8.1)
- ReAsH [17] (T34562, Section 2.2)
- Texas Red®-X succinimidyl ester [18] (T6134, Section 1.6)

1. Science (1982) 217:953; **2.** J Photochem Photobiol B (1991) 11:241; **3.** J Histochem Cytochem (1998) 46:1085; **4.** J Cell Biol (1994) 126:901; **5.** J Cell Biol (1994) 126:877; **6.** Adv Photochem (1993) 18:315; **7.** Photochem Photobiol (1983) 37:271; **8.** J Histochem Cytochem (2001) 49:1351; **9.** Neuroscience Protocols, Wouterlood FG, Ed. (1993) p. 93-050-06; **10.** Cell (1993) 73:1079; **11.** J Neurosci Methods (2002) 117:73; **12.** J Histochem Cytochem (1990) 38:725; **13.** Brain Res (1993) 630:115; **14.** J Histochem Cytochem (1993) 41:777; **15.** J Struct Biol (2008) 161:359; **16.** J Neurosci Methods (1992) 45:87; **17.** Proc Natl Acad Sci U S A (2006) 103:17777; **18.** Nature (2010) 465:478.

IMPORTANT NOTICE: The products described in this manual are covered by one or more Limited Use Label License(s). Please refer to the Appendix on page 971 and Master Product List on page 975. Products are For Research Use Only. Not intended for any animal or human therapeutic or diagnostic use.

Other Cascade Blue® and Alexa Fluor® 405 Derivatives

Cascade Blue® acetyl azide (C2284) and Alexa Fluor® 405 succinimidyl ester (A30000, A30100) are water-soluble, amine-reactive tracers that can be introduced either by microinjection or by fusion of dye-filled liposomes with cells. Once inside the cell, these derivatives will react with the amine groups of intracellular proteins. Cascade Blue® ethylenediamine (C621) and Alexa Fluor® 405 cadaverine (previously called Cascade Blue® cadaverine, A30675) are aldehyde-fixable fluorophores with reactive properties similar to those of the ethylenediamine derivative of lucifer yellow (A1339). A Cascade Blue® dye–labeled 10,000 MW dextran (D1976, Section 14.5) is also available.

Biocytin and Other Biotin Derivatives

Biocytin (ε-biotinoyl-L-lysine, B1592, Figure 14.3.9) and biotin ethylenediamine (A1593, Figure 14.3.10) are microinjectable anterograde and transneuronal tracers.[51–54] Retrograde transport of biocytin and biotin ethylenediamine in neurons has also been reported.[55,56] These water-soluble tracers are often used to label neurons during electrophysiological measurements in order to correlate neuronal function with structure and connectivity.[57,58] Biotin cadaverine (A1594) and biotin-X cadaverine (B1596) have slightly longer spacers than their ethylenediamine counterparts, making the hapten more accessible to the deep biotin-binding site in avidins.[59–61]

Biocytin, biotin ethylenediamine, biotin cadaverine and biotin-X cadaverine all contain primary amines and can therefore be fixed in cells with formaldehyde or glutaraldehyde and subsequently detected using fluorescent- or enzyme-labeled avidin or streptavidin second-step reagents or with NANOGOLD® and Alexa Fluor® FluoroNanogold™ streptavidin[62] (Section 7.6). Biocytin hydrazide (B1603) and DSB-X™ biotin hydrazide (D20653) can serve as aldehyde-fixable tracers[57] and as reactive probes for labeling glycoproteins and nucleic acids (Section 4.2).

As with the reactive lucifer yellow, and Cascade Blue® and Alexa Fluor® 405 derivatives discussed above, amine- or thiol-reactive biotin derivatives are useful for intracellular labeling applications. The succinimidyl esters of biotin and biotin-X (B1513, B1582) have been used to trace retinal axons in avian embryos.[63] Because they are more water soluble, the *sulfo*succinimidyl esters of biotin-X and biotin-XX (B6353, B6352) or the thiol-reactive biocytin maleimide (M1602) may be preferred for these applications.

Fluorescent Biotin Derivatives

Fluorescence of the finer processes of dye-filled neurons may fade rapidly or be obscured by the more intensely stained portions of the neuron, necessitating further amplification of the signal or other ultrastructural detection methods. Lucifer yellow biocytin (L6950), Alexa Fluor® 488 biocytin (A12924), Alexa Fluor® 546 biocytin (A12923), Alexa Fluor® 594 biocytin (A12922), Oregon Green® 488 biocytin (O12920) and tetramethylrhodamine biocytin (T12921) each incorporate a fluorophore, biotin and an aldehyde-fixable primary amine into a single molecule, thus enabling researchers to amplify the signals of these tracers with fluorescent or enzyme-labeled avidin or streptavidin conjugates (Section 7.6). Although our lucifer yellow cadaverine biotin-X (L2601) lacks a primary amine, it was reported that this tracer was well retained in aldehyde-fixed tissues, even after sectioning, extraction with detergents and several washes.[64,65] Because fluorescent biocytin derivatives contain free primary amines, they should be even more efficiently fixed by formaldehyde or glutaraldehyde.

Nonfixable Polar Tracers

Polar fluorescent dyes are commonly used to investigate fusion, lysis and gap-junctional communication and to detect changes in cell or liposome volume. These events are primarily monitored by following changes in the dye's fluorescence caused by interaction with nearby molecules. For example, because the fluorescence of many dyes at high concentrations is quenched, various processes that result in a dilution of the dyes, such as lysis or fusion of fluorescent dye–filled cells or liposomes, can produce an increase in fluorescence, thereby providing an easy method for monitoring these events. Cell–cell and cell–liposome fusion, as well as membrane permeability and transport through gap junctions, can all be monitored using these methods. Furthermore, a fluorogenic substrate such as fluorescein diphosphate (FDP, F2999; Section 10.3) can be incorporated within a cell or vesicle that lacks the enzymatic activity to generate a fluorescent product; subsequent fusion with a cell or vesicle that contains the appropriate enzyme will generate a fluorescent product.[66]

An ultrasensitive fusion assay that can be used to follow fusion of single vesicles utilizes an almost nonfluorescent potassium salt of an ion-sensitive indicator such as fluo-3 (F1240, F3715; Section 19.3) in one vesicle and a polyvalent ion such as Ca^{2+}—or perhaps better La^{3+}, which causes greater enhancement of the fluorescence of fluo-3—in a second vesicle.[67]

Fluorescein Derivatives

The self-quenching of fluorescein derivatives provides a means of determining their concentration in dynamic processes such as lysis or fusion of dye-filled cells or liposomes (Assays of Volume Change, Membrane Fusion and Membrane Permeability—Note 14.3). Calcein

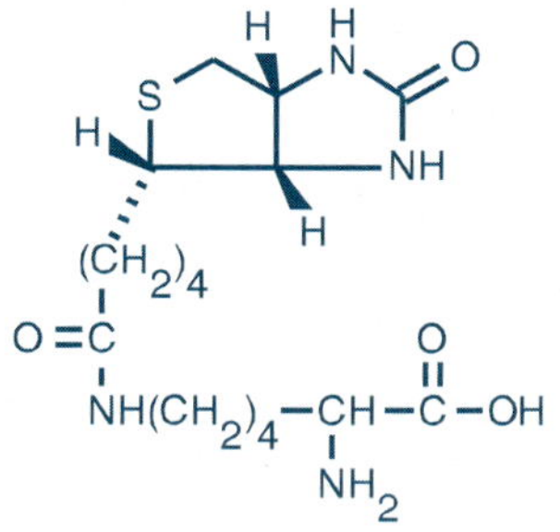

Figure 14.3.9 ε-Biotinoyl-L-lysine (biocytin, B1592).

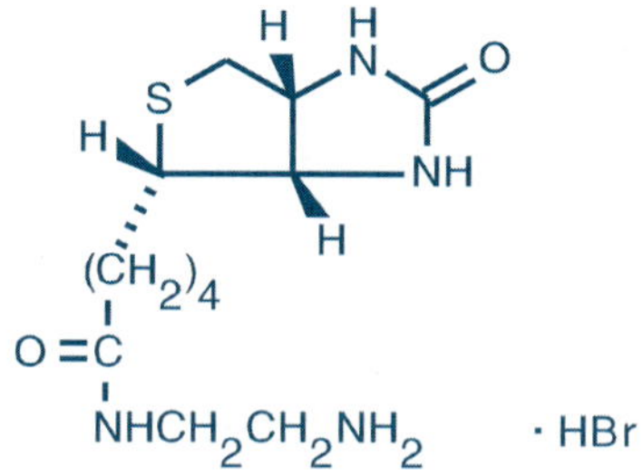

Figure 14.3.10 *N*-(2-aminoethyl)biotinamide, hydrobromide (biotin ethylenediamine, A1593).

IMPORTANT NOTICE: The products described in this manual are covered by one or more Limited Use Label License(s). Please refer to the Appendix on page 971 and Master Product List on page 975. Products are For Research Use Only. Not intended for any animal or human therapeutic or diagnostic use.

NOTE 14.3

Assays of Volume Change, Membrane Fusion and Membrane Permeability Fusion

Assays of membrane fusion report either the mixing or leakage of the aqueous contents of the fused entities (described here) or the mixing of membrane lipids (Lipid-Mixing Assays of Membrane Fusion—Note 13.1). Leakage of aqueous contents from cells or vesicles as a result of lysis, fusion or physiological permeability can be detected fluorometrically using low molecular weight soluble tracers. Assays designed to detect solution mixing often rely on a fluorophore and quencher pair, whereas assays that detect solution leakage into the external medium typically exploit the self-quenching properties of a fluorophore.

Fluorescence Quenching Assays with ANTS/DPX

Originally developed by Smolarsky and co-workers to follow complement-mediated immune lysis,[1] the ANTS/DPX fluorescence quenching assay has since been widely used to detect membrane fusion.[2] This assay is based on the collisional quenching of the polyanionic fluorophore ANTS by the cationic quencher DPX (Figure 1). Separate vesicle populations are loaded with 25 mM ANTS (A350) in 40 mM NaCl and 90 mM DPX (X1525).

Vesicle fusion results in quenching of ANTS fluorescence monitored at 530 nm, with excitation at 360 nm. External leakage of vesicle contents does not cause quenching due to the accompanying high dilution of ANTS and DPX. For assays of vesicle leakage, ANTS (12.5 mM) and DPX (45 mM) can be co-encapsulated into liposomes; upon dilution into surrounding medium, ANTS fluorescence will increase because quenching by DPX will be diminished.

Examples employing ANTS/DPX quenching include studies of the role of amino lipids[3] and phospholipid asymmetry[4] in membrane fusion; glycoprotein-mediated viral fusion with membranes;[5] and the interactions of vesicle membranes with annexins,[6] HIV fusion peptide[7] and apolipoprotein.[8]

Additional Dye/Quencher Pairs

The fluorescence of HPTS and pyrenetetrasulfonic acid is also effectively quenched by DPX.[9–11] Thallium (Tl^+) and cesium (Cs^+) ions quench the fluorescence of ANTS, pyranine (HPTS, H348), pyrenetetrasulfonic acid (P349) and some other polyanionic fluorophores. A review by Garcia[12] describes how fluorescence quenching with a variety of dye/quencher pairs can be used to determine transmembrane ion permeability.

Fluorescence Enhancement Assays with Tb^{3+}/DPA

In the Tb^{3+}/dipicolinic acid (DPA) assay, which was originally described by Wilschut and Papahadjopoulos,[13,14] separate vesicle populations are loaded with 2.5 mM $TbCl_3$ (T1247) in 50 mM sodium citrate, or 50 mM DPA in 20 mM NaCl.

Vesicle fusion results in formation of Tb^{3+}/DPA chelates that are ~10,000 times more fluorescent than free Tb^{3+} (Figure 2). Fluorescence of the chelates is detected at 490 nm or 545 nm, with excitation at 276 nm. Including Ca^{2+} and EDTA in the external medium inhibits formation of the complex outside the fused vesicles. The Tb^{3+}/DPA fluorescence enhancement assay has been used to investigate the role of phospholipid conformation in vesicle fusion[15] and the interaction of cardiotoxin with phospholipid vesicles.[16]

Self-Quenching Assays with Fluorescein Derivatives

Fluorescence of carboxyfluorescein (C194, C1904) or calcein (C481) is >95% self-quenched at concentrations >100 mM. Concentrated solutions of these water-soluble dyes are encapsulated in liposomes, which are then separated from any remaining free dye by gel filtration. Upon addition of a fusogen or other permeabilizing agent, dye release is accompanied by an increase in fluorescence (excitation/emission maxima ~490 nm/520 nm). Complete lysis of the liposomes with 0.1% Triton® X-100 can be used to determine the assay endpoint. Calcein may be preferred over carboxyfluorescein because of higher net charge and lower pH sensitivity. Note that this assay will detect any process that causes leakage of aqueous contents, including fusion, lysis or permeabilization. In a modification of this assay designed to specifically detect vesicle fusion, a nonfluorescent Co^{2+} complex of calcein contained in a vesicle is disrupted upon fusion with a second vesicle that delivers EDTA, a chelator of Co^{2+}.[17,18] Studies employing carboxyfluorescein self-quenching include investigations of interactions of membranes with mycobacterial glycopeptidolipids[19] and with HIV glycoprotein peptide fragments.[20]

1. J Immunol Methods (1977) 15:255; **2.** Biochemistry (1985) 24:3099; **3.** Biochemistry (1994) 33:12573; **4.** Biochemistry (1992) 31:4262; **5.** Biochemistry (1995) 34:1084; **6.** Biochemistry (1993) 32:14194; **7.** Biochemistry (1994) 33:3201; **8.** J Biol Chem (1993) 268:22112; **9.** Biochim Biophys Acta (1993) 1142:277; **10.** Am J Physiol (1992) 262:G30; **11.** Biochim Biophys Acta (1990) 1024:352; **12.** Methods Enzymol (1992) 207:501; **13.** Biochemistry (1980) 19:6011; **14.** Nature (1979) 281:690; **15.** Biochemistry (1994) 33:5805; **16.** J Biol Chem (1994) 269:14473; **17.** Biophys J (1989) 55:973; **18.** J Biol Chem (1982) 257:13892; **19.** Biochemistry (1994) 33:7056; **20.** J Biol Chem (1992) 267:7121.

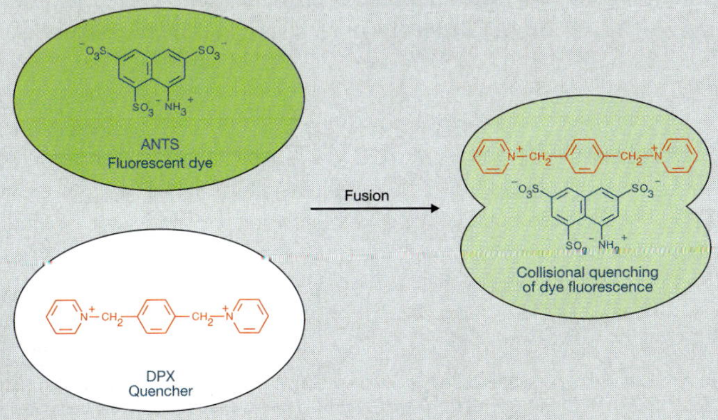

Figure 1 Pictorial representation of the ANTS/DPX vesicle-fusion assay.

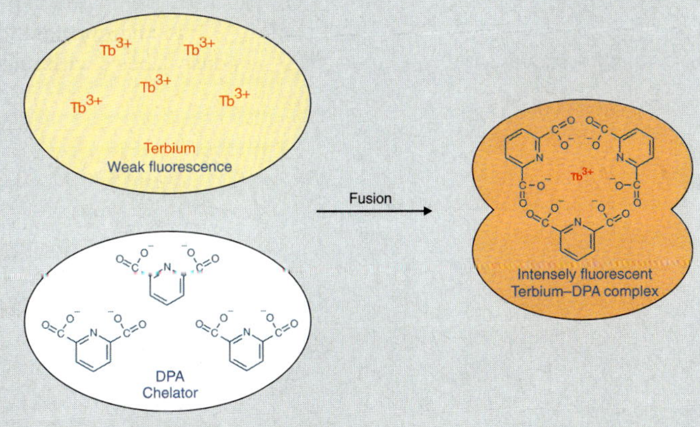

Figure 2 Pictorial representation of the terbium/dipicolinic acid (DPA) fluorescence enhancement assay for vesicle fusion.

IMPORTANT NOTICE: The products described in this manual are covered by one or more Limited Use Label License(s). Please refer to the Appendix on page 971 and Master Product List on page 975. Products are For Research Use Only. Not intended for any animal or human therapeutic or diagnostic use.

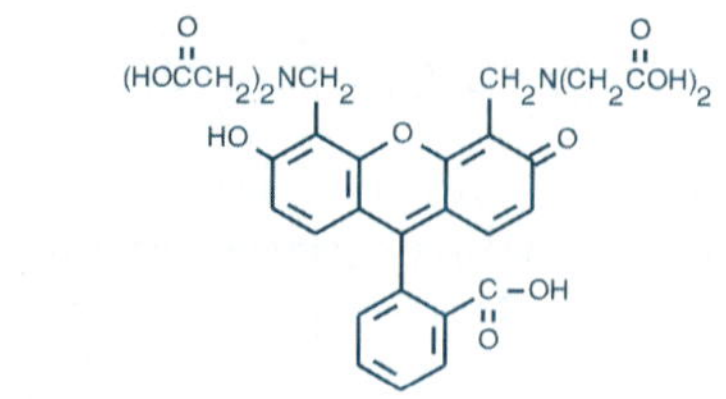

Figure 14.3.11 Calcein (C481).

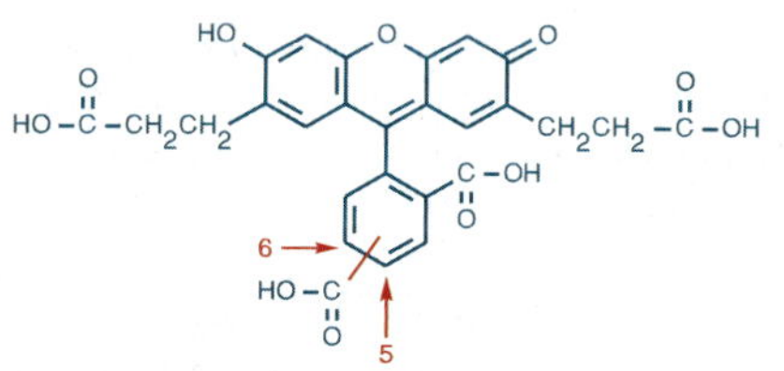

Figure 14.3.12 2′,7′-Bis-(2-carboxyethyl)-5-(and-6)-carboxyfluorescein (BCECF acid, B1151).

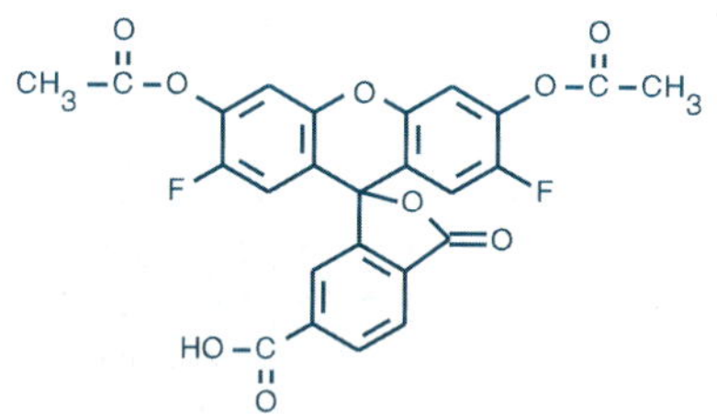

Figure 14.3.13 Oregon Green® 488 carboxylic acid diacetate (carboxy-DFFDA, O6151).

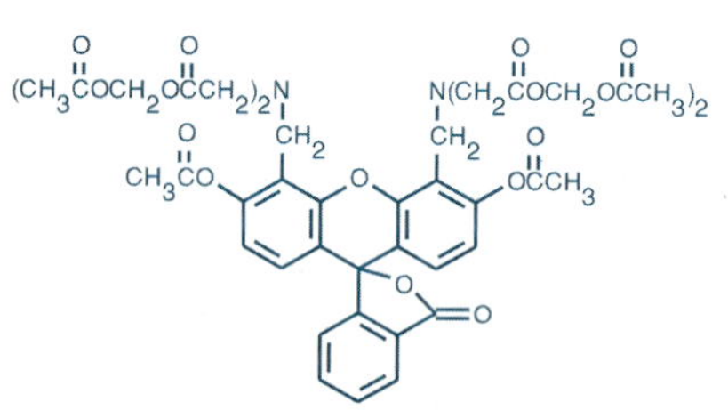

Figure 14.3.14 Calcein, AM (C1430).

(C481)—a polyanionic fluorescein derivative that has about six negative and two positive charges at pH 7 (Figure 14.3.11)—as well as BCECF (B1151, Figure 14.3.12), carboxyfluorescein (C194, C1904), the 5-isomer of Oregon Green® 488 carboxylic acid (O6146) and fluorescein-5-(and 6-) sulfonic acid (F1130) are all soluble in water at >100 mM at pH 7. Unlike the other fluorescein derivatives, both calcein and Oregon Green® 488 carboxylic acid exhibit fluorescence that is essentially independent of pH between 6.5 and 12.

These green-fluorescent polar tracers are widely used for investigating:

- Cell volume changes in neurons and other cells[68–71]
- Gap junctional communication[72–74]
- Liposome formation, fusion and targeting[75–78]
- Membrane integrity and permeability[79–90]

Fluorescence of calcein (but not of carboxyfluorescein or fluorescein sulfonic acid) is strongly quenched by Fe^{3+}, Co^{2+}, Cu^{2+} and Mn^{2+} at physiological pH but not by Ca^{2+} or Mg^{2+} ions.[91] Monitoring the fluorescence level of cells that have been loaded with calcein (or its AM ester, see below) may provide an easy means for following uptake of Fe^{3+}, Co^{2+}, Cu^{2+}, Mn^{2+} and certain other metals through ion channels.[92] Increases in the internal volume of lipid vesicles and virus envelopes cause a decrease in Co^{2+}-induced quenching of calcein, a change that can be followed fluorometrically.[93] In addition, the Co^{2+}-quenched calcein complex is useful for both lysis and fusion assays[94] (Assays of Volume Change, Membrane Fusion and Membrane Permeability—Note 14.3). Calcein is a widely used reagent for following volume changes because its fluorescence is not particularly sensitive to either pH or physiological concentrations of other ions.[69]

We prepare a high-purity grade of calcein (C481) that is generally >97% pure by HPLC. The chemical structure assigned to "calcein" in various literature references and by commercial sources has been inconsistent;[95,96] our structure (Figure 14.3.11) has been confirmed by NMR spectroscopy, and we believe that several past assignments of other structures to calcein were incorrect. We also offer a high-purity grade of 5-(and 6-)carboxyfluorescein (C1904) that contains essentially no polar or nonpolar impurities that might alter transfer rates of the dye between vesicles and cells.[97]

Cell-Permeant Fluorescein Derivatives

Cell-permeant versions of carboxyfluorescein, fluorescein sulfonic acid, calcein and the Oregon Green® dyes permit passive loading of cells (Section 15.2). Acid hydrolysis of nonfluorescent carboxyfluorescein diacetate (CFDA; C195, C1361, C1362; Section 15.2) to fluorescent carboxyfluorescein has been used to detect the fusion of dye-loaded clathrin-coated vesicles with lysosomes.[98] CFDA has also been used to investigate cell–cell communication in plant cells.[99–101] A probenecid-inhibitable anion-transport mechanism permits loading of carboxyfluorescein diacetate and Oregon Green® 488 carboxylic acid diacetate (O6151, Figure 14.3.13; Section 15.2) into hyphal tip-cells of some fungi.[102]

Calcein AM (Figure 14.3.14), but not the AM or acetate esters of BCECF or CFDA, is reported to differentially label lymphocytes, permitting their resolution into two populations based on fluorescence intensity, only one of which is taken up by lymphoid organs. This unique property makes calcein AM a useful probe for determining the lymph node homing potential of lymphocytes.[103]

In an important technique for studying gap junctional communication, cells are simultaneously labeled with calcein AM (C1430, C3099, C3100MP) and DiI (D282, D3911, V22885; Section

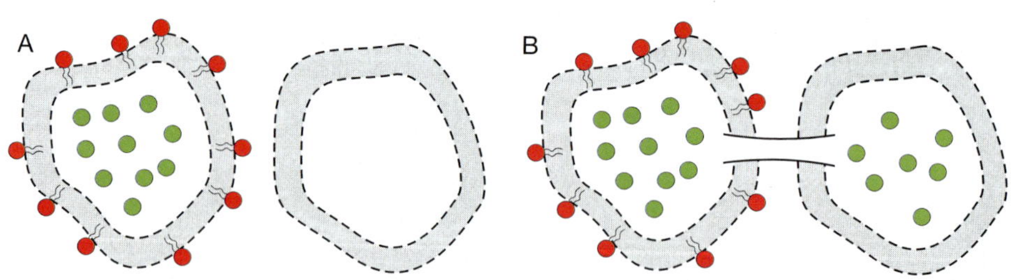

Figure 14.3.15 A simple technique for the study of gap junctional communication. A population of cells is labeled simultaneously with DiI (D282, D3911) and calcein AM (C1430, C3099, C3100MP), and then mixed with an unlabeled cell population (**A**). The formation of gap junctions allows the cytosolic tracer calcein to cross into the unlabeled cell, while the membrane-bound DiI does not (**B**). Cells from the initial unlabeled population that have taken part in gap junctional communication will therefore display the green fluorescence of calcein while lacking the red-fluorescent signal of DiI.

IMPORTANT NOTICE: The products described in this manual are covered by one or more Limited Use Label License(s). Please refer to the Appendix on page 971 and Master Product List on page 975. Products are For Research Use Only. Not intended for any animal or human therapeutic or diagnostic use.
www.invitrogen.com/probes

14.4) and then mixed with unlabeled cells (Figure 14.3.15). When gap junctions are established, only the cytosolic calcein tracer (but not the DiI membrane probe) is transferred from the labeled cell to the unlabeled cell. Thus, after gap-junctional transfer, the initially unlabeled cells exhibit the green fluorescence of calcein but not the red fluorescence of DiI.[73,104–107] This assay can be followed by either imaging or flow cytometry.[108] In addition, calcein AM and DiI have been combined for use in following cell fusion[109] and for analysis of cholesterol processing by macrophages following ingestion of apoptotic cells.[110]

Fluorescein Substitutes

Chapter 1 describes several of our proprietary green-fluorescent dyes that have exceptional optical properties, including our Alexa Fluor® 488 dye (Section 1.3), BODIPY® FL dye (Section 1.4) and Oregon Green® dyes (Section 1.5). Not only do these innovative fluorescein substitutes exhibit high quantum yields in aqueous solution, but the dyes are significantly more photostable than fluorescein (Figure 14.3.16) and their fluorescence is less sensitive to pH (Figure 14.3.17). Their greater photostability makes them the preferred green-fluorescent dyes for fluorescence microscopy. In addition to their membrane-permeant versions, which are described in Section 15.2, highly water-soluble derivatives of these fluorescein substitutes are available for use as polar tracers:

- Oregon Green® 514 carboxylic acid (O6138, Figure 14.3.18), which is highly photostable and has little pH sensitivity at near-neutral pH
- BODIPY® 492/515 disulfonic acid[111] (D3238), which has narrow spectral bandwidths and bright green, pH-independent fluorescence
- Carboxy-2′,7′-dichlorofluorescein (C368), which has a lower pK_a than fluorescein

Sulforhodamines

Sulforhodamine 101 (S359, Figure 14.3.19) and sulforhodamine B (S1307, Figure 14.3.20) are orange- to red-fluorescent, very water-soluble sulfonic acid tracers with strong absorption and good photostability. Sulforhodamine 101—the precursor to reactive Texas Red® derivatives—has been the preferred red-fluorescent polar tracer for use in combination with lucifer yellow CH, carboxyfluorescein or calcein[112–114] (Figure 14.3.6). Activity-dependent uptake of sulforhodamine 101 during nerve stimulation has been reported.[115–119] Sulforhodamine 101 specifically labels astrocytes both *in vivo*[120] and in acute brain slice preparations.[121] Labeling is accomplished by application of a sulforhodamine 101 solution (1–25 µM in artificial cerebrospinal fluid) to the tissue for 1–5 minutes and is stable for several hours.[120] This technique is particularly useful for cellular-context identification in conjunction with calcium imaging using Oregon Green® 488 BAPTA-1, fluo-4 and related fluorescent indicators[122–125] (Section 19.3). Because it is chemically stable, can be prepared in high purity and has a fluorescence quantum yield of nearly 1.0, we have included sulforhodamine 101 in our Reference Dye Sampler Kit (R14782, Section 23.1), along with four other dyes whose spectra cover the visible wavelengths.

Sulforhodamine B is an alternative to sulforhodamine 101 for investigating neuronal morphology,[126,127] preparing fluorescent liposomes,[128] studying cell–cell communications[129,130] and labeling elastic and collagen fibers.[131]

Hydroxycoumarins

7-Hydroxycoumarin-3-carboxylic acid (H185) is a blue-fluorescent polar tracer (excitation/emission maxima of ~388/445 nm) with uses that complement those of calcein and the other green-fluorescent polar tracers.[106] The membrane-permeant AM ester of calcein blue (C1429, Section 15.2), another coumarin-based tracer, can be used for passive loading of cells.[132]

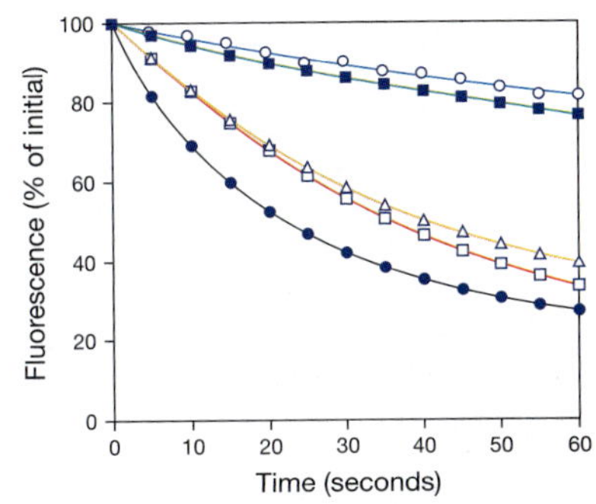

Figure 14.3.16 Comparison of photostability of green-fluorescent antibody conjugates. The following fluorescent goat anti-mouse IgG antibody conjugates were used to detect mouse anti–human IgG antibody labeling of human anti-nuclear antibodies in HEp-2 cells on prefixed test slides (INOVA Diagnostics Corp.): Oregon Green® 514 (O6383, ■), Alexa Fluor® 488 (A11001, ○), BODIPY® FL (B2752, △), Oregon Green® 488 (O6380, □) or fluorescein (F2761, ●). Samples were continuously illuminated and viewed on a fluorescence microscope using a fluorescein longpass filter set. Images were acquired every 5 seconds. For each conjugate, three data sets, representing different fields of view, were averaged and then normalized to the same initial fluorescence intensity value to facilitate comparison.

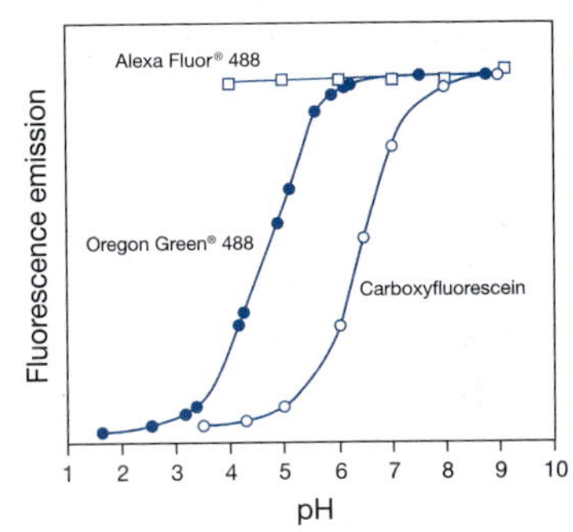

Figure 14.3.17 Comparison of pH-dependent fluorescence of the Oregon Green® 488 (●), carboxyfluorescein (○) and Alexa Fluor® 488 (□) fluorophores. Fluorescence intensities were measured for equal concentrations of the three dyes using excitation/emission at 490/520 nm.

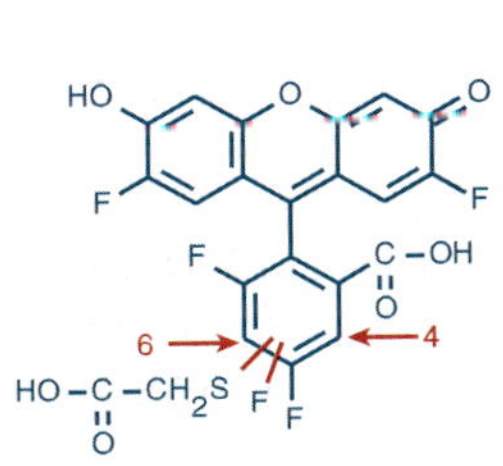

Figure 14.3.18 Oregon Green® 514 carboxylic acid (O6138).

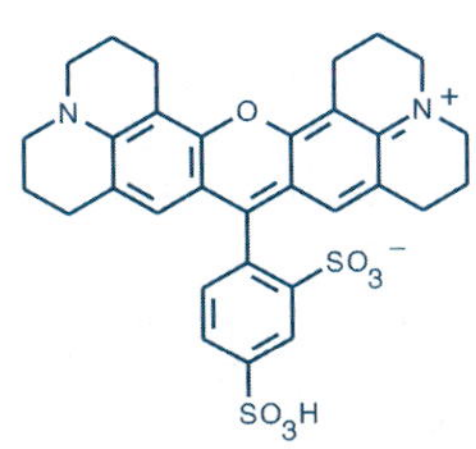

Figure 14.3.19 Sulforhodamine 101 (S359).

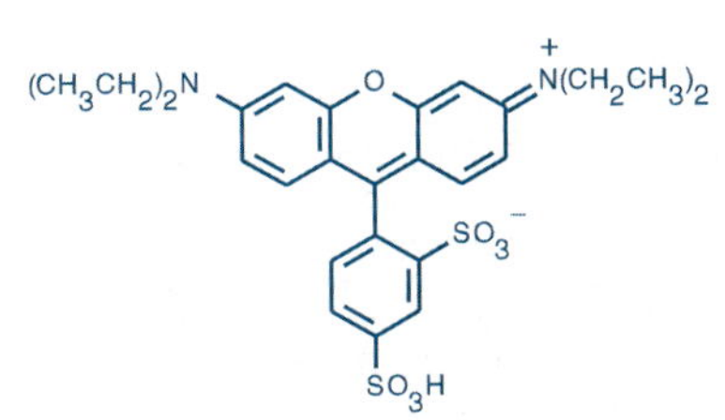

Figure 14.3.20 Sulforhodamine B (S1307).

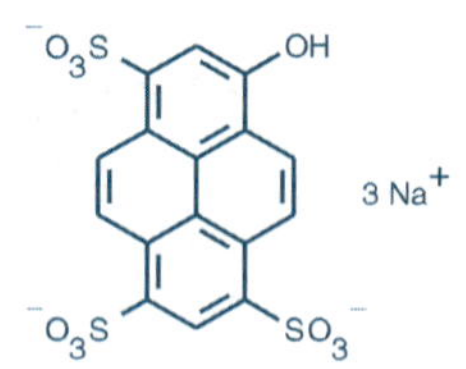

Figure 14.3.21 8-Hydroxypyrene-1,3,6-trisulfonic acid, trisodium salt (HPTS; pyranine; H348).

Figure 14.3.22 Sternomastoid muscle fibers of a living mouse that have been transfected with YOYO®-1 dye–stained DNA (red) containing the *lacZ* reporter gene and then stained with β-galactosidase substrate fluorescein di-β-D-galactopyranoside (FDG, F1179). DNA stained with YOYO®-1 (Y3601) prior to implantation could still be localized 5 days after application. Fluorescence signals were visualized *in situ* by epifluorescence microscopy with a low-light level SIT camera and a computer imaging system. Image contributed by Peter van Mier, Washington University School of Medicine.

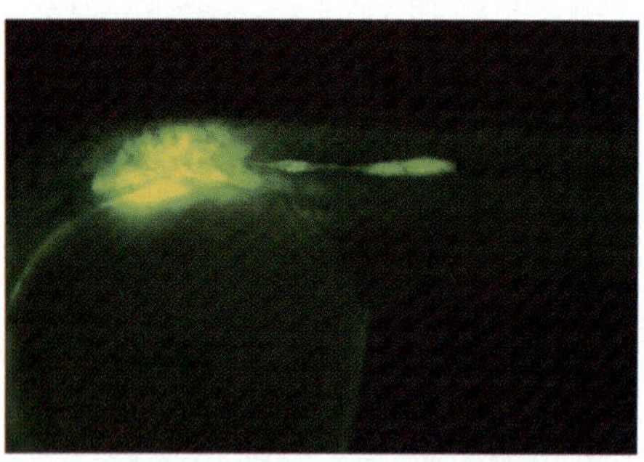

Figure 14.3.23 A zebrafish embryo that was injected with a 1% solution (in 0.2 M KCl) of DMNB-caged fluorescein 10,000 MW dextran at the two-cell stage and then allowed to grow for 19 hours. The posterior lateral line placode was then exposed to a 5-second pulse from an epifluorescence microscope fitted with a DAPI optical filter set. After 6 hours of further development, the labeled primordium has migrated caudally, away from the activation site. Image contributed by Walter K. Metcalfe, Institute of Neuroscience, University of Oregon.

Polysulfonated Pyrenes

HPTS (8-hydroxypyrene-1,3,6-trisulfonic acid, also known as pyranine, H348; Figure 14.3.21) is a unique pH-sensitive tracer. It fluoresces blue in acidic solutions and in acidic organelles,[133–135] but fluoresces green in more basic organelles.[136] In addition to its use as a probe for proton translocation,[137–140] HPTS has been employed for intracellular labeling of neurons[141] and as a fluid-phase endocytic tracer in catecholamine-secreting PC12 rat pheochromocytoma cells.[142] HPTS forms a nonfluorescent complex with the cationic quencher DPX (X1525), and several assays have been described that monitor the increase in HPTS fluorescence that occurs upon lysis or fusion of liposomes or cells containing this quenched complex[135,143–145] (Assays of Volume Change, Membrane Fusion and Membrane Permeability—Note 14.3). HPTS has also been used as a viscosity probe in unilamellar phospholipid vesicles.[146]

The pH-insensitive 8-aminopyrene-1,3,6-trisulfonic acid (APTS, A6257) and 1,3,6,8-pyrenetetrasulfonic acid (P349) are extremely soluble in water (>25%); they have been utilized as blue-fluorescent tracers.[147–150] As with HPTS, the fluorescence of APTS and 1,3,6,8-pyrenetetrasulfonic acid is quenched by DPX[135,144,151] (Assays of Volume Change, Membrane Fusion and Membrane Permeability—Note 14.3) and by the cationic spin label CAT 1[152] (T506, see below). These quenched-fluorophore complexes are useful for following lysis of cells and liposomes.

ANTS–DPX

The polyanionic dye ANTS (A350) is often used in combination with the cationic quencher DPX (X1525) for membrane fusion or permeability assays, including complement-mediated immune lysis[153] (Assays of Volume Change, Membrane Fusion and Membrane Permeability—Note 14.3). Thallium (Tl^+) and cesium (Cs^+) ions quench the fluorescence of ANTS, pyrenetetrasulfonic acid and some other polyanionic fluorophores.[154,155] A review by Garcia[148] describes how this quenching effect can be utilized to determine transmembrane ion permeability. The unusually high Stokes shift of ANTS in water (>150 nm) separates its emission from much of the autofluorescence of biological samples. An approximately 4-fold enhancement of the quantum yield of ANTS is induced by D_2O—a spectral characteristic that has been used to determine water permeability in red blood cell ghosts and kidney collecting tubules.[156,157] ANTS has also been employed as a neuronal tracer.[141]

Lanthanide Chelates

Terbium ion (Tb^{3+} from $TbCl_3$, T1247) forms a chelate with dipicolinic acid (DPA) that is ~10,000 times more fluorescent than free Tb^{3+}. Fusion of vesicles that have been separately loaded with DPA and Tb^{3+} results in enhanced fluorescence, providing the basis for liposome fusion assays (Assays of Volume Change, Membrane Fusion and Membrane Permeability—Note 14.3). The fluorescence emission spectrum of the Tb^{3+}/DPA complex exhibits two sharp spectral peaks at 491 nm and 545 nm and a lifetime of several milliseconds. Because DPA is a major constituent of bacterial spores, Tb^{3+}/DPA complex luminescence provides a straightforward yet sensitive method for their detection.[158–160]

TOTO®, YOYO® and SYTO® Nucleic Acid Stains

Our high-affinity nucleic acid stains, including TOTO®-1 and YOYO®-1 (T3600, Y3601; Chapter 8), form tight complexes with nucleic acids with slow off-rates for release of the dye. Consequently, nucleic acids that have been prelabeled with these dyes can be traced in cells following microinjection or during gene transfer[161–163] (Figure 14.3.22). The cell-to-cell transport via plasmodesmata of TOTO®-1 dye–labeled RNA, single-stranded DNA and double-stranded DNA has been determined following microinjection of the labeled nucleic acids into plant cells.[164–167]

Khoobehi and Peyman have demonstrated the use of our cell-permeant SYTO® nucleic acid stains as ophthalmological tracers of blood flow. White blood cells were passively loaded with the green-fluorescent SYTO® 16 dye (S7578, Section 8.1) or the red-fluorescent SYTO® 59 dye (S11341, Section 8.1). Red blood cell membranes were labeled with DiD (D307, Section 14.4). By using an argon-ion laser to excite SYTO® 16 and a red light–emitting He-Ne laser to excite both SYTO® 59 and DiD, they were able to follow the relative mobility of the two types of blood cells.[168]

IMPORTANT NOTICE: The products described in this manual are covered by one or more Limited Use Label License(s). Please refer to the Appendix on page 971 and Master Product List on page 975. Products are For Research Use Only. Not intended for any animal or human therapeutic or diagnostic use.
www.invitrogen.com/probes
molecular probes® | invitrogen™ by life technologies™

Caged Fluorescent Dye Tracers

UV photolysis of photoactivatable fluorescent dyes provides a means of controlling—both spatially and temporally—the release of fluorescent tracers (Section 5.3). Thus, caged dyes enable researchers to follow the movement of individual molecules and cellular structures,[169–171] as well as to study cell lineage in live organisms (Figure 14.3.23). CMNB-caged fluorescein [172,173] (F7103, Figure 14.3.24) is colorless and nonfluorescent until it is photolyzed at <365 nm to the intensely green-fluorescent free dye. Movement of the liberated fluorophore from the site of photolysis can then be followed. The succinimidyl ester of CMNB-caged carboxyfluorescein (C20050, Section 1.5), which is water soluble at pH 7, is useful for preparation of other caged dye tracers, including those of proteins.

Fluorescent Retrograde Tracers

Anterograde and retrograde tracing of neurons has utilized a wide variety of fluorescent and nonfluorescent probes [174] (Figure 14.3.25). Among the retrograde and anterograde tracers are biotin derivatives and polar fluorescent dyes (described in this section), lipophilic tracers such as DiI (D282, Section 14.4), dextran conjugates (Section 14.5), fluorescent microspheres (Section 14.6) and protein conjugates, including lectins (Section 14.7).

True Blue and Nuclear Yellow

The popular retrograde tracer true blue (T1323) is a UV light–excitable, divalent cationic dye that stains the cytoplasm with blue fluorescence.[175–179] For two-color neuronal mapping, true blue has been combined with longer-wavelength tracers such as nuclear yellow [175,180–182] (Hoechst S769121, N21485) or diamidino yellow,[183–186] which primarily stain the neuronal nucleus with yellow fluorescence (Figure 14.3.26, Figure 14.3.27, Figure 14.3.28). Fluorescent microspheres have also been used as a counterstain with true blue.[187] True blue is reported to be a less cytotoxic retrograde tracer than Fluoro-Gold™ [176] and to be a more efficient retrograde tracer than diamidino yellow.[177] Both true blue and nuclear yellow are stable when subjected to immunohistochemical processing and can be used to photoconvert DAB into an insoluble, electron-dense reaction product.[28,188,189]

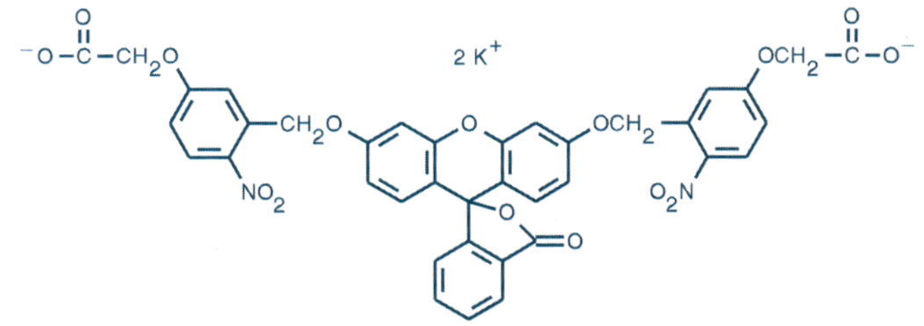

Figure 14.3.24 Fluorescein bis-(5-carboxymethoxy-2-nitrobenzyl) ether, dipotassium salt (CMNB-caged fluorescein, F7103).

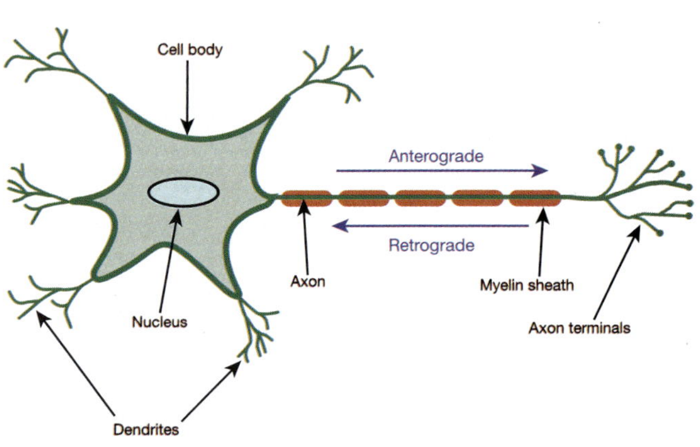

Figure 14.3.25 Structural schematic of a neuron indicating the directions of retrograde and anterograde transport.

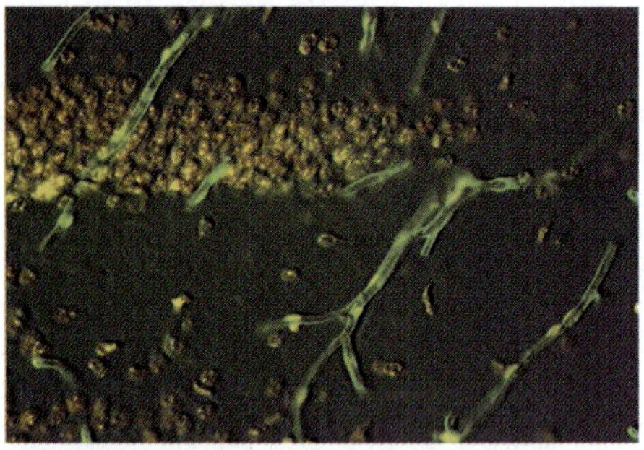

Figure 14.3.26 Capillaries in the hippocampal region of a mouse brain cryosection were visualized with the green-fluorescent Alexa Fluor® 488 conjugate of lectin HPA from *Helix pomatia* (L11271), which specifically binds to type A erythrocytes and *N*-acetylgalactosaminyl residues. The nuclei were counterstained with nuclear yellow (N21485). The multiple-exposure image was acquired using a DAPI longpass filter set and a filter set appropriate for fluorescein.

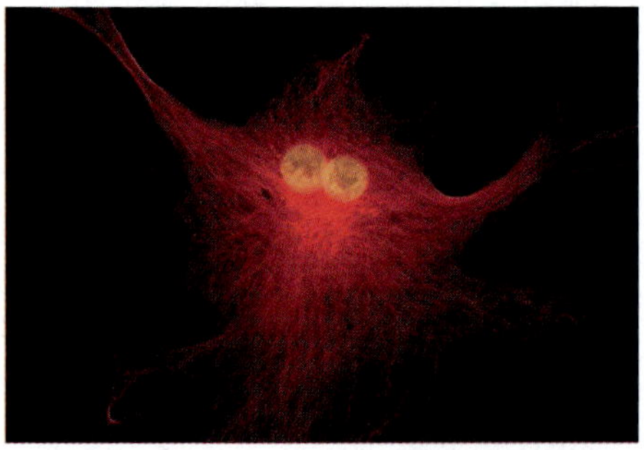

Figure 14.3.27 A binucleate bovine pulmonary artery endothelial cell labeled with the biotin-XX conjugate of anti-α-tubulin antibody (A21371) and Alexa Fluor® 568 streptavidin (S11226), then counterstained with nuclear yellow (N21485).

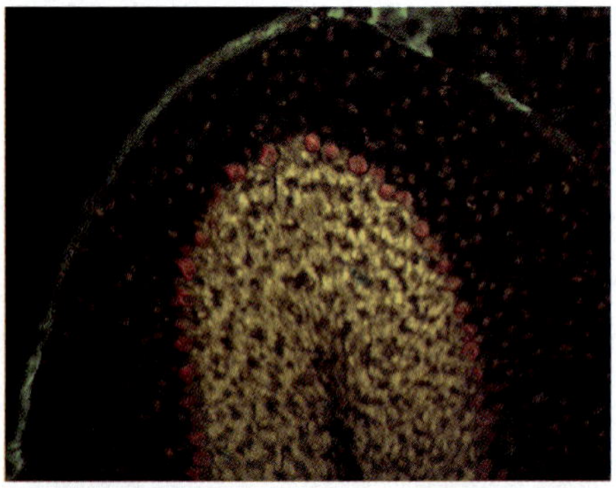

Figure 14.3.28 Neurons in a mouse cerebellum section labeled with NeuroTrace® 530/615 red-fluorescent Nissl stain (N21482). The Nissl substance, ribosomal RNA associated with the rough endoplasmic reticulum, is specific to neuronal cells. Other cells in the sample are identified with the contrasting nuclear counterstain nuclear yellow (N21485). Smaller neurons appear to be labeled primarily with the nuclear yellow stain. This image is a composite of two micrographs acquired using a DAPI longpass filter set and a filter set appropriate for tetramethylrhodamine.

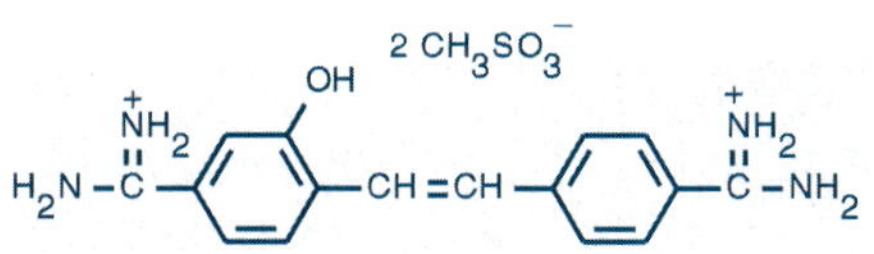

Figure 14.3.29 Hydroxystilbamidine, methanesulfonate (H22845).

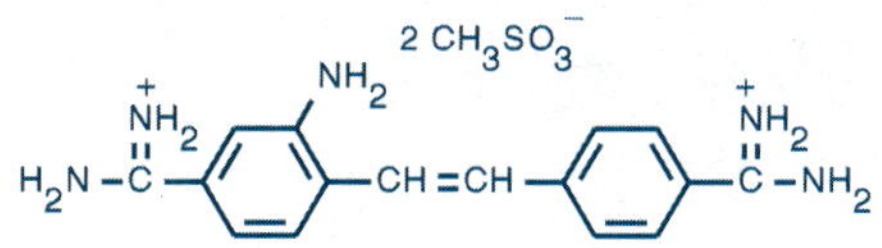

Figure 14.3.30 Aminostilbamidine, methanesulfonate (A22850).

Hydroxystilbamidine and Aminostilbamidine

Hydroxystilbamidine methanesulfonate (H22845, Figure 14.3.29) was originally developed as a trypanocide and has been identified by Wessendorf[190] as the active component of a dye that was named Fluoro-Gold™ by Schmued and Fallon[191] and later sold for retrograde tracing by Fluorochrome, Inc. The use of hydroxystilbamidine as a histochemical stain and as a retrograde tracer for neurons apparently goes back to the work of Snapper and collaborators in the early 1950s who, while studying the effects of hydroxystilbamidine on multiple myeloma, showed that therapeutically administered hydroxystilbamidine gives selective staining of ganglion cells.[192,193] The comprehensive article by Wessendorf[190] also describes several other early applications of hydroxystilbamidine that do not include its therapeutic uses:

- As an AT-selective, nonintercalating nucleic acid stain for discriminating between DNA and RNA[194–196] (Section 8.1)
- For histochemical staining of DNA, mucosubstances and elastic fibers, as well as mast cells[197]
- As a ribonuclease inhibitor[198]
- For lysosomal staining of live cells[199–204] (Section 12.3)

The weakly basic properties of hydroxystilbamidine that result in its uptake by lysosomes are reportedly important for its mechanism of retrograde transport.[190,199] We developed a product that we called hydroxystilbamidine (formerly catalog number H7599) in April 1995; however, we subsequently discovered that the chemical structure of our original "hydroxystilbamidine" corresponded to a novel dye that we now call aminostilbamidine (A22850, Figure 14.3.30). Apparently, aminostilbamidine functions at least as well as authentic hydroxystilbamidine as a tracer.[205] Aminostilbamidine, however, does not show the spectral shifts with DNA and RNA that are observed with authentic hydroxystilbamidine.

Propidium Iodide and DAPI for Retrograde Tracing

A variety of other low molecular weight dyes have been used as fluorescent retrograde neuronal tracers.[206,207] These include propidium iodide[208,209] (P1304MP, P21493) and DAPI.[210,211] Both propidium iodide and DAPI can be used to photoconvert DAB into an insoluble, electron-dense product.[28,212] The lactate salt of DAPI (D3571) has much higher water solubility than the chloride salt (D1306, D21490), making it the preferred form for microinjection.

NeuroTrace® Fluorescent Nissl Stains

The Nissl substance, described by Franz Nissl more than 100 years ago, is unique to neuronal cells.[213] Composed of an extraordinary amount of rough endoplasmic reticulum, the Nissl substance reflects the unusually high protein synthesis capacity of neuronal cells. Various fluorescent or chromophoric "Nissl stains" have been used for this counterstaining, including acridine orange,[214] ethidium bromide,[214] neutral red (N3246, Section 15.2), cresyl violet,[215] methylene blue, safranin-O and toluidine blue-O.[216] We have developed five fluorescent Nissl stains

Table 14.2 Fluorescence characteristics of NeuroTrace® fluorescent Nissl stains.

				Signal Using Various Filter Sets †				
Cat. No.	Fluorescent Color	Ex *	Em *	405 ± 20 445 ≥450	485 ± 11 505 535 ± 17.5	510 ± 11.5 540 ≥550	560 ± 20 590 610 ± 10	640 ± 10 660 682 ± 11
N21479	Blue	435	455	+++	+	–	–	–
N21480	Green	500	525	–	+++	+++	–	–
N21481	Yellow	515	535	–	+++	+++	+++	–
N21482	Red	530	615	–	–	++	+++	–
N21483	Deep red	640	660	–	–	– ‡	–	++

* Approximate fluorescence excitation (Ex) and emission (Em) maxima, in nm. † Spectral characteristics in nm of the excitation filter, dichroic mirror and emission filter are shown top to bottom for each filter set. Relative signal strength for each filter combination is indicated with pluses and minuses. ‡ High background fluorescence.

IMPORTANT NOTICE: The products described in this manual are covered by one or more Limited Use Label License(s). Please refer to the Appendix on page 971 and Master Product List on page 975. Products are For Research Use Only. Not intended for any animal or human therapeutic or diagnostic use.
www.invitrogen.com/probes
molecular probes® | invitrogen™ by life technologies™

(Table 14.2) that not only provide a wide spectrum of fluorescent colors for staining neurons, but also are far more sensitive than the conventional dyes:

- NeuroTrace® 435/455 blue-fluorescent Nissl stain (N21479, Figure 14.3.31)
- NeuroTrace® 500/525 green-fluorescent Nissl stain (N21480; Figure 14.3.32, Figure 14.3.33, Figure 14.3.34)
- NeuroTrace® 515/535 yellow-fluorescent Nissl stain (N21481, Figure 14.3.35)
- NeuroTrace® 530/615 red-fluorescent Nissl stain (N21482; Figure 14.3.28, Figure 14.3.36)
- NeuroTrace® 640/660 deep red–fluorescent Nissl stain (N21483)

In addition, the Nissl substance redistributes within the cell body in injured or regenerating neurons. Therefore, these Nissl stains can also act as markers for the physiological state of the neuron. Staining by the Nissl stains is eliminated by pretreatment of tissue specimens with RNase; however, these dyes are not specific stains for RNA in solutions. The strong fluorescence (emission maximum ~515–520 nm) of NeuroTrace® 500/525 green-fluorescent Nissl stain (N21480) makes it the preferred dye for use as a counterstain in combination with orange- or red-fluorescent neuroanatomical tracers such as DiI [217] (D282, D3911, V22885; Section 14.4; Figure 14.3.34).

Polar Spin Label

The highly water-soluble cationic spin label 4-trimethylammonium-2,2,6,6-tetramethylpiperidine-1-oxyl iodide (CAT 1, T506) has been used to:

- Quench fluorescent dyes in solutions, cells and cell membranes [218]
- Detect oxygen gradients and oxidation–reduction properties of cells [219–223]
- Study liposome permeability [218,223]

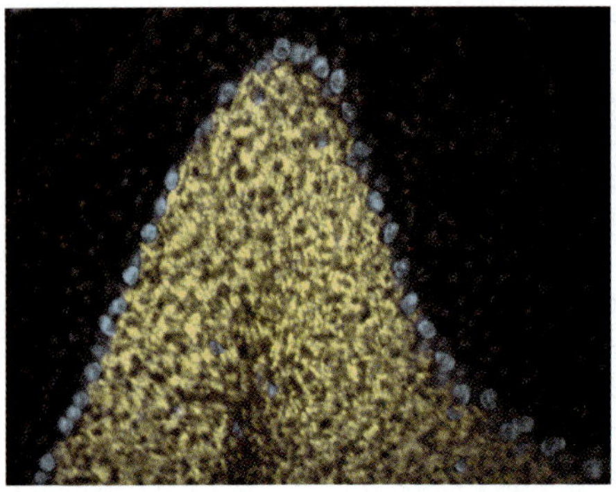

Figure 14.3.31 Mouse brain section stained with NeuroTrace® 435/455 blue-fluorescent Nissl stain (N21479) and counterstained with nuclear yellow (N21485).

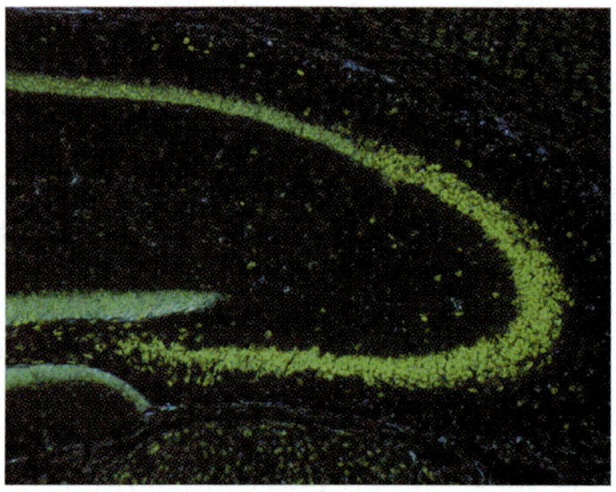

Figure 14.3.32 Pyramidal cells of the hippocampus and dentate gyrus in a transverse cryosection of formaldehyde-fixed mouse brain. NeuroTrace® green fluorescent Nissl stain (N21480) is localized to neuronal somata, while non-neuronal cells can be identified by the presence of DAPI-stained nuclei. This image is a composite of images taken using a 10× objective and filters appropriate for fluorescein and DAPI.

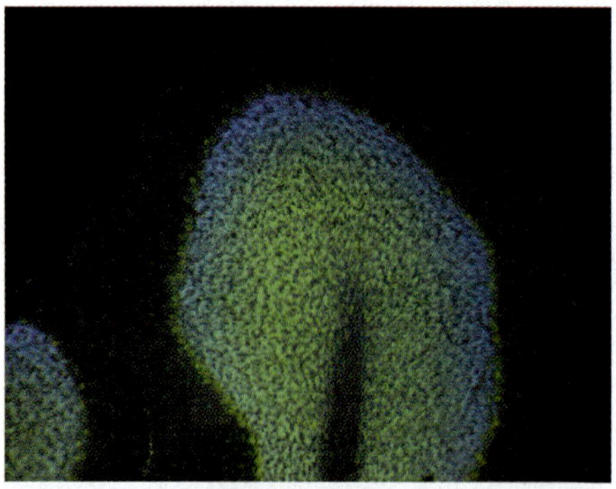

Figure 14.3.33 A mouse brain cryosection stained with the neuron-selective NeuroTrace® 500/525 green-fluorescent Nissl stain (N21480). The nuclei of the non-neuronal cells appear blue after incubation with the cell-permeant DNA counterstain, DAPI (D1306, D3571, D21490). The image is a composite of two micrographs acquired using filter sets appropriate for fluorescein and DAPI.

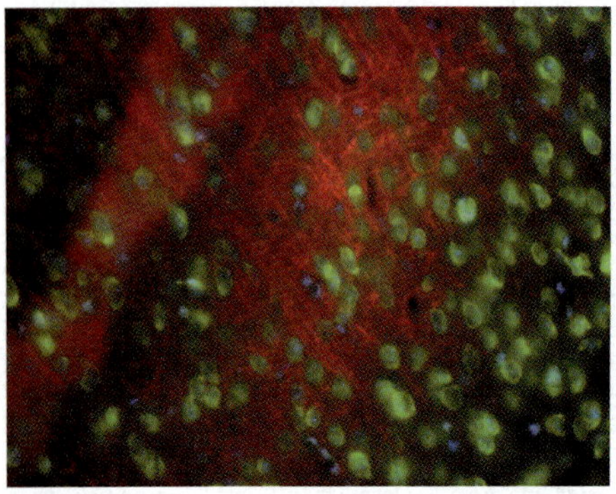

Figure 14.3.34 Neural somata from a mouse brain section labeled with NeuroTrace® 500/525 green-fluorescent Nissl stain (N21480). Nuclei are labeled with DAPI (D1306, D3571, D21490). Myelin and other lipophilic areas are illustrated with the aid of the red-orange fluorescence from CellTracker™ CM-DiI (C7000, C7001). The image is a composite of three micrographs acquired using filters appropriate for fluorescein, tetramethylrhodamine and DAPI.

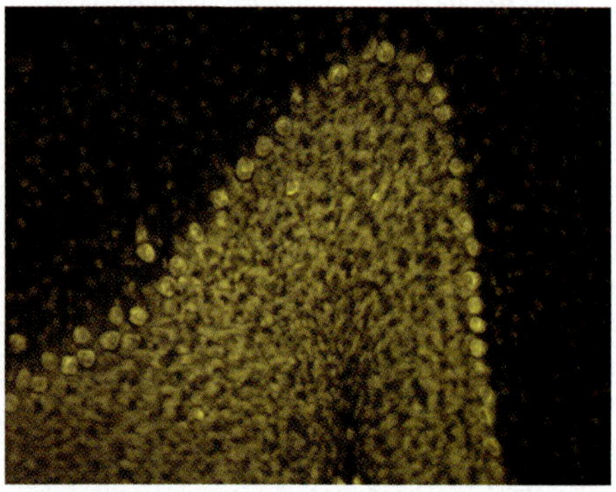

Figure 14.3.35 Mouse brain section stained with NeuroTrace® 515/535 yellow-fluorescent Nissl stain (N21481) and counterstained with nuclear yellow (N21485).

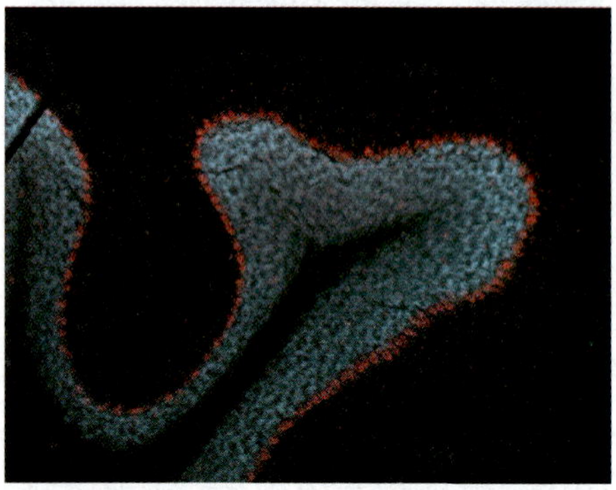

Figure 14.3.36 Neural somata labeled with NeuroTrace® 530/615 red-fluorescent Nissl stain (N21482). Nuclei in this mouse cerebellum section were counterstained with DAPI (D1306, D3571, D21490). The image is a composite of two micrographs acquired using filters appropriate for tetramethylrhodamine and DAPI.

IMPORTANT NOTICE: The products described in this manual are covered by one or more Limited Use Label License(s). Please refer to the Appendix on page 971 and Master Product List on page 975. Products are For Research Use Only. Not intended for any animal or human therapeutic or diagnostic use.

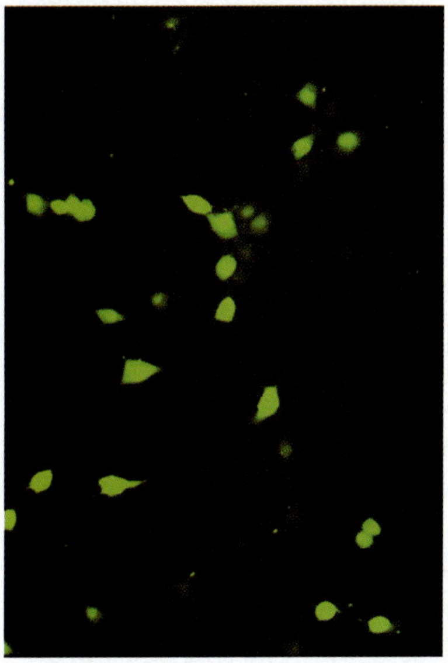

Figure 14.3.38 3T3 cells loaded with calcein (C481) using the Influx™ pinocytic cell-loading reagent (I14402). The image was acquired using a fluorescence microscope equipped with a cooled CCD camera and a bandpass filter set appropriate for fluorescein.

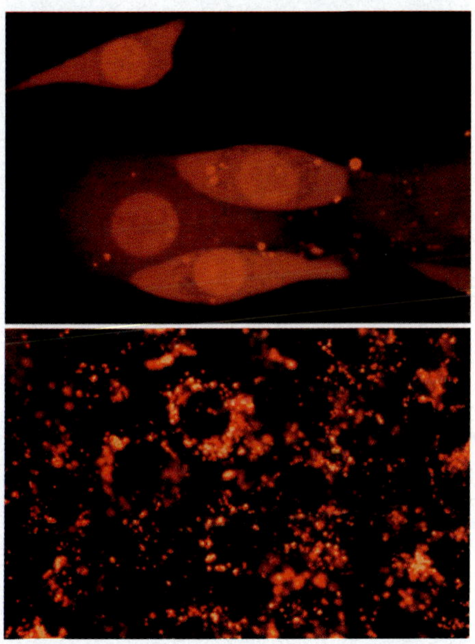

Figure 14.3.39 Adherent CRE BAG 2 cells passively loaded with the membrane-impermeant polar tracer Alexa Fluor® 594 hydrazide (A10438, A10442). The image on top illustrates the relatively uniform, cytoplasmic labeling one can obtain with the Influx™ pinocytic cell-loading reagent (I14402), compared to the punctate labeling that results from pinocytic uptake in normal growth medium (bottom). Both images were acquired using a bandpass filter set appropriate for rhodamine dyes.

Signal Amplification of Polar Tracers

Polar tracers such as Cascade Blue® hydrazide, lucifer yellow CH, Alexa Fluor® 488 hydrazide and biocytin penetrate even the finest structures of neurons and other cells; however, as the thickness of the sample is decreased, the fluorescence signal is reduced. Consequently, it may be necessary to further amplify the signal by secondary detection methods.

We provide rabbit polyclonal antibodies to the Alexa Fluor® 488, Alexa Fluor® 405/Cascade Blue®, lucifer yellow, fluorescein, BODIPY® FL, tetramethylrhodamine and Texas Red® fluorophores (Section 7.4, Table 7.8) and a vast number of dye- or enzyme-labeled anti-rabbit antibodies (Section 7.2, Table 7.1). Our polyclonal and monoclonal antibodies to fluorescein cross-react strongly with the Oregon Green® dyes and somewhat with Rhodamine Green™ fluorophores, and our anti-tetramethylrhodamine and anti–Texas Red® antibodies cross-react with tetramethylrhodamine, Lissamine rhodamine B, Rhodamine red and Texas Red® dyes.

Our Tyramide Signal Amplification (TSA™) Kits (Section 6.2, Table 6.1) can also be utilized to detect aldehyde-fixed biotin derivatives or, in combination with antibodies to aldehyde-fixed fluorophore-labeled polar tracers, to further amplify the signal. Following fixation, biotinylated tracers such as biocytin and biotin ethylenediamine, can be detected using the reagents in our TSA™ Kits that contain horseradish peroxidase streptavidin and a fluorescent tyramide.

The relatively small size and easy penetration into tissues makes the streptavidin conjugates of the NANOGOLD®, Alexa Fluor® 488 FluoroNanogold™ and Alexa Fluor® 594 FluoroNanogold™ 1.4 nm gold clusters (N24918, A24926, A24927; Section 7.6) useful for ultrastructural studies of biotinylated tracers, particularly in combination with the LI Silver (LIS) Enhancement Kit (L24919, Section 7.2). Biotinylated polar tracers can also be detected with light microscopy using the Diaminobenzidine (DAB) Histochemistry Kit #3 (D22187, Section 6.2).

Influx™ Pinocytic Cell-Loading Reagent

Our Influx™ pinocytic cell-loading reagent (I14402) works via a rapid and simple technique based on the osmotic lysis of pinocytic vesicles, an approach introduced by Okada and Rechsteiner.[19] The probe is simply mixed at high concentration with the Influx™ reagent blended into growth medium, then incubated with live cells to allow pinocytic uptake of the surrounding solution. Subsequent transfer of the cells to a slightly hypotonic medium results in bursting of the pinocytic vesicles within the cells and the release of the probe into the cytosol (Figure 14.3.37, Figure 14.3.38, Figure 14.3.39).

The Influx™ pinocytic cell-loading reagent is highly effective for loading a diverse array of probes—including calcein (Figure 14.3.38), Alexa Fluor® hydrazides (Figure 14.3.39), dextran conjugates of fluorophores[224] and ion indicators (Figure 14.3.40), fura-2 salts, Oregon Green® 514 dye–labeled tubulin, Alexa Fluor® 488 dye–labeled actin, heparin, hydroxyurea,[225] DNA, oligonucleotides,[226] and Qdot® nanocrystals[227,228] into a variety of cell lines.

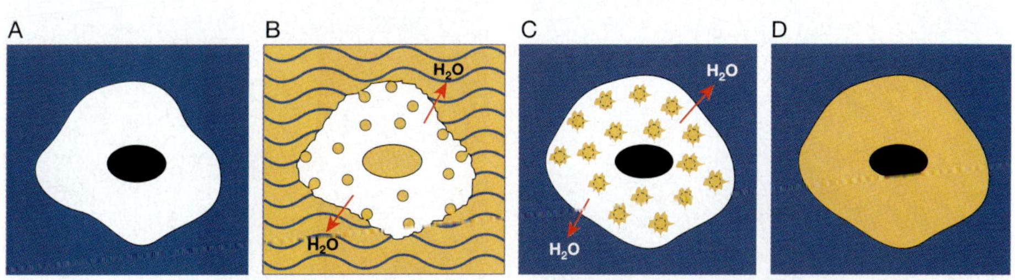

Figure 14.3.37 Principle of the Influx™ reagent pinocytic cell-loading method (I14402). Cultured cells are placed in hypertonic Influx™ loading reagent (**A**), along with the material to be loaded into the cells (yellow fluid, **B**), allowing the material to be carried into the cells via pinocytic vesicles. When the cells are placed in hypotonic medium, the pinocytic vesicles burst (**C**), releasing their contents into the cytosol (**D**).

IMPORTANT NOTICE: The products described in this manual are covered by one or more Limited Use Label License(s). Please refer to the Appendix on page 971 and Master Product List on page 975. Products are For Research Use Only. Not intended for any animal or human therapeutic or diagnostic use.

DATA TABLE 14.3 POLAR TRACERS

Cat. No.	MW	Storage	Soluble	Abs	EC	Em	Solvent	Notes
A350	427.33	L	H_2O	353	7200	520	H_2O	
A1339	491.57	L	H_2O	425	12,000	532	H_2O	
A1340	533.65	L	H_2O	426	11,000	531	H_2O	
A1593	367.30	NC	DMF, DMSO	<300		none		
A1594	442.50	NC	DMF, DMSO	<300		none		
A6257	523.39	D,L	H_2O	424	19,000	505	pH 7	
A10436	570.48	D,L	H_2O	493	71,000	517	pH 7	
A10437	730.74	D,L	H_2O	576	86,000	599	pH 7	1
A10438	758.79	D,L	H_2O	588	97,000	613	pH 7	1
A10439	349.29	L	H_2O, DMSO	345	13,000	445	pH 7	
A10441	730.74	FF,L	H_2O	576	86,000	599	pH 7	2
A10442	758.79	FF,L	H_2O	588	97,000	613	pH 7	2
A12922	1141.31	D,L	DMSO, H_2O	591	80,000	618	pH 7	
A12923	1209.66	D,L	DMSO, H_2O	556	99,000	572	pH 7	
A12924	974.98	D,L	DMSO, H_2O	494	62,000	520	pH 7	
A20501MP	~1150	D,L	H_2O	554	150,000	567	pH 7	
A20502	~1200	D,L	H_2O	649	250,000	666	pH 7	
A22850	471.55	F,D,L	H_2O, DMSO	361	17,000	536	pH 7	
A30000	1028.26	F,DD,L	H_2O, DMSO	400	35,000	424	pH 7	3, 4, 5
A30627	584.52	F,D,L	H_2O, DMSO	353	20,000	437	MeOH	6
A30629	895.07	F,D,L	H_2O, DMSO	494	77,000	518	pH 7	6, 7, 8
A30632	~1220	F,D,L	H_2O, DMSO	651	250,000	672	MeOH	6
A30634	~950	D,L	H_2O, DMSO	624	110,000	643	pH 7	
A30675	666.58	F,D,L	H_2O	399	29,000	422	H_2O	
B1151	520.45	L	pH >6	503	90,000	528	pH 9	9
B1370	831.01	L	DMF, pH >6	494	75,000	518	pH 9	9
B1513	341.38	F,D	DMF, DMSO	<300		none		
B1582	454.54	F,D	DMF, DMSO	<300		none		
B1592	372.48	NC	H_2O	<300		none		
B1596	555.65	NC	DMF, DMSO	<300		none		
B1603	386.51	D	pH >6, DMF	<300		none		
B6352	669.74	F,D	DMF, pH >6	<300		none		4
B6353	556.58	F,D	DMF, pH >6	<300		none		4
B10570	644.70	L	DMSO	494	68,000	523	pH 9	9
C194	376.32	L	pH >6, DMF	492	75,000	517	pH 9	9
C368	445.21	L	pH >6, DMF	504	107,000	529	pH 8	10
C481	622.54	L	pH >5	494	77,000	517	pH 9	11, 12
C621	624.49	L	H_2O	399	30,000	423	H_2O	3
C687	596.44	L	H_2O	399	30,000	421	H_2O	3, 13
C1430	994.87	F,D	DMSO	<300		none		14
C1904	376.32	L	pH >6, DMF	492	78,000	517	pH 9	9, 15
C2284	607.42	F,D,LL	H_2O, MeOH	396	29,000	410	MeOH	3, 16
C3099	994.87	F,D	DMSO	<300		none		2, 14
C3100MP	994.87	F,D	DMSO	<300		none		14
C3221	644.77	L	H_2O	399	31,000	419	H_2O	3, 13
C3239	548.29	L	H_2O	399	29,000	419	H_2O	3, 13
D1306	350.25	L	H_2O, DMF	342	28,000	450	pH 7	
D3238	466.19	F,D,L	H_2O	490	97,000	515	H_2O	17
D3571	457.49	L	H_2O, MeOH	342	28,000	450	pH 7	
D20653	341.45	D	DMSO	<300		none		18
D21490	350.25	L	H_2O, DMF	342	28,000	450	pH 7	15
F1130	478.32	D,L	H_2O, DMF	495	76,000	519	pH 9	9
F7103	826.81	FF,D,LL	H_2O, DMSO	333	15,000	none	DMSO	16, 19, 20
H185	206.15	L	pH >6, DMF	386	29,000	448	pH 10	21
H348	524.37	D,L	H_2O	454	24,000	511	pH 9	22
H22845	472.53	F,D,L	H_2O, DMSO	345	31,000	450	pH 5	23
L453	457.24	L	H_2O	428	12,000	536	H_2O	24, 25
L682	479.44	L	H_2O	428	12,000	533	H_2O	24, 25
L1177	521.56	L	H_2O	427	12,000	535	H_2O	24, 25
L1338	659.51	F,D,L	H_2O	426	11,000	531	pH 7	26
L2601	873.10	D,L	H_2O	428	11,000	531	H_2O	
L6950	850.03	D,L	H_2O	428	11,000	532	pH 7	
L12926	457.24	FF,L	H_2O	428	12,000	536	H_2O	2
M395	538.40	L	H_2O	404	29,000	435	pH 8	27
M1602	523.60	F,D	pH >6, DMF	<300		none		
N21479	see Notes	F,D,L	DMSO	435	see Notes	457	H_2O/RNA	2, 28, 29
N21480	see Notes	F,D,L	DMSO	497	see Notes	524	H_2O/RNA	2, 28, 29
N21481	see Notes	F,D,L	DMSO	515	see Notes	535	H_2O/RNA	2, 28, 29
N21482	see Notes	F,D,L	DMSO	530	see Notes	619	H_2O/RNA	2, 28, 29
N21483	see Notes	F,D,L	DMSO	644	see Notes	663	H_2O/RNA	2, 28, 29
O6138	512.36	L	pH >6, DMF	506	86,000	526	pH 9	30
O6146	412.30	L	pH >6, DMF	492	85,000	518	pH 9	31
O12920	887.39	L	DMSO, H_2O	495	66,000	522	pH 9	31
P349	610.42	L	H_2O	374	51,000	403	H_2O	
P1304MP	668.40	L	H_2O, DMSO	493	5900	636	H_2O	

continued on next page

IMPORTANT NOTICE: The products described in this manual are covered by one or more Limited Use Label License(s). Please refer to the Appendix on page 971 and Master Product List on page 975. Products are For Research Use Only. Not intended for any animal or human therapeutic or diagnostic use.

DATA TABLE 14.3–POLAR TRACERS—*continued*

Cat. No.	MW	Storage	Soluble	Abs	EC	Em	Solvent	Notes
P21493	668.40	L	H_2O, DMSO	493	5900	636	H_2O	15
S359	606.71	L	H_2O	586	108,000	605	H_2O	
S1129	518.43	F,D	DMSO	<300		none		32
S1307	558.66	L	H_2O	565	84,000	586	H_2O	
T506	341.25	F,D	H_2O, MeOH	<300		none		
T1247	373.38	D	H_2O	270	4700	545	H_2O	33, 34
T1323	417.29	L	DMSO	375	68,000	403	H_2O	
T12921	869.09	D,L	DMSO	554	103,000	581	pH 7	
X1525	422.16	D	H_2O	259	8800	none	H_2O	

For definitions of the contents of this data table, see "Using *The Molecular Probes® Handbook*" in the introductory pages.

Notes

1. Maximum solubility in water is ~8% for A10437 and A10438.
2. This product is supplied as a ready-made solution in the solvent indicated under "Soluble."
3. The Alexa Fluor® 405 and Cascade Blue® dyes have a second absorption peak at about 376 nm with EC ~80% of the 395–400 nm peak.
4. This sulfonated succinimidyl ester derivative is water-soluble and may be dissolved in buffer at ~pH 8 for reaction with amines. Long-term storage in water is NOT recommended due to hydrolysis.
5. A30100 is an alternative packaging of A30000 but is otherwise identical.
6. Aqueous stock solutions should be used within 24 hours; long-term storage is NOT recommended.
7. The fluorescence lifetime (τ) of the Alexa Fluor® 488 dye in pH 7.4 buffer at 20°C is 4.1 nanoseconds. Data provided by the SPEX Fluorescence Group, Horiba Jobin Yvon Inc.
8. Abs and Em of the Alexa Fluor® 488 dye are red-shifted by as much as 16 nm and 25 nm respectively on microarrays relative to aqueous solution values. The magnitude of the spectral shift depends on the array substrate material. (Biotechniques (2005) 38:127)
9. Absorption and fluorescence of fluorescein derivatives are pH-dependent. Extinction coefficients and fluorescence quantum yields decrease markedly at pH <7.
10. Absorption and fluorescence of dichlorofluorescein derivatives are pH-dependent. Extinction coefficients and fluorescence quantum yields decrease markedly at pH <5.
11. C481 fluorescence is strongly quenched by micromolar concentrations of Fe^{3+}, Co^{2+}, Ni^{2+} and Cu^{2+} at pH 7. (Am J Physiol (1995) 268:C1354, J Biol Chem (1999) 274:13375)
12. $K_d(Co^{2+})$ for calcein is 120 nM, determined in 10 mM HEPES, 1 µM Ca^{2+}, 1 mM Mg^{2+}, 100 mM KCl. (Anal Biochem (1997) 248:31)
13. Maximum solubility in water is ~1% for C687, ~1% for C3221 and ~8% for C3239.
14. Calcein AM is converted to fluorescent calcein (C481) after acetoxymethyl ester hydrolysis.
15. This product is specified to equal or exceed 98% analytical purity by HPLC.
16. Unstable in water. Use immediately.
17. The absorption and fluorescence spectra of BODIPY® derivatives are relatively insensitive to the solvent.
18. The dissociation constant (K_d) for desthiobiotin binding to streptavidin is 1.9 nM. (Bioconjug Chem (2006) 17:366)
19. All photoactivatable probes are sensitive to light. They should be protected from illumination except when photolysis is intended.
20. This product is colorless and nonfluorescent until it is activated by ultraviolet photolysis. Photoactivation generates a fluorescein derivative with spectral characteristics similar to C1359 .
21. Spectra of hydroxycoumarins are pH-dependent. Below the pK_a (~7.5), Abs shifts to shorter wavelengths (325–340 nm) and fluorescence intensity decreases.
22. H348 spectra are pH-dependent.
23. Hydroxystilbamidine in H_2O has a wide emission bandwidth, with a second peak at ~600 nm.
24. The fluorescence quantum yield of lucifer yellow CH in H_2O is 0.21. (J Am Chem Soc (1981) 103:7615)
25. Maximum solubility in water is ~8% for L453, ~6% for L682 and ~1% for L1177.
26. Iodoacetamides in solution undergo rapid photodecomposition to unreactive products. Minimize exposure to light prior to reaction.
27. Maximum solubility for M395 in water is ~25%.
28. This product is essentially nonfluorescent except when bound to DNA or RNA.
29. The active ingredient of this product is an organic dye with MW <1000. The exact MW and extinction coefficient values for this dye are proprietary.
30. Absorption and fluorescence of Oregon Green® 514 derivatives are pH-dependent only in moderately acidic solutions (pH <5).
31. AbXsorption and fluorescence of Oregon Green® 488 derivatives are pH-dependent only in moderately acidic solutions (pH <5).
32. S1129 is converted to a fluorescent product (F1130) after acetate hydrolysis.
33. Absorption and luminescence of T1247 are extremely weak unless it is chelated. Data are for dipicolinic acid (DPA) chelate. The luminescence spectrum has secondary peak at 490 nm.
34. MW is for the hydrated form of this product.

PRODUCT LIST 14.3 POLAR TRACERS

Cat. No.	Product	Quantity
A12924	Alexa Fluor® 488 biocytin, disodium salt (biocytin Alexa Fluor® 488)	250 µg
A12923	Alexa Fluor® 546 biocytin, sodium salt (biocytin Alexa Fluor® 546)	250 µg
A12922	Alexa Fluor® 594 biocytin, sodium salt (biocytin Alexa Fluor® 594)	250 µg
A30675	Alexa Fluor® 405 cadaverine, trisodium salt	1 mg
A30627	Alexa Fluor® 350 C_5-aminooxyacetamide, trifluoroacetate salt (Alexa Fluor® 350 hydroxylamine)	1 mg
A30629	Alexa Fluor® 488 C_5-aminooxyacetamide, bis(triethylammonium) salt (Alexa Fluor® 488 hydroxylamine)	1 mg
A30632	Alexa Fluor® 647 C_5-aminooxyacetamide, bis(triethylammonium) salt (Alexa Fluor® 647 hydroxylamine)	1 mg
A30000	Alexa Fluor® 405 carboxylic acid, succinimidyl ester	1 mg
A30100	Alexa Fluor® 405 carboxylic acid, succinimidyl ester	5 mg
A10439	Alexa Fluor® 350 hydrazide, sodium salt	5 mg
A10436	Alexa Fluor® 488 hydrazide, sodium salt	1 mg
A20501MP	Alexa Fluor® 555 hydrazide, tris(triethylammonium) salt	1 mg
A10437	Alexa Fluor® 568 hydrazide, sodium salt	1 mg
A10441	Alexa Fluor® 568 hydrazide, sodium salt *for microinjection* *10 mM in 200 mM KCl*	125 µL
A10438	Alexa Fluor® 594 hydrazide, sodium salt	1 mg
A10442	Alexa Fluor® 594 hydrazide, sodium salt *for microinjection* *10 mM in 200 mM KCl*	125 µL
A30634	Alexa Fluor® 633 hydrazide, bis(triethylammonium) salt	1 mg
A20502	Alexa Fluor® 647 hydrazide, tris(triethylammonium) salt	1 mg
A350	8-aminonaphthalene-1,3,6-trisulfonic acid, disodium salt (ANTS)	1 g
A1339	*N*-(2-aminoethyl)-4-amino-3,6-disulfo-1,8-naphthalimide, dipotassium salt (lucifer yellow ethylenediamine)	25 mg
A1593	*N*-(2-aminoethyl)biotinamide, hydrobromide (biotin ethylenediamine)	25 mg

IMPORTANT NOTICE: The products described in this manual are covered by one or more Limited Use Label License(s). Please refer to the Appendix on page 971 and Master Product List on page 975. Products are For Research Use Only. Not intended for any animal or human therapeutic or diagnostic use.
www.invitrogen.com/probes
molecular probes® | invitrogen™ by life technologies™

PRODUCT LIST 14.2 POLAR TRACERS—*continued*

Cat. No.	Product	Quantity
A1340	*N*-(5-aminopentyl)-4-amino-3,6-disulfo-1,8-naphthalimide, dipotassium salt (lucifer yellow cadaverine)	25 mg
A1594	*N*-(5-aminopentyl)biotinamide, trifluoroacetic acid salt (biotin cadaverine)	25 mg
A6257	8-aminopyrene-1,3,6-trisulfonic acid, trisodium salt (APTS)	10 mg
A22850	aminostilbamidine, methanesulfonate	10 mg
B1592	biocytin (ε-biotinoyl-L-lysine)	100 mg
B1603	biocytin hydrazide	25 mg
B10570	biotin-4-fluorescein	5 mg
B1596	biotin-X cadaverine (5-(((*N*-(biotinoyl)amino)hexanoyl)amino)pentylamine, trifluoroacetic acid salt)	10 mg
B1582	6-((biotinoyl)amino)hexanoic acid, succinimidyl ester (biotin-X, SE; biotinamidocaproate, *N*-hydroxysuccinimidyl ester)	100 mg
B6353	6-((biotinoyl)amino)hexanoic acid, sulfosuccinimidyl ester, sodium salt (Sulfo-NHS-LC-Biotin; biotin-X, SSE)	25 mg
B6352	6-((6-((biotinoyl)amino)hexanoyl)amino)hexanoic acid, sulfosuccinimidyl ester, sodium salt (biotin-XX, SSE)	25 mg
B1370	5-((*N*-(5-(*N*-(6-(biotinoyl)amino)hexanoyl)amino)pentyl)thioureidyl)fluorescein (fluorescein biotin)	5 mg
B1513	D-biotin, succinimidyl ester (succinimidyl D-biotin)	100 mg
B1151	2',7'-bis-(2-carboxyethyl)-5-(and-6)-carboxyfluorescein (BCECF acid) *mixed isomers*	1 mg
C481	calcein *high purity*	100 mg
C1430	calcein, AM	1 mg
C3099	calcein, AM *1 mg/mL solution in anhydrous DMSO*	1 mL
C3100MP	calcein, AM *special packaging*	20 x 50 µg
C368	5-(and-6)-carboxy-2',7'-dichlorofluorescein *mixed isomers*	100 mg
C1904	5-(and-6)-carboxyfluorescein (5(6)-FAM) *FluoroPure™ grade* *mixed isomers*	100 mg
C194	5-(and-6)-carboxyfluorescein *mixed isomers*	5 g
C2284	Cascade Blue® acetyl azide, trisodium salt	5 mg
C621	Cascade Blue® ethylenediamine, trisodium salt	10 mg
C3239	Cascade Blue® hydrazide, trilithium salt	10 mg
C3221	Cascade Blue® hydrazide, tripotassium salt	10 mg
C687	Cascade Blue® hydrazide, trisodium salt	10 mg
D1306	4',6-diamidino-2-phenylindole, dihydrochloride (DAPI)	10 mg
D21490	4',6-diamidino-2-phenylindole, dihydrochloride (DAPI) *FluoroPure™ grade*	10 mg
D3571	4',6-diamidino-2-phenylindole, dilactate (DAPI, dilactate)	10 mg
D3238	4,4-difluoro-1,3,5,7,8-pentamethyl-4-bora-3a,4a-diaza-*s*-indacene-2,6-disulfonic acid, disodium salt (BODIPY® 492/515 disulfonate)	10 mg
D20653	DSB-X™ biotin hydrazide	5 mg
F7103	fluorescein bis-(5-carboxymethoxy-2-nitrobenzyl) ether, dipotassium salt (CMNB-caged fluorescein)	5 mg
F1130	fluorescein-5-(and-6)-sulfonic acid, trisodium salt	100 mg
H185	7-hydroxycoumarin-3-carboxylic acid *reference standard*	100 mg
H348	8-hydroxypyrene-1,3,6-trisulfonic acid, trisodium salt (HPTS; pyranine)	1 g
H22845	hydroxystilbamidine, methanesulfonate	10 mg
I14402	Influx™ pinocytic cell-loading reagent *makes 10 x 5 mL*	1 set
L6950	lucifer yellow biocytin, potassium salt (biocytin lucifer yellow)	5 mg
L2601	lucifer yellow cadaverine biotin-X, dipotassium salt	10 mg
L682	lucifer yellow CH, ammonium salt	25 mg
L453	lucifer yellow CH, lithium salt	25 mg
L12926	lucifer yellow CH, lithium salt *for microinjection* *100 mM in water*	100 µL
L1177	lucifer yellow CH, potassium salt	25 mg
L1338	lucifer yellow iodoacetamide, dipotassium salt	25 mg
M1602	N^{α}-(3-maleimidylpropionyl)biocytin	25 mg
N21479	NeuroTrace® 435/455 blue fluorescent Nissl stain *solution in DMSO*	1 mL
N21480	NeuroTrace® 500/525 green fluorescent Nissl stain *solution in DMSO*	1 mL
N21481	NeuroTrace® 515/535 yellow fluorescent Nissl stain *solution in DMSO*	1 mL
N21482	NeuroTrace® 530/615 red fluorescent Nissl stain *solution in DMSO*	1 mL
N21483	NeuroTrace® 640/660 deep-red fluorescent Nissl stain *solution in DMSO*	1 mL
O12920	Oregon Green® 488 biocytin (biocytin Oregon Green® 488)	5 mg
O6146	Oregon Green® 488 carboxylic acid *5-isomer*	5 mg
O6138	Oregon Green® 514 carboxylic acid	5 mg
P1304MP	propidium iodide	100 mg
P21493	propidium iodide *FluoroPure™ grade*	100 mg
P349	1,3,6,8-pyrenetetrasulfonic acid, tetrasodium salt	100 mg
S359	sulforhodamine 101	25 mg
S1307	sulforhodamine B	5 g
T1247	terbium(III) chloride, hexahydrate	1 g
T12921	5-(and-6)-tetramethylrhodamine biocytin (biocytin TMR)	5 mg
T506	4-trimethylammonium-2,2,6,6-tetramethylpiperidine-1-oxyl iodide (CAT 1)	100 mg
T1323	true blue chloride	5 mg
X1525	*p*-xylene-bis-pyridinium bromide (DPX)	1 g

14.4 Tracers for Membrane Labeling

Cell membranes provide a convenient conduit for loading live and fixed cells with lipophilic dyes. Not only can cells tolerate a high concentration of the lipophilic dye, but also lateral diffusion of the dye within the membrane can serve to stain the entire cell, even if the dye is applied locally. These properties have made lipophilic carbocyanine and aminostyryl dyes particularly important for anterograde and retrograde tracing in neuronal cells.[1,2] Lipophilic tracers are used to label cells,[3–5] organelles,[6,7] liposomes,[8] viruses[9] and lipoproteins[10] in a wide variety of long-term tracing applications, including cell transplantation, migration, adhesion and fusion studies. The distinguishing features of these carbocyanine and aminostyryl tracers are summarized in Table 14.3. Other lipophilic probes described in Chapter 13 have also been used as tracers for liposomes. For example, incorporation of a BODIPY® FL cholesterol derivative (C3927MP, Section 13.3) in liposomes is virtually irreversible and the dye-labeled liposomes have been conjugated to antibodies for immunotargeting applications.[11]

Long-Chain Carbocyanines: DiI, DiO and Analogs

DiI, DiO, DiD and DiR

The lipophilic carbocyanines DiI (DiIC$_{18}$(3), Figure 14.4.1), DiO (DiOC$_{18}$(3), Figure 14.4.2), DiD (DiIC$_{18}$(5), Figure 14.4.3) and DiR (DiIC$_{18}$(7), Figure 14.4.4) are weakly fluorescent in water but highly fluorescent and quite photostable when incorporated into membranes (Figure 14.4.5). They have extremely high extinction coefficients

Table 14.3 Summary of our lipophilic carbocyanine and aminostyryl tracers.

Cat. No.	Probe *	Features of Carbocyanine or Aminostyryl Tracers
DiI and Analogs		
D282	DiIC$_{18}$(3) "DiI"	Red-orange–fluorescent lipophilic probe; widely used as a neuronal tracer
D3911	DiIC$_{18}$(3) "DiI"	Large-crystal form of D282 to facilitate direct application of crystals to membranes
N22880	DiIC$_{18}$(3) "DiI"	NeuroTrace® DiI tissue-labeling paste
V22885	DiIC$_{18}$(3) "DiI"	Vybrant® DiI cell-labeling solution
D384	DiIC$_{16}$(3)	Shorter-chain DiI analog that may incorporate into membranes more easily than DiI
D383	DiIC$_{12}$(3)	Shorter-chain DiI analog that may incorporate into membranes more easily than DiI
D3899	*FAST* DiI™ oil	Unsaturated DiI analog that reportedly migrates ~50% faster than DiI within membranes
D7756	*FAST* DiI™	Solid form of D3899 to facilitate direct application of crystals to membranes
D3886	Δ^9-DiI oil	Unsaturated DiI analog that may migrate faster than DiI within membranes
C7000 C7001	CellTracker™ CM-DiI	Chloromethylated DiI analog with enhanced solubility in culture medium; potentially retained after fixation and permeabilization
N22883	CellTracker™ CM-DiI	NeuroTrace® CM-DiI tissue-labeling paste
V22888	CellTracker™ CM-DiI	Vybrant® CM-DiI cell-labeling solution
D7777	SP-DiIC$_{18}$(3)	Anionic DiI analog with enhanced solubility in culture medium; potentially retained after fixation and permeabilization
D7776	DiIC$_{18}$(3)-DS	Anionic DiI analog with enhanced solubility in culture medium; potentially retained after fixation
DiD and Analogs		
D307	DiIC$_{18}$(5) oil "DiD"	Much longer-wavelength DiI analog; useful in autofluorescent samples and as a second tracer in combination with DiI
D7757	DiIC$_{18}$(5) "DiD"	Solid form of D307 to facilitate direct application of crystals to membranes
V22887	DiIC$_{18}$(5) "DiD"	Vybrant® DiD cell-labeling solution
D12730	DiIC$_{18}$(5)-DS	Anionic DiD analog with enhanced solubility in culture medium
DiR		
D12731	DiIC$_{18}$(7) "DiR"	DiI analog with absorption and emission in the near-infrared region, where many tissues are optically transparent
DiO and Analogs		
D275	DiOC$_{18}$(3) "DiO"	Yellow-green–fluorescent lipophilic probe; widely used as a second tracer in combination with DiI
N22881	DiOC$_{18}$(3) "DiO"	NeuroTrace® DiO tissue-labeling paste
V22886	DiOC$_{18}$(3) "DiO"	Vybrant® DiO cell-labeling solution
D1125	DiOC$_{16}$(3)	Shorter-chain DiO analog that may incorporate into membranes more easily than DiO
D3898	*FAST* DiO™	Unsaturated DiO analog that reportedly migrates ~50% faster than DiO within membranes
D7778	SP-DiOC$_{18}$(3)	Anionic DiO analog with enhanced water solubility; potentially retained after fixation
DiA and Analogs		
D3883	4-Di-16-ASP "DiA"	Yellow-green–fluorescent lipophilic probe; useful as a second tracer with DiI
D7758	*FAST* DiA™	Unsaturated DiA analog
D291	4-Di-10-ASP	Shorter-chain DiA analog that may incorporate into membranes more easily than DiA

* Probe is in solid form unless otherwise specified.

IMPORTANT NOTICE: The products described in this manual are covered by one or more Limited Use Label License(s). Please refer to the Appendix on page 971 and Master Product List on page 975. Products are For Research Use Only. Not intended for any animal or human therapeutic or diagnostic use.
www.invitrogen.com/probes
molecular probes® | invitrogen™ by life technologies™

(EC >125,000 $cm^{-1}M^{-1}$ at their longest-wavelength absorption maximum) though modest quantum yields, and short excited-state lifetimes (~1 nanosecond) in lipid environments.[12] Once applied to cells, the dyes diffuse laterally within the plasma membrane, resulting in staining of the entire cell. Transfer of these probes between intact membranes is usually but not always negligible.[13] DiI and its analogs usually exhibit very low cell toxicity; however, moderate inhibition of electron transport chain activity has been reported for some analogs.[14]

DiI, DiO, DiD and DiR exhibit distinct orange, green, red and infrared fluorescence, respectively (Figure 14.4.6), thus facilitating multicolor imaging.[15] DiO (D275) and DiI (D282, D3911) can be used with standard fluorescein and rhodamine optical filters, respectively. Iontophoretic application of these lipophilic dyes to nerve terminals at a single neuromuscular junction in live animals permits tracing of other synaptic terminals of the same motor unit.[16] The He-Ne laser–excitable DiD (D307, D7757) has much longer-wavelength excitation and emission spectra than those of DiI, providing a valuable alternative for labeling cells and tissues that have significant intrinsic fluorescence at wavelengths in the same range as DiI fluorescence (Figure 14.4.7). DiD was used as a population marker in a flow cytometry study that also employed indo-1 to monitor intracellular Ca^{2+} mobilization.[17] Our heptamethine carbocyanine probe $DiIC_{18}(7)$ (DiR, D12731) has even further red-shifted spectra, with absorption and emission maxima in the near-infrared region (absorption/emission maxima = 748/780 nm in methanol, Figure 14.4.6). Because the fluorescence of this dye is invisible to the human eye, it must be detected using a CCD camera or other near-infrared photosensitive device. The high transmission of infrared light through cells and tissues and low level of autofluorescence in the infrared makes DiR particularly useful as an *in vivo* tracer for labeled cells and liposomes in live organisms.[12,13] Photoconversion of diaminobenzidine (DAB) by lipid tracers, including DiI and DiO derivatives, produces an insoluble, electron-dense reaction product[18,19] (Fluorescent Probes for Photoconversion of Diaminobenzidine Reagents—Note 14.2).

Vybrant® DiI, DiO and DiD Cell-Labeling Solutions

The highly lipophilic nature of DiI, DiO and DiD has often posed an obstacle to uniform cellular labeling in aqueous culture media.[20] This technical difficulty has somewhat limited the use of these tracers in cell–cell fusion,[21,22] cellular adhesion[23,24] and migration[25] applications. The Vybrant® DiI cell-labeling solution (V22885) is a dye-delivery solution that can be added directly to normal culture media to uniformly label suspended or attached culture cells. The complementary Vybrant® DiO and DiD cell-labeling solutions (V22886, V22887) allow cell populations to be marked in distinctive fluorescent colors for identification after mixing (Figure 14.4.6). Cells that have fused or that have formed stable clusters can be identified

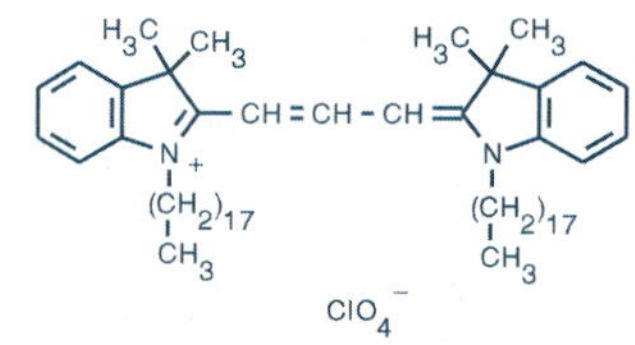

Figure 14.4.1 'DiI'; $DiIC_{18}(3)$ (1,1'-dioctadecyl-3,3,3',3'-tetramethylindocarbocyanine perchlorate; D282).

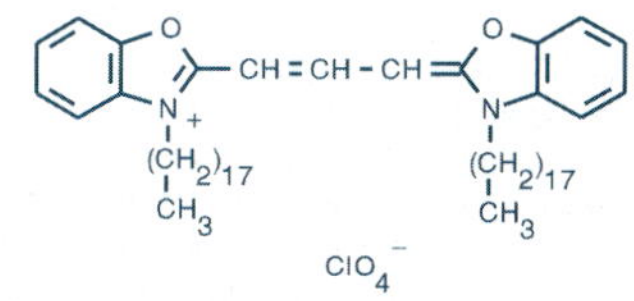

Figure 14.4.2 'DiO'; $DiOC_{18}(3)$ (3,3'-dioctadecyloxacarbocyanine perchlorate; D275).

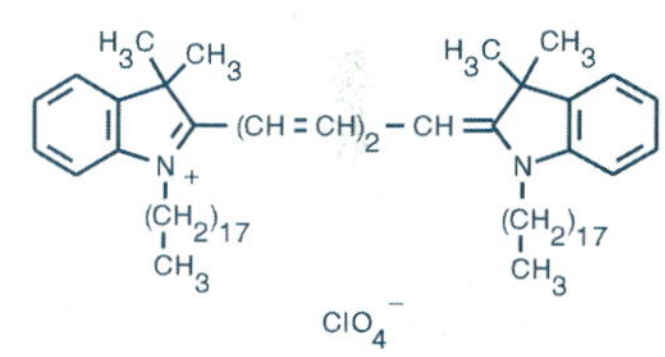

Figure 14.4.3 'DiD'; $DiIC_{18}(5)$ (1,1'-dioctadecyl-3,3,3',3'-tetramethylindodicarbocyanine perchlorate; D307).

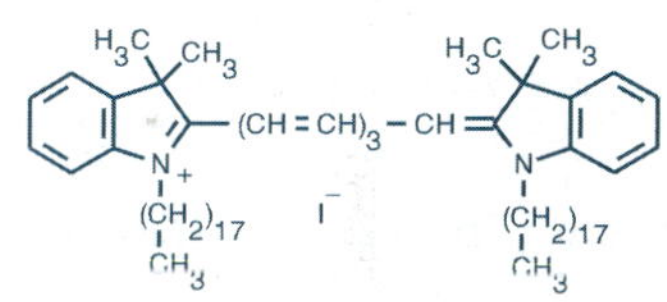

Figure 14.4.4 'DiR'; $DiIC_{18}(7)$ (1,1'-dioctadecyl-3,3,3',3'-tetramethylindotricarbocyanine iodide; D12731).

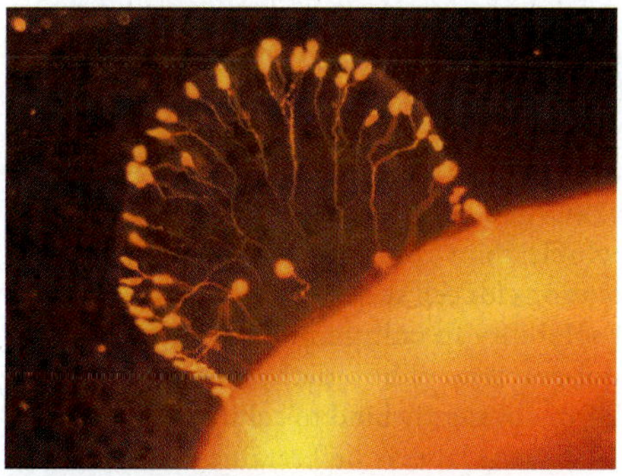

Figure 14.4.5 The rear fin of a squid embryo that has been labeled with $DiIC_{18}(3)$ (DiI; D282, D3911), showing selective staining of a population of ciliated sensory neurons whose cell bodies line the periphery of the fin. The dye was applied by soaking the embryos in a solution; no injection was used. Image contributed by Rachel Fink, Mount Holyoke College.

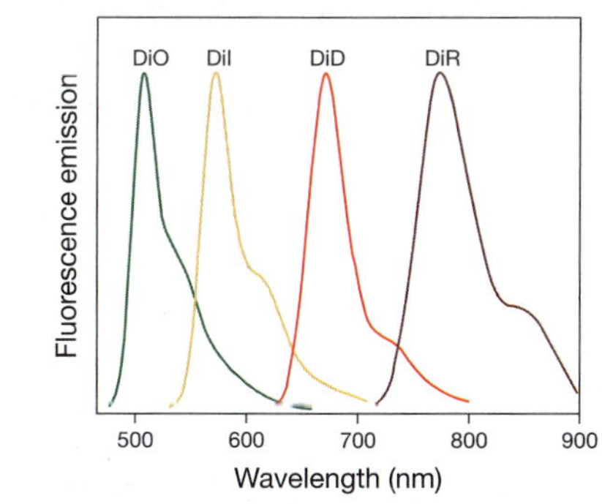

Figure 14.4.6 Normalized fluorescence emission spectra of DiO (D275), DiI (D282), DiD (D307) and DiR (D12731) bound to phospholipid bilayer membranes.

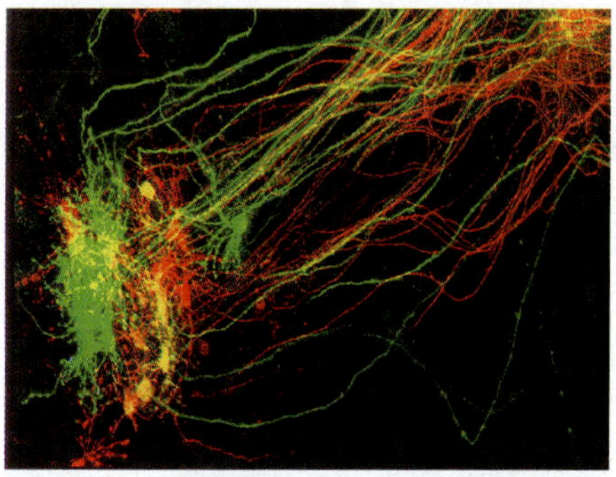

Figure 14.4.7 Pseudocolored confocal microscope image of retrogradely labeled neurons in the ventrobasal nucleus of the thalamus in newborn mouse. Paired dye deposits of DiI ($DiI_{18}(3)$, D282, D3911; pseudocolored green) and DiD ($DiIC_{18}(5)$, D307, D7757; pseudocolored red) were injected into this 500 µm–thick section, which was then fixed with formaldehyde and incubated for two weeks at 37°C to allow retrograde dye transport to proceed. Image contributed by Ariel Agmon, West Virginia University.

IMPORTANT NOTICE: The products described in this manual are covered by one or more Limited Use Label License(s). Please refer to the Appendix on page 971 and Master Product List on page 975. Products are For Research Use Only. Not intended for any animal or human therapeutic or diagnostic use.

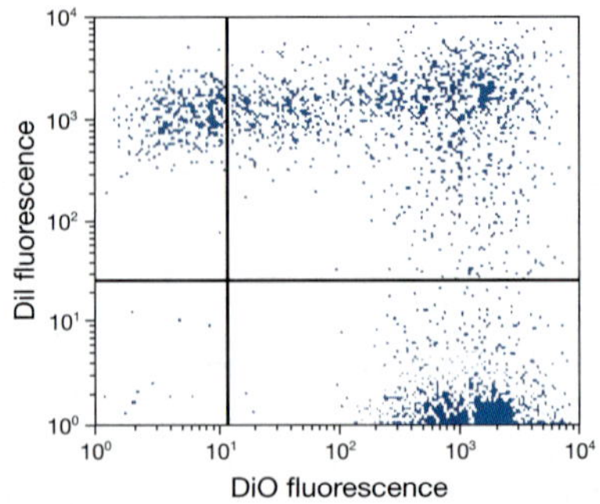

Figure 14.4.8 Poly(ethylene glycol)–induced fusion of Jurkat cells detected by flow cytometry. Two populations of Jurkat cells were separately labeled, one with the Vybrant® DiI cell-labeling solution (V22885) and the other with the Vybrant® DiO cell-labeling solution (V22886). Equal portions (1 mL) of the labeled cell suspensions were combined and treated with poly(ethylene glycol) for 45 seconds to induce fusion. The mixed-cell population was analyzed by flow cytometry. Double-labeled fused cells appear in the upper right quadrant of this bivariate correlation plot.

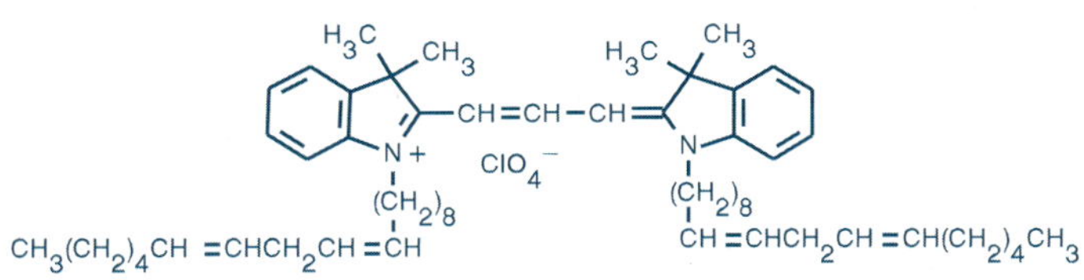

Figure 14.4.9 1,1′-Dilinoleyl-3,3,3′,3′-tetramethylindocarbocyanine perchlorate (*FAST* DiI™; DiI$\Delta^{9,12}$-C_{18}(3), ClO_4; D3899).

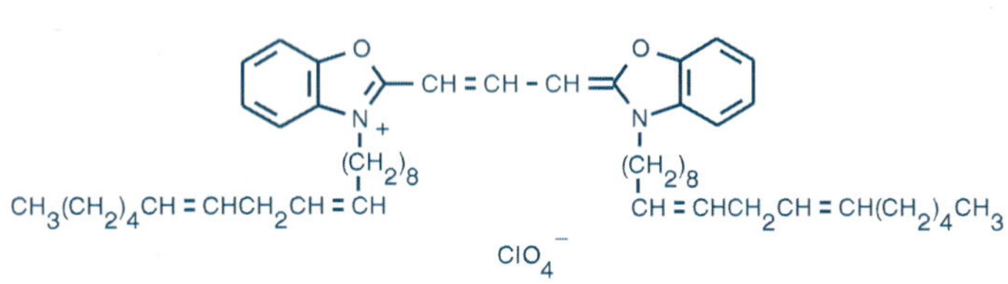

Figure 14.4.10 3,3′-Dilinoleyloxacarbocyanine perchlorate (*FAST* DiO™; DiO$\Delta^{9,12}$-C_{18}(3), ClO_4; D3898).

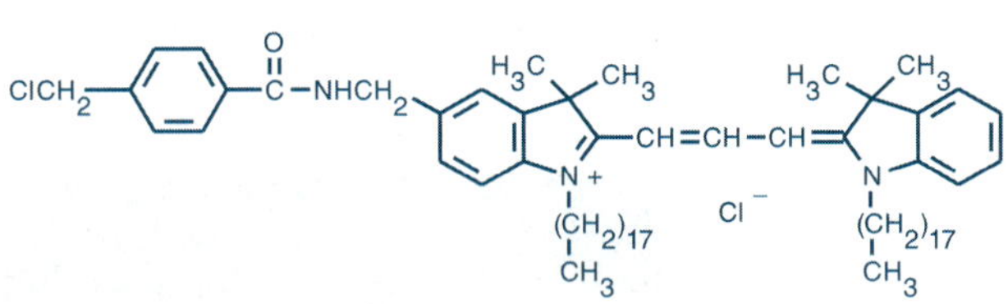

Figure 14.4.11 CellTracker™ CM-DiI (C7000).

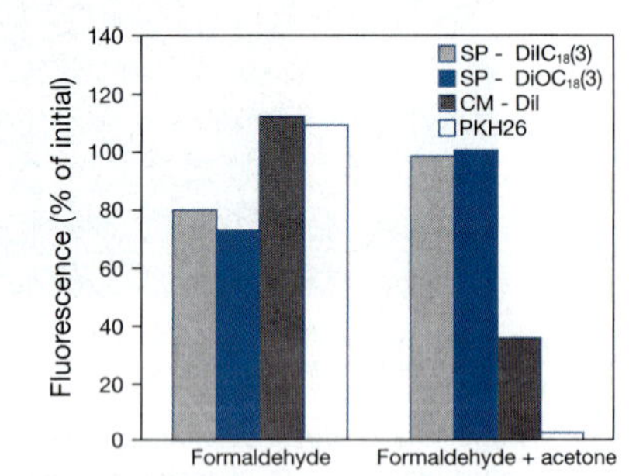

Figure 14.4.12 Persistence of lipophilic tracer fluorescence following fixation. Cultured human B cells were stained with 20 µM SP-DiOC_{18}(3) (D7778), SP-DiIC_{18}(3) (D7777), CM-DiI (C7000, C7001) or PKH26 and then fixed with 3.7% formaldehyde or 3.7% formaldehyde + acetone. The fixed cells were analyzed by flow cytometry to generate a comparison of their fluorescence with that of the original live-cell population.

by double labeling (Figure 14.4.8). Each 1 mM solution of DiI, DiO or DiD has been filtered through a 0.2 µm polycarbonate filter. All three cell-labeling solutions are also available in the Vybrant® Multicolor Cell-Labeling Kit (V22889). A Vybrant® CM-DiI labeling solution is also available (V22888, see below under CM-DiI).

NeuroTrace® DiI, DiO and DiD Tissue-Labeling Pastes

NeuroTrace® DiI and NeuroTrace® DiO tissue-labeling pastes (N22880, N22881) consist of DiI and DiO, respectively, mixed into an inert, water-resistant gel. Three formulations—NeuroTrace® DiI, NeuroTrace® DiO and NeuroTrace® DiD—are available in the convenient NeuroTrace® Multicolor Tissue-Labeling Kit (N22884). A NeuroTrace® CM-DiI tissue-labeling paste is also available (N22883, see below under CM-DiI). These pastes are ready to use as supplied and can be applied directly to live or fixed tissue specimens using the tip of a needle. This method of application improves the penetration of the dye into bundled neurons, labeling axons both on and below the surface. In similar situations, direct application of dye crystals or microinjection of concentrated solutions will only label neurons on the surface. This labeling method has also been found to increase the rate of dye transport by 50–80% (H. Richard Koerber, University of Pittsburgh School of Medicine, personal communication).

FAST DiI™ and *FAST* DiO™

Diffusion of lipophilic carbocyanine tracers from the point of their application to the terminus of a neuron can take several days to weeks.[26] The diffusion process appears to be accelerated by introducing unsaturation in the alkyl tails of the probes. *FAST* DiI™ [27–29] (D3899, D7756; Figure 14.4.9) and *FAST* DiO™ [30] (D3898, Figure 14.4.10) have diunsaturated linoleyl ($C_{18:2}$) tails in place of the saturated octadecyl tails ($C_{18:0}$) of DiI and DiO. Migration of the unsaturated analogs is reported to be at least 50% faster than that of DiI and DiO (Andrea Elberger, University of Tennessee, personal communication). *FAST* DiI™ and the shorter chain DiIC_{12}(3) (D383) internalize during endocytic sorting of lipids and are mainly found in the endocytic recycling compartment, whereas the 16-carbon DiIC_{16}(3) (D384) is delivered to late endosomes.[31] The perchlorate salt of *FAST* DiI™ does not crystallize and is a viscous oil (D3899); however, the 4-chlorobenzenesulfonate salt of *FAST* DiI™ (D7756) is a solid, making its crystals suitable for direct application to cells. We also offer a monounsaturated *cis*-9-octadecenyl ($C_{18:1}$) analog of DiI (Δ^9-DiI, D3886).

CM-DiI

CellTracker™ CM-DiI (C7000, C7001) is a DiI derivative that is somewhat more water-soluble than DiIC_{18}(3), thus facilitating the preparation of staining solutions for cell suspensions and fixed cells. In addition to its improved solubility in culture medium, CellTracker™ CM-DiI contains a thiol-reactive chloromethyl moiety (Figure 14.4.11) that allows the dye to covalently bind to cellular thiols. Thus, unlike other membrane stains, the label is well retained in some cells throughout fixation and permeabilization steps[32–34] (Figure 14.4.12, Figure 14.4.13). Membrane staining with CellTracker™ CM-DiI persists following routine paraffin processing.[33] CellTracker™ CM-DiI is particularly useful in experiments that combine membrane labeling with subsequent immunohistochemical analysis, fluorescence *in situ* hybridization or electron microscopy[35] (Figure 14.4.14). CM-DiI labeling of mesenchymal stem cells has been used to track their

molecular probes® | invitrogen™ by life technologies™

distribution *in vivo* and differentiation following transplantation.[36] Similarly, cationic liposome complexes of DNA have been labeled with CM-DiI to identify their organ and cellular distribution following injection into mice.[37,38]

In addition to providing CM-DiI as a 1 mg solid (C7001) or specially packaged as a set of 20 vials, each containing 50 µg of CM-DiI (C7000), we have available the Vybrant® CM-DiI cell-labeling solution (V22888) and NeuroTrace® CM-DiI tissue-labeling paste (N22883) for efficient labeling of cell suspension and tissues, respectively, with this useful aldehyde-fixable tracer.

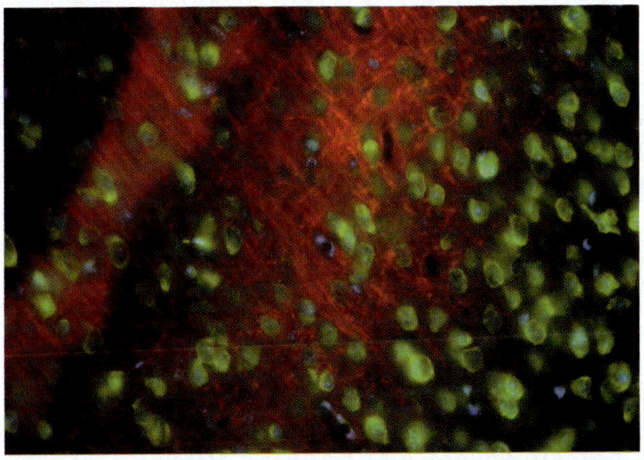

Figure 14.4.13 Neural somata from a mouse brain section labeled with NeuroTrace® 500/525 green-fluorescent Nissl stain (N21480). Nuclei are labeled with DAPI (D1306, D3571, D21490). Myelin and other lipophilic areas are illustrated with the aid of the red-orange fluorescence from CellTracker™ CM-DiI (C7000, C7001). The image is a composite of three micrographs acquired using filters appropriate for fluorescein, tetramethylrhodamine and DAPI.

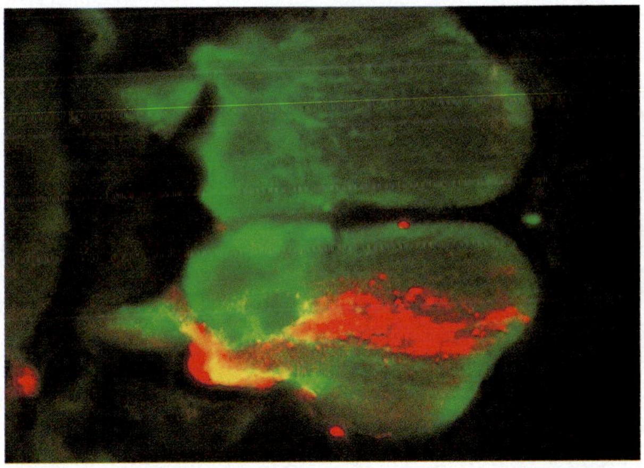

Figure 14.4.14 A 30 µm-thick section of a zebrafish embryo stained with CellTracker™ CM-DiI (C7000, C7001) prior to immunohistochemical analysis. A one-day-old zebrafish embryo was immobilized and impaled in the hindbrain with a microelectrode filled with 1 mg/mL CellTracker™ CM-DiI in 95% ethanol. One day later, the brain was fixed, embedded, frozen in liquid nitrogen and sectioned on a cryostat. After blocking with 0.1% Triton® X-100 in PBS containing 2% BSA and 2% normal goat serum, the 30 µm-thick section was incubated with primary anti-glial antibody in conjunction with fluorescein goat anti-mouse IgG antibody. This section was then viewed sequentially through optical filter sets appropriate for rhodamine and fluorescein, and the resulting images were superimposed. The image was contributed by William Trevarrow, Beckman Institute, California Institute of Technology.

Sulfonated Carbocyanines

Because of poor solubility in water, it is often difficult to load the long-chain carbocyanine dyes such as DiI, DiO and PKH dyes into cells in suspension. To facilitate the staining of cells with long-chain carbocyanine dyes, we have developed sulfonated cyanine dyes[39]—SP-DiIC$_{18}$(3)[40] (D7777, Figure 14.4.15), SP-DiOC$_{18}$(3) (D7778), DiIC$_{18}$(3)-DS[41] (D7776, Figure 14.4.16) and DiIC$_{18}$(5)-DS (D12730)—that retain the 18-carbon lipophilic chains of DiI, DiD and DiO but exhibit improved solubility in culture medium. Cells can be labeled with these dyes by simply diluting a DMSO stock solution of the dye to a labeling concentration of 1–10 µM in unmodified culture medium. We have determined that these sulfonated dyes are completely soluble at 20 µM for at least one hour in Dulbecco's phosphate-buffered saline and Hanks' balanced salt solution; aggregation begins after a few hours in solution. Uptake of the anionic carbocyanine dyes in live cells is usually slower than that of cationic carbocyanine dyes, and the resulting staining patterns may be quite different.

Some applications require retention of the membrane stain in place following aldehyde-mediated fixation and lipid extraction. In contrast to labeling with DiI, DiO or the PKH dyes,[42] cell labeling with some of these sulfonated carbocyanines appears to be compatible with standard aldehyde-based fixation methods and acetone treatment, at least in some cell lines. Although the mechanism for cell retention of the sulfonated carbocyanines has not been determined, we have observed excellent retention of staining of human B cells with SP-DiIC$_{18}$(3) and SP-DiOC$_{18}$(3), whereas virtually all of the PKH26 dye (which is structurally identical to DiIC$_{18}$(3)) is lost during the lipid extraction step (Figure 14.4.12). CM-DiI (C7000, C7001, N22883, V22888; see above) is also well retained during this processing. Furthermore, we have observed that acetone treatment actually enhances the fluorescence of some cells stained with SP-DiIC$_{18}$(3) and especially with SP-DiOC$_{18}$(3), a phenomenon that has not been seen with other carbocyanine-derived membrane stains.

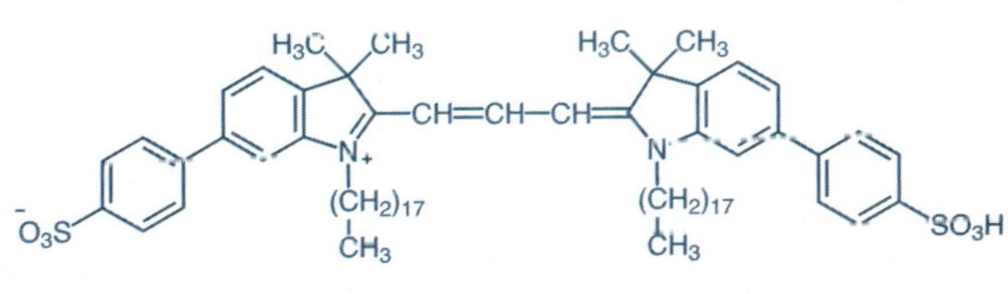

Figure 14.4.15 1,1′-Dioctadecyl-6,6′-di(4-sulfophenyl)-3,3,3′,3′-tetramethylindocarbocyanine (SP-DiIC$_{18}$(3), D7777).

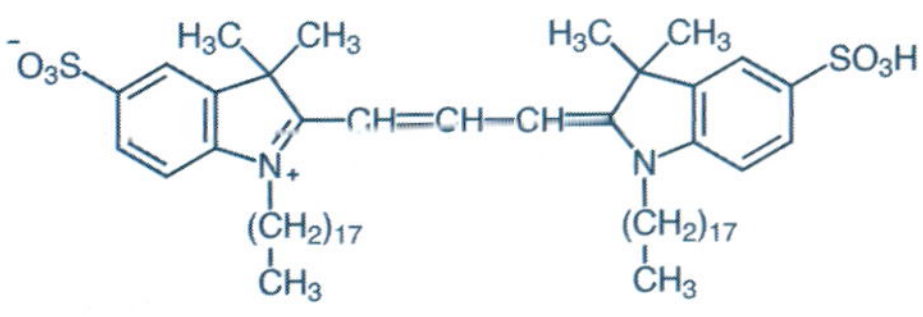

Figure 14.4.16 DiIC$_{18}$(3)-DS (1,1′-dioctadecyl-3,3,3′,3′-tetramethylindocarbocyanine-5,5′-disulfonic acid; D7776).

IMPORTANT NOTICE: The products described in this manual are covered by one or more Limited Use Label License(s). Please refer to the Appendix on page 971 and Master Product List on page 975. Products are For Research Use Only. Not intended for any animal or human therapeutic or diagnostic use.

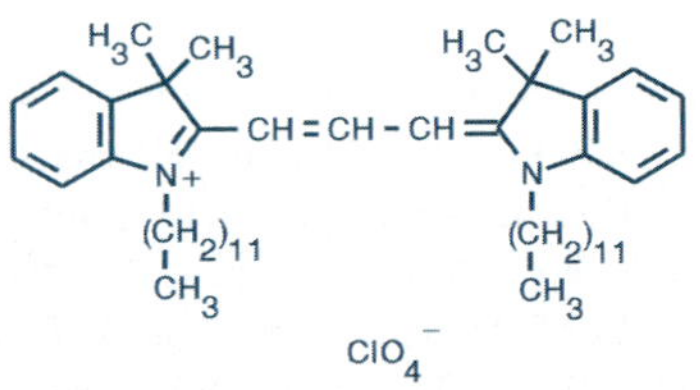

Figure 14.4.17 DiIC$_{12}$(3) (1,1′-didodecyl-3,3,3′,3′-tetramethylindocarbocyanine perchlorate; D383).

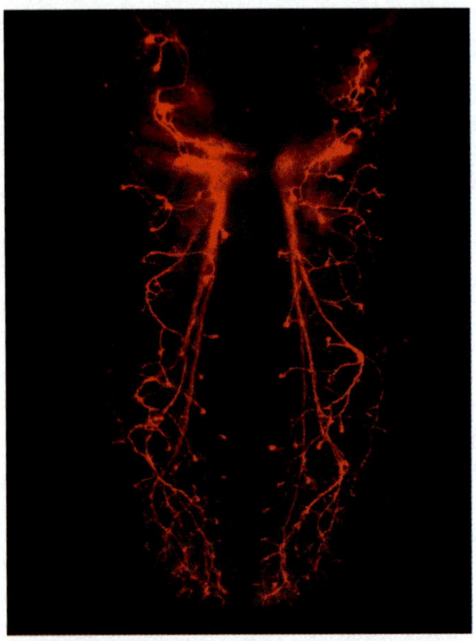

Figure 14.4.18 Visualizing gustatory neurons in a mouse tongue using DiI (D282, D3911). This wild-type mouse embryo was fixed in 4% phosphate-buffered paraformaldehyde, and taste ganglia were labeled with the lipophilic tracer DiI. (J Neurosci (2009) 29:3354) Images were taken with an Olympus confocal microscope. Image provided by Robin F. Krimm, Department of Anatomical Sciences and Neurobiology, University of Louisville School of Medicine.

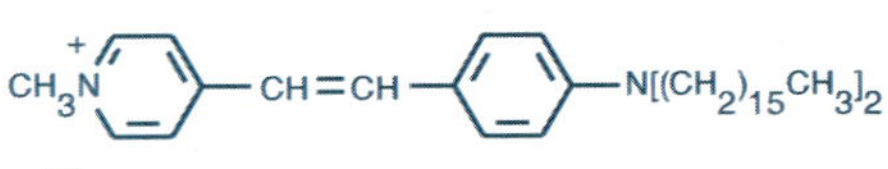

Figure 14.4.19 DiA; 4-Di-16-ASP (4-(4-(dihexadecylamino) styryl)-*N*-methylpyridinium iodide; D3883).

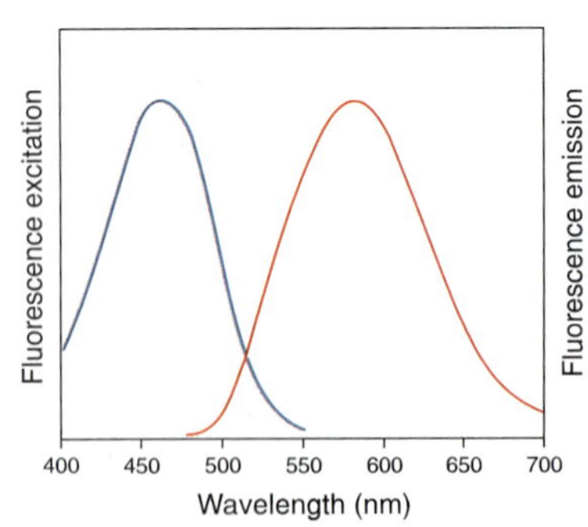

Figure 14.4.20 Fluorescence excitation and emission spectra of DiA bound to phospholipid bilayer membranes.

Other DiO and DiI Analogs

Moderately lipophilic carbocyanines are also useful tracers for labeling cell membranes.[31,43] Some scientists find the slightly less lipophilic DiIC$_{12}$(3) (D383, Figure 14.4.17), DiIC$_{16}$(3) (D384) and DiOC$_{16}$(3) (D1125) easier to load into cell suspensions than their C$_{18}$ homologs.

Neuronal Tracing Studies with DiI and DiO Analogs

DiI, DiO and DiD are widely used to label neuronal projections in live and fixed tissue (Figure 14.4.18, Figure 14.4.5, Figure 14.4.7). The dyes insert into the outer leaflet of the plasma membrane and diffuse laterally, producing detailed labeling of fine neuronal projections. DiI has been reported to diffuse about 6 mm/day in live tissue but more slowly in fixed specimens.[44] The dyes usually do not transfer from labeled to unlabeled cells, unless the membrane of the labeled cell is disrupted, and apparently do not transfer through gap junctions.[45]

Motoneurons labeled with DiI have been reported to remain viable up to four weeks in culture and one year *in vivo*.[46] Staining of neurons in fixed tissue with DiI has been reported to persist for at least two years.[47,48] DiI and DiD can be used simultaneously for two-color tracing[15,49] (Figure 14.4.7) and as a fluorescence donor–acceptor pair in excited-state energy transfer studies.[50,51] Combinations of DiO, DiI, DiD and DiR may permit three- or even four-color measurements (Figure 14.4.6).

Cell labeling with the lipophilic carbocyanines is generally performed by direct application of a dye crystal or dye-coated paper, metal or glass probe[52] or by microinjection onto single cells from a solution in dimethylformamide or dimethylformamide/ethanol.[53,54] A method has been described for iontophoretic loading of cells with DiI and other lipid tracers.[55] We also offer DiI in the form of extra-large crystals (D3911), which many researchers prefer for direct application to tissue. Detailed protocols for DiI labeling and confocal laser-scanning microscopy have been published.[56,57] DiI has also been used to mark sites of microinjection by coating the outside of the glass micropipette with the dye.[58] "DiOlistic labeling" of live or fixed cells and tissues uses a gene gun to deliver pellets coated with lipophilic carbocyanine dyes DiO, DiI and DiD, permitting rapid labeling at high densities with relatively high spatial resolution.[59–62]

Long-Term Cell Tracing with Membrane Tracers

Lipophilic carbocyanine tracers like DiI are also ideal labels for long-term cell tracing *in vivo* and *in vitro*, including studies of cell migration, transplantation,[63] adhesion and fusion.[42] The presence of DiI and DiO in the cell membrane does not appreciably affect cell viability, development or other basic physiological properties.[20,64] Labeling of cell suspensions or perfusion of tissues requires aqueous dispersions of the dyes.[65] Stock solutions of most of the tracers may be prepared in alcohol or dimethylsulfoxide (DMSO) and then added to cell suspensions. Several analogs of DiI that are more soluble in culture medium should facilitate this method of labeling (Table 14.3).

Tissue Processing and Electron Microscopy with Long-Chain Carbocyanines

Methods have been developed to allow immunofluorescent labeling of tissue containing neurons stained with DiI.[32,66,67] Polyacrylamide has been employed to embed DiI-stained brain tissue for vibratome sectioning[68]—a method that has been reported to preserve DiI labeling better than cryostat sectioning.[69] Also, see Table 14.3 for a description of DiI and DiO analogs that may be better retained through fixation and permeabilization procedures.

DiA and *FAST* DiA™

Reports from several researchers indicate that DiA (4-Di-16-ASP, D3883, Figure 14.4.19) is a better neuronal tracer than DiO for multicolor labeling with DiI.[70–72] DiA diffuses more rapidly in membranes and is more soluble than DiO, thus facilitating cell labeling. For example, DiA and DiI have been used together to investigate the interactions between dorsal root axons and their targets[73] and axon outgrowth in the retina.[74] DiA can be excited between 440 nm and 500 nm, and its maximum emission in DOPC vesicles is at ~590 nm (red-orange fluorescence, Figure 14.4.20). Its fluorescence in cells, however, is usually bright green to yellow-green (depending on the optical filter set used). The polyunsaturated *FAST* DiA™, offered as a crystalline solid

IMPORTANT NOTICE: The products described in this manual are covered by one or more Limited Use Label License(s). Please refer to the Appendix on page 971 and Master Product List on page 975. Products are For Research Use Only. Not intended for any animal or human therapeutic or diagnostic use.

(D7758) form, may diffuse faster in membranes than DiA. The C_{10} analog of DiA, 4-Di-10-ASP (D291), has been used as a retrograde tracer to monitor injury-induced degradation of rat neurons *in vivo* and the role of microglial cells in removing debris resulting from natural and axotomy-induced neuronal cell death[75–77] (Figure 14.4.21). 4-Di-10-ASP staining of microglia can persist for at least 12 months and can be used for DAB photoconversion[75] (Fluorescent Probes for Photoconversion of Diaminobenzidine Reagents—Note 14.2). Axonal outgrowth of retinal ganglion cells stained with 4-Di-10-ASP has been observed.[78]

Lipophilic Tracer Sampler Kit

For researchers wishing to investigate the suitability of this group of membrane probes for a particular application, we have prepared a Lipophilic Tracer Sampler Kit (L7781). This kit contains 1 mg samples of nine different membrane stains, including both relatively new and well-established tracers:

- $DiIC_{18}(3)$ (DiI)
- $DiIC_{18}(3)$-DS
- SP-$DiIC_{18}(3)$
- 5,5′-Ph_2-$DiIC_{18}(3)$
- $DiOC_{18}(3)$ (DiO)
- SP-$DiOC_{18}(3)$
- $DiIC_{18}(5)$ (DiD)
- $DiIC_{18}(7)$ (DiR)
- 4-Di-16-ASP (DiA)

Octadecyl Rhodamine B

Octadecyl rhodamine B (R18, O246) is a lipophilic cation that has been extensively used as a membrane probe[79–82] (Lipid-Mixing Assays of Membrane Fusion—Note 13.1). Viral particles that have been labeled with high concentrations of R18 have fluorescence that is highly self-quenched; fusion of the particle with host cell membranes therefore results in a fusion-associated fluorescence increase.[83,84]

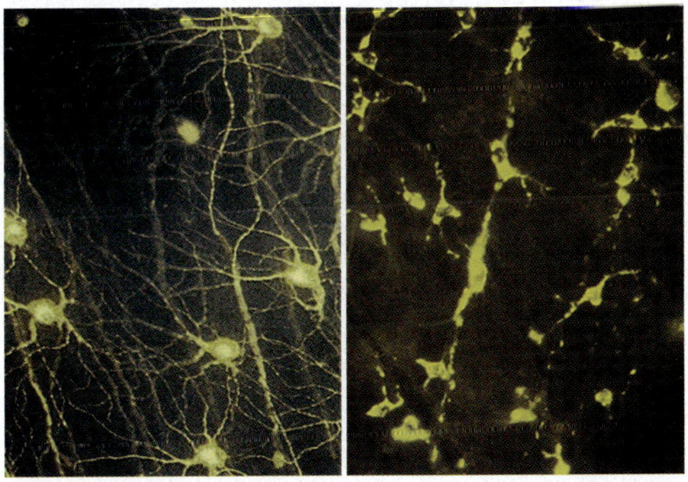

Figure 14.4.21 Rat retinal ganglion cells that have been allowed to regenerate for 2 months after transection into a grafted peripheral nerve piece and then labeled with 4-Di-10-ASP (D291), which is retrogradely transported within the regenerated axons and finest dendrites (left). Microglial cells in the retina were transcellularly labeled with 4-Di-10-ASP after phagocytosing degenerated ganglion cells (right). The images were contributed by Solon Thanos, University of Tübingen School of Medicine.

Plasma Membrane Stains

CellMask™ Plasma Membrane Stains

The CellMask™ plasma membrane stains are designed to generate rapid and uniform cell-surface labeling without the cell type–dependent variability exhibited by lectins.[85] CellMask™ Orange (C10045) and CellMask™ Deep Red (C10046) plasma membrane stains are optimally detected using tetramethylrhodamine (TRITC) and Cy®5 dye detection configurations, respectively. They are particularly important for cell-surface demarcation in experiments using intracellular targets labeled with Green Fluorescent Protein (GFP), Alexa Fluor® 488 dye, BODIPY® FL dye and other green-fluorescent fluorophores.[86–89] Membrane staining persists through fixation, but not permeabilization. Immunofluorescence detection of intracellular targets in cells labeled with CellMask™ plasma membrane stains must therefore be carried out under non-permeabilizing conditions[86]

Fluorescent Protein–Based Plasma Membrane Markers

CellLight® plasma membrane expression vectors (C10606, C10607, C10608) generate cyan-, green- or red-autofluorescent proteins fused to a plasma membrane targeting sequence consisting of the 10 N-terminal amino acids of LcK tyrosine kinase (Lck10). These fusion proteins are lipid raft markers with well established utility,[90] providing alternatives to cholera toxin B conjugates (Section 14.7) or BODIPY® FL C_5-ganglioside G_{M1}[91] (B13950, B34401; Section 13.3) with the inherent advantages of long-lasting and titratable expression conferred by BacMam 2.0 vector technology (BacMam Gene Delivery and Expression Technology—Note 11.1).

FM® and RH Dyes

Although the lipophilic cyanine and styryl dyes described above are useful for staining cell membranes, they are relatively difficult to apply to cells and they do eventually internalize, staining all cell membranes. In contrast, FM® 1-43 (T3163, T35356), FM® 1-43FX (F35355), FM® 4-64 (T3166, T13320), FM® 4-64FX (F34653) and RH 414 (T1111) are easily applied to cells, where they bind rapidly and reversibly to the plasma membrane with strong fluorescence enhancement.[92–94] These probes have large Stokes shifts and can be excited by the argon-ion laser. Because we have found that small differences in the polarity of these probes can play a large role in their rates of uptake and their retention properties, we have introduced FM® 5-95 (T23360), a slightly less lipophilic analog of FM® 4-64 with essentially identical spectroscopic properties.

FM® 1-43 is efficiently excited with standard fluorescein optical filters but poorly excited with standard tetramethylrhodamine optical filters (Figure 14.4.22). FM® 1-43 has been used to outline membranes

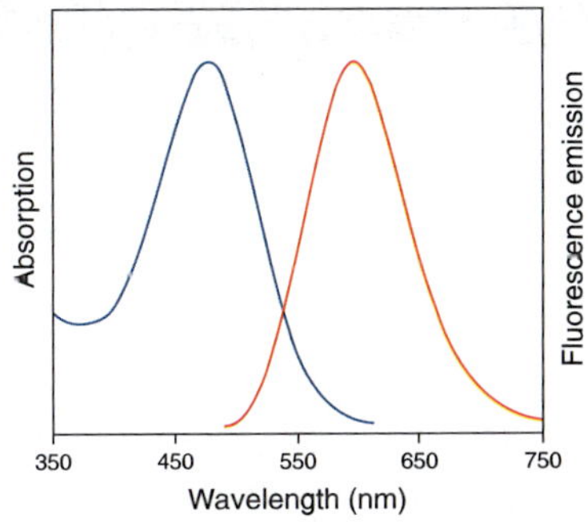

Figure 14.4.22 Absorption and fluorescence emission spectra of FM® 1-43 bound to phospholipid bilayer membranes.

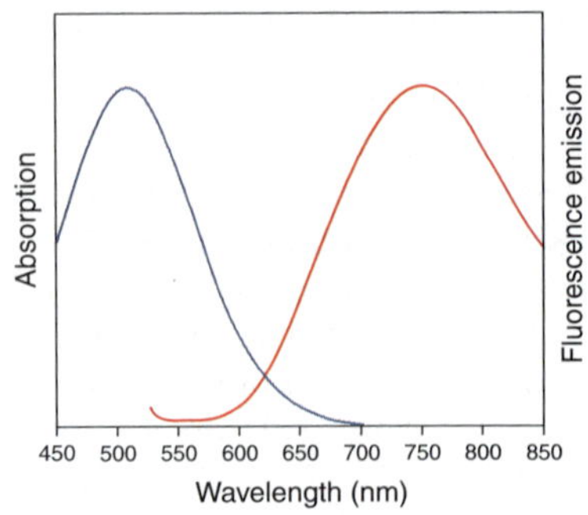

Figure 14.4.23 Absorption and fluorescence emission spectra of FM® 4-64 bound to zwitterionic detergent (CHAPS) micelles.

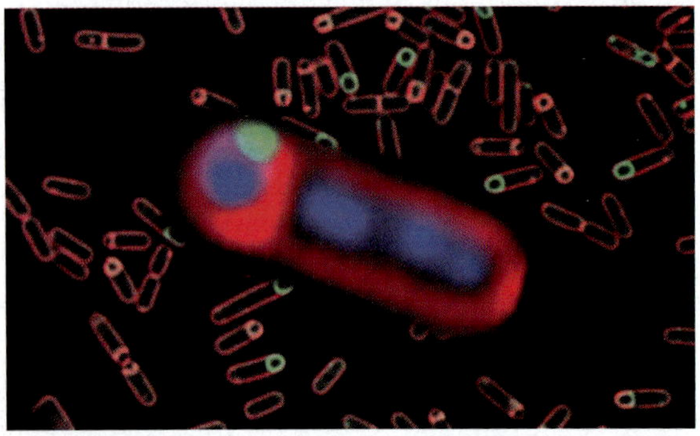

Figure 14.4.24 The morphology of sporulating *Bacillus subtilis* in the early stages of forespore engulfment. The membranes and chromosomes of both the forespore and the larger mother cell are stained with FM® 4-64 (red; T3166, T13320) and DAPI (blue; D1306, D3571, D21490), respectively. The small green-fluorescent patch indicates the localization of a GFP fusion to SPoIIIE, a protein essential for translocation of the forespore chromosome that may also regulate membrane fusion events (see Proc Natl Acad Sci U S A (1999) 96:14553). The background contains sporangia at various stages in the engulfment process stained with MitoTracker® Green FM® (green, M7514) and FM® 4-64 (red).

Figure 14.4.25 Cultured olfactory bulb neuron stained with RH 414 (T1111) and DM-BODIPY®(–)-dihydropyridine (D7443). Left: image at greater than 580 nm; Middle: image at 510–580 nm; Right: ratio of the middle image divided by the left image. Images were acquired with a Leica confocal laser-scanning microscope. The images were contributed by D. Schild, H. Geiling and J. Bischofberger, Physiologisches Institut, Universität Göttingen.

in sea urchin eggs.[95] This styryl dye has also proven extremely valuable for identifying actively firing neurons and for investigating the mechanisms of activity-dependent vesicle cycling in widely different species[96–100] (Section 16.1). We offer FM® 1-43 in a 1 mg vial (T3163) or specially packaged in 10 vials of 100 µg each (T35356).

FM® 1-43FX is an FM® 1-43 analog that has been modified to contain an aliphatic amine. This modification makes the probe fixable with aldehyde-based fixatives including formaldehyde and glutaraldehyde. FM® 1-43FX has been used to study synaptic vesicle cycling in cone photoreceptor terminals[96] and to investigate the functional maturation of glutamatergic synapses.[101] FM® 1-43FX is available specially packaged in 10 vials of 100 µg each (F35355).

Membranes labeled with FM® 4-64 exhibit long-wavelength red fluorescence (Figure 14.4.23) that can be distinguished from the green fluorescence of FM® 1-43 staining (Figure 14.4.22) with the proper optical filter sets, thus permitting two-color observation of membrane recycling in real time.[102,103] FM® 4-64 selectively stains yeast vacuolar membranes and is an important tool for visualizing vacuolar organelle morphology and dynamics and for studying the endocytic pathway in yeast[104] (Section 12.3). In addition, FM® 4-64 staining has been used to visualize membrane migration and fusion during *Bacillus subtilis* sporulation—movements that can be correlated with the translocation of GFP-labeled proteins[105] (Figure 14.4.24). We offer FM® 4-64 in a 1 mg vial (T3166) or specially packaged in 10 vials of 100 µg each (T13320). Additionally, we offer the fixable analog FM® 4-64FX specially packaged in 10 vials of 100 µg each (F34653).

Membranes stained with RH 414 (T1111) exhibit orange fluorescence when observed through a longpass optical filter that permits passage of light beyond 510 nm.[94,106,107] In a confocal laser-scanning microscopy study, the subcellular distribution of L-type Ca^{2+} channels in olfactory bulb neurons was determined using intensity ratio measurements of the green fluorescence of DM-BODIPY® dihydropyridine (D7443, Section 16.3) to the red fluorescence of RH 414–stained plasma membranes[108] (Figure 14.4.25). In this method, staining of the plasma membrane by RH 414 was used to control for optical artifacts and differences in membrane surface area in the optical section. RH 414 has also been used as to follow vacuolization of the transverse tubules of frog skeletal muscle[109] and to measure membrane potential (Chapter 22).

Image-iT® LIVE Plasma Membrane and Nuclear Labeling Kit

The Image-iT® LIVE Plasma Membrane and Nuclear Labeling Kit (I34406) provides two stains—red-fluorescent Alexa Fluor® 594 wheat germ agglutinin (WGA) (excitation/emission maxima ~590/617 nm) and blue-fluorescent Hoechst 33342 dye (excitation/emission maxima when bound to DNA ~350/461 nm)—for highly selective staining of the plasma membrane and nucleus, respectively, of live, Green Fluorescent Protein (GFP)–transfected cells (Figure 14.4.26). These dyes can be combined into one staining solution using the protocol provided, saving labeling time and wash steps while still providing optimal staining. Cell-impermeant Alexa Fluor® 594 WGA binds selectively to *N*-acetylglucosamine and *N*-acetylneuraminic (sialic) acid residues.[110] When used according to the protocol, Alexa Fluor® 594 WGA provides highly selective labeling of the plasma membrane with minimal background, although labeling may not be as distinct for flat cell types when viewed using standard epifluorescence microscopy or

IMPORTANT NOTICE: The products described in this manual are covered by one or more Limited Use Label License(s). Please refer to the Appendix on page 971 and Master Product List on page 975. Products are For Research Use Only. Not intended for any animal or human therapeutic or diagnostic use.
molecular probes® | invitrogen™ by life technologies™

low magnification. Alexa Fluor® 594 WGA is retained after formaldehyde fixation and permeabilization with Triton® X-100. This fluorescent lectin conjugate can also be used to label fixed cells; however, to avoid labeling intracellular components, formaldehyde-fixed cells should not be permeabilized before labeling. It is important to note that Alexa Fluor® 594 WGA can stimulate biological activity, including clustering of glycosylated cell-surface proteins. The kit also includes Hoechst 33342 dye, a cell-permeant nucleic acid stain that is selective for DNA and is spectrally similar to DAPI. This dye does not interfere with GFP fluorescence and is retained after fixation and permeabilization.

The Image-iT® LIVE Plasma Membrane and Nuclear Labeling Kit contains:

- Alexa Fluor® 594 wheat germ agglutinin (WGA)
- Hoechst 33342 nucleic acid stain
- Detailed experimental protocols

Each kit provides enough staining solution for 360 assays using the protocol provided for labeling live, cultured cells that are adhering to coverslips.

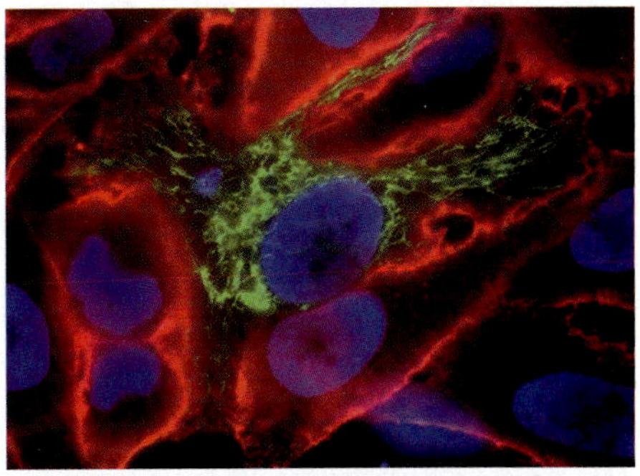

Figure 14.4.26 Live, highly confluent HeLa cells were transfected using pShooter™ vector pCMV/myc/mito/GFP and Lipofectamine® 2000 transfection reagent and stained with the reagents in the Image-iT® LIVE Plasma Membrane and Nuclear Labeling Kit (I34406). Plasma membranes were labeled with Alexa Fluor® 594 wheat germ agglutinin and nuclei were stained with Hoechst 33342. Cells were visualized using epifluorescence microscopy.

Vybrant® Lipid Raft Labeling Kits

The Vybrant® Lipid Raft Labeling Kits (V34403, V34404, V34405) are designed to provide convenient, reliable and extremely bright fluorescent labeling of lipid rafts in live cells. Lipid rafts are detergent-insoluble, sphingolipid- and cholesterol-rich membrane microdomains that form lateral assemblies in the plasma membrane.[111–117] Lipid rafts also sequester glycophosphatidylinositol (GPI)-linked proteins and other signaling proteins and receptors, which may be regulated by their selective interactions with these membrane microdomains.[118–123] Recent research has demonstrated that lipid rafts play a role in a variety of cellular processes—including the compartmentalization of cell-signaling events,[124–131] the regulation of apoptosis[132–134] and the intracellular trafficking of certain membrane proteins and lipids[135–137]—as well as in the infectious cycles of several viruses and bacterial pathogens.[138–143] Examining the formation and regulation of lipid rafts is a critical step in understanding these aspects of eukaryotic cell function.

The Vybrant® Lipid Raft Labeling Kits provide the key reagents for fluorescently labeling lipid rafts *in vivo* with our bright and extremely photostable Alexa Fluor® dyes (Figure 14.4.27). Live cells are first labeled with the green-fluorescent Alexa Fluor® 488, orange-fluorescent Alexa Fluor® 555 or red-fluorescent Alexa Fluor® 594 conjugate of cholera toxin subunit B (CT-B). This CT-B conjugate binds to the pentasaccharide chain of plasma membrane ganglioside G_{M1}, which selectively partitions into lipid rafts.[122,144,145] All Molecular Probes® CT-B conjugates are prepared from recombinant CT-B and are completely free of the toxic subunit A, thus eliminating any concern for toxicity or ADP-ribosylating activity. An antibody that specifically recognizes CT-B is then used to crosslink the CT-B–labeled lipid rafts into distinct patches on the plasma membrane, which are easily visualized by fluorescence microscopy.[146,147]

Each Vybrant® Lipid Raft Labeling Kit contains sufficient reagents to label 50 live-cell samples in a 2 mL assay, including:

- Recombinant cholera toxin subunit B (CT-B) labeled with the Alexa Fluor® 488 (in Kit V34403), Alexa Fluor® 555 (in Kit V34404) or Alexa Fluor® 594 (in Kit V34405) dye
- Anti–cholera toxin subunit B antibody (anti–CT-B)
- Concentrated phosphate-buffered saline (PBS)
- Detailed labeling protocols

Because they are compatible with various multilabeling schemes, the Vybrant® Lipid Raft Labeling Kits can also serve as important tools for identifying physiologically significant membrane proteins that associate with lipid rafts. Cells can be labeled with other live-cell probes during the lipid raft labeling protocol or immediately following the antibody crosslinking step, depending on the specific labeling requirements of the other probes. Alternatively, once the lipid rafts have been labeled and crosslinked, the cells can be fixed for long-term storage or fixed and permeabilized for subsequent labeling with antibodies or other probes that are impermeant to live cells.

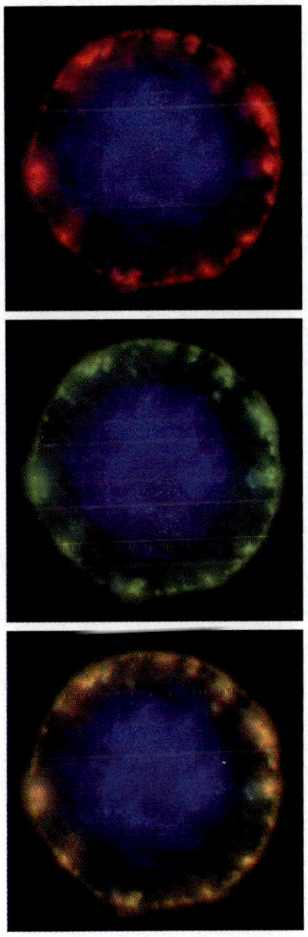

Figure 14.4.27 A J774 mouse macrophage cell sequentially stained with BODIPY® FL ganglioside G_{M1} (B13950) and then with Alexa Fluor® 555 dye–labeled cholera toxin subunit B (C22843, C34776; also available as a component of V34404). The cell was then treated with an anti–CT-B antibody (a component of V34404) to induce crosslinking. Alexa Fluor® 555 dye fluorescence (top, red) and BODIPY® FL dye fluorescence (center, green) were imaged separately and overlaid to emphasize the coincident staining (bottom, yellow). Nuclei were stained with blue-fluorescent Hoechst 33342 (H1399, H3570, H21492).

IMPORTANT NOTICE: The products described in this manual are covered by one or more Limited Use Label License(s). Please refer to the Appendix on page 971 and Master Product List on page 975. Products are For Research Use Only. Not intended for any animal or human therapeutic or diagnostic use.

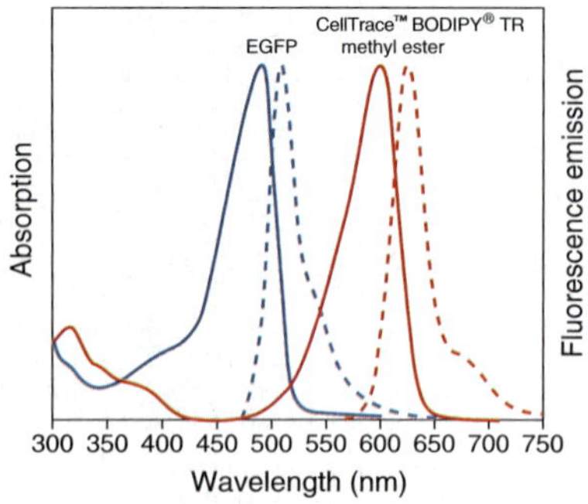

Figure 14.4.28 Normalized absorption (—) and fluorescence emission (- - -) spectra of enhanced Green Fluorescent Protein (EGFP) and CellTrace™ BODIPY® TR methyl ester (C34556).

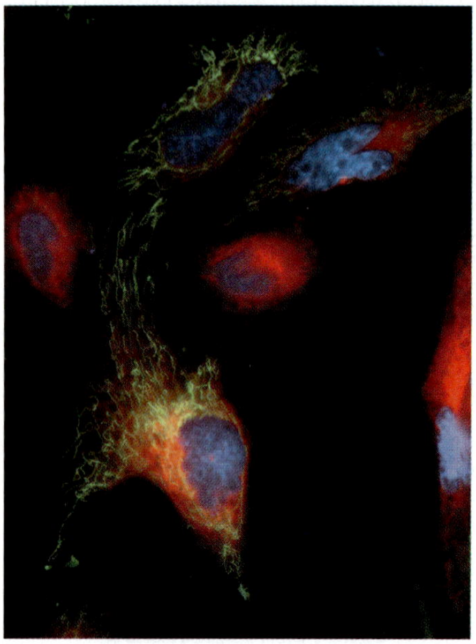

Figure 14.4.29 Live HeLa cells were transfected using pShooter™ vector pCMV/myc/mito/GFP and Lipofectamine® 2000 transfection reagent and stained with the reagents in the Image-iT® LIVE Intracellular Membrane and Nuclear Labeling Kit (I34407). Intracellular membranes were stained with CellTrace™ BODIPY® TR methyl ester, and nuclei were stained with Hoechst 33342. Cells were visualized using epifluorescence microscopy.

Image-iT® LIVE Intracellular Membrane and Nuclear Labeling Kit

Many research and biotechnological applications require detailed three- and four-dimensional visualization of embryonic cells labeled with Green Fluorescent Protein (GFP) within their native tissue environments. Fluorescent counterstains that label all the cells in a live embryo provide a histological context for the GFP-expressing cells in the specimen. The red-fluorescent CellTrace™ BODIPY® TR methyl ester (C34556)—available separately or as part of our Image-iT® LIVE Intracellular Membrane and Nuclear Labeling Kit (I34407)—is an excellent counterstain for cells and tissues that are expressing GFP.[148] This dye readily permeates cell membranes and selectively stains mitochondria and endomembranous organelles such as endoplasmic reticulum and the Golgi apparatus, but does not appear to localize in the plasma membrane. These localization properties make the dye an ideal vital stain that can be used to reveal: (1) the location and shapes of cell nuclei,[149] (2) the shapes of cells within embryonic tissues and (3) the boundaries of organ-forming tissues within the whole embryo. Furthermore, CellTrace™ BODIPY® TR methyl ester staining is retained after formaldehyde fixation and permeabilization with Triton® X-100, and the dye does not appear to produce any teratogenic effects on embryonic development. The emission spectra of enhanced GFP (EGFP) and CellTrace™ BODIPY® TR methyl ester are well separated, with peaks at 508 nm and 625 nm, respectively (Figure 14.4.28), allowing simultaneous dual-channel confocal imaging without significant overspill of GFP fluorescence into the CellTrace™ BODIPY® TR methyl ester detection channel.

The Image-iT® LIVE Intracellular Membrane and Nuclear Labeling Kit provides:

- CellTrace™ BODIPY® TR methyl ester, for staining intracellular membranes
- Hoechst 33342 dye for staining nuclei
- Detailed experimental protocols

These two fluorescent stains were especially chosen for their compatibility with live GFP-expressing cells (Figure 14.4.29), and they can be combined into one staining solution to save labeling time and wash steps while still providing optimal staining. Each kit provides enough staining solution for 250 assays using the protocol provided for labeling live, cultured cells that are adhering to coverslips.

FluoroMyelin™ Fluorescent Myelin Stains

The FluoroMyelin™ Green and FluoroMyelin™ Red fluorescent myelin stains (F34651, F34652) enable quick and selective labeling of myelin in brain cryosections. Visualization of myelin is useful for understanding myelin distribution, for identifying brain structures and relative location of other labels and for mapping a particular section within the brain. Traditional methods require the use of antibodies, such as anti–myelin basic protein, or chromogenic (transmitted-light) methods, such as the Loyez method or Schmued's gold chloride technique, all of which are time consuming, requiring multiple steps over one to three days. The FluoroMyelin™ stains, in contrast, require only a single 20-minute labeling step plus washes. A standard FITC filter set is suitable for imaging the FluoroMyelin™ Green stain, with little or no bleedthrough into standard DAPI, TRITC or far-red filter sets. A standard TRITC or Texas Red® filter set is suitable for imaging the FluoroMyelin™ Red stain, with little or no bleedthrough into standard DAPI, FITC or far-red filter sets. These stains can be used in combination with antibodies and other dyes, and with standard histochemical methods for brain cryosections; they are ideal for studying demyelinating diseases such as multiple sclerosis.[150–152] In addition to being used on their own, FluoroMyelin™ stains can be combined with other dyes in a single labeling step as in our BrainStain™ Imaging Kit described below, which includes NeuroTrace® 530/615 red fluorescent Nissl stain and DAPI nucleic acid stain for three-color visualization.

IMPORTANT NOTICE: The products described in this manual are covered by one or more Limited Use Label License(s). Please refer to the Appendix on page 971 and Master Product List on page 975. Products are For Research Use Only. Not intended for any animal or human therapeutic or diagnostic use.
www.invitrogen.com/probes
molecular probes® | invitrogen™ by Life technologies™

BrainStain™ Imaging Kit

The BrainStain™ Imaging Kit (B34650) enables three-color combinatorial labeling of myelin, neurons and nuclei in brain cryosections in a single 20-minute staining step plus washes (Figure 14.4.30, Figure 14.4.31). This kit contains novel stains that can be used separately or together in one staining solution, replacing traditional methods that can take one to three days. Standard histochemical methods such as immunohistochemistry are compatible with these stains. Each BrainStain™ Kit provides:

- FluoroMyelin™ Green fluorescent myelin stain
- NeuroTrace® 530/615 red-fluorescent Nissl stain
- DAPI nuclear counterstain
- Detailed protocols for staining mouse brain cryosections

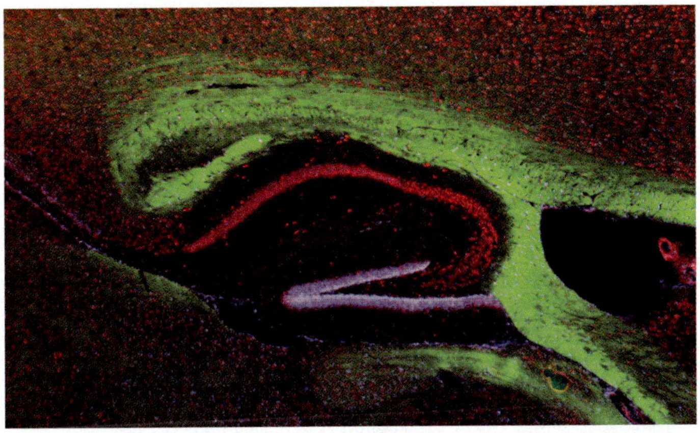

Figure 14.4.30 Mouse hippocampus and corpus callosum sagittal cryosection stained with reagents from the BrainStain™ Imaging Kit (B34650). Neuronal cell bodies were stained with NeuroTrace® red-fluorescent Nissl stain, white matter was stained with FluoroMyelin™ Green dye, and nuclei were stained with DAPI.

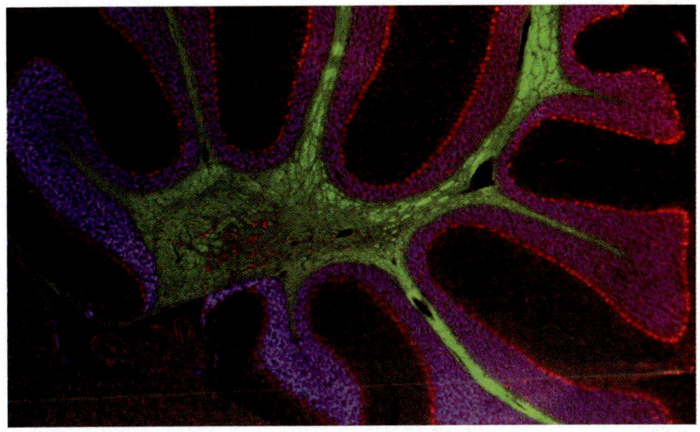

Figure 14.4.31 Mouse cerebellum cryosection stained with reagents from the BrainStain™ Imaging Kit (B34650). Neuronal cell bodies were stained with NeuroTrace® 530/615 red-fluorescent Nissl stain, white matter was stained with FluoroMyelin™ Green dye, and nuclei were stained with DAPI.

REFERENCES

1. J Neurosci Methods (2009) 180:1; **2.** J Neurosci Methods (2007) 163:213; **3.** Neuron (1994) 13:813; **4.** Biotechniques (1992) 13:4 580; **5.** Development (1989) 106:809; **6.** J Cell Biol (1991) 114:929; **7.** J Struct Biol (1990) 105:154; **8.** J Membr Biol (1989) 109:221; **9.** Cytometry (1993) 14:16; **10.** J Cell Biol (1981) 90:595; **11.** Proc Natl Acad Sci U S A (1997) 94:8795; **12.** J Biomed Opt (2009) 14:054005; **13.** Blood (2010) 115:5347; **14.** Biochem Pharmacol (1995) 49:1303; **15.** J Neurosci (1995) 15:549; **16.** J Neurosci Methods (1999) 93:13; **17.** Science (1992) 257:96; **18.** J Neurosci Methods (2002) 117:73; **19.** Microsc Res Tech (1993) 24:2; **20.** J Cell Biol (1986) 103:171; **21.** J Cell Biol (1996) 135:63; **22.** Cytometry (1995) 21:160; **23.** J Biol Chem (1998) 273:33354; **24.** J Cell Biol (1997) 136:1109; **25.** Anticancer Res (1998) 18:4181; **26.** J Neurosci Methods (2005) 141:155; **27.** Neuroreport (1997) 8:3965; **28.** Biophys J (1996) 70:988; **29.** J Cell Biol (1996) 134:1427; **30.** Neuron (1996) 17:91; **31.** J Cell Biol (1999) 144:1271; **32.** J Neurosci Methods (2008) 174:71; **33.** J Immunol Methods (1996) 194:181; **34.** J Cell Biol (1995) 131:1327; **35.** Science (1997) 278:474; **36.** Heart Lung Circ (2008) 17:395; **37.** Am J Physiol (1997) 273:H387; **38.** J Biol Chem (1997) 272:1117; **39.** Bioorg Med Chem Lett (1996) 6:1479; **40.** J Cell Biol (1999) 147:105; **41.** Blood (2000) 96:1180; **42.** Methods Cell Biol (1990) 33:469; **43.** Cancer Res (2006) 66:2346; **44.** J Histochem Cytochem (1998) 46:901; **45.** J Neurosci Methods (1999) 88:27; **46.** J Comp Neurol (1990) 302:729; **47.** J Histochem Cytochem (1990) 38:725; **48.** Exp Neurol (1988) 102:92; **49.** J Cell Biol (1994) 126:519; **50.** Chem Phys Lett (1989) 159:231; **51.** Thin Solid Films (1985) 132:55; **52.** J Neurosci Methods (2006) 154:256; **53.** Neuron (1991) 7:819; **54.** J Neurosci (1990) 10:3947; **55.** Methods Cell Biol (1996) 51:147; **56.** J Neurosci Methods (2007) 162:237; **57.** Neuroscience Protocols, Wouterlood FG, Ed. (1993) p.93-050-16; **58.** Biophys J (1998) 75:2558; **59.** Brain Struct Funct (2008) 213:149; **60.** Trends Biotechnol (2007) 25:530; **61.** J Neurosci Methods (2005) 141:41; **62.** Methods (2003) 30:79; **63.** Cell Transplant (1997) 6:455; **64.** Trends Neurosci (1989) 12:333, 340; **65.** Nat Protoc (2008) 3:1703; **66.** J Neurosci Methods (2009) 184:332; **67.** J Histochem Cytochem (1990) 38:735; **68.** J Neurosci Methods (1992) 42:65; **69.** J Neurosci Methods (1992) 42:45; **70.** J Neurosci (1995) 15:3475; **71.** J Neurosci (1995) 15:990; **72.** Trends Neurosci (1990) 13:14; **73.** J Neurosci (1992) 12:3494; **74.** Vis Neurosci (1993) 10:117; **75.** Trends Neurosci (1994) 17:177; **76.** J Neurosci (1993) 13:455; **77.** Eur J Neurosci (1991) 3:1189; **78.** J Neurosci (1995) 15:5514; **79.** Chem Phys Lipids (2002) 116:3; **80.** Biochemistry (1994) 33:1820; **81.** Biochim Biophys Acta (1994) 1191:375; **82.** J Cell Sci (1977) 28:167; **83.** Chem Phys Lipids (2002) 116:39; **84.** Anal Biochem (2002) 303:145; **85.** J Biol Chem (2005) 280:32811; **86.** J Biol Chem (2010) 285:5931; **87.** J Biol Chem (2009) 284:29798; **88.** Neuroscience (2008) 153:751; **89.** Am J Physiol Cell Physiol (2009) 296:C607; **90.** Cell (2005) 121:937; **91.** J Cell Sci (2009) 122:289; **92.** EMBO J (1993) 12:5209; **93.** J Neurosci (1993) 13:834; **94.** Science (1992) 255:200; **95.** J Cell Biol (1995) 131:1183; **96.** Neuron (2004) 41:755; **97.** J Biol Chem (2000) 275:15279; **98.** J Neurochem (1999) 73:2227; **99.** Annu Rev Neurosci (1999) 22:1; **100.** Proc Biol Sci (1994) 255:61; **101.** Neuron (2001) 29:469; **102.** J Cell Biol (1995) 131:679; **103.** J Neurosci (1995) 15:8246; **104.** J Cell Biol (1995) 128:779; **105.** Proc Natl Acad Sci U S A (1999) 96:14553; **106.** J Neurosci (1995) 15:6327; **107.** J Neurosci (1992) 12:363; **108.** J Neurosci Methods (1995) 59:183; **109.** J Muscle Res Cell Motil (1995) 16:401; **110.** J Mol Biol (1984) 178:91; **111.** J Cell Biol (2003) 162:365; **112.** J Lipid Res (2003) 44:655; **113.** Eur J Biochem (2002) 269:737; **114.** Science (2000) 290:1721; **115.** Mol Membr Biol (1999) 16:145; **116.** Trends Cell Biol (1999) 9:87; **117.** Annu Rev Cell Dev Biol (1998) 14:111; **118.** Proc Natl Acad Sci U S A (2003) 100:5813; **119.** J Immunol (2003) 170:1329; **120.** J Membr Biol (2002) 189:35; **121.** Proc Natl Acad Sci U S A (2001) 98:9098; **122.** J Cell Biol (1999) 147:447; **123.** Mol Biol Cell (1999) 10:3187; **124.** Biochim Biophys Acta (2003) 1610:247; **125.** Annu Rev Immunol (2003) 21:457; **126.** Mol Immunol (2002) 38:1247; **127.** Nat Rev Immunol (2002) 2:96; **128.** Biol Res (2002) 35:127; **129.** Nat Rev Mol Cell Biol (2000) 1:31; **130.** J Exp Med (1999) 190:1549; **131.** J Cell Biol (1998) 143:637; **132.** Immunity (2003) 18:655; **133.** J Biol Chem (2002) 277:39541; **134.** Biochem Biophys Res Commun (2002) 297:876; **135.** Biol Chem (2002) 383:1475; **136.** J Cell Biol (2001) 153:529; **137.** J Cell Sci (2001) 114:3957; **138.** J Virol (2003) 77:9542; **139.** Exp Cell Res (2003) 287:67; **140.** Traffic (2002) 3:705; **141.** J Clin Virol (2001) 22:217; **142.** Curr Biol (2000) 10:R823; **143.** J Virol (2000) 74:3264; **144.** Biochemistry (1996) 35:16069; **145.** Mol Microbiol (1994) 13:745; **146.** J Cell Biol (1998) 141:929; **147.** J Biol Chem (1994) 269:30745; **148.** Dev Dyn (2005) 232:359; **149.** Nature (2008) 452:243; **150.** J Neuroimmunol (2008) 193:94; **151.** J Neurosci (2006) 26:11278; **152.** Am J Physiol Regul Integr Comp Physiol (2006) 290:R1105.

IMPORTANT NOTICE: The products described in this manual are covered by one or more Limited Use Label License(s). Please refer to the Appendix on page 971 and Master Product List on page 975. Products are For Research Use Only. Not intended for any animal or human therapeutic or diagnostic use.

DATA TABLE 14.4 TRACERS FOR MEMBRANE LABELING

Cat. No.	MW	Storage	Soluble	Abs	EC	Em	Solvent	Notes
C7000	1051.50	F,D,L	DMSO, EtOH	553	134,000	570	MeOH	
C7001	1051.50	F,D,L	DMSO, EtOH	553	134,000	570	MeOH	
C10045	~1600	F,D,L	DMSO	555	155,000	572	EtOH	1
C10046	~1600	F,D,L	DMSO	651	250,000	672	EtOH	1
C34556	438.25	F,D,L	DMSO	588	68,000	616	MeOH	
D275	881.72	L	DMSO, DMF	484	154,000	501	MeOH	
D282	933.88	L	DMSO, EtOH	549	148,000	565	MeOH	
D291	618.73	L	DMSO, EtOH	492	53,000	612	MeOH	2
D307	959.92	L	DMSO, EtOH	644	260,000	665	MeOH	3
D383	765.56	L	DMSO, EtOH	549	144,000	565	MeOH	4
D384	877.77	L	DMSO, EtOH	549	148,000	565	MeOH	
D1125	825.61	L	DMSO, DMF	484	156,000	501	MeOH	
D3883	787.05	L	DMSO, EtOH	491	52,000	613	MeOH	2
D3886	925.49	F,L,AA	DMSO, EtOH	549	144,000	564	MeOH	3
D3898	873.65	F,L,AA	DMSO, DMF	484	138,000	499	MeOH	
D3899	925.82	F,L,AA	DMSO, EtOH	549	143,000	564	MeOH	3
D3911	933.88	L	DMSO, EtOH	549	148,000	565	MeOH	
D7756	1017.97	F,L,AA	DMSO, EtOH	549	148,000	564	MeOH	
D7757	1052.08	L	DMSO, EtOH	644	193,000	663	MeOH	
D7758	899.80	F,L,AA	DMSO, EtOH	492	41,000	612	MeOH	2
D7776	993.54	L	DMSO, EtOH	555	144,000	570	MeOH	
D7777	1145.73	L	DMSO, EtOH	556	164,000	573	MeOH	
D7778	1115.55	L	DMSO, EtOH	497	175,000	513	MeOH	
D12730	1019.58	L	DMSO, EtOH	650	247,000	670	MeOH	
D12731	1013.41	L	DMSO, EtOH	748	270,000	780	MeOH	
F34653	788.75	D,L	H_2O, DMSO	505	47,000	725	See Notes	5, 6
F35355	560.09	D,L	H_2O, DMSO	471	38,000	581	See Notes	5, 7
N22880	933.88	L	see Notes	549	148,000	565	MeOH	8
N22881	881.72	L	see Notes	484	154,000	501	MeOH	8
N22883	1051.50	F,L	see Notes	553	134,000	570	MeOH	8
O246	731.50	F,DD,L	DMSO, EtOH	556	125,000	578	MeOH	4
T1111	581.48	D,L	DMSO, EtOH	532	55,000	716	MeOH	9
T3163	611.55	D,L	H_2O, DMSO	471	38,000	581	see Notes	5, 7
T3166	607.51	D,L	H_2O, DMSO	505	47,000	725	see Notes	5, 8
T13320	607.51	D,L	H_2O, DMSO	505	47,000	725	see Notes	5, 8
T23360	565.43	D,L	H_2O, DMSO	560	43,000	734	$CHCl_3$	2, 6
T35356	611.55	D,L	H_2O, DMSO	471	38,000	581	see Notes	5, 7
V22885	933.88	L	see Notes	549	148,000	565	MeOH	10
V22886	881.72	L	see Notes	484	154,000	501	MeOH	10
V22887	1052.08	L	see Notes	644	193,000	663	MeOH	10
V22888	1051.50	F,L	see Notes	553	134,000	570	MeOH	10

For definitions of the contents of this data table, see "Using *The Molecular Probes® Handbook*" in the introductory pages.

Notes

1. This product is supplied as a ready-made solution in the solvent indicated under "Soluble."
2. Abs and Em of styryl dyes are at shorter wavelengths in membrane environments than in reference solvents such as methanol. The difference is typically 20 nm for absorption and 80 nm for emission, but varies considerably from one dye to another. Styryl dyes are generally nonfluorescent in water.
3. This product is intrinsically a liquid or an oil at room temperature.
4. This product is intrinsically a sticky gum at room temperature.
5. Abs, EC and Em determined for dye bound to detergent micelles (20 mg/mL CHAPS in H_2O). These dyes are essentially nonfluorescent in pure water.
6. FM® 4-64 and FM® 5-95 are nonfluorescent in water. For two-color imaging in GFP-expressing cells, these dyes can be excited at 568 nm with emission detection at 690–730 nm. (Am J Physiol Cell Physiol (2001) 281:C624)
7. FM® 1-43 Abs = 479 nm, Em = 598 nm bound to phospholipid bilayer membranes (FM® 1-43/lipid Absorption and fluorescence emission spectra of FM® 1-43 bound to phospholipid bilayer membranes.). Em = 565 nm bound to synaptosomal membranes. (Neuron (1994) 12:1235)
8. This product consists of a tracer dye mixed into an inert gel for direct application to tissues. Molecular weight and spectroscopic data are for the free dye.
9. RH 414 Abs ~500 nm, Em ~635 nm when bound to phospholipid bilayer membranes.
10. This product is supplied as a ready-made staining solution.

IMPORTANT NOTICE: The products described in this manual are covered by one or more Limited Use Label License(s). Please refer to the Appendix on page 971 and Master Product List on page 975. Products are For Research Use Only. Not intended for any animal or human therapeutic or diagnostic use.
www.invitrogen.com/probes
molecular probes® | invitrogen™ by life technologies™

PRODUCT LIST 14.4 TRACERS FOR MEMBRANE LABELING

Cat. No.	Product	Quantity
B34650	BrainStain™ Imaging Kit	1 kit
C10606	CellLight® Plasma Membrane-CFP *BacMam 2.0*	1 mL
C10607	CellLight® Plasma Membrane-GFP *BacMam 2.0*	1 mL
C10608	CellLight® Plasma Membrane-RFP *BacMam 2.0*	1 mL
C10046	CellMask™ Deep Red plasma membrane stain *5 mg/mL solution in DMSO*	100 µL
C10045	CellMask™ Orange plasma membrane stain *5 mg/mL solution in DMSO*	100 µL
C34556	CellTrace™ BODIPY® TR methyl ester *lipophilic counterstain for GFP* *solution in DMSO*	1 mL
C7001	CellTracker™ CM-DiI	1 mg
C7000	CellTracker™ CM-DiI *special packaging*	20 x 50 µg
D291	4-(4-(didecylamino)styryl)-*N*-methylpyridinium iodide (4-Di-10-ASP)	25 mg
D383	1,1′-didodecyl-3,3,3′,3′-tetramethylindocarbocyanine perchlorate (DiIC$_{12}$(3))	100 mg
D3883	4-(4-(dihexadecylamino)styryl)-*N*-methylpyridinium iodide (DiA; 4-Di-16-ASP)	25 mg
D1125	3,3′-dihexadecyloxacarbocyanine perchlorate (DiOC$_{16}$(3))	25 mg
D384	1,1′-dihexadecyl-3,3,3′,3′-tetramethylindocarbocyanine perchlorate (DiIC$_{16}$(3))	100 mg
D7758	4-(4-(dilinoleylamino)styryl)-*N*-methylpyridinium 4-chlorobenzenesulfonate (*FAST* DiA™ solid; DiΔ9,12-C$_{18}$ASP, CBS)	5 mg
D3898	3,3′-dilinoleyloxacarbocyanine perchlorate (*FAST* DiO™ solid; DiO$\Delta^{9,12}$-C$_{18}$(3), ClO4)	5 mg
D7756	1,1′-dilinoleyl-3,3,3′,3′-tetramethylindocarbocyanine, 4-chlorobenzenesulfonate (*FAST* DiI™ solid; DiIΔ9,12-C$_{18}$(3), CBS)	5 mg
D3899	1,1′-dilinoleyl-3,3,3′,3′-tetramethylindocarbocyanine perchlorate (*FAST* DiI™ oil; DiI$\Delta^{9,12}$-C$_{18}$(3), ClO$_4$)	5 mg
D7778	3,3′-dioctadecyl-5,5′-di(4-sulfophenyl)oxacarbocyanine, sodium salt (SP-DiOC$_{18}$(3))	5 mg
D7777	1,1′-dioctadecyl-6,6′-di(4-sulfophenyl)-3,3,3′,3′-tetramethylindocarbocyanine (SP-DiIC$_{18}$(3))	5 mg
D275	3,3′-dioctadecyloxacarbocyanine perchlorate ('DiO'; DiOC$_{18}$(3))	100 mg
D7776	1,1′-dioctadecyl-3,3,3′,3′-tetramethylindocarbocyanine-5,5′-disulfonic acid (DiIC$_{18}$(3)-DS)	5 mg
D282	1,1′-dioctadecyl-3,3,3′,3′-tetramethylindocarbocyanine perchlorate ('DiI'; DiIC$_{18}$(3))	100 mg
D3911	1,1′-dioctadecyl-3,3,3′,3′-tetramethylindocarbocyanine perchlorate *crystalline* ('DiI'; DiIC$_{18}$(3))	25 mg
D7757	1,1′-dioctadecyl-3,3,3′,3′-tetramethylindodicarbocyanine, 4-chlorobenzenesulfonate salt ('DiD' solid; DiIC$_{18}$(5) solid)	10 mg
D12730	1,1′-dioctadecyl-3,3,3′,3′-tetramethylindodicarbocyanine-5,5′-disulfonic acid (DiIC$_{18}$(5)-DS)	5 mg
D307	1,1′-dioctadecyl-3,3,3′,3′-tetramethylindodicarbocyanine perchlorate ('DiD' oil; DiIC$_{18}$(5) oil)	25 mg
D12731	1,1′-dioctadecyl-3,3,3′,3′-tetramethylindotricarbocyanine iodide ('DiR'; DiIC$_{18}$(7))	10 mg
D3886	1,1′-dioleyl-3,3,3′,3′-tetramethylindocarbocyanine methanesulfonate (Δ^9-DiI)	25 mg
F34651	FluoroMyelin™ Green fluorescent myelin stain *solution in water*	1 mL
F34652	FluoroMyelin™ Red fluorescent myelin stain *solution in water*	1 mL
F35355	FM® 1-43FX *fixable analog of FM® 1-43 membrane stain*	10 x 100 µg
F34653	FM® 4-64FX *fixable analog of FM® 4-64 membrane stain*	10 x 100 µg
I34407	Image-iT® LIVE Intracellular Membrane and Nuclear Labeling Kit *counterstains for GFP-expressing cells*	1 kit
I34406	Image-iT® LIVE Plasma Membrane and Nuclear Labeling Kit *counterstains for GFP-expressing cells*	1 kit
L7781	Lipophilic Tracer Sampler Kit	1 kit
N22883	NeuroTrace® CM-DiI tissue-labeling paste	100 mg
N22880	NeuroTrace® DiI tissue-labeling paste	500 mg
N22881	NeuroTrace® DiO tissue-labeling paste	500 mg
N22884	NeuroTrace® Multicolor Tissue-Labeling Kit *DiO, DiI, DiD pastes, 500 mg each*	1 kit
O246	octadecyl rhodamine B chloride (R18)	10 mg
T3163	*N*-(3-triethylammoniumpropyl)-4-(4-(dibutylamino)styryl)pyridinium dibromide (FM® 1-43)	1 mg
T35356	*N*-(3-triethylammoniumpropyl)-4-(4-(dibutylamino)styryl)pyridinium dibromide (FM® 1-43) *special packaging*	10 x 100 µg
T1111	*N*-(3-triethylammoniumpropyl)-4-(4-(4-(diethylamino)phenyl)butadienyl)pyridinium dibromide (RH 414)	5 mg
T3166	*N*-(3-triethylammoniumpropyl)-4-(6-(4-(diethylamino)phenyl)hexatrienyl)pyridinium dibromide (FM® 4-64)	1 mg
T13320	*N*-(3-triethylammoniumpropyl)-4-(6-(4-(diethylamino)phenyl)hexatrienyl)pyridinium dibromide (FM® 4-64) *special packaging*	10 x 100 µg
T23360	*N*-(3-trimethylammoniumpropyl)-4-(6-(4-(diethylamino)phenyl)hexatrienyl)pyridinium dibromide (FM® 5-95)	1 mg
V34403	Vybrant® Alexa Fluor® 488 Lipid Raft Labeling Kit *50 labelings*	1 kit
V34404	Vybrant® Alexa Fluor® 555 Lipid Raft Labeling Kit *50 labelings*	1 kit
V34405	Vybrant® Alexa Fluor® 594 Lipid Raft Labeling Kit *50 labelings*	1 kit
V22888	Vybrant® CM-DiI cell-labeling solution	1 mL
V22887	Vybrant® DiD cell-labeling solution	1 mL
V22885	Vybrant® DiI cell-labeling solution	1 mL
V22886	Vybrant® DiO cell-labeling solution	1 mL
V22889	Vybrant® Multicolor Cell-Labeling Kit *DiO, DiI, DiD solutions, 1 mL each*	1 kit

IMPORTANT NOTICE: The products described in this manual are covered by one or more Limited Use Label License(s). Please refer to the Appendix on page 971 and Master Product List on page 975. Products are For Research Use Only. Not intended for any animal or human therapeutic or diagnostic use.

14.5 Fluorescent and Biotinylated Dextrans

Dextrans are hydrophilic polysaccharides characterized by their moderate to high molecular weight, good water solubility and low toxicity. They are widely used as both anterograde and retrograde tracers in neurons [1,2] and for many other applications. Dextrans are biologically inert due to their uncommon poly-(α-D-1,6-glucose) linkages, which render them resistant to cleavage by most endogenous cellular glycosidases. They also usually have low immunogenicity.

We offer a broad range of fluorescent and biotinylated dextran conjugates in several molecular weight ranges. Because the source and molecular weight of the dextran, as well as the net charge, degree of substitution and nature of the dye may significantly affect the application, references citing the use of Molecular Probes® dextrans may not be directly applicable to dextrans obtained from other sources and should be considered guides rather than definitive protocols. In most cases, Molecular Probes® fluorescent dextrans are much brighter and have higher negative charge than dextrans available from other sources. Furthermore, we use rigorous methods for removing as much unconjugated dye as practical, and then assay our dextran conjugates by thin-layer chromatography to help ensure the absence of low molecular weight contaminants.

Properties of Molecular Probes® Dextran Conjugates

A Wide Selection of Substituents

Molecular Probes® dextrans are conjugated to biotin or a wide variety of fluorophores, including seven of our Alexa Fluor® dyes (Table 14.4). In particular, we would like to highlight the dextran conjugates of Alexa Fluor® 488, Oregon Green® and Rhodamine Green™ dyes, which are significantly brighter and more photostable than most fluorescein dextrans. Dextran-conjugated fluorescent indicators for calcium and magnesium ions (Section 19.4) and for pH (Section 20.4) are described with their corresponding ion indicators in other chapters.

Dextran Size

Molecular Probes® dextrans include those with nominal molecular weights (MW) of 3000; 10,000; 40,000; 70,000; 500,000; and 2,000,000 daltons (Table 14.4). Because unlabeled dextrans are polydisperse—and may become more so during the chemical processes required for their modification and purification—the actual molecular weights present in a particular sample may have a broad distribution. For example, our "3000 MW" dextran preparations contain polymers with molecular weights predominantly in the range of ~1500–3000 daltons, including the dye or other label.

Degree of Substitution of Molecular Probes® Dextrans

Dextrans from other commercial sources usually have a degree of substitution of 0.2 or fewer dye molecules per dextran molecule for dextrans in the 10,000 MW range. Molecular Probes® dextrans, however, typically contain 0.3–0.7 dyes per dextran in the 3000 MW range, 0.5–2 dyes per dextran in the 10,000 MW range, 2–4 dyes in the 40,000 MW range and 3–6 dyes in the 70,000 MW range. The actual degree of substitution is indicated on the product's label. If too many fluorophores are conjugated to the dextran molecule, quenching and undesired interactions with cellular components may occur. We have found our degree of substitution to be optimal for most applications, yielding dextrans that are typically much more fluorescent than the labeled dextrans available from other sources, thus permitting lower quantities to be used for intracellular tracing.

It has been reported that some commercially available fluorescein isothiocyanate (FITC) dextrans yield spurious results in endocytosis studies because of the presence of free dye or metal contamination.[3,4] To overcome this problem, we remove as much of the free dye as possible by a combination of precipitation, dialysis, gel filtration and other techniques. The fluorescent dextran is then assayed by thin-layer chromatography (TLC) to help ensure that it is free of low molecular weight dyes. We prepare several unique products that have two or even three different labels, including the fluoro-ruby, mini-ruby and micro-ruby dextrans, described below. Not all of the individual dextran molecules of these products are expected to have all the substituents, or to be equally fixable, particularly in conjugates of the lowest molecular weight dextrans.

Dextran Net Charge and Method of Substitution

The net charge on the dextran depends on the fluorophore and the method of preparing the conjugate. We prepare most Molecular Probes® dextrans by reacting a water-soluble amino dextran (D1860, D1861, D1862, D3330, D7144) with the succinimidyl ester of the appropriate dye, rather than reacting a native dextran with isothiocyanate derivatives such as FITC. This method provides superior amine selectivity and yields an amide linkage, which is somewhat more stable than the corresponding thioureas formed from isothiocyanates. Except for the Rhodamine Green™ and Alexa Fluor® 488 conjugates, once the dye has been added, the unreacted amines on the dextran are capped to yield a neutral or anionic dextran. In the case of the Rhodamine Green™ and Alexa Fluor® 488 dextrans, the unreacted amines on the dextran are not capped after dye conjugation. Thus, these dextran conjugates may be neutral, anionic or cationic. The Alexa Fluor®, Cascade Blue®, lucifer yellow, fluorescein and Oregon Green® dextrans are intrinsically anionic, whereas most of the dextrans labeled with the zwitterionic rhodamine B, tetramethylrhodamine and Texas Red® dyes are essentially neutral. To produce more highly anionic dextrans, we have developed a proprietary procedure for adding negatively charged groups to the dextran carriers; these products are designated "polyanionic" dextrans.

Dextran Fixability

Some applications require that the dextran tracer be treated with formaldehyde or glutaraldehyde for subsequent analysis.[5,6] For these applications, we offer "lysine-fixable" versions of most of our dextran conjugates of fluorophores or biotin. These dextrans have covalently bound lysine residues that permit dextran tracers to be conjugated to surrounding biomolecules by aldehyde-mediated fixation for subsequent detection by immunohistochemical and ultrastructural techniques. We have also shown that all of our 10,000 MW Alexa Fluor® dextran conjugates can be fixed with aldehyde-based fixatives.

Loading Cells with Dextrans and Subsequent Tissue Processing

Unless taken up by an endocytic process, dextran conjugates are membrane impermeant and usually must be loaded by relatively invasive techniques[7] (Table 14.1). As with the lipophilic tracers in Section 14.4, crystals of the dextran conjugates have been successfully loaded by simply placing them directly on some kinds of samples.[8] We have found the Influx™ pinocytic cell-loading reagent (I14402, Section 19.8) to be useful for loading dextrans into a variety of adherent and nonadherent cells.[9] Sterile filtration of dextran solutions before use with live cells is highly recommended.[10] Biotin and biotinylated biomolecules with molecular weights up to >100,000 daltons are taken up by some plant cells through an endocytic pathway.[11,12]

Our lysine-fixable dextrans and 10,000 MW Alexa Fluor® dextrans can be fixed in place with formaldehyde or glutaraldehyde, permitting subsequent tissue processing, such as sectioning. A protocol has been published for embedding tissues in plastic for high-resolution characterization of neurons filled with lysine-fixable fluorescent dextrans.[13] Fixation of biotinylated or fluorescent dextrans also permits the use of fluorescent- or enzyme-labeled conjugates of avidin and streptavidin (Section 7.6, Table 7.9) or of anti-dye antibodies (Section 7.4, Table 7.8), respectively. These techniques can amplify the signal, which is important for detecting fine structure in sections or for changing the detection mode.[14] We provide antibodies to the Alexa Fluor® 488, Alexa Fluor® 405/Cascade Blue®, lucifer yellow, fluorescein, BODIPY® FL, tetramethylrhodamine and Texas Red® fluorophores and to the 2,4-dinitrophenyl (DNP) and nitrotyrosine haptens (Section 7.4).

Photoconversion of neurons labeled with lysine-fixable fluorescent dextrans in the presence of diaminobenzidine (DAB) using the Diaminobenzidine (DAB) Histochemistry Kits (Section 7.2, Section 7.6) can be used to produce electron-dense products for electron microscopy[15] (Fluorescent Probes for Photoconversion of Diaminobenzidine Reagents—Note 14.2). Electron-dense products can also be generated from peroxidase or colloidal gold conjugates of avidin, streptavidin or anti-dye antibodies.[16] NANOGOLD® and Alexa Fluor® FluoroNanogold™ conjugates of secondary antibodies (Section 7.2) and streptavidin (Section 7.6) can be utilized to allow detection of labeled dextrans in fixed-cell preparations by light microscopy or, following silver enhancement with the LI Silver Enhancement Kit (L24919, Section 7.2), by electron microscopy.

Neuronal Tracing with Dextrans

Fluorescent and biotinylated dextrans are routinely employed to trace neuronal projections. Dextrans can function efficiently as anterograde or retrograde tracers,[2,17–19] depending on the study method and tissue type used. Active transport of dextrans occurs only in live, not fixed tissue.[20] Comparative studies of rhodamine isothiocyanate, rhodamine B dextran (D1824) and lysinated tetramethylrhodamine dextran (fluoro-ruby, D1817) have shown that the dextran conjugates produce less diffusion at injection sites and more permanent labeling

Table 14.4 Molecular Probes® dextran conjugates.

Label(s) (Absorption/Emission Maxima) *	3000 MW	10,000 MW	40,000 MW	70,000 MW	500,000 MW	2,000,000 MW
Cascade Blue® (400/420)	D7132 †	D1976 †				
Lucifer yellow (428/532)		D1825 †				
Alexa Fluor® 488 (494/521)	D34682 ‡	D22910 ‡				
Oregon Green® 488 (496/524)		D7170 D7171 †		D7172 D7173 †		
Fluorescein (494/518)	D3305 D3306 †	D1821 D1820 †	D1844 D1845 †	D1823 D1822 †	D7136 †	D7137 †
Fluorescein + biotin (494/518)	D7156 †	D7178 †				
Rhodamine Green™ (502/527)	D7163	D7153 †				
BODIPY® FL (505/513)		D7168 ‡				
Oregon Green® 514 (511/530)				D7176		
Tetramethylrhodamine (555/580)	D3307 D3308 †	D1816 D1817 † D1868 ‡	D1842	D1819 D1818 †		D7139 †
Tetramethylrhodamine + biotin (555/580)	D7162 †	D3312 †				
Alexa Fluor® 555 (551/565)		D34679 ‡				
Alexa Fluor® 546 (556/573)		D22911 ‡				
Rhodamine B (570/590)		D1824		D1841		
Alexa Fluor® 568 (578/603)		D22912 ‡				
Alexa Fluor® 594 (590/617)		D22913 ‡				
Texas Red® (595/615)	D3329 D3328 †	D1828 D1863 †	D1829	D1830 D1864 †		
Alexa Fluor® 647 (650/668)		D22914 ‡				
Alexa Fluor® 680 (681/704)	D34681 ‡	D34680 ‡				
Biotin (<300/none)	D7135 †	D1956 †		D1957 †	D7142 †	
Amino (<300/none)	D3330	D1860	D1861	D1862	D7144	

* Approximate absorption and emission maxima, in nm, for conjugates. † "Lysine-fixable" dextrans contain lysines and can therefore be fixed in place with formaldehyde or glutaraldehyde. ‡ "Fixable" dextrans contain free amines (but not lysines) and can be fixed in place with formaldehyde or glutaraldehyde.

IMPORTANT NOTICE: The products described in this manual are covered by one or more Limited Use Label License(s). Please refer to the Appendix on page 971 and Master Product List on page 975. Products are For Research Use Only. Not intended for any animal or human therapeutic or diagnostic use.

than do the corresponding free dyes.[5] Dextran conjugates with molecular weights up to 70,000 daltons have been employed as neuronal tracers in a wide variety of species. The availability of fluorescent dextran conjugates with different sizes and charges permitted the analysis of direction and rate of axonal transport in the squid giant axon.[21]

Multilabeled Dextrans

Molecular Probes® fixable dextrans, most of which are lysinated dextrans (see the products marked by a single dagger (†) in Table 14.4), are generally preferred for neuronal tracing because they may transport more effectively and can be fixed in place with aldehydes after labeling. We prepare a number of multilabeled dextrans that are fixable, including some that have acquired the distinction of unique names in various publications:

- Fluoro-ruby[5,15,22–27]—a red-orange–fluorescent, aldehyde-fixable 10,000 MW dextran labeled with both tetramethylrhodamine and lysine (D1817). 3000 MW, 70,000 MW and 2,000,000 MW versions of fluoro-ruby are also available (D3308, D1818, D7139).
- Fluoro-emerald[23,27,28]—a green-fluorescent, aldehyde-fixable 10,000 MW dextran labeled with both fluorescein and lysine (D1820; Figure 14.5.1, Figure 14.5.2, Figure 14.5.3). This labeled dextran is also available in molecular weights from 3000 daltons up to 2,000,000 daltons[29,30] (D3306, D1845, D1822, D7136, D7137).
- Micro-ruby (D7162) and mini-ruby[31–35] (D3312)—red-orange–fluorescent, aldehyde-fixable 3000 MW and 10,000 MW dextrans simultaneously labeled with tetramethylrhodamine, biotin and lysine.
- Micro-emerald (D7156) and mini-emerald (D7178)—green-fluorescent, aldehyde-fixable dextrans simultaneously labeled with fluorescein, biotin and lysine.
- Biotinylated dextran amine (BDA)[1,10,36–43]—nonfluorescent, aldehyde-fixable dextrans simultaneously labeled with biotin and lysine and available in several molecular weights (D1956, D1957, D7135, D7142). A useful review has been published on the BDA products.[44]

Fluoro-ruby and fluoro-emerald (Figure 14.5.3) have been extensively employed for retrograde and anterograde neuronal tracing,[5] transplantation[25] and cell-lineage tracing.[2,24,26,45] Both products can be used to photoconvert DAB into an insoluble, electron-dense reaction product.[15] Like fluoro-ruby and fluoro-emerald, micro-ruby and mini-ruby are brightly fluorescent, making it easy to visualize the electrode during the injection process. DiI (D282, Section 14.4) or other lipophilic probes in Section 14.4 can be used to mark the sites of microinjection.[46] In addition, because these dextrans include a covalently linked biotin, filled cells can be probed with standard enzyme-labeled avidin or streptavidin conjugates or with NANOGOLD® and Alexa Fluor® FluoroNanogold™ streptavidin (Section 7.6) to produce a permanent record of the experiment.[47] Mini-ruby has proven useful for intracellular filling in fixed brain slices[32] and has been reported to produce staining comparable to that achieved with lucifer yellow CH[31] (L453, L682, L1177, L12926; Section 14.3). Moreover, the use of mini-ruby in conjunction with standard peroxidase-mediated avidin–biotin methods does not cause co-conversion of lipofuscin granules found in adult human brain, a common problem during photoconversion of lucifer yellow CH.[31] The lysine-fixable micro-emerald and mini-emerald dextrans (triply labeled with fluorescein, biotin and lysine) provide a contrasting color that is better excited by the argon-ion laser of confocal laser-scanning microscopes; they have uses similar to micro-ruby and mini-ruby, respectively.

3000 MW Dextrans

The nominally 3000 MW dextrans offer several advantages over higher molecular weight dextrans, including faster axonal diffusion and greater access to peripheral cell processes[29] (Figure 14.5.4). Our "3000 MW" dextran preparations contain polymers with molecular weight predominantly in the range of ~1500–3000 daltons, including the dye

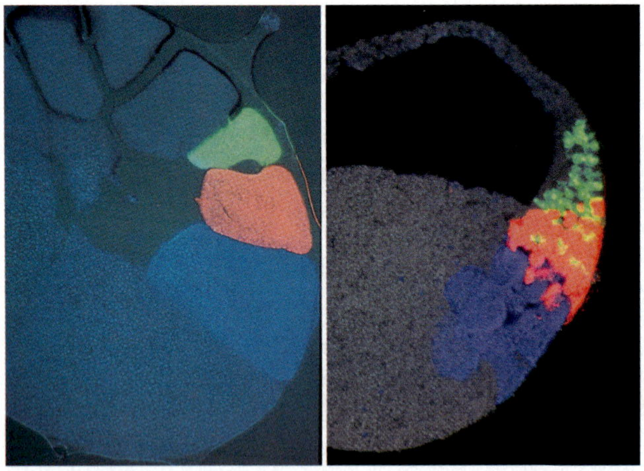

Figure 14.5.1 Lineage tracing of three dorsal blastomeres of 32-cell *Xenopus laevis* embryos injected with Molecular Probes® fluorescent dextran conjugates. The tier 1 dorsal (A1) blastomere was injected with lysine-fixable 10,000 MW fluorescein dextran (D1820), the tier 2 dorsal (B1) blastomere with lysine-fixable 10,000 MW Texas Red® dextran (D1863), and the tier 3 dorsal (C3) blastomere with lysine-fixable 10,000 MW Cascade Blue® dextran (D1976). Embryos were fixed in formaldehyde, embedded and sectioned. The image on the left shows a 13 µm–thick section of a stage 6 (32-cell) embryo fixed right after injection; this section exhibits significant autofluorescence due to the presence of residual yolk. The image on the right is a stage 10 (early gastrula) embryo. Triple-exposure photographs of the sectioned embryos were taken on a Zeiss® Axiophot with a 10× objective. Images contributed by Marie Vodicka, University of California, Berkeley.

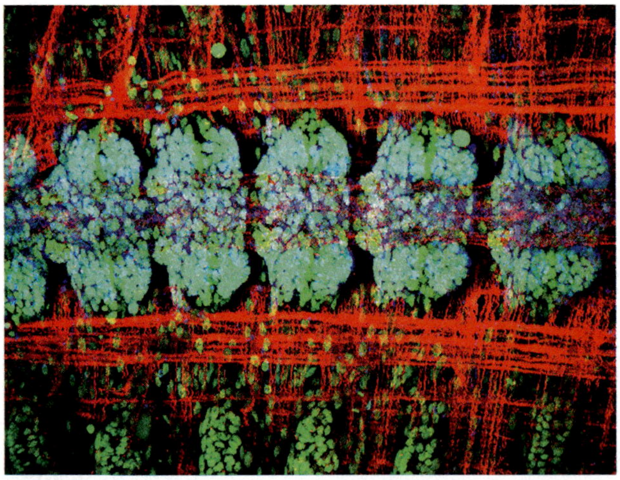

Figure 14.5.2 Development of the leech central nervous system into segments was characterized with the cell lineage tracer lysine-fixable 10,000 MW fluorescein dextran (fluoro-emerald, D1820). Fluoro-emerald was injected into a neuroectodermal cell of a leech embryo. The teloblast eventually differentiated into 5 segmentally iterated ganglia labeled with fluoro-emerald (pseudocolored blue in this image). Associated muscle fibers were identified with a Lan 3-14 monoclonal antibody and visualized with a Cy®3 antibody (pseudocolored red). Nuclei were counterstained with TOTO®-3 dye (T3604, pseudocolored green). Image reprinted from the cover of Development (2000) 127 (4), used with the permission of The Company of Biologists Ltd.

IMPORTANT NOTICE: The products described in this manual are covered by one or more Limited Use Label License(s). Please refer to the Appendix on page 971 and Master Product List on page 975. Products are For Research Use Only. Not intended for any animal or human therapeutic or diagnostic use.
www.invitrogen.com/probes
molecular probes® | invitrogen™ by Life technologies™

or other label. Our selection of 3000 MW dextrans includes Alexa Fluor®, fluorescein, Rhodamine Green™, tetramethylrhodamine, Texas Red® and biotin conjugates. We also offer lysine-fixable 3000 MW dextrans that are simultaneously labeled with both fluorescein and biotin (micro-emerald, D7156) or tetramethylrhodamine and biotin (micro-ruby, D7162).

The 3000 MW fluorescein dextran and tetramethylrhodamine dextran (D3306, D3308; Figure 14.5.5, Figure 14.5.6) have been observed to readily undergo both anterograde and retrograde movement in live cells.[23,29] These 3000 MW dextrans appear to passively diffuse within the neuronal process, as their intracellular transport is not effectively inhibited by colchicine or nocodazole, both of which disrupt active transport by depolymerizing microtubules.[29] Moreover, these small dextrans diffuse at rates equivalent to those of smaller tracers such as sulforhodamine 101 and biocytin (~2 millimeters/hour at 22°C) and about twice as fast as 10,000 MW dextrans. The relatively low molecular weight of the dextrans may result in transport of some labeled probes through gap junctions (see below). The signal from tetramethylrhodamine-conjugated dextrans can be detected in the fine dendrite configuration of cortical projection neurons using anti-tetramethylrhodamine antibodies (A6397, Section 7.4) and peroxidase–anti-peroxidase complex staining.[14]

NeuroTrace® BDA-10,000 Neuronal Tracer Kit

The NeuroTrace® BDA-10,000 Neuronal Tracer Kit (N7167) contains convenient amounts of each of the components required for neuroanatomical tracing using BDA methods,[44] including:

- Lysine-fixable, biotinylated 10,000 MW dextran amine (BDA-10,000)
- Horseradish peroxidase avidin (HRP avidin)
- 3,3′-Diaminobenzidine (DAB)
- Rigorously tested protocols for fast and simple tracing experiments

The neuronal tracer BDA-10,000 is transported over long distances and fills fine processes bidirectionally, including boutons in the anterograde direction and dendritic structures in the retrograde direction.[10,40–43] Two days to two weeks after BDA-10,000 is injected into the

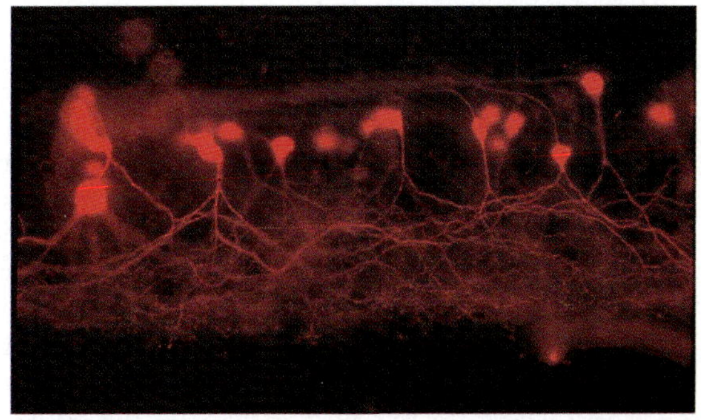

Figure 14.5.4 Secondary motor neurons in a spinal cord whole mount of a male western mosquitofish (*Gambusia affinis affinis*) that have been labeled with lysine-fixable 3000 MW Texas Red® dextran (D3328). The dextran crystals were applied to the bipinnate inclinator muscles of the anal appendicular support fin, and the dye was transported from the axons to cell body and dendrites. Motor neurons were visualized and photographed through a bandpass optical filter appropriate for Texas Red® dye, by epifluorescence microscopy. Image contributed by E. Rosa-Molinar, University of Nebraska Medical Center, and Bernd Fritzsch, Creighton University.

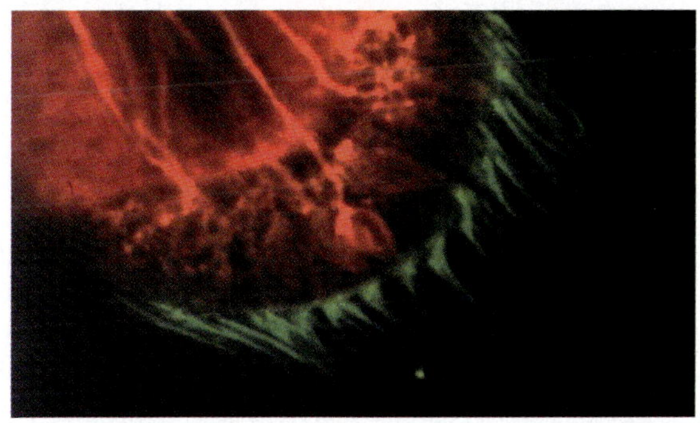

Figure 14.5.5 The attached eighth nerve from the vestibular labyrinth of a turtle, *Pseudemys scripta*, exposed to lysine-fixable 3000 MW tetramethylrhodamine dextran (D3308) and incubated with F-actin–specific BODIPY® FL phallacidin (B607). F-actin–labeled ciliary bundles were stained green by the phallacidin, and the calyceal and bouton endings of their primary afferents were stained red by the dextran. This image is a projection of 40 confocal images. Image contributed by Laura DiCaprio and Ellengene Peterson, Ohio University.

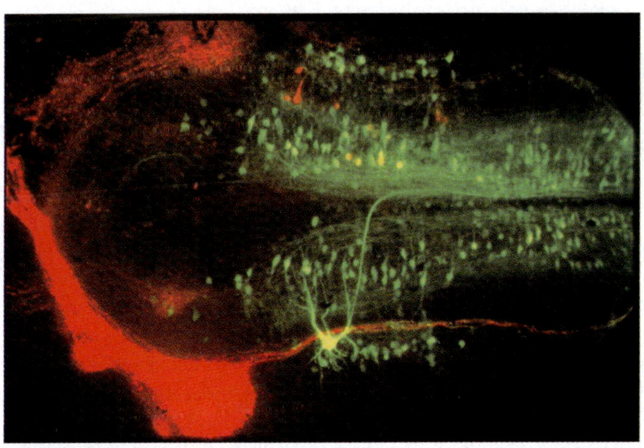

Figure 14.5.3 A whole mount of the embryonic brain of *Xenopus laevis* that has been double-labeled with our lysine-fixable 10,000 MW fluorescein dextran (fluoro-emerald, D1820) and lysine-fixable 10,000 MW tetramethylrhodamine dextran (fluoro-ruby, D1817). The tetramethylrhodamine dextran was used to label the neurons projecting from the retina, whereas the fluorescein dextran was applied to the transected spinal cord, thus allowing the detailed evaluation of the topological relationship of these two populations of neurons. Image contributed by Martina Manns and Bernd Fritzsch, Creighton University.

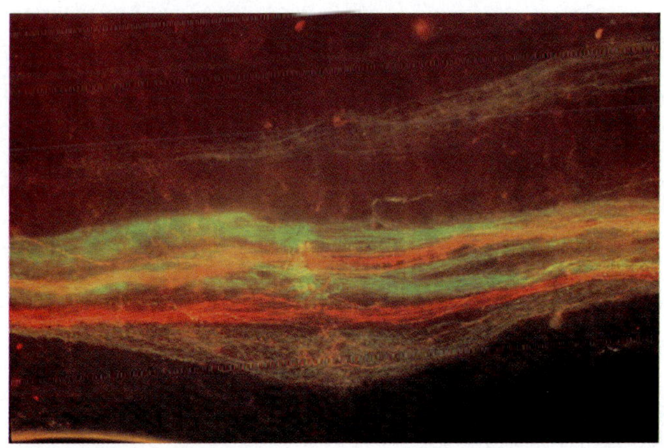

Figure 14.5.6 The antero-ventral and antero-dorsal lateral line nerves in an *Ambystoma mexicanum* whole-mounted brain after labeling with 3000 MW Rhodamine Green™ dextran (D7163) and lysine-fixable 3000 MW tetramethylrhodamine dextran (D3308), respectively. In both cases, the dextran was applied to the respective cut cranial nerves. This photomicrograph shows the four segregated ventral fascicles of the mechanosensory lateral line fibers and the intermingling of the dorsal electrosensory fibers. Image contributed by Bernd Fritzsch, Creighton University.

IMPORTANT NOTICE: The products described in this manual are covered by one or more Limited Use Label License(s). Please refer to the Appendix on page 971 and Master Product List on page 975. Products are For Research Use Only. Not intended for any animal or human therapeutic or diagnostic use.

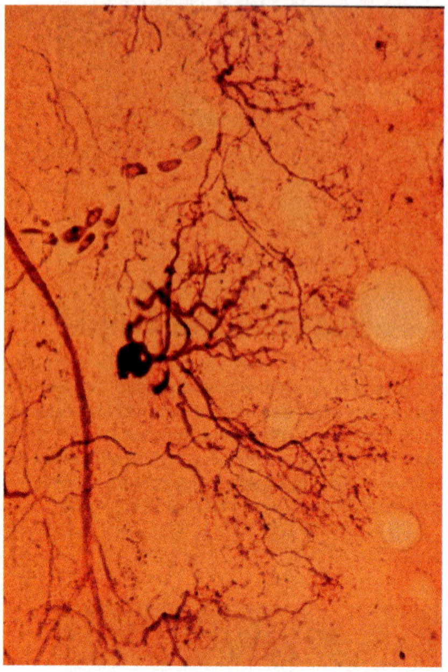

Figure 14.5.7 A mitral cell in the olfactory bulb of a chinook salmon that has been retrogradely labeled using the NeuroTrace® BDA-10,000 Neuronal Tracer Kit (N7167).

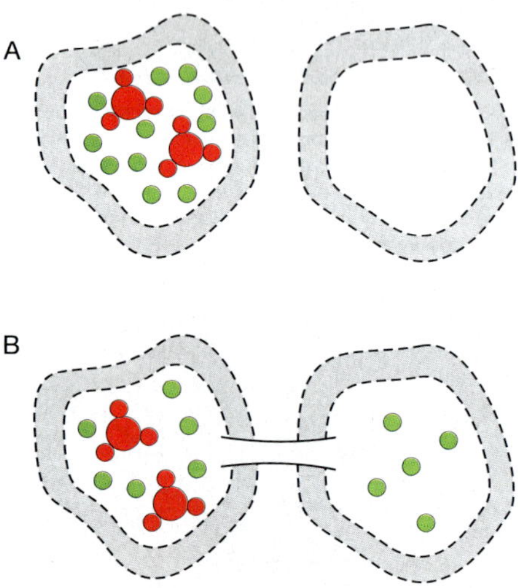

Figure 14.5.8 Dual-tracer technique for identifying gap junction–coupled cells. **A)** Cells are labeled with a mixture of a small polar tracer such as lucifer yellow CH (green circles) and a relatively large tetramethylrhodamine-labeled dextran (red circles). **B)** Adjoining gap junction–coupled cells are accessible to the low molecular weight tracer whereas the much larger dextran conjugate is excluded. Coupled cells with single-color lucifer yellow CH labeling are readily distinguished from initially labeled cells with dual fluorescence.

desired region of the brain, the brain tissue can be fixed and sectioned. BDA-10,000 can also be applied to cut nerves and allowed to transport. Following incubation with HRP avidin and then DAB, the electron-dense DAB reaction product can be viewed by either light or electron microscopy[44,47] (Figure 14.5.7). The NeuroTrace® BDA-10,000 labeling method can be readily combined with other anterograde or retrograde labeling methods or with immunohistochemical techniques. BDA-10,000 is available as a separate product (D1956), as are BDA derivatives with other molecular weights—BDA-3000[48] (D7135), BDA-70,000 (D1957) and BDA-500,000 (D7142). A detailed protocol that utilizes Molecular Probes® BDA-10,000 probe, HRP streptavidin (S911, Section 7.6) and tetramethylbenzidine to anterogradely label fine processes in neurons has been published.[1]

Cell Lineage Tracing with Dextrans

Fluorescent dextrans—particularly the fluorescein and rhodamine conjugates—have been used extensively for tracing cell lineage.[26,49] Our Alexa Fluor® 647 and Alexa Fluor® 680 dextrans (D22914, D34680, D34681) provide longer-wavelength detection options for specimens with high levels of autofluorescence or low transparency. In this technique, the dextran is microinjected into a single cell of the developing embryo, and the fate of that cell and its daughters can be followed *in vivo* (Figure 14.5.1). The lysine-fixable tetramethylrhodamine and Texas Red® dextran conjugates (Table 14.4) are most frequently cited for lineage tracing studies. As a second color, particularly in combination with the Texas Red® dextrans, researchers have most often used Molecular Probes® lysine-fixable fluorescein dextrans[49] (e.g., D3306, D1820, D1822). Although these fixable conjugates can be employed with long-term preservation of the tissue, some researchers prefer to co-inject a fluorescent, nonlysinated dextran along with a nonfluorescent, lysine-fixable biotin dextran (BDA). The nonfluorescent BDA can then be fixed in place with aldehyde-based fixatives and probed with any of our fluorescent or enzyme-labeled streptavidin and avidin conjugates described in Section 7.6 (Table 7.9).

Our 500,000 and 2,000,000 MW fluorescent dextrans (Table 14.4) may be particularly useful for lineage tracing at early stages of development, although these biopolymers have lower water solubility and a greater tendency to precipitate or clog microinjection needles than our lower molecular weight dextrans. Some studies suggest that lower molecular weight dextrans may leak from blastomeres, complicating analysis. Injection of 2,000,000 MW fluorescein dextran and 2,000,000 MW Texas Red® dextran into separate cells of the two-cell stage zebrafish embryo allowed the construction of a fate map.[50] The 500,000 MW and 2,000,000 MW dextrans are labeled with fluorescein, tetramethylrhodamine or Texas Red® dyes or with biotin, and all contain aldehyde-fixable lysine groups. The nonfluorescent 500,000 MW aminodextran (D7144) can be conjugated with the researcher's choice of amine-reactive reagents.[51]

Studying Intercellular Communication with Dextrans

The size of dextrans can be exploited to study connectivity between cells.[52] Examples include studies of the passage of 3000 MW dextrans through plasmodesmata[30] and modulation of gap junctional communication by transforming growth factor–β_1 and forskolin.[53] However, the dispersion of molecular weights in our "3000 MW" dextran preparations, which contain polymers with total molecular weights predominantly in the range of ~1500–3000 daltons but may also contain molecules <1500 daltons, may complicate such analyses.

An important experimental approach to identifying cells that form gap junctions makes use of simultaneous introduction of the polar tracer lucifer yellow CH (~450 daltons) and a tetramethylrhodamine 10,000 MW dextran. Because low molecular weight tracers like lucifer yellow CH (L453, L12926; Section 14.3) pass through gap junctions and dextrans do not, the initially labeled cell exhibits red fluorescence, whereas cells connected through gap junctions have yellow fluorescence[53] (Figure 14.5.8). This technique has been used to follow the loss of intercellular communication in adenocarcinoma cells,[54] to show the re-establishment of communication during wound healing in *Drosophila*[55] and to investigate intercellular communication at different stages in *Xenopus* embryos.[56,57] Simultaneous loading of cells with two (or more) dextrans that differ in both their molecular weight and in the dye's fluorescence properties has been used to assess subcellular heterogeneities in the submicroscopic structure of cytoplasm.[58]

IMPORTANT NOTICE: The products described in this manual are covered by one or more Limited Use Label License(s). Please refer to the Appendix on page 971 and Master Product List on page 975. Products are For Research Use Only. Not intended for any animal or human therapeutic or diagnostic use.
www.invitrogen.com/probes
molecular probes® | invitrogen™ by life technologies™

Probing Membrane Permeability with Dextrans

Labeled dextrans are often used to investigate vascular permeability and blood–brain barrier integrity.[59,60] Fluorescein dextrans with molecular weights ranging from 4000 to 150,000 daltons have been used to determine the effect of electroporation variables—pulse size, shape and duration—on plasma-membrane pore size in chloroplasts,[61] red blood cells[62] and fibroblasts.[63] Fluorescence recovery after photobleaching (FRAP) techniques have been used to monitor nucleocytoplasmic transport of fluorescent dextrans of various molecular weights, allowing the determination of the size-exclusion limit of the nuclear pore membrane,[64–66] as well as to study the effect of epidermal growth factor and insulin on the nuclear membrane and on nucleocytoplasmic transport.[67,68]

Microinjected 3000 MW fluorescent dextrans concentrate in interphase nuclei of *Drosophila* embryos, whereas 40,000 MW dextrans remain in the cytoplasm and enter the nucleus only after breakdown of the nuclear envelope during prophase. This size-exclusion phenomenon was used to follow the cyclical breakdown and reformation of the nuclear envelope during successive cell divisions.[69] Similarly, our 10,000 MW Calcium Green™ dextran conjugate (C3713, Section 19.4) was shown to diffuse across the nuclear membrane of isolated nuclei from *Xenopus laevis* oocytes, but the 70,000 MW and 500,000 MW conjugates could not.[70] Significantly, depletion of nuclear Ca^{2+} stores by inositol 1,4,5-triphosphate (Ins 1,4,5-P_3, I3716; Section 17.2) or by calcium chelators (Section 19.8) blocked nuclear uptake of the 10,000 MW Calcium Green™ dextran conjugate but not entry of lucifer yellow CH. Our 3000 MW Calcium Green™ dextran conjugate (C6765) is actively transported in adult nerve fibers over a significant distance and is retained in presynaptic terminals in a form that allows monitoring of presynaptic Ca^{2+} levels.[71]

Following Endocytosis with Dextrans

Fluorescent Dextrans

Tracing internalization of extracellularly introduced fluorescent dextrans is a standard method for analyzing fluid-phase endocytosis.[72–75] We offer dextrans with nominal molecular weights ranging from 3000 to 2,000,000 daltons, many of which can also be used as pinocytosis or phagocytosis markers (Table 14.4). Discrimination of internalized fluorescent dextrans from dextrans in the growth medium is facilitated by use of reagents that quench the fluorescence of the external probe. For example, most of our anti-fluorophore antibodies (Section 7.4, Table 7.8) strongly quench the fluorescence of the corresponding dyes.[76]

Negative staining produced by fluorescent dextrans that have been intracellularly infused via a patch pipette is indicative of nonendocytic vacuoles in live pancreatic acinar cells; extracellular addition of a second, color-contrasting dextran then allows discrimination of endocytic and nonendocytic vacuoles.[77] An *in vitro* assay for homotypic fusion of early endosomes has been described in which two cell populations are labeled with Alexa Fluor® 488 and Alexa Fluor® 594 10,000 MW dextrans (D22910,D22914) by fluid phase uptake, followed by subcellular fractionation and analysis of endosomal fluorescence colocalization.[78] Intracellular fusion of endosomes has also been monitored with a BODIPY® FL avidin conjugate by following the fluorescence enhancement that occurs when it complexes with a biotinylated dextran.[79]

pH Indicator Dextrans

Some of the dyes we use to prepare Molecular Probes® dextran conjugates exhibit fluorescence that is sensitive to the pH of the medium; pH indicator dextrans and their optical responses are described in detail in Section 20.4. Consequently, internalization of labeled dextrans into acidic organelles of cells can often be tracked by measuring changes in the fluorescence of the dye. The fluorescein dextrans (pK_a ~6.4) are frequently used to investigate endocytic acidification.[80,81] Fluorescence of fluorescein-labeled dextrans is strongly quenched upon acidification; however, fluorescein's lack of a spectral shift in acidic solution makes it difficult to discriminate between an internalized probe that is quenched and residual fluorescence of the external medium. Dextran conjugates that either shift their emission spectra in acidic environments, such as the SNARF® dextrans (Section 20.4), or undergo significant shifts of their excitation spectra, such as BCECF and Oregon Green® dextrans (Section 20.4), provide alternatives to fluorescein. The Oregon Green® 488 and Oregon Green® 514 dextrans exhibit a pK_a of approximately 4.7, facilitating measurements in acidic environments.[80,82] In addition to these pH indicator dextrans, we prepare a dextran that is double-labeled with fluorescein and tetramethylrhodamine (D1951; Section 20.4), which has been used as a ratiometric indicator (Figure 14.5.9) to measure endosomal acidification in Hep G2 cells[83] and murine alveolar macrophages.[84]

In contrast to fluorescein and Oregon Green® 488 dextrans, pHrodo™ 10,000 MW dextran (P10361) exhibits increasing fluorescence in response to acidification[84] (Figure 14.5.10). The minimal fluorescent signal from pHrodo™ dextran at neutral pH prevents the detection of noninternalized and nonspecifically bound conjugates and eliminates

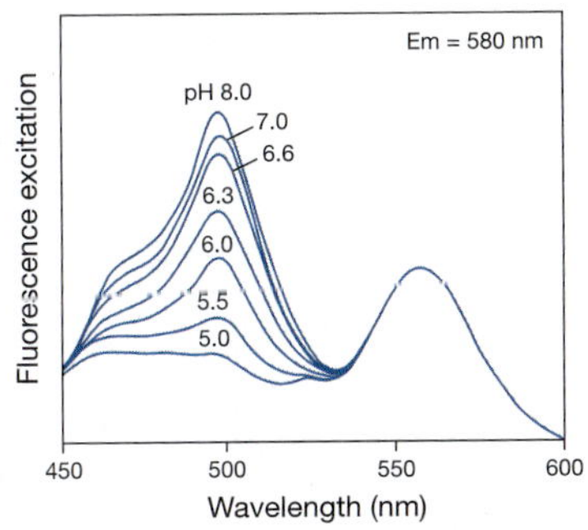

Figure 14.5.9 The excitation spectra of double-labeled fluorescein-tetramethylrhodamine dextran (D1951), which contains pH-dependent (fluorescein) and pH-independent (tetramethylrhodamine) dyes.

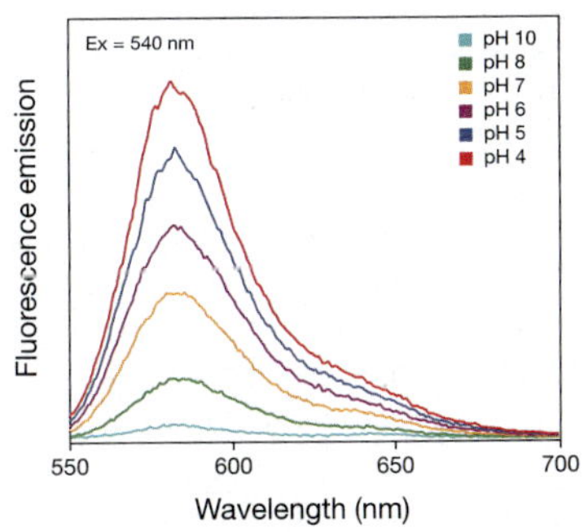

Figure 14.5.10 The pH sensitivity of pHrodo™ dextran. pHrodo™ 10,000 MW dextran (P10361) was reconsitituted in HEPES (20 mM)–buffered PBS and adjusted to pH values from pH 4 to pH 10. The intensity of fluorescence emission increases with increasing acidity, particularly in the pH 5–8 range.

IMPORTANT NOTICE: The products described in this manual are covered by one or more Limited Use Label License(s). Please refer to the Appendix on page 971 and Master Product List on page 975. Products are For Research Use Only. Not intended for any animal or human therapeutic or diagnostic use.

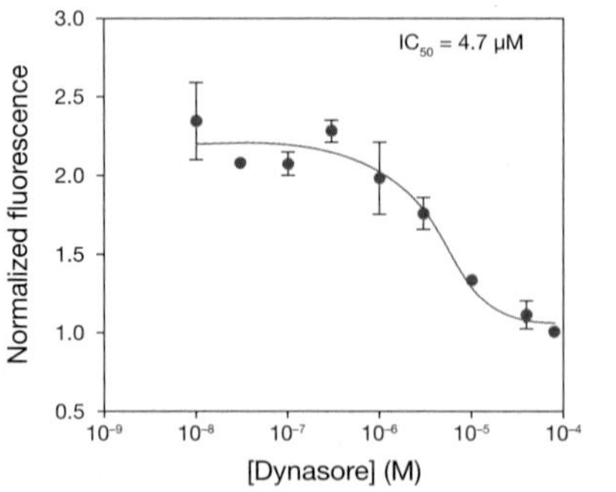

Figure 14.5.11 Tracking endocytosis inhibition with pHrodo™ dextran conjugates. HeLa cells were plated in 96-well format and treated with dynasore for 3 hours at 37°C prior to the pHrodo™ endocytosis assay. Next, 40 µg/mL of pHrodo™ 10,000 MW dextran (P10361) was incubated for 30 minutes at 37°C, and cells were then stained with HCS NuclearMask™ Blue Stain (H10325) for 10 minutes to reveal total cell number and demarcation for image analysis. Images were acquired on the BD Pathway 855 High-Content Bioimager (BD Biosciences).

the need for quenching reagents and extra wash steps, thus providing a simple fluorescent assay for endocytic activity. pHrodo™ dextran's excitation and emission maxima of 560 and 585 nm, respectively, facilitate multiplexing with other fluorophores including blue-, green- and far-red–fluorescent probes. Although pHrodo™ dextran is optimally excited at approximately 560 nm, it is also readily excited by the 488 nm spectral line of the argon-ion laser found on flow cytometers, confocal microscopes and imaging microplate readers (Figure 14.5.11).

Tracing Fluid Transport with Dextrans

Fluorescent dextrans are important tools for studying the hydrodynamic properties of the cytoplasmic matrix. The intracellular mobility of these fluorescent tracers can be investigated using fluorescence recovery after photobleaching (FRAP) techniques. We offer a range of dextran sizes, thus providing a variety of hydrodynamic radii for investigating both the nature of the cytoplasmic matrix and the permeability of the surrounding membrane. Because of their solubility and biocompatibility, fluorescent dextrans have been used to monitor *in vivo* tissue permeability and flow in the uveoscleral tract,[85,86] capillaries[87,88] and proximal tubules,[89] as well as diffusion of high molecular weight substances in the brain's extracellular environment.[90]

REFERENCES

1. J Neurosci Methods (2001) 109:81; **2.** Brain Res Bull (2000) 51:11; **3.** J Cell Sci (1990) 96:721; **4.** J Cell Biol (1987) 105:1981; **5.** Brain Res (1990) 526:127; **6.** Brain Res Bull (1990) 25:139; **7.** J Neurosci Methods (2009) 177:273; **8.** J Neurosci Methods (1998) 81:169; **9.** J Virol (2002) 76:11505; **10.** J Neurosci (1994) 14:5766; **11.** Plant Physiol (1992) 98:673; **12.** Plant Physiol (1990) 93:1492; **13.** Biotech Histochem (1992) 67:153; **14.** J Neurosci Methods (1996) 65:157; **15.** J Histochem Cytochem (1993) 41:777; **16.** Brain Res Bull (1993) 30:115; **17.** J Neurosci Methods (2009) 178:1; **18.** Science (2008) 321:417; **19.** Neuroscience (2007) 146:773; **20.** Trends Neurosci (1990) 13:14; **21.** Proc Natl Acad Sci U S A (1995) 92:11500; **22.** Mol Biol Cell (1995) 6:1491; **23.** J Neurosci Methods (1994) 53:35; **24.** Nature (1993) 363:630; **25.** Proc Natl Acad Sci U S A (1993) 90:1310; **26.** Nature (1990) 344:431; **27.** Dev Biol (1985) 109:509; **28.** J Cell Biol (1995) 128:293; **29.** J Neurosci Methods (1993) 50:95; **30.** Plant J (1993) 4:567; **31.** J Neurosci Methods (1994) 55:105; **32.** Brain Res (1993) 608:78; **33.** J Neurosci Methods (1997) 74:9; **34.** Glia (1997) 20:145; **35.** Hippocampus (1998) 8:57; **36.** J Neurosci (1995) 15:5222; **37.** J Neurosci (1995) 15:5139; **38.** J Neurosci Methods (1994) 53:23; **39.** J Neurosci Methods (1994) 52:153; **40.** Brain Res (1993) 607:47; **41.** J Neurosci Methods (1993) 48:75; **42.** J Neurosci Methods (1992) 45:35; **43.** J Neurosci Methods (1992) 41:239; **44.** J Neurosci Methods (2000) 103:23; **45.** Methods (2006) 39:228; **46.** Biophys J (1998) 75:2558; **47.** Neuroscience Protocols, Wouterlood FG, Ed. (1993) p.93-050-14; **48.** J Neurosci Methods (1997) 76:167; **49.** Methods Mol Biol (2008) 461:351; **50.** Nature (1993) 361:451; **51.** Nat Mater (2005) 4:759; **52.** Biophys J (2008) 94:840; **53.** J Neurosci (1995) 15:262; **54.** Proc Natl Acad Sci U S A (1988) 85:473; **55.** Dev Biol (1988) 127:197; **56.** J Cell Biol (1990) 110:115; **57.** Dev Biol (1988) 129:265; **58.** J Immunol (1996) 157:3396; **59.** Cell (2005) 123:133; **60.** Br J Pharmacol (2003) 140:1201; **61.** Biophys J (1990) 58:823; **62.** Bioelectrochem Bioenerg (1988) 20:57; **63.** Biotechniques (1988) 6:550; **64.** J Cell Biol (1986) 102:1183; **65.** EMBO J (1984) 3:1831; **66.** J Biol Chem (1983) 258:11427; **67.** J Cell Biol (1990) 110:559; **68.** Biochemistry (1987) 26:1546; **69.** Biotechniques (1994) 17: 730; **70.** Science (1995) 270:1835; **71.** Neurosci Lett (1998) 258:124; **72.** Annu Rev Biochem (2009) 78:857; **73.** Cytometry A (2009) 75:941; **74.** J Biol Chem (2008) 283:6764; **75.** Traffic (2008) 9:1801; **76.** Am J Physiol (1990) 258:C309; **77.** Am J Physiol Gastrointest Liver Physiol (2007) 293:G1333; **78.** Proc Natl Acad Sci U S A (2006) 103:2701; **79.** Biophys J (1996) 71:487; **80.** Cell Microbiol (2006) 8:781; **81.** J Cell Sci (2006) 119:1016; **82.** Methods Enzymol (2009) 453:417; **83.** J Cell Biol (1995) 130:821; **84.** J Biol Chem (2009) 284:35926; **85.** Proc Natl Acad Sci U S A (1988) 85:2315; **86.** Arch Ophthalmol (1987) 105:844; **87.** Microvasc Res (1988) 36:172; **88.** Am J Physiol (1983) 245:H495; **89.** Am J Physiol (1987) 253:F366; **90.** Proc Natl Acad Sci U S A (2006) 103:5567.

PRODUCT LIST 14.5 FLUORESCENT AND BIOTINYLATED DEXTRANS

Cat. No.	Product	Quantity
D34682	dextran, Alexa Fluor® 488; 3,000 MW, anionic	2 mg
D22910	dextran, Alexa Fluor® 488; 10,000 MW, anionic, fixable	5 mg
D22911	dextran, Alexa Fluor® 546; 10,000 MW, anionic, fixable	5 mg
D34679	dextran, Alexa Fluor® 555; 10,000 MW, anionic, fixable	5 mg
D22912	dextran, Alexa Fluor® 568; 10,000 MW, anionic, fixable	5 mg
D22913	dextran, Alexa Fluor® 594; 10,000 MW, anionic, fixable	5 mg
D22914	dextran, Alexa Fluor® 647; 10,000 MW, anionic, fixable	2 mg
D34681	dextran, Alexa Fluor® 680; 3,000 MW, anionic	2 mg
D34680	dextran, Alexa Fluor® 680; 10,000 MW, anionic, fixable	5 mg

IMPORTANT NOTICE: The products described in this manual are covered by one or more Limited Use Label License(s). Please refer to the Appendix on page 971 and Master Product List on page 975. Products are For Research Use Only. Not intended for any animal or human therapeutic or diagnostic use.

PRODUCT LIST 14.5 FLUORESCENT AND BIOTINYLATED DEXTRANS—*continued*

Cat. No.	Product	Quantity
D3330	dextran, amino, 3000 MW	100 mg
D1860	dextran, amino, 10,000 MW	1 g
D1861	dextran, amino, 40,000 MW	1 g
D1862	dextran, amino, 70,000 MW	1 g
D7144	dextran, amino, 500,000 MW	100 mg
D7135	dextran, biotin, 3000 MW, lysine fixable (BDA-3000)	10 mg
D1956	dextran, biotin, 10,000 MW, lysine fixable (BDA-10,000)	25 mg
D1957	dextran, biotin, 70,000 MW, lysine fixable (BDA-70,000)	25 mg
D7142	dextran, biotin, 500,000 MW, lysine fixable (BDA-500,000)	10 mg
D7168	dextran, BODIPY® FL, 10,000 MW, fixable	5 mg
D7132	dextran, Cascade Blue®, 3000 MW, anionic, lysine fixable	10 mg
D1976	dextran, Cascade Blue®, 10,000 MW, anionic, lysine fixable	25 mg
D3305	dextran, fluorescein, 3000 MW, anionic	10 mg
D3306	dextran, fluorescein, 3000 MW, anionic, lysine fixable	10 mg
D1821	dextran, fluorescein, 10,000 MW, anionic	25 mg
D1820	dextran, fluorescein, 10,000 MW, anionic, lysine fixable (fluoro-emerald)	25 mg
D1844	dextran, fluorescein, 40,000 MW, anionic	25 mg
D1845	dextran, fluorescein, 40,000 MW, anionic, lysine fixable	25 mg
D1823	dextran, fluorescein, 70,000 MW, anionic	25 mg
D1822	dextran, fluorescein, 70,000 MW, anionic, lysine fixable	25 mg
D7136	dextran, fluorescein, 500,000 MW, anionic, lysine fixable	10 mg
D7137	dextran, fluorescein, 2,000,000 MW, anionic, lysine fixable	10 mg
D7156	dextran, fluorescein and biotin, 3000 MW, anionic, lysine fixable (micro-emerald)	5 mg
D7178	dextran, fluorescein and biotin, 10,000 MW, anionic, lysine fixable (mini-emerald)	10 mg
D1825	dextran, lucifer yellow, 10,000 MW, anionic, lysine fixable	25 mg
D7170	dextran, Oregon Green® 488; 10,000 MW, anionic	5 mg
D7171	dextran, Oregon Green® 488; 10,000 MW, anionic, lysine fixable	5 mg
D7172	dextran, Oregon Green® 488; 70,000 MW, anionic	5 mg
D7173	dextran, Oregon Green® 488; 70,000 MW, anionic, lysine fixable	5 mg
D7176	dextran, Oregon Green® 514; 70,000 MW, anionic	5 mg
D1824	dextran, rhodamine B, 10,000 MW, neutral	25 mg
D1841	dextran, rhodamine B, 70,000 MW, neutral	25 mg
D7163	dextran, Rhodamine Green™, 3000 MW	5 mg
D7153	dextran, Rhodamine Green™, 10,000 MW, lysine fixable	10 mg
D3307	dextran, tetramethylrhodamine, 3000 MW, anionic	10 mg
D3308	dextran, tetramethylrhodamine, 3000 MW, anionic, lysine fixable	10 mg
D1868	dextran, tetramethylrhodamine, 10,000 MW, anionic, fixable	25 mg
D1817	dextran, tetramethylrhodamine, 10,000 MW, lysine fixable (fluoro-ruby)	25 mg
D1816	dextran, tetramethylrhodamine, 10,000 MW, neutral	25 mg
D1842	dextran, tetramethylrhodamine, 40,000 MW, neutral	25 mg
D1818	dextran, tetramethylrhodamine, 70,000 MW, lysine fixable	25 mg
D1819	dextran, tetramethylrhodamine, 70,000 MW, neutral	25 mg
D7139	dextran, tetramethylrhodamine, 2,000,000 MW, lysine fixable	10 mg
D7162	dextran, tetramethylrhodamine and biotin, 3000 MW, lysine fixable (micro-ruby)	5 mg
D3312	dextran, tetramethylrhodamine and biotin, 10,000 MW, lysine fixable (mini-ruby)	10 mg
D3328	dextran, Texas Red®, 3000 MW, lysine fixable	10 mg
D3329	dextran, Texas Red®, 3000 MW, neutral	10 mg
D1863	dextran, Texas Red®, 10,000 MW, lysine fixable	25 mg
D1828	dextran, Texas Red®, 10,000 MW, neutral	25 mg
D1829	dextran, Texas Red®, 40,000 MW, neutral	25 mg
D1864	dextran, Texas Red®, 70,000 MW, lysine fixable	25 mg
D1830	dextran, Texas Red®, 70,000 MW, neutral	25 mg
N7167	NeuroTrace® BDA-10,000 Neuronal Tracer Kit	1 kit
P10361	pHrodo™ dextran, 10,000 MW *for endocytosis*	0.5 mg

IMPORTANT NOTICE: The products described in this manual are covered by one or more Limited Use Label License(s). Please refer to the Appendix on page 971 and Master Product List on page 975. Products are For Research Use Only. Not intended for any animal or human therapeutic or diagnostic use.

14.6 Microspheres and Qdot® Nanocrystals for Tracing

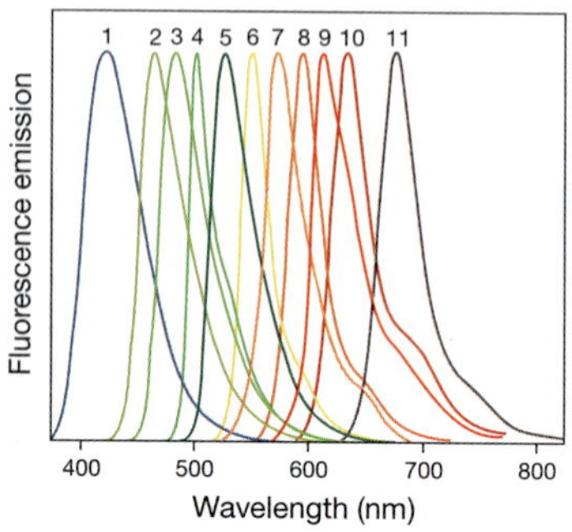

Figure 14.6.1 Normalized fluorescence emission spectra of the dyes contained in FluoSpheres® polystyrene microspheres for blood flow determination, after extraction into 2-ethoxyethyl acetate (Cellosolve® acetate). The eleven colors of fluorescent microspheres represented are: **1)** blue, **2)** blue-green, **3)** green, **4)** yellow-green, **5)** yellow, **6)** orange, **7)** red-orange, **8)** red, **9)** carmine, **10)** crimson and **11)** scarlet.

Fluorescent Microspheres for Regional Blood Flow Studies

FluoSpheres® and TransFluoSpheres® polystyrene microspheres satisfy several prerequisites of ideal long-term biological tracers. Because the dyes in our microspheres are incorporated throughout the microsphere rather than just on its surface, the fluorescence output per microsphere is significantly greater than that obtained from protein or dextran conjugates (Table 6.6) and is relatively immune to photobleaching and other environment-dependent effects. FluoSpheres® and TransFluoSpheres® microspheres are also biologically inert and physically durable, and they are available with a large number of uniform sizes and surface properties. Furthermore, their spectral properties can be freely manipulated during manufacture without altering their surface properties. See Section 6.5 for an extensive discussion of the properties of our FluoSpheres® and TransFluoSpheres® polystyrene beads.

Measuring the effect of various interventions on regional blood flow is an important quantitative application of fluorescent microspheres. Relatively large radiolabeled microspheres (10–15 µm in diameter) have long been used for regional blood flow studies in tissues and organs. Fluorescent microspheres, however, have been shown to be superior to radioactive microspheres in chronic blood flow measurements.[1] In most cases, the microspheres are injected at desired locations in the circulatory system and eventually lodge in the capillaries, where they can later be counted in dissected tissue sections. To eliminate the hazards, expense and disposal problems of the radiolabeled microspheres,[2] researchers have turned to fluorescent and colored microspheres for measuring myocardial and cortical blood flow.[3–6] Blood flow measurements using fluorescent microspheres in other organs, including the kidney,[7,8] lung,[9–15] spleen,[16] adrenal glands,[17] bone[18,19] and teeth,[20] are equally feasible. Our 0.2 µm FluoSpheres® microspheres have been used to trace new blood vessel development, which is important for the study of tumor angiogenesis and microvascular continuity.[9]

FluoSpheres® Microspheres for Blood Flow Determination

We have used Molecular Probes® fluorescent dye technology to produce a range of intensely fluorescent FluoSpheres® microspheres specifically designed for regional blood flow determination (Table 14.5). Regional blood flow studies using our FluoSpheres® polystyrene microspheres for blood flow determination (Figure 14.6.1) have been validated in several side-by-side comparisons with radioactively labeled microspheres.[21–25] The two methods exhibit equivalent detection

Table 14.5 FluoSpheres® microspheres for blood flow determination.

	Bead Diameter	
Color (Ex/Em *)	**15 µm**	**10 µm**
Blue (365/415)	F8837	F8829
Blue-green (430/465)	F8838	F8830
Green (450/480)	F21010	
Yellow-green (505/515)	F8844	F8836
Yellow (515/534)	F21011	
Orange (540/560)	F8841	F8833
Red-orange (565/580)	F21012	
Red (580/605)	F8842	F8834
Carmine (580/620)	F21013	
Crimson (625/645)	F8839	F8831
Scarlet (645/680)	F8843	

* Fluorescence excitation and emission maxima of the encapsulated dye measured in 2-ethoxyethyl acetate. All FluoSpheres® microspheres for blood flow determination are supplied as suspensions containing 1.0×10^6 beads/mL (15 µm microspheres) or 3.6×10^6 beads mL (10 µm microspheres) in a unit size of 10 mL.

Table 14.6 FluoSpheres® blood flow and color kits.

Cat. No.	Application	Bead Size	Concentration	Number of Colors
F8890	Blood flow determination	10 µm	3.6×10^6 beads/mL	7 colors, 10 mL each
F8891	Blood flow determination	15 µm	1.0×10^6 beads/mL	7 colors, 10 mL each
F8892	Blood flow determination	15 µm	1.0×10^6 beads/mL	5 colors, 10 mL each
F21015	Blood flow determination	15 µm	1.0×10^6 beads/mL	4 colors, 10 mL each
F10720	Microinjection/tracing	0.04 µm	5% solids	4 colors, 1 mL each

Table 14.7 FluoSpheres® and TransFluoSpheres® microspheres for tracing.

	Size		
Color (Ex/Em *)	**0.04 µm**	**0.2 µm**	**1.0 µm**
Blue-green (430/465)			F13080 †
Yellow-green (505/515)	F8795 ‡		F13081 †
Orange (540/560)	F8792 ‡		F13082 †
Red-orange (565/580)	F8794 ‡		
Red (580/605)	F8793 ‡		F13083 †
Dark red (660/680)	F8789 ‡		
Europium (365/610)	F20880 §	F20881 §	
Platinum (390/650)	F20886 §		

* Fluorescence or luminescence excitation and emission maxima, in nm. † 1×10^{10} beads/mL, 5 mL unit size. ‡ 5% solids, 1.0 mL unit size. § 0.5% solids, 2.0 mL unit size.

IMPORTANT NOTICE: The products described in this manual are covered by one or more Limited Use Label License(s). Please refer to the Appendix on page 971 and Master Product List on page 975. Products are For Research Use Only. Not intended for any animal or human therapeutic or diagnostic use.

sensitivity, and excellent correlation between the flow measurements has been reported. In addition, techniques have been developed to extract the microspheres and the fluorescent dyes they contain from tissue samples, allowing blood flow quantitation to be performed using readily available instrumentation such as spectrofluorometers and fluorescence microplate readers.[5] Up to seven simultaneously circulating tracers can be discriminated based on uncompensated fluorescent color differences. With spectral spillover corrections, the number of simultaneously detectable tracers can be increased to thirteen.[26] Microspheres that have been perfused or injected into tissues and organs retain their fluorescence following histological serial sectioning.[3,4] Our FluoSpheres® microspheres for blood flow determination, which are available in eleven distinguishable fluorescent colors (Figure 14.6.1), are also compatible with blood flow analyzer systems that perform automated extraction and analysis.[27]

FluoSpheres® Color Kits for Regional Blood Flow Studies

We offer four different FluoSpheres® Blood Flow Determination Fluorescent Color Kits (Table 14.6):

- **Kit #1 (F8890)** contains 10 mL vials of 10 µm microspheres in seven fluorescent colors (blue, blue-green, yellow-green, orange, red, crimson and scarlet).
- **Kit #2 (F8891)** contains 10 mL vials of 15 µm microspheres in seven fluorescent colors (blue, blue-green, yellow-green, orange, red, crimson and scarlet).
- **Kit #3 (F8892)** contains 10 mL vials of 15 µm microspheres in five fluorescent colors (blue-green, yellow-green, orange, red and crimson).
- **Kit #4 (F21015)** contains 10 mL vials of 15 µm microspheres in four additional fluorescent colors (green, yellow, carmine and red-orange), which are spectrally distinguishable from the dyes in Kits #2 and #3 (Figure 14.6.1).

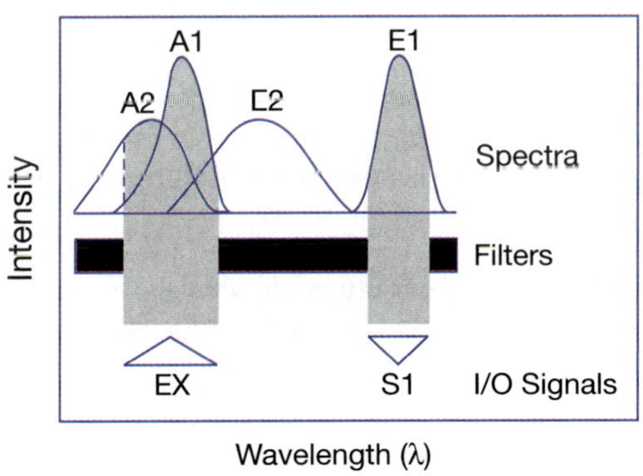

Figure 14.6.2 Schematic diagram of the advantages of the large Stokes shift exhibited by our TransFluoSpheres® beads. A1 and E1 represent the absorption and emission bands of a typical TransFluoSpheres® bead. The large separation of the absorption and emission maxima (Stokes shift) is characteristic of our TransFluoSpheres® beads. Unlike most low molecular weight fluorescent dyes, which show considerable overlap of their absorption and emission spectra, the TransFluoSpheres® beads can be excited (EX) across the entire absorption band A1 and the resulting fluorescence can be detected across the full emission band E1, thereby allowing the researcher to maximize the signal (S1). Moreover, because of the large Stokes shifts of the TransFluoSpheres® beads, researchers can often avoid problems associated with autofluorescence. The absorption and emission bands of a typical autofluorescent component are represented in this figure by A2 and E2. Although the endogenous fluorescent species will be excited simultaneously with the TransFluoSpheres® beads, the resulting emission (E2) does not coincide with E1 and is therefore readily rejected by suitably chosen optical filters.

The aqueous suspensions of 10 µm and 15 µm beads contain 3.6 million and 1 million microspheres per mL, respectively. All kits include a detailed protocol. Each of the colors of FluoSpheres® microspheres for blood flow studies can also be purchased separately (Table 14.5).

Fluorescent Microsphere Resource Center

More information, including a detailed applications manual on the use of fluorescent microspheres for blood flow determination, is available from the Fluorescent Microsphere Resource Center (FMRC) at the University of Washington, Seattle (fmrc.pulmcc.washington.edu).

Particle and Cell Tracking with Fluorescent Microspheres

Availability of intensely fluorescent, highly uniform microspheres in different colors and sizes permits diverse applications in tracking particles and cells, tracing fluid dynamics and amplifying signals. Using a mixture of beads of different sizes, each labeled with a different fluorescent color, researchers can discriminate the size dependence of uptake or transport of microspheres *in vivo* in cells, capillaries, lung or other tissues.[28] In addition to our microspheres specially designed for blood flow studies, we offer a wide range of fluorescent FluoSpheres® carboxylate-modified microspheres in different fluorescent colors, bead diameters and surface functional groups (see Table 6.7 in Section 6.5 for a complete list). Of particular interest are our FluoSpheres® beads with 0.04 µm diameters and FluoSpheres® beads for tracer studies with 1 µm diameters, which have been specially formulated to perform well in cell tracking and particle tracking studies, respectively. Our smallest microspheres can be microinjected into cells (see below) or are actively taken up by phagocytosis (Section 16.1).

FluoSpheres® Beads with 0.04 µm Diameters for Microinjection

Unlike our other fluorescent microspheres, most of which come in aqueous suspensions containing 2% solids and 2 mM sodium azide as a preservative, FluoSpheres® beads with 0.04 µm diameters are prepared as aqueous suspensions containing 5% solids without preservatives. At more than double the concentration of our standard FluoSpheres® microspheres, these carboxylate-modified 0.04 µm beads (F8795, F8792, F8794, F8793, F8789, F8794) are well suited to applications requiring microinjectable tracers. Yellow-green–, orange-, red-orange–, red- and dark red–fluorescent colors are available in the FluoSpheres® Fluorescent Color Kit (F10720). In many biological systems, the concentrated fluorescence and spherical shape of the FluoSpheres® beads permit them to be detected against a relatively high but diffuse background fluorescence. However, our TransFluoSpheres® microspheres (Table 14.7)—microspheres with extremely large Stokes shifts—are preferred for some studies because their fluorescence may be better resolved from the tissue's autofluorescence (Figure 14.6.2).

FluoSpheres® Beads with 1.0 µm Diameters for Tracer Studies

Our FluoSpheres® fluorescent microspheres for tracer studies are 1.0 µm polystyrene beads for analysis by tissue extraction (Table 14.7). Because the dye content of these microspheres is much higher than that of other fluorescent microspheres, stronger signals can be generated using fewer microspheres per tracing experiment. These heavily

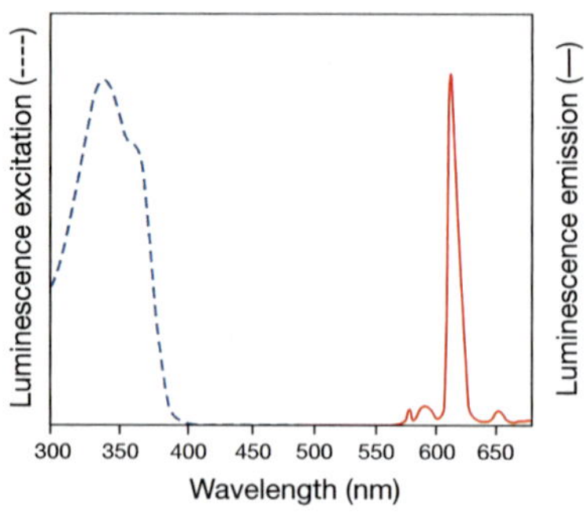

Figure 14.6.3 Fluorescence excitation and emission maxima of the FluoSpheres® europium luminescent microspheres.

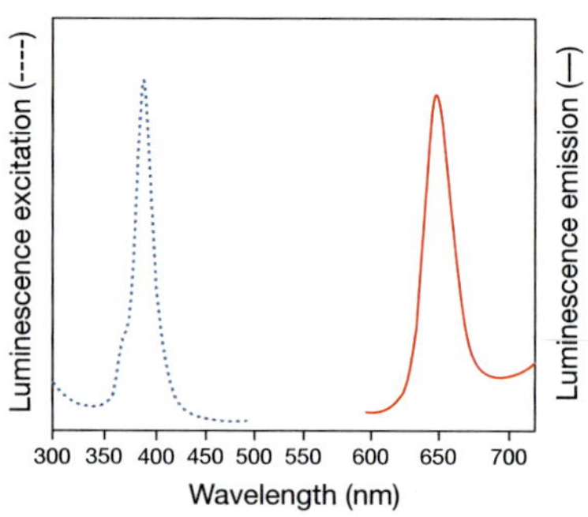

Figure 14.6.4 Luminescence excitation and emission spectra of the FluoSpheres® platinum luminescent microspheres.

dye-loaded beads are available in four well-resolved fluorescent colors: blue-green (F13080), yellow-green (F13081), orange (F13082) and red (F13083), with excitation/emission maxima at approximately 430/465, 505/515, 540/560 and 580/605 nm, respectively (the emission spectra are shown as peaks 2, 4, 6 and 8 of Figure 14.6.1). Aerosols containing these fluorescent microspheres have been used to acquire high-resolution maps of regional pulmonary ventilation.[14] Transport of these fluorescent microspheres through tissues can be determined using methods that have been developed for regional blood flow determination or by confocal laser-scanning microscopy.[29] As with the FluoSpheres® beads for blood flow determination, the microspheres and the fluorescent dyes they contain are first extracted from the tissue sample, and then the fluorescence is quantitated on a spectrofluorometer or fluorescence microplate reader.

SAIVI™ 715 Injectable Contrast Agents with 0.1 µm and 2 µm Diameters

SAIVI™ 715 injectable contrast agents (S31201, S31203) are specially formulated for small-animal *in vivo* imaging of regions of inflammation, blood pooling and wound healing. These contrast agents are polymeric microspheres that have been labeled with a fluorescent dye; each microsphere particle contains many dye molecules protected within the polymer sphere. Their use for *in vivo* imaging offers many advantages over existing contrast agents, including 1) no known intrinsic toxicity, 2) a high degree of localization within diseased vasculature and 3) longer *in vivo* residence times than organic dye–labeled proteins. Formulated to resist liver accumulation, SAIVI™ injectable contrast agents have been observed to circulate throughout the blood and to accumulate in tissues that exhibit damaged, excessive or otherwise abnormal blood vessel development as part of the disease process. These agents are optimized for long-wavelength emission and tested by *in vivo* imaging after injection in disease models established in mice. We also offer a set of complementary contrast agents (SAIVI™ Alexa Fluor® 680 and SAIVI™ Alexa Fluor® 750 conjugates of transferrin and albumin, Section 14.7) for rapid imaging of early-onset events.

Europium and Platinum Luminescent Microspheres for Time-Resolved Fluorometry

Detecting a low level of protein or DNA targets in a tissue sample or on a membrane using classic fluorochromes is sometimes difficult and prone to errors because specific fluorescence signals tend to be low and are usually mixed with nonspecific signals and autofluorescence. One approach to improve detectability is the use of time-resolved luminescence reagents, such as our FluoSpheres® europium luminescent microspheres and FluoSpheres® platinum luminescent microspheres (Table 14.8). The FluoSpheres® europium beads contain Eu^{3+} coordination complexes with luminescence decay times of >600 microseconds—much longer than the <50 nanosecond decay time of conventional fluorophores and autofluorescence. The luminescence of the Pt^{2+} chelate in the FluoSpheres® platinum luminescent microspheres has a decay time of >40 microseconds. Thus, time-gated fluorescence detection using these microspheres results in complete rejection of autofluorescence signals.[30–38] In addition, the europium luminescent microspheres feature long-wavelength emission (610–650 nm) that is well separated from their excitation peak (340–390 nm) (Figure 14.6.3). The platinum luminescent microspheres are maximally excited near 390 nm with narrow emission that is maximal near 650 nm (Figure 14.6.4). Because of these unusually large Stokes shifts, filter combinations can be chosen that effectively isolate the desired luminescence signal. The narrow emissions and different lifetimes permit simultaneous

Table 14.8 Molecular Probes® europium and platinum luminescent FluoSpheres® microspheres.

	Size	
Type of Microsphere*	**0.04 µm**	**0.2 µm**
Carboxylate-Modified Microspheres		
Europium (365/610)	F20880	F20881
Platinum (390/650)	F20886	
NeutrAvidin™-Coated Microspheres †		
Europium (365/610)	F20883	F20884

* Luminescent FluoSpheres® microspheres are supplied as aqueous suspensions containing 0.5% solids. † See Section 7.6 for more information on our avidin and streptavidin conjugates.

IMPORTANT NOTICE: The products described in this manual are covered by one or more Limited Use Label License(s). Please refer to the Appendix on page 971 and Master Product List on page 975. Products are For Research Use Only. Not intended for any animal or human therapeutic or diagnostic use.
www.invitrogen.com/probes
molecular probes® | invitrogen™ by life technologies™

use of the europium and platinum luminescent microspheres as tracers. These microspheres are available with nominal diameters of 0.04 µm or 0.2 µm (Table 14.7).

Microspheres for Monitoring Airflow

A cubic foot of air may contain hundreds of thousands of airborne particles such as viruses, bacteria, pollen, mold spores and gaseous chemicals (Figure 14.6.5). To combat harmful airborne particles, many types of air filters and instruments for detecting airborne particles have been developed for medical and defense purposes. Because most airborne particles are small (typically 10–150 nm, below microscopic resolution) and are difficult to detect, verification of the function of these filters and instruments can be a challenge. Our FluoSpheres® microspheres are specially stained fluorescent polystyrene beads that can mimic airborne particles[14] and be used as unique markers to verify equipment reliability. A single "virus-sized" 20 nm diameter FluoSpheres® microsphere carries 100–200 fluorophore molecules and thus emits sufficiently bright fluorescence for easy detection.

Transplantation and Migration Studies with Fluorescent Microspheres

Labeling cells with fluorescent microspheres prior to transplantation enables researchers to distinguish cell types and analyze graft migration in the host over extended time periods. Unlike other tracers, most of which rapidly diffuse or leach from their site of application in tissues, fluorescent microspheres tend to remain in place for periods of at least months. Their stability and biological inertness give them considerable potential for transplantation studies. Moreover, the intense fluorescence, high uniformity and low debris content of our FluoSpheres® polystyrene microspheres will circumvent many of the problems encountered in use of fluorescent microspheres from other sources. Cells are generally labeled with microspheres by microinjection or other invasive methods (Table 14.1). Potential problems of sterility, unequal uptake and translocation of the beads by the cells need to be considered before using fluorescent microspheres in transplantation and other cell tracing studies.[39] It has been reported that polystyrene microspheres for transplantation studies can be sterilized by heating below their softening temperature.[40]

Neuronal Tracing with Fluorescent Microspheres

Katz, the first scientist to use fluorescent microspheres as neuronal tracers, demonstrated that rhodamine-labeled microspheres could undergo retrograde transport.[41] Although the mechanism of microsphere transport is not completely understood, the process can apparently be facilitated by using a high concentration of particles with small diameters (<0.05 µm) and high negative surface-charge densities.[42–44] Fluorescent microspheres are also suitable for retrograde tracing because they are not cytotoxic and persist for extraordinarily long periods in nerve cells. The fluorescence intensity of microsphere-labeled neuronal perikarya in rats was found to be undiminished one year after injection.[45] Adsorption of neuroactive proteins, including neurotrophins, on small-diameter fluorescent microspheres provides a means of locating their *in vivo* and *in vitro* microinjection sites and of following retrograde transport in the central nervous system.[46] Our fluorescent microspheres are also becoming important tracers in ophthalmological studies[47–49] where they undergo retrograde transport.[50,51]

Our red-fluorescent, carboxylate-modified FluoSpheres® beads[52,53] (F8793) are retrogradely transported in rat lumbosacral ventral root axons that have been subjected to peripheral crush injury.[54] In this study, the fluorescent beads were used to photoconvert diaminobenzidine (DAB) into an insoluble, electron-dense reaction product in order to facilitate ultrastructural analysis by electron microscopy (Fluorescent Probes for Photoconversion of Diaminobenzidine Reagents—Note 14.2).

Following Phagocytosis with Fluorescent Microspheres

Many cell types actively ingest opsonized or nonopsonized fluorescent microspheres (Section 16.1). The preferred microspheres are about 0.5–2 µm in diameter. Confocal laser-scanning microscopy can distinguish between ingested beads and those simply bound to the surface.[55] Flow cytometry, fluorescence microscopy and fluorescence spectrometry have been used to quantitate phagocytosis by macrophage cells[56–60] and protozoan cells.[61] Macrophage cells in primary cultures of rat cerebral cortex have been identified by their ability to selectively phagocytose fluorescent microspheres;[62] macrophage cells have also been sorted based on the absolute number of microspheres phagocytosed.[60] Fluorescent microsphere uptake has been used as a model for alveolar macrophage cell translocation and clearance of inhaled aerosols containing environmental particulates.[29,63,64] As with our fluorescent microspheres for blood flow determination (see above), solvent extraction of the fluorescent dye from the beads can be used to quantitate microsphere uptake by macrophage cells.[65]

For the study of collagen phagocytosis, we manufacture yellow-green–fluorescent FluoSpheres® collagen I–labeled microspheres in either 1.0 µm or 2.0 µm diameters (F20892, F20893; Section 16.1). Fibroblasts phagocytose and then intracellularly digest collagen. These activities play an important role in the remodeling of the extracellular matrix during normal physiological turnover of connective tissues, in development, in wound repair and possibly in aging and various disorders. A well-established procedure for observing collagen phagocytosis by either flow cytometry or fluorescence microscopy involves the use of collagen-coated fluorescent microspheres, which attach to the cell surface and become engulfed by fibroblasts.[66–68]

Tracking the Movement of Proteins and Other Biomolecules with Fluorescent Microspheres

FluoSpheres® microspheres can serve as bright, inert and extremely photostable labels for tracking particle movement and other dynamic processes over extended time periods. The intense fluorescence of our

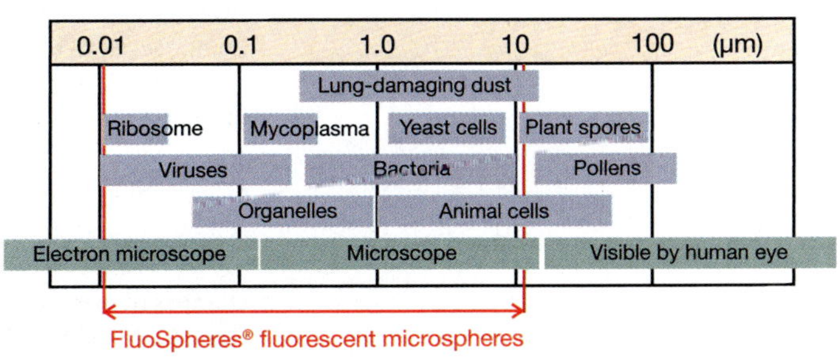

Figure 14.6.5 FluoSpheres® microsphere sizes relative to common airborne particles.

IMPORTANT NOTICE: The products described in this manual are covered by one or more Limited Use Label License(s). Please refer to the Appendix on page 971 and Master Product List on page 975. Products are For Research Use Only. Not intended for any animal or human therapeutic or diagnostic use.

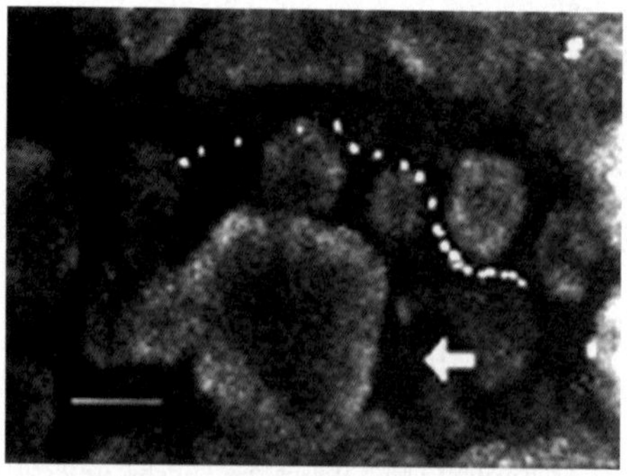

Figure 14.6.6 Superimposed time sequence image (2.3-second intervals) showing a single bead moving through a biofilm channel. Autofluorescence of the cell clusters can be seen as lighter areas relative to the channels. The direction of bulk fluid flow is indicated by the arrow. Photo contributed by Paul Stoodley, Exeter University, and Dirk DeBeer, Max Planck Institute of Marine Biology.

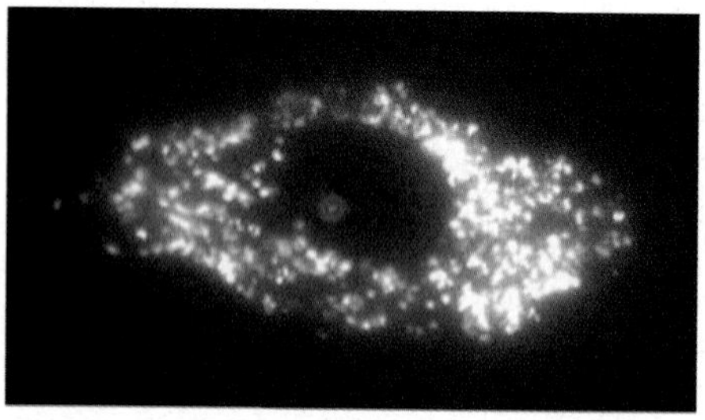

Figure 14.6.7 Distribution of Qdot® nanocrystals in cytoplasmic vesicles after labeling cells with the Qtracker® 655 Cell Labeling Kit (Q25021MP). HeLa cells were labeled with the Qtracker® 655 Cell Labeling Kit and then observed using a Leica TCS SP2 confocal microscope (excitation at 488 nm). This representative image shows the Qdot® nanocrystals distributed in vesicles throughout the cytoplasm.

FluoSpheres® beads permits the detection and tracking of very small single particles in three dimensions.[69–71] Examples of the diverse applications of fluorescent microspheres include studies designed to track or quantitate:

- Binding of kinetochores to microtubules *in vitro*[71,72]
- Brownian motion in protein[73] and dextran solutions[74]
- Exchangeable GTP-binding sites on paclitaxel-stabilized fluorescent microtubules[75]
- Direction and rate of axonal transport in the squid giant axon[76]
- Fluid dynamics and flow in mucus[77] and peripheral lymph[78]
- Injection sites in tissues[79,80]
- Lateral diffusion of lipids and receptors in membranes[70,81–83]
- Movement of microinjected microspheres poleward during mitosis of sea urchin eggs[84] and sand dollar eggs[85]
- Particle flow through a model biofilm containing *Pseudomonas aeruginosa, P. fluorescens* and *Klebsiella pneumoniae* using confocal laser-scanning microscopy[86] (Figure 14.6.6)
- Sentinel lymph node detection[87]
- Three-dimensional motion of microspheres in order to assess water permeability in individual Chinese hamster ovary (CHO) cells expressing CHIP28 water channels[69]

Fluorescent microspheres have been used with optical tweezers[88] to control the movement of single myosin filaments[89] or kinesin molecules[90] and for imaging at suboptical resolution by scanning luminescence X-ray microscopy.[91] In addition, our 0.1 µm red-fluorescent carboxylate-modified microspheres (F8801, Section 6.5) have been used to model particle distribution and penetration of adenovirus-mediated gene transfer in human bronchial submucosal glands using xenografts.[92]

Qdot® Nanocrystal Tracers

Qtracker® Cell Labeling Kits

Qtracker® Cell Labeling Kits provide spectrally distinct Qdot® nanocrystals that have been functionalized on their surface with polyarginine peptides to facilitate spontaneous uptake by live cells.[93] We offer the following eight kits:

- Qtracker® 525 Cell Labeling Kit (Q25041MP)
- Qtracker® 565 Cell Labeling Kit (Q25031MP)
- Qtracker® 585 Cell Labeling Kit (Q25011MP)
- Qtracker® 605 Cell Labeling Kit (Q25001MP)
- Qtracker® 625 Cell Labeling Kit (A10198)
- Qtracker® 655 Cell Labeling Kit (Q25021MP)
- Qtracker® 705 Cell Labeling Kit (Q25061MP)
- Qtracker® 800 Cell Labeling Kit (Q25071MP)

The mechanism whereby such cell penetrating peptides induce the cellular uptake of cargoes including oligonucleotides and proteins as well as Qdot® nanocrystals remains a topic of active investigation and considerable debate.[94,95] It is, however, reasonably well established that the uptake is passive and does not require specific receptors or active transport processes. Therefore, labeling of live cells is a simple matter of adding the Qtracker® cell labeling reagents supplied in the kits to adherent or suspension cells in complete growth medium, incubating at 37°C for 45–60 minutes, followed by a wash step to remove the labeling reagents. The Qdot® nanocrystals are stably incorporated in cytoplasmic vesicles (Figure 14.6.7) and are passed to daughter cells upon cell division but are not transferred to adjacent cells in mixed cultures or host tissues. Fluorescence is not impacted by complex and varying cellular environments including changes in intracellular pH, temperature and metabolic activity. A wide range of applications for Qtracker® Cell Labeling Kits have been reported, including fiducial marking of cell populations for identification during or after downstream analysis,[93,96] labeling of mouse cerebral vascular tissue for visualization of explantation,[97] imaging the assembly of cultured pulmonary artery

IMPORTANT NOTICE: The products described in this manual are covered by one or more Limited Use Label License(s). Please refer to the Appendix on page 971 and Master Product List on page 975. Products are For Research Use Only. Not intended for any animal or human therapeutic or diagnostic use.

adventitial fibroblasts and endothelial cells into three-dimensional structures[98] and tracking embryonic and mesenchymal stem cells.[99–101]

Qtracker® Non-Targeted Quantum Dots

Qtracker® non-targeted quantum dots are designed for small-animal *in vivo* imaging, and especially for studying vascular structure after microinjection[102,103] (Figure 14.6.8). Our selection includes:

- Qtracker® 565 non-targeted quantum dots (Q21031MP)
- Qtracker® 655 non-targeted quantum dots (Q21021MP)
- Qtracker® 705 non-targeted quantum dots (Q21061MP)
- Qtracker® 800 non-targeted quantum dots (Q21071MP)

These nanocrystals exhibit intense red or near-infrared fluorescence emission for maximum transmission through tissues and avoidance of background autofluorescence. Qtracker® non-targeted quantum dots have polyethylene glycol (PEG) surface coatings to minimize nonspecific binding interactions and associated inflammatory responses.[104,105] Because the PEG surface coating does not contain reactive functional groups, the Qtracker® non-targeted quantum dots are retained in circulation longer and can be imaged for up to 3 months without additional injections.[106] Qtracker® non-targeted quantum dots are supplied as 2 µM solutions in 50 mM borate buffer, pH 8.3 in units of 200 µL.

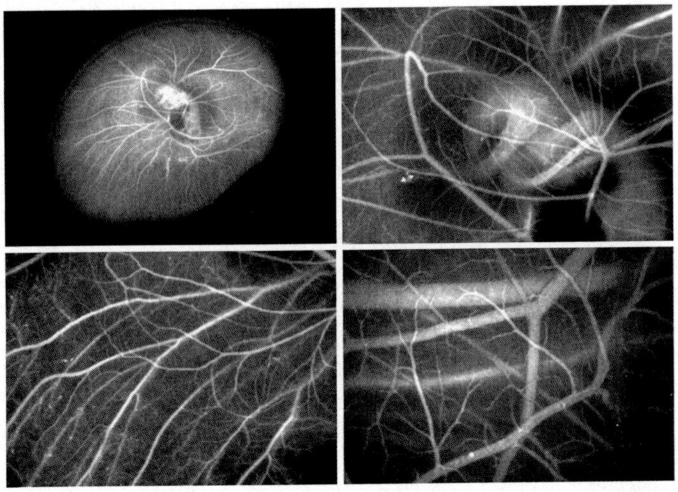

Figure 14.6.8 Chick embryo injected through the major vitelline vein with Qtracker® non-targeted quantum dots. Following a few minutes of circulation of the Qtracker® 705 non-targeted quantum dots (Q21061MP), fluorescence images of the embryo were captured at increasing magnification using 460 nm excitation and a digital imaging system equipped with appropriate emission filters. These Qtracker® reagents revealed highly detailed vascular structure at all levels of magnification. Images contributed by Greg Fisher, Carnegie Mellon University.

REFERENCES

1. Am J Physiol (1998) 275:H110; **2.** Circ Res (1967) 21:163; **3.** Brain Res (2010) 1326:128; **4.** J Neurosci Methods (2003) 122:149; **5.** Am J Physiol (1995) 269:H725; **6.** Cardiovasc Res (1994) 28:1467; **7.** Kidney Int (1981) 20:230; **8.** Nephrol Dial Transplant (1998) 13:594; **9.** Lab Invest (2006) 86:409; **10.** Clin Sci (Lond) (1998) 94:453; **11.** J Appl Physiol (1998) 84:2010; **12.** Anesthesiology (1998) 88:1291; **13.** J Appl Physiol (1998) 84:1278; **14.** J Appl Physiol (1997) 82:943; **15.** J Appl Physiol (1996) 81:1062; **16.** J Appl Physiol (2003) 94:469; **17.** J Appl Physiol (1998) 84:82; **18.** Nat Protoc (2009) 4:1779; **19.** J Appl Physiol (2004) 96:1928; **20.** Acta Odontol Scand (1998) 56:57; **21.** J Appl Physiol (2003) 95:1153; **22.** Am J Physiol (1999) 276:H1150; **23.** Cardiovasc Res (1995) 30:405; **24.** Circ Shock (1993) 41:156; **25.** J Appl Physiol (1993) 74:2585; **26.** Am J Physiol Heart Circ Physiol (2001) 280:H2496; **27.** Can J Physiol Pharmacol (1997) 75:959; **28.** Am J Physiol (1996) 271:H996; **29.** Microsc Res Tech (1993) 26:437; **30.** J Photochem Photobiol B (1995) 27:3; **31.** Clin Chem (1997) 43:1937; **32.** Biophys J (1998) 74:2210; **33.** J Histochem Cytochem (1996) 44:1091; **34.** Histochem J (1999) 31:45; **35.** Biochemistry (1998) 37:2372; **36.** Cytometry (1996) 24:312; **37.** J Histochem Cytochem (1997) 45:1279; **38.** J Histochem Cytochem (1999) 47:183; **39.** J Neurosci Methods (1992) 44:7; **40.** Brain Res (1988) 447:223; **41.** Nature (1984) 310:498; **42.** J Physiol (2004) 554:100; **43.** J Neurosci Methods (1989) 29:1; **44.** J Neurosci Methods (1988) 24:1; **45.** Brain Res (1990) 524:339; **46.** Biotechniques (1997) 23:928; **47.** Ophthalmic Surg Lasers (1997) 28:937; **48.** Ophthalmic Surg Lasers (1998) 29:506; **49.** Ophthalmology (1997) 104:753; **50.** Brain Behav Evol (1996) 48:221; **51.** Nat Med (1997) 3:244; **52.** J Neurosci Methods (2006) 152:163; **53.** Neurosci Lett (1996) 207:53; **54.** Brain Res (1993) 630:115; **55.** J Leukoc Biol (1989) 45:277; **56.** Cell Immunol (1994) 156:508; **57.** Cytometry (1991) 12:677; **58.** J Cell Biol (1991) 115:59; **59.** J Immunol Methods (1986) 88:175; **60.** Science (1982) 215:64; **61.** Eukaryot Cell (2009) 8:1665; **62.** J Neurosci Res (1990) 26:74; **63.** Cytometry (1995) 20:23; **64.** Science (1985) 230:1277; **65.** Anal Biochem (1986) 152:167; **66.** Exp Cell Res (1997) 237:383; **67.** J Cell Physiol (1993) 155:461; **68.** J Cell Biol (1984) 98:1947; **69.** Biophys J (1994) 67:1291; **70.** J Cell Biol (1994) 127:963; **71.** Nature (1992) 359:533; **72.** J Cell Biol (1994) 127:995; **73.** J Cell Biol (1990) 110:1645; **74.** Macromolecules (1991) 24:599; **75.** Science (1993) 261:1044; **76.** Proc Natl Acad Sci U S A (1995) 92:11500; **77.** Proc Natl Acad Sci U S A (2007) 104:1482; **78.** Lymphology (1994) 27:108; **79.** J Neurosci (1995) 15:4209; **80.** J Appl Physiol (1990) 68:1157; **81.** J Membr Biol (1995) 144:231; **82.** Meth Neurosci (1994) 19:320; **83.** J Membr Biol (1993) 135:83; **84.** Cell Motil Cytoskeleton (1987) 8:293; **85.** Dev Growth Differ (1986) 28:461; **86.** Appl Environ Microbiol (1994) 60:2711; **87.** Cancer Sci (2005) 96:353; **88.** Biophys J (2006) 90:2093; **89.** Biophys J (1994) 66:769; **90.** Science (1993) 260:232; **91.** J Microsc (1993) 172:121; **92.** Am J Physiol (1995) 268:L657; **93.** Nano Lett (2004) 4:2019; **94.** Bioconjug Chem (2009) 20:249; **95.** Bioconjug Chem (2008) 19:1785; **96.** J Biomol Screen (2009) 14:845; **97.** Nat Neurosci (2008) 11:429; **98.** Am J Pathol (2006) 168:1793; **99.** Nano Lett (2006) 6:2826; **100.** BMC Biotechnol (2007) 7:67; **101.** Stem Cells (2007) 25:2128; **102.** Methods Mol Biol (2009) 544:393; **103.** Bioconjug Chem (2004) 15:79; **104.** J Invest Dermatol (2007) 127:143; **105.** Nano Lett (2007) 7:2513; **106.** Environ Health Perspect (2007) 115:1339.

IMPORTANT NOTICE: The products described in this manual are covered by one or more Limited Use Label License(s). Please refer to the Appendix on page 971 and Master Product List on page 975. Products are For Research Use Only. Not intended for any animal or human therapeutic or diagnostic use.

PRODUCT LIST 14.6 MICROSPHERES AND QDOT® NANOCRYSTALS FOR TRACING

Cat. No.	Product	Quantity
F8890	FluoSpheres® Blood Flow Determination Fluorescent Color Kit #1, polystyrene microspheres, 10 µm *seven colors, 10 mL each* *3.6 x 10^6 beads/mL*	1 kit
F8891	FluoSpheres® Blood Flow Determination Fluorescent Color Kit #2, polystyrene microspheres, 15 µm *seven colors, 10 mL each* *1.0 x 10^6 beads/mL*	1 kit
F8892	FluoSpheres® Blood Flow Determination Fluorescent Color Kit #3, polystyrene microspheres, 15 µm *five colors, 10 mL each* *1.0 x 10^6 beads/mL*	1 kit
F21015	FluoSpheres® Blood Flow Determination Fluorescent Color Kit #4, polystyrene microspheres, 15 µm *four colors, 10 mL each* *1.0 x 10^6 beads/mL*	1 kit
F8789	FluoSpheres® carboxylate-modified microspheres, 0.04 µm, dark red fluorescent (660/680) *5% solids, azide free*	1 mL
F20880	FluoSpheres® carboxylate-modified microspheres, 0.04 µm, europium luminescent (365/610) *0.5% solids*	2 mL
F8792	FluoSpheres® carboxylate-modified microspheres, 0.04 µm, orange fluorescent (540/560) *5% solids, azide free*	1 mL
F20886	FluoSpheres® carboxylate-modified microspheres, 0.04 µm, platinum luminescent (390/650) *0.5% solids*	2 mL
F8793	FluoSpheres® carboxylate-modified microspheres, 0.04 µm, red fluorescent (580/605) *5% solids, azide free*	1 mL
F8794	FluoSpheres® carboxylate-modified microspheres, 0.04 µm, red-orange fluorescent (565/580) *5% solids, azide free*	1 mL
F8795	FluoSpheres® carboxylate-modified microspheres, 0.04 µm, yellow-green fluorescent (505/515) *5% solids, azide free*	1 mL
F10720	FluoSpheres® Fluorescent Color Kit, carboxylate-modified microspheres, 0.04 µm *four colors, 1 mL each* *5% solids, azide free*	1 kit
F20881	FluoSpheres® carboxylate-modified microspheres, 0.2 µm, europium luminescent (365/610) *0.5% solids*	2 mL
F13080	FluoSpheres® polystyrene microspheres, 1.0 µm, blue-green fluorescent (430/465) *for tracer studies* *1.0 x 10^{10} beads/mL*	5 mL
F13082	FluoSpheres® polystyrene microspheres, 1.0 µm, orange fluorescent (540/560) *for tracer studies* *1.0 x 10^{10} beads/mL*	5 mL
F13083	FluoSpheres® polystyrene microspheres, 1.0 µm, red fluorescent (580/605) *for tracer studies* *1.0 x 10^{10} beads/mL*	5 mL
F13081	FluoSpheres® polystyrene microspheres, 1.0 µm, yellow-green fluorescent (505/515) *for tracer studies* *1.0 x 10^{10} beads/mL*	5 mL
F8829	FluoSpheres® polystyrene microspheres, 10 µm, blue fluorescent (365/415) *for blood flow determination* *3.6 x 10^6 beads/mL*	10 mL
F8830	FluoSpheres® polystyrene microspheres, 10 µm, blue-green fluorescent (430/465) *for blood flow determination* *3.6 x 10^6 beads/mL*	10 mL
F8831	FluoSpheres® polystyrene microspheres, 10 µm, crimson fluorescent (625/645) *for blood flow determination* *3.6 x 10^6 beads/mL*	10 mL
F8833	FluoSpheres® polystyrene microspheres, 10 µm, orange fluorescent (540/560) *for blood flow determination* *3.6 x 10^6 beads/mL*	10 mL
F8834	FluoSpheres® polystyrene microspheres, 10 µm, red fluorescent (580/605) *for blood flow determination* *3.6 x 10^6 beads/mL*	10 mL
F8836	FluoSpheres® polystyrene microspheres, 10 µm, yellow-green fluorescent (505/515) *for blood flow determination* *3.6 x 10^6 beads/mL*	10 mL
F8837	FluoSpheres® polystyrene microspheres, 15 µm, blue fluorescent (365/415) *for blood flow determination* *1.0 x 10^6 beads/mL*	10 mL
F8838	FluoSpheres® polystyrene microspheres, 15 µm, blue-green fluorescent (430/465) *for blood flow determination* *1.0 x 10^6 beads/mL*	10 mL
F21013	FluoSpheres® polystyrene microspheres, 15 µm, carmine fluorescent (580/620) *for blood flow determination* *1.0 x 10^6 beads/mL*	10 mL
F8839	FluoSpheres® polystyrene microspheres, 15 µm, crimson fluorescent (625/645) *for blood flow determination* *1.0 x 10^6 beads/mL*	10 mL
F21010	FluoSpheres® polystyrene microspheres, 15 µm, green fluorescent (450/480) *for blood flow determination* *1.0 x 10^6 beads/mL*	10 mL
F8841	FluoSpheres® polystyrene microspheres, 15 µm, orange fluorescent (540/560) *for blood flow determination* *1.0 x 10^6 beads/mL*	10 mL
F8842	FluoSpheres® polystyrene microspheres, 15 µm, red fluorescent (580/605) *for blood flow determination* *1.0 x 10^6 beads/mL*	10 mL
F21012	FluoSpheres® polystyrene microspheres, 15 µm, red-orange fluorescent (565/580) *for blood flow determination* *1.0 x 10^6 beads/mL*	10 mL
F8843	FluoSpheres® polystyrene microspheres, 15 µm, scarlet fluorescent (645/680) *for blood flow determination* *1.0 x 10^6 beads/mL*	10 mL
F21011	FluoSpheres® polystyrene microspheres, 15 µm, yellow fluorescent (515/534) *for blood flow determination* *1.0 x 10^6 beads/mL*	10 mL
F8844	FluoSpheres® polystyrene microspheres, 15 µm, yellow-green fluorescent (505/515) *for blood flow determination* *1.0 x 10^6 beads/mL*	10 mL
Q25041MP	Qtracker® 525 Cell Labeling Kit	1 kit
Q25031MP	Qtracker® 565 Cell Labeling Kit	1 kit
Q25011MP	Qtracker® 585 Cell Labeling Kit	1 kit
Q25001MP	Qtracker® 605 Cell Labeling Kit	1 kit
A10198	Qtracker® 625 Cell Labeling Kit	1 kit
Q25021MP	Qtracker® 655 Cell Labeling Kit	1 kit
Q25061MP	Qtracker® 705 Cell Labeling Kit	1 kit
Q25071MP	Qtracker® 800 Cell Labeling Kit	1 kit
Q21031MP	Qtracker® 565 non-targeted quantum dots *2 µM solution*	200 µL
Q21021MP	Qtracker® 655 non-targeted quantum dots *2 µM solution*	200 µL
Q21061MP	Qtracker® 705 non-targeted quantum dots *2 µM solution*	200 µL
Q21071MP	Qtracker® 800 non-targeted quantum dots *2 µM solution*	200 µL
S31201	SAIVI™ 715 injectable contrast agent *0.1 µm microspheres*	1 mL
S31203	SAIVI™ 715 injectable contrast agent *2 µm microspheres*	1 mL

IMPORTANT NOTICE: The products described in this manual are covered by one or more Limited Use Label License(s). Please refer to the Appendix on page 971 and Master Product List on page 975. Products are For Research Use Only. Not intended for any animal or human therapeutic or diagnostic use.

www.invitrogen.com/probes

molecular probes® | invitrogen™ by life technologies™

14.7 Protein Conjugates

Unlike the polydisperse dextrans (Section 14.5), fluorescent protein tracers have molecular weights that are reasonably well defined (bovine serum albumin (BSA) ~66,000 daltons; ovalbumin ~45,000; codfish parvalbumin 12,328 daltons, casein ~23,600 daltons; monomeric subunit B of cholera toxin ~12,000 daltons; horseradish peroxidase (HRP) ~40,000 daltons; soybean trypsin inhibitor ~21,500 daltons; *Phaseolus vulgaris* leucoagglutinin (PHA-L) ~126,000 daltons). Some of their applications are similar to those of dextran tracers, although protein conjugates may be more susceptible to proteolysis. Through creative modifications of the proteins, however, researchers have been able to use these conjugates to target receptors, to detect protease activity and to measure intracellular pH.

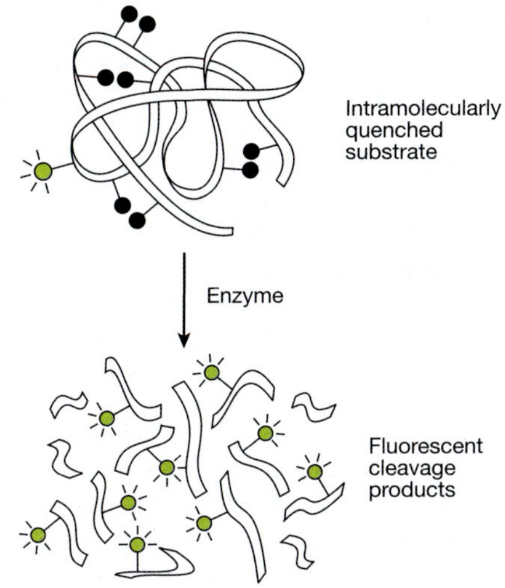

Figure 14.7.1 Principle of enzyme detection via the disruption of intramolecular self-quenching. Enzyme-catalyzed hydrolysis of the heavily labeled and almost totally quenched substrates provided in our EnzChek® Protease Assay Kits (E6638, E6639), EnzChek® *Ultra* Amylase Assay Kit (E33651), EnzChek® Gelatinase/Collagenase Assay Kit (E12055), EnzChek® Elastase Kit (E12056), EnzChek® Lysozyme Assay Kit (E22013)—as well as the stand-alone quenched substrates DQ™ BSA (D12050, D12051), DQ™ collagen (D12052, D12060), DQ™ ovalbumin (D12053) and DQ™ gelatin (D12054)—relieves the intramolecular self-quenching, yielding brightly fluorescent reaction products.

Albumin Conjugates

Fluorescent BSA, Ovalbumin and Parvalbumin

For use as protein tracers, we supply bovine serum albumin (BSA) fluorescently labeled with:

- Alexa Fluor® 488 dye (A13100)
- Alexa Fluor® 555 dye (A34786)
- Alexa Fluor® 594 dye (A13101)
- Alexa Fluor® 647 dye (A34785)
- Alexa Fluor® 680 dye (A34787, S34788)
- Alexa Fluor® 750 dye dye (S34789)
- Fluorescein (A23015)
- BODIPY® FL dye (A2750)
- Tetramethylrhodamine (A23016)
- Texas Red® dye (A23017)

In addition to these BSA conjugates, we prepare fluorescein and Texas Red® conjugates of ovalbumin (O23020, O23021), as well as the intensely fluorescent Alexa Fluor® 488, Alexa Fluor® 555, Alexa Fluor® 594 and Alexa Fluor® 647 conjugates (O34781, O34782, O34783, O34784). Ovalbumin is a protein with a relatively low molecular weight (~45,000 daltons) that is primarily useful as an antigen for activation of dendritic cells, macrophages and B cells.[1,2]

We also offer the Alexa Fluor® 488 conjugate of codfish parvalbumin (P23012), which is the smallest fluorescent protein tracer in this section. Parvalbumin, a stable and protease-resistant protein with a molecular weight of 12,328 daltons, is found in the skeletal muscle of vertebrates and the endocrine glands of mammals.[3]

Dinitrophenylated Bovine Serum Albumin

We offer BSA that has been conjugated to multiple dinitrophenyl hapten molecules (DNP-BSA, A23018), a reagent that is commonly used to study Fc receptor–mediated immune function[4–6] and the IgE- and IgG-mediated responses to crosslinking of DNP-specific antibodies.[7,8] Our dinitrophenylated BSA typically has ~25 dinitrophenyl substituents per molecule of BSA. For detection of DNP-BSA, we offer unlabeled rabbit anti-DNP antibody prepared against DNP-conjugated keyhole limpet hemocyanin (anti–DNP-KLH antibody) and the Alexa Fluor® 488 dye, fluorescein and biotin-XX antibody conjugates (A6430, A11097, A6423, A6435; Section 7.4), as well as a Qdot® 555 rat monoclonal anti-dinitrophenol antibody (Q17421MP).

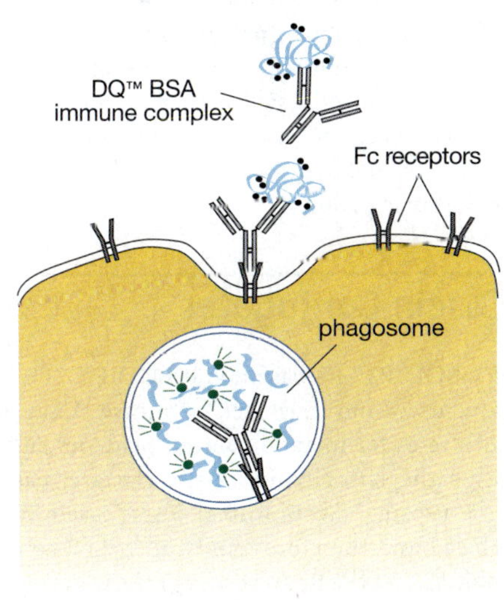

Figure 14.7.2 Immune complex of DQ™ BSA conjugate (D12050, D12051) with an anti–bovine serum albumin (BSA) antibody (A11133) for the fluorescent detection of the Fc receptor–mediated phagocytosis pathway. The DQ™ BSA is a derivative of BSA that is labeled to such a high degree with either the green-fluorescent BODIPY® FL or red-fluorescent BODIPY® TR-X dye that the fluorescence is self-quenched. Upon binding to an Fc receptor, the nonfluorescent immune complex is internalized and the protein is subsequently hydrolyzed to fluorescent peptides within the phagovacuole.

DQ™ BSA, DQ™ Ovalbumin and DQ™ Collagen

DQ™ substrates, including the DQ™ Green BSA and DQ™ Red BSA (D12050, D12051) and DQ™ ovalbumin (D12053) are based on the strong fluorescence quenching effect observed when proteins are heavily conjugated with BODIPY® dyes or, in the case of DQ™ collagen conjugates (D12052, D12060), with multiple fluorescein dyes. Upon hydrolysis of the proteins to single dye-labeled peptides by proteases, this quenching is relieved (Figure 14.7.1). This technology provides an easy continuous assay for numerous proteases (Section 10.4). When complexed with an anti-BSA rabbit IgG antibody (A11133), the DQ™ BSA products are phagocytosed into live cells via the Fc receptor (Section 16.1, Figure 14.7.2), where they are broken down into fluorescent peptides[9,10]

(Figure 14.7.1). DQ™ Green BSA, which gives bright green fluorescence upon proteolysis, and DQ™ Red BSA, which yields red-fluorescent hydrolysis products, are packaged in sets of five vials, each containing 1 mg of the substrate. Our Fc OxyBURST® Green assay reagent (F2902, Section 16.1), which is based on a similar concept, comprises BSA that has been conjugated to a dihydrofluorescein derivative and then complexed with anti-BSA IgG. Upon binding to the Fc receptor, the immune complex is internalized and oxidized to a fluorescent product within the phagovacuole.[11,12]

We have found our highly quenched DQ™ ovalbumin (D12053) to be particularly useful for antigen processing and presentation studies. Ovalbumin conjugates are internalized via the mannose receptor–mediated endocytosis pathway and are recognized by antigen-presenting cells. BSA that was heavily labeled with fluorescein (FITC) has been used for these studies.[13] However, fluorescence of the intracellular hydrolysis products of FITC BSA is significantly decreased by the acidity of the phagovacuole. In contrast, the bright fluorescence of the BODIPY® FL fluorophore in DQ™ ovalbumin is completely independent of the pH. The fluorescence generated after DQ™ collagen (D12052) is hydrolyzed by cellular collagenase has been used to visualize the migratory pathway followed by tumor cells during invasion of a gelatin matrix and to image proteolysis by living breast cancer cells.[14,15]

Injectable Contrast Reagents for Small Animal *in vivo* Imaging Applications

The study of vascularization—often in tumors—requires a probe that will adequately distribute throughout the system under study without being cleared from the tissue prior to the time needed to perform imaging. For this purpose, bovine serum albumin (BSA) and human serum transferrin labeled with near-infrared–fluorescent Alexa Fluor® 680 (S34788, S34790) and Alexa Fluor® 750 (S34789, S34791) dyes are offered as injectable contrast agents in small animal *in vivo* imaging.[16] These SAIVI™ injectable contrast agents have been optimized for emission intensity and tested as reagents for the *in vivo* imaging of regions of blood pooling and vascularization in an inflammatory disease model in mice. Infrared fluorescent microspheres (Section 6.5) can also be used in a similar manner.

Casein Conjugates

BODIPY® FL casein and BODIPY® TR-X casein, which are components of our EnzChek® Protease Assay Kits (E6638, E6639), are substrates for metallo-, serine, acid and sulfhydryl proteases, including cathepsin, chymotrypsin, elastase, papain, pepsin, thermolysin and trypsin[17] (Section 10.4). These casein-based substrates are heavily labeled and therefore highly quenched conjugates; they typically exhibit <3% of the fluorescence of the corresponding free dyes. Protease-catalyzed hydrolysis relieves this quenching, yielding brightly fluorescent peptides (Figure 14.7.1).

The EnzChek® Protease Assay Kit (E6638, E6639) includes:

- BODIPY® FL casein (in Kit E6638) or BODIPY® TR-X casein (in Kit E6639)
- Concentrated digestion buffer
- Detailed protocols

Each kit provides sufficient reagents for ~100 assays using 2 mL assay volumes and standard fluorescence cuvettes or ~1000 assays using 200 µL assay volumes and 96-well microplates.

The EnzChek® Protease Assay Kits, as well as the DQ™ BSA and DQ™ ovalbumin substrates mentioned above, have significant potential as:

- Indicators of protease contamination in culture media
- Fluorogenic substrates for circulating or secreted proteases in extracellular fluids
- Nontoxic, pH-insensitive markers for phagocytic cells, which will ingest and eventually cleave the quenched casein substrates to yield fluorescent BODIPY® FL– or BODIPY® TR-X–labeled peptides
- Microinjectable tracers to detect enhanced protease activity associated with cell activation and fusion
- Nontoxic markers for assessing various cell-loading and cell-transfection techniques, including electroporation (Table 14.1)

The peptide hydrolysis products of BODIPY® FL casein exhibit green fluorescence that is optimally excited by the argon-ion laser at 488 nm, permitting flow sorting of the cells. The red-fluorescent BODIPY® TR-X–labeled peptides, with excitation and emission spectra similar to those of the Texas Red® fluorophore, should be useful for multilabeling experiments or measurements in the presence of green autofluorescence. Following intracellular hydrolysis, some of the lower molecular weight fluorescent peptides may diffuse into other organelles or pass thorough gap junctions.

PHA-L: An Important Anterograde Tracer

Our Alexa Fluor® 488, Alexa Fluor® 594 and Alexa Fluor® 647 conjugates (L11270, L32456, L32457) of *Phaseolus vulgaris* leucoagglutinin (PHA-L) are excellent fluorescent anterograde tracers.[18,19] Iontophoretically injected PHA-L clearly demonstrates the morphological features of the filled neurons at the injection site and the labeled axons and axon terminals. Furthermore, PHA-L that has been transported is not degraded over long periods. PHA-L has been used in combination with our biotin dextran amines[20–23] (BDA dextrans, Section 14.5) or the fluorescent dextrans fluoro-ruby and mini-ruby[22,24,25] (D1817, D3312; Section 14.5) to demonstrate their similar transport properties and to show the distribution of anterograde-labeled fibers.[26–28] PHA-L has been simultaneously injected with the retrograde tracer hydroxystilbamidine (H22845, Section 14.3) for double-labeling experiments.[29,30]

Cholera Toxin Subunit B

Cholera toxin comprises two subunits, A and B, arranged in an AB_5 conformation. Subunit A is an ADP-ribosyltransferase, which disrupts the proper signaling of G proteins and eventually leads to dehydration of the cell.[31] The 12,000-dalton nontoxic subunit B ("choleragenoid"), which is assembled into a 60,000-dalton pentamer above pH 2,[32] allows the protein complex to bind to cellular surfaces via the pentasaccharide chain of ganglioside G_{M1}.[33] All Molecular Probes® cholera toxin subunit B conjugates are prepared from recombinant cholera toxin subunit B, which is completely free of the toxic subunit A, thus eliminating any concern for toxicity or ADP-ribosylating activity.

IMPORTANT NOTICE: The products described in this manual are covered by one or more Limited Use Label License(s). Please refer to the Appendix on page 971 and Master Product List on page 975. Products are For Research Use Only. Not intended for any animal or human therapeutic or diagnostic use.
www.invitrogen.com/probes
molecular probes® | invitrogen™ by life technologies™

Cholera toxin subunit B and its conjugates are established as superior tracers for retrograde labeling of neurons.[34,35] Cholera toxin subunit B conjugates bind to the pentasaccharide chain of ganglioside G_{M1} on neuronal cell surfaces and are actively taken up and transported; alternatively, they can be injected by iontophoresis. Unlike the carbocyanine-based neuronal tracers such as DiI (D282, D3911, V22885; Section 14.4), cholera toxin subunit B conjugates can be used on tissue sections that will be fixed and frozen.[36] Our Alexa Fluor® 488 (C22841, C34775), Alexa Fluor® 555 (C22843, C34776), Alexa Fluor® 594 (C22842, C34777) and Alexa Fluor® 647 (C34778) conjugates of cholera toxin subunit B combine this versatile tracer with the superior brightness of our Alexa Fluor® dyes to provide sensitive and selective receptor labeling and neuronal tracing.[35,37] We also offer biotin-XX (C34779) and horseradish peroxidase (C34780) conjugates of cholera toxin subunit B for use in combination with diaminobenzidine (DAB) oxidation,[38] tyramide signal amplification (TSA™) techniques (Section 6.2) and Qdot® nanocrystal–streptavidin conjugates.[39]

Fluorescent cholera toxins are also markers of lipid rafts—regions of cell membranes high in ganglioside G_{M1} that are thought to be important in cell signaling.[40,41] The Vybrant® Lipid Raft Labeling Kits (V34403, V34404, V34405; Section 14.4) provide the key reagents, including fluorescent cholera toxin subunit B conjugates, for fluorescently labeling lipid rafts *in vivo* with our bright and extremely photostable Alexa Fluor® dyes (Figure 14.7.3).

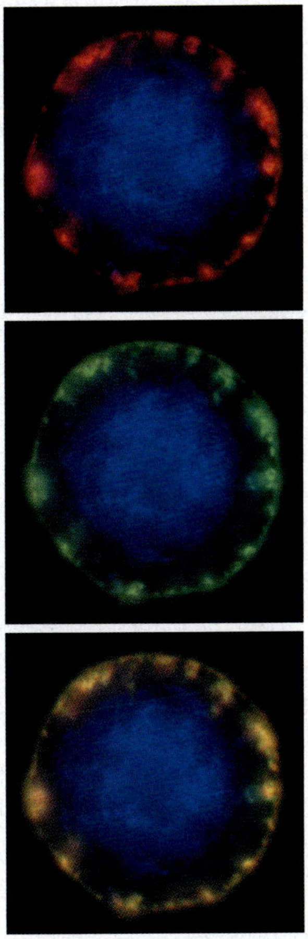

Figure 14.7.3 A J774 mouse macrophage cell sequentially stained with BODIPY® FL ganglioside G_{M1} (B13950) and then with Alexa Fluor® 555 dye–labeled cholera toxin subunit B (C22843, C34776; also available as a component of V34404). The cell was then treated with an anti–CT-B antibody (a component of V34404) to induce crosslinking. Alexa Fluor® 555 dye fluorescence (top, red) and BODIPY® FL dye fluorescence (middle, green) were imaged separately and overlaid to emphasize the coincident staining (bottom, yellow). Nuclei were stained with blue-fluorescent Hoechst 33342 (H1399, H3570, H21492).

Fluorescent Protein–Based Cytosol Marker

Fluorescent protein–based cytosolic marker BacMam GFP transduction control (B10383) incorporates all the generic advantages of BacMam 2.0 delivery technology (BacMam Gene Delivery and Expression Technology—Note 11.1) to produce nonspecific Green Fluorescent Protein (GFP) expression in a wide range of cell types. Once initiated, GFP expression lasts about 5 days in cell lines such as HeLa and CHO. In cells that divide more slowly or show contact inhibition, such as some stem cells, primary cells and neurons, we have observed useful levels of GFP expression for more than 2 weeks. Within this timeframe, labeled cells can be used for any purpose requiring stable incorporation of a passive fluorescent marker in live cells, including wound healing, migration or adhesion assays. In adoptive transfer applications, cells with undetectable levels of GFP fluorescence can be identified immunohistochemically using our anti-GFP antibodies (Section 7.4).

Phycobiliproteins

Phycobiliproteins make effective polar tracers because they are intensely fluorescent, are very soluble in water and have little tendency to bind nonspecifically to cells (Section 6.4). These intrinsically fluorescent biomolecules are stable proteins that are monodisperse and substantially larger than albumins, with molecular weights of 240,000 daltons (B- and R-phycoerythrin, P800, P801) and 104,000 daltons (allophycocyanin, A803, or crosslinked allophycocyanin, A819). Phycoerythrin has been conjugated to nuclear-localization peptide sequences to follow nuclear import processes.[42–45] B-phycoerythrin can pass through pores in erythrocyte membranes created by streptolysin O.[46] Single molecules of R-phycoerythrin have been imaged and tracked within mammalian cells.[47]

Alexa Fluor® 488 Soybean Trypsin Inhibitor

The high water solubility and chemical stability of soybean trypsin inhibitor (SBTI) may make its green-fluorescent Alexa Fluor® 488 conjugate (T23011) a useful polar tracer. This conjugate has been used as a probe for the acrosome reaction of spermatozoa.[48]

REFERENCES

1. Science (2007) 316:612; **2.** Immunology (2007) 120:217; **3.** Eur J Biochem (1994) 222:21; **4.** Cell Regul (1991) 2:181; **5.** FEBS Lett (1990) 270:115; **6.** Cell (1989) 58:317; **7.** J Biol Chem (1995) 270:4013; **8.** Mol Biol Cell (1995) 6:825; **9.** Eur J Cell Biol (1998) 75:192; **10.** J Leukoc Biol (1997) 62:329; **11.** J Leukoc Biol (1988) 43:304; **12.** J Immunol Methods (1990) 130:223; **13.** J Immunol (1997) 159:2177; **14.** FASEB J (2001) 15:932; **15.** Neoplasia (2000) 2:496; **16.** Am J Physiol Cell Physiol (2009) 296:C848; **17.** Anal Biochem (1997) 251:144; **18.** J Comp Neurol (2007) 500:634; **19.** Neuron (2005) 45:119; **20.** J Neurosci Methods (1993) 48:75; **21.** Histochem Cell Biol (1998) 110:509; **22.** J Comp Neurol (1999) 403:158; **23.** J Neurosci (1998) 18:10525; **24.** J Comp Neurol (1990) 302:159; **25.** Anat Embryol (Berl) (1996) 194:559; **26.** Histochem Cell Biol (1997) 107:391; **27.** Brain Res Bull (1996) 40:219; **28.** Neuroscience (1998) 82:1151; **29.** Neurosci Lett (1994) 168:119; **30.** Brain Res Bull (1994) 33:445; **31.** J Biol Chem (1980) 255:1252; **32.** Biochemistry (1996) 35:16069; **33.** Mol Microbiol (1994) 13:745; **34.** PLoS Comput Biol (2009) 5:e1000334; **35.** Brain Struct Funct (2009) 213:367; **36.** J Neurosci (2002) 22:9419; **37.** Nat Protoc (2009) 4:1157; **38.** J Neurosci Methods (2005) 149:101; **39.** BMC Neurosci (2009) 10:80; **40.** Science (2010) 327:46; **41.** FEMS Microbiol Lett (2007) 266:129; **42.** J Membr Biol (1995) 146:239; **43.** Exp Cell Res (1988) 178:318; **44.** Biochim Biophys Acta (1987) 930:419; **45.** J Cell Sci (1995) 108:1849; **46.** J Microsc (1998) 192:114; **47.** Biophys J (2000) 79:2188; **48.** Zygote (2000) 8:127.

PRODUCT LIST 14.7 PROTEIN CONJUGATES

Cat. No.	Product	Quantity
A13100	albumin from bovine serum (BSA), Alexa Fluor® 488 conjugate	5 mg
A34786	albumin from bovine serum (BSA), Alexa Fluor® 555 conjugate	5 mg
A13101	albumin from bovine serum (BSA), Alexa Fluor® 594 conjugate	5 mg
A34785	albumin from bovine serum (BSA), Alexa Fluor® 647 conjugate	5 mg
A34787	albumin from bovine serum (BSA), Alexa Fluor® 680 conjugate	5 mg
A2750	albumin from bovine serum (BSA), BODIPY® FL conjugate	5 mg
A23018	albumin from bovine serum (BSA), 2,4-dinitrophenylated (DNP-BSA)	5 mg
A23015	albumin from bovine serum (BSA), fluorescein conjugate	5 mg
A23016	albumin from bovine serum (BSA), tetramethylrhodamine conjugate	5 mg
A23017	albumin from bovine serum (BSA), Texas Red® conjugate	5 mg
A803	allophycocyanin *4 mg/mL*	0.5 mL
A819	allophycocyanin, crosslinked (APC-XL) *4 mg/mL*	250 µL
A11133	anti-albumin (bovine serum), rabbit IgG fraction (anti-BSA) *2 mg/mL*	0.5 mL
B10383	BacMam GFP transduction control *BacMam 2.0*	1 mL
C34775	cholera toxin subunit B (recombinant), Alexa Fluor® 488 conjugate	100 µg
C22841	cholera toxin subunit B (recombinant), Alexa Fluor® 488 conjugate	500 µg
C34776	cholera toxin subunit B (recombinant), Alexa Fluor® 555 conjugate	100 µg
C22843	cholera toxin subunit B (recombinant), Alexa Fluor® 555 conjugate	500 µg
C34777	cholera toxin subunit B (recombinant), Alexa Fluor® 594 conjugate	100 µg
C22842	cholera toxin subunit B (recombinant), Alexa Fluor® 594 conjugate	500 µg
C34778	cholera toxin subunit B (recombinant), Alexa Fluor® 647 conjugate	100 µg
C34779	cholera toxin subunit B (recombinant), biotin-XX conjugate	100 µg
C34780	cholera toxin subunit B (recombinant), horseradish peroxidase conjugate	100 µg
D12060	DQ™ collagen, type I from bovine skin, fluorescein conjugate	1 mg
D12052	DQ™ collagen, type IV from human placenta, fluorescein conjugate	1 mg
D12050	DQ™ Green BSA *special packaging*	5 x 1 mg
D12053	DQ™ ovalbumin *special packaging*	5 x 1 mg
D12051	DQ™ Red BSA *special packaging*	5 x 1 mg
E6638	EnzChek® Protease Assay Kit *green fluorescence* *100–1000 assays*	1 kit
E6639	EnzChek® Protease Assay Kit *red fluorescence* *100–1000 assays*	1 kit
L11270	lectin PHA-L from *Phaseolus vulgaris* (red kidney bean), Alexa Fluor® 488 conjugate	1 mg
L32456	lectin PHA-L from *Phaseolus vulgaris* (red kidney bean), Alexa Fluor® 594 conjugate	1 mg
L32457	lectin PHA-L from *Phaseolus vulgaris* (red kidney bean), Alexa Fluor® 647 conjugate	1 mg
O34781	ovalbumin, Alexa Fluor® 488 conjugate	2 mg
O34782	ovalbumin, Alexa Fluor® 555 conjugate	2 mg
O34783	ovalbumin, Alexa Fluor® 594 conjugate	2 mg
O34784	ovalbumin, Alexa Fluor® 647 conjugate	2 mg
O23020	ovalbumin, fluorescein conjugate	5 mg
O23021	ovalbumin, Texas Red® conjugate	5 mg
P23012	parvalbumin from codfish, Alexa Fluor® 488 conjugate	1 mg
P800	B-phycoerythrin *4 mg/mL*	0.5 mL
P801	R-phycoerythrin *4 mg/mL*	0.5 mL
S34788	SAIVI™ Alexa Fluor® 680 injectable contrast agent *bovine serum albumin*	1 mL
S34789	SAIVI™ Alexa Fluor® 750 injectable contrast agent *bovine serum albumin*	1 mL
S34790	SAIVI™ Alexa Fluor® 680 injectable contrast agent *human serum transferrin*	1 mL
S34791	SAIVI™ Alexa Fluor® 750 injectable contrast agent *human serum transferrin*	1 mL
T23011	trypsin inhibitor from soybean, Alexa Fluor® 488 conjugate	1 mg

IMPORTANT NOTICE: The products described in this manual are covered by one or more Limited Use Label License(s). Please refer to the Appendix on page 971 and Master Product List on page 975. Products are For Research Use Only. Not intended for any animal or human therapeutic or diagnostic use.

PART V

Probing Cell Function: From Cell Viability to Ion Flux

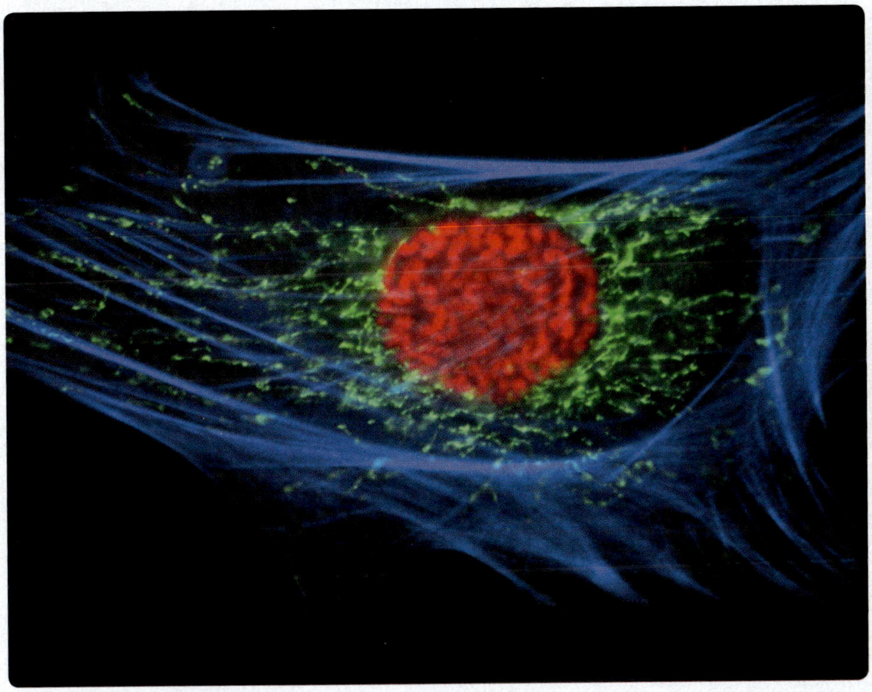

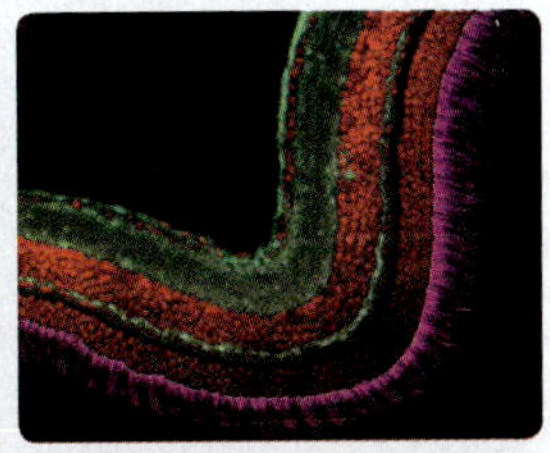

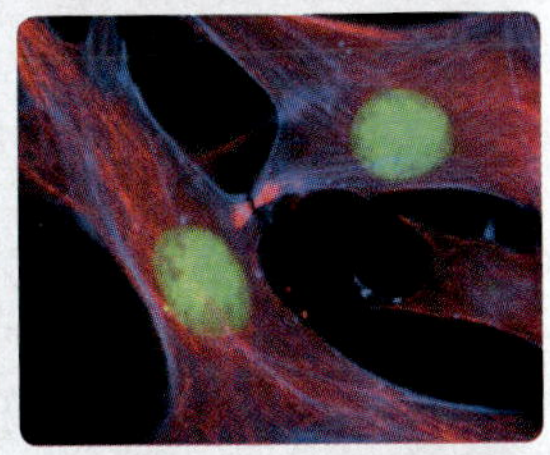

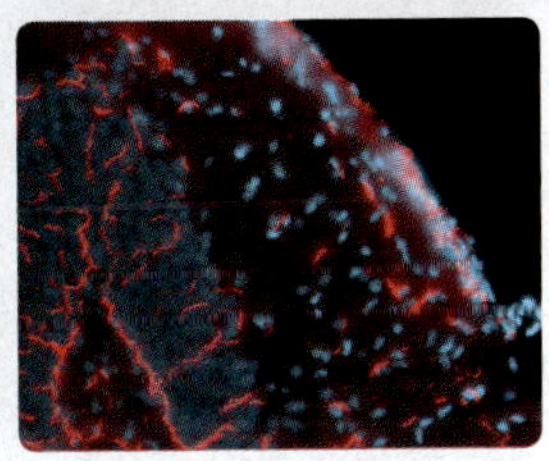

Imaging autophagy with the Premo™ Autophagy Sensor LC3B-GFP and CellLight™ Lysosomes-RFP. HeLa cells were cotransduced with Premo™ Autophagy Sensor LC3B-GFP (P36235, green) and CellLight™ Lysosomes-RFP (C10597, red). The following day, cells were treated with 50 µM chloroquine diphosphate. Twenty-four hours later, cells were incubated with 1 µg/mL Hoechst 33342 (H3570) and 20 µg/mL Alexa Fluor® 647 70,000 MW dextran conjugate (purple) for 40 min at 37°C before imaging. Image contributed by Nick Dolman, Life Technologies Corporation.

This image appeared on the cover of *BioProbes 62*. *BioProbes®* newsletter is published several times each year. This award-winning publication is dedicated to furnishing researchers with the very latest information about cell biology products and their applications, and provides a great way to stay connected with the fluorescence community. You can subscribe and view the latest issues online at **www.invitrogen.com/handbook/bioprobes**.

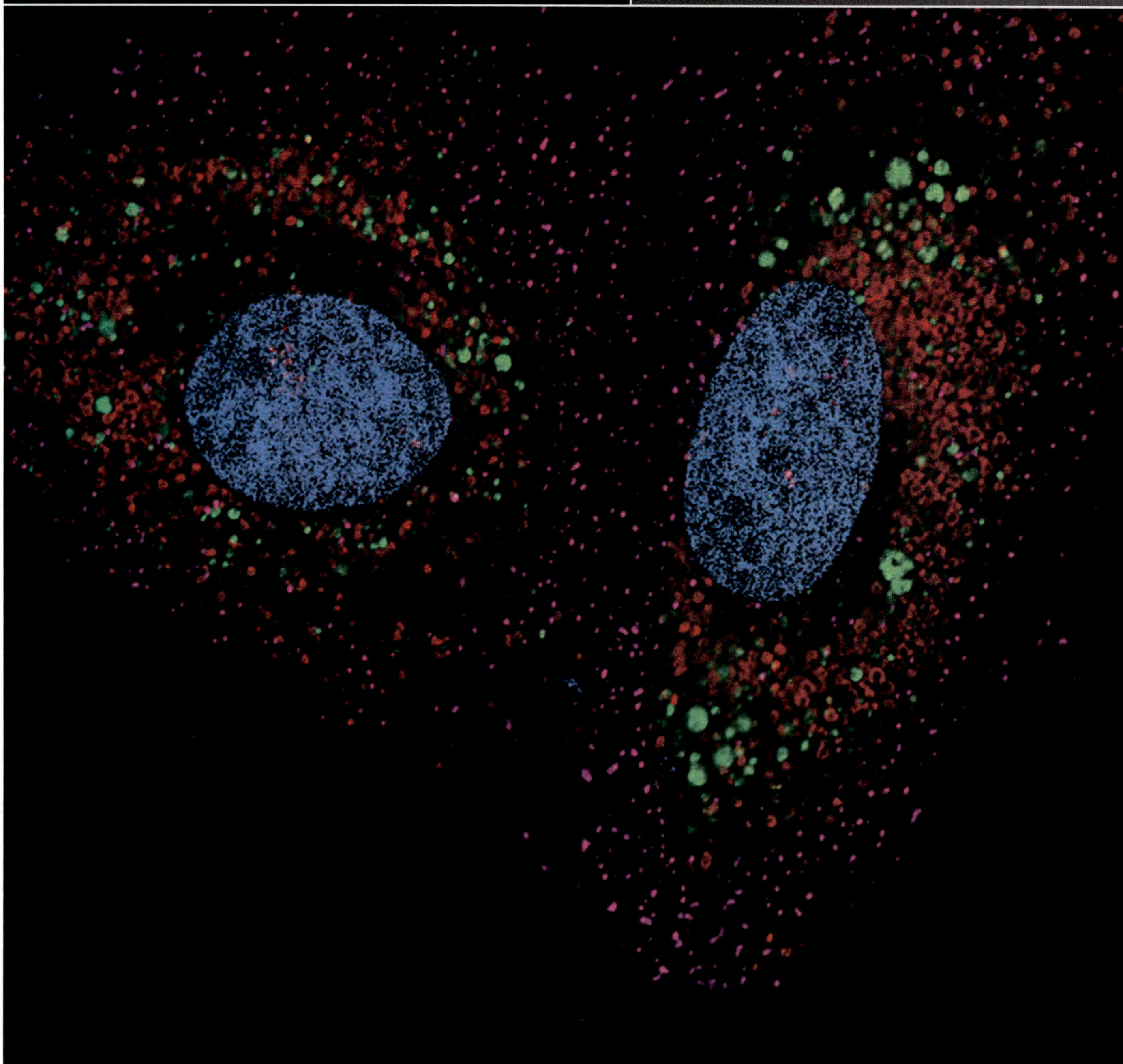

FIFTEEN

CHAPTER 15

Assays for Cell Viability, Proliferation and Function

IMPORTANT NOTICE: The products described in this manual are covered by one or more Limited Use Label License(s). Please refer to the Appendix on page 971 and Master Product List on page 975. Products are For Research Use Only. Not intended for any animal or human therapeutic or diagnostic use.

IMPORTANT NOTICE: The products described in this manual are covered by one or more Limited Use Label License(s). Please refer to the Appendix on page 971 and Master Product List on page 975. Products are For Research Use Only. Not intended for any animal or human therapeutic or diagnostic use.

IMPORTANT NOTICE: The products described in this manual are covered by one or more Limited Use Label License(s). Please refer to the Appendix on page 971 and Master Product List on page 975. Products are For Research Use Only. Not intended for any animal or human therapeutic or diagnostic use.

IMPORTANT NOTICE: The products described in this manual are covered by one or more Limited Use Label License(s). Please refer to the Appendix on page 971 and Master Product List on page 975. Products are For Research Use Only. Not intended for any animal or human therapeutic or diagnostic use.

15.1 Overview of Probes for Cell Viability, Cell Proliferation and Live-Cell Function

Cell viability, cell proliferation and many important live-cell functions—including apoptosis, cell adhesion, chemotaxis, multidrug resistance, endocytosis, secretion and signal transduction—can be stimulated or monitored with various chemical and biological reagents. Many of these processes lead to changes in intracellular radicals (Chapter 18), free-ion concentrations (Chapter 19, Chapter 20, Chapter 21) or membrane potential (Chapter 22) that can be followed with appropriately responsive fluorescent indicators. This chapter discusses Molecular Probes® reagents and assays for detecting these diverse cell processes in live cells. Many of the assays in this chapter can be analyzed on a cell-by-cell basis and some are equally suitable for detection with a fluorescence microscope, flow cytometer or microplate reader. Most of the assays have the capacity for high-throughput analysis.

Our viability and cytotoxicity assay reagents (Section 15.2) and kits (Section 15.3) are principally used to enumerate the proportion of live and dead cells in a population.[1] In contrast, proliferation assays such as the Click-iT® EdU cell proliferation assay (Section 15.4) are primarily designed to monitor the growth rate of a cell population or to detect daughter cells in a growing population. Fluorescence-based cell viability and proliferation assays are generally less hazardous and less expensive than radioisotopic techniques, more sensitive than colorimetric methods and more convenient than animal testing methods. Unlike ^{51}Cr-release assays, fluorescence-based assays of cell-mediated cytotoxicity do not require large samples, which can be difficult to obtain.[2] Furthermore, fluorescence-based protocols are more convenient than the trypan blue–based exclusion assay. This common colorimetric method for determining cell viability must be completed within 3–5 minutes because the number of blue-staining cells increases with time after addition of the dye.[3]

Fluorescent dye–based assays for cell viability and cytotoxicity are reliable and easy to perform. Our stand-alone reagents for these assays are described in Section 15.2, whereas our kits for viability and cytotoxicity analysis are discussed in Section 15.3. Molecular Probes® LIVE/DEAD® Viability Assay Kits (Section 15.3) give researchers a choice of viability/cytotoxicity assays suitable for bacteria, fungi or higher eukaryotic cells. Our LIVE/DEAD® Reduced Biohazard Cell Viability Kit and LIVE/DEAD® Fixable Dead Cell Stain Kits (Section 15.3) permit the original viability status of a mixed-cell population to be determined following aldehyde fixation of the sample in order to kill pathogens or in preparation for antibody staining. In each case, these viability assay kits provide the reagents and a simple protocol for simultaneous quantitation of live and dead cells. We also offer several proliferation assay kits that enable researchers to rapidly monitor numbers of adherent or nonadherent cells based on the presence of newly replicated DNA, total nucleic acid content or total protein content (Section 15.4). Section 15.5 focuses on our probes for monitoring apoptosis, including reagents for selectively detecting apoptotic cells based on their cell-permeability properties, as well as conjugates of annexin V phosphatidylserine-binding protein for detecting phosphatidylserine externalization. Additionally, our Premo™ Autophagy Sensors and other fluorescent probes are useful for examining the role of autophagy in normal and diseased cells.

In addition to our probes for cell viability, cell proliferation, apoptosis and autophagy, several of the reagents for live-cell function described in Section 15.6 can be used to develop assays that measure a particular biochemical parameter of interest. There is a significant overlap between probes for cell viability and probes for live-cell functions. For example, fluorogenic esterase substrates are commonly used to detect viability and proliferation, as well as to monitor cell adhesion, apoptosis and multidrug resistance. Likewise, cell-permeant and cell-impermeant nucleic acid stains are widely applicable to many live-cell function assays. We have organized discussions in this chapter according to several commonly studied cell processes in order to highlight the many published applications for these probes and foster the development of new applications.

The diversity of live cells and their environments (Figure 15.1.1) makes it impossible to devise a single viability or enumeration assay applicable to all cell types. Because viability is not easily defined in terms of a single physiological or morphological parameter,[1] it is often desirable to combine several different measures, such as enzymatic activity, membrane permeability and oxidation–reduction (redox) potential. Each assay method has inherent advantages and limitations and may introduce specific biases into the experiment; thus, different applications often call for different approaches.

REFERENCES

1. Cell Death Differ (2009) 16:3; 2. Hum Immunol (1993) 37:264; 3. J Histochem Cytochem (1985) 33:77.

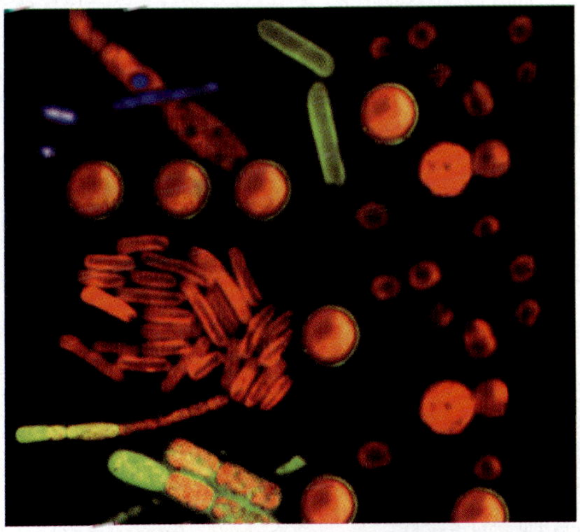

Figure 15.1.1 Collage of images of cyanobacteria stained with various blue- or green-fluorescent probes to complement the natural red autofluorescence from chlorophyll and phycobilisomes. The round cells are *Synechocystis* sp. (strain PCC 6803), and their membranes were labeled with green-fluorescent BODIPY® FL propionic acid (D2183). The cylindrical cells are *Synechococcus* sp. (strain PCC 7942), stained with blue-fluorescent DAPI (D1306, D3571, D21490). The filamentous cyanobacteria, *Anabaena cylindrica*, were labeled with either the green-fluorescent cystosolic stain, CellTracker™ Green BODIPY® (C2102), or with the lipophilic membrane stain BODIPY® FL proprionic acid (D2183). The image was contributed by Mary Sarcina, University College, London.

15.2 Viability and Cytotoxicity Assay Reagents

Fluorometric assays of cell viability and cytotoxicity are easy to perform with the use of a fluorescence microscope, fluorometer, fluorescence microplate reader or flow cytometer,[1] and they offer many advantages over traditional colorimetric and radioactivity-based assays. This section describes our numerous reagents for conducting viability and cytotoxicity assays in a wide variety of cells, including those of animal origin as well as bacteria and yeast. Following this discussion of individual reagents is Section 15.3, which contains a thorough description of each of our viability and cytotoxicity kits, including the:

- LIVE/DEAD® Viability/Cytotoxicity Kit (L3224)
- LIVE/DEAD® Reduced Biohazard Cell Viability Kit (L7013)
- LIVE/DEAD® Fixable Dead Cell Stain Kits (8 different fluorescent stain kits and the Sampler Kit L34960)
- LIVE/DEAD® Cell-Mediated Cytotoxicity Kit (L7010)
- LIVE/DEAD® Sperm Viability Kit (L7011)
- LIVE/DEAD® Cell Vitality Assay Kit (L34951)
- LIVE/DEAD® Violet Viability/Vitality Kit (L34958)
- Vybrant® Cell Metabolic Assay Kit, with C_{12}-resazurin (V23110)
- Vybrant® Cytotoxicity Assay Kit, G6PD release assay (V23111)
- LIVE/DEAD® Yeast Viability Kit (L7009)
- LIVE/DEAD® *Funga*Light™ Yeast Viability Kit (L34952)
- *Funga*Light™ CFDA AM/Propidium Iodide Yeast Vitality Kit (F34953)
- LIVE/DEAD® *Bac*Light™ Bacterial Viability Kits (L7007, L7012, L13152, L34856)
- *Bac*Light™ RedoxSensor™ Vitality Kits (B34954, B34956)
- *Bac*Light™ Bacterial Membrane Potential Kit (B34950)
- ViaGram™ Red+ Bacterial Gram Stain and Viability Kit (V7023)

Also discussed both in this section and Section 15.3 are our unique single-step reagents and kits for assessing gram sign and for simultaneously determining gram sign and viability of bacteria, as well as our novel fluorescent antibiotics. Section 15.4 describes our important probes for quantitating cell proliferation, analyzing the cell cycle and detecting the presence of bacteria and mycoplasma.

Table 15.1 Esterase substrates for cell viability studies.

Esterase Substrate (Cat. No.)	Properties in Cells	pK_a of Product *
BCECF AM (B1150, B1170, B3051)	• Quite well retained • Released during cytolysis • pH-sensitive fluorescence	7.0
Calcein AM (C1430, C3099, C3100MP)	• Quite well retained • Released during cytolysis • Not as pH-sensitive as BCECF	5
Carboxyeosin diacetate, succinimidyl ester (CEDA, SE; C22803; Section 15.4)	• Well retained by reaction with amines • Useful for DAB photoconversion • Phosphorescent	<5 †
Carboxy-2′,7′-dichlorofluorescein diacetate (carboxy-DCFDA, C369)	• Moderately well retained • Not as pH-sensitive as CFDA	4.8
Carboxyfluorescein diacetate (5(6)-CFDA, C195)	• Moderately well retained • pH-sensitive fluorescence	6.4
Carboxyfluorescein diacetate, acetoxymethyl ester (5-CFDA, AM; C1354)	• Easier to load than CFDA yet yields the same product upon hydrolysis • pH-sensitive fluorescence	6.4
Carboxyfluorescein diacetate, succinimidyl ester (5(6)-CFDA, SE; C1157; Section 15.4)	• Well retained by reaction with amines • Not completely released during cytolysis • pH-sensitive fluorescence	6.4 †
CellTracker™ Green CMFDA (CMFDA; C2925, C7025)	• Well retained by reaction with thiols • Not completely released during cytolysis • pH-sensitive fluorescence	6.4 †
Chloromethyl SNARF®-1, acetate (C6826)	• Well retained by reaction with thiols • Not completely released during cytolysis • Long-wavelength, pH-sensitive fluorescence	7.5 †
Fluorescein diacetate (F1303)	• Poorly retained • pH-sensitive fluorescence • Inexpensive	6.4
Oregon Green® 488 carboxylic acid diacetate (carboxy-F_2FDA, O6151)	• Moderately well retained • Not as pH-sensitive as CFDA	4.7

* Approximate pK_a values in aqueous solvents. The actual pK_a of the indicator will vary somewhat depending upon experimental conditions. † pK_a of the unconjugated hydrolysis product; after conjugation to an intracellular amine or thiol, the actual pK_a value may be different.

IMPORTANT NOTICE: The products described in this manual are covered by one or more Limited Use Label License(s). Please refer to the Appendix on page 971 and Master Product List on page 975. Products are For Research Use Only. Not intended for any animal or human therapeutic or diagnostic use.

Viability/Cytotoxicity Assays Using Esterase Substrates

We prepare a wide variety of fluorogenic esterase substrates—including calcein AM, BCECF AM and various fluorescein diacetate derivatives—that can be passively loaded into adherent and nonadherent cells. These cell-permeant esterase substrates serve as viability probes that measure both enzymatic activity, which is required to activate their fluorescence, and cell-membrane integrity, which is required for intracellular retention of their fluorescent products.

As electrically neutral or near-neutral molecules, the esterase substrates freely diffuse into most cells. In general, cell loading of acetate or acetoxymethyl ester derivatives is accomplished by initially preparing a 1–10 mM stock solution of the dye in dimethylsulfoxide (DMSO) and then diluting the stock solution into the cell medium to a final concentration of 1–25 µM (Loading and Calibration of Intracellular Ion Indicators—Note 19.1). Once inside the cell, these nonfluorescent substrates are converted by nonspecific intracellular esterases into fluorescent products that are retained by cells with intact plasma membranes. In contrast, both the unhydrolyzed substrates and their products rapidly leak from dead or damaged cells with compromised membranes, even when the cells retain some residual esterase activity. Low incubation temperatures and highly charged esterase products usually favor retention, although the rate of dye loss from viable cells also depends to a large extent on cell type (see "Multidrug Resistance" in Section 15.6). For example, mast cells and epithelial cells actively secrete many polar products.[2,3] Table 15.1 lists Molecular Probes® esterase substrates that have been used for cell viability studies and compares their cell loading, retention and pH sensitivity. Many of the applications of these esterase substrates—for example, viability, cytotoxicity and adhesion assays—closely parallel those of ^{51}Cr-release assays, except that the fluorescent probes do not carry the risks or the disposal costs associated with the use of radioactive materials.

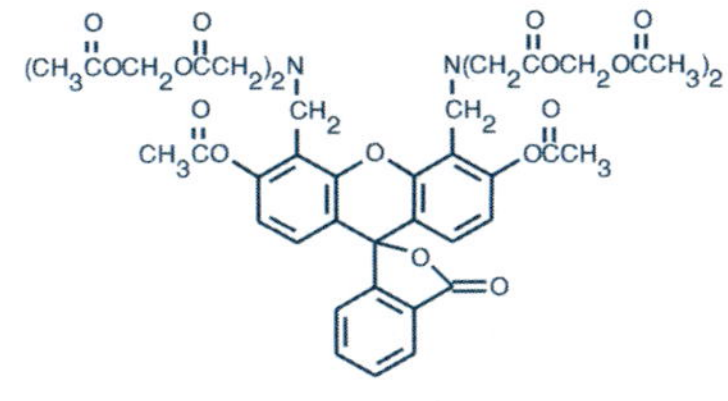

Figure 15.2.1 Calcein, AM (C1430).

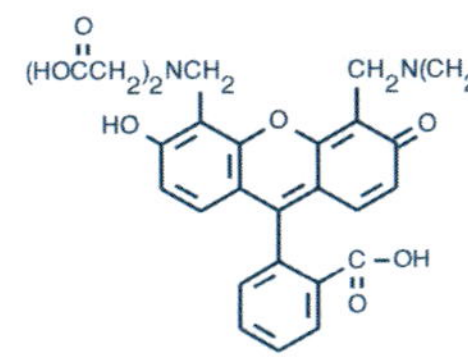

Figure 15.2.2 Calcein (C481).

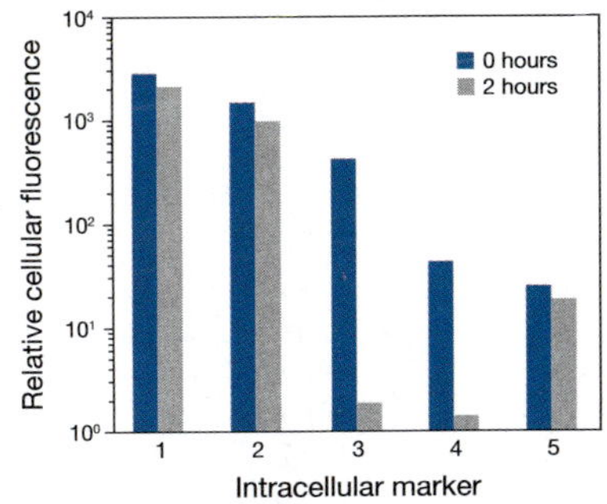

Figure 15.2.3 Loading and retention characteristics of intracellular marker dyes. Cells of a human lymphoid line (GePa) were loaded with the following cell-permeant acetoxymethyl ester (AM) or acetate derivatives of fluorescein. **1)** calcein AM (C1430, C3099, C3100MP), **2)** BCECF AM (B1150), **3)** fluorescein diacetate (FDA, F1303), **4)** carboxyfluorescein diacetate (CFDA, C1354) and **5)** CellTracker™ Green CMFDA (5-chloromethylfluorescein diacetate, C2925, C7025). Cells were incubated in 4 µM staining solutions in Dulbecco's modified eagle medium containing 10% fetal bovine serum (DMEM+) at 37°C. After incubation for 30 minutes, cell samples were immediately analyzed by flow cytometry to determine the average fluorescence per cell at time zero (0 hours). Retained cell samples were subsequently washed twice by centrifugation, resuspended in DMEM+, maintained at 37°C for 2 hours and then analyzed by flow cytometry. The decrease in the average fluorescence intensity per cell in these samples relative to the time zero samples indicates the extent of intracellular dye leakage during the 2-hour incubation period.

CellTrace™ Calceins: Calcein AM, Calcein Blue AM, Calcein Violet AM and Calcein Red-Orange AM

Of the dyes listed in Table 15.1, calcein AM (C1430, C3099, C3100MP; Figure 15.2.1) stands out as the premier indicator of cell viability due to its superior cell retention and the relative insensitivity of its fluorescence to pH in the physiological range.[4–7] Calcein AM, also called CellTrace™ calcein green AM (C34852), is a widely used probe for assays of cell adhesion, chemotaxis and multidrug resistance (Section 15.6). Calcein (C481, Section 14.3), which is the hydrolysis product of calcein AM, is a polyanionic fluorescein derivative (Figure 15.2.2) that has about six negative charges and two positive charges at pH 7.[8] Calcein is better retained in viable cells than are fluorescein, carboxyfluorescein and BCECF (Figure 15.2.3) and tends to have brighter fluorescence in a number of mammalian cell types. Calcein AM has the ability to penetrate intact cornea, revealing cell viability, morphology and organization of living cornea.[9,10] Furthermore, unlike some other dyes—including BCECF AM—calcein AM does not interfere with leukocyte chemotaxis or superoxide production, nor does it affect lymphocyte–target cell conjugation.[6,11–14] Leakage of calcein from calcein AM–loaded cells has been used to measure the increase in membrane permeability that occurs above physiological temperatures,[15] as well as to assay for cytotoxic T lymphocyte activity.[5] Fluorescence of extracellular calcein that has leaked from cells or that has been lost during secretion, lysis or ATP-dependent anion transport can be quenched by 5 µM Co^{2+} ion or by Mn^{2+} ion. Heavy-atom quenching of calcein provides a means of detecting dye leakage and quantitating only the intracellular fluorescence.[16]

Dihydrocalcein AM (Figure 15.2.4) is a reasonably stable, chemically reduced form of calcein AM that requires *both* hydrolysis by intracellular esterases and oxidation within the cell to produce the green-fluorescent calcein dye. Dihydrocalcein AM resembles 2′,7′-dichlorodihydrofluorescein diacetate (D399), the important indicator for oxidative activity in cells (see below), except that its oxidation product (calcein, Figure 15.2.2) should be better retained in cells than is the oxidation product of 2′,7′-dichlorodihydrofluorescein diacetate. Dihydrocalcein AM (D23805) is available as a set of 20 vials, each containing 50 µg of the product.

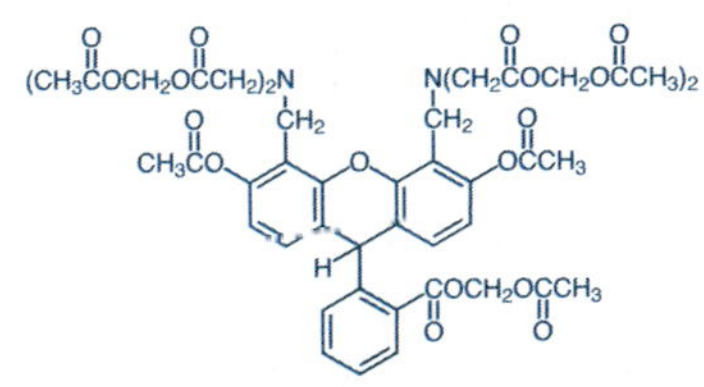

Figure 15.2.4 Dihydrocalcein, AM (D23805).

Calcein blue AM, also called CellTrace™ calcein blue AM (C1429, C34853; Figure 15.2.5), is a viability indicator for use with instruments optimized for the detection of blue fluorescence.[17,18] This tracer possesses AM esters that allow its passive diffusion across cell membranes. Before deesterification, CellTrace™ calcein blue AM is only weakly fluorescent (excitation/emission maxima ~322/435 nm). Upon cleavage of the AM esters by intracellular esterases, however, this tracer becomes relatively polar and is retained by cells for several hours. In addition, its fluorescence intensity increases and its fluorescence spectra shifts to longer wavelengths, with excitation/emission

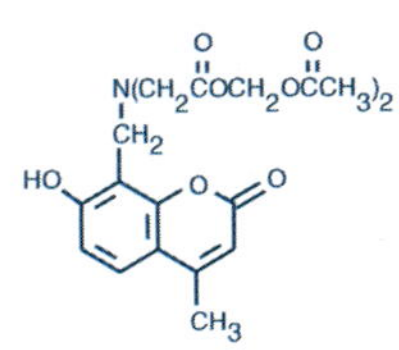

Figure 15.2.5 Calcein blue, AM (C1429).

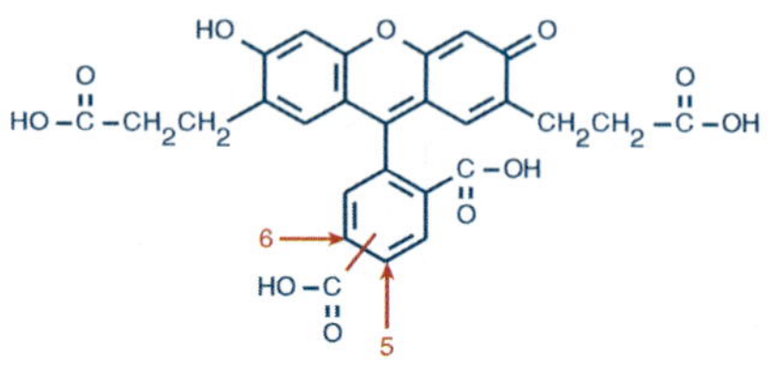

Figure 15.2.6 2′,7′-Bis-(2-carboxyethyl)-5-(and-6)-carboxyfluorescein (BCECF acid, B1151).

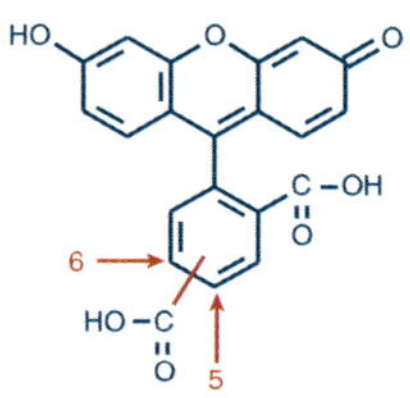

Figure 15.2.7 5-(and-6)-Carboxyfluorescein (C194).

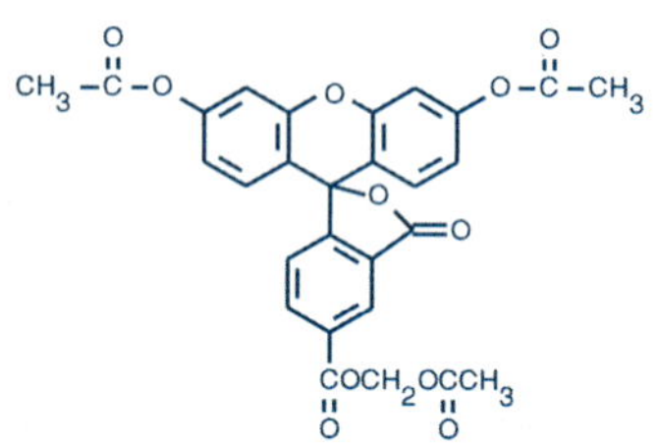

Figure 15.2.8 5-CFDA, AM (5-carboxyfluorescein diacetate, acetoxymethyl ester; C1354).

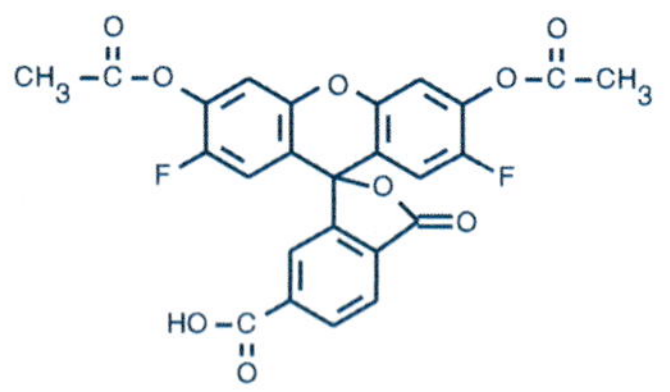

Figure 15.2.9 Oregon Green® 488 carboxylic acid diacetate (carboxy-DFFDA, O6151).

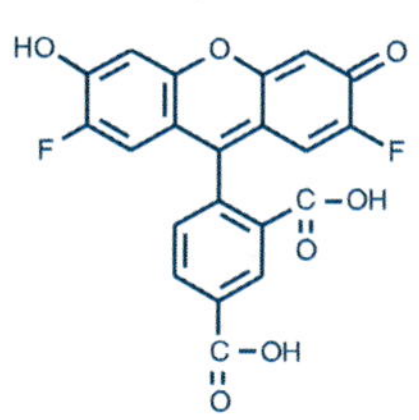

Figure 15.2.10 Oregon Green® 488 carboxylic acid (O6146).

maxima of ~360/449 nm. Calcein blue AM is useful for viability measurements in combination with our SYTOX® Green nucleic acid stain (see below) and other green- or red-fluorescent probes.

Calcein Violet AM (C34858) is optimized for use in flow cytometry. The enzymatic conversion of the virtually nonfluorescent cell-permeant calcein violet AM to the intensely fluorescent calcein violet (excitation/emission maxima ~400/452 nm) is efficiently excited by the 405 nm violet diode laser.

We also offer CellTrace™ calcein red-orange AM (C34851). Upon cleavage by intracellular esterases, CellTrace™ calcein red-orange (excitation/emission maxima ~576/589 nm) is well retained by live cells that possess intact plasma membranes. Unlike calcein AM, CellTrace™ calcein red-orange AM is intrinsically fluorescent; thus, an additional wash step may be necessary to minimize background fluorescence from dye that is not taken up by cells. CellTrace™ calcein red-orange AM is a useful cell tracer and indicator of cell viability and can be used in combination with green-fluorescent probes such as the Fluo-4 Ca^{2+} indicator (Section 19.3).[19]

BCECF AM

BCECF AM (B1150, B1170, B3051) is extensively used for detecting cytotoxicity and for determining the ability of surviving cells to proliferate.[20,21] The intracellular hydrolysis product of BCECF AM, BCECF (B1151, Section 20.2), has 4–5 negative charges (Figure 15.2.6), a property that considerably improves its cell retention in viable cells over that of fluorescein or carboxyfluorescein (Figure 15.2.3). However, because the emission intensity of BCECF is only half-maximal at pH 7.0 ($pK_a = 6.98$)—and is even further reduced in a cell's acidic compartments—the signal intensity of BCECF may be less than optimal in some cell viability and cell adhesion assays.

Using monoclonal antibodies known to either enhance or inhibit natural killer (NK) cell function, researchers found that BCECF AM was at least as effective as ^{51}Cr for measuring NK activity. Furthermore, the fluorescence-based assay could be performed with smaller samples.[22] BCECF AM has also been used to screen for trypanocidal activity[23] and viability of islets.

Fluorescein Diacetate

Fluorescein diacetate (FDA, F1303) was one of the first probes to be used as a fluorescent indicator of cell viability.[24–26] FDA is still occasionally used to detect cell adhesion[27] or, in combination with propidium iodide (P1304MP, P3566, P21493), to determine cell viability.[28,29] However, fluorescein (F1300, Section 20.2), which is formed by intracellular hydrolysis of FDA, rapidly leaks from cells (Figure 15.2.3). Thus, other cell-permeant dyes such calcein AM and BCECF AM are now preferred for cell viability assays.

Carboxyfluorescein Diacetate and Its Derivatives

The high leakage rate of fluorescein from cells[26,30] prompted the development of carboxyfluorescein diacetate (CFDA), which was originally used to measure intracellular pH[31] but was soon adapted for use as a cell viability indicator.[32,33] Upon hydrolysis by intracellular nonspecific esterases, CFDA forms carboxyfluorescein (C194, C1904; Section 14.3). As compared with fluorescein, carboxyfluorescein contains extra negative charges (Figure 15.2.7) and is therefore better retained in cells[6] (Figure 15.2.3). CFDA is moderately permeant to most cell membranes, with uptake greater at pH 6.2 than at pH 7.4.[31] The mixed-isomer preparation of CFDA (5(6)-CFDA, C195) is usually adequate for cell viability measurements; however, we also prepare high-purity single isomers of CFDA (C1361, C1362). In addition, we offer the electrically neutral AM ester of CFDA (5-CFDA, AM; C1354; Figure 15.2.8), which can be loaded into cells at lower concentrations than CFDA. Upon hydrolysis by intracellular esterases, this AM ester also yields carboxyfluorescein.[34–36] CFDA, CFDA AM and sulfofluorescein diacetate (see below) have been proposed for detection of living organisms on Mars.[37] Hemoglobin can be used to quench extracellular fluorescence due to leakage of probes or leakage of products, such as fluorescein or carboxyfluorescein.[38] Alternatively, antibodies directed against the fluorescein hapten (Section 7.4, Table 7.8) or the membrane-impermeant reagent trypan blue can be used to quench low levels of extracellular fluorescence of some fluorescein-based dyes.

CFDA has been used as a viability probe with a variety of cells, including bacteria,[39] fungi (e.g., *Saccharomyces cerevisiae*),[40] spermatozoa,[41] natural killer (NK) cells[19,42] and tumor cells.[43] Cytotoxicity assays using either CFDA or 5-(and 6-)carboxy-2′,7′-dichlorofluorescein diacetate (carboxy-DCFDA, C369) show good correlation with results obtained using the radioisotopic ^{51}Cr-release method.[19,44] With its low pK_a, carboxy-DCFDA is frequently used as a selective

IMPORTANT NOTICE: The products described in this manual are covered by one or more Limited Use Label License(s). Please refer to the Appendix on page 971 and Master Product List on page 975. Products are For Research Use Only. Not intended for any animal or human therapeutic or diagnostic use.
www.invitrogen.com/probes
molecular probes® | invitrogen™ by Life technologies™

probe for the relatively acidic yeast vacuole.[45–47] Oregon Green® 488 carboxylic acid diacetate (carboxy-DFFDA, O6151, Figure 15.2.9) also exhibits a low pK_a (~4.7) and may be similarly useful as a vital stain. Its intracellular hydrolysis product—Oregon Green® 488 carboxylic acid (O6146, Section 1.5, Figure 15.2.10)—is more photostable than carboxyfluorescein.

Sulfofluorescein Diacetate

Sulfofluorescein diacetate (SFDA, S1129), which is converted by intracellular esterases to fluorescein sulfonic acid (F1130, Section 14.3), is more polar than CFDA and consequently may be more difficult to load into some viable cells. However, SFDA's polar hydrolysis product, fluorescein sulfonic acid, is better retained in viable cells than is carboxyfluorescein.[48–51] SFDA was used to stain live bacteria and fungi in soil; little interference from autofluorescence of soil minerals or detritus was observed.[17,52]

CellTracker™ Green CMFDA

The CellTracker™ dyes are thiol-reactive fluorescent dyes that are retained in many live cells through several generations (Figure 15.2.3) and are not transferred to adjacent cells in a population (Figure 15.2.11, Figure 15.2.12, Figure 15.2.13), except possibly through gap junctions. These dyes represent a significant breakthrough in the cellular retention of fluorescent probes and are ideal long-term tracers for transplanted cells or tissues (Section 14.2).

CellTracker™ Green CMFDA (C2925, C7025) freely diffuses into the cell, where its weakly thiol-reactive chloromethyl moieties react with intracellular thiols and their acetate groups are cleaved by cytoplasmic esterases (Figure 15.2.14), generating the fluorescent product (Figure 15.2.15). The other CellTracker™ probes (coumarin, BODIPY® and tetramethylrhodamine derivatives; Section 14.2) do not require enzymatic cleavage to activate their fluorescence. Because the CellTracker™ dyes may react with both glutathione and proteins, cells with membranes that become compromised after staining may retain some residual fluorescent conjugates. However, use of a membrane-impermeant probe such as propidium iodide (P1304MP, P3566, P21493), SYTOX® Blue (S11348, S34857), SYTOX® Orange (S11368), SYTOX® Red (S34859), SYTOX® AADvanced™ (S10274, S10349) or one of our "dimeric" or "monomeric" nucleic acid stains (see below) in combination with CellTracker™ Green CMFDA should permit relatively long-term cytotoxicity assays. CellTracker™ Green CMFDA and ethidium homodimer-1 (EthD-1, E1169) have been used to detect viable and nonviable cells in rat and human coronary and internal thoracic arteries sampled at autopsy[53] and in connective tissue explants.[54]

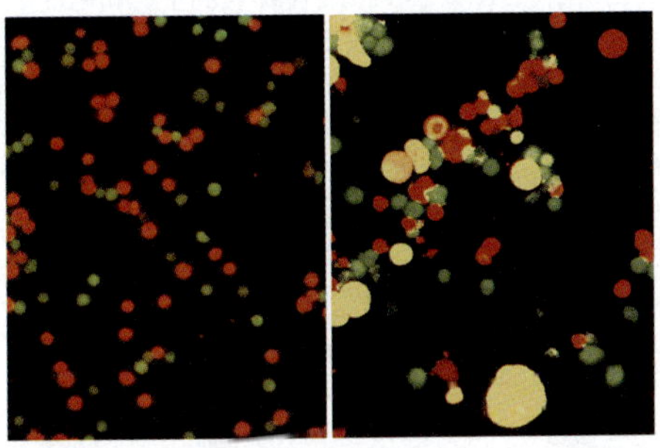

Figure 15.2.11 HL60 cells that have been stained with CellTracker™ Orange CMTMR (C2927) and then mixed with WEHI 7.1 cells stained with CellTracker™ Green CMFDA (C2925, C7025) (left). Several minutes after initiating cell–cell electrofusion, a CMTMR stained HL60 cell is observed fusing with a CMFDA-stained WEHI 7.1 cell; cytoplasmic mixing is evident by the appearance of yellow fluorescence. After electrofusion is complete, dual-fluorescing (yellow) hybrids can be easily distinguished (right). Images contributed by Mark J. Jaroszeski, University of South Florida.

Figure 15.2.12 Individual populations of mouse myeloma (P3X) cells stained with our reactive tracers—CellTracker™ Orange CMTMR (C2927), CellTracker™ Green CMFDA (C2925) and CellTracker™ Blue CMAC (C2110). Each of three cell populations was stained with a different tracer and then the populations were mixed, demonstrating that these tracers allow simultaneous long-term monitoring of different groups of cells in transplantation and other assays.

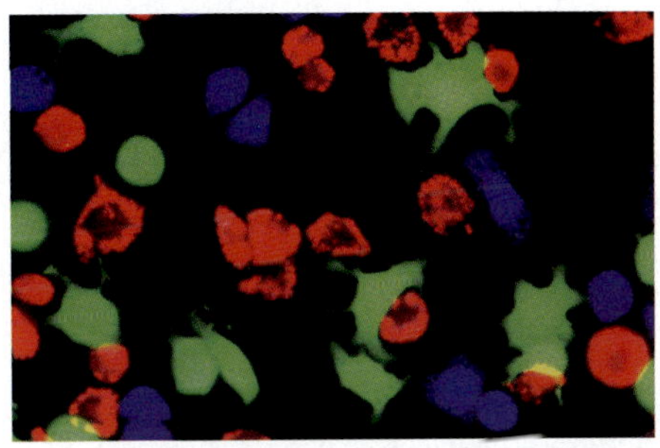

Figure 15.2.13 Rat basophilic leukemia (RBL) cells labeled in suspension with CellTracker™ Blue CMAC (C2110), CellTracker™ Green CMFDA (C2925, C7025) or CellTracker™ Orange CMTMR (C2927) prior to plating. The image was acquired using optical filter sets appropriate for DAPI, fluorescein and the Texas Red® dye.

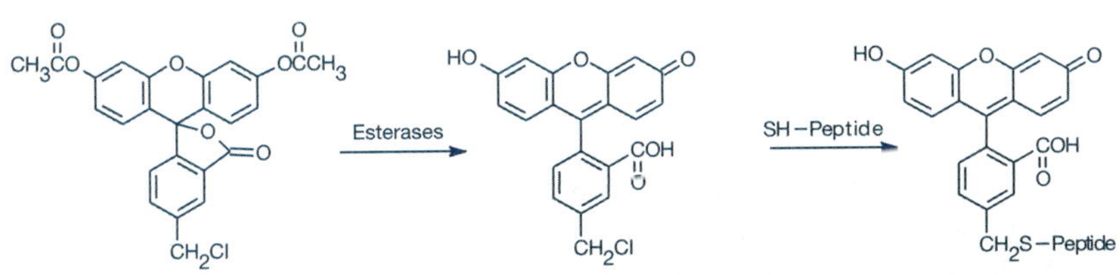

Figure 15.2.14 Intracellular reactions of our fixable CellTracker™ Green CMFDA (5-chloromethylfluorescein diacetate; C2925, C7025). Once this membrane-permeant probe enters a cell, esterase hydrolysis converts nonfluorescent CMFDA to fluorescent 5-chloromethylfluorescein, which can then react with thiols on proteins and peptides to form aldehyde-fixable conjugates. This probe may also react with intracellular thiol-containing biomolecules first, but the conjugate is nonfluorescent until its acetates are removed.

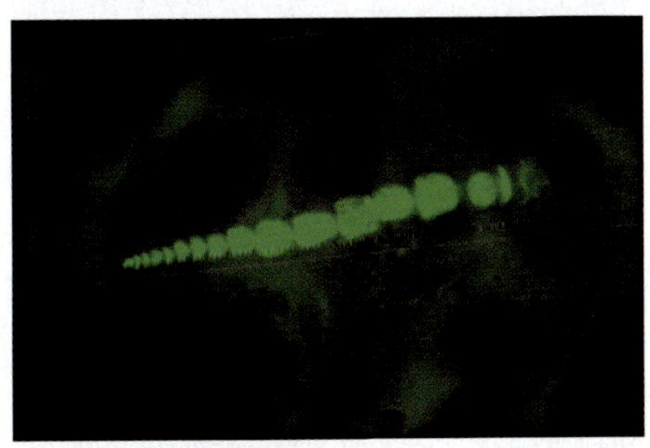

Figure 15.2.15 Detection of organisms in marine sediments by incubating an intact sediment core sample with the fixable, cell-permeant CellTracker™ Green CMFDA (C2925, C7025). The core sample was subsequently embedded, sectioned and examined for fluorescently labeled organisms. The micrograph reveals the microorganism *Leptohalysis scotti*, a marine benthic foraminifera. Image contributed by Joan M. Bernhard, Wadsworth Center, New York State Department of Health.

IMPORTANT NOTICE: The products described in this manual are covered by one or more Limited Use Label License(s). Please refer to the Appendix on page 971 and Master Product List on page 975. Products are For Research Use Only. Not intended for any animal or human therapeutic or diagnostic use.

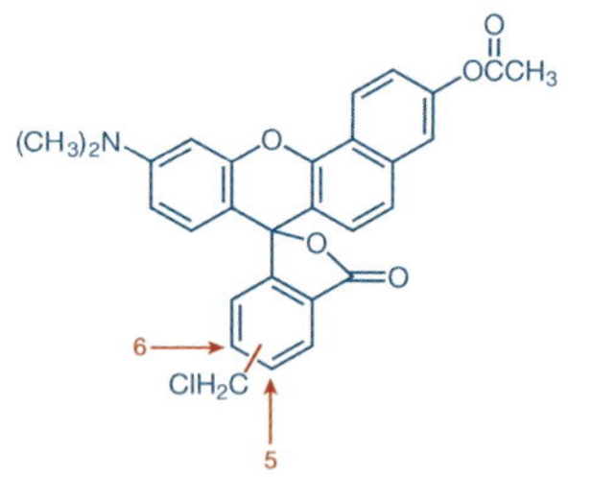

Figure 15.2.16 5-(and-6)-Chloromethyl SNARF®-1, acetate (C6826).

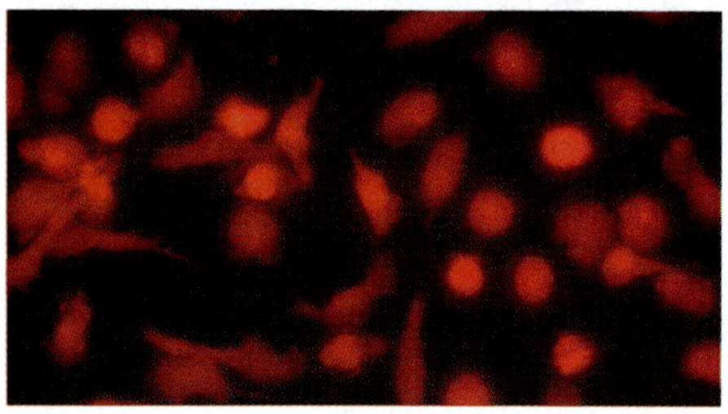

Figure 15.2.17 Human neutrophils loaded with 5-(and-6)-chloromethyl SNARF®-1 acetate (C6826).

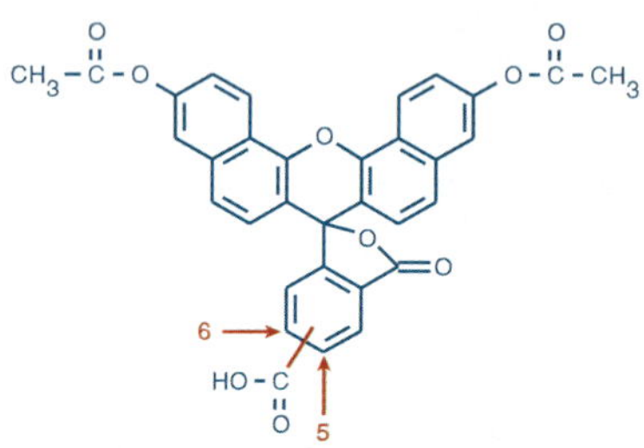

Figure 15.2.18 5-(and-6)-Carboxynaphthofluorescein diacetate (C13196).

Figure 15.2.19 FilmTracer™ FM® 1-43 green biofilm cell stain (F10317) applied to a *Pseudomonas aeruginosa* biofilm. FilmTracer™ FM® 1-43 green biofilm cell stain appears to bind to the cell membrane. This stain has been shown to work equally well on *Staphylococcus epidermidis*, *Pseudomonas aeruginosa* and *Escherichia coli*, exhibiting exceptional cell specificity in each case. The image was obtained using a Leica TCS-SP2 AOBS confocal microscope and a 63×/0.9 NA water immersion objective. Image contributed by Betsey Pitts and Ellen Swogger, Center for Biofilm Engineerig, Montana State University.

Chloromethyl SNARF®-1 Acetate

Chloromethyl SNARF®-1 acetate (C6826, Figure 15.2.16) is the only cell-tracking dye (and pH indicator) that exhibits bright red cytoplasmic fluorescence (Figure 15.2.17) when excited at the same wavelengths used to excite the green-fluorescent hydrolysis product of CMFDA. The spectral characteristics of these two dyes permit simultaneous tracking of two cell populations by either fluorescence microscopy or flow cytometry. The large Stokes shift of the SNARF® fluorophore also makes chloromethyl SNARF®-1 acetate useful as a viability indicator in cells that exhibit green autofluorescence when excited by the 488 nm spectral line of the argon-ion laser.

Carboxynaphthofluorescein Diacetate

Carboxynaphthofluorescein diacetate (C13196, Figure 15.2.18), which is cleaved by intracellular esterases to yield red-fluorescent carboxynaphthofluorescein (excitation/emission maxima ~598/668 nm), is the only long-wavelength tracer of this type that can be passively loaded into live cells.[55] Like chloromethyl SNARF®-1 acetate, carboxynaphthofluorescein diacetate is usually used in combination with a green-fluorescent tracer for detecting cell–cell interactions.

FilmTracer™ Biofilm Stains

Bacterial biofilms present a unique set of challenges for fluorescent staining and subsequent imaging. A typical biofilm not only exhibits heterogeneous thickness throughout the surface, placing stringent restrictions on stain penetration, but also contains regions of widely varying environmental conditions. Evidence suggests that bacterial cells exist in various physiological states within these biofilm microenvironments. Furthermore, biofilms contain many poorly defined components (e.g., the extracellular polymeric matrix) that differ with species and conditions.

FilmTracer™ calcein violet, FilmTracer™ calcein green and FilmTracer™ calcein red-orange biofilm stains (F10320, F10322, F10319) are acetoxymethyl (AM) ester derivatives of fluorescent indicators that provide reliable indication of esterase activity in live cells and are of particular use in biofilm applications. FilmTracer™ calcein violet and FilmTracer™ calcein green biofilm stains are colorless and nonfluorescent until the AM ester is hydrolyzed. FilmTracer™ calcein red-orange biofilm stain is fluorescent prior to cleavage; however, the intracellular fluorescence is much brighter than background fluorescence after rinsing.

FM® dyes are lipophilic styryl compounds useful as general-purpose probes for investigating endocytosis and for identifying cell membrane boundaries (Section 14.4). These styryl dyes are easily applied to cells, where they bind rapidly and reversibly to the plasma membrane with strong fluorescence enhancement.[56–58] FilmTracer™ FM® 1-43 green biofilm cell stain (F10317) has been used successfully to stain the cell bodies in a complex biofilm mileu, including *Pseudomonas aeruginosa* (Figure 15.2.19), *Escherichia coli*, *Staphylococcus* sp., *Acidothiobacillus caldus* and *Vibrio cholerae*.[59]

Our SYPRO® Ruby stain labels most classes of proteins, including glycoproteins, phosphoproteins, lipoproteins, calcium binding proteins, fibrillar proteins and other proteins that are difficult to stain[60] (Section 9.3). FilmTracer™ SYPRO® Ruby Biofilm Matrix Stain (F10318) is a specially formulated version of the SYPRO® Ruby stain and has been found to stain the matrix of *Pseudomonas aeruginosa* (ATCC 15442) and some strains of *Escherichia coli*; it does not stain *E. coli* K-12, which does not produce cellulose. As with all the biofilm stains, staining patterns may vary depending upon the organism and the matrix composition.

Viability/Cytotoxicity and Gram Stain Assays Using Nucleic Acid Stains

Viability assessments of animal cells, bacteria and yeast frequently employ polar and therefore cell-impermeant nucleic acid stains to detect the dead-cell population. Nucleic acid stains are most often used in combination with intracellular esterase substrates (see above), membrane-permeant nucleic acid stains (see below), membrane potential–sensitive probes (Chapter 22), organelle probes (Chapter 12) or cell-permeant indicators (Chapter 19, Chapter 20 and Chapter 21) to simultaneously detect the live-cell population. Although many other cell-impermeant dyes can be used to detect dead cells, the high concentrations of nucleic acids in cells, coupled with the large fluorescence enhancement exhibited by most of our nucleic acid stains upon binding, make cell-impermeant nucleic acid stains the logical candidates for viability probes. See Table 8.2 for

IMPORTANT NOTICE: The products described in this manual are covered by one or more Limited Use Label License(s). Please refer to the Appendix on page 971 and Master Product List on page 975. Products are For Research Use Only. Not intended for any animal or human therapeutic or diagnostic use.
www.invitrogen.com/probes

a list of several cell-impermeant nucleic acid stains and Section 8.1 for a general discussion of dye binding to nucleic acids.

SYTOX® Nucleic Acid Stains

Many polar nucleic acid stains will enter eukaryotic cells with damaged plasma membranes yet will not stain dead bacteria with damaged plasma membranes. SYTOX® Green nucleic acid stain (S7020) is a high-affinity probe that easily penetrates eukaryotic cells and both gram-positive and gram-negative bacteria with compromised plasma membranes, yet is completely excluded from live cells.[61] After brief incubation with the SYTOX® Green nucleic acid stain, dead bacteria fluoresce bright green when excited with the 488 nm spectral line of the argon-ion laser or any other 470–490 nm source (Figure 15.2.20). These properties, combined with its ~1000-fold fluorescence enhancement upon nucleic acid binding, make our SYTOX® Green stain a simple and quantitative dead-cell indicator for use with fluorescence microscopes, fluorometers, fluorescence microplate readers or flow cytometers (Figure 15.2.21). We have taken advantage of the sensitivity of the SYTOX® Green nucleic acid stain in our ViaGram™ Red+ Bacterial Gram Stain and Viability Kit (V7023) and in our Single Channel Annexin V/Dead Cell Apoptosis Kit (V13240, Section 15.5). An important application of the SYTOX® Green nucleic acid stain is the high-throughput screening of bacteria for antibiotic susceptibility by fluorescence microscopy, by flow cytometry or in a fluorescence microplate reader.[62]

The SYTOX® Green nucleic acid stain as a tool for viability assessment is not restricted to bacteria; it is also a very effective cell-impermeant counterstain in eukaryotic systems (Section 12.5). It can be used in conjunction with blue- and red-fluorescent labels for multiparameter analyses in fixed cells and tissue sections (Figure 15.2.22, Figure 15.2.23, Figure 15.2.24). Furthermore, it should be possible to combine the SYTOX® Green nucleic acid stain with one of the membrane-permeant nucleic acid stains in our SYTO® Red-, SYTO® Blue- or SYTO® Orange-Fluorescent Nucleic Acid Stain Sampler Kits (S11340, S11350, S11360) for two-color visualization of dead and live cells.

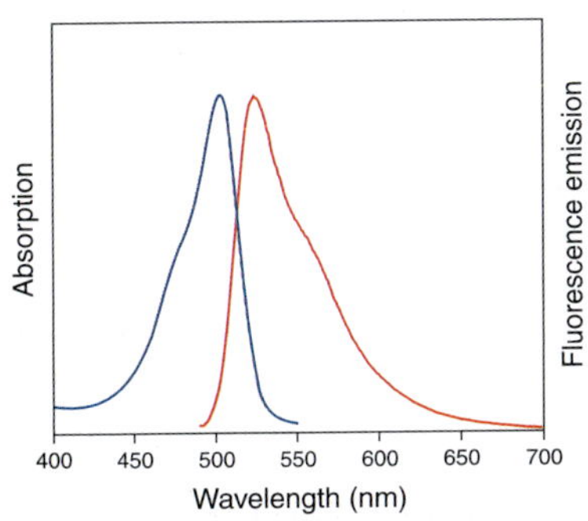

Figure 15.2.20 Absorption and fluorescence emission spectra of the SYTOX® Green nucleic acid stain bound to DNA.

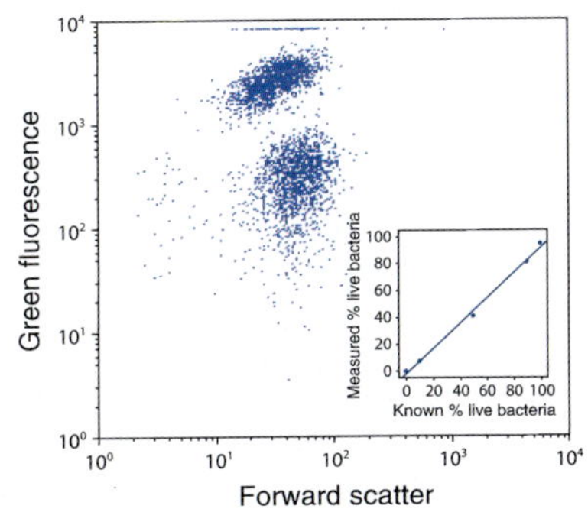

Figure 15.2.21 Quantitative flow cytometric analysis of *Escherichia coli* viability using the SYTOX® Green nucleic acid stain (S7020). A bacterial suspension containing an equal number of live and isopropyl alcohol–killed *E. coli* was stained with SYTOX® Green and analyzed using excitation at 488 nm. A bivariate frequency distribution for forward light scatter versus log fluorescence intensity (collected with a 510 nm longpass optical filter) shows two clearly distinct populations. When live and dead bacteria were mixed in varying proportions, a linear relationship between the population numbers and the actual percentage of live cells in the sample was obtained (see inset).

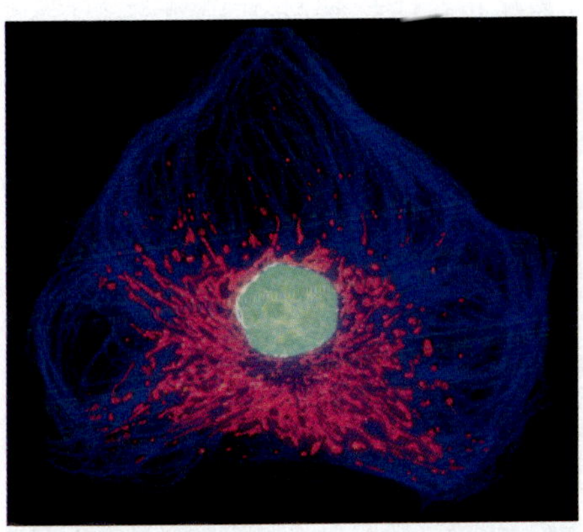

Figure 15.2.22 Bovine pulmonary artery endothelial cells (BPAEC) incubated with the fixable, mitochondrion-selective MitoTracker® Red CMXRos (M7512). After staining, the cells were formaldehyde-fixed, acetone-permeabilized, treated with DNase-free RNase and counterstained using SYTOX® Green nucleic acid stain (S7020). Microtubules were labeled with a mouse monoclonal anti–ß-tubulin antibody, biotin-XX goat anti-mouse IgG antibody (B2763) and Cascade Blue® NeutrAvidin™ biotin-binding protein (A2663). This photograph was taken using multiple exposures through bandpass optical filters appropriate for Texas Red® dye, fluorescein and DAPI using a Nikon® Labophot 2 microscope equipped with a Quadfluor epi-illumination system.

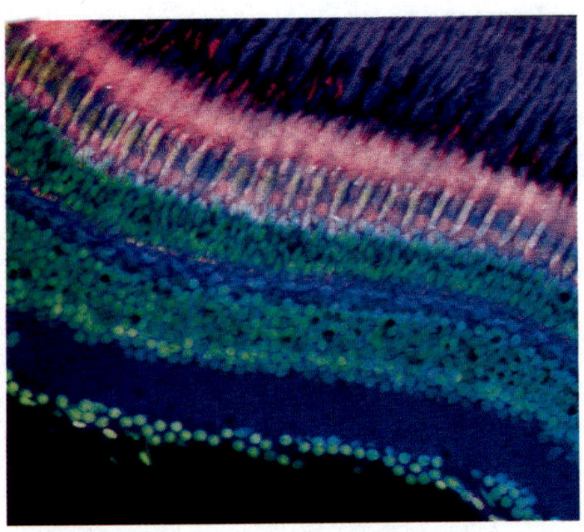

Figure 15.2.23 A frozen section of zebrafish retina stained with mouse monoclonal antibody FRet 43 in conjunction with Texas Red®-X goat anti-mouse IgG (T6390), Alexa Fluor® 350 wheat germ agglutinin (W11263) and SYTOX® Green nucleic acid stain (S7020).

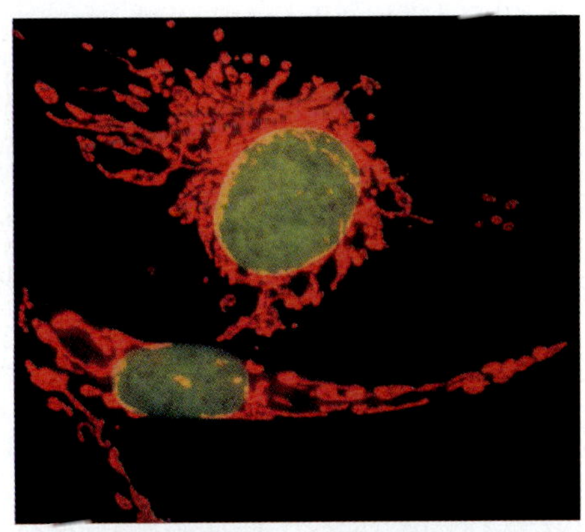

Figure 15.2.24 The mitochondria of bovine pulmonary artery endothelial cells stained with MitoTracker® Red CM-H_2XRos (M7513). The cells were subsequently fixed, permeabilized and treated with RNase. Then the nuclei were stained with SYTOX® Green nucleic acid stain (S7020). The multiple-exposure photomicrograph was acquired using a fluorescence microscope equipped with bandpass filter sets appropriate for fluorescein and Texas Red® dyes.

Like the SYTOX® Green reagent, our SYTOX® Orange (S11368) and SYTOX® Blue nucleic acid stains (5 mM solution in dimethylsulfoxide (DMSO), S11348; 1 mM solution in DMSO, S34857) are high-affinity nucleic acid stains that only penetrate cells with compromised plasma membranes. The SYTOX® Orange nucleic acid stain (S11368, Figure 15.2.25) has absorption/emission maxima of 547/570 nm when bound to DNA and is optimally detected using filters appropriate for rhodamine dyes. As with the other SYTOX® dyes, the SYTOX® Orange stain is virtually nonfluorescent except when bound to nucleic acids and can be used to detect cells that have compromised membranes without a wash step.

Our SYTOX® Blue nucleic acid stain labels both DNA and RNA with extremely bright fluorescence centered near 480 nm (Figure 15.2.26), making it an excellent fluorescent indicator of cell viability (Figure 15.2.27). Unlike many blue-fluorescent dyes, the SYTOX® Blue stain is efficiently excited by the 405 nm violet diode laser. The brightness of the SYTOX® Blue complex with nucleic acids allows sensitive detection of stained cells with fluorometers, fluorescence microplate readers, flow cytometers and epifluorescence microscopes. Quantitation of membrane-compromised bacterial cells carried out with the SYTOX® Blue stain yields results identical to those obtained in parallel assays using the SYTOX® Green stain. Like the SYTOX® Green stain, the SYTOX® Blue stain does not interfere with bacterial cell growth. Because their emission spectra overlap somewhat, we have found that it is not ideal to use SYTOX® Blue stain and green-fluorescent dyes together in the same application, except when the green-fluorescent dye is excited beyond the absorption of the SYTOX® Blue dye (e.g., at >480 nm). However, emission of the SYTOX® Blue complex with nucleic acids permits clear discrimination from red- and orange-fluorescent probes, facilitating development of multicolor assays with minimal spectral overlap between signals.

SYTOX® Red dead cell stain (S34859) is a simple and quantitative single-step dead-cell indicator for use with red laser–equipped flow cytometers. SYTOX® Red dead cell stain is a high-affinity nucleic acid stain that easily penetrates cells with compromised plasma membranes but will not cross uncompromised cell membranes. After brief incubation with SYTOX® Red stain, the nucleic acids of dead cells fluoresce bright red when excited with 633 or 635 nm red laser light. SYTOX® Red nucleic acid stain has absorption/emission maxima of 633/660 nm when bound to DNA and is optimally detected using filters appropriate for Alexa Fluor® 647 dye. SYTOX® Red dead cell stain is distinct from other dead cell probes like 7-AAD and PI, which require 488 nm excitation.

Unlike the other SYTOX® stains, SYTOX® AADvanced™ Dead Cell Stain Kits (S10274, S10349) provide separate vials of dried dye and anhydrous DMSO to help ensure a stable shelf life. SYTOX® AADvanced™ dead cell stain is a high-affinity nucleic acid stain for the detection of dead cells and analysis of cell cycle using the common 488 nm spectral line of the argon-ion laser in flow cytometry. The dye is spectrally similar to 7-AAD but exhibits rapid uptake kinetics and relatively low cellular staining variability. SYTOX® AADvanced™ dead cell stain penetrates cells more efficiently than does 7-AAD, providing better separation of live and dead cell fluorescence signals. SYTOX® AADvanced™ dead cell stain can also be used with fixed cells for DNA content analysis when paired with RNAse treatment.

Dimeric and Monomeric Cyanine Dyes

The dimeric and monomeric cyanine dyes in the TOTO® and TO-PRO® series (Table 8.2) are essentially nonfluorescent unless bound to nucleic acids and have extinction coefficients 10–20 times greater than that of DNA-bound propidium iodide. Spectra of the nucleic acid–bound dyes cover the entire visible spectrum and into the infrared region (Figure 15.2.28). These dyes are typically impermeant to the membranes of live cells [63] but brightly stain dead cells that have compromised membranes. However, YO-PRO®-1 and other polar stains are taken up by some live cells via the $P2X_{1-7}$ receptors [64] as well as by apoptotic cells (Section 15.5), so care needs to be taken when assaying cell viability with nucleic acid stains. The POPO™-1 and BOBO™-1 dyes may be useful blue-fluorescent dead-cell stains, and the YOYO®-3 and TOTO®-3 dyes and the corresponding YO-PRO®-3 and TO-PRO®-3 dyes have excitation maxima beyond 600 nm when bound to DNA. Our JOJO™-1 and JO-PRO™-1 dyes exhibit orange fluorescence (~545 nm) upon binding to nucleic acids and can be excited with a 532 nm Nd:YAG laser. The LOLO™-1 nucleic acid stain has longer-wavelength fluorescence (~580 nm). Our Nucleic Acid Stains Dimer Sampler Kit (N7565)

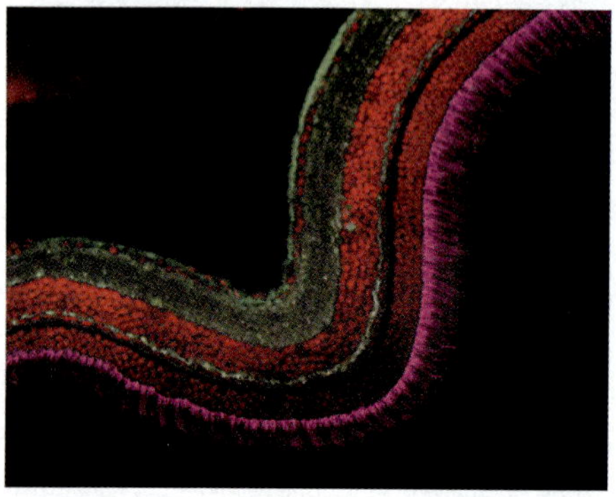

Figure 15.2.25 A zebrafish cryosection incubated with the biotin-XX conjugate of mouse monoclonal anti-α-tubulin antibody (A21371). The signal was amplified with TSA™ Kit #22, which includes HRP–streptavidin and Alexa Fluor® 488 tyramide (T20932). The sample was then incubated with the mouse monoclonal FRet 6 antibody and was visualized with Alexa Fluor® 647 goat anti–mouse IgG (A21235), which is pseudocolored magenta. Finally, the nuclei were counterstained with SYTOX® Orange nucleic acid stain (S11368).

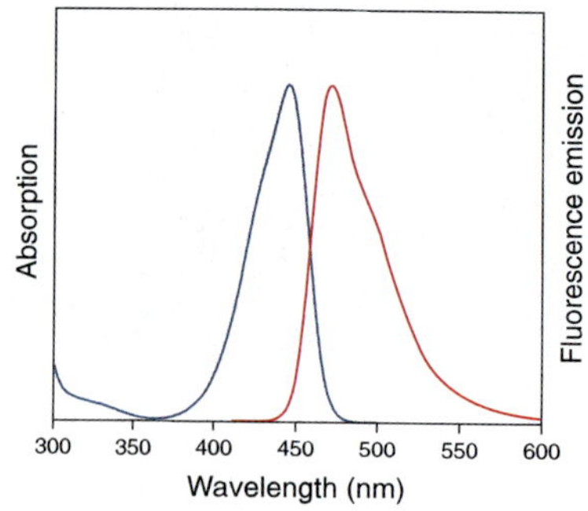

Figure 15.2.26 Absorption and fluorescence emission spectra of SYTOX® Blue nucleic acid stain bound to DNA.

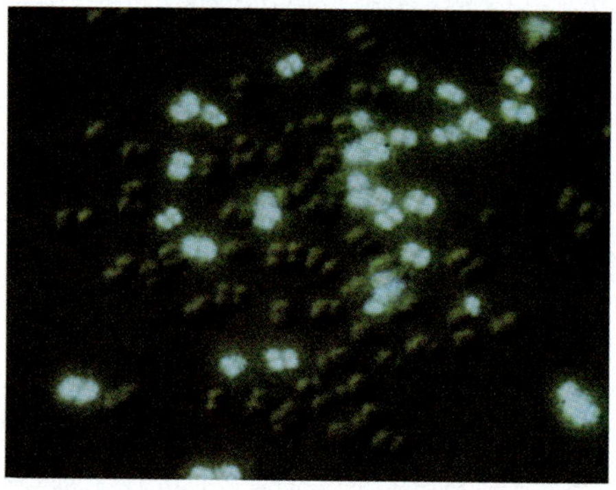

Figure 15.2.27 A mixed population of live and isopropyl alcohol–killed *Micrococcus luteus* stained with SYTOX® Blue nucleic acid stain (S11348), which does not penetrate intact plasma membranes. Dead cells exhibit bright blue-fluorescent staining. The image was acquired using a longpass optical filter set appropriate for the Cascade Blue® dye.

IMPORTANT NOTICE: The products described in this manual are covered by one or more Limited Use Label License(s). Please refer to the Appendix on page 971 and Master Product List on page 975. Products are For Research Use Only. Not intended for any animal or human therapeutic or diagnostic use.
www.invitrogen.com/probes
molecular probes® | invitrogen™ by life technologies™

provides a total of eight cyanine dyes that span the visible spectrum. This kit should be useful when screening dyes for their utility in viability and cytotoxicity assays.

One cell viability assay utilizes YOYO®-1 fluorescence before and after treatment with digitonin as a measure of the dead cells and total cells, respectively, in the sample.[65] The YOYO®-3 dye was used as a stain for dead cells in an assay designed to correlate cell cycle with metabolism in single cells.[66] In addition to their use as dead-cell stains, both the dimeric and monomeric cyanine dyes are also proving useful for staining viruses. TOTO®-1 dye–staining and flow cytometric analysis gave better discrimination of live and dead lactic acid bacteria in several species than did propidium iodide.[67] Viruses stained with the YOYO®-1 and POPO™-1 dyes have been employed to identify and quantitate bacteria and cyanobacteria in marine microbial communities.[68] The YO-PRO®-1 dye has been used to count viruses in marine and freshwater environments by epifluorescence microscopy[69] and is also selectively permeant to apoptotic cells[70] (Section 15.5). TO-PRO®-3 (T3605, Section 15.4) has been utilized to demonstrate transient permeabilization of bacterial cells by sublethal doses of antibiotics.[71]

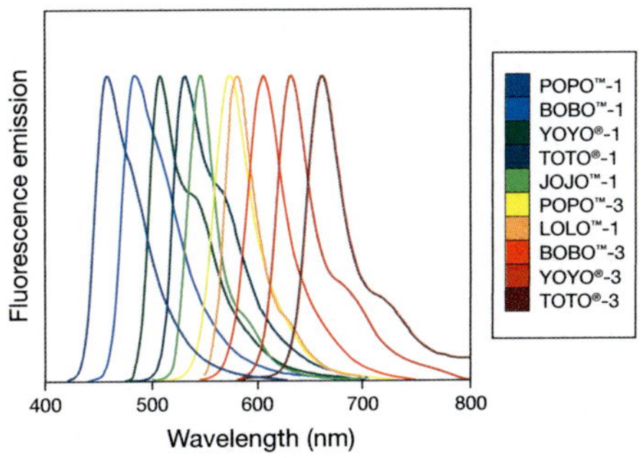

Figure 15.2.28 Normalized fluorescence emission spectra of DNA-bound cyanine dimers, identified by the color key on the sidebar.

Ethidium and Propidium Dyes

The red-fluorescent, cell-impermeant ethidium and propidium dyes—ethidium bromide (15585-011), ethidium homodimer-1 (EthD-1, E1169; Figure 15.2.29) and propidium iodide (P1304MP, P3566, P21493)—can all be excited by the argon-ion laser and are therefore useful for detecting and sorting dead cells by flow cytometry.[72,73] Moreover, these dyes have large Stokes shifts and may be used in combination with fluorescein-based probes (such as calcein, CellTracker™ Green CMFDA or BCECF) or green-fluorescent SYTO® dyes (Table 8.3) for two-color applications (Figure 15.2.30). Both propidium iodide and ethidium bromide have been extensively used to detect dead or dying cells,[74–78] although ethidium bromide may be somewhat less reliable because it is not as highly charged. EthD-1 and propidium iodide are superior to ethidium bromide for two-color flow cytometric viability assays in which either BCECF AM or calcein AM is used as the live-cell stain because their spectra do not overlap as much with those of the green-fluorescent esterase probes.[21]

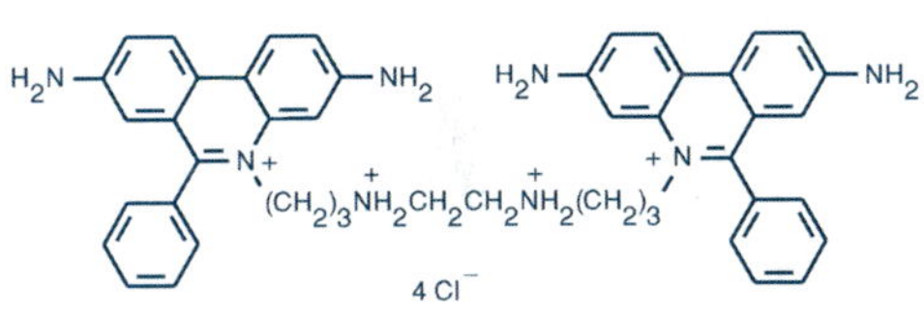

Figure 15.2.29 Ethidium homodimer-1 (EthD-1, E1169).

With its high affinity for DNA and low membrane permeability,[79–82] EthD-1 is often the preferred red-fluorescent dead-cell indicator. EthD-1 binds to nucleic acids 1000 times more tightly than does ethidium bromide and undergoes about a 40-fold enhancement of fluorescence upon binding.[81,83] When used as a viability indicator, EthD-1 typically does not require a wash step. Also, the high affinity of EthD-1 permits the use of very low concentrations to stain dead cells, thus avoiding the use of large quantities of the potentially hazardous ethidium bromide or propidium iodide. EthD-1, the dead-cell indicator in our LIVE/DEAD® Viability/Cytotoxicity Kit (L3224, Section 15.3), has been used alone[84] or in combination with calcein AM[85] to detect tumor necrosis factor activity and to assay neuronal cell death.[75,86,87] Ethidium homodimer-2 (E3599, Figure 15.2.31), which we use as the DEAD Red™ necrotic-cell indicator in our LIVE/DEAD® Reduced Biohazard Cell Viability Kit #1 (L7013, Section 15.3), has a particularly low dissociation rate from cellular nucleic acids, permitting its use for selective marking of dead-cell populations that need to be observed over several hours. Our DEAD Red™ nucleic acid stain has proven useful for determining brain stem lesion size *in vivo* in rats following a neurotoxin injection.[88] Live and dead cells of the yeast-like fungus *Aureobasidium pullulans* have been identified on microscope slides as well as leaf surfaces using CellTracker™ Blue CMAC (C2110, Section 14.2) in conjunction with the DEAD Red™ nucleic acid stain.[89]

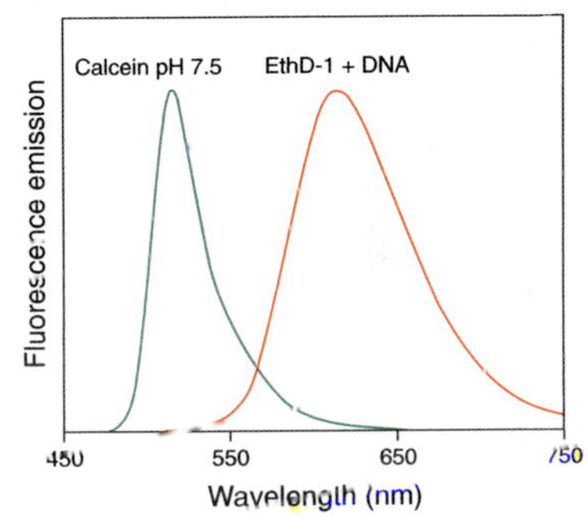

Figure 15.2.30 Normalized fluorescence emission spectra of calcein (C481) and DNA-bound ethidium homodimer-1 (EthD-1, E1169), illustrating the clear spectral separation that allows simultaneous visualization of live and dead eukaryotic cells with Molecular Probes® LIVE/DEAD® Viability/Cytotoxicity Kit (L3224).

Ethidium Monoazide

Ethidium monoazide (E1374) is a fluorescent photoaffinity label that, after photolysis, binds covalently to nucleic acids in solution and in cells with compromised membranes.[90,91] A mixed population of live and dead cells incubated with this membrane-impermeant dye can be illuminated with a visible-light source, washed, fixed and then analyzed in order to determine the viability of the cells at the time of photolysis. Thus, ethidium monoazide reduces some of the hazards inherent in working with pathogenic samples because, once stained, samples can be treated with fixatives before analysis by fluorescence microscopy or flow cytometry.[91] Immunocytochemical analyses requiring fixation are also compatible with this ethidium monoazide–based viability assay. We have developed an alternative two-color fluorescence-based assay for determining the original viability of fixed samples that employs our cell-permeant, green-fluorescent SYTO® 10 and cell-impermeant, red-fluorescent DEAD Red™ nucleic acid stains; the LIVE/DEAD® Reduced Biohazard Cell Viability Kit #1 (L7013) is described in Section 15.3.

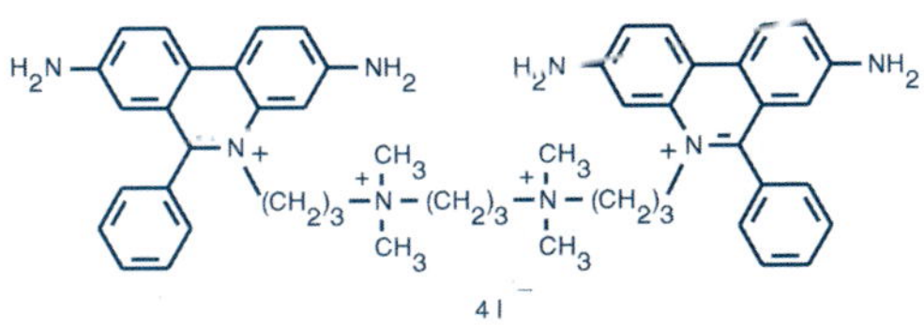

Figure 15.2.31 Ethidium homodimer-2 (EthD-2, E3599).

IMPORTANT NOTICE: The products described in this manual are covered by one or more Limited Use Label License(s). Please refer to the Appendix on page 971 and Master Product List on page 975. Products are For Research Use Only. Not intended for any animal or human therapeutic or diagnostic use.

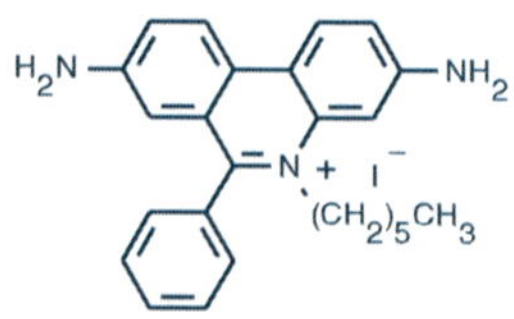

Figure 15.2.32 Hexidium iodide (H7593).

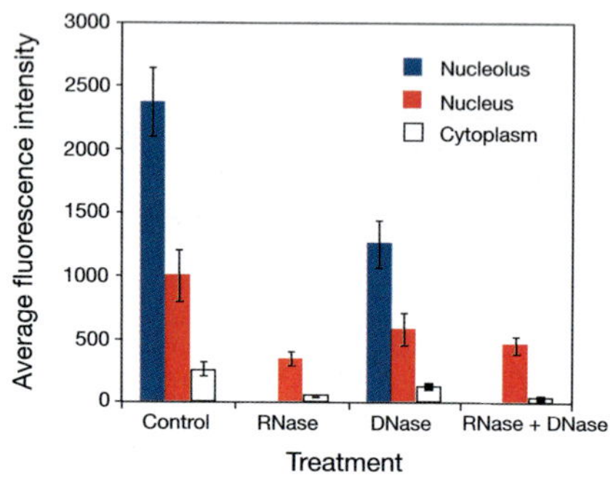

Figure 15.2.33 Methanol-fixed bovine pulmonary artery endothelial cells treated with RNase, DNase or both, and then labeled with SYTO® RNASelect™ Green cell stain (S32703). Removal of RNA with RNase prevented nucleolar labeling and greatly decreased nuclear and cytoplasmic labeling. Use of DNase resulted in less of a loss of label intensity in these cell compartments, reflecting the RNA-selective nature of this dye.

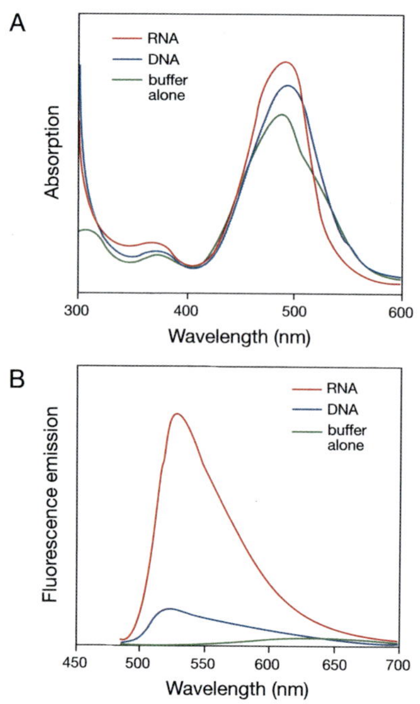

Figure 15.2.34 Absorption (**A**) and fluorescence emission (**B**) spectra of SYTO® RNASelect™ green-fluorescent cell stain (S32703) in the presence of *Escherichia coli* DNA or in buffer alone.

Hexidium Iodide: A Fluorescent Gram Stain

Ethidium bromide is only marginally permeant to cell membranes or bacteria; however, we found that our hexidium iodide stain (H7593) has the right combination of polarity and permeability (Figure 15.2.32) to allow it to rapidly stain most gram-positive bacteria while being excluded by the less-permeant membranes of most gram-negative bacteria.[92] Combining the red-orange–fluorescent hexidium iodide reagent with a green-fluorescent, membrane-permeant nucleic acid stain—as in our LIVE *Bac*Light™ Bacterial Gram Stain Kit (L7005, Section 15.3)—enables taxonomic classification of most bacteria in minutes, using a single staining solution, with no fixatives or wash steps. This rapid gram stain assay should be useful in both clinical and research settings. The validity of using hexidium iodide in combination with the SYTO® 13 green-fluorescent nucleic acid stain to correctly predict the gram sign of 45 clinically relevant organisms, including several known to be gram variable, has been demonstrated.[92] The method of use of hexidium iodide as a gram stain is described further in Section 15.3.

SYTO® Nucleic Acid Stains

Our SYTO® family of dyes, all of which are listed in Table 8.3, are essentially nonfluorescent until they bind to nucleic acids, whereupon their fluorescence quantum yield may increase by 1000-fold or more. These dyes are freely permeant to most cells, although their rate of uptake and ultimate staining pattern may be cell dependent. Their affinity for nucleic acids is moderate and they can be displaced by higher-affinity nucleic acid stains such as SYTOX® Green, propidium iodide, the ethidium dimers and all of the monomeric and dimeric nucleic acid stains described above. Because the membrane of intact cells offers a barrier to entry of these higher-affinity nucleic acid stains, it is common to combine, for instance, a green-fluorescent SYTO® dye with a red-fluorescent, high-affinity nucleic acid stain such as propidium iodide, one of the ethidium homodimers, or TOTO®-3 for simultaneous staining of the live- and dead-cell populations. Although the green-fluorescent SYTO® dye will still bind to nucleic acids in dead cells, it will be displaced or its fluorescence quenched by the red-fluorescent dye, resulting in a yellow-, orange- or red-fluorescent dead-cell population. This principle is the basis of our LIVE/DEAD® *Bac*Light™ Bacterial Viability Kits (L7007, L7012, L13152), our LIVE/DEAD® Sperm Viability Kit (L7011) and our LIVE *Bac*Light™ Bacterial Gram Stain Kit (L7005), which are all discussed in Section 15.3. Four sampler kits of the SYTO® dyes (S7572, S11340, S11350, S11360) provide a total of 27 SYTO® dyes with emission maxima that range from 441 nm to 678 nm. All of the SYTO® dyes in the sampler kits are also available individually (Section 8.1), as well as several other SYTO® green-fluorescent (S32704, S34854, S34855) nucleic acid stains. The SYTO® 13 green-fluorescent nucleic acid stain (S7575) has been used in combination with:

- Ethidium bromide for studies of tissue cryopreservation[93]
- Hexidium iodide for simultaneous viability and gram sign of clinically relevant bacteria[92]
- Ethidium homodimer-1 for quantitation of neurotoxicity[75,94]
- Propidium iodide to detect the effects of surfactants on *Escherichia coli* viability[95]

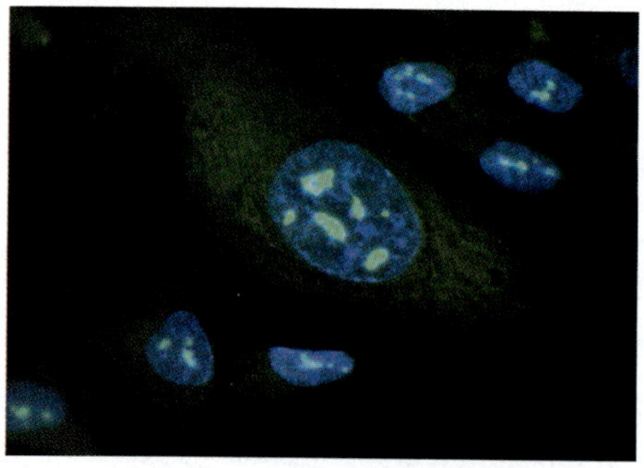

Figure 15.2.35 Methanol-fixed MRC-5 cells stained with SYTO® RNASelect™ green-fluorescent cell stain (S32703). Nuclei were stained with DAPI (D1306, D3571, D21490); the densely stained areas are nucleoli.

Figure 15.2.36 Methanol-fixed MRC-5 cells stained with SYTO® RNASelect™ green-fluorescent cell stain (S32703). Nuclei were stained with DAPI (D1306, D3571, D21490); the densely stained areas are nucleoli.

IMPORTANT NOTICE: The products described in this manual are covered by one or more Limited Use Label License(s). Please refer to the Appendix on page 971 and Master Product List on page 975. Products are For Research Use Only. Not intended for any animal or human therapeutic or diagnostic use.
www.invitrogen.com/probes

SYTO® RNASelect™ Green-Fluorescent Cell Stain

SYTO® RNASelect™ green-fluorescent cell stain (S32703) is a cell-permeant nucleic acid stain that selectively stains RNA (Figure 15.2.33). Although virtually nonfluorescent in the absence of nucleic acids, SYTO® RNASelect™ stain exhibits bright green fluorescence when bound to RNA (absorption/emission maxima ~490/530 nm), but only weak fluorescence when bound to DNA (Figure 15.2.34). Filter sets that are suitable for imaging cells labeled with fluorescein (FITC) will work well for imaging cells stained with SYTO® RNASelect™ stain (Figure 15.2.35, Figure 15.2.36).

Eukaryotic cells stained with the SYTO® RNASelect™ dye show a staining pattern consistent with that of an RNA-selective probe. Maximal fluorescence is observed in the nucleoli, with faint fluorescence throughout the nucleus. Weak fluorescence is also seen throughout the cytoplasm, predominantly associated with mitochondria. The RNA localization of the SYTO® RNASelect™ stain is further supported by RNase and DNase treatments: 1) upon treatment with RNase, the nucleolar and cytoplasmic intensities are significantly reduced, as compared with control cells; 2) upon treatment with DNase, there is no significant loss of fluorescence; and 3) upon treatment with both RNase and DNase, the staining pattern is the same as that observed with RNase treatment alone.

Because the SYTO® RNASelect™ green-fluorescent cell stain is cell permeant, it is suitable for use in live cells. After the cells have been stained, they may be fixed in methanol with minimal loss of the staining pattern. If desired, cells can also be fixed in methanol before staining RNA with the SYTO® RNASelect™ stain. Fixation with formaldehyde alters the staining pattern and is not recommended.

FUN® 1 Dye: A Unique Stain for Assessing Viability of Yeast and Fungi

While structurally related to the SYTO® dyes, FUN® 1 dye (F7030) contains substituents (Figure 15.2.37) that apparently make them chemically reactive with intracellular components of yeast, provided that the yeast are metabolically active. FUN® 1 dye is freely taken up by several species of yeast and fungi and converted from a diffusely distributed pool of yellow-green–fluorescent intracellular stain into compact red-orange– or yellow-orange–fluorescent intravacuolar structures, respectively (Figure 15.2.38). Conversion of FUN® 1 dye to products with longer-wavelength emission (Figure 15.2.39) requires both plasma membrane integrity and metabolic capability. Only metabolically active cells are marked clearly with fluorescent intravacuolar structures, while dead cells exhibit extremely bright, diffuse, yellow-green fluorescence.[61,96] The FUN® 1 cell stain is also available as a component in our LIVE/DEAD® Yeast Viability Kit (L7009, Section 15.3).

7-Aminoactinomycin D

7-AAD (7-aminoactinomycin D, A1310; Section 15.4) is a fluorescent intercalator that undergoes a spectral shift upon association with DNA. 7-AAD/DNA complexes can be excited by the argon-ion laser and emit beyond 610 nm (Figure 15.2.40), making this nucleic acid stain useful for multicolor fluorescence microscopy, confocal laser-scanning microscopy and flow cytometry. 7-AAD appears to be generally excluded from live cells, although it has been reported to label the nuclear region of live cultured mouse L cells and salivary gland polytene chromosomes of *Chironomus thummi thummi* larvae.[97] 7-AAD is also a useful marker for apoptotic cell populations (Section 15.5) and has been utilized to discriminate dead cells from apoptotic and live cells.[98] In addition, 7-AAD can be used in combination with conjugates of R-phycoerythrin (Section 6.4) in three-color flow cytometry protocols.

Viability/Cytotoxicity Assays That Measure Oxidation or Reduction

The generation of reactive oxygen species (ROS) is inevitable for aerobic organisms and, in healthy cells, occurs at a controlled rate. Under conditions of oxidative stress, ROS production is dramatically increased, resulting in subsequent alteration of membrane lipids, proteins and nucleic acids. Oxidative damage of these biomolecules is associated with aging[99,100] and with a variety of pathological events, including atherosclerosis, carcinogenesis, ischemic reperfusion injury and neurodegenerative disorders.[101,102]

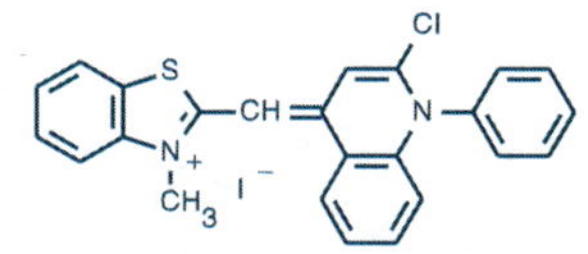

Figure 15.2.37 FUN® 1 cell stain (F7030).

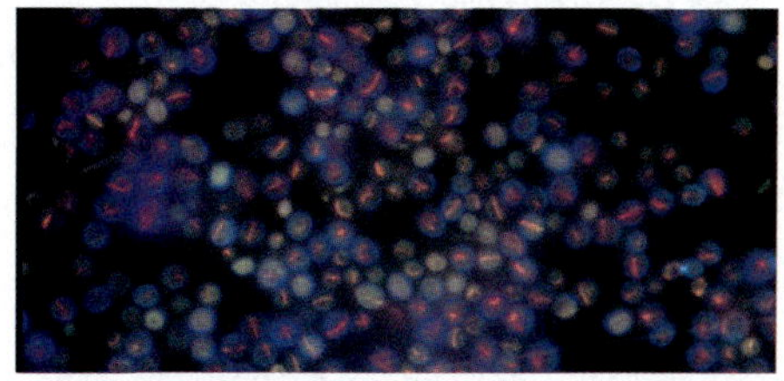

Figure 15.2.38 A culture of *Saccharomyces cerevisiae* incubated in medium containing the FUN® 1 viability indicator (F7030) and the counterstain Calcofluor White M2R, both of which are provided in our LIVE/DEAD® Yeast Viability Kit (L7009). Metabolically active yeast process the FUN® 1 dye, forming numerous red-fluorescent cylindrical structures within their vacuoles. Calcofluor stains the cell walls fluorescent blue, regardless of the yeast's metabolic state. The yeast were photographed in a single exposure through an Omega® Optical triple bandpass filter set.

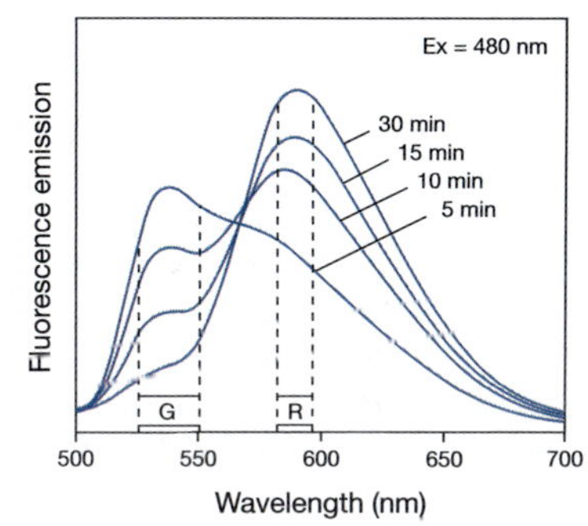

Figure 15.2.39 Fluorescence emission spectra of a *Saccharomyces cerevisiae* suspension that has been stained with the FUN® 1 cell stain, which is available separately (F7030) or in our LIVE/DEAD® Yeast Viability Kit (L7009). After the FUN® 1 reagent was added to the medium, the fluorescence emission spectrum (excited at 480 nm) was recorded in a spectrofluorometer at the indicated times during a 30-minute incubation period. The shift from green (G) to red (R) fluorescence reflects the processing of FUN® 1 by metabolically active yeast cells.

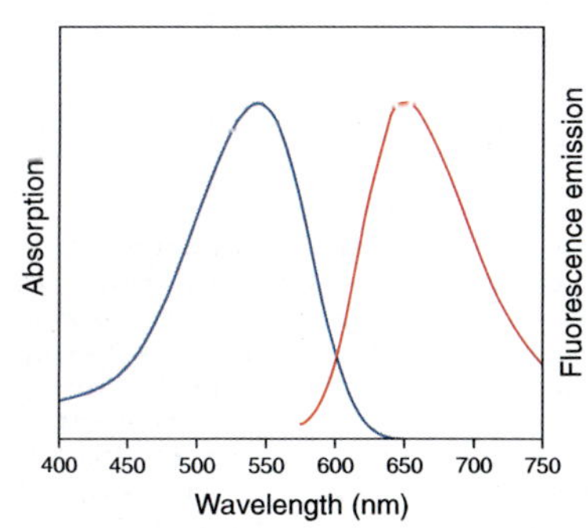

Figure 15.2.40 Absorption and fluorescence emission spectra of 7-aminoactinomycin D (7-AAD) bound to DNA.

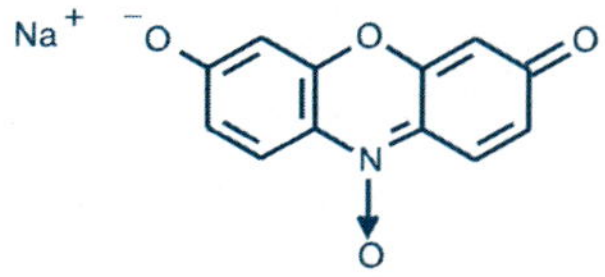

Figure 15.2.41 Resazurin, sodium salt (R12204).

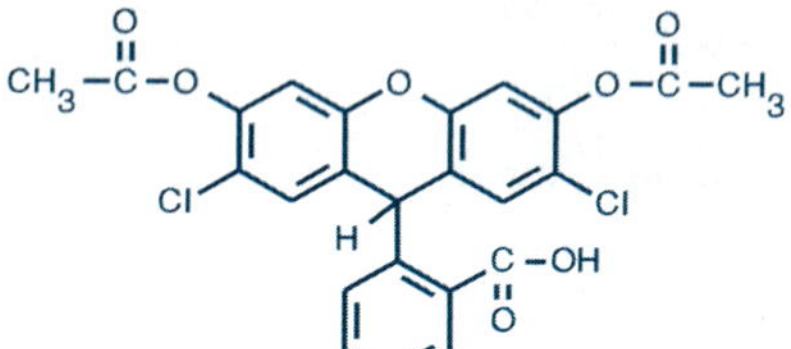

Figure 15.2.42 2′,7′-Dichlorodihydrofluorescein diacetate (2′,7′-dichlorofluorescin diacetate; H_2DCFDA; D399).

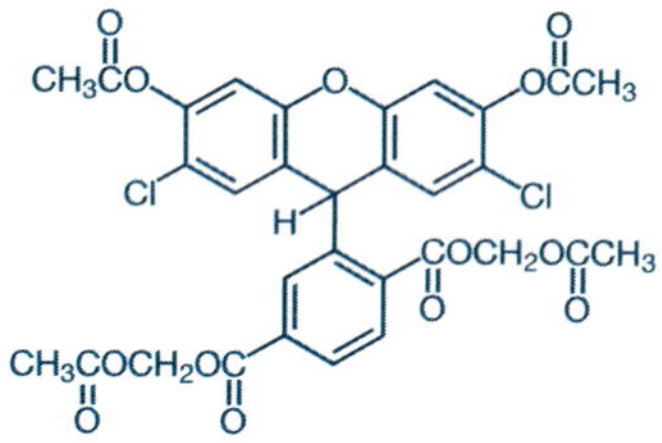

Figure 15.2.43 6-Carboxy-2′,7′-dichlorodihydrofluorescein diacetate, di(acetoxymethyl ester), (C2938).

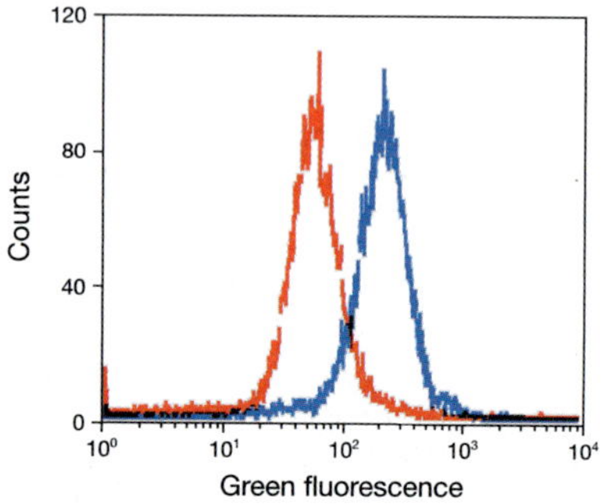

Figure 15.2.44 An oxidative burst was detected by flow cytometry of cells labeled with 5-(and 6-)chloromethyl-2′,7′-dichlorodihydrofluorescein diacetate, acetyl ester (CM-H_2DCFDA, C6827). Jurkat cells were incubated with 100 nM CM-H_2DCFDA. The cells were washed and resuspended in either phosphate-buffered saline (PBS, red) or PBS with 0.03% H_2O_2 (blue). The samples were analyzed on a flow cytometer equipped with a 488 nm argon-ion laser and a 525 ± 10 nm bandpass emission filter.

Metabolically active cells can oxidize or reduce a variety of probes, providing a measure of cell viability and overall cell health. This measure of viability is distinct from that provided by probes designed to detect esterase activity or cell permeability. Detecting oxidative activity and ROS in cells is also discussed in Section 18.2.

High-Purity Resazurin

Resazurin (R12204, Figure 15.2.41) has been extensively used as an oxidation–reduction indicator to detect bacteria and yeast in broth cultures and milk,[103,104] to assess the activity of sperm [105,106] and to assay bile acids [107,108] and triglycerides.[109] Resazurin reduction also occurs with other mammalian cells, including neurons,[110] corneal endothelial cells,[111] lymphocytes, lymphoid tumor cells and hybridoma cells.[112] Furthermore, resazurin has been used in high-throughput screening assays for compounds that act against *Mycobacterium tuberculosis*.[113] However, correlation of the results obtained with resazurin and bioluminescent assays for ATP has been reported to be poor.[114] Resazurin has been reported to be useful for quantitatively measuring cell-mediated cytotoxicity,[115] cell proliferation [116,117] and mitochondrial metabolic activity in isolated neural tissue.[118]

Dodecylresazurin: A Superior Probe for Cell Metabolic Studies

Dodecylresazurin (C_{12}-resazurin), which is available only as a component of our Vybrant® Cell Metabolic Assay Kit (V23110, Section 15.3), LIVE/DEAD® Cell Vitality Assay Kit (L34951, Section 15.3) and Metabolic Activity/Annexin V/Dead Cell Apoptosis Kit (V35114, Section 15.5), has several properties that make it superior to resazurin (and alamarBlue®) for detecting metabolic activity in cells:

- C_{12}-resazurin is freely permeant to the membranes of most cells.
- Less C_{12}-resazurin is required for equivalent sensitivity relative to resazurin.
- Unlike resazurin, which yields a product (resorufin) that rapidly leaks from viable cells, the product of reduction of C_{12}-resazurin—C_{12}-resorufin—is relatively well retained by single cells, permitting flow cytometric assay of cell metabolism and viability on a single-cell basis.
- The fluorescence developed by reduction of C_{12}-resazurin is directly proportional to cell number, and the assay is capable of detecting very low numbers of cells, even in a high-throughput microplate assay.

Dihydrorhodamines and Dihydrofluoresceins

Fluorescein, rhodamine and various other dyes can be chemically reduced to colorless, nonfluorescent leuco dyes. These "dihydro" derivatives are readily oxidized back to the parent dye by some reactive oxygen species (Section 18.2) and thus can serve as fluorogenic probes for detecting oxidative activity in cells and tissues.[119–121] Because reactive oxygen species are produced by live but not dead cells, fluorescent oxidation products that are retained in cells can be used as viability indicators for single cells or cell suspensions. Some probes that are useful for detecting oxidative activity in metabolically active cells include:

- H_2DCFDA [122] (2′,7′-dichlorodihydrofluorescein diacetate, D399, Figure 15.2.42), carboxy-H_2DCFDA (5-(and 6-)carboxy-2′,7′-dichlorodihydrofluorescein diacetate, C400) and the acetoxymethyl ester of H_2DCFDA (C2938, Figure 15.2.43), all of which require both intracellular deacetylation and oxidation to yield green-fluorescent products [123–125]
- CM-H_2DCFDA (chloromethyl-2′,7′-dichlorodihydrofluorescein diacetate, acetyl ester, C6827), which is analogous to H_2DCFDA, except that it forms a mildly thiol-reactive fluorescent product after oxidation by metabolically active cells (Figure 15.2.44), permitting significantly longer-term measurements [126,127]
- Dihydrocalcein AM (D23805), our newest dihydrofluorescein derivative, which is converted intracellularly to calcein, a green-fluorescent dye with superior cell retention (Figure 15.2.3)
- Dihydrorhodamine 123 (D632, D23806; Figure 15.2.45) and dihydrorhodamine 6G (D633), which are oxidized in viable cells to the mitochondrial stains rhodamine 123 [128–131] and rhodamine 6G,[132,133] respectively
- Dihydroethidium (also known as hydroethidine; D1168, D11347, D23107; Figure 15.2.46), which forms the nucleic acid stain ethidium following oxidation [134,135] and has proven useful for detecting the viability of intracellular parasites [136]

IMPORTANT NOTICE: The products described in this manual are covered by one or more Limited Use Label License(s). Please refer to the Appendix on page 971 and Master Product List on page 975. Products are For Research Use Only. Not intended for any animal or human therapeutic or diagnostic use.

- Luminol (L8455), which is useful for chemiluminescence-based detection of oxidative events in cells rich in peroxidases, including granulocytes [137–140] and spermatozoa [141]

These probes are all described in more detail in Section 18.2, which includes products for assaying oxidative activity in live cells and tissues.

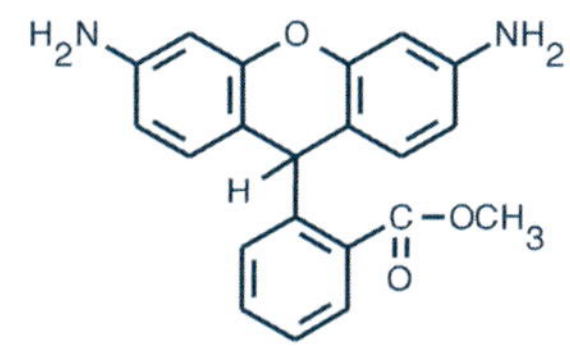

Figure 15.2.45 Dihydrorhodamine 123 (D632).

Image-iT® LIVE Green Reactive Oxygen Species Detection Kit

The Image-iT® LIVE Green Reactive Oxygen Species Detection Kit (I36007) provides the key reagents for detecting reactive oxygen species (ROS) in live cells, including:

- Carboxy-H_2DCFDA (5-(and 6-)carboxy-2′,7′-dichlorodihydrofluorescein diacetate)
- Hoechst 33342
- *tert*-Butyl hydroperoxide (TBHP)
- Dimethylsulfoxide (DMSO)
- Detailed protocols for fluorescence microscopy assays

This assay is based on carboxy-H_2DCFDA (5-(and 6-)carboxy-2′,7′-dichlorodihydrofluorescein diacetate), a reliable fluorogenic marker for ROS in live cells.[142,143] In addition to carboxy-H_2DCFDA, this kit provides the common inducer of ROS production *tert*-butyl hydroperoxide (TBHP) as a positive control [144–147] and the blue-fluorescent, cell-permeant nucleic acid stain Hoechst 33342. Oxidatively stressed and nonstressed cells can be reliably distinguished by fluorescence microscopy using this combination of dyes and the protocol provided (Figure 15.2.47).

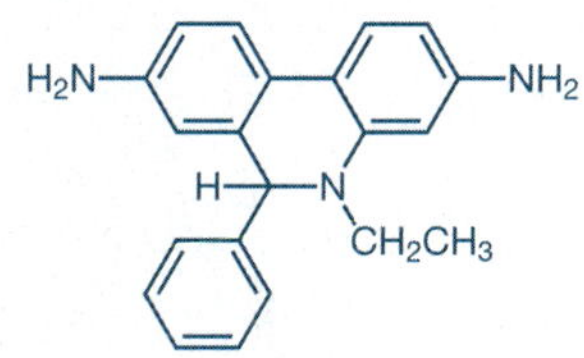

Figure 15.2.46 Dihydroethidium (hydroethidine, D1168).

RedoxSensor™ Red CC-1 Stain

RedoxSensor™ Red CC-1 stain (2,3,4,5,6-pentafluorotetramethyldihydrorosamine, R14060; Figure 15.2.48) passively enters live cells and is subsequently oxidized in the cytosol to a red-fluorescent product (excitation/emission maxima ~540/600 nm), which then accumulates in the mitochondria. Alternatively, this nonfluorescent probe may be transported to the lysosomes where it is oxidized. The differential distribution of the oxidized product between mitochondria and lysosomes appears to depend on the redox potential of the cytosol.[148–150] In proliferating cells, mitochondrial staining predominates; whereas in contact-inhibited cells, the staining is primarily lysosomal (Figure 15.2.49).

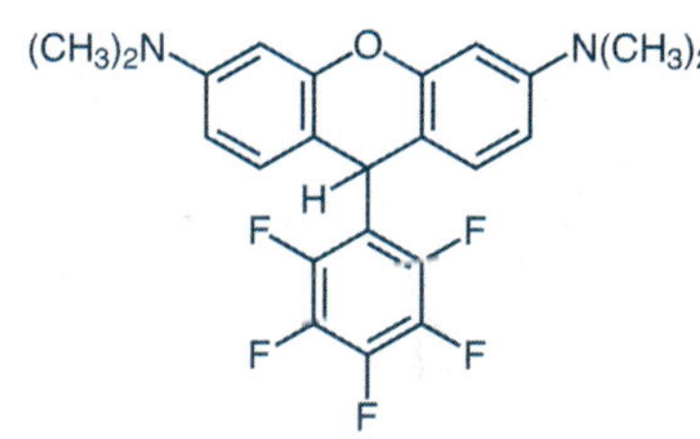

Figure 15.2.48 RedoxSensor™ Red CC-1 (R14060).

Tetrazolium Salts

Tetrazolium salts are widely used for detecting redox potential of cells for viability, cytotoxicity and proliferation assays.[151–156] Following reduction, these water-soluble, colorless compounds form uncharged, brightly colored but nonfluorescent formazans. Several of the formazans precipitate out of solution and are useful for histochemical localization of the site of reduction or, after solubilization in organic solvent, for quantitation by standard spectrophotometric techniques.

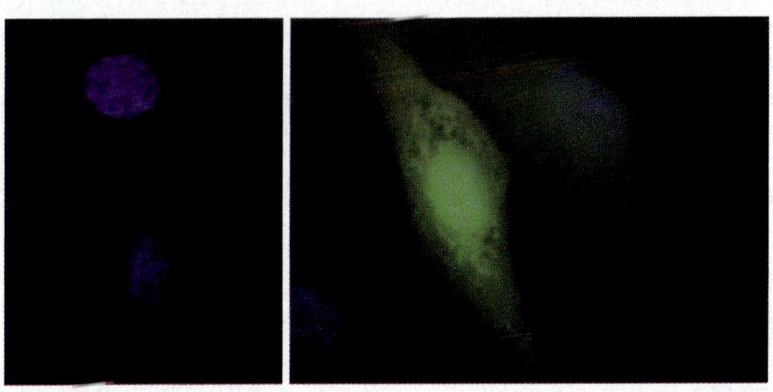

Figure 15.2.47 Detection of oxidative stress in live cells using the Image-iT® LIVE Reactive Oxygen Species (ROS) Kit (I36007). Live bovine pulmonary artery endothelial cells were treated with *tert*-butyl hydroperoxide to induce oxidative stress (right) or were left untreated (left). Cells were then labeled with carboxy-H_2DCFDA, which fluoresces when oxidized by ROS, and nuclei were stained with blue-fluorescent Hoechst 33342. The stressed cells exhibited green fluorescence, signaling an increase in ROS, whereas the untreated cells showed minimal fluorescence.

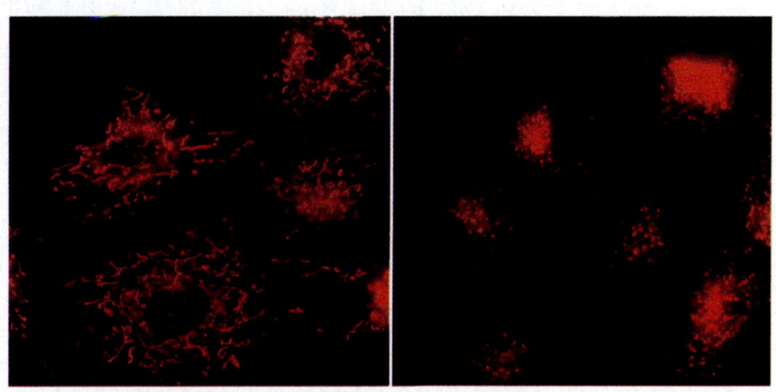

Figure 15.2.49 Cellular proliferation state determines the distribution of the oxidized product of RedoxSensor™ Red CC-1 (R14060). Normal rat kidney (NRK) cells in different growth states were stained with RedoxSensor™ Red CC-1. In proliferating cells (left), the oxidized dye accumulates in mitochondria. In quiescent cells (right), the oxidized product localizes in the lysosomes.

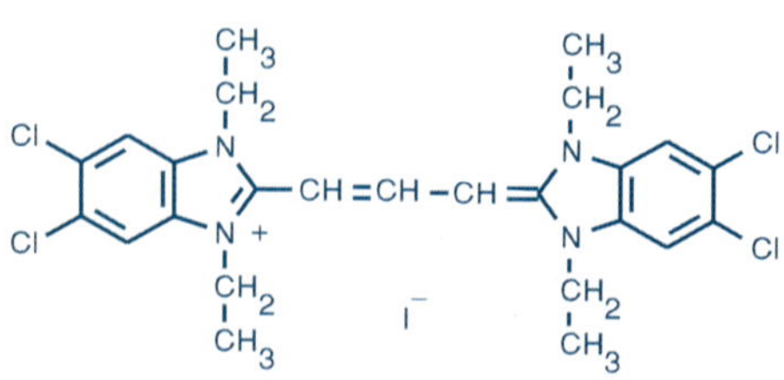

Figure 15.2.50 5,5′,6,6′-Tetrachloro-1,1′,3,3′-tetraethylbenzimidazolylcarbocyanine iodide (JC-1, T3168)

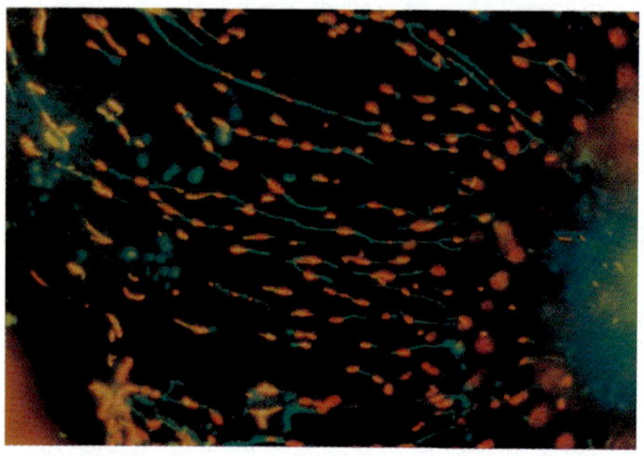

Figure 15.2.51 Potential-dependent staining of mitochondria in CCL64 fibroblasts by JC-1 (T3168). The mitochondria were visualized by epifluorescence microscopy using a 520 nm longpass optical filter. Regions of high mitochondrial polarization are indicated by red fluorescence due to J-aggregate formation by the concentrated dye. Depolarized regions are indicated by the green fluorescence of the JC-1 monomers. The image was contributed by Lan Bo Chen, Dana Farber Cancer Institute, Harvard Medical School.

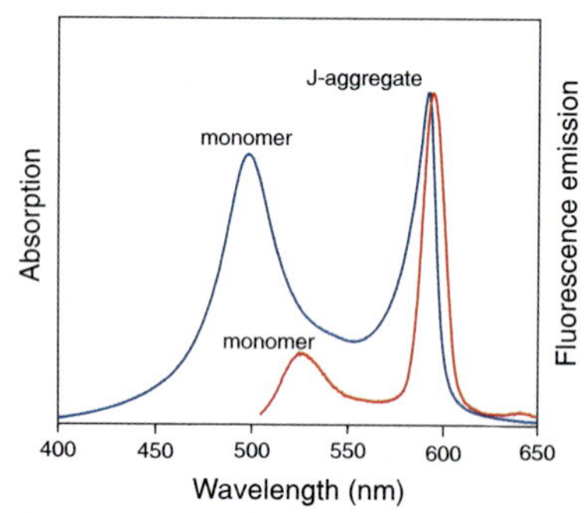

Figure 15.2.52 Absorption and fluorescence emission (excited at 488 nm) spectra of JC-1 in pH 8.2 buffer containing 1% (v/v) DMSO.

Reduction of MTT (M6494) remains the most common assay for tetrazolium salt–based viability testing.[157–159] The Vybrant® MTT Cell Proliferation Assay Kit (V13154, Section 15.4) provides a simple method for determining cell number using standard microplate absorbance readers. MTT has also been used to measure adhesion of HL60 leukemia cells onto endothelial cells.[160] In addition to dehydrogenases, MTT is reduced by glutathione *S*-transferase [161] (GST). Therefore, MTT may not always be a reliable cell viability probe in cells treated with compounds that affect GST activity.

Unlike MTT's purple-colored formazan product, the extremely water-soluble, orange-colored formazan product of XTT (X6493) does not require solubilization prior to quantitation, thereby reducing the assay time in many viability assay protocols. Moreover, the sensitivity of the XTT reduction assay is reported to be similar to or better than that of the MTT reduction assay.[152] The XTT reduction assay is particularly useful for high-throughput screening of antiviral and antitumor agents and for assessing the effect of cytokines on cell proliferation.[156,162–165] NBT (N6495) forms a deep blue–colored precipitate that is commonly used to indicate oxidative metabolism.[166,167]

Other Viability/Cytotoxicity Assay Methods

A viable cell contains an ensemble of ion pumps and channels that maintain both intracellular ion concentrations and transmembrane potentials. Active maintenance of ion gradients ceases when the cell dies, and this loss of activity can be assessed using potentiometric dyes, acidotropic probes, Ca^{2+} indicators [122] (Chapter 19) and pH indicators [122] (Chapter 20).

Potentiometric Dyes

We offer a variety of dyes for detecting transmembrane potential gradients (Chapter 22), including several cationic probes that accumulate in the mitochondria of metabolically active cells (Section 12.2). The mitochondrion-selective rhodamine 123 [122] (R302, R22420) has been used to assess the viability of lymphocytes,[168] human fibroblasts,[169] Simian virus–transformed human cells [170] and bacteria;[171] however, rhodamine 123 is not taken up well by gram-negative bacteria.[171] Rhodamine 123 has also been used in combination with propidium iodide (P1304MP, P3566, P21493) for two-color flow cytometric viability assessment.[172]

The methyl and ethyl esters of tetramethylrhodamine (T668, T669) accumulate in the mitochondria of healthy cells in an amount related to the membrane potential. The dyes are nontoxic and highly fluorescent and do not form aggregates or display binding-dependent changes in their fluorescence efficiency, permitting continuous monitoring of cell heath.[173]

Other potential-sensitive dyes that have proven useful in viability studies include several fast-response styryl dyes and slow-response oxonol and carbocyanine dyes. The fast-response styryl dyes such as di-4-ANEPPS (D1199, Section 22.2) give relatively large fluorescence response to potential changes. Di-4-ANEPPS was used for rapid measurement of toxicity in frog embryos.[174] The symmetrical bis-oxonol dyes [122] (B413, B438; Section 22.3) have been used for viability assessment by flow cytometry [175–177] and imaging. These slow-response dyes have also been employed to determine antibiotic susceptibility of bacteria by flow cytometry,[178,179] and our *Bac*Light™ Bacterial Membrane Potential Kit (B34950, Section 15.3) provides the carbocyanine dye $DiOC_2(3)$ along with the proton ionophore CCCP for detecting membrane potential in both gram-positive and gram-negative bacteria.

The green-fluorescent cyanine dye JC-1 (5,5′,6,6′-tetrachloro-1,1′,3,3′-tetraethylbenzimidazolylcarbocyanine iodide, T3168, M34152; Figure 15.2.50) exists as a monomer at low concentrations or at low membrane potential; however, at higher concentrations (aqueous solutions above 0.1 µM) or higher potentials, JC-1 forms red-fluorescent "J-aggregates" (Figure 15.2.51) that exhibit a broad excitation spectrum and an emission maximum at ~590 nm (Figure 15.2.52). JC-1 has been used to investigate apoptosis,[180,181] as well as mitochondrial poisoning, uncoupling and anoxia.[182] The ability to make ratiometric emission measurements with JC-1 makes this probe particularly useful for monitoring changes in cell health. We have discovered another carbocyanine dye, JC-9 (3,3′-dimethyl-β-naphthoxazolium iodide, D22421; Figure 15.2.53), with potential-dependent spectroscopic properties.

IMPORTANT NOTICE: The products described in this manual are covered by one or more Limited Use Label License(s). Please refer to the Appendix on page 971 and Master Product List on page 975. Products are For Research Use Only. Not intended for any animal or human therapeutic or diagnostic use.

Acidotropic Stains

Membrane-bound proton pumps are used to maintain low pH within the cell's acidic organelles. Our complete selection of stains for lysosomes and other acidic organelles, including LysoTracker® and LysoSensor™ probes, is described in Section 12.3.

The lysosomal stain neutral red (N3246), which was first used for viability measurements by Ehrlich in 1894, has been employed in numerous cytotoxicity, cell proliferation and adhesion assays.[183–188] Although usually used as a chromophoric probe, neutral red also fluoresces at ~640 nm in viable cells and has been detected using a fluorescence microplate reader.[189] Furthermore, the fluorescence of neutral red and BCECF AM (or SYTOX® Green nucleic acid stain) can be measured simultaneously using a single excitation wavelength of 488 nm,[189] suggesting that neutral red may be an effective probe for multicolor flow cytometric determination of cell viability. Our neutral red is highly purified to reduce contaminants that might interfere with these observations.

Acridine orange (A1301, A3568) concentrates in acidic organelles in a pH-dependent manner. The metachromatic green or red fluorescence of acridine orange has been used to assess islet viability[190] and bacterial spore viability[191] and to monitor physiological activity in *Escherichia coli*.[192]

LysoTracker® Green DND-26 (L7526, Section 12.3) was used in a fluorometric assay of cryopreserved sperm to demonstrate both acrosomal integrity and sperm viability.[193] Of several methods used, LysoTracker® Green DND-26 staining was the quickest and easiest to use and gave excellent correlation with the SYBR® 14 staining method used in our LIVE/DEAD® Sperm Viability Kit (L7011, Section 15.3). When observed with a fluorescence microscope, sperm appeared to lose their green-fluorescent LysoTracker® Green DND-26 staining and instead exhibited red-fluorescent propidium iodide staining within about 30 seconds after motility ceased.[194,195]

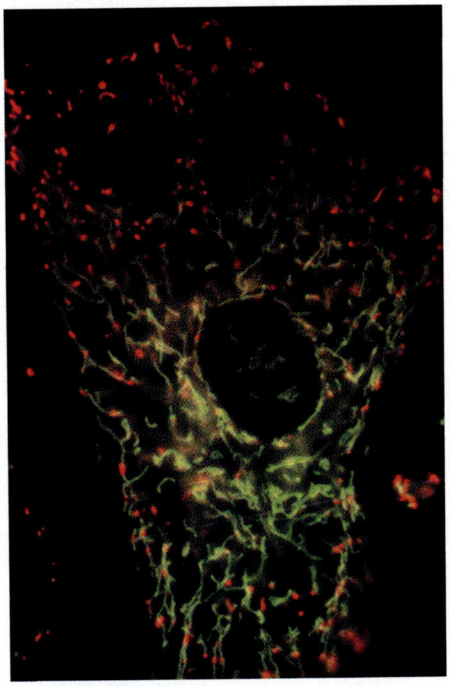

Figure 15.2.53 A viable bovine pulmonary artery endothelial cell incubated with the ratiometric mitochondrial potential indicator, JC-9 (D22421). In live cells, JC-9 exists either as a green-fluorescent monomer at depolarized membrane potentials, or as a red-fluorescent J-aggregate at hyperpolarized membrane potentials.

Fluorescent Glucose Analogs

Measurements of glucose uptake can be used to assess viability in a variety of organisms. 2-NBD-deoxyglucose (2-NBDG, N13195) has been used to monitor glucose uptake in living pancreatic β-cells,[196] the yeast *Candida albicans*[197] and the bacteria *Escherichia coli*.[198–200] We also offer the fluorescent nonhydrolyzable glucose analog 6-NBD-deoxyglucose (6-NBDG, N23106). Using this probe, researchers have studied glucose uptake and transport in isolated cells[201–203] and intact tissues.[204] Although sensitive to its environment, NBD fluorescence typically displays excitation/emission maxima of ~465/540 nm and can be visualized using optical filters designed for fluorescein.

Fluorescent Antibiotics and Related Probes

Fluorescent Polymyxins

Polymyxin B is a cyclic cationic peptide antibiotic that binds to the lipopolysaccharide (LPS) component of the outer membrane of gram-negative bacteria and increases its permeability to lysozyme and hydrophobic compounds. Molecular Probes® fluorescent BODIPY® FL (P13235), Oregon Green® 514 (P13236) and dansyl (P13238) derivatives of polymyxin B are available. Dansyl polymyxin B, which fluoresces weakly when free in solution, becomes highly fluorescent (excitation/emission ~340/485 nm) upon binding to intact cells or LPS.[205] The binding of dansyl polymyxin B to LPS can be displaced by a variety of polycationic antibiotics, as well as by Mg^{2+}. Consequently, dansyl polymyxin B displacement experiments can be used to assess the binding of various compounds, such as antibiotics and macrophage cationic proteins, to LPS and intact bacterial cells.[206–208] Dansyl polymyxin has also been used to localize regions of high anionic lipid content in both sperm[209] and aggregated human platelets,[210] and to analyze the morphology of lipid bilayers in preparations of acetylcholine receptor clusters from rat myotubules.[211] Green-fluorescent BODIPY® FL and Oregon Green® 514 polymyxin B provide additional fluorescent color options for these experiments. Our fluorescent LPS is described in Section 13.3 and Section 16.1.

Fluorescent Penicillin Analogs

BOCILLIN™ FL penicillin and BOCILLIN™ 650/665 penicillin (B13233, B13234) are green- and infrared-fluorescent penicillin analogs, respectively, that bind selectively and with high affinity to penicillin-binding proteins present on the cytoplasmic membranes of

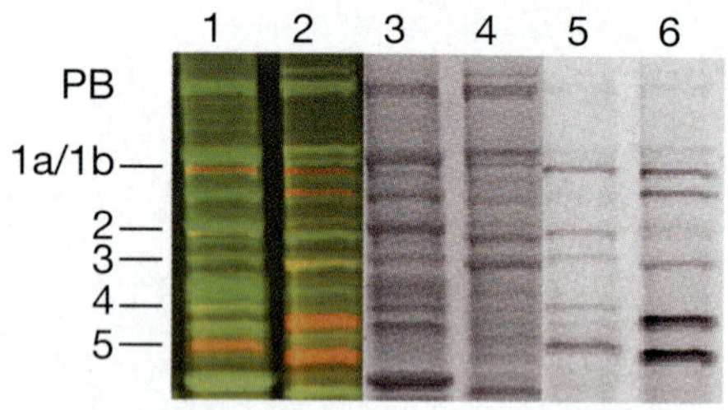

Figure 15.2.54 Detection of penicillin-binding proteins (PBPs) from *Escherichia coli* and *Pseudomonas aeruginosa*. The membrane fractions from *E. coli* and *P. aeruginosa* were prepared as previously described and labeled with BOCILLIN™ 650/665 penicillin (B13234). The labeled membranes were separated on an SDS-polyacrylamide gel, stained with SYPRO® Ruby and visualized using a Typhoon imager. The location of PBPs from *E. coli* are labeled to the left of the gels. Lanes 1, 3 and 5 are *E. coli* membrane preparations; lanes 2, 4 and 6 are *P. aeruginosa* membrane preparations; lanes 1 and 2 are overlays of images obtained from total protein (green) and PBP (red) scans; lanes 3 and 4 are total protein visualized with the SYPRO® Ruby protein gel stain; lanes 5 and 6 are PBPs as detected by BOCILLIN™ 650/665. Image used with permission from Wiley VCH publishers.

eubacteria.[212,213] When electrophoresed under nonreducing conditions, the dye-labeled penicillin-binding proteins are easily visible in the gel with sensitivity in the low nanograms[214] (Figure 15.2.54). BOCILLIN™ FL penicillin, synthesized from penicillin V and the BODIPY® FL dye (spectrally similar to fluorescein), has been used to determine the penicillin-binding protein profiles of *Escherichia coli, Pseudomonas aeruginosa* and *Streptococcus pneumoniae*, and these binding profiles are found to be similar to those reported by researchers using radioactively labeled penicillin V.[212] Fluorescently labeled penicillin has also been used for direct labeling and rapid detection of whole *E. coli* and *Bacillus licheniformis*[215] and of *Enterobacter pneumoniae*.[216] The β-lactam sensor-transducer (BlaR), an integral membrane protein from *Staphylococcus aureus*, covalently and stoichiometrically reacts with β-lactam antibiotics, including BOCILLIN™ FL penicillin, by acylation of its active-site serine residue.[217]

BODIPY® FL Vancomycin

BODIPY® FL vancomycin (V34850), which contains a single BODIPY® FL dye per vancomycin molecule, is a green-fluorescent analog of this important antibiotic, which is active against gram-positive bacteria, including enterrococci. BODIPY® FL vancomycin is useful for detecting vancomycin binding sites[218,219] and for the study and detection of vancomycin-resistant enterococci[220,221] (VRE, Section 15.6).

Antimalarial Agent

The phosphonate antibiotic FR-31564 (fosmidomycin, F23103) is an effective antimalarial agent that functions by blocking a mevalonate-independent methylerythritol phosphate (MEP) pathway of isoprene synthesis.[222–225] The antibiotic activity of fosmidomycin is potentiated by glucose 1-phosphate.[226,227] This antibiotic is also active against several gram-negative bacteria.[228,229]

REFERENCES

1. J Immunol Methods (2000) 243:155; **2.** Biochem Biophys Res Commun (1990) 172:262; **3.** EMBO J (1986) 5:51; **4.** J Immunol Methods (1994) 177:101; **5.** J Immunol Methods (1994) 172:227; **6.** J Immunol Methods (1994) 172:115; **7.** Hum Immunol (1993) 37:264; **8.** Biophys J (1977) 18:3; **9.** J Cell Biol (1994) 125:1077; **10.** Cornea (1996) 15:599; **11.** Biophys J (1995) 68:1207; **12.** Cytometry (1995) 19:366; **13.** Cytometry (1991) 12:666; **14.** J Immunol Methods (1991) 139:281; **15.** Biophys J (1995) 68:2608; **16.** Biochemistry (1998) 37:2243; **17.** Appl Environ Microbiol (2000) 66:4486; **18.** Science (1991) 251:81; **19.** J Neurosci (2008) 28:143; **20.** J Immunol Methods (1994) 172:255; **21.** Cytometry (1990) 11:244; **22.** J Immunol Methods (1989) 122:15; **23.** Trop Med Parasitol (1995) 46:45; **24.** Cryobiology (1977) 14:322; **25.** Vox Sang (1974) 27:13; **26.** Proc Natl Acad Sci U S A (1966) 55:134; **27.** J Immunol Methods (1993) 157:117; **28.** Cancer Res (1989) 49:3776; **29.** J Histochem Cytochem (1985) 33:77; **30.** Cytometry (1986) 7:70; **31.** Biochemistry (1979) 18:2210; **32.** Immunol Lett (1981) 2:187; **33.** J Immunol Methods (1980) 33:33; **34.** Biochemistry (1995) 34:1606; **35.** Cytometry (1992) 13:739; **36.** J Immunol Methods (1990) 130:251; **37.** Proc SPIE-Int Soc Opt Eng (1999) 3755:24; **38.** J Immunol Methods (1987) 100:261; **39.** Cytometry (1994) 15:213; **40.** Appl Environ Microbiol (1994) 60:1467; **41.** Biol Reprod (1986) 34:127; **42.** J Immunol Methods (1986) 86:7; **43.** Anticancer Res (1994) 14:927; **44.** J Immunol Methods (1992) 155:19; **45.** J Cell Biol (1995) 128:779; **46.** Methods Enzymol (1991) 194:644; **47.** Methods Cell Biol (1989) 31:357; **48.** J Cell Biol (1990) 111:3129; **49.** J Immunol Methods (1990) 133:87; **50.** FEBS Lett (1986) 200:203; **51.** Biotechniques (1985) 3:270; **52.** Appl Environ Microbiol (1995) 61:3415; **53.** J Vasc Res (1995) 32:371; **54.** Connect Tissue Res (1996) 33:233; **55.** J Cell Biol (1996) 135:1593; **56.** EMBO J (1993) 12:5209; **57.** J Neurosci (1993) 13:834; **58.** Science (1992) 255:200; **59.** J Bacteriol (2008) 190:311; **60.** Electrophoresis (2000) 21:2509; **61.** Plant Methods (2008) 4:15; **62.** Appl Environ Microbiol (1997) 63:2421; **63.** Appl Environ Microbiol (1994) 60:3284; **64.** Physiol Rev (2002) 82:1013; **65.** Anal Biochem (1994) 221:78; **66.** Cytometry (1999) 37:14; **67.** Appl Environ Microbiol (2001) 67:2326; **68.** Appl Environ Microbiol (1995) 61:3623; **69.** Limnol Oceanogr (1995) 40:1050; **70.** J Immunol Methods (1995) 185:249; **71.** J Microbiol Methods (2000) 42:3; **72.** J Immunol Methods (2010) 358:81; **73.** J Immunol Methods (1982) 52:91; **74.** Appl Environ Microbiol (1995) 61:2521; **75.** Neuron (1995) 15:961; **76.** Anal Biochem (1994) 220:149; **77.** J Immunol Methods (1992) 149:133; **78.** J Immunol Methods (1990) 134:201; **79.** Nucleic Acids Res (1995) 23:1215; **80.** Nucleic Acids Res (1992) 20:2803; **81.** Biochemistry (1978) 17:5078; **82.** Biochemistry (1978) 17:5071; **83.** Proc Natl Acad Sci U S A (1990) 87:3851; **84.** J Immunol Methods (1995) 178:71; **85.** Cytometry (1995) 20:181; **86.** J Neurosci (1995) 15:6239; **87.** J Neurosci (1994) 14:2260; **88.** J Appl Physiol (1998) 85:2370; **89.** Biotechniques (2000) 29:874, 882; **90.** Appl Environ Microbiol (2006) 72:1997; **91.** Methods Mol Biol (2009) 510:415; **92.** Appl Environ Microbiol (1998) 64:2681; **93.** J Physiol (1997) 502:105; **94.** Cytometry (1998) 32:66; **95.** Cytometry (1997) 29:58; **96.** J Cell Biol (1994) 126:1375; **97.** Histochem J (1985) 17:131; **98.** Cytometry (1992) 13:204; **99.** Mitochondrion (2003) 2:361; **100.** Ann N Y Acad Sci (2000) 908:219; **101.** J Cell Mol Med (2002) 6:175; **102.** Free Radic Biol Med (2001) 31:164; **103.** Appl Environ Microbiol (1990) 56:3785; **104.** J Dairy Res (1990) 57:239; **105.** Hum Reprod (1994) 9:1688; **106.** Fertil Steril (1991) 56:743; **107.** Steroids (1981) 38:281; **108.** Clin Chim Acta (1976) 70:79; **109.** Clin Chem (1983) 29:171; **110.** J Neurosci Methods (1996) 70:195; **111.** Invest Ophthalmol Vis Sci (1997) 38:1929; **112.** J Immunol Methods (1994) 170:211; **113.** Antimicrob Agents Chemother (1997) 41:1004; **114.** Gynecol Oncol (1995) 58:101; **115.** J Immunol Methods (1998) 213:157; **116.** J Clin Lab Anal (1995) 9:89; **117.** J Immunol Methods (1994) 175:181; **118.** J Neurosci Res (1996) 45:216; **119.** Arch Toxicol (1994) 68:582; **120.** Brain Res (1994) 635:113; **121.** Chem Res Toxicol (1992) 5:227; **122.** Methods Mol Biol (2010) 594:57; **123.** Biochemistry (1995) 34:7194; **124.** Cell (1993) 75:241; **125.** J Immunol (1983) 130:1910; **126.** J Pharmacol Exp Ther (2002) 300:9; **127.** J Biol Chem (1999) 274:19323; **128.** Cytometry (1994) 18:147; **129.** Methods Enzymol (1994) 233:539; **130.** Naturwissenschaften (1988) 75:354; **131.** J Med Chem (1987) 30:1757; **132.** J Cell Physiol (1993) 156:428; **133.** Biochem Biophys Res Commun (1991) 175:387; **134.** FEMS Microbiol Lett (1994) 122:187; **135.** J Immunol Methods (1994) 170:117; **136.** J Immunol Methods (1991) 140:23; **137.** J Appl Physiol (1994) 76:539; **138.** J Leukoc Biol (1993) 54:300; **139.** J Biochem (Tokyo) (1989) 106:355; **140.** Biochem Biophys Res Commun (1988) 155:106; **141.** J Cell

IMPORTANT NOTICE: The products described in this manual are covered by one or more Limited Use Label License(s). Please refer to the Appendix on page 971 and Master Product List on page 975. Products are For Research Use Only. Not intended for any animal or human therapeutic or diagnostic use.

REFERENCES—*continued*

Physiol (1992) 151:466; **142.** Am J Physiol Heart Circ Physiol (2000) 279:H2424; **143.** J Natl Cancer Inst (1999) 91:1138; **144.** Lipids (2001) 36:57; **145.** Cancer Res (2001) 61:1392; **146.** Histochem Cell Biol (2003) 120:319; **147.** Am J Physiol (1997) 272:C1286; **148.** Am J Pathol (2009) 174:101; **149.** Am J Physiol Renal Physiol (2007) 292:F523; **150.** Free Radic Biol Med (2000) 28:1266; **151.** J Immunol Methods (1995) 179:95; **152.** J Infect Dis (1995) 172:1153; **153.** J Appl Bacteriol (1993) 74:433; **154.** J Immunol Methods (1993) 160:89; **155.** J Immunol Methods (1993) 157:203; **156.** J Immunol Methods (1991) 142:257; **157.** J Cell Biol (1995) 128:201; **158.** Cancer Res (1994) 54:3620; **159.** Cell (1994) 77:817; **160.** J Immunol Methods (1993) 164:255; **161.** Biotechniques (1998) 25:622; **162.** J Immunol Methods (1993) 159:81; **163.** J Immunol Methods (1992) 147:153; **164.** J Natl Cancer Inst (1989) 81:577; **165.** Cancer Res (1988) 48:4827; **166.** Clin Chim Acta (1993) 221:197; **167.** J Leukoc Biol (1993) 53:404; **168.** Clin Bull (1981) 11:47; **169.** J Cell Biol (1981) 91:392; **170.** Somatic Cell Genet (1983) 9:375; **171.** J Appl Bacteriol (1992) 72:410; **172.** Cancer Res (1982) 42:799; **173.** Biophys J (1988) 53:785; **174.** Bull Environ Contam Toxicol (1993) 51:557; **175.** Cytometry (1994) 15:343; **176.** J Immunol Methods (1993) 161:119; **177.** Jpn J Pharmacol (1991) 57:419; **178.** J Appl Bacteriol (1995) 78:309; **179.** J Microsc (1994) 176:8; **180.** J Biol Chem (2008) 283:5188; **181.** Nat Protoc (2007) 2:2719; **182.** Cardiovasc Res (1993) 27:1790; **183.** Clin Chem (1995) 41:1906; **184.** Cell Biol Toxicol (1994) 10:329; **185.** Anal Biochem (1993) 213:426; **186.** Biotech Histochem (1993) 68:29; **187.** Nat Protoc (2008) 3:1125; **188.** Biotechnology (N Y) (1990) 8:1248; **189.** In Vitro Toxicol (1990) 3:219; **190.** In Vitro Cell Dev Biol (1988) 24:266; **191.** Biotech Histochem (1992) 67:27; **192.** J Microbiol Methods (1991) 13:87; **193.** Biol Reprod (1997) 56:991; **194.** Development (1995) 121:825; **195.** J Immunol Methods (1989) 119:45; **196.** J Biol Chem (2000) 275:22278; **197.** Appl Microbiol Biotechnol (1996) 46:400; **198.** J Microbiol Methods (2000) 42:87; **199.** Biochim Biophys Acta (1996) 1289:5; **200.** Biosci Biotechnol Biochem (1996) 60:1899; **201.** Cytometry (1997) 27:262; **202.** Biochemistry (1995) 34:15395; **203.** Biochim Biophys Acta (1992) 1111:231; **204.** Histochem J (1994) 26:207; **205.** Antimicrob Agents Chemother (1986) 29:496; **206.** Antimicrob Agents Chemother (1993) 37:453; **207.** Antimicrob Agents Chemother (1991) 35:1309; **208.** Infect Immun (1988) 56:693; **209.** Proc Natl Acad Sci U S A (1980) 77:6601; **210.** Thromb Res (1988) 50:605; **211.** Exp Cell Res (1991) 195:79; **212.** Antimicrob Agents Chemother (1999) 43:1124; **213.** J Biol Chem (2000) 275:17693; **214.** Electrophoresis (2001) 22:960; **215.** Biochem J (1993) 291:19; **216.** Biochem J (1994) 300:141; **217.** J Biol Chem (2003) 278:18419; **218.** PLos Patholog (2009) 5:e1000285; **219.** Antimicrob Agents Chemother (2007) 51:3267; **220.** Clin Microbiol Infect (2008) 14:766; **221.** N Engl J Med (2001) 344:1427; **222.** J Infect Dis (2004) 189:901; **223.** Proc Natl Acad Sci U S A (2004) 101:7451; **224.** Science (1999) 285:1502; **225.** Science (1999) 285:1573; **226.** Eur J Clin Microbiol (1987) 6:386; **227.** Infection (1987) 15:465; **228.** J Antibiot (Tokyo) (1980) 33:44; **229.** J Antibiot (Tokyo) (1980) 33:24.

DATA TABLE 15.2 VIABILITY AND CYTOTOXICITY ASSAY REAGENTS

Cat. No.	MW	Storage	Soluble	Abs	EC	Em	Solvent	Notes
A1301	301.82	L	H_2O, EtOH	500	53,000	526	H_2O/DNA	1, 2
A3568	301.82	RR,L	H_2O	500	53,000	526	H_2O/DNA	1, 2, 3
B1150	~615	F,D	DMSO	<300		none		4, 5
B1170	~615	F,D	DMSO	<300		none		4, 5
B3051	~615	F,D	DMSO	<300		none		3, 4, 5
B13233	661.46	F,D,L	H_2O, DMSO	504	68,000	511	MeOH	
B13234	653.44	F,D,L	DMSO	646	78,000	659	MeOH	
C195	460.40	F,D	DMSO	<300		none		6
C369	529.29	F,D	DMSO	<300		none		7
C400	531.30	F,D	DMSO, EtOH	290	5600	none	MeCN	8
C1354	532.46	F,D	DMSO	<300		none		9
C1361	460.40	F,D	DMSO	<300		none		6
C1362	460.40	F,D	DMSO	<300		none		6
C1429	465.41	F,D,L	DMSO	322	13,000	437	DMSO	10
C1430	994.87	F,D	DMSO	<300		none		11
C2925	464.86	F,D	DMSO	<300		none		6
C2938	675.43	F,D,AA	DMSO	291	5700	none	MeOH	8
C3099	994.87	F,D	DMSO	<300		none		3, 11
C3100MP	994.87	F,D	DMSO	<300		none		11
C6826	499.95	F,D	DMSO	<350		none		12
C6827	577.80	F,D,AA	DMSO	287	9100	none	MeOH	8
C7025	464.86	F,D	DMSO	<300		none		6
C13196	560.52	F,D	DMSO	<300		none		13
C34851	789.55	F,D,L	DMSO	576	90,000	589	DMSO	14
C34852	994.87	F,D	DMSO	<300		none		11
C34853	465.41	F,D	DMSO	322	13,000	437	DMSO	10
D399	487.29	F,D	DMSO, EtOH	258	11,000	none	MeOH	8
D632	346.38	F,D,L,AA	DMF, DMSO	289	7100	none	MeOH	15, 16
D633	444.57	F,D,L,AA	DMF, DMSO	296	11,000	none	MeOH	15, 16
D1168	315.42	FF,L,AA	DMF, DMSO	355	14,000	see Notes	MeCN	15, 17
D11347	315.42	FF,L,AA	DMF, DMSO	355	14,000	see Notes	MeCN	15, 17
D22421	532.38	D,L	DMSO, DMF	522	143,000	535	$CHCl_3$	18
D23107	315.42	FF,D,L,AA	DMSO	355	14,000	see Notes	MeCN	17, 19
D23805	1068.95	F,D	DMSO	285	5800	none	MeCN	20
D23806	346.38	F,D,L,AA	DMSO	289	7100	none	MeOH	16, 19
E1169	856.77	F,D,L	DMSO	528	7000	617	H_2O/DNA	1, 21, 22
E1374	420.31	F,LL	DMF, EtOH	462	5400	625	pH 7	23
E3599	1292.71	F,D,L	DMSO	535	8000	624	H_2O/DNA	1, 3, 21, 22
F1303	416.39	F,D	DMSO	<300		none		6
F7030	528.84	F,D,L	DMSO	508	71,000	none	pH 7	3, 24
F23103	205.08	F,D,L	H_2O	<300		none		
H7593	497.42	L	DMSO	518	3900	600	H_2O/DNA	1, 25
L8455	177.16	D,L	DMF	355	7500	411	MeOH	26
M6494	414.32	D,L	H_2O, DMSO	375	8300	none	MeOH	27, 28
N3246	288.78	D,L	H_2O, EtOH	541	39,000	640	see Notes	29
N6495	817.65	D,L	H_2O, DMSO	256	64,000	none	MeOH	27
N13195	342.26	F,L	H_2O	466	20,000	540	MeOH	30

continued on next page

IMPORTANT NOTICE: The products described in this manual are covered by one or more Limited Use Label License(s). Please refer to the Appendix on page 971 and Master Product List on page 975. Products are For Research Use Only. Not intended for any animal or human therapeutic or diagnostic use.

DATA TABLE 15.2 VIABILITY AND CYTOTOXICITY ASSAY REAGENTS—*continued*

Cat. No.	MW	Storage	Soluble	Abs	EC	Em	Solvent	Notes
N23106	342.26	F,L	DMSO	475	25,000	552	H_2O	30
O6151	496.38	F,D	DMSO	<300		none		31
P1304MP	668.40	L	H_2O, DMSO	535	5400	617	H_2O/DNA	1, 32
P3566	668.40	RR,L	H_2O	535	5400	617	H_2O/DNA	1, 3, 32
P21493	668.40	L	H_2O, DMSO	535	5400	617	H_2O/DNA	33
R302	380.83	F,D,L	MeOH, DMF	507	101,000	529	MeOH	
R12204	251.17	L	H_2O, MeOH	604	60,000	none	MeOH	34
R14060	434.41	F,D,L,AA	DMSO	239	52,000	none	MeOH	15, 35
R22420	380.83	F,D,L	MeOH, DMF	507	101,000	529	MeOH	33
S1129	518.43	F,D	DMSO	<300		none		36
S7020	~600	F,D,L	DMSO	504	67,000	523	H_2O/DNA	1, 3, 37, 38
S7575	~400	F,D,L	DMSO	488	74,000	509	H_2O/DNA	1, 3, 37, 38, 39
S11348	~400	F,D,L	DMSO	445	38,000	470	H_2O/DNA	1, 3, 37, 38
S11368	~500	F,D,L	DMSO	547	79,000	570	H_2O/DNA	1, 3, 37, 38
S32703	~800	F,D,L	DMSO	491	107,000	532	H_2O/RNA	1, 3, 37, 38
S32704	~350	F,D,L	DMSO	484	67,000	505	H_2O/DNA	1, 3, 37, 38
S34854	~400	F,D,L	DMSO	483	65,000	503	H_2O/DNA	1, 3, 37, 38
S34855	~400	F,D,L	DMSO	480	66,000	502	H_2O/DNA	1, 3, 37, 38
S34859	~450	F,D,L	DMSO	640	92,000	658	H_2O/DNA	1, 3, 37, 38
T668	500.93	F,D,L	DMSO, MeOH	549	115,000	573	MeOH	
T669	514.96	F,D,L	DMSO, EtOH	549	109,000	574	MeOH	
T3168	652.23	D,L	DMSO, DMF	514	195,000	529	MeOH	40
V34850	1723.35	F,D,L	H_2O, DMSO	504	68,000	511	MeOH	
X6493	674.53	F,D	H_2O, DMSO	286	15,000	none	MeOH	41
15585-011(EtBr)	394.31	RR,L	H_2O	518	5200	605	H_2O/DNA	1,3,42

For definitions of the contents of this data table, see "Using *The Molecular Probes® Handbook*" in the introductory pages.

Notes

1. Spectra represent aqueous solutions of nucleic acid–bound dye. EC values are derived by comparing the absorbance of the nucleic acid–bound dye with that of free dye in a reference solvent (H_2O or MeOH).
2. Acridine orange bound to RNA has Abs ~460 nm, Em ~650 nm. (Methods Cell Biol (1994) 41:401, Cytometry (1982) 2:201)
3. This product is supplied as a ready-made solution in the solvent indicated under "Soluble."
4. MW value is approximate. BCECF AM is a mixture of molecular species. Lot-specific average MW values are printed on product labels.
5. BCECF AM is colorless and nonfluorescent until converted to BCECF (B1151) by acetoxymethyl ester hydrolysis.
6. Acetate hydrolysis of this compound yields a fluorescent product with similar pH-dependent spectral characteristics to C1904.
7. C369 is converted to a fluorescent product (C368) after acetate hydrolysis.
8. Dihydrofluorescein diacetates are colorless and nonfluorescent until both of the acetate groups are hydrolyzed and the products are subsequently oxidized to fluorescein derivatives. The materials contain less than 0.1% of oxidized derivative when initially prepared. The oxidation products of C400, C2938, C6827, D399 and D2935 are 2',7'-dichlorofluorescein derivatives with spectra similar to C368.
9. Hydrolysis of the acetate and acetoxymethyl ester groups of C1354 yields C1359.
10. Acetoxymethyl ester hydrolysis of calcein blue, AM yields the corresponding iminodiacetic acid. The iminodiacetic acid product is water soluble and has similar spectroscopic properties to 7-hydroxy-4-methylcoumarin (H189).
11. Calcein AM is converted to fluorescent calcein (C481) after acetoxymethyl ester hydrolysis.
12. C6826 is converted to a fluorescent product with spectra similar to C1270 after acetate hydrolysis.
13. C13196 is converted to a fluorescent product (C652) after acetate hydrolysis.
14. Calcein red-orange, AM is converted to the corresponding nitrilotriacetic acid after acetoxymethyl ester hydrolysis. The nitrilotriacetic acid product is water-soluble and has similar spectroscopic properties to the AM ester form.
15. This compound is susceptible to oxidation, especially in solution. Store solutions under argon or nitrogen. Oxidation may be induced by illumination.
16. D632, D23806 and D633 are essentially colorless and nonfluorescent until oxidized. Oxidation products are R302 (from D632 and D23806) and R634 (from D633).
17. Dihydroethidium has blue fluorescence (Em ~420 nm) until oxidized to ethidium (Ex/Em = 518/605 nm in H_2O/DNA). The reduced dye does not bind to nucleic acids. (FEBS Lett (1972) 26:169)
18. JC-9 exhibits long-wavelength J-aggregate emission at ~635 nm in aqueous solutions and polarized mitochondria.
19. This product is supplied as a ready-made solution in DMSO with sodium borohydride added to inhibit oxidation.
20. D23805 is colorless and nonfluorescent until the AM ester groups are hydrolyzed and the resulting leuco dye is subsequently oxidized. The final product is calcein (C481).
21. Although this compound is soluble in water, preparation of stock solutions in water is not recommended because of possible adsorption onto glass or plastic.
22. E1169 in H_2O: Abs = 493 nm (EC = 9100 $cm^{-1}M^{-1}$). E3599 in H_2O: Abs = 498 nm (EC = 10,800 $cm^{-1}M^{-1}$). Both compounds are very weakly fluorescent in H_2O. QY increases >40-fold on binding to dsDNA.
23. E1374 spectral data are for the free dye. Fluorescence is weak, but intensity increases ~15-fold on binding to DNA. After photocrosslinking to DNA, Abs = 504 nm (EC ~4000 $cm^{-1}M^{-1}$), Em = 600 nm. (Nucleic Acids Res (1978) 5:4891, Biochemistry (1980) 19:3221)
24. F7030 is fluorescent when bound to DNA (Em = 538 nm). Uptake and processing of the dye by live yeast results in red-shifted fluorescence (Em ~590 nm).
25. H7593 in H_2O: Abs = 482 nm (EC = 5500 $cm^{-1}M^{-1}$), Em = 625 nm (weakly fluorescent).
26. This compound emits chemiluminescence upon oxidation in basic aqueous solutions. Emission peaks are at 425 nm (L8455) and 470 nm (L6868).
27. Enzymatic reduction products are water-insoluble formazans with Abs = 505 nm (M6494) and 605 nm (N6495) after solubilization in DMSO or DMF. See literature sources for further information. (Histochemistry (1982) 76:381, Prog Histochem Cytochem (1976) 9:1)
28. M6494 also has Abs = 242 nm (EC = 21,000 $cm^{-1}M^{-1}$) in MeOH.
29. Spectra of N3246 are pH-dependent (pK_a ~6.7). Data reported are for 1:1 (v/v) EtOH/1% acetic acid.
30. Fluorescence of NBD and its derivatives in water is relatively weak. QY and τ increase and Em decreases in aprotic solvents and other nonpolar environments relative to water. (Biochemistry (1977) 16:5150, Photochem Photobiol (1991) 54:361)
31. Acetate hydrolysis of this compound yields a fluorescent product with similar spectral characteristics to O6146.
32. Propidium iodide in H_2O: Abs = 493 nm (EC = 5900 $cm^{-1}M^{-1}$), Em = 636 nm (weakly fluorescent). Fluorescence is enhanced >10-fold on binding to dsDNA.
33. This product is specified to equal or exceed 98% analytical purity by HPLC.
34. Enzymatic reduction of resazurin yields resorufin (R363).
35. R14060 is colorless and nonfluorescent until oxidized. The spectral characteristics of the oxidation product (2,3,4,5,6-pentafluorotetramethylrosamine) are similar to those of T639.
36. S1129 is converted to a fluorescent product (F1130) after acetate hydrolysis.
37. This product is essentially nonfluorescent except when bound to DNA or RNA.
38. MW: The preceding ~ symbol indicates an approximate value, not including counterions.
39. The fluorescence quantum yield (QY) of SYTO® 13 dye bound to dsDNA is 0.56 (measured at 22°C).
40. JC-1 forms J-aggregates with Abs/Em = 585/590 nm at concentrations above 0.1 µM in aqueous solutions (pH 8.0). (Biochemistry (1991) 30:4480)
41. Enzymatic reduction product is a water-soluble formazan, Abs = 475 nm.
42. Ethidium bromide in H_2O: Abs = 480 nm (EC = 5600 $cm^{-1}M^{-1}$), Em = 620 nm (weakly fluorescent). Fluorescence is enhanced >10-fold on binding to dsDNA.

IMPORTANT NOTICE: The products described in this manual are covered by one or more Limited Use Label License(s). Please refer to the Appendix on page 971 and Master Product List on page 975. Products are For Research Use Only. Not intended for any animal or human therapeutic or diagnostic use.

PRODUCT LIST 15.2 VIABILITY AND CYTOTOXICITY ASSAY REAGENTS

Cat. No.	Product	Quantity
A1301	acridine orange	1 g
A3568	acridine orange *10 mg/mL solution in water*	10 mL
B1150	2′,7′-bis-(2-carboxyethyl) -5-(and-6)-carboxyfluorescein, acetoxymethyl ester (BCECF, AM)	1 mg
B3051	2′,7′-bis-(2-carboxyethyl) -5-(and-6)-carboxyfluorescein, acetoxymethyl ester (BCECF, AM) *1 mg/mL solution in anhydrous DMSO*	1 mL
B1170	2′,7′-bis-(2-carboxyethyl) -5-(and-6)-carboxyfluorescein, acetoxymethyl ester (BCECF, AM) *special packaging*	20 x 50 µg
B13234	BOCILLIN™ 650/665 penicillin, sodium salt	1 mg
B13233	BOCILLIN™ FL penicillin, sodium salt	1 mg
C1429	calcein blue, AM	1 mg
C34853	calcein blue, AM *for flow cytometry* *for UV excitation* *special packaging*	20 x 50 µg
C34852	calcein green, AM *for flow cytometry* *for 488 nm excitation* *special packaging*	20 x 50 µg
C34851	calcein red-orange, AM *special packaging*	20 x 50 µg
C34858	calcein violet, AM *for flow cytometry* *for 405 nm excitation* *special packaging*	20 x 25 µg
C1430	calcein, AM	1 mg
C3099	calcein, AM *1 mg/mL solution in anhydrous DMSO*	1 mL
C3100MP	calcein, AM *special packaging*	20 x 50 µg
C2938	6-carboxy-2′,7′-dichlorodihydrofluoresce in diacetate, di(acetoxymethyl ester)	5 mg
C400	5-(and-6)-carboxy-2′,7′-dichlorodihydrofluorescein diacetate (carboxy-H_2DCFDA) *mixed isomers*	25 mg
C369	5-(and-6)-carboxy-2′,7′-dichlorofluorescein diacetate (carboxy-DCFDA) *mixed isomers*	100 mg
C13196	5-(and-6)-carboxynaphthofluorescein diacetate	10 mg
C2925	CellTracker™ Green CMFDA (5-chloromethylfluorescein diacetate)	1 mg
C7025	CellTracker™ Green CMFDA (5-chloromethylfluorescein diacetate) *special packaging*	20 x 50 µg
C1361	5-CFDA (5-carboxyfluorescein diacetate) *single isomer*	100 mg
C1362	6-CFDA (6-carboxyfluorescein diacetate) *single isomer*	100 mg
C195	5(6)-CFDA (5-(and-6)-carboxyfluorescein diacetate) *mixed isomers*	100 mg
C1354	5-CFDA, AM (5-carboxyfluorescein diacetate, acetoxymethyl ester)	5 mg
C6827	5-(and-6)-chloromethyl- 2′,7′-dichlorodihydrofluorescein diacetate, acetyl ester (CM-H_2DCFDA) *mixed isomers* *special packaging*	20 x 50 µg
C6826	5-(and-6)-chloromethyl SNARF®-1, acetate *mixed isomers* *special packaging*	20 x 50 µg
D399	2′,7′-dichlorodihydrofluorescein diacetate (2′,7′-dichlorofluorescin diacetate; H_2DCFDA)	100 mg
D23805	dihydrocalcein, AM *special packaging*	20 x 50 µg
D1168	dihydroethidium (hydroethidine)	25 mg
D23107	dihydroethidium (hydroethidine) *5 mM stabilized solution in DMSO*	1 mL
D11347	dihydroethidium (hydroethidine) *special packaging*	10 x 1 mg
D632	dihydrorhodamine 123	10 mg
D23806	dihydrorhodamine 123 *5 mM stabilized solution in DMSO*	1 mL
D633	dihydrorhodamine 6G	25 mg
D22421	3,3′-dimethyl-α-naphthoxacarbocyanine iodide (JC-9; DiNOC$_1$(3))	5 mg
E1169	ethidium homodimer-1 (EthD-1)	1 mg
E3599	ethidium homodimer-2 (EthD-2) *1 mM solution in DMSO*	200 µL
E1374	ethidium monoazide bromide (EMA)	5 mg
F1303	FDA (fluorescein diacetate)	1 g
F10322	FilmTracer™ calcein green biofilm stain	20 x 50 µg
F10319	FilmTracer™ calcein red-orange biofilm stain	20 x 50 µg
F10320	FilmTracer™ calcein violet biofilm stain	20 x 25 µg
F10317	FilmTracer™ FM® 1-43 green biofilm matrix stain	1 mg
F10318	FilmTracer™ SYPRO® Ruby biofilm matrix stain	200 mL
F23103	fosmidomycin, sodium salt (FR-31564)	25 mg
F7030	FUN® 1 cell stain *10 mM solution in DMSO*	100 µL
H7593	hexidium iodide	5 mg
L8455	luminol (3-aminophthalhydrazide)	25 g
M6494	MTT (3-(4,5-dimethylthiazol- 2-yl)-2,5-diphenyltetrazolium bromide)	1 g
N13195	2-NBDG (2-(*N*-(7-nitrobenz-2-oxa- 1,3-diazol-4-yl)amino)-2-deoxyglucose)	5 mg
N3246	neutral red *high purity*	25 mg
N23106	6-(*N*-(7-nitrobenz-2-oxa- 1,3-diazol-4-yl)amino)-6-deoxyglucose (6-NBDG)	5 mg
N6495	nitro blue tetrazolium chloride (NBT)	1 g
N7565	Nucleic Acid Stains Dimer Sampler Kit	1 kit
O6151	Oregon Green® 488 carboxylic acid diacetate (carboxy-DFFDA) *6-isomer*	5 mg
P13235	polymyxin B, BODIPY® FL conjugate, trifluoroacetic acid salt *mixed species*	100 µg
P13238	polymyxin B, dansyl conjugate, trifluoroacetic acid salt *mixed species*	100 µg
P13236	polymyxin B, Oregon Green® 514 conjugate, trifluoroacetic acid salt *mixed species*	100 µg
P1304MP	propidium iodide	100 mg
P3566	propidium iodide *1.0 mg/mL solution in water*	10 mL

continued on next page

IMPORTANT NOTICE: The products described in this manual are covered by one or more Limited Use Label License(s). Please refer to the Appendix on page 971 and Master Product List on page 975. Products are For Research Use Only. Not intended for any animal or human therapeutic or diagnostic use.

PRODUCT LIST 15.2 VIABILITY AND CYTOTOXICITY ASSAY REAGENTS—*continued*

Cat. No.	Product	Quantity
P21493	propidium iodide *FluoroPure™ grade*	100 mg
R14060	RedoxSensor™ Red CC-1 *special packaging*	10 x 50 µg
R12204	resazurin, sodium salt	10 mg
R302	rhodamine 123	25 mg
R22420	rhodamine 123 *FluoroPure™ grade*	25 mg
S1129	SFDA (5-sulfofluorescein diacetate, sodium salt)	25 mg
S32704	SYTO® 10 green fluorescent nucleic acid stain *5 mM solution in DMSO*	100 µL
S7575	SYTO® 13 green fluorescent nucleic acid stain *5 mM solution in DMSO*	250 µL
S34854	SYTO® 9 green fluorescent nucleic acid stain *5 mM solution in DMSO*	100 µL
S34855	SYTO® BC green fluorescent nucleic acid stain *5 mM solution in DMSO*	100 µL
S11350	SYTO® Blue Fluorescent Nucleic Acid Stain Sampler Kit *SYTO® dyes 40–45* *50 µL each*	1 kit
S7572	SYTO® Green Fluorescent Nucleic Acid Stain Sampler Kit *SYTO® dyes 11–14, 16, 21, 24, and 25* *50 µL each*	1 kit
S11360	SYTO® Orange Fluorescent Nucleic Acid Stain Sampler Kit *SYTO® dyes 80–85* *50 µL each*	1 kit
S11340	SYTO® Red Fluorescent Nucleic Acid Stain Sampler Kit *SYTO® dyes 17 and 59–64* *50 µL each*	1 kit
S32703	SYTO® RNASelect™ green fluorescent cell stain *5 mM solution in DMSO*	100 µL
S10349	SYTOX® AADvanced™ Dead Cell Stain Kit *for flow cytometry* *for 488 nm excitation* *100 tests*	1 kit
S10274	SYTOX® AADvanced™ Dead Cell Stain Kit *for flow cytometry* *for 488 nm excitation* *500 tests*	1 kit
S34857	SYTOX® Blue dead cell stain *for flow cytometry* *1000 assays* *1 mM solution in DMSO*	1 mL
S11348	SYTOX® Blue nucleic acid stain *5 mM solution in DMSO*	250 µL
S7020	SYTOX® Green nucleic acid stain *5 mM solution in DMSO*	250 µL
S11368	SYTOX® Orange nucleic acid stain *5 mM solution in DMSO*	250 µL
S34859	SYTOX® Red dead cell stain *for 633 or 635 nm excitation* *5 µM solution in DMSO*	1 mL
T3168	5,5',6,6'-tetrachloro- 1,1',3,3'-tetraethylbenzimidazolylcarboc yanine iodide (JC-1; $CBIC_2(3)$)	5 mg
T669	tetramethylrhodamine, ethyl ester, perchlorate (TMRE)	25 mg
T668	tetramethylrhodamine, methyl ester, perchlorate (TMRM)	25 mg
15585-011	UltraPure™ ethidium bromide *10 mg/mL*	10 mL
V34850	vancomycin, BODIPY® FL conjugate (BODIPY® FL vancomycin)	100 µg
X6493	XTT (2,3-bis-(2-methoxy-4- nitro-5-sulfophenyl)-2H- tetrazolium-5-carboxanilide)	100 mg

IMPORTANT NOTICE: The products described in this manual are covered by one or more Limited Use Label License(s). Please refer to the Appendix on page 971 and Master Product List on page 975. Products are For Research Use Only. Not intended for any animal or human therapeutic or diagnostic use.
www.invitrogen.com/probes
molecular probes® | invitrogen™ by life technologies™

15.3 Viability and Cytotoxicity Assay Kits for Diverse Cell Types

This section contains a thorough description of Molecular Probes® viability and cytotoxicity kits. Fluorometric assays of cell viability and cytotoxicity are easy to perform with the use of a fluorescence microscope, fluorometer, fluorescence microplate reader or flow cytometer,[1,2] and they offer many advantages over traditional colorimetric and radioactivity-based assays. Also discussed in this section are our unique single-step kits for assessing gram sign and for simultaneously determining gram sign and viability of bacteria, as well as the FilmTracer™ LIVE/DEAD® Biofilm Viability Kit designed specifically for bacterial biofilms.

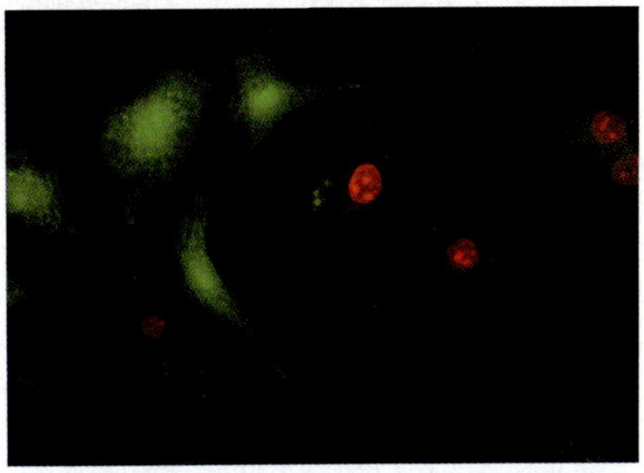

Figure 15.3.2 A mixture of live and ethanol-killed bovine pulmonary artery epithelial cells stained with the reagents in our LIVE/DEAD® Cell Viability/Cytotoxicity Assay Kit (L3224). Live cells fluoresce bright green, whereas dead cells with compromised membranes fluoresce red-orange.

Viability Assay Kits for Animal Cells

To facilitate use of our unique cell viability and cytotoxicity assay technology, we have developed several important products (Table 15.2) that combine fluorescent reagents to yield, in most cases, two-color discrimination of the population of live cells from the dead-cell population by simply adding the reagents, incubating for a brief period and observing the results without any wash steps required. These facile assays are ideal for high-throughput screening applications and, in most cases, for imaging, fluorometry and flow cytometry.

LIVE/DEAD® Viability/Cytotoxicity Kit for Animal Cells

Our LIVE/DEAD® Viability/Cytotoxicity Kit (L3224) for animal cells provides an exceptionally easy fluorescence-based method for determining viability of adherent or nonadherent cells and for assaying cytotoxicity. The kit comprises two probes: calcein AM and ethidium homodimer-1. Calcein AM is a fluorogenic esterase substrate that is hydrolyzed to a green-fluorescent product (calcein); thus, green fluorescence is an indicator of cells that have esterase activity as well as an intact membrane to retain the esterase products. Ethidium homodimer-1 is a high-affinity, red-fluorescent nucleic acid stain that is only able to pass through the compromised membranes of dead cells. The LIVE/DEAD® viability/cytotoxicity assay offers several advantages:

- **Simplicity**. The reagents are simultaneously added to the cell suspension, which is then incubated for 30–45 minutes. No wash steps are required before analysis.
- **Specificity and reliability**. Green-fluorescent cells are live; red-fluorescent cells are dead (Figure 15.3.1, Figure 15.3.2).
- **Versatility**. The LIVE/DEAD® viability/cytotoxicity assay is compatible with adherent cells such as astrocytes,[3] nonadherent cells and certain tissues.[4–6] Results can be analyzed by fluorescence microscopy using standard fluorescein longpass filter sets, as well as by flow cytometry (Figure 15.3.3) or fluorometry. The fluorescence emissions of the two probes are easily resolved (Figure 15.3.4).
- **Simple quantitation**. Flow cytometric measurements yield only two populations; there are rarely any doubly stained cells (Figure 15.3.3). Quantitative assays of bulk cells can be made using a fluorescence microplate reader or fluorometer.
- **Suitability for high-throughput screening**. The ease, reliability and low cost of the LIVE/DEAD® Viability/Cytotoxicity Kit make it an economical assay for high-throughput screening of cytotoxic agents.

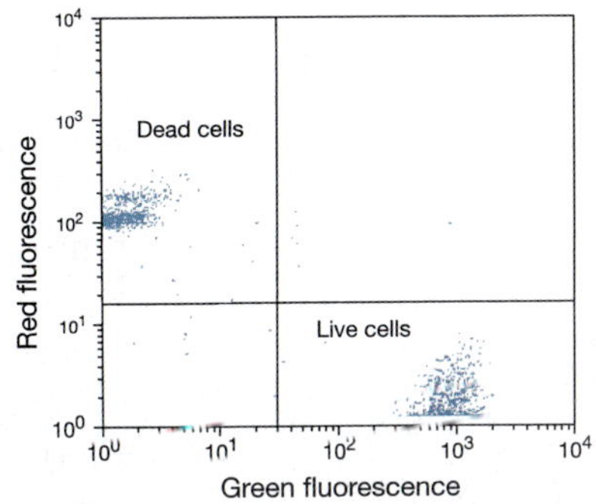

Figure 15.3.3 Flow cytometric viability assay using the LIVE/DEAD® Viability/Cytotoxicity Kit (L3224). A 1:1 mixture of live and ethanol-fixed human B cells was stained with calcein AM and ethidium homodimer-1 according to the kit protocol. After 5 minutes, flow cytometric analysis was carried out with excitation at 488 nm. The resulting bivariate frequency distribution shows the clear separation of the green-fluorescent (530 nm) live-cell population from the red-fluorescent (585 nm) dead-cell population.

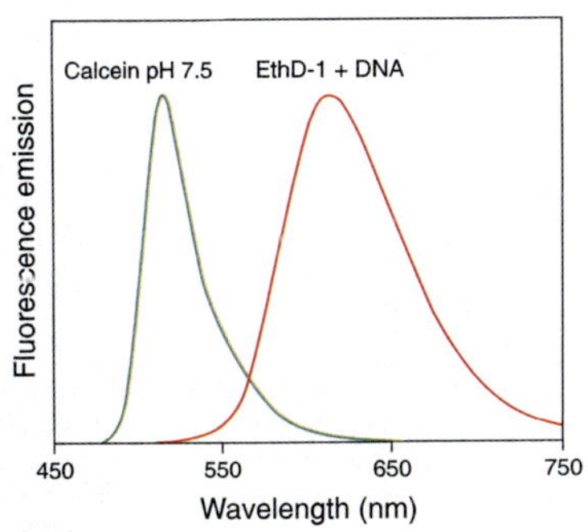

Figure 15.3.4 Normalized fluorescence emission spectra of calcein (C481) and DNA-bound ethidium homodimer-1 (EthD-1, E1169), illustrating the clear spectral separation that allows simultaneous visualization of live and dead eukaryotic cells with the LIVE/DEAD® Viability/Cytotoxicity Kit (L3224).

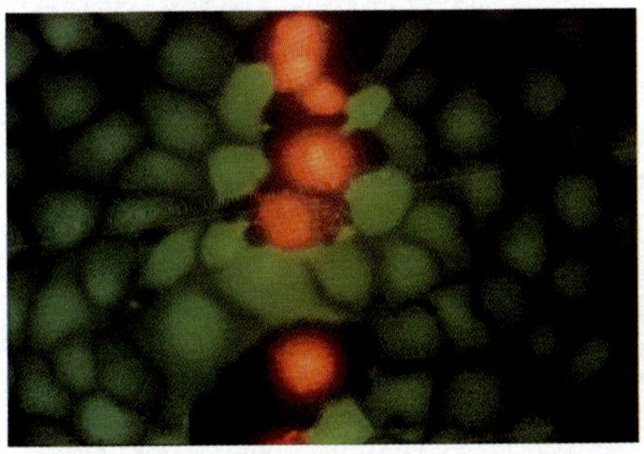

Figure 15.3.1 Live and dead kangaroo rat (PtK2) cells stained with ethidium homodimer-1 and the esterase substrate calcein AM, both of which are provided in our LIVE/DEAD® Viability/Cytotoxicity Kit (L3224). Live cells fluoresce a bright green, whereas dead cells with compromised membranes fluoresce red-orange.

Table 15.2 Molecular Probes® assay kits for cell viability, cell counting and bacterial gram staining.

Cat. No.	Kit Name	Kit Components	# Assays	Assay Principle
Assay Kits for Mammalian Cells				
L3224	LIVE/DEAD® Viability/ Cytotoxicity Kit	• Calcein AM • Ethidium homodimer-1 • Detailed protocols	1000 microscopy assays, 1000 microplate assays or 300 flow cytometry assays	Membrane-permeant calcein AM is cleaved by esterases in live cells to yield cytoplasmic green fluorescence, and membrane-impermeant ethidium homodimer-1 labels nucleic acids of membrane-compromised cells with red fluorescence.
L7013	LIVE/DEAD® Reduced Biohazard Cell Viability Kit #1	• SYTO® 10 nucleic acid stain • DEAD Red™ nucleic acid stain • Detailed protocols	100 microscopy assays or 100 flow cytometry assays	Membrane-permeant SYTO® 10 dye labels the nucleic acids of live cells with green fluorescence, and membrane-impermeant DEAD Red™ dye labels nucleic acids of membrane-compromised cells with red fluorescence. Subsequent fixation inactivates pathogens without distorting the staining pattern.
L23105 L34957 L34959 L34955 L23101 L23102 L10120 L10119 L34960 †	LIVE/DEAD® Fixable Dead Cell Stain Kits *	• Fluorescent reactive dye • DMSO • Detailed protocols	200 flow cytometry assays	Live cells react with the fluorescent reactive dye only on their surface to yield weakly fluorescent cells. Cells with compromised membranes react with the dye throughout their volume, yielding brightly stained cells. Subsequent fixation inactivates pathogens without distorting the staining pattern.
L7010	LIVE/DEAD® Cell-Mediated Cytotoxicity Kit	• $DiOC_{18}(3)$ • Propidium iodide • Detailed protocols	2000 microscopy assays or 200 flow cytometry assays	Target cells are preincubated with the green-fluorescent membrane stain $DiOC_{18}(3)$ and then mixed with effector cells in the presence of the red-fluorescent, membrane-impermeant propidium iodide. Live and dead target cells retain their green-fluorescent membrane stain; target and effector cells with compromised membranes exhibit red-fluorescent nucleic acid staining; live effector cells are nonfluorescent.
L7011	LIVE/DEAD® Sperm Viability Kit	• SYBR® 14 nucleic acid stain • Propidium iodide • Detailed protocols	1000 microscopy assays or 200 flow cytometry assays	Membrane-permeant SYBR® 14 nucleic acid stain labels live sperm with green fluorescence, and membrane-impermeant propidium iodide labels the nucleic acids of membrane-compromised sperm with red fluorescence.
L34951	LIVE/DEAD® Cell Vitality Assay Kit	• Dodecylresazurin (C_{12}-resazurin) • SYTOX® Green nucleic acid stain • DMSO • Phosphate-buffered saline • Detailed protocols	1000 flow cytometry assays	Metabolically active cells reduce C_{12}-resazurin to red-fluorescent C_{12}-resorufin, and cells with compromised membranes (usually late-apoptotic and necrotic cells) are labeled with the green-fluorescent SYTOX® Green nucleic acid stain.
L34958	LIVE/DEAD® Cell Vitality Assay Kit	• CellTrace™ calcein violet AM • Aqua-fluorescent reactive dye • DMSO • Detailed protocols	200 flow cytometry assays	Live cells are labeled with CellTrace™ calcein violet AM, and dead cells with the aqua-fluorescent reactive dye. Both stains are compatible with the 405 nm violet laser diode.
V23110	Vybrant® Cell Metabolic Assay Kit	• Dodecylresazurin (C_{12}-resazurin) • DMSO • Resorufin • Detailed protocols	200 flow cytometry assays or 1000 microplate assays	Nonfluorescent C_{12}-resazurin is reduced to fluorescent C_{12}-resorufin by viable cells; the resulting signal is proportional to the number of cells present.
V23111	Vybrant® Cytotoxicity Assay Kit	• Resazurin • DMSO • Reaction mixture • Reaction buffer • Lysis buffer • Detailed protocols	1000 microplate assays	Damaged and dying cells release glucose 6-phosphate into the surrounding medium, which is detected by an enzymatic process that leads to the reduction of resazurin into red-fluorescent resorufin.
Assay Kits for Yeast				
L7009	LIVE/DEAD® Yeast Viability Kit	• FUN® 1 cell stain • Calcofluor White M2R • Detailed protocols	>1000 microscopy assays or 1000 microplate assays	Plasma membrane integrity and metabolic function of fungi are required to convert the yellow-green–fluorescent intracellular staining of FUN® 1 into red-orange–fluorescent intravacuolar structures; Calcofluor White M2R labels cell-wall chitin with blue fluorescence regardless of metabolic state.
L34952	LIVE/DEAD® *Funga*Light™ Yeast Viability Kit *for flow cytometry*	• SYTO® 9 nucleic acid stain • Propidium iodide • Detailed protocols	~200 flow cytometry assays	Membrane-permeant SYTO® 9 dye generally labels all yeast in a population—those with intact membranes and those with damaged membranes. In contrast, propidium iodide penetrates only yeast with damaged membranes, causing displacement of the SYTO® 9 stain.
F34953	*Funga*Light™ CFDA AM/ Propidium Iodide Yeast Vitality Kit *for flow cytometry*	• 5-carboxyfluorescein diacetate, acetoxymethyl ester (CFDA AM) • Propidium iodide • Dimethylsulfoxide (DMSO), anhydrous • Detailed protocols	~200 flow cytometry assays	The cell-permeant CFDA AM is combined with the membrane integrity indicator propidium iodide to evaluate the vitality of yeast cells by flow cytometry or microscopy. With an appropriate stain mixture, esterase-active yeast with intact cell membranes stain fluorescent green, whereas yeast with damaged membranes stain fluorescent red.

* The LIVE/DEAD® Fixable Dead Cell Stain Kits are available in eight fluorescent colors (excitation/emission maxima): blue (350/450 nm, L23105), aqua (367/526 nm, L34957), yellow (400/575 nm, L34959), violet (416/451 nm, L34955), green (495/520 nm, L23101), red (595/615 nm, L23102), far red (650/665 nm, L10120) and near infrared (750/775 nm, L10119), as well as in the LIVE/DEAD® Fixable Dead Cell Stain Sampler Kit (L34960), which contains all eight fluorescent fixable dead cell stains. † 320 flow cytometry assays.

IMPORTANT NOTICE: The products described in this manual are covered by one or more Limited Use Label License(s). Please refer to the Appendix on page 971 and Master Product List on page 975. Products are For Research Use Only. Not intended for any animal or human therapeutic or diagnostic use.

Table 15.2 Molecular Probes® assay kits for cell viability, cell counting and bacterial gram staining—*continued.*

Cat. No.	Kit Name	Kit Components	# Assays	Assay Principle
Assay Kits for Bacteria				
L7012 L7007 L13152	LIVE/DEAD® *Bac*Light™ Bacterial Viability Kit	• SYTO® 9 nucleic acid stain • Propidium iodide • *Bac*Light™ mounting oil • Detailed protocols	For Kit L7007 and L7012: >1000 microscopy assays, 1000 microplate assays or 200 flow cytometry assays. For Kit L13152: each applicator set allows ~1000 microscopy assays, 50 microplate assays or 10 flow cytometry assays	Membrane-permeant SYTO® 9 dye labels live bacteria with green fluorescence; membrane-impermeant propidium iodide labels membrane-compromised bacteria with red fluorescence. In Kit L7007, the stains are supplied in a mixed, two-component formulation, whereas in Kit L7012 the stains are provided as separate solutions. In Kit L13152, the separate dyes are dry and premeasured into pairs of polyethylene transfer pipettes.
L34856	LIVE/DEAD® *Bac*Light™ Bacterial Viability and Counting Kit	• SYTO® 9 nucleic acid stain • Propidium iodide • Microsphere standard • Detailed protocols	100 flow cytometry assays	Membrane-permeant SYTO® 9 dye labels live bacteria with green fluorescence; membrane-impermeant propidium iodide labels membrane-compromised bacteria with red fluorescence. The calibrated suspension of polystyrene microspheres serves as a standard for the volume of suspension analyzed and is clearly distinguishable from stained bacteria in a fluorescence versus side scatter cytogram.
B34954	*Bac*Light™ RedoxSensor™ Green Vitality Kit	• RedoxSensor™ Green reagent • Propidium iodide, an indicator of membrane integrity • Sodium azide • Carbonyl cyanide 3-chlorophenylhydrazone (CCCP) • Detailed protocols	200 flow cytometry assays	The RedoxSensor™ Green reagent reports reductase activity in both gram-positive and gram-negative bacteria (although differences in signal intensity may be observed based upon cell wall characteristics), providing an accurate view of this important aspect of bacterial vitality. Following reduction, the RedoxSensor™ Green reagent will produce a stable green-fluorescent signal (excitation/emission maxima ~490/520 nm) in 10 minutes that is compatible with formaldehyde fixation techniques.
B34956	*Bac*Light™ RedoxSensor™ CTC Vitality Kit	• 5-cyano-2,3-ditolyl tetrazolium chloride (CTC) • SYTO® 24 green-fluorescent nucleic acid stain • 4′,6-diamidino-2-phenylindole, dihydrochloride (DAPI) • Detailed protocols	50 tests by flow cytometry or microscopy	This kit contains 5-cyano-2,3-ditolyl tetrazolium chloride (CTC), for detection of respiration-linked dehydrogenase activity. SYTO® 24 and DAPI counterstains facilitate the differentiation of cells from debris and the calculation of total cell numbers.
B34950	*Bac*Light™ Bacterial Membrane Potential Kit	• $DiOC_2(3)$ in DMSO • CCCP in DMSO • Phosphate-buffered saline (PBS) • Detailed protocols	100 flow cytometry assays	At low concentrations, $DiOC_2(3)$ exhibits green fluorescence in all bacterial cells, but it becomes more concentrated in healthy cells that are maintaining a membrane potential, causing the dye to self-associate and the fluorescence emission to shift to red. CCCP is included in the kit for use as a control because it eradicates the proton gradient, eliminating bacterial membrane potential.
V7023	ViaGram™ Red$^+$ Bacterial Gram Stain and Viability Kit	• DAPI • SYTOX® Green nucleic acid stain • Texas Red®-X conjugate of wheat germ agglutinin (WGA) • Sodium bicarbonate • *Bac*Light™ mounting oil • Detailed protocols	200 microscopy assays	Membrane-permeant DAPI labels live bacteria with blue fluorescence; membrane-impermeant SYTOX® Green nucleic acid stain labels bacteria with compromised membranes with green fluorescence. Simultaneous Texas Red®-X WGA staining produces red-fluorescent surface labeling of gram-positive bacteria.
L7005	LIVE *Bac*Light™ Bacterial Gram Stain Kit	• SYTO® 9 nucleic acid stain • Hexidium iodide • *Bac*Light™ mounting oil • Detailed protocols	>1000 microscopy assays, 1000 microplate assays or 200 flow cytometry assays	When gram-negative and gram-positive bacteria are simultaneously stained with the membrane-permeant SYTO® 9 dye and hexidium iodide, gram-negative bacteria fluoresce green and gram-positive bacteria fluoresce red.
B7277	Bacteria Counting Kit (Section 15.4)	• SYTO® BC bacteria stain • Suspended microsphere standard • Detailed protocols	100 flow cytometry assays	The membrane-permeant SYTO® BC stain labels both gram-positive and gram-negative bacteria with green fluorescence; the calibrated suspension of polystyrene microspheres serves as a standard for the volume of suspension analyzed.
ATP Assay Kit for All Cell Types				
A22066	ATP Determination Kit	• Luciferin • Luciferase • ATP • Dithiothreitol (DTT) • Reaction buffer • Detailed protocols	1000 microplate assays	Luciferase catalyzes the chemiluminescent reaction of luciferin, ATP and oxygen, producing a photon and other chemical by-products. Stores of ATP are higher in viable cells; therefore, the signal strength correlates with population viability.

* The LIVE/DEAD® Fixable Dead Cell Stain Kits are available in eight fluorescent colors (excitation/emission maxima): blue (350/450 nm, L23105), aqua (367/526 nm, L34957), yellow (400/575 nm, L34959), violet (416/451 nm, L34955), green (495/520 nm, L23101), red (595/615 nm, L23102), far red (650/665 nm, L10120) and near infrared (750/775 nm, L10119), as well as in the LIVE/DEAD® Fixable Dead Cell Stain Sampler Kit (L34960), which contains all eight fluorescent fixable dead cell stains. † 320 flow cytometry assays.

IMPORTANT NOTICE: The products described in this manual are covered by one or more Limited Use Label License(s). Please refer to the Appendix on page 971 and Master Product List on page 975. Products are For Research Use Only. Not intended for any animal or human therapeutic or diagnostic use.

Several laboratories have established the validity of the LIVE/DEAD® viability/cytotoxicity assay for use with animal cells and tissues. Published applications have included measuring the toxic effects of tumor necrosis factor[7] (TNF), β-amyloid protein,[8] adenovirus E1A proteins,[9] tetrodotoxin (TTX) binding to Na^+ channels,[10] methamphetamines,[11] mitogenic sphingolipids[12] and photodynamic therapy.[13] This assay has also been adapted to quantitate lymphocyte-mediated cytotoxicity by flow cytometry,[14] cell-mediated cytotoxicity by fluorescence microscopy[15] and the viability of boar sperm by fluorescence microscopy.[16]

The LIVE/DEAD® Viability/Cytotoxicity Kit is intended for use with animal cells that can be analyzed within about an hour of adding the dyes to the cells. The kit components, number of assays and assay principles are summarized in Table 15.2. This kit's two viability probes—calcein AM (C1430, C3099, C3100MP) and ethidium homodimer-1 (E1169)—are also available separately (see Section 15.2) and may be used in combination with other probes for discrimination of live and dead cells. When assays need to be conducted over longer periods or when hazardous samples are being analyzed, we recommend our LIVE/DEAD® Reduced Biohazard Cell Viability Kit #1 (L7013, see below).

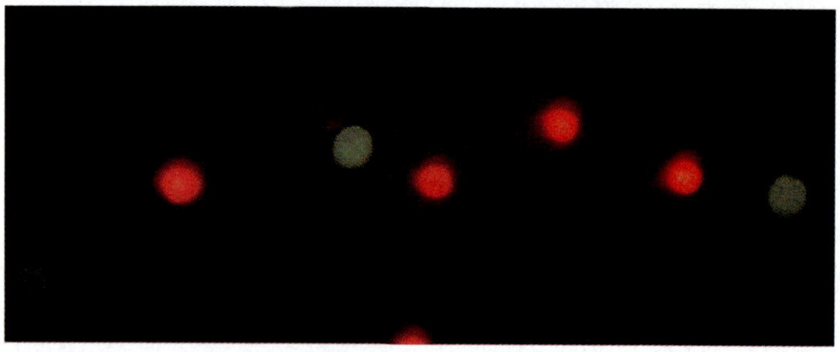

Figure 15.3.5 A mixture of live and dead goat lymphocytes stained with the LIVE/DEAD® Reduced Biohazard Viability/Cytotoxicity Kit (L7013) and subsequently fixed with 4% glutaraldehyde. This image was photographed in a single exposure through an Omega® Optical triple bandpass filter set.

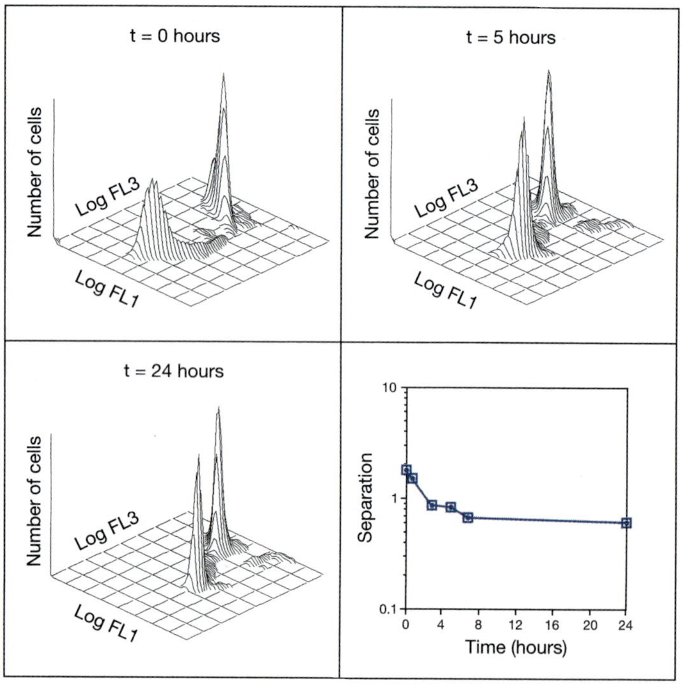

Figure 15.3.6 Flow cytometric analysis of a mixed population of live and complement-treated goat lymphocytes stained using the reagents and protocols provided in our LIVE/DEAD® Reduced Biohazard Cell Viability Kit #1 (L7013) and monitored over a 24-hour period. The panels (left to right, top to bottom) represent the distribution of SYTO® 10 green fluorescence and DEAD Red™ red fluorescence in lymphocytes at 0, 5 and 24 hours after fixation. Data are on a logarithmic scale. The lower right panel is a plot of the separation between the live- and dead-population peaks as a function of time.

LIVE/DEAD® Reduced Biohazard Cell Viability Kit

Rigorous precautions are necessary during analysis of biohazardous specimens.[17,18] Therefore, fixation procedures that inactivate cells yet produce minimal distortion of their characteristics are highly advantageous.[19] The LIVE/DEAD® Reduced Biohazard Cell Viability Kit #1 (L7013) provides a two-color fluorescence assay for animal cell viability that is designed to reduce the risk associated with handling potential biohazards such as viral, bacterial or protozoan pathogens.

Viability analysis with the LIVE/DEAD® Reduced Biohazard Cell Viability Kit #1 is provided by the cell-permeant, green-fluorescent SYTO® 10 and cell-impermeant, red-fluorescent DEAD Red™ (ethidium homodimer-2) nucleic acid stains. The dye concentrations and their relative affinities are balanced so that a cell population exposed simultaneously to both dyes becomes differentially stained—live cells fluoresce green and dead cells fluoresce red (Figure 15.3.5). This assay is simple, fast and can be carried out using a fluorescence microscope, flow cytometer or fluorescence microplate reader. Moreover, the staining pattern of a cell population is retained for several hours after fixation (Figure 15.3.6).

Our LIVE/DEAD® Reduced Biohazard Cell Viability Kit #1 has several unique features:

- **Reduced handling risks.** This kit allows viability staining to take place while the potentially pathogenic sample is well contained. Subsequent treatment with 4% glutaraldehyde (or less effectively with formaldehyde) permits safer handling during analysis, without disrupting the distinctive staining pattern. Glutaraldehyde is known to inactivate cells and viruses, while preserving their overall morphology.[20] In addition, the high sensitivity and specificity of the assay mean that sample sizes can be very small, further reducing potential biohazards.
- **Specificity and reliability.** Live cells initially fluoresce green, and dead cells fluoresce red. With time, this discrimination is reduced but can still be detected, even after 24 hours (Figure 15.3.6).
- **Independence from enzymatic activity.** Because it relies on two nucleic acid stains that differ in their membrane permeability, this assay equates loss of cell viability with loss of membrane integrity. Consequently, the assay is totally independent of variations in enzymatic activity or electrical potential of the cell.
- **Versatility.** The analysis is readily quantitated with a fluorescence microscope or flow cytometer (Figure 15.3.6). This kit's protocol includes methods for analyzing the viability of nonadherent cells, as well as adherent cells on coverslips.
- **Convenience.** Cells can be stained and fixed at various times during the experiment, and the results can be analyzed several hours later, without loss of the discrimination pattern.

The kit components, number of assays and assay principles are summarized in Table 15.2.

IMPORTANT NOTICE: The products described in this manual are covered by one or more Limited Use Label License(s). Please refer to the Appendix on page 971 and Master Product List on page 975. Products are For Research Use Only. Not intended for any animal or human therapeutic or diagnostic use.

LIVE/DEAD® Fixable Dead Cell Stain Kits

The LIVE/DEAD® Fixable Dead Cell Stain Kits (three of which were formerly named LIVE/DEAD® Reduced Biohazard Cell Viability Kits #2, #3 and #4) employ an amine-reactive fluorescent dye to evaluate mammalian cell viability by flow cytometry (Figure 15.3.7).[20] The LIVE/DEAD® Fixable Dead Cell Stain Kits are identical except for the fluorescent color of the amine-reactive dye:

- LIVE/DEAD® Fixable Blue Dead Cell Stain Kit, for UV excitation (L23105, excitation/emission maxima ~350/450 nm)
- LIVE/DEAD® Fixable Aqua Dead Cell Stain Kit, for 405 nm excitation (L34957, excitation/emission maxima ~367/526 nm)
- LIVE/DEAD® Fixable Yellow Dead Cell Stain Kit, for 405 nm excitation (L34959, excitation/emission maxima ~400/575 nm)
- LIVE/DEAD® Fixable Violet Dead Cell Stain Kit, for 405 nm excitation (L34955, excitation/emission maxima ~416/451 nm)
- LIVE/DEAD® Fixable Green Dead Cell Stain Kit, for 488 nm excitation (L23101, excitation/emission maxima ~495/520 nm)
- LIVE/DEAD® Fixable Red Dead Cell Stain Kit, for 488 nm excitation (L23102, excitation/emission maxima ~595/615 nm)
- LIVE/DEAD® Fixable Far Red Dead Cell Stain Kit, for 633 or 635 nm excitation (L10120, excitation/emission maxima ~650/665 nm)
- LIVE/DEAD® Fixable Near-IR Dead Cell Stain Kit, for 633 or 635 nm excitation (L10119, excitation/emission maxima ~750/775 nm)
- LIVE/DEAD® Fixable Dead Cell Stain Sampler Kit, for flow cytometry (L34960), which contains one vial of each of the eight different fluorescent reactive dyes

In cells with compromised membranes, the dye reacts with free amines both in the cell interior and on the cell surface, yielding intense fluorescent staining. In viable cells, the dye's reactivity is restricted to the cell-surface amines, resulting in less intense fluorescence. The difference in intensity between the live and dead cell populations is typically greater than 50 fold (Figure 15.3.8), and this fluorescence intensity discrimination is preserved following formaldehyde fixation, using conditions that inactivate pathogens. These single-color assays use only one channel of a flow cytometer, making the reactive dyes in the LIVE/DEAD® Fixable Dead Cell Stain Kits compatible with multiparameter staining experiments; appropriate flow cytometer channels may vary depending on the instrument. The assays can also be used to detect dead cells by fluorescence microscopy; however, the difference in fluorescence intensity of the live and dead cells can be appreciable, making it relatively difficult to simultaneously photograph the two populations. The kit components, number of assays and assay principles are summarized in Table 15.2.

ArC™ Amine-Reactive Compensation Bead Kit

Optimized for use with the LIVE/DEAD® Fixable Dead Cell Stain Kits, the ArC™ Amine Reactive Compensation Bead Kit (A10346) is a tool designed to remove spectral overlap of the fixable dead-cell stains with other standard fluorophores. This kit provides two polystyrene microsphere samples: ArC™ Reactive Beads, which are reactive to all dyes in the Live/Dead Fixable Dead Cell Stain Kits, and Negative Control Beads, which have no reactivity. The two components provide negative and positive populations that can be used to help accurately set compensation when using the Live/Dead Fixable Dead Cell Stains.

LIVE/DEAD® Cell-Mediated Cytotoxicity Kit

Cytotoxicity triggered by a natural defense mechanism may be much slower than cell lysis triggered by a cytotoxic reagent. Our LIVE/DEAD® Cell-Mediated Cytotoxicity Kit (L7010) is intended for cytotoxicity assessments extending over time periods that are too long for effective use of cytoplasmic markers, such as calcein, which may leak out or become sequestered. This kit is based directly on procedures developed by Kroesen and colleagues for measuring natural killer (NK) cell–mediated, lymphokine-activated killer (LAK) cell–mediated and T cell–mediated cytotoxicity by fluorescence microscopy.[21] The assay has also been adapted for rapid flow cytometric analysis of NK cell activity.[22–24]

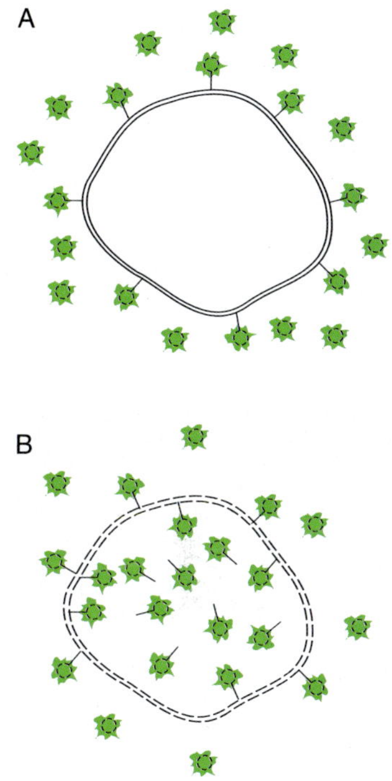

Figure 15.3.7 Principle of our LIVE/DEAD® Fixable Green Dead Cell Stain Kit (L23101). Live cells (**A**) react with the kit's green-fluorescent, amine-reactive dye only on their surface to yield weakly fluorescent cells. Cells with compromised membranes (**B**) react with the dye throughout their volume, yielding brightly stained cells. In both cases, the excess reactive dye is subsequently washed away.

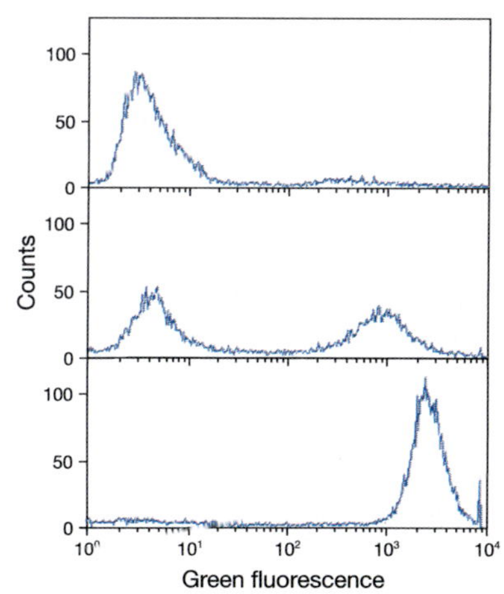

Figure 15.3.8 Live and dead cells distinguished by flow cytometry using the LIVE/DEAD® Fixable Green Dead Cell Stain Kit (L23101). The LIVE/DEAD® Fixable Green Dead Cell Stain Kit was used to differentially stain live and dead Jurkat cells taken from a healthy culture (top panel), an aged culture (middle panel) and a heat-killed culture (bottom panel). Following the staining reaction, the cells were fixed in 3.7% formaldehyde and analyzed by flow cytometry. Nearly identical results were obtained using unfixed cells (data not shown).

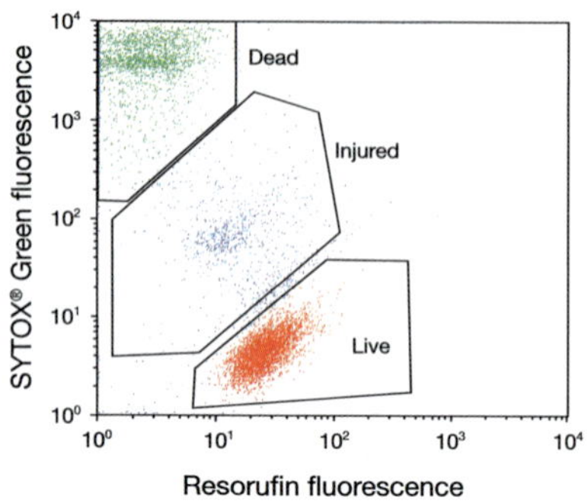

Figure 15.3.10 Flow cytometric analysis of Jurkat cells using the LIVE/DEAD® Cell Vitality Assay Kit (L34951). Jurkat human T-cell leukemia cells were first exposed to 10 µM camptothecin for 4 hours at 37°C, 5% CO_2. The cells were then treated with the reagents in the LIVE/DEAD® Cell Vitality Assay Kit as specified in the kit protocol and analyzed by flow cytometry. This dot plot of SYTOX® Green fluorescence versus resorufin fluorescence shows resolution of live-, injured- and dead-cell populations.

Analysis of cell-mediated cytotoxicity using this kit is easy. In order to distinguish target cells, cultures are labeled overnight with $DiOC_{18}(3)$, a green-fluorescent membrane stain (Section 14.4). Target cells are then washed free of excess $DiOC_{18}(3)$ and combined in various proportions with effector cells. After a suitable incubation period, propidium iodide, a red-fluorescent, membrane-impermeant nucleic acid stain, is added. Propidium iodide labels dead effector cells, as well as dead target cells once their plasma membranes are compromised. Because the target cells retain the green-fluorescent membrane stain, both live and dead effector cells and live and dead target cells can readily be discriminated with a fluorescence microscope. Dead target cells exhibit both green-fluorescent membrane staining and red-fluorescent nuclear staining, whereas dead effector cells show only red-fluorescent nuclear staining. Live target cells have only green-fluorescent membrane staining, and live effector cells are unstained. The kit components, number of assays and assay principles are summarized in Table 15.2.

LIVE/DEAD® Sperm Viability Kit

The LIVE/DEAD® Sperm Viability Kit (L7011), developed in collaboration with Duane L. Garner, provides a novel fluorescence-based method for analyzing the viability of sperm in different species.[25–28] The LIVE/DEAD® Sperm Viability Kit contains the membrane-permeant SYBR® 14 nucleic acid stain, along with the conventional dead-cell stain, propidium iodide. Using this combination of dyes, researchers can rapidly distinguish live and dead cells with visible-light excitation (Figure 15.3.9), thus avoiding the harmful effects of UV exposure and allowing flow cytometric analysis of sperm viability to be performed using an argon-ion laser excitation source. When semen is incubated briefly with these two stains, live sperm with intact membranes fluoresce bright green, whereas sperm cells with damaged membranes fluoresce red. Garner and colleagues assessed bovine sperm viability with flow cytometry and with fluorescence microscopy; both techniques allowed live and dead cells to be visualized simultaneously.[28,29] Furthermore, it was reported that neither the ability to fertilize oocytes nor the development of the embryos was affected by SYBR® 14 staining of porcine sperm.[25] The effect of two-photon illumination on the viability of human sperm stained with these reagents has also been analyzed.[30] This assay is particularly useful for evaluating the viability of cryopreserved sperm.[31–35]

The dyes provided in the LIVE/DEAD® Sperm Viability Kit stain cells more rapidly than conventional stains (within 5–10 minutes), and both label DNA, thereby avoiding the ambiguity that may arise from targeting separate cellular components. The membrane-permeant SYBR® 14 stain provided in the LIVE/DEAD® Sperm Viability Kit should also serve as a valuable tool for labeling and tracking live sperm, thus facilitating analysis of their motility and abundance in

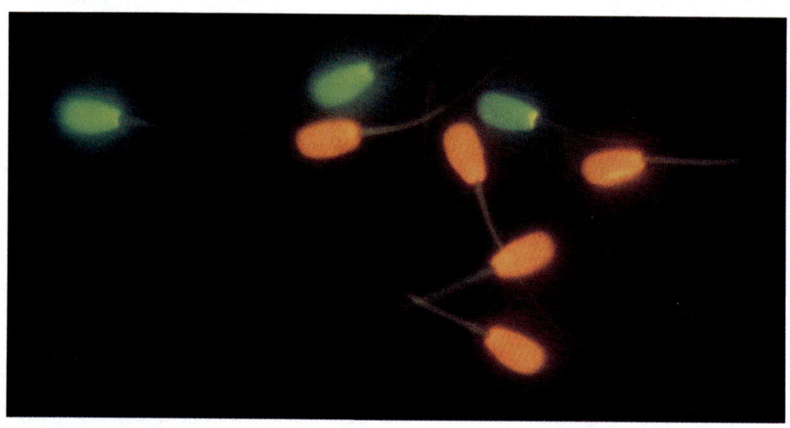

Figure 15.3.9 A mixture of live and dead bovine sperm cells stained with the dyes provided in our LIVE/DEAD® Sperm Viability Kit (L7011). Live sperm with intact membranes are labeled with our proprietary cell-permeant nucleic acid stain, SYBR® 14, and fluoresce green. Dead sperm, which have been killed by unprotected freeze-thawing, are labeled with propidium iodide (P1304MP, P3566, P21493) and fluoresce red-orange. The image was contributed by Duane L. Garner, School of Veterinary Medicine, University of Nevada, Reno, and Lawrence A. Johnson, USDA Agricultural Research Service.

IMPORTANT NOTICE: The products described in this manual are covered by one or more Limited Use Label License(s). Please refer to the Appendix on page 971 and Master Product List on page 975. Products are For Research Use Only. Not intended for any animal or human therapeutic or diagnostic use.

semen samples. Reliable viability measurements with bovine,[28] porcine, ovine, murine,[26] goat, turkey[27] and human sperm[26,30] have been published. The kit components, number of assays and assay principles are summarized in Table 15.2.

Conventional sperm viability assays have employed mixtures of two or three dyes, including fluorescein diacetate derivatives, rhodamine 123 and reduced nucleic acid stains such as dihydroethidium[27,36–39] (hydroethidine, D1168, D11347, D23107; Section 15.2). Acridine orange (A1301, A3568), which fluoresces at different wavelengths when bound to DNA and RNA,[40,41] and the UV light–excitable nucleic acid stains Hoechst 33258 (H1398, H3569, H21491) and Hoechst 33342[42–44] (H1399, H3570, H21492) are frequently used to determine sperm viability and DNA content and to trace sperm–oocyte fusion.[45,46] These nucleic acid stains are described in Section 8.1.

LIVE/DEAD® Violet Viability/Vitality Kit

The LIVE/DEAD® Violet Viability/Vitality Kit (L34958) provides a two-color cell viability and vitality assay that enables the simultaneous identification of live and dead cells. The assay employs two fluorescent probes—calcein violet AM and aqua-fluorescent reactive dye—to indicate recognized parameters of cell health (intracellular esterase activity and plasma membrane integrity, respectively). Both dyes utilize 405 nm violet diode laser excitation, allowing other laser lines to be used for conventional fluorophores. The kit components, number of assays and assay principles are summarized in Table 15.2.

LIVE/DEAD® Cell Vitality Assay Kit

The LIVE/DEAD® Cell Vitality Assay Kit (L34951) provides a simple two-color fluorescence assay that distinguishes metabolically active cells from injured cells and dead cells. The assay is based on the reduction of C_{12}-resazurin to red-fluorescent C_{12}-resorufin in metabolically active cells and the uptake of the cell-impermeant SYTOX® Green nucleic acid stain in cells with compromised plasma membranes (usually late apoptotic and necrotic cells). Dead cells emit mostly green fluorescence, whereas the healthy, metabolically active cells emit mostly red fluorescence (Figure 15.3.10). The injured cells have lower metabolic activity and, consequently, reduced red-fluorescent emission. Because they possess intact membranes, however, injured cells accumulate little SYTOX® Green dye and, therefore, emit very little green fluorescence.

Nonfluorescent resazurin, which can be reduced by viable cells to red-fluorescent resorufin, has been extensively used to detect the metabolic activity of many different cell types, from bacteria to higher eukaryotes.[47–49] Resazurin is nontoxic and stable in culture media, allowing researchers to continuously monitor proliferating cells[50] and to investigate cytotoxicity in both conventional[51] and high-throughput applications.[52] The LIVE/DEAD® Cell Vitality Assay Kit includes a lipophilic version of resazurin, C_{12}-resazurin, which is more permeable to live cells and, after reduction to C_{12}-resorufin, is far better retained than the nonlipophilic resorufin. These characteristics result in brighter signals and better detection limits. The kit components, number of assays and assay principles are summarized in Table 15.2.

Vybrant® Cell Metabolic Assay Kit

One potential drawback in the use of resazurin as a substrate that can detect metabolic activity in live cells (R12204, Section 15.2) is the relatively poor uptake of the substrate and poor cellular retention of the reduced by-product, resorufin. We have found that the lipophilic dodecylresazurin (C_{12}-resazurin), included as a component in the Vybrant® Cell Metabolic Assay Kit (V23110), surpasses resazurin alone in cell permeability and its reduction product (C_{12}-resorufin) is very well retained, permitting single-cell analysis of the substrate's turnover by flow cytometry (Figure 15.3.11). This enhanced uptake, turnover and retention of C_{12}-resazurin by metabolically active cells translates into much brighter signals and far better detection limits when compared with assays using resazurin alone (Figure 15.3.12).

C_{12}-resazurin can be used in any viability/cytotoxicity assay that employs resazurin for both conventional and high-throughput applications. C_{12}-resorufin, which is the reduction product of C_{12}-resazurin, has the same absorption/emission maxima as unmodified resorufin (~571/ 585 nm, Figure 15.3.13); therefore, no changes in instrumentation are required in

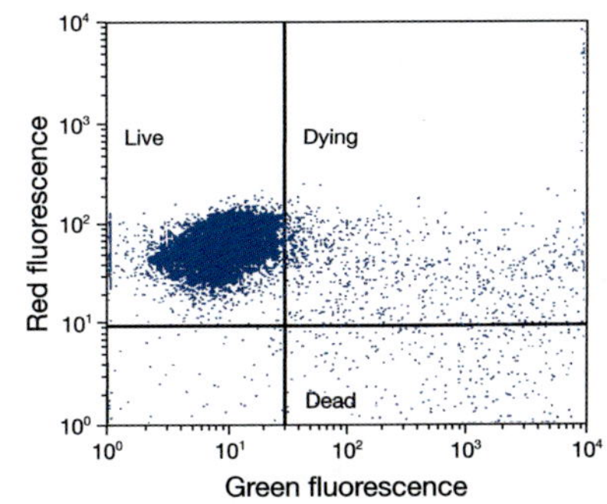

Figure 15.3.11 Flow cytometric analysis of Jurkat cells stained with C_{12}-resazurin. Cells were loaded with 0.1 µM C_{12}-resazurin, a component of the Vybrant® Cell Metabolic Assay Kit (V23110), and 1 mM SYTOX® Green (S7020). After a 15-minute incubation, the cells were analyzed. Healthy (live) cells reduce C_{12}-resazurin into red-fluorescent C_{12}-resorufin and exclude the cell impermeant green-fluorescent SYTOX® Green. Dead cells show little reduction of the C_{12}-resazurin, but strong staining by SYTOX® Green. Cells indicated in the figure as dying are of indeterminate viability, showing both reduction of C_{12}-resazurin and compromised membrane integrity.

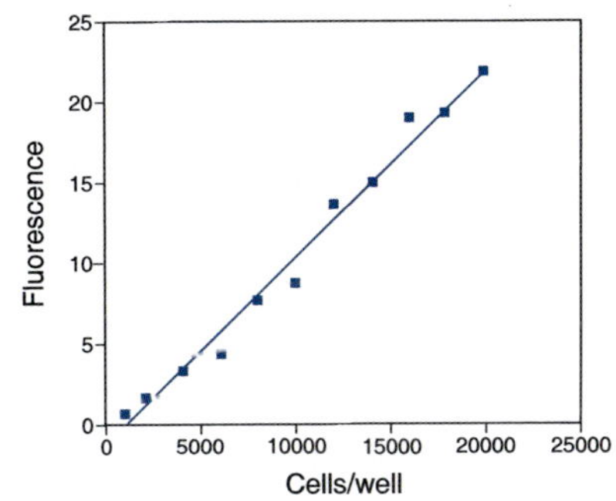

Figure 15.3.12 Detection limit of C_{12}-resazurin and linear response with increasing cell number using our Vybrant® Cell Metabolic Assay Kit (V23110). Jurkat cells were loaded with 5 µM C_{12}-resazurin for 15 minutes. The resulting signal was measured in a fluorescence microplate reader with excitation/emission at 530/590 nm. For comparison, the detection limit for resazurin in a similar experiment was ~8000 cells/well (data not shown).

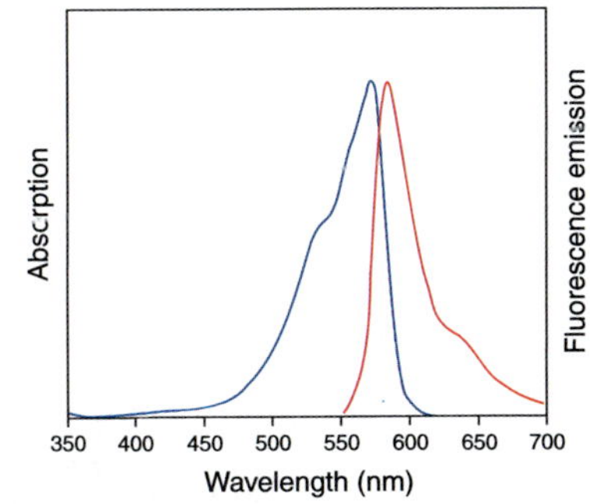

Figure 15.3.13 Absorption and fluorescence emission spectra of resorufin in pH 9.0 buffer.

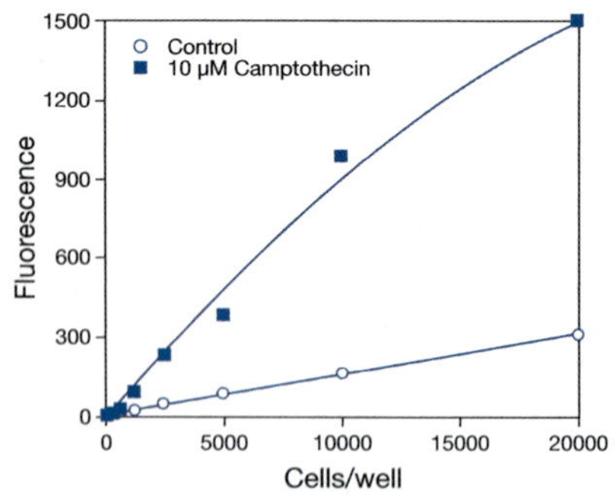

Figure 15.3.15 Detection of dead and dying cells using the Vybrant® Cytotoxicity Assay Kit (V23111). Jurkat cells were treated with 10 µM camptothecin for 6 hours, then assayed for glucose 6-phosphate dehydrogenase release. An untreated control sample is shown for comparison. The fluorescence was measured in a microplate reader (excitation/emission ~530/590 nm). A background of 55 fluorescence units was subtracted from each value.

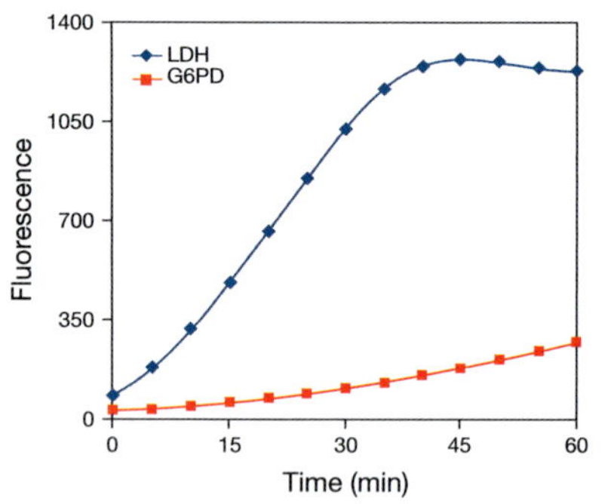

Figure 15.3.16 10% bovine serum was assayed for the presence of lactate dehydrogenase (LDH, blue) and glucose 6-phosphate dehydrogenase (G6PD, red). G6PD was assayed using the Vybrant® Cytotoxicity Assay Kit (V23111); LDH was detected using a similar method, in which LDH reduces lactate to generate NADH. The result clearly shows that, over the time course of the experiment, the serum generates a much lower signal in the G6PD assay than in the LDH assay.

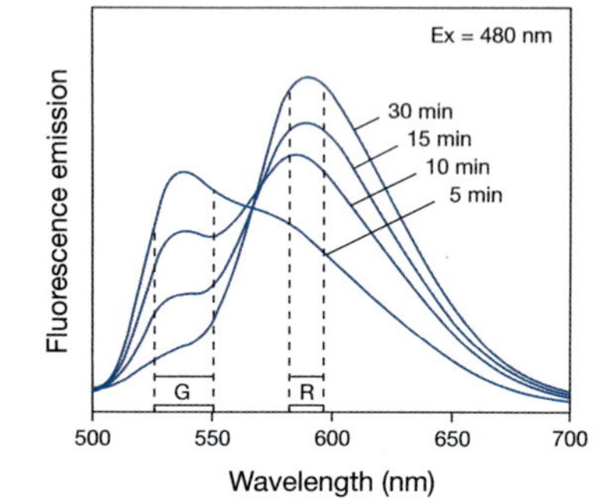

Figure 15.3.17 Fluorescence emission spectra of a *Saccharomyces cerevisiae* suspension that has been stained with the FUN® 1 cell stain, which is available separately (F7030) or in our LIVE/DEAD® Yeast Viability Kit (L7009). After the FUN® 1 reagent was added to the medium, the fluorescence emission spectrum (excited at 480 nm) was recorded in a spectrofluorometer at the indicated times during a 30-minute incubation period. The shift from green (G) to red (R) fluorescence reflects the processing of FUN® 1 by metabolically active yeast cells.

order to use the kit in place of a resazurin-based assay. The Vybrant® Cell Metabolic Assay Kit (V23110) contains:

- Dodecylresazurin (C_{12}-resazurin)
- DMSO
- Resorufin
- Detailed protocols for the assay

Each kit contains sufficient reagents for 500–1000 assays in a fluorescence microplate or about 10,000 flow cytometry assays.

Vybrant® Cytotoxicity Assay Kit

The LIVE/DEAD® kits generally assay cell death via probes that gain entry to the interior of the cell as a result of plasma membrane damage. In contrast, the Vybrant® Cytotoxicity Assay Kit (V23111) monitors the release of the cytosolic enzyme glucose 6-phosphate dehydrogenase (G6PD) from damaged cells into the surrounding medium. G6PD is a ubiquitous enzyme that is part of the pentose phosphate pathway, and is crucial for cellular antioxidant defenses via its production of NADPH.[53,54] Detection of G6PD is by a two-step enzymatic process that leads to the reduction of resazurin into the red-fluorescent resorufin (Figure 15.3.14). The resulting signal is proportional to the amount of G6PD released into the cell medium, which correlates with the number of dead cells in the sample (Figure 15.3.15).

The Vybrant® Cytotoxicity Assay Kit contains all enzymes and substrates needed to detect the release of G6PD from damaged and dying cells. The assay can be completed in less than an hour and is effective with as few as 500 cells per sample. Resorufin, the end product of the G6PD cytotoxicity assay, has absorption and emission maxima at ~571 nm and 585 nm, respectively (Figure 15.3.13), placing the fluorescent signal beyond the autofluorescence of most biological samples. In addition, the levels of G6PD in serum commonly used for cell culture are lower than those of lactate dehydrogenase (LDH), an enzyme often used in similar assays, thus resulting in lower background signals (Figure 15.3.16). The Vybrant® Cytotoxicity Assay Kit (V23111) contains:

- Resazurin (5 vials)
- DMSO
- Reaction mixture (diaphorase, glucose 6-phosphate and $NADP^+$)
- Reaction buffer
- Cell-lysis buffer
- Detailed protocols for the assay

Sufficient reagents are provided for about 1000 assays in a fluorescence microplate reader.

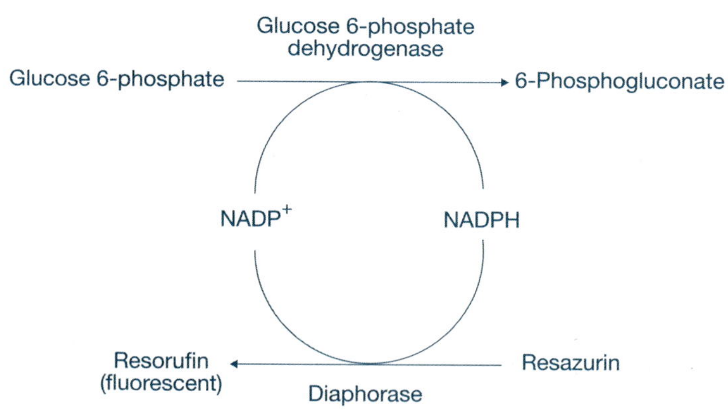

Figure 15.3.14 Principle of the coupled enzymatic assay for detection of glucose 6-phosphate dehydrogenase activity. Oxidation of glucose 6-phosphate by glucose 6-phosphate dehydrogenase results in the generation of NADPH, which in turn leads to the reduction of resazurin by diaphorase to yield fluorescent resorufin.

IMPORTANT NOTICE: The products described in this manual are covered by one or more Limited Use Label License(s). Please refer to the Appendix on page 971 and Master Product List on page 975. Products are For Research Use Only. Not intended for any animal or human therapeutic or diagnostic use.
www.invitrogen.com/probes

Viability Assay Kits for Yeast

LIVE/DEAD® Yeast Viability Kit

Our LIVE/DEAD® Yeast Viability Kit (L7009) provides an extremely simple and sensitive assay for discriminating viable yeast and fungi in complex mixtures or in pure cultures.[55–57] This kit contains our unique two-color fluorescent viability probe, the FUN® 1 dye, which has low intrinsic fluorescence, moderate affinity for nucleic acids and exceptional membrane permeability. Also included is the UV light–excitable counterstain Calcofluor White M2R, which labels the cell walls of yeast and fungi fluorescent blue, regardless of the cell's metabolic state.[58–60]

The FUN® 1 viability probe displays some extraordinary spectral properties when used to stain metabolically active yeast and fungal cells, exploiting normal endogenous biochemical processing mechanisms that appear to be well conserved among different fungal species. The FUN® 1 stain passively diffuses into a variety of cell types and initially stains the cytoplasm with a diffusely distributed green fluorescence. However, in several common species of yeast and fungi, subsequent processing of the dye by live cells results in the formation of distinct vacuolar structures with compact form that exhibit a striking red fluorescence, accompanied by a reduction in the green cytoplasmic fluorescence[57,61] (Figure 15.3.17). Formation of the red-fluorescent intravacuolar structures requires both plasma membrane integrity and metabolic capability. Dead cells fluoresce bright yellow-green, with no discernable red structures.

FUN® 1 stain can be used alone or together with Calcofluor White M2R to determine the metabolic activity of single fungal cells by manual or automated fluorescence microscopy (Figure 15.3.18, Figure 15.3.19). Both live and dead cells may be viewed simultaneously by fluorescence microscopy using a standard fluorescein longpass optical filter set. FUN® 1 dye staining can also be used to assay the viability of suspensions of fungal cells using a fluorescence microplate reader or a fluorometer. The FUN® 1 reagent has been extensively used for flow cytometric analysis of yeast, including of their susceptibility to antifungal agents.[62–66]

The LIVE/DEAD® Yeast Viability Kit has been tested on several fungal species, including *Candida albicans*, *Candida pseudotropicalis* and several strains of *Saccharomyces cerevisiae*, under a variety of experimental conditions. Formation of the red-fluorescent structures was observed not only in logarithmically growing cells but also in nonculturable cells with residual metabolic activity. The LIVE/DEAD® Yeast Viability Kit should be particularly useful for detecting very low numbers of live or dead fungal cells, even in complex mixtures such as blood. The kit components, number of assays and assay principles are summarized in Table 15.2. The FUN® 1 cell stain is also available separately (F7030, Section 15.2).

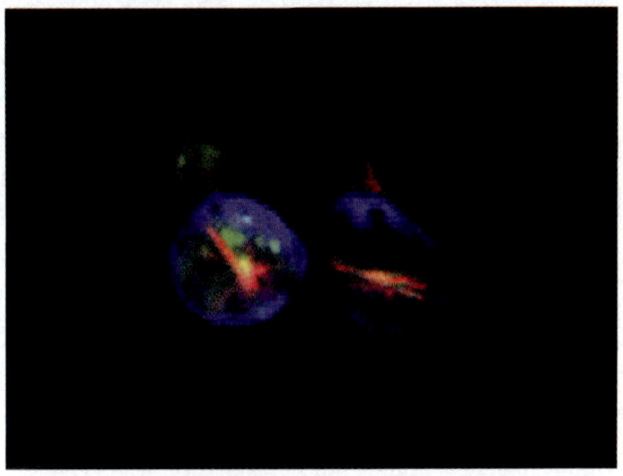

Figure 15.3.18 *Saccharomyces cerevisiae* stained with FUN® 1 cell stain, which generates red-fluorescent intravacuolar structures, and with Calcofluor White M2R, a blue-fluorescent cell wall stain. Both probes are provided in our LIVE/DEAD® Yeast Viability Kit (L7009); FUN® 1 cell stain is also available separately (F7030). A series of z-section images was acquired with a Deltavision wide-field optical sectioning microscope (Applied Precision, Inc.). A three-dimensional projection movie was generated from a deconvolved z-image stack.

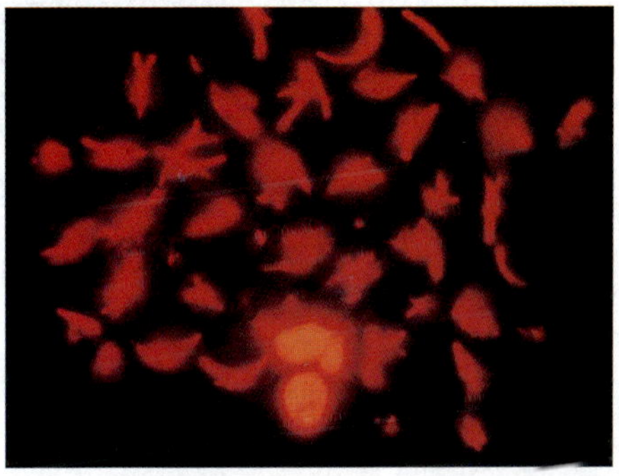

Figure 15.3.19 *Saccharomyces cerevisiae* stained with the FUN® 1 dye, available separately (F7030) or in our LIVE/DEAD® Yeast Viability Kit (L7009). Metabolically active yeast process the FUN® 1 dye, forming numerous red fluorescent cylindrical structures within their vacuoles.

LIVE/DEAD® *Funga*Light™ Yeast Viability Kit

The LIVE/DEAD® *Funga*Light™ Yeast Viability Kit (L34952) allows researchers to easily, reliably and quantitatively distinguish live and dead yeast in minutes. The kit contains solutions of SYTO® 9 green-fluorescent nucleic acid stain and the red-fluorescent nucleic acid stain, propidium iodide. These stains differ both in their spectral characteristics and in their ability to penetrate healthy yeast cells. When used alone, the SYTO® 9 stain generally labels all yeast in a population—those with intact membranes and those with damaged membranes. In contrast, propidium iodide penetrates yeast with damaged membranes, displacing SYTO® 9 stain. Thus, with an appropriate mixture of the SYTO® 9 and propidium iodide stains, yeast with intact cell membranes stain fluorescent green, whereas yeast with damaged membranes stain fluorescent red. The background remains virtually nonfluorescent. Furthermore, this kit also accommodates fine-tuning of the dye combinations so that optimal staining of yeast can be achieved under a variety of experimental conditions. The kit components, number of assays and assay principles are summarized in Table 15.2.

*Funga*Light™ CFDA AM/Propidium Iodide Yeast Vitality Kit

The *Funga*Light™ CFDA AM/Propidium Iodide Yeast Vitality Kit (F34953) combines a cell-permeant esterase substrate with a membrane integrity indicator to evaluate the vitality of yeast cells by flow cytometry or microscopy. The acetoxymethyl ester (AM) of the esterase substrate 5-carboxy-fluorescein diacetate (CFDA) allows this reagent to permeate cell membranes. Once inside the cell, the lipophilic blocking and diacetate groups are cleaved by nonspecific esterases, resulting in a fluorescent, charged form that leaks out of cells very slowly.

IMPORTANT NOTICE: The products described in this manual are covered by one or more Limited Use Label License(s). Please refer to the Appendix on page 971 and Master Product List on page 975. Products are For Research Use Only. Not intended for any animal or human therapeutic or diagnostic use.

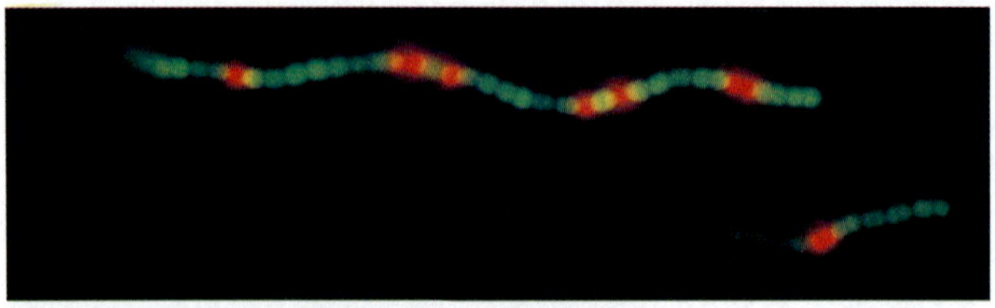

Figure 15.3.20 Use of our LIVE/DEAD® *Bac*Light™ Bacterial Viability Kit (L7007, L7012, L13152) to identify individual live and dead bacteria along a chain of *Streptococcus pyogenes*. This image was photographed in a single exposure through an Omega® Optical triple bandpass filter set.

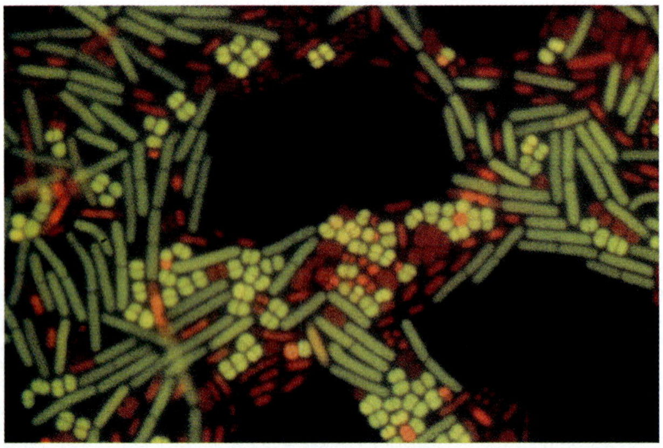

Figure 15.3.21 A mixed population of live and isopropyl alcohol–killed *Micrococcus luteus* and *Bacillus cereus* stained with the LIVE/DEAD® *Bac*Light™ Bacterial Viability Kit (L7007, L7012). Bacteria with intact cell membranes exhibit green fluorescence, whereas bacteria with damaged membranes exhibit red fluorescence. Prior to imaging, the bacteria were placed onto a polycarbonate filter and immersed in *Bac*Light™ mounting oil. This multiple exposure image was acquired with a triple-bandpass optical filter set appropriate for simultaneous imaging of DAPI, fluorescein and Texas Red® dyes.

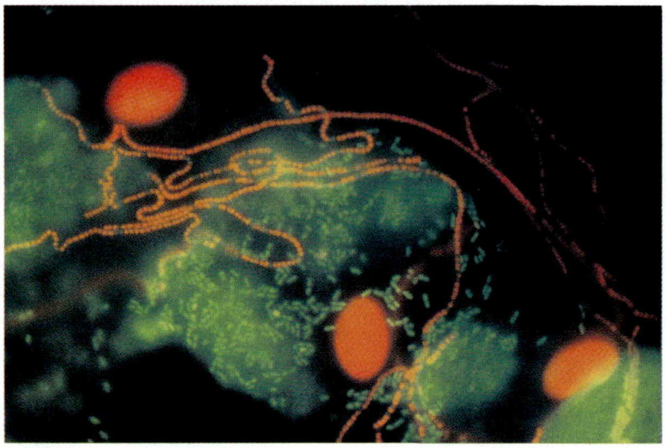

Figure 15.3.22 Live and dead bacteria visualized on freshly isolated human cheek epithelial cells using our LIVE/DEAD® *Bac*Light™ Bacterial Viability Kit (L7007, L7012, L13152). When incubated with the SYTO® 9 and propidium iodide nucleic acid stains provided in this kit, live bacteria with intact cell membranes fluoresce green and dead bacteria with compromised membranes fluoresce red. This image was photographed in a single exposure through an Omega® Optical triple bandpass filter set.

In contrast, the membrane integrity indicator, propidium iodide, penetrates yeast with damaged membranes. With an appropriate mixture of the CFDA AM and propidium iodide stains, esterase-active yeast with intact cell membranes stain fluorescent green, whereas yeast with damaged membranes stain fluorescent red. The excitation/emission maxima for these dyes are 492/517 nm for CFDA AM and 530/635 nm for propidium iodide. The kit components, number of assays and assay principles are summarized in Table 15.2.

Viability Assay and Gram Stain Kits for Bacteria

LIVE/DEAD® *Bac*Light™ Bacterial Viability Kits

Molecular Probes® original LIVE/DEAD® Viability/Cytotoxicity Kit (L3224, see above) is a proven tool for assessing viability of animal cells but is generally not suitable for use with bacterial and yeast cells.[67] Consequently, we have developed the LIVE/DEAD® *Bac*Light™ Bacterial Viability Kits (L7007, L7012, L13152), which provide two different nucleic acid stains—the SYTO® 9 dye and propidium iodide—to rapidly distinguish live bacteria with intact plasma membranes from dead bacteria with compromised membranes[55,57,68] (Figure 15.3.20, Figure 15.3.21).

This assay has several significant features:

- **Ease of use**. The reagents are simultaneously added to the bacterial suspension, which is then incubated for 5–10 minutes. No wash steps are required before analysis.
- **Specificity**. Live bacteria fluoresce green and dead bacteria fluoresce red. Live and dead bacteria can be distinguished and quantitated in minutes, even in a mixed population of bacterial species (Figure 15.3.22).
- **Reliability**. The LIVE/DEAD® *Bac*Light™ Bacterial Viability Kits yield consistent results in tests on a variety of eubacterial genera (Table 15.3). It can also be used to assess the viability of *Eurioplasma eurilytica* and *Mycoplasma hominus* mycoplasma as well as cysts of the protozoan parasite *Giardia muris*[69] (Figure 15.3.23).
- **Validity**. Viability measurements in fresh cultures of bacteria typically correlate well with enumeration techniques involving growth in liquid or solid media. However, variable results have been found using the LIVE/DEAD® *Bac*Light™ reagents to assess viability in some marine bacteria from environmental samples.
- **Versatility**. Bacteria can be stained in suspension or immobilized on microscope slides or filter membranes. Protocols are provided for bacterial viability analysis using a fluorescence microscope, fluorometer (Figure 15.3.24) or fluorescence microplate reader.

The intensities of the fluorescence signals produced by the SYTO® 9 and propidium iodide nucleic acid stains can be adjusted by mixing different proportions of the dye solutions provided in the LIVE/DEAD® *Bac*Light™ Kits. We have balanced the dye concentrations so that, for most bacteria, equal volumes of the two solutions provided give balanced staining of most species. The background remains virtually nonfluorescent, allowing live and dead cells to be easily differentiated in any fluorescence microscope equipped with

IMPORTANT NOTICE: The products described in this manual are covered by one or more Limited Use Label License(s). Please refer to the Appendix on page 971 and Master Product List on page 975. Products are For Research Use Only. Not intended for any animal or human therapeutic or diagnostic use.

a longpass fluorescein or comparable optical filter set. Under certain conditions, bacteria with compromised membranes may recover and reproduce, even though such bacteria may be scored as dead in this assay. Conversely, some bacteria with intact membranes may be unable to reproduce in nutrient medium, yet be scored as live.[70] Combining several different measures of viability, such as membrane permeability, enzyme activity and redox potential, offers a more thorough assessment of bacterial viability and eliminates the inherent limitations of any single viability assay.

The LIVE/DEAD® *Bac*Light™ viability assay has been used to estimate total and viable bacteria in drinking water,[71] to quantitate total and viable concentrations of aerosolized *Pseudomonas fluorescens* by fluorescence microscopy[72] and to measure the reliability of disinfection agents on reducing the viability of *Cryptosporidium parvum* and *Giardia muris* cysts.[69,73,74] The number of live natural planktonic bacteria, as determined with the LIVE/DEAD® *Bac*Light™ Bacterial Viability Kit, reportedly correlated well with the number of bacteria with high DNA content (HDNA), as determined with SYTO® 13 green-fluorescent nucleic acid stain (S7575), leading to the recommendation that %HDNA be used as an index of the percentage of actively growing bacterial cells in marine plankton samples.[75] The reagents in the LIVE/DEAD® *Bac*Light™ Bacterial Viability Kit have been utilized in a high-throughput fluorescence-based screen for bacterial mechanosensitive ion-channel (MscL) activity that replaces otherwise tedious and difficult assays.[76]

Our original packaging of the LIVE/DEAD® *Bac*Light™ Bacterial Viability Kit (L7007), in which the dyes were mixed at different proportions in two solutions, is still available for customers who have already developed protocols using that formulation. However, we recommend use of the LIVE/DEAD® *Bac*Light™ Bacterial Viability Kit (L7012), which is more flexible because it provides separate solutions of the SYTO® 9 and propidium iodide nucleic acid stains, thus facilitating calibration of bacterial fluorescence at each of the two emission wavelengths in quantitative assays. Kit L7007 was designed primarily for use in fluorescence microscopy; Kit L7012 is equally well suited for use in fluorescence microscopy and is better suited than Kit L7007 for use in quantitative analysis with a fluorometer (Figure 15.3.24), fluorescence microplate reader, flow cytometer or other instrumentation.

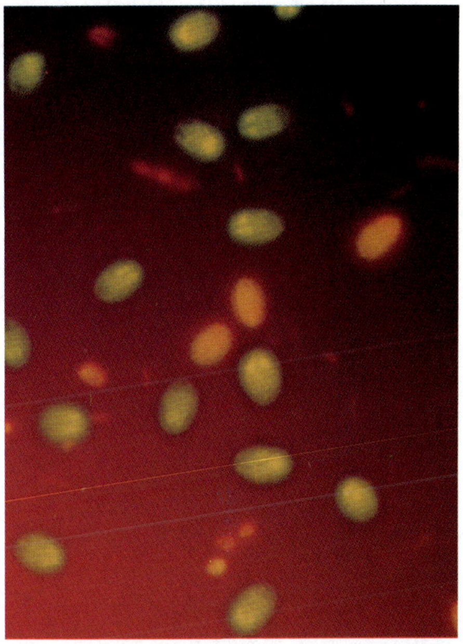

Figure 15.3.23 *Giardia muris* cysts stained with the reagents in the LIVE/DEAD® *Bac*Light™ Bacterial Viability Kit (L7007, L7012). When incubated with the SYTO® 9 and propidium iodide nucleic acid stains provided in this kit, live bacteria with intact cell membranes fluoresce green and dead bacteria with compromised membranes fluoresce red.

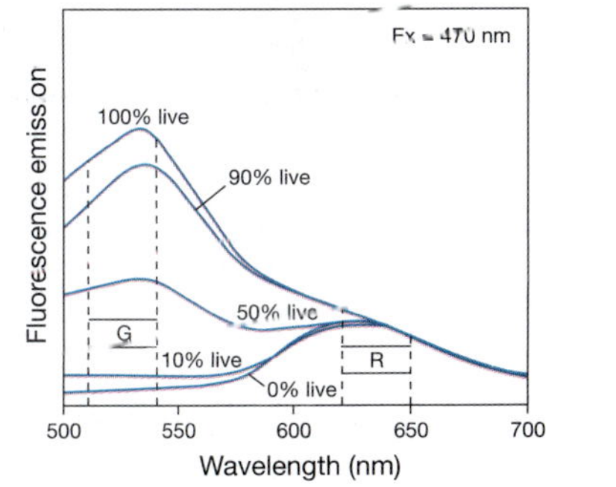

Figure 15.3.24 Viability analysis of bacterial suspensions comprising various proportions of live and isopropyl alcohol–killed *Escherichia coli* using the reagents in the LIVE/DEAD® *Bac*Light™ Bacterial Viability Kit (L7007, L7012, L13152). Live and dead bacteria are stained fluorescent green (G) by SYTO® 9 and fluorescent red (R) by propidium iodide, respectively. Bacterial suspensions that have been incubated in the two stains simultaneously and then excited at 470 nm exhibit a fluorescence spectral shift from red to green as the percentage of live bacteria in the sample is increased.

Table 15.3 Some organisms that have been successfully stained with our LIVE/DEAD® *Bac*Light™ Bacterial Viability Kits.

Bacteria	
Gram-Positive	**Gram-Negative**
• *Bacillus cereus*	• *Agrobacterium tumefaciens*
• *Bacillus subtilis*	• *Edwardsiella ictaluri*
• *Clostridium paerfringens*	• *Escherichia coli*
• *Lactobacillus* sp.	• *Deleya aquamarina*
• *Micrococcus luteus*	• *Helicobacter pylori*
• *Mycobacterium phlei*	• *Klebsiella pneumoniae*
• *Propionibacterium* sp.	• *Legionella pneumophila*
• *Staphylococcus aureus*	• *Pseudomonas aeruginosa*
• *Streptococcus pyogenes*	• *Pseudomonas syringae*
	• *Salmonella oranienburg*
	• *Serratia marcescens*
	• *Shigella sonnei*
	• *Zymomonas* sp.
Mycoplasma	
• *Eurioplasma eurilytica*	
• *Mycoplasma hominus*	
Protozoa	
• *Giardia muris* cysts	

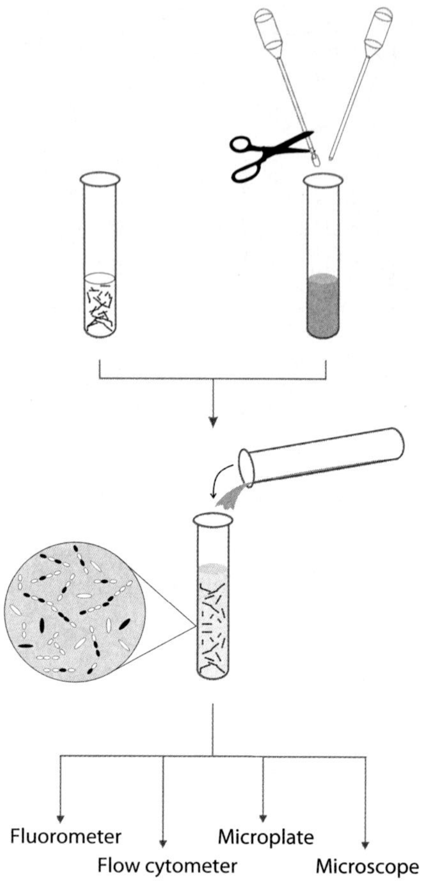

Figure 15.3.25 The convenient and versatile procedure for using the specially packaged LIVE/DEAD® *Bac*Light™ Bacterial Viability Kit (L13152). Simply dissolve the premeasured dyes in buffer, mix with the bacterial sample and observe the fluorescence.

For added convenience, our LIVE/DEAD® *Bac*Light™ Bacterial Viability Kit (L13152) provides the separate stains dry and premeasured into pairs of polyethylene transfer pipettes (Figure 15.3.25). Kit L13152 has several advantages:

- The stains are dry, without DMSO or other potentially harmful solvents, allowing viability determination of solvent-sensitive microorganisms—just dissolve the dyes in virtually any aqueous medium and then add them to the cells.
- The stains are premeasured and supplied in sealed polyethylene transfer pipettes, eliminating the need for pipetting microliter volumes—perfect for educational settings, where there is a need to simplify handling and minimize equipment expenditures.
- This stain formulation does not require refrigeration and is chemically stable, even in poor conditions—storage at 37°C for more than six months produces no detectable changes, making the assay well suited to field testing and other situations where storage or use conditions are less than optimal.

Each of our LIVE/DEAD® *Bac*Light™ Bacterial Viability Kits includes a procedure for mounting bacteria stained with the reagents in the LIVE/DEAD® *Bac*Light™ Bacterial Viability Kit on filter membranes and a proprietary mounting oil that we have found to be useful for the direct epifluorescence filter technique[77] (DEFT). The kit components, number of assays and assay principles are summarized in Table 15.2. The SYTO® 9 nucleic acid stain is also available separately (S34854, Section 15.2).

LIVE/DEAD® *Bac*Light™ Bacterial Viability and Counting Kit

Accurate detection and enumeration of the live and dead bacteria in a sample is an important aspect of many experimental procedures in biotechnology. Because of the marked differences in morphology, cytology and physiology among the many bacterial genera, a universally applicable direct-count viability assay has been very difficult to attain. Conventional direct-count assays of bacterial viability are based on metabolic characteristics or membrane integrity. However, methods relying on metabolic characteristics often only work for a limited subset of bacterial groups,[78] and methods for assessing bacterial membrane integrity commonly have high levels of background fluorescence.[79] Both types of determinations suffer from being very sensitive to growth and staining conditions.[80,81] The LIVE/DEAD® *Bac*Light™ Bacterial Counting and Viability Kit (L34856) allows researchers to reliably distinguish and quantitate live and dead bacteria with the aid of a flow cytometer, even in a mixed population containing a range of bacterial types.

This kit utilizes a mixture of two nucleic acid stains—the green-fluorescent SYTO® 9 dye and red-fluorescent propidium iodide—for viability determinations, as well as a calibrated suspension of beads for accurate sample volume measurements. The SYTO® 9 and propidium iodide stains differ both in their spectral characteristics and in their ability to penetrate healthy bacterial cells. When used alone, the SYTO® 9 stain generally labels all bacteria in a population—those with intact membranes and those with damaged membranes. In contrast, propidium iodide penetrates only bacteria with damaged membranes, causing a reduction in the SYTO® 9 stain fluorescence when both dyes are present. With an appropriate mixture of the SYTO® 9 and propidium iodide stains, bacteria with intact cell membranes fluoresce bright green, whereas bacteria with damaged membranes exhibit significantly less green fluorescence and they often also fluoresce red. The cell type and the gram character influence the amount of red-fluorescent staining exhibited by dead bacteria. Both the SYTO® 9 and propidium iodide stains are efficiently excited by the 488 nm spectral line of the argon-ion laser found in most flow cytometers, and their nucleic acid complexes can be detected in the green and red channels, respectively; the background remains virtually nonfluorescent.

The calibrated suspension of microspheres serves as a reference standard for sample volume. The size and fluorescence of the beads in this microsphere standard have been carefully chosen to ensure that they will be clearly distinguishable from any stained bacteria population in a fluorescence versus side scatter cytogram. A bacterial culture is simply stained with the optimal mixture of SYTO® 9 dye and propidium iodide, and then a fixed number of microspheres are added before analyzing the sample on a flow cytometer. Live and dead bacteria and the microspheres are all easily distinguished in a plot of fluorescence versus side scatter (Figure 15.3.26). The concentration of both the live bacteria and the dead bacteria can then be determined from the ratio of bacteria events to microsphere events in the cytogram (Figure 15.3.27).

The kit components, number of assays and assay principles are summarized in Table 15.2. The SYTO® 9 nucleic acid stain is also available separately (S34854, Section 15.2).

*Bac*Light™ RedoxSensor™ Green Vitality Kit

Bacterial oxidation–reduction activity is an informative parameter for measuring cell vitality. Bacterial oxidases and reductases engage in important functions involving the electron transport chain, catabolic and anabolic pathways and xenobiotic compound metabolism.[82] The RedoxSensor™ Green reagent, available in the *Bac*Light™ RedoxSensor™ Green Vitality Kit (B34954), is an indicator of bacterial reductase activity; this reductase activity is, in turn, a reliable marker for changes in electron transport chain function and for changes in vitality that occur following antibiotic treatment. The RedoxSensor™ Green reagent penetrates both gram-positive and gram-negative bacteria, although differences in signal intensity may be observed based upon cell wall characteristics. Following reduction, the RedoxSensor™ Green reagent will produce a stable green-fluorescent signal (excitation/emission maxima ~490/520 nm) in about 10 minutes that is compatible with formaldehyde fixation techniques. The *Bac*Light™ RedoxSensor™ Green Vitality Kit is useful for measuring the effects of antimicrobial agents and for monitoring cultures in fermenters.

The fluorescence intensity of cells stained with the RedoxSensor™ Green reagent is altered when cells are treated with reagents that disrupt electron transport, such as sodium azide or carbonyl cyanide 3-chlorophenylhydrazone (CCCP), which are both included in this kit. Species differences in responses may be observed. For example, sodium azide disrupts the fluorescence response in *Escherichia coli*, and CCCP disrupts fluorescence response in *Staphylococcus aureus*. The *Bac*Light™ RedoxSensor™ Green Vitality Kit has been tested on logarithmically growing cultures of the following bacterial species: *Micrococcus luteus*, *Staphylococcus aureus*,

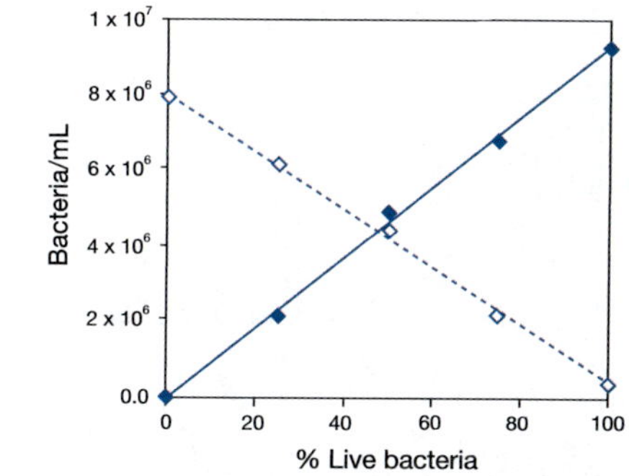

Figure 15.3.27 Best-fit linear regression analysis generated using the LIVE/DEAD® *Bac*Light™ Bacterial Viability and Counting Kit (L34856). Suspensions of live (untreated, (—) (r^2=0.9982)) and dead (alcohol-treated, (- - -),(r^2=0.9974)) *Staphylococcus aureus* were mixed at various live:dead ratios. Mixtures were stained according to the kit protocol and analyzed in triplicate by flow cytometry. Values of bacteria/mL were calculated according to the equation provided in the kit protocol; the mean values are shown above. This experiment may be performed to determine optimal dye concentrations, to practice the cell-staining procedure or to generate a standard curve for unknown samples.

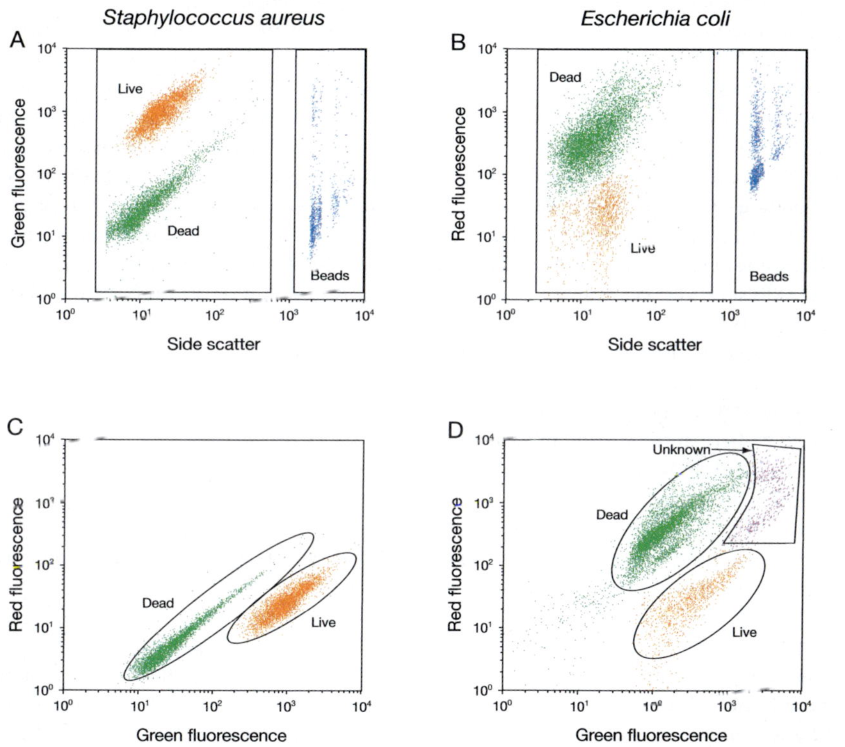

Figure 15.3.26 Analysis of bacterial cultures using the LIVE/DEAD® *Bac*Light™ Bacterial Viability and Counting Kit (L34856). Suspensions of live (untreated) and dead (alcohol-treated) *Staphylococcus aureus* (**A** and **C**) and *Escherichia coli* (**B** and **D**) were stained with the SYTO® 9 nucleic acid stain and propidium iodide and then analyzed by flow cytometry according to the kit protocol. The green or red fluorescence versus side scatter cytogram (**A** or **B**) was used to gate the bacterial population and the bead population (left and right boxes, respectively). Events in the bacteria region of each cytogram are also displayed in red fluorescence versus green fluorescence cytograms (**C** and **D**). Live and dead bacteria/mL can be calculated from either the fluorescence versus side scatter cytogram or the green fluorescence versus red fluorescence cytogram, depending on which one shows the best separation of the live and dead populations. The position of the live and dead populations in these cytograms may be dependent on cell type and gram character. Some samples may exhibit events that fall outside the defined regions and should be evaluated appropriately (e.g., see panel **D**).

IMPORTANT NOTICE: The products described in this manual are covered by one or more Limited Use Label License(s). Please refer to the Appendix on page 971 and Master Product List on page 975. Products are For Research Use Only. Not intended for any animal or human therapeutic or diagnostic use.

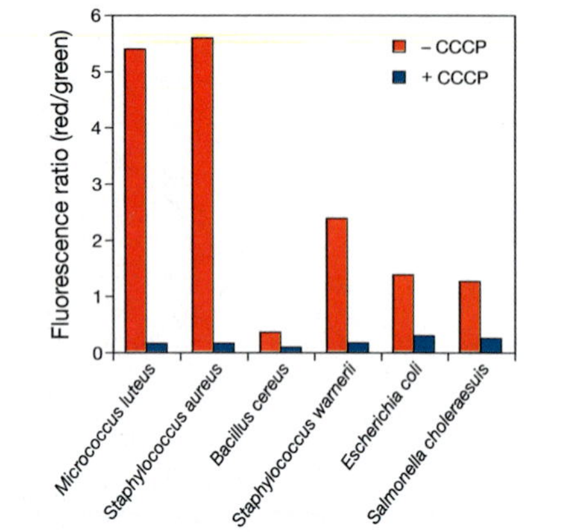

Figure 15.3.28 Detection of membrane potential in various bacteria with the *Bac*Light™ Bacterial Membrane Potential Kit (B34950). Red/green fluorescence ratios were calculated using population mean fluorescence intensities for gram-positive (*Micrococcus luteus, Staphylococcus aureus, Bacillus cereus* and *Staphylococcus warnerii*) and gram-negative (*Escherichia coli* and *Salmonella choleraesuis*) bacteria incubated with 30 µM $DiOC_2(3)$ for 30 minutes in either the presence or absence of 5 µM CCCP, according to the kit protocol.

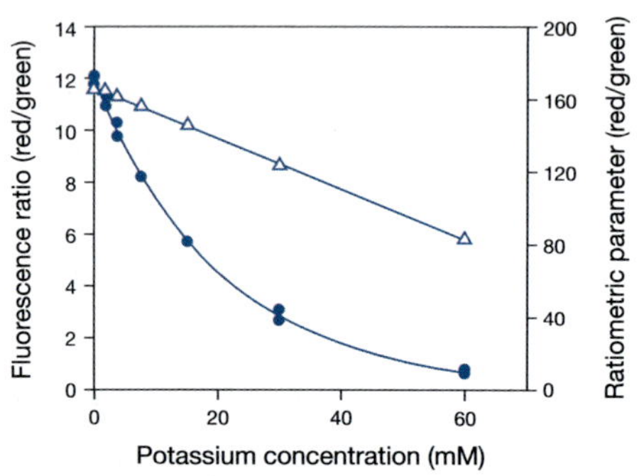

Figure 15.3.29 Response of *Staphylococcus aureus* to valinomycin and external potassium ions, as measured by flow cytometry using the *Bac*Light™ Bacterial Membrane Potential Kit (B34950). Samples containing *S. aureus* were treated with 5 µM valinomycin in different concentrations of potassium buffer, and then stained using 30 µM $DiOC_2(3)$ for 30 minutes, according to the kit protocol. Data are expressed either using a ratiometric parameter based on the formula provided in the kit protocol (Δ, right axis) or as the ratio of population mean red-fluorescence intensity/mean green-fluorescence intensity (●, left axis).

Bacillus cereus, B. subtilis, Klebsiella pneumoniae, Escherichia coli and *Salmonella cholerasuis*. Most gram-positive bacteria stain more efficiently than many gram-negative bacteria; and the response of each bacterial system should be investigated and optimized.

Each *Bac*Light™ RedoxSensor™ Green Vitality Kit contains:

- RedoxSensor™ Green reagent
- Propidium iodide, an indicator of membrane integrity
- Sodium azide
- Carbonyl cyanide 3-chlorophenylhydrazone (CCCP)
- Detailed protocols

At the recommended reagent dilutions and volumes, this kit contains sufficient material to perform ~200 tests by flow cytometry. Although the *Bac*Light™ RedoxSensor™ Green Vitality Kit has been developed for flow cytometric analysis, it may be appropriate for other analysis platforms.

*Bac*Light™ RedoxSensor™ CTC Vitality Kit

The *Bac*Light™ RedoxSensor™ CTC Vitality Kit (B34956) provides effective reagents for evaluating bacterial health and vitality, and these staining reagents can withstand fixation procedures. This kit contains 5-cyano-2,3-ditolyl tetrazolium chloride (CTC), which has been used by researchers to evaluate the respiratory activity of many bacterial populations derived from environmental sources including food,[83,84] soil,[85] stone[86] and marine and fresh water,[87–90] as well as populations undergoing drug efficacy evaluations. Briefly, healthy cells respiring via the electron transport chain will absorb and reduce CTC into an insoluble, red-fluorescent formazan product. Cells not respiring or respiring at slower rates will reduce less CTC, and consequently produce less fluorescent product, giving a semiquantitative estimate of healthy versus unhealthy bacteria. SYTO® 24 green-fluorescent nucleic acid stain and DAPI are provided as counterstains to assist the researcher in differentiating cells from debris and in calculating total cell numbers. Bacteria labeled with CTC and either counterstain may be evaluated immediately or after storage, with or without fixation, using a flow cytometer equipped with appropriate excitation sources (CTC in combination with SYTO® 24 green-fluorescent nucleic acid stain: single 488 nm argon-ion laser; CTC in combination with DAPI: dual UV and 488 nm lasers). Bacteria stained using the *Bac*Light™ RedoxSensor™ CTC Vitality Kit may also be viewed using most standard epifluorescence microscopes equipped with the appropriate filters; for visualizing CTC, SYTO® 24 and DAPI staining, we recommend optical filters optimized for the Texas Red® dye, FITC and DAPI, respectively.

Each *Bac*Light™ RedoxSensor™ CTC Vitality Kit contains:

- 5-cyano-2,3-ditolyl tetrazolium chloride (CTC)
- SYTO® 24 green-fluorescent nucleic acid stain
- 4′,6-diamidino-2-phenylindole, dihydrochloride (DAPI)
- Detailed protocols

At the recommended reagent dilutions and volumes, this kit contains sufficient material to perform at least 50 tests by flow cytometry or microscopy.

*Bac*Light™ Bacterial Membrane Potential Kit

The *Bac*Light™ Bacterial Membrane Potential Kit (B34950) provides a fluorescent membrane-potential indicator dye, $DiOC_2(3)$, along with a proton ionophore (CCCP) and premixed buffer. At low concentrations, $DiOC_2(3)$ exhibits green fluorescence in all bacterial cells, but it becomes more concentrated in healthy cells that are maintaining a membrane potential, causing the dye to self-associate and the fluorescence emission to shift to red. The red- and green-fluorescent bacterial populations are easily distinguished using a flow cytometer. CCCP is included in the kit for use as a control because it eradicates the proton gradient, eliminating bacterial membrane potential.[91,92]

Using the *Bac*Light™ Bacterial Membrane Potential Kit, we have detected membrane potentials in all bacteria tested (including logarithmically growing cultures of *Micrococcus luteus*,

IMPORTANT NOTICE: The products described in this manual are covered by one or more Limited Use Label License(s). Please refer to the Appendix on page 971 and Master Product List on page 975. Products are For Research Use Only. Not intended for any animal or human therapeutic or diagnostic use.
www.invitrogen.com/probes

Staphylococcus aureus, Bacillus cereus, Staphylococcus warnerii, Escherichia coli and *Salmonella choleraesuis*), although the magnitude varies with species (Figure 15.3.28). For many gram-positive species, such as *M. luteus* and *S. aureus*, the DiOC$_2$(3) red:green ratio has been shown to vary with the intensity of the proton gradient (Figure 15.3.29). In gram-negative bacteria, such as *E. coli* and *S. choleraesuis*, a DiOC$_2$(3) response is observed in the presence of a membrane potential but the response does not appear to be proportional to proton gradient intensity.

Each *Bac*Light™ Bacterial Membrane Potential Kit contains:

- DiOC$_2$(3) in dimethylsulfoxide (DMSO)
- CCCP in DMSO
- Phosphate-buffered saline (PBS)
- Detailed protocols

Using the recommended reagent dilutions and volumes, this kit provides sufficient DiOC$_2$(3) to perform approximately 100 individual assays by flow cytometry; sufficient CCCP is provided for 30 depolarized control samples. The *Bac*Light™ Bacterial Membrane Potential Kit is designed to assay bacterial concentrations between 10^5 and 10^7 organisms per mL. Note that DiOC$_2$(3) and CCCP are inhibitors of respiration, rendering the cells nonculturable beyond the brief time window required for staining and analysis.

Figure 15.3.30 Live *Micrococcus luteus* and *Salmonella oranienburg* bacteria stained with the LIVE *Bac*Light™ Bacterial Gram Stain Kit (L7005). Gram-positive *M. luteus* cells fluoresce orange, whereas gram-negative *S. oranienburg* cells fluoresce green. This image was photographed in a single exposure through an Omega® Optical triple bandpass filter set.

LIVE *Bac*Light™ Bacterial Gram Stain Kit

The LIVE *Bac*Light™ Bacterial Gram Stain Kit (L7005) is based on differential nucleic acid staining of *live* gram-negative and gram-positive bacteria. The gram stain is one of the most important and widely used differential stains for the taxonomic classification of bacteria in both clinical and research settings. The original method involves several steps, including heat fixation of the bacteria, a two-step staining protocol, alcohol extraction and counterstaining. Over the years, several improved gram-staining techniques have been developed, but most still involve cell-fixation or cell-permeabilization steps that kill the bacteria being tested. Our single-step LIVE *Bac*Light™ Bacterial Gram Stain Kits can overcome many of the problems inherent in these labor-intensive, fixation-dependent procedures.

Unlike conventional gram stain procedures, the LIVE *Bac*Light™ Bacterial Gram Stain Kit allows researchers to rapidly classify bacteria as either gram-negative or gram-positive in minutes using a single staining solution, no fixatives and no wash steps. The LIVE *Bac*Light™ Bacterial Gram Stain Kit contains our green-fluorescent SYTO® 9 and red-fluorescent hexidium iodide nucleic acid stains. These two dyes differ in both their spectral characteristics and in their ability to label live gram-negative and gram-positive bacteria. When a mixed population of live gram-negative and gram-positive bacteria is simultaneously stained with the membrane-permeant SYTO® 9 dye in combination with hexidium iodide, gram-negative bacteria fluoresce green, and the gram-positive bacteria fluoresce red-orange (Figure 15.3.30, Figure 15.3.31). Dead bacteria do not exhibit predictable staining patterns. Gram-negative and gram-positive organisms can be easily differentiated in any fluorescence microscope equipped with a standard fluorescein longpass optical filter set or by flow cytometry.[93] The assay provides a sensitive indicator system for analyzing low numbers of bacteria in the presence of background material because the unbound reagents exhibit low fluorescence when not bound to nucleic acids. The LIVE *Bac*Light™ Bacterial Gram Stain Kit should be a useful tool for measuring dynamic changes in the composition of bacterial populations.

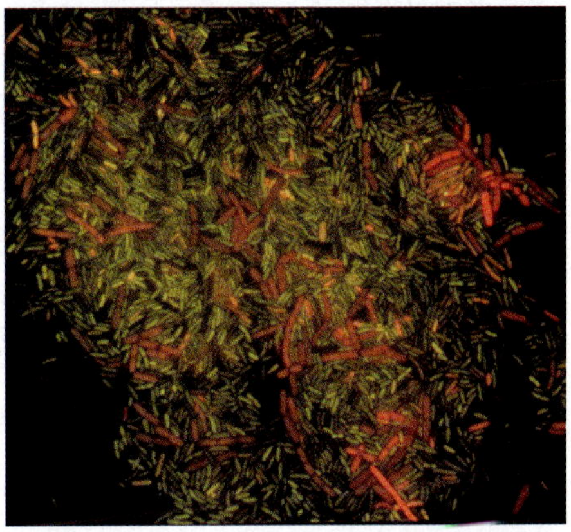

Figure 15.3.31 A mixed population of *Bacillus cereus* and *Pseudomonas aeruginosa* stained with the dye mixture provided in our LIVE *Bac*Light™ Gram Stain Kit (L7005). When live bacteria are incubated with this kit's cell-permeant nucleic acid stains, gram-positive organisms fluoresce orange and gram-negative organisms fluoresce green. The bacteria were photographed through an Omega® Optical triple bandpass filter set.

Our LIVE *Bac*Light™ Bacterial Gram Stain Kit provides separate solutions of the SYTO® 9 and hexidium iodide nucleic acid stains, thus facilitating calibration of bacterial fluorescence at each of the two emission wavelengths in quantitative assays. The kit is well suited for use in fluorescence microscopy, as well as for use in quantitative analysis with a fluorometer (Figure 15.3.32), fluorescence microplate reader, flow cytometer[94] or other instrumentation. The kit includes a procedure for mounting bacteria stained with the LIVE *Bac*Light™ Bacterial Gram Stain Kit on filter membranes, as well as the proprietary *Bac*Light™ mounting oil that we have found to be useful for the direct epifluorescence filter technique[77] (DEFT). The kit components, number of assays and assay principles are summarized in Table 15.2. Hexidium iodide (H7593), the gram-positive bacteria–selective nucleic acid stain, and the SYTO® 9 nucleic acid stain (S34854) are both available separately (Section 15.2). The validity of using hexidium iodide in combination with the SYTO® 13 green-fluorescent nucleic acid stain to correctly predict the

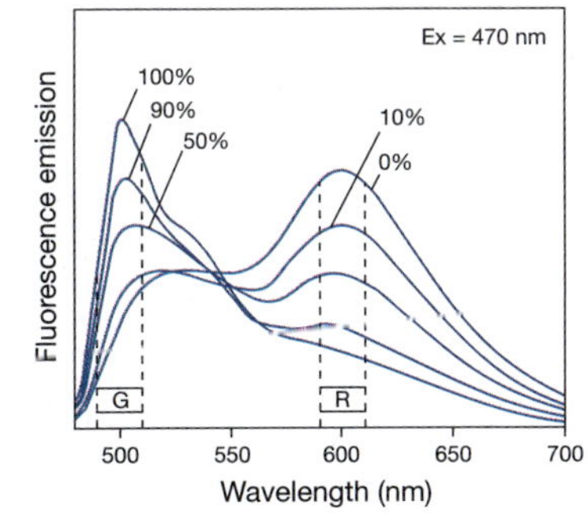

Figure 15.3.32 Analysis of the percentage of gram-negative *Escherichia coli* in mixed suspensions containing gram-positive *Staphylococcus aureus* using the reagents in the LIVE *Bac*Light™ Bacterial Gram Stain Kit (L7005). Live gram-negative and gram-positive bacteria are stained fluorescent green (G) by SYTO® 9 and fluorescent red (R) by hexidium iodide, respectively. Bacterial suspensions that have been incubated in the two stains simultaneously and then excited at 470 nm exhibit a fluorescence spectral shift from red to green as the percentage of gram-negative bacteria in the sample is increased.

IMPORTANT NOTICE: The products described in this manual are covered by one or more Limited Use Label License(s). Please refer to the Appendix on page 971 and Master Product List on page 975. Products are For Research Use Only. Not intended for any animal or human therapeutic or diagnostic use.

gram sign of 45 clinically relevant organisms, including several known to be gram variable, has been demonstrated.[93]

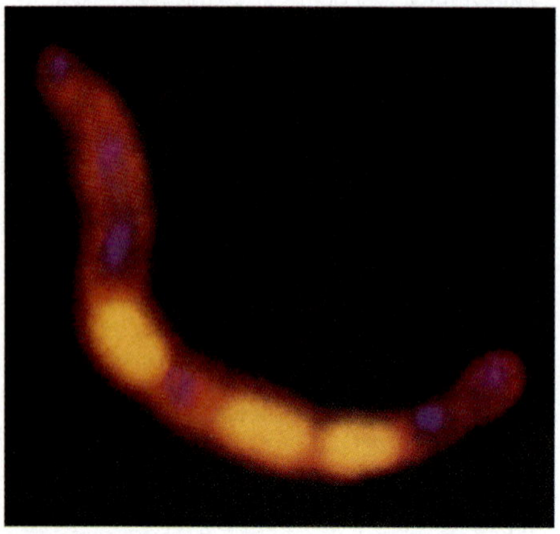

Figure 15.3.33 *Bifidobacterium* sp. bacteria stained with the ViaGram™ Red+ Bacterial Gram Stain and Viability Kit (V7023). While all cells exhibit a red-fluorescent surface stain (gram-positive), live cells exhibit blue-fluorescent internal staining and dead cells with compromised membranes exhibit yellow-green–fluorescent internal staining. This image was obtained by taking multiple exposures through bandpass optical filter sets appropriate for DAPI, fluorescein and Texas Red® dye.

ViaGram™ Red+ Bacterial Gram Stain and Viability Kit

The ViaGram™ Red+ Bacterial Gram Stain and Viability Kit (V7023) provides an easy, three-color fluorescent staining protocol that differentially stains many gram-negative and gram-positive bacterial species and, at the same time, discriminates live from dead cells based on plasma membrane integrity. This kit contains three reagents: two nucleic acid stains for viability determination—the blue-fluorescent cell-permeant DAPI and the green-fluorescent cell-impermeant SYTOX® Green nucleic acid stain—as well as the red-fluorescent Texas Red®-X wheat germ agglutinin (WGA) for gram sign determination. Bacteria with intact cell membranes stain fluorescent blue with DAPI, whereas bacteria with damaged membranes stain fluorescent green with the SYTOX® Green nucleic acid stain. The background remains virtually nonfluorescent. The Texas Red®-X WGA component selectively binds to the surface of gram-positive bacteria, providing a red-fluorescent cell-surface stain that effectively distinguishes them from gram-negative bacteria, even in the presence of the viability stains. Thus, with three fluorescent colors, the four possible combinations of live or dead, gram-negative and gram-positive cells are discriminated with a fluorescence microscope (Figure 15.3.33, Figure 15.3.34). This kit also includes a procedure for mounting bacteria on filter membranes and the *Bac*Light™ mounting oil, which we have found to be useful for the direct epifluorescence filter technique[77] (DEFT). The kit components, number of assays and assay principles are summarized in Table 15.2.

FilmTracer™ LIVE/DEAD® Biofilm Viability Kit

The FilmTracer™ LIVE/DEAD® Biofilm Viability Kit (L10316) provides a two-color fluorescence assay of bacterial viability based on membrane integrity, and has proven useful for a diverse array of bacterial genera, including those growing in biofilm communities. The kit utilizes a mixture of SYTO® 9 green-fluorescent nucleic acid stain and the red-fluorescent nucleic acid stain, propidium iodide, which differ both in their spectral characteristics and in their ability to penetrate healthy bacteria. SYTO® 9 stain generally labels all bacteria in a population—those with intact membranes and those with damaged membranes. In contrast, propidium iodide penetrates only bacteria with damaged membranes; SYTO® 9 stain fluorescence is thereby reduced in cells where both dyes are present. When treated with an appropriate mixture of these stains, bacteria with intact cell membranes display green fluorescence while those with damaged membranes display red fluorescence.

We offer several individual fluorescent FilmTracer™ biofilm stains, which are described in Section 15.2):

- FilmTracer™ calcein red-orange biofilm stain (F10319)
- FilmTracer™ calcein violet biofilm stain (F10320)

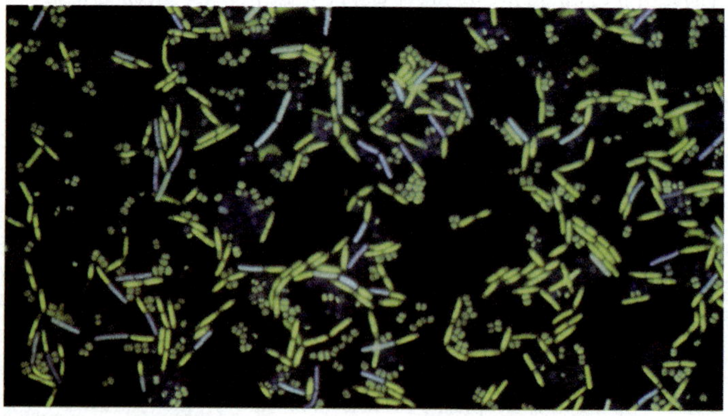

Figure 15.3.34 A mixed population of live and isopropyl alcohol–killed *Micrococcus luteus* and *Bacillus cereus* simultaneously stained with DAPI (D1306, D3571, D21490) and SYTOX® Green nucleic acid stain (S7020), the viability determination components of the ViaGram™ Red+ Bacterial Gram Stain and Viability Kit (V7023). Bacteria with intact cell membranes are stained exclusively with the cell-permeant DAPI nuclear stain and exhibit blue fluorescence, whereas cells with damaged membranes are stained with both fluorophores and exhibit green fluorescence. This image was acquired with a UV longpass optical filter set.

IMPORTANT NOTICE: The products described in this manual are covered by one or more Limited Use Label License(s). Please refer to the Appendix on page 971 and Master Product List on page 975. Products are For Research Use Only. Not intended for any animal or human therapeutic or diagnostic use.

- FilmTracer™ calcein green biofilm stain (F10322)
- FilmTracer™ FM® 1-43 green biofilm cell stain (F10317)
- FilmTracer™ SYPRO® Ruby biofilm matrix stain (F10318)

Wheat Germ Agglutinin Sampler Kit

Fluorescent lectins have proven useful in microbiology applications. Fluorescent wheat germ agglutinin (WGA) conjugates selectively stain chitin in fungal cell walls,[95] as well as the surface of gram-positive but not of gram-negative bacteria. Fluorescent WGA has also been shown to bind to sheathed microfilariae and has been used to detect filarial infection in blood smears.[96] Our Wheat Germ Agglutinin Sampler Kit (W7024) provides 1 mg samples of four of our brightest fluorescent WGA conjugates, spanning the spectrum from blue to red. Included in this kit are conjugates of the blue-fluorescent Alexa Fluor® 350, green-fluorescent Oregon Green® 488, orange-fluorescent tetramethylrhodamine and red-fluorescent Texas Red®-X dyes (Table 7.10). See Section 7.7 for more information on lectins, including additional WGA and concanavalin A (Con A) conjugates.

Reverse Transcriptase Assay for Retrovirus Titering

The EnzChek® Reverse Transcriptase Assay Kit (E22064) is designed to provide a convenient and efficient method for measuring reverse transcriptase activity. This kit has been used to determine HIV-1 reverse transciptase heterodimer activity,[97] to measure SARS-CoV viral titers in 293T host cultures[98] and to quantitate reverse transciptase activity in Friend virus (FV) culture supernatants.[99]

The key to this assay is our PicoGreen® dsDNA quantitation reagent, which preferentially detects dsDNA or RNA–DNA heteroduplexes over single-stranded nucleic acids or free nucleotides. In the assay, the sample to be measured is added to a mixture of a long poly(A) template, an oligo(dT) primer and dTTP. Reverse transcriptase activity in the sample results in the formation of long RNA–DNA heteroduplexes, which are detected by the PicoGreen® reagent at the end of the assay. In less than an hour, samples can be read in a fluorometer or microplate reader with filter sets appropriate for fluorescein (FITC). The assay is sensitive, detecting as little as 0.02 units of HIV reverse transcriptase and has about a 50-fold linear range.

The EnzChek® Reverse Transcriptase Assay Kit contains sufficient reagents for approximately 1000 assays using a fluorescence microplate reader:

- PicoGreen® dsDNA quantitation reagent
- Bacteriophage lambda DNA standard
- Poly(A) ribonucleotide template
- Oligo dT16 primer
- TE buffer, polymerization buffer and an EDTA solution
- Detailed protocols

ATP Determination

The luciferin–luciferase bioluminescence assay is extremely sensitive; most luminometers can detect as little as 1 picomole of pre-existing ATP or ATP as it is generated in kinetic systems. This sensitivity has led to its widespread use for detecting ATP in various enzymatic reactions, as well as for measuring viable cell number,[100] for monitoring ATP release from cells[101–104] and for detecting low-level bacterial contamination in samples such as blood, milk, urine, soil and sludge.[105–109] The luciferin–luciferase bioluminescence assay has also been used to determine cell proliferation and cytotoxicity in both bacterial[110,111] and mammalian cells,[112,113] and to distinguish cytostatic versus cytocidal potential of anticancer drugs on malignant cell growth.[114]

We offer a convenient ATP Determination Kit (A22066) for the sensitive bioluminescence-based detection of ATP with recombinant firefly luciferase and its substrate luciferin. This assay is based on luciferase's absolute requirement for ATP to produce light. In the presence of Mg^{2+}, luciferase catalyzes the reaction of luciferin, ATP and O_2 to form oxyluciferin, AMP, CO_2, pyrophosphate and ~560 nm light (Figure 15.3.35).

The ATP Determination Kit (A22066) contains:

- Luciferin (5 × 3 mg)
- Luciferase
- ATP
- Dithiothreitol (DTT)
- Concentrated reaction buffer
- Detailed protocols for ATP quantitation

Unlike most other commercially available ATP detection kits, our ATP Determination Kit provides the luciferase and luciferin packaged separately, which enables researchers to optimize the reaction conditions for their particular instruments and samples. The ATP Determination Kit provides sufficient reagents to perform 200 ATP assays using 0.5 mL sample volumes or 500 ATP assays using 0.2 mL sample volumes.

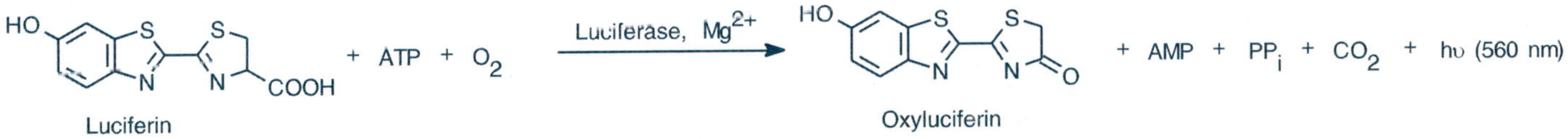

Figure 15.3.35 Reaction scheme for bioluminescence generation via luciferase-catalyzed conversion of luciferin to oxyluciferin.

IMPORTANT NOTICE: The products described in this manual are covered by one or more Limited Use Label License(s). Please refer to the Appendix on page 971 and Master Product List on page 975. Products are For Research Use Only. Not intended for any animal or human therapeutic or diagnostic use.

REFERENCES

1. J Immunol Methods (2000) 243:155; **2.** Prog Neurobiol (2009) 88:221; **3.** J Neurosci (1995) 15:5389; **4.** Biol Bull (1995) 189:218; **5.** Ophthalmologe (1995) 92:452; **6.** J Cell Sci (1993) 106:685; **7.** Cytometry (1995) 20:181; **8.** J Biol Chem (1995) 270:23895; **9.** J Biol Chem (1995) 270:7791; **10.** J Neurosci (1994) 14:2464; **11.** J Neurosci (1994) 14:2260; **12.** J Biol Chem (1994) 269:6803; **13.** Photochem Photobiol (1996) 63:111; **14.** J Immunol Methods (1994) 177:101; **15.** Hum Immunol (1993) 37:264; **16.** Theriogenology (1995) 43:595; **17.** Practical Flow Cytometry, 3rd Ed., Shapiro HM, Ed. (1995) p. 250; **18.** Methods Cell Biol (1994) 42 Pt B:359; **19.** Methods Cell Biol (1994) 42 Pt B:295; **20.** J Immunol Methods (2006) 313:199; **21.** J Immunol Methods (1992) 156:47; **22.** J Immunol Methods (1995) 185:209; **23.** J Immunol Methods (1993) 166:45; **24.** J Immunol Methods (1997) 204:135; **25.** Theriogenology (1996) 45:1103; **26.** Biol Reprod (1995) 53:276; **27.** Poult Sci (1995) 74:1191; **28.** J Androl (1994) 15:620; **29.** Theriogenology (1996) 45:923; **30.** Nature (1995) 377:20; **31.** Biol Reprod (1997) 56:143; **32.** Biol Reprod (1997) 57:1401; **33.** Reprod Dom Anim (1997) 32:279; **34.** Biol Reprod (1998) 58:786; **35.** Biol Reprod (1997) 56:991; **36.** Theriogenology (1993) 39:1009; **37.** Gamete Res (1989) 22:355; **38.** Biol Reprod (1986) 34:127; **39.** J Histochem Cytochem (1982) 30:279; **40.** Methods Cell Biol (1990) 33:401; **41.** J Histochem Cytochem (1977) 25:46; **42.** J Androl (1991) 12:112; **43.** Cytometry (1987) 8:642; **44.** Cytometry (1980) 1:132; **45.** Biol Reprod (1994) 50:987; **46.** J Reprod Fertil (1992) 96:581; **47.** J Neurosci Methods (1996) 70:195; **48.** Appl Environ Microbiol (1990) 56:3785; **49.** J Dairy Res (1990) 57:239; **50.** J Immunol Methods (1997) 210:25; **51.** J Immunol Methods (1998) 213:157; **52.** Antimicrob Agents Chemother (1997) 41:1004; **53.** J Biol Chem (2000) 275:40042; **54.** Nucleic Acids Res (1986) 14:2511; **55.** FEMS Microbiol Lett (1995) 133:1; **56.** J Cell Biol (1994) 126:1375; **57.** Biotechnol Intl (1997) 1:291; **58.** J Cell Biol (1993) 123:1821; **59.** J Cell Biol (1991) 114:111; **60.** J Cell Biol (1991) 114:101; **61.** Appl Environ Microbiol (1997) 63:2897; **62.** J Clin Microbiol (2001) 39:2458; **63.** J Clin Microbiol (1997) 35:5; **64.** Cytometry (1998) 31:307; **65.** Mycoses (1996) 39:45; **66.** Yeast (1998) 14:147; **67.** J Microbiol Methods (1993) 17:1; **68.** Cytometry (1997) 29:58; **69.** Int J Parasitol (1996) 26:637; **70.** Microbiol Rev (1987) 51:365; **71.** J Microbiol Methods (1999) 37:77; **72.** Appl Environ Microbiol (1996) 62:2264; **73.** J Appl Microbiol (2004) 96:684; **74.** Int J Parasitol (1997) 27:787; **75.** Appl Environ Microbiol (1999) 65:4475; **76.** Biochim Biophys Acta (2001) 1514:165; **77.** Appl Environ Microbiol (1980) 39:423; **78.** J Appl Bacteriol (1992) 72:410; **79.** Lett Appl Microbiol (1991) 13:58; **80.** J Microbiol Methods (1991) 13:87; **81.** Current Microbiol (1980) 4:321; **82.** J Microbiol Methods (1997) 29:161; **83.** Cytometry (2003) 54A:27; **84.** Appl Environ Microbiol (2003) 69:2857; **85.** Biodegradation (2002) 13:285; **86.** J Microbiol Methods (2003) 52:75; **87.** Int J Food Microbiol (2004) 92:327; **88.** Antimicrob Agents Chemother (2008) 52:1446; **89.** Appl Environ Microbiol (2003) 69:7462; **90.** J Microbiol Methods (2003) 52:47; **91.** Antimicrob Agents Chemother (2000) 44:827; **92.** Cytometry (1999) 35:55; **93.** Appl Environ Microbiol (1998) 64:2681; **94.** Appl Environ Microbiol (1998) 64:515; **95.** Invest Ophthalmol Vis Sci (1986) 27:500; **96.** Int J Parasitol (1990) 20:1099; **97.** Biochemistry (2009) 48:7646; **98.** Biochem Biophys Res Commun (2008) 371:356; **99.** Science (2008) 321:1343; **100.** J Biolumin Chemilumin (1995) 10:29; **101.** Am J Physiol (1998) 275:C1391; **102.** J Biolumin Chemilumin (1996) 11:149; **103.** Neuroscience (1995) 66:915; **104.** FEBS Lett (1985) 185:323; **105.** Anal Biochem (1988) 175:14; **106.** Bio/Technology (1988) 6:634; **107.** J Clin Microbiol (1984) 20:644; **108.** J Clin Microbiol (1983) 18:521; **109.** Methods Enzymol (1978) 57:3; **110.** Biotechnol Bioeng (1993) 42:30; **111.** J Biolumin Chemilumin (1991) 6:193; **112.** Biochem J (1993) 295:165; **113.** J Immunol Methods (1993) 160:81; **114.** J Natl Cancer Inst (1986) 77:1039.

PRODUCT LIST 15.3 VIABILITY AND CYTOTOXICITY ASSAY KITS FOR DIVERSE CELL TYPES

Cat. No.	Product	Quantity
A10346	ArC™ Amine Reactive Compensation Bead Kit *for use with amine reactive dyes* *for flow cytometry compensation* *100 tests*	1 kit
A22066	ATP Determination Kit *special packaging* *200–1000 assays*	1 kit
B34950	*Bac*Light™ Bacterial Membrane Potential Kit *for flow cytometry* *100 assays*	1 kit
B34956	*Bac*Light™ RedoxSensor™ CTC Vitality Kit *for flow cytometry and microscopy*	1 kit
B34954	*Bac*Light™ RedoxSensor™ Green Vitality Kit *for flow cytometry*	1 kit
E22064	EnzChek® Reverse Transcriptase Assay Kit *1000 assays*	1 kit
L10316	FilmTracer™ LIVE/DEAD® Biofilm Viability Kit	1 kit
F34953	*Funga*Light™ CFDA, AM/propidium iodide Yeast Vitality Kit *for flow cytometry*	1 kit
L7005	LIVE *Bac*Light™ Bacterial Gram Stain Kit *for microscopy and quantitative assays* *1000 assays*	1 kit
L13152	LIVE/DEAD® *Bac*Light™ Bacterial Viability Kit *10 applicator sets* *500 assays*	1 kit
L7007	LIVE/DEAD® *Bac*Light™ Bacterial Viability Kit *for microscopy* *1000 assays*	1 kit
L7012	LIVE/DEAD® *Bac*Light™ Bacterial Viability Kit *for microscopy and quantitative assays* *1000 assays*	1 kit
L34856	LIVE/DEAD® *Bac*Light™ Bacterial Viability and Counting Kit *for flow cytometry* *100 assays*	1 kit
L34951	LIVE/DEAD® Cell Vitality Assay Kit *C_{12}-resazurin/SYTOX® Green* *1000 assays*	1 kit
L7010	LIVE/DEAD® Cell-Mediated Cytotoxicity Kit *for animal cells* *2000 assays*	1 kit
L34957	LIVE/DEAD® Fixable Aqua Dead Cell Stain Kit *for 405 nm excitation* *200 assays*	1 kit
L23105	LIVE/DEAD® Fixable Blue Dead Cell Stain Kit *for UV excitation* *200 assays*	1 kit
L34960	LIVE/DEAD® Fixable Dead Cell Stain Sampler Kit *for flow cytometry* *320 assays*	1 kit
L10120	LIVE/DEAD® Fixable Far Red Dead Cell Stain Kit *for 633 or 635 nm excitation* *200 assays*	1 kit
L23101	LIVE/DEAD® Fixable Green Dead Cell Stain Kit *for 488 nm excitation* *200 assays*	1 kit
L10119	LIVE/DEAD® Fixable Near-IR Dead Cell Stain Kit *for 633 or 635 nm excitation* *200 assays*	1 kit
L23102	LIVE/DEAD® Fixable Red Dead Cell Stain Kit *for 488 nm excitation* *200 assays*	1 kit
L34955	LIVE/DEAD® Fixable Violet Dead Cell Stain Kit *for 405 nm excitation* *200 assays*	1 kit
L34959	LIVE/DEAD® Fixable Yellow Dead Cell Stain Kit *for 405 nm excitation* *200 assays*	1 kit
L34952	LIVE/DEAD® *Funga*Light™ Yeast Viability Kit *for flow cytometry*	1 kit
L7013	LIVE/DEAD® Reduced Biohazard Cell Viability Kit #1 *green and red fluorescence* *100 assays*	1 kit
L7011	LIVE/DEAD® Sperm Viability Kit *200–1000 assays*	1 kit
L3224	LIVE/DEAD® Viability/Cytotoxicity Kit *for mammalian cells*	1 kit
L34958	LIVE/DEAD® Violet Viability/Vitality Kit *for 405 nm excitation* *200 assays*	1 kit
L7009	LIVE/DEAD® Yeast Viability Kit *1000 assays*	1 kit
V7023	ViaGram™ Red+ Bacterial Gram Stain and Viability Kit *200 assays*	1 kit
V23110	Vybrant® Cell Metabolic Assay Kit *with C_{12}-resazurin* *500–1000 assays*	1 kit
V23111	Vybrant® Cytotoxicity Assay Kit *G6PD release assay* *1000 assays*	1 kit
W7024	Wheat Germ Agglutinin Sampler Kit *four fluorescent conjugates, 1 mg each*	1 kit

IMPORTANT NOTICE: The products described in this manual are covered by one or more Limited Use Label License(s). Please refer to the Appendix on page 971 and Master Product List on page 975. Products are For Research Use Only. Not intended for any animal or human therapeutic or diagnostic use.

www.invitrogen.com/probes

molecular probes® | invitrogen™ by life technologies™

15.4 Assays for Cell Enumeration, Cell Proliferation and Cell Cycle

Cell proliferation and the characterization of agents that either promote or retard cell proliferation are extremely important areas of cell biology and drug-discovery research. We offer both traditional reagents for assessing cell proliferation and cell cycle—in particular the Hoechst nucleic acid stains and probes for 5-bromo-2'-deoxyuridine (BrdU) incorporation during cell division—as well as some exceptional tools developed in our laboratories, including the Click-iT® EdU cell proliferation assay. For simply detecting the presence or counting the number of cells, fluorescent stains that identify cells by their characteristic morphology or light-scattering properties may be sufficient.

Cell Enumeration and Cell Proliferation Assays for Animal Cells

Reagents for counting cells and quantitating cell proliferation are valuable research tools. Most cell proliferation assays estimate the number of cells either by incorporating a modified nucleotide into cells during proliferation or by measuring total nucleic acid or protein content of lysed cells.[1,2] Several of our nucleic acid stains (Section 8.1) and nucleotides (Section 8.2) have proven useful in nucleic acid labeling protocols. Here we describe our Click-iT® EdU cell proliferation assay, which provides a superior alternative to bromodeoxyuridine (BrdU) or ^{3}H-thymidine incorporation methods for measuring new DNA synthesis. Alternatively, our CyQUANT® Cell Proliferation Assay Kits use the CyQUANT® GR, CyQUANT® NF or CyQUANT® Direct nucleic acid stains to measure increases in nucleic acid content that accompany cell proliferation

Click-iT® Tools for Cell Proliferation Analysis

Click-iT® labeling technology employs a bioorthogonal reactive chemistry for the *in situ* labeling of specific molecular populations, such as newly synthesized nucleic acids, in an experimental time window of interest. The Click-iT® labeling reaction is based on a copper-catalyzed azide–alkyne cyloaddition[3,4] and derives its high degree of specificity from the fact that the azide and alkyne reaction partners have no endogenous representation in biological molecules, cells, tissues or model organisms.[5–7] Click-iT® labeling technology and the details of the click reaction are discussed in Section 3.1. For a complete list of azide and alkyne derivatives compatible with Click-iT® labeling technology, see Table 3.1. Here we highlight the Click-iT® EdU cell proliferation assay.

The Click-iT® EdU cell proliferation assay provides a superior alternative to bromodeoxyuridine (BrdU) or ^{3}H-thymidine incorporation methods for measuring new DNA synthesis.[8,9] The alkynyl nucleoside analog EdU (5-ethynyl-2'-deoxyuridine; A10044, E10187, E10415) is incorporated into DNA during the synthesis phase (S phase) of the cell cycle and is subsequently detected by copper (I)–catalyzed click coupling to an azide-derivatized fluorophore[10] (Figure 15.4.1). The small size of the click-coupled fluorophore compared to that of antibodies required for immunodetection of BrdU enables efficient penetration of complex samples without the need for harsh cell treatment, simplifying the assay considerably. The Click-iT® EdU assay protocol is compatible with both adherent cells and cell suspensions. From start to finish, the EdU detection assay is complete in as little as 90 minutes, as compared with the antibody-based BrdU method, which takes 6–24 hours to complete. In addition, the Click-iT® EdU cell proliferation assay can be multiplexed with surface and intracellular marker detection using Alexa Fluor® dye–labeled secondary antibodies[11–14] (Section 7.2) (Figure 15.4.2).

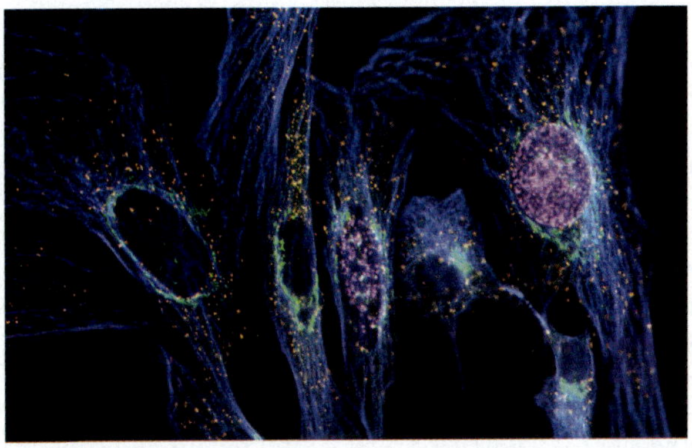

Figure 15.4.2 Multicolor imaging with the Click-iT® EdU Imaging Kits. Muntjac cells were treated with 10 µM EdU for 45 minutes. Cells were then fixed and permeabilized, and EdU that had been incorporated into newly synthesized DNA was detected by the far-red–fluorescent Click-iT® EdU Alexa Fluor® 647 HCS Assay (C10356, C10357). Tubulin was labeled with an anti-tubulin antibody and visualized with an Alexa Fluor® 350 goat anti-mouse IgG antibody (A21049). The Golgi complex was stained with the green-fluorescent Alexa Fluor® 488 conjugate of lectin HPA from *Helix pomatia* (edible snail) (L11271), and peroxisomes were labeled with an anti-peroxisome antibody and visualized with an orange-fluorescent Alexa Fluor® 555 donkey anti-rabbit IgG antibody (A31572).

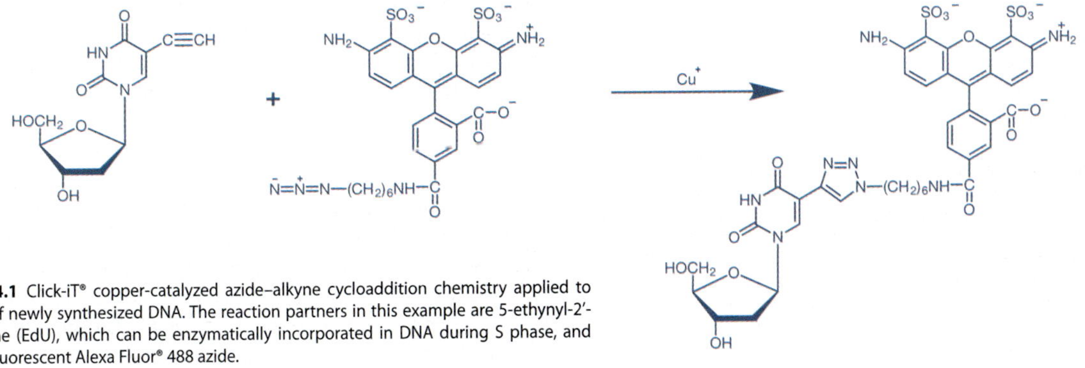

Figure 15.4.1 Click-iT® copper-catalyzed azide–alkyne cycloaddition chemistry applied to detection of newly synthesized DNA. The reaction partners in this example are 5-ethynyl-2'-deoxyuridine (EdU), which can be enzymatically incorporated in DNA during S phase, and the green-fluorescent Alexa Fluor® 488 azide.

IMPORTANT NOTICE: The products described in this manual are covered by one or more Limited Use Label License(s). Please refer to the Appendix on page 971 and Master Product List on page 975. Products are For Research Use Only. Not intended for any animal or human therapeutic or diagnostic use.

Although the majority of applications are in cultured mammalian cells, Click-iT® EdU reagents and methods have also been successfully applied to a wide range of model organisms including:

- *Escherichia coli*[15]
- *Caenorhabditis elegans*[12]
- *Drosophila*[16]
- Zebrafish[17]
- Mouse[13]
- Plants[18,19] (alfalfa, *Arabidopsis*, grape, maize, rice and tobacco)

The Click-iT® EdU Flow Cytometry Assay Kits provide all the reagents needed to perform 50 assays using 0.5 mL reaction buffer per assay, including the nucleoside analog EdU and all components for fixation, permeabilization and labeling whole blood samples, adherent cells or suspension cells.[9,20] Additionally, two cell-cycle stains compatible with the fluorescence excitation and emission characteristics of the fluorescent azide detection reagents are included. We offer three Click-iT® EdU Flow Cytometry Assay Kits:

- Click-iT® EdU Alexa Fluor® 488 Flow Cytometry Assay Kit (C35002)
- Click-iT® EdU Alexa Fluor® 647 Flow Cytometry Assay Kit (A10202)
- Click-iT® EdU Pacific Blue™ Flow Cytometry Assay Kit (A10034)

The Click-iT® EdU Imaging Kits contain all of the components needed to label and detect incorporated EdU on 50 coverslips using 0.5 mL reaction buffer per test, as well as the blue-fluorescent Hoechst 33342 nuclear stain for performing cell-cycle analysis on adherent cell samples. We offer four Click-iT® EdU Imaging Kits:

- Click-iT® EdU Alexa Fluor® 488 Imaging Kit (C10337)
- Click-iT® EdU Alexa Fluor® 555 Imaging Kit (C10338)
- Click-iT® EdU Alexa Fluor® 594 Imaging Kit (C10339)
- Click-iT® EdU Alexa Fluor® 647 Imaging Kit (C10340)

The Click-iT® EdU HCS Assay Kits contain all of the materials needed to label and detect incorporated EdU in adherent cells in 96-well microplates and 100 µL reaction buffer per assay. For cell registration or DNA profiling, these kits also include the blue-fluorescent HCS NuclearMask™ Blue stain (H10325, Section 12.5). We offer four Click-iT® EdU HCS Assay Kits:

- Click-iT® EdU Alexa Fluor® 488 HCS Assay Kit (2-plate size, C10350; 10-plate size, C10351)
- Click-iT® EdU Alexa Fluor® 555 HCS Assay Kit (2-plate size, C10352; 10-plate size, C10353)
- Click-iT® EdU Alexa Fluor® 594 HCS Assay Kit (2-plate size, C10354; 10-plate size, C10355)
- Click-iT® EdU Alexa Fluor® 647 HCS Assay Kit (2-plate size, C10356; 10-plate size, C10357; Figure 15.4.2)

In addition to these kits, our Click-iT® EdU Microplate Assay Kit (C10214) provides a simple and rapid workflow with fewer wash steps resulting in a substantial time-savings advantage over traditional BrdU colorimetric or fluorescent cell proliferation assays. This assay uses Oregon Green® 488 azide for click coupling to synthetically incorporated EdU. The signal is amplified using immunodetection of the Oregon Green® 488 fluorophore by a rabbit anti–Oregon Green® horseradish peroxidase (HRP) conjugate followed by fluorogenic or chromogenic detection with our Amplex® UltraRed HRP substrate. The Click-iT® EdU microplate assay has been successfully tested in HeLa, A549, U2OS and A541 cells with a variety of reagents that modulate DNA synthesis, including the DNA synthesis inhibitor aphidicolin and the mitotic inhibitor paclitaxel. The Click-iT® EdU Microplate Assay Kit contains sufficient reagents for performing 400 individual assays in a 96-well plate format.

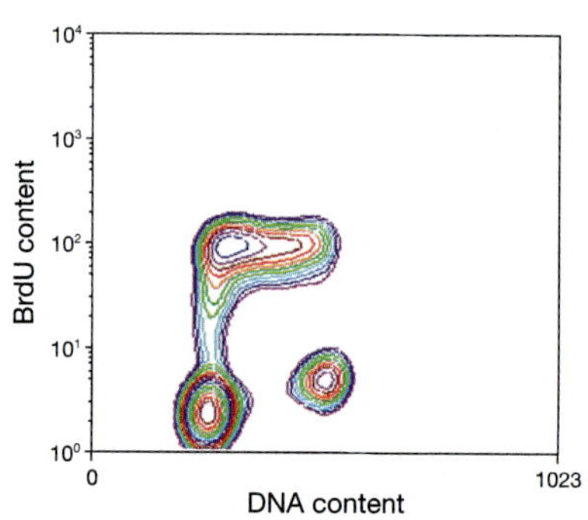

Figure 15.4.3 Detection of proliferation in Wil2S Lymphoma B cells. Cells were treated with 10 µM 5-bromo-2′-deoxyuridine (BrdU, B23151) in culture medium for 1 hour, then pelleted and fixed with cold 70% ethanol. After treatment with RNase and 4 M HCl (to denature the DNA), the cells were labeled with anti-BrdU antibody and detected using green-fluorescent Alexa Fluor® 488 goat anti–mouse IgG antibody (A11001). In addition, the cells were labeled with red-fluorescent propidium iodide (P1304MP, P3566, P21493) to assess the total cellular DNA content. The cells were analyzed by flow cytometry using 488 nm excitation; the fluorescent signals were collected at ~525 nm for the Alexa Fluor® 488 dye and at ~675 nm for propidium iodide. Increased BrdU incorporation is indicative of actively proliferating cells.

Proliferation Assays Using Bromodeoxyuridine Incorporation

Incorporation of 5-bromo-2'-deoxyuridine (BrdU, B23151) into newly synthesized DNA permits indirect detection of rapidly proliferating cells with fluorescently labeled anti-BrdU antibodies or certain nucleic acid stains, thereby facilitating the identification of cells that have progressed through the S phase of the cell cycle during the BrdU labeling period.[2,21,22] We offer fluorescent conjugates of the mouse monoclonal anti-BrdU antibody clone MoBU-1 labeled with our brightest and most photostable dyes:

- Alexa Fluor® 488 anti-BrdU antibody (B35130, B35139)
- Alexa Fluor® 555 anti-BrdU antibody (B35131)
- Alexa Fluor® 594 anti-BrdU antibody (B35132)
- Alexa Fluor® 647 anti-BrdU antibody (B35133, B35140)
- Pacific Blue™ anti-BrdU antibody (B35129)

This anti-BrdU antibody is also available biotinylated (B35138), as well as unlabeled (B35128, B35141). The unlabeled mouse anti-BrdU can be detected with our anti-mouse secondary antibodies (Table 7.1) using either flow cytometry (Figure 15.4.3) or imaging.

Because fluorescence of the Hoechst 33258 (H1398, H3569, H21491) and Hoechst 33342 (H1399, H3570, H21492) dyes bound to DNA is *quenched* at sites where BrdU is incorporated, Hoechst dye fluorescence can also be used to detect BrdU incorporation in single cells.[1,23–26] This technique has been employed to quantitate the noncycling cell fraction, as well as the fraction of cells that are in G_1 and G_2 of two subsequent cycles.[27] The addition of ethidium bromide as a counterstain that is *insensitive* to BrdU incorporation allows the resolution of G_1, S and G_2 compartments of up to three consecutive cell cycles.[28,29]

IMPORTANT NOTICE: The products described in this manual are covered by one or more Limited Use Label License(s). Please refer to the Appendix on page 971 and Master Product List on page 975. Products are For Research Use Only. Not intended for any animal or human therapeutic or diagnostic use.

Unlike the fluorescence of Hoechst dyes, the fluorescence of TO-PRO®-3 (T3605) and LDS 751 (L7595) is considerably *enhanced* by the presence of bromodeoxyuridine in DNA. In conjunction with propidium iodide (P1304MP, P3566, P21493; Section 8.1), these nucleic acid stains have been used to discriminate BrdU-labeled cells from nonproliferating cells by flow cytometry[30,31] and with an imaging system for automated cell proliferation.[32]

Proliferation Assays Using ChromaTide® Nucleotides

In the strand break induction by photolysis (SBIP) technique, proliferating cells that have incorporated BrdU into newly synthesized DNA are subjected to Hoechst 33258 staining, followed by UV photolysis to induce DNA strand breaks[33] (Figure 15.4.4). Once the cells are fixed, strand breaks can be detected *in situ* using mammalian terminal deoxynucleotidyl transferase (TdT), which covalently adds labeled nucleotides to the 3'-hydroxyl ends of these DNA fragments.[34–37] Break sites have traditionally been labeled with biotinylated or haptenylated dUTP conjugates (Section 8.2) in conjunction with antibodies to the hapten (Section 7.4) or conjugates of streptavidin[37,38] (Section 7.6). However, a single-step procedure has been described that uses our ChromaTide® BODIPY® FL-14-dUTP (C7614, Figure 15.4.5) as a TdT substrate for directly detecting DNA strand breaks both in BrdU-labeled cells following SBIP and in apoptotic cells[33,39–42] (Section 15.5; Figure 15.4.6, Figure 15.4.7).

The single-step BODIPY® FL dye–based assay has several advantages over indirect detection of biotinylated or haptenylated nucleotides. With direct detection procedures, no secondary detection reagents are required; fewer protocol steps translate into less chance for error and more immediate results. Moreover, the yield of cells with direct detection procedures is reported to be about three times greater than that of multistep procedures employing biotin- or digoxigenin-conjugated dUTP. Although both BODIPY® FL dye– and fluorescein-labeled nucleotides can be detected with fluorescence microscopy or flow cytometry, the BODIPY® FL dye–labeled nucleotides provide ~40% stronger signal than fluorescein-labeled nucleotides when assaying strand breaks in apoptotic versus nonapoptotic cells. In addition, fading of the fluorescence of the incorporated BODIPY® FL dUTP is less than that of the corresponding fluorescein dUTP analog.[39] Unlike traditional proliferation assays based on BrdU incorporation, no DNA

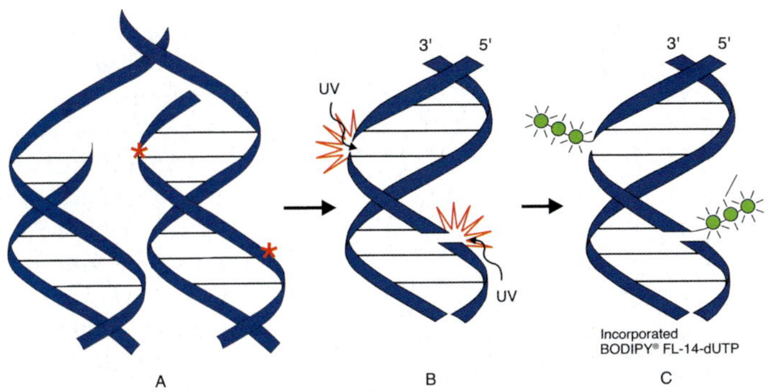

Figure 15.4.4 Schematic diagram showing the sequence of events in the strand break induction by photolysis (SBIP) technique. **A)** Proliferating cells that have incorporated BrdU (*) into newly synthesized DNA are **B)** exposed to UV light in order to induce DNA strand breaks. If the cells are stained with Hoechst 33258 prior to UV illumination, the photolysis efficiency is increased. **C)** Once the cells are fixed, the 3'-hydroxyl ends exposed at these strand breaks can be directly labeled *in situ* using mammalian terminal deoxynucleotidyl transferase (TdT) and our ChromaTide® BODIPY® FL-14-dUTP (C7614).

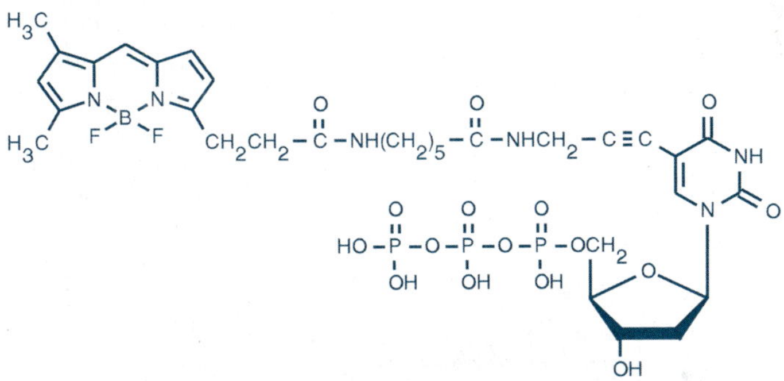

Figure 15.4.5 ChromaTide® BODIPY® FL-14-dUTP (C7614).

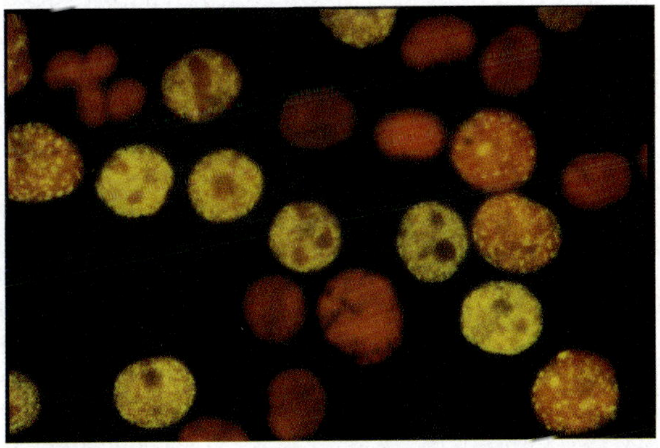

Figure 15.4.6 Detection of BrdU-incorporating cells using the SBIP (Strand Breaks Induced by Photolysis) technique. Exponentially growing human promyelocytic leukemia (HL60) cells were incubated with 20 µM BrdU (B23151) for 40 minutes and then with 20 µg/mL Hoechst 33258 (H1398, H3569, H21491) in the presence of 2% DMSO for an additional 20 minutes. After this incubation, the cells were exposed to 300 nm UV light for 5 minutes to selectively photolyze DNA that contained the incorporated BrdU, and then fixed in 70% ethanol. Subsequent incubation of the permeabilized cells with the ChromaTide® BODIPY® FL-14-dUTP (C7614) in the presence of exogenous terminal deoxynucleotidyl transferase resulted in incorporation of the fluorophore into DNA strand breaks, thereby labeling the S-phase cells. The DNA was counterstained with the red-fluorescent nucleic acid stain propidium iodide (P1304MP, P3566, P21493); therefore, the BODIPY® FL dye–labeled DNA appears yellow. The image was contributed by Zbigniew Darzynkiewicz, Cancer Research Institute, New York Medical College.

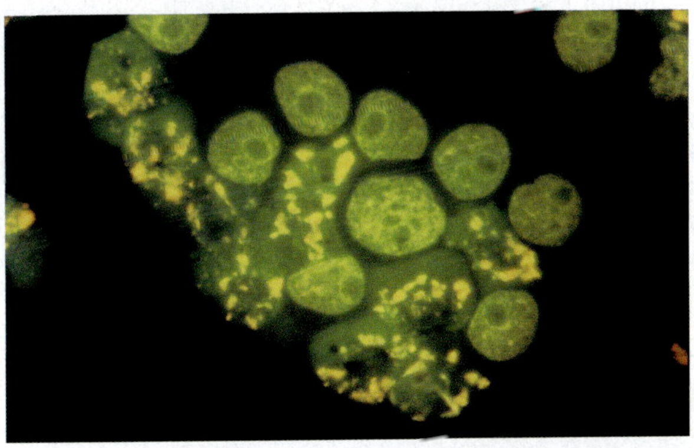

Figure 15.4.7 HL-60 cells treated with camptothecin for 3 hours. The DNA strand nicks characteristic of apoptosis were detected with the TUNEL (terminal deoxynucleotidyl transferase–mediated dUTP nick end-labeling) assay using the fluorescently labeled nucleotide ChromaTide® BODIPY® FL-14-dUTP (C7614). Image contributed by Zbigniew Darzynkiewicz, Cancer Research Institute, New York Medical College.

IMPORTANT NOTICE: The products described in this manual are covered by one or more Limited Use Label License(s). Please refer to the Appendix on page 971 and Master Product List on page 975. Products are For Research Use Only. Not intended for any animal or human therapeutic or diagnostic use.

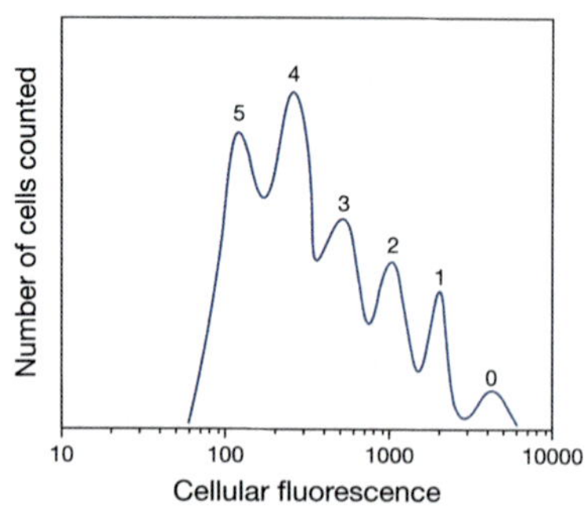

Figure 15.4.8 Tracking of asynchronous cell division using 5-(and 6-)carboxyfluorescein diacetate, succinimidyl ester (5(6)-CFDA SE or CFSE; C1157; V12883) labeling and flow cytometry. Cell division results in sequential halving of the initial fluorescence, resulting in a cellular fluorescence histogram. The peaks labeled 0, 1, 2, 3, 4 and 5 represent successive generations.

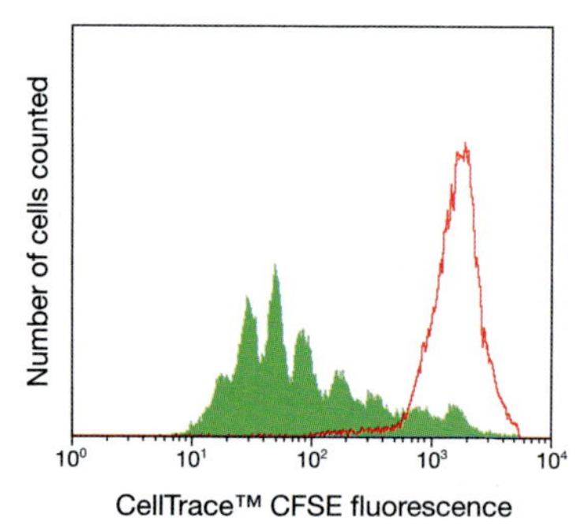

Figure 15.4.9 Following cell proliferation in human peripheral blood lymphocytes using the CellTrace™ CFSE Cell Proliferation Kit (C34554). Human peripheral blood lymphocytes were harvested and stained with CellTrace™ CFSE (carboxyfluorescein diacetate, succinimidyl ester; 5(6)-CFDA, SE) on day 0. A portion of the population was arrested at the parent generation using mitomycin C (red peak). The remainder of the sample was stimulated with phytohemagglutinin and allowed to proliferate for 5 days. Solid green peaks represent successive generations.

heat- or acid-denaturation steps are required with SBIP in order to visualize the labeled strand breaks, allowing simultaneous detection of the morphology of nuclear proteins and other cellular constituents by immunocytochemical analysis. The narrow emission spectrum of the BODIPY® FL dye–labeled nucleotides is especially useful for multicolor labeling experiments.

An elegant technique has been described that permits tracking of labeled chromosomes through mitosis by metabolic incorporation of microinjected fluorescent nucleotides, including our fluorescein-12-dUTP (C7604, Section 8.2), into DNA using endogenous cellular enzymes.[43–45] The procedure does not interfere with subsequent progress through the cell cycle, and fluorescent strands of DNA can be followed as they assemble into chromosomes and segregate into daughters and granddaughters.[43–45] Presumably, injection of 5'-bromo-2'-deoxyuridine triphosphate (BrdUTP, B21550), followed by detection of the incorporated BrdU with one of our Alexa Fluor® conjugates of anti-BrdU would also be suitable for studying mitosis. The corresponding ribonucleotide (BrUTP, B21551) that has been microinjected into cells is incorporated into RNA of a nucleolar compartment,[46,47] a process that should also be detectable with fluorescent anti-BrdU conjugates.

Proliferation Assay Using the Succinimidyl Ester of Carboxyfluorescein Diacetate and Related Probes

The succinimidyl ester of carboxyfluorescein diacetate (5(6)-CFDA, SE or CFSE, C1157) is currently the most widely used probe for generation analysis of cells, although our succinimidyl ester of Oregon Green® 488 carboxylic acid diacetate (O34550, C34555; see below) offers several important advantages over this fluorescein derivative. CFDA SE spontaneously and irreversibly couples to both intracellular and cell-surface proteins by reaction with lysine side chains and other available amine groups. When cells divide, CFDA SE labeling is distributed equally between the daughter cells, which are therefore half as fluorescent as the parents. As a result, each successive generation in a population of proliferating cells is marked by a halving of cellular fluorescence intensity (excitation/emission maxima ~495/525 nm) that is readily detected by a flow cytometer (Figure 15.4.8), fluorescence microscope or fluorescence microplate reader. CFDA SE is available as a single vial containing 25 mg (C1157). CFDA SE is also available conveniently packaged for cell tracing applications in our Vybrant® CFDA SE Cell Tracer Kit (V12883, Figure 15.4.8) and for cell proliferation studies in our CellTrace™ CFSE Cell Proliferation Kit (C34554, Figure 15.4.9). The fluorescent CFDA SE product has excitation/emission maxima of ~492/517 nm and can be detected using a fluorescence microscope, flow cytometer or fluorescence microplate reader. Each kit includes 10 single-use vials of CFDA SE (500 µg each in Kit V12883, 50 µg each in Kit C34554), as well as high-quality anhydrous DMSO and a complete protocol.

CFDA SE produces more homogeneous cellular labeling and, consequently, much better intergenerational resolution than other cell-tracking dyes, such as the membrane marker PKH26. Using flow cytometric analysis of CFDA SE labeling, researchers can resolve 8 to 10 successive generations of lymphocytes.[48,49] In transplanted cells the signal of CFDA SE can be traced *in vivo* for weeks.[50,51] The feasibility of using cell-permeant fluorescent tracers to follow cell division of natural killer (NK) cells, B cells, T cells, thymocytes, lymphocytes, fibroblasts and hematopoietic cells has been demonstrated with CFDA SE.[49,52–62] For instance, researchers have used CFDA SE labeling to show that transplantable hematopoietic cells proliferate *in vitro* in response to stimulation by a growth factor cocktail.[63] These studies helped provide direct evidence that the hematopoietic potential of cultured stem cells is limited by homing activity and not by proliferative capacity. Because the first division results in the largest change in fluorescence intensity, this method is particularly useful for detecting subsets of cells within a population that are resistant to cell division. The method is not limited to mammalian cells; it has also been applied to determine the number of cell divisions in stained *Lactobacillus plantarum.*[64]

Like CFDA SE, the succinimidyl ester of Oregon Green® 488 carboxylic acid diacetate (carboxy-DFFDA SE) should be a useful tool for following proliferating cells. This Oregon Green® 488 probe passively diffuses into cells, where it is colorless and nonfluorescent until its acetate groups are removed by intracellular esterases to yield a highly green-fluorescent, amine-reactive dye. Upon reaction with intracellular amines, the probe forms Oregon Green® 488 conjugates that are well-retained by cells. Unlike fluorescein derivatives, however, Oregon Green® 488

IMPORTANT NOTICE: The products described in this manual are covered by one or more Limited Use Label License(s). Please refer to the Appendix on page 971 and Master Product List on page 975. Products are For Research Use Only. Not intended for any animal or human therapeutic or diagnostic use.
www.invitrogen.com/probes

derivatives exhibit bright green fluorescence that is not pH dependent at typical cellular pH values. Moreover, Oregon Green® 488 probes are usually brighter and more photostable than fluorescein probes. We offer carboxy-DFFDA SE in a 1 mg unit size (O34550) and specially packaged in a set of 20 vials, each containing 50 µg (CellTrace™ Oregon Green® 488 carboxylic acid diacetate succinimidyl ester, C34555).

The intracellular conjugates of 5-(and 6-)carboxyeosin diacetate succinimidyl ester (C22803) have absorption and emission spectra at longer wavelengths than CFDA SE, which may make this probe useful in combination with CFDA SE for studies of proliferation of mixed-cell populations. Eosin conjugates are more effective singlet-oxygen generators than are simple fluorescein derivatives, potentially resulting in their utility for photoablation of cells.

The succinimidyl ester of SNARF®-1 carboxylic acid, acetate (S22801) is also designed to serve as a cell tracer and indicator of cell division. However, unlike the green-fluorescent CFDA SE–labeled cells, cells labeled with the succinimidyl ester of SNARF®-1 carboxylic acid, acetate exhibit red fluorescence when excited near 488 nm. Although the fluorescence intensity of this SNARF® derivative in cells may be weaker than that of cells labeled with CFDA SE, its red fluorescence is easily distinguished from the green fluorescence of CFDA SE–labeled cells. The SNARF® dyes have been predominantly used as indicators of intracellular pH (Chapter 20).

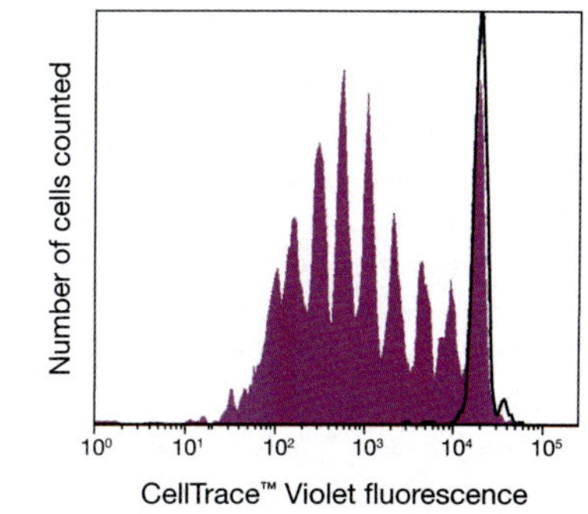

Figure 15.4.10 Human peripheral blood lymphocytes were harvested and stained with CellTrace™ Violet stain (C34557). The violet peaks represent successive generations of cells stimulated with mouse anti–human CD3 and Interleukin-2 and grown in culture for 7 days. The peak outlined in black represents cells that were grown in culture for 7 days with no stimulus.

CellTrace™ Violet Cell Proliferation Kit

CellTrace™ Violet stain is an esterase-activated phenolic fluorophore with a succinimidyl ester substituent for coupling to cell surface and intracellular amines. It is functionally analogous to CFSE, equally partitioning between daughter cells during division resulting in successive 2-fold reductions in cell-associated fluorescence intensity. When analyzed by flow cytometry, this progressive label partitioning provides a direct indication of cell proliferation status (Figure 15.4.10). In contrast to CFSE, CellTrace™ Violet stain is optimally excited by 405 nm violet diode lasers and generates blue fluorescence (emission peak ~455 nm). Consequently, it can be used in combination with CFSE to track cells from different origins after mixing or to analyze proliferation of GFP-expressing cells. The CellTrace™ Violet Cell Proliferation Kit (C34557) includes the CellTrace™ Violet stain together with dimethylsulfoxide (DMSO) for preparation of a stock solution.

CyQUANT® Cell Proliferation Assay Kit

Because cellular DNA content is highly regulated, it is closely proportional to cell number. Therefore, changes in nucleic acid content can serve as a sensitive indicator of overall cell proliferation, as well as of cytotoxic events or pathological abnormalities that affect cell proliferation. Our CyQUANT® Cell Proliferation Assay Kit (C7026) provides an excellent method both for enumerating cells in a population and for measuring their proliferative activity. This assay is an important development for the rapid and quantitative screening of agents that affect cell proliferation. The CyQUANT® assay is based on the use of our green-fluorescent CyQUANT® GR dye, which exhibits strong fluorescence enhancement when bound to cellular nucleic acids.[65] The assay protocol is simple: the culture medium is removed (nonadherent cells require brief centrifugation); the cells are frozen, thawed and lysed by addition of the CyQUANT® cell buffer containing detergent and the CyQUANT® GR dye; and fluorescence is then measured directly in a fluorometer or fluorescence microplate reader (Figure 15.4.11). No washing steps, growth medium changes or long incubations are required.

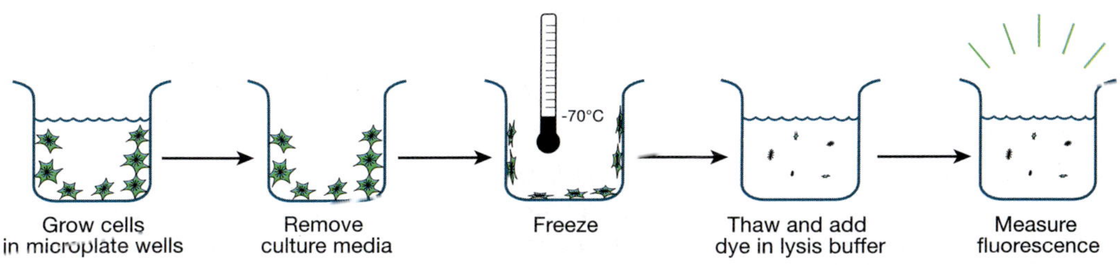

Figure 15.4.11 The simple procedure for using the CyQUANT® Cell Proliferation Assay Kit (C7026).

IMPORTANT NOTICE: The products described in this manual are covered by one or more Limited Use Label License(s). Please refer to the Appendix on page 971 and Master Product List on page 975. Products are For Research Use Only. Not intended for any animal or human therapeutic or diagnostic use.

The CyQUANT® cell proliferation assay has a number of significant advantages over other proliferation assays:

- **Sensitivity and linearity.** The CyQUANT® assay is linear from 50 or fewer cells to at least 50,000 cells in 200 µL volumes (Figure 15.4.12); increasing the dye concentration extends the linear range to at least 250,000 cells. Methods that employ Hoechst 33258 [66] (H1398, H3569, H21491) or Hoechst 33342 [67] (H1399, H3570, H21492) to measure cell number and proliferation are much less sensitive—detection limits of 500 cells for Hoechst 33258 [66] or 2500 cells for Hoechst 33342 [67]—and have much smaller effective ranges.
- **No radioactivity.** Unlike assays that measure ^{3}H-thymidine incorporation, the CyQUANT® assay does not require radioisotopes and thus does not have the hazards or the expense associated with use, storage and disposal of radioisotopes.
- **Quick and easy protocol.** The CyQUANT® assay is a single-step procedure that requires no lengthy incubation steps and can be completed within an hour (Figure 15.4.11).
- **Specificity and reliability.** The assay is specific for total nucleic acids, with essentially no interference from other cell components. No wash steps are required because cellular growth media do not significantly interfere with CyQUANT® GR fluorescence. The CyQUANT® assay is reliable for cell quantitation, even without treatment to eliminate cellular RNA. However, addition of RNase or DNase permits the easy quantitation of DNA or RNA, respectively, in the sample.
- **Convenience.** Unlike assays that use tetrazolium salts, ^{3}H-thymidine, BrdU, neutral red or methylene blue, [67–70] the CyQUANT® procedure is not dependent on cellular metabolism. Thus, cells can be frozen and stored prior to assaying, with no reduction in signal, or they can be assayed immediately after collection. Time-course assays are simplified because data obtained from stored samples taken at widely different time intervals can be assayed together with a single standard curve determination.

We have found the CyQUANT® Cell Proliferation Assay Kit to be useful for assaying widely disparate cell types, including:

- Human neonatal fibroblasts, keratinocytes, melanocytes, umbilical vein endothelial cells (HUVEC) and dermal microvascular endothelial cells (DMVEC)

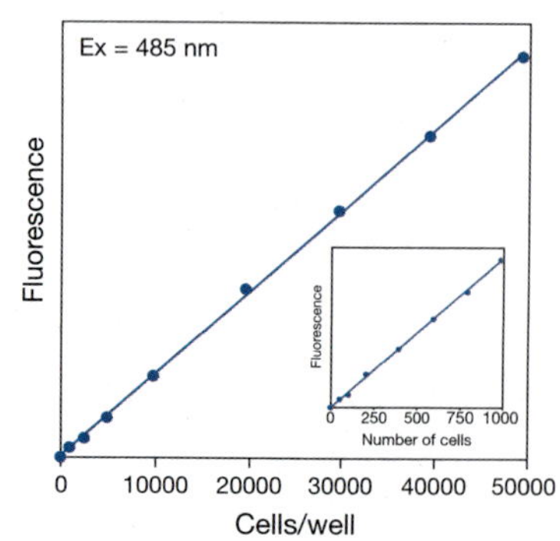

Figure 15.4.12 Quantitation of NIH 3T3 fibroblasts using the CyQUANT® Cell Proliferation Assay Kit (C7026). Fluorescence measurements were made using a microplate reader with excitation at 485 nm and emission detection at 530 nm. The linear range of the assay under these conditions is from 50 to 50,000 cells per 200 µL sample. The inset shows the linearity that can be obtained at very low numbers of cells.

- Murine fibroblasts (NIH 3T3 and CRE BAG 2 cells) and myeloma (P3X63A68) cells
- Madin–Darby canine kidney (MDCK) cells
- Chinook salmon embryo (CHSE) cells
- Rat basophilic leukemia (RBL) and glioma (C6) cells

Determination of total cell number using the CyQUANT® GR reagent is potentially useful for quantitating cell adhesion (see "Cell Adhesion" in Section 15.6) and for determining the total number of cells in a tissue. Each CyQUANT® Cell Proliferation Assay Kit (C7026) includes:

- CyQUANT® GR reagent
- Cell-lysis buffer
- DNA standard for calibration
- Detailed protocols

The kit supplies sufficient materials for performing 1000 assays based on a 200 µL sample volume or a proportionately lower number of assays with a larger sample volume. The CyQUANT® cell-lysis buffer (a 20X concentrate, C7027) is also available separately and has been formulated to produce efficient lysis, to protect nucleic acids from nuclease activity and to dissociate proteins that may interfere with dye binding to nucleic acids. It may prove generally useful in the development of other assays that require cell lysis.

CyQUANT® NF Cell Proliferation Assay Kit

The CyQUANT® NF Cell Proliferation Assay Kit provides a fast and sensitive method for counting cells in a population and measuring proliferation in microplate format. [71] This assay can be completed in 1 hour, with no washes, cell lysis, long incubations or radioactivity required, and it is not dependent on physiological activities that may exhibit cell number–independent variability. The CyQUANT® NF assay eliminates the freeze-thaw cell lysis step of the original CyQUANT® cell proliferation assay by using a cell-permeant DNA-binding dye in combination with a plasma membrane–permeabilization reagent. The CyQUANT® NF assay protocol requires only aspiration of growth medium (for adherent cells), replacement with dye binding solution, incubation for 30–60 minutes and then measurement of fluorescence in a microplate reader. The CyQUANT® NF assay has a linear detection range from at least 100 to 20,000 cells per well in most cell lines using a 96-well microplate format and a 100 µL assay volume.

The CyQUANT® NF Cell Proliferation Assay Kit can be used with either a 96-well or 384-well microplate format and is available in two configurations: a 200-assay kit (C35007) and a 1000-assay kit (C35006) for high-throughput applications.

Each kit contains:

- CyQUANT® NF dye reagent
- Dye delivery reagent
- Concentrated Hank's balanced salt solution (HBSS)
- Detailed protocols

CyQUANT® Direct Cell Proliferation Assay Kit

CyQUANT® Direct Cell Proliferation Assay is a fluorescence-based proliferation and cytotoxicity assay for microplate readers. The no-wash, homogeneous format and fast add-mix-read protocol makes the CyQUANT® Direct assay ideal for high-throughput screening (HTS) applications. The assay can be completed in 1 hour, with no washes, cell

IMPORTANT NOTICE: The products described in this manual are covered by one or more Limited Use Label License(s). Please refer to the Appendix on page 971 and Master Product List on page 975. Products are For Research Use Only. Not intended for any animal or human therapeutic or diagnostic use.

lysis, temperature equilibrations or radioactivity required, and the signal is stable for several hours to provide work-flow convenience. With a dynamic range from less than 50 to more than 20,000 cells of most adherent and suspension cell types, the CyQUANT® Direct assay can be used in 96-, 384- or 1,536-well microplate formats, and is compatible with most HTS and high-content screening (HCS) readers. Because the experimental protocol does not include a lysis step, the assay can conveniently be multiplexed using a spectrally distinct fluorescent or a luminescent readout.

The CyQUANT® Direct Cell Proliferation Assay Kit is available in two configurations: a 10-plate assay kit (C35011) and a 100-plate assay kit (C35012) for high-throughput applications. Each kit contains:

- CyQUANT® Direct nucleic acid stain
- CyQUANT® Direct background suppressor
- Detailed protocols

FluoReporter® Blue Fluorometric Nucleic Acid Assay Kit

The FluoReporter® Blue Fluorometric dsDNA Quantitation Kit (F2962) provides the protocols developed by Rago and colleagues [72] for analyzing cellular DNA with the blue-fluorescent Hoechst 33258 nucleic acid stain. The kit enables researchers to detect ~10 ng of isolated calf thymus DNA or ~1000 mouse NIH 3T3 cells in a 200 µL sample (substantially lower levels are detectable using our CyQUANT® Cell Proliferation Assay Kit described above).

With this kit, quantitation of cellular DNA is rapid, and all manipulations can be carried out in microplate wells. The cells are lysed by freezing them in distilled water, which circumvents the requirement for extraction procedures used in other Hoechst 33258 dye–based protocols.[69,73–76] The diluted dye solution is then added to the lysed cells, and fluorescence is measured. Kit components include:

- Hoechst 33258 in DMSO/H_2O
- TNE buffer
- Detailed protocol

Each kit provides sufficient reagents for assaying approximately 2000 samples using a fluorescence microplate reader.

Vybrant® MTT Cell Proliferation Assay Kit

The convenient Molecular Probes® Vybrant® MTT Cell Proliferation Assay Kit (V13154) simplifies the task of counting cells with a microplate absorbance reader. The colorimetric MTT assay, developed by Mossman, is based on the conversion of the water-soluble MTT to an insoluble purple formazan.[77,78] This formazan is then solubilized, and its concentration determined by optical density at 570 nm. The Vybrant® MTT Cell Proliferation Assay Kit provides a sensitive assay with excellent linearity up to approximately 10^6 cells per well. Each kit includes:

- MTT
- Sodium dodecyl sulfate (SDS)
- Detailed protocol

This kit provides sufficient materials for ~1000 assays using standard 96-well microplates. Numerous variations and modifications of the MTT assay have been published.[79–81] In addition to dehydrogenases, MTT is reduced by glutathione *S*-transferase [82] (GST). Therefore, MTT may not always be a reliable cell viability probe in cells treated with compounds that affect GST activity.

Vybrant® DiI Cell-Labeling Solution

Analysis by mass spectrometry and HPLC indicates that the dye we use in our Vybrant® DiI cell-labeling solution (V22885) is structurally identical to the dye called PKH 26. DiI is a red-fluorescent lipophilic tracer that, in addition to being extensively used for cell tracing (Section 14.4), has been utilized for generational analysis of cells undergoing division.[52,83–86] Unlike the PKH 26 dye, which requires a special cell-labeling medium and low ionic strength for successful cell loading, our Vybrant® DiI cell-labeling solution is simply added to cells in normal growth medium. Dividing cells distribute the lipophilic tracer approximately equally between daughter cells. It is usually possibly to follow at least three or four generations of cells by flow cytometry, although asynchronous division times can quickly complicate the measurements. The dyes in our Vybrant® DiO, Vybrant® DiD and Vybrant® CM-DiI Labeling solutions (V22886, V22887, V22888) may have similar utility for tracing cells through cell division. CM-DiI contains a thiol-reactive chloromethyl that allows the dye to covalently bind to cellular thiols. Thus, unlike other membrane stains, this label is well retained in some cells throughout fixation and permeabilization steps; see Section 14.4 for more information.

CountBright™ Absolute Counting Beads

Flow cytometry provides a rapid method for quantitating cell characteristics, however most flow cytometers cannot directly provide the cell concentration or absolute count of cells in a sample. Absolute cell counts have been widely used in quantitating cell populations and disease progression [87–89] and are generally obtained either by combining a separate cell concentration determination from a hematology analyzer with flow cytometry population data (multiple-platform testing) or by adding an internal microsphere counting standard to the flow cytometry sample (single-platform testing). The single-platform method is preferred as it is technically less complicated and more accurate than multiple-platform testing.[90]

CountBright™ absolute counting beads (C36950) are a calibrated suspension of microspheres that are brightly fluorescent across a wide range of excitation and emission wavelengths and contain a known concentration of microspheres. For absolute counts, a specific volume of the microsphere suspension is added to a specific volume of sample, such that the ratio of sample volume to microsphere volume is known. The volume of sample analyzed can be calculated from the number of microsphere events and then used with cell events to determine cell concentration. In general, at least 1000 bead events should be acquired to assure a statistically significant determination of sample volume. Sufficient reagents are provided for 100 flow cytometry assays, each using 50 µL of counting beads per test.

CountBright™ absolute counting beads are broadly fluorescent and can be used with either a fluorescence or scatter threshold. Fluorescence can be excited by wavelengths from UV to 635 nm; fluorescence emission can be read between 385 nm and 800 nm. The fluorescence intensity of the microspheres has been adjusted to be about 5–50 times brighter than the anticipated intensities of typically stained cells. When using a scatter threshold, the microsphere signal should be above the threshold. The microspheres can be gated by a single parameter, but a combination of parameters can be used to resolve microspheres from cells and other events.

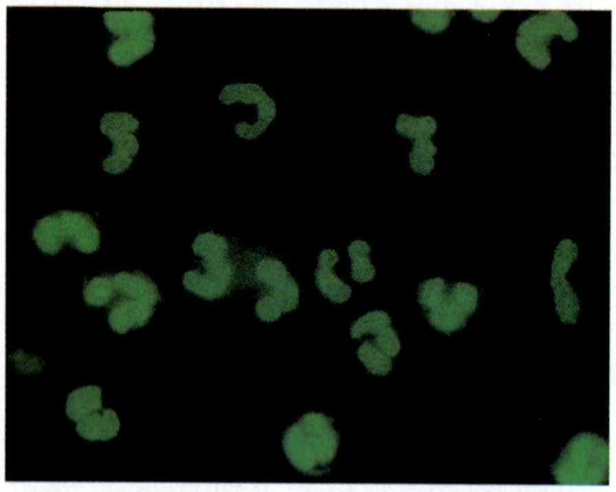

Figure 15.4.13 Adherent cells from human peripheral blood stained with the SYTO® 13 dye (S7575), one of the six visible light–excitable cell-permeant nucleic acid stains in our SYTO® Green Fluorescent Nucleic Acid Stain Sampler Kit #1 (S7572). The multilobed nuclei of these polymorphonuclear leukocytes are particularly striking in this field of view.

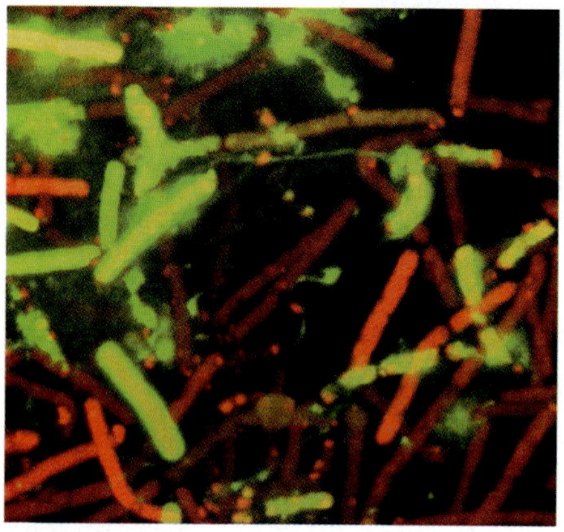

Figure 15.4.14 A mixture of live and heat-killed *Bacillus cereus* cells simultaneously stained with the cell-permeant SYTO® 59 red-fluorescent nucleic acid stain (S11341) and the cell-impermeant SYTOX® Green nucleic acid stain (S7020), each at a concentration of 5 µM. Bacteria with intact cell membranes stain fluorescent red, whereas bacteria with damaged membranes stain fluorescent green. This maximum-projection image was generated from a series of 10 images taken at 0.2 µm increments through the specimen with a Leica confocal laser-scanning microscope.

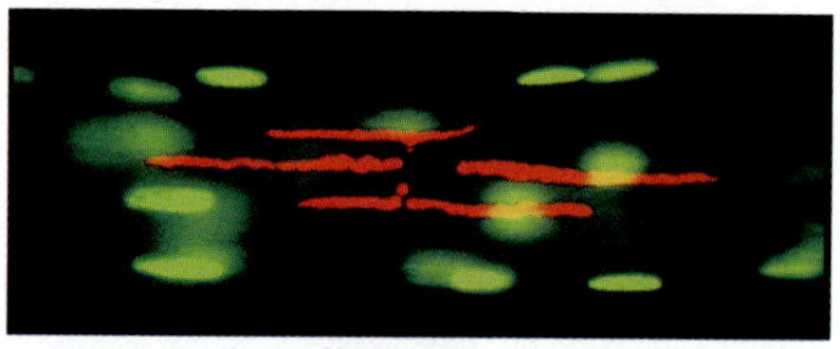

Figure 15.4.15 Pseudocolored photomicrograph of the synaptic region of fluorescently labeled living muscle fibers from the lumbricalis muscle of the adult frog *Rana pipiens*. Six hours after isolation of the muscle fibers, acetylcholine receptors were stained with red-fluorescent tetramethylrhodamine α-bungarotoxin (T1175) and myonuclei were stained with the green-fluorescent SYTO® 13 live-cell nucleic acid stain (S7575). Photo contributed by Christian Brösamle, Brain Research Institute, University of Zurich, and Damien Kuffler, Institute of Neurobiology, University of Puerto Rico.

CountBright™ absolute counting beads can be used with any sample type, including no-wash/lysed whole blood. The microspheres in the reagents are approximately 7 µm in diameter and have settling properties similar to lymphocytes. The accuracy of cell counts based on CountBright™ absolute counting beads depends on sample handling and the precise delivery of the volume of beads. The CountBright™ absolute counting beads must be mixed well to assure a uniform suspension of microspheres. After vortexing for 30 seconds, the microsphere suspension can be pipetted by standard techniques; however, more viscous solutions such as blood require reverse pipetting to facilitate accurate volume delivery. Cell suspensions may be diluted but should be assayed without wash steps. Other sample preparation steps that can lead to cell or microsphere loss should also be avoided. For antibody protocols, CountBright™ absolute counting beads should be used with reagents titered for no-wash staining.

Detection and Enumeration Assays for Microorganisms and Viruses

Detecting Bacteria, Yeast and Plankton Using Nucleic Acid Stains

We recommend our SYTO® nucleic acid stains (see the complete list in Table 8.3) for simple detection of the presence of bacteria, yeast and other microbial cells [91–95] (Figure 15.4.13, Figure 15.4.14, Figure 15.4.15). The SYTO® dyes are essentially nonfluorescent except when bound to nucleic acids, where they become highly fluorescent, often with quantum yields exceeding 0.5. Consequently, it is usually not necessary to remove unbound stains before analysis. SYTO® dyes are available with blue, green, orange or red fluorescence. The SYTO® dyes rapidly penetrate the membranes of almost all cells, including bacteria and yeast. The various cell types can often be identified by their characteristic morphology or, in the case of flow cytometric applications, by their light-scattering properties.

The SYTO® 11 and SYTO® 13 green-fluorescent nucleic acid stains show exceptional ability to penetrate tissues for at least 100 µm, including untreated, unfixed human brain tissue, where they were used to enumerate cells by confocal laser-scanning microscopy.[96] Simultaneous labeling with a green-fluorescent SYTO® dye and a red-fluorescent nucleic acid stain—most often propidium iodide, ethidium homodimer-1 or -2, TOTO®-3 or TO-PRO®-3 (Table 8.2, Table 8.4)—is frequently used to assess cell viability (Section 15.2). Although some of the SYTO® dyes show higher quantum yields on DNA or RNA, they should not be considered specific stains for either of these nucleic acids. We offer four different SYTO® Nucleic Acid Stain Sampler Kits (S11350, S7572, S11360, S11340; Section 8.1).

In addition to its use for cell-cycle analysis (see below) and nuclear staining, DAPI (D1306, D3571, D21490) is frequently employed for DNA content–based counting of bacterial cells [97–99] and for detecting malarial infections by fluorescence microscopy.[100]

*Bac*Light™ Bacterial Stains

The *Bac*Light™ Green and *Bac*Light™ Red bacterial stains (B35000, B35001) are fluorescent, non–nucleic acid labeling reagents for detecting and monitoring bacteria. Bacteria stained with the *Bac*Light™ Green and *Bac*Light™ Red bacterial stains exhibit bright green (excitation/emission maxima ~480/516 nm) and red (excitation/emission maxima ~480/516 nm) fluorescence, respectively, and can be resolved simultaneously using the appropriate flow cytometry channels. Although these dyes were specifically chosen for flow cytometry applications, bacteria stained with these *Bac*Light™ reagents can also be visualized by fluorescence microscopy with only minor, if any, adjustments in the staining concentrations. Furthermore, the *Bac*Light™ bacterial staining patterns are compatible with formaldehyde or alcohol fixation methods.

These *Bac*Light™ bacterial stains efficiently label a variety of different bacteria species. The intensity of the staining appears to depend on several factors, including gram character, outer membrane composition and overall membrane integrity. In the species we tested, gram-positive bacteria generally exhibited brighter fluorescence than gram-negative bacteria, and cells with compromised membranes accumulated more dye than intact cells (Figure 15.4.16).

Bacteria Counting Kit

Accurate enumeration of low numbers of bacteria in samples must be performed daily in many quality-control laboratories.

To facilitate this determination by flow cytometry (Figure 15.4.17), we have developed the Bacteria Counting Kit (B7277), which provides:

- Cell-permeant, green-fluorescent SYTO® BC nucleic acid stain to label bacteria
- Fluorescent polystyrene microspheres to calibrate the volume of bacterial suspension analyzed
- Detailed protocols

SYTO® BC dye, which is also available separately (S34855, Section 8.1), is a high-affinity nucleic acid stain that easily penetrates both gram-negative and gram-positive bacteria, producing an exceptionally bright green-fluorescent signal. The calibrated suspension of polystyrene microspheres contains beads that exhibit a uniform density, low-level fluorescence and optimal size to clearly separate the light scattering of the microspheres from that of most bacteria.

The Bacteria Counting Kit is particularly valuable for monitoring antibiotic sensitivity because it provides a convenient and accurate means for assessing changes in a bacterial population over time. A sample of the population is simply diluted, stained briefly with the SYTO® BC dye, mixed with a fixed number of microspheres and analyzed on a flow cytometer. Signals from both the stained bacteria and the beads are easily detected in the green-fluorescence channel of most

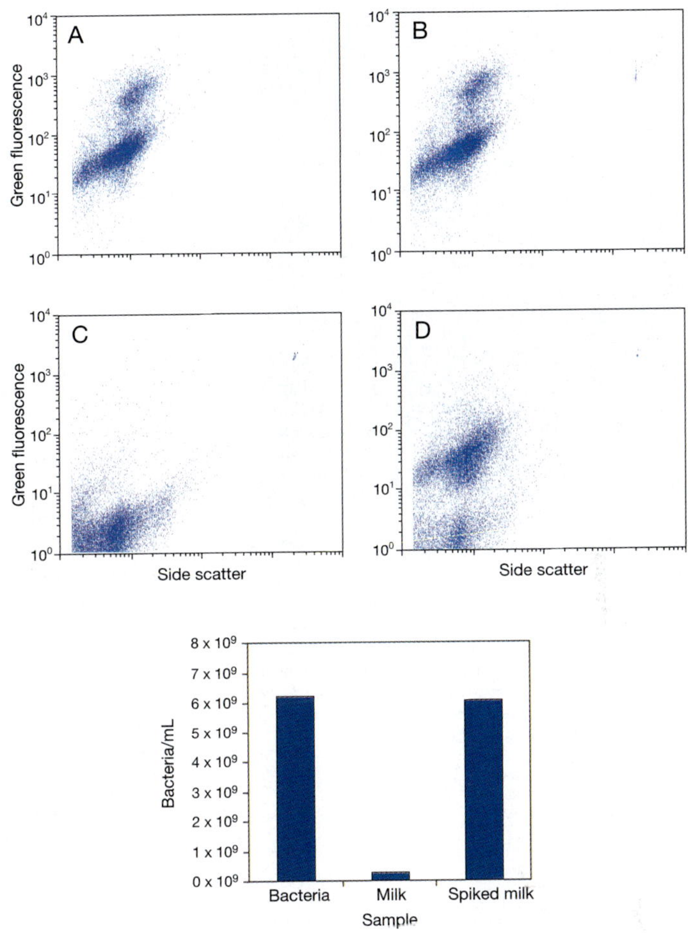

Figure 15.4.17 Detection and counting of bacteria in milk using the Bacteria Counting Kit (B7277). Equal numbers of bacteria were suspended in 150 mM NaCl or a mixture of milk and 150 mM NaCl and assayed using the protocol provided with the kit. As shown in the lower bar chart, the presence of milk does not affect the outcome of the assay. The upper panels plot green fluorescence versus side scatter for: **A)** bacteria alone, **B)** bacteria alone, **C)** milk alone and **D)** bacteria mixed with milk (spiked milk). Panels B–D also contain reference beads, which appear in the upper right corner of the respective plots.

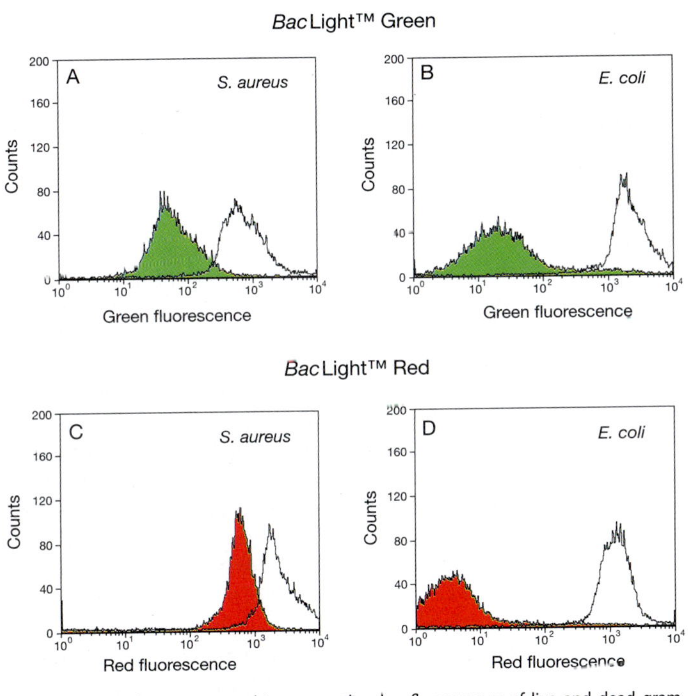

Figure 15.4.16 Flow cytometry histograms showing fluorescence of live and dead gram-positive and gram-negative bacteria stained with the *Bac*Light™ bacterial stains. Untreated and alcohol-treated gram-positive (*Staphylococcus aureus*, (**A** and **C**)) and gram-negative (*Escherichia coli*, (**B** and **D**)) bacteria were each stained separately with 100 nM of either the *Bac*Light™ Green (**A** and **B**) or the *Bac*Light™ Red (**C** and **D**) bacterial stains (B35000, B35001) in 0.85% NaCl buffer and then analyzed by flow cytometry. The histograms for the untreated (colored histogram curve) and alcohol-treated (uncolored histogram curve) bacteria samples were overlaid for each species and *Bac*Light™ bacterial stain.

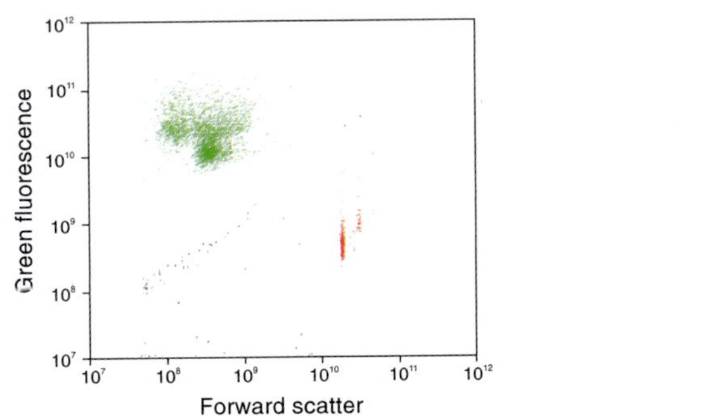

Figure 15.4.18 Flow cytometric enumeration of *Bacillus cereus* using the Bacteria Counting Kit (B7277). In this plot of forward scatter versus fluorescence, signals in the upper lefthand frame represent bacteria stained with SYTO® BC bacteria stain; signals in the lower right-hand frame represent microsphere particles, which serve as a standard used to indicate sample volume.

IMPORTANT NOTICE: The products described in this manual are covered by one or more Limited Use Label License(s). Please refer to the Appendix on page 971 and Master Product List on page 975. Products are For Research Use Only. Not intended for any animal or human therapeutic or diagnostic use.

flow cytometers and can be distinguished on a plot of forward scatter versus fluorescence (Figure 15.4.18). The density of the bacteria in the sample can be determined from the ratio of bacterial signals to microsphere signals in the cytogram. The Bacteria Counting Kit can be used with a variety of gram-negative and gram-positive species of bacteria and provides sufficient reagents for approximately 100 flow cytometry assays.

The fluorescent microspheres in our Bacteria Counting Kit have also been recommended for the enumeration of yeast.[101] We offer a wide selection of labeled beads (Section 6.5, Section 23.1, Section 23.2) that may also prove useful for yeast quantitation and viability assays by flow cytometry.[102]

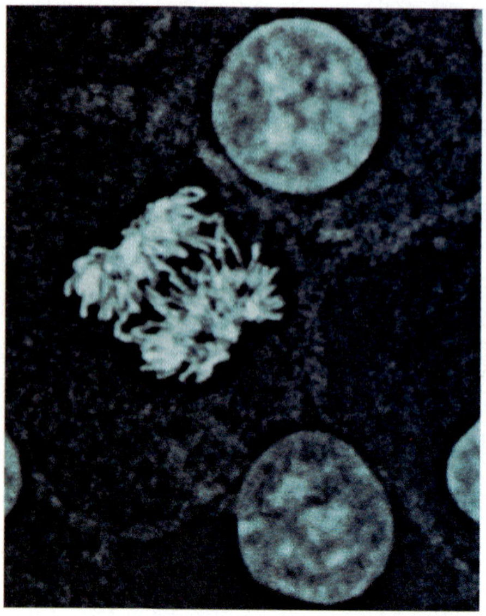

Figure 15.4.19 A mycoplasma-contaminated African green monkey fibroblast cell line was stained with cell-permeant, blue-fluorescent Hoechst 33342 nucleic acid stain (H1399, H3570, H21492), simultaneously illustrating both cell nuclei and mycoplasma. The image was acquired using filters appropriate for DAPI. Image contributed by Heiti Paves, Laboratory of Molecular Genetics, National Institute of Chemical Physics and Biophysics, Estonia.

Cell Culture Contamination Detection Kit

Molecular Probes® Cell Culture Contamination Detection Kit (C7028) uses a simple and effective procedure for visually screening cell cultures for contamination by yeast (and other fungi) or by gram-negative or gram-positive bacteria. This kit not only serves to detect the contaminants, but also identifies the contaminant type, enabling the researcher to choose an appropriate course of action.

A sample of the suspected culture is subjected to two slide-staining protocols. One sample slide is stained with Calcofluor White M2R, a UV light–excitable, blue-fluorescent stain specific for fungal cell walls. A second slide is stained with SYTO® 9 nucleic acid stain to identify all bacteria irrespective of gram signature, and also with the Texas Red®-X conjugate of wheat germ agglutinin (WGA), which selectively stains gram-positive bacteria.[103,104] Gram-positive bacteria on the second slide typically exhibit a green-fluorescent interior with red-fluorescent cell-surface staining, whereas gram-negative bacteria show only green-fluorescent interior staining. Staining and examination of the slides under a fluorescence microscope can be typically performed in less than one hour. Each Cell Culture Contamination Detection Kit contains:

- Green-fluorescent SYTO® 9 nucleic acid stain
- Blue-fluorescent Calcofluor White M2R fungal cell wall stain
- Red-fluorescent Texas Red®-X WGA, for positive identification of gram-positive bacteria
- Buffer for reconstituting Texas Red®-X WGA
- Detailed protocols

This kit provides sufficient material for approximately 200 contamination detection assays. The SYTO® 9 nucleic acid stain, which is also available separately (S34854, Section 15.2), has been used to detect lactic acid–producing bacteria in wine.[105]

MycoFluor™ Mycoplasma Detection Kit

Mycoplasma infections are generally difficult or impossible to detect during routine work with cultured cells because these intracellular pathogens cannot be observed by standard light microscopy. However, mycoplasma infections can be detected with the Hoechst dyes[106,107] (Figure 15.4.19) or with DAPI[108,109] (Section 8.1). Hoechst 33258, either alone[110] or in combination with merocyanine 540[111] (M24571, Section 18.2), has been utilized to eradicate mycoplasma infections from cell cultures. Mycoplasma infections are relatively common. Estimates of contaminated cultures in the United States range from 5% to 35%, whereas the contamination rate is postulated to be much higher in those countries where systematic detection and elimination are not practiced.[112] Not only do mycoplasma cause physiological and morphological distortions that can affect experimental results, but contamination can quickly spread to other cell lines.

The MycoFluor™ Mycoplasma Detection Kit (M7006) provides an extremely rapid and sensitive fluorescence microscopy–based assay for the visual identification of mycoplasma infection in laboratory cell cultures and media. In order to detect mycoplasma, the fluorescent MycoFluor™ reagent is added directly to the culture medium, with or without cells present, and the stained sample is then examined under a fluorescence microscope. The test for the presence of mycoplasma in live or fixed cell cultures takes about 15 minutes from when the reagent is added until when the sample is viewed with a fluorescence microscope equipped with DAPI optical filters. The detection of mycoplasma in cell media requires about 30 minutes, depending on the amount of centrifugation required to concentrate potential contaminants.

IMPORTANT NOTICE: The products described in this manual are covered by one or more Limited Use Label License(s). Please refer to the Appendix on page 971 and Master Product List on page 975. Products are For Research Use Only. Not intended for any animal or human therapeutic or diagnostic use.

Also provided with this kit are mycoplasma MORFS (Microscopic Optical Replicas for Fluorescence assays), which serve as inert positive controls that mimic the size, shape and fluorescence intensity of mycoplasma stained with the blue-fluorescent MycoFluor™ reagent and viewed by fluorescence microscopy. The optical properties of the mycoplasma MORFS enable the researcher to discriminate between stained mycoplasma and other fluorescent material without introducing infectious biological agents into the laboratory environment. No previous experience with mycoplasma testing is required.

Each MycoFluor™ Mycoplasma Detection Kit supplies sufficient materials for at least 100 tests of live cells, fixed cells or culture media. Kit contents include:

- Concentrated MycoFluor™ reagent
- Suspension of mycoplasma MORFS
- Coverslip sealant
- Cotton swab
- Reference micrographs
- Detailed protocols

Alternatively, the LIVE/DEAD® *Bac*Light™ Bacterial Viability Kits (L7007, L7012, L13152; Section 15.3) are useful for detecting mycoplasma infections.[113] Researchers have determined that the reagents in our LIVE/DEAD® *Bac*Light™ Kits can be used for viability determinations in *Eurioplasma eurilytica* and *Mycoplasma hominus* mycoplasma and in the cysts of the protozoan parasite *Giardia muris* (Figure 15.4.20).

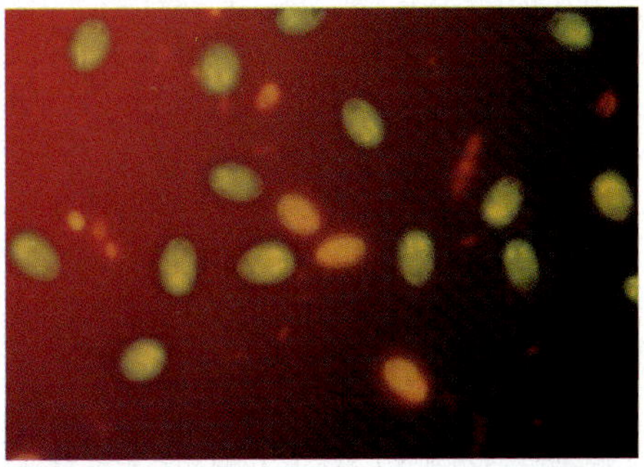

Figure 15.4.20 *Giardia muris* cysts stained with the reagents in the LIVE/DEAD® *Bac*Light™ Bacterial Viability Kit (L7007, L7012). When incubated with the SYTO® 9 and propidium iodide nucleic acid stains provided in this kit, live bacteria with intact cell membranes fluoresce green and dead bacteria with compromised membranes fluoresce red.

Detecting Viruses Using Nucleic Acid Stains

Because of their small size, direct detection of viruses by fluorescence requires highly sensitive reagents or, much more commonly, an amplification scheme. However, direct enumeration of marine viral abundance in seawater using SYBR® Green I nucleic acid gel stain[114–116] (S7563, S7567, S7585; Section 8.3; Figure 15.4.21), YO-PRO®-1 (Y3603, Section 8.1) and DAPI[117] (D1306, D3571, D21490) has been reported.

The slow off-rate of our dimeric nucleic acid stains, such as TOTO®-1 and YOYO®-1 (T3600, Y3601), has permitted their use for labeling nucleic acids, including viral RNA,[118] prior to microinjection into live cells to follow their trafficking. Similar staining techniques may permit tracing of viral uptake and transport by live cells.

Figure 15.4.21 An environmental sample containing marine viruses (smallest dots), bacteria (larger, brighter dots), and a diatom (long thin cell with prominent nucleus) stained with SYBR® Green I nucleic acid stain (S7563, S7567, S7585). Image contributed by Jed Fuhrman, University of Southern California.

Countess® Automated Cell Counter

The Countess® automated cell counter (C10227, C10310, C10311; Figure 15.4.22) uses Trypan Blue staining combined with a sophisticated image analysis algorithm to count all cells in a sample—even mildly clumpy samples—as well as assess viability and measure average cell size. The Countess® instrument allows for accurate cell and viability counts and completes all calculations in about 30 seconds, using as little as 5 µL of your sample; a cell viability percentage value is calculated automatically.

The measurement range of the Countess® automated cell counter extends from 1×10^4 to 1×10^7 cells/mL, with an optimal range from 1×10^5 to 4×10^6 cells/mL, which is broader than that of a hemocytometer. Furthermore, the Countess® instrument counts cell size from 5 to 60 µm and counts beads from 4.5 to 60 µm. Countess® Test Beads (C10284) are designed to provide a reliable and rapid calibration standard for users who choose to operate the instrument outside its default parameters. The Countess® instrument has been validated for use with more than 20 common cell types, including primary cells, blood cells, yeast cells (without viability assessment), insect cells and fish cells.

The Countess® automated cell counter does not require a computer. Disposable chamber slides (C10228, C10312, C10313, C10314, C10315) provide rapid setup, with no cleaning, buffers or maintenance required other than battery replacement after approximately 5 years of use. Trypan Blue stain is also available as a stand-alone reagent (T10282). More information is available at www.invitrogen.com/handbook/countess.

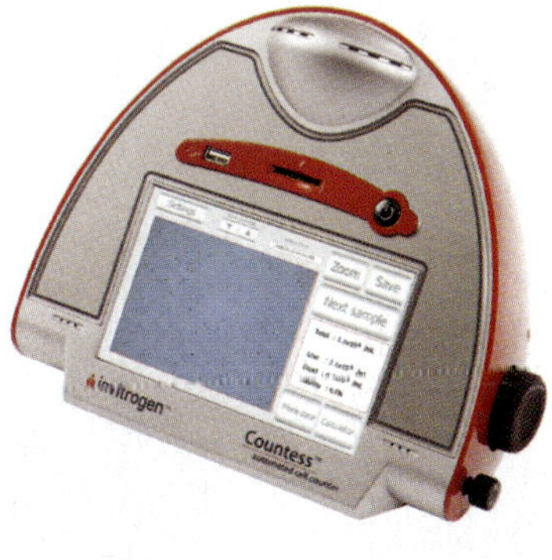

Figure 15.4.22 The Countess® automated cell counter provides fast, easy and accurate cell counting without using a hemocytometer.

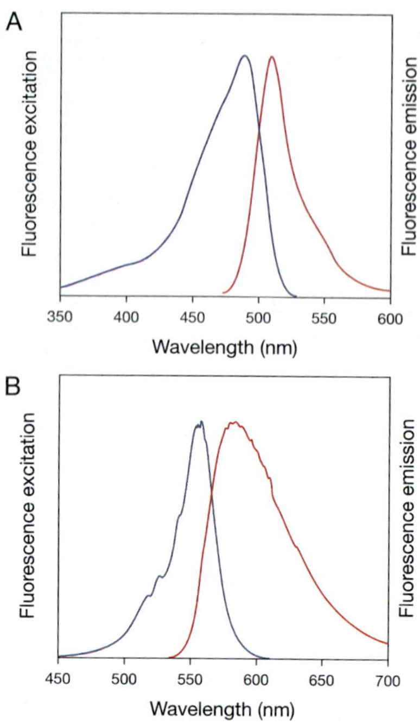

Figure 15.4.24 Fluorescence excitation and emission spectra for **A)** Premo™ geminin–GFP and **B)** Premo™ Cdt1–RFP.

Premo™ FUCCI Cell Cycle Sensor

In 2008, Miyawaki and colleagues developed the Fluorescence Ubiquitination Cell Cycle Indicator (FUCCI), a fluorescent protein (FP)–based sensor that employs a Red Fluorescent Protein (RFP) and a Green Fluorescent Protein (GFP), each of which is fused to one of two different regulators of the cell cycle: Cdt1 and geminin.[119] Ubiquitin E3 ligases add ubiquitin to the Cdt1–GFP and geminin–RFP fusions, thereby targeting these proteins for proteasomal degradation. Temporal regulation of E3 ligase activity results in the biphasic cycling of the geminin and Cdt1 fusions through the cell cycle.

In the G_1 phase, the geminin–GFP fusion is degraded, leaving only the Cdt1–RFP fusion and resulting in red-fluorescent nuclei. During the G_1/S transition, Cdt1 levels decrease and geminin levels increase. Because both proteins are present in the cells, both GFP and RFP fluorescence is visible and the cell nuclei appear yellow when the green and red images are overlaid. In the S, G_2 and M phases, the Cdt1–RFP fusion is degraded, leaving only the geminin–GFP fusion and resulting in green-fluorescent nuclei. This dynamic color change from red to yellow to green serves as an indicator of the progression through cell cycle and division (Figure 15.4.23).

The Premo™ FUCCI Cell Cycle Sensor (P36232) combines the Cdt1 and geminin FP constructs with the powerful BacMam gene delivery and expression system (BacMam Gene Delivery and Expression Technology—Note 11.1). The genetically encoded and pre-packaged reagents are ready for immediate use, eliminating the need to purify plasmid. The BacMam reagent is simply added to cells for 1–2 hours, after which the cells are treated with an enhancer reagent for another 1–2 hours, washed, incubated overnight and then visualized using fluorescence microscopy or high-throughput imaging platforms (Figure 15.4.23). BacMam transduction is efficient and reproducible in most cell types, including primary and stem cells, without apparent cytotoxic effects.

The Premo™ FUCCI Cell Cycle Sensor is designed for live-cell imaging of cell cycle progression and can be used to assess the effect of drugs, siRNA or other factors on the transition of cells through the cell cycle. The fluorescence signals from geminin-GFP and Cdt1-RFP (Figure 15.4.24) have been demonstrated to be resistant to fixation with 4% formaldehyde and permeabilization with 0.1% Triton X-100, thereby enabling processing of labeled cells with antibodies to other cellular targets. Each kit contains all of the components needed to label cells with the Premo™ FUCCI Cell Cycle Sensor using a transduction volume of 2 mL; however, the protocol can easily be adjusted for larger or smaller volumes.

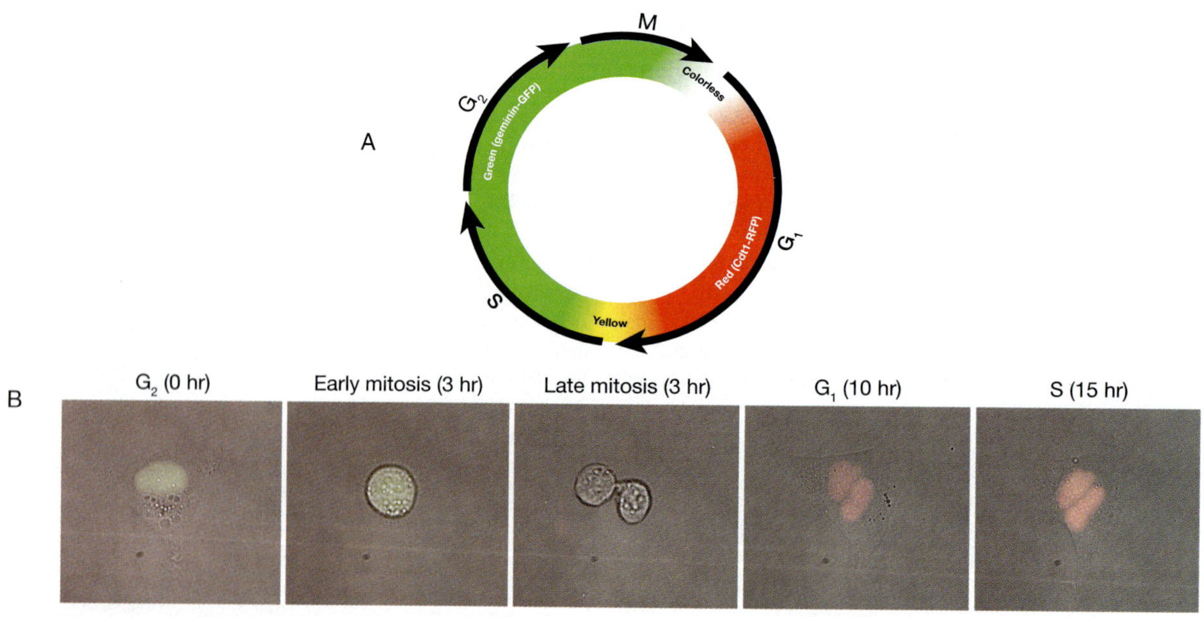

Figure 15.4.23 Imaging cell-cycle progression in live cells with Premo™ FUCCI Cell Cycle Sensor. **A)** Schematic of cell-cycle progression with nuclear fluorescence changes. **B)** U2OS cells were transduced with Premo™ FUCCI Cell Cycle Sensor, and images were collected over 15 hours.

IMPORTANT NOTICE: The products described in this manual are covered by one or more Limited Use Label License(s). Please refer to the Appendix on page 971 and Master Product List on page 975. Products are For Research Use Only. Not intended for any animal or human therapeutic or diagnostic use.
www.invitrogen.com/probes

Nucleic Acid Probes for Cell-Cycle Analysis

Live-cell studies of cellular DNA content and cell-cycle distribution are useful for investigating tumor behavior and suppressor gene mechanisms, for monitoring apoptosis and for detecting variations in growth patterns arising from a variety of physical, chemical and biological means. In a given population, cells will be distributed among three major phases of the cell cycle: G_0/G_1 phase (one set of paired chromosomes per cell), S phase (DNA synthesis with variable amount of DNA) and G_2/M phase (two sets of paired chromosomes per cell, prior to cell division).[120–123] DNA content can be measured using fluorescent, DNA-selective stains that exhibit emission signals proportional to DNA mass. Flow cytometric analysis of these stained populations is then used to produce a frequency histogram that reveals the various phases of the cell cycle. This analysis is typically performed on permeabilized or fixed cells using a cell-impermeant nucleic acid stain,[23,124–126] but is also possible using live cells and a cell-permeant nucleic acid stain.[127,128] While the choices for fixed cell staining are varied, there are only a few examples of useful cell-permeant nucleic acid stains.

Vybrant® DyeCycle™ Stains

All Vybrant® DyeCycle™ stains are DNA-selective, cell membrane–permeant and nonfluorescent until bound to double-stranded DNA. Once bound to DNA, these dyes emit a fluorescence signal that is proportional to DNA mass. Staining cells with the Vybrant® DyeCycle™ stains is simple: suspended cells are incubated in the presence of a Vybrant® DyeCycle™ stain, and fluorescence is measured directly—no additional treatment or centrifugation is required. Fluorescence data can then be used to generate a frequency histogram that reveals the various phases of the cell cycle.

We offer four Vybrant® DyeCycle™ stains for flow cytometry analysis of DNA content in live cells:

- Vybrant® DyeCycle™ Violet stain [129] (V35003)
- Vybrant® DyeCycle™ Green stain (V35004)
- Vybrant® DyeCycle™ Orange stain (V35005)
- Vybrant® DyeCycle™ Ruby stain (V10273, V10309)

The Vybrant® DyeCycle™ stains spectrally match commonly available excitation sources, placing cell-cycle studies within reach of all flow cytometrists and allowing simultaneous staining of the cell population for other parameters. Vybrant® DyeCycle™ Violet stain (excitation/emission maxima ~396/437 nm) is well suited for the 405 nm laser line (Figure 15.4.25) but can also be used with UV excitation. Vybrant® DyeCycle™ Green stain (excitation/emission maxima ~506/534 nm) is efficiently excited with the 488 nm spectral line of the argon-ion laser (Figure 15.4.26). Vybrant® DyeCycle™ Orange stain (excitation/emission maxima ~519/563 nm) can be excited using either the 488 nm or 532 nm laser lines (Figure 15.4.27). Vybrant® DyeCycle™ Ruby stain (excitation/emission maxima ~638/686 nm) can be excited using the 635 nm red diode laser (Figure 15.4.28).

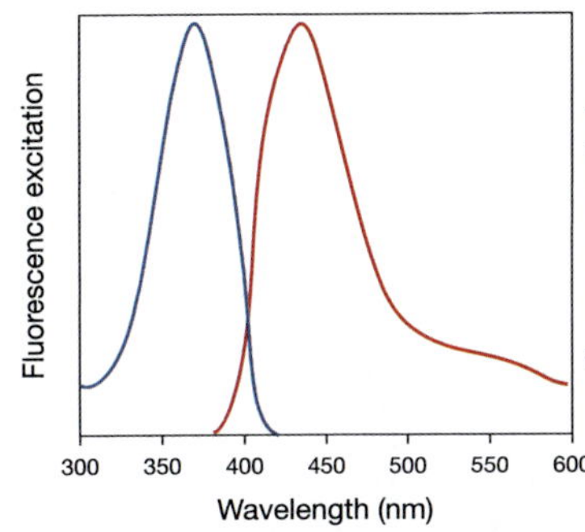

Figure 15.4.25 Fluorescence excitation and emission spectra of Vybrant® DyeCycle™ Violet stain bound to DNA.

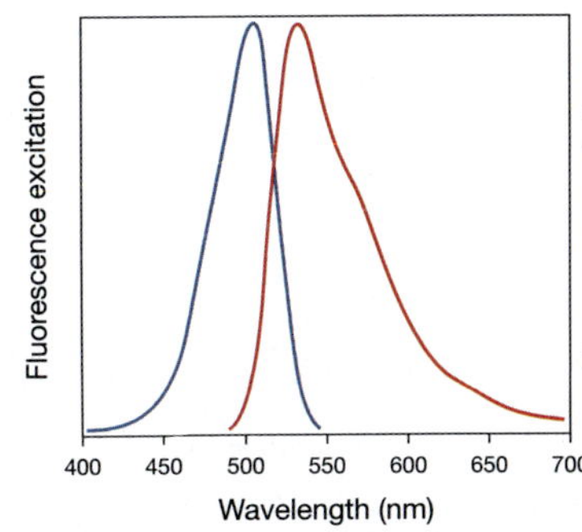

Figure 15.4.26 Fluorescence excitation and emission spectra of Vybrant® DyeCycle™ Green stain bound to DNA.

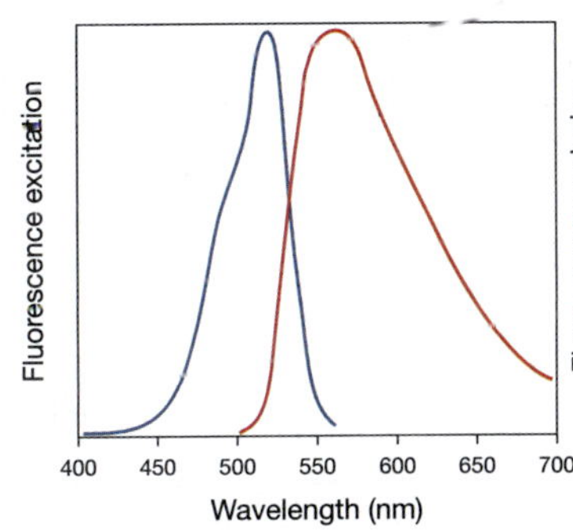

Figure 15.4.27 Fluorescence excitation and emission spectra of Vybrant® DyeCycle™ Orange stain bound to DNA.

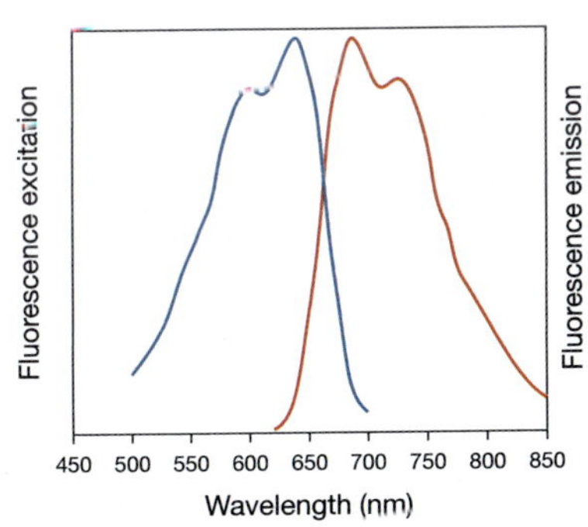

Figure 15.4.28 Fluorescence excitation and emission spectra of Vybrant® DyeCycle™ Ruby stain bound to DNA.

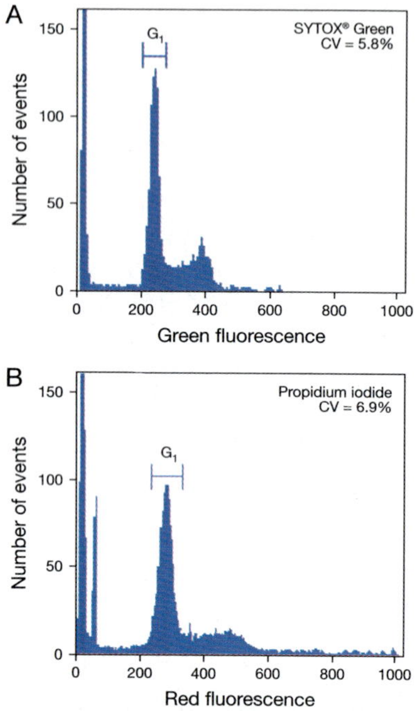

Figure 15.4.31 Comparison of DNA content histograms obtained with **A)** SYTOX® Green nucleic acid stain (S7020) and **B)** propidium iodide (P1304MP, P3566, P21493). Human B cells were suspended in permeabilizing buffer (100 mM Tris, pH 7.4, 154 mM NaCl, 1 mM $CaCl_2$, 0.5 mM $MgCl_2$, 0.1% Nonidet™ P-40) and then stained for 15 minutes with 0.5 µM SYTOX® Green or 5 µM propidium iodide. Flow cytometric analysis of the stained cells was carried out with excitation at 488 nm. SYTOX® Green staining produces a significantly narrower G_1 phase peak, indicated by the smaller coefficient of variation (CV).

FxCycle™ Stains for Cell-Cycle Analysis

FxCycle™ Violet stain (F10347) is a violet laser–excited dye used for cell-cycle analysis in fixed cells. FxCycle™ Violet stain (4',6-diamidino-2-phenylindole, dihydorchloride or DAPI) preferentially stains dsDNA, exhibiting a ~20-fold fluorescence enhancement upon binding. Using FxCycle™ Violet for cell cycle analysis increases the ability to multiplex by freeing up the 488 nm and 633 nm lasers for other cellular analyses, such as immunophenotyping, apoptosis analysis and dead cell discrimination. As compared with several commonly used dyes (propidium iodide and 7-AAD), FxCycle™ Violet stain overlaps less with other fluorescence detection channels, resulting in minimal compensation requirements and more accurate data (Figure 15.4.29).

FxCycle™ Far Red stain (F10348) is useful for flow cytometric analysis of DNA content in fixed cells. Because FxCycle™ Far Red stain binds to both RNA as well as DNA, the addition of RNase A is required for DNA content analysis. This dye takes advantage of the commonly available 633/635 nm excitation sources, with emission around 660 nm (Figure 15.4.30), and is a good choice for DNA content analysis in multicolor cell-cycle studies.

SYTOX® Green Nucleic Acid Stain

SYTOX® Green nucleic acid stain (S7020) is particularly useful for cell-cycle analysis on RNase-treated fixed cells (Figure 15.4.31). In particular, the SYTOX® Green dye produces lower coefficients of variation than propidium iodide (P1304MP, P3566, P21493; Section 8.1), leading to improved resolution of cell phase. Figure 15.4.31 shows a comparison of DNA content histograms obtained with SYTOX® Green nucleic acid stain and propidium iodide after flow cytometric analysis.

Hoechst 33258, Hoechst 33342 and DAPI

The nucleic acid stains most frequently used for cell-cycle analysis—Hoechst 33258 (H1398, H3569, H21491), Hoechst 33342 (H1399, H3570, H21492) and DAPI (D1306, D3571, D21490)—bind to the minor groove of DNA at AT-rich sequences. Hoechst 33342, which more rapidly permeates cells than Hoechst 33258, is commonly used for determining the DNA content of viable cells without detergent treatment or fixation [127,128] (Figure 15.4.32). The Hoechst dyes and DAPI can be excited with a mercury-arc lamp, the UV spectral lines of the argon-ion laser or the 325 nm spectral line of the He-Cd laser. These blue-fluorescent nucleic acid stains preferentially bind to AT-rich sequences and also exhibit higher quantum yields when bound to AT-rich nucleic acids, thus introducing a strong bias into the measurements of nuclear DNA content.[130,131] As a consequence, data obtained with Hoechst 33342 and DAPI correlate very well with each other but less well with data obtained with propidium iodide, a red-fluorescent, cell-impermeant nucleic acid stain [132] (P1304MP, P3566, P21493; Section 8.1). Hoechst 33342 is used in the high-speed sorting of X chromosome– and Y chromosome–bearing sperm based on their DNA content.[133,134]

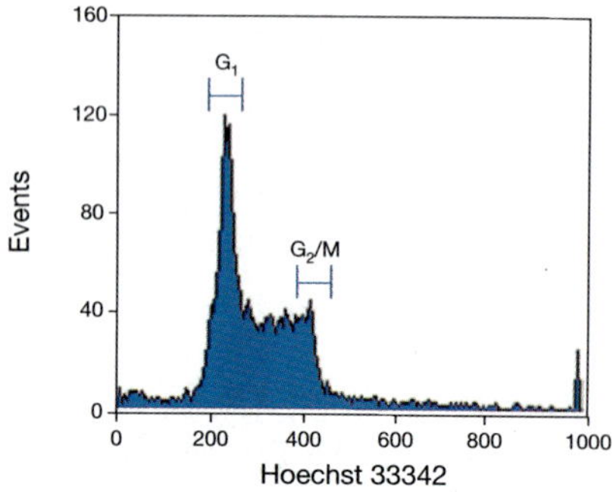

Figure 15.4.32 DNA content histogram for WIL2S cells. The cells were fixed in ethanol, treated with RNase and stained with 2 µg/mL Hoechst 33342 (H1399, H3570, H21492). Flow cytometric analysis of the stained cells was carried out using excitation at 350 nm.

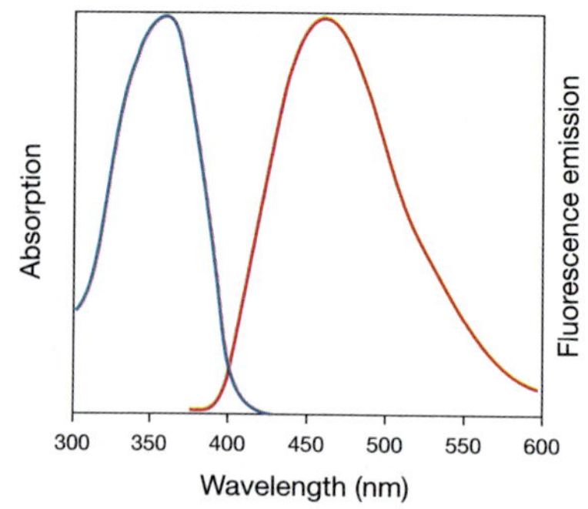

Figure 15.4.29 Absorption and fluorescence and emission spectra of FxCycle™ Violet stain bound to DNA.

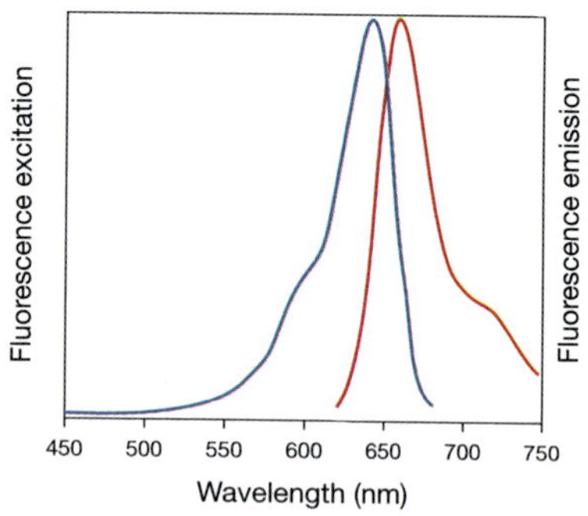

Figure 15.4.30 Fluorescence excitation and emission spectra of FxCycle™ Far Red stain bound to DNA.

IMPORTANT NOTICE: The products described in this manual are covered by one or more Limited Use Label License(s). Please refer to the Appendix on page 971 and Master Product List on page 975. Products are For Research Use Only. Not intended for any animal or human therapeutic or diagnostic use.
www.invitrogen.com/probes

HCS Mitotic Index Kit

Histones are core proteins of DNA in eukaryotic cells. This histones are organized as octamers, around which the DNA is wrapped. The phosphorylation of histone 3 (H3) is involved in condensation of chromatin during mitosis and reaches a peak during mitosis. Mitotic H3 phosphorylation occurs at Ser10 of the amino terminus, and there is a tight correlation between H3 (Ser10) phosphorylation, chromosome condensation and segregation during mitosis. This event can serve as an indication of mitotic progression or inhibition within the context of drug profiling.

The HCS Mitotic Index Kit (H10293) was developed to measure mitotic cells using automated imaging and anlysis and can be combined with other measurements such as DNA profiling, general cytotoxicity or immunocytochemical detection of choice targets. This kit includes sufficient reagents for two 96-well plate assays, including:

- Polyclonal rabbit anti–phospho-H3 antibody
- Alexa Fluor® 488 goat anti–rabbit IgG antibody
- DAPI
- HCS NuclearMask™ Deep Red stain
- Detailed protocols

The primary antibody against phosphorylated histone H3 (Ser10) serves as a sensitive index of mitosis and is detected using the green-fluorescent Alexa Fluor® 488 secondary antibody. The blue-fluorescent DAPI and near-infrared–fluorescent HCS NuclearMask™ Deep Red stain provide two choices for DNA profiling and cell demarcation during image analysis. The HCS Mitotic Index Kit represents a powerful image-based assay for the identification of compounds that affect mitotic progression (Figure 15.4.33).

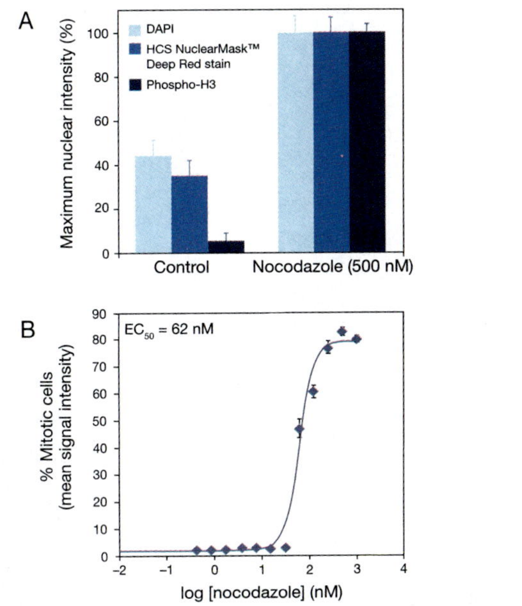

Figure 15.4.33 Imaging of nocodazole-induced mitotic arrest in A549 cells using the HCS Mitotic Index Kit (H10293). **A)** Quantitation of cells treated with 500 nM nocodazole for 24 hr. Nuclear segmentation and DNA content were measured using DAPI or HCS NuclearMask™ Deep Red stain. The strong increase in both phosphohistone 3 and DNA staining is indicative of mitotic cells. **B)** Dose response to 0–1,000 nM nocodazole after 24 hr incubation. The curve generated by nonlinear regression was used to determine the EC_{50} value for nocodazole.

REFERENCES

1. Exp Cell Res (1987) 173:256; **2.** Science (1982) 218:474; **3.** Chem Rev (2009) 109:4207; **4.** Chem Rev (2008) 108:2952; **5.** Angew Chem Int Ed Engl (2009) 48:6974; **6.** Biochemistry (2009) 48:6571; **7.** ACS Chem Biol (2006) 1:644; **8.** J Immunol Methods (2009) 350:29; **9.** Biotechniques (2008) 44:927; **10.** Proc Natl Acad Sci U S A (2008) 105:2415; **11.** J Biol Chem (2010) 285:1544; **12.** Genetics (2009) 183:233; **13.** J Neurosci Methods (2009) 177:122; **14.** Nucleic Acids Res (2009) 37:e31; **15.** Methods (2009) 48:8; **16.** Science (2009) 325:340; **17.** Proc Natl Acad Sci U S A (2009) 106:14896; **18.** Plant Methods (2010) 6:5; **19.** Proc Natl Acad Sci U S A (2009) 106:11806; **20.** Cytometry A (2009) 75:862; **21.** Methods Cell Biol (1994) 41:297; **22.** Proc Natl Acad Sci U S A (1983) 80:5573; **23.** Methods Cell Biol (1994) 41:195; **24.** J Histochem Cytochem (1977) 25:913; **25.** Proc Natl Acad Sci U S A (1973) 70:3395; **26.** Cytometry (2000) 41:89; **27.** Proc Natl Acad Sci U S A (1983) 80:2951; **28.** Methods Cell Biol (1994) 41:327; **29.** Exp Cell Res (1988) 174:319; **30.** Cytometry (1999) 37:221; **31.** Cytometry (1994) 17:310; **32.** Cytometry (2001) 45:13; **33.** Cytometry (1995) 20:172; **34.** Cytometry (1995) 20:257; **35.** Cytometry (1995) 20:245; **36.** J Histochem Cytochem (1993) 41:7; **37.** J Cell Biol (1992) 119:493; **38.** Cancer Res (1994) 54:4289; **39.** Exp Cell Res (1996) 222:28; **40.** Cell Prolif (1995) 28:571; **41.** Cell Biology: A Laboratory Handbook, 2nd Ed., Vol. 1, Celis JE, Ed. (1998) p. 341; **42.** Cytometry (1997) 27:1; **43.** J Biol Chem (2000) 275:16788; **44.** Dev Biol (1999) 206:232; **45.** J Cell Biol (1999) 144:813; **46.** Histochem Cell Biol (2000) 113:181; **47.** Mol Biol Cell (1999) 10:211; **48.** Immunol Cell Biol (1999) 77:509; **49.** J Immunol Methods (1994) 171:131; **50.** J Immunol Methods (1990) 133:87; **51.** Transplant Proc (1992) 24:2820; **52.** Nat Protoc (2007) 2:2049; **53.** J Exp Med (1999) 189:265; **54.** J Immunol (1999) 162:735; **55.** Cell Immunol (1998) 188:1; **56.** Cytometry (1998) 34:143; **57.** Eur J Immunol (1998) 28:1040; **58.** J Immunol (1998) 161:5260; **59.** J Immunol (1998) 160:3666; **60.** Br J Haematol (1997) 98:528; **61.** J Biol Chem (1999) 274:3541; **62.** J Exp Med (1996) 184:277; **63.** Blood (1999) 94:2161; **64.** Lett Appl Microbiol (1997) 25:295; **65.** J Immunol Methods (2001) 254:85; **66.** Anal Biochem (1992) 207:186; **67.** J Immunol Methods (1991) 142:199; **68.** Anal Biochem (1993) 213:426; **69.** In Vitro Toxicol (1990) 3:219; **70.** Exp Cell Res (1979) 124:329; **71.** J Biomol Screen (2008) 13:527; **72.** Anal Biochem (1990) 191:31; **73.** J Immunol Methods (1993) 162:41; **74.** Cancer Res (1989) 49:565; **75.** Anal Biochem (1983) 131:538; **76.** Anal Biochem (1980) 102:344; **77.** J Neurochem (1997) 69:581; **78.** J Immunol Methods (1983) 65:55; **79.** J Immunol Methods (1994) 168:253; **80.** Cancer Res (1987) 47:943; **81.** J Immunol Methods (1986) 93:157; **82.** Biotechniques (1998) 25:622, 626; **83.** J Immunol Methods (1999) 230:99; **84.** Blood (1999) 94:2595; **85.** Leuk Res (1993) 17:873; **86.** Anticancer Res (1998) 18:4243; **87.** J Acquir Immune Defic Syndr (2005) 39:32; **88.** Br J Haematol (2001) 115:953; **89.** Br J Haematol (1999) 106:1059; **90.** MMWR Recomm Rep (2003) 52:1; **91.** J Microbiol Methods (2008) 72:157; **92.** Anal Chem (2007) 79:7510; **93.** J Med Microbiol (2005) 54:843; **94.** J Biol Chem (2007) 282:12112; **95.** J Med Microbiol (2005) 54:77; **96.** Cytometry (1998) 32:66; **97.** J Microbiol Methods (1994) 20:255; **98.** J Microbiol Methods (1994) 19:89; **99.** Microbiol Rev (1994) 58:603; **100.** J Parasitol (1979) 65:421; **101.** Yeast (1998) 14:147; **102.** J Immunol Methods (1998) 211:51; **103.** Appl Environ Microbiol (1990) 56:2245; **104.** J Clin Microbiol (1979) 10:669; **105.** Lett Appl Microbiol (1999) 28:23; **106.** J Immunol Methods (1980) 38:315; **107.** Exp Cell Res (1977) 104:255; **108.** Leukemia (1992) 6:335; **109.** J Immunol Methods (1992) 149:43; **110.** Cytogenet Cell Genet (1983) 36:584; **111.** J Immunol Methods (1994) 168:245; **112.** Nature (1989) 339:487; **113.** Microbiology (2006) 152:913; **114.** Appl Environ Microbiol (1999) 65:45; **115.** Aquat Microbial Ecol (1998) 14:113; **116.** Nature (1999) 399:541; **117.** Aquat Microbial Ecol (1997) 13:225; **118.** Proc Natl Acad Sci U S A (1996) 93:12643; **119.** Cell (2008) 132:487; **120.** Methods Mol Biol (2004) 281:301; **121.** Current Protocols in Cytometry, Robinson JP, Ed. (2004) 7.0.1; **122.** Cytometry A (2004) 58:21; **123.** Lab Chip (2008) 8:68; **124.** Methods Cell Biol (1994) 41:231; **125.** Methods Cell Biol (1994) 41:219; **126.** Methods Cell Biol (1994) 41:211; **127.** Arch Pathol Lab Med (1989) 113:591; **128.** J Histochem Cytochem (1977) 25:585; **129.** Nat Protoc (2008) 3:1187; **130.** Biochemistry (1990) 29:9029; **131.** J Histochem Cytochem (1976) 24:24; **132.** Cytometry (1992) 13:389; **133.** Cytometry (1998) 33:476; **134.** Cytometry (1996) 25:191.

IMPORTANT NOTICE: The products described in this manual are covered by one or more Limited Use Label License(s). Please refer to the Appendix on page 971 and Master Product List on page 975. Products are For Research Use Only. Not intended for any animal or human therapeutic or diagnostic use.

DATA TABLE 15.4 ASSAYS FOR CELL ENUMERATION, CELL PROLIFERATION AND CELL CYCLE

Cat. No.	MW	Storage	Soluble	Abs	EC	Em	Solvent	Notes
B21550	612.99	FF,L	H_2O	<300		none		1
B21551	628.98	FF,L	H_2O	<300		none		1
B23151	307.10	F,L	DMSO	<300		none		
B35000	671.88	F,D,L	DMSO	490	119,000	516	MeOH	
B35001	724.00	F,D,L	DMSO	588	81,000	645	MeOH	
C1157	557.47	F,D	DMF, DMSO	<300		none		2
C7614	~908	FF,L	H_2O	504	68,000	513	pH 8	1, 3, 13
C22803	873.05	F,D	DMSO	<300		none		4
C34555	593.45	F,D,L	DMSO	<300		none		5
D1306	350.25	L	H_2O, DMF	358	24,000	461	H_2O/DNA	6, 7
D3571	457.49	L	H_2O, MeOH	358	24,000	461	H_2O/DNA	
D21490	350.25	L	H_2O, DMF	358	24,000	461	H_2O/DNA	6, 7, 8
H1398	623.96	L	H_2O, DMF	352	40,000	461	H_2O/DNA	6, 9, 10
H1399	615.99	L	H_2O, DMF	350	45,000	461	H_2O/DNA	6, 9, 11
H3569	623.96	RR,L	H_2O	352	40,000	461	H_2O/DNA	1, 6, 9, 10
H3570	615.99	RR,L	H_2O	350	45,000	461	H_2O/DNA	1, 6, 9, 11
H21491	623.96	L	H_2O, DMF	352	40,000	461	H_2O/DNA	6, 8, 9, 10
H21492	615.99	L	H_2O, DMF	350	45,000	461	H_2O/DNA	6, 8, 9, 11
L7595	471.98	L	DMSO, EtOH	543	46,000	712	H_2O/DNA	6
O34550	593.45	F,D,L	DMSO	<300		none		5
S7020	~600	F,D,L	DMSO	504	67,000	523	H_2O/DNA	1, 6, 12, 13
S22801	592.56	F,D	DMSO	<350		none		14
T3600	1302.78	F,D,L	DMSO	514	117,000	533	H_2O/DNA	1, 6, 12, 15
T3605	671.42	F,D,L	DMSO	642	102,000	661	H_2O/DNA	1, 6, 12, 15
V22885	933.88	L	see Notes	549	148,000	565	MeOH	16
V22886	881.72	L	see Notes	484	154,000	501	MeOH	16
V22887	1052.08	L	see Notes	644	193,000	663	MeOH	16
V22888	1051.50	F,L	see Notes	553	134,000	570	MeOH	16
Y3601	1270.65	F,D,L	DMSO	491	99,000	509	H_2O/DNA	1, 6, 12, 15

For definitions of the contents of this data table, see "Using *The Molecular Probes® Handbook*" in the introductory pages.

Notes

1. This product is supplied as a ready-made solution in the solvent indicated under "Soluble."
2. Acetate hydrolysis of this compound yields a fluorescent product with similar pH-dependent spectral characteristics to C1904.
3. The molecular weight (MW) of this product is approximate because the degree of hydration and/or salt form has not been conclusively established.
4. Acetate hydrolysis of C22803 yields a fluorescent product with spectra similar to 5-carboxyeosin, Abs = 519 nm (EC = 100,000 $cm^{-1}M^{-1}$), Em = 542 nm in pH 9 buffer.
5. Acetate hydrolysis of this compound yields a fluorescent product with similar spectral characteristics to O6146.
6. Spectra represent aqueous solutions of nucleic acid-bound dye. EC values are derived by comparing the absorbance of the nucleic acid-bound dye with that of free dye in a reference solvent (H_2O or MeOH).
7. DAPI in H_2O: Abs = 342 nm (EC = 28,000 $cm^{-1}M^{-1}$), Em = 450 nm. The fluorescence quantum yield of DAPI bound to dsDNA is 0.34, representing an ~20-fold increase relative to the free dye in H_2O. (Photochem Photobiol (2001) 73:585)
8. This product is specified to equal or exceed 98% analytical purity by HPLC.
9. MW is for the hydrated form of this product.
10. The fluorescence quantum yield of Hoechst 33258 bound to dsDNA is 0.42, representing an ~30-fold increase relative to the free dye in H_2O. (Photochem Photobiol (2001) 73:585)
11. The fluorescence quantum yield of Hoechst 33342 bound to dsDNA is 0.38, representing an ~10-fold increase relative to the free dye in H_2O. (Photochem Photobiol (2001) 73:585)
12. This product is essentially nonfluorescent except when bound to DNA or RNA.
13. MW: The preceding ~ symbol indicates an approximate value, not including counterions.
14. S22801 is converted to a fluorescent product with spectra similar to C1270 after acetate hydrolysis.
15. Although this compound is soluble in water, preparation of stock solutions in water is not recommended because of possible adsorption onto glass or plastic.
16. This product is supplied as a ready-made staining solution.

PRODUCT LIST 15.4 ASSAYS FOR CELL ENUMERATION, CELL PROLIFERATION AND CELL CYCLE

Cat. No.	Product	Quantity
B35000	*Bac*Light™ Green bacterial stain *special packaging*	20 x 50 µg
B35001	*Bac*Light™ Red bacterial stain *special packaging*	20 x 50 µg
B7277	Bacteria Counting Kit *for flow cytometry*	1 kit
B35130	BrdU, mouse monoclonal antibody (Clone MoBU-1), Alexa Fluor® 488 conjugate *0.2 mg/mL*	350 µL
B35139	BrdU, mouse monoclonal antibody (Clone MoBU-1), Alexa Fluor® 488 conjugate *for flow cytometry* *100 tests*	1 each
B35131	BrdU, mouse monoclonal antibody (Clone MoBU-1), Alexa Fluor® 555 conjugate *0.2 mg/mL*	350 µL
B35132	BrdU, mouse monoclonal antibody (Clone MoBU-1), Alexa Fluor® 594 conjugate *0.2 mg/mL*	350 µL
B35133	BrdU, mouse monoclonal antibody (Clone MoBU-1), Alexa Fluor® 647 conjugate *0.2 mg/mL*	350 µL
B35140	BrdU, mouse monoclonal antibody (Clone MoBU-1), Alexa Fluor® 647 conjugate *for flow cytometry* *100 tests*	1 each
B35138	BrdU, mouse monoclonal antibody (Clone MoBU-1), biotin conjugate *0.2 mg/mL*	350 µL
B35129	BrdU, mouse monoclonal antibody (Clone MoBU-1), Pacific Blue™ conjugate *for flow cytometry* *100 tests*	1 each
B35128	BrdU, mouse monoclonal antibody (Clone MoBU-1), unconjugated *0.1 mg/mL*	350 µL
B35141	BrdU, mouse monoclonal antibody (Clone MoBU-1), unconjugated *for flow cytometry* *100 tests*	1 each
B23151	5-bromo-2′-deoxyuridine (BrdU)	100 mg
B21550	5-bromo-2′-deoxyuridine 5′-triphosphate (BrdUTP) *10 mM in TE buffer*	25 µL
B21551	5-bromouridine 5′-triphosphate (BrUTP) *10 mM in TE buffer*	25 µL
C22803	CEDA, SE (5-(and-6)-carboxyeosin diacetate, succinimidyl ester) *mixed isomers*	5 mg

PRODUCT LIST 15.4 ASSAYS FOR CELL ENUMERATION, CELL PROLIFERATION AND CELL CYCLE—*continued*

Cat. No.	Product	Quantity
C7028	Cell Culture Contamination Detection Kit *200 assays*	1 kit
C34554	CellTrace™ CFSE Cell Proliferation Kit *for flow cytometry*	1 kit
C34555	CellTrace™ Oregon Green® 488 carboxylic acid diacetate, succinimidyl ester (carboxy-DFFDA, SE) *cell permeant* *mixed isomers*	20 x 50 µg
C34557	CellTrace™ Violet Cell Proliferation Kit *for flow cytometry*	1 kit
C1157	5(6)-CFDA, SE; CFSE (5-(and-6)-carboxyfluorescein diacetate, succinimidyl ester) *mixed isomers*	25 mg
C7614	ChromaTide® BODIPY® FL-14-dUTP *1 mM in TE buffer*	25 µL
C35002	Click-iT® EdU Alexa Fluor® 488 Flow Cytometry Assay Kit *50 assays*	1 kit
A10202	Click-iT® EdU Alexa Fluor® 647 Flow Cytometry Assay Kit *50 assays*	1 kit
A10034	Click-iT® EdU Pacific Blue™ Flow Cytometry Assay Kit *50 assays*	1 kit
C10350	Click-iT® EdU Alexa Fluor® 488 HCS Assay *2-plate size*	1 kit
C10351	Click-iT® EdU Alexa Fluor® 488 HCS Assay *10-plate size*	1 kit
C10352	Click-iT® EdU Alexa Fluor® 555 HCS Assay *2-plate size*	1 kit
C10353	Click-iT® EdU Alexa Fluor® 555 HCS Assay *10-plate size*	1 kit
C10354	Click-iT® EdU Alexa Fluor® 594 HCS Assay *2-plate size*	1 kit
C10355	Click-iT® EdU Alexa Fluor® 594 HCS Assay *10-plate size*	1 kit
C10356	Click-iT® EdU Alexa Fluor® 647 HCS Assay *2-plate size*	1 kit
C10357	Click-iT® EdU Alexa Fluor® 647 HCS Assay *10-plate size*	1 kit
C10337	Click-iT® EdU Alexa Fluor® 488 Imaging Kit *for 50 coverslips*	1 kit
C10338	Click-iT® EdU Alexa Fluor® 555 Imaging Kit *for 50 coverslips*	1 kit
C10339	Click-iT® EdU Alexa Fluor® 594 Imaging Kit *for 50 coverslips*	1 kit
C10340	Click-iT® EdU Alexa Fluor® 647 Imaging Kit *for 50 coverslips*	1 kit
C10214	Click-iT® EdU Microplate Assay *400 assays*	1 kit
C36950	CountBright™ absolute counting beads *for flow cytometry* *100 tests*	5 mL
C10227	Countess® automated cell counter *with box of 50 cell counting chambers and Trypan Blue*	1 kit
C10311	Countess® Automated Cell Counter Lab Starter Kit *with 101 boxes of 50 cell counting chamber slides and Trypan Blue*	1 kit
C10310	Countess® Automated Cell Counter Starter Kit *with 11 boxes of 50 cell counting chamber slides and Trypan Blue*	1 kit
C10228	Countess® cell counting chamber slides *for use with Countess® automated cell counter* *box of 50 with Trypan Blue*	1 kit
C10312	Countess® Cell Counting Chamber Slides, 500 Slides (1000 Counts) *for use with Countess® automated cell counter* *10 boxes of 50 slides* *with Trypan Blue*	1 kit
C10313	Countess® Cell Counting Chamber Slides, 1250 Slides (2500 Counts) *for use with Countess® automated cell counter* *25 boxes of 50 slides* *with Trypan Blue*	1 kit
C10314	Countess® Cell Counting Chamber Slides, 2500 Slides (5000 Counts) *for use with Countess® automated cell counter* *50 boxes of 50 slides* *with Trypan Blue*	1 kit
C10315	Countess® Cell Counting Chamber Slides, 5000 Slides (10,000 Counts) *for use with Countess® automated cell counter* *100 boxes of 50 slides* *with Trypan Blue*	1 kit
C10285	Countess® power cord with four adapter cords *for use with Countess® automated cell counter*	1 set
C10284	Countess® test beads *for use with Countess® automated cell counter* *1 x 10^6 beads/mL ±10%*	1 mL
C10286	Countess® USB drive *for use with Countess® automated cell counter* *1 Gbyte*	each
C7027	CyQUANT® cell lysis buffer *20X concentrate*	50 mL
C7026	CyQUANT® Cell Proliferation Assay Kit *for cells in culture* *1000 assays*	1 kit
C35011	CyQUANT® Direct Cell Proliferation Assay *for 10 microplates*	1 kit
C35012	CyQUANT® Direct Cell Proliferation Assay *for 100 microplates*	1 kit
C35006	CyQUANT® NF Cell Proliferation Assay Kit *1000 assays*	1 kit
C35007	CyQUANT® NF Cell Proliferation Assay Kit *200 assays*	1 kit
D1306	4',6-diamidino-2-phenylindole, dihydrochloride (DAPI)	10 mg
D21490	4',6-diamidino-2-phenylindole, dihydrochloride (DAPI) *FluoroPure™ grade*	10 mg
D3571	4',6-diamidino-2-phenylindole, dilactate (DAPI, dilactate)	10 mg
A10044	EdU (5-ethynyl-2'-deoxyuridine)	50 mg
E10187	EdU (5-ethynyl-2'-deoxyuridine)	500 mg
E10415	EdU (5-ethynyl-2'-deoxyuridine)	5 g
F2962	FluoReporter® Blue Fluorometric dsDNA Quantitation Kit *200–2000 assays*	1 kit
F10348	FxCycle™ Far Red stain *for flow cytometry* *500 assays*	1 set
F10347	FxCycle™ Violet stain *for flow cytometry* *500 assays* *DAPI*	1 set
H10293	HCS Mitotic Index Kit *2-plate size*	1 kit
H1398	Hoechst 33258, pentahydrate (bis-benzimide)	100 mg
H3569	Hoechst 33258, pentahydrate (bis-benzimide) *10 mg/mL solution in water*	10 mL
H21491	Hoechst 33258, pentahydrate (bis-benzimide) *FluoroPure™ grade*	100 mg
H1399	Hoechst 33342, trihydrochloride, trihydrate	100 mg
H3570	Hoechst 33342, trihydrochloride, trihydrate *10 mg/mL solution in water*	10 mL
H21492	Hoechst 33342, trihydrochloride, trihydrate *FluoroPure™ grade*	100 mg
L7595	LDS 751	10 mg
M7006	MycoFluor™ Mycoplasma Detection Kit	1 kit
O34550	Oregon Green® 488 carboxylic acid diacetate, succinimidyl ester (carboxy-DFFDA, SE) *mixed isomers*	1 mg

continued on next page

PRODUCT LIST 15.4 ASSAYS FOR CELL ENUMERATION, CELL PROLIFERATION AND CELL CYCLE—*continued*

Cat. No.	Product	Quantity
P36232	Premo™ FUCCI Cell Cycle Sensor	1 kit
S22801	SNARF®-1 carboxylic acid, acetate, succinimidyl ester *special packaging*	10 x 50 µg
S7020	SYTOX® Green nucleic acid stain *5 mM solution in DMSO*	250 µL
T3605	TO-PRO®-3 iodide (642/661) *1 mM solution in DMSO*	1 mL
T3600	TOTO®-1 iodide (514/533) *1 mM solution in DMSO*	200 µL
T10282	Trypan Blue stain 0.4% *for use with Countess® automated cell counter*	2 x 1 mL
V12883	Vybrant® CFDA SE Cell Tracer Kit	1 kit
V22888	Vybrant® CM-DiI cell-labeling solution	1 mL
V22887	Vybrant® DiD cell-labeling solution	1 mL
V22885	Vybrant® DiI cell-labeling solution	1 mL
V22886	Vybrant® DiO cell-labeling solution	1 mL
V35004	Vybrant® DyeCycle™ Green stain *5 mM solution in DMSO* *200 assays*	400 µL
V35005	Vybrant® DyeCycle™ Orange stain *5 mM solution in DMSO* *200 assays*	400 µL
V10309	Vybrant® DyeCycle™ Ruby stain *2.5 mM solution in DMSO* *100 assays*	100 µL
V10273	Vybrant® DyeCycle™ Ruby stain *2.5 mM solution in DMSO* *400 assays*	400 µL
V35003	Vybrant® DyeCycle™ Violet stain *5 mM in water* *200 assays*	200 µL
V13154	Vybrant® MTT Cell Proliferation Assay Kit *1000 assays*	1 kit
Y3601	YOYO®-1 iodide (491/509) *1 mM solution in DMSO*	200 µL

15.5 Assays for Apoptosis and Autophagy

Apoptosis (programmed cell death) is the genetically controlled ablation of cells during normal development. Inappropriately regulated apoptosis is implicated in disease states such as Alzheimer disease, stroke and cancer.[1,2] Apoptosis is distinct from necrosis in both the biochemical and the morphological changes that occur. In contrast to necrotic cells, apoptotic cells are characterized morphologically by compaction of the nuclear chromatin, shrinkage of the cytoplasm and production of membrane-bound apoptotic bodies. Biochemically, apoptosis is distinguished by fragmentation of the genome and cleavage or degradation of several cellular proteins.

As with cell viability, no single parameter fully defines cell death in all systems; therefore, it is often advantageous to use several different approaches when studying apoptosis. Several methods have been developed to distinguish live cells from early and late apoptotic cells and from necrotic cells; these are described below and in a number of review articles.[3–5] Anti-cancer drug candidates failing to induce apoptosis are likely to have decreased clinical efficacy, making apoptosis assays important tools for high-throughput drug screening.[6–9]

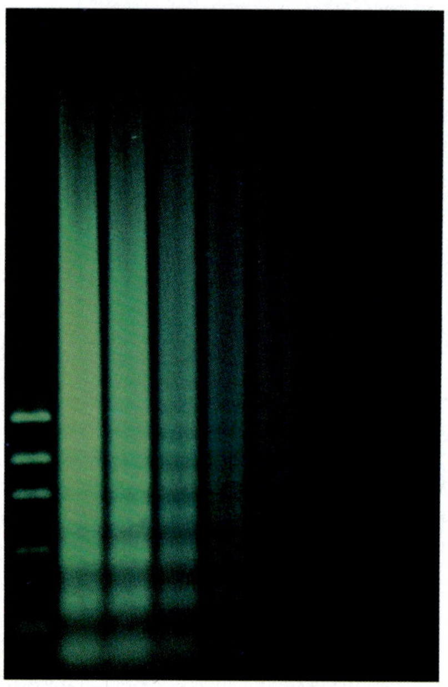

Figure 15.5.1 DNA extracts from camptothecin-treated HL-60 cells separated on an agarose gel and stained with SYBR® Green I nucleic acid gel stain (S7563, S7567, S7585). The 200 to 5000 bp DNA fragments characteristic of apoptotic cells (which appear as "ladders") are clearly visualized with this sensitive nucleic acid stain. Cell preparations were gifts of Zbigniew Darzynkiewicz, Cancer Research Institute, New York Medical College.

Apoptosis Assays Using Nucleic Acid Stains

DNA Stains for Detecting Apoptotic Cells

The characteristic breakdown of the nucleus during apoptosis comprises collapse and fragmentation of the chromatin, degradation of the nuclear envelope and nuclear blebbing, resulting in the formation of micronuclei. Therefore, nucleic acid stains can be useful tools for identifying even low numbers of apoptotic cells in cell populations. Several nucleic acid stains, all of which are listed in Section 8.1, have been used to detect apoptotic cells by fluorescence imaging or flow cytometry.[10–13]

DNA fragmentation can also be detected *in vitro* using electrophoresis. DNA extracted from apoptotic cells and then separated by gel electrophoresis reveals a characteristic ladder pattern of low molecular weight DNA fragments.[14] Our ultrasensitive SYBR® Green I nucleic acid stain (S7567, Section 8.4) allows the detection of even fewer apoptotic cells in these applications (Figure 15.5.1).

Membrane Permeability/Dead Cell Apoptosis Kit

Our Membrane Permeability/Dead Cell Apoptosis Kit (V13243) detects apoptosis based on changes that occur in the permeability of cell membranes. This kit contains ready-to-use solutions of both the YO-PRO®-1 and propidium iodide nucleic acid stains. Our YO-PRO®-1 nucleic acid stain (also available as a stand-alone reagent, Y3603) selectively passes through the plasma membranes of apoptotic cells and labels them with moderate green fluorescence.[15–19]

IMPORTANT NOTICE: The products described in this manual are covered by one or more Limited Use Label License(s). Please refer to the Appendix on page 971 and Master Product List on page 975. Products are For Research Use Only. Not intended for any animal or human therapeutic or diagnostic use.

The dyes included in this kit are effectively excited by the 488 nm spectral line of the argon-ion laser and are useful for both flow cytometry (Figure 15.5.2) and fluorescence microscopy (Figure 15.5.3). The kit components, number of assays and assay principles are summarized in Table 15.4.

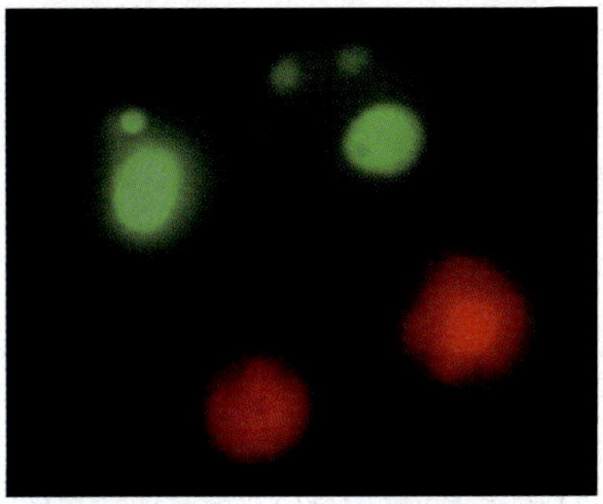

Figure 15.5.3 Apoptosis induced in Jurkat cells with 10 µM camptothecin. The cells were then treated with the reagents in the Membrane Permeability/Dead Cell Apoptosis Kit (V13243). Apoptotic cell nuclei were labeled with YO-PRO®-1 dye (green) (Y3603). Necrotic cells were detected with propidium iodide (red) (P1304MP, P3566, P21493).

Chromatin Condensation/Dead Cell Apoptosis Kit

The Chromatin Condensation/Dead Cell Apoptosis Kit (V13244) provides a rapid and convenient assay for apoptosis based upon fluorescence detection of the compacted state of the chromatin in apoptotic cells. This kit contains ready-to-use solutions of the blue-fluorescent Hoechst 33342 dye (excitation/emission maxima ~350/461 nm when bound to DNA), which stains the condensed chromatin of apoptotic cells more brightly than the chromatin of nonapoptotic cells, and the red-fluorescent propidium iodide (excitation/emission maxima ~535/617 nm when bound to DNA), which is permeant only to dead cells with compromised membranes. The staining pattern resulting from the simultaneous use of these dyes makes it possible to distinguish normal, apoptotic and dead cell populations by flow cytometry or fluorescence microscopy.[20] The 351 nm spectral line of an argon-ion laser or other suitable UV source is required for excitation of the Hoechst 33342 dye, whereas propidium iodide can be excited with the 488 nm spectral line of an argon-ion laser. The kit components, number of assays and assay principles are summarized in Table 15.4.

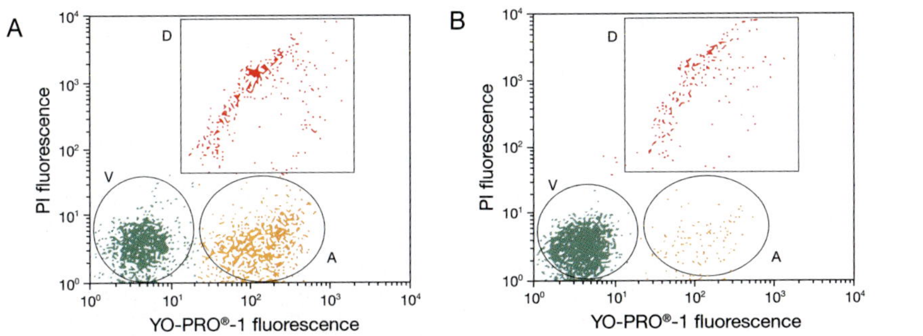

Figure 15.5.2 Flow cytometric analysis of Jurkat cells using the Membrane Permeability/Dead Cell Apoptosis Kit (V13243). Jurkat human T-cell leukemia cells were first exposed to 10 µM camptothecin for 4 hours (**A**) or left untreated (as control, **B**). Cells were then treated with the reagents in the kit and analyzed by flow cytometry. Note that the camptothecin-treated cells (**A**) have a significantly higher percentage of apoptotic cells (indicated by an "A") than the basal level of apoptosis seen in the control cells (**B**). V = viable cells, D = dead cells.

Table 15.4 Molecular Probes® apoptosis assay kits.

Cat. No.	Kit Name	Probe(s) for Apoptotic Cells (Abs/Em) *	Probe for Necrotic or Live Cells (Abs/Em) *	Number of Assays	Kit Features
V13240	Single Channel Annexin V/Dead Cell Apoptosis Kit	Alexa Fluor® 488 annexin V (495/519)	SYTOX® Green nucleic acid stain (504/523)	50 flow cytometry assays, each containing 2×10^5 to 1×10^6 cells in a 1 mL volume	Because this Alexa Fluor® 488 annexin V–based assay uses only the green fluorescence channel on the flow cytometer, other parameters can be measured simultaneously using fluorescent probes that have different emission spectra.
V13241 V13245	Alexa Fluor® 488 Annexin V/Dead Cell Apoptosis Kit	Alexa Fluor® 488 annexin V (495/519)	Propidium iodide (535/617)	50 flow cytometry assays, each containing 2×10^5 to 1×10^6 cells in a 1 mL volume	Apoptotic cells are labeled with annexin V conjugated to our exceptionally bright and photostable green-fluorescent Alexa Fluor® 488 dye. Necrotic cells are labeled with red-fluorescent propidium iodide.
V13242	FITC Annexin V/ Dead Cell Apoptosis Kit	FITC annexin V (494/519)	Propidium iodide (535/617)	50 flow cytometry assays, each containing 2×10^5 to 1×10^6 cells in a 1 mL volume	Similar to Kit V13241 except that it contains the fluorescein conjugate of annexin V.
V13243	Membrane Permeability/Dead Cell Apoptosis Kit	YO-PRO®-1 dye (491/509)	Propidium iodide (535/617)	200 flow cytometry assays, each containing 2×10^5 to 1×10^6 cells in a 1 mL volume	This assay detects changes in cell membrane permeability with YO-PRO®-1 dye, a green-fluorescent nucleic acid stain that is permeant to apoptotic cells but not to live cells. Necrotic cells are labeled with red-fluorescent propidium iodide.

* Approximate absorption and emission maxima, in nm.

continued on next page

IMPORTANT NOTICE: The products described in this manual are covered by one or more Limited Use Label License(s). Please refer to the Appendix on page 971 and Master Product List on page 975. Products are For Research Use Only. Not intended for any animal or human therapeutic or diagnostic use.

Table 15.4 Molecular Probes® apoptosis assay kits—*continued.*

Cat. No.	Kit Name	Probe(s) for Apoptotic Cells (Abs/Em) *	Probe for Necrotic or Live Cells (Abs/Em) *	Number of Assays	Kit Features
V13244	Chromatin Condensation/Dead Cell Apoptosis Kit	Hoechst 33342 (346/460)	Propidium iodide (535/617)	200 flow cytometry assays, each containing 2×10^5 to 1×10^6 cells in a 1 mL volume	This assay uses Hoechst 33342 in combination with propidium iodide to distinguish between the condensed chromatin of apoptotic cells and the looser chromatin structure in live cells.
V23200	Vybrant® Apoptosis Assay Kit #6	Biotin-X annexin V and Alexa Fluor® 350 streptavidin (345/442)	Propidium iodide (535/617)	50 flow cytometry assays, each containing 2×10^5 to 1×10^6 cells in a 1 mL volume	Apoptotic cells are labeled with the biotin-X conjugate of annexin V in conjunction with blue-fluorescent Alexa Fluor® 350 streptavidin. Necrotic cells are labeled with red-fluorescent propidium iodide.
V23201	Chromatin Condensation/ Membrane Permeability/Dead Cell Apoptosis Kit	Hoechst 33342 (346/460) and YO-PRO®-1 dye (491/509)	Propidium iodide (535/617)	200 flow cytometry assays, each containing 2×10^5 to 1×10^6 cells in a 1 mL volume	This kit is a combination of Kits V13243 and V13244. All three dyes can be excited by a UV laser, or by a combination of UV and 488 nm excitation.
V35112	PE Annexin V/Dead Cell Apoptosis Kit	R-phycoerythrin (R-PE) annexin V (496/578)	SYTOX® Green nucleic acid stain (504/523)	50 flow cytometry assays, each containing 2×10^5 to 1×10^6 cells in a 1 mL volume	Apoptotic cells are labeled with annexin V conjugated to the intensely orange-fluorescent R-phycoerythrin (R-PE). Necrotic cells are labeled with the green-fluorescent SYTOX® Green nucleic acid stain. Both probes can be excited by the 488 nm spectral line of the argon-ion laser.
V35113	APC Annexin V/Dead Cell Apoptosis Kit	Allophycocyanin annexin V (650/660)	SYTOX® Green nucleic acid stain (504/523)	50 flow cytometry assays, each containing 2×10^5 to 1×10^6 cells in a 1 mL volume	This kit is similar to Kit V35112 except that it contains the allophycocyanin conjugate of annexin V. Apoptotic cells are labeled with the intensely far-red–fluorescent allophycocyanin annexin V. Necrotic cells are labeled with green-fluorescent SYTOX® Green nucleic acid stain. These populations can easily be distinguished using a flow cytometer equipped with both the 488 nm spectral line of the argon-ion laser and the 633 nm spectral line of the He-Ne laser for excitation.
V35114	Metabolic Activity/ Annexin V/Dead Cell Apoptosis Kit	Allophycocyanin annexin V (650/660)	SYTOX® Green nucleic acid stain (504/523)	50 flow cytometry assays, each containing 2×10^5 to 1×10^6 cells in a 1 mL volume	This kit is identical to Kit V35113 except that it also contains C_{12}-resazurin, which is reduced by viable cells to the orange-fluorescent C_{12}-resorufin (Abs/Em = 571/585 nm). Apoptotic, necrotic and live cell populations can easily be distinguished using a flow cytometer equipped with both the 488 nm spectral line of the argon-ion laser and the 633 nm spectral line of the He-Ne laser for excitation.
V35116	Mitochondrial Membrane Potential/Annexin V Apoptosis Kit	Alexa Fluor® 488 annexin V (495/519)	MitoTracker® Red CMXRos (578/599)	50 flow cytometry assays, each containing 2×10^5 to 1×10^6 cells in a 1 mL volume	Apoptotic cells are labeled with annexin V conjugated to our exceptionally bright and photostable green-fluorescent Alexa Fluor® 488 dye. Live cells are labeled with MitoTracker® Red CMXRos, which exhibits bright red fluorescence in the presence of a mitochondrial transmembrane potential. Apoptotic and live cell populations can easily be distinguished using a flow cytometer and the 488 nm spectral line of the argon-ion laser.
V35123	Violet Membrane Permeability/Dead Cell Apoptosis Kit	PO-PRO™-1 dye (435/455)	7-Aminoactinomycin D (7-AAD) (546/650)	200 flow cytometry assays, each containing 1×10^6 cells in a 1 mL volume	This assay detects changes in cell membrane permeability with PO-PRO™-1 dye, a violet-fluorescent nucleic acid stain that is permeant to apoptotic cells but not to live cells. Necrotic cells are labeled with red-fluorescent 7-AAD. Apoptotic and necrotic cell populations can easily be distinguished using a flow cytometer equipped with a violet laser and an argon-ion laser.
A35135	Violet Chromatin Condensation/Dead Cell Apoptosis Kit	Vybrant® DyeCycle™ Violet dye (370/440)	SYTOX® AADvanced™ dead cell stain (546/647)	200 flow cytometry assays, each containing 1×10^6 cells in a 1 mL volume	This assay uses Vybrant® DyeCycle™ Violet stain to stain the condensed chromatin of apoptotic cells more brightly than the chromatin of normal cells. The SYTOX® AADvanced™ stain labels only necrotic cells, based on membrane integrity. The staining pattern resulting from the simultaneous use of these stains makes it possible to distinguish normal, apoptotic and necrotic cell populations by flow cytometry.
A35136	Violet Annexin V/Dead Cell Apoptosis Kit	Pacific Blue™ annexin V conjugate (415/455)	SYTOX® AADvanced™ dead cell stain (546/647)	50 flow cytometry assays, each containing 1×10^6 cells in a 1 mL volume	The violet-excitable Pacific Blue™ annexin V conjugate stains phosphatidylserine in the exposed cell membrane surface. The SYTOX® AADvanced™ stain labels only necrotic cells, based on membrane integrity. With this assay, apoptotic cells show bright violet fluorescence, dead cells show red fluorescence, and live cells show dim violet fluorescence, allowing users to rapidly and reliably distinguish these populations.
A35137	Violet Ratiometric Membrane Asymmetry Probe/Dead Cell Apoptosis Kit	F2N12S (405/530 for apoptotic and dead cells, 405/585 for live cells)	SYTOX® AADvanced™ dead cell stain (546/647)	100 flow cytometry assays, each containing 1×10^6 cells in a 1 mL volume	This assay detects changes in cell membrane asymmetry using the ratiometric 405 nm–excitable F2N12S dye and SYTOX® AADvanced™ stain for dead cell detection. Apoptotic and necrotic cell populations can easily be distinguished using a flow cytometer equipped with a violet laser and an argon-ion laser.

* Approximate absorption and emission maxima, in nm.

IMPORTANT NOTICE: The products described in this manual are covered by one or more Limited Use Label License(s). Please refer to the Appendix on page 971 and Master Product List on page 975. Products are For Research Use Only. Not intended for any animal or human therapeutic or diagnostic use.
www.invitrogen.com/probes
molecular probes® | invitrogen™ by life technologies™

Chromatin Condensation/Membrane Permeability/Dead Cell Apoptosis Kit

The Chromatin Condensation/Membrane Permeability/Dead Cell Apoptosis Kit (V23201) combines the detection principles used in the two related kits described above. Three nucleic acid stains—Hoechst 33342, YO-PRO®-1 and propidium iodide—are utilized to identify the blue-fluorescent live-cell population, the green-fluorescent apoptotic population and the red-fluorescent dead-cell population by flow cytometry. The stains are provided as separate solutions to facilitate optimization of the assay for the cell line under study and the equipment available. Once optimized, however, the assay can be completed using simultaneous staining with a mixture of the three nucleic acid stains and either UV excitation of all three dyes or with a combination of UV excitation for the Hoechst 33342 dye and excitation by the 488 nm spectral line of the argon-ion laser. The kit components, number of assays and assay principles are summarized in Table 15.4.

Violet Membrane Permeability/Dead Cell Apoptosis Kit

Like the Membrane Permeability/Dead Cell Apoptosis Kit, our Violet Membrane Permeability/Dead Cell Apoptosis Kit (V35123) detects apoptosis based on changes that occur in the permeability of cell membranes (Table 15.4). This kit contains ready-to-use solutions of both PO-PRO™-1 and 7-aminoactinomycin (7-AAD) nucleic acid stains. Our PO-PRO™-1 nucleic acid stain (also available as a stand-alone reagent, P3581) selectively passes through the plasma membranes of apoptotic cells and labels them with violet fluorescence. Furthermore, annexin V labeling of apoptosis yields poor results with trypsinized cells, whereas PO-PRO™-1 dye provides the same efficiency for detecting apoptosis with trypsinized cells as it does with suspension cells. Necrotic cells are stained with the red-fluorescent 7-AAD, a DNA-selective dye that is membrane impermeant but that easily passes through the compromised plasma membranes of necrotic cells. Live cells are not appreciably stained by either PO-PRO™-1 or 7-AAD. The dyes included in this kit are effectively excited by a flow cytometer that uses both the 405 nm spectral line of the violet laser and the 488 nm spectral line of the argon-ion laser for excitation. The kit components, number of assays and assay principles are summarized in Table 15.4.

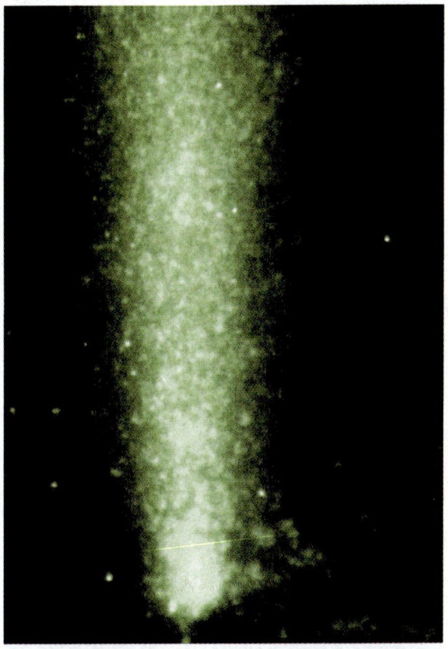

Figure 15.5.4 Comet assay with SYBR® Green I nucleic acid gel stain (S7563, S7567, S7585). DNA fragmentation associated with oxidative DNA damage was visualized using Trevigen's CometAssay® kit. HL-60 cells were treated with H_2O_2 and immobilized onto a Trevigen CometSlide™ for analysis. The cells were gently lysed, washed and treated with endonuclease. Slides were subjected to electrophoresis in alkaline electrophoresis buffer and stained with SYBR® Green I stain.

Comet (Single-Cell Gel Electrophoresis) Assay to Detect Damaged DNA

The Comet assay, or single-cell gel electrophoresis assay, is used for rapid detection and quantitation of DNA damage from single cells.[21,22] The Comet assay is based on the alkaline lysis of labile DNA at sites of damage. Cells are immobilized in a thin agarose matrix on slides and gently lysed. When subjected to electrophoresis, the unwound, relaxed DNA migrates out of the cells. After staining with a nucleic acid stain, cells that have accumulated DNA damage appear as fluorescent comets, with tails of DNA fragmentation or unwinding (Figure 15.5.4). In contrast, cells with normal, undamaged DNA appear as round dots, because their intact DNA does not migrate out of the cell. The ease and sensitivity of the Comet assay has provided a fast and convenient way to measure damage to human sperm DNA,[23] evaluate DNA replicative integrity,[24] monitor the sensitivity of tumor cells to radiation damage[25] and assess the sensitivity of molluscan cells to toxins in the environment.[26] The Comet assay can also be used in combination with FISH to identify specific sequences with damaged DNA.[25] Comet assays have traditionally been performed using ethidium bromide to stain the DNA;[21] however, use of the SYBR® Gold and SYBR® Green I stains[27,28] improves the sensitivity of this assay (Figure 15.5.4).

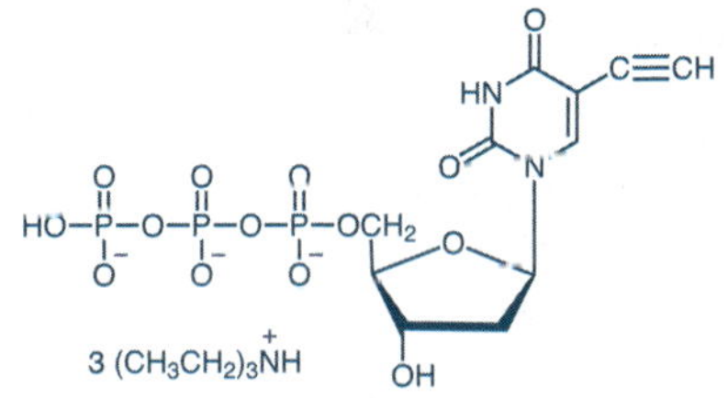

Figure 15.5.5 The EdUTP nucleotide, provided in the Click-iT® TUNEL Imaging Assay Kits.

Apoptosis Assays that Detect DNA Strand Breaks

Click-iT® TUNEL Alexa Fluor® Imaging Assays

The terminal deoxynucleotidyl transferase-dUTP nick end labeling (TUNEL) assay—based on the incorporation of modified dUTPs by terminal deoxynucleotidyl transferase (TdT) at the 3′-OH ends of fragmented DNA—is probably the most widely used *in situ* test for studying apoptotic DNA fragmentation.[29,30] For a sensitive and reliable TUNEL imaging assay, it is vital that the modified nucleotide is an efficient substrate for TdT. We have developed a Click-iT® TUNEL imaging assay that incorporates an alkyne-modified dUTP (Figure 15.5.5) at the 3′-OH ends of fragmented DNA using TdT and then detects the enzymatically

IMPORTANT NOTICE: The products described in this manual are covered by one or more Limited Use Label License(s). Please refer to the Appendix on page 971 and Master Product List on page 975. Products are For Research Use Only. Not intended for any animal or human therapeutic or diagnostic use.

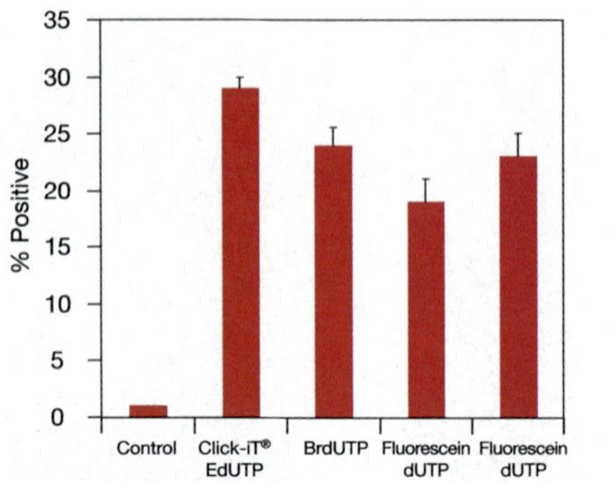

Figure 15.5.7 TUNEL assay comparison—percentage positives detected. HeLa cells were treated with 0.5 µM staurosporine for 4 hours. Following fixation and permeabilization, TUNEL imaging assays were performed according to the manufacturer's instructions. The percent positives were calculated based upon the corresponding negative control. Imaging and analysis was performed using a Thermo Fisher Scientific Cellomics® ArrayScan® II.

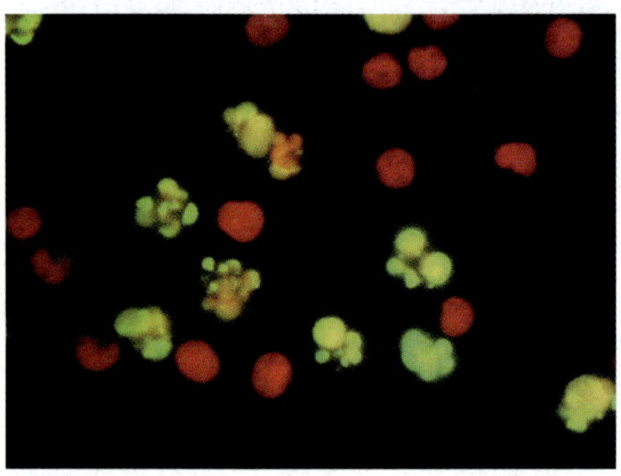

Figure 15.5.8 Human lymphoma cells treated with camptothecin for 4 hours and stained using the APO-BrdU™ TUNEL Assay Kit (A23210). Cells containing DNA strand nicks characteristic of apoptosis are detected by TUNEL and fluoresce green, while necrotic cells are stained with red-fluorescent propidium iodide.

incorporated nucleotide using a copper (I)–catalyzed click reaction with an azide-derivatized fluorophore (Figure 15.5.6).

The Click-iT® labeling reaction is based on a copper-catalyzed azide–alkyne cyloaddition[31,32] and derives its high degree of specificity from the fact that the azide and alkyne reaction partners have no endogenous representation in biological molecules, cells, tissues or model organisms.[33–35] The minimally modified EdUTP nucleotide (Figure 15.5.5) used in the Click-iT® TUNEL imaging assay is rapidly incorporated by TdT, allowing samples to be rapidly fixed in order to preserve late-stage apoptotic cells, thereby lessening the possibility of false-negative results due to cell detachment and subsequent loss. Compared with assays that use one-step incorporation of dye-modified nucleotides, the fast and reliable Click-iT® TUNEL imaging assay can detect a higher percentage of apoptotic cells under identical conditions in two hours or less (Figure 15.5.7). The Click-iT® TUNEL Imaging Assay Kits are available with a choice of azide-derivatized Alexa Fluor® dyes, providing flexibility for combination with other apoptosis detection reagents. They include:

- Click-iT® TUNEL Alexa Fluor® 488 Imaging Assay (C10245)
- Click-iT® TUNEL Alexa Fluor® 594 Imaging Assay (C10246)
- Click-iT® TUNEL Alexa Fluor® 647 Imaging Assay (C10247)

The Click-iT® TUNEL assays have been tested in HeLa, A549 and CHO K1 cells with a variety of reagents that induce apoptosis, including staurosporine, and multiplexed with antibody-based detection of other apoptosis biomarkers such as cleaved poly(ADP-ribose) polymerase (PARP), cleaved caspase-3 and phosphohistone 2B. It has also proven effective for detection of apoptosis induced by siRNA knockdown of the DEC2 transcription factor in human MCF-7 breast cancer cells.[36] Click-iT® labeling technology and the details of the click reaction are discussed in Section 3.1. For a complete list azide and alkyne derivatives compatible with Click-iT® labeling technology, see Table 3.1.

APO-BrdU™ TUNEL Assay Kit

Because DNA fragmentation is one of the most reliable methods for detecting apoptosis, we have collaborated with Phoenix Flow Systems to offer the APO-BrdU™ TUNEL Assay Kit (A23210), which provides all the materials necessary to label and detect the DNA strand breaks of apoptotic cells.[30] When DNA strands are cleaved or nicked by nucleases, a large number of 3′-hydroxyl ends are exposed. In the APO-BrdU™ assay, these ends are labeled with BrdUTP and terminal deoxynucleotidyl transferase (TdT) using the TUNEL technique described above. Once incorporated into the DNA, BrdU is detected using an Alexa Fluor® 488 dye–labeled anti-BrdU monoclonal antibody (Figure 15.5.8). This kit also provides propidium iodide for determining total cellular DNA content, as well as fixed control cells for assessing assay performance.

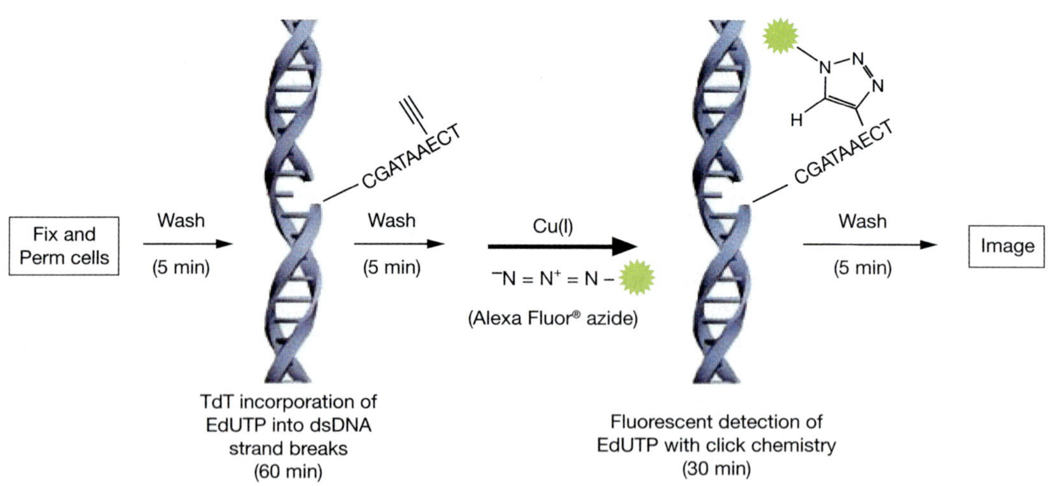

Figure 15.5.6 Detection of apoptosis with the Click-iT® TUNEL imaging assay.

IMPORTANT NOTICE: The products described in this manual are covered by one or more Limited Use Label License(s). Please refer to the Appendix on page 971 and Master Product List on page 975. Products are For Research Use Only. Not intended for any animal or human therapeutic or diagnostic use.

molecular probes® | invitrogen™ by life technologies™

The APO-BrdU™ TUNEL Assay Kit includes complete protocols for use in flow cytometry applications, though it may also be adapted for use with fluorescence microscopy. Each kit contains:

- Terminal deoxynucleotidyl transferase (TdT), for catalyzing the addition of BrdUTP at the break sites
- 5-Bromo-2′-deoxyuridine 5′-triphosphate (BrdUTP)
- Alexa Fluor® 488 dye–labeled anti-BrdU mouse monoclonal antibody PRB-1, for detecting BrdU labels
- Propidium iodide/RNase staining buffer, for quantitating total cellular DNA
- Reaction, wash and rinse buffers
- Positive control cells (a fixed human lymphoma cell line)
- Negative control cells (a fixed human lymphoma cell line)
- Detailed protocols

Sufficient reagents are provided for approximately 60 assays of 1 mL samples, each containing approximately 1 x 10^6 cells/mL.

Detecting DNA Strand Breaks with ChromaTide® Nucleotides

Break sites have traditionally been labeled with biotinylated dUTP, followed by subsequent detection with an avidin or streptavidin conjugate[37–40] (Section 7.6, Table 7.9). However, a more direct approach for detecting DNA strand breaks in apoptotic cells is possible via the use of our ChromaTide® BODIPY® FL-14-dUTP (C7614) as a TdT substrate[41,42] (Figure 15.5.9). The single-step BODIPY® FL dye–based assay has several advantages over indirect detection of biotinylated or haptenylated nucleotides, including fewer protocol steps and increased cell yields. BODIPY® FL dye–labeled nucleotides have also proven superior to fluorescein-labeled nucleotides for detection of DNA strand breaks in apoptotic cells because they provide stronger signals, a narrower emission spectrum and less photobleaching[41] (Figure 15.5.9).

In situ DNA modifications by labeled nucleotides have been used to detect DNA fragmentation in what may be apoptotic cells in autopsy brains of Huntingtons and Alzheimer disease patients.[43–46] DNA fragmentation is also associated with amyotrophic lateral sclerosis.[47] Analogous to TdT's ability to label double-strand breaks, the *E. coli* repair enzyme DNA polymerase I can be used to detect single-strand nicks,[48,49] which appear as a relatively early step in some apoptotic processes.[50–52] Because our ChromaTide® BODIPY® FL-14-dUTP (C7614) and ChromaTide® fluorescein-12-dUTP[53,54] (C7604) are incorporated into DNA by *E. coli* DNA polymerase I, they are also effective for *in situ* labeling with the nick translation method.[55]

High-Content Analysis of Genotoxicity and Cytotoxicity

In mammalian cells, a double-strand break (DSB) in genomic DNA is a potentially lethal lesion. One of the earliest known responses to DSB formation is phosphorylation of H2A histones. Specifically, DNA damaging agents induce phosphorylation of histone variant H2AX at Ser139, leading to the formation of DNA foci at the site of DNA DSBs.[56–58]

The HCS DNA Damage Kit (H10292) was developed to enable simultaneous quantitation of two cell health parameters, genotoxicity and cytotoxicity, by high-content analysis in the same cell (Figure 15.5.10). DNA damage is measured as an indication of genotoxicity and accomplished by specific antibody-based detection of phosphorylated H2AX (Ser139) in the nucleus. Cytotoxicity is measured with the Image-iT® DEAD Green™ viability stain (also available as a stand-alone reagent, I10291),

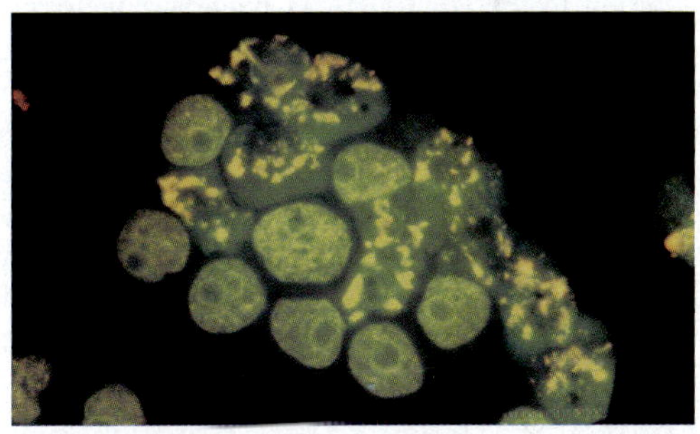

Figure 15.5.9 HL-60 cells treated with camptothecin for 3 hours. The DNA strand nicks characteristic of apoptosis were detected with the TUNEL (terminal deoxynucleotidyl transferase–mediated dUTP nick end-labeling) assay using the fluorescently labeled nucleotide ChromaTide® BODIPY® FL-14-dUTP (C7614). Image contributed by Zbigniew Darzynkiewicz, Cancer Research Institute, New York Medical College.

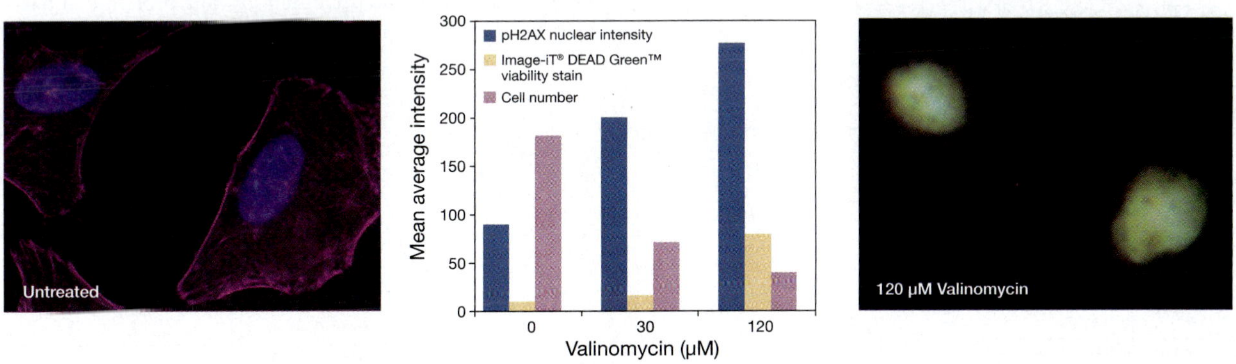

Figure 15.5.10 Detection of genotoxicity and cytotoxicity in valinomycin-treated A549 cells using the HCS DNA Damage Kit (H10292). A549 cells were treated with 30 µM or 120 µM valinomycin for 24 hr before performing the assay. With increasing concentrations of valinomycin, cells showed genotoxic effects as indicated by detection with a pH2AX antibody in conjunction with Alexa Fluor® 555 goat anti–mouse IgG antibody (orange fluorescence), and cytotoxic effects as indicated by staining with the Image-iT® DEAD Green™ viability stain (green fluorescence). Blue-fluorescent Hoechst 33342 was used as a nuclear segmentation tool, and Alexa Fluor® 647 phalloidin was used to visualize F-actin (pseudocolored magenta). The image on the left shows untreated cells with intact F-actin cytoskeletons and no evidence of cytotoxicity or genotoxicity. The image on the right shows cells treated with 120 µM valinomycin, which completely disrupted the actin cytoskeletons, increased levels of DNA damage and compromised plasma membrane integrity.

IMPORTANT NOTICE: The products described in this manual are covered by one or more Limited Use Label License(s). Please refer to the Appendix on page 971 and Master Product List on page 975. Products are For Research Use Only. Not intended for any animal or human therapeutic or diagnostic use.

a cell-impermeant, nonfluorescent, high-affinity DNA stain that forms highly fluorescent and stable dye–nucleic acid complexes when bound to DNA. Thus, staining of nuclear DNA by the Image-iT® DEAD Green™ viability stain cannot occur in live cells due to the impermeability of the plasma membrane to the stain. Drugs and test compounds that lead to serious cell injuries, including plasma membrane permeability, allow entry of the Image-iT® DEAD Green™ viability stain, enabling discrimination of dead cells. Hoechst 33342, which stains nuclear DNA in live and dead cells, is included in the kit as a segmentation tool for automated image analysis.

The HCS DNA Damage Kit contains sufficient material to perform the DNA damage assay on two 96-well plates when used as described in the protocol provided:

- Image-iT® Dead Green viability stain
- pH2AX mouse monoclonal antibody
- Alexa Fluor® 555 goat anti-mouse IgG antibody
- Hoechst 33342 nucleic acid stain
- Detailed protocols

Because no single parameter fully defines cell viability in all systems, it is often useful to use multiple approaches to study cytotoxicity. In addition to the HCS DNA Damage Kit, we offer the HCS LIVE/DEAD® Green Kit (H10290) and the HCS Mitochondrial Health Kit (H10295), both of which employ the Image-iT® DEAD™ Green viability stain. Additionally, the HCS LIVE/DEAD® Kit provides a choice of two nucleic acid stains—the far-red–fluorescent HCS NuclearMask™ Deep Red stain (H10294, Section 12.5) and the blue-fluorescent Hoechst 33342 (H1399, H3570, H21492; Section 12.5)—for use as cell-permeant nuclear segmentation tools. The HCS Mitochondrial Health Kit provides Image-iT® DEAD Green viability stain and Hoechst 33342 stain, as well as the MitoHealth stain, which accumulates in the mitochondria of live cells in proportion to the mitochondrial membrane potential. Both of these kits provide sufficient reagents to perform the assays on two 96-well plates, using the protocol provided.

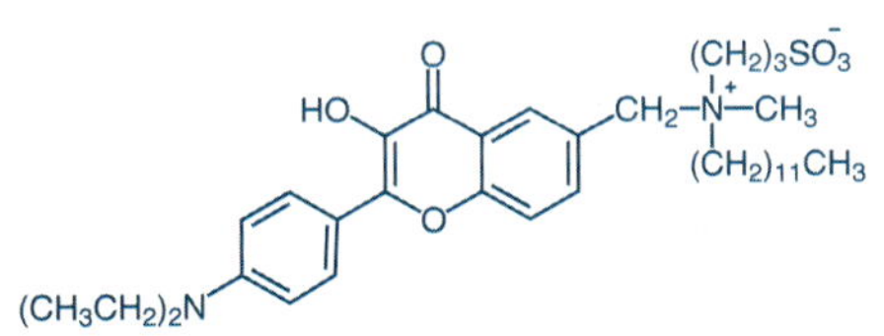

Figure 15.5.11 F2N12S, a component of the Violet Ratiometric Membrane Asymmetry Probe/Dead Cell Apoptosis Kit (A35137).

Apoptosis Assays that Detect Membrane Asymmetry

Violet Ratiometric Membrane Asymmetry Probe/Dead Cell Apoptosis Kit

The Violet Ratiometric Membrane Asymmetry Probe/Dead Cell Apoptosis Kit (A35137) provides a simple and fast method for detecting apoptosis with dead-cell discrimination by flow cytometry. The violet ratiometric membrane asymmetry probe, F2N12S (4′-*N,N*-diethylamino-6-(*N*-dodecyl-*N*-methyl-*N*-(3-sulfopropyl))ammoniomethyl-3-hydroxyflavone, Figure 15.5.11), is a novel violet diode–excitable dye for the detection of membrane phospholipid asymmetry changes during apoptosis. This dye exhibits an excited-state intramolecular proton transfer (ESIPT) reaction, resulting in a dual fluorescence with two emission bands corresponding to 530 nm and 585 nm and producing a two-color ratiometric response to variations in surface charge.[59] This ratiometric probe is therefore a self-calibrating indicator of apoptotic transformation, which is independent of probe concentration, cell size and instrument variation, such as fluctuations of laser intensity or sensitivity of the detectors.

Given that apoptosis modifies the surface charge of the outer leaflet of the plasma membrane, F2N12S can be used to monitor changes in membrane asymmetry that occur during apoptosis through a change in the relative intensity of the two emission bands of the dye[59] (Figure 15.5.12). The F2N12S-based apoptosis assay allows samples to be analyzed after a 5-minute incubation at room temperature and does not require special buffers or wash steps. This kit can be paired with other reagents such as MitoProbe™ DiIC$_1$(5) or annexin V for multiparametric analysis of apoptosis and viability. The kit components, number of assays and assay principles are summarized in Table 15.4.

Annexin V Conjugates

The human vascular anticoagulant annexin V is a 35–36 kilodalton, Ca^{2+}-dependent phospholipid-binding protein that has a high affinity for the anionic phospholipid phosphatidylserine (PS). In normal

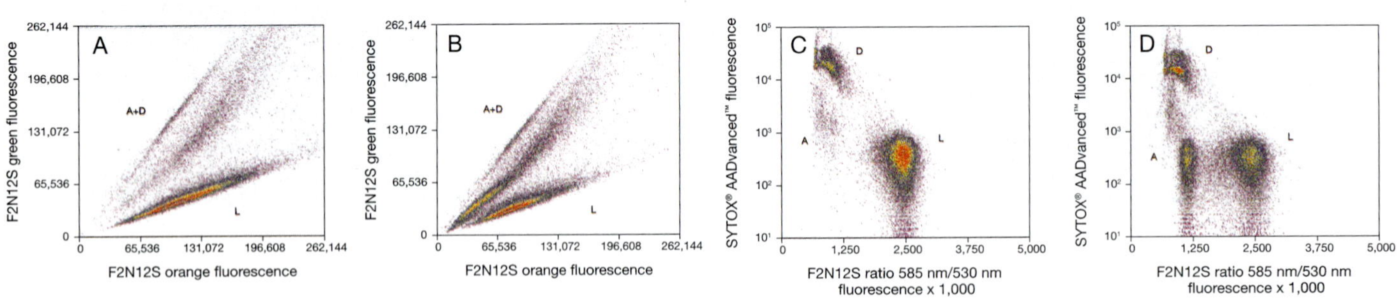

Figure 15.5.12 Jurkat cells (T-cell leukemia, human) treated with 10 µM camptothecin for 4 hours (**B** and **D**) or untreated (**A** and **C**), control. Cells were stained according to the protocol. Samples were analyzed on a flow cytometer with 405 nm excitation using 585 nm and 530 nm bandpass filters for F2N12S and 488 nm excitation for SYTOX® AADvanced™ dead cell stain using a 695 nm bandpass filter. In panels A and B, live cells can be discriminated from apoptotic and dead cells by the relative intensities of the two emission bands from F2N12S. In C and D, SYTOX® AADvanced™ dead cell stain fluorescence is plotted against a derived ratio parameter from the two emission bands (585/530 nm) of F2N12S. A = apoptotic cells, L = live cells, D = dead cells.

IMPORTANT NOTICE: The products described in this manual are covered by one or more Limited Use Label License(s). Please refer to the Appendix on page 971 and Master Product List on page 975. Products are For Research Use Only. Not intended for any animal or human therapeutic or diagnostic use.

viable cells, PS is located on the cytoplasmic surface of the cell membrane. In apoptotic cells, however, PS is translocated from the inner to the outer leaflet of the plasma membrane, exposing PS to the external cellular environment where it can be detected by annexin V conjugates.[60] In leukocyte apoptosis, PS on the outer surface of the cell marks the cell for recognition and phagocytosis by macrophages.[61,62]

Highly fluorescent annexin V conjugates provide quick and reliable detection methods for studying the externalization of phosphatidylserine,[63–65] an indicator of intermediate stages of apoptosis. Nuclear fragmentation, mitochondrial membrane potential flux and caspase-3 activation apparently precede phosphatidylserine "flipping" during apoptosis, whereas permeability to propidium iodide and cytoskeletal collapse occur later. The difference in fluorescence intensity between apoptotic and nonapoptotic cells stained by our fluorescent annexin V conjugates, as measured by flow cytometry, is typically about 100-fold (Figure 15.5.13). Our annexin V conjugates are available as stand-alone reagents, each suitable for 50–100 flow cytometry assays or many more imaging assays, or in several variations of our apoptosis assay kits (Table 15.4). Our annexin V conjugates include:

- Alexa Fluor® 488 annexin V[66] (A13201, Figure 15.5.14)
- Fluorescein (FITC) annexin V (A13199)
- Oregon Green® 488 annexin V (A13200)
- Alexa Fluor® 555 annexin V (A35108)
- R-phycoerythrin (R-PE) annexin V (A35111)
- Alexa Fluor® 568 annexin V (A13202)
- Alexa Fluor® 594 annexin V[67,68] (A13203)
- Alexa Fluor® 647 annexin V[64] (A23204)
- Allophycocyanin (APC) annexin V (A35110)
- Alexa Fluor® 680 annexin V[69] (A35109)
- Alexa Fluor® 350 annexin V (A23202)
- Pacific Blue™ annexin V (A35122)
- Biotin-X annexin V[65] (A13204)

Single Channel Annexin V/Dead Cell Apoptosis Kit

With the Single Channel Annexin V/Dead Cell Apoptosis Kit (V13240), apoptotic cells are detected based on the externalization of phosphatidylserine. This kit contains recombinant annexin V conjugated to the Alexa Fluor® 488 dye, our brightest and most photostable green fluorophore, to permit maximum sensitivity. In addition, the kit includes a ready-to-use solution of the SYTOX® Green nucleic acid stain. The SYTOX® Green dye is impermeant to live cells and apoptotic cells but stains necrotic cells with intense green fluorescence by binding to cellular nucleic acids. After staining a cell population with Alexa Fluor® 488 annexin V and SYTOX® Green dye in the provided binding buffer, apoptotic cells show green fluorescence, dead cells show a higher level of green fluorescence and live cells show little or no fluorescence (Figure 15.5.15). These populations can easily be distinguished using a flow cytometer with the 488 nm spectral line of

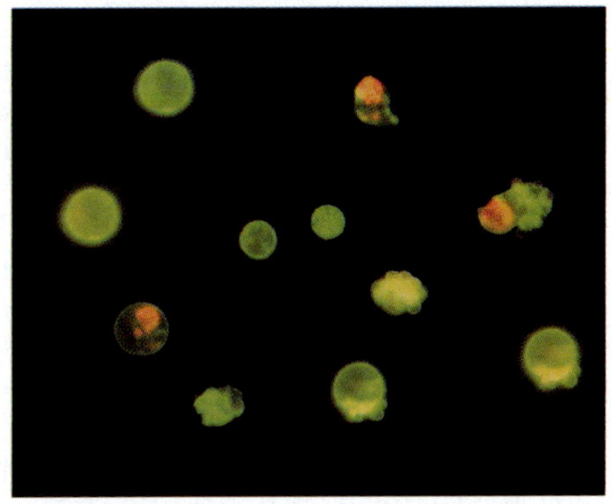

Figure 15.5.14 Jurkat human T-cell leukemia cells treated with 1 µM camptothecin. The externalized phosphatidylserine, a characteristic of early-stage apoptotic cells, was detected with Alexa Fluor® 488 annexin V (A13201). The late-stage apoptotic and necrotic cells were stained with propidium iodide (P1304MP, P3566, P21493). The image was acquired using bandpass filters appropriate for fluorescein.

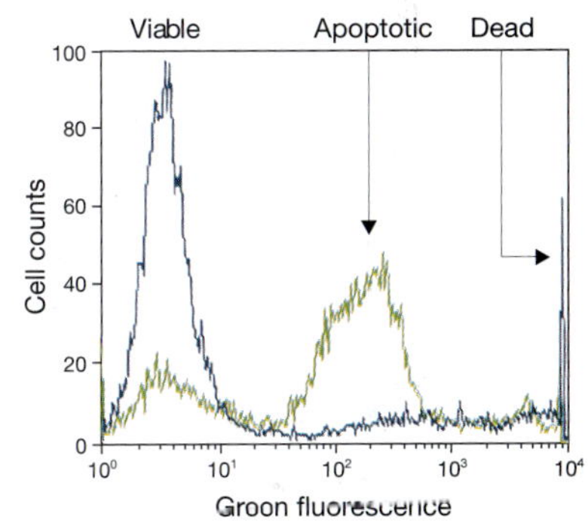

Figure 15.5.15 Flow cytometric analysis of Jurkat cells using the Single Channel Annexin V/Dead Cell Apoptosis Kit (V13240). Jurkat human T-cell leukemia cells were first exposed to 10 µM camptothecin for four hours green line) or left untreated (as control, blue line). Cells were then treated with the reagents in the kit and analyzed by flow cytometry. Note that the camptothecin-treated cells (green line) have a significantly higher percentage of apoptotic cells (intermediate green fluorescence) than the basal level of apoptosis seen in the control cells (blue line).

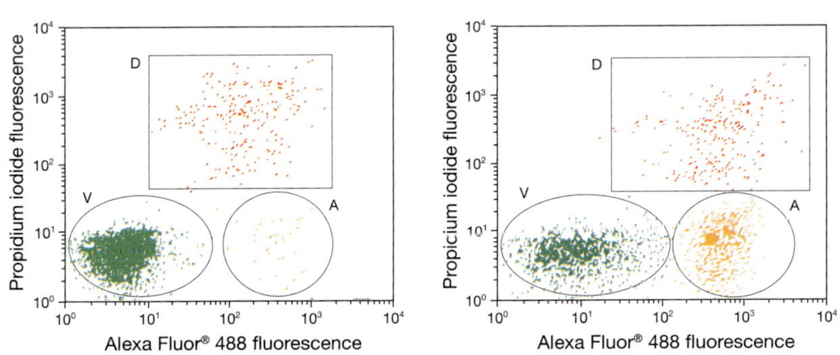

Figure 15.5.13 Flow cytometric analysis of Jurkat cells using the Alexa Fluor® 488 Annexin V/Dead Cell Apoptosis Kit (V13241). Jurkat human T-cell leukemia cells were first exposed to 10 µM camptothecin for 4 hours (right) or left untreated (as control, left). Cells were then treated with the reagents in the kit and analyzed by flow cytometry. Note that the camptothecin-treated cells have a significantly higher percentage of apoptotic cells (indicated by an "A") than the basal level of apoptosis seen in the control cells. V = viable cells, D = dead cells.

IMPORTANT NOTICE: The products described in this manual are covered by one or more Limited Use Label License(s). Please refer to the Appendix on page 971 and Master Product List on page 975. Products are For Research Use Only. Not intended for any animal or human therapeutic or diagnostic use.

an argon-ion laser for excitation. Both Alexa Fluor® 488 annexin and the SYTOX® Green dye emit a green fluorescence that can be detected in the green channel, freeing the other channels for the detection of additional probes in multicolor labeling experiments. The kit components, number of assays and assay principles are summarized in Table 15.4.

Alexa Fluor® 488 Annexin V/Dead Cell Apoptosis Kit

Like the Single Channel Annexin V/Dead Cell Apoptosis Kit, our Alexa Fluor® 488 Annexin V/Dead Cell Apoptosis Kit (V13241) detects the externalization of phosphatidylserine in apoptotic cells.[70] This kit provides a sensitive two-color assay that employs our green-fluorescent Alexa Fluor® 488 annexin and a ready-to-use solution of the red-fluorescent propidium iodide nucleic acid stain. Propidium iodide is impermeant to live cells and apoptotic cells but stains necrotic cells with red fluorescence, binding tightly to the nucleic acids in the cell. After staining a cell population with Alexa Fluor® 488 annexin V and propidium iodide in the provided binding buffer, apoptotic cells show green fluorescence, dead cells show red and green fluorescence, and live cells show little or no fluorescence (Figure 15.5.13). These populations can easily be distinguished using a flow cytometer with the 488 nm spectral line of an argon-ion laser for excitation. The kit components, number of assays and assay principles are summarized in Table 15.4.

FITC Annexin V/Dead Cell Apoptosis Kit

The FITC Annexin V/Dead Cell Apoptosis Kit (V13242) is similar to the Alexa Fluor® 488 Annexin V/Dead Cell Apoptosis Kit, except that it contains fluorescein (FITC) annexin V in place of the Alexa Fluor® 488 conjugate. The kit components, number of assays and assay principles are summarized in Table 15.4.

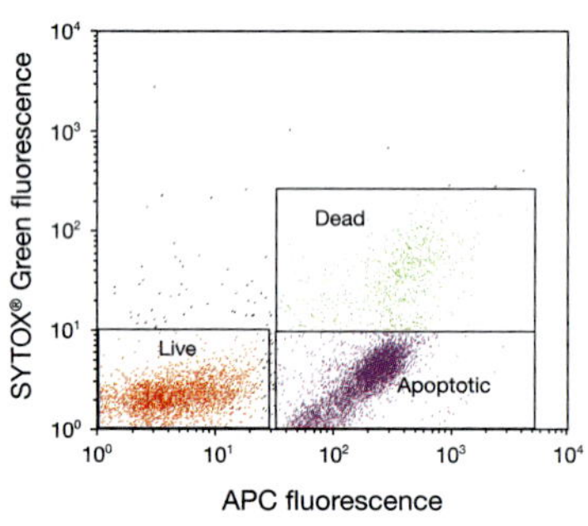

Figure 15.5.16 Flow cytometric analysis of Jurkat cells using the APC Annexin V/Dead Cell Apoptosis Kit (V35113). Jurkat human T-cell leukemia cells were first exposed to 10 µM camptothecin at 37°C, 5% CO_2. The cells were then treated with the reagents in the kit and analyzed by flow cytometry. The SYTOX® Green fluorescence versus allophycocyanin (APC) annexin fluorescence dot plot shows resolution of live, apoptotic and dead cell populations.

Vybrant Apoptosis Assay Kit

The Vybrant Apoptosis Assay Kit #6 (V23200) is similar to the Alexa Fluor® 488 Annexin V/Dead Cell Apoptosis Kit, except that it contains biotin-X annexin V and Alexa Fluor® 350 streptavidin in place of the Alexa Fluor® 488 conjugate. After staining a cell population with biotin-X annexin V in the provided binding buffer, Alexa Fluor® 350 streptavidin is added to fluorescently label the bound annexin V. Finally, propidium iodide is added to detect necrotic cells. Apoptotic cells show blue fluorescence, dead cells show red and blue fluorescence and live cells show little or no fluorescence. These populations can easily be distinguished using a flow cytometer with UV excitation for the Alexa Fluor® 350 fluorophore and 488 nm excitation for the propidium iodide. With this kit, fluorescence in the green channel is minimal. In the same experiment for apoptosis detection, the researcher can apply a green-fluorescent probe, for example an antibody labeled with the Alexa Fluor® 488 dye or with fluorescein. The kit components, number of assays and assay principles are summarized in Table 15.4.

PE Annexin V/ Dead Cell Apoptosis Kit and APC Annexin V/Dead Cell Apoptosis Kit

The PE Annexin V/ Dead Cell Apoptosis Kit and APC Annexin V/Dead Cell Apoptosis Kit (V35112, V35113) are similar to the Single Channel Annexin V/Dead Cell Apoptosis Kit, except that they contain either R-phycoerythrin (R-PE) annexin V or allophycocyanin (APC) annexin V instead of Alexa Fluor® 488 annexin V. In addition to the phycobiliprotein-conjugated annexin V, these kits include the SYTOX® Green nucleic acid stain, which is impermeant to live cells and apoptotic cells but stains necrotic cells with intense green fluorescence. After staining a cell population with R-PE annexin V and SYTOX® Green stain, apoptotic cells show orange fluorescence with very little green fluorescence, late apoptotic cells show a higher level of green and orange fluorescence and live cells show little or no fluorescence; these populations can easily be distinguished using a flow cytometer with the 488 nm spectral line of an argon-ion laser for excitation. After staining a cell population with APC annexin V and the SYTOX® Green stain, apoptotic cells show far-red fluorescence with very little green fluorescence, late apoptotic cells show a higher level of green and far-red fluorescence and live cells show little or no fluorescence (Figure 15.5.16); these populations can easily be distinguished using a flow cytometer

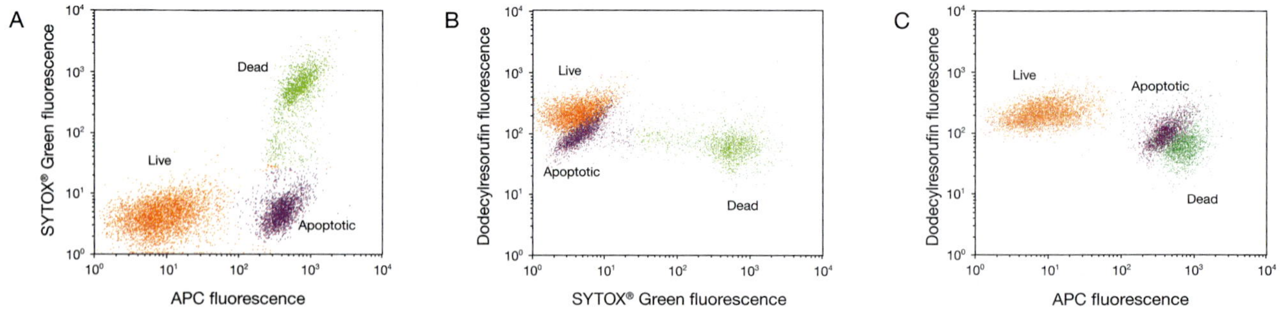

Figure 15.5.17 Flow cytometric analysis of Jurkat cells using the Metabolic Activity/Annexin V/Dead Cell Apoptosis Kit (V35114). Jurkat human T-cell leukemia cells were first exposed to either 10 µM camptothecin or 2 mM hydrogen peroxide for 4 hours at 37°C, 5% CO_2. The cells were then combined, treated with the reagents in the kit and analyzed by flow cytometry. **A)** The SYTOX® Green fluorescence versus allophycocyanin (APC) annexin fluorescence dot plot shows resolution of live, apoptotic and dead cell populations. The cell populations can be evaluated for metabolic activity using **B)** the dodecylresorufin fluorescence versus SYTOX® Green fluorescence dot plot and **C)** the dodecylresorufin fluorescence versus allophycocyanin fluorescence dot plot.

IMPORTANT NOTICE: The products described in this manual are covered by one or more Limited Use Label License(s). Please refer to the Appendix on page 971 and Master Product List on page 975. Products are For Research Use Only. Not intended for any animal or human therapeutic or diagnostic use.

with both the 488 nm spectral line of an argon-ion laser and the 633 nm spectral line of a He-Ne laser for excitation. The kit components, number of assays and assay principles are summarized in Table 15.4.

Metabolic Activity/Annexin V/Dead Cell Apoptosis Kit

The Metabolic Activity/Annexin V/Dead Cell Apoptosis Kit (V35114) is an enhanced version of the APC Annexin V/Dead Cell Apoptosis Kit. Nonfluorescent C_{12}-resazurin is reduced by viable cells to orange-fluorescent C_{12}-resorufin.[71] After staining a cell population with allophycocyanin annexin V, C_{12}-resazurin and the SYTOX® Green stain, apoptotic cells show far-red fluorescence, intermediate orange fluorescence and no green fluorescence; late apoptotic cells show intense far-red and green fluorescence and little orange fluorescence; live cells show little or no green or far-red fluorescence but significant orange fluorescence (Figure 15.5.17). The kit components, number of assays and assay principles are summarized in Table 15.4.

Mitochondrial Membrane Potential/Annexin V Apoptosis Kit

The Mitochondrial Membrane Potential/Annexin V Apoptosis Kit (V35116) provides a rapid and convenient assay for two hallmarks of apoptosis—phosphatidylserine externalization and changes in mitochondrial membrane potential. Recombinant annexin V conjugated to the Alexa Fluor® 488 dye, our brightest and most photostable green fluorophore, provides maximum sensitivity for detecting phosphatidylserine externalization in apoptotic cells. Live cells are labeled with MitoTracker® Red CMXRos, which exhibits bright red fluorescence in the presence of a mitochondrial transmembrane potential. After staining a cell population with Alexa Fluor® 488 annexin V and MitoTracker® Red CMXRos dye in the provided binding buffer, live cells exhibit very little green fluorescence and bright red fluorescence, whereas apoptotic cells exhibit bright green fluorescence and decreased red fluorescence (Figure 15.5.18). These populations can easily be distinguished using a flow cytometer, and the 488 nm line of an argon-ion laser can be used to excite both dyes. The kit components, number of assays and assay principles are summarized in Table 15.4.

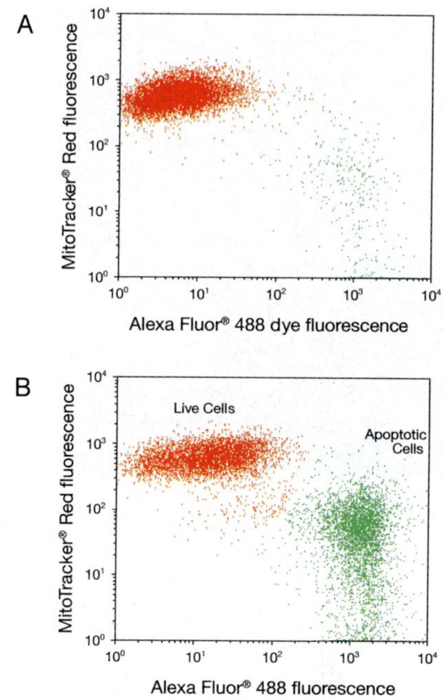

Figure 15.5.18 Flow cytometric analysis of Jurkat cells using the Mitochondrial Membrane Potential/Annexin V Apoptosis Kit (V35116). Jurkat human T-cell leukemia cells in complete medium were **B)** first exposed to 10 µM camptothecin for 4 hours or **A)** left untreated. Both cell populations were then treated with the reagents in the kit and analyzed by flow cytometry. Note that the apoptotic cells show higher reactivity for annexin V and lower MitoTracker® Red dye fluorescence than do live cells.

Violet Chromatin Condensation/Dead Cell Apoptosis Kit

The Violet Chromatin Condensation/Dead Cell Apoptosis Kit (A35135) provides a rapid and convenient assay for apoptosis based upon fluorescence analysis of the compacted state of the chromatin in apoptotic cells. The kit contains the cell-permeant Vybrant® DyeCycle™ Violet stain and the impermeant red-fluorescent SYTOX® AADvanced™ dead cell stain. The condensed chromatin of apoptotic cells are stained more brightly by Vybrant® DyeCycle™ Violet stain than the chromatin of normal cells. The SYTOX® AADvanced™ stain labels only necrotic cells, based on membrane integrity. The staining pattern resulting from the simultaneous use of these stains makes it possible to distinguish normal, apoptotic and necrotic cell populations by flow cytometry (Figure 15.5.19). The Vybrant® DyeCycle™ Violet and SYTOX® AADvanced™ stains are excited with the 405 nm violet diode laser and the 488 nm argon-ion laser, respectively.

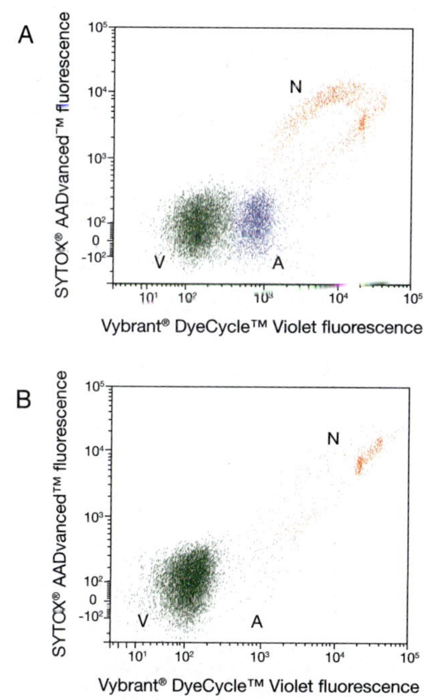

Figure 15.5.19 Jurkat cells (human T-cell leukemia) treated with 10 µM camptothecin for 6 hours (**B**) or untreated (as control, **A**). Cells were then mixed with the reagents in the kit and analyzed by flow cytometry using 405/488 nm dual excitation. Note that the campthothecin-treated cells (**B**) have a higher percentage of apoptotic cells than the basal level of apoptosis seen in the control cells (**A**). A=apoptotic cells, V = viable cells, N = necrotic cells.

Violet Annexin V/Dead Cell Apoptosis Kit

Like the Single Channel Annexin V/Dead Cell Apoptosis Kit, the Violet Annexin V/Dead Cell Apoptosis Kit (A35136) detects the externalization of phosphatidylserine in apoptotic cells.[70] This kit provides a sensitive two-color assay that employs our violet laser–excitable Pacific Blue™ annexin V (excitation/emission maxima ~415/460 nm) and the impermeant red-fluorescent SYTOX® AADvanced™ dead cell stain. The SYTOX® AADvanced™ stain labels only necrotic cells, based on membrane integrity. After staining a cell population with Pacific Blue™ annexin V and SYTOX® AADvanced™ stain in the provided binding buffer, apoptotic cells show blue fluorescence, dead cells show red fluorescence, and live cells show little or no fluorescence. These populations can easily be distinguished with a flow cytometer equipped with both a 405 nm violet diode laser and an argon-ion laser for excitation. The kit components, number of assays and assay principles are summarized in Table 15.4.

IMPORTANT NOTICE: The products described in this manual are covered by one or more Limited Use Label License(s). Please refer to the Appendix on page 971 and Master Product List on page 975. Products are For Research Use Only. Not intended for any animal or human therapeutic or diagnostic use.

Apoptosis Assays Based on Protease Activity

Caspase-3 Substrates

A distinctive feature of the early stages of apoptosis is the activation of caspase enzymes. Members of the caspase (CED-3/ICE) family of cysteine–aspartic acid specific proteases have been identified as crucial mediators of the complex biochemical events associated with apoptosis.[72–75] The recognition site for caspases is marked by three to four amino acids followed by an aspartic acid residue, with the cleavage occurring after the aspartate.[76] The caspase proteases are typically synthesized as inactive precursors. Inhibitor release or cofactor binding activates the caspase through cleavage at internal aspartates, either by autocatalysis or by the action of another protease.[73]

Caspase-3 (CPP32/apopain) is a key effector in the apoptosis pathway, amplifying the signal from initiator caspases (such as caspase-8) and signifying full commitment to cellular disassembly. In addition to cleaving other caspases in the enzyme cascade, caspase-3 has been shown to cleave poly(ADP-ribose) polymerase (PARP), DNA-dependent protein kinase, protein kinase Cδ and actin.[74,77] We offer a selection of fluorogenic caspase substrates (Table 15.5). The Z-DEVD-R110 substrate[78,79]—a component of our EnzChek® Caspase-3 Assay Kit #2 (E13184) and RediPlate™ 96 EnzChek® Caspase-3 Assay Kit (R35100)—is available separately in a 20 mg unit size for high-throughput screening applications (R22120). This nonfluorescent bisamide is first converted by caspase-3 (or a closely related protease) to the fluorescent monoamide and then to the even more fluorescent rhodamine 110 (excitation/emission maxima ~496/520 nm). In addition, the bis-L-aspartic acid amide of R110 (D_2-R110, R22122), which contains rhodamine 110 (R110) flanked by aspartic acid residues, may serve as a substrate for a variety of apoptosis-related proteases, including caspase-3 and caspase-7,[78] and does not appear to require any invasive techniques such as osmotic shock to gain entrance into the cytoplasm (Figure 15.5.20).

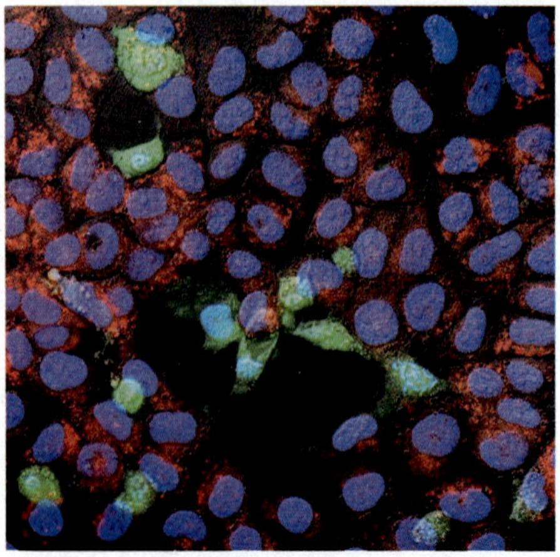

Figure 15.5.20 Detection of apoptosis in SK-N-MC neuroblastoma cells. Following a 6-hour exposure to hydrogen peroxide, cells were labeled with Hoechst 33342 (H1399, H3570, H21492), tetramethylrhodamine ethyl ester (TMRE, T669) and rhodamine 110, bis-L-aspartic acid amide (R22122) for 15 minutes. Apoptotic cells show green cytosolic fluorescence resulting from cleavage of the rhodamine 110, bis-L-aspartic acid amide substrate by active caspase-3. The staining pattern of the Hoechst 33342 dye reveals that the majority of the rhodamine 110–positive cells also contain condensed or fragmented nuclei characteristic of apoptosis. Furthermore, the rhodamine 110–positive cells are also characterized by an absence of polarized mitochondria, as indicated by their failure to load the positively charged mitochondrial indicator TMRE. The image was contributed by A.K. Stout and J.T. Greenamyre, Emory University.

Caspase-8 Substrates

Caspase-8 plays a critical role in the early cascade of apoptosis, acting as an initiator of the caspase activation cascade. Activation of the enzyme itself is accomplished through direct interaction with the death domains of cell-surface receptors for apoptosis-inducing ligands.[80,81] The activated protease has been shown to be involved in a pathway that mediates the release of cytochrome *c* from the mitochondria[82] and is also known to activate downstream caspases, such as caspase-3.[83] A R110-based fluorogenic substrate containing the caspase-8 recognition sequence Ile-Glu-Thr-Asp (IETD) is available (Z-IETD-R110, R22125, A22126; Table 15.5).

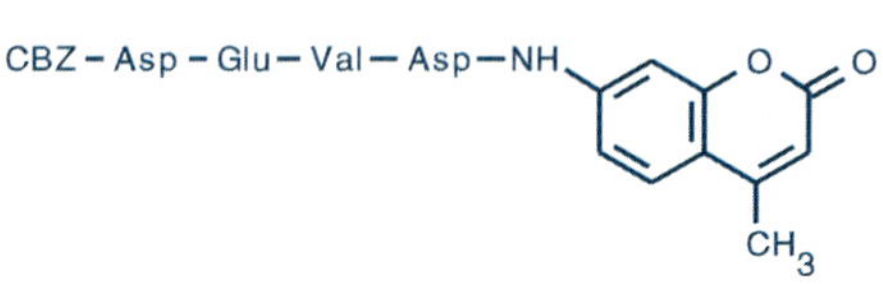

Figure 15.5.21 Z-DEVD-AMC substrate, a component of the EnzChek® Caspase-3 Assay Kit #1 (E13183).

Other Caspase and Granzyme B Substrates

In addition to our R110-derived caspase-3 and -8 substrates, we offer R110-based substrates for caspase-1, -2, -6, -9 and -13, as well as for granzyme B (Table 15.5). Granzyme B, a serine protease contained within cytotoxic T lymphocytes and natural killer cells, is thought to induce apoptosis in target cells by activating caspases and causing mitochondrial cytochrome *c* release.[84]

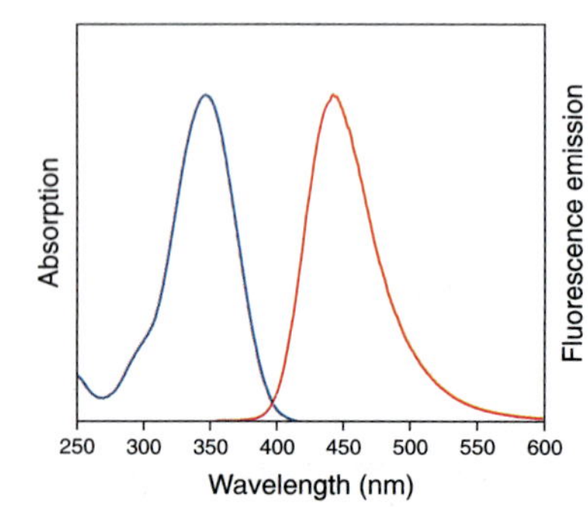

Figure 15.5.22 Absorption and fluorescence emission spectra of 7-amino-4-methylcoumarin in pH 7.0 buffer.

Table 15.5 Fluorogenic substrates for caspase activity.

		Fluorophore †	
Peptide Sequence	**Target for Caspases ***	**R110**	**AMC**
YVAD	1, 4	R33750	
VDVAD	2	R33755	
DEVD	3, 6, 7, 8, 10	E13184 ‡ R22120 § R35100 **	E13183 ‡
VEID	6	R33754	
IETD	8, granzyme B	R22125 R22126 §	
LEED	13	R33753	
AAD	granzyme B	R33752	
Aspartic acid	generic	R22122	

* Caspase substrates can often be cleaved by multiple enzymes. The caspase most often associated with the given peptide sequence is listed first. † R110 = rhodamine 110; AMC = 7-amino-4-methylcoumarin. The absorption and emission maxima for the cleaved fluorophores at pH 7 are: 496/520 nm for R110 and 342/441 nm for AMC. ‡ EnzChek® Caspase-3 Assay Kit. § Bulk packaging. ** RediPlate™ 96 EnzChek® Caspase-3 Assay Kit.

IMPORTANT NOTICE: The products described in this manual are covered by one or more Limited Use Label License(s). Please refer to the Appendix on page 971 and Master Product List on page 975. Products are For Research Use Only. Not intended for any animal or human therapeutic or diagnostic use.
www.invitrogen.com/probes

EnzChek® Caspase-3 Assay Kits

Molecular Probes® EnzChek® Caspase-3 Assay Kits permit the detection of apoptosis by assaying for increases in caspase-3 and caspase-3–like protease activities. Our EnzChek® Caspase-3 Assay Kit #1 (E13183) contains the 7-amino-4-methylcoumarin (AMC)–derived substrate Z-DEVD-AMC (Figure 15.5.21) (where Z represents a benzyloxycarbonyl group). This substrate, which is weakly fluorescent in the UV spectral range (excitation/emission maxima ~330/390 nm), yields the blue-fluorescent product AMC (A191, Section 10.1, Figure 15.5.22), which has excitation/emission maxima of 342/441 nm upon proteolytic cleavage.

The EnzChek® Caspase-3 Assay Kit #2 (E13184) contains the R110-derived substrate, Z-DEVD-R110 [78,79] (Figure 15.5.23). This substrate is a bisamide derivative of R110, containing DEVD peptides covalently linked to each of R110's amino groups, thereby suppressing both the dye's visible absorption and fluorescence. Upon enzymatic cleavage by caspase-3 (or a closely related protease), the nonfluorescent bisamide substrate is converted in a two-step process first to the fluorescent monoamide and then to the even more fluorescent R110 (R6479, Section 10.1, Figure 15.5.24). Both of these hydrolysis products exhibit spectral properties similar to those of fluorescein, with excitation/emission maxima of 496/520 nm. The Z-DEVD-R110 substrate (R22120) is also available separately in a 20 mg unit size for high-throughput screening applications.

Either kit can be used to continuously measure the activity of caspase-3 and closely related proteases in cell extracts and purified enzyme preparations using a fluorescence microplate reader or fluorometer. AMC-based DEVD substrates, which yield blue fluorescence upon proteolytic cleavage, are widely used to monitor caspase-3 activity.[77,85] The longer-wavelength spectra and higher extinction coefficient of the green-fluorescent products of the R110-based substrate in Kit #2 (E13184) should provide even greater sensitivity.[78,79] The reversible aldehyde-based inhibitor Ac-DEVD-CHO can be used to confirm that the observed fluorescence signal in both induced and control cell populations is due to the activity of caspase-3–like proteases.[77] The EnzChek® Caspase-3 Assay Kits contain:

- Z-DEVD-AMC [86,87] (in Kit #1, E13183) or Z-DEVD-R110 [78,88] (in Kit #2, E13184)
- Dimethylsulfoxide (DMSO)
- Concentrated cell-lysis buffer
- Concentrated reaction buffer
- Dithiothreitol (DTT)
- Ac-DEVD-CHO, a reversible aldehyde-based inhibitor
- 7-Amino-4-methylcoumarin (AMC) (in Kit E13183) or rhodamine 110 (in Kit E13184) reference standard to quantitate the amount of fluorophore released in the reaction
- Detailed protocols

Each kit provides sufficient reagents for performing ~500 assays using a volume of 100 µL per assay.

RediPlate™ 96 EnzChek® Caspase-3 Assay Kit

Our EnzChek® Caspase-3 Assay Kit #2 is also available as a convenient RediPlate™ 96 EnzChek® Caspase-3 Assay Kit (R35100), which includes one 96-well microplate, contained in a resealable foil packet to help ensure the integrity of the fluorogenic components, plus all necessary buffers and reagents for performing the assay. The enzyme sample to be assayed is added to the microplate in a suitable buffer, along with any compounds to be tested. Then, after incubation, the resultant fluorescence is quantitated on a fluorescence microplate reader equipped with filters appropriate for the green-fluorescent R110, with excitation/emission maxima of 496/520 nm. The microplate consists of twelve removable strips, each with eight wells, allowing researchers to

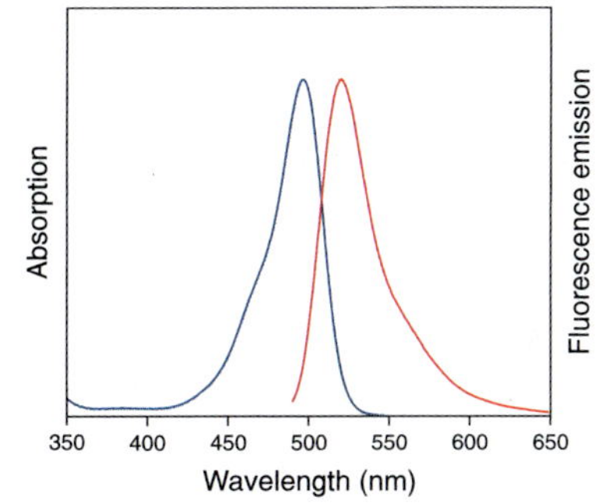

Figure 15.5.24 Absorption and fluorescence emission spectra of rhodamine 110 in pH 7.0 buffer.

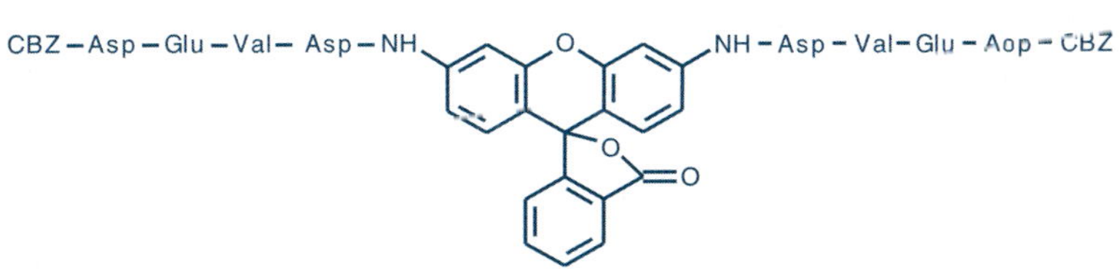

Figure 15.5.23 Z-DEVD-R110 substrate, a component of the EnzChek® Caspase-3 Assay Kit #2 (E13184).

IMPORTANT NOTICE: The products described in this manual are covered by one or more Limited Use Label License(s). Please refer to the Appendix on page 971 and Master Product List on page 975. Products are For Research Use Only. Not intended for any animal or human therapeutic or diagnostic use.

perform only as many assay as required for the experiment. Eleven of the strips (88 wells) are preloaded with the Z-DEVD-R110 substrate. The remaining strip, marked with blackened tabs, contains a dilution series of free R110 that may be used as a fluorescence reference standard. The reversible aldehyde-based inhibitor Ac-DEVD-CHO, which is supplied in a separate vial, can be used to confirm that the observed fluorescence signal in both induced and control cell populations is due to the activity of caspase-3–like proteases.[77] Table 10.3 summarizes our other RediPlate™ 96 and RediPlate™ 384 Assay Kits for protease activity (Section 10.4), phosphatase activity (Section 10.3) and RNA quantitation (Section 8.3).

Image-iT® LIVE Green Caspase Detection Kits for Fluorescence Microscopy

The Image-iT® LIVE Green Caspase-3 and -7 Detection Kit, Image-iT® LIVE Green Caspase-8 Detection Kit and Image-iT® LIVE Green Poly Caspases Detection Kit (I35106, I35105, I35104) employ a novel approach to detect active caspases that is based on a fluorescent inhibitor of caspases (FLICA® methodology). The FLICA® inhibitor comprises a fluoromethyl ketone (FMK) moiety, which can react covalently with a cysteine, a caspase-selective amino acid sequence and a fluorescent carboxyfluorescein (FAM) reporter group. Essentially an affinity label, the FLICA® inhibitor is thought to interact with the enzymatic reactive center of an activated caspase via the recognition sequence, and then to attach covalently to a cysteine through the reactive FMK moiety.[89] The FLICA® inhibitor's recognition sequence is aspartic acid–glutamic acid–valine–aspartic acid (DEVD) for caspase-3 and-7 detection, leucine–glutamic acid–threonine–aspartic acid (LETD) for caspase-8 detection and valine–alanine–aspartic acid (VAD) for detection of most caspases (including caspase-1, -3, -4, -5, -6, -7, -8 and -9). Importantly, the FLICA® inhibitor is cell permeant and not cytotoxic; unbound FLICA® molecules diffuse out of the cell and are washed away. The remaining green-fluorescent signal (excitation/emission maxima ~488/530 nm) can be used as a direct measure of the amount of active caspase that was present at the time the inhibitor was added. FLICA® reagents have been used widely to study apoptosis with flow cytometry and microscopy.[90–94] Recent work indicates that cellular fluorescence from the bound FLICA® reagent is strongly linked to caspase activity in apoptotic cells; however, the interaction of the FLICA® reagent with other cellular sites may contribute to signal intensity in nonapoptotic cells.[95] The Image-iT® LIVE Green Caspase Detection Kit includes:

- FAM-DEVD-FMK caspase-3 and -7 reagent (in Kit I35106), FAM-LETD-FMK caspase-8 reagent (in Kit I35105) or FAM-VAD-FMK poly caspases reagent (in Kit I35104)
- Hoechst 33342
- Propidium iodide
- Dimethylsulfoxide (DMSO)
- Apoptosis fixative solution
- Concentrated apoptosis wash buffer
- Detailed protocols for fluorescence microcscopy assays

In addition to a specific FLICA® reagent, each kit provides Hoechst 33342 and propidium iodide stains, which allow the simultaneous evaluation of caspase activation, nuclear morphology and plasma membrane integrity. Sufficient reagents are provided for 25 assays, based on labeling volumes of 300 µL. These Image-iT® LIVE Green Caspase Detection Kits can also be used in combination with other reagents for multiparametric study of apoptosis.

Image-iT® LIVE Red Caspase Detection Kits for Fluorescence Microscopy

The Image-iT® LIVE Red Caspase-3 and -7 Detection Kit and Image-iT® LIVE Red Poly Caspases Detection Kit (I35102, I35101) are analogous to the Image-iT® LIVE Green Caspase Detection Kits except that the FLICA® reagent contains a red-fluorescent sulforhodamine (SR) reporter group instead of a green-fluorescent carboxyfluorescein (FAM) reporter group. This assay's red-fluorescent signal (excitation/emission maxima ~550/595 nm) can be used as a direct measure of the amount of active caspase that was present at the time the inhibitor was added. The Image-iT® LIVE Red Caspase Detection Kit includes:

- SR-DEVD-FMK caspase-3 and -7 reagent (in Kit I35102) or SR-VAD-FMK poly caspases reagent (in Kit I35101)
- Hoechst 33342
- SYTOX® Green nucleic acid stain
- Dimethylsulfoxide (DMSO)
- Apoptosis fixative solution
- Concentrated apoptosis wash buffer
- Detailed protocols for fluorescence microcscopy assays

In addition to a specific FLICA® reagent, each kit provides Hoechst 33342 and SYTOX® Green nucleic acid stains, which allow the simultaneous evaluation of caspase activation, nuclear morphology and plasma membrane integrity. Sufficient reagents are provided for 25 assays, based on labeling volumes of 300 µL.

Vybrant® FAM Caspase Assay Kits for Flow Cytometry

Like the Image-iT® Kits described above, the Vybrant® FAM Caspase Assay Kits for flow cytometry are based on a fluorescent caspase inhibitor (FLICA® methodology). We offer three different Vybrant® FAM Caspase Assay Kits designed to target different caspases. The Vybrant® FAM Caspase-3 and -7 Assay Kit (V35118) provides a FLICA® inhibitor containing the caspase-3 and -7 recognition sequence DEVD; the Vybrant® FAM Caspase-8 Assay Kit (V35119) provides a FLICA® inhibitor containing the caspase-8 recognition sequence Leu-Glu-Thr-Asp (LETD); and the Vybrant® FAM Poly Caspases Assay Kit (V35117) provides a FLICA® inhibitor containing the caspase recognition sequence Val-Ala-Asp (VAD), which is recognized by caspase-1, -3, -4, -5, -6, -7, -8 and -9. In addition to the selective FLICA® reagent, these kits contain the Hoechst 33342 and propidium iodide nucleic acid stains to permit simultaneous evaluation of caspase activation, membrane permeability and cell cycle. The Vybrant® FAM Caspase Assay Kits include:

- FAM-DEVD-FMK caspase-3 and -7 reagent (in Kit V35118), FAM-LETD-FMK caspase-8 reagent (in Kit V35119) or FAM-VAD-FMK poly caspases reagent (in Kit V35117)
- Hoechst 33342
- Propidium iodide
- Dimethylsulfoxide (DMSO)
- Apoptosis fixative solution
- Concentrated apoptosis wash buffer
- Detailed protocols for flow cytometry assays

Sufficient reagents are provided for 25 assays, based on labeling volumes of 300 µL. These Vybrant® FAM Caspase Assay Kits can be used in combination with other fluorescent probes, such as the

IMPORTANT NOTICE: The products described in this manual are covered by one or more Limited Use Label License(s). Please refer to the Appendix on page 971 and Master Product List on page 975. Products are For Research Use Only. Not intended for any animal or human therapeutic or diagnostic use.
www.invitrogen.com/probes
molecular probes® | invitrogen™ by life technologies™

far-red–fluorescent allophycocyanin annexin V (A35110), for a multiparameter study of apoptosis.

Cathepsins and Calpains

The role of intracellular cathepsins and calpains in apoptosis is unclear, although an upstream role of cathepsin B in activation of some caspases[96,97] and cathepsins during apoptosis has been established.[98] Pepstatin A, which is a selective inhibitor of carboxyl (acid) proteases such as cathepsin D, has been reported to inhibit apoptosis in microglia, lymphoid cells and HeLa cells.[99–101] Consequently, our cell-permeant BODIPY® FL pepstatin derivative (P12271), which we have shown to inhibit cathepsin D *in vitro* (IC_{50} ~10 nM) and to target cathepsin D within lysosomes of live and fixed cells, has demonstrable utility for following the intracellular translocation of cathepsin D.[102,103]

Calpains are a family of ubiquitous calcium-activated thiol proteases that are implicated in a variety of cellular functions including exocytosis, cell fusion, apoptosis and cell proliferation.[101,104] Caspase-dependent downstream processing of calpain has been reported, suggesting that calpain may play a role in the degradation phase of apoptosis that is distinct from that of caspases.[105–107] One mechanism of caspase dependence appears to be processing of the endogenous calpain inhibitor calpastin by caspases.[108] However, calpain activation has also been reported to be upstream of caspases in radiation-induced apoptosis.[109] Our *t*-BOC-Leu-Met-CMAC fluorogenic substrate (A6520) has been used to measure calpain activity in hepatocytes following the addition of extracellular ATP[110] and may be of utility in detecting caspase-activated processing of procalpain in live single cells. Peptidase substrates based on our CMAC fluorophore (7-amino-4-chloromethylcoumarin, C2110; Section 10.1) passively diffuse into several types of cells, where the thiol-reactive chloromethyl group is enzymatically conjugated to glutathione by intracellular glutathione *S*-transferase or reacts with protein thiols, thus transforming the substrate into a membrane-impermeant probe. Subsequent peptidase cleavage results in a bright blue-fluorescent glutathione conjugate; see Section 10.4 for more information on AMC- and CMAC-based peptidase substrates.

Apoptosis Assays Using Mitochondrial Stains

A distinctive feature of the early stages of programmed cell death is the disruption of active mitochondria.[111–113] This mitochondrial disruption includes changes in the membrane potential and alterations to the oxidation–reduction potential of the mitochondria. Changes in the membrane potential are presumed to be due to the opening of the mitochondrial permeability transition pore, allowing passage of ions and small molecules. The resulting equilibration of ions leads in turn to the decoupling of the respiratory chain and then the release of cytochrome *c* into the cytosol.[114,115] These changes can be monitored using our extensive selection of potential-sensitive mitochondrial stains[3,5,64] (Section 12.2). We also offer several kits providing ready-to-use formulations of these reagents in flow cytometry or imaging protocols.

Image-iT® LIVE Mitochondrial Transition Pore Assay Kit for Fluorescence Microscopy

The Image-iT® LIVE Mitochondrial Transition Pore Assay Kit (I35103), based on published experimentation for mitochondrial transition pore opening,[116,117] provides a more direct method of measuring mitochondrial permeability transition pore opening than assays relying on mitochondrial membrane potential alone. This assay employs the acetoxymethyl (AM) ester of calcein, a colorless and nonfluorescent esterase substrate, and $CoCl_2$, a quencher of calcein fluorescence, to selectively label mitochondria. Cells are loaded with calcein AM, which passively diffuses into the cells and accumulates in cytosolic compartments, including the mitochondria. Once inside cells, calcein AM is cleaved by intracellular esterases to liberate the very polar fluorescent dye calcein, which does not cross the mitochondrial or plasma membranes in appreciable amounts over relatively short periods of time. The fluorescence from cytosolic calcein is quenched by the addition of $CoCl_2$, while the fluorescence from the mitochondrial calcein is maintained. As a control, cells that have been loaded with calcein AM and $CoCl_2$ can also be treated with a Ca^{2+} ionophore such as ionomycin (I24222, Section 19.8) to allow entry of excess Ca^{2+} into the cells, which triggers mitochondrial pore activation and subsequent loss of mitochondrial calcein fluorescence. This ionomycin response can be blocked with cyclosporine A, a compound reported to prevent mitochondrial transition pore formation by binding cyclophilin D.

The Image-iT® LIVE Mitochondrial Transition Pore Assay Kit has been tested with HeLa cells and bovine pulmonary artery endothelial cells (BPAEC). Each Image-iT® LIVE Mitochondrial Transition Pore Assay Kit provides:

- Calcein AM
- MitoTracker® Red CMXRos, a red-fluorescent mitochondrial stain (excitation/emission maxima ~579/599 nm)
- Hoechst 33342, a blue-fluorescent nuclear stain (excitation/emission maxima ~350/461 nm)
- Ionomycin
- $CoCl_2$
- Dimethylsulfoxide (DMSO)
- Detailed protocols

Sufficient reagents are provided for 100 assays, based on labeling volumes of 1 mL.

MitoProbe™ Transition Pore Assay Kit for Flow Cytometry

The MitoProbe™ Transition Pore Assay Kit (M34153), based on published experimentation for mitochondrial transition pore opening,[116,117] provides a more direct method of measuring mitochondrial permeability transition pore opening than assays relying on mitochondrial membrane potential alone. As with the Image-iT® LIVE mitochondrial transition pore assay described above, this assay employs the acetoxymethyl (AM) ester of calcein, a colorless and nonfluorescent esterase substrate, and $CoCl_2$, a quencher of calcein fluorescence, to selectively label mitochondria. Cells are loaded with calcein AM, which passively diffuses into the cells and accumulates in cytosolic compartments, including the mitochondria. Once inside cells, calcein AM is cleaved by intracellular esterases to liberate the very polar fluorescent dye calcein, which does not cross the mitochondrial or plasma membranes in appreciable amounts over relatively short periods of time. The fluorescence from cytosolic calcein is quenched by the addition of $CoCl_2$, while the fluorescence from the mitochondrial calcein is maintained. As a control, cells that have been loaded with calcein AM and $CoCl_2$ can also be treated with a Ca^{2+} ionophore such as ionomycin (I24222, Section 19.8) to allow entry of excess Ca^{2+} into the cells, which triggers mitochondrial pore activation and subsequent loss of mitochondrial calcein fluorescence. This ionomycin response can be blocked with cyclosporine A, a compound reported to prevent mitochondrial transition pore formation by binding cyclophilin D.

IMPORTANT NOTICE: The products described in this manual are covered by one or more Limited Use Label License(s). Please refer to the Appendix on page 971 and Master Product List on page 975. Products are For Research Use Only. Not intended for any animal or human therapeutic or diagnostic use.

The MitoProbe™ Transition Pore Assay Kit has been tested with Jurkat cells, MH1C1 cells and bovine pulmonary artery endothelial cells (BPAEC). Each MitoProbe™ Transition Pore Assay Kit provides:

- Calcein AM
- $CoCl_2$
- Ionomycin
- Dimethylsulfoxide (DMSO)
- Detailed protocols

Sufficient reagents are provided for 100 assays, based on labeling volumes of 1 mL.

MitoProbe™ JC-1 Assay Kit for Flow Cytometry

The MitoProbe™ JC-1 Assay Kit (M34152) provides the cationic dye JC-1 and a mitochondrial membrane potential uncoupler, CCCP (carbonyl cyanide 3-chlorophenylhydrazone), for the study of mitochondrial membrane potential. JC-1 (Figure 15.5.25) exhibits potential-dependent accumulation in mitochondria, indicated by a fluorescence emission shift from green (~529 nm) to red (~590 nm), due to concentration-dependent formation of red-fluorescent J-aggregates.[118–120] Consequently, mitochondrial depolarization is indicated by a decrease in the red/green fluorescence intensity ratio, which is dependent only on the membrane potential and not on other factors such as mitochondrial size, shape and density, which may influence single-component fluorescence measurements. Use of fluorescence ratio detection therefore allows researchers to make comparative measurements of membrane potential and to determine the percentage of mitochondria within a population that respond to an applied stimulus. Subtle heterogeneity in cellular responses can be discerned in this way.[119,121] For example, four distinct patterns of mitochondrial membrane potential change in response to glutamate receptor activation in neurons have been identified using confocal ratio imaging of JC-1 fluorescence.[122]

JC-1 can be used as an indicator of mitochondrial potential in a variety of cell types, including myocytes[118] and neurons,[122] as well as in intact tissues[123] and isolated mitochondria.[121] JC-1 is more specific for mitochondrial versus plasma membrane potential and more consistent in its response to depolarization than some other cationic dyes such as $DiOC_6(3)$ and rhodamine 123.[124] The most widely implemented application for JC-1 is the detection of mitochondrial depolarization occurring in apoptosis.[124–127] Each MitoProbe™ JC-1 Assay Kit provides:

- JC-1
- Dimethylsulfoxide (DMSO)
- CCCP
- Concentrated phosphate-buffered saline (PBS)
- Detailed protocols

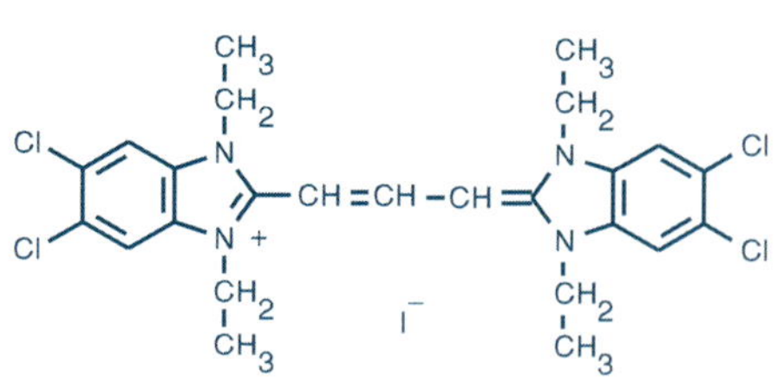

Figure 15.5.25 5,5′,6,6′-Tetrachloro-1,1′,3,3′-tetraethylbenzimidazolylcarbocyanine iodide (JC-1; T3168).

Sufficient reagents are provided for 100 assays, based on a labeling volume of 1 mL.

MitoProbe™ $DiIC_1(5)$ and MitoProbe™ $DiOC_2(3)$ Assay Kits for Flow Cytometry

Cationic carbocyanine dyes have been shown to accumulate in cells in response to membrane potential,[128] and membrane potential changes have been studied in association with apoptosis.[129,130] The MitoProbe™ $DiIC_1(5)$ and MitoProbe™ $DiOC_2(3)$ Assay Kits (M34151, M34150) provide solutions of the far-red–fluorescent $DiIC_1(5)$ (1,1′,3,3,3′,3′-hexamethylindodicarbocyanine iodide) and green-fluorescent $DiOC_2(3)$ (3,3′-diethyloxacarbocyanine iodide) carbocyanine dyes, respectively, along with a mitochondrial membrane potential disrupter, CCCP, for the study of mitochondrial membrane potential. These $DiIC_1(5)$ and $DiOC_2(3)$ carbocyanine dyes penetrate the cytosol of eukaryotic cells and, at concentrations below 100 nM, accumulate primarily in mitochondria with active membrane potentials. In the case of $DiOC_2(3)$, this accumulation is accompanied by a shift from green to red emission due to dye stacking, allowing the use of a ratiometric parameter (red/green fluorescence ratio) that corrects for size differences when measuring membrane potential in bacteria.[131,132] $DiIC_1(5)$ and $DiOC_2(3)$ stain intensities decrease when cells are treated with reagents that disrupt mitochondrial membrane potential, such as CCCP. Each MitoProbe™ $DiIC_1(5)$ and MitoProbe™ $DiOC_2(3)$ Assay Kit provides:

- $DiIC_1(5)$ (in Kit M34151) or $DiOC_2(3)$ (in Kit M34150)
- CCCP
- Detailed protocols for labeling cells with the short-chain carbocyanine dye, as well as with annexin V conjugates (not included)

Cells stained with $DiIC_1(5)$ can be visualized by flow cytometry with red excitation and far-red emission filters; cells stained with $DiOC_2(3)$ can be visualized by flow cytometry with blue excitation and green and red emission filters. $DiIC_1(5)$ can be paired with other reagents, such as propidium iodide and the green-fluorescent Alexa

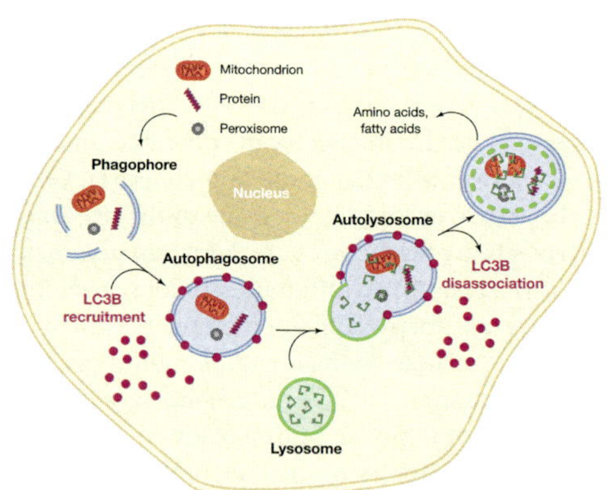

Figure 15.5.26. Schematic depiction of the multistep autophagy pathway in a eukaryotic cell. The first step involves the formation and elongation of isolation membranes, or phagophores. In the second step, which involves the LC3B protein, the cytoplasmic cargo is sequestered, and the double-membrane autophagosome is formed. Fusion of a lysosome with the autophagosome to generate the autolysosome is the penultimate step. In the fourth and final phase, the cargo is degraded.

IMPORTANT NOTICE: The products described in this manual are covered by one or more Limited Use Label License(s). Please refer to the Appendix on page 971 and Master Product List on page 975. Products are For Research Use Only. Not intended for any animal or human therapeutic or diagnostic use.
www.invitrogen.com/probes

Fluor® 488 annexin V (both provided in the Alexa Fluor® 488 Annexin V/Dead Cell Apoptosis Kit, V13241), for multiparameter study of vitality and apoptosis. $DiOC_2(3)$ can be paired with other reagents, such as the far-red–fluorescent allophycocyanin annexin V (A35110), for multiparameter study of vitality and apoptosis. Combining these short-chain carbocyanine dyes with annexin V conjugates results in superior resolution of subpopulations when compared with results obtained with other commonly used dyes.

Assays for Autophagy

Autophagy describes the segregation and delivery of cytoplasmic cargo, including proteins and organelles for degradation by hydrolytic enzymes. The process of autophagy begins with the formation and elongation of isolation membranes, or phagophores (Figure 15.5.26). The cytoplasmic cargo is then sequestered, and the double-membrane autophagosome fuses with a lysosome to generate the autolysosome. Finally, degradation is achieved through the action of hydrolytic enzymes within the autolysosome.

Autophagy was first described in 1963; however, only in the past decade has this pathway become the subject of intense study. Researchers have sought to gain further insight into the role basal autophagy plays in cell homeostasis and development, and to further elucidate the role of induced autophagy in the cell's response to stress, microbial infection, and disease.[1,133,134]

Autophagy Analysis Through Observation of LC3B

The LC3 protein plays a critical role in autophagy. Normally this protein resides in the cytosol, but following cleavage and lipidation with phosphatidylethanolamine, LC3 associates with the phagophore and can be used as a general marker for autophagic membranes (Figure 15.5.26). The new Premo™ Autophagy Sensor Kits (P36235, P36236) combine the selectivity of an LC3B–fluorescent protein chimera[6,135,136] with the transduction efficiency of BacMam technology (BacMam Gene Delivery and Expression Technology—Note 11.1), enabling unambiguous visualization of this protein in live cells (Figure 15.5.27). Recent improvements made to the BacMam system enable efficient transduction in a wider variety of cells, including neurons and neural stem cells (NSCs) with an easy, one-step protocol. To image autophagy, BacMam LC3B-FP is simply added to cells and allowed to incubate overnight for protein expression. Each Premo™ Autophagy Sensor Kit includes:

- BacMam LC3B-FP (GFP fusion, P36235; RFP fusion, P36236)
- Control BacMam LC3B (G120A)-FP
- Chloroquine diphosphate, to artificially induce phagosome formation
- Detailed protocols

Following treatment with chloroquine diphosphate, normal autophagic flux is disrupted, and autophagosomes accumulate as a result of the increase in lysosomal pH. The mutation in the control BacMam LC3B (G120A)-FP prevents cleavage and subsequent lipidation during normal autophagy, and thus protein localization should remain cytosolic and diffuse.

In addition to the Premo™ Autophagy Sensor Kits, we offer the LC3B Antibody Kit for Autophagy (L10382), which includes a rabbit polyclonal anti-LC3B antibody and chloroquine diphosphate for imaging autophagy via LC3B localization in fixed samples. The LC3B Antibody Kit for Autophagy has been validated for use with fluorescence microscopy and high-content imaging and analysis.

Imaging Autophagic Organelles

Two organelles that play a crucial role in auophagy are the mitochondria and lysosomes. Old, damaged and surplus mitochondria are major targets for autophagy, which in this case is sometimes referred to as "mitophagy." Degradation of mitochondria through this process can be used to recover their amino acids and other nutrients, as well as to remove damaged mitochondria from the cell. Fusion of a lysosome with the phagophore to form the autolysosome is the penultimate step of the autophagic pathway (Figure 15.5.26). A variety of reagents including fluorescent

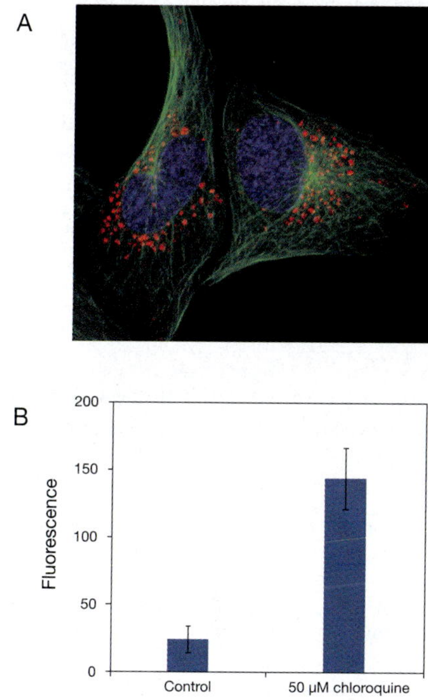

Figure 15.5.27 Detecting autophagy with the Premo™ Autophagy Sensor and fluorescence microscopy (**A**) or high-content imaging and analysis (**B**). (**A**) U2OS cells were co-transduced with the Premo™ Autophagy Sensor LC3B-RFP (P36236) and CellLight® MAP4-GFP (C10598). The following day, cells were treated with 50 µM chloroquine. The following day, cells were incubated with 1 µg/mL Hoechst 33342 before imaging. (**B**) HeLa cells were plated at 5000 cells per well and left to adhere overnight. Cells were then transduced with the Premo™ Autophagy Sensor LC3B-GFP. The following day, cells were incubated with 50 µM chloroquine or left untreated (control) for 16 hr. Quantitative analysis was performed by quantifying fluorescence from vesicular structures in the perinuclear region using the Thermo Scientific Cellomics® ArrayScan® VTI platform.

IMPORTANT NOTICE: The products described in this manual are covered by one or more Limited Use Label License(s). Please refer to the Appendix on page 971 and Master Product List on page 975. Products are For Research Use Only. Not intended for any animal or human therapeutic or diagnostic use.

dyes, fluorescent protein (FP) chimeras and antibodies can be used to image mitochondria and lysosomes during basal and induced autophagy[133,137–139] (Table 15.6). For a description of our mitochondrial and lysosomal-selective organelle probes, see Section 12.2 and Section 12.3.

Visualizing Autolysosome Formation

In conjunction with the Premo™ autophagy sensors LC3B-GFP and LC3B-RFP, the fluorogenic protease substrates DQ™ Green BSA and DQ™ Red BSA can be used to accurately image the formation of the autolysosome in live cells.[140] DQ™ Green BSA and DQ™ Red BSA (D12050, D12051; Section 10.4) are bovine serum albumin (BSA) conjugates that have been labeled to such a high degree that the fluorescence is self-quenched. To visualize autolysosome formation, cells that express a GFP- or RFP-LC3 are incubated with the contrasting color of DQ™ BSA. The convergence of the lysosome with the autophagosome results in dequenching and release of brightly fluorescent fragments. The autolysosomes can then be identified by co-localization of green and red fluorescence.

Table 15.6 Fluorescent detection reagents for imaging mitochondria and lysosomes.

	Organic Dyes (e.g., MitoTracker® and LysoTracker® dyes)	BacMam-Based Fluorescent Proteins (e.g., CellLight® reagents)	Antibodies
How they work	Positively charged MitoTracker® dyes localize to actively respiring mitochondria; weakly basic LysoTracker® dyes accumulate in compartments with low pH.	Combine targeting sequence–fluorescent protein fusion with the transduction efficiency of BacMam to label organelles independently of function (i.e., pH, mitochondrial membrane potential).	Recognize specific target of interest (e.g., LAMP1, a lysosomal protein).
Applications	Live-cell imaging applications; fixable, thus compatible with antibody-based imaging applications.*	Live-cell imaging applications; fixable, thus compatible with antibody-based imaging applications. *†	Imaging fixed cells or tissue. Compatible with labeling with MitoTracker®, LysoTracker® and CellLight® reagents.
Typical workflow	Incubate cells with the MitoTracker® or LysoTracker® reagent for approximately 5–30 min.	The ready-to-use CellLight® reagent is added to live cells, followed by an overnight incubation to allow for protein expression.	Cells are fixed and permeabilized, then incubated with the antibody for labeling, and visualized with a fluorescently labeled secondary antibody. ‡

* Please consult the product manual or contact technical service for additional information on the fixability of these reagents. † With fluorescent protein constructs, anti-GFP or anti-RFP antibodies can be used to amplify fluorescence signals. ‡ Secondary antibody is required if the primary antibody is not directly labeled.

REFERENCES

1. Cell Death Differ (2009) 16:966; **2.** Biotechniques (2002) 32:648; **3.** Methods (2008) 44:222; **4.** Methods Mol Biol (2009) 559:19; **5.** Apoptosis (2007) 12:803; **6.** PLoS One (2009) 4:e7124; **7.** J Biomol Screen (2007) 12:510; **8.** J Biomol Screen (2006) 11:369; **9.** FEBS Lett (2006) 580:885; **10.** J Immunol Methods (2010) 358:81; **11.** Biochemistry (2008) 47:7539; **12.** Cytometry A (2008) 73:563; **13.** Cytometry A (2008) 73:496; **14.** Nat Protoc (2009) 4:125; **15.** J Cell Biol (2000) 150:145; **16.** Cancer Res (1997) 57:3804; **17.** Blood (1996) 87:4959; **18.** J Exp Med (1995) 182:1759; **19.** J Immunol Methods (1995) 185:249; **20.** Cytometry (1994) 17:59; **21.** Exp Cell Res (1988) 175:184; **22.** Int J Radiat Biol (1994) 66:23; **23.** Mol Hum Reprod (1996) 2:613; **24.** Irish J Med Sci (1997) 166:177; **25.** Mutagenesis (1998) 13:1; **26.** Mutat Res (1998) 399:87; **27.** J Histochem Cytochem (2003) 51:873; **28.** Biotechniques (1999) 27:846; **29.** Nucleic Acids Res (2009) 37:3452; **30.** J Histochem Cytochem (1999) 47:1101; **31.** Chem Rev (2009) 109:4207; **32.** Chem Rev (2008) 108:2952; **33.** Angew Chem Int Ed Engl (2009) 48:6974; **34.** Biochemistry (2009) 48:6571; **35.** ACS Chem Biol (2006) 1:644; **36.** Genes Cells (2010) 15:315; **37.** J Histochem Cytochem (1996) 44:959; **38.** Biotechniques (1995) 19:800; **39.** Exp Neurol (1995) 133:225; **40.** Jpn J Cancer Res (1994) 85:939; **41.** Exp Cell Res (1996) 222:28; **42.** Cytometry (1995) 20:172; **43.** Science (1998) 281:1303; **44.** Neuroreport (1995) 6:1053; **45.** Exp Neurol (1995) 133:265; **46.** J Neurosci (1995) 15:3775; **47.** Acta Neuropathol (Berl) (1994) 88:207; **48.** Cancer Res (1993) 53:1945; **49.** Int J Oncol (1992) 1:639; **50.** Trends Cell Biol (1995) 5:21; **51.** Jpn J Cancer Res (1993) 84:566; **52.** Nucleic Acids Res (1993) 21:4206; **53.** Cytometry (1998) 32:95; **54.** Toxicol Appl Pharmacol (1998) 148:176; **55.** J Histochem Cytochem (2001) 49:821; **56.** Nat Protoc (2008) 3:1187; **57.** Cell Prolif (2007) 40:1; **58.** Proc Natl Acad Sci U S A (2006) 103:9891; **59.** J Am Chem Soc (2007) 129:2187; **60.** Cytometry (1998) 31:1; **61.** J Immunol (1992) 148:2207; **62.** J Immunol (1993) 151:4274; **63.** Nat Protoc (2009) 4:1383; **64.** Nat Protoc (2007) 2:2719; **65.** Nano Lett (2006) 6:1863; **66.** J Biol Chem (2000) 275:23807; **67.** J Biol Chem (2001) 276:1127; **68.** J Biol Chem (2000) 275:25336; **69.** BMC Biochem (2009) 10:5; **70.** Proc Natl Acad Sci U S A (2002) 99:10706; **71.** Cancer Res (2008) 68:8384; **72.** Immunol Cell Biol (1998) 76:1; **73.** Science (1998) 281:1312; **74.** Trends Biochem Sci (1997) 22:388; **75.** Cell Death Differ (1999) 6:1067; **76.** J Biol Chem (1997) 272:17907; **77.** Nature (1995) 376:37; **78.** Biochemistry (1999) 38:13906; **79.** Bioorg Med Chem Lett (1999) 9:3231; **80.** Cell Death Differ (1999) 6:821; **81.** Science (1998) 281:1305; **82.** J Biol Chem (1999) 274:17484; **83.** Exp Cell Res (1999) 250:203; **84.** J Biol Chem (2001) 276:6974; **85.** J Neurosci Res (1998) 52:334; **86.** Biochemistry (2000) 39:16056; **87.** Oncogene (1998) 17:1295; **88.** J Biol Chem (2000) 275:288; **89.** Cell Death Differ (1999) 6:1081; **90.** Exp Cell Res (2000) 259:308; **91.** Biotechniques (2001) 31:608; **92.** Leukemia (2002) 16:1589; **93.** Cytometry (2002) 47:143; **94.** J Immunol Methods (2002) 265:111; **95.** Cytometry A (2003) 55:50; **96.** Kidney Int (1987) 31:112; **97.** FEBS Lett (1998) 438:150; **98.** J Biomed Mater Res (1988) 22:837; **99.** Mol Cells (1997) 7:742; **100.** Brain Res (1997) 746:220; **101.** Infect Immun (1994) 62:5126; **102.** J Biochem Biophys Methods (2000) 42:137; **103.** J Neurosci (2008) 28:6926; **104.** Biochem Biophys Res Commun (1997) 236:549; **105.** Cancer Res (1998) 58:1901; **106.** Proc Natl Acad Sci U S A (1999) 96:3200; **107.** J Cell Physiol (1994) 159:229; **108.** Methods Enzymol (1999) 294:117; **109.** Cell Death Differ (1998) 5:1051; **110.** J Biol Chem (1993) 268:23593; **111.** Science (2001) 292:624; **112.** Science (2000) 289:1150; **113.** Trends Cell Biol (2000) 10:369; **114.** Biochim Biophys Acta (1998) 1366:151; **115.** Science (1998) 281:1309; **116.** Biofactors (1998) 8:263; **117.** Biophys J (1999) 76:725; **118.** J Physiol (1995) 486:1; **119.** Proc Natl Acad Sci U S A (1991) 88:3671; **120.** Biochemistry (1991) 30:4480; **121.** Exp Cell Res (1996) 222:84; **122.** J Neurosci (1996) 16:5688; **123.** Methods (1999) 18:104; **124.** FEBS Lett (1997) 411:77; **125.** Exp Cell Res (1998) 245:170; **126.** J Neurosci (1998) 18:932; **127.** J Cell Biol (1997) 138:449; **128.** Proc Natl Acad Sci U S A (1979) 76:5728; **129.** Methods Cell Biol (2001) 63:467; **130.** Exp Cell Res (1994) 214:323; **131.** Methods (2000) 21:271; **132.** Cytometry (1999) 35:55; **133.** Methods Enzymol (2009) 453:417; **134.** J Biol Chem (2008) 283:10892; **135.** Autophagy (2008) 4:770; **136.** J Biol Chem (2009) 284:2383; **137.** Methods Enzymol (2009) 453:111; **138.** Methods Enzymol (2009) 453:217; **139.** Methods Mol Biol (2008) 445:29; **140.** Methods Enzymol (2009) 452:85.

IMPORTANT NOTICE: The products described in this manual are covered by one or more Limited Use Label License(s). Please refer to the Appendix on page 971 and Master Product List on page 975. Products are For Research Use Only. Not intended for any animal or human therapeutic or diagnostic use.
www.invitrogen.com/probes
molecular probes® | invitrogen™ by life technologies™

DATA TABLE 15.5 ASSAYS FOR APOPTOSIS AND AUTOPHAGY

Cat. No.	MW	Storage	Soluble	Abs	EC	Em	Solvent	Notes
A6520	554.10	F,D	DMSO	330	13,000	403	MeOH	1, 2
C7604	~993	FF,L	H_2O	496	68,000	523	pH 8	3, 4
C7614	~908	FF,L	H_2O	504	68,000	513	pH 8	3, 4
E13183	767.74	F,D,L	DMSO	325	16,000	395	pH 7	1, 2, 5
E13184	1515.46	F,D,L	DMSO	232	52,000	none	MeOH	5, 6
P3581	579.26	F,D,L	DMSO	435	50,000	455	H_2O/DNA	4, 7, 8, 9
P12271	1044.14	F,D,L	DMSO	504	86,000	511	MeOH	
R22120	1515.46	F,D	DMSO, DMF	232	52,000	none	MeOH	6
R22122	788.57	F,D	DMSO, DMF	232	55,000	none	MeOH	6
R22125	1515.55	F,D	DMSO, DMF	232	52,000	none	MeOH	6
R22126	1515.55	F,D	DMSO, DMF	232	52,000	none	MeOH	6
R33750	1495.56	F,D	DMSO, DMF	230	76,000	none	MeOH	6
R33752	1113.10	F,D	DMSO, DMF	232	57,000	none	MeOH	6
R33753	1571.57	F,D	DMSO, DMF	232	57,000	none	MeOH	6
R33754	1511.60	F,D	DMSO, DMF	232	57,000	none	MeOH	6
R33755	1597.65	F,D	DMSO, DMF	232	57,000	none	MeOH	6
Y3603	629.32	F,D,L	DMSO	491	52,000	509	H_2O/DNA	4, 7, 8, 9

For definitions of the contents of this data table, see "Using *The Molecular Probes® Handbook*" in the introductory pages.

Notes

1. Peptidase cleavage of this substrate yields 7-amino-4-methylcoumarin (A191).
2. Fluorescence of the unhydrolyzed substrate is very weak.
3. The molecular weight (MW) of this product is approximate because the degree of hydration and/or salt form has not been conclusively established.
4. This product is supplied as a ready-made solution in the solvent indicated under "Soluble."
5. Data represent the substrate component of this kit.
6. Peptidase cleavage of this substrate yields rhodamine 110 (R6479).
7. Spectra represent aqueous solutions of nucleic acid-bound dye. EC values are derived by comparing the absorbance of the nucleic acid–bound dye with that of free dye in a reference solvent (H_2O or MeOH).
8. This product is essentially nonfluorescent except when bound to DNA or RNA.
9. Although this compound is soluble in water, preparation of stock solutions in water is not recommended because of possible adsorption onto glass or plastic.

PRODUCT LIST 15.5 ASSAYS FOR APOPTOSIS AND AUTOPHAGY

Cat. No.	Product	Quantity
V13241	Alexa Fluor® 488 Annexin V/Dead Cell Apoptosis Kit *Alexa Fluor® 488 annexin V/propidium iodide* *50 assays* *for flow cytometry*	1 kit
A6520	7-amino-4-chloromethylcoumarin, *t*-BOC-L-leucyl-L-methionine amide (CMAC, *t*-BOC-Leu-Met)	5 mg
A23202	annexin V, Alexa Fluor® 350 conjugate *100 assays*	500 µL
A13201	annexin V, Alexa Fluor® 488 conjugate *100 assays*	500 µL
A35108	annexin V, Alexa Fluor® 555 conjugate *100 assays*	500 µL
A13202	annexin V, Alexa Fluor® 568 conjugate *100 assays*	500 µL
A13203	annexin V, Alexa Fluor® 594 conjugate *100 assays*	500 µL
A23204	annexin V, Alexa Fluor® 647 conjugate *100 assays*	500 µL
A35109	annexin V, Alexa Fluor® 680 conjugate *100 assays*	500 µL
A35110	annexin V, allophycocyanin conjugate (APC annexin V) *50 assays*	250 µL
A13204	annexin V, biotin-X conjugate *100 assays*	500 µL
A13199	annexin V, fluorescein conjugate (FITC annexin V) *100 assays*	500 µL
A13200	annexin V, Oregon Green® 488 conjugate *100 assays*	500 µL
A35122	annexin V, Pacific Blue™ conjugate *for flow cytometry* *100 assays*	500 µL
A35111	annexin V, R-phycoerythrin conjugate (R-PE annexin V) *50 assays*	250 µL
V35113	APC Annexin V/Dead Cell Apoptosis Kit *with APC annexin V and SYTOX® Green* *50 assays* *for flow cytometry*	1 kit
A23210	APO-BrdU™ TUNEL Assay Kit *with Alexa Fluor® 488 anti-BrdU* *60 assays*	1 kit
C7614	ChromaTide® BODIPY® FL-14-dUTP *1 mM in TE buffer*	25 µL
C7604	ChromaTide® fluorescein-12-dUTP *1 mM in TE buffer*	25 µL
V13244	Chromatin Condensation/Dead Cell Apoptosis Kit *Hoechst 33342/propidium iodide* *200 assays* *for flow cytometry*	1 kit
V23201	Chromatin Condesation/Membrane Permeability/Dead Cell Apoptosis Kit. *Hoechst 33342/YO-PRO®-1/propidium iodide* *200 assays* *for flow cytometry*	1 kit
C10245	Click-iT® TUNEL Alexa Fluor® 488 Imaging Assay *for microscopy and HCS* *50–100 assays*	1 kit
C10246	Click-iT® TUNEL Alexa Fluor® 594 Imaging Assay *for microscopy and HCS* *50–100 assays*	1 kit
C10247	Click-iT® TUNEL Alexa Fluor® 647 Imaging Assay *for microscopy and HCS* *50–100 assays*	1 kit
E13183	EnzChek® Caspase-3 Assay Kit #1 *Z-DEVD-AMC substrate* *500 assays*	1 kit
E13184	EnzChek® Caspase-3 Assay Kit #2 *Z-DEVD-R110 substrate* *500 assays*	1 kit
V13242	FITC Annexin V/ Dead Cell Apoptosis Kit *FITC annexin V/propidium iodide* *50 assays* *for flow cytometry*	1 kit
H10292	HCS DNA Damage Kit *2-plate size*	1 kit
H10290	HCS LIVE/DEAD® Green Kit *2-plate size*	1 kit
H10295	HCS Mitochondrial Health Kit *2-plate size*	1 kit
I10291	Image-iT® DEAD Green™ viability stain *1 mM solution in DMSO*	25 µL
I35106	Image-iT® LIVE Green Caspase-3 and -7 Detection Kit *for microscopy* *25 tests*	1 kit

continued on next page

PRODUCT LIST 15.5 ASSAYS FOR APOPTOSIS AND AUTOPHAGY—*continued*

Cat. No.	Product	Quantity
I35105	Image-iT® LIVE Green Caspase-8 Detection Kit *for microscopy* *25 tests*	1 kit
I35104	Image-iT® LIVE Green Poly Caspases Detection Kit *for microscopy* *25 tests*	1 kit
I35103	Image-iT® LIVE Mitochondrial Transition Pore Assay Kit *for microscopy*	1 kit
I35102	Image-iT® LIVE Red Caspase-3 and -7 Detection Kit *for microscopy*	1 kit
I35101	Image-iT® LIVE Red Poly Caspases Detection Kit *for microscopy*	1 kit
L10382	LC3B Antibody Kit for Autophagy *rabbit polyclonal LC3B* *includes autophagosome inducer*	1 kit
V13243	Membrane Permeability/ Dead Cell Apoptosis Kit *YO-PRO®-1/propidium iodide* *200 assays* *for flow cytometry*	1 kit
V35114	Metabolic Activity/Annexin V/Dead Cell Apoptosis Kit *with C_{12} resazurin, APC annexin V, and SYTOX® Green* *50 assays* *for flow cytometry*	1 kit
V35116	Mitochondrial Membrane Potential/Annexin V Apoptosis Kit *Alexa Fluor® 488 annexin V/MitoTracker® Red CMXRos* *50 assays* *for flow cytometry*	1 kit
M34151	MitoProbe™ $DiIC_1(5)$ Assay Kit *for flow cytometry* *100 assays*	1 kit
M34150	MitoProbe™ $DiOC_2(3)$ Assay Kit *for flow cytometry* *100 assays*	1 kit
M34152	MitoProbe™ JC-1 Assay Kit *for flow cytometry* *100 assays*	1 kit
M34153	MitoProbe™ Transition Pore Assay Kit *for flow cytometry* *100 assays*	1 kit
V35112	PE Annexin V/ Dead Cell Apoptosis Kit *with SYTOX® Green* *50 assays* *for flow cytometry*	1 kit
P12271	pepstatin A, BODIPY® FL conjugate	25 µg
P3581	PO-PRO™-1 iodide (435/455) *1 mM solution in DMSO*	1 mL
P36235	Premo™ Autophagy Sensor LC3B-GFP	1 kit
P36236	Premo™ Autophagy Sensor LC3B-RFP	1 kit
R35100	RediPlate™ 96 EnzChek® Caspase-3 Assay Kit *Z-DEVD-R110 substrate* *one 96-well microplate*	1 kit
R22122	rhodamine 110, bis-(L-aspartic acid amide), trifluoroacetic acid salt	1 mg
R33752	rhodamine 110, bis-(*N*-CBZ-L-alanyl-L- alanyl-L-aspartic acid amide) (Z-AAD-R110)	2 mg
R22120	rhodamine 110, bis-(*N*-CBZ-L-aspartyl- L-glutamyl-L-valyl-L-aspartic acid amide) (Z-DEVD-R110) *bulk packaging*	20 mg
R22125	rhodamine 110, bis-(*N*-CBZ-L-isoleucyl- L-glutamyl-L-threonyl- L-aspartic acid amide) (Z-IETD-R110)	2 mg
R22126	rhodamine 110, bis-(*N*-CBZ-L-isoleucyl- L-glutamyl-L-threonyl- L-aspartic acid amide) (Z-IETD-R110) *bulk packaging*	20 mg
R33753	rhodamine 110, bis-(*N*-CBZ-L-leucyl-L- glutamyl-L-glutamyl-L- aspartic acid amide) (Z-LEED-R110)	2 mg
R33750	rhodamine 110, bis-(*N*-CBZ-L-tyrosinyl- L-valyl-L-alanyl-L-aspartic acid amide) (Z-YVAD-R110)	2 mg
R33755	rhodamine 110, bis-(*N*-CBZ-L-valyl-L-aspartyl- L-valyl-L-alanyl-L-aspartic acid amide) (Z-VDVAD-R110)	2 mg
R33754	rhodamine 110, bis-(*N*-CBZ-L-valyl-L-glutamyl- L-isoleucyl-L-aspartic acid amide) (Z-VEID-R110)	2 mg
V13240	Single Channel Annexin V/ Dead Cell Apoptosis Kit *Alexa Fluor® 488 annexin V/SYTOX® Green* *50 assays* *for flow cytometry*	1 kit
A35136	Violet Annexin V/Dead Cell Apoptosis Kit *Pacific Blue™ annexin V/SYTOX® AADvanced™* *for flow cytometry* *50 assays*	1 kit
A35135	Violet Chromatin Condensation/Dead Cell Apoptosis Kit *Vybrant® DyeCycle™ Violet and SYTOX® AADvanced™* *for flow cytometry* *200 assays*	1 kit
V35123	Violet Membrane Permeability/Dead Cell Apoptosis Kit *with PO-PRO™-1 and 7- aminoactinomycin D* *200 assays* *for flow cytometry*	1 kit
A35137	Violet Ratiometric Membrane Asymmetry Probe/Dead Cell Apoptosis Kit *for flow cytometry* *100 assays*	1 kit
V23200	Vybrant® Apoptosis Assay Kit #6 *biotin-X annexin V/Alexa Fluor® 350 streptavidin/propidium iodide* *50 assays*	1 kit
V35118	Vybrant® FAM Caspase-3 and -7 Assay Kit *for flow cytometry* *25 assays*	1 kit
V35119	Vybrant® FAM Caspase-8 Assay Kit *for flow cytometry* *25 assays*	1 kit
V35117	Vybrant® FAM Poly Caspases Assay Kit *for flow cytometry* *25 assays*	1 kit
Y3603	YO-PRO®-1 iodide (491/509) *1 mM solution in DMSO*	1 mL

IMPORTANT NOTICE: The products described in this manual are covered by one or more Limited Use Label License(s). Please refer to the Appendix on page 971 and Master Product List on page 975. Products are For Research Use Only. Not intended for any animal or human therapeutic or diagnostic use.

15.6 Probes for Cell Adhesion, Chemotaxis, Multidrug Resistance and Glutathione

Cell Adhesion Assays

The fundamental role of cell–cell and cell–matrix adhesion in the morphology and development of organisms, organs and tissues has made identification of molecular mediators of cell adhesion an important research focus in cell biology and immunology.[1–3] The useful review by Löster and Hortstkorte describes a number of different assays that can detect cell adhesion.[4] In a typical fluorescence-based cell adhesion assay, unlabeled cell monolayers in multiwell plates are incubated with fluorescently labeled cells and then washed to separate the adherent and nonadherent populations. Cell adhesion can then be determined simply by correlating the retained fluorescence with cell number. An ideal fluorescent marker will retain proportionality between fluorescence and cell number and introduce minimal interference with the cell adhesion process. Because adhesion is a cell-surface phenomenon, cytoplasmic markers that can be passively loaded are preferable to compounds that label cell-surface molecules, provided they are retained in the cell for the duration of the experiment or their leakage rate can be independently measured. Adhesion of fluorescent dye–labeled cells to matrices such as bone[5] can be directly observed by fluorescence microscopy using cells loaded with the permeant live-cell tracers described in Section 14.2 and Section 15.2 or the lipophilic dyes in Section 14.4. Alternatively, high molecular weight, cell-impermeant, fluorescent dextrans (Section 14.5) have been used to define the area *outside* of adherent cells, with the adherent cells themselves remaining unstained.[6] This same "negative-staining" method can also be used to assess cell spreading and progress toward confluency.

Cell Adhesion Assays Using Enzyme Substrates

Essentially all of the esterase substrates in Table 15.1 useful for monitoring cell viability can also be used for studying cell adhesion. As with cell viability studies, calcein AM (C1430, C3099, C3100MP) appears to best satisfy the criteria for assaying cell adhesion[7–9] and to have the least effect on cell viability and other cell functions.[7] The results obtained with leukocyte adhesion assays using calcein AM correlate well with those obtained with ^{51}Cr assays,[10,11] but the calcein AM protocols take less time and avoid the special handling required when using radioactive material. Calcein AM has been used in numerous cell adhesion assays, including those designed to measure:

- Binding of labeled Jurkat cells to vascular cell adhesion molecule-1 (VCAM1) in cell membrane preparations[12]
- Effects of E-selectin–binding peptides[13] and integrins[14] on neutrophil adhesion
- Integrin-mediated cell adhesion in transfected K562 cells[15,16] and BSC-1 cells[17]
- Leukocyte[7,11,18] and neutrophil[18–20] adhesion to endothelial cells
- Monocyte adhesion in HIV-infected cells[21]

Other fluorogenic esterase substrates that have been used to assess cell adhesion include BCECF AM[5,22–29] (B1150, B1170, B3051), carboxyfluorescein diacetate[7,30–32] (5(6)-CFDA, C195), fluorescein diacetate[33] (F1303), the succinimidyl ester of carboxyfluorescein diacetate[14] (5(6)-CFDA, SE; CFSE; C1157) and CellTracker™ Green CMFDA[23] (C2925, C7025), all of which are discussed in Section 15.2. Because of the possibility of slow calcein leakage from calcein AM–labeled cells, CellTracker™ Green CMFDA and 5(6)-CFDA SE are recommended for quantitative adhesion or aggregation assays that require incubation for more than about an hour.[14,34,35] Use of a combination of a green-fluorescent dye (usually calcein AM, BCECF AM or CMFDA) and a red-fluorescent dye—especially chloromethyl SNARF®-1 acetate (C6826), SNARF®-1 carboxylic acid, acetate, succinimidyl ester (S22801, Section 15.4) or carboxynaphthofluorescein diacetate (C13196, Section 15.2)—enables two-color measurements of adhesion or cell–cell aggregation in mixed cell cultures.[36] Calcein AM has also been used in a quantitative microplate assay of spreading of adherent cells on artificial or biological surfaces.[18]

In their review,[4] Löster and Hortstkorte describe the use of a number of other fluorogenic substrates for quantitating cell adhesion; among these are fluorescein diphosphate (F2999, Section 10.3) and ELF® 97 phosphate (Section 6.3) for lysosomal and membrane phosphatases, as well as various substrates for glycosidases (Section 10.2) and oxidases (Section 10.5).

Vybrant® Cell Adhesion Assay Kit

The Vybrant® Cell Adhesion Assay Kit (V13181) utilizes calcein AM to provide a fast and sensitive method for measuring cell–cell or cell–substratum adhesion.[7,13,14] Calcein AM is nonfluorescent but, once loaded into cells, is cleaved by endogenous esterases to produce calcein, a highly fluorescent and well-retained dye. Calcein provides a bright green-fluorescent, pH-independent, cytoplasmic cell marker that does not appear to affect the cell adhesion process.[9]

To perform this assay, samples of calcein AM–labeled cells are added to monolayers of unlabeled cells in a microplate. Following incubation to allow the labeled cells to adhere to the unlabeled cells, the samples are washed to remove any nonadhering labeled cells (Figure 15.6.1). The calcein fluorescence in this sample, as compared with that of an unlabeled control sample, can then be used to calculate the number of adherent cells. The absorption/emission maxima of calcein (~494/517 nm) are ideally suited for detection by a fluorescence microplate reader equipped with standard fluorescein optical filters.

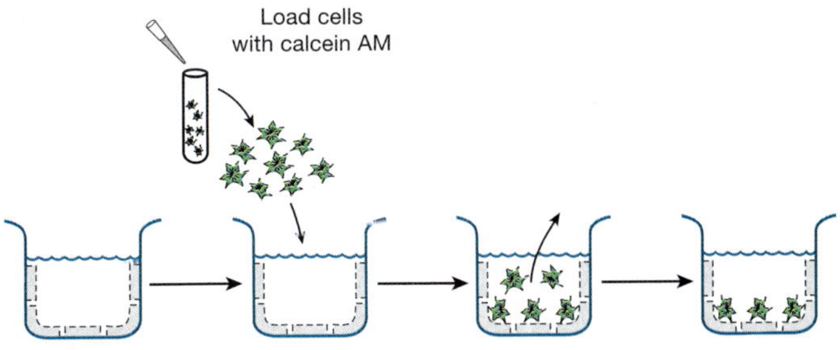

Figure 15.6.1 Principle of the Vybrant® Cell Adhesion Assay Kit (V13181). Microplate wells may be left untreated or precoated with extracellular matrix proteins, antibodies or other reagents.

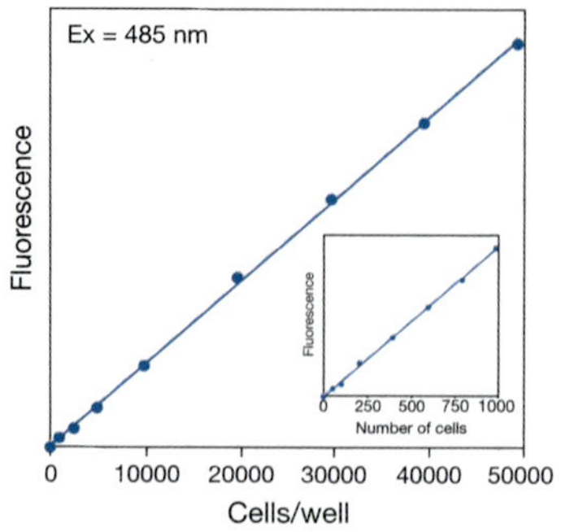

Figure 15.6.2 Quantitation of NIH 3T3 fibroblasts using the CyQUANT® Cell Proliferation Assay Kit (C7026). Fluorescence measurements were made using a microplate reader with excitation at 485 nm and emission detection at 530 nm. The linear range of the assay under these conditions is from 50 to 50,000 cells per 200 µL sample. The inset shows the linearity that can be obtained at very low numbers of cells.

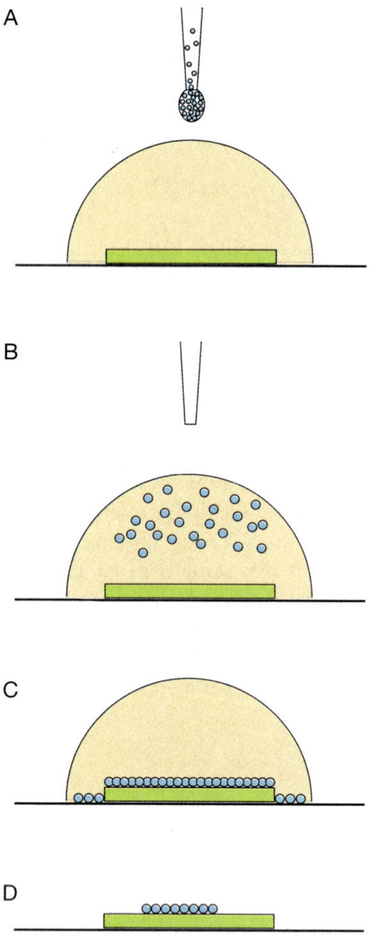

Figure 15.6.3 Schematic view of the microsphere adhesion assay. Membrane-coated microspheres are added to a living tissue slice in a drop of medium (**A**). The microspheres disperse through the medium and eventually cover the tissue slice (**B** and **C**). After incubation, nonadherent microspheres are removed by extensive washing, and the sample is ready for analysis by fluorescence microscopy (**D**).

In addition to calcein AM, the Vybrant® Cell Adhesion Assay Kit includes SYTOX® Green nucleic acid stain, an easy-to-use dead-cell indicator for assessing overall health of cells prior to performing the cell adhesion assay. As a fluorescent substitute for trypan blue, this high-affinity nucleic acid stain easily penetrates cells that have compromised membranes but will not cross the membranes of live cells. Upon binding to nucleic acids, the SYTOX® Green dye exhibits a >500-fold fluorescence enhancement and can be observed with standard fluorescein optical filters (excitation/emission ~504/523 nm).

The Vybrant® Cell Adhesion Assay Kit contains sufficient reagents to perform about 1000 assays using a fluorescence microplate reader, including:

- Calcein AM
- SYTOX® Green nucleic acid stain, an easy-to-use green-fluorescent indicator of the overall health of cells prior to performing the cell adhesion assay
- Detailed protocols

Cell Adhesion Assays Using the CyQUANT® Cell Proliferation Assay Kits

Our CyQUANT® Cell Proliferation Assay Kits (C7026, C35006, C35007, C35011, C35012; Section 15.4) can also serve as important tools for quantitating both cell–cell and cell–surface adhesion. The CyQUANT® GR, CyQUANT® NF and CyQUANT® Direct assays detect total nucleic acids in cells, with a linear detection range from 100 to at least 20,000 cells per well (Figure 15.6.2). To quantitate cell–surface adhesion, cells are simply permitted to adhere to the surface, gently washed to remove nonadherent cells and then analyzed to determine total nucleic acids in the adherent cells using the CyQUANT® assay protocol. As a control, the fluorescence of the total number of cells added to the well before the wash step can be determined by the same assay method to yield the percentage of adherent cells. It should also be possible to extend the CyQUANT® assay to studies of cell–cell adhesion by quantitating both the number of surface-adhering cells originally plated and the number of total cells after a second cell line has been introduced and allowed to adhere. A similar assay for cell adhesion based on DAPI (D1306, D3571, D21490; Section 8.1) has been reported.[37]

Microsphere Adhesion Assays

In a novel method for studying adhesion in live neural tissue slices, including hippocampal slices, fluorescent 4 µm microspheres (F8858, F8859; Section 6.5) are coated with isolated cell membranes from dissociated live cells. The membrane-coated microspheres are then seeded on live tissue slices, and after a short incubation time, the nonadherent microspheres are eliminated by washing (Figure 15.6.3). The pattern of tissue labeling of adherent microspheres can then be visualized by epifluorescence microscopy.[38] A flow cytometry–based microsphere adhesion assay has also been developed to determine integrin-specific adhesion, as well as to analyze integrin-mediated adhesion defects in B-lineage acute lymphoblastic leukemia.[39,40] FluoSpheres® microspheres of different colors (or sizes or intensities) (Section 6.5) can be used for simultaneous labeling of different adhesion factors.

Fluorescent Fibrinogen

Fibrinogen is a key component in the blood clotting process and can support both platelet–platelet and platelet–surface interactions by binding to the glycoprotein IIb-IIIa (GPIIb-IIIa) receptor[41–44] (also called integrin $\alpha_{IIb}\beta_3$). Although the mechanism is not well understood, activation of GPIIb-IIIa is required for fibrinogen binding, which leads to platelet activation, adhesion, spreading and microfilament reorganization of human endothelial cells *in vitro*.[41,44–46] Soluble fibrinogen binds to its receptor with a Ca^{2+}-dependent apparent K_d of 0.18 µM.[47] This binding is apparently mediated by the tripeptide sequence Arg-Gly-Asp (RGD), found both in fibrinogen and fibronectin, as well as some other proteins.[44,48–50]

Fluorescently labeled fibrinogen has proven to be a valuable tool for investigating platelet activation and subsequent fibrinogen binding. Fluorescein fibrinogen has been used to identify activated platelets by flow cytometry.[45,51–53] The binding of fluorescein fibrinogen to activated platelets has been shown to be saturable and can be inhibited completely by underivatized fibrinogen.[52] The preferential binding and accumulation of fluorescein fibrinogen at

IMPORTANT NOTICE: The products described in this manual are covered by one or more Limited Use Label License(s). Please refer to the Appendix on page 971 and Master Product List on page 975. Products are For Research Use Only. Not intended for any animal or human therapeutic or diagnostic use.

the endothelial border of venular blood vessels has been studied by quantitative fluorescence microscopy.[54] We offer five fluorescent conjugates of human fibrinogen, which are useful for investigating platelet activation and subsequent fibrinogen binding using fluorescence microscopy or flow cytometry (Figure 15.6.4):

- Alexa Fluor® 488 human fibrinogen (F13191)
- Oregon Green® 488 human fibrinogen (F7496)
- Alexa Fluor® 546 human fibrinogen (F13192)
- Alexa Fluor® 594 human fibrinogen conjugate (F13193)
- Alexa Fluor® 647 human fibrinogen conjugate (F35200)

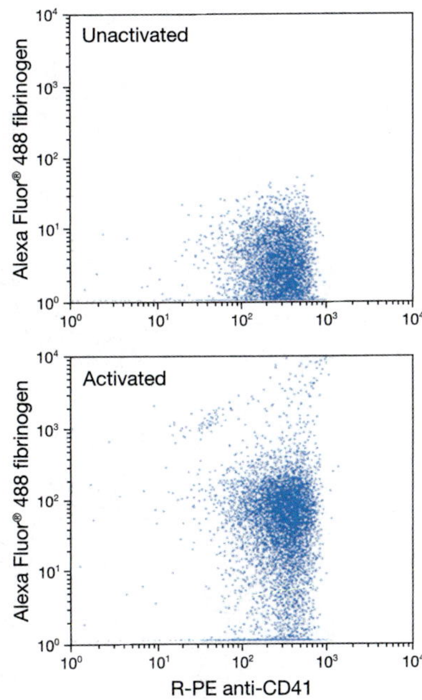

Figure 15.6.4 Interaction of fluorescently labeled fibrinogen with activated platelets. Whole blood was first incubated with an R-phycoerythrin (R-PE)–labeled anti-CD41 antibody to label the platelets. 20 µM adenosine 5′-diphosphate (ADP) was added in order to activate the platelets, then 2 µg/mL Alexa Fluor® 488 fibrinogen (F13191) was added and incubated with the sample for 5 minutes. Cells were analyzed by flow cytometry using excitation at 488 nm. Both activated and unactivated platelets show binding of the anti-CD41 antibody; however, only the activated platelets show strong binding by fibrinogen. A total of 5000 platelets are shown in each experiment.

Fluorescent Gelatin and Collagen

Collagen is a major component of the extracellular matrix and, in vertebrates, constitutes approximately 25% of total cellular protein. This important protein not only serves a structural role, but also is important in cell adhesion and migration. Specific collagen receptors, fibronectin and a number of other proteins involved in cell–cell and cell-surface adhesion have been demonstrated to bind collagen and gelatin[55,56] (denatured collagen). We offer fluorescent conjugates of gelatin and collagen for use in studying collagen-binding proteins and collagen metabolism, as well as gelatinases and collagenases (Section 10.4), which are metalloproteins that digest gelatin and collagen.

We offer two green-fluorescent gelatin conjugates—fluorescein gelatin (G13187) and Oregon Green® 488 gelatin (G13186). When compared with the fluorescein conjugate, the Oregon Green® 488 conjugate exhibits almost identical fluorescence spectra but its fluorescence is much more photostable and less pH dependent. We also offer collagen-coated FluoSpheres® yellow-green–fluorescent microspheres (F20892, F20893; Section 16.1). By analogy to results obtained with fluorescein conjugates of these proteins, these highly fluorescent gelatin conjugates and collagen-coated microspheres are potentially useful for:

- Following integrin-mediated phagocytosis[57]
- Localizing surface fibronectin on cultured cells[58]
- Studying fibronectin–gelatin interactions in solution using fluorescence polarization[56,59]
- Visualizing gelatinase activity using *in situ* gel zymography[60]

Chemotaxis Assays

Direct Detection of Chemotaxis Using Cell Counting

Chemotaxis, defined as directed cell motion toward an extracellular gradient, plays an important role during fertilization, inflammation, wound healing and hematopoiesis.[61] Chemotaxis is typically assayed by determining the number of viable cells that have migrated through a special "chemotaxis chamber." A 96-well disposable chemotaxis chamber is reported to be suitable for fluorescence-based assays that are faster, less labor intensive and more sensitive than visually detected migration assays.[62] The probes used to follow chemotaxis in live cells are often the same esterase substrates that are used for assaying cell viability and cell adhesion (Table 15.1), including calcein AM[63–67] (C1430, C3099, C3100MP), BCECF AM[7,24] (B1150, B1170, B3051) and CellTracker™ Green CMFDA[68] (C2925, C7025). Calcein AM does not interfere with lymphocyte proliferation or with granulocyte or neutrophil chemotaxis or superoxide production,[7,64] and, unlike BCECF AM, calcein AM does not affect chemotaxis in leukocytes.[65]

Because chemotaxis involves translocation of whole cells, assays that simply count cell numbers—such as our CyQUANT® Cell Proliferation Assay Kit (C7026, Section 15.4)—are also quite reliable for following chemotaxis.[69] In addition, the green-fluorescent SYTO® 13 dye (S7575, Section 8.1) has been used to track the co-migration of separately stained populations of neutrophils using opposing gradients of leukotriene B_4 and interleukin 8 as the chemoattractants.[70]

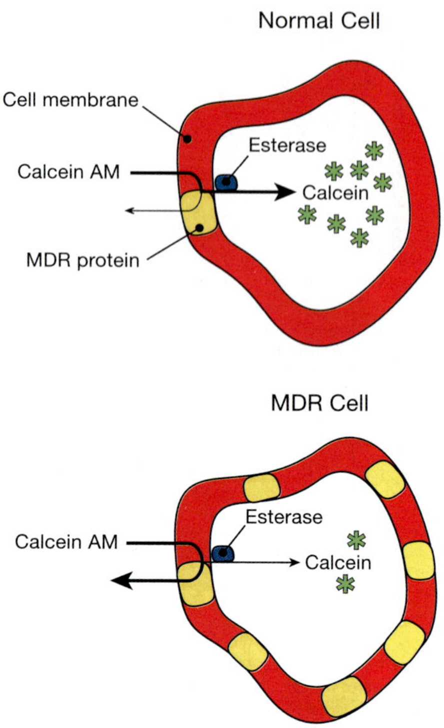

Figure 15.6.5 Principle of the Vybrant® Multidrug Resistance Assay Kit (V13180). In normal cells, nonfluorescent calcein AM readily diffuses across the cell membrane. Fluorescent calcein accumulates in the cytoplasm after cleavage of calcein AM by endogenous esterases. In MDR cells, overexpression of MDR transporter proteins increases expulsion of calcein AM from the cell membrane before enzymatic hydrolysis of its AM esters, thus reducing accumulation of intracellular calcein.

Probes for Chemotaxis Receptors

We prepare the fluorescein conjugate of the chemotactic hexapeptide *N*-formyl-Nle-Leu-Phe-Nle-Tyr-Lys (F1314), which binds to the fMLF receptor[71,72] (Section 16.1). We also offer fluorescein-labeled casein (C2990), which has been used to demonstrate casein-specific chemotaxis receptors in human neutrophils and monocytes with flow cytometry.[73,74] Neutrophils activated with phorbol myristate acetate have been shown to undergo a dose-dependent increase in binding of fluorescein-labeled casein.[75]

Multidrug Resistance Assays

Multidrug resistance (MDR) is a phenomenon representing a complex group of biological processes that are of growing interest in both clinical and experimental oncology.[76–78] The MDR phenotype is characterized by the acquired resistance of tumor cells to structurally and functionally dissimilar anticancer drugs. Among the many mechanisms contributing to this multidrug resistance are the following:

- Amplification of genes encoding drug-metabolizing enzymes
- Elevated levels of glutathione and glutathione-conjugating enzymes
- Mutated DNA topoisomerases
- Overexpression of plasma membrane ATP-dependent drug efflux pumps

Based on substrate and inhibitor profiles, at least four different plasma membrane ATP-dependent drug efflux pumps have been identified.[79,80] The activity of the verapamil-sensitive P-glycoprotein (Pgp) encoded by the *MDR1* gene leads to extrusion of anthracyclins, epipodophyllotoxins, *Vinca* alkaloids, coelenterazine and other cytostatic drugs.[81,82] BODIPY® FL paclitaxel (P7500; Section 11.2) is a substrate for Pgp-mediated transport that has been used as a probe for tumor spheroids.[83]

Many tumor cells do not express Pgp but export daunorubicin via a second, energy-dependent drug export mechanism.[84] A third, energy-dependent drug exporting mechanism is associated with a MDR-associated protein[85] (MRP), which shows high selectivity toward glutathione *S*-conjugates and is inhibited by *Vinca* alkaloids and probenecid.[86,87] A fourth vanadate- and verapamil-resistant but probenecid-sensitive glutathione *S*-conjugate–exporting system in mouse and rat fibroblasts has also been reported.[88] The probes commonly used to follow the transport of glutathione adducts are the same probes used to measure intracellular levels of glutathione (see below): monochlorobimane[89–92] (M1381MP) and CellTracker™ Green CMFDA[91,93,94] (C2925, C7025).

Obviously, the mechanisms of MDR are complex and, in some cases, have overlapping selectivity for the substrates.[80] We offer the Vybrant® Multidrug Resistance Assay Kit, along with a variety of useful fluorescent probes for monitoring various aspects of the MDR phenotype.

MDR Assays Using Acetoxymethyl Esters

The discovery that fluorescent calcium indicators such as indo-1 AM and fluo-3 AM (I1203, F1241, F23915; Section 19.3) and other dyes such as calcein AM (C1430, C3099, C3100MP), are rapidly extruded from cells expressing Pgp[95] presented a new class of highly sensitive probes for functional assays of the *MDR1*-encoded Pgp. Calcein AM—but not calcein—is an activator of Pgp in isolated membranes with a $K_d \leq 1$ µM.[96] Cells expressing the *MDR1*-encoded Pgp rapidly remove the nonfluorescent probe calcein AM, resulting in decreased accumulation of the highly fluorescent calcein in the cytoplasmic compartment.[87,96–98] Calcein AM is also a substrate for the MDR-associated protein (MRP), although in this case the hydrophilic free calcein anion is also exported.[99–101] Because calcein itself is not a substrate for the *MDR1*-encoded Pgp, MDR can be quantitatively assessed by measuring the net accumulation of intracellular fluorescence.[102] Cellular depletion of glutathione does not affect the extrusion of calcein AM by the *MDR1*-encoded Pgp.[99] Fluorescence of intracellular calcein can be distinguished from that of calcein that has leaked into the medium by adding Co^{2+} to totally quench the fluorescence of the extracellular dye.[101] The patented calcein AM assay for Pgp-related MDR is suitable for either flow cytometry[103] or fluorometry and is more rapid and significantly more sensitive than conventional assays based on doxorubicin accumulation.[104] Reduced accumulation of calcein in MDR cells can also be observed in single cells by fluorescence microscopy.[97,100]

IMPORTANT NOTICE: The products described in this manual are covered by one or more Limited Use Label License(s). Please refer to the Appendix on page 971 and Master Product List on page 975. Products are For Research Use Only. Not intended for any animal or human therapeutic or diagnostic use.
molecular probes® | invitrogen™ by life technologies™

Vybrant® Multidrug Resistance Assay Kit

The Vybrant® Multidrug Resistance Assay Kit (V13180), which is based on the fluorescence microplate–based method developed by Tiberghien and Loor,[98] provides a rapid and simple method for large-scale screening of MDR inhibitors. This assay utilizes the fluorogenic dye calcein AM as a substrate for efflux activity of Pgp. Upon hydrolysis by intracellular esterases, calcein is well retained in the cytosol and, unlike the hydrolysis product of other fluorescent Pgp substrates such as BCECF AM or fura-2 AM, its fluorescence is neither pH nor calcium dependent. MDR cells expressing high levels of Pgp rapidly extrude nonfluorescent calcein AM from the plasma membrane, reducing accumulation of fluorescent calcein in the cytosol[87,96,102] (Figure 15.6.5). The amount of Pgp activity is inversely proportional to the accumulation of intracellular calcein fluorescence. This assay is designed for use with fluorescence microplate readers and is particularly useful for rapid and sensitive screening of candidate Pgp inhibitors in MDR cell lines. The absorption/emission maxima of calcein (494/517 nm) are ideally suited for detection by instruments equipped with standard fluorescein filters. The Vybrant® Multidrug Resistance Assay Kit (V13180) contains sufficient reagents to perform approximately 10,000 assays using a fluorescence microplate reader, including:

- Calcein AM
- Cyclosporin A, a competitive inhibitor of drug binding to Pgp
- Verapamil, a calcium channel blocker that noncompetitively inhibits Pgp activity
- Detailed protocols

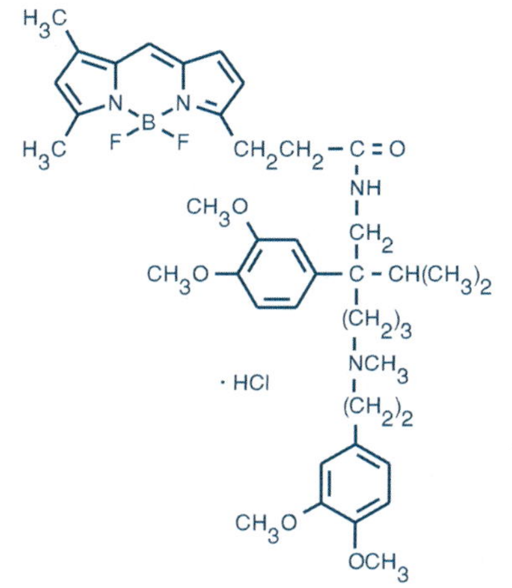

Figure 15.6.6 BODIPY® FL verapamil, hydrochloride (B7431).

MDR Assays Using Mitochondrial Probes

In the classical functional assay for MDR, doxorubicin efflux is measured.[105–107] The principal weakness of this assay is its low sensitivity, which has stimulated the search for other fluorochromes to monitor drug efflux. MDR cells overexpressing the Pgp have been identified using various mitochondrial probes (Section 12.2), including rhodamine 123 (R302, R22420), acridine orange 10-nonyl bromide (nonyl acridine orange, A1372) and rhodamine 6G[79,108–115] (R634). Furthermore, it has been reported that the fluorescence excitation spectrum of rhodamine 123 is different in drug-resistant and drug-sensitive cells.[116,117]

The potential-sensitive carbocyanine dyes, including dipentyl-, dihexyl- and diheptyloxacarbocyanines (D272, D273, D378; Section 22.3), also have advantages over the common doxorubicin assay for MDR.[105,118] Not only are these carbocyanine dyes more fluorescent, permitting use of lower dye concentrations, but their fluorescence increases upon binding to cell membranes,[119] unlike the fluorescence of doxorubicin, which is substantially quenched inside cells.[120] The ratiometric, potential-sensitive di-4-ANEPPS probe (D1199, Section 22.2) has been used to demonstrate that MDR cells have decreased electrical potentials.[121]

MDR Assays Using Nucleic Acid Stains

MDR cells have also been identified by their decreased accumulation of nucleic acid–binding dyes such as Hoechst 33258 (H1398, H3569, H21491; Section 8.1), Hoechst 33342 (H1399, H3570, H21492; Section 8.1) and ethidium bromide[115,122-124] (E1305, 15585-011; Section 8.1). SYTO® 16 green-fluorescent nucleic acid stain (S7578; Section 8.1) was shown to be a useful substrate for detecting Pgp-mediated resistance by flow cytometry and in single cells by confocal laser-scanning microscopy.[125] Our four SYTO® Fluorescent Nucleic Acid Stain Sampler Kits (S7572, S11340, S11350, S11360; Section 8.1) provide samples of SYTO® dyes with fluorescence covering the entire visible spectrum; these dyes may be screened for their utility in detecting MDR cells.

BODIPY® FL Verapamil and Dihydropyridine

The Ca^{2+}-channel blocker verapamil is one of several molecules known to inhibit Pgp-mediated drug efflux.[126] We offer the green-fluorescent BODIPY® FL verapamil probe (B7431, Figure 15.6.6) for the study of Pgp function and localization. Verapamil appears to inhibit drug efflux by acting as a substrate for Pgp, thereby overwhelming the transporter's capacity to expel other drugs. Our BODIPY® FL conjugate of verapamil, with spectral properties similar to fluorescein, also serves as a substrate for Pgp. This fluorescent verapamil derivative preferentially accumulates in the lysosomes of normal, drug-sensitive NIH 3T3 cells but is rapidly transported out of MDR cells, as revealed by fluorescence microscopy.[127–131] The outward transport of BODIPY® FL verapamil from MDR cells is inhibited by underivatized verapamil, as well as by excess vinblastine.[129]

Like verapamil, dihydropyridines are known to inhibit drug efflux. Consequently, our fluorescent dihydropyridines [132,133] labeled with either the green-fluorescent DM-BODIPY® (D7443, Section 16.3) or the orange-fluorescent ST-BODIPY® (S7445, Section 16.3) fluorophores may be useful MDR probes.

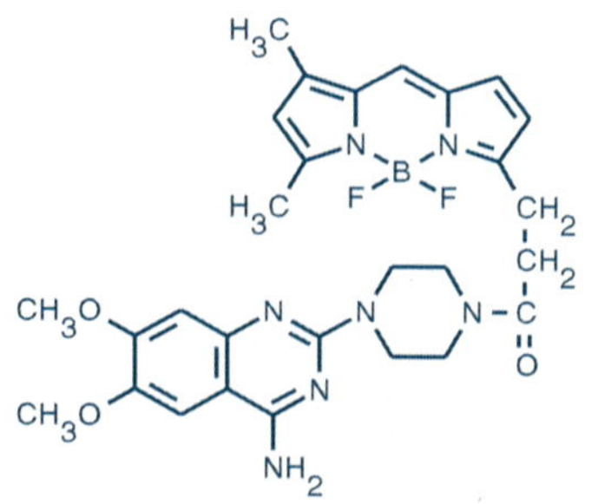

Figure 15.6.7 BODIPY® FL prazosin (B7433).

BODIPY® FL Vinblastine

The antimitotic agent vinblastine inhibits the Pgp-mediated efflux of a number of drugs and other probes from multidrug-resistant cells. Our BODIPY® FL conjugate of vinblastine (V12390) is useful as a highly fluorescent vinblastine analog.[134] A biologically active coumarin dye–labeled vinblastine has previously been described.[135]

BODIPY® FL Prazosin and BODIPY® FL Forskolin

Photoaffinity analogs of prazosin, an α_1-adrenergic receptor antagonist,[136,137] and forskolin, an adenylate cyclase activator, have been shown to selectively photolabel isolated Pgp.[138–141] Our green-fluorescent BODIPY® FL prazosin (B7433, Figure 15.6.7) and BODIPY® FL forskolin (B7469, Figure 15.6.8) are useful tools for probing MDR mechanisms.[82]

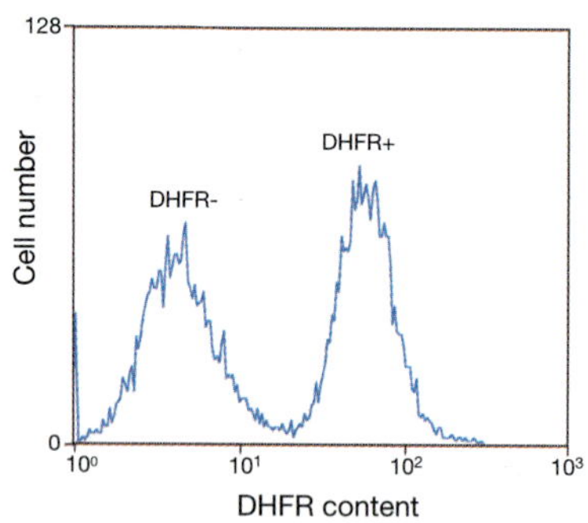

Figure 15.6.9 Analysis of dihydrofolate reductase (DHFR) content in CHO cells. A mixture of DHFR+ and DHFR- cells was stained with 1 µM fluorescein methotrexate (M1198MP) for 2 hours at 37°C. Following incubation with fluorescein methotrexate, cells were trypsinized, washed in phosphate-buffered saline and analyzed by flow cytometry using 488 nm excitation. Emission was collected at 525 nm.

Methotrexate Resistance and Gene Amplification

Tumor cells often undergo gene amplification that leads to overexpression of dihydrofolate reductase (DHFR). This increased DHFR expression (Figure 15.6.9) confers enhanced tolerance to the cytotoxic effects of methotrexate.[76,115] For the study of antimetabolite resistance and spontaneous gene amplification, we offer fluorescent methotrexate conjugates. In addition to green-fluorescent fluorescein methotrexate (M1198MP), which has been used to visualize biochemical networks in living cells,[142] we provide Alexa Fluor® 488 methotrexate (M23271), which exhibits fluorescein-like spectral characteristics. The quantitative binding of fluorescein methotrexate (M1198MP) to dihydrofolate reductase (DHFR) enables researchers to isolate cells based on DHFR expression.[143–146] Our fluorescein cadaverine adduct, which was originally described by Gapski and colleagues,[147] appears to be equivalent to fluorescein lysine methotrexate in its applications.[143]

Glutathione Determination

The tripeptide glutathione (γ-L-glutamyl-L-cysteinylglycine) is the most abundant nonprotein thiol in mammalian cells. Glutathione plays a central role in protecting cells of all organs, including the brain,[149] against damage produced by free radicals, oxidants and electrophiles. A distinct mechanism of MDR involves the overexpression of energy-dependent membrane pumps dedicated to removal of glutathione *S*-conjugates from the cytoplasm by a multidrug resistance–associated protein [85,150] (MRP). An increased rate of efflux of glutathione from Jurkat T lymphocytes during anti-FAS/APO-1 antibody–induced apoptosis has been reported.[151]

Several fluorescent reagents have been proposed for determining cellular levels of glutathione and glutathione *S*-transferase (GST), which catalyzes the formation of glutathione *S*-conjugates, but no probe is without drawbacks in quantitative studies of live cells. The high but variable levels of intracellular glutathione and the multitude of GST isozymes make kinetic measurements under saturating substrate conditions difficult or impossible.[152,153] Isozymes of GST vary both in abundance and activity, further complicating the analysis.[154] Moreover, the fluorescent reagents designed to measure glutathione may react with intracellular thiols other than glutathione, including proteins in glutathione-depleted cells.[155] Therefore, precautions must be

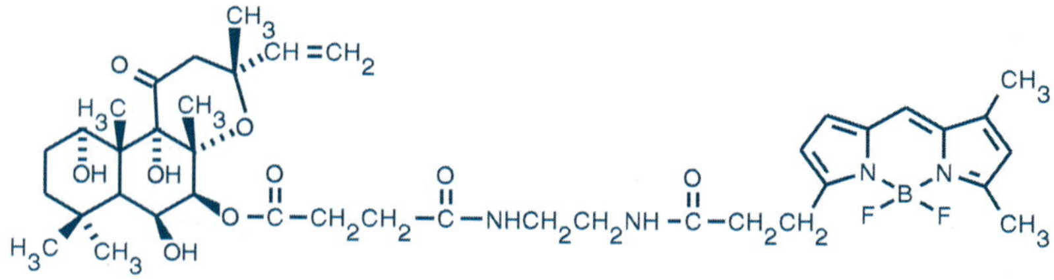

Figure 15.6.8 BODIPY® FL forskolin (B7469).

IMPORTANT NOTICE: The products described in this manual are covered by one or more Limited Use Label License(s). Please refer to the Appendix on page 971 and Master Product List on page 975. Products are For Research Use Only. Not intended for any animal or human therapeutic or diagnostic use.
www.invitrogen.com/probes

taken in applying the reagents mentioned here to quantitate either glutathione or GST in cells. A useful strategy is to test a variety of glutathione-sensitive dyes—those requiring glutathione S-transferase activity, as well as GST-independent fluorophores—under controlled experimental conditions in which glutathione is depleted.[156]

ThiolTracker™ Violet Glutathione Detection Reagent

The ThiolTracker™ Violet reagent (T10095, T10096) reacts with reduced thiols in intact cells and is up to 10-fold brighter than the bimanes traditionally used for gluthathione (GSH) detection. Staining is achieved by applying ThiolTracker™ Violet dye to live cells in thiol-free buffer and then directly imaging labeled cells using excitation with either a 405 nm violet diode laser or conventional xenon or mercury arc lamps (excitation/emission maxima ~404/526 nm, Figure 15.6.10). Alternatively, because this cell-permeant stain survives formaldehyde fixation and detergent extraction, labeled cells can be subjected to immunochemical analysis prior to imaging.

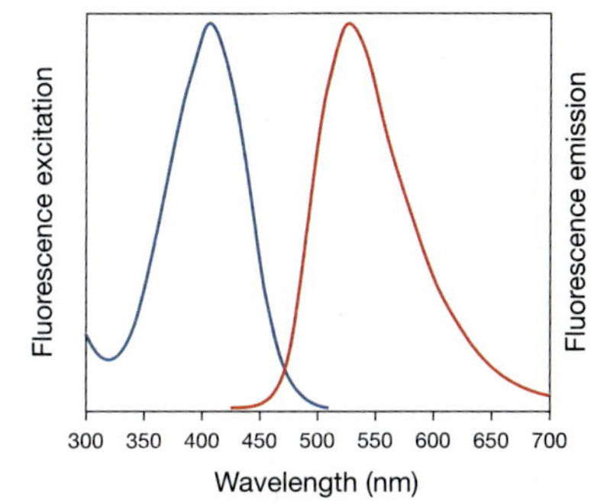

Figure 15.6.10 Fluorescence excitation and emission spectra of ThiolTracker™ Violet dye conjugated to glutathione (GSH) in pH 7.2 buffer.

Glutathione Determination with Monochlorobimane

Cell-permeant monochlorobimane (mBCl, M1381MP), which is essentially nonfluorescent until conjugated to thiols, has long been the preferred thiol-reactive probe for quantitating glutathione levels in cells and for measuring GST activity.[157,158] Because the blue-fluorescent glutathione adduct of monochlorobimane eventually accumulates in the nucleus, it is not a reliable indicator of the nuclear and cytoplasmic distribution of cellular glutathione.[159] Tissue glutathione levels can also be measured fluorometrically by adding both monochlorobimane and glutathione S-transferase to tissue homogenates.[160]

Monochlorobimane is reported to react more selectively with glutathione in whole cells than does monobromobimane (M1378, M20381; Section 2.3; Figure 15.6.11) and has proven useful for assaying drug resistance by flow cytometry[153,158,161] and by fluorescence microscopy.[162,163] Moreover, HPLC analysis has shown that glutathione is the only low molecular weight thiol in hepatocytes that reacts with monochlorobimane.[164] Results from glutathione determination with monochlorobimane have been shown to match those from an independent glutathione-specific assay using glutathione reductase.[165,166] However, although monochlorobimane was shown to be highly selective for glutathione in rodent cells, it was reported to inadequately label glutathione in human cells because of its low affinity for human glutathione S-transferases.[153] The reducing agent tris-(2-carboxyethyl)phosphine (TCEP, T2556; Section 2.1) has been used in place of dithiothreitol (DTT, D1532; Section 2.1) in a simplified monobromobimane-based assay for glutathione[167] and thus may also prove useful for monochlorobimane-based assays. In this monobromobimane-based glutathione assay, an extraction step is reportedly necessary to remove a fluorescent, reductive dehalogenation side product of TCEP and monobromobimane.[168] Probenecid (P36400; Section 19.8) inhibits the ATP-dependent organic anion pump and blocks the loss of the fluorescent bimane–glutathione adduct from rat fibroblasts.[169] Monochlorobimane has also been employed to sort cells based on their expression of recombinant GST.[170]

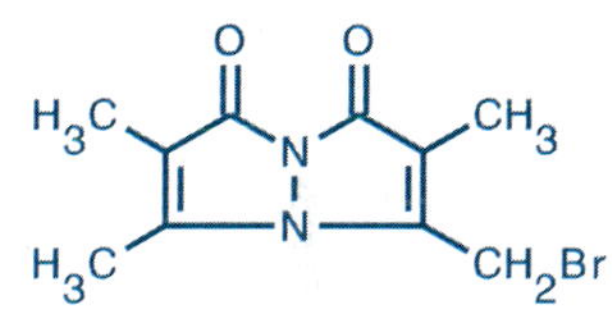

Figure 15.6.11 Monobromobimane (mBBr, M1378).

Glutathione Determination with Visible Light–Excitable Thiol-Reactive Probes

CellTracker™ Green CMFDA (5-chloromethylfluorescein diacetate, C2925, C7025) is a useful alternative to the UV light–excitable monochlorobimane for determining levels of intracellular glutathione.[171,172] CellTracker™ Green CMFDA can be excited by the argon-ion laser, and is compatible with flow cytometry and confocal laser-scanning microscopy applications. CellTracker™ Green CMFDA's enzymatic product has a much higher absorbance and fluorescence quantum yield than that of monochlorobimane. In conjunction with Hoechst 33342 (H1399, H3570, H21492; Section 8.1), CellTracker™ Green CMFDA has also been shown to be effective for analyzing intracellular thiol levels as a function of cell cycle using flow cytometry,[171,173] for following transport of the glutathione adduct to secretory vesicles in multidrug-resistant cells[91] and for detecting apoptotic cells, which have reduced levels of intracellular reduced glutathione.[174] Selectivity of CellTracker™ Green CMFDA for glutathione (versus thiolated proteins) was shown by the isolation of >95% of the intracellular fluorescent products as a mixture of the glutathione adduct and the unconjugated hydrolysis product, chloromethylfluorescein.[153] However, in these experiments, the high fluorescence of unconjugated chloromethylfluorescein resulted in significantly increased background levels. Because glutathione-depleting chemicals may also cause cell death, it has been recommended that

IMPORTANT NOTICE: The products described in this manual are covered by one or more Limited Use Label License(s). Please refer to the Appendix on page 971 and Master Product List on page 975. Products are For Research Use Only. Not intended for any animal or human therapeutic or diagnostic use.

calcein AM be used to independently assess cell viability in assays that use CellTracker™ Green CMFDA to measure changes in the level of intracellular glutathione.[172]

Like CMFDA, chloromethyl SNARF®-1 acetate (C6826, Figure 15.6.12) forms adducts with intracellular thiols that are well retained by viable cells. The glutathione adduct of chloromethyl SNARF®-1 can be excited by the 488 nm spectral line of the argon-ion laser yet emits beyond 630 nm, which may prove advantageous in multicolor applications or when assaying autofluorescent samples. A number of our other CellTracker™ (Section 14.2) and MitoTracker® (Section 12.2) probes have thiol-reactive chloromethyl moieties and may be similarly useful for glutathione determination. All of these probes form glutathione S-conjugates that are likely to be transported from the cytoplasm by an MDR-associated protein[85,150] (MRP).

Glutathiolation Detection with BioGEE

Biotinylated glutathione ethyl ester (BioGEE, G36000; Figure 15.6.13) is a cell-permeant, biotinylated glutathione analog for detecting glutathiolation. Under conditions of oxidative stress, cells may transiently incorporate glutathione into proteins. Stressed cells incubated with BioGEE will also incorporate this biotinylated glutathione derivative into proteins, facilitating the identification of oxidation-sensitive proteins.[175] Once these cells are fixed and permeabilized, glutathiolation levels can be detected with a fluorescent streptavidin conjugate (Section 7.6, Table 7.9) using either flow cytometry or fluorescence microscopy. Proteins glutathiolated with BioGEE can also be extracted and analyzed by mass spectrometry or by western blotting methods in conjunction with fluorophore- or enzyme-labeled streptavidin conjugates.

Glutathione Determination With o-Phthaldialdehyde and Naphthalene-2,3-Dicarboxaldehyde

The reagent o-phthaldialdehyde (OPA, P2331MP) reacts with both the thiol and the amine functions of glutathione, yielding a cyclic derivative that is fluorescent. The spectra of the glutathione adduct of OPA (excitation/emission maxima of 350/420 nm) are shifted from those of its protein adducts.[176] This effect has occasionally been used to estimate glutathione levels in cells.[153,177] OPA has also been used as a derivatization reagent for the chromatographic determination of glutathione in cells, blood and tissues.[178,179]

The membrane-permeant naphthalene-2,3-dicarboxaldehyde (NDA, N1138) has been used to determine glutathione levels in single cells. Cells were treated with the NDA reagent and then analyzed by capillary electrophoresis;[180] glutathione labeling was reported to be complete within two minutes. The glutathione adduct of NDA can be excited by the 458 nm spectral line of the argon-ion laser.

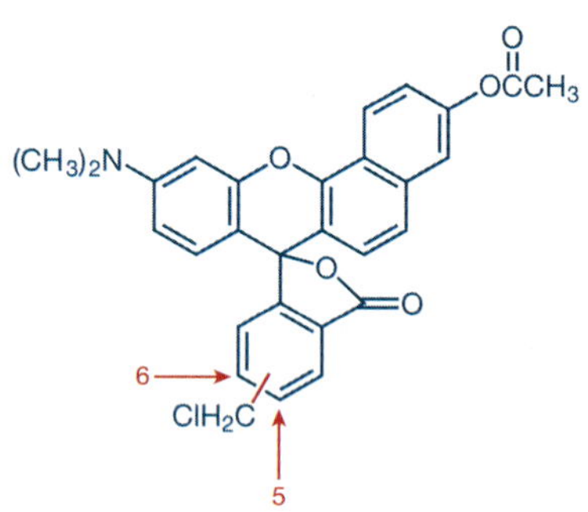

Figure 15.6.12 5-(and-6)-Chloromethyl SNARF®-1, acetate (C6826).

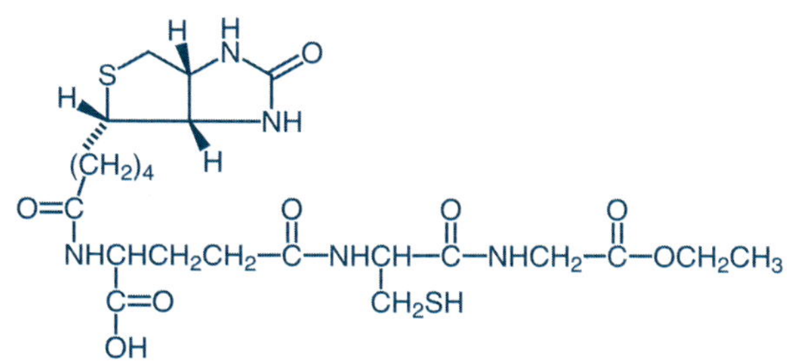

Figure 15.6.13 Glutathione ethyl ester, biotin amide (BioGEE, G36000).

REFERENCES

1. Pharmacol Rev (1998) 50:197; **2.** Cell (1996) 84:345; **3.** Annu Rev Biochem (1991) 60:155; **4.** Micron (2000) 31:41; **5.** Anal Biochem (1999) 267:37; **6.** Cytometry (1998) 33:41; **7.** J Immunol Methods (1994) 172:115; **8.** Cell Biol Toxicol (1994) 10:329; **9.** J Immunol Methods (1995) 178:41; **10.** Biotechniques (1997) 23:1056; **11.** J Immunol Methods (1993) 163:181; **12.** J Immunol Methods (1994) 175:59; **13.** J Biol Chem (1995) 270:21129; **14.** J Immunol Methods (1994) 172:25; **15.** J Cell Biol (1995) 130:745; **16.** J Cell Biol (1994) 127:1129; **17.** Proc Natl Acad Sci U S A (1993) 90:5700; **18.** Anal Biochem (1994) 219:288; **19.** J Immunol (1996) 157:3617; **20.** J Biol Chem (1994) 269:10008; **21.** J Immunol (1996) 156:1638; **22.** J Biol Chem (2000) 275:7052; **23.** J Cell Biol (1996) 133:445; **24.** Mol Biol Cell (1995) 6:661; **25.** J Biol Chem (1994) 269:1033; **26.** J Cell Biol (1994) 125:1395; **27.** J Biol Chem (1993) 268:8835; **28.** J Cell Biol (1993) 123:245; **29.** Biochem Biophys Res Commun (1990) 172:262; **30.** J Cell Biol (1994) 124:609; **31.** Dev Biol (1989) 135:133; **32.** J Cell Biol (1989) 109:3465; **33.** J Immunol Methods (1993) 157:117; **34.** Neuron (1996) 17:1089; **35.** Neuron (1996) 17:353; **36.** J Cell Biol (1996) 135:1593; **37.** J Immunol Methods (1993) 165:93; **38.** Biotechniques (1999) 26:466, 470, 472; **39.** J Immunol (2000) 165:442; **40.** Blood (1999) 94:754; **41.** J Biol Chem (1995) 270:28812; **42.** Thromb Res (1995) 77:543; **43.** Biochem J (1990) 270:149; **44.** Biochem Pharmacol (1987) 36:4035; **45.** J Biol Chem (1995) 270:11358; **46.** J Cell Biol (1987) 104:1403; **47.** J Biol Chem (1983) 258:12582; **48.** J Biol Chem (1998) 273:6821; **49.** Science (1986) 231:1559; **50.** Proc Natl Acad Sci U S A (1985) 82:8057; **51.** J Biol Chem (2000) 275:7795; **52.** Cytometry (1994) 17:287; **53.** J Lab Clin Med (1994) 123:728; **54.** Ann N Y Acad Sci (1983) 416:426; **55.** J Cell Sci (1992) 101:873; **56.** Arch Biochem Biophys (1983) 227:358; **57.** Biophys J (1996) 71:2319; **58.** J Cell Biol (1980) 87:14; **59.** Biochemistry (1993) 32:8168; **60.** FASEB J (1995) 9:974; **61.** Proc Natl Acad Sci U S A (2006) 103:1353; **62.** J Immunol Methods (1998) 213:41; **63.** Anal Biochem (2000) 280:11; **64.** Biophys J (1995) 68:1207; **65.** Cytometry (1995) 19:366; **66.** J Immunol (1996) 156:679; **67.** Immunol Invest (1996) 25:49; **68.** Am J Physiol (1998) 274:C182; **69.** Biotechniques (2000) 29:81; **70.** J Cell Biol (1999) 147:577; **71.** Biochemistry (1990) 29:313; **72.** J Biol Chem (1990) 265:16725; **73.** Inflammation (1983) 7:363; **74.** Inflammation (1977) 2:115; **75.** J Immunol (1987) 139:3028; **76.** Assay Drug Dev Technol (2007) 5:541; **77.** Proc Natl Acad Sci U S A (1994) 91:3497; **78.** Annu Rev Biochem (1993) 62:385; **79.** Biochemistry (1995) 34:32; **80.** Cytometry (1997) 29:279; **81.** Proc Natl Acad Sci U S A (2004) 101:1702; **82.** J Pharmacol Exp Ther (2009) 331:1118; **83.** Br J Cancer (2003) 89:1581; **84.** Cancer Res (1992) 52:17; **85.** Proc Natl

IMPORTANT NOTICE: The products described in this manual are covered by one or more Limited Use Label License(s). Please refer to the Appendix on page 971 and Master Product List on page 975. Products are For Research Use Only. Not intended for any animal or human therapeutic or diagnostic use.
www.invitrogen.com/probes

Acad Sci U S A (1994) 91:13033; **86.** Biochem Pharmacol (1996) 52:967; **87.** FEBS Lett (1995) 368:385; **88.** Int J Biochem Cell Biol (2004) 36:247; **89.** J Pharm Biomed Anal (1998) 18:585; **90.** Int J Radiat Biol (1998) 74:647; **91.** Int J Cancer (1998) 76:55; **92.** Int J Radiat Oncol Biol Phys (1989) 16:1315; **93.** J Biol Chem (1996) 271:30587; **94.** J Cell Sci (1998) 111:1137; **95.** Cytometry (1994) 17:343; **96.** J Biol Chem (1993) 268:21493; **97.** J Biol Chem (1996) 271:13668; **98.** Anticancer Drugs (1996) 7:568; **99.** FEBS Lett (1996) 383:99; **100.** Anticancer Res (1998) 18:2981; **101.** Biochemistry (1998) 37:2243; **102.** Biochim Biophys Acta (1994) 1191:384; **103.** Br J Cancer (1996) 73:849; **104.** Eur J Cancer (1993) 29A:1024; **105.** Cytometry (1987) 8:306; **106.** Cancer Res (1983) 43:5126; **107.** Cancer Res (1980) 40:3895; **108.** Anticancer Res (1995) 15:121; **109.** Cytometry (1994) 17:50; **110.** J Biol Chem (1994) 269:7145; **111.** Mol Pharmacol (1994) 45:1145; **112.** Proc Natl Acad Sci U S A (1994) 91:4654; **113.** FEBS Lett (1993) 329:63; **114.** Blood (1991) 78:1385; **115.** Cancer Res (1991) 51:4665; **116.** Exp Cell Res (1991) 196:323; **117.** Cancer Commun (1989) 1:145; **118.** J Cell Physiol (1986) 126:266; **119.** Biochemistry (1995) 34:3858; **120.** Biochemistry (1992) 31:12555; **121.** Biochemistry (1993) 32:11042; **122.** Anticancer Res (1993) 13:1557; **123.** Cell Biochem Funct (1992) 10:9; **124.** Cytometry A (2009) 75:14; **125.** Br J Cancer (1997) 76:1029; **126.** Cancer Res (1987) 47:1421; **127.** Biochem Biophys Res Commun (1995) 212:494; **128.** Biochim Biophys Acta (1995) 1237:31; **129.** Mol Pharmacol (1991) 40:490; **130.** Blood (1998) 91:4106; **131.** Biochemistry (1999) 38:6630; **132.** Biochemistry (1994) 33:11875; **133.** Proc Natl Acad Sci U S A (1992) 89:3586; **134.** Anticancer Res (1997) 17:3321; **135.** Biochemistry (1994) 33:12665; **136.** Cardiology (1986) 73:164; **137.** Life Sci (1980) 27:1525; **138.** J Biol Chem (1993) 268:11417; **139.** Biochem Pharmacol (1992) 43:89; **140.** J Biol Chem (1991) 266:20744; **141.** J Biol Chem (1989) 264:15483; **142.** Proc Natl Acad Sci U S A (2001) 98:7678; **143.** J Biol Chem (1982) 257:14162; **144.** Cancer Res (1990) 50:4946; **145.** Proc Natl Acad Sci U S A (1983) 80:3711; **146.** J Biol Chem (1978) 253:5852; **147.** J Med Chem (1975) 18:526; **148.** Clin Chem (1996) 42:64; **149.** Prog Neurobiol (2000) 62:649; **150.** J Biol Chem (1994) 269:29085; **151.** J Biol Chem (1996) 271:15420; **152.** Anal Biochem (1994) 217:41; **153.** Cytometry (1994) 15:349; **154.** Cancer Res (1991) 51:1783; **155.** Cytometry (1991) 12:366; **156.** Glia (2000) 30:329; **157.** Ann N Y Acad Sci (1993) 677:341; **158.** Cancer Res (1986) 46:6105; **159.** Biochem J (1993) 294:631; **160.** Anal Biochem (2000) 286:35; **161.** J Biol Chem (1988) 263:14107; **162.** J Biol Chem (2006) 281:17420; **163.** Cytometry (1995) 19:226; **164.** Anal Biochem (1990) 190:212; **165.** Cancer Res (1991) 51:4287; **166.** J Clin Lab Anal (1990) 4:324; **167.** Anal Biochem (1999) 272:107; **168.** Anal Biochem (2003) 318:325; **169.** Cytometry (1996) 23:78; **170.** Cytometry (1991) 12:651; **171.** Cytometry (1991) 12:184; **172.** Toxicol in Vitro (1996) 10:341; **173.** Cytometry (1993) 14:747; **174.** J Immunol (1997) 158:4612; **175.** Biochemistry (2000) 39:11121; **176.** Anal Biochem (1966) 14:434; **177.** Exp Cell Res (1986) 163:518; **178.** Clin Chem (1995) 41:448; **179.** Chromatographia (1993) 36:130; **180.** Anal Chem (1995) 67:4261.

DATA TABLE 15.6 PROBES FOR CELL ADHESION, CHEMOTAXIS, MULTIDRUG RESISTANCE AND GLUTATHIONE

Cat. No.	MW	Storage	Soluble	Abs	EC	Em	Solvent	Notes
B1150	~615	F,D	DMSO	<300		none		1, 2
B1170	~615	F,D	DMSO	<300		none		1, 2
B3051	~615	F,D	DMSO	<300		none		1, 2, 3
B7431	769.18	F,D,L	DMSO, EtOH	504	74,000	511	MeOH	
B7433	563.41	F,D,L	DMSO, EtOH	504	77,000	511	MeOH	
B7469	784.70	F,D,L	DMSO	504	79,000	511	MeOH	
C1430	994.87	F,D	DMSO	<300		none		4
C2925	464.86	F,D	DMSO	<300		none		5
C3099	994.87	F,D	DMSO	<300		none		3, 4
C3100MP	994.87	F,D	DMSO	<300		none		4
C6826	499.95	F,D	DMSO	<350		none		6
C7025	464.86	F,D	DMSO	<300		none		5
F1314	1213.41	F,L	pH >6, DMF	494	72,000	517	pH 9	
G36000	561.67	F,D	DMSO	<300		none		
M1198MP	979.08	F,L	pH >6, DMF	496	67,000	516	pH 9	
M1381MP	226.66	F,L	DMSO	380	6000	see Notes	MeOH	7
M23271	1055.06	F,D,L	DMSO	494	78,000	518	pH 7	
N1138	184.19	L	DMF, MeCN	419	9400	493	see Notes	8
P2331MP	134.13	L	EtOH	334	5700	455	pH 9	9
V12390	1043.02	F,D,L	DMSO, DMF	503	83,000	510	MeOH	

For definitions of the contents of this data table, see "Using *The Molecular Probes® Handbook*" in the introductory pages.

Notes

1. MW value is approximate. BCECF AM is a mixture of molecular species.
2. BCECF AM is colorless and nonfluorescent until converted to BCECF (B1151) by acetoxymethyl ester hydrolysis.
3. This product is supplied as a ready-made solution in the solvent indicated under "Soluble."
4. Calcein AM is converted to fluorescent calcein (C481) after acetoxymethyl ester hydrolysis.
5. Acetate hydrolysis of this compound yields a fluorescent product with similar pH-dependent spectral characteristics to C1904.
6. C6826 is converted to a fluorescent product with spectra similar to C1270 after acetate hydrolysis.
7. Bimanes are almost nonfluorescent until reacted with thiols. For monobromobimane conjugated to glutathione, Abs = 394 nm, Em = 490 nm (QY ~0.1–0.3) in pH 8 buffer. (Methods Enzymol (1987) 143:76, Methods Enzymol (1995) 251:133)
8. Spectral data are for the reaction product with glycine in the presence of cyanide, measured in pH 7.0 buffer/MeCN (40:60). (Anal Chem (1987) 59:1102) Unreacted reagent in MeOH: Abs = 279 nm (EC = 5500 $cm^{-1}M^{-1}$), Em = 330 nm.
9. Spectral data are for the reaction product of P2331MP with alanine and 2-mercaptoethanol. The spectra and stability of the adduct depend on the amine and thiol reactants. (Biochim Biophys Acta (1979) 576:440) Unreacted reagent in H_2O: Abs = 257 nm (EC = 1000 $cm^{-1}M^{-1}$).

IMPORTANT NOTICE: The products described in this manual are covered by one or more Limited Use Label License(s). Please refer to the Appendix on page 971 and Master Product List on page 975. Products are For Research Use Only. Not intended for any animal or human therapeutic or diagnostic use.

PRODUCT LIST 15.6 PROBES FOR CELL ADHESION, CHEMOTAXIS, MULTIDRUG RESISTANCE AND GLUTATHIONE

Cat. No.	Product	Quantity
B1150	2',7'-bis-(2-carboxyethyl)-5-(and-6)-carboxyfluorescein, acetoxymethyl ester (BCECF, AM)	1 mg
B3051	2',7'-bis-(2-carboxyethyl)-5-(and-6)-carboxyfluorescein, acetoxymethyl ester (BCECF, AM) *1 mg/mL solution in anhydrous DMSO*	1 mL
B1170	2',7'-bis-(2-carboxyethyl)-5-(and-6)-carboxyfluorescein, acetoxymethyl ester (BCECF, AM) *special packaging*	20 x 50 µg
B7469	BODIPY® FL forskolin	100 µg
B7433	BODIPY® FL prazosin	100 µg
B7431	BODIPY® FL verapamil, hydrochloride	1 mg
C1430	calcein, AM	1 mg
C3099	calcein, AM *1 mg/mL solution in anhydrous DMSO*	1 mL
C3100MP	calcein, AM *special packaging*	20 x 50 µg
C2990	casein, fluorescein conjugate	25 mg
C2925	CellTracker™ Green CMFDA (5-chloromethylfluorescein diacetate)	1 mg
C7025	CellTracker™ Green CMFDA (5-chloromethylfluorescein diacetate) *special packaging*	20 x 50 µg
C6826	5-(and-6)-chloromethyl SNARF®-1, acetate *mixed isomers* *special packaging*	20 x 50 µg
C7026	CyQUANT® Cell Proliferation Assay Kit *for cells in culture* *1000 assays*	1 kit
C35011	CyQUANT® Direct Cell Proliferation Assay *for 10 microplates*	1 kit
C35012	CyQUANT® Direct Cell Proliferation Assay *for 100 microplates*	1 kit
C35007	CyQUANT® NF Cell Proliferation Assay Kit *200 assays*	1 kit
C35006	CyQUANT® NF Cell Proliferation Assay Kit *1000 assays*	1 kit
F13191	fibrinogen from human plasma, Alexa Fluor® 488 conjugate	5 mg
F13192	fibrinogen from human plasma, Alexa Fluor® 546 conjugate	5 mg
F13193	fibrinogen from human plasma, Alexa Fluor® 594 conjugate	5 mg
F35200	fibrinogen from human plasma, Alexa Fluor® 647 conjugate	5 mg
F7496	fibrinogen from human plasma, Oregon Green® 488 conjugate	5 mg
F1314	formyl-Nle-Leu-Phe-Nle-Tyr-Lys, fluorescein derivative	1 mg
G13187	gelatin from pig skin, fluorescein conjugate	5 mg
G13186	gelatin from pig skin, Oregon Green® 488 conjugate	5 mg
G36000	glutathione ethyl ester, biotin amide (BioGEE) *glutathiolation detection reagent* *special packaging*	10 x 100 µg
M23271	methotrexate, Alexa Fluor® 488, inner salt (Alexa Fluor® 488 methotrexate) *mixed isomers*	500 µg
M1198MP	methotrexate, fluorescein, triammonium salt (fluorescein methotrexate)	1 mg
M1381MP	monochlorobimane (mBCl)	25 mg
N1138	naphthalene-2,3-dicarboxaldehyde (NDA)	100 mg
P2331MP	*o*-phthaldialdehyde (OPA) *high purity*	1 g
T10095	ThiolTracker™ Violet (Glutathione Detection Reagent) *180 assays* *set of 3 vials*	1 set
T10096	ThiolTracker™ Violet (Glutathione Detection Reagent) *for 5 microplates*	each
V12390	vinblastine, BODIPY® FL conjugate (BODIPY® FL vinblastine)	100 µg
V13181	Vybrant® Cell Adhesion Assay Kit	1 kit
V13180	Vybrant® Multidrug Resistance Assay Kit	1 kit

IMPORTANT NOTICE: The products described in this manual are covered by one or more Limited Use Label License(s). Please refer to the Appendix on page 971 and Master Product List on page 975. Products are For Research Use Only. Not intended for any animal or human therapeutic or diagnostic use.
www.invitrogen.com/probes
molecular probes® | invitrogen™ by life technologies™

SIXTEEN

CHAPTER 16

Probes for Endocytosis, Receptors and Ion Channels

IMPORTANT NOTICE: The products described in this manual are covered by one or more Limited Use Label License(s). Please refer to the Appendix on page 971 and Master Product List on page 975. Products are For Research Use Only. Not intended for any animal or human therapeutic or diagnostic use.

16.2 Probes for Neurotransmitter Receptors ... 760

16.3 Probes for Ion Channels and Carriers ... 766

IMPORTANT NOTICE: The products described in this manual are covered by one or more Limited Use Label License(s). Please refer to the Appendix on page 971 and Master Product List on page 975. Products are For Research Use Only. Not intended for any animal or human therapeutic or diagnostic use.
www.invitrogen.com/probes
molecular probes® | invitrogen™ by life technologies™

16.1 Probes for Following Receptor Binding and Phagocytosis

The plasma membrane defines the inside and outside of the cell. It not only encloses the cytosol to maintain the intracellular environment but also serves as a formidable barrier to the extracellular environment. Because cells require input from their surroundings—in the form of hydrated ions, small polar molecules, large biomolecules and even other cells—they have developed strategies for overcoming this barrier. Some of these mechanisms involve initial formation of receptor–ligand complexes, followed by transport of the ligand across the cell membrane.[1–5]

This section focuses on probes for following receptor binding, endocytosis and exocytosis. Section 16.2 describes tools for studying neurotransmitter receptors, which mediate external chemical messenger control over the electrical activity of neurons. Section 16.3 discusses strategies for monitoring ion channels and carriers, which are the molecular centerpiece of neural transmission and bioenergetics.

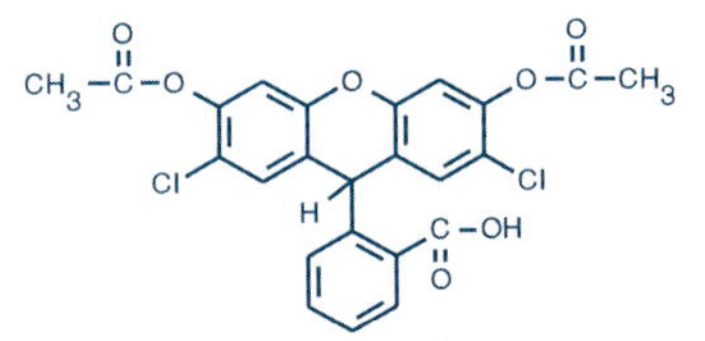

Figure 16.1.1 2',7'-dichlorodihydrofluorescein diacetate (2',7'-dichlorofluorescin diacetate; H_2DCFDA, D399).

Ligands for Studying Receptor-Mediated Endocytosis

We offer a variety of fluorescent and fluorogenic ligands that bind to membrane receptors and are subsequently internalized. In some cases, the bound ligand is released intracellularly and the receptor is then recycled to the plasma membrane. Receptor binding may also result in signal transduction (Chapter 17), Ca^{2+} mobilization (Chapter 19), intracellular pH changes (Chapter 20) and formation of reactive oxygen species (ROS, Chapter 18).

Fc OxyBURST® Green Assay Reagent: Fluorogenic Immune Complex

When soluble or surface-bound IgG immune complexes interact with Fc receptors on phagocytic cells, a number of host defense mechanisms are activated, including phagocytosis and activation of an NADPH oxidase–mediated oxidative burst.[6] Dichlorodihydrofluorescein diacetate (H_2DCFDA, D399; Section 18.2; Figure 16.1.1), a cell-permeant fluorogenic probe that localizes in the cytosol, has frequently been used to monitor this oxidative burst.[7] Its fluorescence response, however, is limited by the diffusion rate of the reactive oxygen species into the cytosol from the phagovacuole where it is generated. In contrast, Fc OxyBURST® assay reagents permit direct measurement of the kinetics of Fc receptor–mediated internalization and the subsequent oxidative burst in the phagovacuole, yielding signals that are many times brighter than those generated by H_2DCFDA (Figure 16.1.2, Figure 16.1.3).

Fc OxyBURST® Green assay reagent (F2902) was developed in collaboration with Elizabeth Simons of Boston University to monitor the oxidative burst in phagocytic cells using fluorescence instrumentation. Fc OxyBURST® Green assay reagent comprises bovine serum albumin (BSA) that has been covalently linked to dichlorodihydrofluorescein (H_2DCF) and then complexed with a purified rabbit polyclonal anti-BSA antibody (A11133). When these immune complexes bind to Fc receptors, the nonfluorescent H_2DCF molecules are internalized within the phagovacuole and subsequently oxidized to green-fluorescent 2',7'-dichlorofluorescein (DCF; Figure 16.1.2, Figure 16.1.3). Unlike H_2DCFDA, Fc OxyBURST® Green assay reagent does not require intracellular esterases for activation, making this reagent particularly suitable for detecting the oxidative burst in cells with low esterase activity such as monocytes.[8] Fc OxyBURST® Green assay reagent reportedly produces >8 times more fluorescence than does H_2DCFDA at 60 seconds and >20 times more at 15 minutes following internalization of the immune complex.[9]

Published reports have described the use of Fc OxyBURST® Green assay reagent to study the oxidative burst in phagovacuoles.[10–12] Neutrophils from patients with chronic granulomatous disease, a genetic deficiency known to disable NADPH oxidase–mediated oxidative bursts, were

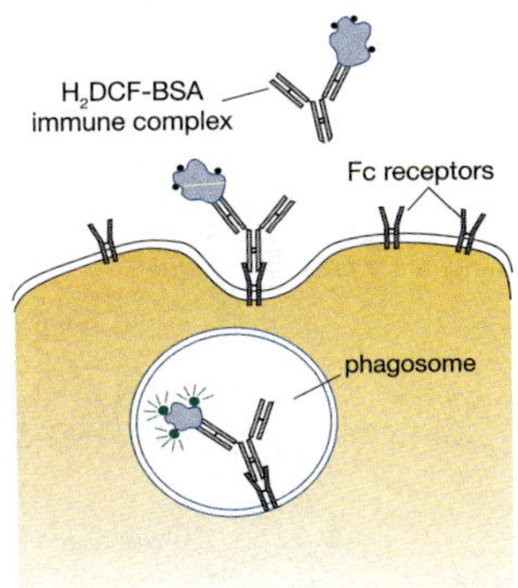

Figure 16.1.2 Fc OxyBURST® Green assay reagent (F2902) for fluorescent detection of the Fc receptor–mediated phagocytosis pathway. Dichlorodihydrofluorescein (H_2DCF) is covalently attached to bovine serum albumin (BSA), then complexed with a rabbit polyclonal anti-BSA antibody (A11133). Upon binding to an Fc receptor, the nonfluorescent immune complex is internalized and subsequently oxidized to the fluorescent DCF.

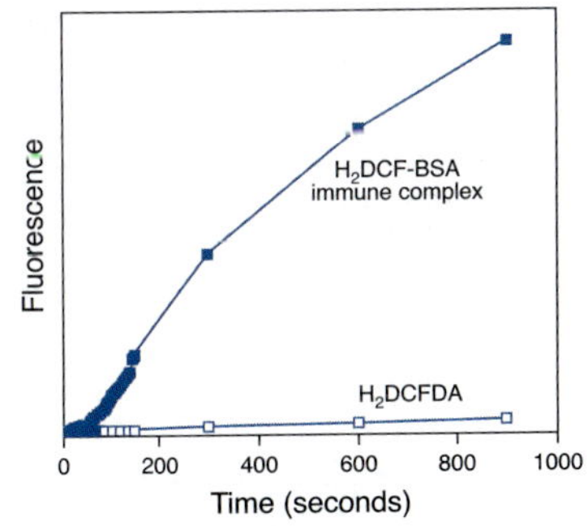

Figure 16.1.3 Fluorescence emission of human neutrophils challenged either with Fc OxyBURST® Green assay reagent (H_2DCF-BSA immune complexes, F2902) or with unlabeled immune complexes in the presence of dichlorodihydrofluorescein diacetate (H_2DCFDA; D399). Fc OxyBURST® Green assay reagent generates significantly more fluorescence than does the more commonly used H_2DCFDA. Flow cytometry data provided by Elizabeth Simons, Boston University.

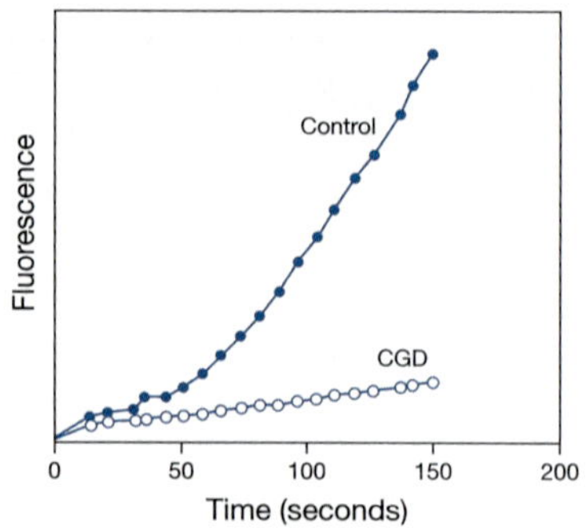

Figure 16.1.4 Oxidative bursts of human neutrophils from a healthy donor (control) compared with those from a patient with chronic granulomatous disease (CGD), as detected using the Fc OxyBURST® Green assay reagent (F2902). Flow cytometry data provided by Elizabeth Simons, Boston University.

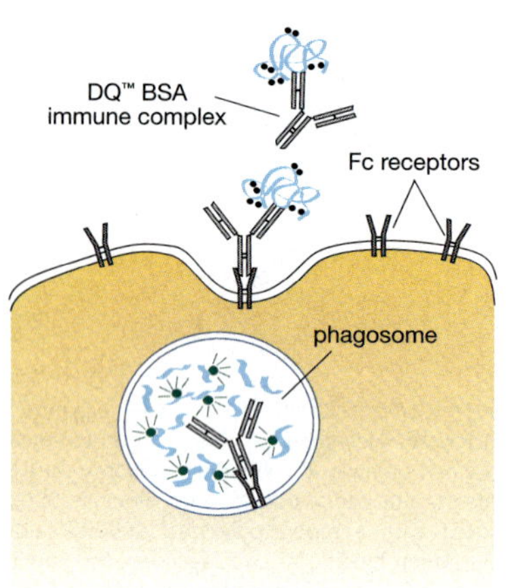

Figure 16.1.5 Immune complex of DQ™ BSA conjugate (D12050, D12051) with an anti–bovine serum albumin (BSA) antibody (A11133) for the fluorescent detection of the Fc receptor–mediated phagocytosis pathway. The DQ™ BSA is a derivative of BSA that is labeled to such a high degree with either the green-fluorescent BODIPY® FL or red-fluorescent BODIPY® TR-X dye that the fluorescence is self-quenched. Upon binding to an Fc receptor, the nonfluorescent immune complex is internalized and the protein is subsequently hydrolyzed to fluorescent peptides within the phagovacuole.

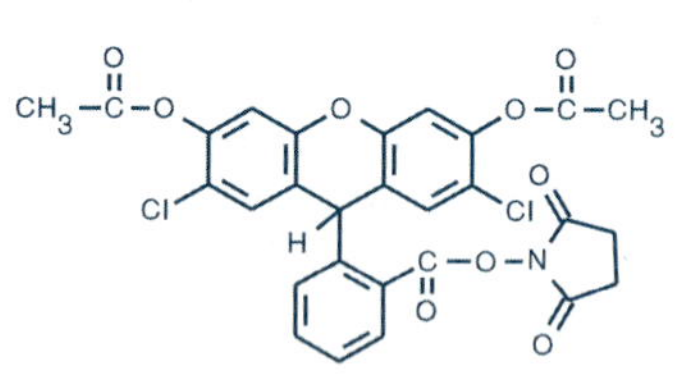

Figure 16.1.6 2',7'-dichlorodihydrofluorescein diacetate, succinimidyl ester (OxyBURST® Green H_2DCFDA, SE, D2935).

observed to bind but not oxidize Fc OxyBURST® Green assay reagent [9] (Figure 16.1.4). Using microfluorometry to detect the Fc OxyBURST® Green signal, researchers were able to simultaneously monitor oxidative activity and membrane currents in voltage-clamped human mononuclear cells.[13]

OxyBURST® Green H_2HFF BSA Reagent

OxyBURST® Green H_2HFF BSA reagent [14–16] (O13291) is similar to Fc OxyBURST® Green assay reagent, except that it is prepared by reacting the succinimidyl ester of a reduced form of our Oregon Green® 488 dye with BSA. The absorption maximum of the oxidation product of this reagent (~492 nm) matches the 488 nm spectral line of the argon-ion laser better than does that of Fc OxyBURST® Green assay reagent (~495 nm). OxyBURST® Green H_2HFF BSA reagent can also be complexed with anti-BSA antibody to form an immune complex that can be utilized like the Fc OxyBURST® Green assay reagent (F2902).

All of the OxyBURST® reagents are slowly oxidized by molecular oxygen and are also susceptible to oxidation catalyzed by illumination in a fluorescence microscope. These reagents are reasonably stable in solution for at least six months when stored under nitrogen or argon in the dark at 4°C. We also offer a purified rabbit polyclonal anti-BSA antibody (A11133), which can bind any of our BSA conjugates (Section 14.7) or fluorogenic DQ™ BSA conjugates (D12050, D12051; Section 10.4) to create immune complexes for analyzing the Fc receptor–mediated phagocytosis pathway. In the case of the anti-BSA antibody complex with DQ™ BSA, initial binding and internalization of the probe is followed by hydrolysis to fluorescent peptides by proteases in the phagovacuole [17] (Figure 16.1.5).

Amine-Reactive OxyBURST® Green Reagent

As an alternative to Fc OxyBURST® Green assay reagent and OxyBURST® Green H_2HFF BSA, we offer amine-reactive OxyBURST® Green H_2DCFDA succinimidyl ester (2',7'-dichlorodihydrofluorescein diacetate, SE; D2935; Figure 16.1.6), which can be used to prepare oxidation-sensitive conjugates of a wide variety of biomolecules and particles, including antibodies, antigens, peptides, proteins, dextrans, bacteria, yeast and polystyrene microspheres.[9,18,19] Following conjugation to amines, the two acetates of OxyBURST® Green H_2DCFDA reagent can be removed by treatment with hydroxylamine at near-neutral pH to yield the oxidant-sensitive dichlorodihydrofluorescein conjugates. Thus, like our Fc OxyBURST® Green assay reagent, they provide a means of detecting the oxidative burst in phagocytic cells.[18]

Several other reagents—dihydrofluoresceins, dihydrorhodamines, dihydroethidium and chemiluminescent probes—that have been used to detect the reactive oxygen species (ROS) produced during phagocytosis are described in Section 18.2.

Fluorescent Low-Density Lipoprotein Complexes

The human LDL complex, which delivers cholesterol to cells by receptor-mediated endocytosis, comprises a core of about 1500 molecules of cholesteryl ester and triglyceride, surrounded by a 20 Å–thick shell of phospholipids, unesterified cholesterol and a single copy of apoprotein B 100 [20] (MW ~500,000 daltons). Once internalized, LDL dissociates from its receptor and eventually appears in lysosomes.[21] In addition to unlabeled LDL (L3486), which has been reported to be an effective vehicle for selectively delivering antitumor drugs to cancer cells,[22] we offer two classes of labeled LDL probes—those containing an unmodified apoprotein, used to study the mechanisms of normal cholesterol delivery and internalization, and those with an acetylated apoprotein, used to study endothelial, microglial and other cell types that express receptors that specifically bind this modified LDL.

For the class of labeled LDL probes containing unmodified apoprotein, we prepare LDL noncovalently labeled with either DiI (DiI LDL, L3482) or the BODIPY® FL fluorophore (BODIPY® FL LDL, L3483), highly fluorescent lipophilic dyes that diffuse into the hydrophobic portion of the LDL complex without affecting the LDL-specific binding of the apoprotein. As compared with DiI LDL, BODIPY® FL LDL is more efficiently excited by the 488 nm spectral line of the argon-ion laser, making it better suited for flow cytometry and confocal laser-scanning microscopy studies. Like our BODIPY® FL C_5-ceramide (D3521, Section 12.4), BODIPY® FL LDL may exhibit concentration-dependent long-wavelength emission (>550 nm), precluding its use for

IMPORTANT NOTICE: The products described in this manual are covered by one or more Limited Use Label License(s). Please refer to the Appendix on page 971 and Master Product List on page 975. Products are For Research Use Only. Not intended for any animal or human therapeutic or diagnostic use.

multicolor labeling with red fluorophores. Both DiI LDL and BODIPY® FL LDL have been used to investigate the binding specificity and partitioning of LDL throughout the *Schistosoma mansoni* parasite[23] (Figure 16.1.7). Fluorescent LDL complexes have also proven useful in a variety of experimental systems to:

- Count the number of cell-surface LDL receptors, analyze their motion and clustering and follow their internalization[24–26]
- Demonstrate that fibroblasts grown continuously in the presence of DiI LDL (L3482) proliferate normally and exhibit normal morphology,[27] making DiI LDL a valuable alternative to ^{125}I-labeled LDL for quantitating LDL receptor activity[28]
- Identify LDL receptor–deficient Chinese hamster ovary (CHO) cell mutants[29]
- Image LDL receptor endocytosis in COS7 cells expressing Green Fluorescent Protein (GFP)–tagged GTPase[30]
- Investigate the modulation of LDL receptor expression by statin drugs[31,32]

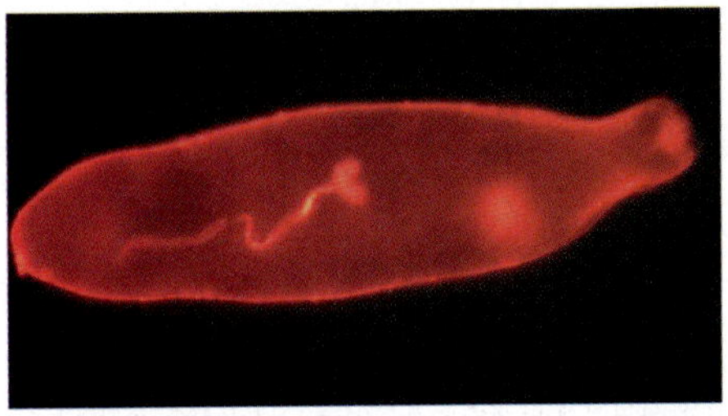

Figure 16.1.7 DiI LDL (L3482) bound to the surface and internalized in the gut of the parasite *Schistosoma mansoni*. The distribution of LDL in the parasite is used to study a putative mechanism by which the parasite may avoid host immune recognition. Image contributed by John P. Caulfield, Harvard School of Public Health.

We prepare fluorescent LDL products from fresh human plasma, and they should be stored refrigerated and protected from light; LDL products must not be frozen. Because preparation of these complexes involves several variables, some batch-to-batch variability in degree of labeling and fluorescence yield is expected.

Fluorescent Acetylated LDL Complexes

If the lysine residues of LDL's apoprotein have been acetylated, the LDL complex no longer binds to the LDL receptor,[33] but rather is taken up by macrophage and endothelial cells that possess "scavenger" receptors specific for the modified LDL.[34,35] Once the acetylated LDL (AcLDL) complexes accumulate within these cells, they assume an appearance similar to that of foam cells found in atherosclerotic plaques.[36–38] We offer unlabeled AcLDL (L35354), as well as AcLDL noncovalently labeled with DiI L(3484) and AcLDL covalently labeled with Alexa Fluor® 488 dye (L23380), Alexa Fluor® 594 dye (L35353) or BODIPY® FL dye (L3485). Fluorescent dye conjugates of high-density lipoproteins, including one prepared using Alexa Fluor® 488 succinimidyl ester (A20000, A20100; Section 1.3), are taken up via the same receptor as acetylated LDL complexes.[39]

Using DiI AcLDL, researchers have discovered that the scavenger receptors on rabbit fibroblasts and smooth muscle cells appear to be up-regulated through activation of the protein kinase C pathway.[40] DiI AcLDL has also been used to show that Chinese hamster ovary (CHO) cells express AcLDL receptors that are distinct from macrophage scavenger receptors.[41,42] Ultrastructural localization of endocytic compartments that maintain a connection to the extracellular space has been achieved by photoconversion of DiI AcLDL in the presence of diaminobenzidine[43] (Fluorescent Probes for Photoconversion of Diaminobenzidine Reagents—Note 14.2). A quantitative assay for LDL- and scavenger-receptor activity in adherent and nonadherent cultured cells that avoids the use of both radioactivity and organic solvents has been described.[44]

It has now become routine to identify endothelial cells and microglial cells in primary cell culture by their ability to take up DiI AcLDL[45,46] (Figure 16.1.8). DiI AcLDL was employed in order to confirm endothelial cell identity in investigations of shear stress[47] and P-glycoprotein expression,[48] as well as to identify blood vessels in a growing murine melanoma.[49] In addition, patch-clamp techniques have been used to investigate membrane currents in mouse microglia, which were identified both in culture and in brain slices by their staining with DiI AcLDL.[50,51] For some applications, Alexa Fluor® 488, Alexa Fluor® 594 and BODIPY® FL AcLDL may be the preferred probes because the dyes are covalently bound to the modified apoprotein portion of the LDL complex and are therefore not extracted during subsequent manipulations of the cells. Furthermore, the green-fluorescent Alexa Fluor® 488 AcLDL has spectral characteristics similar to fluorescein and is useful for analyses with instruments equipped with the 488 nm argon-ion laser excitation sources, including flow cytometers and confocal laser-scanning microscopes. The bright and photostable red-fluorescent Alexa Fluor® 594 AcLDL conjugate is useful for multilabeling experiments with green-fluorescent probes and can be efficiently excited by the 594 nm spectral line of the orange He-Ne laser.[52]

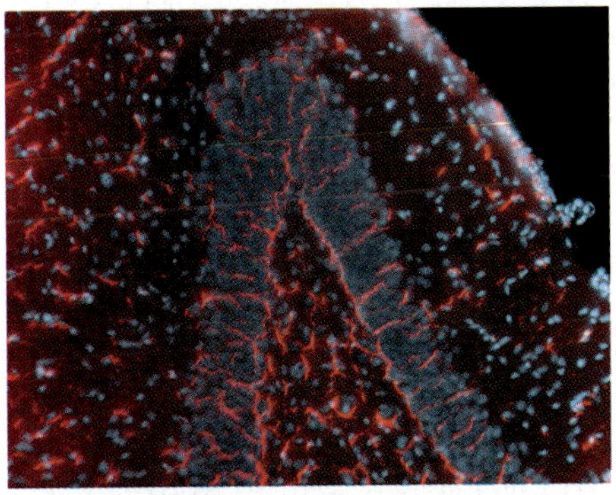

Figure 16.1.8 Microglial cells in a rat hippocampus cryosection labeled with red-orange–fluorescent DiI acetylated low-density lipoprotein (L3484) and stained using blue-fluorescent DAPI (D1306, D3571, D21490).

IMPORTANT NOTICE: The products described in this manual are covered by one or more Limited Use Label License(s). Please refer to the Appendix on page 971 and Master Product List on page 975. Products are For Research Use Only. Not intended for any animal or human therapeutic or diagnostic use.

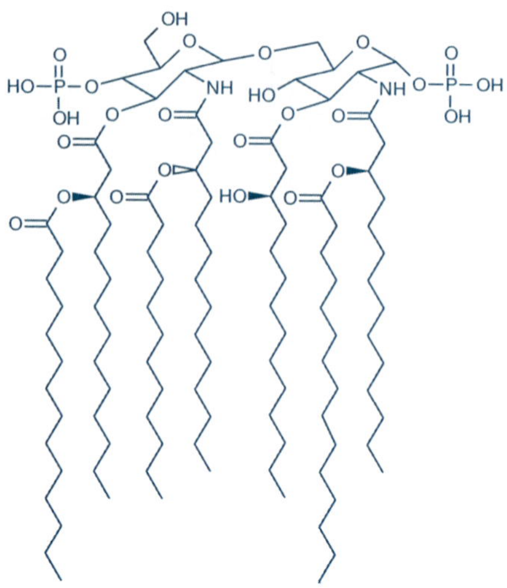

Figure 16.1.9 Structure of the lipid A component of *Salmonella minnesota* lipopolysaccharide.

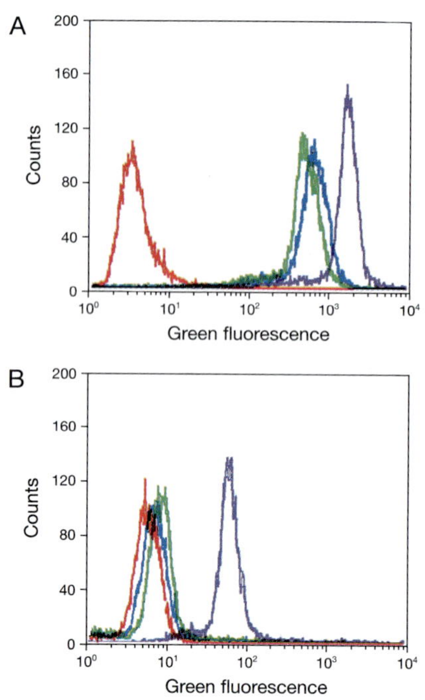

Figure 16.1.10 Detection of epidermal growth factor (EGF) receptors directly or with signal amplification. Cells expressing high (A431 cells, panel **A**) and low (NIH 3T3 cells, panel **B**) levels of EGF receptors were either directly labeled with the preformed Alexa Fluor® 488 complex of biotinylated epidermal growth factor (E13345, blue) or indirectly labeled with biotinylated EGF (E3477) followed by either Alexa Fluor® 488 streptavidin (S11223, green) or HRP-conjugated streptavidin and Alexa Fluor® 488 tyramide (purple), components of our TSA™ Kit #22 (T20932).

Fluorescent Lipopolysaccharides

We offer fluorescent conjugates of lipopolysaccharides (LPS) from *Escherichia coli* and *Salmonella minnesota* (Table 16.1), including:

- Alexa Fluor® 488 LPS from *E. coli* serotype 055:B5 (A23351)
- Alexa Fluor® 488 LPS from *S. minnesota* (A23356)
- Alexa Fluor® 568 LPS from *E. coli* serotype 055:B5 (A23352)
- Alexa Fluor® 594 LPS from *E. coli* serotype 055:B5 (A23353)
- BODIPY® FL LPS from *E. coli* serotype 055:B5 (A23350)

LPS, also known as endotoxins, are a family of complex glycolipid molecules located on the surface of gram-negative bacteria. LPS play a large role in protecting the bacterium from host defense mechanisms and antibiotics. Binding of LPS to the CD14 cell-surface receptor of phagocytes is the key initiation step in the mammalian immune response to infection by gram-negative bacteria.[53] The structural core of LPS, and the primary determinant of its biological activity, is an *N*-acetylglucosamine derivative, lipid A (Figure 16.1.9). In many gram-negative bacterial infections, LPS are responsible for clinically significant symptoms like fever, low blood pressure and tissue edema, which can lead to disseminated intravascular coagulation, organ failure and death.

The fluorescent BODIPY® FL and Alexa Fluor® LPS conjugates, which are labeled with succinimidyl esters of these dyes, allow researchers to follow LPS-elicited inflammatory responses.[53,54] Lipopolysaccharide internalization activates endotoxin-dependent signal transduction in cardiomyocytes.[55] Alexa Fluor® 488 LPS conjugates (L23351, L23356) have been shown to selectively label microglia in a mixed culture containing oligodendrocyte precursors, astrocytes and microglia.[56]

The BODIPY® FL derivative of LPS from *E. coli* strain LCD25 (L23350) was used to measure the transfer rate of LPS from monocytes to high-density lipoprotein[57] (HDL). Another study utilized a BODIPY® FL derivative of LPS from *S. minnesota* to demonstrate transport to the Golgi apparatus in neutrophils,[58,59] although this could have been due to probe metabolism. It has been reported that organelles other than the Golgi are labeled by some fluorescent or nonfluorescent LPS.[60,61] Cationic lipids are reported to assist the translocation of fluorescent lipopolysaccharides into live cells;[62] cell surface–bound LPS can be quenched by trypan blue.[57]

Other probes useful for analyzing lipopolysaccharides include fluorescent analogs of the LPS-binding antibiotic polymyxin B (Section 17.3) and BODIPY® TR cadaverine (D6251, Section 3.4). BODIPY® TR cadaverine binds with high selectivity to lipid A, forming the basis for high-throughput ligand displacement assays for identifying endotoxin antagonists.[63,64]

Epidermal Growth Factor Derivatives

Epidermal growth factor (EGF) is a 53–amino acid polypeptide hormone (MW 6045 daltons) that stimulates division of epidermal and other cells. The EGF receptors include the HER-2/neu receptor (where "HER-2" is an acronym for human epidermal growth factor receptor-2 and "neu" refers to an original mouse origin); HER-2/neu overexpression has evolved as a prognostic/predictive factor in some solid tumors.[65–67] Binding of EGF to its 170,000-dalton receptor protein results in the activation of kinases, phospholipases and Ca^{2+} mobilization and precipitates a wide variety of cellular responses related to differentiation, mitogenesis, organ development and cell motility.

We offer unlabeled mouse submaxillary gland EGF (E3476), as well as the following EGF conjugates, each containing a single fluorophore or biotin on the N-terminal amino acid:

- Fluorescein EGF (E3478)
- Oregon Green® 514 EGF (E7498)
- Tetramethylrhodamine EGF (E3481)
- Biotin-XX EGF (E3477)

The dissociation constant of the EGF conjugates in DMEM-HEPES medium is in the low nanomolar range for human epidermoid carcinoma (A431) cells,[68] a value that approximates that of the unlabeled EGF. Fluorescently labeled EGF has enabled scientists to use fluorescence resonance energy transfer techniques to assess EGF receptor–receptor and receptor–membrane interactions[69–71] (Fluorescence Resonance Energy Transfer (FRET)—Note 1.2). Using fluorescein

IMPORTANT NOTICE: The products described in this manual are covered by one or more Limited Use Label License(s). Please refer to the Appendix on page 971 and Master Product List on page 975. Products are For Research Use Only. Not intended for any animal or human therapeutic or diagnostic use.
www.invitrogen.com/probes
molecular probes® | invitrogen™ by life technologies™

Table 16.1 Fluorescent lipopolysaccharide conjugates.

Fluorophore	Abs *	Em *	*Escherichia coli*	*Salmonella minnesota*
Alexa Fluor® 488	495	519	L23351	L23356
BODIPY® FL	503	513 †	L23350	
Alexa Fluor® 568	578	603	L23352	
Alexa Fluor® 594	590	617	L23353	

* Approximate absorption (Abs) and fluorescence emission (Em) maxima for conjugates, in nm. † At high concentrations, the emission maximum for the BODIPY® FL dye may shift from ~513 nm to ~620 nm.[1,2]
1. J Immunol (1997) 158:3925; 2. J Biol Chem (1996) 271:4100.

EGF as the donor and tetramethylrhodamine EGF as the acceptor, researchers examined temperature-dependent lateral and transverse distribution of EGF receptors in A431 cell plasma membranes.[71] When fluorescein EGF binds to A431 cells, it apparently undergoes a biphasic quenching, which can be attributed first to changes in rotational mobility upon binding and then to receptor–ligand internalization. By monitoring this quenching in real time, the rate constants for the interaction of fluorescein EGF with its receptor were determined.[72] Although fluorescently labeled EGF can be used to follow lateral mobility and endocytosis of the EGF receptor,[73,74] the visualization of fluorescent EGF may require low-light imaging technology or Qdot® nanocrystals, especially in cells that express low levels of the EGF receptor.[75] In cells with few EGF receptors, it can be difficult to detect signal over background fluorescence unless signal amplification methods are employed (Figure 16.1.10).

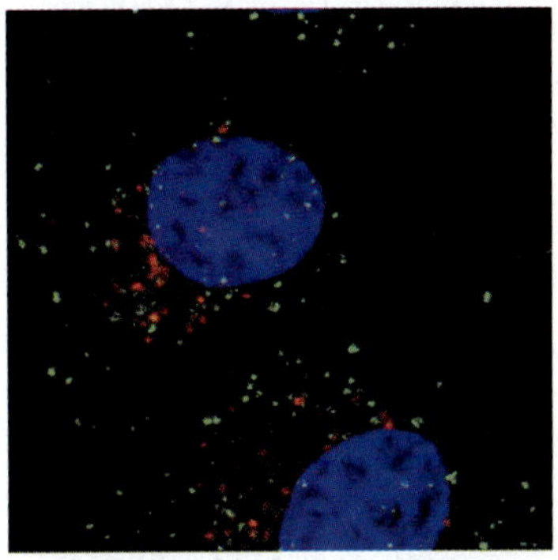

Figure 16.1.11 Early endosomes in live HeLa cells identified after a 10-minute incubation with green-fluorescent Alexa Fluor® 488 epidermal growth factor (E13345). The cells were subsequently fixed with formaldehyde and labeled with an antibody to the late endosomal protein, RhoB. That antibody was visualized with a red-orange–fluorescent secondary antibody. Nuclei were stained with TO-PRO®-3 iodide (T3605, pseudocolored blue). The image was contributed by Harry Mellor, University of Bristol.

Biotin-XX EGF contains a long spacer arm that enhances the probe's affinity for the EGF receptor and facilitates binding of dye-, Qdot® nanocrystal– or enzyme-conjugated streptavidins[75–79] (Section 7.6). Using biotinylated EGF and phycoerythrin-labeled secondary reagents (Section 6.4), researchers were able to detect as few as 10,000 EGF cell-surface receptors by confocal laser-scanning microscopy.[80] Tyramide signal amplification (TSA) technology (Section 6.2) is particularly valuable for detection and localization of low-abundance EGF receptors by both imaging and flow cytometry (Figure 16.1.10). For additional sensitivity, we prepare biotinylated EGF precomplexed to fluorescent streptavidin:

- Biotinylated EGF complexed to Alexa Fluor® 488 streptavidin (E13345, Figure 16.1.11)
- Biotinylated EGF complexed to Alexa Fluor® 555 streptavidin (E35350)
- Biotinylated EGF complexed to Alexa Fluor® 647 streptavidin (E35351)
- Biotinylated EGF complexed to Texas Red® streptavidin[81,82] (E3480)

These products yield several-fold brighter signals per EGF receptor when compared with the direct conjugates. We have found that EGF receptors can easily be detected with these complexes without resorting to low-light imaging technology (Figure 16.1.12). A quantitative high-content screening (HCS) assay for EGF receptor modulators based on imaging the internalization of the Alexa Fluor® 555 EGF complex internalization has been reported.[83]

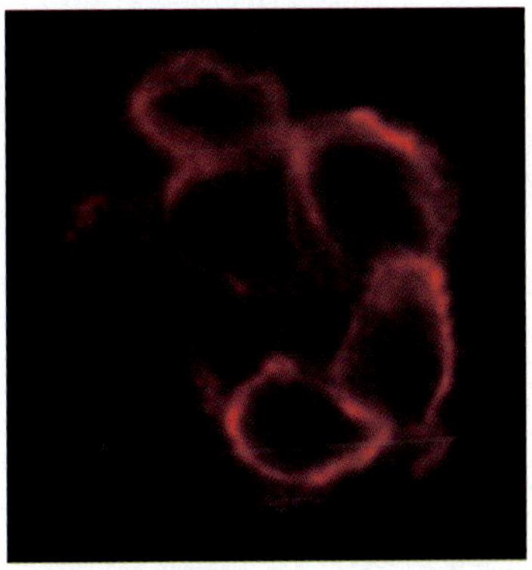

Figure 16.1.12 Lightly fixed human epidermoid carcinoma cells (A431) stained with biotinylated epidermal growth factor (EGF) complexed to Texas Red® streptavidin (E3480). An identical cell preparation stained in the presence of a 100-fold excess of unlabeled EGF (E3476) showed no fluorescent signal.

Transferrin Conjugates

Transferrin is a monomeric serum glycoprotein (MW ~80,000 daltons) that binds up to two Fe^{3+} atoms for delivery to vertebrate cells through receptor-mediated endocytosis. Once iron-carrying transferrin proteins are inside endosomes, the acidic environment favors dissociation of the sequestered iron from the transferrin–receptor complex. Following the release of iron, the apotransferrin is recycled to the plasma membrane, where it is released from its receptor to scavenge more iron. Transferrin uptake is a prototypical and ubiquitous example of clathrin-mediated endocytosis. Although transferrin uptake is widely regarded as a surrogate measure of total clathrin-mediated endocytosis, perturbations that are specific to transferrin endocytosis impel caution in making such extrapolations.[2]

Our fluorescent and biotinylated di-ferric (Fe^{3+}) human transferrin conjugates (Table 16.2) include:

- Fluorescein transferrin (T2871)
- Alexa Fluor® 488 transferrin[84–87] (T13342)
- Alexa Fluor® 546 transferrin[88,89] (T23364)
- Alexa Fluor® 555 transferrin[90] (T35352)

Table 16.2 Transferrin conjugates.

Cat. No.	Label	Abs *	Em *
T2871	Fluorescein	494	518
T13342	Alexa Fluor® 488	495	518
T2872	Tetramethylrhodamine	555	580
T23364	Alexa Fluor® 546	556	575
T35352	Alexa Fluor® 555	555	565
T23365	Alexa Fluor® 568	578	603
T13343	Alexa Fluor® 594	589	616
T2875	Texas Red®	595	615
T23362	Alexa Fluor® 633	632	647
T23366	Alexa Fluor® 647	650	665
T35357	Alexa Fluor® 680	679	702
T23363	Biotin-XX	NA	NA

* Approximate absorption (Abs) and fluorescence emission (Em) maxima for conjugates, in nm. NA = Not applicable.

IMPORTANT NOTICE: The products described in this manual are covered by one or more Limited Use Label License(s). Please refer to the Appendix on page 971 and Master Product List on page 975. Products are For Research Use Only. Not intended for any animal or human therapeutic or diagnostic use.

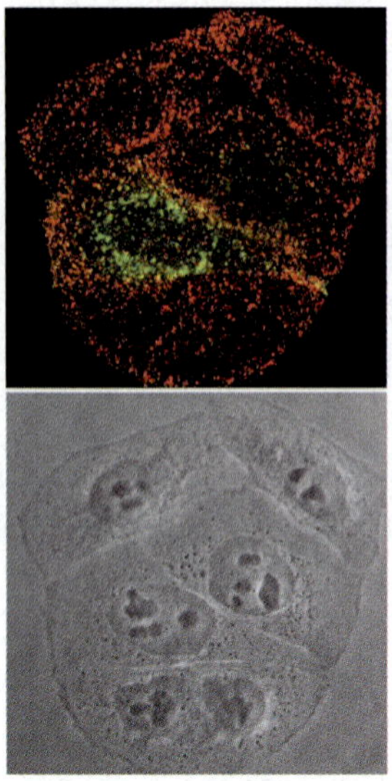

Figure 16.1.13 Live HeLa cells incubated with Alexa Fluor® 594 transferrin (T13343) for 10 minutes to label early endosomes. The cells were subsequently fixed with formaldehyde and labeled with an antibody to the endosomal protein RhoD. That antibody was visualized with a green-fluorescent secondary antibody. Yellow fluorescence indicates regions of co-localization. To illustrate the staining pattern, the cells were imaged by both fluorescence (top panel) and differential interference contrast (DIC) microscopy (bottom panel). The image was contributed by Harry Mellor, University of Bristol.

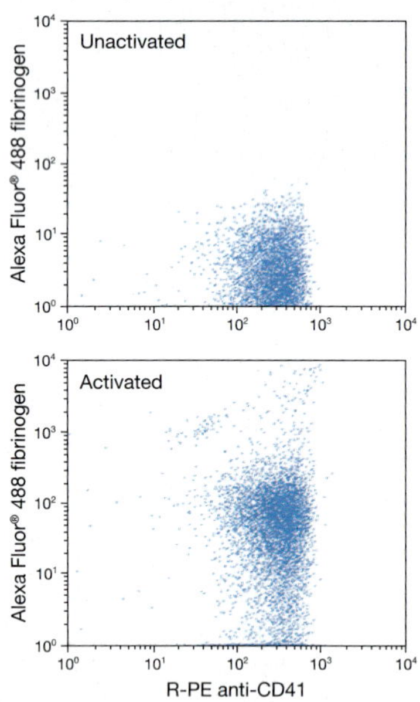

Figure 16.1.14 Interaction of fluorescently labeled fibrinogen with activated platelets. Whole blood was first incubated with an R-phycoerythrin (R-PE)–labeled anti-CD41 antibody to label the platelets. 20 µM adenosine 5′-diphosphate (ADP) was added in order to activate the platelets, then 2 µg/mL Alexa Fluor® 488 fibrinogen (F13191) was added and incubated with the sample for 5 minutes. Cells were analyzed by flow cytometry using excitation at 488 nm. Both activated and unactivated platelets show binding of the anti-CD41 antibody; however, only the activated platelets show strong binding by fibrinogen. A total of 5000 platelets are shown in each experiment.

- Alexa Fluor® 568 transferrin [91] (T23365)
- Alexa Fluor® 594 transferrin [92–94] (T13343, Figure 16.1.13)
- Alexa Fluor® 633 transferrin (T23362)
- Alexa Fluor® 647 transferrin [73,95,96] (T23366)
- Alexa Fluor® 680 transferrin (T35357)
- Tetramethylrhodamine transferrin (T2872)
- Texas Red® transferrin (T2875)
- Biotin-XX transferrin (T23363)

Alexa Fluor® transferrin conjugates are highly recommended because of their brightness, enhanced photostability and lack of sensitivity to pH (Section 1.3). The pH sensitivity of fluorescein-labeled transferrin has been exploited to investigate events occurring during endosomal acidification.[97–100] Fluorescent transferrins have also been used to:

- Analyze the role of the γ-chain of type III IgG receptors in antigen–antibody complex internalization [101]
- Characterize endocytic apparatus phenotypes in drug-resistant cancer cells [85]
- Demonstrate that the fungal metabolite brefeldin A (B7450, Section 12.4) induces an increase in tubulation of transferrin receptors in BHK-21 cells [102] and in the perikaryal–dendritic region of cultured hippocampal neurons [103]
- Image transferrin receptor dynamics using FRET [104]
- Observe receptor trafficking in live cells by confocal laser-scanning microscopy [74]

Uptake of a horseradish peroxidase (HRP) conjugate of transferrin by endosomes has been detected using tyramide signal amplification (TSA, Section 6.2) by catalytic deposition of biotin tyramide and use of fluorescent streptavidin conjugates [105] (Section 7.6).

In addition to fluorescent and biotinylated transferrin conjugates, we offer a mouse monoclonal IgG_1 anti–human transferrin receptor antibody (A11130). This antibody can be used with any of our Zenon® Mouse IgG_1 Labeling Kits (Section 7.3, Table 7.7) for rapid preparation of labeling complexes. Antibodies against transferrin receptors have been used for indirect immunofluorescent staining of the transferrin receptor,[106–108] transport of molecules across the blood–brain barrier,[109] characterization of transferrin in recycling compartments,[106] enzyme-linked immunosorbent assays (ELISAs) [108] and antibody competition with transferrin uptake.[110]

Fluorescent Fibrinogen

Fibrinogen is a key component in the blood clotting process and can support both platelet–platelet and platelet–surface interactions by binding to the glycoprotein IIb-IIIa (GPIIb-IIIa) receptor (also called integrin $\alpha_{IIb}\beta_3$) of activated platelets. Activation of GPIIb-IIIa is required for fibrinogen binding, which leads to platelet activation, adhesion, spreading and microfilament reorganization of human endothelial cells *in vitro*. Bone marrow transplant patients have significantly higher levels of fibrinogen binding, as compared with controls Soluble fibrinogen binds to its receptor with a Ca^{2+}-dependent apparent K_d of 0.18 µM.[111] This binding is mediated by the tripeptide sequence Arg-Gly-Asp (RGD), found in both fibrinogen and fibronectin.

Fluorescently labeled fibrinogen has proven to be a valuable tool for investigating platelet activation and subsequent fibrinogen binding.[112–114] Alexa Fluor® 647 fibrinogen has been used to identify activated platelets by flow cytometry.[115] The binding of fluorescein fibrinogen to activated platelets has been shown to be saturable and can be inhibited completely by underivatized fibrinogen.[116,117]

We offer four conjugates of human fibrinogen in three different fluorescent colors:

- Alexa Fluor® 488 human fibrinogen conjugate (F13191)
- Oregon Green® 488 human fibrinogen conjugate (F7496)
- Alexa Fluor® 546 human fibrinogen conjugate (F13192)
- Alexa Fluor® 647 human fibrinogen conjugate (F35200)

These highly fluorescent fibrinogen conjugates are useful for investigating platelet activation and subsequent fibrinogen binding using fluorescence microscopy or flow cytometry [112,115,118] (Figure 16.1.14).

IMPORTANT NOTICE: The products described in this manual are covered by one or more Limited Use Label License(s). Please refer to the Appendix on page 971 and Master Product List on page 975. Products are For Research Use Only. Not intended for any animal or human therapeutic or diagnostic use.

molecular probes® | invitrogen™ by life technologies™

DQ™ Ovalbumin: A Probe for Antigen Processing and Presentation

Although antigen processing and presentation have been extensively studied, the exact sequence and detailed pathways for generating antigenic peptides have yet to be elucidated. In general, the immunogenic protein is internalized by a macrophage, denatured, reduced and proteolyzed, and then the resulting peptides associate with MHC class II molecules that are expressed at the cell surface.[119] Ovalbumin is efficiently processed through mannose receptor–mediated endocytosis by antigen-presenting cells and is widely used for studying antigen processing.[120–122] DQ™ ovalbumin [123] (D12053), a self-quenched ovalbumin conjugate, is designed specifically for the study of macrophage-mediated antigen processing in flow cytometry and microscopy assays.

Traditionally, fluorescein-labeled bovine serum albumin (FITC-BSA) has been used as a fluorogenic protein antigen for studying the real-time kinetics of antigen processing in live macrophages by flow cytometry,[124] two-photon fluorescence lifetime imaging microscopy (FLIM) [125] and fluorescence polarization.[124,126,127] FITC-ovalbumin has been employed to study antigen uptake in HIV-1–infected monocytic cells.[128] The FITC-ovalbumin and FITC-BSA used in these experiments were heavily labeled with fluorescein such that the intact conjugates were relatively nonfluorescent due to self-quenching. Upon denaturation and proteolysis, however, these FITC conjugates became highly fluorescent, allowing researchers to monitor intracellular trafficking and the processing of ovalbumin and BSA in macrophages.

For studies of antigen processing and presentation, DQ™ ovalbumin offers several advantages when compared with FITC-ovalbumin and FITC-BSA. Like the FITC conjugates, DQ™ ovalbumin is labeled with our pH-insensitive, green-fluorescent BODIPY® FL dye such that the fluorescence is almost completely quenched until the probe is digested by proteases (Figure 16.1.15). Unlike fluorescein, which has greatly reduced fluorescence intensity at acidic pH and is not very photostable, our BODIPY® FL dye exhibits bright, relatively photostable and pH-insensitive fluorescence from pH 3 to 9. Furthermore, the intact DQ™ ovalbumin is more highly quenched than unprocessed FITC-ovalbumin or FITC-BSA at a lower degree of substitution, thereby providing a lower background signal while preserving the protein's antigenic epitopes. Although we offer the green-fluorescent DQ™ Green BSA and red-fluorescent DQ™ Red BSA (D12050, D12051; Section 10.4), which are also self-quenched BODIPY® FL and BODIPY® TR conjugates, we highly recommend DQ™ ovalbumin (D12053) for studying antigen processing and presentation [129,130] because ovalbumin is internalized via the mannose receptor–mediated endocytosis pathway and is thus processed more efficiently by antigen-presenting cells than is BSA.[131]

Fluorescent Gelatin

Collagen is a major component of the extracellular matrix and, in vertebrates, constitutes approximately 25% of total protein. This important protein not only serves a structural role, but also is important in cell adhesion and migration. Specific collagen receptors, fibronectin and a number of other proteins involved in cell–cell and cell–surface adhesion have been demonstrated to bind collagen and gelatin [132,133] (denatured collagen).

We offer highly fluorescent gelatin conjugates for researchers studying collagen-binding proteins and collagen metabolism, as well as gelatinases and collagenases, which are metalloproteins that digest gelatin and collagen. We offer two green-fluorescent gelatin conjugates—fluorescein gelatin and Oregon Green® 488 gelatin (G13187, G13186). Fluorescent gelatin conjugates have been shown to be useful for:

- Assessing gelatinase activity in podosomes of mouse dendritic cells [134]
- Localizing surface fibronectin on cultured cells [135]
- Performing *in situ* gelatinase zymography on canary brain sections [136]
- Studying fibronectin–gelatin interactions in solution using fluorescence polarization [133] (Fluorescence Polarization (FP)—Note 1.4)

We have also developed fluorogenic gelatinase and collagenase substrates—DQ™ gelatin and DQ™ collagen (Figure 16.1.15) (D12054, D12060)—that are described in Section 10.4. In addition, we offer fluorescent microspheres coated with collagen, which are described below.

Fluorescent Casein

Real-time imaging of fluorescein-labeled casein (C2990) and FluoSpheres® fluorescent microspheres has been used to characterize the endocytic apparatus of the protozoan *Giardia lamblia*.[137] The EnzChek® Protease Assay Kits (E6638, E6639; Section 10.4) provide convenient fluorescence-based assays for protease activity and contain either green-fluorescent BODIPY® FL casein or red-fluorescent BODIPY® TR-X casein [138] (Figure 16.1.15). BODIPY® FL casein and BODIPY® TR-X casein have significant utility as nontoxic and pH-insensitive general markers for phagocytic cells in culture.[139,140] Our RediPlate™ 96 (R22132) version of the BODIPY® TR-X casein substrate (Section 10.4) is ideal for high-throughput screening of potential protease inhibitors.

Fluorescent Chemotactic Peptide

A variety of white blood cells containing the formyl-Met-Leu-Phe (fMLF) receptor respond to bacterial *N*-formyl peptides by migrating to the site of bacterial invasion and then initiating an activation pathway to control the spread of infection. Activation involves Ca^{2+} mobilization,[141] transient acidification,[142,143] actin polymerization,[144] phagocytosis [145] and production of oxidative species.[146] We offer the fluorescein conjugate of the hexapeptide formyl-Nle-Leu-Phe-Nle-Tyr-Lys (F1314),

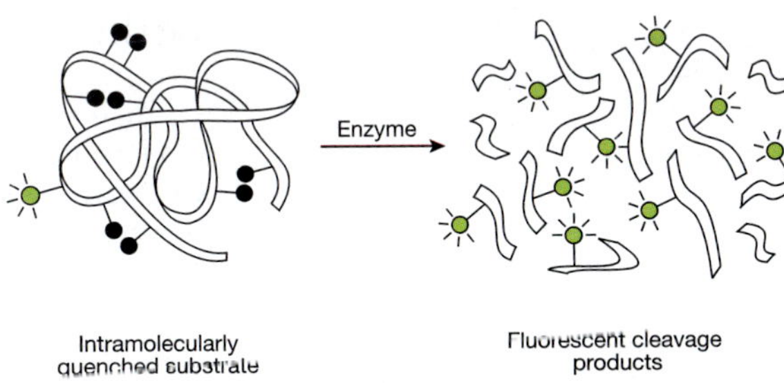

Figure 16.1.15 Principle of enzyme detection via the disruption of intramolecular self-quenching. Enzyme-catalyzed hydrolysis of the heavily labeled and almost totally quenched substrates provided in our EnzChek® Protease Assay Kits (E6638, E6639), EnzChek® *Ultra* Amylase Assay Kit (E33651), EnzChek® Gelatinase/Collagenase Assay Kit (E12055), EnzChek® Elastase Kit (E12056), EnzChek® Lysozyme Assay Kit (E22013)—as well as the stand-alone quenched substrates DQ™ BSA (D12050, D12051), DQ™ collagen (D12052, D12060), DQ™ ovalbumin (D12053) and DQ™ gelatin (D12054)—relieves the intramolecular self-quenching, yielding brightly fluorescent reaction products.

IMPORTANT NOTICE: The products described in this manual are covered by one or more Limited Use Label License(s). Please refer to the Appendix on page 971 and Master Product List on page 975. Products are For Research Use Only. Not intended for any animal or human therapeutic or diagnostic use.

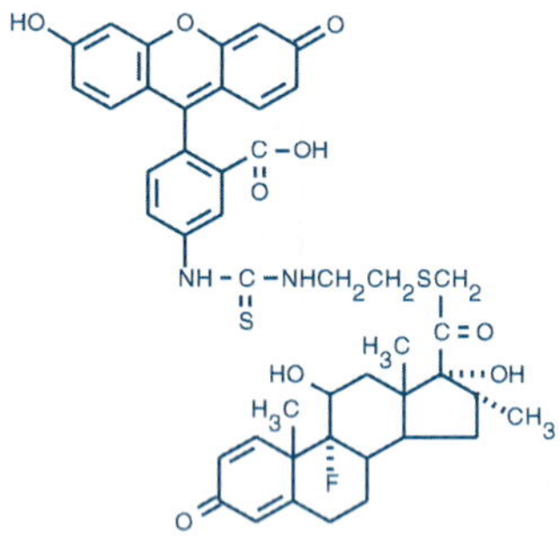

Figure 16.1.16 Dexamethasone fluorescein (D1383).

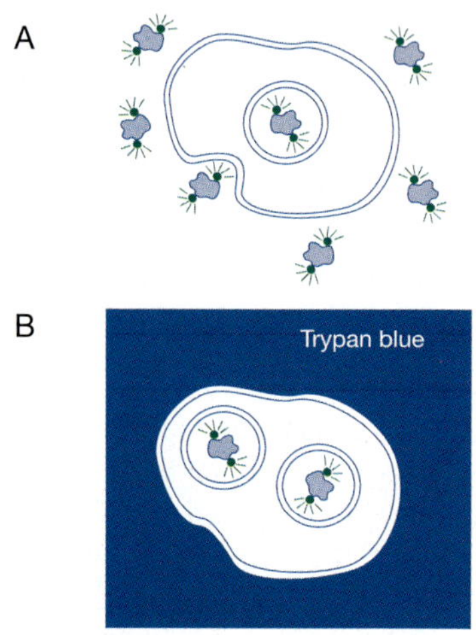

Figure 16.1.17 Principle of the Vybrant® Phagocytosis Assay Kit (V6694) for the simple quantitation of phagocytosis. **A)** Briefly, phagocytic cells are incubated with the green-fluorescent fluorescein-labeled *Escherichia coli* BioParticles® conjugates (E2861). **B)** The fluorescence from any noninternalized BioParticles® product is then quenched by the addition of trypan blue, and the samples are subsequently assayed with a fluorescence microplate reader equipped with filters for the detection of fluorescein (FITC).

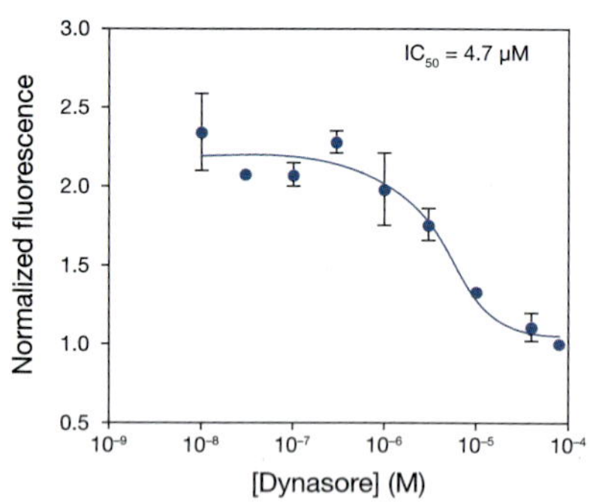

Figure 16.1.18 Tracking endocytosis inhibition with pHrodo™ dextran conjugates. HeLa cells were plated in 96-well format and treated with dynasore for 3 hours at 37°C prior to the pHrodo™ endocytosis assay. Next, 40 µg/mL of pHrodo™ 10,000 MW dextran (P10361) was incubated for 30 minutes at 37°C, and cells were then stained with HCS NuclearMask™ Blue Stain (H10325) for 10 minutes to reveal total cell number and demarcation for image analysis. Images were acquired on the BD Pathway™ 855 High-Content Bioimager (BD Biosciences).

which has been extensively employed to investigate the fMLF receptor.[147–151] The fluorescein-labeled chemotactic peptide has been used to study G-protein coupling and receptor structure,[152–154] expression,[155,156] distribution [157–159] and internalization.[160]

Fluorescent Insulin

We prepare a high-purity, zinc-free fluorescein isothiocyanate conjugate of human insulin (FITC insulin, I13269). Unlike most commercially available preparations, our FITC insulin is purified by HPLC and consists of a singly labeled species of insulin that has been specifically modified at the N-terminus of the B-chain. Because the degree and position of labeling can alter the biological activity of insulin, we have isolated the singly labeled species that has been shown to retain its biological activity in an autophosphorylation assay.[161] Our FITC insulin preparation is useful for imaging insulin and insulin receptor distribution,[162] as well as for conducting insulin-binding assays using microfluidic devices.[163,164]

Fluorescent Dexamethasone Probe for Glucocorticoid Receptors

The synthetic steroid hormone dexamethasone binds to the glucocorticoid receptor, producing a steroid–receptor complex that then localizes in the nucleus and regulates gene transcription. In hepatoma tissue culture (HTC) cells, tetramethylrhodamine-labeled dexamethasone has been shown to have high affinity for the glucocorticoid receptor in a cell-free system and to induce tyrosine aminotransferase (TAT) expression in whole cells, albeit at a much lower rate than unmodified dexamethasone.[165] This labeled dexamethasone also allowed the first observations of the fluorescent steroid–receptor complex in the HTC cell cytosol.[165] Fluorescein dexamethasone (D1383, Figure 16.1.16) should be similarly useful for studying the mechanism of glucocorticoid receptor activation.

Fluorescent Histone H1

The Alexa Fluor® 488 conjugate of the lysine-rich calf thymus histone H1 (H13188) is a useful probe for nuclear protein transport assays.[166] Nuclear-to-mitochondrial translocation of histone H1 is indicative of dsDNA strand breaks. Fluorescent histone H1 conjugates can also be used to detect membrane-surface exposure of acidic phospholipids such as phosphatidylserine.[167]

Fluorescent Probes for the Acrosome Reaction

Soybean trypsin inhibitor (SBTI) inhibits the catalytic activity of serine proteases and binds to acrosin, an acrosomal serine protease associated with binding of spermatozoa to the zona pellucida.[168] Alexa Fluor® 488 dye–labeled trypsin inhibitor from soybean (T23011) is useful for real-time imaging of the acrosome reaction in live spermatozoa.[169] A fluorescent peanut lectin has been combined with ethidium homodimer-1 (EthD-1, E1169; Section 15.2) for a combined acrosome reaction assay and vital staining.[170] Alexa Fluor® 488, Alexa Fluor® 568, Alexa Fluor® 594 and Alexa Fluor® 647 conjugates of *Arachis hypogaea* lectin (PNA) (L21409, L32458, L32459, L32460) have similar utility as acrosomal stains.[171]

Methods for Detecting Internalized Fluorescent Ligands

Many of the fluorescent ligands described in this section first bind to cell-surface receptors, then are internalized and, in some cases, recycled to the cell surface. In most applications, the cell-surface and internalized ligand populations are spatially resolved by imaging. It is often desirable to include noninternalized plasma membrane reference markers in these labeling protocols. CellMask™ Orange and CellMask™ Deep Red plasma membrane stains (C10045, C10046; Section 14.4) are particularly suitable for this purpose.[172,173] Other useful membrane markers include posttranslationally lipidated fluorescent proteins [174] (O36214, O10139; Section 14.4). When spatial resolution is not possible, there are other means by which these signals can be separated and, in some cases, quantitated. These include:

- Use of antibodies to the Alexa Fluor® 488, BODIPY® FL, fluorescein/Oregon Green®, tetramethylrhodamine, Texas Red® and Alexa Fluor® 405/Cascade Blue® dyes (Section 7.4, Table 7.8) to quench most of the fluorescence of surface-bound or exocytosed probes

IMPORTANT NOTICE: The products described in this manual are covered by one or more Limited Use Label License(s). Please refer to the Appendix on page 971 and Master Product List on page 975. Products are For Research Use Only. Not intended for any animal or human therapeutic or diagnostic use.

- Use of a dye such as trypan blue to quench external fluorescent signals but not internalized signals[175,176] (Figure 16.1.17)—a method employed in our Vybrant® Phagocytosis Assay Kit (V6694) described below
- Rapid acidification of the medium to quench the fluorescence of pH-sensitive fluorophores such as fluorescein on the cell surface, thus enabling selective detection of endocytosed probe
- Tagging of proteins, polysaccharides, cells, bacteria, yeast, fungi[177] and other materials to be endocytosed with a pH-sensitive dye—such as our pHrodo™,[178–180] SNARF® or Oregon Green® dyes (Chapter 20)—that undergoes a spectral shift or intensity change in the acidic pH range found in phagovacuoles and late endosomes
- Use of heavily labeled, highly quenched proteins such as our DQ™ BSA and DQ™ gelatin probes, which yield highly fluorescent peptides upon intracellular proteolysis[181] (Section 10.4)

Pathway-specific inhibitors—such as chloropromazine, dynasore (Figure 16.1.18), dansyl cadaverine (D113, Section 3.4), brefeldin A (B7450, Section 12.4), genistein and filipin—are widely used in combination with fluorescently labeled ligands for characterizing endocytic pathways.[182] A critical evaluation[183] highlights some necessary cautions in the application and interpretation of this approach, relating to decreased cell viability caused by some inhibitors as well as cell type–dependent differences in their efficacy.

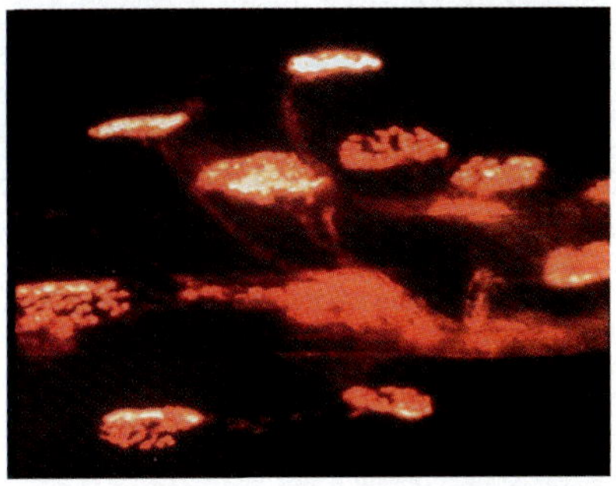

Figure 16.1.19 Live nerve terminals of motor neurons that innervate a rat lumbrical muscle stained with the activity-dependent dye FM® 1-43 (T3163, T35356) and observed under low magnification. The dye molecules, which insert into the outer leaflet of the surface membrane, are captured in recycled synaptic vesicles of actively firing neurons. The image was contributed by William J. Betz, University of Colorado School of Medicine.

Membrane Markers of Endocytosis and Exocytosis

FM® 1-43

FM® dyes—FM® 1-43, FM® 2-10, FM® 4-64, FM® 5-95 and the aldehyde-fixable FM® 1-43FX and FM® 4-64FX—are excellent membrane probes both for identifying actively firing neurons[184] and for investigating the mechanisms of activity-dependent vesicle cycling in widely different species.[185–188] FM® dyes may also be useful as general-purpose probes for investigating endocytosis and for simply identifying cell membrane boundaries.

FM® 1-43 and its analogs, which are nontoxic to cells and virtually nonfluorescent in aqueous medium, are believed to insert into the outer leaflet of the surface membrane, where they become intensely fluorescent. In a neuron that is actively releasing neurotransmitters, these dyes become internalized within the recycled synaptic vesicles and the nerve terminals become brightly stained (Figure 16.1.19, Figure 16.1.20). The nonspecific staining of cell-surface membranes can simply be washed off prior to viewing. Wash removal of noninternalized dye background is more difficult in tissue preparations than in disseminated cell cultures. Extracellular fluorescence quenching[189] and dye adsorption[190] strategies have been developed to address this problem. Alternatively, the optical sectioning capabilities of confocal microscopy, two-photon excitation microscopy (Fluorescent Probes for Two-Photon Microscopy—Note 1.5) and total internal reflection (TIRF) microscopy provide instrument-based solutions for improving the signal-to-background contrast.[191] The amount of FM® 1-43 taken up per vesicle by endocytosis equals the amount of dye released upon exocytosis, indicating that the dye does not transfer from internalized vesicles to an endosome-like compartment during the recycling process.[192] In astrocytes, internalization of FM® 1-43 (and FM® 4-64) is mediated by store-operated calcium channels and not by endocytosis.[193] Like most styryl dyes, the absorption and fluorescence emission spectra of FM® 1-43 are significantly shifted in the membrane environment and are relatively broad (Figure 16.1.21), requiring careful matching with other fluorophores to avoid channel crosstalk in multiplex detection applications (Using the Fluorescence SpectraViewer—Note 23.1). We offer FM® 1-43 in a 1 mg vial (T3163) or specially packaged in 10 vials of 100 µg each (T35356).

FM® 1-43 was employed in a study showing that synaptosomal endocytosis is independent of both extracellular Ca^{2+} and membrane potential in dissociated hippocampal neurons,[194] as well as in a spectrofluorometric assay demonstrating that nitric oxide–stimulated vesicle release is independent of Ca^{2+} in isolated rat hippocampal nerve terminals.[195] FM® 1-43 has been used in combination with fura-2 (Section 19.2) to simultaneously measure intracellular Ca^{2+} and membrane turnover.[196,197] FM® 1-43 dye–mediated photoconversion has been used to visualize recycling vesicles in hippocampal neurons.[198]

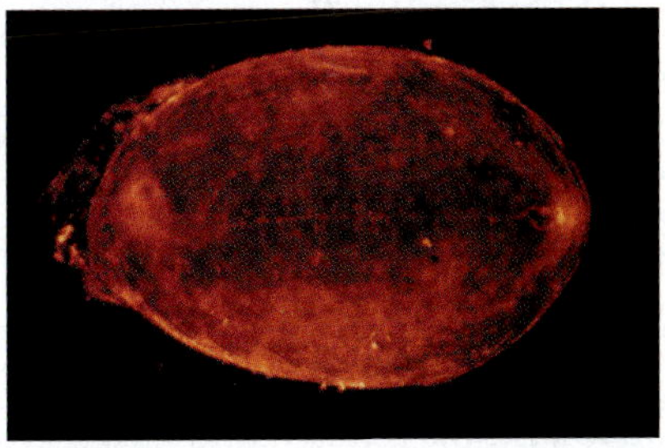

Figure 16.1.20 A feline mesenteric Pacinian corpuscle labeled with FM® 1-43 (T3163, T35356). The image was contributed by Michael Chua, University of North Carolina at Chapel Hill.

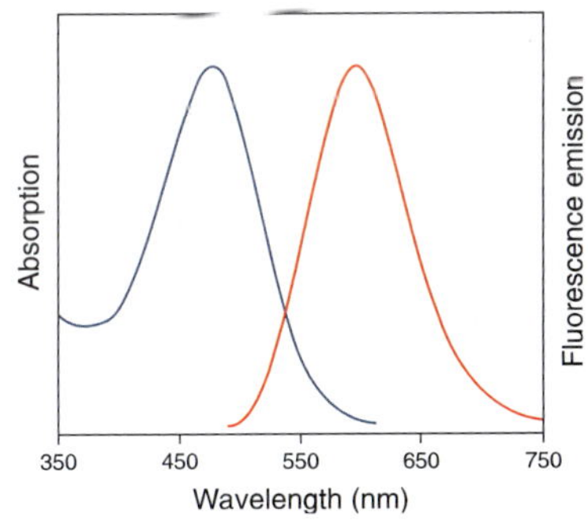

Figure 16.1.21 Absorption and fluorescence emission spectra of FM® 1-43 bound to phospholipid bilayer membranes.

IMPORTANT NOTICE: The products described in this manual are covered by one or more Limited Use Label License(s). Please refer to the Appendix on page 971 and Master Product List on page 975. Products are For Research Use Only. Not intended for any animal or human therapeutic or diagnostic use.

Figure 16.1.22 Feline muscle spindle, a specialized sensory receptor unit that detects muscle length and changes in muscle length and velocity, was labeled with FM® 2-10 (T7508). Image contributed by Michael Chua, University of North Carolina at Chapel Hill.

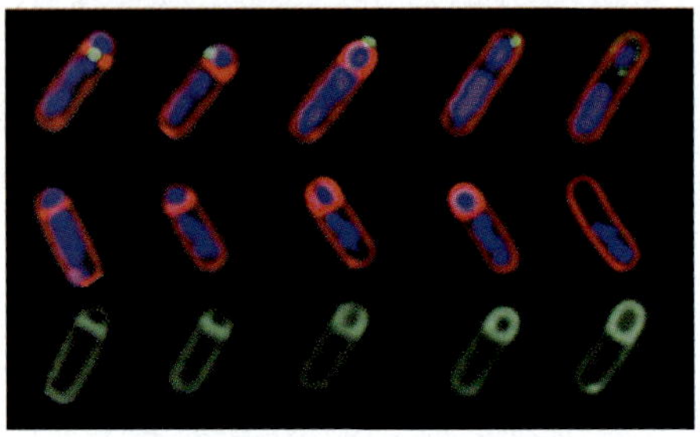

Figure 16.1.23 Correlated fluorescence imaging of membrane migration, protein translocation and chromosome localization during *Bacillus subtilis* sporulation. Membranes were stained with red-fluorescent FM® 4-64 (T3166, T13320). Chromosomes were localized with the blue-fluorescent nuclear counterstain DAPI (D1306, D3571, D21490). The small, green-fluorescent patches (top row) indicate the localization of a GFP fusion to SpoIIIE, a protein essential for both initial membrane fusion and forespore engulfment. Progression of the engulfment is shown from left to right. Green fluorescence in the middle and bottom rows demonstrates fully engulfed sporangia stained with MitoTracker® Green FM® (M7514). Full details of the experimental methods and interpretation are published in Proc Natl Acad Sci U S A 96, 14553 (1999). Image contributed by Kit Pogliano and Marc Sharp, University of California at San Diego. Reproduced from the 7 December 1999 issue of Proc Natl Acad Sci U S A, with permission.

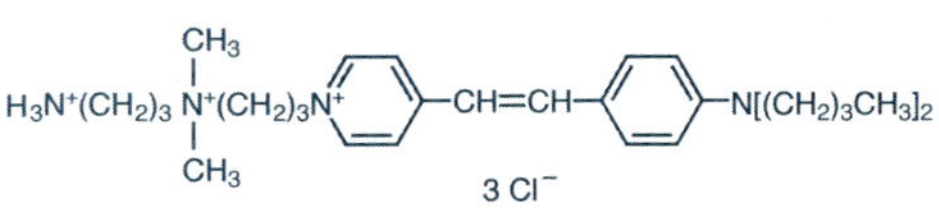

Figure 16.1.24 FM® 1-43FX (F35355).

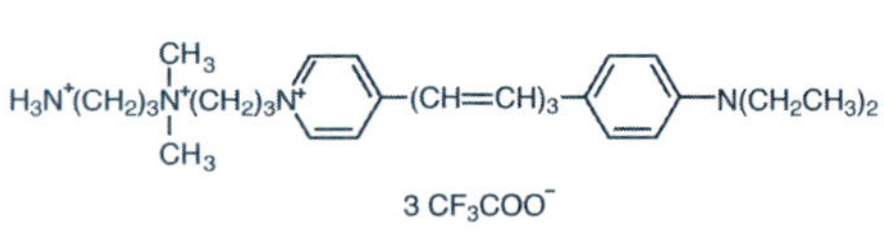

Figure 16.1.25 FM® 4-64FX (F34653).

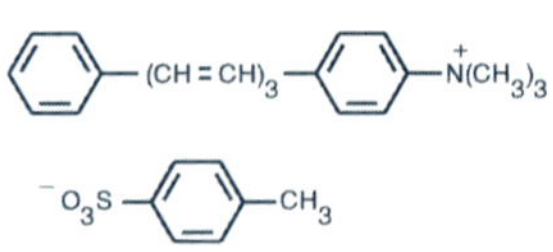

Figure 16.1.26 TMA-DPH (1-(4-trimethylammonium-phenyl)-6-phenyl-1,3,5-hexatriene *p*-toluenesulfonate)(T204).

Other Analogs of FM® 1-43

A comparison of mammalian motor nerve terminals stained with either FM® 1-43 or the more hydrophilic analog FM® 2-10 (T7508, Figure 16.1.22) revealed that lower background staining by FM® 2-10 and its faster destaining rate may make it the preferred probe for quantitative applications.[199,200] However, staining with FM® 2-10 requires much higher dye concentrations[199] (100 µM compared with 2 µM for FM® 1-43). Additionally, it has been shown that both FM® 1-43 and FM® 2-10 are antagonists of muscarinic acetylcholine receptors and may be useful for analyzing receptor distribution and occupancy.[201] This property may be due to the cationic alkylammonium substituent of FM® dyes, which they have in common with choline, and could serve as one of the sources of background FM® dye staining in tissues.

FM® 4-64 (T3166, T13320) and RH 414 (T1111)—both more hydrophobic than FM® 1-43—may also be useful as probes for investigating endocytosis. Because small differences in the polarity of these FM® probes can play a large role in their rates of uptake and their retention properties, we have introduced FM® 5-95 (T23360), a slightly less lipophilic analog of FM® 4-64 with essentially identical spectroscopic properties. FM® 4-64 exhibits long-wavelength red fluorescence that can be distinguished from Green Fluorescent Protein (GFP) with the proper optical filter sets.[202–205]

FM® 4-64 is an endosomal marker and vital stain that persists through cell division,[206,207] as well as a stain for functional presynaptic boutons.[208] In addition, FM® 4-64 staining has been used to visualize membrane migration and fusion during *Bacillus subtilis* sporulation, and these movements can be correlated with the translocation of GFP-labeled proteins[202,209,210] (Figure 16.1.23). Sequential pulse-chase application of FM® 4-64 and FM® 1-43 allows two-color fluorescence discrimination of temporally staged synaptic vesicle populations.[187] FM® 4-64 selectively stains yeast vacuolar membranes and is an important tool for visualizing vacuolar organelle morphology and dynamics and for studying the endocytic pathway and vacuole fusion in yeast[211–213] (Section 12.3). FM® 4-64 and FM® 1-43 also have many applications for visualizing membrane dynamics in plant[204,214–216] and algal[217] cells.

FM® 1-43FX and FM® 4-64FX: Fixable FM® Dyes

FM® 1-43FX and FM® 4-64FX are FM® 1-43 and FM® 4-64 analogs, respectively, that have been modified to contain an aliphatic amine (Figure 16.1.24, Figure 16.1.25). This modification makes the probe fixable with aldehyde-based fixatives, including formaldehyde and glutaraldehyde. FM® 1-43FX has been used to study synaptic vesicle cycling in cone photoreceptor terminals[187] and to investigate the functional maturation of glutamatergic synapses.[218] FM® 1-43FX (F35355) and FM® 4-64FX (F34653) are available specially packaged in 10 vials of 100 µg each.

4-Di-1-ASP and 4-Di-2-ASP

The cationic mitochondrial dyes 4-Di-1-ASP (D288) and 4-Di-2-ASP (D289) stain presynaptic nerve terminals independent of neuronal activity.[219–222] These aminostyrylpyridinium dyes have also been widely used as substrates for functional analysis of biogenic amine transporters[223–227] and renal and hepatic organic cation transporters.[228–230]

TMA-DPH

Also useful as a lipid marker for endocytosis and exocytosis is the cationic linear polyene TMA-DPH (T204, Figure 16.1.26), which readily incorporates in the plasma membrane of live cells.[231,232] TMA-DPH is virtually nonfluorescent in water and is reported to bind to cells in proportion to the available membrane surface.[233] Its fluorescence intensity is therefore sensitive to physiological processes that cause a net change in membrane surface area, making it an excellent probe for monitoring events such as changes in cell volume and exocytosis.[233–236]

Fluorescent Cholera Toxin Subunit B: Markers of Lipid Rafts

Fluorescent cholera toxins, which bind to galactosyl moieties, are markers of lipid rafts—regions of cell membranes high in ganglioside G_{M1} that are thought to be important in cell signaling.[237,238] Lipid rafts are detergent-insoluble, sphingolipid- and cholesterol-rich membrane microdomains that form lateral assemblies in the plasma membrane.[239–245] Lipid rafts also sequester glycophosphatidylinositol (GPI)-linked proteins and other signaling proteins and

The Molecular Probes® Handbook: A Guide to Fluorescent Probes and Labeling Technologies
IMPORTANT NOTICE: The products described in this manual are covered by one or more Limited Use Label License(s). Please refer to the Appendix on page 971 and Master Product List on page 975. Products are For Research Use Only. Not intended for any animal or human therapeutic or diagnostic use.
www.invitrogen.com/probes

receptors, which may be regulated by their selective interactions with these membrane microdomains.[246–251] Recent research has demonstrated that lipid rafts play a role in a variety of cellular processes—including the compartmentalization of cell-signaling events,[252–259] the regulation of apoptosis[260–262] and the intracellular trafficking of certain membrane proteins and lipids[263–265]—as well as in the infectious cycles of several viruses and bacterial pathogens.[266–271]

The Vybrant® Lipid Raft Labeling Kits (V34403, V34404, V34405; Section 14.4) provide the key reagents for fluorescently labeling lipid rafts *in vivo* with our bright and extremely photostable Alexa Fluor® dyes (Figure 16.1.27, Figure 16.1.28). Live cells are first labeled with the green-fluorescent Alexa Fluor® 488, orange-fluorescent Alexa Fluor® 555 or red-fluorescent Alexa Fluor® 594 conjugate of cholera toxin subunit B (CT-B). This CT-B conjugate binds to the pentasaccharide chain of plasma membrane ganglioside G_{M1}, which selectively partitions into lipid rafts.[250,272,273] An antibody that specifically recognizes CT-B is then used to crosslink the CT-B–labeled lipid rafts into distinct patches on the plasma membrane, which are easily visualized by fluorescence microscopy.[274,275] Each Vybrant® Lipid Raft Labeling Kit contains sufficient reagents to label 50 live-cell samples, including:

- Recombinant cholera toxin subunit B (CT-B) labeled with the Alexa Fluor® 488 (in Kit V34403), Alexa Fluor® 555 (in Kit V34404) or Alexa Fluor® 594 (in Kit V34405) dye
- Anti–cholera toxin subunit B antibody (anti–CT-B)
- Concentrated phosphate-buffered saline (PBS)
- Detailed labeling protocol

Cholera toxin subunit B and its conjugates are also established as superior tracers for retrograde labeling of neurons.[276,277] Cholera toxin subunit B conjugates bind to the pentasaccharide chain of ganglioside G_{M1} on neuronal cell surfaces and are actively taken up and transported; alternatively, they can be injected by iontophoresis. Unlike the carbocyanine-based neuronal tracers such as DiI (D282, D3911, V22885; Section 14.4), cholera toxin subunit B conjugates can be used on tissue sections that will be fixed and frozen.[278]

All of our cholera toxin subunit B conjugates are prepared from recombinant cholera toxin subunit B, which is completely free of the toxic subunit A, thus eliminating any concern for toxicity or ADP-ribosylating activity. The Alexa Fluor® 488 (C22841, C34775), Alexa Fluor® 555 (C22843, C34776), Alexa Fluor® 594 (C22842, C34777) and Alexa Fluor® 647 (C34778) conjugates of cholera toxin subunit B combine this versatile tracer with the superior brightness of our Alexa Fluor® dyes to provide sensitive and selective receptor labeling and neuronal tracing. We also offer biotin-XX (C34779) and horseradish peroxidase (C34780) conjugates of cholera toxin subunit B for use in combination with diaminobenzidine (DAB) oxidation,[279] tyramide signal amplification (TSA) and Qdot® nanocrystal–streptavidin conjugates.[280]

Fluorescent Protein–Based Lipid Raft Markers

CellLight® plasma membrane expression vectors (C10606, C10607, C10608; Section 14.4) generate cyan-, green- or red-autofluorescent proteins fused to a plasma membrane targeting sequence consisting of the 10 N-terminal amino acids of LcK tyrosine kinase (Lck10). These fusion proteins are lipid raft markers with well established utility,[90] providing alternatives to cholera toxin B conjugates or BODIPY® FL C_5-ganglioside G_{M1}[91] (B13950, B34401; Section 13.3) with the inherent advantages of long-lasting and titratable expression conferred by BacMam 2.0 vector technology (BacMam Gene Delivery and Expression Technology—Note 11.1).

Fluorescent Protein–Based Synaptic Vesicle Markers

CellLight® Synaptophysin-GFP (C10609) and CellLight® Synaptophysin-RFP (C10610) are valuable counterparts to FM® dyes for visualizing the distribution and density of presynaptic sites in neurons both *in vitro* and *in vivo*. Synaptophysin is a synaptic vesicle membrane glycoprotein that is involved in the biogenesis and fusion of synaptic vesicles but is not essential for neurotransmitter release. It is found in virtually all synaptically active neurons in the brain and spinal cord. These CellLight® reagents incorporate all the customary advantages of BacMam 2.0 delivery technology including high transduction efficiency and long-lasting and titratable expression (BacMam Gene Delivery and Expression Technology—Note 11.1).

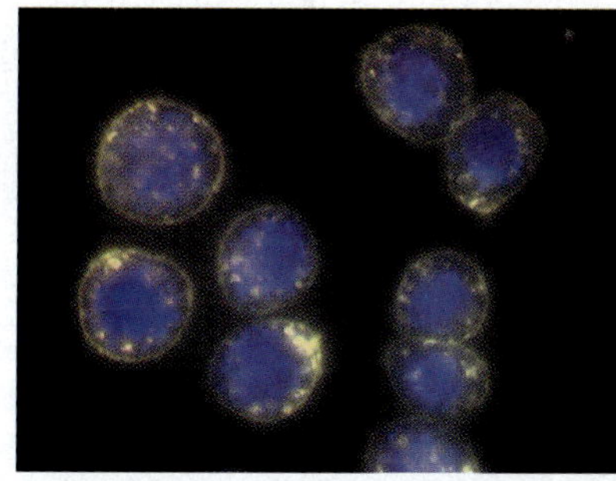

Figure 16.1.27 Live J774 macrophage cells labeled with BODIPY® FL C_5-ganglioside G_{M1} and Alexa Fluor® 555 cholera toxin subunit B conjugate. Live J774 macrophage cells labeled with BODIPY® FL C_5-ganglioside G_{M1} (B13950) and then with Alexa Fluor® 555 cholera toxin subunit B conjugate (C22843; also available as a component of V34404). Cells were then treated with anti–CT-B antibody (a component of V34404) to induce crosslinking. Yellow fluorescence indicates colocalization of the two dyes. Nuclei were stained with the blue-fluorescent Hoechst 33342 dye (H1399, H3570, H21492).

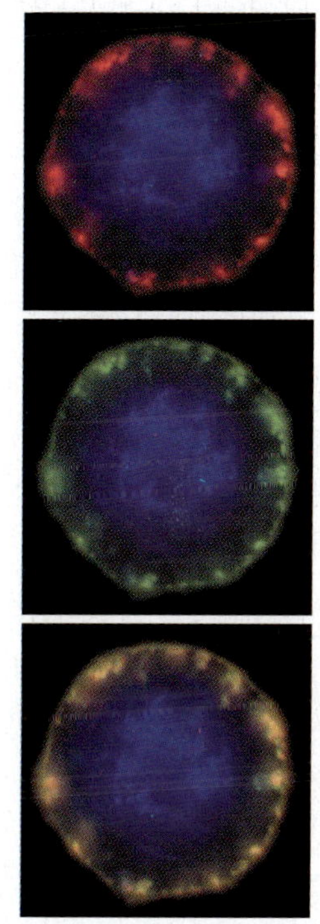

Figure 16.1.28 A J774 mouse macrophage cell stained with BODIPY® FL ganglioside G_{M1} (B13950) and Alexa Fluor® 555 dye–labeled cholera toxin subunit B. A J774 mouse macrophage cell sequentially stained with BODIPY® FL ganglioside G_{M1} (B13950) and then with Alexa Fluor® 555 dye–labeled cholera toxin subunit B (C22843, C34776; also available as a component of V34404). The cell was then treated with an anti–CT-B antibody (a component of V34404) to induce crosslinking. Alexa Fluor® 555 dye fluorescence (top panel, red) and BODIPY® FL dye fluorescence (middle panel, green) were imaged separately and overlaid to emphasize the coincident staining (bottom panel, yellow). Nuclei were stained with blue-fluorescent Hoechst 33258 (H1398, H3569, H21491).

IMPORTANT NOTICE: The products described in this manual are covered by one or more Limited Use Label License(s). Please refer to the Appendix on page 971 and Master Product List on page 975. Products are For Research Use Only. Not intended for any animal or human therapeutic or diagnostic use.

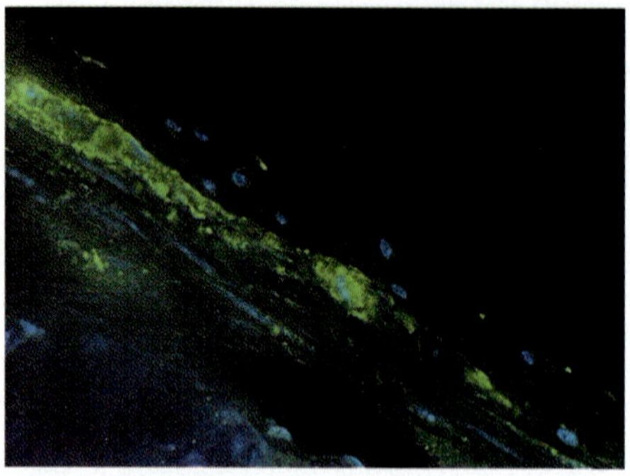

Figure 16.1.29 Peripheral neurons in mouse intestinal cryosections were labeled with rabbit anti–synapsin I antibody (A6442) and detected using Alexa Fluor® 488 goat anti–rabbit IgG antibody (A11008). This tissue was counterstained with DAPI (D1306, D3571, D21490).

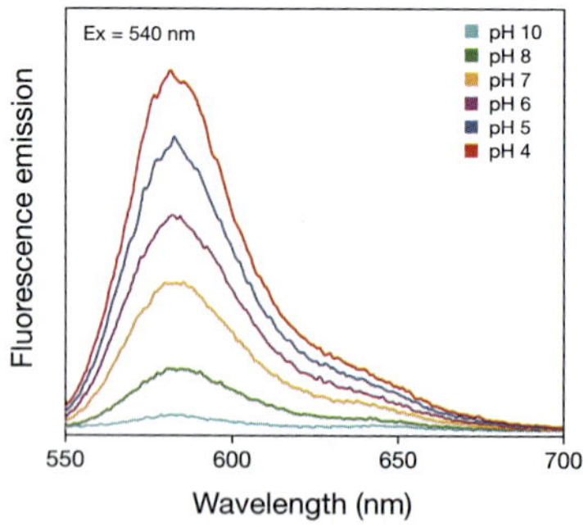

Figure 16.1.30 The pH sensitivity of pHrodo™ dextran. pHrodo™ 10,000 MW dextran (P10361) was reconisitituted in HEPES (20 mM)–buffered PBS and adjusted to pH values from pH 4 to pH 10. The intensity of fluorescence emission increases with increasing acidity, particularly in the pH 5–8 range.

Anti–Synapsin I Antibody

Synapsin I is an actin-binding protein that is localized exclusively to synaptic vesicles and thus serves as an reliable marker for synapses in brain and other neuronal tissues.[283] Synapsin I inhibits neurotransmitter release, an effect that is abolished upon its phosphorylation by Ca^{2+}/calmodulin–dependent protein kinase II[284] (CaM kinase II). Antibodies directed against synapsin I have proven valuable in molecular and neurobiology research, for example, to estimate synaptic density and to follow synaptogenesis.[218,285]

We offer a rabbit polyclonal anti–bovine synapsin I antibody as an affinity-purified IgG fraction (A6442). This antibody was isolated from rabbits immunized against bovine brain synapsin I but is also active against human, rat and mouse forms of the antigen; it has little or no activity against synapsin II. The affinity-purified rabbit polyclonal antibody was fractionated from the serum using column chromatography in which bovine synapsin I was covalently bound to the column matrix. Affinity-purified anti–synapsin I antibody is suitable for immunohistochemistry (Figure 16.1.29), western blots, enzyme-linked immunosorbent assays and immunoprecipitations. Our complete selection of antibodies can be found at www.invitrogen.com/handbook/antibodies.

High Molecular Weight Polar Markers

Fluorescent Protein–Based Endosomal Markers

CellLight® Early Endosomes–GFP (O10104) and CellLight® Early Endosomes–RFP (O36231) provide BacMam expression vectors encoding fusions of GFP or RFP with the small GTPase Rab5a. Rab5a fusions with autofluorescent proteins are sensitive and precise early endosome markers for real-time imaging of clathrin-mediated endocytosis in live cells.[276,286,287] These CellLight® reagents incorporate all the customary advantages of BacMam 2.0 delivery technology, including high transduction efficiency and long-lasting and titratable expression (BacMam Gene Delivery and Expression Technology—Note 11.1).

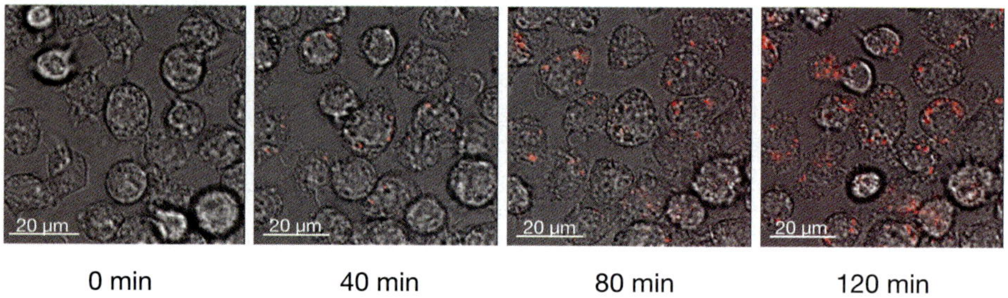

Figure 16.1.31 Time course of pHrodo™ *E. coli* BioParticles® (P35361) uptake by metastatic malignant melanoma cells. Cells were imaged at 37°C in the continued presence of 100 µg/mL pHrodo™ BioParticles®. Uptake of pHrodo™ BioParticles® was observable as early as 20 minutes and reached a plateau within 2 to 3 hours.

Table 16.3 BioParticles® fluorescent bacteria and yeast.

Label (Abs/Em Maxima in nm)	*Escherichia coli* (K-12 strain)	*Staphylococcus aureus* (Wood strain without protein A)	Zymosan A (*Saccharomyces cerevisiae*)
Fluorescein (494/518)	E2861	S2851	Z2841
Alexa Fluor® 488 (495/519)	E13231	S23371	Z23373
BODIPY® FL (505/513)	E2864	S2854	
Tetramethylrhodamine (555/580)	E2862		
pHrodo™ (560/585)	A10025, P35361	A10010	
Alexa Fluor® 594 (590/617)	E23370	S23372	Z23374
Texas Red® (595/615)	E2863		Z2843
Unlabeled		S2859	Z2849

We also offer opsonizing reagents for enhancing the uptake of BioParticles® products. These reagents are derived from purified rabbit polyclonal IgG antibodies that are specific for the *E. coli* (E2870), *S. aureus* (S2860) or zymosan (Z2850) particles. Reconstitution of the lyophilized opsonizing reagents requires only the addition of water, and one unit of opsonizing reagent is sufficient to opsonize ~10 mg of the corresponding BioParticles® product.

IMPORTANT NOTICE: The products described in this manual are covered by one or more Limited Use Label License(s). Please refer to the Appendix on page 971 and Master Product List on page 975. Products are For Research Use Only. Not intended for any animal or human therapeutic or diagnostic use.

BioParticles® Fluorescent Bacteria and Yeast

The BioParticles® product line consists of a series of fluorescently labeled, heat- or chemically killed bacteria and yeast in a variety of sizes, shapes and natural antigenicities. These fluorescent BioParticles® products have been employed to study phagocytosis by fluorescence microscopy,[288,289] quantitative spectrofluorometry[290] and flow cytometry.[288,291] We offer *Escherichia coli* (K-12 strain), *Staphylococcus aureus* (Wood strain without protein A) and zymosan (*Saccharomyces cerevisiae*) BioParticles® products covalently labeled with a variety of fluorophores, including Alexa Fluor®, fluorescein, BODIPY® FL, tetramethylrhodamine, Texas Red® and pHrodo™ dyes (Table 16.3). Special care has been taken to remove any free dye after conjugation. BioParticles® products are freeze-dried and ready for reconstitution in a buffer of choice and are supplied with a general protocol for measuring phagocytosis; we also offer opsonizing reagents for use with each particle type, as described below.

Unlike the fluorescence of fluorescein-labeled BioParticles® bacteria and yeast, which is strongly quenched in acidic environments, the fluorescence of the Alexa Fluor® 488, BODIPY® FL, tetramethylrhodamine and Texas Red® BioParticles® conjugates is uniformly intense between pH 3 and 10. This property is particularly useful for quantitating fluorescent bacteria and zymosan within acidic phagocytic vacuoles.

Fluorescent bacteria and yeast particles are proven tools for studying a variety of phagocytosis parameters. For example, they have been used to:

- Detect the phagocytosis of yeast by murine peritoneal macrophage[292] and human neutrophils[290]
- Determine the effects of different opsonization procedures on the efficiency of phagocytosis of pathogenic bacteria[293] and yeast[290]
- Investigate the kinetics of phagocytosis degranulation and actin polymerization in stimulated leukocytes[290]
- Quantitate the effects of purinergic P2X_7 receptor activation on phagosomal maturation[179]
- Show that *Dictyostelium discoideum* depleted of clathrin heavy chains are still able to undergo phagocytosis of fluorescent zymosans[294]
- Study molecular defects in phagocytic function[178]

Vybrant® Phagocytosis Assay Kit

The Vybrant® Phagocytosis Assay Kit (V6694) provides a convenient set of reagents for quantitating phagocytosis and assessing the effects of certain drugs or conditions on this cellular process. In this assay, cells of interest are incubated first with green-fluorescent fluorescein-labeled *E. coli* BioParticles® conjugates, which are internalized by phagocytosis, and then with trypan blue, which quenches the fluorescence of any extracellular BioParticles® product (Figure 16.1.17). The methodology provided by this kit was developed using the adherent murine macrophage cell line J774;[176] however, researchers have adapted this assay to other phagocytic cell types[295] and other instrument platforms such as flow cytometers.[296] Each kit provides sufficient reagents for 250 tests using a 96-well microplate format and contains:

- BioParticles® fluorescein-labeled *Escherichia coli*
- Hanks' balanced salt solution (HBSS)
- Trypan blue
- Step-by-step instructions for performing the phagocytosis assay

pHrodo™ BioParticles® Fluorescent Bacteria

In contrast to both the fluorescein- and Alexa Fluor® dye–labeled BioParticles® conjugates, the fluorescence of the pHrodo™ *E. coli* and *S. aureus* BioParticles® conjugates (P35361, A10010) *increases* in acidic environments (Figure 16.1.30), providing a continuous positive indicator of phagocytic uptake. With a simple no-cell background subtraction method, a large and specific signal is obtained from cells that ingest the pHrodo™ BioParticles®, providing a specific index of phagocytosis in the context of a variety of pretreatments or conditions (Figure 16.1.31). The optimal absorption and fluorescence emission maxima of the pHrodo™ BioParticles® conjugates are approximately 560 nm and 585 nm, respectively, but the pHrodo™ fluorophore is also readily excited by the 488 nm spectral line of the argon-ion laser used in most flow cytometers.

With each pHrodo™ BioParticles® conjugate, we provide sufficient reagent for 100 microplate wells in a 96-well format, along with step-by-step instructions for performing phagocytosis assays in a fluorescence microplate reader. This methodology has been developed using adherent J774A.1 murine macrophage cells, but can be adapted for use with other adherent cells,[178] primary cells[179,297] or cells in suspension,[298] as well as for *in vivo* applications.[299] Cells assayed for phagocytic activity with pHrodo™ BioParticles® conjugates may be fixed with standard formaldehyde solutions for later analysis, preserving differences in signal between control and experimental samples with high fidelity. pHrodo™ BioParticles® conjugate preparations are also amenable to opsonization (E2870, S2860), which can greatly enhance their uptake and signal strength in the phagocytosis assay.

To facilitate the use of pHrodo™ BioParticles® conjugates for the study of phagocytosis, we offer the pHrodo™ *E. coli* BioParticles® Phagocytosis Kit for Flow Cytometry (A10025), which provides the key reagents for assessing particle ingestion and red blood cell lysis (Figure 16.1.32). Each kit provides sufficient reagents for performing 100 assays when using sample volumes of 100 µL whole blood per assay, including:

- pHrodo™ *E. coli* BioParticles® conjugates
- Lysis and wash buffers
- Detailed protocols

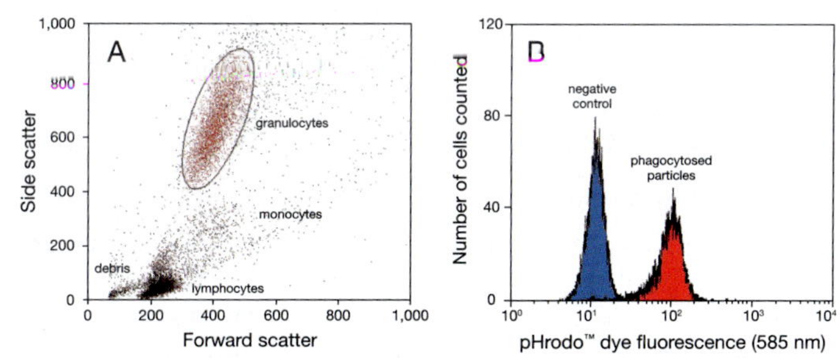

Figure 16.1.32 Flow cytometry analysis showing increased fluorescence of granulocytes treated with pHrodo™ *E. coli* BioParticles® (P35361). A whole blood sample was collected and treated with heparin, and two 100 µL aliquots were prepared. Both aliquots were treated with pHrodo™ BioParticles® and vortexed. One sample was placed in a 37°C water bath, and the other sample (negative control) was placed in an ice bath. After a 15-minute incubation, red blood cells were lysed with an ammonium chloride–based lysis buffer. The samples were centrifuged for 5 minutes at 500 rcf, washed once, and resuspended with HBSS. The samples were then analyzed on a BD FACSCalibur™ cytometer (BD Biosciences) using a 488 nm argon laser and 564–606 nm emission filter. **A)** Granulocytes were gated using forward and side scatter. **B)** The sample incubated at 37°C shows the increased fluorescence of the phagocytosed pHrodo™ BioParticles® (red), in contrast to the negative control sample, which was kept on ice to inhibit phagocytosis (blue).

IMPORTANT NOTICE: The products described in this manual are covered by one or more Limited Use Label License(s). Please refer to the Appendix on page 971 and Master Product List on page 975. Products are For Research Use Only. Not intended for any animal or human therapeutic or diagnostic use.

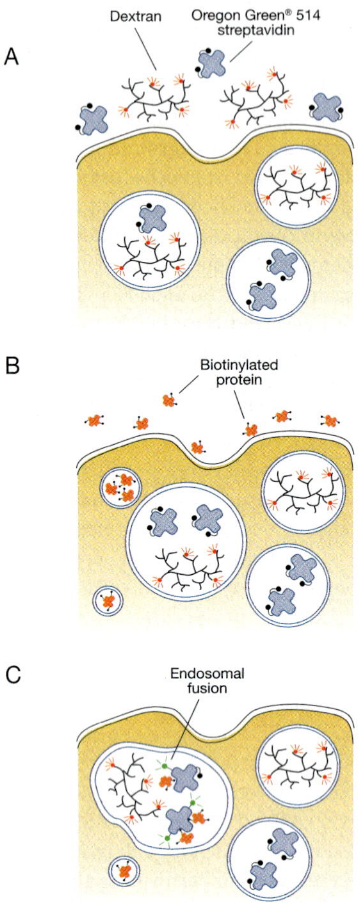

Figure 16.1.33 Detection of endosomal fusion. **A)** Cells are first incubated with a combination of a high molecular weight, red-fluorescent dextran (D1829, D1830, D1864) and the green-fluorescent Oregon Green® 514 streptavidin (S6369), which intrinsically has low fluorescence. **B)** The cells are then incubated with a biotinylated probe, e.g., biotinylated transferrin (T23363), and the excess conjugate is washed. **C)** Endosomal fusion is monitored by an increase in fluorescence of the Oregon Green® 514 dye as it is displaced by the biotinylated protein. The red-fluorescent dextran's fluorescence remains constant and allows for ratiometric measurements of the fused endosomes.

pHrodo™ Phagocytosis Particle Labeling Kit

In addition to the pHrodo™ BioParticles® conjugates, we offer the pHrodo™ Phagocytosis Particle Labeling Kit for Flow Cytometry (A10026), which allows rapid labeling of biological particles, such as bacteria, and subsequent assessment of of phagocytic activity in whole blood samples by flow cytometry. Each kit provides sufficient reagents for performing 100 assays when using sample volumes of 100 µL whole blood per assay, including:

- pHrodo™ succinimidyl ester
- Lysis and wash buffers
- Dimethylsulfoxide (DMSO)
- Sodium bicarbonate
- Detailed protocols

The amine-reactive pHrodo™ succinimidyl ester is also available separately (P36600, Section 20.4) for creating pH-sensitive conjugates for following phagocytosis. pHrodo™ succinimidyl ester was used to label dexamethasone-treated thymocytes for flow cytometry detection of phagocytosis by splenic or peritoneal macrophages.[180]

Opsonizing Reagents and Nonfluorescent BioParticles® Products

Many researchers may want to use autologous serum to opsonize their fluorescent zymosan and bacterial particles; however, we also offer special opsonizing reagents (E2870, S2860, Z2850) for enhancing the uptake of each type of particle, along with a protocol for opsonization. These reagents are derived from purified rabbit polyclonal IgG antibodies that are specific for the *E. coli*, *S. aureus* or zymosan particles. Reconstitution of the lyophilized opsonizing reagents requires only the addition of water, and one unit of opsonizing reagent is sufficient to opsonize ~10 mg of the corresponding BioParticles® product.

In addition, we offer nonfluorescent zymosan (Z2849) and *S. aureus* (S2859) BioParticles® products. These nonfluorescent BioParticles® products are useful either as controls or for custom labeling with the reactive dye or indicator of interest.

Fluorescent Polystyrene Microspheres

Fluorescent polystyrene microspheres with diameters between 0.5 and 2.0 µm have been used to investigate phagocytic processes in murine melanoma cells,[300] human alveolar macrophages,[289] ciliated protozoa[137] and *Dictyostelium discoideum*.[301,302] The phagocytosis of fluorescent microspheres has been quantitated both with image analysis[289,303,304] and with flow cytometry.[305] Section 6.5 includes a detailed description of our full line of FluoSpheres® (Table 6.7) and TransFluoSpheres® (Table 6.9) fluorescent microspheres. Because of their low nonspecific binding, carboxylate-modified microspheres appear to be best for phagocytosis applications. For phagocytosis experiments involving multicolor detection, we particularly recommend our 1.0 µm TransFluoSpheres® fluorescent microspheres[306] (T8880, T8883; Section 6.5). Various opsonizing reagents, such as rabbit serum or fetal calf serum, have been used with the microspheres to facilitate phagocytosis.

Fluorescent Microspheres Coated with Collagen

Fibroblasts phagocytose and subsequently digest collagen. These activities play an important role in the remodeling of the extracellular matrix during normal physiological turnover of connective tissues and wound repair, as well as in development and aging. A well-established procedure for observing collagen phagocytosis by either flow cytometry or fluorescence microscopy entails the use of collagen-coated fluorescent microspheres that attach to the cell surface and become engulfed by fibroblasts.[307] We offer yellow-green–fluorescent FluoSpheres® collagen I–labeled microspheres in either 1.0 µm or 2.0 µm diameter (F20892, F20893) for use in these applications. In the production of these microspheres, collagen I from calf skin is attached covalently to the microsphere's surface.

Fluorescent Dextrans

Tracing internalization of extracellularly introduced fluorescent dextrans is a standard method for analyzing fluid-phase endocytosis.[2,73,308,309] We offer dextrans with nominal molecular weights ranging from 3000 to 2,000,000 daltons, many of which can also be used as pinocytosis or phagocytosis markers (see Section 14.5 and Table 14.4 for further discussion and a complete product list). Discrimination of internalized fluorescent dextrans from dextrans in the growth medium is facilitated by use of reagents that quench the fluorescence of the external probe. For example, most of our anti-fluorophore antibodies (Section 7.4, Table 7.8) strongly quench the fluorescence of the corresponding dyes.

Negative staining produced by fluorescent dextrans that have been intracellularly infused via a patch pipette is indicative of nonendocytic vacuoles in live pancreatic acinar cells.[310] Extracellular addition of a second, color-contrasting dextran then allows discrimination of endocytic and nonendocytic vacuoles. Intracellular fusion of endosomes has been monitored with a BODIPY® FL avidin conjugate by following the fluorescence enhancement that occurs when it complexes with a biotinylated dextran.[311] We have found our Oregon Green® 514 streptavidin (S6369, Section 7.6) to have an over 15-fold increase in fluorescence intensity upon binding free biotin, which may make it the preferred probe for this application (Figure 16.1.33).

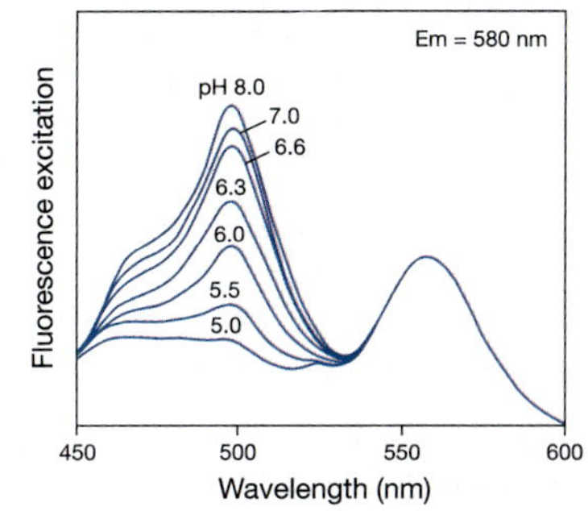

Figure 16.1.34 The excitation spectra of double-labeled fluorescein-tetramethylrhodamine dextran (D1951), which contains pH-dependent (fluorescein) and pH-independent (tetramethylrhodamine) dyes.

pH Indicator Dextrans

The fluorescein dextrans (pK_a ~6.4) are frequently used to investigate endocytic acidification.[312,313] Fluorescence of fluorescein-labeled dextrans is strongly quenched upon acidification; however, fluorescein's lack of a spectral shift in acidic solution makes it difficult to discriminate between an internalized probe that is quenched and residual fluorescence of the external medium. Dextran conjugates that either shift their emission spectra in acidic environments, such as the SNARF® dextrans (Section 20.4), or undergo significant shifts of their excitation spectra, such as BCECF and Oregon Green® dextrans (Section 20.4), provide alternatives to fluorescein. The Oregon Green® 488 and Oregon Green® 514 dextrans exhibit a pK_a of approximately 4.7, facilitating measurements in acidic environments.[312,314] In addition to these pH indicator dextrans, we prepare a dextran that is double-labeled with fluorescein and tetramethylrhodamine (D1951; Section 20.4), which has been used as a ratiometric indicator (Figure 16.1.34) to measure endosomal acidification in Hep G2 cells[315] and murine alveolar macrophages.[178]

In contrast to fluorescein and Oregon Green® 488 dextrans, pHrodo™ 10,000 MW dextran (P10361) exhibits increasing fluorescence in response to acidification[178] (Figure 16.1.30). The minimal fluorescent signal from pHrodo™ dextran at neutral pH prevents the detection of noninternalized and nonspecifically bound conjugates and eliminates the need for quenching reagents and extra wash steps, thus providing a simple fluorescent assay for endocytic activity. pHrodo™ dextran's excitation and emission maxima of 560 and 585 nm, respectively, facilitate multiplexing with other fluorophores including blue-, green- and far-red–fluorescent probes. Although pHrodo™ dextran is optimally excited at approximately 560 nm, it is also readily excited by the 488 nm spectral line of the argon-ion laser found on flow cytometers, confocal microscopes and imaging microplate readers (Figure 16.1.18).

Low Molecular Weight Polar Markers

Hydrophilic fluorescent dyes—including sulforhodamine 101 (S359), lucifer yellow CH (L453), calcein (C481), 8-hydroxypyrene-1,3,6-trisulfonic acid (HPTS, pyranine; H348) and Cascade Blue® hydrazide (C687)—are taken up by actively firing neurons through endocytic recycling of the synaptic vesicles.[316,317] Unlike the fluorescent FM® membrane probes described above, however, the hydrophilic fluorophores appear to work for only a limited number of species in this application. In some tissue preparations, background due to noninternalized polar markers is easier to wash away than that emanating from membrane markers such as FM® 1-43.[316] The same dyes have frequently been used as fluid-phase markers of pinocytosis.[318–321] The highly water-soluble Alexa Fluor® hydrazides and Alexa Fluor® hydroxylamines (Section 14.3, Table 3.2) provide superior spectral properties and can be fixed in cells by aldehyde-based fixatives.[322]

IMPORTANT NOTICE: The products described in this manual are covered by one or more Limited Use Label License(s). Please refer to the Appendix on page 971 and Master Product List on page 975. Products are For Research Use Only. Not intended for any animal or human therapeutic or diagnostic use.

REFERENCES

1. J Cell Sci (2009) 122:1713; **2.** Annu Rev Biochem (2009) 78:857; **3.** Cell Commun Signal (2009) 7:16; **4.** Nat Rev Mol Cell Biol (2006) 7:63; **5.** Nat Rev Drug Discov (2010) 9:29; **6.** Cytometry A (2009) 75:475; **7.** J Immunol (1983) 130:1910; **8.** J Leukoc Biol (1988) 43:304; **9.** J Immunol Methods (1990) 130:223; **10.** J Biol Chem (2008) 283:7983; **11.** Biophys J (1998) 75:2577; **12.** J Leukoc Biol (1998) 64:98; **13.** J Biol Chem (1995) 270:8328; **14.** Proc Natl Acad Sci U S A (2007) 104:20996; **15.** J Biol Chem (2010) 285:1153; **16.** Proc Natl Acad Sci U S A (2003) 100:13326; **17.** Nat Immunol (2009) 10:786; **18.** J Immunol (2008) 180:7497; **19.** J Cell Physiol (1993) 156:428; **20.** Proc Natl Acad Sci U S A (1977) 74:837; **21.** J Cell Biol (1993) 121:1257; **22.** Bioconjug Chem (1994) 5:105; **23.** Am J Pathol (1991) 138:1173; **24.** Biophys J (1994) 66:1301; **25.** Biophys J (1994) 67:1280; **26.** Methods Enzymol (1983) 98:241; **27.** In Vitro Cell Dev Biol (1991) 27A:633; **28.** J Lipid Res (1993) 34:325; **29.** J Biol Chem (1994) 269:20958; **30.** Hum Mutat (2009) 30:1419; **31.** J Clin Pharmacol (2000) 40:421; **32.** Biochem Biophys Res Commun (1997) 235:117; **33.** J Biol Chem (1978) 253:9053; **34.** J Supramol Struct (1980) 13:67; **35.** J Cell Biol (1979) 82:597; **36.** Annu Rev Biochem (1983) 52:223; **37.** Arteriosclerosis (1983) 3:2; **38.** Arteriosclerosis (1981) 1:177; **39.** J Biol Chem (2000) 275:9120; **40.** J Biol Chem (1990) 265:12722; **41.** J Biol Chem (1995) 270:1921; **42.** J Biol Chem (1994) 269:21003; **43.** J Cell Biol (1995) 129:133; **44.** Biochim Biophys Acta (1996) 1303:193; **45.** J Neuroinflammation (2007) 4:26; **46.** Arterioscler Thromb Vasc Biol (2008) 28:1584; **47.** Exp Cell Res (1992) 198:31; **48.** J Biol Chem (1992) 267:20383; **49.** Histochem J (1985) 17:1309; **50.** J Neurosci (1993) 13:4412; **51.** J Neurosci (1993) 13:4403; **52.** Nat Methods (2007) 4:501; **53.** J Biol Chem (2007) 282:4669; **54.** J Biol Chem (2006) 281:21771; **55.** Circ Res (2001) 88:491; **56.** J Neurosci (2002) 22:2478; **57.** J Biol Chem (1999) 274:34116; **58.** J Exp Med (1999) 190:523; **59.** J Exp Med (1999) 190:509; **60.** Electron Microsc Rev (1992) 5:381; **61.** J Periodontol (1985) 56:553; **62.** Biotechniques (2000) 28:510; **63.** Methods Mol Biol (2010) 618:137; **64.** Comb Chem High Throughput Screen (2004) 7:239; **65.** Mol Cancer Ther (2009) 8:232; **66.** Clin Cancer Res (2008) 14:3840; **67.** Clin Cancer Res (2007) 13:2936; **68.** EMBO J (1986) 5:1181; **69.** J Fluorescence (1994) 4:295; **70.** J Biol Chem (1993) 268:23860; **71.** J Membr Biol (1990) 118:215; **72.** Biochemistry (1993) 32:12039; **73.** Cytometry A (2009) 75:941; **74.** Cell (2006) 124:997; **75.** Nat Biotechnol (2004) 22:198; **76.** Methods Mol Biol (2007) 374:69; **77.** J Microsc (2006) 222:22; **78.** J Cell Biol (2005) 170:619; **79.** J Biol Chem (1998) 273:28004; **80.** J Histochem Cytochem (1992) 40:1353; **81.** Mol Biol Cell (2000) 11:747; **82.** Mol Biol Cell (1998) 9:809; **83.** Acta Pharmacol Sin (2007) 28:1698; **84.** Proc Natl Acad Sci U S A (2007) 104:7939; **85.** Cancer Res (2006) 66:2346; **86.** Nat Cell Biol (2006) 8:317; **87.** Proc Natl Acad Sci U S A (2006) 103:10283; **88.** Nat Med (2009) 15:319; **89.** Mol Biol Cell (2007) 18:2667; **90.** J Cell Biol (2009) 185:305; **91.** J Biol Chem (2006) 281:16139; **92.** Cancer Res (2001) 61:7763; **93.** J Biol Chem (2001) 276:13096; **94.** Mol Biol Cell (1999) 10:4403; **95.** Methods Mol Biol (2009) 480:101; **96.** J Biol Chem (2008) 283:6476; **97.** Biochemistry (1992) 31:5820; **98.** J Bioenerg Biomembr (1991) 23:147; **99.** J Biol Chem (1991) 266:3469; **100.** J Biol Chem (1990) 265:6688; **101.** Nature (1992) 358:337; **102.** J Cell Biol (1992) 118:267; **103.** J Cell Biol (1993) 122:1207; **104.** Chem Biol (2005) 12:999; **105.** Eur J Cell Biol (2000) 79:394; **106.** J Biol Chem (1997) 272:13929; **107.** J Cell Biol (1996) 135:1749; **108.** Mol Biol Cell (1996) 7:355; **109.** Pharmacol Toxicol (1992) 71:3; **110.** Nature (1998) 391:499; **111.** J Biol Chem (1983) 258:12582; **112.** Nat Med (2008) 14:325; **113.** Blood (2007) 109:1975; **114.** J Biol Chem (2007) 282:7046; **115.** J Exp Med (2008) 205:1775; **116.** J Lab Clin Med (1994) 123:728; **117.** Cytometry (1994) 17:287; **118.** Proc Natl Acad Sci U S A (2008) 105:19438; **119.** Immunology, 3rd Ed., Roitt IM, Brostoff J, Male DK 1993, p. 6; **120.** Eur J Immunol (1995) 25:1823; **121.** J Immunol (1992) 149:2894; **122.** J Immunol (1990) 145:417; **123.** Proc Natl Acad Sci U S A (1999) 96:15056; **124.** Biol Cell (1996) 87:95; **125.** J Microsc (1997) 185:339; **126.** Biol Cell (1998) 90:169; **127.** Cytometry (1997) 28:25; **128.** J Immunol (1997) 159:2177; **129.** Environ Health Perspect (2006) 114:1154; **130.** J Immunol (2007) 179:5748; **131.** Exp Cell Res (1994) 215:17; **132.** J Cell Sci (1992) 101:873; **133.** Arch Biochem Biophys (1983) 227:358; **134.** J Cell Biol (2008) 182:993; **135.** J Cell Biol (1980) 87:14; **136.** J Neurosci (2008) 28:208; **137.** Eukaryot Cell (2009) 8:1665; **138.** Anal Biochem (1997) 251:144; **139.** J Cell Sci (2000) 113:3093; **140.** J Histochem Cytochem (2001) 49:1473; **141.** J Immunol (2009) 183:6569; **142.** J Leukoc Biol (1993) 53:673; **143.** Biochem Biophys Res Commun (1984) 122:755; **144.** Physiol Rev (1987) 67:285; **145.** J Cell Biol (1979) 82:517; **146.** J Biol Chem (1990) 265:13449; **147.** Biochemistry (1990) 29:313; **148.** J Biol Chem (1990) 265:16725; **149.** J Cell Biol (1989) 109:1133; **150.** Proc Natl Acad Sci U S A (1981) 78:7540; **151.** Science (1979) 205:1412; **152.** J Biol Chem (2003) 278:4041; **153.** J Biol Chem (1995) 270:10686; **154.** J Biol Chem (1994) 269:326; **155.** J Immunol Methods (1992) 149:159; **156.** Biochemistry (1990) 29:11123; **157.** J Cell Biol (1993) 121:1281; **158.** J Cell Sci (1991) 100:473; **159.** Biochem Soc Trans (1990) 18:219; **160.** J Biol Chem (1984) 259:5661; **161.** Anal Chem (1997) 69:4994; **162.** Am J Physiol Endocrinol Metab (2006) 291:E332; **163.** J Biol Chem (2005) 280:23024; **164.** Anal Chem (2003) 75:4711; **165.** J Steroid Biochem (1985) 23:267; **166.** Plant J (2009) 57:680; **167.** Biochemistry (2004) 43:10192; **168.** Mol Reprod Dev (1999) 53:350; **169.** J Cell Sci (2008) 121:2130; **170.** J Androl (1998) 19:542; **171.** J Histochem Cytochem (2005) 53:243; **172.** J Biol Chem (2009) 284:29798; **173.** Neuroscience (2008) 153:751; **174.** Histochemistry (1986) 84:333; **175.** J Insect Physiol (1994) 40:1045; **176.** J Immunol Methods (1993) 162:1; **177.** Infect Immun (1999) 67:885; **178.** J Biol Chem (2009) 284:35926; **179.** J Immunol (2009) 182:2051; **180.** J Immunol Methods (2009) 342:71; **181.** Eur J Cell Biol (1998) 75:192; **182.** J Biol Chem (2005) 280:15300; **183.** Mol Ther (2009) 18:561; **184.** Nat Protoc (2008) 3:835; **185.** Methods Mol Biol (2008) 440:349; **186.** Nat Protoc (2006) 1:2916; **187.** Neuron (2004) 41:755; **188.** J Biol Chem (2000) 275:15279; **189.** Neuron (1999) 24:803; **190.** Neuron (1999) 24:809; **191.** Biotechniques (2006) 40:343; **192.** Nature (1998) 392:497; **193.** Proc Natl Acad Sci U S A (2009) 106:21960; **194.** Neuron (1993) 11:713; **195.** Neuron (1994) 12:1235; **196.** J Neurosci Methods (2007) 160:197; **197.** Cell Calcium (1995) 18:440; **198.** Proc Natl Acad Sci U S A (2001) 98:12748; **199.** Proc Biol Sci (1994) 255:61; **200.** Annu Rev Neurosci (1999) 22:1; **201.** J Physiol (2006) 575:23; **202.** Genes Dev (2008) 22:3475; **203.** Neurobiol Dis (2008) 31:20; **204.** Nature (2007) 448:493; **205.** J Histochem Cytochem (2000) 48:1479; **206.** Science (1998) 282:1516; **207.** J Cell Biol (1998) 141:625; **208.** Neuron (1996) 17:91; **209.** Mol Microbiol (2008) 68:1315; **210.** Proc Natl Acad Sci U S A (1999) 96:14553; **211.** Methods Enzymol (2008) 451:79; **212.** Microbiology (2007) 153:51; **213.** J Cell Biol (2006) 173:615; **214.** J Biol Chem (2007) 282:1916; **215.** Plant Physiol (2006) 141:932; **216.** Proc Natl Acad Sci U S A (2006) 103:10134; **217.** Plant Cell Physiol (2008) 49:1508; **218.** Neuron (2001) 29:469; **219.** J Anat (1998) 192:425; **220.** Cell Tissue Res (1989) 255:125; **221.** J Neurosci (1987) 7:1207; **222.** Nature (1984) 310:53; **223.** J Neurosci Methods (2008) 169:168; **224.** Mol Pharmacol (2007) 71:1222; **225.** J Biomol Screen (2006) 11:1027; **226.** J Neurosci Methods (2005) 143:3; **227.** J Biol Chem (2003) 278:9768; **228.** J Med Chem (2008) 51:5932; **229.** Am J Physiol Renal Physiol (2003) 284:F293; **230.** J Biol Chem (2001) 276:33741; **231.** Neuroscience (2007) 144:135; **232.** Biochemistry (2006) 45:12100; **233.** Biochemistry (1986) 25:2149; **234.** Biol Cell (1993) 79:265; **235.** Biochim Biophys Acta (1991) 1067:71; **236.** Biochim Biophys Acta (1987) 901:138; **237.** Science (2010) 327:46; **238.** FEMS Microbiol Lett (2007) 266:129; **239.** J Cell Biol (2003) 162:365; **240.** J Lipid Res (2003) 44:655; **241.** Eur J Biochem (2002) 269:737; **242.** Science (2000) 290:1721; **243.** Mol Membr Biol (1999) 16:145; **244.** Trends Cell Biol (1999) 9:87; **245.** Annu Rev Cell Dev Biol (1998) 14:111; **246.** Proc Natl Acad Sci U S A (2003) 100:5813; **247.** J Immunol (2003) 170:1329; **248.** J Membr Biol (2002) 189:35; **249.** Proc Natl Acad Sci U S A (2001) 98:9098; **250.** J Cell Biol (1999) 147:447; **251.** Mol Biol Cell (1999) 10:3187; **252.** Biochim Biophys Acta (2003) 1610:247; **253.** Annu Rev Immunol (2003) 21:457; **254.** Mol Immunol (2002) 38:1247; **255.** Nat Rev Immunol (2002) 2:96; **256.** Biol Res (2002) 35:127; **257.** Nat Rev Mol Cell Biol (2000) 1:31; **258.** J Exp Med (1999) 190:1549; **259.** J Cell Biol (1998) 143:637; **260.** Immunity (2003) 18:655; **261.** J Biol Chem (2002) 277:39541; **262.** Biochem Biophys Res Commun (2002) 297:876; **263.** Biol Chem (2002) 383:1475; **264.** J Cell Biol (2001) 153:529; **265.** J Cell Sci (2001) 114:3957; **266.** J Virol (2003) 77:9542; **267.** Exp Cell Res (2003) 287:67; **268.** Traffic (2002) 3:705; **269.** J Clin Virol (2001) 22:217; **270.** Curr Biol (2000) 10:R823; **271.** J Virol (2000) 74:3264; **272.** Biochemistry (1996) 35:16069; **273.** Mol Microbiol (1994) 13:745; **274.** J Cell Biol (1998) 141:929; **275.** J Biol Chem (1994) 269:30745; **276.** Biomaterials (2010) 31:1757; **277.** Brain Struct Funct (2009) 213:367; **278.** J Neurosci (2002) 22:9419; **279.** J Neurosci Methods (2005) 149:101; **280.** BMC Neurosci (2009) 10:80; **281.** Cell (2005) 121:937; **282.** J Cell Sci (2009) 122:289; **283.** Proc Natl Acad Sci U S A (2006) 103:21; **284.** Molecular and Cellular Mechanisms of Neurotransmitter Release, Stjarne L, et al., Eds. 1994, p. 31; **285.** Neuron (2001) 30:737; **286.** J Biol Chem (2009) 284:34296; **287.** J Histochem Cytochem (2009) 57:687; **288.** Proc Natl Acad Sci U S A (2009) 106:1524; **289.** PLoS One (2009) 4:e6209; **290.** J Immunol Methods (1988) 112:99; **291.** Proc Natl Acad Sci U S A (2004) 101:2912; **292.** J Immunol Methods (1989) 123:259; **293.** J Immunol Methods (1989) 116:235; **294.** J Cell Biol (1992) 118:1371; **295.** J Biol Chem (2007) 282:8969; **296.** Cytometry A (2005) 65:93; **297.** PLoS One (2009) 4:e8402; **298.** J Immunol Methods (2009) 340:102; **299.** Methods Mol Biol (2009) 546:255; **300.** Biochem J (2004) 377:159; **301.** Science (2007) 317:678; **302.** J Cell Sci (2002) 115:3703; **303.** Toxicol Sci (2007) 97:398; **304.** Proc Natl Acad Sci U S A (2005) 102:7523; **305.** Brain (2005) 128:1778; **306.** J Biol Chem (2007) 282:34194; **307.** Oncogene (2009) 28:1454; **308.** J Biol Chem (2008) 283:6764; **309.** Traffic (2008) 9:1801; **310.** Am J Physiol Gastrointest Liver Physiol (2007) 293:G1333; **311.** Biophys J (1996) 71:487; **312.** Cell Microbiol (2006) 8:781; **313.** J Cell Sci (2006) 119:1016; **314.** Methods Enzymol (2009) 453:417; **315.** J Cell Biol (1995) 130:821; **316.** J Physiol (2005) 565:743; **317.** Neuron (1993) 11:801; **318.** Am J Pathol (2009) 174:1891; **319.** J Physiol (2005) 568:917; **320.** J Leukoc Biol (2004) 76:1142; **321.** J Cell Biol (2001) 155:649; **322.** J Exp Zool (2002) 292:1.

IMPORTANT NOTICE: The products described in this manual are covered by one or more Limited Use Label License(s). Please refer to the Appendix on page 971 and Master Product List on page 975. Products are For Research Use Only. Not intended for any animal or human therapeutic or diagnostic use.

DATA TABLE 16.1 PROBES FOR FOLLOWING RECEPTOR BINDING AND PHAGOCYTOSIS

Cat. No.	MW	Storage	Soluble	Abs	EC	Em	Solvent	Notes
C481	622.54	L	pH >5	494	77,000	517	pH 9	1
C687	596.44	L	H_2O	399	30,000	421	H_2O	2, 3
D288	366.24	L	DMF	475	45,000	605	MeOH	4
D289	394.30	L	H_2O, DMF	488	48,000	607	MeOH	4
D1383	840.98	L	pH >6, DMF	494	76,000	519	pH 9	
D2935	584.37	F,D,AA	DMF	258	11,000	none	MeOH	5
E3476	~6100	FF,D	H_2O	<300		none		
E3477	~6600	FF,D	H_2O	<300		none		6
E3478	~6500	FF,D,L	H_2O	495	84,000	517	pH 8	6, 7
E3480	see Notes	FF,D,L	H_2O	596	ND	612	pH 7	8, 9
E3481	~6800	FF,D,L	H_2O	555	85,000	581	pH 7	6, 7
E7498	~6600	FF,D,L	H_2O	511	85,000	528	pH 9	6, 7
E13345	see Notes	FF,D,L	H_2O	497	ND	520	pH 8	8, 10
E35350	see Notes	FF,D,L	H_2O	554	ND	568	pH 7	8, 11
E35351	see Notes	FF,D,L	H_2O	653	ND	671	pH 7	8, 12
F1314	1213.41	F,L	pH >6, DMF	494	72,000	517	pH 9	
F2902	see Notes	RR,L,AA	H_2O	<300		none		13, 14, 15
F34653	788.75	D,L	H_2O, DMSO	562	47,000	744	$CHCl_3$	4
F35355	560.09	D,L	H_2O, DMSO	510	50,000	626	MeOH	4
H348	524.37	D,L	H_2O	454	24,000	511	pH 9	16
L453	457.24	L	H_2O	428	12,000	536	H_2O	17, 18
L3482	see Notes	RR,L,AA	see Notes	554	ND	571	see Notes	8, 19, 20, 21
L3483	see Notes	RR,L,AA	see Notes	515	ND	520	see Notes	8, 19, 20, 21
L3484	see Notes	RR,L,AA	see Notes	554	ND	571	see Notes	8, 19, 20, 21
L3485	see Notes	RR,L,AA	see Notes	510	ND	518	see Notes	8, 19, 20, 21
L23380	see Notes	RR,L,AA	see Notes	495	ND	519	see Notes	8, 19, 20, 21
S359	606.71	L	H_2O	586	108,000	605	H_2O	
T204	461.62	D,L	DMF, DMSO	355	75,000	430	MeOH	22
T1111	581.48	D,L	DMSO, EtOH	532	55,000	716	MeOH	4, 23
T3163	611.55	D,L	H_2O, DMSO	471	38,000	581	see Notes	24, 25
T3166	607.51	D,L	H_2O, DMSO	505	47,000	725	see Notes	24, 26
T7508	555.44	D,L	H_2O, DMSO	506	50,000	620	MeOH	4
T13320	607.51	D,L	H_2O, DMSO	505	47,000	725	see Notes	24, 26
T23360	565.43	D,L	H_2O, DMSO	560	43,000	734	$CHCl_3$	26
T35356	611.55	D,L	H_2O, DMSO	471	38,000	581	see Notes	24, 25

For definitions of the contents of this data table, see "Using *The Molecular Probes® Handbook*" in the introductory pages.

Notes

1. C481 fluorescence is strongly quenched by micromolar concentrations of Fe^{3+}, Co^{2+}, Ni^{2+} and Cu^{2+} at pH 7. (Am J Physiol (1995) 268:C1354, J Biol Chem (1999) 274:13375)
2. The Alexa Fluor® 405 and Cascade Blue® dyes have a second absorption peak at about 376 nm with EC ~80% of the 395–400 nm peak.
3. Maximum solubility in water is ~1% for C687, ~1% for C3221 and ~8% for C3239.
4. Abs and Em of styryl dyes are at shorter wavelengths in membrane environments than in reference solvents such as methanol. The difference is typically 20 nm for absorption and 80 nm for emission, but varies considerably from one dye to another. Styryl dyes are generally nonfluorescent in water.
5. Dihydrofluorescein diacetates are colorless and nonfluorescent until both of the acetate groups are hydrolyzed and the products are subsequently oxidized to fluorescein derivatives. The materials contain less than 0.1% of oxidized derivative when initially prepared. The oxidation products of C400, C2938, C6827, D399 and D2935 are 2',7'-dichlorofluorescein derivatives with spectra similar to C368 (see data).
6. α-Bungarotoxin, EGF and phallotoxin conjugates have approximately 1 label per peptide.
7. The value of EC listed for this EGF conjugate is for the labeling dye in free solution. Use of this value for the conjugate assumes a 1:1 dye:peptide labeling ratio and no change of EC due to dye–peptide interactions.
8. ND = not determined.
9. E3480 is a complex of E3477 with Texas Red® streptavidin, which typically incorporates 3 dyes/streptavidin (MW ~52,800).
10. E13345 is a complex of E3477 with Alexa Fluor® 488 streptavidin, which typically incorporates 5 dyes/streptavidin (MW ~52,800).
11. E35350 is a complex of E3477 with Alexa Fluor® 555 streptavidin, which typically incorporates 3 dyes/streptavidin (MW ~52,800).
12. E35351 is a complex of E3477 with Alexa Fluor® 647 streptavidin, which typically incorporates 3 dyes/streptavidin (MW ~52,800).
13. This product is supplied as a ready-made solution in the solvent indicated under "Soluble."
14. F2902 is essentially colorless and nonfluorescent until oxidized. A small amount (~5%) of oxidized material is normal and acceptable for the product as supplied. The oxidation product is fluorescent (Abs = 495 nm, Em = 524 nm). (J Immunol Methods (1990) 130:223)
15. This product consists of a dye–bovine serum albumin conjugate (MW ~66,000) complexed with IgG in a ratio of approximately 1:4 mol:mol (BSA:IgG)
16. H348 spectra are pH-dependent.
17. The fluorescence quantum yield of lucifer yellow CH in H_2O is 0.21. (J Am Chem Soc (1981) 103:7615)
18. Maximum solubility in water is ~8% for L453, ~6% for L682 and ~1% for L1177.
19. LDL complexes must be stored refrigerated BUT NOT FROZEN. The maximum shelf-life under the indicated storage conditions is 4–6 weeks.
20. This LDL complex incorporates multiple fluorescent labels. The number of dyes per apoprotein B (MW ~500,000) is indicated on the product label.
21. LDL complexes are packaged under argon in 10 mM Tris, 150 mM NaCl, 0.3 mM EDTA, pH 8.3 containing 2 mM azide. Spectral data reported are measured in this buffer.
22. Diphenylhexatriene (DPH) and its derivatives are essentially nonfluorescent in water. Absorption and emission spectra have multiple peaks. The wavelength, resolution and relative intensity of these peaks are environment dependent. Abs and Em values are for the most intense peak in the solvent specified.
23. RH 414 Abs ~500 nm, Em ~635 nm when bound to phospholipid bilayer membranes.
24. Abs, EC and Em determined for dye bound to detergent micelles (20 mg/mL CHAPS in H_2O). These dyes are essentially nonfluorescent in pure water.
25. FM® 1-43 Abs = 479 nm, Em = 598 nm bound to phospholipid bilayer membranes. Em = 565 nm bound to synaptosomal membranes. (Neuron (1994) 12:1235)
26. FM® 4-64 and FM® 5-95 are nonfluorescent in water. For two-color imaging in GFP-expressing cells, these dyes can be excited at 568 nm with emission detection at 690–730 nm. (Am J Physiol Cell Physiol (2001) 281:C624)

IMPORTANT NOTICE: The products described in this manual are covered by one or more Limited Use Label License(s). Please refer to the Appendix on page 971 and Master Product List on page 975. Products are For Research Use Only. Not intended for any animal or human therapeutic or diagnostic use.

PRODUCT LIST 16.1 PROBES FOR FOLLOWING RECEPTOR BINDING AND PHAGOCYTOSIS

Cat. No.	Product	Quantity
A6442	anti-synapsin I (bovine), rabbit IgG fraction *affinity purified*	10 µg
A11130	anti-transferrin receptor (human), mouse IgG_1, monoclonal 236-15375	50 µg
C481	calcein *high purity*	100 mg
C687	Cascade Blue® hydrazide, trisodium salt	10 mg
C2990	casein, fluorescein conjugate	25 mg
C10586	CellLight® Early Endosomes-GFP	1 kit
C10587	CellLight® Early Endosomes-RFP	1 kit
C10609	CellLight® Synaptophysin-GFP	1 kit
C10610	CellLight® Synaptophysin-RFP	1 kit
C34775	cholera toxin subunit B (recombinant), Alexa Fluor® 488 conjugate	100 µg
C22841	cholera toxin subunit B (recombinant), Alexa Fluor® 488 conjugate	500 µg
C34776	cholera toxin subunit B (recombinant), Alexa Fluor® 555 conjugate	100 µg
C22843	cholera toxin subunit B (recombinant), Alexa Fluor® 555 conjugate	500 µg
C34777	cholera toxin subunit B (recombinant), Alexa Fluor® 594 conjugate	100 µg
C22842	cholera toxin subunit B (recombinant), Alexa Fluor® 594 conjugate	500 µg
C34778	cholera toxin subunit B (recombinant), Alexa Fluor® 647 conjugate	100 µg
C34779	cholera toxin subunit B (recombinant), biotin-XX conjugate	100 µg
C34780	cholera toxin subunit B (recombinant), horseradish peroxidase conjugate	100 µg
D1383	dexamethasone fluorescein	5 mg
D2935	2',7'-dichlorodihydrofluorescein diacetate, succinimidyl ester (OxyBURST® Green H_2DCFDA, SE)	5 mg
D289	4-(4-(diethylamino)styryl)-*N*-methylpyridinium iodide (4-Di-2-ASP)	1 g
D288	4-(4-(dimethylamino)styryl)-*N*-methylpyridinium iodide (4-Di-1-ASP)	1 g
D12060	DQ™ collagen, type I from bovine skin, fluorescein conjugate	1 mg
D12054	DQ™ gelatin from pig skin, fluorescein conjugate *special packaging*	5 x 1 mg
D12050	DQ™ Green BSA *special packaging*	5 x 1 mg
D12053	DQ™ ovalbumin *special packaging*	5 x 1 mg
D12051	DQ™ Red BSA *special packaging*	5 x 1 mg
E3476	epidermal growth factor (EGF) *from mouse submaxillary glands*	100 µg
E3477	epidermal growth factor, biotin-XX conjugate (biotin EGF)	20 µg
E13345	epidermal growth factor, biotinylated, complexed to Alexa Fluor® 488 streptavidin (Alexa Fluor® 488 EGF complex)	100 µg
E35350	epidermal growth factor, biotinylated, complexed to Alexa Fluor® 555 streptavidin (Alexa Fluor® 555 EGF complex)	100 µg
E35351	epidermal growth factor, biotinylated, complexed to Alexa Fluor® 647 streptavidin (Alexa Fluor® 647 EGF complex)	100 µg
E3480	epidermal growth factor, biotinylated, complexed to Texas Red® streptavidin (Texas Red® EGF complex)	100 µg
E3478	epidermal growth factor, fluorescein conjugate (fluorescein EGF)	20 µg
E7498	epidermal growth factor, Oregon Green® 514 conjugate (Oregon Green® 514 EGF)	20 µg
E3481	epidermal growth factor, tetramethylrhodamine conjugate (rhodamine EGF)	20 µg
E2870	*Escherichia coli* BioParticles® opsonizing reagent	1 U
E13231	*Escherichia coli* (K-12 strain) BioParticles®, Alexa Fluor® 488 conjugate	2 mg
E23370	*Escherichia coli* (K-12 strain) BioParticles®, Alexa Fluor® 594 conjugate	2 mg
E2864	*Escherichia coli* (K-12 strain) BioParticles®, BODIPY® FL conjugate	10 mg
E2861	*Escherichia coli* (K-12 strain) BioParticles®, fluorescein conjugate	10 mg
E2862	*Escherichia coli* (K-12 strain) BioParticles®, tetramethylrhodamine conjugate	10 mg
E2863	*Escherichia coli* (K-12 strain) BioParticles®, Texas Red® conjugate	10 mg
F2902	Fc OxyBURST® Green assay reagent *25 assays* *3 mg/mL*	500 µL
F13191	fibrinogen from human plasma, Alexa Fluor® 488 conjugate	5 mg
F13192	fibrinogen from human plasma, Alexa Fluor® 546 conjugate	5 mg
F13193	fibrinogen from human plasma, Alexa Fluor® 594 conjugate	5 mg
F35200	fibrinogen from human plasma, Alexa Fluor® 647 conjugate	5 mg
F7496	fibrinogen from human plasma, Oregon Green® 488 conjugate	5 mg
F20892	FluoSpheres® collagen I-labeled microspheres, 1.0 µm, yellow-green fluorescent (505/515) *0.5% solids*	0.4 mL
F20893	FluoSpheres® collagen I-labeled microspheres, 2.0 µm, yellow-green fluorescent (505/515) *0.5% solids*	0.4 mL
F35355	FM® 1-43FX *fixable analog of FM® 1-43 membrane stain*	10 x 100 µg
F34653	FM® 4-64FX *fixable analog of FM® 4-64 membrane stain*	10 x 100 µg
F1314	formyl-Nle-Leu-Phe-Nle-Tyr-Lys, fluorescein derivative	1 mg
G13187	gelatin from pig skin, fluorescein conjugate	5 mg
G13186	gelatin from pig skin, Oregon Green® 488 conjugate	5 mg
H13188	histone H1 from calf thymus, Alexa Fluor® 488 conjugate	1 mg
H348	8-hydroxypyrene-1,3,6-trisulfonic acid, trisodium salt (HPTS; pyranine)	1 g
I13269	insulin, human, recombinant from *E. coli*, fluorescein conjugate (FITC insulin) *monolabeled* *zinc free*	100 µg

IMPORTANT NOTICE: The products described in this manual are covered by one or more Limited Use Label License(s). Please refer to the Appendix on page 971 and Master Product List on page 975. Products are For Research Use Only. Not intended for any animal or human therapeutic or diagnostic use.

PRODUCT LIST 16.1 PROBES FOR FOLLOWING RECEPTOR BINDING AND PHAGOCYTOSIS—*continued*

Cat. No.	Product	Quantity
L21409	lectin PNA from *Arachis hypogaea* (peanut), Alexa Fluor® 488 conjugate	1 mg
L32458	lectin PNA from *Arachis hypogaea* (peanut), Alexa Fluor® 568 conjugate	1 mg
L32459	lectin PNA from *Arachis hypogaea* (peanut), Alexa Fluor® 594 conjugate	1 mg
L32460	lectin PNA from *Arachis hypogaea* (peanut), Alexa Fluor® 647 conjugate	1 mg
L23351	lipopolysaccharides from *Escherichia coli* serotype 055:B5, Alexa Fluor® 488 conjugate	100 µg
L23352	lipopolysaccharides from *Escherichia coli* serotype 055:B5, Alexa Fluor® 568 conjugate	100 µg
L23353	lipopolysaccharides from *Escherichia coli* serotype 055:B5, Alexa Fluor® 594 conjugate	100 µg
L23350	lipopolysaccharides from *Escherichia coli* serotype 055:B5, BODIPY® FL conjugate	100 µg
L23356	lipopolysaccharides from *Salmonella minnesota*, Alexa Fluor® 488 conjugate	100 µg
L3486	low-density lipoprotein from human plasma (LDL) *2.5 mg/mL*	200 µL
L35354	low-density lipoprotein from human plasma, acetylated (AcLDL) *2.5 mg/mL*	200 µL
L23380	low-density lipoprotein from human plasma, acetylated, Alexa Fluor® 488 conjugate (Alexa Fluor® 488 AcLDL) *1 mg/mL*	200 µL
L35353	low-density lipoprotein from human plasma, acetylated, Alexa Fluor® 594 conjugate (Alexa Fluor® 594 AcLDL) *1 mg/mL*	200 µL
L3485	low-density lipoprotein from human plasma, acetylated, BODIPY® FL conjugate (BODIPY® FL AcLDL) *1 mg/mL*	200 µL
L3484	low-density lipoprotein from human plasma, acetylated, DiI complex (DiI AcLDL) *1 mg/mL*	200 µL
L3483	low-density lipoprotein from human plasma, BODIPY® FL complex (BODIPY® FL LDL) *1 mg/mL*	200 µL
L3482	low-density lipoprotein from human plasma, DiI complex (DiI LDL) *1 mg/mL*	200 µL
L453	lucifer yellow CH, lithium salt	25 mg
O13291	OxyBURST® Green H_2HFF BSA *special packaging*	5 x 1 mg
P10361	pHrodo™ dextran, 10,000 MW *for endocytosis*	0.5 mg
P35361	pHrodo™ *E. coli* BioParticles® conjugate for phagocytosis	5 x 2 mg
A10025	pHrodo™ *E. coli* BioParticles® Phagocytosis Kit *for flow cytometry* *100 tests*	1 kit
A10026	pHrodo™ Phagocytosis Particle Labeling Kit *for flow cytometry* *100 tests*	1 kit
A10010	pHrodo™ *S. aureus* BioParticles® conjugate for phagocytosis	5 x 2 mg
S2860	*Staphylococcus aureus* BioParticles® opsonizing reagent	1 U
S23371	*Staphylococcus aureus* (Wood strain without protein A) BioParticles®, Alexa Fluor® 488 conjugate	2 mg
S23372	*Staphylococcus aureus* (Wood strain without protein A) BioParticles®, Alexa Fluor® 594 conjugate	2 mg
S2854	*Staphylococcus aureus* (Wood strain without protein A) BioParticles®, BODIPY® FL conjugate	10 mg
S2851	*Staphylococcus aureus* (Wood strain without protein A) BioParticles®, fluorescein conjugate	10 mg
S2859	*Staphylococcus aureus* (Wood strain without protein A) BioParticles®, unlabeled	100 mg
S359	sulforhodamine 101	25 mg
T204	TMA-DPH (1-(4-trimethylammoniumphenyl)-6-phenyl-1,3,5-hexatriene *p*-toluenesulfonate)	25 mg
T13342	transferrin from human serum, Alexa Fluor® 488 conjugate	5 mg
T23364	transferrin from human serum, Alexa Fluor® 546 conjugate	5 mg
T35352	transferrin from human serum, Alexa Fluor® 555 conjugate	5 mg
T23365	transferrin from human serum, Alexa Fluor® 568 conjugate	5 mg
T13343	transferrin from human serum, Alexa Fluor® 594 conjugate	5 mg
T23362	transferrin from human serum, Alexa Fluor® 633 conjugate	5 mg
T23366	transferrin from human serum, Alexa Fluor® 647 conjugate	5 mg
T35357	transferrin from human serum, Alexa Fluor® 680 conjugate	5 mg
T23363	transferrin from human serum, biotin-XX conjugate	5 mg
T2871	transferrin from human serum, fluorescein conjugate	5 mg
T2872	transferrin from human serum, tetramethylrhodamine conjugate	5 mg
T2875	transferrin from human serum, Texas Red® conjugate	5 mg
T3163	*N*-(3-triethylammoniumpropyl)-4-(4-(dibutylamino)styryl)pyridinium dibromide (FM® 1-43)	1 mg
T35356	*N*-(3-triethylammoniumpropyl)-4-(4-(dibutylamino)styryl)pyridinium dibromide (FM® 1-43) *special packaging*	10 x 100 µg
T1111	*N*-(3-triethylammoniumpropyl)-4-(4-(4-(diethylamino)phenyl)butadienyl)pyridinium dibromide (RH 414)	5 mg
T3166	*N*-(3-triethylammoniumpropyl)-4-(6-(4-(diethylamino)phenyl)hexatrienyl)pyridinium dibromide (FM® 4-64)	1 mg
T13320	*N*-(3-triethylammoniumpropyl)-4-(6-(4-(diethylamino)phenyl)hexatrienyl)pyridinium dibromide (FM® 4-64) *special packaging*	10 x 100 µg
T7508	*N*-(3-triethylammoniumpropyl)-4-(4-(diethylamino)styryl)pyridinium dibromide (FM® 2-10)	5 mg
T23360	*N*-(3-trimethylammoniumpropyl)-4-(6-(4-(diethylamino)phenyl)hexatrienyl)pyridinium dibromide (FM® 5-95)	1 mg
T23011	trypsin inhibitor from soybean, Alexa Fluor® 488 conjugate	1 mg
V6694	Vybrant® Phagocytosis Assay Kit *250 assays*	1 kit
Z2850	zymosan A BioParticles® opsonizing reagent	1 U
Z23373	zymosan A (*S. cerevisiae*) BioParticles®, Alexa Fluor® 488 conjugate	2 mg
Z23374	zymosan A (*S. cerevisiae*) BioParticles®, Alexa Fluor® 594 conjugate	2 mg
Z2841	zymosan A (*S. cerevisiae*) BioParticles®, fluorescein conjugate	10 mg
Z2843	zymosan A (*S. cerevisiae*) BioParticles®, Texas Red® conjugate	10 mg
Z2849	zymosan A (*S. cerevisiae*) BioParticles®, unlabeled	100 mg

IMPORTANT NOTICE: The products described in this manual are covered by one or more Limited Use Label License(s). Please refer to the Appendix on page 971 and Master Product List on page 975. Products are For Research Use Only. Not intended for any animal or human therapeutic or diagnostic use.

16.2 Probes for Neurotransmitter Receptors

Fluorescent receptor ligands provide a sensitive means of identifying and localizing various cellular receptors, ion channels and ion carriers. Many of these site-selective fluorescent probes may be used on live or fixed cells, as well as in cell-free extracts. The high sensitivity and selectivity of these fluorescent probes make them especially good candidates for measuring low-abundance receptors.[1–5] Various methods for further amplifying detection of these receptors[6,7] are discussed in Chapter 6 and Chapter 7.

This section is devoted to our probes for neurotransmitter receptors. Additional fluorescently labeled receptor ligands (including low-density lipoproteins, epidermal growth factors, transferrin and fibrinogen conjugates and chemotactic peptides) are described in Section 16.1, along with other probes for studying receptor-mediated endocytosis. Section 16.3 describes a variety of probes for Ca^{2+}, Na^+, K^+ and Cl^- ion channels and carriers. Chapter 17 focuses on reagents for investigating events—such as calcium regulation, kinase, phosphatase and phospholipase activation, and lipid trafficking—that occur downstream from the receptor–ligand interaction (Figure 16.2.1).

α-Bungarotoxin Probes for Nicotinic Acetylcholine Receptors

Fluorescent α-Bungarotoxins

Nicotinic acetylcholine receptors (nAChRs) are neurotransmitter-gated ion channels that produce an increase in Na^+ and K^+ permeability, depolarization and excitation upon activation by acetylcholine[8] (Figure 16.2.1). α-Bungarotoxin is a 74–amino acid (~8000 dalton) peptide containing 5 lysine residues and 10 cysteine residues paired in 5 disulfide bridges. Extracted from *Bungarus multicinctus* venom, α-bungarotoxin binds with high affinity to the α-subunit of the nAChR of neuromuscular junctions.[9] We provide an extensive selection of fluorescent α-bungarotoxin conjugates (Table 16.4) to facilitate visualization of nAChRs with a variety of instrumentation. We attach approximately one fluorophore to each molecule of α-bungarotoxin, thus retaining optimal binding specificity. The labeled bungarotoxins are then chromatographically separated from unlabeled molecules to ensure adequate labeling of the product.

Alexa Fluor® 488 α-bungarotoxin (B13422) has fluorescence spectra similar to those of fluorescein α-bungarotoxin (F1176) and is therefore suitable for use with standard fluorescein optical filter sets. Tetramethylrhodamine α-bungarotoxin[10–12] (T1175) has been the preferred red-orange–fluorescent probe for staining the nAChR (Figure 16.2.2). We not only offer the red-orange–fluorescent Alexa Fluor® 555 α-bungarotoxin (B35451), but also the red-fluorescent Alexa Fluor® 594 α-bungarotoxin (B13423), which has a longer-wavelength emission maximum and therefore offers better spectral separation from green-fluorescent dyes in multicolor experiments. Our two longest-wavelength conjugates—Alexa Fluor® 647 α-bungarotoxin (B35450) and Alexa Fluor® 680 α-bungarotoxin (B35452)—are spectrally separated from both green-fluorescent and orange-fluorescent dyes, allowing researchers to easily perform three- and four-color experiments.

Fluorescent α-bungarotoxins have been used in a variety of informative investigations to:

- Correlate receptor clustering during neuromuscular development with tyrosine phosphorylation of the receptor[13,14]
- Detect reinnervation of adult muscle after nerve damage and to identify and visualize endplates[15,16]
- Document nAChR cluster formation after myoblast fusion.[17]
- Label proteins fused to the BBS expression tag (a 13–amino acid sequence excerpted from the nAChR) *in situ*[18,19]
- Monitor nAChR-mediated responses in neuromuscular damage and degeneration models[20–22]

Biotinylated α-Bungarotoxin

Nicotinic AChRs can also be labeled with biotinylated α-bungarotoxin (B1196), which is then localized using fluorophore- or enzyme-labeled avidin, streptavidin or NeutrAvidin biotin-binding protein conjugates, or NANOGOLD and Alexa Fluor® FluoroNanogold streptavidin[14,23–25] (Section 7.6, Table 7.9). Based on the intracellular dissociation of biotinylated α-bungarotoxin and streptavidin, researchers were able to distinguish new, preexisting and recycled pools of nAChR at the synapses of live mice by sequentially labeling with biotinylated α-bungarotoxin and fluorescent streptavidin conjugates.[26] Complexation of biotinylated α-bungarotoxin with Qdot® nanocrystal–streptavidin conjugates (Section 6.6) enables single-molecule detection of nAChR.[1,2] The nanocrystal labeling methodology allows detection and tracking of diffuse, nonclustered nAChRs, whereas dye-labeled α-bungarotoxin conjugates primarily detect nAChR clusters.[1]

Table 16.4 Labeled and unlabeled α-bungarotoxins.

Cat. No.	Label	Ex/Em (nm)	Notes	Size
F1176	Fluorescein	494/518	Original green-fluorescent conjugate	500 µg
B13422	Alexa Fluor® 488	495/519	Brightest and most photostable green-fluorescent conjugate	500 µg
T1175	Tetramethylrhodamine	553/577	An extensively used red-orange–fluorescent conjugate	500 µg
B35451	Alexa Fluor® 555	555/565	Bright and photostable red-orange–fluorescent conjugate	500 µg
B13423	Alexa Fluor® 594	590/617	Excellent dye to combine with green-fluorescent probes	500 µg
B35450	Alexa Fluor® 647	650/668	Excellent dye to combine with green- and orange-fluorescent probes	500 µg
B35452	Alexa Fluor® 680	679/702	Excellent dye to combine with green-, orange-, and red-fluorescent probes	500 µg
B1196	Biotin-XX	NA	Visualized with labeled avidins and streptavidins (Table 7.9)	500 µg
B1601	Unlabeled	NA	Useful as a control, as well as for radioiodination and for preparation of new conjugates	1 mg

NA = Not applicable.

IMPORTANT NOTICE: The products described in this manual are covered by one or more Limited Use Label License(s). Please refer to the Appendix on page 971 and Master Product List on page 975. Products are For Research Use Only. Not intended for any animal or human therapeutic or diagnostic use.

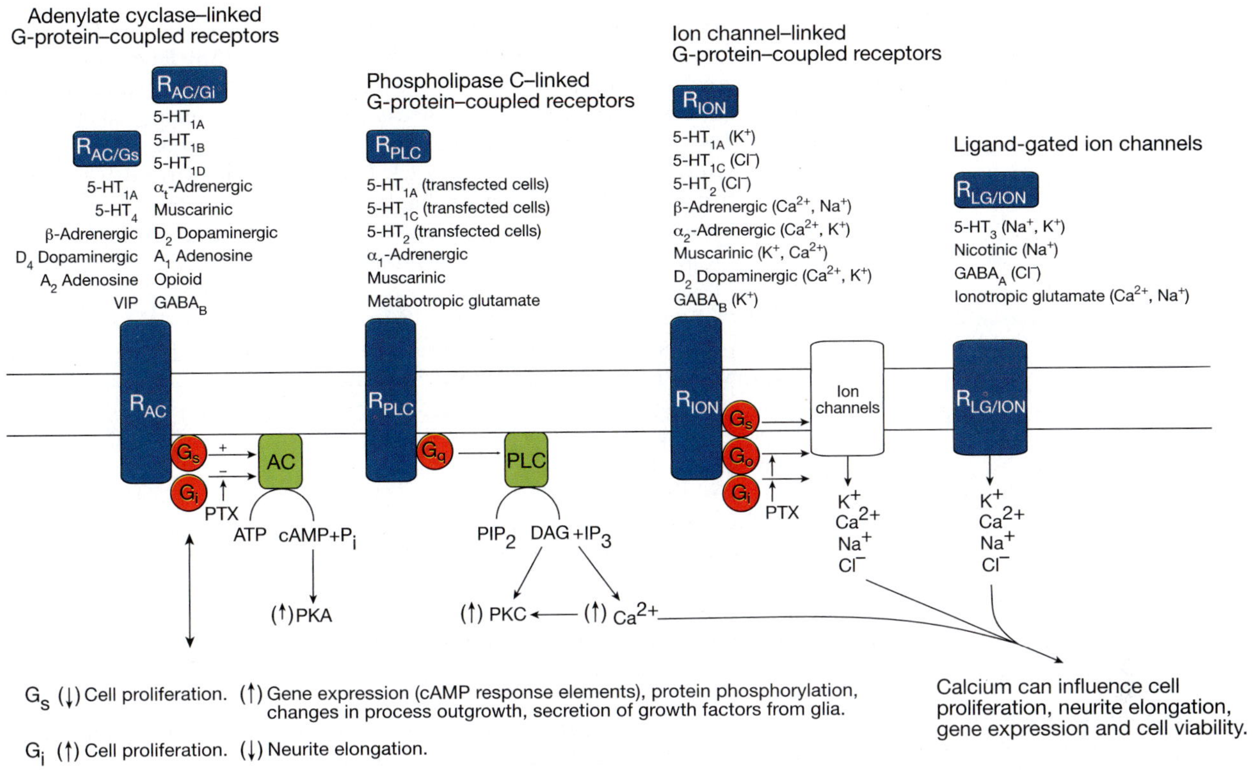

Figure 16.2.1 Neurotransmitter receptors linked to second messengers mediating growth responses in neuronal and nonneuronal cells. Abbreviations: $R_{AC/Gs}$ = Receptors coupled to G-proteins that stimulate adenylate cyclase (AC) activity, leading to cAMP formation and enhanced activity of protein kinase A (PKA). $R_{AC/Gi}$ = Receptors coupled to pertussis toxin (PTX)–sensitive G-proteins that inhibit adenylate cyclase activity. R_{PLC} = Receptors promoting the hydrolysis of phosphatidylinositol 4,5-diphosphate (PIP_2) to inositol 1,4,5-triphosphate (IP_3), which increases intracellular Ca^{2+}, and diacylglycerol (DAG), which activates protein kinase C (PKC). R_{ION} = Receptors indirectly promoting ion fluxes due to coupling to various G-proteins. $R_{LG/ION}$ = Receptors that promote ion fluxes directly because they are structurally linked to ion channels (members of the superfamily of ligand-gated ion channel receptors). Stimulation of proliferation is most often associated with activation of G-proteins negatively coupled to adenylate cyclase (G_i), or positively coupled to phospholipase C (G_q) or to pertussis toxin–sensitive pathways (G_o, G_i). In contrast, activation of neurotransmitter receptors positively coupled to cAMP usually inhibits cell proliferation and causes changes in cell shape indicative of differentiation. Reprinted and modified with permission from J.M. Lauder and Trends Neurosci (1993) 16:233.

In addition, the biotinylated toxin can be employed for affinity isolation of the nAChR using a streptavidin or CaptAvidin™ agarose (S951, C21386; Section 7.6) column.[27,28]

Unlabeled α-Bungarotoxin

In addition to the fluorescent and biotinylated derivatives, we have unlabeled α-bungarotoxin (B1601), which has been shown to be useful for radioiodination.[9,29] Unlabeled α-bungarotoxin has also been employed for ELISA testing of nAChR binding,[30] as well as for investigating the function of the α-bungarotoxin–binding component (α-BgtBC) in vertebrate neurons.[31]

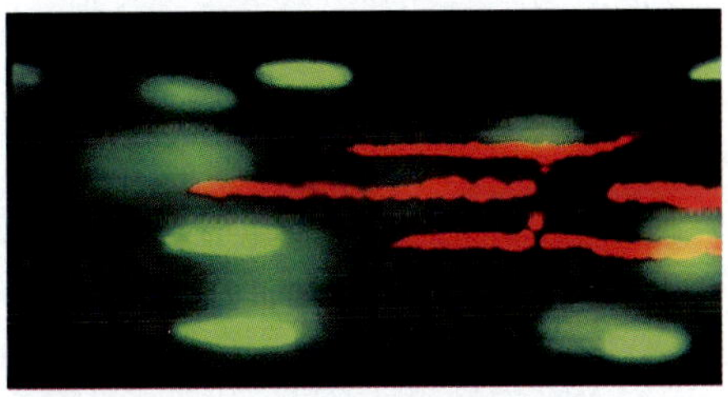

Figure 16.2.2 Pseudocolored photomicrograph of the synaptic region of fluorescently labeled living muscle fibers from the lumbricalis muscle of the adult frog *Rana pipiens.* Six hours after isolation of the muscle fibers, acetylcholine receptors were stained with red-fluorescent tetramethylrhodamine α-bungarotoxin (T1175) and myonuclei were stained with the green-fluorescent SYTO® 13 live-cell nucleic acid stain (S7575). Photo contributed by Christian Brösamle, Brain Research Institute, University of Zurich, and Damien Kuffler, Institute of Neurobiology, University of Puerto Rico.

Amplex® Red Acetylcholine/Acetylcholinesterase Assay Kit

The action of acetylcholine (ACh) at neuromuscular junctions is regulated by acetylcholinesterase (AChE), the enzyme that hydrolyzes ACh to choline and acetate. The Amplex® Red Acetylcholine/Acetylcholinesterase Assay Kit (A12217) provides an ultrasensitive method for continuously monitoring AChE activity and for detecting ACh in a fluorescence microplate reader or fluorometer. Other potential uses for this kit include screening for AChE inhibitors and measuring the release of ACh from synaptosomes. The Amplex® Red Acetylcholine/Acetylcholinesterase Assay Kit can also be used for the ultrasensitive, specific assay of free choline, classified as an essential nutrient in foods.[32]

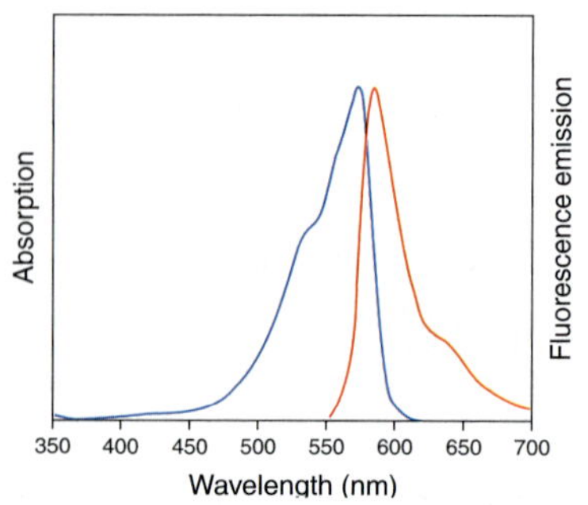

Figure 16.2.3 Absorption and fluorescence emission spectra of resorufin in pH 9.0 buffer.

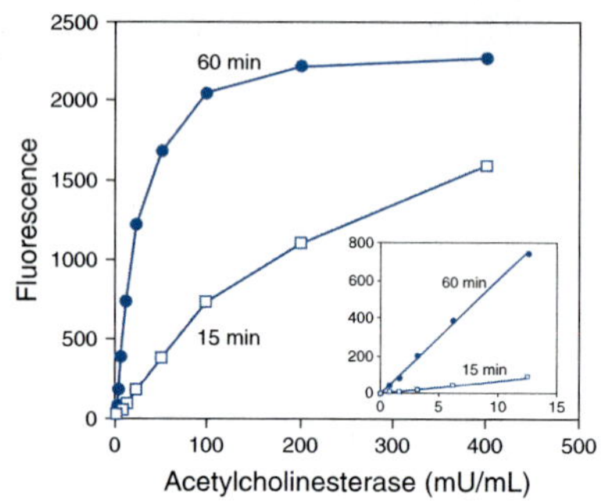

Figure 16.2.4 Detection of electric eel acetylcholinesterase activity using the Amplex® Red Acetylcholine/Acetylcholinesterase Assay Kit (A12217). Each reaction contained 50 µM acetylcholine, 200 µM Amplex® Red reagent, 1 U/mL HRP, 0.1 U/mL choline oxidase and the indicated amount of acetylcholinesterase in 1X reaction buffer. Reactions were incubated at room temperature. After 15 and 60 minutes, fluorescence was measured in a fluorescence microplate reader using excitation at 560 ± 10 nm and fluorescence detection at 590 ± 10 nm. The inset shows the sensitivity of the 15 min (□) and 60 min (●) assays at low levels of acetylcholinesterase activity (0–13 mU/mL).

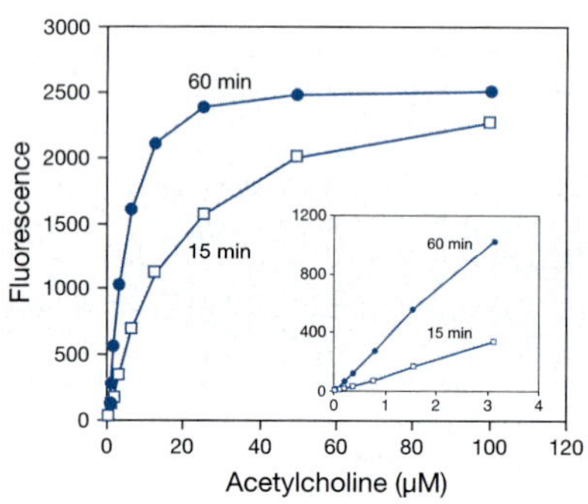

Figure 16.2.5 Detection of acetylcholine using the Amplex® Red Acetylcholine/Acetylcholinesterase Assay Kit (A12217). Each reaction contained 200 µM Amplex® Red reagent, 1 U/mL HRP, 0.1 U/mL choline oxidase, 0.5 U/mL acetylcholinesterase and the indicated amount of acetylcholine in 1X reaction buffer. Reactions were incubated at room temperature. After 15 and 60 minutes, fluorescence was measured with a fluorescence microplate reader using excitation at 560 ± 10 nm and fluorescence detection at 590 ± 10 nm. The inset shows the sensitivity of the 15 min (□) and 60 min (●) assays at low levels of acetylcholine (0–3 µM).

In this assay, AChE activity is monitored indirectly using the Amplex® Red reagent (10-acetyl-3,7-dihydroxyphenoxazine), a highly sensitive and stable fluorogenic probe for H_2O_2 that is also useful in assaying other enzymes and analytes (Section 10.5). First, AChE converts the acetylcholine substrate to choline. Choline is in turn oxidized by choline oxidase to betaine and H_2O_2, the latter of which, in the presence of horseradish peroxidase, reacts with the Amplex® Red reagent to generate the red-fluorescent product resorufin (R363, Section 10.1) with excitation/emission maxima of ~570/585 nm (Figure 16.2.3). Experiments with purified AChE from electric eel indicate that the Amplex® Red Acetylcholine/Acetylcholinesterase Assay Kit can detect AChE levels as low as 0.002 U/mL using a reaction time of only 1 hour (Figure 16.2.4). In our laboratories, we have been able to detect acetylcholinesterase activity from a tissue sample with total protein content as low as 200 ng/mL or 20 ng/well in a microplate assay.[33] By providing an excess of AChE in the assay, the kit can also be used to detect acetylcholine levels as low as 0.3 µM, with a detection range between 0.3 µM and ~100 µM acetylcholine (Figure 16.2.5).

The Amplex® Red Acetylcholine/Acetylcholinesterase Assay Kit contains:

- Amplex® Red reagent
- Dimethylsulfoxide (DMSO)
- Horseradish peroxidase (HRP)
- H_2O_2 for use as a positive control
- Concentrated reaction buffer
- Choline oxidase from *Alcaligenes* sp.
- Acetylcholine (ACh)
- Acetylcholinesterase (AChE) from electric eel
- Detailed protocols

Each kit provides sufficient reagents for approximately 500 assays using a fluorescence microplate reader and a reaction volume of 200 µL per assay.

BODIPY® FL Prazosin for α_1-Adrenergic Receptors

Prazosin is a high-affinity antagonist for the α_1-adrenergic receptor. The green-fluorescent BODIPY® FL prazosin (B7433, Figure 16.2.6) can be used to localize the α_1-adrenergic receptor on cultured cortical neurons[34] and in vascular smooth muscle cells from α_1-adrenergic receptor–knockout mice.[35] BODIPY® FL prazosin has also been successfully employed in multidrug resistance (MDR) transporter activity assays.[36,37]

BODIPY® TMR-X Muscimol for $GABA_A$ Receptors

Muscimol is a powerful agonist of the $GABA_A$ receptor and has been widely used to reversibly inactivate localized groups of neurons.[38,39] Using red-fluorescent BODIPY® TMR-X muscimol (M23400, Figure 16.2.7), researchers can correlate the distribution of muscimol with its pharmacological effects[40] and detect the presence of $GABA_A$ receptors on cell surfaces.[41]

Fluorescent Angiotensin II for AT1 and AT2 Receptors

Angiotensin II (Asp-Arg-Val-Tyr-Ile-His-Pro-Phe) stimulates smooth muscle contraction and plays an important role in blood pressure control and in water and salt homeostasis. These effects are exerted via two G-protein–coupled receptor subtypes, referred to as AT1 and AT2. Our N-terminal–labeled fluorescein and Alexa Fluor® 488 analogs of angiotensin II (A13438, A13439) are useful tools for imaging the distribution of these receptors,[42,43] as well as for flow cytometric analysis of angiotensin II endocytosis.[44] These fluorescent peptides have been characterized for purity by HPLC and mass spectrometry and generally display selectivity for AT1 over AT2 binding.[42]

IMPORTANT NOTICE: The products described in this manual are covered by one or more Limited Use Label License(s). Please refer to the Appendix on page 971 and Master Product List on page 975. Products are For Research Use Only. Not intended for any animal or human therapeutic or diagnostic use.
www.invitrogen.com/probes
molecular probes® | invitrogen™ by life technologies™

Naloxone Fluorescein for μ-Opioid Receptors

The μ-opioid receptor plays a critical role in analgesia. Among the antagonists that have been used to define and characterize these receptors are naloxone, a drug used to counteract the effects of opioid overdose, and naltrexone, a drug used in the treatment of opioid addiction. Naloxone fluorescein (N1384, Figure 16.2.8) has been reported to bind to the μ-opioid binding site with high affinity,[45–47] permitting receptor visualization in transfected Chinese hamster ovary (CHO) cells.[48] Flow cytometry analysis of the binding of naloxone fluorescein to NMDA and μ-opioid receptors (which was displaced by NMDA and met-enkephalin, respectively) has been used to deduce the effects of operant conditioning on visual cortex receptor pattern.[49]

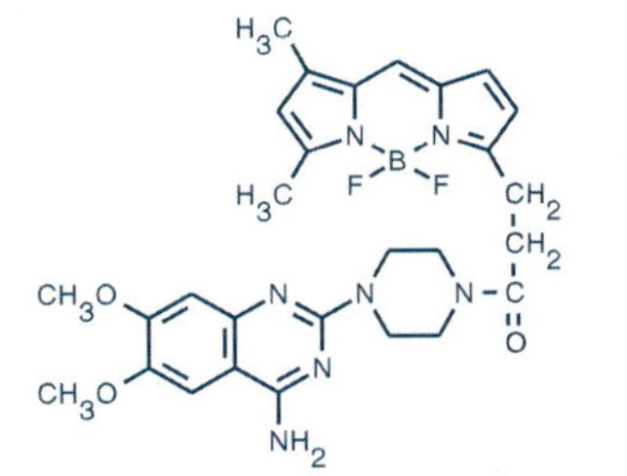

Figure 16.2.6 BODIPY® FL prazosin (B7433).

Probes for Amino Acid Neurotransmitter Receptors

Caged Amino Acid Neurotransmitters

When illuminated with UV light or by multiphoton excitation, caged amino acid neurotransmitters are converted into biologically active amino acids that rapidly initiate neurotransmitter action.[50,51] Thus, these caged probes provide a means of controlling the release—both spatially and temporally—of agonists for kinetic studies of receptor binding or channel opening.

The different caging groups confer special properties on these photoactivatable probes (Table 5.2). We synthesize two caged versions of L-glutamic acid[52–60] (C7122, G7055), as well as caged carbachol[61,62] (*N*-(CNB-caged) carbachol, C13654) and caged γ-aminobutyric acid[56,63–66] (*O*-(CNB-caged) GABA, A7110), all of which are biologically inactive before photolysis.[67] *O*-(CNB-caged) GABA (A7110) and γ-(CNB-caged) L-glutamic acid (G7055), which exhibit fast uncaging rates and high photolysis quantum yields, have been used to investigate the activation kinetics of GABA receptors[66] and glutamate receptors,[55] respectively. *N*-(CNB-caged) L-glutamic acid (C7122) does not hydrolyze in aqueous solution because it is caged on the amino group, thus enabling researchers to use very high concentrations without risk of light-independent glutamic acid production.[55,57]

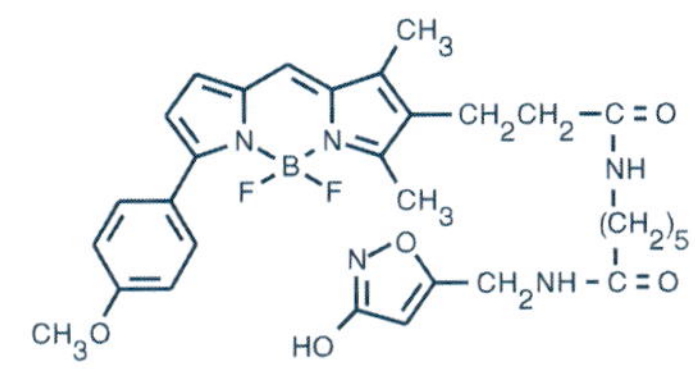

Figure 16.2.7 Muscimol, BODIPY® TMR-X conjugate (M23400).

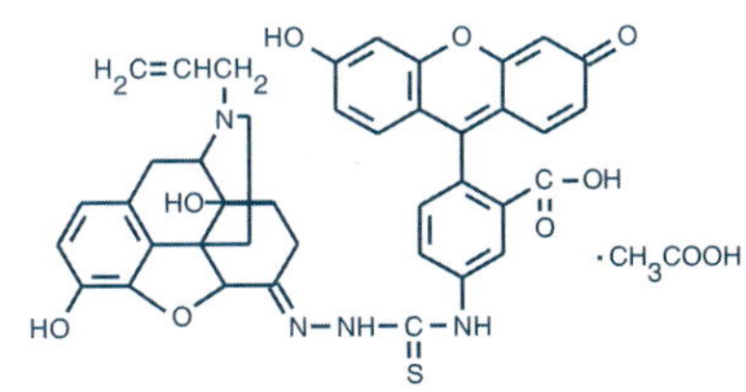

Figure 16.2.8 Naloxone fluorescein (N1384).

Anti–NMDA Receptor Antibodies

N-methyl-D-aspartate (NMDA) receptors constitute cation channels of the central nervous system that are gated by the excitatory neurotransmitter L-glutamate.[68,69] We offer affinity-purified rabbit polyclonal antibodies to NMDA receptor subunits 2A, 2B and 2C (A6473, A6474, A6475). The anti–NMDA receptor subunit 2A and 2B antibodies were generated against fusion proteins containing amino acid residues 1253–1391 of subunit 2A and 984–1104 of subunit 2B, respectively. These two antibodies are active against mouse, rat and human forms of the antigens and are specific for the subunit against which they were generated. In contrast, the anti–NMDA receptor subunit 2C antibody was generated against amino acid residues 25–130 of subunit 2C and recognizes the 140,000-dalton subunit 2C, as well as the 180,000-dalton subunit 2A and subunit 2B from mouse, rat and human. These three affinity-purified antibodies are suitable for immunohistochemistry[70] (Figure 16.2.9), western blots, enzyme-linked immunosorbent assays (ELISAs) and immunoprecipitations.

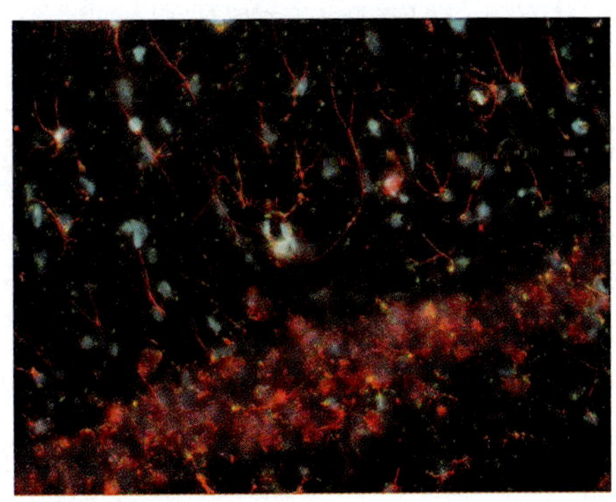

Figure 16.2.9 Rat brain cryosections labeled with anti–NMDA receptor, subunit 2A (rat), rabbit IgG fraction (A6473) and detected using Alexa Fluor® 488 goat anti–rabbit IgG antibody (A11008). The tissue was also labeled with Alexa Fluor® 594 anti–glial fibrillary acidic protein antibody (A21295) and counterstained with TOTO®-3 iodide (T3604), which was pseudocolored light blue in this image.

Amplex® Red Glutamic Acid/Glutamate Oxidase Assay Kit

The Amplex® Red Glutamic Acid/Glutamate Oxidase Assay Kit (A12221) provides an ultrasensitive method for continuously detecting glutamic acid[71] or for monitoring glutamate oxidase activity in a fluorescence microplate reader or a fluorometer.[72] In this assay, L-glutamic acid is oxidized by glutamate oxidase to produce α-ketoglutarate, NH_3 and H_2O_2. L-Alanine and L-glutamate–pyruvate transaminase are also included in the reaction. Thus, the L-glutamic acid is regenerated by transamination of α-ketoglutarate, resulting in multiple cycles of the initial reaction and a significant amplification of the H_2O_2 produced. Hydrogen peroxide reacts with the Amplex® Red reagent in a 1:1 stoichiometry in a reaction catalyzed by horseradish peroxidase (HRP) to generate the highly fluorescent product resorufin[73,74] (R363, Section 10.1).

IMPORTANT NOTICE: The products described in this manual are covered by one or more Limited Use Label License(s). Please refer to the Appendix on page 971 and Master Product List on page 975. Products are For Research Use Only. Not intended for any animal or human therapeutic or diagnostic use.

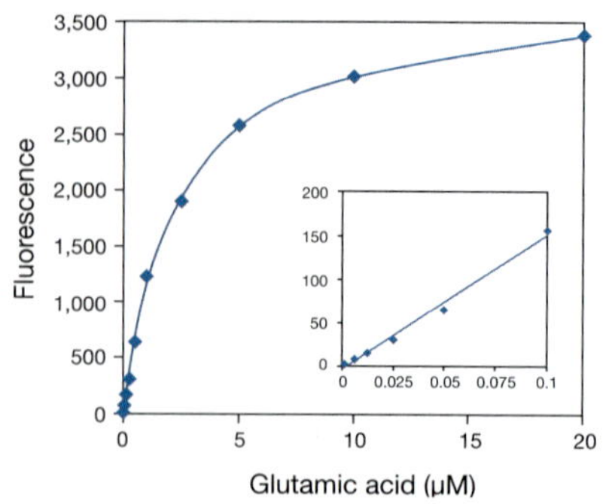

Figure 16.2.10 Detection of L-glutamic acid using the Amplex® Red Glutamic Acid/Glutamate Oxidase Assay Kit (A12221). Each reaction contained 50 µM Amplex® Red reagent, 0.125 U/mL HRP, 0.04 U/mL L-glutamate oxidase, 0.25 U/mL L-glutamate–pyruvate transaminase, 100 µM L-alanine and the indicated amount of L-glutamic acid in 1X reaction buffer. Reactions were incubated at 37°C. After 30 minutes, fluorescence was measured in a fluorescence microplate reader using excitation at 530 ± 12.5 nm and fluorescence detection at 590 ± 17.5 nm.

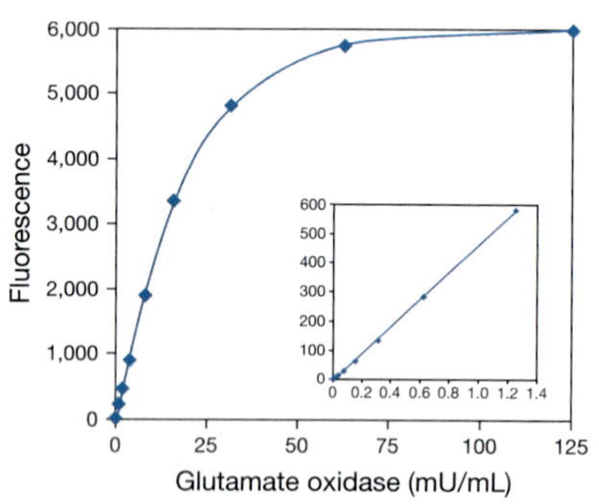

Figure 16.2.11 Detection of L-glutamate oxidase using the Amplex® Red Glutamic Acid/Glutamate Oxidase Assay Kit (A12221). Each reaction contained 50 µM Amplex® Red reagent, 0.125 U/mL HRP, 0.25 U/mL L-glutamate–pyruvate transaminase, 20 µM L-glutamic acid, 100 µM L-alanine and the indicated amount of *Streptomyces* L-glutamate oxidase in 1X reaction buffer. Reactions were incubated at 37°C. After 60 minutes, fluorescence was measured in a fluorescence microplate reader using excitation at 530 ± 12.5 nm and fluorescence detection at 590 ± 17.5 nm. The inset represents data from a separate experiment for lower L-glutamate oxidase concentrations and incubation time of 60 minutes (0–1.25 mU/mL).

Because resorufin has absorption/emission maxima of ~571/585 nm (Figure 16.2.3), there is little interference from autofluorescence in most biological samples.

If the concentration of L-glutamic acid is limiting in this assay, then the fluorescence increase is proportional to the initial L-glutamic acid concentration. The Amplex® Red Glutamic Acid/Glutamate Oxidase Assay Kit allows detection of as little as 10 nM L-glutamic acid in purified systems using a 30-minute reaction time (Figure 16.2.10). If the reaction is modified to include an excess of L-glutamic acid, then this kit can be used to continuously monitor glutamate oxidase activity. For example, purified L-glutamate oxidase from *Streptomyces* can be detected at levels as low as 40 µU/mL (Figure 16.2.11). The Amplex® Red reagent has been used to quantitate the activity of glutamate-producing enzymes in a high-throughput assay for drug discovery.[71] The Amplex® Red Glutamic Acid/Glutamate Oxidase Assay Kit contains:

- Amplex® Red reagent
- Dimethylsulfoxide (DMSO)
- Horseradish peroxidase (HRP)
- H_2O_2
- Concentrated reaction buffer
- L-Glutamate oxidase from *Streptomyces* sp.
- L-Glutamate–pyruvate transaminase from pig heart
- L-Glutamic acid
- L-Alanine
- Detailed protocols

Each kit provides sufficient reagents for approximately 200 assays using a fluorescence microplate reader and a reaction volume of 100 µL per assay.

Probes for Other Receptors

The Molecular Probes® Handbook discusses a diverse array of receptor probes, including fluorescent derivatives of:

- Low-density lipoprotein (LDL)
- Lipopolysaccharides
- Epidermal growth factor (EGF)
- Transferrin
- Fibrinogen
- Gelatin and collagen
- Ovalbumin and bovine serum albumin
- Casein
- Histone H1
- Subunit B of cholera toxin
- Chemotactic peptide
- Insulin

These ligands are all transported into the cell by receptor-mediated endocytosis. Additional information about these probes, as well as membrane and fluid-phase markers, can be found in Section 16.1.

REFERENCES

1. BMC Neurosci (2009) 10:80; **2.** Nano Lett (2008) 8:780; **3.** Proc Natl Acad Sci U S A (2007) 104:13666; **4.** Am J Physiol Cell Physiol (2006) 290:C728; **5.** J Cell Biol (2005) 170:619; **6.** J Immunol Methods (2004) 289:169; **7.** J Histochem Cytochem (2006) 54:817; **8.** Biochemistry (1990) 29:11009; **9.** Meth Neurosci (1992) 8:67; **10.** J Cell Biol (1998) 141:1613; **11.** Proc Natl Acad Sci U S A (1976) 73:4594; **12.** J Physiol (1974) 237:385; **13.** J Cell Biol (1993) 120:197; **14.** J Cell Biol (1993) 120:185; **15.** J Neurosci (1995) 15:520; **16.** J Cell Biol (1994) 124:139; **17.** Biophys J (2006) 90:2192; **18.** Proc Natl Acad Sci U S A (2004) 101:17114; **19.** J Biol Chem (2008) 283:15160; **20.** J Orthop Res (2009) 27:114; **21.** J Neurosci (2006) 26:6873; **22.** J Clin Invest (2004) 113:265; **23.** J Cell Biol (1995) 131:441; **24.** J Biol Chem (1993) 268:25108; **25.** Proc Natl Acad Sci U S A (1980) 77:4823; **26.** Clin Chim Acta (2007) 379:119; **27.** J Neurosci (2008) 28:11468; **28.** Mol Brain (2008) 1:18; **29.** Biochemistry (1979) 18:1875; **30.** Toxicon (1991) 29:503; **31.** Neuron (1992) 8:353; **32.** Science (1998) 281:794; **33.** Proc SPIE-Int Soc Opt Eng (2000) 3926:166; **34.** Brain Res Dev Brain Res (1997) 102:35; **35.** Br J Pharmacol (2009) 158:209; **36.** J Pharmacol Exp Ther (2009) 331:1118; **37.** Br J Pharmacol (2004) 143:899; **38.** J Neurosci Res (1992) 31:166; **39.** Neural Plast (2000) 7:19; **40.** J Neurosci Methods (2008) 171:30; **41.** Proc Natl Acad Sci U S A (2007) 104:335; **42.** Am J Physiol Renal Physiol (2006) 291:F375; **43.** J Neurosci Methods (2005) 143:3; **44.** Am J Physiol Renal Physiol (2005) 288:F420; **45.** Pharm Res (1986) 3:56; **46.** Pharm Res (1985) 6:266; **47.** Life Sci (1983) 33 Suppl 1:423; **48.** J Neurosci Methods (2000) 97:123; **49.** Biol Chem Hoppe Seyler (1995) 376:483; **50.** Nat Methods

IMPORTANT NOTICE: The products described in this manual are covered by one or more Limited Use Label License(s). Please refer to the Appendix on page 971 and Master Product List on page 975. Products are For Research Use Only. Not intended for any animal or human therapeutic or diagnostic use.
www.invitrogen.com/probes

REFERENCES—*continued*

(2007) 4:619; **51.** J Neurosci Methods (2004) 133:153; **52.** Nat Neurosci (1998) 1:119; **53.** Neuroscience (1998) 86:265; **54.** Science (1998) 279:1203; **55.** Proc Natl Acad Sci U S A (1994) 91:8752; **56.** J Org Chem (1996) 61:1228; **57.** Abstr Soc Neurosci (1995) 21:579, abstract 238.11; **58.** J Neurosci Methods (1994) 54:205; **59.** Science (1994) 265:255; **60.** Proc Natl Acad Sci U S A (1993) 90:7661; **61.** J Neurosci (2003) 23:9024; **62.** Proc Natl Acad Sci U S A (2000) 97:13895; **63.** Methods Enzymol (1998) 291:443; **64.** Neuron (1995) 15:755; **65.** J Org Chem (1990) 55:1585; **66.** J Am Chem Soc (1994) 116:8366; **67.** Methods Enzymol (1998) 291:30; **68.** Neuron (1994) 12:529; **69.** Nature (1991) 354:31; **70.** J Neurochem (2000) 75:2040; **71.** Anal Biochem (2000) 284:382; **72.** Anal Chim Acta (1999) 402:47; **73.** Anal Biochem (1997) 253:162; **74.** J Immunol Methods (1997) 202:133.

DATA TABLE 16.2 PROBES FOR NEUROTRANSMITTER RECEPTORS

Cat. No.	MW	Storage	Soluble	Abs	EC	Em	Solvent	Notes
A7110	396.28	F,D,LL	H_2O	262	4500	none	pH 7	1, 2
A13438	1404.50	F,D,L	H_2O, DMSO	494	78,000	522	pH 9	3
A13439	1586.64	F,D,L	H_2O, DMSO	491	78,000	516	pH 7	3
B1196	~8400	F,D	H_2O	<300		none		4
B1601	7984.14	F	H_2O	<300		see Notes		5
B7433	563.41	F,D,L	DMSO, EtOH	504	77,000	511	MeOH	
B13422	9000	F,D,L	H_2O	495	78,000	519	pH 8	4, 6
B13423	9000	F,D,L	H_2O	593	92,000	617	pH 7	4, 6
B35450	9000	F,D,L	H_2O	649	246,000	668	pH 7	4, 6
B35451	9000	F,D,L	H_2O	554	150,000	567	pH 7	4, 6
B35452	9000	F,D,L	H_2O	680	180,000	704	pH 7	4, 6
C7122	326.26	F,D,LL	H_2O	266	4800	none	pH 7	1, 2
C13654	439.34	F,D,LL	H_2O	264	4200	none	H_2O	1, 2
F1176	9000	F,D,L	H_2O	494	84,000	518	pH 8	4, 6
G7055	440.29	F,D,LL	H_2O, DMSO	262	5100	none	pH 7	1, 2
M23400	607.46	F,D,L	DMSO	543	60,000	572	MeOH	
N1384	790.84	D,L	EtOH, DMF	492	79,000	516	pH 9	
T1175	9000	F,D,L	H_2O	553	85,000	577	H_2O	4, 6

For definitions of the contents of this data table, see "Using *The Molecular Probes® Handbook*" in the introductory pages.

Notes

1. All photoactivatable probes are sensitive to light. They should be protected from illumination except when photolysis is intended.
2. This compound has weaker visible absorption at >300 nm but no discernible absorption peaks in this region.
3. The value of EC listed for this peptide conjugate is that of the labeling dye in free solution. Use of this value for the conjugate assumes a 1:1 dye:peptide labeling ratio and no change of EC due to dye–peptide interactions.
4. α-Bungarotoxin, EGF and phallotoxin conjugates have approximately 1 label per peptide.
5. This peptide exhibits intrinsic tryptophan fluorescence (Em ~350 nm) when excited at <300 nm.
6. The value of EC listed for this α-bungarotoxin conjugate is for the labeling dye in free solution. Use of this value for the conjugate assumes a 1:1 dye:peptide labeling ratio and no change of EC due to dye–peptide interactions.

PRODUCT LIST 16.2 PROBES FOR NEUROTRANSMITTER RECEPTORS

Cat. No.	Product	Quantity
A7110	γ-aminobutyric acid, α-carboxy-2-nitrobenzyl ester, trifluoroacetic acid salt (*O*-(CNB-caged) GABA)	5 mg
A12217	Amplex® Red Acetylcholine/Acetylcholinesterase Assay Kit *500 assays*	1 kit
A12221	Amplex® Red Glutamic Acid/Glutamate Oxidase Assay Kit *200 assays*	1 kit
A13439	angiotensin II, Alexa Fluor® 488 conjugate	25 µg
A13438	angiotensin II, fluorescein conjugate	25 µg
A6473	anti-NMDA receptor, subunit 2A (rat), rabbit IgG fraction *affinity purified*	10 µg
A6474	anti-NMDA receptor, subunit 2B (rat), rabbit IgG fraction *affinity purified*	10 µg
A6475	anti-NMDA receptor, subunit 2C (rat), rabbit IgG fraction *affinity purified*	10 µg
B7433	BODIPY® FL prazosin	100 µg
B1601	α-bungarotoxin *from *Bungarus multicinctus**	1 mg
B13422	α-bungarotoxin, Alexa Fluor® 488 conjugate	500 µg
B35451	α-bungarotoxin, Alexa Fluor® 555 conjugate	500 µg
B13423	α-bungarotoxin, Alexa Fluor® 594 conjugate	500 µg
B35450	α-bungarotoxin, Alexa Fluor® 647 conjugate	500 µg
B35452	α-bungarotoxin, Alexa Fluor® 680 conjugate	500 µg
B1196	α-bungarotoxin, biotin-XX conjugate	500 µg
C13654	*N*-(CNB-caged) carbachol (*N*-(α-carboxy-2-nitrobenzyl)carbamylcholine, trifluoroacetic acid salt)	5 mg
C7122	*N*-(CNB-caged) L-glutamic acid (*N*-(α-carboxy-2-nitrobenzyl)-L-glutamic acid)	5 mg
F1176	fluorescein α-bungarotoxin (α-bungarotoxin, fluorescein conjugate)	500 µg
G7055	L-glutamic acid, γ-(α-carboxy-2-nitrobenzyl) ester, trifluoroacetic acid salt (γ-(CNB-caged) L-glutamic acid)	5 mg
M23400	muscimol, BODIPY® TMR-X conjugate	1 mg
N1384	naloxone fluorescein	5 mg
T1175	tetramethylrhodamine α-bungarotoxin (α-bungarotoxin, tetramethylrhodamine conjugate)	500 µg

IMPORTANT NOTICE: The products described in this manual are covered by one or more Limited Use Label License(s). Please refer to the Appendix on page 971 and Master Product List on page 975. Products are For Research Use Only. Not intended for any animal or human therapeutic or diagnostic use.

16.3 Probes for Ion Channels and Carriers

This section describes a variety of probes for Ca^{2+}, Na^{+}, K^{+} and Cl^{-} ion channels and carriers. Chapter 19 and Chapter 21 contain our extensive selection of indicators for these physiologically important ions, providing a means of correlating ion channel activation with subsequent changes in intracellular ion concentration. Ion flux also affects the cell's membrane potential, which can be measured with the probes described in Chapter 22.

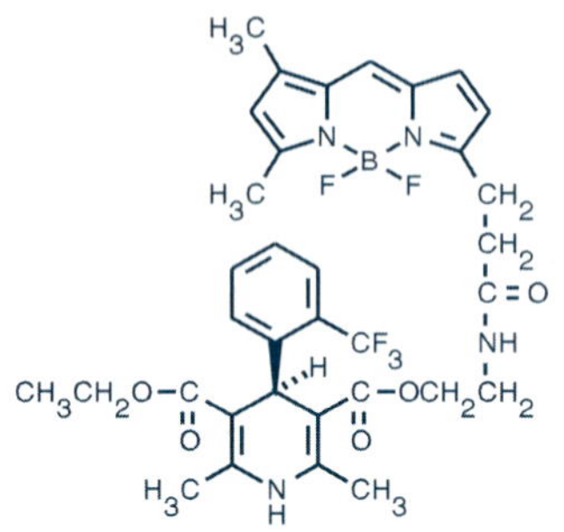

Figure 16.3.1 DM-BODIPY® (–)-dihydropyridine (D7443).

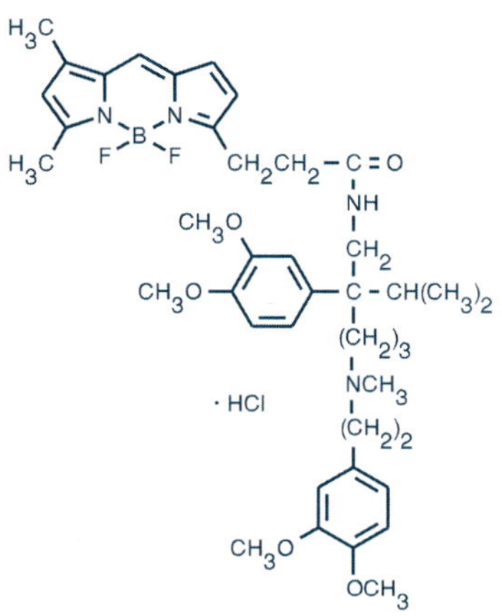

Figure 16.3.2 BODIPY® FL verapamil, hydrochloride (B7431).

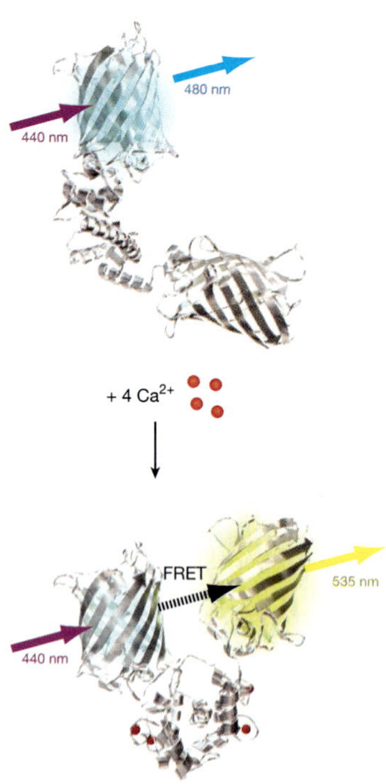

Figure 16.3.4 Schematic of the Premo™ Cameleon Calcium Sensor (P36207, P36208) mechanism.

Probes for Ca^{2+} Channels and Carriers

In both excitable and nonexcitable cells, intracellular Ca^{2+} levels modulate a multitude of vital cellular processes—including gene expression, cell viability, cell proliferation, cell motility and cell shape and volume regulation—and thereby play a key role in regulating cell responses to external activating agents. These dynamic changes in intracellular Ca^{2+} levels are regulated by ligand-gated and G-protein–coupled ion channels in the plasma membrane, as well as by mobilization of Ca^{2+} from intracellular stores. One of the best-studied examples of Ca^{2+}-dependent signal transduction is the depolarization of excitable cells, such as those of neuronal, cardiac, skeletal and smooth muscle tissue, which is mediated by inward Ca^{2+} and Na^{+} currents. The Ca^{2+} current is attributed to the movement of ions through N-, L-, P- and T-type Ca^{2+} channels, which are defined both pharmacologically and by their biophysical properties, including conductance and voltage sensitivity. Here we describe several fluorescent ligands for imaging the spatial distribution and localization of Ca^{2+} channels in cells, as well as Premo™ Cameleon Calcium Sensor, a genetically encoded, protein-based ratiometric sensor for calcium measurements. Our complete selection of Ca^{2+} indicators is described in Chapter 19.

Fluorescent Dihydropyridine for L-Type Ca^{2+} Channels

The L-type Ca^{2+} channel is readily blocked by the binding of dihydropyridine to the channel's pore-forming α_1-subunit. To facilitate the study of channel number and distribution in single cells, we offer fluorescent dihydropyridine derivatives. The high-affinity (–)-enantiomer of dihydropyridine is available labeled with either the green-fluorescent DM-BODIPY® (D7443, Figure 16.3.1) or the orange-fluorescent ST-BODIPY® (S7445) fluorophore. Knaus and colleagues have shown that these BODIPY® dihydropyridines bind to L-type Ca^{2+} channels with high affinity and inhibit the Ca^{2+} influx in GH_3 cells.[1–3] For neuronal L-type Ca^{2+} channels, the (–)-enantiomers of the DM-BODIPY® dihydropyridine and ST-BODIPY® derivatives each exhibit a K_i of 0.9 nM. Their affinities for skeletal muscle L-type Ca^{2+} channels are somewhat lower. Although DM-BODIPY® dihydropyridine exhibits a more intense fluorescence, the particularly high degree of stereoselectivity retained by the ST-BODIPY® derivatives has proven useful for the *in vivo* visualization of L-type Ca^{2+} channels.[4] DM-BODIPY® dihydropyridine has proven effective as a substrate for functional analysis of ABC drug transporters.[5]

BODIPY® FL Verapamil

Like dihydropyridine, phenylalkylamines also bind to the α_1-subunit of L-type Ca^{2+} channels and block Ca^{2+} transport. We offer a green-fluorescent BODIPY® FL derivative (B7431, Figure 16.3.2) of verapamil, a phenylalkylamine known to inhibit P-glycoprotein–mediated drug efflux.

The 170,000-dalton P-glycoprotein is typically overexpressed in tumor cells that have acquired resistance to a variety of anticancer drugs (Section 15.6). P-glycoprotein is thought to mediate the ATP-dependent efflux or sequestration of structurally unrelated molecules, including actinomycin D, anthracyclines, colchicine, epipodophyllotoxins and vinblastine. Verapamil appears to inhibit drug efflux by acting as a substrate of P-glycoprotein, thereby overwhelming the transporter's capacity to expel the drugs. BODIPY® FL verapamil also appears to serve as a substrate for P-glycoprotein. This fluorescent verapamil derivative preferentially accumulates in the lysosomes of normal, drug-sensitive NIH 3T3 cells but is rapidly transported out of multidrug-resistant cells.[6–9]

IMPORTANT NOTICE: The products described in this manual are covered by one or more Limited Use Label License(s). Please refer to the Appendix on page 971 and Master Product List on page 975. Products are For Research Use Only. Not intended for any animal or human therapeutic or diagnostic use.

molecular probes® | invitrogen™ by life technologies™

Eosin Derivatives: Inhibitors of the Calcium Pump

Eosin isothiocyanate (E18) is a potent reversible inhibitor of the erythrocyte plasma membrane calcium pump, with a half-maximal inhibitory concentration of <0.2 µM.[10] Eosin isothiocyanate also reacts irreversibly at the ATP-binding site of this calcium pump. The succinimidyl ester of carboxyeosin diacetate (C22803), a cell membrane–permeant eosin derivative, also inhibits the erythrocyte plasma membrane Ca^{2+} pump.[11,12] Fluorescein isothiocyanate (F143, Section 1.4) is a weaker inhibitor of the erythrocyte plasma membrane calcium pump.

Premo™ Cameleon Calcium Sensor

The Premo™ product line combines genetically encoded ion indicators and environmental sensors with efficient BacMam delivery (BacMam Gene Delivery and Expression Technology—Note 11.1) for intracellular measurements in mammalian cells. Premo™ Cameleon Calcium Sensor (P36207, P36208) is a ratiometric calcium-sensitive fluorescent protein that is delivered by BacMam baculovirus-mediated transduction to a variety of mammalian cell types. This content and delivery system provides an effective and robust technique for measuring Ca^{2+} mobilization in transduced cells using microplate assays or fluorescence microscopy (Figure 16.3.3).

The Premo™ Cameleon Calcium Sensor is based on the YC3.60 version of the fluorescent protein (FP)–based sensor (cameleon) family developed by Tsien, Miyawaki and co-workers, which is reported to have a Ca^{2+} dissociation constant of 240 nM.[13,14] The sensor comprises two fluorescent proteins (Enhanced Cyan Fluorescent Protein or ECFP and Venus variant of Yellow Fluorescent Protein or YFP), linked by the calmodulin-binding peptide M13 and calmodulin. Upon binding four calcium ions, calmodulin undergoes a conformational change by wrapping itself around the M13 peptide, which changes the efficiency of the fluorescence resonance energy transfer (FRET) between the CFP donor and the YFP acceptor fluorophores (Figure 16.3.4). Following this conformational change, there is an increase in YFP emission (525–560 nm) and a simultaneous decrease in CFP emission (460–500 nm) (Figure 16.3.5), making Cameleon an effective reporter of calcium mobilization. The ratiometric readout of the Premo™ Cameleon Calcium Sensor—an increase in YFP emission (535 nm, green-yellow emission) and a decrease in CFP emission (485 nm, blue emission)—reduces assay variations due to compound or cellular autofluorescence, nonuniform cell plating, differences in expression levels between cells, instability of instrument illumination and changes in illumination pathlength.

The Premo™ Cameleon Calcium Sensor is designed to readily and accurately detect intracellular calcium flux from different receptors. Standard pharmacological assays for multiple GPCR agonists and antagonists have been tested. An example of the robustness and reproducibility and accuracy of the system is demonstrated using the endogenous histamine receptor in conjunction with histamine, pyrilamine, and thioperamide in HeLa cells (Figure 16.3.6). Expression levels will be maintained for several days, enabling iterative assays to be run, for instance, when examining agonist/antagonist relationships on the same cells.

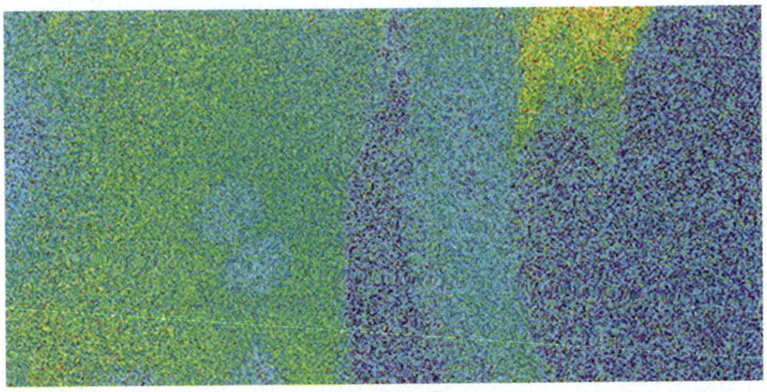

Figure 16.3.3 Porcine left atrial appendage progenitor cells were transfected with Premo™ Cameleon calcium sensor (P36207, P36208); ATP (20 µM final concentration) was applied to the cells the following day and the cells were imaged using a Zeiss 5 Live high-speed confocal system (Carl Zeiss MicroImaging). Excitation was with a 405 nm diode laser (50 mw) operated at 50% power. Emission was collected simultaneously on two linear CCD detectors using a 490 nm dichroic mirror to split the beam through a 415–480 nm bandpass filter for CFP and a 550 nm longpass filter for YFP. Images were collected at a rate of 10 frames per second (512 x 512 pixels) using a 40x Plan-Neofluar 1.3 NA oil immersion objective lens.

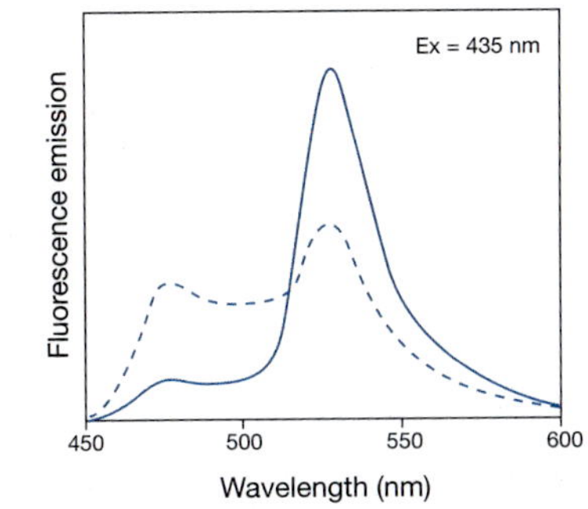

Figure 16.3.5 Fluorescence emission spectra of Premo™ Cameleon Calcium Sensor (P36207, P36208). The dashed line indicates the spectra in the absence of Ca^{2+}; the solid line shows the fluorescence resonance energy transfer (FRET)–based change upon Ca^{2+} binding.

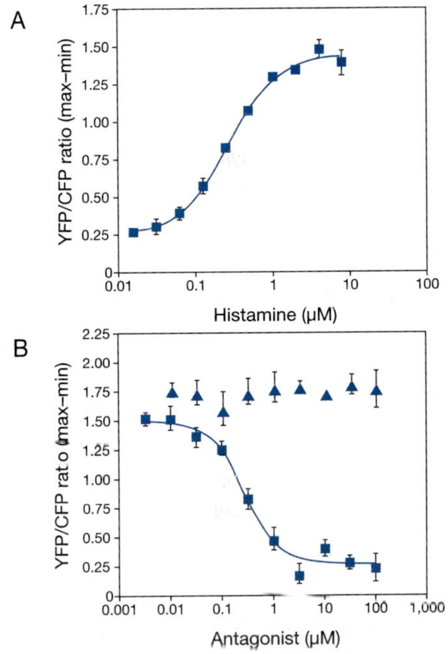

Figure 16.3.6 Agonist and antagonist dose response curves. HeLa cells were plated in a 96-well plate at a density of 15,000 cells/well, transduced with Premo™ Cameleon Calcium Sensor (P36207, P36208), and incubated overnight at 37°C. The following day, a histamine dose response was performed (**A**). A separate plate was used to evaluate an antagonist dose response with pyrilamine (■) and thioperamide (▲) in the presence of an EC_{80} concentration of histamine (**B**). Pyrilamine is a known H1 receptor antagonist that couples through G_q proteins and the second messenger Ca^{2+}. Thioperamide is a known H3 receptor antagonist that couples through G_i proteins and the second messenger cAMP.

IMPORTANT NOTICE: The products described in this manual are covered by one or more Limited Use Label License(s). Please refer to the Appendix on page 971 and Master Product List on page 975. Products are For Research Use Only. Not intended for any animal or human therapeutic or diagnostic use.

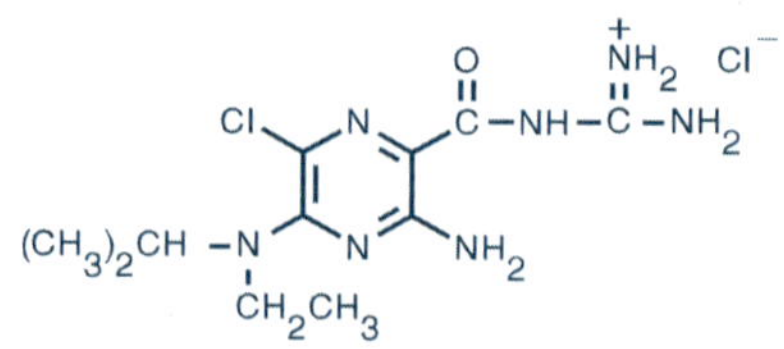

Figure 16.3.7 5-(*N*-ethyl-*N*-isopropyl)amiloride, hydrochloride (E3111).

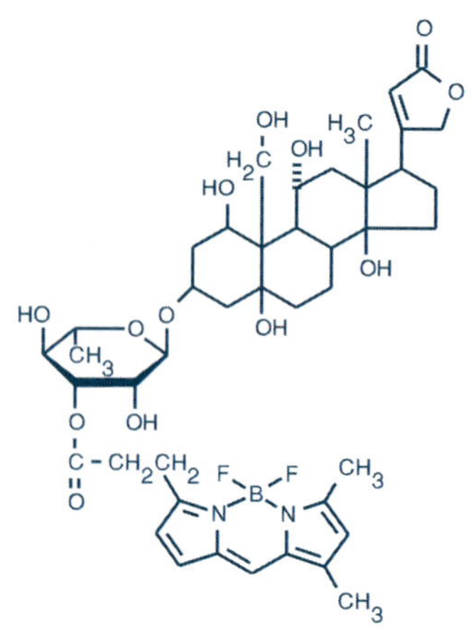

Figure 16.3.8 BODIPY® FL ouabain (B23461).

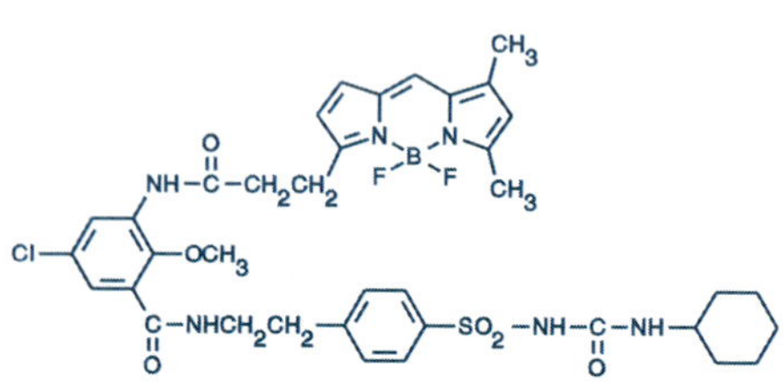

Figure 16.3.9 ER-Tracker™ Green (BODIPY® FL glibenclamide, E34251).

Probes for Na^+ Channels and Carriers

Amiloride Analogs: Probes for the Na^+ Channel and the Na^+/H^+ Antiporter

Amiloride is a compound known to inhibit the Na^+/H^+ antiporter of vertebrate cells by acting competitively at the Na^+-binding site.[15] The antiporter extrudes protons from cells using the inward Na^+ gradient as a driving force, resulting in intracellular alkalinization. In 1967, Cragoe and co-workers reported the synthesis of amiloride and several amiloride analogs, which are pyrazine diuretics that inhibit the Na^+ channel in urinary epithelia.[16] Since then, more than 1000 different amiloride analogs have been synthesized and many of these tested for their specificity and potency in inhibiting the Na^+ channel, Na^+/H^+ antiporter and Na^+/Ca^{2+} exchanger.[17] Unmodified amiloride inhibits the Na^+ channel with an IC_{50} of less than 1 µM. Additionally, amiloride is an important tool for studying the Na^+/H^+ antiporter. Structure–activity relationships have demonstrated that amiloride analogs with hydrophobic groups in the drug are the most potent and specific inhibitors for the Na^+/H^+ antiporter.[17–22] For example, 5-(*N*-ethyl-*N*-isopropyl)amiloride (EIPA, E3111; Figure 16.3.7) is 200-fold more potent than amiloride for inhibiting this antiporter.

Ouabain Probes for Na^+/K^+-ATPase

Ouabain is a member of a class of glycosylated steroids collectively known as cardiac glycosides due to their therapeutic efficacy in the treatment of congestive heart failure. Ouabain achieves this effect by binding to the catalytic α-subunit of Na^+/K^+-ATPase and inhibiting its transport of Na^+ across the plasma membrane. 9-Anthroyl ouabain (A1322) is useful for localizing Na^+/K^+-ATPase and for studying its membrane orientation, mobility and dynamics.[23] Anthroyl ouabain has also been employed to investigate Na^+/K^+-ATPase's active site, inhibition and conformational changes,[24–29] as well as to investigate the kinetics of cardiac glycoside binding.[30–35] BODIPY® FL ouabain (B23461, Figure 16.3.8) has been used in combination with Alexa Fluor® 555 cholera toxin B (C22843, Section 16.1) for visualizing Na^+/K^+-ATPase and ganglioside G_{M1} domain localization in lymphocyte plasma membranes.[36]

Using BacMam Technology to Deliver and Express Sodium Channel cDNA

Sodium channel cDNAs that have been engineered into a baculovirus gene delivery/expression system using BacMam technology (BacMam Gene Delivery and Expression Technology—Note 11.1) are also available, including the Nav1.2 cDNA (B10341) and the Nav1.5 cDNA (B10335).

The BacMam system uses a modified insect cell baculovirus as a vehicle to efficiently deliver and express genes in mammalian cells with minimum effort and toxicity. The use of BacMam delivery in mammalian cells is relatively new, but well described, and has been used extensively in a drug discovery setting.[37] Furthermore, constitutively expressed ion channels and other cell surface proteins have been shown to contribute to cell toxicity in some systems, and may be subject to clonal drift and other inconsistencies that hamper successful experimentation and screening. Thus, transient expression systems such as the BacMam gene delivery and expression system are increasingly methods of choice to decrease variability of expression in such assays.

U2OS cells (ATCC number HTB-96) have been shown to demonstrate highly efficient expression of BacMam delivered targets in a null background ideal for screening in a heterologous expression system. The U2OS cell line is recommended for use if your particular cell line does not efficiently express the BacMam targets. Examples of other cell lines that are efficiently transduced by BacMam technology include HEK 293, HepG2, BHK, Cos-7 and Saos-2.

Probes for K^+ Channels and Carriers

Glibenclamide Probes for the ATP-Dependent K^+ Channel

Glibenclamide blocks the ATP-dependent K^+ channel, thereby eliciting insulin secretion.[38] We have prepared the green-fluorescent BODIPY® FL glibenclamide (BODIPY® FL glyburide, E34251; Figure 16.3.9) and red-fluorescent BODIPY® TR glibenclamide (BODIPY® TR glyburide, E34250) as probes for the ATP-dependent K^+ channel. BODIPY® TR glibenclamide has been used to detect sulfonylurea receptors associated with ATP-dependent K^+ channels in bovine monocytes and in β-cells[39,40] and to label a novel mitochondrial ATP-sensitive potassium channel in brain.[41]

The sulfonylurea receptors of ATP-dependent K^+ channels are prominent on the endoplasmic reticulum (ER). Therefore, because these probes are also effective live-cell stains for ER,

IMPORTANT NOTICE: The products described in this manual are covered by one or more Limited Use Label License(s). Please refer to the Appendix on page 971 and Master Product List on page 975. Products are For Research Use Only. Not intended for any animal or human therapeutic or diagnostic use.

BODIPY® FL glibenclamide and BODIPY® TR glibenclamide are also referred to as ER-Tracker™ Green and ER-Tracker™ Red, respectively; see Section 12.4 for a description of this application. Variable expression of sulfonylurea receptors in some specialized cell types may result in non-ER labeling with these probes.

FluxOR™ Potassium Ion Channel Assay

The FluxOR™ Potassium Ion Channel Assay Kits (F10016, F10017) provide a fluorescence-based assay for high-throughput screening of potassium ion channel and transporter activities.[42,43] The FluxOR™ Potassium Ion Channel Assay Kits take advantage of the well-described permeability of potassium channels to thallium (Tl^+) ions. When thallium is present in the extracellular solution containing a stimulus to open potassium channels, channel activity is detected with a cell-permeant thallium indicator dye that reports large increases in fluorescence emission at 525 nm as thallium flows down its concentration gradient and into the cells (Figure 16.3.10). In this way, the fluorescence reported in the FluxOR™ system becomes a surrogate indicator of activity for any ion channel or transporter that is permeable to thallium, including the human ether-a-go-go–related (hERG) channel, one of the human cardiac potassium channels. The FluxOR™ potassium ion channel assay has been validated for homogeneous high-throughput profiling of hERG channel inhibition using BacMam-mediated transient expression of hERG.[42] The FluxOR™ Potassium Ion Channel Assay Kits can also be used to study potassium co-transport processes that accommodate the transport of thallium into cells.[44] Furthermore, resting potassium channels and inward rectifier potassium channels like Kir2.1 can be assayed by adding stimulus buffer with thallium alone, without any depolarization to measure the signal.

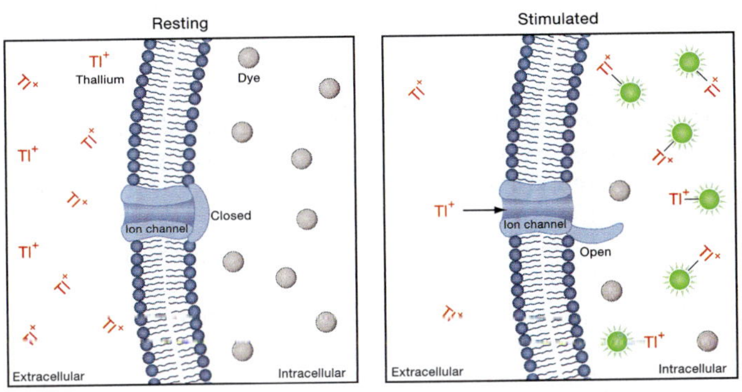

Figure 16.3.10 Thallium redistribution in the FluxOR™ assay. Basal fluorescence from cells loaded with FluxOR™ reagent (provided in the FluxOR™ Potassium Ion Channel Assay Kits; F10016, F10017) is low when potassium channels remain unstimulated, as shown in the left panel. When thallium is added to the assay with the stimulus, the thallium flows down its concentration gradient into the cells, activating the dye as shown in the right panel.

The FluxOR™ reagent, a thallium indicator dye, is loaded into cells as a membrane-permeable AM ester. Loading is assisted by the proprietary PowerLoad™ concentrate, an optimized formulation of nonionic Pluronic® surfactant polyols that act to disperse and stabilize AM ester dyes for optimal loading in aqueous solution. This PowerLoad™ concentrate is also available separately (P10020) to aid the solubilization of water-insoluble dyes and other materials in physiological media.

Once inside the cell, the nonfluorescent AM ester of the FluxOR™ dye is cleaved by endogenous esterases into a weakly fluorescent (basal fluorescence), thallium-sensitive indicator. The thallium-sensitive form is retained in the cytosol, and its extrusion is inhibited by water-soluble probenecid (P36400, Section 19.8), which blocks organic anion pumps. For most applications, cells are loaded with the dye at room temperature. For best results, the dye-loading buffer is then replaced with fresh, dye-free assay buffer (composed of physiological HBSS containing probenecid), and cells are ready for the high-throughput screening assay.

Each FluxOR™ Potassium Ion Channel Assay Kit contains:

- FluxOR™ reagent
- FluxOR™ assay buffer
- PowerLoad™ concentrate
- Probenecid
- FluxOR™ chloride-free buffer
- Potassium sulfate (K_2SO_4) concentrate
- Thallium sulfate (Tl_2SO_4) concentrate
- Dimethylsulfoxide (DMSO)
- Detailed protocols

The FluxOR™ Kits provide a concentrated thallium solution along with sufficient dye and buffers to perform ~4000 (F10016) or ~40,000 (F10017) assays in a 384-well microplate format. These kits allow maximum target flexibility and ease of operation in a homogeneous format. The FluxOR™ potassium ion channel assay has been demonstrated for use with CHO and HEK 293 cells stably expressing hERG, as well as U2OS cells transiently transduced with BacMam hERG reagent[42] (B10019, B10033) (Figure 16.3.11). More information is available at www.invitrogen.com/handbook/fluxorpotassium.

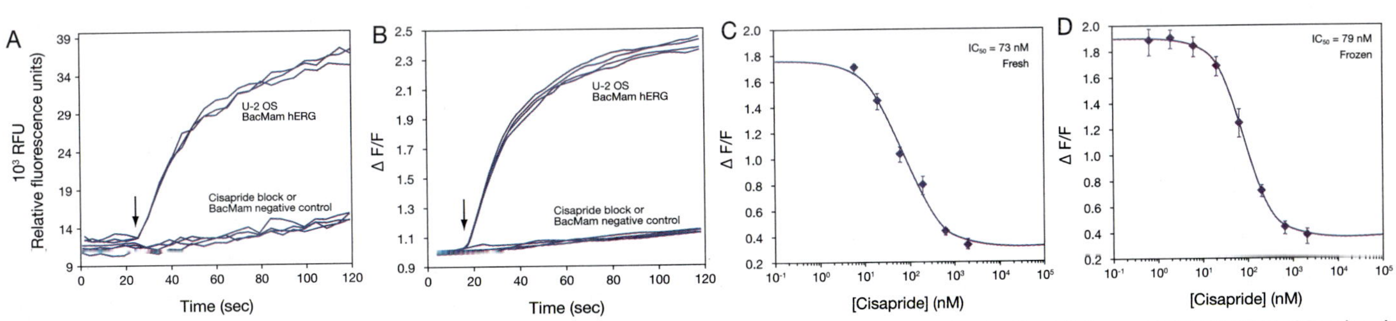

Figure 16.3.11 FluxOR™ potassium ion channel assays (F10016, F10017) performed on fresh and frozen U2OS cells transduced with the BacMam hERG reagent (B10019, B10033). **A)** Raw data obtained in the FluxOR™ assay determination of thallium flux in U2OS cells transduced with BacMam-hERG and kept frozen until the day of use. The arrow indicates the addition of the thallium/potassium stimulus, and upper and lower traces indicate data taken from the minimum and maximum doses of cisapride used in the determination of the dose-response curves. **B)** Raw pre-stimulus peak and baseline values were boxcar averaged and normalized, and indicate the fold increase in fluorescence over time. **C)** Data generated in a dose-response determination of cisapride block on BacMam hERG expressed in U2OS cells freshly prepared from overnight expression after viral transduction. **D)** Parallel data obtained from cells transduced with BacMam-hERG, stored for 2 weeks in liquid nitrogen, thawed, and plated 4 hours prior to running the assay. Error bars indicate standard deviation, n = 4 per determination.

IMPORTANT NOTICE: The products described in this manual are covered by one or more Limited Use Label License(s). Please refer to the Appendix on page 971 and Master Product List on page 975. Products are For Research Use Only. Not intended for any animal or human therapeutic or diagnostic use.

Using BacMam Technology to Deliver and Express Potassium Channel cDNA

Potassium channel cDNAs that have been engineered into a baculovirus gene delivery/expression system using BacMam technology (BacMam Gene Delivery and Expression Technology—Note 11.1) are also available for use with the FluxOR™ Potassium Ion Channel Assay Kits, including the human ether-a-go-go related gene (hERG) (Figure 16.3.12), several members of the voltage-gated K^+ channel (Kv) gene family and two members of the inwardly rectifying K^+ channel (Kir) gene family:

- BacMam hERG[42] (for 10 microplates, B10019; for 100 microplates, B10033)
- BacMam Kv1.1 (for 10 microplates, B10331)
- BacMam Kv1.3 (for 10 microplates, B10332)
- BacMam Kv2.1 (for 10 microplates, B10333)
- BacMam Kv7.2 and Kv7.3 (for 10 microplates, B10147)
- BacMam Kir1.1 (for 10 microplates, B10334)
- BacMam Kir2.1 (for 10 microplates, B10146)

The BacMam system uses a modified insect cell baculovirus as a vehicle to efficiently deliver and express genes in mammalian cells with minimum effort and toxicity. The use of BacMam delivery in mammalian cells is relatively new, but well described, and has been used extensively in a drug discovery setting.[37] Furthermore, constitutively expressed ion channels and other cell surface proteins have been shown to contribute to cell toxicity in some systems, and may be subject to clonal drift and other inconsistencies that hamper successful experimentation and screening. Thus, transient expression systems such as BacMam technology are increasingly methods of choice to decrease variability of expression in such assays.

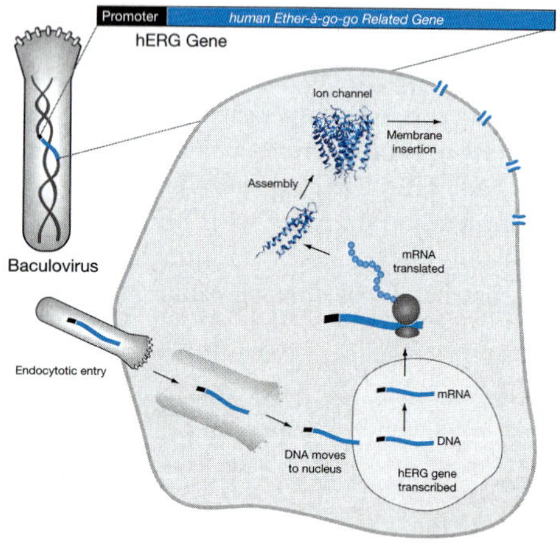

Figure 16.3.12 BacMam-hERG gene delivery and expression. This schematic depicts the mechanism of BacMam-mediated gene delivery into a mammalian cell and expression of the hERG gene (B10019, B10033). The hERG gene resides within the baculoviral DNA, downstream of a CMV promoter that drives its expression when introduced into a mammalian target cell. BacMam viral particles are taken up by endocytic pathways into the cell, and the DNA within them is released for transcription and expression. The translated protein is then folded for insertion into the membrane, forming functional hERG ion channels. This process begins within 4–6 hours and in many cell types is completed after an overnight period.

U2OS cells (ATCC number HTB-96) have been shown to demonstrate highly efficient expression of BacMam-delivered targets in a null background ideal for screening in a heterologous expression system. The U2OS cell line is recommended for use if your particular cell line does not efficiently express the BacMam targets. Examples of other cell lines that are efficiently transduced by BacMam technology include HEK 293, HepG2, BHK, Cos-7 and Saos-2.

Probes for Anion Transporters

Stilbene Disulfonates: Anion-Transport Inhibitors

We offer three stilbene disulfonates that have been employed to inhibit (frequently irreversibly) anion transport[45] in a large number of mammalian cell types:

- DIDS (D337, Figure 16.3.13)
- H_2DIDS (D338)
- DNDS (D673)

Our stilbene disulfonate probes, which are 95–99% pure by HPLC, have significantly higher purity and more defined composition than those available from other commercial sources. DNDS was among the inhibitors used to characterize three different anion exchangers in the membranes of renal brush border cells and to compare these exchangers with the band-3 anion-transport protein of erythrocyte membranes.[46]

These stilbene disulfonates can, in some cases, bind specifically to proteins that are not anion transporters. For example, DIDS and H_2DIDS complex specifically with the CD4 glycoprotein on T-helper lymphocytes and macrophages, blocking HIV type-1 growth at multiple stages of the virus life cycle.[47]

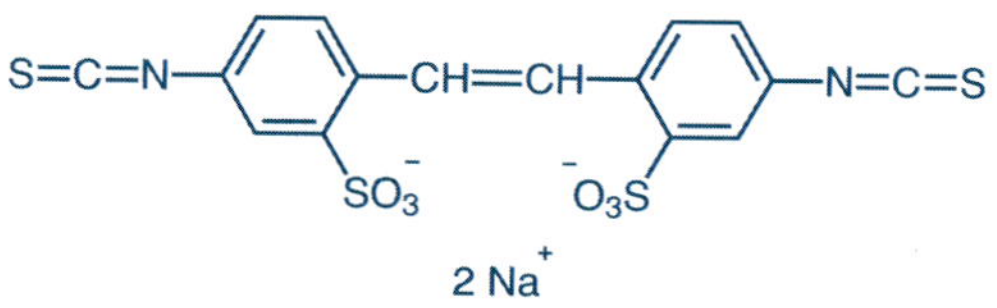

Figure 16.3.13 DIDS (4,4′-diisothiocyanatostilbene-2,2′-disulfonic acid, disodium salt, D337).

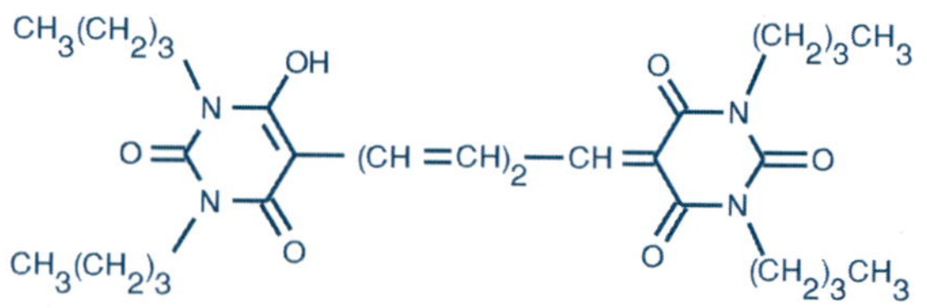

Figure 16.3.14 Bis-(1,3-dibutylbarbituric acid)pentamethine oxonol (DiBAC$_4$(5), B436).

DiBAC$_4$(5)

The membrane potential–sensing dye bis-(1,3-dibutylbarbituric acid)pentamethine oxonol (DiBAC$_4$(5), B436; Figure 16.3.14) initially inhibits Cl^- exchange with an IC_{50} of 0.146 µM. However, this inhibition increases with time to an IC_{50} of 1.05 nM, making DiBAC$_4$(5) a more potent inhibitor than DIDS, which has an IC_{50} of 31 nM under similar conditions.[48]

Eosin Maleimide

Although usually selectively reactive with thiols, eosin-5-maleimide (E118, Section 2.2) is known to react with a specific lysine residue of the band-3 protein in human erythrocytes, inhibiting anion exchange in these cells and providing a convenient tag for observing band-3 behavior in the membrane.[49–51] Eosin-5-isothiocyanate (E18) has similar reactivity with band-3 proteins.[52,53]

Premo™ Halide Sensor

The fluorescent protein–based Premo™ Halide Sensor (P10229) is a pharmacologically relevant sensor for functional studies of ligand- and voltage-gated chloride channels and their modulators in cells. Chloride channels are involved in cellular processes as critical and diverse as transepithelial ion transport, electrical excitability, cell volume regulation and ion homeostasis. Given their physiological significance, it follows that defects in their activity can have severe implications, including such conditions as cystic fibrosis and neuronal degeneration. Thus, chloride channels represent important targets for drug discovery.[54] Other methods for detecting chloride are described in Section 21.2.

Premo™ Halide Sensor combines a Yellow Fluorescent Protein (YFP) variant sensitive to halide ions with the efficient and noncytopathic BacMam delivery and expression technology (BacMam Gene Delivery and Expression Technology—Note 11.1). Premo™ Halide Sensor is based on the Venus variant of *Aequorea victoria* Green Fluorescent Protein (GFP), which displays enhanced fluorescence, increased folding, and reduced maturation time when compared with YFP.[55] Additional mutations H148Q and I152L were made within the Venus sequence to increase the sensitivity of the Venus fluorescent protein to changes in local halide concentration, in particular iodide ions.[56] Because chloride channels are also permeable to the iodide ion (I), iodide can be used as a surrogate for chloride. Upon stimulation, a chloride channel or transporter opens and iodide flows down the concentration gradient into the cells, where it quenches the fluorescence of the expressed Premo™ Halide Sensor protein (Figure 16.3.15). The decrease in Premo™ Halide Sensor fluorescence is directly proportional to the ion flux, and therefore the chloride channel or transporter activity. Premo™ Halide Sensor shows an excitation and emission profile similar to YFP (Figure 16.3.16) and can be detected using standard GFP/FITC or YFP filter sets. Halide-sensitive YFP-based constructs in conjunction with iodide quenching have been used in high-throughput screening (HTS) to identify modulators of calcium-activated chloride channels.[57]

Premo™ Halide Sensor (P10229) is prepackaged and ready for immediate use. It contains all components required for cellular delivery and expression, including baculovirus carrying the genetically encoded biosensor, BacMam enhancer and stimulus buffer. Premo™ Halide Sensor has been demonstrated to transduce multiple cell lines including BHK, U2OS, HeLa, CHO, and primary human bronchial epithelial cells (HBEC), providing the flexibility to assay chloride-permeable channels in a wide range of cellular models. To uncouple cell maintenance and preparation from cell screening, BacMam-transduced cells can be divided into aliquots and frozen for later assay. Both stable cell lines and human primary cells can be prepared frozen and "assay-ready" and can be subsequently plated as little as 4 hours prior to screening. Screening can be conducted in complete medium and without any wash steps. Chloride channel assays with Premo™ Halide Sensor are compatible with standard fluorescence HTS platforms.

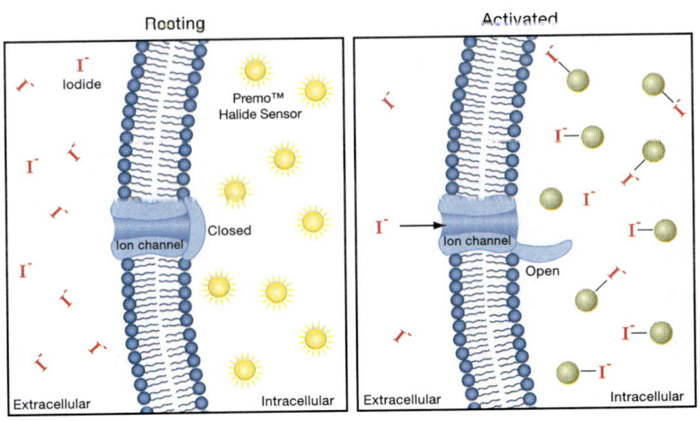

Figure 16.3.15 Principle of Premo™ Halide Sensor Sensor (P10229): Iodide redistribution upon chloride channel activation. Basal fluorescence from Premo™ Halide Sensor is high when chloride channels are closed or blocked. Upon activation (opening) of chloride channels, the iodide ions enter the cell, down its concentration gradient, and quench the fluorescence from Premo™ Halide Sensor.

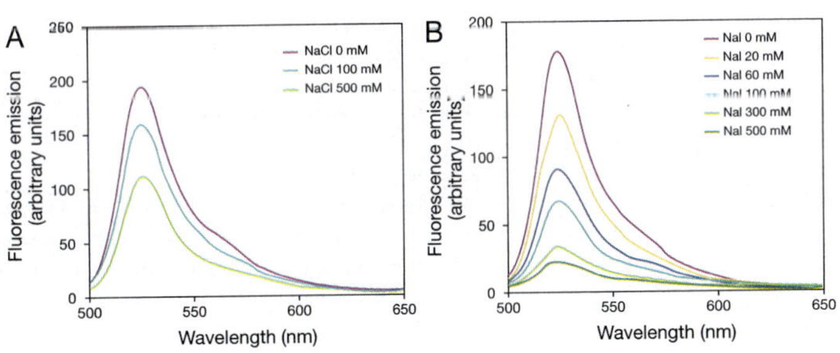

Figure 16.3.16 Quenching of Premo™ Halide Sensor fluorescence by increasing concentrations of iodide and chloride. U2OS cells were transduced with Premo™ Halide Sensor. After 24 hours, cells were trypsinized and lysed by resuspension in sterile distilled water. Fluorescence quenching of the lysate was examined using increasing concentrations of NaCl (**A**) and NaI (**B**). Iodide induces substantially greater quenching of Premo™ Halide Sensor fluorescence than chloride.

REFERENCES

1. J Physiol (2004) 555:251; **2.** Neurosci Lett (2004) 358:75; **3.** Proc Natl Acad Sci U S A (1992) 89:3586; **4.** J Cell Biochem (2007) 100:86; **5.** Biochemistry (2006) 45:8940; **6.** Pharm Res (2003) 20:537; **7.** Biochem Pharmacol (2004) 67:285; **8.** J Histochem Cytochem (2002) 50:731; **9.** Mol Pharmacol (1991) 40:490; **10.** Am J Physiol (1993) 264:C1577; **11.** J Physiol (1999) 515 (Pt 1):109; **12.** Cell Calcium (1997) 22:99; **13.** Proc Natl Acad Sci U S A (2004) 101:10554; **14.** Nature (1997) 388:882; **15.** J Biol Chem (1983) 258:3503; **16.** J Med Chem (1967) 10:66; **17.** J Membr Biol (1988) 105:1; **18.** Biochimie (1988) 70:1285; **19.** Mol Pharmacol (1986) 30:112; **20.** Biochemistry (1984) 23:4481; **21.** J Biol Chem (1984) 259:4313; **22.** Mol Pharmacol (1984) 25:131; **23.** Biochemistry (1977) 16:531; **24.** J Biol Chem (1998) 273:28813; **25.** Cell Biol Int (1994) 18:723; **26.** Physiol Res (1994) 43:33; **27.** Biochemistry (1986) 25:8133; **28.** J Biol Chem (1985) 260:14484; **29.** J Biol Chem (1982) 257:5601; **30.** Biochemistry (1998) 37:6658; **31.** Biophys Chem (1998) 71:245; **32.** Cell Tissue Res (1990) 260:529; **33.** J Cell Biol (1986) 103:1473; **34.** J Biol Chem (1984) 259:11176; **35.** Biochemistry (1980) 19:969; **36.** Biophys J (2008) 94:2654; **37.** Drug Discov Today (2007) 12:396; **38.** Trends Pharmacol Sci (1990) 11:417; **39.** Diabetes (1999) 48:2390; **40.** Pflugers Arch (1997) 434:712; **41.** J Biol Chem (2001) 276:33369; **42.** Anal Biochem (2009) 394:30; **43.** Assay Drug Dev Technol (2008) 6:765; **44.** J Biol Chem (2009) 284:14020; **45.** Am J Physiol (1992) 262:C803; **46.** J Biol Chem (1994) 269:21489; **47.** J Biol Chem (1991) 266:13355; **48.** Am J Physiol (1995) 269:C1073; **49.** Biochemistry (1995) 34:4880; **50.** Biophys J (1994) 66:1726; **51.** Am J Physiol (1993) 264:C1144; **52.** Biochim Biophys Acta (1987) 897:14; **53.** Biochim Biophys Acta (1979) 550:328; **54.** Nat Rev Drug Discov (2009) 8:153; **55.** Nat Biotechnol (2002) 20:87; **56.** FEBS Lett (2001) 499:220; **57.** Mol Pharmacol (2008) 73:758.

DATA TABLE 16.3 PROBES FOR ION CHANNELS AND CARRIERS

Cat. No.	MW	Storage	Soluble	Abs	EC	Em	Solvent	Notes
A1322	788.89	F,D,L	DMSO	362	7500	471	MeOH	
B436	542.67	L	DMSO, EtOH	590	160,000	616	MeOH	1
B7431	769.18	F,D,L	DMSO, EtOH	504	74,000	511	MeOH	
B23461	858.74	F,D,L	DMSO	503	80,000	510	MeOH	
C22803	873.05	F,D	DMSO	<300		none		
D337	498.47	F,DD	H_2O	341	61,000	415	H_2O	2
D338	500.48	F,DD	H_2O	286	41,000	none	MeOH	2
D673	474.32	L	H_2O	352	32,000	none	H_2O	
D7443	686.48	F,D,L,A	DMSO, EtOH	504	83,000	511	MeOH	
E18	704.97	F,DD,L	pH >6, DMF	521	95,000	544	pH 9	2
E3111	336.22	D,L	H_2O, MeOH	378	23,000	423	MeOH	
E34250	915.23	F,D,L	DMSO, H_2O	587	60,000	615	MeOH	
E34251	783.10	F,D,L	DMSO, H_2O	504	76,000	511	MeOH	
S7445	760.57	F,D,L,A	DMSO, EtOH	565	143,000	570	MeOH	

For definitions of the contents of this data table, see "Using *The Molecular Probes® Handbook*" in the introductory pages.

Notes

1. Oxonols may require addition of a base to be soluble.
2. Isothiocyanates are unstable in water and should not be stored in aqueous solution.

PRODUCT LIST 16.3 PROBES FOR ION CHANNELS AND CARRIERS

Cat. No.	Product	Quantity
A1322	9-anthroyl ouabain	5 mg
B10334	BacMam Kir1.1 *for 10 microplates*	1 kit
B10146	BacMam Kir2.1 *for 10 microplates*	1 kit
B10331	BacMam Kv1.1 *for 10 microplates*	1 kit
B10332	BacMam Kv1.3 *for 10 microplates*	1 kit
B10333	BacMam Kv2.1 *for 10 microplates*	1 kit
B10147	BacMam Kv7.2 and Kv7.3 *for 10 microplates*	1 kit
B10341	BacMam Nav1.2 *for 10 microplates*	1 kit
B10335	BacMam Nav1.5 *for 10 microplates*	1 kit
B10019	BacMam-hERG *for 10 microplates*	1 kit
B10033	BacMam-hERG *for 100 microplates*	1 kit
B436	bis-(1,3-dibutylbarbituric acid)pentamethine oxonol (DiBAC$_4$(5))	25 mg
B23461	BODIPY® FL ouabain	100 µg
B7431	BODIPY® FL verapamil, hydrochloride	1 mg
C22803	CEDA, SE (5-(and-6)-carboxyeosin diacetate, succinimidyl ester) *mixed isomers*	5 mg
D337	DIDS (4,4'-diisothiocyanatostilbene-2,2'-disulfonic acid, disodium salt)	100 mg
D338	4,4'-diisothiocyanatodihydrostilbene-2,2'-disulfonic acid, disodium salt (H_2DIDS)	100 mg
D7443	DM-BODIPY® (–)-dihydropyridine *high affinity enantiomer*	25 µg
D673	DNDS (4,4'-dinitrostilbene-2,2'-disulfonic acid, disodium salt)	1 g
E18	eosin-5-isothiocyanate	100 mg
E34251	ER-Tracker™ Green (BODIPY® FL glibenclamide) *for live-cell imaging*	100 µg
E34250	ER-Tracker™ Red (BODIPY® TR glibenclamide) *for live-cell imaging*	100 µg
E3111	5-(*N*-ethyl-*N*-isopropyl)amiloride, hydrochloride	5 mg
F10016	FluxOR™ Potassium Ion Channel Assay *for 10 microplates*	1 kit
F10017	FluxOR™ Potassium Ion Channel Assay *for 100 microplates*	1 kit
P10020	PowerLoad™ concentrate, 100X	5 mL
P36207	Premo™ Cameleon Calcium Sensor *for 10 microplates*	1 kit
P36208	Premo™ Cameleon Calcium Sensor *for 100 microplates*	1 kit
P10229	Premo™ Halide Sensor *for 10 microplates*	1 kit
S7445	ST-BODIPY® (-)-dihydropyridine *high affinity enantiomer*	25 µg

IMPORTANT NOTICE: The products described in this manual are covered by one or more Limited Use Label License(s). Please refer to the Appendix on page 971 and Master Product List on page 975. Products are For Research Use Only. Not intended for any animal or human therapeutic or diagnostic use.
www.invitrogen.com/probes
molecular probes® | invitrogen™ by life technologies™

SEVENTEEN

CHAPTER 17

Probes for Signal Transduction

IMPORTANT NOTICE: The products described in this manual are covered by one or more Limited Use Label License(s). Please refer to the Appendix on page 971 and Master Product List on page 975. Products are For Research Use Only. Not intended for any animal or human therapeutic or diagnostic use.

IMPORTANT NOTICE: The products described in this manual are covered by one or more Limited Use Label License(s). Please refer to the Appendix on page 971 and Master Product List on page 975. Products are For Research Use Only. Not intended for any animal or human therapeutic or diagnostic use.

17.1 Introduction to Signal Transduction

Cells respond to their environment through a complex and interdependent series of signal transduction pathways that frequently begin at the cell membrane. Many cellular receptors are transmembrane proteins with extracellular domains that selectively bind ligands. In response to ligand binding, the receptor's cytoplasmic domain may change conformation and transmit the signal across the membrane, or individual receptors may aggregate and interact with other membrane proteins in order to generate a response. Transmembrane signals trigger a cascade of events in the cell, which can include changes in intracellular Ca^{2+} levels, enzymatic activity and gene expression (Figure 17.1.1).

We offer several important reagents for studying signal transduction mechanisms, including Ca^{2+} regulation and second messenger activities. This chapter focuses on probes for events occurring downstream from the receptor–ligand interaction. These products complement the probes for receptors and ion channels in Chapter 16, as well as the many ion indicators discussed in Chapter 19, Chapter 20 and Chapter 21. Chapter 18 describes our selection of probes for nitric oxide research—including nitric oxide donors, nitric oxide synthase inhibitors and reagents for nitrite detection—as well as for other reactive oxygen species.

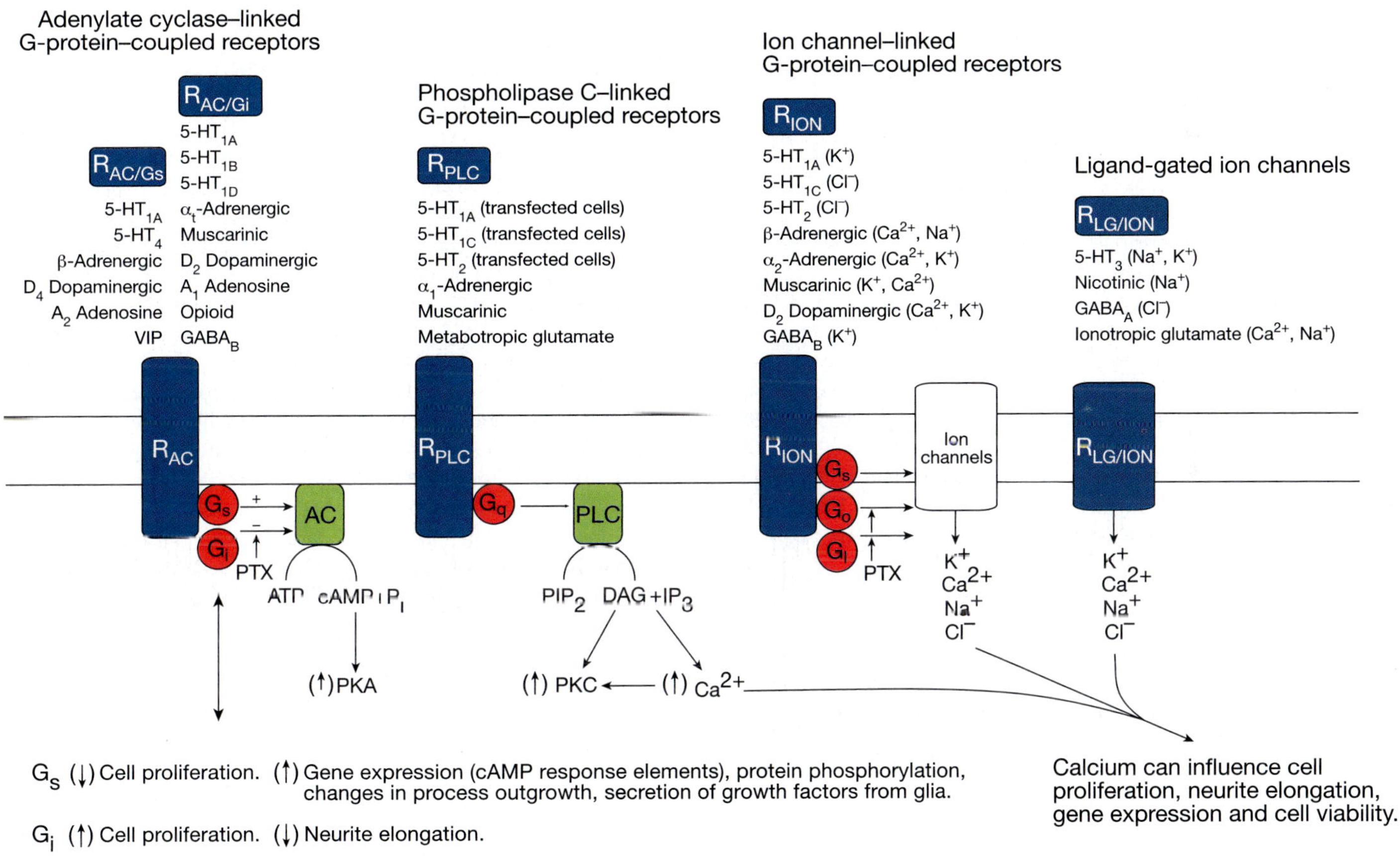

Figure 17.1.1 Neurotransmitter receptors linked to second messengers mediating growth responses in neuronal and nonneuronal cells. Abbreviations: $R_{AC/Gs}$ = Receptors coupled to G-proteins that stimulate adenylate cyclase (AC) activity, leading to cAMP formation and enhanced activity of protein kinase A (PKA). $R_{AC/Gi}$ = Receptors coupled to pertussis toxin (PTX)–sensitive G-proteins that inhibit adenylate cyclase activity. R_{PLC} = Receptors promoting the hydrolysis of phosphatidylinositol 4,5-diphosphate (PIP_2) to inositol 1,4,5-triphosphate (IP_3), which increases intracellular Ca^{2+}, and diacylglycerol (DAG), which activates protein kinase C (PKC). R_{ION} = Receptors indirectly promoting ion fluxes due to coupling to various G-proteins. $R_{LG/ION}$ = Receptors that promote ion fluxes directly because they are structurally linked to ion channels (members of the superfamily of ligand-gated ion channel receptors). Stimulation of proliferation is most often associated with activation of G-proteins negatively coupled to adenylate cyclase (G_i), or positively coupled to phospholipase C (G_q) or to pertussis toxin–sensitive pathways (G_o, G_i). In contrast, activation of neurotransmitter receptors positively coupled to cAMP usually inhibits cell proliferation and causes changes in cell shape indicative of differentiation. Reprinted and modified with permission from J.M. Lauder and Trends Neurosci (1993) 16:233. Learn more about gene specific products for signaling pathways at www.invitrogen.com/handbook/pathways.

IMPORTANT NOTICE: The products described in this manual are covered by one or more Limited Use Label License(s). Please refer to the Appendix on page 971 and Master Product List on page 975. Products are For Research Use Only. Not intended for any animal or human therapeutic or diagnostic use.

17.2 Calcium Regulation

Intracellular Ca^{2+} levels modulate a multitude of vital cellular processes—including gene expression, cell viability, cell proliferation, cell motility, cell shape and volume regulation—thereby playing a key role in regulating cell responses to external signals. These dynamic changes in Ca^{2+} levels are regulated by ligand-gated and G-protein–coupled ion channels in the plasma membrane and by mobilization of Ca^{2+} from intracellular stores. The generation of cytosolic Ca^{2+} spikes and oscillations typically involves the coordinated release and uptake of Ca^{2+} from these stores, mediated by intracellular Ca^{2+} channels and their response to several second messengers such as Ca^{2+} itself, cyclic ADP ribose and inositol triphosphate.[1–3]

This section includes several Molecular Probes® reagents for studying Ca^{2+} regulation in live cells. Fluorescent nucleotides, including analogs of ATP, ADP, AMP, GTP, and GDP, are described in Section 17.3. Our GTP analogs may be particularly useful in the assay of G-protein–coupled receptors. Section 17.4 discusses several selective phosopholipase substrates, as well as labeled ceramide and sphingomyelin probes.

Inositol Triphosphate Pathway

D-*myo*-1,4,5-Inositol Triphosphate and Caged D-*myo*-1,4,5-Inositol Triphosphate

We offer the potassium salt of D-*myo*-inositol 1,4,5-triphosphate (Ins 1,4,5-P_3, I3716) for researchers investigating inositol triphosphate–dependent Ca^{2+} mobilization and signal transduction mechanisms.[1] Cytoplasmic Ins 1,4,5-P_3 is a potent intracellular second messenger that induces Ca^{2+} release from membrane-bound stores in many tissues.

NPE-caged Ins 1,4,5-P_3 can be used to generate rapid and precisely controlled release of Ins 1,4,5-P_3 in intact cells and is widely employed in studies of Ins 1,4,5-P_3–mediated second-messenger pathways.[4–6] Our NPE-caged Ins 1,4,5-P_3 (I23580) is a mixture of the physiologically inert, singly esterified P^4 and P^5 esters (Figure 17.2.1) and does not contain the somewhat physiologically active P^1 ester. NPE-caged Ins 1,4,5-P_3 exhibits essentially no biological activity prior to photolytic release of the biologically active Ins 1,4,5-P_3.

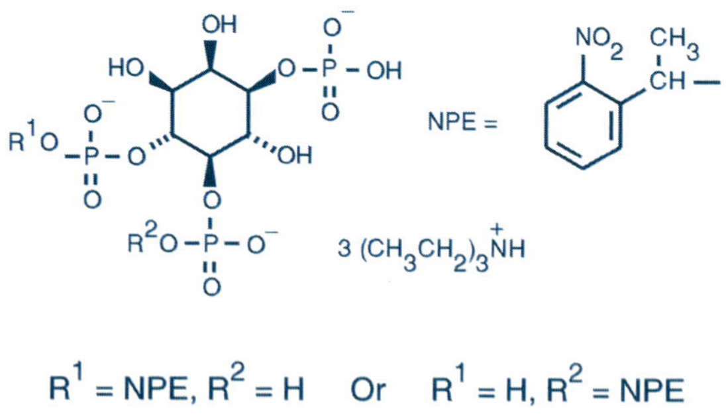

Figure 17.2.1 D-*myo*-inositol 1,4,5-triphosphate, P4(5)-(1-(2-nitrophenyl)ethyl) ester, tris(triethylammonium) salt (NPE-caged Ins 1,4,5-P_3) (I23580).

Fluorescent Heparin

Fluorescein-labeled heparin (H7482) is a useful tool for studying binding of this mucopolysaccharide in cells and tissues.[7,8] In addition to its well-known anticoagulant activity,[9] heparin binds to the Ins 1,4,5-P_3 receptor and inhibits the biological cascade of events mediated by Ins 1,4,5-P_3.[10] Heparin also binds to thrombin [11] and Alzheimer's tau protein,[12] as well as to blood vessel–associated proteins such as laminin and fibronectin.[13] Fluorescence polarization assays using fluorescein-labeled heparin as a tracer provide quantitative assessments of these binding interactions.[14] Fluorescein-labeled heparin has also been used to assess the efficacy of transdermal delivery of heparin by pulsed current iontophoresis as a potential alternative to conventional subcutaneous injections.[15]

Caged Ca^{2+} and Caged Ca^{2+} Chelators

Caged ions and caged chelators can be used to influence the ionic composition of both solutions and cells, particularly for ions such as Ca^{2+} that are present at low concentrations. The properties and uses of caged probes are described in Section 5.3.

NP-EGTA: A Caged Ca^{2+} Reagent

Developed by Ellis-Davies and Kaplan, the photolabile chelator *o*-nitrophenyl EGTA (NP-EGTA) exhibits a high selectivity for Ca^{2+}, a dramatic 12,500-fold *decrease* in affinity for Ca^{2+} upon UV illumination (its K_d increases from 80 nM to >1 mM) and a high photochemical quantum yield [16,17] (~0.2). Furthermore, with a K_d for Mg^{2+} of 9 mM, NP-caged EGTA does not perturb physiological levels of Mg^{2+}. We offer both the potassium salt (N6802) and the acetoxymethyl (AM) ester (N6803) of NP-EGTA. The NP-EGTA salt can be complexed with Ca^{2+} to generate a caged calcium complex that will rapidly deliver Ca^{2+} upon photolysis (Figure 17.2.2). The cell-permeant AM ester of NP-EGTA does not bind Ca^{2+} unless the AM esters are removed. It can potentially serve as a photolabile buffer in cells because, once converted to NP-EGTA by intracellular esterases, it will bind Ca^{2+} with high affinity until photolyzed with UV light. NP-EGTA has been used to measure the calcium buffering capacity of cells.[18]

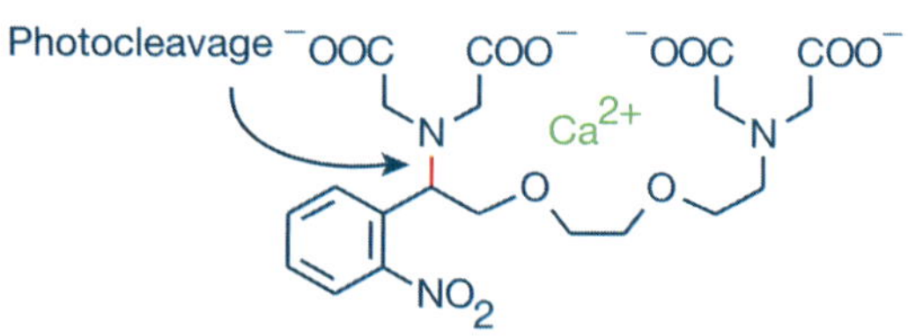

Figure 17.2.2 NP-EGTA (N6802) complexed with Ca^{2+}. Upon illumination, this complex is cleaved to yield free Ca^{2+} and two iminodiacetic acid photoproducts. The affinity of the photoproducts for Ca^{2+} is ~12,500-fold lower than that of NP-EGTA.

DMNP-EDTA: A Caged Ca^{2+} Reagent

The first caged Ca^{2+} reagent described by Ellis-Davies and Kaplan was 1-(4,5-dimethoxy-2-nitrophenyl) EDTA (DMNP-EDTA, D6814), which they named DM-Nitrophen™ [19,20] (now a trademark of Calbiochem-Novabiochem Corp.). Because its structure better resembles that of EDTA than EGTA, we named it as a caged EDTA derivative (Figure 17.2.3). Upon illumination, DMNP-EDTA's K_d for Ca^{2+} increases from 5 nM to 3 mM. Thus, photolysis of DMNP-EDTA complexed with Ca^{2+} results in a pulse of free Ca^{2+}. Furthermore, DMNP-EDTA has significantly higher affinity for Mg^{2+} (K_d = 2.5 µM) than does NP-EGTA [19] (K_d = 9 mM). The photolysis product's K_d for Mg^{2+} is ~3 mM, making DMNP-EDTA an effective caged Mg^{2+} source, in addition to its applications for photolytic Ca^{2+} release.[21,22] Photorelease of Ca^{2+} has been shown to occur in <180 microseconds, with even faster photorelease of Mg^{2+}.[23] Two reviews by Ellis-Davies discuss the uses and limitations of DMNP-EDTA.[24,25]

Diazo-2: A Photoactivatable Ca^{2+} Knockdown Reagent

In contrast to NP-EGTA and DMNP-EDTA, diazo-2 (D3034) is a photoactivatable Ca^{2+} scavenger. Diazo-2 (Figure 17.2.4), which was introduced by Adams, Kao and Tsien,[26,27] is a relatively weak chelator (K_d for Ca^{2+} = 2.2 µM). Following flash photolysis at ~360 nm, however, cytosolic free Ca^{2+} rapidly binds to the diazo-2 photolysis product, which has a high affinity for Ca^{2+} (K_d = 73 nM). Microinjecting a relatively low concentration of fluo-3, fluo-4, or one of the Calcium Green™ or Oregon Green® 488 BAPTA indicators (Section 19.3), along with a known quantity of diazo-2, permits measurement of the extent of depletion of cytosolic Ca^{2+} following photolysis.[27–29] Intracellular loading of NP-EGTA, DMNP-EDTA and diazo-2 is best accomplished by patch pipette infusion with the carboxylate salt form of the caged compound added to the internal pipette solution at 1–10 mM. These reagents are increasingly being applied *in vivo* for controlled intervention in calcium-regulated fundamental processes in neurobiology [30] and developmental biology.[31]

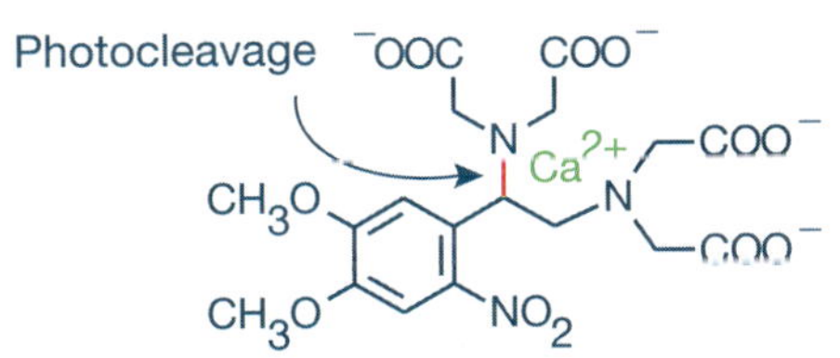

Figure 17.2.3 DMNP-EDTA (D6814) complexed with Ca^{2+}. Upon illumination, this complex is cleaved to yield free Ca^{2+} and two iminodiacetic acid photoproducts. The affinity of the photoproducts for Ca^{2+} is ~600,000-fold lower than that of DMNP-EDTA.

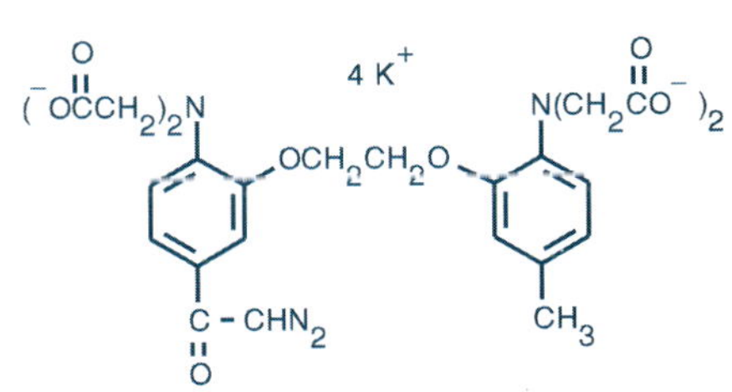

Figure 17.2.4 Diazo-2, tetrapotassium salt (D3034).

Other Probes for Calcium Regulation

Thapsigargin and Fluorescent Thapsigargin

Thapsigargin is a naturally occurring sesquiterpene lactone isolated from the umbelliferous plant *Thapsia garganica*.[32] This tumor promoter releases Ca^{2+} from intracellular stores by specifically inhibiting the sarcoplasmic reticulum Ca^{2+}-ATPase [33,34] (SERCA); it does not directly affect plasma membrane Ca^{2+}-ATPases, Ins 1,4,5-P_3 production or protein kinase C activity.[35,36]

Thapsigargin is available in 1 mg units (T7458) and specially packaged in 20 vials containing 50 µg each (T7459). We have also prepared the green-fluorescent BODIPY® FL thapsigargin (B7487, Figure 17.2.5) and red-fluorescent BODIPY® TR-X thapsigargin (B13800, Figure 17.2.6). BODIPY® FL thapsigargin has proven useful for imaging the intracellular localization of thapsigargin during store-operated calcium entry (SOCE) [37] and for imaging SERCA depletion in injured sensory neurons.[38]

Luminescent Calcium Analog

The trivalent lanthanide terbium (III), which is supplied as its chloride salt (T1247), is a luminescent analog of Ca^{2+} that can be used to study structure–function relationships in Ca^{2+}-binding proteins such as calmodulin, oncomodulin, lactalbumin and ATPases.[39–41] The long-lived luminescence of Tb^{3+} has also been use to probe Ca^{2+}-binding sites of alkaline phosphatase,[42] glutamine synthetase,[43] integrins,[39] protein kinase C [44] and ryanodine-sensitive Ca^{2+} channels.[45] Tb^{3+} reportedly binds most strongly to the I and II sites of calmodulin.[46]

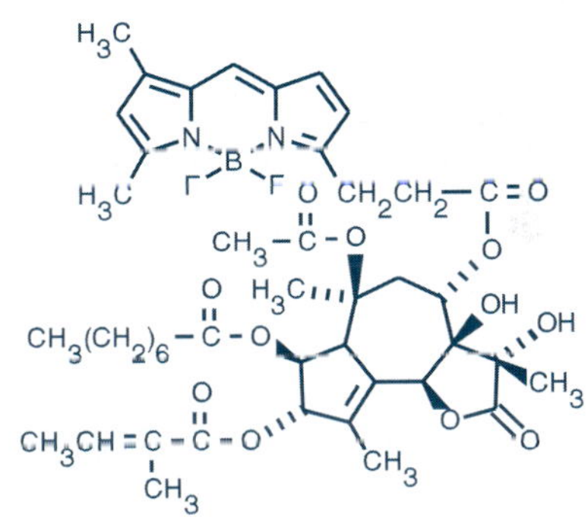

Figure 17.2.5 BODIPY® FL thapsigargin (B7487).

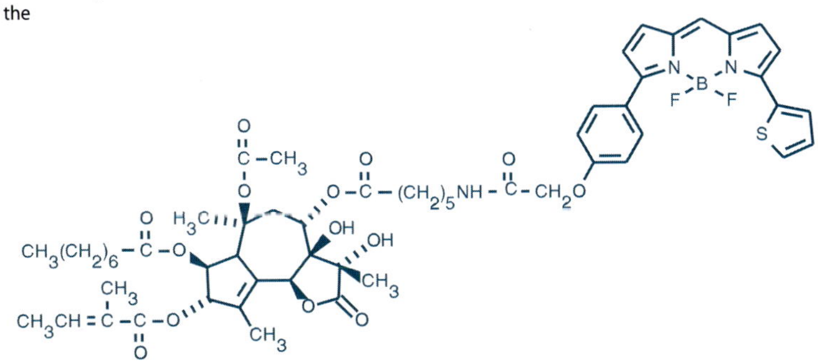

Figure 17.2.6 BODIPY® TR-X thapsigargin (B13800).

REFERENCES

1. Biochim Biophys Acta (2009) 1793:933; **2.** Cell (2007) 131:1047; **3.** Nat Cell Biol (2009) 11:669; **4.** Neuron (2007) 54:611; **5.** J Physiol (1995) 487:343; **6.** Neuron (1995) 15:755; **7.** Biochem Biophys Res Commun (2006) 348:850; **8.** Chem Biol (2004) 11:487; **9.** J Biol Chem (1992) 267:8857; **10.** Biochem J (1994) 302:155; **11.** J Biol Chem (1998) 273:34730; **12.** Biochemistry (2006) 45:6446; **13.** J Biol Chem (1995) 270:18558; **14.** J Biol Chem (2008) 283:19389; **15.** Pharm Res (2006) 23:114; **16.** J Biol Chem (1995) 270:23966; **17.** Proc Natl Acad Sci U S A (1994) 91:187; **18.** Biochem Biophys Res Commun (1998) 250:786; **19.** Proc Natl Acad Sci U S A (1988) 85:6571; **20.** Science (1988) 241:842; **21.** Methods Cell Biol (1994) 40:31; **22.** Neuron (1993) 10:21; **23.** Biochemistry (1992) 31:8856; **24.** Chem Rev (2008) 108:1603; **25.** Nat Methods (2007) 4:619; **26.** Biochim Biophys Acta (1990) 1035:378; **27.** J Am Chem Soc (1989) 111:7957; **28.** Nature (1994) 371:603; **29.** Biophys J (1993) 65:2537; **30.** Science (2009) 325:207; **31.** Dev Growth Differ (2009) 51:617; **32.** Acta Pharm Suec (1978) 15:133; **33.** J Biol Chem (1998) 273:12994; **34.** J Biol Chem (1995) 270:11731; **35.** Proc Natl Acad Sci U S A (1990) 87:2466; **36.** J Biol Chem (1989) 264:12266; **37.** J Biol Chem (2007) 282:12176; **38.** Anesthesiology (2009) 111:393; **39.** Biochemistry (1994) 33:12238; **40.** J Biol Chem (1992) 267:13340; **41.** Photochem Photobiol (1987) 46:1067; **42.** J Photochem Photobiol B (1992) 13:289; **43.** Biochemistry (1991) 30:3417; **44.** J Biol Chem (1988) 263:4223; **45.** J Biol Chem (1994) 269:24864; **46.** Biochem Biophys Res Commun (1986) 138:1243.

DATA TABLE 17.2 CALCIUM REGULATION

Cat. No.	MW	Storage	Soluble	Abs	EC	Em	Solvent	Notes
B7487	854.75	FF,D,L	DMSO	503	85,000	511	MeOH	
B13800	1100.04	FF,D,L	DMSO	589	62,000	616	MeOH	
D3034	710.86	F,D,LL	pH >6	369	18,000	none	pH 7.2	1, 3, 4
D6814	473.39	D,LL	DMSO	348	4200	none	pH 7.2	1, 4, 5
H7482	~18,000	FF,D,L	H_2O	493	ND	514	pH 8	6, 7
I3716	648.64	F,D	H_2O	<250		none		
I23580	872.82	FF,D,LL	H_2O	264	4200	none	H_2O	1, 2, 8
N6802	653.81	FF,D,LL	pH >6	260	3500	none	pH 7.2	1, 2, 4, 9
N6803	789.70	FF,D,LL	DMSO	250	4200	none	MeCN	10, 11
T1247	373.38	D	H_2O	270	4700	545	H_2O	12, 13
T7458	650.76	F,D	DMSO, EtOH	<300		none		
T7459	650.76	F,D	DMSO, EtOH	<300		none		

For definitions of the contents of this data table, see "Using *The Molecular Probes® Handbook*" in the introductory pages.

Notes

1. All photoactivatable probes are sensitive to light. They should be protected from illumination except when photolysis is intended.
2. This compound has weaker visible absorption at >300 nm but no discernible absorption peaks in this region.
3. The Ca^{2+} dissociation constant of diazo-2 is 2200 nM before photolysis and 73 nM after ultraviolet photolysis. The absorption spectrum of the photolysis product is similar to that of BAPTA. (J Am Chem Soc (1989) 111:7957)
4. Abs and EC values determined in Ca^{2+}-free solution (100 mM KCl, 10 mM EGTA, 10 mM MOPS, pH 7.2).
5. K_d (Ca^{2+}) increases from 5 nM to 3 mM after ultraviolet photolysis. K_d values determined in 130 mM KCl, 10 mM HEPES, pH 7.1. (Proc Natl Acad Sci U S A (1988) 85:6571)
6. ND = not determined.
7. This product is a multiply labeled bioconjugate. The number of labels per conjugate is indicated on the vial.
8. Ultraviolet photolysis of I23580 generates I3716.
9. K_d (Ca^{2+}) increases from 80 nM to 1 mM after ultraviolet photolysis. K_d values determined in 100 mM KCl, 40 mM HEPES, pH 7.2. (Proc Natl Acad Sci U S A (1994) 91:187)
10. This product is intrinsically a liquid or an oil at room temperature.
11. N6803 is converted to N6802 via hydrolysis of its acetoxymethyl ester (AM) groups.
12. Absorption and luminescence of T1247 are extremely weak unless it is chelated. Data are for dipicolinic acid (DPA) chelate. The luminescence spectrum has secondary peak at 490 nm.
13. MW is for the hydrated form of this product.

PRODUCT LIST 17.2 CALCIUM REGULATION

Cat. No.	Product	Quantity
A7621	8-amino-cyclic adenosine 5′-diphosphate ribose (8-amino-cADP-ribose)	10 µg
B7487	BODIPY® FL thapsigargin	100 µg
B13800	BODIPY® TR-X thapsigargin	5 µg
C7074	cyclic adenosine 5′-diphosphate ribose, 1-(1-(2-nitrophenyl)ethyl) ester (NPE-caged cADP-ribose) *mixed isomers*	50 µg
D3034	diazo-2, tetrapotassium salt *cell impermeant*	1 mg
D6814	1-(4,5-dimethoxy-2-nitrophenyl)-1,2-diaminoethane-*N,N,N′,N′*-tetraacetic acid (DMNP-EDTA) *cell impermeant*	5 mg
H7482	heparin, fluorescein conjugate	1 mg
I3716	D-*myo*-inositol 1,4,5-triphosphate, hexapotassium salt (Ins 1,4,5-P_3)	1 mg
I23580	D-*myo*-inositol 1,4,5-triphosphate, $P_{4(5)}$-(1-(2-nitrophenyl)ethyl) ester, tris(triethylammonium) salt (NPE-caged Ins 1,4,5-P_3)	25 µg
N6803	*o*-nitrophenyl EGTA, AM (NP-EGTA, AM) *cell permeant* *special packaging*	20 x 50 µg
N6802	*o*-nitrophenyl EGTA, tetrapotassium salt (NP-EGTA) *cell impermeant*	1 mg
T1247	terbium(III) chloride, hexahydrate	1 g
T7458	thapsigargin	1 mg
T7459	thapsigargin *special packaging*	20 x 50 µg

IMPORTANT NOTICE: The products described in this manual are covered by one or more Limited Use Label License(s). Please refer to the Appendix on page 971 and Master Product List on page 975. Products are For Research Use Only. Not intended for any animal or human therapeutic or diagnostic use.

www.invitrogen.com/probes

molecular probes® | invitrogen™ by life technologies™

17.3 Probes for Protein Kinases, Protein Phosphatases and Nucleotide-Binding Proteins

The cascade of cellular events in response to an internal signal or environmental stimulus requires a diversity of molecular participants, ranging from ions to enzymes. Signal transduction pathways frequently activate specific protein kinases, leading to the phosphorylation of particular cellular proteins and subsequent initiation of a multitude of cellular responses. Binding and hydrolysis of nucleotides plays a major role in these activities, and our nucleotide analogs and assays for phosphate-producing enzymes are important tools for signal transduction research and high-throughput screening of compounds that affect signal transduction.

We offer a selection of native and modified biomolecules to aid the researcher in dissecting this highly complex branch of the signal transduction process. In addition to the probes below, we have developed the P_iPer™ and EnzChek® assay kits for quantitation of inorganic phosphate and pyrophosphate that are extremely useful for following hydrolysis of nucleotides by various enzymes and of phosphate esters by protein phosphatases. These kits and other kits to measure ATP by chemiluminescence and protein phosphatase activity are described in Section 10.3.

Protein Kinase Probes and Assays

Protein kinases are critical players in signal transduction pathways. The fluorometric assay of kinases, however, is not straightforward because ATP-dependent phosphorylation of a fluorescent peptide substrate does not directly lead to appreciable changes in the fluorescence of the product.[1,2] We provide an extensive range of assays for protein kinases that utilize a variety of strategies to detect phosphorylation of peptide and protein substrates (Table 17.1).

Antibody Beacon™ Tyrosine Kinase Assay Kit

The Antibody Beacon™ Tyrosine Kinase Assay Kit (A35725) provides a homogeneous solution assay for measuring the activity of tyrosine kinases and the effectiveness of potential inhibitors and modulators.[3] The key to this tyrosine kinase assay is a small-molecule tracer ligand labeled with our bright green-fluorescent Oregon Green® 488 dye. When an anti-phosphotyrosine antibody binds this tracer ligand to form the Antibody Beacon™ detection complex, the fluorescence of the Oregon

Table 17.1 Invitrogen kinase assay platforms.

Assay	Principle	References
Adapta® Universal Kinase Assay	In the absence of an inhibitor, ADP formed by a kinase reaction will displace an Alexa Fluor® 647 dye–labeled ADPtracer from an Eu^{3+}-labeled anti-ADP antibody, resulting in a decrease in the TR-FRET* signal. In the presence of an inhibitor, the amount of ADP formed by the kinase reaction is reduced, and the resulting intact antibody–tracer interaction produces a high TR-FRET signal.	
Antibody Beacon™ Tyrosine Kinase Assay	Peptide substrate phosphorylation is detected via competitive displacement of an Oregon Green® 488 dye–labeled ligand from a phosphospecific antibody.	1
LanthaScreen® Kinase Activity Assays	A terbium (Tb^{3+})– or europium (Eu^{3+})–labeled phosphospecific antibody binds the phosphorylated fluorescein- or Alexa Fluor® 647 dye–labeled peptide substrate, resulting in an increase in the TR-FRET signal.	2, 3
LanthaScreen® Eu Kinase Binding Assay	Binding of an Alexa Fluor® 647 tracer to a kinase is detected by addition of a Eu^{3+}-labeled anti–epitope tag antibody. Binding of the tracer and antibody to a kinase results in a high FRET signal, whereas displacement of the tracer by a kinase inhibitor results in a loss of FRET signal.	4
LanthaScreen® Cellular Assays	Detection of phosphorylation or other protein modification event is measured on a TR-FRET–compatible plate reader. Little or no TR-FRET is observed with unstimulated or inhibited cells, whereas stimulated cell samples display high TR-FRET.	5, 6
NDP Sensor Protein	This fluorescent ADP/ATP biosensor consists of a recombinant bacterial nucleoside diphosphate kinase site-specifically labeled with an environment-sensitive coumarin dye.	7
Omnia® Kinase Assay	Fluorescence enhancement of N- or C-terminal 8-hydroxyquinoline fluorophore (Sox) upon chelation of Mg^{2+} is coupled to phosphorylation of a peptide substrate at an adjacent Ser, Thr or Tyr residue.	8, 9
Z´-LYTE® Kinase Assay	Phosphorylation-dependent protease susceptibility of a double-labeled peptide substrate is detected using FRET.	10
CellSensor® Cell Lines	CellSensor® assays measure pathway-driven activation of transcription factors using GeneBLAzer® β-lactamase reporter technology. Minimal amounts of β-lactamase are expressed in untreated cells or cells treated with a pathway-specific inhibitor. Stimulation of the pathway with a ligand or with a constitutively active mutation in a pathway component leads to activation of downstream transcription factor(s), resulting in β-lactamase reporter gene expression. Cells are loaded with a cell-permeable β-lactamase substrate, and β-lactamase reporter activity is measured on a fluorescence plate reader.	11–13

1. Free Radic Biol Med (2009) 47:983; **2.** J Biomol Screen (2009) 14:121; **3.** Anal Biochem (2006) 356:108; **4.** J Biomol Screen (2009) 14:924; **5.** J Biomol Screen (2009) 14:121; **6.** Anal Biochem (2008) 372:189; **7.** Biochemistry (2001) 40:5087; **8.** Anal Biochem (2006) 352:198; **9.** J Am Chem Soc (2003) 125:14248; **10.** Assay Drug Dev Technol (2002) 1:9; **11.** Current Chemical Genomics (2009) 3:1; **12.** Mol Biosyst (2009) 5:1039; **13.** Mol Cancer (2009) 8:117.

* TR-FRET = Time-resolved fluorescence resonance energy transfer. For further information on these assay technologies and other kinase biology products and services, visit www.invitrogen.com/handbook/kinase.

IMPORTANT NOTICE: The products described in this manual are covered by one or more Limited Use Label License(s). Please refer to the Appendix on page 971 and Master Product List on page 975. Products are For Research Use Only. Not intended for any animal or human therapeutic or diagnostic use.

Green® 488 dye is efficiently quenched. In the presence of a phosphotyrosine-containing peptide, however, this Antibody Beacon™ detection complex is rapidly disrupted, releasing the tracer ligand and relieving its antibody-induced quenching (Figure 17.3.1). Upon its displacement by a phosphotyrosine residue, the Oregon Green® 488 dye–labeled tracer ligand exhibits an approximately 4-fold fluorescence enhancement, enabling the detection of as little as 50 nM phosphotyrosine-containing peptide with excellent signal-to-background discrimination. Key benefits of the Antibody Beacon™ Tyrosine Kinase Assay Kit include:

- Real-time measurements. Unlike many other commercially available tyrosine kinase assays, the Antibody Beacon™ Tyrosine Kinase Assay Kit permits real-time monitoring of kinase activity (Figure 17.3.2). Not only is the Antibody Beacon™ detection complex rapidly dissociated in the presence of phosphotyrosine residues, but the assay components have been designed to be simultaneously combined, reducing any delay in the measurements.
- Simple detection protocol. Tyrosine kinase activity is measured by a simple increase in fluorescence intensity; no special equipment, additional reagents, or extra steps are required. This assay is readily compatible with any fluorescence microplate reader.
- Use of natural substrates. The Antibody Beacon™ tyrosine kinase assay utilizes unlabeled peptide or protein substrates, is compatible wth substrates that are pre-phosphorylated at serine or threonine (but *not* at tyrosine) residues and is applicable to the assay of a wide variety of kinases.

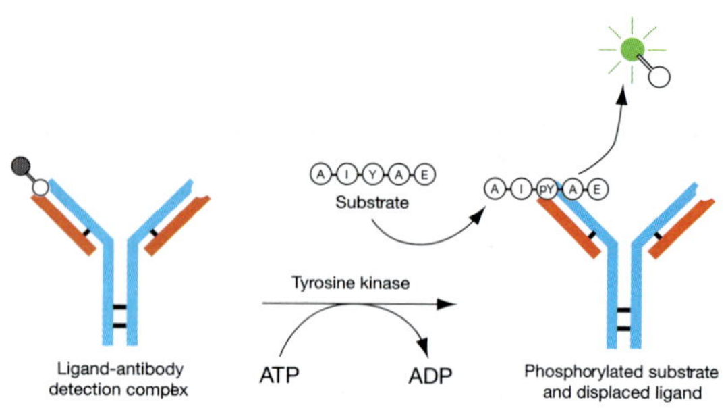

Figure 17.3.1 Reaction scheme for the tyrosine kinase assay used in the Antibody Beacon™ Tyrosine Kinase Assay Kit (A35725). The unlabeled natural substrate (AIYAE) is phosphorylated by the tyrosine kinase to AIY(P)AE, which displaces the quenched Oregon Green® 488 dye–labeled peptide from the anti-phosphotyrosine antibody, resulting in a large increase in its fluorescence that is proportional to the amount of AIY(P)AE formed in the reaction.

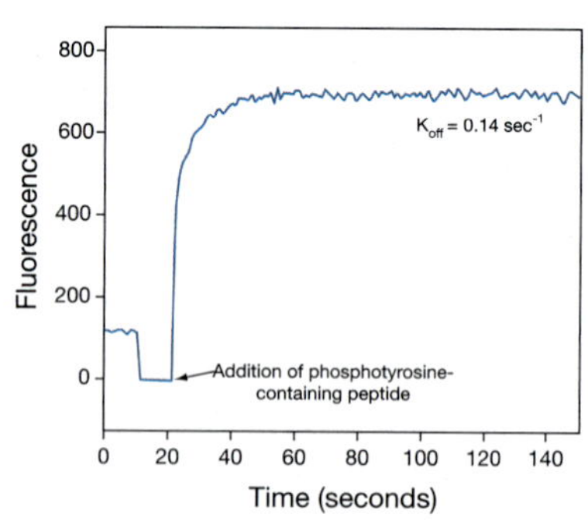

Figure 17.3.2 Real-time detection capability of the Antibody Beacon™ Tyrosine Kinase Assay Kit (A35725). Fluorescence of the Antibody Beacon™ detection complex in tyrosine kinase assay buffer was monitored over time. After ~15 seconds, an excess of phosphotyrosine-containing peptide was added to the Antibody Beacon™ detection complex and the off-rate was calculated.

- Compatibility. The anti-phosphotyrosine antibody provided in the Antibody Beacon™ Tyrosine Kinase Assay Kit is specific for phosphotyrosine residues; assay components such as ATP (up to 1 mM) and reducing agents such as dithiothreitol (DTT, up to 1 mM) do not interfere with this assay. This anti-phosphotyrosine antibody was selected from among several clones to produce the greatest fluorescence enhancement by the kinase-phosphorylated product.
- Reliability. This tyrosine kinase assay has a broad signal window,[4] indicated by a Z′ factor of >0.85.

The Antibody Beacon™ Tyrosine Kinase Assay Kit comes with all the reagents needed to perform this assay, including:

- Oregon Green® 488 dye–labeled tracer ligand
- Anti-phosphotyrosine antibody
- Concentrated tyrosine kinase reaction buffer
- Two generic tyrosine kinase substrate solutions: a poly(Glu:Tyr) solution and a poly(Glu:Ala:Tyr) solution
- Dithiothreitol (DTT)
- Adenosine triphosphate (ATP)
- Phosphotyrosine-containing peptide, phospho-pp60 c-src (521–533), for use as a reference
- Detailed protocols

Each kit provides sufficient reagents to perform ~400 assays using a 50 µL assay volume in a fluorescence microplate reader.

Fluorescent Polymyxin B Analogs

Polymyxin B is a cyclic polycationic peptide antibiotic (Figure 17.3.3) that binds to lipopolysaccharides and anionic lipids.[5] Polymyxin B is also a selective inhibitor of protein kinase C, with an IC_{50} of ~35 µM,[6–9] as well as a potent inhibitor of calmodulin, with an IC_{50} of 80 nM in the presence of 500 µM Ca^{2+}.[10] Our fluorescent polymyxin B analogs include those of the green-fluorescent BODIPY® FL[11] and Oregon Green® 514 fluorophores (P13235, P13236), as well as the ultraviolet light–excitable dansyl polymyxin[5] (P13238).

Hypericin

Hypericin (H7476, Figure 17.3.4), an anthraquinone derivative isolated from plants of the genus *Hypericum*,[12,13] is a potent, selective inhibitor of PKC (IC_{50} = 1.7 µg/mL = 3.4 µM) useful for probing and manipulating PKC in live cells.[14] Hypericin has a variety of pharmacological properties, from antibacterial and antineoplastic activities to antiviral activities[15–18] and induction of apoptosis.[19] Hypericin is also a potent photosensitizer, with a quantum yield of 0.75 for the generation of singlet oxygen.[20]

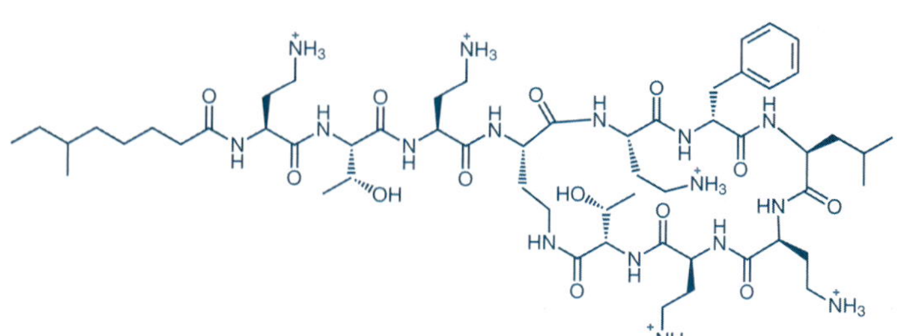

Figure 17.3.3 Polymyxin B.

IMPORTANT NOTICE: The products described in this manual are covered by one or more Limited Use Label License(s). Please refer to the Appendix on page 971 and Master Product List on page 975. Products are For Research Use Only. Not intended for any animal or human therapeutic or diagnostic use.

Protein Phosphatase Assay Kits

RediPlate™ 96 EnzChek® Tyrosine Phosphatase Assay Kits

Protein tyrosine phosphatases (PTP) represent a large family of enzymes that play a very important role in intra- and intercellular signaling. PTPs work antagonistically with protein tyrosine kinases to regulate signal transduction pathways in response to a variety of signals, including hormones and mitogens.[3] Our RediPlate™ 96 EnzChek® Tyrosine Phosphatase Assay Kit (R22067) provides researchers with a sensitive and convenient method to monitor PTP and screen PTP inhibitors in a variety of research areas.[21–24]

The EnzChek® tyrosine phosphatase assay is based on 6,8-difluoro-4-methylumbelliferyl phosphate[25] (DiFMUP, D6567, D22065; Section 10.3). Unlike other end-point tyrosine phosphatase assay kits, the EnzChek® tyrosine phosphatase assay is continuous, allowing researchers to easily measure fluorescence at various time points in order to follow the kinetics of the reaction. Furthermore, the assay is not affected by free phosphate and is compatible with most nonionic detergents, resulting in minimal sample processing before analysis. Most importantly, each assay well contains inhibitors to help ensure that the assay is selective for tyrosine phosphatases; other phosphatases, including serine/threonine phosphatases, will not hydrolyze DiFMUP under our assay conditions (Figure 17.3.5). Unlike phosphopeptide-based assays, this DiFMUP-based assay can be used to monitor a variety of tyrosine phosphatases, including PTP-1B and CD-45 (Figure 17.3.5). Tyrosine phosphatase inhibitors can be evaluated quantitatively in the assay for their effect on tyrosine phosphatase activity.

Each RediPlate™ 96 EnzChek® Tyrosine Phosphatase Assay Kit (R22067) includes:

- One RediPlate™ 96 EnzChek® tyrosine phosphatase assay 96-well microplate
- Reaction buffer
- Detailed assay protocols

RediPlate™ 96 EnzChek® Serine/Threonine Phosphatase Assay Kit

The majority of protein phosphorylation occurs on serine and threonine residues, with <0.01–0.05% on tyrosine residues. Serine/threonine phosphatases represent a large family of enzymes that have been implicated in the regulation of metabolism, transcription, translation, differentiation, cell cycle, cytoskeletal dynamics, oncogenesis and signal transduction. The RediPlate™ 96 EnzChek® Serine/Threonine Phosphatase Assay Kit (R33700) provides a fast, simple and direct fluorescence-based assay for detecting serine/threonine phosphatases and their corresponding modulators and inhibitors[25] (Figure 17.3.6).

As with the RediPlate™ 96 EnzChek® Tyrosine Phosphatase Kit, the substrate incorporated in the RediPlate™ 96 EnzChek® Serine/Threonine Phosphatase Assay Kit is DiFMUP. Inhibitors are included in each assay well to help ensure that the assay is selective for serine/threonine phosphatases; under the prescribed assay conditions, other phosphatases, including tyrosine phosphatases, do not significantly react with the substrate (Figure 17.3.7). Furthermore, unlike phosphopeptide-based assays, this DiFMUP-based assay can be used to monitor a variety of serine/threonine phosphatases including PP-1, PP-2A and PP-2B (Figure 17.3.7). Serine/threonine phosphatase inhibitors can be evaluated quantitatively in the assay for their effect on serine/

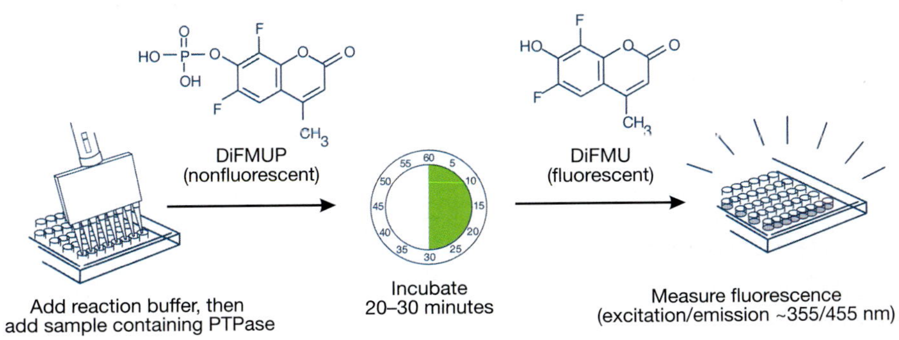

Figure 17.3.6 Schematic diagram of the method used in the RediPlate™ EnzChek® Phosphatase Assay Kits (R22067, R33700).

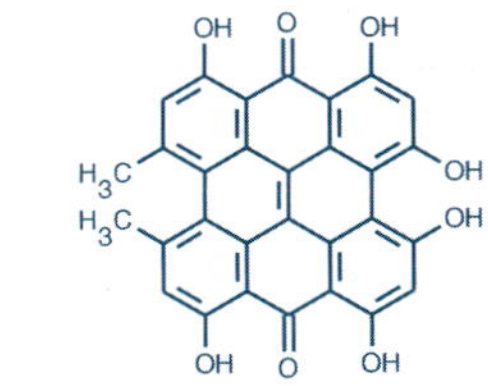

Figure 17.3.4 Hypericin (H7476).

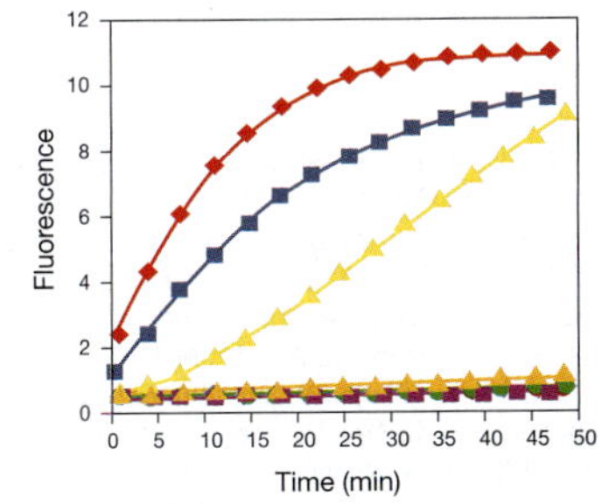

Symbol	Enzyme (Class)	Enzyme Units*
◆	CD-45 (tyrosine phosphatase)	1 U/mL
■	PTP-1B (tyrosine phosphatase)	1 mU/mL
▲	PTPase (tyrosine phosphatase)	1 U/mL
▲	Acid phosphatase	1 U/mL
●	Alkaline phosphatase	1 U/mL
●	PP2A (ser/thr phosphatase)†	1 U/mL
■	PP1 (ser/thr phosphatase)†	1 U/mL
◆	PP-2B (ser/thr phosphatase)†	500 U/mL

* Enzyme unit (U) definitions are standard definitions for each enzyme.
† Serine theonine phosphatase.

Figure 17.3.5 Specificity of the RediPlate™ 96 EnzChek® Tyrosine Phosphatase Assay Kit (R22067). The phosphatases listed in the tables were applied to a RediPlate™ 96 EnzChek® tyrosine phosphatase assay microplate. At the indicated time points, the fluorescence was measured in a fluorescence microplate reader using excitation at 355 ± 20 nm and emission at 460 ± 12.5 nm.

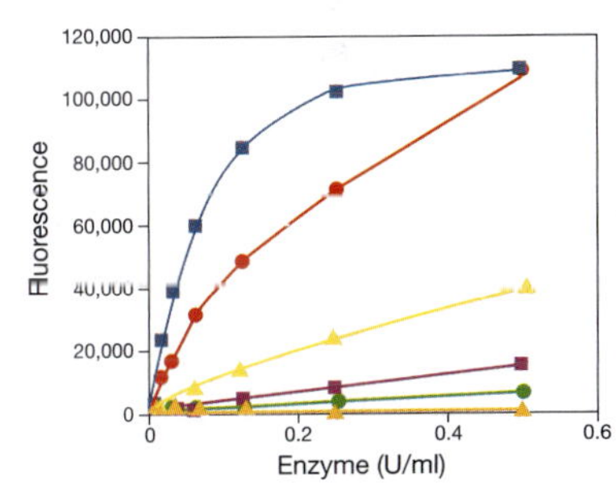

Symbol	Enzyme (Class)
■	PP-2A (Ser/Thr phosphatase)
●	PP-1 (Ser/Thr phosphatase)
▲	PP-2B (Ser/Thr phosphatase)
■	Alkaline phosphatase
●	Acid phosphatase
▲	LAR (tyrosine phosphatase)

Figure 17.3.7 Specificity of the RediPlate™ 96 EnzChek® Serine/Threonine Phosphatase Assay Kit (R33700) for serine/threonine phosphatases. The phosphatases listed in the tables were applied at the indicated concentrations to a RediPlate™ 96 EnzChek® serine/threonine phosphatase assay microplate. Reactions were incubated at 37°C. After 1 hour, fluorescence was measured in a fluorescence microplate reader using excitation at 355 ± 20 nm and emission at 460 ± 12.5 nm.

IMPORTANT NOTICE: The products described in this manual are covered by one or more Limited Use Label License(s). Please refer to the Appendix on page 971 and Master Product List on page 975. Products are For Research Use Only. Not intended for any animal or human therapeutic or diagnostic use.

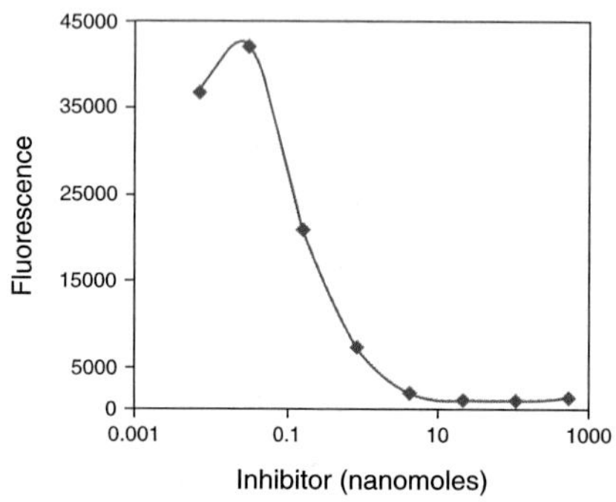

Figure 17.3.8 Detection of PP-2A inhibition by okadaic acid using the RediPlate™ 96 EnzChek® Serine/Threonine Phosphatase Assay Kit (R33700). Each reaction contained 50 µM DiFMUP, 10 mU/mL PP-2A and the indicated concentration (log scale) of okadaic acid in reaction buffer containing 50 mM Tris-HCl, 0.1 mM $CaCl_2$, 1 mM $NiCl_2$, 125 µg/mL bovine serum albumin (BSA) and 0.05% Tween® 20. Reactions were incubated at 37°C. After 30 minutes, fluorescence was measured in a fluorescence microplate reader using excitation at 355 ± 20 nm and emission at 460 ± 12.5 nm.

Figure 17.3.9 A RediPlate™ 96 microplate.

threonine phosphatase activity (Figure 17.3.8). Additional advantages of this RediPlate™ assay include compatibility with nonionic detergents and insensitivity to free phosphate, minimizing sample processing before analysis.

Each RediPlate™ 96 EnzChek® Serine/Threonine Phosphatase Assay Kit includes:

- One RediPlate™ 96 EnzChek® serine/threonine phosphatase assay 96-well microplate
- Concentrated reaction buffer
- $NiCl_2$
- $MnCl_2$
- Dithiothreitol
- Detailed assay protocols

To ensure the integrity of the predispensed reagents, the 96-well microplate provided in both RediPlate™ Protein Phosphatase Assay Kits is packaged in a resealable foil pouch and consists of twelve removable strips, each with eight wells (Figure 17.3.9). Eleven of the strips (88 wells) are preloaded with the fluorogenic substrate DiFMUP; the remaining strip, marked with black tabs, contains a dilution series of the DiFMU reference standard for generating a standard curve.

Pro-Q® Diamond Phosphoprotein/Phosphopeptide Microarray Stain Kit

The Pro-Q® Diamond Phosphoprotein/Phosphopeptide Microarray Stain Kit (P33706) provides a method for selectively staining phosphoproteins or phosphopeptides on microarrays without the use of antibodies or radioactivity. This kit permits direct detection of phosphate groups

NOTE 17.1

G-Proteins and GTP Analogs for Binding Studies

We prepare a wide variety of nucleotide analogs for protein-binding studies; their chemical and spectral properties are described in Section 17.3. These include various fluorescent, photoaffinity and caged versions of adenosine and guanosine triphosphates, diphosphates and cyclic monophosphates. The GTP analogs are among the most important probes for the study of G-proteins and G protein–coupled receptors (GPCR). Heterotrimeric guanine nucleotide–binding regulatory proteins transmit a variety of receptor signals to modulate diverse cellular responses,[1,2] including apoptosis.[3]

G-proteins are composed of α-, β- and γ-subunits. Upon receptor stimulus, the α-subunit of the heterotrimeric G-proteins exchanges GDP for GTP and dissociates from the β-γ-subunit complex. The GTP-bound G protein will interact with various second-messenger systems, either inhibiting (G_i) or stimulating (G_s) their activity. Stimulatory G-proteins are permanently activated by cholera toxin, inhibitory G-proteins by pertussis toxin. The α-subunit has a slow intrinsic rate of GTP hydrolysis, and once the GTP is hydrolyzed it reassociates with the β-γ-subunit complex. The GTP hydrolysis by G-proteins is regulated by interactions with GTPase-activating proteins, or GAPs. There is a large family of GAPs for G-proteins known as regulators of G protein signaling or, RGS proteins.[4] G-proteins are turned off when the α-subunit hydrolyzes the GTP, either spontaneously or upon interaction with a GTPase-activating protein, permitting the heterotrimeric α-β-γ-complex to reassociate.

The GAPs are a diverse group of monomeric GTPases, including ARF, Ran, Ras, Rab, Rac, Rho and Sar, which play an important part in regulating many intracellular processes, such as cytoskeletal organization and secretion. There is less diversity among the β- and γ-subunits, but they may have direct activating effects in their own right. Most β- and γ-subunits are posttranslationally modified by myristoylation or isoprenylation, which may alter their association with membranes.

Our fluorescent GTP analogs include:

- Blue-fluorescent MANT-GTP (M12415) and nonhydrolyzable MANT-GMPPNP (M22353)
- Green-fluorescent BODIPY® FL guanosine 5′-triphosphate (BODIPY® FL GTP, G12411)
- Red-fluorescent BODIPY® TR guanosine 5′-triphosphate (BODIPY® TR GTP, G22351)
- Nonhydrolyzable green-fluorescent BODIPY® FL GTP-γ-*S* thioester (G22183)

1. Annu Rev Biochem (2008) 77:1; **2.** Physiol Rev (1999) 79:1373; **3.** J Biol Chem (2000) 275:20726; **4.** J Biol Chem (1998) 273:1269.

IMPORTANT NOTICE: The products described in this manual are covered by one or more Limited Use Label License(s). Please refer to the Appendix on page 971 and Master Product List on page 975. Products are For Research Use Only. Not intended for any animal or human therapeutic or diagnostic use.
www.invitrogen.com/probes

attached to tyrosine, serine or threonine residues in a microarray environment and has been optimized for microarrays with acrylamide gel surfaces. Each Pro-Q® Diamond Phosphoprotein/Phosphopeptide Microarray Stain Kit provides:

- Pro-Q® Diamond phosphoprotein/phosphopeptide microarray stain
- Pro-Q® Diamond microarray destain solution
- Microarray staining gasket with seal tabs, 10 chambers
- Slide holder tube, 20 tubes
- Detailed protocols

The Pro-Q® Diamond Phosphoprotein/Phosphopeptide Microarray Stain Kit is ideal for identifying kinase targets in signal transduction pathways and for phosphoproteomics studies.[26]

Adenylate Cyclase Assays

3′,5′-Cyclic AMP (cAMP) is an important second messenger in many signal transduction pathways, linking activation of cell-surface membrane receptors to intracellular responses, and, ultimately, to changes in gene expression. cAMP is synthesized by plasma membrane–bound adenylate cyclase, which is coupled to transmembrane receptors for hormones, neurotransmitters and other signaling molecules by heterotrimeric G-proteins. Upon ligand binding, the intracellular receptor domain of a G-protein–coupled receptor (GPCR) interacts with a G-protein, which then dissociates and activates adenylate cyclase, resulting in an increase in the concentration of intracellular cAMP. Subsequently, cAMP activates cAMP-dependent protein kinases (protein kinase A), which phosphorylate specific substrate proteins, including enzymes, structural proteins, transcription factors and ion channels.

Adenylate Cyclase Probe: BODIPY® FL Forskolin

Forskolin, isolated from *Coleus forskohlii*, is a potent activator of adenylate cyclase, the enzyme that catalyzes the formation of cAMP from ATP. Green-fluorescent BODIPY® FL forskolin (B7469, Figure 17.3.10) has been used to visualize adenylyl cyclase internalization and subcellular distribution,[27] as well as for the pharmacological characterization of adenylyl cyclase catalytic subunits.[28]

cAMP Chemiluminescent Immunoassay Kit

The cAMP Chemiluminescent Immunoassay Kit enables ultrasensitive determination of 3′,5′-cyclic AMP (cAMP) levels in cell lysates, providing the highest sensitivity of any commercially available cAMP assay. As few as 60 femtomoles of cAMP can be detected. Furthermore, this assay has a wide dynamic range, detecting from 0.06 to 6000 picomoles without the need for sample dilution or manipulations such as acetylation. This extensive dynamic range is especially important in cell-based assays designed to measure G_s- or G_i-coupled agonist stimulation or inhibition. Intra-assay precision for duplicate samples is typically 5% or less.

This competitive immunoassay is formatted with maximum flexibility to permit either manual assay or automated high-throughput screening. The cAMP immunoassay is based on the highly sensitive CSPD® alkaline phosphate substrate, a chemiluminescent 1,2-dioxetane, with Sapphire-II™ luminescence enhancer. The ready-to-use substrate/enhancer reagent generates sustained glow light emission that is measured 30 minutes after addition. Once the substrate/enhancer reaches the glow signal, the plate can be read for hours with little or no degradation of the signal, facilitating screening protocols in which several plates are compared to each other. In addition, the assay exhibits exceptionally low cross-reactivity with other adenosine-containing or cyclic nucleotides.

The cAMP Chemiluminescent Immunoassay Kit (2-plate size, C10557; 10-plate size, C10558) is designed for the rapid and sensitive quantitation of cAMP in extracts prepared from mammalian cells cultured in microwell plates. Each kit provides all required reagents, including:

- Alkaline phosphate conjugate of cAMP
- Anti-cAMP antibody
- cAMP standard
- CSPD® substrate and Sapphire™-II luminescence enhancer
- Assay and lysis buffer
- Conjugate dilution buffer
- Wash buffer
- Precoated microplates
- Detailed protocols

The cAMP Chemiluminescent Immunoassay Kit is designed for quantitating cellular cAMP for functional assays of receptor activation. It has been used with established cell lines for functional measurements with endogenous receptors,[29–32] with cell lines containing exogenously expressed ligand receptors,[33,34] with primary cells[35,36] and with tissues.[37] It has also been used for receptor characterization,[38] orphan receptor ligand identification[39] and the characterization of novel chimeric receptors.[40] In addition, this assay can be used for high-throughput screening assays[41] of compounds that stimulate or interfere with these signal transduction pathways.

Nucleotide Analogs

Nucleotide analogs that serve as substrates or inhibitors of enzymes, as well as nucleotide derivatives that selectively bind to regulatory sites of nucleotide-binding proteins, have been used as structural and mechanistic probes for isolated proteins, reconstituted membrane-bound enzymes, organelles such as mitochondria, and tissues such as skinned muscle fibers.[42] More recently, however, these analogs have also been employed to study the effects of nucleotides on signal transduction and to screen for compounds that may affect signal transduction, such as G protein inhibitors and activators (G-Proteins and GTP Analogs for Binding Studies—Note 17.1).

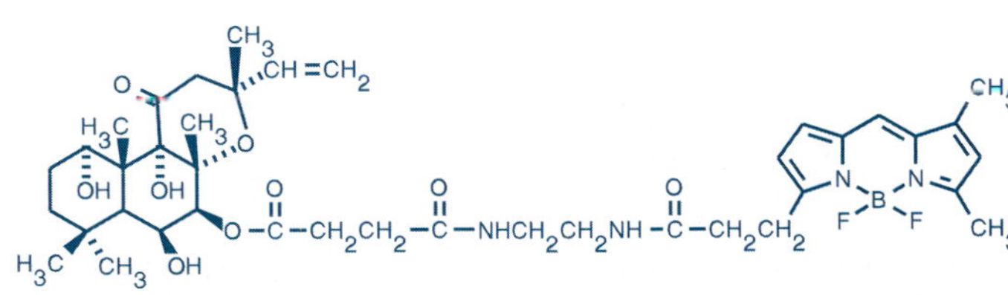

Figure 17.3.10 BODIPY® FL forskolin (B7469).

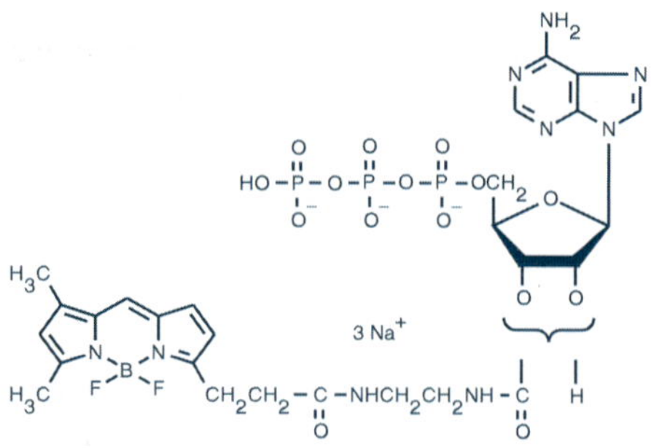

Figure 17.3.12 Adenosine 5′-triphosphate, BODIPY® FL 2′-(or-3′)-*O*-(*N*-(2-aminoethyl)urethane), trisodium salt (BODIPY® FL ATP) (A12410).

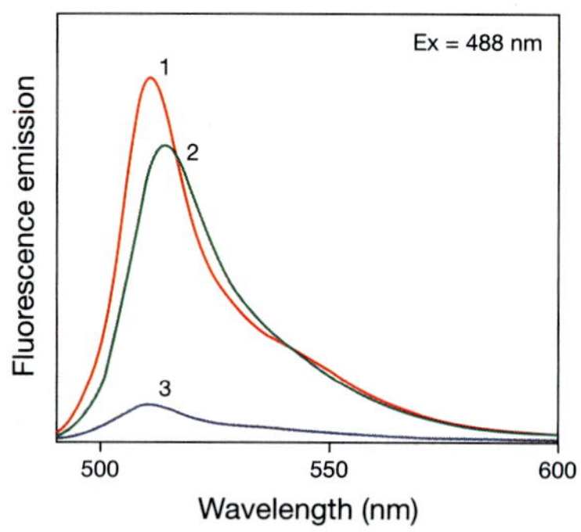

Figure 17.3.13 Fluorescence emission spectra of (**1**) free BODIPY® FL dye in phosphate-buffered saline, pH 7.2; (**2**) BODIPY® FL ATP (A12410); and (**3**) BODIPY® FL GTP (G12411). Samples were prepared with equal absorbance at the excitation wavelength (488 nm). The areas under the curves are therefore proportional to the relative fluorescence quantum yields, clearly showing the quenching effect caused by interaction of the BODIPY® FL fluorophore with the guanine base of GTP.

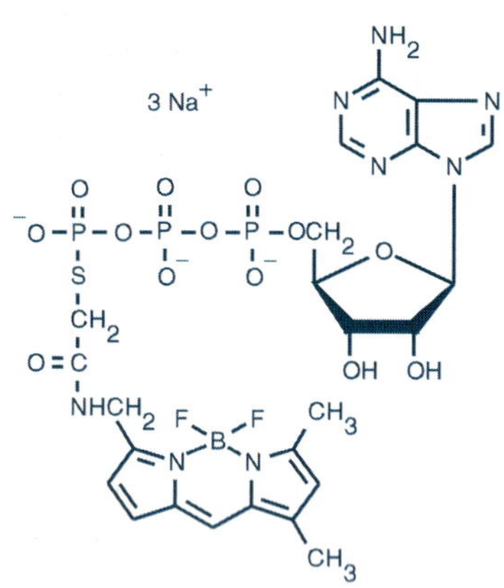

Figure 17.3.14 Adenosine 5′-*O*-(3-thiotriphosphate), BODIPY® FL thioester, sodium salt (BODIPY® FL ATP-γ-*S*, thioester) (A22184).

We prepare a variety of nucleotide analogs, including:

- Alexa Fluor® derivatives of cAMP for use as probes of type I cAMP-dependent protein kinases (PKA I) and Alexa Fluor® 647 ATP (A22362)
- BODIPY® dye–labeled nucleotides for use as enzyme substrates and as long-wavelength probes of nucleotide-binding sites
- Environment-sensitive, blue-fluorescent *N*-methylanthraniloyl (MANT) nucleotides
- Blue-fluorescent ethenoadenosine triphosphate (ε-ATP, E23691)
- Environment-sensitive trinitrophenyl (TNP) nucleotides
- Caged nucleotides, which are important probes for studying the kinetics and mechanism of nucleotide-binding proteins because they allow spatial and temporal control of the release of active nucleotide
- Photoaffinity nucleotides for site-selective covalent labeling
- Fluorescent ChromaTide® nucleotides and aha-dUTP nucleotides, which are primarily used for biosynthetic incorporation into DNA or RNA (Section 8.2)

Alexa Fluor® cAMP and Alexa Fluor® ATP

Our Alexa Fluor® cAMP analogs are 8-(6-aminohexyl)amino derivatives; similar analogs have been shown to exhibit a marked preference for binding to type I cAMP-dependent protein kinases (PKA I). We offer the green-fluorescent Alexa Fluor® 488 cAMP (A35775, Figure 17.3.11) and far-red–fluorescent Alexa Fluor® 647 cAMP (A35777). Alexa Fluor® 488 cAMP was loaded into cells by electroporation and then used to measure intercellular diffusion of cAMP from regulatory to responder T cells via gap junctions.[43]

The Alexa Fluor® 647 conjugate of ATP (A22362) comprises the long-wavelength Alexa Fluor® 647 fluorophore linked to the ribose of ATP by a urethane bridge. Validated applications of this probe include fluorescence resonance energy transfer (FRET) analysis (Fluorescence Resonance Energy Transfer (FRET)—Note 1.2) of nucleotide assocation with Na^+/K^+-ATPase[44] and measurements of the catalytic activity of heavy meromyosin.[45]

BODIPY® Ribonucleotide Di- and Triphosphates

Our selection of BODIPY® dye–modified ribonucleotides includes:

- BODIPY® FL adenosine 5′-triphosphate (BODIPY® FL ATP, A12410)
- BODIPY® TR adenosine 5′-triphosphate (BODIPY® TR ATP, A22352)
- BODIPY® TR adenosine 5′-diphosphate (BODIPY® TR ADP, A22359)
- BODIPY® FL guanosine 5′-triphosphate (BODIPY® FL GTP, G12411)
- BODIPY® TR guanosine 5′-triphosphate (BODIPY® TR GTP, G22351)
- BODIPY® FL guanosine 5′-diphosphate (BODIPY® FL GDP, G22360)

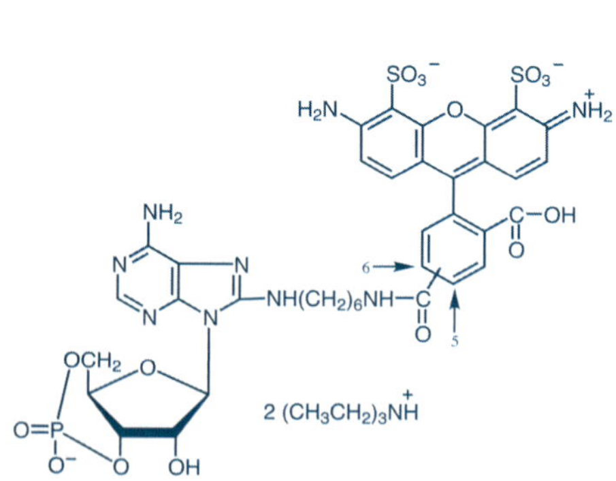

Figure 17.3.11 Alexa Fluor® 488 8-(6-aminohexyl)aminoadenosine 3′,5′-cyclicmonophosphate, bis(triethylammonium) salt (Alexa Fluor® 488 cAMP) (A35775).

IMPORTANT NOTICE: The products described in this manual are covered by one or more Limited Use Label License(s). Please refer to the Appendix on page 971 and Master Product List on page 975. Products are For Research Use Only. Not intended for any animal or human therapeutic or diagnostic use.

These mixed-isomer analogs comprise a BODIPY® fluorophore attached to the 2′ or 3′ position of the ribose ring via an aminoethylcarbamoyl linker (Figure 17.3.12). Interactions between the fluorophore and the purine base are evident from the spectroscopic properties of these nucleotide analogs. The fluorescence quantum yield of BODIPY® FL GTP and BODIPY® FL ATP is significantly quenched in solution (Figure 17.3.13) and increases upon binding to at least some GTP-binding proteins.[46,47] Similar nucleotide analogs incorporating fluorophores such as fluorescein, tetramethylrhodamine and Cy®3 dye have been primarily used for biophysical studies of nucleotide-binding proteins.[48] The BODIPY® dye–labeled nucleotides may be particularly useful for fluorescence polarization–based assays of ATP- or GTP-binding proteins.

Nonhydrolyzable BODIPY® ATP and GTP Analogs

Among the most useful fluorescent nucleotides for protein-binding studies are those that stoichiometrically bind to ATP- or GTP-binding sites but are not metabolized. We offer the following nonhydrolyzable BODIPY® nucleotides:

- BODIPY® FL AMPPNP[49] (B22356)
- BODIPY® FL ATP-γ-*S* (A22184, Figure 17.3.14)
- BODIPY® FL GTP-γ-*S*[46] (G22183)
- BODIPY® 515/530 GTP-γ-*S* (G35779)
- BODIPY® TR GTP-γ-*S* (G35780)
- BODIPY® FL GTP-γ-*NH* amide (G35778)

The fluorescence of the BODIPY® GTP-γ-*S* thioesters is quenched ~90% relative to that of the free dye but is recovered upon protein binding to G-proteins.[46] The green-fluorescent BODIPY® FL GTP-γ-*S* has been used to detect GTP-binding proteins separated by capillary electrophoresis.[50] As compared with BODIPY® FL GTP-γ-*S* thioester, the green-fluorescent BODIPY® 515/530 GTP-γ-*S* thioester has a greater fluorescence increase upon protein binding. The BODIPY® TR GTP-γ-*S* thioester is a red-fluorescent analog with spectral properties similar to the Texas Red® dye.

Although BODIPY® FL GTP-γ-*NH* amide exhibits less fluorescence enhancement upon protein binding, it is reportedly the best of the three green-fluorescent GTP-γ analogs for directly monitoring nucleotide exchange.[51] The different linker lengths of the green-fluorescent GTP-γ analogs (six-carbon for BODIPY® FL GTP-γ-*NH* amide, four-carbon for BODIPY® FL GTP-γ-*S* and one-carbon for BODIPY® 515/530 GTP-γ-*S*) may be useful for understanding protein active-site geometries.

In addition to their potential use for binding studies, BODIPY® FL ATP-γ-*S* and BODIPY® FL GTP-γ-*S* thioesters are important substrates for Fhit (Figure 17.3.15), a member of the histidine triad superfamily of nucleotide-binding proteins that bind and cleave diadenosine polyphosphates.[52–54] Fhit, one of the most frequently inactivated proteins in lung cancer, functions as a tumor suppressor by inducing apoptosis.[53,55,56] These BODIPY® nucleotides should be especially useful for screening potential Fhit inhibitors and activators.

N-Methylanthraniloyl (MANT) Nucleotides

The blue-fluorescent MANT nucleotide analogs of ATP (Figure 17.3.16), AMPPNP, GTP, GMPPNP, ADP and GDP are modified on the ribose moiety, making these probes particularly useful for studying nucleotide-binding proteins that are sensitive to modifications of the

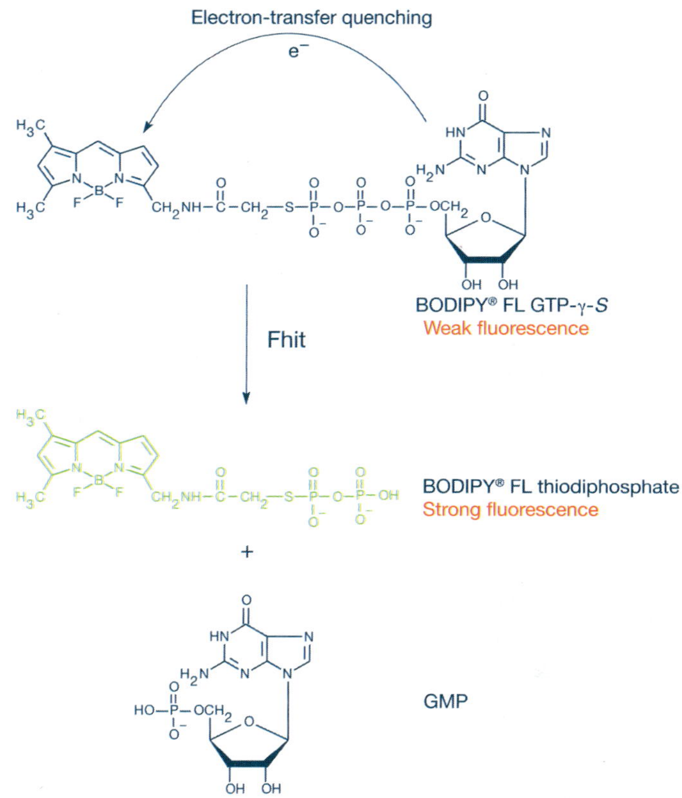

Figure 17.3.15 Principle of fluorescence-based detection of the diadenosine triphosphate hydrolase activity of Fhit using BODIPY® FL GTP-γ-*S* thioester (G22183) as a substrate analog.

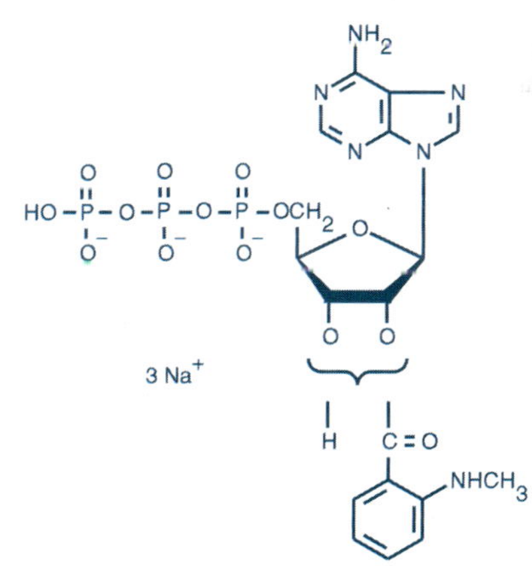

Figure 17.3.16 2′-(or-3′)-*O*-(*N*-methylanthraniloyl)adenosine 5′-triphosphate, trisodium salt (MANT-ATP) (M12417).

IMPORTANT NOTICE: The products described in this manual are covered by one or more Limited Use Label License(s). Please refer to the Appendix on page 971 and Master Product List on page 975. Products are For Research Use Only. Not intended for any animal or human therapeutic or diagnostic use.

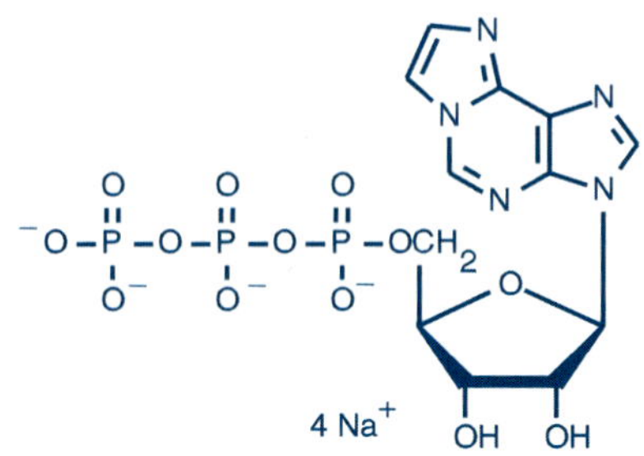

Figure 17.3.17 1,N^6-ethenoadenosine 5′-triphosphate (ε-ATP) (E23691).

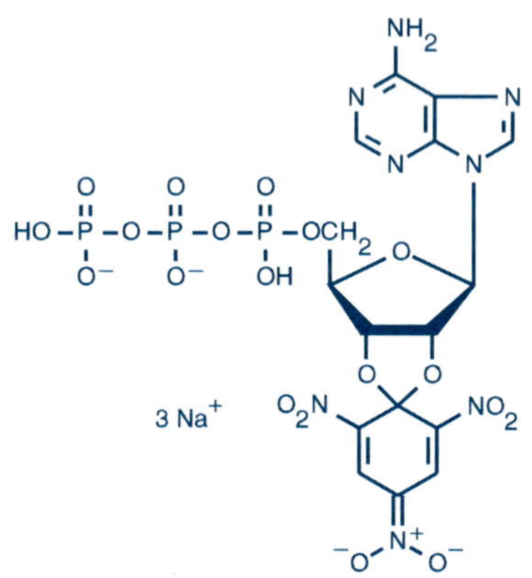

Figure 17.3.18 2′-(or-3′)-*O*-(trinitrophenyl)adenosine 5′-triphosphate, trisodium salt (TNP-ATP) (T7602).

purine base.[57,58] The compact nature of the MANT fluorophore and its attachment position on the ribose ring results in nucleotide analogs that induce minimal perturbation of nucleotide–protein interactions, as confirmed by X-ray crystal structures of MANT nucleotides bound to myosin[59] and H^{-ras} p21.[60] Furthermore, because MANT fluorescence is sensitive to the environment of the fluorophore, nucleotide–protein interactions may be directly detectable. These properties (Table 17.2) make MANT nucleotides valuable probes of the structure and enzymatic activity of nucleotide-binding proteins.[48]

Applications for MANT-ATP (M12417), MANT-ADP (M12416) and MANT-AMPPNP (M22354) include analysis of:

- ATPase kinetics of kinesin[61–63] and other microtubule motor-proteins[64,65] using stopped-flow fluorescence measurements
- Conformation of the myosin subfragment-1 nucleotide-binding site, as indicated by fluorescence quencher accessibility[66,67]
- Interaction of P-glycoprotein ATP-binding sites with drug efflux–modulating steroids[68]
- Myosin ATPase activity in rabbit skeletal muscle[69]
- Structural characteristics of the nucleotide-binding site of *Escherichia coli* DnaB helicase[70,71]

Applications for MANT-GTP (M12415), MANT-GDP (M12414) and MANT-GMPPNP (M22353) include analysis of:

- Activation of protein kinases by Rho subfamily GTP-binding proteins[72]
- Conformational changes during activation of heterotrimeric G-proteins[73]
- Effects of nucleotide structural modifications on binding to H^{-ras} p21[74]
- Nucleotide hydrolysis and dissociation kinetics of H^{-ras} p21 and other low molecular weight GTP-binding proteins[57,75–78]
- GTP-binding proteins Rab5 and Rab7,[79] Raf-1,[80] Rho[81,82] and Rac,[83,84] as well as Ras-related proteins[57,75]

Ethenoadenosine Nucleotide

The ethenoadenosine nucleotides—developed in 1972 by Leonard and collaborators[85,86]—bind like endogenous nucleotides to several proteins. The properties and applications of ethenoadenosine and MANT nucleotides have been comprehensively reviewed.[48] The etheno ATP analog (ε-ATP, E23691; Figure 17.3.17) can often mimic ATP in both binding and function. This probe has been used to replace ATP in actin polymerization reactions[87] and is frequently incorporated in place of the tightly bound actin nucleotide.[88,89] It also supports contraction of actomyosin, facilitates the measurement of nucleotide-exchange kinetics in actin[90] and serves as a substrate for myosin, which converts it to ε-ADP.[91] Sensitized luminescence of Tb^{3+} (T1247, Section 14.3) coordinated to ε-ATP is a sensitive probe of binding to the catalytic site of protein disulfide isomerase.[92]

Trinitrophenyl (TNP) Nucleotides

Unlike the etheno derivatives, the free trinitrophenyl (TNP) nucleotides are essentially nonfluorescent in water. The TNP nucleotides undergo an equilibrium transition to a semiquinoid structure that has relatively long-wavelength spectral properties;[93–95] this form is only fluorescent when bound to the nucleotide-binding site of some proteins. The TNP derivative of ATP frequently exhibits a spectral shift and fluorescence enhancement upon protein binding and actually binds with higher affinity than ATP to some proteins. The broad, long-wavelength absorption of TNP nucleotides makes them useful for FRET studies[96–98] (Fluorescence Resonance

Table 17.2 Spectroscopic properties of MANT-nucleotides in aqueous solution (pH 8).

Parameter	Value	Notes
Absorption maximum	356 nm	Stronger absorption at shorter wavelengths (λ_{max} = 255 nm)
Molar extinction coefficient (EC_{max})	5800 $cm^{-1}M^{-1}$	23,000 $cm^{-1}M^{-1}$ at 255 nm
Fluorescence emission maximum	448 nm	Shifts 10–20 nm shorter in nonpolar solvents and upon binding to most proteins
Fluorescence quantum yield	0.22	Increases in nonpolar solvents and upon binding to most proteins

IMPORTANT NOTICE: The products described in this manual are covered by one or more Limited Use Label License(s). Please refer to the Appendix on page 971 and Master Product List on page 975. Products are For Research Use Only. Not intended for any animal or human therapeutic or diagnostic use.
www.invitrogen.com/probes
molecular probes® | invitrogen™ by life technologies™

Energy Transfer (FRET)—Note 1.2). The TNP derivatives of ATP (TNP-ATP, T7602; Figure 17.3.18), ADP (TNP-ADP, T7601) and AMP (TNP-AMP, T7624) have been used as structural probes for a wide variety of nucleotide-binding proteins.[99–101] We have found that chromatographically purified TNP nucleotides are unstable during lyophilization. Consequently, these derivatives are sold in aqueous solution and should be frozen immediately upon arrival.

Caged Nucleotides

Caged nucleotides are nucleotide analogs in which the terminal phosphate is esterified with a blocking group, rendering the molecule biologically inactive. Photolytic removal of the caging group by UV illumination results in a pulse of the nucleotide—often on a microsecond to millisecond time scale—at the site of illumination. Because photolysis ("uncaging") can be temporally controlled and confined to the area of illumination, the popularity of this technique is growing. We are supporting this development by synthesizing a variety of caged nucleotides, neurotransmitters and Ca^{2+} chelators. Our current selection of caged nucleotides includes:

- NPE-caged ATP (A1048)
- DMNPE-caged ATP (A1049)
- NPE-caged ADP (A7056)
- DMNB-caged c-AMP (D1037)

Section 5.3 discusses our selection of caged probes and the properties of the different caging groups that we use (Table 5.2).

Researchers investigating the cytoskeleton have benefited greatly from advances in caging technology, primarily originating from the work of Trentham, Kaplan and their colleagues.[102] NPE-caged ADP (A7056) is a useful probe for studying the effect of photolytic release of ADP in muscle fibers[103,104] and isolated sarcoplasmic reticulum.[105] Although it is sometimes difficult to properly abstract papers that describe experiments with caged ATP because they could be referring to either NPE-caged ATP (A1048), DMNPE-caged ATP (A1049) or earlier caged versions of this nucleotide, most researchers have used NPE-caged ATP.

Because the caged nucleotides may be added to an experimental system at relatively high concentrations, use of the enzyme apyrase was recommended by Sleep and Burton[106] to eliminate any traces of ATP that may be present in the caged ATP probes.[107–110] Once the caged ATP solutions have been preincubated with apyrase, the enzyme can be removed by centrifugal filtration.[107,109]

These caged nucleotides are generally cell impermeant and must be microinjected into cells or loaded by other techniques (Table 14.1). Permeabilization of cells with staphylococcal α-toxin or the saponin ester β-escin is reported to make the membrane of smooth muscle cells permeable to low molecular weight (<1000 daltons) molecules, while retaining high molecular weight compounds.[111] α-Toxin permeabilization has permitted the introduction of caged nucleotides, including caged ATP (A1048) and caged GTP-γ-S, as well as of caged inositol 1,4,5-triphosphate (NPE-caged Ins 1,4,5; I23580; Section 17.2) into smooth muscle cells.[112] Caged inositol 1,4,5-triphosphate has also been successfully loaded in ECV304 cells using electroporation.[113]

BzBzATP

Functional ion channels can be assembled from both homomeric and heteromeric combinations of the seven P2X receptor subunits so far identified ($P2X_{1-7}$). Due to the lack of specific agonists or antagonists for P2X receptors, it is difficult to determine which receptor subtypes mediate particular cellular responses. We offer one of the most potent and widely used P2X receptor agonists, BzBzATP[114,115] (2´-(or 3´-)*O*-(4-benzoylbenzoyl)adenosine 5´-triphosphate, B22358). BzBzATP has more general applications for site-directed irreversible modification of nucleotide-binding proteins via photoaffinity labeling.[116,117]

REFERENCES

1. Biochim Biophys Acta (2008) 1784:94; **2**. J Am Chem Soc (2009) 131:13286; **3**. Free Radic Biol Med (2009) 47:983; **4**. J Biomol Screen (1999) 4:67; **5**. J Med Chem (2010) 53:1898; **6**. Am J Physiol (1989) 256:C886; **7**. J Neurochem (1986) 47:1405; **8**. Nature (1986) 321:698; **9**. J Neurosci (1985) 5:2672; **10**. Eur J Pharmacol (1991) 207:17; **11**. Antimicrob Agents Chemother (2009) 53:3501; **12**. J Pharm Pharmacol (2001) 53:583; **13**. Med Res Rev (1995) 15:111; **14**. Photochem Photobiol (2006) 82:720; **15**. Anticancer Res (1998) 18:4651; **16**. Photochem Photobiol (1998) 68:593; **17**. Proc Natl Acad Sci U S A (1989) 86:5963; **18**. Pharmacol Ther (1994) 63:1; **19**. J Biol Chem (2002) 277:37718; **20**. Photochem Photobiol (1994) 59:529; **21**. J Biol Chem (2008) 283:3401; **22**. Nat Med (2006) 12:549; **23**. Nature (2005) 437:911; **24**. Nat Cell Biol (2005) 7:78; **25**. Comb Chem High Throughput Screen (2003) 6:341; **26**. Proteomics (2003) 3:1244; **27**. PLoS Biol (2009) 7:e1000172; **28**. J Pharmacol Exp Ther (2008) 325:27; **29**. J Biol Chem (2005) 280:4048; **30**. J Med Chem (2008) 51:1831; **31**. Mol Pharmacol (2008) 73:1371; **32**. J Biomol Screen (2000) 5:239; **33**. J Biochem (2006) 139:543; **34**. Mol Endocrinol (2007) 21:700; **35**. Kidney Int (2007) 71:738; **36**. J Immunol (2005) 174:1073; **37**. J Clin Invest (2003) 112:398; **38**. J Biol Chem (2004) 279:19790; **39**. Nature (2004) 429:188; **40**. Proc Natl Acad Sci U S A (2004) 101:1508; **41**. Nucleic Acids Res (2003) 31:e130; **42**. J Cell Sci (2001) 114:459; **43**. J Exp Med (2007) 204:1303; **44**. Biochim Biophys Acta (2009) 1794:1549; **45**. Biochemistry (2007) 46:7233; **46**. J Biol Chem (2001) 276:29275; **47**. Anal Biochem (2001) 291:109; **48**. Methods Enzymol (1997) 278:363; **49**. J Biol Chem (2004) 279:28402; **50**. Anal Chem (2003) 75:4297; **51**. Proc Natl Acad Sci U S A (2004) 101:2800; **52**. Proc Natl Acad Sci U S A (2003) 100:1592; **53**. Curr Biol (2000) 10:907; **54**. J Biol Chem (2000) 275:4555; **55**. Am J Pathol (2000) 156:419; **56**. J Natl Cancer Inst (2000) 92:338; **57**. Proc Natl Acad Sci U S A (1990) 87:3562; **58**. Biochim Biophys Acta (1983) 742:496; **59**. J Mol Biol (1997) 274:394; **60**. J Mol Biol (1995) 253:132; **61**. Biochemistry (1998) 37:792; **62**. J Biol Chem (1997) 272:717; **63**. Biochemistry (1995) 34:13233; **64**. Biochemistry (1996) 35:2365; **65**. Biochemistry (1995) 34:13259; **66**. Biophys J (1995) 68:142S; **67**. Biochemistry (1990) 29:3309; **68**. Biochemistry (1997) 36:15208; **69**. J Gen Physiol (1995) 106:957; **70**. Biophys J (1996) 71:2075; **71**. J Cell Biol (1997) 139:63; **72**. Biochemistry (1997) 36:1173; **73**. J Biol Chem (1994) 269:13771; **74**. Biochemistry (1995) 34:593; **75**. Biochemistry (1997) 36:4535; **76**. Biochemistry (1995) 34:12543; **77**. Biochemistry (1995) 34:639; **78**. Biochemistry (1993) 32:7451; **79**. J Biol Chem (1996) 271:20470; **80**. J Biol Chem (2000) 275:22172; **81**. Biochemistry (1999) 38:985; **82**. J Biol Chem (1996) 271:10004; **83**. J Biol Chem (1997) 272:18834; **84**. J Biol Chem (1996) 271:19794; **85**. Biochem Biophys Res Commun (1972) 46:597; **86**. Science (1972) 175:646; **87**. Biochemistry (1988) 27:3812; **88**. J Biol Chem (2008) 283:19379; **89**. J Biol Chem (1993) 268:8683; **90**. J Cell Biol (1988) 106:1553; **91**. J Biol Chem (1984) 259:11920; **92**. Am J Physiol Lung Cell Mol Physiol (2000) 278:L1091; **93**. Eur J Biochem (2003) 270:3479; **94**. Biochim Biophys Acta (1973) 320:635; **95**. Biochim Biophys Acta (1976) 453:293; **96**. J Muscle Res Cell Motil (1992) 13:132; **97**. Biochemistry (1992) 31:3930; **98**. Biophys J (1992) 61:553; **99**. Biochemistry (2006) 45:7237; **100**. J Biol Chem (2006) 281:27471; **101**. Br J Pharmacol (2003) 140:202; **102**. Nat Methods (2007) 4:619; **103**. Biophys J (1995) 68:78S-80S; **104**. J Mol Biol (1992) 223:185; **105**. Ann N Y Acad Sci (1982) 402:478; **106**. Biophys J (1994) 67:2436; **107**. Biophys J (1994) 67:1933; **108**. J Biol Chem (1995) 270:23966; **109**. Biophys J (1994) 66:1115; **110**. J Biolumin Chemilumin (1994) 9:29; **111**. Methods Cell Biol (1989) 31:63; **112**. Annu Rev Physiol (1990) 52:857; **113**. J Neurosci Methods (2004) 132:81; **114**. J Physiol (1999) 519 Pt 3:723; **115**. Mol Pharmacol (1999) 56:1171; **116**. J Neurochem (1993) 61:1657; **117**. Biochemistry (1989) 28:3989.

IMPORTANT NOTICE: The products described in this manual are covered by one or more Limited Use Label License(s). Please refer to the Appendix on page 971 and Master Product List on page 975. Products are For Research Use Only. Not intended for any animal or human therapeutic or diagnostic use.

DATA TABLE 17.3 PROBES FOR PROTEIN KINASES, PROTEIN PHOSPHATASES AND NUCLEOTIDE-BINDING PROTEINS

Cat. No.	MW	Storage	Soluble	Abs	EC	Em	Solvent	Notes
A1048	700.30	FF,D,LL	H_2O	259	18,000	none	MeOH	1, 2, 3
A1049	760.35	FF,D,LL	H_2O	351	4400	none	H_2O	1, 2
A7056	614.44	FF,D,LL	H_2O	259	15,000	none	MeOH	1, 2, 3
A12410	933.30	FF,L	H_2O	505	54,000	514	H_2O	4, 5
A12412	1117.18	FF,L	H_2O	323	4200	461	pH 7	4, 5, 6
A22184	878.28	FF,L	H_2O	504	68,000	514	pH 7	4, 5
A22352	1065.43	FF,L	H_2O	591	55,000	620	pH 7	4, 5
A22359	963.47	FF,L	H_2O	592	57,000	621	pH 7	4, 5
A22362	~2050	FF,L	H_2O	648	246,000	667	pH 7	4, 5
A35775	1162.23	FF,L	H_2O	493	71,000	517	pH 7	4, 5
A35777	~1700	FF,L	H_2O	649	246,000	666	pH 7	4, 5
B7469	784.70	F,D,L	DMSO	504	79,000	511	MeOH	
B22356	932.31	FF,L	H_2O	504	68,000	514	H_2O	4, 5
B22358	1018.97	FF,L	H_2O	260	27,000	none	pH 7	
D1037	524.38	F,D,LL	DMSO	338	6100	none	MeOH	1, 2
E23691	619.13	FF	H_2O	265	5000	411	pH 7	5
G12411	949.30	FF,L	H_2O	504	68,000	511	H_2O	4, 5, 7
G22183	894.28	FF,L	H_2O	504	68,000	510	pH 7	4, 5, 7
G22351	1081.43	FF,L	H_2O	591	56,000	620	pH 7	4, 5, 7
G22360	1005.75	FF,L	H_2O	504	68,000	508	pH 7	4, 5, 7
G35778	905.29	FF,L	H_2O	505	68,000	512	pH 7	4, 5, 7
G35779	865.28	FF,L	H_2O	511		520	pH 7	4, 5, 7
G35780	1153.60	FF,L	H_2O	591		621	pH 7	4, 5, 7
H7476	504.45	F,D,L	DMSO, DMF	591	37,000	594	EtOH	
M12414	620.32	FF,L	H_2O	356	5700	447	pH 8	4, 5, 8
M12415	722.28	FF,L	H_2O	356	5700	448	pH 7	4, 5, 8
M12416	604.32	FF,L	H_2O	356	5800	448	pH 7	4, 5, 8
M12417	706.28	FF,L	H_2O	356	5800	447	pH 7	4, 5, 8
M22353	721.29	FF,L	H_2O	357	5700	447	pH 8	4, 5, 8
M22354	705.29	FF,L	H_2O	357	5800	447	pH 8	4, 5, 9
T7601	682.26	FF,L	H_2O	408	26,000	none	pH 8	4, 5, 9
T7602	784.22	FF,L	H_2O	408	26,000	none	pH 8	4, 5, 9
T7624	579.29	F,L	H_2O	408	26,000	none	pH 8	4, 5, 9

For definitions of the contents of this data table, see "Using *The Molecular Probes® Handbook*" in the introductory pages.

Notes

1. Caged nucleotide esters are free of contaminating free nucleotides when initially prepared. However, some decomposition may occur during storage.
2. All photoactivatable probes are sensitive to light. They should be protected from illumination except when photolysis is intended.
3. This compound has weaker visible absorption at >300 nm but no discernible absorption peaks in this region.
4. The molecular weight (MW) of this product is approximate because the degree of hydration and/or salt form has not been conclusively established.
5. This product is supplied as a ready-made solution in the solvent indicated under "Soluble."
6. QY = 0.63 in 50 mM Tris, pH 8.0. Fluorescence shifts to longer wavelengths (Em ~475 nm) on enzymatic cleavage of the α–β phosphoryl bond. (Biochem Biophys Res Commun (1978) 81:35, J Biol Chem (1979) 254:12069)
7. Fluorescence of BODIPY® dye–labeled guanosine derivatives is generally weak due to base-specific intramolecular quenching. (Anal Biochem (2001) 291:109)
8. Fluorescence quantum yields of MANT nucleotides are environment-dependent. In H_2O, QY is ~0.2. (Biochim Biophys Acta (1983) 742:496)
9. Trinitrophenyl nucleotides are in fact very weakly fluorescent in water (Em ~560 nm). Fluorescence is blue-shifted and more intense in organic solvents (DMSO, EtOH) and when bound to proteins (Em ~540 nm). Absorption spectrum also has a second, less intense peak at about 470 nm. (Biochim Biophys Acta (1982) 719:509)

IMPORTANT NOTICE: The products described in this manual are covered by one or more Limited Use Label License(s). Please refer to the Appendix on page 971 and Master Product List on page 975. Products are For Research Use Only. Not intended for any animal or human therapeutic or diagnostic use.

PRODUCT LIST 17.3 PROBES FOR PROTEIN KINASES, PROTEIN PHOSPHATASES AND NUCLEOTIDE-BINDING PROTEINS

Cat. No.	Product	Quantity
A22359	adenosine 5′-diphosphate, BODIPY® TR 2′-(or-3′)-*O*-(*N*-(2-aminoethyl)urethane), disodium salt (BODIPY® TR ADP) *5 mM in buffer*	100 µL
A7056	adenosine 5′-diphosphate, P^2-(1-(2-nitrophenyl)ethyl) ester, monopotassium salt (NPE-caged ADP)	5 mg
A22184	adenosine 5′-*O*-(3-thiotriphosphate), BODIPY® FL thioester, sodium salt (BODIPY® FL ATP-γ-*S*, thioester) *5 mM in buffer*	50 µL
A22362	adenosine 5′-triphosphate, Alexa Fluor® 647 2′-(or-3′)-*O*-(*N*-(2-aminoethyl)urethane), hexa(triethylammonium) salt (Alexa Fluor® 647 ATP) *5 mM in buffer*	100 µL
A12410	adenosine 5′-triphosphate, BODIPY® FL 2′-(or-3′)-*O*-(*N*-(2-aminoethyl)urethane), trisodium salt (BODIPY® FL ATP) *5 mM in buffer*	100 µL
A22352	adenosine 5′-triphosphate, BODIPY® TR 2′-(or-3′)-*O*-(*N*-(2-aminoethyl)urethane), trisodium salt (BODIPY® TR ATP) *5 mM in buffer*	100 µL
A1048	adenosine 5′-triphosphate, P^3-(1-(2-nitrophenyl)ethyl) ester, disodium salt (NPE-caged ATP)	5 mg
A1049	adenosine 5′-triphosphate, P^3-(1-(4,5-dimethoxy-2-nitrophenyl)ethyl) ester, disodium salt (DMNPE-caged ATP)	5 mg
A12412	adenosine 5′-triphosphate, P^3-(5-sulfo-1-naphthylamide), tetra(triethylammonium) salt (ATP γ-AmNS) *5 mM in buffer*	400 µL
A35775	Alexa Fluor® 488 8-(6-aminohexyl)aminoadenosine 3′,5′-cyclicmonophosphate, bis(triethylammonium) salt (Alexa Fluor® 488 cAMP) *5 mM in buffer*	100 µL
A35777	Alexa Fluor® 647 8-(6-aminohexyl)aminoadenosine 3′,5′-cyclicmonophosphate, tetra(triethylammonium) salt (Alexa Fluor® 647 cAMP) *5 mM in buffer*	100 µL
A35725	Antibody Beacon™ Tyrosine Kinase Assay Kit *400 assays*	1 kit
A6442	anti-synapsin I (bovine), rabbit IgG fraction *affinity purified*	10 µg
B22358	2′-(or-3′)-*O*-(4-benzoylbenzoyl)adenosine 5′-triphosphate, tris(triethylammonium) salt (BzBzATP) *5 mM in buffer*	2 mL
B7469	BODIPY® FL forskolin	100 µg
B22356	2′-(or-3′)-*O*-(BODIPY® FL)-β:γ-imidoadenosine 5′-triphosphate, trisodium salt (BODIPY® FL AMPPNP) *5 mM in buffer*	50 µL
C10558	cAMP Chemiluminescent Immunoassay Kit *10-plate size*	1 kit
C10557	cAMP Chemiluminescent Immunoassay Kit *2-plate size*	1 kit
D1037	4,5-dimethoxy-2-nitrobenzyl adenosine 3′,5′-cyclicmonophosphate (DMNB-caged cAMP)	5 mg
E23691	1,N^6-ethenoadenosine 5′-triphosphate (ε-ATP) *5 mM in buffer*	2 mL
G22360	guanosine 5′-diphosphate, BODIPY® FL 2′-(or-3′)-*O*-(*N*-(2-aminoethyl)urethane), bis(triethylammonium) salt (BODIPY® FL GDP) *5 mM in buffer*	100 µL
G35778	guanosine 5′-*O*-(3-iminotriphosphate), BODIPY® FL ethylamide, sodium salt (BODIPY® FL GTP-γ-*NH*, amide) *1 mM in buffer*	100 µL
G35779	guanosine 5′-*O*-(3-thiotriphosphate), BODIPY® 515/530 thioester, sodium salt (BODIPY® 515/530 GTP-γ-*S*, thioester) *1 mM in buffer*	100 µL
G22183	guanosine 5′-*O*-(3-thiotriphosphate), BODIPY® FL thioester, sodium salt (BODIPY® FL GTP-γ-*S*, thioester) *5 mM in buffer*	50 µL
G35780	guanosine 5′-*O*-(3-thiotriphosphate), BODIPY® TR thioester, sodium salt (BODIPY® TR GTP-γ-*S*, thioester) *1 mM in buffer*	100 µL
G12411	guanosine 5′-triphosphate, BODIPY® FL 2′-(or-3′)-*O*-(*N*-(2-aminoethyl)urethane), trisodium salt (BODIPY® FL GTP) *5 mM in water*	100 µL
G22351	guanosine 5′-triphosphate, BODIPY® TR 2′-(or-3′)-*O*-(*N*-(2-aminoethyl)urethane), trisodium salt (BODIPY® TR GTP) *5 mM in water*	100 µL
H7476	hypericin	1 mg
M12416	2′-(or-3′)-*O*-(*N*-methylanthraniloyl)adenosine 5′-diphosphate, disodium salt (MANT-ADP) *5 mM in buffer*	400 µL
M12417	2′-(or-3′)-*O*-(*N*-methylanthraniloyl)adenosine 5′-triphosphate, trisodium salt (MANT-ATP) *5 mM in buffer*	400 µL
M12414	2′-(or-3′)-*O*-(*N*-methylanthraniloyl)guanosine 5′-diphosphate, disodium salt (MANT-GDP) *5 mM in buffer*	400 µL
M12415	2′-(or-3′)-*O*-(*N*-methylanthraniloyl)guanosine 5′-triphosphate, trisodium salt (MANT-GTP) *5 mM in buffer*	400 µL
M22354	2′-(or-3′)-*O*-(*N*-methylanthraniloyl)-β:γ-imidoadenosine 5′-triphosphate, trisodium salt (MANT-AMPPNP) *5 mM in buffer*	50 µL
M22353	2′-(or-3′)-*O*-(*N*-methylanthraniloyl)-β:γ-imidoguanosine 5′-triphosphate, trisodium salt (MANT-GMPPNP) *5 mM in buffer*	50 µL
P13235	polymyxin B, BODIPY® FL conjugate, trifluoroacetic acid salt *mixed species*	100 µg
P13238	polymyxin B, dansyl conjugate, trifluoroacetic acid salt *mixed species*	100 µg
P13236	polymyxin B, Oregon Green® 514 conjugate, trifluoroacetic acid salt *mixed species*	100 µg
P33706	Pro-Q® Diamond Phosphoprotein/Phosphopeptide Microarray Stain Kit	1 kit
R33700	RediPlate™ 96 EnzChek® Serine/Threonine Phosphatase Assay Kit *one 96-well microplate*	1 kit
R22067	RediPlate™ 96 EnzChek® Tyrosine Phosphatase Assay Kit *one 96-well microplate*	1 kit
T7601	2′-(or-3′)-*O*-(trinitrophenyl)adenosine 5′-diphosphate, disodium salt (TNP-ADP) *5 mg/mL in water*	2 mL
T7624	2′-(or-3′)-*O*-(trinitrophenyl)adenosine 5′ monophosphate, sodium salt (TNP-AMP) *5 mg/mL in buffer*	2 mL
T7602	2′-(or-3′)-*O*-(trinitrophenyl)adenosine 5′-triphosphate, trisodium salt (TNP-ATP) *5 mg/mL in buffer*	2 mL

IMPORTANT NOTICE: The products described in this manual are covered by one or more Limited Use Label License(s). Please refer to the Appendix on page 971 and Master Product List on page 975. Products are For Research Use Only. Not intended for any animal or human therapeutic or diagnostic use.

17.4 Probes for Lipid Metabolism and Signaling

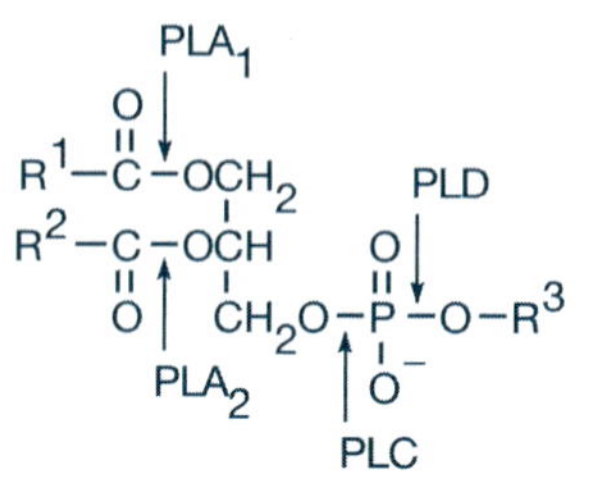

Figure 17.4.1 Cleavage specificities of phospholipases. R^1 and R^2 are typically saturated or unsaturated aliphatic groups. The polar head group R^3 can be choline, ethanolamine, glycerol, inositol, inositol phosphate, serine or other alcohols.

Lipids and lipid metabolites are abundant in cells and have both a structural function and a role in cell regulation. Phospholipases, in particular, play an important part in cellular signaling processes via the generation of second messengers such as diacylglycerols, arachidonate and inositol 1,4,5-triphosphate[1–4] (Ins 1,4,5-P_3, I3716; Section 17.2). In addition, phospholipase A_2 activation is a key step in inflammation processes, and phospholipase A_2 plays major roles in bacterial virulence and in the pathogenesis of acute respiratory distress syndrome[5–7] (ARDS), making this class of enzymes important therapeutic targets.[8,9]

Phospholipases are classified according to the cleavage site on the phospholipid substrate (Figure 17.4.1). There are at least three types of fluorescence-based phospholipase detection methods:[10]

- Continuous methods, which permit direct fluorometric monitoring of enzymatic activity using self-quenching or excimer-forming probes
- Methods that continuously detect nonfluorescent product formation from natural phospholipids, such as detection of fatty acids with our ADIFAB reagent or enzyme-coupled detection of choline with our Amplex® Red Phospholipase Assay Kits
- Discontinuous methods, which require resolution of fluorescent substrates and products by TLC, HPLC or other separation techniques

Table 17.3 summarizes Molecular Probes® products for fluorescence-based phospholipase assays. Other applications for our wide range of fluorescent phospholipids are described in Chapter 13.

Table 17.3 Fluorescence-based phospholipase assays.

Phospholipase *	Probes	Assay Principle	Detection Method
A_1	A10070, E10219, E10221	Intramolecular self-quenching	Fluorescence increase at ~530 nm
A_1, A_2	B7701	Intramolecular self-quenching	Fluorescence increase at ~515 nm [1]
A_1, A_2	B3781, B3782	Intramolecular excimer formation	Emission ratio 380/470 nm [2–4]
A_1, A_2	A3880	Free fatty acid sensor	Emission ratio 432/v505 nm [5–7]
A_2	A10072, E10217, E10218	Intramolecular fluorescence resonance energy transfer (FRET)	Fluorescence increase at ~515 nm or increase in emission ratio at 515/575 nm
A_2	D23739	Intramolecular self-quenching	Fluorescence increase at ~515 nm [8]
A_2	N3786, N3787	Intermolecular self-quenching	Fluorescence increase at ~530 nm [9,10]
A_2	H361, H3809	Intermolecular excimer formation	Emission ratio 380/470 nm [11–13]
A_2	D3803	Release of a fluorescent fatty acid	TLC or fluorescence image scanner [14]
A_2, C, D	D3771	Formation of a fluorescent *O*-alkylglycerol derivative	TLC or HPLC [15–17]
A_2, C, D	H361	Quenching by a disulfide-polymerized lipid matrix	Fluorescence increase at ~380 nm [18,19]
C	A12218	Peroxidase-linked detection of phosphocholine	Conversion of the nonfluorescent Amplex® Red reagent to fluorescent resorufin [20–22]
C	E10215, E10216	Release of dye-labeled diacylglycerol	Fluorescence increase at 516 nm, with potential interference from phospholipase A_2 and phospholipase D activity
D	A12219	Peroxidase-linked detection of choline	Conversion of the nonfluorescent Amplex® Red reagent to fluorescent resorufin [20,22,23]
PAP †	D3805	Release of dye-labeled diacylglycerol	HPLC [24]

* Phospholipase specificity: A_1, A_2, C or D (see Section 17.4 for cleavage specificities). † PAP = phosphatidic acid phosphohydrolase.

1. J Biol Chem (1992) 267:21465; 2. Biochemistry (1993) 32:583; 3. Anal Biochem (1981) 116:553; 4. Biochim Biophys Acta (1994) 1192:132; 5. Anal Biochem (1995) 229:256; 6. Biochem J (1994) 298:23; 7. J Biol Chem (1992) 267:23495; 8. Anal Biochem (1999) 276:27; 9. Lipids (1989) 24:691; 10. Biochem Biophys Res Commun (1984) 118:894; 11. Anal Biochem (2006) 359:280; 12. Chem Phys Lipids (1990) 53:129; 13. Anal Biochem (1989) 177:103; 14. J Biol Chem (1999) 274:19338; 15. Eukaryot Cell (2009) 8:1094; 16. Anal Biochem (1994) 218:136; 17. Biochem J (1995) 307:799; 18. Anal Biochem (1994) 221:152; 19. J Biol Chem (1995) 270:263; 20. Proc Natl Acad Sci U S A (2004) 101:9745; 21. Mol Pharmacol (2000) 57:1142; 22. Proc SPIE-Int Soc Opt Eng (2000) 3926:166; 23. J Biol Chem (2002) 277:45592; 24. Anal Biochem (2008) 374:291.

IMPORTANT NOTICE: The products described in this manual are covered by one or more Limited Use Label License(s). Please refer to the Appendix on page 971 and Master Product List on page 975. Products are For Research Use Only. Not intended for any animal or human therapeutic or diagnostic use.
www.invitrogen.com/probes

Phospholipase A_1 and A_2 Assays

The importance of phospholipases in cellular signaling, lipid metabolism, inflammatory responses and pathological disorders related to these processes has stimulated demand for fluorescence-based enzyme activity monitoring methods. Several of the fluorogenic phospholipase A substrates described here are designed to provide continuous monitoring of phospholipase A activity in purified enzyme preparations, cell lysates and live cells; applications of some of these substrates extend as far as *in vivo* small animal imaging.[11–13] The phospholipase A substrates are generally dye-labeled phospholipids of two types—glycerophosphocholines with BODIPY® dye–labeled *sn*-1 or *sn*-2 (or both) acyl or alkyl chains and glycerophosphoethanoloamines with BODIPY® dye–labeled acyl chains and dinitrophenyl quencher–modified head groups (Figure 17.4.2). These structural variations determine specificity for phospholipase A_1 (which hydrolyzes the *sn*-1 ester linkage between phospholipids and fatty acids) versus phospholipase A_2 (which hydrolyzes the *sn*-2 ester linkage between phospholipids and fatty acids, Figure 17.4.1), and the fluorescence response associated with enzymatic cleavage of the substrate (Table 17.3).

PED-A1 Phospholipase A_1 Substrate

PED-A1 (A10070, Figure 17.4.3) is a fluorogenic substrate designed to provide specific, real-time monitoring of phospholipase A_1 activity in purified enzyme preparations, cell lysates and live cells.[14,15] PED-A1 is comprised of a dinitrophenyl quencher–modified glycerophosphoethanolamine head group and a green-fluorescent BODIPY® FL dye–labeled acyl chain at the *sn*-1 position. Upon cleavage by phospholipase A_1, PED-A1 exhibits an increase in green fluorescence (measured at excitation/emission = 488/530 nm). Phospholipase A_1 specificity is imparted by the placement of the BODIPY® FL acyl chain in the *sn*-1 position and by incorporation of an acyl group with an enzyme-resistant (noncleavable) ether linkage in the *sn*-2 position.

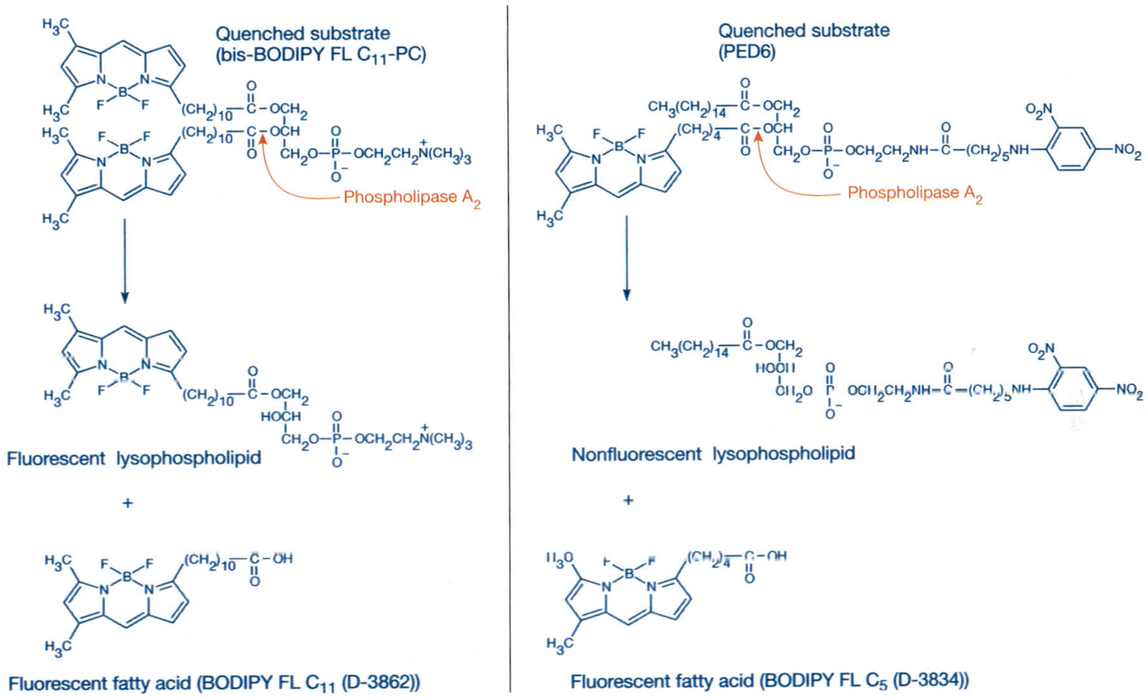

Figure 17.4.2 Mechanism of phospholipase activity-linked fluorescence enhancement responses of bis-BODIPY® FL C_{11}-PC (B7701) and PED6 (D23739). Note that enzymatic cleavage of bis-BODIPY® FL C_{11}-PC yields two fluorescent products, whereas cleavage of PED6 yields only one.

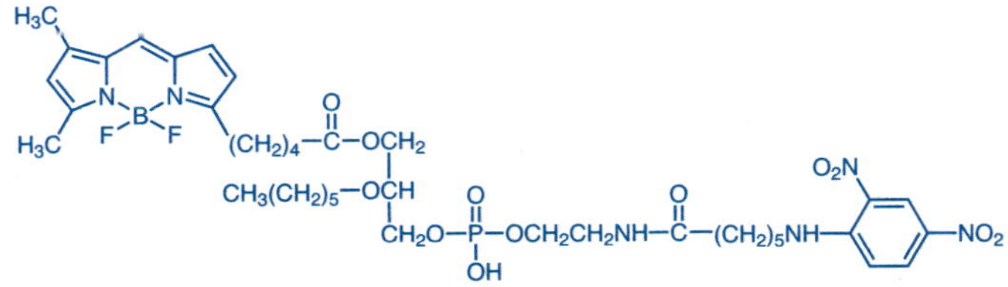

Figure 17.4.3 PED-A1 (*N*-((6-(2,4-DNP)amino)hexanoyl)-1-(BODIPY® FL C5)-2-hexyl-*sn*-glycero-3-phosphoethanolamine, A10070).

IMPORTANT NOTICE: The products described in this manual are covered by one or more Limited Use Label License(s). Please refer to the Appendix on page 971 and Master Product List on page 975. Products are For Research Use Only. Not intended for any animal or human therapeutic or diagnostic use.

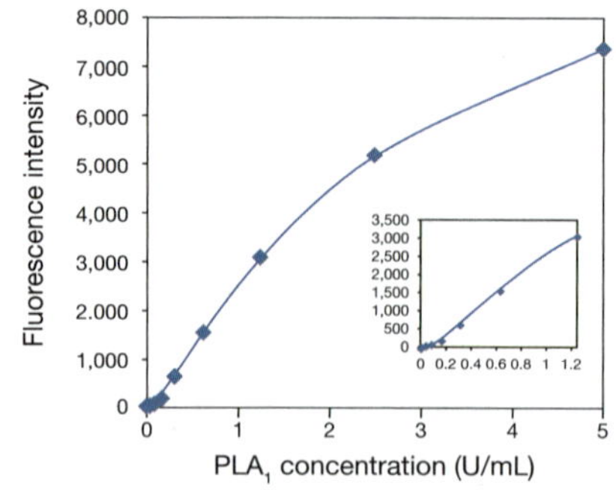

Figure 17.4.4 Detection of phospholipase A$_1$ (PLA$_1$) using the EnzChek® Phospholipase A$_1$ Assay Kit (E10219, E10221). PLA$_1$ reactions were run at ambient temperature with liposomes for 30 minutes according to the assay protocol provided, and fluorescence emission was measured using 460 nm excitation on a Spectra Max M5 (Molecular Devices). Background fluorescence determined for the no-enzyme control reaction has been subtracted.

EnzChek® Phospholipase A$_1$ Assay Kit

The EnzChek® Phospholipase A$_1$ Assay Kit (E10219, E10221) provides a simple, fluorometric method for continuous monitoring of phospholipase A$_1$ activity based on the phospholipase A$_1$–specific PED-A1 substrate (A10070, described above). The EnzChek® Phospholipase A$_1$ Assay Kit can detect phospholipase A$_1$ activity at 0.04 U/mL or lower (Figure 17.4.4). This microplate-based assay is well suited for rapid and direct analysis of phospholipase A$_1$ in purified enzyme preparations and cell lysates using automated instrumentation, as well as for characterizing phospholipase A$_1$ inhibitors.

Each EnzChek® Phospholipase A$_1$ Assay Kit (2-plate size, E10219; 10-plate size, E10221) provides:

- PED-A1 phospholipase A$_1$ substrate
- Phospholipase A$_1$ (Lecitase Ultra)
- Concentrated phospholipase A$_1$ reaction buffer
- Dioleoylphosphatidylcholine (DOPC)
- Dioleoylphosphatidylglycerol (DOPG)
- Dimethylsulfoxide (DMSO)
- Detailed assay protocols

The 2-plate assay kit provides sufficient reagents for 200 reactions in 96-well microplates at a volume of 100 µL per well or 800 reactions using low-volume 384-well microplates at a volume of ≤25 µL per well. The 10-plate assay kit provides sufficient reagents for 1000 reactions in 96-well microplates at a volume of 100 µL per well or 4000 reactions using low-volume 384-well microplates at a volume of ≤25 µL per well.

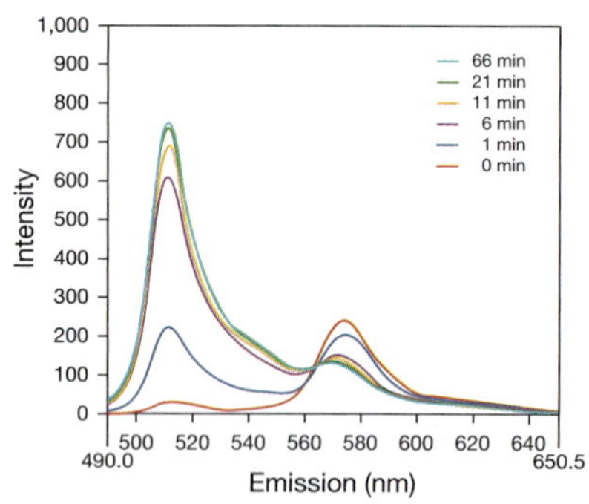

Figure 17.4.6 Fluorescence emission spectra (excitation at 480 nm) of Red/Green BODIPY® PC-A2 phospholipase A$_2$ substrate (A10072) incorporated in liposomes with addition of bee venom phospholipase A$_2$ at ambient temperature.

Red/Green BODIPY® PC-A2 Ratiometric Phospholipase A$_2$ Substrate

Red/Green BODIPY® PC-A2 (A10072, Figure 17.4.5) is a ratiometric fluorogenic substrate designed to provide selective, real-time monitoring of phospholipase A$_2$ activity in purified enzyme preparations, cell lysates and live cells. Cleavage of the BODIPY® FL pentanoic acid substituent at the *sn*-2 position results in decreased quenching by fluorescence resonance energy transfer (FRET) of the BODIPY® 558/568 dye attached at the *sn*-1 position. Thus, upon cleavage by phospholipase A$_2$, Red/Green BODIPY® PC-A2 exhibits an increase in BODIPY® FL fluorescence, detected from 515–545 nm (Figure 17.4.6). The FRET-sensitized BODIPY® 558/568 fluorescence signal is expected to show a reciprocal decrease; in practice, however, this longer-wavelength fluorescence may show a decrease or a slight increase, depending on the formulation of the substrate and the instrument wavelength settings. The ratiometric detection mode of this substrate (emission intensity ratio at 515/575 nm with excitation at ~460 nm) allows measurements of phospholipase A$_2$ activity that are essentially independent of instrumentation and assay conditions. The dual-emission properties of this substrate also provide the capacity to localize the lysophospholipid and fatty acid products of the phospholipase A$_2$ cleavage via their distinct spectroscopic signatures in imaging experiments.

EnzChek® Phospholipase A$_2$ Assay Kit

The EnzChek® Phospholipase A$_2$ Assay Kit (E10217, E10218) provides a simple, fluorometric method for continuous monitoring of phospholipase A$_2$ activity based on the phospholipase A$_2$–selective Red/Green BODIPY® PC-A2 (A10072, described above). This phospholipase A$_2$ assay can be used in an intensity-based detection mode, by following the fluorescence increase at

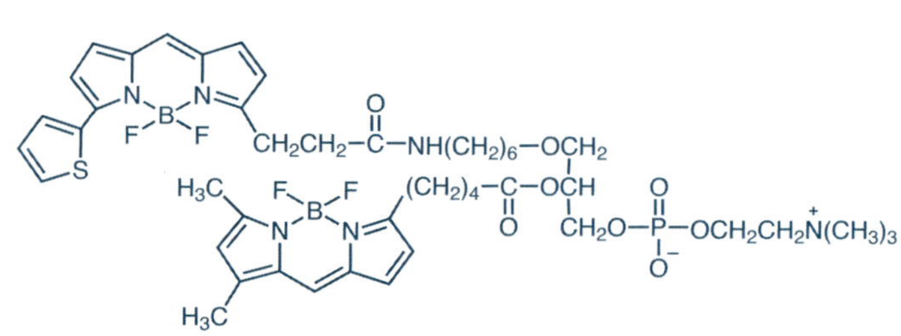

Figure 17.4.5 Red/Green BODIPY® PC-A2 (1-*O*-(6-BODIPY® 558/568-aminohexyl)-2-BODIPY® FL C$_5$-*sn*-glycero-3-phosphocholine, A10072).

IMPORTANT NOTICE: The products described in this manual are covered by one or more Limited Use Label License(s). Please refer to the Appendix on page 971 and Master Product List on page 975. Products are For Research Use Only. Not intended for any animal or human therapeutic or diagnostic use.

~515 nm, or in a ratiometric-based detection mode, by following the changes in the emission intensity ratio at 515/575 nm with excitation at ~460 nm (Figure 17.4.7). The EnzChek® Phospholipase A_2 Assay Kit can detect bee venom phospholipase A_2 activity at 0.05 U/mL or lower (Figure 17.4.7). This microplate-based assay is well suited for rapid and direct analysis of phospholipase A_2 in purified enzyme preparations and cell lysates using automated instrumentation, as well as for characterizing phospholipase A_2 inhibitors.

Each EnzChek® Phospholipase A_2 Assay Kit (2-plate size, E10217; 10-plate size, E10218) provides:

- Red/Green BODIPY® PC-A2 phospholipase A_2 substrate
- Phospholipase A_2 from honey bee venom
- Concentrated phospholipase A_2 reaction buffer
- Dioleoylphosphatidylcholine (DOPC)
- Dioleoylphosphatidylglycerol (DOPG)
- Dimethylsulfoxide (DMSO)
- Detailed assay protocols

The 2-plate assay kit provides sufficient reagents for 200 reactions in 96-well microplates at a volume of 100 µL per well or 800 reactions using low-volume 384-well microplates at a volume of ≤25 µL per well. The 10-plate assay kit provides sufficient reagents for 1000 reactions in 96-well microplates at a volume of 100 µL per well or 4000 reactions using low-volume 384-well microplates at a volume of ≤25 µL per well.

PED6 Phospholipase A_2 Substrate

PED6 (D23739, Figure 17.4.8) is a fluorogenic substrate for phospholipase A_2 incorporating a BODIPY® FL dye–labeled *sn*-2 acyl chain and a dinitrophenyl quencher–labeled head group [16] (Figure 17.4.2). Cleavage of the dye-labeled acyl chain by phospholipase A_2 eliminates the intramolecular quenching effect of the dinitrophenyl group, resulting in a corresponding fluorescence increase. Continuous kinetic assays show PED6 to be a good substrate for both secreted and cytosolic phospholipase A_2 and platelet-activating factor acetylhydrolase.[16] PED6 has been used by Steven Farber and co workers for *in vivo* analysis of intestinal lipid metabolism in zebrafish larvae as a basis for identifying and screening mutant phenotypes [12,13,17] (Figure 17.4.9). PED6 is also useful for high-throughput screening of potential phospholipase A_2 inhibitors or activators.

Other BODIPY® Dye Phospholipase A Substrates

The bis-BODIPY® phospholipase A substrate—bis-BODIPY® FL glycerophosphocholine (bis-BODIPY® FL C_{11}-PC, B7701)—has been specifically designed to allow continuous monitoring of phospholipase A action and to be spectrally compatible with argon-ion laser excitation sources.[18] When this probe is incorporated into cell membranes, the proximity of the BODIPY® FL fluorophores on adjacent phospholipid acyl chains causes fluorescence self-quenching (Figure 17.4.2). Separation of the fluorophores upon hydrolytic cleavage of one of the acyl chains by either phospholipase A_1 or A_2 results in increased fluorescence. Bis-BODIPY® FL C_{11}-PC has been developed in collaboration with Elizabeth Simons, who has successfully employed it for flow cytometric detection of phospholipase A activity in neutrophils.[19] More recently, bis-BODIPY® FL C_{11}-PC has been used to detect phospholipase A_2 activation induced by tumor necrosis factor (TNF) [20] and for high-throughput assays of endothelial lipase, a critical determinant of HDL cholesterol levels.[21,22]

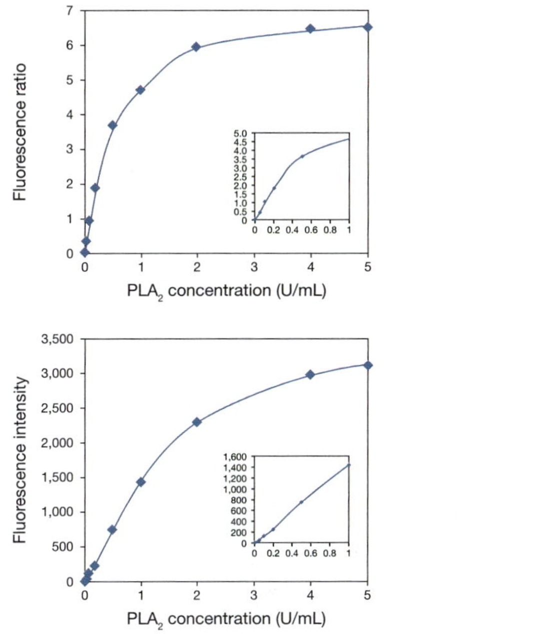

Figure 17.4.7 Detection of phospholipase A_2 (PLA_2) using the EnzChek® Phospholipase A_2 Assay Kit (E10217, E10218). PLA_2 reactions were run at ambient temperature with liposomes for 10 minutes according to the assay protocol provided, and fluorescence emission was measured using 460 nm excitation on a Spectra Max® M5 (Molecular Devices). Background fluorescence determined for the no-enzyme control reaction has been subtracted. Top panel shows ratiometric-based (515/575 nm) detection mode; bottom panel shows intensity-based (515 nm channel) detection mode. Background fluorescence determined for the no-enzyme control reaction has been subtracted for each value.

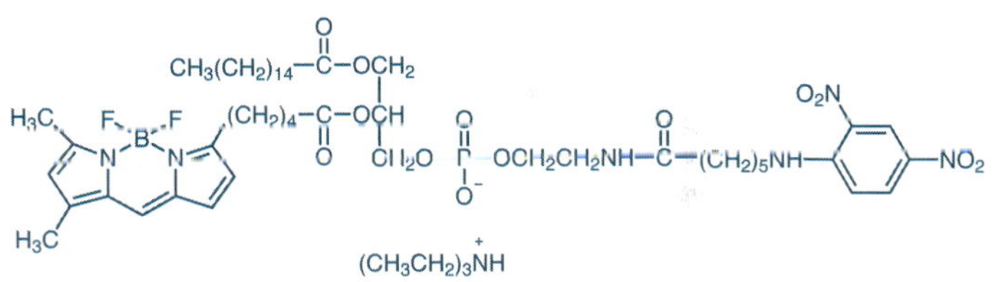

Figure 17.4.8 *N*-((6-(2,4-dinitrophenyl)amino)hexanoyl)-2-(4,4-difluoro-5,7-dimethyl-4-bora-3a,4a-diaza-*s*-indacene-3-pentanoyl)-1-hexadecanoyl-*sn*-glycero-3-phosphoethanolamine, triethylammonium salt (PED6) (D23739).

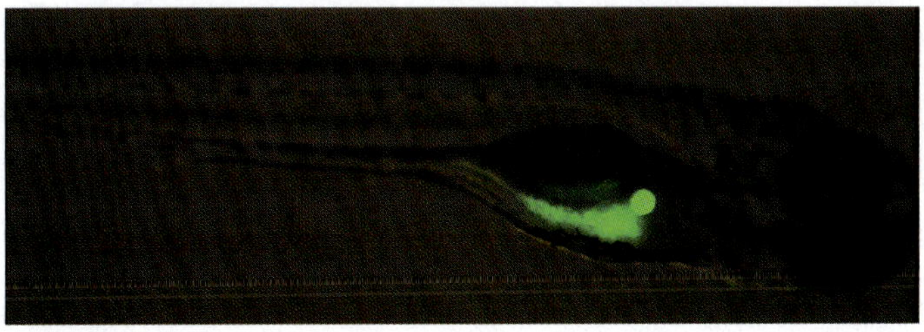

Figure 17.4.9 Imaging of lipid digestion pathways in zebrafish (*Danio rerio*) using the fluorogenic phospholipase A_2 substrate PED6 (D23739). A zebrafish larva (5 days post-fertilization) was incubated with PED6 for 2 hours. Localized fluorescence in the gallbladder and intestinal lumen results from endogenous lipase activity and rapid transport of the substrate cleavage products through the intestinal and hepatobiliary systems. The image was provided by Steven A. Farber, Thomas Jefferson University.

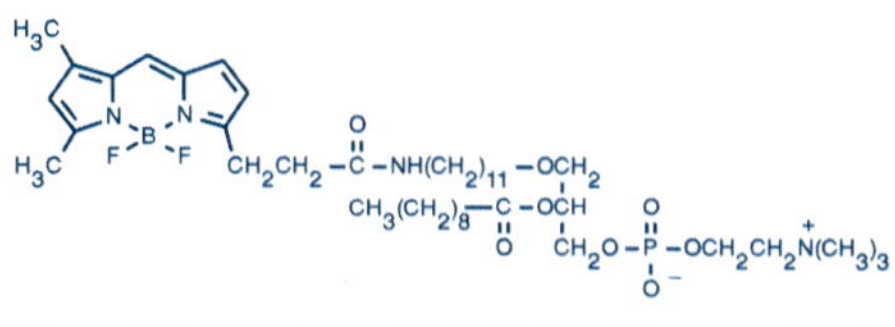

Figure 17.4.10 2-decanoyl-1-(*O*-(11-(4,4-difluoro-5,7-dimethyl-4-bora-3a,4a-diaza-*s*-indacene-3-propionyl)amino)undecyl)-*sn*-glycero-3-phosphocholine (D3771).

Specificity for phospholipase A_2 versus phospholipase A_1 can be obtained using phospholipids with nonhydrolyzable, ether-linked alkyl chains in the *sn*-1 position. A 1-*O*-alkyl–substituted phospholipid containing the BODIPY® FL fluorophore (D3771, Figure 17.4.10) is a useful substrate for a phospholipase A_2–specific chromatographic assay.[23]

The singly labeled BODIPY® phospholipase A_2 substrate—β-BODIPY® FL C_5-HPC (D3803)—has been used to quantitatively delineate a discontinuous increase of Ca^{2+}-dependent cytosolic phospholipase A_2 ($cPLA_2$) activity during zebrafish embryogenesis. The analytical method developed for this study uses a fluorescence image scanner to quantitatively detect the free BODIPY® FL dye–labeled fatty acid generated by the action of $cPLA_2$[24] (Figure 17.4.11).

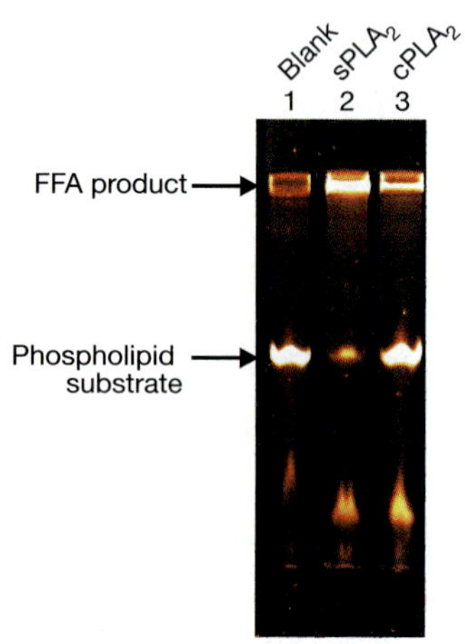

Figure 17.4.11 Assay of cytoplasmic phospholipase A_2 ($cPLA_2$) using β-BODIPY® FL C_5-HPC (D3803) as a substrate. The substrate was incubated in enzyme-free assay buffer (lane 1), with secreted PLA_2 (from *Naja mossambica*; lane 2) or with purified human recombinant $cPLA_2$ (lane 3). Cleavage products were separated by thin-layer chromatography in chloroform/methanol/acetic acid/water (50:30:8:4) and were subsequently analyzed using a fluorescence image scanner. Both phospholipases liberated fluorescent BODIPY® FL dye–labeled fatty acids (FFA) by cleavage of the substrate at the *sn*-2 acyl bond. Figure reproduced with permission from J Biol Chem (1999) 274:19338.

Bis-Pyrenyl Phospholipase A Substrates

Our bis-pyrenyl phospholipase A probes (B3781, B3782) both emit at ~470 nm, indicating that their adjacent pyrene fluorophores (Figure 17.4.12) form excited-state dimers (Figure 17.4.13). Phospholipase A–mediated hydrolysis separates the fluorophores, which then emit as monomers at ~380 nm.[25] These substrates have proven to be effective phospholipase A_2 substrates in model membrane systems (Table 17.3); however, it has been reported that 1,2-bis-(1-pyrenebutanoyl)-*sn*-glycero-3-phosphocholine (B3781) is highly resistant to degradation by phospholipases in human skin fibroblasts.[26] 1,2-bis-(1-pyrenebutanoyl)-*sn*-glycero-3-phosphocholine has been used in a sensitive, continuous assay for lecithin:cholesterol acyltransferase[27,28] (LCAT).

Singly Labeled Pyrenyl and NBD Phospholipase A_2 Substrates

Phospholipase A_2 activity has also been measured using phospholipids labeled with a single pyrene (H361, Figure 17.4.14; H3809, Figure 17.4.15) or NBD (N3786; N3787, Figure 17.4.16) fluorophore (Table 17.3). Because only the *sn*-2 phospholipid acyl chain is labeled, these probes can discriminate between phospholipase A_2 and phospholipase A_1 activity. To obtain a direct fluorescence response to enzymatic cleavage, sufficient phospholipid must be loaded into membranes to cause either intermolecular self-quenching (NBD-acyl phospholipids) or excimer formation[29] (pyreneacyl phospholipids). Pyrene-labeled acidic phospholipids—particularly the phosphoglycerol derivative[30,31] (H3809)—are preferred as substrates by pancreatic and intestinal phospholipase A_2, whereas labeled phosphocholine (H361, Figure 17.4.14) is preferred by phospholipase A_2 from snake venom.[32]

ADIFAB Indicator: A Different View of Phospholipase A Activity

The ADIFAB fatty acid indicator (A3880, Figure 17.4.17) functions as a fluorescent sensor for the free fatty acid cleavage products of phospholipases.[33–36] It does not require membrane loading

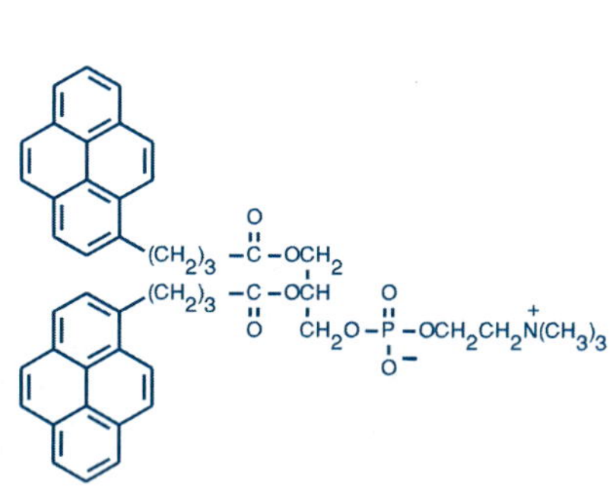

Figure 17.4.12 1,2-bis-(1-pyrenebutanoyl)-*sn*-glycero-3-phosphocholine (B3781).

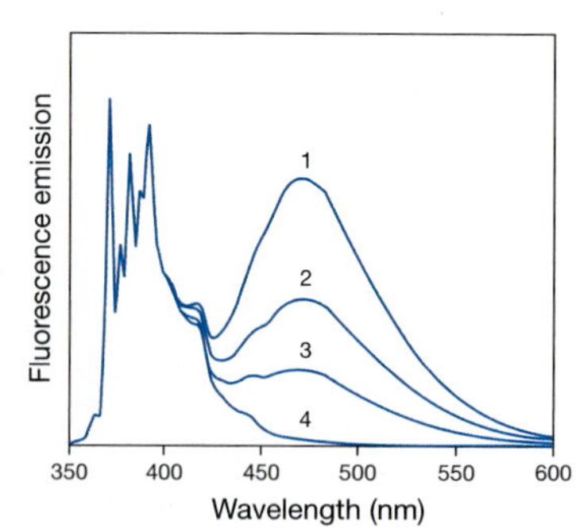

Figure 17.4.13 Excimer formation by pyrene in ethanol. Spectra are normalized to the 371.5 nm peak of the monomer. All spectra are essentially identical below 400 nm after normalization. Spectra are as follows: **1**) 2 mM pyrene, purged with argon to remove oxygen; **2**) 2 mM pyrene, air-equilibrated; **3**) 0.5 mM pyrene (argon-purged); and **4**) 2 µM pyrene (argon-purged). The monomer-to-excimer ratio (371.5 nm/470 nm) is dependent on both pyrene concentration and the excited-state lifetime, which is variable because of quenching by oxygen.

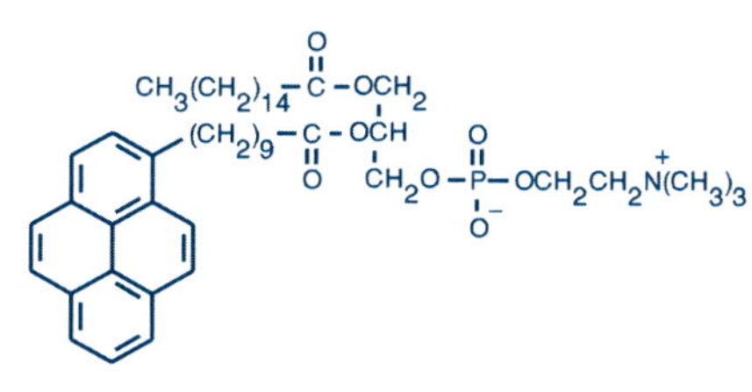

Figure 17.4.14 1-hexadecanoyl-2-(1-pyrenedecanoyl)-*sn*-glycero-3-phosphocholine (β-py-C10-HPC) (H361).

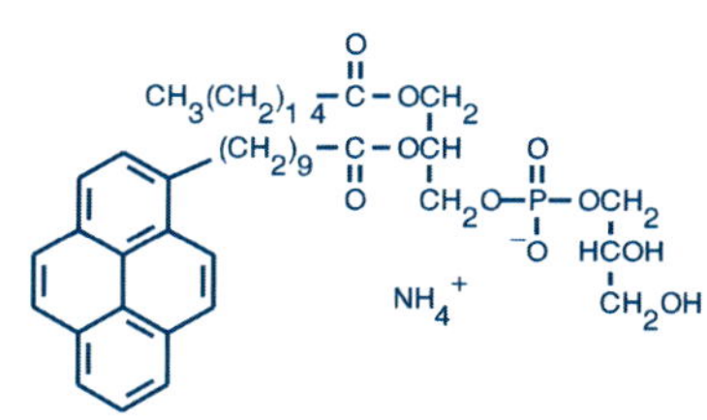

Figure 17.4.15 1-hexadecanoyl-2-(1-pyrenedecanoyl)-*sn*-glycero-3-phosphoglycerol, ammonium salt (β-py-C10-PG) (H3809).

IMPORTANT NOTICE: The products described in this manual are covered by one or more Limited Use Label License(s). Please refer to the Appendix on page 971 and Master Product List on page 975. Products are For Research Use Only. Not intended for any animal or human therapeutic or diagnostic use.
www.invitrogen.com/probes

and can be used to monitor hydrolysis of natural (rather than synthetic) substrates. Assaying lysophospholipase activity with ADIFAB yields sensitivity comparable to radioisotopic methods.[37] Richieri and Kleinfeld have described a methodology for using the ADIFAB reagent to measure the activity of phospholipase A_2 on cell and lipid-vesicle membranes; their assay is capable of detecting hydrolysis rates as low as 10^{-12} mole/minute.[36]

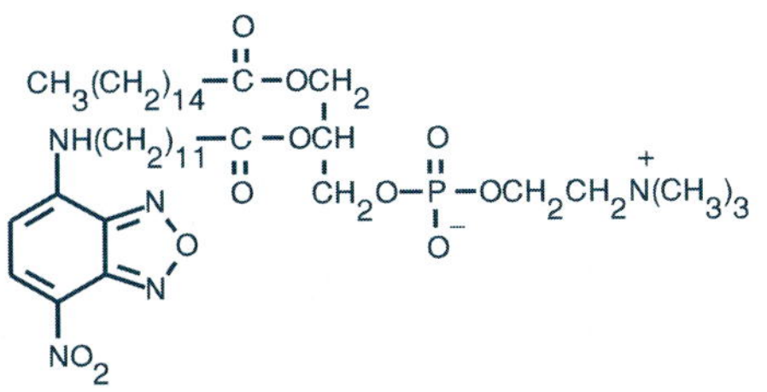

Figure 17.4.16 2-(12-(7-nitrobenz-2-oxa-1,3-diazol-4-yl)amino)dodecanoyl-1-hexadecanoyl-*sn*-glycero-3-phosphocholine (NBD C12-HPC) (N3787).

Phospholipase C Assays

EnzChek® Direct Phospholipase C Assay Kit

The EnzChek® Direct Phospholipase C Assay Kit (E10215, E10216) provides a simple and robust microplate-based method for monitoring phosphatidylcholine-specific phospholipase C (PC-PLC) activity in purified enzyme preparations. PC-PLC plays a crucial role in many cell signaling pathways involved in apoptosis and cell survival, as well as in diseases as diverse as cancer and HIV. This assay uses a proprietary substrate (glycerophosphoethanolamine with a dye-labeled *sn*-2 acyl chain) to detect PC-PLC activity. Substrate cleavage by PC-PLC releases dye-labeled diacylglycerol, which produces a positive fluorescence signal that can be measured continuously using a fluorescence microplate reader. The reaction product has fluorescence excitation and emission maxima of 509 nm and 516 nm, respectively.

The EnzChek® Direct Phospholipase C Assay Kit has been optimized using purified PC-PLC from *Bacillus cereus*. This assay may be amenable for use with cells and cell lysates, although the presence of phospholipase A_2 or phospholipase D activity can potentially result in confounding signal enhancement. Using the EnzChek® Direct Phospholipase C Assay Kit with purified enzyme from *Bacillus cereus*, we can typically detect as little as 10 mU/mL PC-PLC after one hour incubation at room temperature (Figure 17.4.18). This kit is also useful for characterizing PC-PLC inhibition, and because it offers a direct measurement, the potential for false positives in a compound screen is reduced.

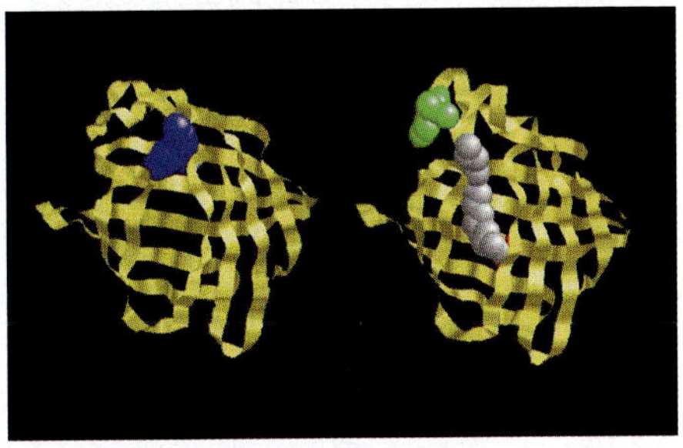

Figure 17.4.17 Ribbon representation of the ADIFAB free fatty acid indicator (A3880). In the left-hand image, the fatty acid binding site of intestinal fatty acid–binding protein (yellow) is occupied by a covalently attached acrylodan fluorophore (blue). In the right-hand image, a fatty acid molecule (gray) binds to the protein, displacing the fluorophore (green) and producing a shift of its fluorescence emission spectrum. Image contributed by Alan Kleinfeld, FFA Sciences LLC, San Diego.

Each EnzChek® Direct Phospholipase C Assay Kit (2-plate size, E10215; 10-plate size, E10216) provides:

- Phosphatidylcholine-specific phospholipase C (PC-PLC) substrate
- Phospholipase C from *Bacillus cereus*
- Concentrated phospholipase C reaction buffer
- Phosphatidylcholine (lecithin)
- Dimethylsulfoxide (DMSO)
- Detailed assay protocols

The 2-plate assay kit provides sufficient reagents for 200 reactions in 96-well microplates at a volume of 200 µL per well or 2000 reactions using low-volume 384-well microplates at a volume of 20 µL per well. The 10-plate assay kit provides sufficient reagents for 1000 reactions in 96-well microplates at a volume of 200 µL per well or 10,000 reactions using low-volume 384-well microplates at a volume of 20 µL per well.

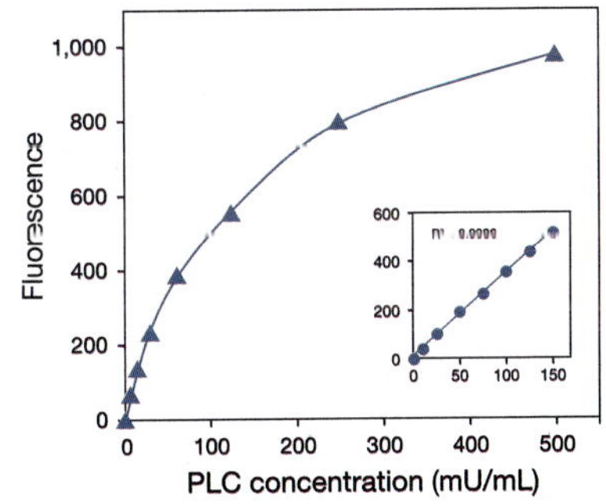

Figure 17.4.18 Detection of phosphatidylcholine-specific phospholipase C (PC-PLC) using the EnzChek® Direct Phospholipase C Assay Kit (E10215, E10216). Triplicate samples of PC-PLC from *Bacillus cereus* were assayed at concentration of 7.8 mU/mL to 500 mU/mL per well in the presence of 1X PLC substrate and 200 µM lecithin in 1X PLC reaction buffer. Reactions were incubated at room temperature for 60 minutes and fluorescence was measured using excitation/emission wavelengths of 490/520 nm. The inset represents a separate experiment and illustrates the linearity of fluorescence response at low levels of PC-PLC. The average variation of replicates (CV) was less than 3%. Background fluorescence determined for the no-enzyme control reaction has been subtracted.

Amplex® Red Phosphatidylcholine-Specific Phospholipase C Assay Kit

The Amplex® Red Phosphatidylcholine-Specific Phospholipase C Assay Kit (A12218) provides a sensitive method for continuously monitoring phosphatidylcholine-specific phospholipase C (PC-PLC) activity *in vitro* using a fluorescence microplate reader or fluorometer.[38–40] In this enzyme-coupled assay, PC-PLC activity is monitored indirectly using the Amplex® Red reagent, a sensitive fluorogenic probe for H_2O_2 (Section 10.5). First, PC-PLC converts the phosphatidylcholine (lecithin) substrate to form phosphocholine and diacylglycerol. After the action of alkaline phosphatase, which hydrolyzes phosphocholine to inorganic phosphate and choline, choline is oxidized by choline oxidase to betaine and H_2O_2. Finally, H_2O_2, in the presence of horseradish peroxidase, reacts with the Amplex® Red reagent in a 1:1 stoichiometry to generate the highly fluorescent product, resorufin. Because resorufin has absorption and fluorescence emission maxima of approximately 571 nm and 585 nm, respectively, there is little interference from autofluorescence in most biological samples.

The Amplex® Red Phosphatidylcholine-Specific Phospholipase C Assay Kit is potentially useful for detecting PC-PLC activity in cell extracts and for screening PC-PLC inhibitors. Experiments

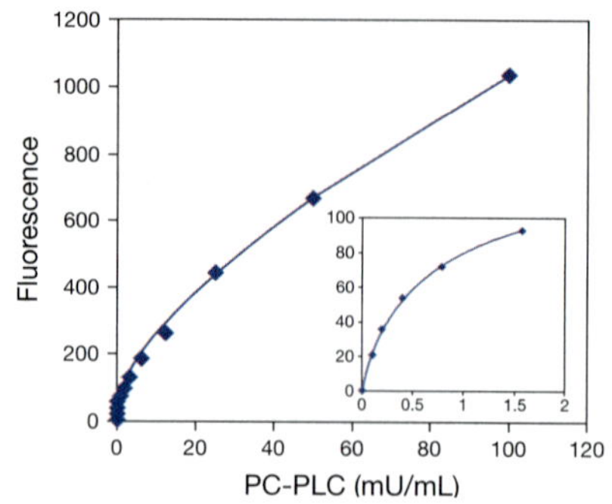

Figure 17.4.19 Detection of phosphatidylcholine-specific phospholipase C using the Amplex® Red Phosphatidylcholine-Specific Phospholipase C Assay Kit (A12218). Fluorescence was measured in a fluorescence microplate reader using excitation at 560 ± 10 nm and fluorescence detection at 590 ± 10 nm. The inset shows the sensitivity at very low enzyme concentrations.

with purified PC-PLC from *Bacillus cereus* indicate that the Amplex® Red Phosphatidylcholine-Specific Phospholipase C Assay Kit can detect PC-PLC levels as low as 0.2 mU/mL using a reaction time of one hour (Figure 17.4.19). One unit of PC-PLC is defined as the amount of enzyme that will liberate 1.0 micromole of water-soluble organic phosphorus from L-α-phosphatidylcholine per minute at pH 7.3 at 37°C.

Each Amplex® Red Phosphatidylcholine-Specific Phospholipase C Assay Kit includes:

- Amplex® Red reagent
- Dimethylsulfoxide (DMSO)
- Horseradish peroxidase (HRP)
- H_2O_2 for use as a positive control
- Concentrated reaction buffer
- Choline oxidase from *Alcaligenes* sp.
- Alkaline phosphatase from calf intestine
- L-α-Phosphatidylcholine (lecithin)
- Phosphatidylcholine-specific phospholipase C from *Bacillus cereus*
- Detailed protocols

Each kit provides sufficient reagents for approximately 500 assays using a fluorescence microplate reader and a reaction volume of 200 µL per assay.

Bacillus cereus PI-PLC

Phosphatidylinositol-specific phospholipase C (PI-PLC, EC 3.1.4.10) from *Bacillus cereus* cleaves phosphatidylinositol (PI), yielding water-soluble D-*myo*-inositol 1,2-cyclic monophosphate and lipid-soluble diacylglycerol.[41] This enzyme also functions to release enzymes that are linked to glycosylphosphatidylinositol (GPI) membrane anchors. We offer highly purified *B. cereus* PI-PLC (P6466), which has been used in studies of PI synthesis and export across the plasma membrane.[42] PI-PLC generates diacylglycerols for PKC-linked signal transduction studies[43] and provides an efficient means of releasing most GPI-anchored proteins from cell surfaces under conditions in which the cells remain viable.[44,45]

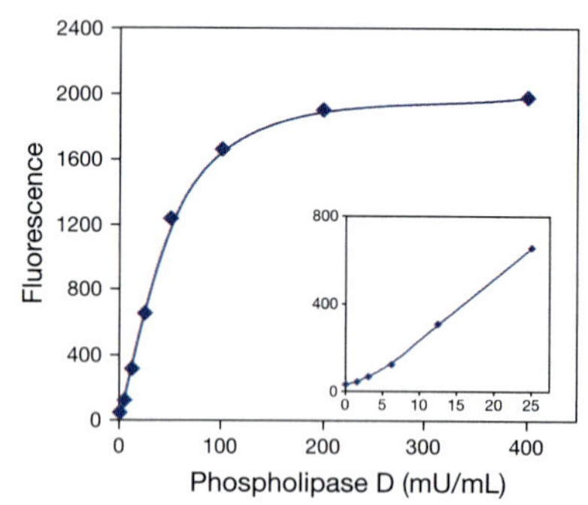

Figure 17.4.20 Quantitation of phospholipase D from *Streptomyces chromofuscus* using the Amplex® Red Phospholipase D Assay Kit (A12219). Fluorescence was measured with a fluorescence microplate reader using excitation at 530 ± 12.5 nm and fluorescence detection at 590 ± 17.5 nm. The inset shows the sensitivity at very low enzyme concentrations (0–25 mU/mL).

Phospholipase D Assays

Amplex® Red Phospholipase D Assay Kit

The Amplex® Red Phospholipase D Assay Kit (A12219) provides a sensitive method for measuring phospholipase D (PLD) activity *in vitro* using a fluorescence microplate reader or fluorometer.[38,40,46] In this enzyme-coupled assay, PLD activity is monitored indirectly using the Amplex® Red reagent (Section 10.5). First, PLD cleaves the phosphatidylcholine (lecithin) substrate to yield choline and phosphatidic acid. Second, choline is oxidized by choline oxidase to betaine and H_2O_2. Finally, H_2O_2, in the presence of horseradish peroxidase, reacts with the Amplex® Red reagent to generate the highly fluorescent product, resorufin (excitation/emission maxima ~571/585 nm).

The Amplex® Red Phospholipase D Assay Kit is designed for detecting PLD activity in cell extracts and for screening PLD inhibitors.This kit can be used to continuously assay PLD enzymes with near-neutral pH optima, whereas PLD enzymes with acidic pH optima can be assayed in a simple two-step procedure. Experiments with purified PLD from *Streptomyces chromofuscus* indicate that the Amplex® Red Phospholipase D Assay Kit can detect PLD levels as low as 10 mU/mL using a reaction time of one hour (Figure 17.4.20). One unit of PLD is defined as the amount of enzyme that will liberate 1.0 micromole of choline from L-α-phosphatidylcholine per minute at pH 8.0 at 30°C. Each Amplex® Red Phospholipase D Assay Kit includes:

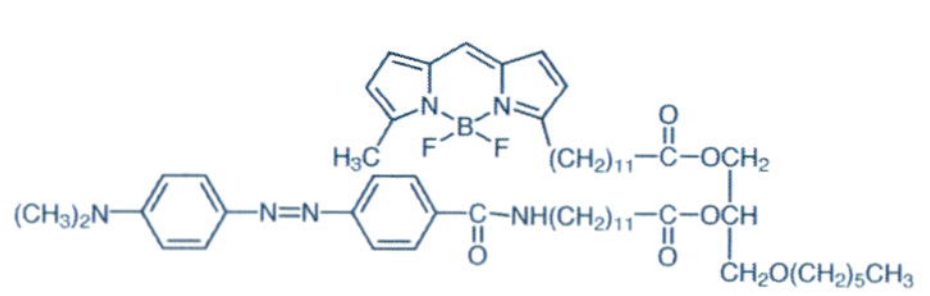

Figure 17.4.21 EnzChek® lipase substrate (E33955).

- Amplex® Red reagent
- Dimethylsulfoxide (DMSO)
- Horseradish peroxidase (HRP)
- H_2O_2 for use as a positive control
- Concentrated reaction buffer
- Choline oxidase from *Alcaligenes* sp.
- L-α-Phosphatidylcholine (lecithin)
- Detailed protocols

IMPORTANT NOTICE: The products described in this manual are covered by one or more Limited Use Label License(s). Please refer to the Appendix on page 971 and Master Product List on page 975. Products are For Research Use Only. Not intended for any animal or human therapeutic or diagnostic use.
www.invitrogen.com/probes

Each kit provides sufficient reagents for approximately 500 assays using a fluorescence microplate reader and a reaction volume of 200 µL per assay.

Fluorescent Substrates for Phospholipase D

The products of phospholipase A_2, C and D cleavage of 1-*O*-alkyl-2-decanoyl-*sn*-glycero-3-phosphocholine labeled with the BODIPY® FL fluorophore (D3771, Figure 17.4.10) can be separated and independently quantitated based on their differential migration on TLC or HPLC.[23,47,48] Our BODIPY® FL analog is preferred for this application because it is relatively photostable and the fluorescence properties of its different enzymatic products are all very similar.[49] Researchers have taken advantage of these features to detect and quantitate phospholipase D activity in vascular smooth muscle cells,[49,50] cultured mammalian cells[51] and yeast.[52–54]

EnzChek® Lipase Substrate

The triacylglycerol-based EnzChek® lipase substrate (E33955, Figure 17.4.21) offers higher throughput and better sensitivity than chromogenic (TLC or HPLC) assays, and a visible light–excitable alternative to 6,8-difluoro-4-methylumbelliferyl octanoate (DiFMU octanoate, D12200; Section 10.6). In the presence of lipases, the nonfluorescent EnzChek® lipase substrate produces a bright, green-fluorescent product (excitation/emission maxima of ~505/515 nm) for the accurate and sensitive detection of lipase activity in solution.

Anti-Phosphoinositide Monoclonal Antibodies

Phosphatidylinositol (PI or PtdIns) and its phosphorylated derivatives represent only a small fraction of eukaryotic cellular phospholipids but are functionally significant in a disproportionately large number of regulatory and signal transduction processes.[55–60] The most familiar of these processes is the phospholipase C–mediated generation of the ubiquitous second messengers inositol 1,4,5-triphosphate ($InsP_3$) and diacylglycerol (DAG) from phosphatidylinositol 4,5-diphosphate (PtdIns(4,5)P_2; Section 17.2). Research has revealed the direct action of phosphatidylinositol 4,5-diphosphate (PtdIns(4,5)P_2) and phosphatidylinositol 3,4,5-triphosphate (PtdIns(3,4,5)P_3) on a diverse array of cellular functions, including actin assembly and cytoskeletal dynamics, vesicular protein trafficking, protein kinase localization and activation, cell proliferation and apoptosis. We offer mouse monoclonal IgM antibodies to PtdIns(4,5)P_2 (A21327) and PtdIns(3,4,5)P_3 (A21328) for immunocytochemical localization of these important lipid metabolites.[61] Both antibodies have been shown to recognize their cognate phosphoinositides in murine and human cells with only slight cross-reactivity with other phosphoinositides or phospholipids.

Sphingolipids

Sphingolipids include sphingomyelins, which are phospholipid analogs, as well as ceramides, glycosyl ceramides (cerebrosides), gangliosides and other derivatives (Figure 17.4.22). Several excellent reviews of the chemistry and biology of sphingolipids and glycosphingolipids and their role in the process of signal transduction are available.[62,63]

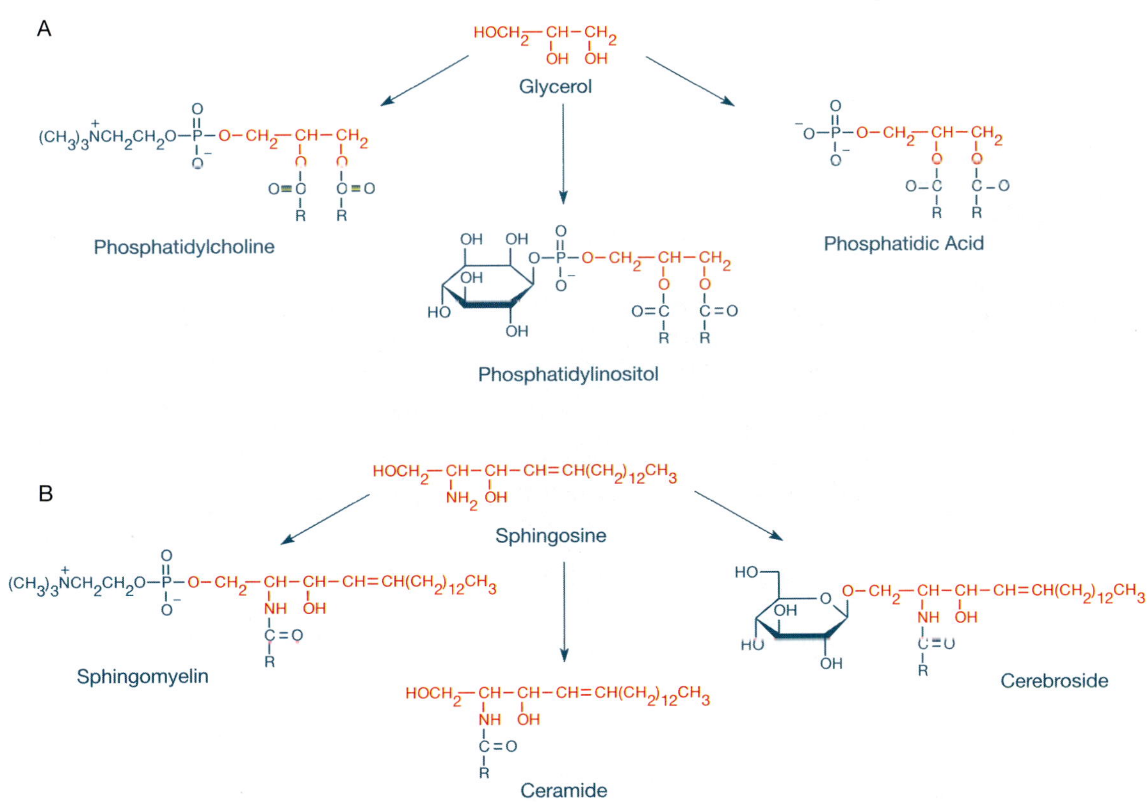

Figure 17.4.22 **A)** Phosphatidylcholines, phosphatidylinositols and phosphatidic acids are examples of glycerolipids derived from glycerol. **B)** Sphingomyelins, ceramides and cerebrosides are examples of sphingolipids derived from sphingosine. In all the structures shown, R represents the hydrocarbon tail portion of a fatty acid residue.

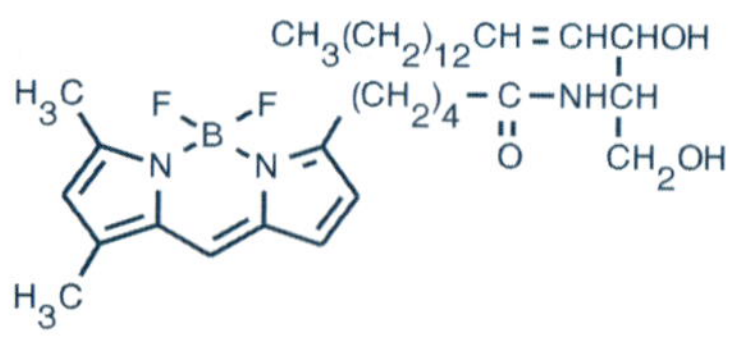

Figure 17.4.23 N-(4,4-difluoro-5,7-dimethyl-4-bora-3a,4a-diaza-s-indacene-3-pentanoyl)sphingosine (BODIPY® FL C_5-ceramide) (D3521).

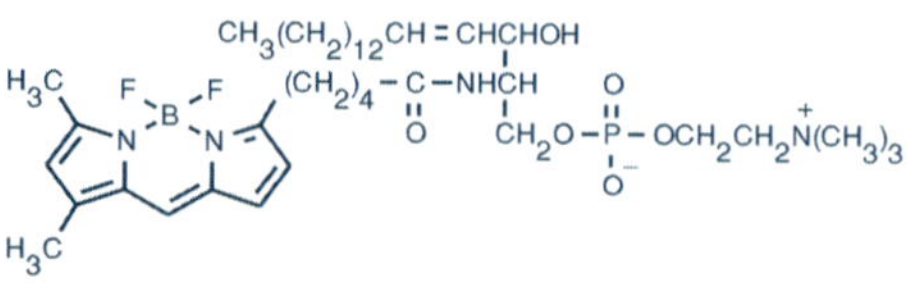

Figure 17.4.24 N-(4,4-difluoro-5,7-dimethyl-4-bora-3a,4a-diaza-s-indacene-3-pentanoyl)sphingosyl phosphocholine (BODIPY® FL C_5-sphingomyelin) (D3522).

BODIPY® Sphingolipids

Ceramides (*N*-acylsphingosines), like diacylglycerols, are lipid second messengers that function in signal transduction processes.[63–65] The concentration-dependent spectral properties of BODIPY® FL C_5-ceramide (D3521, B22650; Figure 17.4.23), BODIPY® FL C_5-sphingomyelin[66–68] (D3522, Figure 17.4.24) and BODIPY® FL C_{12}-sphingomyelin[69] (D7711) make them particularly suitable for investigating sphingolipid transport and metabolism,[68,70–73] in addition to their applications as structural markers for the Golgi complex[74] (Section 12.4). BODIPY® FL C_5-ceramide can be visualized by fluorescence microscopy[75,76] (Figure 17.4.25, Figure 17.4.26) or by electron microscopy following diaminobenzidine (DAB) photoconversion to an electron-dense product.[77] (Fluorescent Probes for Photoconversion of Diaminobenzidine Reagents—Note 14.2).

Our range of BODIPY® sphingolipids also includes the long-wavelength light-excitable BODIPY® TR ceramide[78,79] (D7540, Figure 17.4.27), as well as BODIPY® FL C_5-lactosylceramide[80–85] (D13951), BODIPY® FL C_5-ganglioside G_{M1}[86] (B13950, Figure 17.4.28) and BODIPY® FL C_{12}-galactocerebroside (D7519). All of our sphingolipids are prepared from D-*erythro*-sphingosine and therefore have the same stereochemical conformation as natural biologically active sphingolipids.[87]

Complexing fluorescent lipids with defatted bovine serum albumin (BSA) facilitates cell labeling by eliminating the need for organic solvents to dissolve the lipophilic probe; the

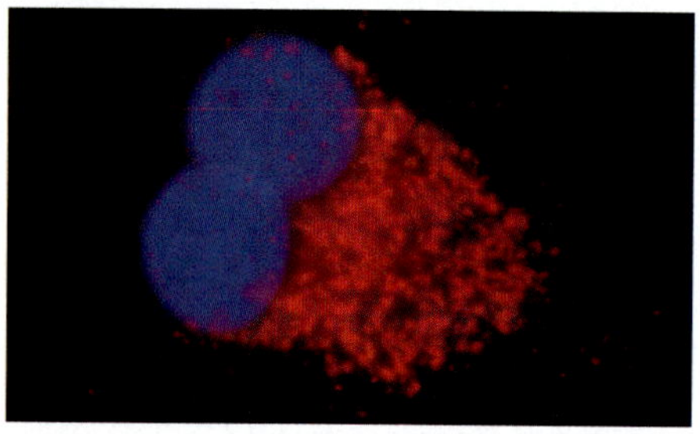

Figure 17.4.25 Nucleus and Golgi apparatus of a bovine pulmonary artery endothelial cell (BPAEC) labeled with Hoechst 33342 (H1399, H3569, H21492) and the BSA complex of BODIPY® FL C_5-ceramide (B22650), respectively.

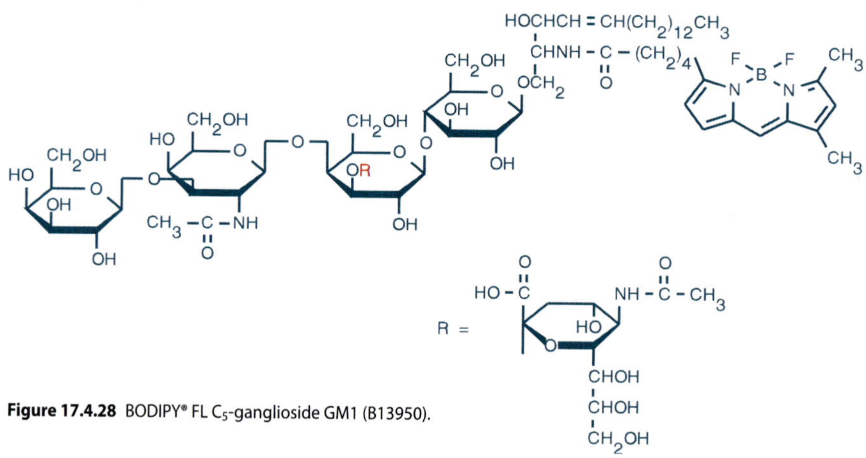

Figure 17.4.28 BODIPY® FL C_5-ganglioside GM1 (B13950).

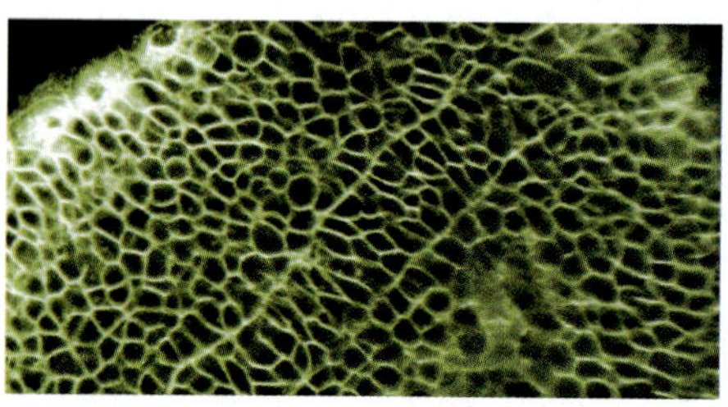

Figure 17.4.26 Cells in the notochord rudiment of a zebrafish embryo undergoing mediolateral intercalation to lengthen the forming notochord. BODIPY® FL C_5-ceramide (D3521) localizes in the interstitial fluid of the zebrafish embryo and freely diffuses between cells, illuminating cell boundaries. This confocal image was obtained using a Bio-Rad® MRC-600 microscope. Image contributed by Mark Cooper, University of Washington.

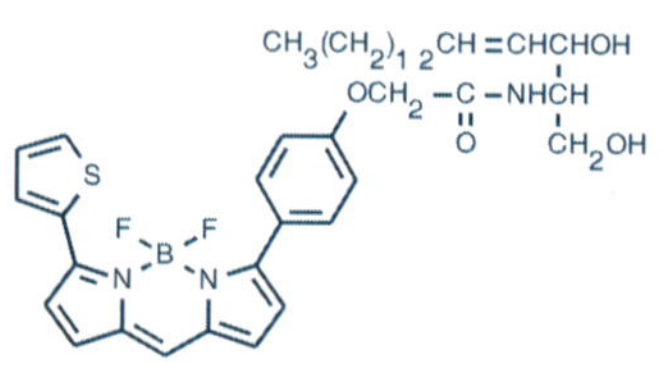

Figure 17.4.27 BODIPY® TR ceramide (*N*-((4-(4,4-difluoro-5-(2-thienyl)-4-bora-3a,4a-diaza-s-indacene-3-yl)phenoxy)acetyl)sphingosine) (D7540).

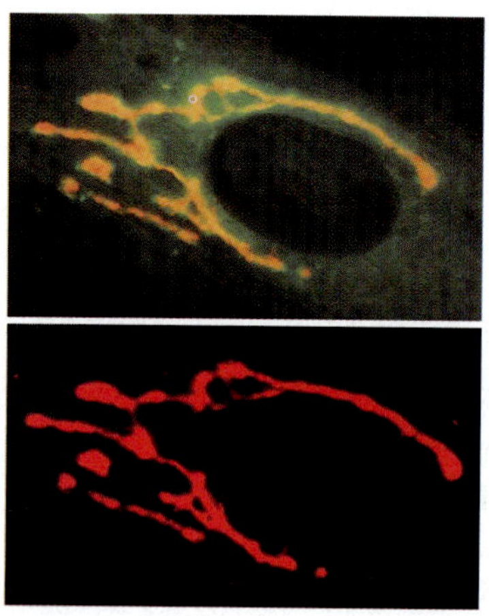

Figure 17.4.29 Selective staining of the Golgi apparatus using the green-fluorescent BODIPY® FL C_5-ceramide (D3521) (top panel). At high concentrations, the BODIPY® FL fluorophore forms excimers that can be visualized using a red longpass optical filter (bottom panel). The BODIPY® FL C_5-ceramide accumulation in the trans-Golgi is sufficient for excimer formation (J Cell Biol (1991) 113:1267). Images contributed by Richard Pagano, Mayo Foundation.

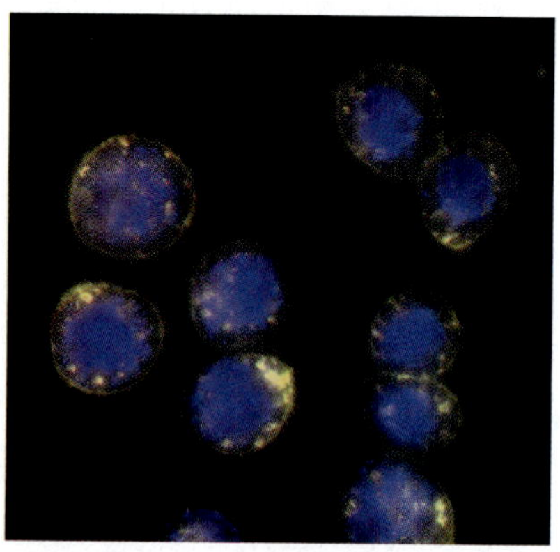

Figure 17.4.30 Live J774 macrophage cells labeled with BODIPY® FL C_5-ganglioside G_{M1} (B13950) and then with Alexa Fluor® 555 cholera toxin subunit B conjugate (C22843). Cells were then treated with anti-CT-B antibody to induce crosslinking. Yellow fluorescence indicates colocalization of the two dyes. Nuclei were stained with the blue fluorescent Hoechst 33342 dye (H1399, H3570, H21492).

IMPORTANT NOTICE: The products described in this manual are covered by one or more Limited Use Label License(s). Please refer to the Appendix on page 971 and Master Product List on page 975. Products are For Research Use Only. Not intended for any animal or human therapeutic or diagnostic use.

BSA-complexed probe can be directly dissolved in water.[88] We offer four BODIPY® sphingolipid–BSA complexes for the study of lipid metabolism and trafficking, including:

- BODIPY® FL C_5-ceramide (B22650)
- BODIPY® TR ceramide (B34400)
- BODIPY® FL C_5-lactosylceramide (B34401)
- BODIPY® FL C_5-ganglioside G_{M1} (B34402)

BODIPY® FL C_5-ceramide has been used to investigate the linkage of sphingolipid metabolism to protein secretory pathways[89–91] and neuronal growth.[83,92] Internalization of BODIPY® FL C_5-sphingomyelin (D3522) from the plasma membrane of human skin fibroblasts results in a mixed population of labeled endosomes that can be distinguished based on the concentration-dependent green (~515 nm) or red (~620 nm) emission of the probe[68] (Figure 17.4.29). BODIPY® C_5-sphingomyelin has also been used to assess sphingomyelinase gene transfer and expression in hematopoietic stem and progenitor cells.[93] BODIPY® FL C_5-lactosylceramide, BODIPY® FL C_5-ganglioside G_{M1} and BODIPY® FL cerebrosides are useful tools for the study of glycosphingolipid transport and signaling pathways in cells.[94–97] BODIPY® FL C_5-ganglioside G_{M1} has been shown to form cholesterol-enhanced clusters in membrane complexes with amyloid β-protein in a model of Alzheimer disease amyloid fibrils.[98] Colocalization of fluorescent cholera toxin B conjugates (Section 7.7) and BODIPY® FL C_5-ganglioside G_{M1} observed by fluorescence microscopy provides a direct indication of the association of these molecules in lipid rafts[99,100] (Figure 17.4.30, Figure 17.4.31).

NBD Sphingolipids

NBD C_6-ceramide (N1154, Figure 17.4.32) and NBD C_6-sphingomyelin (N3524) analogs predate their BODIPY® counterparts and have been extensively used for following sphingolipid metabolism in cells[101–103] and in multicellular organisms.[104] As with BODIPY® FL C_5-ceramide, we also offer NBD C_6-ceramide complexed with defatted BSA (N22651) to facilitate cell loading without the use of organic solvents to dissolve the probe.[88] Elimination of NBD fluorescence at the extracellular surface by dithionite reduction (Figure 17.4.33) can be used to assess endocytosis and recycling of NBD sphingolipids.

Amplex® Red Sphingomyelinase Assay Kit

The Amplex® Red Sphingomyelinase Assay Kit (A12220) is designed for measuring sphingomyelinase activity in solution using a fluorescence microplate reader or fluorometer (Figure 17.4.34). This assay should be useful for screening sphingomyelinase activators or inhibitors or for detecting sphingomyelinase activity in cell and tissue extracts. The assay, which uses natural sphingomyelin as the principal substrate, employs an enzyme-coupled detection scheme in which phosphocholine liberated by the action of sphingomyelinase is cleaved by alkaline phosphatase to generate choline. Choline is, in turn, oxidized to betaine by choline oxidase, generating H_2O_2, which drives the conversion of the Amplex® Red reagent (A12222, A22177; Section 10.5) to red-fluorescent resorufin (Figure 17.4.35). This sensitive assay technique has been employed to detect activation of acid sphingomyelinase associated with ultraviolet radiation–induced apoptosis[105] and to characterize an insecticidal sphingomyelinase C produced by *Bacillus cereus.*[106]

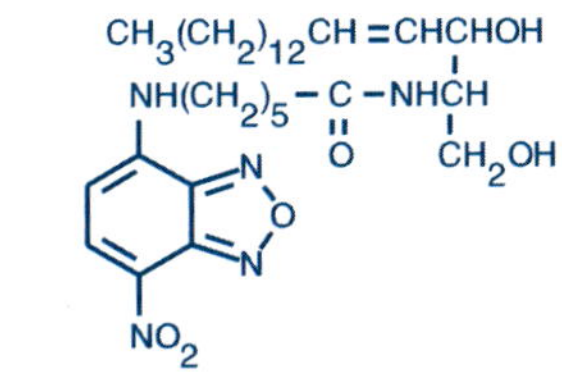

Figure 17.4.32 NBD C_6-ceramide (6-((*N*-(7-nitrobenz-2-oxa-1,3-diazol-4-yl)amino)hexanoyl)sphingosine) (N1154).

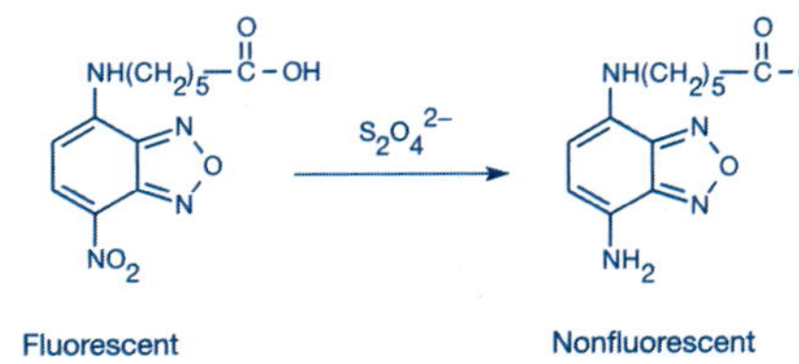

Figure 17.4.33 Dithionite reduction of 6-(*N*-(7-nitrobenz-2-oxa-1,3-diazol-4-yl)amino)hexanoic acid (NBD-X, N316). The elimination of fluorescence associated with this reaction, coupled with the fact that extraneously added dithionite is not membrane permeant, can be used to determine whether the NBD fluorophore is located in the external or internal monolayer of lipid bilayer membranes.

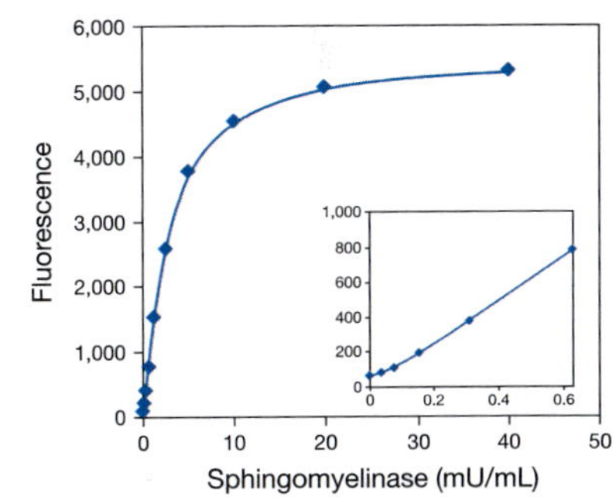

Figure 17.4.34 Measurement of sphingomyelinase activity using the Amplex® Red Sphingomyelinase Assay Kit (A12220). Each reaction contained 50 µM Amplex® Red reagent, 1 U/mL horseradish peroxidase (HRP), 0.1 U/mL choline oxidase, 4 U/mL of alkaline phosphatase, 0.25 mM sphingomyelin and the indicated amount of *Staphylococcus aureus* sphingomyelinase in 1X reaction buffer. Reactions were incubated at 37°C for 1 hour. Fluorescence was measured with a fluorescence microplate reader using excitation at 530 ± 12.5 nm and fluorescence detection at 590 ± 17.5 nm.

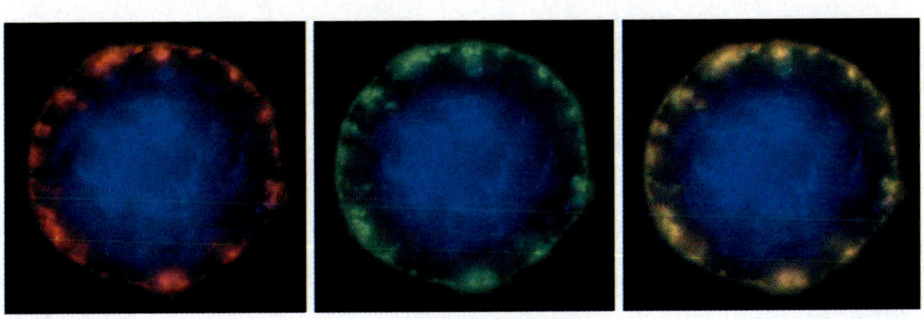

Figure 17.4.31. A J774 mouse macrophage cell sequentially stained with BODIPY® FL ganglioside G_{M1} (B13950) and then with Alexa Fluor® 555 dye–labeled cholera toxin subunit B (C22843, C34776). The cell was then treated with an anti–CT-B antibody to induce crosslinking. Alexa Fluor® 555 dye fluorescence (left panel, red) and BODIPY® FL dye fluorescence (middle panel, green) were imaged separately and overlaid to emphasize the coincident staining (right panel, yellow). Nuclei were stained with blue-fluorescent Hoechst 33258 (H1398, H3569, H21491).

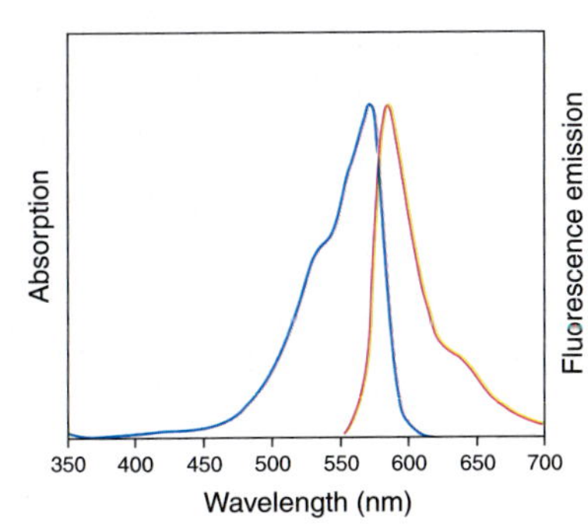

Figure 17.4.35 Absorption and fluorescence emission spectra of resorufin in pH 9.0 buffer.

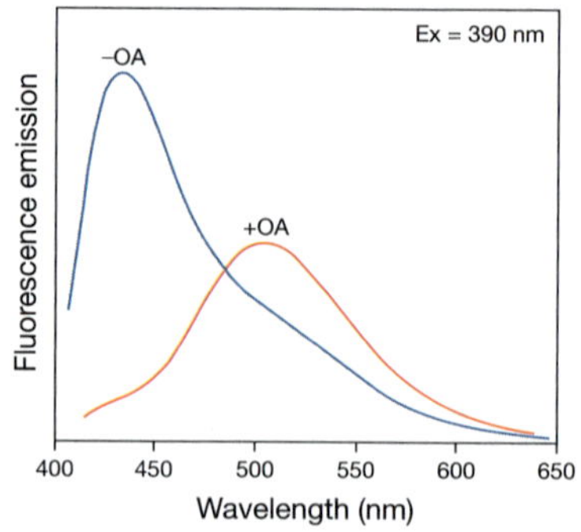

Figure 17.4.36 The free fatty acid–dependent spectral shift of ADIFAB reagent (A3880). Spectra shown represent 0.2 µM ADIFAB in pH 8.0 buffer with (+OA) and without (–OA) addition of 4.7 µM *cis*-9-octadecenoic (oleic) acid (OA). The ratio of fluorescence emission intensities at 505 nm and 432 nm can be quantitatively related to free fatty acid concentrations.

The Amplex® Red Sphingomyelinase Assay Kit contains:

- Amplex® Red reagent
- Dimethylsulfoxide (DMSO)
- Horseradish peroxidase (HRP)
- H_2O_2 for use as a positive control
- Concentrated reaction buffer
- Choline oxidase from *Alcaligenes* sp.
- Alkaline phosphatase from calf intestine
- Sphingomyelin
- Triton X-100
- Sphingomyelinase from *Bacillus* sp.
- Detailed protocols

Each kit provides sufficient reagents for approximately 500 assays using a fluorescence microplate reader and a reaction volume of 200 µL per assay.

ADIFAB Reagent: A Unique Free Fatty Acid Indicator

Elevated levels of free fatty acids (FFA)—which are associated with multiple pathological states, including cancer, diabetes and cardiac ischemia [107]—are generated by inflammatory responses, phospholipase A activity and cytotoxic phenomena.[108] Sensitive techniques are required to detect and quantitate free fatty acids because these important metabolites have low aqueous solubility and are usually found complexed to carriers. ADIFAB (A3880) is a dual-wavelength fluorescent FFA indicator that consists of a polarity-sensitive fluorescent probe (acrylodan, A433; Section 2.3) conjugated to I-FABP, a rat intestinal fatty acid–binding protein with a low molecular weight (15,000 daltons) and a high binding affinity for FFA [109–111] (Figure 17.4.17).

As shown in Figure 17.4.36, titration of the ADIFAB reagent with oleic acid results in a shift of its fluorescence maximum from ~432 nm to ~505 nm. The ratio (R) of these signals (505 nm/432 nm) can be converted to an FFA concentration by using the FFA dissociation constant (K_d) and employing analysis procedures similar to those developed for Ca^{2+} indicators [112] (Chapter 19). Values of K_d vary considerably for different fatty acids; a typical value is 0.28 µM for oleic acid [111] (determined at 37°C). There is little, if any, interference from bile acids, glycerides, sterols or bilirubin. With appropriate precautions, which are described in the product information sheet accompanying this product, ADIFAB can be used to determine FFA concentrations in the range 1 nM to >20 µM.

ADIFAB was used to investigate the physical basis of *cis*-unsaturated fatty acid inhibition of cytotoxic T cells.[113] This effect is due to inhibition of a specific tyrosine phosphorylation event that normally accompanies antigen stimulation.[114,115] Measurements using ADIFAB have also revealed previously undetected differences in FFA binding affinities among fatty acid–binding proteins from different tissues [116,117] and have enabled quantitation of FFA levels in human serum as a potential diagnostic tool.[107,118]

REFERENCES

1. Nat Rev Mol Cell Biol (2001) 2:327; **2.** Curr Opin Struct Biol (2000) 10:737; **3.** Physiol Rev (2000) 80:1291; **4.** Biochim Biophys Acta (1994) 1212:26; **5.** Biochim Biophys Acta (2000) 1488:124; **6.** Mol Med Today (1999) 5:244; **7.** FASEB J (1994) 8:916; **8.** Cardiovasc Drugs Ther (2009) 23:49; **9.** Clin Chim Acta (2010) 411:190; **10.** Methods Enzymol (2007) 434:15; **11.** Circ Res (2009) 104:952; **12.** Am J Physiol Gastrointest Liver Physiol (2009) 296:G445; **13.** Science (2001) 292:1385; **14.** Sci Signal (2009) 2:ra71-ra71; **15.** J Invest Dermatol (2009) 129:2772; **16.** Anal Biochem (1999) 276:27; **17.** Science (2000) 288:1160; **18.** Br J Pharmacol (1998) 124:1675; **19.** J Biol Chem (1992) 267:21465; **20.** J Biol Chem (2001) 276:12035; **21.** J Lipid Res (2007) 48:385; **22.** J Lipid Res (2007) 48:472; **23.** Anal Biochem (1990) 185:80; **24.** J Biol Chem (1999) 274:19338; **25.** Anal Biochem (1981) 116:553; **26.** Biochemistry (1995) 34:2049; **27.** Biochim Biophys Acta (2000) 1486:321; **28.** J Lipid Res (1992) 33:1863; **29.** J Neurosci Methods (2000) 100:127; **30.** Biochemistry (1999) 38:7803; **31.** Biochemistry (1997) 36:14325; **32.** Biochim Biophys Acta (1987) 917:411; **33.** J Biol Chem (2001) 276:22732; **34.** J Biol Chem (1999) 274:11494; **35.** Biochemistry (1999) 38:3867; **36.** Anal Biochem (1995) 229:256; **37.** Biochem J (1994) 298:23; **38.** Proc Natl Acad Sci U S A (2004) 101:9745; **39.** Mol Pharmacol (2000) 57:1142; **40.** Proc SPIE-Int Soc Opt Eng (2000) 3926:166; **41.** J Mol Biol (1998) 275:635; **42.** Biochim Biophys Acta (1994) 1224:247; **43.** J Biol Chem (1994) 269:4098; **44.** J Biol Chem (1991) 266:1926; **45.** J Neurochem (1991) 57:67; **46.** J Biol Chem (2002) 277:45592; **47.** Anal Biochem (2000) 286:277; **48.** J Biol Chem (1997) 272:12909; **49.** Anal Biochem (1994) 218:136; **50.** J Biol Chem (1994) 269:23790; **51.** Mol Biol Cell (1999) 10:3863; **52.** Eukaryot Cell (2009) 8:1094; **53.** Biochem J (1996) 314:15; **54.** Biochem J (1995) 307:799; **55.** Biochem J (2001) 360:513; **56.** Biochem J (2001) 355:249; **57.** Cell (2000) 100:603; **58.** J Biol Chem (1999) 274:8347; **59.** J Biol Chem (1999) 274:9907; **60.** Annu Rev Cell Dev Biol (1998) 14:231; **61.** J Histochem Cytochem (2002) 50:697; **62.** Science (2010) 327:46; **63.** Biochemistry (2001) 40:4893; **64.** Trends Cell Biol (2000) 10:408; **65.** J Biol Chem (1994) 269:3125; **66.** Chem Phys Lipids (1999) 102:55; **67.** Ann N Y Acad Sci (1998) 845:152; **68.** Biophys J (1997) 72:37; **69.** J Cell Biol (1998) 140:39; **70.** Methods Enzymol (2000) 312:293; **71.** Methods Enzymol (2000)

IMPORTANT NOTICE: The products described in this manual are covered by one or more Limited Use Label License(s). Please refer to the Appendix on page 971 and Master Product List on page 975. Products are For Research Use Only. Not intended for any animal or human therapeutic or diagnostic use.

REFERENCES—*continued*

312:523; **72.** Frontiers in Bioactive Lipids, Vanderhoek JV, Ed. 1996, p 203; **73.** J Cell Biol (1996) 134:1031; **74.** J Cell Biol (1991) 113:1267; **75.** Cytometry (1993) 14:251; **76.** J Cell Biol (1993) 120:399; **77.** Eur J Cell Biol (1992) 58:214; **78.** Mol Biochem Parasitol (2000) 106:21; **79.** Infect Immun (2000) 68:5960; **80.** J Cell Biol (2001) 154:535; **81.** Am J Physiol Lung Cell Mol Physiol (2001) 280:L938; **82.** Nat Cell Biol (1999) 1:386; **83.** J Neurochem (1999) 73:1375; **84.** Lancet (1999) 354:901; **85.** Proc Natl Acad Sci U S A (1998) 95:6373; **86.** J Cell Biol (1999) 147:447; **87.** Biophys J (1999) 77:1498; **88.** Cell Biology: A Laboratory Handbook, 2nd Ed., Vol. 2, Celis JE, Ed. 1998, p. 507; **89.** Mol Biol Cell (1995) 6:135; **90.** J Biol Chem (1993) 268:4577; **91.** Biochemistry (1992) 31:3581; **92.** J Biol Chem (1993) 268:14476; **93.** Blood (1999) 93:80; **94.** J Immunol (2007) 179:6770; **95.** Mol Biol Cell (2007) 18:2667; **96.** J Cell Biol (2007) 176:895; **97.** Methods (2005) 36:186; **98.** J Biol Chem (2001) 276:24985; **99.** Mol Biol Cell (2009) 20:3751; **100.** J Cell Sci (2009) 122:289; **101.** Biochim Biophys Acta (1992) 1113:277; **102.** Adv Cell Mol Biol Membranes (1993) 1:199; **103.** Biochim Biophys Acta (1991) 1082:113; **104.** Parasitology (1992) 105:81; **105.** J Biol Chem (2001) 276:11775; **106.** Eur J Biochem (2004) 271:601; **107.** Am J Cardiol (1996) 78:1350; **108.** J Immunol (1991) 147:2809; **109.** J Biol Chem (1995) 270:15076; **110.** Biochemistry (1993) 32:7574; **111.** J Biol Chem (1992) 267:23495; **112.** Mol Cell Biochem (1999) 192:87; **113.** Biochemistry (1993) 32:530; **114.** J Biol Chem (1994) 269:9506; **115.** J Biol Chem (1993) 268:17578; **116.** Biochemistry (2000) 39:7197; **117.** J Biol Chem (1994) 269:23918; **118.** J Lipid Res (1995) 36:229.

DATA TABLE 17.4 PROBES FOR LIPID METABOLISM AND SIGNALING

Cat. No.	MW	Storage	Soluble	Abs	EC	Em	Solvent	Notes
A3880	~15,350	FF,L,AA	H_2O	365	10,500	432	H_2O	1
A10070	880.68	FF,D,L	DMSO	505	92,000	512	MeOH	2, 16
A10072	986.67	FF,D,L	DMSO	505	85,000	567	MeOH	2, 17, 18
B3781	797.88	FF,D,L	see Notes	342	75,000	471	EtOH	2, 3
B3782	966.20	FF,D,L	see Notes	340	62,000	473	EtOH	2, 4
B7701	1029.80	FF,D,L	see Notes	505	123,000	512	MeOH	2, 5
B13950	1582.50	F,D,L	DMSO, EtOH	505	80,000	512	MeOH	6
B22650	~66,000	F,D,L	H_2O	505	91,000	511	MeOH	6, 7
B34400	~66,000	F,D,L	H_2O	589	65,000	616	MeOH	7
B34401	~66,000	F,D,L	H_2O	505	80,000	512	MeOH	6, 7
B34402	~66,000	F,D,L	H_2O	505	80,000	511	MeOH	6, 7
D3521	601.63	FF,D,L	$CHCl_3$, DMSO	505	91,000	511	MeOH	6
D3522	766.75	FF,D,L	see Notes	505	77,000	512	MeOH	2, 6
D3771	854.86	FF,D,L	see Notes	506	71,000	512	EtOH	2
D3803	797.77	FF,D,L	see Notes	503	80,000	512	MeOH	2, 8
D7519	861.96	FF,D,L	DMSO, EtOH	505	85,000	511	MeOH	6
D7540	705.71	FF,D,L	$CHCl_3$, DMSO	589	65,000	616	MeOH	
D7711	864.94	FF,D,L	DMSO	505	75,000	513	MeOH	6, 9
D13951	925.91	FF,D,L	DMSO, EtOH	505	80,000	511	MeOH	6
D23739	1136.13	FF,D,L	DMSO	505	92,000	511	MeOH	2, 10
E33955	1011.15	F,D,L	$CHCl_3$ DMSO	505	94,000	515	MeOH	11
H361	850.13	FF,D,L	see Notes	342	37,000	376	MeOH	2, 12, 13
H3809	856.09	FF,D,L	see Notes	341	38,000	376	MeOH	2, 12, 13
N1154	575.75	FF,D,L	$CHCl_3$, DMSO	466	22,000	536	MeOH	14
N3524	740.88	FF,D,L	see Notes	466	22,000	536	MeOH	2, 14
N3786	771.89	FF,D,L	see Notes	465	21,000	533	EtOH	2, 14, 15
N3787	856.05	FF,D,L	see Notes	465	22,000	534	EtOH	2, 14, 15
N22651	~66,000	F,D,L	H_2O	466	22,000	536	MeOH	7, 14

For definitions of the contents of this data table, see "Using *The Molecular Probes® Handbook*" in the introductory pages.

Notes

1. ADIFAB fatty acid indicator is a protein conjugate with a molecular weight of approximately 15,350. Em shifts from about 432 nm to 505 nm upon binding of fatty acids. (Mol Cell Biochem (1999) 192:87)
2. Chloroform is the most generally useful solvent for preparing stock solutions of phospholipids (including sphingomyelins). Glycerophosphocholines are usually freely soluble in ethanol. Most other glycerophospholipids (phosphoethanolamines, phosphatidic acids and phosphoglycerols) are less soluble in ethanol, but solutions up to 1–2 mg/mL should be obtainable, using sonication to aid dispersion if necessary. Labeling of cells with fluorescent phospholipids can be enhanced by addition of cyclodextrins during incubation. (J Biol Chem (1999) 274:35359)
3. Phospholipase A cleavage generates a fluorescent fatty acid (P1903MP (Section 13.2)) and a fluorescent lysophospholipid.
4. Phospholipase A cleavage generates a fluorescent fatty acid (P31 (Section 13.2)) and a fluorescent lysophospholipid.
5. Phospholipase A cleavage results in increased fluorescence with essentially no wavelength shift. The cleavage products are D3862 (Section 13.2) and a fluorescent lysophospholipid.
6. Em for BODIPY® FL sphingolipids shifts to ~620 nm when high concentrations of the probe (>5 mol %) are incorporated in lipid mixtures. (J Cell Biol (1991) 113:1267)
7. This product is a lipid complexed with bovine serum albumin (BSA). Spectroscopic data are for the free lipid in MeOH.
8. Phospholipase A_2 cleavage generates a fluorescent fatty acid (D3834 (Section 13.2)) and a nonfluorescent lysophospholipid.
9. This product is supplied as a ready-made solution in the solvent indicated under "Soluble."
10. Phospholipase A_2 cleavage results in increased fluorescence with essentially no wavelength shift. The cleavage products are D3834 (Section 13.2) and a dinitrophenylated lysophospholipid.
11. Fluorescence of the intact substrate is weak. Lipase hydrolysis releases a highly fluorescent fatty acid (D3823, Section 13.2).
12. Pyrene derivatives exhibit structured spectra. The absorption maximum is usually about 340 nm with a subsidiary peak at about 325 nm. There are also strong absorption peaks below 300 nm. The emission maximum is usually about 376 nm with a subsidiary peak at 396 nm. Excimer emission at about 470 nm may be observed at high concentrations.
13. Phospholipase A_2 hydrolysis releases a fluorescent fatty acid; P31 (Section 13.2).
14. Fluorescence of NBD and its derivatives in water is relatively weak. QY and τ increase and Em decreases in aprotic solvents and other nonpolar environments relative to water. (Biochemistry (1977) 16:5150, Photochem Photobiol (1991) 54:361)
15. Phospholipase A_2 hydrolysis releases a fluorescent fatty acid; N316 (Section 13.2) from N3786 or N678 (Section 13.2) from N3787.
16. Phospholipase A_1 cleavage results in increased fluorescence with essentially no wavelength shift. The cleavage products are D3834 (Section 13.2) and a dinitrophenylated lysophospholipid.
17. A10072 exhibits dual emission (Em = 510 nm and 567 nm in MeOH, 513 nm and 575 nm when incorporated in phospholipid bilayer membranes). Phospholipase A_2 cleavage results in increased 510–513 nm emission and reciprocally diminshed 567–575 nm emission.
18. A10072 is also soluble at 2 mM in 2-methoxyethanol.

IMPORTANT NOTICE: The products described in this manual are covered by one or more Limited Use Label License(s). Please refer to the Appendix on page 971 and Master Product List on page 975. Products are For Research Use Only. Not intended for any animal or human therapeutic or diagnostic use.

PRODUCT LIST 17.4 PROBES FOR LIPID METABOLISM AND SIGNALING

Cat. No.	Product	Quantity
A3880	ADIFAB fatty acid indicator	200 µg
A12218	Amplex® Red Phosphatidylcholine-Specific Phospholipase C Assay Kit *500 assays*	1 kit
A12219	Amplex® Red Phospholipase D Assay Kit *500 assays*	1 kit
A12220	Amplex® Red Sphingomyelinase Assay Kit *500 assays*	1 kit
A21328	anti-phosphatidylinositol 3,4,5-triphosphate, mouse IgM, monoclonal RC6F8 (anti-PtdIns(3,4,5)P_3) *1 mg/mL*	100 µL
A21327	anti-phosphatidylinositol 4,5-diphosphate, mouse IgM, monoclonal 2C11 (anti-PtdIns(4,5)P_2) *1 mg/mL*	100 µL
B3781	1,2-bis-(1-pyrenebutanoyl)-*sn*-glycero-3-phosphocholine	1 mg
B3782	1,2-bis-(1-pyrenedecanoyl)-*sn*-glycero-3-phosphocholine	1 mg
B7701	1,2-bis-(4,4-difluoro-5,7-dimethyl-4-bora-3a,4a-diaza-*s*-indacene-3-undecanoyl)-*sn*-glycero-3-phosphocholine (bis-BODIPY® FL C_{11}-PC)	100 µg
B22650	BODIPY® FL C_5-ceramide complexed to BSA	5 mg
B13950	BODIPY® FL C_5-ganglioside G_{M1}	25 µg
B34401	BODIPY® FL C_5-ganglioside G_{M1} complexed to BSA	1 mg
B34402	BODIPY® FL C_5-lactosylceramide complexed to BSA	1 mg
B34353	BODIPY® FL phosphatidylinositol(4,5) bisphosphate (BODIPY® FL PtdIns(4,5)P_2)	50 µg
D7540	BODIPY® TR ceramide (*N*-((4-(4,4-difluoro-5-(2-thienyl)-4-bora-3a,4a-diaza-*s*-indacene-3-yl)phenoxy)acetyl)sphingosine)	250 µg
B34400	BODIPY® TR ceramide complexed to BSA	5 mg
D3771	2-decanoyl-1-(*O*-(11-(4,4-difluoro-5,7-dimethyl-4-bora-3a,4a-diaza-*s*-indacene-3-propionyl)amino)undecyl)-*sn*-glycero-3-phosphocholine	1 mg
D7519	*N*-(4,4-difluoro-5,7-dimethyl-4-bora-3a,4a-diaza-*s*-indacene-3-dodecanoyl)sphingosyl 1-β-D-galactopyranoside (BODIPY® FL C_{12}-galactocerebroside)	25 µg
D7711	*N*-(4,4-difluoro-5,7-dimethyl-4-bora-3a,4a-diaza-*s*-indacene-3-dodecanoyl)sphingosyl phosphocholine (BODIPY® FL C_{12}-sphingomyelin) *1 mg/mL in DMSO*	250 µL
D3803	2-(4,4-difluoro-5,7-dimethyl-4-bora-3a,4a-diaza-*s*-indacene-3-pentanoyl)-1-hexadecanoyl-*sn*-glycero-3-phosphocholine (β-BODIPY® FL C_5-HPC)	100 µg
D3521	*N*-(4,4-difluoro-5,7-dimethyl-4-bora-3a,4a-diaza-*s*-indacene-3-pentanoyl)sphingosine (BODIPY® FL C_5-ceramide)	250 µg
D13951	*N*-(4,4-difluoro-5,7-dimethyl-4-bora-3a,4a-diaza-*s*-indacene-3-pentanoyl)sphingosyl 1-β-D-lactoside (BODIPY® FL C_5-lactosylceramide)	25 µg
D3522	*N*-(4,4-difluoro-5,7-dimethyl-4-bora-3a,4a-diaza-*s*-indacene-3-pentanoyl)sphingosyl phosphocholine (BODIPY® FL C_5-sphingomyelin)	250 µg
D23739	*N*-((6-(2,4-dinitrophenyl)amino)hexanoyl)-2-(4,4-difluoro-5,7-dimethyl-4-bora-3a,4a-diaza-*s*-indacene-3-pentanoyl)-1-hexadecanoyl-*sn*-glycero-3-phosphoethanolamine, triethylammonium salt (PED6)	1 mg
E10215	EnzChek® Direct Phospholipase C Assay Kit *phosphatidylcholine specific* *2-plate size*	1 kit
E10216	EnzChek® Direct Phospholipase C Assay Kit *phosphatidylcholine specific* *10-plate size*	1 kit
E33955	EnzChek® lipase substrate *green fluorescent, 505/515*	100 µg
E10219	EnzChek® Phospholipase A_1 Assay Kit *2-plate size*	1 kit
E10221	EnzChek® Phospholipase A_1 Assay Kit *10-plate size*	1 kit
E10217	EnzChek® Phospholipase A_2 Assay Kit *2-plate size*	1 kit
E10218	EnzChek® Phospholipase A_2 Assay Kit *10-plate size*	1 kit
H361	1-hexadecanoyl-2-(1-pyrenedecanoyl)-*sn*-glycero-3-phosphocholine (β-py-C_{10}-HPC)	1 mg
H3809	1-hexadecanoyl-2-(1-pyrenedecanoyl)-*sn*-glycero-3-phosphoglycerol, ammonium salt (β-py-C_{10}-PG)	1 mg
N1154	NBD C_6-ceramide (6-((*N*-(7-nitrobenz-2-oxa-1,3-diazol-4-yl)amino)hexanoyl)sphingosine)	1 mg
N22651	NBD C_6-ceramide complexed to BSA	5 mg
N3524	NBD C_6-sphingomyelin (6-((*N*-(7-nitrobenz-2-oxa-1,3-diazol-4-yl)amino)hexanoyl)sphingosyl phosphocholine)	1 mg
N3787	2-(12-(7-nitrobenz-2-oxa-1,3-diazol-4-yl)amino)dodecanoyl-1-hexadecanoyl-*sn*-glycero-3-phosphocholine (NBD C_{12}-HPC)	5 mg
N3786	2-(6-(7-nitrobenz-2-oxa-1,3-diazol-4-yl)amino)hexanoyl-1-hexadecanoyl-*sn*-glycero-3-phosphocholine (NBD C_6-HPC)	5 mg
A10070	PED-A1 (N-((6-(2,4-DNP)amino)hexanoyl)-1-(BODIPY® FL C_5)-2-hexyl-*sn*-glycero-3-phosphoethanolamine) *phospholipase A_1 selective substrate*	100 µg
P6466	phospholipase C, phosphatidylinositol-specific *from *Bacillus cereus** *100 U/mL*	50 µL
A10072	Red/Green BODIPY® PC-A2 (1-*O*-(6-BODIPY® 558/568-aminohexyl)-2-BODIPY® FL C_5-*sn*-glycero-3-phosphocholine) *ratiometric phospholipase A_2 substrate*	100 µg

IMPORTANT NOTICE: The products described in this manual are covered by one or more Limited Use Label License(s). Please refer to the Appendix on page 971 and Master Product List on page 975. Products are For Research Use Only. Not intended for any animal or human therapeutic or diagnostic use.

www.invitrogen.com/probes

molecular probes® | invitrogen™ by life technologies™

EIGHTEEN

CHAPTER 18

Probes for Reactive Oxygen Species, Including Nitric Oxide

IMPORTANT NOTICE: The products described in this manual are covered by one or more Limited Use Label License(s). Please refer to the Appendix on page 971 and Master Product List on page 975. Products are For Research Use Only. Not intended for any animal or human therapeutic or diagnostic use.

IMPORTANT NOTICE: The products described in this manual are covered by one or more Limited Use Label License(s). Please refer to the Appendix on page 971 and Master Product List on page 975. Products are For Research Use Only. Not intended for any animal or human therapeutic or diagnostic use.

18.1 Introduction to Reactive Oxygen Species

Activated oxygen species are produced during a number of physiological[1–3] and pathological[4–7] processes. Their effects are through reactions with a large variety of easily oxidizable cellular components, including NADH, NADPH, ascorbic acid, histidine, tryptophan, tyrosine, cysteine, glutathione, proteins and nucleic acids.[8–12] Reactive oxygen species can also oxidize cholesterol and unsaturated fatty acids, causing membrane lipid peroxidation.[7,13,14] Several reviews discuss the chemistry of the different reactive oxygen species and their detection.[15–19]

Molecular Probes® products include probes that either generate or detect various reactive oxygen species (Table 18.1), including singlet oxygen ($^1O_2$), superoxide anion ($\bullet O_2^-$), hydroxyl radical (HO•) and various peroxides (ROOR') and hydroperoxides (ROOH). Section 18.2 describes these probes and their applications *in vitro* and *in vivo*.

The importance of the nitric oxide radical (abbreviated NO) and other reactive oxygen species as biological messengers is the focus of intense research.[20–22] Section 18.3 is devoted to our probes for promoting, inhibiting or detecting nitric oxide production in a variety of experimental systems.

Table 18.1 Reactive oxygen species.

Reactive Oxygen Species	Structure	Detection Reagents	
Hydrogen peroxide	H_2O_2	• Carboxy-H_2DCFDA (C400) [1–3] • CM-H_2DCFDA (C6827) [4,5] • Dihydrocalcein AM (D23805) • Dihydrorhodamine 123 (D632, D23806) [6] • Dihydrorhodamine 6G (D633) [7]	• H_2DCFDA (D399) [8–11] • Lucigenin (L6868) [12,13] • Luminol (L8455) [14] • RedoxSensor™ Red CC-1 (R14060) [15]
Hydroxyl radical *	HO•	• 3′-(*p*-Aminophenyl) fluorescein (APF, A36003) • 3′-(*p*-Hydroxyphenyl) fluorescein (HPF, H36004) • CM-H_2DCFDA (C6827) [16]	• Proxyl fluorescamine (C7924) [17] • TEMPO-9-AC (A7923)
Hypochlorous acid	HOCl	• Aminophenyl fluorescein (APF, A36003) • Dihydrorhodamine 123 (D632, D23806) [18]	• Luminol (L8455) [19–21]
Nitric oxide	NO	• DAF-FM (D23841) [22,23] • DAF-FM diacetate (D23842, D23844) [22,23] • DAA (D23840) [24]	• 2,3-Diaminonaphthalene (D7918) [24] • Luminol (L8455) [25]
Peroxyl radical, including both alkylperoxyl and hydroperoxyl [26] radicals (wherein R = H)	ROO•	• BODIPY® FL EDA (D2390) [27] • BODIPY® 665/676 (B3932) [28] • H_2DCFDA (D399) [29–33] • Carboxy-H_2DCFDA (C400) [34] • CM-H_2DCFDA (C6827)	• DPPP (D7894) [35–37] • Luminol (L8455) [38–40] • *cis*-Parinaric acid (P36005) [41,42] • RedoxSensor™ Red CC-1 (R14060) [15]
Peroxynitrite anion †	$ONOO^-$	• 3′-(*p*-Aminophenyl) fluorescein (APF, A36003) • 3′-(*p*-Hydroxyphenyl) fluorescein (HPF, H36004) • H_2DCFDA (D399) [43,44] • Carboxy-H_2DCFDA (C400) • CM-H_2DCFDA (C6827)	• Coelenterazine (C2944) [45] • Dihydrorhodamine 123 (D632, D23806) [43,46–48] • Dihydrorhodamine 6G (D633) • Luminol (L8455) [43,49,50]
Singlet oxygen ‡	$^1O_2$	• Singlet Oxygen Sensor Green reagent (S36002)	• *trans*-1-(2′-methoxyvinyl)pyrene (M7913) [51,52]
Superoxide anion	$\bullet O_2^-$	• Coelenterazine (C2944) [53,54] • Dihydroethidium (D1168, D11347, D23107) [55,56] • Fc OxyBURST® Green assay reagent (F2902) [57,58] • OxyBURST® Green H_2DCFDA SE (D2935) [59,60] • OxyBURST® Green H_2HFF BSA (O13291) [61] • Lucigenin (L6868) [62,63] • Luminol (L8455) [64]	• MCLA (M23800) [65,66] • MTT (M6494) [67] • NBT (N6495) [68] • RedoxSensor™ Red CC-1 (R14060) [15] • TEMPO-9-AC (A7923) • XTT (X6493) [69]

* Hydroxyl radicals can also be photosensitized by malachite green isothiocyanate (M689) or generated by a *N*-(1,10-phenanthrolin-5-yl)iodoacetamide (P6879) metal–ligand complex. † 3-Nitrotyrosine, a product of this potent nitrating reagent, can be detected with an anti-nitrotyrosine antibody (A21285). ‡ Singlet oxygen can also be photosensitized by hypericin (H7476), rose bengal diacetate (R14000) and merocyanine 540 (M24571).

1. Biol Pharm Bull (2000) 23:1153; **2.** J Neurosci (1999) 19:9209; **3.** J Biol Chem (1996) 271:21505; **4.** J Biol Chem (2001) 276:21938; **5.** Proc Natl Acad Sci U S A (1997) 94:11557; **6.** Biochim Biophys Acta (1999) 1454:275; **7.** Proc Natl Acad Sci U S A (2000) 97:8266; **8.** J Biol Chem (2001) 276:514; **9.** J Immunol Methods (1989) 117:53; **10.** Brain Res (1994) 635:113; **11.** J Biol Chem (1999) 274:37111; **12.** Analyst (1986) 3:941; **13.** J Am Chem Soc (1979) 101:5347; **14.** J Bone Miner Res (1992) 7:1139; **15.** Free Radic Biol Med (2000) 28:1266; **16.** Proc Natl Acad Sci U S A (2001) 98:1643; **17.** Anal Chem (1997) 69:4295; **18.** Nitric Oxide (1997) 1:145; **19.** Biochim Biophys Acta (1991) 1097:145; **20.** Luminescence (1999) 14:239; **21.** Am J Physiol (1992) 263:G719; **22.** Anal Biochem (2000) 287:203; **23.** Angew Chem Int Ed Engl (1999) 38:3209; **24.** Neuroreport (1998) 9:4051; **25.** Anal Chem (1993) 65:1794; **26.** DNA Cell Biol (2002) 21:251; **27.** J Biochem Biophys Methods (1997) 35:23; **28.** J Agric Food Chem (2000) 48:1150; **29.** Toxicol Meth (1994) 4:224; **30.** J Biol Chem (2000) 275:40028; **31.** Anal Biochem (1983) 134:111; **32.** Am J Physiol (1989) 257:C347; **33.** Methods Enzymol (1984) 105:352; **34.** J Biol Chem (1998) 273:5294; **35.** J Chromatogr (1993) 628:31; **36.** Anal Lett (1987) 20:731; **37.** Methods Enzymol (1990) 186:157; **38.** Free Radic Biol Med (1995) 18:1; **39.** Biomed Chromatogr (1990) 4:131; **40.** Lipids (1998) 33:1235; **41.** J Biol Chem (1997) 272:12328; **42.** Biochem Biophys Res Commun (1998) 244:647; **43.** Free Radic Biol Med (2001) 30:463; **44.** FEBS Lett (2000) 468:89; **45.** Circ Res (1999) 84:1203; **46.** FASEB J 2001; **47.** Arch Biochem Biophys (2000) 373:302; **48.** FASEB J (2000) 14:1061; **49.** J Biol Chem (1996) 271:29223; **50.** Arch Biochem Biophys (1994) 310:352; **51.** Biochem Biophys Res Commun (1984) 123:869; **52.** Methods Enzymol (1986) 133:569; **53.** Anal Biochem (1992) 206:273; **54.** Free Radic Biol Med (2000) 29:170; **55.** Circ Res (2001) 88:824; **56.** J Biol Chem (2001) 276:17621; **57.** J Leukoc Biol (1997) 62:329; **58.** J Biol Chem (1995) 270:8328; **59.** Immunology (1994) 83:507; **60.** J Immunol Methods (1990) 130:223; **61.** Biophys J (1998) 75:2577; **62.** Free Radic Biol Med (2000) 28:1232; **63.** J Biol Chem (1998) 273:2015; **64.** J Immunol Methods (1992) 155:151; **65.** Free Radic Res (2000) 32:265; **66.** Anal Biochem (1999) 271:53; **67.** Free Radic Res Commun (1993) 18:369; **68.** Arch Biochem Biophys (1997) 342:275; **69.** Plant Physiol (1998) 117:491.

IMPORTANT NOTICE: The products described in this manual are covered by one or more Limited Use Label License(s). Please refer to the Appendix on page 971 and Master Product List on page 975. Products are For Research Use Only. Not intended for any animal or human therapeutic or diagnostic use.

REFERENCES

1. Chem Res Toxicol (2010) 23:568; 2. Cytometry A (2009) 75:475; 3. J Biol Chem (2006) 281:29011; 4. Nat Clin Pract Cardiovasc Med (2008) 5:811; 5. J Neurosci (2009) 29:9090; 6. Mol Cell (2009) 33:627; 7. Physiol Rev (2004) 84:1381; 8. Mitochondrion (2007) 7:106; 9. Electrophoresis (2005) 26:2599; 10. Angew Chem Int Ed Engl (2007) 46:561; 11. Br J Pharmacol (2004) 142:231; 12. Biochem Pharmacol (2003) 66:1527; 13. J Biol Chem (2008) 283:15539; 14. Chem Res Toxicol (2008) 21:432; 15. Methods Mol Biol (2010) 594:57; 16. Methods Cell Biol (2007) 80:355; 17. Free Radic Biol Med (2007) 43:995; 18. J Biochem Biophys Methods (2005) 65:45; 19. Am J Physiol Regul Integr Comp Physiol (2004) 286:R431; 20. Curr Opin Chem Biol (2010) 14:43; 21. Free Radic Biol Med (2009) 47:684; 22. Nitric Oxide (2009) 21:92.

18.2 Generating and Detecting Reactive Oxygen Species

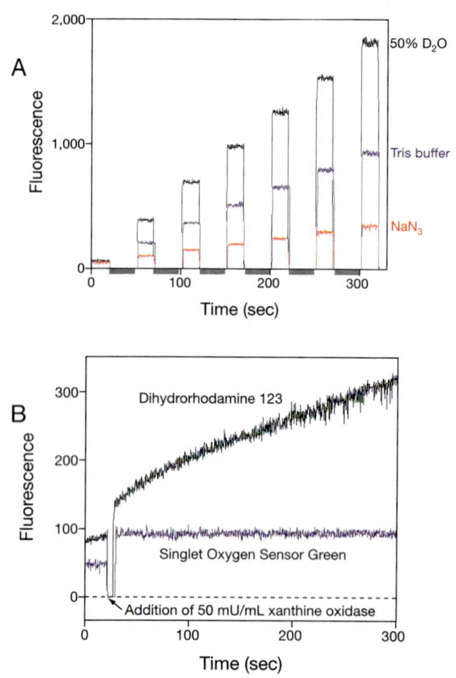

Figure 18.2.1 Fluorescence response and specificity of Singlet Oxygen Sensor Green reagent (S36002) to $^1O_2$. **A)** Fluorescence measurements were made in a spectrofluorometer using excitation/emission wavelengths of 488/525 nm for solutions containing: 1 µM Singlet Oxygen Sensor Green reagent and 10 µM methylene blue in 100 mM pH 7.5 Tris buffer alone, the singlet oxygen scavenger sodium azide (NaN_3, 1 mM), or 50% D_2O, which increases the lifetime of $^1O_2$. Measurements were made for 20-second periods, with 30-second intervals (indicated by grey bars) between each measurement. During the 30-second intervals, the samples were exposed to laser radiation (630–680 nm, <5 mW), resulting in methylene blue–photosensitized generation of $^1O_2$. **B)** Fluorescence measurements were made in a spectrofluorometer using excitation/emission wavelengths of 488/525 nm for solutions of 50 mM pH 7 Tris buffer with 1 mM xanthine containing either 1 µM Singlet Oxygen Sensor Green reagent or dihydrorhodamine 123 (D632, D23806). After ~20 seconds, 50 mU/mL of xanthine oxidase (XO) was added. XO catalzyes the oxidation of xanthine, producing uric acid and superoxide. Superoxide can spontaneously degrade to H_2O_2.

We offer an assortment of Molecular Probes® products for the generation of reactive oxygen species (ROS), including singlet oxygen ($^1O_2$), superoxide ($\bullet O_2^-$), hydroxyl radical (HO•) and various peroxides (ROOR′) and hydroperoxides (ROOH) (Table 18.1), as well as for their fluorometric detection in solution. Although there are no sensors that reversibly monitor the level of reactive oxygen species, this section discusses a number of probes that trap or otherwise react with singlet oxygen, hydroxyl radicals or superoxide. The optical or electron spin properties of the resulting products can be used as a measure of the presence or quantity of the reactive oxygen species and, in certain cases, can report the kinetics and location of their formation.

Generating Singlet Oxygen

Singlet oxygen is responsible for much of the physiological damage caused by reactive oxygen species, including nucleic acid modification through selective reaction with deoxyguanosine to form 8-hydroxydeoxyguanosine [1,2] (8-OHdG). The lifetime of singlet oxygen is sufficiently long (4.4 microseconds in water [3]) to permit significant diffusion in cells and tissues.[4] In the laboratory, singlet oxygen is usually generated in one of three ways: photochemically from dioxygen ($^3O_2$) using a photosensitizing dye (Figure 18.2.1), chemically by thermal decomposition of a peroxide or dioxetane, or by microwave discharge through an oxygen stream. Singlet oxygen can be directly detected by its characteristic weak chemiluminescence at 1270 nm.[4,5]

Hypericin

Among the most efficient reagents for generating singlet oxygen is the photosensitizer hypericin (H7476, Figure 18.2.2), a natural pigment isolated from plants of the genus *Hypericum*. This heat-stable dye exhibits a quantum yield for singlet oxygen generation in excess of 0.7, as well as high photostability, making it an important agent for both anticancer and antiviral research.[6,7]

Rose Bengal Diacetate

Rose bengal diacetate (R14000) is an efficient, cell-permeant generator of singlet oxygen.[8–10] It is an iodinated xanthene derivative that has been chemically modified by the introduction of acetate groups (Figure 18.2.3). These modifications inactivate both its fluorescence and photosensitization properties, while increasing its ability to cross cell membranes. Once inside a live cell, esterases remove the acetate groups, restoring rose bengal to its native structure. Its intracellular localization allows rose bengal diacetate to be a very effective photosensitizer.

Merocyanine 540

Photoirradiation of merocyanine 540 (M24571) produces both singlet oxygen and other reactive oxygen species, including oxygen radicals.[11–14] Merocyanine 540 is often used as a photosensitizer in photodynamic therapy.[15]

IMPORTANT NOTICE: The products described in this manual are covered by one or more Limited Use Label License(s). Please refer to the Appendix on page 971 and Master Product List on page 975. Products are For Research Use Only. Not intended for any animal or human therapeutic or diagnostic use.
www.invitrogen.com/probes

Detecting Singlet Oxygen

Singlet Oxygen Sensor Green Reagent

Unlike other fluorescent and chemiluminescent singlet oxygen detection reagents, the Singlet Oxygen Sensor Green reagent (S36002) is highly selective for singlet oxygen ($^1O_2$); it shows no appreciable response to other reactive oxygen species, including hydroxyl radical (HO•), superoxide ($•O_2^-$) and nitric oxide (NO) (Figure 18.2.1). Before reaction with singlet oxygen, this probe initially exhibits weak blue fluorescence with excitation peaks at 372 and 393 nm and emission peaks at 395 and 416 nm. In the presence of singlet oxygen, however, it emits a green fluorescence similar to that of fluorescein (excitation/emission maxima ~504/525 nm).

We have observed that the fluorescent product of Singlet Oxygen Sensor Green reagent can degrade with time in some solutions and that Singlet Oxygen Sensor Green reagent can become fluorescent at alkaline pH in the absence of singlet oxygen. Nevertheless, with the proper controls the intensity of the green-fluorescent signal can be correlated with singlet oxygen concentration, without significant interference from other reactive oxygen species. The Singlet Oxygen Sensor Green reagent has demonstrated utility for detecting singlet oxygen in solution[16,17] and in plant tissues.[18,19]

trans-1-(2′-Methoxyvinyl)pyrene

trans-1-(2′-Methoxyvinyl)pyrene (M7913) can be used to detect picomole quantities of singlet oxygen in chemical and biological systems (Figure 18.2.4), making this compound one of the most sensitive singlet oxygen probes currently available.[20–22] Furthermore, this highly selective chemiluminescent probe does not react with other activated oxygen species such as hydroxyl radical, superoxide or hydrogen peroxide.

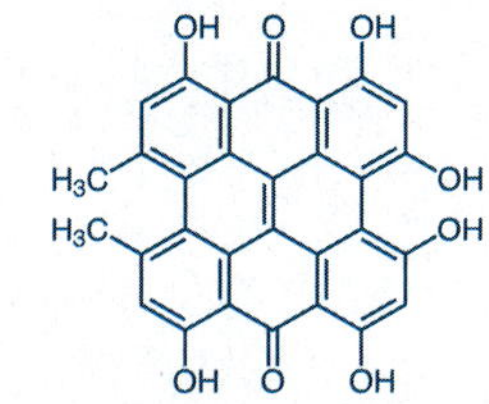

Figure 18.2.2 Hypericin (H7476).

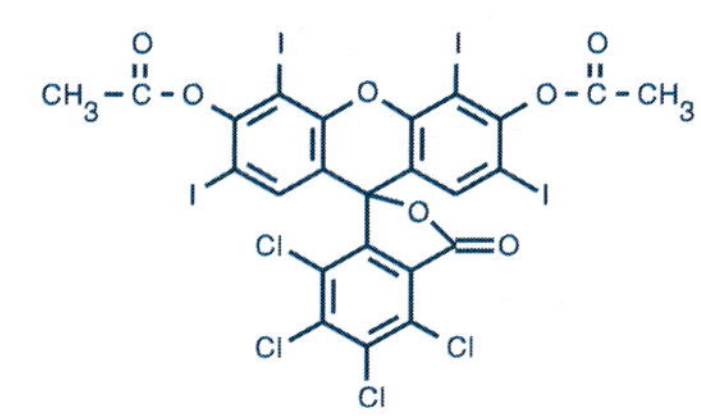

Figure 18.2.3 Rose bengal diacetate (R14000).

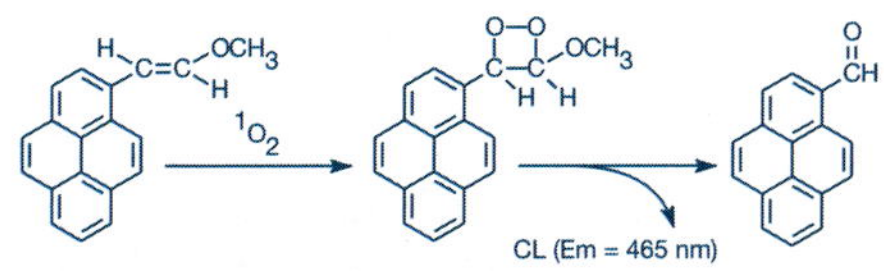

Figure 18.2.4 Reaction of *trans*-1-(2′-methoxyvinyl)pyrene (M7913) with singlet oxygen ($^1O_2$), yielding a dioxetane intermediate that generates chemiluminescence (CL) upon decomposition to 1-pyrenecarboxaldehyde.

Generating Hydroxyl and Superoxide Radicals

Hydroxyl and superoxide radicals have been implicated in a number of pathological conditions, including ischemia, reperfusion and aging. The superoxide anion (Table 18.1) may also play a role in regulating normal vascular function. The hydroxyl radical is a very reactive oxygen species[23] that has a lifetime of about 2 nanoseconds in aqueous solution and a radius of diffusion of about 20 Å. Thus, it induces peroxidation only when it is generated in close proximity to its target. The hydroxyl radical can be derived from superoxide in a Fenton reaction catalyzed by Fe^{2+} or other transition metals, as well as by the effect of ionizing radiation on dioxygen. Superoxide is most effectively generated from a hypoxanthine–xanthine oxidase generating system[24–26] (Figure 18.2.1).

Malachite Green

Malachite green is a nonfluorescent photosensitizer that absorbs at long wavelengths (~630 nm). Its photosensitizing action can be targeted to particular cellular sites by conjugating malachite green isothiocyanate (M689, Figure 18.2.5) to specific antibodies.[27,28] Enzymes and other proteins within ~10 Å of the binding site of the malachite green–labeled antibody can then be selectively destroyed upon irradiation with long-wavelength light. Studies by Jay and colleagues have demonstrated that this photoinduced destruction of enzymes in the immediate vicinity of the chromophore is apparently the result of localized production of hydroxyl radicals, which have short lifetimes that limit their diffusion from the site of their generation.[29]

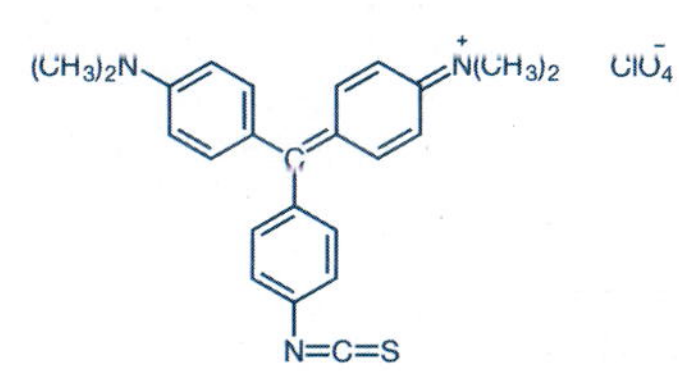

Figure 18.2.5 Malachite green isothiocyanate (M689).

1,10-Phenanthroline Iodoacetamide

Conjugation of the iodoacetamide of 1,10-phenanthroline (P6879, Figure 18.2.6) to thiol-containing ligands confers the metal-binding properties of this important complexing agent on the ligand. For example, the covalent copper–phenanthroline complex of oligonucleotides or nucleic acid–binding molecules in combination with hydrogen peroxide acts as a chemical nuclease to selectively cleave DNA or RNA.[30,31] Hydroxyl radicals or other reactive oxygen species appear to be involved in this cleavage.[32,33]

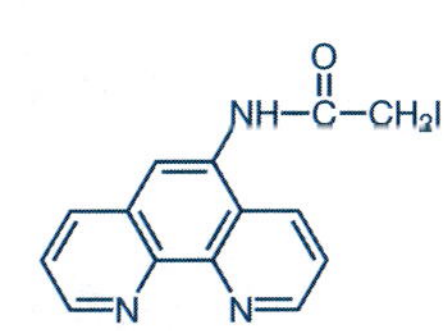

Figure 18.2.6 *N*-(1,10-phenanthrolin-5-yl)iodoacetamide (P6879).

IMPORTANT NOTICE: The products described in this manual are covered by one or more Limited Use Label License(s). Please refer to the Appendix on page 971 and Master Product List on page 975. Products are For Research Use Only. Not intended for any animal or human therapeutic or diagnostic use.

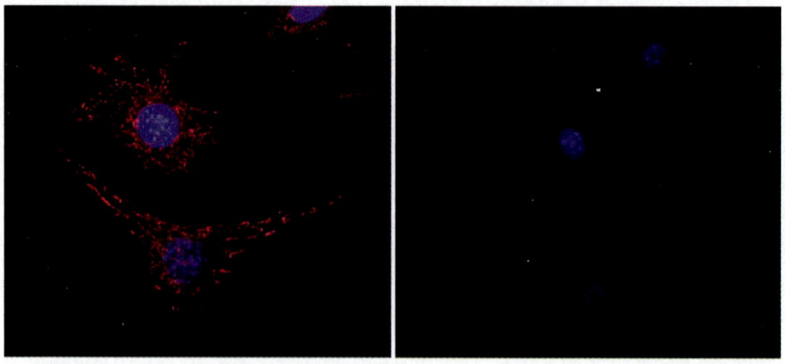

Figure 18.2.7 Detection of superoxide in live cells using MitoSOX™ Red superoxide indicator (M36008). Live 3T3 cells were treated with FeTCPP, a superoxide scavenger (right), or left untreated (left). Cells were then labeled with MitoSOX™ Red reagent, which fluoresces when oxidized by superoxide, and nuclei were stained with blue-fluorescent Hoechst 33342. The mitochondria of untreated cells exhibited red fluorescence, indicating the presence of superoxide, whereas the mitochondria of treated cells showed minimal fluorescence.

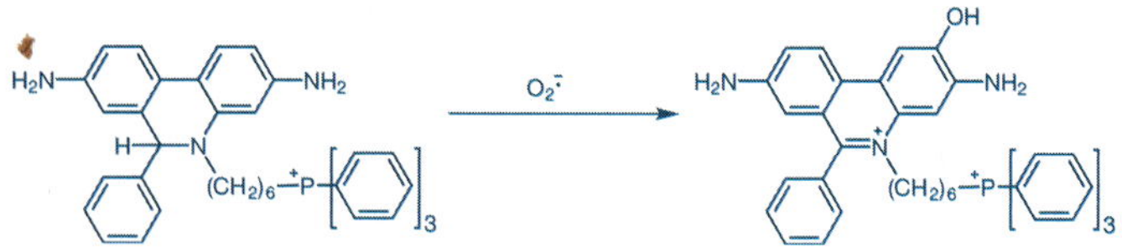

Figure 18.2.8 Oxidation of MitoSOX™ Red mitochondrial superoxide indicator to 2-hydroxy-5-(triphenylphosphonium)hexylethidium by superoxide ($\cdot O_2^-$).

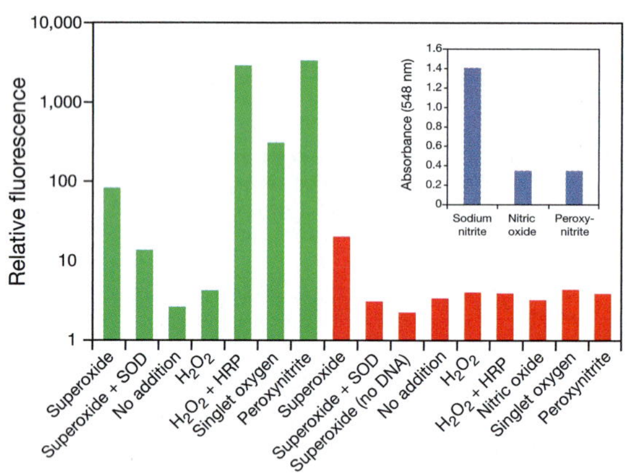

Figure 18.2.9 Selectivity of the MitoSOX™ Red mitochondrial superoxide indicator (M36008). Cell-free systems were used to generate a variety of reactive oxygen species (ROS) and reactive nitrogen species (RNS); each oxidant was then added to a separate 10 µM solution of MitoSOX™ Red reagent and incubated at 37°C for 10 minutes. Excess DNA was added (unless otherwise noted) and the samples were incubated for an additional 15 minutes at 37°C before fluorescence was measured. The Griess Reagent Kit (G7921) (for nitric oxide, peroxynitrite, and nitrite standards only; blue bars) and dihydrorhodamine 123 (DHR 123 (D632); green bars) were employed as positive controls for oxidant generation. Superoxide dismutase (SOD), a superoxide scavenger, was used as a negative control for superoxide. The results show that the MitoSOX™ Red probe (red bars) is readily oxidized by superoxide but not by the other oxidants.

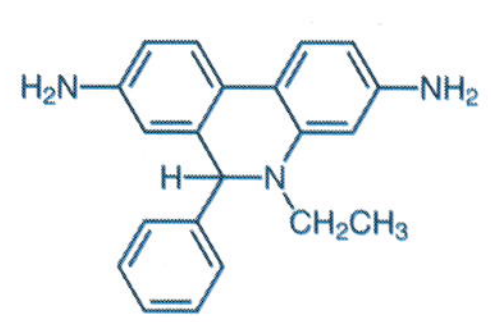

Figure 18.2.10 Dihydroethidium (hydroethidine, D1168).

Detecting Hydroxyl and Superoxide Radicals

MitoSOX™ Red Mitochondrial Superoxide Indicator

Mitochondrial superoxide is generated as a by-product of oxidative phosphorylation. In an otherwise tightly coupled electron transport chain, approximately 1–3% of mitochondrial oxygen consumed is incompletely reduced; these "leaky" electrons can quickly interact with molecular oxygen to form superoxide anion, the predominant reactive oxygen species in mitochondria.[34,35] Increases in cellular superoxide production have been implicated in cardiovascular diseases, including hypertension, atherosclerosis and diabetes-associated vascular injuries,[36] as well as in neurodegenerative diseases such as Parkinson disease, Alzheimer disease and amyotrophic lateral sclerosis (ALS).[35]

MitoSOX™ Red mitochondrial superoxide indicator (M36008) is a cationic derivative of dihydroethidum (also known as hydroethidine; see below) designed for highly selective detection of superoxide in the mitochondria of live cells (Figure 18.2.7). The cationic triphenylphosphonium substituent of MitoSOX™ Red indicator is responsible for the electrophoretically driven uptake of the probe in actively respiring mitochondria. Oxidation of MitoSOX™ Red indicator (or dihydroethidium) by superoxide results in hydroxylation at the 2-position (Figure 18.2.8). 2-hydroxyethidium (and the corresponding derivative of MitoSOX™ Red indicator) exhibits a fluorescence excitation peak at ~400 nm[37] that is absent in the excitation spectrum of the ethidium oxidation product generated by reactive oxygen species other than superoxide. Thus, fluorescence excitation at 400 nm with emission detection at ~590 nm provides optimum discrimination of superoxide from other reactive oxygen species[37–39] (Figure 18.2.9).

Measurements of mitochondrial superoxide generation using MitoSOX™ Red indicator in mouse cortical neurons expressing caspase-cleaved tau microtubule-associated protein have been correlated with readouts from fluorescent indicators of cytosolic and mitochondrial calcium and mitochondrial membrane potential.[40] The relationship of mitochondrial superoxide generation to dopamine transporter activity, measured using the aminostyryl dye substrate 4-Di-1-ASP (D288, Section 12.2), has been investigated in mouse brain astrocytes.[41] MitoSOX™ Red indicator has been used for confocal microscopy analysis of reactive oxygen species (ROS) production by mitochondrial NO synthase (mtNOS) in permeabilized cat ventricular myocytes[42] and, in combination with Amplex® Red reagent, for measurement of mitochondrial superoxide and hydrogen peroxide production in rat vascular endothelial cells.[43] In addition to imaging and microscope photometry measurements, several flow cytometry applications of MitoSOX™ Red indicator have also been reported. Detailed protocols for simultaneous measurements of mitochondrial superoxide generation and apoptotic markers APC annexin V (A35110, Section 15.5) and SYTOX® Green (S7020, Section 8.1) in human coronary artery endothelial cells by flow cytometry have been published by Mukhopadhyay and co-workers.[44]

Dihydroethidium (Hydroethidine)

Although dihydroethidium (Figure 18.2.10), which is also called hydroethidine, is commonly used to analyze respiratory bursts in phagocytes,[45] it has been reported that this probe undergoes significant

IMPORTANT NOTICE: The products described in this manual are covered by one or more Limited Use Label License(s). Please refer to the Appendix on page 971 and Master Product List on page 975. Products are For Research Use Only. Not intended for any animal or human therapeutic or diagnostic use.

oxidation in resting leukocytes, possibly through the uncoupling of mitochondrial oxidative phosphorylation.[46] Cytosolic dihydroethidium exhibits blue fluorescence; however, once this probe is oxidized to ethidium it intercalates within DNA, staining the cell nucleus a bright fluorescent red[47–49] (Figure 18.2.11). The mechanism of dihydroethidium's interaction with lysosomes and DNA has been described.[50] Similar to MitoSOX™ Red mitochondrial superoxide indicator (Figure 18.2.8), dihydroethidium is oxidized by superoxide to 2-hydroxyethidium.[37] It is frequently used for mitochondrial superoxide detection,[51–54] although MitoSOX™ Red indicator provides more specific mitochondrial localization. Indeed, in some cases researchers have used dihydroethidium and MitoSOX™ Red indicator to provide discrete indications of cytosolic and mitochondrial superoxide production, respectively.[55]

Dihydroethidium (hydroethidine) is available in a 25 mg vial (D1168), as a stabilized 5 mM solution in DMSO (D23107) or specially packaged in 10 vials of 1 mg each (D11347); the stabilized DMSO solution or special packaging is recommended when small quantities of the dye will be used over a long period of time.

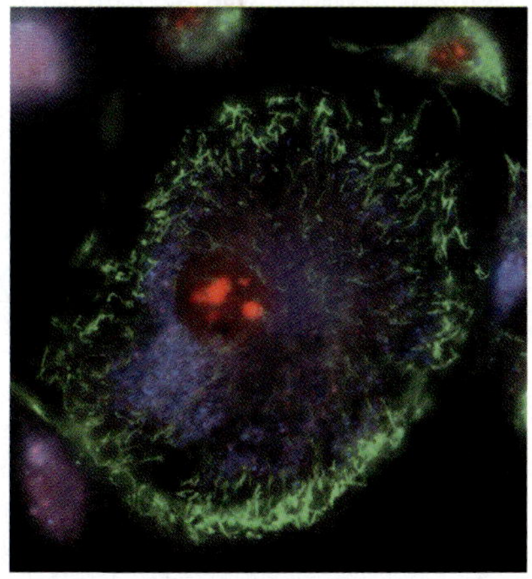

Figure 18.2.11 Live bovine pulmonary artery endothelial cells (BPAEC) were incubated with the cell-permeant, weakly blue-fluorescent dihydroethidium (D1168, D11347, D23107) and the green-fluorescent mitochondrial stain, MitoTracker® Green FM® (M7514). Upon oxidation, red-fluorescent ethidium accumulated in the nucleus.

Fluorogenic Spin Traps

Hydroxyl radicals have usually been detected after reaction with spin traps. We offer TEMPO-9-AC (A7923, Figure 18.2.12) and proxyl fluorescamine[56–59] (C7924, Figure 18.2.13), two fluorogenic probes for detecting hydroxyl radicals[60] and superoxide. Each of these molecules contains a nitroxide moiety that effectively quenches its fluorescence. However, once TEMPO-9-AC or proxyl fluorescamine traps a hydroxyl radical or superoxide, its fluorescence is restored and the radical's electron spin resonance signal is destroyed, making these probes useful for detecting radicals either by fluorescence or by electron spin resonance spectroscopy. TEMPO-9-AC has been reported to detect glutathionyl radicals but not phenoxyl radicals.[61] Proxyl fluorescamine can be used to detect the methyl radicals that are formed by reacting hydroxyl radicals with DMSO.[59] Radical-specific scavengers (Table 18.2)—such as the superoxide-specific *p*-benzoquinone and superoxide dismutase[62] or the hydroxyl radical–specific mannitol and dimethylsulfoxide (DMSO)[56,63,64]—can be used to identify the detected species.

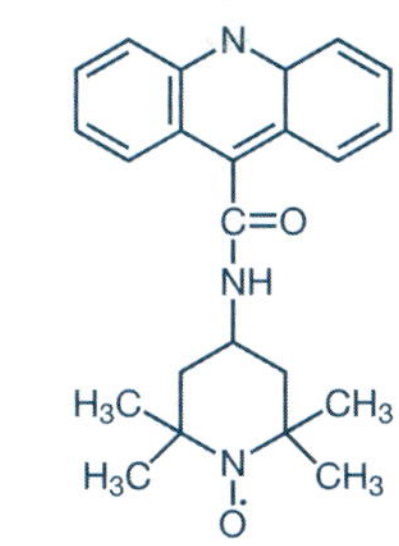

Figure 18.2.12 4-((9-acridinecarbonyl)amino)-2,2,6,6-tetramethylpiperidin-1-oxyl, free radical (TEMPO-9-AC, A7923).

Chemiluminescent and Chromogenic Reagents for Detecting Superoxide

In the absence of apoaequorin, the luminophore coelenterazine (C2944) produces chemiluminescence in response to superoxide generation in cells, organelles, bacteria[65] and tissues.[66] Unlike luminol, coelenterazine exhibits luminescence that does not depend on the activity of cell-derived myeloperoxidase and is not inhibited by azide.[67]

In addition to coelenterazine, we offer MCLA (M23800, Figure 18.2.14) for detecting superoxide.[68] MCLA and coelenterazine are superior alternatives to lucigenin[65] (L6868) for this application because lucigenin can reportedly sensitize superoxide production, leading to false-positive results.[69–73] An additional advantage of MCLA is that its pH optimum for luminescence generation is closer to the physiological near-neutral range than are the pH optima of luminol and lucigenin.[74]

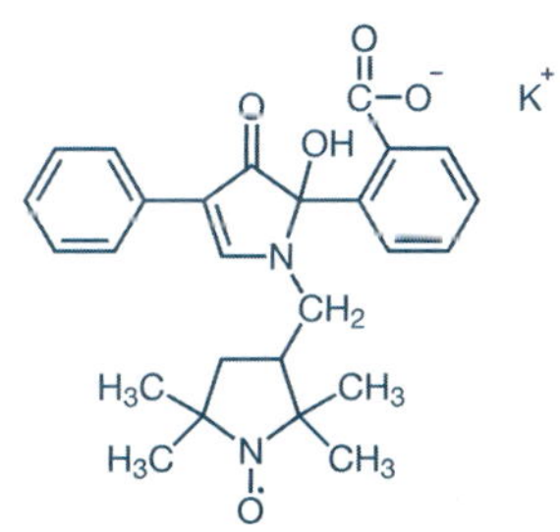

Figure 18.2.13 5-(2-carboxyphenyl)-5-hydroxy-1-((2,2,5,5-tetramethyl-1-oxypyrrolidin-3-yl)methyl)-3-phenyl-2-pyrrolin-4-one, potassium salt (proxyl fluorescamine, C7924).

Table 18.2 Scavengers of reactive oxygen species (ROS).

ROS	Scavenger (Working Concentration)	References
Hydrogen peroxide (H_2O_2)	Sodium pyruvate (10 mM), DMTU* (10 mM)	1
Hydroxyl radical (HO•)	Mannitol (20–100 mM), DMSO (0.28 M) †	1
Nitric oxide (NO)	Carboxy-PTIO (C7912; 100 µM)	1
Peroxyl radical (ROO•)	Trolox ‡ (10–100 µM), α-tocopherol (10–100 µM)	2,3
Peroxynitrite anion ($ONOO^-$)	Ebselen § (10–100 µM), uric acid (100 µM)	1,4
Singlet oxygen ($^1O_2$)	Sodium azide (1–10 mM)	1
Superoxide anion ($•O_2^-$)	MnTBAP ** (100 µM), Tiron †† (10 mM)	5,6

* DMTU = *N,N'*-dimethylthiourea. Disproportionation by catalase is also widely used for suppression of H_2O_2. † 0.28 M DMSO = 2% (v/v). The reactivity of HO• is so high that it can be argued that the actions of these reagents must be indirect. ‡ Trolox = 6-hydroxy-2,5,7,8-tetramethylchroman-2-carboxylic acid. § Ebselen = 2-phenyl-1,2-benzisoselenazol-3(2*H*)-one. ** MnTBAP = manganese(III)-tetrakis(4-benzoic acid)porphyrin. †† Tiron = 4,5-dihydroxybenzene-1,3-disulfonate. **1.** J Biol Chem (2007) 282:30452; **2.** Nitric Oxide (2006) 15:163; **3.** Biochim Biophys Acta (2004) 1636:136; **4.** J Biol Chem (2004) 279:4425; **5.** Circ Res (2004) 94:37; **6.** Bioorg Med Chem (2002) 10:3013.

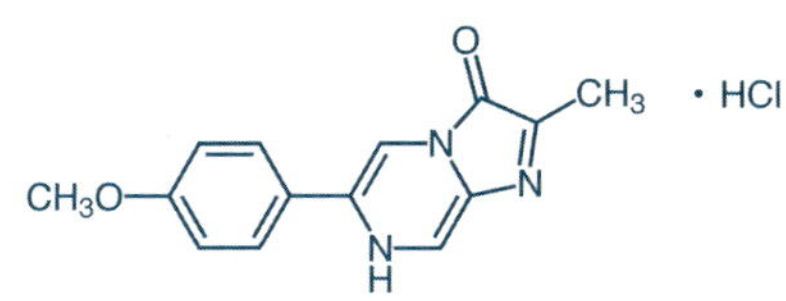

Figure 18.2.14 2-methyl-6-(4-methoxyphenyl)-3,7-dihydroimidazo1,2-apyrazin-3-one, hydrochloride (MCLA, M23800).

IMPORTANT NOTICE: The products described in this manual are covered by one or more Limited Use Label License(s). Please refer to the Appendix on page 971 and Master Product List on page 975. Products are For Research Use Only. Not intended for any animal or human therapeutic or diagnostic use.

Nitro blue tetrazolium salt (NBT, N6495; Table 18.3) and other tetrazolium salts are chromogenic probes useful for superoxide determination.[75,76] The superoxide sensitivity of tetrazolium salts can be a confounding factor in their more common applications for cell viability and proliferation assays.[77]

Detecting Peroxides, Peroxyl Radicals and Lipid Peroxidation

In peroxisomes, H_2O_2 is produced by several enzymes that use molecular oxygen to oxidize organic compounds. This H_2O_2 is then used by catalase to oxidize other substrates, including phenols, formic acid, formaldehyde and alcohol. In liver and kidney cells, these oxidation reactions are important for detoxifying a variety of compounds in the bloodstream.[78–81] However, H_2O_2 also plays a role in neurodegenerative and other disorders through induction of apoptosis[82] and DNA strand breaks,[83] modification of intracellular Ca^{2+} levels and mitochondrial potential, and oxidation of glutathione. In addition, H_2O_2 is released from cells during hypoxia.[84]

Peroxidation of unsaturated lipids affects cell membrane properties,[85] signal transduction pathways,[86,87] apoptosis and the deterioration of foods and other biological compounds.[88] Lipid hydroperoxides have been reported to accumulate in oxidatively stressed individuals, including HIV-infected patients.[89] Lipid peroxidation may also be responsible for aging, as well as for pathological processes such as drug-induced phototoxicity and atherosclerosis,[90] and is often the cause of free radical–mediated damage in cells. To directly assess the extent of lipid peroxidation, researchers either measure the amount of lipid hydroperoxides directly or detect the presence of secondary reaction products[91–93] (e.g., 4-hydroxy-2-nonenal or malonaldehyde; see below).

Peroxyl radicals are formed by the decomposition of various peroxides and hydroperoxides, including lipid hydroperoxides. The hydroperoxyl radical is also the protonated form of superoxide, and approximately 0.3% of the superoxide in the cytosol is present as this protonated radical.[94] Experimentally, peroxyl radicals, including alkylperoxyl (ROO•) and hydroperoxyl (HOO•) radicals, are generated from compounds such as 2,2′-azobis(2-amidinopropane) and from hydroperoxides such as cumene hydroperoxide.

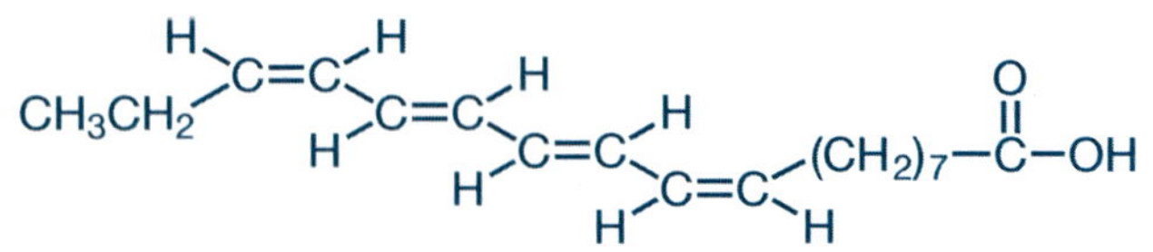

Figure 18.2.15 cis-parinaric acid (P36005).

cis-Parinaric Acid

Fluorescence quenching of the fatty acid analog *cis*-parinaric acid (P36005) has been used in several lipid peroxidation assays,[95–97] including quantitative determinations in live cells.[98,99] Parinaric acid's extensive unsaturation (Figure 18.2.15) makes it quite susceptible to oxidation if not rigorously protected from air.[100] Consequently, we offer *cis*-parinaric acid in a 10 mL unit size of a 3 mM solution in deoxygenated ethanol (P36005); if stored protected from light under an inert argon atmosphere at –20°C, this stock solution should be stable for at least 6 months. During experiments, we advise handling parinaric acid samples under inert gas and preparing solutions using degassed buffers and solvents. Parinaric acid is also somewhat photolabile and undergoes photodimerization when exposed to intense illumination, resulting in loss of fluorescence.[101]

Diphenyl-1-Pyrenylphosphine

Hydroperoxides in lipids, serum, tissues and foodstuffs can be directly detected using the fluorogenic reagent diphenyl-1-pyrenylphosphine[102,103] (DPPP, D7894). DPPP is essentially nonfluorescent until oxidized to a phosphine oxide by peroxides; *in vitro*, DPPP remains nonfluorescent in the presence of hydroxyl radicals generated by the Cu^{2+}-ascorbate method.[104] DPPP has previously been used to detect

Table 18.3 Tetrazolium salts for detecting redox potential in living cells and tissues.

Cat. No.	Tetrazolium Salt	Color of Formazan	Water Solubility of Formazan	Applications
M6494 (MTT)	3-(4,5-Dimethylthiazol-2-yl)-2,5-diphenyltetrazolium bromide	purple	no	• Superoxide generation by fumarate reductase[1] and nitric oxide synthase[2] • Mitochondrial dehydrogenase activity[3] • Cell viability and proliferation[4–9] • Neuronal cell death[10] • Platelet activation[11] • Tumor cell adhesion[12] and invasion[13] • Multidrug resistance[14] • *In vitro* toxicity testing[15–17]
N6495 (NBT)	Nitro blue tetrazolium chloride	deep blue	no	• Superoxide generation by xanthine oxidase[18] • Neutrophil oxidative metabolism[19,20] • NADPH diaphorase activity[21–23] • Succinic dehydrogenase histochemistry[24]
X6493 (XTT)	2,3-Bis-(2-methoxy-4-nitro-5-sulfophenyl)-2*H*-tetrazolium-5-carboxanilide	orange	yes	• Antifungal susceptibility[25] • Drug sensitivity of cells[26] • Parasitic nematode viability[27] • Tumor cell cytotoxicity[28]

1. J Biol Chem (1995) 270:19767; **2.** J Biol Chem (1994) 269:12589; **3.** Cytometry (1992) 13:532; **4.** Biotechniques (1998) 25:622, 626; **5.** J Immunol Methods (1994) 168:253; **6.** Anal Biochem (1993) 214:190; **7.** J Immunol Methods (1993) 164:149; **8.** Anal Biochem (1992) 205:8; **9.** J Immunol Methods (1986) 89:271; **10.** J Cell Biol (1995) 128:201; **11.** J Immunol Methods (1993) 159:253; **12.** J Immunol Methods (1993) 164:255; **13.** Cancer Res (1994) 54:3620; **14.** Leuk Res (1992) 16:1165; **15.** Biosci Biotechnol Biochem (1992) 56:1472; **16.** J Immunol Methods (1991) 144:141; **17.** J Immunol Methods (1990) 131:165; **18.** J Reprod Fertil (1993) 97:441; **19.** Clin Chim Acta (1993) 221:197; **20.** J Leukoc Biol (1993) 53:404; **21.** Neurosci Lett (1993) 155:61; **22.** Proc Natl Acad Sci U S A (1991) 88:7797; **23.** Proc Natl Acad Sci U S A (1991) 88:2811; **24.** Histochemistry (1982) 76:381; **25.** Antimicrob Agents Chemother (1992) 36:1619; **26.** Cancer Res (1988) 48:4827; **27.** Parasitology (1993) 107:175; **28.** J Immunol Methods (1992) 147:153.

IMPORTANT NOTICE: The products described in this manual are covered by one or more Limited Use Label License(s). Please refer to the Appendix on page 971 and Master Product List on page 975. Products are For Research Use Only. Not intended for any animal or human therapeutic or diagnostic use.

picomole levels of hydroperoxides by HPLC.[105,106] Its solubility in lipids makes DPPP quite useful for detecting hydroperoxides in the membranes of live cells[104,107,108] and in low-density lipoprotein particles.[109]

BODIPY® 581/591 C: A Ratiometric Lipid Peroxidation Sensor

The BODIPY® 581/591 C_{11} fatty acid (D3861, Figure 18.2.16) is a sensitive fluorescent reporter for lipid peroxidation, undergoing a shift from red to green fluorescence emission upon oxidation of the phenylbutadiene segment of the fluorophore.[110] This oxidation-dependent emission shift enables fluorescence ratio imaging of lipid peroxidation in live cells.[111–113] Other common applications of BODIPY® 581/591 C_{11} include fluorometric assays of antioxidant efficacy in plasma[114,115] and in lipid vesicles.[116] The oxidation and nitroxidation products of this BODIPY® fatty acid have been characterized by mass spectrometry.[117,118] Based on mass spectrometry analysis of oxidation products, MacDonald and co-workers report that BODIPY® 581/591 C_{11} is more sensitive to oxidation than endogenous lipids, and therefore tends to overestimate oxidative damage and underestimate antioxidant protection effects.[119]

Peroxyl radicals have also been detected in erythrocyte and red blood cell membranes using BODIPY® FL EDA[120] (D2390, Section 3.4), a water-soluble BODIPY® dye, or BODIPY® FL hexadecanoic acid (D3821, Section 13.2). BODIPY® FL hexadecanoic acid exhibits the red shift common to the fluorescence of lipophilic BODIPY® dyes when they are concentrated, permitting ratiometric measurements of hydroxyl radical production and allowing the onset of lipid peroxidation in live cells to be monitored.[121]

Other Scavengers for Peroxyl Radicals

The fluorescence of several other probes is lost following interaction with peroxyl radicals. Lipophilic fluorescein dyes such as hexadecanoylaminofluorescein[122] (H110, Section 13.5) and fluorescein-labeled phosphatidylethanolamine (F362, Section 13.2) have been useful for detecting peroxyl radical formation in membranes and in solution. Phycobiliproteins, such as B-phycoerythrin, R-phycoerythrin and allophycocyanin (P800, P801, A803, A819; Section 6.4), and phenolic dyes such as fluorescein (F1300, F36915; Section 10.1) are extensively used as substrates in total antioxidant capacity assays of plasma and foods.[115,123,124]

Luminol

Although luminol (L8455) is not useful for detecting superoxide in live cells,[125] it is commonly employed to detect peroxidase- or metal ion–mediated oxidative events.[126–128] Used alone, luminol can detect oxidative events in cells rich in peroxidases, including granulocytes[129–132] and spermatozoa.[133] This probe has also been used in conjunction with horseradish peroxidase (HRP) to investigate reoxygenation injury in rat hepatocytes.[134,135] In these experiments, it is thought that the primary species being detected is hydrogen peroxide. In addition, luminol has been employed to detect peroxynitrite generated from the reaction of nitric oxide and superoxide.[136–138]

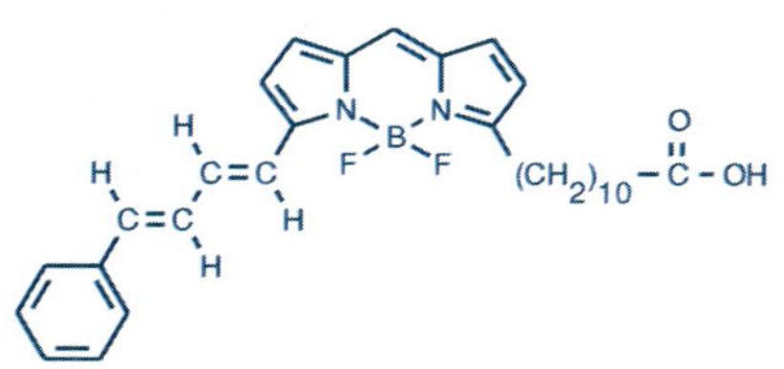

Figure 18.2.16 4,4-difluoro-5-(4-phenyl-1,3-butadienyl)-4-bora-3a,4a-diaza-s-indacene-3-undecanoic acid (BODIPY® 581/591 C_{11}, D3861).

Detecting 4-Hydroxy-2-Nonenal

Formation of 4-hydroxy-2-nonenal (HNE) from linoleic acid is a major cause of lipid peroxidation–induced toxicity. Several reagents for the direct fluorometric detection of aldehydes are described in Section 3.3. The biotinylated hydroxylamine ARP (A10550, Section 4.2) is particularly useful for this purpose.[139] Biotinylation using click chemistry coupling (Section 3.1) enables affinity purification of HNE-modified proteins.[140]

Detecting Peroxides and Peroxidases with Amplex® Red Reagents

Amplex® Red Reagent: Stable Substrate for Peroxidase Detection

In the presence of horseradish peroxidase (HRP), Amplex® Red reagent (10-acetyl-3,7-dihydroxyphenoxazine, A12222, A22177; Figure 18.2.17) reacts with H_2O_2 in a 1:1 stoichiometry to produce highly fluorescent resorufin[141] (R363, Section 10.1, Figure 18.2.18). Amplex® Red reagent has greater stability, yields less background and produces a red-fluorescent product that is more readily detected than the similar reduced methylene blue derivatives commonly used for colorimetric determination of lipid peroxides in plasma, sera, cell extracts and a variety of membrane systems.[142–144]

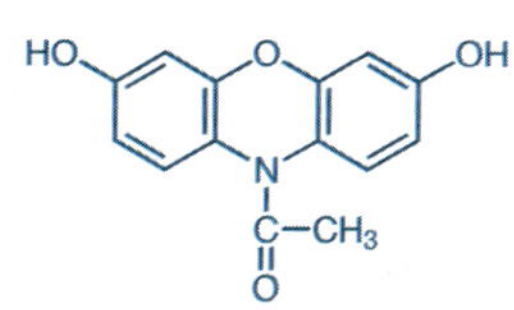

Figure 18.2.17 Amplex® Red reagent (A12222).

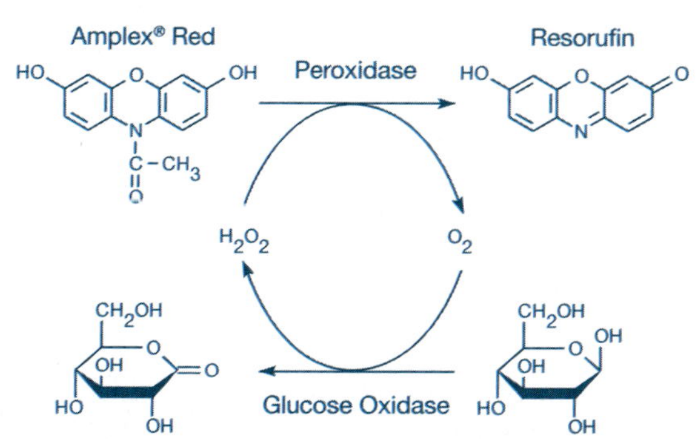

Figure 18.2.18 Principle of coupled enzymatic assays using Amplex® Red reagent. Oxidation of glucose by glucose oxidase results in generation of H_2O_2, which is coupled to conversion of the Amplex® Red reagent to fluorescent resorufin by HRP. The detection scheme shown here is used in the Amplex® Red Glucose/Glucose Oxidase Assay Kit (A22189).

IMPORTANT NOTICE: The products described in this manual are covered by one or more Limited Use Label License(s). Please refer to the Appendix on page 971 and Master Product List on page 975. Products are For Research Use Only. Not intended for any animal or human therapeutic or diagnostic use.

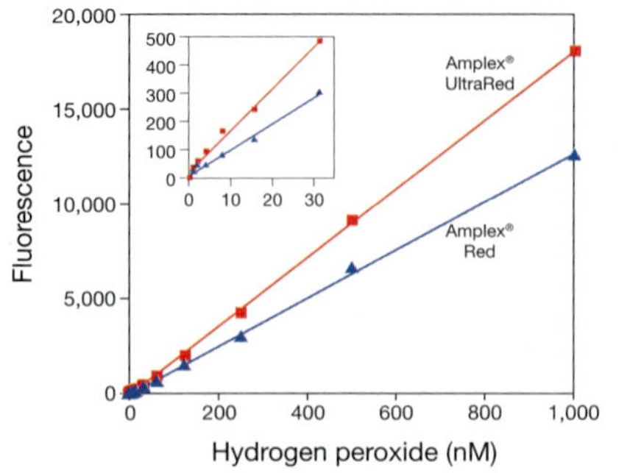

Figure 18.2.19 Detection of H_2O_2 using Amplex® UltraRed reagent (red squares) or Amplex® Red reagent (blue triangles). Reactions containing 50 µM Amplex® UltraRed or Amplex® Red reagent, 1 U/mL HRP and the indicated amount of H_2O_2 in 50 mM sodium phosphate buffer, pH 7.4, were incubated for 30 minutes at room temperature. The inset shows the sensitivity and linearity of the Amplex® UltraRed assay at low levels of H_2O_2.

Amplex® Red reagent has been used to detect the release of H_2O_2 from activated human leukocytes,[141,145] to measure the activity of monoamine oxidase in bovine brain tissue,[146] to demonstrate the extracellular production of H_2O_2 produced by UV light stimulation of human keratinocytes[147–149] and for microplate assays of H_2O_2 and lipid hydroperoxide generation by isolated mitochondria.[51,68,150] Amplex® Red reagent is available in a single 5 mg vial (A12222) or packaged as a set of 10 vials, each containing 10 mg of the substrate, for high-throughput screening applications (A22177).

Amplex® UltraRed Reagent: Brighter and More Sensitive than the Amplex® Red Reagent

Amplex® UltraRed reagent (A36006) improves upon the performance of Amplex® Red reagent, offering brighter fluorescence and enhanced sensitivity on a per-mole basis in horseradish peroxidase or horseradish peroxidase–coupled enzyme assays (Figure 18.2.19). Fluorescence of oxidized Amplex® UltraRed reagent is also less sensitive to pH (Figure 18.2.20), and the substrate and its oxidation product exhibit greater stability than Amplex® Red reagent in the presence of H_2O_2 or thiols such as dithiothreitol (DTT). Like Amplex® Red reagent, nonfluorescent Amplex® UltraRed reagent reacts with H_2O_2 in a 1:1 stoichiometric ratio to produce a brightly fluorescent and strongly absorbing reaction product (excitation/emission maxima ~568/581 nm) (Figure 18.2.21). Although the primary applications of the Amplex® UltraRed reagent are enzyme-linked immunosorbent assays (ELISAs; see Zen™ Myeloperoxidase ELISA Kit below) and *in vitro* antioxidant capacity assays,[151] it is also frequently used (in combination with HRP) to detect H_2O_2 production by isolated mitochondria[152] and cell cultures.[153,154]

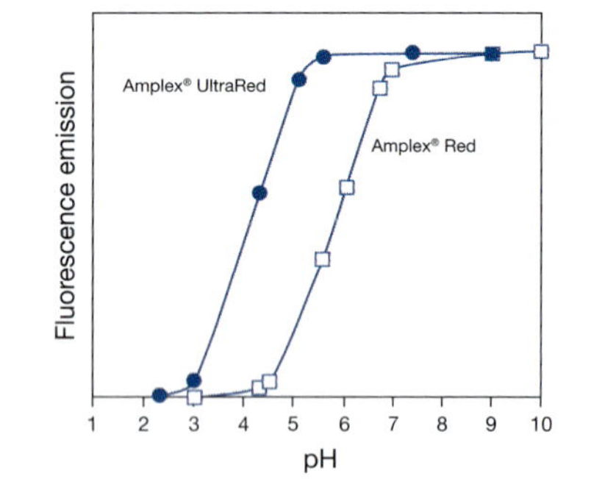

Figure 18.2.20 Comparison of pH-dependent fluorescence of the products derived from oxidation of Amplex® UltraRed reagent (solid blue circles) and Amplex® Red reagent (open blue squares). Fluorescence intensities were measured using excitation/emission of ~570/585 nm.

Amplex® Red Hydrogen Peroxide/Peroxidase Assay Kit

The Amplex® Red Hydrogen Peroxide/Peroxidase Assay Kit (A22188) provides a simple, sensitive, one-step assay for detecting H_2O_2 or the activity of horseradish peroxidase either by measuring fluorescence with a fluorescence-based microplate reader or a fluorometer (Figure 18.2.22) or by measuring absorption with an absorption-based microplate reader or a spectrophotometer. The Amplex® Red peroxidase substrate can detect the presence of active peroxidases and the release of H_2O_2 from biological samples, including cells and cell extracts.[51,141,155,156]

The Amplex® Red Hydrogen Peroxide/Peroxidase Assay Kit contains:

- Amplex® Red reagent
- Dimethylsulfoxide (DMSO)
- Horseradish (HRP)
- H_2O_2 for use as a positive control
- Concentrated reaction buffer
- Detailed protocols

Each kit provides sufficient reagents for approximately 500 assays using a fluorescence- or absorption-based microplate reader and a reaction volume of 100 µL per assay. Several additional kits that utilize the Amplex® Red peroxidase substrate to detect H_2O_2 in coupled enzymatic reactions are described in Section 10.5.

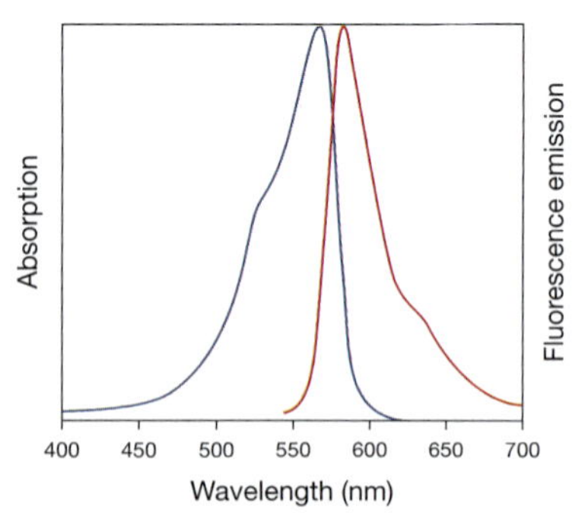

Figure 18.2.21 Absorption and fluorescence emission spectra of the product generated by horseradish peroxidase–mediated oxidation of the Amplex® UltraRed reagent in pH 7.5 buffer.

Amplex® Red Xanthine/Xanthine Oxidase Assay Kit

Xanthine oxidase (E.C. 1.2.3.2) plays a key role in the production of free radicals, including superoxide, in the body. The Amplex® Red Xanthine/Xanthine Oxidase Assay Kit (A22182) provides an ultrasensitive method for detecting xanthine or hypoxanthine or for monitoring xanthine oxidase activity. In the assay, xanthine oxidase catalyzes the oxidation of purine nucleotides, hypoxanthine or xanthine, to uric acid and superoxide. In the reaction mixture, the superoxide spontaneously degrades to H_2O_2, which in the presence of HRP reacts stoichiometrically with Amplex® Red reagent to generate the red-fluorescent oxidation product, resorufin. Resorufin has absorption and fluorescence emission maxima of approximately 571 nm and 585 nm (Figure 18.2.23), respectively, and because the extinction coefficient is high (54,000 $cm^{-1}M^{-1}$), the assay can be performed either fluorometrically or spectrophotometrically.

IMPORTANT NOTICE: The products described in this manual are covered by one or more Limited Use Label License(s). Please refer to the Appendix on page 971 and Master Product List on page 975. Products are For Research Use Only. Not intended for any animal or human therapeutic or diagnostic use.
www.invitrogen.com/probes

The Amplex® Red Xanthine/Xanthine Oxidase Assay Kit (A22182) contains:

- Amplex® Red reagent
- Dimethylsulfoxide (DMSO)
- Horseradish peroxidase (HRP)
- H_2O_2
- Concentrated reaction buffer
- Xanthine oxidase from buttermilk
- Hypoxanthine
- Xanthine
- Detailed protocols

Each kit provides sufficient reagents for approximately 400 assays using either a fluorescence- or absorption-based microplate reader and a reaction volume of 100 µL per assay.

In healthy individuals, xanthine oxidase is present in appreciable amounts only in the liver and jejunum. In various liver disorders, however, the enzyme is released into circulation. Therefore, determination of serum xanthine oxidase levels serves as a sensitive indicator of acute liver damage such as jaundice. The Amplex® Red xanthine/xanthine oxidase assay has been used as a marker of recovery from exercise stress.[157] Previously, researchers have utilized chemiluminescence or absorbance to monitor xanthine oxidase activity. The Amplex® Red Xanthine/Xanthine Oxidase Assay Kit permits the detection of xanthine oxidase in a purified system at levels as low as 0.1 mU/mL by fluorescence (Figure 18.2.24). This kit can also be used to detect as little as 200 nM hypoxanthine or xanthine (Figure 18.2.25), and, when coupled to the purine nucleotide phosphorylase enzyme, to detect inorganic phosphate.[158]

EnzChek® Myeloperoxidase (MPO) Activity Assay Kit

Myeloperoxidase (MPO, EC 1.11.1.7) is a lysosomal hemoprotein located in the azurophilic granules of polymorphonuclear (PMN) leukocytes and monocytes. It is a dimeric protein composed of two 59 kD and two 13.5 kD subunits. MPO is a unique peroxidase that catalyzes the conversion of hydrogen peroxide (H_2O_2) and chloride to hypochlorous acid, a strong oxidant with powerful antimicrobial activity and broad-spectrum reactivity with biomolecules. MPO is considered an important marker for inflammatory diseases, autoimmune diseases and cancer. MPO is also experimentally and clinically important for distinguishing myeloid from lymphoid leukemia and, due to its role in the pathology of atherogenesis, has been advocated as a prognostic marker of cardiovascular disease.

The ferric, or native, MPO reacts with hydrogen H_2O_2 to form the active intermediate MPO-I, which oxidizes chloride (Cl^-) to HOCl; these reactions make up the chlorination cycle (Figure 18.2.26). MPO also oxidizes a variety of substrates, including phenols and anilines, via the classic peroxidation cycle. The relative concentrations of chloride and the reducing substrate determine whether MPO uses hydrogen peroxide for chlorination or peroxidation. Assays based on measurement of chlorination activity are more specific for MPO than those based on peroxidase substrates such as tetramethylbenzidine (TMB).

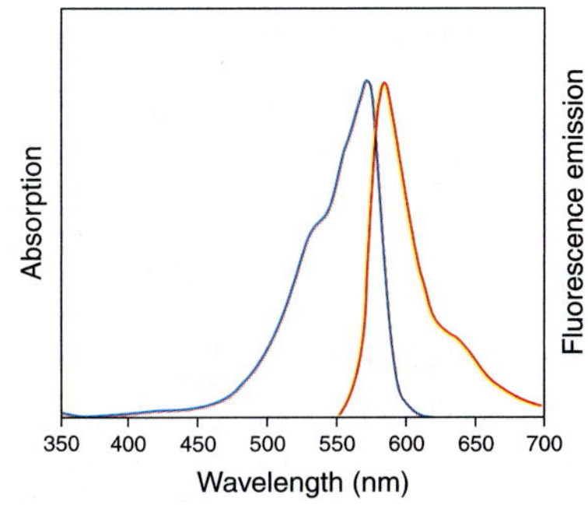

Figure 18.2.23 Absorption and fluorescence emission spectra of resorufin in pH 9.0 buffer.

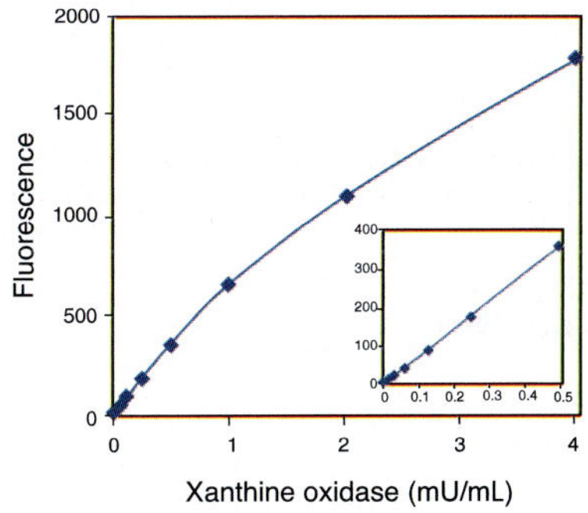

Figure 18.2.24 Detection of xanthine oxidase using the Amplex® Red Xanthine/Xanthine Oxidase Assay Kit (A22182). Each reaction contained 50 µM Amplex® Red reagent, 0.2 U/mL horseradish peroxidase, 0.1 mM hypoxanthine and the indicated amount of xanthine oxidase in 1X reaction buffer. After 30 minutes, fluorescence was measured in a fluorescence microplate reader using excitation at 530 ± 12.5 nm and detection at 590 ± 17.5 nm. A background of 65 fluorescence units was subtracted from each data point. The inset shows the assay's sensitivity and linearity at low hypoxanthine concentrations.

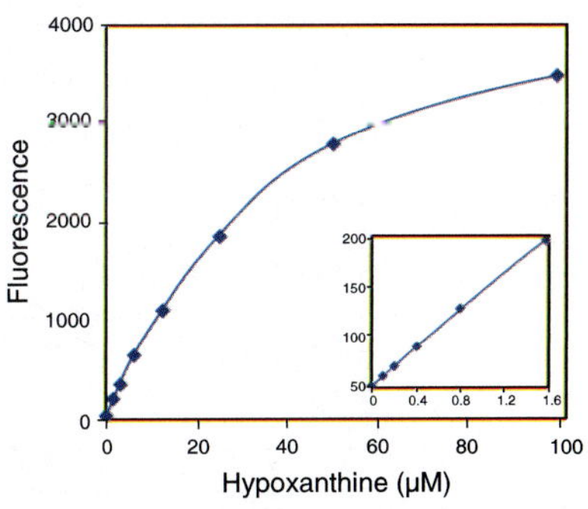

Figure 18.2.25 Detection of hypoxanthine using the Amplex® Red Xanthine/Xanthine Oxidase Assay Kit (A22182). Each reaction contained 50 µM Amplex® Red reagent, 0.2 U/mL horseradish peroxidase, 20 mU/mL xanthine oxidase and the indicated amount of hypoxanthine in 1X reaction buffer. Reactions were incubated at 37°C. After 30 minutes, fluorescence was measured in a fluorescence microplate reader using excitation at 530 ± 12.5 nm and detection at 590 ± 17.5 nm. A background of 54 fluorescence units was subtracted from each data point. The inset shows the assay's sensitivity and linearity at low enzyme concentrations.

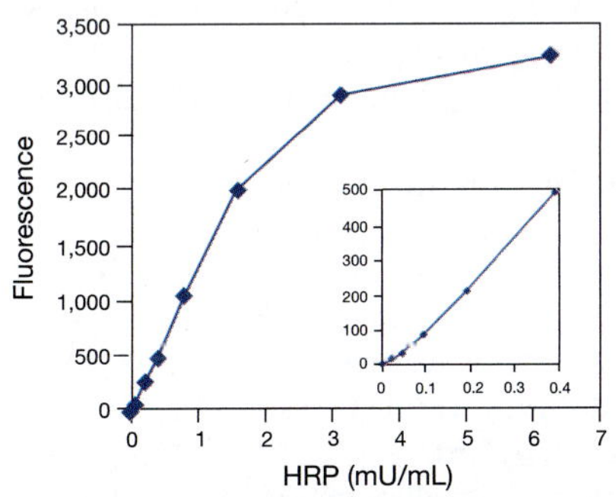

Figure 18.2.22 Detection of HRP using the Amplex® Red Hydrogen Peroxide/Peroxidase Assay Kit (A22188). Reactions containing 50 µM Amplex® Red reagent, 1 mM H_2O_2 and the indicated amount of HRP in 50 mM sodium phosphate buffer, pH 7.4, were incubated for 30 minutes at room temperature. Fluorescence was measured with a fluorescence microplate reader using excitation at 530 ± 12.5 nm and fluorescence detection at 590 ± 17.5 nm. Background fluorescence (3 units), determined for a no-HRP control reaction, was subtracted from each value. The inset shows the sensitivity of the assay at very low levels of HRP.

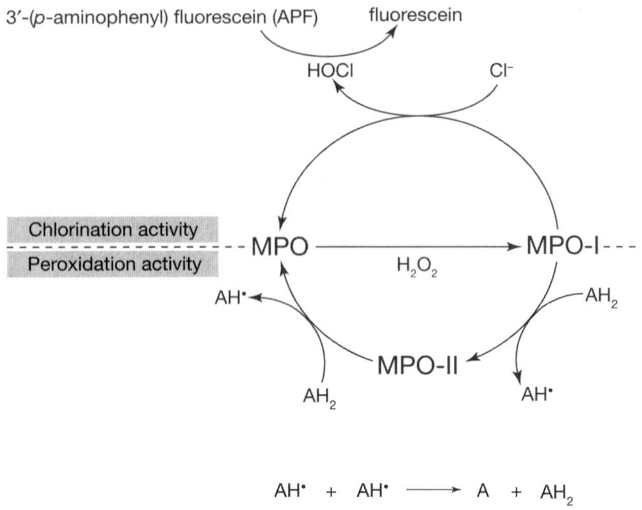

Figure 18.2.26 Schematic diagram for detection of chlorination and peroxidation activity of MPO using the EnzChek® Myeloperoxidase (MPO) Activity Assay Kit (E33856). AH_2 represents the nonfluorescent Amplex® UltraRed substrate, and A represents its fluorescent oxidation product.

The EnzChek® Myeloperoxidase (MPO) Activity Assay Kit (E33856) provides assays for rapid and sensitive determination of both chlorination and peroxidation activities of MPO in solution and in cell lysates [159–161] (Figure 18.2.26). For detection of chlorination, the kit provides nonfluorescent 3′-(*p*-aminophenyl) fluorescein (APF), which is selectively cleaved by hypochlorite (^-OCl) to yield fluorescein. Peroxidation is detected using nonfluorescent Amplex® UltraRed reagent, which is oxidized by the H_2O_2-generated redox intermediates MPO-I and MPO-II to form a fluorescent product. The EnzChek® Myeloperoxidase Activity Assay Kit can be used to continuously detect these activities at room temperature over a broad dynamic range (1.5 to 200 ng/mL) (Figure 18.2.27). The speed (30 minutes), sensitivity and mix-and-read convenience make this kit ideal for measuring MPO activities and for high-throughput screening for MPO-specific inhibitors.

Each EnzChek® Myeloperoxidase (MPO) Activity Assay Kit contains:

- 3′-(*p*-aminophenyl) fluorescein (APF)
- Amplex® UltraRed reagent
- Human myeloperoxidase (MPO) standard
- Chlorination inhibitor
- Peroxidation inhibitor
- Hydrogen peroxide (H_2O_2)
- Phosphate-buffered saline (PBS)
- Dimethylsulfoxide (DMSO)
- Detailed protocols

Sufficient reagents are provided to perform 200 assays for chlorination and 200 assays for peroxidation activity in a 96-well fluorescence microplate format (100 µL per assay).

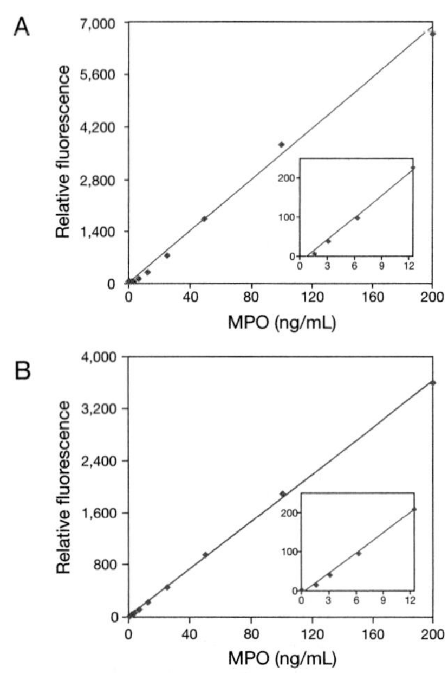

Figure 18.2.27 Typical standard curves for detection of MPO using the APF-based chlorination assay (**A**) and Amplex® UltraRed–based peroxidation assay (**B**) provided in the EnzChek® Myeloperoxidase (MPO) Activity Assay Kit (E33856). Reactions were incubated at room temperature for 30 minutes. Values on the x-axes are concentrations of MPO in the standards prior to adding the detection reagent. Fluorescence was measured with a fluorescence microplate reader using fluorescence excitation and emission at 485 and 530 nm, respectively, for the APF assay, or excitation and emission at 530 and 590 nm, respectively, for the Amplex® UltraRed assay. The background fluorescence measured for each zero-MPO control reaction was subtracted from each fluorescence measurement before plotting.

Zen™ Myeloperoxidase (MPO) ELISA Kit

The Zen™ Myeloperoxidase (MPO) ELISA Kit (Z33857) provides a comprehensive set of components for accurate and sensitive quantitation of human MPO in a variety of biological samples, including human serum. This sandwich immunoassay utilizes the Amplex® UltraRed reagent, a fluorogenic substrate for horseradish peroxidase (HRP) that reacts with H_2O_2 in a 1:1 stoichiometric ratio to produce the brightly fluorescent and strongly absorbing Amplex® UltraRed oxidation product (excitation/ emission maxima ~568/581 nm). Because the Amplex® UltraRed product has long-wavelength emission, there is little interference from the blue or green autofluorescence found in most biological samples. With a high extinction coefficient, good quantum efficiency and resistance to autooxidation, the fluorescence-based Amplex® UltraRed reagent delivers better sensitivity and a broader assay range than colorimetric reagents.

Each Zen™ Myeloperoxidase (MPO) ELISA Kit contains:

- Amplex® UltraRed reagent
- Dimethylsulfoxide (DMSO)
- Concentrated phosphate-buffered saline (PBS)
- Horseradish peroxidase (HRP) labeled goat anti–rabbit IgG antibody
- Amplex® stop reagent
- Hydrogen peroxide (H_2O_2)
- MPO standard
- Bovine serum albumin (BSA)
- Tween® 20
- Mouse anti-MPO antibody (capture antibody)
- Rabbit anti-MPO antibody (detection antibody)
- Zen™ microplates for oriented capture anitbody coating
- Detailed protocols

IMPORTANT NOTICE: The products described in this manual are covered by one or more Limited Use Label License(s). Please refer to the Appendix on page 971 and Master Product List on page 975. Products are For Research Use Only. Not intended for any animal or human therapeutic or diagnostic use.

Sufficient reagents are provided for 200 assays in a microplate format, using a 100 µL per well reaction volume. The Zen™ Myeloperoxidase (MPO) ELISA Kit can be used to detect from 0.2 to 100 ng/mL MPO at room temperature (Figure 18.2.28).

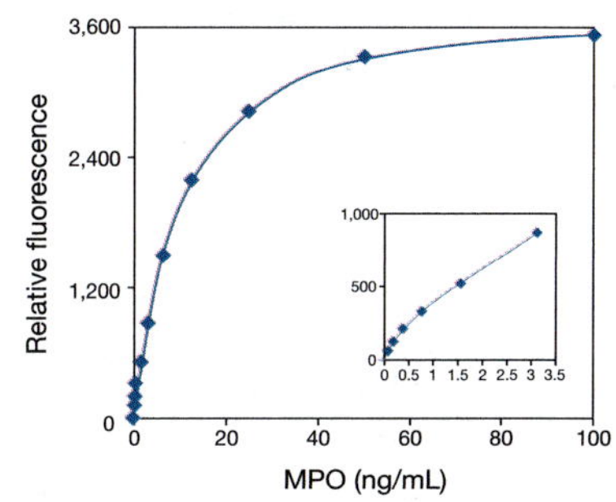

Figure 18.2.28 Typical standard curve for detection of MPO using the Zen™ Myeloperoxidase (MPO) ELISA Kit (Z33857). The sandwich ELISA was carried out as described in the protocol using a mouse anti-MPO primary capture antibody, MPO standards ranging from 0.2 ng/mL to 100 ng/mL, and a rabbit anti-MPO detection antibody.

Assaying Oxidative Activity in Live Cells

The generation of reactive oxygen species (ROS) is inevitable for aerobic organisms and, in healthy cells, occurs at a controlled rate. Under conditions of oxidative stress, however, ROS production is dramatically increased, resulting in subsequent alteration of membrane lipids, proteins and nucleic acids. Oxidative damage of these biomolecules is associated with aging and with a variety of pathological events, including atherosclerosis, carcinogenesis, ischemic reperfusion injury and neurodegenerative disorders.[40,162,163]

Assaying oxidative activity in live cells with fluorogenic, chemiluminescent or chromogenic probes is complicated by the frequent presence of multiple reactive oxygen species in the same cell. Scavengers and enzymes such as superoxide dismutase and catalase are useful knockdown reagents for triaging the optical response of ROS probes (Table 18.2). Quantitative analysis can be further hindered due to: 1) the high intracellular concentration of glutathione, which can form thiyl or sulfinyl radicals or otherwise trap or reduce oxygen species;[164] 2) the variable concentration of metals, which can either catalyze or inhibit radical reactions; and 3) the presence of other free radical–quenching agents such as spermine.[165]

Fluorescein, rhodamine and various other dyes can be chemically reduced to colorless, nonfluorescent leuco dyes. These "dihydro" derivatives are readily oxidized back to the parent dye by reactive oxygen species and thus can serve as fluorogenic probes for detecting oxidative activity in cells and tissues.[166–168] Oxidation also occurs spontaneously, albeit slowly, in air and via photosensitization when illuminated for fluorescence excitation.[169,170] Careful storage and handling, as well as minimizing the duration and intensity of light exposure, are particularly recommended when using these dyes. In general, dihydrofluorescein and dihydrorhodamine do not discriminate between the various reactive oxygen species. It has been reported that dichlorodihydrofluorescein (H_2DCF) and dihydrorhodamine 123 react with intracellular hydrogen peroxide in a reaction mediated by peroxidase, cytochrome *c* or Fe^{2+},[23,171] and these leuco dyes also serve as fluorogenic substrates for peroxidase enzymes (Section 10.5).

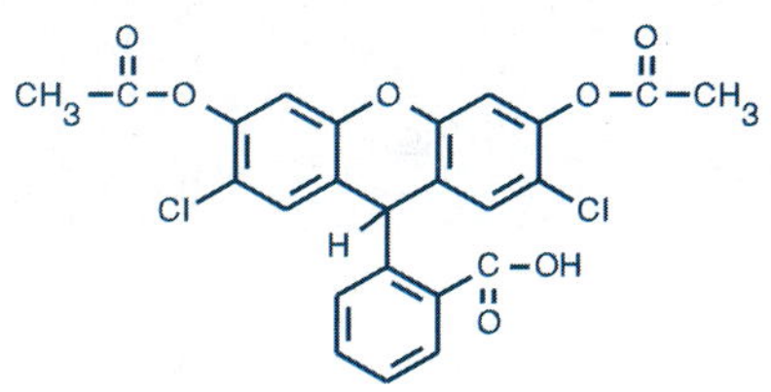

Figure 18.2.29 2′,7′-dichlorodihydrofluorescein diacetate (2′,7′-dichlorofluorescin diacetate; H_2DCFDA, D399).

Dichlorodihydrofluorescein Diacetate

The cell-permeant 2′,7′-dichlorodihydrofluorescein diacetate (H_2DCFDA, D399; Figure 18.2.29), also known as dichlorofluores*cin* diacetate, is commonly used to detect the generation of reactive oxygen intermediates in neutrophils and macrophages.[172–176] Upon cleavage of the acetate groups by intracellular esterases and subsequent oxidation, the nonfluorescent H_2DCFDA is converted to the highly fluorescent 2′,7′-dichlorofluorescein (DCF).

Oxidation of H_2DCFDA is reportedly not sensitive to singlet oxygen directly, but singlet oxygen can indirectly contribute to the formation of DCF through its reaction with cellular substrates that yield peroxy products and peroxyl radicals.[170] In a cell-free system, H_2DCF has been shown to be oxidized to DCF by peroxynitrite anion ($ONOO^-$), by horseradish peroxidase (in the absence of H_2O_2) and by Fe^{2+} (in the absence of H_2O_2).[177] Furthermore, the oxidation of H_2DCF by Fe^{2+} in the presence of H_2O_2 was reduced by the HO• radical scavenger formate and the iron chelator deferoxamine.[177] In addition, DCF itself can act as a photosensitizer for H_2DCFDA oxidation, both priming and accelerating the formation of DCF.[170] Because the oxidation of DCF and H_2DCFDA appears to also generate free radicals, their use for measuring free radical production must be carefully controlled.[178]

A review by Tsuchiya and colleagues outlined methods for visualizing the generation of oxidative species in whole animals. For example, they suggest using propidium iodide (P1304MP, P3566, P21493; Section 8.1) with H_2DCFDA to simultaneously monitor oxidant production and cell injury.[179] H_2DCFDA has been used to visualize oxidative changes in carbon tetrachloride–perfused rat liver[180] and in venular endothelium during neutrophil activation,[181] as well as to examine the effect of ischemia and reperfusion in lung and heart tissue.[182,183] Using H_2DCFDA, researchers characterized hypoxia-dependent peroxide production in *Saccharomyces cerevisiae* as a possible model for ischemic tissue destruction.[184] In neutrophils, H_2DCFDA has proven

IMPORTANT NOTICE: The products described in this manual are covered by one or more Limited Use Label License(s). Please refer to the Appendix on page 971 and Master Product List on page 975. Products are For Research Use Only. Not intended for any animal or human therapeutic or diagnostic use.

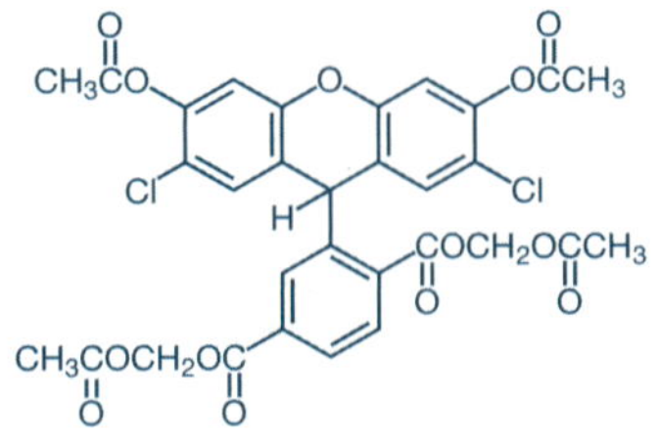

Figure 18.2.30 6-carboxy-2',7'-dichlorodihydrofluorescein diacetate, di(acetoxymethyl ester), C2938.

Figure 18.2.31 Bovine pulmonary artery endothelial (BPAEC) cells were initially stained with the reactive oxygen species (ROS) indicator, 6-carboxy-2',7'-dichlorodihydrofluorescein diacetate, di(acetoxymethyl ester) (C2938). After a 30-minute incubation, the cells were washed and then incubated simultaneously with FM® 5-95 (T23360) and Hoechst 33342 (H1399, H3570, H21492) in phosphate-buffered saline (PBS) for an additional 5 minutes before washing and mounting in PBS. The red-fluorescent FM® 5-95 appears to stain both the plasma membrane and early endosomes; the green-fluorescent, oxidized carboxydichlorofluorescein localizes to the cytoplasm; and the blue-fluorescent Hoechst 33342 dye stains the nucleus.

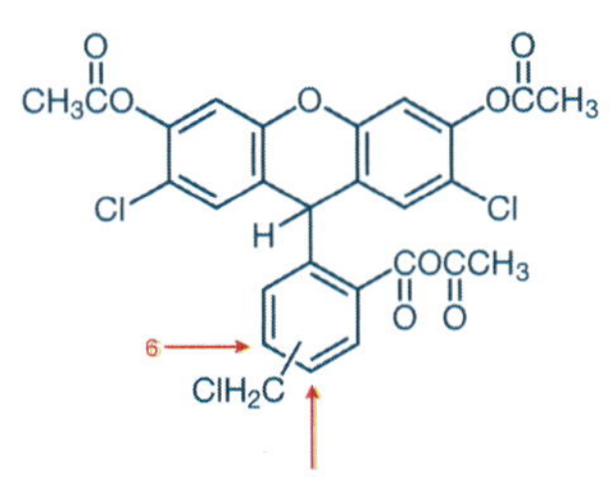

Figure 18.2.32 5-(and-6)-chloromethyl-2',7'-dichlorodihydrofluorescein diacetate, acetyl ester (CM-H_2DCFDA, C6827).

useful for flow cytometric analysis of nitric oxide, forming a product that has spectral properties identical to those produced when it reacts with hydrogen peroxide.[185] In this study, H_2DCFDA's reaction with nitric oxide was blocked by adding the nitric oxide synthase inhibitor N^G-methyl-L-arginine (L-NMMA) to the cell suspension.[185] 2′,7′-Dichlorofluorescein—the oxidation product of H_2DCF—can reportedly be further oxidized to a phenoxyl radical in a horseradish peroxidase–catalyzed reaction, and this reaction may complicate the interpretation of results obtained with this probe in cells undergoing oxidative stress.[186] Although other more specialized ROS probes have been—and continue to be—developed, H_2DCFDA and its chloromethyl derivative CM-H_2DCFDA remain the most versatile indicators of cellular oxidative stress.[187]

Improved Versions of H_2DCFDA

Intracellular oxidation of H_2DCF tends to be accompanied by leakage of the product, 2′,7′-dichlorofluorescein,[188] which may make quantitation or detection of slow oxidation difficult. To enhance retention of the fluorescent product, we offer the carboxylated H_2DCFDA analog[189] (carboxy-H_2DCFDA, C400), which has two negative charges at physiological pH, and its di(acetoxymethyl ester) analog, 6-carboxy-2',7'-dichlorodihydrofluorescein diacetate, di(acetoxymethyl ester)[190] (C2938, Figure 18.2.30, Figure 18.2.31). Upon cleavage of the acetate and ester groups by intracellular esterases and oxidation, both analogs form carboxydichlorofluorescein (C368, Section 14.3), with additional negative charges that impede its leakage out of the cell.

The fluorinated analog 5-(and 6-)carboxy-2′,7′-difluorodihydrofluorescein diacetate (carboxy-H_2DFFDA, C13293) is also useful for visualizing oxidative bursts and inflammatory and infectious processes.[191] As the oxidation potential of deacetylated carboxy-H_2DFFDA is more positive than that of the corresponding chloro compound carboxy-H_2DCFDA, its oxidant sensitivity profile is presumably shifted; however, it is not known if this difference is large enough to have practical utility. The diacetate derivatives of the dichloro- and difluorodihydrofluoresceins are quite stable. When used for intracellular applications, the acetates are cleaved by endogenous esterases, releasing the corresponding dichloro- or difluorodihydrofluorescein derivative. If, however, these nonfluorescent diacetate derivatives are used for *in vitro* assays, they must first be hydrolyzed with mild base to form the colorless probe.[150]

In addition, we have developed 5-(and 6-)chloromethyl-2′,7′-dichlorodihydrofluorescein diacetate, acetyl ester (CM-H_2DCFDA, C6827; Figure 18.2.32; Figure 18.2.33), which is a chloromethyl derivative of H_2DCFDA that exhibits much better retention in live cells.[192,193] As with our other chloromethyl derivatives (see the description of our CellTracker™ probes in Section 14.2), CM-H_2DCFDA passively diffuses into cells, where its acetate groups are cleaved by intracellular esterases and its thiol-reactive chloromethyl group reacts with intracellular glutathione and other thiols. Subsequent oxidation yields a fluorescent adduct that is trapped inside the cell, thus facilitating long-term studies.[192,193] Among its many applications, CM-H_2DCFDA has been used to:

- Analyze FOXO3 transcriptional control of oxidative stress[194,195]
- Assess ROS-mediated cytotoxicity and apoptosis[196–199]
- Detect hydroxyl radicals associated with estrogen-induced DNA damage[200]
- Monitor time courses of ROS generation in neurons and brain slices[192,201]
- Analyze ROS production in chromosomally unstable human–hamster hybrid cells using flow cytometry[202]

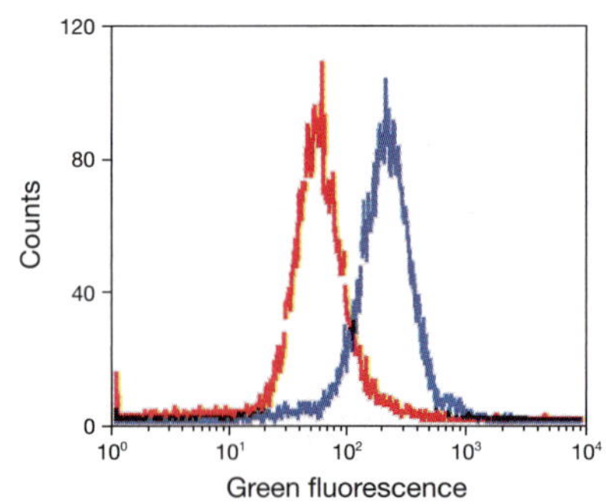

Figure 18.2.33 An oxidative burst was detected by flow cytometry of cells labeled with 5-(and 6-)chloromethyl-2',7'-dichlorodihydrofluorescein diacetate, acetyl ester (CM-H_2DCFDA, C6827). Jurkat cells were incubated with 100 nM CM-H_2DCFDA. The cells were washed and resuspended in either phosphate-buffered saline (PBS, red) or PBS with 0.03% H_2O_2 (blue). The samples were analyzed on a flow cytometer equipped with a 488 nm argon-ion laser and a 525 ± 10 nm bandpass emission filter.

IMPORTANT NOTICE: The products described in this manual are covered by one or more Limited Use Label License(s). Please refer to the Appendix on page 971 and Master Product List on page 975. Products are For Research Use Only. Not intended for any animal or human therapeutic or diagnostic use.

Image-iT® LIVE Green Reactive Oxygen Species Detection Kit

The Image-iT® LIVE Green Reactive Oxygen Species Detection Kit (I36007) provides the key reagents for detecting reactive oxygen species (ROS) in live cells (Figure 18.2.34), including:

- Carboxy-H_2DCFDA (5-(and 6-)carboxy-2′,7′-dichlorodihydrofluorescein diacetate)
- Hoechst 33342
- *tert*-butyl hydroperoxide (TBHP)
- Dimethylsulfoxide (DMSO)
- Detailed protocols for fluorescence microscopy assays

This assay is based on carboxy-H_2DCFDA (5-(and 6-)carboxy-2′,7′-dichlorodihydrofluorescein diacetate), a reliable fluorogenic marker for reactive oxygen species in live cells.[203,204] In addition to carboxy-H_2DCFDA, this kit provides the common inducer of ROS production, *tert*-butyl hydroperoxide (TBHP), as a positive control[205–208] and the blue-fluorescent, cell-permeant nucleic acid stain Hoechst 33342. Oxidatively stressed and nonstressed cells can be effectively distinguished by fluorescence microscopy using this combination of dyes and the protocol provided.[209–211]

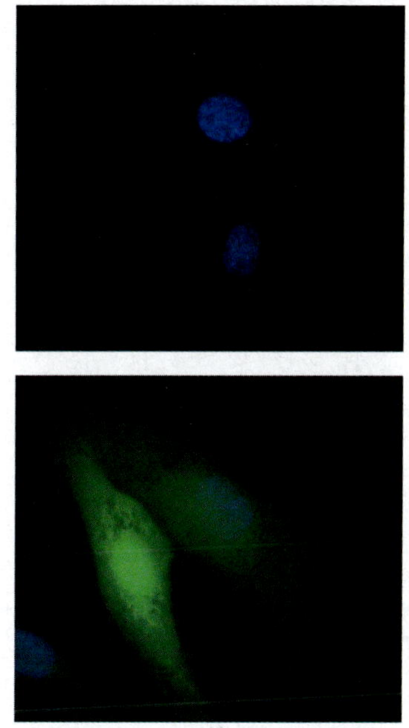

Figure 18.2.34 Detection of oxidative stress in live cells using the Image-iT® LIVE Green Reactive Oxygen Species (ROS) Detection Kit (I36007). Live bovine pulmonary artery endothelial cells were treated with *tert*-butyl hydroperoxide to induce oxidative stress (bottom) or were left untreated (top). Cells were then labeled with carboxy-H_2DCFDA, which fluoresces when oxidized by ROS, and nuclei were stained with blue-fluorescent Hoechst 33342. The stressed cells exhibited green fluorescence, signaling an increase in ROS, whereas the untreated cells showed minimal fluorescence.

Aminophenyl Fluorescein and Hydroxyphenyl Fluorescein

Developed by Nagano, 3′-(*p*-aminophenyl) fluorescein (APF, A36003) and 3′-(*p*-hydroxyphenyl) fluorescein (HPF, H36004) provide greater selectivity and stability than dichlorodihydrofluorescein diacetate (H_2DCFDA, D399) for ROS detection.[212] H_2DCFDA is probably the most commonly used reagent for detecting intracellular reactive oxygen species despite its lack of specificity and tendency to spontaneously photooxidize. The nonfluorescent H_2DCFDA becomes fluorescent in the presence of a wide variety of reactive oxygen species including, but not limited to, peroxyl (ROO•) and hydroxyl (HO•) radicals and the peroxynitrite anion ($ONOO^-$). In contrast, APF and HPF show much more limited reactivity and greater resistance to light-induced oxidation (Table 18.4). Both of these fluorescein derivatives are essentially nonfluorescent until they react with the hydroxyl radical,[60] peroxynitrite anion or singlet oxygen[16] (Figure 18.2.35). APF will also react with the hypochlorite anion ($^-$OCl), making it possible to use APF

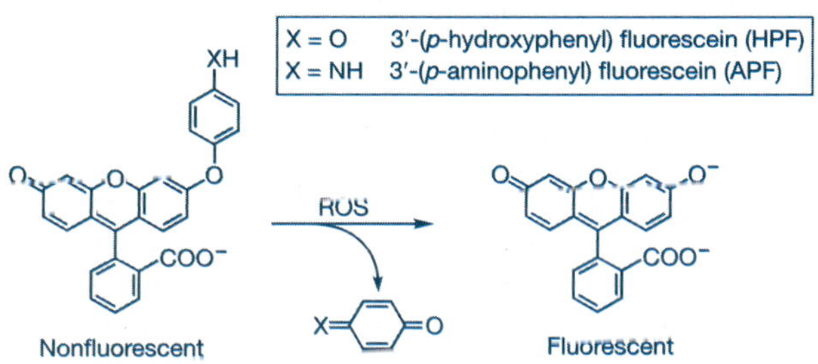

Figure 18.2.35 Detection of reactive oxygen species (ROS) with 3′-(*p*-hydroxyphenyl) fluorescein (HPF, H36004) and 3′-(*p*-aminophenyl) fluorescein (APF, A36003).

Table 18.4 Fluorescence response of APF, HPF and H_2DCFDA to various reactive oxygen species (ROS).

Reactive Oxygen Species (ROS)	ROS Generation Method	APF *	HPF *	H_2DCFDA *
Hydrogen peroxide (H_2O_2)	100 µM H_2O_2	<1	2	190
Hydroxyl radical (HO•)	100 µM ferrous perchlorate (II) and 1 mM of H_2O_2	1200	730	7400
Hypochlorite anion ($^-$OCl)	3 µM (final) $^-$OCl	3600	6	86
Nitric oxide (NO)	100 µM 1-hydroxy-2-oxo-3-(3-aminopropyl)-3-methyl-1-triazene (NOC-7)	<1	6	150
Peroxyl radical (ROO•)	100 µM 2,2′-azobis(2-amidinopropane), dihydrochloride (AAPH)	2	17	710
Peroxynitrite anion ($ONOO^-$)	3 µM (final) $ONOO^-$	560	120	6600
Singlet oxygen ($^1O_2$)	100 µM 3-(1,4-dihydro-1,4-epidioxy-1-naphthyl)propionic acid	9	5	26
Superoxide anion ($\bullet O_2^-$)	100 µM KO_2	6	8	67
Autooxidation	2.5 hours exposure to fluorescent light source	<1	<1	2000

* 10 µm of APF, HPF or DCF (2′,7′-dichlorofluorescein) were added to sodium phosphate buffer (0.1 M, pH 7.4); ROS were generated as indicated; and fluorescence was measured using excitation/emission wavelengths of 490/515 nm (for APF and HPF) or 500/520 nm (for DCF). DCF was obtained by hydrolysis of H_2DCFDA with base as described in J Biol Chem (2003) 278:3170; dihydrofluorescein diacetates are colorless and nonfluorescent until both of the acetate groups are hydrolyzed and the products are subsequently oxidized to fluorescein derivatives.

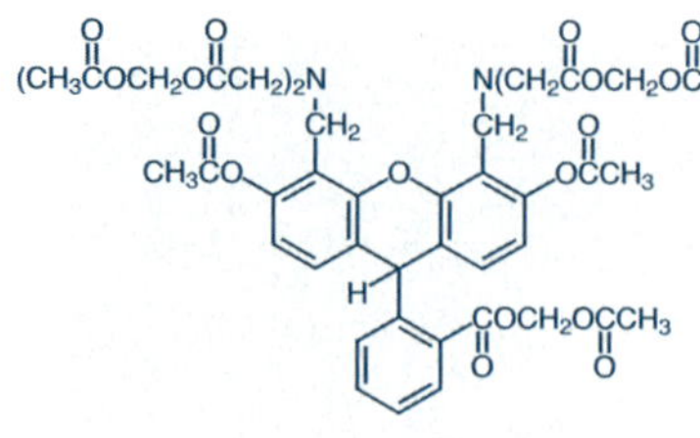

Figure 18.2.36 Dihydrocalcein, AM (D23805).

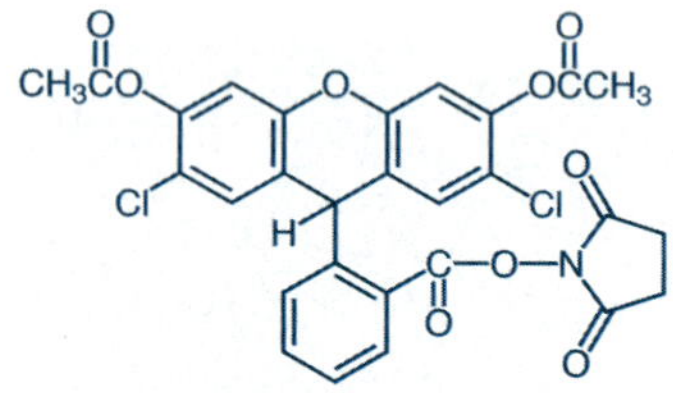

Figure 18.2.37 2′,7′-dichlorodihydrofluorescein diacetate, succinimidyl ester (OxyBURST® Green H_2DCFDA, SE, D2935).

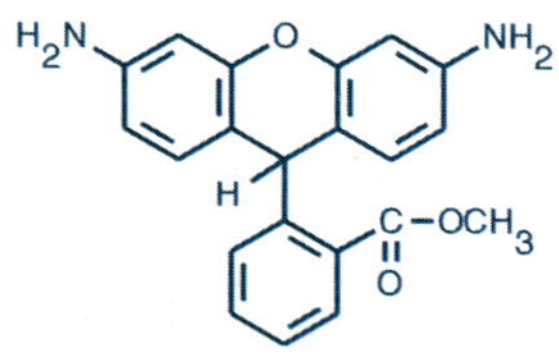

Figure 18.2.38 Dihydrorhodamine 123 (D632).

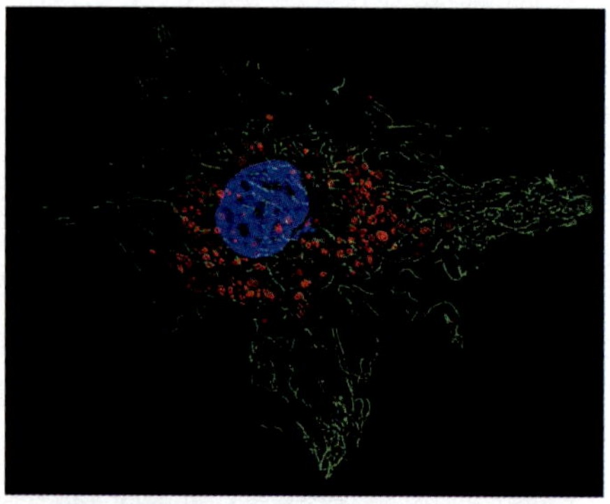

Figure 18.2.39 Live bovine pulmonary artery endothelial cells (BPAEC) were first stained with LysoTracker® Red DND-99 (L7528). Then, a solution of dihydrorhodamine 123 (D632, D23806) and Hoechst 33258 (H1398, H3569, H21491) was added and allowed to incubate with the cells for an additional 10 minutes before the cells were subsequently washed and visualized. The green-fluorescent oxidation product (rhodamine 123, R302) localized primarily to the mitochondria. The red-fluorescent LysoTracker® Red DND-99 stain accumulated in the lysosomes, and the blue-fluorescent Hoechst 33258 dye stained the nuclei. The image was acquired with filters appropriate for DAPI, fluorescein and the Texas Red® dye. The image was deconvolved using Huygens software (Scientific Volume Imaging, www.svi.nl). 3D reconstruction was performed using Imaris software (Bitplane AG, www.bitplane.com).

and HPF together to selectively detect the hypochlorite anion (Section 21.2). In the presence of these specific reactive oxygen species, both APF and HPF yield a bright green-fluorescent product (excitation/emission maxima ~490/515 nm) and are compatible with all fluorescence instrumentation capable of visualizing fluorescein. Using APF, researchers have been able to detect the hypochlorite anion generated by activated neutrophils, a feat that has not been possible with traditional ROS indicators.[212]

Dihydrocalcein AM

We have combined the superior retention of calcein (the intracellular product of calcein AM hydrolysis in viable cells) and the oxidation sensitivity of the dihydrofluoresceins to yield the probe dihydrocalcein AM (D23805, Figure 18.2.36), provided specially packaged as a set of 20 vials, each containing 50 µg. The oxidant sensitivity profile of dihydrocalcein AM has been characterized relative to that of H_2DCFDA[213] and of alkaline elution assays of oxidative DNA modification.[214]

OxyBURST® Green Reagents

Fc OxyBURST® Green assay reagent (F2902) was developed in collaboration with Elizabeth Simons of Boston University to monitor the oxidative burst in phagocytic cells using fluorescence instrumentation. The Fc OxyBURST® Green assay reagent comprises bovine serum albumin (BSA) that has been covalently linked to dichlorodihydrofluorescein (H_2DCF) and then complexed with purified rabbit polyclonal anti-BSA antibodies. When these immune complexes bind to Fc receptors, the nonfluorescent H_2DCF molecules are internalized within the phagovacuole and subsequently oxidized to green-fluorescent dichlorofluorescein (DCF); see Section 16.1 for a more complete description.

OxyBURST® Green H_2HFF BSA (O13291) is a sensitive fluorogenic reagent for detecting extracellular release of oxidative products in a spectrofluorometer or a fluorescence microscope. This reagent comprises BSA that has been covalently linked to dihydro-2′,4,5,6,7,7′-hexafluorofluorescein (H_2HFF), a reduced dye with improved stability. Unlike Fc OxyBURST® Green assay reagent, OxyBURST® Green H_2HFF BSA is not complexed with IgG. OxyBURST® Green H_2HFF BSA provides up to 1000-fold greater sensitivity than conventional methods based on spectrophotometric detection of superoxide dismutase–inhibitable reduction of cytochrome *c*.[215,216]

Amine-Reactive OxyBURST® Green Reagent

As an alternative to Fc OxyBURST® Green assay reagent and OxyBURST® Green H_2HFF BSA, we offer the amine-reactive OxyBURST® Green H_2DCFDA succinimidyl ester (2′,7′-dichlorodihydrofluorescein diacetate, SE; D2935; Figure 18.2.37), which can be used to prepare oxidation-sensitive conjugates of a wide variety of biomolecules and particles, including antibodies, antigens, peptides, proteins, dextrans, bacteria, yeast and polystyrene microspheres.[217,218] Following conjugation to amines, the two acetates of OxyBURST® Green H_2DCFDA can be removed by treatment with hydroxylamine at neutral pH to yield the dihydrofluorescein conjugate. OxyBURST® Green H_2DCFDA conjugates are nonfluorescent until they are oxidized to the corresponding fluorescein derivatives.

Dihydrorhodamine 123

Dihydrorhodamine 123 (D632, D23806; Figure 18.2.38, Figure 18.2.39) is the uncharged and nonfluorescent reduction product of the mitochondrion-selective dye rhodamine 123 (R302, R22420; Section 12.2). This leuco dye passively diffuses across most cell membranes where it is oxidized to cationic rhodamine 123 (Figure 18.2.40), which localizes in the mitochondria. Like H_2DCF, dihydrorhodamine 123 does not directly detect superoxide,[219] but rather reacts with hydrogen peroxide in the presence of peroxidase, cytochrome *c* or Fe^{2+}.[23] However, dihydrorhodamine 123 also reacts with peroxynitrite, the anion formed when nitric oxide reacts with superoxide.[171] Peroxynitrite, which may play a role in many pathological conditions, has been shown to react with sulfhydryl groups, DNA and membrane phospholipids, as well as with tyrosine and other phenolic compounds.[220–223]

Dihydrorhodamine 123 has been used to investigate reactive oxygen intermediates produced by human and murine phagocytes,[45] activated rat mast cells[224] and vascular endothelial

IMPORTANT NOTICE: The products described in this manual are covered by one or more Limited Use Label License(s). Please refer to the Appendix on page 971 and Master Product List on page 975. Products are For Research Use Only. Not intended for any animal or human therapeutic or diagnostic use.

tissues.[225,226] It has also been employed to study the role of the CD14 cell-surface marker in H_2O_2 production by human monocytes.[227]

Dihydrorhodamine 123 is available as a 10 mg vial (D632) or as a stabilized 5 mM solution in DMSO (D23806). Because of the susceptibility of dihydrorhodamine 123 to air oxidation, the DMSO solution is recommended when only small quantities are to be used at a time.

A Longer-Wavelength Reduced Rhodamine

Intracellular oxidation of dihydrorhodamine 6G (D633) yields rhodamine 6G (R634), which localizes in the mitochondria of live cells (Section 12.2). As compared with rhodamine 123, this cationic oxidation product has longer-wavelength spectra, making it especially useful in multicolor applications and in autofluorescent cells and tissues. Dihydrorhodamine 6G has been used for fluorescence microplate assays of granulocyte activation[228] and for analysis of ROS levels in human umbilical vein endothelial (HUVEC) cells by flow cytometry.[194]

Reduced MitoTracker® Probes

Two of our MitoTracker® probes—MitoTracker® Orange CM-H_2TMRos (M7511, Figure 18.2.41) and MitoTracker® Red CM-H_2XRos (M7513, Figure 18.2.42)—are chemically reactive reduced rosamines. Unlike MitoTracker® Orange CMTMRos and MitoTracker® Red CMXRos (M7510, M7512; Section 12.2), the reduced versions of these probes do not fluoresce until they enter an actively respiring cell, where they are oxidized by reactive oxygen species to the fluorescent mitochondrion-selective probe and then sequestered in the mitochondria. Although CM-H_2TMRos and CM-H_2XRos are widely used as indicators of mitochondrial reactive oxygen species,[229–231] their fluorescence cannot be unambiguously associated with the site of oxidant generation, as the cationic charge that drives their electrophoretic sequestration in active mitochondria is only present after the probe has been oxidized. This same caveat also applies to dihydrorhodamine 123 and dihydrorhodamine 6G. Probes such as MitoSOX™ Red mitochondrial superoxide indicator resolve this ambiguity by having their oxidant response and mitochondrial localization functions associated with different structural elements (Figure 18.2.8).

RedoxSensor™ Red CC-1 Stain

RedoxSensor™ Red CC-1 stain (2,3,4,5,6-pentafluorotetramethyldihydrorosamine, R14060; Figure 18.2.43) passively enters live cells and is subsequently oxidized in the cytosol to a red-fluorescent product (excitation/emission maxima ~540/600 nm), which then accumulates in the mitochondria. Alternatively, this nonfluorescent probe may be transported to the lysosomes where it is oxidized. The differential distribution of the oxidized product between mitochondria and lysosomes appears to depend on the redox potential of the cytosol.[232–234] In proliferating cells, mitochondrial staining predominates; whereas in contact-inhibited cells, the staining is primarily lysosomal (Figure 18.2.44).

Glutathiolation Detection with BioGEE

Biotinylated glutathione ethyl ester (BioGEE, G36000; Figure 18.2.45) is a cell-permeant, biotinylated glutathione analog for the detection of glutathiolation. Under conditions of oxidative stress, cells may transiently incorporate glutathione into proteins. Stressed cells incubated with BioGEE will also incorporate this biotinylated glutathione derivative into proteins, facilitating the identification of oxidation-sensitive proteins.[235,236] Once these cells are fixed and

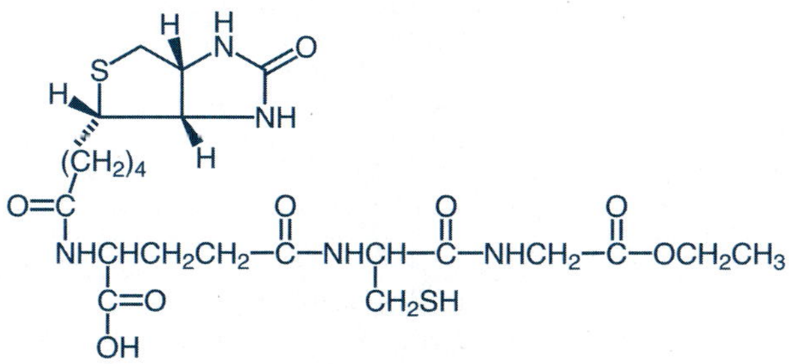

Figure 18.2.45 Glutathione ethyl ester, biotin amide (BioGEE, G36000).

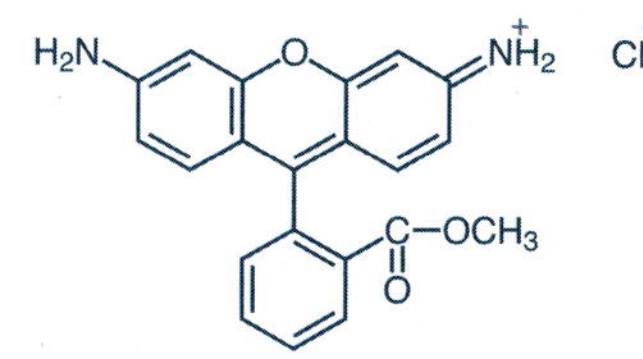

Figure 18.2.40 Rhodamine 123 (R302).

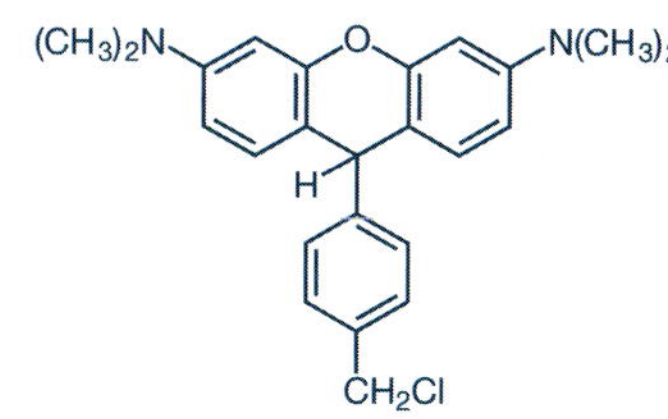

Figure 18.2.41 MitoTracker® Orange CM-H_2TMRos (M7511).

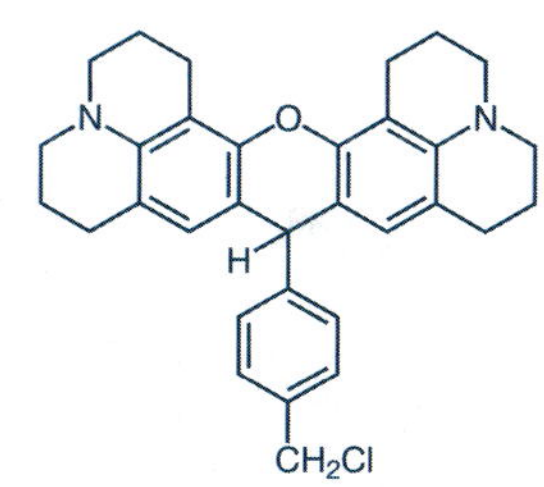

Figure 18.2.42 MitoTracker® Red CM-H_2XRos (M7513).

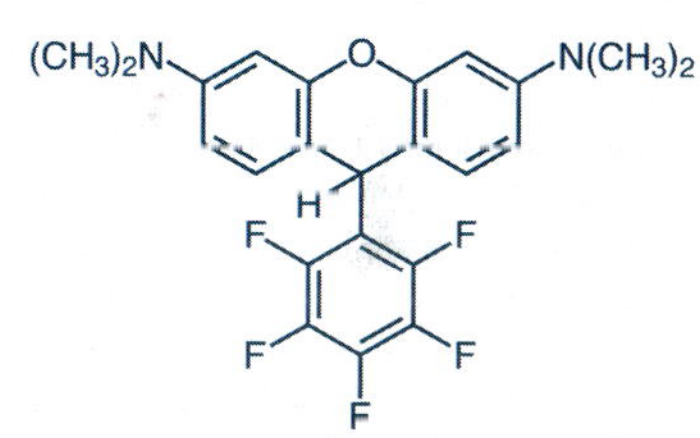

Figure 18.2.43 RedoxSensor™ Red CC-1 (R14060).

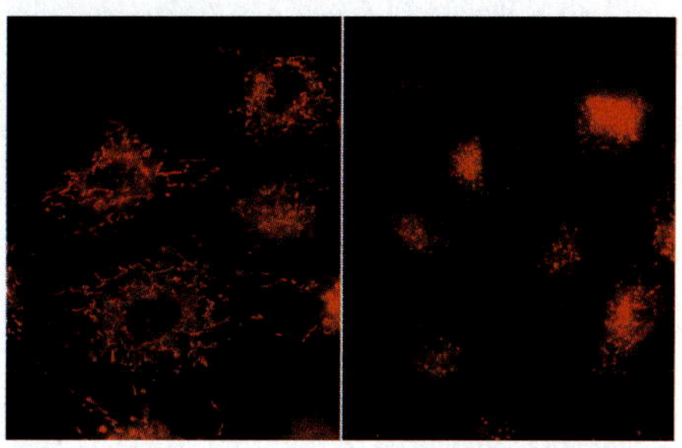

Figure 18.2.44 Cellular proliferation state determines the distribution of the oxidized product of RedoxSensor™ Red CC-1 (R14060). Normal rat kidney (NRK) cells in different growth states were stained with RedoxSensor™ Red CC-1. In proliferating cells (left panel), the oxidized dye accumulates in mitochondria. In quiescent cells (right panel), the oxidized product localizes in the lysosomes.

IMPORTANT NOTICE: The products described in this manual are covered by one or more Limited Use Label License(s). Please refer to the Appendix on page 971 and Master Product List on page 975. Products are For Research Use Only. Not intended for any animal or human therapeutic or diagnostic use.

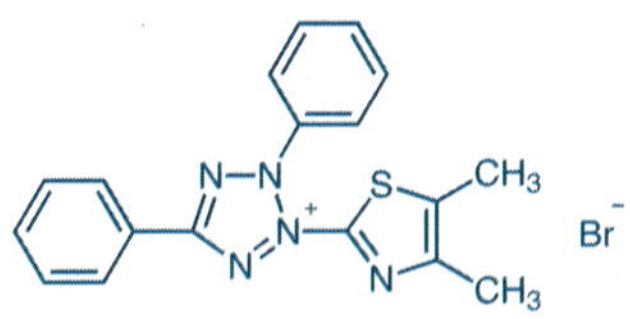

Figure 18.2.46 MTT (3-(4,5-dimethylthiazol-2-yl)-2,5-diphenyltetrazolium bromide), M6494.

permeabilized, glutathiolation levels can be detected with a fluorescent streptavidin conjugate (Section 7.6, Table 7.9) using either flow cytometry or fluorescence microscopy. Proteins glutathiolated with BioGEE can be captured using streptavidin agarose (S951, Section 7.6) and analyzed by mass spectrometry or by western blotting methods.[198,237]

Tetrazolium Salts: Chromogenic Redox Indicators

Tetrazolium salts—especially MTT (M6494, Figure 18.2.46)—are widely used for detecting the redox potential of cells for viability, proliferation and cytotoxicity assays. Upon reduction, these water-soluble colorless compounds form uncharged, brightly colored formazans. Several of the formazans precipitate out of solution and are useful for histochemical localization of the site of reduction or, after solubilization in organic solvent, for quantitation by standard spectrophotometric techniques. The extremely water-soluble formazan product of XTT (X6493) does not require solubilization prior to quantitation.

Selected applications of the tetrazolium salts are listed in Table 18.3. Our Vybrant® MTT Cell Proliferation Assay Kit (V13154, Section 15.4) provides a means of counting metabolically active cells; this Vybrant® MTT assay can detect from 2000 to 250,000 cells, depending on the cell type and conditions. See also Section 15.2 for additional cell applications of tetrazolium salts.

REFERENCES

1. Toxicol Sci (2008) 105:182; **2**. Anal Biochem (2006) 359:151; **3**. J Am Chem Soc (1982) 104:5541; **4**. J Am Chem Soc (2009) 131:332; **5**. Proc Natl Acad Sci USA (1979) 76:6047; **6**. Int J Oncol (2003) 22:933; **7**. Int J Oncol (2002) 21:531; **8**. Histochem Cell Biol (2007) 128:485; **9**. Histochem Cell Biol (2007) 127:263; **10**. Photochem Photobiol (1997) 66:374; **11**. Biochim Biophys Acta (1992) 1105:333; **12**. Arch Biochem Biophys (1991) 291:43; **13**. J Photochem Photobiol A (1991) 58:339; **14**. J Photochem Photobiol B (1988) 1:437; **15**. Biochem Biophys Res Commun (2006) 349:549; **16**. Photochem Photobiol (2009) 85:1177; **17**. Environ Sci Technol (2009) 43:6639; **18**. Methods Mol Biol (2009) 479:109; **19**. J Exp Bot (2006) 57:1725; **20**. Biochim Biophys Acta (1986) 882:210; **21**. Methods Enzymol (1986) 133:569; **22**. Biochem Biophys Res Commun (1984) 123:869; **23**. Free Radic Biol Med (2007) 43:995; **24**. Arch Biochem Biophys (1994) 314:284; **25**. J Reprod Fertil (1993) 97:441; **26**. Methods Enzymol (1967) 12:5; **27**. Curr Biol (1996) 6:1497; **28**. Methods Cell Biol (1994) 44:715; **29**. Proc Natl Acad Sci U S A (1994) 91:2659; **30**. Biochemistry (2000) 39:4068; **31**. Annu Rev Biochem (1990) 59:207; **32**. Biochemistry (1990) 29:8447; **33**. J Am Chem Soc (1987) 109:1990; **34**. J Biol Chem (2004) 279:4127; **35**. J Cell Mol Med (2002) 6:175; **36**. Am J Physiol Heart Circ Physiol (2003) 284:H605; **37**. Nat Protoc (2008) 3:8; **38**. Nat Protoc (2008) 3:941; **39**. Proc Natl Acad Sci U S A (2006) 103:15038; **40**. J Biol Chem (2009) 284:18754; **41**. PLoS ONE (2008) 3:e1616; **42**. J Physiol (2009) 587:851; **43**. Am J Physiol Heart Circ Physiol (2008) 294:H2121; **44**. Nat Protoc (2007) 2:2295; **45**. Cytometry A (2009) 75:475; **46**. J Leukoc Biol (1990) 47:440; **47**. J Leukoc Biol (1994) 55:253; **48**. J Histochem Cytochem (1986) 34:1109; **49**. Biotechniques (1985) 3:270; **50**. Histochemistry (1990) 94:205; **51**. Methods Mol Biol (2009) 476:28; **52**. J Biol Chem (2007) 282:8860; **53**. J Biol Chem (2007) 282:12430; **54**. Biochim Biophys Acta (2007) 1767:989; **55**. Mol Cell (2009) 33:627; **56**. Anal Biochem (1993) 212:85; **57**. Anal Chem (1996) 68:867; **58**. FASEB J (1995) 9:1085; **59**. Anal Chem (1997) 69:4295; **60**. Part Fibre Toxicol (2008) 5:2; **61**. J Biol Chem (2004) 279:23453; **62**. J Biol Chem (2001) 276:35253; **63**. Proc Natl Acad Sci U S A (1990) 87:1620; **64**. Eur J Biochem (1994) 221:695; **65**. Anal Biochem (2004) 324:45; **66**. Am J Physiol Heart Circ Physiol (2009) 296:H840; **67**. Anal Biochem (1992) 206:273; **68**. J Biol Chem (2009) 284:46; **69**. Arch Biochem Biophys (2000) 373:447; **70**. Circ Res (1999) 84:1203; **71**. Biochem Biophys Res Commun (1998) 248:382; **72**. J Biol Chem (1998) 273:33972; **73**. Inflammation (1996) 20:151; **74**. J Biolumin Chemilumin (1997) 12:277; **75**. Anal Biochem (2001) 298:337; **76**. J Biol Chem (1998) 273:6041; **77**. Anal Biochem (2003) 313:338; **78**. Biochim Biophys Acta (1975) 385:232; **79**. Biochem Biophys Res Commun (1989) 163:836; **80**. Biochim Biophys Acta (1989) 981:235; **81**. FEBS Lett (1984) 169:169; **82**. J Cell Biol (1998) 273:26900; **83**. Neurochem Res (1997) 22:333; **84**. Am J Physiol (1996) 271:F209; **85**. Biochim Biophys Acta (2000) 1487:61; **86**. Circ Res (1999) 84:229; **87**. J Biol Chem (2001) 276:15575; **88**. Plant Physiol (2001) 125:1591; **89**. J Biol Chem (1994) 269:798; **90**. Atherosclerosis (2000) 152:307; **91**. J Biol Chem (1999) 274:2234; **92**. Free Radic Biol Med (1991) 11:81; **93**. J Exp Med (1974) 139:208; **94**. DNA Cell Biol (2002) 21:251; **95**. Biochemistry (1998) 37:13781; **96**. Biochim Biophys Acta (1997) 1330:127; **97**. Biochemistry (1995) 34:12755; **98**. Mol Cell Biochem (2003) 247:171; **99**. Cytometry (1992) 13:686; **100**. Biochemistry (1977) 16:819; **101**. Proc Natl Acad Sci USA (1980) 77:26; **102**. Biosci Biotechnol Biochem (1992) 56:605; **103**. Anal Lett (1988) 21:965; **104**. Proc Natl Acad Sci U S A (2002) 99:11599; **105**. Anal Biochem (2007) 360:130; **106**. J Chromatogr A (2000) 881:159; **107**. Free Radic Biol Med (2001) 31:164; **108**. FEBS Lett (2000) 474:137; **109**. Free Radic Res (1998) 29:43; **110**. Methods Enzymol (2000) 319:603; **111**. J Am Soc Nephrol (2002) 13:2990; **112**. Free Radic Biol Med (2002) 33:473; **113**. FEBS Lett (1999) 453:278; **114**. Bioorg Med Chem Lett (2007) 17:2059; **115**. Anal Biochem (2006) 354:290; **116**. Biochemistry (2006) 45:8135; **117**. Nitric Oxide (2006) 15:163; **118**. Free Radic Biol Med (2004) 36:1635; **119**. Free Radic Biol Med (2007) 42:1392; **120**. J Biol Chem (2007) 282:30452; **121**. J Biochem Biophys Methods (1997) 35:23; **122**. Free Radic Biol Med (1997) 22:93; **123**. J Agric Food Chem (2005) 53:4290; **124**. J Agric Food Chem (2002) 50:2772; **125**. Free Radic Biol Med (1993) 15:447; **126**. Biochem Mol Biol Int (1994) 33:1179; **127**. Free Radic Biol Med (1989) 6:623; **128**. J Immunol (1982) 129:1589; **129**. J Appl Physiol (1994) 76:539; **130**. J Leukoc Biol (1993) 54:300; **131**. J Biochem (Tokyo) (1989) 106:355; **132**. Biochem Biophys Res Commun (1988) 155:106; **133**. J Cell Physiol (1992) 151:466; **134**. Am J Physiol (1994) 266:G799; **135**. Am J Physiol (1992) 262:G1015; **136**. Arch Biochem Biophys (1994) 310:352; **137**. Anal Chem (1993) 65:1794; **138**. Biochem J (1993) 290:51; **139**. Anal Chem (2006) 78:6847; **140**. Chem Res Toxicol (2008) 21:432; **141**. Anal Biochem (1997) 253:162; **142**. Proc Soc Exp Biol Med (1994) 206:53; **143**. Free Radic Biol Med (1992) 12:389; **144**. Biochem Int (1985) 10:205; **145**. J Immunol Methods (1997) 202:133; **146**. Anal Biochem (1997) 253:169; **147**. J Invest Dermatol (1999) 112:751; **148**. Free Radic Biol Med (1999) 27:1197; **149**. J Invest Dermatol (1998) 110:966; **150**. J Biol Chem (2006) 281:39766; **151**. J Pharmacol Exp Ther (2008) 324:970; **152**. Methods Enzymol (2009) 456:381; **153**. Am J Physiol Lung Cell Mol Physiol (2007) 292:L1289; **154**. J Biol Chem (2007) 282:14186; **155**. J Immunol (2010) 184:582; **156**. J Neurochem (2001) 79:266; **157**. Am J Physiol Endocrinol Metab (2002) 282:E474; **158**. Anal Biochem (2003) 320:292; **159**. Exp Gerontol (2008) 43:563; **160**. Eur J Immunol (2007) 37:467; **161**. Blood (2007) 109:4716; **162**. J Neurosci (2009) 29:9090; **163**. J Neurosci (2007) 27:1129; **164**. Nat Protoc (2009) 4:1790;

IMPORTANT NOTICE: The products described in this manual are covered by one or more Limited Use Label License(s). Please refer to the Appendix on page 971 and Master Product List on page 975. Products are For Research Use Only. Not intended for any animal or human therapeutic or diagnostic use.

REFERENCES—*continued*

165. Proc Natl Acad Sci U S A (1992) 89:11426; **166**. Arch Toxicol (1994) 68:582; **167**. Brain Res (1994) 635:113; **168**. Chem Res Toxicol (1992) 5:227; **169**. Biochem Biophys Res Commun (2003) 304:619; **170**. Free Radic Biol Med (2002) 33:938; **171**. Methods Enzymol (2008) 441:261; **172**. J Immunol Methods (1993) 159:173; **173**. J Immunol Methods (1993) 159:131; **174**. Exp Cell Res (1993) 209:375; **175**. Cytometry (1992) 13:615; **176**. Cytometry (1992) 13:525; **177**. Biochem Pharmacol (2003) 65:1575; **178**. Free Radic Biol Med (2006) 40:968; **179**. Methods Enzymol (1994) 233:128; **180**. Lab Invest (1991) 64:167; **181**. Am J Physiol (1993) 264:H881; **182**. Lab Invest (1994) 70:579; **183**. Free Radic Res Commun (1992) 16:217; **184**. Cytometry (1993) 14:287; **185**. J Leukoc Biol (1992) 51:496; **186**. J Biol Chem (1999) 274:28161; **187**. Methods Mol Biol (2010) 594:57; **188**. Free Radic Biol Med (1994) 16:509; **189**. Free Radic Biol Med (2007) 43:300; **190**. Oncogene (2009) 28:2690; **191**. Cancer Res (2009) 69:5860; **192**. J Neurosci (2001) 21:1949; **193**. J Neurosci (2007) 27:11315; **194**. J Biol Chem (2009) 284:14476; **195**. J Biol Chem (2008) 283:25692; **196**. Chem Res Toxicol (2010) 23:568; **197**. Toxicol Sci (2008) 103:335; **198**. Nat Immunol (2008) 9:866; **199**. Toxicol In Vitro (2008) 22:1392; **200**. J Biol Chem (2009) 284:8633; **201**. Pflugers Arch (2009) 458:937; **202**. Cancer Res (2003) 63:3107; **203**. Am J Physiol Heart Circ Physiol (2000) 279:H2424; **204**. J Natl Cancer Inst (1999) 91:1138; **205**. Lipids (2001) 36:57; **206**. Cancer Res (2001) 61:1392; **207**. Histochem Cell Biol (2003) 120:319; **208**. Am J Physiol (1997) 272:C1286; **209**. Nature (2007) 447:686; **210**. J Neurochem (2006) 98:1474; **211**. J Biol Chem (2006) 281:6760; **212**. J Biol Chem (2003) 278:3170; **213**. Free Radic Res (2004) 38:1257; **214**. Toxicol In Vitro (2007) 21:1552; **215**. J Biol Chem (1980) 255:1874; **216**. J Clin Invest (1978) 61:1081; **217**. J Cell Physiol (1993) 156:428; **218**. J Immunol Methods (1990) 130:223; **219**. Eur J Biochem (1993) 217:973; **220**. Free Radic Res (2000) 33:771; **221**. Methods Mol Biol (1998) 100:215; **222**. Mol Med (2000) 6:779; **223**. Methods Enzymol (1994) 233:229; **224**. APMIS (1994) 102:474; **225**. Atherosclerosis (2003) 169:19; **226**. Circ Res (2004) 94:239; **227**. J Immunol Methods (2006) 316:27; **228**. Nat Med (2009) 15:300; **229**. Nat Clin Pract Cardiovasc Med (2008) 5:811; **230**. Biochemistry (2006) 45:7237; **231**. Diabetes (2006) 55:120; **232**. Am J Pathol (2009) 174:101; **233**. Am J Physiol Renal Physiol (2007) 292:F523; **234**. Free Radic Biol Med (2000) 28:1266; **235**. Amino Acids (2007) 33:51; **236**. Biochemistry (2000) 39:11121; **237**. J Biol Chem (2009) 284:22213.

DATA TABLE 18.2 GENERATING AND DETECTING REACTIVE OXYGEN SPECIES

Cat. No.	MW	Storage	Soluble	Abs	EC	Em	Solvent	Notes
A7923	376.48	F,D,L	DMSO	358	11,000	424	MeOH	1
A12222	257.25	FF,D,A	DMSO	280	6000	none	pH 8	2
A22177	257.25	FF,D,A	DMSO	280	6000	none	pH 8	
A36003	423.42	RO,L	DMF	454	24,000	515	pH 9	3, 4
A36006	~300	FF,D,A	DMSO	293	11,000	none	pH 8	5
B3932	448.32	F,L	DMSO, $CHCl_3$	665	161,000	676	MeOH	
C400	531.30	F,D	DMSO, EtOH	290	5600	none	MeCN	6
C2938	675.43	F,D,AA	DMSO	291	5700	none	MeOH	6
C2944	423.47	FF,D,LL,AA	MeOH	429	7500	see Notes	pH 7	7, 8, 9
C6827	577.80	F,D,AA	DMSO	287	9100	none	MeOH	6
C7924	487.62	F,D,L	DMSO, H_2O	385	5800	485	pH 7	1
C13293	498.39	F,D	DMSO, EtOH	290	5500	none	MeCN	10
D399	487.29	F,D	DMSO, EtOH	258	11,000	none	MeOH	6
D632	346.38	F,D,L,AA	DMF, DMSO	289	7100	none	MeOH	11, 12
D633	444.57	F,D,L,AA	DMF, DMSO	296	11,000	none	MeOH	11, 12
D1168	315.42	FF,L,AA	DMF, DMSO	355	14,000	see Notes	MeCN	11, 13
D2935	584.37	F,D,AA	DMF	258	11,000	none	MeOH	6
D3861	504.43	F,L	DMSO	582	140,000	591	MeOH	14
D7894	386.43	F,D,LL	MeCN	358	29,000	none	MeOH	15
D11347	315.42	FF,L,AA	DMF, DMSO	355	14,000	see Notes	MeCN	11, 13
D23107	315.42	FF,D,L,AA	DMSO	355	14,000	see Notes	MeCN	13, 16
D23805	1068.95	F,D	DMSO	285	5800	none	MeCN	17
D23806	346.38	F,D,L,AA	DMSO	289	7100	none	MeOH	12, 16
F2902	see Notes	RR,L,AA	H_2O	<300		none		3, 18, 19
G36000	561.67	F,D	DMSO	<300		none		
H7476	504.45	F,D,L	DMSO, DMF	591	37,000	594	EtOH	
H36004	424.41	RO,L	DMF	454	28,000	515	pH 9	3, 4
L6868	510.50	L	H_2O	455	7400	505	H_2O	20, 21
L8455	177.16	D,L	DMF	355	7500	411	MeOH	21
M689	485.98	F,DD,L	DMF, DMSO	629	75,000	none	MeCN	22
M6494	414.32	D,L	H_2O, DMSO	375	8300	none	MeOH	23, 24
M7511	392.93	F,D,L,AA	DMSO	235	57,000	none	MeOH	11, 12
M7513	497.08	F,D,L,AA	DMSO	245	45,000	none	MeOH	11, 12
M7913	258.32	F,L	DMF, DMSO	352	30,000	401	MeOH	25
M23800	291.74	FF,D,LL,AA	DMSO	430	8400	546	MeOH	26
M24571	569.67	D,L	DMSO, EtOH	555	143,000	578	MeOH	
M36008	759.71	FF,L,AA	DMSO	356	10,000	410	MeCN	11, 27
N6495	817.65	D,L	H_2O, DMSO	256	64,000	none	MeOH	23
O13291	~66,000	F,D,L,AA	H_2O	<300		none		28
P800	~240,000	RR,L	see Notes	546	2,410,000	575	pH 7	29
P801	~240,000	RR,L	see Notes	565	1,960,000	578	pH 7	29

continued on next page

IMPORTANT NOTICE: The products described in this manual are covered by one or more Limited Use Label License(s). Please refer to the Appendix on page 971 and Master Product List on page 975. Products are For Research Use Only. Not intended for any animal or human therapeutic or diagnostic use.

DATA TABLE 18.2 GENERATING AND DETECTING REACTIVE OXYGEN SPECIES—*continued*

Cat. No.	MW	Storage	Soluble	Abs	EC	Em	Solvent	Notes
P6879	363.16	F,D,L	DMSO	270	28,000	none	$CHCl_3$	30
P36005	276.42	FF,LL,AA	EtOH	304	77,000	416	MeOH	3, 31
R14000	1057.75	F,D	DMSO	313	9700	none	MeOH	32
R14060	434.41	F,D,L,AA	DMSO	239	52,000	none	MeOH	11, 33
S36002	~600	F,D,L	DMSO	508	105,000	528	pH 7	34, 35
X6493	674.53	F,D	H_2O, DMSO	286	15,000	none	MeOH	36

For definitions of the contents of this data table, see "Using *The Molecular Probes® Handbook*" in the introductory pages.

Notes

1. Fluorescence of A7923 and C7924 is weak. Reaction of the nitroxide moiety with superoxide or hydroxyl radicals results in increased fluorescence without a spectral shift. (Anal Biochem (1993) 212:85)
2. Peroxidase-catalyzed reaction of the Amplex® Red reagent (A12222, A22177) with H_2O_2 produces fluorescent resorufin (R363). Resorufin is unstable in the presence of thiols such as dithiothreitol (DTT) and 2-mercaptoethanol. (Bioorg Chem (1998) 26:63)
3. This product is supplied as a ready-made solution in the solvent indicated under "Soluble."
4. Fluorescence of A36003 and H36004 is extremely weak. Highly fluorescent fluorescein F1300 is generated upon oxidation. (J Biol Chem (2003) 278:3170)
5. Peroxidase-catalyzed reaction of the Amplex® UltraRed reagent (A36006) with H_2O_2 yields a fluorescent product with Abs = 568 nm (EC = 57,000 $cm^{-1}M^{-1}$), Em = 581 nm in pH 7.5 buffer.
6. Dihydrofluorescein diacetates are colorless and nonfluorescent until both of the acetate groups are hydrolyzed and the products are subsequently oxidized to fluorescein derivatives. The materials contain less than 0.1% of oxidized derivative when initially prepared. The oxidation products of C400, C2938, C6827, D399 and D2935 are 2′,7′-dichlorofluorescein derivatives with spectra similar to C368 (Section 14.3).
7. C2944 emits chemiluminescence (Em = 466 nm) on oxidation by superoxide. (Anal Biochem (1992) 206:273)
8. Do NOT dissolve in DMSO.
9. Aqueous solutions of coelenterazine (>1 mM) can be prepared in pH 7 buffer containing 50 mM 2-hydroxypropyl-β-cyclodextrin. (Biosci Biotechnol Biochem (1997) 61:1219)
10. Difluorodihydrofluorescein diacetates are colorless and nonfluorescent. Acetate hydrolysis and subsequent oxidation generate a fluorescent 2′,7′-difluorofluorescein derivative with spectra similar to O6146 (Section 14.3).
11. This compound is susceptible to oxidation, especially in solution. Store solutions under argon or nitrogen. Oxidation may be induced by illumination.
12. These compounds are essentially colorless and nonfluorescent until oxidized. Oxidation products (in parentheses) are as follows: D632 and D23806 (R302); D633 (R634); M7511 (M7510); M7513 (M7512).
13. Dihydroethidium has blue fluorescence (Em ~420 nm) until oxidized to ethidium (Em ~605 nm). The reduced dye does not bind to nucleic acids. (FEBS Lett (1972) 26:169)
14. Oxidation of the polyunsaturated butadienyl portion of the BODIPY® 581/591 dye results in a shift of the fluorescence emission peak from ~590 nm to ~510 nm. (Methods Enzymol (2000) 319:603, FEBS Lett (1999) 453:278)
15. Oxidation of D7894 occurs rapidly in solution when illuminated. The oxidation product is strongly fluorescent. Em = 379 nm.
16. This product is supplied as a ready-made solution in DMSO with sodium borohydride added to inhibit oxidation.
17. D23805 is colorless and nonfluorescent until the AM ester groups are hydrolyzed and the resulting leuco dye is subsequently oxidized. The final product is calcein (C481).
18. F2902 is essentially colorless and nonfluorescent until oxidized. A small amount (~5%) of oxidized material is normal and acceptable for the product as supplied. The oxidation product is fluorescent (Abs = 495 nm, Em = 524 nm). (J Immunol Methods (1990) 130:223)
19. This product consists of a dye–bovine serum albumin conjugate (MW ~66,000) complexed with IgG in a ratio of approximately 1:4 mol:mol (BSA:IgG)
20. L6868 has much stronger absorption at shorter wavelengths (Abs = 368 nm (EC = 36,000 $cm^{-1}M^{-1}$)).
21. This compound emits chemiluminescence upon oxidation in basic aqueous solutions. Emission peaks are at 425 nm (L8455) and 470 nm (L6868).
22. Isothiocyanates are unstable in water and should not be stored in aqueous solution.
23. Enzymatic reduction products are water-insoluble formazans with Abs = 505 nm (M6494) and 605 nm (N6495) after solubilization in DMSO or DMF. See literature sources for further information. (Histochemistry (1982) 76:381, Prog Histochem Cytochem (1976) 9:1)
24. M6494 also has Abs = 242 nm (EC = 21,000 $cm^{-1}M^{-1}$) in MeOH.
25. Generates chemiluminescence (Em = 465 nm in 0.1 M SDS) upon reaction with $^1O_2$. (J Am Chem Soc (1986) 108:4498)
26. Generates chemiluminescence (Em = 455 nm) upon reaction with superoxide.
27. Spectroscopic properties of the product generated by reaction of M36008 with superoxide are described in Nat Protoc (2008) 3:8.
28. Oxidation of O13291 generates a fluorescent protein conjugate (Abs ~508 nm, Em ~528 nm).
29. Phycobiliproteins are packaged as suspensions in 60% ammonium sulfate, pH 7.0. Store refrigerated at 4°C but DO NOT FREEZE.
30. Iodoacetamides in solution undergo rapid photodecomposition to unreactive products. Minimize exposure to light prior to reaction.
31. *Cis*-parinaric acid is readily oxidized to nonfluorescent products. Use under N_2 or Ar except when oxidation is intended. Stock solutions should be prepared in deoxygenated solvents. *Cis*-parinaric acid is appreciably fluorescent in lipid environments and organic solvents but is nonfluorescent in water.
32. Acetate hydrolysis of R14000 yields rose bengal (Abs = 556 nm (EC = 104,000 $cm^{-1}M^{-1}$) Em = 572 nm in MeOH). (Photochem Photobiol (1997) 66:374)
33. R14060 is colorless and nonfluorescent until oxidized. The spectral characteristics of the oxidation product (2,3,4,5,6-pentafluorotetramethylrosamine) are similar to those of T639 (Section 12.2).
34. MW: The preceding ~ symbol indicates an approximate value, not including counterions.
35. The fluorescence of S36002 is relatively weak. Reaction of the dye with singlet oxygen ($^1O_2$) results in fluorescence enhancement with essentially no change in absorption or emission wavelengths.
36. Enzymatic reduction product is a water-soluble formazan, Abs = 475 nm.

IMPORTANT NOTICE: The products described in this manual are covered by one or more Limited Use Label License(s). Please refer to the Appendix on page 971 and Master Product List on page 975. Products are For Research Use Only. Not intended for any animal or human therapeutic or diagnostic use.

PRODUCT LIST 18.2 GENERATING AND DETECTING REACTIVE OXYGEN SPECIES

Cat. No.	Product	Quantity
A7923	4-((9-acridinecarbonyl)amino)-2,2,6,6-tetramethylpiperidin-1-oxyl, free radical (TEMPO-9-AC)	5 mg
A36003	3′-(*p*-aminophenyl) fluorescein (APF) *5 mM solution in DMF*	470 µL
A22188	Amplex® Red Hydrogen Peroxide/Peroxidase Assay Kit *500 assays*	1 kit
A12222	Amplex® Red reagent	5 mg
A22177	Amplex® Red reagent *packaged for high-throughput screening*	10 x 10 mg
A22182	Amplex® Red Xanthine/Xanthine Oxidase Assay Kit *400 assays*	1 kit
A36006	Amplex® UltraRed reagent	5 x 1 mg
B3932	(*E,E*)-3,5-bis-(4-phenyl-1,3-butadienyl)-4,4-difluoro-4-bora-3a,4a-diaza-*s*-indacene (BODIPY® 665/676)	5 mg
C400	5-(and-6)-carboxy-2′,7′-dichlorodihydrofluorescein diacetate (carboxy-H_2DCFDA) *mixed isomers*	25 mg
C2938	6-carboxy-2′,7′-dichlorodihydrofluorescein diacetate, di(acetoxymethyl ester)	5 mg
C13293	5-(and-6)-carboxy-2′,7′-difluorodihydrofluorescein diacetate (carboxy-H_2DFFDA) *mixed isomers*	5 mg
C7924	5-(2-carboxyphenyl)-5-hydroxy-1-((2,2,5,5-tetramethyl-1-oxypyrrolidin-3-yl)methyl)-3-phenyl-2-pyrrolin-4-one, potassium salt (proxyl fluorescamine)	5 mg
C6827	5-(and-6)-chloromethyl-2′,7′-dichlorodihydrofluorescein diacetate, acetyl ester (CM-H_2DCFDA) *mixed isomers* *special packaging*	20 x 50 µg
C2944	coelenterazine	250 µg
D399	2′,7′-dichlorodihydrofluorescein diacetate (2′,7′-dichlorofluorescin diacetate; H_2DCFDA)	100 mg
D2935	2′,7′-dichlorodihydrofluorescein diacetate, succinimidyl ester (OxyBURST® Green H_2DCFDA, SE)	5 mg
D3861	4,4-difluoro-5-(4-phenyl-1,3-butadienyl)-4-bora-3a,4a-diaza-*s*-indacene-3-undecanoic acid (BODIPY® 581/591 C_{11})	1 mg
D23805	dihydrocalcein, AM *special packaging*	20 x 50 µg
D1168	dihydroethidium (hydroethidine)	25 mg
D23107	dihydroethidium (hydroethidine) *5 mM stabilized solution in DMSO*	1 mL
D11347	dihydroethidium (hydroethidine) *special packaging*	10 x 1 mg
D632	dihydrorhodamine 123	10 mg
D23806	dihydrorhodamine 123 *5 mM stabilized solution in DMSO*	1 mL
D633	dihydrorhodamine 6G	25 mg
D7894	diphenyl-1-pyrenylphosphine (DPPP)	5 mg
E33856	EnzChek® Myeloperoxidase (MPO) Activity Assay Kit *400 assays* *for myeloperoxidase chlorination and peroxidation activity*	1 kit
F2902	Fc OxyBURST® Green assay reagent *25 assays* *3 mg/mL*	500 µL
G36000	glutathione ethyl ester, biotin amide (BioGEE) *glutathiolation detection reagent* *special packaging*	10 x 100 µg
H36004	3′-(*p*-hydroxyphenyl) fluorescein (HPF) *5 mM solution in DMF*	470 µL
H7476	hypericin	1 mg
I36007	Image-iT® LIVE Green Reactive Oxygen Species Detection Kit *for microscopy*	1 kit
L6868	lucigenin (bis-*N*-methylacridinium nitrate) *high purity*	10 mg
L8455	luminol (3-aminophthalhydrazide)	25 g
M689	malachite green isothiocyanate	10 mg
M7913	*trans*-1-(2′-methoxyvinyl)pyrene	1 mg
M23800	2-methyl-6-(4-methoxyphenyl)-3,7-dihydroimidazo1,2-apyrazin-3-one, hydrochloride (MCLA)	5 mg
M24571	merocyanine 540	25 mg
M36008	MitoSOX™ Red mitochondrial superoxide indicator *for live-cell imaging*	10 x 50 µg
M7511	MitoTracker® Orange CM-H_2TMRos *special packaging*	20 x 50 µg
M7513	MitoTracker® Red CM-H_2XRos *special packaging*	20 x 50 µg
M6494	MTT (3-(4,5-dimethylthiazol-2-yl)-2,5-diphenyltetrazolium bromide)	1 g
N6495	nitro blue tetrazolium chloride (NBT)	1 g
O13291	OxyBURST® Green H_2HFF BSA *special packaging*	5 x 1 mg
P36005	*cis*-parinaric acid *3 mM in ethanol*	10 mL
P6879	*N*-(1,10-phenanthrolin-5-yl)iodoacetamide	5 mg
P800	B-phycoerythrin *4 mg/mL*	0.5 mL
P801	R-phycoerythrin *4 mg/mL*	0.5 mL
R14060	RedoxSensor™ Red CC-1 *special packaging*	10 x 50 µg
R14000	rose bengal diacetate	5 mg
S36002	Singlet Oxygen Sensor Green *special packaging*	10 x 100 µg
X6493	XTT (2,3-bis-(2-methoxy-4-nitro-5-sulfophenyl)-2*H*-tetrazolium-5-carboxanilide)	100 mg
Z33857	Zen™ Myeloperoxidase (MPO) ELISA Kit *200 assays*	1 kit

IMPORTANT NOTICE: The products described in this manual are covered by one or more Limited Use Label License(s). Please refer to the Appendix on page 971 and Master Product List on page 975. Products are For Research Use Only. Not intended for any animal or human therapeutic or diagnostic use.

18.3 Probes for Nitric Oxide Research

Nitric oxide (NO) plays a critical role as a molecular mediator of a variety of physiological processes, including blood-pressure regulation and neurotransmission.[1–8] In endothelial cells, as well as in neurons and astrocytes, NO is synthesized from L-arginine in a reaction catalyzed by nitric oxide synthase (NOS)[9–12] (Figure 18.3.1). NO that diffuses into smooth muscle cells binds to the heme group of guanylate cyclase.

Because free NO is a transient species with a half-life of about 5 seconds, many investigations of this gaseous molecule have relied largely on studies of NOS. Preparing NO solutions and detecting NO in experimental systems require special precautions to achieve reproducibility.[13,14] NO also reacts at diffusion-controlled rates with superoxide to form a strong oxidant, peroxynitrite anion[15] ($ONOO^-$, Table 18.1). Peroxynitrite is a well-known inflammatory mediator in various cardiovascular pathologies but has more recently been recognized as a modulator of signal transduction pathways due to its ability to nitrate tyrosine residues and thereby influence cellular processes dependent on tyrosine phosphorylation.[15,16] Activated macrophages and neutrophils produce nitric oxide and superoxide, and thus peroxynitrite anion, at similar rates.[17] NO generators are also reported to produce an accumulation of chelatable Zn^{2+} in hippocampal neuronal perikarya, as determined with some of our Zn^{2+} indicators[18] (Section 19.7, Table 19.6).

Spontaneous Nitric Oxide Donors and Antagonist

Spermine NONOate

Spermine NONOate (S7916) solids provide a means of preparing aqueous NO solutions.[19] When dissolved in buffer, cell culture medium or blood, spermine NONOate dissociates to form two molecules of NO and one molecule of the corresponding amine[20] (Figure 18.3.2). The delivery of NO can be easily controlled by preparing moderately basic solutions of this NONOate and then lowering the pH to initiate NO generation. Spermine NONOate releases NO slowly (half-life of 39 minutes at 37°C in pH 7.4 buffer), making it suitable for whole animal infusions and experiments with long incubations,[21] as well as for *in situ* calibration of DAF-FM[22].

SNAP and SIN-1

NO donors SNAP (*S*-nitroso-*N*-acetylpenicillamine; N7892, N7927) and SIN-1 (3-morpholinosydnonimine, hydrochloride; M7891, M7926) spontaneously release NO (and superoxide in the case of SIN-1) under physiological conditions (Figure 18.3.2), thereby stimulating cyclic GMP production.[23–27] SNAP and SIN-1 have been shown to be potent vasodilators *in vivo* and *in vitro* and to inhibit smooth muscle cell mitogenesis and proliferation.[28–31] The relationship between NO generated from SNAP and SIN-1 and intracellular Ca^{2+} has been studied using fluorescent Ca^{2+} indicators[32–34] (Chapter 19). It has also been reported that NO released from SNAP stimulates Ca^{2+}-independent synaptic vesicle release,[34] which can be detected with FM® 1-43 (T3163, T35356; Section 16.1). We offer SNAP and SIN-1 in 25 mg vials (N7892, M7891) and, because of their potential instability in solution, as specially packaged sets of 20 vials, each containing 1 mg (N7927, M7926).

Carboxy-PTIO: A Nitric Oxide Antagonist

Carboxy-PTIO (C7912) is a water-soluble and stable free radical molecule that reacts stoichiometrically with NO.[35–37] Carboxy-PTIO can be used *in vivo* to inhibit the physiological effects mediated by NO[35,37,38] or to quantitate NO levels *in vitro* by ESR spectrometry.[39,40]

SNAP: A Photoactivatable Nitric Oxide Donor

SNAP (*S*-nitroso-*N*-acetylpenicillamine; N7892, N7927) has been shown to release nitric oxide (NO) in response to light stimulation in both aqueous and isopropyl alcohol solutions.[41] The potential spatial and temporal control of NO release made possible by photolysis of NO precursors makes this an attractive approach for generating NO in experimental systems.

Detecting Nitric Oxide, Nitrite and Nitrate

The nitric oxide (NO) radical is short-lived and physiological concentrations are very low,[42] making *in situ* detection a challenging proposition. NO is readily oxidized to the nitrosonium cation (NO^+), which is moderately stable in aqueous solutions but highly reactive with nucleophiles or other nitrogen oxides. Under aerobic conditions, these reactive nitrogen oxides (Table 18.1) can be trapped by various amines, in particular by aromatic amines to form diazonium salts or by aromatic 1,2-diamines to form benzotriazoles (Figure 18.3.3).

DAF-FM Nitric Oxide Indicator

First described in 1998,[43] vicinal diamine derivatives of fluorescein generate stronger fluorescence signals at longer wavelengths than prototypes such as 2,3-diaminonaphthalene. These characteristics result in much enhanced performance for *in situ* nitric oxide detection. DAF-FM (4-amino-5-methylamino-2′,7′-difluorofluorescein) is the foremost example of this class of compounds.[44] We offer DAF-FM (D23841) and its cell-permeant diacetate derivative (D23842, D23844). Like dihydrofluoresein, dihydrorhodamine and dihydroethidium probes (Section 18.2), and in contrast to BAPTA-based Ca^{2+} indicators (Section 19.2, Section 19.3), DAF-FM is an endpoint dosimeter. DAF-FM is not a reversible equilibrium sensor, limiting its ability to track rapid fluctuations of the target analyte (NO) in real time. Extracellularly applied DAF-FM diacetate spontaneously crosses the plasma membrane and is cleaved by esterases to generate intracellular DAF-FM, which is then oxidized by NO to a triazole product accompanied by increased fluorescence (Figure 18.3.3, Figure 18.3.4). The fluorescence quantum yield of DAF-FM is reported to be 0.005 but increases about 160-fold to 0.81 after reacting with NO.[44] The second step of the process as depicted in Figure 18.3.3 is an oversimplification. In fact, DAF-FM must first be nonspecifically oxidized to an anilinyl radical, which then reacts with NO to form the fluorescent triazole product.[45] This mechanistic complication must be borne in mind when interpreting experimental data. Specifically, the question of whether nonspecific pre-oxidation or reaction with NO is the dominant factor controlling observed DAF-FM

IMPORTANT NOTICE: The products described in this manual are covered by one or more Limited Use Label License(s). Please refer to the Appendix on page 971 and Master Product List on page 975. Products are For Research Use Only. Not intended for any animal or human therapeutic or diagnostic use.

fluorescence signals requires critical scrutiny.[45–47] Applications of DAF-FM and DAF-FM diacetate include:

- Assessment of NO production in transaldolase-deficient lymphoblasts by flow cytometry[48]
- Detection of NO accumulation in embryonic cortical neurons following neurotrophin stimulation[49]
- *In vivo* imaging of NO in zebrafish[50]
- Intravital microscopic detection of NO generation associated with angiogenesis in mice[51]
- Quantitation of ATP-induced NO release in rabbit platelets[22]

2,3-Diaminonaphthalene

In a reaction similar to that of DAF-FM (Figure 18.3.3), 2,3-diaminonaphthalene (D7918, Figure 18.3.5) reacts with the nitrosonium cation that forms spontaneously from NO to form the fluorescent product 1*H*-naphthotriazole.[52,53] Using 2,3-diaminonaphthalene, researchers have developed a rapid, quantitative fluorometric assay that can detect from 10 nM to 10 µM nitrite and is compatible with a 96-well microplate format.[54]

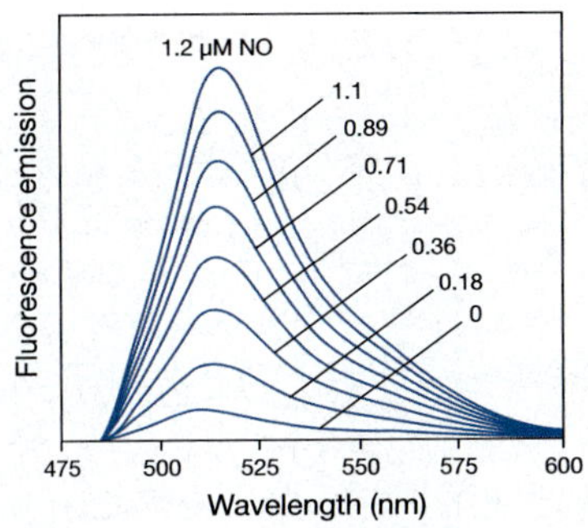

Figure 18.3.4 Fluorescence emission spectra of DAF-FM (D23841, D23842) in solutions containing 0 to 1.2 µM nitric oxide (NO).

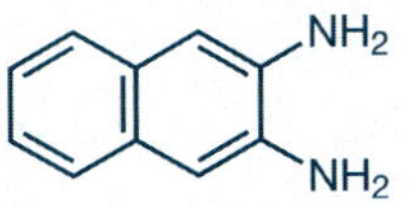

Figure 18.3.5 2,3-diaminonaphthalene (D7918).

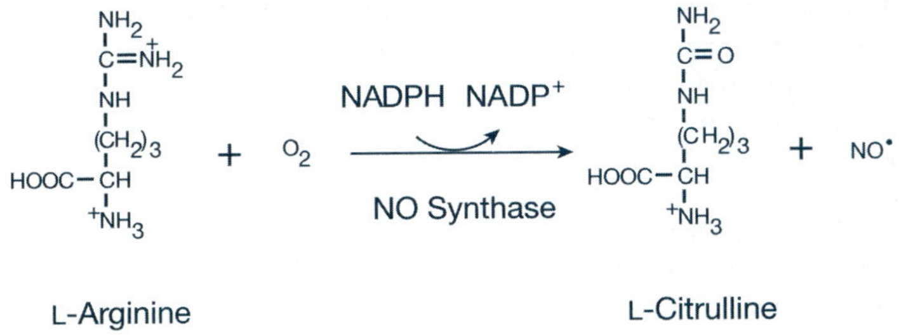

Figure 18.3.1 Nitric oxide synthase production of nitric oxide (NO) and L-citrulline from L-arginine and dioxygen (O_2).

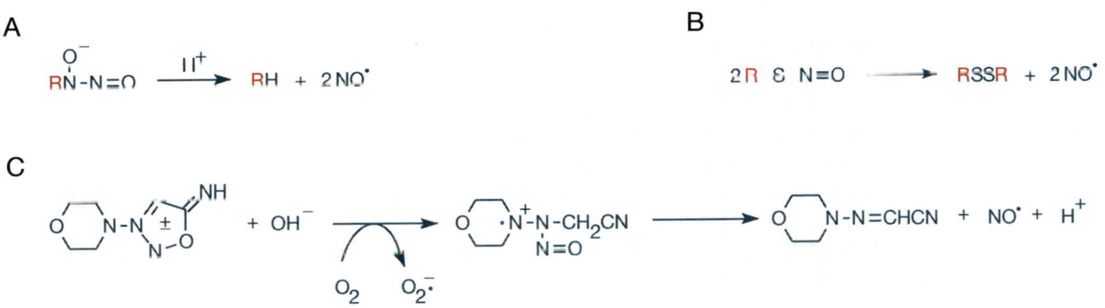

Figure 18.3.2 Mechanisms of spontaneous NO release by: **A)** Spermine NONOate (S7916); **B)** SNAP (N7892, N7927); and **C)** SIN-1 (M7891, M7926).

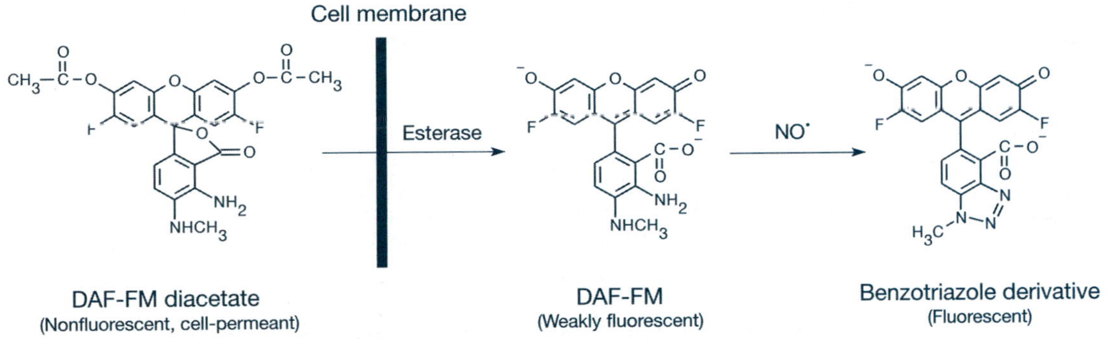

Figure 18.3.3 Reaction scheme for the detection of nitric oxide (NO) by DAF-FM (D23841) and DAF-FM diacetate (D23842, D23844).

IMPORTANT NOTICE: The products described in this manual are covered by one or more Limited Use Label License(s). Please refer to the Appendix on page 971 and Master Product List on page 975. Products are For Research Use Only. Not intended for any animal or human therapeutic or diagnostic use.

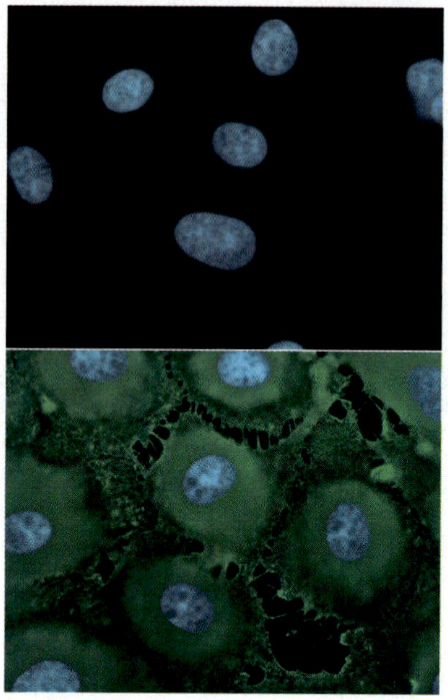

Figure 18.3.7 Fixed and permeabilized bovine pulmonary artery endothelial cells were treated with either degraded peroxynitrite (top panel) or 100 µM peroxynitrite (bottom panel) for 5 minutes at room temperature to induce protein nitration. Nitrated tyrosine residues were detected with our rabbit anti-nitrotyrosine antibody (A21285) and visualized with the green-fluorescent Alexa Fluor® 488 goat anti-rabbit IgG antibody (A11008). Nuclei were counterstained with blue-fluorescent DAPI (D1306, D3571, D21490).

Figure 18.3.8 Specificity of our rabbit anti-nitrotyrosine antibody (A21285) to nitrated proteins. Equal amounts of avidin (A887, lane 1) and CaptAvidin™ biotin-binding protein (C21385, lane 2) were run on an SDS-polyacrylamide gel (4–20%) and blotted onto a PVDF membrane. CaptAvidin™ biotin-binding protein, a derivative of avidin, has nitrated tyrosine residues in the biotin-binding site. On a western blot, nitrated proteins were identified with the anti-nitrotyrosine antibody in combination with an alkaline phosphatase conjugate of goat anti-rabbit IgG antibody (G21079) and the red-fluorescent substrate DDAO phosphate (D6487).

1,2-Diaminoanthraquinone

For directly detecting NO levels *in vivo*, we offer 1,2-diaminoanthraquinone (DAA, D23840). This nitric oxide probe is reported to be nonfluorescent until it reacts with NO to produce a red-fluorescent precipitate. 1,2-Diaminoanthraquinone has been used to detect changes in NO levels in rat retinas after injury to the optic nerve.[55] This methodology may make it possible to test the actions of NO in neurodegeneration, inflammation and other biological processes. The role of NO production in hippocampal long-term potentiation has also been investigated using 1,2-diaminoanthraquinone for spatial imaging of NO in rat brain slices.[56,57]

NBD Methylhydrazine

NBD methylhydrazine (*N*-methyl-4-hydrazino-7-nitrobenzofurazan, M20490) is a unique reagent for the detection of nitrite. Reaction of NBD methylhydrazine with NO_2^- in the presence of mineral acids leads to formation of fluorescent products with excitation/emission maxima of ~468/537 nm. This reaction serves as the principle behind a selective fluorogenic method for the determination of NO_2^- (Figure 18.3.6). Although NBD methylhydrazine has been used to quantitate nitrite in water using a fluorescence microplate reader,[58] it does not seem to have been used yet to detect nitrite formed by spontaneous oxidation of NO.

Dichlorodihydrofluorescein Diacetate and Dihydrorhodamine 123

In addition to their extensive use for detecting other reactive oxygen species such as superoxide, dichlorodihydrofluorescein diacetate (H_2DCFDA) and dihydrorhodamine 123 (D399, D632; Section 18.2) have been reported to be useful for detecting peroxynitrite formation both in solution and in live cells.[59]

Anti-Nitrotyrosine Antibody

High levels of nitrotyrosine are associated with a large number of diseases, including multiple sclerosis, Alzheimer disease and Parkinson disease.[60–62] Increased levels of nitrotyrosine are also indicative of vascular and tissue injury from ischemia–reperfusion and inflammation.[61] Several pathways for the nitration of tyrosine have been suggested. Peroxynitrite ($OONO^-$), formed by spontaneous reaction of nitric oxide (NO) with superoxide ($\bullet O_2^-$), elicits downstream tyrosine nitration.[61,63] Heme peroxidases, such as myeloperoxidase and eosinophil peroxidase, have been shown to utilize hydrogen peroxide (H_2O_2) to oxidize nitrite (NO_2^-) and catalyze tyrosine nitration.[64] In addition, other heme proteins such as hemoglobin and catalase may contribute to tyrosine nitration using NO as a substrate.[65] Tryptophan residues can also be oxidized by peroxynitrite.[66]

We offer a high-activity rabbit polyclonal anti-nitrotyrosine antibody (A21285) for detecting nitrotyrosine-containing proteins and peptides. This antibody is suitable for both immunohistochemical (Figure 18.3.7) and western blotting (Figure 18.3.8) applications and is useful for identifying nitrated proteins and determining the level of protein nitrosylation in tissues.[67,68] Fluorescence of Green Fluorescent Protein (GFP) is extremely sensitive to tyrosine nitration, as confirmed by correlated anti-nitrotyrosine immunoreactivity.[69]

S-Nitrosothiol Detection

S-nitrosylation of thiols, principally in the form of cysteine sidechains or glutathione, is a primary mechanism for downstream propagation of nitric oxide release events. This reversible posttranslational modification regulates enzymatic activity, subcellular localization, chromatin remodeling and protein degradation.[49,70] The primary reactive nitrogen species responsible for

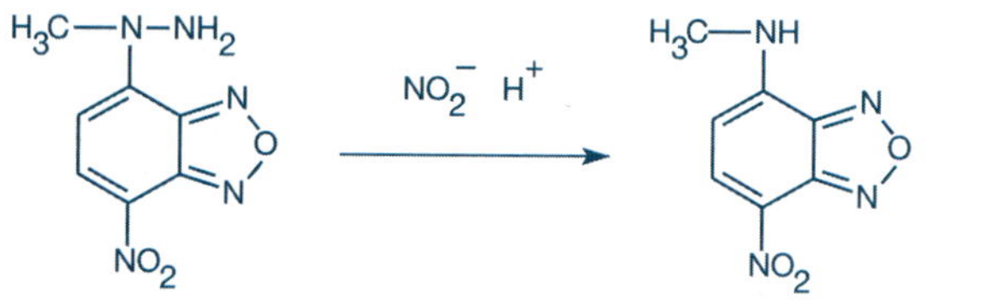

Figure 18.3.6 Reaction scheme illustrating the principle of nitrite detection by NBD methylhydrazine (M20490).

IMPORTANT NOTICE: The products described in this manual are covered by one or more Limited Use Label License(s). Please refer to the Appendix on page 971 and Master Product List on page 975. Products are For Research Use Only. Not intended for any animal or human therapeutic or diagnostic use.

S-nitrosylation of protein thiols is dinitrogen dioxide (N_2O_3) formed from O_2 and NO. Techniques for detecting *S*-nitrosothiol modifications exploit, but are also compromised by, their reversible nature and their susceptibility to photolytic cleavage. The technique with most widespread adoption, often referred to as the biotin switch method,[71–73] consists of three steps: (1) blocking of free thiols with *N*-ethylmaleimide or another alkylating reagent, (2) selective reduction of *S*-nitrosothiols to thiols using ascorbate or TCEP (T2556, Section 2.1) and (3) labeling of thiols created in step 2 with a fluorescent or biotinylated maleimide or iodoacetamide reagent[74,75] (Section 2.2 or Section 4.2, respectively). Streptavidin agarose (S951, Section 7.6) can be used to subsequently pull down biotinylated proteins for further analysis if required. The overall technique is vulnerable to false positives through incomplete blocking of unmodified thiols in step 1 and inadvertent reduction of disulfides in step 2. Other methods take advantage of the fact that *S*-nitrosothiols can be cleaved by heavy metal ions such as Hg^{2+} or by exposure to ultraviolet light, releasing NO and subsequently nitrite (NO_2^-).[76] The NO product of this process can be detected using DAF-FM[77] or the nascent thiol product can be detected using a fluorescent maleimide reagent.[78]

Griess Reagent Kit

Under physiological conditions, NO is readily oxidized to nitrite and nitrate or it is trapped by thiols as an *S*-nitroso adduct. The Griess reagent provides a simple and well characterized colorimetric assay for nitrites, and nitrates that have been reduced to nitrites, with a detection limit of about 100 nM.[52,79,80] Nitrites react with sulfanilic acid in acidic solution to form an intermediate diazonium salt that couples to *N*-(1-naphthyl)ethylenediamine to yield a purple azo derivative that can be monitored by absorbance at 548 nm (Figure 18.3.9).

Our Griess Reagent Kit (G7921) contains all of the reagents required for nitrite quantitation, including:

- *N*-(1-Naphthyl)ethylenediamine dihydrochloride
- Sulfanilic acid in 5% H_3PO_4
- Concentrated nitrite quantitation standard for generating calibration curves
- Detailed protocols for spectrophotometer and microplate reader assays

Both the *N*-(1-naphthyl)ethylenediamine dihydrochloride and the sulfanilic acid in 5% H_3PO_4 are provided in convenient dropper bottles for easy preparation of the Griess reagent. Sample pretreatment with nitrate reductase and glucose 6-phosphate dehydrogenase is reported to reduce nitrate without producing excess NADPH, which can interfere with the Griess reaction.[81] A review of the use of the Griess reagent for nitrite and nitrate quantitation in human plasma describes optimal reaction conditions for minimizing interference from plasma constituents (particularly NADPH).[82] The Griess Reagent Kit can also be used to analyze NO that has been trapped as an *S*-nitroso derivative by a modification that uses mercuric chloride or copper (II) acetate to release the NO from its complex.[83,84]

Measure-iT™ High-Sensitivity Nitrite Assay Kit

The Measure-iT™ High-Sensitivity Nitrite Assay Kit (M36051) provides an easy and accurate method for quantitating nitrite. This kit has an optimal range of 20–500 picomoles nitrite (Figure 18.3.10), making it up to 50 times more sensitive than colorimetric methods utilizing the Griess reagent. Nitrates may be analyzed after quantitative conversion to nitrites through enzymatic reduction;[80] used in this manner, the Measure-iT™ nitrite assay also provides an effective method for quantitating nitric oxide.

Each Measure-iT™ High-Sensitivity Nitrite Assay Kit contains:

- Measure-iT™ nitrite quantitation reagent (100X concentrate in 0.62 M HCl)
- Measure-iT™ nitrite quantitation developer (2.8 M NaOH)
- Measure-iT™ nitrite quantitation standard (11 mM sodium nitrite)
- Detailed protocols

Simply dilute the reagent 1:100, load 100 µL into the wells of a microplate, add 1–10 µL sample volumes and mix. After a 10-minute incubation at room temperature, add 5 µL of developer and read the fluorescence. The assay signal is stable for at least 3 hours, and common contaminants are well tolerated in the assay. The Measure-iT™ High-Sensitivity Nitrite Assay Kit provides sufficient material for 2000 assays, based on a 100 µL assay volume in a 96-well microplate format; this nitrite assay can also be adapted for use in cuvettes or 384-well microplates.

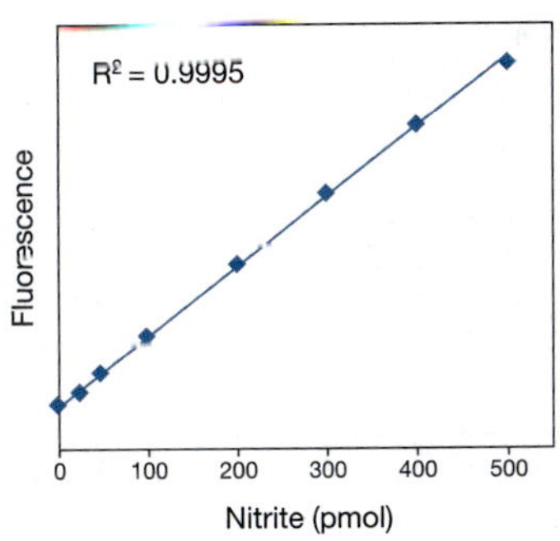

Figure 18.3.10 Linearity and sensitivity of the Measure-iT™ high-sensitivity nitrite assay. Triplicate 10 µL samples of nitrite were assayed using the Measure-iT™ High-Sensitivity Nitrite Assay Kit (M36051). Fluorescence was measured using excitation/emission of 365/450 nm and plotted versus picomoles of nitrite. Background fluorescence was not subtracted. The variation (CV) of replicate samples was <2%.

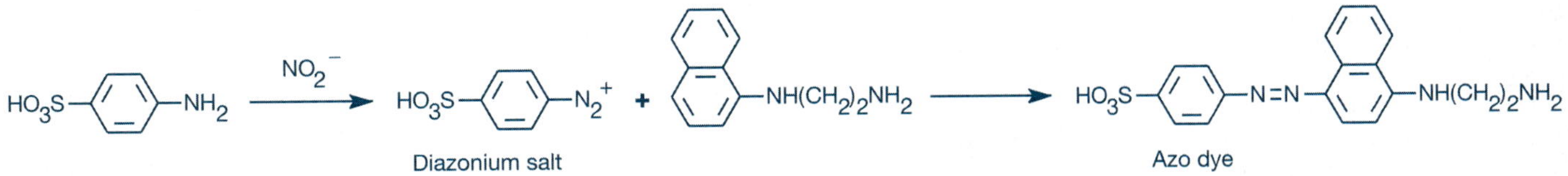

Figure 18.3.9 Principle of nitrite quantitation using the Griess Reagent Kit (G7921). Formation of the azo dye is detected via its absorbance at 548 nm.

IMPORTANT NOTICE: The products described in this manual are covered by one or more Limited Use Label License(s). Please refer to the Appendix on page 971 and Master Product List on page 975. Products are For Research Use Only. Not intended for any animal or human therapeutic or diagnostic use.

REFERENCES

1. Science (2001) 292:2413; **2.** Science (2001) 292:2486; **3.** Curr Biol (1997) 7:R376; **4.** Cell (1994) 78:919; **5.** Am J Physiol (1992) 262:G379; **6.** Annu Rev Biochem (1994) 63:175; **7.** J Med Chem (1995) 38:4343; **8.** Science (1992) 257:494; **9.** Biochem J (2001) 357:593; **10.** Biochem J (1994) 298:249; **11.** Cell (1994) 78:915; **12.** J Med Chem (1994) 37:1899; **13.** Methods Mol Biol (1998) 100:215; **14.** Curr Opin Chem Biol (2010) 14:43; **15.** Free Radic Biol Med (2001) 30:463; **16.** Front Biosci (2009) 14:4809; **17.** Methods Enzymol (1994) 233:229; **18.** Brain Res (1998) 799:118; **19.** Methods Enzymol (1996) 268:281; **20.** J Med Chem (1991) 34:3242; **21.** J Biol Chem (2001) 276:28799; **22.** Anal Chem (2007) 79:2421; **23.** Nature (1993) 364:626; **24.** Nature (1995) 375:68; **25.** Brain Res (1993) 619:344; **26.** FEBS Lett (1993) 315:139; **27.** Thromb Res (1993) 70:405; **28.** J Pharmacol Exp Ther (1992) 260:286; **29.** Eur J Pharmacol (1987) 144:379; **30.** J Pharmacol Exp Ther (1989) 248:762; **31.** J Clin Invest (1989) 83:1774; **32.** Am J Physiol (1994) 266:L9; **33.** Life Sci (1994) 54:1449; **34.** Neuron (1994) 12:1235; **35.** Biochemistry (1995) 34:7177; **36.** Biochem Biophys Res Commun (1994) 202:923; **37.** Biochemistry (1993) 32:827; **38.** Infect Immun (1993) 61:3552; **39.** Photochem Photobiol (1995) 61:325; **40.** Life Sci (1994) 54:185; **41.** Photochem Photobiol (1998) 67:282; **42.** Nitric Oxide (2009) 21:92; **43.** Anal Chem (1998) 70:2446; **44.** Angew Chem Int Ed Engl (1999) 38:3209; **45.** Free Radic Biol Med (2007) 43:995; **46.** Anal Chem (2006) 78:1859; **47.** Free Radic Biol Med (2005) 39:327; **48.** Biochem J (2008) 415:123; **49.** Nature (2008) 455:411; **50.** Free Radic Biol Med (2007) 43:619; **51.** Methods Enzymol (2008) 441:393; **52.** Luminescence (1999) 14:283; **53.** Methods Enzymol (1996) 268:105; **54.** Anal Biochem (1993) 214:11; **55.** Neuroreport (1998) 9:4051; **56.** Neurobiol Dis (2002) 11:96; **57.** Neuroimage (2002) 15:633; **58.** Anal Chem (1999) 71:3003; **59.** Methods Enzymol (2008) 441:261; **60.** J Biol Chem (2003) 278:8380; **61.** Free Radic Res (2000) 33:771; **62.** Methods Enzymol (1999) 301:373; **63.** Proc Natl Acad Sci U S A (1996) 93:11853; **64.** Biochem Biophys Res Commun (2001) 285:273; **65.** Biochim Biophys Acta (2001) 1528:97; **66.** Biochem Biophys Res Commun (1997) 234:82; **67.** Nat Clin Pract Cardiovasc Med (2008) 5:811; **68.** Methods Mol Biol (2008) 477:41; **69.** Proc Natl Acad Sci U S A (2002) 99:3481; **70.** Proc Natl Acad Sci U S A (2005) 102:117; **71.** Free Radic Biol Med (2009) 46:119; **72.** Proteomics (2009) 9:808; **73.** Methods Enzymol (2008) 441:53; **74.** Nitric Oxide (2008) 19:295; **75.** J Bacteriol (2008) 190:4997; **76.** J Biol Chem (1996) 271:18596; **77.** Anal Biochem (2005) 346:69; **78.** J Biol Chem (2006) 281:40354; **79.** FASEB J (1993) 7:349; **80.** Anal Biochem (1982) 126:131; **81.** Anal Biochem (1995) 224:502; **82.** Methods Enzymol (2008) 440:361; **83.** Anal Biochem (1996) 238:150; **84.** Chem Biol (1996) 3:655.

DATA TABLE 18.3 PROBES FOR NITRIC OXIDE RESEARCH

Cat. No.	MW	Storage	Soluble	Abs	EC	Em	Solvent	Notes
C7912	315.39	FF,D	H_2O	367	9300	none	MeOH	
D7915	155.13	FF,DD,A	H_2O, DMSO	248	8000	none	pH 12	1
D7918	158.20	L	DMSO, MeOH	340	5100	377	MeOH	2
D23840	336.32	F,D,L	DMSO	521	6000	none	MeOH	3
D23841	412.35	F,D,L	DMSO	487	84,000	see Notes	pH 8	4
D23842	496.42	F,D,L	DMSO	<300		none		5
D23844	496.42	F,D,L	DMSO	<300		none		5
M7891	206.63	FF,D,LL	DMSO, H_2O	291	11,000	none	pH 7	6
M7926	206.63	FF,D,LL	DMSO, H_2O	291	11,000	none	pH 7	6
M20490	209.16	F,L	MeCN	487	24,000	none	MeOH	7
N7892	220.24	FF,D,LL	DMSO, H_2O	342	700	none	MeOH	6
N7927	220.24	FF,D,LL	DMSO, H_2O	342	700	none	MeOH	6
S7916	262.35	FF,DD,A	H_2O, DMSO	248	8200	none	pH 12	1

For definitions of the contents of this data table, see "Using *The Molecular Probes® Handbook*" in the introductory pages.

Notes

1. Releases nitric oxide upon acid-catalyzed dissociation in solution. Stable in alkaline solutions. (Methods Enzymol (1996) 268:281)
2. Fluorescence of D7918 is weak. Reaction with nitrite yields highly fluorescent 1*H*-naphthotriazole (Abs = 365 nm, Em = 415 nm in H_2O (pH 12)). (Methods Enzymol (1996) 268:105)
3. 1,2-Diaminoanthraquinone reacts with nitrite or nitric oxide to produce 1*H*-anthratriazole-6,11-dione which forms a red-fluorescent (Em >580 nm) precipitate in water. (Neuroreport (1998) 9:4051)
4. DAF-FM fluorescence is very weak. Reaction with nitrite or nitric oxide generates a highly fluorescent benzotriazole derivative with Abs = 495 nm (EC = 73,000 $cm^{-1}M^{-1}$), Em = 515 nm in pH 7.4 buffer. (Angew Chem Int Ed Engl (1999) 38:3209)
5. Acetate hydrolysis and subsequent reaction with nitrite or nitric oxide generate a highly fluorescent benzotriazole derivative with Abs = 495 nm (EC = 73,000 $cm^{-1}M^{-1}$), Em = 515 nm in pH 7.4 buffer. (Angew Chem Int Ed Engl (1999) 38:3209)
6. Spontaneously decomposes in solution.
7. NBD methylhydrazine reacts with nitrite in the presence of strong acid to form fluorescent *N*-methyl-4-amino-7-nitrobenzofurazan (Abs = 459 nm, Em = 537 nm in MeCN). (Anal Chem (1999) 71:3003)

PRODUCT LIST 18.3 PROBES FOR NITRIC OXIDE RESEARCH

Cat. No.	Product	Quantity
A21285	anti-nitrotyrosine, rabbit IgG fraction *1 mg/mL*	0.5 mL
C7912	2-(4-carboxyphenyl)-4,4,5,5-tetramethylimidazoline-1-oxyl-3-oxide, potassium salt (carboxy-PTIO)	25 mg
D23841	DAF-FM (4-amino-5-methylamino-2′,7′-difluorofluorescein)	1 mg
D23842	DAF-FM diacetate (4-amino-5-methylamino-2′,7′-difluorofluorescein diacetate)	1 mg
D23844	DAF-FM diacetate (4-amino-5-methylamino-2′,7′-difluorofluorescein diacetate) *special packaging*	10 x 50 µg
D23840	1,2-diaminoanthraquinone sulfate (DAA) *high purity*	5 mg
D7918	2,3-diaminonaphthalene	100 mg
G7921	Griess Reagent Kit *for nitrite quantitation*	1 kit
M36051	Measure-iT™ High-Sensitivity Nitrite Assay Kit *2000 assays*	1 kit
M20490	*N*-methyl-4-hydrazino-7-nitrobenzofurazan (NBD methylhydrazine)	25 mg
M7891	3-morpholinosydnonimine, hydrochloride (SIN-1)	25 mg
M7926	3-morpholinosydnonimine, hydrochloride (SIN-1) *special packaging*	20 x 1 mg
N7892	*S*-nitroso-*N*-acetylpenicillamine (SNAP)	25 mg
N7927	*S*-nitroso-*N*-acetylpenicillamine (SNAP) *special packaging*	20 x 1 mg
S7916	spermine NONOate	10 mg

IMPORTANT NOTICE: The products described in this manual are covered by one or more Limited Use Label License(s). Please refer to the Appendix on page 971 and Master Product List on page 975. Products are For Research Use Only. Not intended for any animal or human therapeutic or diagnostic use.

molecular probes® | invitrogen™ by life technologies™

NINETEEN

CHAPTER 19

Indicators for Ca^{2+}, Mg^{2+}, Zn^{2+} and Other Metal Ions

IMPORTANT NOTICE: The products described in this manual are covered by one or more Limited Use Label License(s). Please refer to the Appendix on page 971 and Master Product List on page 975. Products are For Research Use Only. Not intended for any animal or human therapeutic or diagnostic use.

IMPORTANT NOTICE: The products described in this manual are covered by one or more Limited Use Label License(s). Please refer to the Appendix on page 971 and Master Product List on page 975. Products are For Research Use Only. Not intended for any animal or human therapeutic or diagnostic use.

www.invitrogen.com/probes

IMPORTANT NOTICE: The products described in this manual are covered by one or more Limited Use Label License(s). Please refer to the Appendix on page 971 and Master Product List on page 975. Products are For Research Use Only. Not intended for any animal or human therapeutic or diagnostic use.

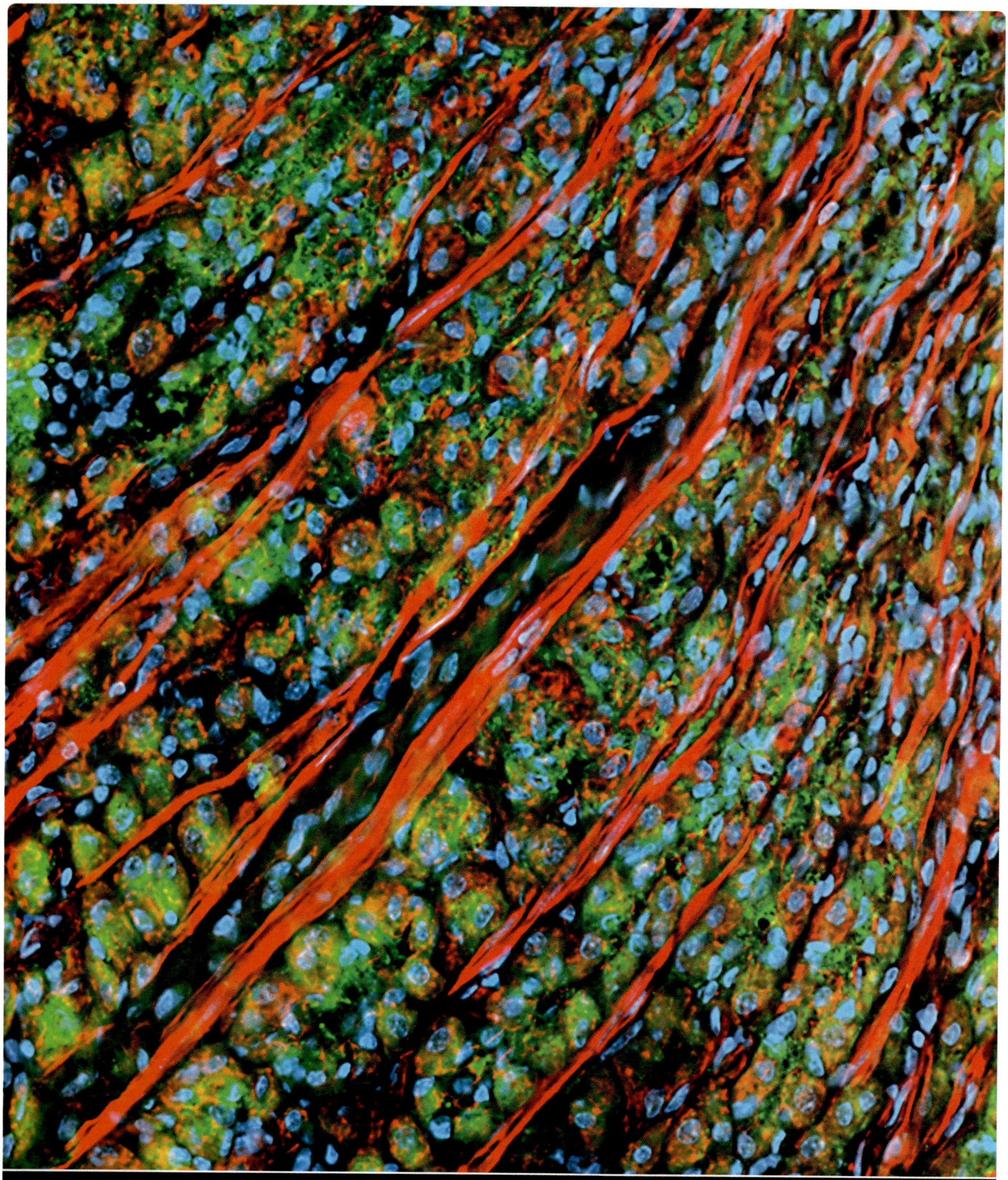

Alexa Fluor® 568 phalloidin, Oregon Green® wheat germ agglutinin and Hoechst 33342.

19.1 Introduction to Ca^{2+} Measurements with Fluorescent Indicators

Fluorescent probes that show a spectral response upon binding Ca^{2+} have enabled researchers to investigate changes in intracellular free Ca^{2+} concentrations using fluorescence microscopy, flow cytometry and fluorescence spectroscopy. The properties and applications of these fluorescent indicators—most of which are derivatives of the Ca^{2+} chelators EGTA, APTRA and BAPTA [1]—have been extensively reviewed.[2–4] Several earlier reviews of these ion indicators also contain useful technical information.[5–8]

We discuss chemical Ca^{2+} indicators according to their excitation requirements in Section 19.2 and Section 19.3, and their high-molecular weight conjugates are described in Section 19.4. Protein-based Ca^{2+} sensors are discussed in Section 19.5.

Selection Criteria for Fluorescent Ca^{2+} Indicators

We offer a wide selection of fluorescent indicators available for detecting changes in intracellular Ca^{2+} over the range of <50 nM to >50 µM (Table 19.1). As the primary suppliers of fura-2, indo-1, fluo-3 (Figure 19.1.1), fluo-4 and rhod-2, we also offer many specialized indicators for intracellular Ca^{2+}. Our fura-4F, fura-6F and fura-FF indicators are designed to provide increased response sensitivity to intracellular Ca^{2+} concentration in the 0.5–5 µM range, as compared with fura-2. The fluo-3, fluo-4, Oregon Green® 488 BAPTA, Calcium Green™, X-rhod-1 and Fura Red™ indicators and their variants enable Ca^{2+} detection in confocal microscopy and high-throughput G protein–coupled receptor (GPCR) screening applications. In addition, we offer indicators that are conjugated to high– or low–molecular weight dextrans for improved cellular retention and less compartmentalization (Section 19.4). We strive to provide the highest-purity indicators available anywhere. The AM ester forms of most of our indicators are typically at least 95% pure by HPLC analysis, although purity often exceeds 98%. Furthermore, the AM esters of many of the Ca^{2+} and Mg^{2+} indicators are available in sets of 50 µg for more convenient handling and reduced risk of deterioration during storage. For high-throughput screening applications, fluo-3 AM and fluo-4 AM are offered in special multi-unit packages (F14242, F14202), as well as in application-specific Fluo-4 NW and Fluo-4 Direct™ Calcium Assay Kits (Section 19.3).

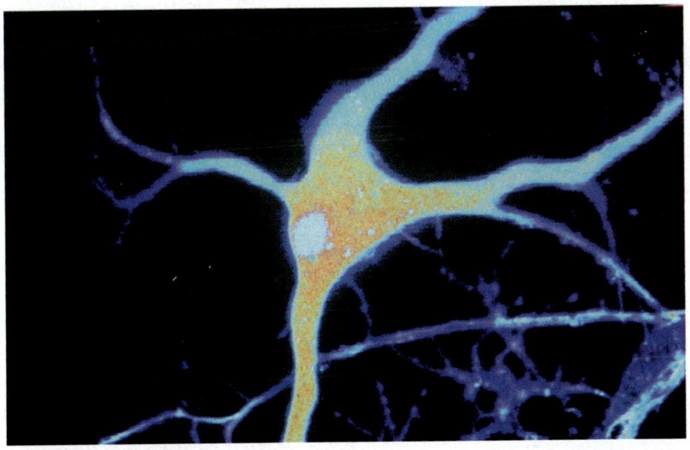

Figure 19.1.1 A pyramidal neuron from rat hippocampus was first exposed to Alzheimer's ß-amyloid peptide and then to the excitatory amino acid glutamate. Confocal laser-scanning microscopy imaging using the intracellular Ca^{2+} indicator fluo-3 (F1241, F1242, F14218, F14242, F23915) shows that ß-amyloid peptide destabilizes the neuron's calcium homeostasis and increases its vulnerability to excitotoxicity. The image was contributed by Mark P. Mattson, Sanders-Brown Center on Aging, University of Kentucky.

A number of factors should be considered when selecting a fluorescent Ca^{2+} indicator, some of which are summarized in Table 19.1 and include the following:

- Indicator form (salt, AM ester or dextran conjugate), which influences the cell-loading method and affects the indicator's intracellular distribution and retention (Loading and Calibration of Intracellular Ion Indicators—Note 19.1). The salt and dextran forms are typically loaded by microinjection, microprojectile bombardment or electroporation or by using the Influx™ pinocytic cell-loading reagent (I14402, Section 19.8) (Table 14.1). In contrast, the cell-permeant acetoxymethyl (AM) esters can be passively loaded into cells, where they are cleaved to cell-impermeant products by intracellular esterases.
- Measurement mode, which is dictated by whether qualitative or quantitative ion concentration data are required. Ion indicators that exhibit spectral shifts upon ion binding can be used for ratiometric measurements of Ca^{2+} concentration, which are essentially independent of uneven dye loading, cell thickness, photobleaching effects and dye leakage (Loading and Calibration of Intracellular Ion Indicators—Note 19.1). Excitation and emission wavelength preferences depend on the type of instrumentation being used, as well as on sample autofluorescence and on the presence of other fluorescent or photoactivatable probes in the experiment.
- Dissociation constant (K_d), which must be compatible with the Ca^{2+} concentration range of interest. Indicators have a detectable response in the concentration range from approximately $0.1 \times K_d$ to $10 \times K_d$. For ratiometric indicators, the Ca^{2+} response range is also somewhat dependent on the measurement wavelengths used.[9–11] The K_d of Ca^{2+} indicators is dependent on many factors, including pH, temperature,[12–15] ionic strength, viscosity, protein binding and the presence of Mg^{2+} and other ions. Consequently, K_d values for intracellular indicators are usually significantly higher than corresponding values measured in cell-free solutions (Table 19.2).

Intracellular calibration of Ca^{2+} indicators may be achieved either by manipulating Ca^{2+} levels inside cells using an ionophore or by releasing the indicator into the surrounding medium of known Ca^{2+} concentration via detergent lysis of the cells. We also offer several compounds and buffers for measuring and manipulating intracellular and extracellular Ca^{2+}. These products, which are discussed in Section 19.8, include caged Ca^{2+} reagents and caged chelators (NP-EGTA, DMNP-EDTA and diazo-2), as well as Calcium Calibration Buffer Kits, BAPTA-derived buffers, ion-selective chelating polymers (Calcium Sponge™) and the important Ca^{2+} ionophores ionomycin, A-23187 and its nonfluorescent analog, 4-bromo A-23187. Reagents for probing Ca^{2+} regulation and second-messenger activity are described in more detail in Chapter 17. Our reagents for the study of Ca^{2+} channels are described in Section 16.3.

IMPORTANT NOTICE: The products described in this manual are covered by one or more Limited Use Label License(s). Please refer to the Appendix on page 971 and Master Product List on page 975. Products are For Research Use Only. Not intended for any animal or human therapeutic or diagnostic use.

Table 19.1 Summary of Molecular Probes® fluorescent Ca^{2+} indicators.

Ca^{2+} Indicator	Water-Soluble Salt *	Cell-Permeant Ester †	Dextran ‡	Mode §	K_d (nM) **	Notes
Bis-fura-2t	B6810			Ex 340/380	370	1
BTC	B6790	B6791		Ex 400/480	7000	2
Calcium Green™-1	C3010MP	C3011MP, C3012	C6765, C3713, C3714	Em 530	190	3, 4
Calcium Green™-2	C3730	C3732		Em 535	550	3, 5
Calcium Green™-5N	C3737	C3739		Em 530	14,000	3
Calcium Orange™	C3013	C3015		Em 575	185	2
Calcium Crimson™		C3018		Em 615	185	2
Fluo-3	F1240, F3715	F1241, F1242, F14218, F14242, F23915		Em 525	390	3, 4
Fluo-4	F14200	F14201, F14202, F14217, F23917	F14240 ††, F36250 ‡‡	Em 520	345	3, 6
Fluo-5F	F14221	F14222		Em 520	2300	3
Fluo-4FF	F23980	F23981		Em 520	9700	3
Fluo-5N	F14203	F14204		Em 520	90,000	3
Fura-2	F1200, F6799	F1201, F1221, F1225, F14185	F3029	Ex 340/380	145	2
Fura-4F	F14174	F14175		Ex 340/380	770	2
Fura-6F	F14178			Ex 340/380	5300	2
Fura-FF	F14180	F14181		Ex 340/380	5500	2
Fura Red™	F14219	F3020, F3021		Ex 420/480	140	2, 7
Indo-1	I1202	I1203, I1223, I1226		Em 405/485	230	2
Mag-fluo-4	M14205	M14206		Em 520	22,000	3
Mag-fura-2	M1290	M1291, M1292		Ex 340/380	25,000	2
Mag-indo-1		M1295		Em 405/485	35,000	2, 8
Magnesium Green™	M3733	M3735		Em 530	6000	3
Oregon Green® 488 BAPTA-1	O6806	O6807	O6798	Em 520	170	3
Oregon Green® 488 BAPTA-2	O6808	O6809		Em 520	580	3, 9
Oregon Green® 488 BAPTA-6F	O23990			Em 520	3000	3
Oregon Green® 488 BAPTA-5N	O6812			Em 520	20,000	3
Quin-2	Q23918			Em 495	60	2, 10
Rhod-2	R14220	R1244, R1245MP	R34676 ‡‡	Em 580	570	3, 11
Rhod-3		R10145		Em 580	570	3
Rhod-FF		R23983		Em 580	19,000	3
Rhod-5N	R14207			Em 580	320,000	3
X-rhod-1		X14210		Em 600	700	3
X-rhod-5F	X23984	X23985		Em 600	1600	3

* Catalog number(s) for the cell-impermeant salt. † Catalog number(s) for the cell-permeant AM ester. ‡ Catalog number(s) for the dextran conjugates. § Measurement wavelengths, in nm, where Ex = fluorescence excitation and Em = fluorescence emission. Indicators for which a pair of wavelengths are listed have dual-wavelength, ratio-measurement capability. ** Ca^{2+} dissociation constant, measured *in vitro* at 22°C in 100 mM KCl, 10 mM MOPS, pH 7.2, unless otherwise noted. K_d values depend on temperature, ionic strength, pH and other factors, and are usually higher *in situ*. Because indicator dextrans are intrinsically polydisperse and have variable degrees of substitution, these values may vary; lot-specific K_d values are printed on the vial in most cases. †† Low-affinity dextran conjugate. ‡‡ High-affinity dextran conjugate.

Notes: 1. Ca^{2+}-dependent fluorescence response similar to fura-2 but ~75% greater molar absorptivity. **2.** The AM ester form is fluorescent (a major potential source of error in Ca^{2+} measurements). **3.** The AM ester form is nonfluorescent. **4.** Calcium Green™-1 is more fluorescent than fluo-3 in both Ca^{2+}-bound and Ca^{2+}-free forms. The magnitude of the Ca^{2+}-dependent fluorescence increase is greater for fluo-3; see Section 19.3. **5.** Larger Ca^{2+}-dependent fluorescence increase than Calcium Green™-1. **6.** The K_d value for the low-affinity fluo-4 dextran (F14240) is ~3 µM, which is much higher than that of the free dye. The K_d value for the high-affinity fluo-4 dextran (F36250) is ~600 nM. **7.** Can also be used in combination with fluo-3 for dual-wavelength ratio measurements, Ex = 488 nm, Em = 530/670 nm (Cell Calcium (1995) 18:377; Cytometry (1994) 17:135; Cell Calcium (1993) 14:359). **8.** K_d determined in 100 mM KCl, 40 mM HEPES, pH 7.0 at 22°C (Biochem Biophys Res Commun (1991) 177:184). **9.** Larger Ca^{2+}-dependent fluorescence increase than Oregon Green® 488 BAPTA-1. **10.** K_d determined in 120 mM KCl, 20 mM NaCl, pH 7.05 at 37°C (Methods Enzymol (1989) 172:230). **11.** The K_d value for the high-affinity rhod dextran (R34676) is ~780 nM.

Table 19.2 Comparison of *in vitro* and *in situ* K_d values for various Ca^{2+} indicators.

Indicator	K_d *in vitro* (nM) *	K_d *in situ* (nM) †	Cell/Tissue Type
Calcium Green™-1	190	930	HeLa cells [1]
Fluo-3	390	2570	Frog skeletal muscle [2]
Fluo-4	345	1000	HeLa cells [1]
Fura-2	145	371	U373-MG astrocytoma cell [3]
Fura-2	145	350	Rabbit gastric gland [4]
Indo-1	230	844	Rabbit cardiac myocyte [5]
Oregon Green® 488 BAPTA-1	170	430	HeLa cells [1]
Rhod-2	570	720	Mouse heart [6]

* Values determined at 22°C in 100 mM KCl, 10 mM MOPS, pH 7.2, 0–10 mM CaEGTA. † Values determined in the cellular environments listed in the adjacent column. **1.** Cell Calcium (2000) 28:213; **2.** Biophys J (1993) 65:865; **3.** Cell Calcium (1997) 21:233; **4.** Methods Enzymol (1990) 192:38; **5.** Biophys J (1995) 68:1453; **6.** Cell Calcium (2001) 29:217.

REFERENCES

1. Biochemistry (1980) 19:2396; **2.** Methods Cell Biol (2007) 81:415; **3.** Trends Mol Med (2008) 14:389; **4.** Methods (2008) 46:143; **5.** Physiol Rev (1999) 79:1089; **6.** Methods Cell Biol (1994) 40:155; **7.** Methods Enzymol (1990) 192:38; **8.** Methods Enzymol (1989) 172:230; **9.** Cell Calcium (1998) 24:17; **10.** Methods Cell Biol (1994) 41:149; **11.** Cell Calcium (1991) 12:29; **12.** J Physiol (2002) 542:843; **13.** Biophys J (2000) 78:2116; **14.** Anal Biochem (1999) 273:60; **15.** Biochem Biophys Res Commun (1990) 171:102.

IMPORTANT NOTICE: The products described in this manual are covered by one or more Limited Use Label License(s). Please refer to the Appendix on page 971 and Master Product List on page 975. Products are For Research Use Only. Not intended for any animal or human therapeutic or diagnostic use.
www.invitrogen.com/probes
molecular probes® | invitrogen™ by life technologies™

NOTE 19.1

Loading and Calibration of Intracellular Ion Indicators

There are two major prerequisites for measuring intracellular ion concentrations using fluorescent indicators:

- **Loading:** The indicator must be localized in the region (most commonly the cytosol but sometimes the mitochondria) where the ion concentration is to be measured.
- **Calibration:** The fluorescence of the indicator must be quantitatively related to the concentration of the free ion.

Loading

Cell loading methods can be divided into two groups. Bulk loading procedures are applicable to large populations of cells and include:

- Acetoxymethyl (AM) ester loading [4–7]
- ATP-induced permeabilization [8]
- Electroporation [9–11]
- Hypoosmotic shock [12]
- Influx™ pinocytic cell-loading reagent [13] (I14402, Section 19.8)
- Coupling to cell-penetrating peptides (CPP) [14–16]
- Ballistic microprojectile delivery [17–20]

Procedures such as microinjection [21] and infusion from whole-cell patch pipettes [22] must be carried out one cell at a time. Reviews of some of these techniques have been published;[23] see also Table 14.1.

The AM Ester Loading Technique

The noninvasive and technically straightforward AM ester technique is by far the most popular method for loading fluorescent ion indicators (Figure 1). The carboxylate groups of indicators for Ca^{2+} and other cations and the phenolic hydroxyl groups of pH indicators are derivatized as acetoxymethyl or acetate esters, respectively, rendering the indicator permeant to membranes and insensitive to ions. Once inside the cell, these derivatized indicators are hydrolyzed by ubiquitous intracellular esterases, releasing the ion-sensitive polyanionic indicator.

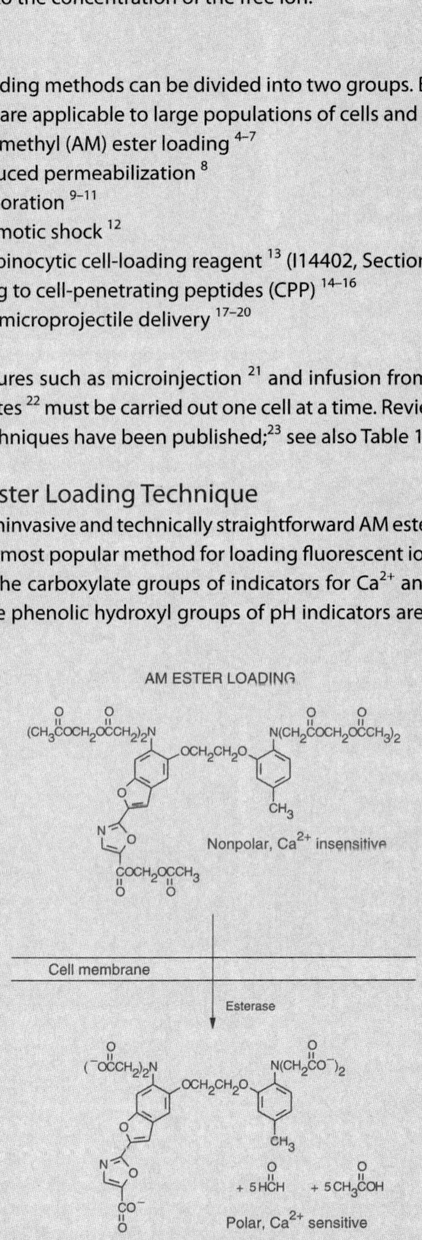

Figure 1 Schematic diagram of the processes involved in loading cells using membrane-permeant acetoxymethyl (AM) ester derivatives of fluorescent indicators, in this case fura-2. Note the generation of potentially toxic by-products (formaldehyde and acetic acid).

In practice, a 1–10 mM stock solution of the ester probe in *anhydrous* dimethylsulfoxide (DMSO) is prepared and divided into appropriately sized aliquots that can be stored desiccated at –20°C. This procedure will curtail the spontaneous ester hydrolysis that can occur in moist environments. Before loading, the DMSO stock solution should be diluted at least 1:200 in *serum-free* culture medium to a final concentration of about 1–10 µM. The nonionic and nondenaturing detergent Pluronic® F-127 or the related PowerLoad™ reagent (P3000MP, P6866, P6867, P10020; Section 19.8) are frequently added to help disperse the indicator in the loading medium.[24] After incubation at 20–37°C for 15–60 minutes, the cells should be washed two to three times with fresh *serum-free* culture medium (serum may contain esterase activity). The loading medium should also be free of amino acids or buffers containing primary or secondary amines because aliphatic amines may cleave the AM esters and prevent loading. The overall loading efficiency is typically 10–40%, depending on the molecular structure of the indicator, the type of cells and the incubation conditions.

Problems with AM Ester Loading

Compartmentalization: For calibration purposes, it is usually assumed that fluorescent indicators are homogeneously distributed in the cytosol and equally responsive to variations of intracellular ion concentration. However, AM esters and their hydrolysis products are capable of accumulating in any membrane-enclosed structure within the cell. In addition, indicators in polyanionic form may be sequestered within organelles via active transport processes.[25] Compartmentalization is usually more pronounced at higher loading temperatures and is particularly acute in plant and fungal cells.[26,27] The extent of compartmentalization can be assessed by image analysis, as well as fluorometrically using membrane permeabilization reagents, such as Triton X-100.[21]

Incomplete AM ester hydrolysis: Residual unhydrolyzed AM esters may be present extracellularly due to incomplete removal by washing. Inside the cell, low levels of intracellular esterase activity, which can vary considerably from one cell type to another, may produce only partial AM ester hydrolysis.[28–30] Because even partially hydrolyzed AM esters are Ca^{2+}-insensitive, detection of their fluorescence as part of the total signal leads to an underestimation of the Ca^{2+} concentration.[31,32] Fluorescence quenching by Mn^{2+}, which only binds with high affinity to completely de-esterified indicators, can be used to quantitate these effects. Note that although some indicators are fluorescent in the AM ester form, others are not (Table 19.1).

Extracellular AM ester hydrolysis: High levels of extracellular esterase activity can make AM ester loading ineffective, particularly for *in vivo* applications.[33]

Leakage: Extrusion of anionic indicators from cells by organic ion transporters can be reduced by cooling the sample or by applying inhibitors such as probenecid (P36400, Section 19.8), sulfinpyrazone and MK571.[25,34] AM esters have been shown to be extruded by the P-glycoprotein multidrug transporter [35] (Section 15.6). Ratiometric measurements help to minimize the impact of indicator leakage on experimental data.[36]

continued on next page

IMPORTANT NOTICE: The products described in this manual are covered by one or more Limited Use Label License(s). Please refer to the Appendix on page 971 and Master Product List on page 975. Products are For Research Use Only. Not intended for any animal or human therapeutic or diagnostic use.

continued from previous page

Calibration

Ion Dissociation Constants

The dissociation constant (K_d) is the key conversion parameter linking fluorescence signals to ion concentrations, assuming that the indicator is operating as an equilibrium sensor. This conventional assumption requires that the concentration of the indicator is close to the K_d value. Because intracellular indicator concentrations can easily reach 10–100 µM, even if the externally applied concentration is only 1–10 µM, this assumption is not always valid.[37] For pH indicators, K_d is conventionally expressed as its negative log (pK_a). The concentration range over which an indicator produces an observable response is approximately $0.1 \times K_d$ to $10 \times K_d$. For ratiometric measurements, the response range also depends on wavelength-dependent parameters.[38,39] For BAPTA-based Ca^{2+} indicators in particular, the K_d is very sensitive to a number of environmental factors, including temperature, pH, ionic strength and interactions of the indicator with proteins.[40–43] Examination of published data shows that values of K_d determined *in situ* within cells can be up to 5-fold higher than values determined *in vitro*[3,40,44–46] (Table 19.2), underscoring the importance of performing calibrations to determine the K_d directly in the system under study.

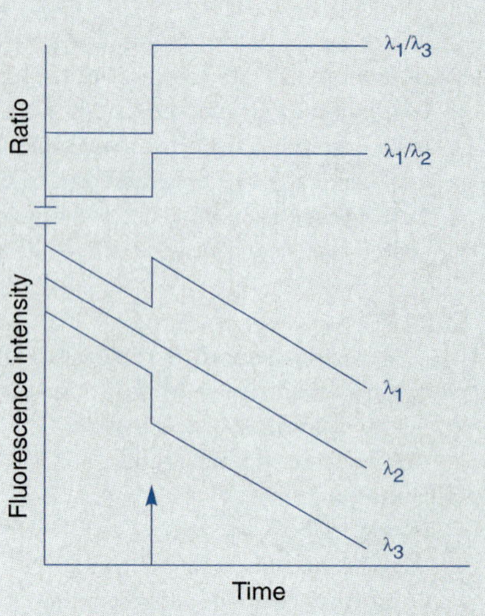

Figure 2 Simulated data demonstrating the practical importance of ratiometric fluorescence techniques. This figure represents an ion indicator that exhibits a fluorescence intensity increase in response to ion binding at wavelength λ_1 and a corresponding decrease at λ_3. Fluorescence measured at an isosbestic point (λ_2) is independent of ion concentration. The intracellular indicator concentration diminishes rapidly due to photobleaching, leakage (assuming the extracellular indicator is not detectable) or some other process. The change of intracellular ion concentration due to a stimulus applied at the time indicated by the arrow is unambiguously identified by recording the fluorescence intensity ratios λ_1/λ_3 or λ_1/λ_2.

Calibration Methodology

Calibration procedures basically consist of recording fluorescence signals corresponding to a series of precisely manipulated ion concentrations. The resulting sigmoidal titration curve is either linearized by means of a Hill plot or analyzed directly by nonlinear regression to yield K_d. For *in vitro* calibrations of Ca^{2+} indicators, EGTA buffering is widely used to produce defined Ca^{2+} concentrations that can be calculated from the K_d of the Ca^{2+}-EGTA complex.[24,47,48] This technique is used in the Calcium Calibration Buffer Kits (Section 19.8). *In situ* calibrations of intracellular indicators generally utilize an ionophore to equilibrate the controlled external ion concentration with the ion concentration within the cell.[4,49] Commonly used ionophores include:

- A-23187 (A1493), 4-bromo A-23187 (B1494) or ionomycin (I24222) for Ca^{2+} and Mg^{2+} (Section 19.8)
- Nigericin (N1495; Section 20.2, Section 21.2) for H^+ and Cl^-
- Gramicidin (G6888, Section 21.1) for Na^+
- Valinomycin (V1644, Section 21.1) for K^+

Ratiometric Calibration

Indicators that show an excitation or emission spectral shift upon ion binding can be calibrated using a ratio of the fluorescence intensities measured at two different wavelengths, resulting in the cancellation of artifactual variations in the fluorescence signal that might otherwise be misinterpreted as changes in ion concentration (Figure 2). Note that background levels must be subtracted from the component fluorescence intensities *before* calculation of the ratio. Examples of indicators exhibiting ion-dependent spectral shifts include the Ca^{2+} indicators fura-2 (Figure 3) and indo-1 (Section 19.2), and the pH indicators BCECF and SNARF®-1 (Section 20.2). The ratio of two intensities with opposite ion-sensitive responses (for example, 340 nm/380 nm in Figure 3) gives the largest possible dynamic range of ratio signals for a particular indicator. Alternatively, the ratio of an ion-sensitive intensity to an ion-insensitive intensity (measured at a spectral isosbestic point, e.g., 360 nm in Figure 3) can be used (Figure 2). Ratiometric measurements reduce or eliminate variations of several determining factors in the measured fluorescence intensity, including indicator concentration, excitation path length, excitation intensity and detection efficiency.[50,51] Artifacts that are eliminated include photobleaching and leakage of the indicator, variable cell thickness, and nonuniform indicator distribution within cells (due to compartmentalization) or among populations of cells (due to loading efficacy variations).

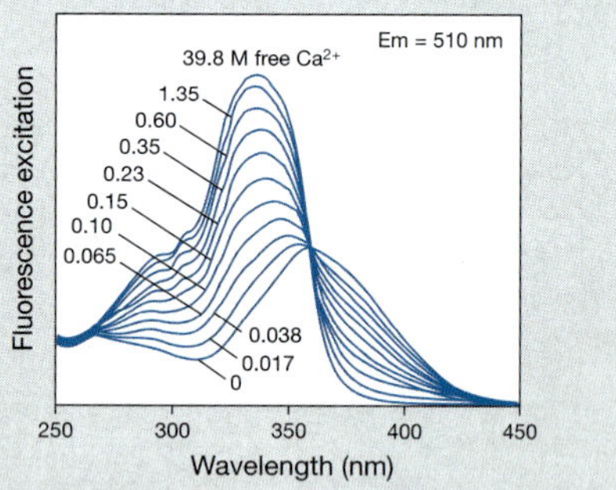

Figure 3 Fluorescence excitation spectra of fura-2 (F1200, F6799) in solutions containing 0–39.8 µM free Ca^{2+}

1. Methods Cell Biol (2007) 81:415; **2.** Methods (2008) 46:143; **3.** Cell Calcium (2000) 28:213; **4.** Methods Mol Biol (2006) 312:229; **5.** Methods Mol Biol (2009) 489:93; **6.** Nat Protoc (2006) 1:380; **7.** Nature (1981) 290:527; **8.** J Biol Chem (1987) 262:8884; **9.** Nat Protoc (2009) 4:862; **10.** Neuron (2007) 53:789; **11.** J Neurophysiol (2005) 93:1793; **12.** Cytometry (1997) 28:316; **13.** Biotechniques (2002) 33:358; **14.** Bioconjug Chem (2009) 20:249; **15.** Org Biomol Chem (2008) 6:4516; **16.** Nano Lett (2004) 4:2019; **17.** Plant J (2006) 46:327; **18.** J Neurosci Methods (2009) 184:332; **19.** J Neurosci Methods (2005) 141:41; **20.** Methods (2003) 30:79; **21.** Biochem J (1994) 302:5; **22.** CSH Protoc (2009) 2009:pdb.prot5201; **23.** Proc Natl Acad Sci U S A (2001) 98:4295; **24.** Methods Cell Biol (1994) 40:155; **25.** Cell Calcium (1990) 11:57; **26.** J Exp Biol (1994) 196:419; **27.** J Microsc (1992) 166:57; **28.** Cell Calcium (1990) 11:63; **29.** Am J Physiol (1988) 255:C304; **30.** Anal Biochem (1988) 169:159; **31.** Biophys J (1994) 67:476; **32.** Biophys J (1993) 65:561; **33.** J Microsc (2007) 226:74; **34.** J Neurosci Methods (2008) 167:140; **35.** J Biol Chem (1993) 268:21493; **36.** Anal Biochem (2009) 390:212; **37.** Mol Pharmacol (2002) 62:618; **38.** Cell Calcium (1991) 12:29; **39.** Cell Calcium (1998) 24:17; **40.** Biophys J (1994) 67:1646; **41.** Biophys J (1992) 63:89; **42.** Biochem Biophys Res Commun (1991) 180:209; **43.** Biochem Biophys Res Commun (1991) 177:184; **44.** Cell Calcium (2003) 34:1; **45.** Biophys J (1995) 68:1453; **46.** Biophys J (1993) 65:865; **47.** Methods Enzymol (1989) 172:230; **48.** Cell Calcium (1991) 12:279; **49.** Cell Calcium (1997) 21:233; **50.** Methods Cell Biol (1989) 30:157; **51.** J Biol Chem (1985) 260:3440.

IMPORTANT NOTICE: The products described in this manual are covered by one or more Limited Use Label License(s). Please refer to the Appendix on page 971 and Master Product List on page 975. Products are For Research Use Only. Not intended for any animal or human therapeutic or diagnostic use.

www.invitrogen.com/probes

molecular probes® | invitrogen™ by life technologies™

19.2 Fluorescent Ca^{2+} Indicators Excited with UV Light

Fura-2, Indo-1 and Related Derivatives

Fura-2 and Indo-1

Fura-2 and indo-1 are UV light–excitable, ratiometric Ca^{2+} indicators. Fura-2 has become the dye of choice for ratio-imaging microscopy (Figure 19.2.1), in which it is more practical to change excitation wavelengths than emission wavelengths.[1,2] Upon binding Ca^{2+}, fura-2 exhibits an absorption shift that can be observed by scanning the excitation spectrum between 300 and 400 nm, while monitoring the emission at ~510 nm (Figure 19.2.2). In contrast, indo-1 is a preferred dye for flow cytometry, where it is more practical to use a single laser for excitation—usually the 351–364 nm spectral lines of the argon-ion laser—and monitor two emissions.[3,4] The emission maximum of indo-1 shifts from ~475 nm in Ca^{2+}-free medium to ~400 nm when the dye is saturated with Ca^{2+} (Figure 19.2.3).

Modern two-photon excitation imaging techniques used with fura-2 and indo-1[5–9] avoid the deleterious effects of conventional ultraviolet illumination on live specimens. Indo-1 may be less subject to compartmentalization than fura-2,[10] whereas fura-2 is more resistant to photobleaching than indo-1.[11,12] Both fura-2 and indo-1 exhibit K_d values that are close to typical basal Ca^{2+} levels in mammalian cells (~100 nM) and display high selectivity for Ca^{2+} binding relative to Mg^{2+}.[13] Nevertheless, Ca^{2+} binding is discernibly perturbed by physiological levels of Mg^{2+}; the K_d for Ca^{2+} of fura-2 is ~135 nM in Mg^{2+}-free Ca^{2+} buffers and ~224 nM in the presence of 1 mM Mg^{2+} (measured at 37°C in 100 mM KCl, 10 mM MOPS, pH 7.0).[13] Fura-2 and indo-1 also exhibit high affinities for other divalent cations such as Zn^{2+} and Mn^{2+}, a property that is discussed further in Section 19.7.

The sodium and potassium salts of fura-2 (F6799, F1200; Figure 19.2.4; Figure 19.2.5) and potassium salt of indo-1 (I1202, Figure 19.2.6) are cell-impermeant probes that can be delivered into cells by microinjection or using our Influx™ pinocytic cell-loading reagent (I14402, Section 19.8). Free acids of fura-2 and indo-1 can also be loaded into some plant cells at pH 4–5.[14–18] In addition, these salts are useful as standards for calibrating Ca^{2+} measurements.

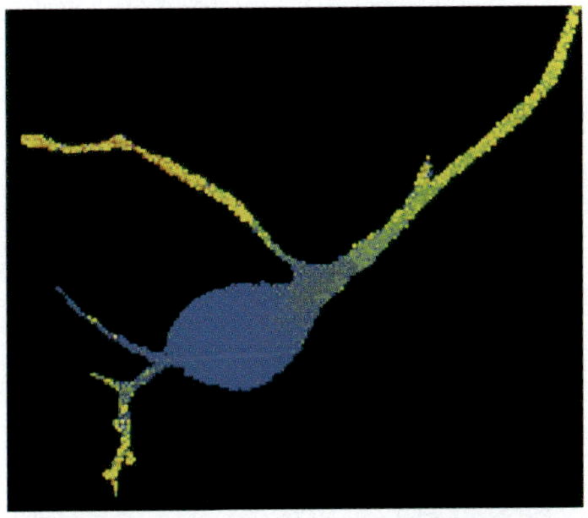

Figure 19.2.1 False-color image of free Ca^{2+} concentration in a Purkinje neuron from embryonic mouse cerebellum. Neurons were grown in dispersed tissue culture for 12 days, loaded with the pentapotassium salt of fura-2 (F1200) using a microelectrode and then challenged with trans-ACPD, an agonist of metabotropic glutamate receptors, in the absence of extracellular Ca^{2+}. The composite image, which represents the ratio of images obtained with excitation at 340 nm and 380 nm, reveals the mobilization of internal Ca^{2+} stores without contribution from Ca^{2+} influx. The image was contributed by D.J. Linden, Johns Hopkins University, and M. Smeyne and J.A. Connor, Roche Institute of Molecular Biology.

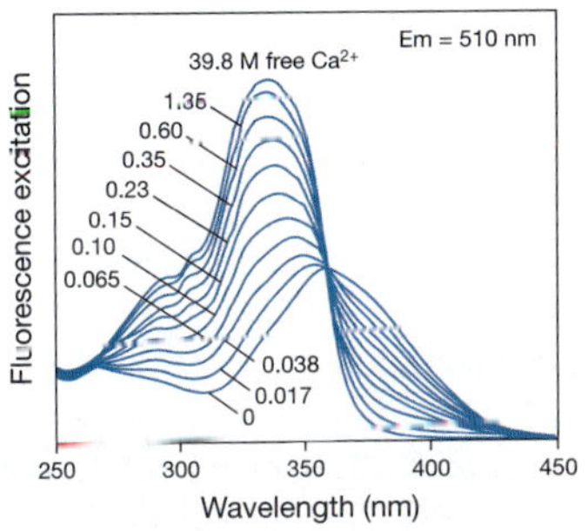

Figure 19.2.2 Fluorescence excitation spectra of fura-2 (F1200, F6799) in solutions containing 0–39.8 µM free Ca^{2+}.

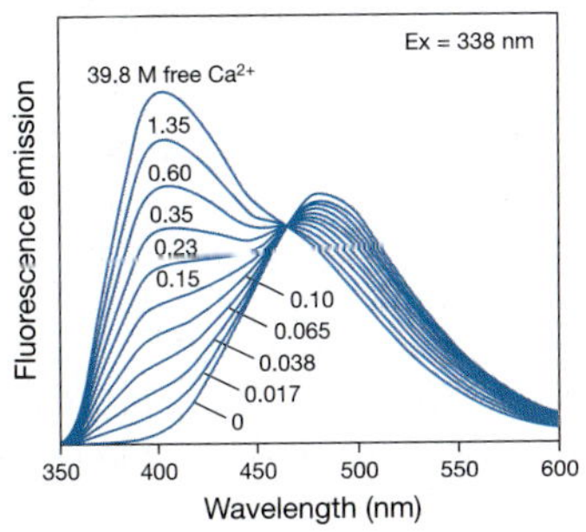

Figure 19.2.3 Fluorescence emission spectra of indo-1 (I1202) in solutions containing 0–39.8 µM free Ca^{2+}.

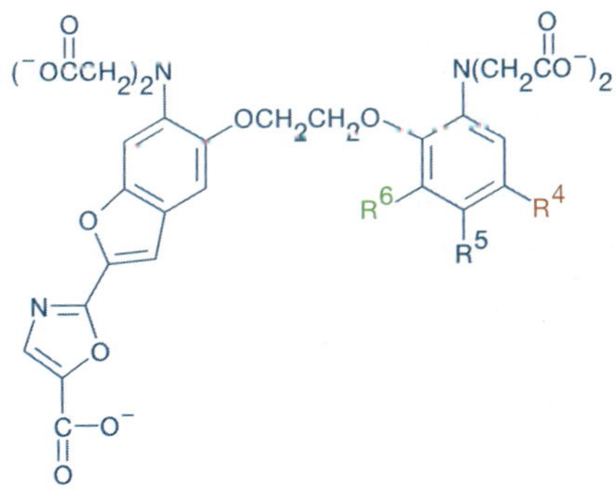

Indicator	$K_d(Ca^{2+})$	R4	R5	R6
Fura-2	0.14 M	H	CH_3	H
Fura-5F	0.40 M	H	F	H
Fura-4F	0.77 M	F	H	H
Fura-6F	5.30 M	H	H	F
Fura-FF	5.50 M	H	F	F

Figure 19.2.4 Fura indicators with varying Ca^{2+} affinities.

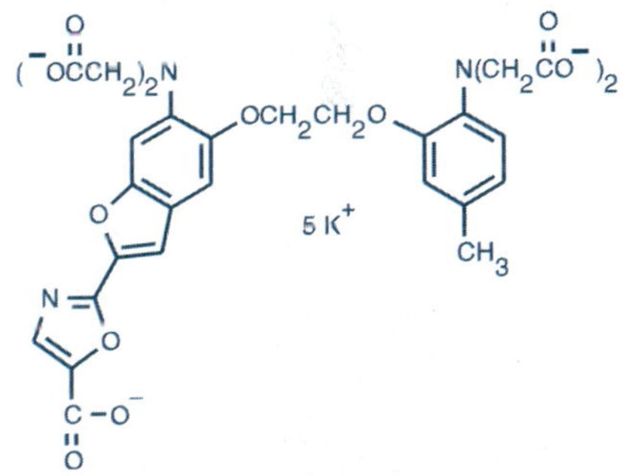

Figure 19.2.5 Fura-2, pentapotassium salt (F1200).

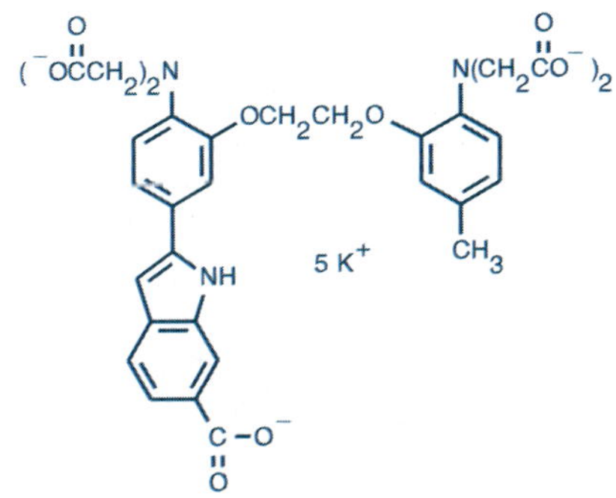

Figure 19.2.6 Indo-1, pentapotassium salt (I1202).

IMPORTANT NOTICE: The products described in this manual are covered by one or more Limited Use Label License(s). Please refer to the Appendix on page 971 and Master Product List on page 975. Products are For Research Use Only. Not intended for any animal or human therapeutic or diagnostic use.

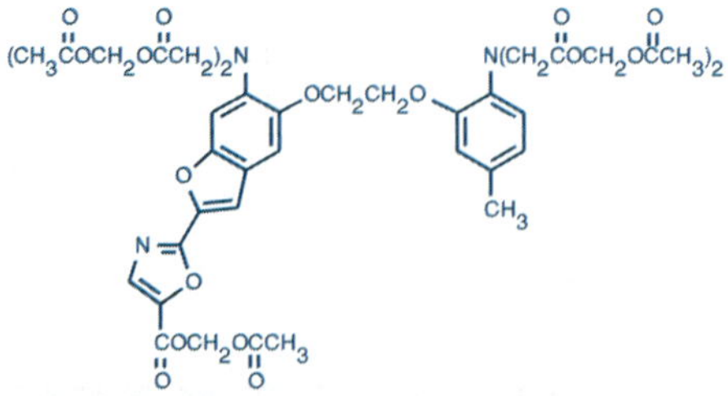

Figure 19.2.7 Fura-2, AM (F1201).

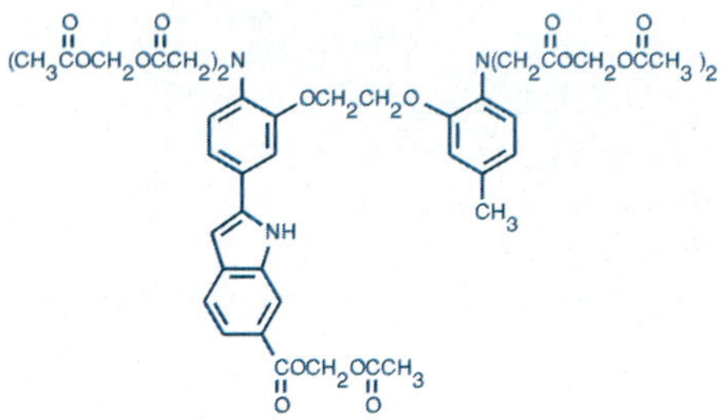

Figure 19.2.8 Indo-1, AM (I1203).

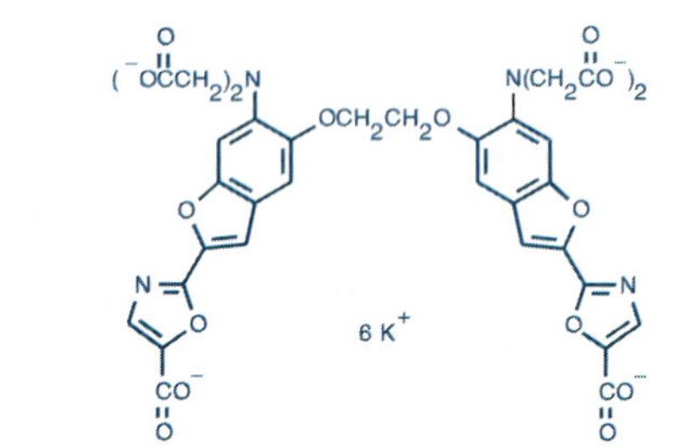

Figure 19.2.9 Bis-fura-2, hexapotassium salt (B6810).

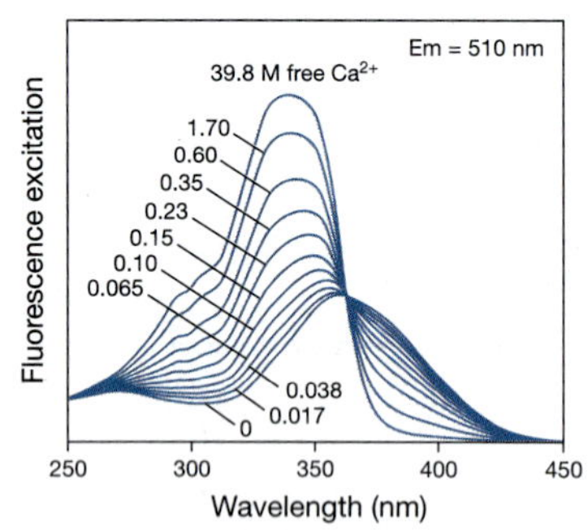

Figure 19.2.10 Fluorescence excitation spectra of bis-fura-2 (B6810) in solutions containing 0–39.8 µM free Ca^{2+}.

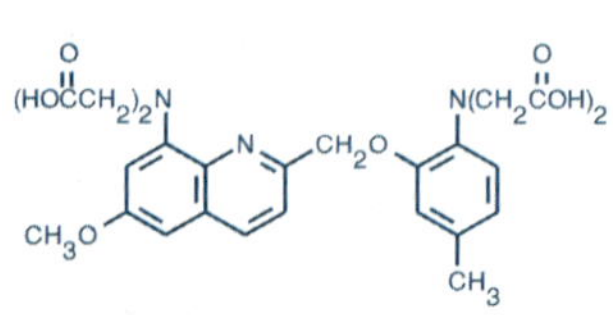

Figure 19.2.11 Quin-2, free acid (Q23918).

Unlike the salt forms, the acetoxymethyl (AM) esters of fura-2 (Figure 19.2.7) and indo-1 (Figure 19.2.8) can passively diffuse across cell membranes, enabling researchers to avoid the use of invasive loading techniques. Once inside the cell, these esters are cleaved by intracellular esterases to yield cell-impermeant fluorescent indicators (Loading and Calibration of Intracellular Ion Indicators—Note 19.1). We offer fura-2 AM and indo-1 AM in 1 mg vials (F1201, I1203) or specially packaged in 20 vials of 50 µg each (F1221, I1223); the special packaging is recommended when small quantities of the dyes are to be used over a long period of time. We also provide stock solutions of fura-2 AM and indo-1 AM in anhydrous DMSO at 1 mg/mL (~1 mM; F1225, I1226). Our standard analytical specifications for fura-2 AM require ≥95% purity by HPLC. We also offer a special packaged high-purity grade of fura-2 AM that is specified to have ≥98% purity by HPLC (as a set of 20 vials, each containing 50 µg; F14185). The 10,000 MW dextran conjugate of fura is described in Section 19.4.

Fura-2 Calcium Imaging Calibration Kit

The Fura-2 Calcium Imaging Calibration Kit (F6774) is designed to facilitate rapid calibration and standardization of digital imaging microscopes.[2,19] This kit provides 11 CaEGTA:EGTA buffer solutions with free Ca^{2+} concentrations from zero (10 mM EGTA) to 39 µM. Each solution also includes 50 µM fura-2, as well as 15 µm unstained polystyrene microspheres to act both as spacers that ensure uniform separation between the slide and the coverslip and as focusing aids. We also provide a twelfth buffer—identical to the 10 mM CaEGTA standard but lacking fura-2—that serves as a control for background fluorescence. Our Calcium Calibration Kits are described further in Section 19.8.

Bis-Fura-2: Brighter Signal with Lower Affinity for Ca^{2+}

By linking two fura fluorophores with one BAPTA chelator (Figure 19.2.9), we have produced bis-fura-2, a Ca^{2+} indicator that exhibits approximately twice the absorptivity of fura-2. Bis-fura-2 has a K_d for Ca^{2+} of ~370 nM and ~525 nM in the absence and presence of 1 mM Mg^{2+}, respectively (measured at ~22°C using our Calcium Calibration Buffer Kits). In other aspects, the quantum yield of bis-fura-2 and its spectral response to Ca^{2+} (Figure 19.2.10) are virtually identical to those of fura-2. Although the difference between the K_d of fura-2 and bis-fura-2 for Ca^{2+} is small, the change in excitation ratio for bis-fura-2 in response to Ca^{2+} concentrations >500 nM is larger than that of fura-2 (Figure 19.2.2); this difference can improve the dynamic range for Ca^{2+} measurements in cells.[20,21] Other potential advantages of bis-fura-2 include:

- Higher fluorescence output per indicator, which may allow the use of lower dye concentrations[21]
- Lower affinity for Ca^{2+}, which decreases the buffering of intracellular Ca^{2+} and produces a faster response to Ca^{2+} spikes[20]
- An additional negative charge, which may facilitate dye retention

The hexapotassium salt of bis-fura-2 (B6810) is available for loading by microinjection[22,23] or by infusion from a patch pipette.[24] We do not currently offer a membrane-permeant AM ester of bis-fura-2.

Quin-2

Quin-2 belongs to the first generation of Ca^{2+} indicators developed by Tsien[25] (Figure 19.2.11). Quin-2 has lower absorptivity and quantum yield values than the fura-2, indo-1, fluo-3, fluo-4 and Calcium Green™ indicators and thus requires higher loading concentrations. The resulting high intracellular concentration of the indicator may buffer intracellular Ca^{2+} transients.[26] Quin-2 AM has been used to intentionally deplete cytosolic free Ca^{2+}[27,28] and to ensure unidirectional Ca^{2+} influx.[29] Measurement of cytosolic free Ca^{2+} with quin-2 has been thoroughly reviewed by Tsien and Pozzan.[30] We offer quin-2 as a high-purity, cell-impermeant free acid (Q23918).

IMPORTANT NOTICE: The products described in this manual are covered by one or more Limited Use Label License(s). Please refer to the Appendix on page 971 and Master Product List on page 975. Products are For Research Use Only. Not intended for any animal or human therapeutic or diagnostic use.
www.invitrogen.com/probes

Indicators with Intermediate Calcium-Binding Affinity

Fura-4F and Fura-6F

Calcium concentrations above 1 µM produce almost complete binding saturation of fura-2 but very low fractional saturation of the low-affinity fura analog mag-fura-2 (M1290). To bridge this gap in the Ca^{2+} measurement range of fura-type indicators, we currently offer two additional ratiometric Ca^{2+} indicators—fura-4F (F14174) and fura-6F (F14178)—as well as the membrane-permeant fura-4F AM [31] (F14175); fura-5F may be available upon request at www.invitrogen.com/handbook/customorganics. Attachment of a single electron-withdrawing fluorine substituent at different positions on the BAPTA chelator moiety of fura-2 (Figure 19.2.4) results in an increase of the K_d value to ~770 nM, ~400 nM and 5.3 µM for fura-4F, fura-5F and fura-6F, respectively (measured at 22°C at 100 mM KCl, 10 mM MOPS, pH 7.2). Except for the change in the Ca^{2+} concentration response range (Figure 19.2.12), the Ca^{2+}-dependent spectral shifts produced by fura-4F, fura-5F and fura-6F are essentially identical to those of fura-2 (Figure 19.2.13) and the probes use the same optical filter sets.

Fura-FF

Fura-FF is a difluorinated derivative of fura-2 (Figure 19.2.4) with a K_d value of ~5.5 µM [32–34] (measured at 22°C in 100 mM KCl, 10 mM MOPS, pH 7.2) and similar spectroscopic properties (Figure 19.2.14). Fura-FF has negligible Mg^{2+} sensitivity, making Ca^{2+} detection less susceptible to interference than with mag-fura-2.[32–34] These properties have made fura-FF particularly useful for spatial and functional characterization of intracellular Ca^{2+} stores [35] and for tracking Ca^{2+} oscillations driven by the inositol 1,4,5-triphosphate receptor.[36,37] The low-affinity indicator fura-FF detected NMDA- and kainate-induced neuronal Ca^{2+} fluxes that were not detectable with the higher-affinity indicator fura-2.[38] Fura-FF has also been used in combination with FluoZin™-3 (Section 19.7) for simultaneous detection of Ca^{2+} and Zn^{2+}.[39] Fura-FF is available in water-soluble potassium salt form (F14180) and as a membrane-permeant AM ester derivative (F14181).

Low-Affinity Calcium Indicators

BTC

The coumarin benzothiazole–based Ca^{2+} indicator BTC (B6790, Figure 19.2.15) and its cell-permeant derivative BTC AM (B6791) were developed in collaboration with Haralambos Katerinopoulos of the University of Crete.[40] This Ca^{2+} indicator exhibits a shift in excitation maximum from about 480 nm to 400 nm upon binding Ca^{2+} (Figure 19.2.16), permitting ratiometric measurements that are essentially independent of uneven dye loading, cell thickness, photobleaching and dye leakage.[41,42] Its high selectivity and moderate affinity for Ca^{2+} (K_d ~7 µM) [42] allows accurate quantitation of high intracellular Ca^{2+} levels that are underestimated by fura-2 measurements.[43,44] When loaded into neurons as its AM ester, BTC exhibits little compartmentalization; however, prolonged excitation appears to cause conversion of the indicator to a calcium-insensitive form.[44]

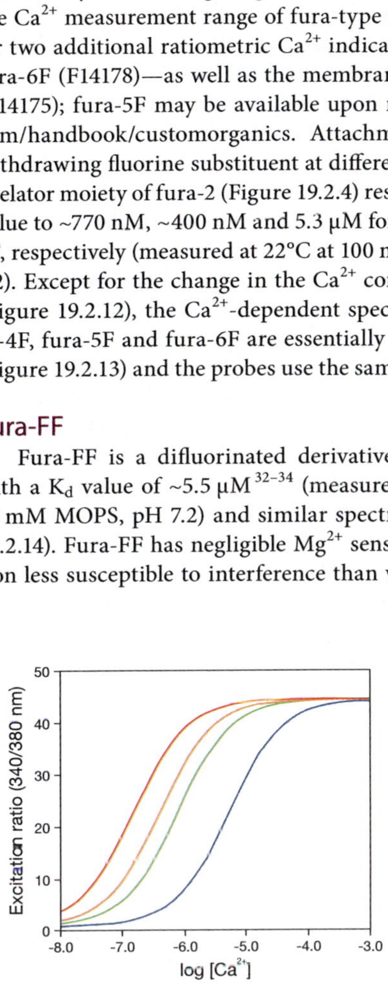

Figure 19.2.12 Fluorescence excitation ratio versus Ca^{2+} concentration curves for fura-2 (red), fura-5F (orange), fura-4F (green) and fura-6F (blue).

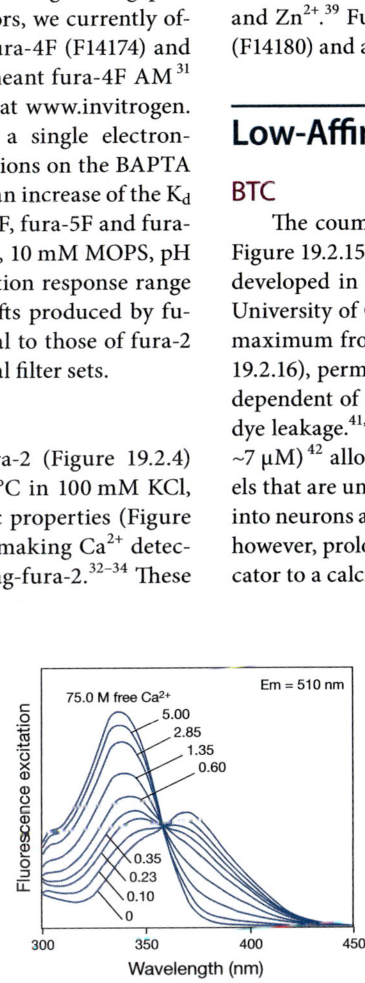

Figure 19.2.13 Ca^{2+}-dependent fluorescence excitation spectra of fura-4F (F14174).

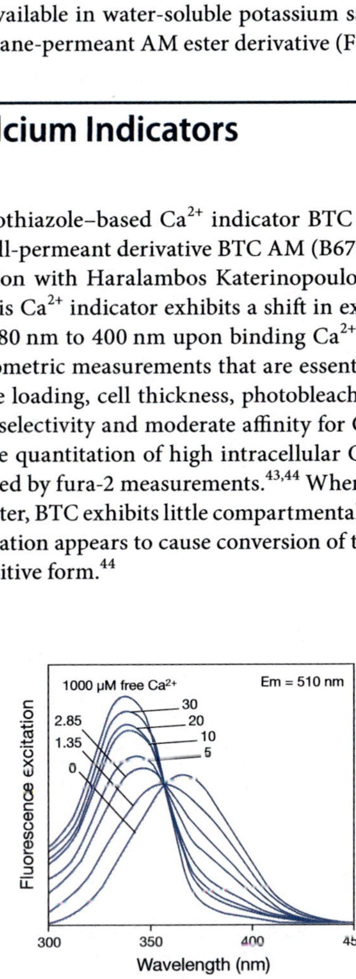

Figure 19.2.14 Ca^{2+}-dependent fluorescence excitation spectra of fura-FF (F14180).

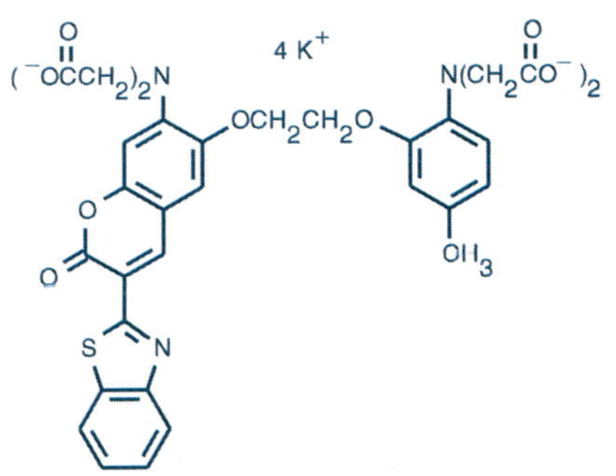

Figure 19.2.15 BTC, tetrapotassium salt (B6790).

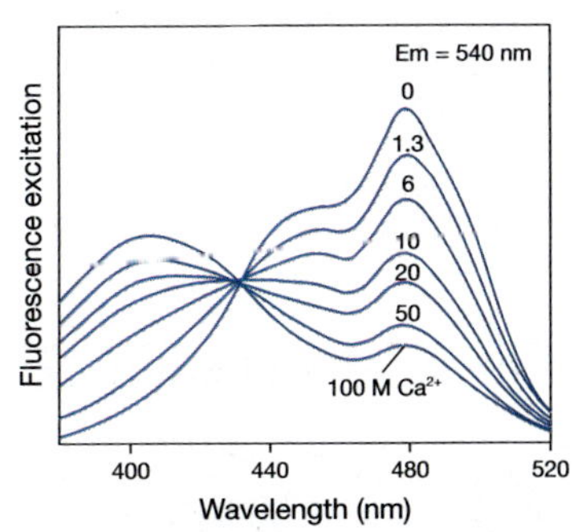

Figure 19.2.16 Fluorescence excitation spectra of BTC (B6790) in solutions containing 0–100 µM free Ca^{2+}.

IMPORTANT NOTICE: The products described in this manual are covered by one or more Limited Use Label License(s). Please refer to the Appendix on page 971 and Master Product List on page 975. Products are For Research Use Only. Not intended for any animal or human therapeutic or diagnostic use.

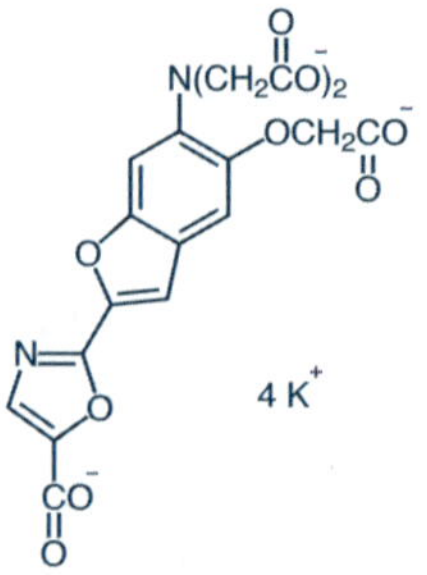

Figure 19.2.17 Mag-fura-2, tetrapotassium salt (M1290).

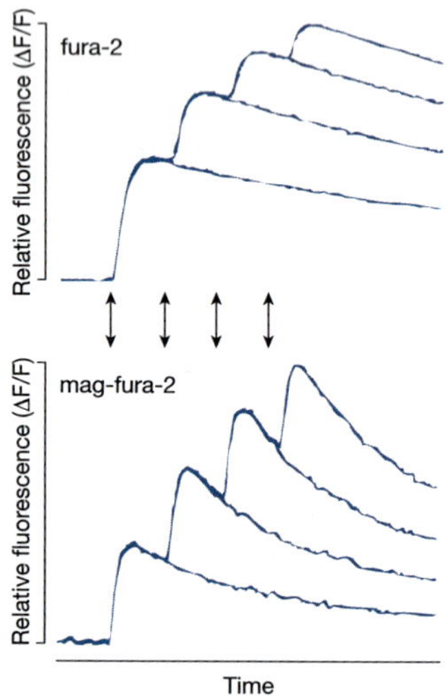

Figure 19.2.18 Ca^{2+} transients evoked by trains of 1–4 action potentials in rat cerebellar granule cells detected by fura-2 (upper panel, F1200) and mag-fura-2 (lower panel, M1290). The stimulus pulses are 50 milliseconds apart (20 Hz); timing is indicated by the double-headed arrows. The amplitude of the transients detected by fura-2 decreases with each successive stimulus due to Ca^{2+} saturation. Mag-fura-2 avoids saturation due to its lower Ca^{2+} binding affinity (K_d for Ca^{2+} = 25 µM), recording transients of approximately equal amplitude from successive action potentials. Adapted with permission from Biophys J (1995) 68:2165.

BTC has been employed in investigations of Ca^{2+}-dependent exocytosis in pancreatic β-cells,[45] CHO fibroblasts[46] and phaeochromocytoma cells.[47,48] Neuronal Ca^{2+} transients detected by the low-affinity Ca^{2+} indicators BTC and mag-fura-2 are significantly more rapid than those reported by the higher-affinity indicators fura-2 and Calcium Green™-2.[44,49]

Mag-Fura-2 and Mag-Indo-1

Mag-fura-2 (also called furaptra, Figure 19.2.17) and mag-indo-1 were originally designed to report intracellular Mg^{2+} levels (Section 19.6); however, these indicators actually have much higher affinity for Ca^{2+} than for Mg^{2+}. Although Ca^{2+} binding by these indicators may complicate analysis when they are employed to measure intracellular Mg^{2+},[50,51] their increased effective range and improved linearity for Ca^{2+} measurements has been exploited for measuring intracellular Ca^{2+} levels between 1 µM and 100 µM.[35]

The spectral shifts of mag-fura-2 and mag-indo-1 are very similar to those of fura-2 and indo-1 but occur at higher Ca^{2+} concentrations. Because the off-rates for Ca^{2+} binding of these indicators are much faster than those of fura-2 and indo-1, these dyes have been used to monitor action potentials in skeletal muscle and nerve terminals with little or no kinetic delay[52–55] (Figure 19.2.18). The spectral properties, kinetics and selectivity of several of our low-affinity Ca^{2+} indicators have been reviewed by Zhao,[56] Hyrc[32] and their co-workers.

The moderate Ca^{2+} affinity of mag-fura-2 and the tendency of its AM ester to accumulate in subcellular compartments have proven useful for *in situ* monitoring of inositol 1,4,5-triphosphate–sensitive Ca^{2+} stores.[35,57] Mag-fura-2 has also been employed to follow Ca^{2+} transients in presynaptic nerve terminals,[49,58–60] gastric epithelial cells[61] and cultured myocytes.[62] Mag-indo-1 has been used to detect gonadotropin-releasing hormone–induced Ca^{2+} oscillations in gonadotropes[63] and to investigate the role of Ca^{2+}/K^+ exchange in intracellular Ca^{2+} storage and release processes.[64] Mag-fura-2 (M1290, M1291, M1292) and mag-indo-1 (M1295) are available as cell-impermeant potassium salts or as cell-permeant AM esters.

REFERENCES

1. Methods Cell Biol (2007) 81:415; **2.** J Vis Exp (2009) doi: 10.3791/1067; **3.** Methods Mol Biol (2010) 612:149; **4.** Methods (2000) 21:221; **5.** Nature (2009) 460:264; **6.** J Neurophysiol (2007) 97:3118; **7.** Nat Protoc (2006) 1:380; **8.** Biophys J (2004) 86:1726; **9.** Nat Neurosci (2000) 3:452; **10.** Cell Calcium (1990) 11:487; **11.** Chem Biol (1996) 3:765; **12.** Am J Physiol (1987) 253:C613; **13.** J Biol Chem (1985) 260:3440; **14.** Proc Natl Acad Sci U S A (1992) 89:3591; **15.** Plant Physiol (1990) 93:841; **16.** Plant Sci (1990) 67:125; **17.** Cell Calcium (1987) 8:455; **18.** Eur J Cell Biol (1988) 46:466; **19.** J Biol Chem (2005) 280:31936; **20.** Pflugers Arch (2008) 456:267; **21.** Biophys J (1998) 75:1635; **22.** J Mol Cell Cardiol (2010) 48:1023; **23.** Am J Physiol Heart Circ Physiol (2007) 292:H2212; **24.** J Neurosci (2006) 26:3482; **25.** Biochemistry (1980) 19:2396; **26.** J Biol Chem (1983) 258:4876; **27.** J Biol Chem (1998) 273:8203; **28.** Brain Res (1990) 528:48; **29.** Biochemistry (1987) 26:6995; **30.** Methods Enzymol (1989) 172:230; **31.** Bioorg Med Chem Lett (2000) 10:1515; **32.** Cell Calcium (2000) 27:75; **33.** Biophys J (1999) 76:2029; **34.** Am J Physiol (1994) 266:C1313; **35.** Methods Mol Biol (2006) 312:229; **36.** J Biol Chem (1999) 274:14157; **37.** EMBO J (1997) 16:3533; **38.** J Neurosci (1998) 18:7727; **39.** Cell Calcium (2005) 37:225; **40.** Cell Calcium (1994) 15:190; **41.** Biochem Biophys Res Commun (2004) 317:77; **42.** J Physiol (2001) 533:757; **43.** J Neurosci (1997) 17:6669; **44.** Cell Calcium (1998) 24:165; **45.** J Cell Biol (1997) 138:55; **46.** J Biol Chem (1996) 271:17751; **47.** J Physiol (2001) 533:627; **48.** J Physiol (1996) 494:53; **49.** Biophys J (1995) 68:2156; **50.** Anal Biochem (2001) 290:221; **51.** Am J Physiol (1992) 263:C300; **52.** Proc Natl Acad Sci U S A (1996) 93:8095; **53.** J Physiol (1994) 475:319; **54.** Biochem Biophys Res Commun (1991) 177:184; **55.** J Gen Physiol (1991) 97:271; **56.** Biophys J (1996) 70:896; **57.** J Cell Sci (2006) 119:226; **58.** J Physiol (2000) 527:33; **59.** Biophys J (1997) 72:1458; **60.** Biophys J (1997) 72:637; **61.** Am J Physiol (1994) 267:G442; **62.** Biophys J (2005) 88:1911; **63.** Proc Natl Acad Sci U S A (1994) 91:9750; **64.** Nature (1998) 395:908.

DATA TABLE 19.2 FLUORESCENT Ca^{2+} INDICATORS EXCITED WITH UV LIGHT

				Low Ca^{2+}				High Ca^{2+}						
Cat. No.	MW	Storage	Soluble	Abs	EC	Em	Solvent	Abs	EC	Em	Solvent	Product	K_d	Notes
B6790	844.03	F,D,L	pH >6	464	29,000	533	H_2O	401	20,000	529	H_2O/Ca^{2+}		7.0 µM	1, 2, 3
B6791	979.92	F,D,L	DMSO	433	39,000	504	MeOH					B6790		
B6810	1007.14	F,D,L	pH >6	366	56,000	511	H_2O	338	68,000	504	H_2O/Ca^{2+}		370 nM	1, 2, 4, 5

IMPORTANT NOTICE: The products described in this manual are covered by one or more Limited Use Label License(s). Please refer to the Appendix on page 971 and Master Product List on page 975. Products are For Research Use Only. Not intended for any animal or human therapeutic or diagnostic use.
www.invitrogen.com/probes
molecular probes® | invitrogen™ by Life technologies™

DATA TABLE 19.2 FLUORESCENT Ca^{2+} INDICATORS EXCITED WITH UV LIGHT—*continued*

				Low Ca^{2+}				High Ca^{2+}						
Cat. No.	**MW**	**Storage**	**Soluble**	**Abs**	**EC**	**Em**	**Solvent**	**Abs**	**EC**	**Em**	**Solvent**	**Product**	**K_d**	**Notes**
F1200	832.00	F,D,L	pH >6	363	28,000	512	H_2O	335	34,000	505	H_2O/Ca^{2+}		145 nM	1, 2, 4, 5
F1201	1001.86	F,D,L	DMSO	370	31,000	476	EtOAc					F1200		
F1221	1001.86	F,D,L	DMSO	370	31,000	476	EtOAc					F1200		
F1225	1001.86	F,D,L	DMSO	370	31,000	476	EtOAc					F1200		6
F6799	751.45	F,D,L	pH >6	363	28,000	512	H_2O	335	34,000	505	H_2O/Ca^{2+}		145 nM	1, 2, 4, 5
F14174	835.96	F,D,L	pH >6	366	21,000	511	H_2O	336	23,000	505	H_2O/Ca^{2+}		770 nM	1, 2, 4
F14175	1005.82	F,D,L	DMSO	370	29,000	475	EtOAc					F14174		
F14178	835.96	F,D,L	pH >6	364	25,000	512	H_2O	336	28,000	505	H_2O/Ca^{2+}		5.3 µM	1, 2, 3
F14180	853.95	F,D,L	pH >6	364	25,000	510	H_2O	335	28,000	506	H_2O/Ca^{2+}		5.5 µM	1, 2, 3
F14181	1023.82	F,D,L	DMSO	370	30,000	476	EtOAc					F14180		
F14185	1001.86	F,D,L	DMSO	370	31,000	476	EtOAc					F1200		7
I1202	840.06	F,D,L	pH >6	346	33,000	475	H_2O	330	33,000	401	H_2O/Ca^{2+}		230 nM	1, 2, 4, 5
I1203	1009.93	F,D,L	DMSO	356	39,000	478	MeOH					I1202		
I1223	1009.93	F,D,L	DMSO	356	39,000	478	MeOH					I1202		
I1226	1009.93	F,D,L	DMSO	356	39,000	478	MeOH					I1202		6
M1290	586.68	F,D,L	pH >6	369	22,000	511	H_2O	329	26,000	508	H_2O/Ca^{2+}		25 µM	1, 2, 3
M1291	722.57	F,D,L	DMSO	366	31,000	475	EtOAc					M1290		
M1292	722.57	F,D,L	DMSO	366	31,000	475	EtOAc					M1290		
mag-indo-1	594.74	F,D,L	pH >6	349	38,000	480	H_2O	328	35,000	390	H_2O/Ca^{2+}		35 µM	1, 2, 8, 9
M1295	730.63	F,D,L	DMSO	354	37,000	472	MeOH					mag-indo-1		
Q23918	541.51	D,L	pH >6	353	4000	495	H_2O	333	3900	495	H_2O/Ca^{2+}		60 nM	1, 2, 10

For definitions of the contents of this data table, see "Using *The Molecular Probes® Handbook*" in the introductory pages.

Notes

1. Dissociation constants are known to vary considerably depending on the temperature, pH, ionic strength, viscosity, protein binding, presence of other ions (especially polyvalent ions), instrument setup and other factors. It is strongly recommended that these values be verified under user-specific experimental conditions.
2. Spectra measured in aqueous buffers containing 10 mM EGTA (H_2O) or a >10-fold excess of free Ca^{2+} relative to the K_d (H_2O/Ca^{2+}).
3. Dissociation constant determined by fluorescence measurements in 100 mM KCl, 10 mM MOPS, pH 7.2, 0 to 1 mM free Ca^{2+} at 22°C.
4. Dissociation constant determined by fluorescence measurements in 100 mM KCl, 10 mM MOPS, pH 7.2, 0 to 39 µM free Ca^{2+} at 22°C.
5. $K_d(Ca^{2+})$ for fura-2 and indo-1 from the original reference by Grynkiewicz, Poenie and Tsien (J Biol Chem (1985) 260:3440) are 224 nM and 250 nM, respectively, measured in 1 mM EGTA, 100 mM KCl, 1 mM free Mg^{2+}, 10 mM MOPS, pH 7.0 at 37°C. For bis-fura-2, $K_d(Ca^{2+})$ in presence of Mg^{2+} is 525 nM (determined in our laboratories using 100 mM KCl, 10 mM MOPS, pH 7.2, 1 mM Mg^{2+} at 22°C).
6. This product is supplied as a ready-made solution in the solvent indicated under "Soluble."
7. This product is specified to equal or exceed 98% analytical purity by HPLC.
8. The emission spectrum of Ca^{2+}-bound mag-indo-1 excited at 340 nm has approximately equal peak intensities at ~390 nm and ~480 nm. (Biochemistry (1991) 30:702)
9. Dissociation constant determined in 100 mM KCl, 40 mM HEPES, pH 7.0 at 22°C. (Biochem Biophys Res Commun (1991) 177:184)
10. $K_d(Ca^{2+})$ for quin-2 was measured in 120 mM KCl, 20 mM NaCl, pH 7.05 at 37°C. Under the same conditions with addition of 1 mM Mg^{2+}, K_d = 115 nM. (Methods Enzymol (1989) 172:230)

PRODUCT LIST 19.2 FLUORESCENT Ca^{2+} INDICATORS EXCITED WITH UV LIGHT

Cat. No.	Product	Quantity
B6810	bis-fura-2, hexapotassium salt *cell impermeant*	1 mg
B6791	BTC, AM *cell permeant*	100 µg
B6790	BTC, tetrapotassium salt *cell impermeant*	1 mg
F6774	Fura-2 Calcium Imaging Calibration Kit *zero to 10 mM CaEGTA, 50 µM fura-2 (11 x 1 mL)*	1 kit
F1225	fura-2, AM *1 mM solution in anhydrous DMSO* *cell permeant*	1 mL
F1201	fura-2, AM *cell permeant*	1 mg
F1221	fura-2, AM *cell permeant* *special packaging*	20 x 50 µg
F14185	fura-2, AM *FluoroPure™ grade* *special packaging*	20 x 50 µg
F1200	fura-2, pentapotassium salt *cell impermeant*	1 mg
F6799	fura-2, pentasodium salt *cell impermeant*	1 mg
F14175	fura-4F, AM *cell permeant* *special packaging*	10 x 50 µg
F14174	fura-4F, pentapotassium salt *cell impermeant*	500 µg
F14178	fura-6F, pentapotassium salt *cell impermeant*	500 µg
F14181	fura-FF, AM *cell permeant* *special packaging*	10 x 50 µg
F14180	fura-FF, pentapotassium salt *cell impermeant*	500 µg
I1226	indo-1, AM *1 mM solution in anhydrous DMSO* *cell permeant*	1 mL
I1203	indo-1, AM *cell permeant*	1 mg
I1223	indo-1, AM *cell permeant* *special packaging*	20 x 50 µg
I1202	indo-1, pentapotassium salt *cell impermeant*	1 mg
M1291	mag-fura-2, AM *cell permeant*	1 mg
M1292	mag-fura-2, AM *cell permeant* *special packaging*	20 x 50 µg
M1290	mag-fura-2, tetrapotassium salt *cell impermeant*	1 mg
M1295	mag-indo-1, AM *cell permeant* *special packaging*	20 x 50 µg
Q23918	quin-2, free acid *cell impermeant*	5 mg

IMPORTANT NOTICE: The products described in this manual are covered by one or more Limited Use Label License(s). Please refer to the Appendix on page 971 and Master Product List on page 975. Products are For Research Use Only. Not intended for any animal or human therapeutic or diagnostic use.

19.3 Fluorescent Ca^{2+} Indicators Excited with Visible Light

Fluo-3, Fluo-4, Rhod-2 and Related Derivatives

Fluo-3

The Ca^{2+} indicator fluo-3 (Figure 19.3.1) was developed by Tsien and colleagues for use with visible-light excitation sources in flow cytometry and confocal laser-scanning microscopy.[1] More recently, imaging with fluo-3 and its derivatives has been extended to include two-photon excitation techniques[2–5] (Figure 19.3.2) and total internal reflection fluorescence (TIRF) microscopy.[6] Fluo-3 imaging has revealed the spatial dynamics of many elementary processes in Ca^{2+} signaling[7–11] (Figure 19.3.3, Figure 19.3.4). Since about 1996, fluo-3 has also been extensively used in cell-based high-throughput screening assays for drug discovery.[12,13] Fluo-3 is essentially nonfluorescent unless bound to Ca^{2+} and exhibits a quantum yield at saturating Ca^{2+} of ~0.14 (Figure 19.3.5) and a K_d for Ca^{2+} of 390 nM (measured at 22°C using our Calcium Calibration Buffer Kits). The intact acetoxymethyl (AM) ester derivative of fluo-3 is almost nonfluorescent, unlike the AM esters of fura-2 and indo-1. The green-fluorescent emission (~525 nm) of Ca^{2+}-bound fluo-3 is conventionally detected using optical filter sets designed for fluorescein (FITC).

In a careful study of the spectral properties of highly purified fluo-3, Harkins, Kurebayashi and Baylor characterized the effects of pH and viscosity on Ca^{2+} measurements with fluo-3 and demonstrated that binding of the indicator to proteins has a significant effect on its K_d for Ca^{2+} [14] (Table 19.2). The temperature dependence of the K_d for fluo-3 has also been reported.[15] In addition, the fluorescence output of fluo-3—the product of the molar absorptivity and the fluorescence quantum yield—may also vary significantly in different cellular environments.[16,17]

Fluo-3 exhibits an at least 100-fold Ca^{2+}-dependent fluorescence enhancement.[4,14] However, fluo-3 lacks a significant shift in emission or excitation wavelength upon binding to Ca^{2+}, which precludes the use of ratiometric measurements (Figure 19.3.6). Simultaneous loading of cells with fluo-3 and our Fura Red™ indicator, which exhibit reciprocal shifts in fluorescence intensity upon binding Ca^{2+}, has enabled researchers to make ratiometric measurements of intracellular Ca^{2+} (Figure 19.3.7) using confocal laser-scanning microscopy[18,19] (Figure 19.3.8) or flow cytometry.[20–22] For ratiometric measurements, fluo-3 (or fluo-4) can also be co-loaded into cells with a spectrally distinct Ca^{2+}-insensitive dye, such as CellTracker™ Orange CMRA[23] (C34551, Section 14.2) or CellTrace™ calcein red-orange AM[24] (C34851, Section 15.2). It is common practice when loading neurons in brain slices via patch pipette infusion with green-fluorescent calcium indicators, such as fluo-3, fluo-4, fluo-5F and Oregon Green® 488 BAPTA-1, to add in a Ca^{2+}-insensitive structural marker such as Alexa Fluor® 594 hydrazide (A10438, A10442; Section 14.3).[25–27]

Fluo-3 is available as a cell-impermeant potassium salt (F3715) or ammonium salt (F1240). The cell-permeant fluo-3 AM is available as a 1 mg vial (F1241), as a set of 20 vials each containing 50 µg (F1242), as a set of 10 vials each containing 50 µg of our high-purity grade fluo-3 AM (F23915) and as a 1 mM solution in DMSO (F14218). A set of 40 vials, each containing 1 mg of fluo-3 AM (F14242), is available at a discounted price to provide a larger quantity for high-throughput screening applications.

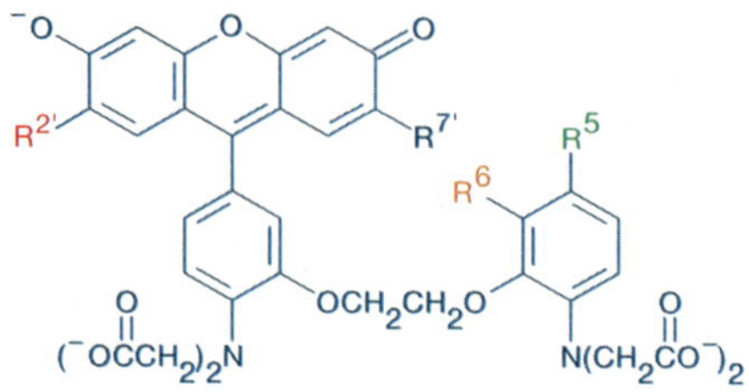

Indicator	$K_d(Ca^{2+})$	$R^{2'}$	$R^{7'}$	R^5	R^6
Fluo-3	0.39 M	Cl	Cl	CH_3	H
Fluo-4	0.35 M	F	F	CH_3	H
Fluo-5F	2.3 M	F	F	F	H
Fluo-5N	90 M	F	F	NO_2	H
Fluo-4FF	9.7 M	F	F	F	F

Figure 19.3.1 Fluo indicators.

Figure 19.3.3 Fluo-3 (F1240) confocal image of a spiral Ca^{2+} wave initiated by injection of a nonhydrolyzable analog of inositol 1,4,5-triphosphate in a *Xenopus laevis* oocyte. The image was contributed by David Clapham, Harvard University, and reproduced with permission from Cell (1992) 69:283.

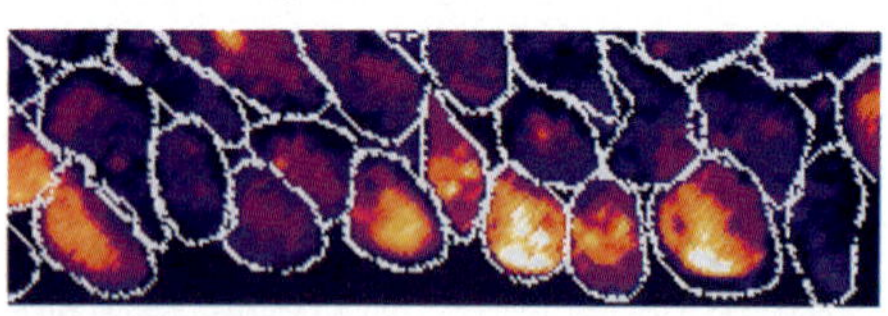

Figure 19.3.4 Spontaneous intracellular Ca^{2+} fluctuations of neurons developing *in vivo*. The spinal cord was dissected from a neurula-stage *Xenopus* embryo and loaded with fluo-3 AM (F1241, F1242, F14218, F14242, F23915). Regions of fluo-3 fluorescence on the ventral side of the spinal cord are pseudocolored in gold and indicate areas of highest intracellular Ca^{2+} (Nature (1995) 375:784). The image was obtained with a Bio-Rad® MRC600 confocal laser-scanning microscope. The image was contributed by Nicholas C. Spitzer, University of California, San Diego.

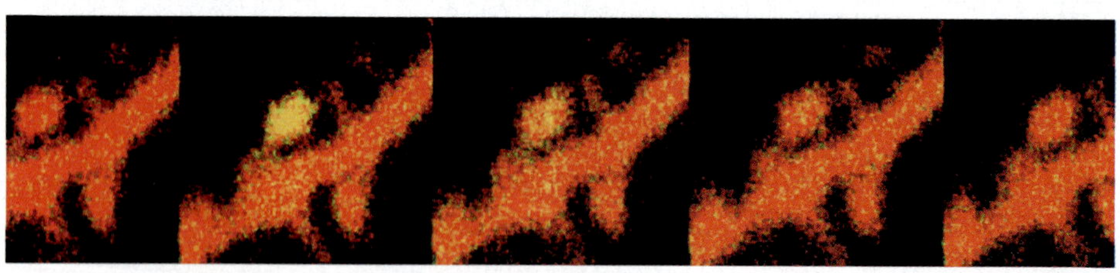

Figure 19.3.2 Two-photon excitation imaging of Ca^{2+} influx in a CA1 pyramidal cell spine. The images are overlays of anatomical images generated by Alexa Fluor® 594 hydrazide (A10438, A10442) and Ca^{2+} signals generated by fluo-5F (F14221). The imaging system uses two-photon excitation at 810 nm and two-channel emission detection (fluo-5F in the green channel, Alexa Fluor® 594 hydrazide in the red channel). The observed Ca^{2+} influx is through NMDA receptors that are activated by glutamate released from the presynaptic terminal following electrode stimulation of a collateral CA3 pyramidal cell axon. The brief (0.2 ms) depolarizing stimulus was applied after the first frame in the image sequence. The frame rate is four frames/second, and each frame represents an area of 5 µm × 5 µm. The image was contributed by Thomas Oertner and Karel Svoboda, Cold Spring Harbor Laboratory.

IMPORTANT NOTICE: The products described in this manual are covered by one or more Limited Use Label License(s). Please refer to the Appendix on page 971 and Master Product List on page 975. Products are For Research Use Only. Not intended for any animal or human therapeutic or diagnostic use.

Fluo-4

Fluo-4 (Figure 19.3.9), an analog of fluo-3 with the two chlorine substituents replaced by fluorine atoms (Figure 19.3.1), exhibits a K_d for Ca^{2+} of 345 nM[28–30] (measured at 22°C using our Calcium Calibration Buffer Kits); the temperature dependence of the K_d for fluo-4 has been reported.[15] The fluorescence quantum yields of Ca^{2+}-bound fluo-3 and fluo-4 are essentially identical. Significantly, however, the absorption maximum of fluo-4 is blue-shifted about 12 nm as compared with fluo-3, resulting in increased fluorescence excitation at 488 nm and consequently higher signal levels for confocal laser-scanning microscopy (Figure 19.3.10), flow cytometry and microplate screening applications (Figure 19.3.11, Figure 19.3.12).

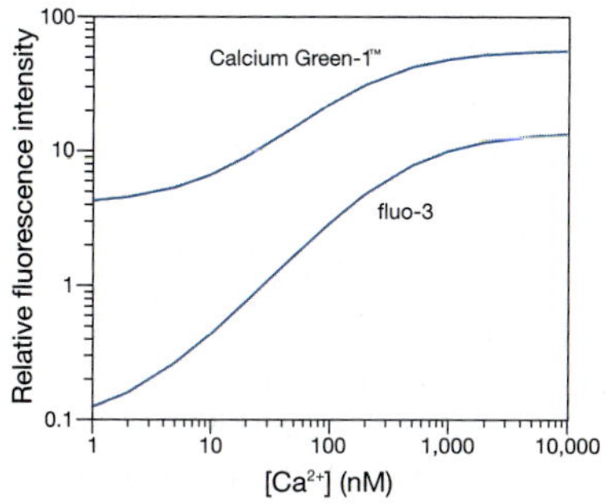

Figure 19.3.5 Comparison of fluorescence intensity responses to Ca^{2+} for the fluo-3 (F1240, F3715) and Calcium Green™-1 (C3010MP) indicators. Responses were calculated from the Ca^{2+} dissociation constants for the two indicators and the extinction coefficients and fluorescence quantum yields of their ion-free and ion-bound forms. They, therefore, represent the relative fluorescence intensities that would be obtained from equal concentrations of the two indicators excited and detected at their peak wavelengths.

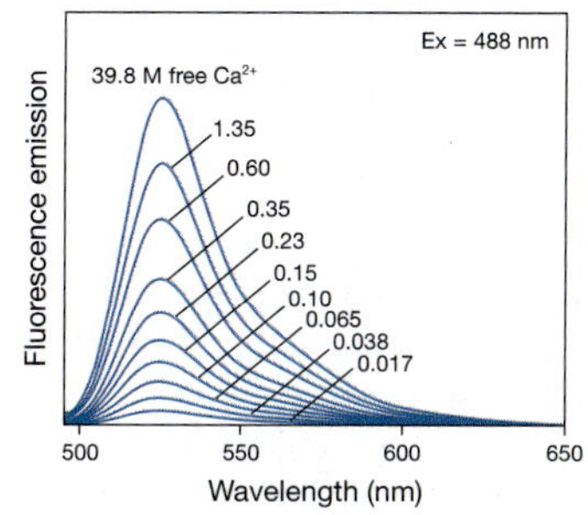

Figure 19.3.6 Ca^{2+}-dependent fluorescence emission spectra of fluo-3 (F1240, F3715). The spectrum for the Ca^{2+}-free solution is indistinguishable from the baseline.

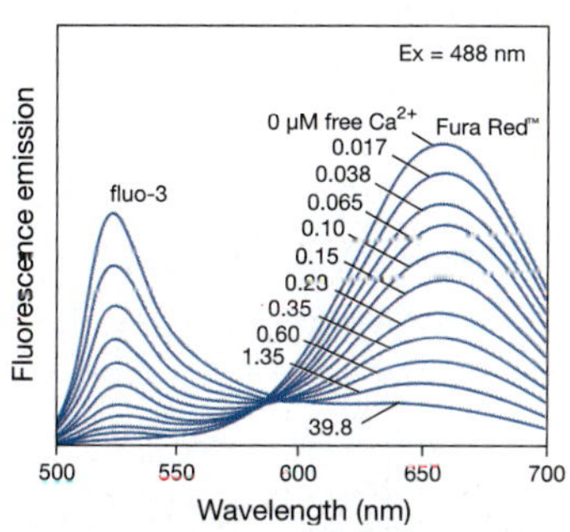

Figure 19.3.7 Fluorescence emission spectra of a 1:10 mole:mole mixture of the fluo-3 (F1240, F3715) and Fura Red™ (F14219) indicators, simultaneously excited at 488 nm, in solutions containing 0–39.8 µM free Ca^{2+}.

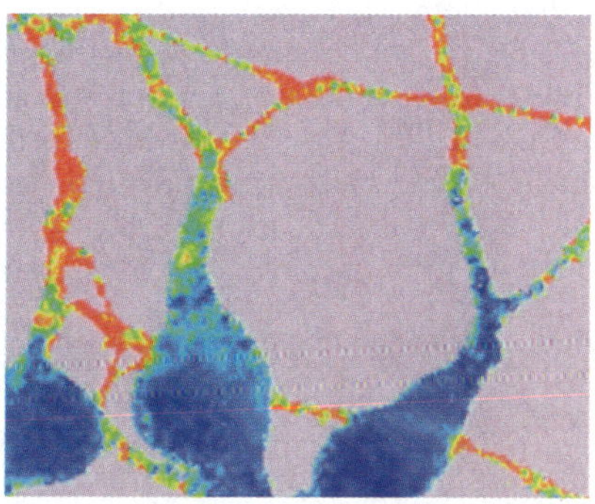

Figure 19.3.8 Frog olfactory bulb neurons labeled with fluo-3 (F1240, F3715) and Fura Red™ (F14219) Ca^{2+} indicators, demonstrating a Ca^{2+} response to treatment with KCl and nifedipine. (J Physiol (1995) 487:305). The image is a ratio of fluo-3 and Fura Red™ fluorescence images acquired with a Leica confocal laser-scanning microscope. The image was contributed by J. Bischofberger and D. Schild, Physiology Institute, University of Göttingen.

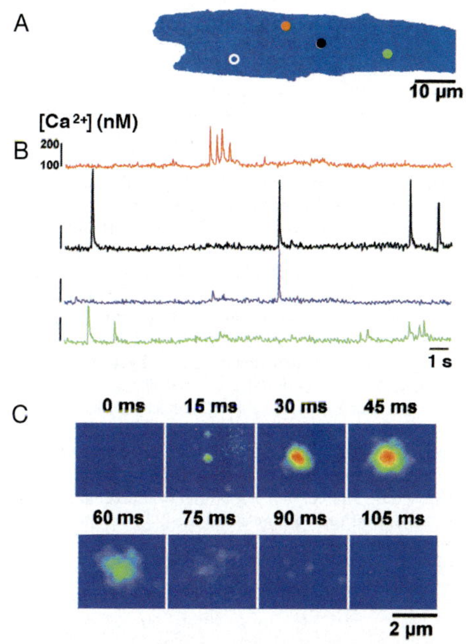

Figure 19.3.10 Fast confocal recording of spontaneous Ca^{2+} sparks in a rat ventricular myocyte. Single myocytes were isolated from rat hearts and loaded with fluo-4 AM (F14201, F14202, F14217, F23917). Images of loaded cells were obtained with a PerkinElmer® UltraVIEW™ LCI confocal imaging system at a constant frame rate of 66 frames/second. A illustrates the outline of a single cell and the position of the four regions of interest that were averaged to produce the kinetic traces shown in B. The traces are color coded in accordance with the regions marked in A and show numerous spontaneous Ca^{2+} sparks. C illustrates a sequence of confocal images taken around the black region in A, indicating the spatiotemporal properties of an individual Ca^{2+} spark. The image was contributed by Peter Lipp, The Babraham Institute, Cambridge, UK, and reproduced with permission from Biomedical Products (2001) 9:52.

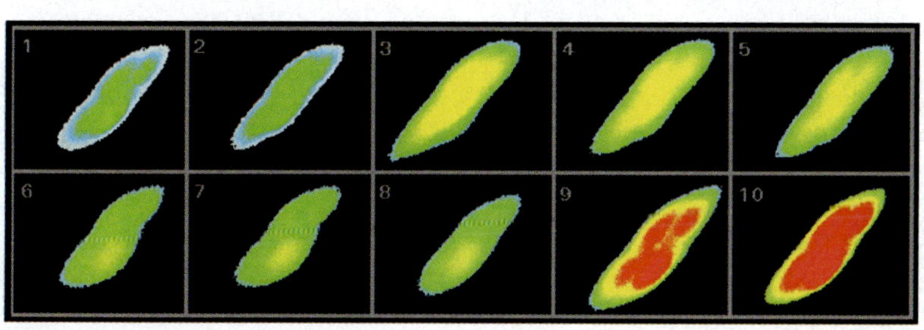

Figure 19.3.9 Pseudocolored images of changes in intracellular free Ca^{2+} in AtT-20/D16v-F2 cells, monitored at 9-second intervals with fluo-4, AM (F14201, F14202, F14217, F23917). In order to induce an influx of Ca^{2+}, the cells were depolarized with 50 mM KCl in frame 2 and exposed to 5 µM ionomycin (I24222, a Ca^{2+} ionophore) in frame 8. The images are pseudocolored according to fluorescence intensity, with red representing high Ca^{2+} concentrations and blue representing low Ca^{2+} concentrations. The images were acquired using a fluorescence microscope equipped with a longpass filter set appropriate for fluorescein and a Photometrics® Quantix cooled CCD camera.

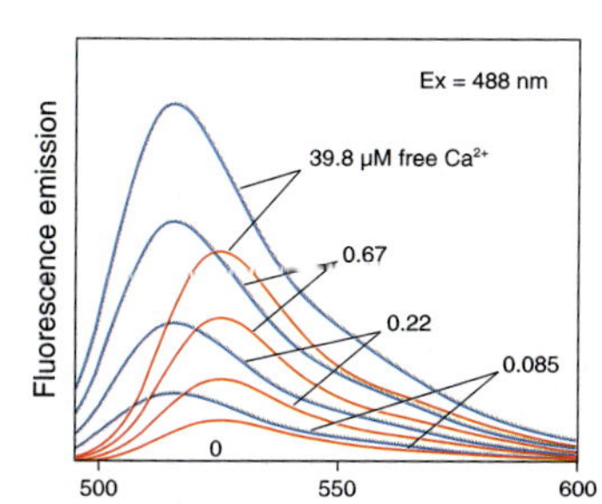

Figure 19.3.11 Fluorescence emission spectra at equal concentrations of fluo-4 (blue, F14200) and fluo-3 (red; F1240, F3715) in solutions containing 0–39.8 µM free Ca^{2+}.

IMPORTANT NOTICE: The products described in this manual are covered by one or more Limited Use Label License(s). Please refer to the Appendix on page 971 and Master Product List on page 975. Products are For Research Use Only. Not intended for any animal or human therapeutic or diagnostic use.

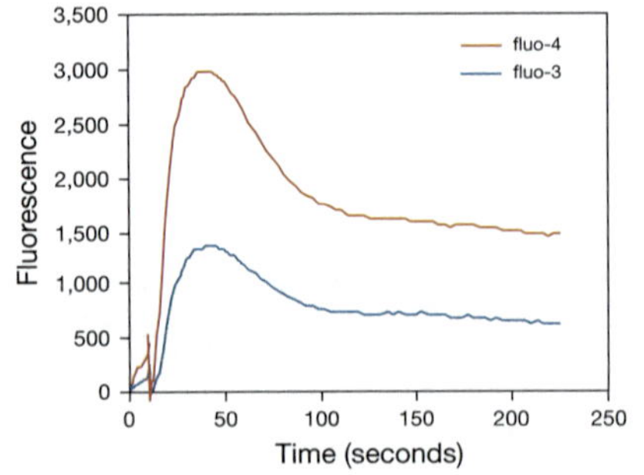

Figure 19.3.12 High-throughput Ca^{2+} influx assays for G-protein–coupled receptor activation. CHO (DHFR-) cells, co-transfected with the orphanin FQ receptor and $G\alpha_{qi3}$ chimeric G-protein, were loaded with 4 µM fluo-3 AM (F1241, F1242, F14218, F14242, F23915) or fluo-4 AM (F14201, F14202, F14217, F23917) for 60 minutes at 37°C. Ca^{2+}-dependent fluorescence traces (1 data point/second, excitation at 488 nm) from loaded cell samples in a 96-well microplate were measured simultaneously by a Fluorometric Imaging Plate Reader (FLIPR®). Ca^{2+} transients were initiated by addition of 25 nM orphanin FQ (nociceptin), as indicated by the baseline discontinuities at approximately 10 seconds. Each trace represents the average of five transient recordings from separate microplate wells. The data were supplied by Sven Merten, Hans-Peter Nothacker and Olivier Civelli, University of California, Irvine.

Intracellular Ca^{2+} measurements using fluo-3 and fluo-4 have become essential for certain types of high-throughput pharmacological screening.[12,13] Applications of this technology include screening for compounds that activate or deactivate G-protein–coupled receptors and identifying receptors for ligands known to be pharmacologically active. Molecular Devices FLIPR® (Fluorometric Imaging Plate Reader) systems have been the leading instrument platforms for these measurements (Figure 19.3.12). Parallel comparisons of fluo-4 with fluo-3, carried out in collaboration with researchers at Molecular Devices, show that the advantages of fluo-4 in microscopy and solution fluorometry measurements are replicated in the FLIPR® system. For example, fluo-4 generates the same fluorescence response as fluo-3 to carbachol-stimulated Ca^{2+} activation in Chinese hamster ovary (CHO) cells using half the AM ester–loading concentration and half the incubation time (Table 19.3). When fluo-4 is substituted for fluo-3 (i.e., using identical loading protocols), fluorescence signals are at least doubled. The stronger fluorescence signals provided by fluo-4 are particularly advantageous in cell types such as human embryonic kidney (HEK 293) cells, which are seeded at low densities for pharmacological screening assays.

Fluo-4 is available as a cell-impermeant potassium salt (F14200) or as its cell-permeant AM ester. The AM ester is available specially packaged as a 1 mM solution in DMSO (F14217), as a set of 10 vials, each containing 50 µg (F14201, F23917) or—for high-throughput screening applications—as a set of five vials, each containing 1 mg (F14202). We also offer both low-affinity and high-affinity 10,000 MW dextran conjugates of the fluo-4 indicator[31] (F14240, F36250; Section 19.4).

We also offer the Fluo-4 Direct™ Calcium Assay Kits (F10471, F10472, F10473) which offer a proprietary assay formulation that allows direct addition to wells containing cells growing in culture media without the requirement of media removal or a wash step. Eliminating the media removal step from the workflow can result in lower variability and higher Z′ values[32] compared with the standard fluo-4 assay, while also providing an easier and faster assay. Contributions to baseline fluorescence by the growth medium are eliminated by the addition of a suppression dye, which reduces background fluorescence. Another source of background fluorescence is extrusion of the indicator out of the cell by organic anion transporters. To address this backgound issue, the Fluo-4 Direct™ Calcium Assay Kits provide a proprietary, water-soluble probenecid, which inhibits this transport and reduces the baseline signal.[33,34] The water-soluble form of probenecid

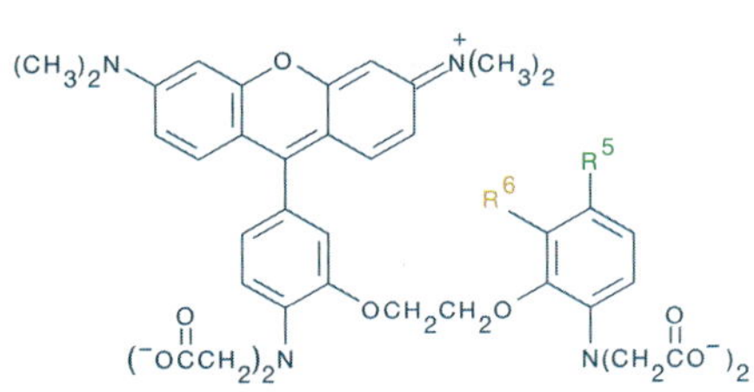

Indicator	$K_d(Ca^{2+})$	R^5	R^6
Rhod-2	0.57 µM	CH_3	H
Rhod-FF	19 µM	F	F
Rhod-5N	320 µM	NO_2	H

Figure 19.3.13 Rhod indicators with varying Ca^{2+} affinities.

Table 19.3 Parallel performance comparison of fluo-3 and fluo-4 on the Molecular Devices FLIPR® system.

			Fluorescence †		
Indicator	µM *	Time (min) *	Basal	Stimulated ‡	Increase §
Fluo-3	4	60	1700	5700	3.4-fold
Fluo-4	4	60	4900	21300	4.3-fold
Fluo-4	2	30	1200	5400	4.5-fold

* CHO cells stably transfected with rat muscarinic M_1 receptors were loaded with fluo-3 AM or fluo-4 AM according to standard protocols, with variations in indicator concentration (µM) and incubation time, as shown. † Relative fluorescence intensities after subtraction of background fluorescence. ‡ Cells were stimulated by addition of the muscarinic agonist carbachol (50 µM). § Ratio of stimulated to basal fluorescence intensities.

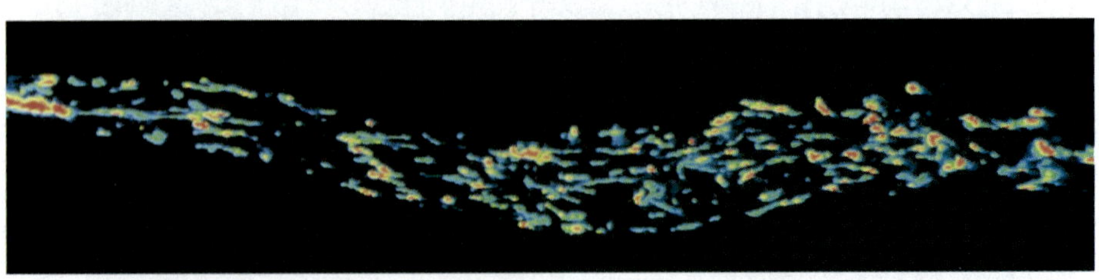

Figure 19.3.14 Pseudocolored micrograph of a single living smooth muscle cell labeled with a reduced form of rhod-2 AM (R1244, R1245MP), dihydrorhod-2. Images were acquired at focal planes spaced at 0.25 µm intervals and then processed using a constrained iterative deconvolution algorithm. This image shows that the rhod-2 fluorescence primarily arises from the mitochondria. The image was contributed by Fredric S. Fay, Program in Molecular Medicine, University of Massachusetts Medical Center.

IMPORTANT NOTICE: The products described in this manual are covered by one or more Limited Use Label License(s). Please refer to the Appendix on page 971 and Master Product List on page 975. Products are For Research Use Only. Not intended for any animal or human therapeutic or diagnostic use.
www.invitrogen.com/probes

is easy to dissolve in physiological buffers, unlike the conventionally used free acid form, which must be initially dissolved in strong base.

Each Fluo-4 Direct™ Calcium Assay Kit provides:

- Fluo-4 calcium assay reagent
- Water-soluble probenecid
- Fluo-4 Direct™ calcium assay buffer (only provided with the Starter pack F10471 and the Surveyor pack F10472)
- Detailed protocols

Sufficient material is supplied for 20 × 96- or 384-well plates (F10471, F10472) or 200 × 96- or 384-well plates (F10473). The Fluo-4 Direct™ Calcium Assay Kits are designed for microplates and high-throughput screening (HTS), and the assay can be performed on adherent as well as nonadherent cells. Water-soluble probenecid is also available separately (P36400, Section 19.8).

The Fluo-4 NW Calcium Assay Kits (F36205, F36206) offer a proprietary assay formulation that requires neither a wash step nor a quencher dye. The fluo-4 NW indicator is nonfluorescent and stable in pH 7–7.5 buffer for several hours, so spontaneous conversion to the Ca^{2+}-sensitive form is not a significant source of background fluorescence. Contributions to baseline fluorescence by the growth medium (e.g., esterase activity, proteins interacting with receptors of interest, or phenol red) are eliminated by removing the medium prior to adding the indicator dye to the wells. A water-soluble form of probenecid is provided to inhibit extrusion of the indicator out of the cell by organic anion transporters.[33,34]

Each Fluo-4 NW Calcium Assay Kit provides:

- Fluo-4 NW dye mix
- Water-soluble probenecid
- Assay buffer (only provided in the Fluo-4 NW Calcium Assay Kit starter pack with buffer, F36206)
- Detailed protocols

The Fluo-4 NW Calcium Assay Kit starter pack with buffer (F36206) contains enough materials for 10 microplates and includes assay buffer; the Fluo-4 NW Calcium Assay Kit for high-throughput assays (F36205) contains enough materials for 100 microplates and does not include the assay buffer. These kits are designed for microplates and HTS, and the assay can be performed on adherent as well as nonadherent cells.

Rhod-2 and X-Rhod-1

The long wavelength Ca^{2+} indicators rhod-2 (Figure 19.3.13, Figure 19.3.14) and X-rhod-1 are valuable for measuring Ca^{2+} in cells and tissues that have high levels of autofluorescence[35,36] and also for detecting Ca^{2+} release generated by photoreceptors and photoactivatable chelators.[37,38] Our chemists have optimized the purification of rhod-2, yielding a highly purified preparation that shows greater than 100-fold enhancement in fluorescence upon binding Ca^{2+} (Figure 19.3.15). The K_d for Ca^{2+} of rhod-2 in the absence of Mg^{2+} has been determined to be 570 nM (measured at 22°C using our Calcium Calibration Buffer Kits), which is considerably lower than that cited in the original paper on rhod-2.[1] Rhod-2 is available as a cell-impermeant potassium salt (R14220) or as a cell-permeant AM ester in either a 1 mg vial (R1244) or specially packaged as a set of 20 vials, each containing 50 µg (R1245MP).

X-rhod-1 is a Ca^{2+} indicator with excitation/emission maxima of ~580/602 nm and a K_d for Ca^{2+} of 700 nM (measured at 22°C using our Calcium Calibration Buffer Kits). It has spectral characteristics that are similar to our Calcium Crimson™ indicator, but the fluorescence response of X-rhod-1 is much more sensitive to Ca^{2+} binding (Figure 19.3.16). The long-wavelength emission characteristics of X-rhod-1 allow simultaneous detection Ca^{2+} transients and Green Fluorescent Protein (GFP) with minimal crosstalk[39] (Using Organic Fluorescent Probes in Combination with GFP—Note 12.1). X-rhod-1 is available as a cell-permeant AM ester (X14210), specially packaged as a set of 10 vials, each containing 50 µg.

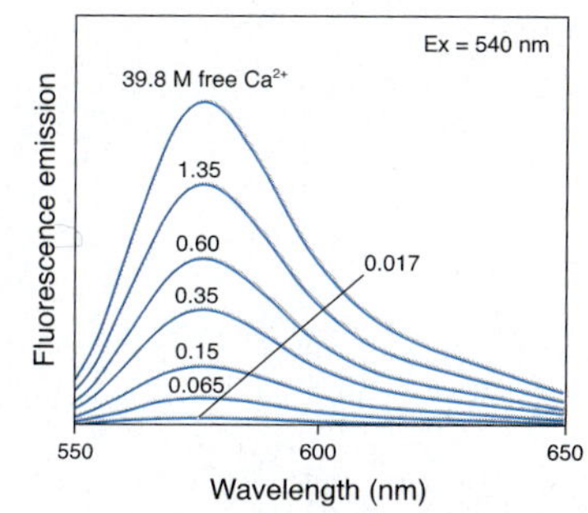

Figure 19.3.15 Fluorescence emission spectra of rhod-2 (R14220) in solutions containing 0–39.8 µM free Ca^{2+}. The spectrum for the Ca^{2+}-free solution is indistinguishable from the baseline.

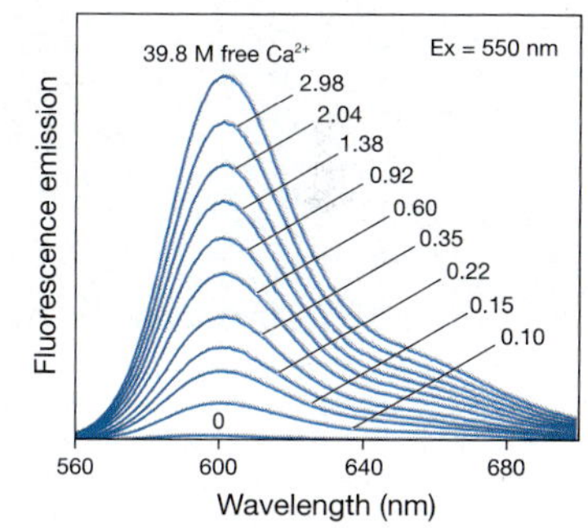

Figure 19.3.16 Fluorescence emission spectra of the tripotassium salt of X-rhod-1 in solutions containing 0–39.8 µM free Ca^{2+}.

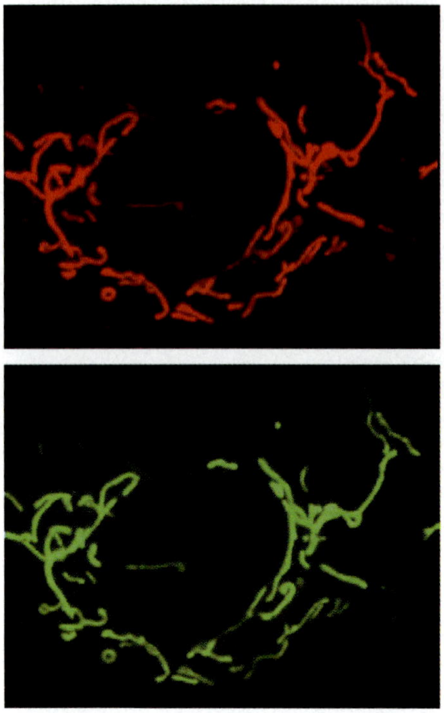

Figure 19.3.17 Colocalization of fluorescent staining by rhod-2 AM (R1244, R1245MP; upper panel) and mitochondrion-selective MitoFluor™ Green stain (lower panel) in an adult rat cortical astrocyte. Cells were simultaneously loaded with rhod-2 AM (4.5 µM) and the MitoFluor™ Green stain (20 nM) for 30 minutes at 22°C. Confocal laser-scanning microscopy using 488 nm excitation and spectrally resolved detection, at 505–530 nm for MitoFluor™ Green stain and ≥585 nm for rhod-2, shows almost identical staining distribution. The images were contributed by Michael Duchen, University College, London. Reproduced with permission from J Cell Biol (1999) 145:795.

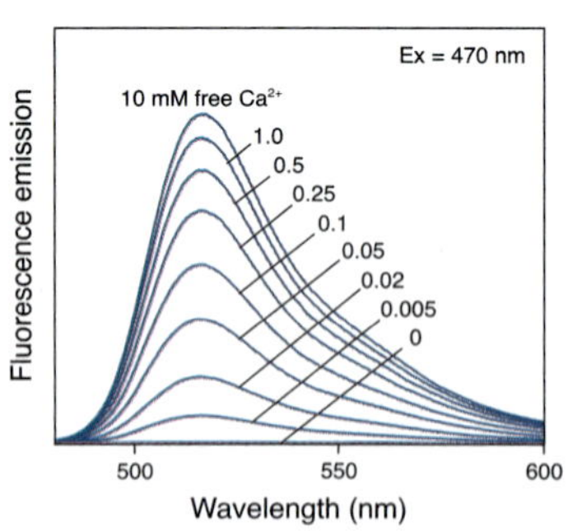

Figure 19.3.18 Ca^{2+}-dependent fluorescence emission spectra of fluo-5N (F14203).

Rhod-3 Imaging Kit

As compared with other red-fluorescent calcium dyes such as rhod-2 AM, rhod-3 AM is an improved red-shifted calcium indicator that displays a more uniform cytosolic distribution and improved signal. The cationic nature of rhod-2 AM results in potential-driven, subcellular localization (Figure 19.3.17). Imaging studies with the rhod-3 AM, however, show minimal subcellular localization. In the presence of PowerLoad™ concentrate and probenecid, the cell-permeant, nonfluorescent rhod-3 AM can be passively loaded into the cells, where intracellular esterases cleave the dye to the cell-impermeant, active form that fluoresces upon Ca^{2+} binding. Rhod-3 exhibits a large increase (>2.5 fold) in fluorescence upon binding Ca^{2+} and very low fluorescence in the absence of Ca^{2+} binding. Rhod-3 has a K_d of 570 nM for Ca^{2+} (determined at 22°C in 30 mM MOPS, pH 7.2 with 100 mM KCl).

We offer the Rhod-3 Imaging Kit (R10145), which provides:

- Rhod-3 AM
- Dimethylsulfoxide (DMSO)
- PowerLoad™ concentrate
- Water-soluble probenecid
- Detailed protocols

Sufficient reagents are supplied for 10 assays using the protocol provided. PowerLoad™ concentrate, an optimized formulation of nonionic, Pluronic® surfactant polyols, aids the solubilization of rhod-3 AM dye in physiological media and is available separately (P10200, Section 19.8). Probenecid, also available separately (P36400, Section 19.8), inhibits organic anion pumps that actively extrude the de-esterified dye.[33,34]

Lower-Affinity Calcium Indicators Based on Fluo-3 and Rhod-2

With Ca^{2+} dissociation constants well above 1 µM, our relatively low-affinity Ca^{2+} indicators can be used to detect intracellular Ca^{2+} levels in the micromolar range—levels that would saturate the response of fluo-3 and rhod-2. Such elevated Ca^{2+} levels are generated by mobilization of intracellular Ca^{2+} stores[40] and by excitatory stimulation of smooth muscle[41,42] and neurons.[43,44] Moreover, low-affinity indicators have faster ion dissociation rates, making them more suitable for tracking the kinetics of rapid Ca^{2+} fluxes than indicators with K_d values <1 µM.[41,42]

Fluo-5F, Fluo-4FF, Fluo-5N and Mag-Fluo-4

Fluo-5F, fluo-4FF, fluo-5N and mag-fluo-4[45] are analogs of fluo-4 with much lower Ca^{2+}-binding affinity, making them suitable for detecting intracellular Ca^{2+} levels in the 1 µM to 1 mM range. Fluo-5F, fluo-4FF, fluo-5N and mag-fluo-4 have K_d values for Ca^{2+} of ~2.3 µM, ~9.7 µM, ~90 µM and ~22 µM, respectively, as compared with fluo-4, which has a K_d for Ca^{2+} of ~345 nM (measured at 22°C using our Calcium Calibration Buffer Kits) (Figure 19.3.1). The temperature dependence of the K_d for fluo-5F has been reported.[15] These low Ca^{2+}-binding affinities are ideal for detecting high concentrations of Ca^{2+} in the endoplasmic reticulum[46,47] and neurons (Figure 19.3.2), as well as for tracking Ca^{2+} flux kinetics.[45] Fluo-5N has been found to be particularly useful for monitoring calcium oscillations in the sarcoplasmic reticulum.[46,48,49] Like fluo-4, these indicators are essentially nonfluorescent in the absence of divalent cations and exhibit strong fluorescence enhancement with no spectral shift upon binding Ca^{2+} (Figure 19.3.18). Because mag-fluo-4 is less Ca^{2+}/Mg^{2+} selective than fluo-5N, it is also useful as an indicator for intracellular Mg^{2+} levels (Section 19.6). Fluo-5F (F14221, F14222), fluo-4FF (F23980, F23981), fluo-5N (F14203, F14204) and mag-fluo-4 (M14205, M14206) are available as cell-impermeant potassium salts or as cell-permeant AM esters. The AM esters are specially packaged as a set of 10 vials, each containing 50 µg.

IMPORTANT NOTICE: The products described in this manual are covered by one or more Limited Use Label License(s). Please refer to the Appendix on page 971 and Master Product List on page 975. Products are For Research Use Only. Not intended for any animal or human therapeutic or diagnostic use.
molecular probes® | invitrogen™ by life technologies™

Rhod-5N

Rhod-5N (Figure 19.3.13) has a lower binding affinity for Ca^{2+} than any other BAPTA-based indicator (K_d = ~320 µM, measured at 22°C using our Calcium Calibration Buffer Kits). These properties confer suitability for Ca^{2+} detection in environmental[50] and food[51] testing applications. Like the parent rhod-2 indicator, rhod-5N is essentially nonfluorescent in the absence of divalent cations and exhibits strong fluorescence enhancement with no spectral shift upon binding Ca^{2+}. Furthermore, rhod-5N has very little detectable response to Mg^{2+} concentrations up to at least 100 mM. Rhod-5N is available as a cell-impermeant potassium salt (R14207).

Rhod-FF and X-Rhod-5F

The fluorinated analogs of rhod-2—rhod-FF (Figure 19.3.13) and X-rhod-5F[52,53]—have intermediate Ca^{2+} sensitivity relative to rhod-2 and rhod-5N. Their Ca^{2+} dissociation constants (K_d) are 19 µM and 1.6 µM, respectively (measured at 22°C using our Calcium Calibration Buffer Kits). X-rhod-5F is available as a water-soluble potassium salt (X23984), and rhod-FF and X-rhod-5F are available as cell-permeant AM esters (R23983, X23985). The AM esters are specially packaged as a set of 10 vials, each containing 50 µg.

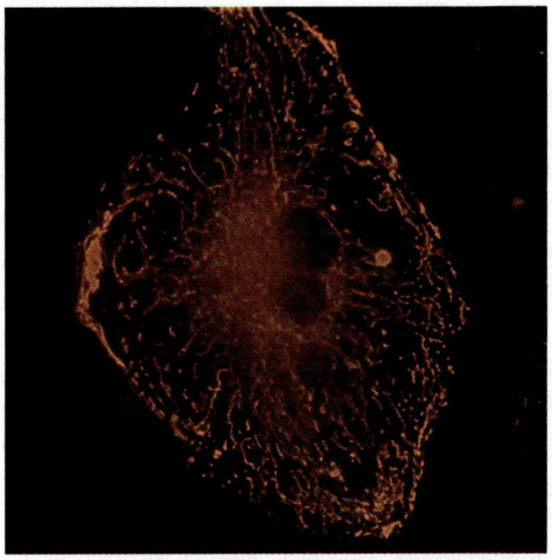

Figure 19.3.19 A live bovine pulmonary artery endothelial cell stained with 500 nM X-rhod-1 AM (X14210) for 30 minutes at 30°C. The image was deconvolved using Huygens software (Scientific Volume Imaging, www.svi.nl).

Indicators of Mitochondrial Ca^{2+} Transients

Our AM esters of rhodamine-based indicators include:

- Rhod-2 (R1244, R1245MP)
- Rhod-FF (R23983)
- X-rhod-1 (X14210)
- X-rhod-5F (X23985)

These rhodamine derivatives form a set of cell-permeant Ca^{2+} indicators with a net positive charge. This property promotes their sequestration into mitochondria in some cells, most likely via membrane potential–driven uptake, and results in a staining pattern that is characteristic of mitochondria (Figure 19.3.14, Figure 19.3.17, Figure 19.3.19). Mitochondria have a high capacity for Ca^{2+} uptake and therefore require low-affinity Ca^{2+} indicators to accurately measure internal Ca^{2+} concentrations. Because the range of Ca^{2+} concentrations that can be detected using rhod-2 AM is limited to within about one order-of-magnitude above and below its Ca^{2+} dissociation constant (K_d = 570 nM), we offer a selection of rhod-2 analogs with Ca^{2+} dissociation constants up to 320 µM (Table 19.1, Figure 19.3.13).

The extent of mitochondrial versus cytosolic localization is influenced by the temperature and incubation time used for AM ester loading.[54–57] By reducing rhod-2 AM to the colorless, nonfluorescent dihydrorhod-2 AM, the discrimination between cytosolic and mitochondrially localized dye can be further improved.[58–61] The AM ester of dihydrorhod-2 exhibits Ca^{2+}-dependent fluorescence only after it is oxidized and its AM esters are cleaved to yield the rhod-2 indicator, processes that occur rapidly in the mitochondrial environment.[62] A detailed protocol for reducing rhod-2 AM to generate dihydrorhod-2 AM is available. The procedure should also be suitable for reduction of the AM esters of the other rhod indicators. Co-loading of cells with rhod-2 AM or X-rhod-1 AM in combination with green-fluorescent Ca^{2+} indicators such as fura-2, fluo-3, fluo-4, Calcium Green™-1 and Oregon Green® 488 BAPTA-1 enables simultaneous two-color imaging of mitochondrial and cytoplasmic Ca^{2+}[63–66] (Figure 19.3.20). Mitochondrial Ca^{2+} and mitochondrial NADH can be simultaneously measured using rhod-2 and intrinsic NADH fluorescence, respectively.[67]

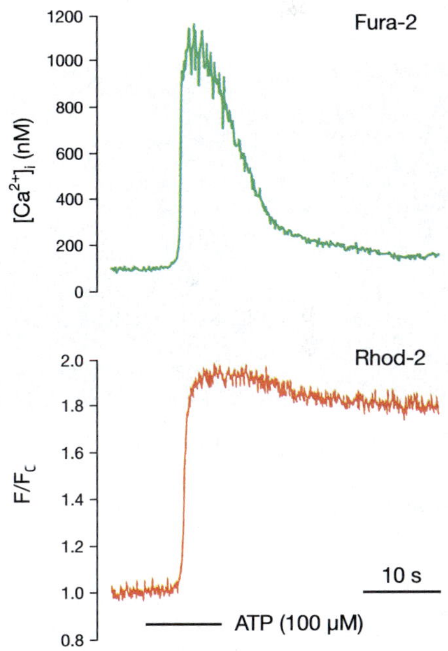

Figure 19.3.20 Simultaneous measurement of intracellular and mitochondrial Ca^{2+} in pulmonary artery smooth muscle cells. Cells were loaded with 2 µM rhod-2 AM (R1244, R1245MP) for 60 minutes at 22°C and were subsequently patch-clamped for electrophysiological measurements. Fura-2 (50 µM; F1200, F6799) was loaded by infusion from the patch pipette; at the same time, residual cytosolic rhod-2 was dialyzed out. Fluorescence excitation was rapidly alternated between 340 nm and 380 nm (for fura-2) and 500 nm (for rhod-2). The Ca^{2+} transient was detected following application of 100 µM ATP for 10 seconds. The observed increases in mitochondrial Ca^{2+} may provide a feedback mechanism for matching ATP supply and demand by activating Ca^{2+}-dependent dehydrogenases in the ATP synthesis pathway. This figure was reproduced with permission from J Physiol (1999) 516:139.

IMPORTANT NOTICE: The products described in this manual are covered by one or more Limited Use Label License(s). Please refer to the Appendix on page 971 and Master Product List on page 975. Products are For Research Use Only. Not intended for any animal or human therapeutic or diagnostic use.

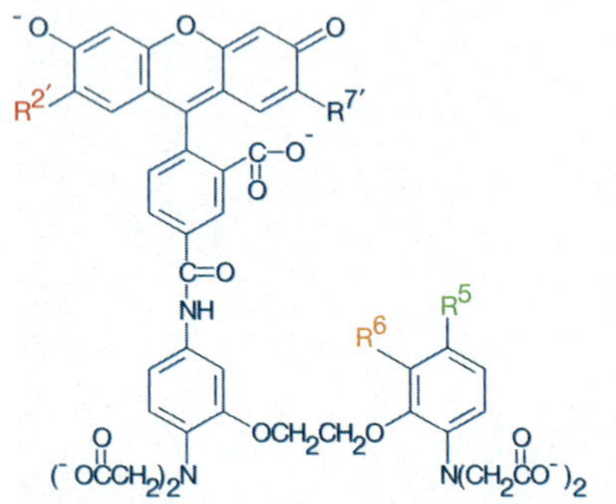

Indicator	$K_d(Ca^{2+})$	R2′	R7′	R5	R6
Calcium Green™-1	0.19 µM	Cl	Cl	H	H
Calcium Green™-5N	14 µM	Cl	Cl	NO_2	H
Oregon Green® 488 BAPTA-1	0.17 µM	F	F	H	H
Oregon Green® 488 BAPTA-6F	3 µM	F	F	H	F
Oregon Green® 488 BAPTA-5N	20 µM	F	F	NO_2	H

Figure 19.3.21 Calcium Green™ and Oregon Green® 488 BAPTA indicators with varying Ca^{2+} affinities.

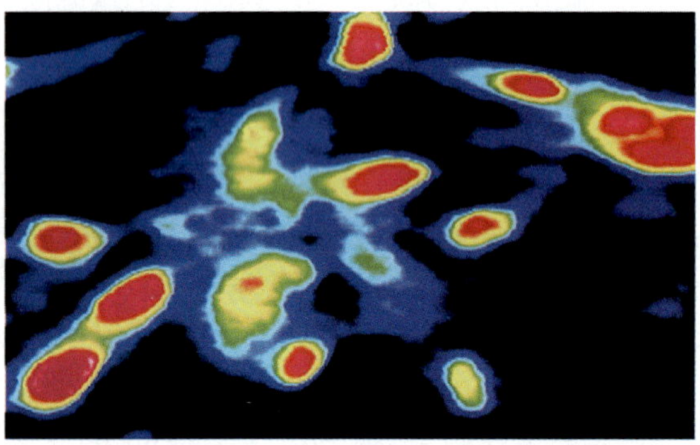

Figure 19.3.22 Brain cells from mediobasal hypothalamus stained with Calcium Green™-1, AM (C3011MP, C3012). The cells are responding to the neurotransmitter, glutamate, with an increase in cytoplasmic Ca^{2+} concentration as shown by this pseudocolored image. The image was contributed by Anthony N. van den Pol, Yale University School of Medicine.

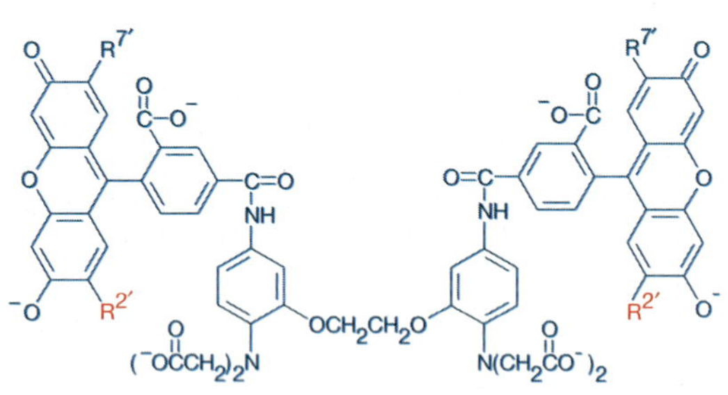

Indicator	$K_d(Ca^{2+})$	R2′	R7′
Calcium Green™-2	0.55 µM	Cl	Cl
Oregon Green® 488 BAPTA-2	0.58 µM	F	F

Figure 19.3.23 Calcium Green™-2 and Oregon Green® 488 BAPTA-2 Ca^{2+} indicators.

Calcium Green™, Calcium Orange™ and Calcium Crimson™ Indicators

Calcium Green™-1 and Calcium Green™-2 Indicators

Molecular Probes® Calcium Green™-1 (Figure 19.3.21, Figure 19.3.22) and Calcium Green™-2 (Figure 19.3.23) indicators, as well as Calcium Orange™ and Calcium Crimson™ indicators (described below), are visible light–excitable indicators developed in our laboratories. Like fluo-3 and fluo-4, the Calcium Green™ indicators exhibit an increase in fluorescence emission intensity upon binding Ca^{2+} with little shift in wavelength[68,69] (Figure 19.3.24, Figure 19.3.25); the fluorescence spectra of the Calcium Green™ indicators are almost identical to those of fluo-3. Further comparison of the Calcium Green™ indicators and fluo-3 reveals that, at high Ca^{2+} levels, Calcium Green™-1 and Calcium Green™-2 are several times brighter than fluo-3 (Figure 19.3.5). Calcium Green™-1 has a quantum yield of 0.75 at saturating Ca^{2+} concentrations,[70] as compared with about 0.14 for fluo-3.[1]

The Calcium Green™ indicators have several other important features:

- Calcium Green™-1 is more fluorescent in resting cells than is fluo-3 (Figure 19.3.5), which increases the visibility of unstimulated cells, facilitates the determination of baseline fluorescence and makes calculations of intracellular Ca^{2+} concentrations more reliable.
- Calcium Green™-1 has been a preferred indicator for multiphoton excitation imaging of Ca^{2+} in living tissues.[4,71–73]
- Calcium Green™-1 is useful for measuring intracellular Ca^{2+} by fluorescence lifetime imaging (FLIM).[74]
- The Ca^{2+} affinity of Calcium Green™-1 in the absence of Mg^{2+} (K_d for Ca^{2+} = 190 nM) is higher than that of fluo-3 (K_d for Ca^{2+} = 390 nM) or Calcium Green™-2 (K_d for Ca^{2+} = 550 nM, measured at 22°C using our Calcium Calibration Buffer Kits).
- Like fluo-3 and fluo-4, Calcium Green™-2 is essentially nonfluorescent in the absence of Ca^{2+} and exhibits an approximately 100-fold increase in emission intensity upon Ca^{2+} binding, which leads to a very large dynamic range.
- Like fluo-3 AM and fluo-4 AM, the AM esters of the Calcium Green™ indicators are nonfluorescent.

Furthermore, the Calcium Green™ indicators are less phototoxic to cells than fluo-3.[75,76] This observation stems at least in part from the fact that the Calcium Green™ indicators are intrinsically more fluorescent than fluo-3, thus requiring lower illumination intensities and lower dye concentrations to achieve the same signal[77] (Figure 19.3.5).

Simultaneous loading of Calcium Green™-2 and carboxy SNARF®-1 AM, acetate (C1271, C1272; Section 20.2) enabled researchers to make ratiometric

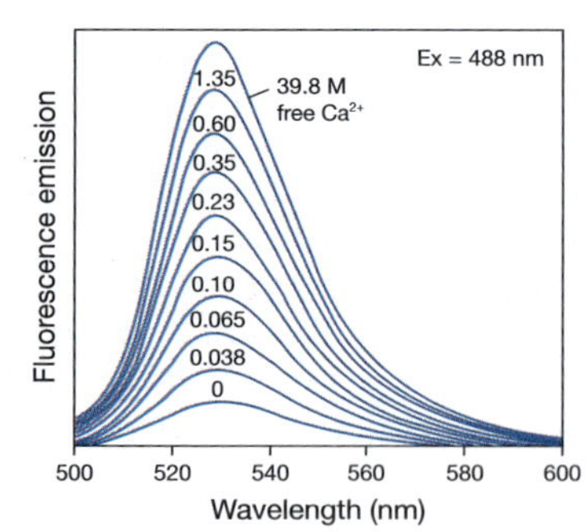

Figure 19.3.24 Ca^{2+}-dependent fluorescence emission spectra of the Calcium Green™-1 indicator (C3010MP).

IMPORTANT NOTICE: The products described in this manual are covered by one or more Limited Use Label License(s). Please refer to the Appendix on page 971 and Master Product List on page 975. Products are For Research Use Only. Not intended for any animal or human therapeutic or diagnostic use.

measurements of intracellular Ca^{2+} in cardiac myocytes.[77] A variation on this approach has been reported by Oheim and co-workers in which Calcium Green™-1 and a Ca^{2+}-insensitive reference dye similar to lucifer yellow CH (L453, L12926; Section 14.3) were loaded into bovine adrenal chromaffin cells by whole-cell patch clamping.[78] Ratio images were obtained using 420/488 nm dual excitation (emission >515 nm) and exhibited superior signal-to-noise characteristics, as compared with conventional UV-excited fura-2 images. This technique should be equally applicable to other 488 nm–excited Ca^{2+} indicators such as fluo-3, fluo-4, fluo-5N and the Oregon Green® 488 BAPTA series. Calcium Green™-1 and Calcium Green™-2 are available as cell-impermeant potassium salts (C3010MP, C3730) or as cell-permeant AM esters (C3011MP, C3012, C3732).

Calcium Orange™ and Calcium Crimson™ Indicators

Like the Calcium Green™ indicators, Calcium Orange™ (C3013, Figure 19.3.26) and Calcium Crimson™ exhibit an increase in fluorescence emission intensity upon binding to Ca^{2+} with little shift in wavelength (Figure 19.3.27, Figure 19.3.28) and can be loaded into cells as their AM esters (C3015, C3018). Both the Calcium Orange™ and Calcium Crimson™ indicators are more photostable than either fluo-3 or the Calcium Green™ indicators.[16]

Calcium Orange™ has an excitation maximum near 550 nm and is compatible with standard tetramethylrhodamine optical filters. Calcium Orange™ has been used to monitor Ca^{2+} in intact photoreceptors containing a genetically altered rhodopsin pigment,[38] as well as to follow Ca^{2+} influx and release in hippocampal astrocytes.[79]

The excitation maximum (~590 nm) of Calcium Crimson™ makes it useful in situations where interference by cellular autofluorescence is problematic.[80] Selective uptake of Calcium Crimson™ AM (C3018) at *Drosophila* motor nerve terminals has been used to obtain images of stimulus-dependent Ca^{2+} influx.[81,82]

Calcium Green™-5N and Magnesium Green™: Low-Affinity Ca^{2+} Indicators

The Ca^{2+} indicators Calcium Green™-5N (Figure 19.3.21) and Magnesium Green™ (Figure 19.3.29) have dissociation constants for Ca^{2+} in the absence of Mg^{2+} of ~14 µM and ~6 µM, respectively (measured at 22°C using our Calcium Calibration Buffer Kits) (Figure 19.3.30). These low-affinity Ca^{2+} indicators exhibit relatively little fluorescence, except in cells in which high-amplitude Ca^{2+} influx or release is occurring.[83,84] Calcium Green™-5N and Magnesium Green™ buffer intracellular Ca^{2+} to a lesser extent than do the higher-affinity Ca^{2+} indicators. Furthermore, the high Ca^{2+} dissociation rates of these indicators are advantageous for tracking rapid Ca^{2+}-release kinetics.[85,86] Use of the low-affinity Ca^{2+} indicator Calcium Green™-5N in combination with the higher-affinity indicator Calcium Green™-2 in the same experimental protocol can give an indication of the absolute magnitude of Ca^{2+} spikes.[87] Furthermore, coinjection of Ca^{2+}-sensitive Calcium Green™-5N and Ca^{2+}-insensitive 8-aminonaphthalene-1,3,6-trisulfonic acid (ANTS, A350; Section 14.3) into *Limulus* ventral nerve photoreceptors permitted ratiometric measurement of Ca^{2+} flux.[88]

Calcium Green™-5N is available as a cell-impermeant potassium salt (C3737) or as a cell-permeant AM ester (C3739). Magnesium Green™, which is also discussed in Section 19.6 with the other Mg^{2+} indicators, is available as a cell-impermeant potassium salt (M3733) or as a cell-permeant AM ester (M3735).

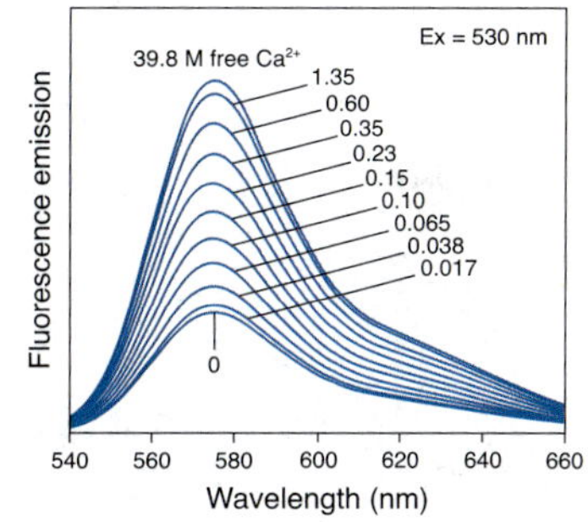

Figure 19.3.27 Ca^{2+}-dependent fluorescence emission spectra of the Calcium Orange™ indicator (C3013).

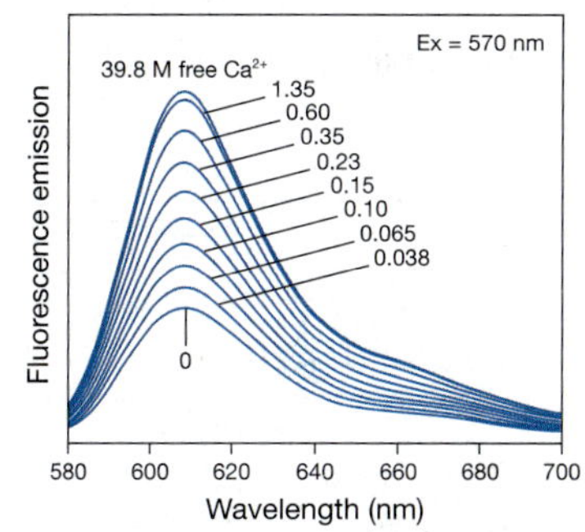

Figure 19.3.28 Ca^{2+}-dependent fluorescence emission spectra of the Calcium Crimson™ indicator (available as the cell-permeant acetoxymethyl (AM) ester, C3018).

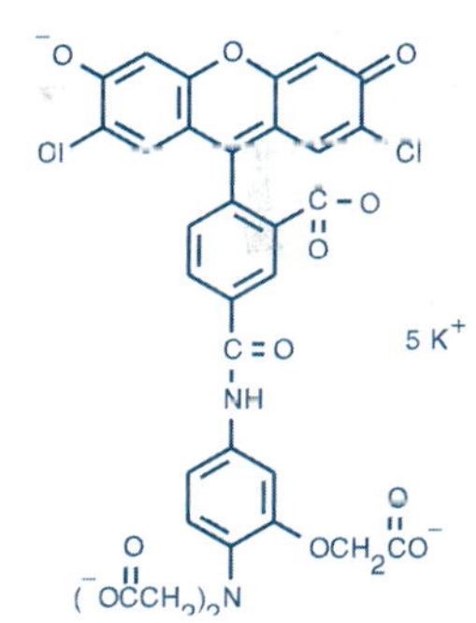

Figure 19.3.29 Magnesium Green™, pentapotassium salt (M3733).

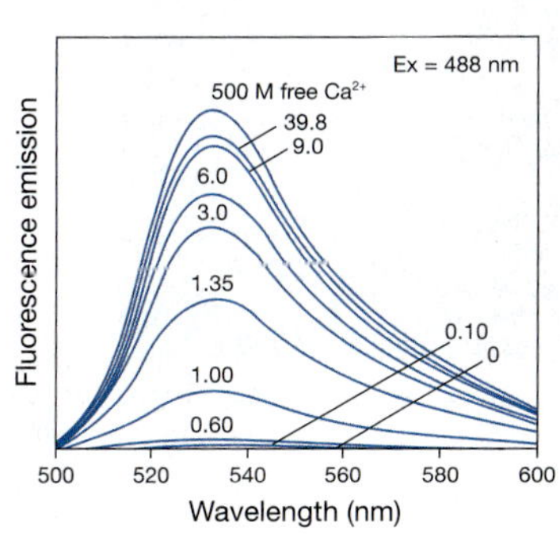

Figure 19.3.25 Ca^{2+}-dependent fluorescence emission spectra of the Calcium Green™-2 indicator (C3730).

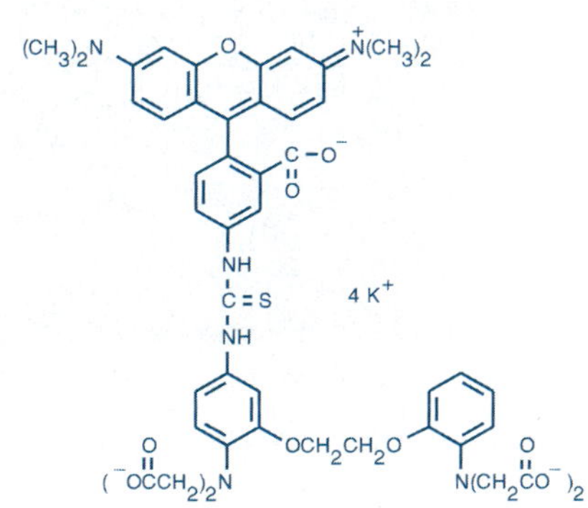

Figure 19.3.26 Calcium Orange™, tetrapotassium salt (C3013).

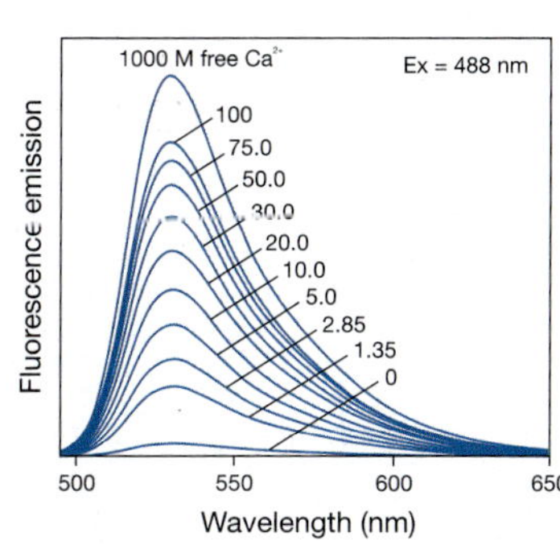

Figure 19.3.30 Ca^{2+}-dependent fluorescence emission spectra of the Calcium Green™-5N indicator (C3737).

IMPORTANT NOTICE: The products described in this manual are covered by one or more Limited Use Label License(s). Please refer to the Appendix on page 971 and Master Product List on page 975. Products are For Research Use Only. Not intended for any animal or human therapeutic or diagnostic use.

Oregon Green® 488 BAPTA Indicators

Oregon Green® 488 BAPTA indicators are based on our exceptionally bright Oregon Green® 488 dyes (Section 1.5). The absorptivity of Oregon Green® 488 BAPTA-1 (Figure 19.3.21) at 488 nm is ~93% of its peak value, whereas the absorptivity of fluo-3 and the Calcium Green™ indicators at 488 nm is only ~45% of their maxima. Consequently, the Oregon Green® 488 BAPTA indicators are more efficiently excited by the 488 nm spectral line of the argon-ion laser than are the fluo-3 and Calcium Green™ indicators.

Oregon Green® 488 BAPTA-1 and Oregon Green® 488 BAPTA-2

The spectral properties of the Oregon Green® 488 BAPTA indicators permit the use of lower dye concentrations when using argon-ion laser excitation sources, making the Oregon Green® 488 BAPTA indicators well suited for intracellular Ca^{2+} measurements by confocal laser-scanning microscopy.[89,90] Oregon Green® 488 BAPTA-1 is also recommended for fluorescence lifetime imaging[91] (FLIM) and applications involving photoactivatable ("caged") probes[92,93] (Figure 19.3.31).

Furthermore, the quantum yields of the Ca^{2+} complexes of Oregon Green® 488 BAPTA-1 and Calcium Green™-1 are ~0.7, as compared with only ~0.14 for fluo-3. As with Calcium Green™-1 (Figure 19.3.5), Oregon Green® 488 BAPTA-1 is moderately fluorescent in Ca^{2+}-free solution, and its fluorescence is enhanced about 14-fold at saturating Ca^{2+} (Figure 19.3.32). In some cases, it may be advantageous to mix Oregon Green® 488 BAPTA-1 and fluo-4 indicators to obtain a calcium response that combines a finite basal signal level with a large stimulus-dependent increase.[94] Oregon Green® 488 BAPTA-1 has a K_d for Ca^{2+} in the absence of Mg^{2+} of about 170 nM. Oregon Green® 488 BAPTA-2 is similar to Calcium Green™-2 in that it contains two dye molecules per BAPTA chelator (Figure 19.3.23) and exhibits very low fluorescence in the absence of Ca^{2+}. The fluorescence of Oregon Green® 488 BAPTA-2 is enhanced at least 37-fold at saturating Ca^{2+}, and it has a K_d for Ca^{2+} in the absence of Mg^{2+} of ~580 nM (Figure 19.3.33) (K_d values for Ca^{2+} are determined at ~22°C using our Calcium Calibration Buffer Kits).

Several research groups are exploiting the power of two-photon laser-scanning microscopy and fluorescent Ca^{2+} indicators for functional imaging of neurons in brain slice preparations[95,96] and intact live brains.[8,97–99] Oregon Green® 488 BAPTA-1 and Calcium Green™-1, excited in the wavelength range 800–850 nm are the indicators of choice in many of these investigations.[100–106] The Oregon Green® 488 BAPTA-2 Ca^{2+} indicator and red-fluorescent, Ca^{2+}-insensitive sulforhodamine 101 reference dye (S359, Section 14.3) have been loaded together into optic nerves to image Ca^{2+} distribution in axons.[90]

Oregon Green® 488 BAPTA-1 and Oregon Green® 488 BAPTA-2 are available as cell-impermeant potassium salts (O6806, O6808) or as cell-permeant AM esters (O6807, O6809), which are specially packaged as a set of 10 vials, each containing 50 µg. A 10,000 MW dextran conjugate of the Oregon Green® BAPTA-1 indicator (O6798) is described in Section 19.4.

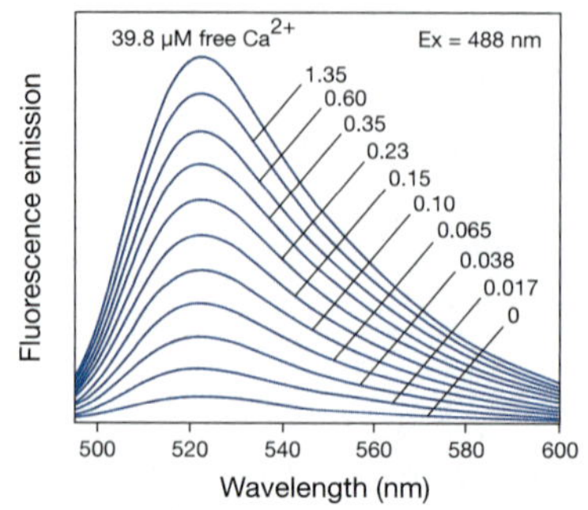

Figure 19.3.32 Ca^{2+}-dependent fluorescence emission spectra of the Oregon Green® 488 BAPTA-1 indicator (O6806).

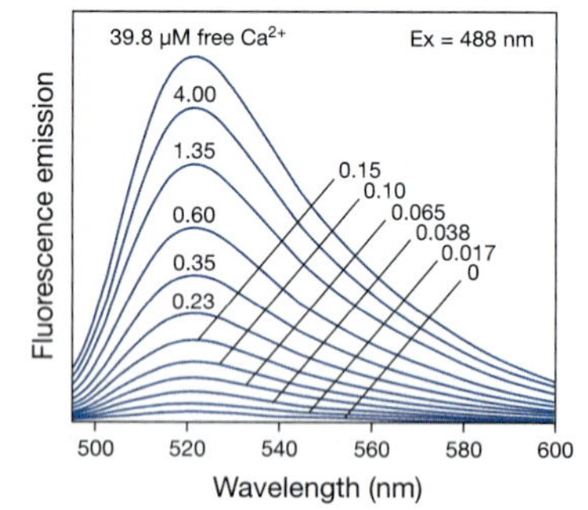

Figure 19.3.33 Ca^{2+}-dependent fluorescence emission spectra of the Oregon Green® 488 BAPTA-2 indicator (O6808).

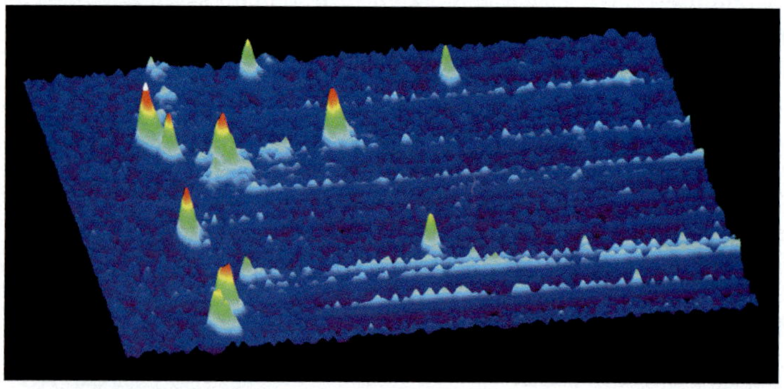

Figure 19.3.31 Confocal linescan image of calcium "puffs" in a *Xenopus* oocyte. Oregon Green® 488 BAPTA-1 (O6806) was used as the calcium indicator and Ca^{2+} liberation was evoked by flash photolysis of NPE-caged Ins 1,4,5-P_3 (I23580). Image contributed by Ian Parker and Nick Callamaras, University of California at Irvine.

IMPORTANT NOTICE: The products described in this manual are covered by one or more Limited Use Label License(s). Please refer to the Appendix on page 971 and Master Product List on page 975. Products are For Research Use Only. Not intended for any animal or human therapeutic or diagnostic use.
www.invitrogen.com/probes

Oregon Green® 488 BAPTA-6F and Oregon Green® 488 BAPTA-5N

Oregon Green® 488 BAPTA-6F (Figure 19.3.21) and Oregon Green® 488 BAPTA-5N (Figure 19.3.21) are low-affinity Ca^{2+} indicators (K_d for Ca^{2+} ~3 µM and ~20 µM, respectively, measured at ~22°C using our Calcium Calibration Buffer Kits) designed for measuring intracellular Ca^{2+} levels above 1 µM.[107–110] By simultaneously imaging Oregon Green® 488 BAPTA-5N and rhod-2, researchers have been able to compare cytosolic and mitochondrial Ca^{2+} responses to action potentials in amphibian motor nerve terminals.[61,111] Oregon Green® 488 BAPTA-6F and Oregon Green® BAPTA-5N are available as cell-impermeant potassium salts (O23990, O6812).

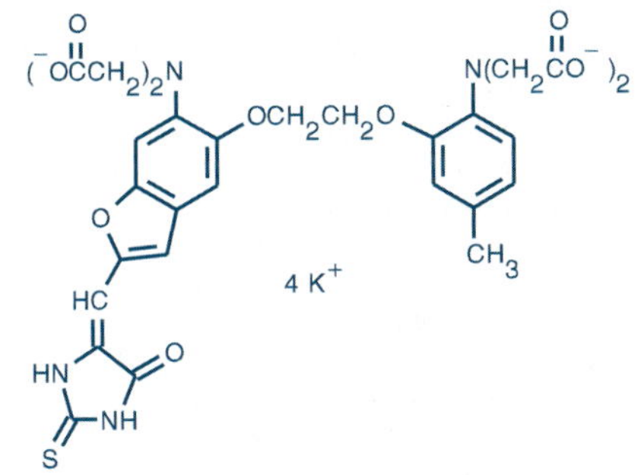

Figure 19.3.34 Fura Red™, tetrapotassium salt (F14219).

Fura Red™ Indicator

Fura Red™ indicator (Figure 19.3.34), a visible light–excitable fura-2 analog, offers unique possibilities for ratiometric measurement of Ca^{2+} in single cells by microphotometry, imaging or flow cytometry. The visible-wavelength excitation (450–500 nm) and very long-wavelength emission maximum (~660 nm) of the Fura Red™ indicator (Figure 19.3.35) minimize interference from autofluorescence and pigmentation in tissues and biological fluids.[112] Fluorescence of the Fura Red™ indicator excited at 488 nm *decreases* once the indicator binds Ca^{2+} (Figure 19.3.35). Even in the absence of Ca^{2+}, fluorescence of the Fura Red™ indicator is much weaker than that of the other visible light–excitable Ca^{2+} indicators, necessitating use of higher concentrations of the indicator in cells to produce equivalent fluorescence. Fura Red™ indicator is available as either a cell-impermeant tetrapotassium salt (F14219) or as a cell-permeant AM ester (F3020, F3021).

Ratiometric measurements of intracellular Ca^{2+} levels with Fura Red™ indicator have been made using excitation wavelengths of 420 nm and 480 nm[113] or 457 nm and 488 nm.[114] A simultaneous assay for Ca^{2+} uptake and ATP hydrolysis by sarcoplasmic reticulum has been developed that uses the large *absorbance* change of Fura Red™ indicator upon Ca^{2+} binding.[115] This assay can measure Ca^{2+} uptake—and probably uptake of heavy metal ions through channels—from the medium in real time without the use of radioactive Ca^{2+}, making it generally useful for measuring Ca^{2+} uptake by cells. In several cell types, simultaneous labeling with Fura Red™ indicator and fluo-3 (Figure 19.3.7) has enabled researchers to use ratiometric measurements for estimating intracellular Ca^{2+} levels using confocal laser-scanning microscopy[19,116–119] or flow cytometry.[20,21]

Furthermore, the huge Stokes shift of Fura Red™ indicator permits multicolor analysis of Fura Red™ indicator fluorescence in combination with fluorescein or fluorescein-like dyes using only a single excitation wavelength (Figure 19.3.7). For example, researchers have been able to simultaneously measure Ca^{2+} fluxes and oxidative bursts in monocytes and granulocytes by simultaneously measuring the fluorescence of Fura Red™ indicator and rhodamine 123—the oxidation product of the probe dihydrorhodamine 123[120] (D632, D23806; Section 18.2). Fura Red™ indicator has also been used in combination with blue-fluorescent protein in transfected cells.[39]

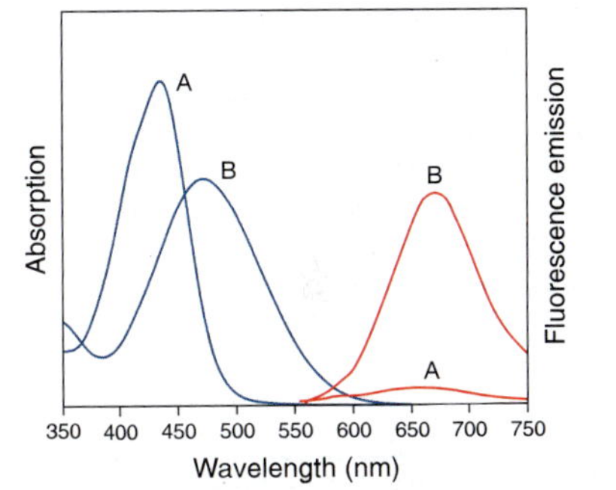

Figure 19.3.35 Absorption and fluorescence emission (excited at 488 nm) spectra of Ca^{2+}-saturated (**A**) and Ca^{2+}-free (**B**) Fura Red™ in pH 7.2 buffer.

Calcein

Calcein (C481, Section 19.7) is a relatively low-affinity Ca^{2+} chelator. The dissociation constants for both the Ca^{2+} and Mg^{2+} complexes of calcein at physiological pH are about 10^{-3} to 10^{-4} M.[121] As a derivative of iminodiacetic acid (Figure 19.3.36), this dye exhibits an ion affinity that increases considerably at higher pH, and thus it is not particularly useful for measuring Ca^{2+} or Mg^{2+} in cells. Instead, it is primarily useful for detecting mineralized Ca^{2+}, particularly in the context of bone[122–124] and bacterial spores. Calcein is also widely used for fluorescence quenching–based detection of Ni^{2+}, Co^{2+}, Cu^{2+} and Fe^{3+} (Section 19.7).

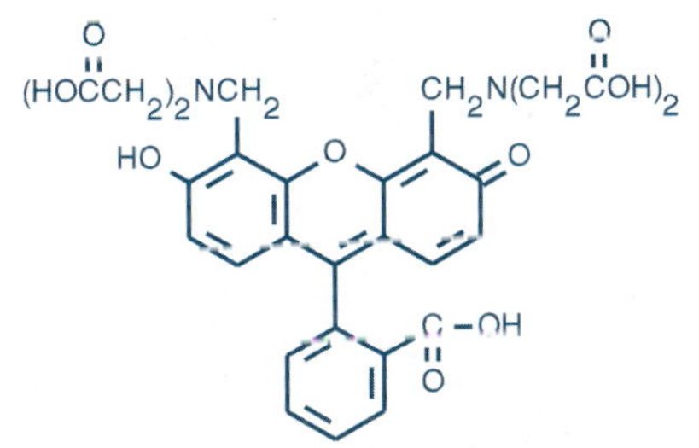

Figure 19.3.36 Calcein (C481).

REFERENCES

1. J Biol Chem (1989) 264:8171; **2.** Methods Mol Biol (2009) 489:93; **3.** J Immunol (2007) 179:6613; **4.** Biophys J (2000) 78:2655; **5.** J Physiol (2000) 526:551; **6.** J Vis Exp (2009) doi: 10.3791/1142; **7.** Nature (2009) 461:930; **8.** Nat Rev Neurosci (2008) 9:195; **9.** Cell (2007) 131:1047; **10.** Proc Natl Acad Sci U S A (2005) 102:9365; **11.** Circ Res (2004) 94:1011; **12.** Methods Mol Biol (2009) 552:269; **13.** Methods Mol Biol (2009) 565:145; **14.** Biophys J (1993) 65:865; **15.** J Physiol (2002) 542:843; **16.** Cell Calcium (2000) 28:213; **17.** Cell Calcium (1997) 21:275; **18.** Nat Med (2008) 14:574; **19.** Cell Calcium (1993) 14:359; **20.** J Immunol Methods (2006) 311:220; **21.** Methods (2000) 21:221; **22.** Cytometry (1994) 17:135; **23.** Nat Immunol (2007) 8:835; **24.** J Neurosci (2008) 28:143; **25.** J Neurosci (2009) 29:7803; **26.** J Neurosci Methods (2009) 179:166; **27.** Proc Natl Acad Sci U S A (2009) 106:3496; **28.** Cell Calcium (2000) 27:97; **29.** J Physiol (1999) 518:815; **30.** Nature (1999) 400:261; **31.** Neuron (2000) 27:25; **32.** J Biomol Screen (1999) 4:67; **33.** J Pharmacol Exp Ther (2003) 304:801; **34.** Cell Calcium (1990) 11:57; **35.** Nat Med (2007) 13:874; **36.** J Physiol (1998) 507:405; **37.** J Neurosci (1997) 17:1701; **38.** Neuron (1994) 13:837; **39.** Biochem J (2001) 356:345; **40.** Cell Calcium (1996) 20:105; **41.** Biophys J (2009) 97:1864; **42.** Circ Res (2004) 94:1589; **43.** Proc Natl Acad Sci U S A (1995) 92:10272; **44.** Neurosci Lett (1993) 162:149; **45.** J Neurosci (2000) 20:6773; **46.** J Physiol (2001) 534:87; **47.** EMBO J (2000) 19:5729; **48.** Circ Res (2008) 103:e105; **49.** Circ Res (2006) 99:740; **50.** J Microbiol Methods (2005) 61:245; **51.** J Dairy Sci (2006) 89:4105; **52.** Neuron (2009) 61:259; **53.** J Biol Chem (2004) 279:16903; **54.** Methods Mol Biol (2007) 372:421; **55.** Methods Enzymol (1999) 302:341; **56.** Mol Pharmacol (1998) 53:974; **57.** Biochem Biophys Res Commun (1997) 236:738; **58.** Methods Enzymol (1999) 307:441; **59.** Biophys J (1998) 75:2004; **60.** Cell (1995) 82:415; **61.** J Physiol (1998) 509:59; **62.** Am J Physiol (1988) 255:C304; **63.** J Biol Chem (2009) 284:18754; **64.** Mol Cell (2009) 33:627; **65.** Proc Natl Acad Sci U S A (2009) 106:2007; **66.** Biochem J (2008) 415:123; **67.** Biophys J (2002) 83:587; **68.** J Fluoresc (2007) 17:739; **69.** Biochem Biophys Res Commun (1991) 180:209; **70.** Proc Natl Acad Sci U S A (1992) 89:3591; **71.** J Neurophysiol (2007) 97:3118; **72.** Proc Natl Acad Sci U S A (2006) 103:17961; **73.** Nature (1997) 385:161; **74.** J Microsc (2007) 225:209; **75.** Three-Dimensional Confocal Microscopy, Stevens JK, Mills LR, Trogadis JE, Eds. (1994) p. 281; **76.** Dev Biol (1992) 149:370; **77.** Pflugers Arch (1995) 430:579; **78.** Cell Calcium (1998) 24:71; **79.** J Neurosci (1996) 16:71; **80.** J Neurosci (1995) 15:5535; **81.** J Neurosci (2004) 24:2496; **82.** J Neurosci (1998) 18:3233; **83.** J Biol Chem (2006) 281:11658; **84.** Biophys J (2005) 88:3946; **85.** Biophys J (2003) 84:2319; **86.** Biophys J (1998) 74:1549; **87.** J Neurosci (1995) 15:4209; **88.** J Gen Physiol (1995) 105:95; **89.** J Vis Exp (2009) doi: 10.3791/1247; **90.** J Neurosci Methods (2000) 102:165; **91.** Science (2009) 323:1211; **92.** EMBO J (2000) 19:3608; **93.** Proc Natl Acad Sci U S A (2000) 97:8635; **94.** J Neurosci Methods (2006) 151:276; **95.** Nature (2008) 452:436; **96.** Nat Neurosci (2008) 11:713; **97.** Nat Protoc (2009) 4:1551; **98.** Proc Natl Acad Sci U S A (2009) 106:19557; **99.** Nat Methods (2008) 5:935; **100.** J Neurosci (2000) 20:2523; **101.** Proc Natl Acad Sci U S A (1999) 96:7035; **102.** J Neurosci (1999) 19:1976; **103.** J Physiol (1999) 520:65; **104.** Nature (1999) 399:151; **105.** Nat Neurosci (1999) 2:65; **106.** Nat Neurosci (1999) 2:989; **107.** Biophys J (2000) 79:202; **108.** J Biol Chem (2001) 276:10655; **109.** J Neurosci (2000) 20:7290; **110.** Nat Neurosci (2001) 4:275; **111.** J Neurosci (1999) 19:7495; **112.** Am J Physiol (1997) 273:H2161; **113.** Biophys J (1993) 64:1934; **114.** Cell Calcium (2003) 34:295; **115.** Anal Biochem (1995) 227:328; **116.** Cell Calcium (1996) 19:255; **117.** Cell Calcium (1995) 18:377; **118.** Cell Calcium (1994) 16:279; **119.** Cell Calcium (1994) 15:341; **120.** Cytometry (1992) 13:693; **121.** Biophys J (1977) 18:3; **122.** Dev Biol (2001) 238:239; **123.** Biochem Biophys Res Commun (2004) 316:943; **124.** Development (2003) 130:1339.

DATA TABLE 19.3 FLUORESCENT Ca^{2+} INDICATORS EXCITED WITH VISIBLE LIGHT

				Low Ca^{2+}				High Ca^{2+}						
Cat. No.	MW	Storage	Soluble	Abs	EC	Em	Solvent	Abs	EC	Em	Solvent	Product	K_d	Notes
C3010MP	1147.19	F,D,L	pH >6	506	81,000	531	H_2O	506	82,000	531	H_2O/Ca^{2+}		190 nM	1, 2, 3, 4
C3011MP	1290.98	F,D,L	DMSO	302	17,000	none	MeOH					C3010MP		
C3012	1290.98	F,D,L	DMSO	302	17,000	none	MeOH					C3010MP		
C3013	1087.33	F,D,L	pH >6	549	80,000	575	H_2O	549	80,000	576	H_2O/Ca^{2+}		185 nM	1, 2, 3, 4
C3015	1223.23	F,D,L	DMSO	540	94,000	566	MeOH					C3013		
Calcium Crimson™	1232.51	F,D,L	pH >6	589	96,000	615	H_2O	589	92,000	615	H_2O/Ca^{2+}		185 nM	1, 2, 3, 4
C3018	1368.40	F,D,L	DMSO	583	113,000	602	MeOH					Calcium Crimson™		
C3730	1665.58	F,D,L	pH >6	506	95,000	536	H_2O	503	147,000	536	H_2O/Ca^{2+}		550 nM	1, 2, 3, 4
C3732	1817.26	F,D	DMSO	302	29,000	none	MeOH					C3730		
C3737	1192.19	F,D,L	pH >6	506	83,000	532	H_2O	506	82,000	532	H_2O/Ca^{2+}		14 µM	1, 2, 4, 5
C3739	1335.98	F,D	DMSO	361	15,000	none	EtOAc					C3737		
F1240	854.70	F,D,L	pH >6	503	92,000	see Notes	H_2O	505	102,000	526	H_2O/Ca^{2+}		390 nM	1, 2, 3, 6
F1241	1129.86	F,D,L	DMSO	464	26,000	see Notes	MeOH					F1240		7
F1242	1129.86	F,D,L	DMSO	464	26,000	see Notes	MeOH					F1240		7
F3020	1089.00	F,D,L	DMSO	458	43,000	597	MeOH					F14219		
F3021	1089.00	F,D,L	DMSO	458	43,000	597	MeOH					F14219		
F3715	960.00	F,D,L	pH >6	506	90,000	see Notes	H_2O	506	100,000	526	H_2O/Ca^{2+}		390 nM	1, 2, 3, 6
F14200	927.09	F,D,L	pH >6	491	82,000	see Notes	H_2O	494	88,000	516	H_2O/Ca^{2+}		345 nM	1, 2, 3, 6
F14201	1096.95	F,D,L	DMSO	456	26,000	see Notes	MeOH					F14200		7
F14202	1096.95	F,D,L	DMSO	456	26,000	see Notes	MeOH					F14200		7
F14203	958.06	F,D,L	pH >6	491	72,000	see Notes	H_2O	493	74,000	518	H_2O/Ca^{2+}		90 µM	1, 2, 5, 6
F14204	1127.92	F,D,L	DMSO	456	26,000	see Notes	MeOH					F14203		7
F14217	1096.95	F,D,L	DMSO	456	26,000	see Notes	MeOH					F14200		7, 8
F14218	1129.86	F,D,L	DMSO	464	26,000	see Notes	MeOH					F1240		7, 8
F14219	808.98	F,D,L	pH >6, MeOH	473	29,000	670	H_2O	436	41,000	655	H_2O/Ca^{2+}		140 nM	1, 2, 3, 9
F14221	931.05	F,D,L	pH >6	491	71,000	see Notes	H_2O	494	74,000	518	H_2O/Ca^{2+}		2.3 µM	1, 2, 5, 6
F14222	1100.92	F,D,L	DMSO	456	24,000	see Notes	MeOH					F14221		7
F14242	1129.86	F,D,L	DMSO	464	26,000	see Notes	MeOH					F1240		7
F23915	1129.86	F,D,L	DMSO	464	26,000	see Notes	MeOH					F1240		7, 10
F23917	1096.95	F,D,L	DMSO	456	26,000	see Notes	MeOH					F14200		7, 10
F23980	949.04	F,D,L	pH >6	491	72,000	see Notes	H_2O	494	75,000	516	H_2O/Ca^{2+}		9.7 µM	1, 2, 5, 6

IMPORTANT NOTICE: The products described in this manual are covered by one or more Limited Use Label License(s). Please refer to the Appendix on page 971 and Master Product List on page 975. Products are For Research Use Only. Not intended for any animal or human therapeutic or diagnostic use.

DATA TABLE 19.3 FLUORESCENT Ca^{2+} INDICATORS EXCITED WITH VISIBLE LIGHT—*continued*

				Low Ca^{2+}				High Ca^{2+}						
Cat. No.	MW	Storage	Soluble	Abs	EC	Em	Solvent	Abs	EC	Em	Solvent	Product	K_d	Notes
F23981	1118.91	F,D,L	DMSO	456	25,000	see Notes	MeOH					F23980		7
F36201	1055.26	F,D,L	pH >6	491	74,000	see Notes	H_2O	494	78,000	518	H_2O/Ca^{2+}		950 nM	1, 2, 3, 6
M3733	915.90	F,D,L	pH >6	506	77,000	531	H_2O	506	77,000	531	H_2O/Ca^{2+}		6 µM	1, 2, 4, 5
M3735	1025.71	F,D	DMSO	302	16,000	none	MeOH					M3733		
M14205	681.77	F,D,L	pH >6	490	74,000	see Notes	H_2O	493	75,000	517	H_2O/Ca^{2+}		22 µM	1, 2, 5, 6
M14206	817.66	F,D,L	DMSO	457	25,000	see Notes	MeOH					M14205		7
O6806	1114.28	F,D,L	pH >6	494	76,000	523	H_2O	494	78,000	523	H_2O/Ca^{2+}		170 nM	1, 2, 3, 4
O6807	1258.07	F,D	DMSO	299	19,000	none	EtOAc					O6806		
O6808	1599.77	F,D,L	pH >6	494	105,000	523	H_2O	494	140,000	523	H_2O/Ca^{2+}		580 nM	1, 2, 3, 4
O6809	1751.45	F,D,L	DMSO	299	31,000	none	MeOH					O6808		
O6812	1159.28	F,D,L	pH >6	494	72,000	521	H_2O	494	76,000	521	H_2O/Ca^{2+}		20 µM	1, 2, 4, 5
O23990	1132.27	F,D,L	pH >6	494	75,000	523	H_2O	494	77,000	523	H_2O/Ca^{2+}		3 µM	1, 2, 4, 5
R1244	1123.96	F,D,L	DMSO	550	125,000	571	see Notes					R14220		11
R1245MP	1123.96	F,D,L	DMSO	550	125,000	571	see Notes					R14220		11
R14207	900.03	F,D,L	pH >6	549	64,000	see Notes	H_2O	551	63,000	577	H_2O/Ca^{2+}		320 µM	1, 2, 6, 12
R14220	869.06	F,D,L	pH >6	548	91,000	see Notes		552	96,000	578	H_2O/Ca^{2+}		570 nM	1, 2, 3, 6
rhod-FF	891.02	F,D,L	pH >6	548	73,000	see Notes	H_2O	552	78,000	577	H_2O/Ca^{2+}		19 µM	1, 2, 5, 6
R23983	1145.91	F,D,L	DMSO	551	104,000	566	$CHCl_3$					rhod-FF		
X-rhod-1	973.21	F,D,L	pH >6	576	88,000	see Notes	H_2O	580	92,000	602	H_2O/Ca^{2+}		700 nM	1, 2, 3, 6
X14210	1228.11	F,D,L	DMSO	578	133,000	596	$CHCl_3$					X-rhod-1		
X23984	977.18	F,D,L	pH >6	576	73,000	see Notes	H_2O	580	73,000	602	H_2O/Ca^{2+}		1.6 µM	1, 2, 5, 6
X23985	1232.07	F,D,L	DMSO	578	92,000	590	$CHCl_3$					X23984		

For definitions of the contents of this data table, see "Using *The Molecular Probes® Handbook*" in the introductory pages.

Notes

1. Dissociation constants are known to vary considerably depending on the temperature, pH, ionic strength, viscosity, protein binding, presence of other ions (especially polyvalent ions), instrument setup and other factors. It is strongly recommended that these values be verified under user-specific experimental conditions.
2. Spectra measured in aqueous buffers containing 10 mM EGTA (H_2O) or a >10-fold excess of free Ca^{2+} relative to the K_d (H_2O/Ca^{2+}).
3. Dissociation constant determined by fluorescence measurements in 100 mM KCl, 10 mM MOPS, pH 7.2, 0 to 39 µM free Ca^{2+} at 22°C.
4. This indicator exhibits fluorescence enhancement in response to ion binding, with essentially no change in absorption or emission wavelengths.
5. Dissociation constant determined by fluorescence measurements in 100 mM KCl, 10 mM MOPS, pH 7.2, 0 to 1 mM free Ca^{2+} at 22°C.
6. Fluorescence of the free indicator is very weak and is enhanced >100-fold on binding Ca^{2+}.
7. Fluorescence of this AM ester derivative is very weak and is enhanced only after hydrolytic cleavage followed by binding of divalent cations to the anionic indicator.
8. This product is supplied as a ready-made solution in the solvent indicated under "Soluble."
9. The fluorescence quantum yield of Fura Red™ is low (~0.013 in Ca^{2+}-free solution). (Methods Cell Biol (1994) 40:155)
10. This product is specified to equal or exceed 98% analytical purity by HPLC.
11. Spectra measured in 90:10 (v/v) $CHCl_3$:MeOH.
12. Dissociation constant determined by fluorescence measurements in 100 mM KCl, 10 mM MOPS, pH 7.2, 0 to 10 mM free Ca^{2+} at 22°C.

PRODUCT LIST 19.3 FLUORESCENT Ca^{2+} INDICATORS EXCITED WITH VISIBLE LIGHT

Cat. No.	Product	Quantity
C3018	Calcium Crimson™, AM *cell permeant* *special packaging*	10 x 50 µg
C3011MP	Calcium Green™-1, AM *cell permeant*	500 µg
C3012	Calcium Green™-1, AM *cell permeant* *special packaging*	10 x 50 µg
C3010MP	Calcium Green™-1, hexapotassium salt *cell impermeant*	500 µg
C3732	Calcium Green™-2, AM *cell permeant* *special packaging*	10 x 50 µg
C3730	Calcium Green™-2, octapotassium salt *cell impermeant*	500 µg
C3739	Calcium Green™-5N, AM *cell permeant* *special packaging*	10 x 50 µg
C3737	Calcium Green™-5N, hexapotassium salt *cell impermeant*	500 µg
C3015	Calcium Orange™, AM *cell permeant* *special packaging*	10 x 50 µg
C3013	Calcium Orange™, tetrapotassium salt *cell impermeant*	500 µg
F14218	fluo-3, AM *1 mM solution in DMSO* *cell permeant*	1 mL
F1241	fluo-3, AM *cell permeant*	1 mg
F1242	fluo-3, AM *cell permeant* *special packaging*	20 x 50 µg
F23915	fluo-3, AM *FluoroPure™ grade* *special packaging*	10 x 50 µg
F14242	fluo-3, AM *packaged for high-throughput screening*	40 x 1 mg
F1240	fluo-3, pentaammonium salt *cell impermeant*	1 mg
F3715	fluo-3, pentapotassium salt *cell impermeant*	1 mg
F10473	Fluo-4 Direct™ Calcium Assay Kit, High-Throughput Pack	1 kit
F10471	Fluo-4 Direct™ Calcium Assay Kit, Starter Pack	1 kit
F10472	Fluo-4 Direct™ Calcium Assay Kit, Surveyor Pack	1 kit
F36205	Fluo-4 NW Calcium Assay Kit (high-throughput) *for 100 microplates*	1 kit

continued on next page

IMPORTANT NOTICE: The products described in this manual are covered by one or more Limited Use Label License(s). Please refer to the Appendix on page 971 and Master Product List on page 975. Products are For Research Use Only. Not intended for any animal or human therapeutic or diagnostic use.

PRODUCT LIST 19.3 FLUORESCENT Ca^{2+} INDICATORS EXCITED WITH VISIBLE LIGHT—*continued*

Cat. No.	Product	Quantity
F36206	Fluo-4 NW Calcium Assay Kit (starter pack with buffer) *for 10 microplates*	1 kit
F14217	fluo-4, AM *1 mM solution in DMSO* *cell permeant*	500 µL
F14201	fluo-4, AM *cell permeant* *special packaging*	10 x 50 µg
F23917	fluo-4, AM *FluoroPure™ grade* *special packaging*	10 x 50 µg
F14202	fluo-4, AM *packaged for high-throughput screening*	5 x 1 mg
F14200	fluo-4, pentapotassium salt *cell impermeant*	500 µg
F23981	fluo-4FF, AM *cell permeant* *special packaging*	10 x 50 µg
F23980	fluo-4FF, pentapotassium salt *cell impermeant*	500 µg
F14222	fluo-5F, AM *cell permeant* *special packaging*	10 x 50 µg
F14221	fluo-5F, pentapotassium salt *cell impermeant*	500 µg
F14204	fluo-5N, AM *cell permeant* *special packaging*	10 x 50 µg
F14203	fluo-5N, pentapotassium salt *cell impermeant*	500 µg
F3020	Fura Red™, AM *cell permeant*	500 µg
F3021	Fura Red™, AM *cell permeant* *special packaging*	10 x 50 µg
F14219	Fura Red™, tetrapotassium salt *cell impermeant*	500 µg
M14206	mag-fluo-4, AM *cell permeant* *special packaging*	10 x 50 µg
M14205	mag-fluo-4, tetrapotassium salt *cell impermeant*	500 µg
M3735	Magnesium Green™, AM *cell permeant* *special packaging*	20 x 50 µg
M3733	Magnesium Green™, pentapotassium salt *cell impermeant*	1 mg
O6807	Oregon Green® 488 BAPTA-1, AM *cell permeant* *special packaging*	10 x 50 µg
O6806	Oregon Green® 488 BAPTA-1, hexapotassium salt *cell impermeant*	500 µg
O6809	Oregon Green® 488 BAPTA-2, AM *cell permeant* *special packaging*	10 x 50 µg
O6808	Oregon Green® 488 BAPTA-2, octapotassium salt *cell impermeant*	500 µg
O6812	Oregon Green® 488 BAPTA-5N, hexapotassium salt *cell impermeant*	500 µg
O23990	Oregon Green® 488 BAPTA-6F, hexapotassium salt *cell impermeant*	500 µg
R1244	rhod-2, AM *cell permeant*	1 mg
R1245MP	rhod-2, AM *cell permeant* *special packaging*	20 x 50 µg
R14220	rhod-2, tripotassium salt *cell impermeant*	1 mg
R10145	Rhod-3 Imaging Kit	1 kit
R14207	rhod-5N, tripotassium salt *cell impermeant*	500 µg
R23983	rhod-FF, AM *cell permeant* *special packaging*	10 x 50 µg
X14210	X-rhod-1, AM *cell permeant* *special packaging*	10 x 50 µg
X23985	X-rhod-5F, AM *cell permeant* *special packaging*	10 x 50 µg
X23984	X-rhod-5F, tripotassium salt *cell impermeant*	500 µg

19.4 Fluorescent Ca^{2+} Indicator Conjugates

When ion indicators are loaded into cells as their acetoxymethyl (AM) esters, they may translocate to intracellular compartments, where they are still fluorescent but no longer respond to changes in cytosolic ion levels.[1,2] This problem frequently limits the experiment's duration because sequestration of the indicator into organelles will cause errors in the estimated cytosolic ion levels. Furthermore, Ca^{2+} indicators such as fura-2 and indo-1 may bind to cellular proteins,[1–5] which can markedly alter the indicator's response to Ca^{2+} (Table 19.2).

To overcome these limitations, we have prepared dextran conjugates of some Molecular Probes® ion indicators[6] (Table 19.1). Dextrans are hydrophilic polysaccharides characterized by their moderate to high molecular weight, good water solubility and low toxicity. They are also biologically inert due to their uncommon poly-(α-D-1,6-glucose) linkages, which render them resistant to cleavage by most endogenous cellular glycosidases. Indicator dextrans must be loaded into cells by microinjection, whole-cell patch clamping, microprojectile bombardment[7–9] or electroporation,[10,11] or by using our Influx™ pinocytic cell-loading reagent (I14402, Section 19.8; Figure 19.4.1). Calcium indicator dextrans are actively transported in adult nerve fibers over a significant distance and are retained in presynaptic terminals in a form that allows monitoring of presynaptic Ca^{2+} levels.[12,13] Dextran conjugates above ~2000 MW are well retained in viable cells, will not readily pass through gap junctions[13] and are less likely to become compartmentalized.[14,15] Also, fluorescence photobleaching measurements have shown that, as compared with low molecular weight dyes, dextran conjugates are much less likely to bind to proteins.[16] Because dextran conjugates are intrinsically polydisperse and their degree of substitution may vary with the production lot, the Ca^{2+} dissociation constant of each lot of these indicators should be calibrated independently using one of our Calcium Calibration Buffer Kits (Section 19.8).

IMPORTANT NOTICE: The products described in this manual are covered by one or more Limited Use Label License(s). Please refer to the Appendix on page 971 and Master Product List on page 975. Products are For Research Use Only. Not intended for any animal or human therapeutic or diagnostic use.

Fura Dextran

Fura dextran (F3029) tends to remain in the cytosol without compartmentalization or leakage and is less likely to bind to cellular proteins, making it useful for long-term Ca^{2+} measurements.[17] Although the spectral response curves of fura dextran are very similar to those of the free dye, its affinity for Ca^{2+} is somewhat weaker. The dissociation constant for Ca^{2+} of fura dextran in the absence of Mg^{2+} varies between 200 nM and 400 nM (measured at ~22°C using our Calcium Calibration Buffer Kits), depending on the molecular weight of the dextran and individual batch characteristics.

Dye compartmentalization has been especially problematic for measurements of ions in plant cells, where the dye is frequently transported out of the cytosol in minutes. Unlike microinjected fura-2 salt, fura dextran is retained for hours in the cytosol of stamen hair cells and *Lilium* pollen tubes[18] (Figure 19.4.2). A comparison of the dextran conjugates of fura and Calcium Green™ with conventional indicators for imaging Ca^{2+} levels in plant and fungal cells has been published.[19]

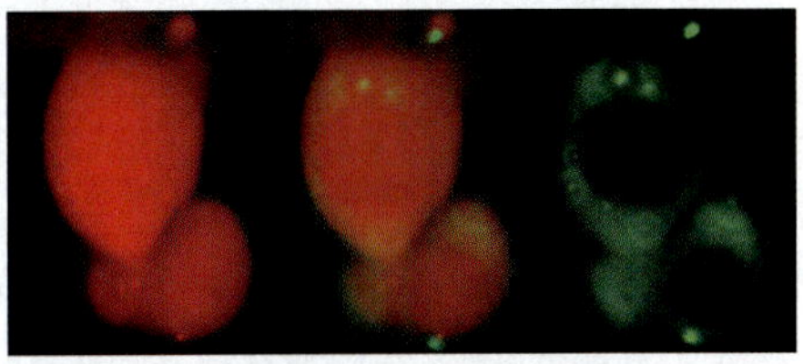

Figure 19.4.1 Adherent 3T3 fibroblasts labeled with 3000 MW Texas Red® dextran (D3329) and 70,000 MW Calcium Green™-1 dextran (C3714). Both dextrans were loaded into the fibroblasts using the Influx™ pinocytic cell-loading reagent (I14402). The images were acquired using a fluorescence microscope equipped with a cooled CCD camera and band-pass filter sets appropriate for either the Texas Red® dye (left image) or fluorescein (right image). The center image represents a composite of the cells labeled with both dextrans. The 3000 MW Texas Red® dextran is distributed throughout the cell, but the 70,000 MW Calcium Green™-1 dextran is excluded from the nucleus by the nuclear membrane.

Visible-Excitation Ca^{2+} Indicator Dextrans

We offer 3000 MW; 10,000 MW; and 70,000 MW dextran conjugates of the Calcium Green™-1 indicator (C6765, C3713, C3714), as well as the 10,000 MW dextran conjugate of the Oregon Green® 488 BAPTA-1 indicator (O6798). We have also developed both low-affinity and high-affinity 10,000 MW dextran conjugates of the fluo-4 indicator (F14240, F36250) and low-affinity and high-affinity 10,000 MW dextran conjugates of our rhod indicator[6] (R34677, R34676). Table 19.1 provides a complete list of our Ca^{2+} indicator dextran conjugates.

A review by Read and co-workers compares the Calcium Green™-1 and fura dextrans with conventional indicators for imaging Ca^{2+} levels in plant and fungal cells.[19] The Calcium Green™ dextran conjugates (and the other green-fluorescent Ca^{2+}-indicating dextrans) can be co-injected with red-fluorescent Texas Red® dextrans (Section 14.5, Table 14.4) to permit ratiometric measurements of Ca^{2+} flux in cells that can be microinjected, including oocytes[20,21] (Figure 19.4.3). Spectra

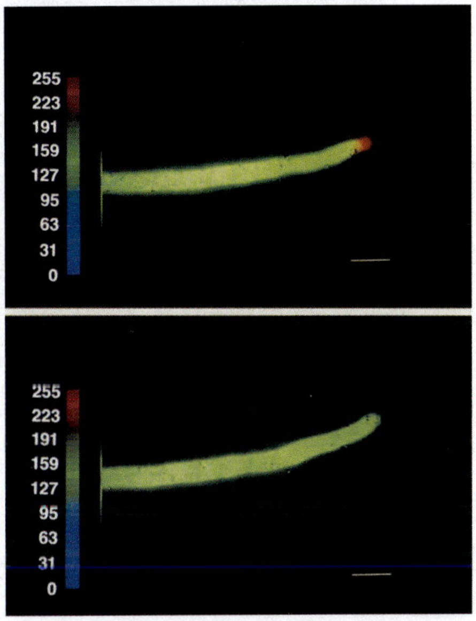

Figure 19.4.2 Top panel: Pseudocolored image of a pollen tube of *Lillium longiflorum* injected with fura dextran (F3029). The cell continues elongating and clearly shows a Ca^{2+} gradient. Bottom panel: The same pollen tube after injection with dibromo BAPTA (D1211) remains healthy but is no longer elongating. The images were contributed by Debra Miller, Dale Callaham, David Gross and Peter Hepler, University of Massachusetts.

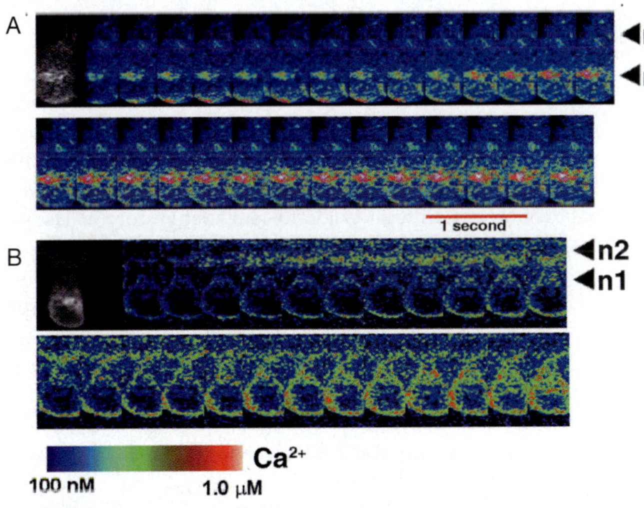

Figure 19.4.3 Rhizoid cells from the marine alga *Fucus serratus* were pressure-microinjected with Ca^{2+}-sensitive 10,000 MW Calcium Green™-1 dextran (C3713) and Ca^{2+}-insensitive 10,000 MW Texas Red® dextran (D1828). Patterns of Ca^{2+} elevation following hypoosmotic treatment in dividing cells were visualized by confocal ratio imaging of Calcium Green™-1 (excited at 488 nm) and Texas Red® (excited at 568 nm). Emission ratio values were pseudocolored to represent calcium concentrations according to the scale bar in the lower left-hand corner. **A)** Sequential ratio images during the onset of a hypoosmotically induced (100% seawater to 50% seawater) Ca^{2+} wave shows an initial elevation of Ca^{2+} in the rhizoid apex, which declines before the onset of Ca^{2+} elevation arising in the apical nucleus (n1) region. **B)** A variation in this pattern was evident in a minority of cells where Ca^{2+} elevations were observed to arise in the subapical nucleus (n2) simultaneously with the apical Ca^{2+} elevation. The image was contributed by Colin Brownlee, Marine Biological Association of the United Kingdom, Plymouth, UK, and reproduced with permission from Proc Natl Acad Sci U S A (2000) 97:1932.

IMPORTANT NOTICE: The products described in this manual are covered by one or more Limited Use Label License(s). Please refer to the Appendix on page 971 and Master Product List on page 975. Products are For Research Use Only. Not intended for any animal or human therapeutic or diagnostic use.

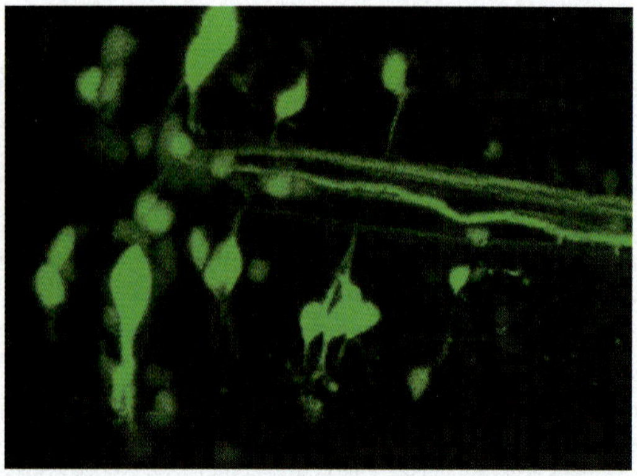

Figure 19.4.4 Reticulospinal neurons in the hindbrain of a live zebrafish labeled with 10,000 MW Calcium Green™-1 dextran (C3713) and optically sectioned in a Bio-Rad® MRC 600 confocal laser-scanning microscope. Image contributed by D.M. O'Malley and J.R. Fetcho, SUNY at Stony Brook.

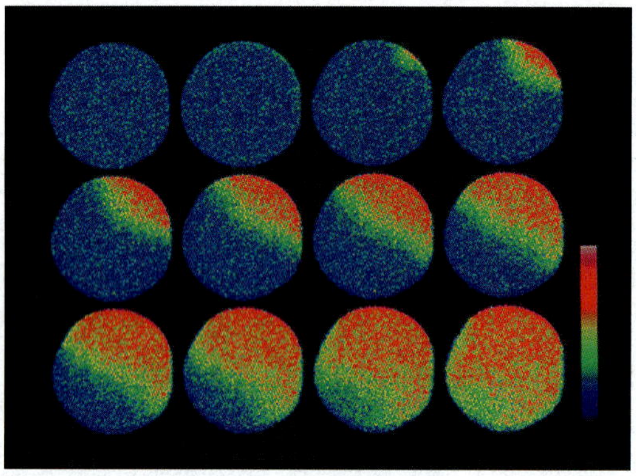

Figure 19.4.5 Confocal images taken at five-second intervals of a fertilization-induced calcium wave in a *Pisaster ochraceus* starfish oocyte that was microinjected with 10,000 MW Calcium Green™-1 dextran (C3713). The image was contributed by Stephen A. Stricker, University of New Mexico.

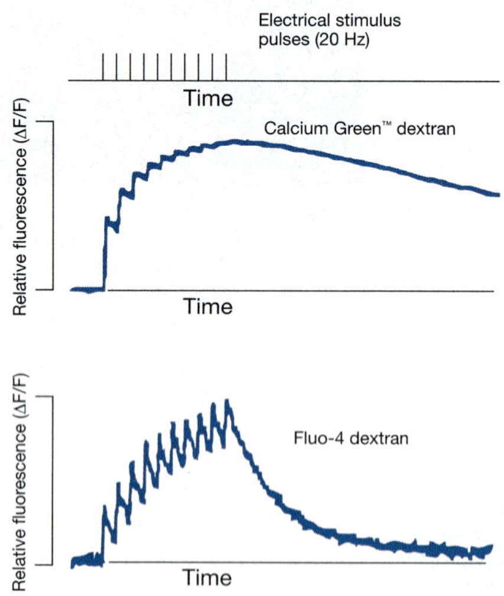

Figure 19.4.6 Ca^{2+} transients in rat climbing fiber pre-synaptic terminals evoked by sequences of 10 applied electrical stimulus pulses (20 Hz) monitored by 10,000 MW Calcium Green™-1 dextran (C3713) (upper panel) and the low-affinity version of fluo-4 dextran (F14240) (lower panel). Each trace represents an average of five recordings. Adapted with permission from Neuron (2000) 27:25.

of the Oregon Green® 488 BAPTA-1 dextrans and fluo-4 dextrans match the 488 nm spectral line of the argon-ion laser and standard fluorescein optical filters better than do dextran conjugates of the Calcium Green™-1 indicator, which should permit the use of lower probe concentrations to achieve the same signal. The red-orange–fluorescent rhod dextran, which exhibits a 50-fold fluorescence enhancement upon Ca^{2+} binding, is valuable for multicolor applications and for experiments in cells and tissues that have high levels of autofluorescence.

The visible light–excitable Calcium Green™-1 and Oregon Green® 488 BAPTA-1 indicator dextrans have been used to:

- Assay Ca^{2+} influx in taste receptors responding to bitter stimuli [22]
- Detect changes in Ca^{2+} levels during cell cleavage in mouse [23] and zebrafish [24] embryos
- Follow transmission of changes in cytosolic Ca^{2+} levels to the nucleus in permeabilized cardiomyocytes [25] and airway epithelial cells [26]
- Label neurons via retrograde or anterograde transport, allowing real-time imaging of neuronal activity [27,28] (Figure 19.4.4)
- Monitor presynaptic Ca^{2+} dynamics [12,29]
- Perform functional imaging of neurons and myocytes during development [30–32]
- Visualize cADP ribose–induced Ca^{2+} waves in starfish eggs [33] (Figure 19.4.5)
- Detect depletion of Ca^{2+} stores in organelles of yeast and mammalian cells [34]
- Examine the regulation of size-specific entry of molecules into the *Xenopus* oocyte nucleus by the nuclear Ca^{2+} store [35]
- Monitor changes in intracellular free Ca^{2+} accompanying photolysis of diazo-2, caged EGTA, caged EDTA or other caged Ca^{2+} probes [36]
- Perform multiphoton excitation functional imaging of brain activity [37,38]

In mammalian presynaptic terminals, the response of Calcium Green™-1 dextran to successive electrical stimuli progressively weakens due to Ca^{2+}-binding saturation (Figure 19.4.6). To solve this problem, we have synthesized fluo-4 dextran (F14240) with a lower Ca^{2+} binding affinity (a batch-dependent K_d for Ca^{2+} ~3 µM). Developed in collaboration with Wade Regehr's laboratory at Harvard University, the low-affinity fluo-4 dextran is a valuable tool for recording Ca^{2+} transients in presynaptic terminals of long axonal projections in heterogeneous fiber tracts.[29,39] We also offer a high-affinity fluo-4 dextran (K_d for Ca^{2+} ~600 nM, F36250) and high-affinity rhod dextran (K_d for Ca^{2+} ~780 nM, R34676). Coinjection of fluo-4 dextran together with a reference marker (e.g., 10,000 MW Texas Red® dextran, D1828; Section 14.5) may be necessary for initial identification of labeled cells due to the intrinsically weak fluorescence of the fluo-4 indicator in the absence of Ca^{2+}.[7,29]

The lack of a ratiometric Ca^{2+} response (Loading and Calibration of Intracellular Ion Indicators—Note 19.1) among the visible light–excitable indicator dextrans can be partially circumvented by co-loading Ca^{2+}-insensitive reference markers. Mixtures of Calcium Green™ dextran and Ca^{2+}-insensitive dextrans (such as the tetramethylrhodamine or Texas Red® dextrans in Section 14.5) have been co-loaded into cells for use in ratio-imaging microscopy [24,40–43] (Figure 19.4.3).

IMPORTANT NOTICE: The products described in this manual are covered by one or more Limited Use Label License(s). Please refer to the Appendix on page 971 and Master Product List on page 975. Products are For Research Use Only. Not intended for any animal or human therapeutic or diagnostic use.
www.invitrogen.com/probes

REFERENCES

1. Biophys J (1990) 58:1491; **2.** Cell Calcium (1990) 11:63; **3.** Biophys J (1994) 67:1646; **4.** Biophys J (1993) 65:865; **5.** Cell Calcium (1992) 13:59; **6.** J Neurophysiol (2004) 92:591; **7.** Plant J (2006) 46:327; **8.** J Neurosci Methods (2005) 141:41; **9.** J Neurosci Methods (2002) 119:37; **10.** Neuron (2007) 53:789; **11.** J Neurophysiol (2005) 93:1793; **12.** J Vis Exp (2007) doi: 10.3791/250; **13.** J Neurophysiol (2004) 91:1025; **14.** J Exp Biol (1994) 196:419; **15.** Methods Cell Biol (1989) 29:59; **16.** J Cell Physiol (1989) 141:410; **17.** Eur J Cell Biol (1992) 58:172; **18.** J Cell Sci (1992) 101:7; **19.** J Microsc (1992) 166:57; **20.** Development (1998) 125:4099; **21.** Proc Natl Acad Sci U S A (2000) 97:1932; **22.** Science (2001) 291:1557; **23.** J Cell Biol (1996) 132:915; **24.** J Cell Biol (1995) 131:1539; **25.** J Physiol (2007) 584:601; **26.** Biophys J (2005) 88:3946; **27.** J Neurosci Methods (2009) 179:166; **28.** J Neurosci Methods (2009) 180:1; **29.** Neuron (2000) 27:25; **30.** J Cell Sci (2005) 118:5181; **31.** Cell Calcium (2005) 37:443; **32.** Cell Calcium (2004) 35:393; **33.** Pflugers Arch (2003) 446:541; **34.** Proc Natl Acad Sci U S A (1999) 96:121; **35.** Science (1995) 270:1835; **36.** Nat Neurosci (2005) 8:305; **37.** Methods Mol Biol (2009)489:43; **38.** Glia (2004) 46:95; **39.** J Neurosci Methods (2006) 150:47; **40.** Biotechniques (2000) 29:492, 496, 498; **41.** Proc Natl Acad Sci U S A (1992) 89:3591; **42.** Biol Bull (1994) 187:234; **43.** Dev Biol (1995) 170:496.

DATA TABLE 19.4 FLUORESCENT Ca^{2+} INDICATOR CONJUGATES

				Low Ca^{2+}				High Ca^{2+}					
Cat. No.	MW	Storage	Soluble	Abs	EC	Em	Solvent	Abs	EC	Em	Solvent	K_d	Notes
C3713	see Notes	F,D,L	H_2O	508	ND	533	H_2O	508	ND	533	H_2O/Ca^{2+}	260 nM	1, 2, 3, 4, 5, 6, 7
C3714	see Notes	F,D,L	H_2O	510	ND	535	H_2O	510	ND	535	H_2O/Ca^{2+}	240 nM	1, 2, 3, 4, 5, 6, 7
C6765	see Notes	F,D,L	H_2O	510	ND	535	H_2O	510	ND	535	H_2O/Ca^{2+}	540 nM	1, 2, 3, 4, 5, 6, 7
F3029	see Notes	F,D,L	H_2O	364	ND	501	H_2O	338	ND	494	H_2O/Ca^{2+}	240 nM	1, 2, 3, 4, 5, 7
F14240	see Notes	F,D,L	H_2O	493	ND	see Notes	H_2O	495	ND	518	H_2O/Ca^{2+}	3.0 µM	1, 2, 3, 5, 7, 8, 9
F36250	see Notes	F,D,L	H_2O	493	ND	see Notes	H_2O	496	ND	518	H_2O/Ca^{2+}	600 nM	1, 2, 3, 4, 5, 7, 9
O6798	see Notes	F,D,L	H_2O	496	ND	524	H_2O	497	ND	524	H_2O/Ca^{2+}	265 nM	1, 2, 3, 4, 5, 6, 7
R34676	see Notes	F,D,L	H_2O	549	ND	see Notes	H_2O	556	ND	578	H_2O/Ca^{2+}	780 nM	1, 2, 3, 4, 5, 7, 9

For definitions of the contents of this data table, see Using *The Molecular Probes® Handbook*" in the introductory pages.

Notes

1. Dissociation constants are known to vary considerably depending on the temperature, pH, ionic strength, viscosity, protein binding, presence of other ions (especially polyvalent ions), instrument setup and other factors. It is strongly recommended that these values be verified under user-specific experimental conditions.
2. Spectra measured in aqueous buffers containing 10 mM EGTA (H_2O) or a >10-fold excess of free Ca^{2+} relative to the K_d (H_2O/Ca^{2+}).
3. Because indicator dextran conjugates are polydisperse both in molecular weight and degree of substitution, dissociation constants and spectra may vary between batches.
4. Dissociation constant determined by fluorescence measurements in 100 mM KCl, 10 mM MOPS, pH 7.2, 0 to 39 µM free Ca^{2+} at 22°C.
5. The molecular weight is nominally as specified in the product name but may have a broad distribution.
6. This indicator exhibits fluorescence enhancement in response to ion binding, with essentially no change in absorption or emission wavelengths.
7. ND = not determined.
8. Dissociation constant determined by fluorescence measurements in 100 mM KCl, 10 mM MOPS, pH 7.2, 0 to 1 mM free Ca^{2+} at 22°C.
9. Fluorescence of the free indicator is very weak and is enhanced >100-fold on binding Ca^{2+}.

PRODUCT LIST 19.4 FLUORESCENT Ca^{2+} INDICATOR CONJUGATES

Cat. No.	Product	Quantity
C6765	Calcium Green™-1 dextran, potassium salt, 3000 MW, anionic	5 mg
C3713	Calcium Green™-1 dextran, potassium salt, 10,000 MW, anionic	5 mg
C3714	Calcium Green™-1 dextran, potassium salt, 70,000 MW, anionic	5 mg
F36250	fluo-4 dextran, potassium salt, 10,000 MW, anionic (high-affinity version)	5 mg
F14240	fluo-4 dextran, potassium salt, 10,000 MW, anionic (low-affinity version)	5 mg
F3029	fura dextran, potassium salt, 10,000 MW, anionic	5 mg
O6798	Oregon Green® 488 BAPTA-1 dextran, potassium salt, 10,000 MW, anionic	5 mg
R34676	rhod dextran, potassium salt, 10,000 MW, anionic (high-affinity version)	5 mg

IMPORTANT NOTICE: The products described in this manual are covered by one or more Limited Use Label License(s). Please refer to the Appendix on page 971 and Master Product List on page 975. Products are For Research Use Only. Not intended for any animal or human therapeutic or diagnostic use.

19.5 Protein-Based Ca^{2+} Indicators

Premo™ Cameleon Calcium Sensor

Premo™ Cameleon Calcium Sensor (P36207, P36208) is a ratiometric calcium-sensitive fluorescent protein that is delivered by BacMam baculovirus-mediated transduction (BacMam Gene Delivery and Expression Technology—Note 11.1) to a variety of mammalian cell types (Figure 19.5.1). This content and delivery system provides an effective and robust technique for measuring Ca^{2+} mobilization in transduced cells using microplate assays or fluorescence microscopy.[1–3]

The Premo™ Cameleon Calcium Sensor is based on the YC3.60 version of the fluorescent protein–based sensor (cameleon) family developed by Tsien, Miyawaki and co-workers,[4] which is reported to have a Ca^{2+} dissociation constant of 250 nM.[5] The sensor comprises two fluorescent proteins (enhanced cyan-fluorescent protein or ECFP and Venus variant of yellow-fluorescent protein or YFP), linked by the calmodulin-binding peptide M13 and calmodulin. Upon binding four calcium ions, calmodulin undergoes a conformational change by wrapping itself around the M13 peptide, which changes the efficiency of the fluorescence resonance energy transfer (FRET) between the CFP donor and the YFP acceptor fluorophores (Figure 19.5.2). Following this conformational change, there is an increase in YFP emission (525–560 nm) and a simultaneous decrease in CFP emission (460–500 nm) (Figure 19.5.3), making Cameleon an effective reporter of calcium mobilization.[6] This Ca^{2+}-dependent emission ratio response reduces assay variations due to compound or cellular autofluorescence, nonuniform cell plating, differences in expression levels between cells, instability of instrument illumination and changes in illumination pathlength.

The Premo™ Cameleon Calcium Sensor is designed to readily and accurately detect intracellular calcium flux from different receptors. An example of the robustness and reproducibility and accuracy of the system is demonstrated using the endogenous histamine receptor in conjunction with histamine, pyrilamine, and thioperamide in HeLa cells (Figure 19.5.4). The no-wash, no-dye format and ratiometric readout eliminates wash steps that can dislodge cells, reduces data variability and increases data integrity. Expression levels will be maintained for several days, enabling iterative assays to be run; for instance, when examining agonist/antagonist

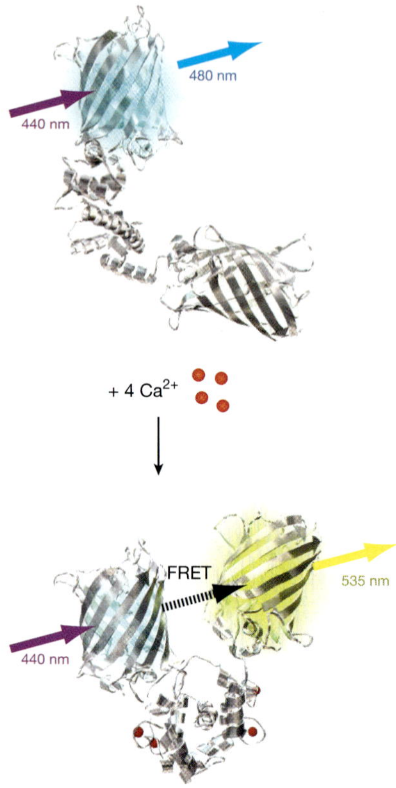

Figure 19.5.2 Schematic of the Premo™ Cameleon Calcium Sensor (P36207, P36208) mechanism.

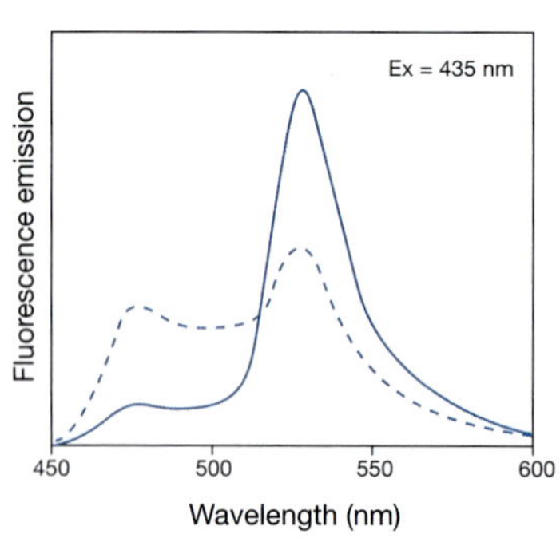

Figure 19.5.3 Fluorescence emission spectra of Premo™ Cameleon Calcium Sensor (P36207, P36208). The dashed line indicates the spectra in the absence of Ca^{2+}; the solid line shows the fluorescence resonance energy transfer (FRET)–based change upon Ca^{2+} binding.

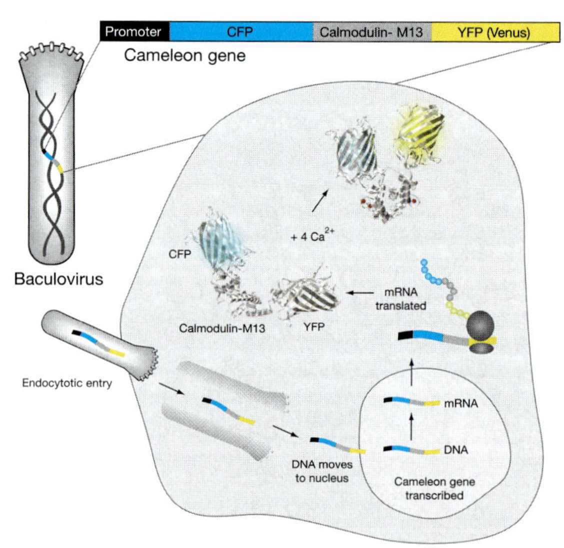

Figure 19.5.1 Schematic representation of BacMam-mediated transduction and expression of the Premo™ Cameleon Calcium Sensor (P36207, P36208). The combination of fluorescent protein biosensors and BacMam delivery technology yields easy-to-use, genetically encoded sensors for cell-based assays. A BacMam virus carrying the Cameleon fusion gene transduces a cell and traffics to the nucleus where only the Cameleon gene is transcribed; baculovirus promoters are not recognized by the mammalian transcriptional machinery, hence no virus replication occurs. Following transcription, the Cameleon mRNA is expressed in the cytosol, or in a specific compartment, depending on the presence of targeting tags.

IMPORTANT NOTICE: The products described in this manual are covered by one or more Limited Use Label License(s). Please refer to the Appendix on page 971 and Master Product List on page 975. Products are For Research Use Only. Not intended for any animal or human therapeutic or diagnostic use.
www.invitrogen.com/probes
molecular probes® | invitrogen™ by life technologies™

relationships in the same cells. Premo™ Cameleon Calcium Sensor is provided as a ready-to-use baculovirus stock suspension containing the Cameleon DNA, which is efficiently delivered to target cells, including primary and stem cells, prior to cell plating. If required, immunolocalization of Premo™ Cameleon Calcium Sensor in fixed specimens can be accomplished using our anti–Green Fluorescent Protein (anti-GFP) antibodies[7] (Section 7.5). Both stable cell lines and human primary cells can be prepared frozen and "assay-ready" and can be subsequently plated as little as four hours prior to screening. Cell-based assays or imaging experiments can be conducted in complete medium without any intervening wash steps. More information is available at www.invitrogen.com/handbook/premocalcium.

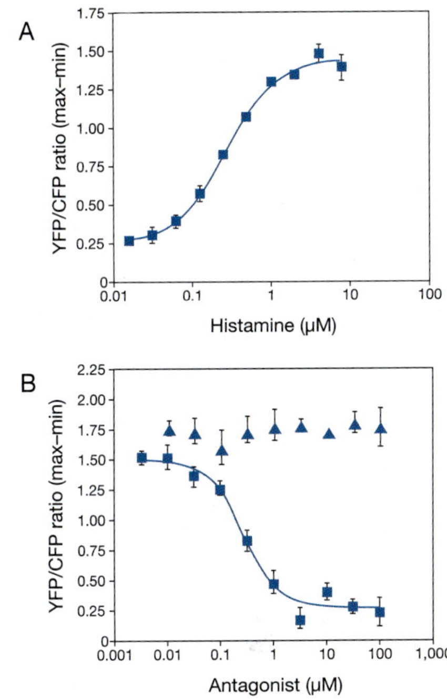

Figure 19.5.4 Agonist and antagonist dose response curves. HeLa cells were plated in a 96-well plate at a density of 15,000 cells/well, transduced with Premo™ Cameleon Calcium Sensor (P36207, P36208), and incubated overnight at 37°C. The following day, a histamine dose response was performed (**A**). A separate plate was used to evaluate an antagonist dose response with pyrilamine (closed squares) and thioperamide (closed triangles) in the presence of an EC_{80} concentration of histamine (**B**). Pyrilamine is a known H1 receptor antagonist that couples through G_q proteins and the second messenger Ca^{2+}. Thioperamide is a known H3 receptor antagonist that couples through G_i proteins and the second messenger cAMP.

Aequorin: A Bioluminescent Calcium Sensor

Bioluminescence is defined as the production of light by biological organisms. Because light is produced by a chemical reaction of specific photoproteins within the organism and does not require illumination, bioluminescence-based assays can be extremely sensitive and free of background.[8] However, the intensity of light produced by bioluminescent cells is often very low, necessitating the use of image enhancement to obtain sufficient signals.

Properties and Applications of Aequorin

We offer recombinant aequorin as well as a variety of synthetic coelenterazine analogs for quantitative Ca^{2+} measurements with aequorin, a photoprotein originally isolated from luminescent jellyfish and other marine organisms. The aequorin complex comprises a 22,000-dalton apoaequorin protein,[9] molecular oxygen and the luminophore coelenterazine[10–14] (Figure 19.5.5). When three Ca^{2+} ions bind to this complex, coelenterazine is oxidized to coelenteramide, with a concomitant release of carbon dioxide and blue light[15–17] (Figure 19.5.6, Figure 19.5.7). The approximately third-power dependence of aequorin's bioluminescence on Ca^{2+} concentration

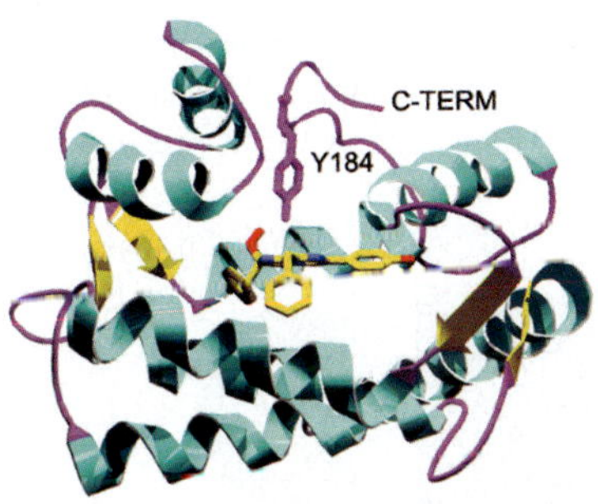

Figure 19.5.5 Ribbon representation of the aequorin/coelenterazine complex showing the secondary structural elements in the protein. Coelenterazine and the side chain of Tyr 184 are shown as stick representations. Reproduced with permission from Nature (2000) 405:372.

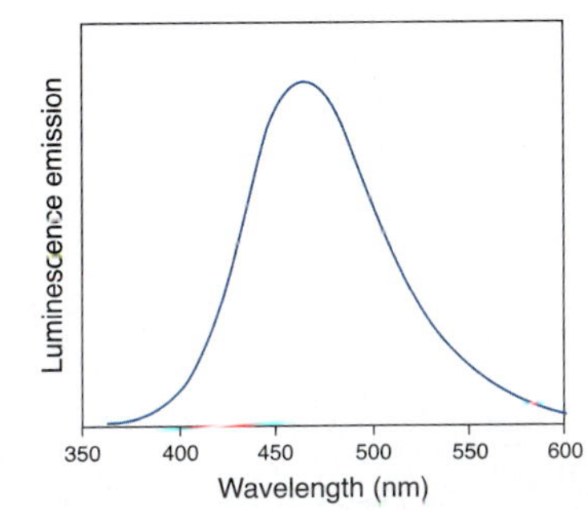

Figure 19.5.6 The Ca^{2+}-induced luminescence emission spectrum of native aequorin incorporating the coelenterazine luminophore (C2944).

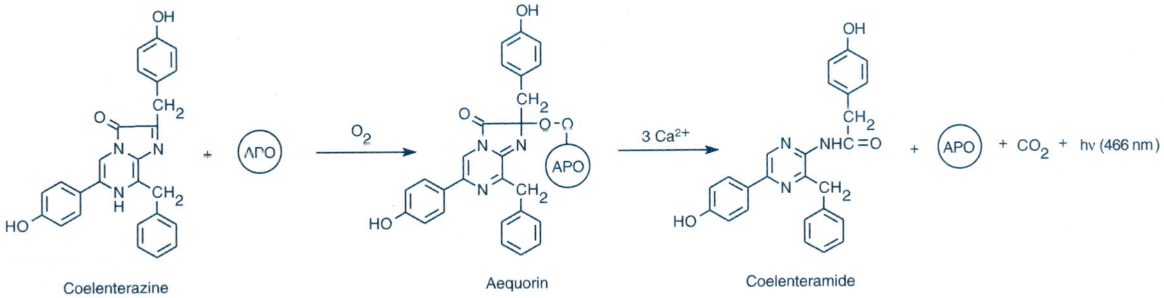

Figure 19.5.7 Ca^{2+}-dependent generation of luminescence by the aequorin complex, which contains apoaequorin (APO) and coelenterazine (C2944).

gives it a broad detection range, allowing the measurement of Ca^{2+} concentrations from ~0.1 µM to >100 µM.[18,19]

Unlike fluorescent Ca^{2+} indicators, Ca^{2+}-bound aequorin [20] can be detected without illuminating the sample, thereby eliminating interference from autofluorescence and allowing simultaneous labeling with caged probes [21] (Section 5.3). Moreover, aequorin that has been microinjected into eggs usually reports higher wave amplitudes (3–30 µM) than do fluorescent ion indicators.[22–25] Aequorin is not exported or secreted, nor is it compartmentalized or sequestered within cells; thus, aequorin measurements can be used to detect Ca^{2+} changes that occur over relatively long periods. In several experimental systems, aequorin's luminescence was detectable many hours to days after cell loading.[18,26,27] Aequorin also does not disrupt cell functions or embryo development [18] (Figure 19.5.8).

Recombinant Aequorin

Conventional purification of aequorin from the jellyfish *Aequorea victoria* requires laborious extraction procedures and sometimes yields preparations that are substantially heterogeneous or that are toxic to the organisms under study.[28,29] Two tons of jellyfish typically yield ~125 mg of the purified photoprotein.[30] In contrast, recombinant AquaLite® aequorin (A6785) is produced by purifying apoaequorin from genetically engineered *Escherichia coli*, followed by reconstitution of the aequorin complex *in vitro* with pure coelenterazine.[31] This method of preparation yields a pure, nontoxic, fully charged aequorin complex that is suitable for measuring intracellular Ca^{2+} by microinjection or other loading techniques, as well as for calibrating aequorin-based assays. Pressure injection is a commonly cited loading method, despite the fact that only large cells can be loaded in this way. Pressure injection has been employed to study the effects of caffeine on mouse diaphragm muscle fibers [32] and the role of Ca^{2+} in the fertilization of sea urchin eggs.[33] Alternatively, human platelets have been transiently permeabilized to the aequorin complex with DMSO,[34] and monkey kidney cells have been loaded by hypoosmotic shock.[35] A method based on the osmotic lysis of pinocytic vesicles—a technique that can be conveniently implemented using our Influx™ pinocytic cell-loading reagent (I14402, Section 19.8)—has been successfully used for cellular loading of aequorin and the related photoprotein obelin.[36]

Because of its Ca^{2+}-dependent luminescence, the aequorin complex has been extensively used as an intracellular Ca^{2+} indicator. *Aequorea victoria* aequorin has been used to:

- Analyze the secretion response of single adrenal chromaffin cells to nicotinic cholinergic agonists [37]
- Calibrate micropipettes with injection volumes of as little as 3 picoliters [38]
- Clarify the role of Ca^{2+} release in heart muscle damage [39]
- Demonstrate the massive release of Ca^{2+} during fertilization [40]
- Study the regulation of the sarcoplasmic reticulum Ca^{2+} pump expression in developing chick myoblasts [41]

Coelenterazine and Its Synthetic Analogs

We offer coelenterazine and several synthetic coelenterazine analogs for reconstituting aequorin in cells that have been transfected with apoaequorin cDNA (Table 19.4). Cell permeation of coelenterazine, which has been demonstrated in organisms as diverse as *Escherichia coli*,[42] yeast,[43,44] *Dictyostelium* cells,[45] fish eggs,[46] mammalian cells [47–49] and plants,[21,50,51] is the rate-limiting step in the reconstitution process.[52] Coelenterazine is also required for generating the bioluminescent aequorin complex when using chimeric aequorin constructs.[53] Furthermore, coelenterazine and its analogs are substrates for the bioluminescent *Renilla* luciferase.[54,55]

In addition to native coelenterazine (C2944), we have synthesized three derivatives of coelenterazine that confer different Ca^{2+} affinities and spectral properties on the aequorin complex [56–58] (Table 19.4). Like native coelenterazine, these derivatives can be used to reconstitute the

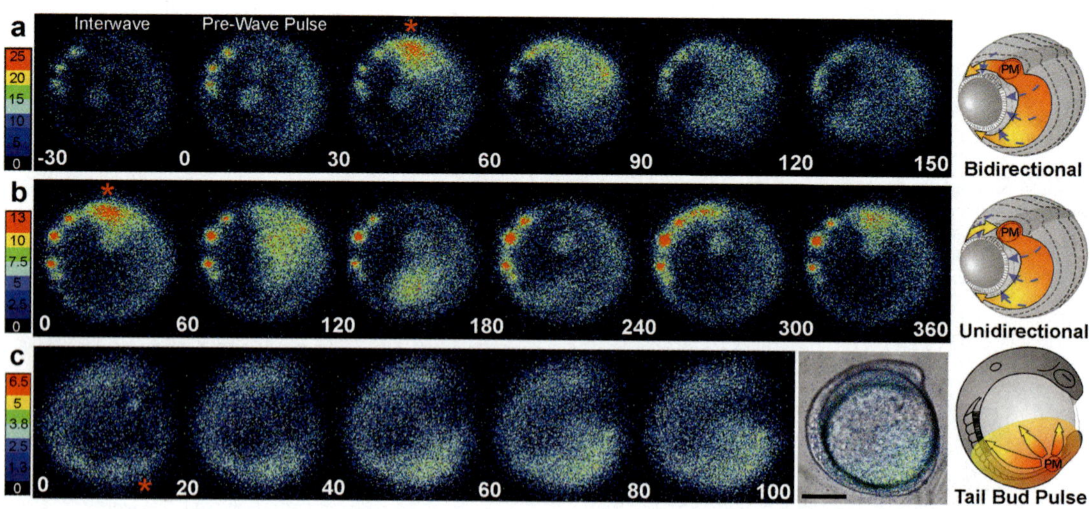

Figure 19.5.8 Images of Ca^{2+} waves in gastrulating zebrafish embryos detected by microinjected f aequorin (recombinant aequorin reconstituted with the coelenterazine f luminophore, C6779). The images are pseudocolored to represent Ca^{2+}-dependent luminescent flux in (photons/pixel/second × 10^{-2}) according to the color scales shown at the left of each of the three time-lapse image sequences (a,b,c). Time in seconds is indicated in the lower left-hand corner of each frame. The sequences depict three different spatial wave types that are represented schematically at the end of each sequence. PM indicates the dorsal midline pacemaker; its position in the luminescence images is marked by a red asterisk. The image was contributed by Edwin Gilland, Marine Biological Laboratory, Woods Hole, MA, and reproduced with permission from Proc Natl Acad Sci U S A (1999) 96:157.

IMPORTANT NOTICE: The products described in this manual are covered by one or more Limited Use Label License(s). Please refer to the Appendix on page 971 and Master Product List on page 975. Products are For Research Use Only. Not intended for any animal or human therapeutic or diagnostic use.

aequorin complex both *in vivo* and *in vitro*. However, intracellular reconstitution of aequorin from coelenterazine analogs can be relatively slow.[57] Aequorins containing the *cp*, *f* or *h* form of coelenterazine (C14260, C6779, C6780) exhibit relative intensities that are reported to be 10–20 times that of apoaequorin reconstituted with native coelenterazine.[21,58,59] Coelenterazine *cp* has been used in an automated high-throughput screening assay for G-protein–coupled receptors.[60] Coelenterazine is readily solubilized in aqueous solutions containing 50 mM hydroxypropyl-β-cyclodextrin.[61]

Table 19.4 Coelenterazines and their properties.

Cat. No.	Coelenterazine Analog	Em (nm) *	RLC †	Relative Intensity ‡	Half-Rise Time (msec) §
C2944	native	466	1.00	1	6–30
C14260	*cp*	442	0.63	28	2–5
C6779	*f*	472	0.80	20	6–30
C6780	*h*	466	0.75	16	6–30

* Emission maxima.

† Relative luminescence capacity = total time-integrated emission of aequorin in saturating Ca^{2+} relative to native aequorin = 1.0.

‡ Relative intensity at 100 nM Ca^{2+}.

§ Half-rise time = time for the luminescence signal to reach 50% of the maximum after addition of 1 mM Ca^{2+} to a standard of aequorin reconstituted with the coelenterazine analog of interest. All data are from Cell Calcium (1993) 14:373.

REFERENCES

1. Trends Mol Med (2008) 14:389; **2.** J Biol Chem (2008) 283:9377; **3.** J Biomed Opt (2007) 12:034017; **4.** Nat Protoc (2006) 1:1057; **5.** Proc Natl Acad Sci U S A (2004) 101:10554; **6.** J Immunol (2009) 183:1759; **7.** J Neurosci Methods (2009) 181:212; **8.** Nat Protoc (2006) 1:337; **9.** Nature (2000) 405:372; **10.** Trends Biotechnol (1999) 17:477; **11.** J Biochem (Tokyo) (1989) 105:473; **12.** J Chem Soc Chem Comm (1986) 21:1566; **13.** Methods Enzymol (1978) 57:271; **14.** Symp Soc Exp Biol (1976) 30:41; **15.** Proc Natl Acad Sci U S A (2006) 103:9500; **16.** J Biol Chem (2004) 279:33647; **17.** Chem Biol (1996) 3:337; **18.** Methods Cell Biol (1994) 40:305; **19.** Methods Enzymol (1989) 172:164; **20.** Trends Biotechnol (1998) 16:216; **21.** Cell Biol Int (1993) 17:111; **22.** Cell Calcium (1993) 14:736; **23.** Ann N Y Acad Sci (1991) 639:112; **24.** Dev Biol (1989) 135:182; **25.** Dev Biol (1986) 118:259; **26.** Biochem J (2001) 355:1; **27.** J Cell Biol (1991) 115:1259; **28.** Biochem J (1990) 270:309; **29.** J Gen Physiol (1985) 85:189; **30.** Biochemistry (1972) 11:1602; **31.** Biochem Biophys Res Commun (1985) 126:1259; **32.** Neurosci Lett (1991) 127:28; **33.** J Cell Biol (1985) 100:1522; **34.** Biochem Biophys Res Commun (1991) 177:888; **35.** Am J Physiol (1984) 247:C396; **36.** Cell Calcium (1985) 6:69; **37.** FEBS Lett (1987) 211:44; **38.** Anal Chem (1997) 69:3115; **39.** Nature (1984) 312:444; **40.** Proc Natl Acad Sci U S A (1977) 74:623; **41.** Am J Physiol (1986) 251:C512; **42.** FEBS Lett (1991) 282:405; **43.** Biochem Biophys Res Commun (1991) 174:115; **44.** Proc Natl Acad Sci U S A (1991) 88:6878; **45.** FEBS Lett (1994) 337:43; **46.** J Neurochem (1998) 71:1298; **47.** Nature (1992) 358:325; **48.** Anal Biochem (1993) 209:343; **49.** Cell Calcium (1993) 14:663; **50.** Nature (1991) 352:524; **51.** J Cell Biol (1993) 121:83; **52.** Cell Calcium (1997) 22:439; **53.** J Biolumin Chemilumin (1989) 4:346; **54.** Nat Methods (2007) 4:641; **55.** Mol Imaging (2004) 3:43; **56.** Biochem J (1995) 306:537; **57.** Biochem J (1993) 296:549; **58.** Biochem J (1989) 261:913; **59.** Cell Calcium (1991) 12:635; **60.** Anal Biochem (1999) 272:34; **61.** Biosci Biotechnol Biochem (1997) 61:1219.

DATA TABLE 19.5 PROTEIN-BASED Ca^{2+} INDICATORS

Cat. No.	MW	Storage	Soluble	Abs	EC	Em	Solvent	Notes
C2944	423.47	FF,D,LL,AA	MeOH	429	7500	see Notes	pH 7	1, 2, 3
C6779	425.46	FF,D,LL,AA	MeOH	437	8700	see Notes	MeOH	1, 2
C6780	407.47	FF,D,LL,AA	MeOH	437	9500	see Notes	MeOH	1, 2
C14260	415.49	FF,D,LL,AA	MeOH	430	7000	see Notes	MeOH	1, 2

For definitions of the contents of this data table, see "Using *The Molecular Probes® Handbook*" in the introductory pages.

Notes

1. Coelenterazine complexes with apoaequorin emit calcium-dependent bioluminescence. Bioluminescence emission maxima (relative intensity at 100 nM Ca^{2+}) are as follows: C2944, 466 nm (1); C6779, 472 nm (20); C6780, 466 nm (16); C14260, 442 nm (28). (Cell Calcium (1993) 14:373)
2. Do NOT dissolve in DMSO.
3. Aqueous solutions of coelenterazine (>1 mM) can be prepared in pH 7 buffer containing 50 mM 2-hydroxypropyl-β-cyclodextrin. (Biosci Biotechnol Biochem (1997) 61:1219)

PRODUCT LIST 19.5 PROTEIN-BASED Ca^{2+} INDICATORS

Cat. No.	Product	Quantity
A6785	AquaLite® aequorin (aequorin) *recombinant*	25 µg
C2944	coelenterazine	250 µg
C14260	coelenterazine *cp*	250 µg
C6779	coelenterazine *f*	250 µg
C6780	coelenterazine *h*	250 µg
P36207	Premo™ Cameleon Calcium Sensor *for 10 microplates*	1 kit
P36208	Premo™ Cameleon Calcium Sensor *for 100 microplates*	1 kit

IMPORTANT NOTICE: The products described in this manual are covered by one or more Limited Use Label License(s). Please refer to the Appendix on page 971 and Master Product List on page 975. Products are For Research Use Only. Not intended for any animal or human therapeutic or diagnostic use.

19.6 Fluorescent Mg^{2+} Indicators

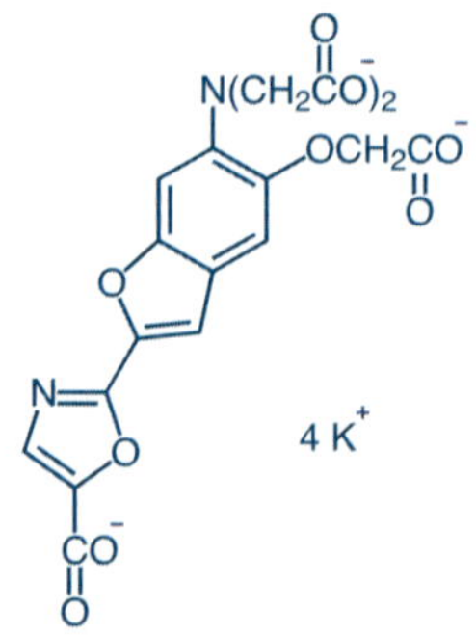

Figure 19.6.1 Mag-fura-2, tetrapotassium salt (M1290).

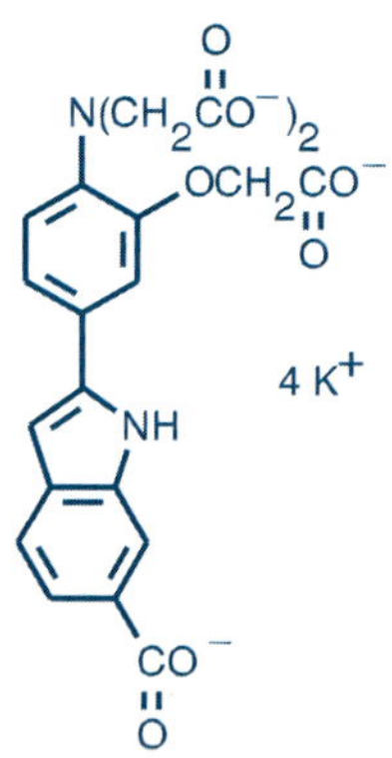

Figure 19.6.2 Mag-indo-1, tetrapotassium salt.

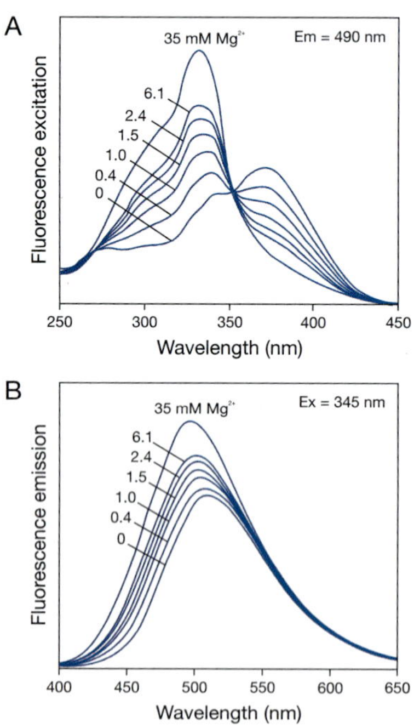

Figure 19.6.3 A) Fluorescence excitation and B) fluorescence emission spectra of mag-fura-2 (M1290) in solutions containing 0–35 mM Mg^{2+}.

Intracellular Mg^{2+} is important for mediating enzymatic reactions, DNA synthesis, hormonal secretion and muscular contraction. To facilitate the investigation of magnesium's role in these and other cellular functions, we offer several different fluorescent indicators for measuring intracellular Mg^{2+} concentration. They include furaptra,[1,2] which we refer to as mag-fura-2 to denote the similarity of its structure (Figure 19.6.1) and spectral response with the Ca^{2+} indicator fura-2; and mag-indo-1, with a structure (Figure 19.6.2) and spectral response similar to that of indo-1. For applications such as confocal laser-scanning microscopy and flow cytometry, we offer the Magnesium Green™ and mag-fluo-4 indicators. The various methods for measuring intracellular Mg^{2+} have been reviewed.[3]

Mg^{2+} indicators are generally designed to maximally respond to the Mg^{2+} concentrations commonly found in cells, typically ranging from about 0.1 mM to 6 mM. Intracellular free Mg^{2+} levels have been reported to be ~0.3 mM in synaptosomes,[4] 0.37 mM in hepatocytes[5] and 0.5–1.2 mM in cardiac cells,[6] whereas the concentration of Mg^{2+} in normal serum is ~0.44–1.5 mM.[7] Measurements using fluorescent Mg^{2+} indicators are somewhat more demanding than intracellular Ca^{2+} determinations because physiological changes in Mg^{2+} concentration are relatively small. Compartmentalization and binding to proteins can also be a problem in use of these indicators in cells.[8] Mg^{2+} indicators also bind Ca^{2+}; however, typical physiological Ca^{2+} concentrations (10 nM–1 µM) usually do not interfere with Mg^{2+} measurements because the affinity of these indicators for Ca^{2+} is low. For ultrasensitive Mg^{2+} measurement applications, intracellular Ca^{2+} background can be suppressed using BAPTA AM (B1205, B6769; Section 19.8).[9] Although Ca^{2+} binding by Mg^{2+} indicators can be a complicating factor in Mg^{2+} measurements,[10,11] this property can also be exploited for measuring high Ca^{2+} concentrations (1–100 µM);[12] see Section 19.2 and Section 19.3 for further examples.

For intracellular calibration of Mg^{2+} indicators, we offer the ionophores A-23187 and the nonfluorescent 4-bromo A-23187 (A1493, B1494; Section 19.8), which are preferred over ionomycin (I24222) because they transport Mg^{2+} more effectively.[2,13] Solutions used to calibrate Mg^{2+} indicators should be initially free of heavy metals such as Mn^{2+} that can interact with the indicators. These metals can be removed by treating the solution with the divalent cation chelator TPEN (T1210, Section 19.8).

Magnesium Indicators Excited by UV Light

Mag-Fura-2 and Mag-Indo-1

The dissociation constant for Mg^{2+} of mag-indo-1 is 2.7 mM, slightly higher than that of mag-fura-2, which is 1.9 mM. The lower-affinity mag-indo-1 indicator is sensitive to somewhat higher spikes in intracellular Mg^{2+}.[14,15] The affinities of mag-fura-2 and mag-indo-1 for Mg^{2+} are reported to be essentially invariant at pH values between 5.5 and 7.4 and at temperatures between 22°C and 37°C.[16] A detailed study of the photophysics of mag-fura-2 has been published.[17] Comparisons of intracellular and solution dissociation constants for mag-fura-2 have been published by Günther[3] and by Tashiro and Konishi.[13]

As with their Ca^{2+} indicator analogs, mag-fura-2 undergoes an appreciable shift in excitation wavelength upon Mg^{2+} binding (Figure 19.6.3), and mag-indo-1 exhibits a shift in both its excitation and emission wavelengths (Figure 19.6.4). Equipment, optical filters and calibration methods are very similar to those required for the Ca^{2+} indicators. The excitation-ratioable mag-fura-2 indicator is most useful for fluorescence microscopy, whereas the emission-ratioable mag-indo-1 indicator is preferred for flow cytometry.[18] Many applications of mag-fura-2 involve estimation of the affinity and selectivity of Mg^{2+} binding to proteins.[19,20] Displacement of bound Mg^{2+} by Li^+ provides a surrogate assay for Li^+ transport, a process for which few direct detection methods exist.[21,22] Researchers have used mag-fura-2 to measure intracellular Mg^{2+} in a wide variety of cells, organelles and tissues, including:

- Cortical neurons[23]
- Isolated mitochondria[24,25]
- Platelets[26,27]
- Rat hepatocytes[28]
- Rat ventricular myocytes[29]
- *Xenopus* oocytes[30]

IMPORTANT NOTICE: The products described in this manual are covered by one or more Limited Use Label License(s). Please refer to the Appendix on page 971 and Master Product List on page 975. Products are For Research Use Only. Not intended for any animal or human therapeutic or diagnostic use.

In addition to the cell-impermeant potassium salt of mag-fura-2 (M1290), we offer the cell-permeant AM esters of mag-fura-2 and mag-indo-1 as a set of 20 vials, each containing 50 µg (M1292, M1295). The special packaging is recommended when small quantities of the dyes are to be used over a long period of time. Mag-fura-2 AM is also available in a single vial containing 1 mg (M1291).

Magnesium Indicators Excited by Visible Light

We also offer visible light–excitable Mg^{2+} indicators, including the Magnesium Green™ and mag-fluo-4 indicators. As with mag-fura-2 and mag-indo-1, these visible light–excitable Mg^{2+} indicators can also be used as low-affinity Ca^{2+} indicators (Section 19.3) and may be useful as indicators for Zn^{2+} and other metals (Section 19.7).

Magnesium Green™ Indicator

The Magnesium Green™ indicator (Figure 19.6.5) exhibits a higher affinity for Mg^{2+} (K_d ~1.0 mM) than does mag-fura-2 (K_d ~1.9 mM) or mag-indo-1 (K_d ~2.7 mM); this indicator also binds Ca^{2+} with moderate affinity (K_d for Ca^{2+} in the absence of Mg^{2+} is ~6 µM, measured at 22°C using our Calcium Calibration Buffer Kits). The spectral properties of the Magnesium Green™ indicator are similar to those of the Calcium Green™ indicators. Upon binding Mg^{2+}, Magnesium Green™ exhibits an increase in fluorescence emission intensity without a shift in wavelength (Figure 19.6.6). The Magnesium Green™ indicator has been used to investigate the binding of free Mg^{2+} by the bacterial SecA protein [31] and by protein tyrosine kinases.[32] By exploiting the fact that ATP has greater Mg^{2+}-binding affinity than ADP, researchers have used Magnesium Green™ to indirectly measure ATP in pancreatic acinar cells,[33] in bullfrog hair cells,[34] in cultured *Xenopus* spinal neurons [35] and in isolated mitochondria.[36] Magnesium Green™ is available as a cell-impermeant potassium salt (M3733) or as a cell-permeant AM ester (M3735).

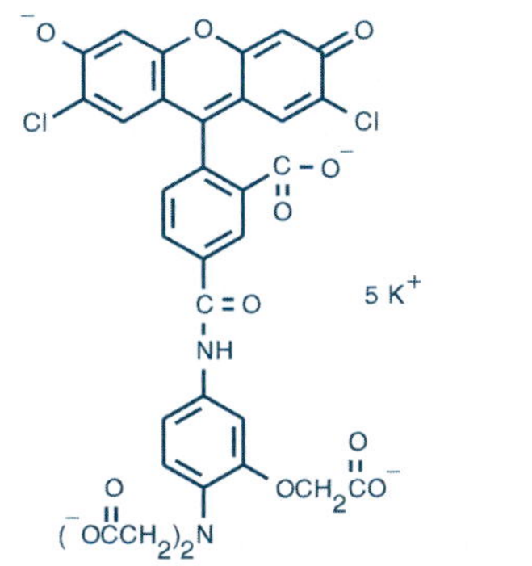

Figure 19.6.5 Magnesium Green™, pentapotassium salt (M3733).

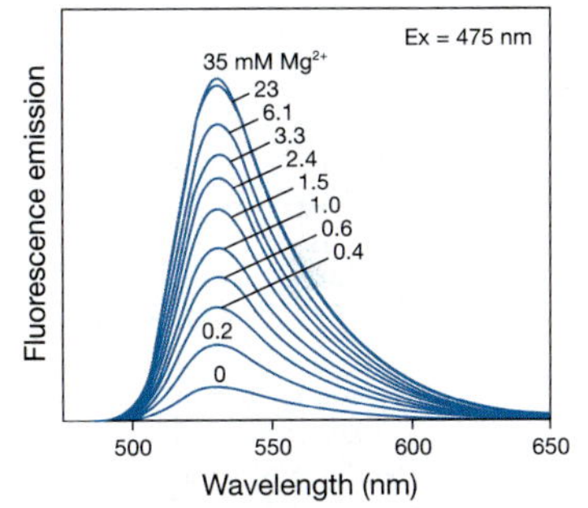

Figure 19.6.6 Mg^{2+}-dependent fluorescence emission spectra of Magnesium Green™ (M3733).

Mag-Fluo-4

Mag-fluo-4 (Figure 19.6.7) is an analog of fluo-4 with a K_d for Mg^{2+} of 4.7 mM and a K_d for Ca^{2+} of 22 µM (measured at 22°C using our Calcium Calibration Buffer Kits), making it useful as an intracellular Mg^{2+} indicator as well as a low-affinity Ca^{2+} indicator (Section 19.3). Mag-fluo-4 has a much more sensitive fluorescence response to Mg^{2+} binding than does our Magnesium Green™ indicator. Because physiological fluctuations of intracellular Mg^{2+} concentration are typically small, this increased sensitivity is a considerable advantage.[9] Like fluo-4, mag-fluo-4 is essentially nonfluorescent in the absence of divalent cations and exhibits strong fluorescence enhancement with no spectral shift upon binding Mg^{2+} (Figure 19.6.8). Mag-fluo-4 is available as a cell-impermeant potassium salt (M14205) or as a cell-permeant AM ester (M14206).

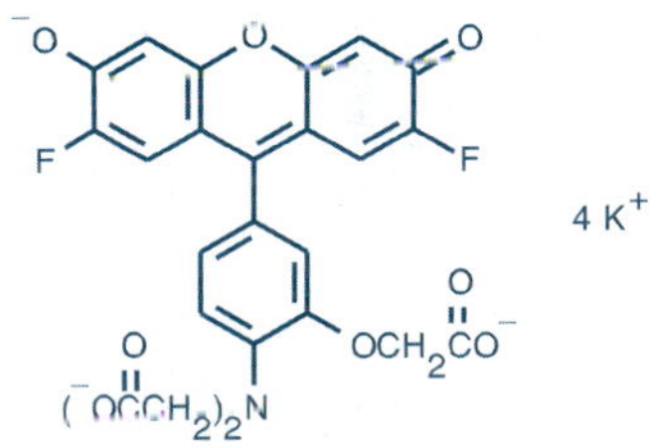

Figure 19.6.7 Mag-fluo-4, tetrapotassium salt. (M14205).

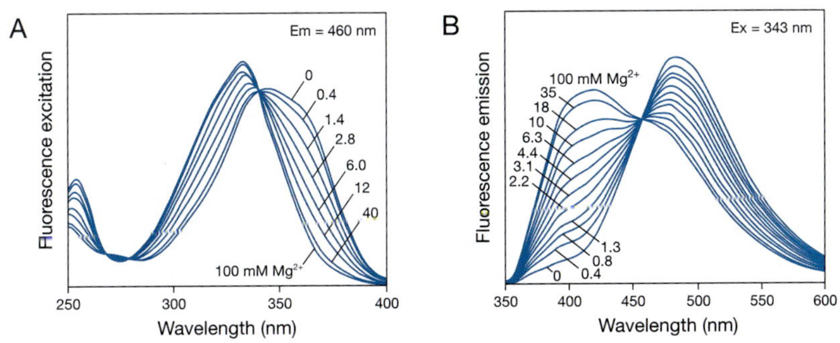

Figure 19.6.4 A) Fluorescence excitation and B) fluorescence emission spectra of mag-indo-1 in solutions containing 0–100 mM Mg^{2+}.

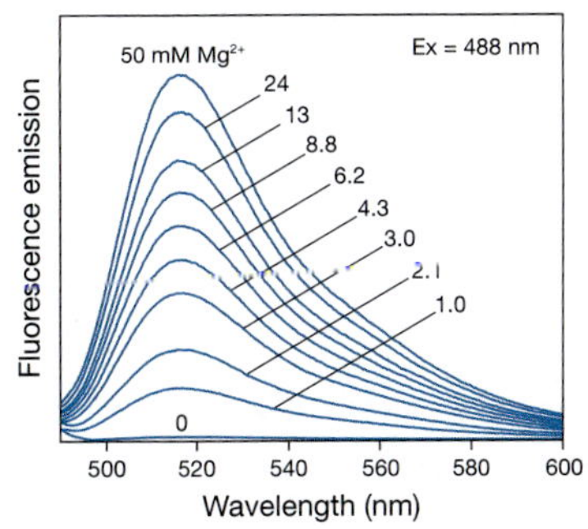

Figure 19.6.8 Fluorescence emission spectra of mag-fluo-4 (M14205) in solutions containing 0–50 mM Mg^{2+}.

REFERENCES

1. Proc Natl Acad Sci U S A (1989) 86:2981; **2.** Am J Physiol (1989) 256:C540; **3.** Magnes Res (2006) 19:225; **4.** Biochim Biophys Acta (1987) 898:331; **5.** J Biol Chem (1986) 261:2567; **6.** Annu Rev Physiol (1991) 53:273; **7.** Clin Chem (1989) 35:1492; **8.** Anal Biochem (2001) 290:221; **9.** Anal Chem (2009) 81:538; **10.** Am J Physiol (1992) 263:C300; **11.** Mol Cell Biochem (1994) 136:11; **12.** Biophys J (2009) 97:1864; **13.** Biophys J (1997) 73:3358; **14.** J Physiol (1994) 475:319; **15.** Biophys J (1998) 75:957; **16.** Biochem Biophys Res Commun (1991) 177:184; **17.** Chem Phys Lett (1998) 287:412; **18.** J Mol Biol (2010) 396:858; **19.** Biochemistry (2007) 46:3692; **20.** Biochemistry (2006) 45:763; **21.** Biochim Biophys Acta (2004) 1691:79; **22.** J Inorg Biochem (2004) 98:691; **23.** Neuron (1993) 11:751; **24.** Biochim Biophys Acta (1997) 1320:310; **25.** J Biol Chem (1998) 273:7850; **26.** Clin Chem (1996) 42:744; **27.** Am J Physiol (1998) 274:R548; **28.** Biochem J (1997) 326:823; **29.** Biophys J (2005) 88:1911; **30.** J Biol Chem (2007) 282:8060; **31.** J Biol Chem (1995) 270:18975; **32.** Biochemistry (1997) 36:2139; **33.** Am J Physiol Cell Physiol (2008) 295:C1247; **34.** Neuron (2007) 53:371; **35.** Mol Biol Cell (2008) 19:150; **36.** Biophys J (2009) 96:2490.

DATA TABLE 19.6 FLUORESCENT Mg^{2+} INDICATORS

				Low Mg^{2+}				High Mg^{2+}						
Cat. No.	**MW**	**Storage**	**Soluble**	**Abs**	**EC**	**Em**	**Solvent**	**Abs**	**EC**	**Em**	**Solvent**	**Product**	**K_d**	**Notes**
M1290	586.68	F,D,L	pH >6	369	22,000	511	H_2O	330	24,000	491	H_2O/Mg^{2+}		1.9 mM	1, 2, 3, 4
M1291	722.57	F,D,L	DMSO	366	31,000	475	EtOAc					M1290		
M1292	722.57	F,D,L	DMSO	366	31,000	475	EtOAc					M1290		
mag-indo-1	594.74	F,D,L	pH >6	349	38,000	480	H_2O	330	33,000	417	H_2O/Mg^{2+}		2.7 mM	1, 2, 3, 4
M1295	730.63	F,D,L	DMSO	354	37,000	472	MeOH					mag-indo-1		
M3733	915.90	F,D,L	pH >6	506	77,000	531	H_2O	506	75,000	531	H_2O/Mg^{2+}		1.0 mM	1, 2, 3, 4, 5
M3735	1025.71	F,D	DMSO	302	16,000	none	MeOH					M3733		
M14205	681.77	F,D,L	pH >6	490	74,000	see Notes	H_2O	493	75,000	517	H_2O/Mg^{2+}		4.7 mM	1, 2, 3, 4, 6
M14206	817.66	F,D,L	DMSO	457	25,000	see Notes	MeOH					M14205		7

For definitions of the contents of this data table, see "Using *The Molecular Probes® Handbook*" in the introductory pages.

Notes

1. Dissociation constants are known to vary considerably depending on the temperature, pH, ionic strength, viscosity, protein binding, presence of other ions (especially polyvalent ions), instrument setup and other factors. It is strongly recommended that these values be verified under user-specific experimental conditions.
2. This indicator binds Ca^{2+} with higher affinity than Mg^{2+}, producing a similar spectral response.
3. $K_d(Mg^{2+})$ values have been determined in 115 mM KCl, 20 mM NaCl, 10 mM Tris, pH 7.05, 0 to 35 mM Mg^{2+} at 22°C.
4. Spectra measured in aqueous buffers containing zero or 35 mM Mg^{2+}, indicated as H_2O and H_2O/Mg^{2+}, respectively.
5. This indicator exhibits fluorescence enhancement in response to ion binding, with essentially no change in absorption or emission wavelengths.
6. Fluorescence of the free indicator is very weak and is enhanced >100-fold on binding Mg^{2+}.
7. Fluorescence of this AM ester derivative is very weak and is enhanced only after hydrolytic cleavage followed by binding of divalent cations to the anionic indicator.

PRODUCT LIST 19.6 FLUORESCENT Mg^{2+} INDICATORS

Cat No.	Product	Quantity
M14206	mag-fluo-4, AM *cell permeant* *special packaging*	10 x 50 µg
M14205	mag-fluo-4, tetrapotassium salt *cell impermeant*	500 µg
M1291	mag-fura-2, AM *cell permeant*	1 mg
M1292	mag-fura-2, AM *cell permeant* *special packaging*	20 x 50 µg
M1290	mag-fura-2, tetrapotassium salt *cell impermeant*	1 mg
M1295	mag-indo-1, AM *cell permeant* *special packaging*	20 x 50 µg
M3735	Magnesium Green™, AM *cell permeant* *special packaging*	20 x 50 µg
M3733	Magnesium Green™, pentapotassium salt *cell impermeant*	1 mg

IMPORTANT NOTICE: The products described in this manual are covered by one or more Limited Use Label License(s). Please refer to the Appendix on page 971 and Master Product List on page 975. Products are For Research Use Only. Not intended for any animal or human therapeutic or diagnostic use.
www.invitrogen.com/probes
molecular probes® | invitrogen™ by life technologies™

19.7 Fluorescent Indicators for Zn^{2+} and Other Metal Ions

Not only do transition metal ions play an important role in biological structure and activity,[1] but they can also serve as surrogates and blockers of ion transport through Ca^{2+} channels[2,3] (Figure 19.7.1). Measuring heavy metal ion concentrations in cells and environmental samples with indicators originally designed for detection of Ca^{2+} and Mg^{2+} has been hampered by competitive binding of other, more abundant, cations.[4] Detection methods rely both on novel fluorescent ion sensors specifically designed for metal detection,[5] as well as on new applications for indicators originally designed for detection of Ca^{2+} and Mg^{2+}[6] (Table 19.5, Figure 19.7.2, Figure 19.7.3).

Several of our fluorescent indicators can be used to selectively determine polyvalent cation concentrations inside cells or to follow metal ion transport through ion channels. Other indicators are primarily useful for measurements in solutions or in extracts of environmental samples.[7] In most cases, the high affinity of the indicators for metal ions allows interference from other compounds to be minimized by diluting the sample with deionized water. As with Ca^{2+} and Mg^{2+} detection, spectroscopic responses to metal ion binding and dissociation constant (K_d) values are dependent on many factors, including pH, temperature, viscosity, protein binding and the presence of other ions. These responses may vary significantly in complex environments such as seawater and the cytosol.[8]

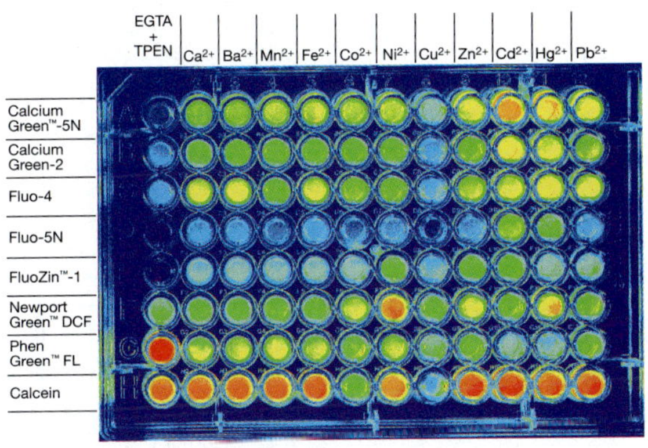

Figure 19.7.2 Visual screening of fluorescent indicator responses to metal ions. The image shows a 96-well microplate containing various combinations of ions and indicators in 50 mM MOPS pH 7.0. Each row of wells represents a different indicator; each column of wells represents a different ion. The indicators (top to bottom) are: Calcium Green™-5N (0.5 µM), Calcium Green™-2 (0.2 µM), fluo-4 (2 µM), fluo-5N (2 µM), FluoZin™-1 (2 µM), Newport Green™ DCF (2 µM), Phen Green™ FL (2 µM; this indicator salt is no longer routinely available, please inquire) and calcein (0.5 µM). The left-hand column of wells contains 10 mM EGTA + 10 µM TPEN (ion-free reference solution). Subsequent columns (left to right) represent 1 µM concentrations of Ca^{2+}, Ba^{2+}, Mn^{2+}, Fe^{2+}, Co^{2+}, Ni^{2+}, Cu^{2+}, Zn^{2+}, Cd^{2+}, Hg^{2+} and Pb^{2+}, respectively. The microplate was scanned using a FLA3000G laser scanner (Fuji® Photo Film Co.) with excitation at 473 nm and fluorescence emission detected at 520 nm. The image is pseudocolored according to fluorescence intensity (high = red > orange > yellow > green > blue = low).

Table 19.5 Response of fura-2 and indo-1 to some divalent cations other than Ca^{2+} and Mg^{2+}.

Ion	Indicator	K_d *	Spectroscopic Effect	Reference
Ba^{2+}	fura-2	1.4×10^{-6} M	Spectra similar to Ca^{2+} complex; no quenching	1
Cd^{2+}	fura-2	1.0×10^{-12} M	Spectra similar to Ca^{2+} complex; higher intensity	2
Co^{2+}	fura-2	8.6×10^{-9} M	>99% Quenched relative to Ca^{2+} complex	1
Mn^{2+}	fura-2	2.8×10^{-9} M	>99% Quenched relative to Ca^{2+} complex	1
Pb^{2+}	fura-2	4.2×10^{-12} M	Spectra shifted relative to Ca^{2+} complex; no quenching	3
Pb^{2+}	indo-1	3.5×10^{-11} M	Fluorescence quenched	4
Sr^{2+}	fura-2	7.6×10^{-6} M	Spectra similar to Ca^{2+} complex; no quenching	1
Zn^{2+}	fura-2	3.0×10^{-9} M	Spectra similar to Ca^{2+} complex; no quenching	5
Zn^{2+}	indo-1	1.6×10^{-10} M	Spectra similar to Ca^{2+} complex; no quenching	6

* Dissociation constant reported in cited literature reference.
1. J Biol Chem (1990) 265:678; **2.** J Biol Chem (1992) 267:25553; **3.** Am J Physiol (1990) 259:C762; **4.** J Biol Chem (1997) 272:8346; **5.** J Biol Chem (1995) 270:2473; **6.** Anal Biochem (1990) 187:328.

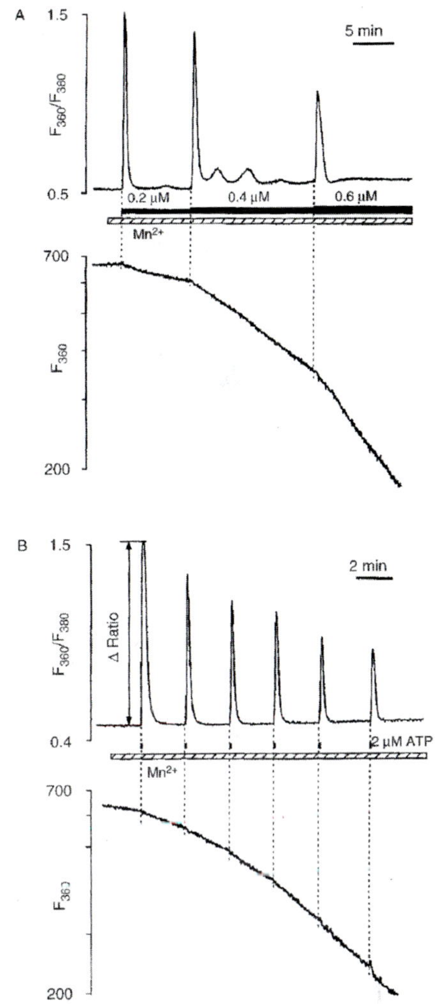

Figure 19.7.1 Discrimination of ATP-induced intracellular Ca^{2+} release from capacitative Ca^{2+} entry in bovine vascular endothelial cells using quenching of fura-2 fluorescence by Mn^{2+}. Intracellular Ca^{2+} release was detected by fura-2 fluorescence excited alternately at 360 nm and 380 nm (F_{360}/F_{380}). Capacitative entry of extracellular Mn^{2+} was detected by its quenching effect on the Ca^{2+}–insensitive fura-2 fluorescence signal excited at 360 nm (F_{360}). Successive Ca^{2+} spikes induced by sustained applications of low ATP concentrations (**A**) or transient 5-second applications of 2 µM ATP (**B**) were followed by accelerated entry of Mn^{2+}. Experiments were conducted in Ca^{2+}-free extracellular medium containing 100 µM Mn^{2+}. Under these conditions, Mn^{2+} functions as an ionic surrogate for Ca^{2+} with respect to transmembrane conductance, and Ca^{2+} signals can be unambiguously attributed to release from intracellular stores. Figure reproduced with permission from Sedova M, Klishin A, Huser J, Blatter LA, J Physiol (2000) 523:549.

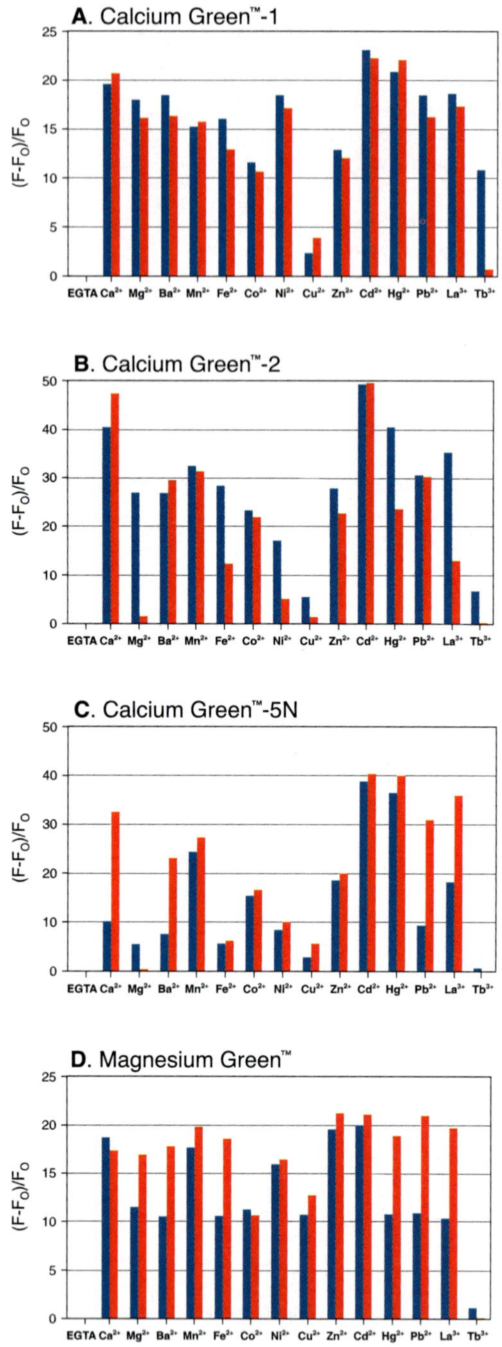

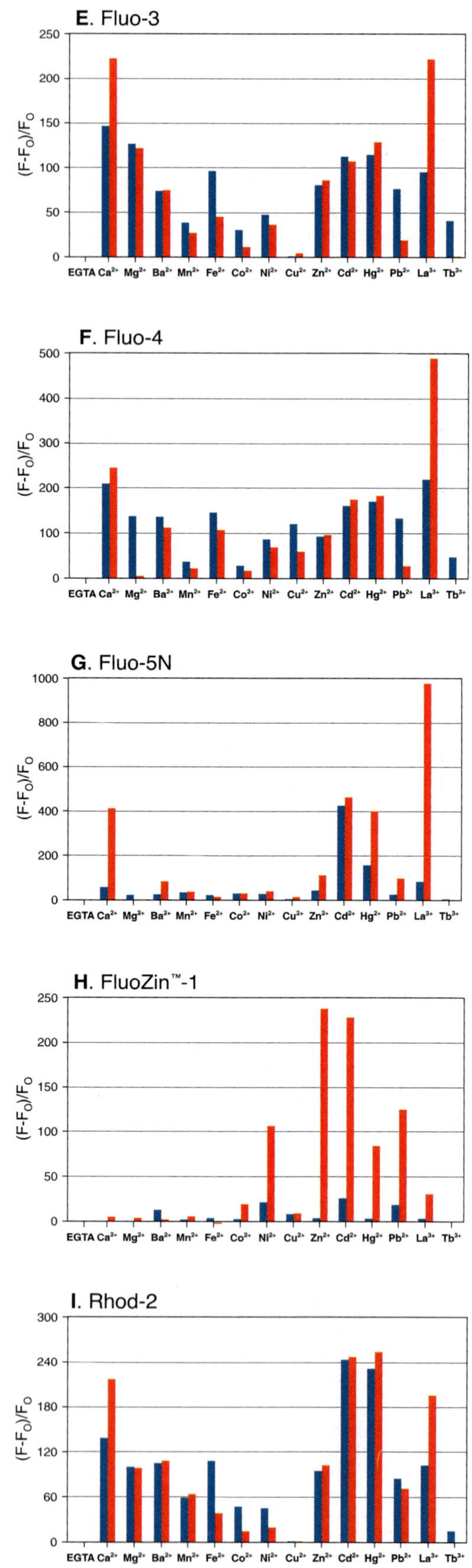

Figure 19.7.3 Metal-ion response screening for various fluorescent indicators. The maximum relative fluorescence intensity was measured for identical indicator concentrations in solutions containing 10 mM EGTA + 10 µM TPEN, 1 µM ion (100 µM for Mg^{2+}) and 100 µM ion (10 mM for Mg^{2+}). Results are plotted as fluorescence changes relative to the ion-free (10 mM EGTA + 10 µM TPEN) reference solution expressed as $(F-F_O)/F_O$, where F is the fluorescence intensity of ion-containing solutions and F_O is the fluorescence intensity of the reference solution. Blue bars indicate the response to 1 µM ion (100 µM for Mg^{2+}), and red bars indicate the response to 100 µM ion (10 mM for Mg^{2+}). The indicators represented are: **A)** Calcium Green™-1, **B)** Calcium Green™-2, **C)** Calcium Green™-5N, **D)** Magnesium Green™, **E)** fluo-3, **F)** fluo-4, **G)** fluo-5N, **H)** FluoZin™-1, **I)** rhod-2, **J)** rhod-5N, **K)** X-rhod-1, **L)** fura-2, **M)** mag-fura-2, **N)** FuraZin™-1 (no longer routinely available; please inquire), **O)** Fura Red™, **P)** fura-FF, **Q)** indo-1, **R)** FluoZin™-3 and **S)** BTC.

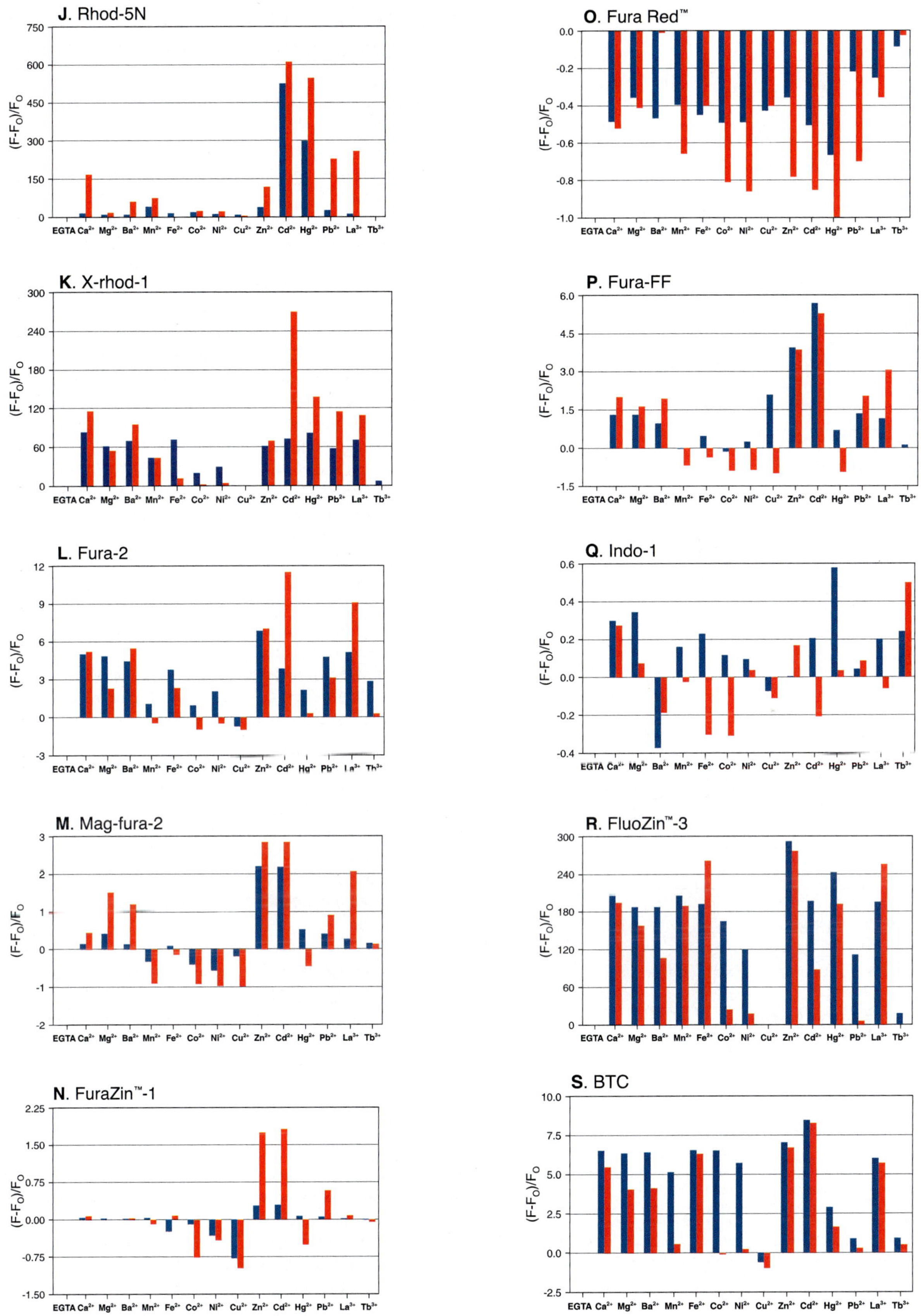

IMPORTANT NOTICE: The products described in this manual are covered by one or more Limited Use Label License(s). Please refer to the Appendix on page 971 and Master Product List on page 975. Products are For Research Use Only. Not intended for any animal or human therapeutic or diagnostic use.

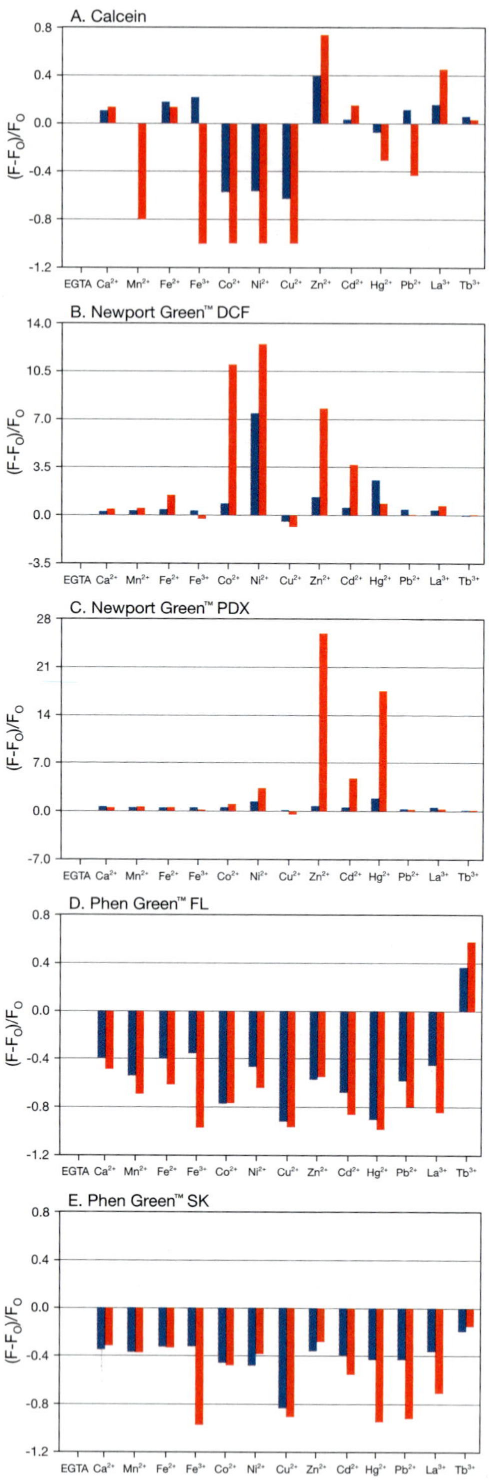

Figure 19.7.4 Metal-ion response screening for calcein, Newport Green™ and Phen Green™ indicators. The maximum relative fluorescence intensity was measured for identical indicator concentrations in solutions containing 10 mM EGTA + 10 µM TPEN, 1 µM ion and 100 µM ion. Results are plotted as fluorescence changes relative to the ion-free (10 mM EGTA + 10 µM TPEN) reference solution expressed as $(F-F_o)/F_o$, where F is the fluorescence intensity of ion-containing solutions and F_o is the fluorescence intensity of the reference solution. Blue bars indicate the response to 1 µM ion, and red bars indicate the response to 100 µM ion. The indicators represented are: **A)** calcein, **B)** Newport Green™ DCF, **C)** Newport Green™ PDX, **D)** Phen Green™ FL and **E)** Phen Green™ SK. The water-soluble salt forms of the Newport Green™ PDX and Phen Green™ FL indicators are no longer routinely available; please inquire.

Applications of Ca^{2+} and Mg^{2+} Indicators for Detection of Zn^{2+} and Other Metals

We have tested the responses of many of the Ca^{2+} and Mg^{2+} indicators (described in Section 19.2 through Section 19.6) to a series of polyvalent metal ions (Figure 19.7.2, Figure 19.7.3). The BAPTA-based indicators, including the Calcium Green™ dyes, fluo-3 and fluo-4, exhibit the highest emission intensities upon binding La^{3+}, Hg^{2+} and Cd^{2+}. The response of the Calcium Green™-1 and Magnesium Green™ indicators shows relatively little variability amongst the ions tested. Indicators nominally designed for detecting low Ca^{2+} concentrations, such as Calcium Green™-5N, fluo-5N and rhod-5N, show little response to transition metal ions such as Fe^{2+}, Co^{2+}, Ni^{2+} and Co^{2+} and much stronger responses to heavier ions such as Cd^{2+}, Hg^{2+} and Pb^{2+}.[9] Note that the responses shown in Figure 19.7.3 are not necessarily saturated or even uniform in some cases. Some indication of the overall response pattern can be obtained by comparing the effects of the 1 µM and 100 µM ion concentrations sampled in these experiments. Also, note that our selection of excitation and emission wavelengths for the measurements may significantly affect the absolute and relative magnitudes of the changes observed.

Fura-2 and indo-1, like other BAPTA-based Ca^{2+} indicators, are highly selective for Ca^{2+} over Mg^{2+}; however, they bind other divalent and trivalent cations with significantly higher affinity. In terms of spectroscopic detection, some of these ions (e.g., Ba^{2+}, Cd^{2+} and Sr^{2+}) mimic the effects of Ca^{2+};[10–12] others (e.g., Mn^{2+}, Co^{2+} and Ni^{2+}) exert strong fluorescence quenching effects (Table 19.5). Although heavy metal cations are a potentially serious source of interference in Ca^{2+} measurements using fluorescent indicators, intracellular concentrations are fortunately very low in most cases. When interference occurs, it can be identified and controlled using the selective heavy metal ion chelator TPEN[13–16] (T1210, Section 19.8). Quenching of indicator fluorescence by Mn^{2+} has several useful applications:

- Calibrating the intracellular Ca^{2+} response of fluo-3, fluo-4, rhod-2 and related indicators[17–19]
- Discriminating fluorescence signals from mitochondrial and cytosolic indicator populations[20,21]
- Estimating the fraction of residual Ca^{2+}-insensitive indicators after AM ester loading[2,22]
- Estimating indicator leakage in measurements on cell suspensions in cuvettes[23]

Mn^{2+} is also extensively used as an ionic surrogate for measuring Ca^{2+} influx through ion channels[3,24–27] (Figure 19.7.1). Because the effect of Mn^{2+} on fura-2 fluorescence is quite different from that of Ca^{2+} (Table 19.5), influx can be clearly distinguished from Ca^{2+} elevation due to mobilization of intracellular stores,[2,28] although such clarity may be impaired by the existence of alternative Mn^{2+}-influx pathways.[29] A similar analysis applies when using fura-2 and indo-1 to study the antagonistic effects of ions such as Ni^{2+}, La^{3+} and Co^{2+} on voltage-activated and Ca^{2+} release–activated Ca^{2+} channels. In this case, the Ca^{2+}-dependent fluorescence signal is of primary interest and permeability of the antagonist ion, resulting in direct interactions with the indicator, will usually invalidate the measurement.[30–32] For the Calcium Green™, Oregon Green® 488 BAPTA and Magnesium Green™ indicator series, the response to Mn^{2+} is not markedly different from the Ca^{2+} response (Figure 19.7.3), making these indicators generally unsuitable for the applications based on Ca^{2+}/Mn^{2+} discrimination described above. A variety of fluorescent

IMPORTANT NOTICE: The products described in this manual are covered by one or more Limited Use Label License(s). Please refer to the Appendix on page 971 and Master Product List on page 975. Products are For Research Use Only. Not intended for any animal or human therapeutic or diagnostic use.
www.invitrogen.com/probes
molecular probes® | invitrogen™ by life technologies™

indicators, including fura-2, fura-FF and our Calcium Orange™ and Magnesium Green™ indicators, have been utilized to detect Sr^{2+}, an ion that can replace Ca^{2+} in triggering neurotransmitter release[33,34] and also serves as a blocker of mitochondrial permeability transition pore opening.[35]

Fluorescence of calcein (C481) is quenched strongly by Co^{2+}, Ni^{2+} and Cu^{2+} and appreciably by Fe^{3+} and Mn^{2+} at physiological pH[36] (panel A, Figure 19.7.4). This fluorescence quenching response can be exploited for detecting the opening of the mitochondrial permeability transition pore,[37–39] for assaying membrane fusion (Assays of Volume Change, Membrane Fusion and Membrane Permeability—Note 14.3) and for monitoring cellular iron transport and the cellular labile iron pool.[36,40–43]

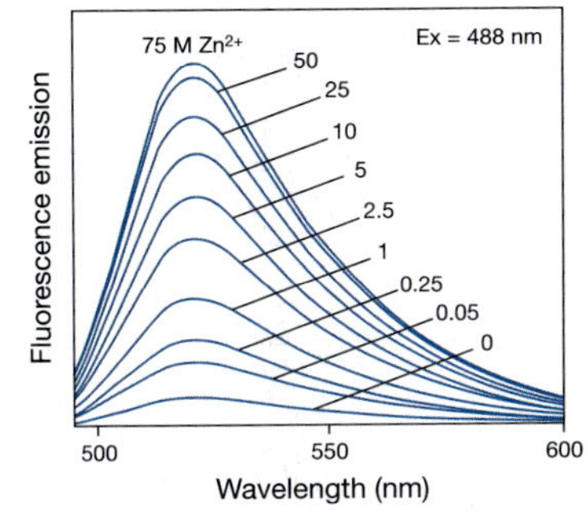

Figure 19.7.5 Zn^{2+} dependence of the fluorescence emission spectra of FluoZin™-2.

Indicators for Zinc

Zinc is the second most abundant transition metal in living organisms after iron. It is of particular importance in the regulation of gene expression, as Zn^{2+} binding proteins account for nearly 50% of the transcription regulatory proteins in the human genome. Zn^{2+} is also functionally active in pancreatic insulin secretion[44,45] and is a contributory factor in neurological disorders including epilepsy and Alzheimer disease.[46–48] Free Zn^{2+} is released from metalloprotein complexes during oxidative stress.[49,50] The intracellular concentration of free Zn^{2+} is extremely low in most cells (<1 nM), with the remainder being bound to proteins or nucleic acids.[51] One calculation[52] yields an intracellular free Zn^{2+} concentration of six orders of magnitude less than one atom per cell. We prepare a variety of Zn^{2+} indicators (Table 19.6) to help elucidate the role of Zn^{2+} release and the localization of free or chelatable Zn^{2+} in cells.[53]

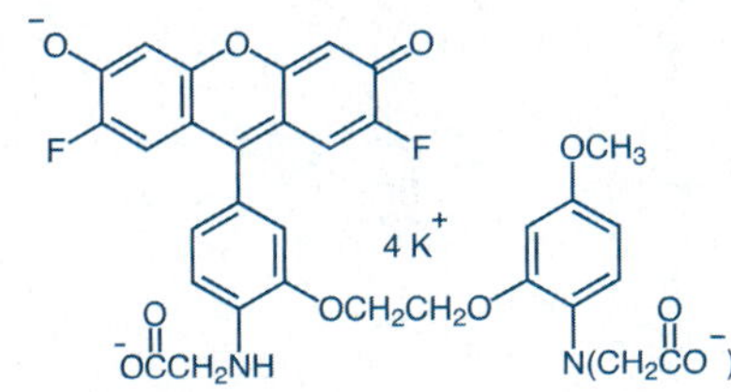

Figure 19.7.6 FluoZin™-3, tetrapotassium salt (F24194).

FluoZin™ Indicators

Zinc concentrations in the 1–100 nM range can be measured using fluorescent indicators nominally designed for Ca^{2+} detection such as fura-2, or more recently developed indicators with greater Zn^{2+} selectivity.[6,54] We have focused our development efforts on probes for detection of higher Zn^{2+} concentrations that are present in synaptic vesicles and released in response to electrical stimulation or excitotoxic agonists.[48] Peak concentrations of synaptically released Zn^{2+} may exceed 100 µM. We have developed FluoZin™-1 (F24180, F24181) and FluoZin™-2 (F24189), a series of unique indicators designed for detection of Zn^{2+} in the 0.1–100 µM range with minimal interfering Ca^{2+} sensitivity[53] (panel H, Figure 19.7.3; Figure 19.7.5). In our laboratories, we determined a $K_d(Zn^{2+})$ of 8.2 µM for FluoZin™-1; however, dissociation constants are known to vary considerably depending on the experimental conditions and a $K_d(Zn^{2+})$ of 0.4 µM for this same indicator has been reported elsewhere,[55] underscoring the importance of calibrating the indicator directly in the system under study. Published applications of FluoZin™-1 have primarily involved characterization of Zn^{2+}-binding proteins.[55–57]

The FluoZin™-3 indicator[53,58] (F24194, Figure 19.7.6; F24195) is a Zn^{2+}-selective indicator with a structure that resembles fluo-4. FluoZin™-3 exhibits high Zn^{2+}-binding affinity (K_d for Zn^{2+} ~15 nM) that is unperturbed by Ca^{2+} concentrations up to at least 1 µM.[15,54,59] The Zn^{2+}-specificity of FluoZin™-3 evident from measurements in calibration solutions is reproduced in cell-based experiments. The responses of FluoZin™-3 to transition metals, including Zn, Mn, Fe, Co, Cu(I), Cu(II), Ni and Cd, have been extensively characterized.[5] In addition, FluoZin™-3 exhibits a large increase in fluorescence in response to saturating levels of Zn^{2+} (greater than 50 fold, Figure 19.7.7). FluoZin™-3 has been found to be a brighter alternative to Zinquin for measuring

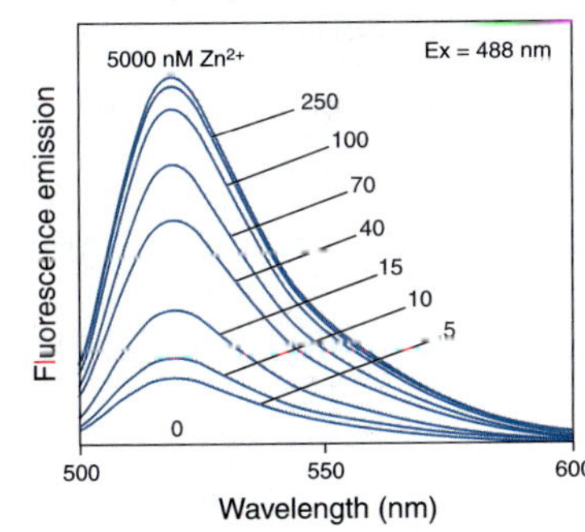

Figure 19.7.7 Zn^{2+} dependence of the fluorescence emission spectra of FluoZin™-3 (F24194). The spectrum for the Zn^{2+}-free solution is indistinguishable from the baseline.

Table 19.6 Fluorescent indicators for Zn^{2+}.

Indicator	Water-Soluble Salt	Cell-Permeant Ester	Ex/Em *	K_d for Zn^{2+} †
FluoZin™-1	F24180	F24181	495/515	8 µM
FluoZin™-2	NA	F24189	495/525	2 µM
FluoZin™-3	F24194	F24195	494/516	15 nM**
RhodZin™-3	R36350	R36351	550/575	65 nM
Newport Green™ DCF	N7990	N7991	505/535	1 µM
Newport Green™ PDX	NA	N24191	495/520	30 µM

* Excitation (Ex) and Emission (Em) maxima, in nm. † Dissociation constant of the indicator Zn^{2+} complex measured in 50 mM MOPS, pH 7.0, at 22°C. ‡ Excitation peak shifts upon binding Zn^{2+}. § Emission peak shifts upon binding Zn^{2+}. ** K_d determined in 135 mM NaCl, 1.1 mM EGTA, 20 mM HEPES, pH 7.4, 0–10 µM free Zn^{2+} at 22°C. NA = not available.

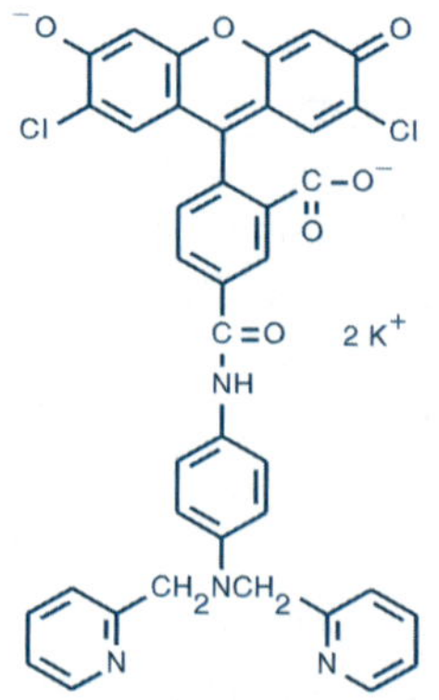

Figure 19.7.8 Newport Green™ DCF, dipotassium salt (N7990).

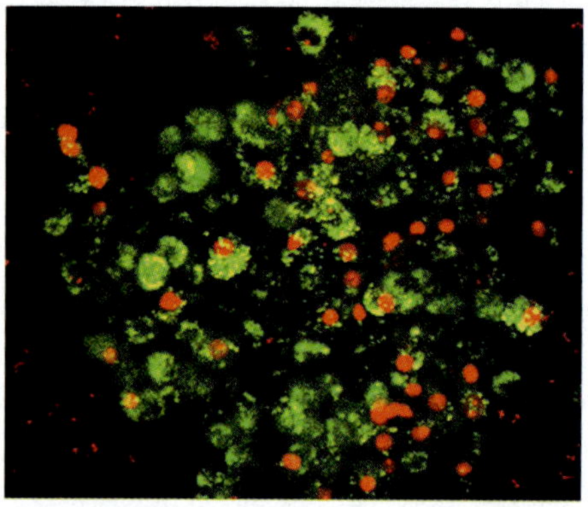

Figure 19.7.9 A human pancreatic islet stained with Newport Green™ DCF diacetate (N7991) and propidium iodide (P1304MP, P3566, P21493). The image represents a projection of 60 optical sections acquired at 2 µm intervals by confocal laser-scanning microscopy. The green-fluorescent Newport Green™ DCF indicator identifies viable insulin-producing ß-cells via binding to intracellular zinc. Dead cells are identified by red-fluorescent propidium iodide staining. Image contributed by Brigitte Vandewalle, University of Lille, and reproduced with permission from J Histochem Cytochem (2001) 49:519.

the exocytotic release of Zn^{2+} from pancreatic β-cells after stimulation with glucose or potassium.[60,61] Although imaging applications of FluoZin™-3 are predominant, cell-based microplate assays[47,62] and flow cytometry protocols[63] have also been developed.

RhodZin™-3 Indicator

The orange-fluorescent RhodZin™-3 zinc indicator exhibits a dramatic 75-fold increase in fluorescence at saturating levels of Zn^{2+} and also possesses a K_d for Zn^{2+} of ~65 nM.[64] The cell-permeant AM ester form of RhodZin™-3 (R36351) effectively localizes into mitochondria and is a valuable tool for investigating the physiological consequences of mitochondrial Zn^{2+} sequestration.[65,66] The cell-impermeant RhodZin™-3 salt (R36350) can be used to measure extracellular Zn^{2+} concentrations.[67]

Newport Green™ DCF and Newport Green™ PDX Indicators

The Newport Green™ DCF indicator (N7990, Figure 19.7.8; N7991) has moderate zinc-binding affinity (K_d for Zn^{2+} ~1 µM) but is essentially insensitive to Ca^{2+} (K_d for Ca^{2+} >100 µM), making this a valuable probe for detecting Zn^{2+} influx into neurons through voltage- or glutamate-gated channels.[46,68,69] When used alongside dyes with dual Ca^{2+}/Zn^{2+} sensitivity such as fura-2 and mag-fura-2, Newport Green™ DCF provides confirmation that changes in Zn^{2+} levels, and not Ca^{2+} or Mg^{2+}, are being detected.[70,71] Newport Green™ DCF has been used to identify insulin-producing β-cells from human pancreatic islets based on their high intracellular Zn^{2+} content[45] (Figure 19.7.9). Newport Green™ DCF has also been used in conjunction with Texas Red® 10,000 MW dextran to create a ratiometric fluorescent PEBBLE (Probe Encapsulated By Biologically Localized Embedding) nanosensor for real-time measurements of intra- and intercellular free zinc.[72] The responses of Newport Green™ DCF to transition metals, including Zn, Mn, Fe, Co, Cu(I), Cu(II), Ni and Cd, have been characterized.[5]

Newport Green™ PDX[53] (N24191) incorporates the same di-(2-picolyl)amine chelator as Newport Green™ DCF but has a higher Zn^{2+} dissociation constant (K_d for Zn^{2+} ~30 µM) and a larger Zn^{2+}-free to Zn^{2+}-saturated fluorescence intensity increase.

TSQ

Use of the membrane-permeant probe *N*-(6-methoxy-8-quinolyl)-*p*-toluenesulfonamide (TSQ, M688; Figure 19.7.10) in cells was first described by Fredrickson.[73] TSQ is selective for Zn^{2+} in the presence of physiological concentrations of Ca^{2+} and Mg^{2+} ions. The complex of TSQ with free Zn^{2+} apparently has a stoichiometry of two dye molecules per metal atom,[74] but a 1:1 complex may be formed with metalloproteins. The intracellular Zn^{2+} chelator dithizone blocks TSQ binding of Zn^{2+}.[75]

Several reports suggest that TSQ can be used to localize Zn^{2+} pools in the central nervous system.[76–78] Histochemical localization using TSQ identified a broad distribution of Zn^{2+} in neonatal mice, particularly associated with rapidly proliferating tissues, such as skin and gastrointestinal epithelium.[79] TSQ has also been used to detect Zn^{2+} translocation from presynaptic nerve terminals into postsynaptic nerve terminals when blood flow is constricted in the brain during ischemic events.[80] TSQ (like Newport Green™ DCF) is a selective nontoxic stain for pancreatic islet cells, which have a high content of Zn^{2+}, and may be useful for their flow cytometric isolation.[81]

TSQ-based assays for Zn^{2+} in seawater and other biological systems exhibit a detection limit of ~0.1 nM.[82,83] The simultaneous determination of Zn^{2+} and Cd^{2+} by spectrofluorometry using TSQ in an SDS micelle has also been reported.[74] TSQ has been used to measure Zn^{2+} levels in artificial lipid vesicles and live sperm cells by flow cytometry.[84] In this latter study, the fluorescence yield of the TSQ–Zn^{2+} complex was shown to be much higher when bound to lipids than in aqueous solution, indicating that quantitative cell assays for Zn^{2+} based on the fluorescence intensity of TSQ may not be accurate because of uncertainty in the quantum yield of the dye when bound to membranes.

Traditional Ca^{2+} and Mg^{2+} Indicators as Zn^{2+} Indicators

Zn^{2+} binds to most BAPTA-based Ca^{2+} indicators with substantially higher affinity than Ca^{2+}.[85] For example, fura-2 (F1200, F6799; Section 19.2) exhibits a K_d for Zn^{2+} in the absence of Ca^{2+} of 3 nM (Table 19.5). The lack of saturation of fura-2 fluorescence in resting cells is indicative of the low intracellular concentration of free Zn^{2+}. Fura-2 remains sensitive to nanomolar

IMPORTANT NOTICE: The products described in this manual are covered by one or more Limited Use Label License(s). Please refer to the Appendix on page 971 and Master Product List on page 975. Products are For Research Use Only. Not intended for any animal or human therapeutic or diagnostic use.

Zn^{2+} levels in the presence of 25–100 nM free Ca^{2+}, allowing the use of fura-2 AM (F1201, F1221, F1225, F14185; Section 19.2) for detecting intracellular Zn^{2+} influx via voltage-gated Ca^{2+} channels [86] and for examining Zn^{2+} levels in live neurons [87–89] and myocytes.[90] Fura-2 has also been used as an indicator for nanomolar levels of free Zn^{2+} and other ions in solution.[10]

Mag-fura-2 (M1290, Section 19.6) exhibits slightly altered spectral characteristics upon binding Zn^{2+} (K_d ~20 nM at pH 7.0–7.8 and 37°C), allowing Zn^{2+} to be measured in the presence of Ca^{2+}.[91] A review by Dineley and co-workers [92] evaluates the performance of currently available fluorescent indicators for neuronal Zn^{2+} and identifies artifacts associated with their use. As usual, the key to indicator selection is matching the ion binding affinity (K_d) to the prevailing range of ion concentrations. In the case of neuronal free Zn^{2+}, physiological concentrations apparently are about 1–50 nM. Under these conditions, indicators nominally designed for detection of intracellular magnesium, such as mag-fura-2 with K_d for Zn^{2+} around 20 nM, appear to be the most suitable.[89] Based on their similar ion-binding properties, the Magnesium Green™ indicator (M3733, Section 19.6) and our mag-fluo-4 indicator (M14205, Section 19.6) are suitable for confocal imaging of Zn^{2+} influx.[93]

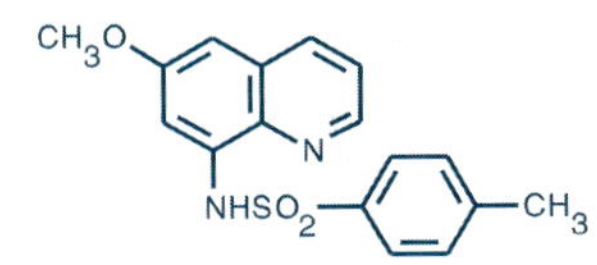

Figure 19.7.10 *N*-(6-methoxy-8-quinolyl)-*p*-toluenesulfonamide (TSQ, M688).

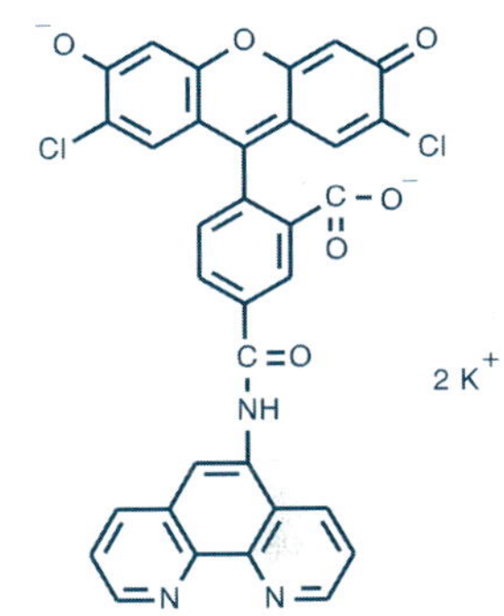

Figure 19.7.11 Phen Green™ SK, dipotassium salt (P14312).

Indicators for Copper

Copper is third in abundance (after Fe^{3+} and Zn^{2+}) among the essential heavy metals in the human body. Dietary copper is required for normal hemoglobin synthesis, for prevention of anemia and for redox enzyme activity.[1] The redox activity of copper (i.e., reversible reduction of Cu^{2+} to Cu^{+}) is both a key to its biological activity and a complicating factor in its detection. Reduction of Cu^{2+} to Cu^{+} is catalyzed by the β-amyloid precursor protein, which is converted to plaques that are characteristic of Alzheimer disease.[94–97]

Phen Green™ FL for Cu^{2+}

The phenanthroline-based Phen Green™ FL indicator, available only as the cell-permant diacetate (P6763), is an excellent general-purpose heavy metal sensor capable of detecting a broad range of metal ions, including both Cu^{2+} and Cu^{+}.[94] The use of Phen Green™ FL for detecting Fe^{2+}, Hg^{2+}, Pb^{2+}, Cd^{2+} and Ni^{2+} at submicromolar concentrations is described below. Uncomplexed Phen Green™ FL is brightly fluorescent, with a fluorescence quantum yield of ~0.8. Binding of certain heavy metal ions is registered by strong fluorescence quenching (Figure 19.7.4). The emission intensity of Phen Green™ FL depends both on metal ion concentration and on the indicator's concentration. Phen Green™ FL diacetate has allowed researchers to discern significant differences in intracellular Cu^{2+} levels for four types of lobster hepatopancreatic epithelial cells.[98,99]

Phen Green™ SK for Cu^{2+}

The phenanthroline-based Phen Green™ SK indicator (P14312) has been reported to be a selective indicator for Cu^{+}, using conditions that minimize the oxidation of Cu^{+} to Cu^{2+}.[100] These researchers then used Phen Green™ SK to characterize the dissociation of Cu^{+} from glutathione.

Traditional Ca^{2+} and Mg^{2+} Indicators as Cu^{2+} Indicators

Fluorescence of calcein (C481) is strongly quenched by Cu^{2+} at neutral pH [36] (Figure 19.7.4), although this spectroscopic response does not appear to have been exploited for Cu^{2+} detection. Cu^{2+} has also been measured in solution using fura-2.[10] Indicators designed for detection of Ca^{2+}, Mg^{2+} and Zn^{2+} generally exhibit relatively weak responses to Cu^{2+}. FluoZin™-3 and Newport Green™ DCF bind Cu^{2+} with extremely high affinity [5] (K_d = 0.09 nM and 0.8 nM, respectively). Cu^{2+} binding is registered as a fluorescence decrease due to competitive displacement of Zn^{2+}. For both indicators, binding affinity for Cu^{+} is more than 1000-fold weaker than for Cu^{2+}.[5]

Indicators for Iron

Phen Green™ FL and Phen Green™ SK Indicators

The intracellular pool of chelatable iron is considered to be a decisive pathogenic factor for various kinds of cell injury.[101–103] Fluorescence of Phen Green™ FL (available only as the cell-permeant diacetate, P6763) and Phen Green™ SK (P14312, Figure 19.7.11; cell-permeant diacetate,

IMPORTANT NOTICE: The products described in this manual are covered by one or more Limited Use Label License(s). Please refer to the Appendix on page 971 and Master Product List on page 975. Products are For Research Use Only. Not intended for any animal or human therapeutic or diagnostic use.

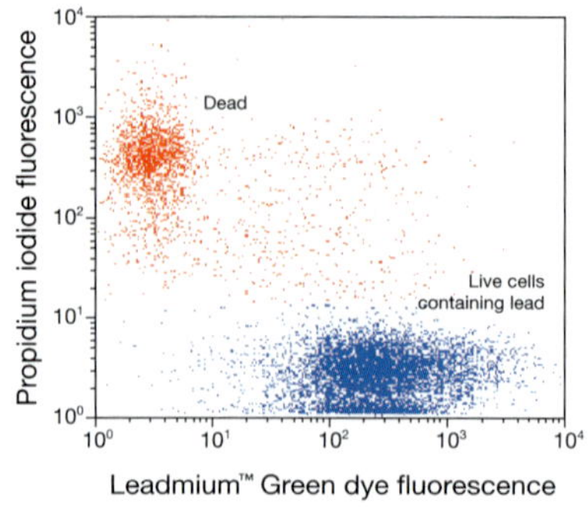

Figure 19.7.12 Dual-color scatter plot showing two populations of Jurkat cells. Jurkat cells were loaded with Leadmium™ Green AM (A10024) and washed. The sample was then incubated in the presence of 1 µM $PbCl_2$ (in saline) and 1 µM ionomycin. After washing the sample, it was incubated in the presence of PI. Dual-color fluorescence was collected using 488 nm excitation and 525/10 nm and 610/10 nm bandpass filters. This results in visualization of two populations: dead cells (positive for PI) and live cells that contain lead (positive for Leadmium™ Green dye and negative for PI).

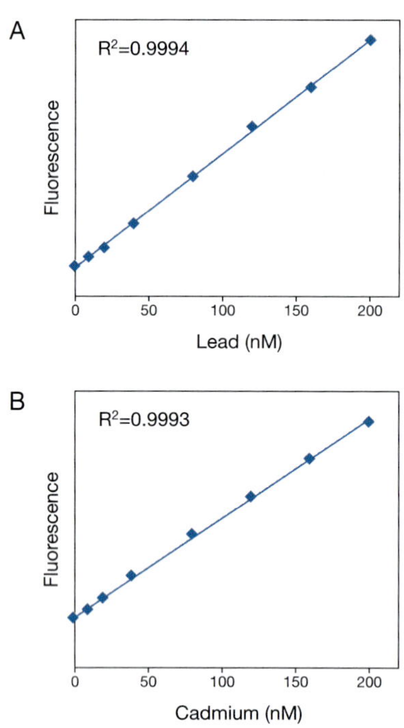

Figure 19.7.13 Linearity and sensitivity of the Measure-iT™ Lead and Cadmium Assay Kit (M36353) for lead (**A**) and cadmium (**B**). Triplicate 10 µL samples of lead and cadmium were assayed; fluorescence was measured at 490/520 nm and plotted versus lead or cadmium concentration. The variation (CV) of replicate samples was <2%.

P14313) is quenched upon binding Fe^{2+} (Figure 19.7.2) and Fe^{3+} (Figure 19.7.4). The emission intensity of the Phen Green™ FL indicator depends on both the metal ion's concentration and the indicator's concentration.[103] Phen Green™ SK diacetate has been successfully used to quantitate the intracellular pool of chelatable iron in rat hepatocytes,[102–104] to directly measure Fe^{2+} transport across chloroplast membranes[105] and to monitor iron uptake in relation to fatigue in mouse skeletal muscle.[101]

Calcein

Cabantchik and co-workers have exploited the fluorescence quenching of calcein at neutral pH to follow intracellular release of Fe^{2+} from transferrin.[36,106,107] Using passively loaded calcein AM (C1430, C3099, C3100MP), they were able to measure cytosolic Fe^{2+} concentrations from about 0.1 µM to 1.0 µM. However, more recent studies on the mechanism of iron chelation indicate that quenching of calcein fluorescence is primarily due to Fe^{3+} and is relatively insensitive to Fe^{2+} [108] (Figure 19.7.4). Nontransferrin-bound iron (NTBI) occurs in the serum of individuals with iron overload and in a variety of other pathological conditions, including thalassemia. A microplate assay based on the fluorescence quenching of calcein by iron has been developed to measure NTBI.[109] The capacity of transition metals such as iron to catalyze the generation of oxidative radicals can be used as a basis for indirect detection using fluorescent ROS sensors (Section 18.2) under conditions where the metal ion is the limiting species. Dihydrorhodamine 123 (D632, D23806; Section 18.2) has been used in this way to detect NTBI in the plasma of healthy and thalassemic patients.[110]

Indicators for Lead, Cadmium and Mercury

Leadmium™ Green Indicator

Leadmium™ Green dye is a fluorescent indicator supplied in cell-permeant AM ester form (A10024) for selectively detecting lead or cadmium in cells. The calcium-insensitive Leadmium™ Green dye can detect nanomolar levels of lead and micromolar levels of cadmium (Figure 19.7.12). In a typical application, Leadmium™ Green AM was used to assess the effects of *N*-acetylcysteine on intracellular lead levels in oligodendrocyte progenitor cells.[111]

Measure-iT™ Lead and Cadmium Assay Kit

The Measure-iT™ Lead and Cadmium Assay Kit (M36353) provides a fluorescence microplate assay for quantitation of lead or cadmium in solution, with a linear detection range from 5 to 200 nM.[112] (Figure 19.7.13). The Measure-iT™ Leadmium™ reagent working solution is mixed with 1–20 µL sample volumes in the wells of a microplate and then immediately analyzed with a fluorescence microplate reader using fluorescein/FITC wavelength settings. The assay is performed at room temperature, and the signal is stable for at least 30 minutes.

Each Measure-iT™ Lead and Cadmium Assay Kit contains:

- Measure-iT™ Leadmium™ reagent
- Concentrated Measure-iT™ Leadmium™ buffer
- Lead standard
- Cadmium standard
- Detailed protocol

Sufficient reagents are provided for 1000 assays using a fluorescence microplate reader.

Phen Green™ FL and Phen Green™ SK Indicators

Our phenanthroline-based Phen Green™ FL and Phen Green™ SK indicators (P14312), which can be passively loaded into cells as their membrane-permeant diacetates (P6763, P14313), are excellent general-purpose heavy metal sensors that are capable of detecting a broad range of metal ions—including Cu^{2+}, Cu^{+} and Fe^{2+}—as well as micromolar concentrations of Hg^{2+}, Pb^{2+}, Cd^{2+}, Zn^{2+} and Ni^{2+} (Figure 19.7.4). We have used these versatile sensors

IMPORTANT NOTICE: The products described in this manual are covered by one or more Limited Use Label License(s). Please refer to the Appendix on page 971 and Master Product List on page 975. Products are For Research Use Only. Not intended for any animal or human therapeutic or diagnostic use.
www.invitrogen.com/probes

to detect metal ions in a variety of matrices, including seawater and various contaminated solids such as paint sludge and soil. In such samples, Phen Green™ FL and Phen Green™ SK detect only the readily soluble (bioavailable) fraction of the total metal ions. Phen Green™ FL and Phen Green™ SK are well suited for initial field testing of metal ion contamination in aqueous samples; the large fluorescence changes produced by micromolar ion concentrations of these heavy metals are easily visible upon illuminating the sample with a hand-held light source (Figure 19.7.2).

Traditional Ca^{2+} and Mg^{2+} Indicators as Pb^{2+}, Cd^{2+} and Hg^{2+} Indicators

The fluorescence of several of our traditional Ca^{2+} and Mg^{2+} indicators is strongly affected by binding of Pb^{2+}, Cd^{2+} and Hg^{2+} (Figure 19.7.3). Fura-2 and quin-2 bind Cd^{2+} with extremely high affinity (Table 19.5). The excitation response of fura-2 to Cd^{2+}—almost identical to its Ca^{2+} response—has been used to monitor Cd^{2+} uptake by cells and to image intracellular free Cd^{2+}.[113] The response is reversed by TPEN (T1210, Section 19.8), which complexes many heavy metals but not Ca^{2+} or Mg^{2+}.[114] Heavy metal binding by rhod-5N (R14207, Section 19.3) has also been reported; the $K_d(Cd^{2+})$ of rhod-5N was determined to be 1.4 nM.[9]

Mag-fura-2 AM (M1291, M1292; Section 19.6) has proven useful as an intracellular Cd^{2+} indicator,[115] and indo-1 AM can be used to simultaneously determine the intracellular concentrations of Ca^{2+} and Cd^{2+} or Ca^{2+} and Ba^{2+}.[116] Use of fura-2 AM (F1201, F1221, F1225, F14185; Section 19.2) to measure the concentration of cytosolic Ba^{2+} has also been reported,[11] as has fluorometric detection of Cd^{2+} at pH 13.3 using calcein [117] (C481). Pb^{2+} entry into cells has been monitored using fura-2 or indo-1 as intracellular indicators.[118–122] Pb^{2+} is efficiently transported into cells with high selectivity by ionomycin [123] (I24222, Section 19.8).[19] F NMR can also be used to detect Pb^{2+} uptake by platelets using fluorinated BAPTA derivatives [124] (Section 19.8).

Indicators for Nickel and Cobalt

Newport Green™ Indicators

Much of the recent interest in Ni^{2+} comes from its use in the detection and isolation of oligohistidine fusion proteins by metal-chelate affinity chromatography. Newport Green™ DCF indicator (N7990) is an exceptionally sensitive probe for Ni^{2+} in solution.[5] 100 µM Ni^{2+} enhances the fluorescence of the Newport Green™ DCF reagent approximately 13-fold without a spectral shift; Zn^{2+} and Co^{2+} enhance this indicator's fluorescence to a lesser extent (Figure 19.7.4). Newport Green™ DCF diacetate (N7991) has been used to measure the cellular uptake of Ni^{2+} in human monocyte–derived dendritic cells.[125] Newport Green™ DCF and Newport Green™ DCF diacetate have also been used with flow cytometry to detect Ni^{2+}-binding metalloproteins involved in human nickel allergy, the most common form of human contact hypersensitivity.[126] The AM ester of Newport Green™ PDX (N24191) incorporates the same di-(2-picolyl)amine chelator as Newport Green™ DCF and should prove to have similar utility for detection of Ni^{2+}.

Traditional Ca^{2+} and Mg^{2+} Indicators as Ni^{2+} and Co^{2+} Indicators

Co^{2+} enhances the fluorescence of the Calcium Green™-2 indicator at least 20-fold and strongly quenches fluorescence of the Fura Red™ indicator (Figure 19.7.3). Co^{2+} and Ni^{2+}, as well as Cu^{2+} and Fe^{3+} (Figure 19.7.4), strongly quench the fluorescence of calcein, even at pH 7.[127–129] Consequently, it should be possible to follow the kinetics of uptake of these ions into cells loaded with calcein AM (C1430, C3099, C3100MP). Note that the 1:1 stoichiometry of calcein–metal binding will require the use of low loading levels of this probe to achieve significant quenching by limited amounts of metal transport. The efficient quenching of calcein fluorescence by Co^{2+} or Ni^{2+} [36] has been used to detect liposome fusion [128] (Assays of Volume Change, Membrane Fusion and Membrane Permeability—Note 14.3). Fura-2 has been used to measure low levels of Ni^{2+} and Co^{2+} in solution.[10]

Indicators for Aluminum

Al^{3+} binding has little effect on the fluorescence of most of the traditional Ca^{2+} and Mg^{2+} indicators. However, Al^{3+} is reported to selectively form a fluorescent complex with calcein (C481) at acidic pH that can be detected with micromolar sensitivity.[130] Quenching of the Al^{3+}-calcein complex fluorescence has been used as the basis of a method for fluoride determination, with a detection limit of 0.2 ng/mL.[131]

Newport Green™ DCF diacetate (N7991) has also been used to detect the uptake and distribution of several metal ions—including Al^{3+} and Ti^{3+}—in human monocyte–derived dendritic cells by flow cytometry and confocal microscopy.[125] Although the intensities varied, intracellular Cr^{3+}, Mo^{2+}, Ni^{2+}, Ti^{4+} and Zr^{4+} also produced a fluorescence response with Newport Green™ DCF diacetate.

Indicators for Lanthanides

La^{3+} has a strong effect on the fluorescence of indicators designed for detection of Ca^{2+}, Mg^{2+} and Zn^{2+} (Figure 19.7.3). For example, fluo-4 fluorescence increases more than 400-fold in the presence of 100 µM La^{3+}, a response that is about twice as large as that generated by Ca^{2+} saturation. Detection of La^{3+} by the prototypical BAPTA-based Ca^{2+} indicator quin-2 has been used to investigate the ion transport selectivity of 4-bromo A-23187 and ionomycin [132] (B1494, I24222; Section 19.8). These indicators generally exhibit relatively weak responses to Tb^{3+}; fluo-3 and fluo-4 are apparently the most sensitive (~40-fold fluorescence enhancement with 1 µM Tb^{3+}) (Figure 19.7.3). Long-lived luminescence of Tb^{3+} (from $TbCl_3$, T1247) is also used to probe Ca^{2+} binding sites of proteins.[133] Fluo-5N (F14203; Section 19.3) is a low-affinity Ca^{2+} indicator with subnanomolar affinity for gadolinium (Gd^{3+}), enabling its use for determining Gd^{3+}-binding affinities of magnetic resonance imaging (MRI) contrast agents via competition titrations.[134]

IMPORTANT NOTICE: The products described in this manual are covered by one or more Limited Use Label License(s). Please refer to the Appendix on page 971 and Master Product List on page 975. Products are For Research Use Only. Not intended for any animal or human therapeutic or diagnostic use.

REFERENCES

1. Chem Rev (2008) 108:1517; **2.** Methods (2008) 46:204; **3.** Circ Res (2003) 92:286; **4.** Cell Calcium (2006) 40:393; **5.** Anal Biochem (2009) 384:34; **6.** Nat Chem Biol (2008) 4:168; **7.** Fresenius J Anal Chem (2000) 368:182; **8.** Cell Calcium (2000) 28:225; **9.** J Fluoresc (2008) 18:1077; **10.** Anal Biochem (2000) 284:307; **11.** J Biol Chem (2000) 275:20274; **12.** J Biol Chem (2000) 275:6980; **13.** Cell Calcium (2009) 45:185; **14.** Methods Mol Biol (2006) 312:229; **15.** Am J Physiol Gastrointest Liver Physiol (2006) 290:G250; **16.** Biophys J (1996) 71:1048; **17.** Cell Calcium (2001) 29:217; **18.** Methods Cell Biol (1994) 40:155; **19.** J Cell Biol (1997) 136:833; **20.** J Physiol (1998) 507:379; **21.** Cell Calcium (1994) 16:87; **22.** Cell Calcium (1995) 18:420; **23.** Cell Calcium (1989) 10:171; **24.** J Physiol (2000) 523:549; **25.** Neurosci Lett (1997) 229:109; **26.** Cell Calcium (1997) 22:157; **27.** Methods Enzymol (1997) 288:301; **28.** J Biol Chem (1989) 264:1522; **29.** Pflugers Arch (1993) 423:225; **30.** Cell Calcium (1997) 22:385; **31.** Endocrinology (1992) 131:1936; **32.** J Biol Chem (1989) 264:197; **33.** J Neurosci (2000) 20:4414; **34.** Biophys J (1999) 76:2029; **35.** Cell Calcium (2008) 43:602; **36.** Am J Physiol (1995) 268:C1354; **37.** J Biol Chem (2009) 284:15117; **38.** J Biol Chem (2005) 280:715; **39.** Biophys J (1999) 76:725; **40.** J Biol Chem (2000) 275:35738; **41.** Mol Biochem Parasitol (1999) 101:43; **42.** Blood (1999) 94:2128; **43.** J Biol Chem (2000) 275:1651; **44.** Stem Cells (2006) 24:2858; **45.** J Histochem Cytochem (2001) 49:519; **46.** J Neurosci Methods (2009) 177:1; **47.** Brain Res Bull (2007) 74:183; **48.** Trends Pharmacol Sci (2000) 21:395; **49.** J Neurosci (2008) 28:3114; **50.** J Biol Chem (2008) 283:15349; **51.** J Biol Inorg Chem (2006) 11:1049; **52.** Science (2001) 292:2488; **53.** Cell Calcium (2002) 31:245; **54.** Cell Calcium (2008) 44:422; **55.** J Am Chem Soc (2008) 130:8847; **56.** Protein Sci (2008) 17:760; **57.** J Biol Chem (2005) 280:33716; **58.** J Am Chem Soc (2002) 124:776; **59.** Cell Calcium (2005) 37:225; **60.** Biotechniques (2004) 37:922; **61.** Anal Chem (2003) 75:3136; **62.** Proc Natl Acad Sci U S A (2009) 106:8374; **63.** Anal Biochem (2006) 352:222; **64.** Cell Calcium (2003) 34:281; **65.** J Neurosci (2006) 26:6851; **66.** Proc Natl Acad Sci U S A (2003) 100:6157; **67.** Diabetes (2006) 55:600; **68.** J Neurosci (2002) 22:1273; **69.** Eur J Neurosci (2000) 12:3813; **70.** Proc Natl Acad Sci U S A (1999) 96:2414; **71.** J Neurochem (2000) 75:1878; **72.** Analyst (2002) 127:11; **73.** J Neurosci Methods (1987) 20:91; **74.** J Fluorescence (1991) 1:267; **75.** Science (1996) 272:1013; **76.** Pain (2000) 86:177; **77.** J Neurosci (1999) 19:2288; **78.** Eur J Neurosci (2000) 12:8; **79.** J Histochem Cytochem (2004) 52:529; **80.** J Neurosci Methods (2001) 110:57; **81.** Anal Biochem (2003) 314:38; **82.** Anal Chem (1994) 66:2732; **83.** Biol Res (1994) 27:49; **84.** Cytometry (1995) 21:153; **85.** Cell Calcium (2000) 27:75; **86.** J Biol Chem (1995) 270:2473; **87.** J Neurochem (1998) 71:2401; **88.** Neurobiol Dis (1997) 4:275; **89.** J Neurosci (1997) 17:9554; **90.** Am J Physiol (1997) 272:H2095; **91.** J Biochem Biophys Methods (1993) 27:25; **92.** Mol Pharmacol (2002) 62:618; **93.** J Physiol (2000) 529:83; **94.** J Neurochem (2008) 104:1249; **95.** J Biol Chem (2006) 281:15145; **96.** J Mol Biol (2006) 356:759; **97.** Science (1996) 271:1406; **98.** J Exp Biol (2002) 205:405; **99.** J Exp Biol (2001) 204:1433; **100.** Anal Chim Acta (2006) 575:223; **101.** J Physiol (2009) 587:4705; **102.** Biochem J (2001) 356:61; **103.** Hepatology (1999) 29:1171; **104.** Arch Biochem Biophys (2000) 376:74; **105.** Plant Physiol (2002) 128:1022; **106.** J Biol Chem (1995) 270:24209; **107.** FEBS Lett (1996) 382:304; **108.** J Biol Chem (1999) 274:13375; **109.** Blood (2000) 95:2975; **110.** Blood (2003) 102:2670; **111.** PLoS Biol (2007) 5:e35; **112.** Langmuir (2007) 23:1974; **113.** J Biol Chem (1992) 267:25553; **114.** J Biol Chem (1985) 260:2719; **115.** Kidney Int (1992) 41:1237; **116.** Mol Cell Biochem (1995) 151:91; **117.** Anal Chem (1974) 46:2036; **118.** Toxicol Sci (1998) 46:90; **119.** Neurotoxicology (2000) 21:365; **120.** J Biol Chem (1997) 272:8346; **121.** Toxicol Appl Pharmacol (1997) 146:127; **122.** Am J Physiol (1990) 259:C762; **123.** J Biol Chem (2000) 275:7071; **124.** Biochim Biophys Acta (1991) 1092:341; **125.** J Neurosci Methods (2009) 178:182; **126.** J Immunol Methods (2007) 328:14; **127.** Chem Pharm Bull (Tokyo) (1991) 39:227; **128.** Biochim Biophys Acta (1982) 691:332; **129.** Anal Chem (1984) 56:810; **130.** Anal Chem (1963) 35:1035; **131.** Fresenius J Anal Chem (2000) 368:501; **132.** Biophys J (1998) 75:1244; **133.** J Am Chem Soc (2006) 128:7346; **134.** J Am Chem Soc (2008) 130:9260.

DATA TABLE 19.7 FLUORESCENT INDICATORS FOR Zn^{2+} AND OTHER METAL IONS

				Low Zn^{2+}				High Zn^{2+}						
Cat. No.	**MW**	**Storage**	**Soluble**	**Abs**	**EC**	**Em**	**Solvent**	**Abs**	**EC**	**Em**	**Solvent**	**Product**	**K_d**	**Notes**
C481	622.54	L	pH >5	494	77,000	517	pH 9			see Notes			see Notes	1, 2
C1430	994.87	F,D	DMSO	<300		none						C481		
C3099	994.87	F,D	DMSO	<300		none						C481		3
C3100MP	994.87	F,D	DMSO	<300		none						C481		
F24180	599.67	F,D,L	pH >6	490	68,000	see Notes	H_2O	495	58,000	517	H_2O/Zn^{2+}		8.2 µM	4, 5, 6, 7
F24181	701.59	F,D,L	DMSO	456	26,000	see Notes	MeOH					F24180		8
FluoZin™-2	800.89	D,L	pH >6	494	74,000	522	H_2O	494	75,000	521	H_2O/Zn^{2+}		2.0 µM	4, 5, 6
F24189	876.73	F,D,L	DMSO	299	16,000	none	MeOH					FluoZin™-2		
F24194	846.96	F,D,L	pH >6	491	82,000	see Notes	H_2O	494	88,000	516	H_2O/Zn^{2+}		15 nM	4,5,6,7,9
F24195	982.85	F,D,L	DMSO	455	26,000	see Notes	MeOH					F24194		8
M688	328.38	L	EtOH	334	4200	385	MeOH			see Notes				10
N7990	793.74	D,L	pH >6	506	82,000	535	H_2O	506	82,000	535	H_2O/Zn^{2+}		1.0 µM	4, 5, 6
N7991	801.64	F,D	DMSO	302	15,000	none	MeOH					N7990		
Newport Green™ PDX	521.52	D,L	pH >6	490	76,000	518	H_2O	491	77,000	518	H_2O/Zn^{2+}	30 µM		4, 5, 6
N24191	593.59	F,D	DMSO	457	21,000	538	MeOH					Newport Green™ PDX		
P6763	668.68	F,D	DMSO	<300		none						Phen Green™ FL		
Phen Green™ FL	660.79	D,L	pH >6	492	68,000	517	pH 9			see Notes				11
P14312	698.60	D,L	pH >6	507	86,000	532	pH 9			see Notes				11
P14313	706.49	F,D	DMSO	<300		none						P14312		
R36350	788.94	F,D,L	pH >6	549	80,000	see Notes	H_2O	552	82,000	576	H_2O/Zn^{2+}		65 nM	4,5,6,7,9
R36351	1009.86	F,D,L	DMSO	546	110,000	570	see Notes					R36350		12
T1247	373.38	D	H_2O	270	4700	545	H_2O							13, 14

For definitions of the contents of this data table, see "Using *The Molecular Probes® Handbook*" in the introductory pages.

Notes

1. C481 fluorescence is strongly quenched by micromolar concentrations of Fe^{3+}, Co^{2+}, Ni^{2+} and Cu^{2+} at pH 7. (Am J Physiol (1995) 268:C1354, J Biol Chem (1999) 274:13375)
2. $K_d(Co^{2+})$ for calcein is 120 nM, determined in 10 mM HEPES, 1 µM Ca^{2+}, 1 mM Mg^{2+}, 100 mM KCl. (Anal Biochem (1997) 248:31)
3. This product is supplied as a ready-made solution in the solvent indicated under "Soluble."

IMPORTANT NOTICE: The products described in this manual are covered by one or more Limited Use Label License(s). Please refer to the Appendix on page 971 and Master Product List on page 975. Products are For Research Use Only. Not intended for any animal or human therapeutic or diagnostic use.

www.invitrogen.com/probes

molecular probes® | invitrogen™ by Life technologies™

DATA TABLE 19.7 FLUORESCENT INDICATORS FOR Zn^{2+} AND OTHER METAL IONS—*continued*

4. Dissociation constants are known to vary considerably depending on the temperature, pH, ionic strength, viscosity, protein binding, presence of other ions (especially polyvalent ions), instrument setup and other factors. It is strongly recommended that these values be verified under user-specific experimental conditions.
5. Spectra measured in aqueous buffers containing zero (H_2O) or a >10-fold excess of free cation X (H_2O/X) relative to the listed dissociation constant (K_d) for cation X.
6. This indicator exhibits fluorescence enhancement in response to ion binding, with essentially no change in absorption or emission wavelengths.
7. Fluorescence of the free indicator is very weak and is enhanced >100-fold on binding Zn^{2+}.
8. Fluorescence of this AM ester derivative is very weak and is enhanced only after hydrolytic cleavage followed by binding of divalent cations to the anionic indicator.
9. $K_d(Zn^{2+})$ determined in 135 mM NaCl, 1.1 mM EGTA, 20 mM HEPES, pH 7.4, 0 to 10 µM free Zn^{2+} at 22°C. $K_d(Zn^{2+})$ is unchanged in the presence of 1 µM free Ca^{2+}. (J Am Chem Soc (2002) 124:776, Cell Calcium (2003) 34:281)
10. Fluorescence of M688 is very sensitive to solvent polarity. Em = 495 nm in aqueous buffer (pH 7.4). Fluorescence is enhanced on binding Zn^{2+} (0.01–10 µM) with no change in emission wavelength. (Cytometry (1995) 21:153)
11. Fluorescence of Phen Green™ indicators is quenched by Fe^{2+}, Fe^{3+}, Cu^{2+}, Cd^{2+}, Hg^{2+} and Pb^{2+} with no change in emission wavelength. The extent of quenching is dependent on the concentration of indicator as well as the ion concentration.
12. Spectra measured in 90:10 (v/v) $CHCl_3$:MeOH.
13. Absorption and luminescence of T1247 are extremely weak unless it is chelated. Data are for dipicolinic acid (DPA) chelate. The luminescence spectrum has secondary peak at 490 nm.
14. MW is for the hydrated form of this product.

PRODUCT LIST 19.7 FLUORESCENT INDICATORS FOR Zn^{2+} AND OTHER METAL IONS

Cat. No	Product	Quantity
C481	calcein *high purity*	100 mg
C1430	calcein, AM	1 mg
C3099	calcein, AM *1 mg/mL solution in anhydrous DMSO*	1 mL
C3100MP	calcein, AM *special packaging*	20 x 50 µg
F24181	FluoZin™-1, AM *cell permeant*	50 µg
F24180	FluoZin™-1, tripotassium salt *cell impermeant*	500 µg
F24189	FluoZin™-2, AM *cell permeant*	50 µg
F24195	FluoZin™-3, AM *cell permeant*	100 µg
F24194	FluoZin™-3, tetrapotassium salt *cell impermeant*	500 µg
A10024	Leadmium™ Green, AM dye *for intracellular detection of lead and cadmium* *special packaging*	5 x 50 µg
M36353	Measure-iT™ Lead and Cadmium Assay Kit *1000 assays*	1 kit
M688	*N*-(6-methoxy-8-quinolyl)-*p*-toluenesulfonamide (TSQ)	25 mg
N7991	Newport Green™ DCF diacetate *cell permeant*	1 mg
N7990	Newport Green™ DCF, dipotassium salt *cell impermeant*	1 mg
N24191	Newport Green™ PDX acetoxymethyl ether	1 mg
P6763	Phen Green™ FL, diacetate *cell permeant*	1 mg
P14313	Phen Green™ SK, diacetate	1 mg
P14312	Phen Green™ SK, dipotassium salt	1 mg
R36351	RhodZin™-3, AM *cell permeant* *special packaging*	10 x 50 µg
R36350	RhodZin™-3, dipotassium salt *cell impermeant*	500 µg
T1247	terbium(III) chloride, hexahydrate	1 g

IMPORTANT NOTICE: The products described in this manual are covered by one or more Limited Use Label License(s). Please refer to the Appendix on page 971 and Master Product List on page 975. Products are For Research Use Only. Not intended for any animal or human therapeutic or diagnostic use.

19.8 Chelators, Calibration Buffers, Ionophores and Cell-Loading Reagents

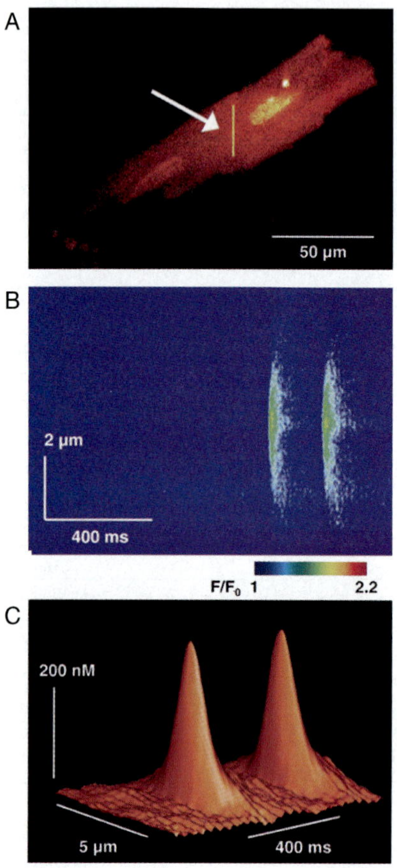

Figure 19.8.3 A) Confocal image of a guinea pig ventricular myocyte filled with fluo-3, pentapotassium salt (F3715). The vertical yellow line (arrow) indicates the region scanned repetitively to generate the line-scan image shown in panel B. **B)** Two precisely localized Ca^{2+} transients were elicited by two-photon photolysis of DMNP-EDTA (D6814) using ultrafast high-repetition (120 fs, 80 MHz) laser pulses at 700 nm. The color scale bar represents fluo-3 fluorescence intensities corresponding to resting (blue) to high (red) Ca^{2+} concentrations. **C)** Surface plot of Ca^{2+} concentration plotted against space and time. The spatial spread of the Ca^{2+} transients was about 2 µm at half-maximum amplitude. Images contributed by Ernst Niggli, University of Bern, and reproduced with permission from Nat Cell Biol (1999) 1:323.

Caged Calcium and Caged Calcium Chelators

Nitrophenyl EGTA: A Superior Caged Calcium Reagent

As an alternative to solely monitoring Ca^{2+} changes using fluorescent indicators, scientists may want to rapidly raise or lower the intracellular Ca^{2+} concentration and study the physiological response that results. Ellis-Davies and Kaplan have developed a photolabile chelator, *o*-nitrophenyl EGTA (NP-EGTA, N6802) that exhibits a high selectivity for Ca^{2+}, a dramatic 12,500-fold decrease in affinity for Ca^{2+} upon UV illumination (its K_d increases from 80 nM to >1 mM) and a high photochemical quantum yield[1,2] (~0.2). Photolysis of NP-EGTA is slightly faster than that of DMNP-EDTA,[3] another "caged Ca^{2+}" reagent that is frequently called DM-Nitrophen™. Furthermore, with a K_d for Mg^{2+} of 9 mM, NP-caged EGTA does not bind physiological levels of Mg^{2+} and thus reduces interference from this abundant cation.[4] Skinned muscle fibers equilibrated with NP-EGTA were shown to contract maximally upon irradiation with a single flash from a frequency-doubled ruby laser[5] (347 nm illumination). Other suitable ultraviolet light sources for photoactivation of NP-EGTA include mercury arc lamps (365 nm), frequency-tripled Nd:YAG lasers[6] (355 nm), UV argon-ion laser lines[7] (351/364 nm) and light-emitting diodes[8] (LED). NP-EGTA is not suitable for two-photon photoactivation due to its extremely low absorption cross section in the wavelength range accessible with femtosecond pulsed Ti:sapphire laser sources.[9]

We offer the tetrapotassium salt (N6802) and the acetoxymethyl (AM) ester (NP-EGTA AM, N6803) of NP-EGTA. The NP-EGTA salt can be complexed with Ca^{2+} to generate a caged Ca^{2+} reagent that will rapidly deliver Ca^{2+} upon photolysis (Figure 19.8.1). The cell-permeant AM ester of NP-EGTA does not bind Ca^{2+} unless its AM ester groups are removed. NP-EGTA AM can serve as a photolabile buffer in cells because, once converted to NP-EGTA by intracellular esterases, it will bind Ca^{2+} with high affinity until photolyzed with UV light.[10]

DMNP-EDTA

The first caged Ca^{2+} reagent to be described by Kaplan and Ellis-Davies was 1-(4,5-dimethoxy-2-nitrophenyl)-EDTA (DMNP-EDTA, D6814), which they named DM-Nitrophen™[11,12] (now a trademark of Calbiochem-Novabiochem Corp.). Because its structure more closely resembles that of EDTA than EGTA, we named it as a caged EDTA derivative (Figure 19.8.2). Upon illumination, DMNP-EDTA's dissociation constant for Ca^{2+} increases from 5 nM to 3 mM. Thus, photolysis of DMNP-EDTA complexed with Ca^{2+} results in a pulse of free Ca^{2+}. Furthermore, DMNP-EDTA has significantly higher affinity for Mg^{2+} (K_d = 2.5 µM) than does NP-EGTA[11] (K_d = 9 mM). Because the photolysis product's K_d for Mg^{2+} is ~3 mM, DMNP-EDTA is an effective caged Mg^{2+} source, in addition to its applications for photolytic Ca^{2+} release.[13,14] Photorelease of Ca^{2+} has been shown to occur in <180 µsec, with even faster photorelease of Mg^{2+}.[15] Moreover, DMNP-EDTA is also useful for photolytic release of other divalent cations such as Sr^{2+}, Ba^{2+}, Mn^{2+}, Co^{2+} and Cd^{2+}.[16] Unlike NP-EGTA, DMNP-EDTA has a low but finite two-photon absorption cross-section in the wavelength range accessible with femtosecond pulsed Ti:sapphire laser sources[9] and can therefore be used to produce two-photon activated calcium release with high spatial precision (Figure 19.8.3). Two reviews by Ellis-Davies discuss the uses and limitations of DMNP-EDTA.[1,17]

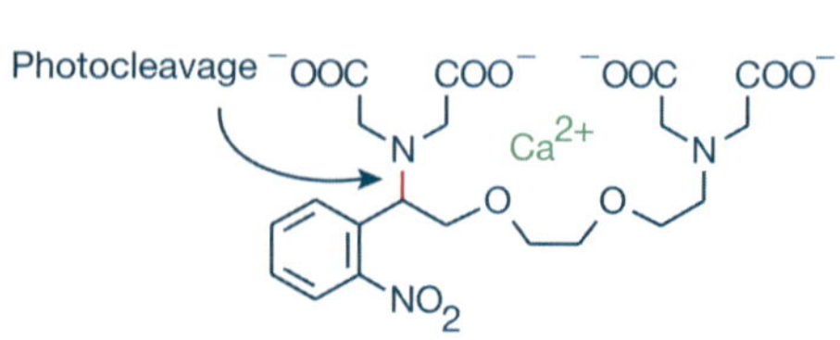

Figure 19.8.1 NP-EGTA (N6802) complexed with Ca^{2+}. Upon illumination, this complex is cleaved to yield free Ca^{2+} and two iminodiacetic acid photoproducts. The affinity of the photoproducts for Ca^{2+} is ~12,500-fold lower than that of NP-EGTA.

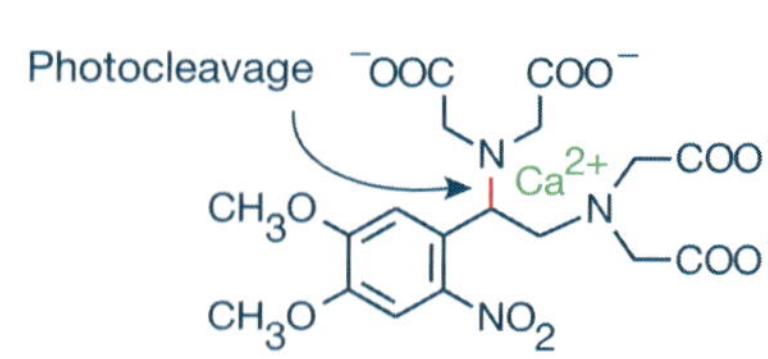

Figure 19.8.2 DMNP-EDTA (D6814) complexed with Ca^{2+}. Upon illumination, this complex is cleaved to yield free Ca^{2+} and two iminodiacetic acid photoproducts. The affinity of the photoproducts for Ca^{2+} is ~600,000-fold lower than that of DMNP-EDTA.

IMPORTANT NOTICE: The products described in this manual are covered by one or more Limited Use Label License(s). Please refer to the Appendix on page 971 and Master Product List on page 975. Products are For Research Use Only. Not intended for any animal or human therapeutic or diagnostic use.
www.invitrogen.com/probes
molecular probes® | invitrogen™ by life technologies™

Diazo-2: A Photoactivatable Calcium Knockdown Reagent

In contrast to NP-EGTA and DMNP-EDTA, diazo-2 is a photoactivatable Ca^{2+} scavenger. Diazo-2 (D3034, Figure 19.8.4), which was introduced by Adams, Kao and Tsien,[18,19] is a relatively weak chelator (K_d for Ca^{2+} = 2.2 µM). Following flash photolysis at ~360 nm, however, cytosolic free Ca^{2+} rapidly binds to the high-affinity photolysis product of diazo-2 (K_d = 73 nM). Microinjecting a relatively low concentration of a visible light–excitable Ca^{2+} indicator—such as fluo-3, fluo-4 or one of our Calcium Green™ or Oregon Green® 488 BAPTA indicators—along with a known quantity of diazo-2 permits measurement of the extent of depletion of cytosolic Ca^{2+} following photolysis. Diazo-2 can be microinjected into larger cells[4] or electroporated into smaller cells[20] as its potassium salt (D3034). Intracellular loading of NP-EGTA, DMNP-EDTA and diazo-2 is best accomplished by patch pipette infusion with the carboxylate salt form of the caged compound added to the internal pipette solution at 1–10 mM.[21,22]

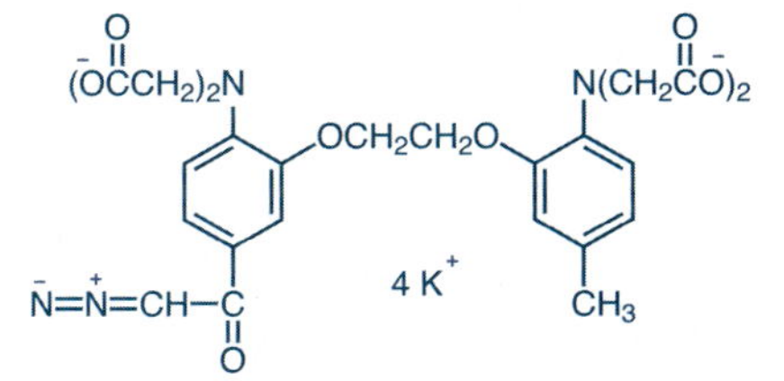

Figure 19.8.4 Diazo-2, tetrapotassium salt (D3034).

Nonfluorescent Chelators

BAPTA and BAPTA AM

The bis(2-aminophenoxy)ethane tetraacetic acid (BAPTA, Figure 19.8.5) buffers developed by Tsien[23] are highly selective for Ca^{2+} over Mg^{2+} and can be used to control the level of both intracellular and extracellular Ca^{2+} (Table 19.7, Figure 19.8.6). The BAPTA buffers are more selective for Ca^{2+} than EDTA and EGTA, and their metal binding is also much less pH sensitive. Furthermore, BAPTA buffers bind and release Ca^{2+} ions about 50–400 times faster than EGTA. Both BAPTA and its membrane-permeant AM ester are extensively used to clamp intracellular Ca^{2+} concentrations, providing insights on the role of free cytosolic Ca^{2+} in a number of important cell systems.[24–26] BAPTA AM (or the higher-affinity 5,5'-dimethyl BAPTA AM) is also useful for establishing an intracellular zero free calcium level for *in situ* calibrations of fluorescent indicators (Loading and Calibration of Intracellular Ion Indicators—Note 19.1). BAPTA and its derivatives also exert physiological effects that are somewhat independent of their calcium-binding activity.[27,28]

BAPTA is available as a cell-impermeant potassium, cesium[29] or sodium salt (B1204, B1212, B1214); the Cs^+ salt of BAPTA has frequently been used for patch-clamp experiments.[29] In addition, we offer the cell-permeant BAPTA AM ester in two packaging formats (B1205, B6769). In addition to these products, we offer a polystyrene conjugate of BAPTA for selective removal of Ca^{2+} from solutions, which we call Calcium Sponge™ S (C3047).

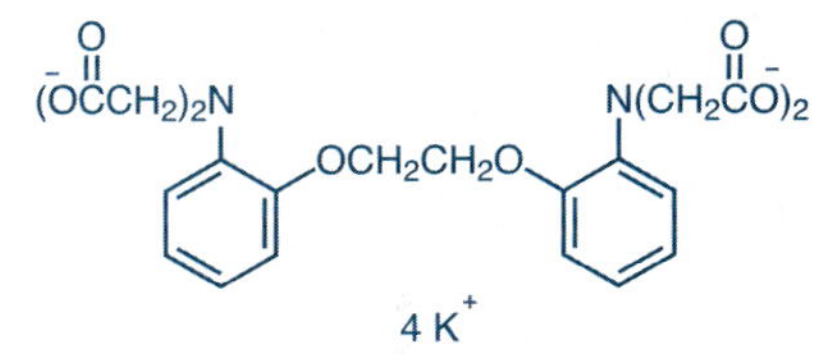

Figure 19.8.5 BAPTA, tetrapotassium salt (B1204).

Other BAPTA Derivatives

Other BAPTA derivatives are listed in Table 19.7, along with their dissociation constants for Ca^{2+}. The most powerful Ca^{2+} chelator among these is 5,5'-dimethyl BAPTA,[30–33] available as its cell-permeant AM ester (D1207).

BAPTA derivatives with intermediate affinity for Ca^{2+}, such as 5,5'-dibromo BAPTA (D1211), have been extensively used to study Ca^{2+} mobilization, spatial Ca^{2+} buffering and Ca^{2+} shuttling in a variety of cells and animal models.[34–36] 5,5'-Dibromo BAPTA and other lower-affinity chelators protect neurons against excitotoxic and ischemic injury, without markedly attenuating intracellular Ca^{2+} levels.[37]

Fluorinated BAPTA derivatives, such as the AM ester of 5,5'-difluoro BAPTA (D1209), have been employed for optical imaging studies[38,39] but are most widely used for NMR analysis of Ca^{2+} in live cells and tissues. The ^{19}F NMR shifts of the 5,5'-difluoro BAPTA have been reported to correlate with intracellular Ca^{2+} in BALB/c thymocytes,[40] normal[41] and sickle[42] erythrocytes and ferret hearts.[43,44]

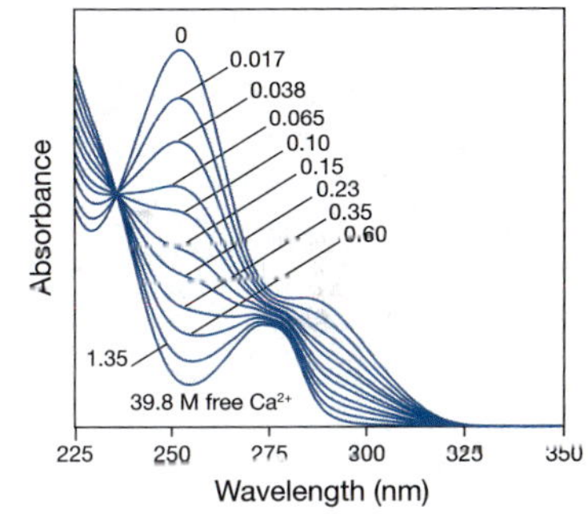

Figure 19.8.6 Absorption spectra of BAPTA (B1204) in solutions containing 0–39.8 µM free Ca^{2+}.

Table 19.7 Ca^{2+} affinities of BAPTA chelators.

Chelator	Water-Soluble Salt	Cell-Permeant Ester	K_d for Ca^{2+}
BAPTA	B1204, B1212, B1214	B1205, B6769	No Mg^{2+}: 160 nM; * 1 mM Mg^{2+}: 700 nM †
5,5'-Dibromo BAPTA	D1211	NA	No Mg^{2+}: 1.6 µM ‡
5,5'-Difluoro BAPTA	NA	D1209	No Mg^{2+}: 635 nM; § 1 mM Mg^{2+}: 705 nM §
5,5'-Dimethyl BAPTA	NA	D1207	No Mg^{2+}: 40 nM ‡

* Measured in 10 mM MOPS, 100 mM KCl, pH 7.2 at 22°C. † Measured in 10 mM HEPES, 300 mM KCl, pH 7.0 at 22°C.[1] ‡ Measured in 10 mM MOPS, 100 mM KCl, pH 7.3 at 22°C.[2] § Measured in 10 mM HEPES, 115 mM KCl, 20 mM NaCl, pH 7.05.[3] NA = Not available. 1. Cell Calcium (1989) 10:491; 2. Biochemistry (1980) 19:2396; 3. Am J Physiol (1987) 252:C441.

NOTE 19.2

MAXC Computer Program for Calculating Free Ca^{2+} Concentrations

In solutions that contain several chelating species, or that contain ions such as Zn^{2+} that compete with Ca^{2+} for binding to BAPTA or EGTA, calculating the free Ca^{2+} concentration requires simultaneously solving multiple equilibrium expressions. Except for the simplest cases where analytical solutions are feasible, researchers must use an iterative computer program to complete these calculations. MAXC, a program developed for this purpose,[1] is freely accessible on the Web at www.stanford.edu/~cpatton/maxc.html.

1. Methods Cell Biol (1994) 40:3.

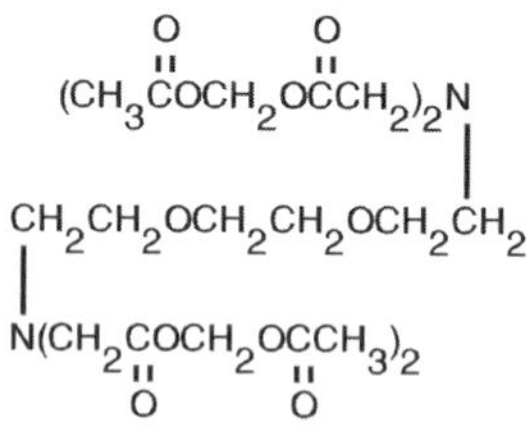

Figure 19.8.7 EGTA, tetra (acetoxymethyl ester) (EGTA, AM; E1219).

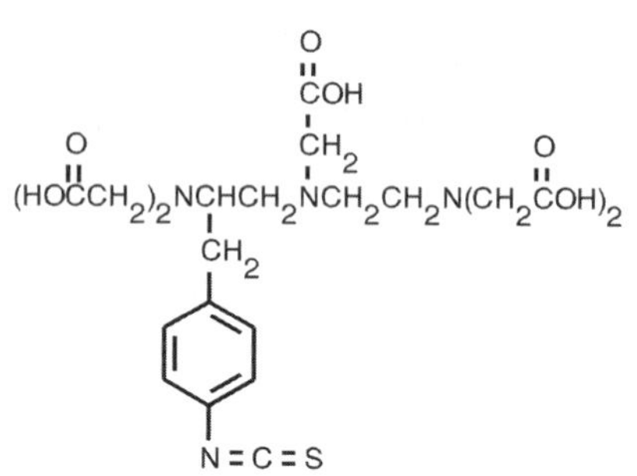

Figure 19.8.8 (*S*)-1-*p*-isothiocyanatobenzyldiethylenetriaminepentaacetic acid (DTPA isothiocyanate, I24221).

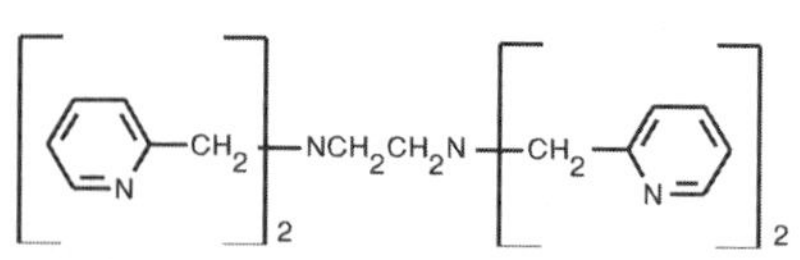

Figure 19.8.9 Tetrakis-(2-pyridylmethyl)ethylenediamine (TPEN, T1210).

EGTA AM

The AM ester derivative of EGTA (E1219, Figure 19.8.7) can be passively loaded into cells to generate intracellular EGTA. The slower on-rate of EGTA relative to the BAPTA-based buffers reduces its ability to inhibit Ca^{2+} diffusion in cells.[45] Because Ca^{2+} binding by intracellular EGTA is relatively slow it is possible to distinguish between buffering of rapid Ca^{2+} transients, which can occur with BAPTA-derived buffers, and the slower effects of general Ca^{2+} buffering.[37,46,47]

DTPA Isothiocyanate

DTPA isothiocyanate (I24221) can be coupled to antibodies and other biomolecules by conventional amine-reactive chemistry, thereby introducing a high-affinity binding site for lanthanides and other metal ions. Unlike reactive anhydride forms of DTPA, the isothiocyanate derivative yields conjugates that retain all five carboxylate groups (Figure 19.8.8), resulting in more stable metal complexation.[48] Antibodies and ligands labeled with the fluorescent lanthanides europium and terbium (T1247, Section 19.7) are widely utilized in time-resolved fluorescence–based assays.[49] DTPA-gadolinium (Gd^{3+}) complexes are extensively used as contrast agents for magnetic resonance imaging.[50] DTPA isothiocyanate–labeled antibodies also have potentially important therapeutic applications for targeted delivery of radionuclides such as indium-111 and yttrium-90.[51]

Calcium Sponge™ Polymer

We offer a biologically compatible conjugate of the BAPTA chelator to selectively remove specific polyvalent ions from solution, as well as from the binding sites of indicators, proteins and polynucleotides. This BAPTA polystyrene conjugate—Calcium Sponge™ S—is selective for Ca^{2+} and certain other ions, including Zn^{2+} and some heavy metals, in the presence of relatively high levels of Mg^{2+}. Many contaminating polycations can be selectively removed from aqueous solutions simply by stirring a solution with the water-insoluble Calcium Sponge™ S polymer[52,53] (BAPTA polystyrene, C3047). For example, free Ca^{2+} can be reduced to less than 40 nM (measured with fura-2) by passing 3 mL of a 100 µM $CaCl_2$ solution through one gram of Calcium Sponge™ S. The polymer can be regenerated several times by washing it with pH 4 buffer, then readjusting to neutral pH with base.

TPEN

TPEN (T1210, Figure 19.8.9) selectively chelates intracellular heavy metal ions such as Zn^{2+}, Cu^{2+} and Fe^{2+} without disturbing Ca^{2+} and Mg^{2+} concentrations, revealing distortions in intracellular Ca^{2+} measurements caused by high-affinity binding of these ions to fluorescent indicators.[54–57] TPEN has been used to show that the effects of BAPTA on mitotic progression and nuclear assembly are specifically Ca^{2+}-mediated and are not attributable to binding of essential heavy metal ions.[58,59] TPEN has also been used to modulate the effects of Zn^{2+} on enzymatic activity[60] and protein conformation.[61]

Calcium Calibration Buffer Kits

Calibration of fluorescent Ca^{2+} indicators is a prerequisite for accurate Ca^{2+} measurements. We offer kits designed to facilitate this calibration using a laboratory fluorometer or quantitative imaging system. These kits contain buffers and detailed protocols—including methods for calculating K_d, a sample response curve and tables to help determine the exact concentration of free Ca^{2+} under conditions of varying pH, temperature and ionic strength. A discussion of methods to correct the fura-2 dissociation constants for differences in temperature and ionic strength has been published.[62] A computer program is available online for calculating the free Ca^{2+} concentrations in solutions that contain several chelating species, or that contain ions such as Zn^{2+} that compete with Ca^{2+} for binding to BAPTA or EGTA (MAXC Computer Program for Calculating Free Ca^{2+} Concentrations—Note 19.2).

Calcium Calibration Buffer Kit #1

Because cells contain very low levels of *free* Ca^{2+}, it is essential to use Ca^{2+} buffers such as EGTA to precisely calibrate Ca^{2+} indicators under specific experimental conditions. When the concentrations of Ca^{2+} and EGTA are very close to each other, the only free Ca^{2+} available is the

IMPORTANT NOTICE: The products described in this manual are covered by one or more Limited Use Label License(s). Please refer to the Appendix on page 971 and Master Product List on page 975. Products are For Research Use Only. Not intended for any animal or human therapeutic or diagnostic use.
www.invitrogen.com/probes

Ca^{2+} that is in equilibrium with EGTA. Thus, the concentration of free Ca^{2+} is determined by the K_d of CaEGTA at a controlled pH, temperature and ionic strength.[63]

Calcium Calibration Buffer Kit #1 (C3008MP) contains:

- 10 mM K_2EGTA buffered solution ("zero" free Ca^{2+})
- 10 mM CaEGTA buffered solution (40 µM free Ca^{2+})
- Detailed protocols for calibrating Ca^{2+} indicators

When used according to the protocol provided, each kit provides sufficient reagents for five complete calibrations using 2 mL samples and a standard fluorometer cuvette. Many more calibrations can be done by digital imaging microscopy.[64,65] This kit employs a reciprocal dilution method—an equal amount of dye is added to a portion of the zero and 40 µM free Ca^{2+} solutions, and the two are then cross-mixed to give a series of solutions with equal dye and varying free Ca^{2+} concentrations. With ratiometric indicators, this method yields a series of curves that exhibit an accurate isosbestic point (see, for example, Figure 19.8.10); it is the method regularly used in our laboratories to determine Ca^{2+} affinities. *In situ* calibrations of intracellular calcium indicators can be carried out using CaEGTA buffer solutions in combination with a suitable ionophore such as ionomycin (I24222).[66]

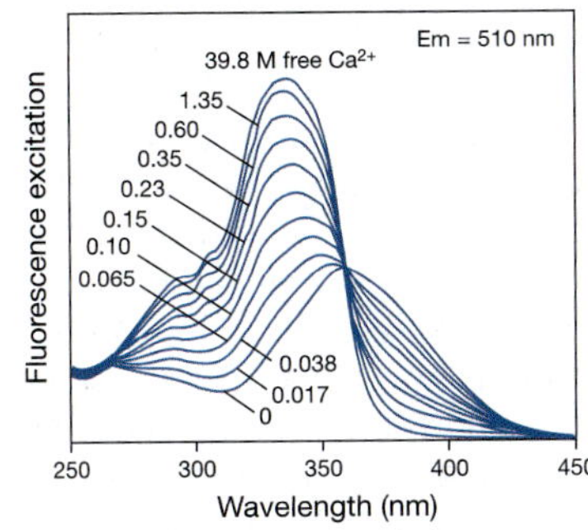

Figure 19.8.10 Fluorescence excitation spectra of fura-2 (F1200, F6799) in solutions containing 0–39.8 µM free Ca^{2+}.

Fura-2 Calcium Imaging Calibration Kit

The Fura-2 Calcium Imaging Calibration Kit (F6774) is designed to facilitate rapid calibration and standardization of digital imaging microscopes.[67,68] This kit provides 11 CaEGTA:EGTA buffer solutions with free Ca^{2+} concentrations from zero to 40 µM. Each solution also includes 50 µM fura-2, as well as 15 µm unstained polystyrene microspheres to act both as spacers that ensure uniform separation between the slide and the coverslip and as focusing aids. We also provide a twelfth buffer, identical to the 10 mM CaEGTA standard but lacking fura-2, that serves as a control for background fluorescence.

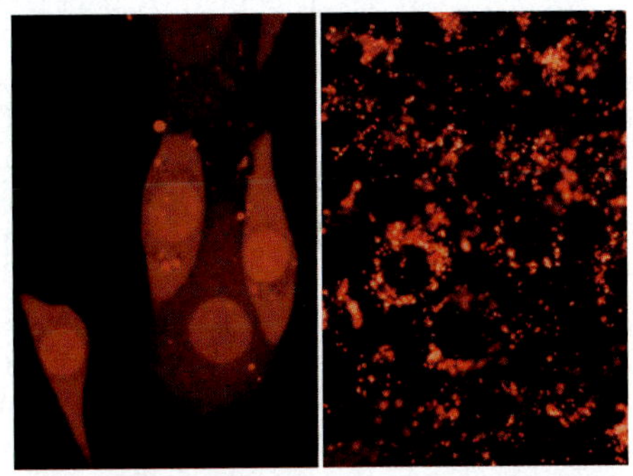

Figure 19.8.12 Adherent CRE BAG 2 cells passively loaded with the membrane-impermeant polar tracer Alexa Fluor® 594 hydrazide (A10438, A10442). The image on the left illustrates the relatively uniform, cytoplasmic labeling one can obtain with the Influx™ pinocytic cell-loading reagent (I14402), compared to the punctate labeling that results from pinocytic uptake in normal growth medium (right panel). Both images were acquired using a bandpass filter set appropriate for rhodamine dyes.

Influx™ Pinocytic Cell-Loading Reagent

The Influx™ pinocytic cell-loading reagent (I14402) facilitates the loading of water-soluble materials into live cells via a rapid and simple technique based on the osmotic lysis of pinocytic vesicles.[69] Simply mix the water-soluble probe at high concentration with the Influx™ reagent blended into growth medium, then incubate the cells in the medium to allow pinocytic uptake of the surrounding solution. When the cells are subsequently transferred to a slightly hypotonic medium, pinocytic vesicles within the cells release the trapped material and fill the cytosol with the probe (Figure 19.8.11).

The Influx™ pinocytic cell-loading reagent is effective for loading a diverse array of probes—including calcein, Alexa Fluor® hydrazides (Figure 19.8.12), dextran conjugates of fluorophores[70] and ion indicators (Figure 19.8.13), fura-2 salts, Oregon Green® 514 dye–labeled tubulin, Alexa Fluor® 488 dye–labeled actin, heparin,[71] DNA, siRNAs,[72] antisense oligonucleotides[73,74] and Qdot® nanocrystals[75,76]—into a variety of cell lines.

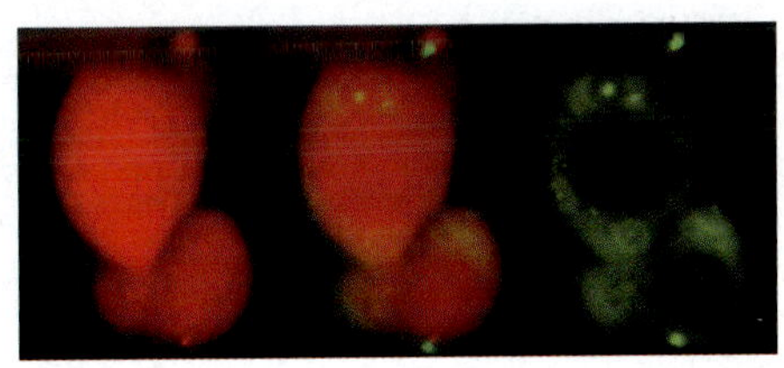

Figure 19.8.13 Adherent 3T3 fibroblasts labeled with 3000 MW Texas Red® dextran (D3329) and 70,000 MW Calcium Green™-1 dextran (C3714). Both dextrans were loaded into the fibroblasts using the Influx™ pinocytic cell-loading reagent (I14402). The images were acquired using a fluorescence microscope equipped with a cooled CCD camera and bandpass filter sets appropriate for either the Texas Red® dye (left image) or fluorescein (right image). The center image represents a composite of the cells labeled with both dextrans. The 3000 MW Texas Red® dextran is distributed throughout the cell, but the 70,000 MW Calcium Green™-1 dextran is excluded from the nucleus by the nuclear membrane.

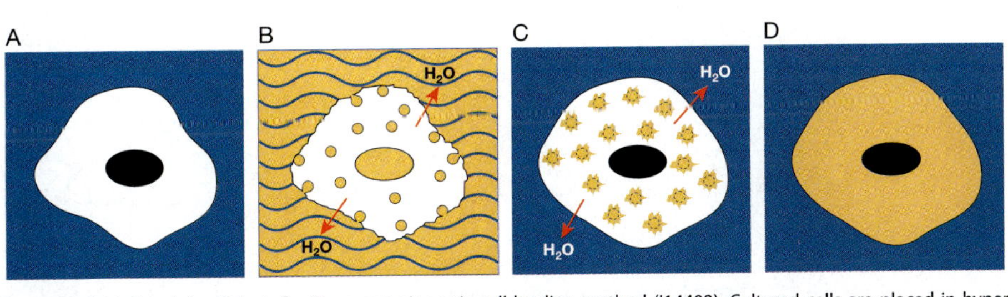

Figure 19.8.11 Principle of the Influx™ reagent pinocytic cell-loading method (I14402). Cultured cells are placed in hypertonic Influx™ loading reagent (**A**), along with the material to be loaded into the cells (yellow fluid, **B**), allowing the material to be carried into the cells via pinocytic vesicles. When the cells are placed in hypotonic medium, the pinocytic vesicles burst (**C**), releasing their contents into the cytosol (**D**).

IMPORTANT NOTICE: The products described in this manual are covered by one or more Limited Use Label License(s). Please refer to the Appendix on page 971 and Master Product List on page 975. Products are For Research Use Only. Not intended for any animal or human therapeutic or diagnostic use.

In addition to the Influx™ pinocytic cell-loading reagent and cell growth medium, all that is required to perform the loading procedure is sterile deionized water and the fluorescent probe or other polar molecule of interest. Cell labeling can be accomplished in a single 30-minute loading cycle and may be enhanced by repetitive loading. Although most types of cells load quickly and easily, optimal conditions for loading must be determined for each cell type. It is also important to note that cell-to-cell variability in the degree of loading is typical and that higher variability is generally observed when using large compounds, such as >10,000 MW dextrans and proteins.

The Influx™ pinocytic cell-loading reagent is packaged as a set of 10 tubes (I14402), each containing sufficient material to load 50 samples of cells grown on coverslips following the standard protocol supplied. Cells in suspension or in culture flasks may also be easily loaded; however, the number of possible cell loadings will depend on the cell suspension volume or size of culture flask used. The information provided with the Influx™ reagent includes general guidelines and detailed suggestions for optimizing cell loading. Use of the coverslip mini-rack or coverslip maxi-rack (C14784, C24784; Section 23.1) facilitates cell loading and slide handling when using the Influx™ reagent.

Loading P2X Receptor–Expressing Cells

$P2X_7$ receptor–expressing cells such as macrophages and thymocytes exhibit reversible pore opening that can be exploited to provide an entry pathway for intracellular loading of both cationic and anionic fluorescent dyes with molecular weights of up to 900 daltons.[77] Pore opening is induced by treatment with 5 mM ATP for five minutes and subsequently reversed by addition of divalent cations (Ca^{2+} or Mg^{2+}). Dyes that have been successfully loaded into macrophage cells by this method include:

- Ca^{2+} indicator: fura-2 [78] (F1200, F6799; Section 19.2)
- pH indicator: HPTS [79] (H348, Section 20.2)
- Aqueous tracers: lucifer yellow CH (L453, L682, L1177, L12926; Section 14.3) and sulforhodamine B [80] (S1307, Section 14.3)
- Nucleic acid stain: YO-PRO®-1 [80,81] (Y3603, Section 8.1)

We offer one of the most potent and widely used P2X receptor agonists, BzBzATP [82–84] (2′-(or 3′-)O-(4-benzoylbenzoyl)adenosine 5′-triphosphate, B22358, Section 17.3). BzBzATP has more general applications for site-directed irreversible modification of nucleotide-binding proteins via photoaffinity labeling;[85,86] see Section 17.3 for more information on nucleotide analogs.

Ionophores

A-23187 and 4-Bromo A-23187

The Ca^{2+} ionophore A-23187 (A1493, Figure 19.8.14) is commonly used for *in situ* calibrations of fluorescent Ca^{2+} indicators, to equilibrate intracellular and extracellular Ca^{2+} concentrations and to permit Mn^{2+} to enter the cell to quench intracellular dye fluorescence. Although the intrinsic fluorescence of A-23187 is too high for use with fura-2, indo-1 and quin-2, it is suitable for use with the visible light–excitable indicators, including Calcium Green™, Magnesium Green™, Calcium Orange™, Calcium Crimson™, Oregon Green® 488 BAPTA, fluo-3, fluo-4, rhod-2, X-rhod-1 and Fura Red™ indicator. Brominated A-23187 (4-bromo A-23187, B1494; Figure 19.8.14), which is essentially nonfluorescent, is the best ionophore for use with fura-2, indo-1 and other UV light–excited Ca^{2+} indicators. Like A-23187, 4-bromo A-23187 rapidly transports both Ca^{2+} and Mn^{2+} into cells.[87] Both A-23187 and 4-bromo A-23187 can also be used to equilibrate intracellular and extracellular Mg^{2+} concentrations, making them useful for calibrating Mg^{2+} indicators (Section 19.6). Furthermore, 4-bromo A-23187 has occasionally been used to equilibrate intracellular Zn^{2+} with controlled extracellular levels for *in situ* calibration of fluorescent indicators.[88] The zero reference level for intracellular Zn^{2+} calibrations is usually set by addition of TPEN [89] (K_d for $Zn^{2+} = 2.6 \times 10^{-16}$ M).

Ionomycin

Ionomycin (I24222, Figure 19.8.15) is an effective Ca^{2+} ionophore that is commonly used both to modify intracellular Ca^{2+} concentrations and to calibrate fluorescent Ca^{2+} indicators when studying the regulatory properties of Ca^{2+} in cellular processes. Ionomycin also transports Pb^{2+} and some other divalent cations, as well as several lanthanide series trivalent cations, at efficiencies that are greater than or equal to those for Ca^{2+}.[90,91]

Probenecid: Inhibitor of Organic-Anion Transporters

Probenecid (P36400) is commonly used to inhibit organic-anion transporters located in the cell membrane. Such transporters can extrude dyes and indicators and thus contribute to poor loading or a high background signal in assays based on retention of the dyes or indicators inside cells.[92] The use of probenecid to block the efflux of intracellular dyes was first reported by Di Virgilio and co-workers.[93] Wash steps or masking dyes may be incorporated into fluorescent assays in order to minimize baseline fluorescence. However, washing introduces an extra

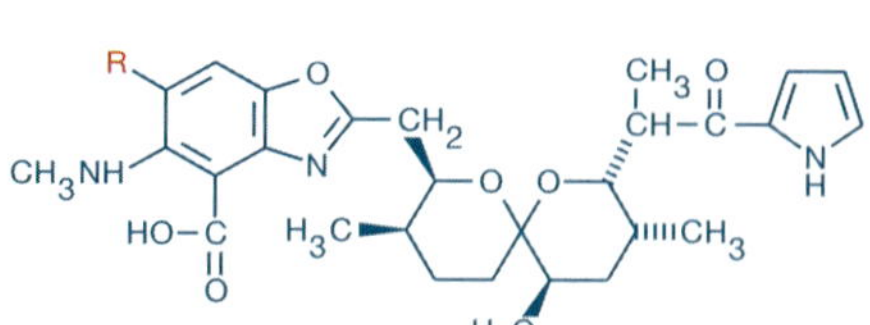

Figure 19.8.14 Chemical structures of the ionophores A-23187 (R = H, A1493) and 4-bromo A-23187 (R = Br, B1494).

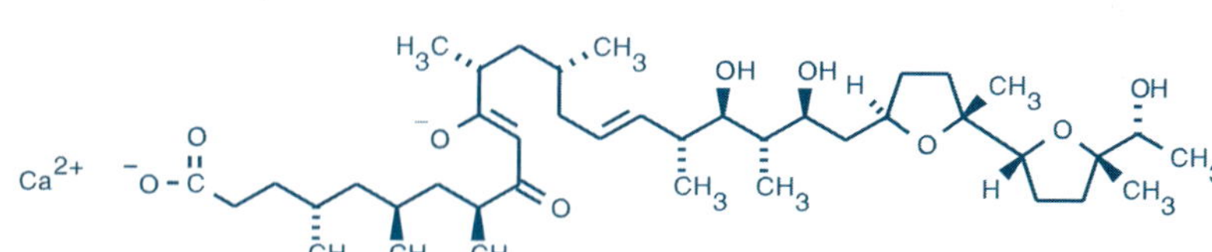

Figure 19.8.15 Ionomycin, calcium salt (I24222).

IMPORTANT NOTICE: The products described in this manual are covered by one or more Limited Use Label License(s). Please refer to the Appendix on page 971 and Master Product List on page 975. Products are For Research Use Only. Not intended for any animal or human therapeutic or diagnostic use.

step that is undesirable for high-throughput applications and that may also risk loss of nonadherent cells. Masking dyes, while offering the advantage of homogeneous (one-step, mix-and-read) assays, may interact negatively with some receptor systems of interest.[94] Our water-soluble probenecid (P36400) has the advantage of being easy to dissolve in physiological buffers, unlike the conventionally used free acid form, which must be initially dissolved in strong base.

Pluronic® F-127 and PowerLoad™ Solutions

Pluronic® F-127

Because acetoxymethyl (AM) esters have low aqueous solubility,[95] dispersing agents—typically fetal calf serum, bovine serum albumin or Pluronic® F-127—are often used to facilitate cell loading.[96,97] Pluronic® F-127 has also proven useful as a blocking reagent to prevent cell adhesion to PDMS (polydimethylsiloxane, a silicon-based organic polymer) microfluidic channels.[98] We provide Pluronic® F-127 in three forms, all of which have low UV absorbance (OD_{280} nm <0.02 at 10 mg/mL):

- Powder (P6867)
- 10% Solution in water (P6866), filtered through a 0.2 µm–pore size membrane filter for use in tissue culture and other applications
- 20% Solution in DMSO (P3000MP)

Cautioning that Pluronic® F-127 is not necessarily physiologically benign, a recent paper shows a Pluronic® F-127–dependent modulation of depolarization-evoked Ca^{2+} transients in rat dorsal ganglion (DRG) neurons, as detected with fura-2 AM[99] (F1201, F1221, F1225, F14185; Section 19.2).

PowerLoad™ Concentrate

PowerLoad™ concentrate (P10020) is an optimized formulation of nonionic Pluronic® surfactants that act to disperse and stabilize water-insoluble dyes in aqueous solution, thereby facilitating cellular uptake from aqueous dispersions.[100] The PowerLoad™ reagent is supplied as a ready-to-use concentrate in sterile water that is mixed with the dye stock solution (typically prepared in DMSO) and then diluted 1:100 into serum-free medium for direct application to adherent cells on coverslips, in microplate wells or in other suitable incubation chambers (Section 23.1).

REFERENCES

1. Nat Methods (2007) 4:619; **2.** Proc Natl Acad Sci U S A (1994) 91:187; **3.** Biophys J (1996) 70:1006; **4.** Cell Calcium (1996) 19:185; **5.** Biophys J (2004) 86:978; **6.** Pflugers Arch (2007) 454:663; **7.** Proc Natl Acad Sci U S A (2007) 104:5674; **8.** Cell Calcium (2005) 37:565; **9.** Biophys J (1999) 76:489; **10.** Am J Physiol (1999) 277:L893; **11.** Proc Natl Acad Sci U S A (1988) 85:6571; **12.** Science (1988) 241:842; **13.** Methods Cell Biol (1994) 40:31; **14.** Neuron (1993) 10:21; **15.** Biochemistry (1992) 31:8856; **16.** J Physiol (2001) 533:627; **17.** Chem Rev (2008) 108:1603; **18.** Biochim Biophys Acta (1990) 1035:378; **19.** J Am Chem Soc (1989) 111:7957; **20.** Proc Natl Acad Sci U S A (1992) 89:11804; **21.** Science (2009) 325:207; **22.** CSH Protoc (2009):pdb.prot5201; **23.** Biochemistry (1980) 19:2396; **24.** Science (2009) 325:207; **25.** Anal Chem (2009) 81:538; **26.** J Physiol (2003) 553:775; **27.** Eur J Biochem (2004) 271:3255; **28.** J Physiol (2000) 522:231; **29.** J Neurosci (1995) 15:903; **30.** J Neurosci (1995) 15:2867; **31.** Exp Physiol (1994) 79:269; **32.** J Biol Chem (1993) 268:6511; **33.** Mol Biol Cell (1993) 4:293; **34.** Int J Dev Biol (2003) 47:411; **35.** Mol Biol Cell (2002) 13:1263; **36.** J Biol Chem (2003) 278:12247; **37.** J Neurophysiol (1994) 72:1973; **38.** Neuron (1993) 11:221; **39.** Proc Natl Acad Sci U S A (1986) 83:6179; **40.** Proc Natl Acad Sci U S A (1983) 80:7178; **41.** Am J Physiol (1987) 252:C441; **42.** Blood (1987) 69:1469; **43.** Am J Physiol (1990) 258:H9; **44.** Proc Natl Acad Sci U S A (1987) 84:6005; **45.** Biophys J (1995) 69:1683; **46.** Proc Natl Acad Sci U S A (2006) 103:13232; **47.** J Neurosci (1994) 14:523; **48.** Bioconjug Chem (1991) 2:180; **49.** Anal Biochem (2004) 328:187; **50.** Chem Rev (1987) 87:901; **51.** J Med Chem (1989) 32:236; **52.** Nat Struct Mol Biol (2010) 17:112; **53.** J Biol Chem (2000) 275:20572; **54.** Cell Calcium (2009) 45:185; **55.** Methods Mol Biol (2006) 312:229; **56.** Am J Physiol Gastrointest Liver Physiol (2006) 290:G250; **57.** Biophys J (1996) 71:1048; **58.** J Cell Biol (1990) 111:183; **59.** Cell Calcium (1998) 23:151; **60.** Exp Cell Res (1998) 239:393; **61.** Mol Carcinog (1998) 21:205; **62.** Cell Calcium (1991) 12:279; **63.** Methods Enzymol (1989) 172:230; **64.** J Neurochem (1994) 62:890; **65.** Cell Calcium (1990) 11:75; **66.** Cell Calcium (1997) 21:233; **67.** J Vis Exp (2009) doi: 10.3791/1067; **68.** J Biol Chem (2005) 280:31936; **69.** Cell (1982) 29:33; **70.** J Virol (2002) 76:11505; **71.** FEBS Lett (2001) 508:484; **72.** Biotechniques (2004) 37:96; **73.** Antisense Nucleic Acid Drug Dev (2000) 10:263; **74.** EMBO J (2002) 21:1743; **75.** Methods Enzymol (2006) 414:211; **76.** Nano Lett (2006) 6:1491; **77.** Physiol Rev (2002) 82:1013; **78.** J Biol Chem (1987) 262:8884; **79.** Am J Physiol (1998) 275:C1158; **80.** J Cell Sci (2008) 121:3261; **81.** Br J Pharmacol (2000) 130:513; **82.** J Immunol (2009) 182:2051; **83.** J Immunol (2008) 180:300; **84.** Br J Pharmacol (2001) 132:1501; **85.** J Neurochem (1993) 61:1657; **86.** Biochemistry (1989) 28:3989; **87.** Anal Biochem (1985) 146:349; **88.** J Neurochem (1998) 71:2401; **89.** J Biol Chem (1985) 260:2719; **90.** J Biol Chem (2000) 275:7071; **91.** Biophys J (1998) 75:1244; **92.** J Pharmacol Exp Ther (2003) 304:801; **93.** Cell Calcium (1990) 11:57; **94.** Biotechniques (2003) 34:164; **95.** Methods Cell Biol (1994) 40:155; **96.** Proc Natl Acad Sci U S A (1987) 84:7793; **97.** Science (1986) 233:886; **98.** Anal Chem (2008) 80:9840; **99.** Brain Res (2006) 1068.131, **100.** J Biomol Screen (2010) 15:441.

DATA TABLE 19.8 CHELATORS, CALIBRATION BUFFERS, IONOPHORES AND CELL-LOADING REAGENTS

Cat. No.	MW	Storage	Soluble	Abs	EC	Em	Solvent	Product	K_d	Notes
A1493	523.63	F,L	DMSO, EtOH	378	8900	438	MeOH			
B1204	628.80	D	pH >6	284	5100	see Notes	pH 7.2		160 nM	1, 2, 3, 4, 5
B1205	764.69	F,D	DMSO	287	5900	ND	$CHCl_3$	B1204		6
B1212	1004.03	D	pH >6	285	5200	see Notes	pH 7.2		160 nM	1, 2, 3, 4, 5
B1214	564.37	D	pH >6	285	5100	see Notes	pH 7.2		160 nM	1, 2, 3, 4, 5
B1494	602.52	F,D	DMSO, EtOH	289	20,000	none	MeOH			7
B6769	764.69	F,D	DMSO	287	5900	ND	$CHCl_3$	B1204		6
5,5′-dimethyl BAPTA	656.85	D	pH >6	290	5100	ND	pH 7.2		40 nM	1, 4, 5, 6, 8
D1207	792.75	F,D	DMSO	291	5900	ND	$CHCl_3$	5,5′-dimethyl BAPTA	6	
5,5′-difluoro BAPTA	664.78	D	pH >6	289	5100	ND	pH 7.2		635 nM	1, 4, 5, 6, 9
D1209	800.67	F,D	DMSO	290	5700	ND	EtOAc	5,5′-difluoro BAPTA	6	
D1211	786.59	D	pH >6	263	18,000	ND	pH 7.2		1.6 µM	1, 4, 5, 6, 8
D3034	710.86	F,D,LL	pH >6	369	18,000	none	pH 7.2		2.2 µM	1, 5, 10, 11

continued on next page

IMPORTANT NOTICE: The products described in this manual are covered by one or more Limited Use Label License(s). Please refer to the Appendix on page 971 and Master Product List on page 975. Products are For Research Use Only. Not intended for any animal or human therapeutic or diagnostic use.

DATA TABLE 19.8 CHELATORS, CALIBRATION BUFFERS, IONOPHORES AND CELL-LOADING REAGENTS—*continued*

Cat. No.	MW	Storage	Soluble	Abs	EC	Em	Solvent	Product	K_d	Notes
D6814	473.39	D,LL	DMSO	348	4200	none	pH 7.2		5 nM	1, 5, 11, 12
E1219	668.60	F,D	DMSO	<300		none				13
I24221	540.54	F,DD	DMSO	<300		none				14
I24222	747.08	F,D	DMSO, EtOH	300	22,000	none	MeOH			
N6802	653.81	FF,D,LL	pH >6	260	3500	none	pH 7.2		80 nM	1, 5, 7, 11, 15
N6803	789.70	FF,D,LL	DMSO	250	4200	none	MeCN	N6802		13, 16
T1210	424.55	D	EtOH	261	14,000	ND	MeOH		see Notes	17

For definitions of the contents of this data table, see "Using *The Molecular Probes® Handbook*" in the introductory pages.

Notes

1. Dissociation constants are known to vary considerably depending on the temperature, pH, ionic strength, viscosity, protein binding, presence of other ions (especially polyvalent ions), instrument setup and other factors. It is strongly recommended that these values be verified under user-specific experimental conditions.
2. Dissociation constant determined from absorption measurements in 100 mM KCl, 10 mM MOPS, pH 7.2 at 22°C. Ca^{2+} concentrations below 5 µM were controlled using CaEGTA buffering (Calcium Calibration Buffer Kit #1, C3008MP).
3. BAPTA is weakly fluorescent in aqueous solutions (Em = 363 nm, QY = 0.03). (Biochemistry (1980) 19:2396)
4. Absorption spectra of BAPTA and its derivatives are Ca^{2+} dependent.
5. Abs and EC values determined in Ca^{2+}-free solution (100 mM KCl, 10 mM EGTA, 10 mM MOPS, pH 7.2).
6. ND = not determined.
7. This compound has weaker visible absorption at >300 nm but no discernible absorption peaks in this region.
8. Dissociation constant determined in CaEGTA buffers in 100 mM KCl, 10 mM MOPS, pH 7.3. (Biochemistry (1980) 19:2396)
9. 5,5′-Difluoro BAPTA dissociation constant determined in 115 mM KCl, 20 mM NaCl, 10 mM MOPS, pH 7.3. (Am J Physiol (1994) 266:C1313, Am J Physiol (1987) 252:C441)
10. The Ca^{2+} dissociation constant of diazo-2 is 2200 nM before photolysis and 73 nM after ultraviolet photolysis. The absorption spectrum of the photolysis product is similar to that of B1204. (J Am Chem Soc (1989) 111:7957)
11. All photoactivatable probes are sensitive to light. They should be protected from illumination except when photolysis is intended.
12. $K_d(Ca^{2+})$ increases from 5 nM to 3 mM after ultraviolet photolysis. K_d values determined in 130 mM KCl, 10 mM HEPES, pH 7.1. (Proc Natl Acad Sci U S A (1988) 85:6571)
13. This product is intrinsically a liquid or an oil at room temperature.
14. Isothiocyanates are unstable in water and should not be stored in aqueous solution.
15. $K_d (Ca^{2+})$ increases from 80 nM to 1 mM after ultraviolet photolysis. K_d values determined in 100 mM KCl, 40 mM HEPES, pH 7.2. (Proc Natl Acad Sci U S A (1994) 91:187)
16. N6803 is converted to N6802 via hydrolysis of its acetoxymethyl ester (AM) groups.
17. TPEN has very high affinity for Zn^{2+} ($K_d = 2.6 \times 10^{-16}$ M), Fe^{2+} ($K_d = 2.4 \times 10^{-15}$ M) and Mn^{2+} ($K_d = 5.4 \times 10^{-11}$ M) but relatively low affinity for Ca^{2+} ($K_d = 4.0 \times 10^{-5}$ M) and Mg^{2+} ($K_d = 2.0 \times 10^{-2}$ M). (J Biol Chem (1985) 260:2719)

PRODUCT LIST 19.8 CHELATORS, CALIBRATION BUFFERS, IONOPHORES AND CELL-LOADING REAGENTS

Cat. No.	Product	Quantity
A1493	A-23187 free acid (calcimycin)	10 mg
B1205	BAPTA, AM *cell permeant*	25 mg
B6769	BAPTA, AM *cell permeant* *special packaging*	20 x 1 mg
B1212	BAPTA, tetracesium salt *cell impermeant*	1 g
B1204	BAPTA, tetrapotassium salt *cell impermeant*	1 g
B1214	BAPTA, tetrasodium salt *cell impermeant*	1 g
B1494	4-bromo A-23187, free acid	1 mg
C3008MP	Calcium Calibration Buffer Kit #1 *zero and 10 mM CaEGTA (2 x 50 mL)*	1 kit
C3047	Calcium Sponge™ S (BAPTA polystyrene)	1 g
D3034	diazo-2, tetrapotassium salt *cell impermeant*	1 mg
D1211	5,5′-dibromo BAPTA, tetrapotassium salt *cell impermeant*	100 mg
D1209	5,5′-difluoro BAPTA, AM *cell permeant*	25 mg
D6814	1-(4,5-dimethoxy-2-nitrophenyl)-1,2-diaminoethane-*N,N,N′,N′*-tetraacetic acid (DMNP-EDTA) *cell impermeant*	5 mg
D1207	5,5′-dimethyl BAPTA, AM *cell permeant*	25 mg
E1219	EGTA, tetra(acetoxymethyl ester) (EGTA, AM)	10 mg
F6774	Fura-2 Calcium Imaging Calibration Kit *zero to 10 mM CaEGTA, 50 µM fura-2 (11 x 1 mL)*	1 kit
I14402	Influx™ pinocytic cell-loading reagent *makes 10 x 5 mL*	1 set
I24222	ionomycin, calcium salt	1 mg
I24221	(*S*)-1-*p*-isothiocyanatobenzyldiethylenetriaminepentaacetic acid (DTPA isothiocyanate)	1 mg
N6803	*o*-nitrophenyl EGTA, AM (NP-EGTA, AM) *cell permeant* *special packaging*	20 x 50 µg
N6802	*o*-nitrophenyl EGTA, tetrapotassium salt (NP-EGTA) *cell impermeant*	1 mg
P6866	Pluronic® F-127 *10% solution in water* *0.2 µm filtered*	30 mL
P3000MP	Pluronic® F-127 *20% solution in DMSO*	1 mL
P6867	Pluronic® F-127 *low UV absorbance*	2 g
P10020	PowerLoad™ concentrate, 100X	5 mL
P36400	probenecid, water soluble	10 x 77 mg
T1210	tetrakis-(2-pyridylmethyl)ethylenediamine (TPEN)	100 mg

IMPORTANT NOTICE: The products described in this manual are covered by one or more Limited Use Label License(s). Please refer to the Appendix on page 971 and Master Product List on page 975. Products are For Research Use Only. Not intended for any animal or human therapeutic or diagnostic use.
www.invitrogen.com/probes

TWENTY

CHAPTER 20

pH Indicators

IMPORTANT NOTICE: The products described in this manual are covered by one or more Limited Use Label License(s). Please refer to the Appendix on page 971 and Master Product List on page 975. Products are For Research Use Only. Not intended for any animal or human therapeutic or diagnostic use.

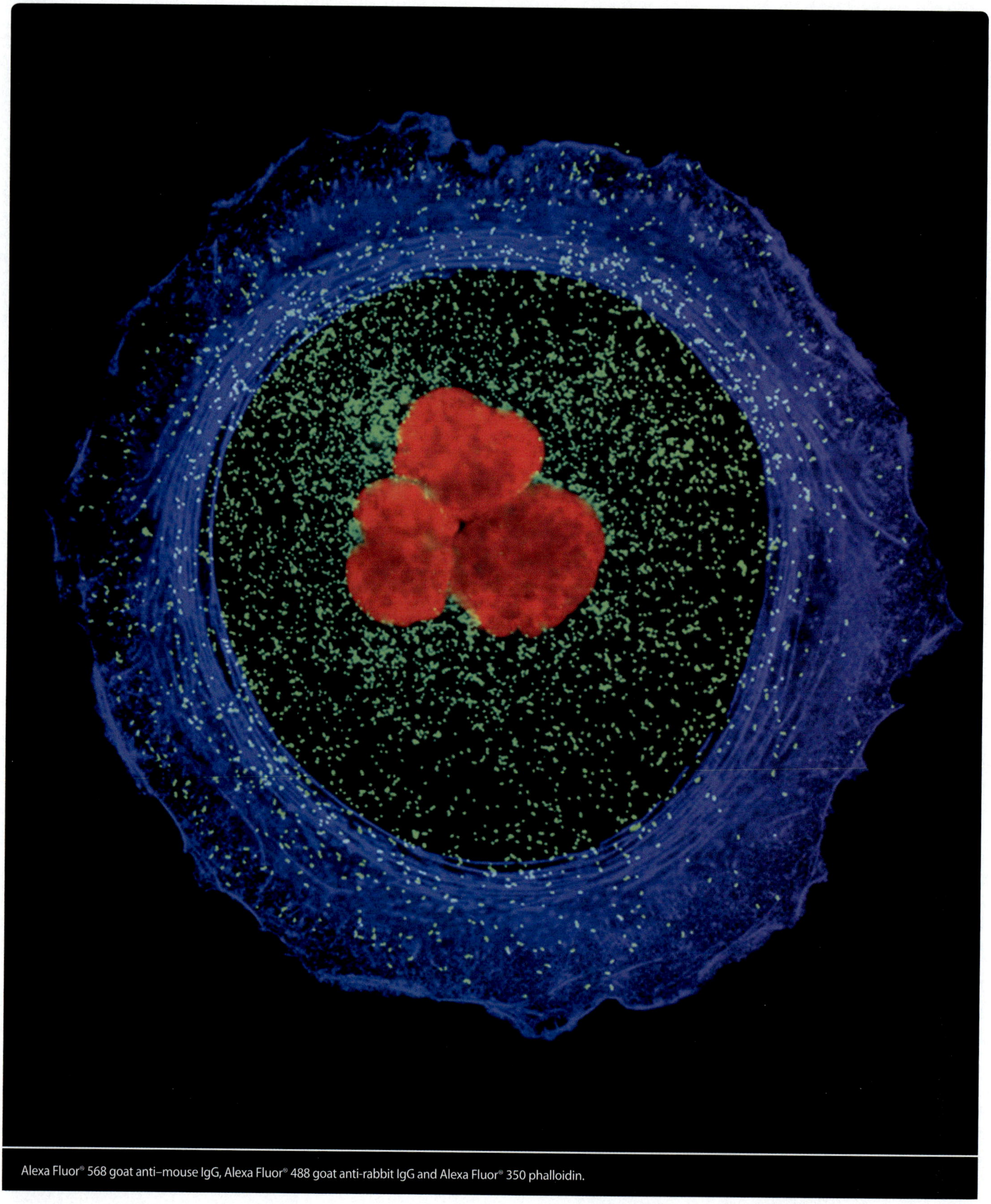

Alexa Fluor® 568 goat anti-mouse IgG, Alexa Fluor® 488 goat anti-rabbit IgG and Alexa Fluor® 350 phalloidin.

IMPORTANT NOTICE: The products described in this manual are covered by one or more Limited Use Label License(s). Please refer to the Appendix on page 971 and Master Product List on page 975. Products are For Research Use Only. Not intended for any animal or human therapeutic or diagnostic use.
www.invitrogen.com/probes

20.1 Overview of pH Indicators

The ability of dyes—notably litmus, phenolphthalein and phenol red—to change their color in response to a pH change has found widespread application in research and industry. Fluorescent dyes, however, provide the increased sensitivity required for optical pH measurements *inside* live cells. They also offer much greater spatial sampling capability when compared with microelectrode techniques.[1] These advantages have spurred the development of improved fluorescent dyes that can sense pH changes within physiological ranges. Of course, many of the same fluorescent pH indicators can also be used as pH sensors in cell-free media.

To quantitatively measure pH, it is essential to match the indicator's pK_a to the pH of the experimental system. Consequently, the following two sections of this chapter are divided into pH indicators for use in environments with near-neutral pH (Section 20.2) and pH indicators for use in relatively acidic environments (Section 20.3). Intracellular pH is generally between ~6.8 and 7.4 in the cytosol and ~4.5 and 6.0 in the cell's acidic organelles. Unlike intracellular free Ca^{2+} concentrations, which can rapidly change by perhaps 100-fold, the pH inside a cell varies by only fractions of a pH unit, and such changes may be quite slow.

We offer a variety of fluorescent pH indicators, pH indicator conjugates and other reagents for pH measurements in biological systems. Among these are several probes with unique optical responses and specialized localization characteristics:

- Visible light–excitable SNARF® pH indicators enable researchers to determine intracellular pH in the physiological range using dual-emission or dual-excitation ratiometric techniques (Section 20.2), thus providing important tools for confocal laser-scanning microscopy and flow cytometry.
- pHrodo™ dye and LysoSensor™ probes, for use in acidic environments such as lysosomes (Section 20.3).
- Fluorescent pH indicators coupled to dextrans are extremely well retained in cells, do not bind to cellular proteins and have a reduced tendency to compartmentalize[2] (Section 20.4).

Families of Molecular Probes® pH indicators are listed in Table 20.1 in approximate order of decreasing pK_a value.

REFERENCES

1. Chem Rev (2010) 110:2709; **2.** Methods Cell Biol (1989) 29:59.

Table 20.1 Molecular Probes® pH indicator families, in order of decreasing pK_a.

Parent Fluorophore	pH Range	Typical Measurement
SNARF® indicators	6.0–8.0	Emission ratio 580/640 nm
HPTS (pyranine)	7.0–8.0	Excitation ratio 450/405 nm
BCECF	6.5–7.5	Excitation ratio 490/440 nm
Fluoresceins and carboxyfluoresceins	6.0–7.2	Excitation ratio 490/450 nm
LysoSensor™ Green DND-189	4.5–6.0	Single emission 520 nm
Oregon Green® dyes	4.2–5.7	Excitation ratio 510/450 nm or excitation ratio 490/440 nm
LysoSensor™ Yellow/Blue DND-160	3.5–6.0	Emission ratio 450/510 nm
pHrodo™ dye	See below *	Single emission 585 nm

* pHrodo™ succinimidyl ester exhibits a complex pH titration profile. Decreasing pH (from pH 9 to pH 2) produces a continuous (but nonlinear) fluorescence increase. This pH response profile typically changes upon conjugation of the dye to proteins and other biomolecules.

IMPORTANT NOTICE: The products described in this manual are covered by one or more Limited Use Label License(s). Please refer to the Appendix on page 971 and Master Product List on page 975. Products are For Research Use Only. Not intended for any animal or human therapeutic or diagnostic use.

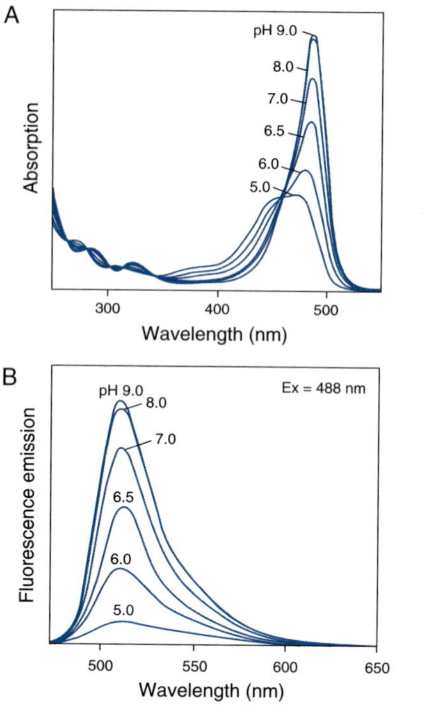

Figure 20.2.2 The pH-dependent spectra of fluorescein (F1300): **A)** absorption spectra, **B)** emission spectra.

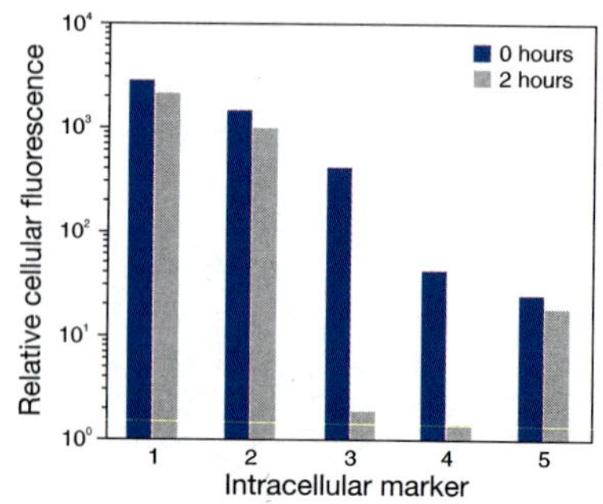

Figure 20.2.3 Loading and retention characteristics of intracellular marker dyes. Cells of a human lymphoid line (GePa) were loaded with the following cell-permeant acetoxymethyl ester (AM) or acetate derivatives of fluorescein: **1)** calcein AM (C1430, C3099, C3100MP), **2)** BCECF AM (B1150), **3)** fluorescein diacetate (FDA, F1303), **4)** carboxyfluorescein diacetate (CFDA, C1354) and **5)** CellTracker™ Green CMFDA (5-chloromethylfluorescein diacetate, C2925, C7025). Cells were incubated in 4 µM staining solutions in Dulbecco's modified eagle medium containing 10% fetal bovine serum (DMEM+) at 37°C. After incubation for 30 minutes, cell samples were immediately analyzed by flow cytometry to determine the average fluorescence per cell at time zero (0 hours). Retained cell samples were subsequently washed twice by centrifugation, resuspended in DMEM+, maintained at 37°C for 2 hours and then analyzed by flow cytometry. The decrease in the average fluorescence intensity per cell in these samples relative to the time zero samples indicates the extent of intracellular dye leakage during the 2-hour incubation period.

20.2 Probes Useful at Near-Neutral pH

Fluorescein and Fluorescein Derivatives

Fluorescein and many of its derivatives exhibit multiple, pH-dependent ionic equilibria.[1–5] Both the phenol and carboxylic acid functional groups of fluorescein are almost totally ionized in aqueous solutions above pH 9 (Figure 20.2.1). Acidification of the fluorescein dianion first protonates the phenol (pK_a ~6.4) to yield the fluorescein monoanion, then the carboxylic acid (pK_a <5) to produce the neutral species of fluorescein. Further acidification generates a fluorescein cation (pK_a ~2.1).

Only the monoanion and dianion of fluorescein are fluorescent, with quantum yields of 0.37 and 0.93, respectively, although excitation of either the neutral or cationic species is reported to produce emission from the anion with effective quantum yields of 0.31 and 0.18, respectively.[2] A further equilibrium involves formation of a colorless, nonfluorescent lactone (Figure 20.2.1). The lactone is not formed in aqueous solution above pH 5 but may be the dominant form of neutral fluorescein in solvents such as acetone. The pH-dependent absorption spectra of fluorescein (Figure 20.2.2) clearly show the blue shift and decreased absorptivity indicative of the formation of protonated species. However, the fluorescence emission spectrum of most fluorescein derivatives, even in acidic solution, is dominated by the dianion, with only small contributions from the monoanion. Consequently, the wavelength and shape of the emission spectra resulting from excitation close to the dianion absorption peak at 490 nm are relatively independent of pH, but the fluorescence intensity is dramatically reduced at acidic pH (Figure 20.2.2).

We offer a broad variety of fluorescein-derived reagents and fluoresceinated probes that can serve as sensitive fluorescent pH indicators in a wide range of applications. Chemical substitutions of fluorescein may shift absorption and fluorescence maxima and change the pK_a of the dye; however, the effects of acidification on the spectral characteristics illustrated in Figure 20.2.2 are generally maintained in all fluorescein derivatives.

Fluorescein and Its Diacetate

The cell-permeant fluorescein diacetate (FDA, F1303) is still occasionally used to measure intracellular pH,[6] as well as to study cell adhesion[7] or, in combination with propidium iodide (P1304MP, P3566, P21493; Section 8.1), to determine cell viability.[8,9] However, fluorescein (F1300), which is formed by intracellular hydrolysis of FDA, rapidly leaks from cells (Figure 20.2.3). Thus, other cell-permeant dyes such as the acetoxymethyl (AM) esters of BCECF and calcein are now preferred for intracellular pH measurements and cell viability assays (Section 15.2).

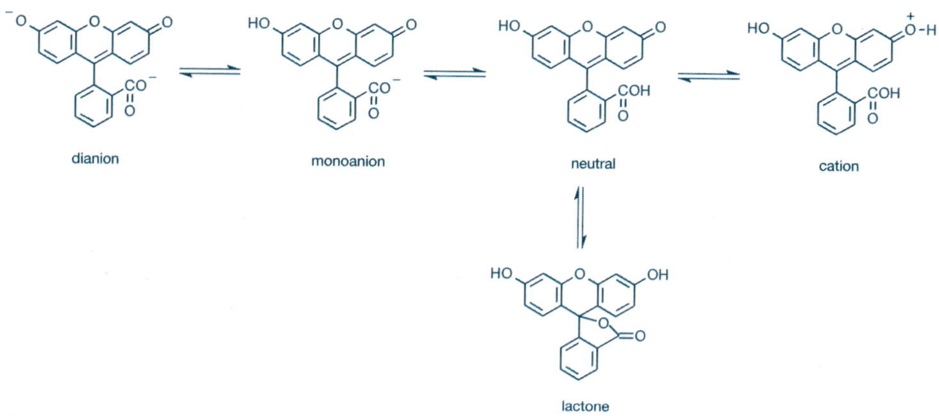

Figure 20.2.1 Ionization equilibria of fluorescein.

Carboxyfluorescein and Its Cell-Permeant Esters

Fluorescein's high leakage rate out of cells makes it very difficult to quantitate intracellular pH, because the decrease in the cell's fluorescence due to dye leakage cannot be easily distinguished from that due to acidification. The use of carboxyfluorescein diacetate (CFDA, C195) for intracellular pH measurements partially addresses this problem.[10,11] CFDA is moderately permeant to most cell membranes and, upon hydrolysis by intracellular nonspecific esterases, forms carboxyfluorescein (5(6)-FAM, C194, C1904), which has a pH-dependent spectral response very similar to that of fluorescein. As compared with fluorescein, carboxyfluorescein contains an extra negative charge and is therefore better retained in cells[12] (Figure 20.2.3). The mixed-isomer preparation of CFDA (C195) is usually adequate for intracellular pH measurements because the single isomers of carboxyfluorescein exhibit essentially identical pH-dependent spectra with a pK_a ~6.5. For experiments requiring a pure isomer, the single-isomer preparations of carboxyfluorescein (C1359, C1360; Section 1.5) and CFDA (C1361, C1362; Section 15.2) are available. In addition, we offer the AM ester of CFDA (5-CFDA, AM, C1354), which is electrically neutral and facilitates cell loading. Upon hydrolysis by intracellular esterases, this AM ester also yields carboxyfluorescein.[13–15]

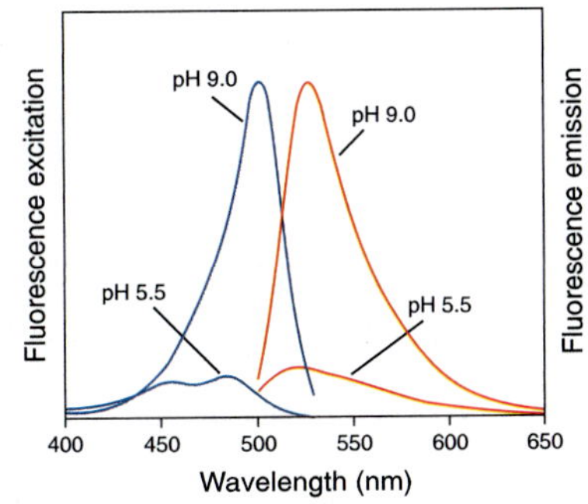

Figure 20.2.5 Fluorescence excitation (detected at 535 nm) and emission (excited at 490 nm) spectra of BCECF in pH 9.0 and pH 5.5 buffers.

BCECF and Its AM Ester

Although carboxyfluorescein is better retained in cells than is fluorescein, its pK_a of ~6.5 is lower than the cytosolic pH of most cells (pH ~6.8–7.4). Consequently, its fluorescence change is less than optimal for detecting small pH changes above pH 7. Since its introduction by Roger Tsien in 1982,[16,17] the polar fluorescein derivative BCECF (B1151) and its membrane-permeant AM ester (B1150, B1170, B3051) have become the most widely used fluorescent indicators for estimating intracellular pH. Also, a flow cytometric assay has been developed that uses BCECF to estimate the concentration of intracellular K^+.[18] BCECF's four to five negative charges at pH 7–8 improve its retention in cells (Figure 20.2.3), and its pK_a of 6.98 is ideal for typical intracellular pH measurements.

As with fluorescein and carboxyfluorescein, absorption of the phenolate anion (basic) form of BCECF is red-shifted and has increased molar absorptivity relative to the protonated (acidic) form (Figure 20.2.4); there is little pH-dependent shift in the fluorescence emission spectrum of BCECF upon excitation at 505 nm. BCECF is typically used as a dual-excitation ratiometric pH indicator. Signal errors caused by variations in concentration, path length, leakage and photobleaching are greatly reduced with ratiometric methods (Loading and Calibration of Intracellular Ion Indicators—Note 19.1). Intracellular pH measurements with BCECF are made by determining the pH-dependent ratio of emission intensity (detected at 535 nm) when the dye is excited at ~490 nm versus the emission intensity when excited at its isosbestic point of ~440 nm (Figure 20.2.4, Figure 20.2.5). Because BCECF's absorption at 440 nm is quite weak, increasing the denominator wavelength to ~450 nm provides improved signal-to-noise characteristics for ratio imaging applications.[19–21] As with other intracellular pH indicators, *in situ* calibration of BCECF's fluorescence response is usually accomplished using 10–50 µM nigericin (N1495) in the presence of 100–150 mM K^+ to equilibrate internal and external pH.[22,23] Alternative calibration methods have also been reported.[24–26]

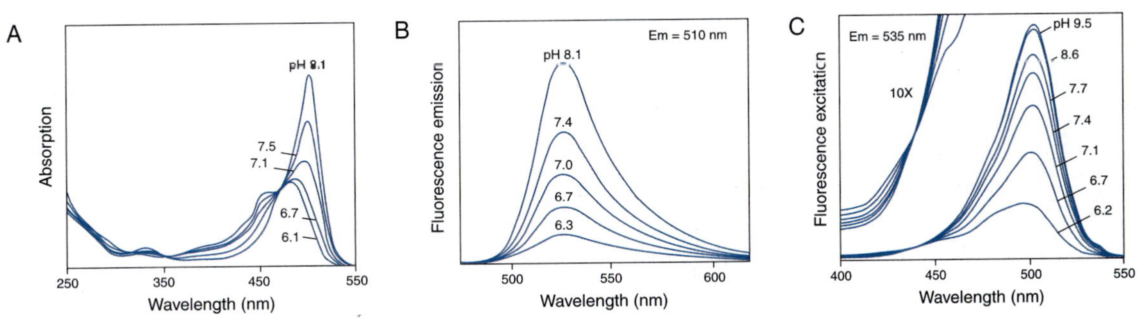

Figure 20.2.4 The pH-dependent spectra of BCECF (B1151): **A)** absorption spectra, **B)** emission spectra and **C)** excitation spectra. The fluorescence excitation spectra on the left in panel C have been enlarged 10X to reveal BCECF's 439 nm isosbestic point. Note that the isosbestic point of the excitation spectra of BCECF is different from that of the absorption spectra (compare panels A and **C**).

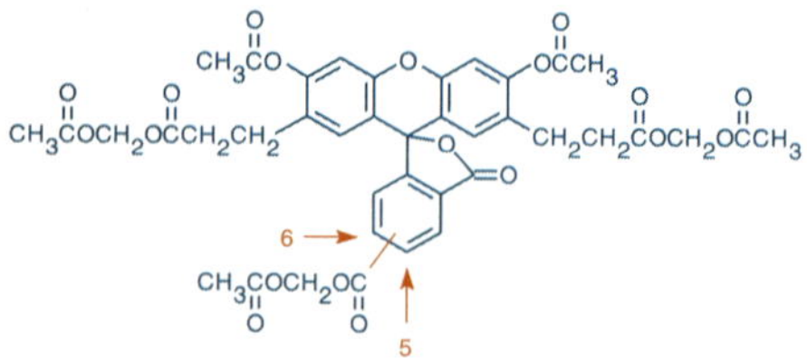

I (Molecular Weight = 820.7)

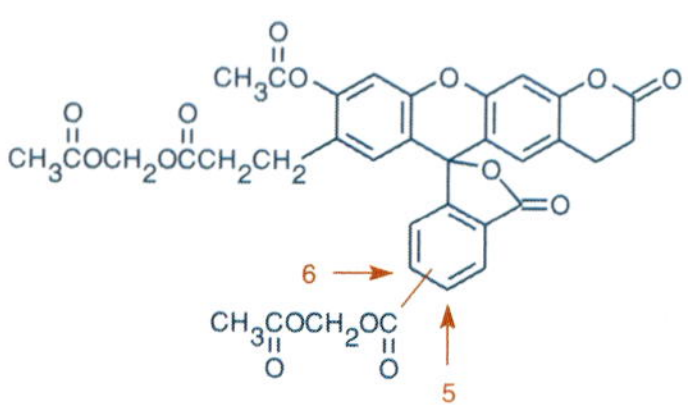

II (Molecular Weight = 688.6)

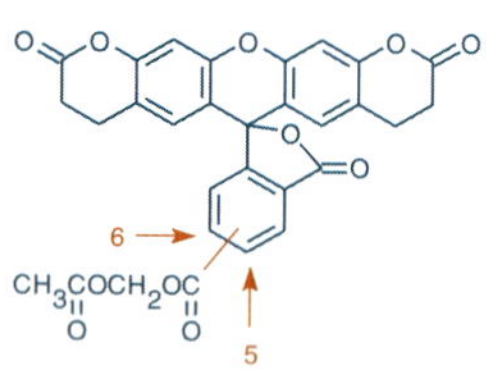

III (Molecular Weight = 556.5)

Figure 20.2.6 Structures of the AM esters of BCECF (B1150, B1170, B3051).

Loading of live cells for measurement of intracellular pH is readily accomplished by incubating cell suspensions or adherent cells in a 1–10 µM solution of the AM ester of BCECF. At least three different molecular species can be obtained in synthetic preparations of the AM ester of BCECF; however, all three forms shown in Figure 20.2.6 appear to be converted to the same product—BCECF acid (B1151, Figure 20.2.7)—by intracellular esterase hydrolysis. Although we can readily prepare the pure tri(acetoxymethyl) ester form (Form I in Figure 20.2.6), some researchers have found that cell loading with a mixture of the lactone Forms II and III is more efficient. Consequently, we produce BCECF AM predominantly as a mixture of Forms II and III with a typical percentage composition ratio of 45:55, as determined by HPLC, NMR and mass spectrometry. The AM ester of BCECF is available in a single 1 mg vial (B1150), specially packaged as a set of 20 vials that each contains 50 µg (B1170) and as a 1 mg/mL solution (~1.6 mM) in anhydrous dimethylsulfoxide (DMSO) (B3051). We highly recommend purchasing the set of 20 vials in order to reduce the potential for product deterioration caused by exposure to moisture.

Our bibliography for BCECF AM lists more than 1200 journal citations, including references for the use of BCECF AM to investigate:

- Cl^-/HCO_3^- exchange [27–30]
- K^+/H^+ exchange [31,32]
- Na^+/H^+ exchange [33–35]
- Na^+/Ca^{2+} exchange [36]
- NH_4^+ transport [37,38]
- Lactate transport and metabolism [39–41]
- Apoptosis [42–44] (Section 15.5)
- Phagocytosis [45–47] (Section 16.1)
- Regulation of pancreatic insulin secretion [48]
- Voltage-activated H^+ conductance in neurons [49]

The cell-impermeant BCECF acid (B1151) is useful for pH measurements in intercellular spaces of epithelial cell monolayers,[50] interstitial spaces of normal and neoplastic tissue [51,52] and isolated cell fractions.[53] BCECF has also been employed for two-photon fluorescence lifetime imaging of the skin stratum corneum to detect aqueous acid pockets within the lipid-rich extracellular matrix.[54] The free acid of BCECF can be loaded into cells by microinjection [26] or electroporation or by using our Influx™ pinocytic cell-loading reagent (I14402, Section 19.8). It has also been loaded into bacterial cells by brief incubation at pH ~2.[55,56] In addition to the cell-permeant BCECF AM and cell-impermeant BCECF acid, we offer dextran conjugates of BCECF (D1878, D1880; Section 20.4).

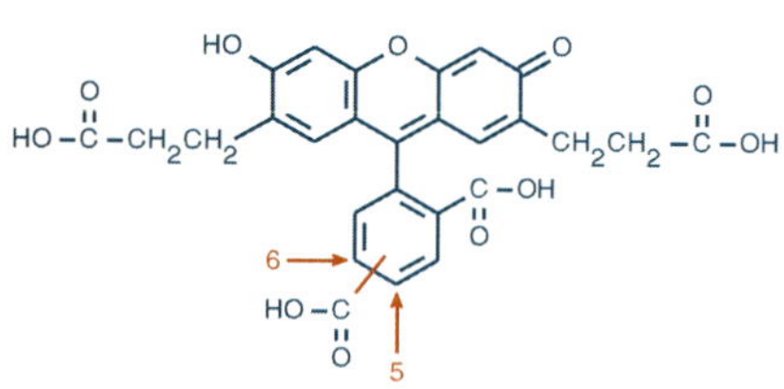

Figure 20.2.7 2',7'-bis-(2-carboxyethyl)-5-(and-6)-carboxyfluorescein (BCECF acid, B1151).

Fluorescein Sulfonic Acid and Its Diacetate

The fluorescein-5-(and 6-)sulfonic acid (F1130, Figure 20.2.8) is much more polar than carboxyfluorescein. Consequently, once inside cells or liposomes, it is relatively well retained. Some cells can be loaded directly with 5-sulfofluorescein diacetate [57–60] (SFDA, S1129). Direct ratiometric measurement of the pH in the trans-Golgi of live human fibroblasts was achieved by simultaneously microinjecting liposomes loaded with both fluorescein sulfonic acid and sulforhodamine 101 [61] (S359, Section 14.3). Fluorescein-5-(and 6-)sulfonic acid is more commonly used to measure barrier permeability of membranes[62,63] (Section 14.3).

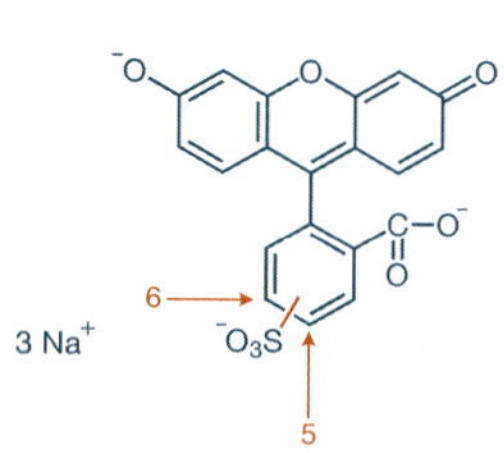

Figure 20.2.8 Fluorescein-5-(and-6)-sulfonic acid, trisodium salt (F1130).

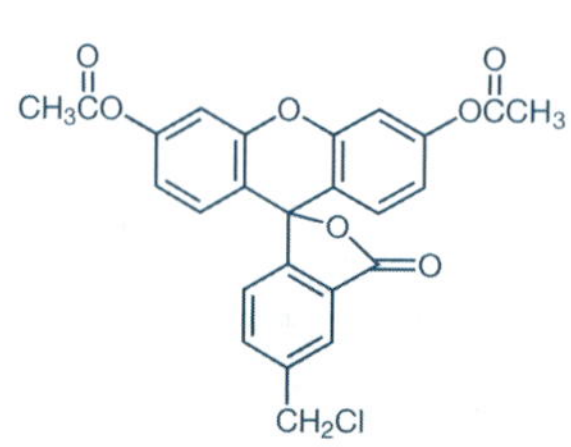

Figure 20.2.9 CellTracker™ Green CMFDA (5-chloromethylfluorescein diacetate, C2925).

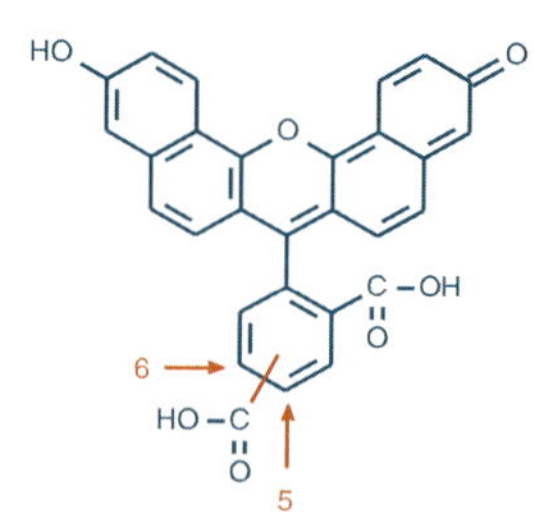

Figure 20.2.10 5-(and-6)-carboxynaphthofluorescein (C652).

IMPORTANT NOTICE: The products described in this manual are covered by one or more Limited Use Label License(s). Please refer to the Appendix on page 971 and Master Product List on page 975. Products are For Research Use Only. Not intended for any animal or human therapeutic or diagnostic use.

Chemically Reactive Fluorescein Diacetates

One means for overcoming the cell leakage problem common to the above pH indicators, including BCECF, is to trap the indicator inside the cell via conjugation to intracellular constituents. CellTracker™ Green CMFDA (C2925, C7025; Figure 20.2.9) and chloromethyl SNARF®-1 (C6826) incorporate a thiol-reactive chloromethyl moiety that reacts with intracellular thiols, including glutathione and proteins, to yield well-retained products (Figure 20.2.3). Cleavage of the acetate groups of the CMFDA conjugate by intracellular esterases yields a conjugate that retains the pH-dependent spectral properties of fluorescein. Because of its superior retention as compared with SNARF® AM and BCECF AM, CellTracker™ Green CMFDA was employed to monitor the intracellular pH response to osmotic stress in CHO, HEK 293 and Caco-2 cells.[64] Similarly, the amine-reactive succinimidyl ester of CFDA (CFSE, C1157) can be used for long-term pH studies of live cells, producing a conjugate with the pH-sensitive properties of carboxyfluorescein.[65]

Carboxynaphthofluorescein

Carboxynaphthofluorescein (C652, Figure 20.2.10) has pH-dependent red fluorescence (excitation/emission maxima ~598/668 nm at pH >9) with a relatively high pK_a of ~7.6. The long-wavelength pH-dependent spectra of carboxynaphthofluorescein have been exploited in the construction of fiber-optic pH sensors.[66,67] This long-wavelength pH indicator is also available in membrane-permeant diacetate form (C13196) for passive intracellular loading and as an amine-reactive succinimidyl ester (C653, Section 20.4) for preparing pH-sensitive conjugates.

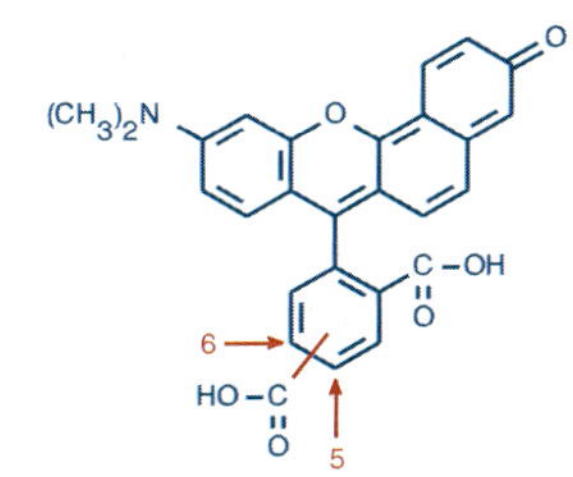

Figure 20.2.12 5-(and-6)-carboxy SNARF®-1 (C1270).

SNARF® pH Indicator

The seminaphthorhodafluors (SNARF® dyes) are visible light–excitable fluorescent pH indicators.[68] The SNARF® indicators have both dual-emission and dual-excitation properties, making them particularly useful for confocal laser-scanning microscopy[69–72] (Figure 20.2.11), flow cytometry[21,73–75] and microplate reader–based measurements.[76] The dual-emission properties of the SNARF® indicators make them preferred probes for use in fiber-optic pH sensors.[77–79] These pH indicators can be excited by the 488 or 514 nm spectral lines of the argon-ion laser and are sensitive to pH values within the physiological range. Dextran conjugates of the SNARF® dyes are described in Section 20.4.

Carboxy SNARF®-1 Dye and Its Cell-Permeant Ester

The carboxy SNARF®-1 dye (C1270, Figure 20.2.12), which is easily loaded into cells as its cell-permeant AM ester acetate (C1271, C1272), has a pK_a of about 7.5 at room temperature and between 7.3 and 7.4 at 37°C. Thus, carboxy SNARF®-1 is useful for measuring pH changes

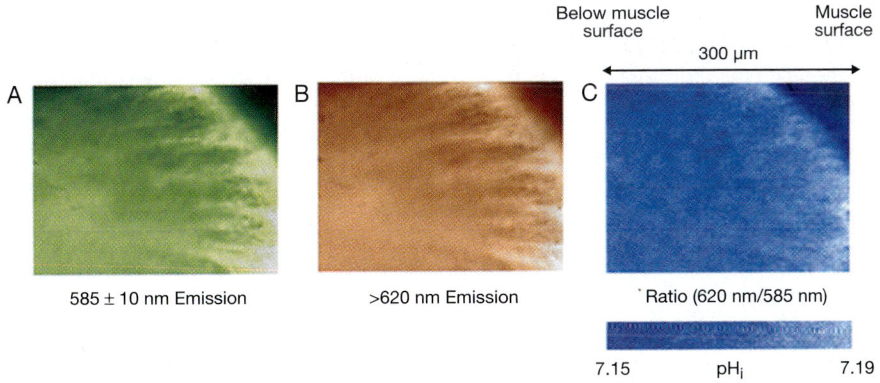

Figure 20.2.11 Confocal fluorescence images of rabbit papillary muscle loaded by perfusion with carboxy SNARF®-1 AM acetate (C1271, C1272). The first two images (A and B) were acquired through 585 ± 10 nm bandpass and >620 nm longpass emission filters, respectively. The 620 nm/585 nm fluorescence ratio image in the third image (C) is more uniform than the component images A and B due to cancellation of intensity variations resulting from heterogeneous uptake of the fluorescent indicator. Images contributed by Barbara Muller-Borer and John Lemasters, University of North Carolina, and reprinted with permission from Am J Physiol (1998) 275:H1937.

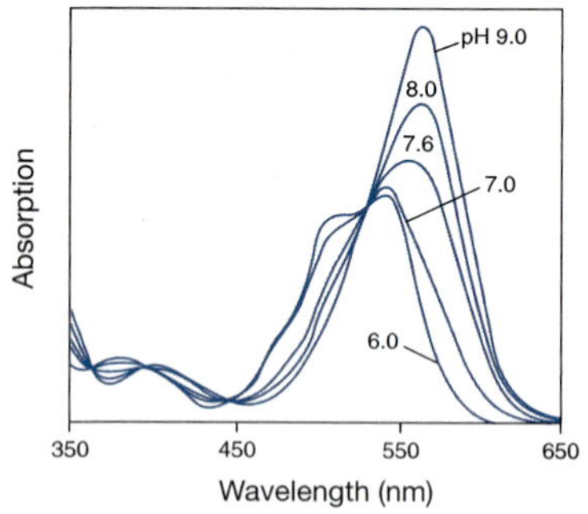

Figure 20.2.13 The pH-dependent absorption spectra of carboxy SNARF®-1 (C1270).

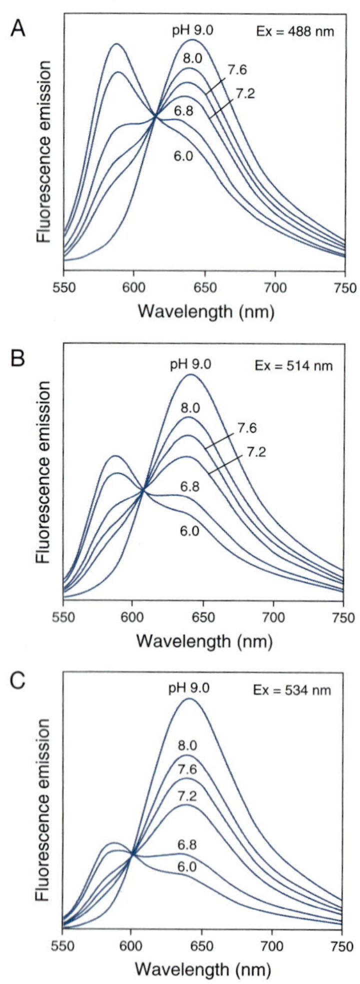

Figure 20.2.14 The pH-dependent emission spectra of carboxy SNARF®-1 (C1270) when excited at **A**) 488 nm, **B**) 514 nm and **C**) 534 nm.

between pH 7 and pH 8. Like fluorescein and BCECF, the absorption spectrum of the carboxy SNARF®-1 pH indicator undergoes a shift to longer wavelengths upon deprotonation of its phenolic substituent (Figure 20.2.13). In contrast to the fluorescein-based indicators, however, carboxy SNARF®-1 also exhibits a significant pH-dependent emission shift from yellow-orange to deep-red fluorescence as conditions become more basic (Figure 20.2.14, Figure 20.2.15). This pH dependence allows the ratio of the fluorescence intensities from the dye at two emission wavelengths—typically 580 nm and 640 nm—to be used for quantitative determinations of pH (Loading and Calibration of Intracellular Ion Indicators—Note 19.1) (Figure 20.2.11). For practical purposes, it is often desirable to bias the detection of carboxy SNARF®-1 fluorescence towards the less fluorescent acidic form by using an excitation wavelength between 488 nm and the excitation isosbestic point at ~530 nm, yielding balanced signals for the two emission ratio components (Figure 20.2.14, Figure 20.2.16). When excited at 488 nm, carboxy SNARF®-1 exhibits an emission isosbestic point of ~610 nm and a lower fluorescent signal than obtained with 514 nm excitation.[72] Alternatively, when excited by the 568 nm spectral line of the Ar-Kr laser found in some confocal laser-scanning microscopes, carboxy SNARF®-1 exhibits a fluorescence increase at 640 nm as the pH increases and an emission isosbestic point at 585 nm.[72] As with other ion indicators, intracellular environments may cause significant changes to both the spectral properties and pK_a of carboxy SNARF®-1,[80–83] and the indicator should always be calibrated in the system under study.

The spectra of carboxy SNARF®-1 are well resolved from those of fura-2[84,85] and indo-1[86] (Section 19.2), as well as those of the fluo-3,[85,87,88] fluo-4, Calcium Green™ and Oregon Green® 488 BAPTA Ca^{2+} indicators (Section 19.3), permitting simultaneous measurements of intracellular pH and Ca^{2+} (Figure 20.2.17). Carboxy SNARF®-1 has also been used in combination with the Na^+ indicator SBFI (S1262, S1263, S1264; Section 21.1) to simultaneously detect pH and Na^+ changes.[89] The relatively long-wavelength excitation and emission characteristics of carboxy SNARF®-1 facilitate studies in autofluorescent cells[90] and permit experiments that employ ultraviolet light–photoactivated caged probes[91] (Section 5.3). Incubation of cells for several hours after loading with carboxy SNARF®-1 AM ester acetate results in compartmentally selective retention of the dye, allowing *in situ* measurements of mitochondrial pH[92] (Figure 20.2.18).

SNARF®-4F and SNARF®-5F Dyes and Their Cell-Permeant Esters

Although the carboxy SNARF®-1 indicator possesses excellent spectral properties, its pK_a of ~7.5 may be too high for measurements of intracellular pH in some cells. For quantitative measurements of pH changes in the typical cytosolic range (pH ~6.8–7.4), we now recommend SNARF®-5F carboxylic acid (Figure 20.2.19), which has a pK_a value of ~7.2, as the indicator with the optimal spectral properties for estimating cytosolic pH (Figure 20.2.20). SNARF®-4F carboxylic acid (Figure 20.2.21) has a somewhat more acidic pH sensitivity maximum (pK_a ~6.4) but retains its dual-emission spectral properties (Figure 20.2.22). SNARF®-4F has been used for pH imaging in kidney tissues using two-photon excitation (780 nm) microscopy; the pH-dependent emission shift response was observed to be essentially the same as seen with one-photon

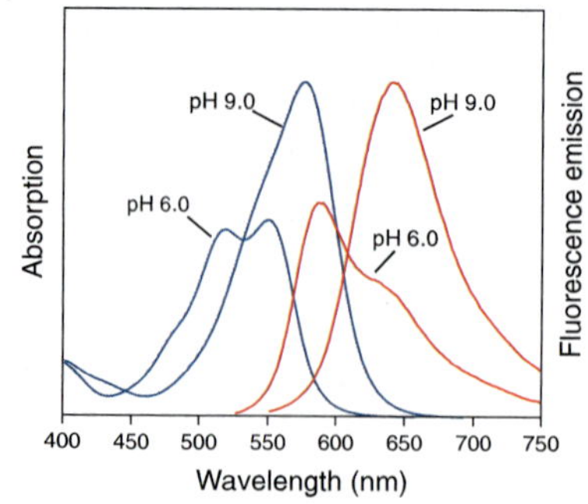

Figure 20.2.15 Absorption and fluorescence emission (excited at 514 nm) spectra of carboxy SNARF®-1 in pH 9.0 and pH 6.0 buffers.

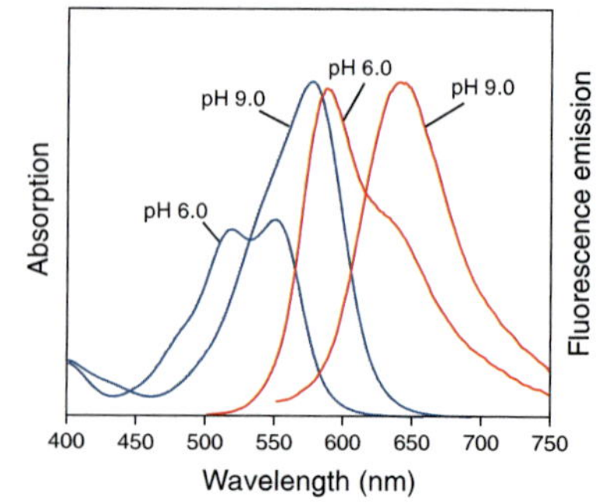

Figure 20.2.16 Absorption and fluorescence emission (excited at 488 nm) spectra of carboxy SNARF®-1 in pH 9.0 and pH 6.0 buffers.

excitation.[93] This study also reported nigericin calibrations that yielded different pK_a values (6.8 versus 7.4) in the kidney cortex and kidney ileum, respectively, emphasizing the importance of performing *in situ* calibrations. Both SNARF®-4F and SNARF®-5F[94] allow dual-excitation and dual-emission ratiometric pH measurements, making them compatible with the same instrument configurations used for carboxy SNARF®-1 in ratio imaging and flow cytometry applications. SNARF®-4F and SNARF®-5F are available as free carboxylic acids (S23920, S23922) and as cell-permeant AM ester acetate derivatives (S23921, S23923).

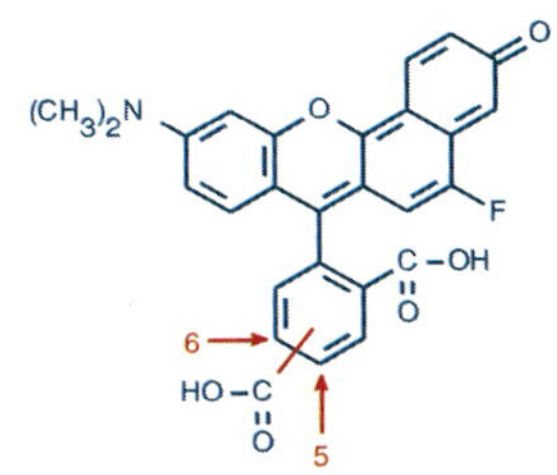

Figure 20.2.19 SNARF®-5F 5-(and-6)-carboxylic acid (S23922).

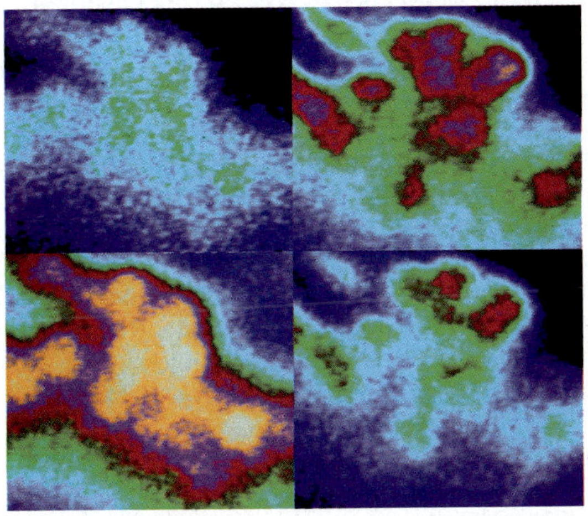

Figure 20.2.17 Rat pituitary intermediate lobe melanotropes labeled with the indo-1 AM (I1203, I1223, I1226) and carboxy SNARF®-1, AM, acetate (C1271, C1272) indicators. Pseudocolored fluorescence from the dual-emission Ca^{2+} indicator indo-1 is shown at 405 and 475 nm (left panels). Pseudocolored fluorescence from the dual-emission pH indicator carboxy SNARF®-1 is shown at 575 and 640 nm (right panels). Image contributed by Stephen J. Morris, University of Missouri-Kansas City, and Diane M. Beatty, Molecular Probes, Inc.

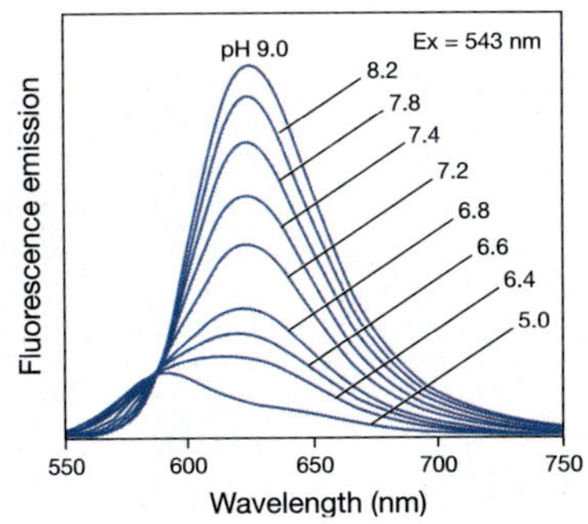

Figure 20.2.20 Fluorescence emission spectra of SNARF®-5F 5-(and 6-)carboxylic acid (S23922) as a function of pH.

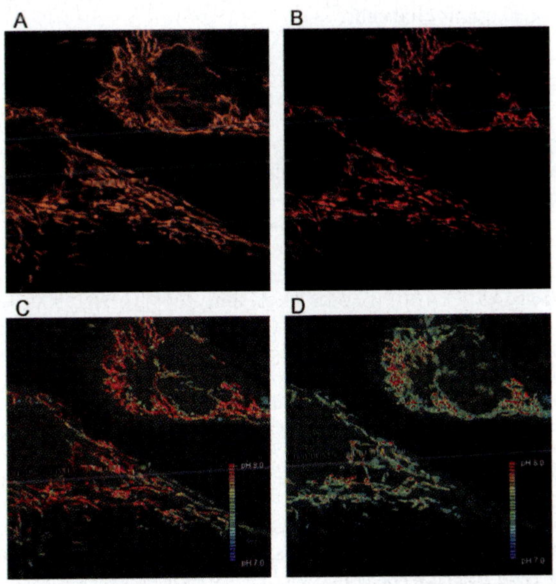

Figure 20.2.18 Selective loading of carboxy SNARF®-1 into mitochondria. BHK cells were loaded with 10 µM carboxy SNARF®-1, AM, acetate (C1271, C1272) for 10 minutes, followed by incubation for 4 hours at room temperature. **A)** Confocal image (488 nm excitation) of mitochondrial-selective loading of carboxy SNARF®-1 visualized through a 560–600 nm bandpass filter. **B)** Confocal image of the same cells as in A, but using a 605 nm dichroic mirror and a 610 nm longpass filter. **C)** Ratio image (A and **B**) of mitochondria in cells pseudocolored to represent different pH levels. **D)** Change in mitochondrial pH following the addition of 10 µM carbonyl cyanide *m*-chlorophenylhydrazone (CCCP), resulting in a decrease (acidification) of mitochondrial pH. Image contributed by Brian Herman, University of Texas Health Science Center, San Antonio, and reprinted with permission from Biotechniques 30, 804 (2001).

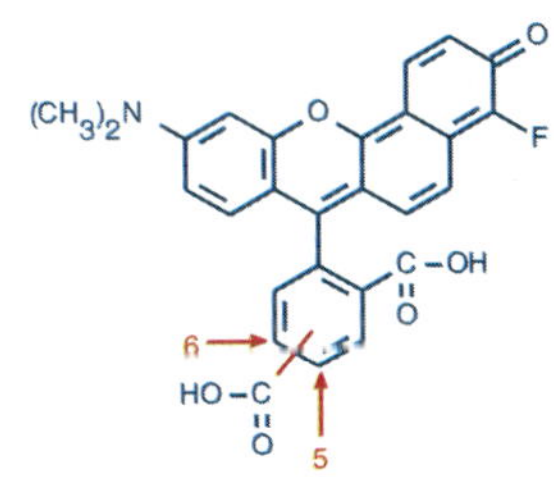

Figure 20.2.21 SNARF®-4F 5-(and-6)-carboxylic acid (S23920).

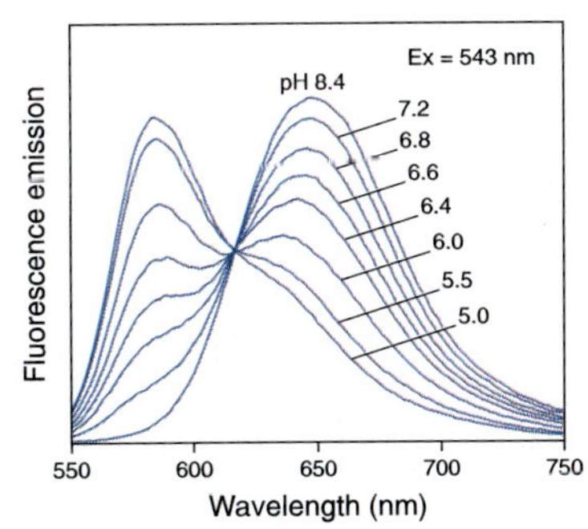

Figure 20.2.22 Fluorescence emission spectra of SNARF®-4F 5-(and 6-)carboxylic acid (S23920) showing the pH-dependent spectral shift that is characteristic of this and other SNARF® pH indicators.

IMPORTANT NOTICE: The products described in this manual are covered by one or more Limited Use Label License(s). Please refer to the Appendix on page 971 and Master Product List on page 975. Products are For Research Use Only. Not intended for any animal or human therapeutic or diagnostic use.

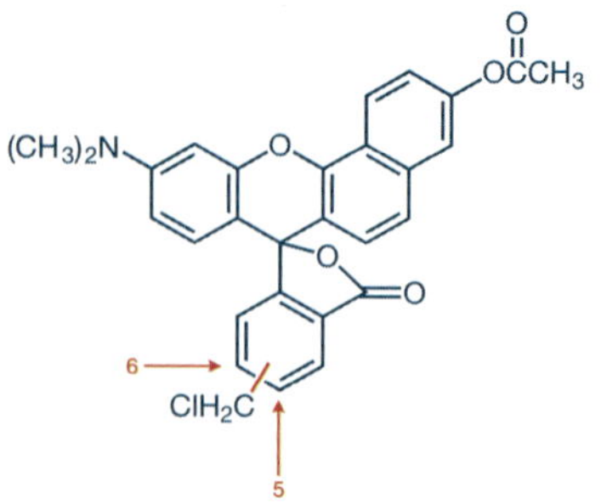

Figure 20.2.23 5-(and-6)-chloromethyl SNARF®-1, acetate (C6826).

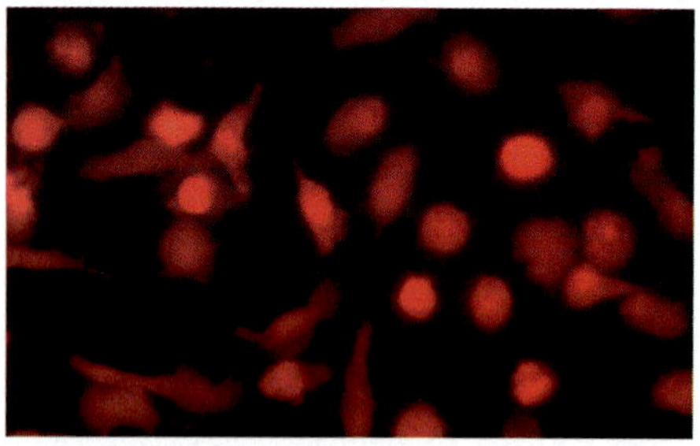

Figure 20.2.24 Human neutrophils loaded with 5-(and-6)-chloromethyl SNARF®-1 acetate (C6826).

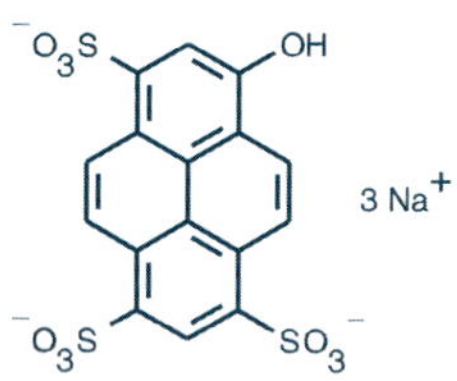

Figure 20.2.25 8-hydroxypyrene-1,3,6-trisulfonic acid, trisodium salt (HPTS; pyranine, H348).

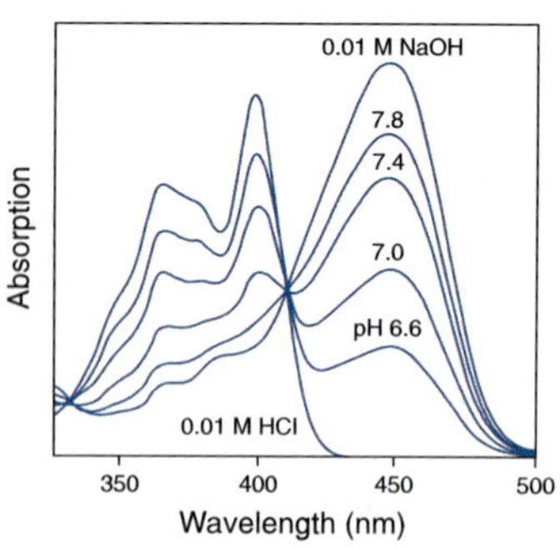

Figure 20.2.26 The pH-dependent absorption spectra of 8-hydroxypyrene-1,3,6-trisulfonic acid (HPTS, H348).

Amine- and Thiol-Reactive SNARF® Dyes

Our 5-(and 6-)chloromethyl SNARF®-1 acetate (C6826, Figure 20.2.23) contains a chloromethyl group that is mildly reactive with intracellular thiols, forming adducts that improve cellular retention of the SNARF® fluorophore (Figure 20.2.24). As with CellTracker™ Green CMFDA, improved retention of this conjugate in cells may permit monitoring of intracellular pH over longer time periods than is possible with other intracellular pH indicators. Similarly, amine-reactive SNARF®-1 succinimidyl ester (S22801, Section 20.4) is useful as an intracellular pH indicator [95] in addition to its more common application as a cell tracer.[96]

8-Hydroxypyrene-1,3,6-Trisulfonic Acid (HPTS)

8-Hydroxypyrene-1,3,6-trisulfonic acid (HPTS, also known as pyranine; H348; Figure 20.2.25) is an inexpensive, highly water-soluble, membrane-impermeant pH indicator with a pK_a of ~7.3 in aqueous buffers.[97] The pK_a of HPTS is reported to rise to 7.5–7.8 in the cytosol of some cells.[98] Unlike indicators based on the SNARF® and fluorescein dyes, there is no membrane-permeant form of HPTS available. Consequently, HPTS must be introduced into cells by microinjection, electroporation [99] or liposome-mediated delivery,[100–102] through ATP-gated ion channels [103] or by other relatively invasive means (Table 14.1, Section 14.1). HPTS exhibits a pH-dependent absorption shift (Figure 20.2.26), allowing ratiometric measurements using an excitation ratio of 450/405 nm.[49,104] Because the excited state of HPTS is much more acidic than the ground state,[105] it is frequently used as a photoactivated source of H^+ in mechanistic studies of bacteriorhodopsin and other proton pumps.[106–108]

Nigericin

Intracellular calibration of the fluorescence response of cytosolic pH indicators is typically performed using the K^+/H^+ ionophore nigericin (N1495), which causes equilibration of intracellular and extracellular pH in the presence of a depolarizing concentration of extracellular K^+ [22,23] (Loading and Calibration of Intracellular Ion Indicators—Note 19.1). Nett and Deitmer have compared this technique with calibrations performed by direct insertion of pH-sensitive microelectrodes in leech giant glial cells.[26]

REFERENCES

1. J Fluorescence (1996) 6:147; **2.** Spectrochim Acta A (1995) 51:7; **3.** Photochem Photobiol (1994) 60:435; **4.** J Luminescence (1975) 10:381; **5.** J Phys Chem (1971) 75:245; **6.** FEBS Lett (1994) 341:125; **7.** J Immunol Methods (1993) 157:117; **8.** Methods Mol Biol (1995) 43:211; **9.** J Histochem Cytochem (1985) 33:77; **10.** Cytometry (1995) 19:235; **11.** Photochem Photobiol (1994) 60:274; **12.** J Immunol Methods (1994) 172:115; **13.** Biochemistry (1995) 34:1606; **14.** Cytometry (1992) 13:739; **15.** J Immunol Methods (1990) 130:251; **16.** Proc Natl Acad Sci U S A (1984) 81:7436; **17.** J Cell Biol (1982) 95:189; **18.** Cytometry (1997) 28:42; **19.** Methtods Cell Biol (2007) 81:415; **20.** Biophys J (2006) 90:608; **21.** Methods Cell Biol (1994) 41:135; **22.** Biochemistry (1979) 18:2210; **23.** Methods Enzymol (1990) 192:38; **24.** J Cell Physiol (1992) 151:596; **25.** J Fluorescence (1992) 2:191; **26.** Biophys J (1996) 71:394; **27.** J Gen Physiol (2009) 133:315; **28.** Kidney Int (1998) 53:432; **29.** Am J Physiol (1998) 274:F358; **30.** J Membr Biol (1997) 159:253; **31.** Biochemistry (1995) 34:15157; **32.** J Biol Chem (1997) 272:25668; **33.** J Biol Chem (2009) 284:19437; **34.** Pflugers Arch (2008) 455:799; **35.** Proc Natl Acad Sci U S A (2007) 104:9325; **36.** J Biol Chem (2007) 282:3720; **37.** Am J Physiol (1997) 273:F817; **38.** Plant Physiol (1997) 113:451; **39.** J Biol Chem (1998) 273:15920; **40.** J Biol Chem (1998) 273:27162; **41.** J Biol Chem (1996) 271:861; **42.** J Biol Chem (2007) 282:2880; **43.** J Biol Chem (2001) 276:514; **44.** J Biol Chem (1996) 271:16260; **45.** J Biol Chem (1997) 272:29810; **46.** J Leukoc Biol (1997) 62:329; **47.** J Biol Chem (1996) 271:2005; **48.** J Biol Chem (2006) 281:22142; **49.** Neuroscience (2008) 151:1084; **50.** J Membr Biol (1994) 140:89; **51.** Proc Natl Acad Sci U S A (2002) 99:9439; **52.** Cancer Res (1994) 54:5670; **53.** Biochemistry (1996) 35:13419; **54.** Biophys J (2002) 83:1682; **55.** Mol Membr Biol (1996) 13:173; **56.** Biochim Biophys Acta (1991) 1115:75; **57.** J Cell Biol (1990) 111:3129; **58.** J Immunol Methods (1990) 133:87; **59.** FEBS Lett (1986) 200:203; **60.** Biotechniques (1985) 3:270; **61.** J Biol Chem (1995) 270:4967; **62.** Am J Physiol Gastrointest Liver Physiol (1995) 268:361; **63.** Am J Physiol Gastrointest Liver Physiol (2001) 281:G833; **64.** AAPS PharmSci (2002) 4:E21; **65.** J Appl Microbiol (2000) 88:809; **66.** Mikrochim Acta (1992) 108:133; **67.** Anal Chem (1997) 69:863; **68.** Anal Biochem (1991) 194:330; **69.** Methods Enzymol (1999) 302:341; **70.** Micron (1993) 24:573; **71.** Am J Physiol (1998) 275:H1937; **72.** Biophys J (1994) 66:942; **73.** Cytometry (1993) 14:916; **74.** J Immunol Methods (1998) 221:43; **75.** J Cell Physiol (1998) 177:109; **76.** Am J Physiol (1997) 273:C1783;

IMPORTANT NOTICE: The products described in this manual are covered by one or more Limited Use Label License(s). Please refer to the Appendix on page 971 and Master Product List on page 975. Products are For Research Use Only. Not intended for any animal or human therapeutic or diagnostic use.

REFERENCES—*continued*

77. J Biomed Mater Res (1998) 39:9; **78.** J Immunol Methods (1993) 159:145; **79.** Anal Chem (1993) 65:2329; **80.** J Photochem Photobiol B (1997) 37:18; **81.** Pflugers Arch (1994) 427:332; **82.** Anal Biochem (1992) 204:65; **83.** J Fluorescence (1992) 2:75; **84.** J Cell Physiol (1994) 161:129; **85.** Cell Calcium (1996) 19:337; **86.** Endocrinology (1993) 133:972; **87.** J Physiol (2000) 528:25; **88.** Cytometry (1996) 24:99; **89.** Pflugers Arch (2004) 449:307; **90.** Proc Natl Acad Sci U S A (2009) 106:16574; **91.** Biophys J (2007) 92:641; **92.** Biotechniques (2001) 30:804; **93.** J Biol Chem (2007) 282:25141; **94.** Bioorg Med Chem Lett (2001) 11:2903; **95.** Exp Cell Res (2004) 298:521; **96.** Cytometry B Clin Cytom (2007) 72:458; **97.** Fresenius Z Anal Chem (1983) 314:119; **98.** Anal Biochem (2005) 347:34; **99.** J Bacteriol (1995) 177:1017; **100.** Pharm Res (1997) 14:1203; **101.** Proc Natl Acad Sci U S A (1997) 94:8795; **102.** Curr Eye Res (1997) 16:1073; **103.** Am J Physiol (1998) 275:C1158; **104.** Proc Natl Acad Sci U S A (1995) 92:3156; **105.** J Phys Chem A (2007) 111:230; **106.** Angew Chem Int Ed Engl (2009) 48:8523; **107.** Biophys J (2003) 84:671; **108.** Biochemistry (1996) 35:6604.

DATA TABLE 20.2 PROBES USEFUL AT NEAR-NEUTRAL pH

				Acidic Solution				Basic Solution						
Cat. No.	MW	Storage	Soluble	Abs	EC	Em	Solvent	Abs	EC	Em	Solvent	pK_a	Product*	Notes
B1150	~615	F,D	DMSO	<300		none							B1151	1
B1151	520.45	L	pH >6	482	35,000	520	pH 5	503	90,000	528	pH 9	7.0		2, 3
B1170	~615	F,D	DMSO	<300		none							B1151	1
B3051	~615	F,D	DMSO	<300		none							B1151	1, 4
C194	376.32	L	pH >6, DMF	475	28,000	517	pH 5	492	75,000	517	pH 9	6.4		2, 3
C195	460.40	F,D	DMSO	<300		none							C194	
C652	476.44	L	pH >6, DMF	512	11,000	563	pH 6	598	49,000	668	pH 10	7.6		2, 3, 5
C1157	557.47	F,D	DMF, DMSO	<300		none							C1311	
C1270	453.45	L	pH >6	548	27,000	587	pH 6	576	48,000	635	pH 10	7.5		2, 3, 6
C1271	567.55	F,D	DMSO	<350		none							C1270	
C1272	567.55	F,D	DMSO	<350		none							C1270	
C1354	532.46	F,D	DMSO	<300		none							C1359	
C1904	376.32	L	pH >6, DMF	475	29,000	517	pH 5	492	78,000	517	pH 9	6.4		2, 3, 7
C2925	464.86	F,D	DMSO	<300		none							see Notes	8
C6826	499.95	F,D	DMSO	<350		none							see Notes	9
C7025	464.86	F,D	DMSO	<300		none							see Notes	8
C13196	560.52	F,D	DMSO	<300		none							C652	
F1130	478.32	D,L	H_2O, DMF	476	31,000	519	pH 5	495	76,000	519	pH 9	6.4		2, 3
F1300	332.31	L	pH >6, DMF	473	34,000	514	pH 5	490	93,000	514	pH 9	6.4		2, 3
F1303	416.39	F,D	DMSO	<300		none							F1300	
H348	524.37	D,L	H_2O	403	20,000	511	pH 4	454	24,000	511	pH 9	7.3		2, 3, 10
N1495	724.97	F,D	MeOH	<300		none								
S1129	518.43	F,D	DMSO	<300		none							F1130	
S23920	471.44	L	pH >6	552	27,000	589	pH 5	581	48,000	652	pH 9	6.4		2, 3
S23921	585.54	F,D	DMSO	<350		none							S23920	
S23922	471.44	L	pH >6	555	27,000	590	pH 5	579	49,000	630	pH 9	7.2		2, 3
S23923	585.54	F,D	DMSO	<350		none							S23922	

For definitions of the contents of this data table, see "Using *The Molecular Probes® Handbook*" in the introductory pages.

* Cat. No. of product generated *in situ* in typical intracellular applications.

Notes

1. MW value is approximate. BCECF, AM is a mixture of molecular species.
2. pK_a values may vary considerably depending on the temperature, ionic strength, viscosity, protein binding and other factors. Unless otherwise noted, values listed have been determined from pH-dependent fluorescence measurements at 22°C.
3. Spectra are in aqueous buffers adjusted to >1 pH unit above and >1 pH unit below the pK_a.
4. This product is supplied as a ready-made solution in the solvent indicated under "Soluble."
5. Data on pH dependence of C652 spectra obtained in our laboratories. Additional relevant data are reported elsewhere. (Mikrochim Acta (1992) 108:133)
6. Values of pK_a for these SNARF® indicators are as reported in published references. (Anal Biochem (1991) 194:330)
7. This product is specified to equal or exceed 98% analytical purity by HPLC.
8. Acetate hydrolysis of this compound yields a fluorescent product with pH-dependent spectral characteristics similar to C1904.
9. C6826 and S22801 are converted to fluorescent products with spectra similar to C1270 after acetate hydrolysis.
10. The pK_a for H348 was determined in 0.066 M phosphate buffers at 22°C. (Fresenius Z Anal Chem (1983) 314:119)

IMPORTANT NOTICE: The products described in this manual are covered by one or more Limited Use Label License(s). Please refer to the Appendix on page 971 and Master Product List on page 975. Products are For Research Use Only. Not intended for any animal or human therapeutic or diagnostic use.

PRODUCT LIST 20.2 PROBES USEFUL AT NEAR-NEUTRAL pH

Cat. No.	Product	Quantity
B1151	2′,7′-bis-(2-carboxyethyl)-5-(and-6)-carboxyfluorescein (BCECF acid) *mixed isomers*	1 mg
B1150	2′,7′-bis-(2-carboxyethyl)-5-(and-6)-carboxyfluorescein, acetoxymethyl ester (BCECF, AM)	1 mg
B3051	2′,7′-bis-(2-carboxyethyl)-5-(and-6)-carboxyfluorescein, acetoxymethyl ester (BCECF, AM) *1 mg/mL solution in anhydrous DMSO*	1 mL
B1170	2′,7′-bis-(2-carboxyethyl)-5-(and-6)-carboxyfluorescein, acetoxymethyl ester (BCECF, AM) *special packaging*	20 x 50 µg
C1270	5-(and-6)-carboxy SNARF®-1	1 mg
C1271	5-(and-6)-carboxy SNARF®-1, acetoxymethyl ester, acetate	1 mg
C1272	5-(and-6)-carboxy SNARF®-1, acetoxymethyl ester, acetate *special packaging*	20 x 50 µg
C194	5-(and-6)-carboxyfluorescein *mixed isomers*	5 g
C1904	5-(and-6)-carboxyfluorescein (5(6)-FAM) *FluoroPure™ grade* *mixed isomers*	100 mg
C195	5(6)-CFDA (5-(and-6)-carboxyfluorescein diacetate) *mixed isomers*	100 mg
C1354	5-CFDA, AM (5-carboxyfluorescein diacetate, acetoxymethyl ester)	5 mg
C1157	5(6)-CFDA, SE; CFSE (5-(and-6)-carboxyfluorescein diacetate, succinimidyl ester) *mixed isomers*	25 mg
C652	5-(and-6)-carboxynaphthofluorescein *mixed isomers*	100 mg
C13196	5-(and-6)-carboxynaphthofluorescein diacetate	10 mg
C6826	5-(and-6)-chloromethyl SNARF®-1, acetate *mixed isomers* *special packaging*	20 x 50 µg
H348	8-hydroxypyrene-1,3,6-trisulfonic acid, trisodium salt (HPTS; pyranine)	1 g
C2925	CellTracker™ Green CMFDA (5-chloromethylfluorescein diacetate)	1 mg
C7025	CellTracker™ Green CMFDA (5-chloromethylfluorescein diacetate) *special packaging*	20 x 50 µg
F1303	FDA (fluorescein diacetate)	1 g
F1300	fluorescein *reference standard*	1 g
F1130	fluorescein-5-(and-6)-sulfonic acid, trisodium salt	100 mg
N1495	nigericin, free acid	10 mg
S1129	SFDA (5-sulfofluorescein diacetate, sodium salt)	25 mg
S23920	SNARF®-4F 5-(and-6)-carboxylic acid	1 mg
S23921	SNARF®-4F 5-(and-6)-carboxylic acid, acetoxymethyl ester, acetate *special packaging*	20 x 50 µg
S23922	SNARF®-5F 5-(and-6)-carboxylic acid	1 mg
S23923	SNARF®-5F 5-(and-6)-carboxylic acid, acetoxymethyl ester, acetate *special packaging*	20 x 50 µg

20.3 Probes Useful at Acidic pH

Eukaryotic cells contain compartments with different degrees of acidity. For example, biomolecules brought into cells by receptor-mediated endocytosis or phagocytosis (Section 16.1) are initially processed through organelles of decreasing pH, and specialized organelles such as plant vacuoles and the acrosome of spermatozoa are intrinsically acidic. A low intracompartmental pH activates enzymes and other protein functions—such as iron release from transferrin—that would be too slow at neutral pH, thereby facilitating cellular metabolism. Abnormal lysosomal or endosomal acidification is associated with various pathological conditions. For example, lysosomes in some tumor cells have been reported to have a lower pH than normal lysosomes,[1] whereas other tumor cells contain lysosomes with higher pH.[2]

The fluorescent pH indicators used to detect acidic organelles and to follow trafficking through acidic organelles must have a lower pK_a than those described in Section 20.2. Also, unlike most pH indicators for cytosolic measurements, pH indicators for acidic organelles need not be intrinsically permeant to membranes. Often they are covalently attached to large biomolecules that are actively taken up and processed through acidic organelles by the cell's own endocytic mechanisms (Chapter 16).

Table 20.2 Summary of the pH response of our LysoSensor™ probes.

Cat. No.	LysoSensor™ Probe	Abs/Em * (nm)	pK_a †	Useful pH Range †
L7533	LysoSensor™ Blue DND-167	373/425	5.1	4.5–6.0
L7534	LysoSensor™ Green DND-153	442/505	7.5	6.5–8.0
L7535	LysoSensor™ Green DND-189	443/505	5.2	4.5–6.0
L7545	LysoSensor™ Yellow/Blue DND-160	384/540 ‡; 329/440 §	4.2	3.5–6.0

* Absorption (Abs) and fluorescence emission (Em) maxima at pH 5; values may vary somewhat in cellular environments. † All pK_a values were determined *in vitro*; values are likely to be different in cells. ‡ At pH 3. § At pH 7.

LysoSensor™ Probes

LysoSensor™ probes are weak bases that are selectively concentrated in acidic organelles as a result of protonation (Table 20.2). This protonation also relieves the fluorescence quenching of the dye that results from photoinduced electron transfer (PET) by its weak-base side chain. Thus, unlike most other pH indicators in this chapter, the LysoSensor™ dyes become *more* fluorescent in acidic environments. Because accumulation of LysoSensor™ probes also appears to cause lysosomal alkalinization, pH measurements should be made rapidly using the lowest practicable concentration of dye.

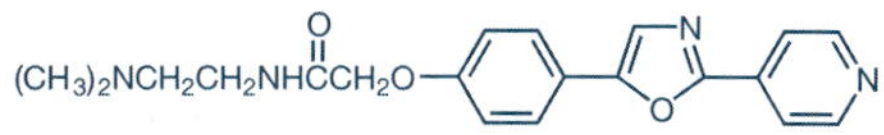

Figure 20.3.1 LysoSensor™ Yellow/Blue DND-160 (PDMPO, L7545).

LysoSensor™ Yellow/Blue Dye

LysoSensor™ Yellow/Blue DND-160 (L7545, Figure 20.3.1) undergoes a pH-dependent emission shift to longer wavelengths in acidic environments when illuminated near its excitation isosbestic point (~360 nm) (Figure 20.3.2). It also undergoes a pH-dependent excitation shift when detected near its emission isosbestic point (~490 nm) (Figure 20.3.2). These properties can be exploited for dual-emission ratio imaging of lysosomal pH[3] (Figure 20.3.3) (emission ratio ~450/510 nm, excitation ~365 nm). Yellow-fluorescent staining by LysoSensor™ Yellow/Blue DND-160 has been used to identify lysosomes as the accumulation site for anthracyclines in a drug-resistant cell line.[4] LysoSensor™ Yellow/Blue DND-160, frequently referred to by the acronym PDMPO, has been widely used as a tracer of silica deposition and transport in marine diatoms.[5–8] We also offer a 10,000 MW dextran conjugate of the LysoSensor™ Yellow/Blue dye (L22460, Section 20.4).

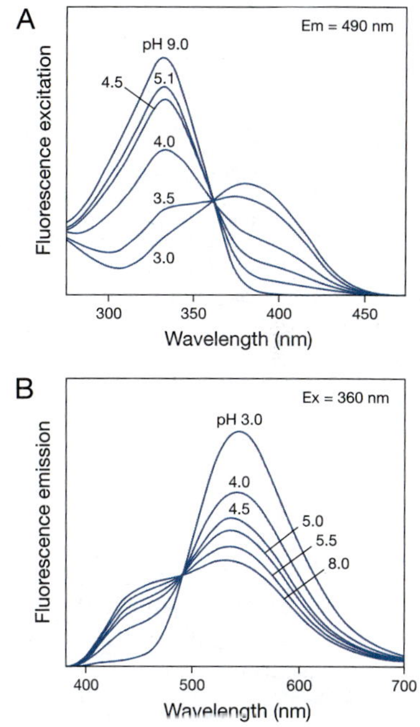

Figure 20.3.2 The pH-dependent spectral response of LysoSensor™ Yellow/Blue DND-160 (L7545): **A)** fluorescence excitation spectra and **B)** fluorescence emission spectra.

LysoSensor™ Green and LysoSensor™ Blue Dyes

The green-fluorescent LysoSensor™ Green dyes are available with optimal pH sensitivity in either the acidic or neutral range (pK_a ~5.2 or ~7.5). The blue-fluorescent LysoSensor™ Blue DND-167 has a pK_a of ~5.1 (Figure 20.3.4). With their low pK_a values, LysoSensor™ Green DND-189 (L7535, Figure 20.3.5) and LysoSensor™ Blue DND-167 (L7533) are almost nonfluorescent except when inside acidic compartments, whereas LysoSensor™ Green DND-153 (L7534, Figure 20.3.6) is brightly fluorescent, even at neutral pH. LysoSensor™ Green DND-189 has been used to monitor the dissipation of neurite vesicle transmembrane pH gradients by bafilomycin A1.[9]

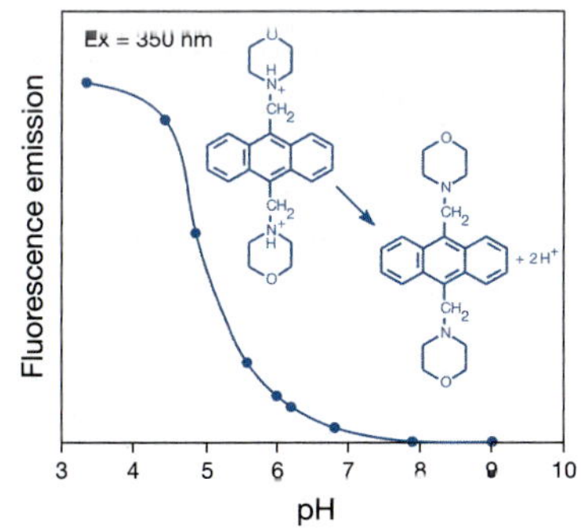

Figure 20.3.4 The pH titration curve of LysoSensor™ Blue DND-167 (L7533), which exhibits a pK_a ~5.1.

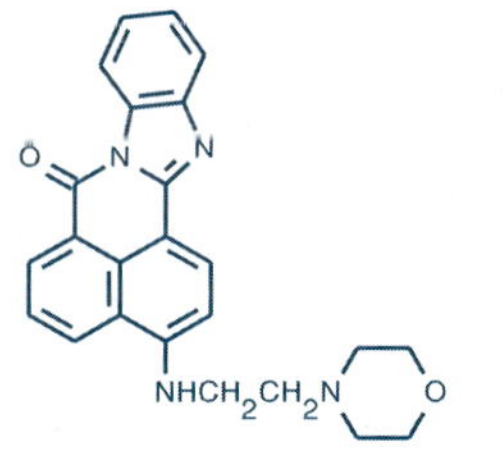

Figure 20.3.5 LysoSensor™ Green DND-189 (L7535).

Figure 20.3.6 LysoSensor™ Green DND-153 (L7534).

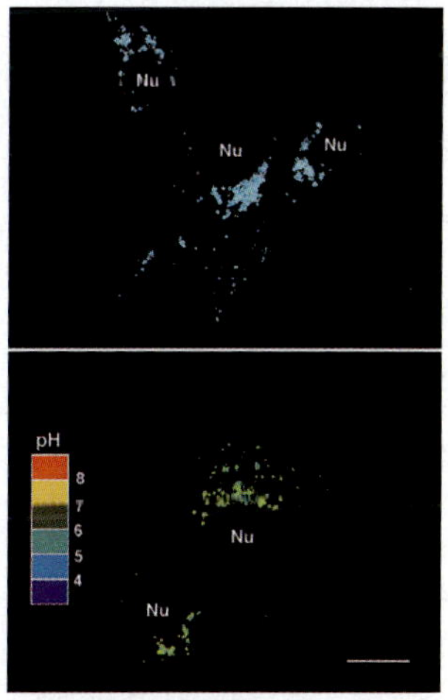

Figure 20.3.3 Dual-emission ratiometric measurement of lysosomal pH using LysoSensor™ Yellow/Blue DND-160 (L7545). Madin-Darby canine kidney cells were exposed to pH-calibration buffers (pH 4.5 or 7.0) in the presence of nigericin (N1495) and monensin. These pseudocolored images were constructed from two emission images at 450 ± 33 nm and 510 ± 20 nm, both excited at 365 ± 8 nm.

IMPORTANT NOTICE: The products described in this manual are covered by one or more Limited Use Label License(s). Please refer to the Appendix on page 971 and Master Product List on page 975. Products are For Research Use Only. Not intended for any animal or human therapeutic or diagnostic use.

pHrodo™ pH Indicator

pHrodo™ pH indicator is an aminorhodamine dye that exhibits increasing fluorescence as the pH of its surroundings becomes more acidic [10] (Figure 20.3.7). Consequently, pHrodo™ fluorescence (excitation/emission ~560/585 nm) provides a positive indication of processes such as phagocytic ingestion and lysosomal sequestration, in contrast to the negative indication generated by fluorescein and Oregon Green® dyes.[11–13] Because the pHrodo™ dye has more than one protolytically ionizable substituent, it exhibits a complex pH titration profile that typically changes upon conjugation to proteins and other biomolecules. We currently offer an amine-reactive succinimidyl ester form of the pHrodo™ dye (P36600, Section 20.4) and various bioconjugates (Section 20.4, Section 16.1). In cases where the nonreactive carboxylic acid form of the dye is required, pHrodo™ carboxylic acid can be obtained by mild alkaline hydrolysis of the succinimidyl ester (e.g., 12-hour incubation in aqueous solution at pH 8 in the dark will yield >99.9% hydrolysis; higher pH will give a more rapid reaction).

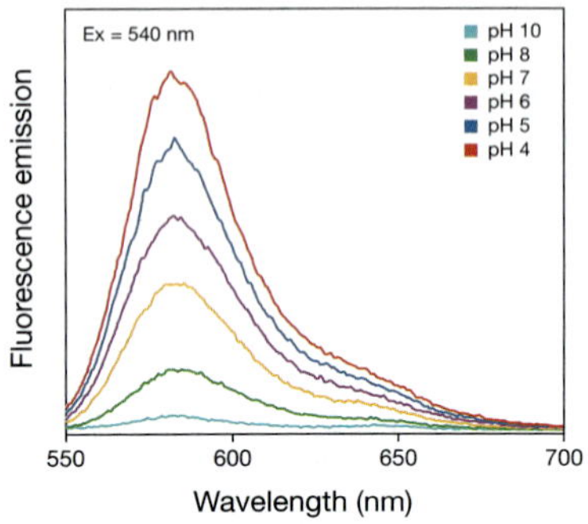

Figure 20.3.7 The pH sensitivity of pHrodo™ dextran. pHrodo™ dextran (P10361) was reconstituted in HEPES (20 mM)–buffered PBS and adjusted to pH values from pH 4 to pH 10. The intensity of fluorescence emission increases with increasing acidity, particularly in the pH 5–8 range.

Oregon Green® and Dichlorofluorescein Derivatives

Introduction of electron-withdrawing groups into fluorescein dyes lowers the pK_a of the phenolic group to 5 or below, as exemplified by our Oregon Green® dyes and their insensitivity to pH changes in the near-neutral pH range.[14] However, these fluorinated fluorescein dyes are still pH sensitive in moderately acidic solutions, with pK_a values of ~4.7 (Figure 20.3.8). With the exception of their lower pK_a values, the pH-dependent spectral characteristics of the Oregon Green® dyes closely parallel those of other fluorescein-based dyes, allowing dual-excitation ratiometric measurements with the same general configuration used for BCECF (Section 20.2).

Oregon Green® 514 carboxylic acid (O6138, Figure 20.3.9), Oregon Green® 488 carboxylic acid (O6146, Figure 20.3.10) and 5-(and 6-)carboxy-2′,7′-dichlorofluorescein (C368) do not readily enter cells but may be useful as fluid phase markers for endocytosis. These fluorinated and chlorinated fluorescein derivatives will likely be most useful in the form of conjugates that are endocytosed and processed through acidic organelles (Section 20.4); dextran conjugates of the Oregon Green® 488 and Oregon Green® 514 dyes are described in Section 20.4.

Cell-permeant diacetate derivatives of carboxydichlorofluorescein (carboxy-DCFDA, C369) and carboxydifluorofluorescein (carboxy-DFFDA, O6151) are also available. Carboxy-DCFDA has been used to measure the pH in acidic organelles,[15] as well as in the cytosol and vacuoles of plants and yeast.[16,17] Furthermore, the mechanism of vacuolar pH rectification following exposure to ammonia has been investigated in rice and maize root hair cells loaded with carboxy-DFFDA.[18]

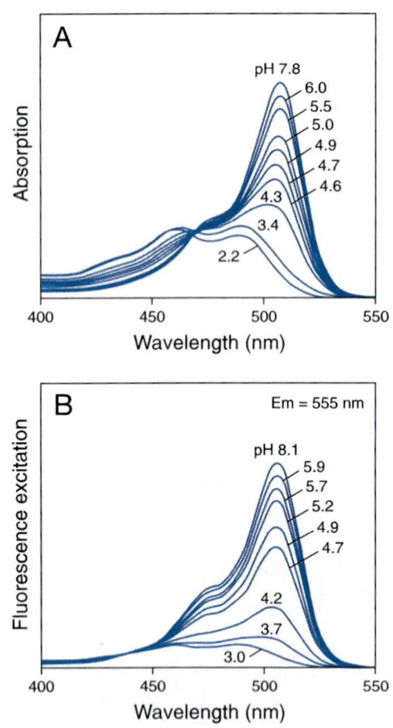

Figure 20.3.8 The pH-dependent spectra of Oregon Green® 514 carboxylic acid (O6138): **A)** absorption spectra and **B)** fluorescence excitation spectra.

Other pH Indicators for Acidic Environments

9-Amino-6-Chloro-2-Methoxyacridine (ACMA)

The nucleic acid stain 9-amino-6-chloro-2-methoxyacridine (ACMA, A1324; Figure 20.3.11) apparently binds to membranes in the energized state, and its fluorescence becomes quenched if a pH gradient forms.[19] Mechanistically, this probe resembles the membrane potential–sensitive carbocyanines (Section 22.3) more than the other probes in this chapter. ACMA is primarily employed to detect the proton-translocating activity of *Escherichia coli* ATP synthase [20–23] and yeast vacuolar H^+-ATPase.[24–26]

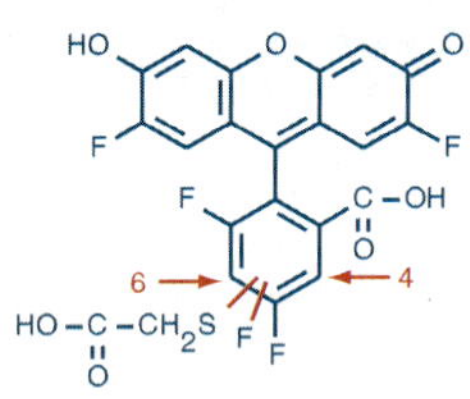

Figure 20.3.9 Oregon Green® 514 carboxylic acid (O6138).

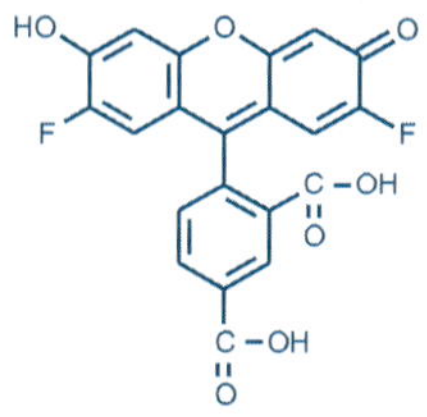

Figure 20.3.10 Oregon Green® 488 carboxylic acid (O6146).

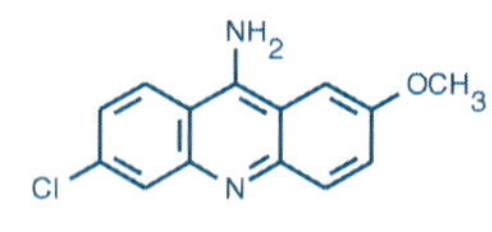

Figure 20.3.11 9-amino-6-chloro-2-methoxyacridine (ACMA, A1324).

IMPORTANT NOTICE: The products described in this manual are covered by one or more Limited Use Label License(s). Please refer to the Appendix on page 971 and Master Product List on page 975. Products are For Research Use Only. Not intended for any animal or human therapeutic or diagnostic use.
www.invitrogen.com/probes
molecular probes® | invitrogen™ by life technologies™

8-Hydroxypyrene-1,3,6-Trisulfonic Acid (HPTS)

Although 8-hydroxypyrene-1,3,6-trisulfonic acid (HPTS, H348; Section 20.2) has a pK_a ~7.3 and is primarily used as a pH indicator in the near-neutral range, it offers several advantages for monitoring intraorganelle pH in endosomal/lysosomal pathways:[27]

- The highly polar character that results from its three sulfonic acid groups prevents leakage across intracellular membranes.
- Uptake by fluid-phase endocytosis is efficient and easily accomplished.
- Determination of excitation ratios allows pH measurements that are independent of vesicular size and indicator concentration.
- Precise calibration permits pH values as low as 4.4 to be accurately measured.
- HPTS is nontoxic and does not perturb normal physiological function.

HPTS can be introduced into cells by microinjection, electroporation or liposome-mediated delivery or through ATP-gated ion channels.[28] HPTS has also been loaded into cells using a patch pipette.[29] HPTS has been reported to be less phototoxic than BCECF.[28]

REFERENCES

1. J Cell Sci (1994) 107:2381; **2.** Biochemistry (1996) 35:2811; **3.** Chem Biol (1999) 6:411; **4.** Blood (1997) 89:3745; **5.** Proc Natl Acad Sci U S A (2008) 105:1579; **6.** Eukaryot Cell (2007) 6:271; **7.** Limnol Oceanogr Meth (2005) 3:462; **8.** Chem Biol (2001) 8:1051; **9.** J Neurochem (1997) 69:1927; **10.** Mol Biosyst (2010) 6:888; **11.** J Biol Chem (2009) 284:35926; **12.** J Immunol Methods (2009) 342:71; **13.** J Immunol (2008) 181:4043; **14.** J Org Chem (1997) 62:6469; **15.** Anal Biochem (1990) 187:109; **16.** Plant Physiol (1992) 98:680; **17.** Methods Enzymol (1991) 194:644; **18.** Planta (1998) 206:154; **19.** Biochim Biophys Acta (1983) 722:107; **20.** J Biol Chem (1998) 273:16229; **21.** Biochemistry (1998) 37:10846; **22.** J Biol Chem (1998) 273:16241; **23.** Biochim Biophys Acta (1993) 1183:161; **24.** J Biol Chem (1996) 271:22487; **25.** J Biol Chem (1996) 271:2018; **26.** J Biol Chem (1994) 269:13224; **27.** Proc Natl Acad Sci U S A (1995) 92:3156; **28.** Am J Physiol (1998) 275:C1158; **29.** J Physiol (2001) 530:405.

DATA TABLE 20.3 PROBES USEFUL AT ACIDIC pH

				Acidic Solution				Neutral/Basic Solution					
Cat. No.	MW	Storage	Soluble	Abs	EC	Em	Solvent	Abs	EC	Em	Solvent	pK_a	Notes
A1324	258.71	L	DMF, DMSO	412	8200	471	MeOH	see Notes		see Notes		8.6	1, 2
C368	445.21	L	pH >6, DMF	495	38,000	529	pH 4	504	107,000	529	pH 8	4.8	3, 4
C369	529.29	F,D	DMSO	<300		none							5
L7533	376.50	F,D,L	DMSO	373	11,000	425	pH 3	373	11,000	425	pH 7	5.1	3, 4, 6, 7
L7534	356.43	F,D,L	DMSO	442	17,000	505	pH 5	442	17,000	505	pH 9	7.5	3, 4, 6, 7
L7535	398.46	F,D,L	DMSO	443	16,000	505	pH 3	443	16,000	505	pH 7	5.2	3, 4, 6, 7
L7545	366.42	F,D,L	DMSO	384	21,000	540	pH 3	329	23,000	440	pH 7	4.2	3, 4, 6, 8
O6138	512.36	L	pH >6, DMF	489	26,000	526	pH 3	506	86,000	526	pH 9	4.7	3, 4, 9
O6146	412.30	L	pH >6, DMF	478	27,000	518	pH 3	492	85,000	518	pH 9	4.7	3, 4, 10
O6151	496.38	F,D	DMSO	<300		none							11

For definitions of the contents of this data table, see "Using *The Molecular Probes® Handbook*" in the introductory pages.

Notes

1. Absorption and fluorescence spectra of the protonated and deprotonated forms of A1324 are quite similar. This observation, and the listed pK_a value, are reported in published references. (Eur Biophys J (1986) 13:251) Accumulation of the probe in the presence of transmembrane pH gradients results in fluorescence quenching. (Biochim Biophys Acta (1983) 722:107)
2. Spectra of this compound are in methanol acidified with a trace of HCl.
3. pK_a values may vary considerably depending on the temperature, ionic strength, viscosity, protein binding and other factors. Unless otherwise noted, values listed have been determined from pH-dependent fluorescence measurements at 22°C.
4. Spectra are in aqueous buffers adjusted to >1 pH unit above and >1 pH unit below the pK_a.
5. C369 is converted to a fluorescent product (C368) after acetate hydrolysis.
6. This product is supplied as a ready-made solution in the solvent indicated under "Soluble."
7. This LysoSensor™ dye exhibits increasing fluorescence as pH decreases with no spectral shift. L7533 has additional absorption and fluorescence emission peaks at Abs = 394 nm and Em = 401 nm.
8. The pK_a value for this product is determined from the pH-dependent variation of the absorption spectrum.
9. The fluorescence lifetime (τ) of the Oregon Green® 514 dye in pH 9.0 buffer at 20°C is 4.2 nanoseconds. Data provided by the SPEX Fluorescence Group, Horiba Jobin Yvon, Inc.
10. The fluorescence lifetime (τ) of the Oregon Green® 488 dye in pH 9.0 buffer at 20°C is 4.1 nanoseconds. Data provided by the SPEX Fluorescence Group, Horiba Jobin Yvon, Inc.
11. Acetate hydrolysis of this compound yields a fluorescent product with spectral characteristics similar to O6146.

IMPORTANT NOTICE: The products described in this manual are covered by one or more Limited Use Label License(s). Please refer to the Appendix on page 971 and Master Product List on page 975. Products are For Research Use Only. Not intended for any animal or human therapeutic or diagnostic use.

PRODUCT LIST 20.3 PROBES USEFUL AT ACIDIC pH

Cat. No.	Product	Quantity
A1324	9-amino-6-chloro-2-methoxyacridine (ACMA)	100 mg
C368	5-(and-6)-carboxy-2',7'-dichlorofluorescein *mixed isomers*	100 mg
C369	5-(and-6)-carboxy-2',7'-dichlorofluorescein diacetate (carboxy-DCFDA) *mixed isomers*	100 mg
L7533	LysoSensor™ Blue DND-167 *1 mM solution in DMSO* *special packaging*	20 x 50 µL
L7534	LysoSensor™ Green DND-153 *1 mM solution in DMSO* *special packaging*	20 x 50 µL
L7535	LysoSensor™ Green DND-189 *1 mM solution in DMSO* *special packaging*	20 x 50 µL
L7545	LysoSensor™ Yellow/Blue DND-160 (PDMPO) *1 mM solution in DMSO*	20 x 50 µL
O6146	Oregon Green® 488 carboxylic acid *5-isomer*	5 mg
O6151	Oregon Green® 488 carboxylic acid diacetate (carboxy-DFFDA) *6-isomer*	5 mg
O6138	Oregon Green® 514 carboxylic acid	5 mg

20.4 pH Indicator Conjugates

This section includes our selection of pH indicators conjugated to dextrans and lipids, as well as our chemically reactive pH indicators for preparing new pH-sensitive conjugates. The pH indicator conjugates described below include both those useful at near-neutral pH and those useful in acidic environments.

pH Indicator Dextrans

The pH-sensitive properties of the pH indicators described in Section 20.2 and Section 20.3 are usually not significantly affected upon conjugation to dextrans. However, coupling of pH indicators to these inert polysaccharides changes several other properties of the dyes:

- Conjugates have high water solubility and therefore must be loaded into cells by microinjection, whole-cell patch-clamping, endocytosis or liposome fusion or by using the Influx™ pinocytic cell loading reagent (I14402, Section 19.8).
- Once loaded, dextrans are retained in viable cells for long periods and (at least those dextrans with average molecular weights above 3000) will not pass through gap junctions.
- Attachment to a dextran significantly decreases the likelihood that the indicator will become compartmentalized, thereby avoiding a substantial problem associated with cell-permeant acetoxymethyl (AM) ester derivatives.[1]

The properties of some of the most useful pH indicator dextrans are listed in Table 20.3 in approximate order of decreasing pK_a value. Our numerous labeled dextrans are discussed in Section 14.5 and listed in Table 14.4.

pHrodo™ Dextran

The 10,000 MW pHrodo™ dextran (P10361) is a superior alternative to other fluorescent dextran conjugates (e.g., BCECF and tetramethylrhodamine) for live-cell imaging of endocytosis[2] (Section 16.1). pHrodo™ dextran possesses a pH-sensitive fluorescence emission that increases in intensity with increasing acidity (Figure 20.4.1). pHrodo™ dextran is essentially nonfluorescent in the extracellular environment; however, upon internalization, the acidic environment of the endosomes elicits a bright, red-fluorescent signal from this dextran conjugate. The minimal fluorescent signal from the pHrodo™ dextran conjugate at neutral pH prevents the detection of noninternalized and nonspecifically bound

Table 20.3 Molecular Probes® pH indicator dextrans, in order of decreasing pK_a.

Dye	Cat. No.	pK_a *	Measurement Wavelengths	Application Notes
SNARF®	D3303, D3304	~7.5	Emission ratio 580/640 nm excited at 514 or 488 nm	• Best conjugate for ratiometric emission measurements
BCECF	D1878, D1880	~7.0	Excitation ratio 490/440 nm detected at 530 nm	• Best conjugate for ratiometric excitation measurements
Fluorescein	D1821, D1823, D1844, D3305	~6.4	Excitation ratio 490/450 nm detected at 520 nm	• Fluorescence is strongly quenched upon uptake into acidic organelles
Fluorescein and tetramethylrhodamine	D1951	~6.4	Excitation ratio 495/555 nm detected at 580 nm †	• Conjugate incorporating both pH-sensitive and pH-insensitive fluorescent dyes
pHrodo™	P10361	NA ‡	Emission at 585 nm when excited at 540–560 nm	• Best conjugate for measurements requiring broad-range pH sensitivity
Oregon Green® 488	D7170, D7172	~4.7	Excitation ratio 490/440 nm detected at 520 nm	• Good photostability • Optimum pH sensitivity for measurements in lysosomes and late endosomes
Oregon Green® 514	D7176	~4.7	Excitation ratio 510/450 nm detected at 530 nm	• Excellent photostability • Optimum pH sensitivity for measurements in lysosomes and late endosomes
LysoSensor™ Yellow/Blue	L22460	~4.2	Excitation ratio 340/400 nm detected at 520 nm; emission ratio 450/510 nm excited at 365 nm	• Options for dual excitation or dual emission ratio measurements • Optimum pH sensitivity for measurements in lysosomes and late endosomes

* pK_a values are those determined for the free dyes. Actual values for dextran conjugates may differ by up to ± 0.3 pH units and may vary with production lots. † Ratiometric emission measurements at 520/580 nm (with excitation at 495 nm) are also possible in principle; however, the response may be complicated by fluorescence resonance energy transfer.
NA = Not applicable. ‡ pHrodo™ pH indicator exhibits a complex pH titration profile. Decreasing pH (from pH 9 to pH 2) produces a continuous (but nonlinear) fluorescence increase.

IMPORTANT NOTICE: The products described in this manual are covered by one or more Limited Use Label License(s). Please refer to the Appendix on page 971 and Master Product List on page 975. Products are For Research Use Only. Not intended for any animal or human therapeutic or diagnostic use.

conjugates and eliminates the need for quenching reagents and extra wash steps, thus providing a simple fluorescent assay for endocytic activity (Figure 20.4.2). pHrodo™ dextran can be used to study or monitor endocytosis on a variety of platforms including fluorescence microscopy, flow cytometry and automated imaging and analysis (also known as high-content imaging or HCS). pHrodo™ dextran's excitation and emission maxima of 560 and 585 nm, respectively, facilitate multiplexing with other fluorophores including Alexa Fluor® 488 and Alexa Fluor® 647 dyes as well as Green Fluorescent Protein (GFP).

BCECF and SNARF® Indicator Dextrans for Measuring Near-Neutral pH

Our 10,000 MW and 70,000 MW BCECF dextrans (D1878, D1880) are important dual-excitation pH indicator conjugates for pH measurements near pH 7.0. BCECF dextran–labeled Swiss 3T3 cells have been shown to produce much more stable fluorescent signals, reduced probe compartmentalization and 10-fold greater resistance to light-induced damage when compared with BCECF AM–labeled cells.[3] The 10,000 MW BCECF dextran (D1878) has been used to monitor intracellular pH increases during developmental processes[4–6] and to measure pH in submucosal gland secretions from human lung tissues.[7] Cytoplasmic Ca^{2+}/H^+ buffering in green algae has been investigated using BCECF dextran in combination with fura dextran[8] (F3029, Section 19.4).

A dextran conjugate of the carboxy SNARF®-1 pH indicator (D3303, D3304) has been microinjected into rhizoid cells of the alga *Pelvetia fastigata* and used with ratiometric imaging to measure pH gradients associated with polar tip growth.[9] SNARF®-1 dextran has also been used to detect cytosolic alkalinization associated with multidrug transporter activity[10] and to investigate pH regulation of connexin 43 channels.[11] SNARF® dextran conjugates have been scrape-loaded into the cytosol of MDF-7/ADR cells. It was found that the 70,000 MW SNARF® dextran conjugate remained exclusively cytosolic, whereas the 10,000 MW conjugate reported the pH of both cytosolic and nuclear compartments.[12]

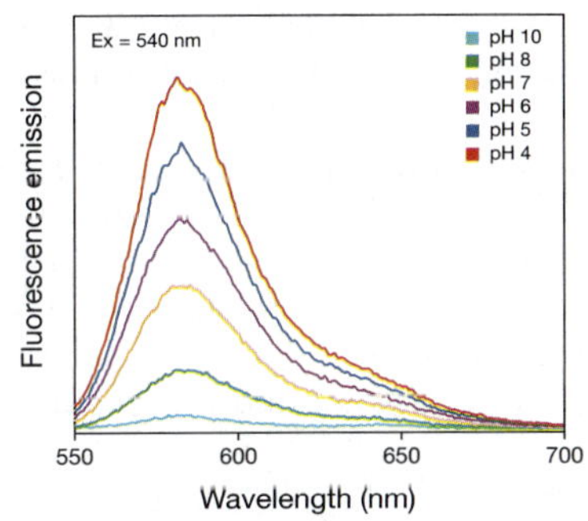

Figure 20.4.1 The pH sensitivity of pHrodo™ dextran. pHrodo™ dextran (P10361) was reconstituted in HEPES (20 mM)–buffered PBS and adjusted to pH values from pH 4 to pH 10. The intensity of fluorescence emission increases with increasing acidity, particularly in the pH 5–8 range.

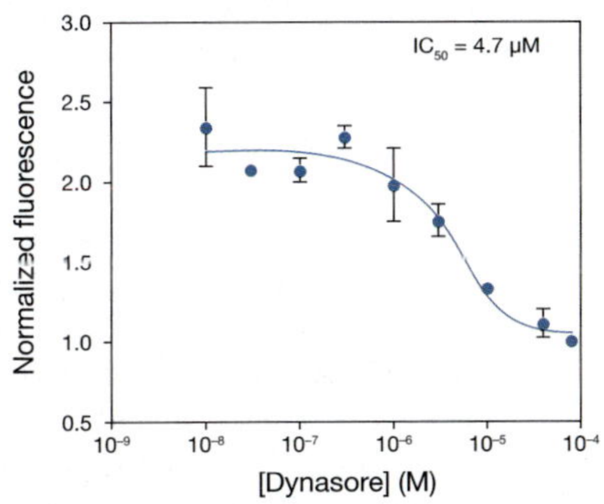

Figure 20.4.2 Tracking endocytosis inhibition with pHrodo™ dextran conjugates. HeLa cells were plated in 96-well format and treated with dynasore for 3 hr at 37°C prior to the pHrodo™ assay. Then, 40 µg/mL pHrodo™ dextran synthesized from pHrodo™ succinimidyl ester (P36600) and amine-reactive 10,000 MW dextran (D1860) was incubated for 30 min at 37°C. Cells were stained with HCS NuclearMask™ Blue stain (H10325) for 10 min to reveal total cell number and demarcation for image analysis. Images were acquired on the BD Pathway™ 855 High-Content Bioimager (BD Biosciences).

Oregon Green® and LysoSensor™ Yellow/Blue Dextrans

Although the fluorescein, BCECF and SNARF® dextrans are intended for pH measurements between pH ~6 and 8, these dextrans are also useful for detecting uptake into acidic organelles, such as occurs during endocytosis. In particular, when these indicator dextrans enter moderately acidic compartments (pH <5.5):

- Fluorescence of the fluorescein and BCECF dextrans is strongly quenched[13–15] (Section 20.2).
- The 520/570 nm emission intensity ratio of the double-labeled fluorescein–tetramethylrhodamine dextran (D1951) decreases[16] (Figure 20.4.3).
- The 580/640 nm emission ratio of the SNARF®-1 dextrans increases (Section 20.2).
- The fluorescein, BCECF and SNARF® dextrans are useful for detecting translocation into compartments that have an acidic pH; however, the relative insensitivity of their fluorescence below pH ~6 limits quantitative pH estimation. The lower pK_a values of the Oregon Green® 488 and Oregon Green® 514 dextran conjugates (Table 20.3) make them more suitable indicators for estimating the pH of relatively acidic lysosomal environments. Moreover, the shift in their excitation spectra in acidic media permits ratiometric pH measurements.

We have also developed a 10,000 MW dextran conjugate of the LysoSensor™ Yellow/Blue dye (L22460), which can be used to quickly and accurately estimate the pH of lysosomes. As this labeled dextran is taken up by the cells and moves through the endocytic pathway, the fluorescence of the LysoSensor™ probe changes from blue in the near-neutral endosomes to yellow in the acidic lysosomes.[17] The greatest change in

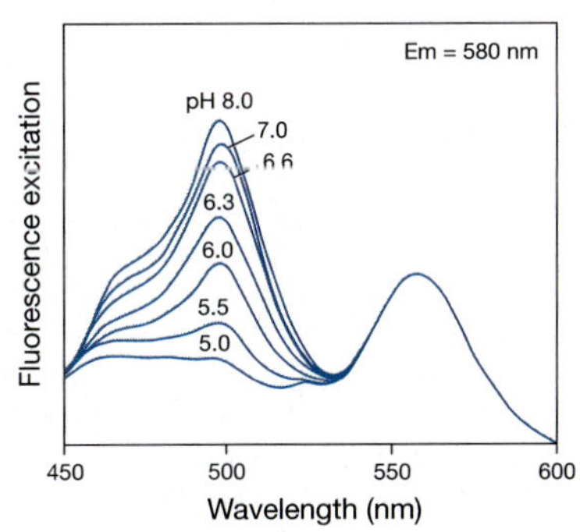

Figure 20.4.3 The excitation spectra of double-labeled fluorescein–tetramethylrhodamine dextran (D1951), which contains pH-dependent (fluorescein) and pH-independent (tetramethylrhodamine) dyes.

IMPORTANT NOTICE: The products described in this manual are covered by one or more Limited Use Label License(s). Please refer to the Appendix on page 971 and Master Product List on page 975. Products are For Research Use Only. Not intended for any animal or human therapeutic or diagnostic use.

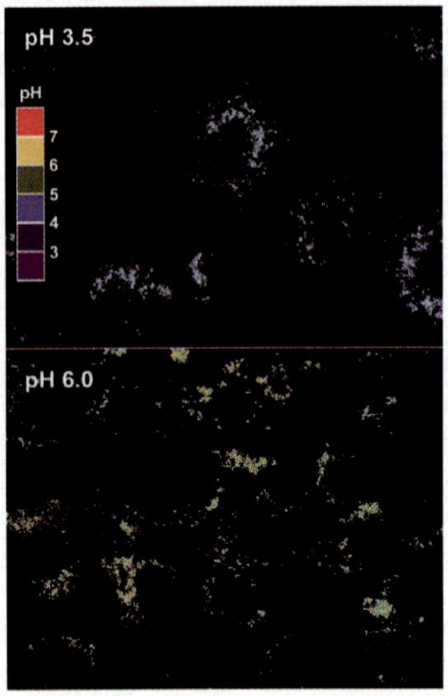

Figure 20.4.4 Dual-emission ratiometric measurement of lysosomal pH using LysoSensor™ Yellow/Blue dextran (L22460). MDCK cells labeled with the fluorescent dextran were exposed to pH-calibration buffers (pH 3.5 or pH 6.0) in the presence of nigericin (N1495) and monensin. Pseudocolored images were constructed from two emission images at 450 ± 33 nm and 510 ± 20 nm, both excited at 365 ± 8 nm. (Chem Biol (1999) 6:411).

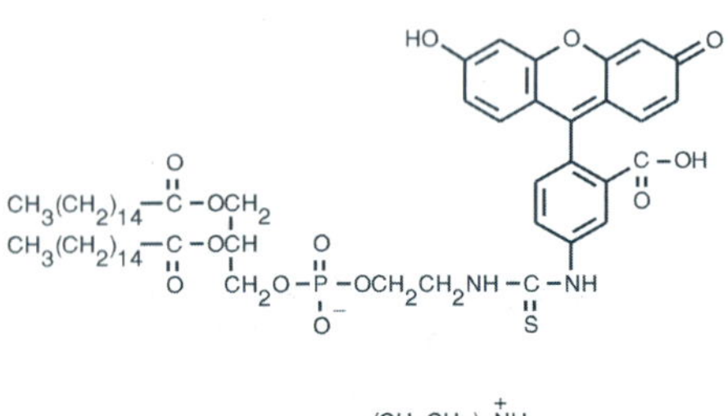

Figure 20.4.5 Fluorescein DHPE (*N*-(fluorescein-5-thio-carbamoyl)-1,2-dihexadecanoyl-*sn*-glycero-3-phosphoethanolamine, triethylammonium salt, F362).

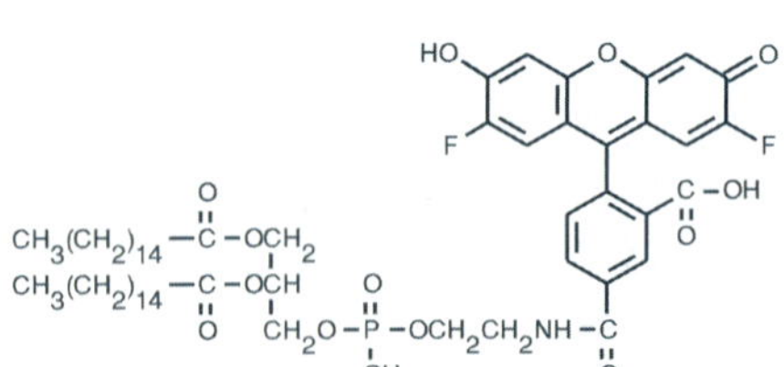

Figure 20.4.6 Oregon Green® 488 DHPE (Oregon Green®488 1,2-dihexadecanoyl-*sn*-glycero-3-phosphoethanolamine, O12650).

fluorescence emission occurs near the pK_a of the dye at pH ~4.2. The pH in lysosomes can be measured with LysoSensor™ Yellow/Blue dextran using fluorescence microscopy (Figure 20.4.4) or flow cytometry.

Lipophilic pH Indicators

Measurement of the pH adjacent to membrane surfaces is often complicated by electrostatic charge and solvation effects on the pK_a of surface-bound indicators.[18,19] The pK_a of membrane-intercalated fluorescein DHPE (F362, Figure 20.4.5) is ~6.2, quite close to that of free fluorescein.[18] Researchers have used the pH-dependent fluorescence of fluorescein DHPE to measure lateral proton conduction along lipid monolayers.[20–23] This fluorescein-labeled phospholipid has also been used to follow proton translocation from internal compartments in phospholipid vesicles.[24] For more acidic environments, Oregon Green® 488 DHPE (O12650, Figure 20.4.6) has potentially similar applications. Other related lipophilic fluorescein derivatives, including 5-dodecanoylaminofluorescein (D109) and 5-hexadecanoylaminofluorescein (H110), are described in Section 13.5.

Reactive Dyes for Preparing pH-Sensitive Conjugates

Amine-Reactive pHrodo™ pH Indicator

pHrodo™ dye is an aminorhodamine pH indicator that increases in fluorescence as the pH of its surroundings becomes more acidic (Figure 20.4.1). The amine-reactive succinimidyl ester form of the dye (P36600) provides access to a wide variety of user-defined bioconjugates by following the detailed protocols provided in the accompanying product information sheet. pHrodo™ dye is extremely sensitive to its local environment; therefore, the pH response of each bioconjugate must be individually determined. pHrodo™ succinimidyl ester has been used to label dexamethansone-treated thymocytes for flow cytometric analysis of phagocytosis by splenic or peritoneal macrophages[25] and for live-cell confocal imaging of antigen transfer from human B lymphocytes to macrophages.[26]

To study endocytosis in a high-throughput format, we have used pHrodo™ conjugates to create dose response curves for the inhibition of endocytosis by dynasore, a dynamin-specific inhibitor. pHrodo™ succinimidyl ester was conjugated to an amine-derivatized 10,000 MW dextran, and cells were treated with serial dilutions of dynasore to create dose response curves for endocytosis inhibition (Figure 20.4.2). The loss of pHrodo™ dextran accumulation in punctuate structures within the cell was observed as a function of dynasore concentration.

Other Amine-Reactive Dyes

Many of the pH indicators described in Section 20.2 and Section 20.3 can be conjugated to biological molecules in order to generate pH-sensitive tracers. The resulting conjugates are useful for following endocytosis, phagocytosis, organelle trafficking and other processes, as described in Section 16.1. For example, the pH sensitivity of fluorescein-labeled transferrin (T2871, Section 16.1) has frequently been exploited to detect pH changes associated with the endocytic processing of this important iron-transporting glycoprotein.[27–30]

For these types of applications, fluorescein conjugates are less than optimal because they have little sensitivity in the pH range below pH 5.5. The response range can be extended using conjugates of our Oregon Green® 488 dye, which has a much lower pK_a than fluorescein (4.7 compared with 6.4, Figure 20.4.7) but essentially identical spectra. Our collaborators, Dr. Elizabeth Simons and her co-workers, have labeled fungi (*Cryptococcus neoformans*) with FITC (F143, F1906, F1907; Section 1.5) or Oregon Green® 488 isothiocyanate (O6080, Section 1.5) and used them to study the influence of phagosomal pH on the fungicidal and fungistatic activity of human monocyte-derived macrophages.[31] At pH levels evoked by phagocytosis of live and heat-killed fungi (pH 4.7–5.7), standard curves of the 498/450 nm fluorescence excitation ratio as a function of pH (Figure 20.4.8) illustrate the greater sensitivity of the Oregon Green® 488 conjugates. Vergne and co-workers have used zymosan (heat-killed yeast) double-labeled with the succinimidyl esters of Oregon Green® 488 carboxylic acid (O6147) and carboxytetramethylrhodamine (C1171, Section 1.6) to measure phagosomal pH in J774 macrophages using dual-emission (530/585 nm) flow cytometry; they were able to estimate pH values as low as 4.0.[32]

IMPORTANT NOTICE: The products described in this manual are covered by one or more Limited Use Label License(s). Please refer to the Appendix on page 971 and Master Product List on page 975. Products are For Research Use Only. Not intended for any animal or human therapeutic or diagnostic use.

Many of the reagents or methods required to prepare conjugates have been described in Chapter 1. The most common method of producing a useful conjugate is the reaction of amines with succinimidyl ester or isothiocyanate derivatives of the pH indicator. Examples of amine-reactive pH indicators include the succinimidyl esters of the carboxy SNARF®-1 and carboxynaphthofluorescein dyes (S22801, C653). The succinimidyl ester of the chlorinated fluorescein derivative 6-JOE (pK_a ~11.5, C6171MP; Section 1.5) can be used to prepare conjugates that are responsive to alkaline pH levels. When the amine-reactive pH indicator is not available, sulfosuccinimidyl esters can generally be prepared *in situ* simply by dissolving the carboxylic acid dye in a buffer that contains *N*-hydroxysulfosuccinimide and 1-ethyl-3-(3-dimethylaminopropyl) carbodiimide (NHSS, H2249; EDAC, E2247; Section 3.4); this method was used to activate SNARF®-4F carboxylic acid (S23920, Section 20.2) for labeling 5′-amine–modified oligonucleotides.[33] Addition of NHSS to the buffer has been shown to enhance the yield of carbodiimide-mediated conjugations.[34] Suitable amine-reactive pH indicators or dyes that can be made reactive using EDAC/NHSS are listed in Table 20.4.

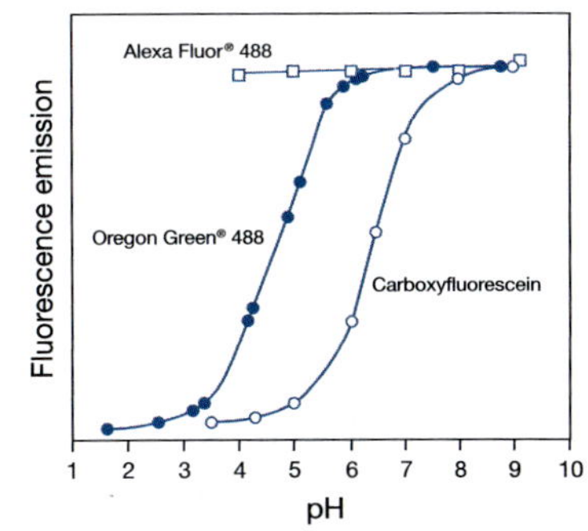

Figure 20.4.7 Comparison of pH-dependent fluorescence of the Oregon Green® 488 (●), carboxyfluorescein (○) and Alexa Fluor® 488 (□) fluorophores. Fluorescence intensities were measured for equal concentrations of the three dyes using excitation/emission at 490/520 nm.

Table 20.4 Reactive pH indicator dyes.

pH Indicator	Preferred Reactive Form
BCECF	BCECF (B1151, Section 20.2) *
Carboxyfluorescein	5-(and 6-)carboxyfluorescein, succinimidyl ester (C1311, Section 1.5)
6-JOE	6-carboxy-4′,5′-dichloro-2′,7′-dimethoxyfluorescein, succinimidyl ester (C6171MP, Section 1.5)
Naphthofluorescein	5-(and 6-)carboxynaphthofluorescein, succinimidyl ester (C653)
Oregon Green® 488	Oregon Green® 488 carboxylic acid, succinimidyl ester (O6147, O6149)
Oregon Green® 514	Oregon Green® 514 carboxylic acid, succinimidyl ester (O6139)
pHrodo™	pHrodo™ succinimidyl ester (P36600)
SNARF®-1	SNARF®-1 carboxylic acid, acetate, succinimidyl ester (S22801) 5-(and 6-)carboxy SNARF®-1 (C1270, Section 20.2) *
SNARF®-4F	SNARF®-4F carboxylic acid (S23920, Section 20.2) *
SNARF®-5F	SNARF®-5F carboxylic acid (S23922, Section 20.2) *

* Carboxylic acids require activation with EDAC/NHSS before reaction with amines.

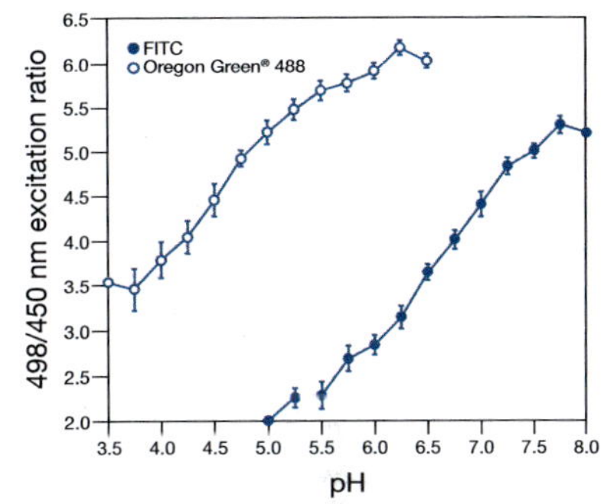

Figure 20.4.8 Calibration curves for intraphagosomal pH measurements using fungi (*Cryptococcus neoformans*) labeled with Oregon Green® 488 isothiocyanate (O6080) or fluorescein isothiocyanate (FITC; F143, F1906, F1907). Human monocyte–derived macrophages laden with phagocytosed *C. neoformans* were exposed to pH-controlled buffers in the presence of the K+/H+ ionophore nigericin (N1495). The 498/450 nm fluorescence excitation ratios corresponding to different pH levels were measured in a spectrofluorometer. The data were provided by Elizabeth Simons, Boston University.

REFERENCES

1. J Exp Biol (1994) 196:419; **2.** J Biol Chem (2009) 284:35926; **3.** J Cell Physiol (1989) 141:410; **4.** Dev Biol (1997) 191:53; **5.** Development (1994) 120:433; **6.** Dev Biol (1993) 156:176; **7.** Proc Natl Acad Sci U S A (2001) 98:8119; **8.** Protoplasma (1997) 198:107; **9.** Science (1994) 263:1419; **10.** J Histochem Cytochem (1990) 38:685; **11.** Biophys J (1996) 70:1294; **12.** Proc Natl Acad Sci U S A (1999) 96:4432; **13.** Proc Natl Acad Sci U S A (1994) 91:4811; **14.** J Biol Chem (1993) 268:25320; **15.** J Cell Sci (1993) 105:861; **16.** J Cell Biol (1995) 130:821; **17.** Nucleic Acids Res (2002) 30:1338; **18.** Biochim Biophys Acta (1988) 939:289; **19.** J Phys Chem (1977) 81:1755; **20.** Biochemistry (1990) 29:59; **21.** J Am Chem Soc (1991) 113:8818; **22.** Eur J Biochem (1987) 162:379; **23.** Nature (1986) 322:756; **24.** Biochim Biophys Acta (1984) 766:161; **25.** J Immunol Methods (2009) 342:71; **26.** J Immunol (2008) 181:4043; **27.** Biochemistry (1992) 31:5820; **28.** Cell (1984) 37:789; **29.** J Biol Chem (1991) 266:3469; **30.** J Biol Chem (1990) 265:6688; **31.** Infect Immun (1999) 67:885; **32.** Anal Biochem (1998) 255:127; **33.** Bioconjug Chem (2005) 16:986; **34.** Anal Biochem (1986) 156:220.

DATA TABLE 20.4 pH INDICATOR CONJUGATES

				Acidic Solution				Basic Solution					
Cat. No.	MW	Storage	Soluble	Abs	EC	Em	Solvent	Abs	EC	Em	Solvent	pK_a	Notes
C653	573.51	F,D,L	DMF, DMSO	515	10,000	565	pH 6	602	42,000	672	pH 10	7.6	1, 2
D1821	see Notes	F,L	H_2O	473	ND	514	pH 5	490	ND	513	pH 9	6.4	1, 3, 4, 5
D1823	see Notes	F,L	H_2O	473	ND	514	pH 5	490	ND	514	pH 9	6.4	1, 3, 4, 5
D1844	see Notes	F,L	H_2O	473	ND	514	pH 5	490	ND	514	pH 9	6.4	1, 3, 4, 5
D1878	see Notes	F,L	H_2O	482	ND	520	pH 5	503	ND	528	pH 9	7.0	1, 3, 4, 5
D1880	see Notes	F,L	H_2O	482	ND	520	pH 5	503	ND	528	pH 9	7.0	1, 3, 4, 5
D1951	see Notes	F,L	H_2O	see Notes		see Notes						ND	1, 3, 4, 6
D3303	see Notes	F,L	H_2O	548	ND	587	pH 6	576	ND	635	pH 10	7.5	1, 3, 4, 5
D3304	see Notes	F,L	H_2O	548	ND	587	pH 6	576	ND	635	pH 10	7.5	1, 3, 4, 5
D3305	see Notes	F,L	H_2O	473	ND	514	pH 5	490	ND	514	pH 9	6.4	1, 3, 4, 5
D7170	see Notes	F,L	H_2O	478	ND	518	pH 3	492	ND	518	pH 9	4.7	1, 3, 4, 5

continued on next page

DATA TABLE 20.4 pH INDICATOR CONJUGATES—*continued*

				Acidic Solution				Basic Solution					
Cat. No.	MW	Storage	Soluble	Abs	EC	Em	Solvent	Abs	EC	Em	Solvent	pK_a	Notes
D7172	see Notes	F,L	H_2O	478	ND	518	pH 3	492	ND	518	pH 9	4.7	1, 3, 4, 5
D7176	see Notes	F,L	H_2O	489	ND	526	pH 3	506	ND	526	pH 9	4.7	1, 3, 4, 5
F362	1182.54	FF,D,L	see Notes	476	32,000	519	MeOH/H^+	496	88,000	519	MeOH/OH^-	6.2	7, 8
L22460	see Notes	F,D,L	H_2O	384	ND	540	pH 3	329	ND	440	pH 7.5	4.2	1, 3, 4, 5
O6139	609.43	F,D,L	DMF, DMSO	489	26,000	526	pH 3	506	85,000	526	pH 9	4.7	2
O6147	509.38	F,D,L	DMF, DMSO	480	24,000	521	pH 3	495	76,000	521	pH 9	4.7	2
O6149	509.38	F,D,L	DMF, DMSO	480	26,000	516	pH 3	496	82,000	516	pH 9	4.7	2
O12650	1086.25	FF,D,L	see Notes	485	26,000	526	MeOH/H^+	501	85,000	526	MeOH/OH^-	ND	4, 7, 8
P10361	see Notes	F,L	H_2O	560	ND	587	pH 2	553	ND	587	pH 9	see Notes	3, 4, 9
P36600	~650	F,D,L	DMSO	560	95,000	587	pH 2	553	90,000	587	pH 9	see Notes	9
S22801	592.56	F,D	DMSO	<350		none		<350		none			10

For definitions of the contents of this data table, see "Using *The Molecular Probes® Handbook*" in the introductory pages.

Notes

1. Spectra are in aqueous buffers adjusted to >1 pH unit above and >1 pH unit below the pK_a.
2. Spectral data for this product represents the unreacted succinimidyl ester. The pK_a value and the spectral data for acidic solutions have been estimated based on the spectra of the parent carboxylic acid.
3. The molecular weight is nominally as specified in the product name but may have a broad distribution.
4. ND = not determined.
5. Abs, Em and pK_a values listed for this dextran conjugate are those obtained for the free dye. Values for actual conjugates are typically very similar, with slight variations between different production lots.
6. These conjugates contain both pH-sensitive fluorescein (Abs = 495, Em = 520 nm) and pH-insensitive tetramethylrhodamine (Abs = 555 nm, Em = 575 nm) fluorophores.
7. The pK_a values of lipophilic pH indicators may vary considerably depending on the electrostatic properties of membrane surfaces. (Biochemistry (1993) 32:10057, J Phys Chem (1977) 81:1755, Biochim Biophys Acta (1988) 939:289) The pK_a values listed are for electrostatically neutral liposomes or micelles. Spectra are in MeOH containing a trace of HCl (MeOH/H^+) or a trace of KOH (MeOH/OH^-).
8. Chloroform is the most generally useful solvent for preparing stock solutions of phospholipids (including sphingomyelins). Glycerophosphocholines are usually freely soluble in ethanol. Most other glycerophospholipids (phosphoethanolamines, phosphatidic acids and phosphoglycerols) are less soluble in ethanol, but solutions up to 1–2 mg/mL should be obtainable, using sonication to aid dispersion if necessary. Labeling of cells with fluorescent phospholipids can be enhanced by addition of cyclodextrins during incubation. (J Biol Chem (1999) 274:35359)
9. pHrodo™ succinimidyl ester (P36600) exhibits a complex pH titration profile. Decreasing pH (from pH 9 to pH 2) produces a continuous (but nonlinear) fluorescence increase. This pH response profile typically changes upon conjugation of the dye to proteins and other biomolecules.
10. S22801 is converted to fluorescent products with spectra similar to C1270 (Section 20.2) after acetate hydrolysis.

PRODUCT LIST 20.4 pH INDICATOR CONJUGATES

Cat. No.	Product	Quantity
C653	5-(and-6)-carboxynaphthofluorescein, succinimidyl ester *mixed isomers*	25 mg
D1878	dextran, BCECF, 10,000 MW, anionic	10 mg
D1880	dextran, BCECF, 70,000 MW, anionic	10 mg
D1951	dextran, fluorescein and tetramethylrhodamine, 70,000 MW, anionic	10 mg
D3305	dextran, fluorescein, 3000 MW, anionic	10 mg
D1821	dextran, fluorescein, 10,000 MW, anionic	25 mg
D1844	dextran, fluorescein, 40,000 MW, anionic	25 mg
D1823	dextran, fluorescein, 70,000 MW, anionic	25 mg
D7170	dextran, Oregon Green® 488; 10,000 MW, anionic	5 mg
D7172	dextran, Oregon Green® 488; 70,000 MW, anionic	5 mg
D7176	dextran, Oregon Green® 514; 70,000 MW, anionic	5 mg
D3303	dextran, SNARF®-1, 10,000 MW, anionic	5 mg
D3304	dextran, SNARF®-1, 70,000 MW, anionic	5 mg
F362	fluorescein DHPE (*N*-(fluorescein-5-thiocarbamoyl)-1,2-dihexadecanoyl-*sn*-glycero-3-phosphoethanolamine, triethylammonium salt)	5 mg
L22460	LysoSensor™ Yellow/Blue dextran, 10,000 MW, anionic, fixable	5 mg
O6147	Oregon Green® 488 carboxylic acid, succinimidyl ester *5-isomer*	5 mg
O6149	Oregon Green® 488 carboxylic acid, succinimidyl ester *6-isomer*	5 mg
O12650	Oregon Green® 488 DHPE (Oregon Green® 488 1,2-dihexadecanoyl-*sn*-glycero-3-phosphoethanolamine)	1 mg
O6139	Oregon Green® 514 carboxylic acid, succinimidyl ester	5 mg
P10361	pHrodo™ dextran, 10,000 MW *for endocytosis*	0.5 mg
P36600	pHrodo™, succinimidyl ester (pHrodo™, SE)	1 mg
S22801	SNARF®-1 carboxylic acid, acetate, succinimidyl ester *special packaging*	10 x 50 µg

IMPORTANT NOTICE: The products described in this manual are covered by one or more Limited Use Label License(s). Please refer to the Appendix on page 971 and Master Product List on page 975. Products are For Research Use Only. Not intended for any animal or human therapeutic or diagnostic use.

molecular probes® | invitrogen™ by life technologies™

TWENTY-ONE

CHAPTER 21

Indicators for Na^+, K^+, Cl^- and Miscellaneous Ions

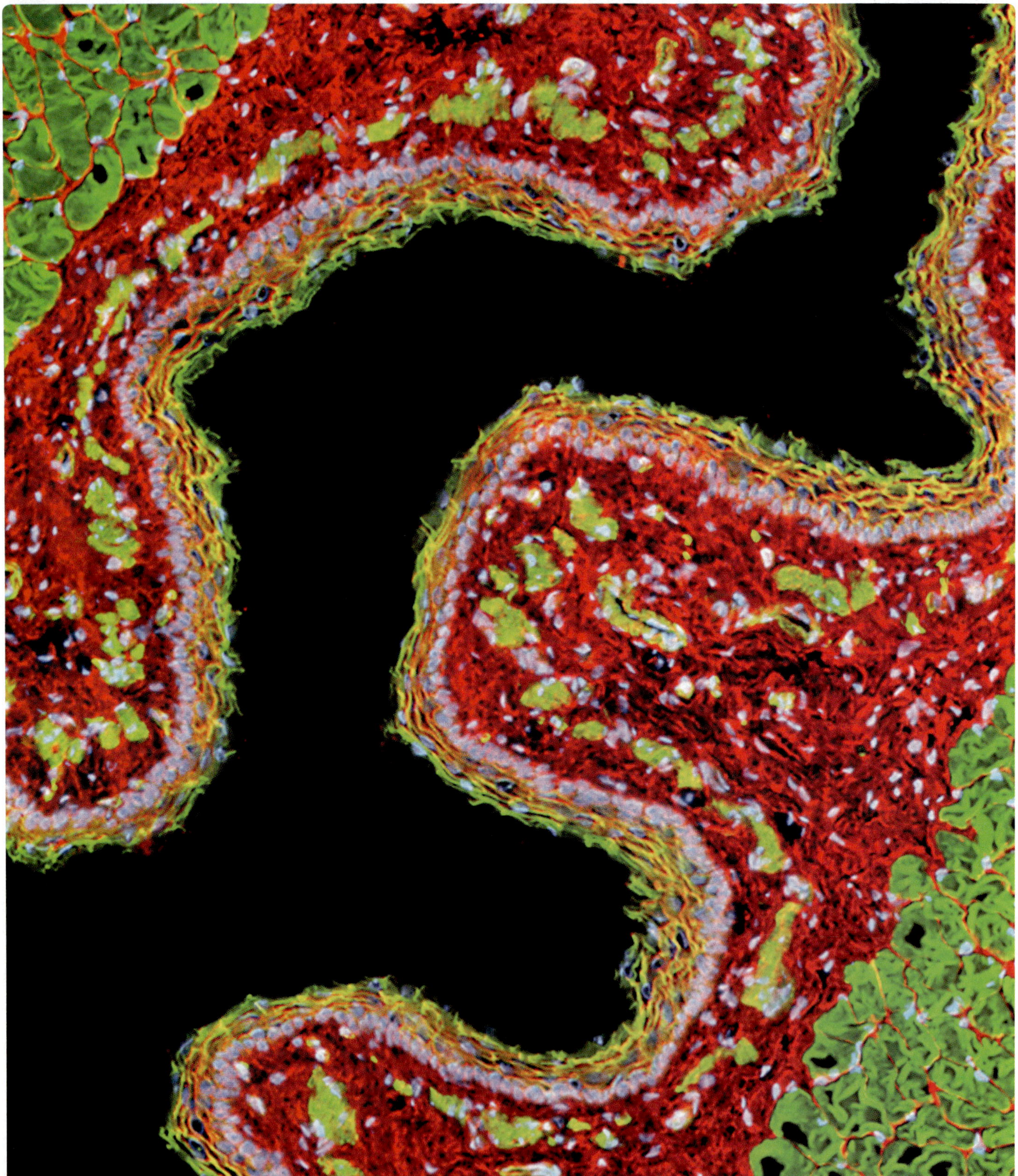

Alexa Fluor® 568 phalloidin, Oregon Green® wheat germ agglutinin and Hoechst 33342.

IMPORTANT NOTICE: The products described in this manual are covered by one or more Limited Use Label License(s). Please refer to the Appendix on page 971 and Master Product List on page 975. Products are For Research Use Only. Not intended for any animal or human therapeutic or diagnostic use.

21.1 Fluorescent Na$^+$ and K$^+$ Indicators

Sodium and potassium channels are ion-selective protein pores that span the cell's plasma membrane and serve to establish and regulate membrane potential. They are typically classified according to their response mechanism: voltage-gated channels open or close in response to changes in membrane potential,[1] whereas ligand-gated or ion-activated channels are triggered by ligand or ion binding.[2] In excitable cells such as neurons and myocytes, these channels function both to create the action potential and to reset the cell's resting membrane potential.

In this section, we describe sodium- and potassium-selective fluorescent indicators, as well as the FluxOR™ Thallium Detection Kits, which provide a fluorescence-based method for assaying potassium ion channel and transporter activities. The next section describes fluorescent indicators for intracellular and extracellular chloride, together with an assortment of analytical reagents and methods for direct or indirect quantitation of other inorganic anions.

SBFI and PBFI

Properties of SBFI and PBFI

SBFI[3] and PBFI[3,4] are fluorescent indicators for sodium and potassium, respectively. Although the selectivity of SBFI and PBFI for their target ions is less than that of calcium indicators such as fura-2, it is sufficient for the detection of physiological concentrations of Na$^+$ and K$^+$ in the presence of other monovalent cations.[3] Furthermore, the spectral responses of SBFI and PBFI upon ion binding permit excitation ratio measurements (Loading and Calibration of Intracellular Ion Indicators—Note 19.1), and these indicators can be used with the same optical filters and equipment used for fura-2.[5,6]

SBFI (Figure 21.1.1) and PBFI (Figure 21.1.2) comprise benzofuranyl fluorophores linked to a crown ether chelator. The cavity size of the crown ether confers selectivity for Na$^+$ versus K$^+$ (or *vice versa* in the case of PBFI). When an ion binds to SBFI or PBFI, the indicator's fluorescence quantum yield increases, its excitation peak narrows and its excitation maximum shifts to shorter wavelengths (Figure 21.1.3), causing a significant change in the ratio of fluorescence intensities excited at 340/380 nm (Figure 21.1.4, Figure 21.1.5). This fluorescence signal is slightly sensitive to changes in pH between 6.5 and 7.5,[7,8] but it is strongly affected by ionic strength[9] and viscosity.[10] Researchers have described the use of SBFI for emission ratio detection[11] (410/590 nm, excited at

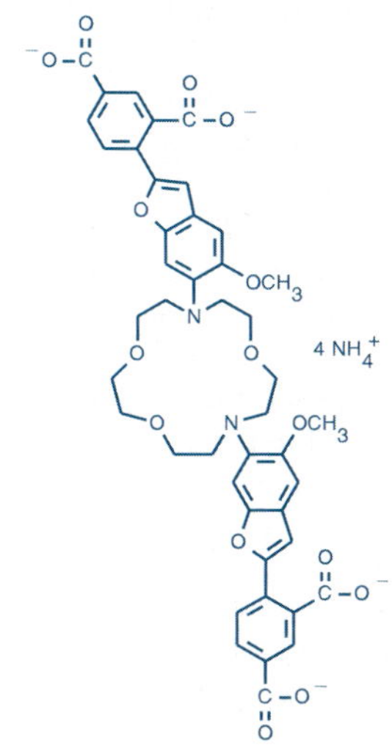

Figure 21.1.1 SBFI, tetraammonium salt (S1262).

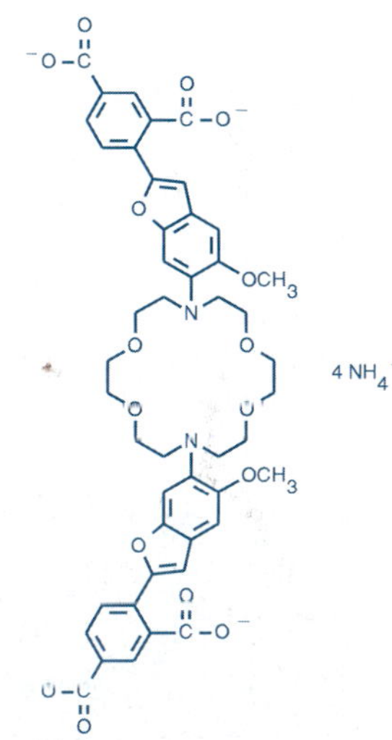

Figure 21.1.2 PBFI, tetraammonium salt (P1265MP).

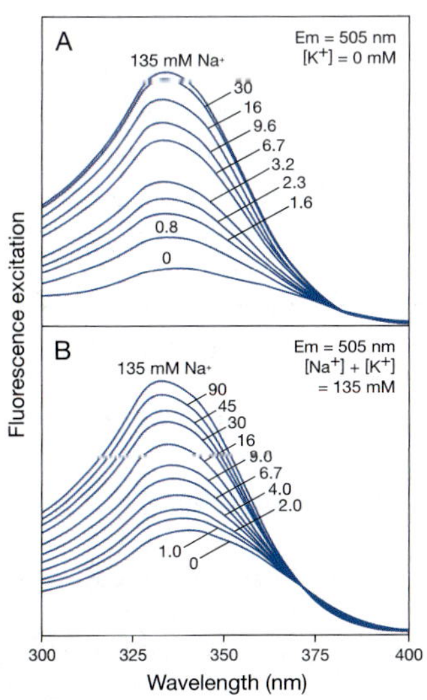

Figure 21.1.4 The excitation spectral response of SBFI (S1262) to Na$^+$: **A)** in K$^+$-free solution and **B)** in solutions containing K$^+$ with the combined Na$^+$ and K$^+$ concentration equal to 135 mM. The scale on the vertical axis is the same for both panels.

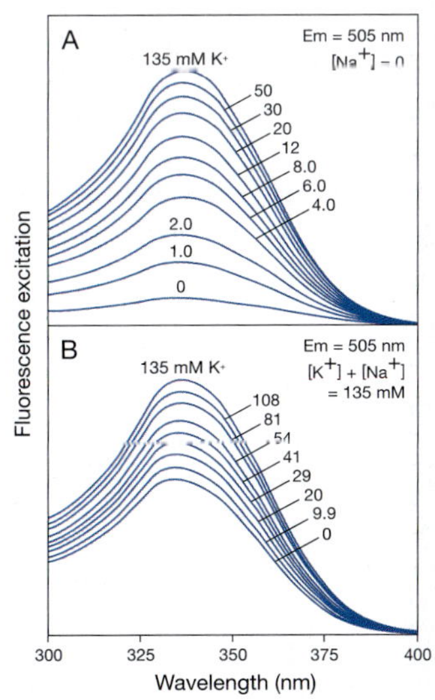

Figure 21.1.5 The excitation spectral response of PBFI (P1265MP) to K$^+$: **A)** in Na$^+$-free solution and **B)** in solutions containing Na$^+$ with the combined K$^+$ and Na$^+$ concentration equal to 135 mM. The scale on the vertical axis is the same for both panels.

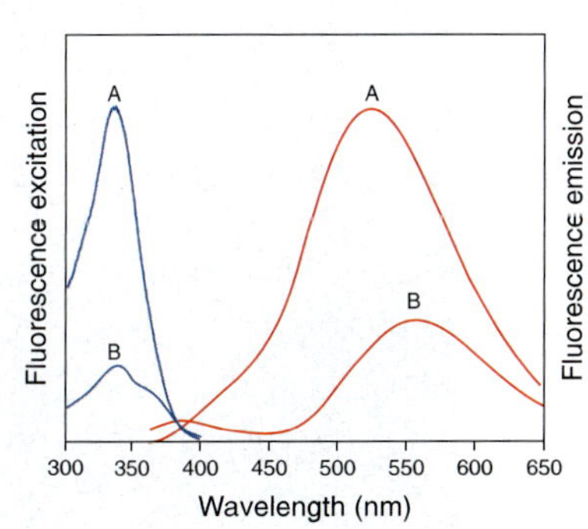

Figure 21.1.3 Fluorescence excitation (detected at 505 nm) and emission (excited at 340 nm) spectra of SBFI in pH 7.0 buffer containing 135 mM **(A)** or 0 mM **(B)** Na$^+$.

IMPORTANT NOTICE: The products described in this manual are covered by one or more Limited Use Label License(s). Please refer to the Appendix on page 971 and Master Product List on page 975. Products are For Research Use Only. Not intended for any animal or human therapeutic or diagnostic use.

340 nm). More recently, the implementation of two-photon excitation of SBFI with infrared light has been reported for Na^+ imaging in spines and fine dendrites of central neurons [12,13] (Figure 21.1.6).

Although SBFI is quite selective for the Na^+ ion, K^+ has some effect on the native affinity of SBFI for Na^+ (Figure 21.1.4). The dissociation constant (K_d) of SBFI for Na^+ is 3.8 mM in the absence of K^+, and 11.3 mM in solutions with a combined Na^+ and K^+ concentration of 135 mM, which approximates physiological ionic strength. SBFI is ~18-fold more selective for Na^+ than for K^+. Likewise, the K_d of PBFI for K^+ is strongly dependent on whether Na^+ is present (Figure 21.1.5), with a value of 5.1 mM in the absence of Na^+ and 44 mM in solutions with a combined Na^+ and K^+ concentration of 135 mM. In buffers in which the Na^+ is replaced by tetramethylammonium chloride, the K_d of PBFI for K^+ is 11 mM; choline chloride and *N*-methylglucamine are two other possible replacements for Na^+ in the medium. Although PBFI is only 1.5-fold more selective for K^+ than for Na^+, this selectivity is often sufficient because intracellular K^+ concentrations are normally about 10 times higher than Na^+ concentrations.

The K_d of all ion indicators depends on factors such as pH, temperature, ionic strength, concentrations of other ions and dye–protein interactions. Due to these environmental factors, the K_d determined *in situ* for intracellular SBFI is substantially higher than that determined in cell-free buffer solutions. K_d (Na^+) values of 29 mM, 26.6 mM and 18.0 mM have been determined for SBFI in lizard peripheral axons, porcine adrenal chromaffin cells and rat hippocampal neurons, respectively.[7,14] Consequently, intracellular SBFI should be calibrated using the pore-forming antibiotic gramicidin [5] (G6888). Palytoxin, an ionophoric toxin isolated from marine coelenterates, is much more effective than gramicidin for equilibrating intracellular and extracellular Na^+.[14] Intracellular PBFI should be calibrated using the K^+ ionophore valinomycin [15] (V1644).

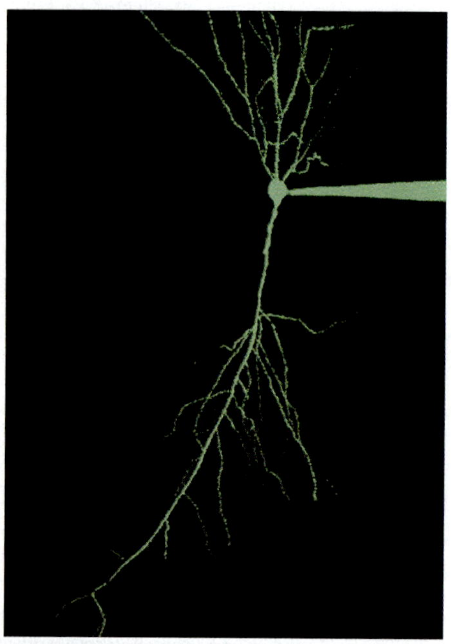

Figure 21.1.6 CA1 pyramidal neuron in a hippocampal slice filled with SBFI (S1262) delivered from a patch pipette (visible on the right). The image was obtained using two-photon excitation of SBFI at 790 nm. Image contributed by Christine R. Rose, Physiological Institute, University of Munich.

Cell Loading with SBFI and PBFI

SBFI and PBFI are available both as cell-impermeant acid salts (S1262, P1265MP) and as cell-permeant acetoxymethyl (AM) esters (S1263, S1264, P1267MP). The anionic acid forms can be loaded into cells using our Influx™ pinocytic cell-loading reagent (I14402, Section 19.8), or by microinjection, patch-pipette infusion or electroporation. For AM ester loading (Loading and Calibration of Intracellular Ion Indicators—Note 19.1), addition of the Pluronic® F-127 (P3000MP, P6866, P6867) or PowerLoad™ (P10020) dispersing agents as well as relatively long incubation times—up to four hours—are typically necessary.[5] ATP-induced permeabilization reportedly produces increased uptake of SBFI AM by bovine pulmonary arterial endothelial cells [16] (BPAEC). Somewhat higher working concentrations of PBFI and SBFI than those used for fura-2 may be required because of the lower fluorescence quantum yields of these indicators. AM ester loading sometimes produces intracellular compartmentalization of SBFI.[10,17] As with other AM esters, reducing the incubation temperature below 37°C may inhibit compartmentalization. Other practical aspects of loading and calibrating SBFI have been reviewed by Negulescu and Machen.[5]

Applications of SBFI

SBFI has been employed to estimate Na^+ gradients in isolated mitochondria,[17–19] as well as to measure intracellular Na^+ levels or Na^+ efflux in cells from a variety of tissues:

- Blood—platelets,[20] monocytes [21] and lymphocytes [22]
- Brain—astrocytes,[23] neurons,[12,24] and presynaptic terminals [25,26]
- Muscle—perfused heart,[27,28] cardiomyocytes [29–32] and smooth muscle [33,34]
- Secretory epithelia [35–37]
- Plants [38]

SBFI has also been used in combination with other fluorescent indicators to correlate changes in intracellular Na^+ with Ca^{2+} and Mg^{2+} concentrations,[24,39,40] intracellular pH and membrane potential.[21]

Applications of PBFI

PBFI [4] has fewer documented applications than SBFI. Renewed interest has been prompted by the observation that intracellular K^+ levels appear to be a controlling factor in apoptotic cell death pathways.[41] Flow cytometric measurements using UV argon-ion laser excitation (351 nm and 364 nm) of PBFI indicate that K^+ efflux induces shrinkage of apoptotic cells and is a trigger for caspase activation.[42–45] Furthermore, PBFI provides a potential alternative to radiometric ^{86}Rb efflux assays for quantitating K^+ transport.[15] Other applications of PBFI include:

- Detecting adrenoceptor-stimulated decreases of intracellular K^+ concentration in astrocytes and neurons [46]
- Evaluating the mediating effects of K^+ depletion on monocytic cell necrosis [47]
- Investigating the relationship between cytoplasmic K^+ concentrations and NMDA excitotoxicity [48]
- Measuring intracellular K^+ fluxes associated with apoptotic cell shrinkage [49,50]
- Monitoring mitochondrial K_{ATP} channel activation [51–53]
- Quantitating K^+ in isolated cochlear outer hair cells [54] and in mammalian ventricles using patch-clamp techniques [55]

IMPORTANT NOTICE: The products described in this manual are covered by one or more Limited Use Label License(s). Please refer to the Appendix on page 971 and Master Product List on page 975. Products are For Research Use Only. Not intended for any animal or human therapeutic or diagnostic use.

- Detecting elevated intracellular K^+ levels associated with HIV-induced cytopathology [56]
- Measuring K^+ levels in plant cells and vacuoles [57]

Sodium Green™ Na+ Indicator

The Sodium Green™ indicator can be excited at 488 nm (Figure 21.1.7), providing a valuable alternative to the UV light–excitable SBFI for use with confocal laser-scanning microscopes [58] and flow cytometers. [59] We offer the cell-impermeant tetra(tetramethylammonium) salt of the Sodium Green™ indicator (S6900), as well as its cell-permeant tetraacetate (S6901).

The Sodium Green™ indicator comprises two 2′,7′-dichlorofluorescein dyes linked to the nitrogen atoms of a crown ether (Figure 21.1.8) with a cavity size that confers selectivity for the Na^+ ion. Upon binding Na^+, the Sodium Green™ indicator exhibits an increase in fluorescence emission intensity with little shift in wavelength (Figure 21.1.9). Although the Sodium Green™ indicator lacks the direct ratiometric readout capability of SBFI, fluorescence intensity fluctuations due to cell size variability can be compensated to some extent by using forward light scatter as a reference signal in flow cytometry. [59]

As compared with SBFI, the Sodium Green™ indicator shows greater selectivity for Na^+ than K^+ (~41-fold versus ~18-fold) and displays a much higher fluorescence quantum yield (0.2 versus 0.08) in Na^+-containing solutions. The longer-wavelength absorption of the Sodium Green™ indicator results in reduction of the potential for photodamage to the cell because the energy of the excitation light is lower than that of the UV light required for excitation of SBFI. The K_d of the Sodium Green™ indicator for Na^+ is about 6 mM in K^+-free solution and about 21 mM in solutions with combined Na^+ and K^+ concentration of 135 mM, approximating physiological ionic strength. Because its K_d may be shifted due to intracellular interactions, the Sodium Green™ indicator should be calibrated *in situ* using the pore-forming antibiotic gramicidin [59] (G6888). In some cases, dye–protein interactions may cause severe dampening or even complete elimination of the Na^+-dependent fluorescence response of intracellular Sodium Green™ indicator. Nevertheless, flow cytometric measurements in Chinese hamster ovary (CHO) cells are well correlated with spectrofluorometric measurements using SBFI. [59] Other applications include:

- Assessing the regulation of Na^+/K^+-ATPase by persistent Na^+ accumulation in rat thalamic neurons [60]
- Confocal imaging of Na^+ transport in rat colonic mucosa [61] and cochlear hair cells by flow cytometry [62]
- Detecting anoxia-induced Na^+ influx in neurons [63]
- Determining intracellular Na^+ concentration in crayfish presynaptic terminals using an area-ratio method [64]
- Fluorescence lifetime imaging of intracellular Na^+ [65–67]
- Measuring intracellular Na^+ concentration in bacterial cells [68] and green algae [69]
- Determining voltage-gated sodium channel NaV1.5–driven endosomal Na^+ levels in macrophages [21]

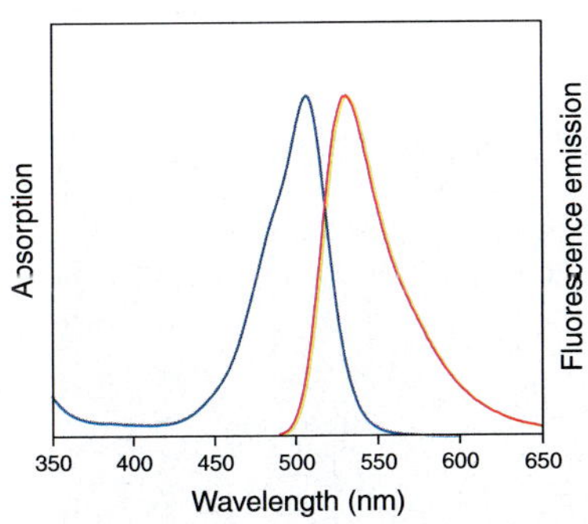

Figure 21.1.7 Absorption and fluorescence emission spectra of Sodium Green™ indicator in pH 7.0 buffer containing 135 mM Na^+.

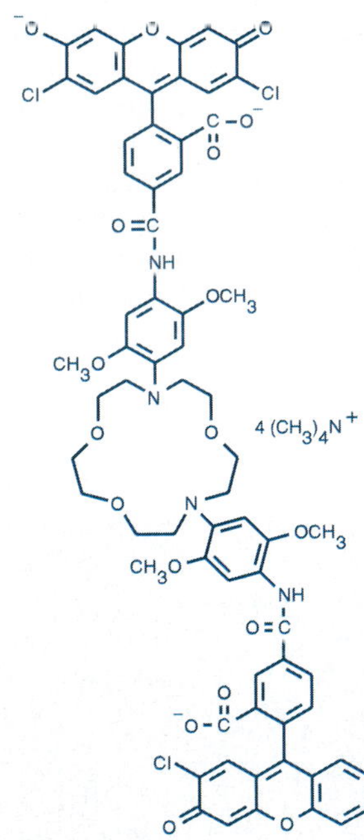

Figure 21.1.8 Sodium Green™, tetra (tetramethylammonium) salt (S6900).

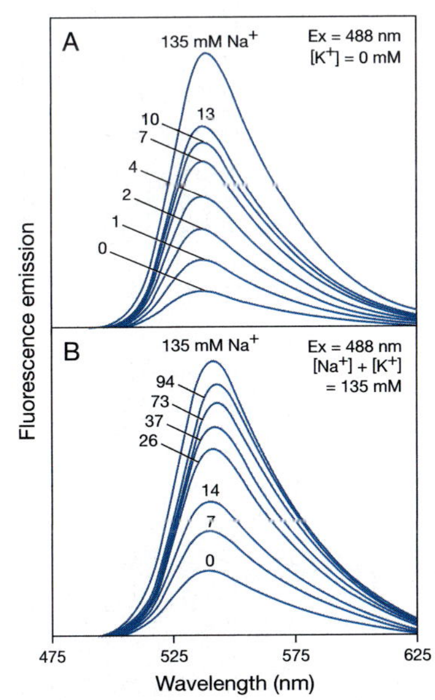

Figure 21.1.9 Emission spectral response of the Sodium Green™ indicator (S6900) to Na^+: **A)** in K^+-free solution and **B)** in solutions containing K^+ with the combined Na^+ and K^+ concentration equal to 135 mM. The scale on the vertical axis is the same for both panels.

CoroNa™ Na^+ Indicators

CoroNa™ Green Na^+ Indicator

The CoroNa™ Green dye is a green-fluorescent Na^+ indicator that exhibits an increase in fluorescence emission intensity upon binding Na^+ (excitation/emission = 492/516 nm), with little shift in wavelength (Figure 21.1.10). Similar to our SBFI and Sodium Green™ Na^+ indicators, the CoroNa™ Green indicator allows spatial and temporal resolution of Na^+ concentrations in the presence of physiological concentrations of other monovalent cations.[70–73] CoroNa™ Green Na^+ indicator has been co-loaded with Alexa Fluor® 594 dextran (an ion-insensitive reference) via suction pipettes into live rat optic nerves for confocal imaging of intracellular Na^+ levels; calcium measurements were also made using fluo-4 dextran and Alexa Fluor® 594 dextran.[74]

Comprising a fluorescein molecule linked to a crown ether with a cavity size that confers selectivity for the Na^+ ion (Figure 21.1.11), the CoroNa™ Green indicator is less than half the size of the Sodium Green™ indicator[75] (molecular weight 586 and 1668, respectively). This smaller size appears to help the cell-permeant CoroNa™ Green AM (Figure 21.1.12) load cells more effectively than the Sodium Green™ tetraacetate. Furthermore, the CoroNa™ Green indicator responds to a broader range of Na^+ concentration, with a K_d of ~80 mM. The cell-impermeant CoroNa™ Green indicator (C36675) is supplied in a unit size of 1 mg. The cell-permeant AM ester of the CoroNa™ Green indicator (C36676) is supplied as a set of 20 vials, each containing 50 µg of the indicator.

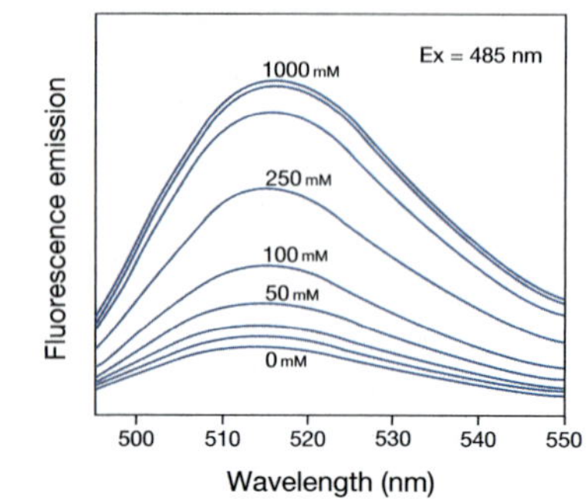

Figure 21.1.10 Fluorescence emission spectra of the CoroNa™ Green indicator (C36675, C36676) in 50 mM MOPS, pH 7.0 (adjusted with tetramethylammonium hydroxide), containing 100 mM K^+ and variable concentrations of Na^+ as indicated.

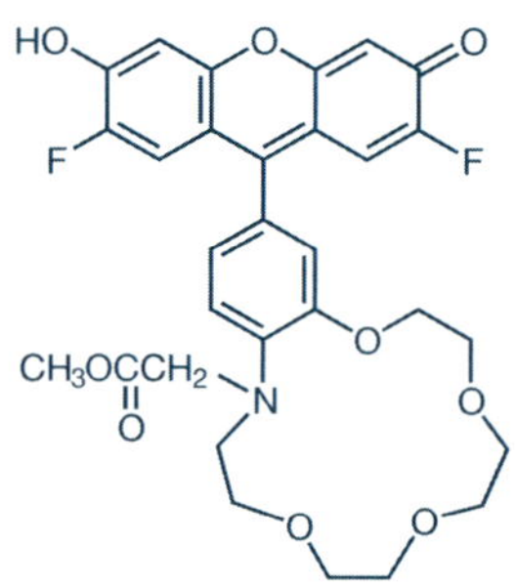

Figure 21.1.11 CoroNa™ Green (C36675).

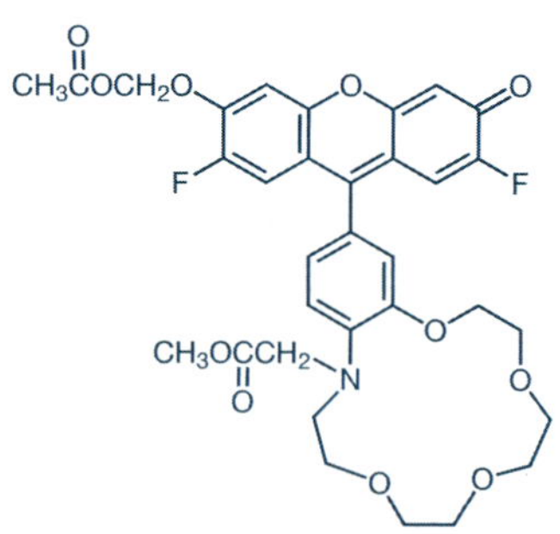

Figure 21.1.12 CoroNa™ Green, AM (C36676).

CoroNa™ Red Na^+ Indicator

CoroNa™ Red chloride is based on a crown ether that has structural similarity to the Ca^{2+} chelator BAPTA (Figure 21.1.13). Unlike SBFI and the Sodium Green™ indicator, the net positive charge of CoroNa™ Red chloride targets the indicator to mitochondria (Figure 21.1.14), and therefore loading of cells does not require use of a permeant ester derivative of the dye. Cells are typically loaded by adding 0.5–1.0 µM CoroNa™ Red chloride from a 1 mM stock solution in DMSO, incubating for 10–30 minutes at 37°C and finally washing with dye-free medium before commencing fluorescence analysis. The CoroNa™ Red indicator is only weakly fluorescent in the absence of Na^+ and its fluorescence increases ~15-fold upon binding Na^+ (Figure 21.1.15). Despite its relatively high K_d for Na^+ of ~200 mM, the CoroNa™ Red indicator exhibits sensitive responses to cellular Na^+ influxes through voltage-gated channels and ATP-gated cation pores. Verkman and co-workers have immobilized the CoroNa™ Red indicator on polystyrene microspheres and used this complex to measure Na^+ concentrations around 100 mM in the tracheal airway–surface liquid (ASL) of cultured epithelial cells and human lung tissues.[76,77] The CoroNa™ Red indicator has also been employed to investigate the Na^+ channel permeation pathway using polyhistidine-tagged and pore-only constructs of a voltage-dependent Na^+ channel.[78] The CoroNa™ Red indicator is available as a single 1 mg vial (C24430) or as a set of 20 vials, each containing 50 µg of the indicator (C24431).

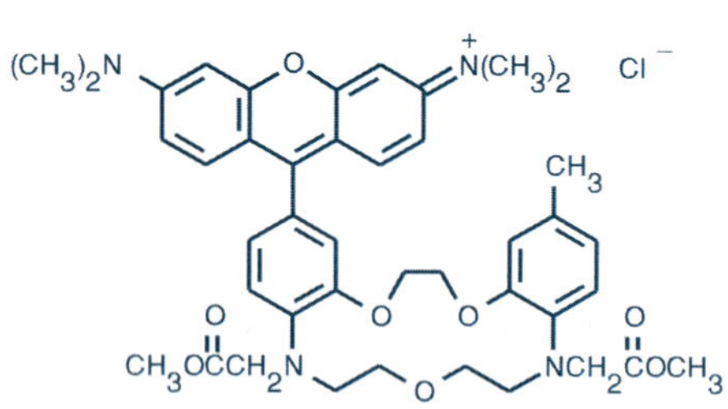

Figure 21.1.13 CoroNa™ Red chloride (C24430).

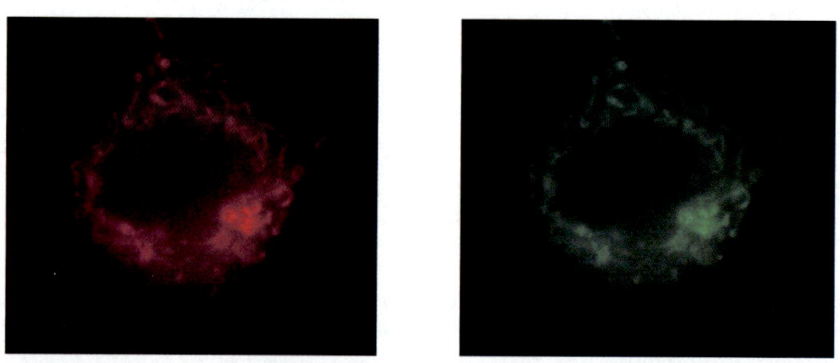

Figure 21.1.14 Images of an NIH 3T3 cell showing colocalization of the CoroNa™ Red sodium indicator (left panel; C24430, C24431) with the MitoTracker® Green FM® mitochondrial marker (right panel, M7514). A cell loaded with both dyes was imaged consecutively using Omega® Optical bandpass filter set XF41 for CoroNa™ Red sodium indicator and set XF23 for MitoTracker® Green FM®.

IMPORTANT NOTICE: The products described in this manual are covered by one or more Limited Use Label License(s). Please refer to the Appendix on page 971 and Master Product List on page 975. Products are For Research Use Only. Not intended for any animal or human therapeutic or diagnostic use.

FluxOR™ Potassium Ion Channel Assay

Assaying K^+ Channels with the FluxOR™ Potassium Ion Channel Assay Kit

The FluxOR™ Potassium Ion Channel Assay Kits (F10016, F10017) provide a fluorescence-based assay for high-throughput screening (HTS) of potassium ion channel and transporter activities.[79–81] The FluxOR™ Potassium Ion Channel Assay Kits take advantage of the well-described permeability of potassium channels to thallium (Tl^+) ions. When thallium is present in the extracellular solution containing a stimulus to open potassium channels, channel activity is detected with a cell-permeant thallium indicator dye that reports large increases in fluorescence emission at 525 nm as thallium flows down its concentration gradient and into the cells (Figure 21.1.16). In this way, the fluorescence reported in the FluxOR™ system becomes a surrogate indicator of activity for any ion channel or transporter that is permeable to thallium, including the human ether-a-go-go–related gene (hERG) channel, one of the human cardiac potassium channels. The FluxOR™ potassium ion channel assay has been validated for homogeneous high-throughput profiling of hERG channel inhibition using BacMam-mediated transient expression of hERG.[80] The FluxOR™ Potassium Ion Channel Assay Kits can also be used to study potassium co-transport processes that accommodate the transport of thallium into cells.[82] Furthermore, resting potassium channels and inward rectifier potassium channels like Kir2.1 can be assayed by adding stimulus buffer with thallium alone, without any depolarization to measure the signal.

The FluxOR™ reagent, a thallium indicator dye, is loaded into cells as a membrane-permeable AM ester. The FluxOR™ dye is dissolved in DMSO and further diluted with FluxOR™ assay buffer, a physiological HBSS (Hank's balanced salt solution), for loading into cells. Loading is assisted by the proprietary PowerLoad™ concentrate, an optimized formulation of nonionic Pluronic® surfactant polyols that act to disperse and stabilize AM ester dyes for optimal loading in aqueous solution. This PowerLoad™ concentrate is also available separately (P10020) to aid the solubilization of water-insoluble dyes and other materials in physiological media.

Once inside the cell, the nonfluorescent AM ester of the FluxOR™ dye is cleaved by endogenous esterases into a weakly fluorescent (basal fluorescence), thallium-sensitive indicator. The thallium-sensitive form is retained in the cytosol, and its extrusion is inhibited by water-soluble probenecid (P36400, Section 19.8), which blocks organic anion pumps. For most applications, cells are loaded with the dye at room temperature. For best results, the dye-loading buffer is then replaced with fresh, dye-free assay buffer (composed of physiological HBSS containing probenecid), and cells are ready for the HTS assay.

Each FluxOR™ Potassium Ion Channel Assay Kit contains:

- FluxOR™ reagent
- FluxOR™ assay buffer
- PowerLoad™ concentrate
- Probenecid
- FluxOR™ chloride-free buffer
- Potassium sulfate (K_2SO_4) concentrate
- Thallium sulfate (Tl_2SO_4) concentrate
- Dimethylsulfoxide (DMSO)
- Detailed protocols

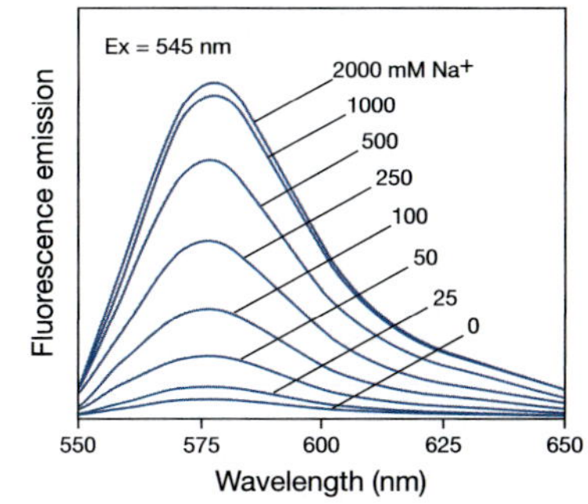

Figure 21.1.15 Fluorescence emission spectra of the CoroNa™ Red indicator (C24430, C24431) in 50 mM MOPS (pH 7.0, adjusted with tetramethylammonium hydroxide) containing 100 mM K^+ and variable concentrations of Na^+ as indicated.

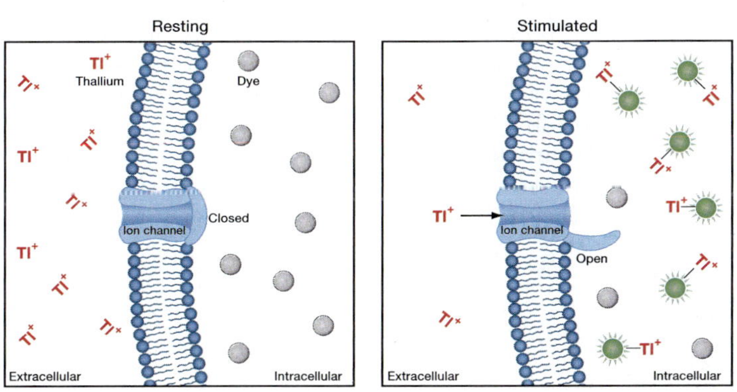

Figure 21.1.16 Thallium redistribution in the FluxOR™ assay. Basal fluorescence from cells loaded with FluxOR™ reagent (provided in the FluxOR™ Potassium Ion Channel Assay Kits; F10016, F10017) is low when potassium channels remain unstimulated, as shown in the left panel. When thallium is added to the assay with the stimulus, the thallium flows down its concentration gradient into the cells, activating the dye as shown in the right panel.

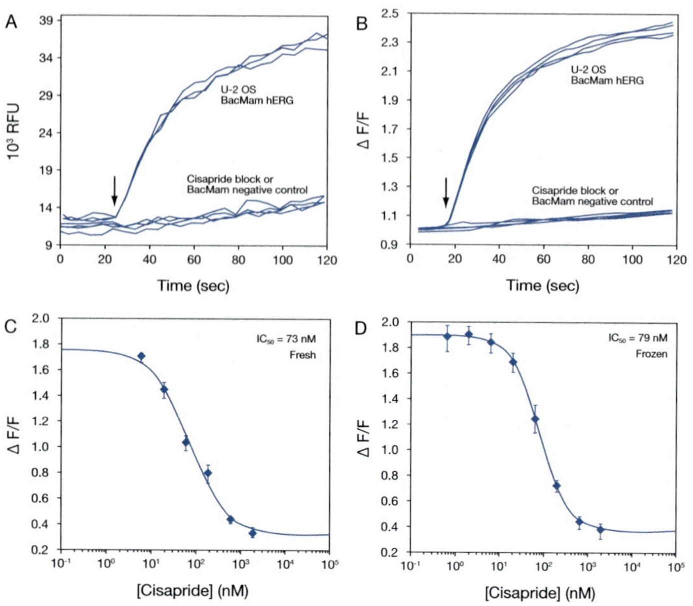

Figure 21.1.17 FluxOR™ potassium ion channel assays (F10016, F10017) performed on fresh and frozen U2OS cells transduced with the BacMam hERG reagent (B10019, B10033). **A)** Raw data (RFU = relative fluorescence units) obtained in the FluxOR™ assay determination of thallium flux in U2OS cells, which had been transduced with BacMam-hERG and kept frozen until the day of use. The arrow indicates the addition of the thallium/potassium stimulus, and upper and lower traces indicate data taken from the minimum and maximum doses of cisapride used in the determination of the dose-response curves. **B)** Raw pre-stimulus peak and baseline values were boxcar averaged and normalized to indicate the fold increase in fluorescence over time. **C)** Data generated in a dose-response determination of cisapride block on BacMam hERG expressed in U2OS cells freshly prepared from overnight expression after viral transduction. **D)** Parallel data obtained from cells transduced with BacMam-hERG, stored for 2 weeks in liquid nitrogen, thawed and plated 4 hours prior to running the assay. Error bars indicate standard deviation, n = 4 per determination.

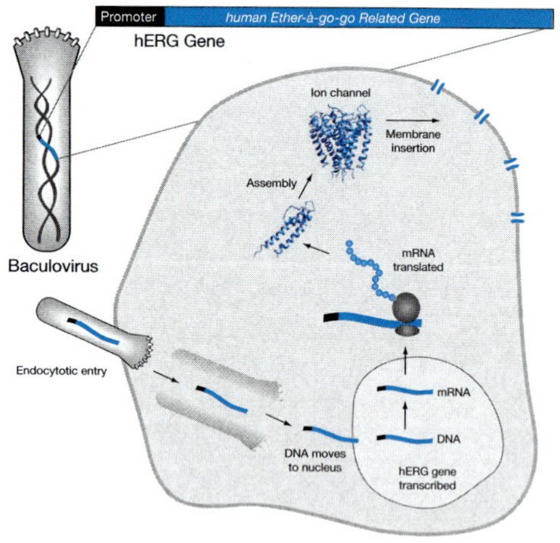

Figure 21.1.18 BacMam-hERG gene delivery and expression. This schematic depicts the mechanism of BacMam-mediated gene delivery into a mammalian cell and expression of the hERG gene (B10019, B10033). The hERG gene resides within the baculoviral DNA, downstream of a CMV promoter that drives its expression when introduced into a mammalian target cell. BacMam viral particles are taken up by endocytic pathways into the cell, and the DNA within them is released for transcription and expression. The translated protein is then folded for insertion into the membrane, forming functional hERG ion channels. This process begins within 4–6 hours and in many cell types is completed after an overnight period.

The FluxOR™ Kits provide a concentrated thallium solution along with sufficient dye and buffers to perform ~4000 (F10016) or ~40,000 (F10017) assays in a 384-well microplate format. These kits allow maximum target flexibility and ease of operation in a homogeneous format. The FluxOR™ potassium ion channel assay has been demonstrated for use with CHO and HEK 293 cells stably expressing hERG, as well as U2OS cells transiently transduced with the BacMam hERG reagent [80] (B10019, B10033; see below) (Figure 21.1.17). More information is available at www.invitrogen.com/handbook/fluxorpotassium.

Using BacMam Technology for Transient Expression of K^+ Channels

Potassium channel cDNAs that have been engineered into a baculovirus gene delivery/expression system using BacMam technology (BacMam Gene Delivery and Expression Technology—Note 11.1) are also available for use with the FluxOR™ Potassium Ion Channel Assay Kits, including the human ether-a-go-go related gene [80] (hERG) (Figure 21.1.18), several members of the voltage-gated K^+ channel (Kv) gene family and two members of the inwardly rectifying K^+ channel (Kir) gene family:

- BacMam hERG (for 10 microplates, B10019; for 100 microplates, B10033)
- BacMam Kv1.1 (for 10 microplates, B10331)
- BacMam Kv1.3 (for 10 microplates, B10332)
- BacMam Kv2.1 (for 10 microplates, B10333)
- BacMam Kv7.2 and Kv7.3 (for 10 microplates, B10147)
- BacMam Kir1.1 (for 10 microplates, B10334)
- BacMam Kir2.1 (for 10 microplates, B10146)

The BacMam system uses a modified insect cell baculovirus as a vehicle to efficiently deliver and express genes in mammalian cells with minimum effort and toxicity. The use of BacMam delivery in mammalian cells is relatively new, but well described, and has been used extensively in a drug discovery setting.[83] Constitutively expressed ion channels and other cell surface proteins have been shown to contribute to cell toxicity in some systems, and may be subject to clonal drift and other inconsistencies that hamper successful experimentation and screening. Thus, inducible, division-arrested or transient expression systems such as BacMam technology are increasingly methods of choice to decrease variability of expression in such assays.

U2OS cells (ATCC number HTB-96) have been shown to demonstrate highly efficient expression of BacMam-delivered targets in a null background ideal for screening in a heterologous expression system. The U2OS cell line is recommended for use if your particular cell line does not efficiently express the BacMam targets. Examples of other cell lines that are efficiently transduced by BacMam technology include HEK 293, HepG2, BHK, Cos-7 and Saos-2.

IMPORTANT NOTICE: The products described in this manual are covered by one or more Limited Use Label License(s). Please refer to the Appendix on page 971 and Master Product List on page 975. Products are For Research Use Only. Not intended for any animal or human therapeutic or diagnostic use.
www.invitrogen.com/probes
molecular probes® | invitrogen™ by life technologies™

REFERENCES

1. Nat Rev Drug Discov (2009) 8:982; **2.** Neuropharmacology (2009) 56:2; **3.** J Biol Chem (1989) 264:19449; **4.** Biophys J (1995) 68:2469; **5.** Methods Enzymol (1990) 192:38; **6.** Proc Natl Acad Sci U S A (1997) 94:7053; **7.** Am J Physiol Cell Physiol (2001) 280:C1623; **8.** J Physiol (1997) 498:295; **9.** Cell Regul (1990) 1:259; **10.** J Biol Chem (1989) 264:19458; **11.** J Mol Cell Cardiol (1997) 29:3375; **12.** J Neurosci (2001) 21:4207; **13.** Pflugers Arch (1999) 439:201; **14.** J Neurosci Methods (1997) 75:21; **15.** J Biol Chem (1990) 265:10522; **16.** J Appl Physiol (1996) 81:509; **17.** J Physiol (1992) 448:493; **18.** J Biol Chem (1995) 270:672; **19.** Am J Physiol (1992) 262:C1047; **20.** J Biol Chem (2004) 279:19421; **21.** J Immunol (2007) 178:7822; **22.** J Biol Chem (2006) 281:2232; **23.** J Physiol (2009) 587:5859; **24.** J Neurosci (2009) 29:7803; **25.** Biophys J (1997) 73:2476; **26.** J Neurochem (1998) 70:1513; **27.** Am J Physiol Heart Circ Physiol (2001) 280:H280; **28.** Am J Physiol (1997) 273:H1246; **29.** Am J Physiol Heart Circ Physiol (2000) 279:H1661; **30.** Am J Physiol Heart Circ Physiol (2000) 279:H2143; **31.** J Mol Cell Cardiol (1997) 29:2653; **32.** Am J Physiol (1996) 270:H2149; **33.** Biophys J (1997) 73:3371; **34.** Proc Natl Acad Sci U S A (1993) 90:8058; **35.** J Biol Chem (2008) 283:4602; **36.** Am J Physiol Cell Physiol (2000) 279:C1648; **37.** Am J Physiol Gastrointest Liver Physiol (2000) 278:G400; **38.** J Exp Bot (2005) 56:3149; **39.** J Pharmacol Exp Ther (2002) 300:9; **40.** J Biol Chem (2001) 276:13657; **41.** Curr Opin Cell Biol (2001) 13:405; **42.** J Biol Chem (2000) 275:19609; **43.** J Biol Chem (1999) 274:21953; **44.** J Biol Chem (1997) 272:32436; **45.** J Biol Chem (1997) 272:30567; **46.** Neurosci Lett (1997) 238:33; **47.** Am J Physiol (1999) 276:C717; **48.** Mol Pharmacol (1999) 56:619; **49.** J Biol Chem (2008) 283:36071; **50.** J Biol Chem (2008) 283:7219; **51.** J Biol Chem (2004) 279:32562; **52.** J Biol Chem (1998) 273:13578; **53.** J Biol Chem (1996) 271:8796; **54.** Brain Res (1994) 636:153; **55.** Circ Res (1994) 74:829; **56.** J Virol (1996) 70:5447; **57.** J Exp Bot (2003) 54:2035; **58.** Methods Enzymol (1999) 307:119; **59.** Cytometry (1995) 21:248; **60.** J Physiol (2000) 525:343; **61.** J Physiol (1999) 514:211; **62.** Hear Res (1998) 119:1; **63.** Brain Res (1994) 663:329; **64.** J Neurosci Methods (2002) 118:163; **65.** Anal Biochem (2000) 281:159; **66.** Anal Biochem (1997) 250:131; **67.** Methods Enzymol (1994) 240:723; **68.** Biophys J (2006) 90:357; **69.** J Biol Chem (2008) 283:15122; **70.** Neuron (2009) 61:259; **71.** J Neurochem (2009) 108:126; **72.** J Biol Chem (2008) 283:9377; **73.** J Neurosci Methods (2006) 155:251; **74.** J Neurosci (2009) 29:1796; **75.** Bioorg Med Chem Lett (2005) 15:1851; **76.** Proc Natl Acad Sci U S A (2001) 98:8119; **77.** J Clin Invest (2001) 107:317; **78.** J Biol Chem (2002) 277:24653; **79.** J Biomol Screen (2010) 15:441; **80.** Anal Biochem (2009) 394:30; **81.** Assay Drug Dev Technol (2008) 6:765; **82.** J Biol Chem (2009) 284:14020; **83.** Drug Discov Today (2007) 12:396.

DATA TABLE 21.1 FLUORESCENT Na^+ AND K^+ INDICATORS

				Low Ion *				High Ion *						
Cat. No.	MW	Storage	Soluble	Abs	EC	Em	Solvent	Abs	EC	Em	Solvent	Product †	K_d	Notes
C24430	773.32	L	DMSO	547	92,000	570	H_2O	551	92,000	576	H_2O/Na^+		200 mM	1, 2, 3, 4
C24431	773.32	L	DMSO	547	92,000	570	H_2O	551	92,000	576	H_2O/Na^+		200 mM	1, 2, 3, 4
C36675	585.56	F,D,L	pH >6	492	68,000	516	H_2O	492	68,000	516	H_2O/Na^+		80 mM	1, 2, 5, 6
C36676	657.62	F,D,L	DMSO	454	23,000	516	pH 7					C36675		
G6888	~1880	D	MeOH	<300	none									
P1265MP	950.99	L	pH >6	336	33,000	557	H_2O	338	41,000	507	H_2O/K^+		5.1 mM	1, 5, 7
P1267MP	1171.13	F,D,L	DMSO	369	37,000	see Notes	MeOH					P1265MP		8
S1262	906.94	L	pH >8	339	45,000	565	H_2O	333	52,000	539	H_2O/Na^+		3.8 mM	1, 5, 9
S1263	1127.07	F,D,L	DMSO	379	32,000	see Notes	MeOH					S1262		8
S1264	1127.07	F,D,L	DMSO	379	32,000	see Notes	MeOH					S1262		8
S6900	1667.57	L	pH >6	506	117,000	532	H_2O	507	133,000	532	H_2O/Na^+		6.0 mM	1, 2, 5, 9
S6901	1543.17	F,D,L	DMSO	302	21,000	none	MeOH					S6900		
V1644	1111.33	F,L	EtOH	<300		none	MeCN							

For definitions of the contents of this data table, see "Using *The Molecular Probes® Handbook*" in the introductory pages.

* For "Low Ion" spectra, the concentration of Na^+ or K^+ is zero. For "High Ion" spectra, the concentration of Na^+ or K^+ is in excess of that required to saturate the response of the indicator.

† Cat. No. of product generated *in situ* in typical intracellular application.

Notes

1. Dissociation constant values vary considerably depending on presence of other ions, temperature, pH, ionic strength, viscosity, protein binding and other factors. It is essential that the spectral response of the probe be calibrated in your system.
2. This indicator exhibits fluorescence enhancement in response to ion binding, with essentially no change in absorption or emission wavelengths.
3. K_d determined in 50 mM MOPS, pH 7.0 (adjusted with tetramethylammonium hydroxide) containing 40% DMSO at 22°C.
4. Spectra measured in aqueous buffers containing 40% DMSO and 0 M (H_2O) or 1 M Na^+ (H_2O/Na^+).
5. Spectra measured in aqueous buffers containing 0 M (H_2O) or a >10-fold excess of free cation X (H_2O/X) relative to the listed dissociation constant (K_d) for cation X.
6. $K_d(Na^+)$ determined in 50 mM MOPS, pH 7.0 (adjusted with tetramethylammonium hydroxide) at 22°C.
7. $K_d(K^+)$ has been determined in 10 mM MOPS, pH 7.0 (adjusted with tetramethylammonium hydroxide) at 22°C. $K_d(K^+)$ is strongly dependent on the concentration of Na^+. In solutions with $Na^+ + K^+$ = 135 mM, $K_d(K^+)$ = 44 mM.
8. Fluorescence of SBFI AM and PBFI AM is very weak.
9. $K_d(Na^+)$ has been determined in 10 mM MOPS, pH 7.0 (adjusted with tetramethylammonium hydroxide) at 22°C. Na^+ dissociation constants for these indicators are dependent on K^+ concentration. In solutions with total $Na^+ + K^+$ = 135 mM, $K_d(Na^+)$ = 11.3 mM (S1262) and 21 mM (S6900).

PRODUCT LIST 21.1 FLUORESCENT Na^+ AND K^+ INDICATORS

Cat. No.	Product	Quantity
B10019	BacMam-hERG *for 10 microplates*	1 kit
B10033	BacMam-hERG *for 100 microplates*	1 kit
B10334	BacMam Kir1.1 *for 10 microplates*	1 kit
B10146	BacMam Kir2.1 *for 10 microplates*	1 kit
B10331	BacMam Kv1.1 *for 10 microplates*	1 kit
B10332	BacMam Kv1.3 *for 10 microplates*	1 kit
B10333	BacMam Kv2.1 *for 10 microplates*	1 kit
B10147	BacMam Kv7.2 and Kv7.3 *for 10 microplates*	1 kit
C36675	CoroNa™ Green *cell impermeant*	1 mg
C36676	CoroNa™ Green, AM *cell permeant* *special packaging*	20 x 50 µg
C24430	CoroNa™ Red chloride	1 mg
C24431	CoroNa™ Red chloride *special packaging*	20 x 50 µg
F10016	FluxOR™ Potassium Ion Channel Assay *for 10 microplates*	1 kit
F10017	FluxOR™ Potassium Ion Channel Assay *for 100 microplates*	1 kit
G6888	gramicidin	100 mg
P1267MP	PBFI, AM *cell permeant* *special packaging*	20 x 50 µg
P1265MP	PBFI, tetraammonium salt *cell impermeant*	1 mg
P6866	Pluronic® F-127 *10% solution in water* *0.2 µm filtered*	30 mL
P3000MP	Pluronic® F-127 *20% solution in DMSO*	1 mL
P6867	Pluronic® F-127 *low UV absorbance*	2 g
P10020	PowerLoad™ concentrate, 100X	5 mL
S1263	SBFI, AM *cell permeant*	1 mg
S1264	SBFI, AM *cell permeant* *special packaging*	20 x 50 µg
S1262	SBFI, tetraammonium salt *cell impermeant*	1 mg
S6901	Sodium Green™ tetraacetate *cell permeant* *special packaging*	20 x 50 µg
S6900	Sodium Green™, tetra(tetramethylammonium) salt *cell impermeant*	1 mg
V1644	valinomycin	25 mg

IMPORTANT NOTICE: The products described in this manual are covered by one or more Limited Use Label License(s). Please refer to the Appendix on page 971 and Master Product List on page 975. Products are For Research Use Only. Not intended for any animal or human therapeutic or diagnostic use.

molecular probes® | invitrogen™ by life technologies™

21.2 Detecting Chloride, Phosphate, Nitrite and Other Anions

This section describes fluorescent indicators for intracellular and extracellular chloride together with an assortment of analytical reagents and methods for direct or indirect quantitation of other inorganic anions, including bromide, iodide, hypochlorite, cyanide, nitrite, nitrate, phosphate, pyrophosphate and selenide.[1]

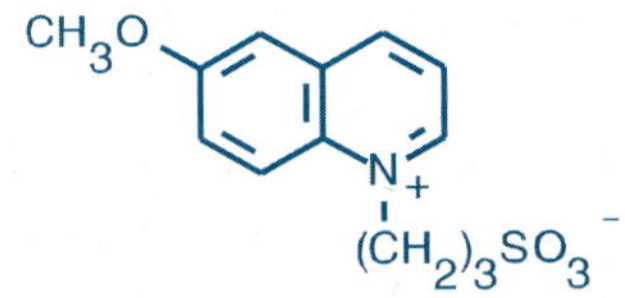

Figure 21.2.1 6-methoxy-*N*-(3-sulfopropyl)quinolinium, inner salt (SPQ, M440).

Fluorescent Chloride Indicators

Most of the fluorescent chloride indicators are 6-methoxyquinolinium derivatives, the prototype of which is 6-methoxy-*N*-(3-sulfopropyl)quinolinium[2,3] (SPQ, Figure 21.2.1). Cl^- detection sensitivity has been improved by modifications of the quinolinium *N* substituent.[4,5] Our current range of Cl^- indicators consists of:

- 6-Methoxy-*N*-(3-sulfopropyl)quinolinium (SPQ, M440)
- *N*-(Ethoxycarbonylmethyl)-6-methoxyquinolinium bromide (MQAE, E3101)
- 6-Methoxy-*N*-ethylquinolinium iodide (MEQ, M6886)
- Lucigenin (L6868)

All of these indicators detect Cl^- via diffusion-limited collisional quenching.[6] This detection mechanism is different from that of fluorescent indicators for Ca^{2+}, Mg^{2+}, Zn^{2+}, Na^+ and K^+. It involves a transient interaction between the excited state of the fluorophore and a halide ion—no ground-state complex is formed. Quenching is not accompanied by spectral shifts (Figure 21.2.2) and, consequently, ratio measurements are not directly feasible. Quenching by other halides, such as Br^- and I^-, and other anions, such as thiocyanate, is more efficient than Cl^- quenching.[6] Fortunately, physiological concentrations of non-choloride ions do not significantly affect the fluorescence of SPQ and other methoxyquinolinium-based Cl^- indicators. With some exceptions,[7] fluorescence of these indicators is not pH sensitive in the physiological range.[4] Because Cl^--dependent fluorescence quenching is a diffusional process, it is quite sensitive to solution viscosity and volume. Exploiting this property, SPQ has been used to measure intracellular volume changes.[8]

The efficiency of collisional quenching is characterized by the Stern–Volmer constant (K_{SV}), defined as the reciprocal of the ion concentration that produces 50% of maximum quenching. For SPQ, K_{SV} is reported to be 118 M^{-1} in aqueous solution and 12 M^{-1} inside cells.[9] For MQAE, *in situ* K_{SV} values of 25–28 M^{-1} have been determined in various cell types,[10,11] compared with the solution value of 200 M^{-1}. Intracellular Cl^- indicators are generally calibrated using high-K^+ buffers and the K^+/H^+ ionophore nigericin (N1495) in conjunction with tributyltin chloride, an organometallic compound that acts as a Cl^-/OH^- antiporter.[4,12] With the exception of diH-MEQ, Cl^- indicators must be loaded into cells by long-term incubation (up to eight hours) in the presence of a large excess of dye or by brief hypotonic permeabilization. Because membranes are slightly permeable to the indicator, rapid leakage may occur. Experimentally determined estimates of leakage vary quite widely.[10–12]

Measurement of intracellular Cl^- concentrations and the study of Cl^- channels have been stimulated by the discovery that cystic fibrosis is caused by mutations in a gene encoding a Cl^- transport channel, which is known as the cystic fibrosis transmembrane conductance regulator[13] (CFTR). Cl^- permeability assays are used to detect activity of the CFTR and other anion transporters.[14–17] In these assays, SPQ- or MQAE-loaded cells are successively perfused with chloride-containing extracellular medium followed by medium in which the Cl^- content is replaced by nitrate (NO_3^-). NO_3^- is used in this assay protocol because it produces no fluorescence quenching of the indicator, yet its channel permeability is essentially the same as that of Cl^-[14,15] (Figure 21.2.3).

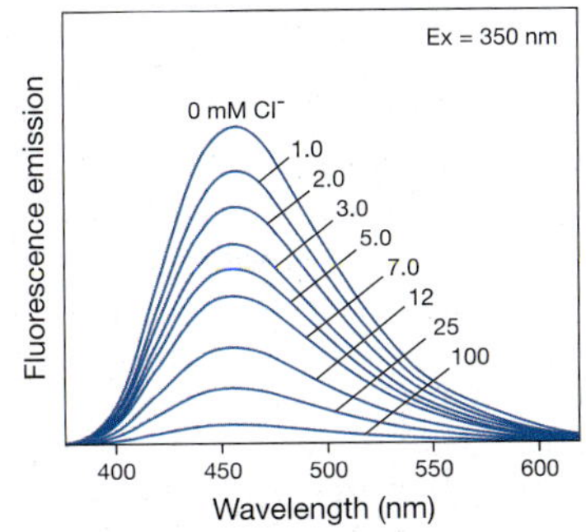

Figure 21.2.2 Fluorescence emission spectra of MQAE (E3101) in increasing concentrations of Cl^-.

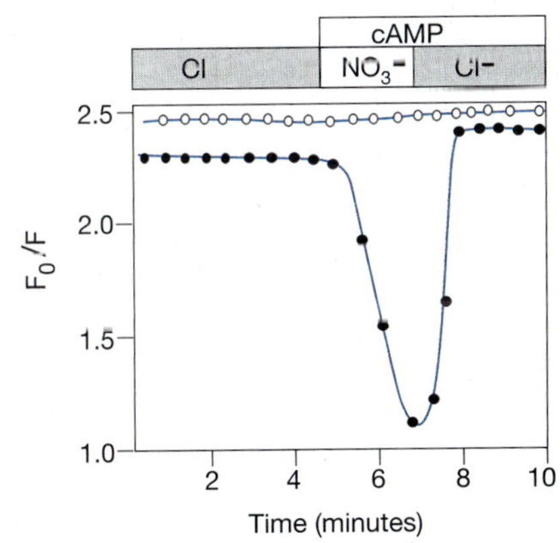

Figure 21.2.3 Detection of cystic fibrosis transmembrane conductance regulator (CFTR) activity using 6-methoxy-*N*-(3-sulfopropyl)quinolinium, inner salt (SPQ, M440). Fluorescence of intracellular SPQ is quenched by collision with chloride ions, indicated by $F_0/F>1$ (F_0 = fluorescence intensity in absence of chloride, F = fluorescence intensity at time points indicated on the x-axis). Upon addition of cyclic AMP to initiate channel opening, and exchange of extracellular Cl^- (135 mM) for nitrate (NO_3^-), SPQ quenching decreases in CFTR-expressing cells (filled circles) as CFTR-mediated anion transport results in replacement of intracellular Cl^- with nonquenching NO_3^-. Control cells with no CFTR expression (open circles) show no response.

IMPORTANT NOTICE: The products described in this manual are covered by one or more Limited Use Label License(s). Please refer to the Appendix on page 971 and Master Product List on page 975. Products are For Research Use Only. Not intended for any animal or human therapeutic or diagnostic use.

SPQ

SPQ (M440, Figure 21.2.1) is currently in widespread use for detecting CFTR activity using the Cl^-/NO_3^- exchange technique described above.[16,18–24] SPQ has also has been employed to investigate Cl^- fluxes through several other transporters such as the $GABA_A$ receptor,[25,26] erythrocyte Cl^-/HCO_3^- exchangers[27,28] and the mitochondrial uncoupling protein.[29–31] Although SPQ requires UV excitation (as do MQAE and MEQ), techniques for flow cytometric detection and calibration of the indicator using argon-ion laser excitation at 351 nm and 364 nm have been successfully demonstrated.[12]

MQAE

MQAE (E3101, Figure 21.2.4) has greater sensitivity to Cl^-[4,5] and a higher fluorescence quantum yield than SPQ; consequently, it is currently the more widely used of the two indicators. However, the ester group of MQAE may slowly hydrolyze inside cells, resulting in a change in its fluorescence response.[32] MQAE has been used in a fluorescence-based microplate assay that has potential for screening compounds that modify Cl^- ion-channel activity.[10] Other applications have included Cl^- measurements in cytomegalovirus-infected fibroblasts,[33] smooth muscle cells[32] and salivary glands,[17] as well as in reconstituted membranes containing the $GABA_A$ receptor[26] or the mitochondrial-uncoupling protein[34,35] (UCP-1).

MEQ and Cell-Permeant Dihydro-MEQ

The Cl^- indicator 6-methoxy-*N*-ethylquinolinium iodide (MEQ) can be rendered cell-permeant by masking its positively charged nitrogen to create a lipophilic, Cl^--insensitive compound, 6-methoxy-*N*-ethyl-1,2-dihydroquinoline[36] (dihydro-MEQ). This reduced quinoline derivative can then be loaded noninvasively into cells, where it is rapidly reoxidized in most cells to the cell-impermeant, Cl^--sensitive MEQ (Figure 21.2.5). Using this technique, researchers have loaded live brain slices and hippocampal neurons with MEQ for confocal imaging of Cl^- responses to $GABA_A$ receptor activation and glutamatergic excitotoxicity.[37–41] Quenching of intracellular MEQ fluorescence by Cl^- has a K_{SV} of 19 M^{-1}, a value that is slightly higher than that reported for SPQ in fibroblasts. MEQ is available in solid form (M6886) and is supplied with a simple protocol for reducing it to dihydro-MEQ with sodium borohydride (not supplied) just prior to cell loading.

Lucigenin

The fluorescence of lucigenin (L6868, Figure 21.2.6) is quantitatively quenched by high levels of Cl^- with a reported $K_{SV} = 390\ M^{-1}$.[42] Lucigenin absorbs maximally at both 368 nm (EC = 36,000 $cm^{-1}M^{-1}$) and 455 nm (EC = 7400 $cm^{-1}M^{-1}$), with an emission maximum at 505 nm. Its fluorescence emission has a quantum yield of ~0.6 and is insensitive to nitrate, phosphate and sulfate. Lucigenin is a useful Cl^- indicator in liposomes and reconstituted membrane vesicles; however, because its fluorescence is reported to be unstable in the cytoplasm, it may not always be suitable for determining intracellular Cl^-.[42] Lucigenin has been used to detect chloride uptake in tonoplast vesicles[43] and to measure Cl^- influx across the pleural surface in perfused mouse lungs.[44]

Alternative Detection Techniques for Halides

As mentioned above, the fluorescence of SPQ and related Cl^- indicators is quenched by collision with a variety of anions, including (in order of increasing quenching efficiency) Cl^-, Br^-, I^- and thiocyanate[45] (SCN^-). For example, fluorescence of SPQ is partially quenched by the anionic pH buffer TES (*N-tris*(hydroxymethyl)methyl-2-aminoethanesulfonic acid) but not by the protonated TES zwitterion, a property that has been exploited to measure proton efflux from proteoliposomes.[31,46] Anion detectability using diffusional fluorescence quenching of these fluorophores is typically limited to the millimolar range. I^- quenches many other fluorophores and is commonly used to determine the accessibility of fluorophores to quenching in proteins and membranes.[47,48]

In addition, halides can be oxidized to hypohalites (^-OCl, ^-OBr, ^-OI), which react with rhodamine 6G (R634, Section 12.2) to yield chemiluminescent products.[49,50] A cell produces ^-OCl by oxidizing Cl^- within

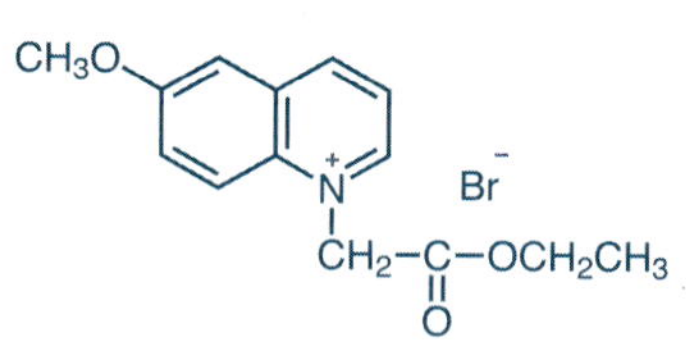

Figure 21.2.4 *N*-(ethoxycarbonylmethyl)-6-methoxyquinolinium bromide (MQAE, E3101).

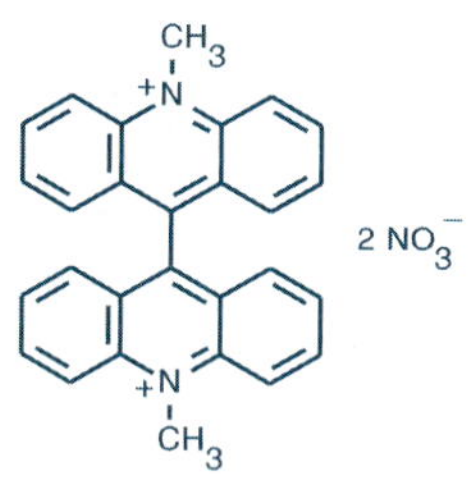

Figure 21.2.6 Lucigenin (bis-*N*-methylacridinium nitrate, L6868).

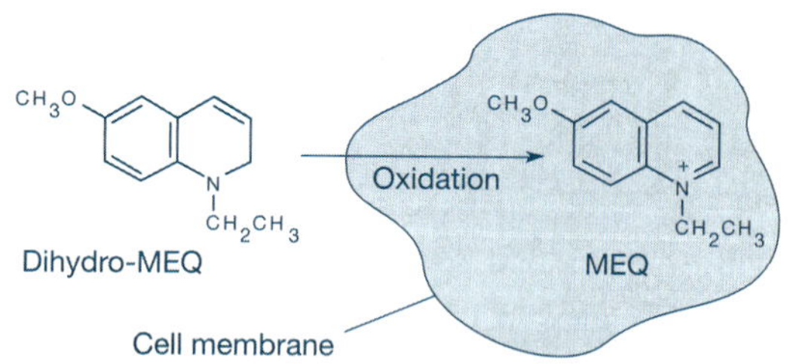

Figure 21.2.5 Intracellular delivery of the fluorescent chloride indicator 6-methoxy-*N*-ethylquinolinium iodide (MEQ, M6886), via oxidation of the membrane-permeant precursor dihydro-MEQ.

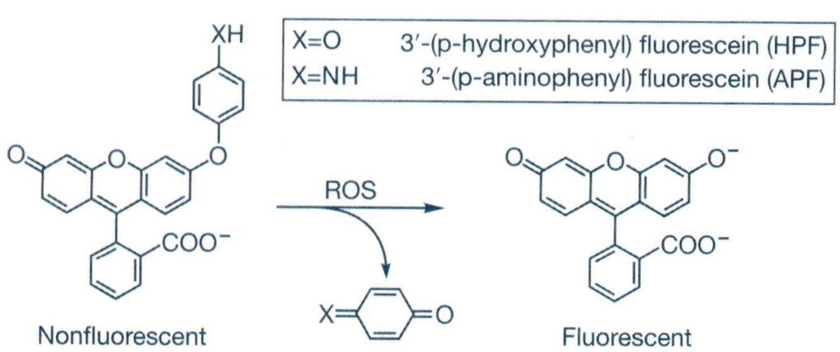

Figure 21.2.7 Detection of reactive oxygen species (ROS) with 3′-(*p*-hydroxyphenyl) fluorescein (HPF, H36004) and 3′-(*p*-aminophenyl) fluorescein (APF, A36003).

IMPORTANT NOTICE: The products described in this manual are covered by one or more Limited Use Label License(s). Please refer to the Appendix on page 971 and Master Product List on page 975. Products are For Research Use Only. Not intended for any animal or human therapeutic or diagnostic use.
www.invitrogen.com/probes
molecular probes® | invitrogen™ by life technologies™

the phagovacuole.[51,52] $^-$OCl also reacts with fluorescein (F1300, Section 1.5) to yield fluorescent products,[53] permitting analysis of $^-$OCl levels in water.

Alternatively, 3′-(*p*-aminophenyl) fluorescein (APF) and 3′-(*p*-hydroxyphenyl) fluorescein (HPF) (A36003, H36004; Section 18.2) can be used for the selective detection of $^-$OCl. Both of these fluorescein derivatives are essentially nonfluorescent until they react with the hydroxyl radical (HO·) or peroxynitrite anion ($ONOO^-$) (Figure 21.2.7). APF will also react with the hypochlorite anion ($^-$OCl), making it possible to use APF and HPF together to selectively detect hypochlorite anion. In the presence of these specific ROS, both APF and HPF yield a bright green-fluorescent product (excitation/emission maxima ~490/515 nm) and are compatible with all fluorescence instrumentation capable of visualizing fluorescein. Using APF, researchers have been able to detect the $^-$OCl generated by activated neutrophils, a feat that has not been possible with traditional ROS indicators.[54]

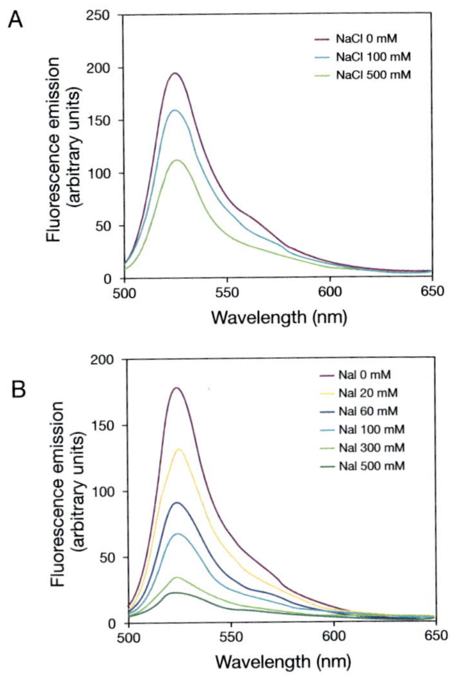

Figure 21.2.9 Quenching of Premo™ Halide Sensor fluorescence by increasing concentrations of iodide and chloride. U2OS cells were transduced with Premo™ Halide Sensor. After 24 hours, cells were trypsinized and lysed by resuspension in sterile distilled water. Fluorescence quenching of the lysate was examined using increasing concentrations of NaCl (**A**) and NaI (**B**). Iodide induces substantially greater quenching of Premo™ Halide Sensor fluorescence than chloride.

Premo™ Halide Sensor

The fluorescent protein–based Premo™ Halide Sensor (P10229) is a pharmacologically relevant sensor for functional studies of ligand- and voltage-gated chloride channels and their modulators in cells. Chloride channels are involved in cellular processes as critical and diverse as transepithelial ion transport, electrical excitability, cell volume regulation and ion homeostasis. Given their physiological significance, it follows that defects in their activity can have severe implications, including such conditions as cystic fibrosis and neuronal degeneration. Thus, chloride channels represent important targets for drug discovery.[55]

The Premo™ Halide Sensor combines a Yellow Fluorescent Protein (YFP) variant sensitive to halide ions with the efficient and noncytopathic BacMam delivery and expression technology (BacMam Gene Delivery and Expression Technology—Note 11.1), yielding a highly sensitive, robust and easy-to-use tool for efficiently screening halide ion channels and transporter modulators in their cellular models of choice. The Premo™ Halide Sensor is based on the Venus variant of *Aequorea victoria* Green Fluorescent Protein (GFP), which displays enhanced fluorescence, increased folding, and reduced maturation time when compared with YFP.[56] Additional mutations (H148Q and I152L) were made within the Venus sequence to increase the sensitivity of the Venus fluorescent protein to changes in local halide concentration, in particular iodide ions.[57] Because chloride channels are also permeable to the iodide ion (I), iodide can be used as a surrogate of chloride. Upon stimulation, a chloride channel or transporter opens and iodide flows down the concentration gradient into the cells, where it quenches the fluorescence of the expressed Premo™ Halide Sensor protein (Figure 21.2.8). The decrease in Premo™ Halide Sensor fluorescence is directly proportional to the ion flux, and therefore the chloride channel or transporter activity. The Premo™ Halide Sensor shows a similar excitation and emission profile to YFP (Figure 21.2.9) and can be detected using standard GFP/FITC or YFP filter sets. Halide-sensitive YFP-based constructs in conjunction with iodide quenching have been used in high-throughput screening (HTS) to identify modulators of calcium-activated chloride channels.[58]

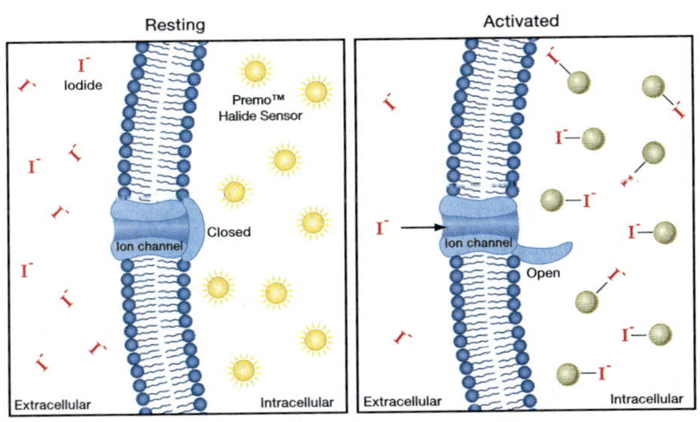

Figure 21.2.8 Principle of Premo™ Halide Sensor Sensor (P10229): Iodide redistribution upon chloride channel activation. Basal fluorescence from Premo™ Halide Sensor is high when chloride channels are low. Upon activation (opening) of chloride channels, the iodide ions enter the cell, down its concentration gradient, and quench the fluorescence from Premo™ Halide Sensor.

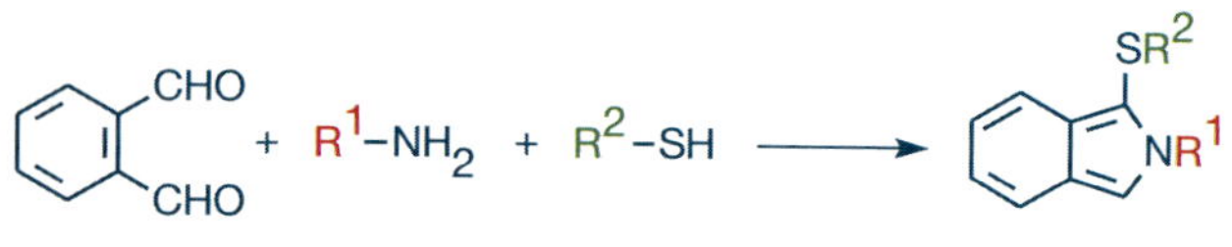

OPA

Figure 21.2.10 Fluorogenic amine-derivatization reaction of *o*-phthaldialdehyde (OPA, P2331MP).

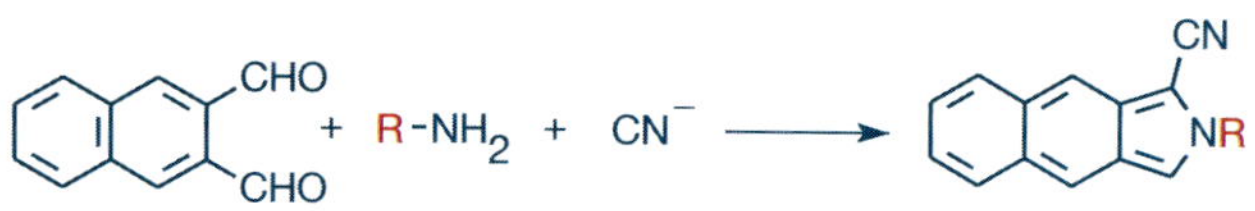

NDA

Figure 21.2.11 Fluorogenic amine-derivatization reaction of naphthalene-2,3-dicarboxaldehyde (NDA, N1138).

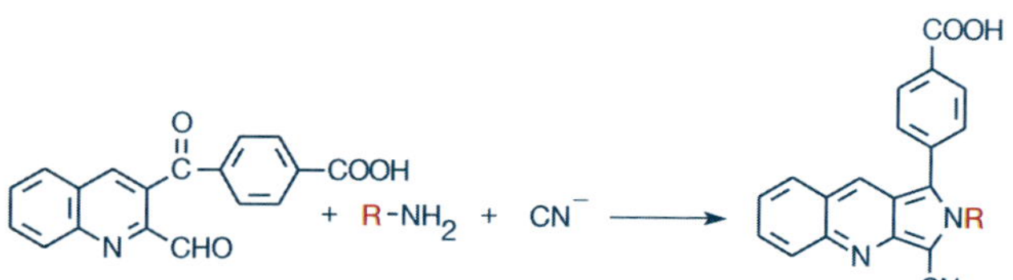

CBQCA

Figure 21.2.12 Fluorogenic amine-derivatization reaction of CBQCA (A6222, A2333).

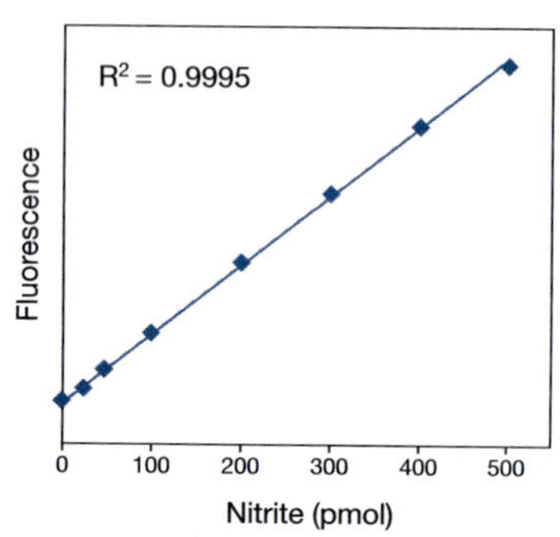

Figure 21.2.13 Linearity and sensitivity of the Measure-iT™ high-sensitivity nitrite assay. Triplicate 10 µL samples of nitrite were assayed using the Measure-iT™ High-Sensitivity Nitrite Assay Kit (M36051). Fluorescence was measured using excitation/emission of 365/450 nm and plotted versus picomoles of nitrite. Background fluorescence was not subtracted. The variation (CV) of replicate samples was <2%.

The Premo™ Halide Sensor (P10229) is pre-packaged and ready for immediate use. It contains all components required for cellular delivery and expression—including the baculovirus carrying the genetically encoded biosensor, BacMam enhancer and stimulus buffer containing iodide—in ten 96- or 384-well plates. The Premo™ Halide Sensor has been demonstrated to transduce multiple cell lines including BHK, U2OS, HeLa, CHO, and primary human bronchial epithelial cells (HBEC), providing the flexibility to assay chloride-permeable channels in a wide range of cellular models. More information is available at www.invitrogen.com/handbook/premohalide.

Cyanide Detection

The homologous aromatic dialdehydes, *o*-phthaldialdehyde [59] (OPA, P2331MP) and naphthalene-2,3-dicarboxaldehyde [60] (NDA, N1138), are essentially nonfluorescent until reacted with a primary amine in the presence of excess cyanide or a thiol, such as 2-mercaptoethanol, 3-mercaptopropionic acid or the less obnoxious sulfite,[61] to yield a fluorescent isoindole (Figure 21.2.10, Figure 21.2.11). Modified protocols that use an excess of an amine and limiting amounts of other nucleophiles permit the determination of cyanide in blood, urine and other samples.[62–65]

We also offer the ATTO-TAG™ CBQCA (A6222) and ATTO-TAG™ FQ (A10192) reagents, which are similar to OPA and NDA in that they react with primary amines in the presence of cyanide or thiols to form highly fluorescent isoindoles [66–74] (Figure 21.2.12). The ATTO-TAG™ CBQCA and ATTO-TAG™ FQ reagents should also be useful for detecting cyanide in a variety of biological samples.

We have found that our Thiol and Sulfide Quantitation Kit (T6060, Section 2.1) also provides an ultrasensitive enzymatic assay for cyanide, with a detection limit of ~5 nanomoles. In this case, interference would be expected from thiols, sulfides, sulfites and other reducing agents.

Nitrite, Nitrate and Nitric Oxide Detection

With the discovery of the role of nitric oxide in signal transduction (Section 18.3), assays for nitrite (NO_2^-) have assumed new importance. Because inorganic nitrite is spontaneously produced by air oxidation of nitric oxide, the same reagents that have been utilized to detect nitric oxide production in cells should be useful for detecting nitrite in aqueous samples. Furthermore, inorganic nitrate (NO_3^-) can be reduced to NO_2^- by both chemical and enzymatic means, permitting the quantitative analysis of NO_3^- in samples.

Measure-iT™ High-Sensitivity Nitrite Assay Kit

The Measure-iT™ High-Sensitivity Nitrite Assay Kit (M36051) provides an easy and accurate method for quantitating nitrite. This kit has an optimal range of 20–500 picomoles nitrite (Figure 21.2.13), making it up to 50 times more sensitive than colorimetric methods utilizing the Griess reagent. Nitrates may be analyzed after quantitative conversion to nitrites through enzymatic reduction.[75]

Each Measure-iT™ High-Sensitivity Nitrite Assay Kit contains:

- Measure-iT™ nitrite quantitation reagent (100X concentrate in 0.62 M HCl)
- Measure-iT™ nitrite quantitation developer (2.8 M NaOH)
- Measure-iT™ nitrite quantitation standard (11 mM sodium nitrite)
- Detailed protocols

IMPORTANT NOTICE: The products described in this manual are covered by one or more Limited Use Label License(s). Please refer to the Appendix on page 971 and Master Product List on page 975. Products are For Research Use Only. Not intended for any animal or human therapeutic or diagnostic use.
www.invitrogen.com/probes
molecular probes® | invitrogen™ by life technologies™

Simply dilute the reagent 1:100, load 100 µL into the wells of a microplate, add 1–10 µL sample volumes and mix. After a 10-minute incubation at room temperature, add 5 µL of developer and read the fluorescence. The assay signal is stable for at least 3 hours, and common contaminants are well tolerated in the assay. The Measure-iT™ High-Sensitivity Nitrite Assay Kit provides sufficient material for 2000 assays, based on a 100 µL assay volume in a 96-well microplate format; this nitrite assay can also be adapted for use in cuvettes or 384-well microplates.

Griess Reagent Kit

Under physiological conditions, NO is readily oxidized to NO_2^- and NO_3^- or it is trapped by thiols as an *S*-nitroso adduct. The Griess reagent provides a simple and well-characterized colorimetric assay for nitrites—and nitrates that have been reduced to nitrites—with a detection limit of about 100 nM.[75–77] The Griess assay is suitable for measuring the activity of nitrate reductase in a microplate.[78] Nitrite reacts with the Griess reagent to form a purple azo derivative that can be monitored by absorbance at 548 nm (Figure 21.2.14).

The Griess Reagent Kit (G7921) contains all of the reagents required for NO_2^- quantitation, including:

- *N*-(1-Naphthyl)ethylenediamine dihydrochloride
- Sulfanilic acid in 5% H_3PO_4
- A concentrated nitrite quantitation standard for generating calibration curves
- Detailed protocols for spectrophotometer- and microplate reader–based assays

Both the *N*-(1-naphthyl)ethylenediamine dihydrochloride and the sulfanilic acid in 5% H_3PO_4 are provided in convenient dropper bottles for easy preparation of the Griess reagent. Sample pretreatment with nitrate reductase and glucose 6-phosphate dehydrogenase is reported to reduce NO_3^- without producing excess NADPH, which can interfere with the Griess reaction.[79] NO that has been trapped as an *S*-nitroso derivative can also be analyzed with the Griess Reagent Kit after first releasing the NO from its complex using mercuric chloride or copper (II) acetate.[80,81]

DAF-FM Reagent

DAF-FM[82] (4-amino-5-methylamino-2′,7′-difluorofluorescein, D23841; Figure 21.2.15) and its diacetate derivative (DAF-FM diacetate, D23842, D23844; Section 18.3) have significant utility for measuring nitric oxide and nitrite production in live cells and solutions. The fluorescence quantum yield of DAF-FM is reported to be 0.005 but increases about 160-fold to 0.81 after reacting with nitrite[82] (Figure 21.2.16). DAF-FM has some important advantages over the similar nitric oxide sensor, DAF-2, and other aromatic diamines:

- Spectra of the NO (NO_2^-) adduct of DAF-FM are independent of pH above pH 5.5.[83]
- NO_2^- adduct of DAF-FM is significantly more photostable than that of DAF-2.[82]
- DAF-FM is a more sensitive reagent for NO_2^- than is DAF-2; the NO and NO_2^- detection limit for DAF-FM is ~3 nM[82] versus ~5 nM for DAF-2.[83]
- The higher absorptivity and greater water solubility of the NO_2^- adduct of DAF-FM should make this assay much more sensitive than detection with 2,3-diaminonaphthalene or other aromatic diamines.

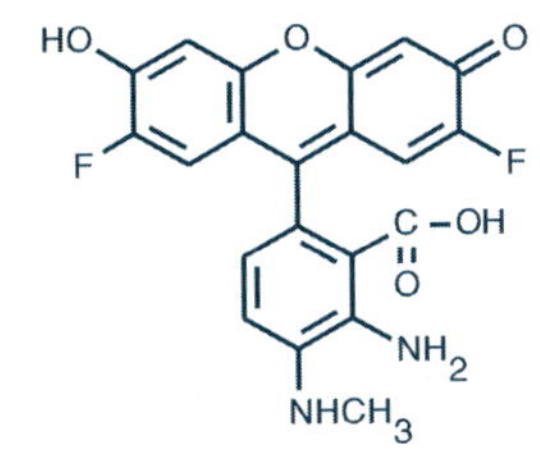

Figure 21.2.15 DAF-FM (4-amino-5-methylamino-2′,7′-difluorofluorescein, D23841).

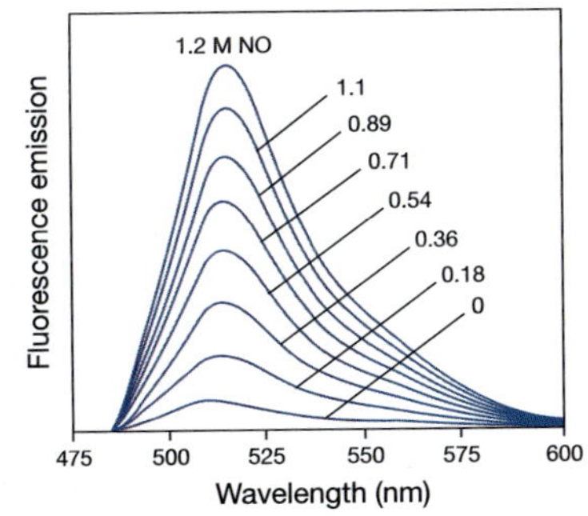

Figure 21.2.16 Fluorescence emission spectra of DAF-FM (D23841, D23842, D23844) in solutions containing 0–1.2 µM nitric oxide (NO).

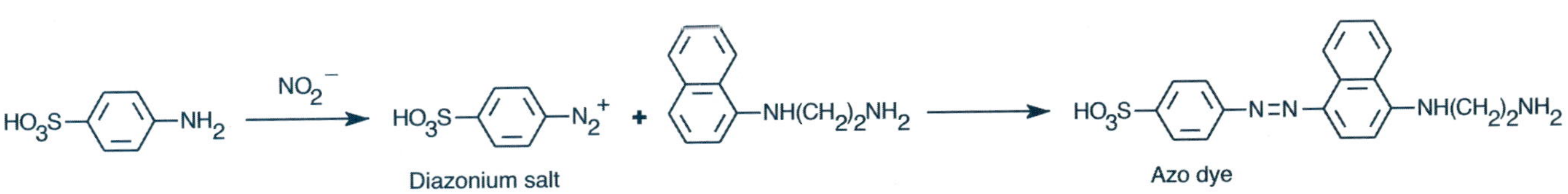

Figure 21.2.14 Principle of nitrite quantitation using the Griess Reagent Kit (G7921). Formation of the azo dye is detected via its absorbance at 548 nm.

IMPORTANT NOTICE: The products described in this manual are covered by one or more Limited Use Label License(s). Please refer to the Appendix on page 971 and Master Product List on page 975. Products are For Research Use Only. Not intended for any animal or human therapeutic or diagnostic use.

Because the reaction of DAF-FM with NO requires a preliminary nonspecific oxidation step, it is important to also perform control experiments with nitric oxide synthase inhibitors to confirm the source of the fluorescent species.[84]

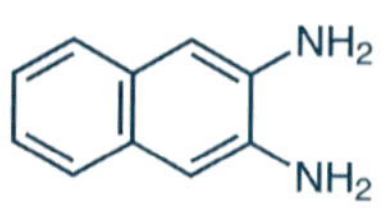

Figure 21.2.17 2,3-diaminonaphthalene (D7918).

2,3-Diaminonaphthalene

We also offer 2,3-diaminonaphthalene (D7918, Figure 21.2.17), which reacts with NO_2^- to form the fluorescent product 1*H*-naphthotriazole. A rapid, quantitative fluorometric assay that employs 2,3-diaminonaphthalene can reportedly detect from 10 nM to 10 µM NO_2^-, and is compatible with a 96-well microplate format.[85] Nitrate (NO_3^-) does not interfere with this assay; however, NO_3^- can be reduced to NO_2^- by bacterial nitrate reductase and then detected using the same reagent.[86] A detailed protocol for measuring the stable products of the nitric oxide pathway (NO_2^- and NO_3^-) using 2,3-diaminonaphthalene has been published and is shown to be approximately 50 times more sensitive than the Griess assay.[86]

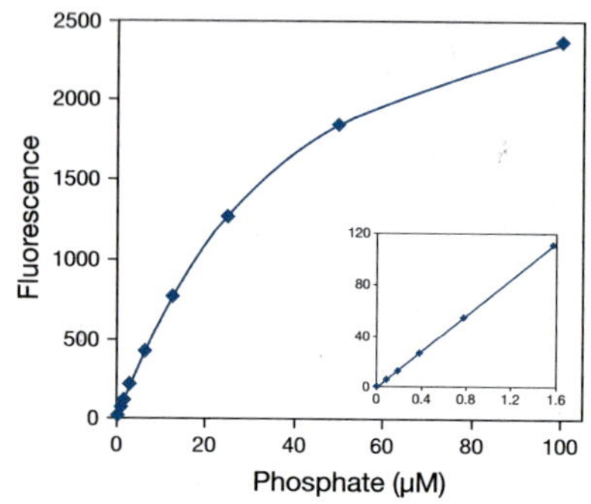

Figure 21.2.19 Detection of inorganic phosphate using the P_iPer™ Phosphate Assay Kit (P22061). Each reaction contained 50 µM Amplex® Red reagent, 2 U/mL maltose phosphorylase, 1 mM maltose, 1 U/mL glucose oxidase and 0.2 U/mL HRP in 1X reaction buffer. Reactions were incubated at 37°C. After 60 minutes, fluorescence was measured in a fluorescence microplate reader using excitation at 530 ± 12.5 nm and fluorescence detection at 590 ± 17.5 nm. Data points represent the average of duplicate reactions, and a background value of 43 (arbitrary units) was subtracted from each reading.

NBD Methylhydrazine

NBD methylhydrazine (*N*-methyl-4-hydrazino-7-nitrobenzofurazan, M20490) has been used to measure NO_2^- in water.[87] Reaction of NBD methylhydrazine with NO_2^- in the presence of mineral acids leads to formation of fluorescent products with excitation/emission maxima of ~468/537 nm. This reaction serves as the principle behind a selective fluorogenic method for the determination of NO_2^-. The assay is suitable for measurements by absorption or fluorescence spectroscopy or by fluorescence-detected HPLC.[87]

Other Nitrate Detection Reagents

Rhodamine 110 (R6479) has proven useful in a fluorescence quenching method for determining trace nitrite.[88] This sensitive assay takes advantage of the reaction of the green-fluorescent rhodamine 110 with nitrite at acidic pH to form a nitroso product that exhibits much weaker fluorescence. With a linear range of 1×10^{-8} to 3×10^{-7} moles/L and a detection limit of 7×10^{-10} moles/L, this assay has been used to measure nitrite in tap water and lake water without any prior extraction procedures.

Efficient quenching of SPQ or MQAE fluorescence (M440, E3101; see above) by nitrite (but not nitrate) has been used for direct measurement of NO_2^- transport across erythrocyte membranes[89] and for functional assays of bacterial nitrite extrusion transporters.[90]

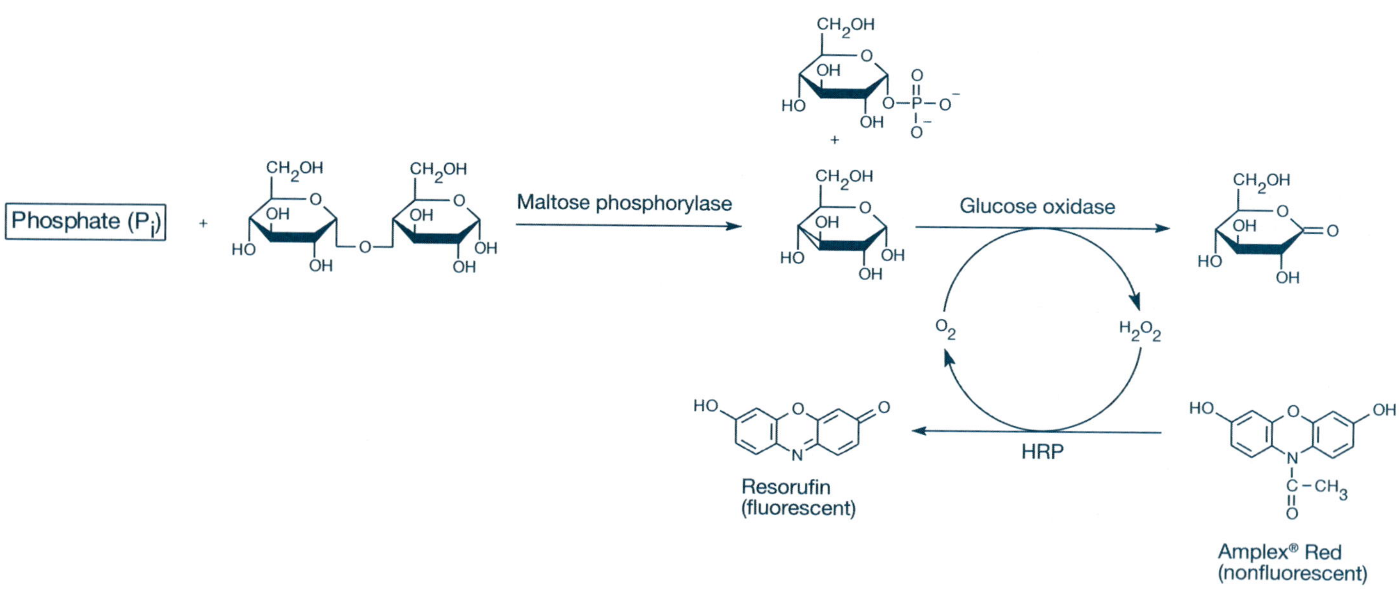

Figure 21.2.18 Principle of the P_iPer™ Phosphate Assay Kit (P22061). In the presence of inorganic phosphate, maltose phosphorylase converts maltose to glucose 1-phosphate and glucose. Then, glucose oxidase converts the glucose to gluconolactone and H_2O_2. Finally, with horseradish peroxidase (HRP) as a catalyst, the H_2O_2 reacts with the Amplex® Red reagent to generate the highly fluorescent resorufin. The resulting increase in fluorescence or absorption is proportional to the amount of P_i in the sample.

IMPORTANT NOTICE: The products described in this manual are covered by one or more Limited Use Label License(s). Please refer to the Appendix on page 971 and Master Product List on page 975. Products are For Research Use Only. Not intended for any animal or human therapeutic or diagnostic use.
www.invitrogen.com/probes

Phosphate and Pyrophosphate Detection

P$_i$Per™ Phosphate Assay Kit

The P$_i$Per™ Phosphate Assay Kit (P22061) provides an ultrasensitive assay that detects free phosphate in solution through formation of the fluorescent product resorufin. Because resorufin also has strong absorption, the assay can be performed either fluorometrically or spectrophotometrically. This kit can be used to detect inorganic phosphate (P_i) in a variety of samples or to monitor the kinetics of phosphate release by a variety of enzymes, including ATPases, GTPases, 5′-nucleotidase, protein phosphatases, acid and alkaline phosphatases and phosphorylase kinase. Furthermore, the assay can be modified to detect virtually any naturally occurring organic phosphate molecule by including an enzyme that can specifically digest the organic phosphate to liberate inorganic phosphate.

In the P$_i$Per™ phosphate assay (Figure 21.2.18), maltose phosphorylase converts maltose (in the presence of P_i) to glucose 1-phosphate and glucose. Then glucose oxidase converts the glucose to gluconolactone and H_2O_2. Finally, with horseradish peroxidase as a catalyst, the H_2O_2 reacts with the Amplex® Red reagent (10-acetyl-3,7-dihydroxyphenoxazine) to generate resorufin, which has absorption/emission maxima of ~571/585 nm.[91,92] The resulting increase in fluorescence or absorption is proportional to the amount of P_i in the sample. This kit can be used to detect as little as 0.2 µM P_i by fluorescence (Figure 21.2.19) or 0.4 µM P_i by absorption.

The P$_i$Per™ Phosphate Assay Kit contains:

- Amplex® Red reagent
- Dimethylsulfoxide (DMSO)
- Concentrated reaction buffer
- Recombinant maltose phosphorylase from *Escherichia coli*
- Maltose
- Glucose oxidase from *Aspergillus niger*
- Horseradish peroxidase
- Phosphate standard
- Hydrogen peroxide
- Detailed protocols for detecting phosphatase activity

Each kit provides sufficient reagents for approximately 1000 assays using a reaction volume of 100 µL per assay and either a fluorescence or absorbance microplate reader.

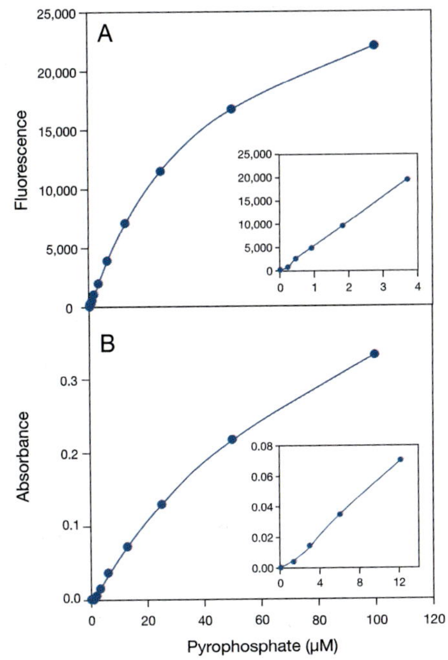

Figure 21.2.20 Detection of pyrophosphate using the P$_i$Per™ Pyrophosphate Assay Kit (P22062). Each reaction contained 50 µM Amplex® Red reagent, 0.01 U/mL inorganic pyrophosphatase, 2 U/mL maltose phosphorylase, 0.2 mM maltose, 1 U/mL glucose oxidase and 0.2 U/mL HRP in 1X reaction buffer. Reactions were incubated at 37°C. After 60 minutes, A) fluorescence was measured in a fluorescence-based microplate reader using excitation at 530 ± 12.5 nm and fluorescence detection at 590 ± 17.5 nm or B) absorbance was measured in an absorption-based microplate reader at 576 ± 5 nm. Data points represent the average of duplicate reactions. In panel A, a background value of 78 (arbitrary units) was subtracted from each reading; in panel B, a background absorbance of 0.011 was subtracted from each reading.

P$_i$Per™ Pyrophosphate Assay Kit

The P$_i$Per™ Pyrophosphate Assay Kit (P22062) provides a sensitive fluorometric or colorimetric method for measuring the inorganic pyrophosphate (PP_i) in experimental samples or for monitoring the kinetics of PP_i release by a variety of enzymes, including DNA and RNA polymerases, adenylate cyclase and S-acetyl coenzyme A synthetase. In the P$_i$Per™ pyrophosphate assay, inorganic pyrophosphatase hydrolyzes PP_i to two molecules of inorganic phosphate (P_i). The P_i then enters into the same cascade of reactions as it does in the P$_i$Per™ Phosphate Assay Kit (Figure 21.2.18). In this case, the resulting increase in fluorescence or absorption is proportional to the amount of PP_i in the sample. This kit can be used to detect as little as 0.1 µM PP_i by fluorescence or 0.2 µM PP_i by absorption (Figure 21.2.20).

The P$_i$Per™ Pyrophosphate Assay Kit contains:

- Amplex® Red reagent
- Dimethylsulfoxide (DMSO)
- Concentrated reaction buffer
- Recombinant maltose phosphorylase from *Escherichia coli*
- Maltose
- Glucose oxidase from *Aspergillus niger*
- Horseradish peroxidase
- Inorganic pyrophosphatase from baker's yeast
- Pyrophosphate standard
- Detailed protocols for detecting pyrophosphatase activity

Each kit provides sufficient reagents for approximately 1000 assays using a reaction volume of 100 µL per assay and either a fluorescence or absorbance microplate reader.

IMPORTANT NOTICE: The products described in this manual are covered by one or more Limited Use Label License(s). Please refer to the Appendix on page 971 and Master Product List on page 975. Products are For Research Use Only. Not intended for any animal or human therapeutic or diagnostic use.

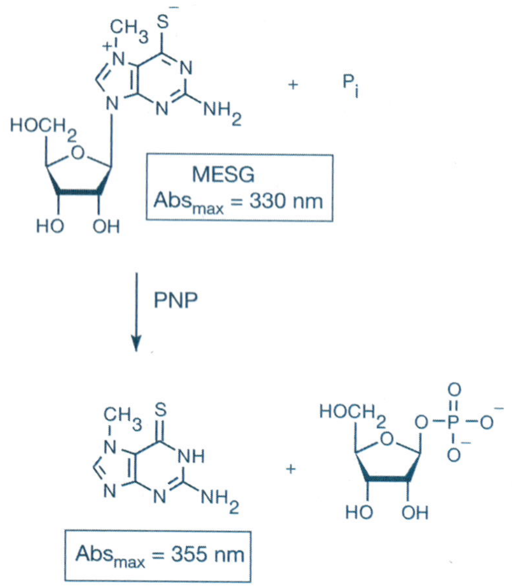

Figure 21.2.21 Enzymatic conversion of 2-amino-6-mercapto-7-methylpurine riboside (MESG) to ribose 1-phosphate and 2-amino-6-mercapto-7-methylpurine by purine nucleoside phosphorylase (PNP), reagents supplied in the EnzChek® Phosphate Assay Kit (E6646). The accompanying change in the absorption maximum (Abs max) allows quantitation of inorganic phosphate (P_i) consumed in the reaction.

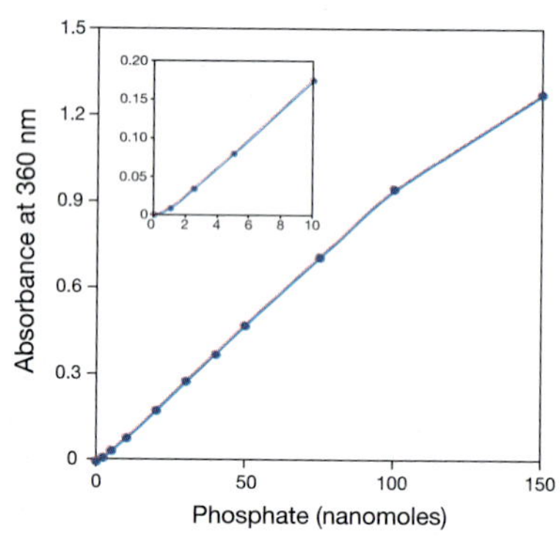

Figure 21.2.22 Quantitative analysis of inorganic phosphate using the EnzChek® Phosphate Assay Kit (E6646). KH_2PO_4 was used as the source for the inorganic phosphate, and the absorbance at 360 nm was corrected for background absorbance. The inset shows an enlargement of the standard curve, demonstrating the lower range of the assay; the units are the same.

EnzChek® Phosphate Assay Kit

The EnzChek® Phosphate Assay Kit (E6646), which is based on a method originally described by Webb,[93,94] provides an easy enzymatic assay for detecting P_i from multiple sources through formation of a chromophoric product (Figure 21.2.21). Although this kit is usually used to determine the P_i produced by a wide variety of enzymes such as ATPases, kinases and phosphatases (Section 10.3), it can also be used to specifically quantitate P_i with a sensitivity of ~2 µM P_i (~0.2 µg/mL) (Figure 21.2.22). Moreover, this colorimetric assay has proven useful for determining the level of P_i contamination in the presence of high concentrations of acid-labile phosphates using a microplate reader.[95] Because the sulfate anion competes with P_i for binding to purine nucleoside phosphorylase (PNP), this kit can be adapted for measurement of sulfate concentrations between 0.1 and 10 mM in the presence of a low (<100 µM) fixed P_i concentration.[96]

The EnzChek® Phosphate Assay Kit contains:

- 2-Amino-6-mercapto-7-methylpurine riboside (MESG)
- Purine nucleoside phosphorylase (PNP)
- Concentrated reaction buffer
- KH_2PO_4 standard
- Detailed protocols for detecting and quantitating P_i

Each kit provides sufficient reagents for about 100 phosphate assays using 1 mL assay volumes and standard cuvettes.

EnzChek® Pyrophosphate Assay Kit

In the EnzChek® Pyrophosphate Assay Kit (E6645), we have adapted the method provided in the EnzChek® Phosphate Assay Kit to permit the sensitive spectrophotometric detection of PP_i, which is converted by the enzyme pyrophosphatase to P_i.[97] Because two moles of P_i are released per mole of PP_i consumed, the sensitivity limit of the EnzChek® Pyrophosphate Assay Kit is 1 µM PP_i (~0.2 µg/mL). This assay has been modified to continuously detect several enzymes that liberate PP_i[98–101] such as aminoacyl-tRNA synthetase,[102] luciferase, cytidylyl transferase[103] and *S*-acetyl coenzyme A synthetase[97] and potentially DNA and RNA polymerases, adenylate cyclase and guanylyl cyclase.[104]

The EnzChek® Pyrophosphate Assay Kit contains:

- Inorganic pyrophosphatase
- 2-Amino-6-mercapto-7-methylpurine riboside (MESG)
- Purine nucleoside phosphorylase (PNP)
- Concentrated reaction buffer
- $Na_2P_2O_7$ standard
- Detailed protocols for detecting and quantitating PP_i

Each kit provides sufficient reagents for about 100 PP_i assays using standard 1 mL assay volumes and standard cuvettes.

IMPORTANT NOTICE: The products described in this manual are covered by one or more Limited Use Label License(s). Please refer to the Appendix on page 971 and Master Product List on page 975. Products are For Research Use Only. Not intended for any animal or human therapeutic or diagnostic use.

REFERENCES

1. Analyst (1982) 107:465; **2.** J Heterocyclic Chem (1982) 19:841; **3.** Biochemistry (1987) 26:1215; **4.** Anal Biochem (1989) 178:355; **5.** Am J Physiol (1990) 259:C375; **6.** Biophys Chem (2000) 85:49; **7.** Am J Physiol (1993) 264:C27; **8.** Am J Physiol (1997) 272:C1405; **9.** Biophys J (1988) 53:955; **10.** Anal Biochem (1996) 241:51; **11.** Neuroscience (1989) 28:725; **12.** Cytometry (1997) 28:316; **13.** Annu Rev Physiol (1998) 60:689; **14.** Proc Natl Acad Sci U S A (1991) 88:7500; **15.** Biochim Biophys Acta (1993) 1152:83; **16.** Am J Physiol (1998) 274:C310; **17.** Pflugers Arch (1994) 427:24; **18.** Hum Gene Ther (1999) 10:861; **19.** Biophys J (2000) 78:1293; **20.** Am J Physiol Cell Physiol (2000) 279:C1088; **21.** Biochemistry (1998) 37:15222; **22.** J Biol Chem (1997) 272:27830; **23.** J Biol Chem (1995) 270:17033; **24.** Biochemistry (1997) 36:1287; **25.** Pflugers Arch (1999) 437:289; **26.** Biochemistry (1994) 33:755; **27.** J Membr Biol (1997) 159:197; **28.** Anal Biochem (1995) 230:1; **29.** J Biol Chem (1999) 274:26003; **30.** FEBS Lett (1997) 408:166; **31.** J Biol Chem (1994) 269:7435; **32.** Am J Physiol (1994) 267:H2114; **33.** Am J Physiol (1998) 275:C1330; **34.** Biochemistry (2001) 40:5243; **35.** J Biol Chem (1998) 273:24368; **36.** Biochemistry (1991) 30:7879; **37.** Methods Enzymol (1999) 307:469; **38.** Methods (1999) 18:197; **39.** J Neurochem (1998) 71:1396; **40.** J Neurochem (1998) 70:2500; **41.** J Neurosci Methods (1997) 75:127; **42.** Anal Biochem (1994) 219:139; **43.** J Membr Biol (2000) 177:199; **44.** J Appl Physiol (2003) 94:343; **45.** Fresenius Z Anal Chem (1983) 314:577; **46.** Biophys J (2008) 94:4493; **47.** Chem Phys Lipids (1991) 60:127; **48.** Anal Biochem (1981) 114:199; **49.** Analyst (1989) 114:1275; **50.** Anal Lett (1988) 21:1887; **51.** J Biol Chem (1984) 259:4812; **52.** Chem Res Toxicol (1997) 10:1080; **53.** Chem Pharm Bull (1984) 32:3702; **54.** J Biol Chem (2003) 278:3170; **55.** Nat Rev Drug Discov (2009) 8:153; **56.** Nat Biotechnol (2002) 20:87; **57.** FEBS Lett (2001) 499:220; **58.** Mol Pharmacol (2008) 73:758; **59.** Proc Natl Acad Sci U S A (1975) 72:619; **60.** Anal Chem (1987) 59:1096; **61.** J Chroma*togr A (1994) 668:323; **62.** J Chroma*togr (1992) 582:131; **63.** Anal Chim Acta (1989) 225:351; **64.** Biomed Chroma*togr (1989) 3:209; **65.** Anal Sci (1986) 2:491; **66.** Anal Chem (1994) 66:3512; **67.** Anal Chem (1994) 66:3477; **68.** Electrophoresis (1993) 14:373; **69.** Anal Chem (1992) 64:973; **70.** Anal Chem (1991) 63:413; **71.** Anal Chem (1991) 63:408; **72.** J Chroma*togr (1991) 559:223; **73.** Proc Natl Acad Sci U S A (1991) 88:2302; **74.** J Chroma*togr (1990) 499:579; **75.** Anal Biochem (1982) 126:131; **76.** Luminescence (1999) 14:283; **77.** FASEB J (1993) 7:349; **78.** Anal Biochem (2000) 282:1; **79.** Anal Biochem (1995) 224:502; **80.** Chem Biol (1996) 3:655; **81.** Anal Biochem (1996) 238:150; **82.** Angew Chem Int Ed Engl (1999) 38:3209; **83.** Anal Chem (1998) 70:2446; **84.** J Biol Chem (2002) 277:48472; **85.** Anal Biochem (1993) 214:11; **86.** Nat Protoc (2006) 1:2223; **87.** Anal Chem (1999) 71:3003; **88.** Spectrochim Acta A (2003) 59:1667; **89.** J Bioenerg Biomembr (1997) 29:611; **90.** Mol Microbiol (1994) 12:579; **91.** Anal Biochem (1997) 253:162; **92.** J Immunol Methods (1997) 202:133; **93.** Anal Biochem (1994) 218:449; **94.** Proc Natl Acad Sci U S A (1992) 89:4884; **95.** Anal Biochem (1995) 230:173; **96.** Biophys Chem (1997) 63:107; **97.** Anal Biochem (1996) 243:41; **98.** J Biol Chem (2000) 275:17962; **99.** Biochemistry (2000) 39:2297; **100.** J Biol Chem (1998) 273:16555; **101.** J Biol Chem (1998) 273:22151; **102.** Nucleic Acids Res (1995) 23:2886; **103.** Biochemistry (2001) 40:5041; **104.** Biochemistry (1996) 35:11013.

DATA TABLE 21.2 DETECTING CHLORIDE, PHOSPHATE, NITRITE AND OTHER ANIONS

Cat. No.	MW	Storage	Soluble	Abs	EC	Em	Solvent	K_{SV}	Notes
A6222	305.29	F,D,L	MeOH	465	ND	560	MeOH		1, 2, 3
A10192	251.24	F,L	EtOH	486	ND	591	MeOH		2, 4
D7918	158.20	L	DMSO, MeOH	340	5100	377	MeOH		5
D23841	412.35	F,D,L	DMSO	487	84,000	see Notes	pH 8		6
E3101	326.19	F,D,L	H_2O	350	2800	460	H_2O	200 M^{-1}	7, 8, 9, 10
E6645	313.33	FF,D	H_2O	332	16,000	none	pH 7		11, 12
E6646	313.33	F,D	H_2O	332	16,000	none	pH 7		11, 12
L6868	510.50	L	H_2O	455	7400	505	H_2O	390 M^{-1}	7, 8, 10, 13, 14
M440	281.33	L	H_2O	344	3700	443	H_2O	118 M^{-1}	7, 8, 9, 10
M6886	315.15	L	H_2O	344	3900	442	H_2O	145 M^{-1}	7, 8, 9, 10, 15
M20490	209.16	F,L	MeCN	487	24,000	none	MeOH		16
N1138	184.19	L	DMF, MeCN	419	9400	493	see Notes		17
N1495	724.97	F,D	MeOH	<300		none			
P2331MP	134.13	L	EtOH	334	5700	455	pH 9		18
R6479	366.80	L	DMSO	499	92,000	521	MeOH		

For definitions of the contents of this data table, see "Using *The Molecular Probes® Handbook*" in the introductory pages.

Notes

1. Spectral data are for the reaction product with glycine in the presence of cyanide. Unreacted reagent in MeOH: Abs = 254 nm (EC = 46,000 $cm^{-1}M^{-1}$), nonfluorescent.
2. ND = not determined.
3. Solubility in methanol is improved by addition of base (e.g., 1–5% (v/v) 0.2 M KOH).
4. Spectral data are for the reaction product with glycine in the presence of cyanide. Unreacted reagent in MeOH: Abs = 282 nm (EC = 21,000 $cm^{-1}M^{-1}$), nonfluorescent.
5. Fluorescence of D7918 is weak. Reaction with nitrite yields highly fluorescent 1*H*-naphthotriazole (Abs = 365 nm, Em = 415 nm in H_2O (pH 12)). (Methods Enzymol (1996) 268:105)
6. DAF-FM fluorescence is very weak. Reaction with nitrite or nitric oxide generates a highly fluorescent benzotriazole derivative with Abs = 495 nm (EC = 73,000 $cm^{-1}M^{-1}$), Em = 515 nm in pH 7.4 buffer. (Angew Chem Int Ed Engl (1999) 38:3209)
7. Values of K_{SV} are taken from published references. (Anal Biochem (1994) 219:139, Am J Physiol (1992) 262:C242, Biochemistry (1991) 30:7879, Anal Biochem (1989) 178:355
8. K_{SV} is the Stern-Volmer quenching constant for Cl^- ions (units are M^{-1}), representing the reciprocal of the ion concentration that produces 50% fluorescence quenching. The quenching constant is very dependent on viscosity and is usually significantly lower in cells. These indicators are quenched more effectively by bromide, iodide and certain other anions.
9. This quinolinium dye also has a slightly stronger (~50%) absorption peak 25–30 nm shorter than the listed Abs wavelength.
10. This product undergoes Cl^--dependent fluorescence quenching with essentially no change in absorption or emission wavelengths.
11. Data represent the substrate component of this kit.
12. Enzymatic phosphorylation of this substrate yields 2-amino-6-mercapto-7-methylpurine (Abs = 355 nm). (Proc Natl Acad Sci U S A (1992) 89:4884)
13. L6868 has much stronger absorption at shorter wavelengths (Abs = 368 nm (EC = 36,000 $cm^{-1}M^{-1}$)).
14. This compound emits chemiluminescence upon oxidation in basic aqueous solutions. Emission peaks are at 425 nm (L8455) and 470 nm (L6868).
15. M6886 may be chemically reduced to cell-permeant diH-MEQ. (Biochemistry (1991) 30:7879)
16. NBD methylhydrazine reacts with nitrite in the presence of strong acid to form fluorescent *N*-methyl-4-amino-7-nitrobenzofurazan (Abs = 459 nm, Em = 537 nm in MeCN). (Anal Chem (1999) 71:3003)
17. Spectral data are for the reaction product with glycine in the presence of cyanide, measured in pH 7.0 buffer/MeCN (40:60). (Anal Chem (1987) 59:1102) Unreacted reagent in MeOH: Abs = 279 nm (EC = 5500 $cm^{-1}M^{-1}$), Em = 330 nm.
18. Spectral data are for the reaction product of P2331MP with alanine and 2-mercaptoethanol. The spectra and stability of the adduct depend on the amine and thiol reactants. (Biochim Biophys Acta (1979) 576:440) Unreacted reagent in H_2O: Abs = 257 nm (EC = 1000 $cm^{-1}M^{-1}$).

PRODUCT LIST 21.2 DETECTING CHLORIDE, PHOSPHATE, NITRITE AND OTHER ANIONS

Cat. No.	Product	Quantity
A6222	ATTO-TAG™ CBQCA derivatization reagent (CBQCA; 3-(4-carboxybenzoyl)quinoline-2-carboxaldehyde)	10 mg
A10192	ATTO-TAG™ FQ derivatization reagent (FQ; 3-(2-furoyl)quinoline-2-carboxaldehyde)	10 mg
D23841	DAF-FM (4-amino-5-methylamino-2',7'-difluorofluorescein)	1 mg
D7918	2,3-diaminonaphthalene	100 mg
E6646	EnzChek® Phosphate Assay Kit *100 assays*	1 kit
E6645	EnzChek® Pyrophosphate Assay Kit *100 assays*	1 kit
E3101	*N*-(ethoxycarbonylmethyl)-6-methoxyquinolinium bromide (MQAE)	100 mg
G7921	Griess Reagent Kit *for nitrite quantitation*	1 kit
L6868	lucigenin (bis-N-methylacridinium nitrate) *high purity*	10 mg
M36051	Measure-iT™ High-Sensitivity Nitrite Assay Kit *2000 assays*	1 kit
M6886	6-methoxy-*N*-ethylquinolinium iodide (MEQ)	100 mg
M440	6-methoxy-*N*-(3-sulfopropyl)quinolinium, inner salt (SPQ)	100 mg
M20490	*N*-methyl-4-hydrazino-7-nitrobenzofurazan (NBD methylhydrazine)	25 mg
N1138	naphthalene-2,3-dicarboxaldehyde (NDA)	100 mg
N1495	nigericin, free acid	10 mg
P2331MP	*o*-phthaldialdehyde (OPA) *high purity*	1 g
P22061	P_iPer™ Phosphate Assay Kit *1000 assays*	1 kit
P22062	P_iPer™ Pyrophosphate Assay Kit *1000 assays*	1 kit
P10229	Premo™ Halide Sensor *for 10 microplates*	1 kit
R6479	rhodamine 110 (R110) *reference standard*	25 mg

IMPORTANT NOTICE: The products described in this manual are covered by one or more Limited Use Label License(s). Please refer to the Appendix on page 971 and Master Product List on page 975. Products are For Research Use Only. Not intended for any animal or human therapeutic or diagnostic use.
www.invitrogen.com/probes
molecular probes® | invitrogen™ by life technologies™

TWENTY-TWO

CHAPTER 22

Probes for Membrane Potential

IMPORTANT NOTICE: The products described in this manual are covered by one or more Limited Use Label License(s). Please refer to the Appendix on page 971 and Master Product List on page 975. Products are For Research Use Only. Not intended for any animal or human therapeutic or diagnostic use.

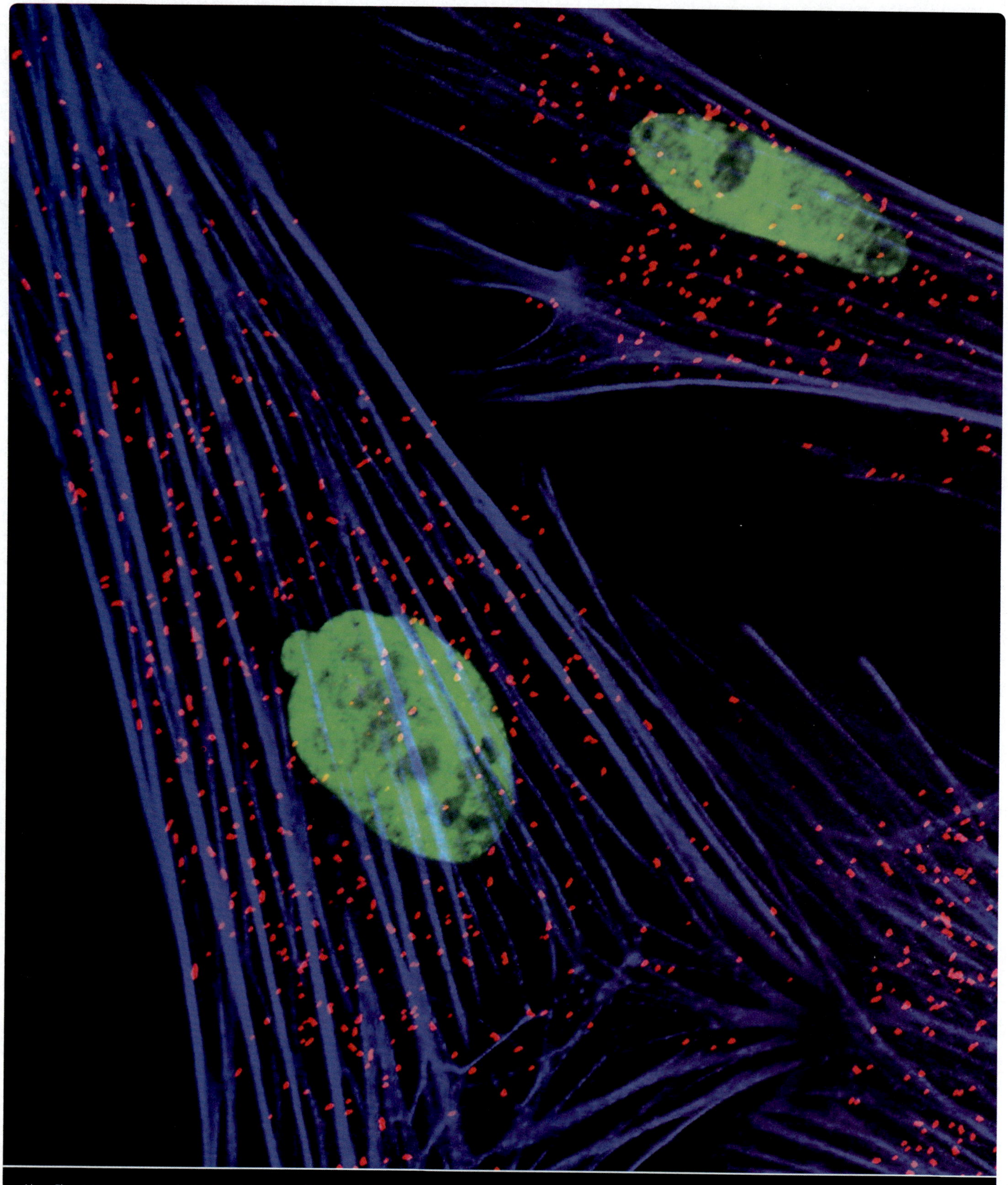

Alexa Fluor® 568 goat anti-rabbit IgG and Alexa Fluor® 350 phalloidin.

IMPORTANT NOTICE: The products described in this manual are covered by one or more Limited Use Label License(s). Please refer to the Appendix on page 971 and Master Product List on page 975. Products are For Research Use Only. Not intended for any animal or human therapeutic or diagnostic use.

22.1 Introduction to Potentiometric Probes

Potentiometric optical probes enable researchers to perform membrane potential measurements in organelles and in bacterial cells that are too small for microelectrodes. Moreover, in conjunction with imaging techniques, these probes can be employed to map variations in membrane potential across excitable cells, in perfused organs[1] and ultimately in the brain *in vivo*,[2–5] with spatial resolution and sampling frequency that cannot be obtained using microelectrodes.

Potentiometric probes include the cationic or zwitterionic styryl dyes, the cationic carbocyanines and rhodamines, the anionic and hybrid oxonols and merocyanine 540. The class of dye determines factors such as accumulation in cells, response mechanism and toxicity. Surveys of techniques and applications using membrane potential probes can be found in several reviews.[2,4,6–8]

Applications for Potentiometric Probes

The plasma membrane of a cell typically has a transmembrane potential of approximately –70 mV (negative inside) as a consequence of K^+, Na^+ and Cl^- concentration gradients that are maintained by active transport processes. Potentiometric probes offer an indirect method of detecting the translocation of these ions, whereas the fluorescent ion indicators discussed in Chapter 21 can be used to directly measure changes in specific ion concentrations.

Increases and decreases in membrane potential—referred to as membrane hyperpolarization and depolarization, respectively—play a central role in many physiological processes, including nerve-impulse propagation, muscle contraction, cell signaling and ion-channel gating. Potentiometric probes are important tools for studying these processes, as well as for visualizing mitochondria (which exhibit transmembrane potentials of approximately –150 mV, negative inside matrix) (Section 12.2), for assessing cell viability (Section 15.2) and for high-throughput screening of new drug candidates.

Selecting a Potentiometric Probe

Selecting the best potentiometric probe for a particular application can be complicated by the substantial variations in their optical responses, phototoxicity and interactions with other molecules. Probes can be divided into two categories based on their response mechanism:

- Fast-response probes (usually styrylpyridinium dyes, Section 22.2) operate by means of a change in their electronic structure, and consequently their fluorescence properties, in response to a change in the surrounding electric field (Figure 22.1.1). Their optical response is sufficiently fast to detect transient (millisecond) potential changes in excitable cells, including single neurons, cardiac cells and intact brains. However, the magnitude of their potential-dependent fluorescence change is often small; fast-response probes typically show a 2–10% fluorescence change per 100 mV.
- Slow-response probes (Section 22.3) exhibit potential-dependent changes in their transmembrane distribution that are accompanied by a fluorescence change (Figure 22.1.1). The magnitude of their

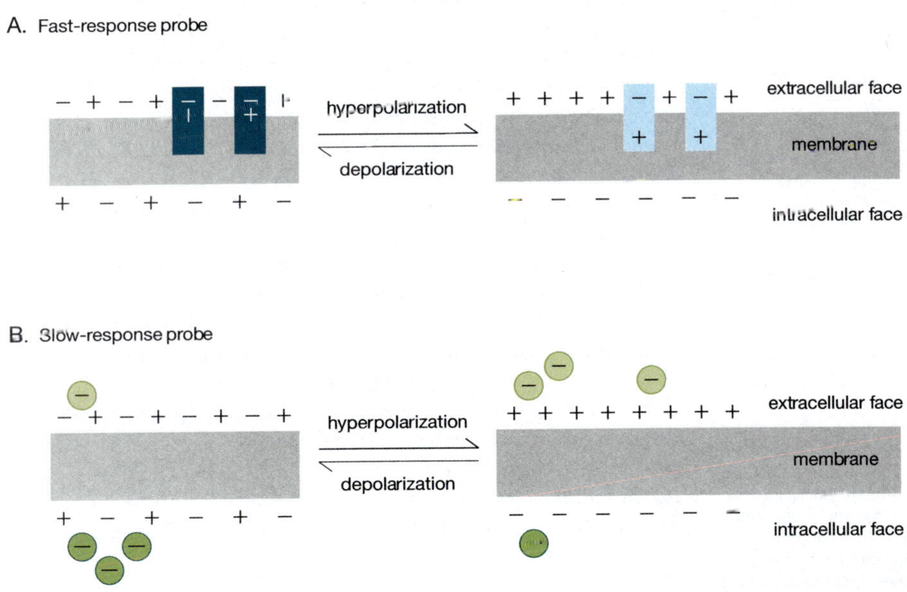

Figure 22.1.1 Response mechanisms of membrane potential–sensitive probes. Fast-response probes undergo electric field–driven changes of intramolecular charge distribution that produce corresponding changes in the spectral profile or intensity of their fluorescence (represented by color changes in the illustration). Slow-response probes are lipophilic anions (in this illustration) or cations that are translocated across membranes by an electrophoretic mechanism. Fluorescence changes associated with transmembrane redistribution (represented by color changes in the illustration) result from sensitivity of the probe to intracellular and extracellular environments. Thus, potentiometric response speeds directly reflect the time constants of the underlying processes—fast intramolecular redistribution of electrons versus relatively slow transmembrane movement of entire molecules.

optical responses is much larger than that of fast-response probes (typically a 1% fluorescence change per mV). Slow-response probes, which include cationic carbocyanines and rhodamines and anionic oxonols, are suitable for detecting changes in average membrane potentials of nonexcitable cells caused by respiratory activity, ion-channel permeability, drug binding and other factors.

Calibration of potentiometric probes can be accomplished by imposing a transmembrane potential using valinomycin or gramicidin (V1644, G6888; Section 21.1) in conjunction with externally applied K^+ solutions.[9,10] The ultimate test of calibration veracity is quantitative agreement with electrophysiological measurements.[5,11]

REFERENCES

1. Nature (1998) 392:78; **2.** Nature (2009) 461:930; **3.** Nat Rev Neurosci (2008) 9:195; **4.** Methods Mol Biol (2009) 489:43; **5.** Nat Protoc (2008) 3:249; **6.** J Physiol Paris (2010) 104:40; **7.** Trends Neurosci (2000) 23:166; **8.** Methods (2000) 21:271; **9.** Cytometry A (2009) 75:593; **10.** Methods Mol Biol (1998) 91:85; **11.** Circ Res (2009) 104:670.

22.2 Fast-Response Probes

Our fast-response potential-sensitive probes (see Figure 22.1.1A in Section 22.1) are listed in Table 22.1, along with their charges, optical responses and selected applications.

ANEP Dyes

Di-4-ANEPPS and Di-8-ANEPPS

The ANEP (AminoNaphthylEthenylPyridinium) dyes developed by Leslie Loew and colleagues[1] are among the most sensitive of the fast-response probes. Zwitterionic di-4-ANEPPS (D1199, Figure 22.2.1) and di-8-ANEPPS (D3167, Figure 22.2.2) exhibit fairly uniform 10% per 100 mV changes in fluorescence intensity in a variety of tissue, cell and model membrane systems.[2,3] The millisecond-range temporal characteristics of the ANEP dyes compensate for this modest response amplitude (Figure 22.2.3). Di-4-ANEPPS is internalized in the cell rather rapidly, precluding its use in all but very short-term experiments, whereas di-8-ANEPPS is better retained in the outer leaflet of the plasma membrane. In addition, although both ANEP dyes exhibit good photostability and low toxicity, di-8-ANEPPS is reported to be slightly more photostable and significantly less phototoxic than di-4-ANEPPS.[4–6]

Like other styryl dyes, the ANEP dyes are essentially nonfluorescent in aqueous solutions and exhibit spectral properties that are strongly dependent on their environment.[7] When bound to phospholipid vesicles, di-8-ANEPPS has absorption/emission maxima of ~467/631 nm (Figure 22.2.4), as compared with ~498/713 nm in methanol. The fluorescence excitation/emission maxima of di-4-ANEPPS bound to neuronal membranes are ~475/617 nm.[8]

Both di-4-ANEPPS and di-8-ANEPPS respond to increases in membrane potential (hyperpolarization) with a decrease in fluorescence excited at approximately 440 nm and an increase in fluorescence excited at 530 nm.[7,9] These spectral shifts permit the use of ratiometric methods (Loading and Calibration of Intracellular Ion Indicators—Note 19.1) to correlate the change in fluorescence signal with membrane potential.[2] Using di-8-ANEPPS, Loew and colleagues were able to follow changes in membrane potential along the surface of a single mouse neuroblastoma cell in their study of the mechanisms underlying cathode-directed neurite elongation[10] and to define differences between transmembrane potentials of neurites and somata.[11] Potential-dependent fluorescence emission ratio measurements (ratio of emission intensities at 560 nm and 620 nm following excitation at 475 nm) have also been reported using both di-4-ANEPPS and di-8-ANEPPS[12–15] (Figure 22.2.3). Some other applications are listed in Table 22.1.

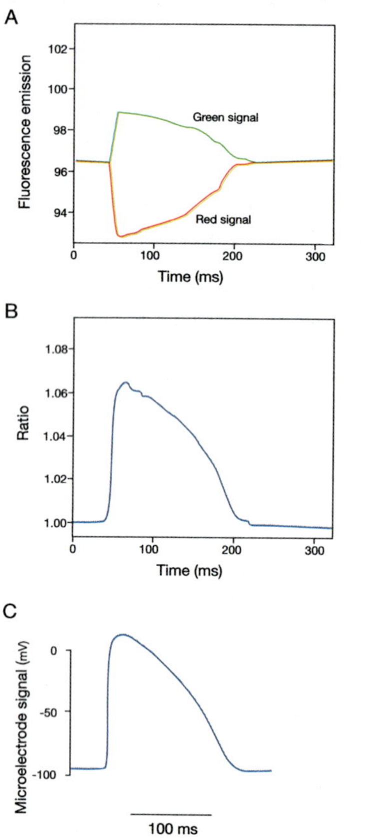

Figure 22.2.3 Detection of action potentials in intact rabbit hearts using the fast potentiometric probe di-4-ANEPPS (D1199). Excised rabbit hearts were loaded with di-4-ANEPPS by perfusion with dye-containing medium. Fluorescence was excited at 488 nm by an argon-ion laser. Emission components at 540 ± 6 nm (green) and >610 nm (red) were detected simultaneously by two photomultipliers (panel **A**). The ratio of the green to red signals (panel **B**) displayed a larger fractional change during action potential cycles than either of the component signals; it also followed transmembrane voltage contours recorded simultaneously by an intracellular microelectrode (panel **C**). In addition, fluorescence ratio measurements reduce the motion artifacts that typically distort optical signals detected from contracting hearts. Figure reproduced with permission from Am J Physiol Heart Circ Physiol (2000) 279:H1421.

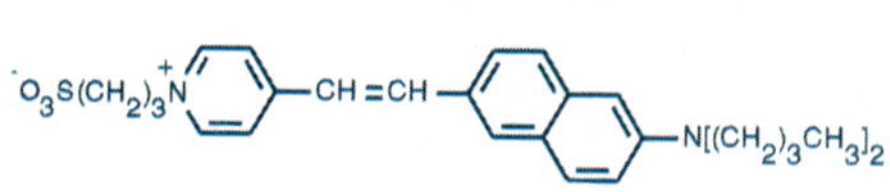

Figure 22.2.1 di-4-ANEPPS (D1199).

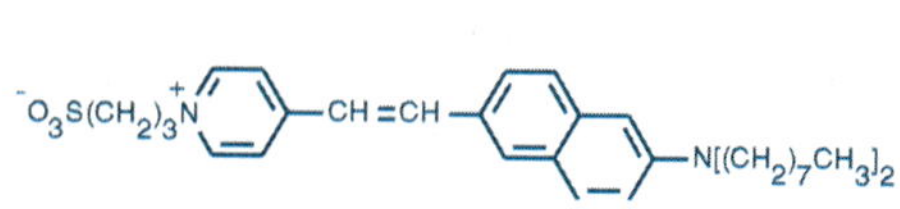

Figure 22.2.2 di-8-ANEPPS (D3167).

IMPORTANT NOTICE: The products described in this manual are covered by one or more Limited Use Label License(s). Please refer to the Appendix on page 971 and Master Product List on page 975. Products are For Research Use Only. Not intended for any animal or human therapeutic or diagnostic use.
www.invitrogen.com/probes

Cationic ANEP Dyes

In collaboration with Leslie Loew and Joe Wuskell of the University of Connecticut, we offer a series of potential-sensitive cationic ANEP dyes.[16–18] The water-soluble di-2-ANEPEQ[19] (JPW 1114, D6923; Figure 22.2.5) can be either microinjected into cells, a mode of delivery that intensifies the staining of remote neuronal processes, or applied topically to deeply stain brain tissue.[20] Microinjection of di-2-ANEPEQ into neurons in ganglia of the snail *Helix aspersa* produced an approximately 50-fold improvement in voltage-sensitive signals from distal processes over that obtained with conventional absorption– and fluorescence-based staining methods.[21] Di-12-ANEPPQ (D6927) is useful for potential-sensitive retrograde labeling of neurons[17,22] using techniques similar to those employed for lipophilic carbocyanine and aminostyryl tracers (Section 14.4). Di-3-ANEPPDHQ (D36801, Figure 22.2.6) and di-4-ANEPPDHQ (D36802, Figure 22.2.7) both exhibit very low rates of internalization and good signal-to-noise ratios, and are useful for neural network analysis.[23] Di-4-ANEPPDHQ has proven useful for visualizing cholesterol-enriched lipid domains in model membranes.[24]

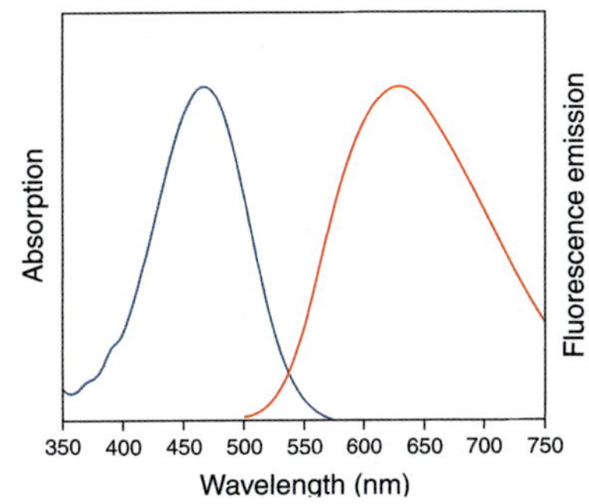

Figure 22.2.4 Absorption and fluorescence emission spectra of di-8-ANEPPS bound to phospholipid bilayer membranes.

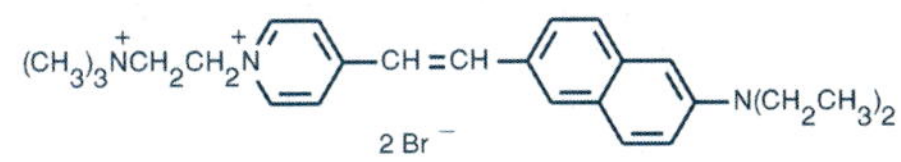

Figure 22.2.5 Di-2-ANEPEQ (JPW 1114, D6923).

RH Dyes

Originally synthesized by Rina Hildesheim, the RH dyes include an extensive series of dialkylaminophenylpolyenylpyridinium dyes that are principally used for functional imaging of neurons (Table 22.1). The existence of numerous RH dye analogs reflects the observation that no single dye provides the optimal response under all experimental conditions.[5,25,26] Currently, the most widely used RH dyes are RH 237 (S1109, Figure 22.2.8), RH 414 (T1111, Figure 22.2.9), RH 421 (S1108) and RH 795 (R649, Figure 22.2.10). Physiological effects of staining with different analogs are not equivalent. For example, staining of the cortex with RH 414 causes arterial constriction, whereas staining with RH 795 does not.[27] RH 795 produced negligible side effects

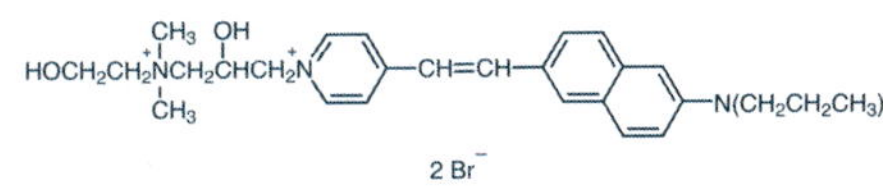

Figure 22.2.6 Di-3-ANEPPDHQ (D36801).

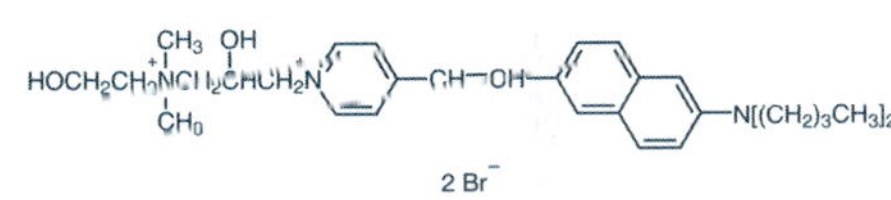

Figure 22.2.7 Di-4-ANEPPDHQ (D36802).

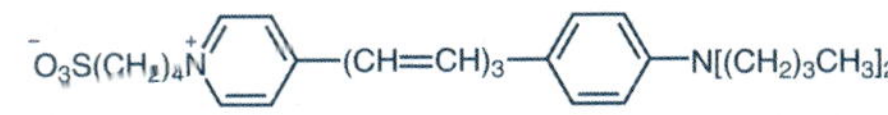

Figure 22.2.8 RH 237 *N*-(4-sulfobutyl)-4-(6-(4-(dibutylamino)-phenyl)hexatrienyl)pyridinium, inner salt (S1109).

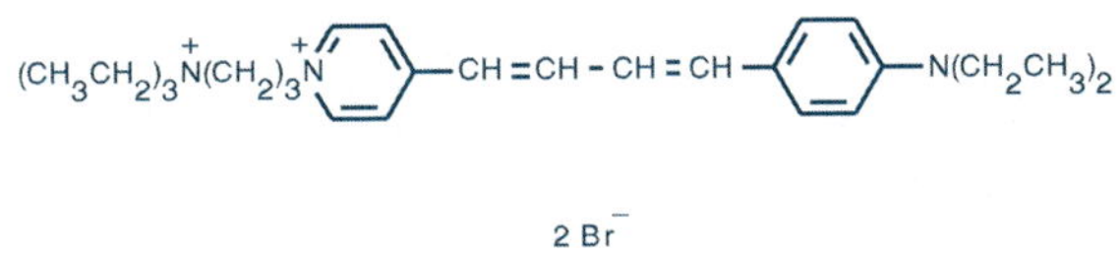

Figure 22.2.9 *N*-(3-triethylammoniumpropyl)-4-(4-(4-(diethylamino)phenyl)butadienyl)pyridinium dibromide (RH 414, T1111).

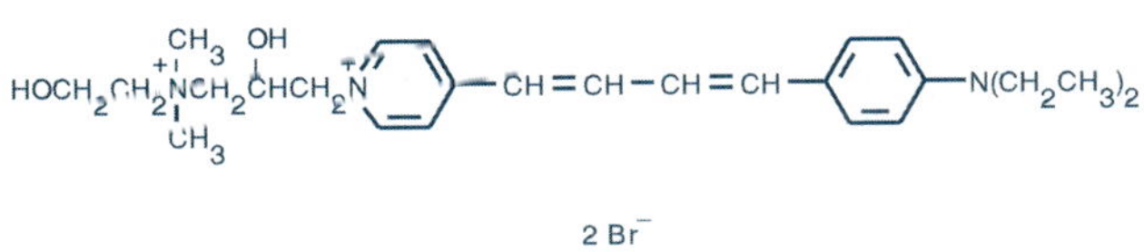

Figure 22.2.10 RH 795 (R649).

Table 22.1 Characteristics and selected applications of Molecular Probes® fast-response probes.

Dyes (Cat. No.)	Structure (Charge)	Optical Response	Selected Applications
Di-4-ANEPPS (D1199) Di-8-ANEPPS (D3167) Di-2-ANEPEQ (D6923) Di-12-ANEPPQ (D6927) Di-3-ANEPPDHQ (D36801) Di-4-ANEPPDHQ (D36802)	Styryl (cationic or zwitterionic)	FAST; fluorescence excitation ratio 440/505 nm decreases upon membrane hyperpolarization	• Combined optical potentiometric and electrophysiological measurements[1] • Combined potentiometric and Ca^{2+} measurements[2,3] • Imaging electrical activity from intact heart tissues[4–6] • Mapping of membrane potentials along neurons[7–9] and muscle fibers[10] • Membrane potential changes in response to pharmacological stimuli[11,12] • Two-photon excitation microscopy[13–16]
RH 237 (S1109) RH 414 (T1111) RH 421 (S1108) RH 795 (R649)	Styryl (cationic or zwitterionic)	FAST; fluorescence decreases upon membrane depolarization	• Electrical activity of cardiomyocytes and cardiac tissue[17,18] • Functional tracing of neurons[19,20] • Membrane potentials evoked by visual[21] and auditory[22,23] stimuli

1. Biophys J (1998) 74:48; **2.** J Biol Chem (2006) 281:40302; **3.** Cardiovasc Res (2005) 65:83; **4.** Am J Physiol Heart Circ Physiol (2005) 289:H2602; **5.** Circ Res (2004) 95:21; **6.** Am J Physiol Heart Circ Physiol (2004) 287:H985; **7.** Neuron (2008) 58:763; **8.** Science (2007) 317:819; **9.** Proc Natl Acad Sci U S A (2006) 103:16550; **10.** J Membr Biol (2005) 208:141; **11.** Am J Physiol (1998) 274:H60; **12.** Pharmacol Res (1996) 34:125; **13.** Methods Mol Biol (2009) 489:43; **14.** J Neurophysiol (2008) 99:1545; **15.** J Gen Physiol (2006) 127:623; **16.** J Neurosci Methods (2005) 148:94; **17.** Am J Physiol Heart Circ Physiol (2008) 294:H1417; **18.** Biophys J (2007) 92:448; **19.** Nat Protoc (2008) 3:249; **20.** J Neurosci Methods (1994) 54:151; **21.** Proc Natl Acad Sci U S A (2006) 103:12586; **22.** Front Neuroengineering (2009) 2:2.1-; **23.** Proc Natl Acad Sci U S A (2006) 103:1918.

IMPORTANT NOTICE: The products described in this manual are covered by one or more Limited Use Label License(s). Please refer to the Appendix on page 971 and Master Product List on page 975. Products are For Research Use Only. Not intended for any animal or human therapeutic or diagnostic use.

REFERENCES

1. Biochemistry (1985) 24:5749; 2. Biophys J (1998) 74:48; 3. J Membr Biol (1992) 130:1; 4. Biophys J (1999) 76:2272; 5. Biophys J (1994) 67:1301; 6. Pflugers Arch (1994) 426:548; 7. Biochemistry (1989) 28:4536; 8. Biochim Biophys Acta (1993) 1150:111; 9. Biophys J (1994) 67:208; 10. Neuron (1992) 9:393; 11. Neuron (1994) 13:1187; 12. Am J Physiol Heart Circ Physiol (2000) 279:H1421; 13. Biophys J (2001) 81:1163; 14. Am J Physiol (1996) 270:H2216; 15. Am J Physiol (1998) 274:H60; 16. Neuron (2003) 37:85; 17. J Neurosci Methods (1996) 70:121; 18. Pure Appl Chem (1996) 68:1405; 19. Biol Bull (2000) 198:1; 20. Proc Natl Acad Sci U S A (1997) 94:7621; 21. J Neurosci (1995) 15:1392; 22. J Neurosci Methods (1996) 70:111; 23. J Neurosci Methods (2004) 134:179; 24. Biophys J (2006) 90:2563; 25. Physiol Rev (1988) 68:1285; 26. Annu Rev Neurosci (1985) 8:263; 27. Nature (1986) 324:361; 28. J Neurosci (1994) 14:2545; 29. Biophys J (1995) 69:299; 30. Biophys J (1995) 68:1406; 31. Am J Physiol Heart Circ Physiol (2008) 294:H1417; 32. Chem Biol (1997) 4:269; 33. Biophys J (1995) 69:1272; 34. J Neurosci (2009) 29:9197.

when tested *in vitro* using hippocampal slices and *in vivo* using single-unit recordings in cat and monkey visual cortices.[28] Electrophysiological measurements indicate a broadening of action potentials that is attributable to the staining of cultured neurons with RH 237.[29]

Like the ANEP dyes, the RH dyes exhibit varying degrees of fluorescence excitation and emission spectral shifts in response to membrane potential changes.[8] Their absorption and fluorescence spectra are also strongly dependent on the environment.[30] Spectra of RH 414 bound to phospholipid vesicles are similar to those obtained on neuronal plasma membranes.[8] Using the RH dyes in conjunction with fluorescent Ca^{2+} indicators allowed the simultaneous optical mapping of membrane potential (with RH 237) and intracellular calcium (with rhod-2 AM; R1245MP, R1244; Section 19.3) in cardiomyocyte monolayers; rhod-FF (R23983, Section 19.3) was used to check for buffering of calcium dynamics by the high-affinity rhod-2 indicator.[31]

FRET-Pair Membrane Potential Sensors

Fluorescence resonance energy transfer (FRET) between a mobile lipophilic anion in the membrane interior and a static donor fluorophore on the membrane surface provides a potential-sensing mechanism that generates a more sensitive fluorescence response than electrochromic dyes and a more rapid temporal response than intracellular–extracellular ion translocation.[32,33] A particularly effective implementation of this concept uses $DiOC_{18}(3)$ or $DiOC_{16}(3)$ (D275, V22886, D1125; Section 14.4) as the static reference marker in combination with the mobile anion dipicrylamine. Characterization of this approach by Bradley and co-workers[34] demonstrated depolarization-induced fluorescence changes of >50% per 100 mV with submillisecond time constants in whole-cell patch-clamped HEK^s293 cells. In neuronal cultures and brain slices, action potentials generated fluorescence increases relative to the resting baseline signal of >25% per 100 mV.

DATA TABLE 22.2 FAST-RESPONSE PROBES

Cat. No.	MW	Storage	Soluble	Abs	EC	Em	Solvent	Notes
D1199	480.66	D,L	DMSO, EtOH	497	42,000	705	MeOH	1
D3167	592.88	D,L	DMSO, EtOH	498	37,000	713	MeOH	1
D6923	549.39	D,L	DMSO, EtOH	517	36,000	721	EtOH	
D6927	843.95	D,L	DMSO, EtOH	519	36,000	719	EtOH	1
D36801	637.50	F,D,L	DMSO, EtOH	512	36,000	712	EtOH	1
D36802	665.55	F,D,L	DMSO, EtOH	512	36,000	712	EtOH	1, 2
R649	585.42	D,L	DMSO, EtOH	530	47,000	712	MeOH	1
S1108	498.72	D,L	DMSO, EtOH	515	50,000	704	MeOH	1, 3
S1109	496.71	D,L	DMSO, EtOH	528	53,000	782	MeOH	1, 3
T1111	581.48	D,L	DMSO, EtOH	532	55,000	716	MeOH	1

For definitions of the contents of this data table, see "Using *The Molecular Probes® Handbook*" in the introductory pages.

Notes

1. Abs and Em of styryl dyes are at shorter wavelengths in membrane environments than in reference solvents such as methanol. The difference is typically 20 nm for absorption and 80 nm for emission, but varies considerably from one dye to another. Styryl dyes are generally nonfluorescent in water.
2. The fluorescence excitation/emission maxima of di-4-ANEPPDHQ (D36802) bound to dioleoylphosphatidylcholine (DOPC) bilayer membranes are 468/635 nm. Em is sensitive to membrane cholesterol content. (Biophys J (2006) 90:2563)
3. Abs/Em for these dyes adsorbed on neuronal plasma membranes are 493/638 nm (S1108) and 506/687 nm (S1109). (Biochim Biophys Acta (1993) 1150:111)

PRODUCT LIST 22.2 FAST-RESPONSE PROBES

Cat. No.	Product	Quantity
D6923	di-2-ANEPEQ (JPW 1114)	5 mg
D36801	di-3-ANEPPDHQ	1 mg
D36802	di-4-ANEPPDHQ	1 mg
D6927	di-12-ANEPPQ	5 mg
D1199	di-4-ANEPPS	5 mg
D3167	di-8-ANEPPS	5 mg
R649	RH 795	1 mg
S1109	*N*-(4-sulfobutyl)-4-(6-(4-(dibutylamino)phenyl)hexatrienyl)pyridinium, inner salt (RH 237)	5 mg
S1108	*N*-(4-sulfobutyl)-4-(4-(4-(dipentylamino)phenyl)butadienyl)pyridinium, inner salt (RH 421)	25 mg
T1111	*N*-(3-triethylammoniumpropyl)-4-(4-(4-(diethylamino)phenyl)butadienyl)pyridinium dibromide (RH 414)	5 mg

IMPORTANT NOTICE: The products described in this manual are covered by one or more Limited Use Label License(s). Please refer to the Appendix on page 971 and Master Product List on page 975. Products are For Research Use Only. Not intended for any animal or human therapeutic or diagnostic use.
www.invitrogen.com/probes

22.3 Slow-Response Probes

Our slow-response potential-sensitive probes[1] (see Figure 22.1.1B in Section 22.1) are listed in Table 22.2, along with their charges, optical responses and selected applications.

Table 22.2 Characteristics and selected applications of Molecular Probes® slow-response probes.

Dyes (Cat. No.)	Structure (Charge)	Optical Response	Selected Applications
• $DiOC_2(3)$ (D14730) • $DiOC_5(3)$ (D272) • $DiOC_6(3)$ (D273) • $DiSC_3(5)$ (D306) • $DiIC_1(5)$ (H14700)	Carbocyanine (cationic)	SLOW; fluorescence response to depolarization depends on the staining concentration and detection method	• Bacterial infection [1–3] • Flow cytometry assays of membrane potential [4] • Membrane potential in intact yeast cells [5] • Mitochondrial activity [6–8] • Oxidative stress [9,10]
• JC-1 (T3168) • JC-9 (D22421)	Carbocyanine (cationic)	SLOW; fluorescence emission ratio 585/520 nm increases upon membrane hyperpolarization	• Apoptotic mitochondrial depolarization [11–13] • Ca^{2+} regulation by mitochondria [14–16] • Mitochondrial function in brain slices [17] and cultured neurons [18,19] • Mitochondrial response to glutamate excitotoxicity [20,21] • Mitochondrial response to oxidative stress [22]
• Tetramethylrhodamine methyl and ethyl esters (T668, T669) • Rhodamine 123 (R302, R22420)	Rhodamine (cationic)	SLOW; used to obtain unbiased images of potential-dependent dye distribution	• Ca^{2+} regulation by mitochondria [23–25] • Mitochondrial permeability transition [26–28] • Oxidative stress [29,30] • Stem cells [31]
• Oxonol V (O266) • Oxonol VI (O267)	Oxonol (anionic)	SLOW; fluorescence decreases upon membrane hyperpolarization	• Ion channels and electrogenic pumps [32–36] • Liposomes [37–39] • Plant physiology [40–42]
• $DiBAC_4(3)$ (B438, B24570) • $DiBAC_4(5)$ (B436) • $DiSBAC_2(3)$ (B413)	Oxonol (anionic)	SLOW; fluorescence decreases upon membrane hyperpolarization	• ATP-sensitive K^+ channel activation [43–46] • Combined potentiometric and Ca^{2+} measurements [47–49] • Confocal imaging of membrane potential [50] • Flow cytometry assays of cell viability [51–54] • Primary vascular smooth muscle cells [47]
• Merocyanine 540 (M24571)	Merocyanine	FAST/SLOW (biphasic response)	• Membrane lipid asymmetry [55,56] • Membrane potentials in mitochondria [57,58] and skeletal muscle [59] • Photodynamic therapy • Structure of membrane surfaces [60–62]

1. J Biol Chem (2007) 282:7742; **2.** Antimicrob Agents Chemother (2005) 49:1127; **3.** Appl Environ Microbiol (2002) 68:37; **4.** J Biol Chem (2007) 282:18069; **5.** Anal Biochem (2001) 293:269; **6.** J Biol Chem (2006) 281:13990; **7.** J Biol Chem (1998) 273:33942; **8.** Cytometry (1998) 33:333; **9.** J Biol Chem (2006) 281:6726; **10.** Science (2003) 299:1751; **11.** Methods Cell Biol (2001) 63:467; **12.** J Cell Biol (1997) 138:449; **13.** FEBS Lett (1997) 411:77; **14.** J Biol Chem (2000) 275:38680; **15.** J Physiol (1998) 509:81; **16.** Biophys J (1998) 75:2004; **17** Methods (1999) 18:104; **18.** J Biol Chem (2009) 284:9540; **19.** Mol Biol Cell (2008) 19:150; **20.** J Neurosci (1996) 16:5688; **21.** Neuron (1995) 15:961; **22.** J Biol Chem (2007) 282:24146; **23.** Methods Mol Biol (2007) 372:421; **24.** J Biol Chem (2001) 276:23329; **25.** J Neurosci (2000) 20:7290; **26.** J Biol Chem (2009) 284:15117; **27** Biochem J (1999) 343; **28.** Biophys J (1998) 74:2129; **29.** J Biol Chem (2009) 284:14476; **30.** J Biol Chem (2009) 284:18754; **31.** Nat Methods (2010) 7:61; **32.** J Biol Chem (1993) 268:23122; **33.** Biochemistry (1990) 29:3859; **34.** Biochim Biophys Acta (1990) 1023:81; **35.** Biochim Biophys Acta (1990) 1017:221; **36.** Biochim Biophys Acta (1989) 980:139; **37.** Biochim Biophys Acta (1993) 1146:87; **38.** Biochem Biophys Res Commun (1990) 173:1008; **39.** Biophys Chem (1989) 34:225; **40.** J Biol Chem (1992) 267:21850; **41.** Plant J (1992) 2:97; **42.** Biophys J (1999) 76:360; **43.** Br J Pharmacol (2000) 129:1323; **44.** J Physiol (1999) 517:781; **45.** J Physiol (1997) 502:397; **46.** Pflugers Arch (1997) 434:712; **47.** Circ Res (2009) 104:670; **48.** J Neurosci (2007) 27:8238; **49.** Methods (2000) 21:335; **50.** Exp Cell Res (1997) 231:260; **51.** Methods Cell Biol (2001) 64:553; **52.** J Appl Microbiol (1998) 84:988; **53.** Yeast (1998) 14:147; **54.** J Microbiol Methods (1998) 32:45; **55.** Biochim Biophys Acta (1991) 1062:24; **56.** Proc Natl Acad Sci U S A (1986) 83:3311; **57.** J Biol Chem (1991) 266:803; **58.** J Membr Biol (1991) 123:23; **59.** J Gen Physiol (1990) 95:147; **60.** Biochim Biophys Acta (1993) 1146:169; **61.** Biochim Biophys Acta (1992) 1107:245; **62.** J Cell Physiol (1989) 138:61.

Carbocyanines

DiI, DiS and DiO Derivatives

Indo- (DiI), thia- (DiS) and oxa- (DiO) carbocyanines with short alkyl tails (<7 carbon atoms) were among the first potentiometric fluorescent probes developed.[2] These cationic dyes accumulate on hyperpolarized membranes and are translocated into the lipid bilayer.[3] Aggregation within the confined membrane interior usually results in decreased fluorescence, although the magnitude and even the direction of the fluorescence response is strongly dependent on the concentration of the dye and its structural characteristics.[1,4] While the distribution of extracellularly applied dye is dependent on both the plasma and mitochondrial membrane potentials, the primary determinant is the latter. Very low applied concentrations (<100 nM) are required to obtain mitochondrial signal specificity and avoid potential-independent background derived from staining of the endoplasmic reticulum and other intracellular membranes.

$DiOC_6(3)$ (D273, Figure 22.3.1) has been the most widely used carbocyanine dye for membrane potential measurements,[5] followed closely by $DiOC_5(3)$ (D272, Figure 22.3.2); see Table 22.2 for selected references. In flow cytometry measurements, the detected intensity of carbocyanine

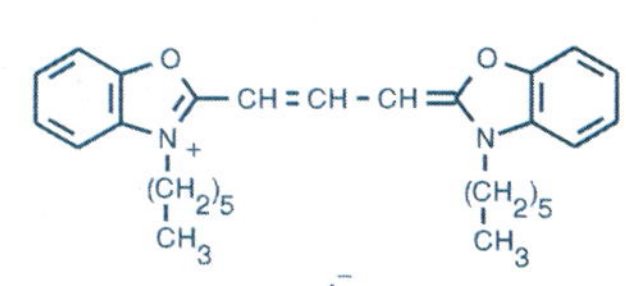

Figure 22.3.1 3,3′-dihexyloxacarbocyanine iodide ($DiOC_6(3)$, D273).

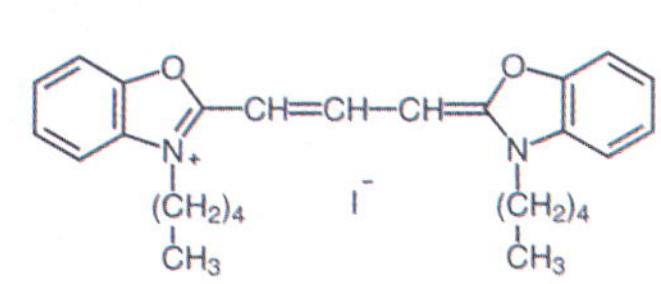

Figure 22.3.2 3,3′-dipentyloxacarbocyanine iodide ($DiOC_5(3)$, D272).

IMPORTANT NOTICE: The products described in this manual are covered by one or more Limited Use Label License(s). Please refer to the Appendix on page 971 and Master Product List on page 975. Products are For Research Use Only. Not intended for any animal or human therapeutic or diagnostic use.

fluorescence is dependent not only on the membrane potential, but also on cell size. In some cases, measurements of forward light scatter have been used to normalize the optical changes for cell size variability. A fluorescence ratio method (Figure 22.3.3) that exploits a potential-dependent red shift in the emission spectrum of $DiOC_2(3)$ (D14730, Figure 22.3.4) has been developed for membrane potential measurements in bacteria [6,7] (B34950). The 633 nm light–excitable indodicarbocyanine $DiIC_1(5)$ [8] (H14700, Figure 22.3.5) enables analysis of mitochondrial potential in apoptotic cells in combination with fluoresceinated annexin V [9] (A13199, Section 15.5), a method we have utilized in two of our MitoProbe™ Assay Kits for flow cytometry (M34150, M34151; see below). Carbocyanine dyes, particularly thiacyanines such as $DiSC_3(5)$ (D306, Figure 22.3.6), can inhibit respiration [4,10] and may therefore be relatively cytotoxic.[11]

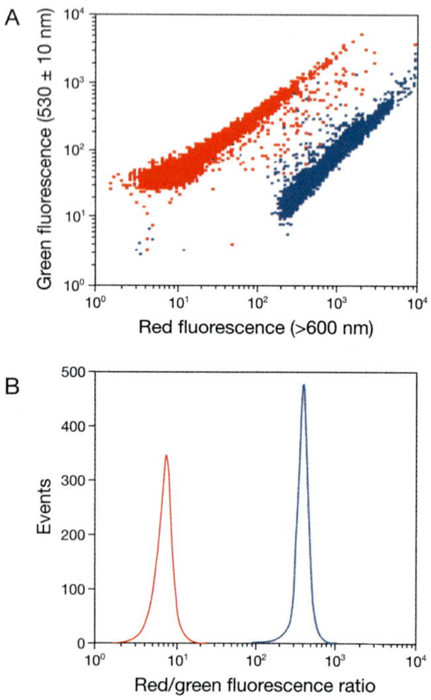

Figure 22.3.3 A) Two-color flow cytometric analysis of *Staphylococcus aureus* populations stained with 30 µM $DiOC_2(3)$ (D14730) in the presence (red) or absence (blue) of the metabolic uncoupler carbonyl cyanide *m*-chlorophenylhydrazone (CCCP). Note the variability (~100-fold range) of the green and red fluorescence intensities. **B)** The same data expressed as red/green fluorescence intensity ratios. Ratio values are calculated by subtracting the logarithmic green fluorescence channel value from the corresponding logarithmic red fluorescence channel value. Figure supplied by Howard M. Shapiro, Harvard Medical School, Boston, MA.

JC-1 and JC-9

JC-1 (5,5′,6,6′-tetrachloro-1,1′,3,3′-tetraethylbenzimidazolylcarbocyanine iodide, T3168; Figure 22.3.7) exists as a green-fluorescent monomer at low concentrations or at low membrane potential. However, at higher concentrations (aqueous solutions above 0.1 µM) or higher potentials, JC-1 forms red-fluorescent "J-aggregates," which exhibit a broad excitation spectrum and a very narrow emission spectrum (Figure 22.3.8). Because J-aggregate formation increases linearly with applied membrane potential over the range of 30–180 mV, this phenomenon can be exploited for potentiometric measurements [12,13] (Table 22.2). JC-1 is more specific for mitochondrial versus plasma membrane potential and more consistent in its response to depolarization than some other cationic dyes such as $DiOC_6(3)$ and rhodamine 123.[14]

Various types of ratio measurements are possible by combining signals from the green-fluorescent JC-1 monomer (absorption/emission maxima ~514/529 nm) and the red-fluorescent J-aggregate (Figure 22.3.9) (absorption/emission maxima ~585/590 nm), which can be effectively excited anywhere between 485 nm and its absorption maximum. Optical filters designed for fluorescein and tetramethylrhodamine can be used to separately visualize the monomer and J-aggregate forms, respectively. Alternatively, both forms can be observed simultaneously using a fluorescein longpass optical filter set. For flow cytometry, JC-1 can be excited at 488 nm and detected in bivariate mode using the green channel for the monomer and the red channel for the J-aggregate form [15,16] (Figure 22.3.10).

JC-1 is widely used for detecting mitochondrial depolarization in apoptotic cells [14,15,17–19] (MitoProbe™ JC-1 Assay Kit, M34152, described below) and for assaying multidrug-resistant cells [20] (Section 15.6). It is also frequently employed for mitochondrial function assessment in cell-based high-throughput assays.[21,22] We have discovered another carbocyanine dye, JC-9 (3,3′-dimethyl-α-naphthoxacarbocyanine iodide, D22421; Figure 22.3.11), with potential-dependent spectroscopic properties (Figure 22.3.12) similar to those of JC-1 for detecting mitochondrial depolarization in apoptotic cells.[23,24]

MitoProbe™ JC-1 Assay Kit for Flow Cytometry

The MitoProbe™ JC-1 Assay Kit (M34152) provides the cationic dye JC-1 and a mitochondrial membrane potential disrupter, CCCP (carbonyl cyanide 3-chlorophenylhydrazone), for the study of mitochondrial membrane potential. Use of JC-1 fluorescence ratio detection allows researchers to make comparative measurements of membrane potential and to determine the percentage of mitochondria within a population that respond to an applied stimulus (Figure 22.3.13). Subtle heterogeneity in cellular responses can be discerned in this way.[13,15,18,25] For example, four distinct patterns of mitochondrial membrane potential change in response to glutamate receptor activation in neurons have been identified using confocal ratio imaging of JC-1 fluorescence.[26]

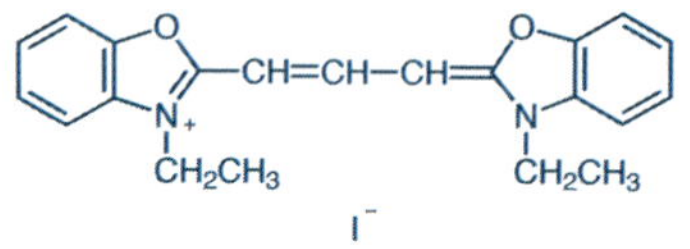

Figure 22.3.4 3,3′-diethyloxacarbocyanine iodide ($DiOC_2(3)$, D14730).

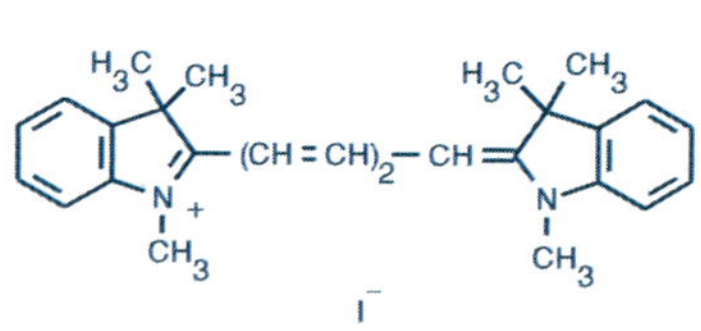

Figure 22.3.5 1,1′,3,3,3′,3′-hexamethylindodicarbocyanine iodide ($DiIC_1(5)$, H14700).

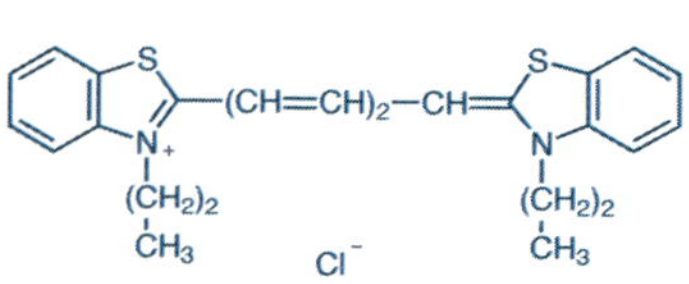

Figure 22.3.6 3,3′-dipropylthiadicarbocyanine iodide ($DiSC_3(5)$, D306).

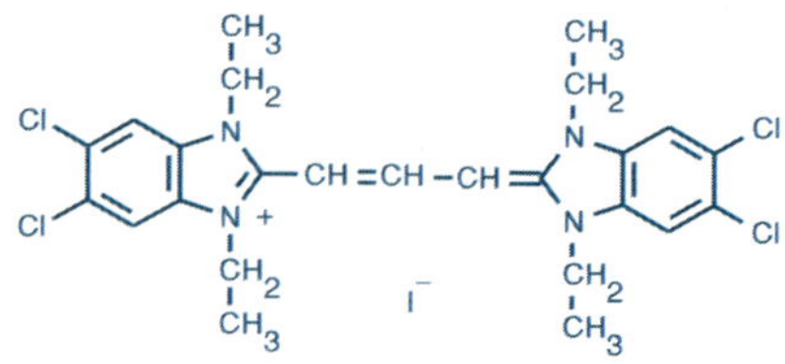

Figure 22.3.7 5,5′,6,6′-tetrachloro-1,1′,3,3′-tetraethylbenzimidazolylcarbocyanine iodide (JC-1; $CBIC_2(3)$, T3168).

IMPORTANT NOTICE: The products described in this manual are covered by one or more Limited Use Label License(s). Please refer to the Appendix on page 971 and Master Product List on page 975. Products are For Research Use Only. Not intended for any animal or human therapeutic or diagnostic use.

Each MitoProbe™ JC-1 Assay Kit provides:

- JC-1
- Dimethylsulfoxide (DMSO)
- CCCP
- Concentrated phosphate-buffered saline (PBS)
- Detailed protocols

Sufficient reagents are provided for 100 assays, based on a labeling volume of 1 mL.

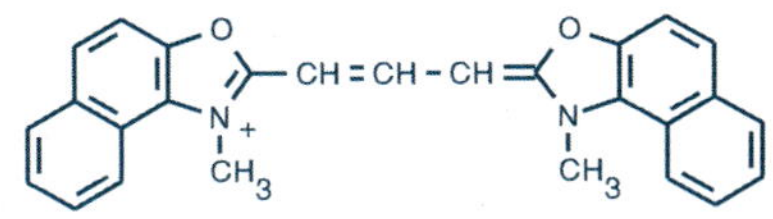

Figure 22.3.11 3,3′-dimethyl-α-naphthoxacarbocyanine iodide (JC-9; $DiNOC_1(3)$, D22421).

MitoProbe™ DiIC(5) and MitoProbe™ DiOC(3) Assay Kits for Flow Cytometry

The MitoProbe™ $DiIC_1(5)$ and MitoProbe™ $DiOC_2(3)$ Assay Kits (M34151, M34150) provide solutions of the far-red–fluorescent $DiIC_1(5)$ (1,1′,3,3,3′,3′-hexamethylindodicarbocyanine

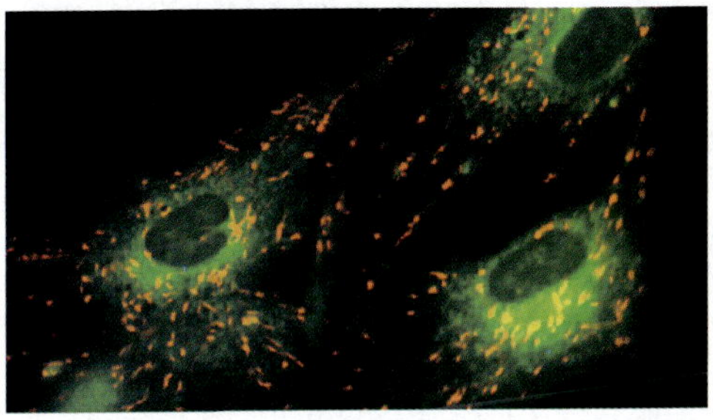

Figure 22.3.8 Cultured human pre-adipocytes loaded with the ratiometric mitochondrial potential indicator JC-1 (T3168) at 5 µM for 30 minutes at 37°C. In live cells, JC-1 exists either as a green-fluorescent monomer at depolarized membrane potentials or as an orange-fluorescent J-aggregate at hyperpolarized membrane potentials. Cells were then treated with 50 nM FCCP, a protonophore, to depolarize the mitochondrial membrane. Approximately 10 minutes after the addition of the uncoupler, the cells were illuminated at 488 nm and the emission was collected between 515–545 nm and 575–625 nm. Image contributed by Bob Terry, BioImage A/S, Denmark.

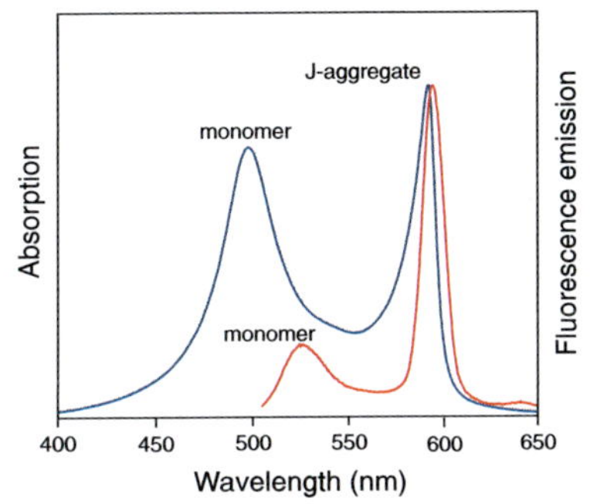

Figure 22.3.9 Absorption and fluorescence emission (excited at 488 nm) spectra of JC-1 in pH 8.2 buffer containing 1% (v/v) DMSO.

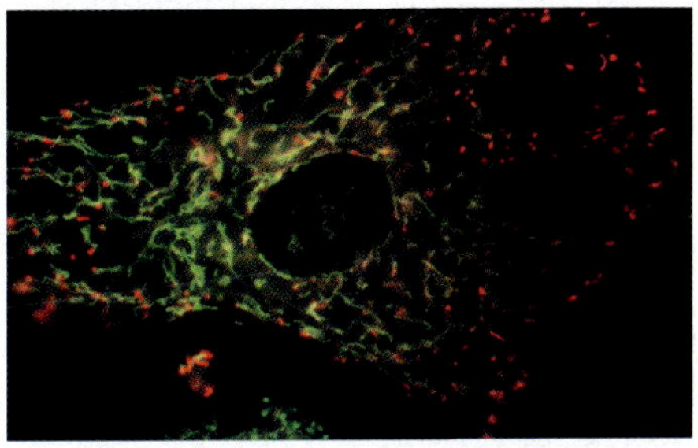

Figure 22.3.12 A viable bovine pulmonary artery endothelial cell incubated with the ratiometric mitochondrial potential indicator, JC-9 (D22421). In live cells, JC-9 exists either as a green-fluorescent monomer at depolarized membrane potentials, or as a red-fluorescent J-aggregate at hyperpolarized membrane potentials.

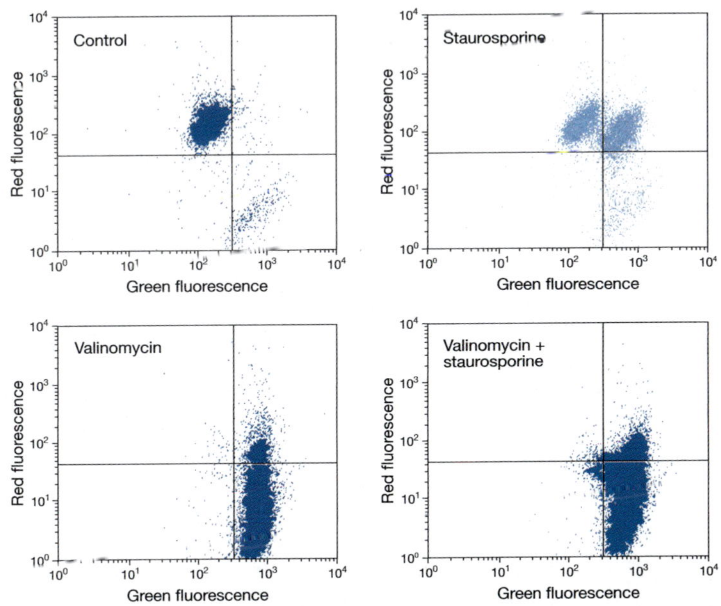

Figure 22.3.10 Bivariate JC-1 (T3168) analysis of mitochondrial membrane potential in HL60 cells by flow cytometry. The sensitivity of this technique is demonstrated by the response to depolarization using K+/valinomycin (V1644) (bottom two panels). Distinct populations of cells with different extents of mitochondrial depolarization are detectable following apoptosis-inducing treatment with 5 µM staurosporine for 2 hours (top right panel). Figure courtesy of Andrea Cossarizza, University of Modena and Reggio Emilia, Italy.

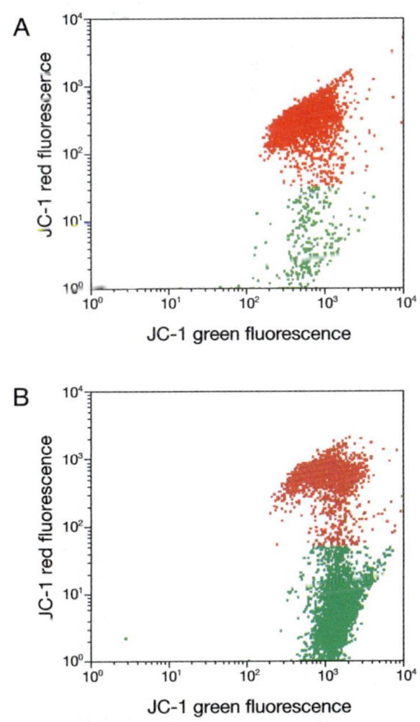

Figure 22.3.13 Flow cytometric analysis of Jurkat cells using the MitoProbe™ JC-1 Assay Kit (M34152). Jurkat cells were stained with 2 µM JC-1 for 15 minutes at 37°C, 5% CO_2, and then washed with phosphate-buffered saline (PBS) and analyzed on a flow cytometer using 488 nm excitation with 530 nm and 585 nm bandpass emission filters. Untreated cultured cells are shown in panel **A**. Panel **B** shows cells induced to apoptosis with 10 µM camptothecin for 4 hours at 37°C.

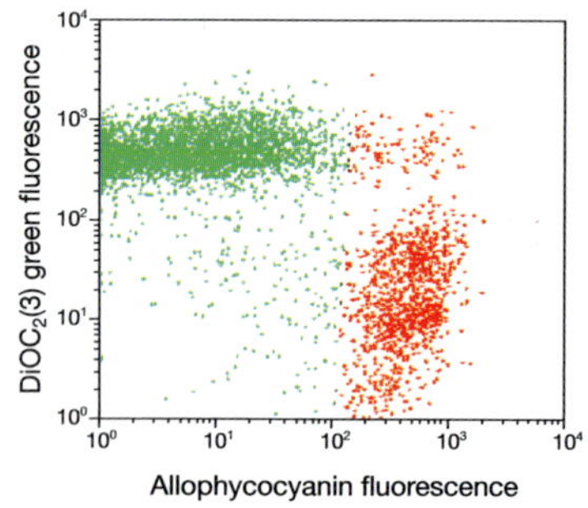

Figure 22.3.14 Flow cytometric analysis of camptothecin-treated Jurkat cells stained with $DiOC_2(3)$ (D14730, M34150) and allophycocyanin annexin V (A35110). Jurkat cells were incubated for 4 hours with camptothecin at 37°C, 5% CO_2, then stained with 50 nM $DiOC_2(3)$ and allophycocyanin annexin V. Cells were analyzed on a flow cytometer using 488 nm and 633 nm excitations with 530 nm and 660 nm bandpass emission filters.

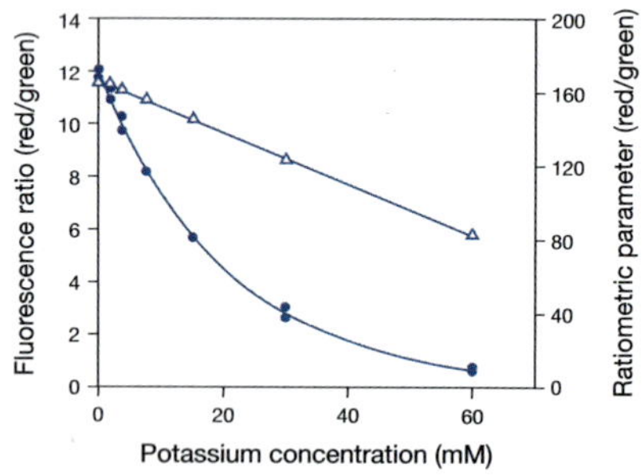

Figure 22.3.15 Response of *Staphylococcus aureus* to valinomycin and external potassium ions, as measured by flow cytometry using the *Bac*Light™ Bacterial Membrane Potential Kit (B34950). Samples containing *S. aureus* were treated with 5 µM valinomycin in different concentrations of potassium buffer, and then stained using 30 µM $DiOC_2(3)$ for 30 minutes, according to the kit protocol. Data are expressed either using a ratiometric parameter based on the formula provided in the kit protocol (Δ, right axis) or as the ratio of population mean red-fluorescence intensity/mean green-fluorescence intensity (●, left axis).

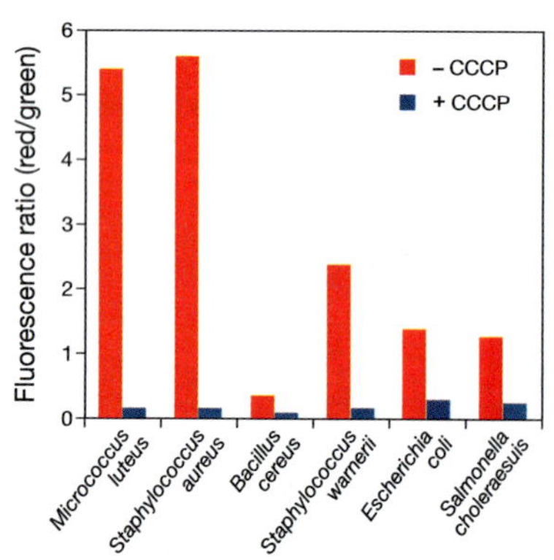

Figure 22.3.16 Detection of membrane potential in various bacteria with the *Bac*Light™ Bacterial Membrane Potential Kit (B34950). Red/green fluorescence ratios were calculated using population mean fluorescence intensities for gram-positive (*Micrococcus luteus, Staphylococcus aureus, Bacillus cereus* and *Staphylococcus warnerii*) and gram-negative (*Escherichia coli* and *Salmonella choleraesuis*) bacteria incubated with 30 µM $DiOC_2(3)$ for 30 minutes in either the presence or absence of 5 µM CCCP, according to the kit protocol.

iodide) and green-fluorescent $DiOC_2(3)$ (3,3′-diethyloxacarbocyanine iodide) carbocyanine dyes, respectively, along with a mitochondrial membrane potential uncoupler, CCCP, for the study of mitochondrial membrane potential. These $DiIC_1(5)$ and $DiOC_2(3)$ carbocyanine dyes penetrate the cytosol of eukaryotic cells and, at concentrations below 100 nM, accumulate primarily in mitochondria with active membrane potentials. In the case of $DiOC_2(3)$, this accumulation is accompanied by a shift from green to red emission due to dye stacking (Figure 22.3.3), allowing the use of a ratiometric parameter (red/green fluorescence ratio) that corrects for size differences when measuring membrane potential in bacteria.[6,7] $DiOC_2(3)$ can be paired with other reagents, such as the far-red–fluorescent allophycocyanin annexin V (A35110, Section 15.5) or TO-PRO®-3 nucleic acid stain (T3605, Section 8.1) for multiparameter study of vitality and apoptosis[27] (Figure 22.3.14).

The MitoProbe™ $DiIC_1(5)$ and MitoProbe™ $DiOC_2(3)$ Assay Kits provide:

- $DiIC_1(5)$ (in Kit M34151) or $DiOC_2(3)$ (in Kit M34150)
- CCCP
- Detailed protocols for labeling cells

Each kit provides sufficient reagents for 100 assays, based on a labeling volume of 1 mL.

*Bac*Light™ Bacterial Membrane Potential Kit

The *Bac*Light™ Bacterial Membrane Potential Kit (B34950) provides a fluorescent membrane-potential indicator dye, $DiOC_2(3)$, along with a proton ionophore (CCCP) and premixed buffer. At low concentrations, $DiOC_2(3)$ exhibits green fluorescence in all bacterial cells, but it becomes more concentrated in healthy cells that are maintaining a membrane potential, causing the dye to self-associate and the fluorescence emission to shift to red. The red- and green-fluorescent bacterial populations are easily distinguished using a flow cytometer (Figure 22.3.15). CCCP is included in the kit for use as a control because it eradicates the proton gradient, eliminating bacterial membrane potential.[7,27]

The *Bac*Light™ Bacterial Membrane Potential Kit contains:

- $DiOC_2(3)$ in dimethylsulfoxide (DMSO)
- CCCP in DMSO
- Phosphate-buffered saline (PBS)
- Detailed protocols

Using the recommended reagent dilutions and volumes, this kit provides sufficient $DiOC_2(3)$ to perform approximately 100 individual assays by flow cytometry; sufficient CCCP is provided for 30 depolarized control samples. The *Bac*Light™ Bacterial Membrane Potential Kit is designed to assay bacterial concentrations between 10^5 and 10^7 organisms per mL. Note that $DiOC_2(3)$ and CCCP are inhibitors of respiration, rendering the cells nonculturable beyond the brief time window required for staining and analysis.

Using the *Bac*Light™ Bacterial Membrane Potential Kit, we have detected membrane potentials in all bacteria tested (including logarithmically growing cultures of *Micrococcus luteus, Staphylococcus aureus, Bacillus cereus, Staphylococcus warnerii, Escherichia coli* and *Salmonella choleraesuis*), although the magnitude varies with species (Figure 22.3.16).

IMPORTANT NOTICE: The products described in this manual are covered by one or more Limited Use Label License(s). Please refer to the Appendix on page 971 and Master Product List on page 975. Products are For Research Use Only. Not intended for any animal or human therapeutic or diagnostic use.

Rhodamine 123, TMRM and TMRE

Rhodamine 123 (R302, R22420; Figure 22.3.17) is widely used as a structural marker for mitochondria (Section 12.2) and as an indicator of mitochondrial activity (Section 15.2, Figure 22.3.18). Highly selective, potential-dependent staining of mitochondria is obtained by setting the extracellular K^+ concentration close to intracellular values (~137 mM), thereby depolarizing the plasma membrane.[28]

TMRM (T668, Figure 22.3.19) and TMRE (T669, Figure 22.3.20), the methyl and ethyl esters of tetramethylrhodamine, are closely related to rhodamine 123. They are primarily mitochondrial membrane potential sensors.[29,30] As with rhodamine 123, accumulation of these cationic dyes in mitochondria results in diminished fluorescence due to self-quenching (Figure 22.3.18).[30] TMRM and TMRE cross the plasma membrane more rapidly than rhodamine 123, and their strong fluorescence allows the use of low probe concentrations, thus avoiding aggregation. Because their fluorescence is relatively insensitive to the environment, spatially resolved fluorescence of TMRM and TMRE presents an unbiased profile of their transmembrane distribution that can be directly related to the plasma membrane potential via the Nernst equation.[30–34] TMRE has been successfully used in a high-throughput screening assay for drugs that affect mitochondrial membrane potential in live cells.[35] TMRM has been used in conjunction with X-rhod-1 AM, (X14210, Section 19.3) for simultaneous confocal imaging of mitochondrial membrane potential and calcium in rat cardiomyocytes.[36]

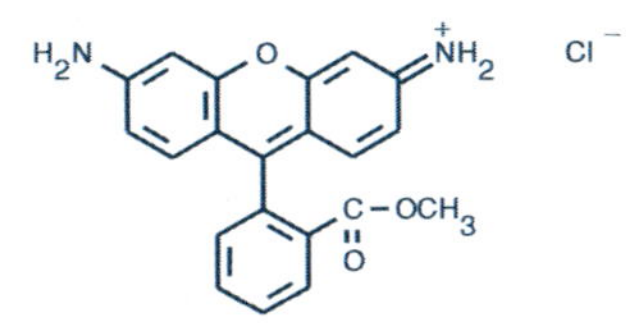

Figure 22.3.17 Rhodamine 123 (R302).

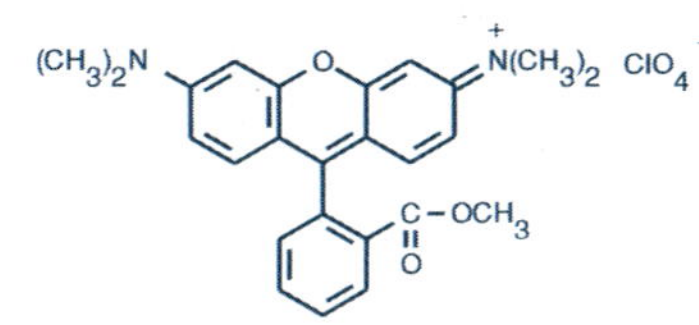

Figure 22.3.19 Tetramethylrhodamine, methyl ester, perchlorate (TMRM, T668).

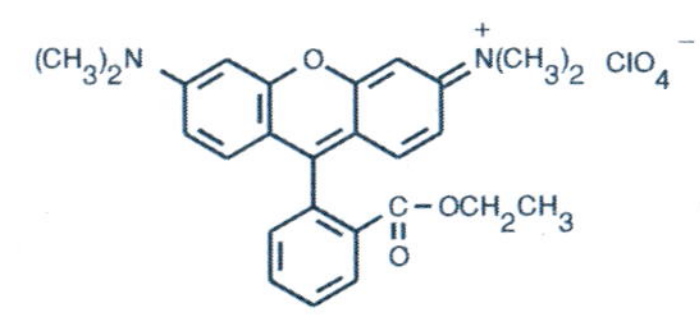

Figure 22.3.20 Tetramethylrhodamine, ethyl ester, perchlorate (TMRE, T669).

Oxonols

Oxonol V and Oxonol VI

The anionic bis-isoxazolone oxonols (O266, Figure 22.3.21; O267, Figure 22.3.22) accumulate in the cytoplasm of depolarized cells by a Nernst equilibrium–dependent uptake from the extracellular solution.[37] Their voltage-dependent partitioning between water and membranes is often measured by absorption rather than fluorescence. Of the oxonols studied by Smith and Chance,[38] oxonol VI (O267) gave the largest spectral shifts, with an isosbestic point at 603 nm. In addition, oxonol VI responds to changes in potential more rapidly than oxonol V and is therefore considered to be the better probe for measuring fast potential changes.[39]

DiBAC (Bis-Oxonol) Dyes

The three bis-barbituric acid oxonols, often referred to as DiBAC dyes, form a family of spectrally distinct potentiometric probes with excitation maxima at approximately 490 nm ($DiBAC_4(3)$; B438, B24570; Figure 22.3.23), 530 nm ($DiSBAC_2(3)$, B413; Figure 22.3.24) and

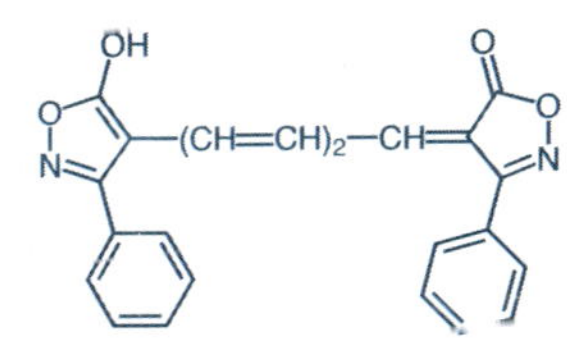

Figure 22.3.21 Oxonol V (bis-(3-phenyl-5-oxoisoxazol-4-yl) pentamethine oxonol, O266).

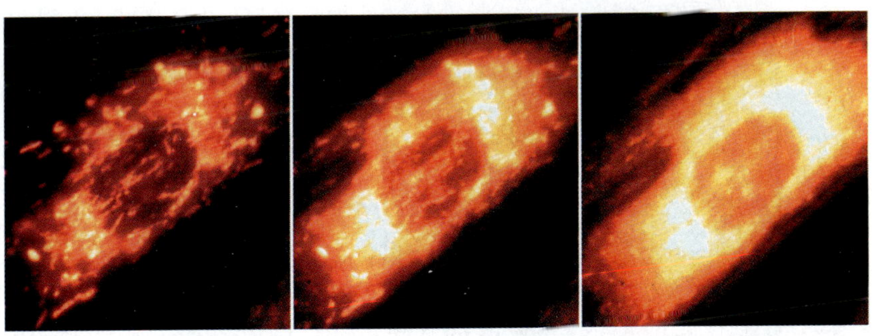

Figure 22.3.18 Staining of rat cortical astrocytes by rhodamine 123 (R302, R22420). Potential-dependent accumulation of the cationic dye in mitochondria results in a relatively weak fluorescence signal due to self-quenching (left panel). Dissipation of the mitochondrial membrane potential by the uncoupler FCCP is marked by increasing fluorescence (middle panel) and subsequent redistribution of the dye throughout the cell (right panel). Images courtesy of Michael Duchen, University College, London.

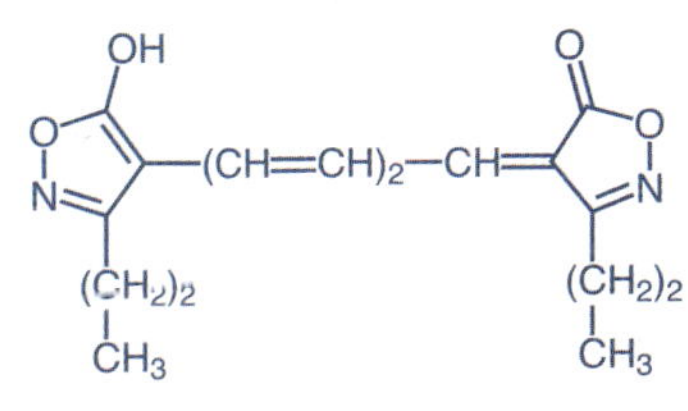

Figure 22.3.22 Oxonol VI (bis-(3-propyl-5-oxoisoxazol-4-yl) pentamethine oxonol, O267).

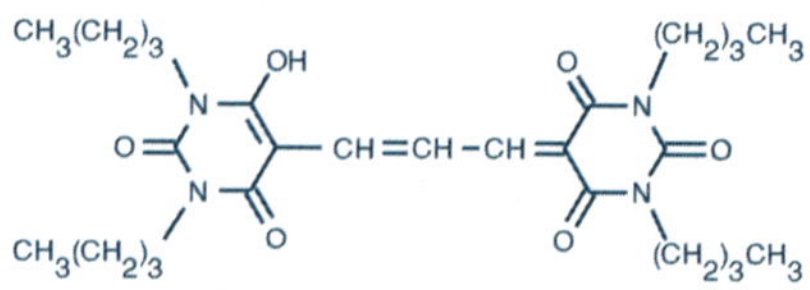

Figure 22.3.23 bis-(1,3-dibutylbarbituric acid)trimethine oxonol (DiBAC$_4$(3), B438).

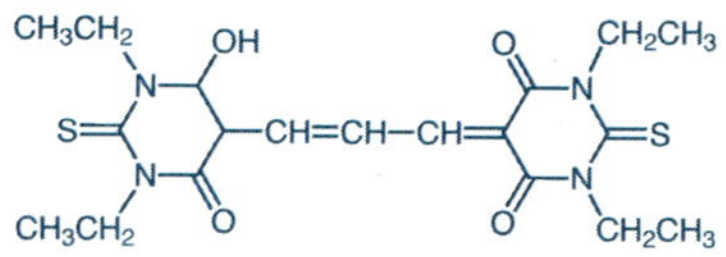

Figure 22.3.24 bis-(1,3-diethylthiobarbituric acid)trimethine oxonol (DiSBAC$_2$(3), B413).

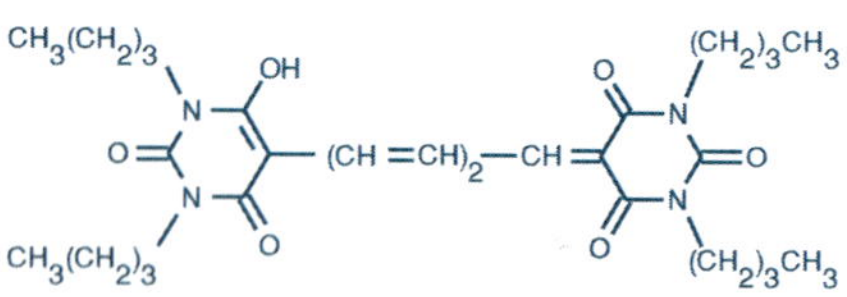

Figure 22.3.25 bis-(1,3-dibutylbarbituric acid)pentamethine oxonol (DiBAC$_4$(5), B436).

590 nm (DiBAC$_4$(5), B436; Figure 22.3.25). Several papers have referred to these dyes simply as "bis-oxonol" and it is not always possible to determine which of the dyes was employed; however, DiBAC$_4$(3) (Figure 22.3.23) has been used in the majority of publications that cite a "bis-oxonol."

These dyes enter depolarized cells where they bind to intracellular proteins or membranes and exhibit enhanced fluorescence and red spectral shifts.[40] Increased depolarization results in more influx of the anionic dye and thus an increase in fluorescence (Figure 22.3.26). Conversely, hyperpolarization is indicated by a decrease in fluorescence (Figure 22.3.27). In contrast to cationic carbocyanines, anionic bis-oxonols are largely excluded from mitochondria and are primarily sensitive to plasma membrane potential. Potential-dependent fluorescence changes generated by DiBAC$_4$(3) are typically ~1% per mV.[41,42] Interactions between anionic oxonols and the cationic K^+-valinomycin complex complicate the use of this ionophore when calibrating potentiometric responses.[43] Oxonol dyes have known pharmacological activity against various ion channels and receptors.[44,45] It is therefore important to establish, as is the case in any experiment using fluorescent probes, that experimental observations with implied physiological significance are independent of the externally applied probe concentration.

Merocyanine 540

Although merocyanine 540 (M24571) was among the first fluorescent dyes to be used as a potentiometric probe,[46] its use for this application has declined with the advent of superior probes. A significant disadvantage of merocyanine 540 is its extreme phototoxicity; consequently, it is now more commonly used as a photosensitizer.[47–56]

Merocyanine 540 exhibits a biphasic kinetic response to membrane polarization changes. It binds to the surface of polarized membranes in a perpendicular orientation, reorienting as the membrane depolarizes to form nonfluorescent dimers with altered absorption spectra.[57,58] This fast (microseconds) reorientation is followed by a slower response caused by an increased dye uptake.

Merocyanine 540 is also a useful probe of lipid packing because it binds preferentially to membranes with highly disordered lipids.[59,60] Fluorescence of merocyanine 540 is sensitive to heat-induced changes in the organization of membrane lipids.[61]

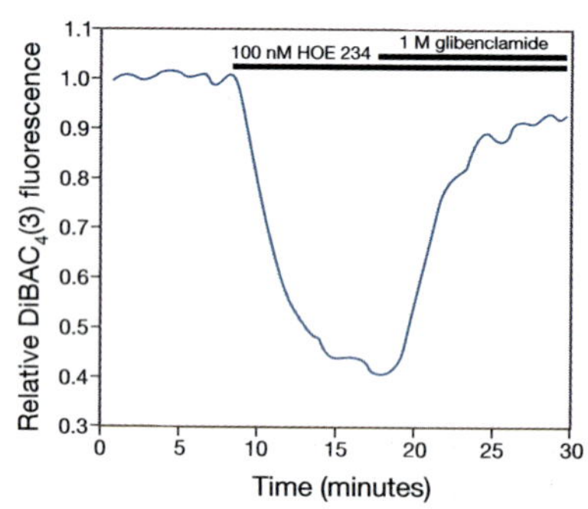

Figure 22.3.27 Detection of ATP-sensitive potassium (K_{ATP}) channel activation in isolated capillaries from guinea pig hearts using DiBAC$_4$(3) (B438, B24570), a slow potentiometric probe. Application of a K^+ channel opener (HOE 234) induced membrane hyperpolarization, resulting in a net efflux of intracellular DiBAC$_4$(3), which is registered as a decrease of fluorescence intensity. These effects were reversed by subsequent treatment with the channel blocker glibenclamide. Figure reproduced with permission from J Physiol (1997) 502:397.

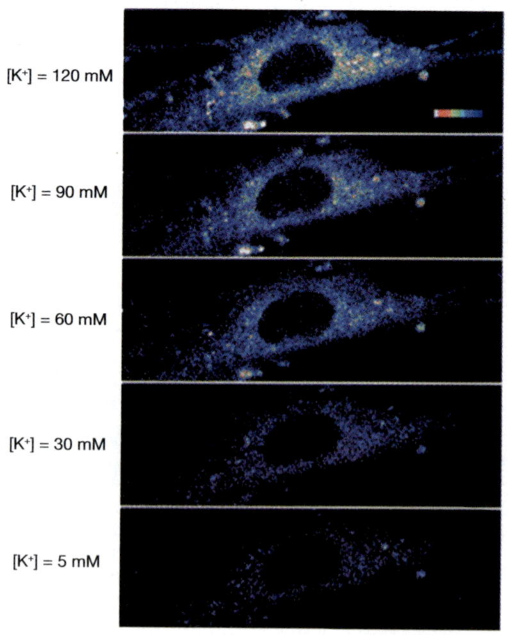

Figure 22.3.26 NIH 3T3 fibroblast undergoing progressive depolarization induced by the stepwise increase of KCl concentration from 5 mM (bottom panel) to 120 mM (top panel). The cell culture was loaded with DiSBAC$_2$(3) (B413) and examined under a confocal laser-scanning microscope after each appropriate medium change, keeping a constant plane of section (Exp Cell Res 231, 260 (1997)). The image was contributed by Rita Gatti, Institute of Histology and General Embryology, University of Parma, Parma, Italy.

IMPORTANT NOTICE: The products described in this manual are covered by one or more Limited Use Label License(s). Please refer to the Appendix on page 971 and Master Product List on page 975. Products are For Research Use Only. Not intended for any animal or human therapeutic or diagnostic use.
www.invitrogen.com/probes

REFERENCES

1. J Photochem Photobiol B (1996) 33:101; **2.** Biochemistry (1974) 13:3315; **3.** J Membr Biol (1986) 92:171; **4.** Biophys J (1989) 56:979; **5.** Biochim Biophys Acta (1998) 1404:393; **6.** Methods (2000) 21:271; **7.** Cytometry (1999) 35:55; **8.** Cytometry (2000) 41:245; **9.** J Biol Chem (2007) 282:18069; **10.** J Biol Chem (1981) 256:1108; **11.** Biochem Pharmacol (1993) 45:691; **12.** Biochemistry (1991) 30:4480; **13.** Proc Natl Acad Sci U S A (1991) 88:3671; **14.** FEBS Lett (1997) 411:77; **15.** Cytometry A (2005) 68:28; **16.** Biochem Biophys Res Commun (1993) 197:40; **17.** J Biol Chem (2008) 283:5188; **18.** Nat Protoc (2007) 2:2719; **19.** Methods Cell Biol (2001) 63:467; **20.** J Biomol Screen (2008) 13:185; **21.** Proc Natl Acad Sci U S A (2008) 105:7387; **22.** Nat Biotechnol (2008) 26:343; **23.** Toxicol Lett (2009) 191:246; **24.** J Am Chem Soc (2005) 127:8686; **25.** Exp Cell Res (1996) 222:84; **26.** J Neurosci (1996) 16:5688; **27.** Antimicrob Agents Chemother (2000) 44:827; **28.** Methods Cell Biol (1989) 29:103; **29.** Trends Neurosci (2000) 23:166; **30.** Biophys J (1999) 76:469; **31.** Methods Enzymol (1999) 302:341; **32.** Neuron (1995) 15:961; **33.** Biophys J (1998) 74:2129; **34.** Am J Physiol (1997) 272:C1286; **35.** J Biomol Screen (2002) 7:383; **36.** Methods Mol Biol (2007) 372:421; **37.** Biochim Biophys Acta (1987) 903:480; **38.** J Membr Biol (1979) 46:255; **39.** Biophys Chem (1989) 34:225; **40.** Chem Phys Lipids (1994) 69:137; **41.** J Physiol (1997) 502:397; **42.** Biochim Biophys Acta (1984) 771:208; **43.** Cytometry A (2009) 75:593; **44.** J Neurosci (2010) 30:2871; **45.** Mol Pharmacol (2007) 71:1075; **46.** Annu Rev Biophys Bioeng (1979) 8:47; **47.** J Immunol Methods (1994) 168:245; **48.** Pharmacol Ther (1994) 63:1; **49.** Arch Biochem Biophys (1993) 300:714; **50.** Bone Marrow Transplant (1993) 12:191; **51.** Cancer Res (1993) 53:806; **52.** Free Radic Biol Med (1992) 12:389; **53.** Biochim Biophys Acta (1991) 1075:28; **54.** J Infect Dis (1991) 163:1312; **55.** Photochem Photobiol (1991) 53:1; **56.** Cancer Res (1989) 49:3637; **57.** Biochemistry (1985) 24:7117; **58.** Biochemistry (1978) 17:5228; **59.** Biochim Biophys Acta (1993) 1146:136; **60.** Biochim Biophys Acta (1983) 732:387; **61.** Biochim Biophys Acta (1990) 1030:269.

DATA TABLE 22.3 SLOW-RESPONSE PROBES

Cat. No.	MW	Storage	Soluble	Abs	EC	Em	Solvent	Notes
B413	436.54	L	DMSO, EtOH	535	170,000	560	MeOH	1
B436	542.67	L	DMSO, EtOH	590	160,000	616	MeOH	1
B438	516.64	L	DMSO, EtOH	493	140,000	516	MeOH	1, 2
B24570	516.64	L	DMSO, EtOH	493	140,000	516	MeOH	1, 2, 3
D272	544.47	D,L	DMSO	484	155,000	500	MeOH	
D273	572.53	D,L	DMSO	484	154,000	501	MeOH	
D306	546.53	D,L	DMSO	651	258,000	675	MeOH	
D14730	460.31	D,L	DMSO	482	165,000	497	MeOH	4
D22421	532.38	D,L	DMSO, DMF	522	143,000	535	$CHCl_3$	5
H14700	510.46	L	DMSO	638	230,000	658	MeOH	6
M24571	569.67	D,L	DMSO, EtOH	555	143,000	578	MeOH	
O266	384.39	L	DMSO, EtOH	610	135,000	639	MeOH	1
O267	316.36	L	DMSO, EtOH	599	136,000	634	MeOH	1
R302	380.83	F,D,L	MeOH, DMF	507	101,000	529	MeOH	
R22420	380.83	F,D,L	MeOH, DMF	507	101,000	529	MeOH	3
T668	500.93	F,D,L	DMSO, MeOH	549	115,000	573	MeOH	
T669	514.96	F,D,L	DMSO, EtOH	549	109,000	574	MeOH	
T3168	652.23	D,L	DMSO, DMF	514	195,000	529	MeOH	7

For definitions of the contents of this data table, see "Using *The Molecular Probes® Handbook*" in the introductory pages.

Notes

1. Oxonols may require addition of a base to be soluble.
2. Fluorescence of DiBAC$_4$(3) increases about 3-fold relative to H_2O on binding to proteins (Abs = 499 nm, Em = 519 nm). (Chem Phys Lipids (1994) 69:137)
3. This product is specified to equal or exceed 98% analytical purity by HPLC.
4. QY for DiOC$_2$(3) in MeOH = 0.04. Abs = 478 nm, Em = 496 nm in H_2O. (Biochemistry (1974) 13:3315)
5. JC-9 exhibits long-wavelength J-aggregate emission at ~635 nm in aqueous solutions and polarized mitochondria.
6. DiIC$_1$(5) in H_2O; Abs = 636 nm, Em = 658 nm.
7. JC-1 forms J-aggregates with Abs/Em = 585/590 nm at concentrations above 0.1 µM in aqueous solutions (pH 8.0). (Biochemistry (1991) 30:4480)

IMPORTANT NOTICE: The products described in this manual are covered by one or more Limited Use Label License(s). Please refer to the Appendix on page 971 and Master Product List on page 975. Products are For Research Use Only. Not intended for any animal or human therapeutic or diagnostic use.

PRODUCT LIST 22.3 SLOW-RESPONSE PROBES

Cat. No.	Product	Quantity
B34950	*Bac*Light™ Bacterial Membrane Potential Kit *for flow cytometry* *100 assays*	1 kit
B436	bis-(1,3-dibutylbarbituric acid)pentamethine oxonol (DiBAC$_4$(5))	25 mg
B438	bis-(1,3-dibutylbarbituric acid)trimethine oxonol (DiBAC$_4$(3))	25 mg
B24570	bis-(1,3-dibutylbarbituric acid)trimethine oxonol (DiBAC$_4$(3)) *FluoroPure™ grade*	5 mg
B413	bis-(1,3-diethylthiobarbituric acid)trimethine oxonol (DiSBAC$_2$(3))	100 mg
D273	3,3′-dihexyloxacarbocyanine iodide (DiOC$_6$(3))	100 mg
D22421	3,3′-dimethyl-α-naphthoxacarbocyanine iodide (JC-9; DiNOC$_1$(3))	5 mg
D272	3,3′-dipentyloxacarbocyanine iodide (DiOC$_5$(3))	100 mg
D14730	DiOC$_2$(3) (3,3′-diethyloxacarbocyanine iodide)	100 mg
D306	DiSC$_3$(5) (3,3′-dipropylthiadicarbocyanine iodide)	100 mg
H14700	1,1′,3,3,3′,3′-hexamethylindodicarbocyanine iodide (DiIC$_1$(5))	100 mg
M24571	merocyanine 540	25 mg
M34151	MitoProbe™ DiIC$_1$(5) Assay Kit *for flow cytometry* *100 assays*	1 kit
M34150	MitoProbe™ DiOC$_2$(3) Assay Kit *for flow cytometry* *100 assays*	1 kit
M34152	MitoProbe™ JC-1 Assay Kit *for flow cytometry* *100 assays*	1 kit
O266	oxonol V (bis-(3-phenyl-5-oxoisoxazol-4-yl)pentamethine oxonol)	100 mg
O267	oxonol VI (bis-(3-propyl-5-oxoisoxazol-4-yl)pentamethine oxonol)	100 mg
R302	rhodamine 123	25 mg
R22420	rhodamine 123 *FluoroPure™ grade*	25 mg
T3168	5,5′,6,6′-tetrachloro-1,1′,3,3′-tetraethylbenzimidazolylcarbocyanine iodide (JC-1; CBIC$_2$(3))	5 mg
T669	tetramethylrhodamine, ethyl ester, perchlorate (TMRE)	25 mg
T668	tetramethylrhodamine, methyl ester, perchlorate (TMRM)	25 mg

IMPORTANT NOTICE: The products described in this manual are covered by one or more Limited Use Label License(s). Please refer to the Appendix on page 971 and Master Product List on page 975. Products are For Research Use Only. Not intended for any animal or human therapeutic or diagnostic use.
www.invitrogen.com/probes
molecular probes® | invitrogen™ by life technologies™

PART VI

Integrating Technology and Instrumentation

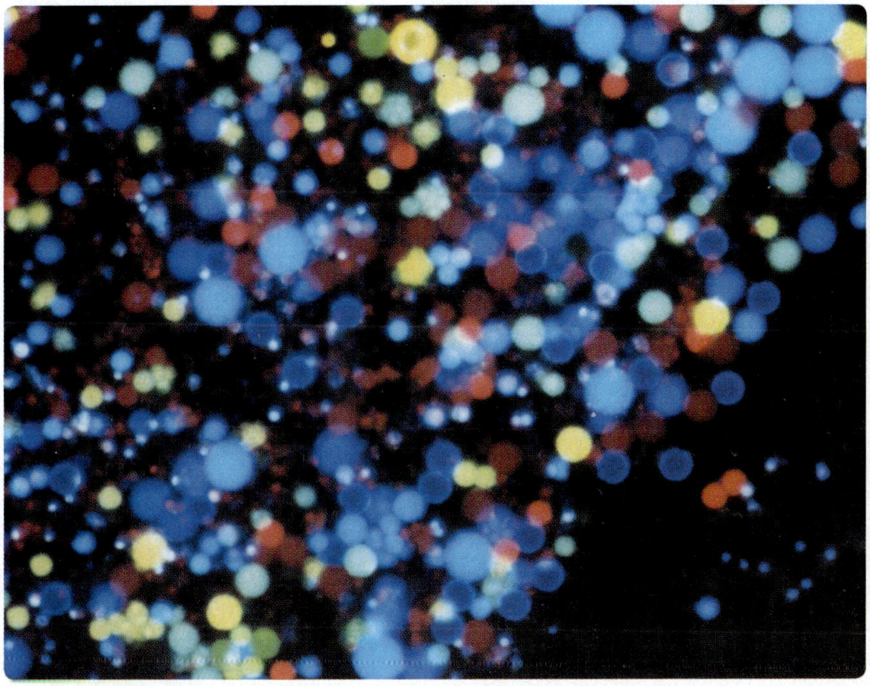
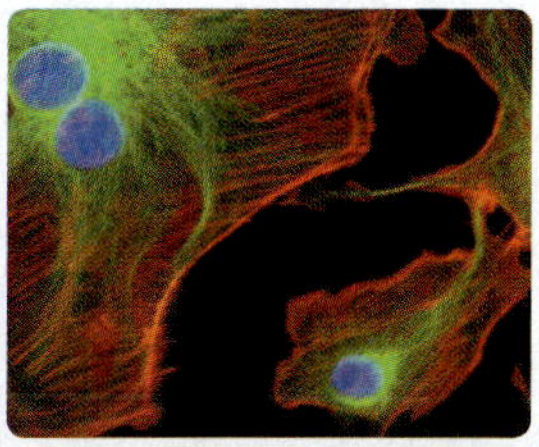
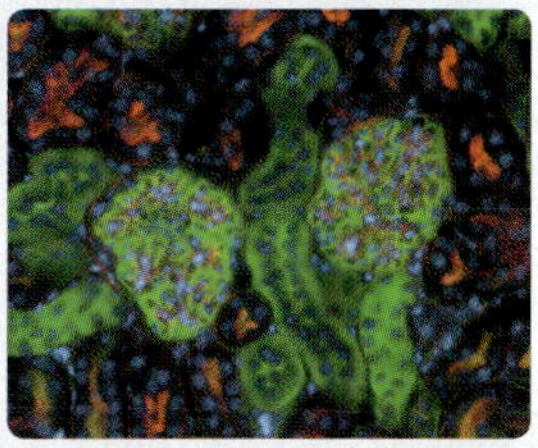
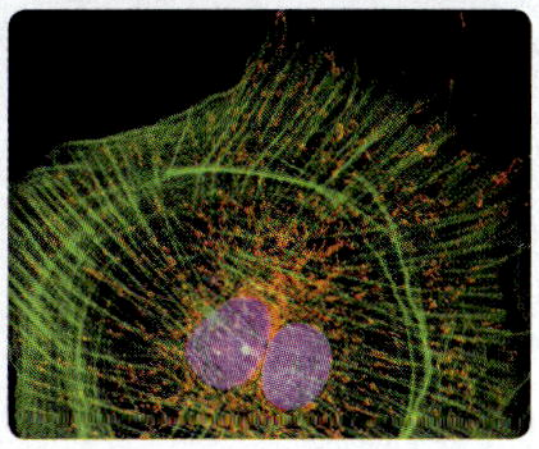

Multiplex imaging of human carcinoma (HeLa) cell labeled with Qdot® nanocrystals and mounted with Qmount® media. Mitochondria were detected with anti–OxPhos Complex V inhibitor protein IgG (A21355) and labeled using Qdot® 625 goat F(ab')$_2$ anti–mouse IgG (A10195, red fluorescence). The Golgi apparatus was detected with rabbit anti-giantin and labeled using Qdot® 585 goat F(ab')$_2$ anti–rabbit IgG conjugate (Q11411MP, yellow fluorescence). Tubulin was detected with rat anti-tubulin and labeled using DSB-X™ biotin goat anti–rat IgG (D20697) and Qdot® 525 streptavidin conjugate (Q10141MP, green fluorescence). Nuclei were labeled with Qnuclear™ Deep Red Stain (Q10363, purple fluorescence), and the slide was mounted using Qmount® Qdot® Mounting Media (Q10336). Image contributed by Kary Oakleaf, Life Technologies Corporation.

This image appeared on the cover of *BioProbes 61*. *BioProbes®* newsletter is published several times each year. This award-winning publication is dedicated to furnishing researchers with the very latest information about cell biology products and their applications, and provides a great way to stay connected with the fluorescence community. You can subscribe and view the latest issues online at **www.invitrogen.com/handbook/bioprobes**.

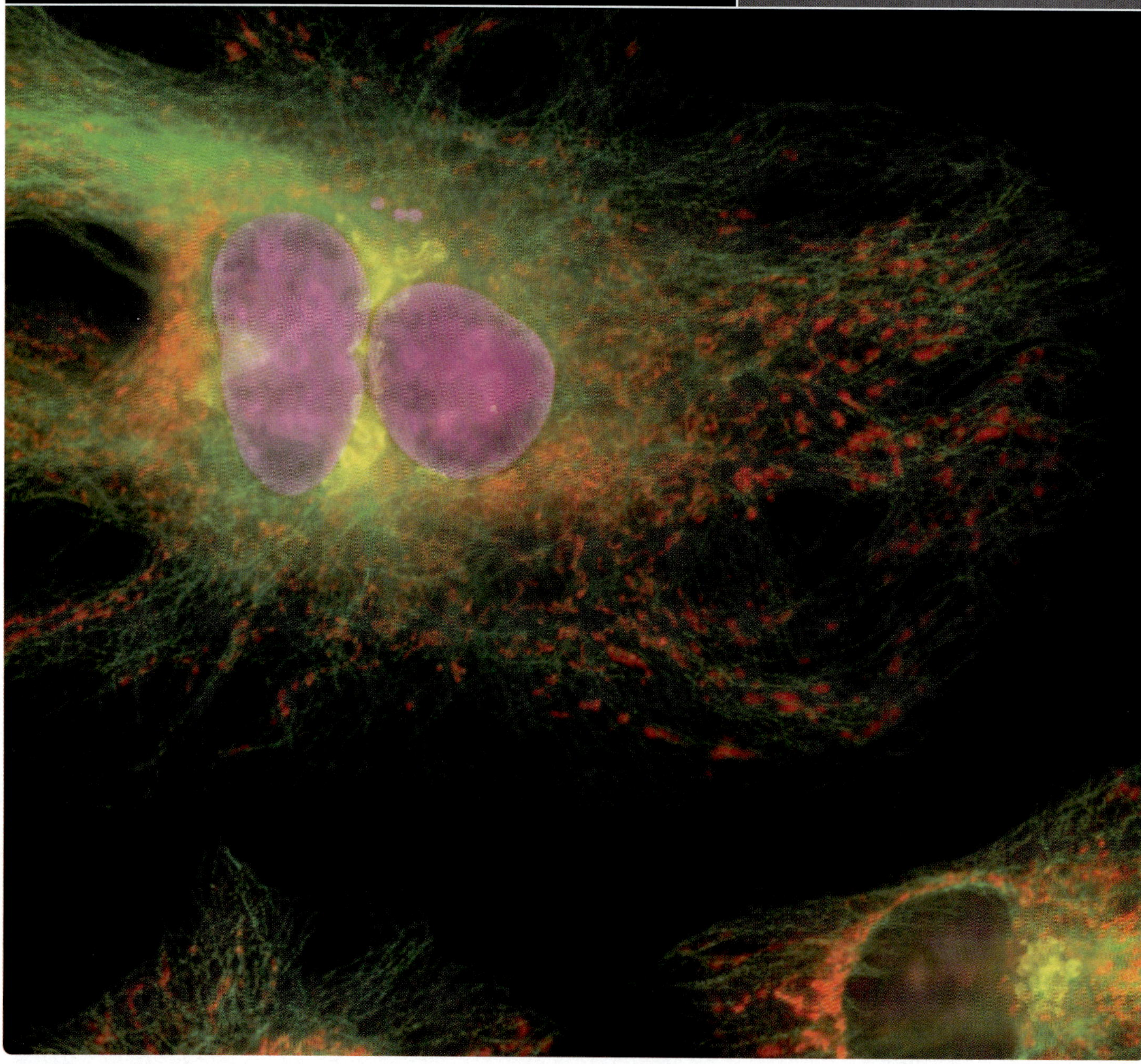

TWENTY-THREE

CHAPTER 23
Antifades and Other Tools for Fluorescence

IMPORTANT NOTICE: The products described in this manual are covered by one or more Limited Use Label License(s). Please refer to the Appendix on page 971 and Master Product List on page 975. Products are For Research Use Only. Not intended for any animal or human therapeutic or diagnostic use.

IMPORTANT NOTICE: The products described in this manual are covered by one or more Limited Use Label License(s). Please refer to the Appendix on page 971 and Master Product List on page 975. Products are For Research Use Only. Not intended for any animal or human therapeutic or diagnostic use.

23.1 Fluorescence Microscopy Accessories and Reference Standards

To obtain accurate and reproducible results from fluorescence imaging applications, it is essential to maximize the intensity and stability of the fluorescence signal in the experimental sample. We have developed several effective antifade reagents that minimize photobleaching of fluorescently labeled specimens. The Image-iT® FX signal enhancer (I36933) blocks nonspecific binding of dye-labeled antibodies, resulting in improved signal:background characteristics in images of immunolabeled cells and tissues. Our FluoCells® prepared microscope slides provide ready-to-use, multicolor-labeled cell or tissue preparations for educational and commercial fluorescence microscopy demonstrations. In collaboration with Grace Bio-Labs, we also offer a wide selection of microscopy accessories, including sample chambers, slides and coverslips.

Likewise, accurate and reproducible results depend on optimal performance of the optical system. The spectral compatibility of dyes and probes with excitation light sources (Table 23.1) and emission wavelength filters must be carefully evaluated. Downloadable reference spectra are available through our online Fluorescence SpectraViewer tool (www.invitrogen.com/handbook/spectraviewer, Using the Fluorescence SpectraViewer—Note 23.1), and guidance in choosing optical filters can be found in Selecting Optical Filters for Fluorescence Microscopy—Note 23.2. Careful calibration and instrumentation adjustment are also required for high-precision imaging of fluorescent probes, particularly in multicolor applications that involve multiple exposures, repetitive scans or three-dimensional sectioning. We offer a variety of microsphere reference standards designed to facilitate adjustment and calibration of both conventional fluorescence microscopes and confocal laser-scanning microscopes. In addition, the Reference Dye Sampler Kit (R14782) provides ready-made stock solutions of five extensively characterized fluorescence standards for use in spectrofluorometers and fluorescence microplate readers. We are also the source of the NIST-traceable fluorescein standard (F36915), which is directly traceable to the fluorescein standard adopted and maintained by the National Institute of Standards and Technology.

Antifade Reagents and Mounting Media

Loss of fluorescence through irreversible photobleaching processes can lead to a significant reduction in sensitivity, particularly when target molecules are of low abundance or when excitation light is of high intensity or long duration.[1–3] To minimize photobleaching of experimental samples, we have developed the ProLong®, ProLong® Gold, *SlowFade®* and *SlowFade®* Gold Antifade Kits and reagents, which have been shown to increase the photostability of many of our fluorophores in fixed cells, fixed tissues and cell-free preparations. The primary function of any antifade reagent is to sustain dye fluorescence, usually by inhibiting the generation and diffusion of reactive oxygen species. Other strategies for avoiding photobleaching include reducing the excitation light intensity by using high-numerical aperture objectives and low magnification as well as hardware control of the excitation light's spatial[3] and temporal[4] distribution. Loss of fluorescence signal due to attenuated excitation can be compensated to some extent by use of high-quality optical filters and high-efficiency photodetectors.

ProLong® Gold Antifade Reagent

ProLong® Gold antifade reagent is an improved version of the ProLong® antifade reagent, a component of the ProLong® Antifade Kit described below. The ProLong® Gold antifade reagent causes little or no quenching of the fluorescent signal while protecting the sample from

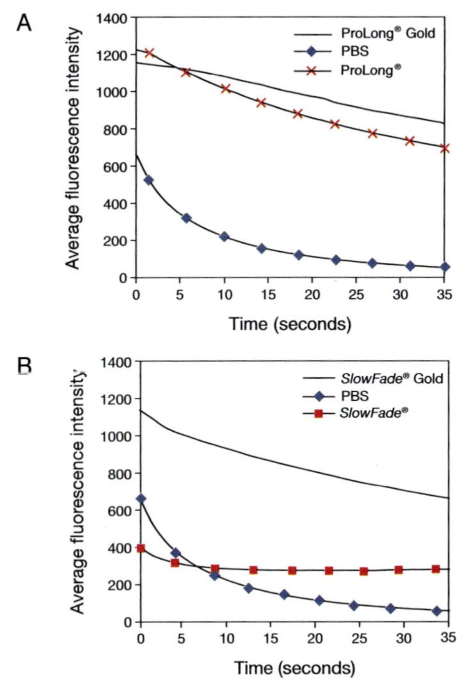

Figure 23.1.1 A comparison of the performance of Molecular Probes® antifade reagents. Bovine pulmonary arterial endothelial cells were labeled with fluorescein phalloidin and mounted using A) PBS, ProLong® Gold antifade reagent (P36930, P36934) or ProLong® Antifade Kit (P7481); B) PBS, SlowFade® Gold antifade reagent (S36936, S36937) or SlowFade® Antifade Kit (S2828). Samples were imaged using a 1.3 NA 40× oil immersion lens, Omega® XF100-2 filter set and frame-capture rate of 1 image/second. Images were acquired with a Hamamatsu Orca ER camera using the same exposure time for all samples. Y-axis values represent averages of the top 10% of the intensity-binned pixel values.

Table 23.1 Fluorescence excitation sources.

Source	Principal Lines (nm)
Mercury-arc lamp	366, 405, 436, 546, 578 *
Xenon-arc lamp	250–1000 *
Tungsten–halogen lamp	350–1000 *
Violet diode laser	405
Helium–cadmium laser	325, 442
Argon-ion laser	457, 488, 514
Nd:YAG laser	532 †
Helium–neon laser	543, 594, 633
Yellow diode laser	561
Krypton-ion laser	568, 647
Red diode laser	635

* Continuous white-light source. † Frequency-doubled principal line output.

NOTE 23.1

Using the Fluorescence SpectraViewer

The Fluorescence SpectraViewer (www.invitrogen.com/handbook/spectraviewer) is an online tool that allows researchers to assess the spectral compatibility of dyes and probes in the course of designing experiments that utilize fluorescence detection techniques. This note outlines the functionality of the SpectraViewer (Figure 1A) and examples of its utility in the experimental design process (Figure 1B, Figure 1C).

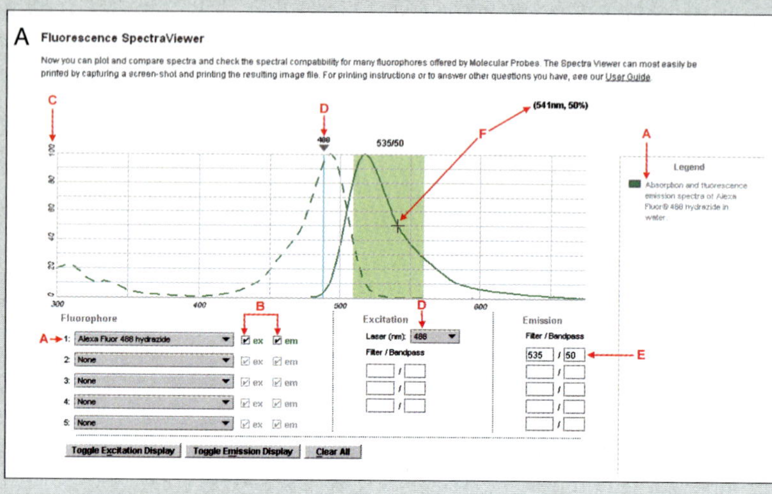

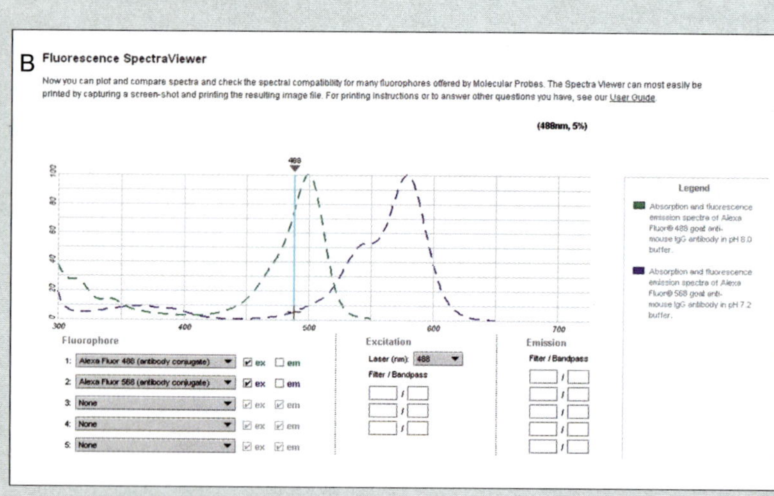

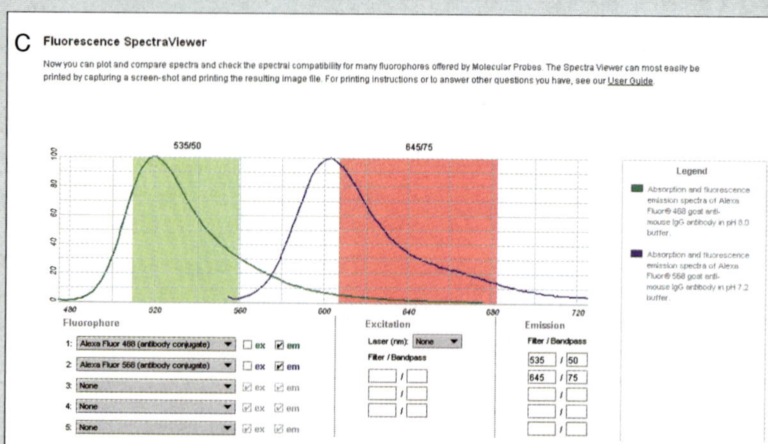

Figure 1A SpectraViewer features: **A.** Fluorophore selection menu. Up to 5 fluorophore data sets may be displayed simultaneously. Open the drop-down menu (▼) and select from the list of available excitation/emission data sets. The number of available data sets (as of July 2010) is 300, encompassing organic dyes, fluorescent proteins and Qdot® nanocrystals. Note that data sets may be added, deleted or modified at our editorial discretion and without notice, although in practice we aim to keep such changes to a minimum to maintain stability of the database. These data sets may be downloaded from our website as 4-column text files for importing into other plotting and calculation utilities. For each selected fluorophore data set, a legend containing sample context information is displayed on the panel to the right of the plot. **B.** For each fluorophore data set, excitation (ex) and emission (em) data may be displayed or hidden by checking or unchecking the respective boxes. **C.** Y-axis scaling for excitation and emission spectra is in terms of percentage of peak intensity value. For Qdot® nanocrystals, which exhibit quasi-continuous excitation profiles, the 100% intensity value has been arbitrarily defined as that at 300 nm. X-axis values on all plots are wavelengths in nanometers (nm). **D.** Excitation source spectral characteristics may be superimposed on the plot in the form of laser lines selected from a drop-down menu or filter characteristics input as numeric center wavelength (CWL) and bandpass (BP) values in nm in the boxes provided. In this example, laser excitation at 488 nm is indicated. **E.** Emission filter spectral characteristics may be superimposed on the plot in the form of numeric CWL and BP values in nm entered in the boxes provided. In this example, a typical FITC emission filter with CWL = 535 nm and BP = 50 nm is indicated. The transmission window of the filter is shown on the plot as a green-shaded rectangle. **F.** Mouse-controlled X,Y cursor. The crosshairs may be moved to any user-selected location within the plot window and are coupled to a numeric display of the corresponding (X,Y) values.

Figure 1B Evaluating fluorophores for multiplex detection experiments: Excitation spectra. This overlay of the fluorescence excitation spectra of Alexa Fluor® 488 and Alexa Fluor® 568 dyes provides a useful initial assessment of their suitability for use in a multiplex detection experiment. It also serves to highlight scaling factors that are critical determinants of the final detected signal levels but that are excluded from the SpectraViewer comparison. The plot indicates that excitation of Alexa Fluor® 568 at 488 nm is relatively inefficient (~5% of maximum indicated by the cursor). However, this consideration takes no account of the molar absorptivities (extinction coefficient, EC) of the fluorophores. The maximum EC values of Alexa Fluor® 488 and Alexa Fluor® 568 dyes are actually quite similar (75,000 cm^{-1} M^{-1} and 93,000 cm^{-1} M^{-1}, respectively, as listed in *The Molecular Probes® Handbook* data tables). An even more significant weighting factor is the relative abundances of the molecular targets of the fluorophores in the detector field of view; in general, cellular abundances of proteins vary over more orders of magnitude than the extinction coefficients of fluorophores. In the situation illustrated with excitation limited to 488 nm, the preferred course would be to use Alexa Fluor® 568 to detect the more abundant of the two molecular targets and Alexa Fluor® 488 to detect the less abundant target, thereby offsetting the absorptivity and target abundance factors.

Figure 1C Evaluating fluorophores for multiplex detection experiments: Emission spectra. The fluorescence emission spectra of Alexa Fluor® 488 and Alexa Fluor® 568 dyes are shown with the spectral characteristics of typical emission filters superimposed. In this case, the main practical concern is the extent of overspill of Alexa Fluor® 488 fluorescence into the Alexa Fluor® 568 detection channel (CWL = 645 nm, BP = 75 nm), which can lead to false indications of molecular target colocalization in imaging applications. As in the excitation spectra comparison, this peak-normalized overlay provides an initial assessment of the suitability of fluorophore combinations for multiplex detection, but weighting of fluorescence signals by the relative abundances of molecular targets and other factors will heavily influence the final experimentally observed outcome.

IMPORTANT NOTICE: The products described in this manual are covered by one or more Limited Use Label License(s). Please refer to the Appendix on page 971 and Master Product List on page 975. Products are For Research Use Only. Not intended for any animal or human therapeutic or diagnostic use.

photobleaching (Figure 23.1.1, Figure 23.1.2). Furthermore, unlike the reagents in the ProLong® Antifade Kit, the ProLong® Gold antifade reagent is premixed and ready to use—just add a drop to the preparation and mount. As with our original ProLong® antifade reagent, ProLong® Gold reagent cures within 24 hours and the sample can be saved for months after mounting. ProLong® Gold reagent offers excellent compatibility with a multitude of dyes and dye complexes, making it an especially valuable tool for multicolor applications[5–7] (Figure 23.1.3). The ProLong® Gold antifade reagent is available in a single 10 mL bottle (P36930), as well as in a set of five 2 mL bottles (P36934).

As an added convenience, we also offer ProLong® Gold antifade reagent containing DAPI, the popular nuclear and chromosome stain that emits blue fluorescence upon binding to DNA. The addition of DAPI in the mounting media eliminates the need for a separate counterstaining step. ProLong® Gold antifade reagent with DAPI is available in a single 10 mL bottle (P36931), as well as in a set of five 2 mL bottles (P36935).

ProLong® Antifade Kit

The ProLong® Antifade Kit (P7481, Figure 23.1.4) contains our original ProLong® antifade reagent, which has proven to effectively enhance the resistance of many different fluorophores to photobleaching. Furthermore, specimens mounted using the ProLong® Antifade Kit exhibit little or no quenching of the fluorescent signal of most dyes.

Each ProLong® Antifade Kit contains:

- ProLong® antifade reagent powder
- ProLong® mounting medium
- Detailed protocols for mounting samples

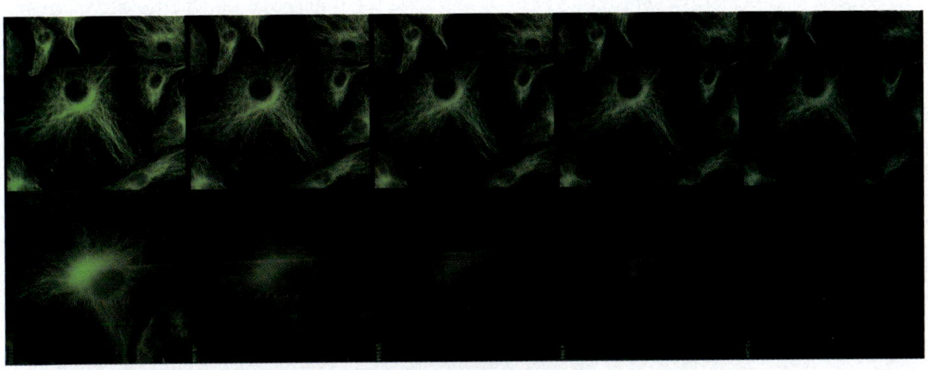

Figure 23.1.2 A 20-second time series showing enhanced resistance to photobleaching afforded by ProLong® Gold antifade reagent. Fixed bovine pulmonary artery endothelial cells were labeled with anti–α-tubulin (A11126) and visualized with fluorescein goat anti-mouse IgG (F2761). The samples were mounted in ProLong® Gold antifade reagent (P36930; top) or phosphate-buffered saline (bottom). Images were acquired at 5-second intervals using a 40x/1.3 NA oil immersion objective with continuous illumination from a standard 100 watt mercury-arc lamp.

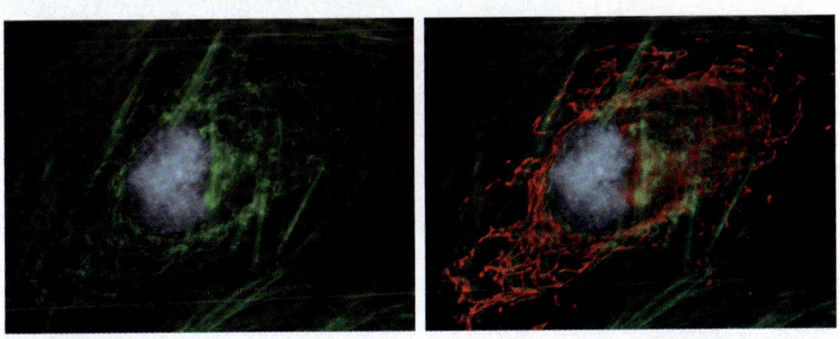

Figure 23.1.3 HeLa cell transfected with pShooter pCMV/myc/mito/GFP, then fixed and permeabilized. Green Fluorescent Protein (GFP) localized in the mitochondria was labeled with anti-GFP mouse IgG_{2a} (A11120) and detected with orange-fluorescent Alexa Fluor® 555 goat anti–mouse IgG (A21422), which colocalized with the dim GFP fluorescence. F-actin was labeled with green-fluorescent Alexa Fluor® 488 phalloidin (A12379), and the nucleus was stained with blue-fluorescent DAPI (D1306, D3571, D21490). The sample was mounted using ProLong® Gold antifade reagent (P36930). Some GFP fluorescence is retained in the mitochondria after fixation (left), but immunolabeling and detection greatly improve visualization (right).

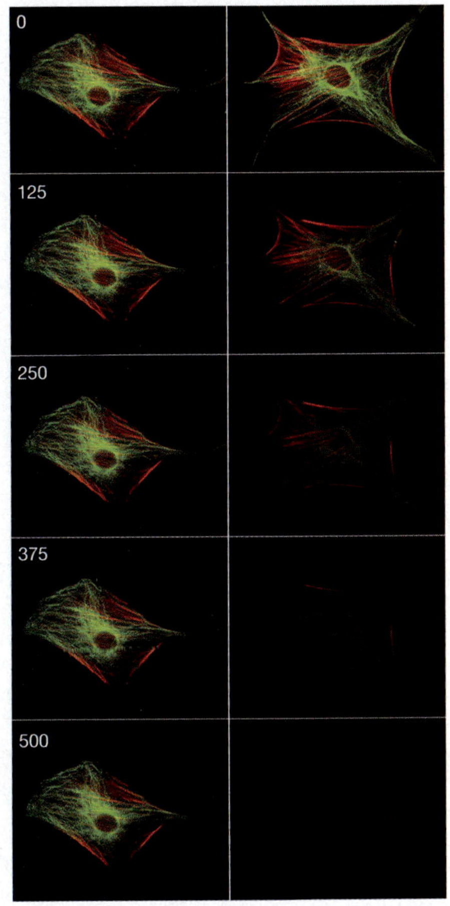

Figure 23.1.4 The relative photobleaching rates of fluorescein and Texas Red® fluorophores with buffer alone (right series) or after treatment with the ProLong® antifade reagent (left series). Bovine pulmonary artery endothelial cells were fixed, permeabilized and labeled with Texas Red®-X phalloidin (T7471), which labels F-actin, and with a mouse monoclonal anti–ß-tubulin antibody and fluorescein goat anti–mouse IgG antibody (F2761), which label microtubules. Images were acquired at appropriate wavelengths using a cooled CCD camera. The numbers on the left of each pair of frames represent the duration of continuous excitation in seconds.

IMPORTANT NOTICE: The products described in this manual are covered by one or more Limited Use Label License(s). Please refer to the Appendix on page 971 and Master Product List on page 975. Products are For Research Use Only. Not intended for any animal or human therapeutic or diagnostic use.

NOTE 23.2

Selecting Optical Filters for Fluorescence Microscopy

Sensitive and versatile fluorescence detection techniques are of ever-increasing importance and popularity in biological research microscopy. In the now-standard epi-illuminated microscope configuration, the optical filter set performs a critical function in separating the fluorescence emission photons that will form the final image from the more-intense excitation light field. For practical purposes, it is necessary to reduce the excitation light intensity in the detection path by a factor of 10^6–10^7. This design objective has to be achieved in parallel with capturing as many of the available fluorescence photons as possible. High capture efficiency allows compensating reductions in overall excitation light levels, with accompanying reductions in dye photobleaching and cellular phototoxicity.

This technical note provides some basic fluorescence microscopy information, including discussions of:

- The Optical Filter Set
- The Trade-Off in Optical Filter Design
- Selecting an Optical Filter Set

The Fluorescence SpectraViewer (www.invitrogen.com/handbook/spectraviewer) is an online tool that allows researchers to assess the spectral compatibility of dyes, probes and optical filters in the course of designing experiments (Using the Fluorescence SpectraViewer—Note 23.1).

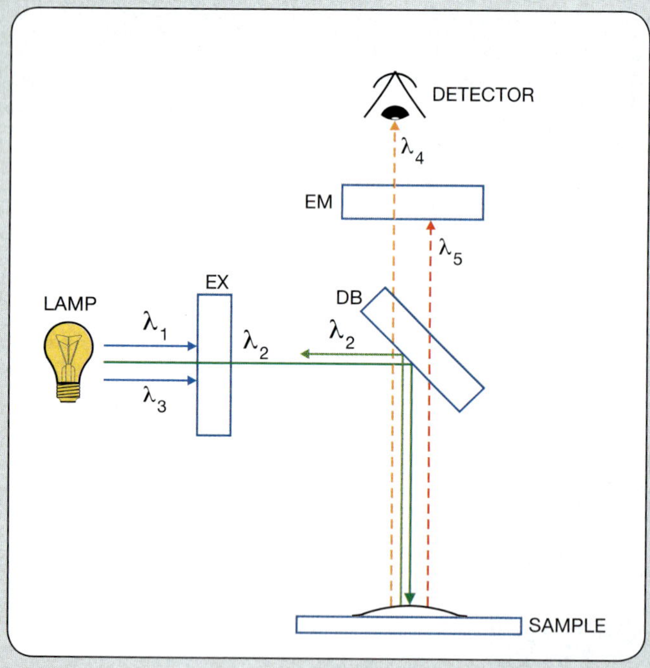

Figure 1 Functions of fluorescence microscope filter set components. The desired excitation wavelength (λ_2) is selected from the spectral output of the lamp by the excitation filter (EX) and directed to the sample via the dichroic beamsplitter (DB). The beamsplitter separates emitted fluorescence (- - -) from scattered excitation light (—). The emission filter (EM) selectively transmits a portion of the sample's fluorescence emission (λ_4) for detection and blocks other emission components (λ_5).

The Optical Filter Set

A set of optical filters for selective excitation and detection of fluorescence typically consists of a minimum of three components: an excitation filter, a dichroic beamsplitter ("dichroic mirror") and an emission filter ("barrier filter") (Figure 1). The excitation filter selectively transmits a portion of the spectral output from the light source (Table 23.1). The dichroic beamsplitter then reflects the selected light, directing it to the sample. Fluorescence emission photons traveling from the sample towards the detector are transmitted by the dichroic beamsplitter, while excitation light reflected back from the sample is diverted out of the detection light path. The emission filter blocks unwanted spectral components of the emitted fluorescence (e.g., sample autofluorescence) as well as any residual excitation light. An interactive Java tutorial demonstrating these functions is available online at the Molecular Expressions website of Florida State University (http://micro.magnet.fsu.edu/primer/java/fluorescence/filtersetprofiles/index.html).

The Trade-Off in Optical Filter Set Design

For optimal fluorescence detection when using a single dye, the excitation and emission filters should be centered on the dye's absorption and emission peaks. To maximize the signal, one can choose excitation and emission filters with wide bandwidths. However, this strategy may result in unacceptable overlap of the emission signal with the excitation signal, resulting in poor resolution. To minimize spectral overlap, one can instead choose excitation and emission filters that are narrow in bandwidth and are spectrally well separated to increase signal isolation. This approach will reduce optical noise but may also reduce the signal strength to unacceptable levels. When overlapping signals from multiple fluorophores in the same sample are being differentiated, both the spectra of the dyes and their expected intensities must be considered before choosing an

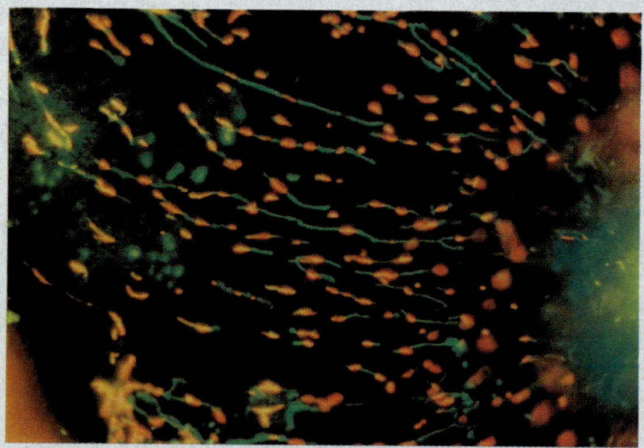

Figure 2 Potential-dependent staining of mitochondria in CCL64 fibroblasts by JC-1 (T3168). The mitochondria were visualized by epifluorescence microscopy using a 520 nm longpass optical filter. Regions of high mitochondrial polarization are indicated by red fluorescence due to J-aggregate formation by the concentrated dye. Depolarized regions are indicated by the green fluorescence of the JC-1 monomers. The image was contributed by Lan Bo Chen, Dana Farber Cancer Institute, Harvard Medical School.

IMPORTANT NOTICE: The products described in this manual are covered by one or more Limited Use Label License(s). Please refer to the Appendix on page 971 and Master Product List on page 975. Products are For Research Use Only. Not intended for any animal or human therapeutic or diagnostic use.

optical filter. Complete spectral data for Molecular Probes® fluorophores can be found at www.invitrogen.com/handbook/spectraviewer using our interactive Fluorescence SpectraViewer utility (Using the Fluorescence SpectraViewer—Note 23.1). An interactive Java tutorial illustrating the trade-off among these parameters is available online at the Molecular Expressions website of Florida State University (http://micro.magnet.fsu.edu/primer/java/fluorescence/fluorocubes/index.html).

Selecting an Optical Filter Set

Filter set selection may involve a straightforward recommendation or a complex analysis of the spectral relationships of dyes and optical filters.[1] Emission filters are available with either longpass or bandpass wavelength transmission profiles. A typical longpass emission filter might transmit all wavelengths ≥530 nm, whereas a typical bandpass filter might transmit only wavelengths between 515 and 545 nm. Longpass filters should be used when the application requires maximum emission collection and when spectral discrimination is not desirable or necessary, which is generally the case for probes that generate a single emitting species in specimens with relatively low levels of background autofluorescence. Longpass filters are also useful for simultaneous detection of spectrally distinct dual emissions such as the monomer and J-aggregate forms of JC-1 (T3168, Section 12.2, Figure 2) and the monomer and excimer forms of BODIPY® FL C_5-ceramide (D3521, B22650; Section 12.4; Figure 3).

Bandpass filters are designed to maximize the signal-to-noise ratio for applications where discrimination of signal components is more important than overall image brightness. The spectral sensitivity of the detection system should also be considered in order to achieve optimum detector signal-to-noise or accurate color rendition. Some applications, such as confocal laser-scanning microscopy, may require the use of sensitive photomultiplier (PMT) detectors. Alternatively, an electron-multiplying charge-coupled device (EMCCD) may be employed for quantitative imaging or microspectrofluorometry. Dual-, triple- and quadruple-band filter sets enable microscopists to excite and detect two, three or four fluorophores simultaneously instead of performing sequential image acquisitions with intervening filter changes (Figure 4).

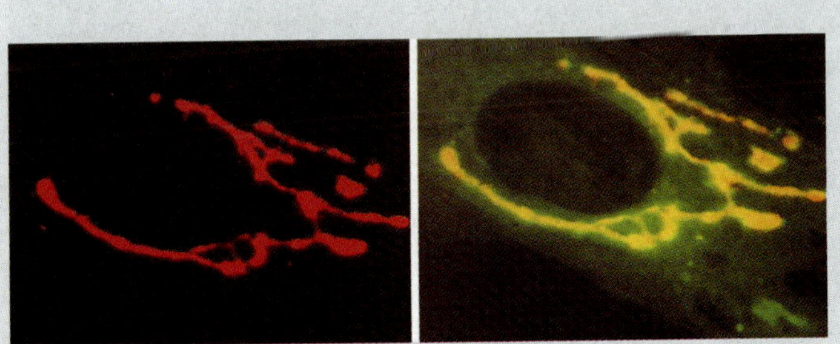

Figure 3 Selective staining of the Golgi apparatus using the green-fluorescent BODIPY® FL C_5-ceramide (D3521) (right panel). At high concentrations, the BODIPY® FL fluorophore forms excimers that can be visualized using a red longpass optical filter (left panel). The BODIPY® FL C_5-ceramide accumulation in the trans-Golgi is sufficient for excimer formation (J Cell Biol (1991) 113:1267). Images contributed by Richard Pagano, Mayo Foundation.

Selecting optimal filter sets for fluorescence microscopy applications requires matching optical filter specifications to the spectral characteristics of dyes. Comparisons should be made with care because some dyes have significantly different spectral properties in a particular application than those reported for the dye in solution. For example, the spectral characteristics of many nucleic acid stains depend on whether the dyes are in aqueous solution or bound to DNA or RNA. Similarly, styryl dyes such as FM® 1-43 (T3163, T35356; Section 14.4, Section 16.1) and di-4-ANEPPS (D1199, Section 22.2) have emission maxima that depend on whether they are dissolved in solvent or associated with membranes. To provide selection guidelines, we have compiled excitation and emission spectra for many of the most widely used probes in fluorescence microscopy in an online tool, the Fluorescence SpectraViewer (www.invitrogen.com/handbook/spectraviewer).

Technical Support

We invite customers to call our Technical Assistance Department for help in selecting the correct optical filter for a specific application. When calling, please be prepared to describe the dye(s), instrumentation and method of detection being used. A technical support scientist will then offer advice on the most effective filter configuration for the specified purposes. Alternatively, we recommend contacting Chroma Technology Corp., Omega Optical, Inc. or Semrock, Inc. or the microscope manufacturer for this information. Chroma Technology, Omega Optical and Semrock provide complete transmission curves and information on their specialty filters for ratio imaging, uncaging, multiphoton and other applications at their respective web sites (www.semrock.com, www.omegafilters.com, www.chroma.com).

1. Biophotonics Int (1999) 6:54.

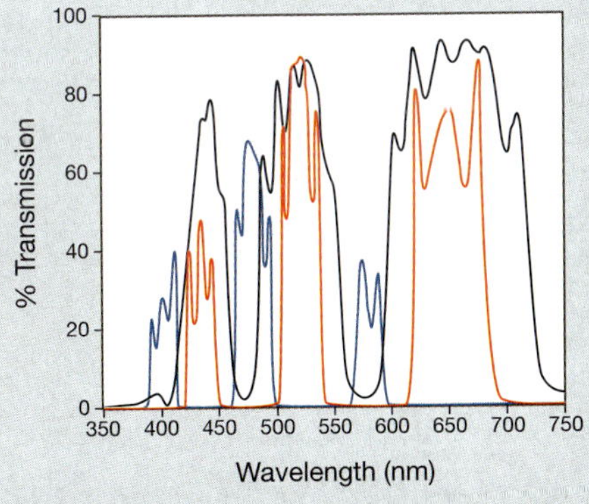

Figure 4 Optical transmission characteristics of a triple-band filter set (XF63, Omega Optical Inc.) designed for simultaneous imaging of DAPI, fluorescein and Texas Red® dyes. Transmission curves for the individual filter set components are shown in blue (excitation filter), black (dichroic beamsplitter) and red (emission filter). Graphic supplied by and used with permission of Omega Optical Inc., Brattleboro, VT.

Bovine pulmonary arterial epithelial cells (BPAEC) labeled with fluorescein phalloidin (F432) photobleached to about 12% of the initial value in 30 seconds in PBS, while staying at the initial value under the same illumination conditions when mounted using the ProLong® Antifade Kit (Figure 23.1.5). As shown in Figure 23.1.6, the ProLong® Antifade Kit provides more fluorescence output than a popular *p*-phenylenediamine–containing antifade reagent[8] when used to mount fluorescein-stained HEp-2 cells. The ProLong® antifade reagent also inhibits the fading of tetramethylrhodamine, as well as the fading of DNA-bound nucleic acid stains such as DAPI, propidium iodide and YOYO®-1 (Section 8.1), again without significantly quenching the fluorescence of these dyes.[9] The compatibility of the ProLong® antifade reagent with a multitude of dyes and dye complexes makes it an especially valuable tool for multicolor analysis procedures such as multiplexed fluorescence *in situ* hybridization.[10]

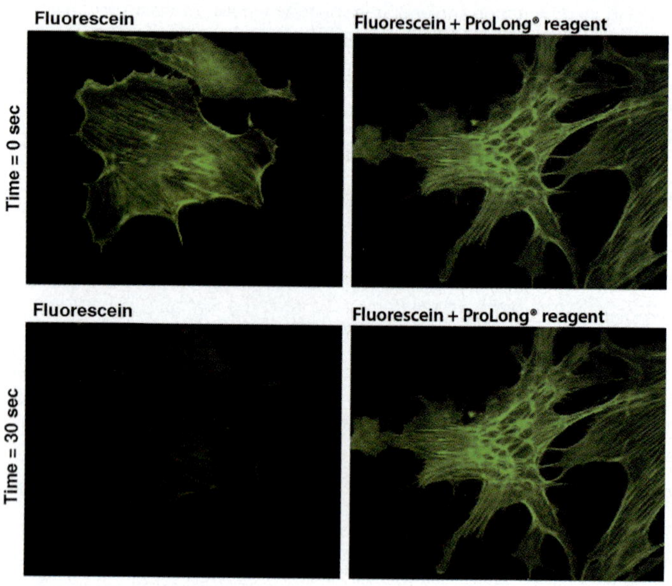

Figure 23.1.5 Bovine pulmonary artery endothelial cells were labeled with fluorescein phalloidin (F432), which labels filamentous actin, and placed under constant illumination on the microscope with a FITC filter set using a 60× objective. Images were acquired at 1-second intervals for 30 seconds. Under these illumination conditions, fluorescein photobleached to about 12% of its initial value in 30 seconds in PBS (left), but stayed at the initial value under the same illumination conditions when mounted using the reagents in the ProLong® Antifade Kit (right, P7481).

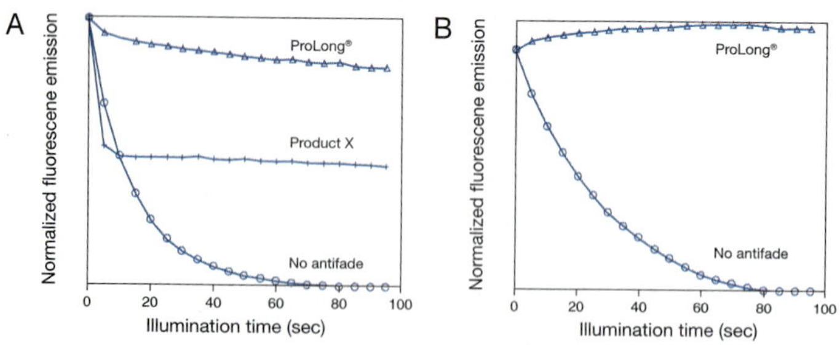

Figure 23.1.6 Bleaching profiles of **A**) fluorescein and **B**) Texas Red® dye conjugates in cell samples. In these photobleaching experiments, human epithelial (HEp-2) cells were probed with human anti-nuclear antibodies and then developed for visualization with fluorophore-labeled secondary reagents. Identical samples were mounted in ProLong® antifade reagent (△), Product X (+) or medium containing no antifade reagent (○). Although these data were normalized, we observed little or no quenching of samples mounted with the ProLong® mounting medium.

SlowFade® Antifade Kit

Our original *SlowFade®* antifade formulation (S2828) was designed to reduce the fading rate of fluorescein to almost zero. Because it provides nearly constant emission intensity from fluorescein, this *SlowFade®* antifade reagent is especially useful for quantitative measurements and applications that employ confocal laser-scanning microscopy.[11] However, this original *SlowFade®* formulation substantially quenches the fluorescence of fluorescein and almost completely quenches that of the Alexa Fluor® 350, Alexa Fluor® 405 and Cascade Blue® fluorophores.

Each *SlowFade®* Antifade Kit contains:

- *SlowFade®* antifade reagent in 50% (v/v) glycerol, ready to use and sufficient for at least 200 coverslip-size experiments
- Concentrated *SlowFade®* antifade reagent solution, provided for those applications in which glycerol may not be compatible
- Equilibration buffer, which raises the pH of the sample, increasing the protection afforded by the *SlowFade®* antifade formulation
- Detailed protocols for mounting samples

SlowFade® Gold Antifade Reagent

To overcome the limitations of the original *SlowFade®* antifade reagent, especially with blue fluorophores, we have developed the *SlowFade®* Gold antifade reagent. The *SlowFade®* Gold antifade formulation slows fluorescein's fading rate by about fivefold without significantly reducing its initial fluorescence intensity, thereby dramatically increasing the signal-to-noise ratio in photomicroscopy. Moreover, quenching of the Alexa Fluor® 350, Alexa Fluor® 405, Cascade Blue®, tetramethylrhodamine and Texas Red® dyes is minimal. In fact, the *SlowFade®* Gold antifade reagent reduces the fading rate of the Cascade Blue® fluorophore to almost zero, while decreasing its emission intensity by only about 30%.

The *SlowFade®* Gold antifade reagent is available in a single 10 mL bottle (S36936), as well as in a set of five 2 mL bottles (S36937). As with the ProLong® Gold antifade reagents, we also offer *SlowFade®* Gold antifade reagent containing the blue-fluorescent nuclear counterstain DAPI. *SlowFade®* Gold antifade reagent with DAPI is available

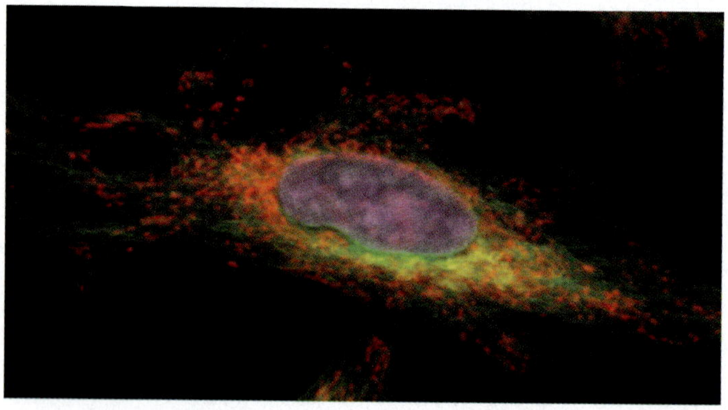

Figure 23.1.7 Human carcinoma (HeLa) cell labeled with Qdot® nanocrystals and mounted with Qmount® media. Mitochondria were detected with anti–OxPhos complex V inhibitor protein IgG (A21355) and labeled with Qdot® 625 goat F(ab')$_2$ anti–mouse IgG antibody (A10195, red fluorescence); the Golgi apparatus was detected with rabbit anti-giantin antibody and labeled with Qdot® 585 goat F(ab')$_2$ anti–rabbit IgG antibody (Q11411MP, yellow fluorescence); tubulin was detected with rat anti-tubulin antibody and labeled with DSB-X™ biotin goat anti–rat IgG antibody (D20697) and Qdot® 525 streptavidin (Q10141MP, green fluorescence). The nucleus was labeled with Qnuclear™ Deep Red Stain (Q10363, purple fluorescence), and the slide was mounted with Qmount® Qdot® mounting media (Q10336).

in a single 10 mL bottle (S36938), as well as in a set of five 2 mL bottles (S36939). These reagents permit simultaneous nuclear staining and protection of the stained sample from photobleaching.[12]

Unlike the ProLong® and ProLong® Gold antifade reagents, the *SlowFade®* and *SlowFade®* Gold antifade reagents do not cure over time so samples can be viewed immediately; however, *SlowFade®* Gold reagents are intended for short-term use (3–4 weeks) only and mounted samples may degrade over longer time periods. Secondary sealing of coverslips with wax or nail polish is recommended when working with high magnification objectives or preparing specimens for storage and subsequent imaging.

Qmount® Qdot® Mounting Media

The unique physical properties of Qdot® nanocrystals (Section 6.6) make them largely incompatible with ProLong® Gold, *SlowFade®* Gold and other mounting media designed primarily for use with organic dyes. Qmount® Qdot® mounting media (Q10336) is a specialized mountant that preserves the fluorescence signal of Qdot® nanocrystals with little to no quenching of the signal's initial intensity. The formulation cures within 12 hours and is provided in a convenient and easy-to-use dropper bottle. This mounting medium offers excellent compatibility with all eight Qdot® nanocrystal spectral types (Section 6.6) and their conjugates, as well as the nuclear counterstain Qnuclear™ Deep Red stain (Q10363, Section 12.5), making it an especially valuable tool for multicolor Qdot® nanocrystal imaging applications (Figure 23.1.7). Qmount® Qdot® mounting media is not recommended for use with Alexa Fluor® dyes or fluorescent proteins.

Image-iT® FX Signal Enhancer

By efficiently blocking nonspecific electrostatic interactions of anionic fluorescent dyes with cationic cell and tissue constituents, the Image-iT® FX signal enhancer (I36933) dramatically improves the signal-to-noise ratio of immunolabeled cells and tissues,[13–15] allowing clear visualization of targets that would normally be indistinguishable due to background fluorescence (Figure 23.1.8, Figure 23.1.9, Figure 23.1.10). Background staining seen with fluorescent conjugates of streptavidin, goat anti–mouse IgG antibody or goat anti–rabbit IgG antibody is largely eliminated when Image-iT® FX signal enhancer is applied to fixed and permeabilized cells prior to staining.

FluoCells® Prepared Microscope Slides

Ideal for educators and instrument manufacturers, our popular FluoCells® prepared microscope slides contain multilabeled cell preparations for observation by epifluorescence or confocal laser-scanning microscopy. The multicolor staining in these cell and tissue preparations can deliver publication-quality images and lasts through repeated viewings. These slides are especially useful for setting up microscopes and camera systems and for assessing the capabilities of optical filter sets.[16–18] When stored properly, these permanently mounted specimens will retain their bright and specific staining patterns for at least six months from the date of purchase. We currently offer five different FluoCells® prepared microscope slides:

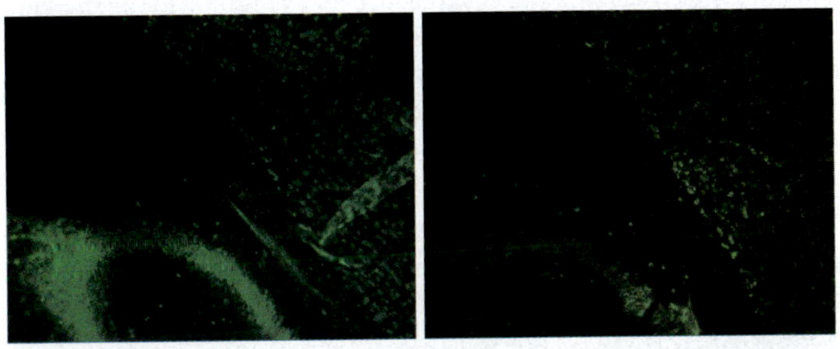

Figure 23.1.8 Reduced background staining afforded by Image-iT® FX signal enhancer. Mouse brain cryosections were permeabilized and antigen retrieval was carried out. The sections were then treated for 30 minutes with Image-iT® FX signal enhancer (I36933, left) or left untreated (right). Sections were labeled with the neural cell body selective antibody anti–HuC/D (A21271) and visualized using TSA™ Kit #2 (T20912) with the HRP conjugate of goat anti–mouse IgG and Alexa Fluor® 488 tyramide. Sections were mounted using the reagents in the ProLong® Antifade Kit (P7481).

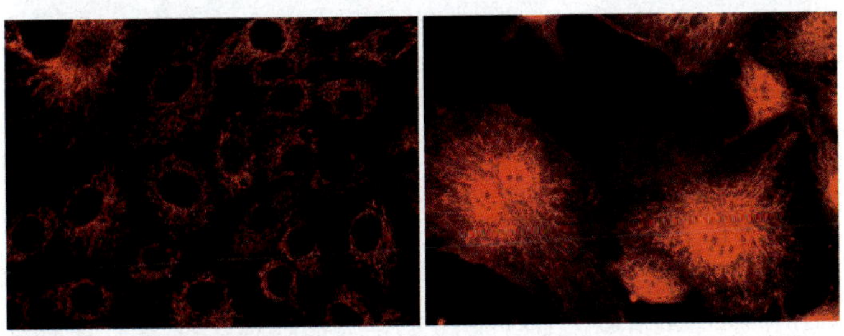

Figure 23.1.9 Increased label specificity and resolution provided by Image-iT® FX signal enhancer. Fixed and permeabilized bovine pulmonary artery endothelial cells were treated with Image-iT® FX signal enhancer (I36933, left) or left untreated (right) and then labeled with tetramethylrhodamine streptavidin (S870).

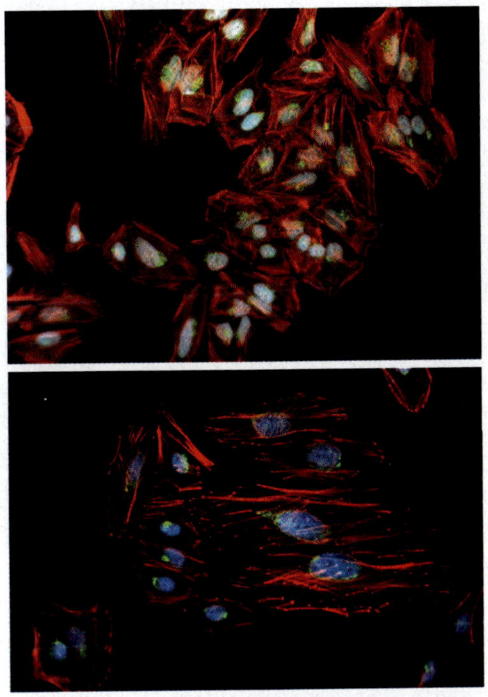

Figure 23.1.10 Golgi complexes in fixed and permeabilized HeLa cells labeled with anti–golgin-97 antibody (A21270) and visualized with green-fluorescent Alexa Fluor® 488 goat anti–mouse IgG antibody (A11001). Actin was stained with red-fluorescent Alexa Fluor® 594 phalloidin (A12381); nuclei were stained with blue-fluorescent DAPI (D1306, D3571, D21490). Treatment with Image-iT® FX signal enhancer (I36933) largely eliminates nonspecific dye binding (bottom panel) as compared with untreated slide (top panel).

IMPORTANT NOTICE: The products described in this manual are covered by one or more Limited Use Label License(s). Please refer to the Appendix on page 971 and Master Product List on page 975. Products are For Research Use Only. Not intended for any animal or human therapeutic or diagnostic use.

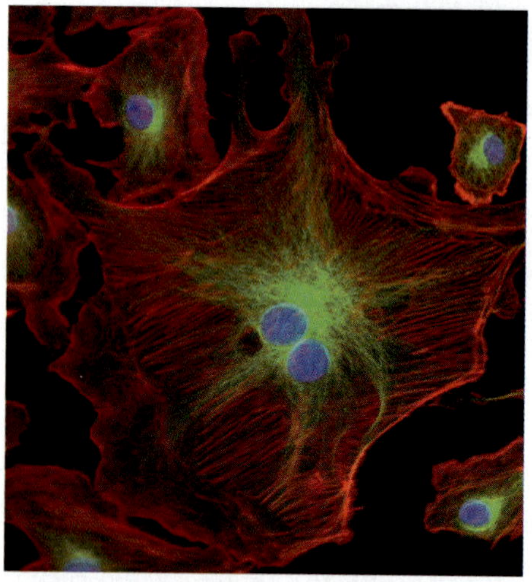

Figure 23.1.11 FluoCells® prepared slide #2 (F14781), which shows bovine pulmonary artery endothelial cells (BPAEC) that have been stained with an anti–ß-tubulin mouse monoclonal antibody in conjunction with BODIPY® FL goat anti–mouse IgG (B2752) for labeling microtubules, Texas Red®-X phalloidin (T7471) for labeling F-actin and DAPI (D1306, D3571, D21490) for labeling nuclei. This multiple-exposure image was acquired using bandpass optical filter sets appropriate for DAPI, fluorescein and Texas Red® dye.

- FluoCells® prepared slide #1 (F36924) shows bovine pulmonary artery endothelial cells (BPAEC) stained with MitoTracker® Red CMXRos for labeling mitochondria, Alexa Fluor® 488 phalloidin for labeling F-actin and DAPI for labeling the nucleus.
- FluoCells® prepared slide #2[19] (F14781, Figure 23.1.11) again contains BPAEC, but this time stained with Texas Red®-X phalloidin for labeling F-actin, mouse monoclonal anti–α-tubulin antibody in conjunction with BODIPY® FL goat anti–mouse IgG antibody for labeling microtubules and DAPI for labeling the nucleus.
- FluoCells® prepared slide #3 (F24630; Figure 23.1.12, Figure 23.1.13, Figure 23.1.14) contains a 16 µm cryostat section of a mouse kidney. Green-fluorescent Alexa Fluor® 488 wheat germ agglutinin stains the glomeruli and convoluted tubules; red-fluorescent Alexa Fluor® 568 phalloidin labels actin, which is especially prevalent in glomeruli and the brush border of proximal convoluted tubules; finally, DAPI stains the nuclei with blue fluorescence.
- FluoCells® prepared slide #4 (F24631, Figure 23.1.15) contains a 16 µm cryostat section of a mouse intestine. Alexa Fluor® 350 wheat germ agglutinin labels the mucus of goblet cells with blue fluorescence; the red-fluorescent Alexa Fluor® 568 phalloidin labels actin filaments, which are especially prevalent in the brush border of the intestinal mucosa; and SYTOX® Green nucleic acid stain labels nuclei with green fluorescence.
- FluoCells® prepared slide #6 (F36925, Figure 23.1.16) contains muntjac skin fibroblast cells stained with a combination of fluorescent stains. Green-fluorescent Alexa Fluor® 488 phalloidin labels the prominent filamentous actin in these cells; a mouse monoclonal anti–OxPhos Complex V inhibitor protein antibody in conjunction with the orange-fluorescent Alexa Fluor® 555 goat anti–mouse IgG antibody labels mitochondria; far-red–fluorescent TO-PRO®-3 nucleic acid stain labels nuclei. Because it contains no blue-fluorescent dyes, this slide is ideal for use with confocal laser-scanning microscopes that rely on non–UV laser light sources.

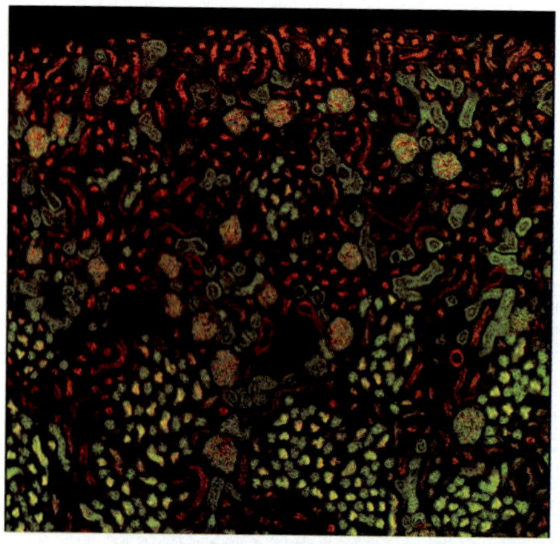

Figure 23.1.12 FluoCells® prepared slide #3 (F24630) contains a mouse kidney section stained with a combination of fluorescent dyes. Alexa Fluor® 488 wheat germ agglutinin (W11261) is a green-fluorescent lectin that was used to label elements of the glomeruli and convoluted tubules. The filamentous actin prevalent in glomeruli and the brush border were stained with red-orange–fluorescent Alexa Fluor® 568 phalloidin (A12380). Finally, the nuclei were stained with the blue-fluorescent DNA stain DAPI (D1306, D3571, D21490). This pseudocolored image was acquired on a Zeiss® confocal microscope located at the Institute of Neuroscience, University of Oregon.

FocalCheck™ Fluorescent Microspheres

FocalCheck™ Ring-Stained Fluorescent Microspheres

FocalCheck™ fluorescent microspheres are specifically designed for examining the alignment, sensitivity and stability of confocal laser-scanning microscopes.[17,20] They are particularly useful for confirming the optical sectioning thickness (Z-resolution) in three-dimensional imaging applications. These polystyrene beads—available in 6 µm and 15 µm diameters—have been treated using a proprietary method in which a fluorescent dye is allowed to penetrate to only a limited depth within the microsphere. The resulting beads have a well-defined dye layer that, when viewed in

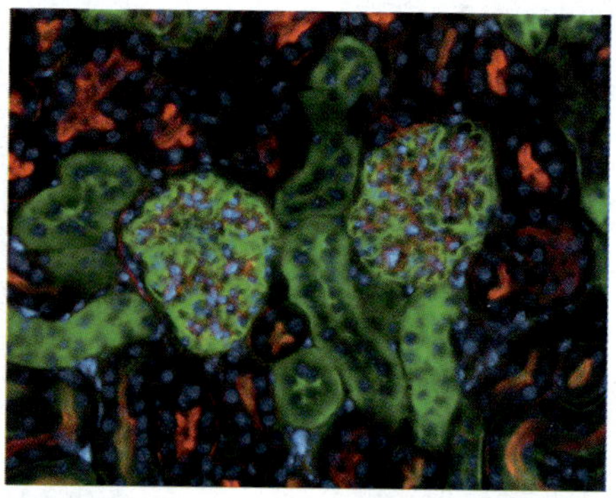

Figure 23.1.13 FluoCells® prepared slide #3 (F24630) contains a section of mouse kidney stained with a combination of fluorescent dyes. Alexa Fluor® 488 wheat germ agglutinin (W11261), a green-fluorescent lectin, was used to label elements of the glomeruli and convoluted tubules. The filamentous actin prevalent in glomeruli and the brush border were stained with red-fluorescent Alexa Fluor® 568 phalloidin (A12380). Finally, the nuclei were counterstained with the blue-fluorescent DNA stain DAPI (D1306, D3571, D21490). This image is a composite of three micrographs acquired using filter sets appropriate for fluorescein, tetramethylrhodamine and DAPI.

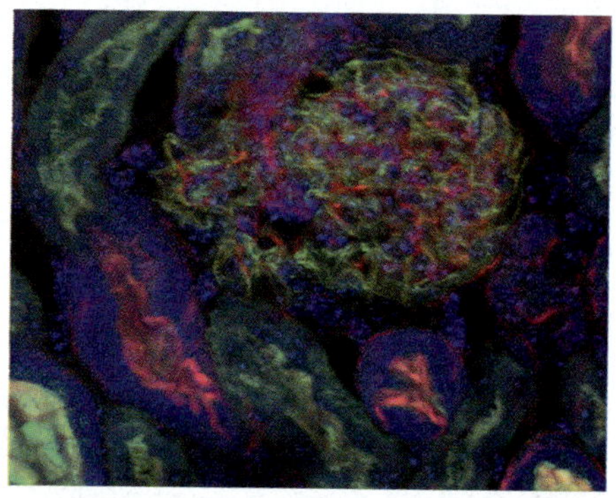

Figure 23.1.14 FluoCells® prepared slide #3 (F24630) containing a 16 µm cryostat section of mouse kidney stained with green-fluorescent Alexa Fluor® 488 wheat germ agglutinin (W11261), red-orange–fluorescent Alexa Fluor® 568 phalloidin (A12380) and blue-fluorescent DAPI (D1306, D3571, D21490). The image represents an optical section obtained by simultaneous two-photon excitation of all three dyes at 797 nm using a Bio-Rad® Radiance 2100 multiphoton microscope system. The image was acquired at the 2001 3D Microscopy of Living Cells Course, University of British Columbia, Vancouver, Canada, by John Jordan, Bio-Rad® Laboratories; and Iain Johnson, Life Technologies.

IMPORTANT NOTICE: The products described in this manual are covered by one or more Limited Use Label License(s). Please refer to the Appendix on page 971 and Master Product List on page 975. Products are For Research Use Only. Not intended for any animal or human therapeutic or diagnostic use.

molecular probes® | invitrogen™ by life technologies™

cross section in the confocal laser-scanning microscope, appears as a fluorescent ring of varying dimensions depending on the focal plane (Figure 23.1.17, Figure 23.1.18). We refer to this proprietary staining procedure as ring staining to differentiate it from routine staining throughout the bead.

FocalCheck™ microspheres are available in a variety of color configurations provided by five different fluorescent stains:

- Blue (365/430 nm)
- Green (505/515 nm)
- Orange (560/580 nm)
- Red (575/600 nm)
- Dark red (660/680 nm)

The excitation/emission maxima of the different stains are well matched to the laser sources and optical filters commonly used in confocal laser-scanning microscopy and are especially useful in testing and aligning confocal laser-scanning microscopes with multiple laser lines and detection channels (Figure 23.1.19). Moreover, because the dyes are localized *within* the bead and therefore protected from environmental factors, the FocalCheck™ microspheres are brighter and much more photostable than conventional surface-stained beads.

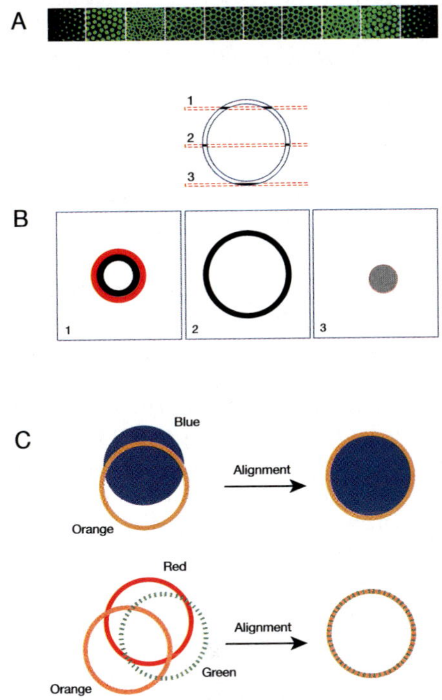

Figure 23.1.17 Confocal laser-scanning microscope optical cross-sectioning and alignment with FocalCheck™ microspheres. **A)** Serial optical sectioning from top to bottom along the z-axis of ring-stained microspheres reveals a continuous pattern of disc-to-ring-to-disc images. **B)** The diameter of the fluorescent ring (or disc) seen is dependent on the depth of the optical focal plane. **C)** In the confocal laser-scanning microscope, separate light paths exist for UV and visible wavelengths. Also, emitted fluorescence is detected by different photomultipliers. Proper optical alignment may be obtained with either of two types of FocalCheck™ microspheres. For example, the microspheres with an orange ring stain that are blue-fluorescent throughout the bead allow UV/visible wavelength alignment in three dimensions upon aligning the orange ring with the blue disc. Focal alignment is also possible simultaneously in three colors by aligning the green, orange and dark red rings of the FocalCheck™ microspheres containing fluorescent green/orange/dark red ring stains.

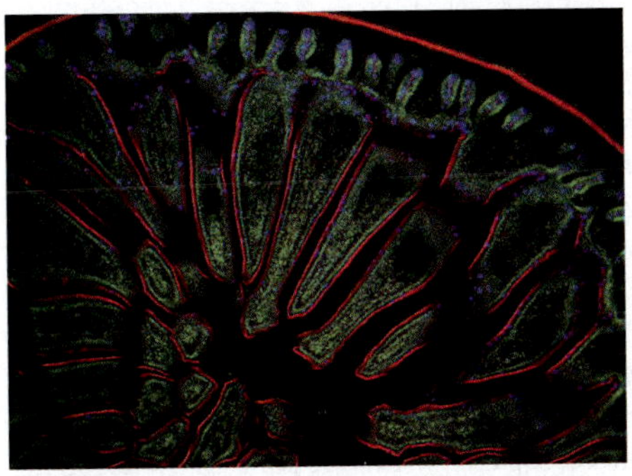

Figure 23.1.15 FluoCells® prepared slide #4 (F24631) contains a section of mouse intestine stained with a combination of fluorescent stains. Alexa Fluor® 350 wheat germ agglutinin (W11263) is a blue-fluorescent lectin that was used to stain the mucus of goblet cells. The filamentous actin prevalent in the brush border was stained with red-orange–fluorescent Alexa Fluor® 568 phalloidin (A12380). Finally, the nuclei were stained with SYTOX® Green nucleic acid stain (S7020). This image is a composite of three digitized images obtained with filter sets appropriate for fluorescein, DAPI and tetramethylrhodamine.

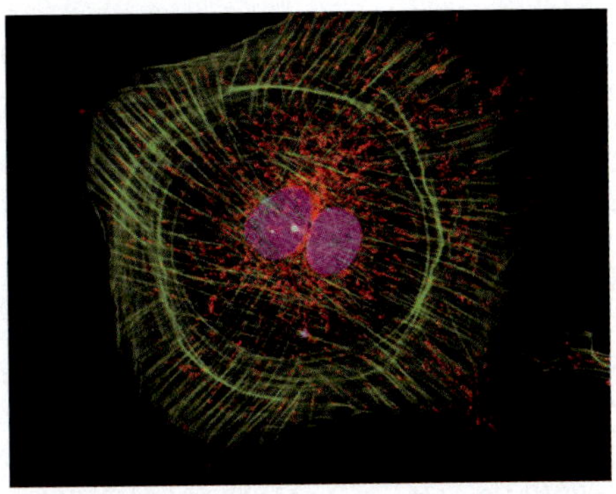

Figure 23.1.16 FluoCells® prepared slide #6 (F36925) showing a fixed, permeabilized and labeled muntjac skin fibroblast. Mitochondria were labeled with mouse anti–OxPhos Complex V inhibitor protein antibody and visualized using orange-fluorescent Alexa Fluor® 555 goat anti–mouse IgG antibody (A21422). F-actin was labeled with green-fluorescent Alexa Fluor® 488 phalloidin (A12379), and the nucleus was stained with TO-PRO®-3 iodide (T3605, pseudocolored magenta).

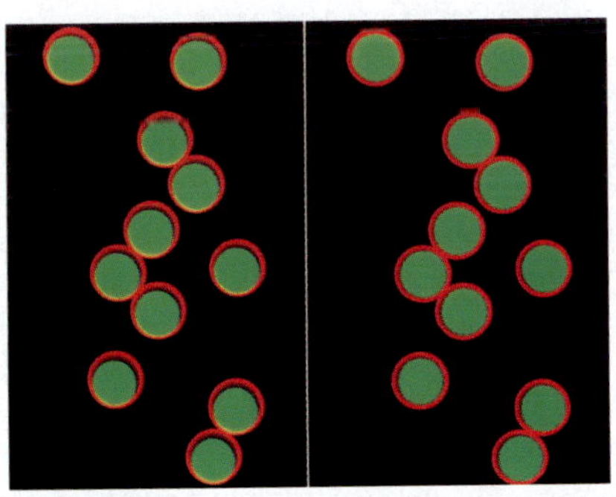

Figure 23.1.18 Images from confocal laser-scanning microscope optical cross-sectioning of our 15 µm FocalCheck™ microspheres that have a dark red–fluorescent ring stain with a green-fluorescent stain throughout the bead (F7239). The left panel provides a clear visual representation of poor instrument alignment. Correct image registration has been achieved in the right panel, where the dark red ring is aligned with the green disk. The image was contributed by Paulette Brunner, Keck Imaging Center, University of Washington, and Yu-Zhong Zhang, Life Technologies.

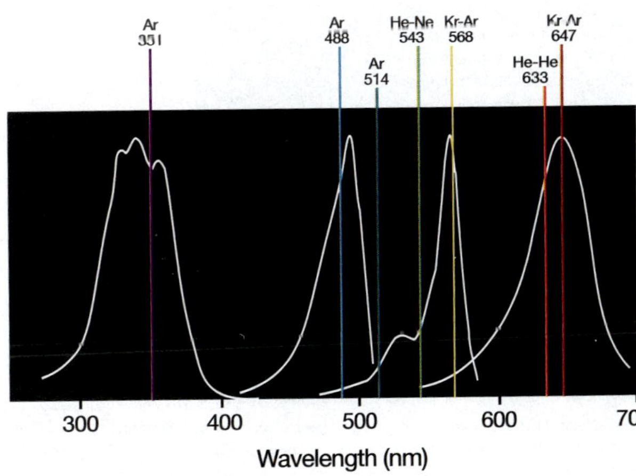

Figure 23.1.19 Normalized excitation spectra of the dyes contained in the FocalCheck™ microspheres. Emission lines of several commonly used laser excitation sources are superimposed on the dyes' excitation spectra to illustrate the wide range of usage of these beads as calibration references. Ar = Argon-ion laser. Kr-Ar = Krypton-argon laser. He-Ne = Helium-neon laser.

IMPORTANT NOTICE: The products described in this manual are covered by one or more Limited Use Label License(s). Please refer to the Appendix on page 971 and Master Product List on page 975. Products are For Research Use Only. Not intended for any animal or human therapeutic or diagnostic use.

FocalCheck™ products are available in various different color configurations, including three suspensions that contain microspheres exhibiting ring stains of two or three different fluorescent colors:

- Blue-fluorescent and orange-fluorescent ring stains (15 µm, F7234)
- Green-fluorescent and dark red–fluorescent ring stains (15 µm, F7240)
- Green-fluorescent, orange-fluorescent and dark red–fluorescent ring stains (6 µm, F14806; 15 µm, F7235)

We also supply four suspensions that contain microspheres exhibiting a ring stain of one fluorescent color combined with a stain of a second fluorescent color throughout the bead:

- Green-fluorescent ring stain with blue-fluorescent stain throughout (6 µm, F14808; 15 µm, F7237)
- Green-fluorescent ring stain with dark red–fluorescent stain throughout (15 µm, F7238)
- Orange-fluorescent ring stain with blue-fluorescent stain throughout (15 µm, F7236)
- Dark red–fluorescent ring stain with green-fluorescent stain throughout (6 µm, F14807; 15 µm, F7239, Figure 23.1.18)

The sharp ring stains exhibited by the FocalCheck™ microspheres produce a striking visual representation of instrument misalignment or other aberrations, making them ideal as reference standards for confocal laser-scanning microscopy. Correct image registration is indicated when the multiple ring images of the ring-stained FocalCheck™ beads (or the ring and disk images of the combination ring-stained and stained-throughout FocalCheck™ beads) are perfectly coincident in all dimensions (Figure 23.1.17).

FocalCheck™ Thin-Ring Fluorescent Microspheres

Our FocalCheck™ Thin-Ring Fluorescent Microspheres Kit (F14791) contains smaller-diameter microspheres that have spectral and physical features similar to those of our 6 µm and 15 µm FocalCheck™ microspheres. Because we prepare these 1.0 µm beads using fluorescent stains that are restricted to the surface only, they exhibit sharper and thinner fluorescent ring patterns when viewed in cross section with a confocal laser-scanning microscope. The FocalCheck™ Thin-Ring Fluorescent Microspheres Kit contains three 200 µL suspensions of 1.0 µm beads. Each suspension contains beads with a different color configuration:

- Green-fluorescent (495/515 nm) and red-fluorescent (575/600 nm) ring stains
- Green-fluorescent (495/515 nm) ring stain with dark red–fluorescent (660/680 nm) stain throughout the bead
- Red-fluorescent (575/600 nm) ring stain with blue-fluorescent (365/430 nm) stain throughout the bead

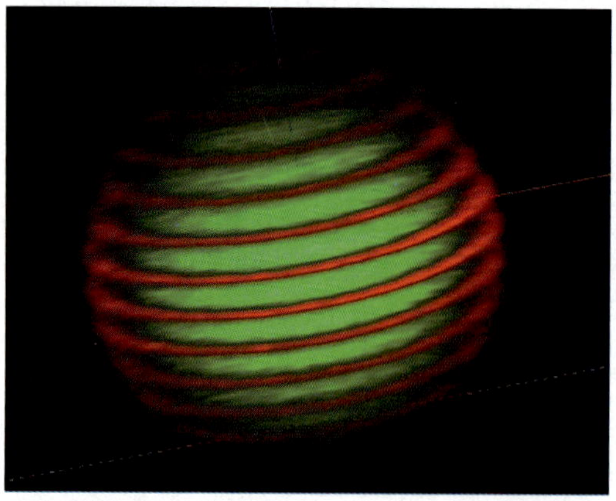

Figure 23.1.20 A double-labeled microsphere from the FocalCheck™ DoubleGreen Fluorescent Microsphere Kit (F36905). The bead was imaged as a z-series using a Carl Zeiss® LSM 510 META system. The two green-fluorescent dyes were separated by spectral unmixing, and one of the dyes was pseudocolored red. In this composite image, the complete z-series is shown prior to software rendering. Rendering fills in the missing information between the slices by interpolation to create a solid object.

FocalCheck™ Microspheres Pre-Mounted on Microscope Slides

In addition to the bead suspensions described above, we offer FocalCheck™ microspheres pre-mounted on microscope slides. The FocalCheck™ Fluorescent Microsphere Kits feature mounted samples of three different color configurations, in either the 6 µm (F24633) or the 15 µm (F24634) bead size:

- FocalCheck™ beads with green-fluorescent, orange-fluorescent and dark red–fluorescent ring stains
- FocalCheck™ beads with green-fluorescent ring stain and blue-fluorescent stain throughout the bead
- FocalCheck™ beads with dark red–fluorescent ring stain and green-fluorescent stain throughout the bead

FocalCheck™ Fluorescence Microscope Test Slides

FocalCheck™ fluorescence microscope test slides #1, #2 and #3 are specifically designed for calibrating fluorescence microscope systems and evaluating system and filter performance:

- FocalCheck™ fluorescence microscope test slide #1 (F36909) is ideal for routine checking and calibration of confocal and widefield fluorescence microscopes[21]
- FocalCheck™ fluorescence microscope test slide #2 (F36913) provides a robust, reproducible method of evaluating the performance of spectral imaging systems,[22] as well as the ability to discriminate closely overlapping spectra. The slide consists of 6 µm–diameter microspheres labeled with a series of spectrally overlapping dyes
- FocalCheck™ fluorescence microscope test slide #3 (F36914) is useful for basic evaluation of filter performance and as a general practice slide for fluorescence microscopy and digital imaging

The slides each contain 10 sample areas arranged in two rows coated with proprietary fluorescent microspheres designed specifically for microscopy applications. The microspheres are mounted in optical cement (refractive index ~1.52) for maximal stability. The optical thickness of the mounted bead specimens may result in spherical aberration or inability to focus when using certain types of objectives. Users who encounter these problems are encouraged to contact our Technical Assistance Department.

FocalCheck™ Fluorescent Microspheres for Spectral Unmixing

We have prepared two FocalCheck™ Fluorescent Microspheres Kits for testing spectral separation on spectral imaging systems.[23] These microspheres are stained with two different fluorescent dyes that appear similar in color by eye but are sufficiently different to be resolved by linear-unmixing techniques. When linear-unmixing data-processing algorithms are applied, the dyes are shown to be spectrally distinct and spatially separated—one appears only within the outer ring and the other appears throughout the microsphere (Figure 23.1.20). These 6 µm, dual-stained microspheres are provided mounted on a microscope slide in each of the following kits:

- FocalCheck™ DoubleGreen Fluorescent Microspheres Kit (F36905), with a green-fluorescent (500/512 nm) ring dye and a slightly longer-wavelength green-fluorescent (512/525 nm) core dye
- FocalCheck™ DoubleOrange Fluorescent Microspheres Kit (F36906), with an orange-fluorescent (532/552 nm) ring dye and a slightly longer-wavelength orange-fluorescent (545/565 nm) core dye

For generating reference spectra, these kits also contain two additional slides containing microspheres stained uniformly with each of the individual dyes. Downloadable reference spectra are also available through our online Fluorescence SpectraViewer tool (www.invitrogen.com/handbook/spectraviewer; Using the Fluorescence SpectraViewer—Note 23.1).

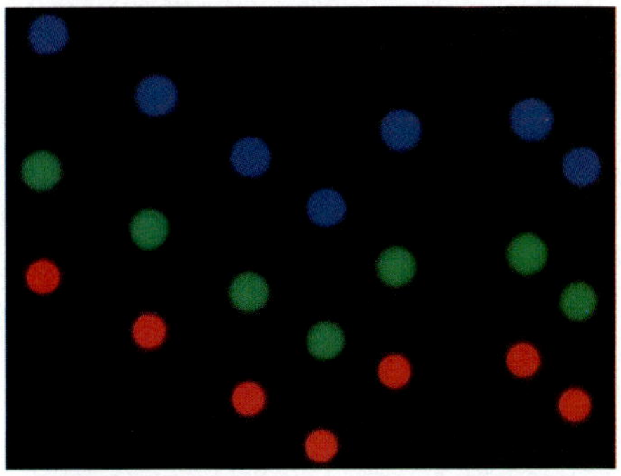

Figure 23.1.21 Multiple exposures of a prepared slide from our MultiSpeck™ Multispectral Fluorescence Microscopy Standards Kit (M7901). This kit provides fluorescent MultiSpeck™ microspheres that exhibit three relatively distinct excitation/emission bands—blue, green and red—all in the same particle. Thus, the same microsphere can appear a different color depending on the optical filter set used. This photograph was taken through bandpass optical filter sets appropriate for DAPI, fluorescein and Texas Red® dyes, with the field of view shifted slightly between exposures.

MultiSpeck™, TetraSpeck™ and Constellation™ Fluorescent Microspheres

Our MultiSpeck™ and TetraSpeck™ fluorescent microspheres greatly facilitate the adjustment and calibration of conventional fluorescence microscopes, confocal laser-scanning microscopes and associated image-processing equipment for multicolor applications. These uniform, multiply stained microspheres are useful for colocalizing and focusing different wavelengths of light in the same optical plane, as well as for checking multicolor image resolution, magnification and sensitivity.

MultiSpeck™ Multispectral Fluorescence Microscopy Standards Kit

The 4 µm MultiSpeck™ microspheres in our MultiSpeck™ Multispectral Fluorescence Microscopy Standards Kit (M7901) exhibit three relatively distinct emission bands—blue, green and red—throughout every particle (Figure 23.1.21). The spectral characteristics (excitation/emission peaks at 365/405 nm, 520/525 nm and 580/600 nm) are compatible with optical filter sets designed for commonly used blue, green and red fluorophores (e.g., DAPI, fluorescein and rhodamine or Texas Red® dyes or their Alexa Fluor® dye counterparts). The MultiSpeck™ beads can be used as external references for assessing image registration in two and three dimensions, allowing the researcher to accurately determine the spatial relationships of different labels in multiparameter experiments.

The MultiSpeck™ Multispectral Fluorescence Microscopy Standards Kit contains:

- Suspension of MultiSpeck™ multispectral microspheres
- Mixed suspension of separately stained blue-, green- and red-fluorescent microspheres, which exhibit the same three excitation/emission bands as the multispectral microspheres but in separate beads
- Mounting medium
- Slide-mounting protocol

Both suspensions are provided at a ready-to-use density and can be mounted on slides or incorporated into an experimental sample. Each kit supplies a sufficient amount of material for ~50 slide preparations using either of the two bead suspensions provided.

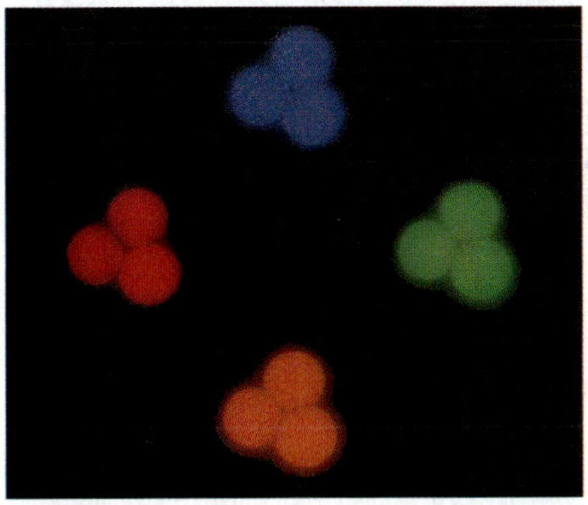

Figure 23.1.22 Four separate exposures of three TetraSpeck™ beads (T7283) photographed using optical filter sets appropriate for DAPI, fluorescein, tetramethylrhodamine and the Texas Red® dye. The stage was shifted after each exposure. Note that the same beads appear blue, green, orange or red, depending on the filters used.

TetraSpeck™ Fluorescent Microspheres and Sampler Kits

Our TetraSpeck™ fluorescent microspheres [17,19,20] expand the multispectral strategy introduced with the MultiSpeck™ beads in two important ways. First, the TetraSpeck™ beads have been stained throughout with a mixture of four different fluorescent dyes, yielding four well-separated excitation and emission peaks (Figure 23.1.22). The excitation/emission maxima of the dyes are

Figure 23.1.23 Luminescent microsphere products provide an extensive range of sizes and fluorescent colors, illustrated by a sample of our Constellation™ microspheres for imaging (C14837).

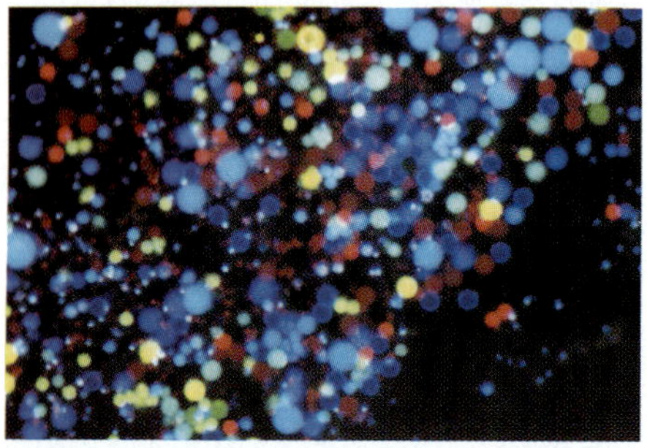

Figure 23.1.24 Constellation™ microspheres (C14837) provide an assortment of sizes and colors for use in fluorescence microscopy demonstrations.

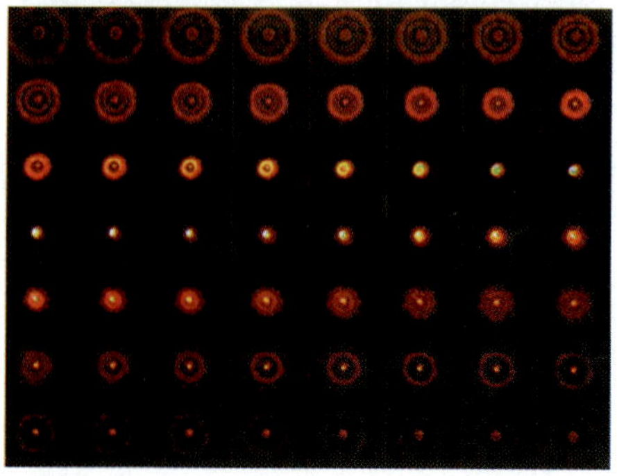

Figure 23.1.25 One PS-Speck™ green-fluorescent microsphere (P7220) used for point-spread function analysis. Shown are individual images, in pseudocolor, taken in x-y planes with 0.25 µm increments in the z-axis between planes. The image was contributed by Regina Armstrong, Uniformed Services University of the Health Sciences, Bethesda, Maryland.

365/430 nm (blue), 505/515 nm (green), 560/580 nm (orange) and 660/680 nm (dark red). Second, these microspheres are available in five nominal sizes (actual bead diameters are indicated on the product labels), spanning the range from subresolution to nearly cell-size particles:

- 0.1 µm (T7279)
- 0.2 µm (T7280)
- 0.5 µm (T7281)
- 1.0 µm (T7282)
- 4.0 µm (T7283)

Each of these products provides a 0.5 mL suspension sample of TetraSpeck™ microspheres that is sufficient for about 100 slides. We offer the TetraSpeck™ Fluorescent Microspheres Sampler Kit (T7284) and the TetraSpeck™ Fluorescent Microspheres Size Kit (T14792). The TetraSpeck™ Fluorescent Microspheres Sampler Kit consists of separate suspension samples of our 0.1 µm, 0.5 µm and 4.0 µm TetraSpeck™ beads, each sufficient for preparing about 20 slides. The TetraSpeck™ Fluorescent Microspheres Size Kit contains six microscope slides; five slides with a mounted sample of the 0.1 µm, 0.2 µm, 0.5 µm, 1.0 µm or 4.0 µm diameter TetraSpeck™ microspheres, and a sixth slide with a mixture of all five sizes. TetraSpeck™ microspheres have been used to calibrate the spatial distribution of illumination for high-content screening (HCS),[24,25] and as reference markers for image alignment in high-resolution immunofluorescence colocalization analysis.[24,26] In addition, various FluoSpheres® and TetraSpeck™ beads have been used as reference standards for two-photon excitation microscopy, and in particular for the *in situ* determination of the two-photon excitation point-spread function (PSF).[27–29]

Constellation™ Microspheres for Imaging

Constellation™ microspheres for imaging (C14837; Figure 23.1.23, Figure 23.1.24) are 3 mL suspensions of assorted fluorescent microspheres with a variety of sizes and colors. Designed for use in laboratory tutorials and customer training sessions, they provide inexpensive and robust test samples for demonstrating filter switching, focus adjustment and other functional capabilities of fluorescence microscopes. The Constellation™ microspheres can be stored at room temperature, protected from light.

PS-Speck™ Microscope Point Source Kit

The fluorescent microspheres in the PS-Speck™ Microscope Point Source Kit (P7220) have a diameter of 0.175 ± 0.005 µm, making them ideal subresolution fluorescent sources for calibrating instrument optics. They are particularly useful for measuring the point-spread function (PSF) required for computational image deconvolution procedures[17,30–32] (Figure 23.1.25, Figure 23.1.26). This kit's four ready-to-use suspensions contain bright monodisperse particles in the following fluorescent colors (and excitation/emission peaks):

- Blue (360/440 nm)
- Green (505/515 nm)
- Orange (540/560 nm)
- Deep red (633/660 nm)

The kit also includes mounting medium and a mounting protocol for the user's convenience. Each suspension provides sufficient material to mount about 100 slides. PS-Speck™ microspheres are too large to represent point source objects for microscopes equipped with high numerical aperture (e.g., NA 1.4) oil immersion objectives. In such cases, we recommend our 0.1 µm TetraSpeck™ fluorescent microspheres (T7279) or 0.1 µm and 0.04 µm FluoSpheres® carboxylate-modified microspheres (Section 6.5) for PSF determinations.

InSpeck™ Microscopy Image Intensity Calibration Kits

InSpeck™ Microscope Image Intensity Calibration Kits provide microsphere standards that generate a series of well-defined fluorescence intensity levels (Figure 23.1.27) for constructing calibration curves and evaluating sample brightness.[6,25,33] InSpeck™ microspheres have been used to estimate the global background and signal response for high-content screening (HCS).[25]

IMPORTANT NOTICE: The products described in this manual are covered by one or more Limited Use Label License(s). Please refer to the Appendix on page 971 and Master Product List on page 975. Products are For Research Use Only. Not intended for any animal or human therapeutic or diagnostic use.

Most of the kits are offered in a choice of two different microsphere sizes (2.5 µm or 6 µm) and five different fluorescent colors:

- InSpeck™ Blue (350/440 nm) Kit, (2.5 µm, I7221)
- InSpeck™ Green (505/515 nm) Kits (2.5 µm, I7219; 6 µm, I14785)
- InSpeck™ Orange (540/560 nm) Kits (2.5 µm, I7223; 6 µm, I14786)
- InSpeck™ Red (580/605 nm) Kits (2.5 µm, I7224; 6 µm, I14787)
- InSpeck™ Deep Red (633/660 nm) Kit (2.5 µm, I7225)

Each kit includes six separate suspensions of InSpeck™ fluorescent microspheres with relative fluorescence intensities of 100%, 30%, 10%, 3%, 1% and 0.3% (Figure 23.1.27), covering the range of intensities commonly encountered in microscopy applications. Unstained control beads and mounting medium are also supplied. The aqueous suspensions of microspheres may be applied directly to the sample for calibrating fluorescence intensities or mounted separately in an adjacent well or on another slide. Each suspension provides sufficient material to prepare about 100 slides.

Fluorescence Reference Standards

Fluorescein NIST-Traceable Standard

The National Institute of Standards and Technology (NIST) chose high-grade Molecular Probes® fluorescein to create Standard Reference Material 1932 (SRM® 1932), a certified fluorescein solution. We now offer a NIST-traceable fluorescein standard (F36915) that not only meets the stringent criteria established by NIST, but is also directly traceable to SRM® 1932. We supply our NIST-traceable fluorescein standard as a calibrated 50 µM solution of fluorescein in 100 mM sodium borate buffer, pH 9.5; under these conditions, fluorescein is completely ionized [34] and is therefore in its most fluorescent form (Figure 23.1.28, Figure 23.1.29), exhibiting an extremely high quantum yield of 0.93.

Academic researchers and industry scientists alike can use our NIST-traceable fluorescein standard to assess day-to-day or experiment-to-experiment variation in fluorescence-based instrumentation, as well as to determine the Molecules of Equivalent Soluble Fluorophore (MESF) value for an experimental solution. The MESF value is defined not as the actual number of dye molecules present, but rather as the number of fluorophores that would yield a fluorescence intensity equivalent to that of the experimental solution when analyzed on the same instrument under the same conditions.[35–38] Consequently, the MESF value is an important tool for characterizing the fluorescence intensity of a solution containing spectrally similar dye molecules attached to antibodies, nucleic acids, microspheres or other substrates that might enhance or diminish the fluorescence. When its pH is carefully matched with that of the experimental solution, our NIST-traceable fluorescein standard can be used for accurate MESF determinations of a wide range of green-fluorescent dye solutions and on an assortment of fluorescence-based instruments.

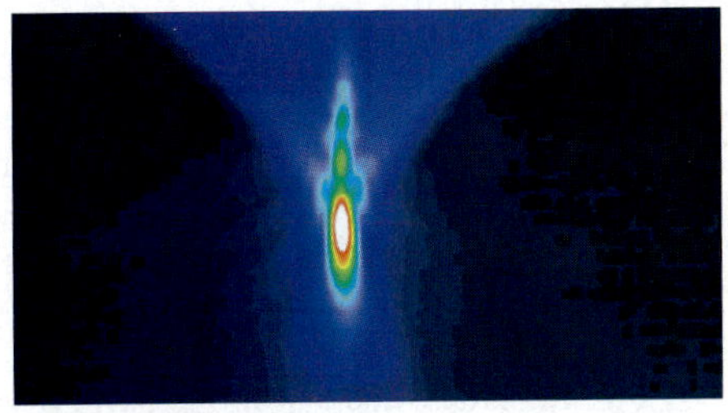

Figure 23.1.26 An orthogonal (x-z) display representing a point-spread function. The microsphere used is a component of the PS-Speck™ Microscope Point Source Kit (P7220). This pseudocolored image was generated electronically from a series of microsphere images taken in x-y planes. The image was contributed by Jennifer Kramer, Scanalytics.

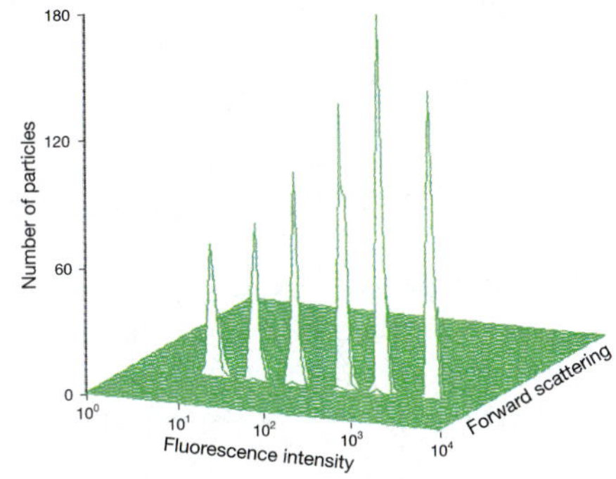

Figure 23.1.27 Flow cytometric analysis of the beads in the 6 µm InSpeck™ Green Microscope Image Intensity Calibration Kit (I14785). The microspheres have nominal relative fluorescence intensities of 100%, 30%, 10%, 3%, 1%, 0.3%. For each lot, actual relative intensities are determined by flow cytometry and printed on the product labels.

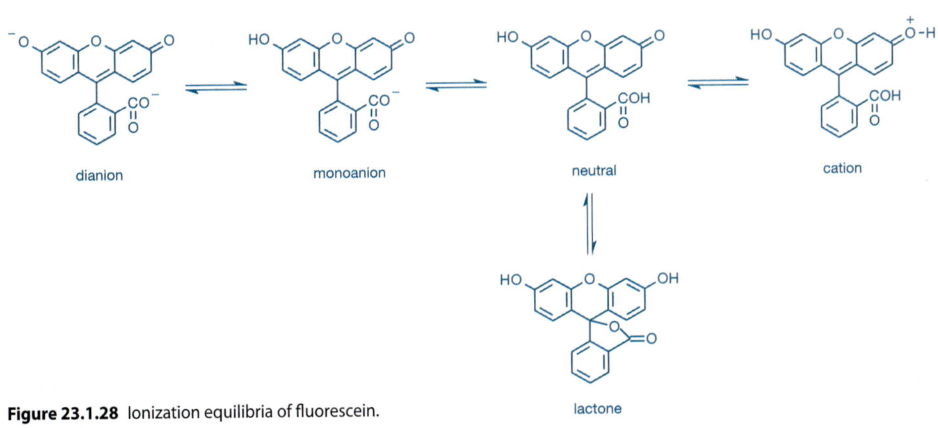

Figure 23.1.28 Ionization equilibria of fluorescein.

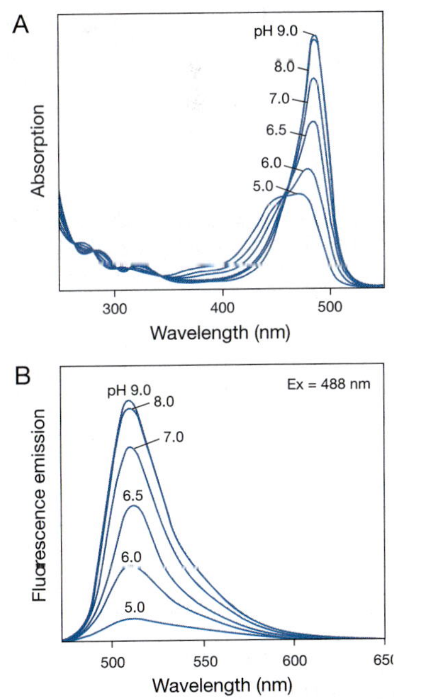

Figure 23.1.29 The pH-dependent spectra of fluorescein (F1300): A) absorption spectra, B) emission spectra.

Reference Dye Sampler Kit

Our Reference Dye Sampler Kit (R14782) provides samples of five extensively characterized fluorescence standards with emission spectra covering the entire visible wavelength range.[39] All five fluorescent standards are supplied as 1 mM stock solutions in 1 mL units, sufficient to prepare approximately 500 diluted working samples for spectrofluorometry. The compositions of the stock solutions are as follows:

- Quinine sulfate in 0.1 M sulfuric acid (H_2SO_4) (Figure 23.1.30)
- Fluorescein in dimethylsulfoxide (DMSO) (Figure 23.1.31)
- 5-Carboxytetramethylrhodamine in DMSO (Figure 23.1.32)
- Sulforhodamine 101 in DMSO (Figure 23.1.33)
- Nile blue perchlorate in DMSO (Figure 23.1.34)

Spectroscopic data for the five standards are summarized in Table 23.2. Reference spectra for all five fluorescent standard solutions are provided through our online Fluorescence SpectraViewer tool (www.invitrogen.com/handbook/spectraviewer).

Sample Chambers, Slides and Coverslips

In collaboration with Grace Bio-Labs, we offer a collection of accessories for imaging and microscopy. These accessories make slide preparation easy, facilitate sample perfusion and simplify sample manipulation during *in situ* hybridization and other procedures that involve multiple wash steps.

CultureWell™ Cell Culture Systems

The CultureWell™ cell culture systems provide an integrated set of tools for preparing cultured cells for staining and imaging. Each system uses medical-grade silicone gaskets preassembled with standard optical-quality coverslips into convenient inserts that fit into matching cell culture plates (Figure 23.1.35). The entire system is provided sterile and ready to use. Cell culture, treatment and staining are performed on the coverslip, which adheres securely to the culture plate via a silicone backing. The samples can then be imaged with or without the silicone gaskets. Two types of systems are available in several configurations to suit a variety of needs (Table 23.3).

The CultureWell™ multiwell cell culture systems use precut silicone gaskets to form convenient no-leak wells on 24 × 50 mm coverslips. The wells are spaced for compatibility with microfluidic handling robots. Low numbers of wells are ideal for titering antibody dilutions or other staining conditions, whereas the higher numbers of wells facilitate high-throughput screening. Each insert includes four coverslips with gaskets, preassembled into a culture plate. Silicone dividers (C24770, Figure 23.1.36) are also available for separating portions of the coverslip into leak-proof wells, for different treatment and washing conditions.

The CultureWell™ multislip cell culture systems comprise multiple coverslips arrayed on a sheet of silicone and assembled into a convenient insert. The silicone backing adheres the coverslips to the tissue culture plate, preventing movement during plating, cell culture and washing steps. Each coverslip can be removed separately for individual

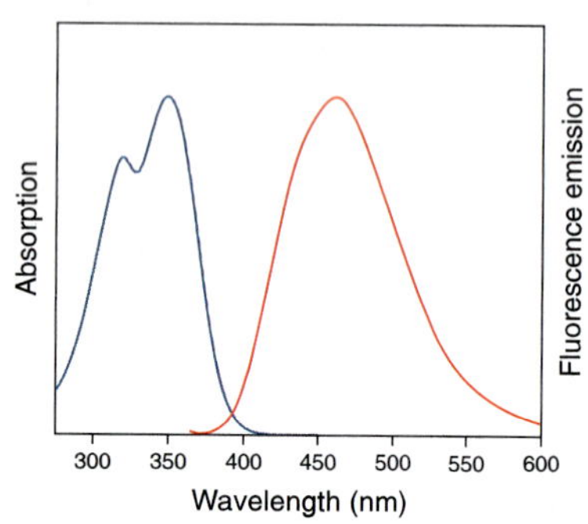

Figure 23.1.30 Absorption and fluorescence emission spectra of quinine sulfate, dihydrate in 0.5 M sulfuric acid.

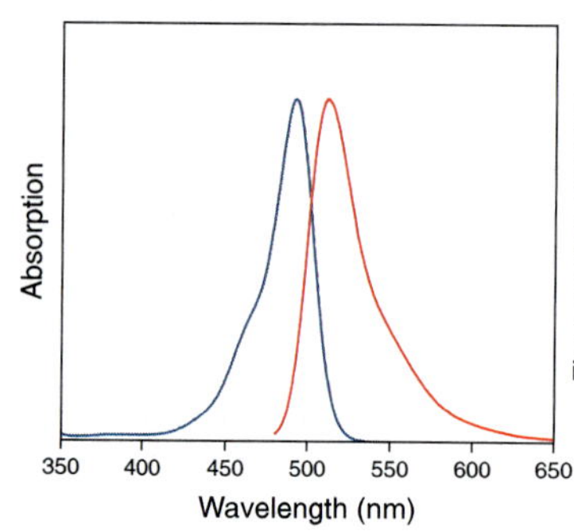

Figure 23.1.31 Absorption and fluorescence emission spectra of fluorescein in 0.1 M sodium hydroxide.

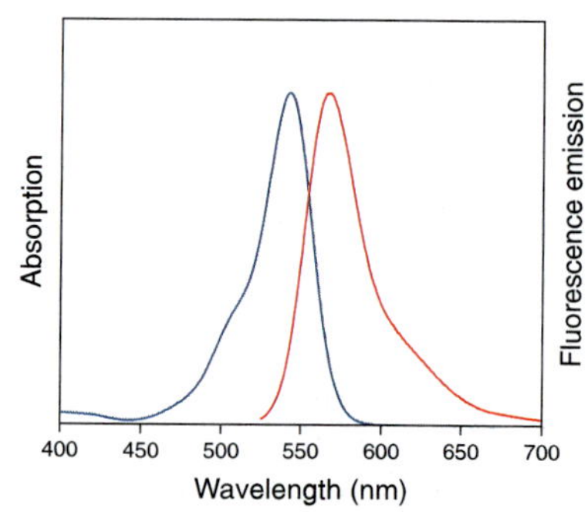

Figure 23.1.32 Absorption and fluorescence emission spectra of 5-carboxytetramethylrhodamine (5-TAMRA) in methanol.

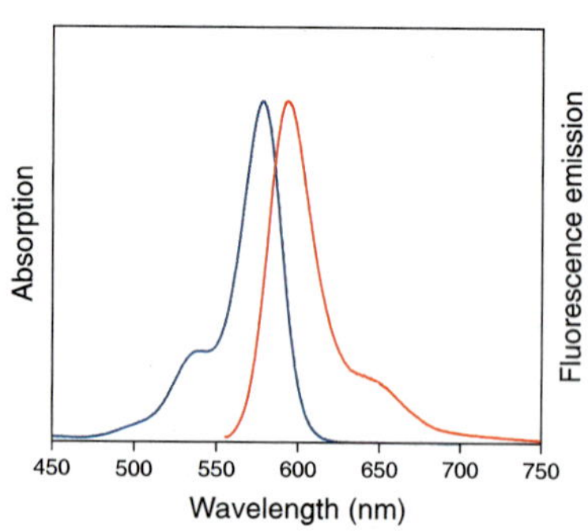

Figure 23.1.33 Absorption and fluorescence emission spectra of sulforhodamine 101 in ethanol.

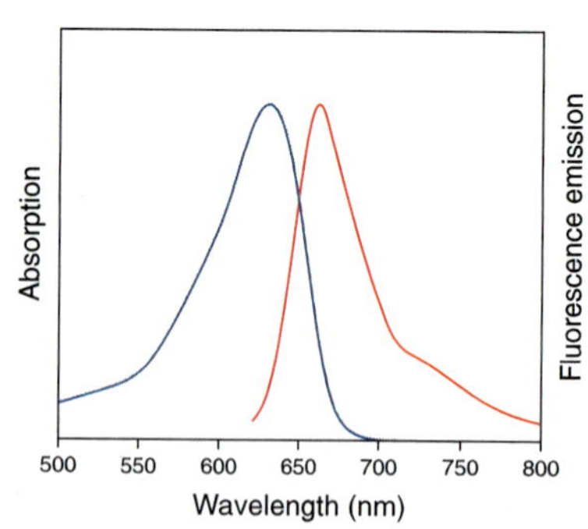

Figure 23.1.34 Absorption and fluorescence emission spectra of nile blue in ethanol.

Table 23.2 Spectroscopic data for components of the Reference Dye Sampler Kit.

Component	Solvent	Abs (nm) *	Em (nm) *	QY †
Quinine sulfate	Water	347 ‡	455 ‡	0.55 ‡
Fluorescein	Water	490 §	514 §	0.92 §
5-CTMR **	Methanol	542	568	0.68
Sulforhodamine 101	Ethanol	578	605	0.90
Nile blue	Ethanol	636	665	0.27

* Approximate absorbance (Abs) and fluorescence emission (Em) maxima. † Fluorescence quantum yield at 22°C. ‡ Standard values in 0.5 M H_2SO_4. § Standard values in 0.1 M NaOH. ** CTMR = Carboxytetramethylrhodamine.

IMPORTANT NOTICE: The products described in this manual are covered by one or more Limited Use Label License(s). Please refer to the Appendix on page 971 and Master Product List on page 975. Products are For Research Use Only. Not intended for any animal or human therapeutic or diagnostic use.

staining experiments. The CultureWell™ cell culture systems are provided in a set of 10 preassembled inserts in plates.

CultureWell™ Chambered Coverslips

The CultureWell™ chambered coverslips (Table 23.3, Figure 23.1.37) are the same gasketed coverslips provided in our CultureWell™ cell culture systems, but they are not preassembled into inserts. The chambered coverslip can be placed in CultureWell™ plates (C24769) or other cell culture dishes for cell culture and staining. The silicone gasket can be easily removed and the coverslip placed on a slide for microscopy. The CultureWell™ chambered coverslips provide maximum versatility for designing smaller scale cell culture applications. They are provided in sets of five sterile pouches, with four chambered coverslips per pouch.

CultureWell™ Chambered Coverglasses

The CultureWell™ chambered coverglass—provided sterile and ready to use—contains 16 wells that can each hold up to 250 µL, allowing cells to be cultured in a number of different conditions on a single slide (Figure 23.1.38). A silicone gasket forms a leakproof seal between the polystyrene upper structure and the coverglass. When the cells are ready to be imaged, coverglass removal is made easy by the use of a simple tool that separates the parts without the need for excessive force, eliminating the risk of coverglass breakage (Figure 23.1.39). Removal of the

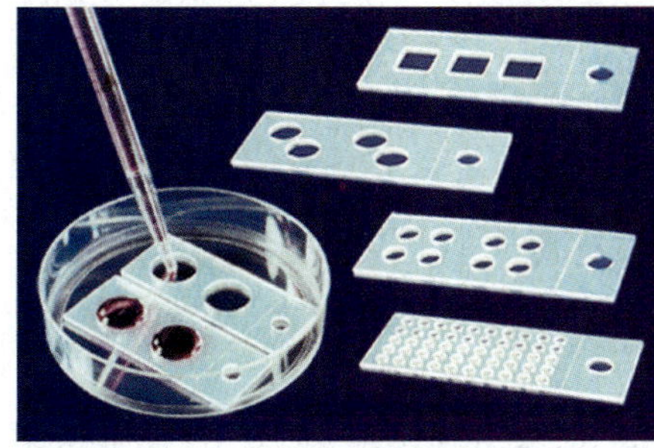

Figure 23.1.37 CultureWell™ chambered coverslips.

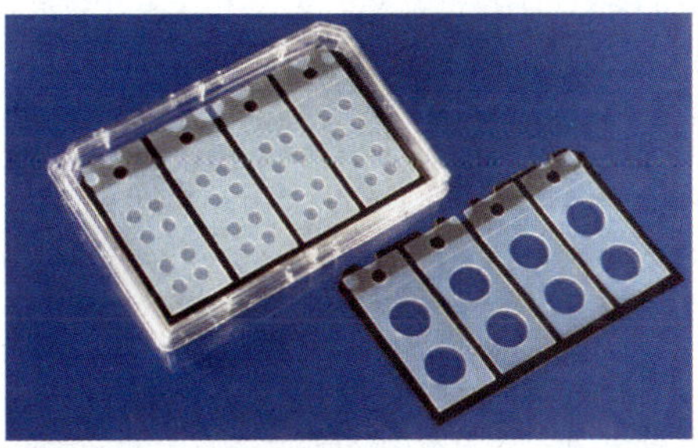

Figure 23.1.35 CultureWell™ cell culture system.

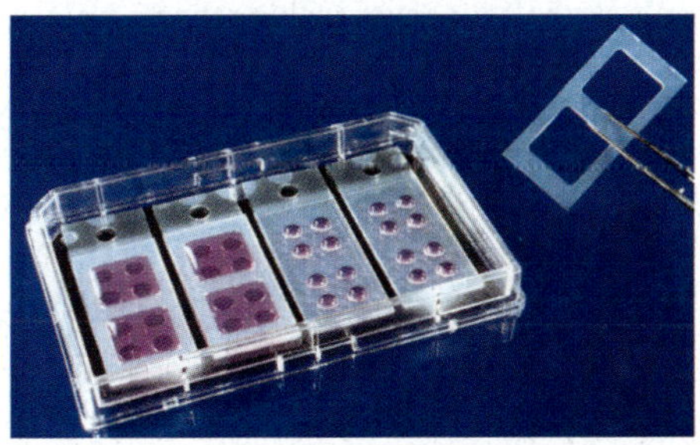

Figure 23.1.36 CultureWell™ silicone dividers (C24770).

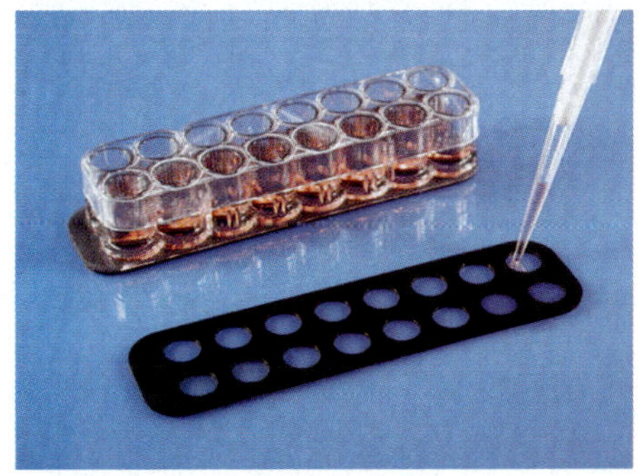

Figure 23.1.38 The CultureWell™ removable chambered coverglass for cell culture (C37000).

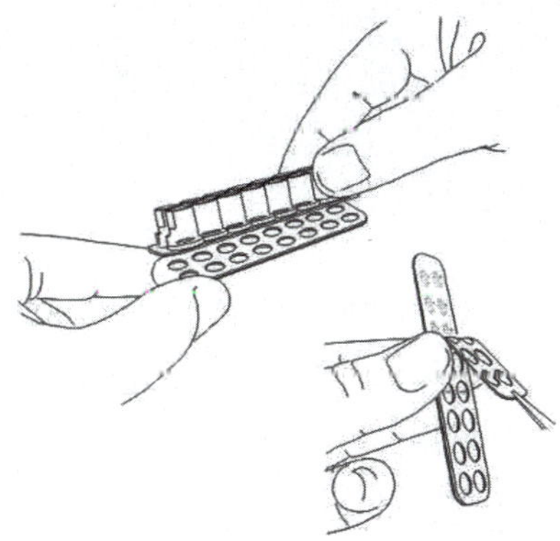

Figure 23.1.39 Using the CultureWell™ removable chambered coverglass for cell culture (C37000). When the cells are ready to be imaged, coverglass removal is made easy by the use of a simple tool (included with the coverglass) that separates the parts without the need for excessive force, eliminating the risk of coverglass breakage.

Table 23.3 CultureWell™ cell culture systems and chambered coverslips.

Cat. No.	Size of Coverslip (mm)	Number of Wells per Coverslip	Well Dimensions	Depth	Volume per Well	Number of Coverslips per Insert
CultureWell™ multiwell cell culture systems *†						
C24762	24 × 50	2	15 mm diameter	1 mm	250–400 µL	4
C24763	24 × 50	2	15 mm diameter	2 mm	300–500 µL	4
C24764	24 × 50	3	9.5 mm square	1 mm	300–500 µL	4
C24765	24 × 50	4	9 mm diameter	1 mm	50–100 µL	4
C24766 C24767	24 × 50	8	6 mm diameter	1 mm	15–30 µL	4
C24768	24 × 50	50	3 mm diameter	1 mm	3–10 µL	4
CultureWell™ chambered coverslips ‡						
C24775	24 × 50	2	15 mm diameter	1 mm	250–400 µL	NA
C24776	24 × 50	2	15 mm diameter	2 mm	300–500 µL	NA
C24777	24 × 50	3	9.5 mm diameter	1 mm	300–500 µL	NA
C24778	24 × 50	4	9 mm diameter	1 mm	50–100 µL	NA
C24779	24 × 50	8	6 mm diameter	1 mm	15–30 µL	NA
C24780	24 × 50	50	3 mm diameter	1 mm	3–10 µL	NA
CultureWell™ multislip cell culture systems *						
C24761	18 × 18	NA	NA	NA	NA	8
C24760	12 × 12	NA	NA	NA	NA	15

* Each system is supplied in packages of 10 inserts, each preassembled in an 86 mm × 128 mm plate. † A trial size is also available (C24767) that includes two inserts — four coverslips with eight 6 mm wells each — in two plates. ‡ Chambered coverslips are supplied in sets, in five pouches of four coverslips each. Packs of 10 sterile plates are also available separately (C24769). NA = Not applicable.

IMPORTANT NOTICE: The products described in this manual are covered by one or more Limited Use Label License(s). Please refer to the Appendix on page 971 and Master Product List on page 975. Products are For Research Use Only. Not intended for any animal or human therapeutic or diagnostic use.

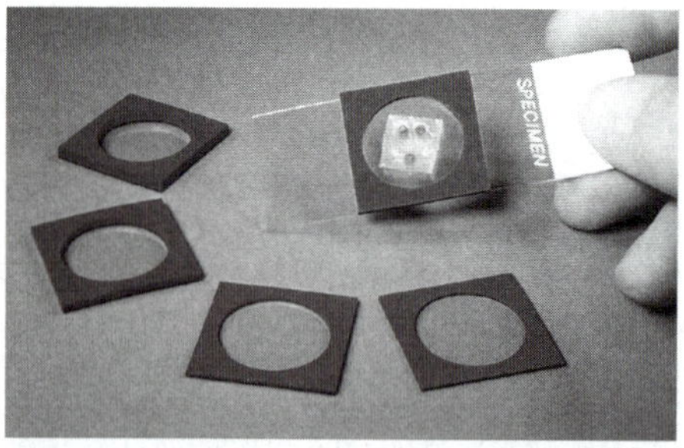

Figure 23.1.40 CoverWell™ imaging chamber gaskets.

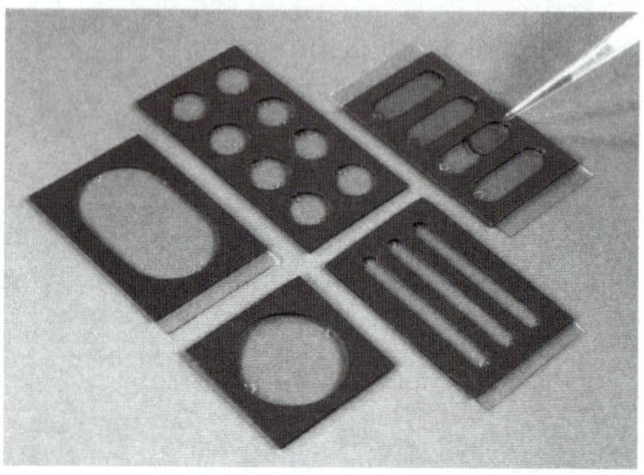

Figure 23.1.41 CoverWell™ perfusion chamber gaskets.

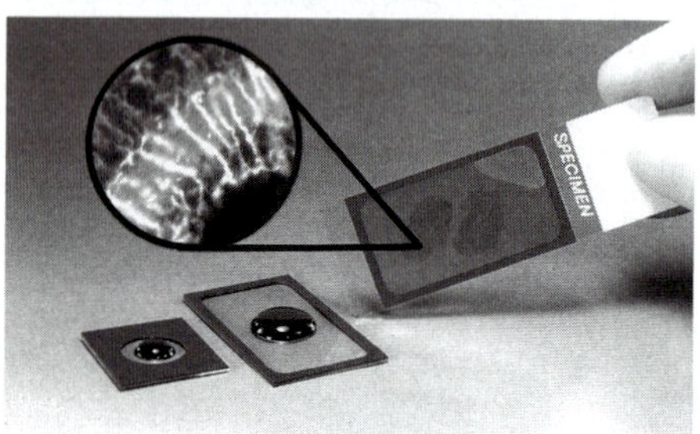

Figure 23.1.42 CoverWell™ incubation chamber gaskets.

silicone gasket leaves no residue. The components of the chambered coverglass are manufactured and assembled with special orientation features to allow easy location of a specific specimen after the coverglass is mounted. Frosted microscope slides are also provided for mounting. The CultureWell™ removable chambered coverglass has several important features:

- Black silicone gasket reduces light scatter, enhancing fluorescence applications.
- Inert, non-cytotoxic silicone permits edge-to-edge growth of cells in the wells.
- Wells are spaced to allow the use of multichannel pipettes for fast and easy cell culturing.
- Silicone gasket remains attached to the coverglass after separation, allowing the wells to be used as reagent reservoirs.
- Gasket design is ideally suited for small-volume incubations, *in situ* hybridization and immunostaining.
- Chambered coverglass is provided sterile and ready to use.

We offer the CultureWell™ chambered coverglass in a package containing eight chambered coverglasses and the removal tool (C37000), as well as in a sample size containing a pair of chambered coverglasses (C37005).

CoverWell™ Imaging Chamber Gaskets

CoverWell™ imaging chamber gaskets (Table 23.4, Figure 23.1.40) incorporate a thin, optically clear plastic cover, making them ideal for light, epifluorescence and confocal laser-scanning microscopy.[40] By simply pressing an imaging chamber gasket to a microscope slide or coverslip, a sealed chamber is formed to contain mounting medium. The watertight chamber supports and stabilizes thick and free-floating specimens, permitting resolution of fine internal structures and analysis of markers without the compression or movement artifacts that affect observations made using an ordinary coverslip.

CoverWell™ Perfusion Chamber Gaskets

CoverWell™ perfusion chamber gaskets (Table 23.4, Figure 23.1.41) are designed for live-cell imaging and manipulation. With the same silicone gasket technology as the CoverWell™ imaging chamber gaskets, these gaskets form watertight "press-to-seal" chambers with dual-access ports for addition and removal of perfusing media. The access ports can be covered using

Table 23.4 CoverWell™ chamber gaskets.

Cat. No.	Number of Wells	Well Dimensions	Depth	Approximate Volume per Chamber	Quantity per Package
CoverWell™ imaging chamber gaskets					
C18160	1	20 mm diameter	0.5 mm	180 µL	40
C18161	1	20 mm diameter	1.0 mm	300 µL	40
C24726 *	1	20 mm diameter	0.5 mm	180 µL	40
C24727 *	1	20 mm diameter	1.0 mm	300 µL	40
CoverWell™ perfusion chamber gaskets					
C18120	1	32 mm × 19 mm	0.5 mm	350 µL	40
C18121	1	32 mm × 19 mm	1.0 mm	550 µL	40
C18128	4	19 mm × 6 mm	0.5 mm	70 µL	40
C18136	1	20 mm diameter	1.0 mm	300 µL	40
C18139	8	9 mm diameter	0.5 mm	35 µL	20
C18140	8	9 mm diameter	1.0 mm	60 µL	20
C18141	8	9 mm diameter	2.0 mm	100 µL	20
C18142	8	9 mm diameter	2.5 mm	150 µL	20
CoverWell™ incubation chamber gaskets					
C18150	1	40 mm × 22 mm	0.2 mm	200 µL	25
C18151	1	40 mm × 22 mm	0.5 mm	500 µL	50
C18155	1	13 mm diameter	0.2 mm	20 µL	25
C18156	1	13 mm diameter	0.5 mm	20 µL	50

* With adhesive on one side.

IMPORTANT NOTICE: The products described in this manual are covered by one or more Limited Use Label License(s). Please refer to the Appendix on page 971 and Master Product List on page 975. Products are For Research Use Only. Not intended for any animal or human therapeutic or diagnostic use.
www.invitrogen.com/probes
molecular probes® | invitrogen™ by life technologies™

adhesive seal tabs (A18211), which are available separately. The heat-resistant gaskets can be sterilized and used for direct culturing of cells and tissues. CoverWell™ perfusion chamber gaskets are available in single- or multiwell configurations, allowing multiple experiments to be performed on a single microscope slide or coverslip.

CoverWell™ Incubation Chamber Gaskets

CoverWell™ incubation chamber gaskets (Table 23.4, Figure 23.1.42) are silicone gaskets with a clear plastic cover that are expressly designed for immunocytochemistry and *in situ* hybridization.[41,42] The gasket is simply pressed onto a wet or dry microscope slide to form a watertight chamber that holds reactants in place and prevents evaporation. The chambers improve the uniformity and sensitivity of staining by enclosing a large sample area while minimizing the reagent volume required. The incubation chamber gaskets are easily removed and reapplied for multiple-step procedures. These chamber gaskets are heat resistant, autoclavable and nuclease free.

Press-to-Seal Silicone Isolators and Secure-Seal™ Adhesive Spacers

For the ultimate in utility and flexibility, Press-to-Seal silicone isolators (Table 23.5, Figure 23.1.43) are removable hydrophobic barriers that can be customized to meet specific experimental requirements. They may be used to isolate cells grown in culture dishes or to separate specimens on microscopy slides during staining procedures. The silicone material can be autoclaved and adheres to any smooth surface. Isolators without adhesive can be easily removed and reapplied for multiple incubation steps. Isolators are also available with adhesive on one side for added security or permanent mounting. In addition, we offer uncut silicone sheets that can easily be trimmed to prepare customized enclosures.

Similar to the Press-to-Seal silicone isolators, Secure-Seal™ adhesive spacers are ultra-thin gaskets with adhesive that can be stacked to any depth desired. For high-resolution microscopy, the spacer and specimen can be sandwiched between two No. 0 glass coverslips. The spacers are available in several configurations (Table 23.5).

Table 23.5 Press-to-Seal gaskets and Secure-Seal™ spacers.

Cat. No.	Number of Wells	Well Dimensions	Depth	Quantity per Package
Press-to-Seal silicone isolators				
P18174	1	20 mm diameter	0.5 mm	50
P18175	1	20 mm diameter	1.0 mm	50
P24740 *	1	20 mm diameter	0.5 mm	50
P24741 *	1	20 mm diameter	1.0 mm	50
P24742 *	24	4.5 mm diameter	2.0 mm	25
P24743 *	8	9 mm diameter	0.5 mm	25
P24744 *	8	9 mm diameter	1.0 mm	25
Secure-Seal™ adhesive spacers				
S24735 *	1	13 mm diameter	0.12 mm	100
S24736 *	1	20 mm diameter	0.12 mm	100
S24737 *	8	9 mm diameter	0.12 mm	100
Press-to-Seal silicone sheets (13 cm × 18 cm)				
P18178	NA	NA	0.5 mm	5
P18179	NA	NA	1.0 mm	5
P24745 *	NA	NA	0.5 mm	5

* With adhesive on one side. NA = Not applicable.

HybriSlip™ Hybridization Covers

HybriSlip™ hybridization covers (Table 23.6, Figure 23.1.44) are nuclease free, ready-to-use and designed specifically for *in situ* hybridization.[43] These hydrophobic coverslips do not require lengthy preparation or blocking procedures to prevent probe trapping or binding. They are heat resistant and do not curl, even at high temperatures, making them ideal for denaturation steps or *in situ* PCR incubations. HybriSlip™ covers are available in three sizes.

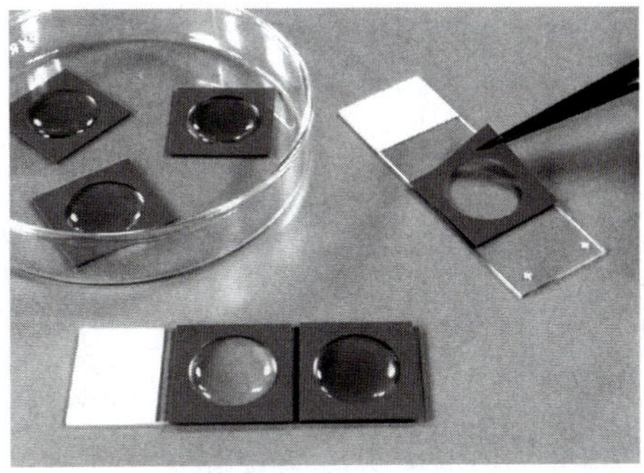

Figure 23.1.43 Press-to-Seal silicone isolators.

Figure 23.1.44 HybriSlip™ hybridization covers.

Table 23.6 Tools for hybridization experiments.

Cat. No.	Chamber Dimensions	Depth	Usable Volume	Quantity per Package
HybriWell™ hybridization sealing system				
H24720	13 mm diameter	0.25 mm	30 µL	100
H24721	20 mm diameter	0.15 mm	30 µL	100
H24723	22 mm × 22 mm	0.15 mm	30–50 µL	100
H18210	40 mm × 21 mm	0.15 mm	50–100 µL	100
H24722	40 mm × 22 mm	0.25 mm	180–200 µL	100
Secure-Seal™ hybridization chambers				
S24734	22 mm × 22 mm	0.8 mm	250 µL	50
S24730	20 mm diameter	0.8 mm	200 µL	40
S24731	20 mm diameter	1.3 mm	280 µL	40
S24732	9 mm diameter	0.8 mm	20 µL	20
S24733	9 mm diameter	1.3 mm	40 µL	20
HybriSlip™ hybridization covers				
H18200	22 mm × 22 mm	NA	NA	500
H18201	40 mm × 22 mm	NA	NA	500
H18202	60 mm × 22 mm	NA	NA	500
Seal tabs				
A18211	Adhesive seal tab	NA	NA	400

NA = Not applicable.

IMPORTANT NOTICE: The products described in this manual are covered by one or more Limited Use Label License(s). Please refer to the Appendix on page 971 and Master Product List on page 975. Products are For Research Use Only. Not intended for any animal or human therapeutic or diagnostic use.

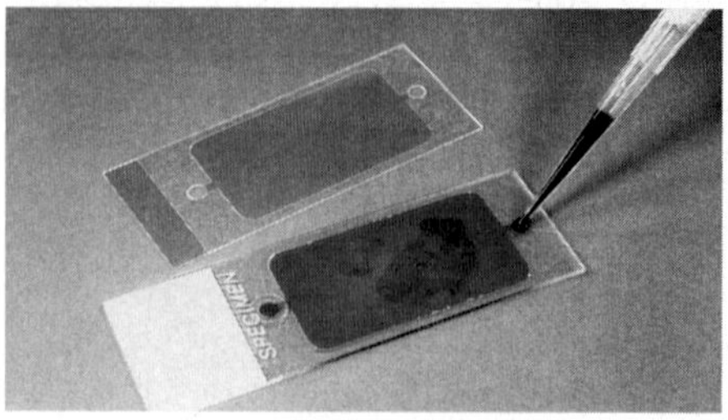

Figure 23.1.45 HybriWell™ hybridization sealing systems.

HybriWell™ Hybridization Sealing Systems

HybriWell™ hybridization sealing systems (Table 23.6, Figure 23.1.45) are coverslip–seal combinations that attach to microscope slides to form microwells optimized for carrying out *in situ* hybridization procedures. These ready-to-use hybridization gaskets have a special adhesive that bonds to glass slides in seconds, creating a water-tight seal that is temperature resistant but can also be removed cleanly and easily after hybridization. Solutions are easily added or removed through dual-access ports without disrupting the specimen. The hydrophobic coverslips are nuclease free and will not trap or bind probes, allowing uniform distribution of the reagent over the specimen. The HybriWell™ sealing systems also include a quick-seal tool to secure the hydrophobic cover to the microscope slide, as well as nuclease-free adhesive seal tabs to cover the access ports. Adhesive seal tabs (A18211) are also available separately in sets of 400.

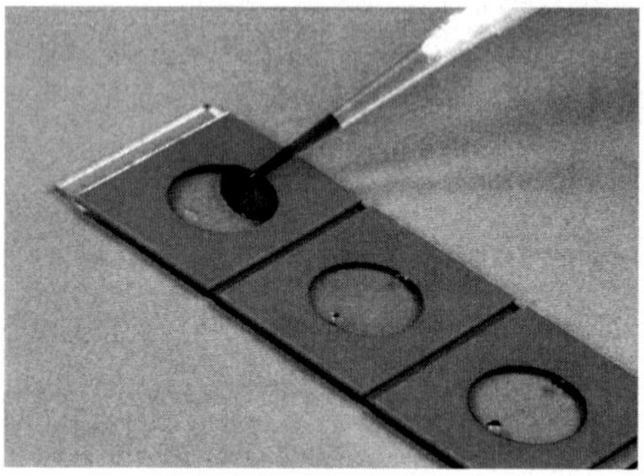

Figure 23.1.46 Secure-Seal™ hybridization chambers.

Secure-Seal™ Hybridization Chambers

Like the HybriWell™ hybridization sealing systems, Secure-Seal™ hybridization chambers (Table 23.6, Figure 23.1.46) are designed to isolate single or multiple specimens on a slide during *in situ* hybridization procedures. Access ports in the chamber surface allow for the addition or removal of solutions and are easily sealed using adhesive seal tabs (A18211), available separately. Because they are deeper than the HybriWell™ chambers, the Secure-Seal™ chambers provide optimum surface-to-volume fluid dynamics, which facilitate more uniform hybridization. However, the shallower chambers created by the HybriWell™ sealing systems hold a smaller reagent volume, minimizing the amount of probe required.

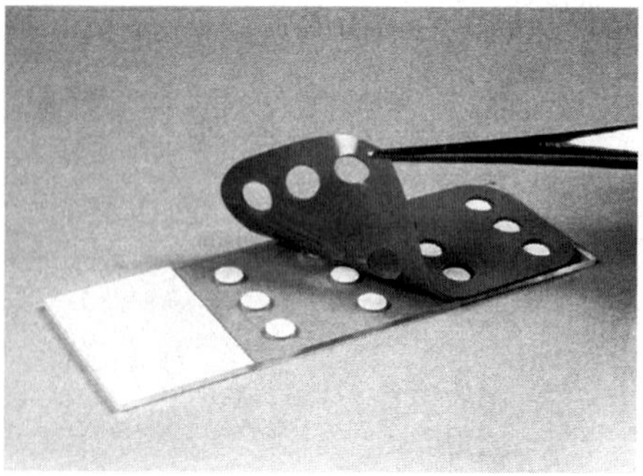

Figure 23.1.47 ONCYTE® MultiWells.

ONCYTE® MultiWells with Slide and Matching Gasket

The versatile ONCYTE® MultiWells (Figure 23.1.47) consist of a two-piece set that includes a slide printed with nitrocellulose circles and a matching removable gasket to enclose and isolate each sample. The nitrocellulose coating on the slide is specially formulated for fluorescence imaging. This ultra-thin microporous coating ensures uniform binding of tissue prints, cells or macromolecules and becomes transparent in a variety of mounting media. The matching press-to-seal silicone gaskets adhere easily to the surface of the slide to isolate specimens and reagents and prevent cross contamination. A coverslip can be added to create enclosed chambers for long incubations. Gaskets can be removed and cleaned simply by peeling them off. ONCYTE® MultiWells are available in two configurations: a set of 20 slides and gaskets, each with 12 wells, 5 mm in diameter (O24750), or a set of 20 slides and gaskets, each with a single well, 13 mm in diameter (O24751).

Figure 23.1.48 ProPlate™ multi-array system (P37004). Individual modules (P37001), covered with seal strips (P37002), fit into a tray (P37003), producing a modular plate with a standard microplate footprint and well spacing.

ProPlate™ Multi-Array System

The ProPlate™ multi-array system (P37004) from Grace Bio-Labs allows integration of microscope array technology with automated microplate processing (Figure 23.1.48). Individual modules (P37001), covered with seal strips (P37002), fit into a tray (P37003), producing a modular plate with a standard microplate footprint and well spacing. The modular design allows loading of 1–4 slides per tray for plate washing and reading. Individual ProPlate™ modules may also be processed by hand without the use of the tray. The large well volumes and isolation of 16 individual arrays (2 × 8) on a single microscope slide is particularly well suited for proteomics applications, including protein expression analysis, protein-protein interactions, antibody profiling and high-throughput automated analysis of multiple proteins. The ProPlate™ multi-array system can also be used to process cDNA or oligonucleotide arrays.[44]

Attofluor Cell Chamber

The Attofluor cell chamber (A7816, Figure 23.1.49) is a durable and practical coverslip holder designed for viewing live-cell specimens on upright or inverted microscopes; spare O-rings for the Attofluor cell chamber are available in sets of 10 (O14804). Features of the Attofluor cell chamber include:

- Surgical stainless steel construction
- Autoclavable, allowing cells to be grown directly in the chamber
- O-ring seal design that prevents sample contamination by oil and leakage of media from the coverslip
- Accepts 25 mm–diameter round coverslips and mounts in a standard 35 mm–diameter stage holder
- Thin 0.5 mm base, allowing clearance for the objective when focusing

Figure 23.1.49 Attofluor cell chamber (A7816).

Coverslip Mini-Rack and Coverslip Maxi-Rack

Our unique coverslip mini-rack (C14784, Figure 23.1.50) is a miniature support designed to vertically hold eight standard round or square coverslips. The mini-rack fits easily into a standard 50 mL beaker and can accommodate a small stir bar beneath the rack. Use of the mini-rack eliminates the necessity for repeatedly moving coverslips between solutions with forceps. Because it is constructed of Teflon®, the mini-rack does not adsorb biopolymers, withstands strong acids and bases, is not damaged by heat and may be sterilized by a variety of methods such as autoclaving, organic solvent treatment or ethylene oxide exposure. The mini-rack is easily disassembled for cleaning and storage. The mini-rack is particularly useful in immunocytochemical and *in situ* hybridization procedures involving sequential wash steps where thorough and consistent removal staining, fixation, permeabilization or blocking reagents from the coverslips is critical. The coverslip maxi-rack (C24784, Figure 23.1.51) provides efficient support for the simultaneous staining and washing of up to 50 samples on 18 mm square or circular coverslips in a self-contained covered container. The maxi-rack includes a convenient handle to remove the rack from the staining solution.

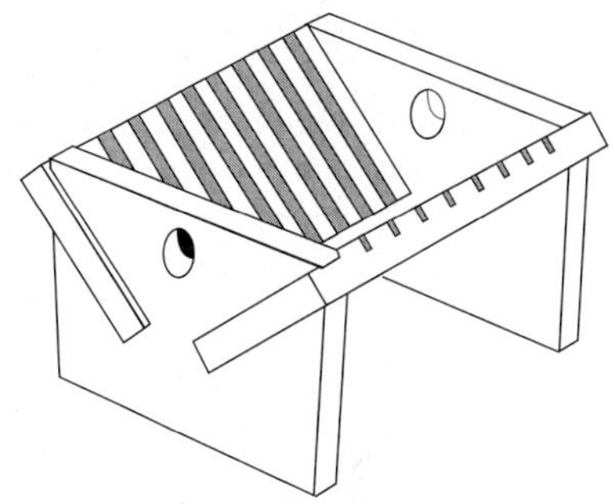

Figure 23.1.50 Coverslip mini-rack (C14784).

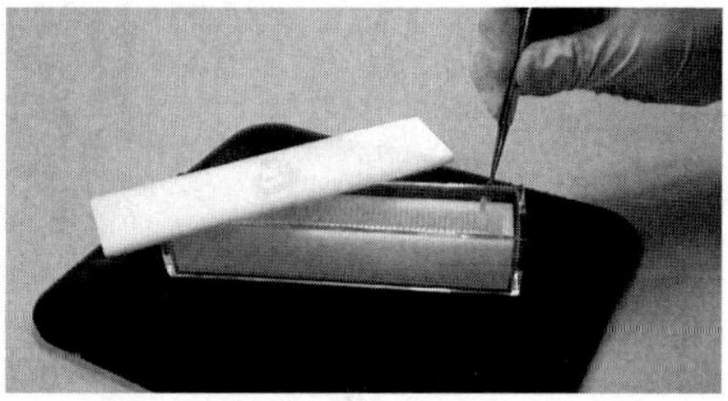

Figure 23.1.51 Coverslip maxi-rack (C24784).

REFERENCES

1. J Phys Chem A (2007) 111:429; **2**. J Am Chem Soc (2007) 129:4643; **3**. Nat Biotechnol (2007) 25:249; **4**. Nat Methods (2008) 5:197; **5**. Methods Mol Biol (2010) 611:151; **6**. Cancer Epidemiol Biomarkers Prev (2007) 16:1371; **7**. J Neurosci Methods (2008) 171:239; **8**. J Histochem Cytochem (1993) 41:1833; **9**. Methods Mol Biol (2006) 315:363; **10**. Cytometry (1998) 32:163; **11**. Dev Dyn (2009) 238:944; **12**. Nat Protoc (2006) 1:2110; **13**. Mol Pharm (2009) 6:1170; **14**. J Immunol (2009) 182:4056; **15**. J Cell Sci (2007) 120:101; **16**. Microsc Res Tech (2005) 68:307; **17**. Cytometry A (2006) 69:659; **18**. Cytometry A (2006) 69:677; **19**. Methods (1999) 18:447; **20**. Appl Immunohistochem Mol Morphol (1999) 7:156; **21**. Environ Sci Technol (2009) 43:6844; **22**. Cytometry A (2007) 71:174; **23**. PLoS One (2009) 4:e4418; **24**. Nat Protoc (2008) 3:619; **25**. Cytometry A (2008) 73:904; **26**. J Neurosci Methods (2009) 176:78; **27**. Biophys J (2007) 93:2519; **28**. Nat Neurosci (2008) 11:713; **29**. Chembiochem (2006) 7:268; **30**. Biochemistry (2006) 45:12411; **31**. Biophys J (2004) 86:2517; **32**. Biotechniques (2001) 31:1076; **33**. J Microsc (2005) 218:148; **34**. J Fluorescence (1996) 6:147; **35**. J Microsc (2007) 228:390; **36**. Anal Biochem (2007) 364:180; **37**. J Res Natl Inst Stand Technol (2001) 106:381; **38**. J Res Natl Inst Stand Technol (2002) 107:83; **39**. J Fluoresc (2004) 14:465; **40**. Appl Environ Microbiol (2009) 75:5952; **41**. PLoS One (2009) 4:e7637; **42**. Cell Vision (1995) 2:165; **43**. CSH Protoc (2010) 2010:pdb.prot5382; **44**. Methods Mol Biol (2010) 632:141.

IMPORTANT NOTICE: The products described in this manual are covered by one or more Limited Use Label License(s). Please refer to the Appendix on page 971 and Master Product List on page 975. Products are For Research Use Only. Not intended for any animal or human therapeutic or diagnostic use.

PRODUCT LIST 23.1 FLUORESCENCE MICROSCOPY ACCESSORIES AND REFERENCE STANDARDS

Cat. No.	Product	Quantity
A18211	Adhesive seal-tab, for HybriWell™ hybridization sealing system *set of 400*	1 set
A7816	Attofluor cell chamber *for microscopy*	each
C14837	Constellation™ microspheres for imaging *mixture of assorted sizes and colors*	3 mL
C24784	coverslip maxi-rack *for 50 coverslips*	each
C14784	coverslip mini-rack *for 8 coverslips*	each
C18160	CoverWell™ imaging chamber gasket, one chamber, 20 mm diameter, 0.5 mm deep *set of 40*	1 set
C18161	CoverWell™ imaging chamber gasket, one chamber, 20 mm diameter, 1.0 mm deep *set of 40*	1 set
C24726	CoverWell™ imaging chamber gasket with adhesive, one chamber, 20 mm diameter, 0.5 mm deep *set of 40*	1 set
C24727	CoverWell™ imaging chamber gasket with adhesive, one chamber, 20 mm diameter, 1.0 mm deep *set of 40*	1 set
C18155	CoverWell™ incubation chamber gasket, one chamber, 13 mm diameter, 0.2 mm deep *set of 25*	1 set
C18156	CoverWell™ incubation chamber gasket, one chamber, 13 mm diameter, 0.5 mm deep *set of 50*	1 set
C18150	CoverWell™ incubation chamber gasket, one chamber, 40 mm x 22 mm, 0.2 mm deep *set of 25*	1 set
C18151	CoverWell™ incubation chamber gasket, one chamber, 40 mm x 22 mm, 0.5 mm deep *set of 50*	1 set
C18139	CoverWell™ perfusion chamber gasket, eight chambers, 9 mm diameter, 0.5 mm deep *set of 20*	1 set
C18140	CoverWell™ perfusion chamber gasket, eight chambers, 9 mm diameter, 1.0 mm deep *set of 20*	1 set
C18141	CoverWell™ perfusion chamber gasket, eight chambers, 9 mm diameter, 2.0 mm deep *set of 20*	1 set
C18142	CoverWell™ perfusion chamber gasket, eight chambers, 9 mm diameter, 2.5 mm deep *set of 20*	1 set
C18128	CoverWell™ perfusion chamber gasket, four chambers, 19 mm x 6 mm, 0.5 mm deep *set of 40*	1 set
C18136	CoverWell™ perfusion chamber gasket, one chamber, 20 mm diameter, 1.0 mm deep *set of 40*	1 set
C18120	CoverWell™ perfusion chamber gasket, one chamber, 32 mm x 19 mm, 0.5 mm deep *set of 40*	1 set
C18121	CoverWell™ perfusion chamber gasket, one chamber, 32 mm x 19 mm, 1.0 mm deep *set of 40*	1 set
C24769	CultureWell™ cell culture plate *set of 10*	1 set
C37005	CultureWell™ chambered coverglass for cell culture *sixteen wells per coverglass* *set of 2*	1 pack
C37000	CultureWell™ chambered coverglass for cell culture *sixteen wells per coverglass* *set of 8*	1 set
C24770	CultureWell™ coverslip divider *set of 4*	1 set
C24760	CultureWell™ multislip cell culture system MSI-12 *plate and insert, fifteen 12 mm coverslips per insert* *set of 10*	1 set
C24761	CultureWell™ multislip cell culture system MSI-18 *plate and insert, eight 18 mm coverslips per insert* *set of 10*	1 set
C24762	CultureWell™ multiwell cell culture system CWI 2R-1.0 *plate and insert, four 24 mm x 50 mm coverslips per insert, two 1 mm-deep wells per coverslip* *set of 10*	1 set
C24763	CultureWell™ multiwell cell culture system CWI 2R-2.0 *plate and insert, four 24 mm x 50 mm coverslips per insert, two 2 mm-deep wells per coverslip* *set of 10*	1 set
C24764	CultureWell™ multiwell cell culture system CWI 3S-1.0 *plate and insert, four 24 mm x 50 mm coverslips per insert, three 1 mm-deep wells per coverslip* *set of 10*	1 set
C24765	CultureWell™ multiwell cell culture system CWI 4R-1.0 *plate and insert, four 24 mm x 50 mm coverslips per insert, four 1 mm-deep wells per coverslip* *set of 10*	1 set
C24766	CultureWell™ multiwell cell culture system CWI 8R-1.0 *plate and insert, four 24 mm x 50 mm coverslips per insert, eight 1 mm-deep wells per coverslip* *set of 10*	1 set
C24767	CultureWell™ multiwell cell culture system CWI 8R-1.0 TS *plate and insert, four 24 mm x 50 mm coverslips per insert, eight 1 mm-deep wells per coverslip* *set of 2*	1 set
C24768	CultureWell™ multiwell cell culture system CWI 50R-1.0 *plate and insert, four 24 mm x 50 mm coverslips per insert, fifty 1 mm-deep wells per coverslip* *set of 10*	1 set
C24775	CultureWell™ multiwell chambered coverslip CWCS 2R-1.0 *24 mm x 50 mm coverslips, two 1 mm-deep wells per coverslip* *set of 20*	1 set
C24776	CultureWell™ multiwell chambered coverslip CWCS 2R-2.0 *24 mm x 50 mm coverslips, two 2 mm-deep wells per coverslip* *set of 20*	1 set
C24777	CultureWell™ multiwell chambered coverslip CWCS 3S-1.0 *24 mm x 50 mm coverslips, three 1 mm-deep wells per coverslip* *set of 20*	1 set
C24778	CultureWell™ multiwell chambered coverslip CWCS 4R-1.0 *24 mm x 50 mm coverslips, four 1 mm-deep wells per coverslip* *set of 20*	1 set
C24779	CultureWell™ multiwell chambered coverslip CWCS 8R-1.0 *24 mm x 50 mm coverslips, eight 1 mm-deep wells per coverslip* *set of 20*	1 set
C24780	CultureWell™ multiwell chambered coverslip CWCS 50R-1.0 *24 mm x 50 mm coverslips, fifty 1 mm-deep wells per coverslip* *set of 20*	1 set
F36924	FluoCells® prepared slide #1 *BPAE cells with MitoTracker® Red CMXRos, Alexa Fluor® 488 phalloidin, DAPI*	each
F14781	FluoCells® prepared slide #2 *BPAE cells with mouse anti-α-tubulin, BODIPY® FL goat anti-mouse IgG, Texas Red®-X phalloidin, DAPI*	each
F24630	FluoCells® prepared slide #3 *mouse kidney section with Alexa Fluor® 488 WGA, Alexa Fluor® 568 phalloidin, DAPI*	each
F24631	FluoCells® prepared slide #4 *mouse intestine section with Alexa Fluor® 350 WGA, Alexa Fluor® 568 phalloidin, SYTOX® Green*	each
F36925	FluoCells® prepared slide #6 *muntjac cells with mouse anti-OxPhos Complex V inhibitor protein, Alexa Fluor® 555 goat anti-mouse IgG, Alexa Fluor® 488 phalloidin, TO-PRO®-3*	each
F36915	fluorescein *NIST-traceable standard* *nominal concentration 50 µM* *special packaging*	5 x 1 mL
F36905	FocalCheck™ DoubleGreen Fluorescent Microspheres Kit, 6 µm *mounted on slides*	1 kit
F36906	FocalCheck™ DoubleOrange Fluorescent Microspheres Kit, 6 µm *mounted on slides*	1 kit
F36909	FocalCheck™ fluorescence microscope test slide #1 *for alignment, intensity, and calibration*	each
F36913	FocalCheck™ fluorescence microscope test slide #2 *for spectral imaging systems*	each
F36914	FocalCheck™ fluorescence microscope test slide #3 *5 colors, high and low intensities*	each
F24633	FocalCheck™ Fluorescent Microspheres Kit, 6 µm *mounted on slides*	1 kit
F24634	FocalCheck™ Fluorescent Microspheres Kit, 15 µm *mounted on slides*	1 kit
F14807	FocalCheck™ microspheres, 6 µm, fluorescent dark-red ring stain/green throughout	0.5 mL
F14808	FocalCheck™ microspheres, 6 µm, fluorescent green ring stain/blue throughout	0.5 mL
F14806	FocalCheck™ microspheres, 6 µm, fluorescent green/orange/dark-red ring stains	0.5 mL
F7234	FocalCheck™ microspheres, 15 µm, fluorescent blue/orange ring stains	0.5 mL
F7239	FocalCheck™ microspheres, 15 µm, fluorescent dark-red ring stain/green throughout	0.5 mL

IMPORTANT NOTICE: The products described in this manual are covered by one or more Limited Use Label License(s). Please refer to the Appendix on page 971 and Master Product List on page 975. Products are For Research Use Only. Not intended for any animal or human therapeutic or diagnostic use.

PRODUCT LIST 23.1 FLUORESCENCE MICROSCOPY ACCESSORIES AND REFERENCE STANDARDS—*continued*

Cat. No.	Product	Quantity
F7237	FocalCheck™ microspheres, 15 µm, fluorescent green ring stain/blue throughout	0.5 mL
F7238	FocalCheck™ microspheres, 15 µm, fluorescent green ring stain/dark red throughout	0.5 mL
F7240	FocalCheck™ microspheres, 15 µm, fluorescent green/dark-red ring stains	0.5 mL
F7235	FocalCheck™ microspheres, 15 µm, fluorescent green/orange/dark-red ring stains	0.5 mL
F7236	FocalCheck™ microspheres, 15 µm, fluorescent orange ring stain/blue throughout	0.5 mL
F14791	FocalCheck™ Thin-Ring Fluorescent Microspheres Kit, 1.0 µm *three suspensions*	1 kit
H18200	HybriSlip™ hybridization cover, 22 mm x 22 mm *RNase free* *set of 500*	1 set
H18201	HybriSlip™ hybridization cover, 40 mm x 22 mm *RNase free* *set of 500*	1 set
H18202	HybriSlip™ hybridization cover, 60 mm x 22 mm *RNase free* *set of 500*	1 set
H24720	HybriWell™ hybridization sealing system, 13 mm diameter chamber, 0.25 mm deep *set of 100*	1 set
H24721	HybriWell™ hybridization sealing system, 20 mm diameter chamber, 0.15 mm deep *set of 100*	1 set
H24723	HybriWell™ hybridization sealing system, 22 mm x 22 mm chamber, 0.15 mm deep *set of 100*	1 set
H18210	HybriWell™ hybridization sealing system, 40 mm x 21 mm chamber, 0.15 mm deep *set of 100*	1 set
H24722	HybriWell™ hybridization sealing system, 40 mm x 22 mm chamber, 0.25 mm deep *set of 100*	1 set
I36933	Image-iT® FX signal enhancer	10 mL
I7221	InSpeck™ Blue (350/440) Microscope Image Intensity Calibration Kit, 2.5 µm	1 kit
I7225	InSpeck™ Deep Red (633/660) Microscope Image Intensity Calibration Kit, 2.5 µm	1 kit
I7219	InSpeck™ Green (505/515) Microscope Image Intensity Calibration Kit, 2.5 µm	1 kit
I14785	InSpeck™ Green (505/515) Microscope Image Intensity Calibration Kit, 6 µm	1 kit
I7223	InSpeck™ Orange (540/560) Microscope Image Intensity Calibration Kit, 2.5 µm	1 kit
I14786	InSpeck™ Orange (540/560) Microscope Image Intensity Calibration Kit, 6 µm	1 kit
I7224	InSpeck™ Red (580/605) Microscope Image Intensity Calibration Kit, 2.5 µm	1 kit
I14787	InSpeck™ Red (580/605) Microscope Image Intensity Calibration Kit, 6 µm	1 kit
M7901	MultiSpeck™ Multispectral Fluorescence Microscopy Standards Kit *in suspension*	1 kit
O24751	ONCYTE® MultiWells, one well, 13 mm diameter, with slide and matching gasket *set of 20*	1 set
O24750	ONCYTE® MultiWells, 12 wells, 5 mm diameter, with slide and matching gasket *set of 20*	1 set
O14804	O-rings for Attofluor cell chamber, set of 10	each
P18174	Press-to-Seal silicone isolator, one well, 20 mm diameter, 0.5 mm deep *set of 50*	1 set
P18175	Press-to-Seal silicone isolator, one well, 20 mm diameter, 1.0 mm deep *set of 50*	1 set
P24743	Press-to-Seal silicone isolator with adhesive, eight wells, 9 mm diameter, 0.5 mm deep *set of 25*	1 set
P24744	Press-to-Seal silicone isolator with adhesive, eight wells, 9 mm diameter, 1.0 mm deep *set of 25*	1 set
P24740	Press-to-Seal silicone isolator with adhesive, one well, 20 mm diameter, 0.5 mm deep *set of 50*	1 set
P24741	Press-to-Seal silicone isolator with adhesive, one well, 20 mm diameter, 1.0 mm deep *set of 50*	1 set
P24742	Press-to-Seal silicone isolator with adhesive, 24 wells, 2.5 mm diameter, 2.0 mm deep *set of 25*	1 set
P24745	Press-to-Seal silicone sheet with adhesive, 13 cm x 18 cm, 0.5 mm thick *set of 5*	1 set
P18178	Press-to-Seal silicone sheet, 13 cm x 18 cm, 0.5 mm thick *set of 5*	1 set
P18179	Press-to-Seal silicone sheet, 13 cm x 18 cm, 1.0 mm thick *set of 5*	1 set
P7481	ProLong® Antifade Kit	1 kit
P36930	ProLong® Gold antifade reagent	10 mL
P36934	ProLong® Gold antifade reagent *special packaging*	5 x 2 mL
P36931	ProLong® Gold antifade reagent with DAPI	10 mL
P36935	ProLong® Gold antifade reagent with DAPI *special packaging*	5 x 2 mL
P37002	ProPlate™ adhesive seal-strips *set of 50 seal-strips and one applicator*	1 set
P37001	ProPlate™ multi-array slide module *set of 2*	1 set
P37004	ProPlate™ multi-array system *includes four 16-well slide modules, one tray and cover, ten seal-strips and one applicator*	1 set
P37003	ProPlate™ tray and cover *includes one tray and cover*	1 set
P7220	PS-Speck™ Microscope Point Source Kit *blue, green, orange and deep red fluorescent beads*	1 kit
Q10336	Qmount® Qdot® mounting media	3 x 2 mL
R14782	Reference Dye Sampler Kit *five 1 mM solutions, 1 mL each*	1 kit
S24732	Secure-Seal™ hybridization chamber gasket, eight chambers, 9 mm diameter, 0.8 mm deep *set of 20*	1 set
S24733	Secure-Seal™ hybridization chamber gasket, eight chambers, 9 mm diameter, 1.3 mm deep *set of 20*	1 set
S24730	Secure-Seal™ hybridization chamber gasket, one chamber, 20 mm diameter, 0.8 mm deep *set of 40*	1 set
S24731	Secure-Seal™ hybridization chamber gasket, one chamber, 20 mm diameter, 1.3 mm deep *set of 40*	1 set
S24734	Secure-Seal™ hybridization chamber gasket, one chamber, 22 mm x 22 mm, 0.8 mm deep *set of 50*	1 set
S24737	Secure-Seal™ spacer, eight wells, 9 mm diameter, 0.12 mm deep *set of 100*	1 set

IMPORTANT NOTICE: The products described in this manual are covered by one or more Limited Use Label License(s). Please refer to the Appendix on page 971 and Master Product List on page 975. Products are For Research Use Only. Not intended for any animal or human therapeutic or diagnostic use.

23.2 Flow Cytometry Reference Standards

Flow cytometers are designed to perform quantitative measurements on individual cells and other particles with speed, accuracy and precision. As with all high-performance instrumentation, flow cytometers must be calibrated frequently to ensure accuracy and reliability. The stability, uniformity and reproducibility of our fluorescent microsphere products make them ideal reference standards for flow cytometry. However, because of the high variability in quantum yields of bound dyes and the heterogeneity of protein labeling, as well as problems with stoichiometry and accessibility in binding to targets, bead standards containing a known number of fluorophores per bead do not necessarily provide accurate information about the number of ligands bound to a cell.[1–4]

AlignFlow™ and AlignFlow™ Plus Flow Cytometry Alignment Beads

In order to ensure accurate and reproducible quantitative results, flow cytometers should be checked at least daily for proper performance. AlignFlow™ and AlignFlow™ Plus flow cytometry alignment beads permit the calibration of a flow cytometer's laser(s), optics and stream flow without wasting valuable and sensitive experimental material.[5–7] These fluorescently stained polystyrene microspheres are highly uniform with respect to both size and fluorescence intensity, and they are designed to approximately replicate the size, emission wavelength and intensity of biological samples. Because the dyes are contained inside the microsphere's matrix instead of on the surface, AlignFlow™ beads have excellent photochemical and physical stability, providing reliable reference signals for aligning, focusing and calibrating flow cytometers. The fluorescent dyes have been carefully selected for optimal excitation by laser sources commonly used in flow cytometry.

The 2.5 µm AlignFlow™ flow cytometry alignment beads are available in four versions: for UV (350–370 nm) excitation (A7304), for 488 nm excitation (A7302), for 633 nm excitation (A7312) and for 630–660 nm excitation (A14835); the 6 µm AlignFlow™ Plus beads are available for the same four excitation-wavelength ranges: for UV (350–370 nm) excitation (A7305), for 488 nm excitation (A7303), for 633 nm excitation (A7313) and for 630–660 nm excitation (A14836). The UV light–excitable beads emit from 400 nm to 470 nm, the 488 nm light–excitable beads emit broadly from 515 nm to 660 nm (Figure 23.2.1), the 633 nm light–excitable beads emit from 645 nm to 680 nm, and the 630–660 nm light–excitable beads emit from 670 nm to 720 nm. The AlignFlow™ and AlignFlow™ Plus flow cytometry alignment beads are supplied as suspensions packaged in dropper vials for convenient dispensation.

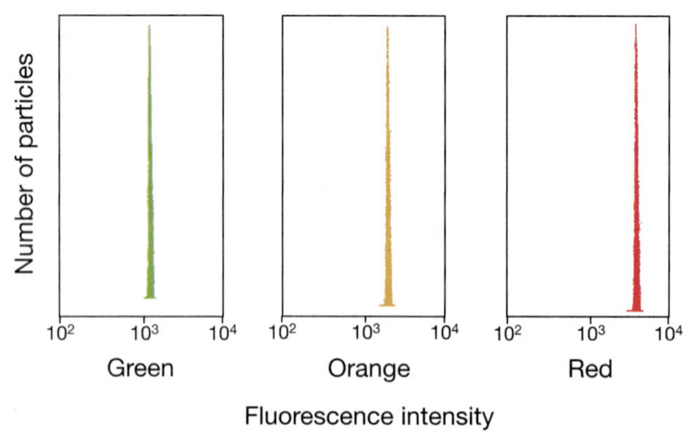

Figure 23.2.1 AlignFlow™ Plus (A7303) flow cytometry alignment beads excited at 488 nm by an argon-ion laser and monitored in three emission channels. The broad fluorescence emission is detected in all three channels. Note the exceptionally small variation of fluorescence intensity of the beads. Contributed by Carleton Stewart, Roswell Park Cancer Institute.

LinearFlow™ Flow Cytometry Intensity Calibration Kits

LinearFlow™ Flow Cytometry Intensity Calibration Kits provide flow cytometer operators with intensity references for generating calibration curves, establishing photomultiplier settings and evaluating sample brightness.[8–10] Each kit contains fluorescent microspheres in which the degree of staining has been carefully controlled to provide precisely determined intensity levels when excited in a flow cytometer (Figure 23.2.2). The microspheres are supplied as suspensions packaged in dropper vials for convenient dispensation. The LinearFlow™ Flow Cytometry Intensity Calibration Kits are available in two different bead sizes (2.5 µm or 6 µm) and five different fluorescent colors covering the spectral ranges commonly encountered in flow cytometry:

- Blue (for UV excitation/430 nm emission). Available in 2.5 µm (L14812) and 6 µm (L14813) sizes; both kits contain microspheres stained at 100%, 20%, 4.0% and 0.8% relative fluorescence intensity levels.
- Green (for 488 nm excitation/515 emission). Available in 2.5 µm (L14821) and 6 µm (L14822) sizes; both kits contain microspheres stained at 100%, 10%, 2.0%, 0.4%, 0.1% and 0.02% relative fluorescence intensity levels. The LinearFlow™ Green Flow Cytometry Low-Intensity Calibration Kits contain 2.5 µm (L14823) or 6 µm (L14824) diameter beads stained at 0.1%, 0.02%, 0.004% and 0.001% relative fluorescence intensity levels. The fluorescence from the beads with the lowest intensity level is approximately half that of the autofluorescence typically observed from unstained cells.
- Orange (for 488 nm excitation/575 nm emission). Available in 2.5 µm (L14814) and 6 µm (L14815) sizes; both kits contain microspheres stained at 100%, 10%, 2.0%, 0.4%, 0.1% and 0.02% relative fluorescence intensity levels.
- Carmine (for 488 nm excitation/620 nm emission). Available in 2.5 µm (L14816) and 6 µm (L14817) sizes; both kits contain microspheres stained at 100%, 10%, 2.0%, 0.4%, 0.1% and 0.02% relative fluorescence intensity levels.
- Deep Red (for 633 nm excitation/660 nm emission). Available in 2.5 µm (L14818) and 6 µm (L14819, Figure 23.2.2) sizes; both kits contain microspheres stained at 100%, 20%, 4.0%, 0.8%, 0.2% and 0.04% relative fluorescence intensity levels.

The 365/430 nm fluorescence excitation/emission maxima of the microspheres in the LinearFlow™ Blue Kits provide a close spectral match to samples stained with DAPI, Hoechst 33258 or Hoechst 33342 nucleic acid stains. These kits are ideal for intensity calibration of flow cytometers equipped with UV laser excitation. The microspheres in the

IMPORTANT NOTICE: The products described in this manual are covered by one or more Limited Use Label License(s). Please refer to the Appendix on page 971 and Master Product List on page 975. Products are For Research Use Only. Not intended for any animal or human therapeutic or diagnostic use.
www.invitrogen.com/probes

LinearFlow™ Green Kits are designed for calibrating the green (FL1) detection channel. Although the microspheres actually have an excitation maximum of ~505 nm, they are effectively excited by the 488 nm spectral line of the argon-ion laser. Their emission maximum of ~515 nm closely matches that of samples labeled with fluorescein, Oregon Green® 488 or Alexa Fluor® 488 dyes or with SYTOX® Green nucleic acid stain. Microspheres in the LinearFlow™ Orange Kit are spectrally similar to phycoerythrin and tetramethylrhodamine conjugates, making this kit useful for calibrating the orange (FL2) channel. Although these microspheres actually have an excitation maximum of ~570 nm, they are effectively excited by the 488 nm spectral line of the argon-ion laser. Microspheres in the LinearFlow™ Carmine Kit exhibit excitation and emission spectra similar to the spectra of the propidium iodide complex with DNA or the spectra of Texas Red® or Alexa Fluor® 594 dyes and are suitable for calibrating the red (FL3) channel. The microspheres in the LinearFlow™ Carmine Kit have an excitation maximum of ~580 nm, but they can also be excited by the 488 nm spectral line of the argon-ion laser. The microspheres in the LinearFlow™ Deep Red Kit have maximal emission at ~660 nm, closely matching that of Alexa Fluor® 647 dye, Cy®5 dye and allophycocyanin, and they are useful for calibrating flow cytometers equipped with 633 nm He-Ne laser excitation. Although primarily intended for 633 nm excitation, the LinearFlow™ Deep Red microspheres can be adequately excited at 488 nm and will provide accurate relative intensity readings with this excitation.

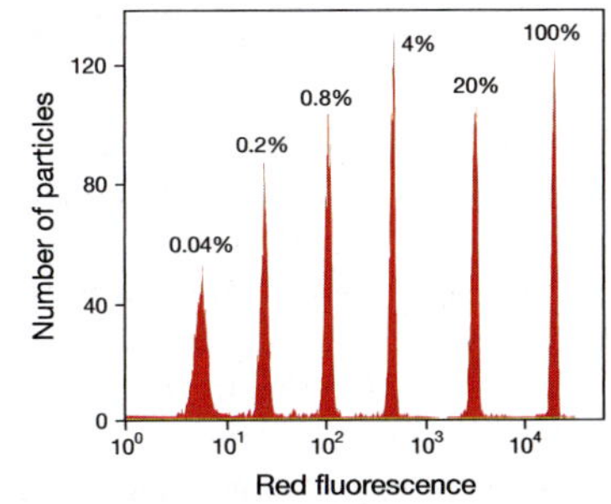

Figure 23.2.2 Fluorescence intensity histogram of the six different 6 µm polystyrene bead samples supplied in the LinearFlow™ Deep Red Flow Cytometry Intensity Calibration Kit (L14819). Fluorescence measurements were performed with a flow cytometer using excitation at 633 nm. This histogram is a composite of two graphs; the same mixture of microspheres was sampled and analyzed using two distinct PMT voltage settings in order to cover the full intensity range.

PeakFlow™ Flow Cytometry Reference Beads

PeakFlow™ flow cytometry reference beads are stained with fluorescent dyes that have been carefully selected to produce emission peaks coincident with labeled cells used in typical flow cytometry applications. The emission profiles for these standards are intentionally narrow in comparison to fluorescein-labeled cells (Figure 23.2.3). Consequently, PeakFlow™ beads serve as reference sources with emissions centered upon the expected fluorescence of the experimental sample. Because PeakFlow™ beads are highly uniform with respect to both size and fluorescence intensity, and because they approximate the size, emission wavelength and intensity of many biological samples, they can be used to calibrate a flow cytometer's laser source, optics, stream flow and cell sorting system without wasting valuable and sensitive experimental material. Furthermore, due to their narrow emission profiles, PeakFlow™ beads of two different fluorescent colors exhibit minimal spectral overlap, and little or no color compensation is needed when setting up for multicolor experiments.

As with all of our flow cytometry standard microspheres, PeakFlow™ beads are stained internally rather than on the surface. The dyes are therefore insulated from environmental interactions that could cause variable fluorescence output, resulting in excellent signal stability. These fluorescent polystyrene microspheres are supplied as suspensions packaged in dropper vials for convenient dispensation, with a choice of seven fluorescent colors and, for most products, two different sizes (Table 23.7).

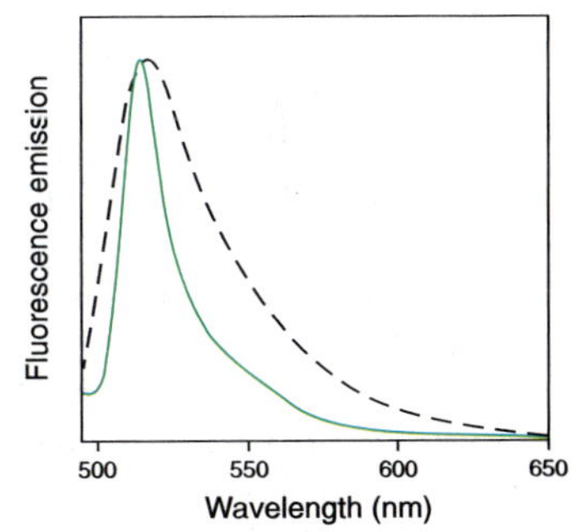

Figure 23.2.3 Normalized emission spectra of PeakFlow™ Green flow cytometry reference beads (P14827, solid line) and fluorescein-labeled cells (dashed line). The narrow emission spectrum of PeakFlow™ beads is approximately centered on the broader emission spectrum of fluorescein.

Table 23.7 Spectral characteristics of PeakFlow™ flow cytometry reference beads.

Cat. No.	Size (µm)	Nominal Color	Abs * (nm)	Em * (nm)	Emission Matches Cells Stained with:
P14825	2.5	Blue	400 †	460	DAPI, Hoechst dyes
P14826	6.0				
P14827	2.5	Green	505 ‡	515	Fluorescein, Alexa Fluor® 488 dye, Oregon Green® 488 dye, DiOC$_{18}$(3) ("DiO")
P14828	6.0				
P14829	2.5	Orange	570 ‡	575	R-phycoerythrin, tetramethylrhodamine, Alexa Fluor® 568 dye, DiIC$_{18}$(3) ("DiI")
P14830	6.0				
P14831	2.5	Carmine	580 ‡	620	Propidium iodide, Texas Red® dye, Alexa Fluor® 594 dye
P14832	6.0				
P24670	6.0	Claret	645 §	680	TOTO®-3, Alexa Fluor® 647 dye, Cy®5 dye, DiIC$_{18}$(5) ("DiD")
P24671	6.0	Ultra red	665 §	695	Alexa Fluor® 660 dye, Cy®5.5 dye
P24672	6.0	Infrared	735	770	Alexa Fluor® 750 dye, Cy®7 dye, DiIC$_{18}$(7) ("DiR")

* Approximate absorption and emission maxima for beads in suspension. † Suitable for excitation by the UV (351–364 nm) spectral line of argon-ion lasers. ‡ Suitable for excitation by the 488 nm spectral line of argon-ion lasers. § Suitable for excitation by the 633 nm spectral line of He-Ne lasers.

AbC™ and ArC™ Bead Kits for Flow Cytometry Compensation

AbC™ Anti-Mouse and AbC™ Anti-Rat/Hamster Bead Kits

The AbC™ Anti-Mouse Bead Kit (A10344) provides a consistent, accurate and simple-to-use technique for the setting of flow cytometry compensation when using fluorophore-conjugated mouse antibodies. The kit contains two types of specially modified polystyrene microspheres, the AbC™ capture beads, that bind all isotypes of mouse immunoglobulin, and the negative beads that have no antibody binding capacity. After incubation with a fluorophore-conjugated mouse antibody, the two bead components provide distinct positive and negative populations of beads that can be used to set compensation (Figure 23.2.4). Because of the consistent nature of bead scatter and high surface antibody–binding capacity, more consistent and accurate compensation settings for any combination of fluorophore-labeled mouse antibodies can be achieved. The AbC™ Anti-Rat/Hamster Bead Kit (A10389) provides parallel functionality for protocols using fluorophore-labeled rat or hamster antibodies. The AbC™ capture beads and negative beads have a diameter of approximately 6 µm (actual size for each lot is listed on the component vial). The bead suspensions are supplied in dropper vials for convenient sample application.

ArC™ Amine-Reactive Compensation Bead Kit

The ArC™ Amine-Reactive Compensation Bead Kit (A10346) provides a consistent, accurate and simple-to-use technique for the setting of flow cytometry compensation when using any of the LIVE/DEAD® fixable dead cell stains. The LIVE/DEAD® Fixable Dead Cell Stain Kits (Section 15.3) use an amine-reactive dye labeling method to evaluate the viability of mammalian cells by flow cytometry.[11] The ArC™ Amine-Reactive Compensation Bead Kit includes two types of specially modified polystyrene microspheres to allow easy compensation of the LIVE/DEAD® fixable stains: the ArC™ reactive beads (Component A), which bind any of the amine-reactive dyes, and the ArC™ negative beads (Component B), which have no reactivity. After incubation with any amine-reactive dye, the two kit components provide distinct positive and negative populations of beads that can be used to set compensation.

Flow Cytometry Size Calibration Kit

The Flow Cytometry Size Calibration Kit (F13838) provides nonfluorescent particle-size calibration standards for use in forward light scattering measurements of cell size by flow cytometry.[12,13] This kit contains suspensions of six different nonfluorescent microspheres packaged in convenient dropper vials. The individual standards contain highly uniform polystyrene microspheres with nominal diameters of 1.0 µm, 2.0 µm, 4.0 µm, 6 µm, 10 µm and 15 µm.

CountBright™ Absolute Counting Beads

Flow cytometry provides a rapid method for quantitating cell characteristics; however, most flow cytometers cannot directly provide the cell concentration or absolute count of cells in a sample. Absolute cell counts have been widely used in quantitating cell populations and disease progression[14–18] and are generally obtained either by combining a separate cell concentration determination from a hematology analyzer with flow cytometry population data (multiple-platform testing) or by adding an internal microsphere counting standard to the flow cytometry sample (single-platform testing). The single-platform method is preferred as it is technically less complicated and more accurate than multiple-platform testing.[19] To facilitate this single-platform method, we offer CountBright™ absolute counting beads (C36950), a calibrated suspension of microspheres that are brightly fluorescent across a wide range of excitation and emission wavelengths and contain a known concentration of microspheres. For absolute counts, a specific volume of the CountBright™ microsphere suspension is added to a specific volume of sample, such that the ratio of sample volume to microsphere volume is known.[20] The volume of sample analyzed can be calculated from the number of microsphere events and then used with cell events to determine cell concentration. In general, at least 1000 bead events should be acquired to assure a statistically significant determination of sample volume. Sufficient reagents are provided for 100 flow cytometry assays, each using 50 µL of counting beads per test.

CountBright™ absolute counting beads are broadly fluorescent and can be used with either a fluorescence or scatter threshold. Fluorescence can be excited by wavelengths from UV to 635 nm; fluorescence

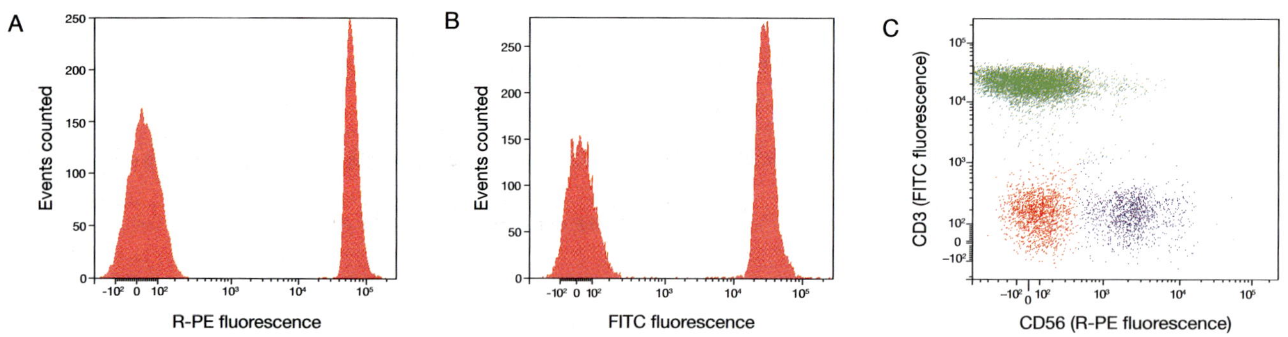

Figure 23.2.4 Compensation using the AbC™ Anti-Mouse Bead Kit (A10344). **(A)** R-Phycoerythrin (R-PE)–conjugated mouse anti–human CD56 antibodies (MHCD56044) label the AbC™ capture beads for a positive signal, and negative beads provide a negative signal. **(B)** FITC-conjugated mouse anti–human CD3 antibodies (MHCD03014) label the AbC™ capture beads for a positive signal, and negative beads provide a negative signal. **(C)** Dual-parameter plot showing gated human lymphocytes labeled with R-PE–conjugated mouse anti–human CD56 and FITC-conjugated mouse anti–human CD3 antibodies using compensation settings obtained with the AbC™ Anti-Mouse Bead Kit.

IMPORTANT NOTICE: The products described in this manual are covered by one or more Limited Use Label License(s). Please refer to the Appendix on page 971 and Master Product List on page 975. Products are For Research Use Only. Not intended for any animal or human therapeutic or diagnostic use.

emission can be read between 385 nm and 800 nm. The fluorescence intensity of the microspheres has been adjusted to be about 5–50 times brighter than the anticipated intensities of typically stained cells. When using a scatter threshold, the microsphere signal should be above the threshold. The microspheres can be gated by a single parameter, but a combination of parameters can be used to resolve microspheres from cells and other events.

CountBright™ absolute counting beads can be used with any sample type, including no-wash/lysed whole blood. The microspheres in the reagents are approximately 7 µm in diameter and have sedimentation properties similar to lymphocytes. The accuracy of cell counts based on CountBright™ absolute counting beads depends on sample handling and the precise delivery of the volume of beads. The CountBright™ absolute counting beads must be mixed well to assure a uniform suspension of microspheres. After vortexing for 30 seconds, the microsphere suspension can be pipetted by standard techniques; however, more viscous solutions such as blood require reverse pipetting for accurate volume delivery. Cell suspensions may be diluted but should be assayed without wash steps. Other sample preparation steps that can lead to cell or microsphere loss should also be avoided. For antibody protocols, CountBright™ absolute counting beads should be used with reagents titered for no-wash staining.

REFERENCES

1. Cytometry B Clin Cytom (2007) 72:442; **2.** Anal Biochem (2007) 364:180; **3.** J Res Natl Inst Stand Technol (2002) 107:83; **4.** Cytometry (1987) 8:632; **5.** J Biol Chem (2008) 283:7219; **6.** Nat Protoc (2007) 2:2233; **7.** Appl Environ Microbiol (2002) 68:37; **8.** Cytometry A (2004) 60:135; **9.** Cytometry A (2005) 68:36; **10.** Appl Environ Microbiol (2000) 66:4258; **11.** J Immunol Methods (2006) 313:199; **12.** Am J Physiol Heart Circ Physiol (2009) 296:H359; **13.** Appl Environ Microbiol (2010) 76:1480; **14.** Br J Haematol (2001) 115:953; **15.** J Acquir Immune Defic Syndr (2005) 39:32; **16.** Clin Diagn Lab Immunol (2000) 7:336; **17.** Br J Haematol (1999) 106:1059; **18.** Cytotherapy (2003) 5:55; **19.** MMWR Recomm Rep (2003) 52:1; **20.** Blood (2009) 114:5081.

PRODUCT LIST 23.2 FLOW CYTOMETRY REFERENCE STANDARDS

Cat. No.	Product	Quantity
A10344	AbC™ Anti-Mouse Bead Kit *for mouse antibody capture* *for flow cytometry compensation* *100 tests*	1 kit
A10389	AbC™ Anti-Rat/Hamster Bead Kit *for rat/hamster antibody capture* *for flow cytometry compensation* *100 tests*	1 kit
A7304	AlignFlow™ flow cytometry alignment beads, 2.5 µm *for UV excitation*	3 mL
A7302	AlignFlow™ flow cytometry alignment beads, 2.5 µm *for 488 nm excitation*	3 mL
A14835	AlignFlow™ flow cytometry alignment beads, 2.5 µm *for 630–660 nm excitation*	3 mL
A7312	AlignFlow™ flow cytometry alignment beads, 2.5 µm *for 633 nm excitation*	3 mL
A7305	AlignFlow™ Plus flow cytometry alignment beads, 6 µm *for UV excitation*	3 mL
A7303	AlignFlow™ Plus flow cytometry alignment beads, 6 µm *for 488 nm excitation*	3 mL
A14836	AlignFlow™ Plus flow cytometry alignment beads, 6 µm *for 630–660 nm excitation*	3 mL
A7313	AlignFlow™ Plus flow cytometry alignment beads, 6 µm *for 633 nm excitation*	3 mL
A10346	ArC™ Amine-Reactive Compensation Bead Kit *for use with amine reactive dyes* *for flow cytometry compensation* *100 tests*	1 kit
C36950	CountBright™ absolute counting beads *for flow cytometry* *100 tests*	5 mL
F13838	Flow Cytometry Size Calibration Kit *nonfluorescent microspheres*	1 kit
L14812	LinearFlow™ Blue Flow Cytometry Intensity Calibration Kit, 2.5 µm *for UV excitation/430 nm emission*	1 kit
L14813	LinearFlow™ Blue Flow Cytometry Intensity Calibration Kit, 6 µm *for UV excitation/430 nm emission*	1 kit
L14816	LinearFlow™ Carmine Flow Cytometry Intensity Calibration Kit, 2.5 µm *for 488 nm excitation/620 nm emission*	1 kit
L14817	LinearFlow™ Carmine Flow Cytometry Intensity Calibration Kit, 6 µm *for 488 nm excitation/620 nm emission*	1 kit
L14818	LinearFlow™ Deep Red Flow Cytometry Intensity Calibration Kit, 2.5 µm *for 633 nm excitation/660 nm emission*	1 kit
L14819	LinearFlow™ Deep Red Flow Cytometry Intensity Calibration Kit, 6 µm *for 633 nm excitation/660 nm emission*	1 kit
L14821	LinearFlow™ Green Flow Cytometry Intensity Calibration Kit, 2.5 µm *for 488 nm excitation/515 nm emission*	1 kit
L14822	LinearFlow™ Green Flow Cytometry Intensity Calibration Kit, 6 µm *for 488 nm excitation/515 nm emission*	1 kit
L14823	LinearFlow™ Green Flow Cytometry Low Intensity Calibration Kit, 2.5 µm *for 488 nm excitation/515 nm emission*	1 kit
L14824	LinearFlow™ Green Flow Cytometry Low Intensity Calibration Kit, 6 µm *for 488 nm excitation/515 nm emission*	1 kit
L14814	LinearFlow™ Orange Flow Cytometry Intensity Calibration Kit, 2.5 µm *for 488 nm excitation/575 nm emission*	1 kit
L14815	LinearFlow™ Orange Flow Cytometry Intensity Calibration Kit, 6 µm *for 488 nm excitation/575 nm emission*	1 kit
P14825	PeakFlow™ Blue flow cytometry reference beads, 2.5 µm *460 nm emission*	3 mL
P14826	PeakFlow™ Blue flow cytometry reference beads, 6 µm *460 nm emission*	3 mL
P14831	PeakFlow™ Carmine flow cytometry reference beads, 2.5 µm *620 nm emission*	3 mL
P14832	PeakFlow™ Carmine flow cytometry reference beads, 6 µm *620 nm emission*	3 mL
P24670	PeakFlow™ Claret flow cytometry reference beads, 6 µm *680 nm emission*	3 mL
P14827	PeakFlow™ Green flow cytometry reference beads, 2.5 µm *515 nm emission*	3 mL
P14828	PeakFlow™ Green flow cytometry reference beads, 6 µm *515 nm emission*	3 mL
P24672	PeakFlow™ Infrared flow cytometry reference beads, 6 µm *770 nm emission*	3 mL
P14829	PeakFlow™ Orange flow cytometry reference beads, 2.5 µm *575 nm emission*	3 mL
P14830	PeakFlow™ Orange flow cytometry reference beads, 6 µm *575 nm emission*	3 mL
P24671	PeakFlow™ Ultra Red flow cytometry reference beads, 6 µm *695 nm emission*	3 mL

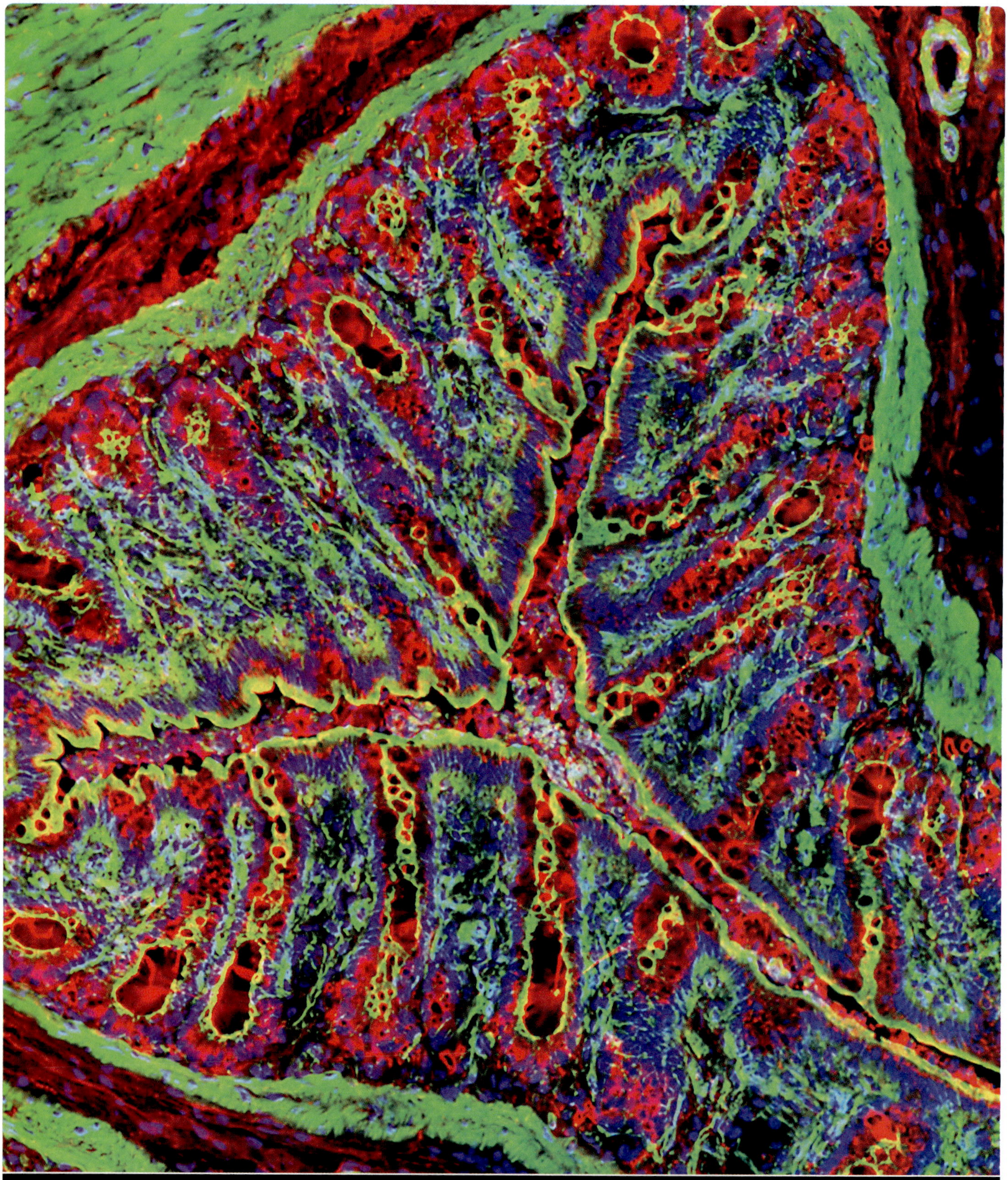

Texas Red® wheat germ agglutinin, Alexa Fluor® 488 phalloidin and Hoechst 33342.

IMPORTANT NOTICE: The products described in this manual are covered by one or more Limited Use Label License(s). Please refer to the Appendix on page 971 and Master Product List on page 975. Products are For Research Use Only. Not intended for any animal or human therapeutic or diagnostic use.

Appendix

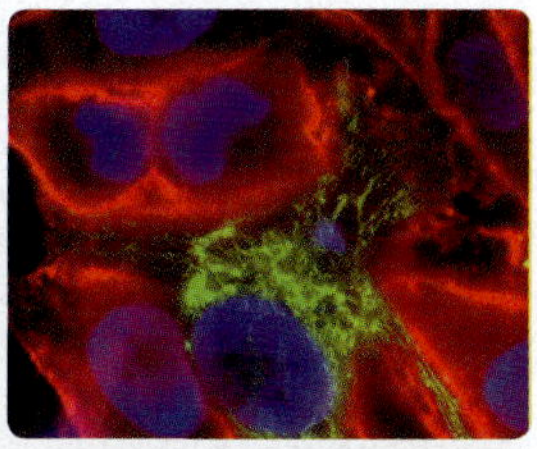
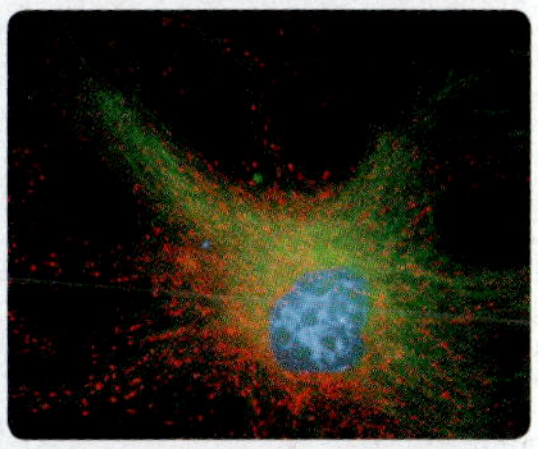
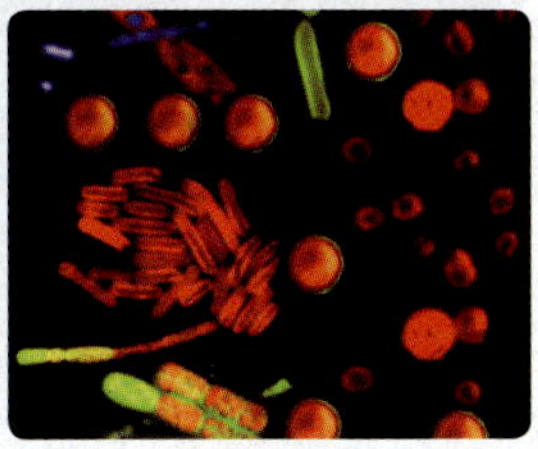

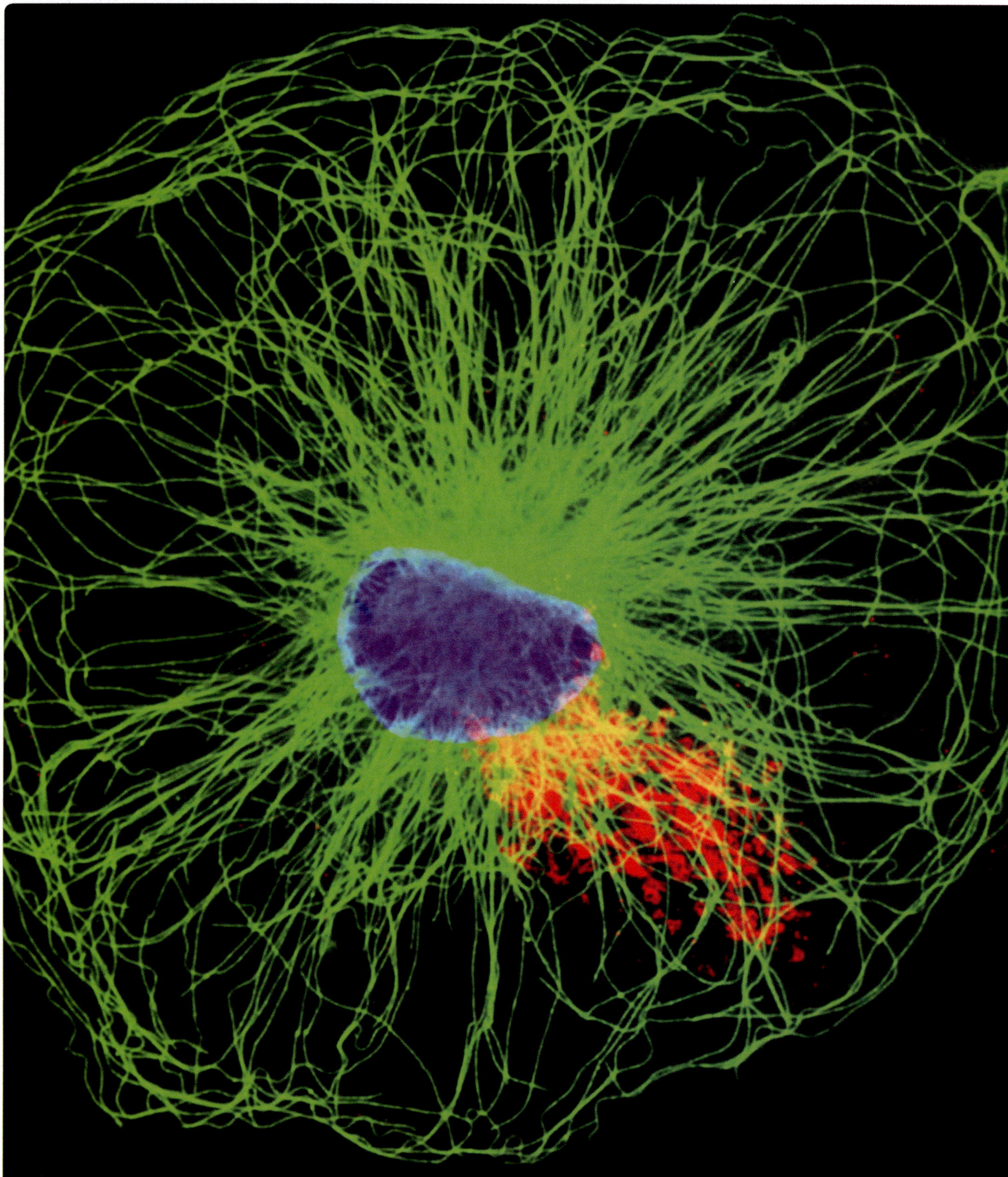

Texas Red® goat anti–rabbit IgG, Alexa Fluor® 488 goat anti–mouse IgG and Hoechst 33342.

IMPORTANT NOTICE: The products described in this manual are covered by one or more Limited Use Label License(s). Please refer to the Appendix on page 971 and Master Product List on page 975. Products are For Research Use Only. Not intended for any animal or human therapeutic or diagnostic use.

Trademark Information

Life Technologies has achieved recognition of its trademarks worldwide. Please respect our trademarks; we will vigorously protect their proper usage. All names containing the designation ® are registered wtih the U.S. Patent and Trademark Office.

Trademarks of Life Technologies Corporation:

AADvanced™
AbC™
ABfinity™
Adapta®
Alexa Fluor®
AlignFlow™
Amplex®
Antibody Beacon™
APEX®
APO-Brdu™
Applied Biosystems®
ArC™
ARES™
ATTO-TAG™
Attune™
*Bac*Light™
BenchMark™
BioParticles®
Biotective™
BlockAid™
BOBO™
BOCILLIN™
BODIPY®
BO-PRO™
BrainStain™
Calcium Crimson™
Calcium Green™
Calcium Orange™
Calcium Sponge™
CandyCane™
CaptAvidin™
Cascade Blue®
Cascade Yellow™
CDP-Star®
CellLight®
CellMask™
CellSensor®
CellTrace™
CellTracker™
Cellular Lights™
ChromaTide®
Click-iT®
Coomassie Fluor™
Constellation™
CoroNa™
CountBright™
Countess®
CSPD®
CyQUANT®
Dapoxyl®
DEAD Red™
DEAD™
DetectaGene™
DQ™
DSB-X™
DyeChrome™
DyeCycle™
Dynabeads®
ELF®
Emerald-III™
Emerald-II™
EnzChek®
ER-Tracker™
EZQ®
FAM™
FAST CAT®
FAST DiA™
FAST DiI™
FAST DiO™
FilmTracer™
FISH Tag™
Fluo-4 Direct™
FluoCells®
FluoReporter®
Fluorocillin™
FluoroMyelin™
FluoroPure™
FluoSpheres®
FluoZin™
FluxOR™
FM®
FocalCheck™
FUN®
*Funga*Light™
Fura Red™
FuraZin™
FxCycle™
Galacton-Star®
Gateway®
GeneBLAzer®
GIBCO®
HEX™
Image-iT®
ImaGene Green™
ImaGene Red™
Influx™
InSpeck™
Invitrogen™
ITK™
JOE™
JOJO™
JO-PRO™
LanthaScreen®
Leadmium™
LinearFlow™
LipidTOX™
Lipofectamine®
LIVE/DEAD®
LOLO™
LO-PRO™
LysoSensor™
LysoTracker®
Magnesium Green™
Marina Blue®
Measure-iT™
MitoFluor™
MitoProbe™
MitoSOX™
MitoTracker®
Molecular Probes®
Multiplexed Proteomics®
MultiSpeck™
MycoFluor™
NanoOrange®
Neon™
NeuroTrace®
Newport Green™
NovaBright™
NuclearMask™
OliGreen®
Omnia®
One Shot®
Oregon Green®
Organelle Lights™
OxyBURST®
Pacific Blue™
Pacific Orange™
Panomer™
PeakFlow™
PeppermintStick™
Phen Green™
pHrodo™
PicoGreen®
P_iPer™
POPO™
PO-PRO™
PowerLoad™
Premo™
ProLong®
Pro-Q®
pShooter™
PS-Speck™
PureLink™
Qdot®
Qmount®
Qnuclear™
QSY®
Qtracker®
Quant-iT™
Qubit®
Qubit™
RediPlate™
RedoxSensor™
Rhinohide™
Rhodamine Green™
Rhodamine Red™
RhodZin™
RiboGreen®
RNASelect™
RNaseOUT™
ROX™
Safe Imager™
SAIVI™
Sapphire-II™
Sapphire™
SelectFX®
SensiFlex™
SlowFade®
SNARF®
Sodium Green™
SYBR®
SYPRO®
SYTO®
SYTOX®
TAMRA™
TC-FlAsH™
TC-ReAsH™
TetraSpeck™
TET™
Texas Red®
ThiolTracker™
TO-PRO®
TOTO®
TransFluoSpheres®
TS-Link™
TubulinTracker™
UltraPure™
ViaGram™
Vybrant®
WesternDot™
YO-PRO®
YOYO®
Z′-LYTE®
Zenon®
Zen™

Trademarks of Other Companies:

alamarBlue® (Trek Diagnostic Systems, Inc.)
Amersham® (GE Healthcare)
AquaLite® (Millipore Corporation)
ArrayScan® (Thermo-Fisher Scientific)
BD FACSCalibur™ (BD Biosciences)
BD FACScan™ (BD Biosciences)
BD Pathway™ (BD Biosciences)
Bio-Rad® (Bio-Rad Laboratories, Inc.)
Celite® (Celite Corp.)
Cellomics (Thermo-Fisher Scientific)
Cellosolve® (Union Carbide Corp.)
Chroma® (Chroma Technology)
CometAssay® (Trevigen, Inc.)
CometSlide™ (Trevigen, Inc.)
Corrositex® (InVitro International, Inc.)
Coulter® (Beckman Coulter, Inc.)
CoverWell™ (Grace Bio-Labs, Inc.)
CultureWell™ (Grace Bio-Labs, Inc.)
Cy® (GE Healthcare)
DM-Nitrophen™ (Calbiochem-Novabiochem Corp.)
FLICA® (Immunochemistry Technologies, LLC)
FLIPR® (Molecular Devices, Inc.)
Fluoro-Gold™ (Fluorochrome, LLC)
FluoroNanogold™ (Nanoprobes, Inc.)
Fuji® (FujiFilm Corporation)
HybriSlip™ (Grace Bio-Labs, Inc.)
HybriWell™ (Grace Bio-Labs, Inc.)
Immobilon® (Millipore Corp.)
iPhone® (Apple, Inc.)
IRDye® (LI-COR, Inc.)
MaxiSorp™ (Thermo-Fisher Scientific)
MetaMorph® (Molecular Devices, Inc.)
Microsoft® (Microsoft Corporation)
Molecular Imager® (Bio-Rad Laboratories)
NANOGOLD® (Nanoprobes, Inc.)
NeutrAvidin® (Pierce Biotechnology, Inc.)
Nikon® (Nikon Corporation)
Nonidet™ (Shell International Petroleum Company Limited, U.K.)
Nunc-Immuno™ (Thermo-Fisher Scientific)
Odyssey® (LI-COR, Inc.)
Omega® (Omega Optical, Inc.)
ONCYTE® (Grace Bio-Labs, Inc.)
PerkinElmer™ (PerkinElmer, Inc.)
Pharmacia® (Pfizer Health AB)
Photometrics® (Roper Scientific, Inc.)
Pluronic® (BASF Corp.)
Polaroid® (Polaroid Corp.)
ProPlate™ (Grace Bio-Labs, Inc.)
ScanArray® (PerkinElmer LAS, Inc.)
SecureSeal™ (Grace Bio-Labs, Inc.)
SpectraMax® (Molecular Devices, Inc.)
SRM® (National Institute of Standards and Technology)
TaqMan® (Roche Molecular Systems, Inc.)
Taxol® (Bristol-Myers Squibb Co.)
Teflon® (E.I. DuPont de Nemours & Co., Inc.)
Triton® (Union Carbide Corp.)
TSA™ (PerkinElmer LAS, Inc.)
Tween® (Uniqema Americas, LLC)
ULS® (KREATECH Biotechnology B.V.)
UltraVIEW® (PerkinElmer LAS, Inc.)
ULYSIS® (KREATECH Biotechnology B.V.)
Vivaspin® (Sartorius Stedim-Biotech Gmbh)
Zeiss® (Carl Zeiss AG)

Limited Use Label Licenses

The products described in this manual are covered by one or more Limited Use Label License(s). Note that all products are covered by LIMITED USE LABEL LICENSE #223. To determine whether a specific product is covered by additional Limited Use Label License(s), please refer to the LULL column of the Master Product List which starts on page 975.

Limited Use Label License No. 21: Bac-to-Bac® and Bac-to-Bac® HT

This product is sold under patent license from Monsanto for research purposes only and no license for commercial use is included. Requests for licenses for commercial manufacture or use should be directed to Director, Monsanto Corporate Research, 800 N. Lindbergh, St. Louis, Missouri 63167.

Limited Use Label License No. 54: ULB ccdB Selection Technology

This product is the subject of one or more of U.S. Patents and is sold under license from the Université Libre de Bruxelles for research purposes only. ccdB selection technology is described in Bernard et al., "Positive Selection Vectors Using the F Plasmid ccdB Killer Gene" Gene 148 (1994) 71-74. The purchase of this product conveys to the buyer the non-transferable right to use the purchased amount of the product and components of the product in research conducted by the buyer (whether the buyer is an academic or for-profit entity). For licensing information for use in other than research, please contact: Out Licensing, Life Technologies Corporation, 5791 Van Allen Way, Carlsbad, California 92008; Phone (760) 603-7200 or e-mail; outlicensing@lifetech.com..

Limited Use Label License No. 127: GFP with Heterologous Promoter

This product and its use is the subject of one or more of U.S. Patents. This product is sold under license from Columbia University. Rights to use this product are limited to research use only, and expressly exclude the right to manufacture, use, sell or lease this product for use for measuring the level of toxicity for chemical agents and environmental samples in cells and transgenic animals. No other rights are conveyed. Not for human use or use in diagnostic or therapeutic procedures. Inquiry into the availability of a license to broader rights or the use of this product for commercial purposes should be directed to Columbia Innovation Enterprise, Columbia University, Engineering Terrace-Suite 363, New York, New York 10027.

Limited Use Label License No. 198: Fluorescent proteins and stable cell lines expressing such proteins (but not for vectors that contain the genes for such fluorescent proteins)

This product and its use is the subject of U.S. Patents. The purchase of this product conveys to the buyer the nontransferable right to use the purchased amount of the product and components of the product in research conducted by the buyer (whether the buyer is an academic or for-profit entity). No rights are conveyed to modify or clone the gene encoding GFP contained in this product. The buyer cannot sell or otherwise transfer (a) this product, (b) its components or (c) materials made by the employment of this product or its components to a third party or otherwise use this product or its components or materials made by the employment of this product or its components for Commercial Purposes. The buyer may transfer information or materials made through the employment of this product to a scientific collaborator, provided that such transfer is not for any Commercial Purpose, and that such collaborator agrees in writing (a) not to transfer such materials to any third party, and (b) to use such transferred materials and/or information solely for research and not for Commercial Purposes. Commercial Purposes means any activity by a party for consideration and may include, but is not limited to: (1) use of the product or its components in manufacturing; (2) use of the product or its components to provide a service, information, or data; (3) use of the product or its components for therapeutic, diagnostic or prophylactic purposes; or (4) resale of the product or its components, whether or not such product or its components are resold for use in research. Life Technologies Corporation will not assert a claim against the buyer of infringement of the above patents based upon the manufacture, use or sale of a therapeutic, clinical diagnostic, vaccine or prophylactic product developed in research by the buyer in which this product or its components was employed, provided that none of this product, or any of its components was used in the manufacture of such product. If the purchaser is not willing to accept the limitations of this limited use statement, Life Technologies Corporation is willing to accept return of the product with a full refund. For information on purchasing a license to use this product for purposes other than those permitted above, contact Licensing Department, Life Technologies Corporation, 5791 Van Allen Way, Carlsbad, California 92008; Phone (760) 603-7200; outlicensing@lifetech.com.

Limited Use Label License No. 204: Metal Ion Indicator Dyes

This product is licensed to Molecular Probes, Inc. For research use only.

Limited Use Label License No. 214: Anti-BrdU Antibodies

This product is provided under an agreement between Phoenix Flow Systems, Inc. and Molecular Probes, Inc. For research use only.

Limited Use Label License No. 215: Multi Drug Resistance

This product is provided under an agreement between Solvo Biotechnology, Inc and Molecular Probes, Inc. For research use only.

Limited Use Label License No. 223: Labeling and Detection Technology

The purchase of this product conveys to the buyer the non-transferable right to use the purchased amount of the product and components of the product in research conducted by the buyer (whether the buyer is an academic or for-profit entity) in a manner consistent with the accompanying product literature. The buyer cannot sell or otherwise transfer (a) this product, (b) its components or (c) materials made using this product or its components to a third party or otherwise use this product or its components or materials made using this product or its components for Commercial Purposes. The buyer may transfer information or materials made through the use of this product to a scientific collaborator, provided that such transfer is not for any Commercial Purpose, and that such collaborator agrees in writing (a) to not transfer such materials to any third party, and (b) to use such transferred materials and/or information solely for research and not for Commercial Purposes. Commercial Purposes means any activity by a party for consideration and may include, but is not limited to: (1) use of the product or its components in manufacturing; (2) use of the product or its components to provide a service, information or data; (3) use of the product or its components for therapeutic, diagnostic or prophylactic purposes; or (4) resale of the product or its components, whether or not such product or its components are resold for use in research. For products that are subject to multiple limited use label licenses, the most restrictive terms apply. Life Technologies Corporation will not assert a claim against the buyer of infringement of patents that are owned or controlled by Life Technologies Corporation and/or Molecular Probes, Inc. which cover this product based upon the manufacture, use or sale of a therapeutic, clinical diagnostic, vaccine or prophylactic product developed in research by the buyer in which this product or its components was employed, provided that neither this product nor any of its components was used in the manufacture of such product. If the purchaser is not willing to accept the limitations of this limited use statement, Life Technologies is willing to accept return of the product with a full refund. For information on purchasing a license to this product for purposes other than research, contact Molecular Probes, Inc., Business Development, 29851 Willow Creek Road, Eugene, OR 97402; Phone: (541) 465-8300; Fax: (541) 335-0354.

Limited Use Label License No. 225: ChromaTide® Nucleotides

The dUTP and UTP nucleotides are NEN-brand products that are distributed and sold under an agreement between ENZO DIAGNOSTICS, INC. and PERKINELMER LIFE SCIENCES, INC. (previously NEN LIFE SCIENCES, INC.) for research purposes only by the end-user in the research market and are not intended for diagnostic or therapeutic use. Purchase does not include or carry any right or license to use, develop or otherwise exploit this product commercially. Any commercial use, development or exploitation of these products without the express prior written authorization of ENZO DIAGNOSTICS, INC. and PERKINELMER LIFE SCIENCES, INC. is strictly prohibited.

Limited Use Label License No. 231: Nanogold® Technology

This product is prepared for Molecular Probes, Inc by Nanoprobes, Inc. For research use only.

Limited Use Label License No. 235: Tyramide Signal Amplification Technology

This product is distributed and sold to the End-User pursuant to a license from PerkinElmer Life Sciences Inc., for use of its tyramide technology by the End-User for life science research applications in the fields of cytochemistry, flow cytometry, histochemistry and in situ hybridization, but not for diagnostic applications. Purchase does not include or carry any right to resell or transfer this product either as a stand-alone product or as a component of another product. Any use of this product other than the licensed use without the express written authorization of PerkinElmer Life Sciences, Inc. and Molecular Probes, Inc., is strictly prohibited.

Limited Use Label License No. 244: ATTO-TAG™ Reagents and CBQCA Reagents

This product is provided under an agreement with Molecular Probes, Inc.

Limited Use Label License No. 267: Mutant GFP Products

This product and its use are the subject of one or more U.S. Patents.

Limited Use Label License No. 272: Humanized GFP

This product is the subject of one or more U.S. Patents licensed by Life Technologies Corporation. This product is sold for research use only. Not for therapeutic or diagnostic use in humans.

Limited Use Label License No. 289: HCS Disclaimer

Purchase of this product does not convey a license to use this product in high content screening ("HCS") methods. Licenses from third parties may be required depending on the intended use of this product.

Limited Use Label License No. 306: Baculovirus Vectors

Certain methods that utilize the product associated with this limited use label license are covered by U.S. Patents. This product is for research use only by those researchers in laboratories of academic, government, industrial and/or clinical institutions engaged in the investigation of biological or biochemical processes, or research and development of biological products. This product is not to be used in the manufacture, use or sale of human or animal diagnostic, therapeutic or prophylactic products.

Limited Use Label License No. 308: WPRE Element

This product contains the Woodchuck Post-transcriptional Regulatory Element ("WPRE") which is the subject of intellectual property owned by The Salk Institute for Biological Studies, and licensed to Life Technologies Corporation. The purchase of this product conveys to the buyer the non-transferable right to use the purchased amount of the product and components of the product in research conducted by the buyer (whether the buyer is an academic or for-profit entity). The buyer cannot sell or otherwise transfer (a) this product, (b) its components or (c) materials made using this product or its components to a third party or otherwise use this product or its components or materials made using this product or its components for Commercial Purposes. The buyer may transfer information or materials made through the use of this product to a scientific collaborator, provided that such transfer is not for any Commercial Purpose, and that such collaborator agrees in writing (a) not to transfer such materials to any third party, and (b) to use such transferred materials and/or information solely for research and not for Commercial Purposes. Commercial Purposes means any activity by a party for consideration and may include, but is not limited to: (1) use of the product or its components in manufacturing; (2) use of the product or its components to provide a service, information, or data; (3) use of the product or its components for therapeutic, diagnostic or prophylactic purposes; and/or (4) resale of the product or its components, whether or not such product or its components are resold for use in research. In addition, any use of WPRE outside of this product or the product's authorized use requires a separate license from the Salk Institute. Life Technologies will not assert a claim against the buyer of infringement of patents owned by Life Technologies and claiming this product based upon the manufacture, use or sale of a therapeutic, clinical diagnostic, vaccine or prophylactic product developed in research by the buyer in which this product or its components was employed, provided that neither this product nor any of its components was used in the manufacture of such product or for a Commercial Purpose. If the purchaser is not willing to accept the limitations of this limited use statement, Life Technologies is willing to accept return of the product with a full refund. For information on purchasing a license to this product for purposes other than research, contact Licensing Department, Life Technologies Corporation, 5791 Van Allen Way, Carlsbad, California 92008, Phone (760) 603-7200; Fax (760) 602-6500; or The Salk Institute for Biological Studies, 10010 North Torrey Pines Road, La Jolla, CA 92037, Attn.: Office of Technology Management, Phone: (858) 453-4100 extension 1275, Fax: (858) 546-8093.

Limited Use Label License No. 321: TagRFP

This product is for the buyer's internal research use only and may not be used for commercial purposes. No rights are conveyed to modify or clone the gene encoding fluorescent protein contained in this product. The right to use this product specifically excludes the right to validate or screen compounds for commercial purposes. For information on commercial licensing, contact Licensing Department, Evrogen JSC, email: license@evrogen.com

continued on next page

Limited Use Label License No. 327: Recombinant Antibody Technology

The purchase of this product conveys to the buyer the non-transferable right to use the purchased amount of the product and components of the product in research and manufacturing conducted by the buyer (whether the buyer is an academic or for-profit entity). The buyer cannot sell or otherwise transfer (a) this product, (b) its components or (c) materials made using this product or its components to a third party or otherwise use this product or its components or materials made using this product or its components for Commercial Purposes. The buyer may transfer information or materials made through the use of this product to a scientific collaborator, provided that such transfer is not for any Commercial Purpose, and that such collaborator agrees in writing (a) not to transfer such materials to any third party, and (b) to use such transferred materials and/or information solely for research and not for Commercial Purposes. Commercial Purposes means any activity by a party for consideration and may include, but is not limited to: (1) use of the product or its components for contract manufacturing services; (2) use of the product or its components to provide a service, information or data; (3) use of the product itself or its components as a therapeutic, diagnostic or prophylactic; or (4) resale of the product or its components, whether or not such product or its components are resold for use in research or manufacturing. If the purchaser is not willing to accept the limitations of this limited use statement, Life Technologies is willing to accept return of the product with a full refund. For information on purchasing a license to this product for purposes other than research, contact Licensing Department, Life Technologies Corporation, 5791 Van Allen Way, Carlsbad, California 92008. Phone (760) 603-7200. Fax (760) 602-6500. Email: outlicensing@lifetech.com.

Limited Use Label License No. 332: BacMam Virus Use

The purchase of this product conveys to the buyer the non-transferable right to use the purchased amount of the product and components of the product in research conducted by the buyer solely in accordance with the accompanying product literature or manual. Purchase of this product does not convey a license to expand, amplify or otherwise propagate the provided viral particles or to otherwise modify or alter the virus by any means.

Limited Use Label License No. 353: Cy® dye antibodies

This product is for research use only as applied to the fields of biotechnology, life science and medicine. Use of this product specifically excludes all uses in therapeutic, diagnostic, in vivo applications, sequencing, drug screening and microarrays. For further information on purchasing additional licenses, please contact Amersham Biosciences Corp, 800 Centennial Avenue, Piscataway, NJ 08855-1327.

IMPORTANT NOTICE: The products described in this manual are covered by one or more Limited Use Label License(s). Please refer to the Appendix on page 971 and Master Product List on page 975. Products are For Research Use Only. Not intended for any animal or human therapeutic or diagnostic use.

Master Product List

Cat. No.	Product	Quantity	Sections	LULL*
A47	1,8-ANS (1-anilinonaphthalene-8-sulfonic acid) *high purity*	100 mg	13.5	
A50	2,6-ANS (2-anilinonaphthalene-6-sulfonic acid)	100 mg	13.5	
A1493	A-23187 free acid (calcimycin)	10 mg	19.8	
A10344	AbC™ Anti-Mouse Bead Kit *for mouse antibody capture* *for flow cytometry compensation* *100 tests*	1 kit	23.2	
A10389	AbC™ anti-Rat/Hamster Bead Kit *for rat/hamster antibody capture* *for flow cytometry* *100 tests*	1 kit	23.2	
A484	4-acetamido-4′-((iodoacetyl)amino)stilbene-2,2′-disulfonic acid, disodium salt	25 mg	2.3	
A485	4-acetamido-4′-maleimidylstilbene-2,2′-disulfonic acid, disodium salt	25 mg	2.3	
A22010	5-acetylaminofluorescein di-β-D-galactopyranoside (C_2FDG)	5 mg	10.2	
A685	5-((2-(and-3)-*S*-(acetylmercapto)succinoyl)amino)fluorescein (SAMSA fluorescein) *mixed isomers*	25 mg	5.2	
A7923	4-((9-acridinecarbonyl)amino)-2,2,6,6-tetramethylpiperidin-1-oxyl, free radical (TEMPO-9-AC)	5 mg	18.2	
A666	acridine homodimer (bis-(6-chloro-2-methoxy-9-acridinyl)spermine)	10 mg	8.1	
A1301	acridine orange	1 g	8.1, 12.3, 15.2	
A3568	acridine orange *10 mg/mL solution in water*	10 mL	8.1, 12.3, 15.2	
A1372	acridine orange 10-nonyl bromide (nonyl acridine orange)	100 mg	12.2	
A20770	6-((acryloyl)amino)hexanoic acid, succinimidyl ester (acryloyl-X, SE)	5 mg	5.2, 9.4	
A433	6-acryloyl-2-dimethylaminonaphthalene (acrylodan)	25 mg	2.3	
A12375	actin from rabbit muscle	1 mg	11.1	
A12373	actin from rabbit muscle, Alexa Fluor® 488 conjugate *in solution*	200 µg	11.1	
A12374	actin from rabbit muscle, Alexa Fluor® 568 conjugate *in solution*	200 µg	11.1	
A34050	actin from rabbit muscle, Alexa Fluor® 594 conjugate *in solution*	200 µg	11.1	
A34051	actin from rabbit muscle, Alexa Fluor® 647 conjugate *in solution*	200 µg	11.1	
A7592	actinomycin D	10 mg	8.1	
A22359	adenosine 5′-diphosphate, BODIPY® TR 2′-(or-3′)-*O*-(*N*-(2-aminoethyl)urethane), disodium salt (BODIPY® TR ADP) *5 mM in buffer*	100 µL	17.3	
A7056	adenosine 5′-diphosphate, P^2-(1-(2-nitrophenyl)ethyl) ester, monopotassium salt (NPE-caged ADP)	5 mg	5.3, 17.3	
A22184	adenosine 5′-*O*-(3-thiotriphosphate), BODIPY® FL thioester, sodium salt (BODIPY® FL ATP-γ-S, thioester) *5 mM in buffer*	50 µL	10.3, 17.3	
A22362	adenosine 5′-triphosphate, Alexa Fluor® 647 2′-(or-3′)-*O*-(*N*-(2-aminoethyl)urethane), hexa(triethylammonium) salt (Alexa Fluor® 647 ATP) *5 mM in buffer*	100 µL	17.3	
A12410	adenosine 5′-triphosphate, BODIPY® FL 2′-(or-3′)-*O*-(*N*-(2-aminoethyl)urethane), trisodium salt (BODIPY® FL ATP) *5 mM in buffer*	100 µL	17.3	
A22352	adenosine 5′-triphosphate, BODIPY® TR 2′-(or-3′)-*O*-(*N*-(2-aminoethyl)urethane), trisodium salt (BODIPY® TR ATP) *5 mM in buffer*	100 µL	17.3	
A1049	adenosine 5′-triphosphate, P^3-(1-(4,5-dimethoxy-2-nitrophenyl)ethyl) ester, disodium salt (DMNPE-caged ATP)	5 mg	5.3, 17.3	
A1048	adenosine 5′-triphosphate, P^3-(1-(2-nitrophenyl)ethyl) ester, disodium salt (NPE-caged ATP)	5 mg	5.3, 17.3	
A12412	adenosine 5′-triphosphate, P^3-(5-sulfo-1-naphthylamide), tetra(triethylammonium) salt (ATP γ-AmNS) *5 mM in buffer*	400 µL	17.3	
A18211	Adhesive seal-tab, for HybriWell™ hybridization sealing system *set of 400*	1 set	23.1	
A3880	ADIFAB fatty acid indicator	200 µg	13.2, 17.4	
A32768	aha-dCTP (5-aminohexylacrylamido-dCTP) *2 mM in TE buffer*	500 µL	8.2	
A32769	aha-dCTP (5-aminohexylacrylamido-dCTP) *50 mM in TE buffer*	50 µL	8.2	
A32761	aha-dUTP (5-aminohexylacrylamido-dUTP) *50 mM in TE buffer*	50 µL	8.2	
A34786	albumin from bovine serum (BSA), Alexa Fluor® 555 conjugate	5 mg	14.7	
A34787	albumin from bovine serum (BSA), Alexa Fluor® 680 conjugate	5 mg	14.7	
A13100	albumin from bovine serum (BSA), Alexa Fluor® 488 conjugate	5 mg	14.7	
A13101	albumin from bovine serum (BSA), Alexa Fluor® 594 conjugate	5 mg	14.7	
A34785	albumin from bovine serum (BSA), Alexa Fluor® 647 conjugate	5 mg	14.7	
A2750	albumin from bovine serum (BSA), BODIPY® FL conjugate	5 mg	14.7	
A23018	albumin from bovine serum (BSA), 2,4-dinitrophenylated (DNP-BSA)	5 mg	14.7	
A23015	albumin from bovine serum (BSA), fluorescein conjugate	5 mg	14.7	
A23016	albumin from bovine serum (BSA), tetramethylrhodamine conjugate	5 mg	14.7	
A23017	albumin from bovine serum (BSA), Texas Red® conjugate	5 mg	14.7	
A32770	Alexa Fluor® 555-aha-dCTP *1 mM in TE buffer*	50 µL	8.2	
A32771	Alexa Fluor® 647-aha-dCTP *1 mM in TE buffer*	50 µL	8.2	
A32762	Alexa Fluor® 555-aha-dUTP *1 mM in TE buffer*	50 µL	8.2	

*Note that all products listed in the above Master Product List are covered by **Limited Use Label License #223**. If a specific product is covered by additional Limited Use Label License(s), the Limited Use Label License #'s are provided in the LULL column above. Please refer to page 971 for the complete text of the Limited Use Label License(s).

IMPORTANT NOTICE: The products described in this manual are covered by one or more Limited Use Label License(s). Please refer to the Appendix on page 971 and Master Product List on page 975. Products are For Research Use Only. Not intended for any animal or human therapeutic or diagnostic use.

Cat. No.	Product	Quantity	Sections	LULL*
A32763	Alexa Fluor® 647-aha-dUTP *1 mM in TE buffer*	50 µL	8.2	
A10267	Alexa Fluor® 488 alkyne (Alexa Fluor® 488 5-carboxamido-(propargyl), bis(triethylammonium salt)) *5-isomer*	0.5 mg	3.1	
A10275	Alexa Fluor® 594 alkyne (Alexa Fluor® 594 carboxamido-(5-(and 6-)propargyl), bis(triethylammonium salt)) *mixed isomers*	0.5 mg	3.1	
A20013	Alexa Fluor® 555 alkyne, triethylammonium salt	0.5 mg	3.1	
A10278	Alexa Fluor® 647 alkyne, triethylammonium salt	0.5 mg	3.1	
A21000	Alexa Fluor® 680–allophycocyanin goat anti-mouse IgG (H+L) *1 mg/mL*	100 µL	6.4, 7.2	
A21006	Alexa Fluor® 750–allophycocyanin goat anti-mouse IgG (H+L) *1 mg/mL*	100 µL	6.4, 7.2	
A21001MP	Alexa Fluor® 680–allophycocyanin goat anti-rabbit IgG (H+L) *1 mg/mL*	100 µL	6.4, 7.2	
A35775	Alexa Fluor® 488 8-(6-aminohexyl)aminoadenosine 3',5'-cyclicmonophosphate, bis(triethylammonium) salt (Alexa Fluor® 488 cAMP) *5 mM in buffer*	100 µL	17.3	
A35777	Alexa Fluor® 647 8-(6-aminohexyl)aminoadenosine 3',5'-cyclicmonophosphate, tetra(triethylammonium) salt (Alexa Fluor® 647 cAMP) *5 mM in buffer*	100 µL	17.3	
V13241	Alexa Fluor® 488 Annexin V/Dead Cell Apoptosis Kit *Alexa Fluor® 488 annexin V/propidium iodide* *50 assays* *for flow cytometry*	1 kit	15.5	
A10266	Alexa Fluor® 488 azide (Alexa Fluor® 488 5-carboxamido-(6-azidohexanyl), bis(triethylammonium salt)) *5-isomer*	0.5 mg	3.1	
A10270	Alexa Fluor® 594 azide (Alexa Fluor® 594 carboxamido-(6-azidohexanyl), triethylammonium salt) *mixed isomers*	0.5 mg	3.1	
A20012	Alexa Fluor® 555 azide, triethylammonium salt	0.5 mg	3.1	
A10277	Alexa Fluor® 647 azide, triethylammonium salt	0.5 mg	3.1	
A12924	Alexa Fluor® 488 biocytin, disodium salt (biocytin Alexa Fluor® 488)	250 µg	4.3, 14.3	
A12923	Alexa Fluor® 546 biocytin, sodium salt (biocytin Alexa Fluor® 546)	250 µg	4.3, 14.3	
A12922	Alexa Fluor® 594 biocytin, sodium salt (biocytin Alexa Fluor® 594)	250 µg	4.3, 14.3	
A20344	Alexa Fluor® 680 C_2-maleimide	1 mg	2.2	
A30459	Alexa Fluor® 750 C_5-maleimide	1 mg	2.2	
A30505	Alexa Fluor® 350 C_5-maleimide	1 mg	2.3	
A10254	Alexa Fluor® 488 C_5-maleimide	1 mg	2.2	
A10255	Alexa Fluor® 532 C_5-maleimide	1 mg	2.2	
A10258	Alexa Fluor® 546 C_5-maleimide	1 mg	2.2	
A20346	Alexa Fluor® 555 C_2-maleimide	1 mg	2.2	
A20341	Alexa Fluor® 568 C_5-maleimide	1 mg	2.2	
A10256	Alexa Fluor® 594 C_5-maleimide	1 mg	2.2	
A20342	Alexa Fluor® 633 C_5-maleimide	1 mg	2.2	
A20347	Alexa Fluor® 647 C_2-maleimide	1 mg	2.2	
A20343	Alexa Fluor® 660 C_2-maleimide	1 mg	2.2	
A30674	Alexa Fluor® 350 cadaverine	1 mg	3.4	
A30675	Alexa Fluor® 405 cadaverine, trisodium salt	1 mg	3.4, 14.3	
A30676	Alexa Fluor® 488 cadaverine, sodium salt	1 mg	3.4	
A30677	Alexa Fluor® 555 cadaverine, disodium salt	1 mg	3.4	
A30680	Alexa Fluor® 568 cadaverine, diammonium salt	1 mg	3.4	
A30678	Alexa Fluor® 594 cadaverine	1 mg	3.4	
A30679	Alexa Fluor® 647 cadaverine, disodium salt	1 mg	3.4	
A30627	Alexa Fluor® 350 C_5-aminooxyacetamide, trifluoroacetate salt (Alexa Fluor® 350 hydroxylamine)	1 mg	3.3, 14.3	
A30629	Alexa Fluor® 488 C_5-aminooxyacetamide, bis(triethylammonium) salt (Alexa Fluor® 488 hydroxylamine)	1 mg	3.3, 14.3	
A30632	Alexa Fluor® 647 C_5-aminooxyacetamide, bis(triethylammonium) salt (Alexa Fluor® 647 hydroxylamine)	1 mg	3.3, 14.3	
A10168	Alexa Fluor® 350 carboxylic acid, succinimidyl ester	5 mg	1.3, 1.7	
A30000	Alexa Fluor® 405 carboxylic acid, succinimidyl ester	1 mg	1.3, 1.7, 4.2, 14.3	
A30100	Alexa Fluor® 405 carboxylic acid, succinimidyl ester	5 mg	1.3, 1.7, 4.2, 14.3	
A10169	Alexa Fluor® 430 carboxylic acid, succinimidyl ester	5 mg	1.3, 1.7	
A20000	Alexa Fluor® 488 carboxylic acid, succinimidyl ester *mixed isomers*	1 mg	1.3, 4.2	
A20100	Alexa Fluor® 488 carboxylic acid, succinimidyl ester *mixed isomers*	5 mg	1.3, 4.2	
A30002	Alexa Fluor® 514 carboxylic acid, succinimidyl ester *mixed isomers*	1 mg	1.3	
A20001	Alexa Fluor® 532 carboxylic acid, succinimidyl ester	1 mg	1.3	
A20101MP	Alexa Fluor® 532 carboxylic acid, succinimidyl ester	5 mg	1.3	
A20002	Alexa Fluor® 546 carboxylic acid, succinimidyl ester	1 mg	1.3	
A20102	Alexa Fluor® 546 carboxylic acid, succinimidyl ester	5 mg	1.3	
A20009	Alexa Fluor® 555 carboxylic acid, succinimidyl ester	1 mg	1.3	
A20109	Alexa Fluor® 555 carboxylic acid, succinimidyl ester	5 mg	1.3	
A20003	Alexa Fluor® 568 carboxylic acid, succinimidyl ester *mixed isomers*	1 mg	1.3	

*Note that all products listed in the above Master Product List are covered by **Limited Use Label License #223**. If a specific product is covered by additional Limited Use Label License(s), the Limited Use Label License #'s are provided in the LULL column above. Please refer to page 971 for the complete text of the Limited Use Label License(s).

Cat. No.	Product	Quantity	Sections	LULL*
A20103	Alexa Fluor® 568 carboxylic acid, succinimidyl ester *mixed isomers*	5 mg	1.3	
A20004	Alexa Fluor® 594 carboxylic acid, succinimidyl ester *mixed isomers*	1 mg	1.3	
A20104	Alexa Fluor® 594 carboxylic acid, succinimidyl ester *mixed isomers*	5 mg	1.3	
A20005	Alexa Fluor® 633 carboxylic acid, succinimidyl ester	1 mg	1.3	
A20105	Alexa Fluor® 633 carboxylic acid, succinimidyl ester	5 mg	1.3	
A20006	Alexa Fluor® 647 carboxylic acid, succinimidyl ester	1 mg	1.3	
A20106	Alexa Fluor® 647 carboxylic acid, succinimidyl ester	5 mg	1.3	
A20007	Alexa Fluor® 660 carboxylic acid, succinimidyl ester	1 mg	1.3	
A20008	Alexa Fluor® 680 carboxylic acid, succinimidyl ester	1 mg	1.3	
A20108	Alexa Fluor® 680 carboxylic acid, succinimidyl ester	5 mg	1.3	
A20010	Alexa Fluor® 700 carboxylic acid, succinimidyl ester	1 mg	1.3	
A20110	Alexa Fluor® 700 carboxylic acid, succinimidyl ester	5 mg	1.3	
A20011	Alexa Fluor® 750 carboxylic acid, succinimidyl ester	1 mg	1.3	
A20111	Alexa Fluor® 750 carboxylic acid, succinimidyl ester	5 mg	1.3	
A30051	Alexa Fluor® 790 carboxylic acid, succinimidyl ester, penta(triethylammonium) salt	100 µg	1.3	
A30005	Alexa Fluor® 488 carboxylic acid, 2,3,5,6-tetrafluorophenyl ester (Alexa Fluor® 488 5-TFP) *5-isomer*	1 mg	1.3, 4.2	
A21467	Alexa Fluor® 488 chicken anti-goat IgG (H+L) *2 mg/mL*	0.5 mL	7.2	
A21468	Alexa Fluor® 594 chicken anti-goat IgG (H+L) *2 mg/mL*	0.5 mL	7.2	
A21469	Alexa Fluor® 647 chicken anti-goat IgG (H+L) *2 mg/mL*	0.5 mL	7.2	
A21200	Alexa Fluor® 488 chicken anti-mouse IgG (H+L) *2 mg/mL*	0.5 mL	7.2	
A21201	Alexa Fluor® 594 chicken anti-mouse IgG (H+L) *2 mg/mL*	0.5 mL	7.2	
A21463	Alexa Fluor® 647 chicken anti-mouse IgG (H+L) *2 mg/mL*	0.5 mL	7.2	
A21441	Alexa Fluor® 488 chicken anti-rabbit IgG (H+L) *2 mg/mL*	0.5 mL	7.2	
A21442	Alexa Fluor® 594 chicken anti-rabbit IgG (H+L) *2 mg/mL*	0.5 mL	7.2	
A21443	Alexa Fluor® 647 chicken anti-rabbit IgG (H+L) *2 mg/mL*	0.5 mL	7.2	
A21470	Alexa Fluor® 488 chicken anti-rat IgG (H+L) *2 mg/mL*	0.5 mL	7.2	
A21471	Alexa Fluor® 594 chicken anti-rat IgG (H+L) *2 mg/mL*	0.5 mL	7.2	
A21472	Alexa Fluor® 647 chicken anti-rat IgG (H+L) *2 mg/mL*	0.5 mL	7.2	
A21081	Alexa Fluor® 350 donkey anti-goat IgG (H+L) *2 mg/mL*	0.5 mL	7.2	
A11055	Alexa Fluor® 488 donkey anti-goat IgG (H+L) *2 mg/mL*	0.5 mL	7.2	
A11056	Alexa Fluor® 546 donkey anti-goat IgG (H+L) *2 mg/mL*	0.5 mL	7.2	
A21432	Alexa Fluor® 555 donkey anti-goat IgG (H+L) *2 mg/mL*	0.5 mL	7.2	
A11057	Alexa Fluor® 568 donkey anti-goat IgG (H+L) *2 mg/mL*	0.5 mL	7.2	
A11058	Alexa Fluor® 594 donkey anti-goat IgG (H+L) *2 mg/mL*	0.5 mL	7.2	
A21082	Alexa Fluor® 633 donkey anti-goat IgG (H+L) *2 mg/mL*	0.5 mL	7.2	
A21447	Alexa Fluor® 647 donkey anti-goat IgG (H+L) *2 mg/mL*	0.5 mL	7.2	
A21083	Alexa Fluor® 660 donkey anti-goat IgG (H+L) *2 mg/mL*	0.5 mL	7.2	
A21084	Alexa Fluor® 680 donkey anti-goat IgG (H+L) *2 mg/mL*	0.5 mL	7.2	
A10035	Alexa Fluor® 350 donkey anti-mouse IgG (H + L) *2 mg/mL*	0.5 mL	7.2	
A21202	Alexa Fluor® 488 donkey anti-mouse IgG (H+L) *2 mg/mL*	0.5 mL	7.2	
A10036	Alexa Fluor® 546 donkey anti-mouse IgG (H + L) *2 mg/mL*	0.5 mL	7.2	
A31570	Alexa Fluor® 555 donkey anti-mouse IgG (H+L) *2 mg/mL*	0.5 mL	7.2	
A10037	Alexa Fluor® 568 donkey anti-mouse IgG (H + L) *2 mg/mL*	0.5 mL	7.2	
A21203	Alexa Fluor® 594 donkey anti-mouse IgG (H+L) *2 mg/mL*	0.5 mL	7.2	
A31571	Alexa Fluor® 647 donkey anti-mouse IgG (H+L) *2 mg/mL*	0.5 mL	7.2	
A10038	Alexa Fluor® 680 donkey anti-mouse IgG (H + L) *2 mg/mL*	0.5 mL	7.2	
A10039	Alexa Fluor® 350 donkey anti-rabbit IgG (H + L) *2 mg/mL*	0.5 mL	7.2	
A21206	Alexa Fluor® 488 donkey anti-rabbit IgG (H+L) *2 mg/mL*	0.5 mL	7.2	
A10040	Alexa Fluor® 546 donkey anti-rabbit IgG (H + L) *2 mg/mL*	0.5 mL	7.2	
A31572	Alexa Fluor® 555 donkey anti-rabbit IgG (H+L) *2 mg/mL*	0.5 mL	7.2	
A10042	Alexa Fluor® 568 donkey anti-rabbit IgG (H + L) *2 mg/mL*	0.5 mL	7.2	
A21207	Alexa Fluor® 594 donkey anti-rabbit IgG (H+L) *2 mg/mL*	0.5 mL	7.2	
A31573	Alexa Fluor® 647 donkey anti-rabbit IgG (H+L) *2 mg/mL*	0.5 mL	7.2	
A10043	Alexa Fluor® 680 donkey anti-rabbit IgG (H + L) *2 mg/mL*	0.5 mL	7.2	
A21208	Alexa Fluor® 488 donkey anti-rat IgG (H+L) *2 mg/mL*	0.5 mL	7.2	
A21209	Alexa Fluor® 594 donkey anti-rat IgG (H+L) *2 mg/mL*	0.5 mL	7.2	
A21097	Alexa Fluor® 350 donkey anti-sheep IgG (H+L) *2 mg/mL*	0.5 mL	7.2	
A11015	Alexa Fluor® 488 donkey anti-sheep IgG (H+L) *2 mg/mL*	0.5 mL	7.2	
A21098	Alexa Fluor® 546 donkey anti-sheep IgG (H+L) *2 mg/mL*	0.5 mL	7.2	

*Note that all products listed in the above Master Product List are covered by **Limited Use Label License #223**. If a specific product is covered by additional Limited Use Label License(s), the Limited Use Label License #'s are provided in the LULL column above. Please refer to page 971 for the complete text of the Limited Use Label License(s).

IMPORTANT NOTICE: The products described in this manual are covered by one or more Limited Use Label License(s). Please refer to the Appendix on page 971 and Master Product List on page 975. Products are For Research Use Only. Not intended for any animal or human therapeutic or diagnostic use.

Cat. No.	Product	Quantity	Sections	LULL*
A21436	Alexa Fluor® 555 donkey anti-sheep IgG (H+L) *2 mg/mL*	0.5 mL	7.2	
A21099	Alexa Fluor® 568 donkey anti-sheep IgG (H+L) *2 mg/mL*	0.5 mL	7.2	
A11016	Alexa Fluor® 594 donkey anti-sheep IgG (H+L) *2 mg/mL*	0.5 mL	7.2	
A21100	Alexa Fluor® 633 donkey anti-sheep IgG (H+L) *2 mg/mL*	0.5 mL	7.2	
A21448	Alexa Fluor® 647 donkey anti-sheep IgG (H+L) *2 mg/mL*	0.5 mL	7.2	
A21102	Alexa Fluor® 680 donkey anti-sheep IgG (H+L) *2 mg/mL*	0.5 mL	7.2	
A10684	Alexa Fluor® 488 F(ab')$_2$ fragment of goat anti-mouse IgG, IgM (H+L) *2 mg/mL*	250 µL	7.2	
A11068	Alexa Fluor® 350 F(ab')$_2$ fragment of goat anti-mouse IgG (H+L) *2 mg/mL*	250 µL	7.2	
A11017	Alexa Fluor® 488 F(ab')$_2$ fragment of goat anti-mouse IgG (H+L) *2 mg/mL*	250 µL	7.2	
A11018	Alexa Fluor® 546 F(ab')$_2$ fragment of goat anti-mouse IgG (H+L) *2 mg/mL*	250 µL	7.2	
A21425	Alexa Fluor® 555 F(ab')$_2$ fragment of goat anti-mouse IgG (H+L) *2 mg/mL*	250 µL	7.2	
A11019	Alexa Fluor® 568 F(ab')$_2$ fragment of goat anti-mouse IgG (H+L) *2 mg/mL*	250 µL	7.2	
A11020	Alexa Fluor® 594 F(ab')$_2$ fragment of goat anti-mouse IgG (H+L) *2 mg/mL*	250 µL	7.2	
A21053	Alexa Fluor® 633 F(ab')$_2$ fragment of goat anti-mouse IgG (H+L) *2 mg/mL*	250 µL	7.2	
A21237	Alexa Fluor® 647 F(ab')$_2$ fragment of goat anti-mouse IgG (H+L) *2 mg/mL*	250 µL	7.2	
A21059	Alexa Fluor® 680 F(ab')$_2$ fragment of goat anti-mouse IgG (H+L) *2 mg/mL*	250 µL	7.2	
A11069	Alexa Fluor® 350 F(ab')$_2$ fragment of goat anti-rabbit IgG (H+L) *2 mg/mL*	250 µL	7.2	
A11070	Alexa Fluor® 488 F(ab')$_2$ fragment of goat anti-rabbit IgG (H+L) *2 mg/mL*	250 µL	7.2	
A11071	Alexa Fluor® 546 F(ab')$_2$ fragment of goat anti-rabbit IgG (H+L) *2 mg/mL*	250 µL	7.2	
A21430	Alexa Fluor® 555 F(ab')$_2$ fragment of goat anti-rabbit IgG (H+L) *2 mg/mL*	250 µL	7.2	
A21069	Alexa Fluor® 568 F(ab')$_2$ fragment of goat anti-rabbit IgG (H+L) *2 mg/mL*	250 µL	7.2	
A11072	Alexa Fluor® 594 F(ab')$_2$ fragment of goat anti-rabbit IgG (H+L) *2 mg/mL*	250 µL	7.2	
A21072	Alexa Fluor® 633 F(ab')$_2$ fragment of goat anti-rabbit IgG (H+L) *2 mg/mL*	250 µL	7.2	
A21246	Alexa Fluor® 647 F(ab')$_2$ fragment of goat anti-rabbit IgG (H+L) *2 mg/mL*	250 µL	7.2	
A21077	Alexa Fluor® 680 F(ab')$_2$ fragment of goat anti-rabbit IgG (H+L) *2 mg/mL*	250 µL	7.2	
A21222	Alexa Fluor® 488 F(ab')$_2$ fragment of rabbit anti-goat IgG (H+L) *2 mg/mL*	250 µL	7.2	
A21223	Alexa Fluor® 594 F(ab')$_2$ fragment of rabbit anti-goat IgG (H+L) *2 mg/mL*	250 µL	7.2	
A21204	Alexa Fluor® 488 F(ab')$_2$ fragment of rabbit anti-mouse IgG (H+L) *2 mg/mL*	250 µL	7.2	
A21205	Alexa Fluor® 594 F(ab')$_2$ fragment of rabbit anti-mouse IgG (H+L) *2 mg/mL*	250 µL	7.2	
A24920	Alexa Fluor® 488 FluoroNanogold™ Fab' fragment of goat anti-mouse IgG *80 µg protein/mL*	1 mL	7.2	231
A24921	Alexa Fluor® 594 FluoroNanogold™ Fab' fragment of goat anti-mouse IgG *80 µg protein/mL*	1 mL	7.2	231
A24922	Alexa Fluor® 488 FluoroNanogold™ Fab' fragment of goat anti-rabbit IgG *80 µg protein/mL*	1 mL	7.2	231
A24923	Alexa Fluor® 594 FluoroNanogold™ Fab' fragment of goat anti-rabbit IgG *80 µg protein/mL*	1 mL	7.2	231
A24926	Alexa Fluor® 488 FluoroNanogold™ streptavidin *80 µg protein/mL*	1 mL	7.6	231
A24927	Alexa Fluor® 594 FluoroNanogold™ streptavidin *80 µg protein/mL*	1 mL	7.6	231
A11039	Alexa Fluor® 488 goat anti-chicken IgG (H+L) *2 mg/mL*	0.5 mL	7.2	
A11040	Alexa Fluor® 546 goat anti-chicken IgG (H+L) *2 mg/mL*	0.5 mL	7.2	
A21437	Alexa Fluor® 555 goat anti-chicken IgG (H+L) *2 mg/mL*	0.5 mL	7.2	
A11041	Alexa Fluor® 568 goat anti-chicken IgG (H+L) *2 mg/mL*	0.5 mL	7.2	
A11042	Alexa Fluor® 594 goat anti-chicken IgG (H+L) *2 mg/mL*	0.5 mL	7.2	
A21103	Alexa Fluor® 633 goat anti-chicken IgG (H+L) *2 mg/mL*	0.5 mL	7.2	
A21449	Alexa Fluor® 647 goat anti-chicken IgG (H+L) *2 mg/mL*	0.5 mL	7.2	
A11073	Alexa Fluor® 488 goat anti-guinea pig IgG (H+L) *highly cross-adsorbed* *2 mg/mL*	0.5 mL	7.2	
A11074	Alexa Fluor® 546 goat anti-guinea pig IgG (H+L) *highly cross-adsorbed* *2 mg/mL*	0.5 mL	7.2	
A21435	Alexa Fluor® 555 goat anti-guinea pig IgG (H+L) *highly cross-adsorbed* *2 mg/mL*	0.5 mL	7.2	
A11075	Alexa Fluor® 568 goat anti-guinea pig IgG (H+L) *highly cross-adsorbed* *2 mg/mL*	0.5 mL	7.2	
A11076	Alexa Fluor® 594 goat anti-guinea pig IgG (H+L) *highly cross-adsorbed* *2 mg/mL*	0.5 mL	7.2	
A21105	Alexa Fluor® 633 goat anti-guinea pig IgG (H+L) *highly cross-adsorbed* *2 mg/mL*	0.5 mL	7.2	
A21450	Alexa Fluor® 647 goat anti-guinea pig IgG (H+L) *highly cross-adsorbed* *2 mg/mL*	0.5 mL	7.2	
A21110	Alexa Fluor® 488 goat anti-hamster IgG (H+L) *2 mg/mL*	0.5 mL	7.2	
A21111	Alexa Fluor® 546 goat anti-hamster IgG (H+L) *2 mg/mL*	0.5 mL	7.2	
A21112	Alexa Fluor® 568 goat anti-hamster IgG (H+L) *2 mg/mL*	0.5 mL	7.2	
A21113	Alexa Fluor® 594 goat anti-hamster IgG (H+L) *2 mg/mL*	0.5 mL	7.2	
A21451	Alexa Fluor® 647 goat anti-hamster IgG (H+L) *2 mg/mL*	0.5 mL	7.2	
A11013	Alexa Fluor® 488 goat anti-human IgG (H+L) *2 mg/mL*	0.5 mL	7.2	
A21089	Alexa Fluor® 546 goat anti-human IgG (H+L) *2 mg/mL*	0.5 mL	7.2	
A21433	Alexa Fluor® 555 goat anti-human IgG (H+L) *2 mg/mL*	0.5 mL	7.2	
A21090	Alexa Fluor® 568 goat anti-human IgG (H+L) *2 mg/mL*	0.5 mL	7.2	
A11014	Alexa Fluor® 594 goat anti-human IgG (H+L) *2 mg/mL*	0.5 mL	7.2	

*Note that all products listed in the above Master Product List are covered by **Limited Use Label License #223**. If a specific product is covered by additional Limited Use Label License(s), the Limited Use Label License #'s are provided in the LULL column above. Please refer to page 971 for the complete text of the Limited Use Label License(s).

IMPORTANT NOTICE: The products described in this manual are covered by one or more Limited Use Label License(s). Please refer to the Appendix on page 971 and Master Product List on page 975. Products are For Research Use Only. Not intended for any animal or human therapeutic or diagnostic use.
www.invitrogen.com/probes

Cat. No.	Product	Quantity	Sections	LULL*
A21091	Alexa Fluor® 633 goat anti-human IgG (H+L) *2 mg/mL*	0.5 mL	7.2	
A21445	Alexa Fluor® 647 goat anti-human IgG (H+L) *2 mg/mL*	0.5 mL	7.2	
A21215	Alexa Fluor® 488 goat anti-human IgM (μ chain) *2 mg/mL*	250 μL	7.2	
A21216	Alexa Fluor® 594 goat anti-human IgM (μ chain) *2 mg/mL*	250 μL	7.2	
A21249	Alexa Fluor® 647 goat anti-human IgM (μ chain) *2 mg/mL*	250 μL	7.2	
A31562	Alexa Fluor® 680 goat anti-mouse IgG_1 (γ1) *2 mg/mL*	250 μL	7.2	
A31560	Alexa Fluor® 488 goat anti-mouse IgG, 5 nm colloidal gold conjugate *30 μg protein/mL*	0.5 mL	7.2	
A31561	Alexa Fluor® 488 goat anti-mouse IgG, 10 nm colloidal gold conjugate *30 μg protein/mL*	0.5 mL	7.2	
A21120	Alexa Fluor® 350 goat anti-mouse IgG_1 (γ1) *2 mg/mL*	250 μL	7.2	
A21121	Alexa Fluor® 488 goat anti-mouse IgG_1 (γ1) *2 mg/mL*	250 μL	7.2	
A21123	Alexa Fluor® 546 goat anti-mouse IgG_1 (γ1) *2 mg/mL*	250 μL	7.2	
A21127	Alexa Fluor® 555 goat anti-mouse IgG_1 (γ1) *2 mg/mL*	250 μL	7.2	
A21124	Alexa Fluor® 568 goat anti-mouse IgG_1 (γ1) *2 mg/mL*	250 μL	7.2	
A21125	Alexa Fluor® 594 goat anti-mouse IgG_1 (γ1) *2 mg/mL*	250 μL	7.2	
A21126	Alexa Fluor® 633 goat anti-mouse IgG_1 (γ1) *2 mg/mL*	250 μL	7.2	
A21240	Alexa Fluor® 647 goat anti-mouse IgG_1 (γ1) *2 mg/mL*	250 μL	7.2	
A21130	Alexa Fluor® 350 goat anti-mouse IgG_{2a} (γ2a) *2 mg/mL*	250 μL	7.2	
A21131	Alexa Fluor® 488 goat anti-mouse IgG_{2a} (γ2a) *2 mg/mL*	250 μL	7.2	
A21133	Alexa Fluor® 546 goat anti-mouse IgG_{2a} (γ2a) *2 mg/mL*	250 μL	7.2	
A21137	Alexa Fluor® 555 goat anti-mouse IgG_{2a} (γ2a) *2 mg/mL*	250 μL	7.2	
A21134	Alexa Fluor® 568 goat anti-mouse IgG_{2a} (γ2a) *2 mg/mL*	250 μL	7.2	
A21135	Alexa Fluor® 594 goat anti-mouse IgG_{2a} (γ2a) *2 mg/mL*	250 μL	7.2	
A21136	Alexa Fluor® 633 goat anti-mouse IgG_{2a} (γ2a) *2 mg/mL*	250 μL	7.2	
A21241	Alexa Fluor® 647 goat anti-mouse IgG_{2a} (γ2a) *2 mg/mL*	250 μL	7.2	
A21140	Alexa Fluor® 350 goat anti-mouse IgG_{2b} (γ2b) *2 mg/mL*	250 μL	7.2	
A21141	Alexa Fluor® 488 goat anti-mouse IgG_{2b} (γ2b) *2 mg/mL*	250 μL	7.2	
A21143	Alexa Fluor® 546 goat anti-mouse IgG_{2b} (γ2b) *2 mg/mL*	250 μL	7.2	
A21147	Alexa Fluor® 555 goat anti-mouse IgG_{2b} (γ2b) *2 mg/mL*	250 μL	7.2	
A21144	Alexa Fluor® 568 goat anti-mouse IgG_{2b} (γ2b) *2 mg/mL*	250 μL	7.2	
A21145	Alexa Fluor® 594 goat anti-mouse IgG_{2b} (γ2b) *2 mg/mL*	250 μL	7.2	
A21146	Alexa Fluor® 633 goat anti-mouse IgG_{2b} (γ2b) *2 mg/mL*	250 μL	7.2	
A21242	Alexa Fluor® 647 goat anti-mouse IgG_{2b} (γ2b) *2 mg/mL*	250 μL	7.2	
A21151	Alexa Fluor® 488 goat anti-mouse IgG_3 (γ3) *2 mg/mL*	250 μL	7.2	
A21155	Alexa Fluor® 594 goat anti-mouse IgG_3 (γ3) *2 mg/mL*	250 μL	7.2	
A31563	Alexa Fluor® 680 goat anti-mouse IgG_{2a} (γ2a) *2 mg/mL*	250 μL	7.2	
A31564	Alexa Fluor® 680 goat anti-mouse IgG_{2b} (γ2b) *2 mg/mL*	250 μL	7.2	
A11045	Alexa Fluor® 350 goat anti-mouse IgG (H+L) *2 mg/mL*	0.5 mL	7.2	
A31553	Alexa Fluor® 405 goat anti-mouse IgG (H+L) *2 mg/mL*	0.5 mL	7.2	
A11063	Alexa Fluor® 430 goat anti-mouse IgG (H+L) *2 mg/mL*	0.5 mL	7.2	
A11001	Alexa Fluor® 488 goat anti-mouse IgG (H+L) *2 mg/mL*	0.5 mL	7.2	
A31555	Alexa Fluor® 514 goat anti-mouse IgG (H+L) *2 mg/mL*	0.5 mL	7.2	
A11002	Alexa Fluor® 532 goat anti-mouse IgG (H+L) *2 mg/mL*	0.5 mL	7.2	
A11003	Alexa Fluor® 546 goat anti-mouse IgG (H+L) *2 mg/mL*	0.5 mL	7.2	
A21422	Alexa Fluor® 555 goat anti-mouse IgG (H+L) *2 mg/mL*	0.5 mL	7.2	
A11004	Alexa Fluor® 568 goat anti-mouse IgG (H+L) *2 mg/mL*	0.5 mL	7.2	
A11005	Alexa Fluor® 594 goat anti-mouse IgG (H+L) *2 mg/mL*	0.5 mL	7.2	
A21050	Alexa Fluor® 633 goat anti-mouse IgG (H+L) *2 mg/mL*	0.5 mL	7.2	
A31574	Alexa Fluor® 635 goat anti-mouse IgG (H+L) *2 mg/mL*	0.5 mL	7.2	
A21235	Alexa Fluor® 647 goat anti-mouse IgG (H+L) *2 mg/mL*	0.5 mL	7.2	
A21054	Alexa Fluor® 660 goat anti-mouse IgG (H+L) *2 mg/mL*	0.5 mL	7.2	
A21057	Alexa Fluor® 680 goat anti-mouse IgG (H+L) *2 mg/mL*	0.5 mL	7.2	
A21036	Alexa Fluor® 700 goat anti-mouse IgG (H+L) *2 mg/mL*	0.5 mL	7.2	
A21037	Alexa Fluor® 750 goat anti-mouse IgG (H+L) *2 mg/mL*	0.5 mL	7.2	
A21049	Alexa Fluor® 350 goat anti-mouse IgG (H+L) *highly cross-adsorbed* *2 mg/mL*	0.5 mL	7.2	
A11029	Alexa Fluor® 488 goat anti-mouse IgG (H+L) *highly cross-adsorbed* *2 mg/mL*	0.5 mL	7.2	
A11030	Alexa Fluor® 546 goat anti-mouse IgG (H+L) *highly cross-adsorbed* *2 mg/mL*	0.5 mL	7.2	
A21424	Alexa Fluor® 555 goat anti-mouse IgG (H+L) *highly cross-adsorbed* *2 mg/mL*	0.5 mL	7.2	
A11031	Alexa Fluor® 568 goat anti-mouse IgG (H+L) *highly cross-adsorbed* *2 mg/mL*	0.5 mL	7.2	
A11032	Alexa Fluor® 594 goat anti-mouse IgG (H+L) *highly cross-adsorbed* *2 mg/mL*	0.5 mL	7.2	

*Note that all products listed in the above Master Product List are covered by **Limited Use Label License #223**. If a specific product is covered by additional Limited Use Label License(s), the Limited Use Label License #'s are provided in the LULL column above. Please refer to page 971 for the complete text of the Limited Use Label License(s).

IMPORTANT NOTICE: The products described in this manual are covered by one or more Limited Use Label License(s). Please refer to the Appendix on page 971 and Master Product List on page 975. Products are For Research Use Only. Not intended for any animal or human therapeutic or diagnostic use.

Cat. No.	Product	Quantity	Sections	LULL*
A21052	Alexa Fluor® 633 goat anti-mouse IgG (H+L) *highly cross-adsorbed* *2 mg/mL*	0.5 mL	7.2	
A31575	Alexa Fluor® 635 goat anti-mouse IgG (H+L) *highly cross-adsorbed* *2 mg/mL*	0.5 mL	7.2	
A21236	Alexa Fluor® 647 goat anti-mouse IgG (H+L) *highly cross-adsorbed* *2 mg/mL*	0.5 mL	7.2	
A21055	Alexa Fluor® 660 goat anti-mouse IgG (H+L) *highly cross-adsorbed* *2 mg/mL*	0.5 mL	7.2	
A21058	Alexa Fluor® 680 goat anti-mouse IgG (H+L) *highly cross-adsorbed* *2 mg/mL*	0.5 mL	7.2	
A10667	Alexa Fluor® 488 goat anti-mouse IgG, IgA, IgM (H+L) *2 mg/mL*	250 µL	7.2	
A10680	Alexa Fluor® 488 goat anti-mouse IgG, IgM (H+L) *2 mg/mL*	250 µL	7.2	
A31552	Alexa Fluor® 350 goat anti-mouse IgM (µ chain) *2 mg/mL*	250 µL	7.2	
A21042	Alexa Fluor® 488 goat anti-mouse IgM (µ chain) *2 mg/mL*	250 µL	7.2	
A21045	Alexa Fluor® 546 goat anti-mouse IgM (µ chain) *2 mg/mL*	250 µL	7.2	
A21426	Alexa Fluor® 555 goat anti-mouse IgM (µ chain) *2 mg/mL*	250 µL	7.2	
A21043	Alexa Fluor® 568 goat anti-mouse IgM (µ chain) *2 mg/mL*	250 µL	7.2	
A21044	Alexa Fluor® 594 goat anti-mouse IgM (µ chain) *2 mg/mL*	250 µL	7.2	
A21046	Alexa Fluor® 633 goat anti-mouse IgM (µ chain) *2 mg/mL*	250 µL	7.2	
A21238	Alexa Fluor® 647 goat anti-mouse IgM (µ chain) *2 mg/mL*	250 µL	7.2	
A21048	Alexa Fluor® 680 goat anti-mouse IgM (µ chain) *2 mg/mL*	250 µL	7.2	
A31619	Alexa Fluor® 488 Goat Anti-Mouse SFX Kit	1 kit	7.2	
A31620	Alexa Fluor® 488 Goat Anti-Mouse SFX Kit *highly cross-adsorbed*	1 kit	7.2	
A31621	Alexa Fluor® 555 Goat Anti-Mouse SFX Kit	1 kit	7.2	
A31622	Alexa Fluor® 555 Goat Anti-Mouse SFX Kit *highly cross-adsorbed*	1 kit	7.2	
A31623	Alexa Fluor® 594 Goat Anti-Mouse SFX Kit	1 kit	7.2	
A31624	Alexa Fluor® 594 Goat Anti-Mouse SFX Kit *highly cross-adsorbed*	1 kit	7.2	
A31625	Alexa Fluor® 647 Goat Anti-Mouse SFX Kit	1 kit	7.2	
A31626	Alexa Fluor® 647 Goat Anti-Mouse SFX Kit *highly cross-adsorbed*	1 kit	7.2	
A31565	Alexa Fluor® 488 goat anti-rabbit IgG, 5 nm colloidal gold conjugate *30 µg protein/mL*	0.5 mL	7.2	
A31566	Alexa Fluor® 488 goat anti-rabbit IgG, 10 nm colloidal gold conjugate *30 µg protein/mL*	0.5 mL	7.2	
A11046	Alexa Fluor® 350 goat anti-rabbit IgG (H+L) *2 mg/mL*	0.5 mL	7.2	
A31556	Alexa Fluor® 405 goat anti-rabbit IgG (H+L) *2 mg/mL*	0.5 mL	7.2	
A11064	Alexa Fluor® 430 goat anti-rabbit IgG (H+L) *2 mg/mL*	0.5 mL	7.2	
A11008	Alexa Fluor® 488 goat anti-rabbit IgG (H+L) *2 mg/mL*	0.5 mL	7.2	
A31558	Alexa Fluor® 514 goat anti-rabbit IgG (H+L) *2 mg/mL*	0.5 mL	7.2	
A11009	Alexa Fluor® 532 goat anti-rabbit IgG (H+L) *2 mg/mL*	0.5 mL	7.2	
A11010	Alexa Fluor® 546 goat anti-rabbit IgG (H+L) *2 mg/mL*	0.5 mL	7.2	
A21428	Alexa Fluor® 555 goat anti-rabbit IgG (H+L) *2 mg/mL*	0.5 mL	7.2	
A11011	Alexa Fluor® 568 goat anti-rabbit IgG (H+L) *2 mg/mL*	0.5 mL	7.2	
A11012	Alexa Fluor® 594 goat anti-rabbit IgG (H+L) *2 mg/mL*	0.5 mL	7.2	
A21070	Alexa Fluor® 633 goat anti-rabbit IgG (H+L) *2 mg/mL*	0.5 mL	7.2	
A31576	Alexa Fluor® 635 goat anti-rabbit IgG (H+L) *2 mg/mL*	0.5 mL	7.2	
A21244	Alexa Fluor® 647 goat anti-rabbit IgG (H+L) *2 mg/mL*	0.5 mL	7.2	
A21073	Alexa Fluor® 660 goat anti-rabbit IgG (H+L) *2 mg/mL*	0.5 mL	7.2	
A21076	Alexa Fluor® 680 goat anti-rabbit IgG (H+L) *2 mg/mL*	0.5 mL	7.2	
A21038	Alexa Fluor® 700 goat anti-rabbit IgG (H+L) *2 mg/mL*	0.5 mL	7.2	
A21039	Alexa Fluor® 750 goat anti-rabbit IgG (H+L) *2 mg/mL*	0.5 mL	7.2	
A21068	Alexa Fluor® 350 goat anti-rabbit IgG (H+L) *highly cross-adsorbed* *2 mg/mL*	0.5 mL	7.2	
A11034	Alexa Fluor® 488 goat anti-rabbit IgG (H+L) *highly cross-adsorbed* *2 mg/mL*	0.5 mL	7.2	
A11035	Alexa Fluor® 546 goat anti-rabbit IgG (H+L) *highly cross-adsorbed* *2 mg/mL*	0.5 mL	7.2	
A21429	Alexa Fluor® 555 goat anti-rabbit IgG (H+L) *highly cross-adsorbed* *2 mg/mL*	0.5 mL	7.2	
A11036	Alexa Fluor® 568 goat anti-rabbit IgG (H+L) *highly cross-adsorbed* *2 mg/mL*	0.5 mL	7.2	
A11037	Alexa Fluor® 594 goat anti-rabbit IgG (H+L) *highly cross-adsorbed* *2 mg/mL*	0.5 mL	7.2	
A21071	Alexa Fluor® 633 goat anti-rabbit IgG (H+L) *highly cross-adsorbed* *2 mg/mL*	0.5 mL	7.2	
A31577	Alexa Fluor® 635 goat anti-rabbit IgG (H+L) *highly cross-adsorbed* *2 mg/mL*	0.5 mL	7.2	
A21245	Alexa Fluor® 647 goat anti-rabbit IgG (H+L) *highly cross-adsorbed* *2 mg/mL*	0.5 mL	7.2	
A21074	Alexa Fluor® 660 goat anti-rabbit IgG (H+L) *highly cross-adsorbed* *2 mg/mL*	0.5 mL	7.2	
A21109	Alexa Fluor® 680 goat anti-rabbit IgG (H+L) *highly cross-adsorbed* *2 mg/mL*	0.5 mL	7.2	
A31627	Alexa Fluor® 488 Goat Anti-Rabbit SFX Kit	1 kit	7.2	
A31628	Alexa Fluor® 488 Goat Anti-Rabbit SFX Kit *highly cross-adsorbed*	1 kit	7.2	
A31629	Alexa Fluor® 555 Goat Anti-Rabbit SFX Kit	1 kit	7.2	
A31630	Alexa Fluor® 555 Goat Anti-Rabbit SFX Kit *highly cross-adsorbed*	1 kit	7.2	
A31631	Alexa Fluor® 594 Goat Anti-Rabbit SFX Kit	1 kit	7.2	

*Note that all products listed in the above Master Product List are covered by **Limited Use Label License #223**. If a specific product is covered by additional Limited Use Label License(s), the Limited Use Label License #'s are provided in the LULL column above. Please refer to page 971 for the complete text of the Limited Use Label License(s).

Cat. No.	Product	Quantity	Sections	LULL*
A31632	Alexa Fluor® 594 Goat Anti-Rabbit SFX Kit *highly cross-adsorbed*	1 kit	7.2	
A31633	Alexa Fluor® 647 Goat Anti-Rabbit SFX Kit	1 kit	7.2	
A31634	Alexa Fluor® 647 Goat Anti-Rabbit SFX Kit *highly cross-adsorbed*	1 kit	7.2	
A21093	Alexa Fluor® 350 goat anti-rat IgG (H+L) *2 mg/mL*	0.5 mL	7.2	
A11006	Alexa Fluor® 488 goat anti-rat IgG (H+L) *2 mg/mL*	0.5 mL	7.2	
A11081	Alexa Fluor® 546 goat anti-rat IgG (H+L) *2 mg/mL*	0.5 mL	7.2	
A21434	Alexa Fluor® 555 goat anti-rat IgG (H+L) *2 mg/mL*	0.5 mL	7.2	
A11077	Alexa Fluor® 568 goat anti-rat IgG (H+L) *2 mg/mL*	0.5 mL	7.2	
A11007	Alexa Fluor® 594 goat anti-rat IgG (H+L) *2 mg/mL*	0.5 mL	7.2	
A21094	Alexa Fluor® 633 goat anti-rat IgG (H+L) *2 mg/mL*	0.5 mL	7.2	
A21247	Alexa Fluor® 647 goat anti-rat IgG (H+L) *2 mg/mL*	0.5 mL	7.2	
A21096	Alexa Fluor® 680 goat anti-rat IgG (H+L) *2 mg/mL*	0.5 mL	7.2	
A21212	Alexa Fluor® 488 goat anti-rat IgM (μ chain) *2 mg/mL*	250 μL	7.2	
A21213	Alexa Fluor® 594 goat anti-rat IgM (μ chain) *2 mg/mL*	250 μL	7.2	
A21248	Alexa Fluor® 647 goat anti-rat IgM (μ chain) *2 mg/mL*	250 μL	7.2	
A10439	Alexa Fluor® 350 hydrazide, sodium salt	5 mg	3.3, 14.3	
A10436	Alexa Fluor® 488 hydrazide, sodium salt	1 mg	3.3, 14.3	
A20501MP	Alexa Fluor® 555 hydrazide, tris(triethylammonium) salt	1 mg	3.3, 14.3	
A10437	Alexa Fluor® 568 hydrazide, sodium salt	1 mg	3.3, 14.3	
A10438	Alexa Fluor® 594 hydrazide, sodium salt	1 mg	3.3, 14.3	
A30634	Alexa Fluor® 633 hydrazide, bis(triethylammonium) salt	1 mg	3.3, 14.3	
A20502	Alexa Fluor® 647 hydrazide, tris(triethylammonium) salt	1 mg	3.3, 14.3	
A10441	Alexa Fluor® 568 hydrazide, sodium salt *for microinjection* *10 mM in 200 mM KCl*	125 μL	14.3	
A10442	Alexa Fluor® 594 hydrazide, sodium salt *for microinjection* *10 mM in 200 mM KCl*	125 μL	14.3	
A30006	Alexa Fluor® 488 Microscale Protein Labeling Kit *for 20–100 μg protein* *3 labelings*	1 kit	1.2, 1.3	
A30007	Alexa Fluor® 555 Microscale Protein Labeling Kit *for 20–100 μg protein* *3 labelings*	1 kit	1.2, 1.3	
A30008	Alexa Fluor® 594 Microscale Protein Labeling Kit *for 20–100 μg protein* *3 labelings*	1 kit	1.2, 1.3	
A30009	Alexa Fluor® 647 Microscale Protein Labeling Kit *for 20–100 μg protein* *3 labelings*	1 kit	1.2, 1.3	
A20180	Alexa Fluor® 350 Monoclonal Antibody Labeling Kit *5 labelings*	1 kit	1.2, 1.3, 1.7	
A20181	Alexa Fluor® 488 Monoclonal Antibody Labeling Kit *5 labelings*	1 kit	1.2, 1.3	
A20182	Alexa Fluor® 532 Monoclonal Antibody Labeling Kit *5 labelings*	1 kit	1.2, 1.3	
A20183	Alexa Fluor® 546 Monoclonal Antibody Labeling Kit *5 labelings*	1 kit	1.2, 1.3	
A20187	Alexa Fluor® 555 Monoclonal Antibody Labeling Kit *5 labelings*	1 kit	1.2, 1.3	
A20184	Alexa Fluor® 568 Monoclonal Antibody Labeling Kit *5 labelings*	1 kit	1.2, 1.3	
A20185	Alexa Fluor® 594 Monoclonal Antibody Labeling Kit *5 labelings*	1 kit	1.2, 1.3	
A20186	Alexa Fluor® 647 Monoclonal Antibody Labeling Kit *5 labelings*	1 kit	1.2, 1.3	
A10631	Alexa Fluor® 488 mouse anti-human IgG_1 *1 mg/mL*	250 μL	7.2	
A20191	Alexa Fluor® 488 Oligonucleotide Amine Labeling Kit *3 labelings*	1 kit	1.2, 1.3, 8.2	
A20196	Alexa Fluor® 647 Oligonucleotide Amine Labeling Kit *3 labelings*	1 kit	1.2, 1.3, 8.2	
A34054	Alexa Fluor® 635 phalloidin	300 U	11.1	
A34055	Alexa Fluor® 555 phalloidin	300 U	11.1	
A22281	Alexa Fluor® 350 phalloidin	300 U	11.1	
A12379	Alexa Fluor® 488 phalloidin	300 U	11.1	
A22282	Alexa Fluor® 532 phalloidin	300 U	11.1	
A22283	Alexa Fluor® 546 phalloidin	300 U	11.1	
A12380	Alexa Fluor® 568 phalloidin	300 U	11.1	
A12381	Alexa Fluor® 594 phalloidin	300 U	11.1	
A22284	Alexa Fluor® 633 phalloidin	300 U	11.1	
A22287	Alexa Fluor® 647 phalloidin	300 U	11.1	
A22285	Alexa Fluor® 660 phalloidin	300 U	11.1	
A22286	Alexa Fluor® 680 phalloidin	300 U	11.1	
A10170	Alexa Fluor® 350 Protein Labeling Kit *3 labelings*	1 kit	1.2, 1.3, 1.7	
A10171	Alexa Fluor® 430 Protein Labeling Kit *3 labelings*	1 kit	1.2, 1.3, 1.7	
A10235	Alexa Fluor® 488 Protein Labeling Kit *3 labelings*	1 kit	1.2, 1.3	
A10236	Alexa Fluor® 532 Protein Labeling Kit *3 labelings*	1 kit	1.2, 1.3	
A10237	Alexa Fluor® 546 Protein Labeling Kit *3 labelings*	1 kit	1.2, 1.3	
A20174	Alexa Fluor® 555 Protein Labeling Kit *3 labelings*	1 kit	1.2, 1.3	
A10238	Alexa Fluor® 568 Protein Labeling Kit *3 labelings*	1 kit	1.2, 1.3	
A10239	Alexa Fluor® 594 Protein Labeling Kit *3 labelings*	1 kit	1.2, 1.3	

*Note that all products listed in the above Master Product List are covered by **Limited Use Label License #223**. If a specific product is covered by additional Limited Use Label License(s), the Limited Use Label License #'s are provided in the LULL column above. Please refer to page 971 for the complete text of the Limited Use Label License(s).

IMPORTANT NOTICE: The products described in this manual are covered by one or more Limited Use Label License(s). Please refer to the Appendix on page 971 and Master Product List on page 975. Products are For Research Use Only. Not intended for any animal or human therapeutic or diagnostic use.

Cat. No.	Product	Quantity	Sections	LULL*
A20170	Alexa Fluor® 633 Protein Labeling Kit *3 labelings*	1 kit	1.2, 1.3	
A20173	Alexa Fluor® 647 Protein Labeling Kit *3 labelings*	1 kit	1.2, 1.3	
A20171	Alexa Fluor® 660 Protein Labeling Kit *3 labelings*	1 kit	1.2, 1.3	
A20172	Alexa Fluor® 680 Protein Labeling Kit *3 labelings*	1 kit	1.2, 1.3	
A11078	Alexa Fluor® 488 rabbit anti-goat IgG (H+L) *2 mg/mL*	0.5 mL	7.2	
A21085	Alexa Fluor® 546 rabbit anti-goat IgG (H+L) *2 mg/mL*	0.5 mL	7.2	
A21431	Alexa Fluor® 555 rabbit anti-goat IgG (H+L) *2 mg/mL*	0.5 mL	7.2	
A11079	Alexa Fluor® 568 rabbit anti-goat IgG (H+L) *2 mg/mL*	0.5 mL	7.2	
A11080	Alexa Fluor® 594 rabbit anti-goat IgG (H+L) *2 mg/mL*	0.5 mL	7.2	
A21086	Alexa Fluor® 633 rabbit anti-goat IgG (H+L) *2 mg/mL*	0.5 mL	7.2	
A21446	Alexa Fluor® 647 rabbit anti-goat IgG (H+L) *2 mg/mL*	0.5 mL	7.2	
A21088	Alexa Fluor® 680 rabbit anti-goat IgG (H+L) *2 mg/mL*	0.5 mL	7.2	
A21062	Alexa Fluor® 350 rabbit anti-mouse IgG (H+L) *2 mg/mL*	0.5 mL	7.2	
A11059	Alexa Fluor® 488 rabbit anti-mouse IgG (H+L) *2 mg/mL*	0.5 mL	7.2	
A11060	Alexa Fluor® 546 rabbit anti-mouse IgG (H+L) *2 mg/mL*	0.5 mL	7.2	
A21427	Alexa Fluor® 555 rabbit anti-mouse IgG (H+L) *2 mg/mL*	0.5 mL	7.2	
A11061	Alexa Fluor® 568 rabbit anti-mouse IgG (H+L) *2 mg/mL*	0.5 mL	7.2	
A11062	Alexa Fluor® 594 rabbit anti-mouse IgG (H+L) *2 mg/mL*	0.5 mL	7.2	
A21063	Alexa Fluor® 633 rabbit anti-mouse IgG (H+L) *2 mg/mL*	0.5 mL	7.2	
A21239	Alexa Fluor® 647 rabbit anti-mouse IgG (H+L) *2 mg/mL*	0.5 mL	7.2	
A21065	Alexa Fluor® 680 rabbit anti-mouse IgG (H+L) *2 mg/mL*	0.5 mL	7.2	
A21210	Alexa Fluor® 488 rabbit anti-rat IgG (H+L) *2 mg/mL*	0.5 mL	7.2	
A21211	Alexa Fluor® 594 rabbit anti-rat IgG (H+L) *2 mg/mL*	0.5 mL	7.2	
A32750	Alexa Fluor® 488 reactive dye decapack *for microarrays* *set of 10 vials*	1 set	1.3, 8.2	
A32756	Alexa Fluor® 555 reactive dye decapack *for microarrays* *set of 10 vials*	1 set	1.3, 8.2	
A32757	Alexa Fluor® 647 reactive dye decapack *for microarrays* *set of 10 vials*	1 set	1.3, 8.2	
A32755	Alexa Fluor® 555 and Alexa Fluor® 647 reactive dye decapacks *for microarrays* *set of 2 x 10 vials* *includes A32756 and A32757 decapacks*	1 set	1.3, 8.2	
A20980	Alexa Fluor® 610–R-phycoerythrin goat anti-mouse IgG (H+L) *1 mg/mL*	100 µL	6.4, 7.2	
A20990	Alexa Fluor® 647–R-phycoerythrin goat anti-mouse IgG (H+L) *1 mg/mL*	100 µL	6.4, 7.2	
A20983	Alexa Fluor® 680–R-phycoerythrin goat anti-mouse IgG (H+L) *1 mg/mL*	100 µL	6.4, 7.2	
A20981	Alexa Fluor® 610–R-phycoerythrin goat anti-rabbit IgG (H+L) *1 mg/mL*	100 µL	6.4, 7.2	
A20991	Alexa Fluor® 647–R-phycoerythrin goat anti-rabbit IgG (H+L) *1 mg/mL*	100 µL	6.4, 7.2	
A20984	Alexa Fluor® 680–R-phycoerythrin goat anti-rabbit IgG (H+L) *1 mg/mL*	100 µL	6.4, 7.2	
A30052	Alexa Fluor® 488 5-SDP ester (Alexa Fluor® 488 sulfodichlorophenol ester)	1 mg	1.3, 4.2	
A11053	Alexa Fluor® 488 Signal-Amplification Kit for Fluorescein- and Oregon Green® Dye–Conjugated Probes *60–120 assays*	1 kit	7.2, 7.4	
A11054	Alexa Fluor® 488 Signal-Amplification Kit for Mouse Antibodies *60–300 assays*	1 kit	7.2	
A11066	Alexa Fluor® 568 Signal-Amplification Kit for Mouse Antibodies *60–300 assays*	1 kit	7.2	
A11067	Alexa Fluor® 594 Signal-Amplification Kit for Mouse Antibodies *60–300 assays*	1 kit	7.2	
A32360	Alexa Fluor® 488 streptavidin, 5 nm colloidal gold conjugate *30 µg protein/mL*	0.5 mL	7.6	
A32361	Alexa Fluor® 488 streptavidin, 10 nm colloidal gold conjugate *30 µg protein/mL*	0.5 mL	7.6	
A30050	Alexa Fluor® 610-X, succinimidyl ester, bis(triethylammonium salt) *6-isomer*	1 mg	1.3	
A7304	AlignFlow™ flow cytometry alignment beads, 2.5 µm *for UV excitation*	3 mL	23.2	
A7302	AlignFlow™ flow cytometry alignment beads, 2.5 µm *for 488 nm excitation*	3 mL	23.2	
A14835	AlignFlow™ flow cytometry alignment beads, 2.5 µm *for 630–660 nm excitation*	3 mL	23.2	
A7312	AlignFlow™ flow cytometry alignment beads, 2.5 µm *for 633 nm excitation*	3 mL	23.2	
A7305	AlignFlow™ Plus flow cytometry alignment beads, 6 µm *for UV excitation*	3 mL	23.2	
A7303	AlignFlow™ Plus flow cytometry alignment beads, 6 µm *for 488 nm excitation*	3 mL	23.2	
A14836	AlignFlow™ Plus flow cytometry alignment beads, 6 µm *for 630–660 nm excitation*	3 mL	23.2	
A7313	AlignFlow™ Plus flow cytometry alignment beads, 6 µm *for 633 nm excitation*	3 mL	23.2	
A10279	alkyne, succinimidyl ester (3-propargyloxypropanoic acid, succinimidyl ester)	1 mg	3.1	
A803	allophycocyanin *4 mg/mL*	0.5 mL	6.4, 14.7	
A819	allophycocyanin, crosslinked (APC-XL) *4 mg/mL*	250 µL	6.4, 14.7	
A10539	allophycocyanin, crosslinked, F(ab')$_2$ fragment of goat anti-mouse (H+L) *1 mg/mL*	250 µL	7.2	
A10541	allophycocyanin, crosslinked, goat anti-mouse IgG$_1$ (γ1) *1 mg/mL*	250 µL	7.2	
A10686	allophycocyanin, crosslinked, goat anti-mouse IgG$_{2a}$ (γ2a) *1 mg/mL*	250 µL	7.2	
A865	allophycocyanin, crosslinked, goat anti-mouse IgG (H+L) *1 mg/mL*	0.5 mL	6.4, 7.2	
A10688	allophycocyanin, crosslinked, goat anti-mouse IgM (µ chain) *1 mg/mL*	250 µL	7.2	

*Note that all products listed in the above Master Product List are covered by **Limited Use Label License #223**. If a specific product is covered by additional Limited Use Label License(s), the Limited Use Label License #'s are provided in the LULL column above. Please refer to page 971 for the complete text of the Limited Use Label License(s).

Cat. No.	Product	Quantity	Sections	LULL*
A10931	allophycocyanin, crosslinked, goat anti-rabbit IgG (H+L) *1 mg/mL*	0.5 mL	6.4, 7.2	
A10540	allophycocyanin, crosslinked, goat anti-rat IgG (H+L) *1 mg/mL*	250 µL	7.2	
A1363	5-(aminoacetamido)fluorescein (fluoresceinyl glycine amide)	10 mg	3.4	
A6289	2-aminoacridone, hydrochloride	25 mg	3.3	
A1310	7-aminoactinomycin D (7-AAD)	1 mg	8.1, 12.5	
A21664	aminoallyl dUTP (5-(3-aminoallyl)-2′-deoxyuridine 5′-triphosphate, trisodium salt) *2 mM in TE buffer*	500 µL	8.2	
A7110	γ-aminobutyric acid, α-carboxy-2-nitrobenzyl ester, trifluoroacetic acid salt (*O*-(CNB-caged) GABA)	5 mg	5.3, 16.2	
A1324	9-amino-6-chloro-2-methoxyacridine (ACMA)	100 mg	8.1, 12.5, 20.3	
A6520	7-amino-4-chloromethylcoumarin, *t*-BOC-L-leucyl-L-methionine amide (CMAC, *t*-BOC-Leu-Met)	5 mg	10.4, 15.5	
A7621	8-amino-cyclic adenosine 5′-diphosphate ribose (8-amino-cADP-ribose)	10 µg	17.2	
A2952	3-amino-3-deoxydigoxigenin hemisuccinamide, succinimidyl ester	5 mg	4.2	
A1339	*N*-(2-aminoethyl)-4-amino-3,6-disulfo-1,8-naphthalimide, dipotassium salt (lucifer yellow ethylenediamine)	25 mg	3.4, 14.3	
A91	5-((2-aminoethyl)amino)naphthalene-1-sulfonic acid, sodium salt (EDANS)	1 g	3.4, 9.5	
A1593	*N*-(2-aminoethyl)biotinamide, hydrobromide (biotin ethylenediamine)	25 mg	4.2, 14.3	
A191	7-amino-4-methylcoumarin *reference standard*	100 mg	3.3, 3.4, 10.1	
A6118	6-((7-amino-4-methylcoumarin-3-acetyl)amino)hexanoic acid, succinimidyl ester (AMCA-X, SE)	10 mg	1.7	
A6521	7-amino-4-methylcoumarin, CBZ-L-phenylalanyl-L-arginine amide, hydrochloride	25 mg	10.4	
A1351	4′-(aminomethyl)fluorescein, hydrochloride	25 mg	3.4	
A1353	5-(aminomethyl)fluorescein, hydrochloride	10 mg	3.4	
A350	8-aminonaphthalene-1,3,6-trisulfonic acid, disodium salt (ANTS)	1 g	3.3, 14.3	
A10550	*N*-(aminooxyacetyl)-*N*′-(D-biotinoyl) hydrazine, trifluoroacetic acid salt (ARP)	10 mg	4.2, 8.2	
A1318	5-(and-6)-((*N*-(5-aminopentyl)amino)carbonyl)tetramethylrhodamine (tetramethylrhodamine cadaverine) *mixed isomers*	10 mg	3.4	
A1340	*N*-(5-aminopentyl)-4-amino-3,6-disulfo-1,8-naphthalimide, dipotassium salt (lucifer yellow cadaverine)	25 mg	3.4, 14.3	
A1594	*N*-(5-aminopentyl)biotinamide, trifluoroacetic acid salt (biotin cadaverine)	25 mg	4.2, 14.3	
A10466	5-((5-aminopentyl)thioureidyl)fluorescein, dihydrobromide salt (fluorescein cadaverine)	25 mg	3.4	
A36003	3′-(*p*-aminophenyl) fluorescein (APF) *5 mM solution in DMF*	470 µL	18.2	
A6257	8-aminopyrene-1,3,6-trisulfonic acid, trisodium salt (APTS)	10 mg	3.3, 14.3	
A22850	aminostilbamidine, methanesulfonate	10 mg	14.3	
A33851	Amplex® ELISA Development Kit for Mouse IgG *with Amplex® UltraRed reagent* *500 assays*	1 kit	6.2, 10.5	
A33852	Amplex® ELISA Development Kit for Rabbit IgG *with Amplex® UltraRed reagent* *500 assays*	1 kit	6.2, 10.5	
A12217	Amplex® Red Acetylcholine/Acetylcholinesterase Assay Kit *500 assays*	1 kit	10.5, 16.2	
A22180	Amplex® Red Catalase Assay Kit *400 assays*	1 kit	10.5	
A12216	Amplex® Red Cholesterol Assay Kit *500 assays*	1 kit	10.5, 13.3	
A22179	Amplex® Red Galactose/Galactose Oxidase Assay Kit *400 assays*	1 kit	10.2, 10.5	
A22189	Amplex® Red Glucose/Glucose Oxidase Assay Kit *500 assays*	1 kit	10.2, 10.5	
A12221	Amplex® Red Glutamic Acid/Glutamate Oxidase Assay Kit *200 assays*	1 kit	10.5, 16.2	
A22188	Amplex® Red Hydrogen Peroxide/Peroxidase Assay Kit *500 assays*	1 kit	10.5, 18.2	
A12214	Amplex® Red Monoamine Oxidase Assay Kit *500 assays*	1 kit	10.5	
A22178	Amplex® Red Neuraminidase (Sialidase) Assay Kit *400 assays*	1 kit	10.2, 10.5	
A12218	Amplex® Red Phosphatidylcholine-Specific Phospholipase C Assay Kit *500 assays*	1 kit	10.5, 17.4	
A12219	Amplex® Red Phospholipase D Assay Kit *500 assays*	1 kit	10.5, 17.4	
A12222	Amplex® Red reagent	5 mg	10.5, 18.2	
A22177	Amplex® Red reagent *packaged for high-throughput screening*	10 x 10 mg	10.5, 18.2	
A12220	Amplex® Red Sphingomyelinase Assay Kit *500 assays*	1 kit	10.5, 13.3, 17.4	
A33855	Amplex® Red/UltraRed stop reagent *500 tests* *set of 5 vials*	1 set	10.5	
A22181	Amplex® Red Uric Acid/Uricase Assay Kit *400 assays*	1 kit	10.5	
A22182	Amplex® Red Xanthine/Xanthine Oxidase Assay Kit *400 assays*	1 kit	10.5, 18.2	
A36006	Amplex® UltraRed reagent	5 x 1 mg	10.5, 18.2	
A13439	angiotensin II, Alexa Fluor® 488 conjugate	25 µg	16.2	
A13438	angiotensin II, fluorescein conjugate	25 µg	16.2	
A23202	annexin V, Alexa Fluor® 350 conjugate *100 assays*	500 µL	15.5	
A13201	annexin V, Alexa Fluor® 488 conjugate *100 assays*	500 µL	15.5	
A35108	annexin V, Alexa Fluor® 555 conjugate *100 assays*	500 µL	15.5	
A13202	annexin V, Alexa Fluor® 568 conjugate *100 assays*	500 µL	15.5	
A13203	annexin V, Alexa Fluor® 594 conjugate *100 assays*	500 µL	15.5	
A23204	annexin V, Alexa Fluor® 647 conjugate *100 assays*	500 µL	15.5	
A35109	annexin V, Alexa Fluor® 680 conjugate *100 assays*	500 µL	15.5	
A35110	annexin V, allophycocyanin conjugate (APC annexin V) *50 assays*	250 µL	6.4, 15.5	

*Note that all products listed in the above Master Product List are covered by **Limited Use Label License #223**. If a specific product is covered by additional Limited Use Label License(s), the Limited Use Label License #'s are provided in the LULL column above. Please refer to page 971 for the complete text of the Limited Use Label License(s).

IMPORTANT NOTICE: The products described in this manual are covered by one or more Limited Use Label License(s). Please refer to the Appendix on page 971 and Master Product List on page 975. Products are For Research Use Only. Not intended for any animal or human therapeutic or diagnostic use.

Cat. No.	Product	Quantity	Sections	LULL*
A13204	annexin V, biotin-X conjugate *100 assays*	500 µL	4.3, 15.5	
A13199	annexin V, fluorescein conjugate (FITC annexin V) *100 assays*	500 µL	15.5	
A13200	annexin V, Oregon Green® 488 conjugate *100 assays*	500 µL	15.5	
A35122	annexin V, Pacific Blue™ conjugate *for flow cytometry* *100 assays*	500 µL	15.5	
A35111	annexin V, R-phycoerythrin conjugate (R-PE annexin V) *50 assays*	250 µL	6.4, 15.5	
A1440	9-anthroylnitrile	25 mg	3.2	
A1322	9-anthroyl ouabain	5 mg	16.3	
A1400	9-anthryldiazomethane (ADAM)	25 mg	3.4	
A11133	anti-albumin (bovine serum), rabbit IgG fraction (anti-BSA) *2 mg/mL*	0.5 mL	14.7	
A5760	anti-Alexa Fluor® 405/Cascade Blue®, rabbit IgG fraction *3 mg/mL*	0.5 mL	7.4	
A11094	anti-Alexa Fluor® 488, rabbit IgG fraction *1 mg/mL*	0.5 mL	7.4	
A5770	anti-BODIPY® FL, rabbit IgG fraction *3 mg/mL*	0.5 mL	7.4	
A35725	Antibody Beacon™ Tyrosine Kinase Assay Kit *400 assays*	1 kit	10.3, 17.3	
A21281	anti-c-myc, chicken IgY fraction *1 mg/mL*	100 µL	7.5	
A21280	anti-c-myc, mouse IgG_1, monoclonal 289-19510 *1 mg/mL*	100 µL	7.5	
A6398	anti-dansyl, rabbit IgG fraction *1 mg/mL*	0.5 mL	7.4	
A21283	anti-desmin, mouse IgG_1, monoclonal 131-15014 *1 mg/mL*	100 µL	11.2	
A6430	anti-dinitrophenyl-KLH, rabbit IgG fraction *2 mg/mL*	0.5 mL	7.4	
A11097	anti-dinitrophenyl-KLH, rabbit IgG fraction, Alexa Fluor® 488 conjugate *2 mg/mL*	0.5 mL	7.4	
A6435	anti-dinitrophenyl-KLH, rabbit IgG fraction, biotin-XX conjugate *2 mg/mL*	0.5 mL	4.3, 7.4	
A6423	anti-dinitrophenyl-KLH, rabbit IgG fraction, fluorescein conjugate *2 mg/mL*	0.5 mL	7.4	
A11095	anti-fluorescein/Oregon Green®, goat IgG fraction *1 mg/mL*	0.5 mL	7.4	
A11096	anti-fluorescein/Oregon Green®, goat IgG fraction, Alexa Fluor® 488 conjugate *1 mg/mL*	0.5 mL	7.4	
A6421	anti-fluorescein/Oregon Green®, mouse IgG_{2a}, monoclonal 4-4-20	0.5 mg	7.4	
A6413	anti-fluorescein/Oregon Green®, rabbit IgG Fab fragment *0.5 mg/mL*	0.5 mL	7.4	
A889	anti-fluorescein/Oregon Green®, rabbit IgG fraction *1 mg/mL*	0.5 mL	7.4	
A11090	anti-fluorescein/Oregon Green®, rabbit IgG fraction, Alexa Fluor® 488 conjugate *1 mg/mL*	0.5 mL	7.4	
A11091	anti-fluorescein/Oregon Green®, rabbit IgG fraction, Alexa Fluor® 594 conjugate *1 mg/mL*	0.5 mL	7.4	
A982	anti-fluorescein/Oregon Green®, rabbit IgG fraction, biotin-XX conjugate *1 mg/mL*	0.5 mL	4.3, 7.4	
A21253	anti-fluorescein/Oregon Green®, rabbit IgG fraction, horseradish peroxidase conjugate	0.5 mg	7.4	
A21250	anti-fluorescein/Oregon Green®, rabbit IgG fraction, R-phycoerythrin conjugate *2 mg/mL*	250 µL	6.4, 7.4	
A11132	anti-β-galactosidase, rabbit IgG fraction *2 mg/mL*	0.5 mL	7.5, 10.2	
A21282	anti-GFAP (anti-glial fibrillary acidic protein, mouse IgG_1, monoclonal 131-17719) *1 mg/mL*	100 µL	11.2	
A21294	anti-GFAP, Alexa Fluor® 488 conjugate (anti-glial fibrillary acidic protein, mouse IgG_1, monoclonal 131-17719, Alexa Fluor® 488 conjugate) *1 mg/mL*	50 µL	11.2	
A21295	anti-GFAP, Alexa Fluor® 594 conjugate (anti-glial fibrillary acidic protein, mouse IgG_1, monoclonal 131-17719, Alexa Fluor® 594 conjugate) *1 mg/mL*	50 µL	11.2	
A5790	anti-β-glucuronidase, rabbit IgG fraction *2 mg/mL*	0.5 mL	7.5, 10.2	
A5800	anti-glutathione *S*-transferase, rabbit IgG fraction *3 mg/mL*	0.5 mL	7.5, 9.4	
A11131	anti-glutathione *S*-transferase, rabbit IgG fraction, Alexa Fluor® 488 conjugate *2 mg/mL*	0.5 mL	7.5, 9.4	
A10262	anti-green fluorescent protein, chicken IgY fraction (anti-GFP, IgY) *2 mg/mL*	100 µL	7.5	
A10263	anti-green fluorescent protein, chicken IgY fraction, biotin-XX conjugate (anti-GFP, IgY, biotin-XX conjugate) *2 mg/mL*	100 µL	4.3, 7.5	
A11120	anti-green fluorescent protein, mouse IgG_{2a}, monoclonal 3E6 (anti-GFP, mAb 3E6)	100 µg	7.5	
A11121	anti-green fluorescent protein, mouse IgG_1, monoclonal 11E5 (anti-GFP, mAb 11E5)	100 µg	7.5	
A11122	anti-green fluorescent protein, rabbit IgG fraction (anti-GFP, IgG) *2 mg/mL*	100 µL	7.5	
A21311	anti-green fluorescent protein, rabbit IgG fraction, Alexa Fluor® 488 conjugate (anti-GFP, IgG, Alexa Fluor® 488 conjugate) *2 mg/mL*	100 µL	7.5	
A31851	anti-green fluorescent protein, rabbit IgG fraction, Alexa Fluor® 555 conjugate (anti-GFP, IgG, Alexa Fluor® 555 conjugate) *2 mg/mL*	100 µL	7.5	
A21312	anti-green fluorescent protein, rabbit IgG fraction, Alexa Fluor® 594 conjugate (anti-GFP, IgG, Alexa Fluor® 594 conjugate) *2 mg/mL*	100 µL	7.5	
A31852	anti-green fluorescent protein, rabbit IgG fraction, Alexa Fluor® 647 conjugate (anti-GFP, IgG, Alexa Fluor® 647 conjugate) *2 mg/mL*	100 µL	7.5	
A10259	anti-green fluorescent protein, rabbit IgG fraction, biotin-XX conjugate (anti-GFP, IgG, biotin-XX conjugate) *2 mg/mL*	100 µL	4.3, 7.5	
A10260	anti-green fluorescent protein, rabbit IgG fraction, horseradish peroxidase conjugate (anti-GFP, IgG, HRP)	200 µg	7.5	
A6455	anti-green fluorescent protein, rabbit serum (anti-GFP, serum)	100 µL	7.5	
A21287	anti-hemagglutinin, mouse IgG_1, monoclonal 16B12, Alexa Fluor® 488 conjugate (anti-HA, Alexa Fluor® 488 conjugate) *1 mg/mL*	100 µL	7.5	

*Note that all products listed in the above Master Product List are covered by **Limited Use Label License #223**. If a specific product is covered by additional Limited Use Label License(s), the Limited Use Label License #'s are provided in the LULL column above. Please refer to page 971 for the complete text of the Limited Use Label License(s).

IMPORTANT NOTICE: The products described in this manual are covered by one or more Limited Use Label License(s). Please refer to the Appendix on page 971 and Master Product List on page 975. Products are For Research Use Only. Not intended for any animal or human therapeutic or diagnostic use.
www.invitrogen.com/probes

Cat. No.	Product	Quantity	Sections	LULL*
A21288	anti-hemagglutinin, mouse IgG_1, monoclonal 16B12, Alexa Fluor® 594 conjugate (anti-HA, Alexa Fluor® 594 conjugate) *1 mg/mL*	100 µL	7.5	
A21272	anti-HuC/HuD neuronal protein (human), mouse IgG_{2b}, monoclonal 16A11, biotin-XX conjugate	100 µg	4.3	
A5750	anti-lucifer yellow, rabbit IgG fraction *3 mg/mL*	0.5 mL	7.4	
A5751	anti-lucifer yellow, rabbit IgG fraction, biotin-XX conjugate *3 mg/mL*	0.5 mL	4.3, 7.4	
A21285	anti-nitrotyrosine, rabbit IgG fraction *1 mg/mL*	0.5 mL	7.4, 18.3	
A6473	anti-NMDA receptor, subunit 2A (rat), rabbit IgG fraction *affinity purified*	10 µg	16.2	
A6474	anti-NMDA receptor, subunit 2B (rat), rabbit IgG fraction *affinity purified*	10 µg	16.2	
A6475	anti-NMDA receptor, subunit 2C (rat), rabbit IgG fraction *affinity purified*	10 µg	16.2	
A21327	anti-phosphatidylinositol 4,5-diphosphate, mouse IgM, monoclonal 2C11 (anti-PtdIns(4,5)P_2) *1 mg/mL*	100 µL	17.4	
A21328	anti-phosphatidylinositol 3,4,5-triphosphate, mouse IgM, monoclonal RC6F8 (anti-PtdIns(3,4,5)P_3) *1 mg/mL*	100 µL	17.4	
A6442	anti-synapsin I (bovine), rabbit IgG fraction *affinity purified*	10 µg	11.2, 16.1, 17.3	
A6397	anti-tetramethylrhodamine, rabbit IgG fraction *1 mg/mL*	0.5 mL	7.4	
A6399	anti-Texas Red®, rabbit IgG fraction *1 mg/mL*	0.5 mL	7.4	
A11130	anti-transferrin receptor (human), mouse IgG_1, monoclonal 236-15375	50 µg	16.1	
A11126	anti-α-tubulin (bovine), mouse IgG_1, monoclonal 236-10501	50 µg	11.2	
A21371	anti-α-tubulin (bovine), mouse IgG_1, monoclonal 236-10501, biotin-XX conjugate	50 µg	4.3, 11.2	
V35113	APC Annexin V/Dead Cell Apoptosis Kit *with APC annexin V and SYTOX® Green* *50 assays* *for flow cytometry*	1 kit	15.5	
A10468	APEX® Alexa Fluor® 488 Antibody Labeling Kit	1 kit	1.2, 1.3	
A10470	APEX® Alexa Fluor® 555 Antibody Labeling Kit	1 kit	1.2, 1.3	
A10474	APEX® Alexa Fluor® 594 Antibody Labeling Kit	1 kit	1.2, 1.3	
A10475	APEX® Alexa Fluor® 647 Antibody Labeling Kit	1 kit	1.2, 1.3	
A10476	APEX® Oregon Green® 488 Antibody Labeling Kit	1 kit	1.2, 1.5	
A10478	APEX® Pacific Blue™ Antibody Labeling Kit	1 kit	1.2, 1.7	
A23210	APO-BrdU™ TUNEL Assay Kit *with Alexa Fluor® 488 anti-BrdU* *60 assays*	1 kit	15.5	214
A6785	AquaLite® aequorin (aequorin) *recombinant*	25 µg	19.5	
A10346	ArC™ Amine Reactive Compensation Bead Kit *for use with amine reactive dyes* *for flow cytometry compensation* *100 tests*	1 kit	15.3, 23.2	
A21665	ARES™ Alexa Fluor® 488 DNA Labeling Kit *10 labelings*	1 kit	1.2, 1.3, 8.2	
A21667	ARES™ Alexa Fluor® 546 DNA Labeling Kit *10 labelings*	1 kit	1.2, 1.3, 8.2	
A21677	ARES™ Alexa Fluor® 555 DNA Labeling Kit *10 labelings*	1 kit	1.2, 1.3, 8.2	
A21669	ARES™ Alexa Fluor® 594 DNA Labeling Kit *10 labelings*	1 kit	1.2, 1.3, 8.2	
A21676	ARES™ Alexa Fluor® 647 DNA Labeling Kit *10 labelings*	1 kit	1.2, 1.3, 8.2	
A22066	ATP Determination Kit *special packaging* *200–1000 assays*	1 kit	10.3, 15.3	
A7816	Attofluor cell chamber *for microscopy*	each	23.1	
A2333	ATTO-TAG™ CBQCA Amine-Derivatization Kit	1 kit	1.8, 9.3	244
A6222	ATTO-TAG™ CBQCA derivatization reagent (CBQCA; 3-(4-carboxybenzoyl)quinoline-2-carboxaldehyde)	10 mg	1.8, 9.2, 9.3, 21.2	244
A2334	ATTO-TAG™ FQ Amine-Derivatization Kit	1 kit	1.8, 9.3	244
A10192	ATTO-TAG™ FQ derivatization reagent (FQ; 3-(2-furoyl)quinoline-2-carboxaldehyde)	10 mg	1.8, 9.3, 21.2	244
A21370	avidin, Alexa Fluor® 488 conjugate	1 mg	7.6	
A2667	avidin, egg white	5 mg	7.6	
A887	avidin, egg white	100 mg	7.6	
A821	avidin, fluorescein conjugate	5 mg	7.6	
A11236	avidin, NeutrAvidin®, Alexa Fluor® 350 conjugate	1 mg	7.6	
A2666	avidin, NeutrAvidin® biotin-binding protein	5 mg	7.6	
A2663	avidin, NeutrAvidin®, Cascade Blue® conjugate	1 mg	7.6	
A2662	avidin, NeutrAvidin®, fluorescein conjugate	1 mg	7.6	
A2664	avidin, NeutrAvidin®, horseradish peroxidase conjugate	1 mg	7.6	
A6374	avidin, NeutrAvidin®, Oregon Green® 488 conjugate	1 mg	7.6	
A6378	avidin, NeutrAvidin®, Rhodamine Red™-X conjugate	1 mg	7.6	
A2660	avidin, NeutrAvidin®, R-phycoerythrin conjugate *1 mg/mL*	1 mL	6.4, 7.6	
A6373	avidin, NeutrAvidin®, tetramethylrhodamine conjugate	1 mg	7.6	
A2665	avidin, NeutrAvidin®, Texas Red® conjugate	1 mg	7.6	
A820	avidin, Texas Red® conjugate	5 mg	7.6	
A10280	azido $(PEO)_4$ propionic acid, succinimidyl ester (3-(azidotetra(ethyleneoxy))propionic acid, succinimidyl ester)	1 mg	3.1	
A2522	4-azido-2,3,5,6-tetrafluorobenzoic acid, succinimidyl ester (ATFB, SE)	25 mg	5.3	
B35000	*Bac*Light™ Green bacterial stain *special packaging*	20 x 50 µg	15.4	
B34950	*Bac*Light™ Bacterial Membrane Potential Kit *for flow cytometry* *100 assays*	1 kit	15.3, 22.3	
B35001	*Bac*Light™ Red bacterial stain *special packaging*	20 x 50 µg	15.4	

*Note that all products listed in the above Master Product List are covered by **Limited Use Label License #223**. If a specific product is covered by additional Limited Use Label License(s), the Limited Use Label License #'s are provided in the LULL column above. Please refer to page 971 for the complete text of the Limited Use Label License(s).

IMPORTANT NOTICE: The products described in this manual are covered by one or more Limited Use Label License(s). Please refer to the Appendix on page 971 and Master Product List on page 975. Products are For Research Use Only. Not intended for any animal or human therapeutic or diagnostic use.

Cat. No.	Product	Quantity	Sections	LULL*
B34956	*Bac*Light™ RedoxSensor™ CTC Vitality Kit *for flow cytometry and microscopy*	1 kit	15.3	
B34954	*Bac*Light™ RedoxSensor™ Green Vitality Kit *for flow cytometry*	1 kit	15.3	
B10383	BacMam GFP transduction control *BacMam 2.0*	1 mL	14.7	21, 127, 198, 267, 272, 306, 308, 332
B10033	BacMam-hERG *for 100 microplates*	1 kit	16.3, 21.1	21, 306, 332
B10019	BacMam-hERG *for 10 microplates*	1 kit	16.3, 21.1	21, 306, 332
B10334	BacMam Kir1.1 *for 10 microplates*	1 kit	16.3, 21.1	21, 306, 332
B10146	BacMam Kir2.1 *for 10 microplates*	1 kit	16.3, 21.1	21, 306, 332
B10331	BacMam Kv1.1 *for 10 microplates*	1 kit	16.3, 21.1	21, 306, 332
B10332	BacMam Kv1.3 *for 10 microplates*	1 kit	16.3, 21.1	21, 306, 332
B10333	BacMam Kv2.1 *for 10 microplates*	1 kit	16.3, 21.1	21, 306, 332
B10147	BacMam Kv7.2 and Kv7.3 *for 10 microplates*	1 kit	16.3, 21.1	21, 306, 332
B10341	BacMam Nav1.2 *for 10 microplates*	1 kit	16.3	21, 306, 332
B10335	BacMam Nav1.5 *for 10 microplates*	1 kit	16.3	21, 306, 332
B7277	Bacteria Counting Kit *for flow cytometry*	1 kit	15.4	
B1205	BAPTA, AM *cell permeant*	25 mg	19.8	
B6769	BAPTA, AM *cell permeant* *special packaging*	20 x 1 mg	19.8	
B1212	BAPTA, tetracesium salt *cell impermeant*	1 g	19.8	
B1204	BAPTA, tetrapotassium salt *cell impermeant*	1 g	19.8	
B1214	BAPTA, tetrasodium salt *cell impermeant*	1 g	19.8	
B1526	benzophenone-4-isothiocyanate	100 mg	5.3	
B1508	benzophenone-4-maleimide	100 mg	5.3	
B1577	4-benzoylbenzoic acid, succinimidyl ester	100 mg	5.3	
B22358	2′-(or-3′)-*O*-(4-benzoylbenzoyl)adenosine 5′-triphosphate, tris(triethylammonium) salt (BzBzATP) *5 mM in buffer*	2 mL	5.3, 17.3	
R441	benzyloxyresorufin (resorufin benzyl ether)	10 mg	10.6	
B30633	bimane amine	5 mg	3.4	
B30600	bimane azide	5 mg	5.3, 13.5	
B30501	bimane C_3-maleimide	5 mg	2.3	
B30500	bimane iodoacetamide	5 mg	2.3	
B30250	bimane mercaptoacetic acid (carboxymethylthiobimane)	5 mg	1.7	
B1592	biocytin (ε-biotinoyl-L-lysine)	100 mg	4.2, 14.3	
B1603	biocytin hydrazide	25 mg	4.2, 14.3	
B1595	D-biotin	1 g	4.2, 7.6	
B20656	D-biotin *50 mM aqueous solution*	10 mL	4.2, 7.6	
B32772	biotin-aha-dCTP *1 mM in TE buffer*	25 µL	4.3, 8.2	
B32766	biotin-aha-dUTP *1 mM in TE buffer*	25 µL	4.3, 8.2	
B10185	biotin alkyne (PEG_4 carboxamide-propargyl biotin)	1 mg	3.1, 4.2	
B10184	biotin azide (PEG_4 carboxamide-6-azidohexanyl biotin)	1 mg	3.1, 4.2	
B1550	biotin DHPE (*N*-(biotinoyl)-1,2-dihexadecanoyl-*sn*-glycero-3-phosphoethanolamine, triethylammonium salt)	10 mg	4.3, 13.2	
B10570	biotin-4-fluorescein	5 mg	4.3, 14.3	
B1582	6-((biotinoyl)amino)hexanoic acid, succinimidyl ester (biotin-X, SE; biotinamidocaproate, *N*-hydroxysuccinimidyl ester)	100 mg	4.2, 14.3	
B6353	6-((biotinoyl)amino)hexanoic acid, sulfosuccinimidyl ester, sodium salt (Sulfo-NHS-LC-Biotin; biotin-X, SSE)	25 mg	4.2, 14.3	
B1606	6-((6-((biotinoyl)amino)hexanoyl)amino)hexanoic acid, succinimidyl ester (biotin-XX, SE)	100 mg	4.2	
B6352	6-((6-((biotinoyl)amino)hexanoyl)amino)hexanoic acid, sulfosuccinimidyl ester, sodium salt (biotin-XX, SSE)	25 mg	4.2, 14.3	
B1370	5-((*N*-(5-(*N*-(6-(biotinoyl)amino)hexanoyl)amino)pentyl)thioureidyl)fluorescein (fluorescein biotin)	5 mg	4.3, 14.3	
B1591	*N*-(biotinoyl)-*N*′-(iodoacetyl)ethylenediamine	25 mg	4.2	
B1513	D-biotin, succinimidyl ester (succinimidyl D-biotin)	100 mg	4.2, 14.3	
B1596	biotin-X cadaverine (5-(((*N*-(biotinoyl)amino)hexanoyl)amino)pentylamine, trifluoroacetic acid salt)	10 mg	4.2, 14.3	
B1616	biotin-X DHPE (*N*-((6-(biotinoyl)amino)hexanoyl)-1,2-dihexadecanoyl-*sn*-glycero-3-phosphoethanolamine, triethylammonium salt)	5 mg	4.3, 13.2	
B2604	biotin-X 2,4-dinitrophenyl-X-L-lysine, succinimidyl ester (DNP-X-biocytin-X, SE)	5 mg	4.2	
B11027	biotin-XX F(ab′)$_2$ fragment of goat anti-mouse IgG (H+L) *2 mg/mL*	250 µL	4.3, 7.2	
B21078	biotin-XX F(ab′)$_2$ fragment of goat anti-rabbit IgG (H+L) *2 mg/mL*	250 µL	4.3, 7.2	
A10519	biotin-XX goat anti-mouse IgG_1 (γ1) *2 mg/mL*	250 µL	7.2	
B2763	biotin-XX goat anti-mouse IgG (H+L) *2 mg/mL*	0.5 mL	4.3, 7.2	
A10676	biotin-XX goat anti-mouse IgG, IgA, IgM (H+L) *2 mg/mL*	250 µL	7.2	
B2770	biotin-XX goat anti-rabbit IgG (H+L) *2 mg/mL*	0.5 mL	4.3, 7.2	
A10517	biotin-XX goat anti-rat IgG (H+L) *2 mg/mL*	0.5 mL	7.2	
B2600	biotin-XX hydrazide (6-((6-((biotinoyl)amino)hexanoyl)amino)hexanoic acid, hydrazide)	25 mg	4.2	

*Note that all products listed in the above Master Product List are covered by **Limited Use Label License #223**. If a specific product is covered by additional Limited Use Label License(s), the Limited Use Label License #'s are provided in the LULL column above. Please refer to page 971 for the complete text of the Limited Use Label License(s).

Cat. No.	Product	Quantity	Sections	LULL*
B30010	Biotin-XX Microscale Protein Labeling Kit *for 20–100 µg protein* *3 labelings*	1 kit	1.2, 4.2	
B30756	Biotin-XX Microscale Protein Labeling Kit with FluoReporter® Biotin Quantitation Assay Kit *includes B30010 and F30751*	1 kit	1.2, 4.2	
A10650	biotin-XX mouse anti-human IgG_1 *1 mg/mL*	250 µL	7.2	
A10663	biotin-XX mouse anti-human IgG_4 *1 mg/mL*	250 µL	7.2	
B7474	biotin-XX phalloidin	50 U	4.3, 11.1	
A10518	biotin-XX rabbit anti-goat IgG (H+L) *2 mg/mL*	0.5 mL	7.2	
B153	bis-ANS (4,4′-dianilino-1,1′-binaphthyl-5,5′-disulfonic acid, dipotassium salt)	10 mg	11.2, 13.5	
B1151	2′,7′-bis-(2-carboxyethyl)-5-(and-6)-carboxyfluorescein (BCECF acid) *mixed isomers*	1 mg	14.3, 20.2	
B1150	2′,7′-bis-(2-carboxyethyl)-5-(and-6)-carboxyfluorescein, acetoxymethyl ester (BCECF, AM)	1 mg	15.2, 15.6, 20.2	
B1170	2′,7′-bis-(2-carboxyethyl)-5-(and-6)-carboxyfluorescein, acetoxymethyl ester (BCECF, AM) *special packaging*	20 x 50 µg	15.2, 15.6, 20.2	
B3051	2′,7′-bis-(2-carboxyethyl)-5-(and-6)-carboxyfluorescein, acetoxymethyl ester (BCECF, AM) *1 mg/mL solution in anhydrous DMSO*	1 mL	15.2, 15.6, 20.2	
B436	bis-(1,3-dibutylbarbituric acid)pentamethine oxonol ($DiBAC_4(5)$)	25 mg	16.3, 22.3	
B24570	bis-(1,3-dibutylbarbituric acid)trimethine oxonol ($DiBAC_4(3)$) *FluoroPure™ grade*	5 mg	22.3	
B438	bis-(1,3-dibutylbarbituric acid)trimethine oxonol ($DiBAC_4(3)$)	25 mg	22.3	
B413	bis-(1,3-diethylthiobarbituric acid)trimethine oxonol ($DiSBAC_2(3)$)	100 mg	22.3	
B7701	1,2-bis-(4,4-difluoro-5,7-dimethyl-4-bora-3a,4a-diaza-*s*-indacene-3-undecanoyl)-*sn*-glycero-3-phosphocholine (bis-BODIPY® FL C_{11}-PC)	100 µg	13.2, 17.4	
B6810	bis-fura-2, hexapotassium salt *cell impermeant*	1 mg	19.2	
B10621	bis-((*N*-iodoacetyl)piperazinyl)sulfonerhodamine	5 mg	2.2, 5.2	
B3932	(*E,E*)-3,5-bis-(4-phenyl-1,3-butadienyl)-4,4-difluoro-4-bora-3a,4a-diaza-*s*-indacene (BODIPY® 665/676)	5 mg	13.5, 18.2	
B3781	1,2-bis-(1-pyrenebutanoyl)-*sn*-glycero-3-phosphocholine	1 mg	13.2, 17.4	
B3782	1,2-bis-(1-pyrenedecanoyl)-*sn*-glycero-3-phosphocholine	1 mg	13.2, 17.4	
B311	1,3-bis-(1-pyrenyl)propane	25 mg	13.5	
B10710	BlockAid™ blocking solution *for use with microspheres*	50 mL	6.5, 7.6	
B3582	BOBO™-1 iodide (462/481) *1 mM solution in DMSO*	200 µL	8.1, 12.5	
B3586	BOBO™-3 iodide (570/602) *1 mM solution in DMSO*	200 µL	8.1, 12.5	
B30300	*N*-(*t*-BOC)-aminooxyacetic acid, tetrafluorophenyl ester	25 mg	1.8, 3.3	
B13233	BOCILLIN™ FL penicillin, sodium salt	1 mg	9.4, 15.2	
B13234	BOCILLIN™ 650/665 penicillin, sodium salt	1 mg	9.4, 15.2	
B3824	BODIPY® 500/510 C_4, C_9 (5-butyl-4,4-difluoro-4-bora-3a,4a-diaza-*s*-indacene-3-nonanoic acid)	1 mg	13.2	
B10250	BODIPY® FL *N*-(2-aminoethyl)maleimide	5 mg	2.2	
B22650	BODIPY® FL C_5-ceramide complexed to BSA	5 mg	12.4, 13.3, 17.4	
B13950	BODIPY® FL C_5-ganglioside G_{M1}	25 µg	13.3, 17.4	
B34401	BODIPY® FL C_5-ganglioside G_{M1} complexed to BSA	1 mg	13.3, 17.4	
B34402	BODIPY® FL C_5-lactosylceramide complexed to BSA	1 mg	10.2, 12.4, 13.3, 17.4	
B20340	BODIPY® FL L-cystine	1 mg	2.2	
B7469	BODIPY® FL forskolin	100 µg	15.6, 17.3	
B2752	BODIPY® FL goat anti-mouse IgG (H+L)	1 mg	7.2	
B2766	BODIPY® FL goat anti-rabbit IgG (H+L)	1 mg	7.2	
B22461	BODIPY® FL histamine	1 mg	12.3	
B22356	2′-(or-3′)-*O*-(BODIPY® FL)-β:γ-imidoadenosine 5′-triphosphate, trisodium salt (BODIPY® FL AMPPNP) *5 mM in buffer*	50 µL	17.3	
B23461	BODIPY® FL ouabain	100 µg	16.3	
B607	BODIPY® FL phallacidin	300 U	11.1	
B34353	BODIPY® FL phosphatidylinositol(4,5) bisphosphate (BODIPY® FL PtdIns(4,5)P_2)	50 µg	17.4	
B7433	BODIPY® FL prazosin	100 µg	15.6, 16.2	
B10006	BODIPY® FL, STP ester, sodium salt	5 mg	1.4, 4.2	
B7487	BODIPY® FL thapsigargin	100 µg	17.2	
B7431	BODIPY® FL verapamil, hydrochloride	1 mg	15.6, 16.3	
B2103	BODIPY® 493/503 methyl bromide (8-bromomethyl-4,4-difluoro-1,3,5,7-tetramethyl-4-bora-3a,4a-diaza-*s*-indacene)	5 mg	2.2, 3.4, 14.2	
B22802	BODIPY® 630/650 methyl bromide (8-bromomethyl-4,4-difluoro-3,5-bis-(2-thienyl)-4-bora-3a,4a-diaza-*s*-indacene)	1 mg	2.2	
B3475	BODIPY® 558/568 phalloidin	300 U	11.1	
B12382	BODIPY® 650/665 phalloidin	300 U	11.1	
B30466	BODIPY® TMR C_5-maleimide	1 mg	2.2	
D7540	BODIPY® TR ceramide (*N*-((4-(4,4-difluoro-5-(2-thienyl)-4-bora-3a,4a-diaza-*s*-indacene-3-yl)phenoxy)acetyl) sphingosine)	250 µg	12.4, 13.3, 17.4	
B34400	BODIPY® TR ceramide complexed to BSA	5 mg	12.4, 13.3, 17.4	
B13800	BODIPY® TR-X thapsigargin	25 µg	17.2	

*Note that all products listed in the above Master Product List are covered by **Limited Use Label License #223**. If a specific product is covered by additional Limited Use Label License(s), the Limited Use Label License #'s are provided in the LULL column above. Please refer to page 971 for the complete text of the Limited Use Label License(s).

IMPORTANT NOTICE: The products described in this manual are covered by one or more Limited Use Label License(s). Please refer to the Appendix on page 971 and Master Product List on page 975. Products are For Research Use Only. Not intended for any animal or human therapeutic or diagnostic use.

Cat. No.	Product	Quantity	Sections	LULL*
B34650	BrainStain™ Imaging Kit	1 kit	14.4	
B35130	BrdU, mouse monoclonal antibody (Clone MoBU-1), Alexa Fluor® 488 conjugate *0.2 mg/mL*	350 µL	15.4	
B35131	BrdU, mouse monoclonal antibody (Clone MoBU-1), Alexa Fluor® 555 conjugate *0.2 mg/mL*	350 µL	15.4	
B35132	BrdU, mouse monoclonal antibody (Clone MoBU-1), Alexa Fluor® 594 conjugate *0.2 mg/mL*	350 µL	15.4	
B35133	BrdU, mouse monoclonal antibody (Clone MoBU-1), Alexa Fluor® 647 conjugate *0.2 mg/mL*	350 µL	15.4	
B35138	BrdU, mouse monoclonal antibody (Clone MoBU-1, biotin conjugate *0.2 mg/mL*	350 µL	4.3, 15.4	
B35129	BrdU, mouse monoclonal antibody (Clone MoBU-1), Pacific Blue™ conjugate *for flow cytometry* *100 tests*	1 each	15.4	
B35128	BrdU, mouse monoclonal antibody (Clone MoBU-1), unconjugated *0.1 mg/mL*	350 µL	15.4	
B35141	BrdU, mouse monoclonal antibody (Clone MoBU-1), unconjugated *for flow cytometry* *100 tests*	1 each	15.4	
B35139	BrdU, mouse monoclonal antibody (Clone MoBU-1), Alexa Fluor® 488 conjugate *for flow cytometry* *100 tests*	1 each	15.4	
B35140	BrdU, mouse monoclonal antibody (Clone MoBU-1), Alexa Fluor® 647 conjugate *for flow cytometry* *100 tests*	1 each	15.4	
B7450	brefeldin A *from *Penicillium brefeldianum**	5 mg	12.4	
B1690	5-bromo-4-chloro-3-indolyl β-D-galactopyranoside (X-Gal)	1 g	10.2	
B1691	5-bromo-4-chloro-3-indolyl β-D-glucuronide, cyclohexylammonium salt (X-GlcU, CHA)	100 mg	10.2	
B6057	6-bromoacetyl-2-dimethylaminonaphthalene (badan)	10 mg	2.3, 3.4	
B1494	4-bromo A-23187, free acid	1 mg	19.8	
B23151	5-bromo-2′-deoxyuridine (BrdU)	100 mg	15.4	
B21550	5-bromo-2′-deoxyuridine 5′-triphosphate (BrdUTP) *10 mM in TE buffer*	25 µL	15.4	
B1355	5-(bromomethyl)fluorescein	10 mg	2.2, 3.4	
B21551	5-bromouridine 5′-triphosphate (BrUTP) *10 mM in TE buffer*	25 µL	15.4	
B6791	BTC, AM *cell permeant*	100 µg	19.2	
B6790	BTC, tetrapotassium salt *cell impermeant*	1 mg	19.2	
B1601	α-bungarotoxin *from *Bungarus multicinctus**	1 mg	16.2	
B35451	α-bungarotoxin, Alexa Fluor® 555 conjugate	500 µg	16.2	
B35452	α-bungarotoxin, Alexa Fluor® 680 conjugate	500 µg	16.2	
B13422	α-bungarotoxin, Alexa Fluor® 488 conjugate	500 µg	16.2	
B13423	α-bungarotoxin, Alexa Fluor® 594 conjugate	500 µg	16.2	
B35450	α-bungarotoxin, Alexa Fluor® 647 conjugate	500 µg	16.2	
B1196	α-bungarotoxin, biotin-XX conjugate	500 µg	4.3, 16.2	
C481	calcein *high purity*	100 mg	14.3, 16.1, 19.7	
C1430	calcein, AM	1 mg	14.3, 15.2, 15.6, 19.7	
C3099	calcein, AM *1 mg/mL solution in anhydrous DMSO*	1 mL	14.3, 15.2, 15.6, 19.7	
C3100MP	calcein, AM *special packaging*	20 x 50 µg	14.3, 15.2, 15.6, 19.7	
C1429	calcein blue, AM	1 mg	15.2	
C34853	calcein blue, AM *for flow cytometry* *for UV excitation* *special packaging*	20 x 50 µg	15.2	
C34852	calcein green, AM *for flow cytometry* *for 488 nm excitation* *special packaging*	20 x 50 µg	15.2	
C34851	calcein red-orange, AM *special packaging*	20 x 50 µg	15.2	
C34858	calcein violet, AM *for flow cytometry* *for 405 nm excitation* *special packaging*	20 x 25 µg	15.2	
C3008MP	Calcium Calibration Buffer Kit #1 *zero and 10 mM CaEGTA (2 x 50 mL)*	1 kit	19.8	
C3018	Calcium Crimson™, AM *cell permeant* *special packaging*	10 x 50 µg	19.3	
C3011MP	Calcium Green™-1, AM *cell permeant*	500 µg	19.3	
C3012	Calcium Green™-1, AM *cell permeant* *special packaging*	10 x 50 µg	19.3	
C6765	Calcium Green™-1 dextran, potassium salt, 3000 MW, anionic	5 mg	19.4	
C3713	Calcium Green™-1 dextran, potassium salt, 10,000 MW, anionic	5 mg	19.4	
C3714	Calcium Green™-1 dextran, potassium salt, 70,000 MW, anionic	5 mg	19.4	
C3010MP	Calcium Green™-1, hexapotassium salt *cell impermeant*	500 µg	19.3	
C3732	Calcium Green™-2, AM *cell permeant* *special packaging*	10 x 50 µg	19.3	
C3730	Calcium Green™-2, octapotassium salt *cell impermeant*	500 µg	19.3	
C3739	Calcium Green™-5N, AM *cell permeant* *special packaging*	10 x 50 µg	19.3	
C3737	Calcium Green™-5N, hexapotassium salt *cell impermeant*	500 µg	19.3	
C3015	Calcium Orange™, AM *cell permeant* *special packaging*	10 x 50 µg	19.3	
C3013	Calcium Orange™, tetrapotassium salt *cell impermeant*	500 µg	19.3	
C3047	Calcium Sponge™ S (BAPTA polystyrene)	1 g	19.8	
C10558	cAMP Chemiluminescent Immunoassay Kit *10-plate size*	1 kit	17.3	
C10557	cAMP Chemiluminescent Immunoassay Kit *2-plate size*	1 kit	17.3	
C21852	CandyCane™ glycoprotein molecular weight standards *200 gel lanes*	400 µL	9.4	
C21386	CaptAvidin™ agarose *sedimented bead suspension*	5 mL	7.6	

*Note that all products listed in the above Master Product List are covered by **Limited Use Label License #223**. If a specific product is covered by additional Limited Use Label License(s), the Limited Use Label License #'s are provided in the LULL column above. Please refer to page 971 for the complete text of the Limited Use Label License(s).

IMPORTANT NOTICE: The products described in this manual are covered by one or more Limited Use Label License(s). Please refer to the Appendix on page 971 and Master Product List on page 975. Products are For Research Use Only. Not intended for any animal or human therapeutic or diagnostic use.

www.invitrogen.com/probes

molecular probes® | invitrogen™ by life technologies™

Cat. No.	Product	Quantity	Sections	LULL*
C21385	CaptAvidin™ biotin-binding protein	1 mg	7.6	
C356	5-(((2-(carbohydrazino)methyl)thio)acetyl)aminofluorescein	25 mg	3.3	
C400	5-(and-6)-carboxy-2',7'-dichlorodihydrofluorescein diacetate (carboxy-H_2DCFDA) *mixed isomers*	25 mg	10.5, 15.2, 18.2	
C2938	6-carboxy-2',7'-dichlorodihydrofluorescein diacetate, di(acetoxymethyl ester)	5 mg	15.2, 18.2	
C6171MP	6-carboxy-4',5'-dichloro-2',7'-dimethoxyfluorescein, succinimidyl ester (6-JOE, SE)	5 mg	1.5	
C368	5-(and-6)-carboxy-2',7'-dichlorofluorescein *mixed isomers*	100 mg	14.3, 20.3	
C369	5-(and-6)-carboxy-2',7'-dichlorofluorescein diacetate (carboxy-DCFDA) *mixed isomers*	100 mg	15.2, 20.3	
C13293	5-(and-6)-carboxy-2',7'-difluorodihydrofluorescein diacetate (carboxy-H_2DFFDA) *mixed isomers*	5 mg	10.5, 18.2	
C1359	5-carboxyfluorescein (5-FAM) *single isomer*	100 mg	1.5	
C1360	6-carboxyfluorescein (6-FAM) *single isomer*	100 mg	1.5	
C1904	5-(and-6)-carboxyfluorescein (5(6)-FAM) *FluoroPure™ grade* *mixed isomers*	100 mg	1.5, 14.3, 20.2	
C194	5-(and-6)-carboxyfluorescein *mixed isomers*	5 g	14.3, 20.2	
C20050	5-carboxyfluorescein-bis-(5-carboxymethoxy-2-nitrobenzyl) ether, β-alanine-carboxamide, succinimidyl ester (CMNB-caged carboxyfluorescein, SE)	1 mg	1.5, 5.3	
C2210	5-carboxyfluorescein, succinimidyl ester (5-FAM, SE) *single isomer*	10 mg	1.5	
C6164	6-carboxyfluorescein, succinimidyl ester (6-FAM, SE) *single isomer*	10 mg	1.5	
C1311	5-(and-6)-carboxyfluorescein, succinimidyl ester (5(6)-FAM, SE) *mixed isomers*	100 mg	1.5	
C652	5-(and-6)-carboxynaphthofluorescein *mixed isomers*	100 mg	1.6, 20.2	
C13196	5-(and-6)-carboxynaphthofluorescein diacetate	10 mg	15.2, 20.2	
C653	5-(and-6)-carboxynaphthofluorescein, succinimidyl ester *mixed isomers*	25 mg	1.6, 20.4	
C7924	5-(2-carboxyphenyl)-5-hydroxy-1-((2,2,5,5-tetramethyl-1-oxypyrrolidin-3-yl)methyl)-3-phenyl-2-pyrrolin-4-one, potassium salt (proxyl fluorescamine)	5 mg	18.2	
C7912	2-(4-carboxyphenyl)-4,4,5,5-tetramethylimidazoline-1-oxyl-3-oxide, potassium salt (carboxy-PTIO)	25 mg	18.3	
C6128	6-carboxyrhodamine 6G, succinimidyl ester (6-CR 6G, SE) *single isomer*	5 mg	1.6	
C6127	5-carboxyrhodamine 6G, succinimidyl ester (5-CR 6G, SE) *single isomer*	5 mg	1.6	
C6157	5-(and-6)-carboxyrhodamine 6G, succinimidyl ester (5(6)-CR 6G, SE) *mixed isomers*	5 mg	1.6	
C1270	5-(and-6)-carboxy SNARF®-1	1 mg	20.2	
C1271	5-(and-6)-carboxy SNARF®-1, acetoxymethyl ester, acetate	1 mg	20.2	
C1272	5-(and-6)-carboxy SNARF®-1, acetoxymethyl ester, acetate *special packaging*	20 x 50 µg	20.2	
C6121	5-carboxytetramethylrhodamine (5-TAMRA) *single isomer*	10 mg	1.6	
C6122	6-carboxytetramethylrhodamine (6-TAMRA) *single isomer*	10 mg	1.6	
C300	5-(and-6)-carboxytetramethylrhodamine (5(6)-TAMRA) *mixed isomers*	100 mg	1.6	
C2211	5-carboxytetramethylrhodamine, succinimidyl ester (5-TAMRA, SE) *single isomer*	5 mg	1.6	
C6123	6-carboxytetramethylrhodamine, succinimidyl ester (6-TAMRA, SE) *single isomer*	5 mg	1.6	
C1171	5-(and-6)-carboxytetramethylrhodamine, succinimidyl ester (5(6)-TAMRA, SE) *mixed isomers*	25 mg	1.6	
C6124	5-carboxy-X-rhodamine, triethylammonium salt (5-ROX) *single isomer*	10 mg	1.6	
C6156	6-carboxy-X-rhodamine (6-ROX) *single isomer*	10 mg	1.6	
C6125	5-carboxy-X-rhodamine, succinimidyl ester (5-ROX, SE) *single isomer*	5 mg	1.6	
C6126	6-carboxy-X-rhodamine, succinimidyl ester (6-ROX, SE) *single isomer*	5 mg	1.6	
C1309	5-(and-6)-carboxy-X-rhodamine, succinimidyl ester (5(6)-ROX, SE) *mixed isomers*	25 mg	1.6	
C2284	Cascade Blue® acetyl azide, trisodium salt	5 mg	1.7, 4.2, 14.3	
C621	Cascade Blue® ethylenediamine, trisodium salt	10 mg	3.4, 14.3	
C962	Cascade Blue® goat anti-mouse IgG (H+L) *2 mg/mL*	0.5 mL	7.2	
C2764	Cascade Blue® goat anti-rabbit IgG (H+L) *2 mg/mL*	0.5 mL	7.2	
C3239	Cascade Blue® hydrazide, trilithium salt	10 mg	14.3	
C3221	Cascade Blue® hydrazide, tripotassium salt	10 mg	14.3	
C687	Cascade Blue® hydrazide, trisodium salt	10 mg	3.3, 14.3, 16.1	
C10164	Cascade Yellow™ succinimidyl ester	5 mg	1.7	
C2990	casein, fluorescein conjugate	25 mg	10.4, 15.6, 16.1	
C6667	CBQCA Protein Quantitation Kit *300–800 assays*	1 kit	9.2	244
C1165	CDCFDA, SE (5-(and-6)-carboxy-2',7'-dichlorofluorescein diacetate, succinimidyl ester) *mixed isomers*	25 mg	14.2	
Q10154	CD10, mouse anti-human, Qdot® 800 conjugate (clone MEM-78) *1 µM solution*	100 µL	6.6	
Q10013	CD14, mouse anti-human, Qdot® 605 conjugate (clone TьK4) *2 µM solution*	100 µL	6.6	
Q10056	CD14, mouse anti-human, Qdot® 655 conjugate (clone TьK4) *1 µM solution*	100 µL	6.6	
Q10064	CD14, mouse anti-human, Qdot® 800 conjugate (clone TьK4) *1 µM solution*	100 µL	6.6	
Q10306	CD19, mouse anti-human, Qdot® 605 conjugate (clone SJ25-C1) *1 µM solution*	100 µL	6.6	
Q10305	CD20, mouse anti-human, Qdot® 655 conjugate (clone HI47) *1 µM solution*	100 µL	6.6	
Q10065	CD27, mouse anti-human, Qdot® 605 conjugate (clone CLB-27/1) *2 µM solution*	100 µL	6.6	
Q10066	CD27, mouse anti-human, Qdot® 655 conjugate (clone CLB-27/1) *2 µM solution*	100 µL	6.6	

*Note that all products listed in the above Master Product List are covered by **Limited Use Label License #223**. If a specific product is covered by additional Limited Use Label License(s), the Limited Use Label License #'s are provided in the LULL column above. Please refer to page 971 for the complete text of the Limited Use Label License(s).

IMPORTANT NOTICE: The products described in this manual are covered by one or more Limited Use Label License(s). Please refer to the Appendix on page 971 and Master Product List on page 975. Products are For Research Use Only. Not intended for any animal or human therapeutic or diagnostic use.

Cat. No.	Product	Quantity	Sections	LULL*
Q10012	CD3, mouse anti-human, Qdot® 655 conjugate (clone S4.1) *1 µM solution*	100 µL	6.6	
Q10053	CD38, mouse anti-human, Qdot® 605 conjugate (clone HIT2) *2 µM solution*	100 µL	6.6	
Q10057	CD38, mouse anti-human, Qdot® 655 conjugate (clone HIT2) *2 µM solution*	100 µL	6.6	
Q10051	CD45, mouse anti-human, Qdot® 605 conjugate (clone HI30) *1 µM solution*	100 µL	6.6	
Q10062	CD45, mouse anti-human, Qdot® 705 conjugate (clone HI30) *1 µM solution*	100 µL	6.6	
Q10156	CD45, mouse anti-human, Qdot® 800 conjugate (clone HI30) *1 µM solution*	100 µL	6.6	
Q10060	CD4, mouse anti-human, Qdot® 705 conjugate (clone S3.5) *1 µM solution*	100 µL	6.6	
Q10008	CD4, mouse anti-human, Qdot® 605 conjugate (clone S3.5) *1 µM solution*	100 µL	6.6	
Q10007	CD4, mouse anti-human, Qdot® 655 conjugate (clone S3.5) *1 µM solution*	100 µL	6.6	
Q10009	CD8, mouse anti-human, Qdot® 605 conjugate (clone 3B5) *1 µM solution*	100 µL	6.6	
Q10055	CD8, mouse anti-human, Qdot® 655 conjugate (clone 3B5) *1 µM solution*	100 µL	6.6	
Q10059	CD8, mouse anti-human, Qdot® 705 conjugate (clone 3B5) *1 µM solution*	100 µL	6.6	
Q10152	CD8, mouse anti-human, Qdot® 565 conjugate (clone 3B5) *1 µM solution*	100 µL	6.6	
Q10153	CD10, mouse anti-human, Qdot® 605 conjugate (clone MEM-78) *2 µM solution*	100 µL	6.6	
Q10307	CD56, mouse anti-human, Qdot® 605 conjugate (clone MEM-188) *2 µM solution*	100 µL	6.6	
Q10172	CD2, mouse anti-human, Qdot® 605 conjugate (clone S5.5) *1 µM solution*	100 µL	6.6	
Q10179	CD19, mouse anti-human, Qdot® 655 conjugate (clone SJ25-C1) *1 µM solution*	100 µL	6.6	
Q10069	CD45RA, mouse anti-human, Qdot® 655 conjugate (clone MEM-56) *1 µM solution*	100 µL	6.6	
Q10047	CD45RA, mouse anti-human, Qdot® 605 conjugate (clone MEM-56) *1 µM solution*	100 µL	6.6	
Q10054	CD3, mouse anti-human, Qdot® 605 conjugate (clone UCHT1) *1 µM solution*	100 µL	6.6	
C22803	CEDA, SE (5-(and-6)-carboxyeosin diacetate, succinimidyl ester) *mixed isomers*	5 mg	14.2, 15.4, 16.3	
C7028	Cell Culture Contamination Detection Kit *200 assays*	1 kit	15.4	
C10582	CellLight® Actin-GFP *BacMam 2.0*	1 mL	11.1	21, 127, 198, 267, 272, 306, 308, 332
C10583	CellLight® Actin-RFP *BacMam 2.0*	1 mL	11.1	21, 306, 308, 321, 332
C10586	CellLight® Early Endosomes-GFP *BacMam 2.0*	1 mL	12.3	21, 127, 198, 267, 272, 306, 308, 332
C10587	CellLight® Early Endosomes-RFP *BacMam 2.0*	1 mL	12.3	21, 306, 308, 321, 332
C10590	CellLight® ER-GFP *BacMam 2.0*	1 mL	12.4	21, 127, 198, 267, 272, 306, 308, 332
C10591	CellLight® ER-RFP *BacMam 2.0*	1 mL	12.4	21, 306, 308, 321, 332
C10592	CellLight® Golgi-GFP *BacMam 2.0*	1 mL	12.4	21, 127, 198, 267, 272, 306, 308, 332
C10593	CellLight® Golgi-RFP *BacMam 2.0*	1 mL	12.4	21, 306, 308, 321, 332
C10594	CellLight® Histone 2B-GFP *BacMam 2.0*	1 mL	12.5	21, 127, 198, 267, 272, 306, 308, 332
C10595	CellLight® Histone 2B-RFP *BacMam 2.0*	1 mL	12.5	21, 306, 308, 321, 332
C10596	CellLight® Lysosomes-GFP *BacMam 2.0*	1 mL	12.3	21, 127, 198, 267, 272, 306, 308, 332
C10597	CellLight® Lysosomes-RFP *BacMam 2.0*	1 mL	12.3	21, 306, 308, 321, 332
C10598	CellLight® MAP4-GFP *BacMam 2.0*	1 mL	11.2	21, 127, 198, 267, 272, 306, 308, 332
C10599	CellLight® MAP4-RFP *BacMam 2.0*	1 mL	11.2	21, 306, 308, 321, 332
C10600	CellLight® Mitochondria-GFP *BacMam 2.0*	1 mL	12.2	21, 127, 198, 267, 272, 306, 308, 332
C10601	CellLight® Mitochondria-RFP *BacMam 2.0*	1 mL	12.2	21, 306, 308, 321, 332
C10602	CellLight® Nucleus-GFP *BacMam 2.0*	1 mL	12.5	21, 127, 198, 267, 272, 306, 308, 332

*Note that all products listed in the above Master Product List are covered by **Limited Use Label License #223**. If a specific product is covered by additional Limited Use Label License(s), the Limited Use Label License #'s are provided in the LULL column above. Please refer to page 971 for the complete text of the Limited Use Label License(s).

IMPORTANT NOTICE: The products described in this manual are covered by one or more Limited Use Label License(s). Please refer to the Appendix on page 971 and Master Product List on page 975. Products are For Research Use Only. Not intended for any animal or human therapeutic or diagnostic use.

Cat. No.	Product	Quantity	Sections	LULL*
C10603	CellLight® Nucleus-RFP *BacMam 2.0*	1 mL	12.5	21, 306, 308, 321, 332
C10615	CellLight® Null (control) *BacMam 2.0*	1 mL	11.1	
C10604	CellLight® Peroxisome-GFP *BacMam 2.0*	1 mL	12.3	21, 127, 198, 267, 272, 306, 308, 332
C10606	CellLight® Plasma Membrane-CFP *BacMam 2.0*	1 mL	14.4	21, 127, 198, 267, 272, 306, 308, 332
C10607	CellLight® Plasma Membrane-GFP *BacMam 2.0*	1 mL	14.4	21, 127, 198, 267, 272, 306, 308, 332
C10608	CellLight® Plasma Membrane-RFP *BacMam 2.0*	1 mL	14.4	21, 306, 308, 321, 332
C10611	CellLight® Talin-GFP *BacMam 2.0*	1 mL	11.2	21, 127, 198, 267, 272, 306, 308, 332
C10612	CellLight® Talin-RFP *BacMam 2.0*	1 mL	11.2	21, 306, 308, 321, 332
C10613	CellLight® Tubulin-GFP *BacMam 2.0*	1 mL	11.2	21, 127, 198, 267, 272, 306, 308, 332
C10614	CellLight® Tubulin-RFP *BacMam 2.0*	1 mL	11.2	21, 306, 308, 321, 332
C10046	CellMask™ Deep Red plasma membrane stain *5 mg/mL solution in DMSO*	100 µL	14.4	
C10045	CellMask™ Orange plasma membrane stain *5 mg/mL solution in DMSO*	100 µL	14.4	
C34554	CellTrace™ CFSE Cell Proliferation Kit *for flow cytometry*	1 kit	14.2, 15.4	
C34553	CellTrace™ Far Red DDAO-SE *special packaging*	20 x 50 µg	14.2	
C34555	CellTrace™ Oregon Green® 488 carboxylic acid diacetate, succinimidyl ester (carboxy-DFFDA, SE) *cell permeant* *mixed isomers*	20 x 50 µg	14.2, 15.4	
C34557	CellTrace™ Violet Cell Proliferation Kit *for flow cytometry*	1 kit	14.2, 15.4	
C2110	CellTracker™ Blue CMAC (7-amino-4-chloromethylcoumarin)	5 mg	10.1, 14.2	
C12881	CellTracker™ Blue CMF_2HC (4-chloromethyl-6,8-difluoro-7-hydroxycoumarin)	5 mg	10.1, 14.2	
C2111	CellTracker™ Blue CMHC (4-chloromethyl-7-hydroxycoumarin)	5 mg	14.2	
C7001	CellTracker™ CM-DiI	1 mg	13.4, 14.4	
C7000	CellTracker™ CM-DiI *special packaging*	20 x 50 µg	13.4, 14.4	
C2102	CellTracker™ Green BODIPY® (8-chloromethyl-4,4-difluoro-1,3,5,7-tetramethyl-4-bora-3a,4a-diaza-*s*-indacene)	5 mg	14.2	
C2925	CellTracker™ Green CMFDA (5-chloromethylfluorescein diacetate)	1 mg	14.2, 15.2, 15.6, 20.2	
C7025	CellTracker™ Green CMFDA (5-chloromethylfluorescein diacetate) *special packaging*	20 x 50 µg	14.2, 15.2, 15.6, 20.2	
C34551	CellTracker™ Orange CMRA *special packaging*	20 x 50 µg	14.2	
C2927	CellTracker™ Orange CMTMR (5-(and-6)-(((4-chloromethyl)benzoyl)amino)tetramethylrhodamine) *mixed isomers*	1 mg	14.2	
C34552	CellTracker™ Red CMTPX *special packaging*	20 x 50 µg	14.2	
C10094	CellTracker™ Violet BMQC (2,3,6,7-tetrahydro-9-bromomethyl-1*H*,5*H*-quinolizino(9,1-*gh*)coumarin)	5 x 100 µg	14.2	
C1361	5-CFDA (5-carboxyfluorescein diacetate) *single isomer*	100 mg	15.2	
C1362	6-CFDA (6-carboxyfluorescein diacetate) *single isomer*	100 mg	15.2	
C195	5(6)-CFDA (5-(and-6)-carboxyfluorescein diacetate) *mixed isomers*	100 mg	15.2, 20.2	
C1354	5-CFDA, AM (5-carboxyfluorescein diacetate, acetoxymethyl ester)	5 mg	15.2, 20.2	
C1157	5(6)-CFDA, SE; CFSE (5-(and-6)-carboxyfluorescein diacetate, succinimidyl ester) *mixed isomers*	25 mg	14.2, 15.4, 20.2	
C10552	Chemiluminescent Alkaline Phosphatase ELISA Kit #1 *with CSPD® Substrate/ Sapphire-II™ enhancer* *1000 assays*	1 kit	6.3, 10.3	
C10553	Chemiluminescent Alkaline Phosphatase ELISA Kit #2 *with CSPD® Substrate/ Emerald-II™ enhancer* *1000 assays*	1 kit	6.3, 10.3	
C10554	Chemiluminescent Alkaline Phosphatase ELISA Kit #3 *with CDP-*Star*® Substrate/ Sapphire-II™ enhancer* *1000 assays*	1 kit	6.3, 10.3	
C10555	Chemiluminescent Alkaline Phosphatase ELISA Kit #4 *with CDP-*Star*® Substrate/ Emerald-II™ enhancer* *1000 assays*	1 kit	6.3, 10.3	
C10556	Chemiluminescent Alkaline Phosphatase ELISA Sampler Kit *1000 assays*	1 kit	6.3, 10.3	
C6827	5-(and-6)-chloromethyl-2',7'-dichlorodihydrofluorescein diacetate, acetyl ester (CM-H_2DCFDA) *mixed isomers* *special packaging*	20 x 50 µg	14.2, 15.2, 18.2	
C6826	5-(and-6)-chloromethyl SNARF®-1, acetate *mixed isomers* *special packaging*	20 x 50 µg	14.2, 15.2, 15.6, 20.2	
C20260	4-chloro-7-nitrobenz-2-oxa-1,3-diazole (NBD chloride; 4-chloro-7-nitrobenzofurazan) *FluoroPure™ grade*	100 mg	1.8, 2.2	

*Note that all products listed in the above Master Product List are covered by **Limited Use Label License #223**. If a specific product is covered by additional Limited Use Label License(s), the Limited Use Label License #'s are provided in the LULL column above. Please refer to page 971 for the complete text of the Limited Use Label License(s).

Cat. No.	Product	Quantity	Sections	LULL*
C22220	6-chloro-9-nitro-5-oxo-5*H*-benzo[a]phenoxazine (CNOB)	1 mg	10.6	
C22841	cholera toxin subunit B (recombinant), Alexa Fluor® 488 conjugate	500 µg	7.7, 14.7, 16.1	
C34775	cholera toxin subunit B (recombinant), Alexa Fluor® 488 conjugate	100 µg	7.7, 14.7, 16.1	
C22843	cholera toxin subunit B (recombinant), Alexa Fluor® 555 conjugate	500 µg	7.7, 14.7, 16.1	
C34776	cholera toxin subunit B (recombinant), Alexa Fluor® 555 conjugate	100 µg	7.7, 14.7, 16.1	
C22842	cholera toxin subunit B (recombinant), Alexa Fluor® 594 conjugate	500 µg	7.7, 14.7, 16.1	
C34777	cholera toxin subunit B (recombinant), Alexa Fluor® 594 conjugate	100 µg	7.7, 14.7, 16.1	
C34778	cholera toxin subunit B (recombinant), Alexa Fluor® 647 conjugate	100 µg	7.7, 14.7, 16.1	
C34779	cholera toxin subunit B (recombinant), biotin-XX conjugate	100 µg	4.3, 7.7, 14.7, 16.1	
C34780	cholera toxin subunit B (recombinant), horseradish peroxidase conjugate	100 µg	7.7, 14.7, 16.1	
C12680	cholesteryl BODIPY® 542/563 C_{11} (cholesteryl 4,4-difluoro-5-(4-methoxyphenyl)-4-bora-3a,4a-diaza-*s*-indacene-3-undecanoate)	1 mg	13.3	
C12681	cholesteryl BODIPY® 576/589 C_{11} (cholesteryl 4,4-difluoro-5-(2-pyrrolyl)-4-bora-3a,4a-diaza-*s*-indacene-3-undecanoate)	1 mg	13.3	
C3927MP	cholesteryl BODIPY® FL C_{12} (cholesteryl 4,4-difluoro-5,7-dimethyl-4-bora-3a,4a-diaza-*s*-indacene-3-dodecanoate)	1 mg	13.3	
C11397	ChromaTide® Alexa Fluor® 488-5-dUTP *1 mM in TE buffer*	25 µL	8.2	
C11403	ChromaTide® Alexa Fluor® 488-5-UTP *1 mM in TE buffer*	25 µL	8.2	
C11398	ChromaTide® Alexa Fluor® 532-5-dUTP *1 mM in TE buffer*	25 µL	8.2	
C11401	ChromaTide® Alexa Fluor® 546-14-dUTP *1 mM in TE buffer*	25 µL	8.2	
C11404	ChromaTide® Alexa Fluor® 546-14-UTP *1 mM in TE buffer*	25 µL	8.2	
C11399	ChromaTide® Alexa Fluor® 568-5-dUTP *1 mM in TE buffer*	25 µL	8.2	
C11400	ChromaTide® Alexa Fluor® 594-5-dUTP *1 mM in TE buffer*	25 µL	8.2	
C21555	ChromaTide® Alexa Fluor® 488-7-OBEA-dCTP *1 mM in TE buffer*	50 µL	8.2	
C21556	ChromaTide® Alexa Fluor® 546-16-OBEA-dCTP *1 mM in TE buffer*	50 µL	8.2	
C21559	ChromaTide® Alexa Fluor® 647-12-OBEA-dCTP *1 mM in TE buffer*	50 µL	8.2	
C7614	ChromaTide® BODIPY® FL-14-dUTP *1 mM in TE buffer*	25 µL	8.2, 15.4, 15.5	
C7604	ChromaTide® fluorescein-12-dUTP *1 mM in TE buffer*	25 µL	8.2, 15.5	225
C7606MP	ChromaTide® tetramethylrhodamine-6-dUTP *1 mM in TE buffer*	25 µL	8.2	225
C7631	ChromaTide® Texas Red®-12-dUTP *1 mM in TE buffer*	25 µL	8.2	
V13244	Chromatin Condensation/Dead Cell Apoptosis Kit *Hoechst 33342/propidium iodide* *200 assays* *for flow cytometry*	1 kit	15.5	
V23201	Chromatin Condesation/Membrane Permeability/Dead Cell Apoptosis Kit. *Hoechst 33342/YO-PRO®-1/propidium iodide* *200 assays* *for flow cytometry*	1 kit	15.5	
C10102	Click-iT® AHA (L-azidohomoalanine) *for nascent protein synthesis*	5 mg	3.1, 9.4	
C10289	Click-iT® AHA Alexa Fluor® 488 Protein Synthesis HCS Assay *2-plate size*	1 kit	3.1, 9.4	
C33372	Click-iT® Biotin Protein Analysis Detection Kit *10 reactions*	1 kit	3.1, 9.4	
C10269	Click-iT® Cell Reaction Buffer Kit	1 kit	3.1, 9.4	
C33371	Click-iT® Dapoxyl® Protein Analysis Detection Kit *for UV excitation* *10 reactions*	1 kit	3.1, 9.4	
A10202	Click-iT® EdU Alexa Fluor® 647 Flow Cytometry Assay Kit *50 assays*	1 kit	3.1, 15.4	
C35002	Click-iT® EdU Alexa Fluor® 488 Flow Cytometry Assay Kit *50 assays*	1 kit	3.1, 15.4	
C10351	Click-iT® EdU Alexa Fluor® 488 HCS Assay *10-plate size*	1 kit	3.1, 15.4	289
C10350	Click-iT® EdU Alexa Fluor® 488 HCS Assay *2-plate size*	1 kit	3.1, 15.4	289
C10355	Click-iT® EdU Alexa Fluor® 594 HCS Assay *10-plate size*	1 kit	3.1, 15.4	289
C10353	Click-iT® EdU Alexa Fluor® 555 HCS Assay *10-plate size*	1 kit	3.1, 15.4	289
C10352	Click-iT® EdU Alexa Fluor® 555 HCS Assay *2-plate size*	1 kit	3.1, 15.4	289
C10354	Click-iT® EdU Alexa Fluor® 594 HCS Assay *2-plate size*	1 kit	3.1, 15.4	289
C10357	Click-iT® EdU Alexa Fluor® 647 HCS Assay *10-plate size*	1 kit	3.1, 15.4	289
C10356	Click-iT® EdU Alexa Fluor® 647 HCS Assay *2-plate size*	1 kit	3.1, 15.4	289
C10337	Click-iT® EdU Alexa Fluor® 488 Imaging Kit *for 50 coverslips*	1 kit	3.1, 15.4	
C10338	Click-iT® EdU Alexa Fluor® 555 Imaging Kit *for 50 coverslips*	1 kit	3.1, 15.4	
C10339	Click-iT® EdU Alexa Fluor® 594 Imaging Kit *for 50 coverslips*	1 kit	3.1, 15.4	
C10340	Click-iT® EdU Alexa Fluor® 647 Imaging Kit *for 50 coverslips*	1 kit	3.1, 15.4	
A10034	Click-iT® EdU Pacific Blue™ Flow Cytometry Assay Kit *50 assays*	1 kit	3.1, 15.4	
C10248	Click-iT® farnesyl alcohol, azide *mixed isomers*	1 mg	3.1, 9.4	
C10264	Click-iT® fucose alkyne (tetraacetylfucose alkyne)	5 mg	3.1, 9.4	
C33365	Click-iT® GalNAz metabolic glycoprotein labeling reagent (tetraacetylated *N*-azidoacetylgalactosamine) *for *O*-linked glycoproteins* *5.2 mg*	each	3.1, 9.4	
C10249	Click-iT® geranylgeranyl alcohol, azide *mixed isomers*	1 mg	3.1, 9.4	

*Note that all products listed in the above Master Product List are covered by **Limited Use Label License #223**. If a specific product is covered by additional Limited Use Label License(s), the Limited Use Label License #'s are provided in the LULL column above. Please refer to page 971 for the complete text of the Limited Use Label License(s).

IMPORTANT NOTICE: The products described in this manual are covered by one or more Limited Use Label License(s). Please refer to the Appendix on page 971 and Master Product List on page 975. Products are For Research Use Only. Not intended for any animal or human therapeutic or diagnostic use.

Cat. No.	Product	Quantity	Sections	LULL*
C33367	Click-iT® GlcNAz metabolic glycoprotein labeling reagent (tetraacetylated *N*-azidoacetylglucosamine) *for *O*-GlcNAc-modified proteins* *5.2 mg*	each	3.1, 9.4	
C10186	Click-iT® HPG (L-homopropargylglycine) *for nascent protein synthesis*	5 mg	3.1, 9.4	
C33366	Click-iT® ManNAz metabolic glycoprotein labeling reagent (tetraacetylated *N*-azidoacetyl-d-mannosamine) *for sialic acid glycoproteins* *5.2 mg*	each	3.1, 9.4	
C10268	Click-iT® myristic acid, azide (12-azidododecanoic acid)	1 mg	3.1, 9.4	
C33368	Click-iT® *O*-GlcNAc Enzymatic Labeling System *for O-linked GlcNAc glycoproteins* *10 labelings*	1 kit	3.1, 9.4	
C33373	Click-iT® *O*-GlcNAc peptide and phosphopeptide LC/MS standards *5 nmol each*	1 set	9.4	
C33374	Click-iT® *O*-GlcNAc peptide LC/MS standard (H-Thr-Ala-Pro-Thr-(O-GlcNAc)Ser-Thr-Ile-Ala-Pro-Gly-OH) *Theoretical Mass (M+H): 1118.50*	5 nmol	9.4	
C10265	Click-iT® palmitic acid, azide (15-azidopentadecanoic acid)	1 mg	3.1, 9.4	
C10276	Click-iT® Protein Reaction Buffer Kit	1 kit	3.1, 9.4	
C10327	Click-iT® RNA Alexa Fluor® 488 HCS Assay *2-plate size*	1 kit	3.1	289
C10328	Click-iT® RNA Alexa Fluor® 594 HCS Assay *2-plate size*	1 kit	3.1	289
C10329	Click-iT® RNA Alexa Fluor® 488 Imaging Kit *for 25 coverslips*	1 kit	3.1	
C10330	Click-iT® RNA Alexa Fluor® 594 Imaging Kit *for 25 coverslips*	1 kit	3.1	
C33370	Click-iT® Tetramethylrhodamine (TAMRA) Protein Analysis Detection Kit *UV/532 nm excitation* *10 reactions*	1 kit	3.1, 9.4	
C10245	Click-iT® TUNEL Alexa Fluor® 488 Imaging Assay *for microscopy and HCS* *50–100 assays*	1 kit	3.1, 15.5	
C10246	Click-iT® TUNEL Alexa Fluor® 594 Imaging Assay *for microscopy and HCS* *50–100 assays*	1 kit	3.1, 15.5	
C10247	Click-iT® TUNEL Alexa Fluor® 647 Imaging Assay *for microscopy and HCS* *50–100 assays*	1 kit	3.1, 15.5	
C10214	Click-iT® EdU Microplate Assay *400 assays*	1 kit	3.1, 15.4	
C13654	*N*-(CNB-caged) carbachol (*N*-(α-carboxy-2-nitrobenzyl)carbamylcholine, trifluoroacetic acid salt)	5 mg	5.3, 16.2	
C7122	*N*-(CNB-caged) L-glutamic acid (*N*-(α-carboxy-2-nitrobenzyl)-L-glutamic acid)	5 mg	5.3, 16.2	
C2944	coelenterazine	250 µg	10.6, 18.2, 19.5	
C14260	coelenterazine cp	250 µg	10.6, 19.5	
C6779	coelenterazine f	250 µg	10.6, 19.5	
C6780	coelenterazine h	250 µg	10.6, 19.5	
C11254	concanavalin A, Alexa Fluor® 350 conjugate	5 mg	7.7	
C11252	concanavalin A, Alexa Fluor® 488 conjugate	5 mg	7.7	
C11253	concanavalin A, Alexa Fluor® 594 conjugate	5 mg	7.7	
C21402	concanavalin A, Alexa Fluor® 633 conjugate	5 mg	7.7	
C21421	concanavalin A, Alexa Fluor® 647 conjugate	5 mg	7.7	
C827	concanavalin A, fluorescein conjugate	10 mg	7.7	
C21401	concanavalin A, succinylated, Alexa Fluor® 488 conjugate	5 mg	7.7	
C860	concanavalin A, tetramethylrhodamine conjugate	10 mg	7.7	
C825	concanavalin A, Texas Red® conjugate	10 mg	7.7	
C14837	Constellation™ microspheres for imaging *mixture of assorted sizes and colors*	3 mL	6.5, 23.1	
C33250	Coomassie Fluor™ Orange protein gel stain *ready-to-use solution*	1 L	9.3	
C33251	Coomassie Fluor™ Orange protein gel stain *ready-to-use solution* *bulk packaging*	5 L	9.3	
C36675	CoroNa™ Green *cell impermeant*	1 mg	21.1	
C36676	CoroNa™ Green, AM *cell permeant* *special packaging*	20 x 50 µg	21.1	
C24430	CoroNa™ Red chloride	1 mg	21.1	
C24431	CoroNa™ Red chloride *special packaging*	20 x 50 µg	21.1	
C606	coumarin phallacidin	300 U	11.1	
C36950	CountBright™ absolute counting beads *for flow cytometry* *100 tests*	5 mL	15.4, 23.2	
C10227	Countess® automated cell counter *with box of 50 cell counting chambers and Trypan Blue*	1 kit	15.4	
C10311	Countess® Automated Cell Counter Lab Starter Kit *with 101 boxes of 50 cell counting chamber slides and Trypan Blue*	1 kit	15.4	
C10310	Countess® Automated Cell Counter Starter Kit *with 11 boxes of 50 cell counting chamber slides and Trypan Blue*	1 kit	15.4	
C10228	Countess® cell counting chamber slides *for use with Countess® automated cell counter* *box of 50 with Trypan Blue*	1 kit	15.4	
C10312	Countess® Cell Counting Chamber Slides, 500 Slides (1000 Counts) *for use with Countess® automated cell counter* *10 boxes of 50 slides* *with Trypan Blue*	1 kit	15.4	
C10313	Countess® Cell Counting Chamber Slides, 1250 Slides (2500 Counts) *for use with Countess® automated cell counter* *25 boxes of 50 slides* *with Trypan Blue*	1 kit	15.4	
C10314	Countess® Cell Counting Chamber Slides, 2500 Slides (5000 Counts) *for use with Countess® automated cell counter* *50 boxes of 50 slides* *with Trypan Blue*	1 kit	15.4	
C10315	Countess® Cell Counting Chamber Slides, 5000 Slides (10,000 Counts) *for use with Countess® automated cell counter* *100 boxes of 50 slides* *with Trypan Blue*	1 kit	15.4	
C10285	Countess® power cord with four adapter cords *for use with Countess® automated cell counter*	1 set	15.4	
C10284	Countess® test beads *for use with Countess® automated cell counter* *1x10^6 beads/mL ±10%*	1 mL	15.4	

*Note that all products listed in the above Master Product List are covered by **Limited Use Label License #223**. If a specific product is covered by additional Limited Use Label License(s), the Limited Use Label License #'s are provided in the LULL column above. Please refer to page 971 for the complete text of the Limited Use Label License(s).

Cat. No.	Product	Quantity	Sections	LULL*
C10286	Countess® USB drive *for use with Countess® automated cell counter* *1Gbyte*	each	15.4	
C24784	coverslip maxi-rack *for 50 coverslips*	each	23.1	
C14784	coverslip mini-rack *for 8 coverslips*	each	23.1	
C18160	CoverWell™ imaging chamber gasket, one chamber, 20 mm diameter, 0.5 mm deep *set of 40*	1 set	23.1	
C18161	CoverWell™ imaging chamber gasket, one chamber, 20 mm diameter, 1.0 mm deep *set of 40*	1 set	23.1	
C24726	CoverWell™ imaging chamber gasket with adhesive, one chamber, 20 mm diameter, 0.5 mm deep *set of 40*	1 set	23.1	
C24727	CoverWell™ imaging chamber gasket with adhesive, one chamber, 20 mm diameter, 1.0 mm deep *set of 40*	1 set	23.1	
C18155	CoverWell™ incubation chamber gasket, one chamber, 13 mm diameter, 0.2 mm deep *set of 25*	1 set	23.1	
C18156	CoverWell™ incubation chamber gasket, one chamber, 13 mm diameter, 0.5 mm deep *set of 50*	1 set	23.1	
C18150	CoverWell™ incubation chamber gasket, one chamber, 40 mm x 22 mm, 0.2 mm deep *set of 25*	1 set	23.1	
C18151	CoverWell™ incubation chamber gasket, one chamber, 40 mm x 22 mm, 0.5 mm deep *set of 50*	1 set	23.1	
C18139	CoverWell™ perfusion chamber gasket, eight chambers, 9 mm diameter, 0.5 mm deep *set of 20*	1 set	23.1	
C18140	CoverWell™ perfusion chamber gasket, eight chambers, 9 mm diameter, 1.0 mm deep *set of 20*	1 set	23.1	
C18141	CoverWell™ perfusion chamber gasket, eight chambers, 9 mm diameter, 2.0 mm deep *set of 20*	1 set	23.1	
C18142	CoverWell™ perfusion chamber gasket, eight chambers, 9 mm diameter, 2.5 mm deep *set of 20*	1 set	23.1	
C18128	CoverWell™ perfusion chamber gasket, four chambers, 19 mm x 6 mm, 0.5 mm deep *set of 40*	1 set	23.1	
C18136	CoverWell™ perfusion chamber gasket, one chamber, 20 mm diameter, 1.0 mm deep *set of 40*	1 set	23.1	
C18120	CoverWell™ perfusion chamber gasket, one chamber, 32 mm x 19 mm, 0.5 mm deep *set of 40*	1 set	23.1	
C18121	CoverWell™ perfusion chamber gasket, one chamber, 32 mm x 19 mm, 1.0 mm deep *set of 40*	1 set	23.1	
C24769	CultureWell™ cell culture plate *set of 10*	1 set	23.1	
C37000	CultureWell™ chambered coverglass for cell culture *sixteen wells per coverglass* *set of 8*	1 set	23.1	
C37005	CultureWell™ chambered coverglass for cell culture *sixteen wells per coverglass* *set of 2*	1 pack	23.1	
C24770	CultureWell™ coverslip divider *set of 4*	1 set	23.1	
C24760	CultureWell™ multislip cell culture system MSI-12 *plate and insert, fifteen 12 mm coverslips per insert* *set of 10*	1 set	23.1	
C24761	CultureWell™ multislip cell culture system MSI-18 *plate and insert, eight 18 mm coverslips per insert* *set of 10*	1 set	23.1	
C24762	CultureWell™ multiwell cell culture system CWI 2R-1.0 *plate and insert, four 24 mm x 50 mm coverslips per insert, two 1 mm-deep wells per coverslip* *set of 10*	1 set	23.1	
C24763	CultureWell™ multiwell cell culture system CWI 2R-2.0 *plate and insert, four 24 mm x 50 mm coverslips per insert, two 2 mm-deep wells per coverslip* *set of 10*	1 set	23.1	
C24764	CultureWell™ multiwell cell culture system CWI 3S-1.0 *plate and insert, four 24 mm x 50 mm coverslips per insert, three 1 mm-deep wells per coverslip* *set of 10*	1 set	23.1	
C24765	CultureWell™ multiwell cell culture system CWI 4R-1.0 *plate and insert, four 24 mm x 50 mm coverslips per insert, four 1 mm-deep wells per coverslip* *set of 10*	1 set	23.1	
C24766	CultureWell™ multiwell cell culture system CWI 8R-1.0 *plate and insert, four 24 mm x 50 mm coverslips per insert, eight 1 mm-deep wells per coverslip* *set of 10*	1 set	23.1	
C24767	CultureWell™ multiwell cell culture system CWI 8R-1.0 TS *plate and insert, four 24 mm x 50 mm coverslips per insert, eight 1 mm-deep wells per coverslip* *set of 2*	1 set	23.1	
C24768	CultureWell™ multiwell cell culture system CWI 50R-1.0 *plate and insert, four 24 mm x 50 mm coverslips per insert, fifty 1 mm-deep wells per coverslip* *set of 10*	1 set	23.1	
C24775	CultureWell™ multiwell chambered coverslip CWCS 2R-1.0 *24 mm x 50 mm coverslips, two 1 mm-deep wells per coverslip* *set of 20*	1 set	23.1	
C24776	CultureWell™ multiwell chambered coverslip CWCS 2R-2.0 *24 mm x 50 mm coverslips, two 2 mm-deep wells per coverslip* *set of 20*	1 set	23.1	
C24777	CultureWell™ multiwell chambered coverslip CWCS 3S-1.0 *24 mm x 50 mm coverslips, three 1 mm-deep wells per coverslip* *set of 20*	1 set	23.1	
C24778	CultureWell™ multiwell chambered coverslip CWCS 4R-1.0 *24 mm x 50 mm coverslips, four 1 mm-deep wells per coverslip* *set of 20*	1 set	23.1	
C24779	CultureWell™ multiwell chambered coverslip CWCS 8R-1.0 *24 mm x 50 mm coverslips, eight 1 mm-deep wells per coverslip* *set of 20*	1 set	23.1	
C24780	CultureWell™ multiwell chambered coverslip CWCS 50R-1.0 *24 mm x 50 mm coverslips, fifty 1 mm-deep wells per coverslip* *set of 20*	1 set	23.1	
A10521	Cy3® goat anti-mouse IgG (H+L) *2 mg/mL*	0.5 mL	7.2	353
A10520	Cy3® goat anti-rabbit IgG (H+L) *2 mg/mL*	0.5 mL	7.2	353
A10522	Cy3® goat anti-rat IgG (H+L) *2 mg/mL*	0.5 mL	7.2	353
A10524	Cy5® goat anti-mouse IgG (H+L) *2 mg/mL*	0.5 mL	7.2	353
A10523	Cy5® goat anti-rabbit IgG (H+L) *2 mg/mL*	0.5 mL	7.2	353
A10525	Cy5® goat anti-rat IgG (H+L) *2 mg/mL*	0.5 mL	7.2	353
C684	3-cyano-7-ethoxycoumarin	10 mg	10.6	
C183	3-cyano-7-hydroxycoumarin	100 mg	10.1	
C7074	cyclic adenosine 5′-diphosphate ribose, 1-(1-(2-nitrophenyl)ethyl) ester (NPE-caged cADP-ribose) *mixed isomers*	50 µg	5.3, 17.2	
C7027	CyQUANT® cell lysis buffer *20X concentrate*	50 mL	15.4	
C7026	CyQUANT® Cell Proliferation Assay Kit *for cells in culture* *1000 assays*	1 kit	15.4, 15.6	

*Note that all products listed in the above Master Product List are covered by **Limited Use Label License #223**. If a specific product is covered by additional Limited Use Label License(s), the Limited Use Label License #'s are provided in the LULL column above. Please refer to page 971 for the complete text of the Limited Use Label License(s).

IMPORTANT NOTICE: The products described in this manual are covered by one or more Limited Use Label License(s). Please refer to the Appendix on page 971 and Master Product List on page 975. Products are For Research Use Only. Not intended for any animal or human therapeutic or diagnostic use.

Cat. No.	Product	Quantity	Sections	LULL*
C35012	CyQUANT® Direct Cell Proliferation Assay *for 100 microplates*	1 kit	15.4, 15.6	
C35011	CyQUANT® Direct Cell Proliferation Assay *for 10 microplates*	1 kit	15.4, 15.6	
C35006	CyQUANT® NF Cell Proliferation Assay Kit *1000 assays*	1 kit	15.4, 15.6	
C35007	CyQUANT® NF Cell Proliferation Assay Kit *200 assays*	1 kit	15.4, 15.6	
A10691	Cy5®–R-phycoerythrin, F(ab')$_2$ fragment of goat anti-rat IgG (H+L) *1 mg/mL*	100 µL	7.2	353
D23841	DAF-FM (4-amino-5-methylamino-2',7'-difluorofluorescein)	1 mg	18.3, 21.2	
D23842	DAF-FM diacetate (4-amino-5-methylamino-2',7'-difluorofluorescein diacetate)	1 mg	18.3	
D23844	DAF-FM diacetate (4-amino-5-methylamino-2',7'-difluorofluorescein diacetate) *special packaging*	10 x 50 µg	18.3	
D2281	*m*-dansylaminophenylboronic acid	100 mg	3.2	
D10460	Dapoxyl® (2-aminoethyl)sulfonamide	10 mg	3.4, 12.3	
D10161	Dapoxyl® carboxylic acid, succinimidyl ester	5 mg	1.7	
D12800	Dapoxyl® sulfonic acid, sodium salt	10 mg	13.5	
D10160	Dapoxyl® sulfonyl chloride	10 mg	1.7	
D3923	DCVJ (4-(dicyanovinyl)julolidine)	25 mg	11.2, 13.5	
D3771	2-decanoyl-1-(*O*-(11-(4,4-difluoro-5,7-dimethyl-4-bora-3a,4a-diaza-*s*-indacene-3-propionyl)amino)undecyl)-*sn*-glycero-3-phosphocholine	1 mg	13.2, 17.4	
D12371	deoxyribonuclease I, Alexa Fluor® 488 conjugate	5 mg	11.1	
D12372	deoxyribonuclease I, Alexa Fluor® 594 conjugate	5 mg	11.1	
D20657	D-desthiobiotin *50 mM aqueous solution*	10 mL	4.2, 7.6	
D2920	DetectaGene™ Green CMFDG *lacZ* Gene Expression Kit	1 kit	10.2	
D1383	dexamethasone fluorescein	5 mg	16.1	
D34682	dextran, Alexa Fluor® 488; 3,000 MW, anionic	2 mg	14.5	
D34681	dextran, Alexa Fluor® 680; 3,000 MW, anionic	2 mg	14.5	
D22910	dextran, Alexa Fluor® 488; 10,000 MW, anionic, fixable	5 mg	14.5	
D22911	dextran, Alexa Fluor® 546; 10,000 MW, anionic, fixable	5 mg	14.5	
D34679	dextran, Alexa Fluor® 555; 10,000 MW, anionic, fixable	5 mg	14.5	
D22912	dextran, Alexa Fluor® 568; 10,000 MW, anionic, fixable	5 mg	14.5	
D22913	dextran, Alexa Fluor® 594; 10,000 MW, anionic, fixable	5 mg	14.5	
D22914	dextran, Alexa Fluor® 647; 10,000 MW, anionic, fixable	2 mg	14.5	
D34680	dextran, Alexa Fluor® 680; 10,000 MW, anionic, fixable	5 mg	14.5	
D3330	dextran, amino, 3000 MW	100 mg	14.5	
D1860	dextran, amino, 10,000 MW	1 g	14.5	
D1861	dextran, amino, 40,000 MW	1 g	14.5	
D1862	dextran, amino, 70,000 MW	1 g	14.5	
D7144	dextran, amino, 500,000 MW	100 mg	14.5	
D1878	dextran, BCECF, 10,000 MW, anionic	10 mg	20.4	
D1880	dextran, BCECF, 70,000 MW, anionic	10 mg	20.4	
D7135	dextran, biotin, 3000 MW, lysine fixable (BDA-3000)	10 mg	4.3, 14.5	
D1956	dextran, biotin, 10,000 MW, lysine fixable (BDA-10,000)	25 mg	4.3, 14.5	
D1957	dextran, biotin, 70,000 MW, lysine fixable (BDA-70,000)	25 mg	4.3, 14.5	
D7142	dextran, biotin, 500,000 MW, lysine fixable (BDA-500,000)	10 mg	4.3, 14.5	
D7168	dextran, BODIPY® FL, 10,000 MW, fixable	5 mg	14.5	
D7132	dextran, Cascade Blue®, 3000 MW, anionic, lysine fixable	10 mg	14.5	
D1976	dextran, Cascade Blue®, 10,000 MW, anionic, lysine fixable	25 mg	14.5	
D3305	dextran, fluorescein, 3000 MW, anionic	10 mg	14.5, 20.4	
D3306	dextran, fluorescein, 3000 MW, anionic, lysine fixable	10 mg	14.5	
D1821	dextran, fluorescein, 10,000 MW, anionic	25 mg	14.5, 20.4	
D1820	dextran, fluorescein, 10,000 MW, anionic, lysine fixable (fluoro-emerald)	25 mg	14.5	
D1844	dextran, fluorescein, 40,000 MW, anionic	25 mg	14.5, 20.4	
D1845	dextran, fluorescein, 40,000 MW, anionic, lysine fixable	25 mg	14.5	
D1823	dextran, fluorescein, 70,000 MW, anionic	25 mg	14.5, 20.4	
D1822	dextran, fluorescein, 70,000 MW, anionic, lysine fixable	25 mg	14.5	
D7136	dextran, fluorescein, 500,000 MW, anionic, lysine fixable	10 mg	14.5	
D7137	dextran, fluorescein, 2,000,000 MW, anionic, lysine fixable	10 mg	14.5	
D7156	dextran, fluorescein and biotin, 3000 MW, anionic, lysine fixable (micro-emerald)	5 mg	4.3, 14.5	
D7178	dextran, fluorescein and biotin, 10,000 MW, anionic, lysine fixable (mini-emerald)	10 mg	4.3, 14.5	
D1951	dextran, fluorescein and tetramethylrhodamine, 70,000 MW, anionic	10 mg	20.4	
D1825	dextran, lucifer yellow, 10,000 MW, anionic, lysine fixable	25 mg	14.5	
D7170	dextran, Oregon Green® 488; 10,000 MW, anionic	5 mg	14.5, 20.4	

*Note that all products listed in the above Master Product List are covered by **Limited Use Label License #223**. If a specific product is covered by additional Limited Use Label License(s), the Limited Use Label License #'s are provided in the LULL column above. Please refer to page 971 for the complete text of the Limited Use Label License(s).

IMPORTANT NOTICE: The products described in this manual are covered by one or more Limited Use Label License(s). Please refer to the Appendix on page 971 and Master Product List on page 975. Products are For Research Use Only. Not intended for any animal or human therapeutic or diagnostic use.

Cat. No.	Product	Quantity	Sections	LULL*
D7171	dextran, Oregon Green® 488; 10,000 MW, anionic, lysine fixable	5 mg	14.5	
D7172	dextran, Oregon Green® 488; 70,000 MW, anionic	5 mg	14.5, 20.4	
D7173	dextran, Oregon Green® 488; 70,000 MW, anionic, lysine fixable	5 mg	14.5	
D7176	dextran, Oregon Green® 514; 70,000 MW, anionic	5 mg	14.5, 20.4	
D1824	dextran, rhodamine B, 10,000 MW, neutral	25 mg	14.5	
D1841	dextran, rhodamine B, 70,000 MW, neutral	25 mg	14.5	
D7163	dextran, Rhodamine Green™, 3000 MW	5 mg	14.5	
D7153	dextran, Rhodamine Green™, 10,000 MW, lysine fixable	10 mg	14.5	
D3303	dextran, SNARF®-1, 10,000 MW, anionic	5 mg	20.4	
D3304	dextran, SNARF®-1, 70,000 MW, anionic	5 mg	20.4	
D3307	dextran, tetramethylrhodamine, 3000 MW, anionic	10 mg	14.5	
D3308	dextran, tetramethylrhodamine, 3000 MW, anionic, lysine fixable	10 mg	14.5	
D1868	dextran, tetramethylrhodamine, 10,000 MW, anionic, fixable	25 mg	14.5	
D1817	dextran, tetramethylrhodamine, 10,000 MW, lysine fixable (fluoro-ruby)	25 mg	14.5	
D1816	dextran, tetramethylrhodamine, 10,000 MW, neutral	25 mg	14.5	
D1842	dextran, tetramethylrhodamine, 40,000 MW, neutral	25 mg	14.5	
D1818	dextran, tetramethylrhodamine, 70,000 MW, lysine fixable	25 mg	14.5	
D1819	dextran, tetramethylrhodamine, 70,000 MW, neutral	25 mg	14.5	
D7139	dextran, tetramethylrhodamine, 2,000,000 MW, lysine fixable	10 mg	14.5	
D7162	dextran, tetramethylrhodamine and biotin, 3000 MW, lysine fixable (micro-ruby)	5 mg	4.3, 14.5	
D3312	dextran, tetramethylrhodamine and biotin, 10,000 MW, lysine fixable (mini-ruby)	10 mg	4.3, 14.5	
D3328	dextran, Texas Red®, 3000 MW, lysine fixable	10 mg	14.5	
D3329	dextran, Texas Red®, 3000 MW, neutral	10 mg	14.5	
D1863	dextran, Texas Red®, 10,000 MW, lysine fixable	25 mg	14.5	
D1828	dextran, Texas Red®, 10,000 MW, neutral	25 mg	14.5	
D1829	dextran, Texas Red®, 40,000 MW, neutral	25 mg	14.5	
D1864	dextran, Texas Red®, 70,000 MW, lysine fixable	25 mg	14.5	
D1830	dextran, Texas Red®, 70,000 MW, neutral	25 mg	14.5	
D3883	4-(4-(dihexadecylamino)styryl)-N-methylpyridinium iodide (DiA; 4-Di-16-ASP)	25 mg	13.4, 14.4	
D1306	4',6-diamidino-2-phenylindole, dihydrochloride (DAPI)	10 mg	8.1, 11.2, 12.5, 14.3, 15.4	
D21490	4',6-diamidino-2-phenylindole, dihydrochloride (DAPI) *FluoroPure™ grade*	10 mg	8.1, 11.2, 12.5, 14.3, 15.4	
D3571	4',6-diamidino-2-phenylindole, dilactate (DAPI, dilactate)	10 mg	8.1, 11.2, 12.5, 14.3, 15.4	
D23840	1,2-diaminoanthraquinone sulfate (DAA) *high purity*	5 mg	18.3	
D22185	Diaminobenzidine (DAB) Histochemistry Kit #1 *with goat anti-mouse IgG–HRP*	1 kit	6.2, 7.2, 10.5	
D22187	Diaminobenzidine (DAB) Histochemistry Kit #3 *with streptavidin–HRP*	1 kit	6.2, 7.6, 10.5	
D1463	1,2-diamino-4,5-dimethoxybenzene, dihydrochloride (DDB)	100 mg	3.3	
D7918	2,3-diaminonaphthalene	100 mg	3.3, 18.3, 21.2	
D6923	di-2-ANEPEQ (JPW 1114)	5 mg	22.2	
D36801	di-3-ANEPPDHQ	1 mg	22.2	
D36802	di-4-ANEPPDHQ	1 mg	22.2	
D6927	di-12-ANEPPQ	5 mg	22.2	
D1199	di-4-ANEPPS	5 mg	22.2	
D3167	di-8-ANEPPS	5 mg	22.2	
D3034	diazo-2, tetrapotassium salt *cell impermeant*	1 mg	5.3, 17.2, 19.8	
D1211	5,5'-dibromo BAPTA, tetrapotassium salt *cell impermeant*	100 mg	19.8	
D1379	dibromobimane (bBBr)	25 mg	2.3, 5.2	
D399	2',7'-dichlorodihydrofluorescein diacetate (2',7'-dichlorofluorescin diacetate; H_2DCFDA)	100 mg	10.5, 15.2, 18.2	
D2935	2',7'-dichlorodihydrofluorescein diacetate, succinimidyl ester (OxyBURST® Green H_2DCFDA, SE)	5 mg	16.1, 18.2	
D6488	9*H*-(1,3-dichloro-9,9-dimethylacridin-2-one-7-yl) β-D-galactopyranoside (DDAO galactoside)	5 mg	10.2	
D6487	9*H*-(1,3-dichloro-9,9-dimethylacridin-2-one-7-yl) phosphate, diammonium salt (DDAO phosphate)	5 mg	10.3	
D16	5-(4,6-dichlorotriazinyl)aminofluorescein (5-DTAF) *single isomer*	100 mg	1.5, 3.2	
D291	4-(4-(didecylamino)styryl)-N-methylpyridinium iodide (4-Di-10-ASP)	25 mg	13.4, 14.4	
D307	1,1'-dioctadecyl-3,3,3',3'-tetramethylindodicarbocyanine perchlorate ('DiD' oil; DiIC_{18}(5) oil)	25 mg	13.4, 14.4	
D337	DIDS (4,4'-diisothiocyanatostilbene-2,2'-disulfonic acid, disodium salt)	100 mg	16.3	
D7757	1,1'-dioctadecyl-3,3,3',3'-tetramethylindodicarbocyanine, 4-chlorobenzenesulfonate salt ('DiD' solid; DiIC_{18}(5) solid)	10 mg	13.4, 14.4	
D1446	7-diethylaminocoumarin-3-carbonyl azide	25 mg	3.2	
D1421	7-diethylaminocoumarin-3-carboxylic acid	100 mg	1.7	

*Note that all products listed in the above Master Product List are covered by **Limited Use Label License #223**. If a specific product is covered by additional Limited Use Label License(s), the Limited Use Label License #'s are provided in the LULL column above. Please refer to page 971 for the complete text of the Limited Use Label License(s).

IMPORTANT NOTICE: The products described in this manual are covered by one or more Limited Use Label License(s). Please refer to the Appendix on page 971 and Master Product List on page 975. Products are For Research Use Only. Not intended for any animal or human therapeutic or diagnostic use.

Cat. No.	Product	Quantity	Sections	LULL*
D355	7-diethylaminocoumarin-3-carboxylic acid, hydrazide (DCCH)	25 mg	3.3	
D1412	7-diethylaminocoumarin-3-carboxylic acid, succinimidyl ester	25 mg	1.7	
D404	7-diethylamino-3-((4′-(iodoacetyl)amino)phenyl)-4-methylcoumarin (DCIA)	25 mg	2.3	
D10253	7-diethylamino-3-(((2-maleimidyl)ethyl)amino)carbonyl)coumarin (MDCC)	5 mg	2.3	
D346	7-diethylamino-3-(4′-maleimidylphenyl)-4-methylcoumarin (CPM)	25 mg	2.3	
D289	4-(4-(diethylamino)styryl)-*N*-methylpyridinium iodide (4-Di-2-ASP)	1 g	16.1	
D1209	5,5′-difluoro BAPTA, AM *cell permeant*	25 mg	19.8	
D20351	4,4-difluoro-3,5-bis(4-methoxyphenyl)-8-(4-maleimidylphenyl)-4-bora-3a,4a-diaza-*s*-indacene (BODIPY® 577/618 maleimide)	5 mg	2.2	
D3822	4,4-difluoro-5,7-dimethyl-4-bora-3a,4a-diaza-*s*-indacene-3-dodecanoic acid (BODIPY® FL C_{12})	1 mg	13.2	
D3792	2-(4,4-difluoro-5,7-dimethyl-4-bora-3a,4a-diaza-*s*-indacene-3-dodecanoyl)-1-hexadecanoyl-*sn*-glycero-3-phosphocholine (β-BODIPY® FL C_{12}-HPC)	100 µg	13.2	
D7519	*N*-(4,4-difluoro-5,7-dimethyl-4-bora-3a,4a-diaza-*s*-indacene-3-dodecanoyl)sphingosyl 1-β-D-galactopyranoside (BODIPY® FL C_{12}-galactocerebroside)	25 µg	10.2, 13.3, 17.4	
D7711	*N*-(4,4-difluoro-5,7-dimethyl-4-bora-3a,4a-diaza-*s*-indacene-3-dodecanoyl)sphingosyl phosphocholine (BODIPY® FL C_{12}-sphingomyelin) *1 mg/mL in DMSO*	250 µL	12.4, 13.3, 17.4	
D3821	4,4-difluoro-5,7-dimethyl-4-bora-3a,4a-diaza-*s*-indacene-3-hexadecanoic acid (BODIPY® FL C_{16})	1 mg	13.2	
D3834	4,4-difluoro-5,7-dimethyl-4-bora-3a,4a-diaza-*s*-indacene-3-pentanoic acid (BODIPY® FL C_5)	1 mg	1.4, 13.2	
D6184	4,4-difluoro-5,7-dimethyl-4-bora-3a,4a-diaza-*s*-indacene-3-pentanoic acid, succinimidyl ester (BODIPY® FL C_5, SE)	5 mg	1.4	
D3805	2-(4,4-difluoro-5,7-dimethyl-4-bora-3a,4a-diaza-*s*-indacene-3-pentanoyl)-1-hexadecanoyl-*sn*-glycero-3-phosphate, diammonium salt (β-BODIPY® FL C_5-HPA)	100 µg	13.2	
D3803	2-(4,4-difluoro-5,7-dimethyl-4-bora-3a,4a-diaza-*s*-indacene-3-pentanoyl)-1-hexadecanoyl-*sn*-glycero-3-phosphocholine (β-BODIPY® FL C_5-HPC)	100 µg	13.2, 17.4	
D3521	*N*-(4,4-difluoro-5,7-dimethyl-4-bora-3a,4a-diaza-*s*-indacene-3-pentanoyl)sphingosine (BODIPY® FL C_5-ceramide)	250 µg	12.4, 13.3, 17.4	
D13951	*N*-(4,4-difluoro-5,7-dimethyl-4-bora-3a,4a-diaza-*s*-indacene-3-pentanoyl)sphingosyl 1-β-D-lactoside (BODIPY® FL C_5-lactosylceramide)	25 µg	10.2, 12.4, 13.3, 17.4	
D3522	*N*-(4,4-difluoro-5,7-dimethyl-4-bora-3a,4a-diaza-*s*-indacene-3-pentanoyl)sphingosyl phosphocholine (BODIPY® FL C_5-sphingomyelin)	250 µg	12.4, 13.3, 17.4	
D2183	4,4-difluoro-5,7-dimethyl-4-bora-3a,4a-diaza-*s*-indacene-3-propionic acid (BODIPY® FL)	5 mg	1.4	
D2371	4,4-difluoro-5,7-dimethyl-4-bora-3a,4a-diaza-*s*-indacene-3-propionic acid, hydrazide (BODIPY® FL hydrazide)	5 mg	3.3	
D2184	4,4-difluoro-5,7-dimethyl-4-bora-3a,4a-diaza-*s*-indacene-3-propionic acid, succinimidyl ester (BODIPY® FL, SE)	5 mg	1.4	
D6140	4,4-difluoro-5,7-dimethyl-4-bora-3a,4a-diaza-*s*-indacene-3-propionic acid, sulfosuccinimidyl ester, sodium salt (BODIPY® FL, SSE)	5 mg	1.4	
D6102	6-((4,4-difluoro-5,7-dimethyl-4-bora-3a,4a-diaza-*s*-indacene-3-propionyl)amino)hexanoic acid, succinimidyl ester (BODIPY® FL-X, SE)	5 mg	1.4, 4.2, 9.3	
D3800	*N*-(4,4-difluoro-5,7-dimethyl-4-bora-3a,4a-diaza-*s*-indacene-3-propionyl)-1,2-dihexadecanoyl-*sn*-glycero-3-phosphoethanolamine, triethylammonium salt (BODIPY® FL DHPE)	100 µg	13.2	
D2390	4,4-difluoro-5,7-dimethyl-4-bora-3a,4a-diaza-*s*-indacene-3-propionyl ethylenediamine, hydrochloride (BODIPY® FL EDA)	5 mg	3.4	
D3862	4,4-difluoro-5,7-dimethyl-4-bora-3a,4a-diaza-*s*-indacene-3-undecanoic acid (BODIPY® FL C_{11})	1 mg	13.2	
D6003	*N*-(4,4-difluoro-5,7-dimethyl-4-bora-3a,4a-diaza-*s*-indacene-3-yl)methyl)iodoacetamide (BODIPY® FL C_1-IA)	5 mg	2.2	
D6117	6-((4,4-difluoro-1,3-dimethyl-5-(4-methoxyphenyl)-4-bora-3a,4a-diaza-*s*-indacene-2-propionyl)amino)hexanoic acid, succinimidyl ester (BODIPY® TMR-X, SE)	5 mg	1.4	
D3832	4,4-difluoro-5,7-diphenyl-4-bora-3a,4a-diaza-*s*-indacene-3-dodecanoic acid (BODIPY® 530/550 C_{12})	1 mg	13.2	
D3815	2-(4,4-difluoro-5,7-diphenyl-4-bora-3a,4a-diaza-*s*-indacene-3-pentanoyl)-1-hexadecanoyl-*sn*-glycero-3-phosphocholine (β-BODIPY® 530/550 C_5-HPC)	100 µg	13.2	
D2187	4,4-difluoro-5,7-diphenyl-4-bora-3a,4a-diaza-*s*-indacene-3-propionic acid, succinimidyl ester (BODIPY® 530/550, SE)	5 mg	1.4	
D6145	2′,7′-difluorofluorescein (Oregon Green® 488)	10 mg	1.5	
D6566	6,8-difluoro-7-hydroxy-4-methylcoumarin (DiFMU) *reference standard*	10 mg	10.1	
D3823	4,4-difluoro-5-methyl-4-bora-3a,4a-diaza-*s*-indacene-3-dodecanoic acid (BODIPY® 500/510 C_1, C_{12})	1 mg	13.2	
D3793	2-(4,4-difluoro-5-methyl-4-bora-3a,4a-diaza-*s*-indacene-3-dodecanoyl)-1-hexadecanoyl-*sn*-glycero-3-phosphocholine (β-BODIPY® 500/510 C_{12}-HPC)	100 µg	13.2	
D12200	6,8-difluoro-4-methylumbelliferyl octanoate (DiFMU octanoate)	10 mg	10.6	
D6567	6,8-difluoro-4-methylumbelliferyl phosphate (DiFMUP)	5 mg	10.3	
D22065	6,8-difluoro-4-methylumbelliferyl phosphate (DiFMUP) *packaged for high-throughput screening*	10 x 10 mg	10.3	
D3825	4,4-difluoro-5-octyl-4-bora-3a,4a-diaza-*s*-indacene-3-pentanoic acid (BODIPY® 500/510 C_8, C_5)	1 mg	13.2	
D3922	4,4-difluoro-1,3,5,7,8-pentamethyl-4-bora-3a,4a-diaza-*s*-indacene (BODIPY® 493/503)	10 mg	13.5	
D3238	4,4-difluoro-1,3,5,7,8-pentamethyl-4-bora-3a,4a-diaza-*s*-indacene-2,6-disulfonic acid, disodium salt (BODIPY® 492/515 disulfonate)	10 mg	14.3	
D6180	4,4-difluoro-5-phenyl-4-bora-3a,4a-diaza-*s*-indacene-3-propionic acid, succinimidyl ester (BODIPY® R6G, SE)	5 mg	1.4	
D2228	4,4-difluoro-5-(4-phenyl-1,3-butadienyl)-4-bora-3a,4a-diaza-*s*-indacene-3-propionic acid, succinimidyl ester (BODIPY® 581/591, SE)	5 mg	1.4	
D3861	4,4-difluoro-5-(4-phenyl-1,3-butadienyl)-4-bora-3a,4a-diaza-*s*-indacene-3-undecanoic acid (BODIPY® 581/591 C_{11})	1 mg	13.2, 18.2	

*Note that all products listed in the above Master Product List are covered by **Limited Use Label License #223**. If a specific product is covered by additional Limited Use Label License(s), the Limited Use Label License #'s are provided in the LULL column above. Please refer to page 971 for the complete text of the Limited Use Label License(s).

Cat. No.	Product	Quantity	Sections	LULL*
D2225	4,4-difluoro-5-(2-pyrrolyl)-4-bora-3a,4a-diaza-*s*-indacene-3-propionic acid, succinimidyl ester (BODIPY® 576/589, SE)	5 mg	1.4	
D10001	6-(((4,4-difluoro-5-(2-pyrrolyl)-4-bora-3a,4a-diaza-*s*-indacene-3-yl)styryloxy)acetyl)aminohexanoic acid, succinimidyl ester (BODIPY® 650/665-X, SE)	5 mg	1.4	
D2222	4,4-difluoro-5-styryl-4-bora-3a,4a-diaza-*s*-indacene-3-propionic acid, succinimidyl ester (BODIPY® 564/570, SE)	5 mg	1.4	
D3921	4,4-difluoro-1,3,5,7-tetramethyl-4-bora-3a,4a-diaza-*s*-indacene (BODIPY® 505/515)	10 mg	13.5	
D2191	4,4-difluoro-1,3,5,7-tetramethyl-4-bora-3a,4a-diaza-*s*-indacene-8-propionic acid, succinimidyl ester (BODIPY® 493/503, SE)	5 mg	1.4	
D6004	*N*-(4,4-difluoro-1,3,5,7-tetramethyl-4-bora-3a,4a-diaza-*s*-indacene-2-yl)iodoacetamide (BODIPY® 507/545 IA)	5 mg	2.2	
D20350	4,4-difluoro-1,3,5,7-tetramethyl-8-(4-maleimidylphenyl)-4-bora-3a,4a-diaza-*s*-indacene (BODIPY® 499/508 maleimide)	5 mg	2.2	
D3835	4,4-difluoro-5-(2-thienyl)-4-bora-3a,4a-diaza-*s*-indacene-3-dodecanoic acid (BODIPY® 558/568 C_{12})	1 mg	13.2	
D2219	4,4-difluoro-5-(2-thienyl)-4-bora-3a,4a-diaza-*s*-indacene-3-propionic acid, succinimidyl ester (BODIPY® 558/568, SE)	5 mg	1.4	
D6116	6-(((4-(4,4-difluoro-5-(2-thienyl)-4-bora-3a,4a-diaza-*s*-indacene-3-yl)phenoxy)acetyl)amino)hexanoic acid, succinimidyl ester (BODIPY® TR-X, SE)	5 mg	1.4, 9.3	
D6251	5-(((4-(4,4-difluoro-5-(2-thienyl)-4-bora-3a,4a-diaza-*s*-indacene-3-yl)phenoxy)acetyl)amino)pentylamine, hydrochloride (BODIPY® TR cadaverine)	5 mg	3.4	
D10000	6-(((4,4-difluoro-5-(2-thienyl)-4-bora-3a,4a-diaza-*s*-indacene-3-yl)styryloxy)acetyl)aminohexanoic acid, succinimidyl ester (BODIPY® 630/650-X, SE)	5 mg	1.4	
D378	3,3′-diheptyloxacarbocyanine iodide ($DiOC_7(3)$)	100 mg	12.2	
D273	3,3′-dihexyloxacarbocyanine iodide ($DiOC_6(3)$)	100 mg	12.2, 12.4, 22.3	
D23805	dihydrocalcein, AM *special packaging*	20 x 50 µg	10.5, 15.2, 18.2	
D11347	dihydroethidium (hydroethidine) *special packaging*	10 x 1 mg	8.1, 15.2, 18.2	
D1168	dihydroethidium (hydroethidine)	25 mg	8.1, 15.2, 18.2	
D23107	dihydroethidium (hydroethidine) *5 mM stabilized solution in DMSO*	1 mL	8.1, 15.2, 18.2	
D23806	dihydrorhodamine 123 *5 mM stabilized solution in DMSO*	1 mL	12.2, 15.2, 18.2	
D632	dihydrorhodamine 123	10 mg	12.2, 15.2, 18.2	
D633	dihydrorhodamine 6G	25 mg	12.2, 15.2, 18.2	
D3886	1,1′-dioleyl-3,3,3′,3′-tetramethylindocarbocyanine methanesulfonate (Δ9-DiI)	25 mg	13.4, 14.4	
D282	1,1′-dioctadecyl-3,3,3′,3′-tetramethylindocarbocyanine perchlorate ('DiI'; $DiIC_{18}(3)$)	100 mg	12.4, 13.4, 14.4	
D3911	1,1′-dioctadecyl-3,3,3′,3′-tetramethylindocarbocyanine perchlorate *crystalline* ('DiI'; $DiIC_{18}(3)$)	25 mg	13.4, 14.4	
D383	1,1′-didodecyl-3,3,3′,3′-tetramethylindocarbocyanine perchlorate ($DiIC_{12}(3)$)	100 mg	13.4, 14.4	
D384	1,1′-dihexadecyl-3,3,3′,3′-tetramethylindocarbocyanine perchlorate ($DiIC_{16}(3)$)	100 mg	12.4, 13.4, 14.4	
D12730	1,1′-dioctadecyl-3,3,3′,3′-tetramethylindodicarbocyanine-5,5′-disulfonic acid ($DiIC_{18}(5)$-DS)	5 mg	13.4, 14.4	
D7776	1,1′-dioctadecyl-3,3,3′,3′-tetramethylindocarbocyanine-5,5′-disulfonic acid ($DiIC_{18}(3)$-DS)	5 mg	13.4, 14.4	
D338	4,4′-diisothiocyanatodihydrostilbene-2,2′-disulfonic acid, disodium salt (H_2DIDS)	100 mg	16.3	
D7758	4-(4-(dilinoleylamino)styryl)-N-methylpyridinium 4-chlorobenzenesulfonate (FAST DiA™ solid; $DiΔ^{9,12}$-C_{18}ASP, CBS)	5 mg	13.4, 14.4	
D3898	3,3′-dilinoleyloxacarbocyanine perchlorate (FAST DiO™ solid; $DiOΔ^{9,12}$-$C_{18}(3)$, ClO4)	5 mg	13.4, 14.4	
D7756	1,1′-dilinoleyl-3,3,3′,3′-tetramethylindocarbocyanine, 4-chlorobenzenesulfonate (FAST DiI™ solid; $DiIΔ^{9,12}$-$C_{18}(3)$, CBS)	5 mg	13.4, 14.4	
D3899	1,1′-dilinoleyl-3,3,3′,3′-tetramethylindocarbocyanine perchlorate (FAST DiI™ oil; $DiIΔ^{9,12}$-$C_{18}(3)$, ClO4)	5 mg	13.4, 14.4	
D1037	4,5-dimethoxy-2-nitrobenzyl adenosine 3′,5′-cyclicmonophosphate (DMNB-caged cAMP)	5 mg	5.3, 17.3	
D6814	1-(4,5-dimethoxy-2-nitrophenyl)-1,2-diaminoethane-*N,N,N′,N′*-tetraacetic acid (DMNP-EDTA) *cell impermeant*	5 mg	5.3, 17.2, 19.8	
D2516	1-(4,5-Dimethoxy-2-nitrophenyl)diazoethane Generation Kit	1 kit	5.3	
D126	7-dimethylaminocoumarin-4-acetic acid (DMACA)	100 mg	1.7	
D374	7-dimethylaminocoumarin-4-acetic acid, succinimidyl ester (DMACA, SE)	25 mg	1.7	
D10166	7-dimethylamino-4-methylcoumarin-3-isothiocyanate (DACITC)	10 mg	1.7	
D10252	*N*-(7-dimethylamino-4-methylcoumarin-3-yl)iodoacetamide (DACIA)	10 mg	2.3	
D10251	*N*-(7-dimethylamino-4-methylcoumarin-3-yl)maleimide (DACM)	10 mg	2.3	
D112	5-dimethylaminonaphthalene-1-(*N*-(2-aminoethyl))sulfonamide (dansyl ethylenediamine)	100 mg	3.4	
D113	5-dimethylaminonaphthalene-1-(*N*-(5-aminopentyl))sulfonamide (dansyl cadaverine)	100 mg	3.4, 12.3	
D6104	6-((5-dimethylaminonaphthalene-1-sulfonyl)amino)hexanoic acid, succinimidyl ester (dansyl-X, SE)	25 mg	1.7, 4.2	
D94	11-((5-dimethylaminonaphthalene-1-sulfonyl)amino)undecanoic acid (DAUDA)	100 mg	13.2	
D21	5-dimethylaminonaphthalene-1-sulfonyl chloride (dansyl chloride)	1 g	1.7, 9.5	
D22	2-dimethylaminonaphthalene-5-sulfonyl chloride	100 mg	1.7	
D23	2-dimethylaminonaphthalene-6-sulfonyl chloride	100 mg	1.7	
D57	*N*-(5-dimethylaminonaphthalene-1-sulfonyl)-1,2-dihexadecanoyl-*sn*-glycero-3-phosphoethanolamine, triethylammonium salt (dansyl DHPE)	25 mg	13.2	
D100	5-dimethylaminonaphthalene-1-sulfonyl hydrazine (dansyl hydrazine)	100 mg	3.3, 9.5	
D2245	4-((4-(dimethylamino)phenyl)azo)benzoic acid, succinimidyl ester (dabcyl, SE)	100 mg	1.8, 9.5	
D6216	ε-(4-((4-(dimethylamino)phenyl)azo)benzoyl)-α-9-fluorenylmethoxycarbonyl-L-lysine (ε-dabcyl-α-FMOC-L-lysine)	100 mg	9.5	
D1521	4-dimethylaminophenylazophenyl-4′-maleimide (DABMI)	100 mg	2.2	

*Note that all products listed in the above Master Product List are covered by **Limited Use Label License #223**. If a specific product is covered by additional Limited Use Label License(s), the Limited Use Label License #'s are provided in the LULL column above. Please refer to page 971 for the complete text of the Limited Use Label License(s).

Cat. No.	Product	Quantity	Sections	LULL*
D426	2-(4-(dimethylamino)styryl)-*N*-ethylpyridinium iodide (DASPEI)	1 g	12.2	
D288	4-(4-(dimethylamino)styryl)-*N*-methylpyridinium iodide (4-Di-1-ASP)	1 g	12.2, 16.1	
D1207	5,5′-dimethyl BAPTA, AM *cell permeant*	25 mg	19.8	
D2004	*N,N′*-dimethyl-*N*-(iodoacetyl)-*N′*-(7-nitrobenz-2-oxa-1,3-diazol-4-yl)ethylenediamine (IANBD amide)	25 mg	2.2	
D22421	3,3′-dimethyl-α-naphthoxacarbocyanine iodide (JC-9; $DiNOC_1(3)$)	5 mg	12.2, 15.2, 22.3	
D2248	6-(2,4-dinitrophenyl)aminohexanoic acid, succinimidyl ester (DNP-X, SE)	25 mg	4.2	
D23739	*N*-((6-(2,4-dinitrophenyl)amino)hexanoyl)-2-(4,4-difluoro-5,7-dimethyl-4-bora-3a,4a-diaza-*s*-indacene-3-pentanoyl)-1-hexadecanoyl-*sn*-glycero-3-phosphoethanolamine, triethylammonium salt (PED6)	1 mg	13.2, 17.4	
D1552	*N*-(3-((2,4-dinitrophenyl)amino)propyl)-*N*-(3-aminopropyl)methylamine, dihydrochloride (DAMP)	100 mg	12.3	
D275	3,3′-dioctadecyloxacarbocyanine perchlorate ('DiO'; $DiOC_{18}(3)$)	100 mg	13.4, 14.4	
D1125	3,3′-dihexadecyloxacarbocyanine perchlorate ($DiOC_{16}(3)$)	25 mg	13.4, 14.4	
D14730	$DiOC_2(3)$ (3,3′-diethyloxacarbocyanine iodide)	100 mg	22.3	
D7778	3,3′-dioctadecyl-5,5′-di(4-sulfophenyl)oxacarbocyanine, sodium salt (SP-$DiOC_{18}(3)$)	5 mg	13.4, 14.4	
D7777	1,1′-dioctadecyl-6,6′-di(4-sulfophenyl)-3,3,3′,3′-tetramethylindocarbocyanine (SP-$DiIC_{18}(3)$)	5 mg	13.4, 14.4	
D6562	1,2-dioleoyl-3-(1-pyrenedodecanoyl)-*rac*-glycerol	1 mg	13.3	
D272	3,3′-dipentyloxacarbocyanine iodide ($DiOC_5(3)$)	100 mg	12.4, 22.3	
D7894	diphenyl-1-pyrenylphosphine (DPPP)	5 mg	18.2	
D12731	1,1′-dioctadecyl-3,3,3′,3′-tetramethylindotricarbocyanine iodide ('DiR'; $DiIC_{18}(7)$)	10 mg	13.4, 14.4	
D306	$DiSC_3(5)$ (3,3′-dipropylthiadicarbocyanine iodide)	100 mg	22.3	
D8451	5,5′-dithiobis-(2-nitrobenzoic acid) (DTNB; Ellman's reagent)	10 g	2.1, 5.2	
D1532	dithiothreitol (DTT)	1 g	2.1, 5.2	
D7443	DM-BODIPY® (-)-dihydropyridine *high affinity enantiomer*	25 µg	16.3	
D673	DNDS (4,4′-dinitrostilbene-2,2′-disulfonic acid, disodium salt)	1 g	16.3	
D109	5-dodecanoylaminofluorescein	100 mg	13.5	
D2893	5-dodecanoylaminofluorescein di-β-D-galactopyranoside ($C_{12}FDG$)	5 mg	10.2	
D250	6-dodecanoyl-2-dimethylaminonaphthalene (laurdan)	100 mg	13.5	
D202	DPH (1,6-diphenyl-1,3,5-hexatriene)	100 mg	13.5	
D476	β-DPH HPC (2-(3-(diphenylhexatrienyl)propanoyl)-1-hexadecanoyl-*sn*-glycero-3-phosphocholine)	1 mg	13.2	
D12060	DQ™ collagen, type I from bovine skin, fluorescein conjugate	1 mg	10.4, 14.7, 16.1	
D12052	DQ™ collagen, type IV from human placenta, fluorescein conjugate	1 mg	10.4, 14.7	
D12054	DQ™ gelatin from pig skin, fluorescein conjugate *special packaging*	5 x 1 mg	10.4, 16.1	
D12050	DQ™ Green BSA *special packaging*	5 x 1 mg	10.4, 14.7, 16.1	
D12053	DQ™ ovalbumin *special packaging*	5 x 1 mg	10.4, 14.7, 16.1	
D12051	DQ™ Red BSA *special packaging*	5 x 1 mg	10.4, 14.7, 16.1	
D20658	DSB-X™ Bioconjugate Isolation Kit #1 *with streptavidin agarose* *5 isolations*	1 kit	4.3	
D30753	DSB-X™ biotin C_2 iodoacetamide (desthiobiotin-X C_2-iodoacetamide)	5 mg	4.2	
D20698	DSB-X™ biotin donkey anti-goat IgG (H+L) *2 mg/mL*	0.5 mL	4.3, 7.2	
D20701	DSB-X™ biotin goat anti-chicken IgG (H+L) *2 mg/mL*	0.5 mL	4.3, 7.2	
D20691	DSB-X™ biotin goat anti-mouse IgG (H+L) *highly cross-adsorbed* *2 mg/mL*	0.5 mL	4.3, 7.2	
D20693	DSB-X™ biotin goat anti-mouse IgM (µ chain) *2 mg/mL*	250 µL	4.3, 7.2	
D20697	DSB-X™ biotin goat anti-rat IgG (H+L) *2 mg/mL*	0.5 mL	4.3, 7.2	
D20653	DSB-X™ biotin hydrazide	5 mg	4.2, 14.3	
D20655	DSB-X™ Biotin Protein Labeling Kit *5 labelings*	1 kit	1.2, 4.2	
D21887	DyeChrome™ Double Western Blot Stain Kit *for mouse IgG, rabbit IgG and total protein detection* *20 minigel blots*	1 kit	7.2	
E10187	EdU (5-ethynyl-2′-deoxyuridine)	500 mg	3.1, 15.4	
A10044	EdU (5-ethynyl-2′-deoxyuridine)	50 mg	3.1, 15.4	
E10415	EdU (5-ethynyl-2′-deoxyuridine)	5 g	3.1, 15.4	
E1219	EGTA, tetra(acetoxymethyl ester) (EGTA, AM)	10 mg	19.8	
E33075	Electrophoretic Mobility-Shift Assay (EMSA) Kit *with SYBR® Green and SYPRO® Ruby EMSA stains* *10 minigel assays*	1 kit	8.4, 9.3	
E6578	ELF® 97 alcohol *1 mM solution in DMSO*	1 mL	6.3, 10.1	
E6603	ELF® 97 Cytological Labeling Kit *with streptavidin, alkaline phosphatase conjugate* *50 assays*	1 kit	6.3, 10.3	
E6601	ELF® 97 Endogenous Phosphatase Detection Kit	1 kit	6.3, 10.3	
E6600	ELF® 97 Immunohistochemistry Kit	1 kit	6.3, 10.3	
E6604	ELF® 97 mRNA *In Situ* Hybridization Kit #1 *50 assays*	1 kit	6.3, 10.3	
E6605	ELF® 97 mRNA *In Situ* Hybridization Kit #2 *with streptavidin, alkaline phosphatase conjugate* *50 assays*	1 kit	6.3, 10.3	
E6589	ELF® 97 phosphatase substrate (ELF® 97 phosphate) *5 mM in water* *contains 2 mM azide*	1 mL	6.3, 10.3	
E6588	ELF® 97 phosphatase substrate (ELF® 97 phosphate) *5 mM in water* *0.2 µm filtered*	1 mL	6.3, 10.3	
E6606	ELF® spin filters *20 filters*	1 box	6.3, 10.3	

*Note that all products listed in the above Master Product List are covered by **Limited Use Label License #223**. If a specific product is covered by additional Limited Use Label License(s), the Limited Use Label License #'s are provided in the LULL column above. Please refer to page 971 for the complete text of the Limited Use Label License(s).

IMPORTANT NOTICE: The products described in this manual are covered by one or more Limited Use Label License(s). Please refer to the Appendix on page 971 and Master Product List on page 975. Products are For Research Use Only. Not intended for any animal or human therapeutic or diagnostic use.

Cat. No.	Product	Quantity	Sections	LULL*
E21390	Endogenous Biotin-Blocking Kit *100 assays*	1 kit	7.6	
E13183	EnzChek® Caspase-3 Assay Kit #1 *Z-DEVD-AMC substrate* *500 assays*	1 kit	10.4, 15.5	
E13184	EnzChek® Caspase-3 Assay Kit #2 *Z-DEVD-R110 substrate* *500 assays*	1 kit	10.4, 15.5	
E33953	EnzChek® cellulase substrate *blue fluorescent, 339/452*	1 mg	10.2	
E10215	EnzChek® Direct Phospholipase C Assay Kit *phosphatidylcholine specific* *2-plate size*	1 kit	17.4	
E10216	EnzChek® Direct Phospholipase C Assay Kit *phosphatidylcholine specific* *10-plate size*	1 kit	17.4	
E12056	EnzChek® Elastase Assay Kit *600 assays*	1 kit	10.4	
E33956	EnzChek® epoxide hydrolase substrate	100 µg	10.6	
E12055	EnzChek® Gelatinase/Collagenase Assay Kit *250–2000 assays*	1 kit	10.4	
E33955	EnzChek® lipase substrate *green fluorescent, 505/515*	100 µg	10.6, 17.4	
E22013	EnzChek® Lysozyme Assay Kit *400 assays*	1 kit	10.2	
E33856	EnzChek® Myeloperoxidase (MPO) Activity Assay Kit *400 assays* *for myeloperoxidase chlorination and peroxidation activity*	1 kit	10.5, 18.2	
E33702	EnzChek® Paraoxonase Assay Kit *100 assays*	1 kit	10.3	
E33758	EnzChek® Peptidase/Protease Assay Kit *100 assays*	1 kit	10.4	
E12020	EnzChek® Phosphatase Assay Kit *1000 assays*	1 kit	10.3	
E6646	EnzChek® Phosphate Assay Kit *100 assays*	1 kit	10.3, 21.2	
E10217	EnzChek® Phospholipase A2 Assay Kit *2-plate size*	1 kit	17.4	
E10218	EnzChek® Phospholipase A2 Assay Kit *10-plate size*	1 kit	17.4	
E10221	EnzChek® Phospholipase A1 Assay Kit *10-plate size*	1 kit	17.4	
E10219	EnzChek® Phospholipase A1 Assay Kit *2-plate size*	1 kit	17.4	
E6658	EnzChek® Polarization Assay Kit for Proteases *green fluorescence* *100–1000 assays*	1 kit	10.4	
E6638	EnzChek® Protease Assay Kit *green fluorescence* *100–1000 assays*	1 kit	10.4, 14.7	
E6639	EnzChek® Protease Assay Kit *red fluorescence* *100–1000 assays*	1 kit	10.4, 14.7	
E6645	EnzChek® Pyrophosphate Assay Kit *100 assays*	1 kit	10.3, 21.2	
E22064	EnzChek® Reverse Transcriptase Assay Kit *1000 assays*	1 kit	15.3	
E33651	EnzChek® *Ultra* Amylase Assay Kit *500 assays*	1 kit	10.2	
E33701	EnzChek® *Ultra* Phytase Assay Kit *500 assays*	1 kit	10.3, 10.5	
E33650	EnzChek® *Ultra* Xylanase Assay Kit *500 assays*	1 kit	10.2	
E18	eosin-5-isothiocyanate	100 mg	1.5, 16.3	
E118	eosin-5-maleimide	25 mg	2.2	
E3476	epidermal growth factor (EGF) *from mouse submaxillary glands*	100 µg	16.1	
E3477	epidermal growth factor, biotin-XX conjugate (biotin EGF)	20 µg	4.3, 16.1	
E13345	epidermal growth factor, biotinylated, complexed to Alexa Fluor® 488 streptavidin (Alexa Fluor® 488 EGF complex)	100 µg	4.3, 16.1	
E35350	epidermal growth factor, biotinylated, complexed to Alexa Fluor® 555 streptavidin (Alexa Fluor® 555 EGF complex)	100 µg	4.3, 16.1	
E35351	epidermal growth factor, biotinylated, complexed to Alexa Fluor® 647 streptavidin (Alexa Fluor® 647 EGF complex)	100 µg	4.3, 16.1	
E3480	epidermal growth factor, biotinylated, complexed to Texas Red® streptavidin (Texas Red® EGF complex)	100 µg	4.3, 16.1	
E3478	epidermal growth factor, fluorescein conjugate (fluorescein EGF)	20 µg	16.1	
E7498	epidermal growth factor, Oregon Green® 514 conjugate (Oregon Green® 514 EGF)	20 µg	16.1	
E3481	epidermal growth factor, tetramethylrhodamine conjugate (rhodamine EGF)	20 µg	16.1	
E12353	ER-Tracker™ Blue-White DPX *for live-cell imaging* *1 mM solution in DMSO*	20 x 50 µL	12.4	
E34251	ER-Tracker™ Green (BODIPY® FL glibenclamide) *for live-cell imaging*	100 µg	12.4, 16.3	
E34250	ER-Tracker™ Red (BODIPY® TR glibenclamide) *for live-cell imaging*	100 µg	12.4, 16.3	
E2870	*Escherichia coli* BioParticles® opsonizing reagent	1 U	16.1	
E13231	*Escherichia coli* (K-12 strain) BioParticles®, Alexa Fluor® 488 conjugate	2 mg	16.1	
E23370	*Escherichia coli* (K-12 strain) BioParticles®, Alexa Fluor® 594 conjugate	2 mg	16.1	
E2864	*Escherichia coli* (K-12 strain) BioParticles®, BODIPY® FL conjugate	10 mg	16.1	
E2861	*Escherichia coli* (K-12 strain) BioParticles®, fluorescein conjugate	10 mg	16.1	
E2862	*Escherichia coli* (K-12 strain) BioParticles®, tetramethylrhodamine conjugate	10 mg	16.1	
E2863	*Escherichia coli* (K-12 strain) BioParticles®, Texas Red® conjugate	10 mg	16.1	
E23691	1,N^6-ethenoadenosine 5′-triphosphate (ε-ATP) *5 mM in buffer*	2 mL	17.3	
E1169	ethidium homodimer-1 (EthD-1)	1 mg	8.1, 8.4, 15.2	
E3599	ethidium homodimer-2 (EthD-2) *1 mM solution in DMSO*	200 µL	8.1, 8.4, 15.2	
E1374	ethidium monoazide bromide (EMA)	5 mg	5.3, 8.1, 15.2	
E3101	*N*-(ethoxycarbonylmethyl)-6-methoxyquinolinium bromide (MQAE)	100 mg	21.2	
R352	ethoxyresorufin (resorufin ethyl ether)	5 mg	10.6	
E2882	7-ethoxy-4-trifluoromethylcoumarin	25 mg	10.6	
E2247	1-ethyl-3-(3-dimethylaminopropyl)carbodiimide, hydrochloride (EDAC)	100 mg	3.4, 5.2	
E3111	5-(*N*-ethyl-*N*-isopropyl)amiloride, hydrochloride	5 mg	16.3	

*Note that all products listed in the above Master Product List are covered by **Limited Use Label License #223**. If a specific product is covered by additional Limited Use Label License(s), the Limited Use Label License #'s are provided in the LULL column above. Please refer to page 971 for the complete text of the Limited Use Label License(s).

IMPORTANT NOTICE: The products described in this manual are covered by one or more Limited Use Label License(s). Please refer to the Appendix on page 971 and Master Product List on page 975. Products are For Research Use Only. Not intended for any animal or human therapeutic or diagnostic use.

Cat. No.	Product	Quantity	Sections	LULL*
E10345	5-ethynyl uridine (EU)	5 mg	3.1	
R33200	EZQ® Protein Quantitation Kit *2000 assays*	1 kit	9.2	
A10534	F(ab')$_2$ fragment of goat anti-mouse IgG (H+L) *2 mg/mL*	250 µL	7.2	
F21452	F(ab')$_2$ fragment of goat anti-mouse IgG (H+L), alkaline phosphatase conjugate	0.5 mg	7.2	
F21453	F(ab')$_2$ fragment of goat anti-mouse IgG (H+L), horseradish peroxidase conjugate	0.5 mg	7.2	
F21456	F(ab')$_2$ fragment of goat anti-rabbit IgG (H+L), alkaline phosphatase conjugate	0.5 mg	7.2	
A10547	F(ab')$_2$ fragment of goat anti-rabbit IgG (H+L), horseradish peroxidase conjugate	0.5 mg	7.2	
A10548	F(ab')$_2$ fragment of goat anti-rat IgG (H+L), horseradish peroxidase conjugate	0.5 mg	7.2	
F2900	*FAST* CAT® Chloramphenicol Acetyltransferase Assay Kit *100 assays*	1 kit	10.6	
F6616	*FAST* CAT® Green (deoxy) Chloramphenicol Acetyltransferase Assay Kit *100 assays*	1 kit	10.6	
F6617	*FAST* CAT® Yellow (deoxy) Chloramphenicol Acetyltransferase Assay Kit *100 assays*	1 kit	10.6	
F2902	Fc OxyBURST® Green assay reagent *25 assays* *3 mg/mL*	500 µL	16.1, 18.2	
F1303	FDA (fluorescein diacetate)	1 g	15.2, 20.2	
F13191	fibrinogen from human plasma, Alexa Fluor® 488 conjugate	5 mg	15.6, 16.1	
F13192	fibrinogen from human plasma, Alexa Fluor® 546 conjugate	5 mg	15.6, 16.1	
F13193	fibrinogen from human plasma, Alexa Fluor® 594 conjugate	5 mg	15.6, 16.1	
F35200	fibrinogen from human plasma, Alexa Fluor® 647 conjugate	5 mg	15.6, 16.1	
F7496	fibrinogen from human plasma, Oregon Green® 488 conjugate	5 mg	15.6, 16.1	
F10322	FilmTracer™ calcein green biofilm stain	20 x 50 µg	15.2	
F10319	FilmTracer™ calcein red-orange biofilm stain	20 x 50 µg	15.2	
F10320	FilmTracer™ calcein violet biofilm stain	20 x 25 µg	15.2	
F10317	FilmTracer™ FM® 1-43 green biofilm matrix stain	1 mg	15.2	
L10316	FilmTracer™ LIVE/DEAD® Biofilm Viability Kit	1 kit	15.3	
F10318	FilmTracer™ SYPRO® Ruby biofilm matrix stain	200 mL	15.2	
F32950	FISH Tag™ DNA Far Red Kit *with Alexa Fluor® 647 dye* *10 reactions*	1 kit	1.3, 8.2	
F32947	FISH Tag™ DNA Green Kit *with Alexa Fluor® 488 dye* *10 reactions*	1 kit	1.3, 8.2	
F32951	FISH Tag™ DNA Multicolor Kit *Alexa Fluor® dye combination* *10 reactions*	1 kit	1.3, 8.2	
F32948	FISH Tag™ DNA Orange Kit *with Alexa Fluor® 555 dye* *10 reactions*	1 kit	1.3, 8.2	
F32949	FISH Tag™ DNA Red Kit *with Alexa Fluor® 594 dye* *10 reactions*	1 kit	1.3, 8.2	
F32955	FISH Tag™ RNA Far Red Kit *with Alexa Fluor® 647 dye* *10 reactions*	1 kit	1.3, 8.2	
F32952	FISH Tag™ RNA Green Kit *with Alexa Fluor® 488 dye* *10 reactions*	1 kit	1.3, 8.2	
F32956	FISH Tag™ RNA Multicolor Kit *Alexa Fluor® dye combination* *10 reactions*	1 kit	1.3, 8.2	
F32953	FISH Tag™ RNA Orange Kit *with Alexa Fluor® 555 dye* *10 reactions*	1 kit	1.3, 8.2	
F32954	FISH Tag™ RNA Red Kit *with Alexa Fluor® 594 dye* *10 reactions*	1 kit	1.3, 8.2	
V13242	FITC Annexin V/ Dead Cell Apoptosis Kit *FITC annexin V/propidium iodide* *50 assays* *for flow cytometry*	1 kit	15.5	
F13838	Flow Cytometry Size Calibration Kit *nonfluorescent microspheres*	1 kit	23.2	
F1241	fluo-3, AM *cell permeant*	1 mg	19.3	204
F1242	fluo-3, AM *cell permeant* *special packaging*	20 x 50 µg	19.3	204
F23915	fluo-3, AM *FluoroPure™ grade* *special packaging*	10 x 50 µg	19.3	204
F14218	fluo-3, AM *1 mM solution in DMSO* *cell permeant*	1 mL	19.3	204
F14242	fluo-3, AM *packaged for high-throughput screening*	40 x 1 mg	19.3	204
F1240	fluo-3, pentaammonium salt *cell impermeant*	1 mg	19.3	204
F3715	fluo-3, pentapotassium salt *cell impermeant*	1 mg	19.3	204
F14201	fluo-4, AM *cell permeant* *special packaging*	10 x 50 µg	19.3	204
F23917	fluo-4, AM *FluoroPure™ grade* *special packaging*	10 x 50 µg	19.3	204
F14217	fluo-4, AM *1 mM solution in DMSO* *cell permeant*	500 µL	19.3	204
F14202	fluo-4, AM *packaged for high-throughput screening*	5 x 1 mg	19.3	204
F36250	fluo-4 dextran, potassium salt, 10,000 MW, anionic (high-affinity version)	5 mg	19.4	
F14240	fluo-4 dextran, potassium salt, 10,000 MW, anionic (low-affinity version)	5 mg	19.4	
F14200	fluo-4, pentapotassium salt *cell impermeant*	500 µg	19.3	204
F23981	fluo-4FF, AM *cell permeant* *special packaging*	10 x 50 µg	19.3	204
F23980	fluo-4FF, pentapotassium salt *cell impermeant*	500 µg	19.3	204
F14222	fluo-5F, AM *cell permeant* *special packaging*	10 x 50 µg	19.3	204
F14221	fluo-5F, pentapotassium salt *cell impermeant*	500 µg	19.3	204
F14204	fluo-5N, AM *cell permeant* *special packaging*	10 x 50 µg	19.3	204
F14203	fluo-5N, pentapotassium salt *cell impermeant*	500 µg	19.3	204
F36924	FluoCells® prepared slide #1 *BPAE cells with MitoTracker® Red CMXRos, Alexa Fluor® 488 phalloidin, DAPI*	each	23.1	
F14781	FluoCells® prepared slide #2 *BPAE cells with mouse anti-α-tubulin, BODIPY® FL goat anti-mouse IgG, Texas Red®-X phalloidin, DAPI*	each	23.1	

*Note that all products listed in the above Master Product List are covered by **Limited Use Label License #223**. If a specific product is covered by additional Limited Use Label License(s), the Limited Use Label License #'s are provided in the LULL column above. Please refer to page 971 for the complete text of the Limited Use Label License(s).

Cat. No.	Product	Quantity	Sections	LULL*
F24630	FluoCells® prepared slide #3 *mouse kidney section with Alexa Fluor® 488 WGA, Alexa Fluor® 568 phalloidin, DAPI*	each	23.1	
F24631	FluoCells® prepared slide #4 *mouse intestine section with Alexa Fluor® 350 WGA, Alexa Fluor® 568 phalloidin, SYTOX® Green*	each	23.1	
F36925	FluoCells® prepared slide #6 *muntjac cells with mouse anti-OxPhos Complex V inhibitor protein, Alexa Fluor® 555 goat anti-mouse IgG, Alexa Fluor® 488 phalloidin, TO-PRO®-3*	each	23.1	
F10472	Fluo-4 Direct™ Calcium Assay Kit, Surveyor Pack	1 kit	19.3	
F10473	Fluo-4 Direct™ Calcium Assay Kit, High-Throughput Pack	1 kit	19.3	
F10471	Fluo-4 Direct™ Calcium Assay Kit, Starter Pack	1 kit	19.3	
F36205	Fluo-4 NW Calcium Assay Kit (high-throughput) *for 100 microplates*	1 kit	19.3	
F36206	Fluo-4 NW Calcium Assay Kit (starter pack with buffer) *for 10 microplates*	1 kit	19.3	
F11830	*N*$^{\alpha}$-(9-fluorenylmethoxycarbonyl)-*N*$^{\varepsilon}$-tetramethylrhodamine-(5-carbonyl)-L-lysine (α-FMOC-ε-TMR-L-lysine)	25 mg	9.5	
F6348	FluoReporter® Biotin/DNP Protein Labeling Kit *5–10 labelings*	1 kit	1.2, 4.2	
F30751	FluoReporter® Biotin Quantitation Assay Kit *for biotinylated proteins* *5 determinations*	1 kit	1.2, 4.2	
F30755	FluoReporter® Biotin Quantitation Assay Kit *for biotinylated nucleic acids* *10 determinations*	1 kit	1.2, 4.2, 8.2	
F2610	FluoReporter® Biotin-XX Protein Labeling Kit *5 labelings of 5–20 mg protein each*	1 kit	1.2, 4.2	
F2962	FluoReporter® Blue Fluorometric dsDNA Quantitation Kit *200–2000 assays*	1 kit	8.3, 15.4	
F20650	FluoReporter® Cell-Surface Biotinylation Kit	1 kit	4.2	
F6434	FluoReporter® FITC Protein Labeling Kit *5–10 labelings*	1 kit	1.2, 1.5	
F6433	FluoReporter® Fluorescein-EX Protein Labeling Kit *5–10 labelings*	1 kit	1.2, 1.5	
F1930	FluoReporter® *lacZ* Flow Cytometry Kit *50 assays*	1 kit	10.2	
F2905	FluoReporter® *lacZ*/Galactosidase Quantitation Kit *1000 assays*	1 kit	10.2	
F6347	FluoReporter® Mini-biotin-XX Protein Labeling Kit *5 labelings of 0.1–3 mg protein each*	1 kit	1.2, 4.2	
F6153	FluoReporter® Oregon Green® 488 Protein Labeling Kit *5–10 labelings*	1 kit	1.2, 1.5	
F6161	FluoReporter® Rhodamine Red™-X Protein Labeling Kit *5–10 labelings*	1 kit	1.2, 1.6	
F6162	FluoReporter® Texas Red®-X Protein Labeling Kit *5–10 labelings*	1 kit	1.2, 1.6	
F20261	fluorescamine *FluoroPure™ grade*	100 mg	1.8, 9.2, 9.3	
F2332	fluorescamine	100 mg	1.8, 9.2, 9.3	
F1300	fluorescein *reference standard*	1 g	1.5, 10.1, 20.2	
F32767	fluorescein-aha-dUTP *1 mM in TE buffer*	25 µL	8.2	
F7103	fluorescein bis-(5-carboxymethoxy-2-nitrobenzyl) ether, dipotassium salt (CMNB-caged fluorescein)	5 mg	5.3, 14.3	
F1176	fluorescein α-bungarotoxin (α-bungarotoxin, fluorescein conjugate)	500 µg	16.2	
F6218	fluorescein-5-carbonyl azide, diacetate	10 mg	3.2	
F6106	6-(fluorescein-5-carboxamido)hexanoic acid, succinimidyl ester (5-SFX) *single isomer*	5 mg	1.5	
F2181	6-(fluorescein-5-(and-6)-carboxamido)hexanoic acid, succinimidyl ester (5(6)-SFX) *mixed isomers*	10 mg	1.5, 4.2	
F6129	6-(fluorescein-5-(and-6)-carboxamido)hexanoic acid, succinimidyl ester (5(6)-SFX) *mixed isomers* *special packaging*	10 x 1 mg	1.5	
F362	fluorescein DHPE (*N*-(fluorescein-5-thiocarbamoyl)-1,2-dihexadecanoyl-*sn*-glycero-3-phosphoethanolamine, triethylammonium salt)	5 mg	13.2, 20.4	
F1179	fluorescein di-β-D-galactopyranoside (FDG)	5 mg	10.2	
F2881	fluorescein di-β-D-glucopyranoside (FDGlu)	5 mg	10.2	
F2915	fluorescein di-β-D-glucuronide (FDGlcU)	5 mg	10.2	
F2999	fluorescein diphosphate, tetraammonium salt (FDP)	5 mg	10.3	
F10240	Fluorescein-EX Protein Labeling Kit *3 labelings*	1 kit	1.2, 1.5	
F6130	fluorescein-5-EX, succinimidyl ester	10 mg	1.5, 4.2	
F11021	fluorescein F(ab')$_2$ fragment of goat anti-mouse IgG (H+L) *2 mg/mL*	250 µL	7.2	
A10683	fluorescein F(ab')$_2$ fragment of goat anti-mouse IgG, IgM (H+L) *2 mg/mL*	250 µL	7.2	
A10526	fluorescein F(ab')$_2$ fragment of goat anti-rabbit IgG (H+L) *2 mg/mL*	250 µL	7.2	
A10527	fluorescein F(ab')$_2$ fragment of goat anti-rat IgG (H+L) *2 mg/mL*	250 µL	7.2	
A10530	fluorescein goat anti-mouse IgG$_1$ (γ1) *2 mg/mL*	250 µL	7.2	
F2761	fluorescein goat anti-mouse IgG (H+L) *2 mg/mL*	0.5 mL	7.2	
A10679	fluorescein goat anti-mouse IgG, IgM (H+L) *2 mg/mL*	250 µL	7.2	
F2765	fluorescein goat anti-rabbit IgG (H+L) *2 mg/mL*	0.5 mL	7.2	
A10528	fluorescein goat anti-rat IgG (H+L) *2 mg/mL*	0.5 mL	7.2	
F143	fluorescein-5-isothiocyanate (FITC 'Isomer I')	1 g	1.5	
F1906	fluorescein-5-isothiocyanate (FITC 'Isomer I') *special packaging*	10 x 10 mg	1.5	
F1907	fluorescein-5-isothiocyanate (FITC 'Isomer I') *special packaging*	10 x 100 mg	1.5	
F150	fluorescein-5-maleimide	25 mg	2.2	
F36915	fluorescein *NIST-traceable standard* *nominal concentration 50 µM* *special packaging*	5 x 1 mL	1.5, 10.1, 23.1	
F3857	fluorescein octadecyl ester	10 mg	13.5	
F432	fluorescein phalloidin	300 U	11.1	
A10529	fluorescein rabbit anti-goat (H+L) *2 mg/mL*	0.5 mL	7.2	

*Note that all products listed in the above Master Product List are covered by **Limited Use Label License #223**. If a specific product is covered by additional Limited Use Label License(s), the Limited Use Label License #'s are provided in the LULL column above. Please refer to page 971 for the complete text of the Limited Use Label License(s).

IMPORTANT NOTICE: The products described in this manual are covered by one or more Limited Use Label License(s). Please refer to the Appendix on page 971 and Master Product List on page 975. Products are For Research Use Only. Not intended for any animal or human therapeutic or diagnostic use.
www.invitrogen.com/probes

Cat. No.	Product	Quantity	Sections	LULL*
F1130	fluorescein-5-(and-6)-sulfonic acid, trisodium salt	100 mg	14.3, 20.2	
F121	fluorescein-5-thiosemicarbazide	100 mg	3.3	
F6053	7-fluorobenz-2-oxa-1,3-diazole-4-sulfonamide (ABD-F)	10 mg	2.2	
F33951	Fluorocillin™ Green 345/530 β-lactamase substrate *precipitating product*	5 mg	10.6	
F33952	Fluorocillin™ Green 495/525 β-lactamase substrate *soluble product*	5 x 100 µg	10.6	
F34651	FluoroMyelin™ Green fluorescent myelin stain *solution in water*	1 mL	14.4	
F34652	FluoroMyelin™ Red fluorescent myelin stain *solution in water*	1 mL	14.4	
F486	4-fluoro-7-nitrobenz-2-oxa-1,3-diazole (NBD fluoride; 4-fluoro-7-nitrobenzofurazan)	25 mg	1.8, 2.2	
F8760	FluoSpheres® aldehyde-sulfate microspheres, 0.02 µm, yellow-green fluorescent (505/515) *2% solids*	10 mL	6.5	
F8762	FluoSpheres® aldehyde-sulfate microspheres, 1.0 µm, yellow-green fluorescent (505/515) *2% solids*	10 mL	6.5	
F8763	FluoSpheres® amine-modified microspheres, 0.2 µm, red fluorescent (580/605) *2% solids*	5 mL	6.5	
F8764	FluoSpheres® amine-modified microspheres, 0.2 µm, yellow-green fluorescent (505/515) *2% solids*	5 mL	6.5	
F8765	FluoSpheres® amine-modified microspheres, 1.0 µm, yellow-green fluorescent (505/515) *2% solids*	5 mL	6.5	
F8766	FluoSpheres® biotin-labeled microspheres, 0.04 µm, yellow-green fluorescent (505/515) *1% solids*	0.4 mL	4.3, 6.5, 7.6	
F8767	FluoSpheres® biotin-labeled microspheres, 0.2 µm, yellow-green fluorescent (505/515) *1% solids*	0.4 mL	4.3, 6.5, 7.6	
F8769	FluoSpheres® biotin-labeled microspheres, 1.0 µm, nonfluorescent *1% solids*	0.4 mL	4.3, 6.5, 7.6	
F8768	FluoSpheres® biotin-labeled microspheres, 1.0 µm, yellow-green fluorescent (505/515) *1% solids*	0.4 mL	4.3, 6.5, 7.6	
F8890	FluoSpheres® Blood Flow Determination Fluorescent Color Kit #1, polystyrene microspheres, 10 µm *seven colors, 10 mL each* *3.6x10^6 beads/mL*	1 kit	14.6	
F21015	FluoSpheres® Blood Flow Determination Fluorescent Color Kit #4, polystyrene microspheres, 15 µm *four colors, 10 mL each* *1.0x10^6 beads/mL*	1 kit	14.6	
F8891	FluoSpheres® Blood Flow Determination Fluorescent Color Kit #2, polystyrene microspheres, 15 µm *seven colors, 10 mL each* *1.0x10^6 beads/mL*	1 kit	14.6	
F8892	FluoSpheres® Blood Flow Determination Fluorescent Color Kit #3, polystyrene microspheres, 15 µm *five colors, 10 mL each* *1.0x10^6 beads/mL*	1 kit	14.6	
F8781	FluoSpheres® carboxylate-modified microspheres, 0.02 µm, blue fluorescent (365/415) *2% solids*	10 mL	6.5	
F8782	FluoSpheres® carboxylate-modified microspheres, 0.02 µm, crimson fluorescent (625/645) *2% solids*	2 mL	6.5	
F8783	FluoSpheres® carboxylate-modified microspheres, 0.02 µm, dark red fluorescent (660/680) *2% solids*	2 mL	6.5	
F8784	FluoSpheres® carboxylate-modified microspheres, 0.02 µm, nile red fluorescent (535/575) *2% solids*	10 mL	6.5	
F8786	FluoSpheres® carboxylate-modified microspheres, 0.02 µm, red fluorescent (580/605) *2% solids*	10 mL	6.5	
F8787	FluoSpheres® carboxylate-modified microspheres, 0.02 µm, yellow-green fluorescent (505/515) *2% solids*	10 mL	6.5	
F8789	FluoSpheres® carboxylate-modified microspheres, 0.04 µm, dark red fluorescent (660/680) *5% solids, azide free*	1 mL	6.5, 14.6	
F8792	FluoSpheres® carboxylate-modified microspheres, 0.04 µm, orange fluorescent (540/560) *5% solids, azide free*	1 mL	6.5, 14.6	
F8793	FluoSpheres® carboxylate-modified microspheres, 0.04 µm, red fluorescent (580/605) *5% solids, azide free*	1 mL	6.5, 14.6	
F8794	FluoSpheres® carboxylate-modified microspheres, 0.04 µm, red-orange fluorescent (565/580) *5% solids, azide free*	1 mL	6.5, 14.6	
F8795	FluoSpheres® carboxylate-modified microspheres, 0.04 µm, yellow-green fluorescent (505/515) *5% solids, azide free*	1 mL	6.5, 14.6	
F8797	FluoSpheres® carboxylate-modified microspheres, 0.1 µm, blue fluorescent (350/440) *2% solids*	10 mL	6.5	
F8799	FluoSpheres® carboxylate-modified microspheres, 0.1 µm, infrared fluorescent (715/755) *2% solids*	1 mL	6.5	
F8800	FluoSpheres® carboxylate-modified microspheres, 0.1 µm, orange fluorescent (540/560) *2% solids*	10 mL	6.5	
F8801	FluoSpheres® carboxylate-modified microspheres, 0.1 µm, red fluorescent (580/605) *2% solids*	10 mL	6.5	
F8803	FluoSpheres® carboxylate-modified microspheres, 0.1 µm, yellow-green fluorescent (505/515) *2% solids*	10 mL	6.5	
F8805	FluoSpheres® carboxylate-modified microspheres, 0.2 µm, blue fluorescent (365/415) *2% solids*	10 mL	6.5	
F8806	FluoSpheres® carboxylate-modified microspheres, 0.2 µm, crimson fluorescent (625/645) *2% solids*	2 mL	6.5	
F8807	FluoSpheres® carboxylate-modified microspheres, 0.2 µm, dark red fluorescent (660/680) *2% solids*	2 mL	6.5	
F8809	FluoSpheres® carboxylate-modified microspheres, 0.2 µm, orange fluorescent (540/560) *2% solids*	10 mL	6.5	
F8810	FluoSpheres® carboxylate-modified microspheres, 0.2 µm, red fluorescent (580/605) *2% solids*	10 mL	6.5	
F8811	FluoSpheres® carboxylate-modified microspheres, 0.2 µm, yellow-green fluorescent (505/515) *2% solids*	10 mL	6.5	
F8812	FluoSpheres® carboxylate-modified microspheres, 0.5 µm, red fluorescent (580/605) *2% solids*	10 mL	6.5	
F8813	FluoSpheres® carboxylate-modified microspheres, 0.5 µm, yellow-green fluorescent (505/515) *2% solids*	10 mL	6.5	
F8814	FluoSpheres® carboxylate-modified microspheres, 1.0 µm, blue fluorescent (365/415) *2% solids*	10 mL	6.5	
F8815	FluoSpheres® carboxylate-modified microspheres, 1.0 µm, blue fluorescent (350/440) *2% solids*	10 mL	6.5	
F8816	FluoSpheres® carboxylate-modified microspheres, 1.0 µm, crimson fluorescent (625/645) *2% solids*	2 mL	6.5	
F8819	FluoSpheres® carboxylate-modified microspheres, 1.0 µm, nile red fluorescent (535/575) *2% solids*	10 mL	6.5	
F8820	FluoSpheres® carboxylate-modified microspheres, 1.0 µm, orange fluorescent (540/560) *2% solids*	10 mL	6.5	
F8821	FluoSpheres® carboxylate-modified microspheres, 1.0 µm, red fluorescent (580/605) *2% solids*	10 mL	6.5	
F8823	FluoSpheres® carboxylate-modified microspheres, 1.0 µm, yellow-green fluorescent (505/515) *2% solids*	10 mL	6.5	
F8824	FluoSpheres® carboxylate-modified microspheres, 2.0 µm, blue fluorescent (365/415) *2% solids*	2 mL	6.5	
F8825	FluoSpheres® carboxylate-modified microspheres, 2.0 µm, nile red fluorescent (535/575) *2% solids*	2 mL	6.5	
F8826	FluoSpheres® carboxylate-modified microspheres, 2.0 µm, red fluorescent (580/605) *2% solids*	2 mL	6.5	

*Note that all products listed in the above Master Product List are covered by **Limited Use Label License #223**. If a specific product is covered by additional Limited Use Label License(s), the Limited Use Label License #'s are provided in the LULL column above. Please refer to page 971 for the complete text of the Limited Use Label License(s).

IMPORTANT NOTICE: The products described in this manual are covered by one or more Limited Use Label License(s). Please refer to the Appendix on page 971 and Master Product List on page 975. Products are For Research Use Only. Not intended for any animal or human therapeutic or diagnostic use.

Cat. No.	Product	Quantity	Sections	LULL*
F8827	FluoSpheres® carboxylate-modified microspheres, 2.0 µm, yellow-green fluorescent (505/515) *2% solids*	2 mL	6.5	
F20880	FluoSpheres® carboxylate-modified microspheres, 0.04 µm, europium luminescent (365/610) *0.5% solids*	2 mL	6.5, 14.6	
F20881	FluoSpheres® carboxylate-modified microspheres, 0.2 µm, europium luminescent (365/610) *0.5% solids*	2 mL	6.5, 14.6	
F20886	FluoSpheres® carboxylate-modified microspheres, 0.04 µm, platinum luminescent (390/650) *0.5% solids*	2 mL	6.5, 14.6	
F20892	FluoSpheres® collagen I-labeled microspheres, 1.0 µm, yellow-green fluorescent (505/515) *0.5% solids*	0.4 mL	6.5, 16.1	
F20893	FluoSpheres® collagen I-labeled microspheres, 2.0 µm, yellow-green fluorescent (505/515) *0.5% solids*	0.4 mL	6.5, 16.1	
F10720	FluoSpheres® Fluorescent Color Kit, carboxylate-modified microspheres, 0.04 µm *four colors, 1 mL each* *5% solids, azide free*	1 kit	6.5, 14.6	
F8772	FluoSpheres® NeutrAvidin® labeled microspheres, 0.04 µm, nonfluorescent *1% solids*	0.4 mL	6.5, 7.6	
F8777	FluoSpheres® NeutrAvidin® labeled microspheres, 1.0 µm, nonfluorescent *1% solids*	0.4 mL	6.5, 7.6	
F8770	FluoSpheres® NeutrAvidin® labeled microspheres, 0.04 µm, red fluorescent (580/605) *1% solids*	0.4 mL	6.5, 7.6	
F8775	FluoSpheres® NeutrAvidin® labeled microspheres, 1.0 µm, red fluorescent (580/605) *1% solids*	0.4 mL	6.5, 7.6	
F8771	FluoSpheres® NeutrAvidin® labeled microspheres, 0.04 µm, yellow-green fluorescent (505/515) *1% solids*	0.4 mL	6.5, 7.6	
F8774	FluoSpheres® NeutrAvidin® labeled microspheres, 0.2 µm, yellow-green fluorescent (505/515) *1% solids*	0.4 mL	6.5, 7.6	
F8776	FluoSpheres® NeutrAvidin® labeled microspheres, 1.0 µm, yellow-green fluorescent (505/515) *1% solids*	0.4 mL	6.5, 7.6	
F20883	FluoSpheres® NeutrAvidin® labeled microspheres, 0.04 µm, europium luminescent (365/610) *0.5% solids*	0.4 mL	6.5, 7.6	
F20884	FluoSpheres® NeutrAvidin® labeled microspheres, 0.2 µm, europium luminescent (365/610) *0.5% solids*	0.4 mL	6.5, 7.6	
F13080	FluoSpheres® polystyrene microspheres, 1.0 µm, blue-green fluorescent (430/465) *for tracer studies* *1.0x10^{10} beads/mL*	5 mL	6.5, 14.6	
F13082	FluoSpheres® polystyrene microspheres, 1.0 µm, orange fluorescent (540/560) *for tracer studies* *1.0x10^{10} beads/mL*	5 mL	6.5, 14.6	
F13083	FluoSpheres® polystyrene microspheres, 1.0 µm, red fluorescent (580/605) *for tracer studies* *1.0x10^{10} beads/mL*	5 mL	6.5, 14.6	
F13081	FluoSpheres® polystyrene microspheres, 1.0 µm, yellow-green fluorescent (505/515) *for tracer studies* *1.0x10^{10} beads/mL*	5 mL	6.5, 14.6	
F8829	FluoSpheres® polystyrene microspheres, 10 µm, blue fluorescent (365/415) *for blood flow determination* *3.6x10^{6} beads/mL*	10 mL	14.6	
F8830	FluoSpheres® polystyrene microspheres, 10 µm, blue-green fluorescent (430/465) *for blood flow determination* *3.6x10^{6} beads/mL*	10 mL	14.6	
F8831	FluoSpheres® polystyrene microspheres, 10 µm, crimson fluorescent (625/645) *for blood flow determination* *3.6x10^{6} beads/mL*	10 mL	14.6	
F8833	FluoSpheres® polystyrene microspheres, 10 µm, orange fluorescent (540/560) *for blood flow determination* *3.6x10^{6} beads/mL*	10 mL	14.6	
F8834	FluoSpheres® polystyrene microspheres, 10 µm, red fluorescent (580/605) *for blood flow determination* *3.6x10^{6} beads/mL*	10 mL	14.6	
F8836	FluoSpheres® polystyrene microspheres, 10 µm, yellow-green fluorescent (505/515) *for blood flow determination* *3.6x10^{6} beads/mL*	10 mL	14.6	
F8837	FluoSpheres® polystyrene microspheres, 15 µm, blue fluorescent (365/415) *for blood flow determination* *1.0x10^{6} beads/mL*	10 mL	14.6	
F8838	FluoSpheres® polystyrene microspheres, 15 µm, blue-green fluorescent (430/465) *for blood flow determination* *1.0x10^{6} beads/mL*	10 mL	14.6	
F8839	FluoSpheres® polystyrene microspheres, 15 µm, crimson fluorescent (625/645) *for blood flow determination* *1.0x10^{6} beads/mL*	10 mL	14.6	
F8841	FluoSpheres® polystyrene microspheres, 15 µm, orange fluorescent (540/560) *for blood flow determination* *1.0x10^{6} beads/mL*	10 mL	14.6	
F8842	FluoSpheres® polystyrene microspheres, 15 µm, red fluorescent (580/605) *for blood flow determination* *1.0x10^{6} beads/mL*	10 mL	14.6	
F8843	FluoSpheres® polystyrene microspheres, 15 µm, scarlet fluorescent (645/680) *for blood flow determination* *1.0x10^{6} beads/mL*	10 mL	14.6	
F8844	FluoSpheres® polystyrene microspheres, 15 µm, yellow-green fluorescent (505/515) *for blood flow determination* *1.0x10^{6} beads/mL*	10 mL	14.6	
F21013	FluoSpheres® polystyrene microspheres, 15 µm, carmine fluorescent (580/620) *for blood flow determination* *1.0x10^{6} beads/mL*	10 mL	14.6	
F21010	FluoSpheres® polystyrene microspheres, 15 µm, green fluorescent (450/480) *for blood flow determination* *1.0x10^{6} beads/mL*	10 mL	14.6	
F21012	FluoSpheres® polystyrene microspheres, 15 µm, red-orange fluorescent (565/580) *for blood flow determination* *1.0x10^{6} beads/mL*	10 mL	14.6	
F21011	FluoSpheres® polystyrene microspheres, 15 µm, yellow fluorescent (515/534) *for blood flow determination* *1.0x10^{6} beads/mL*	10 mL	14.6	
F8887	FluoSpheres® Size Kit #1, carboxylate-modified microspheres, red fluorescent (580/605) *six sizes, 1 mL each* *2% solids*	1 kit	6.5	
F8888	FluoSpheres® Size Kit #2, carboxylate-modified microspheres, yellow-green fluorescent (505/515) *six sizes, 1 mL each* *2% solids*	1 kit	6.5	
F8780	FluoSpheres® streptavidin-labeled microspheres, 0.04 µm, yellow-green fluorescent (505/515) *0.5% solids*	0.4 mL	6.5, 7.6	
F8845	FluoSpheres® sulfate microspheres, 0.02 µm, yellow-green fluorescent (505/515) *2% solids*	10 mL	6.5	

*Note that all products listed in the above Master Product List are covered by **Limited Use Label License #223**. If a specific product is covered by additional Limited Use Label License(s), the Limited Use Label License #'s are provided in the LULL column above. Please refer to page 971 for the complete text of the Limited Use Label License(s).

Cat. No.	Product	Quantity	Sections	LULL*
F8848	FluoSpheres® sulfate microspheres, 0.2 µm, yellow-green fluorescent (505/515) *2% solids*	10 mL	6.5	
F8851	FluoSpheres® sulfate microspheres, 1.0 µm, red fluorescent (580/605) *2% solids*	10 mL	6.5	
F8852	FluoSpheres® sulfate microspheres, 1.0 µm, yellow-green fluorescent (505/515) *2% solids*	10 mL	6.5	
F8853	FluoSpheres® sulfate microspheres, 2.0 µm, yellow-green fluorescent (505/515) *2% solids*	2 mL	6.5	
F8854	FluoSpheres® sulfate microspheres, 4.0 µm, blue fluorescent (365/415) *2% solids*	2 mL	6.5	
F8858	FluoSpheres® sulfate microspheres, 4.0 µm, red fluorescent (580/605) *2% solids*	2 mL	6.5	
F8859	FluoSpheres® sulfate microspheres, 4.0 µm, yellow-green fluorescent (505/515) *2% solids*	2 mL	6.5	
F24181	FluoZin™-1, AM *cell permeant*	50 µg	19.7	
F24180	FluoZin™-1, tripotassium salt *cell impermeant*	500 µg	19.7	
F24189	FluoZin™-2, AM *cell permeant*	50 µg	19.7	
F24195	FluoZin™-3, AM *cell permeant*	100 µg	19.7	
F24194	FluoZin™-3, tetrapotassium salt *cell impermeant*	500 µg	19.7	
F10017	FluxOR™ Potassium Ion Channel Assay *for 100 microplates*	1 kit	16.3, 21.1	
F10016	FluxOR™ Potassium Ion Channel Assay *for 10 microplates*	1 kit	16.3, 21.1	
F11831	5-((2-(FMOC)-γ-L-glutamylaminoethyl)amino)naphthalene-1-sulfonic acid (γ-EDANS-α-FMOC-L-glutamic acid)	100 mg	9.5	
F35355	FM® 1-43FX *fixable analog of FM® 1-43 membrane stain*	10 x 100 µg	14.4, 16.1	
F34653	FM® 4-64FX *fixable analog of FM® 4-64 membrane stain*	10 x 100 µg	14.4, 16.1	
F36905	FocalCheck™ DoubleGreen Fluorescent Microspheres Kit, 6 µm *mounted on slides*	1 kit	23.1	
F36906	FocalCheck™ DoubleOrange Fluorescent Microspheres Kit, 6 µm *mounted on slides*	1 kit	23.1	
F36909	FocalCheck™ fluorescence microscope test slide #1 *for alignment, intensity, and calibration*	each	23.1	
F36913	FocalCheck™ fluorescence microscope test slide #2 *for spectral imaging systems*	each	23.1	
F36914	FocalCheck™ fluorescence microscope test slide #3 *5 colors, high and low intensities*	each	23.1	
F24633	FocalCheck™ Fluorescent Microspheres Kit, 6 µm *mounted on slides*	1 kit	23.1	
F24634	FocalCheck™ Fluorescent Microspheres Kit, 15 µm *mounted on slides*	1 kit	23.1	
F7234	FocalCheck™ microspheres, 15 µm, fluorescent blue/orange ring stains	0.5 mL	23.1	
F14807	FocalCheck™ microspheres, 6 µm, fluorescent dark-red ring stain/green throughout	0.5 mL	23.1	
F7239	FocalCheck™ microspheres, 15 µm, fluorescent dark-red ring stain/green throughout	0.5 mL	23.1	
F7240	FocalCheck™ microspheres, 15 µm, fluorescent green/dark-red ring stains	0.5 mL	23.1	
F14806	FocalCheck™ microspheres, 6 µm, fluorescent green/orange/dark-red ring stains	0.5 mL	23.1	
F7235	FocalCheck™ microspheres, 15 µm, fluorescent green/orange/dark-red ring stains	0.5 mL	23.1	
F14808	FocalCheck™ microspheres, 6 µm, fluorescent green ring stain/blue throughout	0.5 mL	23.1	
F7237	FocalCheck™ microspheres, 15 µm, fluorescent green ring stain/blue throughout	0.5 mL	23.1	
F7238	FocalCheck™ microspheres, 15 µm, fluorescent green ring stain/dark red throughout	0.5 mL	23.1	
F7236	FocalCheck™ microspheres, 15 µm, fluorescent orange ring stain/blue throughout	0.5 mL	23.1	
F14791	FocalCheck™ Thin-Ring Fluorescent Microspheres Kit, 1.0 µm *three suspensions*	1 kit	23.1	
F1314	formyl-Nle-Leu-Phe-Nle-Tyr-Lys, fluorescein derivative	1 mg	15.6, 16.1	
F23103	fosmidomycin, sodium salt (FR-31564)	25 mg	15.2	
F34953	*Funga*Light™ CFDA, AM/propidium iodide Yeast Vitality Kit *for flow cytometry*	1 kit	15.3	
F7030	FUN® 1 cell stain *10 mM solution in DMSO*	100 µL	12.3, 15.2	
F1201	fura-2, AM *cell permeant*	1 mg	19.2	
F1221	fura-2, AM *cell permeant* *special packaging*	20 x 50 µg	19.2	
F1225	fura-2, AM *1 mM solution in anhydrous DMSO* *cell permeant*	1 mL	19.2	
F14185	fura-2, AM *FluoroPure™ grade* *special packaging*	20 x 50 µg	19.2	
F6774	Fura-2 Calcium Imaging Calibration Kit *zero to 10 mM CaEGTA, 50 µM fura-2 (11 x 1 mL)*	1 kit	19.2, 19.8	
F1200	fura-2, pentapotassium salt *cell impermeant*	1 mg	19.2	
F6799	fura-2, pentasodium salt *cell impermeant*	1 mg	19.2	
F14175	fura-4F, AM *cell permeant* *special packaging*	10 x 50 µg	19.2	
F14174	fura-4F, pentapotassium salt *cell impermeant*	500 µg	19.2	
F14178	fura-6F, pentapotassium salt *cell impermeant*	500 µg	19.2	
F3029	fura dextran, potassium salt, 10,000 MW, anionic	5 mg	19.4	
F14181	fura-FF, AM *cell permeant* *special packaging*	10 x 50 µg	19.2	
F14180	fura-FF, pentapotassium salt *cell impermeant*	500 µg	19.2	
F3020	Fura Red™, AM *cell permeant*	500 µg	19.3	
F3021	Fura Red™, AM *cell permeant* *special packaging*	10 x 50 µg	19.3	
F14219	Fura Red™, tetrapotassium salt *cell impermeant*	500 µg	19.3	
F10348	FxCycle™ Far Red stain *for flow cytometry* *500 assays*	1 set	15.4	
F10347	FxCycle™ Violet stain *for flow cytometry* *500 assays* *DAPI*	1 set	15.4	
G13187	gelatin from pig skin, fluorescein conjugate	5 mg	15.6, 16.1	
G13186	gelatin from pig skin, Oregon Green® 488 conjugate	5 mg	15.6, 16.1	

*Note that all products listed in the above Master Product List are covered by **Limited Use Label License #223**. If a specific product is covered by additional Limited Use Label License(s), the Limited Use Label License #'s are provided in the LULL column above. Please refer to page 971 for the complete text of the Limited Use Label License(s).

IMPORTANT NOTICE: The products described in this manual are covered by one or more Limited Use Label License(s). Please refer to the Appendix on page 971 and Master Product List on page 975. Products are For Research Use Only. Not intended for any animal or human therapeutic or diagnostic use.

Cat. No.	Product	Quantity	Sections	LULL*
G10362	GFP, ABfinity™ recombinant rabbit monoclonal antibody, unconjugated (anti-GFP, rabbit mAb)	100 µg	7.5	327
G7055	L-glutamic acid, γ-(α-carboxy-2-nitrobenzyl) ester, trifluoroacetic acid salt (γ-(CNB-caged) L-glutamic acid)	5 mg	5.3, 16.2	
G2879	glutathione agarose, linked through sulfur *sedimented bead suspension*	10 mL	9.4	
G36000	glutathione ethyl ester, biotin amide (BioGEE) *glutathiolation detection reagent* *special packaging*	10 x 100 µg	4.3, 9.4, 15.6, 18.2	
A10538	goat anti-mouse IgG_1 (γ1) *2 mg/mL*	250 µL	7.2	
A10685	goat anti-mouse IgG_{2a} (γ2a), horseradish peroxidase conjugate	200 µg	7.2	
A10535	goat anti-mouse IgG (H+L) *2 mg/mL*	0.5 mL	7.2	
A10516	goat anti-mouse IgG (H+L) agarose *sedimented bead suspension*	2 mL	7.2	
G21060	goat anti-mouse IgG (H+L), alkaline phosphatase conjugate	1 mg	7.2	
G21040	goat anti-mouse IgG (H+L), horseradish peroxidase conjugate	1 mg	7.2, 10.5	
G31567	goat anti-mouse IgG (H+L), β-lactamase TEM-1 conjugate *0.5 mg net protein*	0.5 mg	7.2	
A10551	goat anti-mouse IgG_1 (γ1), horseradish peroxidase conjugate	0.5 mg	7.2	
A10666	goat anti-mouse IgG, IgA, IgM (H+L) *2 mg/mL*	250 µL	7.2	
A10668	goat anti-mouse IgG, IgA, IgM (H+L), horseradish peroxidase conjugate	200 µg	7.2	
A10677	goat anti-mouse IgG, IgM (H+L), horseradish peroxidase conjugate	200 µg	7.2	
A10533	goat anti-rabbit IgG (H+L) *2 mg/mL*	0.5 mL	7.2	
G21079	goat anti-rabbit IgG (H+L), alkaline phosphatase conjugate	1 mg	7.2	
G21234	goat anti-rabbit IgG (H+L), horseradish peroxidase conjugate	1 mg	7.2	
G31568	goat anti-rabbit IgG (H+L), β-lactamase TEM-1 conjugate *0.5 mg net protein*	0.5 mg	7.2	
A10536	goat anti-rat IgG (H+L) *2 mg/mL*	0.5 mL	7.2	
A10546	goat anti-rat IgG (H+L), alkaline phosphatase conjugate	0.5 mg	7.2	
A10549	goat anti-rat IgG (H+L), horseradish peroxidase conjugate	0.5 mg	7.2	
G6888	gramicidin	100 mg	21.1	
G7921	Griess Reagent Kit *for nitrite quantitation*	1 kit	18.3, 21.2	
G22360	guanosine 5′-diphosphate, BODIPY® FL 2′-(or-3′)-*O*-(*N*-(2-aminoethyl)urethane), bis(triethylammonium) salt (BODIPY® FL GDP) *5 mM in buffer*	100 µL	17.3	
G35779	guanosine 5′-*O*-(3-thiotriphosphate), BODIPY® 515/530 thioester, sodium salt (BODIPY® 515/530 GTP-γ-S, thioester) *1 mM in buffer*	100 µL	17.3	
G22183	guanosine 5′-*O*-(3-thiotriphosphate), BODIPY® FL thioester, sodium salt (BODIPY® FL GTP-γ-S, thioester) *5 mM in buffer*	50 µL	10.3, 17.3	
G35780	guanosine 5′-*O*-(3-thiotriphosphate), BODIPY® TR thioester, sodium salt (BODIPY® TR GTP-γ-S, thioester) *1 mM in buffer*	100 µL	17.3	
G12411	guanosine 5′-triphosphate, BODIPY® FL 2′-(or-3′)-*O*-(*N*-(2-aminoethyl)urethane), trisodium salt (BODIPY® FL GTP) *5 mM in water*	100 µL	17.3	
G22351	guanosine 5′-triphosphate, BODIPY® TR 2′-(or-3′)-*O*-(*N*-(2-aminoethyl)urethane), trisodium salt (BODIPY® TR GTP) *5 mM in water*	100 µL	17.3	
G35778	guanosine 5′-*O*-(3-iminotriphosphate), BODIPY® FL ethylamide, sodium salt (BODIPY® FL GTP-γ-NH, amide) *1 mM in buffer*	100 µL	17.3	
H32720	HCS CellMask™ Blue stain *for 10 X 96-well plates*	1 set	12.5	289
H32721	HCS CellMask™ Deep Red stain *for 10 X 96-well plates*	1 set	12.5	289
H32714	HCS CellMask™ Green stain *for 10 X 96-well plates*	1 set	12.5	289
H32713	HCS CellMask™ Orange stain *for 10 X 96-well plates*	1 set	12.5	289
H32712	HCS CellMask™ Red stain *for 10 X 96-well plates*	1 set	12.5	289
H10292	HCS DNA Damage Kit *2-plate size*	1 kit	15.5	289
H34477	HCS LipidTOX™ Deep Red neutral lipid stain *solution in DMSO* *for cellular imaging*	each	13.5	289
H34475	HCS LipidTOX™ Green neutral lipid stain *solution in DMSO* *for cellular imaging*	each	13.5	289
H34350	HCS LipidTOX™ Green phospholipidosis detection reagent *1000X aqueous solution* *for cellular imaging* *10-plate size*	each	13.2	289
H34476	HCS LipidTOX™ Red neutral lipid stain *solution in DMSO* *for cellular imaging*	each	13.5	289
H34158	HCS LipidTOX™ Phospholipidosis and Steatosis Detection Kit *for high content screening* *for cellular imaging* *10-plate size*	1 kit	13.2	289
H34157	HCS LipidTOX™ Phospholipidosis and Steatosis Detection Kit *for high content screening* *for cellular imaging* *2-plate size*	1 kit	13.2	289
H34351	HCS LipidTOX™ Red phospholipidosis detection reagent *1000X aqueous solution* *for cellular imaging* *10-plate size*	each	13.2	289
H10290	HCS LIVE/DEAD® Green Kit *2-plate size*	1 kit	15.5	289
H10295	HCS Mitochondrial Health Kit *2-plate size*	1 kit	15.5	289
H10293	HCS Mitotic Index Kit *2-plate size*	1 kit	15.4	289
H10325	HCS NuclearMask™ Blue stain *for 10 X 96-well plates* *2000X concentrate*	65 µL	12.5	289
H10294	HCS NuclearMask™ Deep Red stain *250X concentrate in DMSO*	400 µL	12.5	289
H10326	HCS NuclearMask™ Red stain *for 10 X 96-well plates* *1000X concentrate*	125 µL	12.5	289

*Note that all products listed in the above Master Product List are covered by **Limited Use Label License #223**. If a specific product is covered by additional Limited Use Label License(s), the Limited Use Label License #'s are provided in the LULL column above. Please refer to page 971 for the complete text of the Limited Use Label License(s).

IMPORTANT NOTICE: The products described in this manual are covered by one or more Limited Use Label License(s). Please refer to the Appendix on page 971 and Master Product List on page 975. Products are For Research Use Only. Not intended for any animal or human therapeutic or diagnostic use.
www.invitrogen.com/probes

molecular probes® | invitrogen™ by life technologies™

Cat. No.	Product	Quantity	Sections	LULL*
H7482	heparin, fluorescein conjugate	1 mg	17.2	
H22730	4-heptadecyl-7-hydroxycoumarin	10 mg	13.5	
H110	5-hexadecanoylaminofluorescein	100 mg	13.5	
H361	1-hexadecanoyl-2-(1-pyrenedecanoyl)-*sn*-glycero-3-phosphocholine (β-py-C_{10}-HPC)	1 mg	13.2, 17.4	
H3809	1-hexadecanoyl-2-(1-pyrenedecanoyl)-*sn*-glycero-3-phosphoglycerol, ammonium salt (β-py-C_{10}-PG)	1 mg	13.2, 17.4	
H14700	1,1',3,3,3',3'-hexamethylindodicarbocyanine iodide (DiIC_1(5))	100 mg	22.3	
H7593	hexidium iodide	5 mg	8.1, 15.2	
H13188	histone H1 from calf thymus, Alexa Fluor® 488 conjugate	1 mg	12.5, 16.1	
H2930	HIV Protease Substrate 1 (Arg-Glu(EDANS)-Ser-Gln-Asn-Tyr-Pro-Ile-Val-Gln-Lys(dabcyl)-Arg)	1 mg	10.4	
Q10052	HLA-DR, mouse anti-human, Qdot® 605 conjugate (clone Tu36) *1 µM solution*	100 µL	6.6	
Q10063	HLA-DR, mouse anti-human, Qdot® 800 conjugate (clone Tu36) *1 µM solution*	100 µL	6.6	
H1398	Hoechst 33258, pentahydrate (bis-benzimide)	100 mg	8.1, 8.3, 12.5, 15.4	
H21491	Hoechst 33258, pentahydrate (bis-benzimide) *FluoroPure™ grade*	100 mg	8.1, 8.3, 12.5, 15.4	
H3569	Hoechst 33258, pentahydrate (bis-benzimide) *10 mg/mL solution in water*	10 mL	8.1, 8.3, 12.5, 15.4	
H1399	Hoechst 33342, trihydrochloride, trihydrate	100 mg	8.1, 12.5, 15.4	
H21492	Hoechst 33342, trihydrochloride, trihydrate *FluoroPure™ grade*	100 mg	8.1, 12.5, 15.4	
H3570	Hoechst 33342, trihydrochloride, trihydrate *10 mg/mL solution in water*	10 mL	8.1, 12.5, 15.4	
H21486	Hoechst 34580	5 mg	8.1	
H18200	HybriSlip™ hybridization cover, 22 mm x 22 mm *RNase free* *set of 500*	1 set	23.1	
H18201	HybriSlip™ hybridization cover, 40 mm x 22 mm *RNase free* *set of 500*	1 set	23.1	
H18202	HybriSlip™ hybridization cover, 60 mm x 22 mm *RNase free* *set of 500*	1 set	23.1	
H24720	HybriWell™ hybridization sealing system, 13 mm diameter chamber, 0.25 mm deep *set of 100*	1 set	23.1	
H24721	HybriWell™ hybridization sealing system, 20 mm diameter chamber, 0.15 mm deep *set of 100*	1 set	23.1	
H24723	HybriWell™ hybridization sealing system, 22 mm x 22 mm chamber, 0.15 mm deep *set of 100*	1 set	23.1	
H18210	HybriWell™ hybridization sealing system, 40 mm x 21 mm chamber, 0.15 mm deep *set of 100*	1 set	23.1	
H24722	HybriWell™ hybridization sealing system, 40 mm x 22 mm chamber, 0.25 mm deep *set of 100*	1 set	23.1	
H185	7-hydroxycoumarin-3-carboxylic acid *reference standard*	100 mg	1.7, 14.3	
H1193	7-hydroxycoumarin-3-carboxylic acid, succinimidyl ester	25 mg	1.7	
H6482	7-hydroxy-9*H*-(1,3-dichloro-9,9-dimethylacridin-2-one) (DDAO) *reference standard*	10 mg	10.1	
H189	7-hydroxy-4-methylcoumarin *reference standard*	1 g	10.1	
H1428	7-hydroxy-4-methylcoumarin-3-acetic acid	100 mg	1.7	
H36004	3'-(*p*-hydroxyphenyl) fluorescein (HPF) *5 mM solution in DMF*	470 µL	18.2	
H348	8-hydroxypyrene-1,3,6-trisulfonic acid, trisodium salt (HPTS; pyranine)	1 g	14.3, 16.1, 20.2	
H22845	hydroxystilbamidine, methanesulfonate	10 mg	8.1, 14.3	
H2249	*N*-hydroxysulfosuccinimide, sodium salt (NHSS)	100 mg	3.4, 5.2	
H7476	hypericin	1 mg	17.3, 18.2	
I10291	Image-iT® DEAD Green™ viability stain *1 mM solution in DMSO*	25 µL	15.5	
I36933	Image-iT® FX signal enhancer	10 mL	7.2, 7.6, 23.1	
I35106	Image-iT® LIVE Green Caspase-3 and -7 Detection Kit *for microscopy* *25 tests*	1 kit	10.4, 15.5	
I35105	Image-iT® LIVE Green Caspase-8 Detection Kit *for microscopy* *25 tests*	1 kit	10.4, 15.5	
I35104	Image-iT® LIVE Green Poly Caspases Detection Kit *for microscopy* *25 tests*	1 kit	10.4, 15.5	
I36007	Image-iT® LIVE Green Reactive Oxygen Species Detection Kit *for microscopy*	1 kit	18.2	
I34407	Image-iT® LIVE Intracellular Membrane and Nuclear Labeling Kit *counterstains for GFP-expressing cells*	1 kit	14.4	
I34202	Image-iT® LIVE Lysosomal and Nuclear Labeling Kit *counterstains for GFP-expressing cells*	1 kit	12.3	
I34154	Image-iT® LIVE Mitochondrial and Nuclear Labeling Kit *counterstains for GFP-expressing cells*	1 kit	12.2	
I35103	Image-iT® LIVE Mitochondrial Transition Pore Assay Kit *for microscopy*	1 kit	12.2, 15.5	
I34406	Image-iT® LIVE Plasma Membrane and Nuclear Labeling Kit *counterstains for GFP-expressing cells*	1 kit	7.7, 14.4	
I35102	Image-iT® LIVE Red Caspase-3 and -7 Detection Kit *for microscopy*	1 kit	10.4, 15.5	
I35101	Image-iT® LIVE Red Poly Caspases Detection Kit *for microscopy*	1 kit	10.4, 15.5	
I2904	ImaGene Green™ C_{12}FDG *lacZ* Gene Expression Kit	1 kit	10.2	
I2908	ImaGene Green™ C_{12}FDGlcU GUS Gene Expression Kit	1 kit	10.2	
I2906	ImaGene Red™ C_{12}RG *lacZ* Gene Expression Kit	1 kit	10.2	
I1203	indo-1, AM *cell permeant*	1 mg	19.2	
I1223	indo-1, AM *cell permeant* *special packaging*	20 x 50 µg	19.2	
I1226	indo-1, AM *1 mM solution in anhydrous DMSO* *cell permeant*	1 mL	19.2	
I1202	indo-1, pentapotassium salt *cell impermeant*	1 mg	19.2	
I14402	Influx™ pinocytic cell-loading reagent *makes 10 x 5 mL*	1 set	14.3, 19.8	

*Note that all products listed in the above Master Product List are covered by **Limited Use Label License #223**. If a specific product is covered by additional Limited Use Label License(s), the Limited Use Label License #'s are provided in the LULL column above. Please refer to page 971 for the complete text of the Limited Use Label License(s).

IMPORTANT NOTICE: The products described in this manual are covered by one or more Limited Use Label License(s). Please refer to the Appendix on page 971 and Master Product List on page 975. Products are For Research Use Only. Not intended for any animal or human therapeutic or diagnostic use.

Cat. No.	Product	Quantity	Sections	LULL*
I3716	D-*myo*-inositol 1,4,5-triphosphate, hexapotassium salt (Ins 1,4,5-P_3)	1 mg	17.2	
I23580	D-*myo*-inositol 1,4,5-triphosphate, $P_{4(5)}$-(1-(2-nitrophenyl)ethyl) ester, tris(triethylammonium) salt (NPE-caged Ins 1,4,5-P_3)	25 µg	5.3, 17.2	
I7221	InSpeck™ Blue (350/440) Microscope Image Intensity Calibration Kit, 2.5 µm	1 kit	23.1	
I7225	InSpeck™ Deep Red (633/660) Microscope Image Intensity Calibration Kit, 2.5 µm	1 kit	23.1	
I7219	InSpeck™ Green (505/515) Microscope Image Intensity Calibration Kit, 2.5 µm	1 kit	23.1	
I14785	InSpeck™ Green (505/515) Microscope Image Intensity Calibration Kit, 6 µm	1 kit	23.1	
I7223	InSpeck™ Orange (540/560) Microscope Image Intensity Calibration Kit, 2.5 µm	1 kit	23.1	
I14786	InSpeck™ Orange (540/560) Microscope Image Intensity Calibration Kit, 6 µm	1 kit	23.1	
I7224	InSpeck™ Red (580/605) Microscope Image Intensity Calibration Kit, 2.5 µm	1 kit	23.1	
I14787	InSpeck™ Red (580/605) Microscope Image Intensity Calibration Kit, 6 µm	1 kit	23.1	
I13269	insulin, human, recombinant from *E. coli*, fluorescein conjugate (FITC insulin) *monolabeled* *zinc free*	100 µg	16.1	
I10189	iodoacetamide alkyne	1 mg	3.1	
I10188	iodoacetamide azide	1 mg	3.1	
I7	2-(4'-(iodoacetamido)anilino)naphthalene-6-sulfonic acid, sodium salt (IAANS)	100 mg	2.3	
I30451	5-iodoacetamidofluorescein (5-IAF)	25 mg	2.2	
I30452	6-iodoacetamidofluorescein (6-IAF)	25 mg	2.2	
I9	*N*-((2-(iodoacetoxy)ethyl)-*N*-methyl)amino-7-nitrobenz-2-oxa-1,3-diazole (IANBD ester)	100 mg	2.2	
I14	5-((((2-iodoacetyl)amino)ethyl)amino)naphthalene-1-sulfonic acid (1,5-IAEDANS)	100 mg	2.3	
I24222	ionomycin, calcium salt	1 mg	19.8	
I21411	isolectin GS-IB_4 from *Griffonia simplicifolia*, Alexa Fluor® 488 conjugate	500 µg	7.7	
I21412	isolectin GS-IB_4 from *Griffonia simplicifolia*, Alexa Fluor® 568 conjugate	500 µg	7.7	
I21413	isolectin GS-IB_4 from *Griffonia simplicifolia*, Alexa Fluor® 594 conjugate	500 µg	7.7	
I32450	isolectin GS-IB_4 from *Griffonia simplicifolia*, Alexa Fluor® 647 conjugate	500 µg	7.7	
I21414	isolectin GS-IB_4 from *Griffonia simplicifolia*, biotin-XX conjugate	500 µg	4.3, 7.7	
I24221	(*S*)-1-*p*-isothiocyanatobenzyldiethylenetriaminepentaacetic acid (DTPA isothiocyanate)	1 mg	19.8	
J7473	jasplakinolide	100 µg	11.1	
J11372	JOJO™-1 iodide (529/545) *1 mM solution in DMSO*	200 µL	8.1	
J11373	JO-PRO™-1 iodide (530/546) *1 mM solution in DMSO*	1 mL	8.1	
L12370	latrunculin A	100 µg	11.1	
L22290	latrunculin B	100 µg	11.1	
L10382	LC3B Antibody Kit for Autophagy *rabbit polyclonal LC3B* *includes autophagosome inducer*	1 kit	15.5	
L7595	LDS 751	10 mg	8.1, 15.4	
A10024	Leadmium™ Green, AM dye *for intracellular detection of lead and cadmium* *special packaging*	5 x 50 µg	19.7	
L21415	lectin GS-II from *Griffonia simplicifolia*, Alexa Fluor® 488 conjugate	500 µg	7.7, 12.4	
L21416	lectin GS-II from *Griffonia simplicifolia*, Alexa Fluor® 594 conjugate	500 µg	7.7, 12.4	
L32451	lectin GS-II from *Griffonia simplicifolia*, Alexa Fluor® 647 conjugate	500 µg	7.7, 12.4	
L11271	lectin HPA from *Helix pomatia* (edible snail), Alexa Fluor® 488 conjugate	1 mg	7.7, 12.4	
L32454	lectin HPA from *Helix pomatia* (edible snail), Alexa Fluor® 647 conjugate	1 mg	7.7, 12.4	
L11270	lectin PHA-L from *Phaseolus vulgaris* (red kidney bean), Alexa Fluor® 488 conjugate	1 mg	7.7, 14.7	
L32456	lectin PHA-L from *Phaseolus vulgaris* (red kidney bean), Alexa Fluor® 594 conjugate	1 mg	7.7, 14.7	
L32457	lectin PHA-L from *Phaseolus vulgaris* (red kidney bean), Alexa Fluor® 647 conjugate	1 mg	7.7, 14.7	
L21409	lectin PNA from *Arachis hypogaea* (peanut), Alexa Fluor® 488 conjugate	1 mg	7.7, 16.1	
L32458	lectin PNA from *Arachis hypogaea* (peanut), Alexa Fluor® 568 conjugate	1 mg	7.7, 16.1	
L32459	lectin PNA from *Arachis hypogaea* (peanut), Alexa Fluor® 594 conjugate	1 mg	7.7, 16.1	
L32460	lectin PNA from *Arachis hypogaea* (peanut), Alexa Fluor® 647 conjugate	1 mg	7.7, 16.1	
L11272	lectin SBA from *Glycine max* (soybean), Alexa Fluor® 488 conjugate	1 mg	7.7	
L32462	lectin SBA from *Glycine max* (soybean), Alexa Fluor® 594 conjugate	1 mg	7.7	
L32463	lectin SBA from *Glycine max* (soybean), Alexa Fluor® 647 conjugate	1 mg	7.7	
L6543	leupeptin hemisulfate	10 mg	10.4	
L14812	LinearFlow™ Blue Flow Cytometry Intensity Calibration Kit, 2.5 µm *for UV excitation/430 nm emission*	1 kit	23.2	
L14813	LinearFlow™ Blue Flow Cytometry Intensity Calibration Kit, 6 µm *for UV excitation/430 nm emission*	1 kit	23.2	
L14816	LinearFlow™ Carmine Flow Cytometry Intensity Calibration Kit, 2.5 µm *for 488 nm excitation/620 nm emission*	1 kit	23.2	
L14817	LinearFlow™ Carmine Flow Cytometry Intensity Calibration Kit, 6 µm *for 488 nm excitation/620 nm emission*	1 kit	23.2	
L14818	LinearFlow™ Deep Red Flow Cytometry Intensity Calibration Kit, 2.5 µm *for 633 nm excitation/660 nm emission*	1 kit	23.2	
L14819	LinearFlow™ Deep Red Flow Cytometry Intensity Calibration Kit, 6 µm *for 633 nm excitation/660 nm emission*	1 kit	23.2	
L14821	LinearFlow™ Green Flow Cytometry Intensity Calibration Kit, 2.5 µm *for 488 nm excitation/515 nm emission*	1 kit	23.2	
L14822	LinearFlow™ Green Flow Cytometry Intensity Calibration Kit, 6 µm *for 488 nm excitation/515 nm emission*	1 kit	23.2	
L14823	LinearFlow™ Green Flow Cytometry Low Intensity Calibration Kit, 2.5 µm *for 488 nm excitation/515 nm emission*	1 kit	23.2	

*Note that all products listed in the above Master Product List are covered by **Limited Use Label License #223**. If a specific product is covered by additional Limited Use Label License(s), the Limited Use Label License #'s are provided in the LULL column above. Please refer to page 971 for the complete text of the Limited Use Label License(s).

IMPORTANT NOTICE: The products described in this manual are covered by one or more Limited Use Label License(s). Please refer to the Appendix on page 971 and Master Product List on page 975. Products are For Research Use Only. Not intended for any animal or human therapeutic or diagnostic use.

Cat. No.	Product	Quantity	Sections	LULL*
L14824	LinearFlow™ Green Flow Cytometry Low Intensity Calibration Kit, 6 µm *for 488 nm excitation/515 nm emission*	1 kit	23.2	
L14814	LinearFlow™ Orange Flow Cytometry Intensity Calibration Kit, 2.5 µm *for 488 nm excitation/575 nm emission*	1 kit	23.2	
L14815	LinearFlow™ Orange Flow Cytometry Intensity Calibration Kit, 6 µm *for 488 nm excitation/575 nm emission*	1 kit	23.2	
L7781	Lipophilic Tracer Sampler Kit	1 kit	13.4, 14.4	
L23351	lipopolysaccharides from *Escherichia coli* serotype 055:B5, Alexa Fluor® 488 conjugate	100 µg	13.3, 16.1	
L23352	lipopolysaccharides from *Escherichia coli* serotype 055:B5, Alexa Fluor® 568 conjugate	100 µg	13.3, 16.1	
L23353	lipopolysaccharides from *Escherichia coli* serotype 055:B5, Alexa Fluor® 594 conjugate	100 µg	13.3, 16.1	
L23350	lipopolysaccharides from *Escherichia coli* serotype 055:B5, BODIPY® FL conjugate	100 µg	13.3, 16.1	
L23356	lipopolysaccharides from *Salmonella minnesota*, Alexa Fluor® 488 conjugate	100 µg	13.3, 16.1	
L24919	LI Silver (LIS) Enhancement Kit	1 kit	7.2, 7.6	
L1392	Lissamine rhodamine B 1,2-dihexadecanoyl-*sn*-glycero-3-phosphoethanolamine, triethylammonium salt (rhodamine DHPE)	5 mg	13.2	
L2424	Lissamine rhodamine B ethylenediamine	10 mg	3.4	
L1908	Lissamine rhodamine B sulfonyl chloride *mixed isomers* *special packaging*	10 x 10 mg	1.6	
L20	Lissamine rhodamine B sulfonyl chloride *mixed isomers*	1 g	1.6	
L7005	LIVE *Bac*Light™ Bacterial Gram Stain Kit *for microscopy and quantitative assays* *1000 assays*	1 kit	15.3	
L34856	LIVE/DEAD® *Bac*Light™ Bacterial Viability and Counting Kit *for flow cytometry* *100 assays*	1 kit	15.3	
L13152	LIVE/DEAD® *Bac*Light™ Bacterial Viability Kit *10 applicator sets* *500 assays*	1 kit	15.3	
L7007	LIVE/DEAD® *Bac*Light™ Bacterial Viability Kit *for microscopy* *1000 assays*	1 kit	15.3	
L7012	LIVE/DEAD® *Bac*Light™ Bacterial Viability Kit *for microscopy and quantitative assays* *1000 assays*	1 kit	15.3	
L7010	LIVE/DEAD® Cell-Mediated Cytotoxicity Kit *for animal cells* *2000 assays*	1 kit	15.3	
L34951	LIVE/DEAD® Cell Vitality Assay Kit *C_{12}-resazurin/SYTOX® Green* *1000 assays*	1 kit	15.3	
L34957	LIVE/DEAD® Fixable Aqua Dead Cell Stain Kit *for 405 nm excitation* *200 assays*	1 kit	15.3	
L23105	LIVE/DEAD® Fixable Blue Dead Cell Stain Kit *for UV excitation* *200 assays*	1 kit	15.3	
L34960	LIVE/DEAD® Fixable Dead Cell Stain Sampler Kit *for flow cytometry* *320 assays*	1 kit	15.3	
L10120	LIVE/DEAD® Fixable Far Red Dead Cell Stain Kit *for 633 or 635 nm excitation* *200 assays*	1 kit	15.3	
L23101	LIVE/DEAD® Fixable Green Dead Cell Stain Kit *for 488 nm excitation* *200 assays*	1 kit	15.3	
L10119	LIVE/DEAD® Fixable Near-IR Dead Cell Stain Kit *for 633 or 635 nm excitation* *200 assays*	1 kit	15.3	
L23102	LIVE/DEAD® Fixable Red Dead Cell Stain Kit *for 488 nm excitation* *200 assays*	1 kit	15.3	
L34955	LIVE/DEAD® Fixable Violet Dead Cell Stain Kit *for 405 nm excitation* *200 assays*	1 kit	15.3	
L34959	LIVE/DEAD® Fixable Yellow Dead Cell Stain Kit *for 405 nm excitation* *200 assays*	1 kit	15.3	
L34952	LIVE/DEAD® *Funga*Light™ Yeast Viability Kit *for flow cytometry*	1 kit	15.3	
L7013	LIVE/DEAD® Reduced Biohazard Cell Viability Kit #1 *green and red fluorescence* *100 assays*	1 kit	15.3	
L7011	LIVE/DEAD® Sperm Viability Kit *200–1000 assays*	1 kit	15.3	
L3224	LIVE/DEAD® Viability/Cytotoxicity Kit *for mammalian cells*	1 kit	15.3	
L34958	LIVE/DEAD® Violet Viability/Vitality Kit *for 405 nm excitation* *200 assays*	1 kit	15.3	
L7009	LIVE/DEAD® Yeast Viability Kit *1000 assays*	1 kit	15.3	
L11376	LOLO™-1 iodide (565/579) *1 mM solution in DMSO*	200 µL	8.1	
L3486	low-density lipoprotein from human plasma (LDL) *2.5 mg/mL*	200 µL	16.1	
L35354	low-density lipoprotein from human plasma, acetylated (AcLDL) *2.5 mg/mL*	200 µL	16.1	
L23380	low-density lipoprotein from human plasma, acetylated, Alexa Fluor® 488 conjugate (Alexa Fluor® 488 AcLDL) *1 mg/mL*	200 µL	16.1	
L35353	low-density lipoprotein from human plasma, acetylated, Alexa Fluor® 594 conjugate (Alexa Fluor® 594 AcLDL) *1 mg/mL*	200 µL	16.1	
L3485	low-density lipoprotein from human plasma, acetylated, BODIPY® FL conjugate (BODIPY® FL AcLDL) *1 mg/mL*	200 µL	16.1	
L3484	low-density lipoprotein from human plasma, acetylated, DiI complex (DiI AcLDL) *1 mg/mL*	200 µL	16.1	
L3483	low-density lipoprotein from human plasma, BODIPY® FL complex (BODIPY® FL LDL) *1 mg/mL*	200 µL	16.1	
L3482	low-density lipoprotein from human plasma, DiI complex (DiI LDL) *1 mg/mL*	200 µL	16.1	
L7085	D-luciferin, 1-(4,5-dimethoxy-2-nitrophenyl)ethyl ester (DMNPE-caged luciferin)	5 mg	5.3, 10.6	
L2911	D-luciferin, free acid	25 mg	10.6	
L2916	D-luciferin, potassium salt	25 mg	10.6	
L2912	D-luciferin, sodium salt	25 mg	10.6	
L6950	lucifer yellow biocytin, potassium salt (biocytin lucifer yellow)	5 mg	4.3, 14.3	
L2601	lucifer yellow cadaverine biotin-X, dipotassium salt	10 mg	4.3, 14.3	
L682	lucifer yellow CH, ammonium salt	25 mg	14.3	
L453	lucifer yellow CH, lithium salt	25 mg	3.3, 14.3, 16.1	
L12926	lucifer yellow CH, lithium salt *for microinjection* *100 mM in water*	100 µL	14.3	
L1177	lucifer yellow CH, potassium salt	25 mg	14.3	
L1338	lucifer yellow iodoacetamide, dipotassium salt	25 mg	2.2, 4.2, 14.3	
L6868	lucigenin (bis-*N*-methylacridinium nitrate) *high purity*	10 mg	12.2, 18.2, 21.2	

*Note that all products listed in the above Master Product List are covered by **Limited Use Label License #223**. If a specific product is covered by additional Limited Use Label License(s), the Limited Use Label License #'s are provided in the LULL column above. Please refer to page 971 for the complete text of the Limited Use Label License(s).

Cat. No.	Product	Quantity	Sections	LULL*
L8455	luminol (3-aminophthalhydrazide)	25 g	6.2, 10.5, 15.2, 18.2	
L7533	LysoSensor™ Blue DND-167 *1 mM solution in DMSO* *special packaging*	20 x 50 µL	12.3, 20.3	
L7534	LysoSensor™ Green DND-153 *1 mM solution in DMSO* *special packaging*	20 x 50 µL	12.3, 20.3	
L7535	LysoSensor™ Green DND-189 *1 mM solution in DMSO* *special packaging*	20 x 50 µL	12.3, 20.3	
L22460	LysoSensor™ Yellow/Blue dextran, 10,000 MW, anionic, fixable	5 mg	12.3, 20.4	
L7545	LysoSensor™ Yellow/Blue DND-160 (PDMPO) *1 mM solution in DMSO*	20 x 50 µL	12.3, 20.3	
L7525	LysoTracker® Blue DND-22 *1 mM solution in DMSO* *special packaging*	20 x 50 µL	12.3	
L7526	LysoTracker® Green DND-26 *1 mM solution in DMSO* *special packaging*	20 x 50 µL	12.3	
L7528	LysoTracker® Red DND-99 *1 mM solution in DMSO* *special packaging*	20 x 50 µL	12.3	
L12491	LysoTracker® Yellow HCK-123 *1 mM solution in DMSO* *special packaging*	20 x 50 µL	12.3	
M14206	mag-fluo-4, AM *cell permeant* *special packaging*	10 x 50 µg	19.3, 19.6	
M14205	mag-fluo-4, tetrapotassium salt *cell impermeant*	500 µg	19.3, 19.6	
M1291	mag-fura-2, AM *cell permeant*	1 mg	19.2, 19.6	
M1292	mag-fura-2, AM *cell permeant* *special packaging*	20 x 50 µg	19.2, 19.6	
M1290	mag-fura-2, tetrapotassium salt *cell impermeant*	1 mg	19.2, 19.6	
M1295	mag-indo-1, AM *cell permeant* *special packaging*	20 x 50 µg	19.2, 19.6	
M3735	Magnesium Green™, AM *cell permeant* *special packaging*	20 x 50 µg	19.3, 19.6	
M3733	Magnesium Green™, pentapotassium salt *cell impermeant*	1 mg	19.3, 19.6	
M689	malachite green isothiocyanate	10 mg	1.6, 18.2	
M8	2-(4'-maleimidylanilino)naphthalene-6-sulfonic acid, sodium salt (MIANS)	100 mg	2.3	
M6026	1-(2-maleimidylethyl)-4-(5-(4-methoxyphenyl)oxazol-2-yl)pyridinium methanesulfonate (PyMPO maleimide)	5 mg	2.2	
M1602	N^{α}-(3-maleimidylpropionyl)biocytin	25 mg	4.2, 14.3	
M12652	Marina Blue® 1,2-dihexadecanoyl-*sn*-glycero-3-phosphoethanolamine (Marina Blue® DHPE)	1 mg	13.2	
M10991	Marina Blue® goat anti-mouse IgG (H+L) *2 mg/mL*	0.5 mL	7.2	
M10992	Marina Blue® goat anti-rabbit IgG (H+L) *2 mg/mL*	0.5 mL	7.2	
M10165	Marina Blue® succinimidyl ester	5 mg	1.7	
M36051	Measure-iT™ High-Sensitivity Nitrite Assay Kit *2000 assays*	1 kit	18.3, 21.2	
M36353	Measure-iT™ Lead and Cadmium Assay Kit *1000 assays*	1 kit	19.7	
M30550	Measure-iT™ Thiol Assay Kit *500 assays*	1 kit	2.1, 5.2	
V13243	Membrane Permeability/ Dead Cell Apoptosis Kit *YO-PRO®-1/propidium iodide* *200 assays* *for flow cytometry*	1 kit	15.5	
M24571	merocyanine 540	25 mg	18.2, 22.3	
V35114	Metabolic Activity/Annexin V/Dead Cell Apoptosis Kit *with C_{12} resazurin, APC annexin V, and SYTOX® Green* *50 assays* *for flow cytometry*	1 kit	15.5	
M23271	methotrexate, Alexa Fluor® 488, inner salt (Alexa Fluor® 488 methotrexate) *mixed isomers*	500 µg	15.6	
M1198MP	methotrexate, fluorescein, triammonium salt (fluorescein methotrexate)	1 mg	15.6	
M1445	7-methoxycoumarin-3-carbonyl azide	25 mg	3.2	
M1420MP	7-methoxycoumarin-3-carboxylic acid	100 mg	1.7	
M1410	7-methoxycoumarin-3-carboxylic acid, succinimidyl ester	25 mg	1.7	
M6886	6-methoxy-*N*-ethylquinolinium iodide (MEQ)	100 mg	21.2	
M440	6-methoxy-*N*-(3-sulfopropyl)quinolinium, inner salt (SPQ)	100 mg	21.2	
M688	*N*-(6-methoxy-8-quinolyl)-*p*-toluenesulfonamide (TSQ)	25 mg	19.7	
M7913	*trans*-1-(2'-methoxyvinyl)pyrene	1 mg	18.2	
M12416	2'-(or-3')-*O*-(*N*-methylanthraniloyl)adenosine 5'-diphosphate, disodium salt (MANT-ADP) *5 mM in buffer*	400 µL	17.3	
M12417	2'-(or-3')-*O*-(*N*-methylanthraniloyl)adenosine 5'-triphosphate, trisodium salt (MANT-ATP) *5 mM in buffer*	400 µL	17.3	
M12414	2'-(or-3')-*O*-(*N*-methylanthraniloyl)guanosine 5'-diphosphate, disodium salt (MANT-GDP) *5 mM in buffer*	400 µL	17.3	
M12415	2'-(or-3')-*O*-(*N*-methylanthraniloyl)guanosine 5'-triphosphate, trisodium salt (MANT-GTP) *5 mM in buffer*	400 µL	17.3	
M22354	2'-(or-3')-*O*-(*N*-methylanthraniloyl)-β:γ-imidoadenosine 5'-triphosphate, trisodium salt (MANT-AMPPNP) *5 mM in buffer*	50 µL	17.3	
M22353	2'-(or-3')-*O*-(*N*-methylanthraniloyl)-β:γ-imidoguanosine 5'-triphosphate, trisodium salt (MANT-GMPPNP) *5 mM in buffer*	50 µL	17.3	
M20490	*N*-methyl-4-hydrazino-7-nitrobenzofurazan (NBD methylhydrazine)	25 mg	3.3, 10.5, 18.3, 21.2	
M25	*N*-methylisatoic anhydride *high purity*	1 g	1.8, 3.2	
M23800	2-methyl-6-(4-methoxyphenyl)-3,7-dihydroimidazo[1,2-a]pyrazin-3-one, hydrochloride (MCLA)	5 mg	6.2, 10.5, 18.2	
M1489MP	4-methylumbelliferyl β-D-galactopyranoside (MUG)	1 g	10.2	
M1490	4-methylumbelliferyl β-D-glucuronide (MUGlcU)	100 mg	10.2	
M6491	4-methylumbelliferyl phosphate, free acid (MUP)	1 g	10.3	
V35116	Mitochondrial Membrane Potential/Annexin V Apoptosis Kit *Alexa Fluor® 488 annexin V/MitoTracker® Red CMXRos* *50 assays* *for flow cytometry*	1 kit	15.5	

*Note that all products listed in the above Master Product List are covered by **Limited Use Label License #223**. If a specific product is covered by additional Limited Use Label License(s), the Limited Use Label License #'s are provided in the LULL column above. Please refer to page 971 for the complete text of the Limited Use Label License(s).

IMPORTANT NOTICE: The products described in this manual are covered by one or more Limited Use Label License(s). Please refer to the Appendix on page 971 and Master Product List on page 975. Products are For Research Use Only. Not intended for any animal or human therapeutic or diagnostic use.
www.invitrogen.com/probes

molecular probes® | invitrogen™ by life technologies™

Cat. No.	Product	Quantity	Sections	LULL*
M34151	MitoProbe™ DilC$_1$(5) Assay Kit *for flow cytometry* *100 assays*	1 kit	15.5, 22.3	
M34150	MitoProbe™ DiOC$_2$(3) Assay Kit *for flow cytometry* *100 assays*	1 kit	15.5, 22.3	
M34152	MitoProbe™ JC-1 Assay Kit *for flow cytometry* *100 assays*	1 kit	15.5, 22.3	
M34153	MitoProbe™ Transition Pore Assay Kit *for flow cytometry* *100 assays*	1 kit	12.2, 15.5	
M36008	MitoSOX™ Red mitochondrial superoxide indicator *for live-cell imaging*	10 x 50 µg	12.2, 18.2	
M22426	MitoTracker® Deep Red FM *special packaging*	20 x 50 µg	12.2	
M7514	MitoTracker® Green FM *special packaging*	20 x 50 µg	12.2	
M7511	MitoTracker® Orange CM-H$_2$TMRos *special packaging*	20 x 50 µg	12.2, 18.2	
M7510	MitoTracker® Orange CMTMRos *special packaging*	20 x 50 µg	12.2	
M7513	MitoTracker® Red CM-H$_2$XRos *special packaging*	20 x 50 µg	12.2, 18.2	
M7512	MitoTracker® Red CMXRos *special packaging*	20 x 50 µg	12.2	
M22425	MitoTracker® Red FM *special packaging*	20 x 50 µg	12.2	
M6248	mono-*N*-(*t*-BOC)-propylenediamine	1 g	3.4	
M1378	monobromobimane (mBBr)	25 mg	2.3, 14.2	
M20381	monobromobimane (mBBr) *FluoroPure™ grade*	25 mg	2.3, 14.2	
M1381MP	monochlorobimane (mBCl)	25 mg	2.3, 14.2, 15.6	
M7891	3-morpholinosydnonimine, hydrochloride (SIN-1)	25 mg	18.3	
M7926	3-morpholinosydnonimine, hydrochloride (SIN-1) *special packaging*	20 x 1 mg	18.3	
03-3700	anti-Biotin, Mouse IgG monoclonal	0.5 mg	7.4	
A10630	mouse anti-human IgG$_1$ *1 mg/mL*	250 µL	7.2	
A10648	mouse anti-human IgG$_1$, horseradish peroxidase conjugate	200 µg	7.2	
A10651	mouse anti-human IgG$_4$ *1 mg/mL*	250 µL	7.2	
A10654	mouse anti-human IgG$_4$, horseradish peroxidase conjugate	200 µg	7.2	
Q10076	Mouse IgG$_{2a}$, Qdot® 705 conjugate (isotype control) *1 µM solution*	100 µL	6.6	
Q10075	Mouse IgG$_{2a}$, Qdot® 800 conjugate (isotype control) *1 µM solution*	100 µL	6.6	
Q10014	Mouse IgG$_{2a}$, Qdot® 605 conjugate (isotype control) *1 µM solution*	100 µL	6.6	
Q10015	Mouse IgG$_{2a}$, Qdot® 655 conjugate (isotype control) *1 µM solution*	100 µL	6.6	
Q10074	Mouse IgG$_{2b}$, Qdot® 605 conjugate (isotype control) *1 µM solution*	100 µL	6.6	
Q10073	Mouse IgG$_1$, Qdot® 605 conjugate (isotype control) *1 µM solution*	100 µL	6.6	
Q10298	Mouse IgG$_1$, Qdot® 800 conjugate (isotype control) *1 µM solution*	100 µL	6.6	
M6494	MTT (3-(4,5-dimethylthiazol-2-yl)-2,5-diphenyltetrazolium bromide)	1 g	15.2, 18.2	
M33307	Multiplexed Proteomics® Glycoprotein Gel Stain Kit *with 1 L each of Pro-Q® Emerald 300 and SYPRO® Ruby (S12000) gel stains*	1 kit	9.4	
MPM33305	Multiplexed Proteomics® Phosphoprotein Gel Stain Kit *includes MPP33300 and S12000*	1 kit	9.4	204
MPM33306	Multiplexed Proteomics® Phosphoprotein Gel Stain Kit *includes MPP33301 and S12001*	1 kit	9.4	204
M33305	Multiplexed Proteomics® Phosphoprotein Gel Stain Kit #1 *with 1 L each of Pro-Q® Diamond (P33300) and SYPRO® Ruby (S12000) gel stains*	1 set	9.4	204
M33306	Multiplexed Proteomics® Phosphoprotein Gel Stain Kit #2 *with 200 mL each of Pro-Q® Diamond (P33301) and SYPRO® Ruby (S12001) gel stains*	1 set	9.4	204
M7901	MultiSpeck™ Multispectral Fluorescence Microscopy Standards Kit *in suspension*	1 kit	23.1	
M23400	muscimol, BODIPY® TMR-X conjugate	1 mg	16.2	
M7006	MycoFluor™ Mycoplasma Detection Kit	1 kit	15.4	
N1384	naloxone fluorescein	5 mg	16.2	
N24915	NANOGOLD® Fab' fragment of goat anti-mouse IgG *80 µg protein/mL*	1 mL	7.2	231
N24916	NANOGOLD® Fab' fragment of goat anti-rabbit IgG *80 µg protein/mL*	1 mL	7.2	231
N20345	NANOGOLD® monomaleimide *special packaging*	5 x 6 nmol	2.2	231
N24918	NANOGOLD® streptavidin (streptavidin, NANOGOLD® conjugate) *80 µg protein/mL*	1 mL	7.6	231
N6666	NanoOrange® Protein Quantitation Kit *200–2000 assays*	1 kit	9.2	
N1138	naphthalene-2,3-dicarboxaldehyde (NDA)	100 mg	1.8, 9.3, 15.6, 21.2	
N2461	2-(2,3-naphthalimino)ethyl trifluoromethanesulfonate	100 mg	3.4	
N1154	NBD C$_6$-ceramide (6-((*N*-(7-nitrobenz-2-oxa-1,3-diazol-4-yl)amino)hexanoyl)sphingosine)	1 mg	12.4, 13.3, 17.4	
N22651	NBD C$_6$-ceramide complexed to BSA	5 mg	12.4, 13.3, 17.4	
N1148	NBD cholesterol (22-(*N*-(7-nitrobenz-2-oxa-1,3-diazol-4-yl)amino)-23,24-bisnor-5-cholen-3β-ol)	10 mg	13.3	
N3524	NBD C$_6$-sphingomyelin (6-((*N*-(7-nitrobenz-2-oxa-1,3-diazol-4-yl)amino)hexanoyl)sphingosyl phosphocholine)	1 mg	12.4, 13.3, 17.4	
N13195	2-NBDG (2-(*N*-(7-nitrobenz-2-oxa-1,3-diazol-4-yl)amino)-2-deoxyglucose)	5 mg	15.2	
N360	NBD-PE (*N*-(7-nitrobenz-2-oxa-1,3-diazol-4-yl)-1,2-dihexadecanoyl-*sn*-glycero-3-phosphoethanolamine, triethylammonium salt)	10 mg	13.2	
N316	NBD-X (6-(*N*-(7-nitrobenz-2-oxa-1,3-diazol-4-yl)amino)hexanoic acid)	100 mg	1.8, 13.2	
N6547	NBT/BCIP Reagent Kit	1 kit	6.3, 10.3	

*Note that all products listed in the above Master Product List are covered by **Limited Use Label License #223**. If a specific product is covered by additional Limited Use Label License(s), the Limited Use Label License #'s are provided in the LULL column above. Please refer to page 971 for the complete text of the Limited Use Label License(s).

Cat. No.	Product	Quantity	Sections	LULL*
N21479	NeuroTrace® 435/455 blue fluorescent Nissl stain *solution in DMSO*	1 mL	8.1, 12.5, 14.3	
N21480	NeuroTrace® 500/525 green fluorescent Nissl stain *solution in DMSO*	1 mL	8.1, 12.5, 14.3	
N21481	NeuroTrace® 515/535 yellow fluorescent Nissl stain *solution in DMSO*	1 mL	8.1, 12.5, 14.3	
N21482	NeuroTrace® 530/615 red fluorescent Nissl stain *solution in DMSO*	1 mL	8.1, 12.5, 14.3	
N21483	NeuroTrace® 640/660 deep-red fluorescent Nissl stain *solution in DMSO*	1 mL	8.1, 12.5, 14.3	
N7167	NeuroTrace® BDA-10,000 Neuronal Tracer Kit	1 kit	14.5	
N22883	NeuroTrace® CM-DiI tissue-labeling paste	100 mg	14.4	
N22880	NeuroTrace® DiI tissue-labeling paste	500 mg	14.4	
N22881	NeuroTrace® DiO tissue-labeling paste	500 mg	14.4	
N22884	NeuroTrace® Multicolor Tissue-Labeling Kit *DiO, DiI, DiD pastes, 500 mg each*	1 kit	14.4	
N3246	neutral red *high purity*	25 mg	12.3, 15.2	
N7991	Newport Green™ DCF diacetate *cell permeant*	1 mg	19.7	
N7990	Newport Green™ DCF, dipotassium salt *cell impermeant*	1 mg	19.7	
N24191	Newport Green™ PDX acetoxymethyl ether	1 mg	19.7	
N1495	nigericin, free acid	10 mg	20.2, 21.2	
N1142	nile red	25 mg	11.2, 13.5	
N23106	6-(*N*-(7-nitrobenz-2-oxa-1,3-diazol-4-yl)amino)-6-deoxyglucose (6-NBDG)	5 mg	15.2	
N678	12-(*N*-(7-nitrobenz-2-oxa-1,3-diazol-4-yl)amino)dodecanoic acid	100 mg	13.2	
N3787	2-(12-(7-nitrobenz-2-oxa-1,3-diazol-4-yl)amino)dodecanoyl-1-hexadecanoyl-*sn*-glycero-3-phosphocholine (NBD C_{12}-HPC)	5 mg	13.2, 17.4	
N3786	2-(6-(7-nitrobenz-2-oxa-1,3-diazol-4-yl)amino)hexanoyl-1-hexadecanoyl-*sn*-glycero-3-phosphocholine (NBD C_6-HPC)	5 mg	13.2, 17.4	
N354	*N*-(7-nitrobenz-2-oxa-1,3-diazol-4-yl)phallacidin (NBD phallacidin)	300 U	11.1	
N6495	nitro blue tetrazolium chloride (NBT)	1 g	6.3, 10.3, 15.2, 18.2	
N6803	*o*-nitrophenyl EGTA, AM (NP-EGTA, AM) *cell permeant* *special packaging*	20 x 50 µg	5.3, 17.2, 19.8	
N6802	*o*-nitrophenyl EGTA, tetrapotassium salt (NP-EGTA) *cell impermeant*	1 mg	5.3, 17.2, 19.8	
N7892	*S*-nitroso-*N*-acetylpenicillamine (SNAP)	25 mg	18.3	
N7927	*S*-nitroso-*N*-acetylpenicillamine (SNAP) *special packaging*	20 x 1 mg	18.3	
N6356	norbiotinamine, hydrochloride	10 mg	4.2	
N10561	NovaBright™ β-galactosidase and Firefly Luciferase Dual Enzyme Reporter Gene Chemiluminescent Detection Kit *200 assays*	1 kit	10.6	
N10562	NovaBright™ β-galactosidase and Firefly Luciferase Dual Enzyme Reporter Gene Chemiluminescent Detection Kit *600 assays*	1 kit	10.6	
N10564	NovaBright™ β-galactosidase Enzyme Reporter Gene Chemiluminescent Detection Kit *for mammalian cells* *1000 assays*	1 kit	10.2	
N10566	NovaBright™ β-galactosidase Enzyme Reporter Gene Chemiluminescent Detection Kit *for yeast cells* *1000 assays*	1 kit	10.2	
N10563	NovaBright™ β-galactosidase Enzyme Reporter Gene Chemiluminescent Detection Kit *for mammalian cells* *200 assays*	1 kit	10.2	
N10565	NovaBright™ β-galactosidase Enzyme Reporter Gene Chemiluminescent Detection Kit *for yeast cells* *200 assays*	1 kit	10.2	
N10560	NovaBright™ Secreted Placental Alkaline Phosphatase (SEAP) Enzyme Reporter Gene Chemiluminescent Detection Kit *1200 assays*	1 kit	10.3	
N10577	NovaBright™ Secreted Placental Alkaline Phosphatase (SEAP) Enzyme Reporter Gene Chemiluminescent Detection Kit 2.0 *192 assays*	1 kit	10.3	
N10578	NovaBright™ Secreted Placental Alkaline Phosphatase (SEAP) Enzyme Reporter Gene Chemiluminescent Detection Kit 2.0 *960 assays*	1 kit	10.3	
N10559	NovaBright™ Secreted Placental Alkaline Phosphatase (SEAP) Enzyme Reporter Gene Chemiluminescent Detection Kit *400 assays*	1 kit	10.3	
N21485	nuclear yellow (Hoechst S769121, trihydrochloride, trihydrate)	10 mg	8.1, 12.5	
N7565	Nucleic Acid Stains Dimer Sampler Kit	1 kit	8.1, 15.2	
O7703	1-octacosanyl-2-(1-pyrenehexanoyl)-*sn*-glycero-3-phosphomethanol, ammonium salt (C_{28}-O-PHPM)	250 µg	17.4	
O246	octadecyl rhodamine B chloride (R18)	10 mg	13.5, 14.4	
O24751	ONCYTE® MultiWells, one well, 13 mm diameter, with slide and matching gasket *set of 20*	1 set	23.1	
O24750	ONCYTE® MultiWells, 12 wells, 5 mm diameter, with slide and matching gasket *set of 20*	1 set	23.1	
O10181	Oregon Green® 488 alkyne *6-isomer*	0.5 mg	3.1	
O10180	Oregon Green® 488 azide (Oregon Green® 6-carboxamido-(6-azidohexanyl), triethylammonium salt) *6-isomer*	0.5 mg	3.1	
O6807	Oregon Green® 488 BAPTA-1, AM *cell permeant* *special packaging*	10 x 50 µg	19.3	
O6798	Oregon Green® 488 BAPTA-1 dextran, potassium salt, 10,000 MW, anionic	5 mg	19.4	
O6806	Oregon Green® 488 BAPTA-1, hexapotassium salt *cell impermeant*	500 µg	19.3	
O6809	Oregon Green® 488 BAPTA-2, AM *cell permeant* *special packaging*	10 x 50 µg	19.3	
O6808	Oregon Green® 488 BAPTA-2, octapotassium salt *cell impermeant*	500 µg	19.3	
O6812	Oregon Green® 488 BAPTA-5N, hexapotassium salt *cell impermeant*	500 µg	19.3	

*Note that all products listed in the above Master Product List are covered by **Limited Use Label License #223.** If a specific product is covered by additional Limited Use Label License(s), the Limited Use Label License #'s are provided in the LULL column above. Please refer to page 971 for the complete text of the Limited Use Label License(s).

IMPORTANT NOTICE: The products described in this manual are covered by one or more Limited Use Label License(s). Please refer to the Appendix on page 971 and Master Product List on page 975. Products are For Research Use Only. Not intended for any animal or human therapeutic or diagnostic use.

Cat. No.	Product	Quantity	Sections	LULL*
O23990	Oregon Green® 488 BAPTA-6F, hexapotassium salt *cell impermeant*	500 µg	19.3	
O12920	Oregon Green® 488 biocytin (biocytin Oregon Green® 488)	5 mg	4.3, 14.3	
O10465	Oregon Green® 488 cadaverine *5-isomer*	5 mg	3.4	
O6146	Oregon Green® 488 carboxylic acid *5-isomer*	5 mg	1.5, 14.3, 20.3	
O6151	Oregon Green® 488 carboxylic acid diacetate (carboxy-DFFDA) *6-isomer*	5 mg	15.2, 20.3	
O34550	Oregon Green® 488 carboxylic acid diacetate, succinimidyl ester (carboxy-DFFDA, SE) *mixed isomers*	1 mg	14.2, 15.4	
O6147	Oregon Green® 488 carboxylic acid, succinimidyl ester *5-isomer*	5 mg	1.5, 20.4	
O6149	Oregon Green® 488 carboxylic acid, succinimidyl ester *6-isomer*	5 mg	1.5, 20.4	
O11033	Oregon Green® 488 goat anti-mouse IgG (H+L) *highly cross-adsorbed* *2 mg/mL*	0.5 mL	7.2	
O6380	Oregon Green® 488 goat anti-mouse IgG (H+L) *2 mg/mL*	0.5 mL	7.2	
O11038	Oregon Green® 488 goat anti-rabbit IgG (H+L) *highly cross-adsorbed* *2 mg/mL*	0.5 mL	7.2	
O6381	Oregon Green® 488 goat anti-rabbit IgG (H+L) *2 mg/mL*	0.5 mL	7.2	
O6382	Oregon Green® 488 goat anti-rat IgG (H+L) *2 mg/mL*	0.5 mL	7.2	
O6010	Oregon Green® 488 iodoacetamide *mixed isomers*	5 mg	2.2	
O6080	Oregon Green® 488 isothiocyanate (F_2FITC) *mixed isomers*	5 mg	1.5	
O6034	Oregon Green® 488 maleimide	5 mg	2.2	
O7466	Oregon Green® 488 phalloidin	300 U	11.1	
O10241	Oregon Green® 488 Protein Labeling Kit *3 labelings*	1 kit	1.2, 1.5	
O6185	Oregon Green® 488-X, succinimidyl ester *6-isomer*	5 mg	1.5, 4.2	
O6138	Oregon Green® 514 carboxylic acid	5 mg	1.5, 14.3, 20.3	
O6139	Oregon Green® 514 carboxylic acid, succinimidyl ester	5 mg	1.5, 20.4	
O6383	Oregon Green® 514 goat anti-mouse IgG (H+L) *2 mg/mL*	0.5 mL	7.2	
O7465	Oregon Green® 514 phalloidin	300 U	11.1	
O12650	Oregon Green® 488 DHPE (Oregon Green® 488 1,2-dihexadecanoyl-*sn*-glycero-3-phosphoethanolamine)	1 mg	13.2, 20.4	
O10104	Organelle Lights™ Endosomes-GFP	1 kit	16.1	21, 127, 198, 267, 272, 306, 332
O36231	Organelle Lights™ Endosomes-RFP	1 kit	16.1	21, 306, 321, 332
C10080	Organelle Lights™ Synaptophysin-GFP	1 kit	16.1	21, 127, 198, 267, 272, 306, 332
O10138	Organelle Lights™ Synaptophysin-RFP	1 kit	16.1	21, 306, 321, 332
O14804	O-rings for Attofluor cell chamber, set of 10	each	23.1	
O34781	ovalbumin, Alexa Fluor® 488 conjugate	2 mg	14.7	
O34782	ovalbumin, Alexa Fluor® 555 conjugate	2 mg	14.7	
O34783	ovalbumin, Alexa Fluor® 594 conjugate	2 mg	14.7	
O34784	ovalbumin, Alexa Fluor® 647 conjugate	2 mg	14.7	
O23020	ovalbumin, fluorescein conjugate	5 mg	14.7	
O23021	ovalbumin, Texas Red® conjugate	5 mg	14.7	
O266	oxonol V (bis-(3-phenyl-5-oxoisoxazol-4-yl)pentamethine oxonol)	100 mg	22.3	
O267	oxonol VI (bis-(3-propyl-5-oxoisoxazol-4-yl)pentamethine oxonol)	100 mg	22.3	
O13291	OxyBURST® Green H_2HFF BSA *special packaging*	5 x 1 mg	16.1, 18.2	
P30506	Pacific Blue™ C_5-maleimide	1 mg	2.3	
P22652	Pacific Blue™ DMPE (Pacific Blue™ 1,2-ditetradecanoyl-*sn*-glycero-3-phosphoethanolamine, triethylammonium salt)	1 mg	13.2	
P31581	Pacific Blue″ F(ab')$_2$ fragment of goat anti-mouse IgG (H+L) *2 mg/mL*	250 µL	7.2	
P10993	Pacific Blue™ goat anti-mouse IgG (H+L) *2 mg/mL*	0.5 mL	7.2	
P31582	Pacific Blue™ goat anti-mouse IgG (H+L) *highly cross-adsorbed* *2 mg/mL*	0.5 mL	7.2	
P10994	Pacific Blue™ goat anti-rabbit IgG (H+L) *2 mg/mL*	0.5 mL	7.2	
P30013	Pacific Blue™ Monoclonal Antibody Labeling Kit *5 labelings*	1 kit	1.2, 1.7	
P30012	Pacific Blue™ Protein Labeling Kit *3 labelings*	1 kit	1.2, 1.7	
P10163	Pacific Blue™ succinimidyl ester	5 mg	1.7	
P31585	Pacific Orange™ F(ab')$_2$ fragment of goat anti-mouse IgG (H+L) *2 mg/mL*	250 µL	7.2	
P31584	Pacific Orange™ goat anti-rabbit IgG (H+L) *2 mg/mL*	0.5 mL	7.2	
P30507	Pacific Orange™ C_5-maleimide	1 mg	2.3	
P30014	Pacific Orange™ Monoclonal Antibody Labeling Kit *5 labelings*	1 kit	1.2, 1.7	
P30016	Pacific Orange™ Protein Labeling Kit *3 labelings*	1 kit	1.2, 1.7	
P30253	Pacific Orange™ succinimidyl ester *triethylammonium salt*	1 mg	1.7	
P3456	paclitaxel (Taxol equivalent) *for use in research only*	5 mg	11.2	

*Note that all products listed in the above Master Product List are covered by **Limited Use Label License #223**. If a specific product is covered by additional Limited Use Label License(s), the Limited Use Label License #'s are provided in the LULL column above. Please refer to page 971 for the complete text of the Limited Use Label License(s).

Cat. No.	Product	Quantity	Sections	LULL*
P7501	paclitaxel, BODIPY® 564/570 conjugate (BODIPY® 564/570 Taxol)	10 µg	11.2	
P7500	paclitaxel, BODIPY® FL conjugate (BODIPY® FL Taxol)	10 µg	11.2	
P22310	paclitaxel, Oregon Green® 488 conjugate (Oregon Green® 488 Taxol; Flutax-2)	100 µg	11.2	
P21680	Panomer™ 9 random oligodeoxynucleotide, Alexa Fluor® 488 conjugate	10 nmol	8.2, 8.4	
P21681	Panomer™ 9 random oligodeoxynucleotide, Alexa Fluor® 546 conjugate	10 nmol	8.2, 8.4	
P21687	Panomer™ 9 random oligodeoxynucleotide, Alexa Fluor® 555 conjugate	10 nmol	8.2, 8.4	
P21686	Panomer™ 9 random oligodeoxynucleotide, Alexa Fluor® 647 conjugate	10 nmol	8.2, 8.4	
P36005	*cis*-parinaric acid *3 mM in ethanol*	10 mL	13.2, 18.2	
P23012	parvalbumin from codfish, Alexa Fluor® 488 conjugate	1 mg	14.7	
P1267MP	PBFI, AM *cell permeant* *special packaging*	20 x 50 µg	21.1	
P1265MP	PBFI, tetraammonium salt *cell impermeant*	1 mg	21.1	
P14825	PeakFlow™ Blue flow cytometry reference beads, 2.5 µm *460 nm emission*	3 mL	23.2	
P14826	PeakFlow™ Blue flow cytometry reference beads, 6 µm *460 nm emission*	3 mL	23.2	
P14831	PeakFlow™ Carmine flow cytometry reference beads, 2.5 µm *620 nm emission*	3 mL	23.2	
P14832	PeakFlow™ Carmine flow cytometry reference beads, 6 µm *620 nm emission*	3 mL	23.2	
P24670	PeakFlow™ Claret flow cytometry reference beads, 6 µm *680 nm emission*	3 mL	23.2	
P14827	PeakFlow™ Green flow cytometry reference beads, 2.5 µm *515 nm emission*	3 mL	23.2	
P14828	PeakFlow™ Green flow cytometry reference beads, 6 µm *515 nm emission*	3 mL	23.2	
P24672	PeakFlow™ Infrared flow cytometry reference beads, 6 µm *770 nm emission*	3 mL	23.2	
P14829	PeakFlow™ Orange flow cytometry reference beads, 2.5 µm *575 nm emission*	3 mL	23.2	
P14830	PeakFlow™ Orange flow cytometry reference beads, 6 µm *575 nm emission*	3 mL	23.2	
P24671	PeakFlow™ Ultra Red flow cytometry reference beads, 6 µm *695 nm emission*	3 mL	23.2	
V35112	PE Annexin V/ Dead Cell Apoptosis Kit *with SYTOX® Green* *50 assays* *for flow cytometry*	1 kit	15.5	
A10070	PED-A1 (N-((6-(2,4-DNP)amino)hexanoyl)-1-(BODIPY® FL C_5)-2-hexyl-*sn*-glycero-3-phosphoethanolamine) *phospholipase A_1 selective substrate*	100 µg	13.2, 17.4	
P11948	5-(pentafluorobenzoylamino)fluorescein di-β-D-galactopyranoside (PFB-FDG)	5 mg	10.2	
P11947	5-(pentafluorobenzoylamino)fluorescein di-β-D-glucopyranoside (PFB-FDGlu)	5 mg	10.2	
P11949	5-(pentafluorobenzoylamino)fluorescein di-β-D-glucuronide (PFB-FDGlcU)	5 mg	10.2	
P21315	Penta·His mouse IgG_1, monoclonal antibody (anti-pentahistidine) *BSA free*	100 µg	7.5, 9.4	
R1147	pentoxyresorufin (resorufin pentyl ether)	5 mg	10.6	
P27167	PeppermintStick™ phosphoprotein molecular weight standards	40 µL	9.4	
P33350	PeppermintStick™ phosphoprotein molecular weight standards *200 gel lanes*	400 µL	9.4	
P12271	pepstatin A, BODIPY® FL conjugate	25 µg	10.4, 15.5	
P917	peroxidase from horseradish, biotin-XX conjugate	10 mg	4.3, 7.6	
P3457	phalloidin	1 mg	11.1	
P6879	*N*-(1,10-phenanthrolin-5-yl)iodoacetamide	5 mg	2.3, 18.2	
P6763	Phen Green™ FL, diacetate *cell permeant*	1 mg	19.7	
P14313	Phen Green™ SK, diacetate	1 mg	19.7	
P14312	Phen Green™ SK, dipotassium salt	1 mg	19.7	
P1692	phenylethyl β-D-thiogalactopyranoside (PETG)	10 mg	10.2	
P6466	phospholipase C, phosphatidylinositol-specific *from *Bacillus cereus** *100 U/mL*	50 µL	17.4	
P33357	phosphopeptide standard mixture *400 pmol of each peptide*	2800 pmol	9.4	
P10361	pHrodo™ dextran, 10,000 MW *for endocytosis*	0.5 mg	14.5, 16.1, 20.4	
P35361	pHrodo™ *E. coli* BioParticles® conjugate for phagocytosis	5 X 2 mg	16.1	
A10025	pHrodo™ *E. coli* BioParticles® Phagocytosis Kit *for flow cytometry* *100 tests*	1 kit	16.1	
A10026	pHrodo™ Phagocytosis Particle Labeling Kit *for flow cytometry* *100 tests*	1 kit	16.1	
A10010	pHrodo™ *S. aureus* Bioparticles® conjugate for phagocytosis	5 x 2 mg	16.1	
P36600	pHrodo™, succinimidyl ester (pHrodo™, SE)	1 mg	20.4	
P2331MP	*o*-phthaldialdehyde (OPA) *high purity*	1 g	1.8, 9.2, 9.3, 15.6, 21.2	
P800	B-phycoerythrin *4 mg/mL*	0.5 mL	6.4, 14.7, 18.2	
P801	R-phycoerythrin *4 mg/mL*	0.5 mL	6.4, 14.7, 18.2	
P811	R-phycoerythrin, biotin-XX conjugate *4 mg/mL*	0.5 mL	4.3, 6.4	
A10690	R-phycoerythrin $F(ab')_2$ fragment of goat anti-mouse IgM (µ chain) *1 mg/mL*	250 µL	7.2	
A10543	R-phycoerythrin $F(ab')_2$ fragment of goat anti-mouse IgG (H+L) *1 mg/mL*	250 µL	7.2	
A10542	R-phycoerythrin $F(ab')_2$ fragment of goat anti-rabbit IgG (H+L) *1 mg/mL*	250 µL	7.2	
A10544	R-phycoerythrin $F(ab')_2$ fragment of goat anti-rat IgG (H+L) *1 mg/mL*	250 µL	7.2	
P21129	R-phycoerythrin goat anti-mouse IgG_1 (γ1) conjugate *1 mg/mL*	250 µL	6.4, 7.2	
P21139	R-phycoerythrin goat anti-mouse IgG_{2a} (γ2a) conjugate *1 mg/mL*	250 µL	6.4, 7.2	

*Note that all products listed in the above Master Product List are covered by **Limited Use Label License #223**. If a specific product is covered by additional Limited Use Label License(s), the Limited Use Label License #'s are provided in the LULL column above. Please refer to page 971 for the complete text of the Limited Use Label License(s).

IMPORTANT NOTICE: The products described in this manual are covered by one or more Limited Use Label License(s). Please refer to the Appendix on page 971 and Master Product List on page 975. Products are For Research Use Only. Not intended for any animal or human therapeutic or diagnostic use.

Cat. No.	Product	Quantity	Sections	LULL*
P21149	R-phycoerythrin goat anti-mouse IgG_{2b} (γ2b) conjugate *1 mg/mL*	250 µL	6.4, 7.2	
P852	R-phycoerythrin goat anti-mouse IgG (H+L) *1 mg/mL*	1 mL	6.4, 7.2	
A10689	R-phycoerythrin goat anti-mouse IgM (µ chain) *1 mg/mL*	250 µL	7.2	
P2771MP	R-phycoerythrin goat anti-rabbit IgG (H+L) *1 mg/mL*	0.5 mL	6.4, 7.2	
A10545	R-phycoerythrin goat anti-rat IgG (H+L) *1 mg/mL*	250 µL	7.2	
P806	R-phycoerythrin, pyridyldisulfide derivative *2 mg/mL*	1 mL	6.4	
P22061	P_iPer™ Phosphate Assay Kit *1000 assays*	1 kit	10.3, 21.2	
P22062	P_iPer™ Pyrophosphate Assay Kit *1000 assays*	1 kit	10.3, 21.2	
P35901	*O*-pivaloyloxymethyl umbelliferone (C-POM) *lipase substrate* *special packaging*	5 x 100 µg	10.6	
P6867	Pluronic® F-127 *low UV absorbance*	2 g	19.8, 21.1	
P3000MP	Pluronic® F-127 *20% solution in DMSO*	1 mL	19.8, 21.1	
P6866	Pluronic® F-127 *10% solution in water* *0.2 µm filtered*	30 mL	19.8, 21.1	
P13235	polymyxin B, BODIPY® FL conjugate, trifluoroacetic acid salt *mixed species*	100 µg	15.2, 17.3	
P13238	polymyxin B, dansyl conjugate, trifluoroacetic acid salt *mixed species*	100 µg	15.2, 17.3	
P13236	polymyxin B, Oregon Green® 514 conjugate, trifluoroacetic acid salt *mixed species*	100 µg	15.2, 17.3	
P3580	POPO™-1 iodide (434/456) *1 mM solution in DMSO*	200 µL	8.1	
P3584	POPO™-3 iodide (534/570) *1 mM solution in DMF*	200 µL	8.1, 8.4	
P3581	PO-PRO™-1 iodide (435/455) *1 mM solution in DMSO*	1 mL	8.1, 15.5	
P3585	PO-PRO™-3 iodide (539/567) *1 mM solution in DMSO*	1 mL	8.1	
P10020	PowerLoad™ concentrate, 100X	5 mL	16.3, 19.8, 21.1	
P36235	Premo™ Autophagy Sensor LC3B-GFP	1 kit	15.5	
P36236	Premo™ Autophagy Sensor LC3B-RFP	1 kit	15.5	
P36207	Premo™ Cameleon Calcium Sensor *for 10 microplates*	1 kit	16.3, 19.5	
P36208	Premo™ Cameleon Calcium Sensor *for 100 microplates*	1 kit	16.3, 19.5	
P36232	Premo™ FUCCI Cell Cycle Sensor	1 kit	15.4	
P10229	Premo™ Halide Sensor *for 10 microplates*	1 kit	16.3, 21.2	
P18174	Press-to-Seal silicone isolator, one well, 20 mm diameter, 0.5 mm deep *set of 50*	1 set	23.1	
P18175	Press-to-Seal silicone isolator, one well, 20 mm diameter, 1.0 mm deep *set of 50*	1 set	23.1	
P24743	Press-to-Seal silicone isolator with adhesive, eight wells, 9 mm diameter, 0.5 mm deep *set of 25*	1 set	23.1	
P24744	Press-to-Seal silicone isolator with adhesive, eight wells, 9 mm diameter, 1.0 mm deep *set of 25*	1 set	23.1	
P24740	Press-to-Seal silicone isolator with adhesive, one well, 20 mm diameter, 0.5 mm deep *set of 50*	1 set	23.1	
P24741	Press-to-Seal silicone isolator with adhesive, one well, 20 mm diameter, 1.0 mm deep *set of 50*	1 set	23.1	
P24742	Press-to-Seal silicone isolator with adhesive, 24 wells, 2.5 mm diameter, 2.0 mm deep *set of 25*	1 set	23.1	
P18178	Press-to-Seal silicone sheet, 13 cm x 18 cm, 0.5 mm thick *set of 5*	1 set	23.1	
P18179	Press-to-Seal silicone sheet, 13 cm x 18 cm, 1.0 mm thick *set of 5*	1 set	23.1	
P24745	Press-to-Seal silicone sheet with adhesive, 13 cm x 18 cm, 0.5 mm thick *set of 5*	1 set	23.1	
P36400	probenecid, water soluble	10 x 77 mg	19.8	
P248	prodan (6-propionyl-2-dimethylaminonaphthalene)	100 mg	11.2, 13.5	
P7481	ProLong® Antifade Kit	1 kit	23.1	
P36930	ProLong® Gold antifade reagent	10 mL	23.1	
P36934	ProLong® Gold antifade reagent *special packaging*	5 x 2 mL	23.1	
P36931	ProLong® Gold antifade reagent with DAPI	10 mL	23.1	
P36935	ProLong® Gold antifade reagent with DAPI *special packaging*	5 x 2 mL	23.1	
P1304MP	propidium iodide	100 mg	8.1, 12.5, 14.3, 15.2	
P21493	propidium iodide *FluoroPure™ grade*	100 mg	8.1, 12.5, 14.3, 15.2	
P3566	propidium iodide *1.0 mg/mL solution in water*	10 mL	8.1, 12.5, 15.2	
P37001	ProPlate™ multi-array slide module *set of 2*	1 set	23.1	
P37004	ProPlate™ multi-array system *includes four 16-well slide modules, one tray and cover, ten seal-strips and one applicator*	1 set	23.1	
P37002	ProPlate™ adhesive seal-strips *set of 50 seal-strips and one applicator*	1 set	23.1	
P37003	ProPlate™ tray and cover *includes one tray and cover*	1 set	23.1	
P33203	Pro-Q® Diamond LC Phosphopeptide Detection Kit	1 kit	9.4	204
P33356	Pro-Q® Diamond Phosphoprotein Blot Stain Kit *20 minigel blots*	1 kit	9.4	204
P33359	Pro-Q® Diamond Phosphoprotein Enrichment and Detection Kit	1 kit	9.4	204
P33358	Pro-Q® Diamond Phosphoprotein Enrichment Kit	1 kit	9.4	
P33310	Pro-Q® Diamond phosphoprotein gel destaining solution	1 L	9.4	
P33311	Pro-Q® Diamond phosphoprotein gel destaining solution *bulk packaging*	5 L	9.4	
P33300	Pro-Q® Diamond phosphoprotein gel stain	1 L	9.4	204

*Note that all products listed in the above Master Product List are covered by **Limited Use Label License #223**. If a specific product is covered by additional Limited Use Label License(s), the Limited Use Label License #'s are provided in the LULL column above. Please refer to page 971 for the complete text of the Limited Use Label License(s).

IMPORTANT NOTICE: The products described in this manual are covered by one or more Limited Use Label License(s). Please refer to the Appendix on page 971 and Master Product List on page 975. Products are For Research Use Only. Not intended for any animal or human therapeutic or diagnostic use.

Cat. No.	Product	Quantity	Sections	LULL*
P33301	Pro-Q® Diamond phosphoprotein gel stain	200 mL	9.4	204
P33302	Pro-Q® Diamond phosphoprotein gel stain *bulk packaging*	5 L	9.4	204
MPP33300	Pro-Q® Diamond Phosphoprotein Gel Staining Kit *includes 1 L stain and 40 μL standard*	1 kit	9.4	204
MPP33301	Pro-Q® Diamond Phosphoprotein Gel Staining Kit *includes 200 mL stain and 40 μL standard*	1 kit	9.4	204
MPP33302	Pro-Q® Diamond Phosphoprotein Gel Staining Kit *includes 5 L stain and 400 μL standard*	1 kit	9.4	204
P33706	Pro-Q® Diamond Phosphoprotein/Phosphopeptide Microarray Stain Kit	1 kit	9.4, 17.3	204
P21857	Pro-Q® Emerald 300 Glycoprotein Gel and Blot Stain Kit *10 minigels or minigel blots*	1 kit	9.4	
P21855	Pro-Q® Emerald 300 Glycoprotein Gel Stain Kit *with SYPRO® Ruby protein gel stain* *10 minigels*	1 kit	9.4	
P20495	Pro-Q® Emerald 300 Lipopolysaccharide Gel Stain Kit *10 minigels*	1 kit	13.3	
P21875	Pro-Q® Emerald 488 Glycoprotein Gel and Blot Stain Kit *10 minigels or minigel blots*	1 kit	9.4	
P11047	protein A, Alexa Fluor® 488 conjugate	1 mg	7.2	
P11049	protein A, Alexa Fluor® 546 conjugate	1 mg	7.2	
P21462	protein A, Alexa Fluor® 647 conjugate	1 mg	7.2	
P11065	protein G, Alexa Fluor® 488 conjugate	1 mg	7.2	
P21041	protein G, horseradish peroxidase conjugate	1 mg	7.2, 10.5	
P6649	protein molecular weight standards *broad range* *200 gel lanes*	400 μL	9.3	
P6305	Protein-Protein Crosslinking Kit *3 conjugations*	1 kit	5.2	
P7220	PS-Speck™ Microscope Point Source Kit *blue, green, orange and deep red fluorescent beads*	1 kit	23.1	
P1903MP	1-pyrenebutanoic acid *high purity*	100 mg	13.2	
P101	1-pyrenebutanoic acid, hydrazide	100 mg	3.3	
P130	1-pyrenebutanoic acid, succinimidyl ester	100 mg	1.7	
P6114	*N*-(1-pyrenebutanoyl)cysteic acid, succinimidyl ester, potassium salt	5 mg	1.7	
P31	1-pyrenedecanoic acid	25 mg	13.2	
P96	1-pyrenedodecanoic acid	25 mg	13.2	
P243	1-pyrenehexadecanoic acid	5 mg	13.2	
P29	*N*-(1-pyrene)iodoacetamide	100 mg	2.3, 11.1	
P28	*N*-(1-pyrene)maleimide	100 mg	2.3	
P2007MP	*N*-(1-pyrenemethyl)iodoacetamide (PMIA amide)	25 mg	2.3	
P24	1-pyrenesulfonyl chloride	100 mg	1.7	
P349	1,3,6,8-pyrenetetrasulfonic acid, tetrasodium salt	100 mg	14.3	
P1405	1-pyrenyldiazomethane (PDAM)	25 mg	3.4	
P6317	*N*-((2-pyridyldithio)ethyl)-4-azidosalicylamide (PEAS; AET)	10 mg	5.3	
Q22041MP	Qdot® 525 Antibody Conjugation Kit	1 kit	6.6	
Q22031MP	Qdot® 565 Antibody Conjugation Kit	1 kit	6.6	
Q22011MP	Qdot® 585 Antibody Conjugation Kit	1 kit	6.6	
Q22001MP	Qdot® 605 Antibody Conjugation Kit	1 kit	6.6	
A10197	Qdot® 625 Antibody Conjugation Kit	1 kit	6.6	
Q22021MP	Qdot® 655 Antibody Conjugation Kit	1 kit	6.6	
Q22061MP	Qdot® 705 Antibody Conjugation Kit	1 kit	6.6	
Q22071MP	Qdot® 800 Antibody Conjugation Kit	1 kit	6.6	
Q10301MP	Qdot® 605 Biotin Conjugate Kit	1 kit	4.3, 6.6	
Q10321MP	Qdot® 655 Biotin Conjugate Kit	1 kit	4.3, 6.6	
Q15431MP	Qdot® 565 goat anti-fluorescein conjugate *1 μM solution* *whole IgG*	200 μL	6.6, 7.4	
Q15421MP	Qdot® 655 goat anti-fluorescein conjugate *1 μM solution* *whole IgG*	200 μL	6.6, 7.4	
Q11201MP	Qdot® 605 goat $F(ab')_2$ anti-human IgG conjugate (H+L) *1 μM solution*	200 μL	6.6, 7.2	
Q11221MP	Qdot® 655 goat $F(ab')_2$ anti-human IgG conjugate (H+L) *1 μM solution*	200 μL	6.6, 7.2	
Q11041MP	Qdot® 525 goat $F(ab')_2$ anti-mouse IgG conjugate (H+L) *1 μM solution*	200 μL	6.6, 7.2	
Q11031MP	Qdot® 565 goat $F(ab')_2$ anti-mouse IgG conjugate (H+L) *1 μM solution*	200 μL	6.6, 7.2	
Q11032MP	Qdot® 565 goat $F(ab')_2$ anti-mouse IgG conjugate (H+L) *1 μM solution* *in 1 M betaine, 50 mM borate, pH 8.3 with 0.05% sodium azide*	100 μL	6.6, 7.2	
Q11011MP	Qdot® 585 goat $F(ab')_2$ anti-mouse IgG conjugate (H+L) *1 μM solution*	200 μL	6.6, 7.2	
Q11001MP	Qdot® 605 goat $F(ab')_2$ anti-mouse IgG conjugate (H+L) *1 μM solution*	200 μL	6.6, 7.2	
Q11002MP	Qdot® 605 goat $F(ab')_2$ anti-mouse IgG conjugate (H+L) *1 μM solution* *in 1 M betaine, 50 mM borate, pH 8.3 with 0.05% sodium azide*	100 μL	6.6, 7.2	
A10195	Qdot® 625 goat $F(ab')_2$ anti-mouse IgG conjugate (H+L) *1 μM solution*	100 μL	6.6, 7.2	
Q11021MP	Qdot® 655 goat $F(ab')_2$ anti-mouse IgG conjugate (H+L) *1 μM solution*	200 μL	6.6, 7.2	
Q11022MP	Qdot® 655 goat $F(ab')_2$ anti-mouse IgG conjugate (H+L) *1 μM solution* *in 1 M betaine, 50 mM borate, pH 8.3 with 0.05% sodium azide*	100 μL	6.6, 7.2	
Q11061MP	Qdot® 705 goat $F(ab')_2$ anti-mouse IgG conjugate (H+L) *1 μM solution*	200 μL	6.6, 7.2	

*Note that all products listed in the above Master Product List are covered by **Limited Use Label License #223**. If a specific product is covered by additional Limited Use Label License(s), the Limited Use Label License #'s are provided in the LULL column above. Please refer to page 971 for the complete text of the Limited Use Label License(s).

IMPORTANT NOTICE: The products described in this manual are covered by one or more Limited Use Label License(s). Please refer to the Appendix on page 971 and Master Product List on page 975. Products are For Research Use Only. Not intended for any animal or human therapeutic or diagnostic use.

Cat. No.	Product	Quantity	Sections	LULL*
Q11062MP	Qdot® 705 goat F(ab')$_2$ anti-mouse IgG conjugate (H+L) *1 µM solution* *in 1 M betaine, 50 mM borate, pH 8.3 with 0.05% sodium azide*	100 µL	6.6, 7.2	
Q11071MP	Qdot® 800 goat F(ab')$_2$ anti-mouse IgG conjugate (H+L) *1 µM solution*	200 µL	6.6, 7.2	
Q11441MP	Qdot® 525 goat F(ab')$_2$ anti-rabbit IgG conjugate (H+L) *1 µM solution*	200 µL	6.6, 7.2	
Q11431MP	Qdot® 565 goat F(ab')$_2$ anti-rabbit IgG conjugate (H+L) *1 µM solution*	200 µL	6.6, 7.2	
Q11432MP	Qdot® 565 goat F(ab')$_2$ anti-rabbit IgG conjugate (H+L) *1 µM solution* *in 1 M betaine, 50 mM borate, pH 8.3 with 0.05% sodium azide*	100 µL	6.6, 7.2	
Q11411MP	Qdot® 585 goat F(ab')$_2$ anti-rabbit IgG conjugate (H+L) *1 µM solution*	200 µL	6.6, 7.2	
Q11401MP	Qdot® 605 goat F(ab')$_2$ anti-rabbit IgG conjugate (H+L) *1 µM solution*	200 µL	6.6, 7.2	
Q11402MP	Qdot® 605 goat F(ab')$_2$ anti-rabbit IgG conjugate (H+L) *1 µM solution* *in 1 M betaine, 50 mM borate, pH 8.3 with 0.05% sodium azide*	100 µL	6.6, 7.2	
A10194	Qdot® 625 goat F(ab')$_2$ anti-rabbit IgG conjugate (H+L) *1 µM solution*	100 µL	6.6, 7.2	
Q11421MP	Qdot® 655 goat F(ab')$_2$ anti-rabbit IgG conjugate (H+L) *1 µM solution*	200 µL	6.6, 7.2	
Q11422MP	Qdot® 655 goat F(ab')$_2$ anti-rabbit IgG conjugate (H+L) *1 µM solution* *in 1 M betaine, 50 mM borate, pH 8.3 with 0.05% sodium azide*	100 µL	6.6, 7.2	
Q11461MP	Qdot® 705 goat F(ab')$_2$ anti-rabbit IgG conjugate (H+L) *1 µM solution*	200 µL	6.6, 7.2	
Q11462MP	Qdot® 705 goat F(ab')$_2$ anti-rabbit IgG conjugate (H+L) *1 µM solution* *in 1 M betaine, 50 mM borate, pH 8.3 with 0.05% sodium azide*	100 µL	6.6, 7.2	
Q11471MP	Qdot® 800 goat F(ab')$_2$ anti-rabbit IgG conjugate (H+L) *1 µM solution*	200 µL	6.6, 7.2	
Q11631MP	Qdot® 565 goat F(ab')$_2$ anti-rat IgG conjugate (H+L) *1 µM solution*	200 µL	6.6, 7.2	
Q11601MP	Qdot® 605 goat F(ab')$_2$ anti-rat IgG conjugate (H+L) *1 µM solution*	200 µL	6.6, 7.2	
Q11621MP	Qdot® 655 goat F(ab')$_2$ anti-rat IgG conjugate (H+L) *1 µM solution*	200 µL	6.6, 7.2	
Q14421MP	Qdot® 655 goat whole IgG anti-chicken IgY (H+L) conjugate *1 µM solution* *in 1 M betaine, 50 mM borate, pH 8.3 with 0.05% sodium azide*	200 µl	6.6, 7.2	
Q20001MP	Qdot® incubation buffer	30 mL	6.6	
Q21541MP	Qdot® 525 ITK™ amino (PEG) quantum dots *8 µM solution*	250 µL	6.6	
Q21591MP	Qdot® 545 ITK™ amino (PEG) quantum dots *8 µM solution*	250 µL	6.6	
Q21531MP	Qdot® 565 ITK™ amino (PEG) quantum dots *8 µM solution*	250 µL	6.6	
Q21511MP	Qdot® 585 ITK™ amino (PEG) quantum dots *8 µM solution*	250 µL	6.6	
Q21501MP	Qdot® 605 ITK™ amino (PEG) quantum dots *8 µM solution*	250 µL	6.6	
Q21521MP	Qdot® 655 ITK™ amino (PEG) quantum dots *8 µM solution*	250 µL	6.6	
Q21561MP	Qdot® 705 ITK™ amino (PEG) quantum dots *8 µM solution*	250 µL	6.6	
Q21571MP	Qdot® 800 ITK™ amino (PEG) quantum dots *8 µM solution*	250 µL	6.6	
Q21341MP	Qdot® 525 ITK™ carboxyl quantum dots *8 µM solution*	250 µL	6.6	
Q21391MP	Qdot® 545 ITK™ carboxyl quantum dots *8 µM solution*	250 µL	6.6	
Q21331MP	Qdot® 565 ITK™ carboxyl quantum dots *8 µM solution*	250 µL	6.6	
Q21311MP	Qdot® 585 ITK™ carboxyl quantum dots *8 µM solution*	250 µL	6.6	
Q21301MP	Qdot® 605 ITK™ carboxyl quantum dots *8 µM solution*	250 µL	6.6	
A10200	Qdot® 625 ITK™ carboxyl quantum dots *8 µM solution*	250 µL	6.6	
Q21321MP	Qdot® 655 ITK™ carboxyl quantum dots *8 µM solution*	250 µL	6.6	
Q21361MP	Qdot® 705 ITK™ carboxyl quantum dots *8 µM solution*	250 µL	6.6	
Q21371MP	Qdot® 800 ITK™ carboxyl quantum dots *8 µM solution*	250 µL	6.6	
Q21791MP	Qdot® 545 ITK™ organic quantum dots *1 µM solution*	4 mL	6.6	
Q21731MP	Qdot® 565 ITK™ organic quantum dots *1 µM solution*	4 mL	6.6	
Q21711MP	Qdot® 585 ITK™ organic quantum dots *1 µM solution*	4 mL	6.6	
Q21701MP	Qdot® 605 ITK™ organic quantum dots *1 µM solution*	4 mL	6.6	
Q21721MP	Qdot® 655 ITK™ organic quantum dots *1 µM solution*	4 mL	6.6	
Q21761MP	Qdot® 705 ITK™ organic quantum dots *1 µM solution*	4 mL	6.6	
Q21771MP	Qdot® 800 ITK™ organic quantum dots *1 µM solution*	4 mL	6.6	
Q10041MP	Qdot® 525 ITK™ Streptavidin Conjugate Kit *2 µM solution*	1 kit	6.6	
Q10091MP	Qdot® 545 ITK™ Streptavidin Conjugate Kit *2 µM solution*	1 kit	6.6	
Q10031MP	Qdot® 565 ITK™ Streptavidin Conjugate Kit *2 µM solution*	1 kit	6.6	
Q10011MP	Qdot® 585 ITK™ Streptavidin Conjugate Kit *2 µM solution*	1 kit	6.6	
Q10001MP	Qdot® 605 ITK™ Streptavidin Conjugate Kit *2 µM solution*	1 kit	6.6	
Q10021MP	Qdot® 655 ITK™ Streptavidin Conjugate Kit *2 µM solution*	1 kit	6.6	
Q10061MP	Qdot® 705 ITK™ Streptavidin Conjugate Kit *2 µM solution*	1 kit	6.6	
Q10071MP	Qdot® 800 ITK™ Streptavidin Conjugate Kit *2 µM solution*	1 kit	6.6	
Q11821MP	Qdot® 655 rabbit F(ab')$_2$ anti-goat IgG conjugate (H+L) *1 µM solution* *in 1 M betaine, 50 mM borate, pH 8.3 with 0.05% sodium azide*	200 µL	6.6, 7.2	
Q17421MP	Qdot® 655 rat anti-dinitrophenol conjugate *1 µM solution* *whole monoclonal IgG*	200 µL	6.6, 7.4	

*Note that all products listed in the above Master Product List are covered by **Limited Use Label License #223**. If a specific product is covered by additional Limited Use Label License(s), the Limited Use Label License #'s are provided in the LULL column above. Please refer to page 971 for the complete text of the Limited Use Label License(s).

Cat. No.	Product	Quantity	Sections	LULL*
Q10141MP	Qdot® 525 streptavidin conjugate *1 µM solution*	200 µL	6.6, 7.6	
Q10131MP	Qdot® 565 streptavidin conjugate *1 µM solution*	200 µL	6.6, 7.6	
Q10111MP	Qdot® 585 streptavidin conjugate *1 µM solution*	200 µL	6.6, 7.6	
Q10101MP	Qdot® 605 streptavidin conjugate *1 µM solution*	200 µL	6.6, 7.6	
A10196	Qdot® 625 streptavidin conjugate *1 µM solution*	200 µL	6.6, 7.6	
Q10121MP	Qdot® 655 streptavidin conjugate *1 µM solution*	200 µL	6.6, 7.6	
Q10161MP	Qdot® 705 streptavidin conjugate *1 µM solution*	200 µL	6.6, 7.6	
Q10171MP	Qdot® 800 streptavidin conjugate *1 µM solution*	200 µL	6.6, 7.6	
Q10151MP	Qdot® Streptavidin Sampler Kit *1 µM solutions*	1 kit	6.6, 7.6	
Q12021MP	Qdot® 655 wheat germ agglutinin conjugate *1 µM solution*	200 µL	6.6, 7.7	
Q10336	Qmount™ Qdot® mounting media	3 x 2 mL	6.6, 23.1	
Q10363	Qnuclear™ Deep Red stain	100 µL	12.5	
Q10464	QSY® 7 amine, hydrochloride	5 mg	3.4, 9.5	
Q10193	QSY® 7 carboxylic acid, succinimidyl ester	5 mg	1.6, 9.5	
Q10257	QSY® 7 C_5-maleimide	5 mg	2.2, 9.5	
Q21930	$N^ε$-(QSY® 7)-$N^α$-(9-fluorenylmethoxycarbonyl)-L-lysine (α-FMOC-ε-QSY® 7-L-lysine)	5 mg	9.5	
Q20131	QSY® 9 carboxylic acid, succinimidyl ester	5 mg	1.6, 9.5	
Q30457	QSY® 9 C_5-maleimide	5 mg	2.2	
Q20132	QSY® 21 carboxylic acid, succinimidyl ester	5 mg	1.6, 9.5	
Q20133	QSY® 35 acetic acid, succinimidyl ester	5 mg	1.8, 9.5	
Q21931	$N^β$-(QSY® 35)-$N^α$-(9-fluorenylmethoxycarbonyl)-L-alanine (α-FMOC-β-QSY® 35-L-alanine)	5 mg	9.5	
Q20348	QSY® 35 iodoacetamide	5 mg	2.2, 9.5	
Q25041MP	Qtracker® 525 Cell Labeling Kit	1 kit	6.6, 14.6	
Q25031MP	Qtracker® 565 Cell Labeling Kit	1 kit	6.6, 14.6	
Q25011MP	Qtracker® 585 Cell Labeling Kit	1 kit	6.6, 14.6	
Q25001MP	Qtracker® 605 Cell Labeling Kit	1 kit	6.6, 14.6	
A10198	Qtracker® 625 Cell Labeling Kit	1 kit	6.6, 14.6	
Q25021MP	Qtracker® 655 Cell Labeling Kit	1 kit	6.6, 14.6	
Q25061MP	Qtracker® 705 Cell Labeling Kit	1 kit	6.6, 14.6	
Q25071MP	Qtracker® 800 Cell Labeling Kit	1 kit	6.6, 14.6	
Q21031MP	Qtracker® 565 non-targeted quantum dots *2 µM solution*	200 µL	6.6, 14.6	
Q21021MP	Qtracker® 655 non-targeted quantum dots *2 µM solution*	200 µL	6.6, 14.6	
Q21061MP	Qtracker® 705 non-targeted quantum dots *2 µM solution*	200 µL	6.6, 14.6	
Q21071MP	Qtracker® 800 non-targeted quantum dots *2 µM solution*	200 µL	6.6, 14.6	
Q33120	Quant-iT™ dsDNA Assay Kit, High Sensitivity, 1000 assays *0.2–100 ng*	1 kit	8.3	
Q33130	Quant-iT™ dsDNA Assay Kit, Broad Range, 1000 assays *2–1000 ng*	1 kit	8.3	
Q32850	Quant-iT™ dsDNA BR Assay Kit, 100 assays *for use with the Qubit® fluorometer* *2–1000 ng*	1 kit	8.3	
Q32853	Quant-iT™ dsDNA BR Assay Kit, 500 assays *for use with the Qubit® fluorometer* *2–1000 ng*	1 kit	8.3	
Q32851	Quant-iT™ dsDNA HS Assay Kit, 100 assays *0.2–100 ng* *for use with the Qubit® fluorometer*	1 kit	8.3	
Q32854	Quant-iT™ dsDNA HS Assay Kit, 500 assays *0.2–100 ng* *for use with the Qubit® fluorometer*	1 kit	8.3	
O11492	Quant-iT™ OliGreen® ssDNA Assay Kit *2000 assays*	1 kit	8.1, 8.3	
O7582	Quant-iT™ OliGreen® ssDNA reagent *2000 assays*	1 mL	8.1, 8.3	
P11496	Quant-iT™ PicoGreen® dsDNA Assay Kit *2000 assays* *10 x 100 µL*	1 kit	8.1, 8.3	
P7589	Quant-iT™ PicoGreen® dsDNA Assay Kit *2000 assays*	1 kit	8.1, 8.3	
P11495	Quant-iT™ PicoGreen® dsDNA reagent *2000 assays* *10 x 100 µL*	10 x 100 µL	8.1, 8.3	
P7581	Quant-iT™ PicoGreen® dsDNA reagent *2000 assays*	1 mL	8.1, 8.3	
Q33211	Quant-iT™ Protein Assay Kit, 100 assays *0.25–5 µg* *for use with the Qubit® fluorometer*	1 kit	9.2	
Q33210	Quant-iT™ Protein Assay Kit, 1000 assays *0.25–5 µg*	1 kit	9.2	
Q33212	Quant-iT™ Protein Assay Kit, 500 Assays *0.25–5 µg* *for use with the Qubit® fluorometer*	1 kit	9.2	
R11490	Quant-iT™ RiboGreen® RNA Assay Kit *2000 assays*	1 kit	8.1, 8.3	
Q32852	Quant-iT™ RNA Assay Kit, 100 assays *5–100 ng* *for use with the Qubit® fluorometer*	1 kit	8.3	
Q32855	Quant-iT™ RNA Assay Kit, 500 assays *5–100 ng* *for use with the Qubit® fluorometer*	1 kit	8.3	
Q33140	Quant-iT™ RNA Assay Kit, 1000 assays *5–100 ng*	1 kit	8.3	
Q10213	Quant-iT™ RNA Assay Kit, Broad Range, 1000 assays *20–1000 ng*	1 kit	8.3	
Q10212	Quant-iT™ ssDNA Assay Kit, 100 assays *1–200 ng* *for use with the Qubit® fluorometer*	1 kit	8.3	
Q10210	Quant-iT™ RNA BR Assay Kit, 100 assays *20–1000 ng* *for use with the Qubit® fluorometer*	1 kit	8.3	
Q10211	Quant-iT™ RNA BR Assay Kit, 500 assays *20–1000 ng* *for use with the Qubit® fluorometer*	1 kit	8.3	
Q32857	Qubit® fluorometer	each	8.3, 9.2	
Q32856	Qubit™ assay tubes *set of 500*	1 set	8.3, 9.2	

*Note that all products listed in the above Master Product List are covered by **Limited Use Label License #223**. If a specific product is covered by additional Limited Use Label License(s), the Limited Use Label License #'s are provided in the LULL column above. Please refer to page 971 for the complete text of the Limited Use Label License(s).

Cat. No.	Product	Quantity	Sections	LULL*
Q32859	Qubit™ fluorometer international power cord *replacement*	each	8.3, 9.2	
Q32858	Qubit™ fluorometer USB cable	each	8.3, 9.2	
Q32860	Qubit™ Quantitation Starter Kit	1 kit	8.3, 9.2	
Q32861	Qubit™ Quantitation Lab Starter Kit	1 kit	8.3	
Q23918	quin-2, free acid *cell impermeant*	5 mg	19.2	
R11491	Quant-iT™ RiboGreen® RNA reagent *2000 assays*	1 mL	8.1, 8.3	
A10537	rabbit anti-goat IgG (H+L) *2 mg/mL*	0.5 mL	7.2	
R21458	rabbit anti-goat IgG (H+L), alkaline phosphatase conjugate	1 mg	7.2	
R21459	rabbit anti-goat IgG (H+L), horseradish peroxidase conjugate	1 mg	7.2	
R21455	rabbit anti-mouse IgG (H+L), horseradish peroxidase conjugate	1 mg	7.2	
A10072	Red/Green BODIPY® PC-A2 (1-*O*-(6-BODIPY® 558/568-aminohexyl)-2-BODIPY® FL C_5-*sn*-glycero-3-phosphocholine) *ratiometric phospholipase A_2 substrate*	100 µg	13.2, 17.4	
R35100	RediPlate™ 96 EnzChek® Caspase-3 Assay Kit *Z-DEVD-R110 substrate* *one 96-well microplate*	1 kit	10.4, 15.5	
R22132	RediPlate™ 96 EnzChek® Protease Assay Kit *red fluorescence* *one 96-well microplate*	1 kit	10.4	
R33700	RediPlate™ 96 EnzChek® Serine/Threonine Phosphatase Assay Kit *one 96-well microplate*	1 kit	10.3, 17.3	
R22067	RediPlate™ 96 EnzChek® Tyrosine Phosphatase Assay Kit *one 96-well microplate*	1 kit	10.3, 17.3	
R32700	RediPlate™ 96 RiboGreen® RNA Quantitation Kit *one 96-well microplate*	1 kit	8.1, 8.3	
R14060	RedoxSensor™ Red CC-1 *special packaging*	10 x 50 µg	12.2, 12.3, 15.2, 18.2	
R14782	Reference Dye Sampler Kit *five 1 mM solutions, 1 mL each*	1 kit	23.1	
R2931	Renin Substrate 1 (Arg-Glu(EDANS)-Ile-His-Pro-Phe-His-Leu-Val-Ile-His-Thr-Lys(dabcyl)-Arg)	1 mg	10.4	
R12204	resazurin, sodium salt	10 mg	10.6, 15.2	
R1159	resorufin β-D-galactopyranoside	25 mg	10.2	
R363	resorufin, sodium salt *reference standard*	100 mg	10.1	
R10367	RFP, rabbit polyclonal antibody, unconjugated (anti-RFP)	100 µg	7.5	
R649	RH 795	1 mg	22.2	
R33400	Rhinohide™ polyacrylamide gel strengthener concentrate *sufficient additive for 1 L of 30% acrylamide/bis-acrylamide (37.5:1)*	200 mL	9.3	
R1244	rhod-2, AM *cell permeant*	1 mg	19.3	204
R1245MP	rhod-2, AM *cell permeant* *special packaging*	20 x 50 µg	19.3	204
R14220	rhod-2, tripotassium salt *cell impermeant*	1 mg	19.3	204
R6479	rhodamine 110 (R110) *reference standard*	25 mg	10.1, 21.2	
R22122	rhodamine 110, bis-(L-aspartic acid amide), trifluoroacetic acid salt	1 mg	10.4, 15.5	
R6506	rhodamine 110, bis-(CBZ-L-alanyl-L-alanyl-L-alanyl-L-alanine amide)	5 mg	10.4	
R6508	rhodamine 110, bis-(CBZ-L-alanyl-L-arginine amide), dihydrochloride	5 mg	10.4	
R6501	rhodamine 110, bis-(CBZ-L-arginine amide), dihydrochloride (BZAR)	5 mg	10.4	
R22120	rhodamine 110, bis-(*N*-CBZ-L-aspartyl-L-glutamyl-L-valyl-L-aspartic acid amide) (Z-DEVD-R110) *bulk packaging*	20 mg	10.4, 15.5	
R22125	rhodamine 110, bis-(*N*-CBZ-L-isoleucyl-L-glutamyl-L-threonyl-L-aspartic acid amide) (Z-IETD-R110)	2 mg	10.4, 15.5	
R22126	rhodamine 110, bis-(*N*-CBZ-L-isoleucyl-L-glutamyl-L-threonyl-L-aspartic acid amide) (Z-IETD-R110) *bulk packaging*	20 mg	10.4, 15.5	
R6505	rhodamine 110, bis-(CBZ-L-isoleucyl-L-prolyl-L-arginine amide), dihydrochloride (BZiPAR)	5 mg	10.4	
R6502	rhodamine 110, bis-(CBZ-L-phenylalanyl-L-arginine amide), dihydrochloride	5 mg	10.4	
R33752	rhodamine 110, bis-(*N*-CBZ-L-alanyl-L-alanyl-L-aspartic acid amide) (Z-AAD-R110)	2 mg	10.4, 15.5	
R33753	rhodamine 110, bis-(*N*-CBZ-L-leucyl-L-glutamyl-L-glutamyl-L-aspartic acid amide) (Z-LEED-R110)	2 mg	10.4, 15.5	
R33750	rhodamine 110, bis-(*N*-CBZ-L-tyrosinyl-L-valyl-L-alanyl-L-aspartic acid amide) (Z-YVAD-R110)	2 mg	10.4, 15.5	
R33755	rhodamine 110, bis-(*N*-CBZ-L-valyl-L-aspartyl-L-valyl-L-alanyl-L-aspartic acid amide) (Z-VDVAD-R110)	2 mg	10.4, 15.5	
R33754	rhodamine 110, bis-(*N*-CBZ-L-valyl-L-glutamyl-L-isoleucyl-L-aspartic acid amide) (Z-VEID-R110)	2 mg	10.4, 15.5	
R22124	rhodamine 110, bis-(*p*-tosyl-L-glycyl-L-prolyl-L-arginine amide)	2 mg	10.4	
R302	rhodamine 123	25 mg	12.2, 15.2, 22.3	
R22420	rhodamine 123 *FluoroPure™ grade*	25 mg	12.2, 15.2, 22.3	
R634	rhodamine 6G chloride	1 g	12.2	
R648MP	rhodamine B, hexyl ester, perchlorate (R 6)	10 mg	12.2	
R6107	Rhodamine Green™ carboxylic acid, succinimidyl ester, hydrochloride (5(6)-CR 110, SE) *mixed isomers*	5 mg	1.5	
R6113	Rhodamine Green™-X, succinimidyl ester, hydrochloride *mixed isomers*	5 mg	1.5	
R415	rhodamine phalloidin	300 U	11.1	
R6029	Rhodamine Red™ C_2-maleimide	5 mg	2.2	
R6393	Rhodamine Red™-X goat anti-mouse IgG (H+L) *2 mg/mL*	0.5 mL	7.2	
R6394	Rhodamine Red™-X goat anti-rabbit IgG (H+L) *2 mg/mL*	0.5 mL	7.2	
R6160	Rhodamine Red™-X, succinimidyl ester *5-isomer*	5 mg	1.6, 4.2	
R34676	rhod dextran, potassium salt, 10,000 MW, anionic (high-affinity version)	5 mg	19.4	
R23983	rhod-FF, AM *cell permeant* *special packaging*	10 x 50 µg	19.3	204

*Note that all products listed in the above Master Product List are covered by **Limited Use Label License #223**. If a specific product is covered by additional Limited Use Label License(s), the Limited Use Label License #'s are provided in the LULL column above. Please refer to page 971 for the complete text of the Limited Use Label License(s).

IMPORTANT NOTICE: The products described in this manual are covered by one or more Limited Use Label License(s). Please refer to the Appendix on page 971 and Master Product List on page 975. Products are For Research Use Only. Not intended for any animal or human therapeutic or diagnostic use.

Cat. No.	Product	Quantity	Sections	LULL*
R10145	Rhod-3 Imaging Kit	1 kit	19.3	
R14207	rhod-5N, tripotassium salt *cell impermeant*	500 µg	19.3	204
R36351	RhodZin™-3, AM *cell permeant* *special packaging*	10 x 50 µg	19.7	
R36350	RhodZin™-3, dipotassium salt *cell impermeant*	500 µg	19.7	
R14000	rose bengal diacetate	5 mg	18.2	
S37102	Safe Imager™ blue-light transilluminator	each	8.4	
S37104	Safe Imager™ international power cord *replacement*	each	8.4	
S37103	Safe Imager™ viewing glasses	each	8.4	
S30044	SAIVI™ Alexa Fluor® 647 Antibody/Protein 1 mg-Labeling Kit *3 labelings*	1 kit	1.2, 1.3	
S34788	SAIVI™ Alexa Fluor® 680 injectable contrast agent *bovine serum albumin*	1 mL	14.7	
S34790	SAIVI™ Alexa Fluor® 680 injectable contrast agent *human serum transferrin*	1 mL	14.7	
S34789	SAIVI™ Alexa Fluor® 750 injectable contrast agent *bovine serum albumin*	1 mL	14.7	
S34791	SAIVI™ Alexa Fluor® 750 injectable contrast agent *human serum transferrin*	1 mL	14.7	
S31201	SAIVI™ 715 injectable contrast agent *0.1 µm microspheres*	1 mL	14.6	
S31203	SAIVI™ 715 injectable contrast agent *2 µm microspheres*	1 mL	14.6	
S30045	SAIVI™ Rapid Antibody Labeling Kit, Alexa Fluor® 680 *3 labelings*	1 kit	1.2, 1.3	
S30046	SAIVI™ Rapid Antibody Labeling Kit, Alexa Fluor® 750 *3 labelings*	1 kit	1.2, 1.3	
S1263	SBFI, AM *cell permeant*	1 mg	21.1	
S1264	SBFI, AM *cell permeant* *special packaging*	20 x 50 µg	21.1	
S1262	SBFI, tetraammonium salt *cell impermeant*	1 mg	21.1	
S24732	Secure-Seal™ hybridization chamber gasket, eight chambers, 9 mm diameter, 0.8 mm deep *set of 20*	1 set	23.1	
S24733	Secure-Seal™ hybridization chamber gasket, eight chambers, 9 mm diameter, 1.3 mm deep *set of 20*	1 set	23.1	
S24730	Secure-Seal™ hybridization chamber gasket, one chamber, 20 mm diameter, 0.8 mm deep *set of 40*	1 set	23.1	
S24731	Secure-Seal™ hybridization chamber gasket, one chamber, 20 mm diameter, 1.3 mm deep *set of 40*	1 set	23.1	
S24734	Secure-Seal™ hybridization chamber gasket, one chamber, 22 mm x 22 mm, 0.8 mm deep *set of 50*	1 set	23.1	
S24737	Secure-Seal™ spacer, eight wells, 9 mm diameter, 0.12 mm deep *set of 100*	1 set	23.1	
S24735	Secure-Seal™ spacer, one well, 13 mm diameter, 0.12 mm deep *set of 100*	1 set	23.1	
S24736	Secure-Seal™ spacer, one well, 20 mm diameter, 0.12 mm deep *set of 100*	1 set	23.1	
S34200	SelectFX® Alexa Fluor® 488 Endoplasmic Reticulum Labeling Kit *for fixed cells*	1 kit	12.4	
S34201	SelectFX® Alexa Fluor® 488 Peroxisome Labeling Kit *for fixed cells*	1 kit	12.3	
S33025	SelectFX® Nuclear Labeling Kit *DAPI, SYTOX® Green, 7-AAD, TO-PRO®-3 iodide* *for fixed cells*	1 kit	12.5	
S1129	SFDA (5-sulfofluorescein diacetate, sodium salt)	25 mg	15.2, 20.2	
V13240	Single Channel Annexin V/ Dead Cell Apoptosis Kit *Alexa Fluor® 488 annexin V/SYTOX® Green* *50 assays* *for flow cytometry*	1 kit	15.5	
S36002	Singlet Oxygen Sensor Green *special packaging*	10 x 100 µg	18.2	
S2828	*SlowFade®* Antifade Kit	1 kit	23.1	
S36936	*SlowFade®* Gold antifade reagent	10 mL	23.1	
S36937	*SlowFade®* Gold antifade reagent *special packaging*	5 x 2 mL	23.1	
S36938	*SlowFade®* Gold antifade reagent with DAPI	10 mL	23.1	
S36939	*SlowFade®* Gold antifade reagent with DAPI *special packaging*	5 x 2 mL	23.1	
S22801	SNARF®-1 carboxylic acid, acetate, succinimidyl ester *special packaging*	10 x 50 µg	14.2, 15.4, 20.4	
S23920	SNARF®-4F 5-(and-6)-carboxylic acid	1 mg	20.2	
S23921	SNARF®-4F 5-(and-6)-carboxylic acid, acetoxymethyl ester, acetate *special packaging*	20 x 50 µg	20.2	
S23922	SNARF®-5F 5-(and-6)-carboxylic acid	1 mg	20.2	
S23923	SNARF®-5F 5-(and-6)-carboxylic acid, acetoxymethyl ester, acetate *special packaging*	20 x 50 µg	20.2	
S6901	Sodium Green™ tetraacetate *cell permeant* *special packaging*	20 x 50 µg	21.1	
S6900	Sodium Green™, tetra(tetramethylammonium) salt *cell impermeant*	1 mg	21.1	
S7916	spermine NONOate	10 mg	18.3	
S2860	*Staphylococcus aureus* BioParticles® opsonizing reagent	1 U	16.1	
S23371	*Staphylococcus aureus* (Wood strain without protein A) BioParticles®, Alexa Fluor® 488 conjugate	2 mg	16.1	
S23372	*Staphylococcus aureus* (Wood strain without protein A) BioParticles®, Alexa Fluor® 594 conjugate	2 mg	16.1	
S2854	*Staphylococcus aureus* (Wood strain without protein A) BioParticles®, BODIPY® FL conjugate	10 mg	16.1	
S2851	*Staphylococcus aureus* (Wood strain without protein A) BioParticles®, fluorescein conjugate	10 mg	16.1	
S2859	*Staphylococcus aureus* (Wood strain without protein A) BioParticles®, unlabeled	100 mg	16.1	
S7445	ST-BODIPY® (-)-dihydropyridine *high affinity enantiomer*	25 µg	16.3	
S888	streptavidin	5 mg	7.6	
S21379	streptavidin acrylamide	1 mg	7.6, 9.4	
S951	streptavidin agarose *sedimented bead suspension*	5 mL	7.6	

*Note that all products listed in the above Master Product List are covered by **Limited Use Label License #223**. If a specific product is covered by additional Limited Use Label License(s), the Limited Use Label License #'s are provided in the LULL column above. Please refer to page 971 for the complete text of the Limited Use Label License(s).

Cat. No.	Product	Quantity	Sections	LULL*
S21002	streptavidin, Alexa Fluor® 680–allophycocyanin conjugate (Alexa Fluor® 680–allophycocyanin streptavidin) *1 mg/mL*	100 µL	6.4, 7.6	
S21005	streptavidin, Alexa Fluor® 700–allophycocyanin conjugate (Alexa Fluor® 700–allophycocyanin streptavidin) *1 mg/mL*	100 µL	6.4, 7.6	
S21008	streptavidin, Alexa Fluor® 750–allophycocyanin conjugate (Alexa Fluor® 750–allophycocyanin streptavidin) *1 mg/mL*	100 µL	6.4, 7.6	
S11249	streptavidin, Alexa Fluor® 350 conjugate	1 mg	7.6	
S32351	streptavidin, Alexa Fluor® 405 conjugate	1 mg	7.6	
S11237	streptavidin, Alexa Fluor® 430 conjugate	1 mg	7.6	
S11223	streptavidin, Alexa Fluor® 488 conjugate	1 mg	7.6	
S32354	streptavidin, Alexa Fluor® 488 conjugate *2 mg/mL*	0.5 mL	7.6	
S32353	streptavidin, Alexa Fluor® 514 conjugate	1 mg	7.6	
S11224	streptavidin, Alexa Fluor® 532 conjugate	1 mg	7.6	
S11225	streptavidin, Alexa Fluor® 546 conjugate	1 mg	7.6	
S21381	streptavidin, Alexa Fluor® 555 conjugate	1 mg	7.6	
S32355	streptavidin, Alexa Fluor® 555 conjugate *2 mg/mL*	0.5 mL	7.6	
S11226	streptavidin, Alexa Fluor® 568 conjugate	1 mg	7.6	
S11227	streptavidin, Alexa Fluor® 594 conjugate	1 mg	7.6	
S32356	streptavidin, Alexa Fluor® 594 conjugate *2 mg/mL*	0.5 mL	7.6	
S21375	streptavidin, Alexa Fluor® 633 conjugate	1 mg	7.6	
S32364	streptavidin, Alexa Fluor® 635 conjugate	1 mg	7.6	
S21374	streptavidin, Alexa Fluor® 647 conjugate	1 mg	7.6	
S32357	streptavidin, Alexa Fluor® 647 conjugate *2 mg/mL*	0.5 mL	7.6	
S21377	streptavidin, Alexa Fluor® 660 conjugate	1 mg	7.6	
S21378	streptavidin, Alexa Fluor® 680 conjugate	1 mg	7.6	
S32358	streptavidin, Alexa Fluor® 680 conjugate *2 mg/mL*	0.5 mL	7.6	
S21383	streptavidin, Alexa Fluor® 700 conjugate	1 mg	7.6	
S21384	streptavidin, Alexa Fluor® 750 conjugate	1 mg	7.6	
S20982	streptavidin, Alexa Fluor® 610–R-phycoerythrin conjugate (Alexa Fluor® 610–R-phycoerythrin streptavidin) *1 mg/mL*	100 µL	6.4, 7.6	
S20992	streptavidin, Alexa Fluor® 647–R-phycoerythrin conjugate (Alexa Fluor® 647–R-phycoerythrin streptavidin) *1 mg/mL*	100 µL	6.4, 7.6	
S20985	streptavidin, Alexa Fluor® 680–R-phycoerythrin conjugate (Alexa Fluor® 680–R-phycoerythrin streptavidin) *1 mg/mL*	100 µL	6.4, 7.6	
S32363	streptavidin, Alexa Fluor® 750–R-phycoerythrin conjugate (Alexa Fluor® 750–R-phycoerythrin streptavidin) *1 mg/mL*	100 µL	6.4, 7.6	
S921	streptavidin, alkaline phosphatase conjugate *2 mg/mL*	0.5 mL	7.6	
S32362	streptavidin, allophycocyanin conjugate *premium grade* *1 mg/mL*	250 µL	6.4, 7.6	
S868	streptavidin, allophycocyanin, crosslinked, conjugate *1 mg/mL*	0.5 mL	6.4, 7.6	
S11228	streptavidin, Cascade Yellow™ conjugate	1 mg	7.6	
S869	streptavidin, fluorescein conjugate	1 mg	7.6	
S931	streptavidin, β-galactosidase conjugate	1 mg	7.6, 10.2	
S911	streptavidin, horseradish peroxidase conjugate	1 mg	7.6	
S31569	streptavidin, β-lactamase TEM-1 conjugate *0.5 mg net protein*	0.5 mg	7.6	
S11221	streptavidin, Marina Blue® conjugate	1 mg	7.6	
S6368	streptavidin, Oregon Green® 488 conjugate	1 mg	7.6	
S6369	streptavidin, Oregon Green® 514 conjugate	1 mg	7.6	
S11222	streptavidin, Pacific Blue™ conjugate	1 mg	7.6	
S32365	streptavidin, Pacific Orange™ conjugate	1 mg	7.6	
S32350	streptavidin, B-phycoerythrin conjugate *1 mg/mL*	1 mL	6.4, 7.6	
S866	streptavidin, R-phycoerythrin conjugate (SAPE) *1 mg/mL*	1 mL	6.4, 7.6	
S871	streptavidin, rhodamine B conjugate	1 mg	7.6	
S6366	streptavidin, Rhodamine Red™-X conjugate	1 mg	7.6	
S21388	streptavidin, R-phycoerythrin conjugate (SAPE) *premium grade* *1 mg/mL*	1 mL	6.4, 7.6	
S870	streptavidin, tetramethylrhodamine conjugate	1 mg	7.6	
S872	streptavidin, Texas Red® conjugate	1 mg	7.6	
S6370	streptavidin, Texas Red®-X conjugate	1 mg	7.6	
S1553	succinimidyl acetylthioacetate (SATA)	100 mg	5.2	
S1534	succinimidyl *trans*-4-(maleimidylmethyl)cyclohexane-1-carboxylate (SMCC)	100 mg	5.2	
S1167	succinimidyl 6-(*N*-(7-nitrobenz-2-oxa-1,3-diazol-4-yl)amino)hexanoate (NBD-X, SE)	25 mg	1.8	

*Note that all products listed in the above Master Product List are covered by **Limited Use Label License #223**. If a specific product is covered by additional Limited Use Label License(s), the Limited Use Label License #'s are provided in the LULL column above. Please refer to page 971 for the complete text of the Limited Use Label License(s).

Cat. No.	Product	Quantity	Sections	LULL*
S6110	1-(3-(succinimidyloxycarbonyl)benzyl)-4-(5-(4-methoxyphenyl)oxazol-2-yl)pyridinium bromide (PyMPO, SE)	5 mg	1.7	
S1531	succinimidyl 3-(2-pyridyldithio)propionate (SPDP)	100 mg	5.2	
S1109	*N*-(4-sulfobutyl)-4-(6-(4-(dibutylamino)phenyl)hexatrienyl)pyridinium, inner salt (RH 237)	5 mg	22.2	
S1108	*N*-(4-sulfobutyl)-4-(4-(4-(dipentylamino)phenyl)butadienyl)pyridinium, inner salt (RH 421)	25 mg	22.2	
S359	sulforhodamine 101	25 mg	14.3, 16.1	
S1307	sulforhodamine B	5 g	14.3	
S10490	4-sulfo-2,3,5,6-tetrafluorophenol, sodium salt (STP)	100 mg	3.4	
S21500	SYBR® 101, succinimidyl ester	1 mg	8.1, 8.2	
S11494	SYBR® Gold nucleic acid gel stain *10,000X concentrate in DMSO*	500 µL	8.1, 8.4	
S7563	SYBR® Green I nucleic acid gel stain *10,000X concentrate in DMSO*	500 µL	8.1, 8.3, 8.4	
S7567	SYBR® Green I nucleic acid gel stain *10,000X concentrate in DMSO*	1 mL	8.1, 8.3, 8.4	
S7585	SYBR® Green I nucleic acid gel stain *10,000X concentrate in DMSO* *special packaging*	20 x 50 µL	8.1, 8.3, 8.4	
S7564	SYBR® Green II RNA gel stain *10,000X concentrate in DMSO*	500 µL	8.1, 8.4	
S7568	SYBR® Green II RNA gel stain *10,000X concentrate in DMSO*	1 mL	8.1, 8.4	
S7586	SYBR® Green II RNA gel stain *10,000X concentrate in DMSO* *special packaging*	20 x 50 µL	8.1, 8.4	
S7580	SYBR® Green Nucleic Acid Gel Stain Starter Kit	1 kit	8.1, 8.4	
S7569	SYBR® photographic filter	each	8.4	
S33102	SYBR® Safe DNA gel stain *10,000X concentrate in DMSO*	400 µL	8.1, 8.4	
S33111	SYBR® Safe DNA gel stain in 1X TAE	1 L	8.1, 8.4	
S33112	SYBR® Safe DNA gel stain in 1X TAE	4 L	8.1, 8.4	
S33100	SYBR® Safe DNA gel stain in 0.5X TBE	1 L	8.1, 8.4	
S33101	SYBR® Safe DNA gel stain in 0.5X TBE	4 L	8.1, 8.4	
S33110	SYBR® Safe DNA Gel Stain Starter Kit *with 1 L of SYBR® Safe DNA gel stain in 0.5X TBE (S33100) and one photographic filter (S37100)*	1 kit	8.1, 8.4	
S37100	SYBR® Safe photographic filter	each	8.4	
S6650	SYPRO® Orange protein gel stain *5000X concentrate in DMSO*	500 µL	9.3	
S6651	SYPRO® Orange protein gel stain *5000X concentrate in DMSO* *special packaging*	10 x 50 µL	9.3	
S6656	SYPRO® photographic filter	each	9.3	
S12012	SYPRO® Protein Gel Stain Starter Kit	1 kit	9.3	
S6653	SYPRO® Red protein gel stain *5000X concentrate in DMSO*	500 µL	9.3	
S6654	SYPRO® Red protein gel stain *5000X concentrate in DMSO* *special packaging*	10 x 50 µL	9.3	
S11791	SYPRO® Ruby protein blot stain *10–40 blots*	200 mL	9.3	
S12000	SYPRO® Ruby protein gel stain	1 L	9.3	
S12001	SYPRO® Ruby protein gel stain	200 mL	9.3	
S21900	SYPRO® Ruby protein gel stain *bulk packaging*	5 L	9.3	
S12010	SYPRO® Tangerine protein gel stain *5000X concentrate in DMSO*	500 µL	9.3	
S11351	SYTO® 40 blue fluorescent nucleic acid stain *5 mM solution in DMSO*	250 µL	8.1	
S11352	SYTO® 41 blue fluorescent nucleic acid stain *5 mM solution in DMSO*	250 µL	8.1	
S11353	SYTO® 42 blue fluorescent nucleic acid stain *5 mM solution in DMSO*	250 µL	8.1	
S11356	SYTO® 45 blue fluorescent nucleic acid stain *5 mM solution in DMSO*	250 µL	8.1	
S11350	SYTO® Blue Fluorescent Nucleic Acid Stain Sampler Kit *SYTO® dyes 40-45* *50 µL each*	1 kit	8.1, 12.5, 15.2	
S34854	SYTO® 9 green fluorescent nucleic acid stain *5 mM solution in DMSO*	100 µL	8.1, 15.2	
S32704	SYTO® 10 green fluorescent nucleic acid stain *5 mM solution in DMSO*	100 µL	8.1, 15.2	
S7573	SYTO® 11 green fluorescent nucleic acid stain *5 mM solution in DMSO*	250 µL	8.1	
S7574	SYTO® 12 green fluorescent nucleic acid stain *5 mM solution in DMSO*	250 µL	8.1	
S7575	SYTO® 13 green fluorescent nucleic acid stain *5 mM solution in DMSO*	250 µL	8.1, 15.2	
S7576	SYTO® 14 green fluorescent nucleic acid stain *5 mM solution in DMSO*	250 µL	8.1	
S7578	SYTO® 16 green fluorescent nucleic acid stain *1 mM solution in DMSO*	250 µL	8.1	
S7556	SYTO® 21 green fluorescent nucleic acid stain *5 mM solution in DMSO*	250 µL	8.1	
S7559	SYTO® 24 green fluorescent nucleic acid stain *5 mM solution in DMSO*	250 µL	8.1	
S7560	SYTO® 25 green fluorescent nucleic acid stain *5 mM solution in DMSO*	250 µL	8.1	
S34855	SYTO® BC green fluorescent nucleic acid stain *5 mM solution in DMSO*	100 µL	8.1, 15.2	
S7572	SYTO® Green Fluorescent Nucleic Acid Stain Sampler Kit *SYTO® dyes 11-14,16,21,24, and 25* *50 µL each*	1 kit	8.1, 12.5, 15.2	
S11361	SYTO® 80 orange fluorescent nucleic acid stain *5 mM solution in DMSO*	250 µL	8.1	
S11362	SYTO® 81 orange fluorescent nucleic acid stain *5 mM solution in DMSO*	250 µL	8.1	
S11363	SYTO® 82 orange fluorescent nucleic acid stain *5 mM solution in DMSO*	250 µL	8.1	
S11364	SYTO® 83 orange fluorescent nucleic acid stain *5 mM solution in DMSO*	250 µL	8.1	
S11365	SYTO® 84 orange fluorescent nucleic acid stain *5 mM solution in DMSO*	250 µL	8.1	
S11366	SYTO® 85 orange fluorescent nucleic acid stain *5 mM solution in DMSO*	250 µL	8.1	

*Note that all products listed in the above Master Product List are covered by **Limited Use Label License #223**. If a specific product is covered by additional Limited Use Label License(s), the Limited Use Label License #'s are provided in the LULL column above. Please refer to page 971 for the complete text of the Limited Use Label License(s).

IMPORTANT NOTICE: The products described in this manual are covered by one or more Limited Use Label License(s). Please refer to the Appendix on page 971 and Master Product List on page 975. Products are For Research Use Only. Not intended for any animal or human therapeutic or diagnostic use.

www.invitrogen.com/probes

molecular probes® | invitrogen™ by life technologies™

Cat. No.	Product	Quantity	Sections	LULL*
S11360	SYTO® Orange Fluorescent Nucleic Acid Stain Sampler Kit *SYTO® dyes 80-85* *50 µL each*	1 kit	8.1, 12.5, 15.2	
S7579	SYTO® 17 red fluorescent nucleic acid stain *5 mM solution in DMSO*	250 µL	8.1	
S11341	SYTO® 59 red fluorescent nucleic acid stain *5 mM solution in DMSO*	100 µL	8.1	
S11342	SYTO® 60 red fluorescent nucleic acid stain *5 mM solution in DMSO*	250 µL	8.1	
S11343	SYTO® 61 red fluorescent nucleic acid stain *5 mM solution in DMSO*	250 µL	8.1	
S11344	SYTO® 62 red fluorescent nucleic acid stain *5 mM solution in DMSO*	250 µL	8.1	
S11345	SYTO® 63 red fluorescent nucleic acid stain *5 mM solution in DMSO*	250 µL	8.1	
S11346	SYTO® 64 red fluorescent nucleic acid stain *5 mM solution in DMSO*	100 µL	8.1	
S11340	SYTO® Red Fluorescent Nucleic Acid Stain Sampler Kit *SYTO® dyes 17 and 59-64* *50 µL each*	1 kit	8.1, 12.5, 15.2	
S32703	SYTO® RNASelect™ green fluorescent cell stain *5 mM solution in DMSO*	100 µL	8.1, 15.2	
S10349	SYTOX® AADvanced™ Dead Cell Stain Kit *for flow cytometry* *for 488 nm excitation* *100 tests*	1 kit	12.5, 15.2	
S10274	SYTOX® AADvanced™ Dead Cell Stain Kit *for flow cytometry* *for 488 nm excitation* *500 tests*	1 kit	12.5, 15.2	
S34857	SYTOX® Blue dead cell stain *for flow cytometry* *1000 assays* *1 mM solution in DMSO*	1 mL	8.1, 12.5, 15.2	
S11348	SYTOX® Blue nucleic acid stain *5 mM solution in DMSO*	250 µL	8.1, 12.5, 15.2	
S7020	SYTOX® Green nucleic acid stain *5 mM solution in DMSO*	250 µL	8.1, 12.5, 15.2, 15.4	
S11368	SYTOX® Orange nucleic acid stain *5 mM solution in DMSO*	250 µL	8.1, 12.5, 15.2	
S34859	SYTOX® Red dead cell stain *for 633 or 635 nm excitation* *5 µM solution in DMSO*	1 mL	8.1, 12.5, 15.2	
A10067	TC-FlAsH™ Expression Analysis Detection Kit - Orange *fluorescent in-gel detection of TC-tagged and total protein*	1 kit	9.4	
A10068	TC-FlAsH™ Expression Analysis Detection Kit - Red *fluorescent in-gel detection of TC-tagged and total protein*	1 kit	9.4	
T11493	20X TE buffer *RNase free*	100 mL	8.3	
T1247	terbium(III) chloride, hexahydrate	1 g	9.4, 14.3, 17.2, 19.7	
T3168	5,5',6,6'-tetrachloro-1,1',3,3'-tetraethylbenzimidazolylcarbocyanine iodide (JC-1; CBIC$_2$(3))	5 mg	12.2, 15.2, 22.3	
T1210	tetrakis-(2-pyridylmethyl)ethylenediamine (TPEN)	100 mg	19.8	
T12921	5-(and-6)-tetramethylrhodamine biocytin (biocytin TMR)	5 mg	4.3, 14.3	
T1175	tetramethylrhodamine α-bungarotoxin (α-bungarotoxin, tetramethylrhodamine conjugate)	500 µg	16.2	
T6219	tetramethylrhodamine-5-carbonyl azide	5 mg	3.2	
T6105	6-(tetramethylrhodamine-5-(and-6)-carboxamido)hexanoic acid, succinimidyl ester (5(6)-TAMRA-X, SE) *mixed isomers*	10 mg	1.6, 4.2	
T669	tetramethylrhodamine, ethyl ester, perchlorate (TMRE)	25 mg	12.2, 15.2, 22.3	
T2762	tetramethylrhodamine goat anti-mouse IgG (H+L) *2 mg/mL*	0.5 mL	7.2	
T2769	tetramethylrhodamine goat anti-rabbit IgG (H+L) *2 mg/mL*	0.5 mL	7.2	
A10531	tetramethylrhodamine goat anti-rat IgG (H+L) *2 mg/mL*	0.5 mL	7.2	
T6006	tetramethylrhodamine-5-iodoacetamide dihydroiodide (5-TMRIA) *single isomer*	5 mg	2.2	
T1480	tetramethylrhodamine-5-isothiocyanate (5-TRITC; G isomer)	5 mg	1.6	
T1481	tetramethylrhodamine-6-isothiocyanate (6-TRITC; R isomer)	5 mg	1.6	
T490	tetramethylrhodamine-5-(and-6)-isothiocyanate (5(6)-TRITC) *mixed isomers*	10 mg	1.6	
T6027	tetramethylrhodamine-5-maleimide *single isomer*	5 mg	2.2	
T6028	tetramethylrhodamine-6-maleimide *single isomer*	5 mg	2.2	
T668	tetramethylrhodamine, methyl ester, perchlorate (TMRM)	25 mg	12.2, 15.2, 22.3	
A10532	tetramethylrhodamine rabbit anti-goat IgG (H+L) *2 mg/mL*	0.5 mL	7.2	
T10183	tetramethylrhodamine (TAMRA) alkyne (5-carboxytetramethylrhodamine, propargylamide) *5-isomer*	0.5 mg	3.1	
T10182	tetramethylrhodamine (TAMRA) azide (tetramethylrhodamine 5-carboxamido-(6-azidohexanyl)) *5-isomer*	0.5 mg	3.1	
T639	tetramethylrosamine chloride	25 mg	12.2	
T7284	TetraSpeck™ Fluorescent Microspheres Sampler Kit	1 kit	23.1	
T14792	TetraSpeck™ Fluorescent Microspheres Size Kit *mounted on slide*	1 kit	23.1	
T7279	TetraSpeck™ microspheres, 0.1 µm, fluorescent blue/green/orange/dark red	0.5 mL	23.1	
T7280	TetraSpeck™ microspheres, 0.2 µm, fluorescent blue/green/orange/dark red	0.5 mL	23.1	
T7281	TetraSpeck™ microspheres, 0.5 µm, fluorescent blue/green/orange/dark red	0.5 mL	23.1	
T7282	TetraSpeck™ microspheres, 1.0 µm, fluorescent blue/green/orange/dark red	0.5 mL	23.1	
T7283	TetraSpeck™ microspheres, 4.0 µm, fluorescent blue/green/orange/dark red	0.5 mL	23.1	
T6008	Texas Red® C_2-maleimide	5 mg	2.2	
T6009	Texas Red® C_5-bromoacetamide	5 mg	2.2	
T30200	Texas Red® C_2-dichlorotriazine	5 mg	1.6, 3.2	
T2425	Texas Red® cadaverine (Texas Red® C_5)	5 mg	3.4	
T1395MP	Texas Red® DHPE (Texas Red® 1,2-dihexadecanoyl-*sn*-glycero-3-phosphoethanolamine, triethylammonium salt)	1 mg	13.2	
T862	Texas Red® goat anti-mouse IgG (H+L) *2 mg/mL*	0.5 mL	7.2	
T2767	Texas Red® goat anti-rabbit IgG (H+L) *2 mg/mL*	0.5 mL	7.2	
T6256	Texas Red® hydrazide *>90% single isomer*	5 mg	3.3	

*Note that all products listed in the above Master Product List are covered by **Limited Use Label License #223**. If a specific product is covered by additional Limited Use Label License(s), the Limited Use Label License #'s are provided in the LULL column above. Please refer to page 971 for the complete text of the Limited Use Label License(s).

IMPORTANT NOTICE: The products described in this manual are covered by one or more Limited Use Label License(s). Please refer to the Appendix on page 971 and Master Product List on page 975. Products are For Research Use Only. Not intended for any animal or human therapeutic or diagnostic use.

Cat. No.	Product	Quantity	Sections	LULL*
T1905	Texas Red® sulfonyl chloride *mixed isomers* *special packaging*	10 x ~1 mg	1.6	
T353	Texas Red® sulfonyl chloride *mixed isomers*	10 mg	1.6	
T6390	Texas Red®-X goat anti-mouse IgG (H+L) *2 mg/mL*	0.5 mL	7.2	
T6391	Texas Red®-X goat anti-rabbit IgG (H+L) *2 mg/mL*	0.5 mL	7.2	
T6392	Texas Red®-X goat anti-rat IgG (H+L) *2 mg/mL*	0.5 mL	7.2	
T7471	Texas Red®-X phalloidin	300 U	11.1	
T10244	Texas Red®-X Protein Labeling Kit *3 labelings*	1 kit	1.2, 1.6	
T20175	Texas Red®-X, succinimidyl ester *single isomer*	2 mg	1.6, 4.2	
T6134	Texas Red®-X, succinimidyl ester *mixed isomers*	5 mg	1.6, 4.2	
T7458	thapsigargin	1 mg	17.2	
T7459	thapsigargin *special packaging*	20 x 50 µg	17.2	
T6060	Thiol and Sulfide Quantitation Kit *50–250 assays*	1 kit	2.1, 5.2	
T10095	ThiolTracker™ Violet (Glutathione Detection Reagent) *180 assays* *set of 3 vials*	1 set	15.6	
T10096	ThiolTracker™ Violet (Glutathione Detection Reagent) *for 5 microplates*	each	15.6	
T204	TMA-DPH (1-(4-trimethylammoniumphenyl)-6-phenyl-1,3,5-hexatriene *p*-toluenesulfonate)	25 mg	13.5, 16.1	
T53	2,6-TNS (2-(*p*-toluidinyl)naphthalene-6-sulfonic acid, sodium salt)	100 mg	13.5	
T3602	TO-PRO®-1 iodide (515/531) *1 mM solution in DMSO*	1 mL	8.1, 8.4	
T3605	TO-PRO®-3 iodide (642/661) *1 mM solution in DMSO*	1 mL	8.1, 8.4, 12.5, 15.4	
T7596	TO-PRO®-5 iodide (745/770) *1 mM solution in DMSO*	1 mL	8.1	
T3600	TOTO®-1 iodide (514/533) *1 mM solution in DMSO*	200 µL	8.1, 8.4, 12.5, 15.4	
T3604	TOTO®-3 iodide (642/660) *1 mM solution in DMSO*	200 µL	8.1, 8.4, 12.5	
T35357	transferrin from human serum, Alexa Fluor® 680 conjugate	5 mg	16.1	
T13342	transferrin from human serum, Alexa Fluor® 488 conjugate	5 mg	16.1	
T23364	transferrin from human serum, Alexa Fluor® 546 conjugate	5 mg	16.1	
T35352	transferrin from human serum, Alexa Fluor® 555 conjugate	5 mg	16.1	
T23365	transferrin from human serum, Alexa Fluor® 568 conjugate	5 mg	16.1	
T13343	transferrin from human serum, Alexa Fluor® 594 conjugate	5 mg	16.1	
T23362	transferrin from human serum, Alexa Fluor® 633 conjugate	5 mg	16.1	
T23366	transferrin from human serum, Alexa Fluor® 647 conjugate	5 mg	16.1	
T23363	transferrin from human serum, biotin-XX conjugate	5 mg	4.3, 16.1	
T2871	transferrin from human serum, fluorescein conjugate	5 mg	16.1	
T2872	transferrin from human serum, tetramethylrhodamine conjugate	5 mg	16.1	
T2875	transferrin from human serum, Texas Red® conjugate	5 mg	16.1	
T8864	TransFluoSpheres® carboxylate-modified microspheres, 0.04 µm (488/560) *2% solids*	0.5 mL	6.5	
T8870	TransFluoSpheres® carboxylate-modified microspheres, 0.04 µm (633/720) *2% solids*	0.5 mL	6.5	
T8872	TransFluoSpheres® carboxylate-modified microspheres, 0.1 µm (488/560) *2% solids*	0.5 mL	6.5	
T8880	TransFluoSpheres® carboxylate-modified microspheres, 1.0 µm (488/560) *2% solids*	0.5 mL	6.5	
T8883	TransFluoSpheres® carboxylate-modified microspheres, 1.0 µm (488/645) *2% solids*	0.5 mL	6.5	
T8860	TransFluoSpheres® NeutrAvidin® labeled microspheres, 0.04 µm (488/605) *1% solids*	0.4 mL	6.5, 7.6	
T8861	TransFluoSpheres® NeutrAvidin® labeled microspheres, 0.1 µm (488/605) *1% solids*	0.4 mL	6.5, 7.6	
T10711	TransFluoSpheres® streptavidin-labeled microspheres, 0.04 µm (488/645) *0.5% solids*	0.4 mL	6.5, 7.6	
T3163	*N*-(3-triethylammoniumpropyl)-4-(4-(dibutylamino)styryl)pyridinium dibromide (FM® 1-43)	1 mg	14.4, 16.1	
T35356	*N*-(3-triethylammoniumpropyl)-4-(4-(dibutylamino)styryl)pyridinium dibromide (FM® 1-43) *special packaging*	10 x 100 µg	14.4, 16.1	
T1111	*N*-(3-triethylammoniumpropyl)-4-(4-(4-(diethylamino)phenyl)butadienyl)pyridinium dibromide (RH 414)	5 mg	14.4, 16.1, 22.2	
T13320	*N*-(3-triethylammoniumpropyl)-4-(6-(4-(diethylamino)phenyl)hexatrienyl)pyridinium dibromide (FM® 4-64) *special packaging*	10 x 100 µg	12.3, 14.4, 16.1	
T3166	*N*-(3-triethylammoniumpropyl)-4-(6-(4-(diethylamino)phenyl)hexatrienyl)pyridinium dibromide (FM® 4-64)	1 mg	12.3, 14.4, 16.1	
T7508	*N*-(3-triethylammoniumpropyl)-4-(4-(diethylamino)styryl)pyridinium dibromide (FM® 2-10)	5 mg	16.1	
T23360	*N*-(3-trimethylammoniumpropyl)-4-(6-(4-(diethylamino)phenyl)hexatrienyl)pyridinium dibromide (FM® 5-95)	1 mg	12.3, 14.4, 16.1	
T506	4-trimethylammonium-2,2,6,6-tetramethylpiperidine-1-oxyl iodide (CAT 1)	100 mg	14.3	
T7601	2′-(or-3′)-*O*-(trinitrophenyl)adenosine 5′-diphosphate, disodium salt (TNP-ADP) *5 mg/mL in water*	2 mL	17.3	
T7624	2′-(or-3′)-*O*-(trinitrophenyl)adenosine 5′-monophosphate, sodium salt (TNP-AMP) *5 mg/mL in buffer*	2 mL	17.3	
T7602	2′-(or-3′)-*O*-(trinitrophenyl)adenosine 5′-triphosphate, trisodium salt (TNP-ATP) *5 mg/mL in buffer*	2 mL	17.3	
T2556	tris-(2-carboxyethyl)phosphine, hydrochloride (TCEP)	1 g	2.1, 5.2	
T1391	TRITC DHPE (*N*-(6-tetramethylrhodaminethiocarbamoyl)-1,2-dihexadecanoyl-*sn*-glycero-3-phosphoethanolamine, triethylammonium salt)	1 mg	13.2	
T1323	true blue chloride	5 mg	14.3	
T10282	Trypan Blue stain 0.4% *for use with Countess® automated cell counter*	2 x 1 mL	15.4	

*Note that all products listed in the above Master Product List are covered by **Limited Use Label License #223**. If a specific product is covered by additional Limited Use Label License(s), the Limited Use Label License #'s are provided in the LULL column above. Please refer to page 971 for the complete text of the Limited Use Label License(s).

Cat. No.	Product	Quantity	Sections	LULL*
T23011	trypsin inhibitor from soybean, Alexa Fluor® 488 conjugate	1 mg	14.7, 16.1	
T20911	TSA™ Kit #1 *with HRP–goat anti-mouse IgG and biotin-XX tyramide* *50–150 slides*	1 kit	6.2	235
T20912	TSA™ Kit #2 *with HRP–goat anti-mouse IgG and Alexa Fluor® 488 tyramide* *50–150 slides*	1 kit	6.2	235
T20913	TSA™ Kit #3 *with HRP–goat anti-mouse IgG and Alexa Fluor® 546 tyramide* *50–150 slides*	1 kit	6.2	235
T20914	TSA™ Kit #4 *with HRP–goat anti-mouse IgG and Alexa Fluor® 568 tyramide* *50–150 slides*	1 kit	6.2	235
T20915	TSA™ Kit #5 *with HRP–goat anti-mouse IgG and Alexa Fluor® 594 tyramide* *50–150 slides*	1 kit	6.2	235
T20916	TSA™ Kit #6 *with HRP–goat anti-mouse IgG and Alexa Fluor® 647 tyramide* *50–150 slides*	1 kit	6.2	235
T20917	TSA™ Kit #7 *with HRP–goat anti-mouse IgG and Alexa Fluor® 350 tyramide* *50–150 slides*	1 kit	6.2	235
T20921	TSA™ Kit #11 *with HRP–goat anti-rabbit IgG and biotin-XX tyramide* *50–150 slides*	1 kit	6.2	235
T20922	TSA™ Kit #12 *with HRP–goat anti-rabbit IgG and Alexa Fluor® 488 tyramide* *50–150 slides*	1 kit	6.2	235
T20923	TSA™ Kit #13 *with HRP–goat anti-rabbit IgG and Alexa Fluor® 546 tyramide* *50–150 slides*	1 kit	6.2	235
T20924	TSA™ Kit #14 *with HRP–goat anti-rabbit IgG and Alexa Fluor® 568 tyramide* *50–150 slides*	1 kit	6.2	235
T20925	TSA™ Kit #15 *with HRP–goat anti-rabbit IgG and Alexa Fluor® 594 tyramide* *50–150 slides*	1 kit	6.2	235
T20926	TSA™ Kit #16 *with HRP–goat anti-rabbit IgG and Alexa Fluor® 647 tyramide* *50–150 slides*	1 kit	6.2	235
T20927	TSA™ Kit #17 *with HRP–goat anti-rabbit IgG and Alexa Fluor® 350 tyramide* *50–150 slides*	1 kit	6.2	235
T20931	TSA™ Kit #21 *with HRP–streptavidin and biotin-XX tyramide* *50–150 slides*	1 kit	6.2	235
T20932	TSA™ Kit #22 *with HRP–streptavidin and Alexa Fluor® 488 tyramide* *50–150 slides*	1 kit	6.2	235
T20933	TSA™ Kit #23 *with HRP–streptavidin and Alexa Fluor® 546 tyramide* *50–150 slides*	1 kit	6.2	235
T20934	TSA™ Kit #24 *with HRP–streptavidin and Alexa Fluor® 568 tyramide* *50–150 slides*	1 kit	6.2	235
T20935	TSA™ Kit #25 *with HRP–streptavidin and Alexa Fluor® 594 tyramide* *50–150 slides*	1 kit	6.2	235
T20936	TSA™ Kit #26 *with HRP–streptavidin and Alexa Fluor® 647 tyramide* *50–150 slides*	1 kit	6.2	235
T20937	TSA™ Kit #27 *with HRP–streptavidin and Alexa Fluor® 350 tyramide* *50–150 slides*	1 kit	6.2	235
T20939	TSA™ Kit #29 *with HRP–streptavidin and Oregon Green® 488 tyramide* *50–150 slides*	1 kit	6.2	235
T30953	TSA™ Kit #40 *with HRP–goat anti-mouse IgG and Alexa Fluor® 555 tyramide* *50–150 slides*	1 kit	6.2	235
T30954	TSA™ Kit #41 *with HRP–goat anti-rabbit IgG and Alexa Fluor® 555 tyramide* *50–150 slides*	1 kit	6.2	235
T30955	TSA™ Kit #42 *with HRP–streptavidin and Alexa Fluor® 555 tyramide* *50–150 slides*	1 kit	6.2	235
T30456	TS-Link™ BODIPY® 630/650 C_5-thiosulfate, sodium salt	5 mg	2.2	
T30453	TS-Link™ BODIPY® FL C_2-thiosulfate, sodium salt	5 mg	2.2	
T30454	TS-Link™ BODIPY® TMR C_5-thiosulfate, sodium salt	5 mg	2.2	
T30455	TS-Link™ BODIPY® TR C_5-thiosulfate, sodium salt	5 mg	2.2	
T30754	TS-Link™ DSB-X™ biotin C_5-thiosulfate (TS-Link™ desthiobiotin-X C_5-thiosulfate, sodium salt)	5 mg	4.2	
T34075	TubulinTracker™ Green (Oregon Green® 488 Taxol, bis-acetate) *for live-cell imaging*	1 set	11.2	
15585-011	UltraPure™ ethidium bromide *10 mg/mL*	10 mL	8.1, 8.4, 15.2	
U21650	ULYSIS® Alexa Fluor® 488 Nucleic Acid Labeling Kit *20 labelings*	1 kit	1.2, 1.3, 8.2	
U21651	ULYSIS® Alexa Fluor® 532 Nucleic Acid Labeling Kit *20 labelings*	1 kit	1.2, 1.3, 8.2	
U21652	ULYSIS® Alexa Fluor® 546 Nucleic Acid Labeling Kit *20 labelings*	1 kit	1.2, 1.3, 8.2	
U21653	ULYSIS® Alexa Fluor® 568 Nucleic Acid Labeling Kit *20 labelings*	1 kit	1.2, 1.3, 8.2	
U21654	ULYSIS® Alexa Fluor® 594 Nucleic Acid Labeling Kit *20 labelings*	1 kit	1.2, 1.3, 8.2	
U21660	ULYSIS® Alexa Fluor® 647 Nucleic Acid Labeling Kit *20 labelings*	1 kit	1.2, 1.3, 8.2	
U21659	ULYSIS® Oregon Green® 488 Nucleic Acid Labeling Kit *20 labelings*	1 kit	1.2, 1.5, 8.2	
V1644	valinomycin	25 mg	21.1	
V34850	vancomycin, BODIPY® FL conjugate (BODIPY® FL vancomycin)	100 µg	15.2	
V7023	ViaGram™ Red+ Bacterial Gram Stain and Viability Kit *200 assays*	1 kit	15.3	
V12390	vinblastine, BODIPY® FL conjugate (BODIPY® FL vinblastine)	100 µg	11.2, 15.6	
A35136	Violet Annexin V/Dead Cell Apoptosis Kit *Pacific Blue™ annexin V/SYTOX® AADvanced™* *for flow cytometry* *50 assays*	1 kit	15.5	
A35135	Violet Chromatin Condensation/Dead Cell Apoptosis Kit *Vybrant® DyeCycle™ Violet and SYTOX® AADvanced™* *for flow cytometry* *200 assays*	1 kit	15.5	
V35123	Violet Membrane Permeability/Dead Cell Apoptosis Kit *with PO-PRO®-1 and 7- aminoactinomycin D* *200 assays* *for flow cytometry*	1 kit	15.5	
A35137	Violet Ratiometric Membrane Asymmetry Probe/Dead Cell Apoptosis Kit *for flow cytometry* *100 assays*	1 kit	15.5	
V34403	Vybrant® Alexa Fluor® 488 Lipid Raft Labeling Kit *50 labelings*	1 kit	13.3, 14.4	
V34404	Vybrant® Alexa Fluor® 555 Lipid Raft Labeling Kit *50 labelings*	1 kit	13.3, 14.4	
V34405	Vybrant® Alexa Fluor® 594 Lipid Raft Labeling Kit *50 labelings*	1 kit	13.3, 14.4	
V23200	Vybrant® Apoptosis Assay Kit #6 *biotin-X annexin V/Alexa Fluor® 350 streptavidin/propidium iodide* *50 assays*	1 kit	15.5	
V13181	Vybrant® Cell Adhesion Assay Kit	1 kit	15.6	
V23110	Vybrant® Cell Metabolic Assay Kit *with C_{12}-resazurin* *500–1000 assays*	1 kit	15.3	
V12883	Vybrant® CFDA SE Cell Tracer Kit	1 kit	14.2, 15.4	
V22888	Vybrant® CM-DiI cell-labeling solution	1 mL	14.4, 15.4	
V23111	Vybrant® Cytotoxicity Assay Kit *G6PD release assay* *1000 assays*	1 kit	15.3	

*Note that all products listed in the above Master Product List are covered by **Limited Use Label License #223**. If a specific product is covered by additional Limited Use Label License(s), the Limited Use Label License #'s are provided in the LULL column above. Please refer to page 971 for the complete text of the Limited Use Label License(s).

Cat. No.	Product	Quantity	Sections	LULL*
V22887	Vybrant® DiD cell-labeling solution	1 mL	14.4, 15.4	
V22885	Vybrant® DiI cell-labeling solution	1 mL	14.4, 15.4	
V22886	Vybrant® DiO cell-labeling solution	1 mL	14.4, 15.4	
V35004	Vybrant® DyeCycle™ Green stain *5 mM solution in DMSO* *200 assays*	400 µL	15.4	
V35005	Vybrant® DyeCycle™ Orange stain *5 mM solution in DMSO* *200 assays*	400 µL	15.4	
V10309	Vybrant® DyeCycle™ Ruby stain *2.5 mM solution in DMSO* *100 assays*	100 µL	15.4	
V10273	Vybrant® DyeCycle™ Ruby stain *2.5 mM solution in DMSO* *400 assays*	400 µL	15.4	
V35003	Vybrant® DyeCycle™ Violet stain *5 mM in water* *200 assays*	200 µL	15.4	
V35118	Vybrant® FAM Caspase-3 and -7 Assay Kit *for flow cytometry* *25 assays*	1 kit	10.4, 15.5	
V35119	Vybrant® FAM Caspase-8 Assay Kit *for flow cytometry* *25 assays*	1 kit	10.4, 15.5	
V35117	Vybrant® FAM Poly Caspases Assay Kit *for flow cytometry* *25 assays*	1 kit	10.4, 15.5	
V13154	Vybrant® MTT Cell Proliferation Assay Kit *1000 assays*	1 kit	15.4	
V22889	Vybrant® Multicolor Cell-Labeling Kit *DiO, DiI, DiD solutions, 1 mL each*	1 kit	14.4	
V13180	Vybrant® Multidrug Resistance Assay Kit	1 kit	15.6	215
V6694	Vybrant® Phagocytosis Assay Kit *250 assays*	1 kit	16.1	
W10132	WesternDot™ 625 Goat Anti-Mouse IgG Western Blot Kit *20 minigel blots*	1 kit	6.6, 7.2	
W10142	WesternDot™ 625 Goat Anti-Rabbit IgG Western Blot Kit *20 minigel blots*	1 kit	6.6, 7.2	
W11263	wheat germ agglutinin, Alexa Fluor® 350 conjugate	5 mg	7.7, 12.4	
W11261	wheat germ agglutinin, Alexa Fluor® 488 conjugate	5 mg	7.7, 12.4	
W32464	wheat germ agglutinin, Alexa Fluor® 555 conjugate	5 mg	7.7, 12.4	
W11262	wheat germ agglutinin, Alexa Fluor® 594 conjugate	5 mg	7.7, 12.4	
W21404	wheat germ agglutinin, Alexa Fluor® 633 conjugate	5 mg	7.7, 12.4	
W32466	wheat germ agglutinin, Alexa Fluor® 647 conjugate	5 mg	7.7, 12.4	
W32465	wheat germ agglutinin, Alexa Fluor® 680 conjugate	5 mg	7.7, 12.4	
W834	wheat germ agglutinin, fluorescein conjugate	5 mg	7.7, 12.4	
W6748	wheat germ agglutinin, Oregon Green® 488 conjugate	5 mg	7.7, 12.4	
W7024	Wheat Germ Agglutinin Sampler Kit *four fluorescent conjugates, 1 mg each*	1 kit	7.7, 12.4, 15.3	
W849	wheat germ agglutinin, tetramethylrhodamine conjugate	5 mg	7.7, 12.4	
W21405	wheat germ agglutinin, Texas Red®-X conjugate	1 mg	7.7, 12.4	
X14210	X-rhod-1, AM *cell permeant* *special packaging*	10 x 50 µg	19.3	204
X23985	X-rhod-5F, AM *cell permeant* *special packaging*	10 x 50 µg	19.3	204
X23984	X-rhod-5F, tripotassium salt *cell impermeant*	500 µg	19.3	204
X491	X-rhodamine-5-(and-6)-isothiocyanate (5(6)-XRITC) *mixed isomers*	10 mg	1.6	
X6493	XTT (2,3-bis-(2-methoxy-4-nitro-5-sulfophenyl)-2*H*-tetrazolium-5-carboxanilide)	100 mg	15.2, 18.2	
X1525	*p*-xylene-bis-pyridinium bromide (DPX)	1 g	14.3	
Y7530	Yeast Mitochondrial Stain Sampler Kit	1 kit	12.2	
Y7531	Yeast Vacuole Marker Sampler Kit	1 kit	12.3	
Y7536	yeast vacuole membrane marker MDY-64	1 mg	12.3	
Y3603	YO-PRO®-1 iodide (491/509) *1 mM solution in DMSO*	1 mL	8.1, 8.4, 12.5, 15.5	
Y3607	YO-PRO®-3 iodide (612/631) *1 mM solution in DMSO*	1 mL	8.1, 12.5	
Y3601	YOYO®-1 iodide (491/509) *1 mM solution in DMSO*	200 µL	8.1, 8.4, 12.5, 15.4	
Y3606	YOYO®-3 iodide (612/631) *1 mM solution in DMSO*	200 µL	8.1, 12.5	
Z25302	Zenon® Alexa Fluor® 488 Rabbit IgG Labeling Kit *50 labelings*	1 kit	1.2, 1.3, 7.3	
Z25452	Zenon® Biotin-XX Human IgG Labeling Kit *50 labelings*	1 kit	1.2, 4.2, 7.3	
Z25400	Zenon® Alexa Fluor® 350 Human IgG Labeling Kit *50 labelings*	1 kit	1.2, 1.3, 1.7, 7.3	
Z25000	Zenon® Alexa Fluor® 350 Mouse IgG_1 Labeling Kit *50 labelings*	1 kit	1.2, 1.3, 1.7, 7.3	
Z25100	Zenon® Alexa Fluor® 350 Mouse IgG_{2a} Labeling Kit *50 labelings*	1 kit	1.2, 1.3, 1.7, 7.3	
Z25200	Zenon® Alexa Fluor® 350 Mouse IgG_{2b} Labeling Kit *50 labelings*	1 kit	1.2, 1.3, 1.7, 7.3	
Z25300	Zenon® Alexa Fluor® 350 Rabbit IgG Labeling Kit *50 labelings*	1 kit	1.2, 1.3, 1.7, 7.3	
Z25013	Zenon® Alexa Fluor® 405 Mouse IgG_1 Labeling Kit *50 labelings*	1 kit	1.2, 1.3, 1.7, 7.3	
Z25113	Zenon® Alexa Fluor® 405 Mouse IgG_{2a} Labeling Kit *50 labelings*	1 kit	1.2, 1.3, 1.7, 7.3	
Z25213	Zenon® Alexa Fluor® 405 Mouse IgG_{2b} Labeling Kit *50 labelings*	1 kit	1.2, 1.3, 1.7, 7.3	
Z25313	Zenon® Alexa Fluor® 405 Rabbit IgG Labeling Kit *50 labelings*	1 kit	1.2, 1.3, 1.7, 7.3	
Z25001	Zenon® Alexa Fluor® 430 Mouse IgG_1 Labeling Kit *50 labelings*	1 kit	1.2, 1.3, 1.7, 7.3	
Z25301	Zenon® Alexa Fluor® 430 Rabbit IgG Labeling Kit *50 labelings*	1 kit	1.2, 1.3, 1.7, 7.3	
Z25402	Zenon® Alexa Fluor® 488 Human IgG Labeling Kit *50 labelings*	1 kit	1.2, 1.3, 7.3	
Z25002	Zenon® Alexa Fluor® 488 Mouse IgG_1 Labeling Kit *50 labelings*	1 kit	1.2, 1.3, 7.3	

*Note that all products listed in the above Master Product List are covered by **Limited Use Label License #223**. If a specific product is covered by additional Limited Use Label License(s), the Limited Use Label License #'s are provided in the LULL column above. Please refer to page 971 for the complete text of the Limited Use Label License(s).

Cat. No.	Product	Quantity	Sections	LULL*
Z25102	Zenon® Alexa Fluor® 488 Mouse IgG_{2a} Labeling Kit *50 labelings*	1 kit	1.2, 1.3, 7.3	
Z25202	Zenon® Alexa Fluor® 488 Mouse IgG_{2b} Labeling Kit *50 labelings*	1 kit	1.2, 1.3, 7.3	
Z25003	Zenon® Alexa Fluor® 532 Mouse IgG_1 Labeling Kit *50 labelings*	1 kit	1.2, 1.3, 7.3	
Z25303	Zenon® Alexa Fluor® 532 Rabbit IgG Labeling Kit *50 labelings*	1 kit	1.2, 1.3, 7.3	
Z25004	Zenon® Alexa Fluor® 546 Mouse IgG_1 Labeling Kit *50 labelings*	1 kit	1.2, 1.3, 7.3	
Z25104	Zenon® Alexa Fluor® 546 Mouse IgG_{2a} Labeling Kit *50 labelings*	1 kit	1.2, 1.3, 7.3	
Z25204	Zenon® Alexa Fluor® 546 Mouse IgG_{2b} Labeling Kit *50 labelings*	1 kit	1.2, 1.3, 7.3	
Z25304	Zenon® Alexa Fluor® 546 Rabbit IgG Labeling Kit *50 labelings*	1 kit	1.2, 1.3, 7.3	
Z25405	Zenon® Alexa Fluor® 555 Human IgG Labeling Kit *50 labelings*	1 kit	1.2, 1.3, 7.3	
Z25005	Zenon® Alexa Fluor® 555 Mouse IgG_1 Labeling Kit *50 labelings*	1 kit	1.2, 1.3, 7.3	
Z25105	Zenon® Alexa Fluor® 555 Mouse IgG_{2a} Labeling Kit *50 labelings*	1 kit	1.2, 1.3, 7.3	
Z25205	Zenon® Alexa Fluor® 555 Mouse IgG_{2b} Labeling Kit *50 labelings*	1 kit	1.2, 1.3, 7.3	
Z25305	Zenon® Alexa Fluor® 555 Rabbit IgG Labeling Kit *50 labelings*	1 kit	1.2, 1.3, 7.3	
Z25006	Zenon® Alexa Fluor® 568 Mouse IgG_1 Labeling Kit *50 labelings*	1 kit	1.2, 1.3, 7.3	
Z25106	Zenon® Alexa Fluor® 568 Mouse IgG_{2a} Labeling Kit *50 labelings*	1 kit	1.2, 1.3, 7.3	
Z25206	Zenon® Alexa Fluor® 568 Mouse IgG_{2b} Labeling Kit *50 labelings*	1 kit	1.2, 1.3, 7.3	
Z25306	Zenon® Alexa Fluor® 568 Rabbit IgG Labeling Kit *50 labelings*	1 kit	1.2, 1.3, 7.3	
Z25407	Zenon® Alexa Fluor® 594 Human IgG Labeling Kit *50 labelings*	1 kit	1.2, 1.3, 7.3	
Z25007	Zenon® Alexa Fluor® 594 Mouse IgG_1 Labeling Kit *50 labelings*	1 kit	1.2, 1.3, 7.3	
Z25107	Zenon® Alexa Fluor® 594 Mouse IgG_{2a} Labeling Kit *50 labelings*	1 kit	1.2, 1.3, 7.3	
Z25207	Zenon® Alexa Fluor® 594 Mouse IgG_{2b} Labeling Kit *50 labelings*	1 kit	1.2, 1.3, 7.3	
Z25307	Zenon® Alexa Fluor® 594 Rabbit IgG Labeling Kit *50 labelings*	1 kit	1.2, 1.3, 7.3	
Z25020	Zenon® Alexa Fluor® 610–R-Phycoerythrin Mouse IgG_1 Labeling Kit *10 labelings*	1 kit	1.2, 1.3, 6.4, 7.3	
Z25408	Zenon® Alexa Fluor® 647 Human IgG Labeling Kit *50 labelings*	1 kit	1.2, 1.3, 7.3	
Z25008	Zenon® Alexa Fluor® 647 Mouse IgG_1 Labeling Kit *50 labelings*	1 kit	1.2, 1.3, 7.3	
Z25108	Zenon® Alexa Fluor® 647 Mouse IgG_{2a} Labeling Kit *50 labelings*	1 kit	1.2, 1.3, 7.3	
Z25208	Zenon® Alexa Fluor® 647 Mouse IgG_{2b} Labeling Kit *50 labelings*	1 kit	1.2, 1.3, 7.3	
Z25308	Zenon® Alexa Fluor® 647 Rabbit IgG Labeling Kit *50 labelings*	1 kit	1.2, 1.3, 7.3	
Z25021	Zenon® Alexa Fluor® 647–R-Phycoerythrin Mouse IgG_1 Labeling Kit *10 labelings*	1 kit	1.2, 1.3, 6.4, 7.3	
Z25121	Zenon® Alexa Fluor® 647–R-Phycoerythrin Mouse IgG_{2a} Labeling Kit *10 labelings*	1 kit	1.2, 1.3, 6.4, 7.3	
Z25221	Zenon® Alexa Fluor® 647–R-Phycoerythrin Mouse IgG_{2b} Labeling Kit *10 labelings*	1 kit	1.2, 1.3, 6.4, 7.3	
Z25009	Zenon® Alexa Fluor® 660 Mouse IgG_1 Labeling Kit *50 labelings*	1 kit	1.2, 1.3, 7.3	
Z25010	Zenon® Alexa Fluor® 680 Mouse IgG_1 Labeling Kit *50 labelings*	1 kit	1.2, 1.3, 7.3	
Z25110	Zenon® Alexa Fluor® 680 Mouse IgG_{2a} Labeling Kit *50 labelings*	1 kit	1.2, 1.3, 7.3	
Z25210	Zenon® Alexa Fluor® 680 Mouse IgG_{2b} Labeling Kit *50 labelings*	1 kit	1.2, 1.3, 7.3	
Z25310	Zenon® Alexa Fluor® 680 Rabbit IgG Labeling Kit *50 labelings*	1 kit	1.2, 1.3, 7.3	
Z25022	Zenon® Alexa Fluor® 680–R-Phycoerythrin Mouse IgG_1 Labeling Kit *10 labelings*	1 kit	1.2, 1.3, 6.4, 7.3	
Z25011	Zenon® Alexa Fluor® 700 Mouse IgG_1 Labeling Kit *50 labelings*	1 kit	1.2, 1.3, 7.3	
Z25030	Zenon® Alexa Fluor® 700–Allophycocyanin Mouse IgG_1 Labeling Kit *10 labelings*	1 kit	1.2, 1.3, 6.4, 7.3	
Z25312	Zenon® Alexa Fluor® 750 Rabbit IgG Labeling Kit *50 labelings*	1 kit	1.2, 1.3, 7.3	
Z25031	Zenon® Alexa Fluor® 750–Allophycocyanin Mouse IgG_1 Labeling Kit *10 labelings*	1 kit	1.2, 1.3, 6.4, 7.3	
Z25090	Zenon® Alexa Fluor® 488 Mouse IgG_1 Labeling Kit *enhanced with TSA™ technology* *25 labelings*	1 kit	1.2, 1.3, 6.2, 7.3	235
Z25350	Zenon® Alkaline Phosphatase Rabbit IgG Labeling Kit *25 labelings*	1 kit	1.2, 7.3	
Z25451	Zenon® Allophycocyanin Human IgG Labeling Kit *25 labelings*	1 kit	1.2, 6.4, 7.3	
Z25051	Zenon® Allophycocyanin Mouse IgG_1 Labeling Kit *25 labelings*	1 kit	1.2, 6.4, 7.3	
Z25151	Zenon® Allophycocyanin Mouse IgG_{2a} Labeling Kit *25 labelings*	1 kit	1.2, 6.4, 7.3	
Z25251	Zenon® Allophycocyanin Mouse IgG_{2b} Labeling Kit *25 labelings*	1 kit	1.2, 6.4, 7.3	
Z25351	Zenon® Allophycocyanin Rabbit IgG Labeling Kit *25 labelings*	1 kit	1.2, 6.4, 7.3	
Z25052	Zenon® Biotin-XX Mouse IgG_1 Labeling Kit *50 labelings*	1 kit	1.2, 4.2, 7.3	
Z25152	Zenon® Biotin-XX Mouse IgG_{2a} Labeling Kit *50 labelings*	1 kit	1.2, 4.2, 7.3	
Z25252	Zenon® Biotin-XX Mouse IgG_{2b} Labeling Kit *50 labelings*	1 kit	1.2, 4.2, 7.3	
Z25352	Zenon® Biotin-XX Rabbit IgG Labeling Kit *50 labelings*	1 kit	1.2, 4.2, 7.3	
Z25042	Zenon® Fluorescein Mouse IgG_1 Labeling Kit *50 labelings*	1 kit	1.2, 1.5, 7.3	
Z25342	Zenon® Fluorescein Rabbit IgG Labeling Kit *50 labelings*	1 kit	1.2, 1.5, 7.3	
Z25454	Zenon® Horseradish Peroxidase Human IgG Labeling Kit *25 labelings*	1 kit	1.2, 6.2, 7.3	
Z25054	Zenon® Horseradish Peroxidase Mouse IgG_1 Labeling Kit *25 labelings*	1 kit	1.2, 6.2, 7.3	
Z25154	Zenon® Horseradish Peroxidase Mouse IgG_{2a} Labeling Kit *25 labelings*	1 kit	1.2, 6.2, 7.3	
Z25254	Zenon® Horseradish Peroxidase Mouse IgG_{2b} Labeling Kit *25 labelings*	1 kit	1.2, 6.2, 7.3	

*Note that all products listed in the above Master Product List are covered by **Limited Use Label License #223**. If a specific product is covered by additional Limited Use Label License(s), the Limited Use Label License #'s are provided in the LULL column above. Please refer to page 971 for the complete text of the Limited Use Label License(s).

IMPORTANT NOTICE: The products described in this manual are covered by one or more Limited Use Label License(s). Please refer to the Appendix on page 971 and Master Product List on page 975. Products are For Research Use Only. Not intended for any animal or human therapeutic or diagnostic use.

Cat. No.	Product	Quantity	Sections	LULL*
Z25354	Zenon® Horseradish Peroxidase Rabbit IgG Labeling Kit *25 labelings*	1 kit	1.2, 6.2, 7.3	
Z25041	Zenon® Pacific Blue™ Mouse IgG_1 Labeling Kit *50 labelings*	1 kit	1.2, 1.7, 7.3	
Z25341	Zenon® Pacific Blue™ Rabbit IgG Labeling Kit *50 labelings*	1 kit	1.2, 1.7, 7.3	
Z25257	Zenon® Pacific Orange™ Mouse IgG_{2a} Labeling Kit *50 labelings*	1 kit	1.2, 1.7, 7.3	
Z25256	Zenon® Pacific Orange™ Mouse IgG_1 Labeling Kit *50 labelings*	1 kit	1.2, 1.7, 7.3	
Z25455	Zenon® R-Phycoerythrin Human IgG Labeling Kit *25 labelings*	1 kit	1.2, 6.4, 7.3	
Z25055	Zenon® R-Phycoerythrin Mouse IgG_1 Labeling Kit *25 labelings*	1 kit	1.2, 6.4, 7.3	
Z25155	Zenon® R-Phycoerythrin Mouse IgG_{2a} Labeling Kit *25 labelings*	1 kit	1.2, 6.4, 7.3	
Z25255	Zenon® R-Phycoerythrin Mouse IgG_{2b} Labeling Kit *25 labelings*	1 kit	1.2, 6.4, 7.3	
Z25355	Zenon® R-Phycoerythrin Rabbit IgG Labeling Kit *25 labelings*	1 kit	1.2, 6.4, 7.3	
Z25045	Zenon® Texas Red®-X Mouse IgG_1 Labeling Kit *50 labelings*	1 kit	1.2, 1.6, 7.3	
Z25460	Zenon® Tricolor Human IgG Labeling Kit #1 *for green, orange and deep red fluorescence imaging* *3 x 10 labelings*	1 kit	1.2, 7.3	
Z25470	Zenon® Tricolor Human IgG Labeling Kit #2 *for blue, green and red fluorescence imaging* *3 x 10 labelings*	1 kit	1.2, 7.3	
Z25060	Zenon® Tricolor Mouse IgG_1 Labeling Kit #1 *for green, orange and deep red fluorescence imaging* *3 x 10 labelings*	1 kit	1.2, 7.3	
Z25070	Zenon® Tricolor Mouse IgG_1 Labeling Kit #2 *for blue, green and red fluorescence imaging* *3 x 10 labelings*	1 kit	1.2, 7.3	
Z25080	Zenon® Tricolor Mouse IgG_1 Labeling Kit #3 *for flow cytometry, 488 nm excitation* *3 x 10 labelings*	1 kit	1.2, 7.3	
Z25160	Zenon® Tricolor Mouse IgG_{2a} Labeling Kit #1 *for green, orange and deep red fluorescence imaging* *3 x 10 labelings*	1 kit	1.2, 7.3	
Z25170	Zenon® Tricolor Mouse IgG_{2a} Labeling Kit #2 *for blue, green and red fluorescence imaging* *3 x 10 labelings*	1 kit	1.2, 7.3	
Z25180	Zenon® Tricolor Mouse IgG_{2a} Labeling Kit #3 *for flow cytometry, 488 nm excitation* *3 x 10 labelings*	1 kit	1.2, 7.3	
Z25260	Zenon® Tricolor Mouse IgG_{2b} Labeling Kit #1 *for green, orange and deep red fluorescence imaging* *3 x 10 labelings*	1 kit	1.2, 7.3	
Z25270	Zenon® Tricolor Mouse IgG_{2b} Labeling Kit #2 *for blue, green and red fluorescence imaging* *3 x 10 labelings*	1 kit	1.2, 7.3	
Z25280	Zenon® Tricolor Mouse IgG_{2b} Labeling Kit #3 *for flow cytometry, 488 nm excitation* *3 x 10 labelings*	1 kit	1.2, 7.3	
Z25360	Zenon® Tricolor Rabbit IgG Labeling Kit #1 *for green, orange and deep red fluorescence imaging* *3 x 10 labelings*	1 kit	1.2, 7.3	
Z25370	Zenon® Tricolor Rabbit IgG Labeling Kit #2 *for blue, green and red fluorescence imaging* *3 x 10 labelings*	1 kit	1.2, 7.3	
Z25380	Zenon® Tricolor Rabbit IgG Labeling Kit #3 *for flow cytometry, 488 nm excitation* *3 x 10 labelings*	1 kit	1.2, 7.3	
Z25156	Zenon® Pacific Blue™ Mouse IgG_{2a} Labeling Kit *50 labelings*	1 kit	1.2, 7.3	
Z25602	Zenon® Alexa Fluor® 488 Goat IgG Labeling Kit *50 labelings*	1 kit	1.2, 1.3, 7.3	
Z25605	Zenon® Alexa Fluor® 555 Goat IgG Labeling Kit *50 labelings*	1 kit	1.2, 1.3, 7.3	
Z25607	Zenon® Alexa Fluor® 594 Goat IgG Labeling Kit *50 labelings*	1 kit	1.2, 1.3, 7.3	
Z25608	Zenon® Alexa Fluor® 647 Goat IgG Labeling Kit *50 labelings*	1 kit	1.2, 1.3, 7.3	
Z25606	Zenon® Alexa Fluor® 568 Goat IgG Labeling Kit *50 labelings*	1 kit	1.2, 1.3, 7.3	
Z33857	Zen™ Myeloperoxidase (MPO) ELISA Kit *200 assays*	1 kit	10.5, 18.2	
Z2850	zymosan A BioParticles® opsonizing reagent	1 U	16.1	
Z23373	zymosan A (*S. cerevisiae*) BioParticles®, Alexa Fluor® 488 conjugate	2 mg	16.1	
Z23374	zymosan A (*S. cerevisiae*) BioParticles®, Alexa Fluor® 594 conjugate	2 mg	16.1	
Z2841	zymosan A (*S. cerevisiae*) BioParticles®, fluorescein conjugate	10 mg	16.1	
Z2843	zymosan A (*S. cerevisiae*) BioParticles®, Texas Red® conjugate	10 mg	16.1	
Z2849	zymosan A (*S. cerevisiae*) BioParticles®, unlabeled	100 mg	16.1	

*Note that all products listed in the above Master Product List are covered by **Limited Use Label License #223**. If a specific product is covered by additional Limited Use Label License(s), the Limited Use Label License #'s are provided in the LULL column above. Please refer to page 971 for the complete text of the Limited Use Label License(s).

IMPORTANT NOTICE: The products described in this manual are covered by one or more Limited Use Label License(s). Please refer to the Appendix on page 971 and Master Product List on page 975. Products are For Research Use Only. Not intended for any animal or human therapeutic or diagnostic use.

Master Catalog Number List

This alphabetical listing of catalog numbers references all *Handbook* sections associated with each product. If you know the name of the product, consult the Master Product List on page 975.

Cat. No.	Sections
03--3700	7.4
15585-011	8.1, 8.4, 15.2
A10010	16.1
A10024	19.7
A10025	16.1
A10026	16.1
A10034	3.1, 15.4
A10035	7.2
A10036	7.2
A10037	7.2
A10038	7.2
A10039	7.2
A10040	7.2
A10042	7.2
A10043	7.2
A10044	3.1, 15.4
A10067	9.4
A10068	9.4
A10070	13.2, 17.4
A10072	13.2, 17.4
A10168	1.3, 1.7
A10169	1.3, 1.7
A10170	1.2, 1.3, 1.7
A10171	1.2, 1.3, 1.7
A10192	1.8, 9.3, 21.2
A10194	6.6, 7.2
A10195	6.6, 7.2
A10196	6.6, 7.6
A10197	6.6
A10198	6.6, 14.6
A10200	6.6
A10202	3.1, 15.4
A10235	1.2, 1.3
A10236	1.2, 1.3
A10237	1.2, 1.3
A10238	1.2, 1.3
A10239	1.2, 1.3
A10254	2.2
A10255	2.2
A10256	2.2
A10258	2.2
A10259	4.3, 7.5
A10260	7.5
A10262	7.5
A10263	4.3, 7.5
A10266	3.1
A10267	3.1
A10270	3.1
A10275	3.1
A10277	3.1
A10278	3.1
A10279	3.1
A10280	3.1
A10344	23.2

Cat. No.	Sections
A10346	15.3, 23.2
A10389	23.2
A10436	3.3, 14.3
A10437	3.3, 14.3
A10438	3.3, 14.3
A10439	3.3, 14.3
A10441	14.3
A10442	14.3
A10466	3.4
A10468	1.2, 1.3
A10470	1.2, 1.3
A10474	1.2, 1.3
A10475	1.2, 1.3
A10476	1.2, 1.5
A10478	1.2, 1.7
A1048	5.3, 17.3
A1049	5.3, 17.3
A10516	7.2
A10517	7.2
A10518	7.2
A10519	7.2
A10520	7.2
A10521	7.2
A10522	7.2
A10523	7.2
A10524	7.2
A10525	7.2
A10526	7.2
A10527	7.2
A10528	7.2
A10529	7.2
A10530	7.2
A10531	7.2
A10532	7.2
A10533	7.2
A10534	7.2
A10535	7.2
A10536	7.2
A10537	7.2
A10538	7.2
A10539	7.2
A10540	7.2
A10541	7.2
A10542	7.2
A10543	7.2
A10544	7.2
A10545	7.2
A10546	7.2
A10547	7.2
A10548	7.2
A10549	7.2
A10550	4.2, 8.2
A10551	7.2
A10630	7.2

Cat. No.	Sections
A10631	7.2
A10648	7.2
A10650	7.2
A10651	7.2
A10654	7.2
A10663	7.2
A10666	7.2
A10667	7.2
A10668	7.2
A10676	7.2
A10677	7.2
A10679	7.2
A10680	7.2
A10683	7.2
A10684	7.2
A10685	7.2
A10686	7.2
A10688	7.2
A10689	7.2
A10690	7.2
A10691	7.2
A10931	6.4, 7.2
A11001	7.2
A11002	7.2
A11003	7.2
A11004	7.2
A11005	7.2
A11006	7.2
A11007	7.2
A11008	7.2
A11009	7.2
A11010	7.2
A11011	7.2
A11012	7.2
A11013	7.2
A11014	7.2
A11015	7.2
A11016	7.2
A11017	7.2
A11018	7.2
A11019	7.2
A11020	7.2
A11029	7.2
A11030	7.2
A11031	7.2
A11032	7.2
A11034	7.2
A11035	7.2
A11036	7.2
A11037	7.2
A11039	7.2
A11040	7.2
A11041	7.2
A11042	7.2

Cat. No.	Sections
A11045	7.2
A11046	7.2
A11053	7.2, 7.4
A11054	7.2
A11055	7.2
A11056	7.2
A11057	7.2
A11058	7.2
A11059	7.2
A11060	7.2
A11061	7.2
A11062	7.2
A11063	7.2
A11064	7.2
A11066	7.2
A11067	7.2
A11068	7.2
A11069	7.2
A11070	7.2
A11071	7.2
A11072	7.2
A11073	7.2
A11074	7.2
A11075	7.2
A11076	7.2
A11077	7.2
A11078	7.2
A11079	7.2
A11080	7.2
A11081	7.2
A11090	7.4
A11091	7.4
A11094	7.4
A11095	7.4
A11096	7.4
A11097	7.4
A11120	7.5
A11121	7.5
A11122	7.5
A11126	11.2
A11130	16.1
A11131	7.5, 9.4
A11132	7.5, 10.2
A11133	14.7
A11236	7.6
A12214	10.5
A12216	10.5, 13.3
A12217	10.5, 16.2
A12218	10.5, 17.4
A12219	10.5, 17.4
A12220	10.5, 13.3, 17.4
A12221	10.5, 16.2
A12222	10.5, 18.2
A12373	11.1

Cat. No.	Sections
A12374	11.1
A12375	11.1
A12379	11.1
A12380	11.1
A12381	11.1
A12410	17.3
A12412	17.3
A12922	4.3, 14.3
A12923	4.3, 14.3
A12924	4.3, 14.3
A1301	8.1, 12.3, 15.2
A1310	8.1, 12.5
A13100	14.7
A13101	14.7
A1318	3.4
A13199	15.5
A13200	15.5
A13201	15.5
A13202	15.5
A13203	15.5
A13204	4.3, 15.5
A1322	16.3
A1324	8.1, 12.5, 20.3
A1339	3.4, 14.3
A1340	3.4, 14.3
A13438	16.2
A13439	16.2
A1351	3.4
A1353	3.4
A1363	3.4
A1372	12.2
A1400	3.4
A1440	3.2
A14835	23.2
A14836	23.2
A1493	19.8
A1593	4.2, 14.3
A1594	4.2, 14.3
A18211	23.1
A191	3.3, 3.4, 10.1
A20000	1.3, 4.2
A20001	1.3
A20002	1.3
A20003	1.3
A20004	1.3
A20005	1.3
A20006	1.3
A20007	1.3
A20008	1.3
A20009	1.3
A20010	1.3
A20011	1.3
A20012	3.1
A20013	3.1

IMPORTANT NOTICE: The products described in this manual are covered by one or more Limited Use Label License(s). Please refer to the Appendix on page 971 and Master Product List on page 975. Products are For Research Use Only. Not intended for any animal or human therapeutic or diagnostic use.

Cat. No.	Sections
A20100	1.3, 4.2
A20101MP	1.3
A20102	1.3
A20103	1.3
A20104	1.3
A20105	1.3
A20106	1.3
A20108	1.3
A20109	1.3
A20110	1.3
A20111	1.3
A20170	1.2, 1.3
A20171	1.2, 1.3
A20172	1.2, 1.3
A20173	1.2, 1.3
A20174	1.2, 1.3
A20180	1.2, 1.3, 1.7
A20181	1.2, 1.3
A20182	1.2, 1.3
A20183	1.2, 1.3
A20184	1.2, 1.3
A20185	1.2, 1.3
A20186	1.2, 1.3
A20187	1.2, 1.3
A20191	1.2, 1.3, 8.2
A20196	1.2, 1.3, 8.2
A20341	2.2
A20342	2.2
A20343	2.2
A20344	2.2
A20346	2.2
A20347	2.2
A20501MP	3.3, 14.3
A20502	3.3, 14.3
A20770	5.2, 9.4
A20980	6.4, 7.2
A20981	6.4, 7.2
A20983	6.4, 7.2
A20984	6.4, 7.2
A20990	6.4, 7.2
A20991	6.4, 7.2
A21000	6.4, 7.2
A21001MP	6.4, 7.2
A21006	6.4, 7.2
A21036	7.2
A21037	7.2
A21038	7.2
A21039	7.2
A21042	7.2
A21043	7.2
A21044	7.2
A21045	7.2
A21046	7.2
A21048	7.2
A21049	7.2
A21050	7.2
A21052	7.2
A21053	7.2
A21054	7.2
A21055	7.2
A21057	7.2

Cat. No.	Sections
A21058	7.2
A21059	7.2
A21062	7.2
A21063	7.2
A21065	7.2
A21068	7.2
A21069	7.2
A21070	7.2
A21071	7.2
A21072	7.2
A21073	7.2
A21074	7.2
A21076	7.2
A21077	7.2
A21081	7.2
A21082	7.2
A21083	7.2
A21084	7.2
A21085	7.2
A21086	7.2
A21088	7.2
A21089	7.2
A21090	7.2
A21091	7.2
A21093	7.2
A21094	7.2
A21096	7.2
A21097	7.2
A21098	7.2
A21099	7.2
A21100	7.2
A21102	7.2
A21103	7.2
A21105	7.2
A21109	7.2
A21110	7.2
A21111	7.2
A21112	7.2
A21113	7.2
A21120	7.2
A21121	7.2
A21123	7.2
A21124	7.2
A21125	7.2
A21126	7.2
A21127	7.2
A21130	7.2
A21131	7.2
A21133	7.2
A21134	7.2
A21135	7.2
A21136	7.2
A21137	7.2
A21140	7.2
A21141	7.2
A21143	7.2
A21144	7.2
A21145	7.2
A21146	7.2
A21147	7.2
A21151	7.2

Cat. No.	Sections
A21155	7.2
A21200	7.2
A21201	7.2
A21202	7.2
A21203	7.2
A21204	7.2
A21205	7.2
A21206	7.2
A21207	7.2
A21208	7.2
A21209	7.2
A21210	7.2
A21211	7.2
A21212	7.2
A21213	7.2
A21215	7.2
A21216	7.2
A21222	7.2
A21223	7.2
A21235	7.2
A21236	7.2
A21237	7.2
A21238	7.2
A21239	7.2
A21240	7.2
A21241	7.2
A21242	7.2
A21244	7.2
A21245	7.2
A21246	7.2
A21247	7.2
A21248	7.2
A21249	7.2
A21250	6.4, 7.4
A21253	7.4
A21272	4.3
A21280	7.5
A21281	7.5
A21282	11.2
A21283	11.2
A21285	7.4, 18.3
A21287	7.5
A21288	7.5
A21294	11.2
A21295	11.2
A21311	7.5
A21312	7.5
A21327	17.4
A21328	17.4
A21370	7.6
A21371	4.3, 11.2
A21422	7.2
A21424	7.2
A21425	7.2
A21426	7.2
A21427	7.2
A21428	7.2
A21429	7.2
A21430	7.2
A21431	7.2
A21432	7.2

Cat. No.	Sections
A21433	7.2
A21434	7.2
A21435	7.2
A21436	7.2
A21437	7.2
A21441	7.2
A21442	7.2
A21443	7.2
A21445	7.2
A21446	7.2
A21447	7.2
A21448	7.2
A21449	7.2
A21450	7.2
A21451	7.2
A21463	7.2
A21467	7.2
A21468	7.2
A21469	7.2
A21470	7.2
A21471	7.2
A21472	7.2
A21664	8.2
A21665	1.2, 1.3, 8.2
A21667	1.2, 1.3, 8.2
A21669	1.2, 1.3, 8.2
A21676	1.2, 1.3, 8.2
A21677	1.2, 1.3, 8.2
A22010	10.2
A22066	10.3, 15.3
A22177	10.5, 18.2
A22178	10.2, 10.5
A22179	10.2, 10.5
A22180	10.5
A22181	10.5
A22182	10.5, 18.2
A22184	10.3, 17.3
A22188	10.5, 18.2
A22189	10.2, 10.5
A22281	11.1
A22282	11.1
A22283	11.1
A22284	11.1
A22285	11.1
A22286	11.1
A22287	11.1
A22352	17.3
A22359	17.3
A22362	17.3
A22850	14.3
A23015	14.7
A23016	14.7
A23017	14.7
A23018	14.7
A23202	15.5
A23204	15.5
A23210	15.5
A2333	1.8, 9.3
A2334	1.8, 9.3
A24920	7.2
A24921	7.2

Cat. No.	Sections
A24922	7.2
A24923	7.2
A24926	7.6
A24927	7.6
A2522	5.3
A2660	6.4, 7.6
A2662	7.6
A2663	7.6
A2664	7.6
A2665	7.6
A2666	7.6
A2667	7.6
A2750	14.7
A2952	4.2
A30000	1.3, 1.7, 4.2, 14.3
A30002	1.3
A30005	1.3, 4.2
A30006	1.2, 1.3
A30007	1.2, 1.3
A30008	1.2, 1.3
A30009	1.2, 1.3
A30050	1.3
A30051	1.3
A30052	1.3, 4.2
A30100	1.3, 1.7, 4.2, 14.3
A30459	2.2
A30505	2.3
A30627	3.3, 14.3
A30629	3.3, 14.3
A30632	3.3, 14.3
A30634	3.3, 14.3
A30674	3.4
A30675	3.4, 14.3
A30676	3.4
A30677	3.4
A30678	3.4
A30679	3.4
A30680	3.4
A31552	7.2
A31553	7.2
A31555	7.2
A31556	7.2
A31558	7.2
A31560	7.2
A31561	7.2
A31562	7.2
A31563	7.2
A31564	7.2
A31565	7.2
A31566	7.2
A31570	7.2
A31571	7.2
A31572	7.2
A31573	7.2
A31574	7.2
A31575	7.2
A31576	7.2
A31577	7.2
A31619	7.2

IMPORTANT NOTICE: The products described in this manual are covered by one or more Limited Use Label License(s). Please refer to the Appendix on page 971 and Master Product List on page 975. Products are For Research Use Only. Not intended for any animal or human therapeutic or diagnostic use.

Cat. No.	Sections
A31620	7.2
A31621	7.2
A31622	7.2
A31623	7.2
A31624	7.2
A31625	7.2
A31626	7.2
A31627	7.2
A31628	7.2
A31629	7.2
A31630	7.2
A31631	7.2
A31632	7.2
A31633	7.2
A31634	7.2
A31851	7.5
A31852	7.5
A32360	7.6
A32361	7.6
A32750	1.3, 8.2
A32755	1.3, 8.2
A32756	1.3, 8.2
A32757	1.3, 8.2
A32761	8.2
A32762	8.2
A32763	8.2
A32768	8.2
A32769	8.2
A32770	8.2
A32771	8.2
A33851	6.2, 10.5
A33852	6.2, 10.5
A33855	10.5
A34050	11.1
A34051	11.1
A34054	11.1
A34055	11.1
A34785	14.7
A34786	14.7
A34787	14.7
A350	3.3, 14.3
A35108	15.5
A35109	15.5
A35110	6.4, 15.5
A35111	6.4, 15.5
A35122	15.5
A35135	15.5
A35136	15.5
A35137	15.5
A3568	8.1, 12.3, 15.2
A35725	10.3, 17.3
A35775	17.3
A35777	17.3
A36003	18.2
A36006	10.5, 18.2
A3880	13.2, 17.4
A433	2.3
A47	13.5
A484	2.3
A485	2.3
A50	13.5

Cat. No.	Sections
A5750	7.4
A5751	4.3, 7.4
A5760	7.4
A5770	7.4
A5790	7.5, 10.2
A5800	7.5, 9.4
A6118	1.7
A6222	1.8, 9.2, 9.3, 21.2
A6257	3.3, 14.3
A6289	3.3
A6373	7.6
A6374	7.6
A6378	7.6
A6397	7.4
A6398	7.4
A6399	7.4
A6413	7.4
A6421	7.4
A6423	7.4
A6430	7.4
A6435	4.3, 7.4
A6442	11.2, 16.1, 17.3
A6455	7.5
A6473	16.2
A6474	16.2
A6475	16.2
A6520	10.4, 15.5
A6521	10.4
A666	8.1
A6785	19.5
A685	5.2
A7056	5.3, 17.3
A7110	5.3, 16.2
A7302	23.2
A7303	23.2
A7304	23.2
A7305	23.2
A7312	23.2
A7313	23.2
A7592	8.1
A7621	17.2
A7816	23.1
A7923	18.2
A803	6.4, 14.7
A819	6.4, 14.7
A820	7.6
A821	7.6
A865	6.4, 7.2
A887	7.6
A889	7.4
A91	3.4, 9.5
A982	4.3, 7.4
B10006	1.4, 4.2
B10019	16.3, 21.1
B10033	16.3, 21.1
B10146	16.3, 21.1
B10147	16.3, 21.1
B10184	3.1, 4.2
B10185	3.1, 4.2
B10250	2.2

Cat. No.	Sections
B10331	16.3, 21.1
B10332	16.3, 21.1
B10333	16.3, 21.1
B10334	16.3, 21.1
B10335	16.3
B10341	16.3
B10383	14.7
B10570	4.3, 14.3
B10621	2.2, 5.2
B10710	6.5, 7.6
B11027	4.3, 7.2
B1150	15.2, 15.6, 20.2
B1151	14.3, 20.2
B1170	15.2, 15.6, 20.2
B1196	4.3, 16.2
B1204	19.8
B1205	19.8
B1212	19.8
B1214	19.8
B12382	11.1
B13233	9.4, 15.2
B13234	9.4, 15.2
B13422	16.2
B13423	16.2
B1355	2.2, 3.4
B1370	4.3, 14.3
B13800	17.2
B13950	13.3, 17.4
B1494	19.8
B1508	5.3
B1513	4.2, 14.3
B1526	5.3
B153	11.2, 13.5
B1550	4.3, 13.2
B1577	5.3
B1582	4.2, 14.3
B1591	4.2
B1592	4.2, 14.3
B1595	4.2, 7.6
B1596	4.2, 14.3
B1601	16.2
B1603	4.2, 14.3
B1606	4.2
B1616	4.3, 13.2
B1690	10.2
B1691	10.2
B20340	2.2
B20656	4.2, 7.6
B2103	2.2, 3.4, 14.2
B21078	4.3, 7.2
B21550	15.4
B21551	15.4
B22356	17.3
B22358	5.3, 17.3
B22461	12.3
B22650	12.4, 13.3, 17.4
B22802	2.2
B23151	15.4
B23461	16.3
B24570	22.3
B2600	4.2

Cat. No.	Sections
B2604	4.2
B2752	7.2
B2763	4.3, 7.2
B2766	7.2
B2770	4.3, 7.2
B30010	1.2, 4.2
B30250	1.7
B30300	1.8, 3.3
B30466	2.2
B30500	2.3
B30501	2.3
B3051	15.2, 15.6, 20.2
B30600	5.3, 13.5
B30633	3.4
B30756	1.2, 4.2
B311	13.5
B32766	4.3, 8.2
B32772	4.3, 8.2
B34353	17.4
B34400	12.4, 13.3, 17.4
B34401	13.3, 17.4
B34402	10.2, 12.4, 13.3, 17.4
B34650	14.4
B3475	11.1
B34950	15.3, 22.3
B34954	15.3
B34956	15.3
B35000	15.4
B35001	15.4
B35128	15.4
B35129	15.4
B35130	15.4
B35131	15.4
B35132	15.4
B35133	15.4
B35138	4.3, 15.4
B35139	15.4
B35140	15.4
B35141	15.4
B35450	16.2
B35451	16.2
B35452	16.2
B3582	8.1, 12.5
B3586	8.1, 12.5
B3781	13.2, 17.4
B3782	13.2, 17.4
B3824	13.2
B3932	13.5, 18.2
B413	22.3
B436	16.3, 22.3
B438	22.3
B6057	2.3, 3.4
B607	11.1
B6352	4.2, 14.3
B6353	4.2, 14.3
B6769	19.8
B6790	19.2
B6791	19.2
B6810	19.2
B7277	15.4

Cat. No.	Sections
B7431	15.6, 16.3
B7433	15.6, 16.2
B7450	12.4
B7469	15.6, 17.3
B7474	4.3, 11.1
B7487	17.2
B7701	13.2, 17.4
C10045	14.4
C10046	14.4
C10080	16.1
C10094	14.2
C10102	3.1, 9.4
C10164	1.7
C10186	3.1, 9.4
C10214	3.1, 15.4
C10227	15.4
C10228	15.4
C10245	3.1, 15.5
C10246	3.1, 15.5
C10247	3.1, 15.5
C10248	3.1, 9.4
C10249	3.1, 9.4
C10264	3.1, 9.4
C10265	3.1, 9.4
C10268	3.1, 9.4
C10269	3.1, 9.4
C10276	3.1, 9.4
C10284	15.4
C10285	15.4
C10286	15.4
C10289	3.1, 9.4
C10310	15.4
C10311	15.4
C10312	15.4
C10313	15.4
C10314	15.4
C10315	15.4
C10327	3.1
C10328	3.1
C10329	3.1
C10330	3.1
C10337	3.1, 15.4
C10338	3.1, 15.4
C10339	3.1, 15.4
C10340	3.1, 15.4
C10350	3.1, 15.4
C10351	3.1, 15.4
C10352	3.1, 15.4
C10353	3.1, 15.4
C10354	3.1, 15.4
C10355	3.1, 15.4
C10356	3.1, 15.4
C10357	3.1, 15.4
C10552	6.3, 10.3
C10553	6.3, 10.3
C10554	6.3, 10.3
C10555	6.3, 10.3
C10556	6.3, 10.3
C10557	17.3
C10558	17.3
C10582	11.1

Cat. No.	Sections
C10583	11.1
C10586	12.3
C10587	12.3
C10590	12.4
C10591	12.4
C10592	12.4
C10593	12.4
C10594	12.5
C10595	12.5
C10596	12.3
C10597	12.3
C10598	11.2
C10599	11.2
C10600	12.2
C10601	12.2
C10602	12.5
C10603	12.5
C10604	12.3
C10606	14.4
C10607	14.4
C10608	14.4
C10611	11.2
C10612	11.2
C10613	11.2
C10614	11.2
C10615	11.1
C11252	7.7
C11253	7.7
C11254	7.7
C11397	8.2
C11398	8.2
C11399	8.2
C11400	8.2
C11401	8.2
C11403	8.2
C11404	8.2
C1157	14.2, 15.4, 20.2
C1165	14.2
C1171	1.6
C12680	13.3
C12681	13.3
C1270	20.2
C1271	20.2
C1272	20.2
C12881	10.1, 14.2
C1309	1.6
C1311	1.5
C13196	15.2, 20.2
C13293	10.5, 18.2
C1354	15.2, 20.2
C1359	1.5
C1360	1.5
C1361	15.2
C1362	15.2
C13654	5.3, 16.2
C14260	10.6, 19.5
C1429	15.2
C1430	14.3, 15.2, 15.6, 19.7
C14784	23.1
C14837	6.5, 23.1

Cat. No.	Sections
C18120	23.1
C18121	23.1
C18128	23.1
C18136	23.1
C18139	23.1
C18140	23.1
C18141	23.1
C18142	23.1
C18150	23.1
C18151	23.1
C18155	23.1
C18156	23.1
C18160	23.1
C18161	23.1
C183	10.1
C1904	1.5, 14.3, 20.2
C194	14.3, 20.2
C195	15.2, 20.2
C20050	1.5, 5.3
C20260	1.8, 2.2
C2102	14.2
C2110	10.1, 14.2
C2111	14.2
C21385	7.6
C21386	7.6
C21401	7.7
C21402	7.7
C21421	7.7
C21555	8.2
C21556	8.2
C21559	8.2
C21852	9.4
C2210	1.5
C2211	1.6
C22220	10.6
C22803	14.2, 15.4, 16.3
C2284	1.7, 4.2, 14.3
C22841	7.7, 14.7, 16.1
C22842	7.7, 14.7, 16.1
C22843	7.7, 14.7, 16.1
C24430	21.1
C24431	21.1
C24726	23.1
C24727	23.1
C24760	23.1
C24761	23.1
C24762	23.1
C24763	23.1
C24764	23.1
C24765	23.1
C24766	23.1
C24767	23.1
C24768	23.1
C24769	23.1
C24770	23.1
C24775	23.1
C24776	23.1
C24777	23.1
C24778	23.1
C24779	23.1
C24780	23.1

Cat. No.	Sections
C24784	23.1
C2764	7.2
C2925	14.2, 15.2, 15.6, 20.2
C2927	14.2
C2938	15.2, 18.2
C2944	10.6, 18.2, 19.5
C2990	10.4, 15.6, 16.1
C300	1.6
C3008MP	19.8
C3010MP	19.3
C3011MP	19.3
C3012	19.3
C3013	19.3
C3015	19.3
C3018	19.3
C3047	19.8
C3099	14.3, 15.2, 15.6, 19.7
C3100MP	14.3, 15.2, 15.6, 19.7
C3221	14.3
C3239	14.3
C33250	9.3
C33251	9.3
C33365	3.1, 9.4
C33366	3.1, 9.4
C33367	3.1, 9.4
C33368	3.1, 9.4
C33370	3.1, 9.4
C33371	3.1, 9.4
C33372	3.1, 9.4
C33373	9.4
C33374	9.4
C34551	14.2
C34552	14.2
C34553	14.2
C34554	14.2, 15.4
C34555	14.2, 15.4
C34556	13.5, 14.4
C34557	14.2, 15.4
C34775	7.7, 14.7, 16.1
C34776	7.7, 14.7, 16.1
C34777	7.7, 14.7, 16.1
C34778	7.7, 14.7, 16.1
C34779	4.3, 7.7, 14.7, 16.1
C34780	7.7, 14.7, 16.1
C34851	15.2
C34852	15.2
C34853	15.2
C34858	15.2
C35002	3.1, 15.4
C35006	15.4, 15.6
C35007	15.4, 15.6
C35011	15.4, 15.6
C35012	15.4, 15.6
C356	3.3
C36675	21.1
C36676	21.1
C368	14.3, 20.3
C369	15.2, 20.3

Cat. No.	Sections
C36950	15.4, 23.2
C37000	23.1
C37005	23.1
C3713	19.4
C3714	19.4
C3730	19.3
C3732	19.3
C3737	19.3
C3739	19.3
C3927MP	13.3
C400	10.5, 15.2, 18.2
C481	14.3, 16.1, 19.7
C606	11.1
C6121	1.6
C6122	1.6
C6123	1.6
C6124	1.6
C6125	1.6
C6126	1.6
C6127	1.6
C6128	1.6
C6156	1.6
C6157	1.6
C6164	1.5
C6171MP	1.5
C621	3.4, 14.3
C652	1.6, 20.2
C653	1.6, 20.4
C6667	9.2
C6765	19.4
C6779	10.6, 19.5
C6780	10.6, 19.5
C6826	14.2, 15.2, 15.6, 20.2
C6827	14.2, 15.2, 18.2
C684	10.6
C687	3.3, 14.3, 16.1
C7000	13.4, 14.4
C7001	13.4, 14.4
C7025	14.2, 15.2, 15.6, 20.2
C7026	15.4, 15.6
C7027	15.4
C7028	15.4
C7074	5.3, 17.2
C7122	5.3, 16.2
C7604	8.2, 15.5
C7606MP	8.2
C7614	8.2, 15.4, 15.5
C7631	8.2
C7912	18.3
C7924	18.2
C825	7.7
C827	7.7
C860	7.7
C962	7.2
D100	3.3, 9.5
D10000	1.4
D10001	1.4
D10160	1.7
D10161	1.7

Cat. No.	Sections
D10166	1.7
D10251	2.3
D10252	2.3
D10253	2.3
D1037	5.3, 17.3
D10460	3.4, 12.3
D109	13.5
D112	3.4
D1125	13.4, 14.4
D113	3.4, 12.3
D11347	8.1, 15.2, 18.2
D1168	8.1, 15.2, 18.2
D1199	22.2
D12050	10.4, 14.7, 16.1
D12051	10.4, 14.7, 16.1
D12052	10.4, 14.7
D12053	10.4, 14.7, 16.1
D12054	10.4, 16.1
D12060	10.4, 14.7, 16.1
D1207	19.8
D1209	19.8
D1211	19.8
D12200	10.6
D12371	11.1
D12372	11.1
D126	1.7
D12730	13.4, 14.4
D12731	13.4, 14.4
D12800	13.5
D1306	8.1, 11.2, 12.5, 14.3, 15.4
D1379	2.3, 5.2
D1383	16.1
D13951	10.2, 12.4, 13.3, 17.4
D1412	1.7
D1421	1.7
D1446	3.2
D1463	3.3
D14730	22.3
D1521	2.2
D1532	2.1, 5.2
D1552	12.3
D16	1.5, 3.2
D1816	14.5
D1817	14.5
D1818	14.5
D1819	14.5
D1820	14.5
D1821	14.5, 20.4
D1822	14.5
D1823	14.5, 20.4
D1824	14.5
D1825	14.5
D1828	14.5
D1829	14.5
D1830	14.5
D1841	14.5
D1842	14.5
D1844	14.5, 20.4
D1845	14.5

IMPORTANT NOTICE: The products described in this manual are covered by one or more Limited Use Label License(s). Please refer to the Appendix on page 971 and Master Product List on page 975. Products are For Research Use Only. Not intended for any animal or human therapeutic or diagnostic use.

Cat. No.	Sections
D1860	14.5
D1861	14.5
D1862	14.5
D1863	14.5
D1864	14.5
D1868	14.5
D1878	20.4
D1880	20.4
D1951	20.4
D1956	4.3, 14.5
D1957	4.3, 14.5
D1976	14.5
D2004	2.2
D202	13.5
D20350	2.2
D20351	2.2
D20653	4.2, 14.3
D20655	1.2, 4.2
D20657	4.2, 7.6
D20658	4.3
D20691	4.3, 7.2
D20693	4.3, 7.2
D20697	4.3, 7.2
D20698	4.3, 7.2
D20701	4.3, 7.2
D21	1.7, 9.5
D21490	8.1, 11.2, 12.5, 14.3, 15.4
D2183	1.4
D2184	1.4
D2187	1.4
D21887	7.2
D2191	1.4
D22	1.7
D22065	10.3
D22185	6.2, 7.2, 10.5
D22187	6.2, 7.6, 10.5
D2219	1.4
D2222	1.4
D2225	1.4
D2228	1.4
D22421	12.2, 15.2, 22.3
D2245	1.8, 9.5
D2248	4.2
D2281	3.2
D22910	14.5
D22911	14.5
D22912	14.5
D22913	14.5
D22914	14.5
D23	1.7
D23107	8.1, 15.2, 18.2
D2371	3.3
D23739	13.2, 17.4
D23805	10.5, 15.2, 18.2
D23806	12.2, 15.2, 18.2
D23840	18.3
D23841	18.3, 21.2
D23842	18.3
D23844	18.3
D2390	3.4

Cat. No.	Sections
D250	13.5
D2516	5.3
D272	12.4, 22.3
D273	12.2, 12.4, 22.3
D275	13.4, 14.4
D282	12.4, 13.4, 14.4
D288	12.2, 16.1
D289	16.1
D2893	10.2
D291	13.4, 14.4
D2920	10.2
D2935	16.1, 18.2
D3034	5.3, 17.2, 19.8
D306	22.3
D307	13.4, 14.4
D30753	4.2
D3167	22.2
D3238	14.3
D3303	20.4
D3304	20.4
D3305	14.5, 20.4
D3306	14.5
D3307	14.5
D3308	14.5
D3312	4.3, 14.5
D3328	14.5
D3329	14.5
D3330	14.5
D337	16.3
D338	16.3
D346	2.3
D34679	14.5
D34680	14.5
D34681	14.5
D34682	14.5
D3521	12.4, 13.3, 17.4
D3522	12.4, 13.3, 17.4
D355	3.3
D3571	8.1, 11.2, 12.5, 14.3, 15.4
D36801	22.2
D36802	22.2
D374	1.7
D3771	13.2, 17.4
D378	12.2
D3792	13.2
D3793	13.2
D3800	13.2
D3803	13.2, 17.4
D3805	13.2
D3815	13.2
D3821	13.2
D3822	13.2
D3823	13.2
D3825	13.2
D383	13.4, 14.4
D3832	13.2
D3834	1.4, 13.2
D3835	13.2
D384	12.4, 13.4, 14.4
D3861	13.2, 18.2

Cat. No.	Sections
D3862	13.2
D3883	13.4, 14.4
D3886	13.4, 14.4
D3898	13.4, 14.4
D3899	13.4, 14.4
D3911	13.4, 14.4
D3921	13.5
D3922	13.5
D3923	11.2, 13.5
D399	10.5, 15.2, 18.2
D404	2.3
D426	12.2
D476	13.2
D57	13.2
D6003	2.2
D6004	2.2
D6102	1.4, 4.2, 9.3
D6104	1.7, 4.2
D6116	1.4, 9.3
D6117	1.4
D6140	1.4
D6145	1.5
D6180	1.4
D6184	1.4
D6216	9.5
D6251	3.4
D632	12.2, 15.2, 18.2
D633	12.2, 15.2, 18.2
D6487	10.3
D6488	10.2
D6562	13.3
D6566	10.1
D6567	10.3
D673	16.3
D6814	5.3, 17.2, 19.8
D6923	22.2
D6927	22.2
D7132	14.5
D7135	4.3, 14.5
D7136	14.5
D7137	14.5
D7139	14.5
D7142	4.3, 14.5
D7144	14.5
D7153	14.5
D7156	4.3, 14.5
D7162	4.3, 14.5
D7163	14.5
D7168	14.5
D7170	14.5, 20.4
D7171	14.5
D7172	14.5, 20.4
D7173	14.5
D7176	14.5, 20.4
D7178	4.3, 14.5
D7443	16.3
D7519	10.2, 13.3, 17.4
D7540	12.4, 13.3, 17.4
D7711	12.4, 13.3, 17.4
D7756	13.4, 14.4
D7757	13.4, 14.4

Cat. No.	Sections
D7758	13.4, 14.4
D7776	13.4, 14.4
D7777	13.4, 14.4
D7778	13.4, 14.4
D7894	18.2
D7918	3.3, 18.3, 21.2
D8451	2.1, 5.2
D94	13.2
E10187	3.1, 15.4
E10215	17.4
E10216	17.4
E10217	17.4
E10218	17.4
E10219	17.4
E10221	17.4
E10345	3.1
E10415	3.1, 15.4
E1169	8.1, 8.4, 15.2
E118	2.2
E12020	10.3
E12055	10.4
E12056	10.4
E1219	19.8
E12353	12.4
E13183	10.4, 15.5
E13184	10.4, 15.5
E13231	16.1
E13345	4.3, 16.1
E1374	5.3, 8.1, 15.2
E18	1.5, 16.3
E21390	7.6
E22013	10.2
E22064	15.3
E2247	3.4, 5.2
E23370	16.1
E23691	17.3
E2861	16.1
E2862	16.1
E2863	16.1
E2864	16.1
E2870	16.1
E2882	10.6
E3101	21.2
E3111	16.3
E33075	8.4, 9.3
E33650	10.2
E33651	10.2
E33701	10.3, 10.5
E33702	10.3
E33758	10.4
E33856	10.5, 18.2
E33953	10.2
E33955	10.6, 17.4
E33956	10.6
E34250	12.4, 16.3
E34251	12.4, 16.3
E3476	16.1
E3477	4.3, 16.1
E3478	16.1
E3480	4.3, 16.1
E3481	16.1

Cat. No.	Sections
E35350	4.3, 16.1
E35351	4.3, 16.1
E3599	8.1, 8.4, 15.2
E6578	6.3, 10.1
E6588	6.3, 10.3
E6589	6.3, 10.3
E6600	6.3, 10.3
E6601	6.3, 10.3
E6603	6.3, 10.3
E6604	6.3, 10.3
E6605	6.3, 10.3
E6606	6.3, 10.3
E6638	10.4, 14.7
E6639	10.4, 14.7
E6645	10.3, 21.2
E6646	10.3, 21.2
E6658	10.4
E7498	16.1
F10016	16.3, 21.1
F10017	16.3, 21.1
F10240	1.2, 1.5
F10317	15.2
F10318	15.2
F10319	15.2
F10320	15.2
F10322	15.2
F10347	15.4
F10348	15.4
F10471	19.3
F10472	19.3
F10473	19.3
F10720	6.5, 14.6
F11021	7.2
F1130	14.3, 20.2
F1176	16.2
F1179	10.2
F11830	9.5
F11831	9.5
F1200	19.2
F1201	19.2
F121	3.3
F1221	19.2
F1225	19.2
F1240	19.3
F1241	19.3
F1242	19.3
F1300	1.5, 10.1, 20.2
F1303	15.2, 20.2
F13080	6.5, 14.6
F13081	6.5, 14.6
F13082	6.5, 14.6
F13083	6.5, 14.6
F1314	15.6, 16.1
F13191	15.6, 16.1
F13192	15.6, 16.1
F13193	15.6, 16.1
F13838	23.2
F14174	19.2
F14175	19.2
F14178	19.2
F14180	19.2

IMPORTANT NOTICE: The products described in this manual are covered by one or more Limited Use Label License(s). Please refer to the Appendix on page 971 and Master Product List on page 975. Products are For Research Use Only. Not intended for any animal or human therapeutic or diagnostic use.

Cat. No.	Sections
F14181	19.2
F14185	19.2
F14200	19.3
F14201	19.3
F14202	19.3
F14203	19.3
F14204	19.3
F14217	19.3
F14218	19.3
F14219	19.3
F14221	19.3
F14222	19.3
F14240	19.4
F14242	19.3
F143	1.5
F14781	23.1
F14791	23.1
F14806	23.1
F14807	23.1
F14808	23.1
F150	2.2
F1906	1.5
F1907	1.5
F1930	10.2
F20261	1.8, 9.2, 9.3
F20650	4.2
F20880	6.5, 14.6
F20881	6.5, 14.6
F20883	6.5, 7.6
F20884	6.5, 7.6
F20886	6.5, 14.6
F20892	6.5, 16.1
F20893	6.5, 16.1
F21010	14.6
F21011	14.6
F21012	14.6
F21013	14.6
F21015	14.6
F21452	7.2
F21453	7.2
F21456	7.2
F2181	1.5, 4.2
F23103	15.2
F2332	1.8, 9.2, 9.3
F23915	19.3
F23917	19.3
F23980	19.3
F23981	19.3
F24180	19.7
F24181	19.7
F24189	19.7
F24194	19.7
F24195	19.7
F24630	23.1
F24631	23.1
F24633	23.1
F24634	23.1
F2610	1.2, 4.2
F2761	7.2
F2765	7.2
F2881	10.2

Cat. No.	Sections
F2900	10.6
F2902	16.1, 18.2
F2905	10.2
F2915	10.2
F2962	8.3, 15.4
F2999	10.3
F3020	19.3
F3021	19.3
F3029	19.4
F30751	1.2, 4.2
F30755	1.2, 4.2, 8.2
F32767	8.2
F32947	1.3, 8.2
F32948	1.3, 8.2
F32949	1.3, 8.2
F32950	1.3, 8.2
F32951	1.3, 8.2
F32952	1.3, 8.2
F32953	1.3, 8.2
F32954	1.3, 8.2
F32955	1.3, 8.2
F32956	1.3, 8.2
F33951	10.6
F33952	10.6
F34651	14.4
F34652	14.4
F34653	14.4, 16.1
F34953	15.3
F35200	15.6, 16.1
F35355	14.4, 16.1
F362	13.2, 20.4
F36205	19.3
F36206	19.3
F36250	19.4
F36905	23.1
F36906	23.1
F36909	23.1
F36913	23.1
F36914	23.1
F36915	1.5, 10.1, 23.1
F36924	23.1
F36925	23.1
F3715	19.3
F3857	13.5
F432	11.1
F486	1.8, 2.2
F6053	2.2
F6106	1.5
F6129	1.5
F6130	1.5, 4.2
F6153	1.2, 1.5
F6161	1.2, 1.6
F6162	1.2, 1.6
F6218	3.2
F6347	1.2, 4.2
F6348	1.2, 4.2
F6433	1.2, 1.5
F6434	1.2, 1.5
F6616	10.6
F6617	10.6
F6774	19.2, 19.8

Cat. No.	Sections
F6799	19.2
F7030	12.3, 15.2
F7103	5.3, 14.3
F7234	23.1
F7235	23.1
F7236	23.1
F7237	23.1
F7238	23.1
F7239	23.1
F7240	23.1
F7496	15.6, 16.1
F8760	6.5
F8762	6.5
F8763	6.5
F8764	6.5
F8765	6.5
F8766	4.3, 6.5, 7.6
F8767	4.3, 6.5, 7.6
F8768	4.3, 6.5, 7.6
F8769	4.3, 6.5, 7.6
F8770	6.5, 7.6
F8771	6.5, 7.6
F8772	6.5, 7.6
F8774	6.5, 7.6
F8775	6.5, 7.6
F8776	6.5, 7.6
F8777	6.5, 7.6
F8780	6.5, 7.6
F8781	6.5
F8782	6.5
F8783	6.5
F8784	6.5
F8786	6.5
F8787	6.5
F8789	6.5, 14.6
F8792	6.5, 14.6
F8793	6.5, 14.6
F8794	6.5, 14.6
F8795	6.5, 14.6
F8797	6.5
F8799	6.5
F8800	6.5
F8801	6.5
F8803	6.5
F8805	6.5
F8806	6.5
F8807	6.5
F8809	6.5
F8810	6.5
F8811	6.5
F8812	6.5
F8813	6.5
F8814	6.5
F8815	6.5
F8816	6.5
F8819	6.5
F8820	6.5
F8821	6.5
F8823	6.5
F8824	6.5
F8825	6.5

Cat. No.	Sections
F8826	6.5
F8827	6.5
F8829	14.6
F8830	14.6
F8831	14.6
F8833	14.6
F8834	14.6
F8836	14.6
F8837	14.6
F8838	14.6
F8839	14.6
F8841	14.6
F8842	14.6
F8843	14.6
F8844	14.6
F8845	6.5
F8848	6.5
F8851	6.5
F8852	6.5
F8853	6.5
F8854	6.5
F8858	6.5
F8859	6.5
F8887	6.5
F8888	6.5
F8890	14.6
F8891	14.6
F8892	14.6
G10362	7.5
G12411	17.3
G13186	15.6, 16.1
G13187	15.6, 16.1
G21040	7.2, 10.5
G21060	7.2
G21079	7.2
G21234	7.2
G22183	10.3, 17.3
G22351	17.3
G22360	17.3
G2879	9.4
G31567	7.2
G31568	7.2
G35778	17.3
G35779	17.3
G35780	17.3
G36000	4.3, 9.4, 15.6, 18.2
G6888	21.1
G7055	5.3, 16.2
G7921	18.3, 21.2
H10290	15.5
H10292	15.5
H10293	15.4
H10294	12.5
H10295	15.5
H10325	12.5
H10326	12.5
H110	13.5
H1193	1.7
H13188	12.5, 16.1

Cat. No.	Sections
H1398	8.1, 8.3, 12.5, 15.4
H1399	8.1, 12.5, 15.4
H1428	1.7
H14700	22.3
H18200	23.1
H18201	23.1
H18202	23.1
H18210	23.1
H185	1.7, 14.3
H189	10.1
H21486	8.1
H21491	8.1, 8.3, 12.5, 15.4
H21492	8.1, 12.5, 15.4
H2249	3.4, 5.2
H22730	13.5
H22845	8.1, 14.3
H24720	23.1
H24721	23.1
H24722	23.1
H24723	23.1
H2930	10.4
H32712	12.5
H32713	12.5
H32714	12.5
H32720	12.5
H32721	12.5
H34157	13.2
H34158	13.2
H34350	13.2
H34351	13.2
H34475	13.5
H34476	13.5
H34477	13.5
H348	14.3, 16.1, 20.2
H3569	8.1, 8.3, 12.5, 15.4
H3570	8.1, 12.5, 15.4
H36004	18.2
H361	13.2, 17.4
H3809	13.2, 17.4
H6482	10.1
H7476	17.3, 18.2
H7482	17.2
H7593	8.1, 15.2
I10188	3.1
I10189	3.1
I10291	15.5
I1202	19.2
I1203	19.2
I1223	19.2
I1226	19.2
I13269	16.1
I14	2.3
I14402	14.3, 19.8
I14785	23.1
I14786	23.1
I14787	23.1
I21411	7.7
I21412	7.7

IMPORTANT NOTICE: The products described in this manual are covered by one or more Limited Use Label License(s). Please refer to the Appendix on page 971 and Master Product List on page 975. Products are For Research Use Only. Not intended for any animal or human therapeutic or diagnostic use.

Cat. No.	Sections
I21413	7.7
I21414	4.3, 7.7
I23580	5.3, 17.2
I24221	19.8
I24222	19.8
I2904	10.2
I2906	10.2
I2908	10.2
I30451	2.2
I30452	2.2
I32450	7.7
I34154	12.2
I34202	12.3
I34406	7.7, 14.4
I34407	14.4
I35101	10.4, 15.5
I35102	10.4, 15.5
I35103	12.2, 15.5
I35104	10.4, 15.5
I35105	10.4, 15.5
I35106	10.4, 15.5
I36007	18.2
I36933	7.2, 7.6, 23.1
I3716	17.2
I7	2.3
I7219	23.1
I7221	23.1
I7223	23.1
I7224	23.1
I7225	23.1
I9	2.2
J11372	8.1
J11373	8.1
J7473	11.1
L10119	15.3
L10120	15.3
L10316	15.3
L10382	15.5
L11270	7.7, 14.7
L11271	7.7, 12.4
L11272	7.7
L11376	8.1
L1177	14.3
L12370	11.1
L12491	12.3
L12926	14.3
L13152	15.3
L1338	2.2, 4.2, 14.3
L1392	13.2
L14812	23.2
L14813	23.2
L14814	23.2
L14815	23.2
L14816	23.2
L14817	23.2
L14818	23.2
L14819	23.2
L14821	23.2
L14822	23.2
L14823	23.2
L14824	23.2

Cat. No.	Sections
L1908	1.6
L20	1.6
L21409	7.7, 16.1
L21415	7.7, 12.4
L21416	7.7, 12.4
L22290	11.1
L22460	12.3, 20.4
L23101	15.3
L23102	15.3
L23105	15.3
L23350	13.3, 16.1
L23351	13.3, 16.1
L23352	13.3, 16.1
L23353	13.3, 16.1
L23356	13.3, 16.1
L23380	16.1
L2424	3.4
L24919	7.2, 7.6
L2601	4.3, 14.3
L2911	10.6
L2912	10.6
L2916	10.6
L3224	15.3
L32451	7.7, 12.4
L32454	7.7, 12.4
L32456	7.7, 14.7
L32457	7.7, 14.7
L32458	7.7, 16.1
L32459	7.7, 16.1
L32460	7.7, 16.1
L32462	7.7
L32463	7.7
L3482	16.1
L3483	16.1
L3484	16.1
L3485	16.1
L34856	15.3
L3486	16.1
L34951	15.3
L34952	15.3
L34955	15.3
L34957	15.3
L34958	15.3
L34959	15.3
L34960	15.3
L35353	16.1
L35354	16.1
L453	3.3, 14.3, 16.1
L6543	10.4
L682	14.3
L6868	12.2, 18.2, 21.2
L6950	4.3, 14.3
L7005	15.3
L7007	15.3
L7009	15.3
L7010	15.3
L7011	15.3
L7012	15.3
L7013	15.3
L7085	5.3, 10.6
L7525	12.3

Cat. No.	Sections
L7526	12.3
L7528	12.3
L7533	12.3, 20.3
L7534	12.3, 20.3
L7535	12.3, 20.3
L7545	12.3, 20.3
L7595	8.1, 15.4
L7781	13.4, 14.4
L8455	6.2, 10.5, 15.2, 18.2
M10165	1.7
M10991	7.2
M10992	7.2
M1198MP	15.6
M12414	17.3
M12415	17.3
M12416	17.3
M12417	17.3
M12652	13.2
M1290	19.2, 19.6
M1291	19.2, 19.6
M1292	19.2, 19.6
M1295	19.2, 19.6
M1378	2.3, 14.2
M1381MP	2.3, 14.2, 15.6
M1410	1.7
M14205	19.3, 19.6
M14206	19.3, 19.6
M1420MP	1.7
M1445	3.2
M1489MP	10.2
M1490	10.2
M1602	4.2, 14.3
M20381	2.3, 14.2
M20490	3.3, 10.5, 18.3, 21.2
M22353	17.3
M22354	17.3
M22425	12.2
M22426	12.2
M23271	15.6
M23400	16.2
M23800	6.2, 10.5, 18.2
M24571	18.2, 22.3
M25	1.8, 3.2
M30550	2.1, 5.2
M33305	9.4
M33306	9.4
M33307	9.4
M34150	15.5, 22.3
M34151	15.5, 22.3
M34152	15.5, 22.3
M34153	12.2, 15.5
M36008	12.2, 18.2
M36051	18.3, 21.2
M36353	19.7
M3733	19.3, 19.6
M3735	19.3, 19.6
M440	21.2
M6026	2.2
M6248	3.4

Cat. No.	Sections
M6491	10.3
M6494	15.2, 18.2
M688	19.7
M6886	21.2
M689	1.6, 18.2
M7006	15.4
M7510	12.2
M7511	12.2, 18.2
M7512	12.2
M7513	12.2, 18.2
M7514	12.2
M7891	18.3
M7901	23.1
M7913	18.2
M7926	18.3
M8	2.3
MPM33305	9.4
MPM33306	9.4
MPP33300	9.4
MPP33301	9.4
MPP33302	9.4
N10559	10.3
N10560	10.3
N10561	10.6
N10562	10.6
N10563	10.2
N10564	10.2
N10565	10.2
N10566	10.2
N10577	10.3
N10578	10.3
N1138	1.8, 9.3, 15.6, 21.2
N1142	11.2, 13.5
N1148	13.3
N1154	12.4, 13.3, 17.4
N13195	15.2
N1384	16.2
N1495	20.2, 21.2
N20345	2.2
N21479	8.1, 12.5, 14.3
N21480	8.1, 12.5, 14.3
N21481	8.1, 12.5, 14.3
N21482	8.1, 12.5, 14.3
N21483	8.1, 12.5, 14.3
N21485	8.1, 12.5
N22651	12.4, 13.3, 17.4
N22880	14.4
N22881	14.4
N22883	14.4
N22884	14.4
N23106	15.2
N24191	19.7
N2461	3.4
N24915	7.2
N24916	7.2
N24918	7.6
N316	1.8, 13.2
N3246	12.3, 15.2
N3524	12.4, 13.3, 17.4
N354	11.1

Cat. No.	Sections
N360	13.2
N3786	13.2, 17.4
N3787	13.2, 17.4
N6356	4.2
N6495	6.3, 10.3, 15.2, 18.2
N6547	6.3, 10.3
N6666	9.2
N678	13.2
N6802	5.3, 17.2, 19.8
N6803	5.3, 17.2, 19.8
N7167	14.5
N7565	8.1, 15.2
N7892	18.3
N7927	18.3
N7990	19.7
N7991	19.7
O10104	16.1
O10138	16.1
O10180	3.1
O10181	3.1
O10241	1.2, 1.5
O10465	3.4
O11033	7.2
O11038	7.2
O11492	8.1, 8.3
O12650	13.2, 20.4
O12920	4.3, 14.3
O13291	16.1, 18.2
O14804	23.1
O23020	14.7
O23021	14.7
O23990	19.3
O246	13.5, 14.4
O24750	23.1
O24751	23.1
O266	22.3
O267	22.3
O34550	14.2, 15.4
O34781	14.7
O34782	14.7
O34783	14.7
O34784	14.7
O36231	16.1
O6010	2.2
O6034	2.2
O6080	1.5
O6138	1.5, 14.3, 20.3
O6139	1.5, 20.4
O6146	1.5, 14.3, 20.3
O6147	1.5, 20.4
O6149	1.5, 20.4
O6151	15.2, 20.3
O6185	1.5, 4.2
O6380	7.2
O6381	7.2
O6382	7.2
O6383	7.2
O6798	19.4
O6806	19.3
O6807	19.3

IMPORTANT NOTICE: The products described in this manual are covered by one or more Limited Use Label License(s). Please refer to the Appendix on page 971 and Master Product List on page 975. Products are For Research Use Only. Not intended for any animal or human therapeutic or diagnostic use.

Cat. No.	Sections
O6808	19.3
O6809	19.3
O6812	19.3
O7465	11.1
O7466	11.1
O7582	8.1, 8.3
O7703	17.4
P10020	16.3, 19.8, 21.1
P101	3.3
P10163	1.7
P10229	16.3, 21.2
P10361	14.5, 16.1, 20.4
P10993	7.2
P10994	7.2
P11047	7.2
P11049	7.2
P11065	7.2
P11495	8.1, 8.3
P11496	8.1, 8.3
P11947	10.2
P11948	10.2
P11949	10.2
P12271	10.4, 15.5
P1265MP	21.1
P1267MP	21.1
P130	1.7
P1304MP	8.1, 12.5, 14.3, 15.2
P13235	15.2, 17.3
P13236	15.2, 17.3
P13238	15.2, 17.3
P1405	3.4
P14312	19.7
P14313	19.7
P14825	23.2
P14826	23.2
P14827	23.2
P14828	23.2
P14829	23.2
P14830	23.2
P14831	23.2
P14832	23.2
P1692	10.2
P18174	23.1
P18175	23.1
P18178	23.1
P18179	23.1
P1903MP	13.2
P2007MP	2.3
P20495	13.3
P21041	7.2, 10.5
P21129	6.4, 7.2
P21139	6.4, 7.2
P21149	6.4, 7.2
P21315	7.5, 9.4
P21462	7.2
P21493	8.1, 12.5, 14.3, 15.2
P21680	8.2, 8.4
P21681	8.2, 8.4
P21686	8.2, 8.4
P21687	8.2, 8.4
P21855	9.4
P21857	9.4
P21875	9.4
P22061	10.3, 21.2
P22062	10.3, 21.2
P22310	11.2
P22652	13.2
P23012	14.7
P2331MP	1.8, 9.2, 9.3, 15.6, 21.2
P24	1.7
P243	13.2
P24670	23.2
P24671	23.2
P24672	23.2
P24740	23.1
P24741	23.1
P24742	23.1
P24743	23.1
P24744	23.1
P24745	23.1
P248	11.2, 13.5
P27167	9.4
P2771MP	6.4, 7.2
P28	2.3
P29	2.3, 11.1
P3000MP	19.8, 21.1
P30012	1.2, 1.7
P30013	1.2, 1.7
P30014	1.2, 1.7
P30016	1.2, 1.7
P30253	1.7
P30506	2.3
P30507	2.3
P31	13.2
P31581	7.2
P31582	7.2
P31584	7.2
P31585	7.2
P33203	9.4
P33300	9.4
P33301	9.4
P33302	9.4
P33310	9.4
P33311	9.4
P33350	9.4
P33356	9.4
P33357	9.4
P33358	9.4
P33359	9.4
P33706	9.4, 17.3
P3456	11.2
P3457	11.1
P349	14.3
P35361	16.1
P3566	8.1, 12.5, 15.2
P3580	8.1
P3581	8.1, 15.5
P3584	8.1, 8.4
P3585	8.1
P35901	10.6
P36005	13.2, 18.2
P36207	16.3, 19.5
P36208	16.3, 19.5
P36232	15.4
P36235	15.5
P36236	15.5
P36400	19.8
P36600	20.4
P36930	23.1
P36931	23.1
P36934	23.1
P36935	23.1
P37001	23.1
P37002	23.1
P37003	23.1
P37004	23.1
P6114	1.7
P6305	5.2
P6317	5.3
P6466	17.4
P6649	9.3
P6763	19.7
P6866	19.8, 21.1
P6867	19.8, 21.1
P6879	2.3, 18.2
P7220	23.1
P7481	23.1
P7500	11.2
P7501	11.2
P7581	8.1, 8.3
P7589	8.1, 8.3
P800	6.4, 14.7, 18.2
P801	6.4, 14.7, 18.2
P806	6.4
P811	4.3, 6.4
P852	6.4, 7.2
P917	4.3, 7.6
P96	13.2
Q10001MP	6.6
Q10007	6.6
Q10008	6.6
Q10009	6.6
Q10011MP	6.6
Q10012	6.6
Q10013	6.6
Q10014	6.6
Q10015	6.6
Q10021MP	6.6
Q10031MP	6.6
Q10041MP	6.6
Q10047	6.6
Q10051	6.6
Q10052	6.6
Q10053	6.6
Q10054	6.6
Q10055	6.6
Q10056	6.6
Q10057	6.6
Q10059	6.6
Q10060	6.6
Q10061MP	6.6
Q10062	6.6
Q10063	6.6
Q10064	6.6
Q10065	6.6
Q10066	6.6
Q10069	6.6
Q10071MP	6.6
Q10073	6.6
Q10074	6.6
Q10075	6.6
Q10076	6.6
Q10091MP	6.6
Q10101MP	6.6, 7.6
Q10111MP	6.6, 7.6
Q10121MP	6.6, 7.6
Q10131MP	6.6, 7.6
Q10141MP	6.6, 7.6
Q10151MP	6.6, 7.6
Q10152	6.6
Q10153	6.6
Q10154	6.6
Q10156	6.6
Q10161MP	6.6, 7.6
Q10171MP	6.6, 7.6
Q10172	6.6
Q10179	6.6
Q10193	1.6, 9.5
Q10210	8.3
Q10211	8.3
Q10212	8.3
Q10213	8.3
Q10257	2.2, 9.5
Q10298	6.6
Q10301MP	4.3, 6.6
Q10305	6.6
Q10306	6.6
Q10307	6.6
Q10321MP	4.3, 6.6
Q10336	6.6, 23.1
Q10363	12.5
Q10464	3.4, 9.5
Q11001MP	6.6, 7.2
Q11002MP	6.6, 7.2
Q11011MP	6.6, 7.2
Q11021MP	6.6, 7.2
Q11022MP	6.6, 7.2
Q11031MP	6.6, 7.2
Q11032MP	6.6, 7.2
Q11041MP	6.6, 7.2
Q11061MP	6.6, 7.2
Q11062MP	6.6, 7.2
Q11071MP	6.6, 7.2
Q11201MP	6.6, 7.2
Q11221MP	6.6, 7.2
Q11401MP	6.6, 7.2
Q11402MP	6.6, 7.2
Q11411MP	6.6, 7.2
Q11421MP	6.6, 7.2
Q11422MP	6.6, 7.2
Q11431MP	6.6, 7.2
Q11432MP	6.6, 7.2
Q11441MP	6.6, 7.2
Q11461MP	6.6, 7.2
Q11462MP	6.6, 7.2
Q11471MP	6.6, 7.2
Q11601MP	6.6, 7.2
Q11621MP	6.6, 7.2
Q11631MP	6.6, 7.2
Q11821MP	6.6, 7.2
Q12021MP	6.6, 7.7
Q14421MP	6.6, 7.2
Q15421MP	6.6, 7.4
Q15431MP	6.6, 7.4
Q17421MP	6.6, 7.4
Q20001MP	6.6
Q20131	1.6, 9.5
Q20132	1.6, 9.5
Q20133	1.8, 9.5
Q20348	2.2, 9.5
Q21021MP	6.6, 14.6
Q21031MP	6.6, 14.6
Q21061MP	6.6, 14.6
Q21071MP	6.6, 14.6
Q21301MP	6.6
Q21311MP	6.6
Q21321MP	6.6
Q21331MP	6.6
Q21341MP	6.6
Q21361MP	6.6
Q21371MP	6.6
Q21391MP	6.6
Q21501MP	6.6
Q21511MP	6.6
Q21521MP	6.6
Q21531MP	6.6
Q21541MP	6.6
Q21561MP	6.6
Q21571MP	6.6
Q21591MP	6.6
Q21701MP	6.6
Q21711MP	6.6
Q21721MP	6.6
Q21731MP	6.6
Q21761MP	6.6
Q21771MP	6.6
Q21791MP	6.6
Q21930	9.5
Q21931	9.5
Q22001MP	6.6
Q22011MP	6.6
Q22021MP	6.6
Q22031MP	6.6
Q22041MP	6.6
Q22061MP	6.6
Q22071MP	6.6
Q23918	19.2
Q25001MP	6.6, 14.6
Q25011MP	6.6, 14.6
Q25021MP	6.6, 14.6
Q25031MP	6.6, 14.6
Q25041MP	6.6, 14.6

IMPORTANT NOTICE: The products described in this manual are covered by one or more Limited Use Label License(s). Please refer to the Appendix on page 971 and Master Product List on page 975. Products are For Research Use Only. Not intended for any animal or human therapeutic or diagnostic use.
www.invitrogen.com/probes
molecular probes® | invitrogen™ by life technologies™

Cat. No.	Sections
Q25061MP	6.6, 14.6
Q25071MP	6.6, 14.6
Q30457	2.2
Q32850	8.3
Q32851	8.3
Q32852	8.3
Q32853	8.3
Q32854	8.3
Q32855	8.3
Q32856	8.3, 9.2
Q32857	8.3, 9.2
Q32858	8.3, 9.2
Q32859	8.3, 9.2
Q32860	8.3, 9.2
Q32861	8.3
Q33120	8.3
Q33130	8.3
Q33140	8.3
Q33210	9.2
Q33211	9.2
Q33212	9.2
R10145	19.3
R10367	7.5
R1147	10.6
R11490	8.1, 8.3
R11491	8.1, 8.3
R1159	10.2
R12204	10.6, 15.2
R1244	19.3
R1245MP	19.3
R14000	18.2
R14060	12.2, 12.3, 15.2, 18.2
R14207	19.3
R14220	19.3
R14782	23.1
R21455	7.2
R21458	7.2
R21459	7.2
R22067	10.3, 17.3
R22120	10.4, 15.5
R22122	10.4, 15.5
R22124	10.4
R22125	10.4, 15.5
R22126	10.4, 15.5
R22132	10.4
R22420	12.2, 15.2, 22.3
R23983	19.3
R2931	10.4
R302	12.2, 15.2, 22.3
R32700	8.1, 8.3
R33200	9.2
R33400	9.3
R33700	10.3, 17.3
R33750	10.4, 15.5
R33752	10.4, 15.5
R33753	10.4, 15.5
R33754	10.4, 15.5
R33755	10.4, 15.5
R34676	19.4
R35100	10.4, 15.5

Cat. No.	Sections
R352	10.6
R363	10.1
R36350	19.7
R36351	19.7
R415	11.1
R441	10.6
R6029	2.2
R6107	1.5
R6113	1.5
R6160	1.6, 4.2
R634	12.2
R6393	7.2
R6394	7.2
R6479	10.1, 21.2
R648MP	12.2
R649	22.2
R6501	10.4
R6502	10.4
R6505	10.4
R6506	10.4
R6508	10.4
S10274	12.5, 15.2
S10349	12.5, 15.2
S10490	3.4
S1108	22.2
S1109	22.2
S11221	7.6
S11222	7.6
S11223	7.6
S11224	7.6
S11225	7.6
S11226	7.6
S11227	7.6
S11228	7.6
S11237	7.6
S11249	7.6
S1129	15.2, 20.2
S11340	8.1, 12.5, 15.2
S11341	8.1
S11342	8.1
S11343	8.1
S11344	8.1
S11345	8.1
S11346	8.1
S11348	8.1, 12.5, 15.2
S11350	8.1, 12.5, 15.2
S11351	8.1
S11352	8.1
S11353	8.1
S11356	8.1
S11360	8.1, 12.5, 15.2
S11361	8.1
S11362	8.1
S11363	8.1
S11364	8.1
S11365	8.1
S11366	8.1
S11368	8.1, 12.5, 15.2
S11494	8.1, 8.4
S1167	1.8
S11791	9.3

Cat. No.	Sections
S12000	9.3
S12001	9.3
S12010	9.3
S12012	9.3
S1262	21.1
S1263	21.1
S1264	21.1
S1307	14.3
S1531	5.2
S1534	5.2
S1553	5.2
S20982	6.4, 7.6
S20985	6.4, 7.6
S20992	6.4, 7.6
S21002	6.4, 7.6
S21005	6.4, 7.6
S21008	6.4, 7.6
S21374	7.6
S21375	7.6
S21377	7.6
S21378	7.6
S21379	7.6, 9.4
S21381	7.6
S21383	7.6
S21384	7.6
S21388	6.4, 7.6
S21500	8.1, 8.2
S21900	9.3
S22801	14.2, 15.4, 20.4
S23371	16.1
S23372	16.1
S23920	20.2
S23921	20.2
S23922	20.2
S23923	20.2
S24730	23.1
S24731	23.1
S24732	23.1
S24733	23.1
S24734	23.1
S24735	23.1
S24736	23.1
S24737	23.1
S2828	23.1
S2851	16.1
S2854	16.1
S2859	16.1
S2860	16.1
S30044	1.2, 1.3
S30045	1.2, 1.3
S30046	1.2, 1.3
S31201	14.6
S31203	14.6
S31569	7.6
S32350	6.4, 7.6
S32351	7.6
S32353	7.6
S32354	7.6
S32355	7.6
S32356	7.6
S32357	7.6

Cat. No.	Sections
S32358	7.6
S32362	6.4, 7.6
S32363	6.4, 7.6
S32364	7.6
S32365	7.6
S32703	8.1, 15.2
S32704	8.1, 15.2
S33025	12.5
S33100	8.1, 8.4
S33101	8.1, 8.4
S33102	8.1, 8.4
S33110	8.1, 8.4
S33111	8.1, 8.4
S33112	8.1, 8.4
S34200	12.4
S34201	12.3
S34788	14.7
S34789	14.7
S34790	14.7
S34791	14.7
S34854	8.1, 15.2
S34855	8.1, 15.2
S34857	8.1, 12.5, 15.2
S34859	8.1, 12.5, 15.2
S359	14.3, 16.1
S36002	18.2
S36936	23.1
S36937	23.1
S36938	23.1
S36939	23.1
S37100	8.4
S37102	8.4
S37103	8.4
S37104	8.4
S6110	1.7
S6366	7.6
S6368	7.6
S6369	7.6
S6370	7.6
S6650	9.3
S6651	9.3
S6653	9.3
S6654	9.3
S6656	9.3
S6900	21.1
S6901	21.1
S7020	8.1, 12.5, 15.2, 15.4
S7445	16.3
S7556	8.1
S7559	8.1
S7560	8.1
S7563	8.1, 8.3, 8.4
S7564	8.1, 8.4
S7567	8.1, 8.3, 8.4
S7568	8.1, 8.4
S7569	8.4
S7572	8.1, 12.5, 15.2
S7573	8.1
S7574	8.1
S7575	8.1, 15.2

Cat. No.	Sections
S7576	8.1
S7578	8.1
S7579	8.1
S7580	8.1, 8.4
S7585	8.1, 8.3, 8.4
S7586	8.1, 8.4
S7916	18.3
S866	6.4, 7.6
S868	6.4, 7.6
S869	7.6
S870	7.6
S871	7.6
S872	7.6
S888	7.6
S911	7.6
S921	7.6
S931	7.6, 10.2
S951	7.6
T10095	15.6
T10096	15.6
T10182	3.1
T10183	3.1
T10244	1.2, 1.6
T10282	15.4
T10711	6.5, 7.6
T1111	14.4, 16.1, 22.2
T11493	8.3
T1175	16.2
T1210	19.8
T1247	9.4, 14.3, 17.2, 19.7
T12921	4.3, 14.3
T1323	14.3
T13320	12.3, 14.4, 16.1
T13342	16.1
T13343	16.1
T1391	13.2
T1395MP	13.2
T14792	23.1
T1480	1.6
T1481	1.6
T1905	1.6
T20175	1.6, 4.2
T204	13.5, 16.1
T20911	6.2
T20912	6.2
T20913	6.2
T20914	6.2
T20915	6.2
T20916	6.2
T20917	6.2
T20921	6.2
T20922	6.2
T20923	6.2
T20924	6.2
T20925	6.2
T20926	6.2
T20927	6.2
T20931	6.2
T20932	6.2
T20933	6.2

IMPORTANT NOTICE: The products described in this manual are covered by one or more Limited Use Label License(s). Please refer to the Appendix on page 971 and Master Product List on page 975. Products are For Research Use Only. Not intended for any animal or human therapeutic or diagnostic use.

Cat. No.	Sections
T20934	6.2
T20935	6.2
T20936	6.2
T20937	6.2
T20939	6.2
T23011	14.7, 16.1
T23360	12.3, 14.4, 16.1
T23362	16.1
T23363	4.3, 16.1
T23364	16.1
T23365	16.1
T23366	16.1
T2425	3.4
T2556	2.1, 5.2
T2762	7.2
T2767	7.2
T2769	7.2
T2871	16.1
T2872	16.1
T2875	16.1
T30200	1.6, 3.2
T30453	2.2
T30454	2.2
T30455	2.2
T30456	2.2
T30754	4.2
T30953	6.2
T30954	6.2
T30955	6.2
T3163	14.4, 16.1
T3166	12.3, 14.4, 16.1
T3168	12.2, 15.2, 22.3
T34075	11.2
T353	1.6
T35352	16.1
T35356	14.4, 16.1
T35357	16.1
T3600	8.1, 8.4, 12.5, 15.4
T3602	8.1, 8.4
T3604	8.1, 8.4, 12.5
T3605	8.1, 8.4, 12.5, 15.4
T490	1.6
T506	14.3
T53	13.5
T6006	2.2
T6008	2.2
T6009	2.2
T6027	2.2
T6028	2.2
T6060	2.1, 5.2
T6105	1.6, 4.2
T6134	1.6, 4.2
T6219	3.2
T6256	3.3
T639	12.2
T6390	7.2
T6391	7.2
T6392	7.2
T668	12.2, 15.2, 22.3

Cat. No.	Sections
T669	12.2, 15.2, 22.3
T7279	23.1
T7280	23.1
T7281	23.1
T7282	23.1
T7283	23.1
T7284	23.1
T7458	17.2
T7459	17.2
T7471	11.1
T7508	16.1
T7596	8.1
T7601	17.3
T7602	17.3
T7624	17.3
T862	7.2
T8860	6.5, 7.6
T8861	6.5, 7.6
T8864	6.5
T8870	6.5
T8872	6.5
T8880	6.5
T8883	6.5
U21650	1.2, 1.3, 8.2
U21651	1.2, 1.3, 8.2
U21652	1.2, 1.3, 8.2
U21653	1.2, 1.3, 8.2
U21654	1.2, 1.3, 8.2
U21659	1.2, 1.5, 8.2
U21660	1.2, 1.3, 8.2
V10273	15.4
V10309	15.4
V12390	11.2, 15.6
V12883	14.2, 15.4
V13154	15.4
V13180	15.6
V13181	15.6
V13240	15.5
V13241	15.5
V13242	15.5
V13243	15.5
V13244	15.5
V1644	21.1
V22885	14.4, 15.4
V22886	14.4, 15.4
V22887	14.4, 15.4
V22888	14.4, 15.4
V22889	14.4
V23110	15.3
V23111	15.3
V23200	15.5
V23201	15.5
V34403	13.3, 14.4
V34404	13.3, 14.4
V34405	13.3, 14.4
V34850	15.2
V35003	15.4
V35004	15.4
V35005	15.4
V35112	15.5
V35113	15.5

Cat. No.	Sections
V35114	15.5
V35116	15.5
V35117	10.4, 15.5
V35118	10.4, 15.5
V35119	10.4, 15.5
V35123	15.5
V6694	16.1
V7023	15.3
W10132	6.6, 7.2
W10142	6.6, 7.2
W11261	7.7, 12.4
W11262	7.7, 12.4
W11263	7.7, 12.4
W21404	7.7, 12.4
W21405	7.7, 12.4
W32464	7.7, 12.4
W32465	7.7, 12.4
W32466	7.7, 12.4
W6748	7.7, 12.4
W7024	7.7, 12.4, 15.3
W834	7.7, 12.4
W849	7.7, 12.4
X14210	19.3
X1525	14.3
X23984	19.3
X23985	19.3
X491	1.6
X6493	15.2, 18.2
Y3601	8.1, 8.4, 12.5, 15.4
Y3603	8.1, 8.4, 12.5, 15.5
Y3606	8.1, 12.5
Y3607	8.1, 12.5
Y7530	12.2
Y7531	12.3
Y7536	12.3
Z23373	16.1
Z23374	16.1
Z25000	1.2, 1.3, 1.7, 7.3
Z25001	1.2, 1.3, 1.7, 7.3
Z25002	1.2, 1.3, 7.3
Z25003	1.2, 1.3, 7.3
Z25004	1.2, 1.3, 7.3
Z25005	1.2, 1.3, 7.3
Z25006	1.2, 1.3, 7.3
Z25007	1.2, 1.3, 7.3
Z25008	1.2, 1.3, 7.3
Z25009	1.2, 1.3, 7.3
Z25010	1.2, 1.3, 7.3
Z25011	1.2, 1.3, 7.3
Z25013	1.2, 1.3, 1.7, 7.3
Z25020	1.2, 1.3, 6.4, 7.3
Z25021	1.2, 1.3, 6.4, 7.3
Z25022	1.2, 1.3, 6.4, 7.3
Z25030	1.2, 1.3, 6.4, 7.3
Z25031	1.2, 1.3, 6.4, 7.3
Z25041	1.2, 1.7, 7.3
Z25042	1.2, 1.5, 7.3
Z25045	1.2, 1.6, 7.3
Z25051	1.2, 6.4, 7.3

Cat. No.	Sections
Z25052	1.2, 4.2, 7.3
Z25054	1.2, 6.2, 7.3
Z25055	1.2, 6.4, 7.3
Z25060	1.2, 7.3
Z25070	1.2, 7.3
Z25080	1.2, 7.3
Z25090	1.2, 1.3, 6.2, 7.3
Z25100	1.2, 1.3, 1.7, 7.3
Z25102	1.2, 1.3, 7.3
Z25104	1.2, 1.3, 7.3
Z25105	1.2, 1.3, 7.3
Z25106	1.2, 1.3, 7.3
Z25107	1.2, 1.3, 7.3
Z25108	1.2, 1.3, 7.3
Z25110	1.2, 1.3, 7.3
Z25113	1.2, 1.3, 1.7, 7.3
Z25121	1.2, 1.3, 6.4, 7.3
Z25151	1.2, 6.4, 7.3
Z25152	1.2, 4.2, 7.3
Z25154	1.2, 6.2, 7.3
Z25155	1.2, 6.4, 7.3
Z25156	1.2, 7.3
Z25160	1.2, 7.3
Z25170	1.2, 7.3
Z25180	1.2, 7.3
Z25200	1.2, 1.3, 1.7, 7.3
Z25202	1.2, 1.3, 7.3
Z25204	1.2, 1.3, 7.3
Z25205	1.2, 1.3, 7.3
Z25206	1.2, 1.3, 7.3
Z25207	1.2, 1.3, 7.3
Z25208	1.2, 1.3, 7.3
Z25210	1.2, 1.3, 7.3
Z25213	1.2, 1.3, 1.7, 7.3
Z25221	1.2, 1.3, 6.4, 7.3
Z25251	1.2, 6.4, 7.3
Z25252	1.2, 4.2, 7.3
Z25254	1.2, 6.2, 7.3
Z25255	1.2, 6.4, 7.3
Z25256	1.2, 1.7, 7.3
Z25257	1.2, 1.7, 7.3
Z25260	1.2, 7.3
Z25270	1.2, 7.3
Z25280	1.2, 7.3
Z25300	1.2, 1.3, 1.7, 7.3
Z25301	1.2, 1.3, 1.7, 7.3
Z25302	1.2, 1.3, 7.3
Z25303	1.2, 1.3, 7.3
Z25304	1.2, 1.3, 7.3
Z25305	1.2, 1.3, 7.3
Z25306	1.2, 1.3, 7.3
Z25307	1.2, 1.3, 7.3
Z25308	1.2, 1.3, 7.3
Z25310	1.2, 1.3, 7.3
Z25312	1.2, 1.3, 7.3
Z25313	1.2, 1.3, 1.7, 7.3
Z25341	1.2, 1.7, 7.3
Z25342	1.2, 1.5, 7.3
Z25350	1.2, 7.3
Z25351	1.2, 6.4, 7.3
Z25352	1.2, 4.2, 7.3

Cat. No.	Sections
Z25354	1.2, 6.2, 7.3
Z25355	1.2, 6.4, 7.3
Z25360	1.2, 7.3
Z25370	1.2, 7.3
Z25380	1.2, 7.3
Z25400	1.2, 1.3, 1.7, 7.3
Z25402	1.2, 1.3, 7.3
Z25405	1.2, 1.3, 7.3
Z25407	1.2, 1.3, 7.3
Z25408	1.2, 1.3, 7.3
Z25451	1.2, 6.4, 7.3
Z25452	1.2, 4.2, 7.3
Z25454	1.2, 6.2, 7.3
Z25455	1.2, 6.4, 7.3
Z25460	1.2, 7.3
Z25470	1.2, 7.3
Z25602	1.2, 1.3, 7.3
Z25605	1.2, 1.3, 7.3
Z25606	1.2, 1.3, 7.3
Z25607	1.2, 1.3, 7.3
Z25608	1.2, 1.3, 7.3
Z2841	16.1
Z2843	16.1
Z2849	16.1
Z2850	16.1
Z33857	10.5, 18.2

IMPORTANT NOTICE: The products described in this manual are covered by one or more Limited Use Label License(s). Please refer to the Appendix on page 971 and Master Product List on page 975. Products are For Research Use Only. Not intended for any animal or human therapeutic or diagnostic use.

molecular probes® | invitrogen™ by life technologies™

Index

A

IMPORTANT NOTICE: The products described in this manual are covered by one or more Limited Use Label License(s). Please refer to the Appendix on page 971 and Master Product List on page 975. Products are For Research Use Only. Not intended for any animal or human therapeutic or diagnostic use.

IMPORTANT NOTICE: The products described in this manual are covered by one or more Limited Use Label License(s). Please refer to the Appendix on page 971 and Master Product List on page 975. Products are For Research Use Only. Not intended for any animal or human therapeutic or diagnostic use.

C

IMPORTANT NOTICE: The products described in this manual are covered by one or more Limited Use Label License(s). Please refer to the Appendix on page 971 and Master Product List on page 975. Products are For Research Use Only. Not intended for any animal or human therapeutic or diagnostic use.
www.invitrogen.com/probes
molecular probes® | invitrogen™ by life technologies™

IMPORTANT NOTICE: The products described in this manual are covered by one or more Limited Use Label License(s). Please refer to the Appendix on page 971 and Master Product List on page 975. Products are For Research Use Only. Not intended for any animal or human therapeutic or diagnostic use.

D

IMPORTANT NOTICE: The products described in this manual are covered by one or more Limited Use Label License(s). Please refer to the Appendix on page 971 and Master Product List on page 975. Products are For Research Use Only. Not intended for any animal or human therapeutic or diagnostic use.

H

IMPORTANT NOTICE: The products described in this manual are covered by one or more Limited Use Label License(s). Please refer to the Appendix on page 971 and Master Product List on page 975. Products are For Research Use Only. Not intended for any animal or human therapeutic or diagnostic use.
www.invitrogen.com/probes

I

J

K

L

IMPORTANT NOTICE: The products described in this manual are covered by one or more Limited Use Label License(s). Please refer to the Appendix on page 971 and Master Product List on page 975. Products are For Research Use Only. Not intended for any animal or human therapeutic or diagnostic use.

www.invitrogen.com/probes

molecular probes® | invitrogen™ by life technologies™

N

IMPORTANT NOTICE: The products described in this manual are covered by one or more Limited Use Label License(s). Please refer to the Appendix on page 971 and Master Product List on page 975. Products are For Research Use Only. Not intended for any animal or human therapeutic or diagnostic use.

IMPORTANT NOTICE: The products described in this manual are covered by one or more Limited Use Label License(s). Please refer to the Appendix on page 971 and Master Product List on page 975. Products are For Research Use Only. Not intended for any animal or human therapeutic or diagnostic use.
www.invitrogen.com/probes

IMPORTANT NOTICE: The products described in this manual are covered by one or more Limited Use Label License(s). Please refer to the Appendix on page 971 and Master Product List on page 975. Products are For Research Use Only. Not intended for any animal or human therapeutic or diagnostic use.

IMPORTANT NOTICE: The products described in this manual are covered by one or more Limited Use Label License(s). Please refer to the Appendix on page 971 and Master Product List on page 975. Products are For Research Use Only. Not intended for any animal or human therapeutic or diagnostic use.
www.invitrogen.com/probes
molecular probes® | invitrogen™ by life technologies™

IMPORTANT NOTICE: The products described in this manual are covered by one or more Limited Use Label License(s). Please refer to the Appendix on page 971 and Master Product List on page 975. Products are For Research Use Only. Not intended for any animal or human therapeutic or diagnostic use.

IMPORTANT NOTICE: The products described in this manual are covered by one or more Limited Use Label License(s). Please refer to the Appendix on page 971 and Master Product List on page 975. Products are For Research Use Only. Not intended for any animal or human therapeutic or diagnostic use.
www.invitrogen.com/probes

molecular probes® | invitrogen™ by *life* technologies™

S

IMPORTANT NOTICE: The products described in this manual are covered by one or more Limited Use Label License(s). Please refer to the Appendix on page 971 and Master Product List on page 975. Products are For Research Use Only. Not intended for any animal or human therapeutic or diagnostic use.

IMPORTANT NOTICE: The products described in this manual are covered by one or more Limited Use Label License(s). Please refer to the Appendix on page 971 and Master Product List on page 975. Products are For Research Use Only. Not intended for any animal or human therapeutic or diagnostic use.

IMPORTANT NOTICE: The products described in this manual are covered by one or more Limited Use Label License(s). Please refer to the Appendix on page 971 and Master Product List on page 975. Products are For Research Use Only. Not intended for any animal or human therapeutic or diagnostic use.

Contact Information

United States Headquarters
5791 Van Allen Way
Carlsbad, California 92008
United States
Tel: +1 760 603 7200
Toll-Free Tel: 800 955 6288
Fax: +1 760 603 7229
Email: techsupport@invitrogen.com

European Headquarters
Inchinnan Business Park
3 Fountain Drive
Paisley, Scotland PA4 9RF
United Kingdom
Tel: +44 141 814 6100
Toll-Free Tel: 0800 269 210
Toll-Free Tech: 0800 838 380
Fax: +44 141 814 6260
Email: euroinfo@invitrogen.com
Email Tech: eurotech@invitrogen.com

Asia Pacific Headquarters
Shanghai Representative Office
Room 4010, Grand Gateway Tower 1
No. 1 Hongqiao Road
Shanghai 200030
People's Republic of China
Tel: +86 21 6145 2000
Fax: +86 21 6448 4077
Email: sales@invitrogen.com.cn

Japan Headquarters
LOOP-X Bldg. 6F
3-9-15, Kaigan, Minato-ku
Tokyo 108-0022
Tel: +81 3 5730 6509
Fax: +81 3 5730 6519

For distributors and branch offices in regions not listed above, please visit www.invitrogen.com/contactus.

On the cover:
Mink uterus endometrial cells were fixed and permeabilized, then labeled with a mouse anti-vimentin primary antibody and visualized using a green-fluorescent Alexa Fluor® 488 goat anti–mouse IgG. Filamentous actin was labeled with blue-fluorescent coumarin phalloidin; nuclei were stained with SYTOX® Orange nucleic acid stain. Image contributed by John D. Griffin and Michael W. Davidson, National High Magnetic Field Laboratory, Florida State University.

Selected full-page feature images throughout this publication were also graciously provided by Michael Davidson and his laboratory.

IMPORTANT NOTICE: The products described in this manual are covered by one or more Limited Use Label License(s). Please refer to the Appendix on page 971 and Master Product List on page 975. Products are For Research Use Only. Not intended for any animal or human therapeutic or diagnostic use.